CHILTON'S
TRUCK AND VAN MANUAL 1993-97

Publisher & Editor-In-Chief	Kerry A. Freeman, S.A.E.
Executive Editors	Dean F. Morgantini, S.A.E., W. Calvin Settle Jr., S.A.E.
Managing Editor	Nick D'Andrea
Senior Editors	Jacques Gordon, Michael L. Grady, Kevin M. G. Maher, Debra McCall, Richard J. Rivele, S.A.E., Richard T. Smith, Jim Taylor, Ron Webb
Project Managers	Larry Braun, S.A.E., A.S.C., Thomas P. Browne III, Joseph L. DeFrancesco, A.S.E., Robert E. Doughten, Ben Greisler, S.A.E., Martin J. Gunther, Craig P. Nangle, A.S.E., S.A.E., Ernest H. Ralph, A.S.E., S.A.E., Richard Schwartz
Editorial Staff	Jaffer A. Ahmad, Robert Chabot, William Cottman, A.S.E., Leonard Davis, A.S.E., Michael DiFurio Jr., S.A.E., Sam Fiorani, Matthew E. Frederick, William C. Friedauer, Edward Giacomucci, A.S.E., S.A.E., Al Gibbs, Herbert Guie Jr, George B. Heinrich III, Dawn M. Hoch, Daniel Howells, A.S.E., David E. Jester, A.S.E., Lori L. Johnson, A.S.E., Will Kessler, A.S.E., Kenneth F. Konzelman, Neil J. Leonard, A.S.E., James R. Marotta, Robert McAnally, Thomas A. Mellon, Raymond K. Moore, A.S.E., Norman D. Norville, A.S.E., Christine L. Nuckowski, Eric S. Peterson, A.S.E., Charles Ramsey, A.S.E., Roy Ripple, A.S.E., George E. Ritter, Robert Saxton, A.S.E., S.A.E., Paul Shanahan, Larry E. Stiles, Gordon L. Tobias, S.A.E., Albert A. Wood, A.S.E.
Production Manager	Andrea M. Steiger
Assistant Production Manager	Marsha Park-Herman
Production Specialists	Christina Davis, Kimberly T. Hayes, Joseph C. McGinty, Elizabeth E. Thompson
Director of Manufacturing	Mike D'Imperio
Manufacturing Manager	Robin Norman
OFFICERS	
Senior Vice President	Ronald A. Hoxter

CHILTON BOOK COMPANY
ONE OF THE **DIVERSIFIED PUBLISHING COMPANIES,**
A PART OF **CAPITAL CITIES/ABC,INC.**

Manufactured in
© 1996 Chilton Book Company
Chilton Way, Radnor, PA 19089
ISBN 0-8019-7921-1
ISSN

TRUCK MODELS

Table of Contents

Truck Sections

HOW TO USE THIS MANUAL

Car Section

Car sections are grouped by manufacturer and arranged in alphabetical order. The text and illustrations that comprise the service procedures in each Car Section are arranged in the following order of systems and components: Firing Orders, Engine Electrical, Chassis Electrical, Engine Cooling, Fuel System, Emission Controls, Engine Mechanical, Engine Lubrication, Transmission/Transaxle, Drive Axle, Steering, Front Suspension, Rear Suspension, Brakes.

All illustrations are located as close as possible to the pertinent text. Procedures are for all models in the particular section unless specifically noted otherwise.

Locating Information

The Table of Contents, at the front of the book, lists the beginning of each Car Section in the manual.

To find where a particular Car Section is located in the book, you need only look in the Table of Contents. Once you have found the proper section, you may wish to find where specific procedures are located in that section. Turn to the Index at the front of the section. At the upper left-hand side is a listing of the main topics within the section and the page number they will be found on. Following the main topics is an alphabetical listing of all the procedures within the section and their page numbers.

Safety Notice

Proper service and repair procedures are vital to the safe, reliable operation of all motor vehicles, as well as the personal safety of those performing repairs. This manual outlines procedures for servicing and repairing vehicles using safe effective methods. The procedures contain many NOTES and CAUTIONS which should be followed along with standard safety procedures to eliminate the possibility of personal injury or improper service which could damage the vehicle or compromise its safety.

It is important to note that repair procedures and techniques, tools and parts for servicing vehicles, as well as the skill and experience of the individual performing the work vary widely. It is not possible to anticipate all of the conceivable ways or conditions under which vehicles may be serviced, or to provide cautions as to all of the possible hazards that may result. Standard and accepted safety precautions and equipment should be used when handling toxic or flammable fluids, and safety goggles or other protection should be used during cutting, grinding, chiseling, prying, or any other process that can cause material removal or projectiles.

Some procedures require the use of tools specially designed for a specific purpose. Before substituting another tool or procedure, you must be completely satisfied that neither your personal safety, nor the performance of the vehicle will be endangered.

Part Numbers

Part numbers listed in this book are not recommendations by Chilton for any product by brand name. They are references that can be used with interchange manuals and aftermarket supplier catalogs to locate each brand supplier's discrete part number.

Although information in this manual is based on industry sources and is as complete as possible at the time of publication, the possibility exists that some car manufacturers made later changes which could not be included here. Information on very late models may not be available in some circumstances. While striving for total accuracy, Chilton Book Company cannot assume responsibility for any errors, changes, or omissions that may occur in the compilation of this data.

Copyright Notice

ACURA and ISUZU

ACURA-SLX ISUZU-Trooper

FIRING ORDERS

NOTE: To avoid confusion, always replace spark plug wires one at a time.

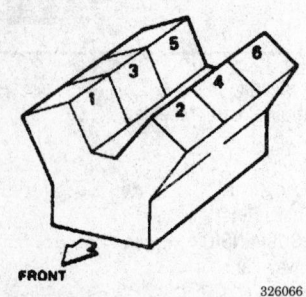

1993–95 3.2L (VIN W) DOHC and
3.2L (VIN V) SOHC Engines
Firing Order: 1-2-3-4-5-6
Distributorless Ignition

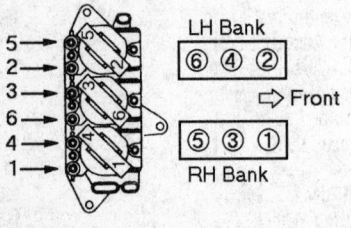

1996–97 3.2L (VIN V) Engine
Firing Order: 1-2-3-4-5-6
Distributorless Ignition

ENGINE ELECTRICAL

NOTE: Disconnecting the negative battery cable on some vehicles may interfere with the functions of the on board computer systems and may require the computer to undergo a relearning process, once the negative battery cable is reconnected.

Ignition Timing

ADJUSTMENT

These engines are equipped with a distributorless ignition system. The ignition timing is controlled by the Powertrain Control Module through the input of engine control system sensors. The ignition timing is set at 5 degrees BTDC for vehicles equipped with manual or automatic transmissions. The ignition timing cannot be adjusted.

Alternator

PRECAUTIONS

Several precautions must be observed with alternator equipped vehicles to avoid damage to the unit.

• If the battery is removed for any reason, make sure it is reconnected with the correct polarity. Reversing the battery connections may result in damage to the 1-way rectifiers.
• When utilizing a booster battery as a starting aid, always connect the positive to positive terminals and the negative terminal from the booster battery to a good engine ground on the vehicle being started.
• Never use a fast charger as a booster to start vehicles.
• Disconnect the battery cables when charging the battery with a fast charger.
• Never attempt to polarize the alternator.
• Do not use test lights of more than 12 volts when checking diode continuity.
• Do not short across or ground any of the alternator terminals.
• The polarity of the battery, alternator and regulator must be matched and considered before making any electrical connections within the system.
• Never separate the alternator on an open circuit. Make sure all connections within the circuit are clean and tight.
• Disconnect the battery ground terminal when performing any service on electrical components.
• Disconnect the battery if arc welding is to be done on the vehicle.

REMOVAL AND INSTALLATION

NOTE: The voltage regulator is an internal part of the alternator. If the voltage regulator is found to be faulty, the alternator must be rebuilt or replaced.

1. Disconnect negative battery cable.
2. If necessary, raise and support the vehicle safely.
3. Remove the radiator skid plate and front lower cover.
4. Loosen the alternator mounting bolt.

5. Loosen the lock and adjusting bolts to slip the drive belt off the pulley.
6. Remove the alternator belt tension adjuster bolt.
7. Remove the alternator mounting bolt.
8. Hold the alternator and disconnect the electrical harness and the B terminal connector.
9. Remove the alternator from the vehicle.
   To install:
10. Install the alternator. Connect the electrical connector to terminal B.
11. Connect the electrical harness.
12. Install the alternator mounting bolts but hand–tighten them only.
13. Install the belt tension adjuster bolt.
14. Fit the belt back onto the pulleys and tighten the mounting bolt to 16 ft. lbs. (22 Nm).
15. Adjust the belt tension by turning the adjuster bolt.
16. Tighten lockbolt to 17 ft. lbs. (24 Nm).
17. Install the front lower cover and skid plate. Tighten the skid plate bolts to 27 ft. lbs. (37 Nm).
18. Connect the negative battery cable.

Drive Belt

REMOVAL AND INSTALLATION

Power Steering Pump

1. Loosen the lockbolt and the pivot bolt.
2. Pivot the power steering pump toward the crankshaft to relieve the belt tension. Remove the power steering belt.
   To install:
3. Install the belt on the proper pulleys. Pivot the power steering pump outward until the proper belt tension (90 ±20 lbs.) is reached and tighten the lockbolt. Tighten the lockbolt to 30 ft. lbs. (46 Nm).

Air Conditioning Compressor

1. Remove the power steering belt.
2. Loosen the lockbolt in the center of the A/C compressor tensioner pulley.
3. Turn the adjusting bolt to relieve the belt tension.
4. Remove the belt.
   To install:
5. Install the A/C compressor belt on the crankshaft and compressor pulleys.
6. Turn the adjusting bolt until the proper belt tension (120 ±20 lbs.

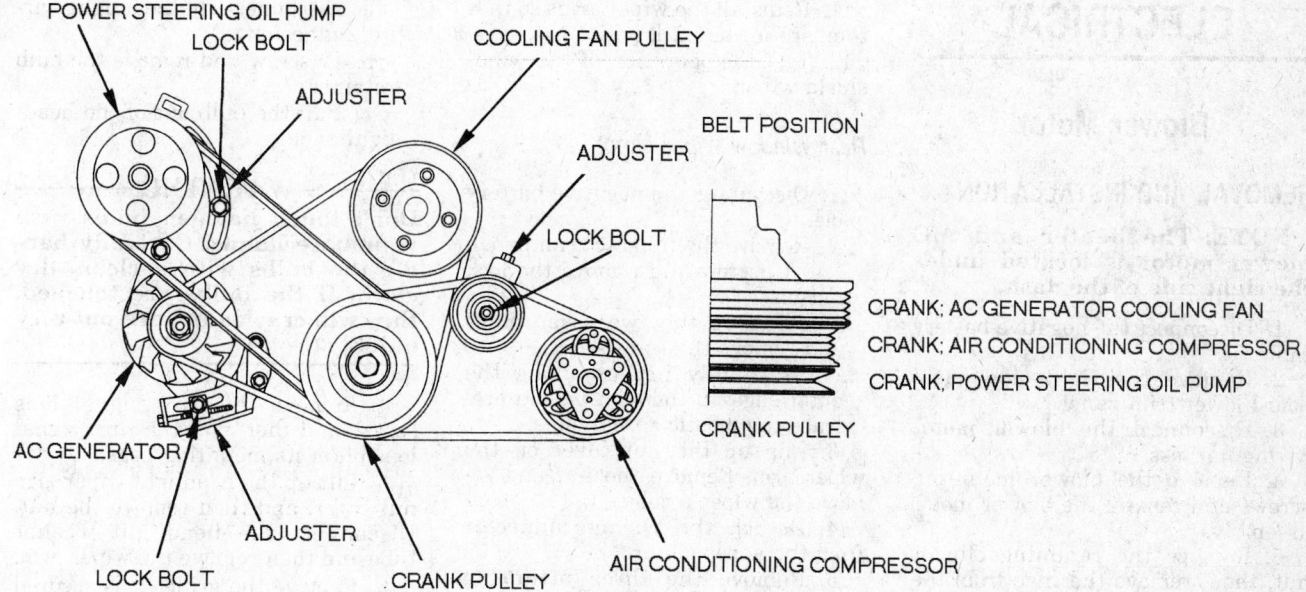

Power steering pump belt routing

322248

for a new belt or 100 ±20 lbs. for a readjustment) is reached and tighten the lockbolt to 37 ft. lbs. (50 Nm).

7. Install the power steering belt and adjust its tension.

Alternator and Cooling Fan

1. Remove the power steering pump belt and the A/C compressor belt.
2. Loosen the lockbolt.
3. Turn the adjusting bolt to relieve the belt tension.
4. Remove the belt.
To install:
5. Install the belt on the alternator, cooling fan, and crankshaft pulleys.
6. Turn the adjusting bolt on the alternator until the proper belt tension is reached:
• 1993 vehicles: 120 ±20 lbs. for a new belt or 100 ±20 lbs. for a readjustment
• 1994–97 vehicles: 180 ±10 lbs. for a new belt or 110 ±10 lbs. for a readjustment
7. Tighten the adjusting bolt to 16 ft. lbs. (22 Nm). Tighten the lockbolt to 17 ft. lbs. (24 Nm).
8. Install the A/C compressor belt and the power steering belt.

Starter

REMOVAL AND INSTALLATION

1. Disconnect negative battery cable.
2. Raise and safely support the vehicle.
3. Remove the transfer case skid plates.
4. Label and disconnect the heated oxygen sensor connectors.

NOTE: The suspension cross-member may be removed for extra working room. Make sure the transmission is securely supported with a jack.

5. Unbolt and separate the front exhaust pipe section from the crossover pipe or catalytic convertor. Remove the front exhaust pipe section.
6. Remove the starter motor heat shield.
7. Disconnect the electrical connectors from terminals B and S.
8. Remove the two starter mounting bolts.
9. Remove the starter from the bottom of the engine.

To install:
10. Install the starter from the bottom of the engine.
11. Install the two starter mounting bolts and tighten to 30 ft. lbs. (40 Nm).
12. Connect the electrical connector to terminals B and S. Tighten the nut at terminal B to 6.3 ft. lbs. (8.5 Nm).
13. Install the heat shield to the engine.
14. Install the front exhaust pipe and reconnect the crossover pipe. Tighten the mounting nuts to 49 ft. lbs. (67 Nm). On 1993–94 vehicles, tighten the bolt to 20 ft. lbs. (27 Nm). On 1995–97 vehicles, tighten the bolt to 32 ft. lbs. (43 Nm).
15. Reconnect the heated oxygen sensor connectors.
16. If removed, install the suspension crossmember and tighten the bolts to 58 ft. lbs. (78 Nm).
17. Install the transfer case skid plates and tighten their bolts to 27 ft. lbs. (37 Nm).
18. Lower the vehicle.
19. Connect the negative battery cable.
20. Verify that the starter works properly.

CHASSIS ELECTRICAL

Blower Motor

REMOVAL AND INSTALLATION

NOTE: The heater and A/C blower motor is located under the right side of the dash.

1. Disconnect the negative battery cable.
2. Remove the right–side dashboard lower trim panel.
3. Disconnect the blower motor wiring harness.
4. Remove the blower mounting screws and remove the blower motor assembly.
5. Remove the retaining clip or nut; then, remove the cage from the motor.

To install:

6. Install the cage onto the motor, and install the retaining clip or nut.
7. Install the blower motor into its housing and install the mounting screws.
8. Connect the wiring harness to the blower motor.
9. Install the right–side dashboard lower trim panel.
10. Connect the negative battery cable.
11. Test the blower motor operation at all speeds.

Wiper Motor

REMOVAL AND INSTALLATION

Front Windshield Wiper Motor

1. Disconnect the negative battery cable.
2. Detach the wiring connector from the wiper motor.
3. Remove the nut securing the wiper motor shaft to the linkage.
4. Remove the four mounting bolts.
5. Remove the wiper motor.

To install:

6. Connect the motor shaft to the linkage. Tighten the nut to 10 ft. lbs. (14 Nm).
7. Install the wiper motor and the four mounting nuts.
8. Connect the wiring connector to the wiper motor.
9. Remove both wiper arms and connect the negative battery cable.

10. Operate the wiper motor to set the wiper arm shafts into the park position.
11. Reinstall the wiper arms so that they are in the park position.
12. Test the operation of the windshield wipers.

Rear Window Wiper Motor

1. Disconnect the negative battery cable.
2. Remove the tailgate trim panel.
   a. Unscrew and remove the lock button.
   b. Pop up the two cover plugs and remove the screws.
   c. Carefully pry up along the outer edges of the trim panel to release the plastic spring clips.
3. Flip up the nut cover on the wiper arm. Remove the nut and remove the wiper arm.
4. Detach the wiring connector from the wiper motor.
5. Remove the three mounting bolts and remove the wiper motor.

To install:

6. Install the wiper motor to the tailgate. Install the mounting nuts.
7. Install the wiper motor shaft nut. Tighten the nut to 5 ft. lbs. (6 Nm).
8. Connect the wiring connector to the wiper motor.
9. Connect the negative battery cable.
10. Operate the rear wiper motor to set the wiper arm shaft in the park position.
11. Install the wiper arm. Tighten the nut to 7 ft. lbs. (9 Nm).
12. In the parked position, the wiper blade should rest 0.79 in. (20mm) from the bottom of the glass.
13. Install the tailgate trim panel.

Headlight Wiper Motor

1. Make sure the headlight wipers are in the park position.

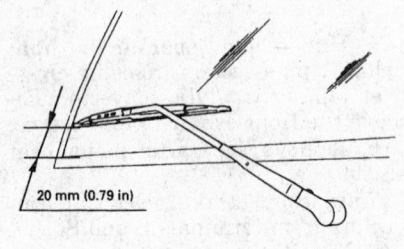

326360

Rear wiper arm in the park position

2. Disconnect the negative battery cable.
3. Remove the headlight bulb:
   a. Disconnect the headlight wiring connectors.
   b. Unscrew and remove the bulb retainer.
   c. Pull the bulb out of the headlight shell.

WARNING

Don't touch halogen bulbs with your bare hands. Carefully handle the bulbs with a clean, dry cloth. If the bulbs are touched, they will crack and burn out very quickly.

---

4. Remove the turn signal lens screw, and then pull the turn signal lens off of its mounting pegs.
5. Lift up the headlight wiper arm nut cover, and then remove the nut. Disconnect the headlight washer tube and then remove the wiper arm.
6. Remove the grille. It is secured with five clips and two screws.
7. Remove the nose panel section from below the headlight. It is held in place by two screws.
8. Remove the two screws and two nuts which attach the headlight assembly bracket to the nose panel. Disconnect the adjustment spring. Remove the headlight shell and mounting bracket as an assembly.
9. Uncouple the headlight wiper motor wiring harness.
10. Remove the headlight wiper motor.

To install:

11. Install the headlight wiper motor. Reconnect the wiring harness.
12. Install the headlight assembly. Make sure the adjustment spring is installed correctly.
13. Install the nose panel section and its two screws.
14. Install the grille and its clips and screws.
15. Install the turn signal lenses.
16. Install the headlight wiper arm. Reconnect the washer tube. Install the wiper arm nut and tighten it to 4 ft. lbs. (5 Nm).
17. Install the headlight bulb and retainer. Reconnect the wiring connector.
18. Reconnect the negative battery cable.
19. Test the operation of the headlight wipers and washers.
20. Test the operation of the headlights and turn signals.
21. Check the headlight aim with an aim tester. Adjust the headlight aim if necessary.

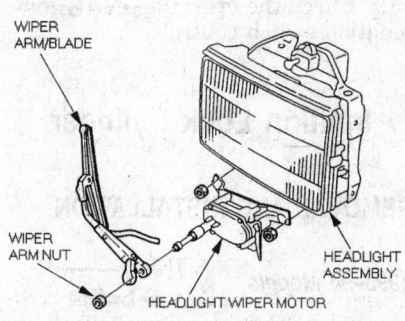

Headlight wiper motor and arm

Headlight Switch

REMOVAL AND INSTALLATION

1993–94 Models

1. Disconnect the negative battery cable.
2. Remove the eight instrument panel mounting screws and the instrument panel.
3. Pull the knob from the headlight switch.
4. Release the two locking tabs for the switch bezel from the rear of the instrument panel.
5. Remove the two mounting screws and two locking tabs.
6. Remove the headlight switch from the instrument panel.

To install:
7. Install the headlight switch into the instrument panel.
8. Install the switch bezel and the knob.
9. Install the instrument panel.
10. Connect the negative battery cable.

Combination Switch

REMOVAL AND INSTALLATION

1993–94 Models

1. Disconnect the negative battery cable.
2. Remove the steering wheel.
3. Remove the steering column upper and lower covers.
4. Disconnect the combination switch electrical connector near the base of the steering column.
5. Remove the combination switch mounting screws and remove the combination switch.

To install:
6. Install the combination switch to the steering column and secure it with screws.
7. Connect the combination switch electrical connector.
8. Install the steering column cover.
9. Install the contact ring.
10. Install the steering wheel.
11. Connect the negative battery cable.

1995–97 Models

> **CAUTION**
> *The SRS/air bag system must be disabled before the steering wheel and combination switch are removed. Failure to disable the air bag system may result in unnecessary system repairs and possible personal injury. Do not damage, cut, or attempt to alter the yellow SRS wiring harness.*

1. Remove the air bag module from the vehicle and place it face–up on a bench away from your work area.

NOTE: For information regarding the air bag and steering wheel, refer to the appropriate procedures in this section.

> **CAUTION**
> *Always carry a live air b... its trim cover and c... pointed away from your... When placing a live air bag... bench or other surface, al... point the trim cover up, a... from the surface. Following th... precautions will lessen t... chance of personal injury if th... air bag accidentally deploys.*

2. Remove the steering wheel.

> **WARNING**
> **The puller bolts may damage the air bag cable reel if they are threaded too deeply into the steering wheel.**

3. Remove the shift lever trim console.
4. Remove the radio trim piece. It is secured by two screws and four clips.
5. Unbolt the hood latch bracket from the lower dashboard cover.
6. Remove the lower dashboard cover and driver's knee bolster.
7. Remove the upper and lower steering column covers. Be careful with the wiring harness that runs through the lower cover — it contains the SRS wiring.
8. Remove the retaining screws and then remove the combination switch and SRS cable reel assembly from the steering column.

NOTE: The combination switch and cable reel assembly cannot be disassembled. They are serviced and replaced as one complete assembly.

To install:
9. Install the combination switch and cable reel assembly onto the steering column shaft.
10. Reconnect the switch wiring harnesses.
11. Install the lower steering column cover, making sure that the SRS

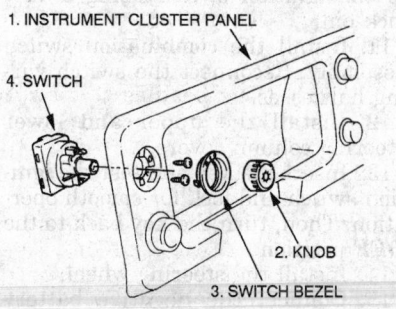

1. INSTRUMENT CLUSTER PANEL
4. SWITCH
2. KNOB
3. SWITCH BEZEL

183141

Headlight switch mounting — 1993–94 vehicles

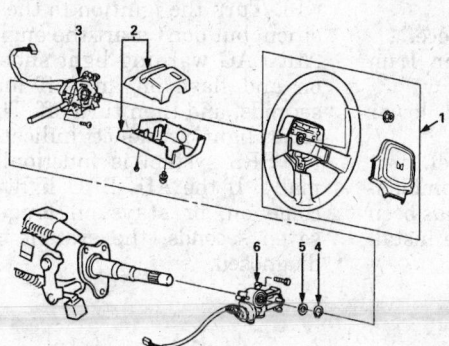

1. Steering wheel
2. Steering cowl
3. Combination switch
4. Snap ring
5. Bushing
6. Steering lock and bearing

328549

Ignition switch and combination switch — 1993–94 vehicles (without air bags)

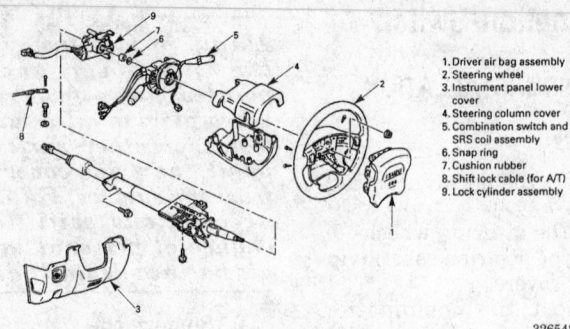

1. Driver air bag assembly
2. Steering wheel
3. Instrument panel lower cover
4. Steering column cover
5. Combination switch and SRS coil assembly
6. Snap ring
7. Cushion rubber
8. Shift lock cable (for A/T)
9. Lock cylinder assembly

326548

Ignition switch and combination switch — 1995–97 vehicles (with air bags)

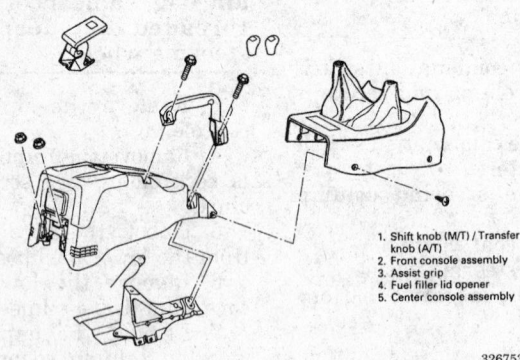

1. Shift knob (M/T) / Transfer knob (A/T)
2. Front console assembly
3. Assist grip
4. Fuel filler lid opener
5. Center console assembly

326753

Shift console components — 1995–97 vehicles

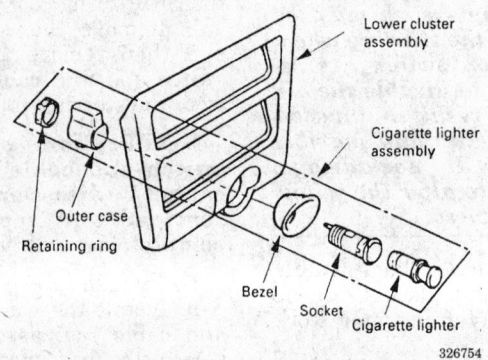

Lower cluster assembly

Cigarette lighter assembly

Outer case

Retaining ring

Bezel

Socket

Cigarette lighter

326754

Radio trim piece — 1995–97 vehicles

wiring is routed correctly and not pinched. Install the upper steering column cover.

12. Install the radio trim piece.

13. Install the shift lever trim console.

14. Install the dashboard lower cover.

15. Install the steering wheel.

16. Verify that none of the combination switch or air bag wiring has been pinched or kinked during the installation process.

17. Install the air bag onto the steering wheel.

18. Turn the ignition to the **ON** position, but don't start the engine. The AIR BAG warning light should turn on and flash on and off for seven seconds, and then turn off. This indicator light sequence indicates that the SRS system is functioning normally. If the AIR BAG light doesn't come on, or stays on longer than seven seconds, the system must be diagnosed.

19. Check the operation of the combination switch controls.

Ignition Lock Cylinder

REMOVAL AND INSTALLATION

1993–94 Models

1. Disconnect the negative battery cable.

2. Remove the steering wheel.

3. Remove the upper and lower steering column covers.

4. Unscrew, disconnect, and remove the combination switch assembly.

5. If equipped with an automatic transmission, disconnect the shift lock cable from the ignition lock cylinder.

NOTE: For information regarding the steering wheel, refer to the appropriate procedures in this section.

6. Unscrew the starter switch from the lock cylinder. Disconnect the starter switch wiring harness.

7. Unbolt the ignition lock cylinder from the steering column. If equipped with shear bolts, use a center punch to mark the bolt heads before drilling them out. Remove the snapring and the bushing to remove the lock cylinder from the steering column.

To install:

8. Install the lock cylinder onto the steering column. Install and evenly tighten the bolts. Tighten shear bolts until their heads snap off. Install a new bushing and snapring.

9. Reconnect the starter switch to the lock cylinder. Connect the wiring harness.

10. Connect the A/T shift lock cable to the ignition switch using a new lock pin.

11. Install the combination switch assembly. Reconnect the switch wiring harnesses.

12. Install the upper and lower steering column covers.

13. Insert the key and test the ignition switch and lock for smooth operation. Then, turn the key back to the **OFF** position.

14. Install the steering wheel.

15. Connect the negative battery cable.

16. Check the ignition switch and the combination switch controls.

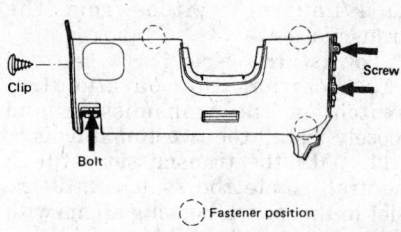

Dashboard lower trim panel — 1995-97 vehicles

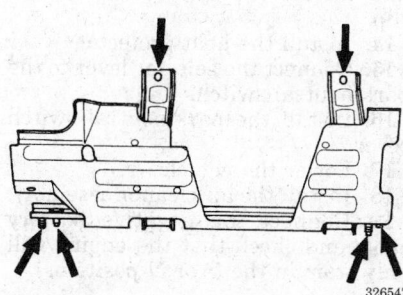

Driver's knee bolster — 1995-97 vehicles

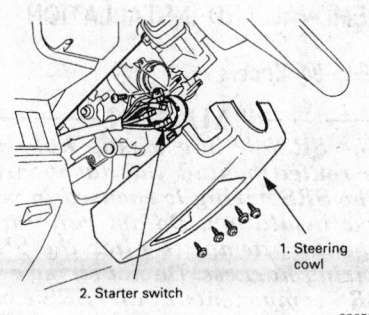

Ignition switch component assembly — 1993-94 vehicles

1995-97 Models

— CAUTION —

The SRS/air bag system must be disabled before the steering wheel and ignition switch are removed. Failure to disable the air bag system may result in unnecessary system repairs and possible personal injury. Do not damage, cut, or attempt to alter the yellow SRS wiring harness.

NOTE: For information regarding the air bag and steering wheel, refer to the appropriate procedures in this section.

1. Remove the air bag module from the vehicle and place it face-up on a bench away from your work area. (Refer to the air bag removal and installation procedure later in this section.

2. Remove the steering wheel.

NOTE: The puller bolts may damage the air bag cable reel if they are threaded too deeply into the steering wheel.

3. Remove the shift lever center console.

4. Remove the radio trim piece. It is held in place by two screws and four clips.

5. Remove the lower dashboard cover and disconnect the dimmer switch.

6. Remove the upper and lower steering column covers. Be careful with the wiring harness that runs through the lower cover: it contains the SRS wiring.

7. Unscrew and remove the combination switch and SRS cable reel assembly. The switch and cable reel assembly cannot be disassembled.

8. If equipped with an automatic transmission, disconnect the shift lock cable from the ignition lock cylinder.

9. Unscrew the starter switch from the ignition lock body and disconnect its wiring harness from the fuse box.

10. Unbolt the ignition lock cylinder from the steering column. If the lock cylinder is equipped with shear bolts, mark the bolt heads with a center punch before drilling them out. Remove the snapring and the

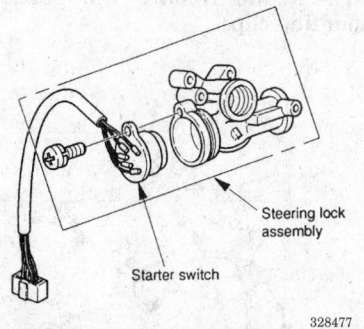

Ignition starter switch and lock cylinder — 1995-97 vehicles

bushing to remove the lock cylinder from the steering column.

To install:

11. Install the lock cylinder onto the steering column. Install and evenly tighten the bolts. Tighten shear bolts until their heads snap off. Install a new bushing and snapring.

12. Reconnect the starter switch to the lock cylinder. Connect the wiring harness.

13. Connect the A/T shift lock cable to the ignition switch using a new lock pin.

14. Install the combination switch and cable reel assembly. Reconnect the switch wiring harnesses.

15. Install the lower steering column cover, making sure that the SRS wiring is routed correctly and not pinched. Install the upper steering column cover.

16. Reconnect the dimmer switch and install the dashboard lower cover.

17. Install the radio trim piece.

18. Install the shift lever center console.

19. Check the alignment of the SRS cable reel.

   a. Turn the cable reel clockwise to its fully locked position. Don't turn the cable reel past the point at which you begin to feel resistance to its rotation.

   b. Turn the cable reel about three turns in the opposite direction until the pointer on the cable reel is in alignment with the neutral mark.

20. Insert the key and test the ignition switch and lock for smooth operation. Then, turn the key back to the **OFF**.

21. Install the steering wheel.

22. Connect the air bag lead and install the air bag module.

23. Reconnect the positive and negative battery cables.

24. Turn the ignition to the **ON** position, but don't start the engine. The AIR BAG warning light should turn on and flash on and off for seven seconds, and then turn off. This indicator light sequence indicates that the SRS system is functioning normally. If the AIR BAG light doesn't come on, or stays on longer than seven seconds, the system must be diagnosed.

25. Check the operation of the ignition switch and combination switch controls.

Ignition Switch

REMOVAL AND INSTALLATION

1993–94 Models

1. Disconnect the negative battery cable from the battery.
2. Remove the lower steering column cover by removing the five screws.
3. Disconnect the connector to the ignition switch.
4. Remove the ignition switch by removing the screw.

To install:

5. Install the ignition switch to the steering column with the screw.
6. Connect the electrical connector.
7. Install the lower steering column cover and install the five screws.
8. Connect the negative battery cable to the battery.

1995–97 Models

——————— CAUTION ———————

The SRS/air bag system must be disabled before the ignition switch is removed. Failure to disable the air bag system may result in unnecessary system repairs and possible personal injury. Do not damage, cut, or attempt to alter the yellow SRS wiring harness.

1. Turn the steering wheel so that the vehicle's front wheels are pointing straight ahead.
2. Turn the ignition switch to the **LOCK** position. Remove the key.
3. Disconnect the negative and positive battery cables. Wait at least five minutes before working around the air bags.
4. Remove the shift lever center console.
5. Remove the radio trim piece, it is secured by two screws.
6. Remove the lower dashboard cover and disconnect the dimmer switch.
7. Remove the upper and lower steering column covers. Be careful with the wiring harness that runs through the lower cover: it contains the SRS wiring.
8. Unscrew the starter switch from the ignition lock body and disconnect its wiring harness from the fuse box.

To install:

9. Reconnect the starter switch to the lock cylinder. Connect the wiring harness.
10. Install the lower steering column cover, making sure that the SRS wiring is routed correctly and not pinched. Install the upper steering column cover.
11. Install the lower dashboard cover and reconnect the dimmer switch.
12. Install the radio trim piece.
13. Install the shift lever center console.
14. Reconnect the yellow 3–way SRS harness connector at the base of the steering column.
15. Reconnect the positive and negative battery cables.
16. Turn the ignition to the **ON** position, but don't start the engine. The AIR BAG warning light should turn on and flash on and off for seven seconds, and then turn off. This indicator light sequence indicates that the SRS system is functioning normally. If the AIR BAG light doesn't come on, or stays on longer than seven seconds, the system must be diagnosed.
17. Check the operation of the ignition switch.

Park/Neutral Safety Switch

REMOVAL AND INSTALLATION

1. Disconnect the negative battery cable.
2. Place the gear selector lever in the **N** position.
3. Remove the air cleaner assembly.
4. Raise and support the vehicle safely.
5. Remove the park/neutral switch cover.
6. Disconnect the selector lever from the park/neutral switch.
7. Remove the heat protector from the transmission case.
8. Disconnect the black switch electrical connector from the engine harness and remove the harness mounting clip.

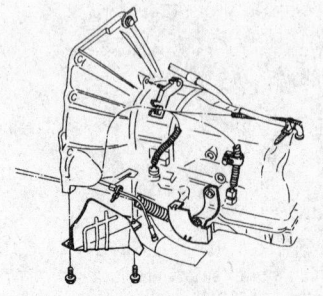

320650

Park / Neutral switch and heat shield

9. Remove the two 10mm mounting bolts and remove the park/neutral switch from the transmission.

To install:

10. Position the park/neutral switch on the transmission and loosely install the two 10mm bolts.
11. With the transmission still in neutral, rotate the switch until the slot in the switch housing aligns with the selector shaft bushing, and insert a drill bit or punch into the slot. Tighten the bolts to 10 ft. lbs. (13 Nm).
12. Remove the drill bit or punch.
13. Connect the electrical connector and install the harness mounting clip.
14. Install the heat protector.
15. Connect the selector lever to the park/neutral switch.
16. Install the park/neutral switch cover.
17. Lower the vehicle.
18. Install the air cleaner assembly.
19. Connect the negative battery cable and check that the engine will only start in the **P** or **N** positions.

Powertrain Control Module

REMOVAL AND INSTALLATION

1993–94 Models

——————— CAUTION ———————

The SRS/air bag wiring harness is routed behind the dashboard. The SRS wiring is encased in yellow insulation. Do not cut, damage, or attempt to alter the SRS wiring harness. Do not strike any SRS components. If any SRS component must be disconnected, the system must first be disabled.

1. Disconnect the negative and positive battery cables. Wait at least five minutes before working around the air bags.
2. Remove the shift lever knobs. If equipped with an automatic transmission, only the transfer case shift lever knob must be removed.
3. Remove the four shift lever console screws. Lift the console up and disconnect the electrical switches.
4. Remove the two screws under the top edge of the radio trim panel. Pull the trim panel out to detach the four clips.
5. Unbolt the hood latch bracket from the driver's side dashboard lower cover. Remove the lower cover.
6. Remove the air conditioner/heater transfer duct.

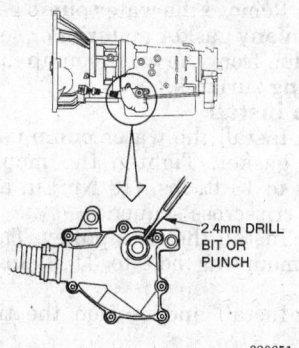

2.4mm DRILL
BIT OR
PUNCH

320651

Park / Neutral switch adjustment

7. If equipped, unbolt the anti-lock brake controller and move it out of the way.

8. Remove the four mounting bolts on the control module mounting bracket.

9. Disconnect the electrical connectors at the control module.

10. Remove the control module from under the dash panel with the bracket still attached.

To install:

11. Install the control module and bracket under the dash.

12. Connect the electrical connectors to the control module.

13. Install the anti-lock brake controller into its position if it was removed.

14. Install the air conditioner/heater transfer duct.

15. Install the driver's side dashboard lower cover. Reconnect the hood latch bracket.

16. Reconnect the cigarette lighter if it was disconnected. Install the radio trim panel and its two retaining screws.

17. Fit the shift lever console back into place and reconnect any switches. Install the four retaining screws and the shift knobs.

18. Connect the positive and negative battery cables.

19. Turn the ignition to the **ON** position, but don't start the engine. The AIR BAG indicator light should turn on, flash on and off for seven seconds, and then turn off. This sequence indicates that the SRS system is functioning normally. If the light doesn't follow this sequence, the system fault must be diagnosed.

1995–97 Models

────── **CAUTION** ──────
The SRS/air bag wiring harness is routed through the dashboard. The SRS wiring is encased in yellow insulation. Do not cut, damage, or strike the SRS wiring or any SRS components. If any SRS component must be disconnected, the system must first be disabled.

1. Disconnect the negative and positive battery cables. Wait at least five minutes before working around the air bags.

2. If equipped with an automatic transmission, engage the parking brake and shift the selector lever into the **N** position.

3. Remove the shift lever knobs. If equipped with an automatic transmission, only the transfer case shift lever knob must be removed.

4. Remove the four shift lever console screws. Lift the console up and disconnect the electrical switches.

5. Label and disconnect the red, white, and blue wiring harness connectors from the back of the Powertrain Control Module.

6. Remove the three screws from the Powertrain Control Module mounting brackets.

7. Remove the Powertrain Control Module.

To install:

8. Install the Powertrain Control Module and its three mounting screws.

9. Reconnect the red, white, and blue connectors.

10. Verify that no dashboard wiring has been pinched or kinked during Powertrain Control Module installation.

11. Fit the shift lever console back into place and reconnect any switches. Install the for retaining screws and the shift knobs.

12. If equipped with an automatic transmission, shift the selector lever back into the **P** position.

13. Reconnect the positive and negative battery cables.

14. Turn the ignition to the **ON** position, but don't start the engine. The AIR BAG indicator light should turn on, flash on and off for seven seconds, and then turn off. This sequence indicates that the SRS system is functioning normally. If the light doesn't follow this sequence, the system fault must be diagnosed.

15. Check and diagnose any stored trouble codes.

ENGINE COOLING

Radiator

REMOVAL AND INSTALLATION

1993–94 Models

1. Disconnect the negative battery cable.

2. Drain the cooling system into a clean container so it can be reused.

3. Remove the upper, lower, and reservoir hoses from the radiator.

4. Disconnect the oil cooling lines from the radiator if equipped with an automatic transmission.

5. Remove the fan shroud attaching bolts and the shroud.

6. Remove the radiator attaching bolts and the radiator.

To install:

7. Install the radiator and its attaching bolts.

8. Install the fan shroud and the shroud attaching bolts.

9. Reconnect the radiator hoses.

10. Connect the oil cooling lines, if removed.

11. Refill the cooling system.

12. Connect the negative battery cable and check for coolant system leaks.

1995–97 Models

1. Disconnect the negative battery cable.

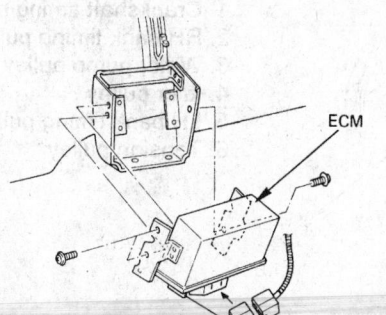

Removal steps
1. Negative battery cable
2. Lower trim panel at console
3. A/C Heater transfer tube
4. ABS controller electrical controller
5. ABS controller
6. Four fasteners for ECM bracket
7. ECM electrical connectors
8. ECM from under dash bracket still attached

ECM

327593

Engine control module removal — 1993–94 vehicles

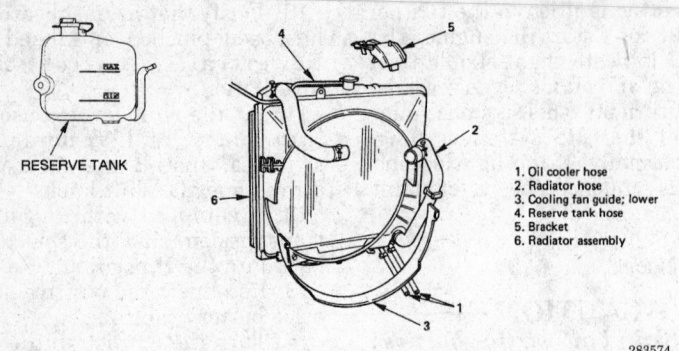

RESERVE TANK

1. Oil cooler hose
2. Radiator hose
3. Cooling fan guide; lower
4. Reserve tank hose
5. Bracket
6. Radiator assembly

283574

Radiator assembly — 1993–94 vehicles

2. Loosen the drain plug on the bottom of the radiator and drain the coolant.

3. If equipped with an automatic transmission, disconnect and plug the ATF cooler hoses.

4. Disconnect the upper and lower radiator hoses from the engine.

5. Disconnect the lower fan guide clips and the bottom lock, then remove the lower fan guide from the vehicle.

6. Disconnect the reserve tank hose from the radiator.

7. Remove the radiator bracket.

8. Lift the radiator from the vehicle with the hoses attached. Remove the cushions from the bottom of the radiator.

9. Remove the upper and lower hoses from the radiator.

To install:

10. Install the rubber cushions to the bottom of the radiator. Install the upper and lower radiator hoses to the radiator.

11. Carefully lower the radiator into the vehicle, take care not to damage the radiator or the fan blades.

12. Install the radiator bracket.

13. Connect the reserve tank hose to the radiator.

14. Install the lower fan guide, make sure that the clips are fully engaged.

15. Connect the upper and lower radiator hoses to the engine.

16. If equipped with a automatic transmission, connect the ATF cooler hoses.

17. Refill and bleed the air from the cooling system.

18. Connect the negative battery cable.

19. Warm up the vehicle and check the operation of the cooling system and heater. Check the transmission fluid level on automatic–equipped vehicles and add ATF if necessary.

Water Pump

REMOVAL AND INSTALLATION

1. Disconnect the negative battery cable.

2. Drain the engine coolant into a clean container.

3. Remove the upper radiator hose.

4. Remove the timing belt and the idler pulley. The timing belt must be replaced if it has been contaminated by oil or coolant.

5. Unbolt and remove the water pump.

6. Remove the water pump gasket. Clean any gasket material or sealant residue from the water pump mating sealing surfaces.

To install:

7. Install the water pump using a new gasket. Tighten the mounting bolts to 13 ft. lbs. (18 Nm) in a two-step crisscross sequence.

8. Install the idler pulley. Tighten the mounting bolt to 31 ft. lbs. (42 Nm).

9. Install and tension the timing belt.

10. Install the upper radiator hose.

11. Refill and bleed the cooling system.

12. Connect the negative battery cable. Start the engine and check for coolant leaks.

Thermostat

REMOVAL AND INSTALLATION

1. Disconnect the negative battery cable.

2. Drain the coolant into a clean container so it can be reused.

3. Disconnect the upper radiator hose from the thermostat housing.

4. Unbolt and remove the thermostat housing. Remove the thermostat and its gasket.

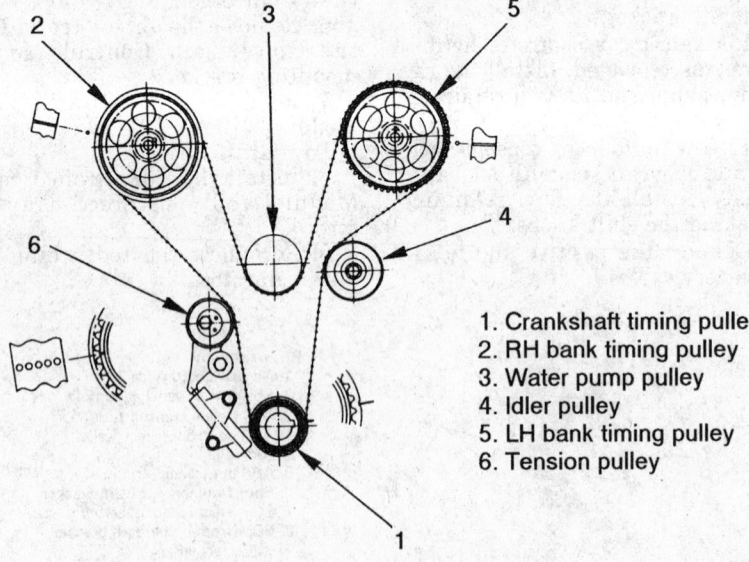

1. Crankshaft timing pulley
2. RH bank timing pulley
3. Water pump pulley
4. Idler pulley
5. LH bank timing pulley
6. Tension pulley

323878

Belt routing

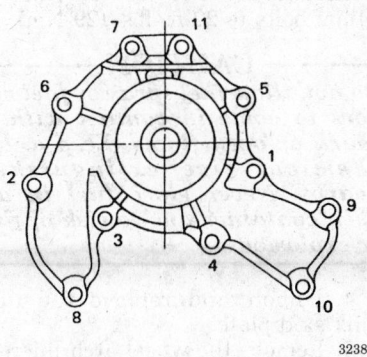

1. Timing belt
2. Idler pulley
3. Water pump assembly
4. Gasket

323876

Water pump assembly

Water pump bolt tightening sequence

323877

Cooling Fan

REMOVAL AND INSTALLATION

1. Disconnect the negative battery cable.

2. Remove the nuts attaching the cooling fan to the fan pulley assembly.

3. Remove the fan from the fan clutch.

To install:

4. Install the fan on the fan clutch.

5. Install the fan clutch to the cooling fan pulley assembly. Torque the nuts to 6 ft. lbs. (8 Nm).

6. Connect the negative battery cable.

Cooling System Bleeding

PROCEDURE

1. Fill the radiator with a 50/50 mixture of coolant and water to just below the radiator filler neck. Do not use more than 60% coolant, or the efficiency of the cooling system will be impaired.

2. Fill the reserve tank to the MAX line with the coolant mixture.

3. With the radiator cap off, warm the engine to operating temperature. The heater controls should be set to maximum heat. Add coolant as needed while the engine is warming up.

4. When the thermostat opens (upper hose hot and visible coolant flow in radiator) top off the coolant and install the radiator cap.

5. Refill the reserve tank to the proper level.

6. Shut off the engine and allow it to cool. As the engine cools it will draw coolant from the reserve tank.

NOTE: Any air in the cooling system will be released through the reservoir bottle during normal engine operation.

7. Refill the reserve tank to the MAX line after the engine has cooled.

To install:

5. Clean the sealing surface. Use a new gasket and install the thermostat with the spring toward the engine. Torque the thermostat housing mounting bolts to 14 ft. lbs. (19 Nm).

6. Connect the radiator hose to the thermostat housing.

7. Refill and bleed the cooling system.

8. Connect the negative battery cable.

9. Start the engine and add coolant as needed until the level stabilizes below the radiator cap.

10. Install the radiator cap and check for leaks.

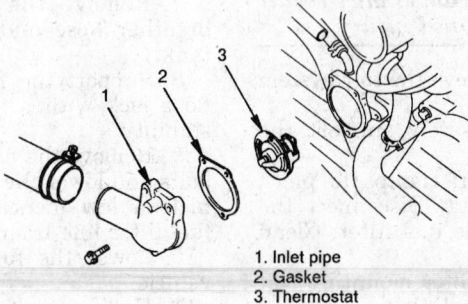

1. Inlet pipe
2. Gasket
3. Thermostat

320577

Thermostat assembly

FUEL SYSTEM

Fuel System Service Precautions

Safety is the most important factor when performing not only fuel system maintenance but any type of maintenance. Failure to conduct maintenance and repairs in a safe manner may result in serious personal injury or death. Maintenance and testing of the vehicle's fuel system components can be accomplished safely and effectively by adhering to the following rules and guidelines.

• To avoid the possibility of fire and personal injury, always disconnect the negative battery cable unless the repair or test procedure requires that battery voltage be applied.

• Always relieve the fuel system pressure prior to disconnecting any fuel system component (injector, fuel rail, pressure regulator, etc.), fitting or fuel line connection. Exercise extreme caution whenever relieving fuel system pressure to avoid exposing skin, face and eyes to fuel spray. Please be advised that fuel under pressure may penetrate the skin or any part of the body that it contacts.

• Always place a shop towel or cloth around the fitting or connection prior to loosening to absorb any excess fuel due to spillage. Ensure that all fuel spillage (should it occur) is quickly removed from engine surfaces. Ensure that all fuel soaked cloths or towels are deposited into a suitable waste container.

• Always keep a dry chemical (Class B) fire extinguisher near the work area.

• Do not allow fuel spray or fuel vapors to come into contact with a spark or open flame.

• Always use a backup wrench when loosening and tightening fuel line connection fittings. This will prevent unnecessary stress and torsion to fuel line piping. Always follow the proper torque specifications.

• Always replace worn fuel fitting O-rings with new. Do not substitute fuel hose or equivalent, where fuel pipe is installed.

Fuel System Pressure

RELIEVING

1. Remove the fuel filler cap.
2. Remove the fuel pump relay from the underhood relay box.
3. Start the engine and let it run until it stalls, then crank the engine for an additional 30 seconds.
4. Turn the ignition switch to the **OFF** position and remove the key. Disconnect the negative battery cable and then install the fuel pump relay.

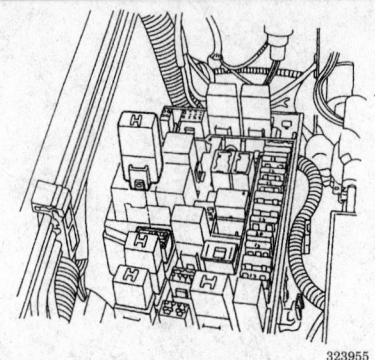

323955

Fuel pump relay location — 1995–97 vehicles

Idle Speed

ADJUSTMENT

The idle speed is controlled by the Powertrain Control Module. It is not adjustable. The correct idle speed is 750 RPM for vehicles equipped with manual or automatic transmissions.

Mixture

ADJUSTMENT

The air/fuel mixture is computer controlled according to the needs of the engine and is not adjustable. If the air/fuel mixture is too lean or too rich, other problems with the engine and/or engine control system exist.

Fuel Filter

REMOVAL AND INSTALLATION

——— **CAUTION** ———
Fuel injection systems remain under pressure even after the engine has been turned off. The fuel system pressure must be relieved before disconnecting any fuel lines. Failure to do so may result in fire and personal injury.

1. Properly relieve the fuel system pressure.
2. Raise and safely support the vehicle.
3. Use block–off clamps to pinch the fuel lines shut. Disconnect the fuel lines from the fuel filter. Clean up any fuel spills.
4. Loosen the filter mounting bolt and remove the fuel filter.
   To install:
5. Install the fuel filter with the label, arrow, or manufacturer's mark

facing toward the front of the vehicle and tighten the bracket bolt.
6. Connect the fuel lines to the fuel filter.
7. Lower the vehicle to the floor and install the filler cap.
8. Reconnect the negative battery cable.
9. Start the engine and inspect the fuel filter connections for leaks.

Fuel Pump

REMOVAL AND INSTALLATION

——— **CAUTION** ———
Fuel injection systems remain under pressure after the engine has been turned OFF. Properly relieve fuel pressure before disconnecting any fuel lines. Failure to do so may result in fire or personal injury.

1. Properly relieve the fuel pressure.
2. Raise and support the vehicle safely.
3. Drain the fuel into an approved sealable container. Install the drain plug with a new washer and tighten 8mm bolts to 14 ft. lbs. (20 Nm), or 14mm bolts to 22 ft. lbs. (29 Nm).

——— **CAUTION** ———
Do not allow fuel spray or fuel vapors to come in contact with a spark or open flame. Keep a dry chemical fire extinguisher nearby. Never store fuel in an open container due to risk of fire or explosion.

4. Unbolt and remove the fuel tank skid plate.
5. Remove the wheel arch liner.
6. Disconnect all fuel lines and wiring connectors from the fuel pump. Plug the fuel lines to prevent leakage. Immediately clean up any fuel spills.
7. Remove the filler neck and breather hose and clamp from the tank.
8. Support the fuel tank using a floor jack with a wooden plank for stability.
9. Remove the fuel tank mounting bolts and lower the tank from the vehicle a few inches. Disconnect the fuel filter line from the pump.
10. Lower the fuel tank from the vehicle.
11. Unbolt the fuel pump bracket plate. Lift the fuel pump assembly out of the tank and allow any fuel to drain into the tank.

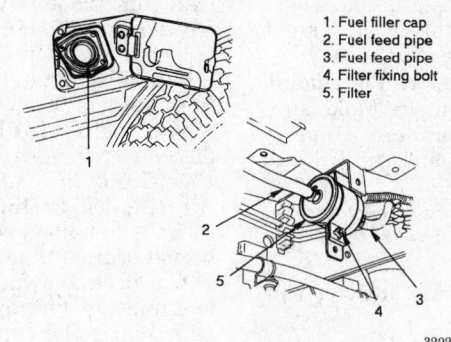

1. Fuel filler cap
2. Fuel feed pipe
3. Fuel feed pipe
4. Filter fixing bolt
5. Filter

320201

Fuel filter mounting

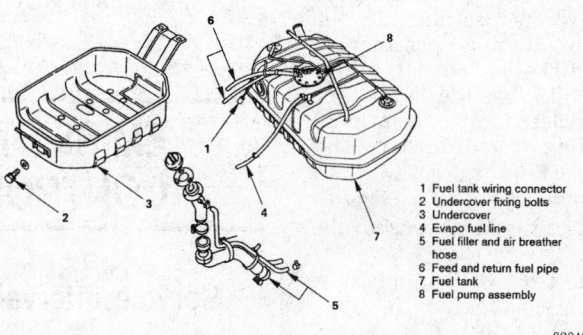

1. Fuel tank wiring connector
2. Undercover fixing bolts
3. Undercover
4. Evapo fuel line
5. Fuel filler and air breather hose
6. Feed and return fuel pipe
7. Fuel tank
8. Fuel pump assembly

320486

Fuel tank and pump components

12. Remove the fuel pump. If the pump is out of the tank for a long period of time, cover the opening to keep out moisture and foreign objects.

To install:

13. Install the fuel pump assembly into the tank with a new gasket. Evenly tighten the bracket plate bolts to prevent leakage.

14. Raise the tank and connect the hoses.

15. Connect the fuel lines to their ports on the pump.

16. Connect the breather hose and clamp.

17. Connect the wiring connectors.

18. Connect the fuel filler line and breather hose.

19. Raise the tank into position. Install the fuel tank mounting bolts and tighten them to 27–30 ft. lbs. (36–39 Nm). Lower the jack.

20. Install the fuel tank skid plate and tighten its bolts to 27 ft. lbs. (37 Nm).

21. Install the wheel arch liner.

22. Lower the vehicle safely.

23. Refill the fuel tank.

24. Install the fuel pump relay. Reconnect the negative battery cable.

25. Turn the ignition to the **ON** position to pressurize the fuel system. Then, check for leaks and proper fuel system operation.

Fuel Injector

REMOVAL AND INSTALLATION

1995–97 Models

──────── **CAUTION** ────────

Fuel injection systems remain under pressure even after the engine has been turned off. The fuel system pressure must be relieved

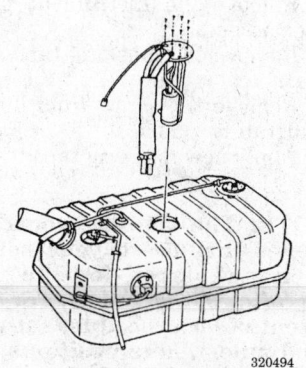

320494

Fuel pump assembly

before disconnecting any fuel lines. Failure to do so may result in fire and personal injury.

1. Properly relieve the fuel system pressure.

2. Disconnect the negative battery cable and reinstall the fuel pump relay.

3. Remove the air cleaner and air intake duct.

4. Disconnect the throttle cable from the throttle body linkage.

5. Label and disconnect the following vacuum hoses from the intake manifold chamber:
   a. PCV hose
   b. EVAP canister vacuum hose
   c. Brake booster hose

6. Label and disconnect the following sensor connectors from the rear of the intake manifold chamber:
   a. Ignition control module connectors.
   b. Linear EGR valve.
   c. MAP sensor.
   d. EVAP purge valve.
   e. Throttle position sensor.
   f. Idle air control valve.
   g. Intake air temperature sensor.

7. Disconnect the EGR valve supply tube and bracket.

8. Remove the throttle body.

9. Disconnect the MAP sensor tube, and then unbolt the MAP sensor bracket.

10. Unbolt the intake manifold chamber from its brackets which are located at its front and rear edges.

11. Loosen the six manifold mounting bolts and two nuts in a crisscross sequence.

12. Remove the bolts and nuts, and then lift the chamber off of the base of the intake manifold. Note the positions of the long and short bolts.

13. Cover the intake manifold with a sheet of plastic or clean shop towels to keep out dirt and foreign objects.

14. Carefully clean any dirt from the fuel rail and fuel fittings.

15. Disconnect the fuel feed and return lines from the front of the fuel rail. Clean up any spilled fuel.

16. Label and disconnect the fuel injector wiring harness.

17. Unbolt the fuel return line bracket from the front of the intake manifold.

18. Unbolt the fuel rail from the intake manifold.

19. Carefully lift the fuel rail and injectors off of the intake manifold as an assembly.

20. Cover the fuel injector ports on the intake manifold to keep dirt out.

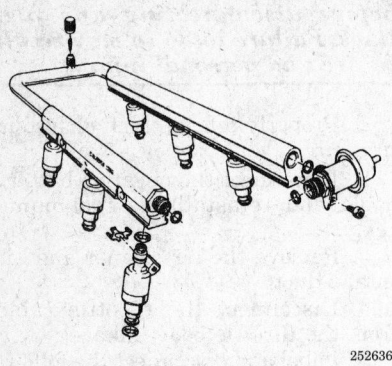

Fuel rail assembly

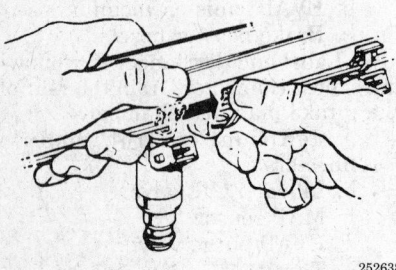

Slide the retainer to remove the fuel injector — 1993–94 vehicles

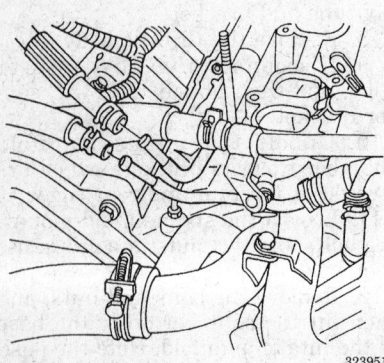

Fuel feed and return lines — 1995–97 vehicles

21. Remove the fuel injector retaining clips.

22. Remove the fuel injectors from the fuel rail.

23. Remove the O–rings from the fuel injectors.

To install:

24. Lubricate new O–rings with clean engine oil. Install them onto the fuel injectors.

25. Install the fuel injectors onto the fuel rail. Install new injector retaining clips.

26. Coat the intake manifold end of each injector with a thin coat of clean engine oil before installation. Don't submerge the injector in oil, or get any oil inside the injector.

27. Install the fuel rail assembly onto the intake manifold. Make sure all the fuel injectors are properly seated. Tighten the fuel rail bolts to 5.2 ft. lbs. (7 Nm).

28. Reconnect the fuel feed and return lines. Install the return line bracket bolt.

29. Reconnect the fuel injector wiring harness.

30. Install a new intake manifold chamber gasket. Install the intake manifold chamber.

31. Install the six intake manifold chamber bolts and two nuts. Torque the nuts and bolts in a crisscross sequence. For 1993–94 vehicles, torque them to 17 ft. lbs. (24 Nm). For 1995–97 vehicles, torque them to 15 ft. lbs. (20 Nm).

32. Install the intake manifold chamber bracket bolts.

33. Reconnect the MAP sensor tube and bracket.

34. Reconnect the EGR supply tube and bracket.

35. Install the throttle body with a new gasket. Tighten the bolts in a crisscross sequence. For 1993–94 vehicles, tighten the bolts to 16 ft. lbs. (22 Nm). For 1995–97 vehicles, tighten the bolts to 10 ft. lbs. (13.5 Nm).

36. Reconnect the following sensor connectors to the rear of the intake manifold chamber:

   a. Ignition control module connectors.

   b. Linear EGR valve.

   c. MAP sensor.

   d. EVAP purge valve.

   e. Throttle position sensor.

   f. Idle air control valve.

   g. Intake air temperature sensor.

37. Reconnect the following vacuum hoses to the intake manifold chamber:

   a. PCV hose.

   b. EVAP canister vacuum hose.

   c. Brake booster hose.

38. Reconnect the throttle cable to the throttle body linkage.

39. Install the air cleaner and air intake duct.

40. Check the condition of the engine oil. Leaky fuel injectors may have contaminated the engine oil with fuel. If necessary, install a new oil filter and refill the engine with fresh oil.

41. Reconnect the negative battery cable.

42. For 1993–94 vehicles, turn the ignition key to the **ON** position for two seconds, then **OFF**. Turn the ignition **ON** again to pressurize the fuel system and check for leaks where the fuel system was disconnected.

43. Crank the engine until it starts. Air trapped in the fuel lines may cause the engine to crank for a longer period of time than normal.

44. Check the fuel lines, fuel rail, and injectors for any signs of leakage.

45. Warm the engine up to normal operating temperature and check the operation of the throttle cable and linkage. Adjust if necessary.

46. Check the engine oil level.

EMISSIONS CONTROLS

Service Interval Lamp

RESETTING

Oxygen Sensor Life Indicator Light

1993–95

The oxygen sensor must be replaced after 90,000 miles of vehicle operation. When the odometer reading reaches 90,000 miles, the oxygen sensor life indicator light (O_2) will illuminate to remind the driver to change the oxygen sensor.

After replacing the oxygen sensor, the oxygen sensor life indicator light must be reset to remind the driver to replace the oxygen sensor after the next 90,000 miles. The reset screw is located in the back of the instrument cluster. Perform the reset procedure as follows:

1. Remove the Instrument panel cluster assembly.

2. Remove the masking tape from hole **B**.

3. Remove the screw from hole **A** and install it to hole **B**.

4. Apply new masking tape to hole **A**.

NOTE: The above procedure assumes that the oxygen sensor is being replaced for the first time (after 90,000 miles). For subsequent reset procedure (at next 90,000 miles), hole positions will be the opposite of the above procedure.

ENGINE MECHANICAL

Engine Assembly

REMOVAL AND INSTALLATION

NOTE: The transmission must be removed from the vehicle before the engine is removed. The transmission must safely supported.

1. Relieve the fuel pressure:
   a. Remove the fuel filler cap.
   b. Remove the fuel pump relay from the underhood relay box.
   c. Start the engine and let it run until it stalls. Then, crank the engine for an additional 30 seconds.
   d. Turn the ignition switch to the **OFF** position.
2. Disconnect the negative and positive battery cables. Remove the battery. Reinstall the fuel pump relay.
3. Use a felt–tipped marker to matchmark the hood hinges to the hood. Remove the hood.
4. Remove the radiator skid plates.
5. Drain the coolant into a sealable container.
6. Remove the air cleaner duct and hose. Use a clean shop cloth to cover the air cleaner port to prevent dirt from entering the engine.
7. Label and disconnect the necessary hoses, electrical connectors, control cables, and control rods from the engine.
8. Label and disconnect the following items:
   a. Air switch valve hose.
   b. Oxygen sensor harness.
   c. Vacuum switch valve hose.
   d. Thermal Vacuum Switching Valve (VSV) hose.

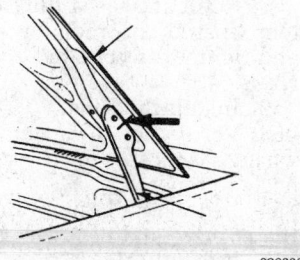

328330

Matchmark the hood and hinges

   e. Pressure regulator vacuum hose.
   f. Canister hose.
   g. Powertrain Control Module harness.
   h. Fuel inlet and return hoses.

— **CAUTION** —

Fuel injection systems remain under pressure after the engine has been turned OFF. Properly relieve fuel pressure before disconnecting any fuel lines. Failure to do so may result in fire or personal injury.

9. Remove the grille from the deflector panel.
10. Disconnect the upper and lower radiator hoses and the reservoir tank hose.
11. Remove the fan shroud, fan blade assembly, and the radiator.
12. If equipped with air conditioning, remove the compressor from the engine and move it aside; do not disconnect the pressure hoses.
13. If equipped with a manual transmission, remove the gear shift lever by performing the following procedures:
   a. Place the gear shift lever in **N**.
   b. Remove the front shift console.
   c. Pull the shift lever boot and grommet upward.
   d. Remove the shift lever cover bolts and the shift lever.
14. Remove the transfer shift lever by performing the following procedures:
   a. Place the transfer shift lever in **2H**.
   b. Pull the shift lever boot and dust cover upward.
   c. Remove the shift lever retaining bolts.
   d. Pull the shift lever from the transfer case.
15. Raise and safely support the vehicle. Remove the front wheels. Drain the oil from the engine.
16. Drain the transmission and transfer case fluid.
17. If equipped with an automatic transmission, perform the following procedures:
   a. Remove the dipstick and its tube.
   b. Disconnect the shift select control link rod from the select lever.
   c. Disconnect the downshift cable and vehicle speed sensor cable from the transmission.
   d. Disconnect and plug the fluid coolant lines from the transmission.

18. Matchmark the front and rear driveshaft flanges. Unbolt and remove the front and rear driveshafts.
19. Remove the starter.
20. If equipped with a clutch slave cylinder, remove it from the transmission and move it aside.
21. Remove the exhaust pipe-to-exhaust manifold nuts, the exhaust pipe bracket-to-transmission bolts, the front exhaust pipe-to-second exhaust pipe bolts and the front exhaust pipe from the vehicle.
22. Attach an engine hangers to the rear of the exhaust manifolds.
23. Connect a chain hoist to the engine hangers and support the engine.
24. Remove the transmission/transfer case assembly by performing the following procedures:
   a. Place a transmission jack under the transmission for support.
   b. Remove the rear transmission mount nuts.
   c. Remove the rear mount crossmember member and then remove crossmember and mount.
   d. Unbolt the transmission from the engine bolts.
   e. Move the transmission assembly rearward.
   f. Carefully lower the transmission from the vehicle.
25. Raise the chain hoist to support the weight of the engine. Unbolt the engine mounts and separate them from the engine.
26. Verify that all vacuum lines, hoses, and wiring harnesses have been disconnected.
27. Slowly lift the engine out of the vehicle with the chain hoist. Hold the front of the engine higher than the rear.
28. Place the engine on a work stand.

To install:

NOTE: Note the position of the transmission dowel pins before installation. If the dowels are installed in the incorrect position, the transmission case may crack.

29. Using the hoist, slowly lower the engine into the vehicle; be sure to hold the front of the engine higher than the rear.
30. Install the engine mount nuts bolts. Tighten the engine mount nuts to 30 ft. lbs. (41 Nm), and the bolts to 37 ft. lbs. (50 Nm).
31. Install the transmission/transfer assembly by performing the following procedures:
   a. Raise the transmission into position.
   b. Move the transmission forward and engage it with the en-

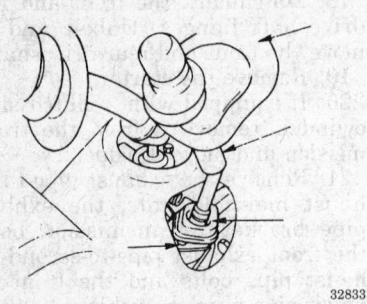

328331

Remove the shift levers from the transmission and transfer case

gine. Make sure the dowel pins engage with the holes on the engine block.

c. Install the transmission mounting bolts. Tighten the upper six bolts to 56 ft. lbs. (76 Nm). Tighten the two bolts on the lower right to 35 ft. lbs. (48 Nm), and the bolt on the lower left (A/T only) to 20 ft. lbs. (27 Nm). Tighten the remaining bolts to 4.4 ft. lbs. (6 Nm).

d. Install the transmission mount and crossmember. Tighten the mount bolts to 37 ft. lbs. (50 Nm). Tighten the front crossmember bolts to 58 ft. lbs. (78 Nm).

Tighten the third crossmember bolts to 37 ft. lbs. (50 Nm).

e. Remove the transmission jack.

32. Remove the engine hoist and the engine hanger from the rear of the exhaust manifold.

33. Install the front exhaust pipe, exhaust pipe-to-exhaust manifold nuts, the exhaust pipe bracket-to-transmission bolts, the front exhaust pipe-to-second exhaust pipe bolts. Tighten the manifold nuts to 49 ft. lbs. (67 Nm). Tighten the flange bolts to 32–37 ft. lbs. (43–50 Nm).

34. Install the clutch slave cylinder, if equipped. Tighten the slave cylinder bolts to 32 ft. lbs. (43 Nm).

35. Install the starter and tighten the mounting bolts to 30 ft. lbs. (40 Nm).

36. Install the front and rear driveshafts and tighten the flange bolts to 46 ft. lbs. (63 Nm).

37. If equipped with an automatic transmission, perform the following procedures:

a. Connect the fluid coolant lines to the transmission.

b. Connect the downshift cable to the transmission.

c. Connect the shift select control link rod to the select lever.

d. Install the dipstick and tube.

38. Connect the backup light switch connector and the vehicle speed sensor cable to the transmission.

39. Install the front wheels and lower the vehicle.

40. Install the transfer shift lever by performing the following procedures:

a. Position the shift lever into the transfer case.

b. Install the shift lever retaining bolts.

c. Push the dust cover and the shift lever boot downward.

41. If equipped with manual transmission, install the gear shift lever by performing the following procedures:

a. Install the shift lever and the shift lever cover bolts.

b. Push the grommet and shift lever boot downward.

c. Install the front console to the floor panel.

42. Install the power steering pump and bracket to the engine.

43. If equipped with air conditioning, install the compressor to the engine.

44. Install the radiator, the fan blade assembly, and the fan shroud.

45. Connect the upper and lower radiator hoses and the reservoir tank hose.

46. Install the grille to the deflector panel.

47. Connect the following items:

a. Air switch valve hose.

b. Oxygen sensor harness.

c. Vacuum switch valve hose.

d. Thermal Vacuum Switching Valve (VSV) hose.

e. Pressure regulator vacuum hose.

f. Canister hose.

g. Powertrain Control Module harness.

h. Fuel hose(s).

48. Connect the necessary hoses, electrical connectors, control cables and control rods to the engine.

49. Install the air cleaner duct and hose.

50. Refill the engine, the transmission, and the transfer case.

51. Refill and bleed the cooling system. Install the radiator skid plate and tighten its bolts to 27 ft. lbs. (37 Nm).

52. Install the hood.

53. Verify that all vacuum lines and wiring harnesses have been reconnected properly.

54. Install the battery and connect the positive and negative cables.

55. Adjust the accessory belt tensions.

56. Start the engine and check for fluid and fuel leaks.

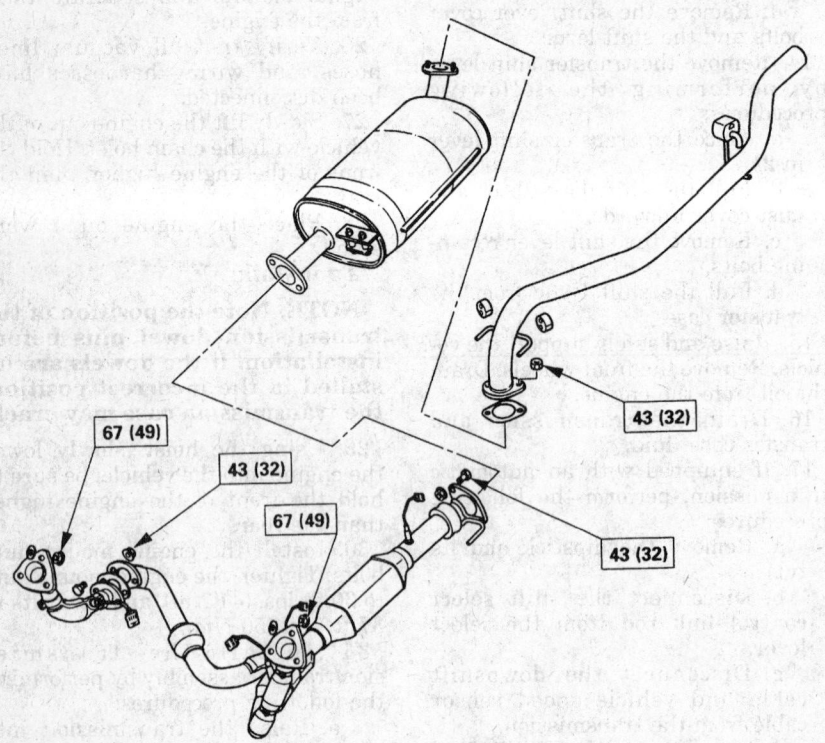

67 (49)   43 (32)   67 (49)   43 (32)   43 (32)

328373

Exhaust system components

Engine Mounts

REMOVAL AND INSTALLATION

Engine Mounts

1. Disconnect the negative battery cable.
2. Attach a chain hoist to the engine lifting hooks; or use a floor jack with a block of wood on its pad to lift the engine and remove the weight from its mounts.
3. Loosen the engine mount fasteners and remove the mounts from the engine block and frame rails.

------ **WARNING** ------
The engine may need to be moved with a pry bar to get the bolt holes to line up. Be careful when prying around the engine components. The engine must be securely supported.

To install:
4. Install the mount and replace the fasteners.
5. Tighten the mount bracket bolts and nuts to 30 ft. lbs. (41 Nm).
6. Remove the engine lifting equipment.
7. Reconnect the negative battery cable.
8. Run the engine and check the mounts for excess motion and any signs of looseness.

Transmission Mounts

1. Disconnect the negative battery cable.
2. Raise and safely support the vehicle.
3. Remove the transmission and transfer case skid plates.
4. Use a suitable transmission jack with a wooden plank on its pad to raise the transmission enough to take the weight off its mounts.
5. Remove the rear mount and its bracket.
To install:
6. Install the rear mount bracket and tighten its bolts to 30 ft. lbs. (41 Nm).
7. Install the rear mount and tighten its bolts to 37 ft. lbs. (50 Nm).
8. Remove the jack.
9. Install the skid plates and tighten their bolts to 27 ft. lbs. (37 Nm).

Cylinder Head

REMOVAL AND INSTALLATION

1993–95 3.2L (VIN V) Engines

Single Overhead Camshaft (SOHC) Engine

NOTE: Isuzu has issued a recall notice for 1993–1994 Troopers equipped with SOHC engines. This is campaign number 94V-094, and involves faulty camshaft end plugs. The plugs may dislodge from the cylinder heads and can cause rapid oil loss. Remember this when ordering parts: a service kit is available for affected vehicles.

------ **CAUTION** ------
Fuel injection systems remain under pressure even after the engine has been turned off. The fuel system pressure must be relieved before disconnecting any fuel lines. Failure to do so may result in fire and personal injury.

1. Matchmark the hood hinge to the hood and remove the hood.
2. Properly relieve the fuel system pressure.
3. Disconnect the negative battery cable.
4. Remove the air cleaner assembly.
5. Remove the upper cooling fan shroud.
6. Remove the cooling fan assembly.
7. Disconnect the accelerator cable from the throttle body and the bracket.
8. Disconnect the canister vacuum hose from the vacuum pipe.
9. Disconnect the air vacuum hose from the common chamber.
10. Disconnect the vacuum booster hose from the common chamber.
11. Disconnect the MAP sensor; Canister Vacuum Switching Valve (VSV); Exhaust gas recirculation VSV; Intake air temperature sensor and ground connectors.
12. Remove the spark plug wires from the cylinder head cover.
13. Remove the ignition control module assembly.
14. Remove the four bolts and the throttle body from the common chamber.
15. Disconnect the vacuum hoses from the throttle body.
16. Disconnect the positive crankcase ventilation hose from the common chamber.

17. Disconnect the fuel pressure control valve vacuum hose from the common chamber.
18. Disconnect the evaporative emission canister purge hose from the common chamber.
19. Remove the EGR valve assembly.
20. Remove the common chamber from the intake manifold.
21. Disconnect the fuel feed and return hoses from the fuel rail assembly.
22. Disconnect the connectors to the fuel injectors and the thermo sensor.
23. Remove the intake manifold.
24. Remove the engine coolant manifold by removing the heater hose and four mounting bolts.
25. Remove the accessory drive belts.
26. Remove the power steering pump.
27. Remove the fan pulley assembly.
28. Remove the crankshaft pulley and damper.
29. Remove the oil cooler hoses and bracket on the timing belt cover.
30. Remove the timing belt cover.
31. Align the timing marks.
32. Remove the timing belt auto tensioner (pusher). The pusher prevents air from entering the oil chamber. Its rod must always be facing upward.
33. Remove the timing belt.
34. Remove the cylinder head cover.
35. Remove the power steering pump bracket.
36. Remove the front exhaust pipes from the exhaust manifolds.
37. Remove the dipstick tube bracket from the cylinder head.

------ **WARNING** ------
The cylinder head and engine block must be at room temperature before removing the cylinder head.

38. Remove the cylinder head bolts in the reverse of the installation sequence, gradually and in two steps.
39. Remove the cylinder head.
To install:
40. Install new camshaft seals and retaining plates onto the cylinder. Tighten the right camshaft seal retaining plate 6mm bolts to 65 inch lbs. Tighten the left camshaft seal retaining plate 8mm bolts to 191 inch lbs.
41. Thoroughly clean the cylinder head and engine block sealing surfaces.
42. Place a new head gasket on the engine block and carefully position

the cylinder head on top of the new gasket.

NOTE: Do not reuse or apply oil to the cylinder head bolts.

43. Install new cylinder head bolts and torque them in sequence to 47 ft. lb. (64 Nm) for the M11 bolts and 15 ft. lbs. (21 Nm) for the M8 bolts.

44. Install the dipstick tube bracket to the cylinder head.

45. Connect the front exhaust pipes to the exhaust manifolds. Tighten the exhaust bolts to 48 ft. lbs. (67 Nm).

46. Install the power steering pump bracket. Torque the mounting bolts to 34 ft. lbs. (46 Nm).

47. Install the cylinder head covers.

48. Install the timing belt and the auto tensioner (pusher). Torque the mounting bolt to 14 ft. lbs. (19 Nm).

49. Install the timing belt cover and oil cooler hoses and bracket.

50. Install the crankshaft pulley. Torque the center bolt to 123 ft. lbs. (167 Nm).

51. Install the fan pulley assembly.

52. Install the power steering pump.

53. Install the accessory drive belts.

54. Install the engine coolant manifold and the heater hose.

55. Install the intake manifold. Torque the nuts and bolts to 17 ft. lbs. (24 Nm) in a crisscross pattern.

56. Install the fuel injector connectors and the fuel hoses to the fuel rail.

57. Install the common chamber. Torque the nuts and bolts to 17 ft. lbs. (24 Nm).

58. Install the EGR valve assembly. Torque the mounting bolts on the valve side to 69 inch lbs. (8 Nm) and the bolts on the exhaust side to 21 ft. lbs. (28 Nm).

59. Connect the evaporative emission canister purge hose.

60. Connect the fuel pressure control valve vacuum hose.

61. Connect the positive crankcase ventilation hose.

62. Install the throttle body assembly. Torque the mounting bolts to 14 ft. lbs. (19 Nm).

63. Connect the vacuum hoses to the throttle body.

64. Install the ignition control module and the spark plug wires.

65. Connect the MAP sensor; Canister Vacuum Switching Valve (VSV); Exhaust gas recirculation VSV; Intake air temperature sensor and ground connectors.

66. Connect the vacuum booster hose.

67. Connect the air vacuum hose.

68. Connect the accelerator cable. Adjust the accelerator cable by pulling the cable housing while closing the throttle valve and tightening the adjusting nut and screw cap by hand temporarily. Now loosen the adjusting nut by three turns and then tightening the screw cap. Make sure the throttle valve reaches the screw stop when the throttle is closed.

69. Install the cooling fan assembly and the upper fan shroud.

70. Install the air cleaner assembly.

71. Connect the negative battery cable.

72. Refill and bleed the cooling system.

73. Refill and bleed the power steering pump if necessary.

74. Refill the engine with fresh oil.

75. Run the engine and check for leaks and proper compression.

Double Overhead Camshaft (DOHC) Engine

NOTE: Isuzu has issued a recall notice for 1993–1994 Troopers equipped with DOHC engines. This is campaign number 94V-094, and involves faulty camshaft end plugs. The plugs may dislodge from the cylinder heads and can cause rapid oil loss. A service kit is available for affected vehicles.

1. Use a felt–tipped marker to matchmark the hood hinge to the hood. Remove the hood.

2. Properly relieve the fuel system pressure.

3. Disconnect the negative battery cable.

4. Drain the engine coolant into a sealable container.

5. Remove the air cleaner assembly.

6. Remove the upper cooling fan shroud from the radiator.

7. Remove the cooling fan assembly.

8. Disconnect the accelerator cable from the throttle body and the bracket.

9. Disconnect the canister vacuum hose from the vacuum pipe.

10. Disconnect the air vacuum hose from the common chamber.

11. Disconnect the vacuum booster hose from the common chamber.

12. Disconnect the MAP sensor; Canister Vacuum Switching Valve (VSV); Exhaust gas recirculation VSV; Intake air temperature sensor; and ground connectors.

13. Remove the spark plug wires from the cylinder head cover.

14. Remove the ignition control module assembly.

15. Remove the four bolts and the throttle body from the common chamber.

16. Disconnect the vacuum hoses from the throttle body.

17. Disconnect the positive crankcase ventilation hose from the common chamber.

18. Disconnect the fuel pressure control valve vacuum hose from the common chamber.

19. Disconnect the evaporative emission canister purge hose from the common chamber.

20. Remove the EGR valve assembly.

21. Remove the common chamber air duct.

22. Remove the common chamber from the intake manifold.

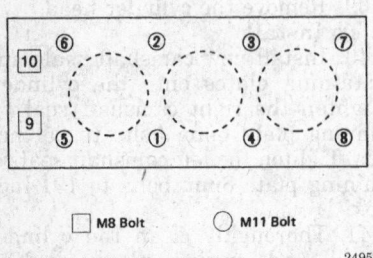

Cylinder head bolt torque sequence — 1993–95 SOHC and DOHC engine

□ M8 Bolt ○ M11 Bolt

249520

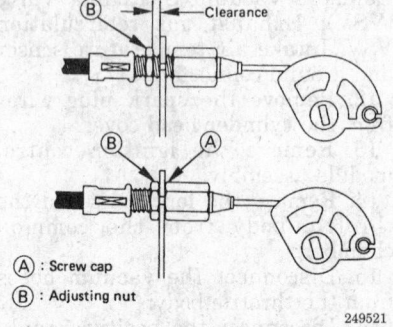

Ⓐ : Screw cap
Ⓑ : Adjusting nut

249521

Accelerator cable adjustment — 1993–95 SOHC engine, DOHC engine is similar

23. Disconnect the fuel feed and return hoses from the fuel rail assembly.

24. Disconnect the connectors to the fuel injectors and the thermo sensor.

25. Remove the intake manifold.

26. Remove the engine coolant manifold by removing the heater hose and four mounting bolts.

27. Remove the accessory drive belts.

28. Remove the power steering pump.

29. Remove the fan pulley assembly.

30. Remove the crankshaft pulley and damper.

31. Remove the oil cooler hoses and bracket on the timing belt cover.

32. Remove the timing belt cover.

33. Align the timing marks.

34. Remove the timing belt auto tensioner (pusher). The pusher prevents air from entering the oil chamber. Its rod must always be facing upward.

35. Remove the timing belt.

36. Remove the cylinder head cover.

37. Remove the power steering pump bracket.

38. Remove the front exhaust pipes from the exhaust manifolds.

39. Remove the dipstick tube bracket from the cylinder head.

40. Remove the cylinder head bolts in sequence, gradually in two steps.

41. Remove the cylinder head.

To install:

42. Thoroughly clean the cylinder head and engine block sealing surfaces.

43. Place a new head gasket on the engine block and carefully position the cylinder head on top of the new gasket.

NOTE: Do not reuse or apply oil to the cylinder head bolts.

44. Install new cylinder head bolts and torque them in sequence to 47 ft. lb. (64 Nm) for the M11 bolts and 15 ft. lbs. (21 Nm) for the M8 bolts.

45. Install new camshaft seals and retaining plates onto the cylinder. Tighten the rear camshaft seal retaining plate 6mm bolts to 70 inch lbs. Tighten the front camshaft seal retaining bolts to 65 inch lbs.

46. Install the dipstick tube bracket to the cylinder head.

47. Connect the front exhaust pipes to the exhaust manifolds.

48. Install the power steering pump bracket. Torque the mounting bolts to 34 ft. lbs. (46 Nm).

49. Install the cylinder head covers.

50. Install the timing belt and the auto tensioner (pusher). Torque the mounting bolt to 14 ft. lbs. (19 Nm).

51. Install the timing belt cover and oil cooler hoses and bracket.

52. Install the crankshaft pulley. Torque the center bolt to 123 ft. lbs. (167 Nm).

53. Install the fan pulley assembly and tighten the bolts to 16 ft. lbs. (22 Nm).

54. Install the power steering pump.

55. Install the accessory drive belts.

56. Install the engine coolant manifold and the heater hose.

57. Install the intake manifold. Torque the nuts and bolts to 17 ft. lbs. (24 Nm).

58. Install the fuel injector connectors and the fuel hoses to the fuel rail.

59. Install the common chamber. Torque the nuts and bolts to 17 ft. lbs. (24 Nm).

60. Install the common chamber air duct. Torque the nuts and bolts to 17 ft. lbs. (24 Nm).

61. Install the EGR valve assembly. Torque the mounting bolts on the valve side to 69 inch lbs. (8 Nm) and the bolts on the exhaust side to 21 ft. lbs. (28 Nm).

62. Connect the evaporative emission canister purge hose.

63. Connect the fuel pressure control valve vacuum hose.

64. Reconnect the fuel lines using new washers.

65. Connect the positive crankcase ventilation hose.

66. Install the throttle body assembly. Torque the mounting bolts to 14 ft. lbs. (19 Nm).

67. Connect the vacuum hoses to the throttle body.

68. Install the ignition control module and the spark plug wires.

69. Connect the MAP sensor; Canister Vacuum Switching Valve (VSV); Exhaust gas recirculation VSV; Intake air temperature sensor and ground connectors.

70. Connect the vacuum booster hose.

71. Connect the air vacuum hose.

72. Connect the accelerator cable. Adjust the accelerator cable by pulling the cable housing while closing the throttle valve and tightening the adjusting nut and screw cap by hand temporarily. Now loosen the adjusting nut by three turns and then tightening the screw cap. Make sure the throttle valve reaches the screw stop when the throttle is closed.

73. Install the cooling fan assembly and the upper fan shroud.

74. Install the air cleaner assembly.

75. Align the matchmarks and install the hood.

76. Connect the negative battery cable.

77. Change the engine oil and filter.

78. Refill the engine with the proper amount of coolant, start the engine and check for leaks.

1996–97 3.2L (VIN V) Engine

NOTE: There was (only) an SOHC engine available for the 1996–97 vehicle years.

———— WARNING ————
The cylinder head should be cool to the touch before it is removed. If the head bolts are loosened on a hot engine, the cylinder head may warp.

1. Properly relieve the fuel system pressure.

———— CAUTION ————
Fuel injection systems remain under pressure even after the engine has been turned off. The fuel system pressure must be relieved before disconnecting any fuel lines. Failure to do so may result in fire and personal injury.

2. Disconnect the negative battery cable and reinstall the fuel pump relay.

3. Raise and support the vehicle safely.

4. Disconnect the front exhaust pipes from the exhaust manifolds. If necessary, separate the front exhaust pipes from the crossover pipe. Label and disconnect the oxygen sensors.

5. Lower the vehicle.

6. Drain the coolant into a sealable container.

7. Use a felt–tipped marker to matchmark the hood hinge plates. Remove the hood.

8. Remove the air intake duct and the air cleaner box.

9. Disconnect and remove the upper and lower radiator hoses. Catch the coolant that runs out.

10. Loosen and remove the power steering pump, A/C compressor, and alternator drive belts.

11. Remove the cooling fan and its pulley assembly.

12. Unbolt the power steering pump mounting bracket. Move the pump and bracket out of the way without disconnecting the hydraulic lines.

13. Disconnect the throttle cable from the throttle body linkage.

14. Label and disconnect the following vacuum hoses from the intake manifold chamber:

   a. PCV hose.

   b. EVAP canister vacuum hose.

   c. Brake booster hose.

15. Label and disconnect the following sensor connectors from the rear of the intake manifold chamber:

   a. Ignition control module connectors.

   b. Linear EGR valve.

   c. MAP sensor.

   d. EVAP purge valve.

   e. Throttle position sensor.

   f. Idle air control valve.

   g. Intake air temperature sensor.

16. Disconnect the EGR valve supply tube and bracket.

17. First, remove the throttle body, and then remove the intake manifold chamber.

18. Carefully clean any dirt from the fuel rail and fuel fittings.

19. Disconnect the fuel feed and return lines from the front of the fuel rail. Clean up any spilled fuel.

20. Label and disconnect the fuel injector wiring harness.

21. Remove the fuel injectors and lower intake manifold as an assembly. If desired, the fuel rail and injectors may be removed separately as an assembly.

22. Remove the intake manifold gaskets. Be careful not to drop any pieces of the gaskets into the engine. Don't scratch or gouge the machined aluminum mating surfaces of the intake manifold and engine block.

23. Cover the intake openings with a sheet of plastic or clean shop towels to keep out dirt and foreign objects.

24. Label the ignition coil assemblies and disconnect them from the wiring harness. Remove the coil assemblies so they won't be damaged.

25. Unbolt the oil cooler line brackets from the timing belt covers.

26. Rotate the crankshaft to align the camshaft timing marks with the pointer dots on the back covers. When the timing marks are aligned, the No. 2 cylinder is at TDC/compression.

27. Remove the crankshaft pulley. Remove the lower timing belt cover.

28. Remove the pusher assembly (tensioner) from below the timing belt tensioner pulley. The pusher rod must always be facing upward to prevent oil leakage. Push the pusher rod in, and insert a wire pin into the hole to keep the pusher rod retracted.

29. Remove the timing belt.

WARNING

If the timing belt is worn, damaged, or shows signs of oil or coolant contamination, it must be replaced.

---

30. Disconnect the heater hoses from the engine; then, unbolt and remove the engine coolant manifold.

31. Unbolt the dipstick tube from the cylinder head.

32. Loosen the valve cover bolts in a crisscross sequence in the reverse of the installation sequence. Remove the valve covers.

33. Loosen the cylinder head bolts in a two-step crisscross pattern working from the outer bolts to those at the center of the head. First, partially loosen the 11mm bolts, then partially loosen the 8mm bolts. Finally loosen all the bolts and then remove them.

34. Remove the cylinder head. If it sticks, tap it with a wooden or plastic–faced mallet.

35. Remove the head gasket.

36. Inspect the cylinder head for cracking or warpage. Inspect the engine block for any signs of damage. Carefully clean the head gasket mating surfaces, don't scratch or gouge the machined aluminum surfaces.

37. Cover the engine block with a sheet of plastic or clean shop towels to keep any dirt and foreign objects out of the combustion chambers.

   To install:

NOTE: Use new head bolts when installing the cylinder head. Do not apply oil to the head bolt threads.

38. Make sure all mating surfaces are clean and free of oil, coolant, or gasket residue.

39. Install new cylinder head gaskets.

40. Install the cylinder head. Install the new head bolts and tighten them by hand only.

41. Follow these steps to tighten the cylinder head bolts to their final torque specification:

   a. Use a two-step crisscross pattern to tighten the 11mm bolts to 47 ft. lbs. (64 Nm). Start tightening with the center bolts, and work toward the outer bolts.

   b. Tighten the 8mm bolts to 15 ft. lbs. (21 Nm). Start with the bolt closest to the exhaust side of the head and work toward the intake side.

42. Apply a 2–3mm bead of sealant to the joint were the camshaft holders meet the cylinder head. Install the valve cover with a new gasket before the sealant cures.

---

43. Tighten the valve cover bolts to 6 ft. lbs. (8 Nm) in crisscross pattern.

44. Verify that the camshaft and crankshaft timing marks are properly aligned.

45. Install and tension the timing belt. Tighten the pusher bolts to 14 ft. lbs. (19 Nm).

46. Install the lower timing belt covers and tighten the bolts to 13 ft. lbs. (18 Nm). Install the crankshaft pulley. Tighten the pulley bolt to 123 ft. lbs. (167 Nm).

47. Install the upper timing belt covers and tighten the bolts to 13 ft. lbs. (18 Nm).

48. Fit the oil cooler line brackets onto the timing cover and tighten the bolts to 13 ft. lbs. (18 Nm).

49. Install the engine coolant manifold and reconnect the heater hoses. Tighten the bolts to 16 ft. lbs. (22 Nm).

50. Install the dipstick tube bracket.

51. Raise and safely support the vehicle. Install and reconnect the front exhaust pipes. Reconnect the oxygen sensors. Lower the vehicle.

52. Install the intake manifold with a new gasket . Tighten the bolts and nuts to 17 ft. lbs. (24 Nm).

53. Reconnect the fuel injector wiring harness. Reconnect the fuel feed and return lines.

54. Install and reconnect the ignition coil assemblies.

55. Install the intake manifold chamber and throttle body with new gaskets. Tighten the nuts and bolts to 17 ft. lbs. (24 Nm).

56. Reconnect the throttle cable to the throttle body linkage.

57. Reconnect the EGR valve supply tube and bracket.

58. Reconnect the following vacuum to the intake manifold chamber:

   a. PCV hose.

   b. EVAP canister vacuum hose.

   c. Brake booster hose.

59. Reconnect the following sensor connectors to the rear of the intake manifold chamber:

   a. Ignition control module connectors

   b. Linear EGR valve.

   c. MAP sensor.

   d. EVAP purge valve.

   e. Throttle position sensor.

   f. Idle air control valve.

   g. Intake air temperature sensor.

60. Install the power steering pump and mounting bracket.

61. Install the cooling fan and its pulley assembly.

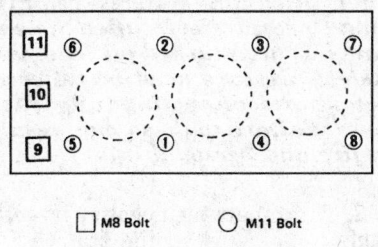

M8 Bolt   M11 Bolt

324741

Cylinder head bolt torque sequence — 1996–97 3.2L (VIN V) engine

62. Install and tension the alternator, A/C compressor, and power steering pump drive belts.

63. Install and reconnect the upper and lower radiator hoses.

64. Install the air cleaner box and air intake duct.

65. Install the hood.

66. Verify that all fuel lines, vacuum and coolant hoses, and wiring harness have been reconnected.

67. Refill the engine with fresh coolant.

68. Drain the engine oil. Install a new oil filter and refill the engine with fresh oil. If the engine oil was severely contaminated with coolant, a second oil and filter change may be necessary.

69. Crank the engine until it starts. A longer–than–normal starting time may be necessary due to air in the fuel lines. Check all fuel line connections for leaks.

70. Bleed any air from the cooling system.

71. Bleed the power steering system if necessary.

72. Warm the engine up to normal operating temperature and check the operation of the thermostat and water pump.

73. Check the throttle cable operation and adjustment.

74. Check the engine oil level and add if necessary.

Lash Adjusters

BLEEDING

Single Overhead Camshaft (SOHC) Engine

NOTE: The hydraulic lash adjusters are self–bleeding. The lash adjusters must be primed before installation to purge excess air. If the lash adjusters are worn, they should be replaced.

1. Disassemble the rocker arms and shafts.

2. Remove the old hydraulic lash adjuster from the rocker arm. The rocker arms and lash adjusters are a set. Label them so they aren't confused upon reassembly.

3. Inspect the hydraulic lash adjusters for excess movement and replace them if necessary. The hydraulic lash adjusters are designed to be self–bleeding, but new lash adjusters must be primed before installation.

   a. Use a small–diameter rod (.08 in. or 2mm) to push in the adjuster's check ball.

   b. Submerge the lash adjuster in a tub of clean engine oil.

   c. Hold the lash adjuster and pump the plunger with your finger to fill the adjuster with oil and displace any air.

   d. Keep pumping the plunger until it's hard and no more air bubbles come out. Then, remove the rod to release the check ball.

   e. Leave the lash adjuster in the tub of oil.

4. Submerge the rocker arm into the tub of engine oil. Install the hydraulic lash adjuster into the rocker arm.

5. Reassemble the rocker arm and shaft assemblies and install them.

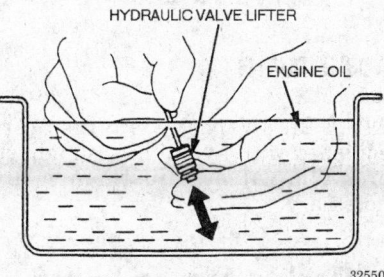

325506

Priming the valve lash adjuster — SOHC engine component shown, DOHC component similar

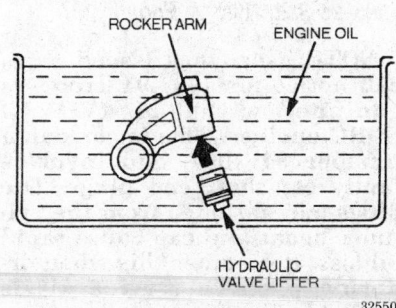

325507

Installing the valve lash adjuster into the rocker arm — SOHC engine

Double Overhead Camshaft (DOHC) Engine

NOTE: The hydraulic lash adjuster is a precision component that relies on a clean operating environment. When handling the lash adjuster, make certain that no dirt or foreign particles are allowed to get inside the unit. Do not try to take the unit apart: it can't be rebuilt. The adjuster is filled with oil. Hold the adjuster in the upright position so that the oil does not spill out. Use only clean engine oil when bleeding the lifters.

To bleed the hydraulic lash adjusters when disassembling the rocker arms or as part of an engine overhaul, perform the following:

1. Hold the adjuster upright in a container of clean engine oil.

2. Insert a 0.08 in. (2mm) diameter wire into the adjuster. Lightly press down on the steel check ball.

3. Hold the check ball in with the wire. Manually move the plunger up and down at one second intervals until all air bubbles disappear.

4. Remove the wire from the adjuster and push down firmly on the plunger. If the plunger moves even slightly, repeat the previous steps until the plunger stops moving. If the plunger continues to move, the lifter must be replaced.

Valve Swing Arm and Hydraulic Lifter

REMOVAL AND INSTALLATION

1993–95 3.2L (VIN W) Engine

NOTE: This procedure pertains only to the DOHC engine. Isuzu has issued a recall notice for 1993–1994 Troopers equipped with DOHC engines. This is campaign number 94V–094, and involves faulty camshaft end plugs. The plugs may dislodge from the cylinder heads and can cause rapid oil loss. Remember this when ordering parts: a service kit is available for affected vehicles.

1. Disconnect the negative battery cable.

2. Remove the valve covers and front timing covers.

3. Align the timing marks (the keyway on the camshaft sprockets will align with the front plate mark). When the timing marks are aligned, no pistons are at TDC/compression.

The engine must not be disturbed once it is in this position.

4. Remove the timing belt.

5. Remove the camshafts and camshaft timing chains as an assembly.

6. Label and remove the swing arms. Inspect them for any signs of wear and replace if necessary.

7. Remove the hydraulic lifters from their ports in the cylinder head. The lifters cannot be disassembled.

8. Bleed any air from the hydraulic lifters:

a. Place the lifter upright in a container filled with clean engine oil.

b. Insert a 2mm punch through the hole in the lifter's plunger. Press the punch against the internal check ball.

c. Push the plunger up and down for a few seconds until no more air bubbles come from the lifter.

d. After the air bubbles disappear, remove the punch and push the plunger in to lock it. If the plunger won't lock, the lifter must be replaced.

e. Store the lifters in engine oil until they are to be installed.

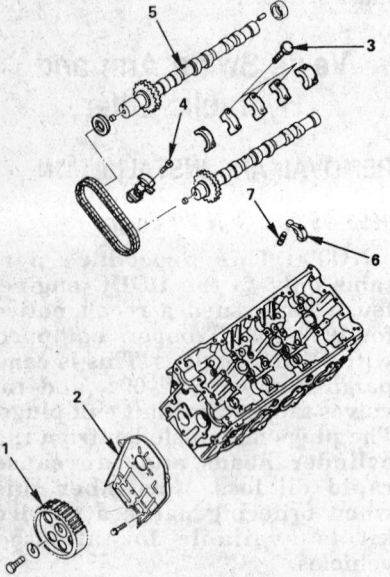

1. Camshaft timing pulley
2. Front plate
3. Camshaft bracket fixing bolts
4. Chain tensioner fixing bolts
5. Camshaft, oil seal and camshaft end plug
6. Swing arm
7. Hydraulic lash adjusters

253616

Camshaft and valve components — 1993–94 DOHC engines

To install:

9. Coat the lifters with clean engine oil and install them into their ports.

10. Install the swing arms, making sure that they fit correctly onto the valve stems and lifters.

11. Lubricate the camshaft journals and lobes.

12. Install the camshafts and camshaft timing chains. Make sure the sprocket and chain timing marks align. Apply sealant to the camshaft holder contact surfaces and tighten the camshaft holder bolts from to 87 inch lbs. (10 Nm). Tighten the chain tensioner bolts to 14 ft. lbs. (19 Nm).

13. Install new camshaft seals and retaining plates onto the cylinder. Tighten the rear camshaft seal retaining plate 6mm bolts to 70 inch lbs. Tighten the front camshaft seal retaining bolts to 65 inch lbs.

14. Make sure the sprocket and timing belt alignment marks are aligned. Install the timing belt.

15. Install the valve covers.

16. Verify that all vacuum hoses and wiring harnesses have been reconnected.

17. Connect the negative battery cable.

18. Check the engine oil and add as necessary.

Valve Lash

ADJUSTMENT

All 3.2L (VIN V) engines are equipped with hydraulic lash adjusters. No valve adjustments are necessary.

Rocker Arm and Shaft

REMOVAL AND INSTALLATION

1993–95 3.2L (VIN V) Engine

NOTE: Isuzu has issued a recall notice for 1993–94 Troopers equipped with 3.2L (VIN V) SOHC engines. This is campaign number 94V–094, and involves faulty camshaft end plugs. The plugs may dislodge from the cylinder heads and can cause rapid oil loss. Remember this when ordering parts: a service kit is available for affected vehicles.

1. Properly relieve the fuel system pressure.

— **CAUTION** —

Fuel injection systems remain under pressure even after the engine has been turned off. The fuel system pressure must be relieved before disconnecting any fuel lines. Failure to do so may result in fire and personal injury.

2. Disconnect the negative battery cable.

3. Remove the air cleaner assembly.

4. Disconnect the accelerator pedal cable from the throttle body and cable brackets.

5. Disconnect the canister vacuum hose from the Vacuum Switch Valve (VSV).

6. Disconnect the vacuum booster hose from the common chamber duct.

7. Disconnect the electrical connectors from the Idle Air Control Valve, Throttle Position sensor, Manifold Absolute Pressure sensor, canister VSV, EGR VSV, Intake Air Temperature sensor and VSV.

8. Remove the high tension cable from the cylinder heads.

9. Disconnect the connectors from the ignition module.

10. Remove the three bolts from the electronic ignition bracket and assembly.

11. Remove the four bolts from the throttle body and remove the throttle body.

12. Disconnect the canister VSV and the EGR VSV vacuum hose from the throttle body.

13. Disconnect the fuel pressure control valve vacuum hose from the common chamber duct.

14. Disconnect the PCV hose from the common chamber duct.

15. Disconnect the evaporative emission canister purge hose from the common chamber duct.

16. Remove the four bolts from the EGR valve assembly common chamber duct and remove the exhaust manifold.

17. Remove the four bolts, four nuts and three manifold bracket fixing bolts from the common chamber duct.

18. Remove the ground cable fixing bolt from the rear of the common chamber duct.

19. Remove the six bolts and two nuts from the common chamber duct.

20. Remove the common chamber duct bracket fixing bolts from the rear of the common chamber duct.

21. Remove the following components, and then remove the timing belt:

a. Remove the upper fan shroud from the radiator.

b. Remove the four nuts retaining the cooling fan assembly. Remove the cooling fan.

c. Remove the power steering drive belt.

d. Remove the air conditioning compressor drive belt.

e. Remove the generator drive belt.

f. Remove the fan pulley assembly.

g. Make sure the timing belt is set at TDC: all the timing marks must align. On 1994–1995 vehicles, the engine will be at TDC/compression for the No. 2 cylinder. On earlier engines, no cylinders are at TDC/compression when the marks are aligned.

h. Remove the crankshaft pulley center bolt. Remove the crankshaft pulley.

i. Remove the two oil cooler hose bracket fixing bolts on the timing cover. Remove the oil cooler hose.

j. Remove the timing belt cover.

k. Remove the pusher. The rod must always be facing upward.

l. Mark the timing belt, cam pulley and crankshaft pulley. Remove the timing belt.

22. Remove the cylinder head cover.

23. Remove the camshaft holders and camshaft.

24. Remove the rocker arm shaft bolts. Lift the rocker arm assembly from the cylinder head.

25. The hydraulic lifters are attached to the rocker arms. Remove them and inspect, bleed, or replace as necessary.

To install:

26. Install new camshaft seals and retaining plates onto the cylinder head. Tighten the right camshaft seal retaining plate 6mm bolts to 65 inch lbs. Tighten the left camshaft seal retaining plate 8mm bolts to 191 inch lbs.

27. Install the rocker arm assembly and tighten the bolts in sequence to 13 ft. lbs. (18 Nm).

28. Oil the camshaft bearing journals, camshaft lobes, and rocker arm contact areas.

29. Install the camshaft. Apply sealant to the contact edges of the camshaft holders. Tighten the camshaft 6mm holder bolts to 69 inch lbs. (8 Nm). Tighten the remaining bolts to 13 ft. lbs. (18 Nm).

30. Install the cylinder head covers and carefully tighten the bolts to 69 inch lbs. (8 Nm). Do not overtighten the head cover bolts: they crack very easily

31. Install the timing belt:

a. Align the groove on the crankshaft timing pulley with mark on the oil pump.

b. Align the marks on the camshaft timing pulleys with the dots on the front plate.

c. Install the timing belt. Align the dotted marks on the timing belt with the mark on the crankshaft gear.

d. Align the white line on the timing belt with the alignment mark on the right bank camshaft timing pulley. Secure the belt with a double clip.

e. Turn the crankshaft counterclockwise to remove the slack between the crankshaft pulley and the right camshaft timing pulley.

f. Install the belt on the water pump pulley.

g. Install the belt on the idler pulley.

h. Align the white alignment mark on the timing belt with the alignment mark on the left bank camshaft timing pulley.

i. Install the crankshaft pulley and tighten the center bolt by hand. Turn the crankshaft pulley clockwise to give slack between the crankshaft timing pulley and the right bank camshaft timing pulley.

j. Install the pusher while pushing the tension pulley to the belt.

k. Pull the pin out from the pusher.

l. Remove the double clips from the pulleys. Turn the crankshaft pulley clockwise 2 turns. Measure the rod protrusion to be sure it is between 0.16–0.24 in. (4–6mm).

m. Tighten the adjusting bolt to 31 ft. lbs. (42 Nm).

n. Tighten the pusher bolt to 14 ft. lbs. (19 Nm).

o. Remove crankshaft pulley. Install the timing belt cover and tighten bolts to 12 ft. lbs. (17 Nm).

p. Install the oil cooler hose and tighten brackets to 16 ft. lbs. (22 Nm).

q. Install the crankshaft pulley and tighten the bolt to 123 ft. lbs. (167 Nm).

r. Install the fan pulley assembly and tighten the bolts to 16 ft. lbs. (22 Nm).

s. Engage and adjust the alternator drive belt.

t. Engage and adjust the air conditioning drive belt.

u. Engage and adjust the power steering pump drive belt.

v. Install the cooling fan assembly and tighten bolts to 69 inch lbs. (8 Nm).

w. Install the upper fan shroud to the radiator.

32. Install the common chamber duct bracket bolts to the rear of the common chamber duct.

33. Install the six bolts and two nuts to the common chamber duct.

34. Install the ground cable fixing bolt to the rear of the common chamber duct.

35. Install the four bolts, four nuts and three manifold bracket bolts to the common chamber duct.

36. Install the exhaust manifold and install the four bolts to the EGR valve assembly and common chamber duct.

37. Connect the evaporative emission canister purge hose to the common chamber duct.

38. Connect the PCV hose to the common chamber duct.

39. Connect the fuel pressure control valve vacuum hose to the common chamber duct.

40. Connect the canister VSV and the EGR VSV vacuum hose to the throttle body.

41. Install the throttle body and install the four bolts to the throttle body and remove the throttle body.

42. Install the electronic ignition assembly and its bracket.

43. Connect the three connectors to the electronic ignition module.

44. Connect the high tension cable to the cylinder head cover clips.

45. Connect the electrical connectors to the Idle Air Control Valve, Throttle Position sensor, Manifold Absolute Pressure sensor, canister VSV, EGR VSV, Intake Air Temperature sensor and VSV.

46. Connect the vacuum booster hose to the common chamber duct.

47. Connect the canister vacuum hose to the Vacuum Switch Valve (VSV).

48. Connect the accelerator pedal cable to the throttle body and cable brackets.

49. Verify that all vacuum hoses, lines, and wiring harnesses are reconnected.

50. Install the air cleaner assembly.

51. Connect the negative battery cable.

1996–97 3.2L (VIN V) Engine

1. Properly relieve the fuel system pressure.

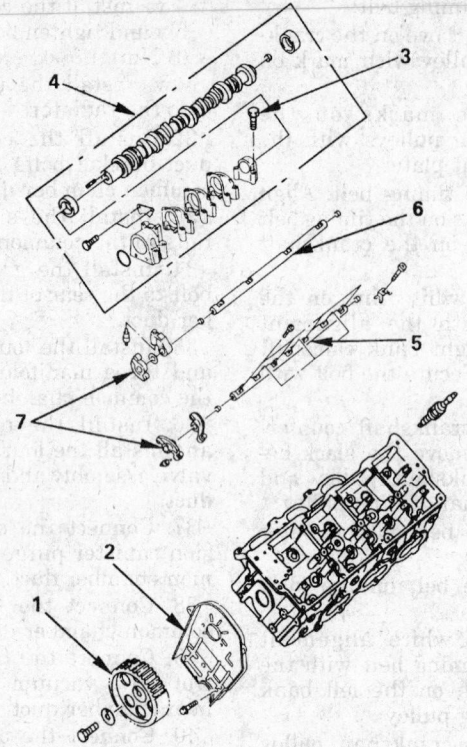

1. Camshaft timing pulley
2. Front plate
3. Camshaft bracket housing fixing bolts
4. Camshaft assembly
5. Rockershaft assembly (Exhaust side)
6. Rockershaft assembly (Intake side)
7. Rocker arm

249528

Rocker arm assembly — 1993–95 3.2L (VIN V) SOHC engines

Intake side Alignment mark / Wave washer

Exhaust side / Wave washer

249529

Rocker arms and shafts — 1993–95 3.2L (VIN V) SOHC engines, others similar

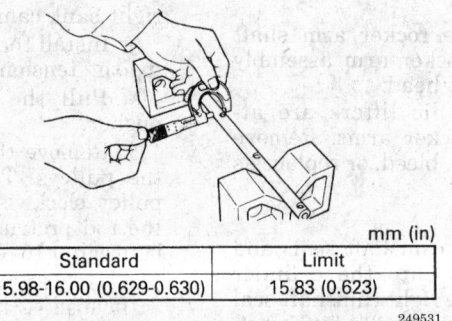

mm (in)	
Standard	Limit
15.98-16.00 (0.629-0.630)	15.83 (0.623)

249531

Measure the shaft at the point where the rocker arm moves on the shaft — 1993–95 3.2L (VIN V) SOHC engines

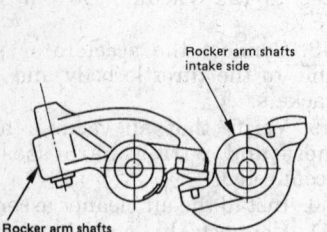

Rocker arm shafts intake side

Rocker arm shafts exhaust side

249532

Exhaust valve rocker arms — 1993–95 3.2L (VIN V) SOHC engines

—— CAUTION ——
Fuel injection systems remain under pressure even after the engine has been turned off. The fuel system pressure must be relieved before disconnecting any fuel lines. Failure to do so may result in fire and personal injury.

2. Disconnect the negative battery cable.
3. Drain the coolant into a sealable container.
4. Support the hood as far open as possible.

5. Remove the air intake duct and the air cleaner box.
6. Disconnect and remove the upper and lower radiator hoses. Catch any coolant that runs out.
7. Loosen and remove the power steering pump, A/C compressor, and alternator drive belts.
8. Remove the cooling fan and its pulley assembly.
9. Unbolt the power steering pump mounting bracket. Move the pump and bracket out of the way without disconnecting the hydraulic lines.

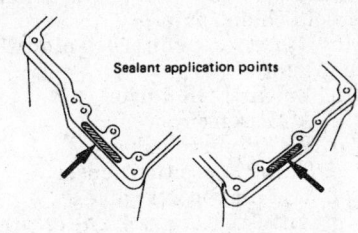

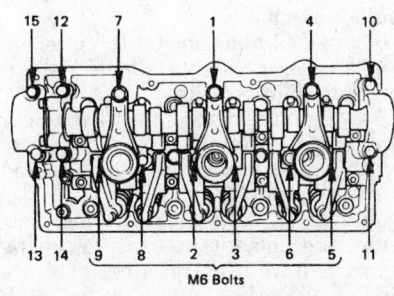

Camshaft holder bolt tightening sequence and sealant application points — 1993–95 3.2L (VIN V) SOHC engines

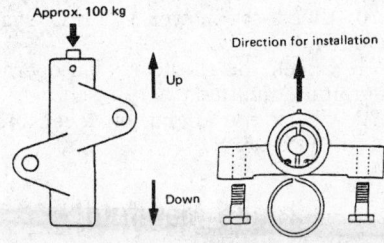

Timing belt auto tensioner — 1993–95 3.2L (VIN V) SOHC engines

10. Disconnect the throttle cable from the throttle body linkage.

11. Label and disconnect the following vacuum hoses from the intake manifold chamber:

a. PCV hose.

b. EVAP canister vacuum hose.

c. Brake booster hose.

12. Label and disconnect the following sensor connectors from the rear of the intake manifold chamber:

a. Ignition control module connectors.

b. Linear EGR valve.

c. MAP sensor.

d. EVAP purge valve.

e. Throttle position sensor.

f. Idle air control valve.

g. Intake air temperature sensor.

13. Disconnect the EGR valve supply tube and bracket.

14. First, remove the throttle body, and then remove the intake manifold chamber.

15. Carefully clean any dirt from the fuel rail and fuel fittings.

16. Disconnect the fuel feed and return lines from the front of the fuel rail. Clean up any spilled fuel.

17. Remove the intake manifold gaskets. Be careful not to drop any pieces of the gaskets into the engine. Don't scratch or gouge the machined aluminum mating surfaces of the intake manifold and engine block.

18. Cover the intake openings with a sheet of plastic or clean shop towels to keep out dirt and foreign objects.

19. Label the ignition coil assemblies and disconnect them from the wiring harness. Remove the coil assemblies so they won't be damaged.

20. Unbolt the oil cooler line brackets from the timing belt covers.

21. Remove the upper timing belt covers.

22. Rotate the crankshaft to align the camshaft timing marks with the pointer dots on the back covers. When the timing marks are aligned, the No. 2 cylinder is at TDC/compression.

23. Remove the crankshaft pulley. Remove the lower timing belt cover.

24. Remove the pusher assembly (tensioner) from below the timing belt tensioner pulley. The pusher rod must always be facing upward to prevent oil leakage. Push the pusher rod in, and insert a wire pin into the hole to keep the pusher rod retracted.

25. Remove the timing belt.

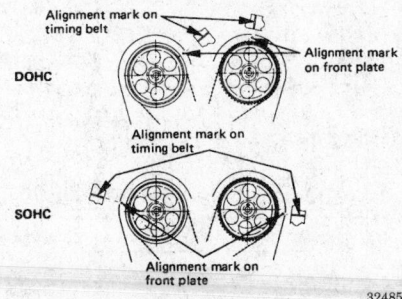

Camshaft sprocket alignment marks — 1996–97 3.2L (VIN V) SOHC engines

WARNING

If the timing belt is worn, damaged, or shows signs of oil or coolant contamination, it must be replaced.

26. Loosen the valve cover bolts in a crisscross sequence. Remove the valve covers.

27. Remove the camshaft sprockets and back covers.

28. Loosen the camshaft holder bolts in a crisscross sequence to prevent warping.

29. Remove the camshaft and camshaft holders from the cylinder head.

30. Inspect the camshaft lobes and journals for signs of wear or damage.

31. Loosen the exhaust and intake rocker shaft bolts in a crisscross sequence to prevent warping.

32. Remove the intake and exhaust rocker shafts from the cylinder head.

33. If the rocker arms and shafts must be disassembled, label the parts and wave washers so that they can be reassembled in the same positions.

34. If necessary, remove the hydraulic valve lash adjusters from the rocker arms.

35. Inspect the hydraulic lash adjusters for excess movement and replace if necessary. The hydraulic lash adjusters are designed to be self–bleeding, but new adjusters must be primed before installation.

a. Use a small–diameter rod (.08 in. or 2mm) to push in the adjuster's check ball.

b. Submerge the adjuster in a tub of clean engine oil.

c. Pump the plunger with your finger to fill the adjuster with oil and displace any air.

d. Keep pumping the plunger until it's hard and no more air bubbles come out. Then, remove the rod to release the check ball.

To install:

36. Reassemble the rocker arm, shaft, and hydraulic lash adjuster components. Assemble the hydraulic lash adjusters to the rockers before removing them from the tub of oil. The intake rocker arms all face the same direction when installed.

37. Lubricate the rocker arms and shafts with clean engine oil.

38. Install the intake and exhaust rocker arms and shaft assemblies. Tighten the rocker shaft holder bolts to 13 ft. lbs. (18 Nm), starting with the intake shaft and then moving to the exhaust shaft. Make sure the intake and exhaust rockers contact each other properly.

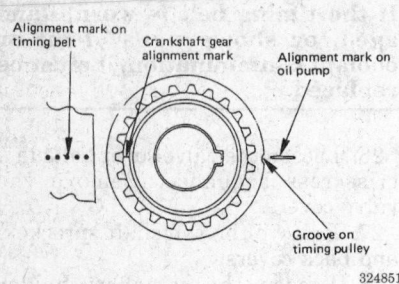

Timing belt dotted mark alignment — 1996-97
3.2L (VIN V) SOHC engines

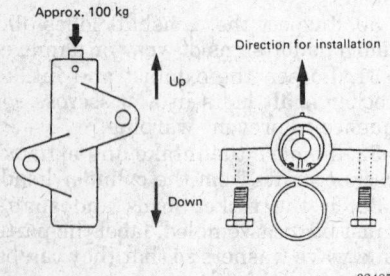

Timing belt pusher (tensioner) with pin
installed — 1996-97 3.2L (VIN V) SOHC engines

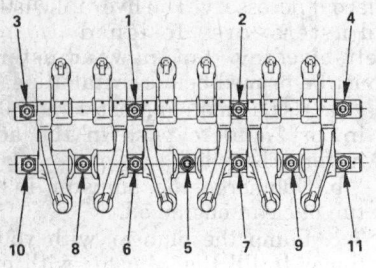

Rocker shaft bolt tightening pattern — 1996-97
3.2L (VIN V) SOHC engines

39. Make sure all mating surfaces are clean and free of oil, coolant, or gasket residue.

40. Lubricate the camshaft lobes and journals with clean engine oil.

41. Apply a bead of sealant to the front and rear camshaft holder mating surfaces on the cylinder head.

42. Install the camshaft and holder assembly onto the cylinder head before the sealant cures. Install the camshaft holder bolts, but don't tighten them yet.

43. Use a crisscross sequence to tighten the camshaft holder bolts. Tighten the 8mm bolts to 13 ft. lbs.

(18 Nm). Tighten the 6mm bolts to 6 ft. lbs. (8 Nm).

44. Use a seal driver to install a new camshaft seal.

45. Install the camshaft sprocket back covers and tighten their bolts to 12 ft. lbs. (17 Nm).

46. Install the camshaft sprockets so that the timing marks are aligned. Tighten the bolts to 46 ft. lbs. (64 Nm).

47. Apply a 2-3mm bead of sealant to the joint were the camshaft holders meet the cylinder head. Install the valve cover with a new gasket before the sealant cures.

48. Tighten the valve cover bolts to 6 ft. lbs. (8 Nm) in crisscross pattern.

49. Verify that the camshaft and crankshaft timing marks are properly aligned.

50. Install and tension the timing belt. Tighten the pusher bolts to 14 ft. lbs. (19 Nm).

51. Install the lower timing belt covers and tighten the bolts to 13 ft. lbs. (18 Nm). Install the crankshaft pulley. Tighten the pulley bolt to 123 ft. lbs. (167 Nm).

52. Install the upper timing belt covers and tighten the bolts to 13 ft. lbs. (18 Nm).

53. Fit the oil cooler line brackets onto the timing cover and tighten the bolts to 13 ft. lbs. (18 Nm).

54. Reconnect the fuel feed and return lines.

55. Install and reconnect the ignition coil assemblies.

56. Install the intake manifold chamber and throttle body with new gaskets. Tighten the nuts and bolts to 17 ft. lbs. (24 Nm).

57. Reconnect the throttle cable to the throttle body linkage.

58. Reconnect the EGR valve supply tube and bracket.

59. Reconnect the following vacuum to the intake manifold chamber:

  a. PCV hose.
  b. EVAP canister vacuum hose.
  c. Brake booster hose.

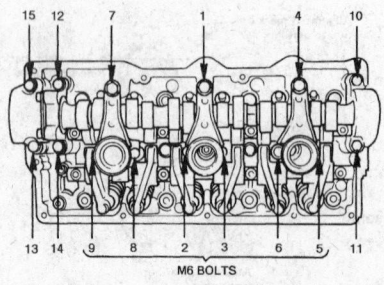

Camshaft mounting bolts tightening
sequence — 1996-97 3.2L (VIN V) SOHC
engines

60. Reconnect the following sensor connectors to the rear of the intake manifold chamber:

  a. Ignition control module connectors.
  b. Linear EGR valve.
  c. MAP sensor.
  d. EVAP purge valve.
  e. Throttle position sensor.
  f. Idle air control valve.
  g. Intake air temperature sensor.

61. Install the power steering pump and mounting bracket.

62. Install the cooling fan and its pulley assembly.

63. Install and tension the alternator, A/C compressor, and power steering pump drive belts.

64. Install and reconnect the upper and lower radiator hoses.

65. Install the air cleaner box and air intake duct.

66. Verify that all fuel lines, vacuum and coolant hoses, and wiring harness have been reconnected.

67. Refill the engine with fresh coolant.

68. Crank the engine until it starts. A longer-than-normal starting time may be necessary due to air in the fuel lines. Check all fuel line connections for leaks.

69. Bleed any air from the cooling system.

70. Bleed the power steering system if necessary.

71. Check the throttle cable operation and adjustment.

72. Check the engine oil level and add if necessary.

Intake Manifold

REMOVAL AND INSTALLATION

1993-95 3.2L (VIN V) Engines

Single Overhead Camshaft (SOHC) Engine

——— **CAUTION** ———
Fuel injection systems remain under pressure even after the engine has been turned off. The fuel system pressure must be relieved before disconnecting any fuel lines. Failure to do so may result in fire and personal injury.

1. Properly relieve the fuel system pressure.

2. Disconnect the negative battery cable.

3. Remove the air cleaner assembly.

4. Disconnect the accelerator pedal cable from the throttle body and bracket.

5. Disconnect the charcoal canister vacuum hose from the vacuum pipe.

6. Disconnect the air vacuum hose and the vacuum booster hose from the common chamber.

7. Disconnect the following electrical connectors:

    a. MAP sensor.

    b. Charcoal canister Vacuum Switching Valve (VSV).

    c. Exhaust Gas Recirculation VSV.

    d. Intake Air Temperature sensor.

    e. Engine ground cable.

    f. Fuel injector connectors.

    g. Thermo sensor connector.

8. Tag and then disconnect the spark plug wires.

9. Remove the ignition module assembly with the spark plug wires attached.

10. Tag and then disconnect the vacuum hoses from the throttle body.

11. Remove the four throttle body mounting bolts. Then, remove the throttle body.

12. Disconnect the PCV hose from the common chamber.

13. Disconnect the fuel pressure control valve vacuum hose from the common chamber.

14. Disconnect the Evaporative Emission Canister Purge hose from the common chamber.

15. Remove the EGR valve assembly from the common chamber.

16. Remove the common chamber (six bolts, two nuts, and three brackets).

17. Disconnect the fuel feed and return hoses from the fuel rail. Remove the bracket mounting bolts from the cylinder head cover.

18. Remove the two bolts and four nuts to remove the intake manifold from the engine.

To install:

NOTE: Use new self-locking nuts when installing the intake manifold. Use new manifold gaskets. Use new sealing washers when reconnecting the fuel lines.

19. Install the intake manifold on the engine. Torque the bolts and nuts to 17 ft. lbs. (24 Nm). Tighten the bolts from the center towards the ends.

20. Connect the electrical connectors to the fuel injectors and the thermo sensor.

21. Connect the fuel return and feed hoses to the fuel rail.

22. Install the common chamber. Torque the bolts and nuts to 17 ft. lbs. (24 Nm).

23. Install the EGR assembly. Torque the bolts to 78 inch lbs. (9 Nm).

24. Connect the charcoal canister purge and fuel pressure regulator vacuum hoses to the common chamber.

25. Connect the PCV hose to the common chamber.

26. Install the throttle body assembly and connect the vacuum hoses to the throttle body. Torque the throttle body mounting bolts to 16 ft. lbs. (22 Nm).

27. Install the ignition module assembly. Torque the mounting bolts to 16 ft. lbs. (22 Nm).

28. Reconnect the spark plug wires.

29. Reconnect the following electrical connectors:

    a. MAP sensor.

    b. Charcoal canister Vacuum Switching Valve (VSV).

    c. Exhaust Gas Recirculation VSV.

    d. Intake Air Temperature sensor.

    e. Engine ground cable.

30. Connect the air vacuum hose and the vacuum booster hose to the common chamber.

31. Connect the charcoal canister vacuum hose to the vacuum pipe.

32. Connect the accelerator cable to the throttle body and bracket. Adjust the cable so that the linkage rests against the stop when moved by hand.

33. Install the air cleaner assembly.

34. Verify that all electrical connectors and vacuum lines have been reconnected.

35. Connect the negative battery cable.

36. Turn the ignition key to the ON position for two seconds, then off. Turn the ignition ON again to pressurize the fuel system and check for leaks.

Double Overhead Camshaft (DOHC) Engine

———— **CAUTION** ————
Fuel injection systems remain under pressure even after the engine has been turned off. The fuel system pressure must be relieved before disconnecting any fuel lines. Failure to do so may result in fire and personal injury.

1. Properly relieve the fuel pressure.

2. Disconnect the negative battery cable.

3. Remove the air cleaner assembly.

4. Disconnect the accelerator pedal cable from the throttle body and bracket.

5. Disconnect the charcoal canister vacuum hose from the vacuum pipe.

6. Disconnect the air vacuum hose and the vacuum booster hose from the common chamber.

7. Disconnect the following electrical connectors:

    a. MAP sensor.

    b. charcoal canister Vacuum Switching Valve (VSV).

    c. Exhaust Gas Recirculation VSV.

    d. Intake Air Temperature sensor.

    e. Engine ground cable.

    f. Fuel injector connectors.

    g. Thermo sensor connector

8. Label and disconnect the spark plug wires.

9. Remove the ignition module assembly with the spark plug wires attached.

10. Disconnect the vacuum hoses from the throttle body.

11. Remove the four throttle body mounting bolts to remove the throttle body.

12. Disconnect the PCV hose from the common chamber.

13. Disconnect the fuel pressure control valve vacuum hose from the common chamber.

14. Disconnect the Evaporative Emission Canister Purge hose from the common chamber.

15. Remove the EGR valve assembly from the common chamber.

16. Remove the common chamber (six bolts, two nuts, and three brackets).

17. Disconnect the fuel feed and return hoses from the fuel rail and the bracket mounting bolts from the cylinder head cover.

18. Remove the two bolts and four nuts and remove the intake manifold from the engine.

To install:

NOTE: Use new self-locking nuts when installing the intake manifold. Use new manifold gaskets. Use new sealing washers when reconnecting the fuel lines.

19. Install the intake manifold onto the engine. Torque the bolts and nuts to 17 ft. lbs. (24 Nm).

20. Connect the electrical connectors to the fuel injectors and the thermo sensor.

21. Connect the fuel return and feed hoses to the fuel rail.

22. Install the common chamber. Torque the bolts and nuts to 17 ft. lbs. (24 Nm).

23. Install the EGR assembly. Torque the valve side bolts to 78 inch lbs. (9 Nm) and the manifold side bolts to 21 ft. lbs. (28 Nm).

24. Connect the charcoal canister purge and fuel pressure regulator vacuum hoses to the common chamber.

25. Connect the PCV hose to the common chamber.

26. Install the throttle body assembly and connect the vacuum hoses to the throttle body. Torque the throttle body mounting bolts to 14–16 ft. lbs. (19–22 Nm).

27. Install the ignition module assembly. Torque the mounting bolts to 14–16 ft. lbs. (19–22 Nm).

28. Reconnect the spark plug wires.

29. Reconnect the following electrical connectors:
   a. MAP sensor.
   b. Charcoal canister Vacuum Switching Valve (VSV).
   c. Exhaust Gas Recirculation VSV.
   d. Intake Air Temperature sensor.
   e. Engine ground cable.

30. Connect the air vacuum hose and the vacuum booster hose to the common chamber.

31. Connect the charcoal canister vacuum hose to the vacuum pipe.

32. Connect the accelerator cable to the throttle body and bracket. Adjust the cable so that the linkage rests against the stop when moved by hand.

33. Install the air cleaner and the negative battery cable.

34. Verify that all electrical connectors and vacuum lines have been reconnected.

35. Turn the ignition key to the ON position for two seconds, then off. Turn the ignition ON again to pressurize the fuel system and check for leaks.

1996–97 3.2L (VIN V) Engine

1. Properly relieve the fuel system pressure.

2. Disconnect the negative battery cable.

3. Remove the air cleaner and air intake duct.

4. Drain the engine coolant to a level below the upper radiator hose. Catch the coolant in a clean drain pan if it is to be reused.

5. Disconnect the throttle cable from the throttle body linkage.

6. Label and disconnect the following vacuum hoses from the intake manifold chamber:
   a. PCV hose.
   b. EVAP canister vacuum hose.
   c. Brake booster hose.

7. Label and disconnect the following sensor connectors from the rear of the intake manifold chamber:
   a. Ignition control module connectors.
   b. Linear EGR valve.
   c. MAP sensor.
   d. EVAP purge valve.
   e. Throttle position sensor.
   f. Idle air control valve.
   g. Intake air temperature sensor.

8. Disconnect the EGR valve supply tube and bracket.

9. Remove the throttle body.

— WARNING —

Don't use solvent of any type when cleaning the gasket mating surfaces of the throttle body and intake manifold. Solvent may damage the machined surfaces of these components. Be careful not to scratch the mating surfaces.

10. Disconnect the MAP sensor tube, and then unbolt the MAP sensor bracket.

11. Unbolt the intake manifold chamber from its brackets which are located at its front and rear edges.

12. Loosen the six manifold mounting bolts and two nuts in a crisscross sequence.

13. Remove the bolts and nuts, and then lift the chamber off of the base of the intake manifold. Note the positions of the long and short bolts.

14. Cover the intake manifold with a sheet of plastic, or clean shop towels to keep out dirt and foreign objects.

15. Carefully clean any dirt from the fuel rail and fuel fittings.

— CAUTION —

Fuel injection systems remain under pressure even after the engine has been turned off. The fuel system pressure must be relieved before disconnecting any fuel lines. Failure to do so may result in fire and personal injury.

16. Disconnect the fuel feed and return lines from the front of the fuel rail. Clean up any spilled fuel.

17. Label and disconnect the fuel injector wiring harness.

18. Unbolt the fuel return line bracket from the front of the intake manifold.

19. Unbolt the fuel rail from the intake manifold.

20. Carefully lift the fuel rail and injectors off of the intake manifold as an assembly. Move the fuel rail out of the work are so it won't be damaged; then, clean up any spilled fuel.

21. Remove the fuel rail spacer grommets from the sides of the intake manifold. Replace the spacer grommets if they are cracked or ripped.

22. Loosen the intake manifold nuts and bolts in a crisscross sequence working from the outer edges of the manifold toward the center.

23. Push the engine wiring harnesses aside and lift the intake manifold up and off of the engine block.

24. Remove the intake manifold gaskets. Be careful not to drop any pieces of the gaskets into the engine. Don't scratch or gouge the machined aluminum mating surfaces of the intake manifold and engine block.

25. Cover the intake openings with a sheet of plastic or clean shop towels to keep out dirt and foreign objects.

To install:

26. Remove the covers from the intake openings. Install new intake manifold gaskets.

27. Fit the intake manifold into position. Move the wiring harness back into position.

28. Install the intake manifold nuts and bolts. Tighten them in a two-step crisscross pattern to 15 ft. lbs. (20.5 Nm) working from the center of the manifold toward the outer edges.

29. Install the fuel rail spacer grommets.

30. Install the fuel rail assembly onto the intake manifold. Make sure all the fuel injectors are properly seated. Tighten the fuel rail bolts to 5.2 ft. lbs. (7 Nm).

31. Reconnect the fuel feed and return lines. Install the return line bracket bolt.

32. Reconnect the fuel injector wiring harness.

33. Install a new intake manifold chamber gasket. Install the intake manifold chamber.

34. Install the six intake manifold chamber bolts and two nuts. Tighten the nuts and bolts to 15 ft. lbs. (20.5 Nm) in a crisscross sequence.

35. Install the intake manifold chamber bracket bolts.

36. Reconnect the MAP sensor tube and bracket.

37. Reconnect the EGR supply tube and bracket.

38. Install the throttle body with a new gasket. Tighten the bolts to 10 ft. lbs. (13.5 Nm) in a crisscross sequence.

39. Reconnect the following sensor connectors to the rear of the intake manifold chamber:

    a. Ignition control module connectors.

    b. Linear EGR valve.

    c. MAP sensor.

    d. EVAP purge valve.

    e. Throttle position sensor.

    f. Idle air control valve.

    g. Intake air temperature sensor.

40. Reconnect the following vacuum hoses to the intake manifold chamber:

    a. PCV hose.

    b. EVAP canister vacuum hose.

    c. Brake booster hose.

41. Reconnect the throttle cable to the throttle body linkage.

42. Install the air cleaner and air intake duct.

43. Reconnect the negative battery cable.

44. Refill and bleed the cooling system.

45. Crank the engine until it starts. Air trapped in the fuel lines may cause the engine to crank for a longer period of time than normal.

46. Check the fuel lines, fuel rail, and injectors for any signs of leakage.

47. Warm the engine up to normal operating temperature and check the operation of the throttle cable and linkage. Adjust if necessary.

48. Check the manifold and throttle body mating surfaces for vacuum leaks.

Exhaust Manifold

REMOVAL AND INSTALLATION

Driver's Side Exhaust Manifold

NOTE: Allow the engine to cool completely before removing the exhaust manifolds.

1. Disconnect the negative battery cable.

2. Remove the air duct.

3. Remove the EGR pipe mounting bolts from the exhaust manifold.

4. Raise and safely support the vehicle.

5. If necessary to gain extra working room, remove the transfer case skid plate.

6. Label and disconnect the oxygen sensor connectors.

7. Remove the two stud nuts and two bolts and nuts and separate the front exhaust pipes from the exhaust manifold. Be careful not to damage the oxygen sensors when working around the exhaust pipes.

8. Lower the vehicle.

9. Remove the engine hanger and the heat shield.

10. Remove the seven nuts and then remove the exhaust manifold from the cylinder head.

    To install:

NOTE: Use new self–locking nuts and new gaskets when installing the exhaust manifolds.

11. Install the exhaust manifold and gasket to the cylinder head using new nuts. Tighten the new nuts to 42 ft. lbs. (57 Nm) in a crisscross sequence.

12. Install the heat shield and the engine hanger.

13. Raise and safely support the vehicle.

14. Install the front exhaust pipes and reconnect the exhaust system. Tighten the exhaust fasteners to the following specifications:

    a. Stud nuts: 49 ft. lbs. (67 Nm).

    b. Flange nuts and bolts: 32–37 ft. lbs. (43–50 Nm).

15. Reconnect the oxygen sensor connectors.

16. Install the skid plate and tighten the bolts to 27 ft. lbs. (37 Nm).

17. Lower the vehicle.

18. Install the EGR pipe to the exhaust manifold. Tighten the mounting bolts to 21 ft. lbs. (28 Nm).

19. Install the air duct and connect the negative battery cable. Verify that all wires and vacuum lines have been reconnected.

20. Start the engine and check for exhaust leaks.

Passenger's Side Exhaust Manifold

NOTE: Allow the engine to cool completely before removing the exhaust manifolds.

1. Disconnect the negative battery cable.

2. Raise and safely support the vehicle.

3. If necessary to gain extra working room, remove the transfer case skid plate.

4. Label and disconnect the oxygen sensor connectors.

5. Remove the two stud nuts and two bolts and nuts and separate the front exhaust pipes from the exhaust manifold. Be careful not to damage the oxygen sensors when working around the exhaust pipes.

6. Lower the vehicle.

7. Remove the engine hanger.

8. Remove the five heat shield mounting bolts and then remove the heat shield.

9. Remove the seven nuts and then separate the exhaust manifold from the cylinder head.

    To install:

NOTE: Use new self–locking nuts and new gaskets when installing the exhaust manifolds.

10. Install the exhaust manifold and gasket to the cylinder head using new nuts. Tighten the new nuts to 42 ft. lbs. (57 Nm) in a crisscross sequence.

11. Install the heat shield and the engine hanger.

12. Raise and safely support the vehicle.

13. Install the front exhaust pipes and reconnect the exhaust system. Tighten the exhaust fasteners to the following specifications:

    a. Stud nuts: 49 ft. lbs. (67 Nm).

    b. Flange nuts and bolts: 32–37 ft. lbs. (43–50 Nm).

14. Reconnect the oxygen sensor connectors.

15. Install the skid plate and tighten the bolts to 27 ft. lbs. (37 Nm).

16. Lower the vehicle.

17. Verify that all wires and vacuum hoses have been reconnected.

18. Reconnect the negative battery cable.

19. Start the engine and check for exhaust leaks.

Front Crankshaft Seal

REMOVAL AND INSTALLATION

1. Disconnect the negative battery cable.

2. Loosen and remove the accessory drive belts.

3. Drain the coolant to a level below the upper radiator hose. Disconnect the upper radiator hose from the coolant inlet.

4. Remove the cooling fan and pulley assembly.

5. Remove the upper timing covers.

6. Rotate the crankshaft to align the timing marks.

7. Remove the crankshaft pulley and the lower timing belt cover.

8. Remove the timing belt.

9. Remove the crankshaft timing sprocket and key.

10. Use a seal puller and remove the crankshaft oil seal. Be careful not to damage the crankshaft or the oil pump sealing surface.

    To install:

11. Apply engine oil to the lip of the seal and install the oil seal using seal

installer tool No. J–39202 or an equivalent seal driver.

12. Install the crankshaft timing sprocket and key.

13. Install the timing belt.

14. Install the lower timing belt covers and tighten the bolts to 12 ft. lbs. (17 Nm).

15. Install the crankshaft pulley and tighten it to 123 ft. lbs. (167 Nm).

16. Verify that the timing belt has been installed correctly.

17. Install the upper timing covers.

18. Install the cooling fan and pulley assembly and tighten the bolts to 16 ft. lbs. (22 Nm).

19. Install and adjust the accessory drive belts.

20. Connect the upper radiator hose. Refill and bleed the cooling system.

21. Check the engine oil level and top up if necessary.

22. Reconnect the negative battery cable.

Timing Belt, Sprockets, Tensioner and Front Cover

ADJUSTMENT

———— WARNING ————

It is very highly recommended that a timing belt be replaced any time its tension is released. A timing belt should not be viewed as an adjustable component. The timing belt's tension cannot be increased to compensate for wear. If the engine has been disassembled for mechanical work, a new timing belt should be installed. The small cost of a new timing belt is cheap insurance against expensive engine damage which can be caused by the failure of a re-used timing belt.

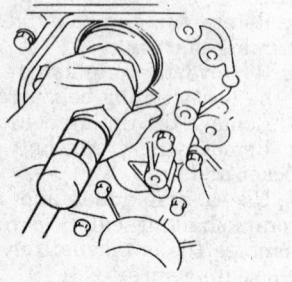

Oil pump seal installation

Setting the Tension of a New Timing Belt

1. Verify that the camshaft sprocket timing marks are aligned. The groove and the keyway on the crankshaft timing sprocket align with mark on the oil pump. The white pointers on the camshaft timing sprockets align with the dots on the front plate.

2. Install the timing belt. Use clips to secure the belt onto each sprocket until the installation is complete. Align the dotted marks on the timing belt with the timing mark opposite the groove on the crankshaft sprocket.

NOTE: The arrows on the timing belt must follow the belt's direction rotation. The manufacturer's trademark on the belt's spine should be readable left-to-right when the belt is installed.

3. Align the white line on the timing belt with the alignment mark on the right bank camshaft timing pulley. Secure the belt with a clip.

4. Rotate the crankshaft counterclockwise to remove the slack between the crankshaft sprocket and the right camshaft timing sprocket.

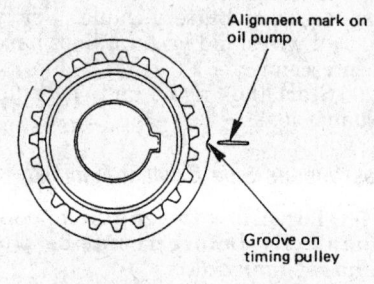

Crankshaft sprocket groove and alignment marks

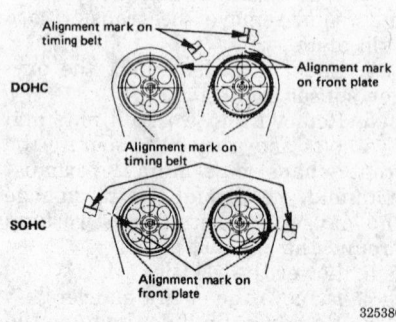

Camshaft sprocket alignment marks

5. Install the belt around the water pump pulley.

6. Install the belt on the idler pulley.

7. Align the white alignment mark on the timing belt with the alignment mark on the left bank camshaft timing sprocket.

8. Install the crankshaft pulley and tighten the center bolt by hand. Rotate the crankshaft pulley clockwise to give slack between the crankshaft timing pulley and the right bank camshaft timing pulley.

9. Insert a 1.4mm piece of wire through the hole in the pusher to hold the rod in. Install the pusher assembly while pushing the tension pulley toward the belt.

10. Pull the pin out from the pusher to release the rod.

11. Remove the clamps from the sprockets. Rotate the crankshaft pulley clockwise two turns. Measure the rod protrusion to be sure it is between 0.16–0.24 in. (4–6mm).

12. If the tensioner pulley bracket pivot bolt was removed, tighten it to 31 ft. lbs. (42 Nm).

13. Tighten pusher bolts to 14 ft. lbs. (19 Nm).

14. Remove crankshaft pulley. Install the lower and upper timing belt covers and tighten their bolts to 12 ft. lbs. (17 Nm).

15. Install the crankshaft pulley and tighten the pulley bolt to 123 ft. lbs. (167 Nm).

Timing Belt Inspection

1. Disconnect the negative and positive battery cables.

2. Label and disconnect the ignition wires and remove the spark plugs. Remove the upper timing belt cover.

3. Rotate the crankshaft to align the camshaft timing marks with the pointer dots on the back covers. Verify that the pointer on the crankshaft aligns with the mark on the lower timing cover.

NOTE: When the timing marks are aligned on 1994–95 vehicles, no cylinders will be at TDC/compression. When the timing marks are aligned on 1996–97 vehicles, the No. 2 cylinder is at TDC/compression.

4. Rotate the crankshaft pulley to cycle the belt through its entire rotation.

5. Inspect the entire length of the timing belt. Look carefully for any signs of the following conditions:

    a. Cracked, chipped, or broken teeth.

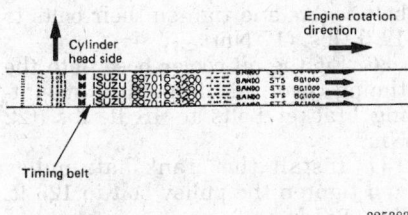

Timing belt direction of installation and rotation

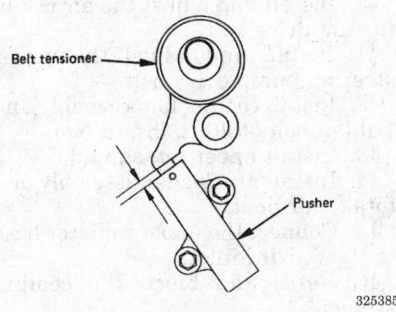

Timing belt tensioner pulley with pusher installed

b. Fraying, separation, or heat damage to the belt's rubber and fiber layers.

c. Oil or coolant leaks which may have contaminated the belt.

d. Make sure the timing marks align. Misaligned marks may indicate that the belt has jumped one or more teeth, or has been improperly tensioned.

6. Check the camshaft and crankshaft oil seals for any signs of leakage. Also check the water pump for leakage. The source of any oil or coolant leaks must be found and corrected before a new timing belt is installed.

7. Replace the timing belt if it's damaged in any way, or if its condition or the vehicle's maintenance history is uncertain. Isuzu's recommended service interval for timing belt replacement is 60,000 miles (96,000 Km). It's also recommended to install a new water pump when the timing belt is replaced

8. After inspection, retighten the crankshaft pulley bolt to 123 ft. lbs. (167 Nm). Install the timing belt covers and valve covers. Install the spark plugs and reconnect the ignition wires. Reconnect the battery cables.

REMOVAL AND INSTALLATION

1993–95 3.2L (VIN V) Engine

1. Disconnect the negative battery cable.

2. Drain the engine coolant into a sealable container.

3. Remove the air cleaner assembly.

4. Remove the upper fan shroud from the radiator.

5. Remove the four nuts retaining the cooling fan assembly. Remove the cooling fan.

6. Remove the power steering drive belt.

7. Remove the air conditioning compressor drive belt.

8. Remove the alternator drive belt.

9. Remove the fan pulley assembly.

10. Set the timing belt so that the timing marks on the sprockets align with those on the oil pump and front plate.

11. Remove the crankshaft pulley center bolt. Remove the crankshaft pulley.

12. Remove the two oil cooler hose bracket fixing bolts on the timing cover. Remove the oil cooler hoses and bracket.

13. Remove the timing belt cover.

--- **WARNING** ---

Set the timing sprockets so that the crankshaft and camshaft sprocket timing marks align with the timing marks on the oil pump and front plate. When the timing marks are align, no pistons are at TDC for the compression stroke. Valve damage may result if the sprockets and belt are out of alignment.

14. Remove the pusher. The rod must always be facing upward.

15. Mark the timing belt, cam pulley and crankshaft pulley. Remove the timing belt.

To install:

16. Align the groove on the crankshaft timing pulley with mark on the oil pump.

17. Align the marks on the camshaft timing pulleys with the dots on the front plate.

18. Install the timing belt. Align the dotted marks on the timing belt with the mark on the crankshaft pulley.

19. Align the white line on the timing belt with the alignment mark on the right bank camshaft timing pulley. Secure the belt with a small clamp.

20. Turn the crankshaft counter-clockwise to remove the slack between the crankshaft pulley and the right camshaft timing pulley.

21. Install the belt on the water pump pulley.

22. Install the belt on the idler pulley.

23. Align the white alignment mark on the timing belt with the alignment mark on the left bank camshaft timing pulley.

24. Install the crankshaft pulley and tighten the center bolt by hand. Turn the crankshaft pulley clockwise to give slack between the crankshaft timing pulley and the right bank camshaft timing pulley.

25. Install the pusher while pushing the tension pulley to the belt.

26. Pull the pin out from the pusher.

27. Remove the clamps from the pulleys. Turn the crankshaft pulley clockwise two turns. Measure the rod protrusion to be sure it is between 0.16–0.24 in. (4–6mm).

28. Check the stick–out of the pusher plunger against the belt tensioner. Clearance should be 0.16–0.24 in. (4–6mm).

29. Tighten adjusting bolt to 31 ft. lbs. (42 Nm).

30. Tighten pusher bolt to 14 ft. lbs. (19 Nm).

31. Remove crankshaft pulley. Install the timing belt cover and tighten bolts to 12 ft. lbs. (17 Nm).

32. Install the oil cooler hose and tighten brackets to 16 ft. lbs. (22 Nm).

33. Install the crankshaft pulley and tighten bolts to 123 ft. lbs. (167 Nm).

34. Install fan pulley assembly and tighten fixing bolts to 16 ft. lbs. (22 Nm).

35. Install and adjust alternator drive belt.

36. Install and adjust air conditioning drive belt.

37. Install and adjust power steering pump drive belt.

38. Install cooling fan assembly and tighten bolts to 69 inch lbs. (8 Nm).

39. Install upper fan shroud to the radiator.

40. Install air cleaner assembly.

41. Refill and bleed the cooling system.

42. Reconnect the negative battery cable.

43. Check the operation of the engine and valve train.

1996–97 3.2L (VIN V) Engine

1. Disconnect the negative battery cable.

2. Drain the engine coolant into a sealable container.

3. Remove the air cleaner assembly and intake air duct.

4. Disconnect the upper radiator hose from the coolant inlet.

5. Remove the upper fan shroud from the radiator.

6. Remove the four nuts retaining the cooling fan assembly. Remove the cooling fan from the fan pulley.

7. Loosen and remove the power steering drive belt.

8. Loosen and remove the air conditioning compressor drive belt.

9. Loosen and remove the alternator drive belt.

10. Remove the upper timing belt covers.

11. Remove the fan pulley assembly.

12. Rotate the crankshaft to align the camshaft timing marks with the pointer dots on the back covers. Verify that the pointer on the crankshaft aligns with the mark on the lower timing cover.

NOTE: When the timing marks are aligned on 1994–95 vehicles, no cylinders will be at TDC/compression. When the timing marks are aligned on 1996–97 vehicles, the No. 2 cylinder is at TDC/compression.

———— WARNING ————
Align the camshaft and crankshaft sprockets with their alignment marks before removing the timing belt. Failure to align the belt and sprocket marks may result in valve damage.

13. Use tool No. J–8614–01, or a suitable pulley holding tool to remove the crankshaft pulley center bolt. Remove the crankshaft pulley.

14. Disconnect the two oil cooler hose bracket bolts on the timing cover. Move the oil cooler hoses and bracket off of the lower timing cover.

15. Remove the lower timing belt cover.

16. Remove the pusher assembly (tensioner) from below the belt tensioner pulley. The pusher rod must always be facing upward to prevent oil leakage. Push the pusher rod in, and insert a wire pin into the hole to keep the pusher rod retracted.

17. Remove the timing belt.

18. Use tool No. J–41472, or a suitable pulley holding tool to loosen and remove the camshaft sprocket bolt. Remove the camshaft sprockets.

19. Inspect the water pump and replace it if there is any doubt about its condition.

20. Repair any oil or coolant leaks before installing a new timing belt. If the timing belt has been contaminated with oil or coolant, or is damaged, it must be replaced.

To install:

21. Install the camshaft sprockets. Use a holding tool, and tighten their bolts to 41 ft. lbs. (55 Nm).

22. Verify that the sprocket timing marks are still aligned. The groove and the keyway on the crankshaft timing sprocket align with mark on the oil pump. The white pointers on the camshaft timing sprockets align with the dots on the front plate.

23. Install the timing belt. Use clips to secure the belt onto each sprocket until the installation is complete. Align the dotted marks on the timing belt with the timing mark opposite the groove on the crankshaft sprocket.

NOTE: The arrows on the timing belt must follow the belt's direction rotation. The manufacturer's trademark on the belt's spine should be readable left-to-right when the belt is installed.

24. Align the white line on the timing belt with the alignment mark on the right bank camshaft timing pulley. Secure the belt with a clip.

25. Rotate the crankshaft counterclockwise to remove the slack between the crankshaft sprocket and the right camshaft timing sprocket.

26. Install the belt around the water pump pulley.

27. Install the belt on the idler pulley.

28. Align the white alignment mark on the timing belt with the alignment mark on the left bank camshaft timing sprocket.

29. Install the crankshaft pulley and tighten the center bolt by hand. Rotate the crankshaft pulley clockwise to give slack between the crankshaft timing pulley and the right bank camshaft timing pulley.

30. Insert a 1.4mm piece of wire through the hole in the pusher to hold the rod in. Install the pusher assembly while pushing the tension pulley toward the belt.

31. Pull the pin out from the pusher to release the rod.

32. Remove the clamps from the sprockets. Rotate the crankshaft pulley clockwise two turns. Measure the rod protrusion to be sure it is between 0.16–0.24 in. (4–6mm).

33. If the tensioner pulley bracket pivot bolt was removed, tighten it to 31 ft. lbs. (42 Nm).

34. Tighten pusher bolts to 14 ft. lbs. (19 Nm).

35. Remove the crankshaft pulley. Install the lower and upper timing belt covers and tighten their bolts to 12 ft. lbs. (17 Nm).

36. Fit the oil cooler hose onto the timing cover and tighten its mounting bracket bolts to 16 ft. lbs. (22 Nm).

37. Install the crankshaft pulley and tighten the pulley bolt to 123 ft. lbs. (167 Nm).

38. Install fan pulley assembly and tighten bolts to 16 ft. lbs. (22 Nm).

39. Install and adjust the alternator drive belt.

40. Install and adjust the air conditioning drive belt.

41. Install and adjust the power steering pump drive belt.

42. Install cooling fan assembly and tighten bolts to 6 ft. lbs. (8 Nm).

43. Install upper fan shroud.

44. Install air cleaner assembly and intake air duct.

45. Connect the upper radiator hose to the coolant inlet.

46. Refill and bleed the cooling system.

47. Connect the negative battery cable.

Camshaft

REMOVAL AND INSTALLATION

1993–95 3.2L (VIN V and W) Engines

Single Overhead Camshaft (SOHC) Engine

NOTE: Isuzu has issued a recall notice for 1993–1994 Troopers equipped with 3.2L (VIN V) 6VD1 SOHC engines. This is campaign number 94V–094, and involves faulty camshaft end plugs. The plugs may dislodge from the cylinder heads and can cause rapid oil loss. Remember this when ordering parts; a service kit is available for affected vehicles.

1. Disconnect the negative battery cable.

2. Remove the air cleaner assembly.

3. Remove the upper cooling fan shroud.

4. Remove the cooling fan assembly.

5. Disconnect the accelerator cable from the throttle body and remove it from the bracket.

6. Disconnect the canister vacuum hose from the vacuum pipe.

7. Disconnect the vacuum booster hose from the common chamber.

8. Disconnect the MAP sensor; Canister Vacuum Switching Valve (VSV); Exhaust gas recirculation VSV; Intake air temperature sensor and ground connectors.

9. Remove the spark plug wires from the cylinder head cover.

10. Remove the ignition control module assembly.

11. Remove the four bolts and the throttle body from the common chamber.

12. Disconnect the vacuum hoses from the throttle body.

13. Disconnect the positive crankcase ventilation hose from the common chamber.

14. Disconnect the fuel pressure control valve vacuum hose from the common chamber.

15. Disconnect the evaporative emission canister purge hose from the common chamber.

16. Remove the EGR valve assembly.

17. Remove the common chamber from the intake manifold.

18. Disconnect the fuel feed and return hoses from the fuel rail assembly.

19. Remove the accessory drive belts.

20. Remove the power steering pump.

21. Remove the fan pulley assembly.

22. Remove the crankshaft pulley and damper.

23. Remove the timing belt cover.

24. Remove the timing belt auto tensioner (pusher). The pusher prevents air from entering the oil chamber. Its rod must always be facing upward.

25. Align the timing marks and remove the timing belt. After the timing marks are aligned, the engine must not be disturbed.

26. Remove the cylinder head cover(s).

27. Remove the camshaft pulley.

28. Remove the camshaft front plate.

29. Remove the camshaft mounting bracket bolts and the camshaft.

To install:

30. Install new camshaft seals and retaining plates onto the cylinder. Tighten the right camshaft seal retaining plate 6mm bolts to 65 inch lbs. Tighten the left camshaft seal retaining plate 8mm bolts to 191 inch lbs.

31. Apply sealant to the mounting surfaces on the cylinder head where the front and rear camshaft mounting brackets attach to the cylinder head.

32. Install the camshaft and mounting brackets. Torque the bolts in sequence to 69 inch lbs. (8 Nm) for the M6 bolts and 13 ft. lbs. (18 Nm) for the M8 bolts.

33. Install the front plate. Torque the bolts to 12 ft. lbs. (17 Nm).

34. Install the camshaft pulley. Torque the mounting bolts to 41 ft. lbs. (55 Nm).

35. Apply sealant to both sides of the front and rear camshaft mounting brackets and install the cylinder head cover(s). Torque the bolts to 69 inch lbs. (8 Nm). Do not overtighten the cylinder head covers: they crack very easily.

36. Install the timing belt.

37. Install the timing belt auto tensioner. Torque the mounting bolts to 13 ft. lbs. (18 Nm).

38. Rotate the crankshaft by hand to verify that the timing belt is aligned properly and there is no piston-to-valve interference.

39. Install the timing belt cover.

40. Install the oil cooler hoses and bracket.

41. Install the crankshaft pulley assembly. Torque the center bolt to 123 ft. lbs. (167 Nm).

42. Install the fan pulley assembly. Torque the mounting bolts to 16 ft. lbs. (22 Nm).

43. Install the power steering pump.

44. Install the accessory drive belts.

45. Connect the fuel hoses to the fuel rail assembly.

46. Install the common chamber. Torque the bolts and nuts to 17 ft. lbs. (24 Nm) in a crisscross pattern.

47. Install the EGR valve assembly. Torque the mounting bolts on the valve side to 69 inch lbs. (8 Nm) and the bolts on the exhaust side to 21 ft. lbs. (28 Nm).

48. Connect the evaporative emission canister purge hose.

49. Connect the fuel pressure control valve vacuum hose.

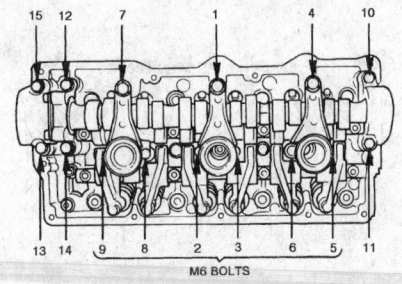

Camshaft mounting bolts torque sequence — 1993–95 SOHC engine

50. Connect the positive crankcase ventilation hose.

51. Install the throttle body assembly. Torque the mounting bolts to 14 ft. lbs. (19 Nm).

52. Connect the vacuum hoses to the throttle body.

53. Install the ignition control module and the spark plug wires.

54. Connect the MAP sensor; Canister Vacuum Switching Valve (VSV); Exhaust gas recirculation VSV; Intake air temperature sensor and ground connectors.

55. Connect the vacuum booster hose.

56. Connect the air vacuum hose.

57. Connect the accelerator cable. Adjust the accelerator cable by pulling the cable housing while closing the throttle valve and tightening the adjusting nut and screw cap by hand temporarily. Now loosen the adjusting nut by three turns and then tightening the screw cap. Make sure the throttle valve reaches the screw stop when the throttle is closed.

58. Install the cooling fan assembly and the upper fan shroud.

59. Install the air cleaner assembly.

60. Connect the negative battery cable.

Double Overhead Camshaft (DOHC) Engine

NOTE: Isuzu has issued a recall notice for 1993–1994 Troopers equipped with 3.2L (VIN V) 6VD1-W DOHC engines. This is campaign number 94V–094, and involves faulty camshaft end plugs. The plugs may dislodge from the cylinder heads and can cause rapid oil loss. Remember this when ordering parts: a service kit is available for affected vehicles.

1. Properly relieve the fuel system pressure.

2. Disconnect the negative battery cable.

3. Remove the air cleaner assembly.

4. Remove the upper cooling fan shroud.

5. Remove the cooling fan assembly.

6. Disconnect the accelerator cable from the throttle body and remove it from the bracket.

7. Disconnect the canister vacuum hose from the vacuum pipe.

8. Disconnect the vacuum booster hose from the common chamber.

9. Disconnect the following connectors:
   a. MAP sensor.

b. Canister Vacuum Switching Valve (VSV).

c. Exhaust gas recirculation (VSV).

d. Intake air temperature sensor.

e. Ground cables.

10. Disconnect and label the spark plug wires.

11. Remove the ignition control module assembly with the spark plug wire attached.

12. Remove the four bolts to separate the throttle body from the common chamber.

13. Disconnect the vacuum hoses from the throttle body.

14. Disconnect the positive crankcase ventilation (PCV) hose from the common chamber.

15. Disconnect the fuel pressure control valve vacuum hose from the common chamber.

16. Disconnect the evaporative emission canister purge hose from the common chamber.

17. Remove the EGR valve assembly.

18. Remove the common chamber air duct.

19. Remove the common chamber from the intake manifold.

20. Disconnect the fuel feed and return hoses from the fuel rail assembly.

21. Remove the accessory drive belts.

22. Remove the power steering pump.

23. Remove the fan pulley assembly.

24. Remove the crankshaft pulley and damper.

25. Remove the oil cooler hoses and bracket.

26. Remove the timing belt cover.

27. Align the timing marks

28. Remove the timing belt auto tensioner (pusher). The pusher prevents air from entering the oil chamber. Its rod must always be facing upward.

29. Remove the timing belt.

30. Remove the cylinder head covers.

31. Remove the camshaft pulley.

32. Remove the camshaft front plate.

33. Remove the camshaft mounting bracket bolts.

34. Remove the camshaft chain tensioner bolts.

35. Remove the camshafts with the timing chain and tensioner attached.

To install:

NOTE: Install a retainer on the chain tensioner to prevent the plunger from moving. Remove the retainer after installation.

36. Apply sealant to the cylinder head mounting surface of the front and rear camshaft mounting brackets.

37. Apply clean engine oil to the camshaft journals, lobes, and sprockets.

38. Install the camshaft assembly with the chain and tensioner. Take care not to install the wrong tensioner, the left and right tensioners are different and they are marked accordingly. Make sure the timing marks on the camshaft chain sprockets are aligned with the timing marks on the chain links. Torque the camshaft holder mounting bolts to 87 inch lbs. (10 Nm).

39. Install the chain tensioner. Torque the bolts to 14 ft. lbs. (19 Nm).

40. Install new camshaft seals and retaining plates onto the cylinder. Tighten the rear camshaft seal retaining plate 6mm bolts to 70 inch lbs. Tighten the front camshaft seal retaining bolts to 65 inch lbs.

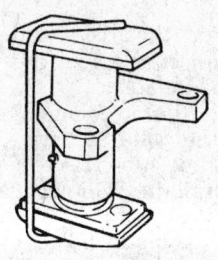

253473

Install a retainer on the chain tensioner and remove it after installation — 1993–95 DOHC engine

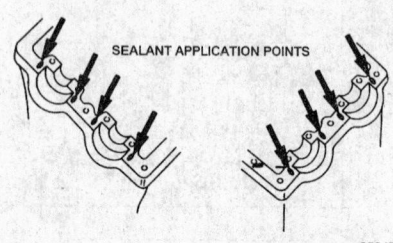

SEALANT APPLICATION POINTS

253477

Apply sealant to the bracket mounting surface on the cylinder head — 1993–95 DOHC engine

41. Install the camshaft pulley. Hold the camshaft with an open ended wrench to prevent it from turning and torque the pulley bolts to 41 ft. lbs. (55 Nm) for 1992–94 vehicles, or 46 ft. lbs. (63 Nm) for 1995 vehicles.

42. Apply sealant to both sides of the front and rear camshaft mounting brackets and install the cylinder head cover(s). Torque the bolts to 69 inch lbs. (8 Nm). Don't over-torque the bolts, as the head covers may crack.

43. Install the timing belt.

44. Install the timing belt auto tensioner. Torque the mounting bolts to 14 ft. lbs. (19 Nm).

45. Install the timing belt cover.

46. Install the oil cooler hoses and bracket.

47. Install the crankshaft pulley assembly. Torque the center bolt to 123 ft. lbs. (167 Nm).

48. Install the fan pulley assembly. Torque the mounting bolts to 16 ft. lbs. (22 Nm).

49. Install the power steering pump.

50. Install the accessory drive belts.

51. Connect the fuel hoses to the fuel rail assembly.

52. Install the common chamber. Torque the bolts and nuts to 17 ft. lbs. (24 Nm).

53. Install the common chamber air duct. Torque the nuts and bolts to 17 ft. lbs. (24 Nm).

54. Connect the ground cable.

55. Install the EGR valve assembly. Torque the mounting bolts on the valve side to 69 inch lbs. (8 Nm) and the bolts on the exhaust side to 21 ft. lbs. (28 Nm).

56. Connect the evaporative emission canister purge hose.

57. Connect the fuel pressure control valve vacuum hose.

58. Reconnect the fuel lines using new washers.

59. Connect the positive crankcase ventilation hose.

60. Install the throttle body assembly. Torque the mounting bolts to 14 ft. lbs. (19 Nm).

61. Connect the vacuum hoses to the throttle body.

62. Install the ignition control module and the spark plug wires.

63. Connect the MAP sensor; Canister Vacuum Switching Valve (VSV); Exhaust gas recirculation VSV, the Intake air temperature sensor connectors; and ground cables.

64. Connect the vacuum booster hose.

65. Connect the air vacuum hose.

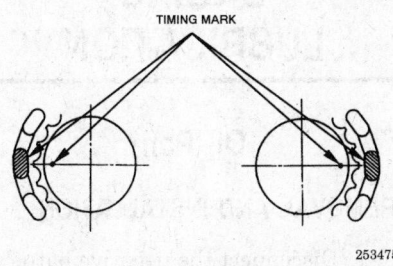

Camshaft and chain timing marks

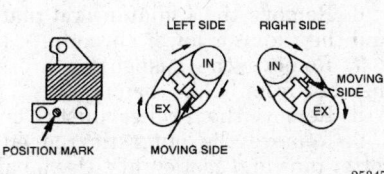

Chain tensioner installed position — 1993–95 DOHC engine

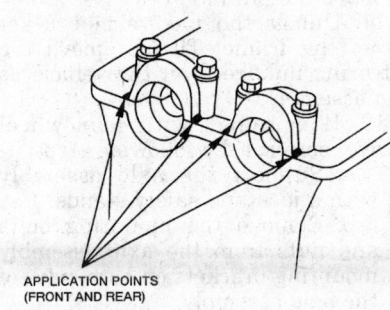

Apply sealant to the camshaft mounting brackets — 1993–95 DOHC engine

66. Connect the accelerator cable. Adjust the accelerator cable by pulling the cable housing while closing the throttle valve and tightening the adjusting nut and screw cap by hand temporarily. Now loosen the adjusting nut by three turns and then tightening the screw cap. Make sure the throttle valve reaches the screw stop when the throttle is closed.

67. Install the cooling fan assembly and the upper fan shroud.

68. Install the air cleaner assembly.

69. Connect the negative battery cable.

70. Start the engine and check for fuel leaks.

1996–97 3.2L (VIN V) Engine

─── CAUTION ───

Fuel injection systems remain under pressure even after the engine has been turned off. The fuel system pressure must be relieved before disconnecting any fuel lines. Failure to do so may result in fire and personal injury.

1. Properly relieve the fuel system pressure.

2. Disconnect the negative battery cable.

3. Drain the coolant into a sealable container.

4. Support the hood as far open as possible.

5. Remove the air intake duct and the air cleaner box.

6. Disconnect and remove the upper and lower radiator hoses. Catch any coolant that runs out.

7. Loosen and remove the power steering pump, A/C compressor, and alternator drive belts.

8. Remove the cooling fan and its pulley assembly.

9. Unbolt the power steering pump mounting bracket. Move the pump and bracket out of the way without disconnecting the hydraulic lines.

10. Disconnect the throttle cable from the throttle body linkage.

11. Label and disconnect the following vacuum hoses from the intake manifold chamber:
   a. PCV hose.
   b. EVAP canister vacuum hose.
   c. Brake booster hose.

12. Label and disconnect the following sensor connectors from the rear of the intake manifold chamber:
   a. Ignition control module connectors.
   b. Linear EGR valve.
   c. MAP sensor.
   d. EVAP purge valve.
   e. Throttle position sensor.
   f. Idle air control valve.
   g. Intake air temperature sensor.

13. Disconnect the EGR valve supply tube and bracket.

14. First, remove the throttle body, and then remove the intake manifold chamber.

15. Carefully clean any dirt from the fuel rail and fuel fittings.

16. Disconnect the fuel feed and return lines from the front of the fuel rail. Clean up any spilled fuel.

17. Remove the intake manifold gaskets. Be careful not to drop any pieces of the gaskets into the engine. Don't scratch or gouge the machined aluminum mating surfaces of the intake manifold and engine block.

18. Cover the intake openings with a sheet of plastic or clean shop towels to keep out dirt and foreign objects.

19. Label the ignition coil assemblies and disconnect them from the wiring harness. Remove the coil assemblies so they won't be damaged.

20. Unbolt the oil cooler line brackets from the timing belt covers.

21. Remove the upper timing belt covers.

22. Rotate the crankshaft to align the camshaft timing marks with the pointer dots on the back covers. When the timing marks are aligned, the No. 2 cylinder is at TDC/compression.

23. Remove the crankshaft pulley. Remove the lower timing belt cover.

24. Remove the pusher assembly (tensioner) from below the timing belt tensioner pulley. The pusher rod must always be facing upward to prevent oil leakage. Push the pusher rod in, and insert a wire pin into the hole to keep the pusher rod retracted.

25. Remove the timing belt.

─── WARNING ───

If the timing belt is worn, damaged, or shows signs of oil or coolant contamination, it must be replaced.

26. Loosen the valve cover bolts in a crisscross sequence. Remove the valve covers.

27. Remove the camshaft sprockets and back covers.

28. Loosen the camshaft holder bolts in a crisscross sequence to prevent warping.

29. Remove the camshaft and camshaft holders from the cylinder head.

30. Inspect the camshaft lobes and journals for signs of wear or damage.

   To install:

31. Make sure all mating surfaces are clean and free of oil, coolant, or gasket residue.

32. Lubricate the camshaft lobes and journals with clean engine oil.

33. Apply a bead of sealant to the front and rear camshaft holder mating surfaces on the cylinder head.

34. Install the camshaft and holder assembly onto the cylinder head before the sealant cures. Install the camshaft holder bolts, but don't tighten them yet.

35. Use a crisscross sequence to tighten the camshaft holder bolts. Tighten the 8mm bolts to 13 ft. lbs.

(18 Nm). Tighten the 6mm bolts to 6 ft. lbs. (8 Nm).

36. Use a seal driver to install a new camshaft seal.

37. Install the camshaft sprocket back covers and tighten their bolts to 12 ft. lbs. (17 Nm).

38. Install the camshaft sprockets so that the timing marks are aligned. Tighten the bolts to 46 ft. lbs. (64 Nm).

39. Apply a 2–3mm bead of sealant to the joint were the camshaft holders meet the cylinder head. Install the valve cover with a new gasket before the sealant cures.

40. Tighten the valve cover bolts to 6 ft. lbs. (8 Nm) in crisscross pattern.

41. Verify that the camshaft and crankshaft timing marks are properly aligned.

42. Install and tension the timing belt. Tighten the pusher bolts to 14 ft. lbs. (19 Nm).

43. Install the lower timing belt covers and tighten the bolts to 13 ft. lbs. (18 Nm). Install the crankshaft pulley. Tighten the pulley bolt to 123 ft. lbs. (167 Nm).

44. Install the upper timing belt covers and tighten the bolts to 13 ft. lbs. (18 Nm).

45. Fit the oil cooler line brackets onto the timing cover and tighten the bolts to 13 ft. lbs. (18 Nm).

46. Reconnect the fuel feed and return lines.

47. Install and reconnect the ignition coil assemblies.

48. Install the intake manifold chamber and throttle body with new gaskets. Tighten the nuts and bolts to 17 ft. lbs. (24 Nm).

49. Reconnect the throttle cable to the throttle body linkage.

50. Reconnect the EGR valve supply tube and bracket.

51. Reconnect the following vacuum to the intake manifold chamber:

   a. PCV hose.
   b. EVAP canister vacuum hose.
   c. Brake booster hose.

52. Reconnect the following sensor connectors to the rear of the intake manifold chamber:

   a. Ignition control module connectors.
   b. Linear EGR valve.
   c. MAP sensor.
   d. EVAP purge valve.
   e. Throttle position sensor.
   f. Idle air control valve.
   g. Intake air temperature sensor.

53. Install the power steering pump and mounting bracket.

54. Install the cooling fan and its pulley assembly.

55. Install and tension the alternator, A/C compressor, and power steering pump drive belts.

56. Install and reconnect the upper and lower radiator hoses.

57. Install the air cleaner box and air intake duct.

58. Verify that all fuel lines, vacuum and coolant hoses, and wiring harness have been reconnected.

59. Refill the engine with fresh coolant.

60. Crank the engine until it starts. A longer–than–normal starting time may be necessary due to air in the fuel lines. Check all fuel line connections for leaks.

61. Bleed any air from the cooling system.

62. Bleed the power steering system if necessary.

63. Check the throttle cable operation and adjustment.

64. Check the engine oil level and add if necessary.

Piston and Connecting Rod

POSITIONING

ENGINE LUBRICATION

Oil Pan

REMOVAL AND INSTALLATION

1. Disconnect the negative battery cable.

2. Drain the engine oil.

3. Raise and safely support the vehicle.

4. Remove the front wheels.

5. Remove the dipstick from the dipstick tube.

6. Remove the radiator skid plate and the radiator lower shroud.

7. Remove the suspension crossmember from below the oil pan.

8. Remove the flywheel dust cover.

9. Remove the outer tie rod end cotter pins and castle nuts. Use a ball joint separator tool to disconnect the tie rod ends from the steering knuckles.

10. Matchmark the pitman arm to the steering shaft and use a puller, (tool No. J–29107, or equivalent) to remove the pitman arm.

11. Unbolt the idler arm bracket from the frame. Then, remove the steering linkage from the vehicle as an assembly.

12. If equipped with four wheel drive, perform the following steps:

   a. Support the axle assembly with a jack and safety stands.

   b. Remove the mounting bolts and nuts from the axle assembly mounting brackets on both sides of the axle assembly.

   c. Lower the axle assembly to gain access to the oil pan. Support the axle assembly so that the half-shafts aren't stressed.

13. Remove the oil pan mounting bolts. Use a sealer cutter to break the seal and separate the oil pan from the engine block.

To install:

14. Thoroughly clean and dry the sealing surface of the oil pan and engine block. Apply a continuous bead of sealant to the oil pan flange and install the oil pan to the engine block. Do not allow the sealant to cure before installation. Tighten the mounting bolts to 74. ft. lbs. (10 Nm) in a two-step crisscross sequence.

15. Install the axle housing assembly. Tighten the axle mounting bracket bolts and nut to 112 ft. lbs. (152 Nm). If the axle housing flange bolts at the mounting brackets were

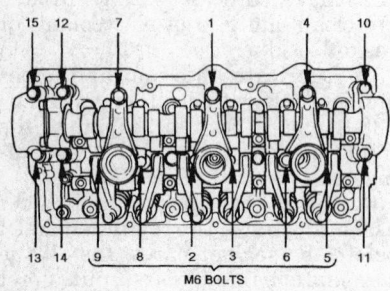

Camshaft mounting bolts tightening sequence — 1996–97 SOHC engine

324837

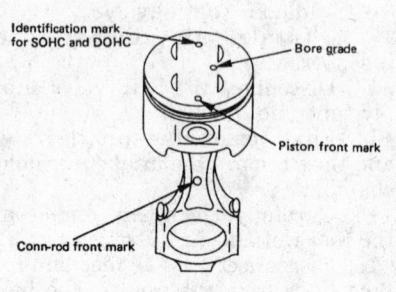

Piston identification marks

324245

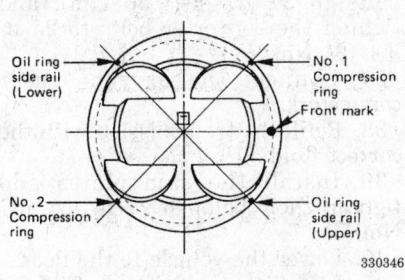

Piston ring end-gap locations

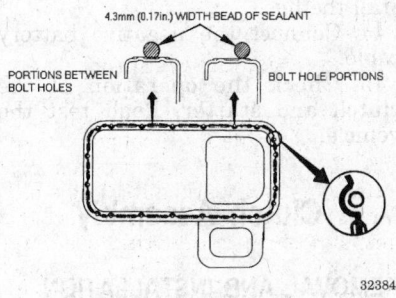

Apply sealant to the oil pan flange

loosened or removed, tighten them to 61 ft. lbs. (82 Nm).

16. Install the idler arm. Tighten the mounting bolts to 33 ft. lbs. (45 Nm).

17. Align the matchmark and install the pitman arm on the sector shaft. Tighten the nut to 160 ft. lbs. (216 Nm).

18. Install the flywheel dust cover.

19. Install the suspension crossmember. Tighten the mounting bolts to 58 ft. lb. (78 Nm).

20. Install the steering linkage, pitman arm, and idler arm assembly.

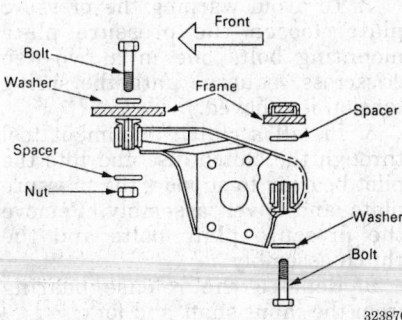

Axle bracket mounting bolt, spacer, and nut locations

Tighten the castle nuts to the following specifications:

   a. Tie rod end castle nuts: 73 ft. lbs. (98 Nm).

   b. Idler arm bracket bolts: 33 ft. lbs. (44 Nm).

   c. Pitman arm nuts: 159 ft. lbs. (216 Nm).

21. Install the lower fan shroud.

22. Install the radiator skid plate. Tighten the bolts to 27 ft. lbs. (37 Nm).

23. Verify that the front axle and any suspension components have been correctly installed.

24. Lower the vehicle to the floor.

25. Install the dipstick and refill the engine with the proper amount of oil.

26. Connect the negative battery cable.

27. Start the engine and check for oil leaks.

Oil Pump

REMOVAL AND INSTALLATION

1. Disconnect the negative battery cable.

2. Remove the timing belt. If the timing belt is damaged or has been contaminated with oil or coolant, it must be replaced.

3. Remove the crankshaft timing pulley.

4. Raise and safely support the vehicle.

5. Drain the engine oil.

6. Remove the oil pan.

7. Remove the oil pipe and O-ring.

8. Remove the oil strainer and O-ring.

9. Remove the oil cooler assembly.

10. Remove the oil pump mounting bolts, and then remove the oil pump from the engine block.

   To install:

11. Install a new oil seal into the oil pump housing.

12. Thoroughly clean the sealing surface of the oil pump and the engine block.

13. Apply sealant to the oil pump. Be careful not to block the oil ports.

14. Apply engine oil to the seal lip and install the oil pump on the engine block. Tighten the mounting bolts to 13 ft. lbs. (18 Nm). Take care not to drop the garter spring from the seal lid during installation.

15. Install the oil cooler assembly.

16. Install the oil pipe and O-ring.

17. Install the oil strainer and O-ring.

18. Install the oil pan.

19. Install the crankshaft timing pulley.

20. Install the timing belt.

21. Install the remaining accessories and drive belts.

22. Lower the vehicle.

23. Refill the engine with oil.

24. Connect the negative battery cable, start the engine and check for proper oil pressure.

NOTE: If the oil pressure does not build up almost immediately, stop the engine and investigate the cause.

25. Check for oil leaks.

TRANSMISSION

Manual Transmission Assembly

REMOVAL AND INSTALLATION

NOTE: The transfer case is an integral part of the transmission housing. Although the two cases can be separated, the transfer case should be removed with the transmission.

1. Shift the transmission into **N**. Shift the transfer case into **2H** and drive the vehicle forward and backward a few feet/meters to make sure the front axle is not engaged.

2. Use a felt–tipped marker to matchmark the hood to the hood hinges. Remove the hood.

3. Disconnect the negative battery cable.

4. Remove the shift knobs and the console and shift boots.

5. Remove the shift lever and the transfer case shift lever, by unbolting their mounting plates from the transmission case.

6. Raise and safely support the vehicle.

7. Drain the transmission fluid.

8. Remove the exhaust and transfer case skid plates.

9. Label and disconnect the oxygen sensor connectors from the transmission harness.

10. Remove the catalytic converter, left front, and center exhaust pipes.

11. Remove the harness heat protector.

12. Remove the slave cylinder heat protector. Unbolt and remove the slave cylinder. Don't disconnect the hydraulic line.

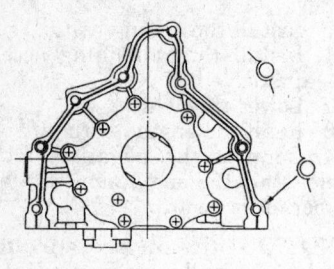

Oil pump sealant application points

324141

13. Remove the slave cylinder dust covers.

14. Matchmark the driveshafts at the their flanges. Unbolt and remove the driveshafts.

15. Label and disconnect the reverse light switch, 4WD indicator switch, and 1–2 and 3–4 indicator switch harness connectors.

16. Disconnect the speed sensor harness connector.

17. Remove the two harness clamps from the transmission case.

18. Using a transmission jack, raise the transmission slightly. Remove the two rear transmission mounting nuts.

19. Remove the center crossmember (8 bolts).

20. Attach a chain hoist to the engine lifting brackets, but don't raise the engine yet.

21. Remove the front crossmember.

NOTE: Make sure the engine assembly is properly supported when removing the front crossmember.

22. Remove the three flywheel inspection cover bolts.

23. Use the clutch release bearing remover tool No. J–39207, or an equivalent, pry tool to release the bearing from the pressure plate. Push the release fork toward the rear of the vehicle. Insert the tool between the release bearing and the pressure plate collar. Move the lever toward the rear to pry.

24. Raise the engine slightly with a chain hoist and remove the bolts and nuts securing the transmission to the engine.

25. Carefully pull the transmission rearward. Lower the transmission from the vehicle.

To install:

NOTE: Make sure the transmission dowel pins are installed in the correct position. If the dowels are in the wrong hole, the transmission case may crack.

26. Apply a thin coating of molybdenum grease to the spline of the input shaft, slowly raise the transmission into position with the rear of the engine. Align the splines of the input shaft with the grooves of the clutch disc hub and install the transmission to the engine.

NOTE: It may be helpful the put the transmission in gear and rotate the driveshaft flange so the input shaft will turn and engage the grooves in the clutch disc hub.

27. Install the transmission case bolts. Tighten the upper six bolts to 56 ft. lbs. (76 Nm). Tighten the two remaining large bolts to 56 ft. lbs. (76 Nm). Tighten the remaining three bolts to 5.2 inch lbs. (6 Nm).

28. Push the release bearing fork rearward with a force of 13–18 lbs. (59–78 N) to engage the release bearing with the pressure plate. A click sound will be heard when the bearing engages the pressure plate properly.

29. Install the flywheel inspection cover.

30. Install the front crossmember and the front driveshaft. Tighten the crossmember bolts to 58 ft. lbs. (78 Nm). Tighten the driveshaft flange bolts to 46 ft. lbs. (60 Nm).

31. Install the rear crossmember and mount. Tighten the crossmember mounting bolts to 37 ft. lbs. (50 Nm) and the transmission mounting nuts to 30 ft. lbs. (41 Nm).

32. Remove the transmission jack and the engine hoist.

33. Connect the transmission harness connectors and install the harness clamps.

34. Install the rear driveshaft and tighten the flange bolts to 46 ft. lbs. 60 (Nm).

35. Apply grease to the dimple on the end of the release bearing fork and install the slave cylinder, heat protector, and dust covers. Tighten the slave cylinder bolts to 32 ft. lbs. (43 Nm). Tighten the dust cover bolts to 5.2 ft. lbs. (6 Nm).

36. Install the exhaust pipes and heat protectors using new self–locking nuts. Tighten the exhaust manifold flange bolts to 49 ft. lbs. (67 Nm). Tighten the center pipe flange bolts to 32 ft. lbs. (43 Nm) for 1995–97 vehicles. On 1993–1994 ve-

hicles: tighten the center pipe flange bolts to 37 ft. lbs. (50 Nm); and tighten the converter bolts to 25 ft. lbs. (34 Nm).

37. Connect the oxygen sensor connectors.

38. Refill the transmission with the correct fluid.

39. Install the skid plates and tighten their bolts to 27 ft. lbs. (37 Nm).

40. Lower the vehicle to the floor.

41. Install the shift levers. Tighten the shift lever mounting plate bolts to 15 ft. lbs. (20 Nm).

42. Install the console, shift boots, and shift knobs.

43. Align the matchmarks and install the hood.

44. Connect the negative battery cable.

45. Check the operation of the clutch and starter. Road test the vehicle.

Clutch Assembly

REMOVAL AND INSTALLATION

1. Shift the transmission into **N** and the transfer case into **2H**. Drive the vehicle forward and backward for a few feet to make sure the front axle and hubs are disengaged.

2. Disconnect the negative battery cable.

3. Remove the shift knobs . Remove the console screws and lift the console and shift boots over the shift levers.

4. Raise and support the vehicle safely.

5. Remove the transmission and transfer case as an assembly.

6. If the pressure plate is going to be reused, matchmark the pressure plate to the flywheel so that it can be reassembled in the same position. Lock the flywheel in place to prevent it from turning.

7. To avoid warping the pressure plate, loosen the pressure plate mounting bolts one in a two-step crisscross sequence until the spring tension is relieved.

8. Install a clutch alignment tool through the clutch disc and into the pilot bearing to support the pressure plate and cover assembly. Remove the pressure plate bolts and the clutch assembly.

9. Remove the release bearing from the input shaft and fork.

10. Remove the snap pin and release–fork fulcrum pin. Remove the release fork.

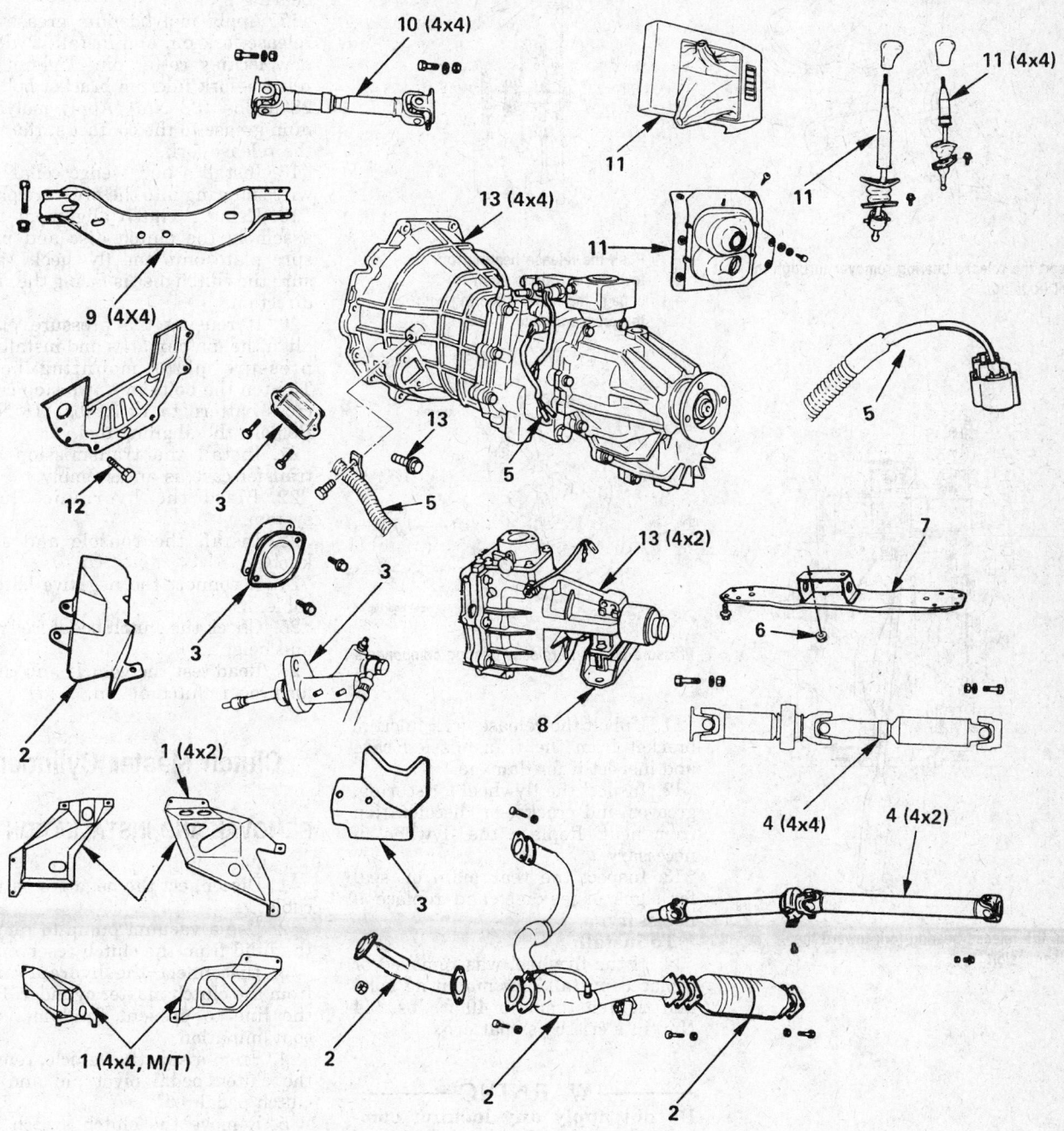

1 Exhaust and transfer protectors
2 Exhaust pipe and heat protector
3 Slave cylinder, heat protector,
  dust cover (1) and (2)
4 Rear propeller shaft
5 Transmission harness connectors
  and clamps
6 Engine rear mount nuts

7 Center crossmember
8 Engine rear mount
9 Front propeller shaft
10 Front propeller shaft
11 Gear control lever
12 Flywheel under cover bolts
13 Transmission assembly

328374

Manual transmission assembly and related components

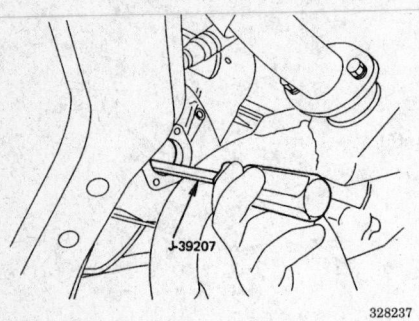

328237

Insert the release bearing remover through the bell housing

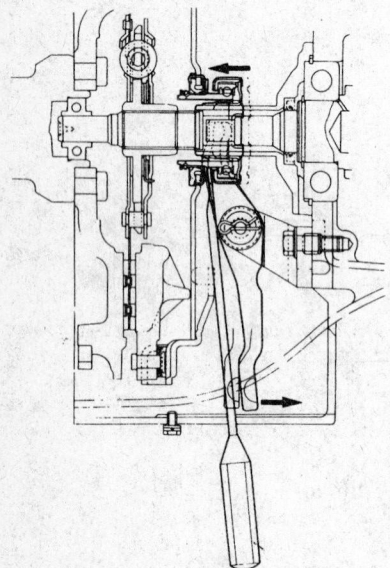

328238

Pull the release bearing fork toward the transmission

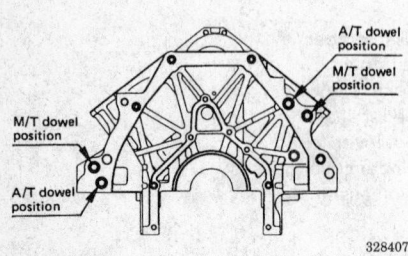

328407

Transmission mounting dowel position

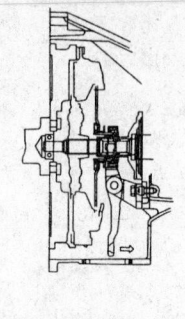

328242

Push the release bearing fork toward the transmission to engage the release bearing with the pressure plate

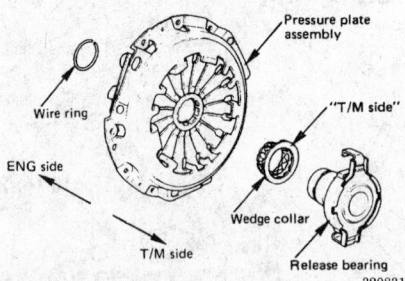

320831

Pressure plate and release bearing components

11. Unbolt the release–fork fulcrum bracket from the transmission case and inspect it for damage.

12. Inspect the flywheel for scoring, grooves and cracks, or discoloration from heat. Replace the flywheel if necessary.

13. Inspect the rear main oil seal for signs of leakage, and replace if necessary.

To install:

14. If the flywheel was removed or replaced, install new mounting bolts and tighten them to 40 ft. lbs. (54 Nm) in a crisscross pattern.

——— **WARNING** ———
Do not apply any locking compound to the flywheel bolts of 1994–97 vehicles. On earlier vehicles, apply only a small amount of locking compound to the flywheel bolts.

15. Apply a thin coat of molybdenum grease to the splines in the clutch disc and the front bearing retainer on the transmission where the release fork and bearing slide.

16. Pack the recess in the release bearing with molybdenum grease and apply grease to the tabs where the clutch fork attaches to the release bearing.

17. Apply molybdenum grease the release fork pin and install it with a new locking cotter pin. Tighten the release–fork fulcrum bracket bolts to 28 ft. lbs. (38 Nm). Apply molybdenum grease to the contact surfaces of the release fork.

18. Install a new wedge collar and wire snapring into the pressure plate.

19. Using a clutch alignment tool, assemble the clutch disc and pressure plate onto the flywheel. Make sure the clutch disc is facing the right direction.

20. If reusing the pressure plate, align the matchmarks and install the pressure plate mounting bolts. Tighten the bolts in a two-step crisscross pattern to 13 ft. lbs. (18 Nm). Remove the aligning tool.

21. Install the transmission and transfer case as an assembly.

22. Bleed the hydraulic clutch system.

23. Install the console and shift knobs.

24. Reconnect the negative battery cable.

25. Check the clutch pedal free play and height.

26. Road test the vehicle and check for proper clutch action.

Clutch Master Cylinder

REMOVAL AND INSTALLATION

1. Disconnect the negative battery cable.

2. Use a vacuum pump to remove the fluid from the clutch reservoir.

3. Disconnect the hydraulic line from the clutch master cylinder. Plug the line to prevent fluid loss and contamination.

4. From inside the vehicle, remove the clutch pedal pivot pin and the clutch pedal.

5. Remove the clutch switch and pedal bracket.

6. Remove the master cylinder mounting nuts and the master cylinder.

To install:

7. Install the clutch master cylinder on the bulkhead with a new gasket. Tighten the mounting nuts to 12 ft. lbs. (16 Nm).

8. Install the clutch pedal bracket and switch. Tighten the mounting nuts to 15 ft. lbs. (21 Nm).

9. Install the clutch pedal and pivot pin. Adjust the clutch pedal height.

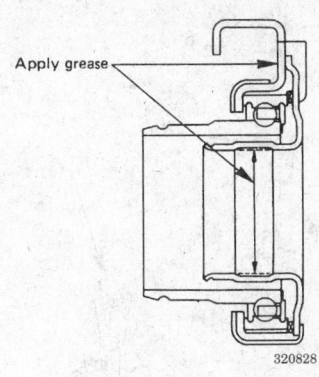

Release bearing grease points

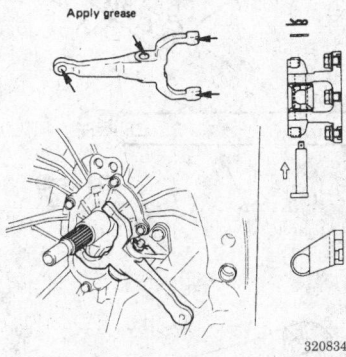

Release fork and fulcrum grease points

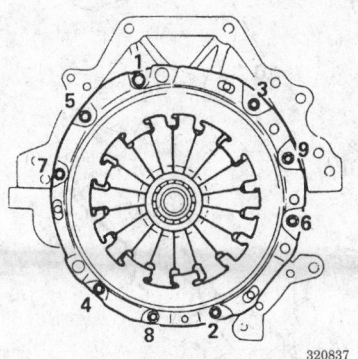

Pressure plate torque sequence

10. Connect the hydraulic line to the master cylinder. Tighten the line fitting nut to 14 ft. lbs. (20 Nm).

11. Bleed the clutch hydraulic system.

a. Fill the clutch master cylinder reservoir with fresh DOT 3 brake fluid. Cap the reservoir.

b. Attach a piece of clear tubing to the slave cylinder bleed screw and route it into a container partially filled with brake fluid.

c. Open the bleed screw.

d. Pump the clutch pedal until the fluid coming out the tube is free

of air bubbles. Then, tighten the bleed screw and remove the hose.

e. Refill the clutch master cylinder as necessary.

12. Reconnect the negative battery cable.

Clutch Slave Cylinder

REMOVAL AND INSTALLATION

1. Disconnect the negative battery cable.

2. Use a suction pump to remove the fluid from the clutch reservoir.

3. Raise and safely support the vehicle.

4. Use a flare wrench to disconnect the hydraulic line from the clutch slave cylinder.

5. Remove the slave cylinder mounting bolts. Remove the slave cylinder from its bracket on the transmission case.

To install:

6. Install the slave cylinder and connect the hydraulic line. Tighten the slave cylinder mounting bolts to 32 ft. lbs. (43 Nm).

7. On slave cylinders with flare fittings, tighten the hydraulic line fitting to 11–17 ft. lbs. (16–25 Nm).

8. On slave cylinders with banjo fittings, use new washers and tighten the hydraulic line fitting to 14 ft. lbs. (20 Nm).

9. Lower the vehicle.

10. Bleed the hydraulic clutch system.

a. Fill the clutch master cylinder reservoir with fresh DOT 3 brake fluid. Cap the reservoir

b. Attach a piece of hose to the slave cylinder bleed screw and route it into a container of brake fluid.

c. Open the bleed screw.

d. Pump the clutch pedal until the fluid coming out the tube is free of air bubbles. Tighten the bleed screw and remove the hose.

e. Refill the clutch master cylinder as necessary.

11. Reconnect the negative battery cable.

Hydraulic Clutch System Bleeding

PROCEDURE

1. Firmly set the parking brake.

2. Check the reservoir fluid level and refill as necessary.

3. Raise and safely support the front of the vehicle. Block the rear wheels.

4. Connect a vinyl tube to the bleeder screw. Submerge the other end of the tube in a container of brake fluid.

5. Have an assistant pump the clutch pedal slowly several times and hold it depressed.

6. Loosen the bleeder screw and allow the air bubbles and fluid to flow into the container. Tighten the bleeder screw.

7. Release the clutch pedal.

8. Add fluid to the reservoir if necessary. Don't let the reservoir run dry.

9. Continue to pump the pedal until the fluid draining from the tube is free of air bubbles.

10. Refill the reservoir to the full mark. Remove the vinyl tube. Tighten the bleeder screw and install its rubber cap.

Automatic Transmission Assembly

REMOVAL AND INSTALLATION

NOTE: The transfer case is an integral part of the transmission housing. Although the two cases can be separated, the transfer case should be removed with the transmission.

1. Use a felt–tipped marker to matchmark the hood to the hood hinges. Remove the hood.

2. Disconnect the negative battery cable.

3. Shift the transmission into the **N** position, and the transfer case into the **2H** position.

4. Remove the air cleaner assembly.

5. Remove the transfer case shift knob. Remove the four console retaining screws.

6. Remove the center console assembly and disconnect the console switch wiring connectors.

7. Disconnect the shift lock cable and the shift control rod from the selector lever assembly.

8. Remove the transfer case control lever.

9. Raise and safely support the vehicle.

10. Remove the transmission and transfer case skid plates.

11. Remove the exhaust pipe protectors.

12. Drain the transmission fluid.

13. Label and disconnect the oxygen sensor connectors.

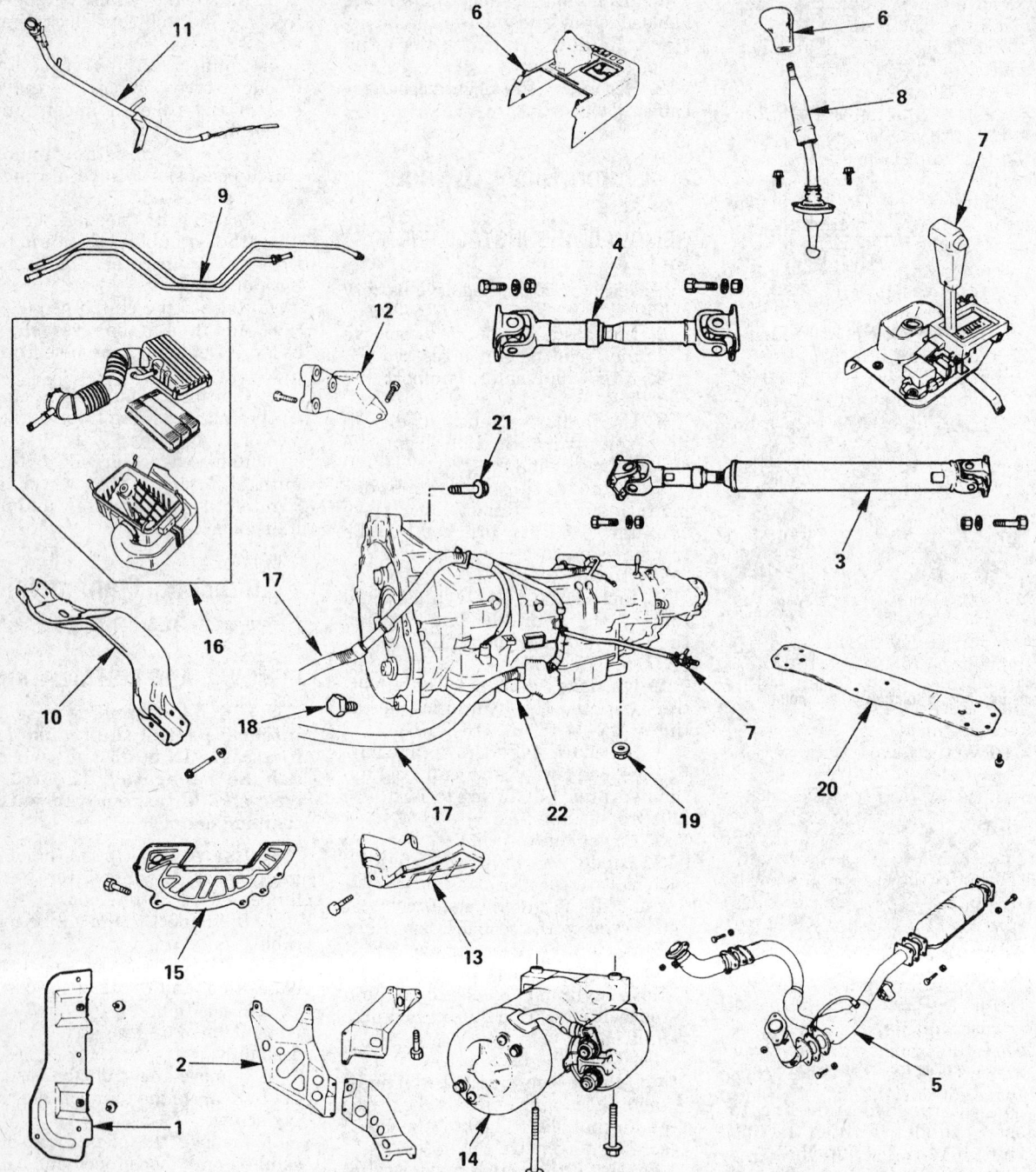

1. Skid plate
2. Exhaust and transfer pretectors
3. Rear propeller shaft
4. Front propeller shaft
5. Exhaust pipe
6. Front console
7. Selector lever assembly
8. Transfer control lever
9. Transmission oil cooler pipe

10. Front crossmember
11. Oil level gauge and guide tube
12. Stiffener
13. Heat protector
14. Starter
15. Flywheel under cover
16. Air cleaner assembly
17. Harness

18. Flywheel-Torque converter bolt
19. Rear mount nut
20. Third crossmember
21. Engine-transmission bolt
22. Transmission assembly

328377

Automatic transmission assembly and related components

14. Remove the catalytic converter, center and the front exhaust pipes.

15. Matchmark the front and rear driveshafts to the differential and transfer case flanges.

16. Unbolt and remove the front driveshaft.

17. Unbolt and remove the rear driveshaft.

18. Disconnect the oxygen sensor connector from the transmission harness.

19. Disconnect the oil cooler lines from the transmission. Plug the lines to keep moisture out.

20. Remove the brackets securing the oil cooler lines to the stiffener.

21. Remove the front crossmember.

22. Remove the dipstick and tube. Disconnect the breather hoses from the tube.

23. Remove the five engine stiffener bracket bolts and the stiffener bracket.

24. Remove the heat protector.

25. Disconnect the transmission harness connectors and the mode switch harness connector from the engine harness.

26. Disconnect the harness clamp from the clamp bracket.

27. Disconnect the ground cable from the engine.

28. Remove the starter.

29. Remove the flexplate inspection cover.

30. Remove the three bolts securing the flexplate to the torque converter.

NOTE: Remove the radiator upper fan shroud and the cooling fan to access the crankshaft center bolt to turn the crankshaft.

31. Place a suitable transmission jack under the transmission and transfer case unit for support.

32. Raise the transmission slightly and remove the eight bolts securing the rear mount and the transmission crossmember.

NOTE: Make sure the engine and transmission assembly is properly supported before removing the rear mount and third crossmember.

33. Raise the engine slightly with a hoist and remove the transmission-to-engine bolts.

34. Separate the transmission from the engine and lower the transmission from the vehicle.

To install:

NOTE: Use new self-locking nuts when reconnecting the exhaust system. Replace any

color-coded self-locking bolts when installing the frame crossmembers.

35. Install the transfer case onto the transmission case if the two were separated.

36. Install a new O-ring on the front pump shaft.

NOTE: Make sure the transmission dowel pins are installed in the correct position. If the dowels are in the wrong hole, the transmission case may crack.

37. Make sure the torque converter is fully seated in the front pump and slowly raise the front of the transmission into position until it is flush with the rear of the engine and install the transmission-to-engine bolts. Tighten the upper six mounting bolts to 56 ft. lbs. (76 Nm). Tighten the lower three mounting bolts to 35 ft. lbs. (48 Nm). Tighten the remaining two bolts to 4.4 ft. lbs. (6 Nm).

38. Install the third crossmember. Tighten the bolts to 37 ft. lbs. (50 Nm).

39. Install the rear mount and lower the engine from the hoist. Tighten the nuts to 30 ft. lbs. (41 Nm).

40. Remove the transmission jack and the engine hoist.

41. Install the flexplate to torque converter bolts. Tighten the bolts in a crisscross pattern to 41 ft. lbs. (55 Nm).

42. Install the flexplate inspection cover.

43. Connect the transmission and mode switch harness connectors to the engine harness.

44. Connect the harness clamp to the clamp bracket and radiator clip.

45. Connect the ground cable to the engine.

46. Install the starter and tighten the mounting bolts to 30 ft. lbs. (40 Nm).

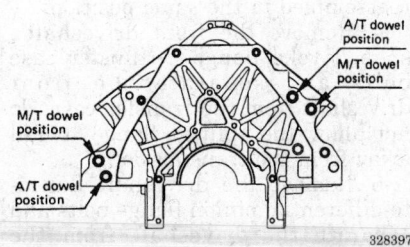

Transmission mounting dowel position.

47. Install the heat protector.

48. Install the stiffener bracket. Tighten the bolts to 35 ft. lbs. (48 Nm).

49. Install the transmission dipstick tube and dipstick. Connect the breather hoses to the tube.

50. Install the front crossmember. Tighten the bolts to 58 ft. lbs. (79 Nm).

51. Connect the transmission fluid cooling lines to the transmission. Tighten the line fittings to the following specifications:

- 1993–94 Trooper: 33 ft. lbs. (44 Nm)
- 1995–97 Trooper and SLX: 40 ft. lbs. (54 Nm)

52. Install the front and rear driveshafts. Tighten the flange bolts to 46 ft. lbs. (64 Nm).

53. Install the exhaust pipes and the catalytic converter. Tighten the manifold nuts to 49 ft. lbs. (67 Nm). On 1995–97 vehicles, tighten the center pipe and converter bolts to 32 ft. lbs. (43 Nm). On 1993–94 vehicles, tighten the center pipe bolts to 37 ft. lbs. (50 Nm), and the converter bolts to 25 ft. lbs. (34 Nm).

54. Reconnect the oxygen sensors.

55. Install the exhaust pipe protectors.

56. Install the skid plates and tighten their mounting bolts to 27 ft. lbs. (37 Nm).

57. Refill the transmission with DEXRON® ATF.

58. Lower the vehicle.

59. Install the transfer case control lever.

60. Connect the shift control rod to the selector lever. Adjust the shift cable so that there is 2.0 ± 0.5mm of slack at the selector lever.

61. Connect the console switch connectors and the shift lock cable to the selector lever.

62. Install the center console assembly and the transfer case lever knob.

63. Install the upper fan shroud and the cooling fan.

64. Install the air cleaner assembly.

65. Align the matchmarks and install the hood.

66. Connect the negative battery cable.

67. Start the engine and shift through the gear range three times to circulate the ATF through the valve body.

68. Recheck the fluid level and top-up if necessary.

69. Road test the vehicle.

Transfer Case Assembly

REMOVAL AND INSTALLATION

NOTE: On manual transmission equipped vehicles, the transfer case is an integral part of the transmission housing. Although the two cases can be separated if the transmission is to be rebuilt, the transfer case should be removed with the transmission.

1. Shift the transfer case into the **2H** position. Drive the vehicle forward and backward for a few feet/meters to make sure the front axle and hubs are disengaged. Shift the transmission to the **N** position.
2. Disconnect the negative battery cable.
3. Remove the center console.
4. Remove the shift knob and boot from the transfer case shift lever. Unbolt the shift lever from the transfer case.
5. Disconnect the shift lock cable from the transmission shift lever.
6. Raise and safely support the vehicle.
7. Remove the transmission and transfer case skid plates.
8. Disconnect the oxygen sensor from the front exhaust pipe.
9. Unbolt the front exhaust pipe from the exhaust manifolds and the catalytic converter. Remove the exhaust pipe and converter assembly.
10. Matchmark the front and rear driveshafts to the differential and transfer case flanges.
11. Unbolt and remove the front driveshaft.
12. Unbolt the rear driveshaft from the rear differential and transfer case flanges.
13. Unbolt the center bearing and remove the rear driveshaft.
14. Drain the transfer case oil.
15. Disconnect the transmission shift linkage from the shift lever rod.
16. Disconnect the two wiring harnesses from the transfer case.
17. Support the transfer case with a transmission jack.
18. Remove the transfer case-to-transmission bolts.
19. Separate the transfer case from the transmission output shaft. Lower the transfer case from the vehicle.

To install:

NOTE: Use new self-locking nuts when installing the exhaust pipe and converter.

20. Apply a thin coating of molybdenum grease to transfer case input shaft splines.

21. Raise the transfer case to the level of the transmission and align the output and input shaft splines.
22. Install the transfer case-to-transmission case bolts, and tighten them to 34 ft. lbs. (46 Nm).
23. Remove the transmission jack.
24. Fill the transfer case with fresh engine oil.
25. Connect the two wiring harnesses. Connect the shift lever rod to the shift linkage.
26. Align the matchmarks and install the front and rear driveshafts. Tighten the flange bolts to 46 ft. lbs. (63 Nm).
27. Install the exhaust pipe and catalytic converter. Tighten the nuts to 32 ft. lbs. (43 Nm).
28. Install the skid plates and tighten their bolts to 27 ft. lbs. (37 Nm).
29. Lower the vehicle.
30. Install the transfer case shift lever.
31. Install the console, shift boot, and shift knob.
32. Reconnect the negative battery cable.
33. Check to make sure that the transmission, transfer case, and front axle engage correctly.

DRIVE AXLE

Driveshaft

REMOVAL AND INSTALLATION

1993–94 Models

1. Raise and safely support the vehicle.
2. Matchmark the driveshaft flange-to-transfer case flange and the driveshaft-to-differential pinion flange.
3. Matchmark the front and rear parts of the driveshaft so it can be reassembled in the same position.
4. Remove the front driveshaft's splined yoke flange-to-transfer case bolts and separate the front driveshaft from the transfer case; do not allow the splined flange to fall away from the transmission.
5. Remove the driveshaft flange-to-differential pinion flange bolts and separate the driveshaft from the front differential.

To install:
6. Align the matchmarks and install the driveshaft.

7. Install the driveshaft flange-to-differential pinion flange bolts.
8. Install the front driveshaft's splined yoke flange-to-transfer case bolts and separate the front driveshaft to the transfer case.
9. Lower the vehicle.

1995–97 Models

Front Driveshaft

1. Shift the transmission into **N**, and the transfer case into **2H**. Make sure the front axle and hubs are disengaged.
2. Raise and safely support the vehicle.
3. Remove the transmission and transfer case skid plates.
4. Matchmark the driveshaft flanges to the transfer case flange and differential pinion flange.
5. Matchmark the front and rear parts of the driveshaft so they can be reassembled in the same position if the sliding yoke is to be separated from the rear part of the driveshaft.
6. Remove the bolts attaching the front driveshaft flange to transfer case flange and separate the front driveshaft from the transfer case.
7. Remove the bolts attaching the driveshaft flange to the differential pinion flange and separate the driveshaft from the front differential.
8. Remove the driveshaft from the vehicle.
9. Clean the flange mounting surfaces to remove any rust or dirt.

To install:
10. Align the matchmarks and install the driveshaft.
11. Install the bolts attaching the driveshaft flange to the differential pinion flange. Tighten the bolts to 46 ft. lbs. (63 Nm).
12. Connect the driveshaft flange to the transfer case flange and install the attaching bolts. Tighten the bolts to 46 ft. lbs. (63 Nm).
13. Repaint the exposed portions of the flanges so they don't rust.
14. Install the transfer case and transmission skid plates and tighten their bolts to 27 ft. lbs. (37 Nm).
15. Lower the vehicle.

Rear Driveshaft

1. Shift the transmission into **N**, and the transfer case into **2H**. Make sure the front axle and hubs are disengaged.
2. Raise and safely support the vehicle.
3. Matchmark the driveshaft flanges to the transfer case flange and differential pinion flange.
4. Matchmark the front and rear parts of the driveshaft so they can be

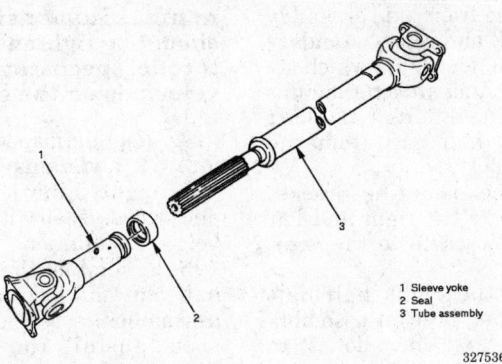

1 Sleeve yoke
2 Seal
3 Tube assembly

327536

Driveshaft assembly — 1995–97 vehicles

reassembled in the same position if the sliding yoke is to be separated from the rear part of the driveshaft.

5. Remove the bolts attaching the driveshaft flange to transfer case flange and separate the driveshaft from the transfer case.

6. Remove the bolts attaching the driveshaft flange to the differential pinion flange and separate the driveshaft from rear differential.

7. Remove the driveshaft from the vehicle.

8. Clean the flange mounting surfaces to remove any rust or dirt.

To install:

9. Align the matchmarks and install the driveshaft.

10. Install the bolts attaching the driveshaft flange to the differential pinion flange. Tighten the bolts to 46 ft. lbs. (63 Nm).

11. Connect the driveshaft flange to the transfer case flange and install the attaching bolts. Tighten the bolts to 46 ft. lbs. (63 Nm).

12. Repaint the exposed portions of the flanges so they don't rust.

13. Lower the vehicle.

U-Joints

REMOVAL AND INSTALLATION

1. Raise and support the vehicle safely.

2. Matchmark and remove the driveshafts.

3. Matchmark the driveshaft yokes so that driveline balance is preserved upon reassembly. Remove the snaprings that retain the bearing caps.

4. Select two press fixtures: one must be small enough to pass through the yoke holes for the bearing caps; the other must be large enough to receive the bearing cap.

5. Use a vise or a press and position the small and large press fixtures on either side of the U-joint. Press in on the smaller press component so it presses the opposite bearing cap out of the yoke and into the larger press component. If the cap does not come all the way out, grasp it with a pair of pliers and carefully work it out.

6. Reverse the position of the press components so that the smaller press component presses on the cross. Press the other bearing cap out of the yoke.

7. Repeat the procedure on the other bearing caps.

To install:

8. Grease the bearing caps and needles thoroughly if they are not pre-greased. Start a new bearing cap into a side of the yoke. Position the cross in the yoke.

NOTE: Some U-joints have a grease fitting that must be installed in the joint before assembly. When installing the fitting, make sure that once the driveshaft is installed in the vehicle the fitting is accessible to be greased during service.

9. Select two press fixtures small enough to pass through the yoke holes. Put the press fixtures against the cross and the cap and press the bearing cap ¼ in. (6mm) below the surface of the yoke. If there is a sudden increase in the force needed to press the cap into place, or if the cross starts to bind, the bearings are cocked. They must be removed and restarted in the yoke. Failure to do so will cause premature bearing failure.

10. Install a new snapring.

11. Start the new bearing cap into the opposite side. Place a press fixture on it and press in until the opposite bearing contacts the snapring.

12. Install a new snapring. It may be necessary to grind the facing surface of the snapring slightly to permit easier installation.

13. Install the other bearings in the same manner.

14. Check the joint for free movement. If binding exists, tap the yoke ears with a brass or plastic faced hammer to seat the bearing needles. If binding still exists, disassemble the joint and check to see if the needles are in place. Do not strike the bearings unless the shaft is supported firmly. Do not install the driveshaft until free movement exists at all joints.

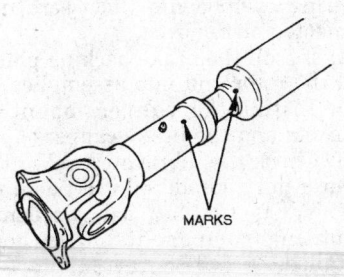

MARKS

327537

Matchmark the two parts of the driveshaft — 1995–97 vehicles

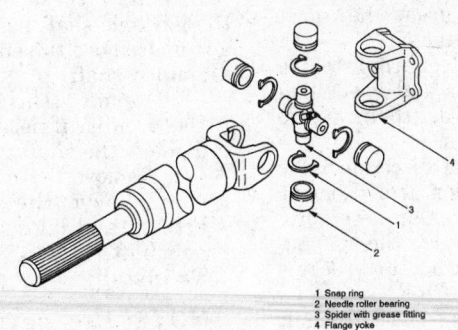

1 Snap ring
2 Needle roller bearing
3 Spider with grease fitting
4 Flange yoke

327572

U-Joint assembly (exploded view)

15. Install the driveshafts into the vehicle.

16. Lower the vehicle. Test drive the vehicle and check for driveline vibrations.

Halfshaft

REMOVAL AND INSTALLATION

1993–95 Models

NOTE: The right axle shaft is an integral part of the right halfshaft assembly. Removal of the left axle shaft involves disassembling the shift-on-the-fly four-wheel drive gearbox. The inboard joints of both halfshafts fit through the axle mounting brackets, which are bolted to the axle housing.

1. Shift the transfer case shift lever into 2H. Drive the vehicle a few feet forward and reverse to verify that the front axle is disengaged.

2. Set the front wheels and steering wheel in the straight–ahead position. Lock the steering column in this position, and remove the key.

3. Raise and safely support the vehicle.

4. Remove the front wheels.

5. Remove the radiator skid plate.

6. Remove the transfer case skid plates.

7. Drain the oil from the differential.

8. Unbolt the calipers from their brackets. Support the calipers out of the way on wire hangers. Don't disconnect the brake hoses.

9. Remove the caliper mounting bracket from the steering knuckle.

10. Remove the automatic hub and brake rotor assemblies. Note the positions of the hub snaprings, shims, and lock washers for reassembly.

11. If equipped with ABS, unbolt the front wheel sensor brackets from the steering knuckles. Move the sensors out of the work area. They don't need to be disconnected.

12. Use a ball joint separator tool to disconnect the upper and lower ball joints and tie rod ends, then remove the steering knuckles.

13. Matchmark and disconnect the pitman arm and idler arm. Remove the steering linkage as an assembly.

14. Unbolt and remove the suspension crossmember from its brackets at the lower control arms.

15. Matchmark the front driveshaft flanges to the differential and transfer case flanges. Unbolt and remove the front driveshaft.

16. Support the front axle assembly with a floor jack and safety stands.

17. Remove the four bolts which secure the right halfshaft axle mounting bracket to the differential. Don't unbolt the left halfshaft from the axle.

18. Remove the mounting bracket bolts which secure the right and left axle mounting brackets to the vehicle's frame.

19. Separate the right halfshaft and axle mounting bracket assembly from the differential and allow it to rest on the lower control arm.

WARNING
Be careful not to damage the CV-joints, boots, or splined shafts when removing the axle.

---

20. Follow these steps to remove the axle assembly:

a. Verify that the axle assembly is securely supported by the floor jack. Remove the safety stands.

b. First, slide the axle to the left to release the splined stub axle of the right halfshaft.

c. Next, lower the axle slightly and slide it to the right so that the left halfshaft clears the left lower control arm.

d. Finally, completely lower the axle from the vehicle.

21. Remove the right halfshaft from the vehicle. Follow these steps to remove the right axle shaft seal and bearing:

a. Remove the snapring from the splined shaft.

b. Remove the shaft bearing. Use a puller if necessary, but don't damage the shaft or splines.

c. Remove the inner snapring.

d. Remove the axle mounting bracket and oil seal from the right halfshaft.

22. With the axle out of the vehicle, unbolt the left halfshaft from the axle case. Remove the halfshaft together with the left axle mounting bracket.

23. Follow these steps to remove the left axle shaft seal and bearing:

a. Remove the snapring from the splined shaft.

b. Remove the shaft bearing. Use a puller if necessary, but don't damage the shaft or splines.

c. Remove the inner snapring.

d. Remove the axle mounting bracket and oil seal from the left halfshaft.

To install:

NOTE: Use new self-locking nuts and color-coded bolts when assembling the axle assembly mounts and suspension compo-

nents. Suspension fasteners should be tightened to their final torque specifications when the vehicle is on the ground.

24. Visually inspect the axle shafts for wear and damage.

25. Lightly lubricate the bearing and grease seal with differential oil before installation.

26. Assemble the left and right halfshaft and axle mounting bracket assemblies

a. Install the axle mounting bracket onto the halfshaft.

b. Lubricate and install a new oil seal.

c. Install a new inner snapring.

d. Install a new bearing.

e. Install a new outer snapring.

27. Place the right halfshaft and axle mounting bracket assembly into position, and rest it on the right lower control arm.

28. Position the axle and left halfshaft assembly on a floor jack. Raise the axle into position.

29. Fit the right halfshaft and axle mounting bracket assembly into the differential. Make sure the splined stub axle shaft is fully seated. Be careful not to distort the oil seal when connecting the right mounting bracket to the differential.

30. Install the mounting brackets and halfshafts to the axle housing. Tighten the mounting bolts to the following specifications:

• 1993–94 vehicles: 61 ft. lbs. (82 Nm)

• 1995 vehicles: 85 ft. lbs. (116 Nm)

31. Raise the axle assembly into its final position and support it with safety stands.

32. Install the mounting bracket bolts, nuts, and spacers. The washer fits under the bolt and the spacer is used with the nut. Tighten the mounting nuts and bolts to 112 ft. lbs. (152 Nm).

33. Install the steering knuckles and assemble any suspension and steering components that were disconnected or removed.

34. Install the brake backing plates and the rotor and hub assemblies.

35. Install the caliper mounting brackets and the brake calipers.

36. Check the lubricant level in the differential and add oil if needed.

37. Verify that all axle assembly mounting components have been installed.

38. Align the front driveshaft matchmarks. Install the driveshaft flange bolts and tighten them to 46 ft. lbs. (63 Nm).

39. Install the suspension crossmember and tighten the bolts to 58 ft. lbs. (78 Nm).

40. Install the steering linkage assembly.

41. If equipped, reconnect the ABS front wheel sensors.

42. Install the radiator skid plate. Tighten the bolts to 58 ft. lbs. (78 Nm).

43. Install the transfer case skid plates and tighten their bolts to 27 ft. lbs. (37 Nm).

44. Install the front wheels.

45. Lower the vehicle and adjust the ride height.

46. Check and adjust the front wheel alignment.

47. Tighten the suspension bushing fasteners to their final torque specifications.

48. If necessary, bleed the brake system.

49. Verify that the front axle and hubs engage and disengage properly.

50. Road test the vehicle.

1996–97 Models

NOTE: The right axle shaft is an integral part of the right halfshaft assembly. Removal of the left axle shaft involves disassembling the shift-on-the-fly four-wheel drive gearbox. The inboard joints of both halfshafts fit through the axle mounting brackets, which are bolted to the axle housing.

1. Shift the transfer case shift lever into **2H**. Drive the vehicle a few feet forward and reverse to verify that the front axle is disengaged.

2. Set the front wheels and steering wheel in the straight-ahead position. Lock the steering column in this position, and remove the key.

3. Raise and safely support the vehicle.

4. Remove the front wheels.

5. Remove the radiator skid plate.

6. Remove the transfer case skid plates.

7. Drain the oil from the differential.

8. Unbolt the calipers from their brackets. Support the calipers out of the way on wire hangers. Don't disconnect the brake hoses.

9. Remove the caliper mounting brackets from the steering knuckle.

10. Remove the automatic hub and brake rotor assemblies. Note the positions of the hub snaprings, shims, and lock washers for reassembly.

11. If equipped with four-wheel ABS, unbolt the front wheel sensor brackets from the steering knuckles.

Move the sensors out of the work area. They don't need to be disconnected.

12. Remove the shift-on-the-fly four-wheel drive actuator:

 a. Remove the skid plate from the shift-on-the-fly gear housing.

 b. Label and disconnect the Vacuum Switching Valve (VSV) hoses and 2P connector from the gear housing.

 c. Unbolt the VSV assembly from the left axle tube and remove it so it won't be damaged. Don't disconnect the two vacuum hoses from the body of the VSV.

13. Use a ball joint separator tool to disconnect the upper and lower ball joints and tie rod ends, and then remove the steering knuckles.

14. Matchmark and disconnect the pitman arm and idler arm. Remove the steering linkage as an assembly.

15. Unbolt and remove the suspension crossmember from its brackets at the lower control arms.

16. Matchmark the front driveshaft flanges to the differential flange and transfer case flange. Unbolt and remove the front driveshaft.

17. Support the front axle assembly with a floor jack and safety stands.

18. Remove the four bolts which secure the right halfshaft axle mounting bracket to the differential. Don't unbolt the left halfshaft from the axle.

19. Remove the mounting bracket bolts which secure the right and left axle mounting brackets to the vehicle's frame.

20. Separate the right halfshaft and axle mounting bracket assembly from the differential and allow it to rest on the lower control arm.

— WARNING —

Be careful not to damage the CV-joints, boots, or splined shafts when removing the axle.

---

21. Follow these steps to remove the axle assembly:

 a. Verify that the axle assembly is securely supported by the floor jack. Remove the safety stands.

 b. First, slide the axle to the left to release the splined stub axle of the right halfshaft.

 c. Next, lower the axle slightly and slide it to the right so that the left halfshaft clears the left lower control arm.

 d. Finally, completely lower the axle from the vehicle.

22. Remove the right halfshaft from the vehicle. Follow these steps to remove the right axle shaft seal and bearing:

 a. Remove the snapring from the splined shaft.

 b. Remove the shaft bearing. Use a puller if necessary, but don't damage the shaft or splines.

 c. Remove the inner snapring.

 d. Remove the axle mounting bracket and oil seal from the right halfshaft.

23. Drain the lubricant from the shift-on-the-fly gearbox.

24. With the axle out of the vehicle, unbolt the left halfshaft from the axle case. Remove the halfshaft together with the left axle mounting bracket.

25. Loosen the shift-on-the-fly actuator mounting bolts in a crisscross pattern. Remove the actuator assembly from the gearbox.

26. Unbolt the shift-on-the-fly actuator gearbox from the axle tube flange.

27. Slowly pull the gearbox straight off of the axle tube. Be careful not to loose the sleeve and clutch gear if they fall out of the gearbox.

28. Remove the outer snapring from the left axle shaft. Draw the axle shaft out of the axle tube.

29. Remove the left axle shaft oil seal from the axle tube. Be careful not to damage the sealing surface.

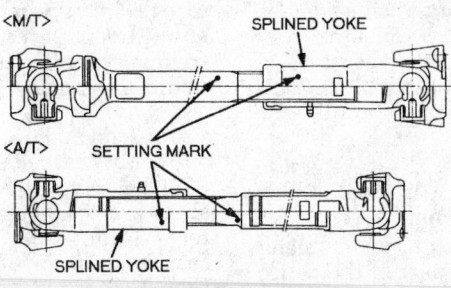

334700

Driveshaft alignment marks — 1996–97 vehicles

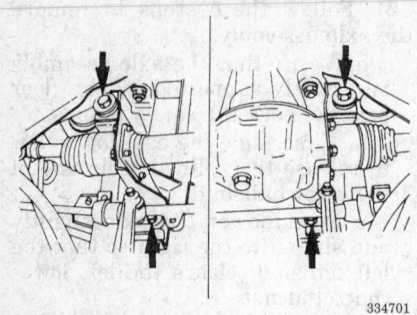

Axle assembly mounting bracket bolts — 1996–97 vehicles

Axle removal and installation method — 1996–97 vehicles

30. Inspect the axle shaft, bearing, and clutch gear components for any damage:

a. Inspect the axle shaft splines for damage, and replace as necessary. Use a dial gauge and center blocks to inspect the shaft run-out. If run-out exceeds 0.02 in. (0.5mm), replace the shaft. Don't try to heat the axle shaft to correct excess run-out.

b. Insert the clutch gear into the axle shaft and inspect the motion and play of the inner bearing and needle bearing. If either bearing exhibits smoothness or play, they should be replaced.

c. Check the clutch sleeve for wear. First, coat the clutch gear with gear oil, and then slide the clutch sleeve back and forth over the gear to simulate operation. If the sleeve and gear exhibit smoothness or play, they must be replaced.

d. Check the clutch sleeve groove width. Groove width shouldn't exceed 0.28 in. (7.1mm).

e. Check the external diameter of the narrowest part of the clutch gear. The diameter shouldn't exceed 1.456 in. (36.98mm).

31. Remove the inner snapring from the axle shaft bearing.

32. Remove the inner shaft bearing:

a. Install tool No. J–37452, or an equivalent bearing remover onto the axle shaft.

b. Place the axle shaft and the bearing remover into a press.

c. Press the bearing from the axle shaft. Don't damage the axle shaft.

33. Remove the needle bearing from the axle shaft clutch gear:

a. Support the axle shaft in a padded vise.

b. Install tool No. J–26941 into the bearing. Install tool No. J–2619–01, or an equivalent slide hammer onto the bearing removal tool.

c. Work the slide hammer to gradually remove the needle bearing from the axle shaft.

34. Inspect the condition and operation of the shift-on-the-fly actuator:

a. Disconnect the shift position switch.

b. Use a vacuum pump to apply 15.7 in. (400mm) Hg of negative pressure to the port on the actuator's vacuum disc (port A). The shift fork should move toward the vacuum disc.

c. Repeat step B with vacuum applied to the port beside the shift position switch (port B). The shift fork should move away from the vacuum disc.

d. If the actuator needs more than 15.7 in. (400mm) Hg of vacuum to function, or if it functions incorrectly, it should be replaced.

e. Inspect the shift fork for damage and wear. The internal distance between each end of the fork should be 2.52 in. (64.1mm). The width of the fork ends should be 0.26 in. (6.7mm) when measured at their narrowest point.

To install:

NOTE: Use new self-locking nuts and color-coded bolts when assembling the axle assembly mounts and suspension components. Suspension fasteners should be tightened to their final torque specifications when the vehicle is on the ground.

35. Clean and dry all axle and shift-on-the-fly gearbox sealing surfaces.

36. Thoroughly lubricate a new axle shaft seal with clean gear oil. Use tool No. J–41693, or an equivalent seal driver to install it into the axle tube.

37. Use tool No. J–41694, or an equivalent bearing driver to install a new needle bearing.

38. Install a new inner snapring.

39. Use tool No. J–4169, or an equivalent press base to press a new bearing onto the axle shaft.

40. Install the axle shaft into the axle tube. Don't damage the new oil seal. Install a new outer snapring.

41. Lubricate the clutch gear and clutch sleeve with SAE 75W–90 GL5–5 gear oil and install them.

42. Apply a 1mm–wide bead of liquid gasket to the axle tube sealing surface. Install the shift-on-the-fly gearbox onto the axle tube before the sealant cures.

43. Tighten the shift-on-the-fly gearbox bolts to 85 ft. lbs. (116 Nm) in a two-step crisscross pattern.

44. Install the shift position switch onto the actuator and tighten it to 29 ft. lbs. (39 Nm).

45. Install the actuator assembly:

a. Apply a 1mm–wide bead of liquid gasket to the actuator sealing surface. Don't allow the sealant to cure before installation.

b. Align the shift fork arms with the groove of the clutch sleeve and install the actuator.

c. Tighten the actuator mounting bolts to 10 ft. lbs. (13 Nm) in a crisscross sequence.

46. Install the spacer onto the left front halfshaft.

47. Install the halfshaft and axle mounting bracket assembly onto the axle. Tighten the bolts to 85 ft. lbs. (116 Nm) in a crisscross sequence.

48. After the sealant has fully cured, refill the shift-on-the-fly gearbox with SAE 75W–90 GL5–gear oil.

49. Assemble the right halfshaft and axle mounting bracket assembly

a. Install the axle mounting bracket onto the halfshaft.

b. Lubricate and install a new oil seal.

c. Install a new inner snapring.

d. Install a new bearing.

e. Install a new outer snapring.

50. Place the right halfshaft and axle mounting bracket assembly into position, and rest it on the right lower control arm.

51. Position the axle and left halfshaft assembly on a floor jack. Raise the axle into position.

52. Fit the right halfshaft and axle mounting bracket assembly into the differential. Make sure the splined stub axle shaft is fully seated. Be careful not to distort the oil seal when connecting the right mounting bracket to the differential.

53. Tighten the right halfshaft and axle mounting bracket assembly mounting bolts to 85 ft. lbs. (116 Nm) in a crisscross sequence.

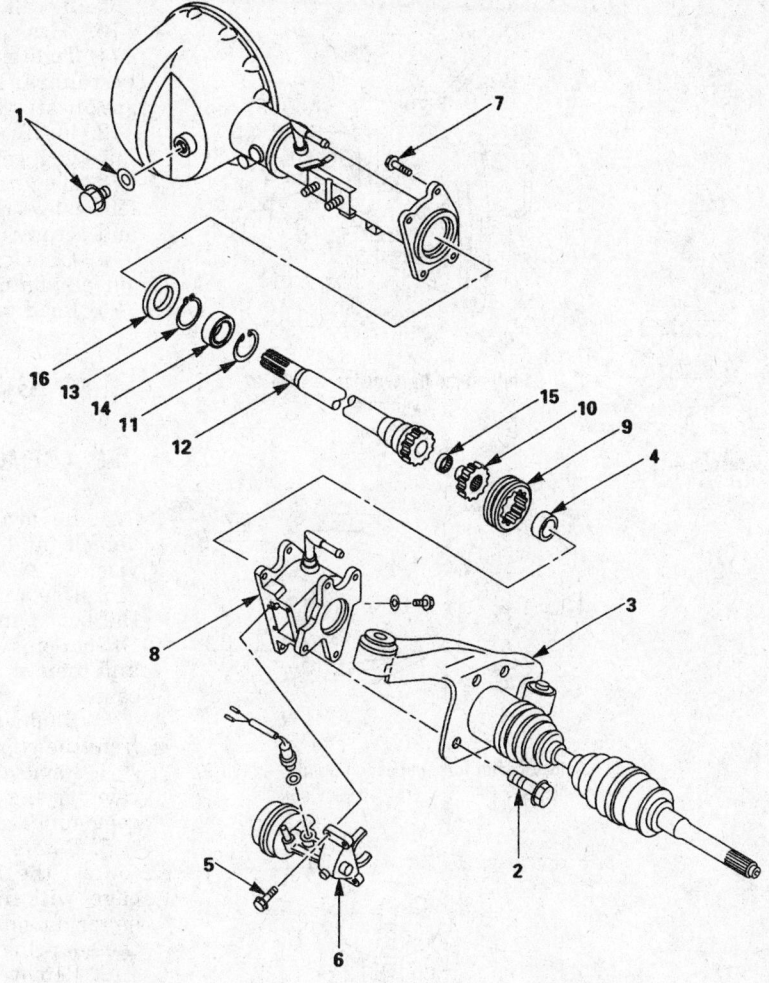

1. Filler plug
2. Bolt
3. Front axle drive shaft (LH side)
4. Spacer
5. Bolt
6. Actuator assembly
7. Bolt
8. Housing

9. Sleeve
10. Clutch gear
11. Snap ring
12. Inner shaft
13. Snap ring
14. Inner shaft bearing
15. Needle bearing
16. Oil seal

334706

Left halfshaft, axle shaft, and gearbox exploded view — 1996–97 vehicles

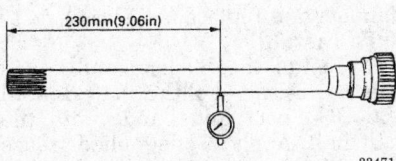

230mm(9.06in)

334711

Axle shaft run-out inspection — 1996–97 vehicles

54. Raise the axle assembly into its final position and support it with safety stands.

55. Install the mounting bracket bolts, nuts, and spacers. The washer fits under the bolt and the spacer is used with the nut. Tighten the mounting nuts and bolts to 112 ft. lbs. (152 Nm).

56. Install the steering knuckles and assemble any suspension components that were disconnected or removed.

57. Install the brake backing plates and the rotor and hub assemblies.

58. Install the caliper mounting brackets and the brake calipers.

59. Install the VSV onto the axle tube. Reconnect the vacuum hoses and the 2P connector. Then, install the VSV skid plate.

60. Refill the differential with gear oil. Check the oil level in the gear case. Use new crush washers and tighten both drain plugs to 58 ft. lbs. (78 Nm).

61. Verify that all axle assembly mounting components have been installed.

62. Align the front driveshaft matchmarks. Install the driveshaft flange bolts and tighten them to 46 ft. lbs. (63 Nm).

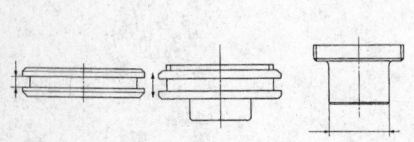

Several views of clutch sleeve inspection — 1996-97 vehicles

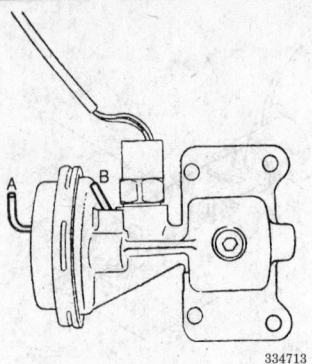

Shift-on-the-fly actuator — 1996-97 vehicles

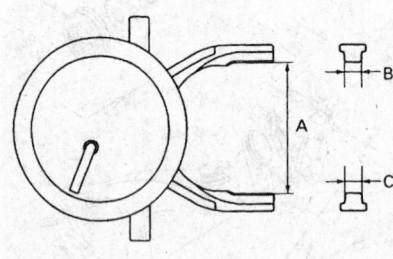

Actuator shift fork inspection points — 1996-97 vehicles

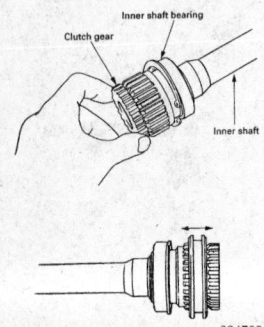

Clutch gear and sleeve inspection — 1996-97 vehicles

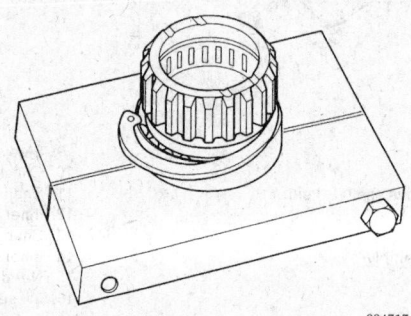

Axle shaft bearing installation press tool — 1996-97 vehicles

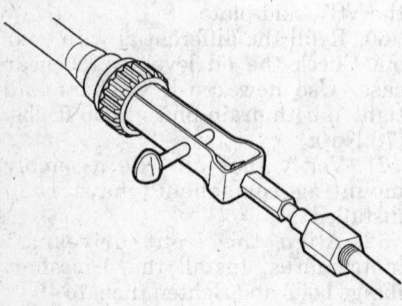

Needle bearing puller — 1996-97 vehicles

69. Install the front wheels.

70. Lower the vehicle.

71. Tighten the suspension bushing fasteners to their final torque specifications.

72. Verify that the front axle and hubs engage and disengage properly.

73. If equipped with shift-on-the-fly four-wheel drive, make sure the VSV and actuator function correctly.

74. Check and adjust the front wheel alignment and ride height.

75. Road test the vehicle.

CV-Joint Boot

REPLACEMENT

1. Remove the halfshaft from the vehicle and secure in a soft-jawed vise.

2. Use a flat bladed tool to remove the boot bands on the inner Double Offset Joint (DOJ). Pull the boot back and remove the circlip from the DOJ case.

3. Remove the DOJ outer race from the shaft.

4. Remove the balls from the DOJ cage prying from the inside of the cage outward.

5. Rotate the cage while pushing toward the Birfield Joint (BJ). The cage will drop down to expose a snapring on the halfshaft. Remove the snapring.

6. Remove the DOJ cage and inner race from the halfshaft.

7. Wrap tape over the splines of the halfshaft to protect the boot if it is to be reused. Remove the boot from the shaft.

8. Remove the dust seal from the BJ.

NOTE: Do not disassemble the (outer) Birfield Joint (BJ).

9. Remove the boot bands. Remove the BJ boot from the DOJ end of the halfshaft.

10. Inspect all parts for wear or damage and replace, as necessary.

To install:

11. Install the BJ boot and bands over the DOJ end of the shaft. Install the DOJ boot and bands onto the halfshaft. Apply ½ of supplied grease into the BJ boot and install the boot to the BJ shaft. Make sure that the bands are set in the proper direction and install. Do not distort the boots during installation of the bands.

12. Install the DOJ cage onto the halfshaft with the smaller diameter side of the cage facing the installed BJ boot.

63. Install the suspension crossmember and tighten the bolts to 58 ft. lbs. (78 Nm).

64. Install the steering linkage assembly.

65. If equipped, reconnect the ABS front wheel sensors.

66. Install the radiator skid plate. Tighten the bolts to 58 ft. lbs. (78 Nm).

67. Install the transfer case skid plates and tighten their bolts to 27 ft. lbs. (37 Nm).

68. If necessary, bleed the brake system.

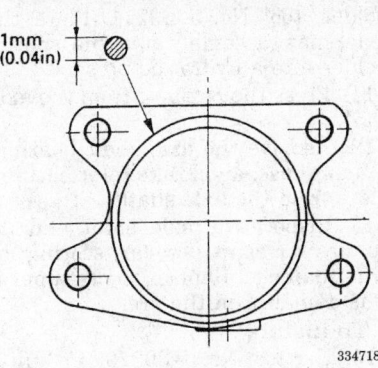

1mm
(0.04in)

Axle tube sealing surface — 1996–97 vehicles

334718

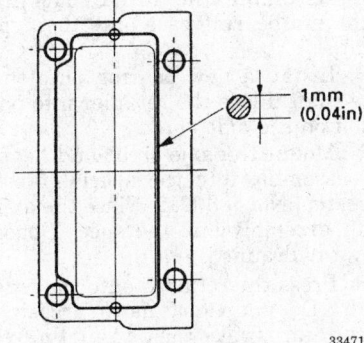

1mm
(0.04in)

Actuator sealing surface — 1996–97
vehicles

334719

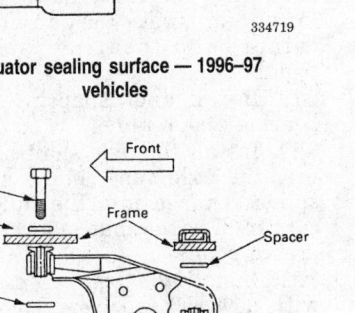

Front

Bolt

Washer

Frame

Spacer

Spacer

Nut

Washer

Bolt

334704

**Axle bracket mounting bolt, spacer, and nut
locations — 1996–97 vehicles**

13. Install the DOJ inner race onto the halfshaft and secure with the snapring.

14. Apply the proper grease to the DOJ inner race, the DOJ cage and the balls. Insert the balls into the cage from the outside pushing in toward the shaft.

15. Apply 1.9 oz. of the proper grease to the DOJ outer race. Install the outer race onto the halfshaft.

16. Apply another 1.9 oz. of the proper grease to the DOJ outer race and install the circlip.

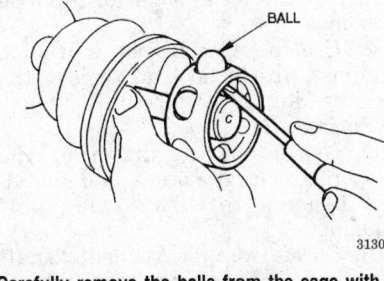

BALL

313092

Carefully remove the balls from the cage with a flat bladed tool — 1993–95 vehicles

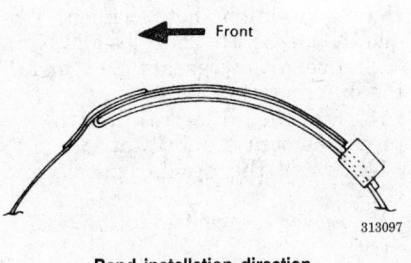

Front

313097

Band installation direction

17. Add to the BJ as much of the proper grease as was wiped away at the time of inspection.

18. Install the BJ boot and secure with new boot bands. To control air in the DOJ boot, set the distance between the boot bands to 3.03–3.27 in. (77–83mm) before securing the boot bands. The joint's extended length from the end of the case to the outer edge of the boot should be 6.98–7.0 in. (171mm).

19. Install a new dust seal on the outer joint.

20. Install the halfshaft into the vehicle.

21. Install new boot bands and clench them so that the crimp faces the front of the vehicle.

22. Assemble any suspension components that were disconnected to remove the halfshaft.

23. Install the front wheels.

24. Lower the vehicle. Tighten any suspension fasteners to their final torque specifications.

Rear Axle Shaft

REMOVAL AND INSTALLATION

1993–94 Models

Dana Rear Axle

1. Raise and safely support the vehicle.

2. Remove the rear wheel assembly and brake drum or rotor.

3. Remove the 4 axle retaining nuts and lock washers.

4. Remove the axle shaft assembly from the axle housing.

5. Remove the snapring and bearing cup.

6. Break the retainer ring with a hammer and chisel.

7. Remove the oil seal, retainer, and the parking brake assembly from the axle.

8. Remove the inner race from the axle shaft using a press and a bearing splitter.

To install:

9. Install the parking brake assembly and the oil seal, in the proper direction, into the bearing holder. Install the bearing holder onto the axle assembly.

10. Press the bearing assembly on the axle shaft using a steel pipe 25 ½ inches long, 2.0 inches O.D and 1.625 inches I.D.

NOTE: Install the bearing with cup towards the inboard side.

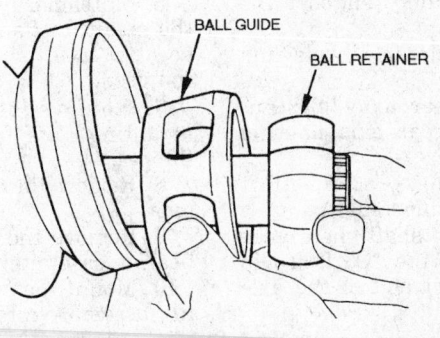

BALL GUIDE

BALL RETAINER

313094

The smaller diameter of the cage faces the outer CV-joint

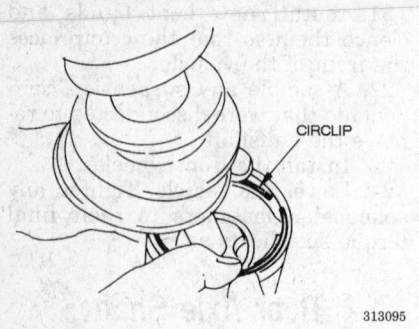

Installing the circlip in the outer race of the inner joint

313095

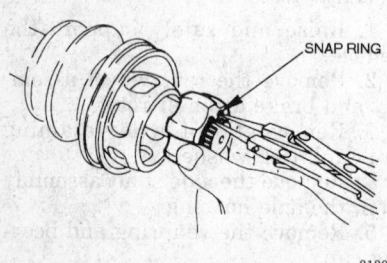

Remove the snapring from the shaft — 1996–97 vehicles

313093

11. Install the retainer ring using a press.

12. Install the snapring.

13. Place the axle assembly into the axle housing and install the lock washers and nuts. Torque the nuts to 55 ft. lbs. (75 Nm).

14. Install the brake rotors and calipers or drums.

15. Install the wheels and lower the vehicle to the floor.

Saginaw Rear Axle

1. Raise and safely support the vehicle.

2. Remove the rear wheels and drums.

3. Drain the rear axle lubricant and remove the rear axle housing cover.

4. Remove the pinion shaft lockbolt and the pinion shaft.

5. Push the axle shaft slightly inward and remove the "C" lock clip from the innermost end of the axle shaft.

6. Remove the axle from the axle housing.

7. Pry the oil seal out of the axle housing.

8. Use the special tool or a slide hammer with a hook to remove the axle bearing.

To install:

9. Use a bearing driver or the proper size large socket and install the bearing into the axle shaft housing.

10. Carefully slide the axle shaft into the axle housing taking care not to damage the oil seal. The axle may have to be rotated a little to properly engage the splines in the differential side gear.

11. Install the "C" lock clip, pinion shaft and pinion shaft lockbolt. Torque the lockbolt to 25 ft. lbs. (34 Nm).

12. Use a new gasket and install the differential cover.

13. Fill the differential with the proper amount of lubricant.

14. Install the drums, wheels and tires.

15. Lower the vehicle to the floor.

1995–97 Models

1. Raise and safely support the vehicle.

2. Remove the rear wheels.

3. Remove the brake caliper and caliper brackets. Support the calipers up and out of the way with wire. Don't disconnect the brake hoses.

4. Remove the rear brake rotors. If equipped with ABS, unbolt the wheel sensor wire brackets from the axle tubes. Be careful not to damage the wheel sensors.

5. Disconnect the parking brake cable from the parking brake shoe assembly. Unbolt the brake cable bracket from the backing plate. Remove the parking brake shoe assembly.

6. Matchmark the bearing case to the axle tube flange. Mark the hubs so the axle shafts will not be confused.

7. Remove the bearing case mounting nuts from the axle tube flanges.

8. Remove the axle shafts from the axle.

9. Remove the snapring from the bearing case retainer.

10. Mount the axle shaft and bearing assembly into the special press holder, tool No. J–39211. Place the axle shaft assembly and the special holder into a hydraulic press.

11. Press the retainer from the axle shaft.

12. Remove the axle shaft bearing, bearing case, and brake rotor backing plate from the axle shaft.

13. Inspect the axle shaft and its splines for signs of wear, scoring, or other damage. Replace the retainer if it is damaged in the press.

To install:

14. Use tool No. J–39379 or a suitably–sized seal installer to install a new shaft seal into the bearing case.

15. Assemble the brake backing plate and bearing case onto the axle shaft.

16. Install a new bearing onto the axle shaft. Place the retainer into position on the axle shaft.

17. Mount the axle shaft and bearing assembly into the special press base, tool No. J–39212. Place the axle shaft assembly and the special base into a hydraulic press.

18. Press the retainer onto the axle shaft. Do not exert more pressing force than is necessary to fit the retainer and bearing assembly together.

19. Install a new snapring onto the bearing case retainer.

20. Install the axle shaft assembly into the axle tube. Make sure the splines engage into the differential carrier. Use the matchmarks as reference points.

21. Install the bearing case nuts with new lock washers. Tighten the nuts to 54 ft. lbs. (74 Nm).

22. Install the parking brake shoes. Reconnect the parking brake cable to the shoes and install its backing plate bracket mounting bolts.

23. Install the ABS wheel sensor wire brackets onto the axle tubes.

24. Install the rear brake rotors.

25. Install the brake caliper brackets and calipers. Tighten the bracket bolts on 1995 vehicles to 109 ft. lbs. (148 Nm), or to 76 ft. lbs. (103 Nm) for 1996–97 vehicles. Tighten the caliper bolts to 32 ft. lbs. (44 Nm).

26. Install the rear wheels.

27. Bleed the brakes if necessary.

28. Check the differential oil level and refill as necessary.

29. Road test the vehicle and check for abnormal bearing and gear noises.

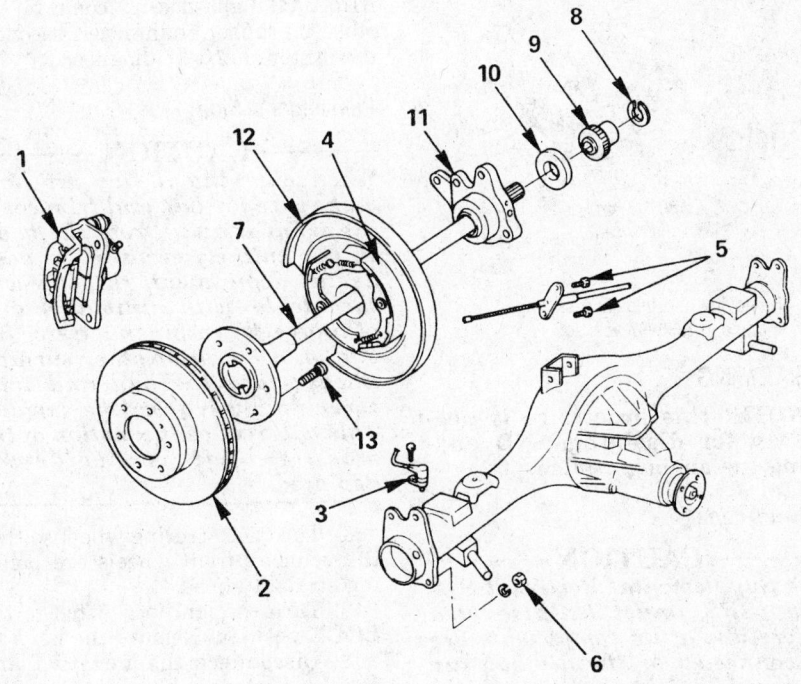

1. Brake caliper
2. Brake disc
3. Antilock brake system (ABS) speed sensor
4. Parking brake assembly
5. Bolt
6. Nut
7. Axle shaft assembly
8. Snap ring
9. Retainer
10. Bearing
11. Bearing holder
12. Back plate
13. Wheel pin

327334

Rear axle components — 1995–97 vehicles

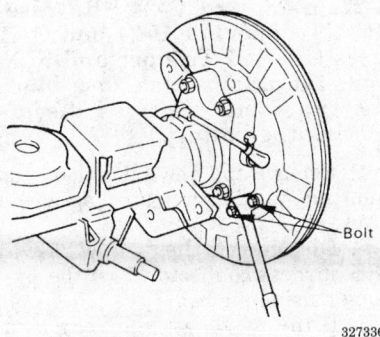

327336

Bearing case nuts and parking brake cable bracket — 1995–97 vehicles

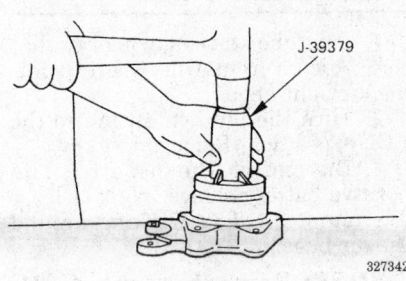

J-39379

327342

Tool No. J-39379: oil seal installer — 1995–97 vehicles

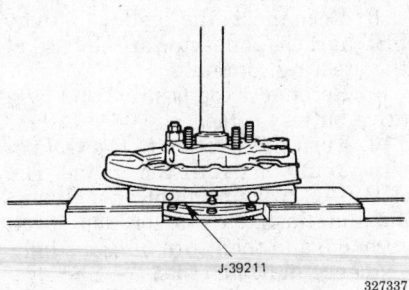

J-39211

327337

Tool No. J-39211: press bearing holder — 1995–97 vehicles

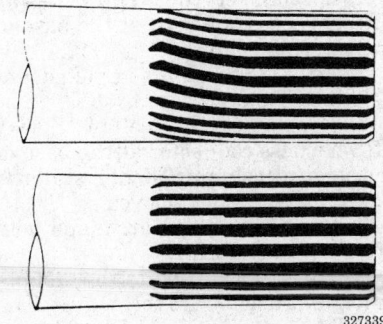

327339

Axle shaft spline wear patterns — 1995–97 vehicles

STEERING

Air Bag

—CAUTION—

Some vehicles are equipped with an air bag system, also known as the Supplemental Restraint System (SRS). The system must be disabled before performing service on or around system components, steering column, instrument panel components, wiring and sensors. Failure to follow safety and disabling procedures could result in accidental air bag deployment, possible personal injury and unnecessary system repairs.

PRECAUTIONS

Several precautions must be observed when handling the inflator module to avoid accidental deployment and possible personal injury.

- Never carry the inflator module by the wires or connector on the underside of the module.
- When carrying a live inflator module, hold securely with both

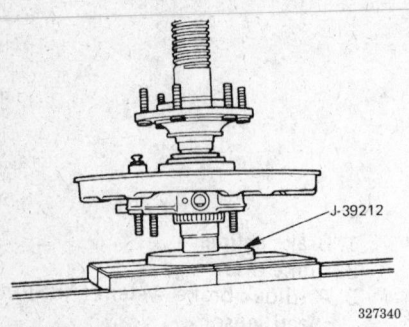

Tool No. J-39379: press bearing base — 1995–97 vehicles

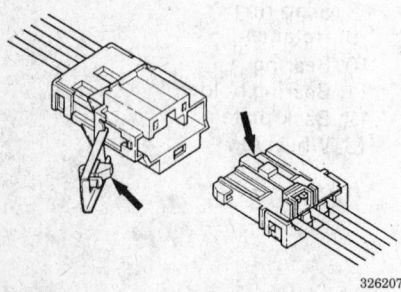

SRS wiring harness connectors — 1995–97 vehicles

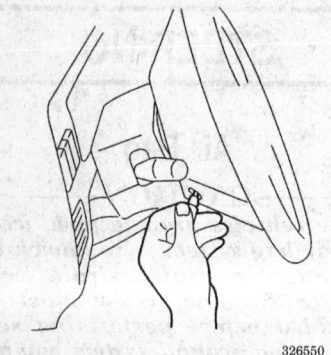

Air bag retaining bolts — 1995–97 vehicles

hands, and ensure that the bag and trim cover are pointed away.

• Place the inflator module on a bench or other surface with the bag and trim cover facing up.

• With the inflator module on the bench, never place anything on or close to the module which may be thrown in the event of an accidental deployment.

Removing and installing air bag module — 1995–97 vehicles

DISARMING

NOTE: This procedure is necessary for disabling AND enabling the air bag system.

Driver's Side

CAUTION

The Supplemental Restraint System (SRS) must be disarmed before any of its components are disconnected or the air bag are removed. Failing to disarm the SRS before servicing its components may cause accidental deployment of the air bag, resulting in unnecessary SRS repairs and possible personal injury.

1. Turn the steering wheel so that the vehicle's front wheels are pointing straight ahead.

2. Turn the ignition switch to the **LOCK** position. Remove the key.

3. Disconnect the negative and positive battery cables. Wait at least five minutes before working around the air bags.

NOTE: Removing the C–21, C–22, C–23, and C–24 SRS fuses (1995 Trooper) or C–21 and C–22 fuses (1996–97 Trooper and SLX) from the under–dash fuse block has the same effect as disconnecting the battery.

4. Disconnect the yellow 3–way SRS harness connector at the base of the steering column.

5. After servicing is completed, enable the SRS.

6. Reconnect the yellow 3–way SRS harness connector at the base of the steering column. Then, install the fuses if they were removed.

7. Reconnect the positive and negative battery cables.

8. Turn the ignition to the **ON** position, but don't start the engine. The AIR BAG warning light should turn on and flash on and off for seven seconds, and then turn off. This light

sequence indicates that the SRS system is functioning normally. If the AIR BAG light doesn't come on, or stays on longer than seven seconds, the system must be diagnosed.

Passenger's Side

CAUTION

When carrying a live air bag, make sure the bag and trim cover are pointed away from the body. In the unlikely event of an accidental deployment, the bag will then deploy with minimal chance of injury. When placing a live air bag on a bench or other surface, always face the bag and trim cover up, away from the surface. This will reduce the motion of the module if it is accidently deployed.

1. Turn the steering wheel so that the vehicle's front wheels are pointing straight ahead.

2. Turn the ignition switch to the **LOCK** position. Remove the key.

3. Disconnect the negative and positive battery cables. Wait at least five minutes before working around the air bags.

NOTE: Removing the C–21, C–22, C–23, and C–24 SRS fuses (1995 Trooper) or C–21 and C–22 fuses (1996–97 Trooper and SLX) from the under–dash fuse block has the same effect as disconnecting the battery.

4. Remove the glove box door and and the passenger's air bag lower cover.

5. Disconnect the yellow 4–way SRS harness connector from the passenger's side air bag.

6. If the passenger's air bag module must be removed; carefully remove the lower mounting bracket and mounting nuts. Lift the air bag out of the dashboard and remove it from the vehicle.

7. After servicing is completed, enable the SRS.

8. Reconnect the yellow 3–way SRS harness connector at the base of the steering column.

9. Reconnect the positive and negative battery cables.

10. Turn the ignition to the **ON** position, but don't start the engine. The AIR BAG warning light should turn on and flash on and off for seven seconds, and then turn off. This light sequence indicates that the SRS system is functioning normally. If the AIR BAG light doesn't come on, or stays on longer than seven seconds, the system must be diagnosed.

Steering Wheel

REMOVAL AND INSTALLATION

1993–94 Models

1. Set the steering wheel so that the front wheels are pointing straight–ahead.
2. Disconnect the negative battery cable.
3. Remove the horn pad screws from the rear of the steering wheel. Pry the horn pad upward from the steering wheel to remove it.
4. Disconnect the horn lead.
5. Remove the steering column nut.
6. Matchmark the steering wheel-to-steering shaft position for reassembly.
7. Using a steering wheel puller, press the steering wheel from the steering shaft.

NOTE: Do not strike the steering column or use air tools to loosen the steering wheel nut. The impact may damage the steering column's energy–absorbing properties

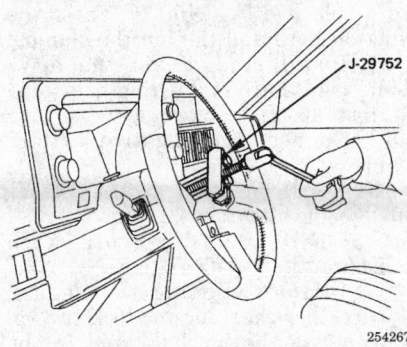

Use a puller to remove the steering wheel — 1993–94 vehicles

To install:

8. Align the matchmarks and install the steering wheel on the shaft. Torque the steering wheel nut to 22–29 ft. lbs. (30–39 Nm).
9. Reconnect the horn lead and install the horn pad.
10. Connect the negative battery cable.
11. Check the steering wheel spoke angle and test the operation of the horn.

1995–97 Models

——— **CAUTION** ———
The SRS/air bag system must be disabled before the steering wheel is removed. Failure to disable the air bag system may result in unnecessary system repairs and possible personal injury. Do not damage, cut, or attempt to alter the yellow SRS wiring harness.

1. Properly remove the air bag module from the vehicle and place it face–up on a bench away from your work area. (Refer to the air bag removal and installation procedure in this section.)

——— **CAUTION** ———
Always carry a live air bag with its trim cover and cushion pointed away from your body. When placing a live air bag on a bench or other surface, always point the trim cover up, away from the surface. Following these precautions will lessen the chance of personal injury if the air bag accidentally deploys.

2. Disconnect the horn leads.
3. Make alignment marks on the steering wheel and steering column shaft.
4. Loosen and remove the steering wheel nut. Use tool No. J–29752 or

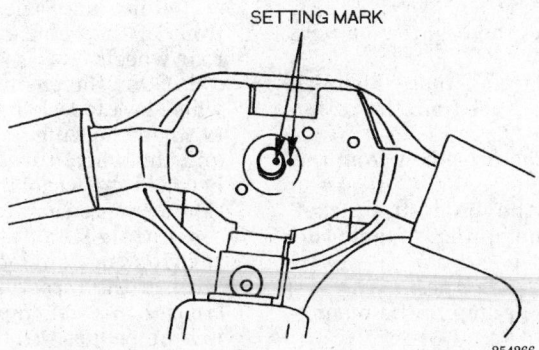

Matchmark the steering wheel to the shaft — 1993–94 vehicles

an equivalent steering wheel puller to remove the steering wheel.

NOTE: The puller bolts may damage the air bag cable reel if they are threaded too deeply into the steering wheel.

To install:
5. Check the alignment of the SRS cable reel.
   a. Turn the cable reel clockwise to its fully locked position. Stop turning the cable reel as soon as you feel resistance. Don't turn the cable reel past the point at which you begin to feel resistance to its rotation.
   b. Turn the cable reel about three turns in the opposite direction until the pointer on the cable reel is in alignment with the neutral mark.
6. Align the steering wheel–to–shaft matchmarks. Install the steering wheel and tighten the nut to 25 ft. lbs. (34 Nm). Reconnect the horn leads.
7. Carefully connect the yellow 2–way connector to the air bag module. Properly install the air bag module onto the steering wheel.
8. Reconnect the positive and negative battery cables.
9. Turn the ignition to the **ON** position, but don't start the engine. The AIR BAG warning light should turn on and flash on and off for seven seconds, and then turn off. This indicator light sequence indicates that the SRS system is functioning normally. If the AIR BAG light doesn't come on, or stays on longer than seven seconds, the system must be diagnosed.
10. Check the operation of the horn buttons.

Tie Rod Ends

REMOVAL AND INSTALLATION

1. Set the front wheels and steering wheel in the straight-ahead position. Remove the ignition key and lock the steering column in this position.
2. Raise and safely support the vehicle.
3. If necessary, remove the radiator skid plate.
4. Matchmark the tie rod ends to the tie rod shaft for reinstallation.
5. Remove the cotter pin and castle nut from the tie rod end. Loosen the tie rod end locknut.

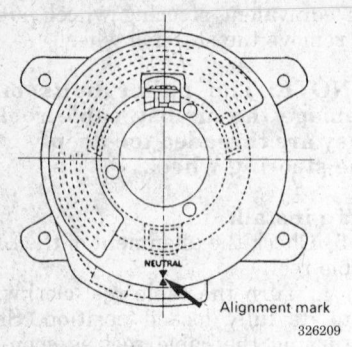

SRS cable reel alignment marks — 1995–97 vehicles

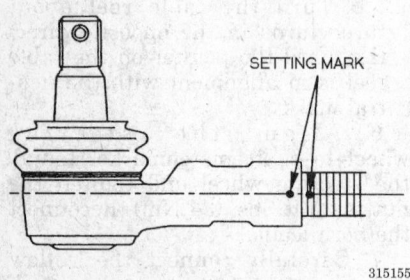

Matchmark the tie rod end to the shaft

6. Using a ball joint separator tool, separate the tie rod ends from the steering knuckle and center link.

7. Remove the tie rod.

8. Unscrew the tie rod ends from the tie rod.

9. Check the tie rod ends for damage and excess play. Replace any damaged or worn parts.

To install:

10. Install the tie rod ends onto the tie rod, then align the matchmarks. Torque the locknut (inner or outer) to 87 ft. lbs. (118 Nm).

11. Install the tie rod end onto the steering knuckle and center link. Install the castle nuts and torque them

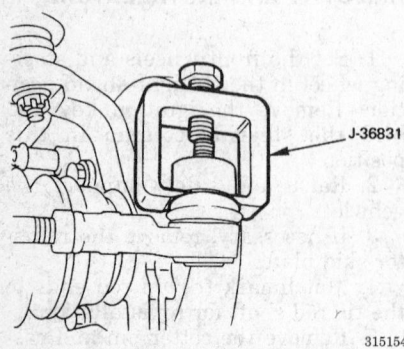

Outer tie rod end removal

to 72 ft. lbs. (98 Nm). Tighten the castle nuts only enough to install a new cotter pin.

12. Install the radiator skid plate and tighten its bolts to 27 ft. lbs. (37 Nm).

13. Check and adjust the front wheel alignment. Tighten the tie rod locknuts to 87 ft. lbs. (118 Nm).

Power Steering Pump

BLEEDING

──────── **WARNING** ────────
Use only Dexron® III ATF in the power steering system. Other types of fluids may cause system damage.

1. Fill the reservoir with fresh ATF. Allow the fluid to settle for a few minutes.

2. Start the engine and let it run for a few minutes. Don't turn the steering wheel. Shut the engine off.

3. Recheck the fluid level and add more if necessary.

4. Raise and support the vehicle so its front wheels are off the ground. Block the rear wheels.

5. Start the engine and slowly turn the steering wheel lock–to–lock three or four more times until any buzzing sound disappears and the wheel turns smoothly. Do not hold the wheel in either lock position for more than 5 seconds.

6. Add fluid if necessary.

7. Lower the vehicle.

8. Straighten the wheels and turn off the engine. If the fluid level in the reservoir increases, or the fluid foams — repeat the bleeding procedure.

9. The bleeding procedure is completed when the fluid level in the reservoir remains constant.

REMOVAL AND INSTALLATION

1993–94 Models

1. Disconnect the negative battery cable.

2. Disconnect and plug the inlet and outlet fluid lines from the power steering pump.

3. Remove the drive belt from the pump.

4. Remove the pump-to-bracket bolts and the pump from the brackets.

To install:

5. Install the pump to the mounting bracket.

6. Connect the hydraulic lines to the pump.

7. Install and adjust the drive belt.

8. Refill the reservoir with Dexron® III ATF.

9. Connect the negative battery cable, start the engine and bleed the power steering system.

1995–97 Models

1. Disconnect the negative battery cable.

2. Drain the fluid from the power steering pump reservoir.

3. Disconnect the inlet and outlet fluid lines from the power steering pump. Plug the lines to prevent fluid loss and contamination.

4. Loosen the adjusting bolt to relieve the belt tension.

5. Remove the drive belt from the pump pulley.

6. Unbolt and remove the pump from its mounting bracket.

7. Unbolt and remove the steering pump pulley. Use a puller if necessary.

To install:

8. Install the pump to the mounting bracket and tighten the mounting bolts to 34 ft. lbs. (46 Nm).

9. Install the pulley and tighten its bolt to 58 ft. lbs. (78 Nm).

10. Connect the hydraulic line banjo bolt to the pump using new crush washers. Tighten the banjo bolt to 40 ft. lbs. (54 Nm). Connect the suction hose and tighten the clamps.

11. Install and adjust the drive belt. Tighten the adjusting bolt to 34 ft. lbs. (46 Nm).

12. Connect the negative battery cable.

13. Refill and bleed the power steering system.

a. Fill the reservoir with Dexron® III ATF.

b. Before starting the engine, turn the wheel lock to lock several times so the level of fluid in the reservoir drops. Refill the reservoir as needed to bring the fluid to the specified level.

c. Raise and safely support the front of the vehicle and block the rear wheels.

d. Start the engine and turn the wheel lock to lock 3 or 4 more times until any buzzing sound disappears and the wheel turns smoothly. Do not hold the wheel in the lock position for more than 5 seconds.

e. Straighten the wheels and turn off the engine. If the fluid level in the reservoir foams or increases in level: repeat the bleeding procedure. If the fluid level stays the same: bleeding is completed.

BRAKES

Anti-Lock Brake System Service

PRECAUTIONS

• Certain components within the Anti-lock Brake System (ABS) are not intended to be serviced or repaired individually. Only those components with removal and installation procedures should be serviced.

• Do not use rubber hoses or other parts not specifically specified for and ABS system. When using repair kits, replace all parts included in the kit. Partial or incorrect repair may lead to functional problems and require the replacement of components.

• Lubricate rubber parts with clean, fresh brake fluid to ease assembly. Do not use lubricated shop air to clean parts; damage to rubber components may result.

• Use only specified brake fluid from an unopened container.

• If any hydraulic component or line is removed or replaced, it may be necessary to bleed the entire system.

• A clean repair area is essential. Always clean the reservoir and cap thoroughly before removing the cap. The slightest amount of dirt in the fluid may plug an orifice and impair the system function. Perform repairs after components have been thoroughly cleaned; use only denatured alcohol to clean components. Do not allow ABS components to come into contact with any substance containing mineral oil; this includes used shop rags.

• The Anti-lock control unit is a microprocessor similar to other computer units in the vehicle. Ensure that the ignition switch is **OFF** before removing or installing controller harnesses. Avoid static electricity discharge at or near the controller.

• If any arc welding is to be done on the vehicle, the control unit should be unplugged before welding operations begin.

DEPRESSURIZING

Before bleeding the brake system, remove the 40A ABS main fuse from the fuse and relay box. Removing the fuse allows air to be thoroughly bled from the hydraulic modulator unit.

1. Set the parking brake and start the engine.

2. Bleed the brake system with the engine running.
3. Shut the engine off.
4. After the brake system has been bled, reinstall the 40A ABS main fuse.

NOTE: If the system is bleed without the fuse removed, the hydraulic valve unit will be damaged.

Master Cylinder

REMOVAL AND INSTALLATION

1. Disconnect the negative battery cable.
2. Disconnect the fluid level indicator connector from the brake fluid reservoir.
3. Use a vacuum pump to remove the brake fluid from the reservoir.
4. Use a flare wrench to disconnect the fluid lines from the master cylinder. Plug the lines to prevent fluid loss and contamination.
5. Disconnect the vacuum hose from the power booster. Releasing the vacuum will relieve the power booster's internal pressure and prevent the piston from popping loose when the master cylinder is removed.
6. Remove the master cylinder mounting nuts.
7. If applicable, remove the proportioning valve bracket off of the power booster's studs.
8. Remove the master cylinder by pulling it straight off of the power booster's studs.
To install:
9. Check the adjustment of the power booster pushrod.
   a. Use a small flat-bladed tool to loosen the seal retainer from the power booster pushrod. Remove the seal by hand to avoid damaging it.
   b. Install tool No. J–39216, or an equivalent pushrod gauge, onto the power booster. Use a vacuum pump to apply 20 in. (500mm) Hg of NEGATIVE pressure to the booster.
   c. Measure the distance between the end of the pushrod gauge's piston and the gauge body. Clearance should be 0 ± 0.004 in. (0 ± 0.1mm).
   d. If the clearance is out of adjustment, use tool No. J–39241, or an equivalent pushrod adjusting tool to bring it into specification.
   e. After adjustment is completed, install the seal assembly and its retainer.
10. Position the master cylinder and its gaskets and spacers onto the power booster.
11. If applicable, fit the proportioning valve bracket onto the mounting studs.
12. Install the mounting nuts and tighten them to 10 ft. lbs. (13 Nm).
13. Connect the brake lines to the master cylinder. Use a flare wrench to tighten the fittings to 9 ft. lbs. (12 Nm).
14. Reconnect the power booster vacuum hose.
15. Connect the fluid level indicator connector to the fluid reservoir.
16. Properly bleed the system (according to the appropriate brake system bleeding procedure.

Brake Caliper

REMOVAL AND INSTALLATION

Front Brake Caliper

1. Raise and safely support the vehicle.
2. Remove some brake fluid from the reservoir.
3. Remove the front wheels.
4. Disconnect the brake fluid line from the caliper. Plug the line to prevent fluid loss.

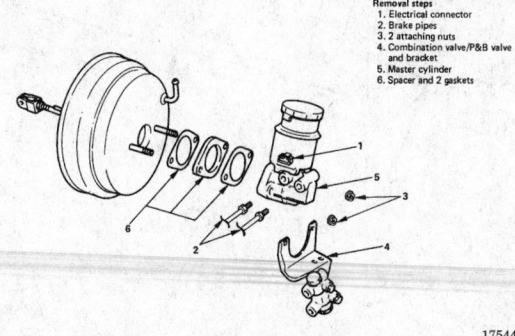

Removal steps
1. Electrical connector
2. Brake pipes
3. 2 attaching nuts
4. Combination valve/P&B valve and bracket
5. Master cylinder
6. Spacer and 2 gaskets

175445

Master cylinder

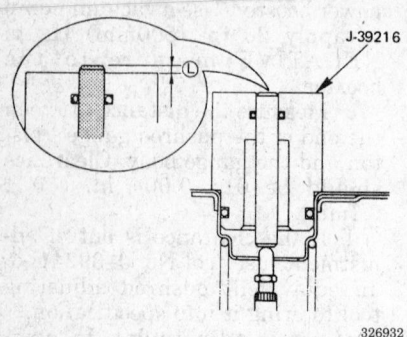

Pushrod adjustment gauge piston clearance

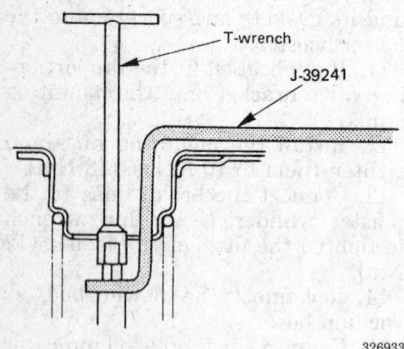

Pushrod adjuster tool

5. Loosen the brake caliper mounting bolt and guide bolt. Remove the caliper from the mount.

6. Remove the brake pads and clips from the caliper. Inspect the brake pads for wear and replace them if necessary.

To install:

7. Fill the brake caliper with clean brake fluid and connect the fluid line to the caliper using new washers. Tighten the brake line banjo fitting to 26 ft. lbs. (35 Nm). Install the brake pads and clips onto the caliper.

8. Install the caliper onto the mounting bracket. Lubricate the caliper bolts and their boots. Then, install the caliper mounting bolts and tighten them to 54 ft. lbs. (74 Nm).

9. Refill and bleed the brake system.

10. Install the front wheels and lower the vehicle.

Rear Brake Caliper

1. Raise and safely support the vehicle.

2. Remove some brake fluid from the reservoir.

3. Remove the rear wheels.

4. Disconnect the brake fluid line from the caliper. Plug the line to prevent fluid loss.

5. Loosen the brake caliper mounting bolt and guide bolt. Remove the caliper from the mount bracket.

6. Remove the brake pads and clips from the caliper. Inspect the brake pads for wear; replace them if necessary.

7. If necessary for servicing, unbolt the caliper mounting bracket from the backing plate.

To install:

8. If removed, install the caliper mounting bracket and tighten its bolts to 76 ft. lbs. (103 Nm).

9. Fill the brake caliper with clean brake fluid and connect the fluid line to the caliper using new washers. Tighten the brake line banjo fitting to 26 ft. lbs. (35 Nm). Install the brake pads and clips onto the caliper.

10. Install the caliper on the mounting bracket. Lubricate the caliper bolts and their boots. Then, install the caliper mounting bolts. For 1993–95 vehicles, tighten them to 32 ft. lbs. (44 Nm). For 1996–97 vehicles, tighten them to 32 ft. lbs. (44 Nm).

11. Refill and bleed the brake system.

12. Install the rear wheels and lower the vehicle.

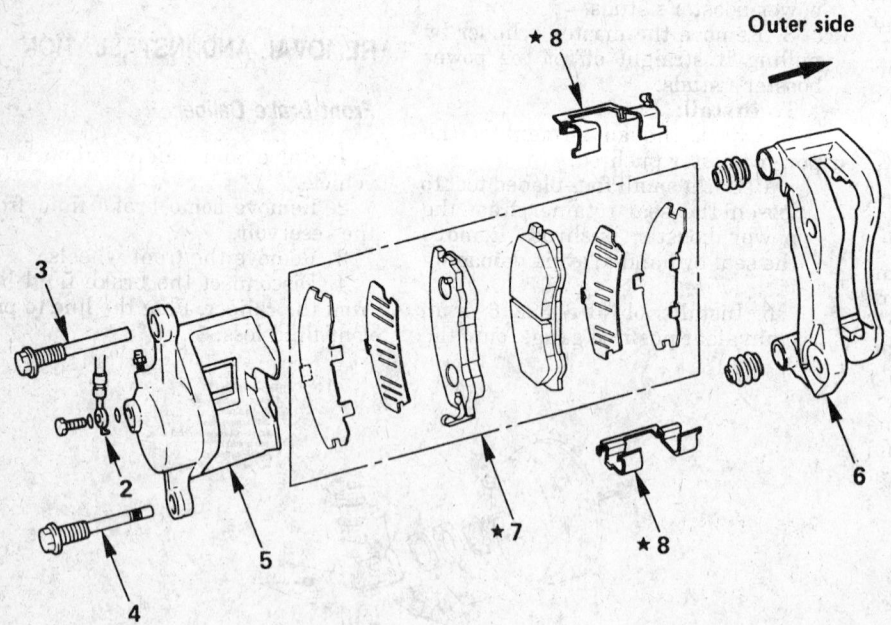
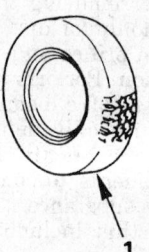

Outer side

1 Wheel and tire assembly
2 Brake flexible hose
3 Guide bolt
4 Lock bolt
5 Caliper assembly
6 Support bracket with pad asssembly
7 Pad assembly with shim
8 Clip

★ ; Repair kit

Front caliper assembly — 1993–95 vehicle shown, later designs are similar

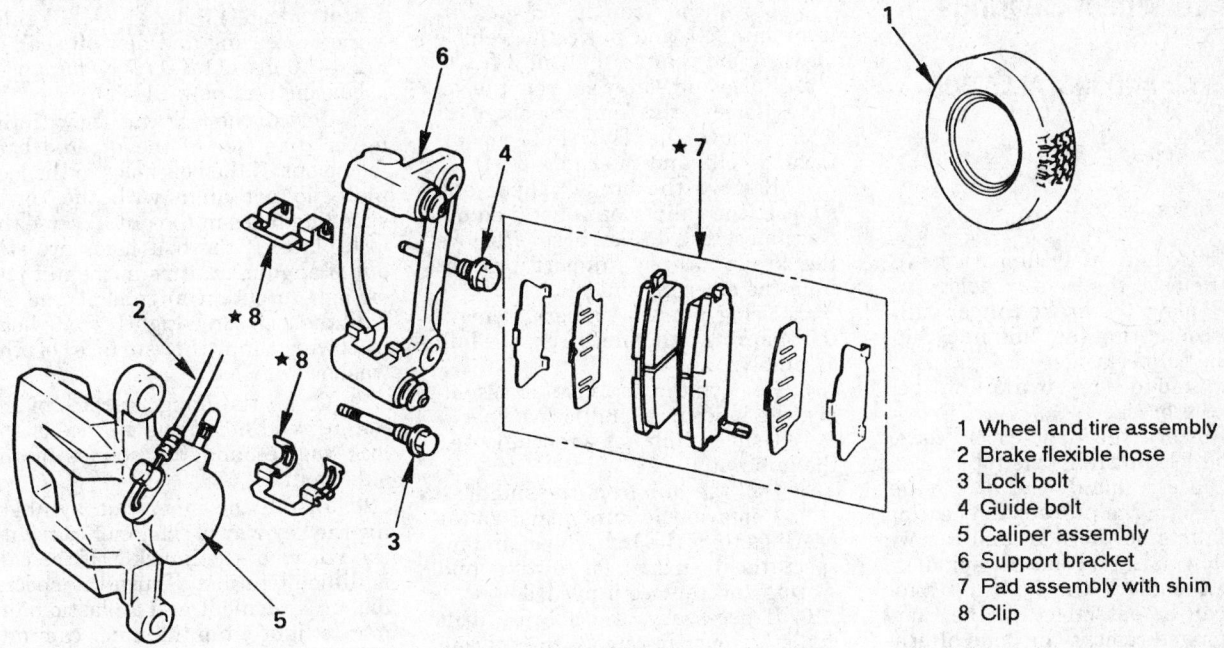

1 Wheel and tire assembly
2 Brake flexible hose
3 Lock bolt
4 Guide bolt
5 Caliper assembly
6 Support bracket
7 Pad assembly with shim
8 Clip

256320

Rear caliper assembly

Disc Brake Pads

REMOVAL AND INSTALLATION

Front Disc Brakes

─────── CAUTION ───────

Brake pads may contain asbestos, which has been determined to be a cancer causing agent. Never clean the brake surfaces with compressed air. Avoid inhaling any dust from any brake surface. Use a commercially available brake cleaning fluid when cleaning disc brake surfaces.

1. Remove about half of the brake fluid from the master cylinder reservoir to prevent overflow when the caliper piston is compressed.
2. Raise and safely support the vehicle.
3. Remove the front wheels.
4. Remove the brake caliper from the caliper bracket without disconnecting the brake line. Support the caliper with a length of wire. Do not let the caliper hang from the brake hose.
5. Remove the brake pads and shims. Inspect the brake rotor and machine or replace as necessary. Check the minimum thickness (speci-

fication is cast into the rotor) before machining.

To install:

6. Use a large C-clamp or brake piston tool to push the caliper piston into its bore.
7. Apply a thin coat of brake grease to both sides of both inner shims. Assemble the pads and shims, then install them into the caliper. The wear indicator on the inner pad must face down.
8. Install the calipers. Clean and lubricate the caliper mounting bolts and lubricate the mounting bolt boots. Install the mounting bolts and tighten them to 54 ft. lbs. (74 Nm).
9. Install the front wheels and lower the vehicle.
10. Apply the brakes several times to seat the pads before moving the vehicle. Check the fluid level in the master cylinder reservoir and add as necessary.

Rear Disc Brakes

1. Use a vacuum pump to remove some brake fluid from the master cylinder reservoir to prevent overflow when the caliper piston is compressed.
2. Raise and safely support the vehicle.
3. Remove the rear wheels.

4. Remove the brake caliper from the caliper bracket without disconnecting the brake line. Support the caliper with a length of wire. Do not let the caliper hang from the brake hose.
5. Remove the brake pads and shims. Inspect the brake rotor and machine or replace as necessary. Check the minimum thickness (specification is cast into the rotor) before machining.

To install:

6. Use a large C-clamp or brake piston tool to push the caliper piston into its bore.
7. Apply a thin coat of brake grease to both sides of both inner shims. Assemble the pads and shims, then install them into the caliper. The wear indicator on the inner pad must face down.
8. Install the calipers. Clean and lubricate the caliper mounting bolts and lubricate the mounting bolt boots. Install the mounting bolts and tighten them to 32 ft. lbs. (44 Nm).
9. Install the rear wheels and lower the vehicle.
10. Apply the brakes several times to seat the pads before moving the vehicle. Check the fluid level in the master cylinder reservoir and add as necessary.

Front Brake Rotor, Hub and Wheel Bearings

REMOVAL AND INSTALLATION

1993–95 Models

2WD Vehicles

1. Raise and safely support the vehicle. Remove the front wheels.

2. Remove the brake caliper without disconnecting the fluid line. Support the caliper aside.

3. Remove the brake caliper mounting bracket.

4. Remove the dust cover, cotter pin and locknut from the rotor.

5. Place a hand over the outer wheel bearing to prevent the bearing from falling on the floor and remove the rotor and hub from the spindle.

6. Matchmark the hub and rotor so it can be assembled in the same position and remove the bolts attaching the hub to the rotor, then pull off the rotor.

7. Thoroughly clean, inspect and repack the bearings with wheel bearing grease. Replace the bearings if needed. Use a brass drift to drive the bearing races out of the hub.

To install:

8. Use a suitably–sized bearing driver to drive new bearing races into the hub.

9. Install the rotor on the hub and torque the bolts to 76 ft. lbs. (103 Nm).

10. Install the inner bearing and the grease seal in the hub if removed and position the hub and rotor assembly on the spindle.

11. Install the outer bearing and spindle nut in the hub. Torque the hub nut to 22 ft. lbs. (29 Nm) to seat the bearings and then loosen the nut. Using a spring gauge, connected to the stud bolt at 90 degrees, retorque the hub nut until the spring gauge measures a bearing preload of 4.4–5.5 lbs. (2.0–2.5 kg) for a new bearing and grease seal, or 2.6–4.0 lbs. (1.2–1.8 kg) for a used bearing and a new grease seal.

12. Install the caliper mounting bracket and torque the bolts to 115 ft. lbs. (156 Nm).

13. Install the caliper and torque the mounting bolts to 20–27 ft. lbs. (28–38 Nm).

14. Install the front wheels and lower the vehicle.

15. Pump the brake pedal several times to test the system. Bleed the brakes if necessary.

4WD Automatic Hub

1. Move the transfer case shift lever into **2H** and move the vehicle forward and rearward about 3 ft.

2. Raise and safely support the vehicle. Remove the front wheels.

3. Remove the 4WD hub cap attaching bolts and the cap.

4. Remove the brake caliper and support the caliper on a wire; do not disconnect the brake hose. Remove the brake caliper support bracket from the steering knuckle.

5. Using snapring pliers, remove the snapring and shims from the hub assembly.

6. Remove the drive clutch assembly, the inner cam and lockwasher.

7. Using a hub nut wrench, loosen the hub nut.

8. Pull the hub from the spindle.

9. Remove the inner and outer bearings from the hub. Clean and inspect the bearings for pitting and scoring and replace if needed.

10. If necessary, use a brass drift and a hammer to remove the bearing races from the hub.

NOTE: Always replace the bearing and race as a matched set if replacement is needed.

11. If equipped with ABS, unbolt and remove the ABS sensor ring.

12. If removing the disc from the hub, scribe matchmarks, remove the bolts attaching the hub to the rotor, then separate the disc from the hub.

To install:

13. Use a suitably–sized bearing driver to install new bearing races into the hub.

14. If the hub and rotor were separated, install the hub to the rotor and torque the bolts to 76 ft. lbs. (103 Nm).

15. If equipped, install the ABS sensor ring and tighten the bolts to 13 ft. lbs. (18 Nm).

16. Pack the bearings with wheel bearing grease and install the inner bearing and a new grease seal in the hub. Position the hub and disc assembly on the spindle, place 1 to 1½ oz. of bearing grease in the hub and install the outer bearing and hub nut on the spindle.

17. When installing the hub nut, perform the following procedures:

a. Torque the hub nut to 22 ft. lbs. (30 Nm) and loosen the nut to seat the bearings.

b. Use a spring gauge connected to the stud bolt at 90° to measure preload.

c. Retorque the hub nut until the spring gauge measures a bear-

ing preload of 4.4–5.5 lbs. (2.0–2.5 kg) for a new bearing and grease seal, or 2.6–4.0 lbs. (1.2–1.8 kg) for a new bearing and new oil seal, or 2.6–4.0 lbs. (11.8–17.7 N) for a used bearing and new oil seal.

18. Install the lock washer with the larger diameter of the tapered bore facing out. If the bolt holes in the lock plate do not align with the corresponding holes in the nut, reverse the lock plate. If the bolt holes are still out of alignment, turn in the nut just enough to obtain alignment and install the lock screw tightly so its head is lower than the surface of the washer.

19. Clean the flange surface of the hub, thread holes, the surface of the lock washer and the splines of the axle shaft.

20. Install the inner cam by aligning the key way of the inner cam with the groove of the knuckle. If the cam is difficult to install, use the special tool or equivalent and a plastic hammer to lightly tap the inner cam into place.

21. Do the following steps to select the proper shim.

a. Lower the vehicle and support the lower control arm with a block of wood and a floor jack to place the axle in the normal horizontal position.

b. Install the special adjusting tools onto the hub until it comes in contact with the lock washer.

c. Pull the axle out as far as possible and using a feeler gauge, measure the clearance "t" between the hub and the snapring groove on the axle shaft.

d. If the clearance is larger than the snapring groove, selected shims must be installed so that clearance "t" is 0–0.039 in. (0– 0.1mm). Shims come in thicknesses of; 0.2, 0.3, 0.5, and 1.0mm.

e. Remove the special tool J-36836 or equivalent.

22. Install the drive clutch assembly.

23. Install the shims selected above by hand on the axle and use the following steps to install a new snapring.

a. Install special tool J-36835-2 or equivalent onto the axle.

b. Install the snapring on the tool.

c. Install tool driver J-36835-1 or equivalent.

d. Pull out the axle shaft by pulling tool J-36835-2 or equivalent. Install the snapring to the axle by pushing on tool J-36835-1.

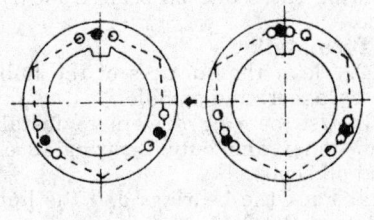

Lock washer installation — 1993–95 4WD automatic hub

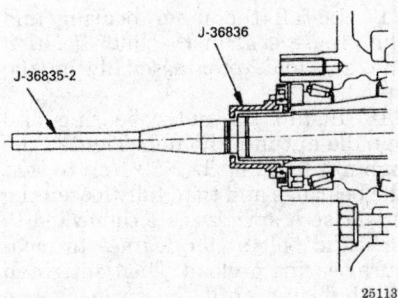

Install special tools to measure shim thickness — 1993–95 4WD automatic hub

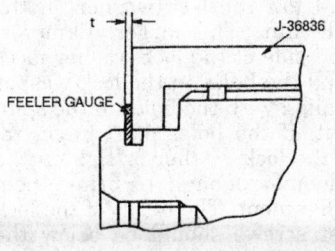

Measure clearance "t" between the special tool and the snapring groove — 1993–95 4WD automatic hub

e. Remove tool driver J-36835-2 or equivalent from the axle and check the fit of the snapring.

24. Install the housing assembly and cap. Torque the cap bolts to 43.3 ft. lbs. (58.9 Nm).

25. Install the front wheels.

26. Lower the vehicle to the floor.

27. Pump the brake pedal several times to test the system. Bleed the brakes if necessary.

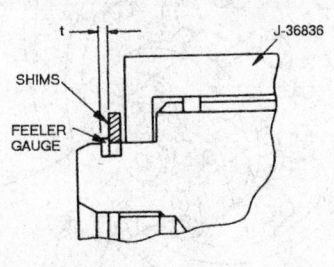

If clearance "t" is larger than snapring groove, selected shims must be used — 1993–95 4WD automatic hub

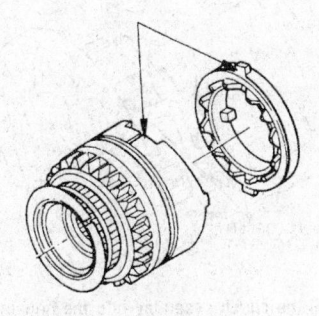

Drive clutch assembly installation — 1993–95 4WD automatic hub

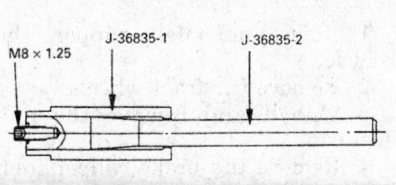

Special tools for snapring installation — 1993–95 4WD automatic hub

4WD Manual Hub

1. Shift the transfer case lever into "2H" and the manual hub into the "FREE" position.

2. Raise and safely support the vehicle.

3. Remove the front wheels.

4. Remove the brake caliper assembly and mounting bracket.

5. Loosen the 6 bolts and remove the housing assembly. Be careful not to loose the detent ball and spring.

6. Use snapring pliers and remove the snapring and shims.

7. Remove 6 bolts and remove the housing assembly from the hub.

8. Remove the lock washer retaining screw and remove the lock washer.

9. Use a special hub nut wrench and remove the hub nut.

10. Remove the hub and rotor assembly from the spindle.

11. To disassembly the clutch assembly, push the follower knob and turn the clutch assembly clockwise and remove the clutch assembly from the knob. Remove the retaining spring from the clutch assembly by turning it counterclockwise.

12. Matchmark the hub and rotor. Place the rotor in a soft jaw vise and remove the bolts attaching the rotor to the hub.

To install:

13. Align the matchmark and assemble the hub and rotor. Torque the mounting bolts to 76 ft. lbs. (103 Nm).

14. Pack the clean bearings with wheel bearing grease and install the inner bearing and a new grease seal in the hub. Position the hub and disc assembly on the spindle, place 1 to 1½ oz. of bearing grease in the hub and install the outer bearing and hub nut on the spindle.

15. When installing the hub nut, perform the following procedures:

  a. Torque the hub nut to 22 ft. lbs. (29 Nm) and loosen the nut to seat the bearings.

  b. Use a spring gauge connected to the stud bolt at 90° to measure preload.

  c. Retorque the hub nut until the spring gauge measures a bearing preload of 4.4–5.5 lbs. (2.0–2.5 kg) a new bearing and oil seal or 2.6–4.0 lbs. (1.2–1.8 kg.) for a used bearing and new oil seal.

16. Install the lock washer with the larger diameter of the tapered bore to the outer side of the vehicle. If the bolt holes in the lock plate do not align with the corresponding holes in the nut, reverse the lock plate. If the bolt holes are still out of alignment, turn in the nut just enough to obtain alignment and install the lock screw tightly so it's head is lower than the surface of the washer.

17. Apply grease to the spacer, ring, snapring and the splined part of the inner assembly.

18. Assemble the inner assembly and the clutch assembly.

19. Install the body assembly on the hub. Torque the bolts to 8.6 ft. lbs. (11.8 Nm).

20. Install the selective shim and a new snapring. The clearance between

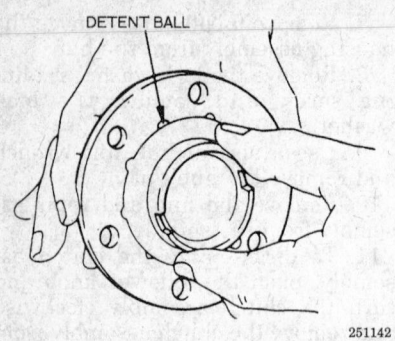

Align the detent ball with the groove cut in the cover — 1993–95 4WD manual hub

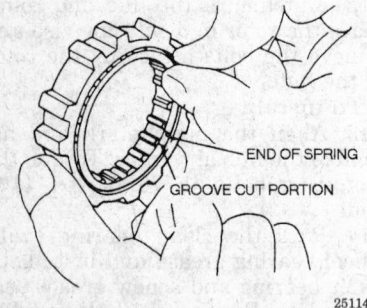

Align the end of the retainer spring to the end of the cut portion of clutch spring groove — 1993–95 4WD manual hub

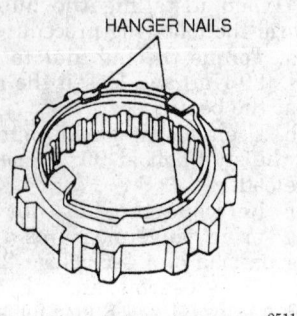

Hook the retainer spring onto the upper portion of hanger nails — 1993–95 4WD manual hub

the free wheeling hub body and the snapring should be 0–0.01 inch, shims are available in selective sizes.

21. Install the housing assembly. Torque the bolts to 43.4 ft. lbs. (58 Nm).

22. Install the front wheels.

23. Lower the vehicle to the floor.

24. Pump the brake pedal several times to test the system. Bleed the brakes if necessary.

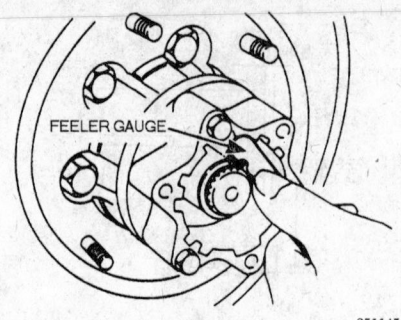

Use a feeler gauge to measure the snapring clearance — 1993–95 4WD manual hub

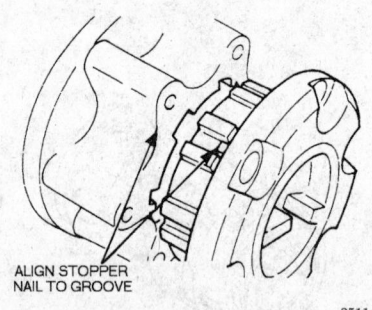

Install the clutch assembly into the housing, align the stopper nails to the grooves of the body — 1993–95 4WD manual hub

1996–97 Models

2WD Vehicles

1. Raise and safely support the vehicle.

2. Remove the front wheels.

3. Unbolt and remove the hub dust cap.

4. Remove the brake caliper from its support bracket and support it with a wire hook. Don't disconnect the brake hose. Remove the brake caliper support bracket from the steering knuckle.

5. Remove the lock washer.

6. Use a hub nut wrench to loosen and remove the spindle nut.

7. Pull the hub from the spindle.

8. Hold the outer wheel bearing with your hand to prevent it from falling out of the hub. Remove the hub and rotor assembly from the spindle.

9. Matchmark the hub and rotor so they can be assembled in the same positions. Place the hub and rotor assembly in a padded vice and then remove the six hub bolts to separate the rotor from the hub.

10. Thoroughly clean, inspect, and repack the bearings with wheel bearing grease. Replace the bearings if necessary.

11. Use a hammer and a brass drift to carefully tap the outer and inner bearing races and oil seals from the hub.

To install:

12. Clean the surfaces of the hub, lock washer, and spindle.

13. Use bearing drivers to install new inner and outer bearing races and oil seals.

14. Pack the bearings and the hub cavity with wheel bearing grease.

15. If a new rotor is being installed, use brake cleaner or alcohol to remove any anti–rust coating.

16. Install the rotor onto the hub and tighten the bolts to 76 ft. lbs. (103 Nm).

17. Install the inner bearing and the grease seal in the hub. Position the hub and rotor assembly on the spindle.

18. Install the outer bearing and spindle nut into the hub. Tighten the hub nut to 22 ft. lbs. (29 Nm) to seat the bearings and then fully loosen the nut. Use a spring scale connected to the stud bolt at 90 degrees to measure bearing preload. Then, retighten the hub nut until the spring gauge measures a bearing preload of 4.4–5.5 lbs. (2.0–2.5 kg) for a new bearing and grease seal, or 2.6–4.0 lbs. (1.2–1.8 kg) for a used bearing and a new grease seal.

19. Install the lock washer on the spindle nut. The larger diameter, tapered side of the lock washer faces out, and the holes in the lock washer must align with the holes in the spindle nut. If the holes don't align, reverse the lock washer or tighten the spindle just enough to bring them into alignment. The heads of the lock washer screws should be below the surface of the washer after tightening.

20. Pack the hub dust cap with fresh grease and install it. Tighten the hub bolts to 43 ft. lbs. (59 Nm).

21. Install the brake caliper mount and the caliper.

22. Install the front wheels.

23. Lower the vehicle.

4WD Vehicles

1. Move the transfer case shift lever into **2H** and move the vehicle forward and rearward about 3 ft. to be sure that the front axle isn't engaged.

2. Raise and safely support the vehicle.

3. Remove the front wheels.

4. Unbolt and remove the hub dust cap.

5. Remove the brake caliper from its support bracket and support it with a wire hook. Don't disconnect

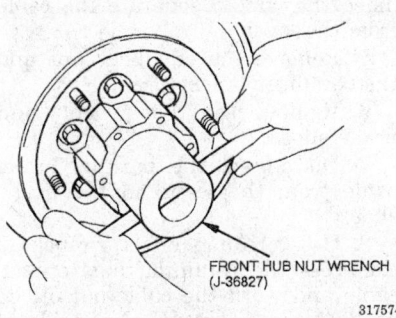

Hub nut wrench — 1996–97 2WD

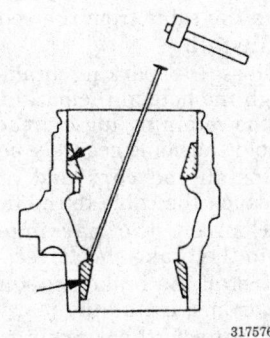

Use a brass drift to remove the inner or outer races — 1996–97 2WD

the brake hose. Remove the brake caliper support bracket from the steering knuckle.

6. Use snapring pliers to remove the snapring. Remove the shim.

7. Remove the hub flange assembly.

8. Remove the lock washer and its screws.

9. Use a hub nut wrench to loosen and remove the spindle nut.

10. Pull the hub from the spindle.

11. Hold the outer wheel bearing with your hand to prevent it from falling out of the hub. Remove the hub and rotor assembly from the spindle.

12. Matchmark the hub and rotor so they can be assembled in the same positions. Place the hub and rotor assembly in a padded vice and then remove the six hub bolts to separate the rotor from the hub. If equipped with ABS, unbolt the sensor ring from the rotor.

13. Thoroughly clean, inspect, and repack the bearings with wheel bearing grease. Replace the bearings if necessary.

14. Use a hammer and a brass drift to carefully tap the outer and inner bearing races and oil seals from the hub.

To install:

15. Clean the flange surface of the hub, the thread holes, the surface of the lock washer, and the splines of the axle shaft.

16. Use bearing drivers to install new inner and outer bearing races and oil seals.

17. Pack the bearings and the hub cavity with wheel bearing grease.

18. If a new rotor is being installed, use brake cleaner or alcohol to remove any anti–rust coating.

19. Install the rotor onto the hub and tighten the bolts to 76 ft. lbs. (103 Nm).

20. If equipped, install the ABS sensor ring and tighten the bolts to 13 ft. lbs. (18 Nm).

21. Install the inner bearing and the grease seal in the hub. Position the hub and rotor assembly on the spindle.

22. Install the outer bearing and spindle nut into the hub. Tighten the hub nut to 22 ft. lbs. (29 Nm) to seat the bearings and then fully loosen the nut. Use a spring scale connected to the stud bolt at 90 degrees to measure bearing preload. Then, retighten the hub nut until the spring gauge measures a bearing preload of 4.4–5.5 lbs. (2.0–2.5 kg) for a new bearing and grease seal, or 2.6–4.0 lbs. (1.2–1.8 kg) for a used bearing and a new grease seal.

23. Install the lock washer on the spindle nut. The larger diameter, tapered side of the lock washer faces out, and the holes in the lock washer must align with the holes in the spindle nut. If the holes don't align, reverse the lock washer, or tighten the spindle just enough to bring them into alignment.

24. Apply sealant to the both mating surfaces of the hub flange assembly and install it.

25. Install the shim and the snapring. There should be 0–0.012 in. (0–0.3mm) of clearance between the hub body and the snapring. Shims of

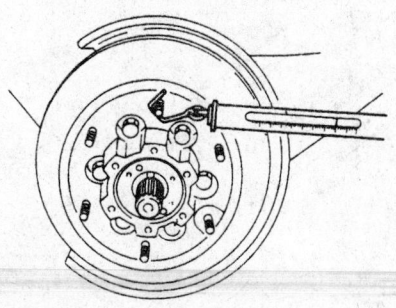

Use a spring scale to measure bearing preload

different thicknesses may be used if necessary.

26. Pack the hub dust cap with fresh grease and install it. Tighten the hub bolts to 43 ft. lbs. (59 Nm).

27. Install the front wheels.

28. Lower the vehicle.

29. Verify that the front axle engages and the shift-on-the-fly 4WD system works correctly.

Rear Brake Rotor

REMOVAL AND INSTALLATION

1. Raise and safely support the vehicle. Remove the rear wheels.

2. Engage the parking brake.

3. Remove the brake caliper without disconnecting the fluid line. Support the caliper aside with wire.

4. Remove the brake caliper from the mounting bracket. Remove the brake caliper mounting bracket.

5. Release the parking brake.

6. Remove the rotor from the hub.

To install:

7. If a new rotor is being installed, use brake cleaner or alcohol to remove any anti–rust coating.

8. Install the brake rotor onto the hub. Install any retaining screws.

9. Install the caliper mounting bracket. Torque the mounting bracket bolts to 69–84 ft. lbs. (95–116 Nm), or 109 ft. lbs. (148 Nm) for 1995 Trooper vehicles.

10. Fit the caliper into place and install the mounting bolts. Torque the caliper mounting bolt to 12–17 ft. lbs. (16–24 Nm), or 32 ft. lbs. (44 Nm) for 1995 Trooper vehicles.

11. Install the rear wheels and lower the vehicle.

12. Pump the brake pedal to test the system. Bleed the brakes if necessary.

Parking Brake Cable

ADJUSTMENT

Rear Drum Brakes

1. Remove the parking brake lever trim console, if equipped. Alternately, pull back the parking brake lever boot to expose the adjusting nut and cable equalizer.

2. Raise and safely support the vehicle.

3. Properly adjust the rear service brakes.

   a. Remove the plug from the drum adjusting hole.

   b. Use a small pry tool to turn the shoe adjuster downward until the wheel cannot be turned by hand.

c. Turn the adjuster upward with the pry tool until the wheel can be turned by hand.

d. Make sure the shoes don't drag or bind against the drum.

4. Adjust the parking brake at the equalizer so the parking brake is fully engaged (wheel cannot be turned by hand) when the handle is pulled nine to eleven notches. On 1996 vehicles, the handle only needs to be pulled up six notches.

Rear Disc Brakes

1. Remove the parking brake lever trim console.
2. Raise the safely support the vehicle.
3. Loosen the equalizer adjusting nut to provide slack in the brake cable.
4. Properly adjust the parking brake shoes.

a. Remove the plug from the drum adjusting hole.

b. Use a small pry tool to turn the shoe adjuster downward until the wheel cannot be turned by hand.

c. Turn the adjuster upward approximately six to eight notches

with the pry tool until the wheel can be turned by hand.

d. Make sure the shoes don't drag or bind against inside of the brake rotor.

5. Remove the rubber plug in the backing plate and turn the adjusting screw downward until the rotor will not turn by hand.
6. Slowly turn the adjuster upward just until the rotor begins to turn and install the rubber plug.
7. Adjust the cable adjusting nut so the lever travels six notches when pulled with a force of 66 lbs. (30 kg) and then tighten the cable locknut. Tighten the locknut to 8 ft. lbs. (11 Nm).

REMOVAL AND INSTALLATION

Rear Drum Brakes

1. If equipped, remove the rear console.
2. Make sure the parking brake lever or handle is in the fully released position.
3. Raise and safely support the vehicle.
4. If equipped with a dash-board–mounted parking brake lever, disconnect the front cable from the relay lever. Next, remove the nut to release the equalizer lever and re-

move the clip to separate the cable from the lever.

5. Remove the equalizer nut and the equalizer bracket.
6. Remove the rear wheels and brake shoes.
7. Disconnect the parking brake cable from the lever on the brake shoe.
8. Use a 13mm offset wrench to compress the retaining lugs on the cable and work the cable out of the backing plate (toward the inside of the vehicle).
9. Remove any clips or brackets attaching the cable to the frame and remove the cable from the vehicle.

To install:

10. Pass the parking brake cable through the hole in the backing plate until the retaining lugs spread apart and hold the cable securely in place.
11. Install the clips and brackets that attach the cable to the frame.
12. Connect the parking brake cable to the brake shoe lever.
13. Install the brake shoes and the brake drums. Install the rear wheels.
14. Connect the parking brake cable at the equalizer bracket.
15. If equipped, reconnect the equalizer to the lever and reconnect the front cable to the relay lever using new clips and washers.

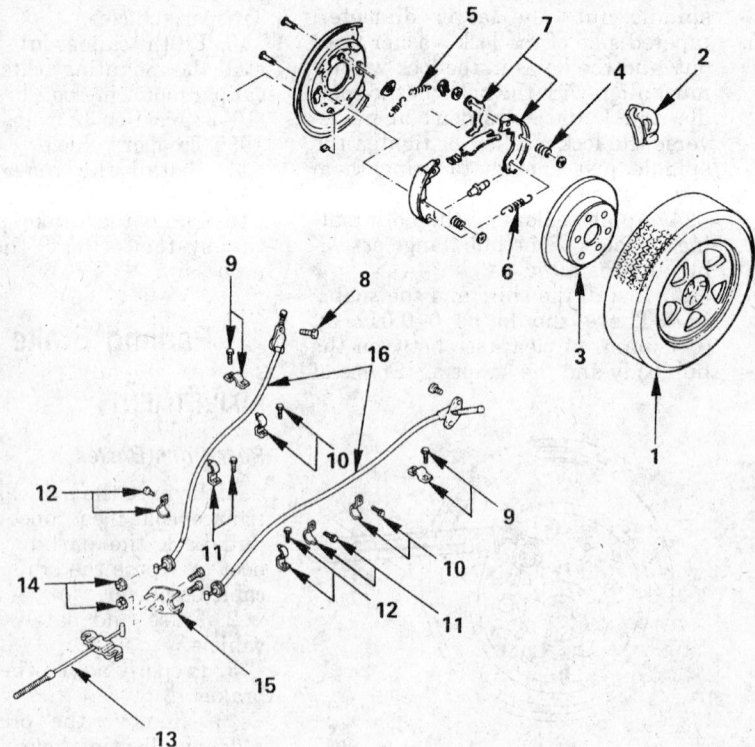

1. Rear wheels
2. Caliper assembly
3. Rotor (Drum)
4. Holding spring
5. Return spring; upper
6. Return spring; lower
7. Shoe assembly
8. Cable fixing bolt
9. Clip
10. Clip
11. Clip
12. Clip
13. Front cable
14. Nuts
15. Retainer assembly
16. Rear cable

324999

Rear disc brake parking brake components

16. Adjust the parking brake and lower the vehicle to the floor.

17. Install the rear console.

Rear Disc Brakes

1. Remove the rear console.

2. Raise and safely support the vehicle.

3. If equipped with a dash-board–mounted parking brake lever, disconnect the front cable from the relay lever. Next, remove the nut to release the equalizer lever and remove the clip to separate the cable from the lever.

4. Remove the rear wheels.

5. Remove the calipers and support them with wire hooks. Don't disconnect the brake line.

6. Remove the rear brake rotors.

7. Remove the parking brake shoes and remove the cable from the lever on the shoe.

8. Remove the cable bracket bolts and the clips attaching the cable to the frame.

9. Remove the cable from the equalizer assembly.

To install:

10. Apply grease to the front and rear linkages of the cable.

11. Attach the cable to the equalizer assembly. Torque the nuts on the retainer assembly to 10 ft. lbs. (13 Nm).

12. Install the clips holding the cable to the frame.

13. Pass the cable through the backing plate and install the bracket bolts. Torque the bolts to 4 ft. lb. (6 Nm).

14. Apply grease to the brake shoe contact areas, adjuster assembly, and brake shoe anchors.

15. Install the parking brake shoes and the rotor.

16. Install the caliper assembly and adjust the parking brake shoes.

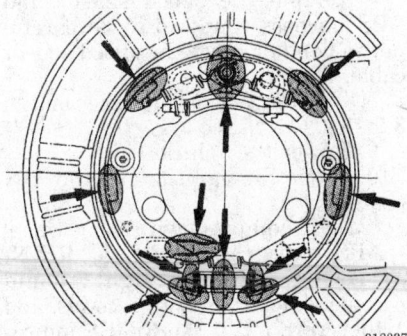

318337

Apply grease lightly to the shaded areas

17. If equipped, reconnect the equalizer to the lever and reconnect the front cable to the relay lever using new clips and washers.

18. Install the wheels and lower the vehicle to the floor.

19. Make the final parking brake cable adjustment at the parking brake lever.

20. Install the rear console.

Parking Brake Shoe

ADJUSTMENT

NOTE: Adjust the parking brake shoes before adjusting the parking brake lever and cable.

1. Remove the parking brake lever console.

2. Raise and safely support the rear of the vehicle. Block the front wheels.

3. Loosen the adjusting nut on the parking brake cable equalizer.

4. Remove the adjusting hole plug from the rear brake backing plate.

5. Use a brake adjusting tool or a small pry bar to turn the adjuster wheel downward until the rotor cannot be rotated by hand.

6. Turn the adjuster wheel upward 7 or 8 notches until the rotor can be turned by hand. Make sure the rotor doesn't drag when turned.

7. Inspect the parking brake shoes for excess wear. The minimum lining thickness is 0.039 in. (1.0mm).

8. Make sure the return springs and areas of metal–to–metal contact are lubricated with brake grease. No lubricant or cleaner must contact the brake shoe friction surfaces.

9. Adjust the parking brake cable so the parking brakes are fully engaged when the lever is pulled upward 6 or 7 notches. Tighten the equalizer nut to 4.4 ft. lbs. 6 Nm).

Brake System Bleeding

PROCEDURE

NOTE: On vehicles equipped with 4-wheel anti-lock brakes, remove the 40A ABS main fuse located at the relay and the fuse block before starting the engine. If the system is bleed without the fuse removed, the hydraulic valve unit will be damaged.

1. Set the parking brake and start the engine.

> ## ─ WARNING ─
> The vacuum booster seal will be damaged if the bleeding operation is performed with the engine off.

2. Remove the master cylinder reservoir cap and fill the reservoir with brake fluid. Keep the reservoir at least half full during the bleeding operation.

3. If the master cylinder is replaced or overhauled, first bleed the air from the master cylinder and then from each caliper or wheel cylinder. Bleed the master cylinder as follows:

 a. Disconnect the rear wheel brake line from the master cylinder.

 b. Have an assistant depress the brake pedal slowly once and hold it depressed.

 c. Seal the delivery port of the master cylinder where the line was disconnected with a finger, then release the brake pedal slowly.

 d. Release the finger from the delivery port after the brake pedal returns completely.

 e. Repeat Steps B through D until the brake fluid (not air) comes out of the delivery port during Step B.

NOTE: Do not let the fluid level in the reservoir drop below the half-way mark.

 f. Reconnect the brake line to the master cylinder.

 g. Have an assistant depress the brake pedal slowly once and hold it depressed.

 h. Loosen the rear wheel brake line at the master cylinder.

 i. Retighten the brake line, then release the brake pedal slowly.

 j. Repeat Steps G through I until no air comes out from the port when the brake line is loosened.

 k. Bleed the air from the front wheel brake line connection by repeating Steps A through J.

4. If equipped with a rear wheel ABS proportioning valve, this valve must be bled before the calipers.

5. Bleed the air from each wheel in the following order: Right rear wheel, Left rear wheel, Right front caliper and Left front caliper. Bleed the air as follows:

 a. Place the proper size box wrench over the bleeder screw.

 b. Cover the bleeder screw with a transparent tube and submerge the free end of the tube in a transparent container containing brake fluid.

c. Have an assistant pump the brake pedal slowly 3 times, then hold it depressed.

d. Remove the air along with the brake fluid by loosening the bleeder screw.

e. Retighten the bleeder screw, then release the brake pedal slowly.

f. Repeat Steps C through E until the air is completely removed. It may be necessary to repeat the bleeding procedure 10 or more times for front wheels and 15 or more times for rear wheels.

g. Go to the next wheel in sequence after each wheel is bled.

6. Depress the brake pedal several times after the air has been removed from all wheel cylinders and calipers. If the pedal feels spongy, the entire bleeding procedure must be repeated.

7. After the bleeding operation is completed on each individual wheel, check the level of brake fluid in the reservoir and refill up to the **MAX** level, if necessary.

8. Install the master cylinder reservoir cap and shut off the engine.

Wheel Speed Sensor

REMOVAL AND INSTALLATION

Front Wheel

1. Make sure the ignition is turned **OFF**. Remove the key.

2. Raise and safely support the vehicle.

3. Unbolt the speed sensor cable brackets.

4. Uncouple the speed sensor connector from the ABS wiring harness.

5. Unbolt the speed sensor from the steering knuckle and remove it.

To install:

6. Inspect the speed sensor. Replace the sensor if it is damaged or shorted:

a. Clean any dirt or corrosion from the speed sensor probe.

b. Check the speed sensor for a short circuit. Bend the cable while checking for continuity.

c. Check the sensor ring for damaged or chipped teeth.

7. Install the speed sensor to the steering knuckle.

8. Install the speed sensor brackets. Be careful not to bend or twist the cable during installation.

9. Couple the speed sensor and ABS harness connectors. Verify that the white line on the cable insulation is not twisted.

10. Tighten the speed sensor bolts to the following specifications:

a. Speed sensor bolt: 8 ft. lbs. (11 Nm)

b. Lower cable bracket: 18 ft. lbs. (24 Nm)

c. Upper cable bracket: 4 ft. lbs. (6 Nm)

11. Lower the vehicle.

12. Turn the ignition to the **ON** position. Verify that both the ABS or ANTILOCK and ABS ACTIVE (if equipped) indicators come on. Start the engine and verify that all ABS indicator lights turn off.

Rear Wheel

Four-Wheel ABS

1. Make sure the ignition is turned **OFF**. Remove the key.

2. Raise and safely support the vehicle.

3. Unbolt the speed sensor cable brackets from the rear axle. Detach any cable–securing clips

4. Uncouple the speed sensor connector from the ABS wiring harness.

5. Unbolt the speed sensors from the brake backing plates and remove them.

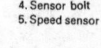
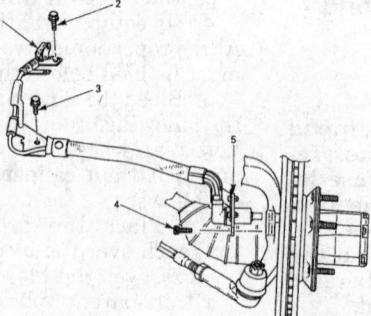

1. Speed sensor connector
2. Sensor cable bolt (Upper side)
3. Sensor cable bolt (Lower side)
4. Sensor bolt
5. Speed sensor

330195

Front wheel speed sensor

To install:

6. Inspect the speed sensors. Replace the sensor if it is damaged or shorted:

a. Clean any dirt or corrosion from the speed sensor probe.

b. Check the speed sensor for a short circuit. Bend or the cable while checking for continuity.

c. Check the sensor ring for damaged or chipped teeth.

7. Install the speed sensors to the backing plates.

8. Install the speed sensor brackets and clips. Be careful not to bend or twist the cable during installation.

9. Couple the speed sensor and ABS harness connectors.

10. Tighten the speed sensor bolts to 8 ft. lbs. (11 Nm)

11. Lower the vehicle.

12. Turn the ignition to the **ON** position. Verify that both the ABS or ANTILOCK and ABS ACTIVE (if equipped) indicators come on. Start the engine and verify that all ABS indicator lights turn off.

Rear-Wheel ABS and 2WD Vehicles Only

1. Make sure the ignition is turned **OFF**. Remove the key.

2. Raise and safely support the vehicle.

3. Uncouple the speed sensor connector from the ABS wiring harness.

4. Unbolt the speed sensor from the differential case.

To install:

NOTE: The speed sensor can only be replaced as any assembly. Do not replace the sensor unless it is diagnosed as faulty.

5. Inspect the speed sensor. Replace the sensor if it is damaged or shorted:

a. Clean any dirt or corrosion from the speed sensor probe.

b. Check the speed sensor for a short circuit. Bend the cable while checking for continuity.

6. Install the speed sensor into the differential case.

7. Couple the speed sensor and ABS harness connectors. Be careful not to bend or twist the speed sensor cable.

8. Tighten the speed sensor bolt to 8 ft. lbs. (11 Nm).

9. Lower the vehicle.

10. Turn the ignition to the **ON** position:

a. If equipped with rear–wheel ABS only, the RrABS or RrANTILOCK indicator light should come on for about two seconds, and then turn off. If the RrABS indicator comes after starting the engine, it has stored a diagnostic code.

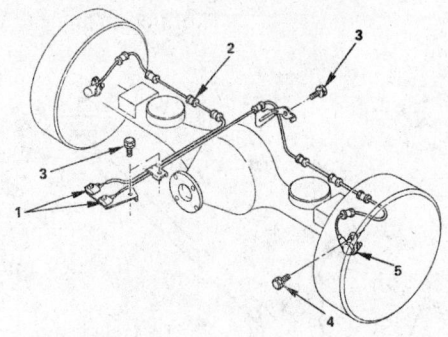

1. Speed sensor connector
2. Clip (11 pieces)
3. Sensor cable bolt
4. Sensor bolt
5. Speed sensor

330196

Rear wheel speed sensors (4WD)

b. If equipped with four–wheel ABS, verify that both the ABS or ANTILOCK and ABS ACTIVE (if equipped) indicators come on. Start the engine and verify that all ABS indicator lights turn off.

FRONT SUSPENSION

Shock Absorber

REMOVAL AND INSTALLATION

1. Raise and support the vehicle safely.
2. Remove the front wheels.
3. Support the lower control arm with a floor jack.
4. Hold the shaft of the shock absorber with a wrench to keep it from turning and remove the upper mounting nut, retainer, and rubber grommet.

NOTE: On some vehicles, it may be necessary to remove the bump stops to gain access to the lower mounting bolt.

5. Unbolt the shock absorber from the lower control arm. Remove the shock absorber from the vehicle.
To install:
6. Install the lower retainer and rubber grommet onto the upper shaft of the shock absorber. Then, fully extend the shock and install the upper shaft into the mounting hole in the frame bracket.
7. Install the upper rubber grommet, retainer and attaching nut onto shock absorber upper shaft. Only hand–tighten the mounting nut at this time.
8. Install the the shock absorber to the lower control arm bracket and install the mounting bolt and nut. Tor-

que the mounting bolt to 60–61 ft. lbs. (82–84 Nm).
9. Tighten the upper mounting nut to 14–15 ft. lbs. (19–20 Nm).
10. Install the bump stop if removed, and tighten the bolts to 30 ft. lbs. (41 Nm).
11. Install the front wheels and lower the vehicle.

Torsion Bar

REMOVAL AND INSTALLATION

1. Raise and safely support the vehicle.
2. Remove the front wheels.
3. Mark the location of the height adjustment bolt and remove it from the height control arm.
4. Mark the location of the torsion bar shaft and the height control arm. Remove the height control arm from the torsion bar and the frame crossmember.
5. Mark the location of the torsion bar on the lower control arm bracket.
6. Unbolt the torsion bar bracket, and remove the torsion bar from the lower control arm. The torsion bars are marked with an **L** or **R** for identification.
To install:
7. Inspect the torsion bar and height control arm components for signs of damage. If the rubber torsion bar seat is damaged, replace it.
8. Apply a generous amount of grease to the serrated ends of the torsion bar.
9. Use a floor jack to raise the lower control arm to help hold the rubber seat in contact with the torsion bar bracket while the torsion bar is being installed.
10. Align the matchmark and insert the front end of the torsion bar into the control arm.
11. Align the matchmark and install the height control arm in position so its end is reaching the adjust-

ing bolt. Be sure to grease the part of the height control arm that fits into the chassis.
12. Turn the adjusting bolt to so that its matchmark aligns with the mark made on the height control arm.
13. Tighten the torsion bar control arm bracket bolts to 86 ft. lbs. (116 Nm).
14. Install the front wheels.
15. Lower the vehicle and check the vehicle height.
16. Check and adjust the front wheel alignment.

Upper Ball Joints

REMOVAL AND INSTALLATION

1. Raise and safely support the vehicle.
2. Remove the front wheels.
3. Mark the position of the torsion bar adjuster. Loosen the adjuster to relieve the torsion bar tension.
4. Support the lower control arm with a floor jack.
5. Remove the upper ball joint castle nut.
6. Using a ball joint separator tool, separate the upper ball joint from the steering knuckle.
7. Unbolt and remove the upper ball joint from the control arm.
To install:
8. Install the ball joint onto the control arm. Tighten the upper ball joint bolts:
   • Four–bolt–style ball joint: 21–25 ft. lbs. (29–35 Nm)
   • 1993–94 vehicles with three–bolt–style ball joints: 51 ft. lbs. (69 Nm)
   • 1995–97 vehicles with three–bolt–style ball joints: 42 ft. lbs. (57 Nm).
9. Install the upper ball joint to the steering knuckle. Torque the upper ball joint castle nut to 72–73 ft. lbs. (96–98 Nm). Then, tighten the castle nut only enough to install a new cotter pin.
10. Install the front wheels.
11. Adjust the tension on the torsion bar to its original position.
12. Lower the vehicle to the floor.

Lower Ball Joints

REMOVAL AND INSTALLATION

Two–Wheel Drive

1. Raise and safely support the vehicle.
2. Remove the front wheels.

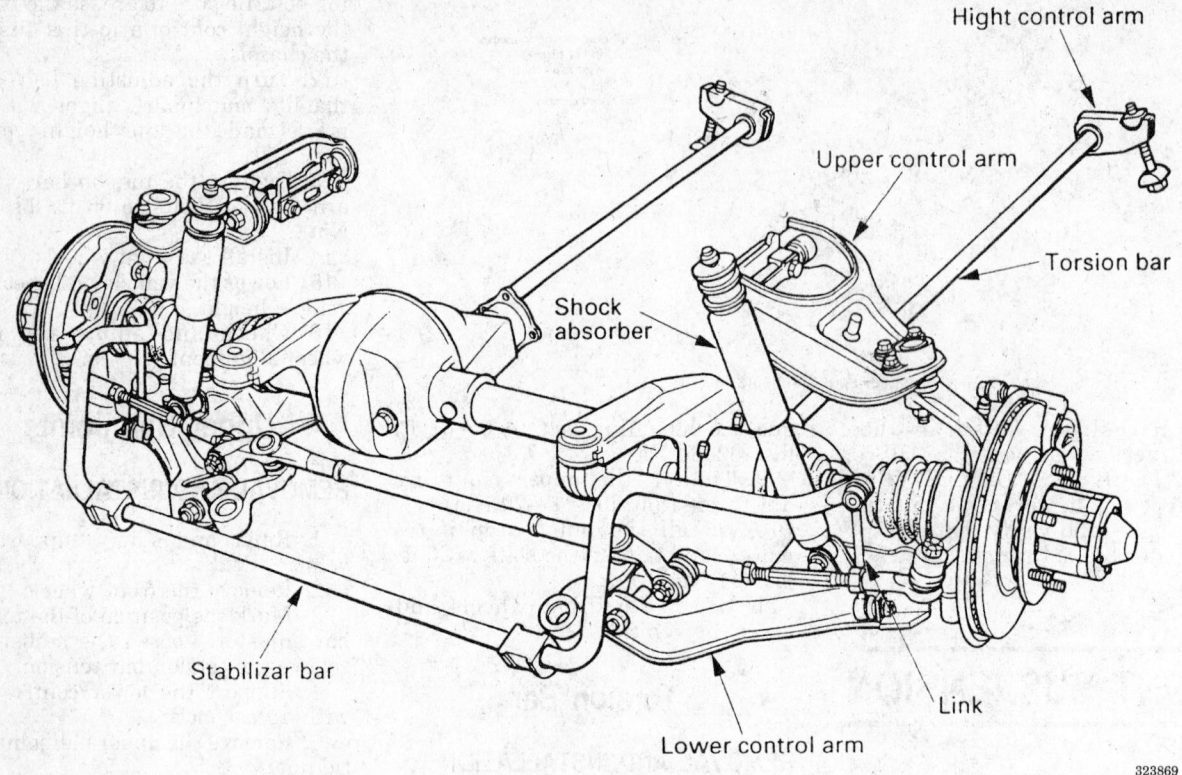

Front suspension components

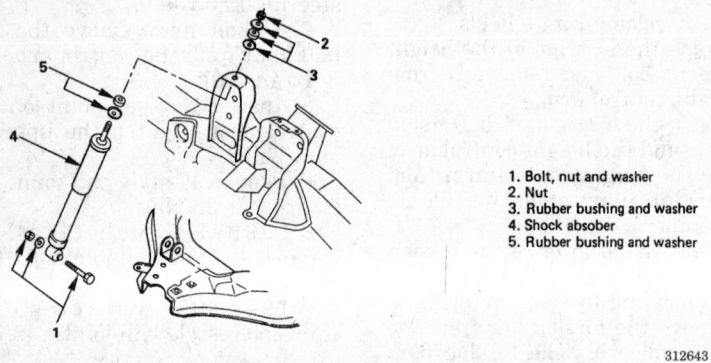

1. Bolt, nut and washer
2. Nut
3. Rubber bushing and washer
4. Shock absober
5. Rubber bushing and washer

312643

Shock absorber and mounting components

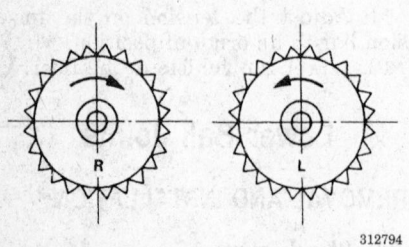

312794

Left and right torsion bar identification

3. Use a ball joint separator tool to disconnect the outer tie rod end from the steering knuckle.

4. Mark the position of the torsion bar adjuster. Loosen the adjuster to release the torsion bar tension.

5. Remove the cotter pin and castle nut from the upper and lower ball joints.

6. Using tool No. J–29107 or an equivalent ball joint separator tool, disconnect the lower ball joint from the knuckle.

7. Unbolt the lower ball joint from the lower control arm.

8. Remove the ball joint. Inspect the ball joint for excess play and damage.

To install:

NOTE: Use new self–locking nuts when installing the ball joint.

9. Install the lower ball joint on the lower control arm and torque the bolts to 68–83 ft. lbs. (93–113 Nm).

10. Install the lower ball joint stud into the steering knuckle and install the castle nut. Torque the nut to 87–111 ft. lbs. (117–137 Nm). Then, tighten the nut just enough to install a new cotter pin.

11. Connect the tie rod end to the steering knuckle. Tighten the tie rod securing nut to 80 ft. lbs. (109 Nm).

12. Lubricate the lower ball joint through the grease fitting.

13. Adjust the torsion bar tension to its original position.

14. Install the front wheels and lower the vehicle.

Four–Wheel Drive

1. Shift the transfer case into the **2H** position. Then, drive the vehicle forward and backward for a few feet to make sure the front axle and hubs are disengaged.

2. Raise and safely support the vehicle.

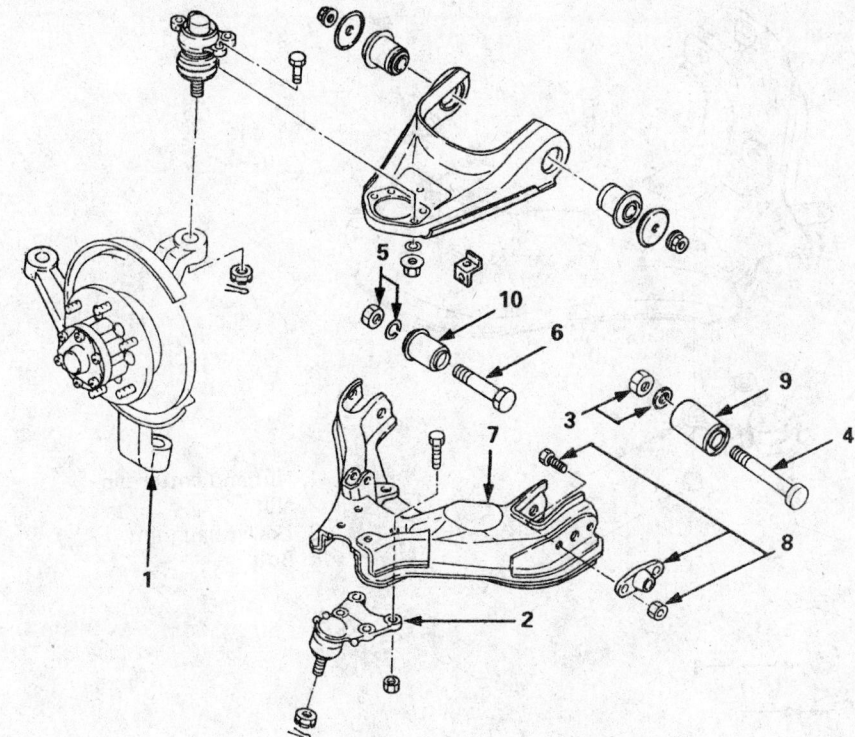

Control arm and ball joint components

1. Knuckle
2. Lower end
3. Nut and washer, rear
4. Bolt, rear
5. Nut and washer, front
6. Bolt, front
7. Lower control arm assembly
8. Torsion bar arm bracket
9. Bushing, rear
10. Bushing, front

312640

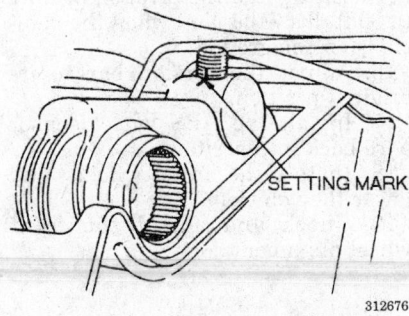

SETTING MARK

312676

Torsion bar adjusting bolt

3. Remove the front wheels.

4. Unbolt and remove the cap from the locking hub assembly. Do not lose any of the internal shims.

5. Remove the snapring to release the axle shaft from the hub.

6. Use a ball joint separator tool to disconnect the outer tie rod end from the steering knuckle.

7. Mark the position of the torsion bar adjuster. Loosen the adjuster to release the torsion bar tension.

8. Remove the cotter pin and castle nut from the upper and lower ball joints.

9. Using tool No. J–29107 or an equivalent ball joint separator tool,

disconnect the lower ball joint from the knuckle.

10. Unbolt the lower ball joint from the lower control arm.

11. Remove the ball joint. Inspect the ball joint for excess play and damage.

To install:

NOTE: Use new self-locking nuts when installing the ball joint.

12. Install the lower ball joint on the lower control arm and torque the bolts to 68–83 ft. lbs. (93–113 Nm).

13. Install the lower ball joint stud into the steering knuckle and install the castellated nut. Torque the nut to 87–111 ft. lbs. (117–147 Nm). Then, tighten the nut just enough to align the cotter pin hole with a hole in the castellated nut. Install a new cotter pin.

14. Connect the tie rod end to the steering knuckle. Tighten the tie rod securing nut to 72–80 ft. lbs. (98–109 Nm). Then, tighten the nut only enough to install a new cotter pin.

15. Reinstall the shims and the snapring on the axle shaft. Repack the hub with heavy-duty multi-purpose grease.

16. Install the hub cover and tighten the bolts to 43 ft. lbs. (59 Nm).

17. Lubricate the lower ball joint through the grease fitting.

18. Adjust the torsion bar tension to its original position.

19. Install the front wheels and lower the vehicle.

20. Make sure the front hubs engage into four-wheel drive correctly.

21. Check the vehicle ride height and front wheel alignment and adjust as necessary.

Upper Control Arms

REMOVAL AND INSTALLATION

1. Raise and safely support the vehicle.

2. Remove the front wheels.

3. If equipped with ABS, unbolt the wheel sensor wire bracket. Move the sensor out of the way, but don't disconnect it.

4. Mark the position of the torsion bar adjuster. Loosen the adjuster to release the torsion bar tension.

5. Support the lower control arm with a floor jack.

6. Remove the cotter pin and castle nut from the upper control arm ball joint. Use a ball joint separator

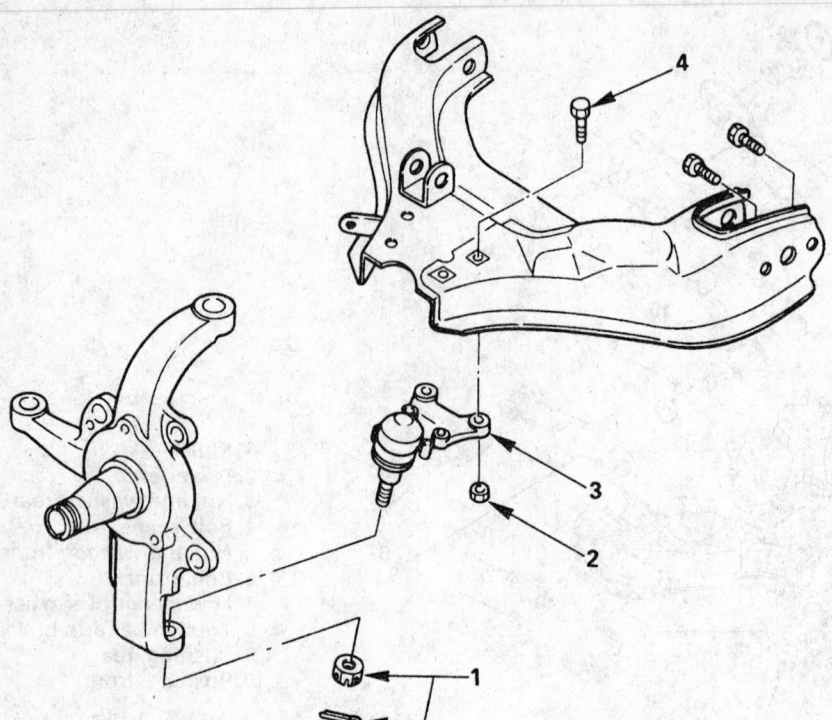

1. Nut and cotter pin
2. Nut
3. Lower ball joint
4. Bolt

312675

Lower ball joint and related components

tool to separate the upper control arm from the steering knuckle.

NOTE: Do not allow the steering knuckle to hang by the flexible brake line. Wire the steering knuckle up to the frame temporarily or support it with a floor jack.

7. Unbolt the upper control arm pivot shaft and remove the upper control arm from the frame bracket. Be sure to note the position and number of shims used for adjusting the camber and caster angles when removing the upper control arm. The shims must be replaced in their original position.

NOTE: It is helpful to wrap each shim pack in electrical tape for assembly. This will keep the appropriate shims together and organized.

8. Evenly loose the pivot shaft nuts to separate the pivot shaft and bushings from the upper control arm assembly. Then, remove the pivot shaft and bushings using tool No. J-29755 or its equivalent and a press.
To install:
9. Press new bushings into the control arm and install the pivot shaft.

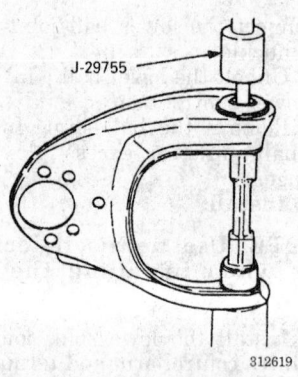

J-29755

312619

Removing the upper control arm bushings

10. Install the pivot shaft and bushing in the upper control arm. Do not tighten the pivot nuts at this time.

NOTE: Tighten the thinner shim pack's nut first for improved shaft-to-frame clamping force and torque retention.

11. Install the control arm to the frame. Replace the shims in the same position that they were removed from. Torque the bolts to 113 ft. lbs. (153 Nm).

12. If the ball joint was removed from the upper control arms, install it and tighten the nuts to 42 ft. lbs. (57 Nm).
13. Install the ball joint stud through the steering knuckle. Install the castle nut and tighten it to 73 ft. lbs. (98 Nm). Then, tighten the castle nut just enough to install a new cotter pin. Remove the wire used to support the knuckle.
14. Raise the lower control arm with a floor jack to adjust the clearance of the control arm to its rubber buffer. Buffer clearance:
• 1993–1994 Trooper: 0.83 in. (21mm)
• 1995–97 Trooper and SLX: 0.79 in. (20mm)
15. After the buffer clearance has been set, tighten the pivot shaft nuts to 80 ft. lbs. (108 Nm) when the vehicle is on the ground.
16. Adjust the torsion bar to its original position.
17. Install the ABS wheel sensor wire back into position.
18. Install the front wheels and lower the vehicle.
19. Check and adjust the front wheel alignment.

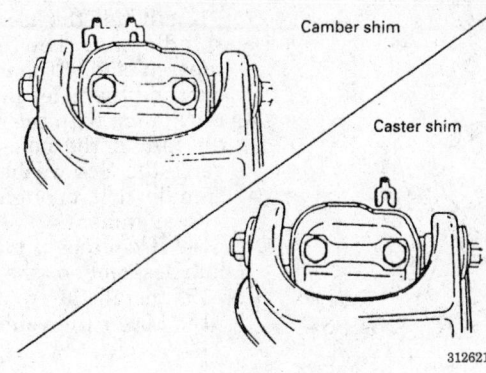

Camber shim

Caster shim

312621

Camber and caster shims

Lower Control Arms

REMOVAL AND INSTALLATION

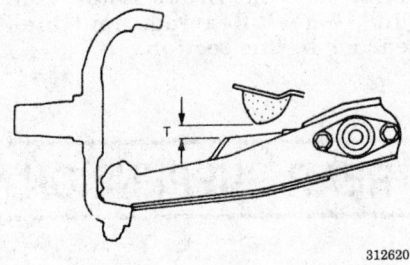

312620

Control arm buffer clearance

1. Raise and safely support the vehicle.
2. Remove the front wheels.
3. Mark the position of the torsion bar adjuster and release the torsion bar tension.
4. Disconnect the stabilizer bar from the lower control arm.
5. Unbolt the torsion bar bracket from the lower control arm. Then, remove the torsion bar from the lower control arm.

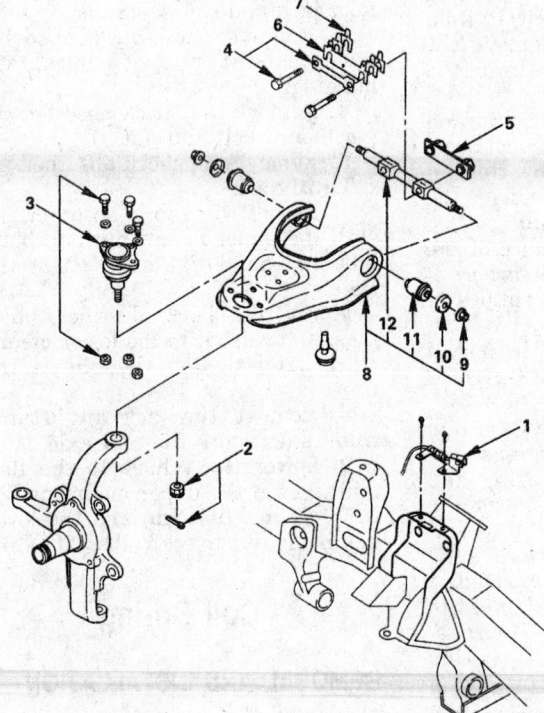

1. Speed sensor cable (If equipped with antilock brake system)
2. Nut and cotter pin
3. Upper ball joint
4. Bolt and plate
5. Nut assembly
6. Camber shims
7. Caster shims
8. Upper control arm assembly
9. Nut
10. Plate
11. Bushing
12. Fulcrum pin

312617

Upper control arm components

6. Disconnect the shock absorber from the lower control arm.
7. Unbolt the lower ball joint from the lower control arm.

NOTE: On four-wheel drive vehicles, the knuckle/hub should be supported with a jack so that the front axle shafts do not bear the weight of the suspension components.

8. Remove the retaining nuts and use a soft metal drift to drive out the bolts holding the lower control arm to the chassis.
9. Remove the lower control arm from the vehicle.
   To install:

NOTE: Use new self-locking nuts when assembling the suspension components.

10. Mount the lower control arm to the frame. Drive the bolts into position carefully to avoid damaging the threads. Install the nuts, but do not tighten them until the vehicle is on the ground.
11. Install the lower ball joint to the lower control arm. Tighten the retaining bolts 76 ft. lbs. (103 Nm).
12. Install the stabilizer bar to the lower control arm. Tighten the nuts to 8 ft. lbs. (10 Nm), or 37 ft. lbs. (50 Nm).
13. Assemble the lower ball joint to the steering knuckle. Torque the castle nut to 94–108 ft. lbs. (128–147 Nm). Then, tighten the nut only enough to install a new cotter pin.
14. Install the torsion bar and adjust the tension to its original position.
15. Install the front wheels and lower the vehicle.
16. With the weight of the vehicle on the suspension tighten the front pivot bolt to 145 ft. lbs. (197 Nm) and the rear pivot bolt to 116 ft. lbs. (157 Nm). Tighten the torsion bar bracket bolts to 86 ft. lbs. (116 Nm). Tighten the shock absorber mounting bolt to 61 ft. lbs. (84 Nm).
17. Check the vehicle ride height and adjust as necessary.
18. Check and adjust the front wheel alignment.

Sway Bar

REMOVAL AND INSTALLATION

1. Raise and safely support vehicle.
2. Remove the sway bar link nuts and separate the links from the sway bar.
3. Remove the radiator skid plate.

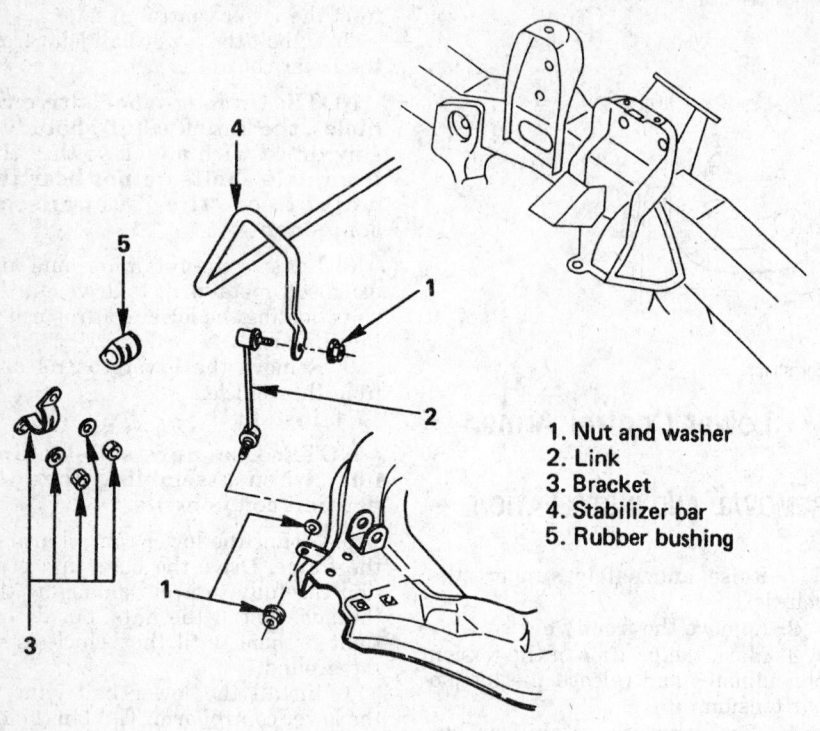

1. Nut and washer
2. Link
3. Bracket
4. Stabilizer bar
5. Rubber bushing

312718

Front sway bar components

4. Unbolt the frame bushing brackets and remove the stabilizer bar.

5. If necessary, remove the linkages from the lower control arms. Note the position of the washer between the linkage joint and the lower control arm flange.

To install:

6. Installation the sway bar with its frame bushings and brackets.

7. Install the linkages and connect them to the sway bar and lower control arms.

8. Torque the bracket bolts to 16 ft. lbs. (22 Nm) and the sway bar link nuts to 37 ft. lbs. (50 Nm).

9. Install the radiator skid plate and tighten its bolts to 27 ft. lbs. (37 Nm).

10. Lower the vehicle.

Front Wheel Bearings

ADJUSTMENT

1. Raise and safely support the vehicle.

2. Remove the front wheels.

3. If equipped, unbolt and remove the hub dust cap.

4. If equipped, remove the locking hub assembly.

5. Use a hub nut wrench to loosen and remove the spindle nut.

6. Pull the hub from the spindle.

7. Hold the outer wheel bearing with your hand to prevent it from falling out of the hub. Remove the hub and rotor assembly from the spindle.

8. Thoroughly clean, inspect and repack the bearings with wheel bearing grease. Replace the bearings if needed. Use a brass drift to drive the bearing races out of the hub.

9. Clean the flange surface of the hub, the thread holes, the surface of the lock washer, and the splines of the axle shaft.

10. Use bearing drivers to install new inner and outer bearing races and oil seals.

11. Pack the bearings and the hub cavity with wheel bearing grease.

12. Install the outer bearing and spindle nut into the hub. Tighten the hub nut to 22 ft. lbs. (29 Nm) to seat the bearings and then fully loosen the nut. Use a spring scale connected to the wheel stud bolt at 90 degrees to measure bearing preload. Then, retighten the hub nut until the spring gauge measures a bearing preload of 4.4–5.5 lbs. (2.0–2.5 kg) for a new bearing and grease seal, or 2.6–4.0 lbs. (1.2–1.8 kg) for a used bearing and a new grease seal.

13. Install the lock washer on the spindle nut. The larger diameter, tapered side of the lock washer faces out, and the holes in the lock washer must align with the holes in the spindle nut. If the holes don't align, reverse the lock washer, or tighten the spindle just enough to bring them into alignment.

14. If equipped, install the locking hub assembly or the hub dust cap.

15. Install the front wheels.

16. Lower the vehicle.

REMOVAL AND INSTALLATION

NOTE: For front wheel bearing replacement procedures, please refer to Front Brake Rotor, Hub and Wheel Bearings, outlined earlier in this section.

REAR SUSPENSION

Shock Absorber

REMOVAL AND INSTALLATION

1. Raise and safely support the vehicle.

2. Support the rear axle with a floor jack and safety stands.

3. Remove the shock absorber lower mount nut, washers, and bushings.

4. Remove the shock absorber upper mount bolt and nut.

5. Remove the shock absorber.

To install:

6. Install the shock absorber. Install the upper mount nut and bolts, but only hand–tighten them at this time.

7. Fit the shock absorber, bushings, and washer to the lower mount. Only hand–tighten the nut at this time.

8. Remove the jack and stands from underneath the rear axle.

9. Lower the vehicle to the floor and tighten the upper mounting bolt to 70 ft. lbs. (95 Nm) and the lower mounting nut to 58 ft. lbs. (78 Nm).

Coil Spring

REMOVAL AND INSTALLATION

1. Raise and safely support the vehicle under the frame.

2. Remove the rear wheels.

13. Lower the vehicle to the floor.

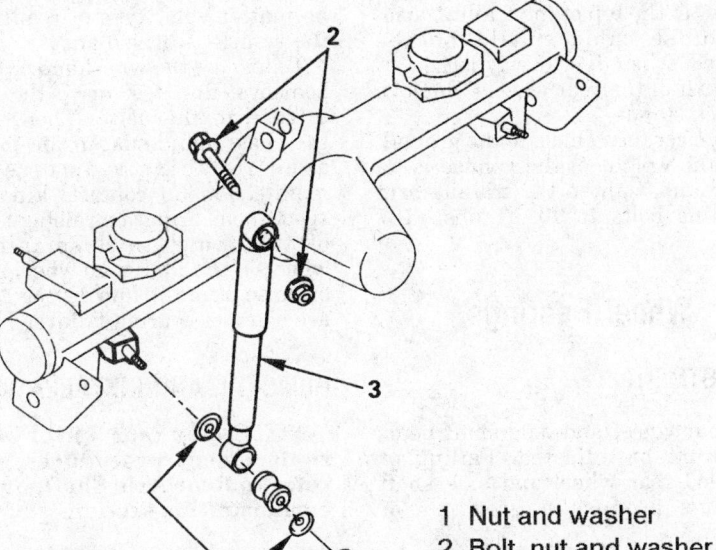

1 Nut and washer
2 Bolt, nut and washer
3 Shock absorber

326164

Rear shock absorber assembly

3. Place a floor jack under the rear axle housing and raise it slightly to compress the spring.

4. Unbolt the parking brake cable bracket from the trailing link.

5. Disconnect the stabilizer bar from its linkage.

6. Unbolt the shock absorber from the axle case mount.

───── **WARNING** ─────
Do not let the brake hose, parking brake cable or the breather hose extend to their full length.

7. Slowly lower the rear axle with the jack to release the coil spring tension. Remove the coil spring and the insulator.

To install:

8. Place the coil spring on the axle assembly and put the insulator on top of the spring.

9. Raise the axle assembly into position and connect the shock absorbers to the axle assembly. Tighten the shock absorber nut and bolt to 58 ft. lbs. (79 Nm).

10. Connect the stabilizer linkage to the bar. Tighten the nut to 37 ft. lbs. (50 Nm).

11. Install the parking brake bracket onto the trailing arm.

12. Install the rear wheels.

Sway Bar

REMOVAL AND INSTALLATION

1. Raise and safely support the vehicle.

2. Remove the rear wheels.

───── **WARNING** ─────
Be careful not to break the linkage ball joint boots.

3. Remove the nut attaching the linkage to the sway bar.

4. Remove the bolts attaching the sway bar brackets to the frame.

5. Remove the sway bar.

6. Inspect the rubber bushings and linkage ball joints for wear and replace them if necessary.

To install:

7. Install the sway bar and brackets to the frame. Tighten the bracket bolts to 16 ft. lbs. (22 Nm).

8. Connect the sway bar to the linkages and install the washer and nut. Tighten the nut to 37 ft. lbs. (50 Nm).

9. Install the rear wheels.

10. Lower the vehicle to the floor.

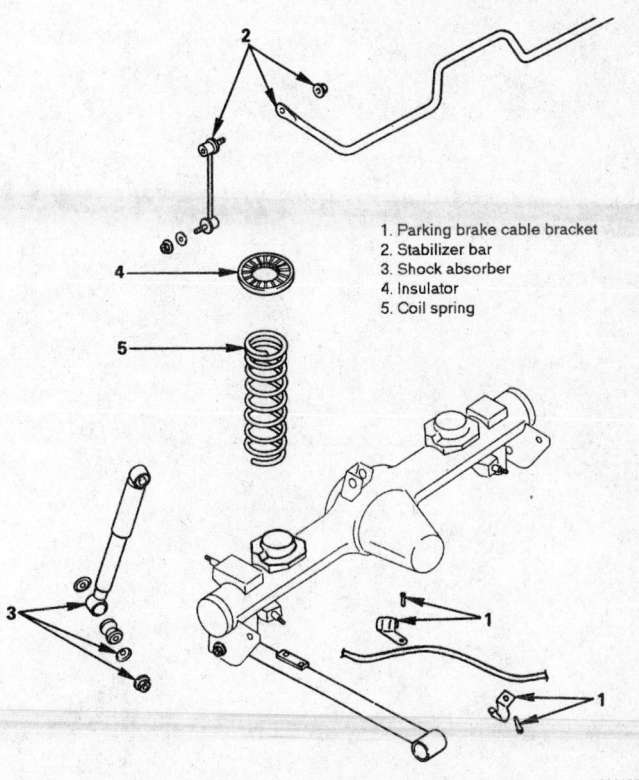

1. Parking brake cable bracket
2. Stabilizer bar
3. Shock absorber
4. Insulator
5. Coil spring

326167

Rear coil spring and related components

Trailing Arms

REMOVAL AND INSTALLATION

1. Raise and safely support the vehicle.
2. Remove the rear wheels.
3. Raise the rear axle slightly and support it with jackstands.
4. Unbolt the parking brake cable and bracket from the trailing arm.
5. Remove the mounting bolts and nuts; then, remove the trailing arm from the vehicle.
6. Use a press and special adapters to remove the bushings and install new ones if necessary.

To install:

NOTE: Do not use grease on or near the rubber bushings.

7. Install the trailing arm to its frame and axle housing brackets. Only hand–tighten the nuts and bolts at this time.
8. Install the parking brake and bracket to the top of the trailing arm. The brake cable should not be strained, twisted, or overly loose.
9. Install the rear wheels. Remove the jack stands.
10. Lower the vehicle to the ground. Once the weight of the vehicle is on the ground, tighten the trailing arm mounting bolts to 101 ft. lbs. (137 Nm).

Wheel Bearings

ADJUSTMENT

The rear wheel and axle shaft bearings can't be adjusted. Failing or damaged rear wheel and axle shaft bearings produce a growling or grating noise that can be heard when the vehicle is coasting at low speeds. Bearing noise will remain constant no matter what type of road surface the vehicle is driven on.

To inspect the wheel and axle shaft bearings, first test drive the vehicle to confirm the noise. Then, inspect the chassis and suspension for damaged bushings or areas of metal–to–metal contact. Finally, inspect the bearings by raising the rear of the vehicle and spinning the wheels by hand. Rough wheel motion, or noise from the hub during rotation are signs of bearing failure.

REMOVAL AND INSTALLATION

NOTE: For rear wheel bearing replacement procedures, please refer to Rear Axle Shaft, outlined earlier in this section.

DODGE 2

B-Series Van • Dakota • Ramcharger • Ram Pick-Up

FIRING ORDERS

NOTE: To avoid confusion, always replace spark plug wires one at a time.

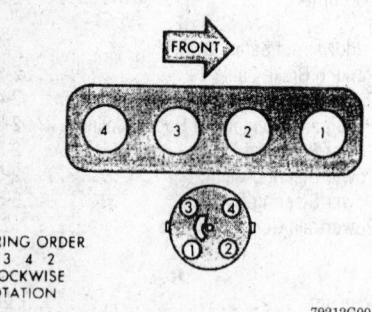

FIRING ORDER
1 3 4 2
CLOCKWISE
ROTATION

79212G00

2.5L (VIN G, K and P) 4-cylinder engine
Engine Firing Order: 1–3–4–2
Distributor Rotation: Clockwise

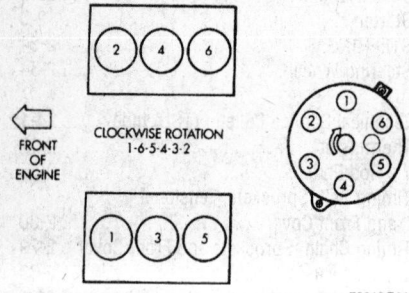

CLOCKWISE ROTATION
1-6-5-4-3-2

FRONT OF ENGINE

79212G01

3.9L (VIN X) engine
Engine Firing Order: 1–6–5–4–3–2
Distributor Rotation: Clockwise

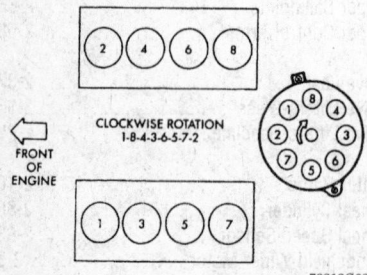

CLOCKWISE ROTATION
1-8-4-3-6-5-7-2

FRONT OF ENGINE

79212G02

5.2L (VIN T and Y) and 5.9L (VIN A, Z and 5) engines
Engine Firing Order: 1–8–4–3–6–5–7–2
Distributor Rotation: Clockwise

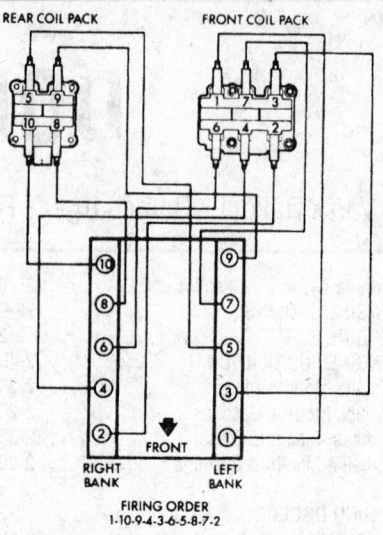

REAR COIL PACK   FRONT COIL PACK

RIGHT BANK   LEFT BANK

FIRING ORDER
1-10-9-4-3-6-5-8-7-2

79212G03

8.0L (VIN W) engine
Engine Firing Order: 1–10–9–4–3–6–5–8–7–2
Distributorless Ignition

ENGINE ELECTRICAL

NOTE: Disconnecting the negative battery cable on some vehicles may interfere with the functions of the on-board computer system(s) and may require the computer to undergo a relearning process, once the negative battery cable is reconnected.

Distributor

REMOVAL AND INSTALLATION

1. Remove the splash shield (if equipped).
2. If necessary, remove the air cleaner assembly and connecting tubes.
3. Unplug the pickup lead wire connector(s) from the wiring harness.
4. Leaving the distributor wires connected, unfasten the clips or screws that retain the distributor cap and lift off the cap.

NOTE: If insufficient clearance is obtained by removing the cap with the wires still connected, all or some of the spark plug wires should be tagged and disconnected from the cap for better access.

5. Bump the engine around until the rotor is pointing at No. 1 cylinder firing position, with the timing marks on the front case and crank pulley are aligned. Disconnect the negative battery cable from the battery.
6. Mark the distributor body and the engine block to indicate the position of the distributor in the block. Mark the distributor body to indicate the rotor position. These marks are used as guides when installing the distributor.
7. Remove the distributor hold-down bolt and bracket. Carefully lift the distributor from the engine. The shaft may rotate slightly as the distributor is removed. Make a note of where the movement stops. That is where the rotor must point when the distributor is reinstalled into the block.

To install:

Ignition Timing Undisturbed

1. If the crankshaft has not been rotated while the distributor was removed from the engine, use the reference marks made before removal to correctly position the distributor in the block. The shaft may have to be rotated slightly to engage the intermediate shaft gear (3.9L (VIN X), 5.2L (VIN T and Y) and 5.9L (VIN A, Z and 5).
2. Install the distributor while holding the rotor in position, allowing it to move only enough to engage the slot in the drive gear.
3. Install pickup coil leads and distributor cap. Make sure all high tension wires are firmly snapped in cap towers. Install distributor hold-down clamp screw and tighten to 200 inch lbs. (22.5 Nm).

Ignition Timing Disturbed

Perform this procedure if the crankshaft was rotated or otherwise disturbed (for example, during engine rebuilding) after the distributor was removed.

1. Rotate the crankshaft until the No. 1 cylinder is at Top Dead Center (TDC) of the compression stroke. The simplest way to do this is to remove the spark plug from the No. 1 cylinder and place your thumb over the hole. Slowly turn the engine by hand

in the normal direction of rotation until compression is felt at the hole.

NOTE: Other ways to check for TDC of the compression stroke: The most fail-safe (and time consuming) way is to remove the cylinder head cover and crank the engine until both the intake and exhaust valves are closed. Or you could use a compression gauge to determine that as you rotate the engine, the piston is coming up on compression. Also there are specialty tools available that screw in the sparkplug hole (like a compression tester) and whistle as you rotate the engine into its compression stroke. In all cases, verify the timing mark is at 0 degrees when you believe you have found TDC.

2. Ensure that the indicating mark on the crankshaft vibration damper is aligned to the 0 degree (TDC) mark on the timing chain cover.

3. Clean the distributor mounting at the engine and distributor base. Lightly oil the rubber O-ring seal on the distributor housing.

4. Hold the distributor over the mounting pad on the cylinder block so that the distributor body flange coincides with the mounting pad and the rotor points to the No. 1 cylinder firing position.

5. Install the distributor while holding the rotor in position, allowing it to move only enough to engage the slot in the drive gear.

6. Install pickup coil leads and distributor cap. Make sure all high tension wires are firmly snapped in cap towers. Install distributor hold-down clamp screw and tighten to 200 inch lbs. (22.5 Nm).

Ignition Timing

ADJUSTMENT

The ignition timing is automatically set by the Powertrain Control Module and is not adjustable.

Alternator

PRECAUTIONS

Several precautions must be observed with alternator equipped vehicles to avoid damage to the unit.

• If the battery is removed for any reason, make sure it is reconnected with the correct polarity. Reversing the battery connections may result in damage to the 1-way rectifiers.

• When utilizing a booster battery as a starting aid, always connect the positive to positive terminals and the negative terminal from the booster battery to a good engine ground on the vehicle being started.

• Never use a fast charger as a booster to start vehicles.

• Disconnect the battery cables when charging the battery with a fast charger.

• Never attempt to polarize the alternator.

• Do not use test lights of more than 12 volts when checking diode continuity.

• Do not short across or ground any of the alternator terminals.

• The polarity of the battery, alternator and regulator must be matched and considered before making any electrical connections within the system.

• Never separate the alternator on an open circuit. Make sure all connections within the circuit are clean and tight.

• Disconnect the battery ground terminal when performing any service on electrical components.

• Disconnect the battery if arc welding is to be done on the vehicle.

REMOVAL AND INSTALLATION

1. Disconnect the negative battery cable.

2. Disconnect and label the alternator output (BATT) and field (FLD) leads, then disconnect the ground wire.

3. You may need to spray a lubricant on the existing nuts and bolts for easy removal. Sometime it may be necessary to allow the lubricant to sit for a few minutes.

4. Loosen the alternator adjusting bolt and swing the alternator in toward the engine. Disengage the alternator drive belt. On models with a serpentine belt, release the automatic belt tensioner.

5. Remove the alternator mounting bolts and remove the alternator from the vehicle.

To install:

6. Install the alternator using the old bolts. Tighten the bolts, but not all the way, you still need to install the belt.

7. Install the old belt, or a replacement if worn.

8. Be sure to connect all ground wires and leads securely.

9. Adjust the belt tension as required.

10. Connect the negative battery cable.

Drive Belt

REMOVAL AND INSTALLATION

On most vehicles, to remove and install a new belt, you will need to loosen the mounting bracket bolts of the component. Using a suitable pry tool wrapped with a towel or rag, carefully pry the component forward slightly to give easy access to slide the belt off and on. You may need to do this on additional component(s) in order to make removal or installation easier. Some models have a few belts running different accessories; if you must replace a belt in the rear, you will first have to remove the belt(s) that are in your way. Look at your truck first to decide how many belts require removal to access the belt in need of replacement.

For those vehicles equipped with an automatic belt tensioner, the same logic in accessing the belt applies. To remove a belt on these vehicles, use a socket or wrench to rotate the tensioner pulley and loosen the belt. To install, place the belt over all pulleys but the tensioner pulley, then back off the tensioner in the same manner and slip the belt over the tensioner pulley. When you release the tensioner, it is spring loaded and will automatically apply the correct tension to the belt.

Starter

REMOVAL AND INSTALLATION

Gasoline Engines

1. Disconnect the negative battery cable.

2. Remove the cable from the starter.

3. Disconnect the solenoid leads at their solenoid terminals. If necessary,

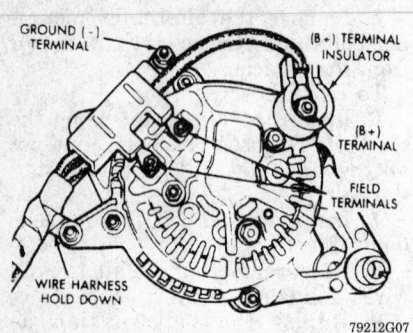

79212G07

Detach the connectors shown as depicted on this Dodge truck alternator

tag then to avoid confusion when re-installing.

NOTE: On 4WD Dakota models with the 3.9L (VIN X) engine, the starter is removed from above. On these models you will have to disconnect the steering gear and position the shaft out of the way to gain working clearance.

4. Remove the starter attachment bolts and withdraw the starter from the engine flywheel housing. On some models with automatic transmissions, the oil cooler tube bracket will interfere with the starter removal. In this case, remove the starter attachment bolts, slide the cooler tube bracket off the stud, and then withdraw the starter.

To install:

5. Be sure that the starter and flywheel housing mating surfaces are free of dirt and oil to make a good electrical contact.

6. Slide the starter in place, secure using the bolts and washers removed. Tighten the bolts to 50 ft. lbs. (68 Nm). Connect wires and battery cable end at the starter.

7. If removed for clearance, reconnect the steering column shaft.

8. Secure the oil cooler bracket if equipped. Connect the negative battery cable on the battery.

NOTE: When tightening the mounting bolt and nut on the starter, hold the starter away from the engine for the correct alignment.

Diesel Engines

1. Disconnect the negative battery cable.

2. Raise the truck and support it safely on jackstands.

3. Disconnect and label the wires at the starter motor.

4. Remove the attaching bolt, nut and washer, then lift the starter and solenoid assembly from the engine.

To install:

5. Before installing the starter motor, be sure the mounting surface on the drive end housing and the flywheel housing are clean, to ensure good electrical contact.

6. When tightening the attaching bolt and nut, hold the starter away from the engine to ensure the proper alignment. Tighten the bolts to 32 ft. lbs. (43 Nm).

7. Attach the wiring at the starter. Tighten the solenoid connection to 44 inch lbs. (5 Nm). Tighten the battery cable at the starter motor to 16 ft. lbs. (22 Nm).

8. Attach the negative battery cable.

CHASSIS ELECTRICAL

Blower Motor

REMOVAL AND INSTALLATION

Van Models

Without Air Conditioning

1. Disconnect the negative battery cable.

2. Remove the air intake duct and top half of the fan shroud, if necessary. Disconnect the blower connector.

3. Remove the 7 screws that fasten the back plate to the heater housing.

4. Remove the blower motor from the vehicle.

5. Remove the spring clip fastening the blower wheel to the blower shaft and pull off the wheel.

6. Remove the vent tube.

7. Remove the nuts fastening the blower motor to the back plate and remove the motor.

To install:

8. Check the seal for breaks or poor adhesion; repair as needed.

9. Install the blower motor to the back plate.

10. Install the vent tube.

11. Install the blower wheel to the shaft and secure the spring clip.

12. Install the assembly to the heater housing and install the 7 screws.

13. Install electrical connector.

14. Install the fan shroud and air duct if removed.

15. Connect the negative battery cable and check the blower motor for proper operation.

With Air Conditioning

1. Disconnect the negative battery cable.

2. Remove the air intake duct and top half of the fan shroud.

3. Disconnect the blower connector.

4. Remove the blower motor cooling tube.

5. Remove the retaining nuts and washers from the studs holding the blower.

6. Pull the air conditioning lines inboard and upward while removing the blower assembly from the vehicle. Remove the spring clip fastening the blower wheel to the blower shaft and pull off the wheel.

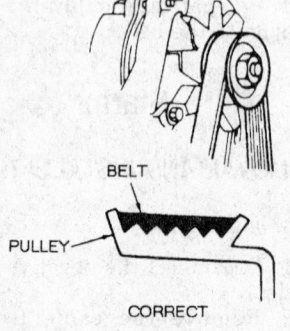
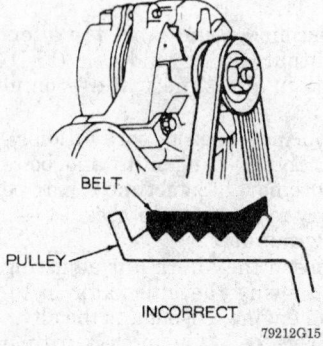

79212G15

Make sure the serpentine belt is correctly placed on the parallel-running teeth on all the pulleys

To install:

7. Install the blower wheel to the shaft and install the spring clip. Inspect the blower mounting plate seal and repair as needed. Apply rubber adhesive to the seal to aid in assembly.

8. Install the blower into the housing and install the washers and nuts.

9. Install the cooling tube.

10. Reconnect the electrical connector.

11. Install the fan shroud and air intake duct.

12. Connect the negative battery cable and check the blower motor for proper operation.

Dakota

1. Disconnect the negative battery cable.

2. Remove the steering column cover, intermittent wiper control and the lower instrument panel module retaining screw to the right of the steering column.

3. Remove the center distribution duct retaining screws and panel support screw at the bottom of the module.

4. Remove the courtesy lamp at the lower right corner of the module and the screw near the ash receiver.

5. Open the glovebox and remove the screws along the top edge.

6. Move the module out and down far enough to unclip the wiring harness and antenna cable and disconnect the speaker wire (if equipped with monaural radio) and glovebox light wire. Remove the module from the vehicle.

7. If the vehicle is equipped with air conditioning, disconnect the two vacuum lines from the recirculating air door actuator and disconnect the blower lead wires.

8. Remove two screws at the top of the blower housing, five screws from around the housing and remove the blower housing from the unit.

9. Remove three screw attaching the blower to the unit and remove the blower from the vehicle.

10. Remove the fan from the blower motor.

To install:

11. Install the fan to the blower motor and secure the clip.

12. Install the blower to the unit and install the blower housing.

13. Connect the two vacuum lines from the recirculating air door actuator, if equipped and connect the blower lead wires.

14. Hold the module in position and clip the wiring harness and antenna

cable in place and connect the speaker wire (if equipped with monaural radio) and glovebox light wire.

15. Install the retaining screws along the top of the inside of the glovebox.

16. Install the courtesy lamp at the lower right corner of the module and the screw near the ash receiver.

17. Install the panel support screw at the bottom of the module and the center distribution duct retaining screws.

18. Install the lower instrument panel module retaining screw to the right of the steering column, intermittent wiper control and the steering column cover.

19. Connect the negative battery cable and check the blower motor for proper operation.

Ram Pick-up and Ramcharger Models

1. Disconnect the negative battery cable.

2. Disconnect the blower wiring.

3. Remove the blower motor cooling tube.

4. Remove the screws or retaining nuts retaining the blower plate to the housing.

5. Remove the assembly from the housing.

6. Remove the spring clip fastening the blower wheel to the blower shaft and pull off the wheel. Remove the blower from the plate.

To install:

7. Inspect the blower mounting plate seal and repair as necessary.

8. Install the blower to the plate. Install the blower wheel to the shaft and install the spring clip.

9. Install the blower into the housing and install the screws or washers and nuts.

10. Install the cooling tube.

11. Connect the wiring.

12. Connect the negative battery cable and check the blower motor for proper operation.

Windshield Wiper Motor

REMOVAL AND INSTALLATION

Dakota Models

1. Disconnect the negative battery cable.

2. Disconnect the connector from the motor.

3. Remove the wiper arms, raise the hood and remove the cowl panel.

4. Hold the drive crank with a wrench while remove the crank nut. Remove the drive crank.

5. Remove the mounting nuts and remove the motor from the vehicle.

6. The installation is the reverse of the removal procedure.

7. Connect the negative battery cable and check the wipers for proper operation.

Ram Pick-up, Ramcharger and Van

1. Disconnect the negative battery cable.

2. Disconnect the wires from the wiper motor.

3. Remove the mounting bolts.

4. Pull the motor out far enough to gain access to the crank arm to motor link retainer bushing.

5. Remove the crank arm from the drive link by prying the retainer bushing from the crank arm pin with a suitable prying tool.

6. Remove the motor from the vehicle.

7. Hold the crank arm with a wrench while removing the crank nut to prevent from overloading the gears.

8. Remove the crank arm from the motor.

To install:

9. Index the slot correctly and position the crank arm on the motor shaft. Start the crank nut making sure the crank arm does not move from its slotted position.

10. Hold the crank arm with a wrench and torque the nut to 95 inch lbs. (11 Nm).

11. Lubricate the drive link retainer bushing and install the crank arm pin to the bushing by snapping them together suitable pliers.

12. Install the motor to the vehicle and tighten the mounting bolts for 1993 models to 65 inch lbs. (7 Nm) and 1994–96 models to 55 inch lbs. (6 Nm).

13. Connect the wires to the motor.

14. Connect the negative battery cable and check the wiper motor for proper operation.

Headlight Switch

REMOVAL AND INSTALLATION

Van

1. Disconnect the negative battery cable.

2. Remove the lower steering column cover.

3. Unscrew the hood release handle and lower.

4. Working under the instrument panel, depress the spring button on

the headlight switch and pull the stem out.

5. Open the glove box. Remove the screws that fasten the dash bezel assembly. Pull the bezel off of the upper retaining clips.

6. Remove the switch bezel and remove the illumination bulb socket.

7. Remove the switch mounting nut from the panel, remove the switch and disconnect the wiring.

To install:

8. Connect the switch, install the switch to the panel and install the mounting nut.

9. Install the illumination bulb socket to the switch bezel and install the bezel.

10. Install the dash bezel and headlight switch stem.

11. Connect the negative battery cable and check the switch for proper operation.

12. Install the hood release handle and lower steering column cover.

Dakota Models

1. Disconnect the negative battery cable.

2. Remove the steering column cover and remove the instrument panel bezel. Two screws are hidden behind the steering column cover.

3. Remove the screws from the headlight switch bezel, pull the assembly out and disconnect the wiring.

4. Remove the nut retaining the bezel to the bracket. Depress the spring button on the right side of the switch and remove the headlight switch knob and stem.

5. Remove the spanner nut and remove the switch.

6. The installation is the reverse of the removal procedure.

1993 Ram Pick-up and Ramcharger Models

1. Disconnect the negative battery cable.

2. Remove the map light.

3. Remove screws which attach the faceplate to the base panel. There is a screw below the heater-air conditioning control panel which is not visible from above.

4. If equipped with automatic transmission, place the shift lever in its lowest position.

5. Remove the faceplate by pulling the top edge rearward to clear the brow and pulling the bottom out, disengaging the attaching clips. If the vehicle is equipped with 4WD, disconnect the indicator wires.

6. Reach under the instrument panel, depress the spring button and remove the headlight switch knob.

7. Remove the wiper and power mirror switch knobs off their levers, if equipped.

8. Remove the bezel.

9. Remove the switch mounting nut from the panel, remove the switch and disconnect the wiring.

To install:

10. Connect the switch, install the switch to the panel and install the mounting nut.

11. Install the bezel.

12. Install the headlight stem and wiper and power mirror switch knobs, if removed.

13. Connect the 4WD indicator if equipped, install the cluster faceplate and map light.

14. Connect the negative battery cable and check the switch for proper operation.

1994-96 Ram Pick-up

1. Remove the cluster bezel (as described in the instrument panel removal procedure).

2. Remove the switch retaining three screws. Remove the switch and bezel from the instrument panel.

3. Unplug the two harness connectors from the headlight switch.

4. Pull the headlight switch knob and stem out to its stop. Depress the button on the bottom of the switch housing, then pull to remove the knob and stem from the switch.

To install:

5. Install the knob and stem into the switch.

6. Plug in the harness connectors, then install the switch.

7. Install the instrument cluster bezel.

Turn Signal Switch

REMOVAL AND INSTALLATION

1993 Column Mounted

Except Tilt Wheel

1. Disconnect the negative battery cable.

2. Remove the lower steering column cover, if equipped.

3. Remove the horn pad mounting screws from behind the steering wheel and remove the horn pad.

4. Remove the steering wheel nut, matchmark the steering wheel to the shaft and remove the steering wheel with a suitable puller.

5. Remove the plastic wiring channel from the underside of the steering column.

6. Disconnect the wiper switch connector, intermittent wipe module connector and cruise control connector, if equipped.

7. Remove the side lock housing cover.

8. Remove the slotted hex-head screw that attaches the wiper switch to the turn signal switch and remove the switch.

To install:

9. Install the turn signal switch, install the screws, the hider and the control knob.

10. Run the wiring through the opening and down the steering column, position the switch and install the hex-head screw. If removed, make sure the dimmer switch rod is properly engaged.

11. Install the side lock housing cover.

12. Connect the wires and install the wiring channel.

13. Install the steering wheel torque the nut to 45 ft. lbs. (61 Nm).

14. Install the horn pad.

15. Connect the negative battery cable and check the wiper and washer, cruise control, turn signal switch and dimmer switch for proper operation.

16. Install the lower column cover, if equipped.

Tilt Wheel

1. Disconnect the negative battery cable.

2. Remove the lower steering column cover, if equipped, and remove the plastic wiring channel from the underside of the steering column.

3. Remove the horn pad mounting screws from behind the steering wheel and remove the horn pad.

4. Remove the steering wheel nut, matchmark the steering wheel to the shaft and remove the steering wheel with a suitable puller.

5. Depress the lock plate with the proper depressing tool, remove the retaining ring from its groove and remove the tool, ring, lock plate, cancelling cam and spring.

6. Remove the switch stalk actuator screw and arm.

7. Remove the hazard switch knob.

8. Disconnect the turn signal switch.

9. Remove the three screws and remove the turn signal switch. Tape the connector to the wires to aid in removal.

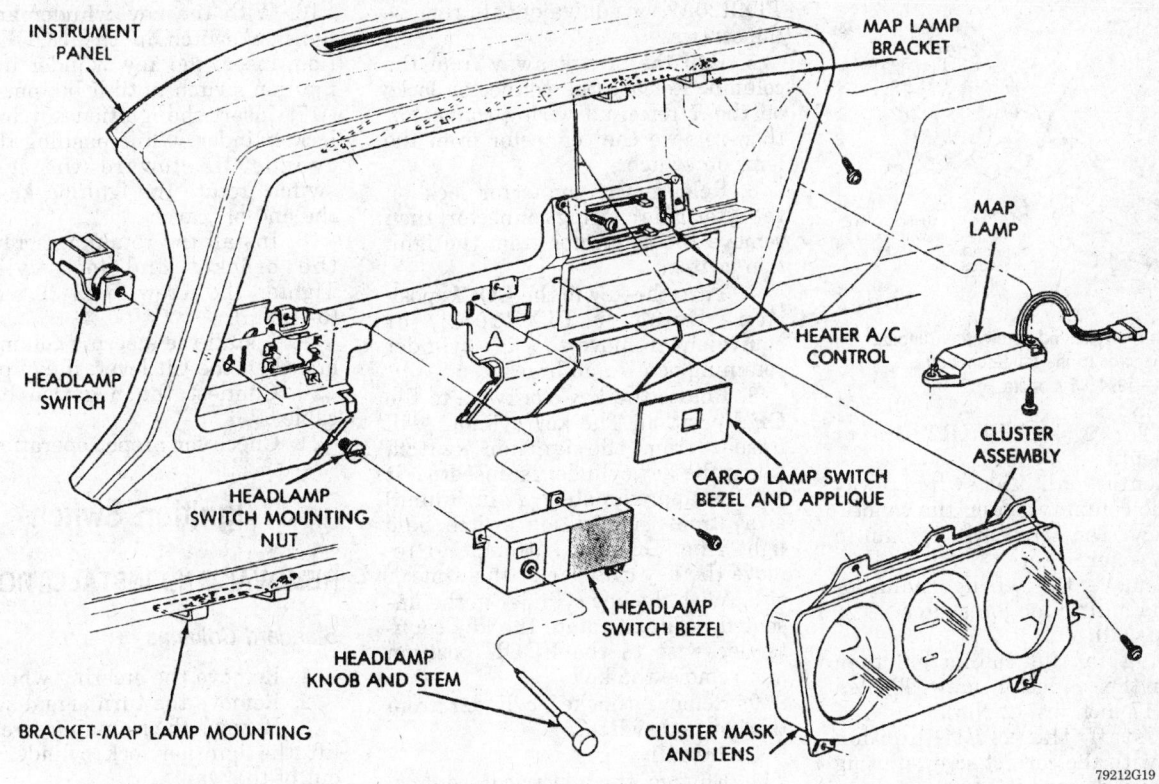

Exploded view of the instrument panel switch mounting — 1993 Ram Pick-up and Ramcharger models

79212G19

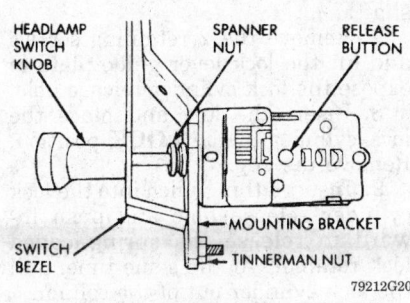

Details of the headlight switch — 1993 Ram Pick-up and Ramcharger models

79212G20

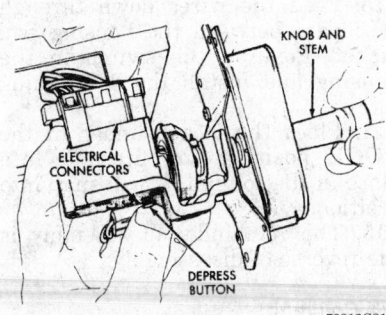

Depress the button on the switch housing, then pull to remove the knob and stem — 1994–96 Ram Pick-up models

79212G21

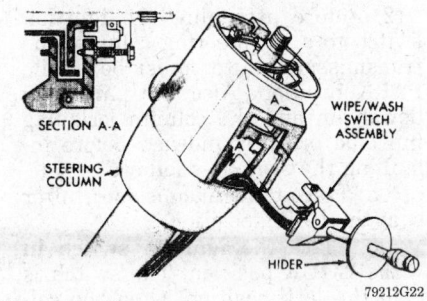

A common windshield wiper/washer switch — Dakota models

79212G22

To install:

10. Install the turn signal switch, switch stalk actuator arm and hazard switch knob.

11. Install the spring, cancelling cam, lock plate and ring on the steering shaft. Depress the plate with the depressing tool and install the ring securely in the groove. Remove the tool slowly.

12. Connect the turn signal switch, wiper switch, intermittent module and cruise control connectors, if equipped. Install the trough.

13. Install the steering wheel and tighten the nut to 45 ft. lbs. (61 Nm).

14. Install the horn pad.

15. Connect the negative battery cable and check the wiper and washer, cruise control, turn signal switch and dimmer switch for proper operation.

16. Install the lower column cover, if equipped.

Combination Switch

REMOVAL AND INSTALLATION

1994–96 Models

NOTE: This switch is referred to (in Mopar terminology) as the "Multi-function Switch." To remove the switch, a tamper resistant Torx® bit, such as the Snap-on® TTXR20B2, or equivalent, is required.

1. Disconnect the negative battery cable.

2. Remove the tilt steering column lever, if equipped.

3. Remove the lower fixed column shroud.

4. Move the upper fixed column shroud to gain access to the rear of the multi-function switch.

5. Remove the multi-function switch tamper resistant mounting screws using the correct tamper resistant, such as a Snap-on® tamper

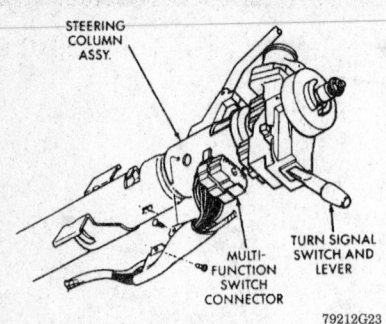

Remove the retaining screw, then unplug the connector to the multi-function switch — 1994-96 models

proof Torx® bit TTXR20B2, or equivalent.

6. Gently pull the switch away from the column. Loosen the connector screw. The screw will remain in the connector.

7. Remove the wiring connector from the multi-function switch.

To install:

8. Plug in the connector, then screw in the connector screw. Tighten this to 17 inch lbs. (2 Nm).

9. Install the multi-function switch with the correct screws, using the correct driver bit.

10. Install the column upper and lower shroud.

11. If equipped, install the tilt steering column lever.

12. Connect the negative battery cable.

Ignition Lock Cylinder

REMOVAL AND INSTALLATION

———— CAUTION ————
Disconnect and isolate the negative (ground) battery cable. This will disable the air bag system. Failure to disconnect the battery could result in accidental deployment and possible personal injury. Allow the system capacitor to discharge for two minutes then begin air bag system component removal. Refer to the air bag system precautions and information in this section.

1. Disconnect the negative battery cable.

2. If equipped with a tilt column, remove the tilt lever by turning it counterclockwise.

3. Remove the upper and lower covers from the column.

4. Remove the ignition switch mounting screws. Use a tamper resistant Torx® bit (Snap-On®

TTXR20A2, or equivalent) to remove the screws.

5. Pull the switch away from the column. Release the connector locks on the 7 terminal wiring connector, then remove the connector from the ignition switch.

6. Release the connector lock on the 4 terminal wiring connector, then remove the connector from the ignition switch.

7. Turn the key to the **LOCK** position. Using a TTXR20A2, or equivalent, remove the key cylinder retaining screw and bracket.

8. Rotate the key clockwise to the **OFF** position. The key cylinder will unseat from the ignition switch. When the key cylinder is unseated, it will be approximately ⅛ in. (3mm) away from the ignition switch halo light ring. DO NOT attempt to remove the key cylinder at this time.

9. With the key cylinder in the unseated position, rotate the key counterclockwise to the **LOCK** position and remove the key.

10. Remove the key cylinder from the ignition switch.

To install:

11. Engage the electrical connectors to the ignition switch and make sure the switch locking tabs are fully seated.

12. Before attaching the ignition switch to a tilt steering column, the transmission shifter must be in the PARK position. Also the park lock dowel pin and the column lock flag must be properly indexed before installing the switch as follows:

   a. Place the transmission shifter in the PARK position

   b. Place the ignition switch in the **LOCK** position. The switch is in the lock position when the column lock flag is parallel to the ignition switch terminals.

   c. Position the ignition switch park lock dowel pin so it will engage the steering column park lock slider linkage.

   d. Apply a light coating of grease to the column lock flag and the park lock dowel pin.

13. Place the ignition switch against the lock housing opening on the steering column. Ensure that the ignition switch park lock dowel pin enters the slot in the park lock slider linkage in the steering column.

14. Install the retaining bracket and ignition switch mounting screws, then tighten the screws to 22–30 inch lbs. (3–4 Nm).

15. If the vehicle is equipped with a tilt steering column, install the tilt lever.

16. With the key cylinder and the ignition switch in the **LOCK** position, insert the key cylinder into the ignition switch until it bottoms.

17. Insert the ignition key into the lock cylinder. While pushing the key cylinder in toward the ignition switch, rotate the ignition key until the end of travel.

18. Install the retaining screw into the bracket and lock cylinder. Tighten the screw to 22–30 inch lbs. (3–4 Nm).

19. Install the steering column covers, then the tilt lever, if equipped.

20. Connect the negative battery cable.

21. Check for proper operation.

Ignition Switch

REMOVAL AND INSTALLATION

Standard Columns

1. Remove the steering wheel.

2. Remove the turn signal switch.

3. Remove the retaining screw and lift the ignition lock cylinder lamp out of the way.

4. Remove the bearing housing.

5. Remove the coil spring.

6. Remove the lock plate from the shaft.

7. Remove the 2 retaining screws and lift the lock lever guide plate to expose the lock cylinder release hole.

8. Insert the key and place the lock cylinder in the **LOCK** position. Remove the key.

9. Insert a thin punch into the lock cylinder release hole and push inward to release the spring-loaded lock retainer. At the same time, pull the lock cylinder out of the column.

10. Remove the 3 retaining screws and lift out the ignition switch.

To install:

11. Position the ignition switch in the center detent position (**OFF**).

12. Place the shift lever in PARK.

13. Feed the wires down through the space between the housing and jacket. Position the switch in the housing and install the 3 retaining screws.

14. Place the lock cylinder in the **LOCK** position and press it into place in the column. It will snap into position.

15. The remainder of assembly is the reverse of disassembly.

Tilt Columns

1. Remove the steering wheel.

2. Remove the tilt lever and turn signal lever.

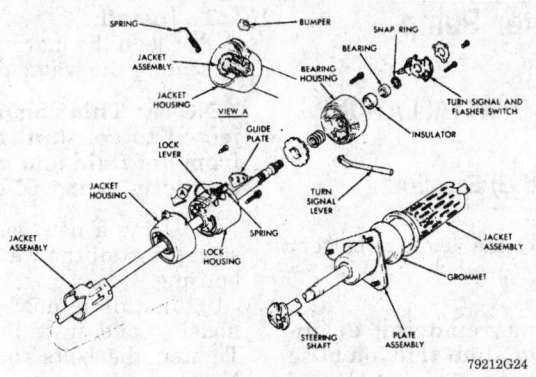

Exploded view of the steering column with floor shift — 1993 standard columns

3. If equipped, remove the turn signal lever.

4. Remove the turn signal switch.

5. Using the key, place the lock cylinder in the **LOCK** position. Remove the key.

6. Insert a thin punch in the slot next to the switch mounting screw boss and depress the spring latch at the bottom of the slot. Hold the spring latch depressed and pull the lock cylinder out of the column.

7. Place the ignition switch in the **ACCESSORY** position and remove the mounting screws. Lift off the switch. The **ACCESSORY** position is the one opposite the spring-loaded end position.

To install:

8. First, install the lock cylinder. Place the cylinder in the **LOCK** position and push it into the housing. It will snap into place.

9. Rotate the lock cylinder to the **ACCESSORY** position.

10. Fit the actuator rod in the slider hole and position the switch on the column. Insert the mounting screws, but don't tighten them yet.

11. Push the switch gently down the column to remove all lash from the actuator rod. Tighten the mounting screws. Make sure that you didn't take the switch out of the **ACCESSORY** detent!

12. The remainder of installation is the reverse of removal.

Neutral Safety Switch

REMOVAL AND INSTALLATION

The neutral safety switch is thread mounted into the transmission case. When the gearshift lever is placed in either the Park or Neutral position, a cam, which is attached to the transmission throttle lever inside the transmission, contacts the neutral safety switch and provides a ground to complete the starter solenoid circuit. The back-up light switch is incorporated into the neutral safety switch. The center terminal is for the neutral safety switch and the two outer terminals are for the back-up lights. There is no adjustment for the switch. If a malfunction occurs, the switch must be removed and replaced. To remove the switch:

1. Disconnect the electrical leads and unscrew the switch. Use a drain pan to catch the transmission fluid.

2. Using a new seal, install the new switch and torque it to 24 ft. lbs. (32.5 Nm).

3. Pour four quarts of Dexron®II fluid through the filler tube.

4. Start the engine and idle it for at least 2 minutes.

5. Set the parking brake and move the selector through each position, ending in Park.

6. Add sufficient fluid to bring the lever to the FULL mark on the dipstick. The level should be checked in Park, with the engine idle at normal operating temperature.

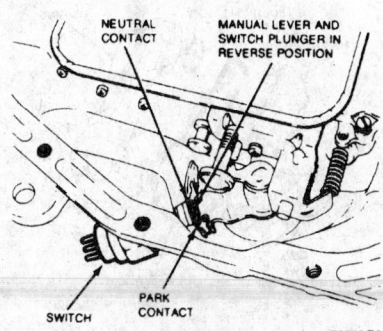

The neutral safety switch is threaded into the transmission case

Powertrain Control Module

REMOVAL AND INSTALLATION

The Powertrain Control Module (PCM) is the computer/microprocessor that regulates all ignition functions. Most models models employ a 60-way connector. In many cases, the powertrain control module is mounted in the engine bay to the firewall or a fender. Some earlier models may mount the powertrain control module in an air duct mounted on the inside left fender. Access to these units hidden in the air duct requires some dismantling of the air duct to get to them.

1. Disconnect the negative battery cable.

2. Remove the air cleaner duct from the powertrain control module, if equipped.

3. Remove the three module retaining screws.

4. Remove the 60-way connectors from the powertrain control module.

To install:

NOTE: Check the 60-way connectors for damaged or bent pins (and correct if any are found) prior to plugging it in.

5. Connect the 60-way connectors to the powertrain control module.

6. Install and tighten the module retaining screws.

7. If applicable, install the air cleaner duct to the powertrain control module

8. Connect the negative battery cable.

ENGINE COOLING

Radiator

REMOVAL AND INSTALLATION

1. Drain the cooling system.

—— **CAUTION** ——

When draining the coolant, keep in mind that cats and dogs are attracted by ethylene glycol antifreeze, and are quite likely to drink any that is left in an uncovered container or in puddles on the ground. This will prove fatal in sufficient quantity. Always drain the coolant into a sealable container.

2. Disconnect the negative battery cable.

3. Detach the upper hose from the radiator.

4. Remove the shroud mounting nuts and position it out of the way.

5. Remove the radiator top mounting screws. If equipped with air conditioning, remove the condenser attaching screws, accessible through the grille. Do not disconnect any air conditioning lines.

6. Raise the vehicle and support it.

7. Disconnect and plug the automatic transmission cooler lines and cap the openings in the cooler.

8. Hold the radiator in place and remove the lower mounting screws. Carefully lift it up and out of the truck.

To install:

9. Hold the radiator in place and install the lower mounting screws.

10. Connect the automatic transmission cooler lines.

11. Install the radiator top mounting screws. If equipped with air conditioning, install the condenser attaching screws, accessible through the grille.

12. Install the shroud mounting nuts.

13. Connect the upper hose from the radiator.

14. Connect the negative battery cable.

15. Fill the cooling system.

16. Check all fluid levels and run the engine, making sure there are no leaks.

Water Pump

REMOVAL AND INSTALLATION

2.5L (VIN K and G) Engine

1. Disconnect the negative battery cable.

NOTE: When removing or installing the constant tension hose clamps from vehicles so equipped, use only the correct clamp tool, such as the Snap-On No. HPC–20, or equivalent.

2. Drain the cooling system.

———— **CAUTION** ————

When draining the coolant, keep in mind that cats and dogs are attracted by ethylene glycol antifreeze, and are quite likely to drink any that is left in an uncovered container or in puddles on the ground. This will prove fatal in sufficient quantity. Always drain the coolant into a sealable container.

3. If the vehicle is equipped with air conditioning, remove the compressor from the bracket and position it to the side.

4. Raise the vehicle and support safely, if necessary and remove the alternator and bracket. Remove the pulley from the water pump.

5. Disconnect the lower radiator hose and heater hose from the water pump.

6. Remove the water pump housing attaching screws and remove the assembly from the vehicle. Discard the O-ring.

7. Remove the water pump from the housing.

To install:

8. Clean the mating surfaces prior to sealing the water pump.

NOTE: This component is subjected to constant high pressure from hot fluid and must be sealed correctly or it will leak.

9. Using a new gasket or silicone sealer, install the water pump to the housing.

10. Install a new O-ring to the housing and install to the engine. Tighten the bolts to 21 ft. lbs. (30 Nm).

11. Install the water pump pulley. Connect the radiator hose and heater hose to the water pump.

12. Install all items removed to gain access to the water pump and adjust the belt(s).

13. Remove the hex-head plug on the top of the thermostat housing (some carbureted engines have a vacuum switching valve at that location). Fill the radiator with coolant until the coolant comes out the plug hole. Install the plug or valve and continue to fill the radiator.

14. Connect the negative battery cable, run the vehicle until the thermostat opens, fill the radiator completely and check for leaks.

15. Once the vehicle has cooled, recheck the coolant level.

3.9L (VIN X) Engine

1. Disconnect the negative battery cable.

2. Drain the coolant.

NOTE: When removing or installing the constant tension hose clamps from vehicles so equipped, use only the correct clamp tool, such as the Snap-On No. HPC–20, or equivalent.

3. Remove the radiator and lower hose.

4. Remove the fan blade, spacer or viscous drive unit, pulley, bolts and shroud together. Remove the air pump belt and power steering pump belt.

NOTE: Do not place the viscous fan in an upright position because the silicone fluid in the drive could drain into the bearing and contaminate its lubricant.

5. Loosen the alternator mounting bolts and remove the alternator/air conditioning or idler pulley belt(s).

6. Remove the alternator bracket. This bracket also supports the air conditioning compressor or idler pul-

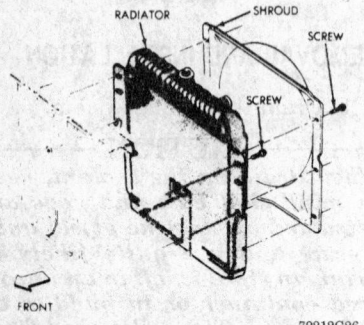

Exploded view of the radiator and fan shroud — 1994–97 Ram Pick-up model shown

79212G26

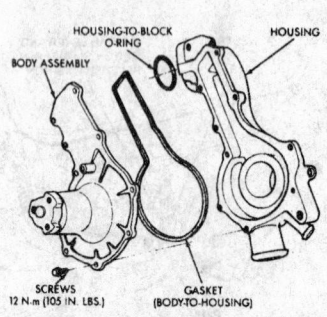

Exploded view of the water pump assembly including the gasket — 2.5L (VIN K and G) engine

79212G27

ley, which will remain supported by its rear mount.

7. Remove the air pump. Remove the air pump bracket and unbolt the power steering pump bracket with the pump still attached and position it out of the way.

8. Disconnect the heater hose and bypass hose.

9. Remove the air conditioning compressor pulley and field coil assembly, if necessary to remove the water pump to compressor front mount bolts and bracket.

10. Remove the remaining pump retaining bolts (if any) and remove the pump from the engine.

To install:

11. Clean and dry the pump mating surfaces. Install a new by-pass hose to the engine.

NOTE: This component is subjected to constant high pressure from hot fluid and must be sealed correctly or it will leak.

12. Install a new gasket and install the water pump to the engine.

13. Install any bolts that do not retain a bracket.

14. Tighten the bypass hose clamps. Install the heater hose.

15. Install the water pump to compressor front mount bolts and bracket, if it was removed.

16. Install all remaining brackets and components that were removed. Torque all water pump retaining bolts that do not go through adjusting slots to 30 ft. lbs. (41 Nm).

17. Install the air conditioning compressor pulley and field coil assembly, if it was removed. Install and adjust the alternator/air conditioning or idler belt(s). If applicable, place the remaining belts over their components.

18. Install the fan blade, spacer or viscous drive unit, pulley and bolts along with the shroud.

19. Install the radiator, lower hose and shroud.

20. Adjust the belt(s) and tighten the remaining water pump retaining bolts to 30 ft. lbs. (41 Nm).

21. Fill the radiator with coolant. This cooling system is self-bleeding, so system bleeding is not required.

22. Connect the negative battery cable, run the vehicle until the thermostat opens, fill the radiator completely and check for leaks.

23. Once the vehicle has cooled, recheck the coolant level.

5.2L (VIN T and Y) and 5.9L (VIN A, Z and 5) Engines

1. Remove the radiator.

NOTE: When removing or installing the constant tension hose clamps from vehicles so equipped, use only the correct clamp tool, such as the Snap-On No. HPC–20, or equivalent.

2. Loosen all accessories that are belt driven and remove the belt(s).

3. On engines without air conditioning, remove the alternator bracket attaching bolts and tie the alternator and bracket out of the way.

4. On engines with air conditioning, remove the idler pulley assembly, alternator, and adjusting bracket.

5. Remove the fan blade, spacer (or fluid unit), pulley, and bolts as an assembly.

NOTE: To prevent silicone fluid from draining into the drive bearing and ruining the lubricant, do not place the thermostatic fan drive unit with the shaft pointing downward.

6. Disconnect all hoses from the water pump.

7. Remove the air conditioning compressor front mounting bolts.

8. Remove the water pump-to-compressor front bracket bolts and the bracket. Remove the water pump.

NOTE: Do not disconnect any refrigerant lines from the compressor.

To install:

NOTE: This component is subjected to constant high pressure from hot fluid and must be sealed correctly or it will leak.

9. Clean the mating surfaces prior to sealing the water pump.

10. Install the water pump on the engine. Use a new gasket coated with sealer.

11. Install the water-pump-to-compressor front bracket mounting bolts. Tighten the bolts to 30 ft. lbs. (41 Nm).

12. Install the air conditioning compressor front mounting bolts. Tighten the bolts to 30 ft. lbs. (41 Nm).

13. Reconnect all hoses to the water pump.

14. Install the fan blade, spacer (or fluid unit), pulley, and bolts as an assembly.

15. On engines with air conditioning, install the idler pulley assembly, alternator, and adjusting bracket.

16. On engines without air conditioning, install the alternator and bracket.

17. Install the drive belt(s).

18. Install the radiator. Fill the cooling system and adjust the tension of the drive belt(s).

8.0L (VIN W) Engine

1. Disconnect the negative battery cable.

2. Drain the cooling system.

3. Remove the washer bottle from the fan shroud and disconnect the fan shroud but do not remove it from the radiator.

4. Remove the upper radiator hose from the radiator.

NOTE: When removing or installing the constant tension hose clamps from vehicles so equipped, use only the correct clamp tool, such as the Snap-On No. HPC–20, or equivalent.

5. Remove the fan blade, spacer (or fluid unit), pulley, and bolts as an assembly.

NOTE: To prevent silicone fluid from draining into the drive bearing and ruining the lubricant, do not place the thermostatic fan drive unit with the shaft pointing downward.

6. Loosen all accessories that are belt driven and remove the belt.

7. Remove the four water pump pulley-to-water pump hub bolts and remove the pulley from the vehicle.

8. Remove the lower radiator hose from the water pump.

9. Remove the heater hose at the water pump fitting.

10. Remove the seven water pump mounting bolts.

11. Loosen the clamp at the water pump end of the bypass hose. Slip the hose from the water pump while removing the pump from the vehicle. Do not remove the clamp from the bypass hose.

12. Discard the water pump-to-timing chain/case/cover O-ring seal.

13. Remove the heater hose fitting from the water pump if the pump replacement is necessary. Note the position (direction) of the fitting before removal. The fitting must be installed in the same position.

NOTE: Do not disconnect any refrigerant lines from the compressor.

To install:

NOTE: This component is subjected to constant high pressure from hot fluid and must be sealed correctly or it will leak.

14. Clean the mating surfaces prior to sealing the water pump.

15. Install the water pump on the engine. Use a new gasket coated with sealer.

16. Install the water-pump-to-compressor front bracket mounting bolts. Tighten the bolts to 30 ft. lbs. (41 Nm).

17. If a new pump has been installed, then install the heater hose fitting to the pump. Tighten the fitting to 144 inch lbs. (16 Nm). After the fitting is tightened, position it as shown in the drawing. When positioning the fitting, do not back it off (rotate counterclockwise). Use a suitable Teflon® containing thread sealant. Refer to the directions on the package.

18. Clean the O-ring groove and install a new O-ring.

19. Apply a small amount of petroleum jelly to the O-ring to help it stay in place on the water pump.

20. Install the water pump to the engine as follows:

   a. Guide the pump fitting bypass hose as the hose is being installed.

   b. Install the water pump and tighten them to 30 ft. lbs. (40 Nm).

21. Position the bypass clamp on the hose.

22. Spin the water pump to be sure that the pump impeller does not rub against the timing chain case/cover.

23. Connect the radiator lower hose to the to the water pump.

24. Connect the heater hose and hose clamp to the heater hose fitting.

25. Install the water pump pulley. Tighten the bolts to 16 ft. lbs. (22 Nm). Place a prybar between the water pump pulley bolts to prevent the pulley from relaxing.

26. Install the serpentine belt.

27. Position the fan shroud assembly and fan blade/viscous fan drive assembly as a complete unit.

28. Install the fan shroud to the radiator. Tighten the bolts to 50 inch lbs. (6 Nm).

29. Install the fan blade and viscous fan drive to the water pump shaft.

30. Fill the cooling system, connect the negative battery cable and check for leaks.

5.9L (VIN C and 8) Engine

1. Disconnect the negative battery cable.

2. Drain the coolant.

3. Use a ⅜ in. drive breaker bar to lift the belt tensioner and remove the belt.

4. Remove the two water pump retaining bolts and remove the pump from the engine.

5. Remove the O-ring from the pump groove.

To install:

6. Clean the O-ring groove and install a new O-ring.

7. Clean the pump mating surfaces and install the pump to the engine.

8. Tighten the mounting bolts to 18 ft. lbs. (24 Nm). Fill the radiator with coolant.

9. Install the drive belt.

10. Connect the negative battery cable, run the vehicle until the thermostat opens, fill the radiator completely and check for leaks.

11. Once the vehicle has cooled, recheck the coolant level.

Thermostat

REMOVAL AND INSTALLATION

Gasoline Engines

NOTE: The thermostat on all gasoline-powered vehicles is located beneath the thermostat housing at the front of the intake manifold.

1. Drain the cooling system to below the level of the thermostat.

CAUTION

When draining the coolant, keep in mind that cats and dogs are attracted by ethylene glycol antifreeze, and are quite likely to drink any that is left in an uncovered container or in puddles on the ground. This will prove fatal in sufficient quantity. Always drain the coolant into a sealable container.

2. For 1994–96 3.9L (VIN X), 5.2L (VIN T and Y) and 5.9L (VIN A, Z and 5) engines equipped with air conditioning:

   a. Remove the support bracket (rod) located near the rear of the alternator.

   b. Remove the drive belt.

   c. Partially remove the alternator by removing the necessary alternator bolts. If equipped with 4WD, unplug the 4WD indicator lamp harness (at the rear of the unit).

   d. Remove the alternator. Position it to gain access for thermostat gasket removal.

3. On 8.0L (VIN W) engines, remove the two support rod mounting bolts and remove the support rod (intake manifold-to-alternator mount).

4. All models: Remove the upper radiator hose from the thermostat housing. Note the positioning of the thermostat. It is important that the thermostat is correctly installed.

5. Withdraw the housing bolts and remove the housing and the thermostat.

6. Check to make sure that the thermostat valve closes tightly. If the valve does not close completely due to foreign material, carefully clean the sealing edge of the valve while being careful not to damage the sealing edge. If the valve does not close tightly after it has been cleaned, a new thermostat must be installed.

7. Immerse the thermostat in a container of warm water so that its pellet is completely covered and does not touch the bottom or sides of the container.

8. Heat the water and, while stirring the water continuously (to ensure uniform temperature), check the water temperature with a thermometer at the point when a 0.001 in. (0.025mm) feeler gauge can be inserted in the valve opening at a water temperature within plus or minus 5° of the standard thermostat temperature. If the thermostat does not open within the temperature range, replace it with a new thermostat.

9. Continue heating the water to a temperature of approximately 20° higher than the standard thermostat opening temperature. At this point, the thermostat should be fully open. If it is not, install a new thermostat.

To install:

10. To install, use a new gasket and position the thermostat so that its pellet end (the part with the spring) is toward the engine block. On the 3.9L engine, the vent hole must be up. Refit the thermostat housing and tighten its securing bolts to 25–30 ft. lbs. (34–41 Nm).

11. Connect the upper radiator hose.

12. On 1994–96 3.9L (VIN X), 5.2L (VIN T and Y) and 5.9L (VIN A, Z and 5) models, install the alternator.

13. On 8.0L (VIN W) engines, install the two support rod mounting bolts and the support rod.

14. On 2.5L (VIN K and G) engines, remove the hex-head plug or vacuum switching valve on the thermostat housing. Fill the radiator with cool-

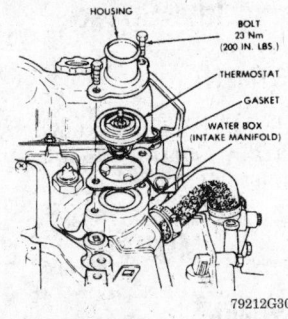

Exploded view of the thermostat installation — 3.9L (VIN X), 5.2L (VIN T and Y) and 5.9L (VIN A, Z and 5) engines

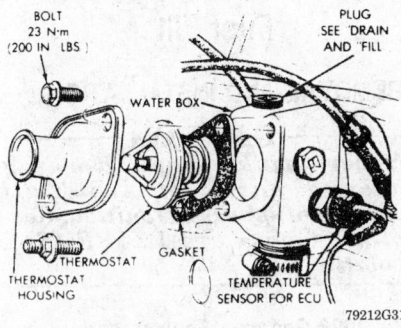

Exploded view of the thermostat and water box — 2.5L (VIN K and G) engine

ant until the coolant come out the plug hole. Install the plug or valve and continue to fill the radiator.

NOTE: Poor heater output and slow engine warm-up are often caused by a thermostat stuck in the open position: occasionally one sticks shut causing immediate overheating. Do not attempt to correct an overheating condition by permanently removing the thermostat. Thermostat flow restriction is designed into the system; without it, localized overheating due to turbulence may occur.

15. On 3.9L (VIN X), 5.2L (VIN T and Y) and 5.9L (VIN A, Z and 5) engines, fill the cooling system to 1.25 in. (32mm) below the filler neck with the correct water and antifreeze mixture. Warm the engine and inspect the upper radiator hose and the thermostat housing for leaks.

Diesel Engine

1. Disconnect the negative battery cable.
2. Drain the coolant.

---- CAUTION ----
When draining the coolant, keep in mind that cats and dogs are attracted by ethylene glycol antifreeze, and are quite likely to drink any that is left in an uncovered container or in puddles on the ground. This will prove fatal in sufficient quantity. Always drain the coolant into a sealable container.

3. Use a ⅜ in. drive breaker bar to lift the belt tensioner and remove the belt.
4. Disconnect the upper radiator hose from the thermostat housing.
5. Loosen the alternator mounting bolts and lower the alternator.
6. Unbolt the thermostat housing and remove the housing engine lifting bracket and thermostat with seal.

To install:

7. Install the new thermostat to the housing, making sure the tang on the thermostat is aligned with the slot in the housing. This will ensure correct positioning of the jiggle pins in the housing. This is important because during filling, air vents through the jiggle pin openings through the upper hose and out the radiator fill neck.
8. Clean the pump mating surfaces.
9. Install the engine lifting bracket, new seal, thermostat and housing. Make sure the seal is installed with the beveled side facing out.
10. Tighten the thermostat housing bolts to 18 ft. lbs. (24 Nm). Connect the upper hose to the housing.
11. Reinstall the alternator into position. Tighten the upper bolt to 18 ft. lbs. (24 Nm) and the lower bolt to 32 ft. lbs. (43 Nm).
12. Install the drive belt.
13. Connect the negative battery cable, run the vehicle until the thermostat opens, fill the radiator completely and check for leaks.

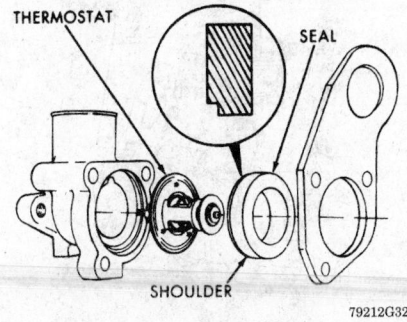

Be careful to install the diesel engine thermostat seal as shown

---

14. Once the vehicle has cooled, recheck the coolant level. With the Thermostat plug on the relief valve removed from the thermostat housing, fill the radiator with coolant until the coolant comes out the plug hole. Install the plug or valve and continue to fill the radiator.

Electric Cooling Fan

REMOVAL AND INSTALLATION

2.5L (VIN K and G) Engine

1. Disconnect the negative battery cable.
2. Unplug the wire connector.
3. Remove the fasteners and remove the fan, shroud and motor from the radiator as an assembly.
4. Support the fan motor and shaft on a bench. Remove the retaining clip to detach the fan blade.

---- WARNING ----
Do not allow the fan blades to contact the bench

To install:
5. Position the fan on the motor shaft, so that it properly engages the roll pin.
6. Support the fan motor and shaft on a bench WITHOUT allowing the fan to contact the bench. This will prevent the fan from being damaged by excessive force.
7. Install the fan retaining clip.
8. Install the complete assembly into the pockets in the radiator lower tank. Tighten the screws to 95 inch lbs. (11 Nm).
9. Plug in the electrical connector to the harness and connect the negative battery cable.

GASOLINE FUEL SYSTEM

Fuel System Service Precautions

Safety is the most important factor when performing not only fuel system maintenance but any type of maintenance. Failure to conduct maintenance and repairs in a safe manner may result in serious personal injury or death. Maintenance and testing of the vehicle's fuel system components can be accomplished safely and effectively by adhering to the following rules and guidelines.

• To avoid the possibility of fire and personal injury, always disconnect the negative battery cable unless the repair or test procedure requires that battery voltage be applied.

• Always relieve the fuel system pressure prior to disconnecting any fuel system component (injector, fuel rail, pressure regulator, etc.), fitting or fuel line connection. Exercise extreme caution whenever relieving fuel system pressure to avoid exposing skin, face and eyes to fuel spray. Please be advised that fuel under pressure may penetrate the skin or any part of the body that it contacts.

• Always place a shop towel or cloth around the fitting or connection prior to loosening to absorb any excess fuel due to spillage. Ensure that all fuel spillage (should it occur) is quickly removed from engine surfaces. Ensure that all fuel soaked cloths or towels are deposited into a suitable waste container.

• Always keep a dry chemical (Class B) fire extinguisher near the work area.

• Do not allow fuel spray or fuel vapors to come into contact with a spark or open flame.

• Always use a backup wrench when loosening and tightening fuel line connection fittings. This will prevent unnecessary stress and torsion to fuel line piping. Always follow the proper torque specifications.

• Always replace worn fuel fitting O-rings with new. Do not substitute fuel hose or equivalent, where fuel pipe is installed.

Fuel System Pressure

RELIEVING

TBI-Equipped Engines

Except 1993–95 2.5L (VIN K and G) Engines

1. Loosen the gas cap to release tank pressure.
2. Remove the injector wiring harness connector from the engine harness.
3. Connect a jumper wire between ground terminal No. 1 of the injector harness and the engine ground.
4. Connect a jumper wire to the second terminal and touch the battery positive post for no longer than five seconds. This releases system pressure.

5. Remove the jumper wire and continue fuel system service.

1993–95 2.5L (VIN K and G) Engines

1. Disconnect the negative battery cable.
2. Remove the fuel tank filler cap to release any fuel tank pressure.
3. Unscrew the plastic cap from the pressure test port on the fuel rail.
4. Obtain a fuel pressure gauge/hose from a fuel pressure gauge tool set No. 5069, or equivalent. Remove the gauge, then place the gauge end of the hose into a suitable gasoline container.
5. Place a shop towel under the test port.
6. Screw the other end of the hose onto the fuel pressure port to relieve the pressure.
7. When the pressure has been relieved, remove the hose and cap the port.

MFI-Equipped Engines

1. Disconnect the negative battery cable.
2. Remove the fuel tank filler cap to release any fuel tank pressure.
3. Unscrew the plastic cap from the pressure test port on the fuel rail. On the 8.0L engine, the test port is found at the front of the engine.
4. Obtain a fuel pressure gauge/hose from a fuel pressure gauge tool set No. 5069, or equivalent. Remove the gauge, then place the gauge end of the hose into a suitable gasoline container.
5. Place a shop towel under the test port.
6. Screw the other end of the hose onto the fuel pressure port to relieve the pressure.
7. When the pressure has been relieved, remove the hose and cap the port.

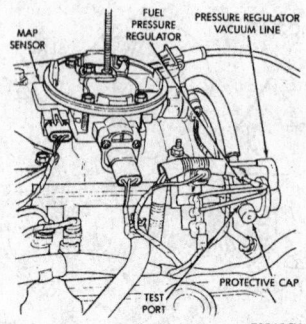

Fuel pressure test port — 1993–96 3.9L (VIN X) engine

Idle Speed

ADJUSTMENT

The idle speed is controlled by the Powertrain Control Module and is not adjustable.

Mixture

ADJUSTMENT

The air fuel mixture is controlled by the Powertrain Control Module and is not adjustable.

Fuel Filter

REMOVAL AND INSTALLATION

——— CAUTION ———
Never smoke when working around gasoline! Avoid all sources of sparks or ignition. Gasoline vapors are EXTREMELY volatile!

All 1993 Gasoline Engines and 1994–96 2.5L (VIN K and G) Engines

The fuel filter is located in the fuel line on the inside of the left frame rail, just ahead of the crossmember.

1. Relieve the fuel system pressure as described in this section.
2. Unfasten the filter retaining screw and remove the filter from the frame rail.
3. Wrap a shop towel around the hose connections to catch fuel spillage. Loosen the fuel hose clamps and disconnect the hoses from the filter.
4. Remove the old fuel filter and replace it with a new one.
5. Connect the fuel hoses and tighten the hose clamps. Position the filter on the frame rail and tighten the retaining screw to 75 inch lbs. (8 Nm).

All 1994–96 Engines — Except 2.5L (VIN K and G) Engine

The fuel filter is integrated into the fuel pressure regulator which is mounted in the fuel tank. This unit is not controlled by the PCM or engine vacuum. It is calibrated to deliver approximately 35–45 psi (241–310 kPa) of fuel pressure to the injectors. If the pressure exceeds the maximum of the specified range, an internal diaphragm closes to route fuel back into the fuel tank. This system eliminates the need for conventional return lines from the engine bay and accounts for

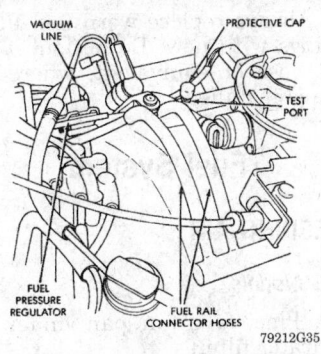

Fuel pressure test port — 1993–96 5.2L (VIN T and Y) engine

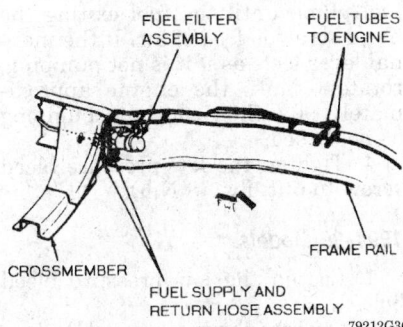

Fuel filter location — 1993 gasoline engines and 1994–97 2.5L (VIN K and G) engines

the name of the "Returnless" fuel injection system employed in these vehicles.

NOTE: Fuel tank removal is required for this procedure. Also needed will be external snapring pliers and proper hose clamp pliers, such as No. C-4124 pliers (available through Plymouth/Dodge dealers), or equivalent.

1. Properly relieve the fuel system pressure.
2. Drain the fuel tank and remove the tank.
3. Remove the fuel filter/regulator (which is pressed into a rubber grommet) by twisting and pulling it straight up.
4. Remove the snapring retaining the cover tube, then slide it down to reveal the clear plastic fuel tube and its retaining clamp.
5. Gently cut off the old clamp without damaging the tube, then discard the clamp.
6. Carefully pull the tube off, then remove the filter/regulator from the fuel pump module.
To install:
7. Install a new clamp over the plastic fuel tube and attach it loosely to the filter/regulator. Rotate the unit

in the line until it is pointed to the driver's side of the vehicle.
8. Tighten the clamp using hose clamp pliers.

NOTE: Do not use conventional side cutters to tighten the clamp.

9. Slide the cover tube up to the bottom of the filter/regulator and install the snapring.
10. Carefully press the assembly back into the rubber grommet by hand. It should be pointed to the driver's side of the vehicle.
11. Install the fuel tank.

Fuel Pump

REMOVAL AND INSTALLATION

1993 3.9L (VIN X) and 5.2L (VIN T and Y) Engines

──── **CAUTION** ────
The fuel injection system is under a constant pressure. Before servicing any part of the fuel injection system, the system pressure must be released. Use a clean shop towel to catch any fuel spray and take precautions to avoid the risk of fire.
───────────────

1. Perform the fuel pressure release procedure described previously in this section.
2. Properly relieve the fuel system pressure and disconnect the negative battery cable.
3. Remove the fuel tank from the truck.
4. Remove the locking ring, then lift out the fuel pump module. Tag any lines for replacement.
5. Remove the sending unit attaching screws from the mounting bracket located on the drain tube.
6. Disconnect the wires from the sending unit, then remove the sending unit.
7. Remove the drain tube from the mounting lug at the bottom of the reservoir.
8. Remove the lower-most coil of the drain tube from the mounting lugs on top of the reservoir. Be careful to avoid unsnapping the return line check valve cover from the bottom of the reservoir.
9. Release the pump mounting bracket from the reservoir. Press the bracket with both thumbs toward the center of the reservoir.
10. Remove the pump mounting bracket and rubber collar from the hose. Cut the hose clamp on the sup-

ply line and discard the clamp. Remove the pump/filter assembly. Pry the filter from the pump.
To install:
11. Press a new filter onto the pump.
12. Using a new clamp, attach the supply hose.
13. Position the pump mounting bracket and rubber collar on the supply hose between the bulge in the hose and the pump.
14. Position the pump in the reservoir so that the filter aligns with the cavity in the reservoir.
15. Snap the pump bracket into the reservoir.
16. Position the coil tube on the reservoir so that the drain tube aligns with the mounting lugs on the reservoir.
17. Snap the lowermost coil into the mounting lugs on top of the reservoir.
18. Snap the drain tube into the lugs on the bottom of the reservoir.
19. Connect the wires to the sending unit.
20. Align the index tab on the level unit with the index hole in the mounting bracket.
21. Install the level unit screws.
22. Install the assembly in the tank.

1993–96 5.9L (VIN A, Z and 5), 1994–96 3.9L (VIN X), 5.2L (VIN T and Y), 8.0L (VIN W) and 1996 2.5L (VIN K and G) Engines

──── **CAUTION** ────
The fuel injection system is under a constant pressure. Before servicing any part of the fuel injection system, the system pressure must be released. Use a clean shop towel to catch any fuel spray and take precautions to avoid the risk of fire.
───────────────

1. Perform the fuel pressure release procedure described previously in this section.
2. Properly relieve the fuel system pressure and disconnect the negative battery cable.
3. Remove the fuel tank from the truck.
4. Note the direction of the fuel filter/fuel pressure regulator, the pressure relief/rollover valve and the pump electrical connector. These should all be pointed to the driver's side. Tag any lines for replacement, if necessary.
5. Remove the locknut threaded into the fuel tank. The fuel pump module will spring up when the locknut is removed.

6. Remove the module from the fuel tank.

To install:

7. Use a new gasket and position the fuel pump module in the opening. The fuel filter/fuel pressure regulator, the pressure relief/rollover valve and the fuel pump electrical connector should all be pointed to the driver's side of the vehicle when this unit is properly installed.

8. Position a new locknut over the top of the fuel pump module, then tighten the locknut.

9. Install the fuel tank.

Fuel Injector

REMOVAL AND INSTALLATION

1. Remove the air cleaner assembly.

2. Relieve the fuel system pressure as previously described in this section.

3. Disconnect the negative battery cable.

4. Remove the fuel pressure regulator.

5. Remove the Torx® screw holding down the injector cap.

6. With two small prytools, lift the cap off the injector using the slots provided.

7. Using a small prytool placed in the hole in the front of the electrical connector, gently pry the injector from pod.

8. Make sure the injector lower O-ring has been removed from the pod.

To install:

9. Place a new lower O-ring on the injector and a new O-ring on the injector cap. The injector will have the upper O-ring already installed.

10. Put the injector cap on the injector. (Injector and cap are keyed). The cap should sit on the injector without interference. Apply a light coating of castor oil or petroleum jelly on the O-rings. Place the assembly in the pod.

11. Rotate the cap and injector to line up the attachment hole.

12. Push down on the cap until it contacts the injector pod.

13. Install the Torx® screw and tighten it to 35–45 inch lbs. (4–5 Nm).

14. Install the fuel pressure regulator.

15. Connect the negative battery cable.

16. Test for leaks using Mopar® ATM tester C–4805, or equivalent. With the ignition in the **RUN** position depress the tester button. This will activate the pump and pressurize the system. Check for leaks.

17. Reinstall the air cleaner assembly.

DIESEL FUEL SYSTEM

Fuel System Service Precautions

Safety is the most important factor when performing not only fuel system maintenance but any type of maintenance. Failure to conduct maintenance and repairs in a safe manner may result in serious personal injury or death. Maintenance and testing of the vehicle's fuel system components can be accomplished safely and effectively by adhering to the following rules and guidelines.

• To avoid the possibility of fire and personal injury, always disconnect the negative battery cable unless the repair or test procedure requires that battery voltage be applied.

• Always relieve the fuel system pressure prior to disconnecting any fuel system component (injector, fuel rail, pressure regulator, etc.), fitting or fuel line connection. Exercise extreme caution whenever relieving fuel system pressure to avoid exposing skin, face and eyes to fuel spray. Please be advised that fuel under pressure may penetrate the skin or any part of the body that it contacts.

• Always place a shop towel or cloth around the fitting or connection prior to loosening to absorb any excess fuel due to spillage. Ensure that all fuel spillage (should it occur) is quickly removed from engine surfaces. Ensure that all fuel soaked cloths or towels are deposited into a suitable waste container.

• Always keep a dry chemical (Class B) fire extinguisher near the work area.

• Do not allow fuel spray or fuel vapors to come into contact with a spark or open flame.

• Always use a backup wrench when loosening and tightening fuel line connection fittings. This will prevent unnecessary stress and torsion to fuel line piping. Always follow the proper torque specifications.

• Always replace worn fuel fitting O-rings with new. Do not substitute fuel hose or equivalent, where fuel pipe is installed.

Fuel System

BLEEDING AIR

1993 Models

1. Place a drain pan under the separator filter.

2. Open the low pressure bleed screw.

3. Manual pump the lever on the fuel pump until the fuel exiting the bleed screw is free of air. If the manual lever feels as if it is not pumping, rotate (crank) the engine approximately 90°, then continue pumping as described.

4. Tighten the low pressure bleed screw to 6 ft. lbs. (8 Nm).

1994–96 Models

1. Loosen the low pressure bleed bolt.

2. Operate the rubber push-button primer on the fuel transfer pump. Do this until the fuel exiting the bleed screw is free of air. If the primer button feels as if it is not pumping, rotate (crank) the engine approximately 90°, then continue pumping as described.

3. Tighten the low pressure bleed screw to 6 ft. lbs. (8 Nm).

IDLE SPEED

ADJUSTMENT

1. Start the engine and run until at normal operating temperature.

2. An optical tachometer must be used to read engine speed; a conventional tachometer connected to the coil is useless in this instance.

3. Turn the air conditioning **ON**, if equipped.

4. Turn the idle speed screw until the desired idle speed is obtained. The specification for a vehicle equipped with automatic transmission is 700 rpm. The specification for a vehicle equipped with manual transmission is 750 rpm.

Fuel Water Separator/Filter

DRAIN WATER

Filtration and separation of water from the fuel is important for trouble

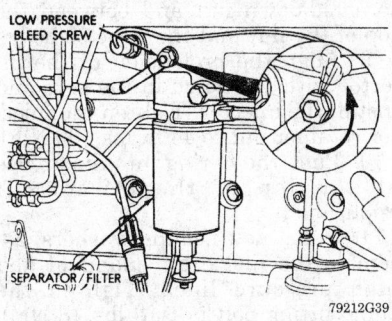

Turn the low pressure bleed screw and work the lever to purge the lines of air — 1993 models

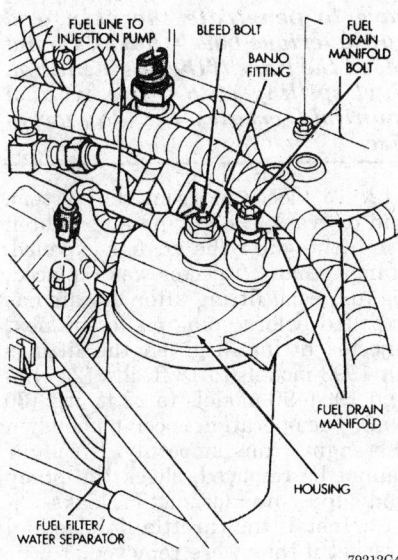

Location of the low pressure bleed bolt — 1994–96 models

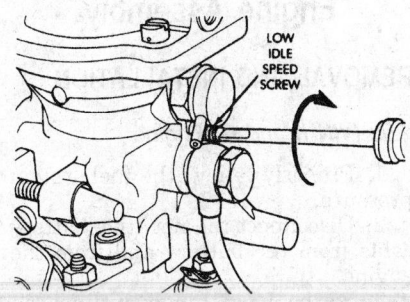

Turn the screw shown to adjust the low idle speed

free operation and long life of the fuel system. Regular maintenance, including draining moisture from the fuel/water separator filter is essential to keep water out of the fuel pump. To remove the collected water, simply unscrew the drain at the bottom of the WIF assembly located at the bottom of the filter separator.

REMOVAL AND INSTALLATION

1. Disconnect the negative battery cable.
2. Disconnect the Water In Filter (WIF) sensor connector.
3. Remove the separator filter assembly from the filter head with a standard oil filter wrench.
4. Remove the square cut O-ring from the filter mounting bushing.
5. Drain the fuel/water separator filter and remove the assembly from the fuel filter.
   To install:
6. Install a new O-ring to the WIF assembly and install to the new separator filter.
7. Install a new square cut O-ring to the mounting bushing.
8. Fill the fuel/water separator filter with clean diesel fuel.
9. Apply a light coat of oil to the sealing surface of the separator filter.
10. Install the assembly and tighten it ½ turn after the seal contacts the filter head.
11. Reconnect the WIF sensor connector.
12. Connect the negative battery cable, start the engine and check for leaks.

Diesel Injection Pump

REMOVAL AND INSTALLATION

NOTE: The Bosch VE lever is indexed to the shaft during pump calibration. Do not remove it from the pump during removal.

1. Disconnect the negative battery cable.
2. Remove the throttle linkage and bracket.
3. Disconnect the fuel drain manifold.
4. Remove the injection pump supply line.
5. Remove the high pressure lines.
6. Disconnect the electrical wire to the fuel shut off valve.
7. Remove the fuel air control tube.
8. Remove the pump support bracket.

9. Remove the oil fill tube bracket and adapter from the front gear cover.
10. Place a shop towel in the gear cover opening in a position that will prevent the nut and washer from falling into the gear housing. Remove the gear retaining nut and washer.
11. Install the turning tool into the flywheel housing opening on the exhaust side of the engine. Place a ½ in. drive universal joint in the turning tool and attach enough extensions to the joint to make it convenient to turn the tool.
12. Using a ratchet to turn the barring tool, turn the engine until the keyway on the fuel pump shaft is pointing approximately in the six o'clock position.
13. Locate TDC for cylinder No. 1 by turning the engine slowly while pushing in on the TDC pin. Stop turning the engine as soon as the pin engages with the gear timing hole. Disengage the pin after locating TDC and remove the turning equipment.
14. Loosen the lockscrew, remove the special washer from the injection pump and wire it to the line above it so it will not get misplaced. Retighten the lockscrew to 22 ft. lbs. (30 Nm) to lock the driveshaft.
15. Using a suitable puller, pull the pump drive gear from the driveshaft.

NOTE: Be careful not to drop the drive gear key into the front cover when removing or installing the pump. If it does drop in, it must be removed before proceeding.

16. Remove the 3 mounting nuts and remove the injection pump from the vehicle.
17. Remove the gasket and clean the mounting surface.
   To install:
18. Install a new gasket.

NOTE: The shaft of a new or reconditioned pump is locked so the key aligns with the drive gear keyway with cylinder No. 1 at TDC.

19. Install the pump and finger-tighten the mounting nuts; the pump must be free to move in the slots.
20. Install the pump drive gear, washer and nut to the driveshaft. The pump will rotate slightly because of gear helix and clearance. This is acceptable providing the pump is free to move on the flange slots and the crankshaft does not move. Tighten the nut to 11–15 ft. lbs. (15–20 Nm). This is not the final torque; do not overtighten.

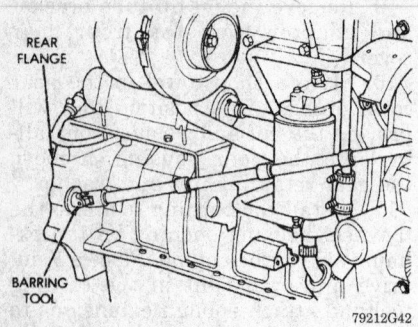

REAR FLANGE

BARRING TOOL

79212G42

Use the barring tool to rotate the diesel engine

21. If installing the original pump, rotate the pump to align the original timing marks and tighten the mounting nuts to 18 ft. lbs. (24 Nm).

22. If installing a replacement pump, take up gear lash by rotating the pump counterclockwise toward the cylinder head, and tighten the mounting nuts to 18 ft. lbs. (24 Nm). Permanently mark the new injection pump flange to match the mark on the gear housing.

23. Loosen the lockscrew and install the special washer under the lockscrew; tighten to 13 ft. lbs. (18 Nm). Disengage the TDC pin.

24. Install the injection pump support bracket. Finger-tighten the bolts initially, then tighten them to 18 ft. lbs. (24 Nm) in the following sequence:

a. Bracket to block bolts.
b. Bracket to injection pump bolts.
c. Throttle support bracket bolts.

25. Now perform the final tighten of the pump drive gear retaining nut to 48 ft. lbs. (65 Nm).

26. Install the oil filler tube assembly and clamp. Tighten the bolts to 32 ft. lbs. (43 Nm).

27. Install all fuel lines and the electrical connector to the fuel shut off valve. Tighten the high pressure lines to 18 ft. lbs. (24 Nm).

28. Install the fuel air control tube. Tighten the banjo fitting bolt to 9 ft. lbs. (12 Nm).

29. Install the throttle bracket and linkage. When connecting the cable to the control lever, adjust the length so the lever has stop-to-stop movement.

30. Connect the negative battery cable.

—— **CAUTION** ——
Do not place any part of the hand near the base of the high pressure line. A fuel leak from a high pres-

sure fuel line has sufficient pressure to penetrate the skin and cause serious bodily harm. Do not bleed the lines if the engine is hot. Fuel spilling onto a hot exhaust manifold creates the danger of fire.

31. To bleed air from the system, run or crank the engine and carefully loosen the high pressure fitting from each injector one at a time. Retighten the fitting after the air has expelled before going on to the next injector fitting. The operation is complete when the engine runs smoothly. If the air cannot be removed, check the pump and supply line for suction leaks.

32. Adjust the idle speed if necessary.

Fuel Injector

REMOVAL AND INSTALLATION

1. Disconnect the negative battery cable. Remove the throttle linkage and bracket if necessary.
2. Disconnect the high pressure fuel supply line to the injector.
3. Disconnect the fuel drain manifold.
4. Clean the area around the injector.

NOTE: Certain types of injectors MAY have an O-ring located above the hold-down nut.

5. Using the correct deepwell socket, remove the injector from the cylinder head.
   To install:
6. Clean the injector bore with a bore brush.
7. Assemble the injector and 1 new copper sealing washer. Never use more than 1 copper washer.
8. Apply a thin coat of anti-seize compound to the threads of the injec-

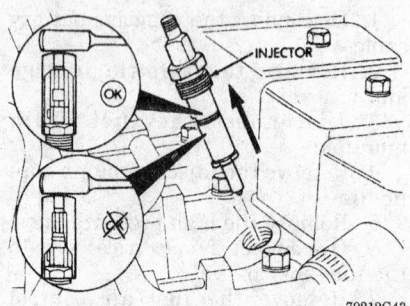

INJECTOR

OK

79212G43

Unscrew and remove the diesel fuel injector

tor hold-down nut and between the top of the nut and the injector body.

9. Align the protrusion in the injector with the notch in the bore and install the injector. Tighten the injector retainer nut to 44 ft. lbs. (60 Nm).

10. Push the O-ring into the groove at the top of the injector, if applicable.

11. Using new sealing washers, assemble the fuel drain manifold and high pressure lines. Tighten the banjo fitting bolt to 6 ft. lbs. (8 Nm). Leave the high pressure line loose temporarily.

—— **CAUTION** ——
Do not place any part of the hand near the base of the high pressure line. A fuel leak from a high pressure fuel line has sufficient pressure to penetrate the skin and cause serious bodily harm. Do not bleed the lines if the engine is hot. Fuel spilling onto a hot exhaust manifold creates the danger of fire.

12. To bleed air from the system, run or crank the engine and tighten the fitting after the air has expelled. If more than 1 injector was replaced, tighten each fitting after the air has expelled before going on to the next injector fitting. Tighten the fittings for 1993 models to 18 ft. lbs. (24 Nm) and 1994–96 models to 22 ft. lbs. (30 Nm). The operation is complete when the engine runs smoothly. If the air cannot be removed, check the pump and supply line for suction leaks.

13. Install the throttle linkage and bracket if they were removed.

GASOLINE ENGINE MECHANICAL

Engine Assembly

REMOVAL AND INSTALLATION

2.5L (VIN K and G) Engine

1. Properly relieve the fuel system pressure.
2. Disconnect the negative battery cable from the battery and from the engine.
3. Scribe hood hinge outlines on the hood and remove the hood from the vehicle. Remove the hood and the oil dipstick.

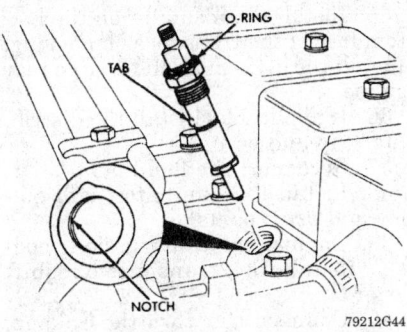

79212G44

Install the diesel injector so that its tab is aligned with the notch in the bore

---

CAUTION

The EPA warns that prolonged contact with used engine oil may cause a number of skin disorders, including cancer! You should make every effort to minimize your exposure to used engine oil. Protective gloves should be worn when changing the oil. Wash your hands and any other exposed skin areas as soon as possible after exposure to used engine oil. Soap and water, or waterless hand cleaner should be used.

---

4. Raise the vehicle and support safely, Drain the engine oil and coolant. Remove the lower radiator hose.

5. Remove the starter. Remove the engine to transmission struts, if equipped.

CAUTION

When draining the coolant, keep in mind that cats and dogs are attracted by ethylene glycol antifreeze, and are quite likely to drink any that is left in an uncovered container or in puddles on the ground. This will prove fatal in sufficient quantity. Always drain the coolant into a sealable container.

---

6. Remove the exhaust pipe from the exhaust manifold(s).

7. If the vehicle is equipped with a manual transmission, remove the transmission and all related parts.

8. If the vehicle is equipped with an automatic transmission, remove the inspection plate, matchmark the flexplate to the converter, remove the torque converter bolts and push the torque converter backwards as far as it will go.

9. Remove the engine mount lower nuts.

10. If the vehicle is equipped with 4WD:

a. On the left side, remove the two bolts attaching the bracket to the transmission bell housing and two brackets to pinion nose adaptor bolts. Separate the engine from the insulator by removing the upper nut washer assembly and through-bolt from the engine support bracket.

b. On the right side, remove the two brackets to axle. Disconnect the housing bolts and one bracket-to-bell housing bolt. Separate the engine from the insulator by removing the upper nut/washer assembly and the through-bolt from the engine support bracket.

11. Disconnect and plug the rubber fuel inlet and return hoses from the fuel lines at the front of the vehicle. Lower the vehicle.

12. Remove the air cleaner assembly and disconnect all linkages and cables. On the 3.9L (VIN X) engine, remove the throttle body. Stuff a clean shop towel into the intake manifold opening to prevent foreign objects from entering.

13. Unbolt the air conditioning compressor from the engine and position it to the side without discharging the system or disconnecting the lines.

14. Remove the radiator and shroud. Remove the fan and all related parts.

15. Unbolt the power steering pump brackets from the engine and position it to the side.

16. Remove the alternator, air pump and all brackets. Disconnect the heater hoses.

17. Remove the distributor cap with all spark plug wires attached. Disconnect all remaining electrical connectors, vacuum hoses and check for any other items preventing engine removal.

18. Attach a hoist or similar engine removal device to the intake manifold or cylinder head.

19. If the vehicle is equipped with an automatic transmission, support the transmission with a floor jack, or equivalent. Remove the remaining bell housing bolts.

20. Remove the engine from the vehicle slowly and carefully.

To install:

21. Lower the engine into position and install the engine mount bolts on the 2.5L (VIN K and G) engine. Install the upper bell housing bolts. Remove the hoist or engine removal device. Install the oil dipstick.

22. Raise the vehicle and support it safely.

23. If the vehicle is equipped with a manual transmission, install the transmission and all related parts.

24. If the vehicle is equipped with an automatic transmission, align the torque converter and flexplate and install the bolts. Install the inspection plate, starter and the engine-to-transmission struts, if equipped.

25. Connect the rubber fuel inlet and return hoses to the fuel lines at the front of the vehicle.

26. Install the exhaust pipe to the exhaust manifold(s). Install the lower radiator hose. Lower the vehicle.

27. Connect the heater hoses.

28. Make sure the negative battery cable is not connected to the battery. Connect the engine side of the negative cable to the engine. Install the alternator, air pump, power steering pump and all brackets.

29. Install the air conditioning compressor.

30. Install the throttle body using a new base gasket and connect all linkages and cables. Connect all electrical connectors, vacuum hoses, etc. that were disconnected during the engine removal procedure.

31. Install the fan and all related parts. Adjust all belt tensions.

32. Install the radiator and shroud.

33. Install the distributor cap with all spark plug wires attached.

34. Install the air cleaner assembly.

35. Install the hood.

36. Fill the engine with the specified amount of oil and fill the radiator with coolant.

37. Connect the negative battery cable and set all adjustments to specifications.

3.9L (VIN X), 5.2L (VIN T and Y), 5.9L (VIN A, Z and 5) and 8.0L (VIN W) Engines

1. Scribe hood hinge outlines on the hood and remove the hood from the vehicle. Disconnect the battery.

2. Drain the coolant from the radiator and engine block.

3. Drain the engine oil.

CAUTION

When draining the coolant, keep in mind that cats and dogs are attracted by ethylene glycol antifreeze, and are quite likely to drink any that is left in an uncovered container or in puddles on the ground. This will prove fatal in sufficient quantity. Always drain the coolant into a sealable container.

---

4. Remove the air cleaner.

5. Remove the starter.

CAUTION

The EPA warns that prolonged contact with used engine oil may cause a number of skin disorders, including cancer! You should make every effort to minimize your exposure to used engine oil. Protective gloves should be worn when changing the oil. Wash your hands and any other exposed skin areas as soon as possible after exposure to used engine oil. Soap and water, or waterless hand cleaner should be used.

6. Depressurize the fuel injection system.

7. Have the air conditioning refrigerant properly discharged at a reputable repair shop.

8. If equipped, remove the transmission oil cooler.

9. Disconnect both radiator and heater hoses. Remove the radiator and set the fan shroud aside.

10. Remove the power steering pump and air pump with the hoses attached and lay them aside.

11. Remove the washer bottle.

12. Disconnect and tag the vacuum lines.

13. Remove the A/C compressor and lay it aside with the hoses connected, if enough slack exists. If not, disconnect the A/C hoses.

14. On manual transmission models, remove the shift lever.

15. Disconnect the throttle linkage and all electrical connection to the engine sensors. Tag all electrical connections before disconnecting.

16. Remove the alternator, fan, pulley and drive belt(s).

17. If applicable, remove the distributor cap and wires.

18. Remove and plug the fuel line.

19. Disconnect the exhaust pipe at the manifold.

20. On automatic transmission models, remove the bell housing bolts and inspection plate. Attach a C-clamp on the front bottom of the transmission torque converter housing to prevent the torque converter from coming out.

21. Remove the retaining bolts from the torque converter drive plate. Mark the converter and drive plate to aid in reassembly.

22. Remove the driveshaft and engine rear support.

23. Support the transmission with a suitable transmission jack.

24. Remove the transmission.

25. Raise the rear of the engine approximately 2 in. (51mm) and remove the clutch or drive plate and the flywheel.

26. Using a boom hoist attached to the engine with the shortest hook-up possible, take up all tension and support the engine.

WARNING

Do not lift the engine by the intake manifold.

27. Remove the engine front mounts and insulators.

28. Carefully remove the engine from the vehicle. Lift slowly and watch for any snagged or missed connections. Mount the engine on a suitable engine stand for further disassembly or service work.

To install:

29. Carefully lower the engine into the vehicle.

30. Install the engine front mounts and insulators.

31. Raise the vehicle.

32. Install the torque converter drive plate bolts.

33. Install the clutch or drive plate and the flywheel.

34. If equipped with an automatic transmission, install the transmission intact with the torque converter. Use the match marks to align the drive plate and torque converter mounting bolts.

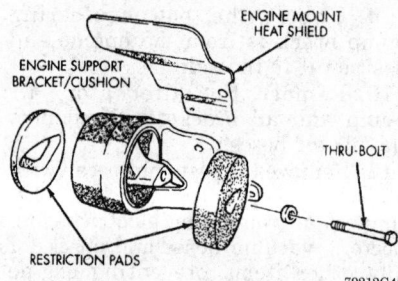

Exploded view of the front engine mounts — all 1994–96 Ram models

79212G45

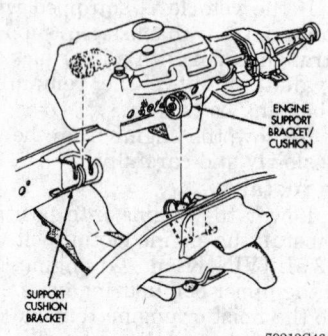

Positioning the front engine mounts — all 1994–96 Ram models

79212G46

35. Install the insulator on the bottom face of the transmission housing. Install the driveshaft and engine rear support.

36. If applicable, install the distributor cap and wires.

37. Reconnect the fuel line.

38. Install the alternator, fan, pulley and drive belt(s).

39. On manual transmission models, install the transmission shift lever.

40. Connect the throttle linkage, heater and vacuum hoses and all electrical connections to the ignition, alternator, and all other electrical connections.

41. Install the A/C compressor with lines attached.

42. Install the power steering and air pumps.

43. Install the radiator and fan shroud.

44. Connect both radiator and heater hoses.

45. If equipped, install the transmission oil cooler.

46. Install the starter.

47. Install the air cleaner.

48. Fill the crankcase with engine oil.

49. On the 8.0L (VIN W) engine, prime the oil pump by filling the J-trap of the front timing cover with oil and quickly installing an oil-filled filter in place just as the oil is pouring out of the J-trap.

50. Fill the cooling system.

51. Connect the battery.

52. Check all fluid levels and perform all tune-up adjustments if the engine was rebuilt.

53. Take the vehicle back to the air conditioning service facility and have them properly evacuate and charge the air conditioning system.

Engine Mounts

REMOVAL AND INSTALLATION

3.9L (VIN X), 5.2L (VIN T and Y), 5.9L (VIN A, Z and 5) Engines — Van models

Front Mounts

1. Disconnect the negative battery cable.

2. If necessary, position the fan to assure clearance for the radiator top tank and hose.

3. Install a suitable engine hoist or use a floor jack to SLIGHTLY raise the engine.

4. Remove the nuts from the brackets and insulators.

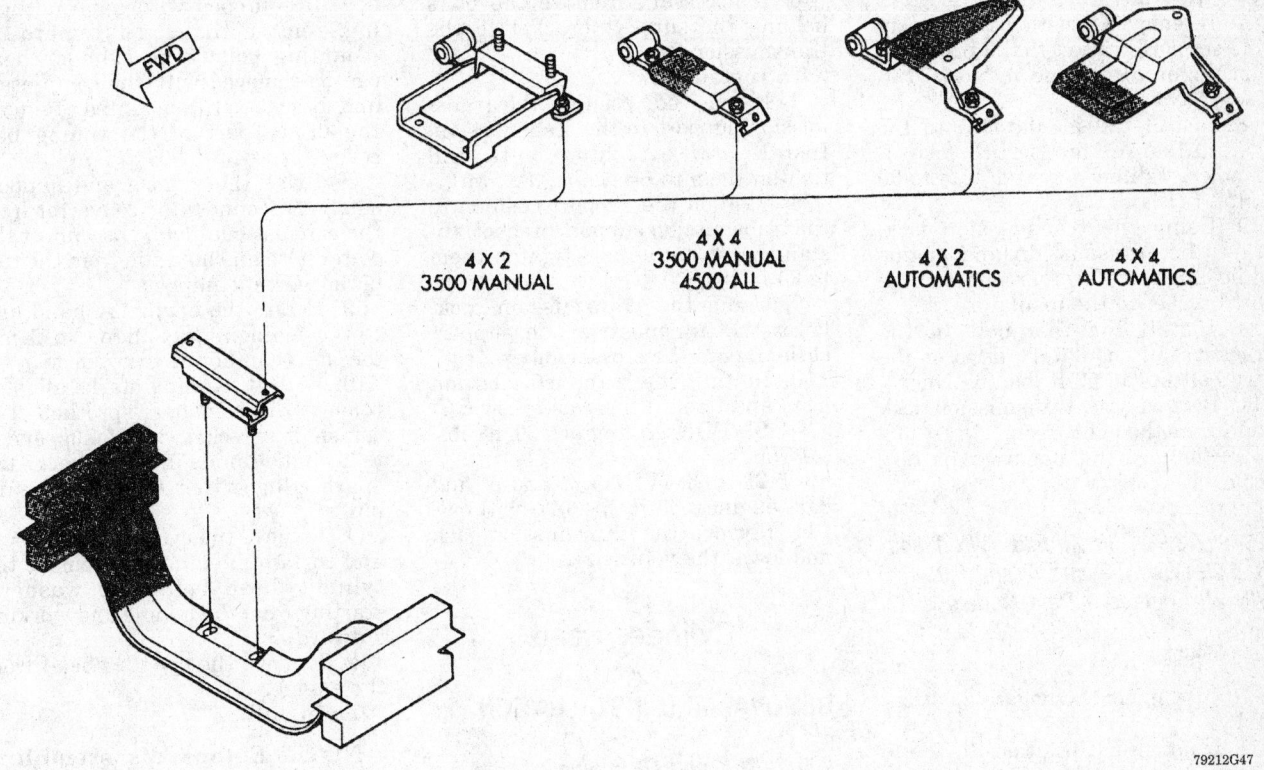

Various rear engine support cushion assemblies — all 1994–96 Ram models

5. Remove the engine mount heat shields.

To install:

6. With the engine raised SLIGHTLY, install the insulators and brackets.

7. Lower the engine while guiding the insulator studs into attaching holes in the crossmember and brackets.

8. Tighten the attaching nuts to 75 ft. lbs. (88 Nm).

9. Lower the vehicle.

10. Remove the engine hoist or jack.

11. Position the fan to its original location.

12. Connect the negative battery cable.

Rear Mount

1. Raise and safely support the vehicle.

2. Position a transmission jack in place.

3. Raise the rear of the transmission and engine SLIGHTLY.

4. Remove the rear mount through-bolt and nut.

5. Remove the U-shaped bracket from the frame crossmember. Remove the insulator from the U-shaped bracket.

6. Remove the rear engine support bracket from the bottom face of the transmission extension housing.

To install:

7. Attach the rear engine support bracket to the bottom face of the transmission extension housing. Tighten the two front bolts to 50 ft. lbs. (68 Nm). Tighten the rear bolt to 17 ft. lbs. (23 Nm)

8. Install the through-bolt through the insulator and U-shaped bracket. Tighten the through-bolt nut to 50 ft. lbs. (68 Nm)

9. Attach the U-shaped bracket to the crossmember. Tighten the attaching bolts to 30 ft. lbs. (41 Nm).

10. Remove the transmission jack and lower the vehicle.

2.5L (VIN K and G), 3.9L (VIN X) and 5.2L (VIN T and Y) Engines — Dakota models

Front Mounts

1. Disconnect the negative battery cable.

2. If necessary, position the fan to assure clearance for the radiator top tank and hose.

3. Install a suitable engine hoist or use a floor jack to SLIGHTLY raise the engine.

4. Remove the thru-bolt and nut.

5. Remove the engine mount heat shields.

To install:

6. With the engine raised SLIGHTLY, position the support bracket/cushion and heat shields to the block. Install new bolts and tighten to 40 ft. lbs. (54 Nm).

7. Install the thru-bolt into the engine support bracket/cushion.

8. Lower the engine while guiding the bracket/cushion and thru-bolt into the support cushion brackets.

9. Install the thru-bolt nuts and tighten them to 50 ft. lbs. (68 Nm).

10. Lower the vehicle and remove the hoist or jack.

Rear Mount

1. Disconnect the negative battery cable.

2. Raise and safely support the vehicle.

3. Position a transmission jack in place.

4. Remove the rear engine support bracket thru-bolt and nut.

5. Raise the rear of the transmission and engine SLIGHTLY.

6. Remove the stud nuts and insulator from the transmission mounting crossmember. Remove the insulator.

To install:

7. If removed, position the rear engine support bracket to the transmission. Tighten the stud nuts to 30 ft. lbs. (41 Nm).

8. Install the insulator onto the transmission mounting crossmember. tighten the stud nuts to 30 ft. lbs. (41 Nm).

9. Using the transmission jack, lower the transmission and engine while aligning the rear engine support bracket to the insulator.

10. Install the thru-bolt in the bracket and insulator. Tighten the thru-bolt nut to 50 ft lbs. (68 Nm).

11. Remove the transmission jack and lower the vehicle.

12. Connect the negative battery cable.

1994–97 3.9L (VIN X), 5.2L (VIN T and Y), 5.9L (VIN A, Z and 5) and 8.0L (VIN W) Engines — Ram Models

Front Mounts

1. Disconnect the negative battery cable.

2. If necessary, position the fan to assure clearance for the radiator top tank and hose.

3. Install a suitable engine hoist or use a floor jack to SLIGHTLY raise the engine.

4. Remove the thru-bolt and nut.

5. Remove the engine mount heat shields.

To install:

6. With the engine raised SLIGHTLY, position the support bracket/cushion and heat shields to the block. Install new bolts and tighten to 60 ft. lbs. (81 Nm).

7. Install the thru-bolt into the engine support bracket/cushion.

8. Lower the engine while guiding the bracket/cushion and thru-bolt into the support cushion brackets.

9. Install the thru-bolt nuts and tighten them to 75 ft. lbs. (102 Nm).

10. Lower the vehicle and remove the hoist or jack.

Rear Mount

1. Raise and safely support the vehicle.

2. Position a transmission jack in place.

3. Remove the support cushion stud nuts.

4. Raise the rear of the transmission and engine SLIGHTLY.

5. Remove the bolts holding the support cushion to the transmission support bracket. Remove the support cushion.

6. If necessary, remove the bolts holding the support bracket to the transmission.

To install:

7. If removed, position the transmission support to the transmission. Install new attaching bolts and tighten them to 65 ft. lbs. (88 Nm).

8. Position the support cushion to the transmission support bracket. Install the stud nuts and tighten them to 30 ft. lbs. (41 Nm).

9. Using the transmission jack, lower the transmission and support cushion onto the crossmember.

10. Install the support cushion bolts and tighten them as follows:
- 3.9L (VIN X) engine: 30 ft. lbs. (41 Nm)
- 5.2L, 5.9L (VIN A, Z and 5) and 8.0L engines: 35 ft. lbs. (47 Nm).

11. Remove the transmission jack and lower the vehicle.

Cylinder Head

REMOVAL AND INSTALLATION

2.5L (VIN K and G) Engine

1. Relieve the fuel system pressure.

2. Disconnect the negative battery cable and unbolt it from the head.

3. Drain the cooling system. Remove the dipstick bracket nut from the thermostat housing.

4. Remove the air cleaner assembly. Remove the upper radiator hose and disconnect the heater hoses.

5. Disconnect and label the vacuum lines, hoses and wiring connectors from the manifold(s), throttle body and from the cylinder head. Remove the air pump, if equipped.

6. Disconnect the all linkages and the fuel line from the throttle body. Unbolt the cable bracket. Remove the ground strap attaching screw from the firewall.

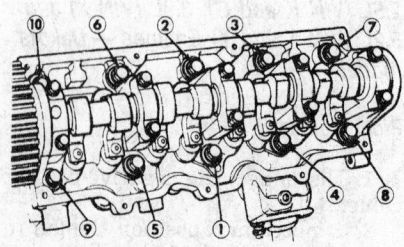

79212G48

Cylinder head tightening sequence — 2.5L (VIN K and G) engine

7. If equipped with air conditioning, remove the upper compressor mounting bolts. The cylinder head can be removed with the compressor and bracket still mounted. Remove the upper part of the timing belt cover.

8. Raise the vehicle and support safely. Disconnect the converter from the exhaust manifold. Disconnect the water hose and oil drain from the turbocharger, if equipped.

9. Rotate the engine by hand, until the timing marks align (No. 1 piston at TDC). Lower the vehicle.

10. With the timing marks aligned, remove the camshaft sprocket. The camshaft sprocket can be suspended to keep the timing intact. Remove the spark plug wires from the spark plugs.

11. Remove the cylinder head cover and curtain, if equipped. Remove the cylinder head bolts and washers, starting from the middle and working outward.

12. Remove the cylinder head from the engine.

To install:

NOTE: Before disassembling or repairing any part of the cylinder head assembly, identify factory installed oversized components. To do so, look for the tops of the bearing caps painted green and O/SJ stamped rearward of the oil gallery plug on the rear of the head. In addition, the barrel of the camshaft is painted green and O/SJ is stamped onto the rear end of the camshaft. Installing standard sized parts in a head equipped with oversized parts-or visa versa-will cause severe engine damage.

13. Clean the cylinder head gasket mating surfaces.

14. Using new gaskets and seals, install the head to the engine.

15. Using new head bolts assembled with the old washers, tighten the cylinder head bolts in sequence, to 45 ft. lbs. (61 Nm). Repeating the sequence, tighten the bolts to 65 ft. lbs. (88 Nm). With the bolts at 65 ft. lbs. (88 Nm), turn each bolt an additional 1/4 turn.

NOTE: Head bolt diameter is 11mm. These bolts are identified with the number 11 of the head of the bolt. The 10mm bolts used on some earlier vehicles will thread into an 11mm bolt hole, but will permanently damage the cylinder block. Make sure the correct bolts are being used when replacing old head bolts.

16. Install the timing belt.

17. Install or connect all items that were removed or disconnected during the removal procedure.

18. Refill the cooling system. Connect the negative battery cable. Start the engine and check for leaks.

3.9L (VIN X) Engine

1. Relieve the fuel pressure. Disconnect the negative battery cable from the battery and drain the cooling system.

2. Raise the vehicle and safely support. Disconnect the exhaust pipe from the manifolds.

3. Remove the alternator if the right head is being removed and the air pump and negative battery cable if the left head is being removed.

4. Remove the air cleaner assembly. Unbolt the air conditioning compressor and lay it to the side, if equipped. Remove the distributor cap with all wires attached.

5. Disconnect all wires, hoses, linkages and cables from the throttle body. Disconnect and plug the fuel line.

6. Disconnect the ignition coil, coolant temperature sending unit wire and all other connectors along the wiring harness connected to items on the intake manifold.

7. Disconnect the heater hose, upper radiator hose and the lower bypass hose clamp.

8. Remove the cylinder head covers.

9. Remove the intake manifold assembly. Remove the exhaust manifolds.

10. Remove the rocker arm and shaft assembly from the heads. Do not disassemble unless service is required.

11. Remove the pushrods and identify them to ensure installation in their original locations.

12. Remove the head bolts and remove the cylinder head(s).

   To install:

13. Clean and dry all gasket surfaces of the cylinder block and head. Inspect all surfaces with a straightedge. If the flatness exceeds 0.0075 times the length of the span measured (in any direction), replace or machine the head gasket surface.

14. Using no sealer whatsoever, install the new head gasket(s) to the block. Clean, dry and lightly oil all head bolts threads. Install the heads and install the head bolts.

15. Tighten the head bolts in sequence to 50 ft. lbs. (68 Nm). Repeat the sequence retightening the bolts to a final torque of 105 ft. lbs. (143 Nm)

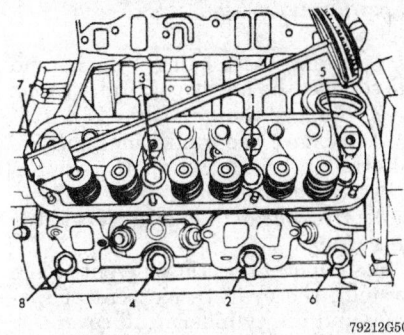

79212G50

Cylinder head tightening sequence — 3.9L (VIN X) engine

and repeat the second step to ensure that all bolts are accurately torqued.

16. Assemble the rocker shaft assembly, if it was serviced. Make sure all rocker arms with an RH stamped on them are installed to the right of those with an LH. Install the pushrods, rocker arms and shaft(s) with the notch on the rear of the right bank or to the font of the left bank. Make sure the long stamped steel retainers are at the number 2 and 4 positions.

17. Tighten the bolts evenly and gradually to 117 ft. lbs. (23 Nm).

18. Clean and dry the intake manifold contact surfaces. Coat the intake manifold side gaskets very lightly with sealer and install the gaskets to the heads. Cutouts at the front of the gaskets differentiate the right and left sides.

19. Apply a thin uniform coat of quick dry cement to the front and rear intake manifold gaskets and mounting surfaces on the block and apply a thin bead of sealer to each of the four corners. Install the front and rear gaskets engaging the hole in the block and the tangs from the head gaskets. Apply a second thin bead of sealer above the gaskets in the four corners.

20. Carefully lower the intake manifold into position engaging the bypass hose; after it is satisfactorily in place, inspect the gasket to make sure they have not become dislodged.

21. Install the intake manifold bolts and tighten in sequence to 25 ft. lbs. (34 Nm). Repeat the sequence retightening the bolts to a final torque of 40 ft. lbs. (54 Nm) and repeat the second step to ensure all bolts are accurately torqued.

22. Install the exhaust manifold(s) and tighten the bolts to 20 ft. lbs. (27 Nm). Tighten the end nuts to 15 ft. lbs. (20 Nm).

23. Clean and dry the cylinder head cover mating surfaces, bolts and bolt holes. Install the cylinder head covers each with a new gasket.

24. Connect the heater hose, upper radiator hose and the lower bypass hose clamp.

25. Connect the ignition coil, coolant temperature sending unit wire and all other connectors that were disconnected along the wiring harness.

26. Install the air conditioning compressor, if equipped. Install the distributor cap and all spark plug wires.

27. Install the alternator, battery ground and air pump, if they were removed.

28. Connect all wires, hoses, cables and the fuel line to the throttle body. Install the air cleaner assembly.

29. Raise the vehicle and safely support. Connect the exhaust pipe to the manifolds.

30. Fill the cooling system.

31. Connect the negative battery cable and set all adjustments to specification.

5.2L (VIN T and Y) and 5.9L (VIN A, Z and 5) Engines

1. Drain the cooling system and disconnect the negative battery cable.

2. Remove the alternator, air cleaner, fuel line and any obstructing heat shields.

3. Disconnect the accelerator linkage.

4. Remove the vacuum lines from the throttle body. On trucks with the 5.9L engine, remove the battery.

5. Remove the distributor cap and wires as an assembly.

6. Disconnect the coil wires, water temperature sending unit, heater hoses, and bypass hose. On the 5.9L engine, remove the distributor.

7. Remove the closed ventilation system, the evaporative control system, and the cylinder head covers.

8. Remove the intake manifold.

9. Remove the exhaust manifolds.

10. On 5.9L engines, tag the center bolts.

11. Remove the tappet chamber cover. Remove the rocker and shaft assemblies.

12. Remove the pushrods and keep them in order to ensure installation in their original locations. On the 5.9L engine, remove the water pump-to-head bolts.

13. Remove the head bolts from each cylinder head and remove the cylinder heads.

14. Clean all the gasket surfaces of the engine block and the cylinder heads. Install the spark plugs.

15. Inspect all surfaces with a straightedge. If warpage is indicated,

measure the amount. This amount must not exceed 0.00075 times the span length in any direction. For example, if a 12 in. (305mm) span is 0.004 in. (0.10mm) warped, the maximum allowable difference is 12 x 0.00075 = 0.009 in. (305 x 0.00075 = 0.22875mm). In this case, the head is within limits. If the warpage exceeds the specified limits, either replace the head or lightly machine the head gasket surface.

To install:

16. Coat new cylinder head gaskets with sealer, install the gaskets and install the cylinder heads.

17. Install the cylinder head bolts. Tighten the cylinder head bolts to 105 ft. lbs. (143 Nm) in the sequence indicated. Repeat this sequence to retorque all the cylinder head bolts to specifications.

18. Install the pushrods.

19. On the 5.9L engine, install the water pump-to-head bolts.

20. Install the rocker and shaft assemblies. Install the tappet chamber cover.

21. Install the exhaust manifolds.

22. Install the intake manifold.

23. Install the PCV system, the evaporative control system and the cylinder head covers.

24. Connect the coil wires, water temperature sending unit, heater hoses, and bypass hose. On the 5.9L engine, install the distributor.

25. Install the distributor cap and wires as an assembly.

26. Install the vacuum lines to the throttle body. On trucks with the 5.9L engine, install the battery.

27. Connect the accelerator linkage.

28. Install the alternator, air cleaner, and fuel line.

29. Fill the cooling system and connect the negative battery cable.

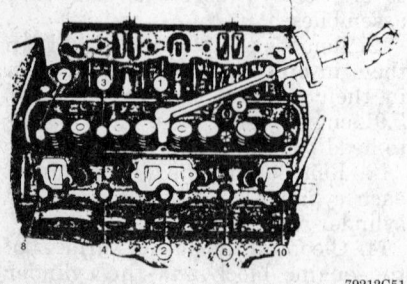

79212G51

Cylinder head tightening sequence — 5.2L (VIN T and Y) and 5.9L (VIN A, Z and 5) engines

8.0L (VIN W) Engine

1. Drain the cooling system and disconnect the negative battery cable.

2. Remove the heat shields.

3. Remove the intake manifold-to-alternator bracket support rod. Remove the alternator.

4. Disconnect the accelerator linkage.

5. Remove the closed ventilation system, the evaporative control system, and the cylinder head covers.

6. Disconnect the evaporation control system.

7. Remove the air cleaner.

8. Relieve the fuel system pressure.

9. Disconnect the throttle linkage and the speed control and transmission kickdown cables, if equipped.

10. Remove the coil pack and bracket.

11. Disconnect the coil wires.

12. Disconnect the heat indicator sending unit wire.

13. Disconnect the heater hoses and bypass hose.

14. Remove the upper intake manifold and throttle body as an assembly.

15. Remove the cylinder head covers and (reusable) gaskets.

16. Remove the EGR tube. Discard the gasket for right side only.

17. Remove the lower intake manifold. Discard the flange side gaskets and the front and rear cross-over gaskets.

18. Disconnect the exhaust pipe from the exhaust manifold.

19. Remove rocker arm assemblies and pushrods. Organize them so as to be able to install them in the exact same location.

20. Remove the head bolts from each cylinder head and remove the cylinder heads. Discard the gasket.

21. Remove the spark plugs.

22. Clean all the gasket surfaces of the engine block and the cylinder heads.

23. Inspect all surfaces with a straightedge. If warpage is indicated, measure the amount. The out-of-flatness specifications are 0.0004 in. per 1 in. (0.0007mm per 1mm) or 0.005 in. per 6 in. (0.127mm per 152mm) in any direction, or 0.010 in. (0.254mm) overall across the head. If exceeded, either replace the head, or machine the surface to skim it flat. The cylinder head surface finish should be 15–80 microinches (1.78–4.57 microns).

To install:

24. Install the gaskets and install the cylinder heads.

25. Install the cylinder head bolts. Tighten the cylinder head bolts in two steps:
- Tighten all cylinder head bolts in sequence to 43 ft. lbs. (58 Nm).
- Tighten all cylinder head bolts in sequence to 105 ft. lbs. (143 Nm).

26. Install the pushrods and rocker arm assemblies in their original positions. Tighten the bolts to 21 ft. lbs. (28 Nm).

27. Install the intake manifold gaskets. Be sure that the locator dowels are positioned in the head.

—————— **WARNING** ——————
Make absolutely sure the block sealing surface is free of oil.

28. Peel off the protective paper (blue — rear, and brown — front).

29. Align the slots in the end seals with the notches in the intake manifold.

30. Insert Mopar® Silicone Rubber Adhesive Sealant, or equivalent, into the four corner pockets. Fill — without overfilling — the pockets.

31. Install the lower intake manifold within NO MORE THAN three minutes of applying the sealant. Carefully lower the intake manifold into position on the cylinder block and heads. After the intake manifold is in place, inspect to make sure the seals and gaskets are in place.

32. Finger-start all bolts while alternating one side to the other.

33. Tighten the lower intake manifold bolts to 40 ft. lbs. (54 Nm).

34. Use a new gasket and position the upper intake manifold onto the lower intake manifold.

35. Tighten the upper intake manifold bolts to 16 ft. lbs. (22 Nm).

36. Install the exhaust pipe to the exhaust manifold. Tighten the bolts to 25 ft. lbs. (34 Nm).

37. Using a new gasket, position the EGR tube to the intake manifold and the exhaust manifold.

38. Tighten the nut to 25 ft. lbs. (34 Nm). Tighten the bolts to 174 inch lbs. (20 Nm).

39. Install the heat shields and the washers. Make sure that the heat shields tabs hook over the exhaust gasket. Tighten the nuts to 132 inch lbs. (15 Nm).

40. Install the sparkplugs. Tighten to 30 ft. lbs. (41 Nm).

41. Install the coil packs and bracket. Tighten the bolts to 190 inch lbs. (21 Nm). Connect the coil wires.

42. Connect the heat indicator sending unit wire.

43. Connect the heater hoses and bypass hose.

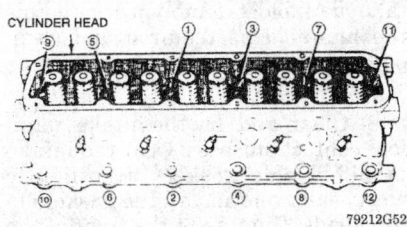

Cylinder head tightening sequence — 8.0L (VIN W) engine

44. Connect the throttle linkage and (if equipped) the speed control and transmission kickdown cables.
45. Install the fuel line.
46. Install the alternator and drive belt. Tighten the mounting bolt to 30 ft. lbs. (41 Nm). Tighten the adjusting strap bolt to 17 ft. lbs. (23 Nm).
47. Install the cylinder head cover onto the gasket. Install the stud bolts and hex head bolts in their proper positions. Tighten the stud bolts to 144 inch lbs. (16 Nm).
48. Install the closed crankcase ventilation system.
49. Connect the evaporation control system.
50. Install the air cleaner.
51. Fill the cooling system.
52. Connect the negative battery cable.
53. Start the truck and check for any leaks.

Valve Lifters

BLEEDING

The hydraulic valve lifters do not require a special bleeding procedure. Lubricate them upon installation. Be sure not to run the engine above fast idle until they are filled with oil.

REMOVAL AND INSTALLATION

2.5L (VIN K and G) Engine

1. Disconnect the negative battery cable.
2. Remove the cylinder head cover.
3. For each rocker arm, rotate the cam until the base circle is in contact with the rocker arm. Depress the valve spring using tool No. MD–998772A. Slide the rocker arm out. Keep all rocker arms in order for installation in the exact same sequence.
4. Remove the hydraulic valve lifter.

To install:
5. Install the hydraulic lifters. Make sure that the adjusters are at least partially full of oil.
6. Rotate the cam until the base circle is in contact with the rocker arm. Depress the valve spring and slide the rocker arm in place.

NOTE: Keep the rockers in the original order. It is possible for the valve spring retainer locks to become dislocated when depressing the valve spring. Check to make sure the locks are in their proper location.

7. Install the vapor curtain, if equipped, and the cylinder head cover.
8. Connect the negative battery cable.

3.9L (VIN X), 5.2L (VIN T and Y) and 5.9L (VIN A, Z and 5) Engines

1. Remove the cylinder head and rocker assembly keeping all parts in original order.
2. Once the rocker assembly and pushrods are removed, use tool C–4129–A, or equivalent, to remove the valve lifters from their bores. Pull the tappet out of the bore with a twisting motion. Keep the tappets in order so they can be installed in their original bores.
To install:
3. Lubricate and install the lifters in their original locations. Make sure the oil feed hole in the side of the lifter body faces up (away from the crankshaft.
4. Install the rocker arm and shaft assembly.
5. Install the vent cover, distributor, start the engine.

8.0L Engine

1. Disconnect the negative battery cable.
2. Remove the air cleaner and cylinder head cover.

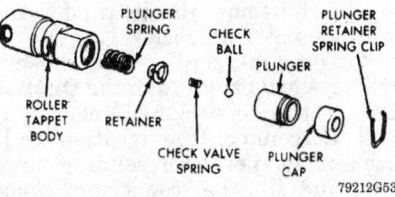

Exploded view of hydraulic tappet assembly assembly — 3.9L (VIN X) engine

3. Remove the rocker arms assembly and pushrods. Label or organize the pushrods to enable installation in the original location.
4. Remove the upper and lower intake manifold.
5. If the end tappets are to be removed, remove the cylinder head.
6. Remove the yoke retainer spider and tappet aligning yokes.
7. Pull the tappets out of the bore with a twisting motion. Label or organize the tappets to enable installation in the original location.
To install:
8. Lubricate the tappets.
9. Install the tappets in their original positions. make sure the oil bleed hole (if equipped) faces FORWARD.
10. Install the alignment yokes and retainer spider. Tighten the bolts to 16 ft. lbs. (22 Nm).
11. If removed, install the cylinder head.
12. Install the pushrods in their original location.
13. Position the rocker arm assembly on the pedestal and align the pushrods. Install the bolts and tighten to 21 ft. lbs. (28 Nm).
14. Install the lower and upper intake manifold.
15. Install the cylinder head cover and air cleaner.

Rocker Arms/Shafts

REMOVAL AND INSTALLATION

2.5L (VIN K and G) Engine

1. Disconnect the negative battery cable.
2. Remove the cylinder head cover.
3. Rotate the crankshaft until the low point of the desired cam lobe is contacting the rocker arm.
4. Use the special valve spring compressor tool, or equivalent, depress the valve spring (without dislodging the keeper) and slide the rocker arm out.
To install:
5. Depress the valve spring with the compressor tool and install in reverse order, turning the camshaft as necessary.
6. Install the cylinder head cover.
7. Connect the negative battery cable.

3.9L (VIN X) Engine

1. Disconnect the negative battery cable.
2. Remove the cylinder head cover and gasket.

3. Note the positioning of the oil notch and remove the rocker arms and shaft assembly from the head.

4. Disassemble the unit as required and replace all worn parts.

NOTE: On engines with exhaust valve rotators, the exhaust rocker arm must have relief for clearance.

To install:

5. Make sure all rocker arms with an RH stamped on them are installed to the right of those with an LH. Install the assembly with the notch on the end of the shaft pointing to the engine centerline and to the rear of the right bank or to the front of the left bank. Make sure the longer stamped steel retainers are at the number 2 and 4 positions. Tighten the bolts evenly and gradually to 17 ft. lbs. (23 Nm).

6. Install the cylinder head cover.

7. Connect the negative battery cable and check for leaks.

5.2L (VIN T and Y), 5.9L (VIN A, Z and 5) and 8.0L (VIN W) Engines

The stamped steel rocker arms are arranged on one rocker arm shaft per cylinder head. Because the angle of the pushrods tend to force the rocker arm pairs toward each other, oilite spacers are fitted to absorb the side thrust at each rocker arm. The shaft is secured by bolts and steel retainers attached to the brackets on the cylinder head.

1. Disconnect the spark plug wires.

2. Disconnect the closed ventilation system and evaporative control system (if so equipped) from the cylinder head cover.

3. Remove each cylinder head cover and gasket.

4. Remove the rocker shaft bolts and retainer.

5. Remove each rocker arm and shaft assembly. Keep everything in

order for installation in the original position.

To install:

6. Install the components in the same order as removed. For 5.2L (VIN T and Y) and 5.9L (VIN A, Z and 5) engines, the notch on the end of both rocker shafts should point to the engine centerline and toward the front of the engine on the left cylinder head and toward the rear on the right side.

7. Tighten the rocker arm bolts to 21 ft. lbs. (28 Nm).

8. Install the cylinder head and cover gasket.

9. Connect the spark plug wires.

Intake Manifold

REMOVAL AND INSTALLATION

2.5L (VIN K and G) Engine

Please refer to the Combination Manifold procedure for the 2.5L (VIN K and G) engine.

3.9L (VIN X) Engine

1. Relieve the fuel pressure. Disconnect the negative battery cable from the battery and drain the cooling system.

— **CAUTION** —

When draining the coolant, keep in mind that cats and dogs are attracted by ethylene glycol antifreeze, and are quite likely to drink any that is left in an uncovered container or in puddles on the ground. This will prove fatal in sufficient quantity. Always drain the coolant into a sealable container.

2. Remove the air pump and bracket. Removal of the bracket will allow for easier installation of the left front corner of the intake manifold.

3. Remove the air cleaner assembly. Unbolt the air conditioning compressor and lay it to the side, if equipped. Remove the distributor cap with all wires attached.

4. Disconnect all wires, hoses, linkages and cables from the throttle body. Disconnect the fuel line.

5. Disconnect the ignition coil, coolant temperature sending unit wire and all other connectors along the wiring harness connected to items on the intake manifold.

6. Disconnect the heater hose, upper radiator hose and the lower bypass hose clamp.

7. Unbolt the intake manifold from the heads and remove the intake manifold assembly. Disassemble the manifold as required and clean out the exhaust crossover passages.

To install:

8. Clean and dry the intake manifold contact surfaces. Coat the intake manifold side gaskets very lightly with sealer and install the gaskets to the heads. Cutouts at the front of the gaskets differentiate the right and left sides.

9. Apply a thin uniform coat of quick dry cement to the front and rear intake manifold gaskets and mounting surfaces on the block and apply a thin bead of sealer to each of the four corners. Install the front and rear gaskets engaging the hole in the block and the tangs from the head gaskets. Apply a second thin bead of sealer above the gaskets in the four corners.

10. Carefully lower the intake manifold into position engaging the bypass hose; after it is satisfactorily in place, inspect the gaskets to make sure they have not become dislodged.

11. Install the intake manifold bolts with the aspirator tube, air pump bracket and kickdown linkage bracket in place, if equipped. Tighten the bolts in sequence to 25 ft. lbs. (34 Nm). Repeat the sequence retightening the bolts to a final torque of 40 ft. lbs. (54 Nm) and repeat the second step to ensure all bolts are accurately torqued.

12. Clean and dry the cylinder head cover mating surfaces, bolts and bolt holes. Install the cylinder head covers with a new gasket. Tighten the screws or nuts to 95 inch lbs. (11 Nm).

13. Connect the heater hose, upper radiator hose and the lower bypass hose clamp.

14. Connect the ignition coil, coolant temperature sending unit wire and all other connectors that were disconnected along the wiring harness.

15. Install the air conditioning compressor, if equipped. Install the distributor cap and all spark plug wires.

16. Install air pump.

17. Connect all wires, hoses, cables and the fuel line to the throttle body. Install the air cleaner assembly.

18. Fill the cooling system.

19. Connect the negative battery cable. Turn the ignition **ON** to repressurize the fuel system by activating the fuel pump. Check for leaks.

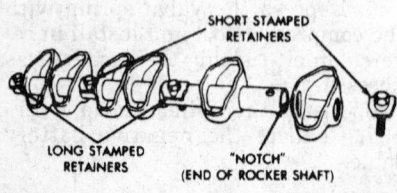

SHORT STAMPED RETAINERS

LONG STAMPED RETAINERS

"NOTCH" (END OF ROCKER SHAFT)

79212G55

Details of the rocker arm assembly — 3.9L (VIN X) engine

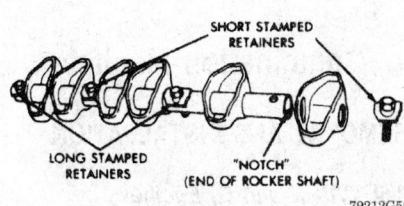

Torque sequence for the 3.9L (VIN X) engine intake manifold

5.2L (VIN T and Y) and 5.9L (VIN A, Z and 5) Engines

1. Drain the cooling system and disconnect the negative battery cable.

—————— **CAUTION** ——————

When draining the coolant, keep in mind that cats and dogs are attracted by ethylene glycol antifreeze, and are quite likely to drink any that is left in an uncovered container or in puddles on the ground. This will prove fatal in sufficient quantity. Always drain the coolant into a sealable container.

2. Remove alternator and air cleaner.
3. Disconnect accelerator linkage.
4. Depressurize the fuel system and disconnect the fuel line.
5. Remove the distributor cap and wires.
6. Disconnect coil wires, temperature sending unit wire, heater hoses and bypass hose.
7. Remove intake manifold and throttle body as an assembly.

To install:

8. Clean the gasket surfaces so they are clean and dry.
9. Position the intake manifold and install. If the fasteners were se-

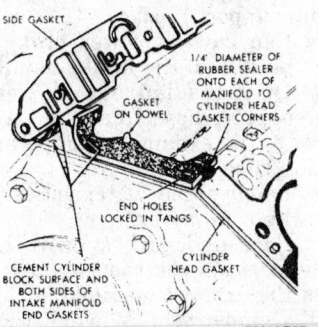

Seal the intake manifold as shown — 5.2L (VIN T and Y) and 5.9L (VIN A, Z and 5) engines

verely rusted when removed, use new fasteners.

10. Tighten the intake manifold to head bolts in the sequence illustrated.

8.0L (VIN W) Engines

1. Drain the cooling system and disconnect the negative battery cable.
2. Remove alternator and air cleaner.
3. Remove the serpentine belt.
4. Remove the alternator and its brace.
5. Remove the air conditioning compressor brace, remove the compressor and set it aside without disconnecting the lines.
6. Remove the air cleaner cover and filter. Remove the filter. Discard the gasket.
7. Depressurize the fuel system and disconnect the fuel line.
8. Disconnect accelerator linkage, and if so equipped, the speed control and transmission kickdown cables.
9. Remove the coil assemblies. Disconnect the vacuum lines.
10. Disconnect the heater hoses and bypass hose.
11. Remove the closed crankcase ventilation and evaporation control systems.

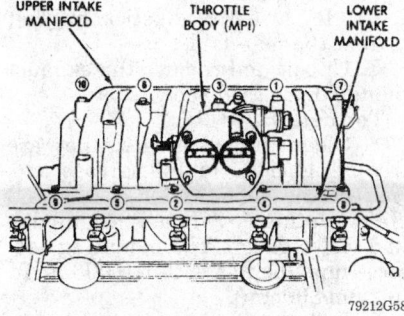

Upper intake manifold tightening sequence — 8.0L (VIN W) engine

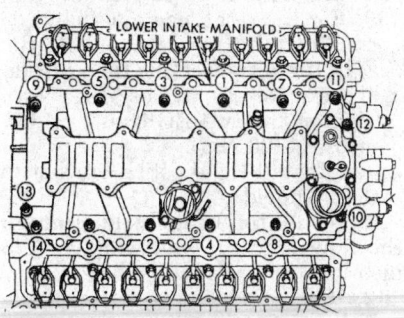

Lower intake manifold tightening sequence — 8.0L (VIN W) engine

12. Remove throttle body and lift it off the upper intake manifold. Discard the gasket.
13. Remove the front upper intake manifold bolts. Retain the three rear bolts in the up position with tape or rubber bands.
14. Lift the upper intake manifold out of the engine bay. Discard the gasket.
15. Remove the lower intake manifold bolts and remove the manifold. Discard the gasket.

To install:

16. Clean the manifolds and mounting surfaces with solvent and let dry or dry completely with compressed air.
17. Inspect the surfaces for cracks and for warpage using a straightedge.
18. With the locator dowels in place on the head, install the intake manifold side gaskets.
19. When sure that the block is oil-FREE, peel off the paper (blue-rear, brown-front) and press firmly onto the block. Align the slots in the end seals with the notches in the intake manifold gaskets.
20. Into each of the four corner pockets, insert Mopar® Silicon Rubber Adhesive Sealant, or equivalent. Do NOT overfill.
21. The lower intake manifold must be installed within three minutes of the sealant's having been applied. After in place, inspect to be sure all the gaskets and the seals are in their proper places. Finger-start all lower intake bolts.
22. Tighten in sequence to 40 ft. lbs. (54 Nm).
23. Install a new gasket and the upper intake manifold and finger-tighten all bolts. Alternate from one side to the other.
24. Tighten the bolts to 16 ft. lbs. (22 Nm).
25. Install a new gasket and the throttle body onto the upper intake manifold. Tighten the bolts to 17 ft. lbs. (23 Nm).
26. Install the closed crankcase ventilation and evaporation control systems.
27. Connect the heater hoses and bypass hose.
28. Connect the vacuum lines.
29. Install the oil assemblies and the ignition cables.
30. Connect the accelerator linkage, and if so equipped, the speed control and transmission kickdown cables.
31. Install the fuel lines.
32. Using a new gasket, install the air cleaner housing. Tighten the nuts

to 96 inch lbs. (11 Nm). Install the air cleaner assembly and its cover.

33. Install the air conditioning compressor. Install the brace, tightening the bolts to 30 ft. lbs. (41 Nm).

34. Install the serpentine belt.

35. Fill the cooling system and connect the negative battery cable.

Exhaust Manifold

REMOVAL AND INSTALLATION

2.5L (VIN K and G) Engine

Please refer to the combination manifold procedure for the 2.5L (VIN K and G) engine.

3.9L (VIN X) Engine

1. Disconnect the negative battery cable.

2. Remove the hot air tube and heat shield, if necessary.

3. Raise the vehicle and safely support it on jackstands.

4. Remove the exhaust pipe from the exhaust manifolds.

5. Lower the vehicle.

6. Take note of all conical washer locations and remove the bolts, nuts and washers attaching the manifold to the head.

7. Remove the manifold.

To install:

8. If either of the end studs came out with the nuts, install a new stud using sealer on the coarse threads.

9. Position the manifold on the end studs. Install conical washers and nuts on the studs.

10. Install the remaining bolts and washers in their proper locations. The inner bolts are not mounted with washers. Working outward from the center, tighten the bolts to 20 ft. lbs. (27 Nm) and the nuts to 15 ft. lbs. (20 Nm).

11. Install the exhaust pipe to the manifolds.

12. Connect the negative battery cable and check for exhaust leaks.

5.2L (VIN T and Y) and 5.9L (VIN A, Z and 5) Engines

1. Disconnect the exhaust manifold at the flange where it mates to the exhaust pipe.

2. If the vehicle is equipped with air injection and/or a heated air stove, remove them.

3. Remove the exhaust manifold by removing the securing bolts and washers. To reach these bolts, it may be necessary to jack the engine slightly off its front mounts. When

the exhaust manifold is removed, sometimes the securing studs will screw out with the nuts. If this occurs, the studs must be replaced with the aid of sealing compound on the coarse thread ends. If this is not done, water leaks may develop at the studs.

To install:

4. Clean the mounting surfaces so they are clean and dry.

5. Replace hardware, as necessary. Install a new gasket and the exhaust manifold. On the center branch of the exhaust manifold, no conical washers are used.

6. Install the air injection and/or heated air stove, if removed.

7. Install the exhaust pipe to the exhaust manifold flange.

8.0L (VIN W) Engine

1. Disconnect the negative battery cable.

2. Raise and safely support the vehicle.

3. Remove the hardware that attaches the exhaust pipe to the exhaust manifold.

4. Lower the vehicle.

5. Remove the exhaust heat shields.

6. To remove the right exhaust manifold:

   a. Remove the EGR tube. Discard the gasket.

   b. Remove the dipstick bracket from the manifold.

7. Unbolt and remove the exhaust manifold. Discard the gasket.

To install:

8. Clean the surfaces so they are free of old material, oil, grease, etc, Inspect the surfaces for cracks and check flatness with a straightedge and feeler gauge. The mounting surfaces must be flat within 0.008 in./ft. (0.2mm/300mm).

9. Install a new gasket and the exhaust manifold(s). Once installed in the correct order, tighten the stud bolts and bolts to 16 ft. lbs. (22 Nm).

10. For the right side manifold:

   a. Install a new gasket and the EGR tube. Tighten the tube assembly nut to 25 ft. lbs. (34 Nm). Tighten the two EGR nuts to 175 inch lbs. (20 Nm).

   b. Install the dipstick bracket to the manifold.

11. Install the heat shield. Tighten the nuts onto the stud bolts to 175 inch lbs. (20 Nm).

12. Raise and safely support the vehicle.

13. Install the exhaust pipe. Tighten the bolts to 25 ft. lbs. (34 Nm).

14. Lower the vehicle and connect the negative battery cable.

Combination Manifold

REMOVAL AND INSTALLATION

2.5L (VIN K and G) Engine

The "combination manifold" is, in fact, the intake and exhaust manifold intermeshed into one unit.

NOTE: On some vehicles, some of the manifold attaching bolts may not be accessible or are too heavily sealed from the factory. Examine the mounting hardware prior to committing yourself to this procedure. If you find this to be the case, it will be necessary to remove the cylinder head, then separate the manifold from the cylinder head on a workbench.

1. Disconnect the negative battery cable.

2. Drain the cooling system.

CAUTION

When draining the coolant, keep in mind that cats and dogs are attracted by ethylene glycol antifreeze, and are quite likely to drink any that is left in an uncovered container or in puddles on the ground. This will prove fatal in sufficient quantity. Always drain the coolant into a sealable container.

3. Depressurize the fuel system. Remove the air cleaner and disconnect all vacuum lines, electrical wiring and fuel lines from the throttle body and/or manifold.

4. Disconnect the throttle linkage.

5. Loosen the power steering pump and remove the drive belt. Remove the power steering and air pump support bracket.

6. Remove the power brake vacuum hose from the intake manifold.

7. On Canadian models, remove the coupling hose from the diverter valve to the exhaust manifold air injection tube assembly.

8. Remove the water hoses from the water crossover.

9. Raise and safely support the vehicle. Disconnect the exhaust pipe from the exhaust manifold.

10. Remove the power steering pump and set it aside.

11. Remove the intake manifold support bracket, if equipped.

12. Remove the EGR tube.

13. If equipped, remove the air injection tube bolts and the air injection tube assembly.

14. Remove the intake manifold bolts.

15. Lower the vehicle and remove the intake manifold.

16. Remove the exhaust manifold nuts.

17. Remove the exhaust manifold.

To install:

18. Install a new combination manifold gasket lightly coated on the manifold side with Mopar® Gasket Sealer, or equivalent.

19. Set the gasket in place and install and tighten the nuts to 17 ft. lbs. (23 Nm). The tightening should begin in the center and progress outwards in both directions.

20. Repeat the tightening sequence until all nuts are at the specified torque.

21. Install the intake manifold strut. Tighten the bolt to 70 ft. lbs. (95 Nm). Tighten the nut to 40 ft. lbs (54 Nm).

22. Install with a new gasket the EGR tube. Tighten to 17 ft. lbs. (23 Nm).

23. Install the exhaust pipe to the exhaust manifold, tightening to 20 ft. lbs. (27 Nm).

24. Connect the water hose.

25. Install the diverter valve assembly and the air injection tube to the exhaust manifold.

26. Install the power brake vacuum hose to the intake manifold.

27. Install the power steering and air pump support bracket.

28. Install the throttle linkage.

29. Install a new gasket and the throttle body onto the intake manifold. Tighten the bolts to 175 inch lbs. (20 Nm).

30. Install the air cleaner. Connect all the vacuum lines, electrical wiring and fuel lines to the throttle body.

31. Fill the cooling system and connect the negative battery cable.

Front Cover Seal

REMOVAL AND INSTALLATION

3.9L (VIN X) Engine

1. Disconnect the negative battery cable.

2. Remove the belt(s) from the crankshaft pulley.

3. Remove the fan and shroud from the vehicle.

4. Remove the crankshaft pulley.

5. Remove the vibration damper using the proper puller.

6. Using a suitable tool behind the lips of the oil seal, pry outward. Take care not to damage the crankshaft seal surface of the cover.

To install:

7. Install the new seal by installing the threaded shaft part of special tool C–4251, or equivalent, into the threads of the crankshaft.

8. Place the seal into the opening with the spring toward the engine. Place the installing adapter C–4251-3, or equivalent, with the thrust bearing and nut on the shaft. Tighten the nut until the tool is flush with the timing chain cover. Remove the tool.

9. Install the damper with tool C–3638, or equivalent, install the bolt and washer and tighten to specification.

10. Apply a small amount of sealer to the bolts and install the crankshaft pulley.

11. Install the fan and shroud.

12. Connect the negative battery cable and check for leaks.

5.2L (VIN T and Y) and 5.9L (VIN A, Z and 5) Engines

NOTE: A seal remover and installer tool is required to prevent seal damage.

1. Using a seal puller, separate the seal from the retainer.

2. Pull the seal from the case.

To install:

3. Place the seal face down in the case with the seal lips downward.

4. Seat the seal tightly against the cover face. There should be a maximum clearance of 0.0014 in (0.035mm) between the seal and the cover. Be careful not to overcompress the seal.

Front Crankshaft Seal

REMOVAL AND INSTALLATION

2.5L (VIN K and G) Engine

1. Remove the timing belt.

2. Remove the crankshaft sprocket using Special Tool No. C–4685, or equivalent, Insert and a 5.9 in. long screw.

3. Install the crankshaft sprocket using plate L–4524, or equivalent, Thrust Bearing/washer and a 5.9 in. long screw.

4. Hold the engine sprocket with Special Tool No. C–4687, or equivalent (with adapter tool No. –4687–1, or equivalent) while removing/installing the screw.

5. Remove the crankshaft seal using Special Tool No. 6341. Remove the intermediate and camshaft seals, if necessary, with Special Tool No. C–4679, or equivalent.

6. Shaft seal lip surface must be free of varnish, dirt and nicks. Polish with 400 grit sandpaper, if necessary.

7. Install engine crankshaft seal into the retainer using Special Tool No. 6342 and 6343. Install the intermediate and camshaft seals, if removed, with Special tool C–4680. Install the seals until they fit flush.

3.9L, 5.2L (VIN T and Y) and 5.9L (VIN A, Z and 5) Engines

1. Disconnect the negative battery cable.

2. Remove the vibration damper.

3. If the front seal is suspected of leaking, check the front seal alignment to the crankshaft. Fit a Seal Installation/Alignment tool No. 6635, or equivalent. the tool should fit with minimum interference. If the tool does not fit, the cover must be removed and installed properly.

4. Place a suitable tool behind the lips of the oil seal to pry the oil seal outward. Be careful not to damage the crankshaft seal surface of the cover.

5. Place the smaller diameter of the oil seal over tool No. 6635, or equivalent.

6. Position the seal and tool onto the crankshaft.

7. Using the vibration damper, tighten the bolt to draw the seal into position on the crankshaft.

8. Remove the vibration damper bolt and seal installation tool.

9. Inspect the seal flange on the vibration damper.

10. Install the vibration damper.

11. Connect the negative battery cable.

Timing Chain, Sprockets and Front Cover

REMOVAL AND INSTALLATION

3.9L (VIN X) Engine

1. If possible, crank the engine around so that the No. 1 cylinder is at TDC on the compression stroke. Remove the distributor cap to confirm and line the timing mark on the damper pulley with "0" on the timing scale. This will aid in aligning timing marks when installing the timing gears. Disconnect the negative battery cable.

2. Drain the cooling system.

3. Remove the radiator, fan and all related parts. Remove the water pump.

4. Remove the crankshaft pulley.

5. Remove the vibration damper using the proper puller.

6. Unbolt the chain cover from the block and remove, using caution to avoid damaging the oil pan gasket. Remove the fuel pump from the cover, if equipped.

7. Remove the camshaft gear retaining bolt and cup washer. Remove the timing chain and gears. If necessary, remove the chain and upper gear, then use a gear puller to remove the lower gears.

To install:

8. Place both camshaft and crankshaft gears on the bench with the timing marks on the exact imaginary center line through both gear bores as they are installed on the engine.

9. Place the timing chain around both sprockets.

10. Turn the crankshaft and camshaft so the keys line up with the keyways in the gears when the timing marks are in proper position.

11. Slide both gears over their respective shafts and use a straightedge to check timing mark alignment.

12. Install the fuel pump eccentric and cup washer, if equipped. Tighten the camshaft gear retaining bolt to 35 ft. lbs. (47 Nm).

13. Clean and dry the mating surfaces of the timing chain cover and block. Apply a thin bead of sealer to the oil pan gasket.

14. Install a new cover gasket and install the cover. Tighten the bolts to 30 ft. lbs. (41 Nm).

15. Install the water pump with a new gasket, if it was removed.

16. Install the damper with tool C–3638, or equivalent, install the bolt and washer and tighten to specifica-

tion. Apply a small amount of sealer to the bolts and install the crankshaft pulley.

17. Install the fuel pump using a new gasket, if equipped and connect the fuel lines. Install the two oil pan bolts if they were removed.

18. Install the radiator, fan and all related parts.

19. Fill the cooling system.

20. Connect the negative battery cable, set all adjustment to specifications and check for leaks.

5.2L (VIN T and Y) and 5.9L (VIN A, Z and 5) Engines

1. Remove the front cover.

2. Remove the camshaft sprocket lockbolt, securing cup washer, and fuel pump eccentric. Remove the timing chain with both sprockets.

To install:

3. To begin the installation procedure, place the camshaft and crankshaft sprockets on a flat surface with the timing indicators on an imaginary centerline through both sprocket boxes. Place the timing chain around both sprockets. Be sure that the timing marks are in alignment.

——— WARNING ———

When installing the timing chain, have an assistant support the camshaft with a suitable tool to prevent it from contacting the plug in the rear of the engine block. Remove the distributor and the oil pump/distributor drive gear. Position the suitable tool against the rear side of the cam gear and be careful not to damage the cam lobes.

4. Turn the crankshaft and camshaft to align them with the keyway location in the crankshaft sprocket and the keyway or dowel hole in the camshaft sprocket.

5. Lift the sprockets and timing chain while keeping the sprockets tight against the chain in the correct position. Slide both sprockets evenly onto their respective shafts.

6. Use a straightedge to measure the alignment of the sprocket timing marks. They must be perfectly aligned.

7. Install the fuel pump eccentric, cup washer, and camshaft sprocket lockbolt and tighten to 35 ft. lbs. (47 Nm). If camshaft end-play exceeds 0.010 in. (0.25mm), install a new thrust plate. It should be 0.002-0.006 in. (0.05-0.15mm) with the new plate.

8.0L (VIN W) Engine

1. Align the camshaft and crankshaft centerline. Remove the camshaft sprocket attaching bolt and remove the timing chain and camshaft sprockets.

2. Use puller No. 6444 and jaws No. 6920, or equivalent, to pull the crankshaft sprocket.

To install:

3. Align the key in the crankshaft sprocket with the sprocket. Press on the crankshaft sprocket using tools No. C–3688, C–3718 and MB–990799, or their equivalents. Seat the sprocket against the crankshaft shoulder.

4. Turn the crankshaft to line up the timing mark with the crankshaft and camshaft centerline.

5. Put the timing chain on the camshaft sprocket.

6. Align the timing marks and install the chain and camshaft sprocket onto the crankshaft sprocket. Check to see that the timing marks are on the centerline of the crankshaft and camshaft centerline.

7. Install the camshaft bolt. Tighten the bolt to 45 ft. lbs. (61 Nm).

8. Check the camshaft end-play. If a new thrust plate was installed, the clearance should be 0.002–0.006 in. (0.051–0.152mm). If the used thrust plate is installed, the end-play clearance may be up to 0.010 in. (0.254mm).

9. If not within these limits, install a new thrust plate.

Timing Belt, Sprockets, Tensioner and Front Cover

ADJUSTMENT

2.5L (VIN K and G) Engine

1. Disconnect the negative battery.

2. Raise the vehicle and support safely. Remove the right front inner splash shield.

3. Remove the tensioner cover.

4. Place the special tensioning tool C–4703, or equivalent, on the hex of the tensioner so the weight is at about the 10 o'clock position and loosen the bolt.

5. The tensioner should drop down to the 9 o'clock position. Reposition the tool as required in order to have it end up at the 9 o'clock position (parallel to the ground, hanging to the left) plus or minus 15°.

6. Hold the tool in position and tighten the bolt. Do not pull the tool past the 9 o'clock position; this will

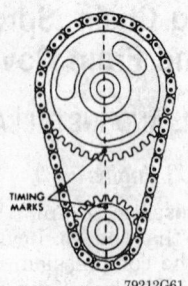

79212G61

Make sure the timing alignment marks are in line as shown — 3.9L (VIN X), 5.2L (VIN T and Y) and 5.9L (VIN A, Z and 5) engines

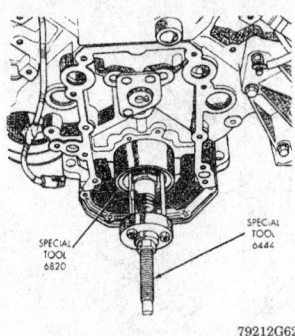

Use the special puller tool to remove the crankshaft sprocket — 8.0L (VIN W) engine

make the belt too tight and will cause it to howl or possibly break.

7. Install the cover and the splash shield.

REMOVAL AND INSTALLATION

2.5L (VIN K and G) Engine

NOTE: If possible, position the engine so that the No. 1 piston is at TDC.

1. Disconnect the negative battery cable.

2. Remove the timing belt covers.

3. Remove the timing belt tensioner and allow the belt to hang free.

4. Remove the air conditioning compressor belt idler pulley, if equipped and remove the mounting stud. Unbolt the compressor/alternator bracket and position it to the side.

5. Remove the timing belt from the vehicle.

To install:

6. Turn the crankshaft sprocket and intermediate shaft sprocket until the marks are in line. Use a straight-edge shaft sprocket until the marks are in line. Use a straightedge from bolt to bolt to confirm alignment.

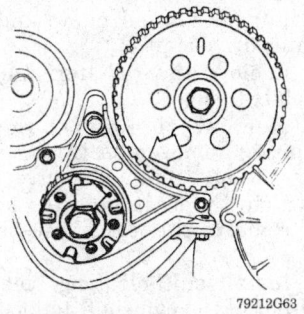

Align the crankshaft and intermediate shaft timing mark as indicated — 2.5L (VIN K and G) engine

7. Turn the camshaft until the small hole in the sprocket is at the top and rows on the hub are in line with the camshaft cap to cylinder head mounting lines. Use a mirror to see the alignment so it is viewed straight on and not at an angle from above. Install the belt, built let it hang free at this point.

8. Install the air conditioning compressor/alternator bracket, idler pulley and motor mount. Remove the floor jack. Raise the vehicle and support safely. Have the tensioner at an arm's reach because the timing belt will have to be held in position with one hand.

9. To properly install the timing belt, reach up and engage it with the camshaft sprocket. Turn the intermediate shaft counterclockwise slightly, then engage the belt with the intermediate shaft sprocket. Hold the belt against the intermediate shaft sprocket and turn clockwise to take up all tension; if the timing marks are out of alignment, repeat until alignment is correct.

10. Using a 13mm wrench, turn the crankshaft sprocket counterclockwise slightly and wrap the belt around it. Turn the sprocket clockwise so there is no slack in the belt between sprockets, if the timing marks are out of alignment, repeat until alignment is correct.

NOTE: If the timing marks are in line but slack exists in the belt between either the camshaft and intermediate shaft sprockets or the intermediate and crankshaft sprockets, the timing will be incorrect when the belt is tensioned. All slack must be only between the crankshaft and camshaft sprockets.

11. Install the tensioner and install the mounting bolt loosely. Place special tensioning tool C–4703, or equivalent, on the hex of the tensioner so the weight is at about the 9 o'clock position (parallel to the ground, hanging to the left) plus or minus 15°.

12. Hold the tool in position and tighten the bolt to 45 ft. lbs. (61 Nm). Do not pull the tool past the 9 o'clock position; this will make the belt too tight and will cause it to howl or possibly break.

13. Lower the vehicle and recheck the camshaft sprocket positioning. If it is correct install the timing belt covers and all related parts.

14. Connect the negative battery cable and road test the vehicle.

Camshaft

REMOVAL AND INSTALLATION

2.5L (VIN K and G) Engine

1. Disconnect the negative battery cable. Relieve the fuel pressure, if equipped with fuel injection.

2. Turn the crankshaft so the No. 1 piston is at the TDC of the compression stroke. Remove the upper timing belt cover. Remove the air pump pulley, if equipped.

3. Remove the camshaft sprocket bolt and the sprocket and suspend tightly so the belt does not lose tension. If it does, the belt timing will have to be reset.

4. Remove the cylinder head cover.

5. If the rocker arms are being reused, mark them for installation identification and loosen the camshaft bearing bolts evenly and gradually.

6. Using a soft mallet, tap the rear of the camshaft a few times to break the bearing caps loose.

7. Remove the bolts, bearing caps and the camshaft with seals.

NOTE: Before replacing the camshaft, identify factory installed oversized components. To do so, look for the tops of the bearing caps painted green and O/SJ stamped rearward of the oil gallery plug on the rear of the head. In addition, the barrel of the camshaft is painted green and O/SJ is stamped onto the rear end of the camshaft. Installing standard sized parts in a head equipped with oversized parts — or vice versa — will cause severe engine damage. Also, take note of the color of the paint stripe on the rear camshaft seal. These stripes differentiate seal sizes. If a seal with a different color stripe is installed, a severe leak will develop if the seal is too small, or the cap will not be able to be fully installed if the seal is too big.

8. Check the oil passages for blockages and the parts for wear and damage and replace parts, as required. Clean the gasket mounting surfaces.

To install:

9. Transfer the sprocket key to the new camshaft. New rocker arms and a new camshaft sprocket bolt are normally included with the camshaft package. Install the rocker arms, lubricate the camshaft and install with end seals installed.

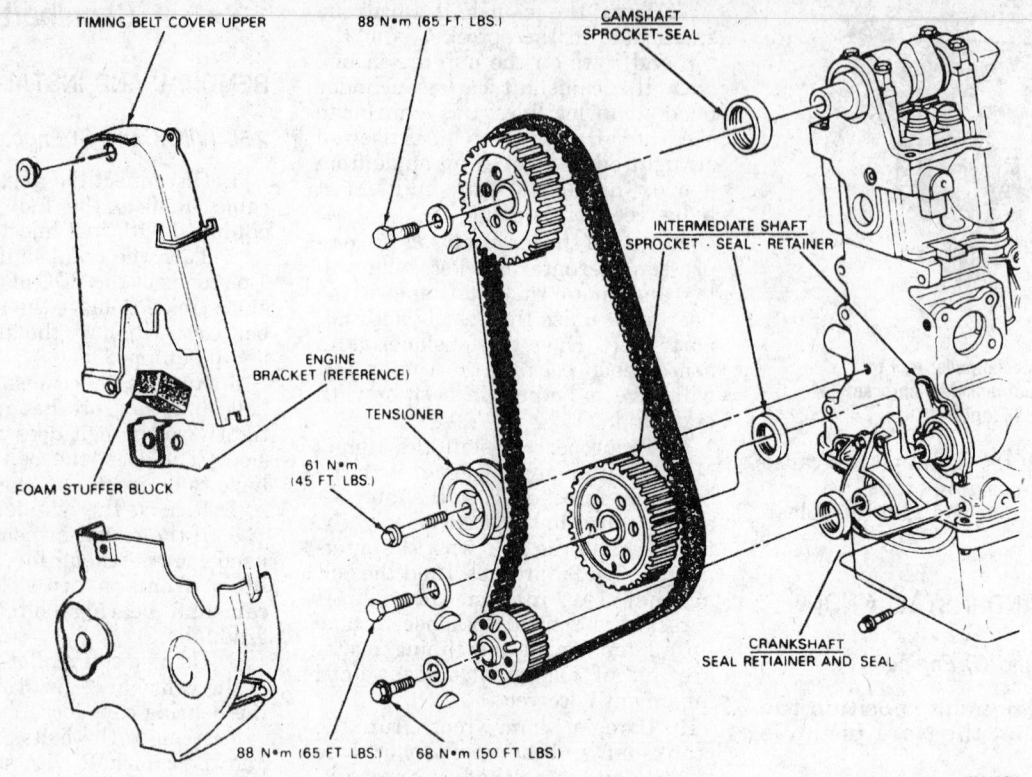

Exploded view of the timing belt, sprockets, cover and seals — 2.5L (VIN K and G) engine

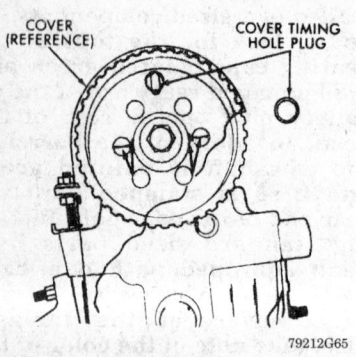

Be careful to properly align the camshaft timing marks — 2.5L (VIN K and G) engine

10. Place the bearing caps with No. 1 at the timing belt end and No. 5 at the transaxle end. The camshaft bearing caps are numbered and have arrows facing forward. Tighten the camshaft bearing bolts evenly and gradually to 18 ft. lbs. (24 Nm).

NOTE: Apply RTV silicone gasket material to the No. 1 and 5 bearing caps. Install the bearing caps before the seals are installed.

11. Mount a dial indicator to the front of the engine and check the

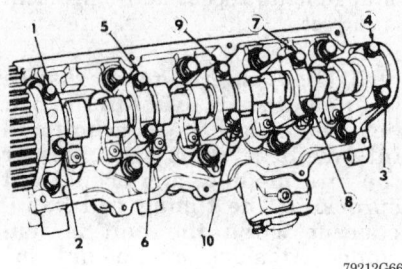

Camshaft cap bolt tightening sequence — 2.5L (VIN K and G) engine

camshaft end-play. Play should not exceed 0.006 in. (0.15mm).

12. Install the camshaft sprocket and the new bolt. Install the air pump pulley, if equipped.

13. Install the cylinder head cover with a new gasket.

14. Connect the negative battery cable and check for leaks.

3.9L (VIN X) Engine

1. If possible, crank the engine around so that the No. 1 cylinder is at TDC on the compression stroke. Remove the distributor cap to confirm

and line the timing mark on the damper pulley with "O" on the timing scale. This will aid in aligning timing marks when installing the timing gears.

2. If the vehicle is equipped with fuel injection, relieve the fuel pressure. Disconnect the negative battery cable.

3. Drain the cooling system.

4. Remove the cylinder head cover(s).

5. Remove the rocker shaft assemblies. Identify and remove the pushrods.

6. Remove the intake manifold. Identify and remove all lifters.

7. Remove the distributor.

8. Lift out the oil pump and distributor driveshaft.

9. Remove the radiator, fan and all related parts.

10. Remove the fuel pump, if equipped. Remove the timing chain cover, timing chain and gears.

11. Note the location of the oil tab and remove the camshaft thrust plate.

12. Install suitable long bolt into the front of the camshaft to facilitate removal. Remove the camshaft, being careful not to damage any of the cam bearings with the cam lobes.

To install:

13. Install the camshaft to within 2 in. (50mm) of its final installation position.

14. Install the camshaft blocking tool C–3509, or equivalent, and bolt it in place with the distributor hold-down bolt. This will prevent the camshaft from being pushed in too far and knocking out the welch plug at the rear of the block. This tool should remain in place until the timing chain installation has been completed.

15. Install the camshaft thrust plate and chain oil tab. Make sure the tang of the oil tab enters the hole in the thrust plate at the lower right. Tighten the bolts to 18 ft. lbs. (24 Nm). Make sure the top edge of the toil tab is flat against the thrust plate or it will not feed oil to the chain.

16. Place both camshaft and crankshaft gears on the bench with the timing marks on the exact imaginary center link through both gear bores as they are installed on the engine. Place the timing chain around both sprockets.

17. Turn the crankshaft and camshaft so the keys line up with the keyways in the gears when the timing marks are in proper position.

18. Slide both gears over their respective shafts and use a straightedge to check timing mark alignment.

19. Install the fuel pump eccentric and cup washer, if equipped. Tighten the camshaft gear retaining bolt to 35 ft. lbs. (47 Nm).

20. Remove the camshaft blocking tool, if it was installed.

21. Measure camshaft end-play, if applicable. Replace the thrust plate if not within specifications.

22. Coat the oil pump and distributor driveshaft with oil. Install the shaft so that when the gear spirals into place and drops into the oil pump, the slot in the top of the gear

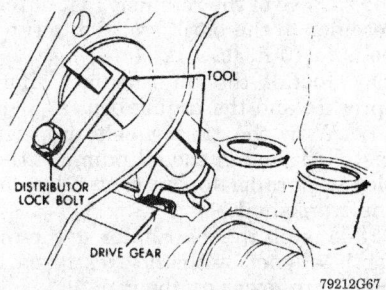

Use the special holding tool to keep the camshaft in position during installation — 3.9L (VIN X), 5.2L (VIN T and Y) and 5.9L (VIN A, Z and 5) engines

is pointing directly to the left front intake manifold bolt hole.

23. If the camshaft was not replaced, lubricate and install the lifters in their original locations. If the camshaft was replaced, new lifters must be used.

24. Install the pushrods and rocker shaft assemblies.

25. Install the intake manifold, if it was removed. Install the cylinder head covers.

26. Install the distributor so the rotor points to the No. 1 spark plug wire position on the cap.

27. Install the timing chain cover and all related parts.

28. Install the fuel pump if equipped and radiator.

29. When everything is bolted in place, change the engine oil and replace the oil filter.

NOTE: If the camshaft or lifters have been replaced, add one pint of Mopar crankcase conditioner, or equivalent when replenishing the oil to aid in break in. This mixture should be left in the engine for a minimum of 500 miles (805 km) and drained at the next normal oil change.

30. Fill the radiator with coolant.

31. Connect the negative battery cable, set all adjustments to specifications and check for leaks.

5.2L (VIN T and Y) and 5.9L (VIN A, Z and 5) Engines

1. Drain the cooling system and disconnect the negative battery cable.

2. Remove the intake manifold, cylinder head covers, rocker arm assemblies, pushrods, and valve tappets, keeping them in order to insure the installation in their original locations.

3. Remove the timing gear cover, the camshaft and the crankshaft sprockets, and the timing chain.

4. Remove the distributor and lift out the oil pump and distributor driveshaft.

5. If the distributor driveshaft bushing is excessively worn, it must be pulled with a threaded inserting puller tool as follows:

    a. Insert distributor driveshaft bushing puller tool C–3052, or equivalent, into the old bushing and thread it in until the fit is tight.

    b. Turn the puller nut while holding the screw with a wrench and pull the bushing out.

6. Remove the camshaft thrust plate.

7. Install a long bolt into the front of the camshaft and remove the camshaft, being careful not to damage the cam bearings with the cam lobes.

To install:

NOTE: Prior to installation, lubricate the camshaft lobes and bearings journals. It is recommended that one pint of Mopar® Crankcase Conditioner be added to the initial crankcase oil fill. Insert the camshaft into the engine block within 2 in. (51mm) of its final position in the block. Have an assistant support the camshaft with a suitable tool to prevent the camshaft from contacting the plug in the rear of the engine block. Position the suitable tool against the rear side of the cam gear and be careful not to damage the cam lobes.

8. Install the camshaft thrust plate. If camshaft end-play exceeds 0.010 in. (0.25mm), install a new thrust plate. It should be 0.002–0.006 in (0.15mm) with the new plate.

9. Install the timing chain and sprockets, timing gear cover, and pulley.

10. Install the tappets, pushrods, rocker arms, and cylinder head covers. Install fuel pump, if removed.

11. If the distributor driveshaft bushing was removed, install a new one as follows:

    a. Slide the new bushing over the burnishing end of the special Driver/Burnisher tool No. C–3053, or equivalent, and insert the tool and bushing into the bore.

    b. Use a hammer to drive the bushing and tool into position.

    c. As the burnisher tool is pulled through the bushing, the bushing will be expanded tight in the block and burnished to the correct size.

——— **WARNING** ———
Bushing installation must be performed as described. Do NOT attempt to ream the bushing. Failure to follow these instructions could result in distributor shaft seizure!

12. Install the distributor and oil pump driveshaft. If necessary, install a new bushing.

13. Install the distributor.

14. Install the remaining components which were removed, in reverse order of their removal.

15. Refill the cooling and reconnect the negative battery cable.

16. After starting the engine, adjust the ignition timing.

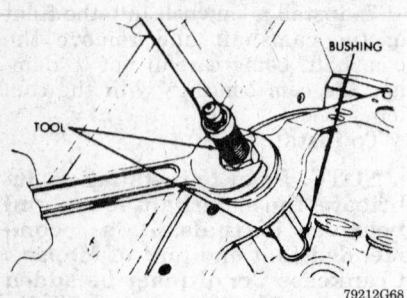

Burnish the bushing using the special tool as shown — 5.2L (VIN T and Y) and 5.9L (VIN A, Z and 5) engines

8.0L (VIN W) Engine

1. Drain the cooling system and disconnect the negative battery cable.
2. Remove the cylinder head covers, rocker arm assemblies, pushrods, and valve tappets, keeping them in order to insure the installation in their original locations.

NOTE: Keep all tappets in order for reassembly in the exact same position as when removed. The four corner tappets cannot be removed without removing the cylinder heads and gaskets. However they can be lifted and retained for camshaft removal.

3. Remove the upper and lower intake manifold.
4. Remove the timing chain cover, timing chain, the timing chain thrust plate and the timing chain.
5. Remove the distributor and lift out the oil pump and distributor driveshaft.
6. Install a long bolt into the front of the camshaft and remove the camshaft, being careful not to damage the cam bearings with the cam lobes.
To install:

NOTE: Prior to installation, lubricate the camshaft lobes and bearings journals. It is recommended that one pint of Mopar® Crankcase Conditioner be added to the initial crankcase oil fill. Insert the camshaft into the engine block within 2 in. (51mm) of its final position in the block. Have an assistant support the camshaft with a suitable tool to prevent the camshaft from contacting the plug in the rear of the engine block. Position the suitable tool against the rear side of the cam gear and be careful not to damage the cam lobes.

7. Install the camshaft thrust plate. If camshaft end-play exceeds 0.010 in. (0.254mm), install a new thrust plate. Normal tolerance for a new plate should be 0.051–0.152 in. (0.002–0.006mm).
8. Line up the key with the keyway in the sprocket and press on the crankshaft timing sprocket using special tool Nos. C–3688, C–3718 and MB990799, or equivalent, to seat the sprocket against the crankshaft shoulder.
9. Align the timing mark on the crankshaft sprocket with the centerline of the crankshaft and camshaft.
10. Put the timing chain on the camshaft sprocket.
11. Take the chain and camshaft sprockets and align the mark with the centerline of the crankshaft and camshaft install camshaft sprocket and chain to camshaft.
12. Install the camshaft bolt. Tighten the bolt to 55 ft. lbs. (75 Nm).
13. Install the timing chain cover.
14. Install the crankshaft pulley/damper using tool C7–3688, or equivalent.
15. Prime the oil pump by squirting oil in the oil filter mounting hole and filling the J-trap of the front timing cover. Fill a new oil filter with fresh engine oil and install quickly just as the oil starts to come out.
16. Each tappet reused must be installed in the same position from which it was removed.

NOTE: When the camshaft is replaced, all of the tappets must be replaced.

17. Install the tappets and pushrods in their original location.
18. Install the rocker arms.

NOTE: The cylinder head cover gasket can be reused. For the left side, the number tab is at the front of the engine with the number up. For the right side the number tab is at the rear of the engine.

——— **WARNING** ———
The cylinder head cover fasteners have a special plating and should not be substituted for another fastener.

19. Position the cylinder head cover onto the gasket. Install the stud bolts and hex head bolts in the proper positions. Tighten the stud bolts and the bolts to 144 inch lbs. (16 Nm).
20. Install the intake manifolds.
21. Refill the cooling and reconnect the negative battery cable.

Intermediate Shaft

REMOVAL AND INSTALLATION

2.5L (VIN K and G) Engine

1. Disconnect the negative battery cable.
2. Crank the engine around until the No. 1 piston is at TDC. Remove the timing belt covers to confirm that all timing marks are lined up.
3. Remove the fuel pump, if equipped. Remove the distributor. Looking down at the oil pump, the slot in the shaft must be paralleled with the center line of the crankshaft. Remove the oil pump.
4. Remove the timing belt and the intermediate shaft sprocket.
5. Remove the shaft retainer bolts and remove the retainer from the block.
6. Remove the intermediate shaft from the engine.
7. If necessary, remove the front bushing using tool C–4697–2 and the rear bushing using tool C–4686–2, or their equivalents.
To install:
8. Install the front bushing using tool C4697–1 until the tool is flush with the block. Install the rear bushing using tool C–4686–1 until the tool is flush with the block.
9. Lubricate the distributor drive gear and install the intermediate shaft.
10. Replace the seal in the retainer and apply silicone sealer to the mating surface of the retainer. Install the retainer to the block and tighten the bolts to 10 ft. lbs. (12 Nm).
11. Install the intermediate shaft sprocket and the timing belt.
12. With the timing belt properly installed, install the oil pump so the slot is parallel to the center line of the crankshaft.
13. Install the distributor so the rotor is aligned with the No. 1 spark plug wire tower on the cap.
14. Install the fuel pump, if equipped.
15. Connect the negative battery cable and road test the vehicle.

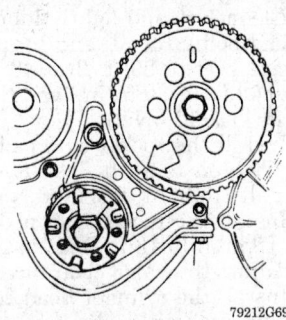

79212G69

Align the crankshaft and intermediate shaft timing marks as shown — 2.5L (VIN K and G) engine

Piston and Connecting Rod

POSITIONING

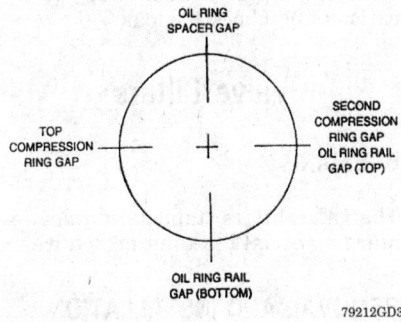

79212GD3

Install the rings so the gaps are spaced as shown — gasoline engines

DIESEL ENGINE MECHANICAL

Engine Assembly

REMOVAL AND INSTALLATION

── CAUTION ──

This engine has a dry weight of 880 lbs. (400 kg). Make sure the engine removal equipment is rated adequately, or serious personal injury could result.

1. Scribe hood hinge outlines on the hood and remove the hood from the vehicle.
2. Disconnect the negative cable from the battery and from the engine.
3. Drain the coolant.

── CAUTION ──

When draining the coolant, keep in mind that cats and dogs are attracted by ethylene glycol antifreeze, and are quite likely to drink any that is left in an uncovered container or in puddles on the ground. This will prove fatal in sufficient quantity. Always drain the coolant into a sealable container.

4. Remove the radiator, shroud, belt, fan and all related parts.
5. Remove the intake and exhaust pipes from the turbocharger.
6. Disconnect the air conditioner connections. Cover the openings on the compressor.
7. Disconnect the alternator and all other electrical connection to the engine.
8. Disconnect the accelerator linkage.
9. Disconnect the throttle linkage from the control lever, but do not remove the control lever from the injection pump.
10. Disconnect all engine driven accessories.
11. Raise the vehicle and support safely. Remove the starter.
12. If equipped with automatic transmission, remove the torque converter bolts and remove the lower bell housing bolts. If equipped with manual transmission, remove the transmission.
13. Drain the oil from the engine.

── CAUTION ──

The EPA warns that prolonged contact with used engine oil may cause a number of skin disorders, including cancer! You should make every effort to minimize your exposure to used engine oil. Protective gloves should be worn when changing the oil. Wash your hands and any other exposed skin areas as soon as possible after exposure to used engine oil. Soap and water, or waterless hand cleaner should be used.

14. Disconnect the transmission oil cooler lines from their brackets, if equipped.
15. Disconnect the exhaust pipe from the turbocharger. Lower the vehicle.
16. Disconnect and plug the fuel lines.
17. Using the proper equipment, hoist the engine slightly using the lifting eyes and support the transmission, if still installed.
18. Remove the motor mounts.

19. Remove the upper bell housing bolts.
20. Remove the engine from the vehicle.

To install:

21. Position the engine in the engine compartment and install the motor mounts. Tighten the nuts and bolts to 57 ft. lbs. (77 Nm).
22. Install the bell housing bolts and torque converter bolts, if equipped. Install the manual transmission, if equipped.
23. Install the starter.
24. Connect the exhaust pipe.
25. Connect the transmission oil cooler lines to their brackets, if equipped.
26. Connect the fuel lines.
27. Connect the power steering lines.
28. Connect all engine driven accessories.
29. Connect the accelerator linkage.
30. Connect the throttle linkage to the control lever.
31. Connect the air conditioner connections.
32. Connect the alternator and all other electrical connections to the engine.
33. Install the intake and exhaust pipes to the turbocharger.
34. Install the fan and all related parts, shroud and radiator.
35. Fill the engine with the proper amount of diesel engine oil.
36. Fill the radiator with coolant.
37. Connect the negative battery cable, set all adjustments to specifications and check for leaks.

Engine Mounts

REMOVAL AND INSTALLATION

1994–97 Engines

Front Mounts

1. Disconnect the negative battery cable.
2. If necessary, position the fan to assure clearance for the radiator top tank and hose.
3. Install a suitable engine hoist or use a floor jack to SLIGHTLY raise the engine.
4. Remove the thru-bolt and nut.
5. Remove the engine mount heat shields.

To install:

6. With the engine raised SLIGHTLY, position the support bracket/cushion and heat shields to the block. Install new bolts and tighten to 140 ft. lbs. (189 Nm).

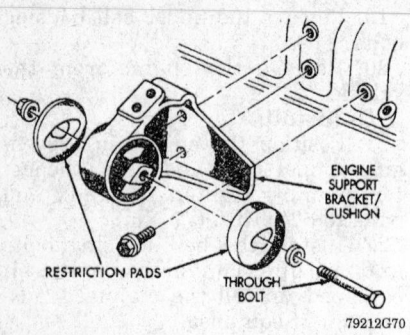

Exploded view of the front engine mounts — 1994–96 diesel engines

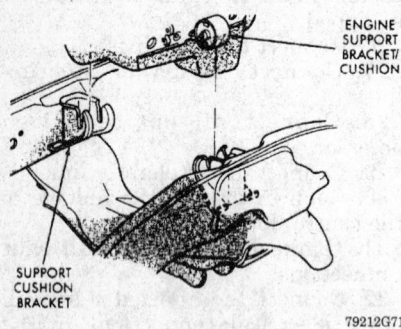

Position the front engine mounts as shown — 1994–96 diesel engines

7. Install the thru-bolt into the engine support bracket/cushion.

8. Lower the engine while guiding the bracket/cushion and thru-bolt into the support cushion brackets.

9. Install the thru-bolt nuts and tighten them to 50 ft. lbs. (68 Nm).

10. Lower the vehicle and remove the hoist or jack.

Rear Mount

1. Raise and safely support the vehicle.

2. Position a transmission jack in place.

3. Remove the support cushion stud nuts.

4. Raise the rear of the transmission and engine SLIGHTLY.

5. Remove the bolts holding the support cushion to the transmission support bracket. Remove the support cushion.

6. If necessary, remove the bolts holding the support bracket to the transmission.

To install:

7. If removed, position the transmission support to the transmission. Install new attaching bolts and tighten them to 65 ft. lbs. (88 Nm).

8. Position the support cushion to the transmission support bracket. In-

stall the stud nuts and tighten them to 30 ft. lbs. (41 Nm).

9. Using the transmission jack, lower the transmission and support cushion onto the crossmember.

10. Install the support cushion bolts and tighten them to 35 ft. lbs. (47 Nm).

11. Remove the transmission jack and lower the vehicle.

Cylinder Head

REMOVAL AND INSTALLATION

1. Disconnect the negative battery cable.

2. Drain the coolant.

3. Disconnect the radiator hose and heater hoses.

4. Remove the turbocharger and air crossover.

5. Remove the exhaust manifold.

6. Remove all fuel lines from the injection pump and injector nozzles. Remove the fuel filter.

7. Remove the cylinder head covers.

8. Remove the rocker arms and pushrods.

9. Unbolt the cylinder head from the block. Remove the cylinder head.

10. Inspect the coolant passages. A large accumulation of rust or lime will require service to the block.

11. Inspect the surface of the head for flatness. The maximum variation is 0.0004 in. (0.010mm) within any 2 in. (50mm) diameter area or 0.012 in. (0.30mm) overall end-to-end or side-to-side.

To install:

12. Thoroughly clean and dry the mating surfaces of the head and block. Position the new head gasket on the dowels.

13. Install the head onto the dowels on the block.

14. Lubricate the pushrod sockets and install the pushrods and rocker arms.

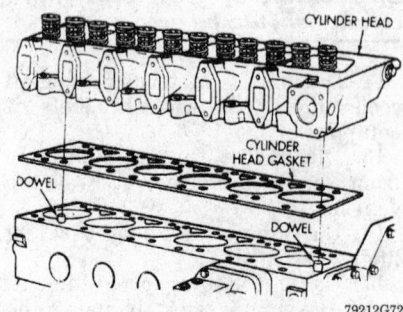

Align the cylinder head gasket with the dowels as shown — 5.9L (VIN C and 8) diesel engine

15. Clean, dry and lightly lubricate the head bolts. Install and torque in sequence first to 29 ft. lbs. (40 Nm), then to 62 ft. lbs. (85 Nm) and finally to 93 ft. lbs. (126 Nm).

16. Install the rocker arm pedestal bolts. Tighten to 18 ft. lbs. (85 Nm) and finally to 93 ft. lbs. (126 Nm).

17. Install the rocker arm pedestal bolts. Tighten to 18 ft. lbs. (24 Nm).

18. Adjust the valve clearance.

19. Install the cylinder head covers with new gaskets. Tighten the bolts to 18 ft. lbs. (24 Nm).

20. Install all fuel lines and the fuel filter.

21. Install the exhaust manifold.

22. Install the turbocharger and air crossover.

23. Connect the radiator hose and heater hoses.

24. Fill the radiator with coolant.

25. Connect the negative battery cable, set all adjustments to specifications and check for leaks.

Valve Lifters

BLEEDING

The valve lifters (tappets) do not require a special bleeding procedure.

REMOVAL AND INSTALLATION

NOTE: The lifters for the Cummins diesel engine are known as "tappets".

1. Remove the camshaft.

2. Insert a trough the full length of the cam bore. Cummins Tappet Changing Tool No. 3822513, or equivalent should be used for this task.

3. Make sure the trough is positioned so it will catch the tappet when the wooden dowel is removed.

4. Identify the location of each tappet as it is removed. The tappets must be installed in their original position.

5. Remove one tappet at a time. Remove the rubber band from the two companion tappets, securing the tappet not to be removed with the rubber band.

6. Pull the wooden dowel from the tappet bore allowing the tappet to fall into the trough.

7. Normally the tappet will fall over when it drops into the trough. Use a flashlight to determine this. If the tappet does not fall over, shake the trough gently to cause it to do so.

8. Take special care when removing the No. 6 cylinder tappets. DO

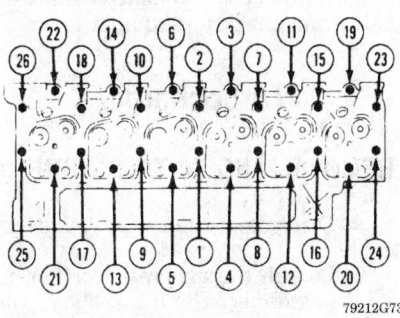

Cylinder head tightening sequence — 5.9L (VIN C and 8) diesel engine

NOT knock or shake the tappet over the end barrier of the trough.

9. Carefully pull the trough and tappet from the cam bore and remove the tappet. Repeat the process until all tappets are removed.

10. Inspect the tappet socket, stem and face for excessive wear, cracks and other damage. The minimum tappet stem diameter is 0.0627 in. (15.925mm). If any tappet is out of limit, replace it.

To install:

11. Insert the trough the full length of the cam bore.

12. Feed the installation tool down the tappet bore and into the trough.

13. Feed the installation tool cord through the cam bores. Carefully pull the trough and installation tool out the front. The barrier at the rear of the trough will assure the tool will be pulled out with it.

14. Lubricate the tappets with a suitable assembly lube, such as Lubriplate 105, or equivalent.

15. Insert the installation tool into the tappet. to aid in removing the installation tool after the tappets are installed, work the tool in and out of the tappet several times before installing the tappets.

16. Press the tappet and tool in the trough and slide the trough back into the cam bore.

17. Pull the tool/tappet through the cam bore and up into the tappet bore.

NOTE: It may be difficult to get the tappet to make the bend from the trough up to the tappet bore (due to the internal webbing in the block). If so, pull the trough out enough to allow the tappet to drop down and align itself. Now pull the tappet up into the bore carefully.

18. After the tappet has been pulled up into position, slide the trough back into the cam bore and rotate it ½ turn. This will position the round

side of the trough up, which will hold the tappet in place.

19. Remove the installation tool from the tappet.

20. Install a wooden dowel into the top of the tappet and secure it with a rubber band.

21. Repeat this process for the rest of the tappets.

22. Install the camshaft.

Valve Lash

ADJUSTMENT

Valve adjustment is required on the diesel engine every 36,000 miles (57,971 km).

NOTE: The timing pin is used in this procedure to locate Top Dead Center (TDC). It is found at the back of the gear housing and below the injection pump. Be sure to disengage the timing pin after locating TDC.

——————— WARNING ———————
Do not set the valve lash closer than specified in an attempt to quiet the lifters. This will only result in burned valves.

1. The engine must be cold for this adjustment (below 140°F/60°C).

2. Remove the valve cover.

3. Manually turn the engine and use the timing pin to locate Top Dead Center (TDC) for cylinder No. 1. Disengage the timing pin after locating TDC.

4. Perform the following two-step procedure to adjust the valves. Refer to the accompanying illustrations to determine which valves to adjust in each of the steps:

a. Valve lash is measured between the rocker arm and the end of the valve. Check the lash by inserting a feeler gauge between the rocker arm and the valve. Adjust

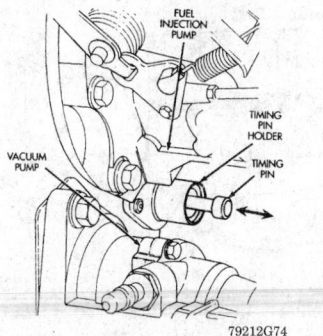

Use the timing pin to locate TDC for diesel engines

the clearance, if necessary. The clearance for the intake valves is 0.010 in. (0.254mm). The clearance for the exhaust valves is 0.20 in. (0.508mm). Adjust the lash by loosening the locknut on the rocker arm and turning the adjusting screw. After the adjustment is made, tighten the locknut to 18 ft. lbs. (24 Nm) and recheck the lash to be sure it did not change as the locknut was being tightened.

b. Double-check that the timing pin is disengaged, then mark the pulley and turn the engine crankshaft 360° in the normal direction of rotation (clockwise). Check and adjust the clearance of the indicated valves following the same specifications as in Step 4a.

5. Install the rocker cover with a new gasket, then attach the fuel line to the injector. Start the engine and check for leaks.

Rocker Arms/Shafts

REMOVAL AND INSTALLATION

1. Disconnect the negative battery cable.

2. Remove the cylinder head cover.

3. Loosen the adjusting screw locknuts. Loosen the screws until they stop.

4. Remove the 8mm bolt and 12mm bead bolt from the pedestal.

5. Remove the pedestal and rocker arm assembly. Remove the pushrods if necessary.

6. Remove the retaining ring and thrust washer.

7. Remove the rocker arm from the pedestal.

NOTE: Do not disassemble the rocker shaft and pedestal; they must be replaced as an assembly.

8. Remove the locknut and adjusting screw from the rocker arm.

To install:

9. Install the adjusting screw and locknut.

10. Lubricate the shaft with oil and install the rocker arm to the shaft. Install the thrust washer and snapring.

11. Install the pushrods to the engine if they were removed.

12. Install the pedestal and rocker arm assembly to the head aligning the dowel in the pedestal with the dowel bore in the head. If the pushrod is holding the pedestal off of the head, turn the engine until the pedestal will set on the head without interference.

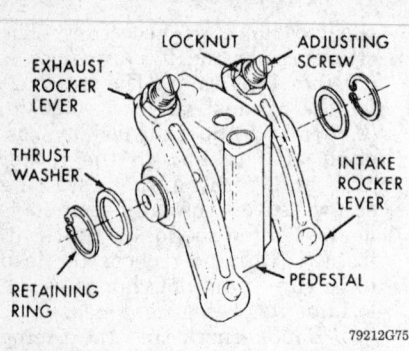

Details of the rocker arm components — 5.9L (VIN C and 8) diesel engine

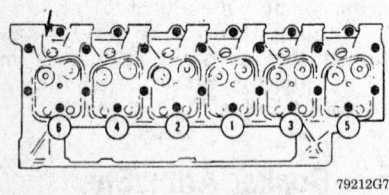

Rocker arm bolt tightening sequence — 5.9L (VIN C and 8) diesel engine

13. Lubricate the threads of the bolt with oil. Install and torque first to 29 ft. lbs. (40 Nm), then to 62 ft. lbs. (85 Nm) and finally to 93 ft. lbs. (126 Nm). If all of the pedestals were removed, follow the entire head bolt torque sequence including those head bolts that were not removed in this procedure.

14. Tighten the 8mm bolts of 18 ft. lbs. (24 Nm).

15. Adjust the valves.

16. Install the cylinder head cover with a new gasket. Tighten the bolts to 18 ft. lbs. (24 Nm).

17. Connect the negative battery cable.

Intake Manifold

REMOVAL AND INSTALLATION

1. Disconnect the negative battery cable.

2. Remove the throttle control bracket and linkage.

3. Remove the high pressure fuel lines.

4. Disconnect the intake manifold heater.

5. Disconnect the fuel heater ground wire from the intake manifold.

6. Remove the air crossover tube and the intake manifold heater.

7. Remove the manifold cover and gasket. Clean the gasket sealing surface.

To install:

8. Install the new gasket and cover. Some of the bolt holes are drilled through. Apply liquid Teflon® sealant to these bolts. Torque all bolts to 18 ft. lbs. (24 Nm).

9. Assemble all intake piping and intake manifold heater with the throttle control bracket and linkage.

10. Connect the fuel heater ground wiring.

11. Install and bleed the high pressure fuel lines.

12. Connect the negative battery cable.

Exhaust Manifold

REMOVAL AND INSTALLATION

1. Disconnect the negative battery cable.

2. Disconnect the air intake hose and exhaust pipe from the turbo.

3. Remove the turbocharger and gasket.

4. Remove the cab heater supply and return lines.

5. Remove the exhaust manifold and gasket. Clean the gasket sealing surface.

To install:

6. Install the new gasket and manifold. Tighten the exhaust manifold bolts in sequence to 32 ft. lbs. (43 Nm).

7. Install the cab heater supply and return lines.

8. Install the turbocharger and gasket.

9. Connect the air intake and exhaust piping.

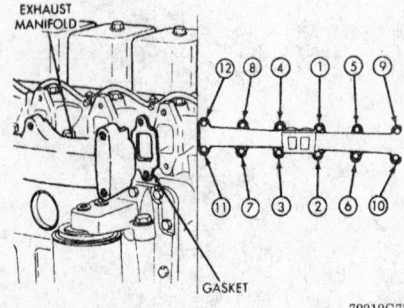

Tightening sequence for the diesel engine exhaust manifold

10. Connect the negative battery cable and check for exhaust leaks.

Turbocharger

REMOVAL AND INSTALLATION

1. Disconnect the negative battery cable.

2. Loosen the air crossover hose.

3. Disconnect the intake hose and exhaust pipe.

4. Remove the oil drain tube bolts.

5. Remove the oil supply line from the turbocharger.

6. Remove the turbocharger mounting nuts and remove the turbocharger.

To install:

7. Inspect the mounting surface for cracks and damage.

8. Install a new gasket and apply anti-seize compound to the mounting studs.

9. Install the turbocharger and tighten the mounting nuts to 24 ft. lbs. (32 Nm).

10. Install the air crossover hose.

11. Install a new gasket and install the oil drain tube. Tighten the bolts to 18 ft. lbs. (24 Nm).

12. New turbochargers must be pre-lubricated with fresh engine oil before operation. To do so, pour about 2 or 3 oz. (65–100 ml) of oil into the supply fitting and rotate the turbine wheel to circulate the oil.

13. Install the oil supply line to the turbocharger.

14. Connect the intake and exhaust piping.

15. Connect the negative battery cable and check for exhaust leaks.

Front Cover Seal

REMOVAL AND INSTALLATION

1. Disconnect the negative battery cable.

2. Remove the drive belt.

3. Remove the crankshaft pulley.

4. Drill two 1/8 in. (3mm) holes into the seal face, 180° apart. Drill two 1/8 in. (3mm) holes into the seal face, 180° apart.

5. Using a slide hammer with a No. 10 sheet metal screw, pull the seal out alternating from side-to-side until the seal is out.

To install:

6. Thoroughly clean and dry the end of the crankshaft.

7. Apply a bead of Loctite® 277 to the outside diameter of the seal.

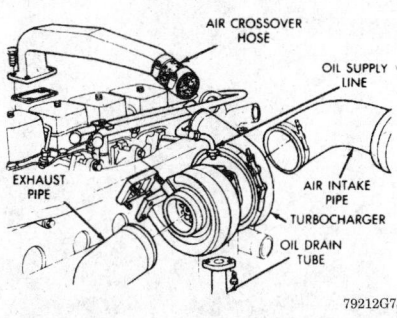

Exploded view of the turbocharger mounting — 5.9L (VIN C and 8) diesel engine

8. Install the pilot from the seal kit onto the crankshaft.

9. Install the seal onto the pilot and start it into the front cover seal bore. Remove the pilot.

10. Use the alignment/installation tool and a plastic hammer to fully install the seal.

11. Install the crankshaft pulley, but do not tighten the bolts at this point.

12. Install the drive belt.

13. Tighten the crankshaft pulley bolts to 92 ft. lbs. (125 Nm).

14. Connect the negative battery cable and check for leaks.

Front Crankshaft Seal

REMOVAL AND INSTALLATION

1. Remove the drivebelt.

2. Remove the vibration damper.

3. Drill two ⅛ in. holes into the seal face, 180° apart.

4. Use a slide hammer tool with a #10 metal screw. Pull alternating from side-to-side until the seal is free.

To install:

5. Make sure the sealing surface on the crankshaft is completely free of all oil residue and debris to prevent seal leaks.

6. If the gear cover was replaced, use the alignment tool from the seal kit to make sure the cover is aligned with the crankshaft.

7. Apply a bead of Loctite® 277, or equivalent, to the outside diameter of the seal.

8. Install the pilot from the seal kit onto the crankshaft.

9. Install the seal onto the pilot and start it into the gear housing cover seal bore.

10. Remove the pilot.

11. Use the alignment/installation tool and a plastic hammer to install the seal to the correct depth.

12. Install the vibration damper, but DO NOT tighten the damper bolt until the belt is installed.

13. Install the drive belt.

14. Tighten the vibration damper bolts to 92 ft. lbs. (125 Nm). Use an engine barring tool to keep the engine from rotating during the tightening procedure.

Timing Gears

REMOVAL AND INSTALLATION

1. Remove the timing gear cover to gain access to the timing gears.

2. Simply unbolt the gears and slide them off their respective shafts. See the illustration for correct timing gear alignment.

Camshaft

REMOVAL AND INSTALLATION

1. Disconnect the negative battery cable.

2. Remove the cylinder head covers.

3. Remove the rocker pedestal and arm assemblies.

4. Remove the pushrods.

5. Remove the drive belt.

6. Drain the cooling system. Remove the fan assembly, radiator and all related parts.

7. Remove the crankshaft pulley.

8. Remove the front gear cover.

9. Remove the fuel pump.

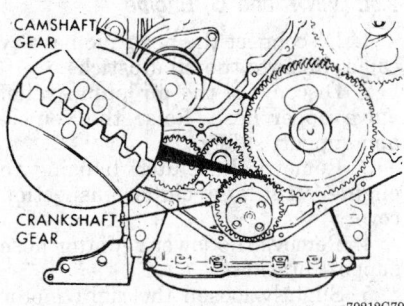

Timing gear alignment — 5.9L (VIN C and 8) diesel engine

10. Insert the special dowels into the pushrod holes and onto the top of each lifter. When properly installed, the dowels can be used to hold the tappets up securely. Wrap rubber bands around the top of the dowels to prevent them from dropping down.

11. Rotate the crankshaft to align the crankshaft to camshaft timing marks.

12. Remove the bolts from the thrust plate.

13. Remove the camshaft and thrust plate.

14. Press the gear from the camshaft and remove the key.

To install:

15. Install the key on the camshaft.

16. Heat the camshaft gear to 250°F (121°C) for 45 minutes. Lubricate the gear mount surface with Lubriplate 105®. Install the gear to the camshaft with the timing marks facing away from the shaft.

17. Lubricate the camshaft bores, lobes, journals and thrust washer with Lubriplate 105®.

NOTE: Do not push the camshaft in too far or it may dislodge the plug in the rear of the camshaft bore, possibly creating a leak.

18. Install the camshaft and thrust washer so the "E" timing mark on the injection pump gear aligns with the "C" timing mark on the camshaft gear and the timing mark on the crankshaft gear align with those on the camshaft gear.

19. Install the thrust washer bolts and tighten to 18 ft. lbs. (24 Nm).

20. Check the end-play of the camshaft. The specification is 0.006-0.010 in. (0.152-0.254mm).

21. Check the backlash of the camshaft gear. The specification is 0.003-0.013 in. (0.080-0.330mm).

22. Install the tappets and pushrods.

23. Install the rocker pedestal and arm assemblies.

24. Install the front cover and crankshaft pulley.

25. Install the drive belt and fan assembly.

26. Install the fuel pump.

27. Adjust the valves.

28. Install the cylinder head covers.

29. Connect the negative battery cable and check for leaks.

Piston and Connecting Rod

POSITIONING

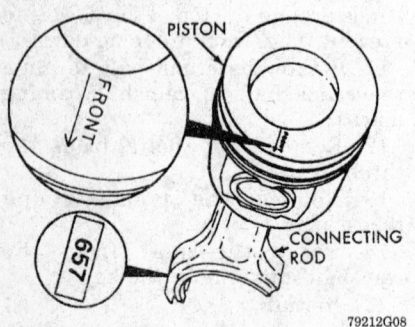

Identify the piston and connecting rod as shown — 5.9L (VIN C and 8) diesel engine

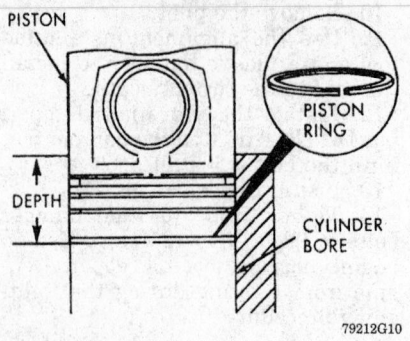

Ring position in the cylinder bore — 5.9L (VIN C and 8) diesel engine

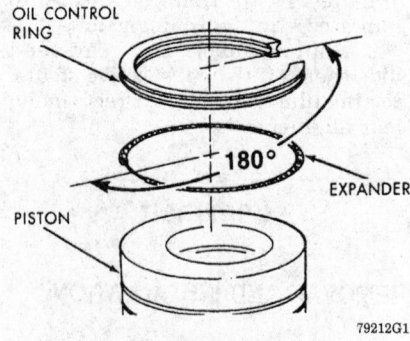

Oil control ring/expander location in the groove — 5.9L (VIN C and 8) diesel engine

Piston ring positioning — 5.9L (VIN C and 8) diesel engine

ENGINE LUBRICATION

Oil Pan

REMOVAL AND INSTALLATION

2.5L (VIN K and G) Engine

1. Disconnect the negative battery cable. Remove the oil dipstick.
2. Disconnect the air pump relief valve upper hose. Raise the vehicle and support safely.
3. Remove the clutch housing to engine strut and clutch inspection cover.
4. Remove the lower radiator hose support bracket.
5. Slightly loosen the right motor mount through-bolt just enough to relieve the tension.
6. Using the proper equipment, support the weight of the engine. Loosen the left motor mount through-bolt enough to clear the bracket.

7. Raise the left side of the engine about 2 in. (51mm).
8. Remove the oil pan retaining screws and remove the oil pan and gasket.

To install:

9. Thoroughly clean and dry all sealing surfaces, bolts and bolt holes.
10. Apply silicone sealer to the four end seal to block corners and install the end seals making sure the corners are not twisted.
11. Apply silicone to the four pan to block corners. Install a new pan gasket or apply a silicone sealer to the sealing surface of the pan and install to the engine making sure not to dislodge the end seals.
12. Install the pan retaining screws and tighten to 17 ft. lbs. (23 Nm).
13. Lower the engine. Tighten the motor mount through-bolts to 50 ft. lbs. (68 Nm).
14. Install the clutch inspection cover and housing to engine strut. Install the lower radiator hose support bracket. Lower the vehicle.
15. Install the dipstick and air pump hose. Fill the engine with the proper amount of oil.

16. Connect the negative battery cable and check for leaks.

3.9L (VIN X) Engine

2-Wheel Drive

1. Disconnect the negative battery cable. Remove the oil dipstick. Disengage the distributor cap and remove it away from the firewall.
2. Raise the vehicle and support safely.
3. Drain the engine oil.
4. Remove the exhaust crossover.
5. Loosen the motor mount bolts. Using the proper equipment, raise the engine. When the engine is high enough, install replacement bolts (similar in size to the motor mount bolts), in the engine mount attaching points on the frame brackets.
6. Lower the engine so the bottom of the motor mounts rest on the two replacement bolts. Remove the torque converter inspection cover, if equipped.
7. Remove the oil pan retaining screws and remove the oil pan and gaskets.

To install:

8. Thoroughly clean and dry all sealing surfaces, bolts and bolt holes.
9. Place a drop of silicone sealer to the timing chain cover to block mating seam.
10. Install the new gaskets to the engine and add a drop of silicone sealer to the corners where the rubber and cork meet. Instal the rubber seals to the pan.
11. Install the pan to the engine and tighten the retaining screws to 17 ft. lbs. (23 Nm). Install the torque converter inspection cover, if equipped.
12. Reinstall the engine to the mount and install the exhaust crossover. Lower the vehicle.
13. Install the distributor cap.
14. Install the dipstick. Fill the engine with the proper amount of oil.
15. Connect the negative battery cable and check for leaks.

4-Wheel Drive

1. Disconnect the negative battery cable. Remove the oil dipstick.
2. Raise the vehicle and support safely.
3. Using the proper equipment, support the weight of the engine. Remove the front driving axle.
4. Remove the exhaust crossover and the lower transmission cover.
5. Remove the oil pan retaining screws and remove the oil pan and gaskets.

To install:

6. Thoroughly clean and dry all sealing surfaces, bolts and bolt holes.
7. Place a drop of silicone sealer to the timing chain cover to block mating seam.
8. Install the new gaskets to the engine and add a drop of silicone sealer to the corners where the rubber and cork meet. Install the rubber seals to the pan.
9. Install the pan to the engine and tighten the retaining screws to 17 ft. lbs. (23 Nm). Install the lower transmission cover, if equipped.
10. Install the exhaust crossover.
11. Install the front driving axle. Lower the vehicle.
12. Install the dipstick. Fill the engine with the proper amount of oil.
13. Connect the negative battery cable and check for leaks.

5.2L (VIN T and Y) and 5.9L (VIN A, Z and 5) Engine

1. Disconnect the negative battery cable.
2. Remove the oil dipstick.
3. Raise and support the front end on jackstands.
4. Drain the oil.
5. Remove the exhaust cross-over pipe.
6. Remove the left engine-to-transmission strut.
7. Remove the bolts and lower the oil pan.
8. Thoroughly clean the gasket mating surfaces.
9. When installing the pan, always use new gaskets coated with sealer. Apply a drop of RTV silicone sealer where the cork and rubber gaskets meet. On the 5.9L engine, make sure the gasket notches are positioned as shown. Tighten the oil pan bolts to 15 ft. lbs. (20 Nm). Tighten the cross-over pipe to 24 ft. lbs. (32 Nm).

8.0L (VIN W) Engine

1. Disconnect the negative battery cable.
2. Remove the oil dipstick.
3. Raise and support the front end on jackstands.
4. Drain the oil.
5. Remove the left engine-to-transmission strut.
6. Remove the bolts and lower the oil pan. Remove the one piece gasket. The engine may have to be raised slightly on 2WD vehicles.
7. Remove the oil pickup tube assembly.

To install:

8. Thoroughly clean the gasket mating surfaces. If present, trim excess gasket sealant from the inside of the engine.
9. Clean the oil pan with solvent and dry thoroughly with a lint-free cloth.
10. Clean the oil screen and pipe. Inspect the condition of the screen.
11. Fabricate four alignment dowels from 5/16 x 1½ in. bolts. Cut the heads off the bolts and cut a slot in the top to allow installation and removal with a screwdriver.
12. Install the dowels into the four corners with a screwdriver.
13. Apply a small quantity of Mopar® Silicon Rubber Adhesive Sealant, or equivalent at the split lines. These are between the cylinder block, timing chain cover and rear crankshaft seal.

NOTE: After the sealant is applied, you have three minutes to install the gasket and oil pan.

14. Slide the one-piece gasket over the alignment dowels and position it on the block.
15. Position the oil pan over the gasket. The engine may have to be raised slightly on 2WD vehicles.
16. Install the bolts. Tighten the ½ in. bolts to 96 inch lbs. (11 Nm). Tighten the stud bolts to 144 inch lbs. (16 Nm). Tighten the 5/16 in. bolts to 144 inch lbs. (16 Nm).
17. Remove the dowels and, in their place, install the four remaining 5/16 in. bolts. Tighten these bolts to 144 inch lbs. (16 Nm).
18. Install the drain plug. Tighten it to 25 ft. lbs. (34 Nm).
19. Install the engine-to-transmission strut.
20. Lower the vehicle. Connect the negative battery cable. Fill the crankcase with oil.

5.9L (VIN C and 8)

1. Remove the engine and place it safely on an engine stand.
2. Drain the engine oil into a suitable container and dispose of it properly.
3. Remove the oil pan and gasket. Be sure to connect the support bracket.
4. If required, remove the suction tube and gasket.

To install:

5. Clean the sealing surface.
6. Install the suction tube and gasket. Tighten the bolts to 18 ft. lbs. (24 Nm).
7. Fill the joint between the pan rail/gear housing and pan rail/rear

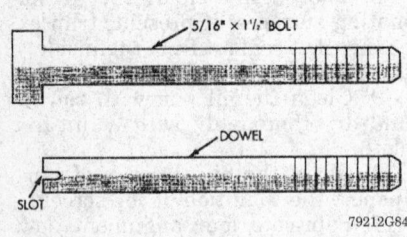

Fabricate four alignment dowels by cutting the bolt heads off and slotting the tops — 8.0L (VIN W) engine

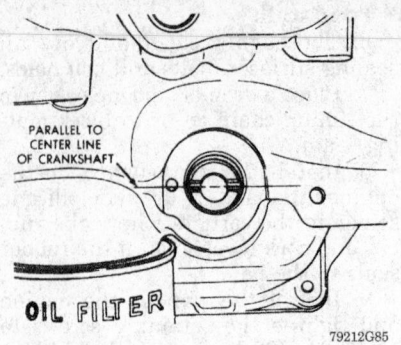

Align the oil pump shaft as shown — 2.5L (VIN K and G) engine

cover with sealant. Use Three Bond 1207–C, or equivalent.

8. Install the pan and gasket. Tighten the bolts to 18 ft. lbs. (24 Nm).

9. Install the drain plug with a new sealing washer. Tighten to 60 ft. lbs. (80 Nm).

10. Install the engine assembly into the vehicle.

11. Fill the engine with clean engine oil. Run the engine and check for leaks.

12. Stop the engine and let it set for five minutes. Check the oil level. Add oil, if necessary.

Oil Pump

REMOVAL AND INSTALLATION

2.5L (VIN K and G) Engine

1. Crank the engine around so that the No. 1 piston is at TDC. Disconnect the negative battery cable.

2. Matchmark the rotor to the block and remove the distributor to confirm that the slot in the oil pump shaft is parallel to the centerline of the crankshaft. Matchmark the slot to the distributor bore, if desired.

3. Remove the dipstick. Raise the vehicle and support safely. Drain the engine oil and remove the pan.

4. Remove the oil pickup.

5. Remove the two mounting bolts and remove the oil pump from the engine.

To install:

6. Prime the pump by pouring fresh oil into the pump intake and turning the driveshaft until oil comes out the pressure port. Repeat a few times until no air bubbles are present.

7. Apply sealer (Loctite® 515, or equivalent) to the pump body to block machined surface interface. Lubri-

cate the oil pump and distributor driveshaft.

8. Align the slot so it will be in the same position as when it was removed. If it is not, the distributor will not be timed correctly. Install the pump fully and rotate back and forth to ensure proper positioning between the pump mounting surface and the machined surface of the block.

9. Install the mounting bolts finger-tight and lower the vehicle to confirm proper slot positioning. If the slot is not properly positioned, raise the vehicle and move the gear as required. If the slot is correct, hold the pump firmly against the block and tighten the mounting bolts to 17 ft. lbs. (23 Nm).

10. Clean out the oil pickup or replace as required. Replace the oil pickup O-ring and install the pickup to the pump.

11. Install the oil pan using new gaskets. Lower the vehicle.

12. Install the distributor.

13. Install the dipstick. Fill the engine with the proper amount of oil.

14. Connect the negative battery cable, check the timing and check the oil pressure.

3.9L (VIN X) Engine

1. Disconnect the negative battery cable.

2. Raise the vehicle and support safely. Drain the oil and remove the oil pan.

3. Remove the screen.

4. Unbolt the oil pump from the rear main bearing cap and remove it from the vehicle.

To install:

5. Prime the pump by pouring fresh oil into the pump intake and turning the driveshaft until oil comes out the pressure port. Repeat a few times until no air bubbles are present. Install the oil pump with a ro-

tating motion to ensure proper pump driveshaft engagement.

6. Hold the pump flush against the main cap and finger-tighten the attaching bolts.

7. Tighten the bolts to 130 ft. lbs. (176 Nm).

8. Install the screen.

9. Install the oil pan with a new gasket.

10. Connect the negative battery cable and check the oil pressure.

5.2L (VIN T and Y) and 5.9L (VIN A, Z and 5) Engines

It is necessary to remove the oil pan, and to remove the oil pump from the rear main bearing cap to service the oil pump.

1. Drain the engine oil and remove the oil pan.

2. Remove the oil pump mounting bolts and remove the oil pump from the rear main bearing cap.

To install:

3. Install the pump and tighten the cover bolts to 95 inch lbs. (11 Nm).

4. Prime the oil pump before installation by filling the rotor cavity with engine oil. Install the oil pump on the engine and tighten attaching bolts to 30 ft. lbs. (41 Nm).

5. Install the oil pan.

6. Fill the engine with the proper grade motor oil. Start the engine and check for leaks.

8.0L (VIN W) Engine

1. Remove the timing chain cover.

2. Remove the relief valve plug, gasket, spring and the valve. Discard the gasket.

3. Remove the oil pump cover.

4. Remove the pump rotors.

To install:

5. Install the oil pump and lubricate the pump rotors with petroleum jelly or Lubriplate®.

6. Install the timing chain cover.

7. Position the oil pump cover onto the timing chain cover and tighten the cover bolts to 125 inch lbs. (14 Nm).

8. After the cover is installed, make sure that the inner ring can still move freely and does not bind in any way.

9. Install the timing chain cover. Squirt oil into the relief valve hole until oil runs out.

10. Using a new pressure relief valve gasket, install the relief valve plug, tightening it to 15 ft. lbs. (20 Nm).

11. Fill the oil filter with oil and install it on the engine.

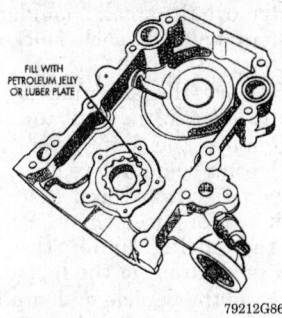

79212G86

Lubricate the rotors as a means of priming the oil pump so it is not dry on start-up — 8.0L (VIN W) engine

5.9L (VIN C and 8)

1. Disconnect the negative battery cable.
2. Remove the drive belt.
3. Remove the radiator.
4. Remove the fan assembly.
5. Remove the oil fill tube and adaptor.
6. Remove the crankshaft pulley.
7. Remove the front cover.
8. Remove the four pump mounting bolts and remove the pump from the block.
   To install:
9. Prime the pump by pouring fresh oil into the pump intake and turning the driveshaft until oil comes out the pressure port. Repeat a few times until no air bubbles are present.
10. Align the idler gear pin with the locating bore in the block and install the pump.
11. Tighten the mounting bolts in the proper sequence to 44 inch lbs. (5 Nm), then repeat the torquing sequence to 18 ft. lbs. (24 Nm).

NOTE: When the pump is correctly installed, the flange on the pump should not touch the block; the back plate on the pump seats against the bottom of the bore.

12. Measure the backlash of the idler to pump drive gears. The specification is 0.003-0.013 in. (0.08-0.33mm).
13. Measure the backlash of the idler to crankshaft gears. The specification is 0.003-0.013 in. (0.08-0.33mm).
14. Install the front cover and crankshaft pulley.
15. Install the oil fill tube and adaptor.
16. Install the fan assembly, radiator and drive belt.
17. Connect the negative battery cable and check the oil pressure.

TRANSMISSION

Manual Transmission Assembly

REMOVAL AND INSTALLATION

4-Speed 2WD Ram Pick-up and Ramcharger Models

1. Disconnect the negative battery cable. Remove the engine cover. Remove the gear shift lever and shift unit. Raise and support the vehicle. Drain the transmission fluid.
2. Mark the rear driveshaft yoke to differential for correct installation reference. Disconnect the rear universal joint and remove the driveshaft.
3. Disconnect the speedometer cable and back-up light switch electrical harness.
4. Support the rear of the engine. Remove the rear mount to crossmember bolts/nuts. Raise the rear of the engine slightly. Take care when raising the engine so nothing in the front of the engine, hoses lines, fan to radiator is strained or damaged.
5. Remove the rear transmission support crossmember. Position a suitable support jack under he transmission. Remove transmission to clutch bell housing retaining bolts and pull transmission rearward until drive pinion clears clutch, then remove transmission.
   To install:
6. Place a small amount of Mopar® Multi-purpose Grease No. 4318063, or equivalent into clutch pilot bushing and on the clutch release bearing sleeve. Do not get any grease on the flywheel face. Do not lubricate the end of the transmission input shaft, clutch disc splines, or clutch release levers.
7. Align clutch disc and backing plate with a spare drive pinion shaft or clutch aligning tool.
8. Raise the transmission into position. Slide the transmission forward until the input shaft is aligned with and starts to enter the clutch disc splines. With the transmission in gear, turn the splines until they are aligned then push the transmission in until it mates to the bell housing.
9. Install transmission to bell housing bolts, tightening to 50 ft. lbs. (68 Nm). Install the crossmember and tighten the mounting bolts to 30 ft. lbs. (41 Nm). Attach the rear

transmission mount to the crossmember and tighten to 50 ft. lbs. (68 Nm).
10. Install gear shift unit and tighten the mounting bolts to 24 ft. lbs. (33 Nm). Install and adjust the shift linkage.
11. Install the driveshaft. Connect the speedometer cable and back-up light.
12. Install the transmission drain plug and fill the transmission with lubrication.
13. Install the shift lever and engine cover. Lower the vehicle. Connect the battery cable.

4-Speed 4WD Ram Pick-up and Ramcharger Models

1. Raise and support the truck.
2. Remove the skid plate, if any. Drain lubricant from the transmission and transfer case.
3. Disconnect the speedometer cable.
4. Disconnect and match-mark the front and rear driveshafts. Suspend each shaft from a convenient place; do not allow them to hang free.
5. Disconnect the shift rods at the transfer case. Remove the shift lever retainer by pressing down and turning it counterclockwise. Remove the shift lever springs and washers.
6. Remove the rear driveshaft. Matchmark the driveshaft and rear U-joints before removing the driveshaft.
7. Support the transfer case.
8. Remove the extension-to-transfer case mounting bolts.
9. Move the transfer case rearward to disengage the front input shaft spline.
10. Lower and remove the transfer case.
11. Disconnect the back-up light switch.
12. Support the engine.
13. Support the transmission.
14. Remove the transmission crossmember.
15. Remove the transmission-to-clutch housing bolts.
16. Slide the transmission rearward until the mainshaft clears the clutch disc.
17. Lower and remove the transmission.
18. The transmission pilot bushing in the end of the crankshaft requires high-temperature grease. Multi-purpose grease should be used. Do not lubricate the end of the mainshaft, clutch splines, or clutch release levers.
19. Raise and position the transmission.

20. Slide the transmission forward until the mainshaft enters the clutch disc, then push it all the way forward.

21. Install the transmission-to-clutch housing bolts. Tighten the bolts to 50 ft. lbs. (68 Nm).

22. Install the transmission crossmember.

23. Connect the back-up light switch.

24. Raise and position the transfer case.

25. Move the transfer case rearward to disengage the front input shaft spline.

26. Install the extension-to-transfer case mounting bolts. Tighten the bolts to 50 ft. lbs. (68 Nm).

27. Install the transmission crossmember.

28. Connect the back-up light switch.

29. Raise and position the transfer case.

30. Move the transfer case rearward to disengage the front input shaft spline.

31. Install the extension-to-transfer case mounting bolts. Tighten the bolts to 50 ft. lbs. (68 Nm).

32. Install the rear driveshaft.

33. Connect the shift rods at the transfer case. Install the shift lever retainer by pressing down and turning it clockwise. Install the shift lever springs and washers.

34. Connect the front driveshafts.

35. Connect the speedometer cable.

36. Install the skid plate, if any. Fill the transmission and transfer case.

37. Lower the truck.

Ram Pick-up and Ramcharger Models with NP2500 Transmission

1. Disconnect the negative battery cable. Remove the engine cover. Remove the upper shift lever. Raise and support the vehicle. Drain the transmission fluid.

2. Mark the rear driveshaft yoke to differential for correct installation reference. Disconnect the rear universal joint and remove the driveshaft.

3. If necessary, remove the exhaust system Y-pipe for clearance.

4. Disconnect the back-up light switch and distance sensor electrical harness. Separate the sensor from the speedometer adapter. Remove the speedometer components.

5. Support the rear of the engine. Remove the rear mount to crossmember bolts/nuts. Raise the rear of the engine slightly. Take care when raising the engine so nothing in the

front of the engine, hoses lines, fan to radiator is strained or damaged.

6. Remove the rear transmission support crossmember. Position a suitable support jack under the transmission. Remove transmission to clutch bell housing retaining bolts and pull transmission rearward until drive pinion clears clutch, then remove transmission.

To install:

7. Place a small amount of Mopar® Multi-purpose Grease No. 4318-63, or equivalent into clutch pilot bushing and on the clutch release bearing sleeve. Do not get any grease on the flywheel face. Do not lubricate the end of the transmission input shaft, clutch disc splines, or clutch release levers.

8. Align clutch disc and backing plate with a spare drive pinion shaft or clutch aligning tool.

9. Raise the transmission into position. Slide the transmission forward until the input shaft is aligned with and starts to enter the clutch disc splines. With the transmission in gear, turn the splines until they are aligned then push the transmission in until it mates to the bell housing.

10. Install transmission to bell housing bolts. Install the crossmember. Attach the rear transmission mount to the crossmember.

11. Install the driveshaft. Connect the speedometer components. Install the distance sensor. Connect the back-up light.

12. Install the transmission drain plug and fill the transmission with lubricant.

13. Install the upper shift lever and engine cover. Lower the vehicle. Connect the battery cable.

2WD Ram Pick-up with Getrag G-360 Transmission

1. Disconnect the negative battery cable.

2. Remove the screws attaching the shift lever upper boot to the metal plate on the lower boot. Slide the boot upward on the lever.

3. Remove the screws attaching the lower boot and plate assembly to the floor pan.

4. Remove the shift lever extension by unthreading it from the shift lever.

5. Slide the lower boot and plate upward and off the shift lever.

6. Remove the transmission shift tower boot. Loosen the clamp securing the boot to the shift tower. Slide the boot up and off the shift lever and tower.

7. Remove the small shoulder-type alignment bolts at each side of the shift tower.

8. Remove the snapring securing the shift lever in the shift tower.

9. Remove the shift lever. If the lever doesn't come out readily, press the lever downward and quickly release it two or three times. Pressure from the spring under the lever should push it out of the housing.

10. Raise the vehicle and support it safely.

11. Mark the propeller shaft U-joints and axle yokes for assembly alignment reference.

12. Remove the propeller shaft. Remove the clamp strap bolts at the front of the shaft first. Remove the center bearing bracket bolts and support the shaft with a jackstand. Remove the rear clamp strap bolts and remove the shaft.

13. Unplug the distance sensor and back-up light switch connectors.

14. Spread the clamps securing the switch wires to the transmission top cover and remove the wires from the clamps. Move the wires aside.

15. Remove the bolts attaching the transmission rear mount to the crossmember.

16. Support the transmission with a suitable jack. Use a safety chain to secure the transmission to the jack.

17. Remove the screws attaching the fuel line clamps to the driver side frame rail. There are three of them.

18. Remove the left and right crossmember-to-frame braces, then remove the crossmember.

19. Remove the clutch slave cylinder shield.

20. Loosen the clutch slave cylinder attaching nuts until the cylinder piston rod is clear of the release lever. The cylinder does not have to be completely removed.

21. Remove the bolts attaching the transmission to the clutch housing and remove the transmission. Rock the transmission slightly to help free the input shaft from the clutch disc and release bearing. Support the engine with a wood block and adjustable jackstand.

To install:

22. Apply a light coating of Mopar® high temperature bearing grease to the transmission input shaft and release bearing slide surface of the front bearing retainer. Apply a light coating of grease to the bore of the release bearing.

23. Align the transmission and clutch housing. Move the transmission forward and start the input

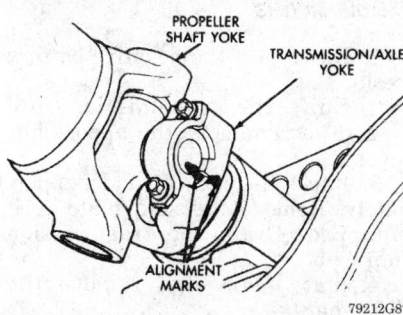

Mark the driveshaft U-joints and axle yokes as shown for an assembly alignment reference

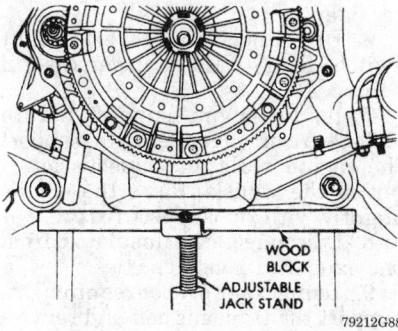

Use a reliable method (as shown) to support the engine during transmission removal

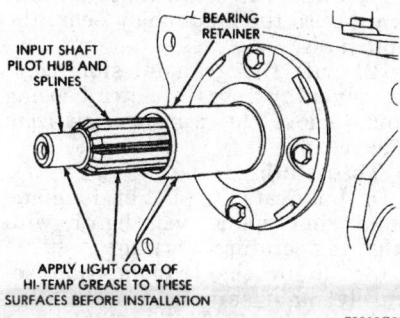

Lubricate the input shaft at the locations indicated when installing the transmission

shaft into the release bearing and clutch disc hub.

24. Move the transmission forward and seat it against the housing.

25. Install and tighten the transmission mounting bolts to 50 ft. lbs. (68 Nm).

26. Install the rear crossmember and crossmember brackets. Start all crossmember bolts before tightening to 50 ft. lbs. (68 Nm).

27. Remove the transmission jack.

28. Install the fuel line clamps back onto the driver side frame rail.

29. Align the clutch slave cylinder in the housing and tighten the cylin-der attaching nuts. Install the clutch slave cylinder shield.

30. Secure the back-up light and distance sensor wires into the clips on top of the transmission. Route the wires across the top of the transmission and reconnect the switch and sensor connectors.

31. Align and install the propeller shaft. Install the shaft front U-joint first. Secure the center bearing bracket to frame bracket, then install the shaft rear U-joint in the axle yoke. Tighten the U-joint clamp bolts to 14 ft. lbs. (19 Nm) and the center bearing bracket bolts to 50 ft. lbs. (68 Nm).

32. Lower the vehicle. Lubricate the shift leverball and collar with Mopar® all-purpose grease. Align the notches in the shift lever ball and collar and insert the shift lever in the shift tower. Press the shift lever downward and install the snapring. Tighten the shift lever alignment bolts completely.

33. Install the lower boot and plate assembly. Thread the shift lever extension onto the shift lever, then install the upper boot and plate. Install the shift knob and connect the negative battery cable.

4WD Ram Pick-up with Getrag G-360 Transmission

1. Disconnect the negative battery cable.

2. Remove the screws attaching the shift lever upper boot to the metal plate on the lower boot. Slide the boot upward on the lever.

3. Remove the screws attaching the lower boot and plate assembly to the floor pan.

4. Remove the shift lever extension by un-threading it from the shift lever.

5. Slide the lower boot and plat upward and off the shift lever.

6. Remove the transmission shift tower boot. Loosen the clamp securing the boot to the shift tower. Slide the boot up and off the shift lever and tower.

7. Remove the small should-type alignment bolts at each side of the shift tower.

8. Remove the snapring securing the shift lever in the shift tower.

9. Remove the shift lever. If the lever doesn't come out readily, press the lever downward and quickly release it two or three times. Pressure from the spring under the lever should push it out of the housing.

10. Raise the vehicle and support it safely.

11. Mark the front and rear propeller shafts for installation reference, then disconnect and remove both shafts.

12. Remove the transfer case skid plate and crossmember.

13. Disconnect the distance sensor and back-up light with connectors.

14. Disconnect the transfer case shift lever from the transfer case.

15. Support the transmission with an adjustable jackstand and support the transfer case with a transmission jack.

16. Remove the nuts attaching the transfer case to the transmission adapter. Lower the transfer case and move it from under the vehicle.

17. Move the adjustable jackstand from under the transmission and position it to support the engine.

18. Spread the clamps securing the switch wires to the transmission top cover and remove the wires from the clamps. Move the wires aside.

19. Remove the bolts attaching the transmission rear mount to the crossmember.

20. Support the transmission with a suitable jack. Use a safety chain to secure the transmission to the jack.

21. Remove the screws attaching the fuel line clamps to the driver side frame rail. There are three of them.

22. Remove the left and right crossmember-to-frame braces, then remove the crossmember.

23. Remove the clutch slave cylinder shield.

24. Loosen the clutch slave cylinder attaching nuts until the cylinder piston rod is clear of the release lever. The cylinder does not have to be completely removed.

25. Remove the bolts attaching the transmission to the clutch housing and remove the transmission. Rock the transmission slightly to help free the input shaft from the clutch disc and release bearing. Support the engine with a wood block and adjustable jackstand.

To install:

26. Apply a light coating of Mopar® high temperature bearing grease to the transmission input shaft and release bearing slide surface of the front bearing retainer. Apply a light coating of grease to the bore of the release bearing.

27. Align the transmission and clutch housing. Move the transmission forward and start the input shaft into the release bearing and clutch disc hub.

28. Move the transmission forward and seat it against the housing.

29. Install and tighten the transmission mounting bolts to 50 ft. lbs. (68 Nm).

30. Install the rear crossmember and crossmember brackets. Start all crossmember bolts before tightening to 50 ft. lbs. (68 Nm).

31. Remove the transmission jack.

32. Install the fuel line clamps back onto the driver side frame rail.

33. Align the clutch slave cylinder in the housing and tighten the cylinder attaching nuts. Install the clutch slave cylinder shield.

34. Secure the back-up light and distance sensor wires into the clips on top of the transmission. Route the wires across the top of the transmission and reconnect the switch and sensor connectors.

35. Install the gasket and transfer case on the transmission adapter. Tighten the transfer case attaching nuts to 35 ft. lbs. (47 Nm).

36. Check the transmission and transfer case lubricant levels and top off, if necessary. Connect the transfer case shift lever to the transfer case.

37. Align and install the propeller shaft. Install the shaft front U-joint first. Secure the center bearing bracket to frame bracket, then install the shaft rear U-joint in the axle yoke. Tighten the U-joint clamp bolts to 14 ft. lbs. (19 Nm) and the center bearing bracket bolts to 50 ft. lbs. (68 Nm).

38. Lower the vehicle. Lubricate the shift lever ball and collar with Mopar® all-purpose grease. Align the notches in the shift lever ball and collar and insert the shift lever in the shift tower. Install the shift lever downward and install the snapring. Tighten the shift lever alignment bolts completely.

39. Install the lower boot and plate assembly. Thread the shift lever extension onto the shift lever, then install the upper boot and plate. Install the shift knob and connect the negative battery cable.

Ram Pick-up and Ramcharger Models with NV3500/4500

1. Disconnect the negative battery cable. Remove the upper shift lever. Raise and support the vehicle. Drain the transmission fluid.

2. Mark the rear driveshaft yoke to differential for correct installation reference. Disconnect the rear universal joint and remove the driveshaft.

3. 4WD vehicles: If necessary, remove the exhaust system Y-pipe for clearance.

4. Disconnect the back-up light switch and distance sensor electrical harness. Separate the sensor from the speedometer adapter. Remove the speedometer components.

5. Support the rear of the engine. Remove the rear mount to crossmember bolts/nuts. Raise the rear of the engine slightly. Take care when raising the engine so nothing in the front of the engine, hoses lines, fan to radiator is strained or damaged.

6. Remove the rear transmission support crossmember. Position a suitable support jack under the transmission. Remove transmission to clutch bell housing retaining bolts and pull transmission rearward until drive pinion clears clutch, then remove transmission.

To install:

7. Place a small amount of Mopar® Multi-purpose Grease No. 4318-63, or equivalent onto the following components:
- Drive gear splines and the pilot bearing hub
- Release bearing slide surface of the front retainer
- Pilot bearing
- Release bearing bore
- Release fork
- Release fork ball stud
- Driveshaft slip yoke

———— WARNING ————
Do not get any grease on the flywheel face. Do not lubricate the end of the transmission input shaft, clutch disc splines, or clutch release levers.

8. Align clutch disc and backing plate with a spare drive pinion shaft or clutch aligning tool.

9. Raise the transmission into position. Slide the transmission forward until the input shaft is aligned with and starts to enter the clutch disc splines. With the transmission in gear, turn the splines until they are aligned then push the transmission in until it mates to the bell housing.

10. Install transmission to bell housing bolts. Install the crossmember. Attach the rear transmission mount to the crossmember.

11. Install the driveshaft. Connect the speedometer components. Install the distance sensor. Connect the back-up light.

12. If removed, install the exhaust system Y-pipe.

13. Install the transmission drain plug and fill the transmission with lubricant.

14. Install the upper shift lever and engine cover. Lower the vehicle. Connect the battery cable.

Dakota Models

1. Disconnect the negative battery cable.

2. Shift the transmission into Neutral and remove the upper shift lever.

3. Raise the vehicle and support safely. Remove the skid plate(s), if equipped. Drain the transmission lubricant.

4. Matchmark and remove the driveshaft(s).

5. Disconnect the wires from the distance sensor, if equipped. Then loosen the sensor coupling and remove the sensor from the speedometer adaptor.

6. Disconnect the reverse light switch. Remove the transfer case, if equipped.

7. Install engine support fixture C-3487-A, or equivalent, to support the engine while the transmission is out of the vehicle. Raise the engine slightly with the support fixture.

8. Disconnect the insulator from the extension housing.

9. Using the proper equipment, support the transmission and remove the crossmember.

10. Remove the transmission to clutch housing bolts.

11. Slide the transmission rearward until the input shaft clears the clutch disc.

12. Pull the transmission completely away from the clutch housing and remove the transmission from the vehicle.

To install:

13. Lubricate the pilot bushing and input shaft splines very lightly with high temperature lubricant.

14. Mount the transmission securely on a suitable transmission jack and lift it is place until the input shaft is centered in the clutch housing opening. Roll the transmission forward until the input shaft splines fully engage with the clutch disc.

15. Install the transmission to clutch housing bolts. Tighten the bolts to 50 ft. lbs. (68 Nm).

16. Install the transmission crossmember. Remove the transmission and engine support fixtures.

17. Install the transfer case, if equipped and connect all linkage, electrical connectors and vacuum lines to the transfer case.

18. Connect the reverse light switch connector and clip all wiring to the transmission case.

19. Connect speedometer cable and distance sensor, if equipped.

20. Install the driveshaft(s).

21. Fill the transmission with the proper amount of 10W-30 engine oil

with an API classification SG/CD. Fill the transfer case, if equipped, with the proper amount of Dexron®II lubricant.

22. Apply silicone sealer to the perimeter of the shifter base and install the shifter assembly.

23. Install the skid plate(s), if equipped.

24. Connect the negative battery cable and check the transmission for proper operation.

Clutch Assembly

REMOVAL AND INSTALLATION

Except Dakota Models

1. Remove the transmission. Remove transfer case, if equipped.
2. Remove the clutch housing.
3. Remove the clutch fork and release bearing assembly.
4. Mark the clutch cover and flywheel, with a suitable tool or paint to assure correct reassembly.
5. Remove the pressure plate retaining bolts, loosening them evenly so the clutch cover will not be distorted.
6. Pull the pressure plate assembly clear of flywheel and, while the supporting pressure plate, slide the clutch disc from between flywheel and plate.

To install:

—————— **CAUTION** ——————
Clutch discs may contain asbestos. Do NOT use compressed air to clean components. Instead, use a commercially available evaporative spray cleaner. Always use proper eye protection when working around spray chemicals.

7. Thoroughly clean all working surfaces of the flywheel and the pressure plate.
8. Grease radius at back of bushing.
9. Rotate clutch cover and pressure plate assembly for maximum clearance between flywheel and frame crossmember (if crossmember was not removed during clutch removal).
10. Tilt top edge of clutch cover and pressure plate assembly back and move it up into the clutch housing. Support clutch cover and pressure plate assembly, then slide clutch disc into position.
11. Position the clutch disc and the plate against the flywheel, then insert spare transmission main drive gear shaft or clutch installing tool

through clutch disc hub and into main drive pilot bearing.

12. Rotate clutch cover until the punch marks on cover and flywheel line up.

13. Bolt the pressure plate loosely to flywheel. Tighten the bolts a few turns at a time, in progression, until tight. Then tighten bolts to:
- 5/16 in. (8mm) bolts: 20 ft. lbs. (27 Nm)
- 3/8 in. (9.5mm) bolts: 30 ft. lbs. (40 Nm)

14. Install transmission.
15. Install frame crossmembers and insulator, tighten all bolts.

Dakota Models

1. Disconnect the negative battery cable.
2. Raise the vehicle and support safely.
3. Remove the transmission and transfer case, if equipped.
4. Remove the inspection cover at the bottom of the bell housing.
5. Rotating the engine with a flywheel turner, remove the clutch cover bolts gradually as they appear.
6. Remove the clutch cover and disc by lowering it through the opening at the bottom of the housing.

To install:

7. Sparingly apply anti-seize compound to the input shaft and clutch disc splines. Install a new release bearing.
8. Raise the clutch cover and disc into place and use a suitable clutch aligning tool or spare input shaft to center the disc. Tighten all of the bolts finger tight.
9. The cover bolts must be turned gradually, evenly and to the proper torque to avoid distorting the cover. Tighten the bolts to 21 ft. lbs. (28 Nm).
10. Install the transmission and transfer case, if equipped.
11. Install the inspection cover.

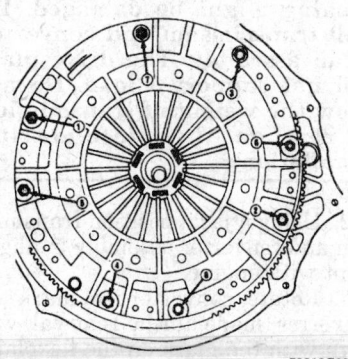
79212G92

Clutch cover bolt tightening/loosening sequence

12. Connect the negative battery cable and check the clutch for proper operation.

Clutch Master Cylinder and Slave Cylinder

REMOVAL AND INSTALLATION

The hydraulic clutch master cylinder is mounted on the firewall and is fed fluid via gravity from a remote reservoir. The master cylinder mounting nuts are inside the cab. The torque for the nuts is 200 inch lbs. (22 Nm). Torque for the reservoir nuts is 95 inch lbs. (10 Nm). The slave cylinder is mounted on a bracket on the left side of the transmission. Mounting nut torque is 200 inch lbs. (22 Nm). The clutch master cylinder, remote reservoir, slave cylinder and connecting lines are serviced as an assembly only. The linkage components cannot be overhauled or serviced separately. The cylinders and connecting lines are filled and factory sealed.

1. Disconnect the negative battery cable.
2. Raise the vehicle and support safely.
3. Remove the nuts attaching the slave cylinder to the bell housing.
4. Remove the slave cylinder and clip from the housing.
5. Lower the vehicle.
6. Remove the locating clip from the clutch master cylinder mounting bracket.
7. Remove the retaining ring, flat washer and wave washer that attach the clutch master cylinder pushrod to the clutch pedal. Slide the pushrod off of the pedal pin. Inspect the bushing on the pedal pin and replace if it is excessively worn.
8. Verify that the cap on the clutch master cylinder reservoir is tight so fluid will not spill during removal.
9. Remove the screws attaching the reservoir and bracket, if equipped, to the dash panel and remove the reservoir.
10. Pull the clutch master cylinder rubber seal from the dash panel.
11. Rotate the clutch master cylinder counterclockwise 45° to unlock it. Remove the cylinder from the dash panel.
12. Remove the clutch master cylinder, remote reservoir, slave cylinder and connecting lines from the vehicle.

To install:

13. Verify that the cap on the fluid reservoir is tight so fluid will not spill during installation.

14. Position the components in the replacement kit in their places on the vehicle.

15. Insert the master cylinder in the dash. Rotate clockwise 45 ° to lock in place.

16. Lubricate the rubber seal with a lubricant to ease installation. Seat the seal around the cylinder in the dash.

17. Install the fluid reservoir and bracket, if equipped, to the dash panel.

18. Install the master cylinder pushrod to the clutch pedal pin. Secure the rod with the wave washer, flat washer and retaining ring. Install the locating clip. Do not remove the plastic shipping stop from the pushrod until the slave cylinder has been installed.

19. Raise the vehicle and support safely.

20. Insert the slave cylinder pushrod through the opening and make sure the cap on the end of the pushrod is securely engaged in the release lever before tightening the attaching nuts. Tighten the nuts to 17 ft. lbs. (23 Nm).

21. Install slave cylinder cover, if equipped. Lower the vehicle. Remove the plastic shipping stop from the master cylinder pushrod. Connect clutch pedal interlock switch wires.

22. Operate the clutch pedal a few times to verify proper system operation. The system will self-bleed any air in and vent through the reservoir.

23. Connect the negative battery cable and road test the vehicle.

Hydraulic Clutch System Bleeding

The system is self-bleeding. Press the clutch pedal repeatedly to release air from the fluid. The air will be vented from the reservoir.

Automatic Transmission Assembly

REMOVAL AND INSTALLATION

Ram Pick-up and Ramcharger Models

1. Remove the transmission and converter as an assembly; otherwise the converter drive plate pump bushing and oil seal will be damaged. The drive plate will not support a load. Therefore, none of the weight of transmission should be allowed to rest on the plate during removal.

2. Disconnect the negative battery cable. Remove the engine cover. Raise and safely support the vehicle.

3. Remove engine to transmission struts, if necessary. You may have to drop the exhaust system on some models.

4. Remove the starter. Remove wire from the neutral safety switch and converter solenoid (if equipped).

5. Remove the gearshift cable or rod from the transmission and the lever.

6. Disconnect the throttle rod from left side of transmission. Disconnect the distance sensor and speedometer.

7. Drain the transmission fluid and reinstall the pan.

8. Disconnect the oil cooler lines at transmission and remove the oil filler tube. Disconnect the speedometer cable.

9. Remove the converter front cover. Mark the converter and flexplate to converter mounting bolts/nuts. Turn the engine in a clockwise direction to gain access to the bolts. Mark for reference and remove the driveshaft.

10. Install engine support fixture or position a jack to hold up the rear of the engine. Care should be taken not to place strain on the radiator hoses, etc. Watch the fan-to-radiator clearance.

11. Position a suitable transmission jack under the transmission. Raise the transmission slightly with the jack to relieve load and remove support bracket or crossmember. Remove all of the bell housing bolts and carefully work transmission and converter rearward off engine dowels and disengage converter hub from end of crankshaft.

— WARNING —

Attach a small C-clamp to edge of bell housing to hold converter in place during transmission removal; otherwise the front pump bushing might be damaged. Install transmission and converter as an assembly. The drive plate will not support a load. Do not allow the weight of transmission to rest on the plate during installation.

12. Using a jack, position transmission and converter assembly in alignment with engine.

13. Rotate converter so mark on converter (made during removal) will align with the mark on the flexplate. Carefully work the transmission assembly forward over engine block dowels with converter hub entering the crankshaft opening and the flexplate to converter mounts lined up.

14. Install converter housing to engine bolts and tighten them.

15. Install the converter to flexplate bolts.

16. Install the rear crossmember and connect the rear transmission mount if loosened. Install the driveshaft.

17. Connect oil cooler lines, install oil filler tube and connect the speedometer and distance sensor.

18. Connect gearshift cable or rod and torque shaft assembly to the transmission case and to the lever.

19. Connect throttle rod to the lever at left side of transmission bell housing.

20. Connect wire to back-up and neutral staring switch. Connect the converter solenoid (if equipped).

21. Install cover plate in front of the converter assembly. Install the starter motor and engine struts.

22. Lower the vehicle and refill transmission with fluid.

23. Adjust linkage.

Dakota Models

1. Disconnect the negative battery cable.

2. Raise the vehicle and support safely. Drain the transmission and transfer case, if equipped.

3. Remove the exhaust crossover pipe.

4. Remove the skid plates, if equipped.

5. Matchmark and remove the driveshaft(s).

6. Disconnect the distance sensor, if equipped and the speedometer cable.

7. If equipped with 4WD, disconnect all linkage, electrical connectors and vacuum lines from the transfer case. Using a suitable jack, support the transfer case, unbolt the transfer case from the transmission and slide it backwards to remove it from the vehicle.

8. Remove the engine to transmission struts.

9. Remove the starter and the fluid cooler lines bracket.

10. Remove the torque converter inspection cover.

11. Matchmark the converter to the flex plate. Remove the torque converter bolts.

12. Disconnect the wires to the neutral safety switch and lock-up solenoid, if equipped.

13. Disconnect and plug the oil cooler lines from the transmission.

14. Disconnect the gearshift rod and torque shaft assembly from the transmission.

15. Disconnect the throttle rod from the lever.

16. Unbolt the oil filler tube brace and lift the oil filler tube out of its bore.

17. Install an appropriate engine support fixture to hold the engine in place when the transmission is out of the vehicle.

18. Raise the transmission slightly using a suitable transmission jack.

19. Remove the transmission crossmember.

20. Remove the oil filter, if necessary. Remove all bell housing bolts and remove the transmission from the vehicle.

To install:

21. Install the transmission securely on the transmission jack. Rotate the converter so it will align with the positioning of the flex plate.

22. Apply a coating of high temperature grease to the torque converter pilot hub.

23. Raise the transmission into place and push it forward until the dowels engage and the bell housing is flush with the block.

24. Install the oil filler tube. Install the bell housing bolts and tighten to 30 ft. lbs. (41 Nm). Install the oil filter, if it was removed.

25. Install the transmission crossmember. Remove the engine support fixture and the transmission jack.

26. install the torque converter bolts and tighten to 23 ft. lbs. (31 Nm).

27. Connect the oil cooler lines.

28. Connect the throttle rod to the lever and adjust if necessary.

29. Connect the gearshift rod and torque shaft assembly to the transmission and adjust if necessary.

30. Connect the wires to the neutral safety switch an lockup solenoid, if equipped. Make sure all wires are routed correctly and clipped in place.

31. Install the torque converter inspection cover, starter and transmission struts.

32. Install the transfer case, if equipped.

33. Connect the distance sensor, if equipped.

34. Connect the speedometer cable.

35. Install the driveshaft(s).

36. Install exhaust parts that were removed in order to remove the transmission.

37. Fill the transfer case, if equipped. Lower the vehicle.

38. Connect the negative battery cable.

39. Fill the transmission with the proper amount of Dexron®II.

40. Road test the vehicle, check for leaks and recheck the fluid level.

Transfer Case Assembly

REMOVAL AND INSTALLATION

Ram Pick-up and Ramcharger Models

1. Raise and support the truck.

2. Remove the skid plate, if equipped.

3. Drain the transfer case by removing the bottom bolt from the front output rear cover.

4. Disconnect the speedometer cable.

5. Disconnect the front and rear output shafts. Suspend these from a convenient location; do not allow them to hang free.

6. Disconnect the shift rods at the transfer case.

7. Support the transfer case.

8. Remove the adapter-to-transfer case mounting bolts and move the transfer case rearward to disengage the front input splines.

9. Lower and remove the transfer case.

10. Installation is the reverse of removal. Adjust the linkage.

Dakota Models

1. Disconnect the negative battery cable.

2. Raise the vehicle and support safely.

3. Remove the skid plates, if equipped. Drain the transfer case fluid.

4. Disconnect the distance sensor, if equipped and disconnect the speedometer cable from the transfer case.

5. Matchmark and remove the driveshafts.

6. Disconnect the Power Take-Off (PTO), if equipped.

7. Disconnect the linkage, electrical connectors and vacuum lines from the transfer case. Using a suitable jack, support the transfer case and remove the crossmember.

8. Unbolt the transfer case from the transmission and slide it backward to remove it from the vehicle.

9. The installation is the reverse of the removal procedure. Tighten the transfer case to transmission case nuts to 26 ft. lbs. (35 Nm). Fill the transfer case with Dexron®II.

DRIVE AXLE

Driveshaft

REMOVAL AND INSTALLATION

Front Driveshaft

1. Remove the four flange retaining bolts and lockwashers from the constant velocity U-joint at the transfer case. Mark the parts to reinstall them in the same position. To prevent the constant velocity joint from turning while remove the nut, use a press bar.

2. Remove the nuts and lockwashers from the U-bolts at the differential flange and remove the U-bolts.

3. Support the driveshaft and separate the U-joint at the front the driveshaft yoke, pulling backward to clear the flange. The driveshaft should never be allowed to hang by either universal joint.

4. Remove the driveshaft.

5. Installation is the reverse of removal.

Rear Driveshaft

One-Piece

1. Raise the vehicle and support safely.

2. Matchmark the driveshaft and the rear axle drive pinion gear shaft yoke.

3. Remove the rear U-joint attaching bolts and both strap clamps from the rear axle drive gear shaft yoke.

4. Fluid may run from the rear of the extension housing or transfer case when the shaft is removed, so position a suitable drain pan under the area.

5. Remove the driveshaft from the transmission or transfer case.

6. The installation is the reversal of the removal procedure. Tighten ¼ – 28 clamp bolts to 14 ft. lbs. (19 Nm) and 5/16 – 24 to 25 ft. lbs. (34 Nm).

Two-Piece

NOTE: This driveshaft has a universal joint at either end, with a third universal joint and a support bearing at the center.

1. Matchmark the shaft and the rear axle pinion hub yoke. Match-

mark the center bearing spline and slip yoke.

NOTE: Do not allow the driveshaft to hang down the during removal. Suspend it from the frame. Raise the rear of the truck to prevent loss of transmission fluid.

2. Remove both rear U-joint roller and bushing assembly clamps from the rear axle pinion yoke. Do not disturb the retaining strap used to hold the bushing assemblies on the U-joint cross.

3. Slide the rear half of the shaft off the front shaft splines at the center bearing. Remove the rear half.

4. At the transmission end of the front half, remove the bushing retaining bolts and clamps, after matchmarking. If there is a driveshaft brake, there will be flange nuts.

5. Unbolt the center bearing mounting nuts and bolts and remove the front half of the shaft.

To install:

6. Align the matchmarks at the transmission and start all the bolts and nuts at the front U-joint and the center support bearing.

7. Tighten ¼ in. clamp bolts to 170 inch lbs. (19 Nm) and ⁵⁄₁₆ in. bolts to 300 inch lbs. (34 Nm). Tighten driveshaft brake flange nuts to 35 ft. lbs. (47 Nm). Leave the center bearing bolts just snug.

8. Align the rear U-joint matchmarks and install the bushing clamps and bolts. Tighten the bolts to the torque given in Step 7. Grease the joints and splines.

9. Jack up the rear wheels and let the engine drive the shaft. The center support bearing will align itself.

10. Tighten the center bearing bolts to 50 ft. lbs. (68 Nm).

U-Joints

REMOVAL AND INSTALLATION

Single Cardan Universal Joint

1. Remove the driveshaft and slip yoke, if equipped from the vehicle.

NOTE: Do not clamp the driveshaft tube in a vise. Clamp only the forged portion of the welded yoke or the slip yoke in a vise. Do not overtighten the vise jaws.

2. Clamp the yoke in a vise and remove the bearing cap retainers.

3. Place a socket which has an inside diameter larger than the outside

diameter of the bearing cap, against the yoke around the perimeter of the first cap to be removed. Place a socket which is slightly smaller than the cap, on the cap, opposite the cap to be removed. Then position the yoke in a vise.

4. Compress the jaws until the smaller socket has driven the other cap into the larger socket.

5. Release the jaws and remove the cap that is partially out of the yoke.

6. Repeat the procedure for the remaining cap(s).

To install:

7. Clean and remove and rust from the yoke bores and lubricate lightly with suitable lithium based grease.

8. Position the spider cylinders in the yoke bores. Insert the seals into the yoke bore and against the spider cylinders. Tap the bearing caps into the yoke bores far enough to keep the spider in place.

9. Place the socket that is slightly smaller than the cap against the first cap and position the assembly in a vise.

10. Compress the jaws to force the bearing caps into the yoke bores far enough so the retainer grooves are visible.

11. Repeat the procedure for the remaining caps if necessary.

12. Install the retaining clips.

13. Install the driveshaft assembly to the vehicle.

Double Cardan Constant Velocity Joint

1. Remove the front driveshaft from the vehicle.

2. Matchmark the yokes before disassembling so they will be installed in their original locations to retain driveshaft balance.

3. To expedite removal, remove the bearing caps in the sequence indicated.

4. Support the driveshaft horizontally and aligned with the base plate of the press. Shear the bearing cap plastic retaining ring and position the first link yoke rear arm over a 1⅛ in. (29mm) socket. Place Spider Press Tool C-4365-1, or equivalent, on the bearing caps in the flange yoke arms. Force the bearing caps out of the yoke with a press.

5. If the bearing cap is not completely removed, insert a suitable spacer between the spider and bearing cap and complete the removal.

6. Rotate the driveshaft 180° and repeat the procedure.

7. Disengage the spider trunnions from the link yoke. Pull the flange yoke and the spider from the centering ball on the ball support tube yoke.

8. To remove the ball socket, separate the CV-joint between the link yoke and the flange yoke by forcing the spider trunnion bushing from the link yoke. Pull the flange yoke and the spider with the ball socket from the centering ball as an assembly.

9. Pry the seal from the ball socket and remove the washers, spring and 3 ball seats.

10. Remove the centering ball from the ball socket using Tool set C-4365, or equivalent.

To install:

11. Install the centering ball in the socket using special tool C-4365-3, or equivalent. Force the ball into the socket until it is seated firmly against the shoulder at the base of the socket.

12. To install the spider, insert one bearing cap partially into on of the yoke bores and then rotate the yoke 180°. Insert the spider into the yoke bore and seat the spider trunnion in the bearing cap. Partially insert the opposite bearing cap in the remaining yoke bore.

13. Force the bearing caps inward while pivoting the spider back and forth to provided free movement of the trunnions in the bearing.

14. When the retainer grooves become visible, install the retainer.

15. Continue to force the caps inward until the opposite retainer can be installed in its groove.

16. Lubricate the centering ball and socket with the lubricant provided in the replacement kit.

17. Repeat the installation procedure with the remaining portion of the assembly.

Halfshaft

REMOVAL AND INSTALLATION

Dakota Models

1. Disconnect the negative battery cable.

2. Raise the vehicle and support safely.

3. Remove the tire and wheel assembly.

4. Remove the cotter pin form the end of the halfshaft. Remove the nut lock, spring washer, axle nut and washer.

5. Remove the ball joint retaining bolt and pry the control arm down to release the ball stud from the steering knuckle.

6. Position a drain pan under the transaxle where the halfshaft enters the differential or extension housing. Remove the halfshaft from the transaxle where the halfshaft enters the differential or extension housing. Remove the halfshaft from the transaxle or center bearing. Unbolt the center bearing from the block and remove the intermediate shaft from the transaxle, if equipped.

To install:

7. Install the halfshaft or intermediate shaft to the transaxle, being careful not to damage the side seals. Make sure the inner joint clicks into place inside the differential. Install the outer shaft to the center bearing if equipped. Install the outer shaft to the center bearing if equipped.

8. Pull the front strut out and insert the outer joint into the front hub.

9. Turn the ball joint stud, if necessary, to position the bolt retaining indent to the inside of the vehicle. Install the ball join stud into the steering knuckle. Install the retaining bolt and nut.

10. Install the axle nut washer and nut and tighten the nut to 180 ft. lbs. (244 Nm). Install the spring washer, nut lock and a new cotter pin.

11. Install the tire and wheel assembly.

Axles

REMOVAL AND INSTALLATION

Model 44

Left Side Axle Shaft

1. Raise and support the front end on jackstands.
2. Remove the wheel.
3. Remove the caliper and support it up and out of the way. DO NOT DISCONNECT THE BRAKE LINE!
4. Remove the inboard brake pad.
5. Remove the hub-rotor assembly.
6. Unbolt and remove the splash shield and spindle from the knuckle. It may be necessary to tap the spindle loose with a soft mallet.
7. Disconnect the vacuum lines and wire at the disconnect housing on the axle.
8. Remove the disconnect housing cover, gasket and shield.
9. Carefully pull the intermediate shaft through the seal and out of the axle housing.
10. Remove the shift collar from the disconnect housing.

11. Drive out the inner axle shaft seal and remove it from the disconnect housing. Discard the seal.

NOTE: Some axles may have a seal guard. Replacement seals do not use a guard.

12. Remove the needle bearing from the intermediate shaft using tool D-330 or equivalent.
13. Remove the differential cover. Allow the oil to drain into a drain pan.
14. While pushing inward on the inner shaft, remove the C-lock from the groove on the shaft.
15. Remove the inner shaft with tools D-354-4 and adapter D-354-3 or equivalents.
16. Remove the bearing from the inner shaft with tools D-354-4, D-354-1 and C-637, or equivalents.
17. Thoroughly clean and inspect all parts. Replace any worn or damaged parts.

To install:

18. Using tools D-354-4, D-354-2 and C-637 or equivalents, install a new bearing on the inner shaft.
19. Position the inner shaft in the axle housing using tools D-354-4 and adapter D-354-3, or equivalents.
20. Carefully slide the shaft into position and install the C-lock.
21. Place a new axle shaft seal on tool D-5041-1 and position the assembly in the disconnect housing. Screw threaded bar 5041-2 through the seal and into tool 5041-1. Place tool 5041-3 and the nut on the end of the bar and tighten the nut until the tool reaches the shoulder of the bar.
22. Install the shift collar onto the splined end of the inner shaft.
23. Using tool D-328 and handle C-4171 or equivalents, install the needle bearing into the intermediate shaft.
24. Coat the splined end of the intermediate shaft with multi-purpose chassis lube and install it through the inner axle seal. Avoid damage to the seal!

25. Install the disconnect housing cover and gasket. Make sure that the shift fork indexes the groove in the shift collar.
26. Tighten the cover bolts to 10 ft. lbs. (14 Nm).
27. Connect the vacuum lines and wire. Install the clip securely around the 4-wheel drive indicator switch connector.
28. Install the splash shield and spindle. Use new nuts tightened to 30 ft. lbs. (40 Nm).
29. Install the hub and bearings.
30. Install the spacer, drive gear and snapring.
31. Apply RTV sealant on the mating edge of the cap and install it.
32. Install the inboard brake pad and caliper.
33. Thoroughly clean the carrier cover mating surfaces and install the cover using a bead of RTV sealant in place of a gasket. Tighten the cover bolts to 40 ft. lbs. (54 Nm).
34. Fill the axle with fluid.
35. Install the wheel.

Right Side Axle Shaft

1. Raise and support the front end on jackstands.
2. Remove the wheel.
3. Remove the caliper and support it up and out of the way. DO NOT DISCONNECT THE BRAKE LINE!
4. Remove the inboard brake pad.
5. Remove the hub-rotor assembly.
6. Unbolt and remove the splash shield and spindle from the knuckle. It may be necessary to tap the spindle loose with a soft mallet.
7. Remove the caliper adapter from the knuckle.
8. Carefully pull the axle shaft from the housing.
9. Thoroughly clean and inspect all parts. Replace any worn or damaged parts. The axle U-joint can be rebuilt in the same manner as a

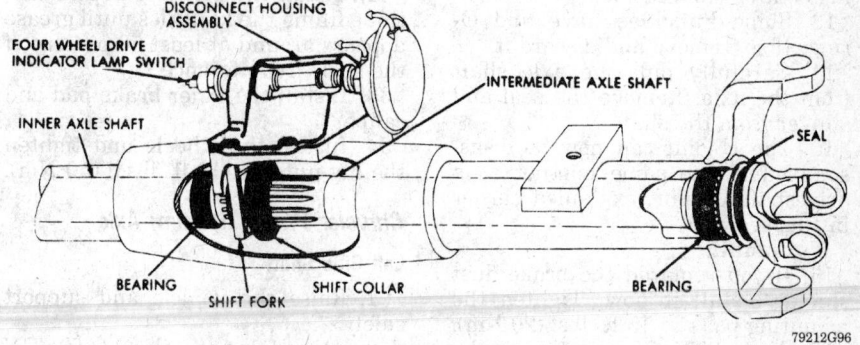

Details of the axle housing assembly

driveshaft U-joint, as previously described.

To install:

10. Install a new seal on the axle shaft stone shield with the lip facing the shaft splines.

11. Carefully slide the shaft into the housing, avoiding damage to the seal at the side gears.

12. Install the spindle and splash shield. Tighten the nuts to 30 ft. lbs. (40 Nm).

13. Install the hub/rotor assembly.

14. Install the spacer, drive gear and snapring.

15. Apply RTV sealer to the mating edge of the grease cap and install the cap.

16. Install the caliper adapter and tighten the bolts to 85 ft. lbs. (116 Nm).

17. Install the inboard pad and caliper.

18. Install the wheel.

Models With Full-Time 4-Wheel Drive

1. Remove the wheel cover.

2. Remove the cotter pin and loosen the axle shaft nut.

3. Raise and support the front end on jackstands.

4. Remove the wheels.

5. Unbolt the caliper and support it out of the way. DO NOT DISCONNECT THE BRAKE LINE!

6. Remove the inboard brake pad.

7. Remove the axle shaft nut and washer.

8. Through the hole provided in the rotor, remove the six retainer bolts.

9. Position puller C-4358, or equivalent, over the wheel lugs and install the lug nuts. Tighten the main screw of the puller to remove the hub, bearings, retainer and outer seal as an assembly.

10. Remove the puller.

11. Remove the caliper adapter from the knuckle.

12. Place a pry bar behind the inner axle shaft yoke and push the bearings out of the knuckle.

13. Some knuckles have and O-ring. If so, remove and discard it.

14. Carefully pull the axle shaft from the axle. Remove the seal and slinger from the shaft.

15. The U-joint can now be disassembled in the same manner as a driveshaft U-joint, explained earlier in this section.

To install:

16. If you removed the brake dust shield, install it now. Tighten the mounting bolts to 15 ft. lbs. (20 Nm).

17. Apply RTV silicone sealer to the sealing surfaces of the axle shaft.

18. Using a driver, install the slinger onto the shaft.

19. Install a new seal on the slinger with the lip toward the splines.

20. Carefully insert the shaft into the axle housing, being careful to avoid damage to the differential side gear seal.

21. Insert a prybar through the U-joint and wedge the shaft inward as far as it will go. Make sure that it is wedged securely and can't be moved.

22. Using adapter tool C-4398-2 and driver, or equivalents, install the seal cup in the knuckle until it is bottomed. A small amount of wheel bearing grease on the adapter will help hold the cup in position. Leave the tool in position for the time being.

23. Apply a ¼ in. (6mm) bead of RTV sealant to the retainer face on the chamfer.

24. Carefully remove the seal installing tool so that the outer shaft remains centered.

NOTE: If the shaft is touched, make sure that the seal lip is still riding inside the cup.

25. Position the bearing retainer on the knuckle so that the lube fitting is facing directly forward. This is extremely important!

26. Install the hub, rotor and bearing assembly and tighten the retainer bolts to 30 ft. lbs. (40 Nm) in a crisscross pattern.

27. Install the brake adapter. Tighten the bolts to 85 ft. lbs. (116 Nm).

28. Remove the prybar from the U-joint.

29. Install the axle shaft nut and tighten it to 100 ft. lbs. (136 Nm), then continue tightening it to align the cotter pin holes. Install the cotter pin.

30. Using the lube fitting, fill the knuckle with NLGI, Grade 2, multipurpose grease until the grease is seen flowing through the inner seal.

31. Remove the grease gun and rotate the hub several times.

32. Install the grease gun and continue filling the knuckles until grease appears around at least 50 percent of the seal circumference.

33. Install the inner brake pad and caliper.

34. Install the wheels and tighten the lug nuts to 110 ft. lbs. (150 Nm).

Chrysler 7¼ In. (184mm) Axle

Left Side Shaft

1. Raise the vehicle and support safely.

2. Disconnect the left CV driveshaft from the axle shaft flange.

3. Remove the differential housing cover. Rotate the differential case so that the differential pinion mate gear shaft lock screw is accessible. Remove the lock screw and the pinion mate gear shaft from the differential case.

4. Force the left axle shaft toward the center of the vehicle and remove the shaft C-clip lock from the recessed groove in the axle shaft.

5. Remove the axle shaft from the differential housing. Inspect the axle shaft bearing contact surface for brinelling, spalling and pitting. If any of these conditions exist, replace the axle shaft and bearing.

6. Remove the axle shaft seal from the end of the housing bore with a suitable pry bar. Remove the axle shaft bearing only if it being replaced; do not reinstall used bearings. Use removal tool C-4167 and slide hammer tool C-637, or their equivalents, to remove the bearing.

To install:

7. Thoroughly clean and dry the bearing bore in the differential housing.

8. Insert the new bearing into the pilot of bearing installation tool C-4198, or equivalent, and attach to the handle. Insert the bearing to the housing until it is seated against the bore shoulder.

9. Install the new seal using tool C-4203, or equivalent, with the flat side of the tool facing the seal. When the installation tool contacts the housing flange face, the seal is installed to the correct depth.

10. Lubricate the bearing bore and seal lip, insert the axle shaft into the housing and engage the splines. With the shaft in place, install the C-clip lock and push the shaft outward to seal the lock.

11. Install the mate shaft, align the hole in the shaft with the lock screw hole in the differential case and install the lock screw. Tighten the lock screw to 8 ft. lbs. (11 Nm).

12. Thoroughly clean and dry the case cover, mating surface, bolts and bolt holes. Apply silicone sealer to the cover and install.

13. Connect the CV driveshaft.

14. Level the vehicle and fill the differential with multi-purpose gear lubricant.

15. Road test the vehicle and check for leaks and correct front axle operation.

Right Side and Intermediate Shaft

1. Raise the vehicle and support safely.

2. Remove the tire and wheel assembly.

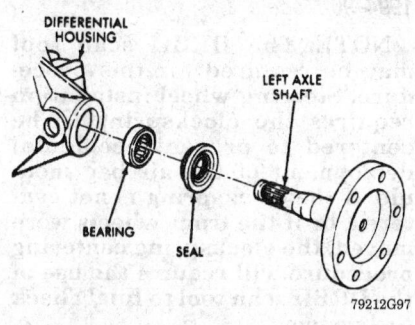

Exploded view of the left axle shaft, seal and bearing on the 7¼ in. axle

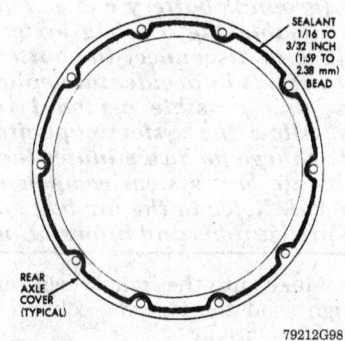

Correct sealant application on the housing cover

3. Remove the cotter pin, nut lock and spring washer from the stub shaft. Remove the hub nut and washer.

4. Remove the bolts that attach the inner CV-joint to the axle shaft flange.

5. Separate the stub shaft splines from the hub bearing splines and remove the CV-joint shaft.

6. Remove the differential housing cover.

7. Disconnect the 4WD indicator lamp switch and the vacuum tube from the shift motor. Remove the shift motor, housing cover and gasket from the shift motor housing.

8. Remove the bearing seal/retainer attaching screws accessible via the holes in the axle shaft flange.

9. Remove the outer axle shaft from the differential housing.

10. Remove the snapring and the splined gear from the outer axle shaft. Separate the bearing and seal from the outer axle shaft.

11. Remove the shift collar from the shift motor housing.

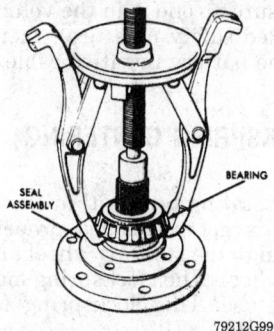

Use a puller to remove the bearing and seal assembly from the axle shaft

12. Rotate the differential case so that the different pinion mate gear shaft lock screw is accessible. Remove the lock screw and the pinion mate gear shaft from the differential case.

13. Force the intermediate shaft toward the center of the vehicle and remove the shaft C-clip lock from the recessed groove in the axle shaft. Remove the intermediate shaft from the differential housing and tube.

To install:

14. Insert the axle shaft into the tube and housing and engage the splines. With the shaft in place, install the C-clip lock and push the shaft outward to seal the lock.

15. Install the mate shaft, align the hole in the shaft with the lock screw hole in the differential case and install the lock screw. Tighten the lock screw to 8 ft. lbs. (11 Nm).

16. Install the shift collar to the intermediate shaft.

17. Install a new seal to the seal retainer and install to the outer shaft. Press on the new bearing, install the splined gear and the snapring.

18. Insert the outer axle shaft into the shift motor housing. Tighten the attaching screws to 17 ft. lbs. (23 Nm).

19. Install the shift motor housing cover and gasket. Ensure that the shift fork is correctly engaged in the collar groove. Install the cover bolts. Connect the vacuum tubes to the shift motor and the connector to the 4WD lamp switch.

20. Thoroughly clean and dry the case cover, mating surface, bolts and bolt holes. Apply silicone sealer to the cover and install.

21. Lubricate the contact surface area of the CV driveshaft wear sleeve with grease and install the CV

driveshaft. Tighten the hub nut to 190 ft. lbs. (258 Nm). Install the spring washer, nut lock and a new cotter pin.

22. Level the vehicle and fill the differential with multi-purpose gear lubricant.

23. Road test the vehicle and check for leaks and correct front axle operation.

Model 60 Axle

1. Raise and support the front end on jackstands.

2. Remove the caliper from the knuckle and wire it out of the way.

3. Remove the locking hub.

4. Remove the front wheel bearing.

5. Remove the hub and rotor assembly.

6. Remove the spindle-to-knuckle bolts. Tap the spindle from the knuckle using a plastic mallet.

7. Remove the splash shield and a caliper support.

8. Pull the axle shaft out through the knuckle.

9. Using a slidehammer and bearing cup puller, remove the needle bearing from the spindle.

10. Clean the spindle bore thoroughly and make sure that it is free of nicks and burrs. If the bore is excessively pitted or scored, the spindle must be replaced.

To install:

11. Insert a new spindle bearing in its bore with the printing facing outward. Drive it into place with a driver. Install a new bearing seal with the lip facing away from the bearing.

12. Pack the bearing with waterproof wheel bearing grease.

13. Pack the thrust face of the seal in the spindle bore and the V-seal on the axle shaft with waterproof wheel bearing grease.

14. Carefully guide the axle shaft through the knuckle and into the housing. Align the splines and fully seat the shaft.

15. Place the bronze spacer on the shaft. The chamfered side of the space must be inboard.

16. Install the splash shield and caliper support.

17. Place the spindle on the knuckle and install the bolts. Tighten the bolts to 50–60 ft. lbs. (68–82 Nm).

18. Install the hub/rotor assembly on the spindle.

19. Assemble the wheel bearings.

20. Assemble the locking hub.

STEERING

Air Bag

—— CAUTION ——

Some vehicles are equipped with an air bag system, also known as the Supplemental Inflatable Restraint (SIR) or Supplemental Restraint System (SRS). The system must be disabled before performing service on or around system components, steering column, instrument panel components, wiring and sensors. Failure to follow safety and disabling procedures could result in accidental air bag deployment, possible personal injury and unnecessary system repairs.

PRECAUTIONS

Several precautions must be observed when handling the inflator module to avoid accidental deployment and possible personal injury.
• Never carry the inflator module by the wires or connector on the underside of the module.
• When carrying a live inflator module, hold securely with both hands, and ensure that the bag and trim cover are pointed away.
• Place the inflator module on a bench or other surface with the bag and trim cover facing up.
• With the inflator module on the bench, never place anything on or close to the module which may be thrown in the event of an accidental deployment.

DISARMING

1. First read the system precautions.
2. Disconnect and isolate the negative battery cable.
3. If the air bag module is undeployed, wait two minutes for the system capacitor to discharge.

ARMING THE SYSTEM

Assuming the system components (air bag control module, sensors, air bag, etc.) are installed correctly and are in good working order, the system is armed whenever the battery positive and negative battery cables are connected.

If you have disarmed the air bag system for any reason, to re-arm,

make sure no one is in the vehicle (as an added safety measure), then connect the battery negative cable.

CLOCKSPRING CENTERING

If the rotating tape within the clockspring is not positioned properly in relation to the steering wheel and the front wheels, the clockspring may fail during use. The clockspring MUST BE CENTERED.

1. Place the front wheels in the straight-ahead position.
2. Disconnect the negative battery cable and wait at least two full minutes.
3. Remove the air bag module. Remove the steering wheel with a suitable puller tool.
4. Unlock the two plastic autolocking tabs.
5. Keep the locking mechanism disengaged while rotating the clockspring rotor CLOCKWISE to the end of its travel. Do not apply excessive torque.
6. From the end of travel, rotate the rotor 2 1/2 turns COUNTER CLOCKWISE. The horn wire should end up at the top, and the air bag wire at the bottom.
7. Install the steering wheel and the air bag module.
8. Perform an air bag system check BEFORE connecting the negative battery cable.

Steering Wheel

REMOVAL AND INSTALLATION

1993 Models

1. Disconnect the battery ground cable.
2. If applicable, unscrew the horn pad from behind the steering wheel.
3. Working through the access holes in the back of the wheel, push the horn pad off. Do NOT pry the pad off!
4. Disconnect the horn wire.
5. Matchmark the steering wheel and shaft.
6. Remove the steering wheel retaining nut.
7. Using a puller, remove the steering wheel from the shaft. NEVER hammer the shaft to free the wheel!
8. Installation is the reverse of removal. Tighten the nut to 45 ft. lbs. (61 Nm).

1994–96 Models

NOTE: The DRBII scan tool may be required for this procedure. Steering wheel installation requires the clockspring to be centered to prevent accidental deployment of the air bag module. If the clockspring is not centered, or if the front wheels were moved, the clockspring centering procedure will require the use of the DRBII scan tool to final check the system.

—— CAUTION ——

Disconnect and isolate the negative (ground) battery cable. This will disable the air bag system. Failure to disconnect the battery could result in accidental deployment and possible personal injury. Allow the system capacitor to discharge for two minutes then begin air bag system component removal. Refer to the air bag system precautions and information.

1. Make sure the front wheels are straight and the steering column is locked in place.
2. Disconnect the negative battery cable at the battery and isolate. Wait 2 minutes for the reserve capacitor to discharge before removing non-deployed module.
3. Remove the 4 nuts attaching the air bag module from the back side of the steering wheel.
4. Lift the module and unplug the connector from the rear of the module.

NOTE: Refer to precautions in this section concerning the handling of a live air bag module.

5. Remove the vehicle speed control switch and connector, if so equipped, or the cover.
6. Remove the steering wheel retaining nut. The steering wheel and shaft are master splined for installation reference.
To install:

—— CAUTION ——

If the clockspring is not properly positioned or if the front wheels were moved, follow the clockspring centering procedure as outlined in this section before installing the steering wheel.

7. Pull the air bag and speed control wires through the lower, larger hole in the steering wheel and horn wire through the smaller hole at the top. Be sure not to pinch the wires.
8. With the front wheels in a straight-ahead position, position the

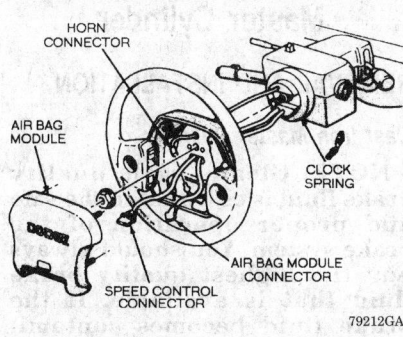

HORN
CONNECTOR

AIR BAG
MODULE

CLOCK
SPRING

AIR BAG MODULE
CONNECTOR

SPEED CONTROL
CONNECTOR

79212GA0

Exploded view of the air bag steering wheel mounting

steering wheel on the steering column by aligning the master splines.

9. Install, then tighten the nut to 45 ft. lbs. (61 Nm). Do not attempt to force the wheel onto the shaft. The force of the nut is all that is needed to properly seat the steering wheel.

10. Install the wires to the horn buttons, speed control and air bag module.

11. Connect the clockspring wiring to the module by pressing straight in on the connector. It should latch securely beneath the module locking clip.

12. Mount the air bag module, install the 4 nuts, tightening them to 80–100 inch lbs. (9–11 Nm).

13. Do not connect the negative battery until you perform an Air Bag System Check as outlined in this section.

Tie Rod Ends

REMOVAL AND INSTALLATION

1. Raise and safely support the vehicle on jackstands.

2. Remove the cotter pin and nut from the tie rod end.

3. Using a suitable puller, remove the tie rod from the steering knuckle of center link.

NOTE: Either count the number of visible threads or paint an alignment reference mark before unscrewing the end from the tie rod. This should enable you to thread the tie rod end back to NEAR the original position. Even if this is done, a front end alignment check should still be performed.

4. Loosen the sleeve clamp nut and bolt and unscrew the tie rod end from the sleeve.

To install:

5. Install the tie rod end in the sleeve. Screw it in as many turns as

the tie rod end was when removed (or to the painted mark).

6. Install the tie rod end into the steering knuckle. Install the castellated nut, tightening it to 45 ft. lbs. (61 Nm). Always install a new cotter pin.

7. Have the alignment checked at a reputable repair facility.

Manual Rack and Pinion

REMOVAL AND INSTALLATION

Dakota Models

NOTE: This procedure applies to 2WD Dakota models only.

1. Disconnect the negative battery cable.

2. Remove the tie rod ends from the steering knuckle.

3. Remove the coupling pin.

4. Remove the bolts attaching the rack to the crossmember.

5. Remove the gear from the vehicle.

6. The installation is the reverse of the removal procedure. Tighten the mounting bolts to 150 ft. lbs. (203 Nm).

Power Rack and Pinion

REMOVAL AND INSTALLATION

Dakota Models

NOTE: This procedure applies to 2WD Dakota models only.

1. Disconnect the negative battery cable.

2. Remove the tie rod ends from the steering knuckle.

3. Disconnect and plug the power steering fluid lines.

4. Remove the coupling pin.

5. Remove the bolts attaching the rack to the crossmember.

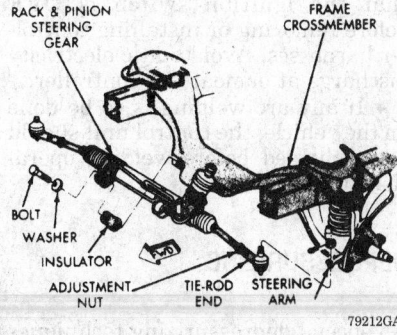

RACK & PINION
STEERING
GEAR

FRAME
CROSSMEMBER

BOLT

WASHER

INSULATOR

ADJUSTMENT
NUT

TIE-ROD
END

STEERING
ARM

79212GA1

Rack-and-pinion mounting location–2WD Dakota models

6. Remove the gear from the vehicle.

7. The installation is the reverse of the removal procedure. Tighten the mounting bolts to 150 ft. lbs. (203 Nm). Refill the power steering pump.

Power Steering Gear

REMOVAL AND INSTALLATION

B-Series Van, Ram Pick-Up, Ramcharger

——— WARNING ———
To avoid accidental air bag deployment and possible personal injury, always disconnect and isolate the negative battery cable. Allow 2 minutes to elapse before beginning any component removal.

1. Place the front wheels in straight-ahead position.

2. Disconnect the negative battery cable.

3. Remove the windshield washer solvent reservoir and the coolant overflow tank, if necessary.

4. Position a drain pan under the steering gear.

5. Disconnect the fluid hoses from the gear and plug them.

6. Disconnect the steering column shaft from the stub shaft.

7. Raise the vehicle and support safely.

8. On 2WD vehicles, matchmark and remove the pitman arm from the center link.

9. On 4WD vehicles, disconnect the drag link from the pitman arm. Remove the pitman arm from the pitman shaft.

10. Remove the retaining bolts and remove the steering gear from the vehicle.

To install:

11. On 2WD vehicles, position steering gear at frame rail and install bolts loosely. Align steering shaft and stub shaft and install bolts to 33 ft. lbs. (45 Nm) torque. Realign gear at frame and torque bolts to 100 ft. lbs. (136 Nm).

12. On 4WD vehicles, install steering gear to reinforcement and tighten screws to 100 ft. lbs. (136 Nm). Position steering gear at frame rail and install bolts loosely. Align steering shaft and stub shaft and install bolts to 33 ft. lbs. (45 Nm) torque. Realign gear at frame and torque bolts to 100 ft. lbs. (136 Nm).

13. Install pitman arm to steering shaft and torque nut to 175 ft. lbs.

(237 Nm). Connect steering linkage to arm. Install replacement cotter pins.

14. Lower the vehicle.

15. Connect the negative battery cable.

16. Road test the vehicle for proper operation.

Power Steering Pump

BLEEDING

1. Fill the reservoir with power steering fluid.

2. Turn the wheels to the full left turn position and add fluid until the reservoir is full.

3. Start the engine and add fluid to bring the level to the correct level.

4. To purge the system of air, turn the steering wheel from side to side without contacting the stops.

5. Return the wheel to the straight-ahead position and operate the engine for 2 minutes before road testing. This should bleed the system completely.

REMOVAL AND INSTALLATION

1. Disconnect the negative battery cable.

2. Position a drain pan under the power steering pump.

3. Disconnect the fluid hoses from the pump and plug them.

4. Remove the front bracket attaching bolts and remove the belt from the pulley.

5. Remove the rear pump to bracket nut and remove the pump.

6. Remove the bracket from the pump.

7. Remove the pulley from the pump with the proper puller. Install the pulley on the new pump using the special installation tools.

8. Installation is the reversal of the removal procedure.

BRAKES

Anti-Lock Brake System Service

PRECAUTIONS

• Certain components within the Anti-Lock Brake System (ABS) are not intended to be serviced or repaired individually. Only those components with removal and installation procedures should be serviced.

• Do not use rubber hoses or other parts not specifically specified for and ABS system. When using repair kits, replace all parts included in the kit. Partial or incorrect repair may lead to functional problems and require the replacement of components.

• Lubricate rubber parts with clean, fresh brake fluid to ease assembly. Do not use lubricated shop air to clean parts; damage to rubber components may result.

• Use only specified brake fluid from an unopened container.

• If any hydraulic component or line is removed or replaced, it may be necessary to bleed the entire system.

• A clean repair area is essential. Always clean the reservoir and cap thoroughly before removing the cap. The slightest amount of dirt in the fluid may plug an orifice and impair the system function. Perform repairs after components have been thoroughly cleaned; use only denatured alcohol to clean components. Do not allow ABS components to come into contact with any substance containing mineral oil; this includes used shop rags.

• The Anti-Lock control unit is a microprocessor similar to other computer units in the vehicle. Ensure that the ignition switch is **OFF** before removing or installing controller harnesses. Avoid static electricity discharge at or near the controller.

• If any arc welding is to be done on the vehicle, the control unit should be unplugged before welding operations begin.

DEPRESSURIZING

No special depressurizing techniques are necessary with the models covered by this manual.

Master Cylinder

REMOVAL AND INSTALLATION

Cast Iron Master Cylinder

NOTE: Clean, high quality brake fluid is essential to the safe and proper operation of the brake system. You should always buy the highest quality brake fluid that is available. If the brake fluid becomes contaminated with dirt or water, drain and flush the system. Never reuse any brake fluid. Any brake fluid that is removed from the system should be discarded.

1. Disconnect the brake lines from the master cylinder.

2. Remove the bolt securing the pushrod to the pedal linkage.

3. Remove the master cylinder attaching nuts.

4. Slide the master cylinder from the truck.

To install:

5. Position the master cylinder to the booster.

6. Install the and tighten the mounting nuts to 16 ft. lbs. (22 Nm).

7. Connect the pushrod.

8. Install all the brake lines.

9. Bleed the master cylinder and brake system.

Aluminum Master Cylinder

1. Disconnect the brake lines from the master cylinder.

2. Remove the master cylinder mounting nuts.

3. Slide the master cylinder out.

To install:

4. Position the master cylinder to the booster.

5. Install the and tighten the mounting nuts to 16 ft. lbs. (22 Nm).

6. Install all the brake lines.

7. Bleed the master cylinder and brake system.

Brake Caliper

REMOVAL AND INSTALLATION

Dakota Models

1. Raise and support the front end on jackstands.

2. Remove the wheels.

3. Disconnect the rubber brake hose from the tubing at the frame mount. If the pistons are to be removed from the caliper, leave the brake hose connected to the caliper. Check the rubber hose for cracks or chafed spots.

4. Plug the brake line to prevent loss of fluid.

5. Remove the caliper slide pins.

6. Remove the caliper and brake pads from the rotor adapter.

To install:

7. Position the outboard shoe in the caliper. The shoe should not rattle in the caliper. If it does, or if any movement is obvious, bend the shoe tabs over the caliper to tighten the fit.

8. Slide the caliper into position on the adaptor and over the rotor.

------- **WARNING** -------

Take great care to avoid dislodging the piston dust boot!

9. Align the caliper and start the pins in by hand.

10. Tighten the pins to 22 ft. lbs. (30 Nm).

11. Connect the brake hose to the caliper. Use new washers to attach the hose fitting if the original washers are scored, worn or damaged.

12. Fill and bleed the brake system.

13. Install the wheels.

14. Lower the vehicle.

Chrysler Sliding Caliper

1. Raise and support the front end on jackstands.

2. Remove the wheels.

3. Disconnect the rubber brake hose from the tubing at the frame mount. If the pistons are to be removed from the caliper, leave the brake hose connected to the caliper. Check the rubber hose for cracks or chafed spots.

4. Plug the brake line to prevent loss of fluid.

5. Remove the retaining screw, clip and anti-rattle spring that attach the caliper to the adaptor.

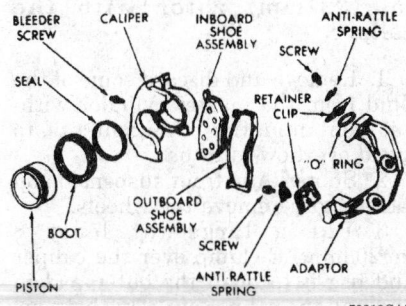

Exploded view of the Chrysler sliding caliper and brake pad assembly

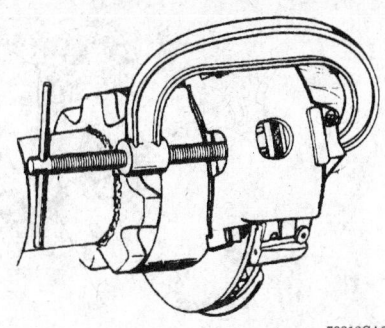

79212GA6

Use a large C-clamp to bottom the piston in the bore

6. Carefully slide the caliper out and away from the disc. Check the pads to be sure that they are reinstalled in the same position.

To install:

7. Position the outboard shoe in the caliper. The shoe should not rattle in the caliper. If it does, or if any movement is obvious, bend the shoe tabs over the caliper to tighten the fit.

8. Install the inboard shoe.

9. Slide the caliper into position on the adaptor and over the rotor.

------- **WARNING** -------

Take great care to avoid dislodging the piston dust boot!

10. Install the anti-rattle springs and retaining clips and tighten the retaining screws to 16 ft. lbs. (22 Nm).

NOTE: The inboard shoe must always be installed on top of the retainer spring plate.

11. Fill the system with fresh fluid and bleed the brakes.

12. Lower the vehicle.

Bendix Sliding Caliper

1. Support the front end on jackstands. Remove the wheel and tire.

2. Disconnect the brake hose. Cap the hose and plug the caliper.

3. Remove the key retaining screw and drive the key out with a brass drift.

4. Rotate the key end of the caliper out and slide the other end out.

5. Thoroughly clean the sliding areas.

To install:

6. Position the caliper rail into the slide on the support and rotate the caliper onto the rotor. Start the key and spring by hand. The spring should be between the key and cali-

per and the spring ends should overlap the key. If necessary, use a screwdriver to hold the caliper against the support assembly. Drive the key and spring into position, aligning the correct notch with the hole in the support. Install the key retaining screw and tighten to 12–20 ft. lbs. (16–27 Nm).

7. Bleed the system of air. Replace the wheel and tire and lower the truck to the floor.

Delco Sliding Caliper

NOTE: Most 1994–96 models use the Delco sliding caliper.

1. Raise and support the front end on jackstands.

2. Remove the wheels.

3. Press the caliper piston back into the bore with a suitable prytool. Use a large C-clamp to drive the piston into the bore of additional force is required.

4. Remove the caliper mounting bolts with a 3/8 in. hex wrench or socket.

5. Loosen the bolt that secures the front brake hose fitting bolt in the caliper.

6. Rotate the caliper rearward off the rotor and out from its mount.

7. Remove the front brake hose fitting bolt completely, then remove the caliper with the pads installed as an assembly. Take care not to drip fluid onto the pad surfaces.

------- **WARNING** -------

DO NOT depress the brake pedal with the caliper removed!

8. Cover the open end of the front brake hose fitting to prevent dirt entry.

To install:

9. Clean the caliper and steering knuckle sliding surfaces with a wire brush. Then, apply a coat of Mopar® multi-mileage grease, or equivalent.

10. Lubricate the caliper mounting bolts, collars, bushings and bores with Dow 111®, or GE 661® silicone grease, or equivalent. (Please refer to illustrations).

11. Install the caliper over the rotor and seat it flush in its original position.

12. Install the mounting bolts by hand, then tighten them to 38 ft. lbs. (51 Nm).

13. Install the wheels.

14. Lower the vehicle.

15. Pump the brakes several times to seat the pads.

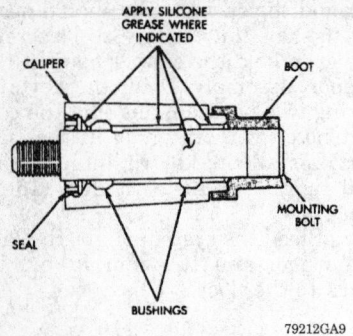

Mounting bolt lubrication points —
1994–96 Ram Pick-up light duty caliper

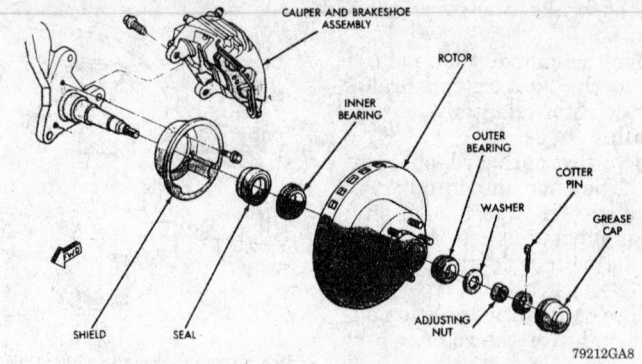

Exploded view of the Dakota disc brake components

Disc Brake Pads

REMOVAL AND INSTALLATION

———— WARNING ————
NEVER replace the pads on one side only! ALWAYS replace pads on both wheels as a set!

Dakota Models

NOTE: Dakota models use a single piston sliding caliper.

1. Raise and support the front end on jackstands.
2. Remove the wheels.
3. Press the caliper piston back into the bore with a suitable prytool. Use a large C-clamp to drive the piston into the bore of additional force is required.
4. Remove the caliper mounting bolts with a ³/₈ in. hex wrench or socket.
5. Rotate the caliper rearward off the rotor and out from its mount.
6. Set the caliper on a crate or sturdy box, then remove the inboard and outboard brake pads. The inboard pad has a spring clip that holds it in the caliper. Tilt this pad out at the top to unseat the clip. The outboard pad has a retaining spring that secures it in the caliper. Unseat one spring end and rotate the pad out of the caliper.

———— WARNING ————
DO NOT depress the brake pedal with the pads removed!

7. Secure the caliper to a chassis or suspension component with a sturdy wire. Do not let it hang from the hose.
To install:
8. Clean the caliper and steering knuckle sliding surfaces with a wire brush. Then, apply a coat of Mopar® multi-mileage grease, or equivalent.

9. Clean the caliper slide pins with brake cleaner or brake fluid. Then apply a light coating of silicone grease to the pins.

NOTE: If there is minor rust or corrosion on the pins, first polish them with a crocus cloth. If they are severely rusted, replace them.

10. Install the inboard pad and its spring clip.
11. Install the outboard brake pad.
12. Install the caliper over the rotor and seat it flush in its original position.
13. Install the mounting bolts by hand, then tighten them to 38 ft. lbs. (51 Nm).
14. Final tighten the caliper slide pins to 18–26 ft. lbs. (25–35 Nm).
15. Install the wheels.
16. Lower the vehicle.
17. Pump the brakes several times to seat the pads.

Chrysler Sliding Caliper

NOTE: The 1993 D100/150, W100/150 and Ramcharger models use the Chrysler sliding caliper with the 11.75 in. (298.5mm) rotor. The 1993 D200/250 and D300/350 and W200/250 with the Dana 44 axle use the Chrysler sliding caliper with a 12.82 in. (325.6mm) rotor.

1. Raise and support the front end on jackstands.
2. Remove the wheels.
3. Remove the caliper retaining clips and anti-rattle springs.
4. Remove the caliper from the disc by slowly sliding the caliper and brake pad assembly out and away from the disc. Do not damage the flexible brake hose.
5. Drain some of the fluid from the master cylinder.
6. Remove the outboard pad from the caliper by prying between the pad and the caliper fingers. Remove the

inboard pad from the caliper support by the same method. DO NOT depress the brake pedal with the pads removed!
To install:
7. Push the pistons to the bottom of their bores. This may be done with a large C-clamp or a pair of large pliers by placing a flat metal bar against the pistons and depressing the pistons with a steady force. This operation will displace some of the fluid in the master cylinder.
8. Slide the new pads into the caliper and caliper support. The ears of the pad should rest on the bridges of the caliper.
9. Install the caliper on the disc and install the caliper retaining clips, pins and anti-rattle springs. Pump the brake pedal until it is firm.
10. Check the fluid level in the master cylinder and add fluid as needed.
11. Install the wheels.
12. Road test the truck. The truck may pull to one side, but the pull should disappear shortly as the pads are seated.

Bendix Sliding Caliper

NOTE: This brake type is utilized by the W200/250 and W300/350 with the Dana or Spicer 60 axle. These models use a 12.88 in. (327mm) rotor with the caliper.

1. Remove and discard some of the fluid from the master cylinder without contaminating the contents to avoid overflow later on.
2. Support the front suspension on jackstands. Remove the wheels.
3. Put a large (at least 8 in./203mm) C-clamp over the caliper and use it to push the outer pad in and pull the caliper out. This bottoms the caliper piston in its bore.
4. Remove the key retaining screw. Drive the caliper support key

and spring out toward the outside, using a brass drift.

5. Push the caliper down and rotate the upper end up and out. Support the caliper, so as not to damage the brake hose.

6. Remove the outer pad from the caliper. You may have to tap it to loosen it. Remove the inner pad, removing the anti-rattle clip from the lower end of the pad.

To install:

7. Thoroughly clean the sliding contact areas on the caliper and adaptor.

8. Put the new anti-rattle clip on the lower end of the new inner pad. Put the pad and clip in the pad abutment with the clip tab against the abutment and the loop-type spring away from the disc. Compress the clip and slide the upper end of the pad into place.

9. If the caliper piston isn't bottomed, press it in with a C-clamp.

10. The replacement outer pad may differ slightly from the original equipment. Put the outer pad in place and press the tabs into place with your fingers. You can press the tabs in with a C-clamp, but be careful of the lining.

11. Position the caliper on the adaptor by pivoting it around the upper mounting surface. Be careful of the boot.

12. Use a suitable prytool to hold the upper machined surface of the caliper against the support assembly. Drive a new key and spring assembly into place with a plastic mallet. Install the retaining screw and tighten to 12–20 ft. lbs. (16–27 Nm).

13. Replace the wheels and tires and lower the truck to the floor. Fill the master cylinder. Depress the brake pedal firmly several times to seat the pads on the disc. Don't drive until you get a firm pedal.

Delco Sliding Caliper

NOTE: The 1994–96 Dakota and Ram Pick-up models use the Delco sliding caliper.

1. Raise and support the front end on jackstands.

2. Remove the wheels.

3. Press the caliper piston back into the bore with a suitable prytool. Use a large C-clamp to drive the piston into the bore of additional force is required.

4. Remove the caliper mounting bolts with a 3/8 in. hex wrench or socket.

5. Rotate the caliper rearward off the rotor and out from its mount.

6. Set the caliper on a crate or sturdy box, then remove the inboard and outboard brake pads. The inboard pad has a spring clip that holds it in the caliper. Tilt this pad out at the top to unseat the clip. The outboard pad has a retaining spring that secures it in the caliper. Unseat one spring end and rotate the pad out of the caliper.

———— **WARNING** ————

DO NOT depress the brake pedal with the pads removed!

7. Secure the caliper to a chassis or suspension component with a sturdy wire. Do not let it hang from the hose.

To install:

8. Clean the caliper and steering knuckle sliding surfaces with a wire brush. Then, apply a coat of Mopar® multi-mileage grease, or equivalent.

NOTE: If there is minor rust or corrosion on the pins, first polish them with a crocus cloth. If they are severely rusted, replace them.

9. Lubricate the caliper mounting bolts, collars, bushings and bores with Dow 111®, or GE 661® silicone grease, or equivalent. (Please refer to illustrations).

10. Install the inboard pad and its spring clip.

11. Install the outboard brake pad.

12. Install the caliper over the rotor and seat it flush in its original position.

13. Install the mounting bolts by hand, then tighten them to 38 ft. lbs. (51 Nm).

14. Install the wheels.

15. Lower the vehicle.

16. Pump the brakes several times to seat the pads.

Brake Rotor

REMOVAL AND INSTALLATION

1. Raise and safely support the front of the truck with jackstands. Remove the front wheel.

2. Remove the caliper assembly and support it to the frame with a piece of wire without disconnecting the brake fluid hose.

3. Remove the hub and rotor assembly.

4. Install the rotor in the reverse order of removal, and adjust the wheel bearing as outlined in this section.

Brake Drums

REMOVAL AND INSTALLATION

Chrysler Servo Type with Single Anchor

1. Raise and safely support the truck.

2. Remove the plug from the brake adjustment access hole.

3. Insert a thin bladed screwdriver through the adjusting hole and hold the adjusting lever away from the starwheel.

4. Release the brake by prying down against the starwheel with a brake spoon.

5. Remove the rear wheel and clips from the wheel studs. Remove the brake drum.

6. Installation is the reverse of removal. Adjust the brakes.

Bendix Duo-Servo Type

1. Raise and safely support the vehicle.

2. Remove the rear wheel and tire.

3. Remove the axle shaft nuts, washers and cones. If the cones do not readily release, rap the axle shaft sharply in the center.

4. Remove the axle shaft.

5. Remove the outer hub nut.

6. Straighten the lockwasher tab and remove it along with the inner nut and bearing.

7. Carefully remove the drum.

To install:

8. Position the drum on the axle housing.

9. Install the bearing and inner nut. While rotating the wheel and tire, tighten the adjusting nut until a slight drag is felt.

10. Back off the adjusting nut 1/6 turn so that the wheel rotates freely without excessive end-play.

11. Install the lockrings and nut. Place a new gasket on the hub and install the axle shaft, cones, lockwashers and nuts.

12. Install the wheel and tire.

13. Road test the vehicle.

Brake Shoes

REMOVAL AND INSTALLATION

Servo Type With Single Anchor

1. Raise and support the vehicle.

2. Remove the rear wheel, drum retaining clips and the brake drum.

3. Remove the brake shoe return springs, noting how the secondary spring overlaps the primary spring.

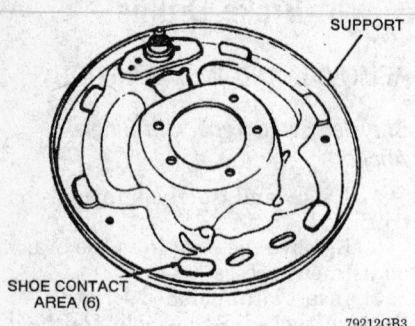

Shoe contact points on the backing plate

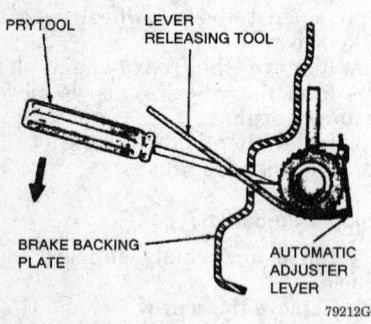

Use a lever releasing tool to depress the adjuster lever while turning the starwheel with a prytool — Bendix brakes

4. Remove the brake shoe retainer, springs and nails.

5. Disconnect the automatic adjuster cable from the anchor and unhook it from the lever. Remove the cable, cable guide, and anchor plate.

6. Remove the spring and lever from the shoe web.

7. Spread the anchor ends of the primary and secondary shoes and remove the parking brake spring and strut.

8. Disconnect the parking brake cable and remove the brake assembly.

9. Remove the primary and secondary brake shoe assemblies and the star adjuster as an assembly. Block the wheel cylinders to retain the pistons.

To install:

10. Measure the drum as described in this section.

11. Apply a thin coat of lubricant to the support platforms.

12. Attach the parking brake lever to the back side of the secondary shoe.

13. Place the primary and secondary shoes in their relative positions on a workbench.

14. Lubricate the adjuster screw threads. Install it between the primary and secondary shoes with the star wheel next to the secondary shoe. The star wheels are stamped with an L (left) and R (right).

15. Overlap the ends of the primary and second brake shoes and install the adjusting spring and lever at the anchor end.

16. Hold the shoes in position and install the parking brake cable into the lever.

17. Install the parking brake strut and spring between the parking brake lever and primary shoe.

18. Place the brake shoes on the support and install the retainer nails and springs.

19. Install the anchor pin plate.

20. Install the eye of the adjusting cable over the anchor pin and install the return spring between the anchor pin and primary shoe.

21. Install the cable guide in the secondary shoe and install the secondary return spring. Be sure that the primary spring overlaps the secondary spring.

22. Position the adjusting cable in the groove of the cable guide and engage the hook of the cable in the adjusting lever.

23. Install the brake drum and retaining clips. Install the wheel and tire.

24. Adjust the brakes and road test the truck.

Bendix Duo-Servo Type

1. Unhook and remove the adjusting lever return spring.

2. Remove the lever from the lever pivot pin.

3. Unhook the adjuster lever from the adjuster cable.

4. Unhook the upper shoe-to-shoe spring.

5. Unhook and remove the shoe hold-down springs.

6. Disconnect the parking brake cable from the parking brake lever.

7. Remove the shoes with the lower shoe-to-shoe spring and star wheel as an assembly.

To install:

8. The pivot screw and adjusting nut on the left side have left-hand threads and right hand threads on the right side.

9. Lubricate and assemble the star wheel assembly. Lubricate the guide pads on the support plates.

10. Assemble the star wheel, lower shoe-to-shoe spring, and the primary and secondary shoes. Position this assembly on the support plate.

11. Install and hook the hold-down springs.

12. Install the upper shoe-to-shoe spring.

13. Install the cable and retaining clip.

14. Position the adjuster lever return spring on the pivot (green springs on left brakes and red springs on right brakes).

15. Install the adjuster lever. Route the adjuster cable and connect it to the adjuster.

16. Install the brake drum and adjust the brakes.

Wheel Cylinder

REMOVAL AND INSTALLATION

When the brake drums are removed, carry out an inspection of the wheel cylinder boots for cuts, tears, cracks, or leaks. If any of these are present, the wheel cylinder should be replaced.

NOTE: Preservative fluid is used during assembly; its presence in small quantities does not indicate a leak.

1. Remove the brake shoes and check them. Replace them if they are soaked with grease or brake fluid.

2. Detach the brake hose.

3. Unfasten the wheel cylinder attachment bolts and slide the wheel cylinder off its support.

To install:

4. Install the wheel cylinder on its support.

5. Attach the jumper tube to the wheel cylinder. Install the brake hose on the frame bracket. Connect the brake line to the hose. Connect the end of the brake hose. Connect the end of the brake hose through the end of the stand-off. Attach the jumper tube to the brake hose and attach the hose to the stand-off.

Parking Brake Cable

ADJUSTMENT

1. Adjust the service brakes by making a few stops in reverse.

2. Raise and support the rear end on jackstands.

3. Release the parking brake lever and loosen the cable adjusting nut to be sure that the cable is slack.

4. Tighten the cable adjusting nut until a slight drag is felt while rotating the wheels.

5. Loosen the cable adjusting nut until the wheels can be rotated freely, then back off the cable adjusting nut two turns.

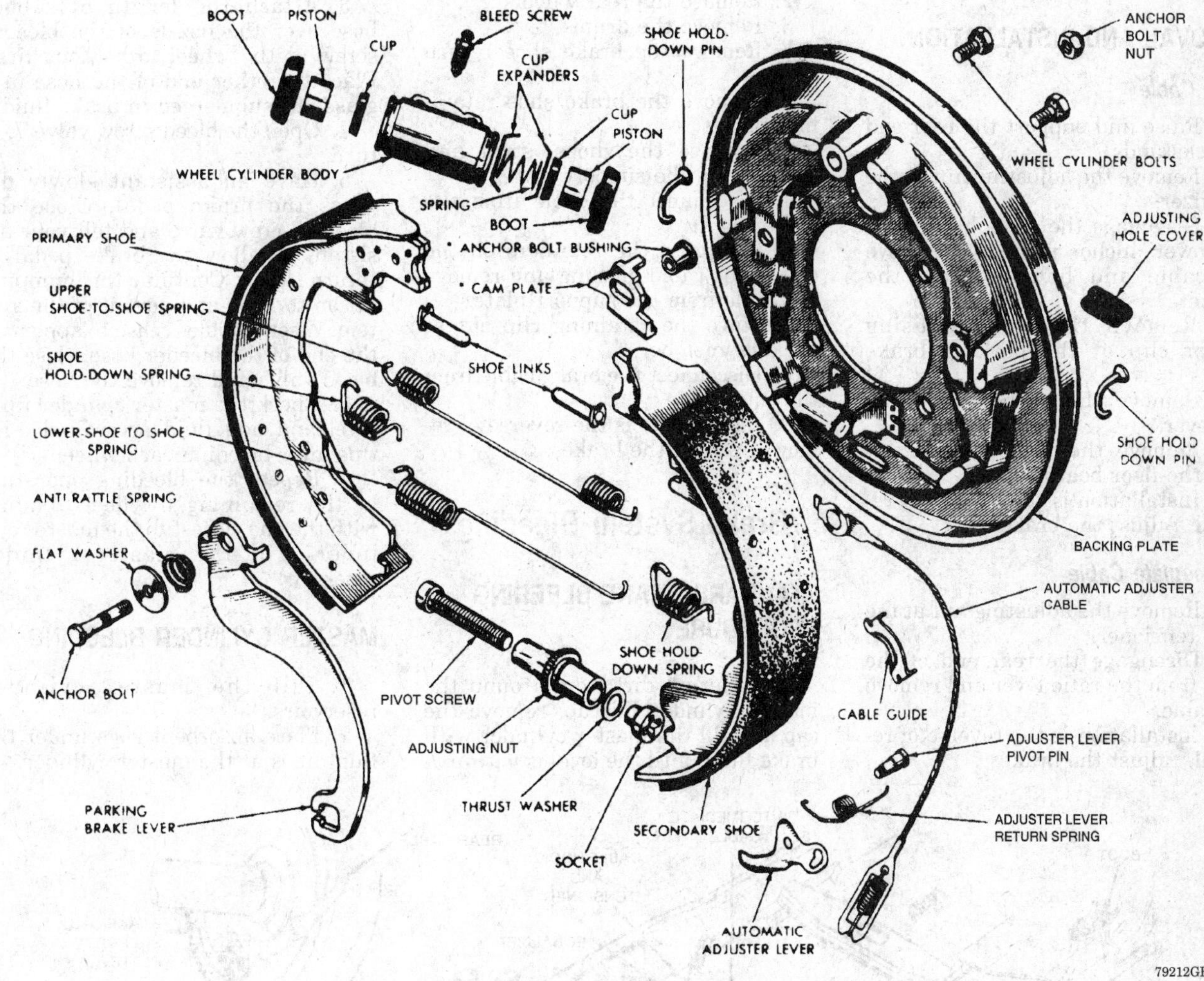

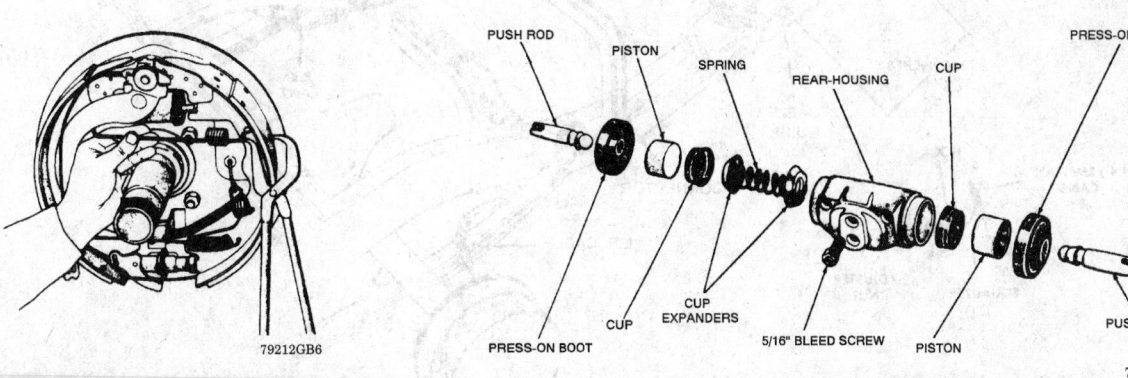

BOOT PISTON
CUP
BLEED SCREW
CUP EXPANDERS
CUP PISTON
SHOE HOLD-DOWN PIN
ANCHOR BOLT NUT
WHEEL CYLINDER BODY
SPRING
BOOT
WHEEL CYLINDER BOLTS
ADJUSTING HOLE COVER
PRIMARY SHOE
ANCHOR BOLT BUSHING
CAM PLATE
UPPER SHOE-TO-SHOE SPRING
SHOE HOLD-DOWN SPRING
SHOE LINKS
LOWER-SHOE TO SHOE SPRING
SHOE HOLD DOWN PIN
BACKING PLATE
ANTI RATTLE SPRING
FLAT WASHER
AUTOMATIC ADJUSTER CABLE
ANCHOR BOLT
PIVOT SCREW
ADJUSTING NUT
SHOE HOLD-DOWN SPRING
CABLE GUIDE
ADJUSTER LEVER PIVOT PIN
PARKING BRAKE LEVER
THRUST WASHER
SECONDARY SHOE
ADJUSTER LEVER RETURN SPRING
SOCKET
AUTOMATIC ADJUSTER LEVER

79212GB5

Exploded view of the Bendix rear brake components

79212GB6

Use a special brake tool as shown to remove the shoe-to-shoe spring

PUSH ROD
PISTON
SPRING
REAR-HOUSING
CUP
PRESS-ON BOOT
PRESS-ON BOOT
CUP
CUP EXPANDERS
5/16" BLEED SCREW
PISTON
PUSH ROD

79212GB7

Exploded view of a common wheel cylinder assembly

6. Apply the parking brake several times, then release it and check to be sure that the rear wheels rotate freely.

REMOVAL AND INSTALLATION

Front Cable

1. Raise and support the rear end on jackstands.
2. Remove the adjusting nut at the equalizer.
3. Disengage the cable housing at the lower anchor point and remove the cable and housing from the bracket.
4. Remove the cable housing anchor clip at the parking brake lever.
5. Remove the anchor clip from the lever.
6. Remove the housing grommet from the floor board.
7. Installation is the reverse of removal. Adjust the brakes.

Intermediate Cable

1. Remove the adjusting nut at the cable tensioner.
2. Disengage the rear end of the cable from the ratio lever and remove the cable.
3. Installation is the reverse of removal. Adjust the brakes.

Rear Cable

1. Raise and support the rear end on jackstands.
2. Remove the rear wheels.
3. Remove the drums.
4. Remove the brake shoe return springs.
5. Remove the brake shoe retaining springs.
6. Remove the shoes, strut and spring from the support plate.
7. Disconnect the cable from the operating arm.
8. Compress the retainers on the end of the cable housing and remove the cable from the support plate.
9. Move the retaining clip out at the crossmember.
10. Disconnect the brake cable from the equalizer.
11. Installation is the reverse of removal. Adjust the brakes.

Brake System Bleeding

STANDARD BRAKE BLEEDING PROCEDURE

1. Clean all dirt from around the master cylinder fill cap, remove the cap and fill the master cylinder with brake fluid until the level is within ¼ in. (6mm) of the top of the edge of the reservoir.
2. Clean off the bleeder screws at the wheel cylinders and calipers.
3. Attach the length of rubber hose over the nozzle of the bleeder screw at the wheel to be done first. Place the other end of the hose in a glass jar, submerged in brake fluid.
4. Open the bleed screw valve ½-¾ turn.
5. Have an assistant slowly depress the brake pedal. Close the bleeder screw valve and tell your assistant to allow the brake pedal to return slowly. Continue this pumping action to force any air out of the system. When bubbles cease to appear at the end of the bleeder hose, close the bleed valve and remove the hose.
6. Check the master cylinder fluid level and add fluid accordingly. Do this after bleeding each wheel.
7. Repeat the bleeding operation at the remaining 3 wheels, ending with the one closest to the master cylinder. Fill the master cylinder reservoir.

MASTER CYLINDER BLEEDING

1. Fill the master cylinder reservoirs.
2. Place absorbent rags under the fluid lines at the master cylinder.

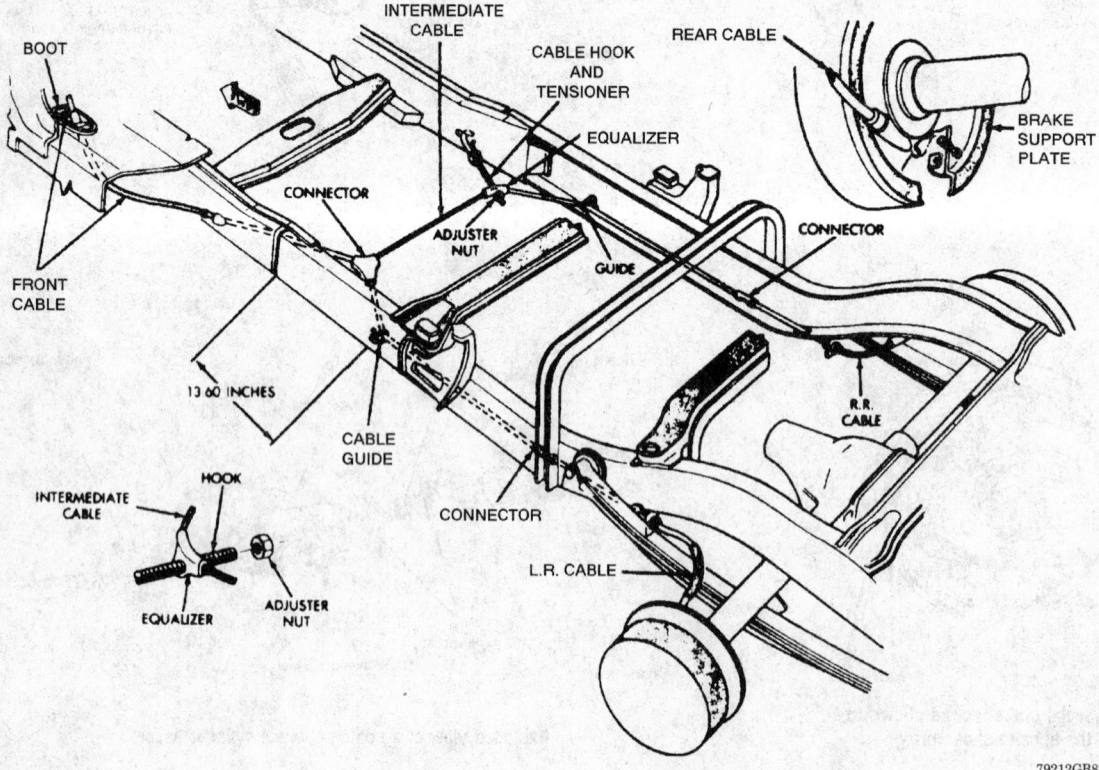

Parking brake cable layout — 2WD Dakota models

79212GB8

3. Have an assistant depress and hold the brake pedal.

4. With the pedal held down, slowly crack open the hydraulic line fitting, allowing the air to escape. Close the fitting and have the pedal released.

5. Repeat Steps 3 and 4 for each fitting until all the air is released.

Wheel Speed Sensor

Three wheel sensors are used in the 1993–96 all-wheel ABS system, one for each front wheel, and one to monitor both rear wheels.

REMOVAL AND INSTALLATION

Front Wheel

2WD Dakota Models

1. Raise and safely support the vehicle.
2. Remove the front wheel.
3. Remove the disc brake caliper.
4. Remove the disc brake hub and rotor assembly.
5. Remove the special bolts attaching the sensor to the splash shield. If these bolts are in good condition, keep them for reuse. If they are in need of replacement, you must replace them with the same type.
6. Remove the clamps securing the sensor wire to the control arm and inner fender.
7. Unseat the grommet securing the sensor wire in the fender panel.
8. In the engine compartment, unplug the sensor wire from the harness connector.
9. Remove the clamps securing the sensor wire to the engine compartment body panels.
10. Work the sensor wire out of the engine compartment and through the fender panel grommet hole. Then slide the wire out of the steering knuckle splash shield and out from the vehicle.
To install:
11. Guide the sensor wire through the splash shield, around the control arm and through the grommet hole in the fender panel. Do not seat the grommet in the fender panel at this time.
12. Position the sensor in the splash shield and install the special sensor attaching bolts. Tighten them to 13–18 ft. lbs. (18–25 Nm).
13. Secure the sensor wire retaining clamps to the control arm and fender panel with screws.
14. Seat the sensor wire grommet in the inner fender panel.

15. In the engine compartment, connect the sensor wire to the harness plug. Make sure the wire is not routed near any hot or rotating components.
16. Install the disc brake hub and rotor assembly.
17. Install the disc brake caliper.
18. Install the wheel.
19. One more time, check the sensor wire routing. Turn the steering wheel back-and-forth to verify the wire is clear of steering and suspension components. If it binds in any way, re-route.
20. Lower the vehicle.

4WD Dakota Models

1. Raise and safely support the vehicle.
2. Remove the front wheel.
3. Remove the special bolts attaching the sensor to the steering knuckle. If the bolts are in good condition, keep them. Otherwise they will have to be replaced with new ones.
4. Remove the clamps that secure the sensor wire to the control arm and inner fender panel.
5. Unseat the grommet securing the sensor wire in the fender panel.
6. In the engine compartment, unplug the sensor wire from the harness connector.
7. Remove the clamps securing the sensor wire to the engine compartment body panels.
8. Work the sensor out of the engine compartment, through the fender panel grommet hole and out of the vehicle.
To install:
9. Guide the sensor wire through the splash shield, around the control arm and through the grommet hole in the fender panel. Do not seat the grommet in the fender panel at this time.
10. Position the sensor in the splash shield and install the special sensor attaching bolts. Tighten them to 13–18 ft. lbs. (18–25 Nm).
11. Secure the sensor wire retaining clamps to the control arm and fender panel with screws.
12. Seat the sensor wire grommet in the inner fender panel.
13. In the engine compartment, connect the sensor wire to the harness plug. Make sure the wire is not routed near any hot or rotating components.
14. Install the wheel.
15. One more time, check the sensor wire routing. Turn the steering wheel back-and-forth to verify the wire is clear of steering and suspen-

sion components. If it binds in any way, re-route.
16. Lower the vehicle.

2WD Ram Pick-up Models

1. Raise and safely support the front end of the vehicle.
2. Remove the wheel.
3. Use a prytool to press the caliper piston back into the bore.
4. Remove the brake caliper bolts, then lift the caliper from the knuckle and rotor. Use a wire to suspend the caliper from the frame or chassis. Do not let it hang freely from its hose.
5. Remove the rotor.
6. Remove the special sensor attaching bolts. If they are in good condition, keep them for reuse. Otherwise, new bolts of the same type will have to be used.
7. Disconnect the sensor wire and remove the sensor from the vehicle.
To install:
8. Position the sensor and install the special bolts. Tighten them to 16–21 ft. lbs. (21–25 Nm).
9. Connect the sensor wire to the harness wire.
10. Install the rotor and brake caliper.
11. Install the wheel.
12. Lower the vehicle.
13. If available, use a scan tool to verify operation.

4WD Ram Pick-up Models

1. Remove the bolt attaching the sensor to the inside of the steering knuckle.
2. In the engine compartment, unplug the sensor wire connector at the harness plug.
3. Remove the sensor and wire assembly.
To install:
4. Position the sensor on the steering knuckle. Seat the sensor locating tab in the hole in the knuckle and install the sensor attaching bolt. Tighten it to 11 ft. lbs. (14 Nm).
5. Route the sensor wire from the steering knuckle to the harness connector.
6. Check sensor wire routing. Make sure the wire is clear of all chassis components and is not twisted or kinked at any spot.
7. Connect the sensor wire harness in the engine compartment.

Rear Wheel

Dakota Models

1. Raise the rear of the vehicle.
2. Unplug the wire from the sensor on the axle housing.

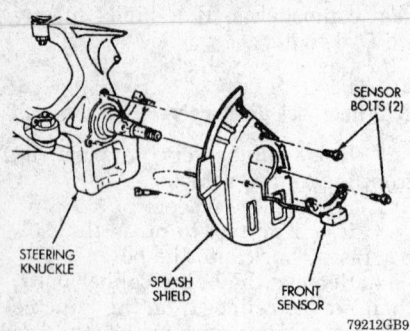

Exploded view of the front wheel speed sensor mounting — 2WD Ram Pick-up design shown

3. Unplug the sensor wire from the harness connector at the driver's side frame rail.

4. Unclip the sensor wire from the rear brake hose, then remove the sensor wire from the vehicle.

To install:

5. Connect the sensor wire to the rear sensor and harness connector at the frame rail.

6. Secure the sensor wire to the rear brake hose with the clips provided.

7. Lower the vehicle.

Ram Pick-up Models

1. Raise and safely support the rear of the vehicle.

2. Locate the sensor, its cover and mounting area on the axle and clean before proceeding.

3. Unplug the harness wires from the sensor.

4. Remove the screw that secures the brake cable, brake line, sensor cover and sensor axle housing.

5. Remove the sensor and cover.

6. Cover the sensor opening in the axle housing to prevent dirt entry.

To install:

7. Insert the sensor in the axle housing opening.

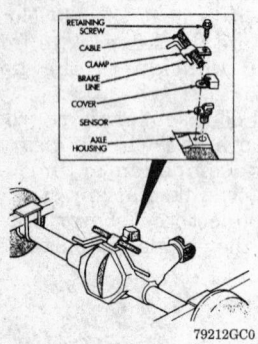

Exploded view of the rear wheel speed sensor mounting — 2WD Ram Pick-up models

8. Position the cover over the sensor and install the cover and cover attaching screw. Tighten to 18 ft. lbs. (24 Nm).

9. Connect the harness wires to the sensor and lower vehicle.

10. Lower the vehicle.

FRONT SUSPENSION

Shock Absorber

REMOVAL AND INSTALLATION

Coil Spring Suspension

1. Raise and support the vehicle with jackstands positioned at the extreme front ends of the frame rails.

2. Remove the wheel.

3. Remove the upper mounting bolt.

4. Remove the two lower mounting bolts.

5. Remove the shock absorber.

6. When installing the shock absorber, make sure the upper busings are in the correct position. Replace any worn or cracked bushings. Tighten the top nut to 25 ft. lbs. (34

TORQUE	
A	52 FT. LBS.
B	25 FT. LBS.
C	175 FT. LBS.
D	200 IN. LBS.

Exploded view of the front shock absorber mounting — Ram Pick-up and Ramcharger models

Nm). Then, tighten the lower bolts to 15 ft. lbs. (20 Nm).

Leaf Spring Suspension

1. Raise and safely support the vehicle on jackstands.

2. Remove the wheel.

3. Remove the two upper shock absorber bracket-to-frame bolts.

4. Remove the lower bracket nut and remove the shock absorber.

5. If new shocks are being installed, remove the upper bracket from the old shock. Replace any worn or cracked bushings.

6. When installing the shock absorbers, tighten all fasteners to 50 ft. lbs. (68 Nm). Installation of the remaining components is the reverse of the removal.

Torsion Bar Suspension

1. Remove the hardware from the shock absorber stud.

2. Raise and safely support the vehicle on jackstands.

3. Remove the wheel.

4. Remove the lower bolts, then remove the shock absorber.

To install:

5. Install the lower retainer and grommet on the shock absorber stud. Insert the shock absorber through the frame hole. Install the lower bolt.

6. Tighten the bolt to 100 ft. lbs. (136 Nm.)

7. Install the upper grommet and retainer on the shock absorber stud. Install the bayonet nut and tighten to 30 ft. lbs. (41 Nm).

Coil Spring

REMOVAL AND INSTALLATION

Except Dakota Models

1993 Models

1. Raise and support the front end on jackstands placed under the frame.

2. Remove the wheels.

3. Remove the brake calipers and suspend them out of the way. DO NOT DISCONNECT THE BRAKE LINE! Remove the inner pad.

4. Remove the shock absorbers.

5. Disconnect the sway bar.

6. Remove the lower control arm strut.

7. Install a spring compressor, such as tool DD-1278, tighten the nut finger-tight and back it off ½ turn.

8. Remove the ball joint nuts.

9. Using ball joint separator C-3564-A, or equivalent, spread the

tool against the lower joint just enough to exert pressure then strike the knuckle sharply with a hammer to free the joint. NEVER ATTEMPT TO FORCE THE BALL JOINT OUT WITH TOOL PRESSURE ALONE! Remove the tool.

10. Slowly loosen the spring compressor until all tension is relieved from the coil.

11. Remove the compressor and spring.

To install:

12. Position the spring on the control arm and install the compressor.

13. Compress the spring until the ball joint is properly positioned.

14. Install the ball joint nuts and tighten them to 135 ft. lbs. (183 Nm).

15. Install the strut. Tighten the mounting bolts to 95 ft. lbs. (129 Nm); the retainer nut to 50 ft. lbs. (68 Nm).

16. Connect the sway bar and tighten the link to 100 inch lbs. (11 Nm).

17. Remove the spring compressor.

18. Install the shock absorber. Tighten the upper end to 25 ft. lbs. (34 Nm); the lower end to 15 ft. lbs. (20 Nm).

19. Install the inboard brake pad and caliper. If equipped, tighten the retaining clips with a suitable torque wrench to 15 ft. lbs. (20 Nm).

20. Install the wheels.

1994–96 Models

1. Raise and safely support the vehicle on jackstands.

2. Remove the wheel.

3. Remove the brake caliper and rotor.

4. Disconnect the tie rod from the steering knuckle.

5. Disconnect the stabilizer bar link from the lower arm.

6. Support the lower arm outboard end with a floor jack. Place the jack under the arm in front of the shock mount.

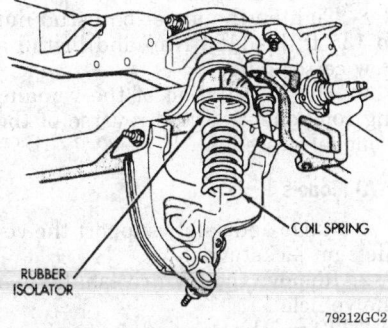

RUBBER ISOLATOR

COIL SPRING

79212GC2

View of the coil spring mounting — 1994–96 models

7. Remove the cotter pin and nut from the lower ball stud. Use remover tool C–4150A, or equivalent, to separate the ball stud.

8. Remove the lower shock bolt from the suspension arm.

9. Lower the jack and suspension arm until spring tension is relieved. Remove the spring and rubber isolator.

To install:

10. Install the rubber isolator on top of the spring. Position the spring into the upper spring seat and lower suspension arm.

11. Raise the suspension arm with the jack and position the shock into the suspension arm mount. Install the shock bolt and tighten to specification.

12. Install the steering knuckle on the lower ball stud. Install the lower ball stud nut and tighten to specification. Install a new cotter pin and remove the jack.

13. Install the stabilizer bar link on the lower suspension arm. Install the grommet, retainer and nut, then tighten to specifications.

14. Install the tie rod on the steering knuckle and tighten the nut to specification.

15. Install the brake caliper and rotor.

16. Install the wheels.

17. Remove the jackstands and lower the vehicle.

Dakota Models

NOTE: For 4WD models, refer to the torsion bar procedure.

1. Raise the vehicle and support safely.

2. Remove the shock absorber.

3. Disconnect the sway bar from the lower control arm, if equipped.

4. Install spring compressor tool DD–1278, or equivalent to the coil spring. Tighten the nut finger-tight, then back off half a turn.

5. Remove the cotter pin and lower ball joint nut.

6. Release the lower ball joint taper using ball stud loosening tool C–3564–A, or equivalent.

7. Remove the tool and remove the ball stud from the control arm. Release the compressor tool from the coil spring.

8. Pull the arm down and remove the spring with the rubber isolation pad from the vehicle.

To install:

9. Install the spring with the rubber isolator. Install the compressor tool and compress it enough so the lower ball joint can be inserted through the knuckle.

10. Tighten the lower ball joint nut to 135 ft. lbs. (183 Nm). Install a new cotter pin. Remove the spring compressor.

11. Connect the sway bar to the lower control arm, if equipped.

12. Install the shock absorber.

Leaf Spring

REMOVAL AND INSTALLATION

1. Raise and support the front end on jackstands placed under the frame. The wheels should still be on the ground, but the weight should be off of the springs.

2. Remove the nuts, lockwashers and U-bolts securing the spring to the axle.

3. Remove the spring shackle bolts, shackles and front eye bolt.

4. Remove the spring.

To install:

5. Position the spring in place and install the eye bolt and nut. Do not tighten the bolt yet.

6. Install the shackles and bolts. Tighten them just enough to make them snug.

7. Make sure that the spring center bolt enters the locating hole in the axle pad.

8. Install the U-bolts, lockwashers and nuts. Make them just snug for now.

9. Lower the truck to its normal position with the weight back on the springs. Now tighten all bolts and nuts as follows:
• U-bolt nuts: 95 ft. lbs. (129 Nm)
• Shackle bolts: 80 ft. lbs. (108 Nm)
• Eye bolt: 80 ft. lbs. (108 Nm)

Torsion Bar

REMOVAL AND INSTALLATION

NOTE: Torsion bars are used on 4WD Dakotas only.

NOTE: The left and right side torsion bars are not interchangeable. The bars are identified by the letter R or L stamped into one end of the bar. The bars do not have a specific front or rear end and can be installed with either end facing forward.

1. Remove the upper control arm jounce bumper.

2. Raise the vehicle with the front suspension hanging free and support it safely with jackstands.

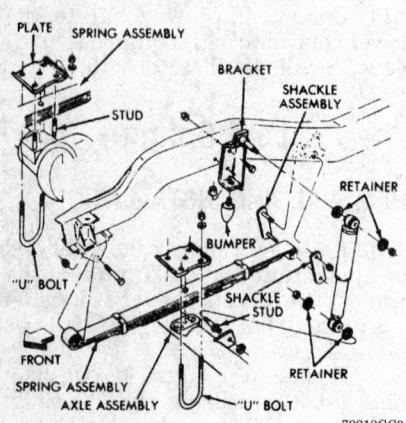

Exploded view of the front leaf spring suspension — 4WD Ram Pick-up and Ramcharger models with Model 60 front axle

79212GC3

3. Release the load from the torsion bar by turning the adjustment bolt counterclockwise.

4. Remove the adjustment bolt from the swivel and then remove the torsion bar and the anchor together from the vehicle. Remove the torsion bar from the anchor.

5. Remove any dirt, rust or pebbles from the hex shaped socket in the anchor and lower control arm.

6. Inspect the anchor and bolt and replace them if any damage or corrosion exists.

To install:

7. Insert the torsion bar ends into the sockets.

8. Position the anchor and the bushing in the crossmember. Insert the adjustment bolt and thread it into the swivel.

9. Turn the adjustment bolt clockwise to apply a load on the bar.

10. Lower the vehicle.

11. Set the front suspension height. The difference between the distance from the surface that the tires are on to the lower control arm inner pivot and the distance that the tires are on to the outer end of the arm is 1–1½ in. (25–38mm).

12. Install the upper control arm jounce bumper.

Upper Ball Joints

REMOVAL AND INSTALLATION

Except Dakota Models

2WD Models

1. Raise and safely support the vehicle on jackstands.

2. Position a support at the outer end of the lower control arm and lower the vehicle so that the support compresses the coil spring.

3. Remove the tire and wheel assembly.

4. Remove the cotter pin and the stud nut.

5. Release the upper ball joint taper using ball stud loosening tool C–3564–A, or equivalent.

6. Unthread the ball joint from the control arm with tool C–3561, or equivalent.

7. The installation is the reversal of the removal procedure. Tighten the ball joint itself to 125 ft. lbs. (169 Nm). Tighten the upper ball stud nut to 135 ft. lbs (183 Nm).

4WD Models

1. Raise and safely support the vehicle on jackstands.

2. Remove the front axle shaft.

3. Disconnect the tie rod end from the steering knuckle. On the left side, disconnect the drag link ball stud from the steering knuckle.

4. On the left side, remove the nuts and washers from the steering knuckle arm and remove the arm and spring, if equipped, from the knuckle.

5. If equipped with a Model 44 front axle, remove the ball joint nuts and discard the lower nut. Use a brass drift and hammer to separate the steering knuckle from the axle tube yoke. Use tool C–4169 to remove the sleeve from the upper yoke arm.

6. Remove the snapring from the ball joint. Install the knuckle in a vise and use tools D–150–1, D–150–3 and C–4212–L to remove the ball joint from the knuckle.

7. If equipped with a Model 60 front axle, remove the bolts from the knuckle lower cap. Dislodge the cap from the steering knuckle and axle tube yoke. Remove the steering knuckle. Use tool D–192 to remove the upper socket pin from the axle tube upper arm bore. Remove the seal.

To install:

8. If equipped with a Model 44 front axle, use tools C–4212–L and C–4288 to force the upper ball joint into the steering knuckle. Install the snapring and install a new rubber

boot. Thread the replacement sleeve into the upper yoke bore so that 2 threads are exposed at the top of the yoke. Position the knuckle on the axle tube yoke and install a new lower ball stud nut, then tighten to 80 ft. lbs. (108 Nm). Using the special socket, tighten the sleeve to 40 ft. lbs. (54 Nm). Install the upper ball stud nut and torque to 100 ft. lbs. (136 Nm) and install a new cotter pin.

9. If equipped with a Model 60 front axle, use tool D–192 to install the upper socket pin to the axle tube upper arm bore. Install a new seal. Tighten to 500–600 ft. lbs. (668–813 Nm). Position the knuckle over the socket pin. Fill the lower socket cavity with grease. Install the lower cap and tighten the bolts to 80 ft. lbs. (108 Nm).

10. On the left side, install the spring, if equipped and the steering knuckle arm to the steering knuckle.

11. Connect the tie rod to the end of the steering knuckle. On the left side, connect the drag link ball stud to the steering knuckle.

12. Install the front axle shaft and all related components.

Dakota Models

2WD Models

1. Raise and safely support the vehicle on jackstands.

2. Position a support at the outer end of the lower control arm and lower the vehicle so that the support compresses the coil spring.

3. Remove the tire and wheel assembly.

4. Remove the cotter pin and stud nut.

5. Release the upper ball joint taper using ball stud loosening tool C–3564–A, or equivalent.

6. Unthread the ball joint from the control arm with tool C–3561, or equivalent.

To install:

7. Tighten the ball joint itself to 125 ft. lbs. (169 Nm).

8. Tighten the upper ball stud nut to 135 ft. lbs (183 Nm) and install a new cotter pin.

9. The installation of the remaining components is the reverse of the removal procedure.

4WD Models

1. Raise and safely support the vehicle on jackstands.

2. Remove the CV driveshaft from the vehicle.

3. Turn the torsion bar adjustment nut counterclockwise to relieve all tension from the torsion bar.

4. Remove the cotter pin from the upper ball stud.

5. Release the upper ball joint taper using ball stud loosening tool C–3564–A, or equivalent. Remove the tool.

6. Remove the upper ball stud seal.

7. Force the ball joint from the arm using the ball stud removal and installation tool C–4212.

To install:

8. Install the new ball joint using the ball stud removal and installation tool C–4212.

9. Install the ball stud seal.

10. Insert the upper ball stud in the steering knuckle arm bore and install the nut. Tighten the nut to 105 ft. lbs. (142 Nm) and install a new cotter pin.

11. Turn the torsion bar adjustment nut clockwise to apply a load on the bar.

12. Install the CV driveshaft and all related parts.

13. Lower the vehicle.

14. Set the front suspension height. The difference between the distance from the surface that the tires are on to the lower control arm inner pivot and the distance that the tires are on to the outer end of the arm is 1–1½ in. (25–38mm).

Lower Ball Joints

REMOVAL AND INSTALLATION

Except Dakota Models

2WD Models

1. Raise and safely support the vehicle on jackstands.

2. Remove the shock absorber.

3. Remove the strut bar and disconnect the sway bar from the lower control arm, if equipped.

4. Install spring compressor tool DD–1278, or equivalent to the coil spring and tighten the nut finger-tight, then back off half a turn.

5. Remove the cotter pin and lower ball joint nut.

6. Release the lower ball joint taper using ball stud loosening tool C–3564–A, or equivalent.

7. Remove the tool and remove the ball stud from the control arm. Release the compressor tool from the coil spring.

8. Pull the arm down and remove the spring with the rubber isolation pad from the vehicle. Remove the ball joint boot. Use tool C–4212, or an appropriate ball joint press to remove the ball joint from the arm.

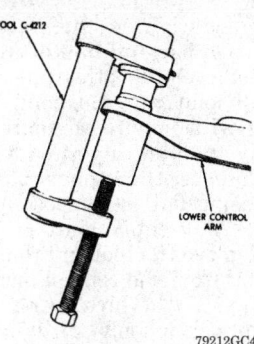

Use the special tool shown, or equivalent to remove/install the lower ball joint

To install:

9. Use the remover tool to press the ball joint into the arm. Install a new rubber boot. Install the spring with the rubber isolators. Install the compressor tool and compress it enough so the lower ball joint can be inserted through the knuckle.

10. Tighten ¹¹/₁₆ lower ball joint nuts to 135 ft. lbs. (183 Nm). Tighten ¾ nuts to 175 ft. lbs. (237 Nm). Install a new cotter pin. Remove the spring compressor.

11. Install the strut bar and connect the sway bar from the lower control arm, if equipped.

12. Install the shock absorber.

4WD Models

1. Raise and safely support the vehicle on jackstands.

2. Remove the front axle shaft.

3. Disconnect the tie rod end from the steering knuckle. On the left side, disconnect the drag link ball stud from the steering knuckle.

4. On the left side, remove the nuts and washers from the steering knuckle arm and remove the arm and spring, if equipped, from the knuckle.

5. If equipped with a Model 44 front axle, remove the ball joint nuts and discard the lower nut. Use a brass drift and hammer to separate the steering knuckle from the axle tube yoke.

6. Remove the snapring from the ball joint. Install the knuckle in a vise and use tools D–150–1, D–150–3 and C–4212–L to remove the ball joint from the knuckle.

7. If equipped with a Dana 60 front axle, use tools C–4212–L, C–4366–1 and C–4366–2 to remove the lower ball joint.

To install:

8. If equipped with a Model 44 front axle, use tools C–4212–L and C–4288 to force the lower ball joint into the steering knuckle. Install the snapring and install a new rubber

boot. Position the knuckle on the axle tube yoke and install a new lower ball stud nut. Tighten to 80 ft. lbs. (108 Nm). Install the upper ball stud nut and tighten to 100 ft. lbs. (136 Nm), then install a new cotter pin.

9. If equipped with a Model 60 front axle, use tools C–4212–L, C–4366–3 and C–4366–4 to install the seal and lower bearing cup into the axle tube yoke lower bore. Reposition the tools and install the lower bearing and seal into the bore. Position the knuckle over the socket pin. Fill the lower socket cavity with grease. Install the lower cap and tighten the bolts to 80 ft. lbs. (108 Nm).

10. On the left side, install the spring, if equipped and the steering knuckle arm to the steering knuckle.

11. Connect the tie rod to the end of the steering knuckle. On the left side, connect the drag link ball stud to the steering knuckle.

12. Install the front axle shaft and all related components.

Dakota Models

2WD Models

1. Raise and safely support the vehicle on jackstands.

2. Remove the shock absorber.

3. Disconnect the sway bar from the lower control arm, if equipped.

4. Install spring compressor tool DD–1278, or equivalent to the coil spring and tighten the nut finger-tight, then back off half a turn.

5. Remove the cotter pin and lower ball joint nut.

6. Release the lower ball joint taper using ball stud loosening tool C–3564–A, or equivalent.

7. Remove the tool and remove the ball stud from the control arm. Release the compressor tool from the coil spring.

8. Pull the arm down and remove the spring with the rubber isolation pad from the vehicle. Remove the ball joint boot. Use tool C–4212, or an appropriate ball joint press to remove the ball joint from the arm.

To install:

9. Use the remover tool to press the ball joint into the arm. Install a new rubber boot. Install the spring with the rubber isolators. Install the compressor tool and compress it enough so the lower ball joint can be inserted through the knuckle.

10. Tighten the lower ball joint nut to 135 ft. lbs. (183 Nm). Install a new cotter pin. Remove the spring compressor.

11. Connect the sway bar to the lower control arm, if equipped.

12. Install the shock absorber.

4WD Models

1. Raise and safely support the vehicle on jackstands.

2. Remove the CV driveshaft.

3. Remove the torsion bar.

4. Remove the shock absorber lower attaching bolt.

5. Disconnect the stabilizer bar from the lower control arm.

6. Remove the cotter pin and the nut from the lower ball stud. Separate the lower ball stud from the steering knuckle.

7. Pry the peened ball joint retainer sections upward from the lower control arm and remove the ball joint from the arm.

To install:

8. Install the new ball joint in the control arm. Peen the ball joint housing retainer over to secure the ball joint.

9. Install the grease seal.

10. Insert the ball stud into the steering knuckle bore. Install the nut. Tighten to 120 ft. lbs. (163 Nm) and install a new cotter pin.

11. Attach the stabilizer bar to the control arm and install the shock mount bolt.

12. Install the torsion bar and turn the adjustment bolt clockwise to apply a load to the bar.

13. Install the CV driveshaft and all related parts.

14. Lower the vehicle.

15. Set the front suspension height so that the difference between the distance from the surface that the tires are on to the lower control arm inner pivot and the distance that the tires are on to the outer end of the arm is 1–1½ in. (25–38mm).

16. Have the alignment checked at a reputable repair facility.

Upper Control Arms

REMOVAL AND INSTALLATION

NOTE: Any time the control arm is removed, it is necessary to have the front suspension aligned.

Ram Pick-up and Ramcharger Models

1. Raise and safely support the vehicle on jackstands.

2. Remove the shock absorber.

3. Remove the strut bar and disconnect the sway bar from the lower control arm, if equipped.

4. Install spring compressor tool DD–1278, or equivalent to the coil spring and tighten the nut finger-tight. Then back off half a turn.

5. Remove the cotter pin and upper ball joint nut. Suspend the assembly with a wire so there is not excessive pull on the brake hose.

6. Release the upper ball joint taper using ball stud loosening tool C–3564–A, or equivalent.

7. Remove the tool and remove the ball stud from the control arm.

8. Remove the pivot bar retaining bolts on Van models or the cam bolt assemblies on Ram Pick-up and Ramcharger models and remove the arm from the vehicle.

To install:

9. Install the arm to the frame rail bracket and install the retaining bolts.

10. Tighten the ball joint nut to 135 ft. lbs. (183 Nm). Install a new cotter pin. Remove the spring compressor.

11. Install the strut bar and connect the sway bar from the lower control arm, if equipped.

12. Install the shock absorber.

13. Align the front end. When all settings are at specifications, torque the pivot bar retaining bolts on Van models to 195 ft. lbs. (264 Nm). Torque the cam bolts to 70 ft. lbs. (95 Nm) on Ram Pick-up and Ramcharger models.

Dakota Models

2WD Models

1. Raise and safely support the vehicle on jackstands.

2. Remove the shock absorber.

3. Disconnect the sway bar from the lower control arm, if equipped.

4. Install spring compressor tool DD–1278, or equivalent to the coil spring and tighten the nut finger-tight, then back off half a turn.

5. Remove the cotter pin and upper ball joint nut. Suspend the rotor assembly with a wire so there is not excessive pull on the brake hose.

6. Release the upper ball joint taper using ball stud loosening tool C–3564–A, or equivalent.

7. Remove the tool and remove the ball stud from the control arm.

8. Remove the pivot bar retaining bolts and remove the arm from the vehicle.

To install:

9. Install the arm to the frame rail bracket and install the retaining bolts.

10. Tighten the ball joint nut to 135 ft. lbs. (183 Nm). Install a new cotter pin. Remove the spring compressor.

11. Connect the sway bar from the lower control arm, if equipped.

12. Install the shock absorber.

13. Roughly align the front end, then tighten the cam bolts to 155 ft. lbs. (210 Nm).

14. Have the alignment checked at a reputable repair facility.

4WD Models

1. Raise and safely support the vehicle on jackstands.

2. Remove the CV driveshaft from the vehicle.

3. Turn the torsion bar adjustment nut counterclockwise to relieve all tension from the torsion bar. Remove the screws that attach the brake hose bracket to the upper arm. Remove the shock absorber.

4. Remove the cotter pin from the upper ball stud, then remove the retaining nut.

5. Release the upper ball joint taper using ball stud loosening tool C–3564–A, or equivalent. Remove the tool. Remove the ball stud from the steering knuckle.

6. Remove the pivot bar retaining bolts.

7. Remove the arm from the vehicle.

To install:

8. Position the arm at the frame rail bracket.

9. Install the pivot arm nuts and bolts.

10. Insert the upper ball stud in the steering knuckle arm bore and install the nut. Tighten the nut to 105 ft. lbs. (142 Nm) and install a new cotter pin. Install the shock absorber and attach the brake hose bracket.

11. Turn the torsion bar adjustment nut clockwise to apply a load on the bar.

12. Install the CV driveshaft and all related parts.

13. Lower the vehicle.

14. Set the front suspension height. The difference between the distance from the surface that the tires are on to the lower control arm inner pivot and the distance that the tires are on to the outer end of the arm is 1–1½ in. (25–38mm).

15. Have the alignment checked at a reputable repair facility.

BUSHING REPLACEMENT

NOTE: 1994–96 Ram Pick-up models do not have replaceable bushings.

1. Remove the upper control arm.

2. Place the upper control arm in a vice and assemble Remover/Installer tool No C–3962, or equivalent and adapter No. SP–3953, or equivalent over the bushing. Tighten the nuts

and force the bushings out of the arm bore.

To install:

NOTE: Be sure the control arm is supported firmly at the area where the bushing is being forced in. Do not use any lubricant to aid in the installation.

3. Position the flange end of the replacement bushing in the Remover/Installer tool No C–3962, or equivalent. Support the control arm firmly and force the bushing in the control arm bore (from outside). Tighten until the flange end is seated on the arm.

4. Install the control arm.

Lower Control Arms

REMOVAL AND INSTALLATION

Ram Pick-up and Ramcharger Models

1. Raise and safely support the vehicle on jackstands.
2. Remove the shock absorber.
3. Remove the strut bar and disconnect the sway bar from the lower control arm, if equipped.
4. Install spring compressor tool DD–1278, or equivalent to the coil spring and tighten the nut finger-tight. Then back off half a turn.
5. Remove the cotter pin and lower ball joint nut.
6. Release the lower ball joint taper using ball stud loosening tool C–3564–A, or equivalent.
7. Remove the tool and remove the ball stud from the control arm. Release the compressor tool from the coil spring.
8. Pull the arm down and remove the spring with the rubber isolation pad from the vehicle. Remove the lower control arm pivot bolt from the crossmember and remove the arm from the vehicle.

To install:

9. Install the arm to the crossmember finger-tight. Install the spring with the rubber isolators. Install the compressor tool and compress it enough so the lower ball joint can be inserted through the knuckle.
10. Tighten $^{11}/_{16}$ lower ball joint nuts to 135 ft. lbs. (183 Nm). Tighten $^{3}/_{4}$ nuts to 175 ft. lbs. (237 Nm). Install a new cotter pin. Remove the spring compressor.
11. Install the strut bar and connect the sway bar from the lower control arm, if equipped.
12. Install the shock absorber.
13. Lower the vehicle completely. When the weight of the vehicle is off

of the lifting apparatus, tighten the lower arm pivot bolts to 225 ft. lbs. (305 Nm).
14. Have the alignment checked at a reputable repair facility.

Dakota Models

2WD Models

1. Raise and safely support the vehicle on jackstands.
2. Remove the shock absorber.
3. Disconnect the sway bar from the lower control arm, if equipped.
4. Install spring compressor tool DD–1278, or equivalent to the coil spring and tighten the nut finger-tight, then back off half a turn.
5. Remove the cotter pin and lower ball joint nut.
6. Release the lower ball joint taper using ball stud loosening tool C–3564–A, or equivalent.
7. Remove the tool and remove the ball stud from the control arm. Release the compressor tool from the coil spring.
8. Pull the arm down and remove the spring with the rubber isolation pad from the vehicle. Remove the lower control arm pivot bolts from the crossmember and remove the arm from the vehicle.

To install:

9. Install the arm to the crossmember finger-tight. Install the spring with the rubber isolators. Install the compressor tool and compress it enough so the lower ball joint can be inserted through the knuckle.
10. Tighten the lower ball joint nut to 135 ft. lbs. (183 Nm). Install a new cotter pin. Remove the spring compressor.
11. Connect the sway bar from the lower control arm, if equipped.
12. Install the shock absorber.
13. Lower the vehicle completely. When the weight of the vehicle is off of the lifting apparatus, tighten the front lower arm pivot bolt to 130 ft.

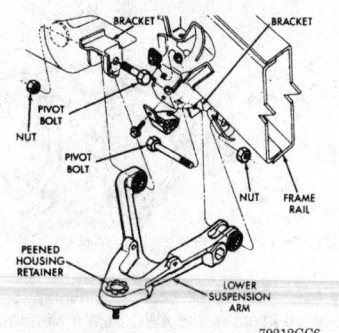

Exploded view of the lower control arm mounting — Dakota models

lbs. (176 Nm) and the rear nut to 80 ft. lbs. (108 Nm).
14. Have the alignment checked at a reputable repair facility.

4WD Models

1. Raise and safely support the vehicle on jackstands.
2. Remove the CV driveshaft.
3. Remove the torsion bar.
4. Remove the shock absorber lower attaching bolt.
5. Disconnect the stabilizer bar from the lower control arm.
6. Remove the cotter pin and the nut from the lower ball stud. Separate the lower ball stud from the steering knuckle.
7. Remove the pivot bolts and remove the arm from the vehicle.

To install:

8. Install the new control arm to the vehicle.
9. Install the pivot bolts.
10. Insert the ball stud into the steering knuckle bore. Install the nut, tighten to 120 ft. lbs. (163 Nm) and install a new cotter pin.
11. Attach the stabilizer bar to the control arm and install the shock mount bolt.
12. Install the torsion bar and turn the adjustment bolt clockwise to apply a load to the bar.
13. Install the CV driveshaft and all related parts.
14. Lower the vehicle so the weight of the vehicle is completely off of the lifting apparatus.
15. Tighten the front pivot nut to 80 ft. lbs. (108 Nm) and the rear nut to 130 ft. lbs. (176 Nm).
16. Set the front suspension height. The difference between the distance from the surface that the tires are on to the lower control arm inner pivot and the distance that the tires are on to the outer end of the arm is 1–1½ in. (25–38mm).
17. Have the alignment checked at a reputable repair facility.

BUSHING REPLACEMENT

NOTE: The 1994–96 Ram Pick-up models do not have replaceable bushings.

1. Remove the lower control arm.
2. Use an arbor press and an appropriate size sleeve to force the original bushing from the lower control arm.

To install:

3. Use an arbor press and an appropriate size sleeve to force the bushing into the lower control arm bore. Make sure that it is fully seated.

4. Install the lower control arm.

Sway Bar

REMOVAL AND INSTALLATION

1. Disconnect the bar at each end link.
2. Remove the bolts from the frame mounting brackets.
3. Remove the sway bar.
4. Installation is the reverse of removal. Tighten the frame bracket bolts to 23 ft. lbs. (31 Nm).; the end links to 100 inch lbs. (11 Nm).

Front Wheel Bearings

ADJUSTMENT

1. Tighten the wheel bearing nut to 20–25 ft. lbs. (27–34 Nm) while turning the rotor.
2. Loosen the wheel bearing adjusting nut completely.
3. Tighten the nut finger-tight.
4. Check the wheel bearing endplay. The specification is 0.001–0.003 in. (0.025–0.076mm).

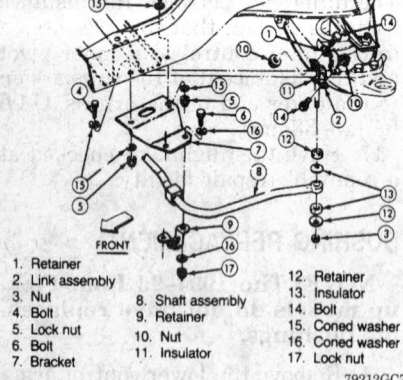

1. Retainer	
2. Link assembly	
3. Nut	
4. Bolt	8. Shaft assembly
5. Lock nut	9. Retainer
6. Bolt	10. Nut
7. Bracket	11. Insulator

12. Retainer
13. Insulator
14. Bolt
15. Coned washer
16. Coned washer
17. Lock nut

79212GC7

Exploded view of the sway bar and related components — Ram Pick-up and Ramcharger models

5. Install the nut lock and cotter pin.

REMOVAL AND INSTALLATION

1. Raise and safely support the vehicle on jackstands.
2. Remove the tire and wheel assembly.
3. Remove the caliper and disc brake pads.
4. Remove the dust cap.
5. Remove the cotter pin, castellated nut lock, wheel bearing nut and washer from the spindle.
6. Remove the outer wheel bearing.
7. Remove the rotor with the inner wheel bearing from the spindle. Remove the grease seal.
   To install:
8. Lubricate and install the inner wheel bearing. Install a new grease seal.
9. Install the rotor to the spindle.
10. Lubricate and install the outer wheel bearing, washer and nut. When the bearing preload is properly set, install the nut lock and a new cotter pin.
11. Install the grease cap.
12. Install the brake pads and caliper.
13. Install the wheel.

REAR SUSPENSION

Shock Absorber

REMOVAL AND INSTALLATION

1. Raise and support the axle.

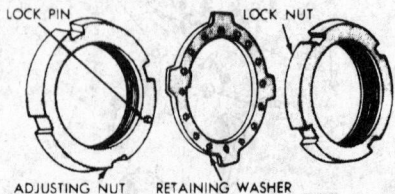

LOCK PIN   LOCK NUT

ADJUSTING NUT   RETAINING WASHER

79212GC8

Exploded view of the bearing retaining hardware used for the 4WD Ram Pick-up and Ramcharger Models

2. Remove the bolt and flag nut from the frame crossmember bracket.
3. Remove the bolt and nut from the axle bracket.
4. Remove the rear shock absorber from the vehicle.
   To install:
5. Install the bolts through the brackets and shock.
6. For Dakota models, tighten the lower bolt and nut to 60 ft. lbs. (81 Nm). Tighten the upper bracket nuts to 20 ft. lbs. (27 Nm).
7. For Ram models, tighten the upper bolt to 70 ft. lbs. (95 Nm). Tighten the lower bolt to 100 ft. lbs. (136 Nm).

Leaf Spring

REMOVAL AND INSTALLATION

1. Raise the truck and support the rear with jackstands under the frame. Be sure that the front wheels are chocked and that the parking brake is set. The wheels should be touching the floor, but the weight must be off of the springs.
2. Remove the nuts, lockwashers and U-bolts that hold the axle to the springs.
3. Remove the front pivot bolt.
4. Remove the rear shackle bolt nuts and the rear shackle plate.
5. Remove the spring.
   To install:
6. Position the spring in place and install the eye bolt and nut. Do not tighten the bolt yet.
7. Install the shackles and bolts. Tighten them just enough to make them snug.
8. Make sure that the spring center bolt enters the locating hole in the axle pad. On headless-type spring bolts, install the bolts with the lock groove lined up with the lockbolt hole in the bracket. Install the lockbolt and tighten the lockbolt nut. Install the lubrication fittings.
9. Install the U-bolts, lockwashers and nuts. Make them just snug for now. Align the auxiliary spring parallel with the main spring.
10. Lower the truck to its normal position with the weight back on the springs. Now tighten all bolts and nuts as follows:
- U-bolt nuts: $\frac{1}{2}$ in. 65 ft. lbs. (88 Nm); $\frac{9}{16}$ in. nuts — 110 ft. lbs. (149 Nm)
- Shackle bolts and eye bolts: $\frac{1}{2}$ 95 ft. lbs. (129 Nm); $\frac{5}{8}$ 125 ft. lbs. (169 Nm); $\frac{3}{4}$ in. 155 ft. lbs. (210 Nm)

CHRYSLER CORP.

CHRYSLER-Town & Country **DODGE**-Caravan **PLYMOUTH**-Voyager

FIRING ORDERS

NOTE: To avoid confusion, always replace spark plug wires one at a time.

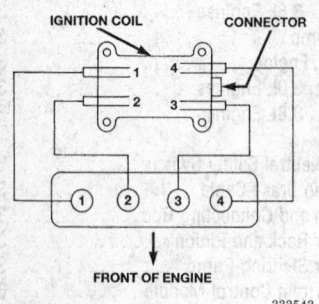

2.4L (VIN B) Engine
Engine Firing Order: 1–3–4–2
Distributorless Ignition System

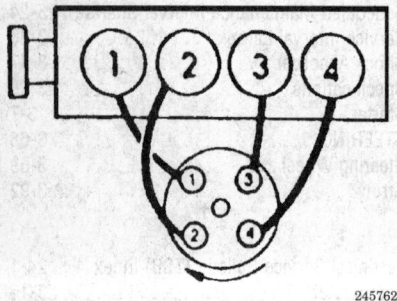

245762

2.5L (VIN K) Engine
Engine Firing Order: 1–3–4–2
Distributor Rotation: Clockwise

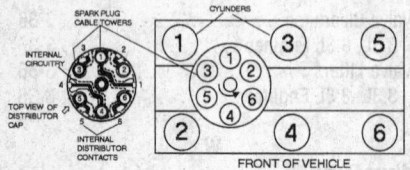

333541

3.0L (VIN 3) Engine
Engine Firing Order: 1–2–3–4–5–6
Distributor Rotation: Counterclockwise

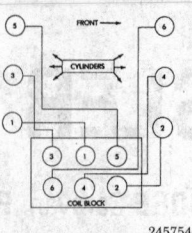

245754

3.3L (VIN R) and 3.8L
(VIN L) Engines
Engine Firing Order:
1–2–3–4–5–6
Distributorless Ignition
System

ENGINE ELECTRICAL

NOTE: Disconnecting the negative battery cable on some vehicles may interfere with the functions of the on board computer systems and may require the computer to undergo a relearning process, once the negative battery cable is reconnected.

Distributor

REMOVAL

2.5L (VIN K) and 3.0L (VIN 3) Engines

1. Disconnect the negative battery cable.
2. On the 2.5L (VIN K) engine, disconnect the distributor lead wires, and vacuum hose, if equipped.
3. On the 2.5L (VIN K) engine, remove the distributor splash shield.
4. On the 3.0L (VIN 3) engine, disconnect the electrical connector from the distributor.
5. Remove the distributor cap.
6. Mark the position of the rotor in relation to the distributor housing and mark the position of the distributor housing in relation to the engine. It is a good idea to take the time to turn the crankshaft to TDC No. 1 cylinder, compression stroke (firing position). This aligns the distributor rotor with the No. 1 spark plug tower in the distributor cap. This is a good reference point and makes installation easier.
7. Remove the distributor hold-down bolt.

8. Carefully lift the distributor from the engine. The shaft will rotate slightly when the drive gear disengages as the distributor is removed.

INSTALLATION

Engine Not Disturbed

2.5L (VIN K) and 3.0L (VIN 3) Engines

1. Lower the distributor into the engine, aligning the marks made during removal. Be sure the O-ring is properly seated on the distributor. Replace the O-ring with a new one if the O-ring is cracked or nicked.
2. Be sure the distributor drive is engaged with the gear on the camshaft.
3. Install the distributor cap.
4. Tighten the hold-down bolt.
5. On the 2.5L (VIN K) engine, install the distributor splash shield, if equipped.
6. On the 2.5L (VIN K) engine, reconnect the distributor pick-up lead connector and vacuum hose, if necessary.
7. On the 3.0L (VIN 3) engine, connect the electrical connector to the distributor.
8. Reconnect the negative battery cable.
9. Check and, if necessary, adjust the ignition timing.

Engine Disturbed

2.5L (VIN K) and 3.0L (VIN 3) Engines

1. Remove the spark plug from No. 1 cylinder. Place a finger over the spark plug hole.
2. Rotate the crankshaft in the normal direction of rotation, until compression is felt at the spark plug hole.
3. On the 2.5L (VIN K) engine, continue rotating the crankshaft until the **O** mark on the flywheel is aligned with the pointer on the bellhousing.
4. On the 3.0L (VIN 3) engine, continue rotating the crankshaft until the No. 1 piston is at the top of the compression stroke.
5. Install the distributor so, with the distributor housing fully seated, the rotor points to the No. 1 spark plug terminal on the distributor cap.
6. Install the distributor cap.
7. Tighten the hold-down bolt. Connect the electrical connector to the distributor.
8. Check and, if necessary, adjust the ignition timing.

Ignition Timing

ADJUSTMENT

2.4L (VIN B), 3.3L (VIN R) and 3.8L (VIN L) Engines

The basic ignition timing cannot be adjusted. The Powertrain Control Module (PCM) regulates the ignition timing.

2.5L (VIN K) and 3.0L (VIN 3) Engines

1993–95

1. Apply the parking brake and/or block the wheels.
2. Place the gearshift in the **P** or **N** position. Be sure to turn all lights and accessories OFF.
3. If a magnetic timing light is being used, place the pickup probe into the open receptacle next to the timing scale window. If a conventional timing light is being used, connect it to the No. 1 cylinder spark plug wire.

NOTE: Do not puncture boots or cables with test probes. This will damage them. Make sure proper adapters are always used.

4. Start the engine and allow it to run until it reaches operating temperature.
5. Once the engine has reached operating temperature, disconnect the electrical connector for the engine coolant temperature sensor. This will turn ON the radiator cooling fan and the malfunction indicator lamp (Check Engine light).
6. Aim the timing light at the timing scale or read the magnetic timing unit. The timing is advanced if a flash occurs when the timing mark is ahead of the specified degree mark. Make the adjustment by turning the distributor housing in the direction of rotor rotation. The timing is retarded if flashing occurs when the timing mark is after the specified degree mark. The adjustment is made by turning the distributor housing against the direction of rotor rotation.
7. Refer to the Vehicle Emission Control Information label, located under the hood, for the correct ignition timing specification.
8. If the timing is within ± 2 degrees of the timing specification:
   a. Turn OFF the engine.
   b. Remove the timing light or magnetic timing unit and tachometer.
   c. Reconnect the engine coolant temperature sensor and erase any

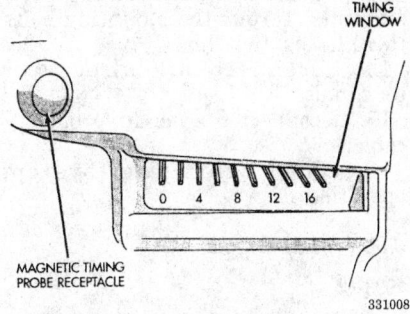

Timing scale at transaxle bell housing — 1993–95 2.5L (VIN K) engine

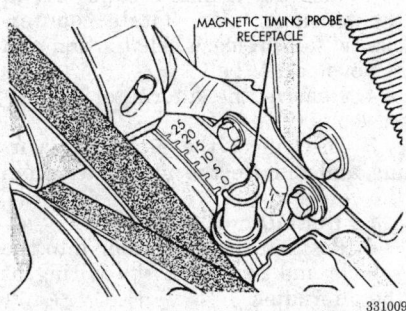

Timing scale — 1993–95 3.0L (VIN 3) engine

diagnostic trouble codes using a DRB or equivalent scan tool.
9. If the timing needs adjustment, proceed as follows:
   a. Loosen the distributor hold-down screw enough to rotate the distributor housing.
   b. Turn the distributor housing to adjust the timing.
   c. Tighten the distributor hold-down screw and recheck the ignition timing.
   d. Turn the engine OFF.
   e. Remove the timing light or magnetic timing unit and tachometer.
   f. Reconnect the engine coolant temperature sensor and erase any diagnostic trouble codes using a DRB or equivalent scan tool.

Alternator

PRECAUTIONS

Several precautions must be observed with alternator equipped vehicles to avoid damage to the unit.

• If the battery is removed for any reason, make sure it is reconnected with the correct polarity. Reversing the battery connections may result in damage to the 1-way rectifiers.
• When utilizing a booster battery as a starting aid, always connect the positive to positive terminals and the negative terminal from the booster battery to a good engine ground on the vehicle being started.
• Never use a fast charger as a booster to start vehicles.
• Disconnect the battery cables when charging the battery with a fast charger.
• Never attempt to polarize the alternator.
• Do not use test lights of more than 12 volts when checking diode continuity.
• Do not short across or ground any of the alternator terminals.
• The polarity of the battery, alternator and regulator must be matched and considered before making any electrical connections within the system.
• Never separate the alternator on an open circuit. Make sure all connections within the circuit are clean and tight.
• Disconnect the battery ground terminal when performing any service on electrical components.
• Disconnect the battery if arc welding is to be done on the vehicle.

REMOVAL AND INSTALLATION

2.4L (VIN B) Engine

1. Disconnect the negative battery cable.
2. Unplug the field circuit from the alternator.
3. Remove the B+ nut and wire.
4. Loosen the adjusting bolt, but do not remove.
5. Loosen the pivot bolt and the adjusting bolt until the drive belt can be removed. Remove the accessory drive belt.
6. Remove the adjusting bolt and pivot bolt.
7. Remove the alternator from the vehicle.
To install:
8. Install the alternator into the bracket on the engine.

9. Install the adjusting and pivot bolts, but do not tighten at this time.

10. Connect the B+ wire and torque the nut to 75 inch lbs. (9 Nm).

11. Reconnect the field circuit to the alternator.

12. Install the accessory drive belt. Be sure the drive belt is correctly routed on the engine and correctly seated on the alternator pulley.

13. Adjust the drive belt and tighten the adjusting bolt to 40 ft. lbs. (54 Nm).

14. Tighten the pivot bolt to 40 ft. lbs. (54 Nm).

15. Reconnect the negative battery cable. Verify the alternator charge rate. Road test the vehicle.

2.5L (VIN K) Engine

1. Disconnect the negative battery cable.

2. Disconnect the wiring connections and label for installation, if necessary.

NOTE: When removing the alternator from some vehicles equipped with air conditioning, clearance may be restricted by the condenser cooling fan assembly or the A/C compressor and mounting bracket assembly. If so, removal of one of these components will be necessary.

3. Remove the air conditioning compressor drive belt, if equipped. Remove the A/C compressor and place it to one side, if necessary.

4. Loosen the alternator adjusting bracket bolt and adjusting bolt. Remove the alternator belt.

5. Remove the bracket bolt and the mounting bolt.

6. Remove the pivot bolt and nut.

7. Lift the alternator from the vehicle.

NOTE: When lifting the alternator out of the vehicle on some years with A/C, clearance may be restricted by the condenser cooling fan assembly or the A/C compressor and mounting bracket assembly. If so, removal of one of these items will be necessary.

To install:

8. Position the alternator against the engine.

9. Install the pivot bolt and nut.

10. Install the mounting bracket bolts, and adjusting bolt.

11. Install the A/C compressor assembly, if necessary.

12. Install the accessory drive belts and adjust to specification.

13. Tighten all the mounting bolts and nuts. Torque the mounting bolts to 40 ft. lbs. (54 Nm).

14. Reconnect all alternator terminals.

15. Reconnect the negative battery cable.

16. Check charging system operation.

3.0L (VIN 3) Engine

1993–95

1. Disconnect the negative battery cable.

2. Remove the air conditioning compressor drive belt.

3. Install a ½ inch breaker bar in the tensioner slot. Rotate counterclockwise to release belt tension and remove poly-V belt.

4. Remove the alternator mounting bolts (2).

5. Disconnect the alternator wiring and remove the alternator from the vehicle.

To install:

6. Position the alternator into the vehicle and reconnect the wiring at the alternator.

7. Set the alternator against the mounting bracket and install the mounting bolts. Torque the mounting bolts to 40 ft. lbs. (54 Nm).

8. Rotate the tensioner counterclockwise and install the alternator belt. Install the A/C compressor drive belt and adjust to the proper tension.

9. Reconnect the negative battery cable.

10. Check charging system operation.

1996–97

1. Disconnect the negative battery cable.

2. Remove the wiper arm and blade assemblies.

3. Remove the cowl cover from the vehicle.

4. Open the hood. Disconnect the positive lock on the wiper unit electrical connector.

5. Disconnect the wiper unit electrical connector from the engine compartment wiring harness.

6. Disconnect the windshield washer hose from the hose coupling inside the wiper unit.

7. Remove the drain tubes from the bottom of the wiper unit.

8. Remove the sound absorbers from the ends of the wiper unit.

9. Remove the attaching nuts securing the wiper unit to the lower windshield fence.

10. Remove the attaching bolts securing the wiper unit to the dash panel.

11. Raise the wiper unit from the weld-studs on the lower windshield fence and remove the wiper unit from the vehicle.

12. Remove the air conditioning compressor drive belt.

13. Install a ½ inch breaker bar in the tensioner slot. Rotate counterclockwise to release belt tension and remove poly-V belt.

14. Remove the alternator mounting bolts (2).

15. Disconnect the alternator wiring and remove the alternator out from the vehicle.

To install:

16. Position alternator into the vehicle and reconnect the wiring at the alternator.

17. Set the alternator against the mounting bracket and install the mounting bolts. Torque the mounting bolts to 40 ft. lbs. (54 Nm).

18. Rotate the tensioner counterclockwise and install the alternator belt. Install the A/C compressor drive belt and adjust to the proper tension.

19. Install the wiper unit into the vehicle engine compartment, making sure the wiper unit is installed properly over the weld-studs on the lower windshield fence. Install and tighten the attaching nuts to the weld-studs.

20. Install and tighten the attaching bolts securing the wiper unit to the dash panel.

21. Install the sound absorbers to each end of the wiper unit.

22. Install the drain tubes to the bottom of the wiper unit and reconnect the windshield washer hose to the hose coupling inside the wiper unit.

23. Reconnect the wiper unit wiring connector to the engine wiring harness.

24. Place the cowl cover onto the vehicle. Reconnect the right side windshield washer hose to the right washer nozzle located on the underside of the cowl cover.

25. Engage the retainers that secure the cowl cover to the front fender edge.

26. Install and tighten the wing nuts that secure the front of the cowl cover to the wiper module.

27. Engage the quarter turn fasteners that secure the outer ends of the cowl cover to the wiper module.

28. Install and tighten the screws that hold the lower area of the cowl cover to the wiper module.

29. Reconnect the negative battery cable and verify that the wiper motor and wiper linkage are in the **PARK** position.

30. Install the wiper arm in correct position over the wiper arm pivot. Align the wiper arm positions as follows:

a. Left side should be no closer than 2.5 in. (65mm) from the lower edge of the windshield.

b. Right side should be no closer than 1.5 in. (40mm) from the lower edge of the windshield.

31. Install the the wiper arm-to-wiper arm pivot retaining nut and torque to 26 ft. lbs. (35 Nm).

32. Push down the wiper arm cap cover and engage the clip that secures the outside end of the cover to the wiper arm.

33. Check charging system operation.

3.3L (VIN R) and 3.8L (VIN L) Engines

1993–95

1. Disconnect the negative battery cable.

2. Remove the alternator drive belt, by relieving the tension on the dynamic tensioner.

3. Loosen the nut on the support bracket at the exhaust manifold, but do not remove it.

4. Remove the alternator tensioner/power steering bracket bolt.

5. Remove the tensioner stud nut and remove the tensioner.

6. Remove the alternator mounting bolts.

7. Remove the power steering reservoir from the mounting bracket, do not disconnect the hoses, and position it out of the way.

8. Remove the alternator support bracket bolts. Remove the intake plenum to alternator bracket bolt and remove the alternator support bracket from the engine.

9. Disconnect the alternator electrical leads and remove the alternator from the engine.

To install:

10. Install the alternator in position on the engine and reconnect the electrical leads.

11. Install the alternator support bracket and tighten the retaining bolts to 40 ft. lbs. (54 Nm).

12. Install the power steering reservoir on the mounting bracket.

13. Install the alternator mounting bolts and tighten to 40 ft. lbs. (54 Nm).

14. Install the tensioner and tensioner mounting stud. Tighten the nut on the support bracket on the exhaust manifold.

15. Install the alternator belt, insert a ½ in. extension into the square hole in the tensioner and turn the tensioner. Tighten the tensioner bolt.

16. Reconnect the negative battery cable.

17. Check charging system operation.

1996–97

1. Disconnect the negative battery cable.

2. Remove the wiper arm and blade assemblies.

3. Remove the cowl cover from the vehicle.

4. Open the hood. Disconnect the positive lock on the wiper unit electrical connector.

5. Disconnect the wiper unit electrical connector from the engine compartment wiring harness.

6. Disconnect the windshield washer hose from the hose coupling inside the wiper unit.

7. Remove the drain tubes from the bottom of the wiper unit.

8. Remove the sound absorbers from the ends of the wiper unit.

9. Remove the attaching nuts securing the wiper unit to the lower windshield fence.

10. Remove the attaching bolts securing the wiper unit to the dash panel.

11. Raise the wiper unit from the weld-studs on the lower windshield fence and remove the wiper unit from the vehicle.

12. Remove the alternator drive belt, by relieving the tension on the dynamic tensioner.

13. Remove the bolt securing the top of the alternator mounting bracket to the air intake plenum.

14. Remove the bolts securing the outside of the alternator mounting bracket to the alternator mounting plate.

15. Remove the upper alternator mounting bolt.

16. Remove the mounting bracket.

17. Disconnect the alternator electrical leads

18. Remove the lower alternator pivot bracket bolt.

19. Remove the alternator from the engine.

To install:

20. Install the alternator in position on the engine and install the lower alternator pivot bracket bolt.

21. Reconnect the electrical leads.

22. Install the alternator mounting bracket into position and tighten the attaching bolts to 40 ft. lbs. (54 Nm).

23. Install the upper alternator mounting bolt and tighten to 40 ft. lbs. (54 Nm).

24. Install the alternator belt, insert a ½ inch extension into the square hole in the tensioner and turn the tensioner. Tighten the tensioner bolt.

25. Install the wiper unit into the vehicle engine compartment, making sure the wiper unit is installed properly over the weld-studs on the lower windshield fence. Install and tighten the attaching nuts to the weld-studs.

26. Install and tighten the attaching bolts securing the wiper unit to the dash panel.

27. Install the sound absorbers to each end of the wiper unit.

28. Install the drain tubes to the bottom of the wiper unit and reconnect the windshield washer hose to the hose coupling inside the wiper unit.

29. Reconnect the wiper unit wiring connector to the engine wiring harness.

30. Place the cowl cover onto the vehicle. Reconnect the right side windshield washer hose to the right washer nozzle located on the underside of the cowl cover.

31. Engage the retainers that secure the cowl cover to the front fender edge.

32. Install and tighten the wing nuts that secure the front of the cowl cover to the wiper module.

33. Engage the quarter turn fasteners that secure the outer ends of the cowl cover to the wiper module.

34. Install and tighten the screws that hold the lower area of the cowl cover to the wiper module.

35. Reconnect the negative battery cable and verify that the wiper motor and wiper linkage are in the **PARK** position.

36. Install the wiper arm in correct position over the wiper arm pivot. Align the wiper arm positions as follows:

a. Left side should be no closer than 2.5 in. (65mm) from the lower edge of the windshield.

b. Right side should be no closer than 1.5 in. (40mm) from the lower edge of the windshield.

37. Install the the wiper arm-to-wiper arm pivot retaining nut and torque to 26 ft. lbs. (35 Nm).

38. Push down the wiper arm cap cover and engage the clip that secures the outside end of the cover to the wiper arm.

39. Check charging system operation.

Drive Belt

REMOVAL AND INSTALLATION

2.4L (VIN B) Engine

Alternator and Air Conditioning Compressor Belt

1. Disconnect the negative battery cable at the battery terminal.

2. Loosen the locking nut and the pivot bolt located on the top and bottom of the alternator (generator).

3. Rotate the adjusting bolt to loosen the the drive belt tension enough to slide it off the pulleys and remove it from the engine.

4. Inspect the belt for wear and/or damage; replace as necessary.

To install:

5. Install the drive belt and properly route it around the alternator (generator), idler, A/C compressor and crankshaft damper pulleys. Be sure the belt is seated correctly in each pulley otherwise, the belt could jump out of position.

6. Rotate the adjusting bolt at the alternator until tension starts to increase on the drive belt. The proper belt tension is as follows:

    a. **NEW** belt: 150 lbs.
    b. **USED** belt: 80 lbs.

7. Tighten the locking nut first, and then the pivot bolt to 40 ft. lbs. (54 Nm) and check the belt tension again. Adjust if necessary.

8. Reconnect the negative battery cable. Start engine and check for proper belt operation.

Power Steering Pump Belt

1. Disconnect the negative battery cable at the negative battery terminal.

2. Working from on top of the vehicle, loosen the locking nuts located in front and behind the power steering pump.

3. Loosen the power steering pump pivot bolt located below the power steering pump.

4. Rotate the adjusting bolt until it loosens the drive belt tension enough to release the drive belt from the pump pulley and crankshaft damper pulley. Remove the drive belt from the engine.

5. Inspect the belt for wear and/or damage; replace as necessary.

To install:

6. Install the drive belt and properly route it around the power steering pump and crankshaft damper pulleys. Be sure the belt is properly seated in each pulley otherwise, the belt could jump out of position.

7. Adjust the belt tension by rotating the adjusting bolt until the proper tension is achieved. The proper belt tension is as follows:

    a. **NEW** belt: 130 lbs.
    b. **USED** belt: 80 lbs.

8. Tighten the locking nuts first and then the pivot bolt to 40 ft. lbs. (54 Nm). Check the belt tension again.

9. Reconnect the negative battery cable. Start the engine and check for proper belt operation.

2.5L (VIN K) Engine

Air Conditioning Compressor

1. Disconnect the negative battery cable.

2. Loosen the idler bracket pivot screw and locking screws. The lower locking screw must be removed to install the drive belt.

3. Move the idler and remove the belt.

4. Inspect the belt for wear and/or damage; replace as necessary.

To install:

5. Install the belt, and, using a breaker bar in the square hole in the idler bracket, apply tension to the belt. Using a suitable belt tension gauge tool, adjust the tension to 135 lbs. for a new compressor belt, or 80 lbs. for a used compressor belt.

6. While holding the tension, tighten the locking screws and then the pivot screw. Tighten each to 40 ft. lbs. (54 Nm).

7. Reconnect the negative battery cable. Start the engine and check for proper drive belt operation.

Power Steering Pump

1. Disconnect the negative battery cable.

2. Loosen the locking screw from the top of the vehicle.

3. Raise and safely support the vehicle, then loosen the pivot screw and pivot nut from under the vehicle.

4. Move the pump and remove the drive belt.

5. Inspect the belt for wear and/or damage; replace as necessary.

To install:

6. Install the belt, and, using a ½ in. breaker bar in the square hole in the bracket, apply tension to the belt. Using a suitable belt tension gauge tool, adjust the tension to 105 lbs. for a new power steering pump belt, or 80 lbs. for a used belt.

7. While holding the tension, tighten the locking screw, then tighten the pivot screw and nut. Torque each to 40 ft. lbs. (54 Nm).

8. Reconnect the negative battery cable. Start the engine and check for proper drive belt operation.

Alternator

1. Disconnect the negative battery cable.

2. Loosen the alternator pivot nut, then the locking nut and adjusting screw.

3. Move the alternator and remove the drive belt.

4. Inspect the belt for wear and/or damage; replace as necessary.

To install:

5. Install the drive belt, and adjust the belt tension using the adjusting screw. Using a suitable belt ten-

POWER STEERING PUMP   GENERATOR

IDLER

CRANKSHAFT DAMPER   AIR CONDITIONING COMPRESSOR

331945

Accessory drive belt routing — 2.4L (VIN B) engine

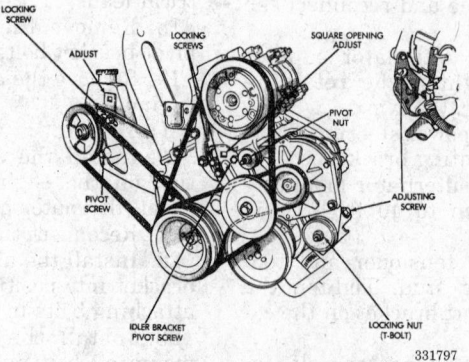

Drive belt routing — 2.5L (VIN K) engine

331797

sion gauge tool, adjust the tension with the adjusting screw to 135 lbs. for a new alternator belt, or 80 lbs. for a used belt.

6. Tighten the locking nut first, and then the pivot nut to 40 ft. lbs. (54 Nm).

7. Reconnect the negative battery cable. Start the engine and check for proper operation.

3.0L (VIN 3) Engine

Air Conditioning Compressor

1. Disconnect the negative battery cable.

2. Loosen the idler pulley locking nut.

3. Turn the adjusting screw to relieve tension and remove the belt.

4. Inspect the belt for wear and/or damage; replace as necessary.

To install:

5. Install the belt, making sure it is properly routed.

6. Using a suitable belt tension gauge tool, adjust the tension with the adjusting screw to 125 lbs. for a new compressor belt, or 80 lbs. for a used belt.

7. Torque the locking nut to 40 ft. lbs. (54 Nm).

8. Reconnect the negative battery cable. Start the engine and check for proper drive belt operation.

Alternator/Power Steering Pump

The Poly-V drive belt uses a dynamic tensioner which maintains the correct drive belt tension.

1. Disconnect the negative battery cable.

2. Using a breaker bar and socket, apply force, in a clockwise direction, to the tensioner pulley bolt.

3. Remove the old belt from the pulleys.

4. Inspect the belt for wear and/or damage; replace as necessary.

To install:

5. Install the new belt and release the tensioner pulley. Be sure the

drive belt is properly seated in each belt pulley or the belt could jump out of position while the engine is running.

6. Reconnect the negative battery cable. Start the engine and check for proper drive belt operation.

3.3L (VIN R) and 3.8L (VIN L) Engines

1. Disconnect the negative battery cable.

2. Raise and safely support the vehicle.

3. Remove the right front splash shield.

4. Rotate the tensioner clockwise and remove the drive belt.

5. Inspect the belt for wear and/or damage; replace as necessary.

To install:

6. Rotate the belt tensioner clockwise and install the accessory drive belt. Release the tensioner. Be sure the drive belt is properly seated on each belt pulley, otherwise the belt could jump out of position.

7. Install the right front splash shield.

8. Reconnect the negative battery cable. Start the engine and check for proper drive belt operation.

Starter

REMOVAL AND INSTALLATION

2.4L (VIN B) Engines

1. Disconnect the negative battery cable.

2. Raise and safely support the vehicle.

3. Disconnect the starter solenoid and **B** positive wiring connectors from the starter terminals.

4. Remove the mounting bolts that attach the starter motor to the transaxle bellhousing.

5. Remove the starter from the vehicle.

To install:

6. Place the starter motor onto the transaxle bellhousing and install mounting bolts.

7. Torque the mounting bolts to 40 ft. lbs. (54 Nm).

NOTE: Before reconnecting the wiring to the starter solenoid, be sure to clean the wiring of any dirt or corrosion.

8. Reconnect the positive battery cable to the solenoid post and torque the retaining nut to 90 inch lbs. (10 Nm).

9. Reconnect the push-on solenoid connector.

10. Lower the vehicle.

11. Reconnect the negative battery cable. Verify starter motor operation.

2.5L (VIN K) Engines

1. Disconnect the negative battery cable.

2. Raise and safely support the vehicle. The starter can be removed by reaching over the crossmember.

3. Remove the heat shield and its clamps, if equipped.

4. On some engines, it may be necessary loosen the air pump tube at the exhaust manifold and move the tube bracket away from the starter.

5. Remove the electrical connections from the starter.

6. Remove the bolts attaching the starter to the flywheel housing and the rear bracket to the engine or transaxle.

7. Remove the starter.

To install:

8. Position the replacement starter against the mounting surface.

9. Install the mounting bolts. Torque the mounting bolts to 40 ft. lbs. (54 Nm).

10. Reconnect the starter wiring.

11. Position the air pump tube toward starter and connect tube bracket the to exhaust manifold, if necessary.

12. Install the heat shield and clamp. Lower the vehicle.

13. Reconnect the negative battery cable.

14. Check starter operation.

3.0L (VIN 3), 3.3L (VIN R) and 3.8L (VIN L) Engines

1. Disconnect the negative battery cable.

2. Raise and safely support the vehicle.

3. Remove the 3 starter mounting bolts.

4. Remove the wiring connector terminal nuts and disconnect the wiring connectors.

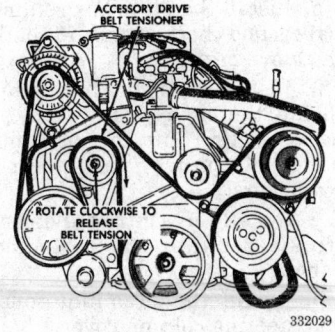

Drive belt routing — 3.0L (VIN 3) engine

Drive belt routing — 3.3L (VIN R) and 3.8L (VIN L) engines

5. Remove the starter motor from the vehicle. If equipped, separate the starter spacer from the transaxle bellhousing.

6. Clean all corrosion and dirt from the wiring terminals before installation.

To install:

7. Install the starter spacer in position on the transaxle bellhousing flange toward the flywheel, if equipped.

8. Place the starter motor into position against the engine/transaxle assembly.

9. Reconnect the wiring connectors to the starter motor. Torque the starter solenoid battery nut to 90 inch lbs. (10 Nm).

10. Install the starter motor mounting bolts. Torque the mounting bolts to 40 ft. lbs. (54 Nm).

11. Lower the vehicle. Reconnect the negative battery cable. Verify starter motor operation.

CHASSIS ELECTRICAL

Blower Motor

REMOVAL AND INSTALLATION

1993–95 Models

—————— CAUTION ——————

The Supplemental Inflatable Restraint (SIR) system must be disarmed before removing the blower motor assembly. Failure to do so may cause accidental deployment of the air bag, resulting in unnecessary SIR system repairs and/or personal injury.

Front Blower Motor

1. Disconnect and isolate the negative battery cable, then wait 2 minutes before proceeding with the removal. This will disable the air bag system, preventing possible personal injury.

2. For 1993 models, perform the following procedures:

   a. Remove the center console bezel and accessory switch carrier from the instrument panel.

   b. Remove the heater control board, cigar lighter and ashtray from the instrument panel. Label

electrical connectors for installation, if necessary.

   c. Remove the center/forward console. Remove the retaining screws and the right lower instrument panel.

3. For 1994–95 models, perform the following procedures:

   a. Remove the steering column cover and premium console.

   b. Remove the center trim bezel and ashtray from the instrument panel.

   c. Remove the A/C-heater control and cigar lighter from the instrument panel.

   d. Remove screw in air duct and retaining screws for the lower right panel.

   e. Remove the glove box retaining screws and check straps. Disconnect the glove box light.

   f. Remove the lower right instrument panel.

4. Remove the blower motor screws attaching the motor to the A/C-heater unit.

5. Allow the blower assembly to drop downward to clear the instrument panel.

To install:

6. Place the blower motor assembly into position on the A/C-heater unit and tighten the mounting screws.

7. Install the lower right instrument panel and tighten the retaining screws.

8. For 1993 models, perform the following procedures:

   a. Install the center/forward console. Install the ashtray, cigar lighter and heater control board into the instrument panel. Reconnect the electrical connectors.

   b. Install the accessory switch carrier and center console bezel into the instrument panel.

9. For 1994–95 models, perform the following procedures:

   a. Install the glove box retaining screws and check straps. Reconnect the glove box light.

   b. Install screw in the air duct. Install the cigar lighter and A/C-heater control into the instrument panel. Reconnect the electrical connectors.

   c. Install the ashtray and center trim bezel into the instrument panel.

   d. Install the premium console and steering column cover.

10. Reconnect the negative battery cable. Check blower motor operation.

Rear Blower Motor

1. Disconnect the negative battery cable.

2. Remove the left lower interior quarter trim panel, as follows:

   a. Remove the ashtray.

   b. Remove the screw that holds the quarter trim insert to the quarter trim inside the ashtray opening.

   c. Pull outward at the bottom of the trim insert to disengage the clips holding the insert to the quarter trim.

   d. Lift the insert upward and separate the quarter trim insert from the vehicle.

   e. Remove the screws holding the lower quarter trim panel to the inner quarter panel.

   f. For 1994–95 models, Remove the rear door opening scuff plate.

   g. Separate the lower quarter trim from the vehicle.

   h. Remove the covers and bolts holding the shoulder harness turning loops to the upper quarter panel.

   i. Remove the bolts holding the outboard seat belt anchor to the floor.

   j. Remove the seat belt retractor covers.

   k. Remove the screws holding the coat hook to the quarter panel.

   l. Remove the screws holding the upper quarter trim panel to the inner quarter panel.

   m. Separate the upper quarter trim panel from the vehicle.

3. Remove one blower cover-to-floor mounting screw and 7 blower cover-to-unit mounting screws.

4. Rotate the blower motor cover from underneath the AC-heater unit cover.

5. Compress the clamp at the center of the blower motor wheel and remove the blower motor wheel from the blower motor shaft.

6. Remove the 3 blower motor mounting screws.

7. Remove the blower motor from the A/C-heater housing unit.

To install:

8. Place the blower motor assembly into position in the A/C-heater housing unit and tighten the 3 mounting screws.

9. Install the blower motor wheel onto the blower motor shaft and secure with the retaining clamp.

10. Rotate the blower motor cover under the A/C-heater unit cover.

11. Install and tighten the 7 blower cover-to-unit mounting screws and one blower cover-to-floor mounting screw.

12. Install the quarter trim panel, as follows:

　a. Replace the upper quarter trim panel to the vehicle. Install the screws that secure the panel to the inner quarter panel.

　b. Replace the coat hook and install the retaining screw.

　c. Install the seat belt retractor covers.

　d. Install the bolts holding the outboard seat belt anchor to the floor.

　e. Install the bolts holding the shoulder harness turning loops to the upper quarter panel. Replace the covers.

　f. Install the lower quarter trim and install the retaining screws.

　g. Install the quarter trim insert.

　h. Install the ashtray.

13. Reconnect the negative battery cable. Test A/C-heater blower operation.

1996–97 Models

CAUTION

The Supplemental Inflatable Restraint (SIR) system must be disarmed before removing the blower motor assembly. Failure to do so may cause accidental deployment of the air bag, resulting in unnecessary SIR system repairs and/or personal injury.

Front Blower Motor

1. Disconnect and isolate the negative battery cable, then wait 2 minutes before proceeding with the removal. This will effectively disable the air bag system preventing possible personal injury.

2. Remove the glove box from the instrument panel.

3. Remove the A/C-heater blower motor cover.

4. Disconnect the blower motor wiring connector.

5. Remove the blower motor wiring grommet and feed the wiring through the blower motor housing.

6. Remove the blower motor mounting screws.

7. Allow the blower motor assembly to drop downward to clear the instrument panel and remove from vehicle.

To install:

8. Place the blower motor assembly into position on the A/C-heater unit and tighten the mounting screws.

9. Route the blower motor wiring through the blower motor housing

and seat the wiring grommet in the hole of the blower motor housing.

10. Reconnect the blower motor electrical connector.

11. Install the A/C-heater blower motor cover.

12. Install the glove box into the instrument panel.

13. Reconnect the negative battery cable. Check blower motor operation.

Rear Blower Motor

1. Disconnect and isolate the negative battery cable.

2. Remove the right quarter trim panel as follows:

　a. Remove first rear seat. Remove the second rear seat, if equipped.

　b. Remove the sliding door sill plate.

　c. Remove the quarter trim bolster.

　d. Remove the upper C-pillar trim and D-pillar trim panel.

　e. Remove the first and second rear seat belt anchors.

　f. Remove the screws securing the quarter trim to quarter panel from the bolster area.

　g. Remove the screws securing the rear edge of the quarter trim to the support bracket.

　h. Carefully disengage the hidden clips securing the front of the quarter trim to the quarter panel just rearward of the sliding door opening.

　i. Disconnect the wiring harness connector from the accessory power outlet, if equipped.

　j. On long wheelbase wagons only, remove the quarter trim from the quarter panel and pull second rear seat belt through access hole.

　k. Pull first rear seat belt through the access hole.

　l. Remove the quarter trim panel from the vehicle.

3. Remove the 5 screws mounting the A/C-heater blower motor housing to the rear HVAC unit (one mounting screw located on evaporator cover).

4. Twist the blower motor out of scroll housing.

5. Disconnect the blower motor wiring connector.

6. Remove the blower motor from vehicle.

To install:

7. Install the blower motor unit into the vehicle.

8. Reconnect the blower motor wiring connector.

9. Secure the blower motor into scroll housing.

10. Install and tighten the 5 screws mounting the A/C-heater blower motor housing to the rear HVAC unit

(one mounting screw located on evaporator cover).

11. Install the right quarter trim panel as follows:

　a. Install the quarter trim panel in position in the vehicle.

　b. Insert and pull the first rear seat belt through the access hole in quarter trim panel.

　c. On long wheelbase wagons only, insert and pull the second rear seat belt through access hole in quarter trim panel.

　d. Reconnect the wiring harness connector to the accessory power outlet, if equipped.

　e. Engage the hidden clips to secure the front of quarter trim to quarter panel rearward of sliding door opening.

　f. Install the retaining screws that secure the rear edge of quarter trim to support bracket.

　g. Install the retaining screws that secure the quarter trim to the quarter panel in bolster area.

　h. Install the second rear seat belt anchor (if equipped), and the first rear seat belt anchor.

　i. Install the D-pillar trim panel and upper C-pillar trim.

　j. Install the quarter trim bolster.

　k. Install the sliding door sill plate.

　l. Install the second rear seat, if equipped, and install the first rear seat.

12. Reconnect the negative battery cable. Check blower motor operation.

Windshield Wiper Motor

REMOVAL AND INSTALLATION

1993–95 Models

1. Disconnect the negative battery cable.

2. Remove the wiper arm and blade assemblies as follows:

　a. Lift up the wiper arm to allow the latch to be pulled out to the holding position. The wiper arm should be positioned off of the windshield.

　b. Using a rocking motion, remove the wiper arm from the wiper pivot.

3. Remove the cowl plenum grille and plastic screen.

4. Remove the hoses from the turret connector. Remove the pivot mounting screws.

5. Disconnect the motor wiring connector from the motor.

6. Remove the retaining nut from the wiper motor shaft-to-linkage

drive crank, and remove the drive crank from the wiper motor shaft.

7. Remove the wiper motor assembly mounting screws and nuts, and remove the wiper motor.

To install:

8. Position the wiper motor against its mounting surface and secure in position with the mounting screws and nuts. Reconnect the wiring harness.

9. Install the linkage drive crank onto the wiper motor shaft and secure it with the retaining nut. Torque the nut to 95 inch lbs. (10 Nm).

10. Reconnect the hoses to the turret connector.

11. Install the cowl plenum grille and plastic screen. Reconnect the negative battery cable. Close the hood.

NOTE: If the original wiper blades are of different lengths, the short windshield wiper is installed on the driver side and the longer wiper is installed on the passenger side of the windshield.

12. Install the windshield wiper arm as follows:

   a. Position the wiper arm on top of the wiper pivot and push down.

   b. Slightly lift up the wiper arm and push in the latch so the wiper arm can now lay on the windshield.

13. Cycle the windshield wipers into the **PARK** position. The ends of the wiper blades should be in the blackout area of the lower windshield.

14. Equal length wiper blades should be positioned no closer than 1 in. (25mm) from the lower edge of the windshield.

15. If equipped with wiper blades of different lengths, they should be positioned as follows:

   a. Left side should be no closer than 2 in. (50mm) from the lower edge of the windshield.

   b. Right side should be no closer than 1 in. (25mm) from the lower edge of the windshield.

16. Operate the windshield wipers and check to make sure they work and park properly.

1996–97 Models

1. Disconnect the negative battery cable.

2. Remove the wiper arm and blade assemblies as follows:

   a. Disengage the wiper arm pivot cover retaining clip from the wiper arm.

   b. Lift up the wiper arm cap.

   c. Remove the wiper arm-to-wiper arm pivot mounting nut.

   d. Separate the wiper arm from the wiper arm pivot using a suitable two-jaw puller tool.

3. Remove the cowl cover from the vehicle as follows:

   a. Close the hood to aid in the removal procedure.

   b. Remove the lower area cowl cover-to-wiper module mounting screws.

   c. Disengage the quarter turn fasteners securing the outer ends of the cowl cover to the wiper module.

   d. Open the hood.

   e. Remove the wing nuts securing the front of the cowl cover to the wiper module.

   f. Using a small flat bladed tool, disengage the retainers securing the cowl cover to the front fender edge.

   g. Raise up the cowl cover enough to gain access to the right side washer hose and disconnect from the right washer nozzle.

   h. Close the hood, but do not latch. Remove the cowl cover from the vehicle.

4. Open the hood. Disconnect the positive lock on the wiper unit electrical connector.

5. Disconnect the wiper unit electrical connector from the engine compartment wiring harness.

6. Disconnect the windshield washer hose from the hose coupling inside the wiper unit.

7. Remove the drain tubes from the bottom of the wiper unit.

8. Remove the sound absorbers from the ends of the wiper unit.

9. Remove the attaching nuts securing the wiper unit to the lower windshield fence.

10. Remove the attaching bolts securing the wiper unit to the dash panel.

11. Raise up the wiper unit from the from the weld-studs on the lower windshield fence and remove the wiper unit from the vehicle.

12. Place the wiper unit on a solid, level working surface. Remove the cowl cover brackets from the wiper unit.

13. Remove the attaching nuts securing the wiper linkage and wiper motor mount plate to the wiper unit.

14. Remove the wiper linkage from the wiper unit.

15. Disconnect the electrical connectors from the wiper motor.

16. Disconnect the wiper linkage from the wiper motor crank, but do

NOT remove the crank from the wiper motor.

17. Remove the attaching bolts that secure the wiper motor to the mount plate and remove the wiper motor.

To install:

18. Place the wiper motor onto the mounting plate and secure in position with the mounting bolts.

19. Install the wiper motor mount plate and wiper linkage into the wiper unit. Install and tighten the attaching nuts.

20. Install the cowl cover brackets onto the wiper unit and tighten the attaching nuts.

21. Reconnect the wiper motor electrical connectors.

22. Install the wiper unit into the vehicle engine compartment and be sure the wiper unit is installed properly over the weld-studs on the lower windshield fence. Install and tighten the attaching nuts to the weld-studs.

23. Install and tighten the attaching bolts securing the wiper unit to the dash panel.

24. Install the sound absorbers to each end of the wiper unit.

25. Install the drain tubes to the bottom of the wiper unit and reconnect the windshield washer hose to the hose coupling inside the wiper unit.

26. Reconnect the wiper unit wiring connector to the engine wiring harness.

27. Place the cowl cover onto the vehicle. Reconnect the right side windshield washer hose to the right washer nozzle located on the underside of the cowl cover.

28. Engage the retainers that secure the cowl cover to the front fender edge.

29. Install and tighten the wing nuts that secure the front of the cowl cover to the wiper module.

30. Engage the quarter turn fasteners that secure the outer ends of the cowl cover to the wiper module.

31. Install and tighten the screws that hold the lower area of the cowl cover to the wiper module.

32. Reconnect the negative battery cable and verify that the wiper motor and wiper linkage are in the **PARK** position.

33. Install the wiper arm in correct position over the wiper arm pivot. Align the wiper arm positions as follows:

   a. Left side should be no closer than 2.5 in. (65mm) from the lower edge of the windshield.

   b. Right side should be no closer than 1.5 in. (40mm) from the lower edge of the windshield.

34. Install the the wiper arm-to-wiper arm pivot retaining nut and torque to 26 ft. lbs. (35 Nm).

35. Push down the wiper arm cap cover and engage the clip that secures the outside end of the cover to the wiper arm.

36. Operate the windshield wipers and check to make sure they work and park properly.

Headlight Switch

REMOVAL AND INSTALLATION

The 1993–95 vehicles use so-called POD switches which clip into the instrument panel. The left POD switch consists of the headlight, parking light and hazard switches which operate with a push-on, push-off action.

1993 Models

1. Disconnect the negative battery cable.
2. Remove the warning indicator grille by prying up with a flat bladed tool.
3. Remove the 3 mounting screws from the warning indicator module assembly.
4. Remove the 8 cluster bezel-to-instrument panel retaining screws.
5. If equipped, tilt the column down to ease removal.
6. Gently pull the cluster bezel out to gain access to the switch assembly snap fingers. Push the switch out through the mounting hole.
7. Disconnect the wiring connector and remove the switch.
   To install:
8. Snap the switch into the cluster bezel.
9. Reconnect the wire connectors and install the cluster bezel into the instrument panel.
10. Install and tighten the 8 cluster bezel-to-instrument panel retaining screws.
11. Install the 3 warning indicator module assembly screws.
12. Install the warning indicator grille.
13. Reconnect the negative battery cable. Test the headlight switch for proper operation.

1994–95 Models

1. Disconnect the negative battery cable.
2. Obtain a ⁵⁄₁₆ in. diameter metal rod approximately 10 in. long. Bend a 90 degree leg 1½ in. long on both ends.

3. Place the shop-made tool in the hole underneath and to the right of the switch in the cluster bezel below the switch.
4. Move the tool to the lower tab on the switch and depress. Pull the switch out to free the lower tab.
5. Move the tool upward along the right side of the switch to place on the top switch tab. Pull down on the tab and pull the switch out of the bezel.
6. Disconnect the switch connector and remove the switch.
   To install:
7. Reconnect the wiring connector and push the switch into position until the tabs lock in place.
8. Reconnect the negative battery cable. Test the headlight switch to be sure it is functioning properly.

1996–97 Models

1. Disconnect the negative battery cable.
2. Remove the instrument cluster bezel.
3. Remove the mounting screws securing the headlight switch bezel to the instrument cluster bezel.
4. Disconnect the electrical connectors from the headlight switch and the power mirror switch.

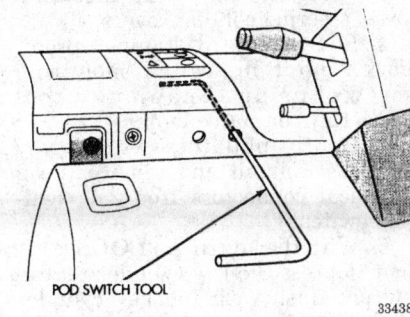

POD SWITCH TOOL 334385

Insert the tool into the cluster bezel left side — 1994–95 models

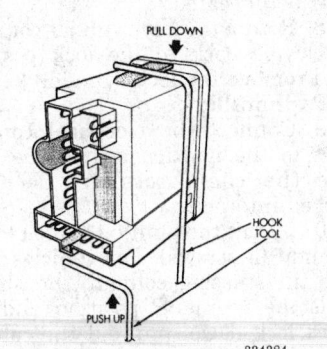

PULL DOWN

HOOK TOOL

PUSH UP

334384

Removing the headlight switch — 1994–95 models

5. Remove the headlight switch bezel from the instrument cluster bezel.
   To install:
6. Install the headlight switch bezel to the instrument cluster bezel and install the mounting screws.
7. Reconnect the electrical connectors to the power mirror switch and the headlight switch.
8. Install the instrument cluster bezel and headlight switch bezel to the instrument panel.
9. Reconnect the negative battery cable. Test the headlight switch for proper operation.

Combination Switch

REMOVAL AND INSTALLATION

— **CAUTION** —
The Supplemental Inflatable Restraint (SIR) system must be disarmed before working around the air bag or SIR system wiring. Failure to do so may cause accidental deployment of the air bag, resulting in unnecessary SIR system repairs and/or personal injury.

1. Disconnect and isolate the negative battery cable from the battery.
2. In order to disarm the air bag system, allow the system capacitor to discharge for at least 2 minutes, before performing any removal procedures.
3. On 1993–95 models, remove the tilt lever, if equipped.
4. Remove the upper and lower steering column shrouds.
5. Disconnect the electrical connector from behind the combination switch.
6. Remove the mounting screws securing the combination switch to the steering column adapter collar.
7. Remove the combination switch from the vehicle.
   To install:
8. On 1996–97 models, install the combination switch to the steering column adapter collar. Tighten the combination switch mounting screws.
9. Reconnect the electrical connector to the combination switch.
10. On 1993–95 models, install the tilt lever, if equipped.
11. Install the upper and lower steering column shrouds and tighten the retaining screws.
12. Reconnect the negative battery cable and check all functions of the combination switch for proper operation.

Ignition Lock Cylinder

REMOVAL AND INSTALLATION

1996–97 Models

──────── CAUTION ────────
The Supplemental Inflatable Restraint (SIR) system must be disarmed before working around the air bag or SIR wiring. Failure to do so may cause accidental deployment of the air bag, resulting in unnecessary SIR system repairs and/or personal injury.

1. Disconnect and isolate the negative battery cable from the battery.
2. In order to disarm the air bag system, allow the SIR system capacitor to discharge for at least 2 minutes, before performing any removal procedures.
3. Remove the screws that secure the parking brake release handle to the instrument panel.
4. Remove the screws that secure the bottom of the lower steering column cover to the instrument panel.
5. Disengage the the retaining clip that holds the right side of the lower steering column cover to the instrument panel.
6. Remove the lower steering column cover from the vehicle.
7. Remove the mounting screws that hold the upper and lower steering column shrouds and remove the lower steering column shroud.
8. Install the ignition key and turn the key cylinder to the **RUN** position. This is important because it will not only aid in removal of the ignition lock, but will also place the socket (in the lock cylinder housing) in proper alignment for the installation of the replacement ignition lock.
9. Press in the lock cylinder retaining tab and remove the key cylinder (ignition lock).

To install:

10. Install the ignition key into the replacement ignition lock key cylinder. Turn the key to the **RUN** position. Depress the ignition lock key cylinder retaining tab.
11. Align the shaft at the end of the ignition lock cylinder with the socket in the end of the ignition lock cylinder housing.
12. Align the ignition lock cylinder with the grooves in the lock cylinder housing. Push the ignition lock cylinder into the housing until the retaining tab sticks through the opening in the lock cylinder housing.
13. Turn the ignition key to the **OFF** position and remove the key.

14. Install the lower steering column shroud and install the mounting screws.
15. Place the lower steering column cover into correct position on the lower instrument panel and engage the clip that secures the right side of the cover to the instrument panel.
16. Install the lower steering column mounting screws.
17. Install the mounting screws that hold the parking brake release handle to the instrument panel.
18. Reconnect the negative battery cable.

Ignition Switch

REMOVAL AND INSTALLATION

1993–95 Models

NOTE: The ignition switch is located on the steering column. The ignition lock key cylinder is located in the ignition switch assembly. If the ignition lock needs replacement, the ignition switch assembly must be removed.

1. Disconnect the negative battery cable.
2. If equipped with tilt steering column, remove the tilt lever.
3. Carefully remove the upper and lower steering column covers.
4. Remove the 3 "tamper proof" Torx® ignition switch mounting screws using Snap-on® tool TTXR15A2 or equivalent.
5. Gently pull the switch away from the column and remove the 2 electrical connectors from the ignition switch.
6. With the key in the **LOCK** position, depress the key cylinder retaining pin flush with the key cylinder surface.
7. Rotate the key clockwise to the OFF position; the key cylinder should now be unseated.
8. Rotate the key cylinder counterclockwise back to the lock position and remove the lock cylinder.

To install:

9. Connect the electrical connectors to the ignition switch. Be sure the the connectors are securely locked into place.
10. Mount the ignition switch to the column (3 screws). For vehicles with the tilt steering column, the shifter must be in the **P** position and the park lock dowel pin and the column lock flag must be indexed before the switch is installed.
11. Place shifter in the **P** position.

12. Place the ignition switch in the **LOCK** position. The switch is in the lock position when the steering column lock flag is parallel to the ignition switch terminals.
13. Position the park lock dowel pin of the ignition switch so it engages the steering column park lock slider linkage.
14. Apply a light coating of grease to the park lock dowel pin and the column lock flag.
15. Position the ignition switch against the lock housing opening on the steering column. Be sure the switch park lock dowel pin enters the slot in the park lock slider linkage. Install the retaining screw and bracket. Torque the screw to 26 inch lbs. (3 Nm).
16. Install the steering column covers and the tilt lever if removed.
17. With the key cylinder and the ignition in the **LOCK** position. install the key cylinder into the switch until it bottoms.
18. Reconnect the negative battery cable and check the ignition switch for proper operation of halo light, shift lock (if equipped) and column lock. Also check for proper operation of the ignition switch accessory, lock, off, run and start functions.

1996–97 Models

──────── CAUTION ────────
The Supplemental Inflatable Restraint (SIR) system must be disarmed before removing any components in the area of the air bag system. Failure to do so may cause accidental deployment of the air bag, resulting in unnecessary SIR system repairs and/or personal injury.

1. Disconnect and isolate the negative battery cable from the battery.
2. Allow the SIR system capacitor to discharge for at least 2 minutes, before performing any removal procedures. The air bag system is now disabled.
3. Remove the key cylinder (ignition lock).
4. Using a No. 10 Torx® tamper proof bit, remove the ignition switch mounting screw.
5. Depress the ignition switch retaining tab and carefully pry the ignition switch from the steering column.
6. Disconnect the electrical connectors from the ignition switch and remove the switch from the vehicle.

To install:

7. Be sure the ignition switch is in the **RUN** position and the actuator

shaft in the ignition lock housing is in the **RUN** position.

8. Reconnect the electrical connectors to the ignition switch.

9. Carefully install the ignition switch, making sure the ignition switch snaps over the retaining tabs.

10. Install and tighten the ignition switch mounting screw.

11. Install the key cylinder (ignition lock). Complete the installation by reversing the removal procedures.

12. Reconnect the negative battery cable.

13. Check switch operation.

Park/Neutral Safety Switch

REMOVAL AND INSTALLATION

1. Disconnect the negative battery cable.

2. Raise and safely support the vehicle. Position a drain pan under the switch.

3. Disconnect the switch electrical connector.

4. Remove the switch from the case.

To install:

5. Verify that the switch operating lever fingers are centered in the switch opening in the case when in **P** and **N**.

6. Install a new seal and screw the switch on the case. Tighten to 24 ft. lbs. (33 Nm). Be sure the new seal has seated properly between the switch and the transaxle housing or a fluid leak could result.

7. Check the continuity of the switch for proper operation. Reconnect the electrical connector.

8. Lower the vehicle and check the transmission fluid level. Add if necessary.

9. Reconnect the negative battery cable and check the switch for proper operation.

Powertrain Control Module

REMOVAL AND INSTALLATION

1993–95 Models

1. Remove the air cleaner duct or air cleaner assembly.

2. Disconnect the battery cables from the battery, negative cable first.

3. Remove the battery.

4. Remove the Powertrain Control Module (PCM) mounting screws.

5. Disconnect the electrical connector.

6. Remove the control module.

To install:

7. Reconnect the engine control electrical connector. Install the control module.

8. Install the module retaining screws.

9. Install the battery and reconnect the battery cables, negative cable last.

10. Install the air cleaner duct or air cleaner assembly.

1996–97 Models

1. Disconnect the battery cables from the battery, negative cable first.

2. Remove the 2 attaching screws that hold the Power Distribution Center (PDC) to its mounting bracket.

3. Remove the battery heat shield.

4. Remove the battery from the vehicle.

5. Remove the PDC from the rear PDC mounting bracket by rotating it toward the center of the vehicle. Remove the PDC from the front mounting bracket by pulling it rearward. Lay the PDC aside, but do not remove it from the vehicle.

6. Disconnect the two, 40-way electrical connectors from the side of the Powertrain Control Module (PCM) by squeezing the connector locking tabs.

7. Remove the 3 PCM mounting screws.

8. Remove the PCM from the vehicle.

To install:

9. Connect the two, 40-way electrical connectors to the PCM.

10. Place the PCM in correct position against the left inner fender and install the PCM mounting screws.

11. Install the PDC into the front and rear PDC mounting brackets. Install and tighten the 2 PDC attaching screws.

12. Install the battery into the vehicle and secure in the battery tray.

13. Install the battery heat shield.

14. Reconnect the battery cables, negative cable last.

15. Road test the vehicle.

ENGINE COOLING

Radiator

REMOVAL AND INSTALLATION

1993–95 Models

1. Disconnect the negative battery cable.

——— **CAUTION** ———

Do not remove the radiator cap or drain with the cooling system hot and under pressure or serious personal injury can occur from hot pressurized coolant.

2. Place a drain pan under the radiator drain. Open the radiator drain and allow the coolant to drain.

3. Disconnect the radiator hoses from the radiator.

4. Remove the coolant reserve overflow tank hose from the radiator.

5. If equipped with automatic transmission and/or auxiliary oil cooler, disconnect the cooler lines and cap the lines.

6. Remove the cooling fan and fan support assembly by disconnecting the fan motor electrical connector. Remove the upper shroud fasteners, and lift the shroud up and out of the bottom shroud attachment clips separating the shroud from the radiator.

7. If equipped with air conditioning, remove the front grille, remove the A/C condenser fasteners and separate the condenser from the radiator.

8. Disconnect the block heater electrical connector, if equipped.

9. Remove the mounting screws and carefully lift the radiator out of the engine compartment.

To install:

10. Lower the radiator into position and install the mounting screws.

11. Connect the block heater electrical connector, if equipped.

12. Install the A/C condenser to the radiator, connect the fasteners and install the front grille, if removed.

13. Install the cooling fan and shroud assembly.

14. Connect the cooling fan electrical connector.

15. If equipped with automatic transmission and/or auxiliary oil cooler, connect the cooler lines.

16. Connect the radiator hoses to the radiator including the coolant reserve overflow tank hose.

17. Refill the system with coolant to the proper level.

18. Reconnect the negative battery cable.

19. Start the engine and run until it reaches normal operating temperature, then check the coolant level and the automatic transmission fluid level, if equipped. Add fluids, if necessary.

1996-97 Models

1. Disconnect the negative battery cable.

----- CAUTION -----
Do not remove the radiator cap or drain with the cooling system hot and under pressure or serious personal injury can occur from hot pressurized coolant.

2. Place a drain pan under the radiator drain. Open the radiator drain plug and allow the coolant to drain.

3. Remove the air intake resonator from the air cleaner assembly.

4. Remove the Coolant Recovery System (CRS) overflow tank filler neck hose.

5. Disconnect the cooling fan electrical connector located on the left side of the cooling fan module.

6. Remove the CRS overflow tank mounting screw from the upper radiator crossmember.

7. Remove the upper radiator-to-crossmember mounting screws.

8. If equipped, disconnect the engine block heater wiring connector.

9. Remove the upper radiator crossmember.

10. Remove the entire air cleaner assembly.

11. Disconnect and plug the automatic transaxle oil cooler lines from the radiator.

12. Disconnect the upper and lower radiator hoses from the radiator. Remove the lower radiator hose clip from the cooling fan module.

13. Remove the A/C condenser mounting fasteners and separate the A/C condenser from the radiator. Be sure the condenser is supported in position.

14. Remove the A/C filter/drier mounting bracket, 2 mounting bolts to the cooling fan module and 2 mounting nuts to the filter/drier. Remove the mounting bracket.

15. Carefully lift the radiator out of the engine compartment. Be careful not to damage the radiator cooling fins or water tubes during removal.

To install:

16. Be sure the air seals are properly positioned before installation of the radiator. Lower the radiator into position and seat the radiator with

the rubber isolators into the mounting holes provided.

17. Install the A/C filter/drier and mounting bracket onto the cooling fan module. Install the bracket mounting fasteners.

18. Install the A/C condenser to the radiator.

19. Unplug and connect the transaxle oil cooler lines to the radiator.

20. Connect the upper and lower radiator hoses to the radiator.

21. Connect the CRS overflow tank filler neck hose to the radiator.

22. Reconnect the cooling fan motor electrical connector.

23. Install the entire air cleaner assembly.

24. Install the upper radiator crossmember.

25. Install the upper radiator mounting screws and torque to 105 inch lbs. (12 Nm).

26. Reconnect the engine block heater electrical connector, if equipped.

27. Install the CRS overflow tank mounting screw to the upper radiator crossmember. Torque the screw to 18 inch lbs. (2 Nm).

28. Install the air intake resonator.

29. Refill the cooling system with a ⁵⁰/₅₀ mixture of clean, fresh ethylene glycol antifreeze and water to the proper level.

30. Reconnect the negative battery cable.

31. Start the engine and run until it reaches normal operating temperature, then check the coolant level and the automatic transmission fluid level. Add fluids, if necessary.

Water Pump

REMOVAL AND INSTALLATION

2.4L (VIN B) Engine

1. Disconnect the negative battery cable.

NOTE: This procedure requires removing the engine timing belt and the auto tensioner. The factory specifies that the timing marks should always be aligned before removing the timing belt. Set the piston in No. 1 cylinder to TDC on the compression stroke. This should align all timing marks on the crankshaft sprocket and both camshaft sprockets.

2. Raise and safely support the vehicle.

3. Remove the right inner splash shield.

4. Remove the accessory drive belts.

5. Place a drain pan under the radiator drain plug. Drain the cooling system.

6. Support the engine and remove the right motor mount.

7. Remove the power steering pump mounting bracket bolts and place the pump/bracket assembly off to one side. Do not disconnect the power steering fluid lines.

8. Remove the right engine mount bracket.

9. Remove the front timing belt upper and lower covers.

10. Loosen the timing belt tensioner screws and remove the belt tensioner and timing belt.

----- CAUTION -----
With the timing belt removed, DO NOT rotate the camshaft or crankshaft or damage to the engine could occur.

11. Remove the camshaft sprockets. With the timing belt removed, remove both camshaft sprocket bolts. Do not allow the camshafts to turn when the camshaft sprockets are being removed.

12. Remove the rear timing belt cover to access the water pump.

13. Remove the water pump attaching bolts.

14. Remove the water pump.

To install:

15. Thoroughly clean all parts. Replace the water pump if there are any cracks, signs of coolant leakage from the shaft seal, loose or rough turning bearing, damaged impeller or sprocket or sprocket flange loose or damaged.

16. Clean all sealing surfaces. Install a new rubber O-ring into the water pump O-ring groove.

NOTE: Make sure the O-ring is properly seated in the water pump groove before tightening the screws. An improperly located O-ring may cause damage to the O-ring and cause a coolant leak.

17. Install the water pump to the engine and torque the bolts to 105 inch lbs. (12 Nm).

18. Pressurize the cooling system to 15 psi and check for leaks. If okay, release the pressure and continue the engine assembly process.

19. Install the rear timing belt cover.

20. Install the camshaft sprockets and tighten the attaching bolts to 75 ft. lbs. (101 Nm). DO NOT allow the camshafts to turn while the sprockets

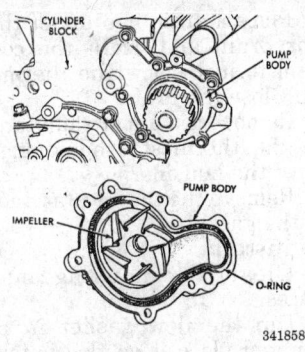

Water pump assembly — 2.4L (VIN B) engine

bolts are being tightened to maintain timing mark alignment.

─────── **CAUTION** ───────
Do not attempt to compress the tensioner plunger with the tensioner assembly installed in the engine. This will cause damage to the tensioner and other related components. The tensioner MUST be compressed in a vise.

21. Install the timing belt tensioner and timing belt. Be sure to properly tension the timing belt.
22. Install the front upper and lower timing belt covers.
23. Install the right engine mount bracket and engine mount.
24. Install the crankshaft damper and torque the center bolt to 105 ft. lbs. (142 Nm).
25. Install the right inner splash shield.
26. Lower the vehicle.
27. Install the power steering pump bracket and power steering pump. Torque the bracket mounting bolts to 40 ft. lbs. (54 Nm).
28. Install and adjust the drive belts.
29. Refill the cooling system using a mixture of $^{50}/_{50}$ water and ethylene glycol antifreeze. Bleed the cooling system.
30. Start the engine and check for proper operation.
31. Check and top off cooling system, if necessary.

2.5L (VIN K) Engine

1. Disconnect the negative battery cable.
2. Drain the cooling system.
3. Remove drive belts.
4. If equipped with air conditioning, remove the compressor from the bracket and position it aside.
5. Remove the alternator and mounting bracket and set it aside.

6. Raise and safely support the vehicle, if necessary. Remove the pulley from the water pump.
7. Disconnect the lower radiator hose and heater hose from the water pump.
8. Remove the water pump housing attaching screws and remove the assembly from the vehicle. Discard the O-ring.
9. Remove the water pump from the housing.
To install:
10. Using a new gasket or silicone sealer, install the water pump on the housing.
11. Install a new O-ring to the housing and install on the engine. Torque the top 3 bolts to 21 ft. lbs. (30 Nm). Torque the lower bolt to 50 ft. lbs. (68 Nm).
12. Install the water pump pulley. Torque the water pump pulley screws to 20 ft. lbs. (28 Nm). Connect the radiator hose and heater hose to the water pump.
13. Install the mounting bracket, alternator and A/C compressor, if removed. Install the drive belts and adjust the belts to the proper tension, if necessary.
14. Remove the hex-head plug or vacuum switching valve on the top of the thermostat housing. Fill the radiator with coolant until the coolant comes out the plug hole. Install the plug or valve and continue to fill the radiator. Torque the vent plug to 15 ft. lbs. (20 Nm).
15. Reconnect the negative battery cable, run the vehicle until the thermostat opens and fill the overflow tank. Check for leaks.
16. Once the vehicle has cooled, recheck the coolant level.

3.0L (VIN 3) Engine

1. Disconnect the negative battery cable.
2. Drain the cooling system.

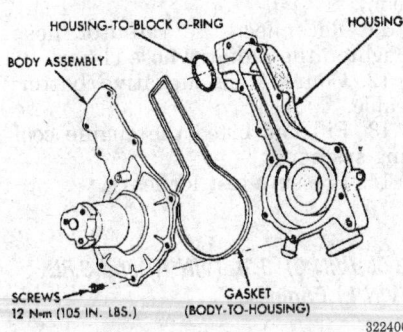

Water pump housing (exploded view) — 2.5L (VIN K) engine

3. Remove the drive belts. Remove the timing belt covers and the timing belt.
4. Remove the water pump mounting bolts.
5. Separate the water pump from the water inlet pipe and remove the water pump.
6. Inspect the water pump and replace as necessary.
To install:
7. Clean all gasket and O-ring surfaces on the water pump and water pipe inlet tube.
8. Wet a new O-ring with water and install it on the water inlet pipe.
9. Install a new gasket on the water pump.
10. Install the pump inlet opening over the water pipe and press until the pipe is completely inserted into the pump housing.
11. Install the water pump-to-block mounting bolts and tighten to 20 ft. lbs. (27 Nm).
12. Install the timing belt and timing belt covers. Install and adjust the drive belts.
13. Reconnect the negative battery cable. Fill the cooling system to the proper level with a $^{50}/_{50}$ mixture of clean, fresh ethylene glycol antifreeze and water.
14. Run the engine and check for leaks. Top off the coolant level, if necessary.

3.3L (VIN R) and 3.8L (VIN L) Engines

1. Disconnect the negative battery cable.
2. Drain the cooling system.
3. Remove the serpentine belt.
4. Raise and safely support the vehicle. Remove the right front wheel and lower fender shield.
5. Remove the water pump pulley.
6. Remove the 5 mounting screws and remove the pump from the engine.
7. Discard the O-ring. Clean the O-ring sealing surface and inspect the water pump for damage, cracks, seal leaks, loose or rough turning bearings.
To install:
8. Install a new O-ring into the water pump groove. Install the pump onto the engine. Tighten the mounting bolts to 108 inch lbs. (12 Nm).
9. Install the water pump pulley. Torque the water pump pulley screws to 20 ft. lbs. (28 Nm).
10. Install the fender shield and wheel. Torque the wheel lug nuts, in sequence, to 95 ft. lbs. (129 Nm). Lower the vehicle.
11. Install the serpentine belt.

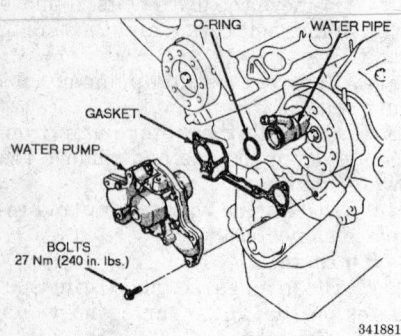

Water pump installation — 3.0L (VIN 3) engine

341881

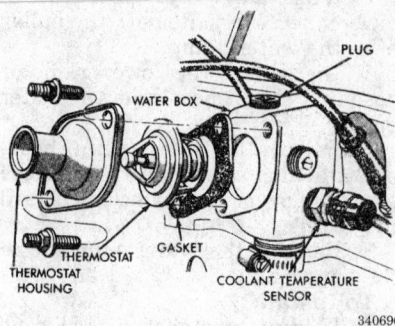

Thermostat installation — 2.5L (VIN K) engine

340696

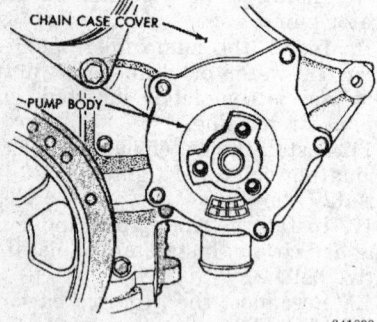

Water pump removal and installation — 3.3L (VIN R) and 3.8L (VIN L) engines

341899

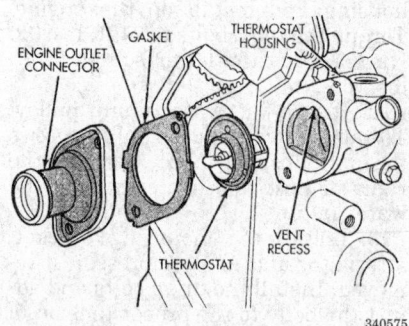

Thermostat installation — 2.4L (VIN B) engine

340575

12. Refill the cooling system to the correct level with a ⁵⁰/₅₀ mixture of clean, fresh ethylene glycol antifreeze and water. Bleed the cooling system.

13. Reconnect the negative battery cable, run the vehicle until the thermostat opens, fill the overflow tank and check for leaks.

14. Once the vehicle has cooled, recheck the coolant level.

Thermostat

REMOVAL AND INSTALLATION

2.4L (VIN B) and 2.5L (VIN K) Engines

1. Disconnect the negative battery cable.

2. Place a large drain pan under the radiator drain plug. Allow the cooling system to sufficiently cool down before opening the drain plug to avoid personal injury. Drain the coolant to below the thermostat level.

3. Disconnect the upper radiator hose at the thermostat housing.

4. Remove the thermostat housing bolts and coolant outlet connector of the thermostat housing.

5. Remove the thermostat assembly from the vehicle and discard. Discard the old thermostat gasket.

To install:

6. Clean all gasket mating surfaces.

7. Install the new thermostat in the correct position. If equipped, align the air bleed valve on top of thermostat to the vent recess in the water box (engine side) of the thermostat housing.

8. Dip the new new gasket in clean water and install on the water box surface of the thermostat housing.

9. Install the thermostat housing over the gasket and thermostat, making sure the thermostat is in correct position.

10. Install the thermostat housing bolts and torque to 20 ft. lbs. (28 Nm).

11. Reconnect the radiator hose. Tighten the radiator hose clamp.

12. Connect the negative battery cable.

13. Fill and bleed the engine cooling system.

14. Pressure test for leaks.

3.0L (VIN 3), 3.3L (VIN R) and 3.8L (VIN L) Engines

1. Disconnect the negative battery cable.

2. Place a drain pan under the radiator drain and drain the cooling system to just below the thermostat level. Close the drain.

3. Remove the upper radiator hose from the thermostat housing, then remove the housing.

4. Remove the thermostat and discard the gasket.

To install:

5. Clean the housing mating surfaces.

6. Dip the new gasket in clean water and place it on the water box surface.

7. Center the thermostat on the gasket, in the water box.

8. Make certain the bolt threads are clean. Threaded bolt holes exposed to coolant are subject to corrosion and should be cleaned with a small wire brush or correct size thread-cutting tap. Install the housing, making sure the thermostat is still in the recess, and tighten the retaining bolts to 105 inch lbs. (12 Nm) for 3.0L (VIN 3) engine or to 21 ft. lbs. (28 Nm) for 3.3L (VIN R) and 3.8L (VIN L) engines. Reconnect the upper radiator hose and tighten the hose clamp.

9. Connect the negative battery cable.

10. Fill and bleed the engine cooling system with a clean ⁵⁰/₅₀ mixture of ethylene glycol antifreeze and water.

11. Make sure the radiator is full and start the vehicle.

CAUTION

Do not remove the radiator cap once the vehicle is warm. Coolant is under pressure and may cause scalding or personal injury.

12. Run the vehicle until the thermostat opens. Check the coolant level in the overflow tank and fill if necessary.

13. Pressure test for leaks.

Electric Cooling Fan

REMOVAL AND INSTALLATION

1993–95 Models

NOTE: The electric cooling fan assembly cannot be disassembled. If the fan is warped, cracked or otherwise damaged, it must be replaced as an assembly.

1. Disconnect the negative battery cable.

2. Disconnect the fan motor electrical connector.

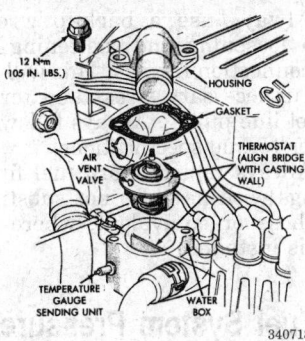

Thermostat installation — 3.0L (VIN 3) engine

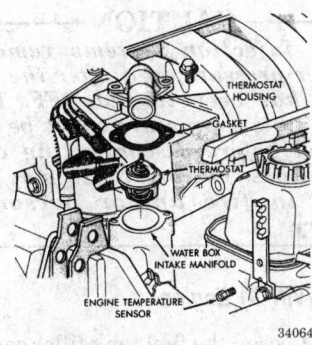

Thermostat installation — 3.3L (VIN R) and 3.8L (VIN L) engines

3. Remove the upper fan shroud retaining nuts and lift out the shroud and motor as an assembly. Be careful not to damage the lower fan shroud retaining clips.

To install:

4. Install the fan, shroud and motor assembly into the retaining brackets at the bottom of the radiator. Install the fan shroud retaining nuts and torque the nuts to 105 inch lbs. (12 Nm).

5. Reconnect the fan motor electrical connector.

6. Reconnect the negative battery cable.

1996–97 Models

NOTE: The electric cooling fan assembly cannot be disassembled. If the fan is warped, cracked or otherwise damaged, it must be replaced as an assembly.

1. Disconnect the negative battery cable.

2. Raise and safely support the vehicle.

3. Remove the radiator outlet hose from the radiator hose retaining clip and remove the retaining clip from the shroud.

4. If equipped, remove the lower auxiliary transaxle cooler lines from

the retaining clips on the cooling fan module shroud.

5. Lower the vehicle.

6. Remove the entire air cleaner assembly and the air intake resonator.

7. Disconnect the fan motor electrical connector.

8. Remove the Coolant Recovery System (CRS) mounting screw from the upper radiator crossmember.

9. Disconnect the mounts to the upper radiator from the radiator crossmember.

10. Remove the upper radiator crossmember.

11. Remove the cooling fan module mounting screws.

12. If equipped, remove the upper auxiliary transaxle cooler lines from the retaining clips on the cooling fan module shroud.

13. Disconnect and plug the transaxle oil cooler line from the radiator fitting on the lower left side.

14. Raise and safely support the vehicle.

15. Remove the filter/drier, cooling fan module and radiator mounting bolts located on the lower right of the cooling fan module.

16. Lower the vehicle. Remove the upper cooling fan module-to-radiator retaining clips.

17. Remove the cooling fan module from the vehicle.

To install:

18. Install the cooling fan module assembly into the retaining clips of the radiator.

19. Install the upper cooling fan module-to-radiator retaining clips.

20. Raise and safely support the vehicle.

21. Install the filter/drier, cooling fan module and radiator mounting bolts located on the lower right of the cooling fan module.

22. Lower the vehicle. Reconnect the transaxle cooler line to the radiator fitting on the lower left side.

23. If equipped, install the upper auxiliary transaxle cooler lines to the retaining clips on the cooling fan module shroud.

24. Install the cooling fan module retaining screws and torque to 105 inch lbs. (12 Nm).

25. Install the entire air cleaner assembly.

26. Install the upper radiator crossmember and connect the upper radiator mounts to the crossmember. Torque the mounting fasteners to 105 inch lbs. (12 Nm).

27. Install the Coolant Recovery System (CRS) mounting screw to the upper radiator crossmember. Torque

the mounting screw to 18 inch lbs. (2 Nm).

28. Reconnect the fan motor electrical connector.

29. Install the air intake resonator to the air cleaner assembly.

30. Raise and safely support the vehicle.

31. If equipped, install the lower auxiliary transaxle cooler lines to the retaining clips on the cooling fan module shroud.

32. Install the radiator outlet hose retainer clip to the fan module shroud.

33. Install the radiator outlet hose to the retaining clip. Lower the vehicle.

34. Reconnect the negative battery cable. Run the engine and check for proper cooling fan operation when the engine reaches operating temperature.

Cooling System Bleeding

PROCEDURE

1. Be sure the radiator drain cock is closed; hand tightened only.

2. If removed, install the cylinder block drain plugs.

WARNING

Do not use well water or the water supply that may already be in the cooling system.

3. Refill the cooling system with a $^{50}/_{50}$ solution of clean, fresh ethylene glycol antifreeze and water.

4. On the 2.4L (VIN B) Engine, perform the following procedure:

a. Vent the cooling system by removing the coolant temperature sensor, located on top of the water outlet connection.

b. When the coolant reaches this hole, install the coolant temperature sensor and torque to 60 inch lbs. (7 Nm) on 2.4L (VIN B) engine.

5. On the 2.5L (VIN K) Engine, perform the following procedure:

a. Vent the cooling system by removing the sealing plug, located just above the thermostat housing.

b. When the coolant reaches this hole, install the vent plug and torque to 15 ft. lbs. (20 Nm).

6. On the 3.3L (VIN R) and 3.8L (VIN L) engines, perform the following procedures:

a. Vent the cooling system by disconnecting and removing the Engine Temperature Sending Unit for 1993–95 engines or the cooling system vent plug for 1996–97 en-

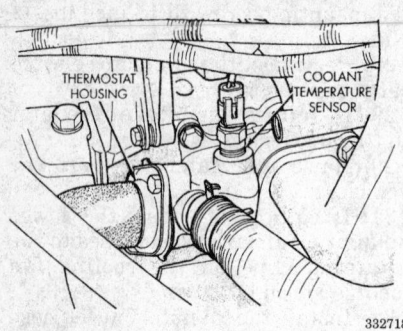

Coolant temperature sensor location — 2.4L (VIN B) engine

gines, located on the front of the cylinder head.

b. When the coolant reaches this hole, install the Engine Temperature Sending Unit or vent plug and torque to 60 inch lbs. (7 Nm). Reconnect the wiring connector to the sending unit.

7. Continue to refill the cooling system until it is full. Be careful not to spill any coolant on drive belts or the alternator.

8. Fill the coolant recovery tank to the "MAX" mark with the same $^{50}/_{50}$ solution of antifreeze and water. Install cap to the coolant recovery tank.

9. Start the engine with the radiator cap off and run until engine reaches full operating temperature.

10. Add coolant as necessary to maintain the level. It may be necessary to add coolant to the coolant recovery tank after 3–4 warm up/cool down cycles to maintain the coolant level between the MIN and MAX mark on the recovery tank. This will ensure that all trapped air will be removed from the system.

NOTE: Air can only be bled from the cooling system by gathering beneath the radiator pressure cap. On the next engine heat

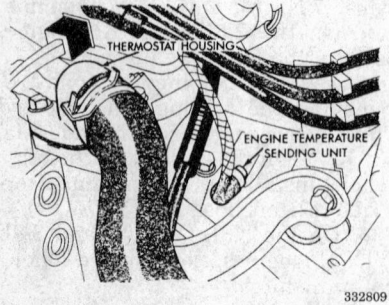

Engine Temperature Sending Unit located on the cylinder head — 1993–95 with 3.3L (VIN R) and 3.8L (VIN L) engines

up, the air will be pushed past the pressure cap into the coolant overflow tank due to thermal expansion of the coolant. It then escapes into the atmosphere and is replaced with solid coolant upon engine cool down.

11. Install the radiator cap.
12. Check the cooling system for leaks and correct fluid level.

FUEL SYSTEM

Fuel System Service Precautions

Safety is the most important factor when performing not only fuel system maintenance but any type of maintenance. Failure to conduct maintenance and repairs in a safe manner may result in serious personal injury or death. Maintenance and testing of the vehicle's fuel system components can be accomplished safely and effectively by adhering to the following rules and guidelines.

• To avoid the possibility of fire and personal injury, always disconnect the negative battery cable unless the repair or test procedure requires that battery voltage be applied.

• Always relieve the fuel system pressure prior to disconnecting any fuel system component (injector, fuel rail, pressure regulator, etc.), fitting or fuel line connection. Exercise extreme caution whenever relieving fuel system pressure to avoid exposing skin, face and eyes to fuel spray. Please be advised that fuel under pressure may penetrate the skin or any part of the body that it contacts.

• Always place a shop towel or cloth around the fitting or connection prior to loosening to absorb any excess fuel due to spillage. Ensure that all fuel spillage (should it occur) is quickly removed from engine surfaces. Ensure that all fuel soaked cloths or towels are deposited into a suitable waste container.

• Always keep a dry chemical (Class B) fire extinguisher near the work area.

• Do not allow fuel spray or fuel vapors to come into contact with a spark or open flame.

• Always use a backup wrench when loosening and tightening fuel line connection fittings. This will prevent unnecessary stress and torsion to fuel line piping. Always follow the proper torque specifications.

• Always replace worn fuel fitting O-rings with new. Do not substitute fuel hose or equivalent, where fuel pipe is installed.

Fuel System Pressure

RELIEVING

— **CAUTION** —
Fuel injection systems remain under pressure even after the engine has been turned OFF. The fuel system pressure must be relieved before disconnecting any fuel lines. Failure to do may result in fire and/or personal injury.

2.5L (VIN K) Engine

1. Loosen the fuel tank filler cap to release the fuel tank pressure.
2. Disconnect the fuel injector wiring harness from the engine harness.
3. Connect one end of a jumper wire (18 gauge or smaller) to ground terminal No. 1 of the injector harness. Connect the other end of the jumper wire to engine ground.
4. Connect one end of a jumper wire (18 gauge or smaller) to the positive terminal No. 2 of the injector harness. Touch the other end of the jumper wire to the positive battery post (for no longer than 5 seconds), to release the fuel system pressure.
5. Remove the jumper wires and disconnect the negative battery cable, then continue with the service procedure.

2.4L (VIN B), 3.3L (VIN R) and 3.8L (VIN L) Engines

NOTE: The following procedure requires fuel pressure release hose C-4799-1 or equivalent. Fuel gauge C-4799-A contains a hose to direct fuel into an approved container.

1. Disconnect the negative battery cable.
2. Remove the fuel tank filler cap to release the pressure in the fuel tank.
3. Remove the cap from the fuel pressure test port on the fuel rail.

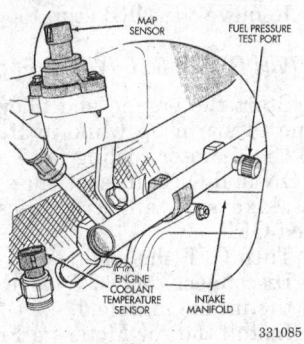

Fuel pressure test port location —
2.4L (VIN B) engine

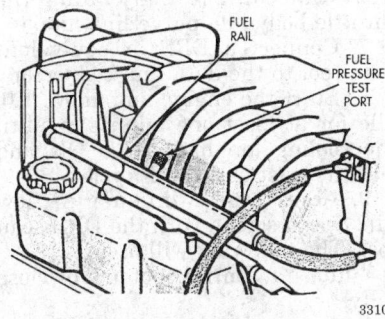

Fuel pressure test port location — 3.3L (VIN R)
and 3.8L (VIN L) engines

—————— CAUTION ——————

Always wear proper eye protection when relieving fuel system pressure. Do not allow fuel to spill on the intake or exhaust manifolds. Place shop towels under and around the pressure test port to absorb fuel when the pressure is released from the fuel rail.

4. Place the open end of fuel pressure release hose, tool No. C-4799-1, or equivalent, into an approved gasoline container. Place a shop towel under the test port.

5. Connect the other end of the hose onto the fuel pressure test port, to release the fuel pressure.

6. After the fuel pressure has been released, remove the hose from the test port and install the cap.

3.0L (VIN 3) Engine

1993–95

1. Loosen the fuel tank filler cap to release the fuel tank pressure.

2. Disconnect the fuel injector wiring harness from the engine harness.

3. Connect one end of a jumper wire to the A142 circuit terminal (dark green/orange colored wire) of the fuel injector harness connector

(black connector located on fuel rail harness). Connect the other end of the jumper wire to a 12 volt power source.

4. Connect one end of a jumper wire to a good ground.

5. Momentarily ground one of the injectors by connecting the other end of the jumper wire to an injector terminal in the harness connector. Repeat the procedure for 2 to 3 injectors.

6. Remove the jumper wires and disconnect the negative battery cable, then continue with the service procedure.

1996–97

1. Remove the fuel pump relay from the Power Distribution Center (PDC) located on the left side in the engine compartment. Location of the relay can be verified by the label located on the underside of the PDC cover.

2. Start the engine and allow it to run until it stalls.

3. Continue to start the engine until it will no longer run.

4. Turn the ignition key to the **OFF** position.

—————— WARNING ——————

Steps 1–4 must be performed to relieve the pressurized fuel from within the fuel rail. Do NOT use the following steps to relieve this fuel pressure as excessive fuel will be forced into a cylinder chamber.

5. Disconnect a wiring connector from any fuel injector.

6. Connect one end of a jumper wire (18 gauge or smaller) to either injector terminal of the fuel injector harness connector (black connector located on fuel rail harness). Connect the other end of the jumper wire to the positive battery terminal.

7. Connect one end of a second jumper wire to the remaining (other) injector terminal.

—————— WARNING ——————

Do NOT supply power to the injector for more than 4 seconds or permanent damage to the injector will result. Do NOT leave the injector connected to power for more than 4 seconds.

8. Momentarily touch the other end of the second jumper wire to the negative battery terminal for **no more than 4 seconds.**

9. Place a shop rag or towel under the fuel line at the quick-connect fitting to the fuel rail.

10. Disconnect the negative battery cable. Squeeze the quick-connect fitting retainer tabs together and pull the fuel tube/quick-connect fitting assembly off the fuel tube nipple. The retainer will remain on the fuel tube.

11. Install the fuel pump relay into the PDC.

12. Removal of the fuel pump relay from the PDC may have caused one or more Diagnostic Trouble Codes (DTC's) to be stored in the Powertrain Control Module (PCM) memory. Erasing any DTC's will require the use of a DRB scan tool or equivalent.

Idle Speed

ADJUSTMENT

2.4L (VIN B) Engine

1. Start the engine and allow it to idle in **P** or **N** until the radiator cooling fan has cycled ON and OFF at least once.

2. Be sure to turn all accessories OFF.

3. Turn OFF the engine.

4. Disconnect the hose from the PCV valve. Plug the nipple on the PCV valve.

5. Disconnect the idle purge hose from the throttle body.

6. Install the Air Metering Fitting tool 6457 to the idle purge hose nipple on the throttle body.

7. Connect a DRB scan tool to the data link connector.

8. Start the engine and allow the engine to idle until the cooling fan has turned ON and OFF at least one 180°F cycle.

9. Access the Minimum Airflow Idle Speed screen on the DRB scan tool. The following will occur:

• Idle air control motor will close fully.

• Idle spark advance will become fixed.

• Engine RPM will be displayed on the DRB scan tool.

10. Check the engine idle RPM. The throttle body minimum air flow is set correctly if it meets the following specifications:

• Below 1000 miles: 500–875 RPM

• Above 1000 miles: 550–875 RPM

—————— CAUTION ——————

Be sure to work in a well ventilated area when cleaning the throttle body. Wear rubber or butyl gloves and do not allow the cleaning solvent to come into contact with eyes or skin. Be sure to wash thoroughly after using the parts cleaner.

11. If the engine idle is not within specifications, turn OFF the engine and clean the throttle body.

12. Turn OFF the engine. Remove the Air Metering Fitting tool 6457. Reconnect the idle purge hose to the purge hose nipple on the throttle body.

13. Reconnect the PCV valve hose to the PCV valve.

14. Remove the DRB scan tool.

2.5L (VIN K) Engine

1. Connect the DRB or equivalent scan tool to the data link connector.

2. Remove the air cleaner assembly and plug the heated air door vacuum hose.

3. Warm up the engine in **P** or **N** until the radiator cooling fan has cycled ON and OFF at least once.

4. Install a timing light and tachometer.

5. Disconnect the electrical connector to the engine coolant temperature sensor. Set the base engine timing to 12 degrees BTDC ± 2 degrees BTDC.

6. Turn OFF the engine. Reconnect the engine coolant temperature sensor.

7. Disconnect the PCV valve hose from the intake manifold.

8. Connect Air Metering Fitting 6457 to the PCV nipple on the intake manifold.

9. Start the engine and allow the engine to idle for at least one minute.

10. Using the DRB scan tool, access the Minimum Airflow Idle Speed in the sensor read test mode. The following will then occur:
• Idle air control motor will close fully.
• Idle spark advance will become fixed.
• Idle fuel will be provided at a set value.
• The DRB scan tool will display the engine RPM.

11. Using the tachometer, check the idle RPM. The throttle body minimum air flow is set correctly if within the following idle specifications:
1993–95 vehicles:
Below 1000 miles: 600–1200 RPM
Above 1000 miles: 800–1200 RPM

12. If the idle is not within the idle specifications, the throttle body needs replacement.

13. Turn the engine OFF. Remove Air Metering Fitting 6457 from the PCV nipple on the intake manifold.

14. Reconnect the PCV valve hose.

15. Remove the DRB scan tool.

16. Install the air cleaner assembly and unplug the heated air door vacuum hose.

17. Remove the tachometer and timing check device.

3.0L (VIN 3) Engine

1. Start the engine and allow it to idle in **P** or **N** until the radiator cooling fan has cycled ON and OFF at least once.

2. Be sure to turn all accessories OFF.

3. Connect a tachometer and timing check device.

4. Disconnect the electrical connector to the engine coolant temperature sensor. Set the base engine timing to 12 degrees BTDC ± 2 degrees BTDC.

5. Turn OFF the engine. Reconnect the engine coolant temperature sensor.

6. Disconnect the hose from the PCV valve. Plug the nipple on the PCV valve.

7. Disconnect the idle purge hose from the engine vacuum harness tee.

8. Install the Air Metering Fitting tool 6457 in the idle purge hose at the intake manifold.

9. Connect a DRB scan tool to the data link connector.

10. Start the engine and allow the engine to idle for at least one minute.

11. Access the Minimum Airflow Idle Speed screen on the DRB scan tool. The following will occur:
• Idle air control motor will close fully.
• Idle spark advance will become fixed.
• Engine RPM will be displayed on the DRB scan tool.

12. Using the tachometer, check the engine idle RPM. The throttle body minimum air flow is set correctly if it meets the following specifications:
1993–96 vehicles:
Below 1000 miles: 560–910 RPM
Above 1000 miles: 610–910 RPM

CAUTION

Be sure to work in a well ventilated area when cleaning the throttle body. Wear rubber or butyl gloves and do not allow the cleaning solvent to come into contact with eyes or skin. Be sure to wash thoroughly after using the parts cleaner.

13. If the engine idle is not within specifications, turn OFF the engine and clean the throttle body.

14. Turn OFF the engine. Remove the Air Metering Fitting tool 6457. Reconnect the vacuum hose to the engine vacuum harness tee.

15. Reconnect the PCV valve hose to the PCV valve.

16. Remove the DRB scan tool.

3.3L (VIN R) and 3.8L (VIN L) Engines

1. Start the engine and allow the engine to warm up while in **P** or **N** until the radiator cooling fan has cycled ON and OFF at least once.

2. Make sure all accessories are turned OFF.

3. Turn OFF the engine.

4. Disconnect the PCV valve hose from the intake manifold.

5. Install the Air Metering Fitting Tool 6457 to the PCV nipple on the intake manifold.

6. Disconnect the idle purge line from the throttle body. Plug the throttle body idle purge line nipple.

7. Connect a DRB or equivalent scan tool to the data link connector.

8. Start the engine and allow it to idle for at least one minute or until the cooling fan has cycled ON and OFF at least one 180°F cycle.

9. Access the Minimum Airflow Idle Speed screen with the DRB scan tool. The following will occur:
• Idle air control motor will close fully.
• Idle spark advance will become fixed.
• Engine RPM will be displayed on the DRB scan tool.

10. The throttle body minimum air flow is set correctly if it meets the following specifications:
1993–96 vehicles:
Below 1000 miles: 525–875 RPM
Above 1000 miles: 575–875 RPM

11. If the RPM is out of specifications, turn OFF the engine and clean the throttle body.

12. Turn OFF the engine. Remove the Air Metering Fitting tool from the PCV nipple on the intake manifold. Reconnect the PCV valve hose.

13. Reconnect the idle purge line to the throttle body.

14. Remove the DRB scan tool.

Mixture

ADJUSTMENT

The Powertrain Control Module (PCM) operates the Fuel Injection System. Under most driving conditions, the PCM maintains an air/fuel ratio of 14.7:1 by constantly adjusting the fuel injector pulse width. The PCM adjusts the fuel injector pulse width by opening and closing the ground path to the fuel injector. There are no service adjustments that can be made to alter the air/fuel mixture; it is regulated by the PCM programming.

Fuel Filter

REMOVAL AND INSTALLATION

——— CAUTION ———

Fuel injection systems remain under pressure, even after the engine has been turned OFF. The fuel system pressure must be relieved before disconnecting any fuel lines. Failure to do so may result in fire and/or personal injury.

1993–95

The Front Wheel Drive (FWD) and the All Wheel Drive (AWD) vans have different fuel delivery systems. The tanks, pumps and fuel lines are different. Both systems use quick connect fittings in some locations. The fuel filter location is similar for both FWD and AWD vehicles.

1. Properly relieve the fuel system pressure.
2. Disconnect the negative battery cable.
3. Raise and safely support the vehicle.
4. On Front Wheel Drive, locate the fuel filter in its mounting along the frame rail. Remove the filter retaining screw and remove the filter from the rail.
5. On All Wheel Drive, remove the converter support bracket and the exhaust pipe heat shield.
6. Loosen the hose clamps. Wrap a shop towel around the hoses to absorb excess fuel.
7. On All Wheel Drive, remove the filter retaining screw and remove the filter from the rail.
8. Disconnect the hoses from the filter and remove the filter from the vehicle. Discard the clamps.
   To install:
9. Connect the hoses and new clamps to the filter.
10. Install the filter to the rail and tighten the retaining screw to 75 inch lbs. (8 Nm).
11. Position and tighten the hose clamps.
12. On All Wheel Drive, install the exhaust pipe heat shield and converter support bracket.
13. Lower the vehicle. Start the engine and check for leaks.

1996–97

NOTE: The fuel delivery system uses quick connect fittings. The fuel filter mounts to the top of the fuel tank.

1. Properly relieve the fuel system pressure.
2. Disconnect the negative battery cable.
3. Raise and safely support the vehicle.
4. Locate the fuel filter in its mounting on top of the fuel tank. Disconnect the quick-connect fittings from the chassis fuel supply tube and fuel pump module by squeezing the quick-connect fitting retainer tabs together and pulling the fitting assembly away from the fuel line nipple. The retainer will remain on the fuel tube.
5. Remove the fuel filter mounting bolt and remove the fuel filter from the fuel tank.
   To install:
6. Install the fuel filter to the top of the fuel tank and tighten the mounting bolt.
7. The fuel supply (to chassis fuel line), return tube (to pump module) and fuel supply (to fuel filter) tube are permanently attached to the fuel filter. The quick-connect fitting ends of the fuel supply and return tubes are of different sizes.
8. Apply a light coating of 30W engine oil to the nipples of the fuel filter.
9. Push the quick-connect fitting over the fuel line until the retainer seats and clicks into place. Be sure the retainer tabs have locked into the case of the quick-connect fitting.
10. Lower the vehicle. Start the engine and check for leaks.

Fuel Pump

REMOVAL AND INSTALLATION

1993–95

Although the Front Wheel Drive (FWD) van and All Wheel Drive (AWD) van have different fuel systems, both have fuel pump modules with an internal fuel reservoir, a fuel level sending unit and a fuel strainer mounted on the pump housing. Both systems use quick connect fittings at the fuel tank.

——— CAUTION ———

Fuel injection systems remain under pressure, even after the engine has been turned OFF. The fuel system pressure must be relieved before disconnecting any fuel lines. Failure to do so may result in fire and/or personal injury.

Front Wheel Drive

1. Properly relieve the fuel system pressure.
2. Disconnect the negative battery cable.
3. Raise the vehicle and support safely.

——— CAUTION ———

Observe all applicable safety precautions when working around fuel. Do not allow fuel spray or fuel vapors to come in contact with a spark or open flame. Keep a dry chemical (Class B) fire extinguisher near the work area. Never drain or store fuel in an open container due to the possibility of fire or explosion.

4. Remove the fuel tank from the vehicle.
5. Clean all dirt and foreign material from the pump mounting area.
6. Using a hammer and brass drift, carefully tap the lock ring counterclockwise to release the fuel pump.
7. Remove the fuel pump from the tank with the O-ring seal. Discard the O-ring seal.
8. Remove the fuel pump inlet strainer and O-ring from the fuel pump and discard.
9. Cover the fuel tank opening to prevent dirt from getting into the fuel tank.
   To install:
10. Lubricate a new strainer O-ring with silicone grease or spray lube and install into the outlet of the strainer. The O-ring must sit evenly on the step of the filter outlet.
11. Push the new strainer onto the inlet of the fuel pump reservoir body, making sure the locking tabs on the reservoir body lock over the locking tangs on the filter.
12. Wipe the seal area of the fuel tank clean. Install a new O-ring seal to the pump.
13. Install the fuel pump to the tank.
14. Install the lock ring with a hammer and brass drift, tapping the ring clockwise to lock the pump in place.

——— CAUTION ———

Over tightening the pump lock ring may result in a leak.

15. Install the fuel tank and lower the vehicle.
16. Reconnect the negative battery cable, start the engine and check for leaks.

All Wheel Drive

1. Properly relieve the fuel system pressure.

2. Disconnect the negative battery cable.

3. Raise the vehicle and support safely.

—————— CAUTION ——————
Observe all applicable safety precautions when working around fuel. Do not allow fuel spray or fuel vapors to come in contact with a spark or open flame. Keep a dry chemical (Class B) fire extinguisher near the work area. Never drain or store fuel in an open container due to the possibility of fire or explosion.

4. Remove the fuel tank from the vehicle.

5. Clean all dirt and foreign material from the pump mounting area.

6. Unclip the fuel vapor hose and fuel drain hose from the fuel tank.

7. While holding down on the fuel pump assembly, remove the band clamp.

NOTE: The fuel pump assembly is spring loaded and may rise up slightly when the band clamp is removed.

8. Remove the fuel pump assembly from the tank and discard the rubber seal.

9. Bend the locking tabs on the fuel inlet strainer to clear the locking tangs on the fuel pump. Remove and discard the strainer.

To install:

10. Align the orientation tabs in the inlet strainer with the slot in the bottom of the fuel pump. Push the strainer onto the fuel pump inlet, making sure the locking tabs on the filter snap over the tangs on the pump module.

11. Clean the seal area of the tank and install a new seal on the pump.

12. Position the fuel pump assembly on the fuel tank, aligning the arrow on the edge of the pump between the 2 lines molded into the fuel tank.

13. Push the module down and install the band clamp. Tighten the clamp to 40 inch lbs. (5 Nm).

14. Install the fuel tank and lower the vehicle.

15. Reconnect the negative battery cable, start the engine and check for leaks.

1996–97

The in-tank fuel pump module contains the fuel pump and pressure regulator which adjusts fuel system pressure to approximately 49 psi. Voltage to the fuel pump is supplied through the fuel pump relay.

The fuel pump is serviced as part of the fuel pump module. The fuel pump module is installed in the top of the fuel tank and contains the electric fuel pump, fuel pump reservoir, inlet strainer fuel gauge sending unit, fuel supply and return line connections and the pressure regulator. The inlet strainer, fuel pressure regulator and level sensor are the only serviceable items. If the fuel pump requires service, replace the fuel pump module. Use the following procedure.

—————— CAUTION ——————
Fuel injection systems remain under pressure, even after the engine has been turned OFF. The fuel system pressure must be relieved before disconnecting any fuel lines. Failure to do so may result in fire and/or personal injury.

1. Remove the fuel filler cap and properly relieve the fuel system pressure.

2. Disconnect and isolate the negative battery cable.

3. Drain and remove the fuel tank.

—————— CAUTION ——————
Observe all applicable safety precautions when working around fuel. Do not allow fuel spray or fuel vapors to come in contact with a spark or open flame. Keep a dry chemical (Class B) fire extinguisher near the work area. Never drain or store fuel in an open container due to the possibility of fire or explosion.

4. Clean the top of the tank to remove any loose dirt.

5. Disconnect fuel lines from the fuel pump module by squeezing the quick-connect fitting with thumb and fore finger.

6. Disconnect the fuel pump module electrical connector from the top of the fuel pump module.

7. Using special tool 6856 or equivalent, remove the fuel pump locknut by turning counterclockwise.

—————— CAUTION ——————
The fuel reservoir of the fuel pump module does not empty out when the tank is drained. The fuel in the reservoir may spill out when the module is removed.

8. Remove the fuel pump and O-ring from the tank. Discard the O-ring.

To install:

9. Thoroughly clean all parts. Wipe the seal area of the tank clean. Place a new O-ring on the ledge between the tank threads and the pump module opening.

10. Position the fuel pump module in the tank. Make sure the alignment tab on the underside of the pump module flange sits in the corresponding notch in the fuel tank.

11. While holding the fuel pump module in place install the locking ring and torque to 40 inch lbs. (5 Nm) using special tool 6856 or equivalent spanner-type tool.

12. Install the fuel tank assembly.

13. Connect fuel pump module electrical connector.

14. Reconnect the negative battery cable.

15. Fill the fuel tank with fuel. Install the fuel filler cap. Turn the ignition switch to the **ON** position to pressurize the system. Check the fuel system for leaks.

Fuel Injector

REMOVAL AND INSTALLATION

—————— CAUTION ——————
Fuel injection systems remain under pressure, even after the engine has been turned OFF. The fuel system pressure must be relieved before disconnecting any fuel lines. Failure to do so may result in fire and/or personal injury.

2.4L (VIN B) Engine

1. Properly relieve the fuel system pressure.

2. Disconnect the negative battery cable.

3. Disconnect the air inlet hose from the throttle body.

4. Disconnect the throttle cable and speed control cable (if equipped), from the throttle lever.

5. Compress the throttle cable retaining tabs and remove the throttle cables from the throttle bracket.

6. Disconnect the electrical connectors to the Throttle Position Sensor (TPS) and idle air control motor.

7. Disconnect vacuum hoses from the fittings on the intake plenum.

8. Disconnect electrical connectors to the intake air temperature sensor and MAP sensor.

9. Disconnect the fuel supply line from the fuel rail. This is a quick connect fitting. Squeeze the fitting retainer tabs together and pull the quick connect fitting assembly apart.

CAUTION

Wrap shop towels around the hose connection to catch any gasoline spillage.

10. Remove the bolt securing the bottom of the intake support bracket.

11. Remove the intake manifold mounting screws.

12. Disconnect the fuel injector electrical connectors.

13. Remove the fuel rail attaching screws.

14. Lift the fuel rail and injectors off the intake manifold. Cover the fuel injector openings in the intake manifold.

15. Remove the fuel injector clip and pull the injector from the fuel rail. Note that whenever a fuel injector is removed, the O-rings must be replaced.

To install:

16. Apply a light coating of clean engine oil to the O-ring on the fuel rail end of each fuel injector.

17. Install the injectors into the fuel rail and install the retaining clips.

18. Apply a light coating of clean engine oil to the O-ring on the nozzle end of each injector.

19. Insert the fuel injector nozzles into the openings in the intake manifold. Seat the injectors and tighten the fuel rail mounting screws to 200 ± 30 inch lbs. (22.5 ± 3 Nm).

20. Reconnect the electrical connectors to the fuel injectors.

21. Install a new intake manifold gasket and position the intake manifold onto the engine.

22. Tighten the intake manifold mounting bolts, starting at the center of the manifold and working outward in both directions. Torque the mounting bolts to 200 inch lbs. (23 Nm).

23. Lightly oil the tube end, then reconnect the fuel supply line quick connect fitting to the fuel rail. Be sure the quick connect fittings are fully engaged.

24. Install and tighten bolt that secures the bottom of the intake support bracket.

25. Reconnect the intake air temperature sensor and MAP sensor electrical connectors.

26. Reconnect the vacuum lines to the intake plenum fittings.

27. Reconnect the idle air control motor and TPS electrical connectors.

28. Secure the throttle cables into the cable bracket by engaging the retaining tabs.

29. Reconnect the throttle cable and speed control cable (if equipped) to the throttle lever.

30. Connect the air cleaner inlet hose to the throttle body.

31. Reconnect negative battery cable.

WARNING

When using the ASD Fuel System Test, the ASD relay and fuel pump relay remain energized for 7 minutes or until the test is stopped, or until the ignition switch is turned to the OFF position.

32. Turn the ignition key to the **ON** position and access the DRB scan tool ASD Fuel System Test to pressurize the fuel system. Check the fuel system for leaks.

2.5L (VIN K) Engine

The engine uses single point (throttle body) type fuel injection. The fuel injector is installed in the top of the throttle body. The injector is covered by a cap.

1. Remove the air cleaner assembly.

2. Properly relieve the fuel system pressure.

3. Disconnect the negative battery cable.

4. Remove the injector cap hold-down screw.

5. Fit 2 small flat-bladed tools in the slots provided and lift the top off the injector.

6. Fit a small flat-bladed tool in the hole in the front of the electrical connector and gently pry the injector from its housing.

7. Make sure the injector lower O-ring has been removed from the housing.

To install:

8. Apply a light coating of clean engine oil on the O-rings and install on the fuel injector and injector cap.

9. Carefully position the injector assembly in its housing, while aligning the injector wiring terminals with the injector cap fastener hole.

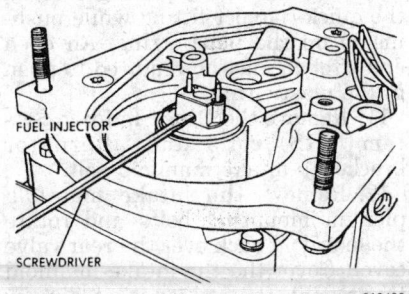

FUEL INJECTOR

SCREWDRIVER

319439

Prying the fuel injector from the housing — 2.5L (VIN K) engine

10. Install the injector cap with the locating notch aligned with the locating lobe on the injector. Push down on the cap to ensure a good seal.

11. Rotate the cap and injector to line up the retaining screw holes. Install the injector cap hold-down screws and tighten to 35–45 inch lbs. (4–5 Nm).

12. Reconnect the negative battery cable. Start the engine and check for leaks and proper engine operation.

13. Install the air cleaner assembly.

3.0L (VIN 3) Engine

The engine uses a sequential Multi-Port Electronic Fuel Injection (MPI) system. The MPI system uses fuel injectors positioned in the intake manifold with the nozzle ends directly above the intake port. The fuel rail assembly must be removed to service the injectors. The system pressure is approximately 48 psi. Perform the fuel pressure release procedure before attempting to service the fuel injectors.

1. Properly relieve the fuel system pressure.

2. Disconnect the negative battery cable.

3. Remove the air cleaner assembly.

4. Remove the throttle cable.

5. Remove the transaxle kickdown linkage, if equipped.

6. Label and disconnect the electrical connectors from the idle air control motor, throttle position sensor and coolant temperature sensor.

7. Label and disconnect the vacuum harness from the throttle body and the vacuum hoses from the intake plenum.

8. Place a shop towel under the fuel hoses to absorb any spilled fuel and remove the fuel hoses from the fuel rail.

9. Remove the intake plenum-to-intake manifold retaining bolts. Remove the ignition coil.

10. Remove the air intake plenum. Cover the the intake manifold to prevent the entrance of dirt or other foreign material.

11. Disconnect the vacuum hoses from the fuel rail.

12. Disconnect the fuel injector wiring harness from the engine wiring harness.

13. Remove the fuel pressure regulator attaching bolts and remove the fuel pressure regulator. Be careful not to damage the regulator O-ring.

14. Remove the fuel rail retaining bolts. Lift the fuel rail assembly off the intake manifold.

15. Label and disconnect the injector electrical connector(s).

16. Remove the fuel injector retaining clip(s).

17. Remove the injector(s) from the fuel rail.

18. Inspect the injector O-rings for damage and replace, as necessary.

To install:

19. Apply a light coating of clean engine oil to the upper injector O-ring.

20. Install the injector(s) on the fuel rail and insert the retaining clip(s).

21. Make sure the injectors are fully seated in the fuel rail and the clips are securely in place. Make sure the injector holes in the manifold are clean.

22. Lubricate the lower injector O-rings with clean engine oil. Start each injector into its hole and push the assembly into place until all injectors are fully seated.

23. Install the fuel rail retaining bolts and tighten to 115 inch lbs. (13 Nm).

24. Lubricate the fuel pressure regulator O-ring with clean engine oil and install the fuel pressure regulator. Tighten the attaching bolts to 90 inch lbs. (10 Nm).

25. Install the fuel supply and return tube hold-down bolt and the vacuum crossover tube hold-down bolt and tighten to 90 inch lbs. (10 Nm).

26. Connect the electrical connectors to the fuel injectors in correct order. Connect the fuel injector wiring harness to the engine harness. Connect the vacuum harness to the fuel rail assembly.

27. Clean the intake manifold mating surfaces. Place the manifold gaskets, beaded sealer side up, on the lower manifold.

28. Place the air intake plenum in position and install the ignition coil. Install the retaining bolts and tighten to 115 inch lbs. (13 Nm).

29. Connect the fuel lines to the fuel rail. Connect the vacuum lines to the intake plenum and fuel pressure regulator.

30. Connect the electrical connectors to the coolant temperature sensor, throttle position sensor and idle air control motor.

31. Reconnect the PCV and power brake booster supply hose to the intake manifold.

32. Reconnect the vacuum vapor harness to the throttle body.

33. Install the throttle cable and transaxle kickdown linkage, if equipped. Install the air inlet hose.

34. Reconnect the negative battery cable. Start the engine and check for leaks and proper engine operation.

3.3L (VIN R) and 3.8L (VIN L) Engines

The engines use a sequential Multi-Port Electronic Fuel Injection (MPI) system. The MPI system uses fuel injectors positioned in the intake manifold with the nozzle ends directly above the intake port. The fuel rail assembly must be removed to service the injectors. The system pressure is approximately 48 psi. Perform the fuel pressure release procedure before attempting to service the fuel injectors.

1993–95

1. Properly relieve the fuel system pressure.

2. Disconnect the negative battery cable.

3. Remove the air cleaner and hose assembly.

4. Remove the throttle cable. Remove the wiring harness from the throttle cable bracket and intake manifold water tube.

5. Disconnect the Idle Air Control (IAC) motor and Throttle Position Sensor (TPS) electrical connectors. Tag and disconnect the vacuum hose harness from the throttle body.

6. Remove the PCV and brake booster vacuum hoses from the air intake plenum.

7. Remove the EGR tube-to-intake manifold flange bolts. Label and remove the vacuum harness connectors from the intake manifold plenum.

8. Remove the cylinder head-to-intake plenum strut.

9. Disconnect the electrical connectors from the Manifold Absolute Pressure (MAP) sensor and oxygen sensor. Remove the engine mounted ground strap.

10. Place a shop towel under the fuel lines to catch any gasoline spillage. Disconnect the fuel lines from the chassis tubes by pulling back on the quick-connect fitting while pushing in on the plastic ring. An open end wrench may be required to push in the plastic ring.

11. Remove the Direct Ignition System (DIS) coils and alternator bracket-to-intake manifold bolt.

12. Remove the intake manifold plenum mounting bolts and rotate the manifold back over the rear valve cover. Cover the intake manifold to prevent the entrance of dirt or other foreign material.

13. Disconnect the vacuum hose from the fuel pressure regulator.

14. Remove the screws from the fuel tube clamp and fuel rail retaining bolts. Spread the retainer bracket to allow fuel tube removal clearance.

15. Remove the fuel rail injector wiring clip from the alternator bracket.

16. Disconnect the camshaft position sensor, engine coolant temperature sensor and engine temperature sensors. Remove the fuel rail.

17. Rotate the fuel injector(s) and pull the injector(s) out of the fuel rail. The clip will stay on the injector.

18. Check the injector O-rings for damage and replace, as necessary. Replace the injector clip if damaged.

To install:

19. Apply a light coating of clean engine oil to the upper O-ring.

20. If removed, install the injector clip by sliding the open end into the top slot of the injector. The edge of the receiver cup will slide into the side slots of the clip. Install the injector into the fuel rail, being careful not to damage the O-ring.

21. Lubricate the injector O-rings with clean engine oil. Install the tip of each fuel injector into their ports and push the assembly into place until fully seated.

22. Install the fuel rail mounting bolts and tighten to 16 ft. lbs. (22 Nm).

23. Install the fuel tube retaining bracket screw and tighten to 35 inch lbs. (4 Nm).

24. Connect the electrical connectors to the camshaft position sensor, coolant temperature sensor and engine temperature sensors.

25. Install the fuel injector wiring harness clips to the alternator bracket and intake manifold water tube.

26. Connect the vacuum line to the fuel pressure regulator.

27. Install the intake manifold gasket onto the lower manifold and install the upper manifold. Hand tighten the bolts.

28. Install the alternator bracket-to-intake manifold bolt and cylinder head-to-intake manifold strut bolts. Do not tighten.

29. Tighten the manifold bolts, in sequence, to 21 ft. lbs. (28 Nm).

30. Tighten the alternator bracket-to-intake manifold bolt and cylinder head-to-intake manifold strut bolts to 40 ft. lbs. (54 Nm).

31. Connect the ground strap, MAP and oxygen sensor electrical connectors.

32. Connect the vacuum harness to the intake plenum. Connect the PCV hoses.

33. Install a new gasket and connect the EGR tube to the intake manifold plenum. Tighten the bolts to 17 ft. lbs. (22 Nm).

34. Clip the wiring harness into the hole in the throttle cable bracket.

35. Connect the electrical connector to the TPS and IAC motor. Connect the vacuum harness to the throttle body.

36. Install the ignition coils and tighten the retainers to 105 inch lbs. (12 Nm).

37. Install the fuel hose quick-connect fittings to the chassis tubes. Push the fittings onto the tubes until they click in place. Pull on the fittings to ensure complete insertion.

38. Install the throttle cable and air cleaner assembly.

39. Reconnect the negative battery cable, start the engine and check for leaks and proper engine operation.

1996-97

1. Properly relieve the fuel system pressure.

2. Disconnect the negative battery cable.

3. Remove the intake manifold cover.

4. Remove the air inlet resonator.

5. Disconnect the throttle cable and speed control cable (if equipped) from the throttle lever and cable bracket.

6. Disconnect the Idle Air Control (IAC) motor and Throttle Position Sensor (TPS) electrical connectors. Tag and disconnect the vacuum hose harness from the throttle body and intake manifold.

7. Remove the EGR tube-to-intake manifold flange bolts.

8. Remove the cylinder head-to-intake plenum strut.

9. Disconnect the electrical connector from the Manifold Absolute Pressure (MAP) sensor. Remove the engine mounted ground strap.

10. Place a shop towel under the fuel lines to catch any gasoline spillage. Disconnect the fuel lines from the chassis tubes by squeezing the retainer tabs together and pulling the quick-connect fitting off the fuel tube nipple.

11. Remove the Direct Ignition System (DIS) coils and alternator bracket-to-intake manifold bolt.

12. Remove the intake manifold mounting bolts and rotate the manifold back over the rear valve cover. Cover the the intake manifold to prevent the entrance of dirt or other foreign material.

13. Remove the screws from the fuel tube clamp and fuel rail retaining bolts. Spread the retainer bracket to allow fuel tube removal clearance.

14. Disconnect the fuel injector wiring connector from the fuel injector.

15. Disconnect the camshaft position sensor and engine coolant temperature sensor. Remove the fuel rail.

16. Rotate the fuel injector(s) and pull the injector(s) out of the fuel rail. The clip will stay on the injector.

17. Check the injector O-rings for damage and replace, as necessary. Replace the injector clip if damaged.

To install:

18. Apply a light coating of clean engine oil to the upper fuel injector O-ring.

19. If removed, install the injector clip by sliding the open end into the top slot of the injector. The edge of the receiver cup will slide into the side slots of the clip. Install the injector into the fuel rail, being careful not to damage the O-ring.

20. Lubricate the injector O-rings with clean engine oil. Install the tip of each fuel injector into their ports and push the assembly into place until fully seated.

21. Install the fuel rail mounting bolts and tighten to 16 ft. lbs. (22 Nm).

22. Install the fuel tube retaining bracket screw and tighten to 35 inch lbs. (4 Nm).

23. Reconnect the fuel injector wiring.

24. Connect the electrical connectors to the camshaft position sensor and engine coolant temperature sensor.

25. Remove the covering on the lower intake manifold and clean the mounting surface.

26. Install the intake manifold gasket onto the lower manifold and install the upper manifold. Hand tighten the bolts only.

27. Install the alternator bracket-to-intake manifold bolt and cylinder head-to-intake manifold strut bolts. Do not tighten.

28. Tighten the manifold bolts, in sequence, to 21 ft. lbs. (28 Nm).

29. Tighten the cylinder head-to-intake manifold strut bolts to 40 ft. lbs. (54 Nm).

30. Connect the ground strap and MAP sensor electrical connectors.

31. Connect the vacuum harness to the intake plenum. Connect the PCV hoses.

32. Install a new gasket and connect the EGR tube to the intake manifold plenum. Tighten the bolts to 17 ft. lbs. (22 Nm).

33. Connect the electrical connector to the TPS and IAC motor. Connect the vacuum harness to the throttle body.

34. Install the ignition coils and tighten the retainers to 105 inch lbs. (12 Nm).

35. Install the fuel hose quick-connect fittings to the chassis tubes. Push the fittings onto the tubes until they click in place. Pull on the fittings to ensure complete insertion.

36. Install the throttle cable and speed control cable, if equipped.

37. Install the air inlet resonator.

38. Reconnect the negative battery cable.

WARNING

When using the ASD Fuel System Test, the ASD relay and fuel pump relay remain energized for 7 minutes or until the test is stopped, or until the ignition switch is turned to the OFF position.

---

39. Turn the ignition key to the **ON** position and access the DRB scan tool ASD Fuel System Test to pressurize the fuel system. Check the fuel system for leaks.

EMISSION CONTROLS

Service Interval Lamp

RESETTING

The system is designed to act as a reminder that scheduled vehicle emission maintenance should be performed. It is not intended to indicate a warning that a state of emergency exists which must be corrected to insure safe vehicle operation.

NOTE: To reset the emission maintenance light, it is necessary to use the Chrysler Digital Read Out Box II Tester (DRB–II) tool C-4805 or equivalent.

1. Connect the Chrysler DRB–II Tester C-4805 or equivalent to the vehicle as directed by the manufacturers instructions.

2. Turn the tester selector switch to the **EMR MEMORY CHECK.** The tester display should read **EMR MEMORY CHECK ARE YOU**

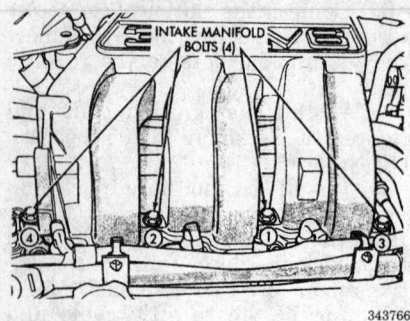

343766

Intake manifold bolt torque sequence — 3.3L (VIN R) and 3.8L (VIN L) engines

SURE?. Press the **YES** key on the tester.

3. The tester should next display the following message; **IS INSTRUMENT PANEL MILEAGE BETWEEN 9953 AND 10051?**. If the odometer mileage on the vehicle is within the specified mileage, press the **YES** key on the tester.

4. The tester should now display, **EMR MEMORY CHECK TEST COMPLETE**. The EMR light is now reset. If the speedometer mileage is not within specifications, go on to Step 5.

5. Press the **NO** key on the tester. The tester will then display **DO YOU WANT TO CORRECT EMR MILEAGE?**. Press the **YES** key on the tester.

6. The tester should now display **ENTER MILEAGE SHOWN ON INSTRUMENT PANEL USE ENTER TO KEY TO END**.

7. Enter the mileage shown on the speedometer. Do not enter the tenths. Press the **ENTER** key on the tester. The tester will now ask for verification of the entry.

8. If the mileage entered was correct, the tester will display **EMR MEMORY CHECK TEST COMPLETE**. The vehicle must now be driven approximately 8–10 miles for the mileage reset to take place.

ENGINE MECHANICAL

Engine Assembly

REMOVAL AND INSTALLATION

— **CAUTION** —

Fuel injection systems remain under pressure, even after the engine has been turned OFF. The fuel system pressure must be relieved before disconnecting any fuel lines. Failure to do so may result in fire and/or personal injury.

1993–95

1. Properly relieve the fuel system pressure.

2. Disconnect the battery cables from the battery, negative cable first.

3. Scribe the hood hinge outlines on the hood, and remove the hood.

4. Drain the cooling system. Remove the radiator hoses from the radiator and engine connections.

5. Remove the radiator and cooling fan assembly.

6. Remove the air cleaner assembly and related hoses.

7. Remove the air conditioner compressor from the engine and mounting brackets, leaving the refrigerant lines connected. Position the assembly aside and secure out of the way.

8. Remove the power steering pump from the engine with mounting brackets and hoses connected. Position the assembly aside and secure out of the way.

9. Label and disconnect all electrical connectors at the alternator, throttle body and engine.

10. Disconnect the fuel lines at the quick-connect fittings by pulling back on the fitting while pushing in on the plastic ring. Disconnect the heater hose from the engine. Disconnect the accelerator cable.

11. Remove the alternator. Disconnect the clutch cable from the clutch lever, if equipped with a manual transaxle.

12. Remove the transaxle case lower cover.

13. If equipped with automatic transaxle, matchmark the flex plate to torque converter position. Remove the bolts that mount the converter to the flex plate. Attach a small C-clamp to the front bottom of the converter

housing to prevent the converter rom falling out of the transaxle.

14. Disconnect the starter motor wiring and remove the starter motor.

15. Disconnect the exhaust pipe from the exhaust manifold.

16. Remove the right inner engine splash shield. Drain the engine oil and remove the oil filter. Disconnect the engine ground strap.

17. Attach a suitable hoist to the engine.

18. Support the transaxle. Apply slight upward pressure with the chain hoist and remove the through bolt from the right (timing case cover) engine mount.

NOTE: If the complete engine mount is to be removed, mark the insulator position on the side rail to insure exact reinstallation location.

19. Remove the transaxle to cylinder block mounting bolts.

20. Remove the front engine mount through bolt. Remove the manual transaxle anti-roll strut, if equipped.

21. Remove the insulator through bolt from the inside wheel house mount, or remove the insulator bracket to transaxle mounting bolt.

22. Raise the engine slowly with the hoist (transaxle supported). Separate the engine and transaxle and remove the engine.

To install:

23. With the hoist attached to the engine, lower the engine into the engine compartment.

24. Align the converter to flex plate, if equipped, and the engine mounts. Install all mounting bolts loosely until all are in position, then tighten to 40 ft. lbs.

25. Install the engine to transaxle mounting bolts. Tighten to 70 ft. lbs. (95 Nm) for the 2.5L (VIN K) and 3.0L (VIN 3) engines or 75 ft. lbs. (102 Nm) for the 3.3L (VIN R) and 3.8L (VIN L) engines.

26. Remove the engine hoist and transaxle support.

27. Secure the engine ground strap.

28. Install the inner splash shield.

29. Install the starter assembly.

30. Install the exhaust system.

31. If equipped with manual transaxle, install the transaxle case lower cover.

32. If equipped with automatic transaxle, remove the C-clamp from the torque converter housing. Align the flex plate and torque converter with mark previously made. Install the convertor to flex plate mounting bolts and torque to 55 ft. lbs. (75 Nm). Install the case lower cover.

33. If equipped with manual transaxle, reconnect the clutch cable.
34. Install the power steering pump.
35. Install the air conditioning compressor.
36. Install the alternator.
37. Reconnect all wiring.
38. Install the radiator, cooling fan and shroud assembly.
39. Reconnect all cooling system hoses, accelerator cable and fuel lines.
40. Install the engine oil filter. Fill the crankcase to proper oil level with clean engine oil.
41. Refill the cooling system.
42. Adjust linkages.
43. Install the air cleaner and hoses.
44. Install the hood.
45. Reconnect the battery cables, positive cable first. Check to make sure all electrical connections, cables, hoses, vacuum and fuel lines have been reconnected.
46. Start the engine and run until normal operation temperature is indicated. Check for leaks and check for proper engine operation.

1996–97

1. Properly relieve the fuel system pressure.
2. Disconnect the battery cables from the battery, negative cable first.
3. Disconnect the fuel line-to-fuel rail quick-connect fittings by squeezing the retainer tabs together and pulling back on the fitting.
4. Scribe the hood hinge outlines on the hood and remove the hood.
5. Remove the wiper arm and blade assemblies.
6. Remove the cowl cover from the vehicle.
7. Disconnect the positive lock on the wiper unit electrical connector.
8. Disconnect the wiper unit electrical connector from the engine compartment wiring harness.
9. Disconnect the windshield washer hose from the hose coupling inside the wiper unit.
10. Remove the drain tubes from the bottom of the wiper unit.
11. Remove the sound absorbers from the ends of the wiper unit.
12. Remove the attaching nuts securing the wiper unit to the lower windshield fence.
13. Remove the attaching bolts securing the wiper unit to the dash panel.
14. Raise up the wiper unit from the from the weld-studs on the lower

windshield fence and remove the wiper unit from the vehicle.
15. Remove air cleaner assembly and related hoses.
16. Remove the battery cover, battery, battery tray and integral vacuum reservoir from the vehicle.
17. If equipped, block off rear heater to the rear heater unit.
18. Drain the cooling system. Remove the upper and lower radiator hoses from the radiator.
19. Disconnect and plug the heater hoses from the engine.
20. Remove the radiator and cooling fan assembly.
21. Disconnect the transaxle shift linkage.
22. Disconnect the throttle body linkage and all vacuum hoses to the throttle body.
23. Remove accessory drive belts.
24. Remove the air conditioner compressor from the engine and mounting brackets. Keep the A/C compressor hoses connected. Position the assembly aside and secure out of the way.
25. Disconnect the wiring harness connector to the alternator.
26. Remove the alternator.
27. Raise and safely support the vehicle. Remove the right and left halfshaft assemblies.
28. Disconnect the starter motor wiring and remove the starter motor.
29. Drain the engine oil and remove the oil filter.
30. Remove the right and left fender inner splash shields.
31. Disconnect the exhaust pipe from the exhaust manifold.
32. Remove the front motor mount and mount bracket as an assembly.
33. Remove the rear transaxle motor mount and bracket.
34. Remove the power steering pump and mounting bracket assembly from the engine.
35. Disconnect, label and remove the wiring harness and connectors from the front of the engine.
36. Remove bending braces.
37. Remove the transaxle case inspection cover.
38. Mark the flex plate to torque converter location.
39. Remove the bolts that mount the converter to the flex plate. Attach a small C-clamp to the front bottom of the converter housing to prevent the converter from falling out of the transaxle, if necessary.
40. Lower the vehicle to the ground.
41. Disconnect the engine ground straps.
42. Attach an engine lifting hoist to the engine.

43. Remove the right engine mount assembly and left transaxle mount through bolt.
44. Using the engine hoist, raise the engine and transaxle assembly slowly out of the vehicle. Separate the engine and transaxle. Secure the engine on an engine stand.

To install:
45. Attach the transaxle to the engine assembly.
46. With the hoist attached to the engine/transaxle assembly, lower the engine/transaxle into the engine compartment.
47. Align the engine and transaxle motor mounts to their attaching points. Install the right engine and left transaxle mount bolts.
48. Remove the engine hoist.
49. Install bending braces.
50. Install the alternator.
51. Install and reconnect the wiring harness connectors on the front of the engine.
52. Install the air conditioning compressor to the engine.
53. Install the power steering pump and mounting bracket to the engine.
54. Install accessory drive belts and adjust to the proper tension, if necessary.
55. Raise and safely support the vehicle.
56. Remove the C-clamp from the torque converter housing, if utilized.
57. Align flex plate and torque converter with marks previously made.
58. Install the convertor to flex plate mounting screws. Torque to 55 ft. lbs. (75 Nm).
59. Install the transaxle case lower inspection cover.
60. Install right and left halfshaft assemblies.
61. Install the engine and transaxle mount and bracket assemblies.
62. Install the exhaust system to the exhaust manifolds.
63. Install the left and right fender inner splash shields.
64. Install the starter assembly and reconnect the starter motor wiring.
65. Reconnect the automatic transaxle shift linkage.
66. Lower the vehicle.
67. Reconnect the fuel line quick-connect fitting to the fuel rail.
68. Reconnect the cooling system heater hoses to the engine.
69. Un-block the heater hoses to the rear heater unit, if equipped.
70. Secure the engine ground straps.
71. Reconnect the engine and throttle body vacuum connections and wiring harness connectors.

72. Reconnect the throttle body linkage.

73. Install the radiator, cooling fan and shroud assembly.

74. Reconnect the upper and lower radiator hoses.

75. Install the battery tray, battery and battery cover into the vehicle.

76. Install the air cleaner assembly and hoses.

77. Install a new engine oil filter. Fill the crankcase to proper oil level with the correct type of clean, fresh engine oil.

78. Refill the cooling system to the proper level with a $^{50}/_{50}$ mixture of clean fresh ethylene glycol antifreeze and water.

79. Install the wiper unit into the vehicle engine compartment, making sure the wiper unit is installed properly over the weld-studs on the lower windshield fence. Install and tighten the attaching nuts to the weld-studs.

80. Install and tighten the attaching bolts securing the wiper unit to the dash panel.

81. Install the sound absorbers to each end of the wiper unit.

82. Install the drain tubes to the bottom of the wiper unit and reconnect the windshield washer hose to the hose coupling inside the wiper unit.

83. Reconnect the wiper unit wiring connector to the engine wiring harness.

84. Place the cowl cover onto the vehicle. Reconnect the right side windshield washer hose to the right washer nozzle located on the underside of the cowl cover.

85. Engage the retainers that secure the cowl cover to the front fender edge.

86. Install and tighten the wing nuts that secure the front of the cowl cover to the wiper module.

87. Engage the quarter turn fasteners that secure the outer ends of the cowl cover to the wiper module.

88. Install and tighten the screws that hold the lower area of the cowl cover to the wiper module.

89. Reconnect the positive battery cable, then the negative battery cable and verify that the wiper motor and wiper linkage are in the **PARK** position.

90. Install the wiper arm in correct position over the wiper arm pivot. Align the wiper arm positions as follows:

 a. Left side should be no closer than 2.5 in. (65mm) from the lower edge of the windshield.

 b. Right side should be no closer than 1.5 in. (40mm) from the lower edge of the windshield.

91. Install the the wiper arm-to-wiper arm pivot retaining nut and torque to 26 ft. lbs. (35 Nm).

92. Push down the wiper arm cap cover and engage the clip that secures the outside end of the cover to the wiper arm.

93. Operate the windshield wipers and check to make sure they work and park properly.

94. Adjust linkages.

95. Install the hood.

96. Check to make sure all electrical connections, cables, hoses, vacuum and fuel lines have been reconnected.

97. Start the engine and run until normal operation temperature is indicated. Check for leaks. Road test the vehicle.

Engine Mounts

REMOVAL AND INSTALLATION

2.5L (VIN K) Engine
1993–95 3.0L (VIN 3), 3.3L (VIN R) and 3.8L (VIN L) Engines

Right Side Engine Mount

1. Disconnect the negative battery cable.

2. Remove the right engine mount insulator vertical retaining bolts from the frame rail.

3. Using a floor jack, take the load off the engine mounts by carefully supporting the engine and transaxle assembly.

4. Remove the yoke bolt and nut from the insulator assembly. Remove the insulator assembly.

 To install:

5. Install the engine mount insulator assembly. Install the yoke bolt and nut, but only finger tight.

6. Lower the floor jack until the engine mount insulator contacts the right frame rail. Torque the engine mount insulator vertical retaining bolts to 50 ft. lbs. (68 Nm). Remove the floor jack.

7. Adjust the engine mount. Torque the insulator yoke nut to 75 ft. lbs. (102 Nm), then torque the yoke bolt to 100 ft. lbs. (133 Nm) on 2.5L (VIN K) and 3.0L (VIN 3) engines or 110 ft. lbs. (149 Nm) on 3.3L (VIN R) and 3.8L (VIN L) engines.

8. Reconnect the negative battery cable.

Front Engine Mount

1. Disconnect the negative battery cable.

2. Using a floor jack, support the engine and transaxle assembly to prevent rotation.

3. Remove the through bolt from the insulator assembly and front mounting bracket.

4. Remove the front engine mount bracket-to-front crossmember bolts and nuts. Remove the insulator assembly.

 To install:

5. Install the insulator assembly to the front body crossmember. Torque the bracket-to-crossmember fasteners to 40 ft. lbs. (54 Nm).

6. Install the insulator assembly-to-front mounting bracket through bolt.

7. Remove the floor jack. Adjust the front engine mount. Torque the through bolt and nut to 50 ft. lbs. (68 Nm).

8. Reconnect the negative battery cable.

Left Side Engine Mount

1. Disconnect the negative battery cable.

2. Raise and safely support the vehicle. Remove the left front wheel.

3. Remove the left inner fender splash shield.

4. Using a transmission jack, support the transaxle.

5. Remove the insulator through bolt from the engine mount.

6. For 2.5L (VIN K) engine with a manual transaxle, remove the damper assembly.

7. Remove the transaxle mount bolts and remove the mount.

 To install:

8. Install the transaxle mount and the mount bolts. Torque the bolts to 40 ft. lbs. (54 Nm) except for 2.5L (VIN K) engine with a manual transaxle.

9. For 2.5L (VIN K) engine with a manual transaxle, torque the transaxle damper bushing retainer nut to 200 inch lbs. (23 Nm).

10. Install the insulator through bolt to the engine mount. Remove the transmission jack.

11. Adjust the engine mount. Torque the insulator through bolt to 55 ft. lbs. (75 Nm) except for 2.5L (VIN K) engine with a manual transaxle.

12. For 2.5L (VIN K) engine with a manual transaxle, torque the transaxle damper assembly through bolt to 40 ft. lbs. (54 Nm).

13. Torque the dampener weight to 200 inch lbs. (23 Nm).

14. Install the inner fender splash shield.

15. Install the left front wheel and lug nuts. Torque the lug nuts to 95 ft. lbs. (129 Nm).

16. Lower the vehicle. Reconnect the negative battery cable.

2.4L (VIN B) Engine
1996–97 3.0L (VIN 3), 3.3L (VIN R) and 3.8L (VIN L) Engines

Right Side Engine Mount

1. Disconnect the negative battery cable.

2. Remove the purge duty solenoid and wiring harness from the engine mount.

3. Remove the 2 right engine mount insulator vertical retaining bolts from the frame rail and loosen the horizontal mounting bolt. **Do not** remove the large nut on the end of the core from the chassis frame rail.

4. Using a floor jack, take the load off the engine mounts by carefully supporting the engine and transaxle assembly.

5. Remove the horizontal and vertical fasteners from the engine side bracket. Remove the right side engine mount assembly.

To install:

6. Install the right side mount assembly. Install the horizontal and vertical fasteners, but only finger tight.

7. Lower the floor jack until the right side engine mount contacts the vehicle frame rail.

8. Install the 2 right engine mount insulator vertical retaining bolts. Torque the engine mount-to-frame rail retaining bolts to 40 ft. lbs. (54 Nm). Remove the floor jack.

9. Torque the vertical engine mount fastener to 110 ft. lbs. (150 Nm), then torque the horizontal engine mount fastener to 75 ft. lbs. (102 Nm).

10. Install the purge duty solenoid and wiring harness to the engine mount.

11. For the 3.0L (VIN 3), 3.3L (VIN R) and 3.8L (VIN L) engines, perform the engine mount insulator adjustment procedure.

12. Reconnect the negative battery cable.

Front Engine Mount

1. Disconnect the negative battery cable.

2. Using a floor jack, support the engine and transaxle assembly to prevent rotation.

3. Remove the thru-bolt from the insulator assembly and front mounting bracket.

4. Remove the 6 air dam mounting screws and air dam unit from the vehicle to provide access to the front engine mount screws.

5. Remove the front engine mount screws. Remove the insulator assembly.

6. If necessary, remove the front engine mounting bracket.

To install:

7. If removed, install the front engine mounting bracket.

8. Install the insulator assembly to the front body crossmember and install the front engine mount screws.

9. Install the front engine mount thru-bolt into the insulator assembly and front mounting bracket.

10. Torque front engine mount bolts 1 and 5 to 40 ft. lbs. (54 Nm). Torque front engine mount bolts 2, 3 and 4 to 80 ft. lbs. (108 Nm).

11. Install the vehicle air dam and mounting screws. Torque the 6 air dam mounting screws to 105 inch lbs. (12 Nm).

12. Remove the floor jack. For the 3.0L (VIN 3), 3.3L (VIN R) and 3.8L (VIN L) engines, perform the engine mount insulator adjustment procedure.

13. Reconnect the negative battery cable.

Left Side Engine Mount

1. Disconnect the negative battery cable.

2. Raise and safely support the vehicle. Remove the left front wheel.

3. Remove the left inner fender splash shield.

4. Using a transmission jack, support the transaxle.

5. Remove the insulator thru-bolt from the engine mount.

6. Remove the transaxle mount bolts and remove the mount.

To install:

7. Install the transaxle mount and the mount bolts. Torque the bolts to 40 ft. lbs. (54 Nm).

8. Install the insulator thru-bolt to the engine mount. Torque the thru-bolt to 55 ft. lbs. (75 Nm).

9. Remove the transmission jack.

10. For the 3.0L (VIN 3), 3.3L (VIN R) and 3.8L (VIN L) engines, perform the engine mount insulator adjustment procedure.

11. Install the inner fender splash shield.

12. Install the left front wheel and lug nuts. Torque the lug nuts, in a star sequence, to 95 ft. lbs. (129 Nm).

13. Lower the vehicle. Reconnect the negative battery cable.

Rear Engine Mount

1. Disconnect the negative battery cable.

2. Raise and safely support the vehicle. Remove the left front wheel.

3. Support the transaxle assembly with a transmission jack to prevent the transaxle from rotating.

4. Remove the insulator thru-bolt from the rear engine mount and rear suspension crossmember.

5. Remove the 4 rear engine mount bolts and remove engine mount.

To install:

6. Install the rear engine mount and tighten the 4 mounting bolts.

7. Install the insulator thru-bolt to the rear engine mount and rear suspension crossmember. Torque the thru-bolt to 55 ft. lbs. (75 Nm).

8. Remove the transmission jack out from under the vehicle.

9. Install the left front wheel and lug nuts. Torque the lug nuts, in a star sequence, to 95 ft. lbs. (129 Nm).

10. Lower the vehicle.

11. Reconnect the negative battery cable.

ADJUSTMENT

2.5L (VIN K) Engine
1993–95 3.0L (VIN 3), 3.3L (VIN R) and 3.8L (VIN L) Engines

1. Using a floor jack, support the engine and transaxle assembly.

2. Loosen, but do not remove, the right side insulator vertical retaining bolts, front bracket-to-crossmember fasteners and the transaxle insulator through bolt.

3. Using a tape measure and measuring at the bottom of the halfshaft assembly, measure the direct distance from the inner edge of the outboard CV-boot to the inner edge of the inboard boot on both halfshafts. The dimension specifications should be as follows:

2.5L (VIN K) Engine
   Manual
      Right halfshaft: 18.7–19.1 in. (476–486mm)
      Left halfshaft: 7.2–7.6 in. (184–194mm)
   Automatic
      Right halfshaft: 18.7–19.1 in. (476–486mm)
      Left halfshaft: 7.3–7.7 in. (185–195mm)
1993–95 3.0L (VIN 3), 3.3L (VIN R) and 3.8L (VIN L) Engines
   Front Wheel Drive

Right halfshaft: 18.7–19.1 in. (476–486mm)

Left halfshaft: 7.2–7.6 in. (184–194mm)

All Wheel Drive

Right halfshaft: 11.5–11.9 in. (292–302mm)

Left halfshaft: 7.2–7.6 in. (184–194mm)

4. Pry the engine assembly to the right or left side to achieve the correct halfshaft assembly length.

5. Torque the right side engine mount insulator vertical bolts to 50 ft. lbs. (68 Nm), then torque the front bracket-to-crossmember fasteners to 40 ft. lbs. (54 Nm) and torque the transaxle insulator through bolt to 55 ft. lbs. (75 Nm).

6. Remove the floor jack. Recheck the halfshaft lengths and re-adjust, if necessary.

1996–97 3.0L (VIN 3), 3.3L (VIN R) and 3.8L (VIN L) Engines

1. Using a floor jack, support the engine and transaxle assembly.

2. Loosen, but do not remove, the right side engine mount insulator vertical bolt, the fore and aft bolts and the front engine mount bracket-to-crossmember fasteners.

3. Pry the engine assembly to the right or left side to achieve the correct halfshaft assembly length.

4. Using a tape measure and measuring at the bottom of the halfshaft assembly, measure the direct distance from the inner edge of the outboard CV-boot to the inner edge of the inboard boot on both halfshafts. The dimension specifications should be as follows:

Right halfshaft: 18.7–19.1 in. (476–486mm)

Left halfshaft: 7.2–7.6 in. (184–194mm)

5. Torque the right side engine mount insulator vertical bolts to 75 ft. lbs. (102 Nm) and the fore and aft bolts to 110 ft. lbs. (150 Nm). Then torque the front engine mount screws to 40 ft. lbs. (54 Nm). The clearance between the engine and the snubbers should be 0.078 inch (2mm) on each side.

6. Torque the left engine mount thru-bolt to 55 ft. lbs. (75 Nm).

7. Remove the floor jack. Recheck the halfshaft lengths and re-adjust, if necessary.

Cylinder Head

REMOVAL AND INSTALLATION

--------- **CAUTION** ---------
Fuel injection systems remain under pressure, even after the engine has been turned OFF. The fuel system pressure must be relieved before disconnecting any fuel lines. Failure to do so may result in fire and/or personal injury.

2.4L (VIN B) Engine

1. Properly relieve the fuel system pressure.

2. Disconnect the negative battery cable.

3. Place a large drain pan under the radiator drain plug. Open up the drain plug and drain the cooling system.

4. Remove the air cleaner assembly and disconnect all vacuum lines, electrical wiring and fuel lines from the throttle body.

5. Disconnect the throttle linkage.

6. Remove the accessory drive belts.

7. Disconnect the power brake vacuum hose from the intake manifold.

8. Raise and safely support the vehicle. Disconnect the exhaust pipe from the exhaust manifold.

9. Lower the vehicle as required to remove the power steering pump. Do not disconnect the fluid lines. Set the pump aside.

10. Label the spark plug wires for correct installation. Disconnect the coil pack wiring connector and remove the coil pack and spark plug wires from the engine.

11. Disconnect the cam sensor and fuel injectors' wiring connectors.

12. Remove the timing belt covers, timing belt and camshaft sprockets.

13. Remove the timing belt idler pulley and rear timing belt cover.

14. Remove the cylinder head cover mounting fasteners and cylinder head cover. Remove the ground strap.

15. Identify the camshafts, if they are to be reused, for later installation. The camshafts are not interchangeable. Remove the camshaft bearing caps and the camshafts.

16. Remove the camshaft followers. Any components that are to be reused must be installed in their original locations. Use care to identify and mark the positions of any removed valve train components so they may be reinstalled correctly.

17. Remove the intake and exhaust manifolds.

18. Remove the cylinder head bolts.

19. Remove the cylinder head from the vehicle, using care not to damage the aluminum gasket surfaces.

20. Remove all gasket material from the cylinder head and engine block. Be careful not to gouge or scratch the sealing surface of the aluminum head. The cylinder head should be checked for flatness using a good straightedge and feeler gauges. The cylinder head must be flat within 0.004 in. (0.1mm).

21. Inspect the camshaft bearing oil feed holes in the cylinder head for clogging. Inspect the camshaft bearing journals for wear or scoring. Check the cam surface for abnormal wear and damage. A visible worn groove in the roller path or on the cam lobes is cause for replacement. Valve service may be performed at this time.

To install:

22. Clean all parts well. Note that the cylinder head bolts are torqued using a new procedure. The cylinder head bolts should be checked carefully BEFORE reuse. If the threads are necked down the bolts should be replaced with new bolts. Necking can be checked by holding a steel scale or straight edge against the threads. If all the threads do not contact the scale, the bolt should be replaced. New cylinder head bolts are recommended for any engine rebuild, especially if known that the engine has been disassembled before.

23. Make sure both the top of the engine block and the bottom of the cylinder head are clean. Install a new gasket making sure all holes align with the openings in the engine block. Carefully set the cylinder head in place.

24. Before installing the bolts, the threads should be oiled with clean engine oil. Install the bolts and torque in sequence in 4 Steps as follows:

a. Tighten all bolts to 25 ft. lbs. (34 Nm).

b. Tighten all bolts to 50 ft. lbs. (68 Nm).

c. Tighten all bolts again to 50 ft. lbs. (68 Nm).

d. Tighten all bolts an additional ¼ turn.

NOTE: Do not use a torque wrench for the fourth step.

25. The camshaft end-play should be checked using the following procedure:

a. Oil the camshaft journals and install the camshaft **WITHOUT** the cam follower assemblies. In-

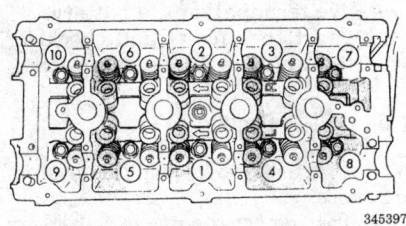

Cylinder head tightening sequence — 2.4L (VIN B) engine

345397

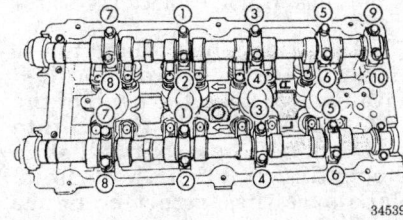

Camshaft bearing cap tightening sequence — 2.4L (VIN B) engine

345399

stall the rear cam caps and torque to specification.

b. Carefully push the camshaft as far rearward as it will go.

c. Set up a dial indicator to bear against the front of the camshaft (the sprocket end). Zero the indicator.

d. Move the camshaft forward as far as it will go. Read the dial indicator. End-play specification is 0.002–0.010 in. (0.05–0.15mm).

26. When satisfied with the fit and condition of the camshafts, remove the camshafts for installation of the cam followers.

27. Lubricate the camshaft followers with clean engine oil. Install the cam followers in their original positions on the hydraulic adjuster and valve stem.

—— WARNING ——
Make sure none of the pistons are at Top Dead Center when installing the camshafts.

28. Lubricate the camshaft bearing journals and cam followers with clean engine oil and install the camshafts. Install right and left camshaft bearing caps No. 2 through No. 5 and right side No. 6. Tighten the M6 fasteners to 105 inch lbs. (12 Nm) in sequence.

29. Apply Mopar Gasket Maker or equivalent sealer to the No. 1 and 6 bearing caps. Install the bearing caps and tighten the M8 fasteners to 250 inch lbs. (28 Nm). The end caps must be installed before the seals may be installed.

30. Apply a light coating of clean engine oil to the lip of the new camshaft seal. Install the camshaft seal until it fits flush with the cylinder head.

31. Install the camshaft sprockets, if removed. Install the rear timing belt cover and timing belt using care to make sure all timing marks are

properly aligned. Install the timing belt cover.

—— WARNING ——
Verify that all timing marks are correct. If the timing belt or sprockets are incorrectly installed, engine damage will occur. Take time to make sure all timing marks are correctly aligned.

32. Install the intake and exhaust manifolds.

33. Clean the cylinder head cover and cylinder head gasket rails (mating surfaces). Make certain the rails are flat.

34. Install new cylinder head cover gaskets. Use care. DO NOT allow oil or solvents to contact the timing belt as they can deteriorate the rubber and cause tooth skipping. Apply Mopar Silicone Rubber Adhesive Sealant, or equivalent, at the camshaft cap corners and at the top edge of the ½ round seal.

35. Install the cylinder head cover assembly to the head and tighten the fasteners in sequence using the following 3 Steps:

a. Tighten all cylinder head cover fasteners to 40 inch lbs. (4.5 Nm).

b. Tighten all fasteners to 80 inch lbs. (9 Nm).

c. Tighten all fasteners to 105 inch lbs. (12 Nm).

36. Install the ground strap.

37. Install the ignition coil pack and reconnect the spark plug wiring.

38. Connect the cam sensor and fuel injectors' wiring.

39. Install the power steering pump assembly.

40. Connect the exhaust pipe to the exhaust manifold.

41. Connect all vacuum lines and remaining wiring. Connect the throttle linkage and fuel lines.

42. Install and adjust the accessory drive belts.

43. Refill the cooling system. An oil and filter change is recommended since coolant can enter the oil system when a head is removed.

44. Connect the remaining air ducting. Connect the negative battery cable and test run the vehicle. Check for leaks and for proper operation.

2.5L (VIN K) Engine

1. Properly relieve the fuel system pressure.

2. Disconnect the negative battery cable and unbolt it from the head.

3. Drain the cooling system. Remove the dipstick bracket nut from the thermostat housing and rotate the bracket from the stud.

4. Remove the air cleaner assembly. Remove the upper radiator hose and disconnect the heater hoses.

5. Remove the accessory drive belts.

6. Label and disconnect the vacuum lines, hoses and wiring connectors from the manifold, throttle body and cylinder head. Remove the air pump, if equipped.

7. Disconnect all linkage and the fuel lines from the throttle body. Disconnect the ground strap.

8. Remove the power steering pump and position aside, without disconnecting the hoses. If equipped with air conditioning, remove the compressor and compressor mounting bracket.

9. Remove the timing belt cover.

10. Raise the vehicle and support safely. Disconnect the exhaust pipe from the exhaust manifold.

11. Rotate the engine by hand, until the piston in No. 1 cylinder is at TDC on the compression stroke (firing position). Make sure all the timing marks are aligned.

12. Lower the vehicle.

13. With the timing marks aligned, remove the camshaft sprocket. Suspend the camshaft sprocket and timing belt under light tension to maintain engine timing.

—— WARNING ——
Do not rotate the crankshaft after the sprocket has been removed from the camshaft. Do not allow the timing belt to become disengaged from the camshaft sprocket. If the engine is disturbed, engine timing will have to be reset.

14. Label and disconnect the spark plug wires from the spark plugs. Disconnect the coil wiring connector and the coil wire from the coil.

15. Remove the valve cover. Remove the air/oil separation curtain, if equipped. Remove the cylinder head bolts and washers, starting from the ends of the cylinder head and working inward.

16. Remove the cylinder head from the engine.

17. If necessary, remove the intake and exhaust manifolds from the cylinder head.

—— WARNING ——

Before disassembling or repairing any part of the cylinder head assembly, check for factory installed oversized components, indicated by green paint and the letters O/SJ. If equipped with oversize components, the tops of the bearing caps would be painted green and O/SJ would be stamped rearward of the oil gallery plug on the rear of the head. In addition, the barrel of the camshaft would be painted green and O/SJ stamped onto the rear end of the camshaft. Installing standard sized parts in a head equipped with oversized parts, or oversize parts in a standard size head will cause severe engine damage.

18. Clean all gasket mating surfaces.

19. Check the cylinder head gasket surface with a straightedge; it must be flat within 0.004 inch (0.1mm). Machine or replace the cylinder head, as necessary.

To install:

20. If removed, assemble the manifolds to the cylinder head.

21. Position a new head gasket on the cylinder block.

22. Install the cylinder head to the engine with the head bolts and washers.

NOTE: Check the head bolts for stretching before they are reused. Check the bolts by holding a straightedge against the threads. If all threads do not contact the straightedge, replace the bolt.

23. Torque the cylinder head bolts in sequence, as follows:

   a. Torque the bolts, in sequence, to 45 ft. lbs. (61 Nm).

   b. Torque the bolts, in sequence, to 65 ft. lbs. (88 Nm).

   c. Repeat Step b, making sure all bolts are torqued to 65 ft. lbs. (88 Nm).

   d. Turn each bolt, in sequence, an additional ¼ turn. Do not use a torque wrench for this step.

   e. Check the torque on each bolt after the ¼ turn. It should be over 90 ft. lbs. If not, replace the bolt.

NOTE: Head bolt diameter is 11mm. These bolts are identified with the No. 11 on the head of the bolt. 10mm bolts will thread into an 11mm bolt hole, but will strip the cylinder block bolt hole. Make sure the correct bolts are being used.

24. Install the camshaft sprocket and tighten the bolt to 65 ft. lbs. (88 Nm). Make sure the timing marks are aligned.

25. Install the valve cover, curtain, timing belt and timing belt cover. Torque the valve cover screws to 105 inch lbs. (12 Nm).

26. Reconnect the spark plug wires to the spark plugs. Reconnect the coil wire and coil wiring connector.

27. Raise and safely support the vehicle.

28. Reconnect the exhaust system to the exhaust manifold. Lower the vehicle.

29. If equipped, install the A/C compressor mounting bracket and compressor.

30. Install the power steering pump.

31. Install the air pump, if equipped.

32. Reconnect all linkages and the fuel lines to the throttle body. Install the ground strap.

33. Connect all vacuum lines, hoses and wiring connectors in their proper locations, as labeled during the removal procedure.

34. Install the accessory drive belts and adjust to the proper tension.

35. Install the radiator hose and connect the heater hoses. Install the air cleaner assembly.

36. Install the dipstick bracket to the thermostat housing.

37. Refill the cooling system to the correct level with clean engine cool-

ant. Reconnect the negative battery cable to the cylinder head and the negative terminal on the battery.

38. Start the engine and check for leaks and proper engine operation.

3.0L (VIN 3) Engine

1. Properly relieve the fuel system pressure.

2. Disconnect the negative battery cable. Drain the cooling system.

3. Remove the accessory drive belts and the air conditioning compressor from its mount and support it aside. Remove the alternator and power steering pump from the brackets and move them aside.

4. Raise the vehicle and support safely. Remove the right front wheel and the right inner splash shield.

5. Remove the crankshaft pulleys and the torsional damper.

6. Lower the vehicle. Using a floor jack and a block of wood positioned under the oil pan, raise the engine slightly. Remove the engine mount bracket from the timing cover end of the engine.

7. Remove the timing belt covers.

8. Remove the timing belt.

9. Hold the camshaft sprocket using a suitable tool and remove the camshaft sprocket retaining bolt. Remove the sprocket and the inner timing belt cover (left bank) and/or alternator bracket (right bank).

10. Label and disconnect the spark plug wires from the spark plugs.

11. If removing the left (front) cylinder head, remove the distributor cap and spark plug wires. Mark the position of the rotor and distributor in relation to the cylinder head and remove the distributor. Remove the distributor drive adaptor.

12. Remove the valve cover.

13. Install the auto lash adjuster retainers on the rocker arms. Remove the camshaft bearing cap-to-cylinder head bolts (do not remove the bolts from the assembly). Remove the rocker arms, rocker shafts and bearing caps as an assembly. Remove the camshaft from the cylinder head.

14. Remove the intake manifold assembly.

15. Remove the exhaust manifold and cross over pipe.

16. Remove the cylinder head bolts starting from the outside and working inward. Remove the cylinder head from the engine.

17. Clean the gasket mounting surfaces and check the head gasket surface for leaks, damage or warpage; the maximum warpage allowed is 0.008 in. (0.20mm).

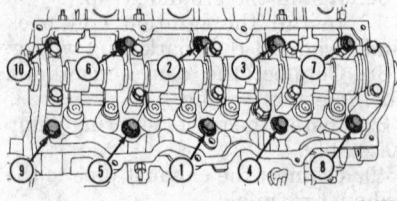

Cylinder head bolt torque sequence — 2.5L (VIN K) engine

314563

To install:

18. Clean all parts well. Install the new cylinder head gasket over the dowels on the engine block.

19. Install the cylinder head on the engine block and torque the 10mm Allen® hex cylinder head bolts in sequence, using 3 even steps, to 80 ft. lbs. (108 Nm).

20. Install the intake and exhaust manifolds.

21. Lubricate the camshaft with clean engine oil and install on the cylinder head.

22. Apply silicone sealant to the cylinder head at the front and rear cam bearing cap contact areas. Install the rocker arm shaft assembly and tighten the bearing cap bolts to 85 inch lbs. (10 Nm) in the following sequence: No. 3, No. 2, No. 1 and No. 4. Retighten to 180 inch lbs. (20 Nm) in the same sequence.

23. If removed, install the distributor drive adaptor. Install a new camshaft seal.

24. Install the valve cover.

25. Install the inner timing belt cover (left bank) and/or alternator bracket (right bank).

26. Position the camshaft sprocket. Hold the sprocket using a suitable tool and install the camshaft sprocket bolt. Tighten the bolt to 70 ft. lbs. (95 Nm).

27. Make sure the timing belt sprocket timing marks are aligned.

28. If removed, install the distributor, aligning the marks made during removal. Install the distributor cap and spark plug wires.

29. Turn the timing belt tensioner counterclockwise full travel and temporarily tighten the bolt.

30. Install the timing belt in the original direction of rotation. Loosen the timing belt tensioner bolt and allow the tensioner spring to tension the belt.

31. Turn the crankshaft 2 turns in clockwise direction. Recheck the timing mark alignment.

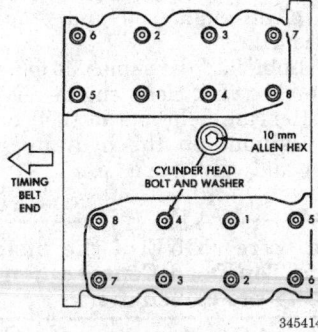

Cylinder head bolt torque sequence — 3.0L (VIN 3) engine

32. If the timing belt alignment is correct, tighten the tensioner bolt to 250 inch lbs. (25 Nm). If the timing belt alignment is incorrect, remove the belt, align the sprockets and reinstall the belt.

33. Install the timing belt covers and the engine support bracket.

34. Install the torsional damper and tighten the bolt to 112 ft. lbs. (151 Nm). Install the crankshaft pulleys.

35. Install the inner splash shield and the right front wheel. Torque the wheel lug nuts to 95 ft. lbs. (129 Nm). Lower the vehicle.

36. Install the alternator, power steering pump and air conditioning compressor.

37. Install the accessory drive belts and adjust to the proper tension.

38. Refill the cooling system. Since coolant can contaminate the engine oil when a cylinder head is removed, an oil and filter change is recommended.

39. Reconnect the negative battery cable.

40. Start the engine and check for leaks. Check the ignition timing.

3.3L (VIN R) and 3.8L (VIN L) Engines

1. Properly relieve the fuel system pressure.

2. Disconnect the negative battery cable and drain the cooling system.

3. Remove the intake manifold with the throttle body.

4. Disconnect the coil wires, sending unit wire, heater hoses and bypass hose.

5. Remove the closed ventilation system, evaporation control system and cylinder head cover(s).

6. Remove the exhaust manifold(s).

7. Remove the rocker arm and shaft assemblies. Remove the pushrods and identify them to ensure installation in their original positions.

8. Loosen the cylinder head bolts in the reverse order of the torque sequence. Remove the cylinder head bolts and remove the cylinder head(s) from the block.

9. Clean all gasket mating surfaces.

To install:

10. Clean the gasket mounting surfaces and install a new head gasket to the block.

NOTE: The cylinder head bolts are torqued using the torque yield method. The bolts should be examined before they are reused. If the threads are stretched, the

bolts should be replaced. Stretching can be checked by holding a straightedge against the threads. If all the threads do not contact the straightedge, the bolt should be replaced.

11. Install the cylinder head to the block with the bolts.

12. Torque cylinder head bolts numbers 1 through 8 in sequence as follows:

   a. Torque each bolt to 45 ft. lbs. (61 Nm).

   b. Repeat the sequence and torque the bolts to 65 ft. lbs. (88 Nm).

   c. Repeat the sequence again, making sure the bolts are at 65 ft. lbs. (88 Nm).

   d. Finally, turn each bolt, in sequence, ¼ turn. Do not use a torque wrench for this step.

   e. Check the torque of each bolt after the ¼ turn. The torque should be over 90 ft. lbs. (122 Nm). If not, replace the bolt.

13. Torque head bolt No. 9 to 25 ft. lbs. (33 Nm) after the other 8 bolts have been properly torqued.

14. Install the pushrods, rocker arms and shafts and torque the bolts to 250 inch lbs. (28 Nm).

15. Place a drop of silicone sealer onto each of the 4 manifold to cylinder head gasket corners.

CAUTION
The intake manifold gasket is composed of very thin and sharp metal. Handle this gasket with care or damage to the gasket or personal injury could result.

16. Install the intake manifold gasket and torque the end retainers to 105 inch lbs. (12 Nm).

17. Install the intake manifold and torque the bolts in sequence to 10 inch lbs. (1 Nm). Repeat the sequence, increasing the torque to 200 inch lbs. (22 Nm). Recheck each bolt, making sure it is at 200 inch lbs. (22 Nm) of torque. As the bolts are torqued, inspect the seals to ensure that they have not become dislodged.

18. Install the valve cover with a new gasket. Torque the valve cover bolts to 105 inch lbs. (12 Nm). Install the exhaust manifold and cross over pipe. Torque the bolts to 20 ft. lbs. (27 Nm) and the nuts to 15 ft. lbs. (20 Nm).

19. Install the closed ventilation system and the evaporation control system.

20. Reconnect the coil wires, sending unit wire, heater hoses and bypass hose.

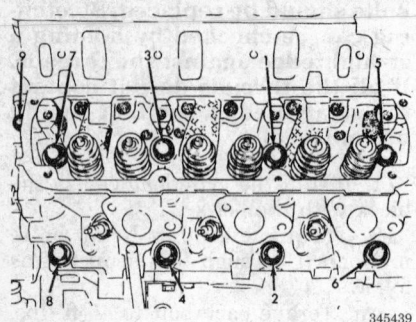

Cylinder head bolt torque sequence — 3.3L (VIN R) and 3.8L (VIN L) engines

21. Check and make sure all wiring connections, cables, hoses, vacuum and fuel lines have been reconnected.

22. Refill the cooling system to the correct level. Coolant can contaminate the engine oil when performing cylinder head service. An oil and filter change is recommended.

23. Reconnect the negative battery cable. Start the engine and check for leaks.

Lash Adjusters

BLEEDING

2.4L (VIN B) and 2.5L (VIN K) Engines

The hydraulic valve lash adjusters do not need to be bled. If any lash adjuster is not functioning properly it must be replaced. When installing a new lash adjuster soak the adjuster in oil until the adjuster has little or no plunger travel when depressed. However, if plunger travel is not reduced, the lash adjuster needs replacement.

3.0L (VIN 3) Engine

The hydraulic lash adjusters are precision units installed in machined openings in the valve actuating ends of the rocker arms, The rocker arm/lash adjuster assembly is not to be disassembled for any reason or damage to the assembly could result. The lash adjuster and rocker arm are serviced as an assembly. Lash adjuster bleeding is not possible.

Perform a free-play check on the lash adjusters by inserting a small wire through the air bleed hole in the rocker arm and push down very lightly on the auto adjuster check ball. While lightly holding down the check ball, move the rocker arm up and down to check for free-play. If there is no free-play, the rocker

arm/lash adjuster assembly must be replaced.

ADJUSTMENT

2.4L (VIN B) and 2.5L (VIN K) Engines

The valves are actuated by roller cam followers which pivot on stationary hydraulic lash adjusters. The hydraulic lash adjusters maintain 0 valve lash during the entire valve opening and closing process; any lash is instantaneously taken up by hydraulic action. The valve lash cannot be manually adjusted.

3.0L (VIN 3) Engine

The hydraulic lash adjusters are precision units installed in machined openings in the valve actuating ends of the rocker arms. The rocker arm/lash adjuster assembly is not to be disassembled for any reason or damage to the assembly could result. The lash adjuster and rocker arm are serviced as an assembly. Lash adjuster bleeding is not possible.

Perform a free-play check on the lash adjusters by inserting a small wire through the air bleed hole in the rocker arm and push down very lightly on the auto adjuster check ball. While lightly holding down the check ball, move the rocker arm up and down to check for free-play. If there is no free-play, the rocker arm/lash adjuster assembly must be replaced.

REMOVAL AND INSTALLATION

2.4L (VIN B) Engine

This engine uses a Dual Over Head Camshaft (DOHC) 4-valves per cylinder cross flow aluminum cylinder head. The valves are actuated by roller cam followers which pivot on stationary hydraulic lash adjusters. Care must be taken to make sure all

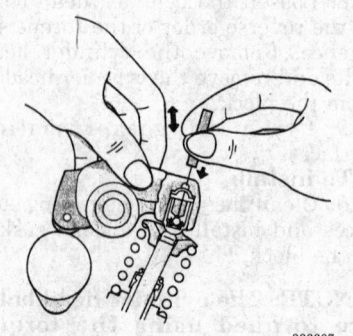

Hydraulic lash adjuster check — 3.0L (VIN 3) engine

valve timing marks align after cylinder head and valvetrain service.

CAUTION

Fuel injection systems remain under pressure, even after the engine has been turned OFF. The fuel system pressure must be relieved before disconnecting any fuel lines. Failure to do so may result in fire and/or personal injury.
---

1. Properly relieve the fuel system pressure.

2. Disconnect the negative battery cable.

3. Label and disconnect the spark plug wires from the spark plugs.

4. Remove the ignition coil pack and spark plug wires.

5. Remove the cylinder head cover retaining fasteners and remove the cylinder head cover from the cylinder head. Discard the old cylinder head cover gasket.

6. Remove the ground strap.

7. Remove the timing belt covers, timing belt and camshaft sprockets.

8. Remove the camshafts.

9. Remove the camshaft follower assemblies from the cylinder head. The cam followers must be installed in their original locations if they are to be reused. Be sure to keep the cam followers in order as they are removed.

10. Remove the lash adjusters. Use care to identify and mark the positions of the hydraulic lash adjusters so they may be reinstalled correctly.

11. Inspect the camshaft follower assemblies and lash adjusters for wear or damage and replace if necessary.

To install:

12. Thoroughly clean all camshaft followers, lash adjusters and related parts.

13. Install the hydraulic lash adjuster assemblies making sure they are at least partially full of oil. Make sure they are clean, well-lubricated with clean engine oil and properly positioned.

14. Lubricate the camshaft follower assemblies with clean engine oil. Install the cam followers in their original positions on the hydraulic adjuster and valve stem.

WARNING

Make sure NONE of the pistons are at Top Dead Center when installing the camshafts.
---

15. Install the camshafts. Install new camshaft end seals.

16. Install the camshaft sprockets, if removed. Install the timing belt using care to make sure all timing marks are properly aligned, using the recommended procedure. Install the timing belt covers.

——————— WARNING ———————
Verify that all timing marks are correct. If the timing belt or sprockets are incorrectly installed, engine damage will occur. Take time to make sure all timing marks are correctly aligned.

17. Clean the cylinder head cover and cylinder head gasket rails (mating surfaces). Make certain the rails are flat.
18. Install new cylinder head cover gaskets. Use care. DO NOT allow oil or solvents to contact the timing belt as they can deteriorate the rubber and cause tooth skipping. Apply Mopar Silicone Rubber Adhesive Sealant, or equivalent, at the camshaft cap corners and at the top edge of the ½ round seal.

NOTE: Inspect the spark plug well seals for cracking and/or swelling and replace if necessary.

19. Install the cylinder head cover assembly to the head and tighten the fasteners in sequence using the following 3 Steps:
   a. Tighten all cylinder head cover fasteners to 40 inch lbs. (4.5 Nm).
   b. Tighten all fasteners to 80 inch lbs. (9 Nm).
   c. Tighten all fasteners to 105 inch lbs. (12 Nm).
20. Install the ignition coil pack and connect the spark plug wiring to the correct spark plugs. Torque the coil pack retaining fasteners to 105 inch lbs. (12 Nm).
21. Reconnect the ground strap.
22. Check to be sure all vacuum lines and remaining wiring have been reconnected.
23. An oil and filter change is recommended.
24. Connect the negative battery cable and test run the vehicle. Check for leaks and for proper operation.

2.5L (VIN K) Engine

1. Disconnect the negative battery cable.
2. Remove the air cleaner assembly.
3. Remove the valve cover and oil vapor curtain.
4. Rotate the camshaft until the lobe base is on the camshaft follower that is to be removed.

5. Slightly depress the valve spring using tool C-4682A or the equivalent. Slide the follower off the lash adjuster and valve tip and remove. Label the camshaft followers for position identification. Proceed to next cam follower and repeat Step 4.
6. Remove the lash adjusters. Label the lash adjusters for position identification.
   To install:
7. Partially fill the lash adjusters with oil and install in their original locations.
8. Rotate the camshaft until lobe base is in position with cam follower. Slightly depress the valve spring using tool C-4682A or equivalent. Slide cam follower into position.

NOTE: When depressing the valve spring with Chrysler tool C-4682A, or the equivalent, the valve locks can become dislocated. Check and make sure both locks are fully seated in the valve grooves and retainer.

9. Install the valve cover and oil vapor curtain. Torque the valve cover retaining bolts to 105 inch lbs. (12 Nm).
10. Install the air cleaner assembly.
11. Reconnect the negative battery cable.
12. Run the engine and check for leaks and proper engine operation.

3.0L (VIN 3) Engine

The automatic lash adjusters are precision units installed in machined openings in the valve actuating ends of the rocker arms. Do not disassembled the auto lash adjusters.
1. Disconnect the negative battery cable. Disconnect and label spark plug wires.
2. Remove the air cleaner assembly. Remove accessory drive belts and disconnect vacuum connections.
3. Remove the valve cover.
4. Before removing the rocker arm shaft assembly, a function check can be made of the auto lash adjusters. Use the following procedure.
   a. Check the adjusters for free-play by inserting a small wire through the air bleed hole in the rocker arm.
   b. **VERY LIGHTLY** push the auto adjuster ball check down.
   c. While lightly holding the check ball down, move the rocker arm up and down to check for free-play. If there is no free-play, replace the adjuster.
5. Install auto lash adjuster retainers MD 998443 or equivalent.

6. Loosen all the camshaft bearing cap bolts. Do not remove the bolts from bearing caps. Remove the rocker arms, rocker shafts and bearing caps as an assembly.
7. Remove the bolts from the camshaft bearing caps and remove the rocker shafts and arms. Keep all parts in order. Note the way the rocker shaft, rocker arms, bearing caps and springs are mounted. The rocker arm shaft on the intake side has a 3mm diameter oil passage hole from the cylinder head. The exhaust side does not have this oil passage.
8. Inspect the rocker arm mounting area and rocker for damage. Replace if worn or heavily damaged. Check oil passages for clogging and clean, if necessary. Replace any lash adjusters that failed the function check made earlier.
   To install:
9. Lubricate the rocker arms and shafts with clean engine oil, prior to installation.
10. Identify No. 1 bearing cap, (No. 1 and 4 caps are similar). Install the rocker shafts into the bearing cap with notches in proper position. Insert the attaching bolts to retain assembly.
11. Install the rocker arms, springs and bearing caps on shafts in numerical sequence.
12. Align the camshaft bearing caps with arrows (depending on cylinder bank).
13. Install the bolts in No. 4 cap to retain assembly.
14. Apply silicone rubber sealant at bearing cap ends.
15. Install the rocker arm shaft assembly.

NOTE: Make sure the arrow mark on the bearing caps and the arrow mark on the cylinder heads are in the same direction. The direction of arrow marks on the front and rear assemblies are opposite to each other.

16. Tighten the bearing caps bolts to 85 inch lbs. (10 Nm) in the following order:
   a. No. 3 Cap
   b. No. 2 Cap
   c. No. 1 Cap
   d. No. 4 Cap
17. Repeat the previous Step, increasing torque to 180 inch lbs. (20 Nm).
18. Remove the auto lash adjuster retainers.
19. Install the distributor drive adapter, if removed.
20. Install the valve cover. Torque the valve cover retaining bolts to 88 inch lbs. (10 Nm).

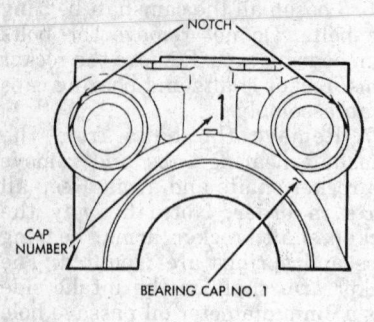

Install the rocker shafts so the end notches are positioned as shown — 3.0L (VIN 3) engine

21. Reconnect vacuum connections. Install accessory drive belts and adjust to proper tension.

22. Install air cleaner assembly. Reconnect the spark plug wires.

23. Reconnect the negative battery cable. Run the engine and check for leaks and proper engine operation.

Valve Clearance

ADJUSTMENT

3.3L (VIN R) and 3.8L (VIN L) Engines

These engines are equipped with hydraulic lifters. The lifters are located in bores in the cylinder block. As the camshaft turns, a camshaft lobe forces the lifter upwards in its bore. The lifter drives a pushrod upwards and acts on a pivoting rocker arm, which in turn opens the valve. When the valve opening event is over and the base circle of the camshaft lobe contacts the hydraulic lifter, the valve spring pulls the valve closed. The function of the hydraulic valve lifter is to maintain zero valve lash during the entire valve opening and closing process; any lash is instantaneously taken up by hydraulic action. The valve lash cannot be manually adjusted.

Valve Lifters

REMOVAL AND INSTALLATION

3.3L (VIN R) and 3.8L (VIN L) Engines

— CAUTION —

Fuel injection systems remain under pressure, even after the engine has been turned OFF. The fuel system pressure must be relieved before disconnecting any

fuel lines. Failure to do so may result in fire and/or personal injury.

1. Properly relieve the fuel system pressure. Disconnect the negative battery cable.

2. Drain the cooling system.

3. Remove the intake manifold and the cylinder heads.

4. Remove the lifter guide (sometimes called the yoke) retainer, aligning yokes and then remove the lifters from their bores.

NOTE: If the lifters are to be reused, be sure to replace them in the same bore.

To install:

5. Coat the roller with clean engine oil and place the lifter in the same bore that it was removed from.

6. Install the aligning yoke, then install the yoke retainer. Torque the yoke retainer screws to 105 inch lbs. (12 Nm).

7. Install the cylinder heads and the intake manifold.

8. Check to make sure all wiring connections, hoses, cables, vacuum and fuel lines have been reconnected properly.

9. Refill the cooling system to the correct level with a $^{50}/_{50}$ mix of clean coolant and water.

10. Reconnect the negative battery cable. Start the engine and allow it to warm up to normal operating temperature. Check the area for leaks.

— WARNING —

To prevent valve mechanism damage, do not run the engine above fast idle until all of the hydraulic lifters have filled up with oil and have become quiet.

Rocker Arm and Shaft

REMOVAL AND INSTALLATION

3.0L (VIN 3) Engine

The rocker arms and shafts are retained by the camshaft bearing journal caps. Four shafts are used, one for each intake and exhaust rocker arm assembly on each cylinder head. The hollow shafts provide a duct for lubricating oil from the cylinder head to the valve mechanisms. The rocker arms are light weight die-cast with roller-type followers operating against the camshaft. The valve actuating end of the rocker arms are machined to retain hydraulic lash ad-

justers, eliminating valve lash adjustment.

1. Disconnect the negative battery cable. Disconnect and label spark plug wires.

2. Remove the air cleaner assembly. Remove accessory drive belts and disconnect vacuum connections.

3. Remove the valve cover.

4. Before removing the rocker arm shaft assembly, a function check can be made of the auto lash adjusters. Use the following procedure.

a. The auto lash adjusters are precision units installed in machined openings in the valve actuating ends of the rocker arms. Do not disassemble the auto lash adjusters.

b. Check the adjusters for free-play by inserting a small wire through the air bleed hole in the rocker arm.

c. **VERY LIGHTLY** push the auto adjuster ball check down.

d. While lightly holding the check ball down, move the rocker arm up and down to check for free-play. If there is no free-play, replace the adjuster.

5. Install lash adjuster retainers MD998443 or equivalent, on the rocker arms.

6. Loosen all the camshaft bearing cap bolts. Do not remove the bolts from bearing caps. Remove the rocker arms, rocker shafts and bearing caps as an assembly.

7. Remove the bolts from the camshaft bearing caps and remove the rocker shafts and arms. Keep all parts in order. Note the way the rocker shaft, rocker arms, bearing caps and springs are mounted. The rocker arm shaft on the intake side has a 3mm diameter oil passage hole from the cylinder head. The exhaust side does not have this oil passage.

8. Inspect the rocker arm mounting area and rocker for damage. Replace if worn or heavily damaged. Check oil passages for clogging and clean, if necessary.

To install:

9. Lubricate the rocker arms and shafts with clean engine oil prior to installation.

10. Identify No. 1 bearing cap, (No. 1 and No. 4 caps are similar). Install the rocker shafts into the bearing cap with notches in proper position. Insert the attaching bolts to retain assembly.

11. Install the rocker arms, springs and bearing caps on shafts in numerical sequence.

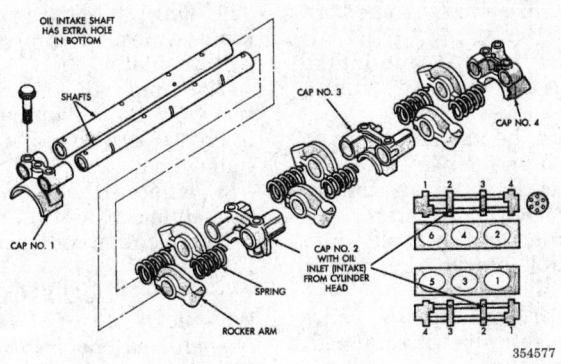

Rocker arms and shafts assembly — 3.0L (VIN 3) engine

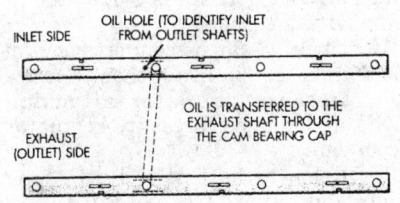

Rocker arm shaft identification — 3.0L (VIN 3) engine

12. Align the camshaft bearing caps with arrows (depending on cylinder bank).

13. Install the bolts in No. 4 cap to retain assembly.

14. Apply silicone rubber sealant at bearing cap ends.

15. Install the rocker arm shaft assembly.

NOTE: Make sure the arrow mark on the bearing caps and the arrow mark on the cylinder heads are in the same direction. The direction of arrow marks on the front and rear assemblies are opposite to each other.

16. Tighten the bearing caps bolts to 85 inch lbs. (10 Nm) in the following order:
    a. No. 3 Cap
    b. No. 2 Cap
    c. No. 1 Cap
    d. No. 4 Cap

17. Repeat the previous Step, increasing torque to 180 inch lbs. (20 Nm).

18. Remove the lash adjuster retainers.

19. Install the distributor drive adapter, if removed.

20. Install the valve cover. Torque the valve cover retaining bolts to 88 inch lbs. (10 Nm).

21. Reconnect vacuum connections. Install accessory drive belts and adjust to proper tension.

22. Install air cleaner assembly. Reconnect the spark plug wires.

23. Reconnect the negative battery cable. Run the engine and check for leaks and proper engine operation.

3.3L (VIN R) and 3.8L (VIN L) Engines

1. Disconnect the negative battery cable.

2. Remove the upper intake manifold assembly. Disconnect and label the spark plug wires.

3. Disconnect the closed ventilation system.

4. Remove the rocker arm cover and gasket.

5. Remove the 4 rocker shaft retaining bolts and retainers.

6. Remove the rocker arms and shaft assembly.

7. If disassembling the rocker shaft, be sure to identify all components so they can be reinstalled in their original locations.

8. Inspect the rocker arms and shafts for wear and/or damage; replace components as necessary.

9. If necessary, remove the pushrods. Identify each pushrod as it is removed, so it can be reinstalled in its original location.

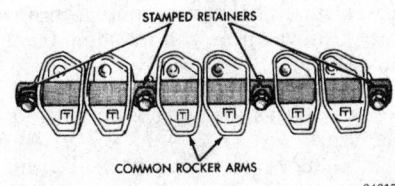

Rocker arms and shaft — 3.3L (VIN R) and 3.8L (VIN L) engines

10. Inspect the pushrods for wear and/or damage. Roll each pushrod on a flat surface to check for a bent condition. Replace pushrods as necessary.

To install:

11. If removed, install the pushrods in their proper locations. Lubricate the pushrod ends with clean engine oil, prior to installation. Make sure the pushrods are seated in the lifters.

12. Lubricate the rocker arms and shafts with clean engine oil, prior to installation.

13. If the rocker shaft was disassembled, reassemble making sure all components are installed in their original locations.

14. Install the rocker arm and shaft assembly, using the 4 retainers. Make sure the pushrods are seated in the rocker arms. Tighten the retaining bolts to 250 inch lbs. (28 Nm).

NOTE: The rocker arm shaft should be torqued down slowly, starting with the center bolts. Allow 20 minutes for tappet bleed down after installation, before engine operation.

15. Install the rocker cover with a new gasket. Be sure the cover gasket mating surface is clean and smooth. Torque the rocker cover retaining bolts to 105 inch lbs. (12 Nm).

16. Install the crankcase ventilation components and reconnect the spark plug wires.

17. Install the upper intake manifold assembly. Reconnect the negative battery cable.

18. Run the engine and check for leaks and proper engine operation.

Intake Manifold

REMOVAL AND INSTALLATION

———— **CAUTION** ————
Fuel injection systems remain under pressure even after the engine has been turned OFF. The fuel system pressure must be relieved before disconnecting any fuel lines. Failure to do so may result in fire and/or personal injury.

2.4L (VIN B) Engine

The 2.4L (VIN B) DOHC (Dual Over Head Camshaft) engine intake manifold is a long branch design made of cast aluminum. It is attached to the cylinder head with 8 fasteners.

1. Properly relieve the fuel system pressure.

2. Disconnect the negative battery cable.

3. Disconnect the air cleaner inlet hose from the throttle body.

4. Disconnect the throttle cable and speed control cable (if equipped) from the throttle lever and cable bracket. The cable(s) can be removed from the bracket by compressing the retaining tabs.

5. Disconnect the idle air control motor and Throttle Position Sensor (TPS) wiring connectors on the throttle body.

6. Disconnect the vacuum hoses from the intake plenum fittings.

7. Disconnect the electrical connectors to the Manifold Absolute Pressure (MAP) and intake air temperature sensors.

8. Disconnect the fuel line quick-connect fitting from the chassis fuel line tube by squeezing the retainer tabs together and pulling the fuel tube/quick connect fitting assembly from the fuel tube nipple. The retainer will remain on the fuel tube. Use shop towels to catch any spilled fuel.

9. Remove the EGR tube and gasket at the EGR valve.

10. Drain the cooling system and remove the accessory drive belt.

11. If necessary, remove the alternator mounting bracket.

12. Remove the mounting bolts that secure the bottom of the intake support bracket.

13. Remove the 8 intake manifold fasteners and washers. Remove the intake manifold from the engine.

To install:

14. Thoroughly clean all parts. Check the mating surfaces for cracks or distortion.

15. Install a new intake manifold gasket and position the manifold on the cylinder head. Install and torque the fasteners to 200 inch lbs. (23 Nm) starting from the center and working outward in both directions.

16. Install the mounting bolt that secures the bottom of the intake support bracket and install the alternator mounting bracket (if removed).

17. Install the accessory drive belt and adjust to the proper tension.

18. Install the EGR gasket and tube to the EGR valve.

19. Inspect the quick-connect fittings for damage and repair as required. Lightly lube the fuel line tube with clean 30W engine oil. Reconnect the fuel hose quick-connect fitting to the chassis fuel tube. Push the quick-connect fitting onto the chassis fuel tube until it clicks into place. Check the connection by pulling on the con-

nector to insure it is locked in position.

20. Reconnect the MAP and intake air temperature sensor wiring connectors.

21. Reconnect the vacuum hoses to the intake plenum fittings.

22. Reconnect the idle air control motor and TPS wiring connectors.

23. Install the throttle cables into the throttle cable bracket. Be sure to engage the retaining tabs.

24. Reconnect the throttle cable and speed control cable (if equipped) to the throttle lever.

25. Install the air cleaner inlet hose to the throttle body.

26. Refill the cooling system with a $^{50}/_{50}$ mixture of clean, fresh ethylene glycol antifreeze and water to the proper level.

27. Reconnect the negative battery cable.

WARNING

When using the ASD Fuel System Test, the ASD relay and fuel pump relay remain energized for 7 minutes or until the test is stopped, or until the ignition switch is turned to the OFF position.

28. Turn the ignition key to the **ON** position and access the DRB scan tool ASD Fuel System Test to pressurize the fuel system. Check the fuel system for leaks.

29. Run the engine and check for leaks and proper operation.

3.0L (VIN 3) Engine

1. Properly relieve the fuel system pressure.

2. Disconnect the negative battery cable.

3. Drain the cooling system.

4. Remove the air cleaner to throttle body hose.

5. Remove the throttle cable and transaxle kickdown cable.

6. Remove the Automatic Idle Speed (AIS) motor and Throttle Position Sensor (TPS) electrical connectors from the throttle body. Disconnect the vacuum connections from throttle body.

7. Remove the PCV and brake booster hoses from the air intake plenum.

8. Remove the ignition coil from the intake plenum.

9. Remove the EGR tube to intake plenum (if equipped).

10. Remove the electrical connection from the coolant temperature sensor.

11. On 1993–95, remove the vacuum connection from the fuel pressure regulator.

12. Remove the air intake connection from the air intake plenum. Remove fuel hoses to fuel rail connection.

13. Remove the air intake plenum to manifold bolts (8) and remove air intake plenum and gasket.

CAUTION

Whenever the air intake plenum is removed, cover the intake manifold properly to keep objects from entering the cylinder head.

14. Label and disconnect the fuel injector wiring harness from the engine wiring harness.

15. On 1993–95, disconnect vacuum hose from the fuel rail; then, remove the pressure regulator attaching bolts and remove pressure regulator from rail.

16. Remove the fuel rail attaching bolts and remove the fuel rail.

17. Remove the radiator hose from the thermostat housing and the heater hose from the pipe.

18. Remove the intake manifold attaching nuts and washers and remove the intake manifold.

19. Clean the gasket material from the cylinder head and manifold gasket surface. Check for cracks or damaged mounting surfaces.

To install:

20. Install a new gasket on the intake surface of the cylinder head and install the intake manifold.

21. Install the intake manifold washers and nuts. Torque the intake manifold attaching nuts, in sequence, to 15 ft. lbs. (20 Nm).

22. Clean the injectors and lubricate the injector O-rings with a drop of clean engine oil.

23. Place the tip of each injector into their ports. Push the assembly into place until the injectors are seated in their ports.

24. Install the fuel rail attaching bolts and tighten to 115 inch lbs. (13 Nm).

25. On 1993–95, install the pressure regulator to rail, the pressure regulator mounting bolts and tighten to 95 inch lbs. (11 Nm).

26. Install the fuel supply and return tube hold-down bolt and the vacuum crossover tube hold-down bolt. Torque to 95 inch lbs. (11 Nm).

27. On 1993–95, torque the fuel pressure regulator hose clamps to 10 inch lbs. (1 Nm).

28. Connect the injector wiring harness to the engine wiring harness.

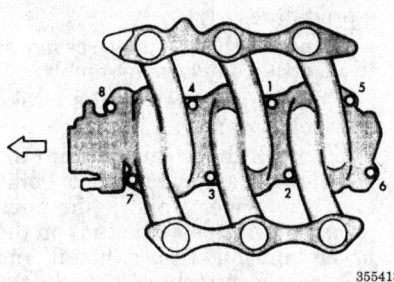

Intake manifold torque sequence — 3.0L (VIN 3) engine

355413

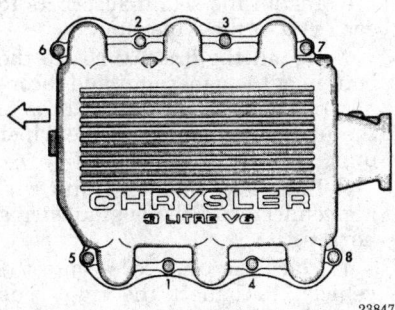

Intake plenum torque sequence — 1993–95 3.0L (VIN 3) engine

238479

29. Remove the covering from the intake manifold.

30. Position the intake manifold gasket, beaded side up, on the intake manifold.

31. Place the air intake plenum in position. Install the attaching bolts and tighten, in sequence, to 115 inch lbs. (13 Nm).

32. Connect the fuel line to the fuel rail. Tighten the clamps to 10 inch lbs. (1 Nm).

33. Connect the vacuum hoses to the intake plenum.

34. Connect the electrical connection to the coolant temperature sensor.

35. Connect the EGR tube flange to the intake plenum (if equipped) and torque to 15 ft. lbs. (20 Nm).

36. Reconnect the PCV hose and the brake booster supply hose to the intake plenum.

37. Reconnect the automatic idle speed control motor and TPS electrical connectors.

38. Connect the throttle body vacuum hoses and electrical connections.

39. Install the throttle cable and transaxle kickdown linkage.

40. Install the air inlet hose assembly.

41. Install the radiator and heater hose.

42. Refill the cooling system with a $^{50}/_{50}$ mixture of clean, fresh ethylene glycol antifreeze and water to the proper level.

43. Reconnect the negative battery cable. An engine oil and filter change is recommended.

— WARNING —

On 1996–97, when using the ASD Fuel System Test, the ASD relay and fuel pump relay remain energized for 7 minutes or until the test is stopped, or until the ignition switch is turned to the OFF position.

44. On 1996–97, turn the ignition key to the **ON** position and access the DRB scan tool ASD Fuel System Test to pressurize the fuel system.

45. Run the vehicle until it reaches operating temperature and check the cooling system, fuel system and engine for fuel, oil or coolant leaks.

3.3L (VIN R) and 3.8L (VIN L) Engines

1. Properly relieve the fuel system pressure.

2. Disconnect the negative battery cable.

3. On 1996–97, perform the following procedures:

  a. Remove the wiper arm and blade assemblies.

  b. Remove the cowl cover from the vehicle.

  c. Open the hood. Disconnect the positive lock on the wiper unit electrical connector.

  d. Disconnect the wiper unit electrical connector from the engine compartment wiring harness.

  e. Disconnect the windshield washer hose from the hose coupling inside the wiper unit.

  f. Remove the drain tubes from the bottom of the wiper unit.

  g. Remove the sound absorbers from the ends of the wiper unit.

  h. Remove the attaching nuts securing the wiper unit to the lower windshield fence.

  i. Remove the attaching bolts securing the wiper unit to the dash panel.

  j. Raise up the wiper unit from the from the weld-studs on the lower windshield fence and remove the wiper unit from the vehicle.

4. Drain the cooling system.

5. If equipped, remove the intake manifold cover.

6. Disconnect the air inlet resonator to throttle body hose assembly.

7. Disconnect the throttle cable and remove the wiring harness from the cable bracket.

8. Remove the Automatic Idle Speed (AIS) motor and Throttle Position Sensor (TPS) wiring connectors from the throttle body.

9. Remove the vacuum hose harness from the throttle body.

10. Remove the PCV and brake booster hoses from the air intake plenum.

11. Remove the EGR tube flange and the vacuum harness connectors from the intake plenum.

12. On 1993–95, disconnect the charge temperature sensor electrical connector. Remove the vacuum harness connectors from the intake plenum.

13. Remove the cylinder head to intake plenum strut.

14. Disconnect the MAP sensor wiring connector. Remove the engine mounted ground strap.

15. Remove the fuel hose quick-connect fittings from the fuel rail by pulling back on the fitting while pushing in the plastic ring. This may require the use of an open end wrench to push in the plastic ring. Be sure to plug the open fuel lines to prevent system contamination. Wrap a shop towel around the fuel hoses to absorb any fuel spill.

16. Remove the DIS coils and the alternator bracket to intake manifold bolt.

17. On 1996–97, remove the alternator wiring harness from the back of the upper intake manifold plenum.

18. Remove the upper intake manifold attaching bolts and remove the upper intake manifold. Cover the intake manifold openings to prevent foreign material from entering the engine.

19. On 1993–95, remove the vacuum harness connector from the fuel pressure regulator.

20. Remove the fuel tube retainer bracket screw and fuel rail attaching bolts. Spread the retainer bracket to allow for clearance when removing the fuel tube.

21. On 1993–95, remove the fuel rail injector wiring clip from the alternator bracket.

22. Disconnect the cam sensor and coolant temperature sensor connectors.

23. Remove the injector wiring clip from the intake manifold water tube.

24. Remove the fuel rail. Be careful not to damage the fuel injector O-rings.

25. Remove the upper radiator hose, bypass hose and rear intake manifold hose.

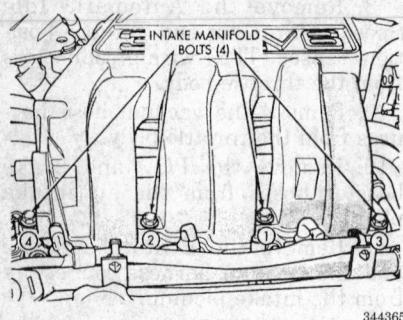

Intake manifold bolts — 1996–97 3.3L (VIN R) and 3.8L (VIN L) engines

26. Remove the intake manifold bolts and remove the manifold from the engine.

27. Remove the intake manifold seal retaining screws and remove the manifold gasket.

28. Clean out clogged end water passages and fuel runners.

To install:

29. Clean and dry all gasket mating surfaces.

30. Place a bead of approximately ¼ inch (6mm) diameter of silicone sealant onto each of the 4 manifold-to-cylinder head gasket corners.

— CAUTION —

The intake manifold gasket is made of very thin material and could cause personal injury. Handle with care.

---

31. Carefully install the intake manifold gasket and torque the end seal retainer screws to 105 inch lbs. (12 Nm).

32. Install the intake manifold and 8 retaining bolts and torque to 10 inch lbs. (1 Nm). Then torque the bolts, in sequence, to 200 inch lbs. (22 Nm).

33. When the bolts are torqued, inspect the seals to ensure that they have not become dislodged.

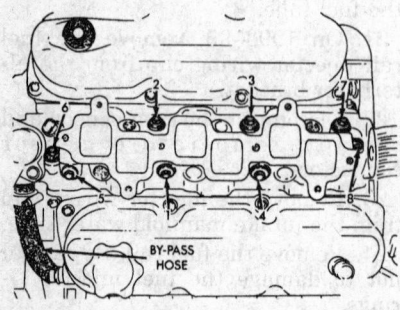

Intake manifold torque sequence — 3.3L (VIN R) and 3.8L (VIN L) engines

34. Lubricate the injector O-rings with clean oil to ease installation. Put the tip of each injector into their ports and place the fuel rail in position. Install the fuel rail mounting bolts and tighten to 200 inch lbs. (22 Nm).

35. Install the fuel tube retaining bracket screw. Torque the screw to 35 inch lbs. (4 Nm).

36. Connect the cam sensor and coolant temperature sensor and engine temperature sensor (1993–95).

37. On 1993–95, perform the following:

a. Install the fuel injector harness wiring clip to the alternator bracket. Install the intake manifold water tube.

b. Install the vacuum harness to the pressure regulator.

38. Install the upper intake manifold with a new intake manifold gasket. Install the bolts only finger tight. Install the alternator bracket to intake manifold bolt and the cylinder head to intake manifold strut bolts. Torque the intake manifold mounting bolts to 250 inch lbs. (28 Nm) starting from the middle and working outward.

39. Torque the alternator bracket and cylinder head-to-intake manifold strut bolts to 40 ft. lbs. (54 Nm).

40. On 1996–97, connect the alternator wiring harness to the alternator and to the rear of the intake manifold.

41. Reconnect the ground strap and MAP sensor connectors. On 1993–95, reconnect the oxygen sensor and charge temperature sensor connectors.

42. Reconnect the intake plenum vacuum harness.

43. Install a new gasket and connect the EGR tube to the intake manifold. Torque the retaining fasteners to 200 inch lbs. (22 Nm).

44. Clip the wiring harness into the throttle cable bracket hole. Reconnect the Throttle Position Sensor (TPS) and Automatic Idle Speed (AIS) control motor wiring connectors.

45. Reconnect the throttle body vacuum harness and install the ignition coils. Torque the ignition coil fasteners to 105 inch lbs. (12 Nm).

46. Lightly lubricate the the ends of the fuel lines with clean, 30W engine oil. Reconnect the fuel hoses to the rail. Push the fittings in until they click in place. Pull back on the quick-connect fitting to ensure that the fuel lines are securely locked in place.

47. Reconnect the throttle cable.

48. Reconnect the fuel injector wiring harness.

49. On 1996–97, perform the following procedures:

a. Install the air inlet resonator to throttle body hose assembly.

b. If equipped, install the intake manifold cover.

c. Install the wiper unit into the vehicle engine compartment. Make sure the wiper unit is installed properly over the weld-studs on the lower windshield fence. Install and tighten the attaching nuts to the weld-studs.

d. Install and tighten the attaching bolts securing the wiper unit to the dash panel.

e. Install the sound absorbers to each end of the wiper unit.

f. Install the drain tubes to the bottom of the wiper unit and reconnect the windshield washer hose to the hose coupling inside the wiper unit.

g. Reconnect the wiper unit wiring connector to the engine wiring harness.

h. Place the cowl cover onto the vehicle. Reconnect the right side windshield washer hose to the right washer nozzle located on the underside of the cowl cover.

i. Engage the retainers that secure the cowl cover to the front fender edge.

j. Install and tighten the wing nuts that secure the front of the cowl cover to the wiper module.

k. Engage the quarter turn fasteners that secure the outer ends of the cowl cover to the wiper module.

l. Install and tighten the screws that hold the lower area of the cowl cover to the wiper module.

m. Reconnect the negative battery cable and verify that the wiper motor and wiper linkage are in the **PARK** position.

n. Install the wiper arm in correct position over the wiper arm pivot.

50. Install the the wiper arm-to-wiper arm pivot retaining nut and torque to 26 ft. lbs. (35 Nm).

51. Push down the wiper arm cap cover and engage the clip that secures the outside end of the cover to the wiper arm.

52. Reconnect the negative battery cable.

53. On 1996–97, operate the windshield wipers and check to make sure they work and park properly.

54. On 1996–97, refill the cooling system with a ⁵⁰⁄₅₀ mixture of clean, fresh ethylene glycol antifreeze and water. Bleed the cooling system.

─────── **WARNING** ───────
When performing the ASD Fuel System Test, the Auto Shutdown Relay (ASD) will stay energized for 7 minutes, or until the ignition switch is turned to the OFF position, or "STOP ALL TEST" is chosen.

55. With the ignition key turned to the **ON** position, select the DRB scan tool ASD Fuel System Test to pressurize the fuel system. Check the fuel system for leaks.

56. Start the engine. Check for leaks and proper engine operation.

Combination Manifold

REMOVAL AND INSTALLATION

2.5L (VIN K) Engine

NOTE: On some vehicles, some of the manifold attaching bolts are not accessible or too heavily sealed from the factory and cannot be removed on the vehicle. Removal of the cylinder head would be necessary in these situations.

─────── **CAUTION** ───────
Fuel injection systems remain under pressure, even after the engine has been turned OFF. The fuel system pressure must be relieved before disconnecting any fuel lines. Failure to do so may result in fire and/or personal injury.

1. Properly relieve the fuel system pressure.
2. Disconnect the negative battery cable.
3. Drain the cooling system.
4. Remove the air cleaner. Label and disconnect all vacuum lines, electrical wiring and fuel lines from the throttle body and intake manifold.
5. Disconnect the throttle linkage.
6. Loosen the power steering pump and remove the power steering drive belt, if necessary.
7. Remove the power brake vacuum hose from the intake manifold.
8. Remove the EGR tube from the intake manifold. Remove the water hoses from the water crossover.
9. Raise and safely support the vehicle. Disconnect the exhaust pipe from the exhaust manifold.
10. Remove the power steering pump and set aside.
11. Remove the intake manifold support bracket, if equipped.

12. If equipped, remove the air injection tube bolts and the air injection tube assembly.
13. Remove the intake manifold screws/bolts.
14. Lower the vehicle and remove the intake manifold.
15. Remove the exhaust manifold nuts and remove the exhaust manifold.

To install:
16. Clean the mating surfaces and install a new combination manifold gasket. Apply a light coat of gasket sealer to the manifold side of the new combination manifold gasket.
17. Install the exhaust manifold. Install the mounting nuts and tighten them to 17 ft. lbs. (23 Nm) starting from the middle and working outward.
18. Install the intake manifold.
19. Raise and safely support the vehicle. Install the mounting screws/bolts and tighten them to 17 ft. lbs. (23 Nm) starting from the middle and working outward.
20. Connect the exhaust pipe to the exhaust manifold.
21. Install the air injection tube assembly, if equipped.
22. Install the EGR tube.
23. Install the intake support bracket.
24. Install the power steering pump and drive belt. Adjust the drive belt to the proper tension.
25. Lower the vehicle and install the water hoses to the water crossover.
26. Install the power brake vacuum hose to the intake manifold.
27. Connect the throttle linkage.
28. Install all vacuum lines, electrical wiring and fuel lines to the throttle body and intake manifold.
29. Install the air cleaner assembly.
30. Refill the cooling system to the proper level.
31. Reconnect the negative battery cable, run the engine and check the manifolds for leaks. Check the coolant level and top off, if necessary.

Exhaust Manifold

REMOVAL AND INSTALLATION

2.4L (VIN B) Engine

1. Disconnect the negative battery cable.
2. Raise and safely support vehicle.
3. Disconnect the exhaust pipe from the exhaust manifold at the flex joint. Apply penetrating oil on the ex-

haust manifold-to-exhaust pipe flange bolts to aid in removal. It may be necessary to remove the entire exhaust system.
4. Disconnect the oxygen sensor wiring connector at the rear of the exhaust manifold.
5. Remove the 8 manifold attaching bolts and remove the manifold from the cylinder head.

To install:
6. Thoroughly clean all parts. Discard the gasket (if equipped) and clean all gasket surfaces of the manifold and cylinder head. Test the manifold gasket surface for flatness with a straightedge and feeler gauge. The surface must be flat within 0.006 inches per foot (0.15mm per 300mm) of manifold length. Inspect the manifold for cracks or distortion. Replace if necessary.
7. Install the manifold to the vehicle with a new gasket. DO NOT APPLY SEALER.
8. Install the 8 manifold bolts and tighten, starting at the center and working outward in both directions. Torque to 17 ft. lbs. (23 Nm).
9. Reconnect the oxygen sensor wiring connector.
10. Reconnect the exhaust pipe to the exhaust manifold and torque fasteners to 20 ft. lbs. (28 Nm).
11. Reconnect the negative battery cable. Start the engine and allow to idle while inspecting the manifold for exhaust leaks.

3.0L (VIN 3) Engine

1. Disconnect the negative battery cable. Raise and safely support the vehicle.
2. Disconnect the exhaust pipe from the rear exhaust manifold, at the flex joint.
3. On 1993–95, disconnect the EGR tube from the rear manifold, if equipped, and disconnect the oxygen sensor wire.
4. Remove the crossover pipe-to-exhaust manifold attaching bolts.
5. On 1996–97, remove the rear exhaust manifold heat shield.
6. Remove the rear exhaust manifold-to-cylinder head nuts and remove the exhaust manifold.
7. Lower the vehicle and remove the heat shield from the front exhaust manifold.
8. Remove the bolts fastening the crossover pipe to the front exhaust manifold. Remove the front exhaust manifold-to-cylinder head nuts and remove the exhaust manifold.
9. Clean the gasket mounting surfaces. Inspect the manifolds for cracks, flatness and/or damage.

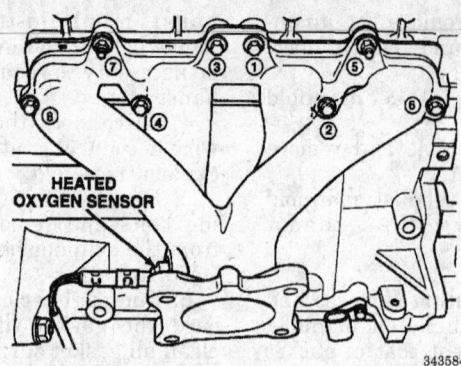

343584

Exhaust manifold torque sequence — 2.4L (VIN B) engine

To install:

NOTE: Install the gasket with the numbers 1-3-5 embossed on the top on the rear bank and those with the numbers 2-4-6 on the front (radiator side) bank.

10. Raise and safely support the vehicle.

11. Install the new gasket and rear exhaust manifold to the cylinder head. Install and torque the rear exhaust manifold-to-cylinder head nuts to 175 inch lbs. (20 Nm).

12. Attach the exhaust pipe to the exhaust manifold and tighten the shoulder bolts to 20 ft. lbs. (28 Nm)

13. Attach the crossover pipe to the exhaust manifold and tighten the bolts to 51 ft. lbs. (69 Nm).

14. On 1996–97, install the rear exhaust manifold heat shield and torque the heat shield mounting fasteners to 130 inch lbs. (15 Nm).

15. Connect the EGR tube to the rear manifold, if removed, and reconnect the oxygen sensor wiring connector.

16. Lower the vehicle.

17. Install the front exhaust manifold and attach the exhaust crossover pipe. Torque the bolts to 51 ft. lbs. (69 Nm).

18. Install the front exhaust manifold heat shield and tighten the screws to 130 inch lbs. (15 Nm).

19. Reconnect the negative battery cable. Start the engine and check for exhaust leaks.

3.3L (VIN R) and 3.8L (VIN L) Engine

1. Disconnect the negative battery cable.

2. On 1996–97, perform the following procedures:

  a. Remove accessory drive belt.

  b. Remove the alternator.

3. Raise and safely support the vehicle.

4. Disconnect the exhaust pipe from the rear exhaust manifold at the flex joint.

5. Separate the EGR tube from the exhaust manifold and disconnect the oxygen sensor.

6. Remove the alternator/power steering support strut.

7. Remove the crossover pipe attaching bolts to the rear exhaust manifold.

8. On 1996–97, remove the rear exhaust manifold heat shield.

9. Remove the rear manifold-to-cylinder head nuts and the remove the rear manifold.

10. Lower the vehicle and remove the front exhaust manifold heat shield.

11. Remove the front exhaust manifold crossover pipe bolts.

12. Remove the front manifold-to-cylinder head nuts and the remove the front exhaust manifold.

13. Clean the mounting surfaces. Inspect the manifolds for cracks or other damage. With a straightedge and feeler gauge, check for flatness. Standard is 0.004 in. with 0.008 in. out-of-flatness the service limit. If distorted beyond these specifications, replace the manifold.

To install:

14. Raise and safely support the vehicle.

15. Install the new gasket and rear manifold. Tighten the manifold to cylinder head nuts to 17 ft. lbs. (23 Nm).

16. On 1996–97, perform tyhe following procedures:

  a. Install the rear exhaust manifold heat shield. Torque the heat shield mounting fasteners to 17 ft. lbs. (23 Nm).

  b. Install the alternator unit.

17. Connect the exhaust pipe to the exhaust manifold and torque the bolt to 20 ft. lbs. (28 Nm).

18. Connect the crossover pipe to the rear manifold and tighten the bolts to 25 ft. lbs. (33 Nm).

19. Connect the oxygen sensor electrical connector.

20. Connect the EGR tube to the exhaust manifold and install the alternator/power steering support strut.

21. Lower the vehicle.

22. Install the new gasket and front exhaust manifold to the cylinder head. Tighten the exhaust manifold-to-cylinder head nuts to 17 ft. lbs. (23 Nm).

23. Connect the crossover pipe to the front manifold.

24. Install the front manifold heat shield. Torque the heat shield mounting fasteners to 200 inch lbs. (23 Nm).

25. On 1996–97, install the accessory drive belt.

26. Reconnect the negative battery cable.

27. Start the engine and check for exhaust leaks.

Front Cover Seal

REMOVAL AND INSTALLATION

3.3L (VIN R) and 3.8L (VIN L) Engines

1. Disconnect the negative battery cable.

2. Raise the vehicle and support safely.

3. Remove the right front wheel and the splash shield.

4. Remove the accessory drive belt.

5. Remove the crankshaft pulley bolt and remove the pulley using a suitable puller.

6. Remove the crankshaft oil seal from the cover using crankshaft seal removal tool C-4991 or equivalent. Be careful not to damage the crankshaft seal surface of the front timing chain cover.

To install:

7. Lubricate the lip of the new crankshaft oil seal with clean engine oil.

8. Use tool C-4992 or equivalent seal driver, to install the new crankshaft oil seal.

9. Install the crankshaft pulley using a bolt approximately 5.9 in. long and thrust bearing and washer plate. Make sure the pulley bottoms out on the crankshaft seal diameter. Install the bolt and torque to 40 ft. lbs. (54 Nm).

10. Install the accessory drive belt.

11. Install the inner splash shield and the wheel. Torque the wheel lug

nuts, in a star pattern, to 95 ft. lbs. (129 Nm).

12. Lower the vehicle.

13. Reconnect the negative battery cable. Check the engine oil level; add oil as necessary.

14. Start the engine and check for leaks.

Front Crankshaft Seal

REMOVAL AND INSTALLATION

2.4L (VIN B) Engine

The timing belt must be removed for this procedure. Use care that all timing marks are aligned after installation or the engine will become damaged.

1. Disconnect the negative battery cable.

2. Remove the accessory drive belts.

3. Raise and safely support the vehicle. Drain the engine oil.

4. Remove the crankshaft damper/pulley using a jaw puller tool.

5. Remove the timing belt cover and timing belt.

6. Remove the crankshaft timing belt sprocket using tool No. 6793 or equivalent.

——— WARNING ———
Do not nick the seal surface of the crankshaft or the seal bore.

7. Remove the front crankshaft seal using tool No. 6771 or equivalent seal puller. Be careful not to damage the seal contact area of the crankshaft.

To install:

8. Apply a light coating of clean engine oil to the lip of the new oil seal. Install the new front crankshaft oil seal using oil seal installer tool No. 6780-1 or equivalent. Install the new oil seal into the opening with the seal spring facing the inside of the engine. Be sure the oil seal is installed flush with the front cover.

9. Install the crankshaft timing belt sprocket using tool No. 6792.

NOTE: Be sure the word "FRONT" on the timing belt sprocket is facing you.

10. Install the timing belt and timing belt cover.

11. Install the crankshaft damper/pulley onto the crankshaft. Use thrust bearing/washer and 12M-1.75 x 150mm bolt from special tool No. 6792. Install the crankshaft

damper/pulley retaining bolt and torque to 105 ft. lbs. (142 Nm).

12. Lower the vehicle.

13. Install the accessory drive belts. Adjust the belts to the proper tension.

14. Refill the engine with the correct amount of clean engine oil.

15. Reconnect the negative battery cable. Start the engine and check for leaks.

2.5L (VIN K) Engine

The timing belt must be removed to service the crankshaft front oil seal. Working on any engine but especially overhead camshaft engines requires much care be given to valve timing. It is good practice to set the engine up to TDC No. 1 cylinder firing position. Verify that all timing marks on the crankshaft and camshaft sprockets are properly aligned before removing the timing belt and beginning camshaft service. This serves as a point of reference for all work that follows. Valve timing is most important and engine damage will result if the work is incorrect.

1. Disconnect the negative battery cable.

2. Remove the accessory drive belts.

3. Raise and safely support the vehicle. Remove the right inner splash shield on front wheel drive models.

4. Loosen and remove the 3 water pump pulley mounting screws and remove the pulley.

5. Remove the 4 crankshaft pulley retaining screws and the crankshaft pulley.

6. Remove the nuts at upper portion of timing cover and screws from lower portion and remove both halves of cover.

7. On front wheel drive, separate the right engine mount. Loosen the timing belt tensioner and remove the timing belt.

8. Using the proper puller tool, remove the crankshaft sprocket.

9. Remove the oil seal using seal removal tool 6341 or equivalent seal puller.

To install:

10. Clean all parts well. Inspect the front of the crankshaft. The seal lip surface must be clean of any scores, dirt or varnish. It may be necessary to polish the surface with 400 grit sandpaper. Clean all debris well or the new seal will wear quickly.

11. Install a new oil seal using tools 6342 and 6343 or equivalent seal driver.

12. Install the crankshaft sprocket and the timing belt. Torque the sprocket bolt to 85 ft. lbs. (115 Nm).

13. Reconnect the right side engine mount, if required.

14. Install the timing belt covers. Secure the upper section to the cylinder head with nuts and lower section to the cylinder block with screws. Torque the timing belt cover fasteners to 40 inch lbs. (4 Nm).

15. Install the crankshaft pulley and tighten the bolt to 20 ft. lbs. (27 Nm).

16. Install the water pump pulley and tighten screws to 105 inch lbs. (12 Nm).

17. Install the right side inner fender splash shield, if required. Lower the vehicle.

18. Install the accessory drive belts.

19. Reconnect the negative battery cable.

20. Check the engine oil level; add oil as required.

21. Run the engine and check for leaks.

3.0L (VIN 3) Engine

1. Disconnect the negative battery cable.

2. Remove the accessory drive belts.

3. Remove the air conditioning compressor from the mounting

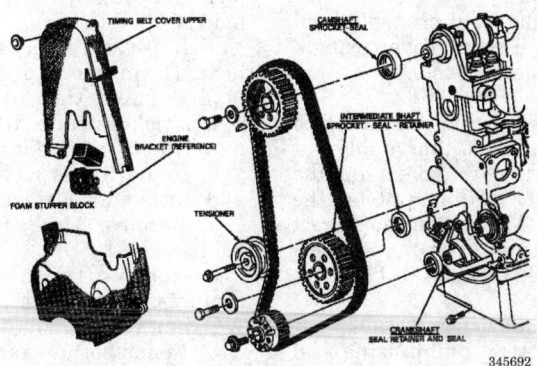

Timing belt and seals — 2.5L (VIN K) engine

345692

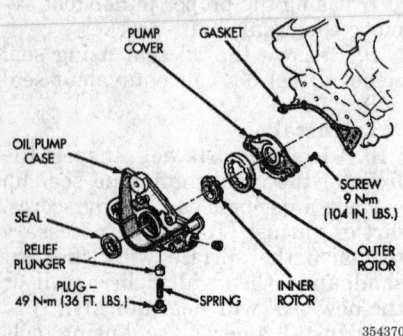

Oil pump assembly — 3.0L (VIN 3) engine

bracket and lay it aside, if equipped. Remove the compressor mounting bracket and adjustable drive belt tensioner from the engine.

4. Remove the power steering pump/alternator belt tensioner mounting bolt and remove the tensioner.

5. Remove the power steering pump mounting bracket bolts, rear support lock nut and set the power steering pump aside.

6. Raise and safely support the vehicle.

7. Drain the engine oil into a suitable container.

8. Remove the right inner fender inner splash shield.

9. Remove the crankshaft drive pulley bolt, drive pulley and torsional damper. Lower the vehicle.

10. Place a floor jack under the engine. Separate the engine mount insulator from the engine mount bracket.

11. Raise the engine slightly and remove the engine mount bracket.

12. Remove the timing belt cover and timing belt.

13. Remove the crankshaft sprocket.

14. Remove the oil pump mounting bolts (5), and remove the oil pump assembly. Mark the mounting bolts for proper installation during reassembly.

15. Remove the front crankshaft oil seal from the oil pump cover using a suitable seal removal tool.

To install:

16. Clean the oil pump and engine block gasket surfaces thoroughly.

17. Position a new gasket on the pump assembly and install on the cylinder block. Make sure the correct length bolts are in their proper locations and torque all bolts to 11 ft. lbs. (15 Nm).

18. Install a new front crankshaft oil seal into the oil pump using seal installer tool MD-998717 or equivalent seal driver tool.

19. Install the crankshaft sprocket and timing belt. Recheck engine timing marks.

20. Install the timing belt covers. Torque the timing belt cover fasteners to 10 ft. lbs. (14 Nm).

21. Raise the engine slightly and install the engine mount bracket. Install the engine mount insulator into the engine mount bracket.

22. Install the torsional damper, crankshaft drive pulley and drive pulley bolt. Torque to 112 ft. lbs. (151 Nm).

23. Install the right fender inner splash shield. Lower the vehicle.

24. Install the power steering mount bracket and power steering pump.

25. Install the power steering/alternator belt tensioner.

26. Install the air conditioning adjustable drive belt tensioner and mounting bracket. Torque the mounting bolts to 40 ft. lbs. (54 Nm).

27. Install the air conditioning compressor. Torque the compressor mounting bolts to 40 ft. lbs. (54 Nm).

28. Install the accessory drive belts and adjust to the proper tension.

29. Reconnect the negative battery cable.

30. Refill the crankcase with the correct amount of clean engine oil and start the engine. Check for leaks.

Timing Chain, Sprockets and Front Cover

REMOVAL AND INSTALLATION

3.3L (VIN R) and 3.8L (VIN L) Engines

1. Turn the crankshaft and position the engine so the No. 1 piston is at TDC on the compression stroke (firing position).

2. Disconnect the negative battery cable. Drain the coolant.

3. Support the engine with a floor jack and remove the right engine mount.

4. Raise and safely support the vehicle. Drain the engine oil.

5. Remove the oil pan and oil pump pick-up tube. If necessary, remove the transaxle inspection cover.

6. Remove the right front wheel and inner fender splash shield.

7. Remove the accessory drive belt.

8. Remove the air conditioning compressor and set aside. Remove A/C compressor mounting bracket.

9. Remove the crankshaft pulley bolt and remove the pulley using a suitable puller.

10. Remove the idler pulley from the engine bracket and remove the bracket.

11. Remove the cam sensor from the timing chain cover.

12. Unbolt and remove the cover from the engine. Make sure the oil pump inner rotor does not fall out. Remove the 3 O-rings from the coolant passages and the oil pump outlet.

13. Remove the camshaft sprocket attaching cup washer and remove the timing chain with both sprockets attached. Remove the timing chain snubber.

14. Remove the crankshaft sprocket using a suitable puller tool.

To install:

15. Assemble the timing chain and sprockets.

16. Turn the crankshaft and camshaft to line up with the key way locations of the sprockets.

17. Slide both sprockets over their respective shafts and use a straightedge to confirm alignment.

18. Install the cup washer and camshaft bolt. Torque the bolt to 40 ft. lbs. (54 Nm).

19. Check camshaft end-play. The specification with a new plate is 0.005–0.012 in. (0.0127–0.304mm) and 0.012 in. (0.310mm) maximum with a used plate. Replace the thrust plate if not within specifications.

20. Install the timing chain snubber.

21. Clean all parts well. Thoroughly clean and dry the gasket mating surfaces. Install new O-rings to the block.

22. Remove the crankshaft oil seal from the cover. The seal must be removed from the cover when installing to ensure proper oil pump engagement. The seal is installed from the front AFTER the timing chain front cover is completely installed as outlined below.

23. Using a new gasket, install the chain case cover to the engine. DO NOT adhere the new gasket to the cover. Make sure the lower edge of the gasket is flush to 0.020 in. past the lower edge of the cover.

24. If necessary, rotate the crankshaft so the oil pump drive flats are vertical and position the oil pump rotor so its mating flats are in the same position. Make certain that the oil pump is engaged onto the crankshaft before proceeding, or severe engine damage will result. Install the attaching bolts and torque to 20 ft. lbs. (27 Nm).

25. Use tool C-4992 or equivalent seal driver to install the crankshaft oil seal.

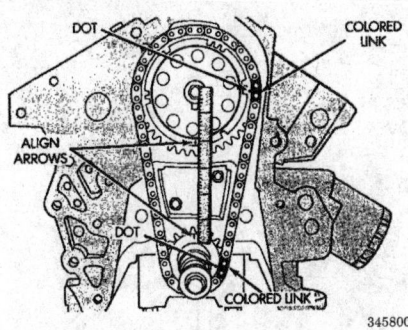

Timing mark alignment — 3.3L (VIN R) and 3.8L (VIN L) engines

26. Install the crankshaft pulley using a bolt approximately 5.9 in. long and thrust bearing and washer plate. Make sure the pulley bottoms out on the crankshaft seal diameter. Install the bolt and torque to 40 ft. lbs. (54 Nm).

27. Install the engine bracket and torque the bolts to 40 ft. lbs. (54 Nm). Install the idler pulley to the engine bracket.

28. To install the cam sensor, first clean off the old spacer from the sensor face completely. Inspect the O-ring for damage and replace if necessary. A new spacer must be attached to the cam sensor, prior to installation. If a new spacer is not used, engine performance will be affected. Oil the O-ring lightly and push the sensor into its bore in the timing case cover until contact is made with the cam timing sprocket. Hold in this position and tighten it to 9 ft. lbs. (12 Nm).

29. Reconnect the sensor connector to the wiring harness connector. Position the wiring harness away from the accessory drive belt.

30. Install the air conditioning compressor and mounting bracket.

31. Install the drive belt.

32. Install the inner splash shield and the wheel. Torque the wheel lug nuts, in a star pattern, to 95 ft. lbs. (129 Nm).

33. Install the oil pan with a new gasket. Torque the oil pan bolts to 105 inch lbs. (12 Nm).

34. Install the motor mount.

35. Refill the cooling system to the correct level with a ⁵⁰/₅₀ mixture of clean, fresh ethylene glycol antifreeze and water.

36. Fill the engine with the proper amount and type of clean engine oil.

37. Reconnect the negative battery cable. Road test the vehicle and check for leaks.

Timing Belt, Sprockets, Tensioner and Front Cover

REMOVAL AND INSTALLATION

2.4L (VIN B) Engine

1. Disconnect the negative battery cable remote connection, located on the left strut tower.

2. Remove the right inner splash shield.

3. Remove the accessory drive belts.

4. Remove the crankshaft damper.

5. Remove the right engine mount.

6. Place a floor jack under the vehicle to support the engine.

7. Remove the engine mount bracket

8. Remove the timing belt cover.

NOTE: Do not rotate the crankshaft or the camshafts after the timing belt has been removed. Damage to the valve components may occur. Before removing the timing belt, always align the timing marks.

9. Align the timing marks of the timing belt sprockets to the timing marks on the rear timing belt cover and oil pump cover. Loosen the timing belt tensioner bolts.

10. Remove the timing belt and the tensioner.

11. Remove the camshaft timing belt sprockets.

12. Remove the crankshaft timing belt sprocket using removal tool No. 6793 or equivalent.

13. Place the tensioner into a soft jawed vise to compress the tensioner.

14. After compressing the tensioner place a pin (⁵/₆₄ in. Allen wrench will work) into the plunger side hole to retain the plunger until installation.

To install:

15. Using tool No. 6792 or equivalent, install the crankshaft timing belt sprocket onto the crankshaft.

16. Install the camshaft sprockets onto the camshafts. Install and torque the camshaft sprocket bolts to 75 ft. lbs. (101 Nm).

17. Set the crankshaft sprocket to Top Dead Center (TDC) by aligning the notch on the sprocket with the arrow on the oil pump housing.

18. Set the camshafts to align the timing marks on the sprockets.

19. Move the crankshaft to ½ notch before TDC.

20. Install the timing belt starting at the crankshaft, then around the water pump sprocket, idler pulley,

camshaft sprockets and then around the tensioner pulley.

21. Move the crankshaft sprocket to TDC to take up the belt slack.

22. Install the tensioner to the block but do not tighten.

23. Using a torque wrench on the tensioner pulley apply 250 inch lbs. (28 Nm) of torque to the tensioner pulley.

24. With torque being applied to the tensioner pulley, move the tensioner up against the tensioner pulley bracket and torque the fasteners to 275 inch lbs. (31 Nm).

25. Remove the tensioner plunger pin, the tension is correct when the plunger pin can be removed and replaced easily.

26. Rotate the crankshaft 2 revolutions and recheck the timing marks. Wait several minutes and then recheck that the plunger pin can easily be removed and installed.

27. Install the front timing belt cover.

28. Install the engine mount bracket.

29. Install the right engine mount.

30. Remove the floor jack from under the vehicle.

31. Install the crankshaft damper and torque to 105 ft. lbs. (142 Nm).

32. Install the accessory drive belts and adjust to the proper tension.

33. Install the right inner splash shield.

34. Reconnect the negative battery cable.

35. Perform the crankshaft and camshaft relearn alignment procedure using the DRB scan tool or equivalent.

2.5L (VIN K) Engine

Working on any engine but especially overhead camshaft engines requires much care be given to valve timing. It is good practice to set the engine up to TDC No. 1 cylinder firing position. Verify that all timing marks on the crankshaft and camshaft sprockets are properly aligned before removing the timing belt. This serves as a point of reference for all work that follows. Valve timing is most important and engine damage will result if the work is incorrect.

1. Disconnect the negative battery cable.

2. Remove the accessory drive belts.

3. Remove the right engine mount yoke screw.

4. Remove the air conditioning compressor and set it aside. Remove the solid mount compressor bracket

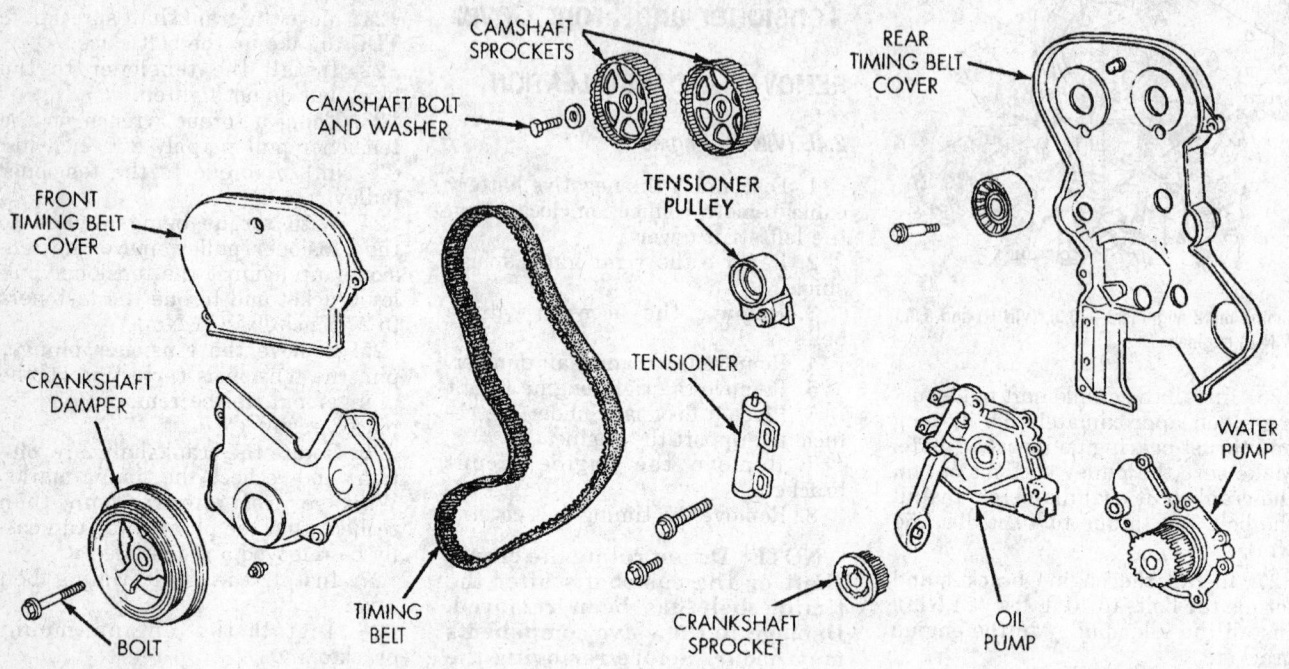

Timing belt and related components — 2.4L (VIN B) engine

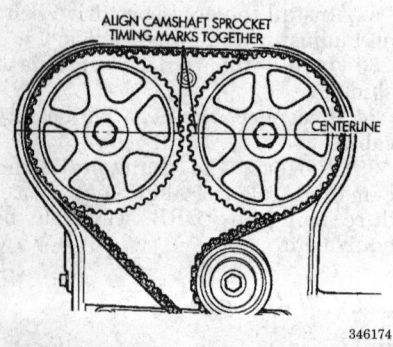

Camshaft sprocket alignment — 2.4L (VIN B) engine

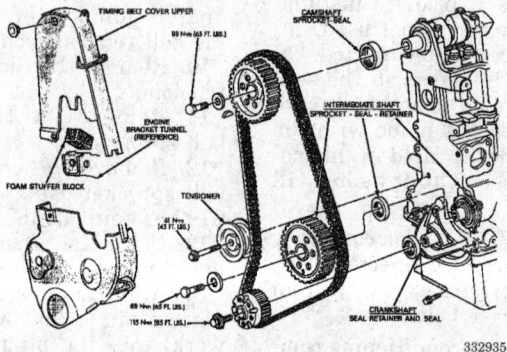

Timing belt and related components — 2.5L (VIN K) engine

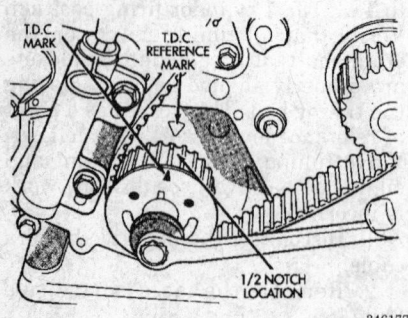

Crankshaft position — 2.4L (VIN B) engine

mounting bolts in the following order: 1-4-5-6-7-2-3.

5. Turn the solid mount bracket away from the engine and slide it on the No. 2 stud until it is free. The front bolt and spacer will be removed with the bracket.

6. Remove the alternator and the drive belt idler.

7. Raise and safely support the vehicle. Remove the right inner fender splash shield.

8. Loosen and remove the 3 water pump pulley mounting bolts and remove the pulley.

9. Remove the 4 crankshaft pulley retaining bolts and the crankshaft pulley.

10. Remove the nuts at upper portion of timing cover and bolts from lower portion and remove both halves of cover.

11. Remove the timing belt covers.

12. Place a jack under the engine.

13. Separate the right engine mount and raise the engine slightly.

14. Loosen the timing belt tensioner bolt, rotate the hex nut, and remove timing belt.

15. Remove timing belt tensioner, if necessary.

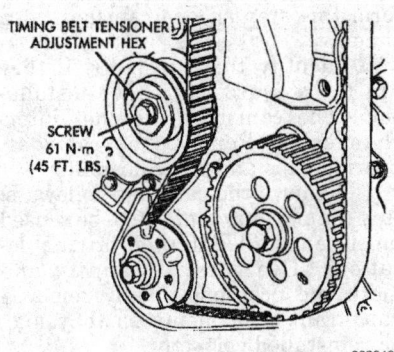

Removing the timing belt — 2.5L (VIN K) engine

16. Remove the crankshaft sprocket with a suitable puller tool and a bolt approximately 6 in. long.

17. Remove the camshaft sprocket and intermediate shaft sprocket, if necessary.

To install:

18. Clean all parts well. A small amount of white paint on the sprocket timing marks may make alignment easier.

19. Install the crankshaft sprocket. Torque the crankshaft sprocket bolt to 85 ft. lbs. (115 Nm).

20. If necessary, turn the crankshaft and intermediate shaft until markings on both sprockets are aligned.

21. Rotate the camshaft so the arrows on the hub are in line with No. 1 camshaft cap to cylinder head line. The small hole in the cam sprocket should be centered, in vertical center line.

22. Install the timing belt tensioner, if removed.

23. Install the timing belt over the drive sprockets and adjust.

24. Tighten the tensioner by turning the tensioner hex to the right. Tension should be correct when the belt can be twisted 90 degrees with the thumb and forefinger, midway between the camshaft and intermediate sprocket.

25. Turn the engine clockwise from TDC, 2 complete revolutions with crankshaft bolt. Check the timing marks for correct alignment.

WARNING

Do not use the camshaft or intermediate shaft to rotate the engine. Do not allow oil or solvent to contact timing belt as they will deteriorate the belt and cause slipping.

---

26. Tighten lock nut on tensioner, while holding weighted wrench (tool C-4503 or equivalent) in position, to 45 ft. lbs. (61 Nm).

27. Lower the engine onto the right engine mount and install the fasteners. Remove the support from the engine.

28. Some engines use a foam stuffer block inside the timing belt housing. Inspect the foam block's condition and position. The stuffer block should be intact and secure within the engine bracket tunnel.

29. Install the timing belt cover. Secure the upper section to the cylinder head with nuts and lower section to the cylinder block with screws. Torque all of the timing belt cover fasteners to 40 inch lbs. (4 Nm).

30. Again check valve timing. With the timing belt cover installed, and with No. 1 cylinder at TDC, the small hole in the sprocket must be centered in the timing belt cover hole. If the hole is not aligned correct, perform the timing belt installation again.

31. Install the water pump pulley and the crankshaft pulley. Torque the water pump pulley bolts to 250 inch lbs. (28 Nm). Torque the crankshaft pulley bolts to 280 inch lbs. (31 Nm).

32. Install the inner fender splash shield. Lower the vehicle.

33. Install the solid mount compressor bracket. The bracket mounting fasteners must be torqued to 40 ft. lbs. (54 Nm) and in the following order: 2-3-1-4-5-6-7.

34. Install the alternator and drive belt idler. Torque mounting bolts to 40 ft. lbs. (54 Nm).

35. Install the right engine mount yoke bolt and torque to 100 ft. lbs. (133 Nm).

36. Install the accessory drive belts and adjust to the proper tension.

NOTE: With timing belt cover installed and the piston in No. 1 cylinder at TDC, the small hole in the cam sprocket should be centered in timing belt cover hole.

37. Reconnect the negative battery cable. Road test the vehicle.

3.0L (VIN 3) Engine

The timing belt can be inspected by removing the upper front outer timing belt cover.

Working on any engine but especially overhead camshaft engines requires much care be given to valve timing. It is good practice to set the engine up at TDC No. 1 cylinder firing position before beginning work. Verify that all timing marks on the crankshaft and camshaft sprockets are properly aligned before removing the timing belt and starting camshaft service. This serves as a point of reference for all work that follows. Valve timing is most important and engine damage will result if the work is incorrect.

1. Disconnect the negative battery cable.

2. Remove the accessory drive belts. Remove the engine mount insulator from the engine support bracket.

3. Remove the engine support bracket. Remove the crankshaft pulleys and torsional damper. Remove the timing belt covers.

4. Rotate the crankshaft until the sprocket timing marks are aligned. The crankshaft sprocket timing mark should align with the oil pump timing mark. The rear camshaft sprocket timing mark should align with the generator bracket timing mark and the front camshaft sprocket timing mark should align with the inner timing belt cover timing mark.

5. If the belt is to be reused, mark the direction of rotation on the belt for installation reference.

6. Loosen the timing belt tensioner bolt and remove the timing belt.

7. If necessary, remove the timing belt tensioner.

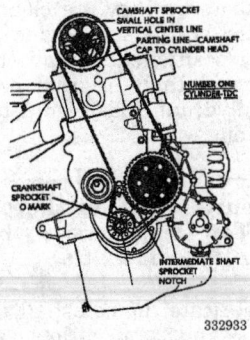

Timing marks — 2.5L (VIN K) engine

Camshaft sprocket timing mark alignment — 2.5L (VIN K) engine

8. Remove the crankshaft sprocket flange shield and crankshaft sprocket.

9. Hold the camshaft sprocket using spanner tool MB990775 or equivalent, and remove the camshaft sprocket bolt and washer. Remove the camshaft sprocket.

To install:

10. Install the camshaft sprocket to the camshaft with the retaining bolt and washer. Hold the camshaft sprocket using spanner tool MB990775 or equivalent, and tighten the bolt to 70 ft. lbs. (95 Nm).

11. Install the crankshaft sprocket.

12. If removed, install the timing belt tensioner and tensioner spring. Hook the spring upper end to the water pump pin and the lower end to the tensioner bracket with the hook out.

13. Turn the timing belt tensioner counterclockwise full travel in the adjustment slot and tighten the bolt to temporarily hold in this position.

14. Rotate the crankshaft sprocket until the timing mark on the crankshaft sprocket is aligned with the oil pump timing mark.

15. Rotate the rear camshaft sprocket until the timing mark on the sprocket is aligned with the timing mark on the generator bracket.

16. Rotate the front (radiator side) camshaft sprocket until the mark on the sprocket is aligned with the timing mark on the inner timing belt cover.

17. Install the timing belt on the crankshaft sprocket while keeping the belt tight on the tension side.

NOTE: If the original belt is being reused, be sure to install it in the same rotational direction.

18. Position the timing belt over the front camshaft sprocket (radiator side). Next, position the belt under the water pump pulley, then over the rear camshaft sprocket and finally over the tensioner.

19. Apply rotating force in the opposite direction to the front camshaft sprocket (radiator side) to create tension on the timing belt tension side. Check that all timing marks are aligned.

20. Install the crankshaft sprocket flange.

21. Loosen the tensioner bolt and allow the tensioner spring to tension the belt.

22. Rotate the crankshaft 2 full turns in a clockwise direction. Turn the crankshaft smoothly and in a clockwise direction only.

23. Again line up the timing marks. If all marks are aligned, tighten the

tensioner bolt to 250 inch lbs. (28 Nm).

24. Install the timing belt covers. Install the engine support bracket. Torque the support bracket mounting bolts to 35 ft. lbs. (47 Nm).

25. Install the engine mount insulator, torsional damper and crankshaft pulleys. Torque the crankshaft pulley bolt to 112 ft. lbs. (151 Nm).

26. Install the accessory drive belts and adjust to the proper tension.

27. Reconnect the negative battery cable.

28. Run the engine and check for proper operation. Road test the vehicle.

Camshaft

REMOVAL AND INSTALLATION

————— **CAUTION** —————
Fuel injection systems remain under pressure, even after the engine has been turned OFF. The fuel system pressure must be relieved before disconnecting any fuel lines. Failure to do so may result in fire and/or personal injury.

2.4L (VIN B) Engine

This engine uses a Dual Over Head Camshaft (DOHC) 4-valves per cylinder cross flow aluminum cylinder head. The valves are actuated by roller cam followers which pivot on stationary hydraulic valve lash adjusters. Care must be taken to make sure all valve timing marks align after cylinder head and valvetrain service.

1. Properly relieve the fuel system pressure.

2. Disconnect the negative battery cable.

3. Label and disconnect the spark plug wires from the spark plugs.

4. Remove the ignition coil pack and spark plug wires.

5. Remove the cylinder head cover retaining fasteners and remove the cylinder head cover from the cylinder head. Discard the old cylinder head cover gasket.

6. Remove the ground strap.

7. Remove the timing belt covers, timing belt and camshaft sprockets.

8. Take note that the camshaft bearing caps are numbered for correct location during installation. Remove the outer bearing caps first.

9. Loosen, but do not remove, the camshaft bearing cap retaining fasteners in the correct sequence. Per-

form this step on one camshaft at a time.

10. Identify the camshafts, if they are to be reused, for later installation. The camshafts are not interchangeable. Remove the camshaft bearing caps and the camshafts.

11. Remove the camshaft followers. Any components that are to be reused must be installed in their original locations. Use care to identify and mark the positions of any removed valvetrain components so they may be reinstalled correctly.

12. Inspect the camshaft bearing oil feed holes in the cylinder head for clogging. Inspect the camshaft bearing journals for wear or scoring. Check the cam surface for abnormal wear and damage. A visible worn groove in the roller path or on the cam lobes is cause for replacement.

To install:

13. Thoroughly clean the camshafts and related parts.

14. The camshaft end play should be checked using the following procedure:

   a. Oil the camshaft journals and install the camshaft **WITHOUT** the cam follower assemblies. Install the rear cam caps and torque to 250 inch lbs. (28 Nm).

   b. Carefully push the camshaft as far rearward as it will go.

   c. Set up a dial indicator to bear against the front of the camshaft (the sprocket end). Zero the indicator.

   d. Move the camshaft forward as far as it will go. Read the dial indicator. End-play specification is 0.002–0.010 in. (0.05–0.15mm).

   e. If excessive end-play is present, inspect the cylinder head and camshaft for wear; replace if necessary.

15. If satisfied with the fit and condition of the camshafts, remove the camshafts for installation of the cam followers.

16. The hydraulic valve lash adjusters are inside the roller cam followers. Make sure they are clean, well-lubricated with clean engine oil and properly positioned. Install the cam followers in their original positions on the hydraulic lash adjuster and valve stem.

————— **WARNING** —————
Make sure NONE of the pistons are at Top Dead Center when installing the camshafts.

17. Lubricate the camshaft bearing journals and cam followers with clean engine oil and install the camshafts. Install right and left camshaft

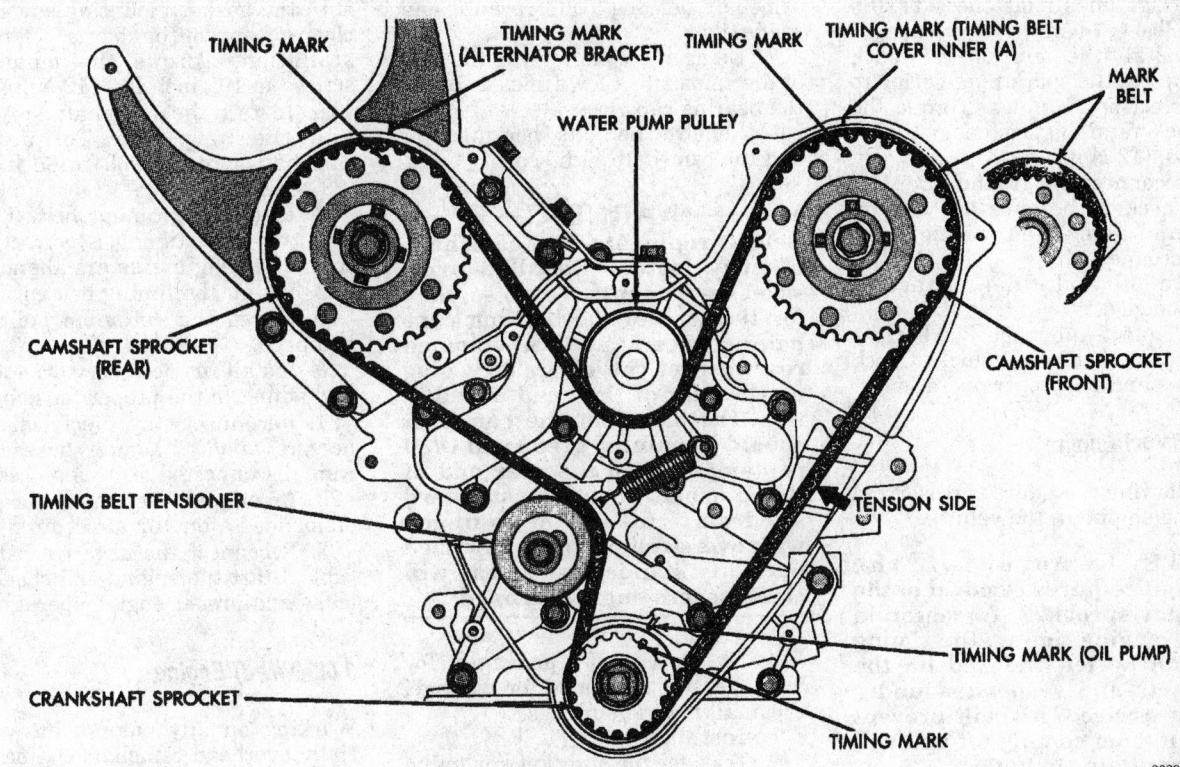

Timing mark alignment — 3.0L (VIN 3) engine

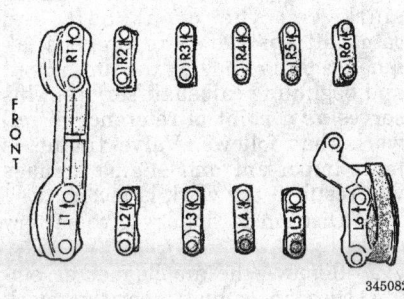

Camshaft bearing cap identification — 2.4L (VIN B) engine

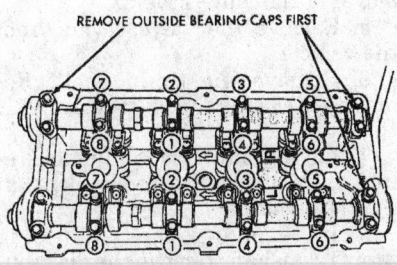

Camshaft bearing cap removal sequence — 2.4L (VIN B) engine

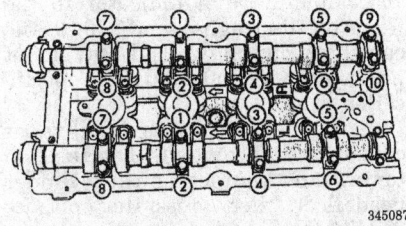

Camshaft bearing cap tightening sequence — 2.4L (VIN B) engine

bearing caps No. 2 through No. 5 and right side No. 6. Tighten the M6 fasteners to 105 inch lbs. (12 Nm) in correct sequence.

18. Apply Mopar® Gasket Maker or equivalent sealer to the No. 1 and left side No. 6 bearing caps. Install the bearing caps and tighten the M8 fasteners to 250 inch lbs. (28 Nm). The end caps must be installed before the seals may be installed.

19. Install the camshaft end seals.

20. Install the camshaft sprockets, if removed. Install the timing belt using care to make sure all timing marks are properly aligned. Install the timing belt covers.

WARNING

Verify that all timing marks are correct. If the timing belt or sprockets are incorrectly installed, engine damage will occur. Take time to make sure all timing marks are correctly aligned.

21. Clean the cylinder head cover and cylinder head gasket rails (mating surfaces). Make certain the rails are flat.

22. Install new cylinder head cover gaskets. Use care. DO NOT allow oil or solvents to contact the timing belt as they can deteriorate the rubber and cause tooth skipping. Apply Mopar Silicone Rubber Adhesive Sealant, or equivalent, at the camshaft cap corners and at the top edge of the 1/2-round seal.

NOTE: Inspect the spark plug well seals for cracking and/or swelling and replace if necessary.

23. Install the cylinder head cover assembly to the head and tighten the fasteners in sequence using the following 3 Steps:

a. Tighten all cylinder head cover fasteners to 40 inch lbs. (4.5 Nm).

b. Tighten all fasteners to 80 inch lbs. (9 Nm).

c. Tighten all fasteners to 105 inch lbs. (12 Nm).

24. Install the ignition coil pack and connect the spark plug wiring to the correct spark plugs. Torque the coil pack retaining fasteners to 105 inch lbs. (12 Nm).

25. Reconnect the ground strap.

26. Check to be sure all vacuum lines and remaining wiring have been reconnected.

27. An oil and filter change is recommended.

28. Connect the negative battery cable and test run the vehicle. Check for leaks and for proper operation.

2.5L (VIN K) Engine

The following procedure is performed with the engine in the vehicle.

NOTE: Removal of the camshaft requires removal of the camshaft sprocket. To maintain proper engine timing, the timing belt can be left indexed on the sprockets and suspended under light pressure. This will prevent the belt from coming off and will help maintain timing.

1. Properly relieve the fuel system pressure.

2. Disconnect the negative battery cable.

3. It is good practice to turn the engine to TDC No. 1 cylinder, compression stroke (firing position) before beginning timing belt and camshaft service. This gives a point of reference to help verify timing mark alignment.

4. Remove the accessory drive belts. Disconnect and label any electrical connectors and/or vacuum hoses that will allow for easier removal of the camshaft.

5. Turn the crankshaft so the No. 1 piston is at TDC on the compression stroke. Remove the upper timing belt cover. Remove the air pump pulley, if equipped.

6. Remove the camshaft sprocket bolt and the sprocket and suspend tightly so the belt does not lose tension. If it does, the belt timing will have to be reset.

7. Remove the valve cover. Under the valve cover may be a large baffle or shroud that Chrysler calls an oil vapor curtain. Remove the curtain. Be careful not to lose the rubber bumpers on the top of the curtain.

8. If the rocker arms are being reused, mark them for installation identification since they must be reinstalled in same location from which they were removed. Loosen the

camshaft bearing bolts, evenly and gradually.

9. Using a soft mallet, rap the rear of the camshaft a few times to break the bearing caps loose.

10. Remove the bolts, bearing caps and the camshaft with seals.

————— WARNING —————

Before replacing the camshaft, identify factory installed oversized components. To do so, look for the tops of the bearing caps painted green and O/SJ stamped rearward of the oil gallery plug on the rear of the head. In addition, the barrel of the camshaft should be painted green and O/SJ is stamped onto the rear end of the camshaft. Installing standard sized parts in a head equipped with oversized parts or oversized parts in a standard size head will cause severe engine damage.

11. Check the oil passages for blockage and the parts for damage. Clean all mating surfaces.

To install:

12. Transfer the sprocket key to the new camshaft. New rocker arms and a new camshaft sprocket bolt are normally included with the camshaft package. Install the rocker arms, lubricate the camshaft and install with end seals installed. NEVER install used rocker arms on a new camshaft. A new camshaft requires new rockers.

13. Position the bearing caps with No. 1 at the timing belt end and No. 5 at the transmission end. The camshaft bearing caps are numbered and have arrows facing forward. Torque the camshaft bearing bolts evenly and gradually to 215 inch lbs. (25 Nm).

NOTE: Apply RTV silicone gasket material to the No. 1 and No. 5 bearing caps. Install the bearing caps before the seals are installed.

14. Mount a dial indicator to the front of the engine and check the camshaft end-play. End-play must not exceed the 0.005–0.013 in. specification range.

15. Install the valve cover oil vapor curtain with the cutouts over the cam towers and contacting the cylinder head floor, then press the opposite distributor side into position below cylinder head rail. Make sure the rubber bumpers are properly positioned on top of the curtain. Install the valve cover and a new gasket. Be

sure the gasket mounting surface is clean of any oil or debris before installation. Torque the mounting screws to 105 inch lbs. (12 Nm).

16. Install the camshaft sprocket and the new bolt. Torque the camshaft sprocket bolt to 65 ft. lbs. (89 Nm).

17. Install the timing belt. Verify that the valve timing is correct and that all timing marks are aligned.

18. Install the timing belt cover.

19. Install the air pump pulley, if equipped.

20. Install the accessory drive belts and adjust to the proper tension.

21. Reconnect any electrical connectors and/or vacuum hoses that were disconnected for this procedure.

22. An oil and filter change are recommended after this procedure.

23. Reconnect the negative battery cable. Start the engine and check for leaks and proper engine operation.

3.0L (VIN 3) Engine

Working on any engine, but especially overhead camshaft engines requires much care be given to valve timing. It is good practice to set the engine up at TDC No. 1 cylinder firing position. Verify that all timing marks on the crankshaft and camshaft sprockets are properly aligned before removing the timing belt and beginning camshaft service. This serves as a point of reference for all work that follows. Valve timing is most important and engine damage will result if the work is incorrect.

1. Disconnect the negative battery cable.

2. Remove the air cleaner assembly. Disconnect and label the spark plug wires.

3. Remove the accessory drive belts, if necessary.

4. Move away any wiring harnesses for easier access to the rear valve cover. Disconnect and label any vacuum hoses required for easier removal of the valve covers.

5. Remove the valve covers from the vehicle.

6. Remove the timing belt. Remove the camshaft timing belt sprockets.

7. Install auto lash adjuster retainers MD998443 or equivalent on the rocker arms.

8. If removing the right side (front) camshaft, remove the distributor adaptor.

9. Remove the camshaft bearing caps but do not remove the bolts from the caps.

10. Remove the rocker arms, rocker shafts and bearing caps, as an assembly.

NOTE: Use care not to mix rocker arms and shafts. The oil holes are different in the shafts and each must be returned to its original location or the engine overhead will not be lubricated.

11. Remove the camshaft from the cylinder head.

12. Inspect the bearing journals on the camshaft, cylinder head and bearing caps.

To install:

13. Clean all parts well. Pay particular attention to the oil feed holes in the cylinder head, checking for clogging. If the camshafts are to be reused, check with a micrometer. Measure the cam height and replace if out of limit. Standard value is 1.624 in. (41.25mm) and wear **limit** is 1.604 in. (40.75mm).

14. Lubricate the camshaft journals and camshaft with clean engine oil and install the camshaft in the cylinder head.

15. Align the camshaft bearing caps with the arrow mark (depending on cylinder numbers) and in numerical order.

16. Apply sealer at the ends of the bearing caps and install the assembly.

17. Torque the bearing cap bolts, in the following sequence: No. 3, No. 2, No. 1 and No. 4 to 85 inch lbs. (10 Nm).

18. Repeat the sequence, increasing torque to 180 inch lbs. (20 Nm).

19. Install the distributor adaptor, if it was removed.

20. Remove the lash adjuster retainers. Install the camshaft timing belt sprockets and timing belt.

21. Install the valve covers and new gaskets. Torque the valve cover retaining screws to 88 inch lbs. (10 Nm).

22. Reconnect all vacuum hoses that were disconnected during the removal procedure. Be sure they are all reconnected to the correct vacuum lines.

23. Position the wiring harnesses back to their original locations. Install the accessory drive belts and adjust to the proper tension.

24. Reconnect the spark plug wires to the correct spark plugs.

25. Install the air cleaner assembly. Reconnect the negative battery cable.

26. Run the engine and check for leaks and proper engine operation.

Intermediate Shaft

REMOVAL AND INSTALLATION

2.5L (VIN K) Engine

This engine is equipped with an intermediate shaft (also called an accessory or auxiliary shaft). It has 2 bearing journals and is housed in the forward facing side of the block. A seal installed in an aluminum housing attached to the block, provides retention, shaft thrust and oil control. The intermediate shaft is driven by the timing belt through a sintered iron sprocket mounted on the nose of the intermediate shaft. The intermediate shaft drives the oil pump and distributor. The following procedures are to be performed with engine removed from vehicle.

1. Disconnect the negative battery cable.

2. Remove the engine from the vehicle.

3. Remove the distributor assembly.

4. Remove the accessory drive belts.

5. Remove the timing belt cover and timing belt from the engine.

6. Remove the intermediate shaft sprocket.

7. Remove the oil pan.

8. Remove the oil pump assembly.

9. Remove the intermediate shaft retainer screws and remove the retainer.

10. Remove the intermediate shaft and inspect the journals and bushing.

11. Remove and discard the intermediate shaft bushings, if necessary.

To install:

12. Install new bushings to the intermediate shaft, if necessary.

13. When installing the shaft, lubricate the oil pump and distributor drive gears. Install the intermediate shaft.

14. Inspect the shaft seal in the retainer; replace if necessary.

15. Lightly lubricate the seal lip with engine oil.

16. Apply a 0.06 in. (1.5mm) diameter bead of RTV (Form-in-Place) gasket material to the retainer sealing surface before installing.

17. Install the intermediate shaft retainer assembly and retainer screws. Tighten the screws to 105 inch lbs. (12 Nm).

18. Install the oil pump assembly and oil pan.

19. Install the intermediate shaft sprocket.

20. Perform timing procedure for timing belt and intermediate shaft sprockets.

21. Install the timing belt and adjust.

22. Install the timing belt cover.

23. Install the accessory drive belts and adjust to the proper tension.

24. Install the distributor.

25. Install the engine into the vehicle.

26. Reconnect the negative battery cable. Be sure all lines, hoses, wiring and linkages have all been properly installed, reconnected and/or tightened.

27. Road test the vehicle.

Balance Shaft

REMOVAL AND INSTALLATION

2.4L (VIN B) and 2.5L (VIN K) Engines

These engines are equipped with 2 balance shafts located in a housing attached to the lower crankcase. These shafts are driven by a chain from the crankshaft at twice crankshaft speed. The balance shafts are interconnected by 2 gears and are designed to counterbalance certain engine reciprocating masses. The engine must be removed from the vehicle to service the balance shafts.

———— **CAUTION** ————

Fuel injection systems remain under pressure, even after the engine has been turned OFF. The fuel system pressure must be relieved before disconnecting any fuel lines. Failure to do so may result in fire and/or personal injury.

1. Properly relieve the fuel system pressure.

2. Disconnect the negative battery cable.

3. Drain the cooling system and engine oil into suitable containers.

4. Remove the engine from vehicle and position on a suitable workstand.

5. Remove the accessory drive belts.

6. Remove the timing case cover, timing belt and sprockets.

7. Remove the engine oil pan.

8. Remove the front crankshaft seal retainer.

9. Remove the balance shafts chain cover.

10. Remove the chain guide and tensioner.

11. Remove the balance shafts sprocket retaining screws and crankshaft chain sprocket Torx screws. Re-

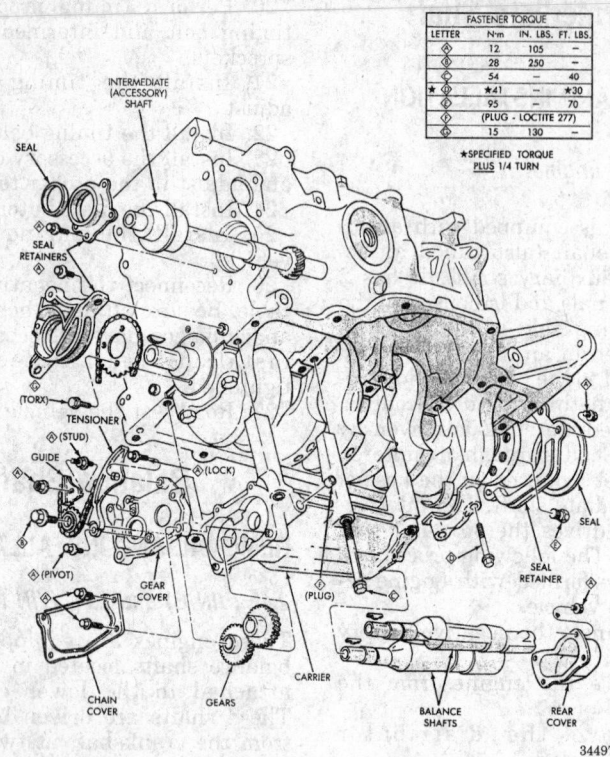

FASTENER TORQUE		
LETTER	N·m	IN. LBS.	FT. LBS.
A	12	105	
B	28	250	
C	54	—	40
★D	★41	—	★30
E	95	—	70
F	(PLUG - LOCTITE 277)		
G	15	130	

★SPECIFIED TORQUE
PLUS 1/4 TURN

Crankshaft, intermediate and balance shaft assemblies and oil seals —
2.5L (VIN K) engine

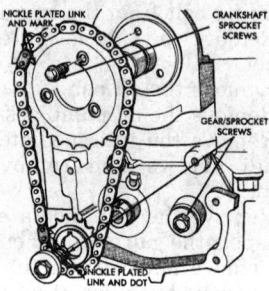

Drive chain and sprockets for
balance shafts — 2.5L (VIN K)
engine

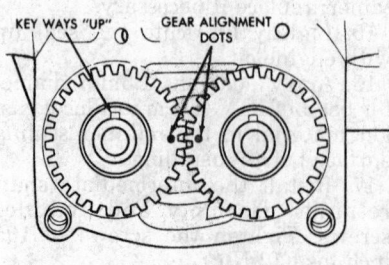

Drive and driven gear timing marks — 2.5L (VIN
K) engine

move the chain and sprocket
assembly.

12. Remove the balance shafts carrier front gear cover retaining double ended stud. Remove the cover and balance shafts gears.

13. Remove the carrier rear cover and balance shafts.

14. To separate the carrier, remove 6 crankcase to carrier attaching bolts and remove the carrier.

To install:

15. Install both shafts into the carrier assembly from the rear of the carrier.

16. Install the rear cover.

17. With the balance shafts installed in the carrier, position the carrier assembly on the crankcase and tighten the 6 bolts to 40 ft. lbs. (54 Nm).

18. Crankshaft-to-Balance Shaft Timing must be established. Rotate both balance shafts until the keyways are in the Up position (parallel to the vertical centerline of the engine). Install the short hub drive gear on sprocket driven shaft and long hub gear on gear driven shaft. After installation, gear and balance shaft keyways must be in the Up position with the drive and driven gear timing marks meshed.

19. Align the balance shaft carrier cover with the carrier housing dowel

pin and install double ended stud. Tighten to 105 inch lbs. (12 Nm).

20. Install the crankshaft sprocket and tighten sprocket Torx® screw to 130 inch lbs. (13 Nm).

21. Turn the crankshaft until the piston in No. 1 cylinder is at TDC. The timing marks on the chain sprocket should line up with the parting line on the left side of No. 1 main bearing cap.

22. Install the chain over the crankshaft sprocket so the nickel plated link of the chain is over the timing mark on the crankshaft sprocket.

23. Install the balance shaft sprocket into the timing chain so the timing mark on the sprocket (yellow dot) mates with the lower nickel plated link (some engines might have a yellow painted link) on the chain.

24. With the balance shaft key ways in the 12 o'clock position, slide the balance shaft chain drive sprocket onto the nose of the balance shaft. The balance shaft may have to be pushed in slightly to allow for clearance.

NOTE: The timing mark on the sprocket, the lower nickel plated link (some engines might have a yellow painted link) and arrow on the side of the gear cover should all line up if the balance shafts are timed correctly.

25. When satisfied that all timing marks are correct, install the balance shaft bolts and tighten to 20 ft. lbs. (28 Nm). Place a wooden block between the crankcase and crankshaft counterbalance to prevent crankshaft from turning.

26. Proper balance shaft Timing Chain Tension must be established. Use the following procedure.

a. Place a shim 1.0mm thick by 70mm long between the chain and tensioner.

b. Apply firm hand pressure behind the adjustment slot and tighten adjustment bolt first, followed by the pivot screw to 105 inch lbs. (12 Nm). Remove the shim.

c. Install the chain guide making sure the tab on the guide fits into slot on the gear cover. Install nut/washer and tighten to 105 inch lbs. (12 Nm).

d. Install the chain cover and tighten screws to 105 inch lbs. (12 Nm).

27. Apply a 1.5mm diameter bead of RTV gasket material to retainer sealing surface. Install retainer assembly.

28. Install the front crankshaft seal retainer.

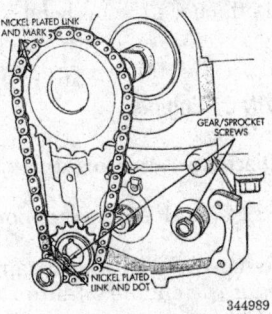

Drive chain and drive chain sprockets — 2.4L (VIN B) DOHC engine

operation. After the engine has been turned off and allowed to cool, check the coolant and oil levels. Add if necessary.

Piston and Connecting Rod

POSITIONING

29. Install the oil pan, new oil pan gasket and mounting bolts.
30. Install the crankshaft sprocket and timing belt.
31. Install the timing cover. Install the engine into the vehicle.
32. Install the accessory drive belts and adjust to the proper tension.
33. Refill the engine with the correct amount of clean engine oil. Refill the cooling system to the level with clean engine coolant.
34. Reconnect the negative battery cable. Start the engine and check the engine for leaks and proper engine

Markings for correct piston installation — 2.4L (VIN B) engine

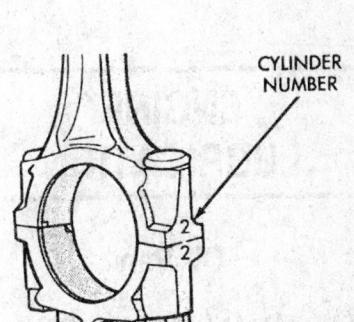

Markings for correct piston installation — 2.5L (VIN K) engine

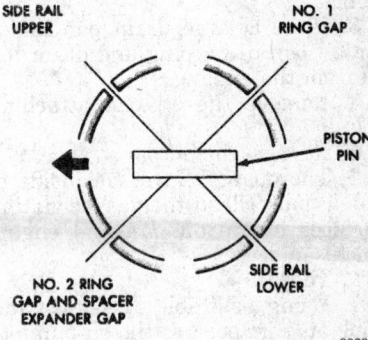

Numbering the connecting rod and bearing cap to the cylinder

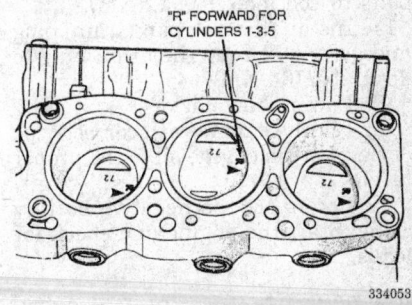

Piston ring end gap positions — all engines

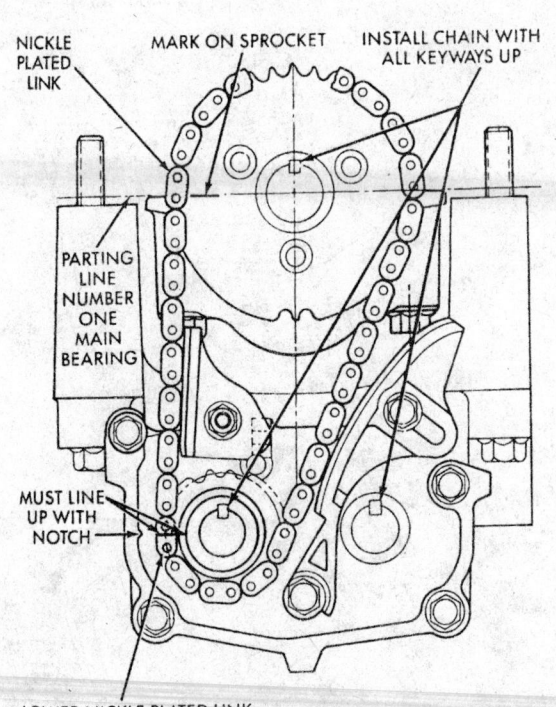

Balance shaft timing — 2.5L (VIN K) and 2.4L (VIN B) engines

Correct position of pistons in the cylinder block — 3.0L (VIN 3) engine

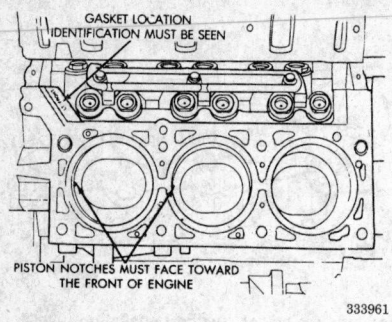

Piston notches must face the front of the engine — 3.3L (VIN R) and 3.8L (VIN L) engines

ENGINE LUBRICATION

Oil Pan

REMOVAL AND INSTALLATION

2.4L (VIN B) Engine

1. Disconnect the negative battery cable.
2. Raise and safely support the vehicle.
3. Place a large drain pan under the oil pan drain plug and drain the oil from the engine.
4. Remove the oil pan attaching bolts.
5. Remove the oil pan.
6. Thoroughly clean the inside of the oil pan. Clean the gasket mating surfaces of the oil pan and engine block.
   To install:
7. Using a suitable gasket sealant, apply a 1/8 in. bead at the oil pump to engine block parting line.
8. Install the new oil pan gasket.
9. Install the oil pan to the engine.
10. Torque the oil pan attaching bolts to 105 inch lbs. (12 Nm).
11. Install the oil pan drain plug and gasket. Torque the drain plug to 25 ft. lbs. (34 Nm).
12. Lower the vehicle.
13. Fill the engine with correct type of clean, fresh engine oil to the proper level.
14. Reconnect the negative battery cable. Start the engine and check for leaks.

2.5L (VIN K) Engine

1. Disconnect the negative battery cable.

2. Raise and safely support the vehicle.
3. Remove the oil pan drain plug and drain the engine oil into a suitable container.
4. Remove the oil pan attaching bolts and remove the oil pan.
   To install:
5. Thoroughly clean the inside of the oil pan. Clean the oil pan and engine block gasket mating surfaces.
6. Apply RTV sealant to the oil pan rail at the front seal retainer parting line.
7. Attach the oil pan side gaskets using RTV to hold the gasket in place.
8. Install the new oil pan seals and apply RTV sealant to the ends of the seals at the junction where the seals and gasket meets.
9. Install the oil pan and tighten the M8 bolts to 17 ft. lbs. (23 Nm), and the M6 bolts to 105 inch lbs. (12 Nm).
10. Install the oil pan drain plug and tighten to 20 ft. lbs. (27 Nm).
11. Lower the vehicle.
12. Refill the engine with the proper type and quantity of engine oil.
13. Connect the negative battery cable.
14. Start the engine and check for leaks. After the engine has been

turned off, check the oil level and top off, if necessary.

3.0L (VIN 3) Engine

1. Disconnect the negative battery cable.
2. Raise and safely support the vehicle.
3. Remove the oil pan drain plug and drain the engine oil into a suitable container.
4. Remove the oil pan attaching bolts and remove the oil pan.
   To install:
5. Thoroughly clean the inside of the oil pan. Clean the oil pan and engine block gasket mating surfaces.
6. Apply RTV sealant to the oil pan.
7. Install the oil pan to the engine and tighten the bolts, in sequence, to 50 inch lbs. (6 Nm).
8. Install the oil pan drain plug and tighten to 30 ft. lbs. (40 Nm).
9. Lower the vehicle.
10. Refill the engine with the proper type and quantity of oil.
11. Connect the negative battery cable.
12. Start the engine and check for leaks. After the engine has been turned off, check the oil level and top off, if necessary.

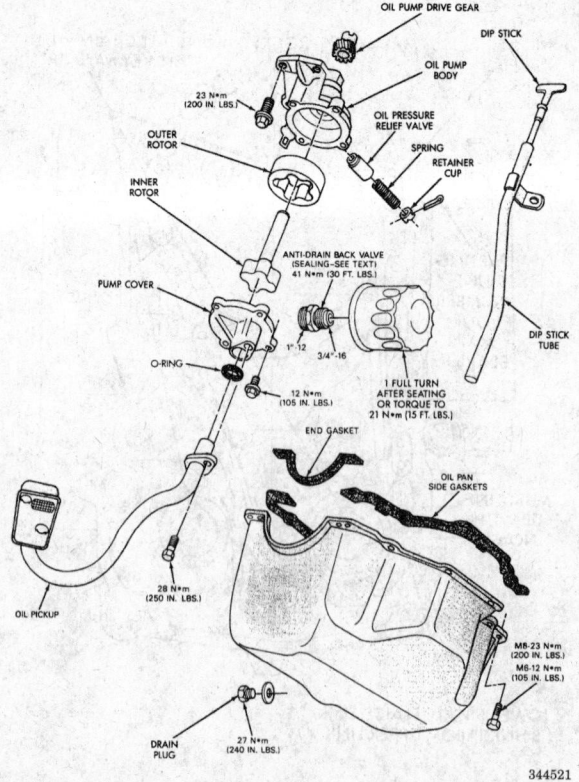

Engine lubrication system components — 2.5L (VIN K) engine

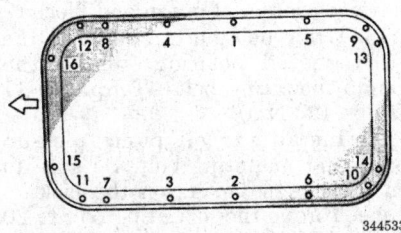

Oil pan bolt torque sequence — 3.0L (VIN 3) engine

3.3L (VIN R) and 3.8L (VIN L) Engines

1. Disconnect the negative battery cable. Remove the engine oil dipstick.
2. Raise the vehicle and support safely.
3. Remove the torque converter bolt access cover, if equipped.
4. Remove the oil pan drain plug and drain the engine oil into a suitable container.
5. Remove the oil pan retaining bolts and remove the oil pan and gasket.

To install:

6. Thoroughly clean the inside of the oil pan. Thoroughly clean and dry all sealing surfaces, bolts and bolt holes.
7. Apply a ⅛ in. bead of silicone sealer to the chain cover-to-block mating seam and the rear main seal retainer-to-block seam.
8. Install a new oil pan gasket and install the oil pan to the engine.
9. Install the retaining bolts and torque to 105 inch lbs. (12 Nm).
10. Install the oil pan drain plug and tighten to 25 ft. lbs. (34 Nm).
11. Install the torque converter bolt access cover, if equipped. Lower the vehicle.
12. Install the oil dipstick. Refill the engine with the proper type and quantity of oil.
13. Reconnect the negative battery cable. Start the engine and check for leaks. After the engine has been turned off, check the oil level and top off, if necessary.

Oil Pump

REMOVAL AND INSTALLATION

2.4L (VIN B) Engine

1. Disconnect the negative battery cable.
2. Raise and safely support the vehicle.

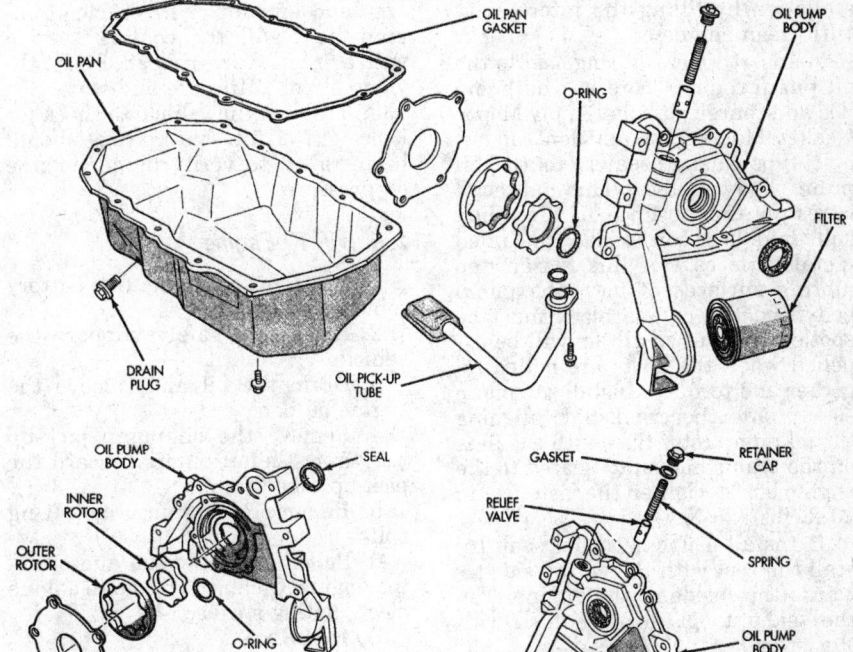

Lubrication system components — 2.4L (VIN B) engine

3. Drain the engine oil and the engine coolant into suitable containers.
4. Lower the vehicle.
5. Remove the accessory drive belts, as required.
6. Take up the weight of the engine with an engine support tool and remove the right engine mount and bracket. Make sure the engine is safely supported.
7. Remove the timing belt cover.
8. Loosen the timing belt tensioner bolts and remove the tensioner and the timing belt.
9. Raise and safely support the vehicle.
10. Remove the oil pan assembly. Remove the oil pump pick-up tube and O-ring.
11. Using a suitable puller, remove the crankshaft damper from the front of the crankshaft.
12. Using a suitable puller, draw the crankshaft sprocket from the front of the crankshaft.
13. Loosen the oil pump bolts and remove. Take note of the location of each bolt for reassembly. Remove the oil pump from the face of the engine block. If necessary, tap lightly with a soft face mallet. Use care working with light alloy parts.
14. Remove the relief valve from the pump body by removing the

threaded plug and gasket and pulling out the spring and relief valve. Note the order of parts removal.

To install:

15. Clean all parts well for inspection. Remove the screws holding the back cover to the pump body. Remove the pump rotors. The mating surface of the oil pump should be smooth. Replace the pump cover if scratched or grooved.
16. The pump should be checked for wear by carefully measuring the components.

NOTE: If oil pressure is low and the pump is within specifications, inspect for worn engine bearings or other reasons for oil pressure loss.

17. Clean all oil pump parts in suitable solvent before assembly. Assemble the pump with new parts as required. **Install the inner rotor with the chamfer facing the cast iron oil pump cover (back of the pump).** Tighten the cover screws to 105 inch lbs. (12 Nm).
18. Install the relief valve first, then the spring, gasket and cover cap into the pump body. Note that installing the spring first will seriously damage the engine. The relief valve goes in first. Tighten the cover cap to 40 ft. lbs. (55 Nm).

19. Prime the oil pump before installation by filling the rotor cavity with clean engine oil.

20. Insert a new oil ring seal to the oil pump counter bore on the pump body discharge passage. Apply Mopar Gasket Maker or equivalent anaerobic type gasket sealer, to the oil pump body flange. This material cures in the absence of air when squeezed between 2 flat machined metal surfaces. For this reason, the mating surfaces of both the pump body and the engine block must be spotlessly clean so all air will be expelled when the parts are bolted together and torqued. Install the pump slowly onto the crankshaft aligning the oil pump rotor flats with the flats on the crankshaft until seated to the engine block. Tighten the fasteners to 20 ft. lbs. (28 Nm).

21. Install a new front oil seal. Install the seal with the spring side towards the inside of the engine. Tap the seal into place until flush with the cover.

22. Install the crankshaft sprocket. A special tool is used to draw the sprocket onto the end of the crankshaft. Use care if using substitutes.

23. Raise and safely support the vehicle.

24. Install the oil pump pick-up tube and O-ring. Torque the oil pump pick-up tube mounting bolt to 20 ft. lbs. (28 Nm).

25. Thoroughly clean the oil pan and make sure the gasket rails are in good condition. Use Mopar Silicone Rubber Adhesive Sealant or equivalent sealer at the oil pump to engine block parting line. Use a new oil pan gasket, install the oil pan and torque the 13 oil pan bolts to 105 inch lbs. (12 Nm).

26. Install a new oil filter.

27. Lower the vehicle.

28. Install the timing belt and covers using the recommended procedures. Use care to make sure all valve timing marks are aligned. This is most important. Failure to properly align the timing marks will result in severe engine damage.

29. Install the crankshaft damper. A special tool making use of a 12mm x 1.75 x 150mm bolt is used to draw the crankshaft damper onto the end of the crankshaft. Use care if using substitutes. Torque the center bolt to 105 ft. lbs. (142 Nm).

30. Install the accessory drive belts, as required. Adjust the accessory drive belts to the proper tension.

31. Install the engine mount and bracket, as required. Remove the engine support fixture from the vehicle.

32. Refill the engine with correct type and amount of fresh, clean engine oil. Refill the cooling system with a $^{50}/_{50}$ mixture of clean, fresh ethylene glycol antifreeze and water.

33. Test run the vehicle to check for leaks. An oil pressure gauge should be installed to verify proper engine oil pressure.

2.5L (VIN K) Engine

1. Disconnect the negative battery cable

2. Raise and safely support the vehicle.

3. Drain the oil and remove the engine oil pan.

4. Remove the oil pump pick-up tube from the oil pump. Discard the pick-up tube O-ring.

5. Remove the pump mounting bolts.

6. Pull the pump down and out of the engine. Clean the pump to engine block mating surface.

To install:

7. Clean all parts well for inspection. Remove the screws holding the pump cover to the pump body. Remove the pump rotors. The mating surface of the oil pump should be smooth. Replace the pump cover if scratched or grooved.

8. The pump should be checked for wear by carefully measuring the components.

NOTE: If oil pressure if low and the pump is within specifications, inspect for worn engine bearings or other reasons for oil pressure loss.

9. Clean all oil pump parts in suitable solvent before assembly. Assemble the pump with new parts as required. **Install the outer rotor with the small chamfered edge facing the oil pump cover.** Tighten the cover screws to 105 inch lbs. (12 Nm).

10. Install the relief valve first, then the spring, cup and cotter key into the pump body. Note that installing the spring first will seriously damage the engine. The relief valve goes in first.

11. Apply gasket sealer to the pump body-to-engine block mating surface.

12. Prime, by filling the pump with fresh oil. Check crankshaft/intermediate shaft timing and oil pump drive alignment. Adjust if necessary.

13. The slot in the oil pump shaft must be positioned parallel to the center line of the crankshaft when the crankshaft and intermediate shaft are timed properly.

14. Install the pump and rotate back and forth slightly to ensure full surface contact of pump and block.

15. While holding the pump in the fully seated position, install the pump mounting bolts. Torque to 17 ft. lbs. (23 Nm).

16. Install the oil pump pick-up tube and mounting bolt. Be sure to install a new O-ring on the pick-up tube. Torque the mounting bolt to 20 ft. lbs. (28 Nm).

17. Install the engine oil pan.

18. Lower the vehicle and connect the negative battery cable

19. Refill the crankcase with the proper type and quantity of clean, fresh engine oil.

20. Start the engine and check for leaks. Check engine oil pressure.

3.0L (VIN 3) Engine

The oil pump assembly is mounted at the front of the crankshaft. The oil pump housing also retains the crankshaft front oil seal. Since the timing belt must be removed to access the front cover, care must be taken. It is good practice to set the engine up to TDC No. 1 cylinder firing position. Verify that all timing marks on the crankshaft and camshaft sprockets are properly aligned before removing the timing belt. This service as a point of reference for all work that follows. Valve timing is most important and engine damage will result if the work is incorrect.

1. Disconnect the negative battery cable.

2. Remove the accessory drive belts.

3. Remove the air conditioning compressor from the mounting bracket and lay it aside, if equipped. Remove the compressor mounting bracket and adjustable drive belt tensioner from the engine.

4. Remove the power steering pump/alternator belt tensioner mounting bolt and remove the tensioner.

5. Remove the power steering pump mounting bracket bolts, rear support lock nut and set the power steering pump aside.

6. Raise and safely support the vehicle.

7. Drain the engine oil into a suitable container.

8. Remove the right inner fender inner splash shield.

9. Remove the crankshaft drive pulley bolt, drive pulley and torsional damper. Lower the vehicle.

10. Place a floor jack under the engine. Separate the engine mount in-

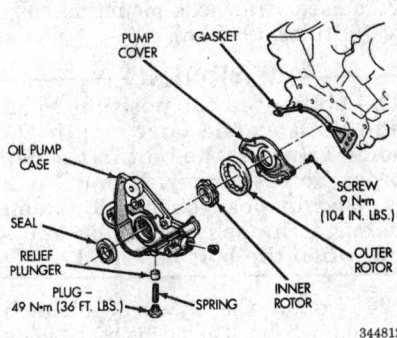

PUMP COVER — GASKET — OIL PUMP CASE — SEAL — RELIEF PLUNGER — PLUG — 49 N•m (36 FT. LBS.) — SPRING — INNER ROTOR — OUTER ROTOR — SCREW 9 N•m (104 IN. LBS.)

344812

Oil pump assembly — 3.0L (VIN 3) engine

sulator from the engine mount bracket.

11. Raise the engine slightly and remove the engine mount bracket.

12. Remove the timing belt cover and timing belt.

13. Remove the crankshaft sprocket.

14. Remove the oil pump mounting bolts (5), and remove the oil pump assembly. Mark the mounting bolts for proper installation during reassembly.

15. Remove the front crankshaft oil seal from the oil pump cover.

To install:

16. Clean all parts well for inspection. Remove the screws holding the rear cover to the oil pump body. Remove the pump rotors. The mating surface of the oil pump should be smooth. Replace the pump cover if scratched or grooved.

17. The pump should be checked for wear by carefully measuring the components.

NOTE: If oil pressure is low and the pump is within specifications, inspect for worn engine bearings or other reasons for oil pressure loss.

18. Clean all oil pump parts in suitable solvent before assembly. Assemble the pump with new parts as required.

19. Clean the oil pump and engine block gasket surfaces thoroughly.

20. Position a new gasket on the pump assembly and install on the cylinder block. Make sure the correct length bolts are in their proper locations and torque all bolts to 11 ft. lbs. (15 Nm).

21. Install a new front crankshaft oil seal into the oil pump using seal installer tool MD-998717 or equivalent seal driver tool.

22. Install the crankshaft sprocket and timing belt. Recheck engine timing marks.

23. Install the timing belt covers. Torque the timing belt cover fasteners to 10 ft. lbs. (14 Nm).

24. Raise the engine slightly and install the engine mount bracket. Install the engine mount insulator into the engine mount bracket.

25. Install the torsional damper, crankshaft drive pulley and drive pulley bolt. Torque to 112 ft. lbs. (151 Nm).

26. Install the right fender inner splash shield. Lower the vehicle.

27. Install the power steering mount bracket and power steering pump.

28. Install the power steering/alternator belt tensioner.

29. Install the air conditioning adjustable drive belt tensioner and mounting bracket. Torque the mounting bolts to 40 ft. lbs. (54 Nm).

30. Install the air conditioning compressor. Torque the compressor mounting bolts to 40 ft. lbs. (54 Nm).

31. Install the accessory drive belts and adjust to the proper tension.

32. Reconnect the negative battery cable.

33. Refill the crankcase with the correct amount of clean engine oil and start the engine. Check for leaks.

34. Check engine oil pressure.

3.3L (VIN R) and 3.8L (VIN L) Engines

1. Disconnect the negative battery cable. Remove the dipstick. Drain the cooling system.

2. Raise the vehicle and support safely. Support the engine and remove the right side engine mount.

3. Drain the oil and remove the oil pan.

4. Remove the oil pickup tube. Remove the transaxle inspection cover, if necessary.

5. Remove the timing chain case cover.

6. Disassemble the oil pump as required.

To install:

7. Clean all parts well for inspection. Remove the screws holding the back cover to the pump body. Remove the pump rotors. The mating surface of the oil pump should be smooth. Replace the pump cover if scratched or grooved.

8. The pump should be checked for wear by carefully measuring the components.

NOTE: If oil pressure is low and the pump is within specifications, inspect for worn engine bearings or other reasons for oil pressure loss.

9. Clean all oil pump parts in suitable solvent before assembly. Assemble the pump with new parts as required. **Install the inner rotor with the chamfer facing the cast iron oil pump cover (back of the pump).** Tighten the cover screws to 105 inch lbs. (12 Nm).

10. Install the relief valve first, then the spring, then the cover cap into the pump body. Note that installing the spring first will seriously damage the engine. The relief valve goes in first.

11. Prime the oil pump by filling the rotor cavity with fresh oil and turning the rotors until oil comes out the pressure port. Repeat a few times until no air bubbles are present.

12. Install the chain case cover. Torque the timing chain case cover screws as follows:
   a. M8 x 1.25 – 20 ft. lbs. (27 Nm)
   b. M10 x 1.5 – 40 ft. lbs. (54 Nm)

13. Clean out the oil pickup or replace as required. Replace the oil pickup O-ring and install the pickup to the pump. Torque the pick-up tube retaining screw to 250 inch lbs. (28 Nm).

14. Install the oil pan. Install the right side engine mount.

15. Lower the vehicle. Install the dipstick. Refill the engine with the proper amount of oil.

16. Refill the cooling system to the proper level.

17. Reconnect the negative battery cable and start the engine.

18. Check the area for leaks and check the oil pressure.

TRANSAXLE

Manual Transaxle Assembly

REMOVAL AND INSTALLATION

1993–94

NOTE: Transaxle removal does not require engine removal.

1. Disconnect the negative battery cable from the battery.

2. Install a sling or lifting bracket to the engine. Place an engine support device across the engine compartment and connect to the sling. Tighten until slight upward pressure is applied to the engine.

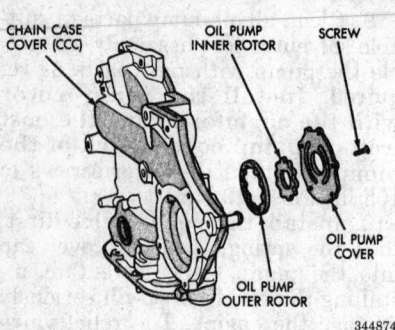

CHAIN CASE
COVER (CCC) OIL PUMP
INNER ROTOR SCREW

OIL PUMP
COVER

OIL PUMP
OUTER ROTOR

344874

Oil pump assembly — 3.3L (VIN R) and 3.8L
(VIN L) engines

3. Disconnect the gearshift cables from the transaxle selector lever. Remove the shifter cables bracket at the transaxle. Disconnect the speedometer.

4. Raise and safely support the vehicle.

5. Remove both front wheels. Remove the left front engine splash shield.

6. Remove the left engine mount from the transaxle.

7. Remove the anti-rotational link (also called anti-hop damper) from the crossmember bracket, if equipped. Do not remove the bracket from the transaxle.

8. Remove both lower ball joint-to-steering knuckle mounting bolts. Using a puller to prevent damage, remove the ball joints from the steering knuckles. Remove right and left side halfshaft assemblies. Drain the fluid from the transaxle.

9. Remove the bell housing inspection cover.

10. Remove the speedometer cable adapter and pinion from the transaxle.

11. Disconnect the front sway bar.

12. Remove the back-up light switch connector.

13. Remove the engine mount bracket from the front crossmember.

14. Remove the front mount insulator through bolt. Place a suitable floor jack or transmission jack under the transaxle and raise to gently support.

15. Remove the drive plate (flywheel)-to-clutch assembly bolts. The heads of the bolts face the engine and the bolts thread into the clutch pressure plate assembly, the reverse of the usual practice on earlier rear-drive type vehicles.

16. Remove the top bell housing bolts. Note that the bolts may be of different lengths. Mark each bolt and its location so that it may be returned to its original location. On some tran-saxles, the wrong bolt (too long) can damage the selector shaft housing.

17. Remove the left engine mount at the rear cover plate. Remove the starter motor. Do not allow the starter motor to hang by the starter wires.

18. When removing or installing the transaxle, it may be helpful to use 2 locating pins in place of the top 2 transaxle-to-engine block bolts. Make 2 locating pins from extra same thread bolts that are slightly longer than the mounting bolts. Cut the heads off with a hacksaw, remove any burrs or sharp edges with a file or grinder. Install the 2 guide bolts into the top bellhousing bolt holes and thread into the rear of the engine.

—————— CAUTION ——————
Use care when pulling the transaxle back away from the engine. Since the flywheel to clutch assembly bolts have already been removed, the entire clutch assembly may slide off the input shaft when removing the transaxle from the vehicle. Damage to the clutch assembly and/or personal injury may occur if the clutch assembly falls out of the bellhousing during transaxle removal.

19. Secure the transaxle to a transmission jack and remove the remaining lower bellhousing bolts. Check that all transaxle support mounts and bolts are removed. Slide the jack and transaxle away from the engine and lower assembly. Use care. Do not pry of the light alloy case.

To install:

20. Clean all parts well. Inspect the clutch assembly and service as required. Inspect for oil leakage through the rear main bearing seal as well as the transaxle drive pinion seal. If leakage is noted, it should be corrected at this time.

21. Inspect the throwout bearing. It should turn smoothly. It is prelubricated and sealed and should not be immersed in solvent. The plastic sleeve should be wiped clean. Refill the cavities with multi-purpose grease. Also lubricate the thrust pads on the release fork.

22. The locating pins made from spare bolts should still be in place. Install the transaxle assembly into position, aligning with the guide pins, and position the transaxle against the engine block.

23. After the transaxle is in position, remove the guide bolts and install the mounting bolts before removing the transmission jack. Torque the transaxle mounting bolts to 70 ft. lbs. (95 Nm).

—————— WARNING ——————
The bolts used for position No. 1 and No. 3 are the same length. On some vehicles, the bolt in the No. 2 position is longer. If bolt No. 2 is used in position No. 3, it can damage the selector shaft housing when the bolt is seated.

24. Install the flywheel-to-clutch assembly bolts. The heads of the bolts face the engine and the bolt threads into the clutch pressure plate assembly. Torque the bolts to 250 inch lbs. (28 Nm).

25. Install the starter motor.

26. Install the left engine mount at the rear cover plate.

27. Install the front mount insulator through bolt.

28. Install the engine mount bracket to the front crossmember.

29. Reconnect the back-up light switch connector.

30. Reconnect the anti-rotational link to the crossmember bracket and reconnect the front sway bar.

31. Install the speedometer cable adapter and pinion to the transaxle.

32. Install the left front engine mount.

33. Install the bell housing inspection cover.

34. Install the halfshaft assemblies and secure the steering knuckle to the lower ball joint.

35. Refill the transaxle to the bottom of the transaxle fill hole with the proper amount of SG or SG-CD SAE 5W-30 engine oil.

36. Install the left front engine splash shield.

37. Install the front wheels and lug nuts. Torque the lug nuts to 95 ft. lbs. (129 Nm).

38. Lower the vehicle.

39. Install the shifter cable bracket at the transaxle and reconnect the gearshift cables to the transaxle selector lever.

40. Remove the engine support bracket from the top of the engine.

41. Reconnect the negative battery cable. Road test the vehicle.

Clutch Assembly

REMOVAL AND INSTALLATION

1993–94

NOTE: The clutch disc and pressure plate are serviced as an assembly. Do NOT disassemble the clutch assembly.

1. Disconnect the negative battery cable.

2. Raise and safely support the vehicle.

3. Remove the bell housing access cover.

4. Matchmark the flywheel to the clutch assembly for installation reference. Remove the flywheel-to-clutch assembly bolts.

5. Remove the transaxle assembly.

6. Remove the the clutch assembly by sliding it out of the transaxle bellhousing.

To install:

7. Inspect for oil leakage through the engine rear main bearing seal and transaxle drive pinion seal. If there is leakage, it should be corrected at this time. Inspect the flywheel for damaged teeth. The clutch release bearing is sealed and should not be washed in solvent. The bearing should turn smoothly when held in the hand under a light thrust load. If the bearing is noisy, rough or dry replace the bearing assembly.

8. If required, assemble the clutch fork to the bearing by sliding the thrust pads under the spring clips. Use care not to distort the spring clips. These clips prevent the thrust plate from rotating with the bearing. A small amount of high-temperature grease may be placed between the release shaft bushing and the shaft. Install the throw out bearing, fork and component parts.

9. Install the clutch assembly into the transaxle bellhousing.

10. Install the transaxle into the vehicle. Torque the transaxle case-to-engine block bolts to 70 ft. lbs. (95 Nm).

11. Align the clutch assembly-to-flywheel matchmarks that were made during the removal procedure. Install the clutch assembly-to-flywheel bolts. Tighten the bolts a few turns at a time in rotation. New bolts are recommended. Torque the bolts to 21 ft. lbs. (28 Nm).

12. Install the bellhousing access cover. Check to be sure all components have been properly reinstalled, tightened and/or reconnected.

13. Reconnect the negative battery cable. Road test vehicle for proper shift operation.

Clutch Cable

REMOVAL AND INSTALLATION

1993–94

The manual transaxle clutch release system has a self-adjusting mechanism to compensate for clutch wear. The adjuster mechanism is located within the clutch pedal. The preload spring maintains tension on the cable. The tension keeps the clutch release bearing continuously loaded against the fingers of the clutch cover assembly. When the pedal is depressed, teeth on the adjuster and the positioner engage and pull the release cable. A spring located behind the adjuster ensures proper tooth engagement. When the pedal is released, the adjuster contacts the bumper. This separates the adjuster and positioner teeth allowing the preload spring to function.

The mechanism which automatically adjusts the clutch may emit a load clicking or pop noise. The cause of this noise in most cases is the clutch cable auto-adjust spring being below design load specifications. To correct this condition, either bend the auto-adjust spring to bring it back to specifications or replace the spring. The auto-adjust spring is located on the back of the clutch pedal.

1. Disconnect the negative battery cable.

2. Remove the clutch cable retainer from the clutch release lever on the transaxle by pulling on the end of the ball stud.

3. Pry out the ball end of the clutch cable from the positioner adjuster on the clutch pedal assembly and remove the clutch cable.

4. Pass the clutch cable through the hoop in the mounting bracket on the strut tower.

5. Inspect the condition of the clutch cable and replace if worn or damaged. Do not lubricate the cable.

6. Inspect the clutch pedal and adjuster mechanism for wear and apply a multipurpose lubricant.

To install:

7. Install the clutch cable by routing it through the hoop in the strut tower mounting bracket.

8. Connect the ball end of the clutch cable to the positioner adjuster in the clutch pedal assembly.

9. Connect the transaxle end of the clutch cable to the clutch release lever on the transaxle assembly.

10. Secure the clutch cable to the clutch release lever on the transaxle by installing the clutch cable retainer.

11. After installation, press and lift the clutch pedal several times to allow the adjuster mechanism to function.

12. Connect the negative battery cable.

13. Check the clutch pedal position switch for proper operation.

Clutch Cable

ADJUSTMENT

1993–94

The manual transaxle clutch release system has a unique self-adjusting mechanism to compensate for clutch disc wear. This adjuster mechanism is located with the clutch cable assembly. The preload spring maintains tension on the cable. This tension keeps the clutch release bearing continuously loaded against the fingers of the clutch cover assembly. There is no clutch freeplay adjustment.

Automatic Transaxle Assembly

REMOVAL AND INSTALLATION

1993–95

The transaxle combines the clutch-type torque converter, 3 or 4 speed transmission, final drive gearing and differential into a front wheel drive,

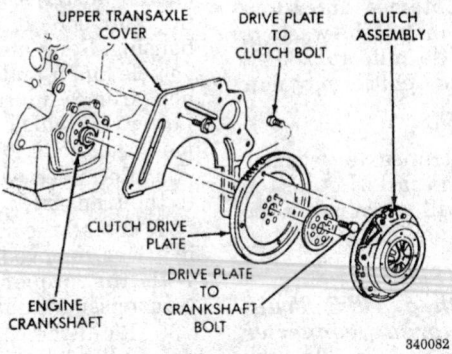

UPPER TRANSAXLE COVER

DRIVE PLATE TO CLUTCH BOLT

CLUTCH ASSEMBLY

CLUTCH DRIVE PLATE

DRIVE PLATE TO CRANKSHAFT BOLT

ENGINE CRANKSHAFT

340082

Clutch assembly — 1993–94

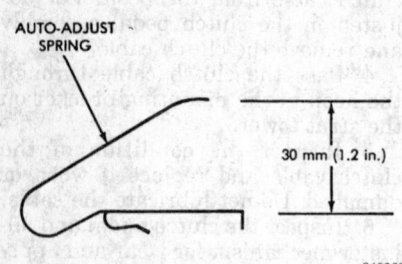

AUTO-ADJUST SPRING

30 mm (1.2 in.)

340069

Clutch cable auto-adjust spring

or depending on option, an all-wheel drive system. The torque converter, transaxle area and differential are housed in an integral aluminum die casting. The differential oil sump is common with the transaxle sump. Separate filling of the differential is not necessary. The engine should be running at idle speed for at least one minute with the vehicle on level ground before checking fluid level with the dipstick. It is important that the proper fluid be used. Chrysler recommends their Mopar ATF Plus Type 7176. Dexron II is not recommended except if Mopar ATF Plus is unavailable. Shudder or shift quality problems may be encountered if using fluid other than Mopar ATF Plus. Chrysler recommends a band adjustment and filter change be performed at the time of fluid change. The magnet inside the oil pan should be cleaned at filter change.

NOTE: If the vehicle is going to be rolled on its own wheels while the transaxle is out of the vehicle, obtain 2 outer CV-joints to install to the hubs. If the vehicle is rolled without the proper torque applied to the front wheel bearings, the bearings will be destroyed.

1. Disconnect the negative battery cable. If equipped with the 3.0L (VIN 3), 3.3L (VIN R) or 3.8L engines, drain the coolant. Remove the dipstick.
2. Use an engine support fixture to support the engine.
3. Remove the air cleaner assembly if preventing access to the upper bell housing bolts. Remove the upper bell housing bolts and water tube, where applicable. Unplug and label all electrical connectors from the transaxle.
4. Disconnect the shifter and kickdown linkages, if equipped.
5. If equipped with 2.5L (VIN K) engine, remove the starter attaching

nut and bolt at the top of the bell housing.
6. Raise and safely support the vehicle. Remove the front wheels. Remove the halfshaft assemblies.
7. Position a drain pan under the transaxle where the halfshafts enter the differential or extension housing.
8. Drain the transaxle. Disconnect and plug the fluid cooler hoses.
9. Remove the speedometer cable adaptor bolt and remove the adaptor from the transaxle.
10. Remove the starter. Remove the torque converter inspection cover, matchmark the torque converter to the flex plate and remove the torque converter bolts.
11. Using the proper equipment, support the weight of the engine.
12. Remove the front motor mount and bracket.
13. If equipped with D.I.S. ignition system, remove the crankshaft position sensor from the bell housing.
14. Position a transmission jack under the transaxle.
15. Remove the lower bell housing bolts.
16. Remove the left side splash shield. Remove the transaxle mount bolts.
17. Carefully pry the engine from the transaxle.
18. Slide the transaxle rearward until the locating dowels disengage from the mating holes in the transaxle.

NOTE: Attach a small C-clamp to the edge of the bell housing. This will hold the torque converter in place during transaxle removal.

19. Pull the transaxle completely away from the engine and carefully lower it from the vehicle.
20. To prepare the vehicle for rolling, support the engine with a suitable support or reinstall the front motor mount to the engine. Then reinstall the ball joints to the steering knuckle and install the retaining bolt. Install the obtained outer CV-joints to the hubs, install the washers and torque the axle nuts to 180 ft. lbs. (244 Nm). The vehicle may now be safely rolled.

To install:

21. Install the transaxle securely on the transmission jack. Rotate the converter so it will align with the positioning of the flex plate.

CAUTION

If equipped with a 41TE Transaxle, and the torque converter has been replaced, a Torque Clutch Break-in Procedure must

be performed. This procedure will reset the transaxle control module break-in status. Failure to perform this procedure may cause transaxle shutter. To properly do this a DRB or equivalent scan tool, is required to read or reset the break-in status.

22. Apply a coating of high temperature grease to the torque converter pilot hub.
23. Raise the transaxle into place and push it forward until the dowels engage and the bell housing is flush with the block.
24. Install the transaxle bell housing bolts.
25. Jack the transaxle up and install the left side mount bolts. Install the torque converter bolts and torque to 55 ft. lbs. (75 Nm). Install the torque converter inspection cover.
26. Install the front motor mount and bracket. Remove the engine and transaxle support fixtures.
27. Install the starter to the transaxle. Install the bolt finger tight if equipped with a 2.5L (VIN K) engine.
28. Install a new O-ring to the speedometer cable adaptor and install to the extension housing; make sure it snaps in place. Install the retaining bolt and torque to 60 inch lbs. (7 Nm).
29. Reconnect the shifter and kickdown linkage to the transaxle, if equipped.
30. Install the halfshaft assemblies and center bearing, if equipped. Install the ball joints to the steering knuckles. Torque the axle nuts to 180 ft. lbs. (244 Nm) and install new cotter pins. Install the splash shield and install the wheels. Lower the vehicle. Install the dipstick.
31. Install the upper bell housing bolts and water pipe, if removed. Torque the transaxle bell housing mounting bolts to 70 ft. lbs. (95 Nm).
32. If equipped with 2.5L (VIN K) engine, install the starter attaching nut and bolt at the top of the bell housing. Raise and safely support the vehicle again and tighten the starter bolt from underneath the vehicle. Lower the vehicle. Torque the starter mounting bolts to 40 ft. lbs. (54 Nm).
33. Reconnect all electrical wiring to the transaxle.
34. Install the air cleaner assembly, if it was removed. Fill the transaxle with the proper amount of clean transmission fluid.
35. Refill the cooling system to the correct level with a 50/50 mix of clean coolant and water, if necessary.

36. Check to make sure all linkages, electrical connectors and fluid lines have been reconnected properly.

37. Reconnect the negative battery cable and check the transaxle for proper operation. On the A-604 transaxle, perform the upshift and kickdown learn procedure.

1996–97

Model 41TE Transaxle

The transaxle combines the clutch-type torque converter 4 speed transmission, final drive gearing and differential into a front wheel drive system. The torque converter, transaxle area and differential are housed in an integral aluminum die casting. The differential oil sump is common with the transaxle sump. Separate filling of the differential is not necessary. The engine should be running at idle speed for at least one minute with the vehicle on level ground before checking fluid level with the dipstick. It is important that the proper fluid be used. Chrysler recommends their Mopar ATF Plus Type 7176. Dexron II is not recommended except if Mopar ATF Plus is unavailable. Shudder or shift quality problems may be encountered if using fluid other than Mopar ATF Plus. Chrysler recommends a filter change be performed at the time of fluid change. The magnet inside the oil pan should be cleaned at filter change.

NOTE: If the vehicle is going to be rolled on its own wheels while the transaxle is out of the vehicle, obtain 2 outer CV-joints to install to the hubs. If the vehicle is rolled without the proper torque applied to the front wheel bearings, the bearings will be destroyed.

1. Disconnect the negative battery cable. If equipped with the 3.0L (VIN 3) engine, drain the coolant.

2. Use an engine support fixture to support the engine.

3. Remove the air cleaner assembly if preventing access to the upper bell housing bolts.

4. Disconnect the transaxle shift linkage at the manual valve lever.

5. Squeeze grommet clips and disconnect cable at the transaxle bracket.

6. Remove the transaxle oil dipstick tube.

7. Disconnect and plug the transaxle fluid cooler lines.

8. Remove the input and output speed sensors.

9. Remove the upper bell housing mounting bolts.

10. Raise and safely support the vehicle. Remove the front wheels.

11. Position a drain pan under the transaxle where the halfshafts enter the differential or extension housing. Remove the right and left halfshaft assemblies.

12. Drain the transmission fluid.

13. Remove the torque converter dust shield (inspection cover), match-mark the torque converter to the flex plate and rotate the engine clockwise to remove the torque converter bolts.

14. Disconnect the wiring harness connections to the transmission range switch and the Park/Neutral Position Switch.

15. Remove the front motor mount insulator and bracket.

16. If equipped with D.I.S. ignition system, remove the crankshaft position sensor from the bell housing.

17. Remove the starter motor mounting bolts and set the starter motor aside. Do not allow the starter motor to hang suspended from the battery cable.

18. Position a transmission jack under the transaxle.

19. With the transaxle mount firmly in position, remove the left transaxle mount.

20. Remove the lower bell housing bolts.

21. Pull the transaxle completely away from the engine and carefully lower it from the vehicle.

22. To prepare the vehicle for rolling, support the engine with a suitable support or reinstall the front motor mount to the engine. Then reinstall the ball joints to the steering knuckle and install the retaining bolt. Install the obtained outer CV-joints to the hubs, install the washers and torque the axle nuts to 180 ft. lbs. (244 Nm). The vehicle may now be safely rolled.

To install:

23. Install the transaxle securely on the transmission jack. Rotate the converter so it will align with the positioning of the flex plate.

——— WARNING ———
If the torque converter has been replaced, a Torque Clutch Break-in Procedure must be performed. This procedure will reset the transaxle control module break-in status. Failure to perform this procedure may cause transaxle shutter. To properly do this, a DRB or equivalent scan tool, is required to read or reset the break-in status.

24. Apply a coating of high temperature grease to the torque converter pilot hub.

25. Raise the transaxle into place and push it forward until the dowels engage and the bell housing is flush with the block.

26. Install the lower transaxle bell housing bolts.

27. Jack the transaxle up and install the left transaxle mount.

28. Install the starter to the transaxle. Torque the starter motor mounting bolts to 40 ft. lbs. (54 Nm).

29. Remove the transaxle jack from under the vehicle.

30. If equipped with D.I.S. ignition system, clean off the old spacer on the crankshaft position sensor and install a new spacer. Install the crankshaft position sensor to the transaxle bell housing and push down until contact is made with the drive plate. Torque the sensor retaining bolts to 105 inch lbs. (12 Nm).

31. Install the front engine mount insulator and bracket.

32. Reconnect the wiring harness connectors to the Park/Neutral Position Switch and the transmission range switch.

33. Align the torque converter to the flex plate mounting bolt holes. Install the torque converter bolts and torque to 55 ft. lbs. (75 Nm). Install the torque converter inspection cover.

34. Install the right and left halfshaft assemblies. Install the ball joints to the steering knuckles. Torque the axle nuts to 180 ft. lbs. (244 Nm) and install new cotter pins.

35. Install the front wheels and lug nuts. Torque the lug nuts, in a star pattern sequence, to 95 ft. lbs. (129 Nm).

36. Lower the vehicle.

37. Install the upper bell housing bolts. Torque the transaxle bell housing mounting bolts to 70 ft. lbs. (95 Nm).

38. Remove the engine support fixture.

39. Install the input and output speed sensors.

40. Reconnect the transaxle oil cooler lines.

41. Install the transaxle oil dipstick tube.

42. Attach the shift cable to the transaxle bracket.

43. Reconnect the transaxle shift linkage to the manual valve lever.

44. Install the air cleaner assembly. Fill the transaxle with the proper amount of clean, fresh MOPAR® ATF Plus 7176 automatic transmission fluid.

45. If equipped with the 3.0L (VIN 3) engine, refill the cooling system to the correct level with a 50/50 mix of clean, fresh ethylene glycol antifreeze and water. Bleed the cooling system.

46. Check to make sure all linkages, electrical connectors and fluid lines have been reconnected properly.

47. Reconnect the negative battery cable and check the transaxle for proper operation. Perform the transaxle quick learn and torque converter clutch break-in procedures.

Model 31TH Transaxle

The transaxle combines the clutch-type torque converter, 3 speed transmission, final drive gearing and differential into a front wheel drive system. The torque converter, transaxle area and differential are housed in an integral aluminum die casting. The differential oil sump is common with the transaxle sump. Separate filling of the differential is not necessary. The engine should be running at idle speed for at least one minute with the vehicle on level ground before checking fluid level with the dipstick. It is important that the proper fluid be used. Chrysler recommends their Mopar ATF Plus Type 7176. Dexron II is not recommended except if Mopar ATF Plus is unavailable. Shudder or shift quality problems may be encountered if using fluid other than Mopar ATF Plus. Chrysler recommends a band adjustment and filter change be performed at the time of fluid change. The magnet inside the oil pan should be cleaned at filter change.

NOTE: If the vehicle is going to be rolled on its own wheels while the transaxle is out of the vehicle, obtain 2 outer CV-joints to install to the hubs. If the vehicle is rolled without the proper torque applied to the front wheel bearings, the bearings will be destroyed.

1. Disconnect the negative battery cable.

2. Remove the entire air cleaner assembly and hoses.

3. Disconnect the shift linkage and throttle linkage from the transaxle.

4. Disconnect the torque converter clutch connector, located near the transaxle fluid dipstick.

5. Disconnect the wiring harness connector from the gear position switch.

6. Disconnect and plug the transaxle fluid cooler lines.

7. Use an engine support fixture to support the engine.

8. Remove the upper bell housing bolts.

9. Raise and safely support the vehicle. Remove the front wheels.

10. Position a drain pan under the transaxle where the halfshafts enter the differential or extension housing. Remove the halfshaft assemblies.

11. Remove the torque converter inspection cover, matchmark the torque converter to the flex plate and remove the torque converter mounting bolts.

12. Remove the front crossmember engine mount bracket.

13. Remove the front mount insulator through-bolt and transaxle belhousing bolts.

14. Position a transmission jack under the transaxle.

15. Remove the left engine mount through-bolt and left engine mount from the transaxle.

16. Remove the starter.

17. Remove the lower bell housing bolts.

18. Carefully pry the transaxle and torque converter assembly rearward off the engine block dowels and disengage the torque converter hub from the end of the crankshaft.

NOTE: Attach a small C-clamp to the edge of the bell housing. This will hold the torque converter in place during transaxle removal.

19. Pull the transaxle completely away from the engine and carefully lower it from the vehicle.

20. Remove the torque converter assembly from the transaxle by removing the C-clamp from the edge of the bellhousing. The torque converter can now be slipped out of the transaxle.

21. To prepare the vehicle for rolling, support the engine with a suitable support or reinstall the front motor mount to the engine. Then reinstall the ball joints to the steering knuckle and install the retaining bolt. Install the obtained outer CV-joints to the hubs, install the washers and torque the axle nuts to 180 ft. lbs. (244 Nm). The vehicle may now be safely rolled.

To install:

22. If installing the torque converter assembly into the transaxle, be sure to align the pump inner gear pilot flats with the torque converter impeller hub flats.

23. Install the transmission securely on the transmission jack. Rotate the converter so it will align with the positioning of the flex plate.

24. Apply a coating of high temperature grease to the torque converter pilot hub.

25. Raise the transaxle into place and push it forward until the dowels engage and the bell housing is flush with the engine block.

26. Install the lower transaxle bell housing bolts.

27. Install the starter motor assembly.

28. Jack the transaxle up and install the left side engine mount and engine mount through-bolt. Torque the left motor mount bolts 40 ft. lbs. (54 Nm).

29. Install the front crossmember engine mount bracket and front mount insulator through-bolt. Torque the bolts to 40 ft. lbs. (54 Nm).

30. Torque the lower bell housing cover bolts to 105 inch lbs. (12 Nm) and the lower bell housing cover screws to 30 ft. lbs. (41 Nm).

31. Install the torque converter bolts and torque to 50 ft. lbs. (68 Nm). Install the torque converter inspection cover.

32. Install the right and left half-shaft assemblies. Install the ball joints to the steering knuckles.

33. Install the front wheels and lug nuts. Torque the lug nuts, in a star pattern sequence, to 95 ft. lbs. (129 Nm).

34. Lower the vehicle to the ground.

35. Install the upper transaxle bell housing bolts and torque the bolts to 105 inch lbs. (12 Nm).

36. Remove the engine support fixture from the vehicle.

37. Reconnect the transaxle fluid cooler lines to the transaxle.

38. Reconnect the gear position switch connector and the torque converter clutch connector.

39. Reconnect the shifter and throttle linkages to the transaxle.

40. Install the air cleaner and hose assembly.

41. Refill the transaxle with the proper amount of clean, fresh MOPAR® ATF Plus Type 7176 automatic transaxle fluid.

42. Check to make sure all linkages, electrical connectors and fluid lines have been reconnected properly. Adjust the gearshift and throttle cables.

43. Reconnect the negative battery cable and check the transaxle for proper operation.

DRIVE AXLE

Driveshaft

REMOVAL AND INSTALLATION

1993–95

All Wheel Drive Vehicles

1. Ensure the vehicle is in **N**. Disconnect the negative battery cable.
2. Raise and safely support the vehicle.

WARNING

Do not allow the end of the driveshaft to hang free during removal or installation or damage to the driveshaft could result.

3. Remove the rear mounting bolts of the driveshaft and support the end of the driveshaft.
4. Remove the mounting bolts for the torque tube bracket.
5. Remove the front mounting bolts of the driveshaft and remove the driveshaft assembly from the vehicle. Be sure to clean the flange mating surfaces before installing the driveshaft.

To install:

6. Install the front part of the driveshaft and the mounting bolts. Support the rear part of the driveshaft while securing the front of the driveshaft. Torque the driveshaft front mounting bolts to 20 ft. lbs. (28 Nm).
7. Install the rear of the driveshaft into position and install the mounting bolts. Torque the driveshaft rear mounting bolts to 20 ft. lbs. (28 Nm).
8. Install the torque tube bracket mounting bolts. Torque the bracket bolts to 40 ft. lbs. (54 Nm).
9. Lower the vehicle and reconnect the negative battery cable. Road test the vehicle.

Halfshaft

REMOVAL AND INSTALLATION

The halfshaft assemblies are 3 piece units. Each halfshaft has a "Tripod" joint, an interconnecting shaft and a "Rzeppa" joint. The "Tripod" joint is splined into the transaxle side gear and the "Rzeppa" joint has a stub shaft that is splined into the wheel hub.

Procedures for the removal and installation of the halfshafts are essentially the same for all front wheel drive vehicles. Please note that boot sealing is vital to retain the special lubricants and to prevent dirt and water from entering the joint. Use care when handling a halfshaft. Do not allow the assemblies to dangle unsupported. Pulling or pushing on the ends can cut the boots and damage the CV-joints. Always support both ends of the halfshaft to prevent damage.

1993–95

Front Halfshaft

1. Disconnect the negative battery cable.
2. Remove the cotter pin from the end of the stub axle. Remove the nut lock and spring washer. Loosen the axle nut and washer with the vehicle still on the ground.
3. Raise and safely support the vehicle. Remove the axle nut and washer.
4. Remove the wheel.
5. If removing the right halfshaft, remove the speedometer pinion retaining bolt and remove the pinion from the transaxle.
6. Remove the ball joint retaining bolt and pry the control arm down to release the ball stud from the steering knuckle.

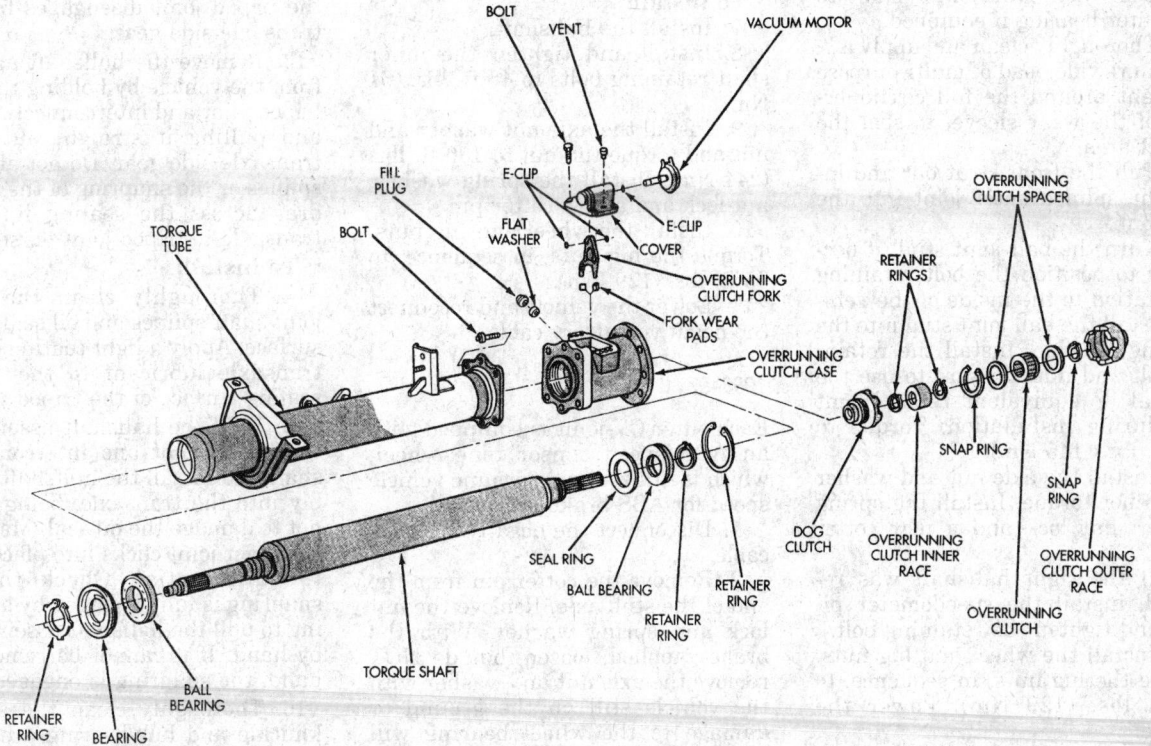

Driveshaft, torque tube and overrunning clutch assembly — All Wheel Drive

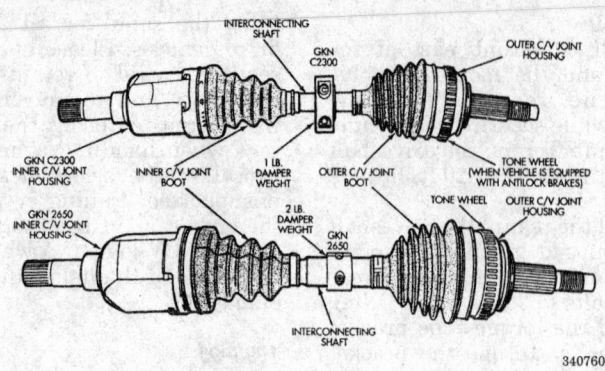

Halfshaft identification — 1993–95

7. Separate the outer CV-joint splined shaft from the hub and bearing assembly by holding the CV-joint housing and pulling the steering knuckle away. Be careful not to pry or damage the outer CV-joint wear sleeve.

8. Support the halfshaft assembly at the CV-joint housing. Remove the halfshaft assembly by by pulling outward on the inner CV-joint housing. Do not pull on the shaft.

To install:

9. Install the halfshaft assembly to the transaxle, being careful not to damage the seal and wear sleeves. Make sure the inner joint clicks into place inside the differential. Install the center bearing retaining bolts, if equipped. Install the outer shaft to the center bearing if equipped.

10. Thoroughly clean and apply a ¼ in. (6mm) wide bead of multi-purpose lubricant around the full circumference of the wear sleeve, to seal the contact area.

11. Pull the front strut out and insert the splined outer joint into the front hub.

12. Turn the ball joint stud, if necessary to position the bolt retaining indentation to the inside of the vehicle. Install the ball joint stud into the steering knuckle. Install the retaining bolt and nut. Be sure to use the original or equivalent replacement bolt during installation. Torque to 105 ft. lbs. (145 Nm).

13. Install the axle nut and washer but do not torque. Install the spring washer, nut lock and a new cotter pin.

14. If the right halfshaft was removed, install the speedometer pinion and tighten the retaining bolt.

15. Install the wheel and lug nuts. Torque the lug nuts, in sequence, to 95 ft. lbs. (129 Nm). Lower the vehicle.

16. Torque the axle nut to 180 ft. lbs. (244 Nm). Install the spring washer, nut lock and a new cotter

pin. Reconnect the negative battery cable.

Rear Halfshaft

1. Disconnect the negative battery cable.

2. Raise and safely support the vehicle.

3. Remove the wheel.

4. Remove the cotter pin from the end of the halfshaft. Remove the nut lock, spring washer, axle nut and washer.

5. Remove the inner shaft retaining bolts. The halfshaft is spring loaded. Compress the inner halfshaft joint slightly and pull downward to clear the differential.

6. Remove the halfshaft.

To install:

7. Install the halfshaft.

8. Install and tighten the inner shaft retaining bolts to 45 ft. lbs. (61 Nm).

9. Install the axle nut washer and nut and torque the nut to 180 ft. lbs. (244 Nm). Install the spring washer, nut lock and a new cotter pin.

10. Install the wheel and lug nuts. Torque the lug nuts, in sequence, to 95 ft. lbs. (129 Nm).

11. Lower the vehicle and reconnect the negative battery cable.

1996–97

Each outer CV-joint is equipped with an ABS speed sensor tone wheel, which is utilized to determine vehicle speed for ABS brake operation.

1. Disconnect the negative battery cable.

2. Remove the cotter pin from the end of the stub axle. Remove the nut lock and spring washer. With the brakes applied, loosen, but do NOT remove the axle nut and washer with the vehicle still on the ground or damage to the wheel bearing will result.

3. Raise and safely support the vehicle. Remove the wheel.

4. Remove the front brake caliper assembly from the steering knuckle assembly and support from the strut assembly using a strong piece of wire.

5. Remove the front brake rotor from the hub/bearing assembly.

6. Remove the retaining nut and washer from the halfshaft stub axle.

7. Separate the outer tie rod from the steering knuckle.

8. Remove the ABS wheel speed sensor from the steering knuckle.

9. Remove the wheel stop from the steering knuckle, if equipped.

10. Remove the nut and bolt that clamps the steering knuckle to the ball joint stud. Using a prybar, pry the control arm down to release the ball stud from the steering knuckle. Be careful not to tear the ball joint grease seal when prying down from the steering knuckle.

11. Separate the outer CV-joint splined shaft from the hub and bearing assembly by holding the CV-joint housing and pulling the steering knuckle away. Be careful not to damage the outer CV-joint wear sleeve or separate the inner CV-joint.

12. Support the halfshaft assembly at the CV-joint housing. Install a prybar between the transaxle housing and the inner tripod joint (CV-joint). Pry against the inner tripod joint until the retainer snapring on the tripod joint disengages from the transaxle side gear.

13. Remove the halfshaft assembly from the vehicle by holding the inner tripod joint and interconnecting shaft and pulling it straight out of the transaxle side gear. Do not allow the splines or the snapring of the shaft to drag across the sealing lip of the transaxle-to-tripod joint oil seal.

To install:

14. Thoroughly clean the tripod joint shaft splines and oil seal sealing surface. Apply a light coating of clean transaxle lubricant to the oil seal sealing surface of the tripod joint.

15. Hold the halfshaft assembly by the tripod joint and interconnecting shaft and install the halfshaft assembly into the transaxle, being careful not to damage the oil seal. Make sure the inner joint clicks into place inside the differential. Check that the snapring is fully engaged by attempting to pull the halfshaft assembly out by hand. If it cannot be removed by hand, the snapring is engaged.

16. Thoroughly clean the steering knuckle and hub/bearing area of all debris and moisture, where the CV-joint will be installed into the steering knuckle. Also thoroughly clean

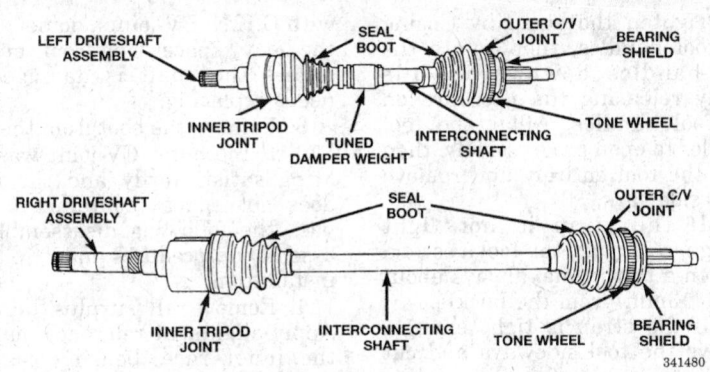

Halfshaft assembly identification — 1996–97

the bearing shield of the outer CV-joint.

17. Pull the front strut out and insert the splined outer CV-joint into the front hub.

18. Insert the ball joint stud into the steering knuckle clamp. Install a **new steering knuckle-to-ball joint stud clamping nut and bolt**. Be sure to use an exact replacement nut and bolt during installation. Torque the bolt to 105 ft. lbs. (145 Nm).

19. Install the tie rod end into the steering knuckle. Torque the tie rod end-to-steering knuckle nut to 45 ft. lbs. (61 Nm).

20. Install the disc brake rotor.

21. Install the brake caliper assembly onto the steering knuckle.

22. Install the axle washer and nut. Tighten but do not torque.

23. Install the ABS wheel speed sensor.

24. Install the wheel and lug nuts. Torque the lug nuts, in sequence, to 95 ft. lbs. (129 Nm).

25. Lower the vehicle. Do NOT roll the vehicle until the axle nut has been properly torqued or damage to the front wheel bearings will result.

26. With the vehicle's brakes applied, torque the axle nut to 180 ft. lbs. (244 Nm). Install the spring washer, nut lock and a new cotter pin. Wrap the cotter pin prongs tightly around the axle nut lock.

27. Reconnect the negative battery cable. Road test the vehicle.

CV-Joint Boot

REPLACEMENT

1993–95

All vehicles use the unequal length halfshaft system which has a short solid interconnecting shaft on the left side of the vehicle for all engine tran-saxle applications. On the right side of the vehicle the interconnecting shafts vary depending on the transaxle used in the vehicle. The manual transaxle equipped vehicles use a tubular shaft and the automatic applications use a solid shaft.

Front Halfshaft
Inner CV-Joint

1. Remove the halfshaft from the vehicle.

2. Secure the halfshaft in a soft-jawed vise.

3. Remove the boot clamps and pull back the boot.

— WARNING —
When removing the housing from the tripod, hold the rollers in place on the trunnion studs to prevent the rollers and needle bearings from falling away. After the tripod is out of the housing, secure the rollers in place with tape.

4. If equipped with S.S.G. CV-joint, use a small prybar to pry the wire ring out of the groove, then slide the tripod from the housing. Be careful not to mangle or destroy the retainer during disassembly.

5. If equipped with G.K.N. CV-joint, separate the CV-joint housing from the tripod as follows:

   a. Clamp the joint housing stub shaft in a vise. Use protective caps on the vise jaws to prevent damage to the stub shaft.

   b. Hold the halfshaft on an angle, while gently pulling on the shaft until one of the tripod bearings is free of the retaining collar.

   c. Continue holding the halfshaft on an angle and gently pull on the shaft until all rollers are free of the retaining collar.

6. Remove the snapring from the end of the shaft, then remove the tri-pod from the shaft. If necessary, tap the tripod from the shaft using a brass drift on the tripod body.

7. If equipped, remove the inner snapring from the shaft.

8. Remove the CV-joint boot from the shaft.

9. Remove as much grease as possible from the assembly and inspect the joint housing ball raceway and tripod components for excessive wear, and replace as necessary.

10. Inspect the spring, spring cup, and the spherical end of the connecting shaft for excessive wear or damage and replace, if necessary.

To install:

NOTE: **Vehicles may be equipped with rubber or plastic CV-joint boots, each requiring different types of boot clamps. As the plastic boots require much more clamping force than the rubber boots, the type clamp specified for rubber boots must never be used on the plastic boots.**

11. If equipped with rubber CV-joint boot, slide the small end of the boot over the shaft and position the boot to the edge of the locating mark or the groove, as necessary. Do not clamp at this time.

12. If equipped with plastic CV-joint boot, install the boot as follows:

   a. Slide the small clamp onto the shaft.

   b. Position the small end of the boot over the shaft with the lip of the boot in the 3rd groove, towards the center of the shaft.

   c. Position the boot clamp evenly over the boot.

   d. Place clamp installer tool C-4975 or equivalent, over the bridge of the clamp and tighten the nut until the jaws of the tool are closed completely, face to face.

13. If equipped, install the inner snapring into the groove on the shaft.

14. Install the tripod onto the shaft until it is past the outer snapring groove. If necessary, tap the tripod assembly onto the shaft using a brass drift on the tripod body. Do not strike the outer tripod bearings.

15. Install a new retaining snapring into the shaft groove.

16. Distribute ½ the amount of grease provided in the boot kit into the housing and the remaining amount in the boot.

17. Position the spring in the housing spring pocket with the spring cup attached to the exposed end of the spring. Place a small amount of

grease on the concave surface of the spring cup.

NOTE: Make sure the spring is positioned properly. The spring must remain centered in the housing spring pocket when the tripod is installed and seated in the spring cup.

18. If equipped with G.K.N. CV-joint, install the tripod into the housing as follows.

a. Clamp the joint housing stub shaft in a vise. Use protective caps on the vise jaws to prevent damage to the stub shaft.

b. Position the shaft and tripod assembly on top of the plastic retaining collar. Carefully insert each of the tripod rollers into the retaining collar, one at a time.

c. Hold the shaft on an angle and carefully push down on the shaft until the rollers are locked into the retaining collar on the housing.

d. Remove the stub shaft from the vise.

19. If equipped with S.S.G. CV-joint, slip the tripod into the housing and install the tripod wire retaining ring into position. Make sure the retaining ring holds the tripod in the housing.

20. If equipped with rubber CV-joint boot, attach the boot to the shaft and joint housing as follows:

a. If not already done, position the small end of the boot to the edge of the locating mark or groove, as necessary.

b. Slide the large end of the boot into the locating groove on the joint housing.

c. Wrap the binding strap around the boot twice, plus 2½ in. (63mm). Pass the strap through the buckle and fold it back about 1⅛ in. (29mm) on the inside of the buckle.

d. Put the strap around the boot with the eye of the buckle toward you. Wrap the strap around the boot once and pass it through the buckle, then wrap it around the second time, also passing it through the buckle.

e. Fold the strap back slightly to prevent it from slipping backwards. Open the tool all the way and place the strap in the narrow slot approximately ½ in. (13mm) from the buckle.

f. Hold the binding strap with the left hand and push the tool forward and slightly upward, and then fit the hook of the tool into the eye of the buckle.

g. Tighten the strap by closing the tool handles, then rotate the tool handles downward while slowly releasing the pressure on the tool handles. Allow the tool handles to open progressively, then open the tool entirely and remove them sideways.

h. If the strap is not tight enough, re-engage the tool a second or even a third time, always about ½ in. (13mm) from the buckle.

i. If the strap is tight enough, remove the tool sideways and cut off the strap ⅛ in. (3mm), so it does not overlap the edge of the buckle. Complete the job by folding the strip back neatly.

j. Repeat the procedure for the other end of the boot.

21. If equipped with plastic CV-joint boot, position the large end of the boot on the joint housing and install the clamp. Crimp the bridge of the clamp using crimper tool C-4975 or equivalent.

NOTE: Some vehicles may be equipped with different type boot clamps requiring different installation tools, such as crimper tool C-4124, clamp locking tool YA3050 or clamp installer 6679. Use only the clamps provided in the boot package, otherwise damage to the boot or CV-joint may occur.

22. Install the halfshaft in the vehicle.

Outer CV-Joint

1. Remove the halfshaft assembly from the vehicle.

2. Clamp the shaft in a vise with soft jaws. Support the outer CV-joint.

3. Remove the boot clamps and slide the boot off the outer joint and down the shaft.

4. Remove the lubricant to expose the outer CV-joint components.

5. If equipped with G.K.N. CV-joint, give a sharp tap to the top of the housing using a soft hammer to dislodge the joint from the internal circlip installed in a groove at the outer end of the shaft.

6. If equipped with S.S.G. CV-joint, loosen the damper weight bolts and slide it and the boot towards the inner joint. Expand the circlip using snapring pliers and slide the joint from the shaft. Install the damper weight.

7. If bent or damaged, carefully pry the wear sleeve from the CV-joint machined ledge.

8. Remove the circlip from the shaft groove and discard. If equipped

with G.K.N. CV-joints, do not remove the heavy spacer ring from the shaft, unless the shaft is damaged and needs replacing.

9. Remove the boot from the shaft.

10. If the outer CV-joint was operating satisfactorily and the grease does not appear contaminated, bypass the following disassembly and assembly procedures and skip to boot installation.

11. Remove all surplus lubricant. Apply alignment reference marks on the inner race, bearing cage and housing with dabs of paint.

12. Clamp the outer CV-joint in a vertical position. Place the stub shaft in a soft-jawed vise, to prevent spline damage.

13. Press down on one side of the inner race to tilt the cage. This will provide access to a ball at the opposite side of the cage. If the CV-joint is tight, use a hammer and brass drift to loosen the inner race. Do not contact the cage with the drift.

14. Remove the ball from the cage. If necessary, use a small prybar to pry the ball loose. Repeat until all 6 balls are removed from the cage.

15. Tilt the cage and inner race assembly vertically and position 2 opposing cage windows in the area between the ball grooves remove the cage and inner race from the housing.

16. Turn the inner race 90 degrees from the bearing cage. Align one pair of the race lands with the cage windows. Raise and insert one of the lands into the adjacent cage window. Remove the inner race by rolling it out of the cage.

17. Clean all components with suitable solvent and allow to dry.

18. Inspect the ball raceways in the housing for excessive wear and scoring. Examine the stub shaft splines and threads for damage. Inspect the balls for pitting, cracks, scoring and excessive wear (a dull exterior surface is normal).

19. Inspect the cage for wear, grooves, ripples, cracks and chipping. Inspect the bearing hub for excessive wear and scoring in ball raceways. Polished contact surface areas on the raceways and on the cage spheres are normal.

20. If any parts are defective, the entire CV-joint assembly must be replaced as a unit.

To install:

21. If removed, install a new wear sleeve on the joint housing using installer tool C-4698 or equivalent.

22. Lightly lubricate the outer CV-joint components with oil before assembly.

23. Align the inner race, cage and housing according to the reference marks made during removal.

24. Insert one of the inner race lands into the cage window. Roll the inner race into the cage. Rotate the inner race 90 degrees to complete the installation.

25. Align the opposing cage windows with the housing land and feed the cage assembly into the housing. Rotate the cage 90 degrees to complete the installation. When properly assembled, the inner race counterbore should be facing outwards from the joint on G.K.N. CV-joints. On S.S.G. CV-joints, the internal circlip in the inner race will be facing outward from the housing.

26. Apply the lubricant supplied with the new boot to the ball raceways. Spread the lubricant equally between all the raceways. One packet of lubricant is enough to lubricate the complete CV-joint.

27. Tilt the inner race and cage and install the balls in the raceways.

28. Install the boot as follows:

   a. Slide the small clamp onto the shaft.

   b. Position the small end of the boot over the shaft with the lip of the boot in the 3rd groove, towards the center of the shaft.

   c. Position the boot clamp evenly over the boot.

   d. Place clamp installer tool C-4975 or equivalent, over the bridge of the clamp and tighten the nut until the jaws of the tool are closed completely, face to face.

29. If equipped with G.K.N. CV-joint, install a new circlip, provided with the kit in the shaft groove. Do not over expand or twist the circlip during assembly. The S.S.G. CV-joint has a reusable circlip retainer that is an integral part of the driver assembly.

30. Position the outer CV-joint on the splined end with the hub nut on the stub shaft. Engage the splines and tap sharply with a mallet.

31. Make sure the circlip is properly seated by attempting to pull the joint from the shaft.

32. Position the large end of the boot on the joint housing and install the clamp. Crimp the bridge of the clamp using crimper tool C-4975 or equivalent.

NOTE: Some vehicles may be equipped with different type boot clamps requiring different installation tools, such as crimper tool C-4124, clamp locking tool YA3050 or clamp installer 6679. Use only the clamps provided in the boot package, otherwise damage to the boot or CV-joint may occur.

33. Install the halfshaft into the vehicle.

Rear Halfshaft

1. Remove the halfshaft from the vehicle.

2. Clamp the shaft in a soft-jawed vise.

3. Cut the boot clamps and discard. Pull the boot back and slide it down the shaft.

4. Remove the housing from the tripod assembly.

5. Remove the snapring from the end of the shaft and remove the tripod assembly from the shaft.

6. Remove the boot from the shaft.

7. Clean all old lubricant from the shaft, tripod assembly and housing. Inspect all components and replace parts, as necessary.

To install:

8. Slide the new boot onto the shaft. Position the small end of the boot and secure with a new clamp.

9. Install the tripod assembly to the shaft and install the snapring.

10. Distribute 1/2 the amount of grease supplied with the boot kit into the CV-joint housing and the rest into the boot.

11. Install the housing on the tripod assembly.

12. Position the large end of the boot on the joint housing and secure with a new clamp.

13. Install the halfshaft into the vehicle.

1996–97

Inner CV-Joint Boot

1. Raise and safely support the vehicle.

2. Remove the halfshaft from the vehicle.

3. Mount the halfshaft assembly firmly in a bench vise with soft jaws.

4. Remove the 2 boot clamps (one large, one small diameter clamp) and discard.

5. Slide the CV-joint boot down the interconnecting shaft.

6. Slide the tripod joint housing off the spider assembly while holding the spider bearings to prevent them from falling off the spider assembly.

7. Remove the spider assembly retaining snapring.

8. Using a brass drift near the shaft, tap the spider assembly toward the end of the shaft to remove.

9. Remove boot from the shaft.

10. Thoroughly clean and inspect the spider assembly, interconnecting shaft and tripod housing for indications of excessive wear. The halfshaft assembly will require replacement if there are signs of excessive wear. Component parts of these halfshaft assemblies are not serviceable separately.

To install:

NOTE: The inner CV-joint boots are made from 2 different types of material. Silicone rubber is used for high temperature applications where hytrel plastic is used for standard temperature applications. The silicone rubber boots are soft and pliable while the hytrel boots are stiff and rigid. The new CV-joint boot MUST be of the same material as the CV-joint boot which was removed.

11. Install the small inner boot retaining clamp onto the halfshaft. Slide the new boot onto the shaft. Be sure the boot is positioned onto the shaft so the raised bead on the inside of the boot is seated on the groove on the shaft.

12. Install the spider assembly.

13. Using a brass drift, tap the spider assembly onto the shaft until the retaining snapring can be installed.

14. Fill the tripod joint housing with 1/2 the amount of grease provided in the boot service package. Fill the CV-joint boot with the remaining grease. Do NOT use any other type of grease in this procedure.

15. Install the large inner CV-joint boot clamp onto the tripod housing.

16. Slide the boot onto the tripod housing and seat it on the tripod housing retaining groove.

17. Correctly position the small diameter sealing clamp over the boot. Using a suitable CV-joint boot clamp crimping tool, crimp the CV-joint boot clamp. Be sure the boot is shaped correctly and if not, shape it by hand.

18. Insert a flat plastic stick between the boot and the joint to vent the joint.

19. Lift the stick slightly and press the joint inward to release any excess air from the boot.

20. Remove the plastic stick.

21. Install the tripod boot clamp and crimp it using the CV-joint boot clamp crimping tool.

22. Install the halfshaft assembly into the vehicle.

Outer CV-Joint Boot

1. Raise and safely support the vehicle.

2. Remove the halfshaft from the vehicle.

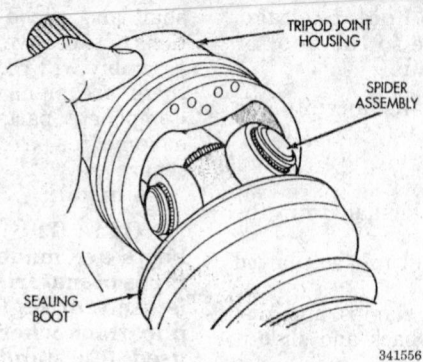

Tripod housing — 1996–97

3. Clamp the shaft in a vise with soft jaws. Support the outer CV-joint.

4. Remove the boot clamps and slide the boot off the outer joint and down the interconnecting shaft.

5. Remove the lubricant to expose the outer CV-joint components.

6. Give a sharp tap to the top of the housing using a soft hammer to dislodge the joint from the internal circlip installed in a groove at the outer end of the shaft.

7. Remove the CV-joint boot from the interconnecting shaft.

8. Clean all components with suitable solvent and allow to dry.

9. Inspect the ball raceways in the housing for excessive wear and scoring. Examine the stub shaft splines and threads for damage. Inspect the balls for pitting, cracks, scoring and excessive wear (a dull exterior surface is normal).

10. Inspect the cage for wear, grooves, ripples, cracks and chipping. Inspect the bearing hub for excessive wear and scoring in ball raceways. Polished contact surface areas on the raceways and on the cage spheres are normal.

11. If any parts are defective, the entire halfshaft assembly must be replaced as a unit.

To install:

12. Install a new small CV-joint boot clamp by sliding the clamp onto the interconnecting shaft.

13. Slide the small end of the new CV-joint boot over the interconnecting shaft with the lip of the boot in the groove on the shaft.

14. Position the outer CV-joint on the splined end of the interconnecting shaft, with the hub nut on the stub shaft. Engage the splines and tap sharply with a soft mallet.

15. Make sure the circlip is properly seated by attempting to pull the joint from the shaft.

16. Fill the outer CV-joint assembly housing with ½ the amount of grease provided in the boot service package. Fill the CV-joint boot with the remaining grease. Do NOT use any other type of grease in this procedure.

17. Correctly position the interconnecting shaft end, small diameter sealing clamp over the CV-joint boot. Using a suitable CV-joint boot clamp crimping tool, crimp the CV-joint boot clamp. Be sure the boot is shaped correctly and if not, shape it by hand.

NOTE: Some vehicles may be equipped with different type boot clamps requiring different installation tools, such as crimper tool C-4124, clamp locking tool YA3050 or clamp installer 6679. Use only the clamps provided in the boot package, otherwise damage to the boot or CV-joint may occur.

18. Position the large end of the CV-joint boot into its retaining groove on the outer CV-joint housing and install the clamp. Crimp the bridge of the clamp using the correct CV-joint boot clamp crimper tool.

19. Install the halfshaft assembly into the vehicle.

STEERING

Air Bag

——— CAUTION ———
Some vehicles are equipped with an air bag system, also known as the Supplemental Inflatable Restraint (SIR) or Supplemental Restraint System (SRS). The system must be disabled before performing service on or around system components, steering column, instrument panel components, wiring and sensors. Failure to follow safety and disabling procedures could result in accidental air bag deployment, possible personal injury and unnecessary system repairs.

PRECAUTIONS

Several precautions must be observed when handling the inflator module to avoid accidental deployment and possible personal injury.

• Never carry the inflator module by the wires or connector on the underside of the module.

• When carrying a live inflator module, hold securely with both hands, and ensure that the bag and trim cover are pointed away.

• Place the inflator module on a bench or other surface with the bag and trim cover facing up.

• With the inflator module on the bench, never place anything on or close to the module which may be thrown in the event of an accidental deployment.

DISARMING

——— CAUTION ———
The Supplemental Inflatable Restraint (SIR) system must be disarmed before working around the air bag or SIR wiring. Failure to do so may cause accidental deployment of the air bag, resulting in unnecessary SIR system repairs and/or personal injury.

1. Disconnect and isolate the negative battery cable from the battery.

2. Allow the SIR system capacitor to discharge for at least 2 minutes, before performing any removal procedures.

3. The air bag system is now disabled.

——— CAUTION ———
When carrying a live air bag, make sure the bag and trim cover are pointed away from the body. In the unlikely event of an accidental deployment, the bag will then deploy with minimal chance of injury. When placing a live air bag on a bench or other surface, always face the bag and trim cover up, away from the surface. This will reduce the motion of the module if accidentally deployed.

Steering Wheel

REMOVAL AND INSTALLATION

—————— CAUTION ——————

The Supplemental Inflatable Restraint (SIR) system must be disarmed before removing the steering wheel. Failure to do so may cause accidental deployment of the air bag, resulting in unnecessary SIR system repairs and/or personal injury.

1993–95

1. Make sure the front wheels are in the straight-ahead position and the steering column is locked in place.
2. Disconnect and isolate the negative (ground) battery cable. Allow system capacitor to discharge for at least 2 minutes before continuing with the removal procedure. This will disable the air bag system.
3. Remove the 4 nuts attaching the air bag module to the back side of the steering wheel.

—————— CAUTION ——————

When carrying a live air bag, make sure the bag and trim cover are pointed away from the body. In the unlikely event of an accidental deployment, the bag will then deploy with minimal chance of injury. When placing a live air bag on a bench or other surface, always face the bag and trim cover up, away from the surface. This will reduce the motion of the module if accidentally deployed.

4. Lift the module and disconnect the connector from the rear of the module.
5. Remove the vehicle speed control switch and connector, if equipped, or cover.
6. Mark the column shaft and wheel for reinstallation reference and remove the steering wheel retaining nut.
7. Remove the steering wheel using a steering wheel puller tool.

To install:

—————— CAUTION ——————

If the clockspring is not properly positioned or if the front wheels were moved, follow the clockspring centering procedure before proceeding.

8. With the front wheels in a straight-ahead position, position the steering wheel on the steering column. Be sure to fit the flats on the hub of the steering wheel with the formations on the inside of the clockspring.
9. Pull the air bag and speed control wires through the lower, larger hole in the steering wheel and horn wire through the smaller hole at the top. Be sure not to pinch the wires.
10. Install and tighten the nut to 45 ft. lbs. (61 Nm).
11. Reconnect the horn wire connector.
12. Reconnect the 4-way connector to the vehicle speed control switch and attach the switch to the steering wheel.
13. Reconnect the air bag lead wire to the air bag module and secure the module to the steering wheel. Tighten to 80–100 inch lbs. (9–11 Nm).
14. Do not reconnect the negative battery until you perform an Air Bag System Check as follows:

   a. Remove the forward console cover as necessary.
   b. Connect a DRB or equivalent scan tool to the ACM data link connector. This is located in the area to the lower right of the steering column.
   c. Turn the ignition key to the **ON** position. Exit vehicle with the scan tool. Be sure to use the latest version of the proper scan tool cartridge.
   d. After making sure no one is inside the vehicle, reconnect the negative battery cable.
   e. Read and record all active diagnostic data using the scan tool.
   f. Using the scan tool, read and record all stored diagnostic trouble codes.
   g. Air Bag System diagnostics will be required if any trouble codes are found.
   h. Erase the stored trouble codes only if there are no active trouble codes. If the problem continues, the codes will not erase.
   i. Turn the ignition key to **OFF** then **ON** and observe the air bag lamp on the dashboard. It should light for 6–8 seconds, then go out. This would indicate a normally functioning system.
15. If the light fails to go ON, goes ON and stays ON or blinks continuously, then there is an air bag system malfunction.

1996–97

1. Disconnect and isolate the negative (ground) battery cable. Allow the system capacitor to discharge for at least 2 minutes before continuing with the removal procedure. This will disable the air bag system.
2. Make sure the front wheels are in the straight-ahead position.
3. Turn the ignition key cylinder to the **LOCK** position and remove the key. Turn the steering wheel a ½ turn toward the left until the steering column becomes locked in position.
4. Remove the 3 bolts attaching the air bag module to the steering wheel. Remove the air bag module.

—————— CAUTION ——————

When carrying a live air bag, make sure the bag and trim cover are pointed away from the body. In the unlikely event of an accidental deployment, the bag will then deploy with minimal chance of injury. When placing a live air bag on a bench or other surface, always face the bag and trim cover up, away from the surface. This will reduce the motion of the module if accidentally deployed.

5. Disconnect the wiring connectors from the air bag module, horn switch and speed control switches.
6. Remove the routing clip for the wiring harness from the air bag module studs.
7. Remove the steering wheel mounting nut from the steering column shaft and remove the steering wheel damper from the steering wheel.
8. Using a steering wheel puller tool, remove the steering wheel from the steering column shaft. Do not thread the steering wheel puller tool bolts into the steering wheel more than a ½ in. or damage to the SRS clockspring will result. Do not hammer or bump the steering column or shaft when removing the steering wheel.

To install:

9. Before installing the steering wheel, the clockspring **must be centered**. Center the clockspring as follows:

   a. Disengage the lock mechanism by depressing the 2 plastic locking pins.
   b. Turn the clockspring rotor clockwise until it stops, while the lock mechanism is disengaged. Do not use excessive force.
   c. From the end of the clockwise travel, turn the clockspring rotor counter-clockwise. The clockspring wires should be at the top. Engage the clockspring locking pins.
   d. Turn the clockspring a ½ additional rotation counter-clockwise from the center locked position.

This should put the clockspring wiring at the bottom.

e. The clockspring is now correctly positioned for installation of the steering wheel.

10. Install the steering wheel onto the steering column shaft. Make sure the master splines of the steering wheel and steering shaft are in correct alignment. Be sure the flats on the steering wheel align with the formations on the clockspring. All wiring leads from the clockspring must be routed correctly.

11. Install the steering wheel damper onto the steering wheel in its original position.

12. Install the steering wheel-to-steering shaft retaining nut. Torque the retaining nut to 45 ft. lbs. (61 Nm).

13. Reconnect the horn switch wiring lead from the clockspring, onto the steering wheel horn switch wiring.

14. If equipped with speed control, reconnect the speed control wiring from the clockspring onto the speed control switch.

15. Install the wiring lead from the clockspring onto the air bag module. Make the wiring connection onto the air bag module, by pressing straight in on the connector. Be sure it is fully seated.

16. Install the air bag module to the steering wheel and then install the 3 air bag module attaching nuts. Torque all 3 air bag module attaching nuts to 100 inch lbs. (11 Nm).

17. When reconnecting the battery on a vehicle that has had the air bag removed, the following procedure should be used:

a. Connect the DRB or equivalent scan tool, to the ASDM diagnostic 6-way connector.

b. Turn the ignition key to the ON position. Exit the vehicle with the DRB, and install the latest version of the proper diagnostic cartridge into the DRB.

c. Make sure there are no occupants in the vehicle and reconnect the negative battery cable.

d. Using the DRB, read and record active or stored fault codes. Take appropriate actions to correct any faults.

e. Erase stored fault codes. If problems remain, fault codes will not erase.

f. From the passenger side of the vehicle, turn the ignition key to **OFF** and then **ON**, observing the instrument cluster air bag lamp. It should go ON for 6–8 seconds, then go out. This will indicate that the air bag system is functioning normally.

WARNING
If the air bag warning lamp fails to light, blinks ON and OFF or goes ON and stays ON, there is an air bag system malfunction.

g. Test the operation of any steering column functions such as the horn, lights or speed control system.

18. Road test vehicle. Be sure the speed control and steering systems are functioning properly.

Tie Rod Ends

REMOVAL AND INSTALLATION

1. Raise and safely support the front of the vehicle on safety stands. Remove the front wheels.

2. Wire brush the tie rod threads and liberally soak with penetrating oil. Loosen the inner tie rod-to-outer tie rod jam nut.

3. Mark the tie rod position on the inner tie rod threads.

4. For 1993–95, remove the tie rod cotter pin and nut.

5. For 1996–97, remove the tie rod-to-steering knuckle nut by holding the tie rod end stud with an $^{11}/_{32}$ socket while loosening and removing the nut with a box wrench.

6. Using a puller tool, separate the tie rod end from the steering knuckle. If the joint is to be reused, do not use a wedge-type tool to hammer the connection apart or the outer tie rod end will be damaged.

NOTE: Count the number of turns when removing tie rod end. Install the new tie rod end the same amount of turns.

7. Unscrew the outer tie rod end from the rack inner tie rod.

To install:

8. Thread the new tie rod end onto the inner tie rod, the same number of turns as was required for removal.

9. Install the tie rod end to the steering knuckle. Install the outer tie rod-to-steering knuckle nut. Torque the nut to 38 ft. lbs. (52 Nm) for 1993–95 or using a crowfoot and $^{11}/_{32}$ socket, torque the nut to 40 ft. lbs. (54 Nm) for 1996–97. Install a new cotter pin.

10. Check the toe setting. Adjust the toe setting to the correct specifications by turning the inner tie rod, taking care not to twist the boot.

11. Tighten the jam nut to 55 ft. lbs. (75 Nm).

12. Install the front wheels and lug nuts. Torque the lug nuts to 95 ft. lbs. (129 Nm).

13. Lower the vehicle.

14. Re-check the wheel alignment and make any adjustments, if necessary. Road test the vehicle.

Power Rack and Pinion

REMOVAL AND INSTALLATION

1993–95

Front Wheel Drive Vehicles

1. Disconnect the negative battery cable.

2. Loosen the wheel lugs slightly. Raise and safely support the front of the vehicle, not on the front crossmember. Use safety stands for support.

3. Remove the front wheels. Remove the tie rod ends from the steering knuckles.

4. Disconnect the engine damper strut from the crossmember, if equipped.

5. If equipped, remove the air diverter valve from the left side of the crossmember.

6. Place a transmission jack, or floor jack with a wide lifting flange, under the front suspension K-crossmember. Support the crossmember and remove the crossmember to frame attaching bolts. Slowly lower the crossmember until enough room is gained to disconnect the steering column from the steering gear assembly. Place safety stands under the crossmember.

7. Remove the splash and boot shields. Disconnect the power steering hoses. Place a drain pan under the power steering pump hoses to catch draining power steering fluid.

8. Remove the bolts that mount the steering gear assembly to the

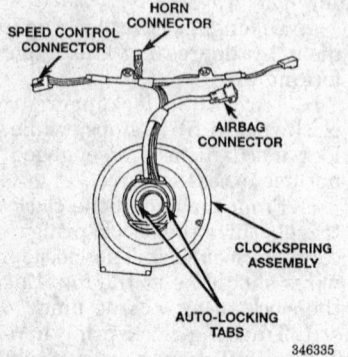

HORN CONNECTOR
SPEED CONTROL CONNECTOR
AIRBAG CONNECTOR
CLOCKSPRING ASSEMBLY
AUTO-LOCKING TABS

346335

Clockspring assembly — 1996–97

crossmember. Remove the assembly from the vehicle.

To install:

9. Install the steering gear rack assembly onto the front crossmember. Install the mounting bolts.

10. Using a transmission jack, raise the front crossmember and steering gear rack and pinion assembly up to the frame rails.

11. Line up the gear pinion with the column. Install the crossmember-to-frame mounting bolts and nut. The right rear crossmember stud is the alignment pilot for re-installation.

12. Torque all crossmember attaching bolts to 90 ft. lbs. (122 Nm). Steering gear rack mounting bolts are torqued to 50 ft. lbs. (68 Nm). To ensure proper alignment of the steering gear assembly, tighten the left front bolt first.

13. Reconnect the engine damper strut to the front crossmember, if equipped. Install the air diverter valve to the left side of the crossmember, if equipped.

14. Reconnect the power steering fluid hoses. Torque the fluid pressure fittings to 275 inch lbs. (31 Nm).

15. Install the outer tie rod ends to the steering knuckles. Torque the tie rod end-to-steering knuckle nuts to 38 ft. lbs. (52 Nm). Install a new cotter pin.

16. Install the front wheels and lug nuts. Torque the lug nuts, in a star sequence, to 95 ft. lbs. (129 Nm).

17. Lower the vehicle. Reconnect the negative battery cable.

18. Refill power steering reservoir with the correct amount of clean power steering fluid and bleed the system.

19. Check the toe setting and adjust, if necessary. Road test the vehicle and check steering operation.

All Wheel Drive Vehicles

Before removing the steering gear on All Wheel Drive vehicles, the steering column must be removed to provide clearance for steering rack removal.

1. Disconnect the negative battery cable.

2. Raise and safely support the vehicle. Remove the front wheels.

3. Remove the steering column assembly from the vehicle.

4. Remove the tie rod ends from the steering knuckle using a suitable puller.

5. Remove the 2 bolts and the 2 nuts that attach the bridge assembly to the crossmember. The bolts and nuts can be reached through the access holes in the top of the bridge assembly.

6. Remove the bracket securing the power steering fluid hoses to the crossmember.

7. Remove the crossmember to frame rail attaching bolts and the nut from the locating stud. Use a jack to lower the crossmember so it is suspended from the lower control arms. It is not necessary to remove the crossmember completely from the vehicle.

8. Disconnect and plug the power steering lines from the steering gear. Place a drain pan under the power steering hoses to catch any spilling fluid.

9. Remove the 4 bolts that retain the steering gear to the bridge assembly.

NOTE: Note the position of each bolt as it is removed, there are different bolts for the left and right sides.

10. Remove the lower steering column coupler from the steering gear. Drive the roll pin from the coupler using a punch. If this is not done, there will not be enough clearance for rack removal.

11. Remove the steering gear from the vehicle by pulling it out through the drivers side wheel well. Rotate the gear to clear the frame rail.

To install:

12. Install the steering gear in the vehicle. Work it in through the left wheel opening, rotating it as needed.

13. Install the steering column coupler; be sure to fully seat the roll pin.

14. Install the steering gear mounting bolts in their proper locations. Do not torque them at this time.

15. Install the steering hose bracket and tighten to 17 ft. lbs. (23 Nm). Install the hoses to the correct fittings on the steering rack and tighten them to 25 ft. lbs. (34 Nm).

16. Raise the crossmember into position, install the bolts and torque as follows:

 a. Crossmember-to-frame rail stud nut — 90 ft. lbs. (122 Nm)

 b. Crossmember-to-frame rail bolt and washer — 90 ft. lbs. (122 Nm)

17. Install the bridge assembly onto the crossmember and tighten the mounting nuts to 50 ft. lbs. (68 Nm).

18. Install the outer tie rod ends on the steering knuckle and tighten the nuts, tighten to 38 ft. lbs. (52 Nm). Be sure to install a new cotter pin.

19. Install the front wheels and lug nuts. Torque the lug nuts, in a star sequence, to 95 ft. lbs. (129 Nm). Lower the vehicle.

20. Reconnect the negative battery cable. Refill the power steering fluid reservoir with the correct amount of clean power steering fluid and bleed the system.

21. Check the toe setting and adjust, if necessary. Road test the vehicle and check steering operation.

1996–97

—— **WARNING** ——
Position the steering column in the locked position to prevent the SRS clockspring from becoming accidentally over-extended, when the steering column is disconnected from the intermediate coupler.

1. With the ignition key in the locked position, turn the steering wheel to the left until the steering wheel locks itself in position.

2. Disconnect the negative battery cable.

3. Disconnect the steering column shaft coupler from the steering gear intermediate coupler.

4. Raise and safely support the vehicle.

5. Remove the front wheels.

6. Place a drain pan under the power steering fluid lines and disconnect the fluid hose from the metal tube portion of the power steering fluid return line and allow the steering fluid to drain into the pan.

7. Remove the tie rod ends from the steering knuckles.

8. Remove the 2 bolts and loosen the third, mounting the ABS Hydraulic Control Unit (HCU) to the front suspension cradle. Rotate the HCU rearward to allow access to the cradle plate mounting nut and bolt just ahead of the HCU.

9. Remove the front suspension cradle plate from the front suspension cradle.

10. Remove the retaining bracket attaching the power steering fluid lines to the front suspension cradle.

11. Using an 18mm crowfoot, disconnect the power steering fluid pressure and return lines from the power steering gear.

12. Remove the 3 bolts and nuts mounting the rack and pinion steering gear to the suspension cradle.

13. Lower the steering gear from the suspension cradle enough to allow access to the steering column intermediate coupler roll pin. Remove the roll pin and separate the intermediate coupler from the steering gear shaft.

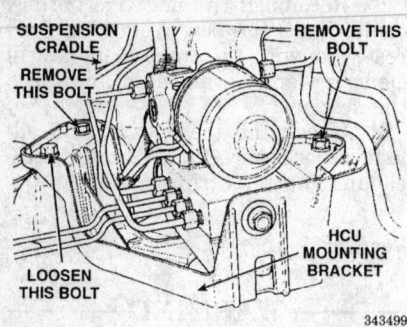

SUSPENSION CRADLE

REMOVE THIS BOLT

REMOVE THIS BOLT

LOOSEN THIS BOLT

HCU MOUNTING BRACKET

343499

HCU bracket mounting bolts — 1996–97

14. Remove the rack and pinion steering gear assembly from the front suspension cradle.

To install:

15. Install the rack and pinion steering gear assembly into the front suspension cradle, providing enough room to install the intermediate coupler.

16. Connect the steering gear shaft to the intermediate coupler and secure with roll pin.

17. Place the steering gear assembly in correct position on the suspension cradle and install the 3 steering gear mounting bolts and nuts. Torque the bolts and nuts to 100 ft. lbs. (136 Nm).

18. Connect the power steering fluid pressure and return lines to the correct fittings on the steering gear. Torque the power steering fluid line tube fittings to 275 inch lbs. (31 Nm).

19. Install the outer tie rod ends to the steering knuckles. Torque the tie rod end-to-steering knuckle nuts to 40 ft. lbs. (54 Nm).

20. Install the front suspension cradle plate to the front suspension cradle. Torque the 10 mounting bolts and nuts to 123 ft. lbs. (165 Nm).

21. Install the bracket mounting the power steering fluid lines to the suspension cradle. Be sure the protective heat shields cover the entire rubber hose hose-to-tube connection of both power steering fluid hoses.

22. Install the hose onto the metal tube portion of the power steering fluid return line. Install hose clamp on the return hose. Be sure the hose clamp is installed past the upset bead on the tube.

23. Install the front wheels and lug nuts. Torque the lug nuts, in a star sequence, to 95 ft. lbs. (129 Nm).

24. Lower the vehicle just enough to access the interior of the vehicle.

25. Using the intermediate coupler, turn the front wheels to the left until the intermediate coupler shaft is correctly aligned with the steering col-

umn coupler. Connect the steering column shaft coupler to the steering gear intermediate coupler. Install the steering column coupler-to-intermediate shaft retaining pinch bolt. Torque the pinch bolt nut to 250 inch lbs. (28 Nm).

26. Reconnect the negative battery cable.

27. Refill power steering reservoir, with the correct amount of clean, fresh MOPAR® Power Steering Fluid or equivalent, and bleed the system.

28. Check the toe setting and adjust, if necessary. Road test the vehicle and check steering operation.

Power Steering Pump

BLEEDING

———— CAUTION ————

The power steering fluid level should be checked with the engine OFF to prevent injury from moving components. Power steering oil, engine components and exhaust system may be extremely hot if the engine has been running. Do not start the engine with any loose or disconnected hoses or allow hoses to touch a hot exhaust manifold or catalyst.

NOTE: In all power steering pumps, use only MOPAR® Power Steering Fluid or equivalent. DO NOT use any type of automatic transmission fluid in the power steering system.

Wipe the filler cap clean, then check the fluid level. The dipstick should indicate FULL COLD when the fluid is at normal room temperature of approximately 70–80°F.

1. Fill the power steering pump fluid reservoir to the proper level. Allow the fluid to settle for at least 2 minutes.

2. Start the engine and let run for a few seconds. Turn the engine OFF.

3. Add fluid if necessary. Repeat this procedure until the fluid level remains constant after running the engine.

4. Raise and safely support the vehicle so the front wheels of the vehicle are off the ground.

5. Start the engine. Slowly turn the steering wheel right and left, lightly contacting the wheel stops. Then turn the engine OFF.

6. Add power steering fluid if necessary.

7. Lower the vehicle and turn the steering wheel slowly from lock to lock.

8. Turn OFF the engine. Check the fluid level and refill as required.

9. If the fluid is extremely foamy, allow the vehicle to stand a few minutes and repeat the above procedure.

REMOVAL AND INSTALLATION

1993–95

2.5L (VIN K) Engine

1. Disconnect and isolate the negative battery cable from the battery.

2. Loosen the pump adjustment bolt and rotate the pump forward in the bracket.

3. Remove the power steering drive belt. It is not necessary to remove the belt from the engine.

4. Raise and safely support the vehicle.

5. Place a drain pan under the vehicle to catch the draining power steering fluid.

6. Disconnect the low pressure fluid hose from the power steering pump and drain the fluid into the drain pan.

7. Remove the fluid pressure line from the power steering pump and drain any excess fluid from the line.

8. Loosen, but do not remove the nut securing the back of the power steering pump to the mounting bracket.

9. Remove the bolt attaching the pulley side of the pump to the mounting bracket.

10. Remove the drain pan. Lower the vehicle. Remove the adjusting bolt from the mounting bracket.

11. Remove the power steering pump from the vehicle. Transfer any parts from the power steering pump to the new replacement power steering pump.

To install:

12. Place the pump in position on the mounting bracket. Install the adjusting bolt into the mounting bracket.

13. Raise and safely support the vehicle. Install the mounting bolt attaching the pulley side of the pump to the mounting bracket. At this time, do not tighten the mounting or adjusting bolts on the power steering pump at the mounting bracket.

14. Install the high and low pressure hoses to the back of the power steering pump. Use new hose clamps on the rubber hoses. Torque the pressure line pump fitting nut to 275 inch lbs. (31 Nm). Inspect the pressure line O-ring for damage and replace, if necessary.

15. Tighten the power steering pump mounting bolts as follows:

- Power steering pump-to-bracket mounting stud M-10: 35 ft. lbs. (48 Nm)
- Power steering pump-to-bracket bolt and nut M-10: 30 ft. lbs. (40 Nm)
- Power steering pump-to-bracket mounting bolts M-8: 21 ft. lbs. (28 Nm)

16. Install the power steering pump drive belt and adjust it to the proper tension.

17. Refill the power steering pump reservoir with the correct amount of clean power steering fluid.

18. Reconnect the negative battery cable. Bleed the power steering system.

19. Run the engine and check the system for leaks and proper steering operation.

3.0L (VIN 3) Engine

1. Disconnect and isolate the negative battery cable.

2. Remove the accessory drive belt. The belt does not have to be removed from the engine.

3. Remove the power steering pump filler tube and dipstick assembly.

4. Raise and safely support the vehicle.

5. Disconnect the exhaust pipe from the exhaust manifold and move out of the way to provide clearance for removal of the power steering pump.

6. Place a drain pan under the power steering pump. Remove the hose clamp and low pressure fluid hose from the power steering gear fluid tube. Allow the excess fluid to drain from the pump and hose.

7. Loosen the fitting for the high pressure fluid line at the power steering pump and remove the high pressure line from the pump.

8. Remove the rear support bracket mounted behind the power steering pump to the engine block.

9. Remove the 2 mounting bolts that secure the front of the pump to the mounting plate.

10. Remove the power steering pump and pulley assembly from the vehicle. Transfer all required parts from the pump to the new replacement pump before installation.

To install:

11. Position the front of the power steering pump up against the mounting plate. Torque the 2 power steering pump-to-mounting plate bolts to 40 ft. lbs. (54 Nm).

12. Install the rear power steering pump-to-engine block support bracket. Torque the 2 support

bracket mounting bolts to 40 ft. lbs. (54 Nm). Install the nut to the mounting stud behind the pump and torque to 40 ft. lbs. (54 Nm).

13. Install the high pressure fluid line to the pump outlet fitting. Torque the high pressure line-to-power steering pump fitting to 275 inch lbs. (31 Nm).

14. Install the low pressure power steering fluid hose to the power steering gear fluid tube. Install a new hose clamp.

15. Reconnect the exhaust pipe to the exhaust manifold. Install the spring assemblies, nuts and bolts and torque to 250 inch lbs. (28 Nm).

16. Remove the drain pan and lower the vehicle.

17. Install the power steering pump filler tube and dipstick assembly. Tighten the hose clamp on the rubber boot of the filler tube. Tighten the mounting bolt for the filler neck to the alternator bracket.

18. Install the accessory drive belt.

19. Refill the power steering pump reservoir with the correct amount of clean power steering fluid.

20. Reconnect the negative battery cable. Bleed the power steering system.

21. Run the engine and check the system for leaks and proper steering operation.

3.3L (VIN R) and 3.8L (VIN L) Engines

1. Remove and isolate the negative battery cable.

2. Remove the accessory drive belt.

3. Raise and safely support the vehicle. Place a drain pan under the power steering pump.

4. Remove the hose clamp and low pressure fluid hose from the power steering pump.

5. Remove the hose clamp and disconnect the hose to the power steering pump from the remote power steering fluid reservoir. Drain off any excess fluid from the hoses.

6. Remove the fluid pressure line from the power steering pump. Drain any excess fluid.

7. Remove the right front wheel for easier access to the pump mounting bolts.

8. Remove the bolts that secure the power steering pump to the alternator, power steering and belt tensioner mounting bracket.

9. Remove the engine block-to-power steering pump support strut.

10. Remove the serpentine drive belt tensioner. Remove the support strut for the alternator/power steering pump bracket.

11. Remove the drain pan. Lower the vehicle.

12. Remove the remote fluid reservoir and tube/hose assembly from the alternator/power steering pump brackets.

13. Remove the engine wiring harness routing clip from the alternator bracket.

14. Loosen, but do not remove, the bolt securing the engine bracket assembly to the engine support assembly.

15. Remove the upper alternator mounting bolt and rotate the alternator unit back toward the firewall.

16. Remove the alternator bracket from the engine.

17. Remove the lower alternator-to-mounting bracket bolt. Without disconnecting the wiring harness from the alternator, remove the alternator from the bracket and lay it on top of the intake manifold.

18. Remove the power steering pump from the vehicle. Transfer all required parts from the power steering pump to the new replacement pump.

To install:

19. Place the power steering pump into the vehicle and lay it down onto the steering gear.

20. Install the alternator onto the lower alternator bracket. Install, but do not tighten, the bolt and nut.

21. Install the alternator bracket onto the engine and intake manifold. Loosely install the engine-to-bracket mounting bolts. Be sure to install the spacer between the engine mounting strut and alternator bracket, if equipped.

22. Temporarily install the serpentine belt tensioner bolt through the both alternator brackets to align the mounting holes. Torque the alternator-to-engine and intake manifold mounting bolts 40 ft. lbs. (54 Nm). Remove the serpentine belt tensioner.

23. Torque the engine bracket assembly-to-engine support assembly bolt to 110 ft. lbs. (150 Nm).

24. Reconnect the engine wiring harness routing clip to the alternator bracket.

25. Install the alternator-to-alternator bracket mounting bolt and torque to 40 ft. lbs. (54 Nm). Torque the alternator pivot bolt to 40 ft. lbs. (54 Nm).

26. Install the fluid reservoir and tube/hose assembly to the power steering pump and alternator brackets. Tighten the bolts.

27. Raise and safely support the vehicle.

28. Install the power steering/alternator bracket to engine support strut. Torque the nut and bolt to 40 ft. lbs. (54 Nm).

29. Install the serpentine belt tensioner. Torque the tensioner mounting nut to 40 ft. lbs. (54 Nm).

30. Install the power steering pump onto the mounting bracket. Install the mounting bolts and torque to 40 ft. lbs. (54 Nm).

31. Install the support strut between the engine block and behind the power steering pump. Torque the nut and bolt to 40 ft. lbs. (54 Nm).

32. Inspect the O-ring on the pressure line for damage and replace, if necessary. Install the fluid pressure line to the output fitting of the power steering pump. Torque the fitting to 275 inch lbs. (31 Nm).

33. Install the low pressure return hose on the power steering pump low pressure fitting. Reconnect the remote fluid reservoir hose to the power steering pump and tighten the hose clamp. Be sure it is properly installed.

34. Install the right front wheel and lug nuts. Torque the lug nuts, in a star sequence, to 95 ft. lbs. (129 Nm).

35. Lower the vehicle.

36. Install the serpentine drive belt.

37. Refill the power steering pump reservoir with the correct amount of clean power steering fluid.

38. Reconnect the negative battery cable. Bleed the power steering system.

39. Run the engine and check the system for leaks and proper steering operation.

1996–97

2.4L (VIN B) Engine

1. Disconnect and isolate the negative battery cable from the battery.

2. Remove the power steering drive belt. It is not necessary to remove the belt from the engine.

3. Loosen, but do not remove the nut securing the front bracket for the power steering pump to the aluminum mounting bracket.

4. Raise and safely support the vehicle.

5. Disconnect the wiring harness connector to the oxygen sensor which is accessible through the oxygen sensor wiring harness grommet in the vehicle floor pan.

6. Remove the catalytic converter from the exhaust manifold and remove all exhaust system hangers and isolators from the exhaust system brackets. Move the exhaust system out of the way as far rearward and to the left as possible to provide access to the power steering pump.

7. Place a drain pan under the power steering pump. Remove the power steering fluid return line hose on the front suspension cradle. Allow the fluid to drain from the pump and hose.

8. Remove the accessory drive belt splash shield.

9. Disconnect the power steering remote reservoir supply hose from the fitting on the power steering pump. Allow fluid to drain from the hose.

10. Remove power steering fluid pressure line from the power steering pump and drain any excess power steering fluid.

11. Remove the power steering fluid return hose from the power steering pump.

12. Remove the nut securing the rear of the power steering pump to the cast mounting bracket.

13. Loosen the 3 bolts securing the power steering pump to the front mounting bracket and then remove the nut and bolt mounting the front of the power steering pump to the cast mounting bracket.

14. Remove the power steering pump and front bracket as an assembly from the cast bracket.

15. Remove the 3 mounting bolts securing the bracket to the power steering pump and separate the bracket from the power steering pump.

16. Remove the power steering pump from the vehicle. Transfer any parts from the power steering pump to the new replacement power steering pump.

To install:

17. Install the power steering pump into the vehicle and position the front of the pump onto the cast mounting bracket. Loosely install mounting nut to secure the pump in place.

18. Install the front mounting bracket on the power steering pump and loosely install the 3 mounting bolts, then install the nut and bolt securing the front bracket to the cast bracket.

19. Torque the 3 power steering pump mounting bracket bolts to 40 ft. lbs. (54 Nm).

20. Install the high pressure fluid line to the pump output fitting. Torque the high pressure line-to-power steering pump fitting to 275 inch lbs. (31 Nm). Be sure to inspect the pressure line O-ring for any damage before connecting the pressure line to the steering pump.

21. Install the low pressure power steering fluid hose to the power steering pump low pressure fitting. Be sure the hose clamps are properly reinstalled and hoses are clear of the accessory drive belts.

22. Install the power steering fluid reservoir supply hose to the power steering pump fluid fitting. Be sure all hoses are clear of any accessory drive belts and hose clamps correctly installed.

23. Install the power steering drive belt.

24. Install the accessory drive belt splash shield.

25. Install the hose on the power steering fluid return line on the front suspension cradle. Be sure the hose clamps and heat shield tubes are correctly reinstalled.

26. Reconnect the exhaust pipe to the exhaust manifold. Install the hangers and isolators onto the exhaust system brackets. Torque the nuts and bolts to 250 inch lbs. (28 Nm).

27. Reconnect the wiring harness connectors to the oxygen sensor. Install the wiring harness grommet into the vehicle floor pan.

28. Remove the drain pan and lower the vehicle.

29. Torque the top mounting nut and bottom mounting bolt on the power steering pump front mounting bracket to 40 ft. lbs. (54 Nm).

30. Refill the power steering pump reservoir with the correct amount of clean power steering fluid.

31. Reconnect the negative battery cable. Bleed the power steering system.

32. Run the engine and check the system for leaks and proper steering operation.

3.0L (VIN 3) Engine

1. Disconnect and isolate the negative battery cable.

2. Remove the accessory drive belt. The belt does not have to be removed from the engine.

3. Raise and safely support the vehicle.

4. Disconnect the wiring harness connector to the oxygen sensor which is accessible through the oxygen sensor wiring harness grommet in the vehicle floor pan.

5. Remove the catalytic converter from the exhaust manifold and remove all exhaust system hangers and isolators from the exhaust system brackets. Move exhaust system out of the way as far rearward and to the left as possible to provide access to the power steering pump.

6. Place a drain pan under the power steering pump. Remove the power steering fluid return line hose on the front suspension cradle. Allow the fluid to drain from the pump and hose.

7. Remove the accessory drive belt splash shield.

8. Disconnect the power steering remote reservoir supply hose from the fitting on the power steering pump. Allow fluid to drain from the hose.

9. Remove power steering fluid pressure line from the power steering pump and drain any excess power steering fluid.

10. Remove the power steering fluid return hose from the power steering pump.

11. Remove the rear support bracket mounted behind the power steering pump to the engine block.

12. Remove the 2 mounting bolts that secure the pump to the alternator/power steering pump and belt tensioner mounting bracket.

13. Remove the power steering pump and pulley assembly out from the vehicle. Transfer all required parts from the pump to the new replacement pump before installation.

To install:

14. Position the front of the power steering pump up onto the mounting bracket. Torque the 2 power steering pump-to-mounting bracket bolts to 40 ft. lbs. (54 Nm).

15. Install the rear power steering pump-to-engine block support bracket. Torque the 2 support bracket mounting bolts to 40 ft. lbs. (54 Nm). Install the nut to the mounting stud behind the pump and torque to 40 ft. lbs. (54 Nm).

16. Install the high pressure fluid line to the pump output fitting. Torque the high pressure line-to-power steering pump fitting to 275 inch lbs. (31 Nm). Be sure to inspect the pressure line O-ring for any damage before connecting the pressure line to the steering pump.

17. Install the low pressure power steering fluid hose to the power steering pump low pressure fitting. Be sure the hose clamps are properly reinstalled and hoses are clear of the accessory drive belts.

18. Install the accessory drive belt.

19. Install the hose on the power steering fluid return line on the front suspension cradle. Be sure the hose clamps and heat shield tubes are correctly reinstalled.

20. Reconnect the exhaust pipe to the exhaust manifold. Install the hangers and isolators onto the exhaust system brackets. torque the nuts and bolts to 250 inch lbs. (28 Nm).

21. Reconnect the wiring harness connectors to the oxygen sensor. Install the wiring harness grommet into the vehicle floor pan.

22. Install the accessory drive belt splash shield.

23. Remove the drain pan and lower the vehicle.

24. Refill the power steering pump reservoir with the correct amount of clean power steering fluid.

25. Reconnect the negative battery cable. Bleed the power steering system.

26. Run the engine and check the system for leaks and proper steering operation.

3.3L (VIN R) and 3.8L (VIN L) Engines

1. Remove and isolate the negative battery cable.

2. Raise and safely support the vehicle. Place a drain pan under the power steering pump.

3. Disconnect the wiring harness connector to the oxygen sensor which is accessible through the oxygen sensor wiring harness grommet in the vehicle floor pan.

4. Remove the catalytic converter from the exhaust manifold and remove all exhaust system hangers and isolators from the exhaust system brackets. Move exhaust system out of the way as far rearward and to the left as possible to provide access to the power steering pump.

5. Remove the power steering fluid return line hose on the front suspension cradle. Allow the fluid to drain from the pump and hose.

6. Remove the accessory drive belt splash shield.

7. Remove accessory drive belt.

8. Disconnect the power steering remote reservoir supply hose from the fitting on the power steering pump. Allow fluid to drain from the hose.

9. Remove power steering fluid pressure line from the power steering pump and drain any excess power steering fluid.

10. Remove the power steering fluid return hose from the power steering pump.

11. Remove the rear support bracket mounted behind the power steering pump to the engine block.

12. Remove the 3 mounting bolts that secure the pump to the alternator/power steering pump and belt tensioner mounting bracket.

13. Remove the power steering pump and pulley assembly from the vehicle. Transfer all required parts from the pump to the new replacement pump before installation.

To install:

14. Position the front of the power steering pump up onto the mounting bracket. Torque the 3 power steering pump-to-mounting bracket bolts to 40 ft. lbs. (54 Nm).

15. Install the rear power steering pump-to-engine block support bracket. Torque the support bracket mounting bolts to 40 ft. lbs. (54 Nm). Install the nut to the mounting stud behind the pump and torque to 40 ft. lbs. (54 Nm).

16. Install the high pressure fluid line to the pump output fitting. Torque the high pressure line-to-power steering pump fitting to 275 inch lbs. (31 Nm). Be sure to inspect the pressure line O-ring for any damage before connecting the pressure line to the steering pump.

17. Install the low pressure power steering fluid hose to the power steering pump low pressure fitting. Be sure the hose clamps are properly reinstalled and hoses are clear of the accessory drive belts.

18. Install the accessory drive belt.

19. Install the hose on the power steering fluid return line on the front suspension cradle. Be sure the hose clamps and heat shield tubes are correctly reinstalled.

20. Reconnect the exhaust pipe to the exhaust manifold. Install the hangers and isolators onto the exhaust system brackets. Torque the nuts and bolts to 250 inch lbs. (28 Nm).

21. Reconnect the wiring harness connectors to the oxygen sensor. Install the wiring harness grommet into the vehicle floor pan.

22. Install the accessory drive belt splash shield.

23. Remove the drain pan and lower the vehicle.

24. Refill the power steering pump reservoir with the correct amount of clean power steering fluid.

25. Reconnect the negative battery cable. Bleed the power steering system.

26. Run the engine and check the system for leaks and proper steering operation.

BRAKES

Anti-Lock Brake System Service

PRECAUTIONS

• Certain components within the Anti-Lock Brake System (ABS) are not intended to be serviced or repaired individually. Only those components with removal and installation procedures should be serviced.

• Do not use rubber hoses or other parts not specifically specified for and ABS system. When using repair kits, replace all parts included in the kit. Partial or incorrect repair may lead to functional problems and require the replacement of components.

• Lubricate rubber parts with clean, fresh brake fluid to ease assembly. Do not use lubricated shop air to clean parts; damage to rubber components may result.

• Use only specified brake fluid from an unopened container.

• If any hydraulic component or line is removed or replaced, it may be necessary to bleed the entire system.

• A clean repair area is essential. Always clean the reservoir and cap thoroughly before removing the cap. The slightest amount of dirt in the fluid may plug an orifice and impair the system function. Perform repairs after components have been thoroughly cleaned; use only denatured alcohol to clean components. Do not allow ABS components to come into contact with any substance containing mineral oil; this includes used shop rags.

• The Anti-Lock control unit is a microprocessor similar to other computer units in the vehicle. Ensure that the ignition switch is **OFF** before removing or installing controller harnesses. Avoid static electricity discharge at or near the controller.

• If any arc welding is to be done on the vehicle, the control unit should be unplugged before welding operations begin.

DEPRESSURIZING

1993 Only

— CAUTION —
The hydraulic accumulator contains brake fluid and nitrogen gas at extremely high pressures on these models. Certain portions of the hydraulic system also contain brake fluid at high pressure. It is mandatory that the system be depressurized before disconnecting any hoses, lines or fittings or personal injury may result.

These vehicles are equipped with the Bendix Anti-Lock 10 brake system. The pump/motor unit keeps the accumulator charged between approximately 1600–2000 psi anytime the ignition switch is in the **ON** position and therefore, must be depressurized before removing any portion of the hydraulic system. Use the following procedure.

1. Disconnect and isolate the negative battery cable from the battery.
2. Pump the brake pedal, at least, 40 times using approximately 50 pounds of pedal force. As the accumulator discharges, a noticeable change in pedal feel should occur.
3. After a definite increase in brake pedal effort is felt, pump the pedal a few more times. This will remove all pressure from the hydraulic system.

— CAUTION —
Even after the system has been depressurized, it is still good practice to use care before disconnecting any part of the hydraulic system. Some residual pressure may be in the system and brake fluid could be released under pressure when loosening hydraulic connections. Wear eye protection and use care to keep brake fluid off painted surfaces.

Master Cylinder

REMOVAL AND INSTALLATION

1993 Models

Without ABS

1. Disconnect the negative battery cable.
2. Disconnect the brake lines from the master cylinder. Plug the outlet ports and brake lines to prevent fluid loss and dirt entry.
3. Remove the master cylinder attaching nuts and remove the master cylinder.

   To install:
4. Bench bleed the replacement master cylinder.
5. Position the master cylinder on the booster studs, aligning the booster pushrod with the master cylinder piston. Secure the master cylinder with the attaching nuts. Torque the attaching nuts to 250 inch lbs. (28 Nm).
6. Reconnect the brake lines to the ports in the master cylinder. Make sure the fluid in the reservoir is at the proper level.
7. Have an assistant apply the brake pedal. With the pedal depressed, loosen the forward brake line fitting until the brake pedal drops to the floor. Tighten the fitting before the brake pedal is allowed to return. Repeat until no more air bubbles are released.
8. Repeat the previous Step at the rear brake line.
9. Final tighten the brake line fittings to 145 inch. lbs. (17 Nm).
10. Reconnect the negative battery cable.
11. Bleed the brake system.
12. Road test and check brake operation.

With ABS

NOTE: The master cylinder and power brake booster are integral components of the ABS hydraulic assembly.

— CAUTION —
The hydraulic accumulator contains brake fluid and nitrogen gas at extremely high pressures. Certain portions of the hydraulic system also contain brake fluid at high pressure. It is mandatory that the system be depressurized before disconnecting any hoses, lines or fittings or personal injury may result.

1. Depressurize the hydraulic accumulator.
2. Disconnect the negative battery cable.
3. Remove the fresh air intake duct, air cleaner and windshield washer fluid bottle.
4. Label and disconnect the wiring connectors from the hydraulic assembly.
5. Remove as much brake fluid as possible from the brake fluid reservoir.
6. Disconnect the pump high pressure hose fitting. Disconnect the pump return hose from the steel tube and cap the end of the steel tube.
7. Disconnect the brake lines from the hydraulic assembly.
8. Working under the instrument panel, position a small screwdriver between the center tang on the retainer clip and the pin in the brake pedal. Rotate the screwdriver enough to allow the retainer clip center tang to pass over the end of the brake pedal pin. Discard the retainer clip.

9. Remove the 4 hydraulic assembly-to-dash panel nuts and remove the hydraulic assembly.

To install:

10. Install the hydraulic assembly into the dash panel mounting holes. Install the nuts and tighten to 20 ft. lbs. (28 Nm).

11. Lubricate the bearing surface of the brake pedal pin. Connect the pushrod to the pedal pin and install a new retainer clip. Make sure the pushrod is installed in the proper position.

12. Connect the brake lines to the hydraulic assembly and tighten the fittings to 12 ft. lbs. (16 Nm). If the proportioning valves were removed from the hydraulic assembly, install and tighten to 30 ft. lbs. (40 Nm).

13. Connect the return hose to the steel tube and tighten the clamp to 10 inch lbs. (1 Nm). Connect the high pressure hose and tighten the fitting to 12 ft. lbs. (16 Nm).

14. Fill the brake fluid reservoir to the top of the screen on the reservoir filter strainer.

15. Reconnect the wiring connectors to the hydraulic assembly.

16. Bleed the brake system.

17. Install the fresh air intake ducts, air cleaner and washer bottle.

18. Reconnect the negative battery cable. Make sure the fluid level in the reservoir is at the correct level before moving the vehicle.

1994–95 Models

1. Disconnect the negative battery cable.

2. Disconnect the brake lines from the master cylinder. Plug the outlet ports and brake lines to prevent fluid loss and dirt entry.

3. Remove the master cylinder attaching nuts and remove the master cylinder.

To install:

4. Bench bleed the replacement master cylinder.

5. Position the master cylinder on the booster studs, aligning the booster pushrod with the master cylinder piston. Secure the master cylinder with the attaching nuts. Torque the attaching nuts to 250 inch lbs. (28 Nm).

6. Reconnect the brake lines to the ports in the master cylinder. Make sure the fluid in the reservoir is at the proper level.

7. Have an assistant apply the brake pedal. With the pedal depressed, loosen the forward brake line fitting until the brake pedal drops to the floor. Tighten the fitting before the brake pedal is allowed to return. Repeat until no more air bubbles are released.

8. Repeat the previous Step at the rear brake line.

9. Final torque the brake line fittings to 145 inch lbs. (17 Nm).

10. Bleed the brake system.

11. Reconnect the negative battery cable. Check the brake fluid level and top off to the reservoir FULL mark, if necessary.

12. Road test vehicle to check brake system operation.

1996–97 Models

1. With the engine turned OFF, pump the brake pedal several times until a firm brake pedal is achieved.

2. Disconnect the negative battery cable.

3. To prevent possible hydraulic system contamination, thoroughly clean all surfaces of the brake fluid reservoir, filler neck and master cylinder. Use MOPAR® Brake Parts Cleaner, or an equivalent solvent.

4. Remove the brake fluid reservoir filler tube by pushing down and turning. Remove the cap from the removed filler tube and install it on the brake fluid reservoir.

5. Disconnect the brake fluid level sensor wiring connector from the side of the brake fluid reservoir.

6. Disconnect the brake lines from the master cylinder. Plug the outlet ports and brake lines to prevent fluid loss and dirt entry.

7. To prevent any dirt particles from falling into the vacuum booster, thoroughly clean the master cylinder and power brake booster using MOPAR® Brake Parts Cleaner, or equivalent. Remove the master cylinder attaching nuts and remove the master cylinder.

8. Remove the vacuum seal located on the mounting flange of the master cylinder by carefully pulling it away from the master cylinder. Discard the old vacuum seal.

To install:

9. Bench bleed the replacement master cylinder.

10. Install a new vacuum seal onto the master cylinder. Be sure the new seal is seated squarely in the groove of the master cylinder casting.

11. Position the master cylinder on the booster studs, aligning the booster pushrod with the master cylinder piston. Secure the master cylinder with the mounting nuts. Torque the mounting nuts to 225 inch lbs. (25 Nm).

12. Reconnect the primary and secondary brake lines to the primary and secondary ports in the master cylinder. Make sure the fluid in the reservoir is at the proper level.

13. Have an assistant apply the brake pedal. With the pedal depressed, loosen the forward brake line fitting until the brake pedal drops to the floor. Tighten the fitting before the brake pedal is allowed to return. Repeat until no more air bubbles are released.

14. Repeat the previous Step at the rear brake line.

15. Final tighten the brake lines at the master cylinder, being sure to hold the brake lines securely during the tightening of the brake line fittings to control orientation of the flex section. Torque the brake line fittings to 145 inch lbs. (17 Nm).

16. Reconnect the wiring connector to the brake fluid level sensor on the side of the master cylinder.

17. Install the filler tube onto the master cylinder brake fluid reservoir.

18. Reconnect the negative battery cable. Refill the master cylinder brake fluid reservoir and bleed the brake system.

19. Road test the vehicle using extreme caution while testing brake system operation. Check the brake fluid level and top off if necessary.

Brake Caliper

REMOVAL AND INSTALLATION

1993–95 Models

1. Raise and safely support the front of the vehicle. Remove the front wheels.

2. Reach to the inside of the caliper assembly and pull it outboard as far as possible. This will push the piston back into the bore of the caliper making removal of the caliper from the adapter easier.

3. If the caliper is only being removed from the bracket (as for a brake pad change), move to Step 4. If the caliper is being removed from the vehicle (as for replacement or overhaul and reseal), remove the brake hose attaching bolt from the caliper. Remove the hose from the caliper and discard the washers. New seal washers will be required at assembly. Plug the brake hose to prevent fluid leakage.

4. Remove the caliper guide pin bolts.

5. Insert a small prybar between the front edge of the caliper and the adapter rail. Be sure to apply a steady, upward pressure.

6. Remove the caliper by slowly sliding it up and off the adapter and

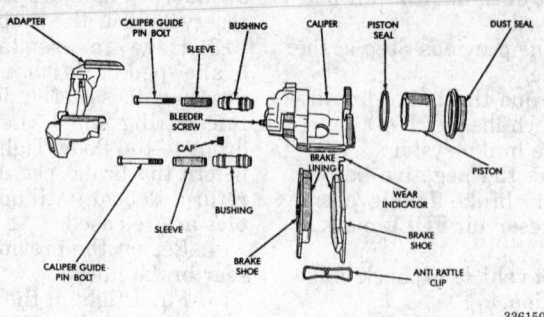

336150

Kelsey-Hayes double pin single caliper and related components — 1993–95 models

disc rotor. Support the brake caliper assembly using a strong piece of wire to suspend it from the strut spring. Do not allow the caliper to hang from the brake fluid hose or damage to the hose will result.

To install:

7. Clean all parts well. Inspect the caliper guide pin bolts and their seals in the caliper. Replace the bolts if corroded. New bushings should be installed if worn or hardened with age. A light coat of silicone grease should make bushing installation easier and will help protect the seals and guide pin bolts.

8. Thoroughly clean both caliper adapter abutment rails of any dirt, grease or corrosion. Do not sand the abutment rails, instead use a steel wire brush.

9. Lubricate both caliper adapter abutments with a liberal amount of MOPAR® Multipurpose Lubricant, or equivalent.

10. Lower the caliper over the brake pads and disc rotor. Be sure to clear the adapter so possible damage to the guide pin bolts, bushings and sleeves does not result. Install the caliper guide pin bolts and torque to 25–35 ft. lbs. (34–47 Nm). Be careful not to cross thread the caliper guide pin bolts.

11. If removed, attach the brake hose to the caliper using new washers. Torque the banjo bolt to 24 ft. lbs. (33 Nm).

12. Bleed the brake system.

13. Install the front wheels and lug nuts. Torque the lug nuts, in a star pattern sequence, to half torque specifications. Then repeat the tightening sequence to the full torque specification of 95 ft. lbs. (129 Nm). Lower the vehicle.

14. Pump the brake pedal several times to insure that the brake pedal is firm. Road test the vehicle and make several stops to clear away any foreign material on the brakes and to seat the brake pads.

1996–97 Models

1. Raise and safely support the front of the vehicle. Remove the front wheels.

2. If the caliper is only being removed from the bracket (as for a brake pad change), move to Step 3. If the caliper is being removed from the vehicle (as for replacement or an overhaul and reseal), remove the brake hose attaching bolt from the caliper. Remove the hose from the caliper and discard the washers. New seal washers will be required at assembly. Plug the brake hose to prevent fluid leakage.

3. Remove the caliper guide pin bolts that secure the caliper to the steering knuckle.

4. Remove the caliper by slowly sliding it away from the steering knuckle. Slide the opposite end of the brake caliper out from under the machined abutment on the steering knuckle.

5. Using a strong piece of wire, support the brake caliper assembly off the strut unit. Do NOT allow the caliper to hang from the brake fluid flex hose or damage to the hose will result.

To install:

6. Clean both steering knuckle abutment surfaces of any dirt, grease

or corrosion. Then lubricate the abutment surfaces with a liberal amount of MOPAR® Multipurpose Lubricant, or equivalent.

7. Properly position the brake caliper over the brake pads and disc rotor. Be careful not to allow the caliper seals or guide pin bushings to get damaged by the steering knuckle bosses. Install the caliper guide pin bolts and torque to 30 ft. lbs. (41 Nm). Be careful not to cross thread the guide pin bolts.

8. If removed, attach the brake hose to the caliper using new washers. Tighten the banjo bolt to 35 ft. lbs. (47 Nm).

9. Bleed the brake system.

10. Install the front wheels and lug nuts. Torque the lug nuts, in a star pattern sequence, to half torque specifications. Then repeat the tightening sequence to the full torque specification of 100 ft. lbs. (135 Nm). Lower the vehicle.

11. Pump the brake pedal several times to insure that the brake pedal is firm. Road test the vehicle.

Disc Brake Pads

REMOVAL AND INSTALLATION

1993–95 Models

1. Remove brake fluid from the master cylinder brake fluid reservoir until the reservoir is approximately ½ full. Discard the removed fluid.

2. Raise and safely support the front of the vehicle. Remove the front wheels.

3. Reach to the inside of the caliper assembly and pull it outboard as far as possible. This will push the piston back into the caliper bore, making caliper removal easier.

4. Remove the caliper guide pin bolts.

5. Insert a small prybar between the front edge of the caliper and the

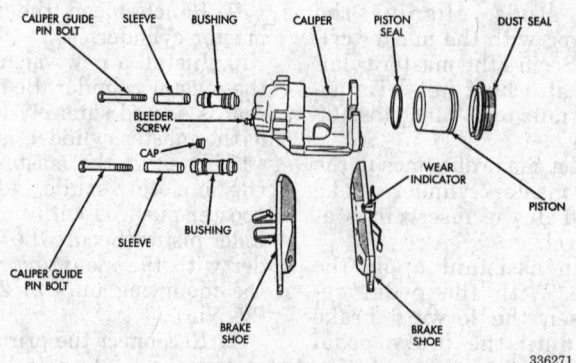

336271

Front brake caliper assembly — 1996–97 models

adapter rail. Apply steady upward pressure to loosen the adhesive seals.

6. Remove the caliper by slowly sliding it up and off the adapter and brake rotor. Support the caliper out of the way with wire. Do not let the caliper hang by the brake hose or damage to the hose will result.

7. If necessary, compress the caliper piston into the bore using a C-clamp. Insert a suitable piece of wood between the C-clamp and caliper piston to protect the piston.

8. Remove the outboard disc brake pad from the caliper adapter.

9. Remove the disc brake rotor.

10. Remove the inboard disc brake pad by sliding it out along the bottom adapter abutment until the pad loosens from the anti-rattle clip. Remove the anti-rattle clip from the top adapter abutment.

To install:

11. Thoroughly clean the adapter abutment rails with a wire brush. Lubricate the abutments with a liberal amount of MOPAR® Multipurpose Lubricant, or equivalent.

12. Install the anti-rattle clip to the upper adapter abutment. If equipped, remove the protective paper from the noise suppression gaskets on the new disc brake pads.

13. Install the inboard disc brake pad, making sure it is properly positioned against the anti-rattle clip. Be careful not to get grease on the pad surface.

14. Install the disc brake rotor.

15. Install the outboard disc brake pad on the adapter abutment.

16. Lower the caliper over the brake pads and disc rotor. Install the caliper guide pins and tighten to 25–35 ft. lbs. (34–47 Nm).

17. Install the front wheels and lug nuts. Torque the lug nuts, in a star pattern sequence, to 95 ft. lbs. (129 Nm). Apply the brake pedal several times until a firm pedal is obtained. Lower the vehicle.

18. Check the fluid level in the master cylinder and add fluid as necessary. Road test the vehicle.

1996–97 Models

1. Remove brake fluid from the master cylinder brake fluid reservoir until the reservoir is approximately ½ full. Discard the removed fluid.

2. Raise and safely support the front of the vehicle. Remove the front wheels.

3. Remove the front brake caliper guide pin bolts.

4. Remove the brake caliper by slowly sliding it up and off the adapter and brake rotor. Support the caliper out of the way with a strong piece of wire. Do not let the caliper hang by the brake hose or damage to the brake hose will result.

5. If necessary, compress the caliper piston into the bore using a C-clamp. Insert a suitable piece of wood between the C-clamp and caliper piston to protect the piston.

6. Remove the outboard disc brake pad from the caliper by prying the brake pad retaining clip over the raised area on the caliper. Slide the brake pad down and off the caliper.

7. Remove the inboard disc brake pad from the caliper by pulling the brake pad away from the caliper piston until the retaining clip on the pad is free from the caliper piston cavity.

To install:

8. Be sure the caliper piston has been completely retracted into the piston bore of the caliper assembly. This is required when installing the brake caliper equipped with new brake pads.

9. If equipped, remove the protective paper from the noise suppression gaskets on the new disc brake pads.

10. Install the new inboard disc brake pad into the caliper piston by pressing the pad firmly into the cavity of the caliper piston. Be sure the new inboard brake pad is seated squarely against the face of the brake caliper piston.

11. Install the outboard disc brake pad by sliding it onto the caliper assembly.

12. Install the brake caliper assembly over the brake rotor and onto the steering knuckle adapter. Install the caliper guide pins and tighten to 30 ft. lbs. (41 Nm).

13. Install the front wheels and lug nuts. Torque the lug nuts, in a star pattern sequence, to 95 ft. lbs. (129 Nm). Apply the brake pedal several times until a firm pedal is obtained.

14. Check the fluid level in the master cylinder and add fluid as necessary. Road test the vehicle.

Brake Rotor

REMOVAL AND INSTALLATION

1. Raise and safely support the vehicle.

2. Remove the wheel.

3. Remove the caliper and brake pads. Do not disconnect the brake hose from the caliper. Support the caliper from the strut using a strong piece of wire; do not let the caliper hang from the brake hose.

4. Remove the factory installed clips, if equipped, from the wheel studs. It is not necessary to reinstall these clips.

5. Chalk an index mark on the brake rotor before removal to show its exact installation on the hub. Remove the rotor from the hub. Be sure to clean the mating surface between the hub and the rotor of any dirt, grease or debris before installing the rotor.

6. Inspect the rotor for thickness, excessive scoring or warpage and resurface, reface or replace, if necessary. Minimum allowable thickness is 0.881 in. (22.4mm).

NOTE: All brake rotors have the specification for minimum allowable thickness cast on the face of the rotor.

To install:

7. Install the rotor on the hub.

8. Install the caliper and brake pads.

9. Install the wheel and lug nuts. Torque the lug nuts, in a star pattern sequence, to 95 ft. lbs. (129 Nm).

10. Lower the vehicle and depress the brake pedal several times to position the caliper piston. Road test the vehicle.

Brake Drums

REMOVAL AND INSTALLATION

1993–95 Models

Front Wheel Drive Vehicles

1. Raise and safely support the vehicle.

2. Remove the wheel.

3. Remove the grease cap.

4. Remove the cotter pin and nut lock.

5. Remove the wheel bearing nut and washer from the spindle.

6. Remove the outer wheel bearing.

7. Remove the drum with the inner wheel bearing from the spindle. If necessary, remove the grease seal and inner wheel bearing.

NOTE: If the drum is difficult to remove, remove the plug from the adjusting hole in the rear of the backing plate. Push the self adjuster lever away from the star wheel using a thin screwdriver. Rotate the star wheel using a brake adjusting tool to retract the shoes.

To install:

8. If removed, lubricate and install the inner wheel bearing. Install a new grease seal.

9. Install the drum to the spindle.

10. Lubricate and install the outer wheel bearing. Install the thrust washer and nut.

11. Tighten the nut to 240–300 inch lbs. (27–34 Nm) while rotating the hub, to seat the bearings. Back off the nut ¼ turn, then tighten the nut finger tight.

12. Install the nut lock with one pair of slots in line with the cotter pin hole, then install a new cotter pin. Install the grease cap.

13. Install the wheel and lug nuts. Torque the lug nuts, in sequence, to 95 ft. lbs. (129 Nm).

14. Adjust the brake shoes and lower the vehicle.

All Wheel Drive Vehicles

1. Raise and safely support the vehicle.

2. Remove the wheel.

3. Remove the brake drum from the hub assembly.

NOTE: If the drum is difficult to remove, push the self adjuster lever away from the star wheel using a thin screwdriver. The lever and star wheel are accessed through the hole on the front of the brake drum. Rotate the star wheel using a brake adjusting tool to retract the shoes.

To install:

4. Install the brake drum to the hub assembly.

5. Adjust the brake shoes.

6. Install the wheel and lug nuts. Torque the lug nuts, in sequence, to 95 ft. lbs. (129 Nm). Lower the vehicle.

1996–97 Models

1. Raise and safely support the vehicle.

2. Remove the rear wheels.

3. Remove the brake drum from the hub assembly by pulling the drum straight off the wheel studs.

NOTE: If the drum is difficult to remove, remove the plug from the rear brake support plate. Push the self adjuster lever away from the star wheel using a thin screwdriver. The lever and star wheel are accessed through the hole on the rear brake support plate. Rotate the star wheel using a brake adjusting tool to retract the shoes.

4. Inspect the brake drum for thickness and runout. Replace or machine as necessary.

To install:

5. Install the brake drum to the hub assembly.

6. Adjust the brake shoes.

7. Install the rear wheel and lug nuts. Torque the lug nuts, in a star pattern sequence, to 95 ft. lbs. (129 Nm). Lower the vehicle.

Brake Shoes

REMOVAL AND INSTALLATION

—— **CAUTION** ——

Be aware that, although factory installed brake shoes are manufactured of asbestos-free materials, some aftermarket brake shoes do contain asbestos. Asbestos can cause serious bodily harm such as asbestosis and/or cancer. This should be taken into account when performing any service on the vehicle's brake system, if aftermarket brake shoes or pads have been installed. When cleaning brake components, always wear a respirator. Never use compressed air to clean brake components, always use an approved vacuum cleaner specifically designed for brake dust removal. If a vacuum cleaner is not available, clean the brake components using water-dampened shop towels. Do not sand brake shoes or pads while servicing the brake system as this will create brake lining dust. Dispose of all dust and dirt containing asbestos materials using sealed containers or air-tight bags. Be sure to follow all recommended safety practices as advised by the Environmental Protection Agency (EPA) and the Occupational Safety and Health Administration (OSHA), for the handling and disposal of products which contain asbestos.

1993–95 Models

1. Raise and safely support the vehicle.

2. Remove the rear wheels. Remove the the brake drums.

3. Remove the primary and secondary (front and rear) shoe return springs from the anchor pin. Be sure to remember that the secondary shoe return spring overlaps the primary shoe return spring.

4. Lift the adjuster lever and disconnect the actuator cable.

5. Remove the actuator cable, overload spring, actuator cable guide and anchor pin plate.

6. Disengage the adjusting lever from the spring by sliding it forward to clear the pivot, then working it out from under the spring. Remove the spring from the pivot.

7. Remove the shoe-to-shoe spring from the secondary shoe (rear brake shoe) and disengage from the primary shoe. Remove the spring.

8. Separate the primary and secondary brake shoes from the adjuster screw assembly. Remove the adjuster screw assembly.

9. Remove the shoe retainers, hold-down springs and nails from the brake support plate using a brake hold-down spring removal/installer tool.

10. Remove the parking brake lever from the secondary brake shoe on the left side of the vehicle and/or primary brake shoe on the right side of the vehicle. Remove the brake shoes from the vehicle.

11. Disconnect the parking brake lever from the parking brake cable.

To install:

12. Thoroughly clean and dry the backing plate. To prepare the backing plate, lubricate the contact pads, anchor pin and parking brake actuating lever pivot surface lightly with MOPAR® Multipurpose Lubricant, or equivalent.

13. Remove, clean and dry all parts still on the old shoes. Lubricate the star wheel shaft threads with anti-seize lubricant and transfer all parts to their proper locations on the new shoes.

14. Place the primary brake shoe into position against the brake support plate. The upper web of the shoe should engage the wheel cylinder piston and anchor pin on the support plate. Install the brake shoe retaining nail, hold-down spring and retainer using a brake hold-down spring removal/installer tool.

15. Install the anchor plate over the anchor pin. Install the eye of the actuating cable over the anchor pin.

16. Install the primary brake shoe return spring into the brake shoe web and engage the return spring to the anchor pin using a brake spring removal/installer tool.

17. Install the parking brake strut into the slot in the parking brake lever and install the anti-rattle spring over the free end of the strut.

18. Connect the parking brake cable to the parking brake lever. Install the parking brake lever into the rectangular hole of the secondary

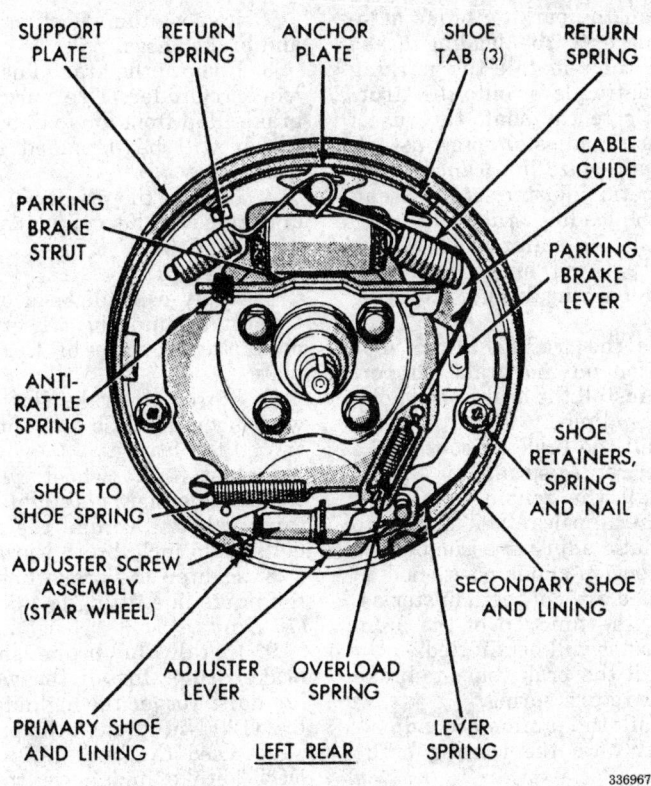

SUPPORT PLATE — RETURN SPRING — ANCHOR PLATE — SHOE TAB (3) — RETURN SPRING — CABLE GUIDE — PARKING BRAKE STRUT — PARKING BRAKE LEVER — ANTI-RATTLE SPRING — SHOE RETAINERS, SPRING AND NAIL — SHOE TO SHOE SPRING — ADJUSTER SCREW ASSEMBLY (STAR WHEEL) — SECONDARY SHOE AND LINING — ADJUSTER LEVER — OVERLOAD SPRING — PRIMARY SHOE AND LINING — **LEFT REAR** — LEVER SPRING

336967

Brake shoe assembly — 1993–95 front-wheel drive (left rear side shown)

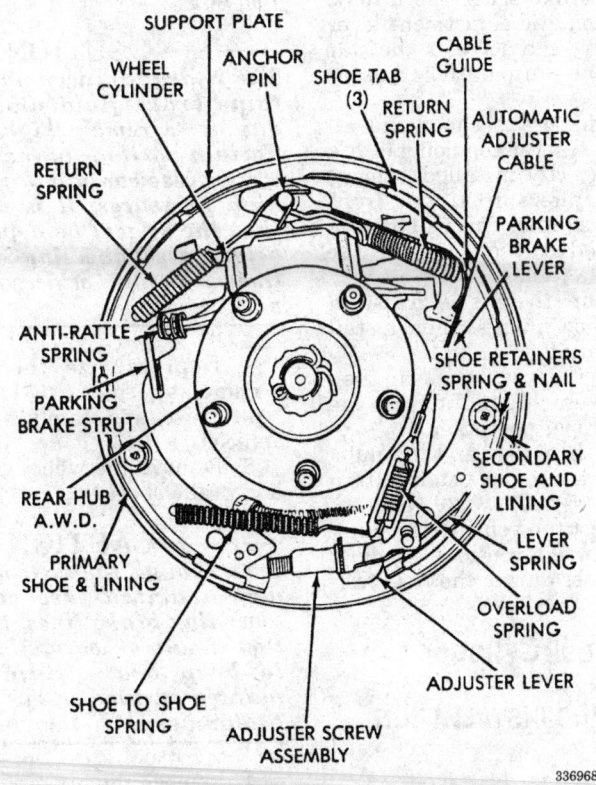

SUPPORT PLATE — WHEEL CYLINDER — ANCHOR PIN — SHOE TAB (3) — CABLE GUIDE — RETURN SPRING — AUTOMATIC ADJUSTER CABLE — RETURN SPRING — PARKING BRAKE LEVER — ANTI-RATTLE SPRING — PARKING BRAKE STRUT — SHOE RETAINERS SPRING & NAIL — REAR HUB A.W.D. — SECONDARY SHOE AND LINING — PRIMARY SHOE & LINING — LEVER SPRING — SHOE TO SHOE SPRING — OVERLOAD SPRING — ADJUSTER SCREW ASSEMBLY — ADJUSTER LEVER

336968

Brake shoe assembly — 1993–95 all-wheel drive

brake shoe on the left side of the vehicle or the primary shoe on the right side of the vehicle.

19. Place the secondary brake shoe into position against the brake support plate. The upper web of the shoe should engage the wheel cylinder piston and the parking brake strut. Install the brake shoe retaining nail, hold-down spring and retainer using a brake hold-down spring removal/installer tool.

20. Insert the protruding hole rim of the actuating cable guide into the hole of the secondary brake shoe web. While holding the cable guide in position, install the secondary brake shoe return spring through the hole of the brake shoe web/actuating cable guide and then over the anchor pin, using a brake spring removal/installer tool.

21. Be sure the cable guide remains flat against the web of the secondary brake shoe and the return spring overlaps the primary return spring. Using pliers, squeeze the ends of the return spring loops around the anchor pin until parallel. Lubricate the sliding surface of the actuator cable plate lightly and make sure the eye of the adjuster cable is still installed over the anchor pin.

22. Install the adjuster screw assembly between the brake shoes. Be sure the star wheel of the adjuster assembly is positioned next to the secondary brake shoe. The left side star wheel adjusting stud is cadmium plated and stamped with the letter **L** to indicate its position on the vehicle. The right side star wheel adjusting stud is black and stamped with the letter **R** to indicate its position on the vehicle.

23. Install the lower shoe-to-shoe spring, engaging the primary brake shoe first.

24. Install the adjusting lever spring over the pivot pin on the web of the brake shoe. Install the adjusting lever under the spring and over the pivot pin. Lock the lever in position by slightly sliding the adjusting lever rearward.

25. Thread the adjuster cable over the cable guide and hook the end of the overload spring to the adjuster lever. Be sure that the cable is pulled tight against the anchor pin and in a straight line with the cable guide.

26. Check the operation of the automatic adjuster mechanism by pulling the adjuster cable rearward which should cause the star wheel to rotate upwards.

27. Make sure there is no grease on the brake shoe linings and install the brake drums.

28. Adjust the brakes, then lower the vehicle and check the brakes for proper operation.

29. Install the wheel and lug nuts. Torque the lug nuts, in a star pattern sequence, to 95 ft. lbs. (129 Nm).

30. Road test the vehicle and check for proper brake operation.

1996–97 Models

1. Raise and safely support the vehicle.

2. Remove the rear wheels and the brake drums.

3. Be sure the parking brake pedal is in the released position. Create slack in the rear parking brake cables by grasping an exposed section of the front parking brake cable, pulling it down and rearward. Maintain the slack in the brake cable by installing a pair of locking pliers onto the parking brake cable just rearward of **only the rear** body outrigger bracket.

4. Remove the adjustment lever spring from the automatic adjustment lever and front brake shoe (leading brake shoe).

5. Remove the automatic adjustment lever from the front brake shoe (leading brake shoe).

6. Remove the brake shoe-to-brake shoe lower return spring.

7. Remove the tension clip that secures the upper return spring to the automatic adjuster assembly.

8. Remove the brake shoe-to-brake shoe upper return spring.

9. Remove the rear brake shoe (trailing brake shoe) hold-down clip and pin.

10. Remove the trailing brake shoe, parking brake actuating lever and parking brake actuating strut from the brake support plate.

11. Remove the automatic adjuster assembly from the leading brake shoe.

12. Remove the leading brake shoe hold-down clip and pin. Remove the leading brake shoe.

13. Remove the parking brake actuator plate from the leading brake shoe and install onto the replacement brake shoe.

To install:

14. Thoroughly clean and dry the backing plate. To prepare the backing plate, lubricate the 8 brake shoe contact areas and brake shoe anchor, using suitable grease.

15. Install the leading brake shoe into position on the brake shoe support plate. Secure the leading brake shoe by installing the brake shoe hold-down clip and pin.

16. Install the parking brake actuating strut onto the leading brake shoe and then install the parking brake actuating lever onto the strut.

17. Lubricate the shaft threads of the automatic adjuster screw assembly with anti-seize lubricant. Install the automatic adjuster screw assembly onto the leading brake shoe.

18. Install the trailing brake shoe onto the parking brake actuating lever and parking brake actuating strut.

19. Place the trailing brake shoe into position on the brake support plate and install the brake shoe hold-down clip and pin.

20. Install the brake shoe-to-brake shoe upper return spring.

21. Install the tension clip that secures the upper return spring to the automatic adjuster assembly. Be sure the tension clip is positioned on the threaded area of the adjuster assembly or the function of the automatic adjuster will be affected.

22. Install the brake shoe-to-brake shoe lower return spring.

23. Install the automatic adjustment lever onto the leading brake shoe.

24. Install the actuating spring onto the automatic adjustment lever and leading brake shoe. Check to be sure the automatic adjustment lever makes positive contact with the star wheel on the automatic adjuster assembly.

25. Once the brake shoes and all other brake system components are fully and correctly installed, remove the locking pliers from the front parking brake cable. This will remove the slack and correctly adjust the parking brake cables.

26. Make sure there is no grease on the brake shoe linings, then install the brake drums.

27. Adjust the rear brakes, then lower the vehicle and check the brakes for proper operation.

28. Install the wheel and lug nuts. Torque the lug nuts, in a star pattern sequence, to 95 ft. lbs. (129 Nm).

29. Road test the vehicle. The automatic adjuster will continue to adjust the brake shoes during the road test.

Wheel Cylinder

REMOVAL AND INSTALLATION

1993 Models

Without ABS

1. Raise and safely support the vehicle.

2. Remove the wheel, brake drum and brake shoes.

3. Remove the brake line from the wheel cylinder. Use care to keep brake fluid from any painted surface which will be damaged by brake fluid.

4. Remove the wheel cylinder bolts and remove the cylinder from the backing plate.

To install:

5. Apply a small bead of silicone sealant around the wheel cylinder mounting surface of the backing plate.

6. Start the brake line into the wheel cylinder. Use care not to cross-thread the fitting.

7. Install the wheel cylinder on the backing plate and tighten the retaining bolts. Torque the retaining bolts to 75 inch lbs. (8 Nm).

8. Tighten the brake line. Torque the brake line fitting to 145 inch lbs. (17 Nm).

9. Install the brake shoes and brake drum. Install the wheel and lug nuts. Torque the lug nuts to 95 ft. lbs. (129 Nm). Adjust the brakes.

10. Bleed the brake system using DOT 3 brake fluid. Lower the vehicle. Test the brakes for proper operation.

With ABS

--- CAUTION ---
The hydraulic accumulator contains brake fluid and nitrogen gas at extremely high pressure. Certain portions of the hydraulic system also contain brake fluid at high pressures. It is mandatory that the system be depressurized before disconnecting any hoses, lines or fittings or personal injury may result.

1. Depressurize the hydraulic accumulator.

2. Raise and safely support the vehicle.

3. Remove the wheel, brake drum and brake shoes.

--- CAUTION ---
Even though the system has been depressurized, use care when loosening brake lines. Eye protection should be worn. Also use care to keep brake fluid off any painted surface or the paint will be damaged.

4. Remove the brake line from the wheel cylinder.

5. Remove the wheel cylinder bolts and remove the cylinder from the backing plate.

To install:

6. Apply a small bead of silicone sealant around the wheel cylinder mounting surface of the backing plate.

7. Start the brake line into the wheel cylinder.

8. Install the wheel cylinder on the backing plate and tighten the retaining bolts. Torque the retaining bolts to 75 inch lbs. (8 Nm).

9. Tighten the brake line. Torque the brake line fitting to 145 inch lbs. (17 Nm).

10. Install the brake shoes and brake drum. Install the wheel and lug nuts. Torque the lug nuts to 95 ft. lbs. (129 Nm). Adjust the brakes.

11. Bleed the brake system and lower the vehicle.

12. Road test the vehicle check the brakes for proper operation.

1994–97 Models

1. Raise and safely support the vehicle.

2. Remove the wheel, brake drum and brake shoes.

3. Using a line wrench, remove the brake line from the wheel cylinder. Use care to keep brake fluid off any painted surfaces which will be damaged by the brake fluid. Plug the brake line opening to prevent fluid spillage and/or system contamination.

4. Remove the wheel cylinder bolts and remove the cylinder from the backing plate.

To install:

5. Apply a small bead of silicone sealant around the wheel cylinder mounting surface of the backing plate.

6. Start the brake line into the wheel cylinder. Use care not to cross-thread the fitting.

7. Install the wheel cylinder on the backing plate and tighten the retaining bolts. Torque the retaining bolts to 75 inch lbs. (8 Nm).

8. Tighten the brake line. Torque the brake line fitting to 145 inch lbs. (17 Nm).

9. Install the brake shoes and brake drum. Install the wheel and lug nuts. Torque the lug nuts, in a star pattern sequence, to 95 ft. lbs. (129 Nm). Adjust the brakes.

10. Bleed the brake system using DOT 3 brake fluid. Lower the vehicle.

11. Road test the vehicle and check the brakes for proper operation.

Parking Brake Cable

ADJUSTMENT

1993–95 Models

1. Raise and safely support the vehicle.

2. Check and adjust the rear brake shoes before adjusting the parking brake cable.

3. Remove the plastic adjuster cover by pressing in the fingers of the cover through the holes in the bracket.

4. Clean the adjuster threads with a brush and lubricate with grease.

—— **CAUTION** ——
Do not use a metal wire brush as it will scratch off the plating and coating.

5. Release the parking brake pedal, then back off the cable adjusting nut so there is slack in the cable.

6. Push the parking brake pedal all the way to the floor.

7. Mark "bent nail" adjuster approximately ¼ inch (6mm) from bracket.

8. Tighten the adjusting nut until the mark moves into alignment with the edge of the bracket.

NOTE: Replace the bent nail with a new part if the nail end hook has moved all the way to the bracket.

9. Install the plastic cover.

10. Apply and release the parking brake. Make sure the rear wheels rotate freely without dragging.

11. Lower the vehicle.

1996–97 Models

—— **CAUTION** ——
The self-adjusting feature of this parking brake lever assembly contains a clockspring loaded to approximately 8 pounds. Care

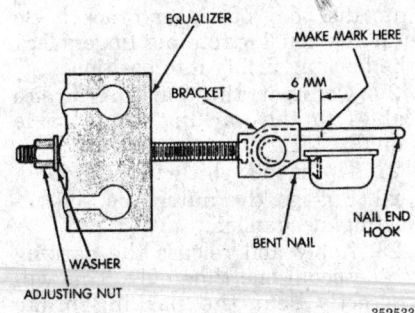

Equalizer/bent nail adjuster assembly — 1993–95 models

must be taken to prevent excessive jarring of the assembly. Do not release the self-adjuster lockout device before installing cables into the equalizer. Keep hands out of the self-adjuster sector and pawl area. Failure to observe this warning in handling this mechanism could lead to serious injury.

Manual Lock-Out of Automatic Self-Adjuster

1. Be sure the parking brake pedal (system) is in the fully released position.

2. From underneath the vehicle, have an assistant grasp the exposed section of the front parking brake cable and pull downward until all free movement is eliminated from the cable.

3. Install a ³⁄₁₆ in. drill bit into the clockspring and position against the parking brake pedal arm.

Engaging the Automatic Self-Adjuster

1. Be sure all the parking brake cables are correctly installed, clipped as required and properly connected.

2. Using a pair of pliers, firmly grasp the lock pin previously installed in the parking brake pedal mechanism.

3. Remove the lock pin from the parking brake pedal mechanism by pulling it firmly and rapidly from the park brake mechanism. This will allow the mechanism to correctly adjust the parking brake cables.

4. Apply and release the parking brake pedal one time. This will seat the parking brake cables. The rear wheels should rotate freely without dragging.

REMOVAL AND INSTALLATION

1993–95 Models

Front Cable

1. Raise and safely support the vehicle.

2. Loosen the adjusting nut and disconnect the front cable from the adjuster. Use a hose clamp to compress the fingers of the cable housing retainer and remove the cable from the anchor bracket.

3. Remove the cable guide brackets from the frame rail and loosen the cable housing at the pedal.

4. Lift the floor mat for access to the floor pan. Force the seal surrounding the cable from the floor.

5. Pull the cable forward and disconnect from the lever clevis. Tap the cable housing end fitting out of the pedal assembly bracket.

6. Pull the cable assembly into the vehicle through the floor pan hole.

To install:

7. Feed the new cable through the floor pan hole. Attach the front end of the cable to the parking brake lever clevis and support.

8. Install the floor pan seal and floor mat.

9. Install the cable and housing into the anchor bracket. Be sure the housing retainer fingers lock the housing firmly into position.

10. Install the cable guide brackets and adjust the parking brake.

11. Lower the vehicle and check for proper brake operation.

Rear Cables

1. Raise and safely support the vehicle.

2. Back off the cable adjustment and disconnect the rear cable that is to be replaced from the equalizer. Remove the cable from the anchor bracket. A 14mm box wrench or small hose clamp can be used to compress the cable lock.

3. Disconnect the cable retaining clips.

4. Remove the rear wheel, brake drum and brake shoes from the side requiring replacement.

5. Disconnect the cable from the rear brake shoe lever. Compress the cable lock with a mini-hose clamp or 14mm box wrench and pull the cable from the backing plate.

To install:

6. Install the new cable through the brake backing plate. Engage the locks.

7. Attach the cable to the brake shoe lever. Install the brake shoes, brake drum, and rear wheel. Torque the lug nuts, in sequence, to 95 ft. lbs. (129 Nm).

8. Install the cable to the anchor bracket. Engage the locks.

9. Install the cable retaining clips. Connect the equalizer bracket.

10. Adjust the brakes and parking brake.

11. Lower the vehicle and check for proper brake operation.

1996–97 Models

Front Cable

1. Raise and safely support the vehicle.

2. Manually lock out the automatic self-adjuster mechanism of the parking brake pedal assembly.

3. Remove the intermediate brake cable and the left rear parking brake cable from the parking brake cable equalizer.

4. Remove the front parking brake cable housing retainer from the body outrigger bracket. Use a hose clamp or a 14mm box wrench to compress the fingers of the cable housing retainer.

5. Lower the vehicle to the ground.

6. Remove the left front door sill molding from the vehicle.

7. Remove the left front kick panel for access to the parking brake cable and parking brake pedal assembly.

8. Lift the floor mat/carpet for access to the parking brake cable and vehicle floor pan. Remove the seal and the cable from the floor pan.

9. Pull the cable forward and disconnect from the lever clevis. Tap the cable housing end fitting out of the pedal assembly bracket.

10. Remove the parking brake cable retainer from the parking brake pedal assembly bracket.

11. Pull the cable assembly out of the vehicle through the floor pan hole.

To install:

12. Feed the new cable through the floor pan hole.

13. Route the brake cable end button through the hole in the parking brake pedal assembly bracket.

14. Install the brake cable retainer onto the parking brake cable. Install the parking brake cable retainer into the park brake pedal assembly bracket.

15. Install the parking brake cable end into the retainer previously installed into the parking brake pedal bracket.

16. Install the front parking brake cable end button into the clevis of the parking brake pedal mechanism.

17. Install the floor pan seal into the hole in the floor pan. Install the seal so the flange on the seal is flush with the floor pan. Place the floor mat/carpet back down on the floor.

18. Raise and safely support the vehicle.

19. Install the cable and housing into the body outrigger bracket. Be sure the housing retainer fingers lock the housing firmly into position.

20. Connect the parking brake cables to the parking brake cable equalizer.

21. Lower the vehicle to the ground.

22. Engage the automatic self-adjusting mechanism.

23. Apply and release the parking brake pedal one time only. This will correctly seat the parking brake cables.

24. Road test the vehicle and check for proper brake operation.

Intermediate Parking Brake Cable

1. Raise and safely support the vehicle.

2. Manually lock out the automatic self-adjuster mechanism of the parking brake pedal assembly.

3. Remove the intermediate brake cable and the left rear parking brake cable from the parking brake cable equalizer.

4. Disconnect the intermediate parking brake cable from the cable connector attaching it to the right rear parking brake cable.

5. Remove the intermediate parking brake cable from the cable guides on the vehicle frame rail.

To install:

6. Install the ends of the parking brake cables through the parking brake cable guides.

7. Connect the intermediate parking brake cable to the cable connector at the right rear parking brake cable.

8. Connect the intermediate parking brake cable to the parking brake cable equalizer.

9. Engage the automatic self-adjusting mechanism.

10. If equipped, install and properly position the foam collar on the parking brake cable to prevent the brake cable from rattling against the floor of the vehicle.

11. Lower the vehicle to the ground.

12. Apply and release the parking brake pedal one time only. This will correctly seat the parking brake cables.

13. Road test the vehicle and check for proper brake operation.

Right Rear Parking Brake Cable

1. Raise and safely support the vehicle. Remove the right rear wheel.

2. Remove the right rear wheel brake drum.

3. Manually lock out the automatic self-adjuster mechanism of the parking brake pedal assembly as follows:

   a. Position the park brake pedal in the fully released position.

   b. From below the vehicle, firmly grasp the exposed section of the front parking brake cable and pull downward until all free movement is removed from the parking brake cable.

   c. Install a pair of locking pliers onto the parking brake cable just rearward of the second body outrigger bracket.

4. Disconnect the right rear parking brake cable from the connector on the intermediate parking brake cable.

5. Remove the right parking brake cable housing from the body bracket by slipping a 14mm box wrench over the end of the brake cable retainer to compress the retaining fingers.

6. Remove the brake shoes from the brake support plate.

7. Disconnect the cable from the parking brake actuator lever in the brake shoe assembly. Compress the cable housing retainer lock with a mini-hose clamp or 14mm box wrench and pull the cable from the brake support plate.

To install:

8. Install the new cable through the brake support plate. Engage the cable housing retainer fingers until they lock the cable housing firmly into place.

9. Attach the cable to the parking brake actuator lever on the brake shoe assembly. Install the brake shoes, brake drum, and rear wheel. Torque the lug nuts, in sequence, to 95 ft. lbs. (129 Nm).

10. Install the brake cable housing retainer into the body bracket, making sure cable housing retainer fingers lock the cable housing firmly into place.

11. Connect the right rear parking brake cable to the connector on the end of the intermediate parking brake cable.

12. Remove the locking pliers from the front parking brake cable. The parking brake cables will automatically adjust.

13. Lower the vehicle.

14. Apply and release the parking brake pedal one time only. This will correctly seat the parking brake cables.

15. Road test the vehicle and check for proper brake operation.

Left Rear Parking Brake Cable

1. Raise and safely support the vehicle. Remove the left rear wheel.

2. Remove the left rear wheel brake drum.

3. Manually lock out the automatic self-adjuster mechanism of the parking brake pedal assembly as follows:

    a. Position the park brake pedal in the fully released position.

    b. From below the vehicle, firmly grasp the exposed section of the front parking brake cable and pull downward until all free movement is removed from the parking brake cable.

    c. Install a pair of locking pliers onto the parking brake cable just rearward of the second body outrigger bracket.

4. Disconnect the left rear parking brake cable from the parking brake cable equalizer.

5. Remove the left parking brake cable housing from the body bracket by slipping a 14mm box wrench over the end of the brake cable retainer to compress the retaining fingers. A small aircraft type hose clamp can also be used.

6. Remove the brake shoes from the brake support plate.

7. Disconnect the cable from the parking brake actuator lever in the brake shoe assembly. Compress the cable housing retainer lock with a mini-hose clamp or 14mm box wrench and pull the cable from the brake support plate.

To install:

8. Install the new cable through the brake support plate. Engage the cable housing retainer fingers until they lock the cable housing firmly into place.

9. Attach the cable to the parking brake actuator lever on the brake shoe assembly. Install the brake shoes, brake drum, and rear wheel. Torque the lug nuts, in sequence, to 95 ft. lbs. (129 Nm).

10. Install the brake cable housing retainer into the body bracket, making sure cable housing retainer fingers lock the cable housing firmly into place.

11. Connect the left rear parking brake cable end to the parking brake cable equalizer bracket.

12. Remove the locking pliers from the front parking brake cable. The parking brake cables will automatically adjust.

13. Lower the vehicle.

14. Apply and release the parking brake pedal one time only. This will correctly seat the parking brake cables.

15. Road test the vehicle and check for proper brake operation.

Brake System

BLEEDING

Without ABS

1993–95 Models
Master Cylinder

If the master cylinder has been removed from the vehicle, it should be bench bled prior to installation, as follows:

1. Firmly mount the master cylinder in a bench vise using only the master cylinder mounting flange. Connect short pieces of brake line to the outlet fittings, bend them until the free end is below the fluid level in the master cylinder reservoir.

2. Fill the reservoir with fresh DOT 3 type brake fluid. Pump the piston slowly until no more air bubbles appear in each chamber of the reservoir.

3. Disconnect the lines, plug the outlet ports, refill the master cylinder and install the fluid reservoir caps. Install the master cylinder onto the power brake booster on the vehicle.

4. Install the brake lines to the master cylinder.

5. Open the brake line(s) slightly with a flare nut wrench while pressure is applied to the brake pedal by a helper inside the vehicle.

6. Tighten the line before the brake pedal is released.

7. Repeat the process with both lines until no air bubbles are released.

Calipers and Wheel Cylinders

1. Fill the master cylinder with fresh DOT 3 brake fluid. Check the level often during the procedure.

2. Starting with the right rear wheel, remove the protective cap from the bleeder and place where it will not be lost. Clean the bleed screw.

--- **CAUTION** ---

When bleeding the brakes, wear eye protection and keep your face away from the brake area. Spewing fluid may cause facial and/or visual injury. Do not allow brake fluid to spill on the vehicle finish. It will remove the paint.

3. If the system is empty, the most efficient way to get fluid down to the wheel is to loosen the bleeder about $1/2$ to $3/4$ turn, place a finger firmly over the bleeder and have a helper pump the brakes slowly until fluid comes out the bleeder. Once fluid is at the bleeder, close it before the pedal is released inside the vehicle.

NOTE: If the pedal is pumped rapidly, the fluid will churn and create small air bubbles, which are almost impossible to remove from the system. These air bubbles will eventually congregate and result in a spongy pedal.

4. Once fluid has been pumped to the caliper or wheel cylinder, open the bleed screw again, have the helper press the brake pedal to the floor, close the bleeder and have the helper slowly release the pedal. Wait 15 seconds and repeat the procedure (including the 15 second wait) until

no more air comes out of the bleeder upon application of the brake pedal. Remember to close the bleeder before the pedal is released inside the vehicle each time the bleeder is opened. If not, air will be induced into the system.

5. If a helper is not available, connect a small hose to the bleeder, place the end in a container of brake fluid and proceed to pump the pedal from inside the vehicle until no more air comes out the bleeder. The hose will prevent air from entering the system. Inspect the master cylinder often and refill as required. If the reservoir runs out of fluid and air is introduced into the system, the entire hydraulic system will need to be bled again.

6. Repeat the procedure on the remaining wheel cylinders/calipers in the following order:
   a. Left rear
   b. Right front
   c. Left front

7. Hydraulic brake systems must be totally flushed if the fluid becomes contaminated with water, dirt or other corrosive chemicals. To flush, bleed the entire system until all fluid has been replaced with new fluid.

8. Install the bleeder cap(s) on the bleeder to keep dirt out. Road test the vehicle and check for proper brake system operation.

With ABS
1993 Models

CAUTION

The hydraulic accumulator contains brake fluid and nitrogen gas at extremely high pressure. Certain portions of the hydraulic system also contain brake fluid at high pressure. It is mandatory that the system be depressurized before disconnecting any hoses, lines or fittings or personal injury may result.

NOTE: Use care when working with brake fluid. It will damage painted surfaces of the vehicle.

1. Disconnect and isolate the negative battery cable from the battery.
2. Depressurize the hydraulic accumulator.
3. Connect a transparent hose to the bleed screw on the wheel cylinder or caliper that is to be bled. Submerge the other end of the hose in a clear glass container partially filled with clean brake fluid.
4. Have an assistant slowly pump the brake pedal several times, using full strokes of the pedal and allowing approximately 5 seconds between

pedal strokes. After 2–3 strokes, continue to hold pressure on the pedal, keeping it at the bottom of its travel.

5. With pressure on the pedal, open the bleed screw ¾–1 full turn. Leave the bleed screw open until fluid no longer flows from the hose. Tighten the bleed screw and have the assistant release the pedal.

NOTE: Make sure the bleed screw is tightened before the brake pedal is released, or air may be drawn back into the hydraulic system.

6. Repeat Steps 3 and 4 at each wheel until clear, bubble-free fluid flows from the hose. Bleed the system in the following sequence:
   a. Left rear
   b. Right rear
   c. Left front
   d. Right front

7. When bleeding is completed, fill the hydraulic assembly reservoir to the proper fill level with clean DOT 3 brake fluid and install the reservoir caps.

8. Reconnect the negative battery cable.

9. Turn the ignition switch to the **RUN** position to allow the ABS pump motor to run and recharge the accumulator.

1994-95 Models

The Bendix Anti-lock 4 brake system must be bled as 2 independent brake systems. The base (non-ABS) portion of the brake system is bled the same as a conventional non-ABS system. The ABS anti-lock portion of the brake system must be bled separately, using the DRB diagnostic tester and the bleeding sequence outlined in this procedure.

Base Brake System
Master Cylinder

If the master cylinder has been removed from the vehicle, it should be bench bled prior to installation, as follows:

1. Connect short piece(s) of brake line to the outlet fitting(s), bend them until the free end is below the fluid level in the master cylinder reservoirs.
2. Fill the reservoir with fresh DOT 3 brake fluid. Pump the piston slowly until no more air bubbles appear in the reservoir(s).
3. Disconnect the lines, plug the outlet ports, refill the master cylinder and securely install the cylinder caps.
4. Install the master cylinder to the vehicle and connect the brake lines.

5. Open the brake line(s) slightly with a flare nut wrench while pressure is applied to the brake pedal by a helper inside the vehicle.
6. Tighten the line before the brake pedal is released.
7. Repeat the process with both lines until no air bubbles are released.

Calipers and Wheel Cylinders

1. Fill the master cylinder with fresh DOT 3 brake fluid. Check the level often during the procedure.
2. Starting with the right rear wheel, remove the protective cap from the bleeder and place where it will not be lost. Clean the bleed screw.

CAUTION

When bleeding the brakes, wear eye protection and keep your face away from the brake area. Spewing fluid may cause facial and/or visual injury. Do not allow brake fluid to spill on the vehicle finish. It will remove the paint.

3. If the system is empty, the most efficient way to get fluid down to the wheel is to loosen the bleeder about ½ to ¾ turn, place a finger firmly over the bleeder and have a helper pump the brakes slowly until fluid comes out the bleeder. Once fluid is at the bleeder, close it before the pedal is released inside the vehicle.

NOTE: If the pedal is pumped rapidly, the fluid will churn and create small air bubbles, which are almost impossible to remove from the system. These air bubbles will eventually congregate and a spongy pedal will result.

4. Once fluid has been pumped to the caliper or wheel cylinder, open the bleed screw again, have the helper press the brake pedal to the floor, lock the bleeder and have the helper slowly release the pedal. Wait 15 seconds and repeat the procedure (including the 15 second wait) until no more air comes out of the bleeder upon application of the brake pedal. Remember to close the bleeder before the pedal is released inside the vehicle each time the bleeder is opened. If not, air will be induced into the system.

5. If a helper is not available, connect a small hose to the bleeder, place the end in a container of brake fluid and proceed to pump the pedal from inside the vehicle until no more air comes out the bleeder. The hose will prevent air from entering the system.

6. Repeat the procedure on the remaining wheel cylinders/calipers in the following order:

   a. Left rear
   b. Right front
   c. Left front

7. Hydraulic brake systems must be totally flushed if the fluid becomes contaminated with water, dirt or other corrosive chemicals. To flush, bleed the entire system until all fluid has been replaced with new fluid.

8. Install the bleeder cap(s) on the bleeder to keep dirt out. Road test the vehicle.

Anti-Lock Brake System

1. Bleed the base brake system, using the proper procedure.

2. To perform the bleeding procedure on the ABS modulator assembly, the battery, battery tray and acid shield must be removed from the vehicle. Then reconnect the vehicle's battery to the vehicle's battery cables, using ONLY approved battery jumper cables.

3. Connect the DRB Diagnostics Tester to the vehicle's diagnostics connector. The vehicle diagnostic connector is located behind the fuse panel access cover on the lower section of the dash panel to the left of the steering column. The diagnostic connector is a blue 6 way connector.

4. Using the DRB, check to make sure the Controller Anti-lock Brake (CAB) does not have any stored fault codes. If it does, remove them using the DRB.

CAUTION

When bleeding the modulator assembly, wear safety glasses. A clear bleed tube must be attached to the modulator bleed screws and submerged in a clear container filled part way with fresh clean brake fluid. Direct the flow of brake fluid away from the painted surfaces of the vehicle. Brake fluid at high pressure may come out of the bleeder screws, when opened.

5. When bleeding the Antilock modulator assembly, the following bleeding sequence MUST be followed to insure a complete bleeding of all air from the Antilock brake, and base brake hydraulic systems. The modulator assembly can ONLY be bled using a manual bleeding procedure to pressurize the hydraulic system.

1 — Modulator Primary Check Valve Circuit

NOTE: To bleed hydraulic circuits of the Bendix Anti-lock 4 Brake System modulator assembly, the aid of an assistant will be required to pump the brake pedal.

1. Install a clear bleed tube on the primary check valve circuit bleed screw. Then install the bleed tube into a clear container partially filled with fresh clean brake fluid.

2. Pump the brake pedal several times, then apply and hold a constant medium to heavy force on the brake pedal.

3. Open the primary check valve circuit bleed screw at least 1 full turn to ensure an adequate flow of brake fluid. Continue bleeding the primary check valve circuit until the brake pedal bottoms.

4. After the brake pedal bottoms, close and tighten the bleed screw. Then release the brake pedal. Do not release the brake pedal prior to closing and tightening the bleed screw.

5. Continue bleeding the modulator assembly, repeating Steps 2 through 4 until a clear, bubble-free flow of brake fluid is evident.

6. When all air is bled from the primary check valve circuit, tighten the bleed screw and remove the bleed hose from the bleed screw. Do not remove the bleed hose before tightening the bleed screw. Air may re-enter the modulator.

7. Torque the modulator assembly primary bleed screw to 80 inch lbs. (9 Nm).

2 — Modulator Secondary Check Valve Circuit

1. Move the clear bleed tube to the secondary check valve circuit bleed screw. Then submerge the other end of the bleed tube into a container partially filled with fresh clean brake fluid.

2. Pump the brake pedal several times, then apply and hold a constant medium to heavy force on the brake pedal.

3. Open the secondary check valve circuit bleeder screw, at least 1 full turn to ensure an adequate flow of brake fluid. Continue to bleed the secondary check valve circuit until the brake pedal bottoms.

4. After the brake pedal bottoms, close and tighten the bleed screw and release the brake pedal. Do not release the brake pedal prior to closing and tightening bleed screw.

5. Continue bleeding the secondary check valve circuit, repeating Steps 2 through 4, until a clear, bubble free flow of brake fluid is evident.

6. When air is bled from the primary check valve circuit, tighten the bleed screw and remove the bleed hose from the bleed screw. Do not remove the bleed hose before tightening the bleed screw. Air may re-enter the modulator.

7. Torque the modulator assembly primary bleed screw to 80 inch lbs. (9 Nm).

3 — Modulator Assembly Primary Sump Circuit

1. Move the clear bleed tube to the primary sump bleed screw. Then submerge the other end of the bleed tube into a container partially filled with fresh clean brake fluid.

2. Pump the brake pedal several times, then apply and hold a constant medium to heavy force on the brake pedal.

3. Open the modulator assembly primary sump circuit bleed screw at least 1 full turn. This will ensure an adequate flow of brake fluid from the primary sump circuit.

4. Using the DRB or equivalent scan tool, select the Bleed ABS Hydraulic Unit Mode. Then select the primary circuit. (The RF and LR solenoids will alternately fire for 5 seconds). Using the DRB, continue to select the primary circuit until an air-free flow of brake fluid from the primary sump bleed screw is maintained or the brake pedal bottoms. If an air-free flow is not maintained before the brake pedal bottoms, close the bleed screw and repeat Steps 2 to 4, until an air free flow is maintained.

5. After an air-free flow of brake fluid is maintained from the primary sump bleed screw, close and lightly tighten the bleeder screw. Then release the brake pedal. Do not release the brake pedal prior to closing and tightening the bleeder screw.

6. After the primary sump bleed screw is closed, remove the bleed hose from the primary sump bleed screw.

7. Torque the modulator assembly primary sump bleed screw to 80 inch lbs. (9 Nm).

4 — Modulator Assembly Primary Accumulator Circuit

1. Transfer the clear bleed tube to the primary accumulator bleed screw. Then submerge the other end of the bleed tube into a container par-

tially filled with fresh clean brake fluid.

2. Pump the brake pedal several times, then apply a constant medium to heavy force on the brake pedal. Using the DRB, select the Bleed ABS Hydraulic Unit Mode. Then select the primary circuit valves. (The RF and LR modulator assembly solenoids will fire for 5 seconds).

3. Open the modulator assembly primary accumulator circuit bleed screw at least one full turn. This will ensure an adequate flow of brake fluid from the primary accumulator circuit. Continue bleeding the primary accumulator circuit until an air-free flow of brake fluid from the bleed screw is maintained or the brake pedal bottoms. If an air-free flow of brake fluid is not maintained before the brake pedal bottoms, close the bleed screw and repeat Steps 1 and 2 until an air free flow is maintained.

4. After an air-free flow of brake fluid is maintained from the primary accumulator bleed screw, close and lightly tighten the bleed screw. Then release the pressure from the brake pedal. Do not release force from the brake pedal prior to closing and tightening the bleed screw.

NOTE: For the next modulator assembly bleeding procedure, use of the DRB is not required. This step of the bleed procedure does not require modulator solenoids to be operated for bleeding to be performed.

5. Pump the brake pedal several times, then apply and hold a constant medium to heavy force on the brake pedal.

6. Again without firing the modulator solenoids, open the primary accumulator circuit bleed screw 1 full turn. This will ensure an adequate flow of brake fluid from the primary accumulator circuit.

7. Bleed the primary accumulator circuit until a clear, air-free flow of brake fluid is maintained from the accumulator bleed screw or the brake pedal bottoms. If an air-free flow of brake fluid is not maintained from the bleed screw before the brake pedal bottoms, first, close the bleed screw and then repeat Steps 5 and 6 of this bleeding procedure until an air-free flow is maintained.

8. After an air-free flow of brake fluid is maintained from the primary accumulator circuit bleed screw, close and lightly tighten the bleed screw. Then release force from the brake pedal. Do not release force from the

brake pedal prior to closing and tightening the bleeder screw.

9. After the primary accumulator bleed screw is closed, remove the bleed hose from the bleed screw.

10. Torque the primary accumulator bleed screw to 80 inch lbs. (9 Nm).

5 — Modulator Assembly Secondary Sump Circuit

1. Transfer the clear bleed tube to the secondary sump bleed screw on the modulator assembly. Then submerge the other end of the bleed tube into a container partially filled with fresh clean brake fluid.

2. Pump the brake pedal several times, then apply and hold a constant medium to heavy force on the brake pedal.

3. Open the secondary sump circuit bleed screw at least 1 full turn. This will ensure an adequate flow of brake fluid is expelled from the secondary sump circuit.

4. Using the DRB, select the bleed ABS hydraulic unit mode. Then select the secondary circuit valves. (The LF and RR solenoids will alternately fire for 5 seconds). Continue bleeding the secondary sump circuit until an air-free flow of brake fluid from the secondary sump bleed screw is maintained or the brake pedal bottoms. If an air-free flow of brake fluid is not maintained before the brake pedal bottoms, close the bleed screw and repeat Steps 2 through 4 until an air-free flow is maintained.

5. After an air-free flow of brake fluid is maintained from the secondary sump bleed screw, close and lightly tighten the bleed screw. Then release force from the brake pedal. Do not release the brake pedal prior to closing and tightening the bleeder screw.

6. After the secondary sump bleed screw is closed, remove the bleed hose from the bleed screw.

7. Torque the secondary sump bleed screw to 80 inch lbs. (9 Nm).

6 — Modulator Assembly Secondary Accumulator Circuit

1. Transfer the bleed tube to the secondary accumulator bleed screw. Then submerge the other end of the bleed tube into a container partially filled with fresh clean brake fluid.

2. Apply constant, medium to heavy force on the brake pedal. Then using the DRB, select the bleed ABS hydraulic unit mode, and then select the secondary circuit valves. (The LF and RR modulator assembly solenoids will fire for 5 seconds).

3. Open the secondary accumulator circuit bleed screw at least one full turn. This will ensure an adequate flow of brake fluid is expelled from the secondary accumulator circuit. Continue to bleed the primary accumulator circuit, until an air-free flow of brake fluid from the bleed screw is maintained or the brake pedal bottoms. If an air-free flow of brake fluid is not maintained from the bleed screw before the brake pedal bottoms, close the bleed screw and then repeat Steps 1 and 2 until an air free flow is maintained.

4. After an air-free flow of brake fluid is maintained from the bleed screw, close and lightly tighten the bleed screw. Then release force from the brake pedal.

5. Do not release force from the brake pedal prior to closing and tightening the bleeder screw.

NOTE: For the next modulator assembly bleeding procedure, use of the DRB is not required. This step of the bleeding procedure does not require the modulator solenoids to be operated for bleeding to be performed.

6. Pump the brake pedal several times, then apply and hold constant medium to heavy force on the brake pedal.

7. Again without firing the modulator assembly solenoids, open the secondary accumulator circuit bleed screw at least 1 full turn. This will ensure an adequate flow of brake fluid is expelled from the secondary accumulator circuit.

8. Bleed the secondary accumulator circuit until a clear, air-free flow of brake fluid is maintained from the secondary accumulator bleed screw or the brake pedal bottoms. If an air-free flow of brake fluid is not maintained from the secondary accumulator bleed screw before brake pedal bottoms, repeat Steps 6 and 7 of this bleeding procedure until an air-free flow is maintained from the bleeder screw.

9. After an air free flow of brake fluid is maintained from the secondary accumulator circuit bleed screw, close and lightly tighten the bleed screw. Then release force from the brake pedal. Do not release force from the brake pedal prior to closing and tightening the bleed screw.

10. After the secondary accumulator bleed screw is closed, remove the bleed hose from the bleed screw.

11. Torque the secondary accumulator bleed screw to 80 inch lbs. (9 Nm).

Wheel Speed Sensor

REMOVAL AND INSTALLATION

1993–95

One of the primary inputs to the ABS system is from the wheel speed sensors. There is a sensor at each wheel that reads magnetic impulses from a toothed gear-like tone wheel. The sensors are easily damaged and must be handled with care. Make sure the wheel sensor surfaces are clean since they are magnetic and attract metal chips and debris. Use care when removing, installing and routing the sensor wiring. The wiring must be correctly installed to avoid ABS problems later.

Front Wheel Sensor

1. Disconnect negative battery cable.
2. Raise and safely support the vehicle. Remove the front wheel.
3. Remove the attaching screw that mounts the grommet retaining clip to the inner fender shield.
4. Gently remove the speed sensor cable grommet from the fender shield. Do NOT pull on the ABS speed sensor wire.
5. Disconnect the vehicle wiring harness connector from the speed sensor connector.
6. Remove the 2 mounting screws that secure the speed sensor wire routing tube to the inner fender well.
7. Remove the 2 speed sensor assembly grommets from the retaining bracket on the strut assembly.
8. Remove the bolt that mounts the speed sensor head to the steering knuckle.
9. Gently remove the speed sensor from the steering knuckle.

— **WARNING** —

Do not remove the speed sensor with pliers for any reason. If the speed sensor has seized, due to

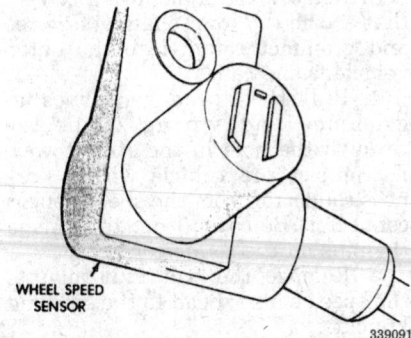

WHEEL SPEED SENSOR

339091

ABS wheel speed sensor — 1993–95

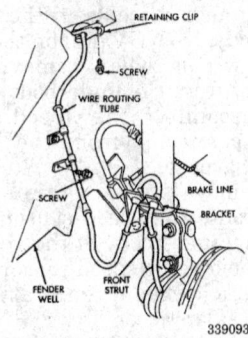

339093

Front wheel speed sensor cable (wire) routing — 1993–95

corrosion, remove it with a small mallet and punch. Lightly tap the edge of the sensor ear, rocking the sensor from side to side until it is free.

To install:

NOTE: Coat the head of the speed sensor with some High Temperature Multi-Purpose E.P. Grease (disc brake-rated wheel bearing grease) before installing into the steering knuckle.

10. Install the speed sensor head into the steering knuckle. Install the mounting screw and torque to 60 inch lbs. (7 Nm).
11. Be sure to check the wheel speed sensor air gap clearance. The allowable clearance range is 0.020–0.065 in. (0.52–1.64mm).

NOTE: Correct system operation depends on the wheel speed sensor cables being installed properly. Be sure the sensor cables are installed in the retainers. Failure to do this could result in cable over extension and/or contact with moving parts. This could result in an open circuit and/or false sensor readings.

12. Install the speed sensor wire routing tube to the inner fender well. Torque the routing tube mounting screws to 35 inch lbs. (4 Nm).
13. Install the speed sensor wiring grommets into the retaining bracket on the strut assembly. Be sure the sensor cable is routed correctly up to the strut assembly.
14. Push the sensor cable grommet into hole in the inner fender shield. Install the grommet retaining clip and screw. Torque the retaining screw to 35 inch lbs. (4 Nm).
15. Connect the wheel speed sensor connector to the vehicle wiring har-

ness connector. Be sure to securely latch the connector locking tab.
16. Install the front wheel and lug nuts. Tighten the wheel lug nuts in a star pattern sequence and torque the nuts to half specification. Then repeat the lug nut torquing sequence to full specified torque of 95 ft. lbs. (129 Nm).
17. Lower the vehicle. Reconnect the negative battery cable.

Rear Wheel Sensor — FWD

1. Raise and safely support the vehicle. Remove the rear wheel.
2. Remove the speed sensor cable grommet retaining bracket. Remove the grommet, then the wiring harness through the hole in the vehicle underbody. Do NOT pull on the speed sensor wiring when removing the grommet from the underbody.
3. Disconnect the speed sensor wiring connector from the vehicle wiring harness.
4. Remove the 4 retaining clips which route the speed sensor wiring along the underbody of the vehicle.
5. Remove the speed sensor wiring-to-vehicle frame rail attaching bracket.
6. Loosen and remove the rear axle U-bolt nuts which secures the rear wheel speed sensor cable wiring assembly mounting bracket. Remove the mounting bracket.
7. Remove the speed sensor head mounting bolt from the rear brake support plate. Remove the wheel speed sensor head from the rear brake support plate assembly.

— **WARNING** —

Do not remove the wheel speed sensor with pliers for any reason. If the speed sensor has seized, due to corrosion, remove it with a hammer and punch. Lightly tap the edge of the sensor ear, rocking the sensor from side to side until it is free.

8. Remove the wheel speed sensor from the vehicle.
To install:
9. Place the rear wheel speed sensor wiring and mounting bracket assembly onto the rear axle U-bolt. Install the rear axle U-bolt nuts and torque to 65 ft. lbs. (88 Nm).

NOTE: Coat the wheel speed sensor, where it slides into the rear brake support plate, with High Temperature Multi-Purpose E.P. Grease (disc brake-rated wheel bearing grease) before installation.

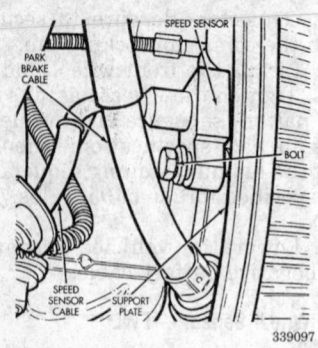

Rear wheel speed sensor location — 1993-95 FWD

339097

10. Install the wheel speed sensor head into the rear brake support plate. The speed sensor wiring on the left side (driver side) is routed between the brake tube and the wheel brake. Install the mounting bolt and torque to 60 inch lbs. (7 Nm).

11. Be sure to check the wheel speed sensor air gap clearance. The allowable clearance range is 0.017–0.047 in. (0.45–1.21mm).

12. Bend the speed sensor cable assembly (rubber coated section) toward the front of the vehicle. Install the anti-rotation tab of the speed sensor assembly frame rail bracket into its respective mounting hole. Install the frame rail bracket-to-frame rail attaching screw and torque to 35 inch lbs. (4 Nm).

13. Connect the speed sensor wiring connector to the vehicle wiring harness and install the sensor cable grommet back into the wiring access hole on the vehicle underbody.

14. Install the cable grommet retaining bracket back on the vehicle underbody, making sure the bracket does not pinch the speed sensor wiring. Install the 2 bracket mounting screws and torque to 35 inch lbs. (4 Nm).

15. Route the speed sensor wiring along the vehicle frame rail and install the 4 routing clips.

16. Install the wheel and lug nuts. Tighten the wheel lug nuts in a star pattern sequence and torque the nuts to half specification. Then repeat the lug nut tightening sequence to full specified torque of 95 ft. lbs. (129 Nm).

17. Lower the vehicle to the ground.

18. Reconnect the negative battery cable.

Rear Wheel Drive — AWD

1. Raise and safely support the vehicle. Remove the rear wheel.

2. Remove the speed sensor cable grommet retaining bracket. Remove the grommet, then the wiring harness through the hole in the vehicle underbody. Do NOT pull on the speed sensor wiring when removing the grommet from the underbody.

3. Disconnect the speed sensor wiring connector from the vehicle wiring harness.

4. Remove the 4 retaining clips which route the speed sensor wiring along the underbody of the vehicle.

5. Remove the speed sensor wiring and vehicle frame rail attaching bracket as an assembly.

6. Remove the 2 attaching bolts which secure the rear wheel speed sensor cable wiring assembly mounting bracket to the rear axle. Remove the mounting bracket assembly from the vehicle by rotating the mounting bracket and removing it from underneath the brake tube.

7. Remove the speed sensor head mounting bolt from the rear axle assembly casting. Remove the wheel speed sensor head from the rear axle assembly casting.

----- **WARNING** -----

Do not remove the wheel speed sensor with pliers for any reason. If the speed sensor has seized, due to corrosion, remove it with a hammer and punch. Lightly tap the edge of the sensor ear, rocking the sensor from side to side until it is free.

---

8. Remove the wheel speed sensor from the vehicle.

To install:

9. Place the rear wheel speed sensor mounting bracket assembly under the brake tube. Begin the outboard bracket attaching bolt by hand. Line up the brake tube clip and speed sensor assembly bracket and secure with attaching bolts. Torque the bolts to 145 inch lbs. (17 Nm).

NOTE: Coat the wheel speed sensor, where it slides into the rear axle assembly casting, with High Temperature Multi-Purpose E.P. Grease (disc brake-rated wheel bearing grease) before installation.

10. Install the wheel speed sensor head into the rear axle assembly casting. Install the mounting bolt and torque to 60 inch lbs. (7 Nm).

11. Be sure to check the wheel speed sensor air gap clearance. The allowable clearance range is 0.017–0.047 in. (0.45–1.21mm).

12. Bend the speed sensor cable assembly (rubber coated section) toward the rear of the vehicle. Install the anti-rotation tab of the speed sensor assembly frame rail bracket into its respective mounting hole. Install the frame rail bracket-to-frame rail attaching screw and torque to 35 inch lbs. (4 Nm).

13. Connect the speed sensor wiring connector to the vehicle wiring harness and install the sensor cable grommet back into the wiring access hole on the vehicle underbody.

14. Install the cable grommet retaining bracket back on the vehicle underbody, making sure the bracket does not pinch the speed sensor wiring. Install the 2 bracket mounting screws and torque to 35 inch lbs. (4 Nm).

15. Route the speed sensor wiring along the vehicle frame rail and install the 4 routing clips.

16. Install the wheel and lug nuts. Tighten the wheel lug nuts in a star pattern sequence and torque the nuts to half specification. Then repeat the lug nut tightening sequence to full specified torque of 95 ft. lbs. (129 Nm).

17. Lower the vehicle to the ground.

18. Reconnect the negative battery cable.

1996–97

One of the primary inputs to the ABS system is from the wheel speed sensors. There is a sensor at each wheel that reads magnetic impulses from a toothed gear-like tone wheel. The sensors are easily damaged and must be handled with care. Make sure the wheel sensor surfaces are clean since they are magnetic and attract metal chips and debris. Use care when removing, installing and routing the sensor wiring. The wiring must be correctly installed to avoid ABS problems later.

Front Wheel Sensor

1. Disconnect negative battery cable.

2. Carefully raise and safely support the vehicle. Remove the front wheel.

3. Remove the 2 mounting screws that attach the front channel bracket and grommet retainer to the outer vehicle frame rail.

4. Pull the speed sensor cable grommet and wiring connector through the hole in the strut tower. Disconnect the vehicle wiring harness connector from the speed sensor connector. Be careful not to damage the pins on the connector.

5. Remove the bolt that mounts the speed sensor head to the steering knuckle.

6. Gently remove the speed sensor head from the steering knuckle.

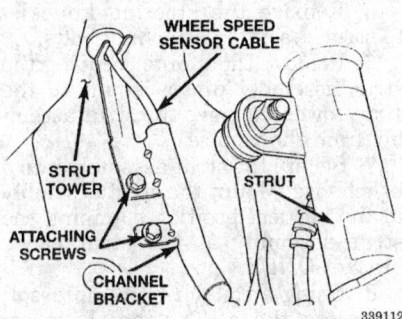

Removal of front speed sensor cable channel bracket — 1996–97

WARNING

Do not remove the speed sensor with pliers for any reason. If the speed sensor has seized, due to corrosion, remove it with a small mallet and punch. Lightly tap the edge of the sensor ear, rocking the sensor from side to side until it is free.

---

7. Remove the speed sensor cable grommets from the retaining bracket.

8. Remove the wheel speed sensor from the vehicle.

To install:

NOTE: Coat the head of the speed sensor with some High Temperature Multi-Purpose E.P. Grease (disc brake-rated wheel bearing grease) before installing into the steering knuckle.

9. Install the speed sensor head into the steering knuckle. Install the mounting screw and torque to 105 inch lbs. (12 Nm).

10. Be sure to check the wheel speed sensor air gap clearance. The allowable clearance range is 0.020–0.065 in. (0.52–1.64mm).

NOTE: Correct system operation depends on the wheel speed sensor cables being installed properly. Be sure the sensor cables are installed in the retainers. Failure to do this could result in cable over extension and/or contact with moving parts. This could result in an open circuit and/or false sensor readings.

11. Install the speed sensor wiring grommets into the intermediate retaining bracket on the strut assembly. Be sure the sensor cable is routed correctly to the strut assembly on the rearward side of the stabilizer bar link.

12. Install the 2 mounting bolts that attach the channel bracket to the vehicle frame. Torque the 2 bolts to 95 inch lbs. (11 Nm).

13. Install the channel bracket and grommet retainer onto the vehicle frame rail. Be careful not to pinch the speed sensor cable under the channel bracket.

14. Connect the wheel speed sensor connector to the vehicle wiring harness connector. Be sure to securely latch the connector locking tab and seat the connector properly.

15. Insert the sensor cable and cable grommet into the hole in the strut tower.

16. Install the front wheel and lug nuts. Tighten the wheel lug nuts in a star pattern sequence and torque the nuts to half specification. Then repeat the lug nut torquing sequence to full specified torque of 95 ft. lbs. (129 Nm).

17. Lower the vehicle. Reconnect the negative battery cable.

18. Road test the vehicle to check the operation of the ABS and base brake systems.

Rear Wheel Sensor

1. Disconnect the negative battery cable.

2. Carefully raise and safely support the vehicle. Remove the rear wheel.

3. Remove the speed sensor cable grommet, then the wiring harness through the hole in the vehicle floor pan. Do NOT pull on the speed sensor wiring when removing the grommet from the underbody.

4. Disconnect the speed sensor wiring connector from the vehicle wiring harness. Be careful not to damage the pins of the wiring connectors. Also inspect the connectors for any signs of previous damage.

5. Remove the speed sensor wiring from the rear brake flex hose routing clips. Be careful to NOT damage the routing clips on the rear brake flex hose. The routing clips are molded to the brake hoses and if they are damaged, the brake flex hoses will require replacement.

6. If removing the right side rear speed sensor, remove the speed sensor cable grommet from the axle flange, brake tube clip and routing clip from the track bar bracket on the axle.

7. Remove the 2 rear speed sensor cable/brake tube routing clips and then unclip the speed sensor cable from the routing clips on the rear brake tube.

8. Remove the speed sensor head mounting bolt from the rear bearing. Remove the wheel speed sensor head from the rear bearing assembly.

WARNING

Do not remove the wheel speed sensor with pliers for any reason. If the speed sensor has seized, due to corrosion, remove it with a hammer and punch. Lightly tap the edge of the sensor ear, rocking the sensor from side to side until it is free.

---

9. Remove the wheel speed sensor from the vehicle.

To install:

NOTE: Coat the wheel speed sensor, where it slides into the rear bearing assembly, with High Temperature Multi-Purpose E.P. Grease (disc brake-rated wheel bearing grease) before installation.

10. Install the wheel speed sensor head into the rear wheel bearing assembly. Be sure the plastic, anti-rotation pin is fully seated in wheel bearing flange before installing the mounting bolt. Install the mounting bolt and torque to 105 inch lbs. (12 Nm).

11. Be sure to check the wheel speed sensor air gap clearance. The allowable clearance range is 0.017–0.047 in. (0.45–1.21mm).

12. Install the 2 routing brackets that secure the speed sensor cable and brake tube to the rear axle. The speed sensor cable must be routed underneath the brake tube. Be careful not to damage the brake hose routing clips.

13. Install the speed sensor cable into the rear brake flex hose routing clips.

14. If installing the right rear speed sensor cable, install the cable grommet onto the rear axle brake flex hose bracket.

WARNING

The left and right rear wheel speed sensor connectors are keyed differently. Therefore, when connecting a speed sensor cable to a vehicle wiring harness, do NOT force the connectors together, or damage to the connectors will result.

---

15. Connect the speed sensor wiring connector to the vehicle wiring harness and install the sensor cable grommet back into the wiring access hole on the vehicle underbody. Be sure the speed sensor connector is fully seated and locked into the vehicle wiring harness. Be sure the speed sensor cable grommet is fully seated into the vehicle underbody wiring access hole.

16. Install the wheel and lug nuts. Tighten the wheel lug nuts in a star pattern sequence and torque the nuts to half specification. Then repeat the lug nut tightening sequence to full specified torque of 95 ft. lbs. (129 Nm).

17. Lower the vehicle to the ground.

18. Reconnect the negative battery cable. Road test the vehicle and check for proper base and ABS braking system operation.

FRONT SUSPENSION

Strut

REMOVAL AND INSTALLATION

1993–95

1. Loosen the front wheel lug nuts slightly. Raise and safely support the front of the vehicle on safety stands.

2. Remove the front wheels.

NOTE: If the original strut assemblies are to be reinstalled, mark the camber eccentric bolt (if used) and strut for installment in same position.

3. Remove the lower camber bolt and nut (at the steering knuckle) and the knuckle bolt and nut. Remove the brake hose to strut bracket mounting bolt.

4. Remove the upper mounting nuts and washers on the fender shield in the engine compartment. Remove the strut assembly from the vehicle.

5. Using a coil spring compressor, compress the coil spring. It is required that 5 spring coils be secured within the jaws of the spring compressor tool.

6. Firmly mount the strut assembly into a bench vise. Using a wrench, hold the end of strut shaft from rotating and remove the strut shaft nut.

7. Remove the upper strut mount from the strut assembly.

8. Remove the coil spring from the strut. If the coil spring is being reused, mark the coil spring for correct re-installation position.

To install:

9. With the strut assembly mounted firmly in the bench vise, install the spring on the strut assembly. Be sure the end of the coil spring is seated in the seat recess of the lower spring mount.

10. Install the upper strut mount and the strut shaft nut. Position the top spring seat alignment tab correctly with respect to the bottom steering knuckle bracket.

11. Torque the strut shaft nut to 55 ft. lbs. (75 Nm) plus a ¼ turn. This step is performed with the spring compressor tool still installed on the spring.

12. Check that the coil spring is aligned correctly to the bottom steering knuckle bracket. Remove the spring compressor tool.

13. Position the strut assembly under the fender well and loosely install the upper washers and nuts. Position the lower mount over the steering knuckle and loosely install the mounting and camber bolts and nuts. Attach the brake hose retaining bracket and tighten the mounting bolts to 10 ft. lbs. (13 Nm).

14. Tighten the upper mount nuts to 20 ft. lbs. (27 Nm). Index the camber bolt to the reference mark and snug the nut. Install the nut on the mounting bolt and tighten slightly.

15. Mount a 4 in. (102mm) C-clamp over the inner edge of the strut and outer edge of the steering knuckle. Tighten the clamp just enough to eliminate any looseness between the knuckle and the strut. Check the alignment of the camber bolt and strut reference marks. Tighten the mounting and camber nuts to 75 ft. lbs. (100 Nm) plus ¼ turn more. Remove the C-clamp.

16. Install the front wheel and lug nuts. Torque the lug nuts to 95 ft. lbs. (129 Nm). Lower the vehicle and check the wheel alignment. Adjust, if necessary.

1996–97

1. Raise and safely support the front of the vehicle on safety stands.

2. Remove the front wheel.

3. Remove the brake hose routing bracket and ABS speed sensor cable routing bracket from the strut damper brackets.

4. Remove the sway bar attaching link from the mounting bracket on the strut assembly. Hold the sway bar attaching link stud using a 6mm hex bit while removing the retaining nut with a box wrench.

——— WARNING ———

The steering knuckle-to-strut assembly attaching bolts are of the serrated type and therefore, cannot be turned during the removal procedure. Hold the bolts stationary in the steering knuckle while removing the nuts.

5. Remove the steering knuckle-to-strut assembly attaching bolts.

6. Remove the 3 nuts securing the strut assembly upper mount to the strut tower. Remove the strut assembly from the vehicle.

7. Secure the strut assembly into a bench vise. Mount the strut assembly in the vertical position clamping the strut assembly at the strut clevis bracket ONLY.

8. Using a coil spring compressor, compress the coil spring. It is required that the upper spring seat and second spring coil from the bottom be secured within the jaws of the spring compressor tool.

9. Using a wrench, hold the end of strut shaft from rotating and remove the strut shaft nut.

10. Remove the upper strut mount from the strut assembly. Carefully remove the coil spring compressor tool from the coil spring.

11. Remove the upper spring seat and pivot bearing assembly from the coil spring.

12. Remove the coil spring from the strut. If the coil spring is being reused, mark the coil spring for correct re-installation position.

To install:

13. With the strut assembly mounted firmly in the bench vise, install the coil spring on the strut assembly. Be sure the end of the coil spring's bottom coil aligns with the strut clevis bracket.

14. Install the upper spring seat onto the coil spring. Position the notch on the top of the spring seat in alignment with the strut clevis bracket.

15. Install compressor tool onto the coil spring and compress spring.

16. Install the pivot bearing onto the top of the upper spring seat with the smaller diameter side of the bearing facing toward the spring seat.

17. Install the strut mount on the upper spring seat of the strut assembly. Loosely install the strut shaft nut.

18. Torque the strut shaft nut to 70 ft. lbs. (94 Nm). This step is performed with the spring compressor tool still installed on the spring.

19. Check that the top of the coil spring is seated correctly against the upper spring seat. Carefully remove the spring compressor tool.

20. Position the strut assembly into the strut tower and loosely install the upper washers and nuts. Torque the 3 attaching upper mount nuts to 250 inch lbs. (28 Nm).

21. Position the lower mount over the steering knuckle and loosely in-

stall the attaching bolts and nuts. If one of the attaching bolts is a cam bolt, the cam bolt must be installed in the lower slotted hole of the strut clevis bracket. Be sure that the attaching nuts face the front of the vehicle. Torque the attaching nuts to 65 ft. lbs. (88 Nm) plus an additional ¼ turn.

22. Install the sway bar attaching link to the bracket on the strut assembly. Using a 6mm hex bit and crowfoot, torque the sway bar link bracket attaching nut to 65 ft. lbs. (88 Nm).

23. Install the brake hose retaining bracket and ABS speed sensor cable routing bracket onto the strut assembly bracket. Torque the mounting bolts to 10 ft. lbs. (13 Nm).

24. Install the front wheels and lug nuts. Torque the lug nuts, in a star pattern sequence, to 95 ft. lbs. (129 Nm). Lower the vehicle and check the wheel alignment. Adjust, if necessary.

Lower Ball Joints

REMOVAL AND INSTALLATION

1993–95

NOTE: Special Chrysler Tools C-4699-1 and C-4699-2 or equivalents, are required to remove and install the lower ball joint.

1. Raise and safely support the vehicle. Remove the front wheel.
2. Remove the lower control arm. Pry off the seal from the ball joint.
3. Position a receiving cup, special tool C-4699-2 or its equivalent to support the lower control arm.
4. Install a 1⅛ inch deep socket over the stud and against the joint upper housing.
5. Press the joint assembly from the lower control arm.

To install:

6. Position the ball joint housing into the control arm cavity. Be sure the ball joint is not cocked in the control arm bore, or this will cause the ball joint to bind.
7. Position the assembly in a press with special tool C-4699-1 or its equivalent, supporting the control arm.
8. Install the receiver cup special tool C-4699-2 over the ball joint stud and down on the lower control arm assembly.
9. Carefully align the ball joint assembly. Using a press, apply pressure against the control arm assem-

bly until the housing ledge of the ball joint assembly stops against the control arm cavity down flange.

10. To install a new seal, support the ball joint housing with tool C-4699-2 and place a new seal over the stud, against the housing.
11. With a 1½ in. socket, press the seal onto the joint housing with the seat against the control arm.
12. Install the control arm in the vehicle.
13. Install the front wheel and lug nuts. Torque the lug nuts to 95 ft. lbs. (129 Nm).
14. Lower the vehicle. Check the wheel alignment.

1996–97

NOTE: Special Chrysler Tools 6758, 6908-4 and 6919 or equivalents, are required to remove and install the lower ball joint. An arbor press is also required to remove and install the ball joint.

1. Raise and safely support the vehicle. Remove the front wheel.
2. Remove the lower control arm. Using a flat blade tool, pry off the seal from the ball joint.
3. Position a receiving cup, special tool 6758 or its equivalent, to support the lower control arm.
4. Install remover, special tool 6919 or its equivalent, over the stud and against the joint upper housing.
5. Using an arbor press, press the joint assembly from the lower control arm.

To install:

6. Position the ball joint housing into the control arm cavity. Be sure the ball joint is not cocked in the control arm bore, or this will cause the ball joint to bind. The notch in the ball joint stud must face inward to the control arm to allow for clearance of the ball joint-to-steering knuckle clamp bolt.

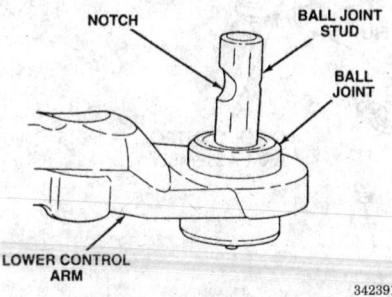

342391

Correct installation position of ball joint in lower control arm — 1996–97

7. Position the assembly in a press with special tool 6758 or its equivalent, supporting the control arm.
8. Position the ball joint installer special tool 6908-4 or its equivalent, on the bottom of the ball joint.
9. Carefully align the ball joint assembly. Using a press, apply pressure against the control arm assembly until the housing ledge of the ball joint assembly seats completely against the lower control arm surface and there is no gap between the lower control arm and ball joint. Be careful not to apply excessive force on the ball joint or the control arm.
10. Install a new seal boot as far as it will go on the ball joint by hand, first making sure the shield of the ball joint seal is facing outward from the end of the control arm.
11. Grease the ball joint using MOPAR® Multi-Mile grease, or equivalent. Do NOT over grease the ball joint or this will prevent the seal boot from being properly installed.
12. To install the new seal, place special tool 6758 or equivalent, over the seal boot and align it squarely with the bottom edge of the seal boot. Apply hand pressure only, until the new seal boot is against the top surface of the control arm.
13. Install the control arm in the vehicle.
14. Install the front wheel and lug nuts. Torque the lug nuts to 95 ft. lbs. (129 Nm).
15. Lower the vehicle. Check the wheel alignment.
16. Road test the vehicle.

Lower Control Arms

REMOVAL AND INSTALLATION

1993–95

1. Raise and safely support the vehicle.
2. Remove the front wheel.
3. Remove the ball joint–to–steering knuckle clamp bolt.
4. Remove the sway bar-to-control arm retainer on both sides of the vehicle. Rotate the sway bar down away from the control arms.
5. Separate the ball joint stud from the steering knuckle by prying between the ball stud retainer on the knuckle and the lower control arm.
6. Remove the front and rear control arm pivot bushing-to-crossmember retaining nuts and bolts. Remove the lower control arm from the vehicle. Inspect the lower control arm for distortion and bushings for exces-

sive deterioration. Replace if necessary.

WARNING

Pulling the steering knuckle out from the vehicle after releasing it from the ball joint can separate the inner CV-Joint. Use care NOT to allow the inner CV-Joint to separate. In addition, DO NOT use substitute fasteners of a lower grade (strength) than those originally used.

To install:

7. Place the lower control arm up into the crossmember and install the front and rear pivot bushing-to-crossmember retaining bolts. Loosely install the nuts to the bolts.

8. Insert the ball joint stud into the steering knuckle and install the clamp bolt. Torque the clamp bolt to 100 ft. lbs. (136 Nm).

9. Position the sway bar and sway bar bushings up against the control arms. Install the sway bar-to-control arm retainers. Install the retainer mounting bolts and torque the nuts to 50 ft. lbs. (70 Nm).

10. Install the front wheels and lug nuts. Torque the lug nuts, in a star pattern, to 95 ft. lbs. (129 Nm).

11. Lower the vehicle, so the weight of the vehicle is resting on the wheels. Torque the lower control arm-to-crossmember pivot bolts to 95 ft. lbs. (129 Nm).

12. Check the wheel alignment and adjust as necessary.

1996–97

1. Disconnect the negative battery cable.

2. Raise and safely support the vehicle on a frame contact hoist.

3. Remove the front wheels.

4. Remove the wheel stop from the steering knuckle.

5. Remove the ball joint-to-steering knuckle clamp bolts.

6. Remove the 10 front suspension cradle plate mounting bolts. Remove the cradle plate from the vehicle.

WARNING

Pulling the steering knuckle out from the vehicle after releasing it from the ball joint can separate the inner CV-Joint. Use care NOT to allow the inner CV-Joint to separate. In addition, DO NOT use substitute fasteners of a lower grade (strength) than those originally used.

7. Separate the ball joint stud from the steering knuckle by prying between the ball stud at the knuckle and the lower control arm. Be careful not to tear the ball joint seal.

8. Loosen, but **do not** remove the front lower control arm-to-suspension cradle pivot bolt.

9. Remove the retainer that secures the rear lower control arm bushing to the front suspension cradle.

NOTE: If the left side lower control arm requires removal, lower the front suspension cradle for the pivot bolt to clear the transaxle.

10. Loosen, but **do not** fully remove the 2 left side suspension cradle mounting bolts and lower the cradle.

11. Lower the left front corner of the suspension cradle until the lower control arm pivot bolt clears end of transaxle. Remove the pivot bolt and lower control arm from the vehicle. Inspect the lower control arm for distortion and bushings for excessive deterioration. Replace if necessary.

To install:

12. Place the lower control arm into the front suspension cradle and install the front lower control arm-to-suspension cradle pivot bolt. **Do not** tighten or torque the pivot bolt at this time.

13. If installing the left side lower control arm, pry down on the left front corner of the suspension cradle to allow the pivot bolt to clear the end of the transaxle. Install the pivot bolt into the suspension cradle and lower control arm.

14. Install the lower control arm rear bushing retainer. Be sure the raised rib of the rear bushing is correctly seated in the groove of the bushing retainer. **Do not** torque the retainer mounting bolts at this time.

15. Insert the ball joint stud into the steering knuckle and install the clamp bolt. Torque the clamp bolt to 105 ft. lbs. (145 Nm).

16. Install the front suspension cradle plate to the suspension cradle and install the 10 mounting bolts. Torque the cradle plate mounting bolts to 123 ft. lbs. (165 Nm).

17. Place jack stands underneath the lower control arms **as close to, but not on**, the ball joints as possible.

18. Lower the vehicle onto the jack stands until the vehicle's total weight is supported by the jack stands.

19. Torque the front lower control arm pivot bolt to 120 ft. lbs. (163 Nm).

20. Torque the lower control arm rear bushing retainer mounting bolts to 50 ft. lbs. (68 Nm).

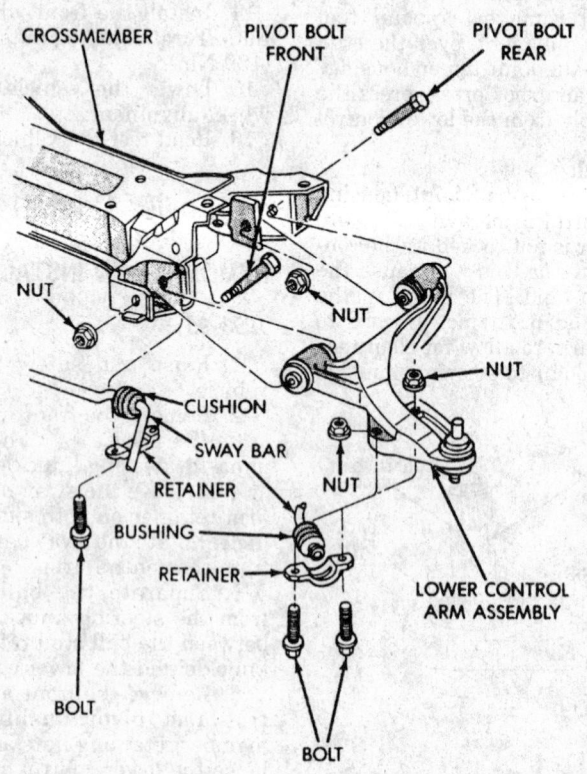

CROSSMEMBER PIVOT BOLT FRONT PIVOT BOLT REAR

NUT NUT NUT CUSHION SWAY BAR RETAINER NUT BUSHING RETAINER BOLT BOLT LOWER CONTROL ARM ASSEMBLY

342437

Lower control arm assembly — 1993–95

21. Install the front wheels and lug nuts. Torque the lug nuts, in a star pattern, to 95 ft. lbs. (129 Nm).

22. Remove the jack stands and lower the vehicle to the ground.

23. Reconnect the negative battery cable. Inspect wheel alignment and adjust, if necessary.

24. Road test the vehicle.

Sway Bar

REMOVAL AND INSTALLATION

1993–95

1. Raise and safely support the front of the vehicle on safety stands.

2. Remove the nuts, bolts and retainers connecting the sway bar to the control arms.

3. Remove the bolts that mount the sway bar to the crossmember. Remove the sway bar and crossmember mounting clamps from the vehicle.

4. Inspect the bushings for wear. Replace as necessary. End bushings are replaced by cutting or driving them from the retainer. Center bushings are split and are removed by opening the split and sliding from the sway bar.

To install:

5. Force the new end bushings into the retainers; allow about ½ in. (13mm) to protrude on the sway bar-to-lower control arm bushings. If required, install the new sway bar-to-crossmember bushings on the sway bar with the external rib up and bushing void facing the rear of the vehicle.

6. Install the sway bar assembly onto the crossmember and install the clamps and mounting bolts.

7. Install the sway bar retainers at the control arms, insert the bolts and install the nuts.

8. Place a jack under the control arm and raise the arm to normal de-

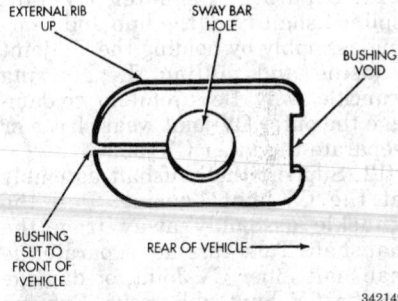

EXTERNAL RIB UP   SWAY BAR HOLE   BUSHING VOID   BUSHING SLIT TO FRONT OF VEHICLE   REAR OF VEHICLE →

342149

Placement of the sway bar bushing — 1993–95

sign height. Torque the sway bar bracket bolts to 50 ft. lbs. (70 Nm). Lower the vehicle.

1996–97

1. Raise and safely support the front of the vehicle on jack stands or a frame contact hoist.

2. Remove the 10 mounting bolts securing the cradle plate to the front suspension cradle. Remove the cradle plate from the vehicle.

3. Remove the nuts connecting the sway bar connecting links to each end of the sway bar. Disconnect the links from the ends of the sway bar.

4. Remove the sway bar bushing retainers from the front suspension cradle and remove the sway bar and bushings as an assembly from the vehicle.

5. Inspect the bushings for wear. Replace as necessary. Center bushings are split, and are removed by opening the split and sliding from the sway bar.

To install:

6. If required, install the new sway bar-to-crossmember bushings on the sway bar with the split in the replacement bushing positioned toward the rear of the vehicle, with the square corner of the bushing facing down toward the ground when installed.

7. Install the sway bar assembly onto the cradle. Be sure the sway bar bushings are aligned with the depressions in the suspension cradle.

8. Install the sway bar bushing retainers onto the cradle, aligning the raised bead of the retainer with the cutouts in the bushings. Install, but do NOT tighten the retainer mounting bolts at this time.

9. Check the position of the sway bar at this time and make sure the center curved section of the sway bar lines up with the center curved section of the suspension cradle.

10. Connect the sway bar links to each end of the sway bar and install the attaching nuts. Torque the sway bar link-to-sway bar attaching nuts to 65 ft. lbs. (88 Nm).

11. Torque the sway bar bushing retainer mounting bolts to 50 ft. lbs. (68 Nm).

12. Install the front suspension cradle plate to the front suspension cradle and install the mounting bolts. Torque the 10 cradle plate mounting bolts to 123 ft. lbs. (165 Nm).

13. Lower the vehicle.

14. Check the wheel alignment.

Front Wheel Bearings

ADJUSTMENT

All Models

The front wheel bearing is designed for the life of the vehicle and requires no type of adjustment or periodic maintenance. The bearing is a sealed unit with the wheel hub and can only be removed and/or replaced as an assembly.

If the wheel bearing is worn or damaged, the vehicle will produce vibration and noise. With the bearings loaded, the noise will generally change. Take the vehicle on a road test to determine the location of the worn/damaged bearing. Driving on a smooth, level road surface, accelerate the vehicle to a constant speed. Once the vehicle has reached a constant speed, swerve the vehicle back and forth from left to right. This will change the noise level by loading and unloading the wheel bearings. If the wheel bearing damage is slight, the noise will usually not be noticeable at speeds above 30 m.p.h.

Front Hub and Bearing

REMOVAL AND INSTALLATION

The hub and wheel bearing unit is serviced as a complete assembly. Use care when selecting the correct replacement wheel hub and bearing assembly. Vehicles equipped with 14 in. wheels have a 4 in. wheel mounting stud pattern. Vehicles equipped with 15 in. wheels have a 4½ inch wheel mounting stud pattern. If the hub and bearing assembly needs to be replaced, be sure the replacement assembly has the same size wheel mounting stud pattern as the original part. The hub and bearing assembly is mounted to the steering knuckle with 4 mounting bolts that are removed from the rear of the steering knuckle. Replacement of the front drive hub and bearing assembly can be done without having to remove the steering knuckle from the vehicle.

1993–95

1. Disconnect the negative battery cable.

2. Remove the cotter pin, nut lock and spring washer from the front wheel.

3. Loosen the front hub nut and wheel lug nuts. Raise the front of the vehicle and support on safety stands.

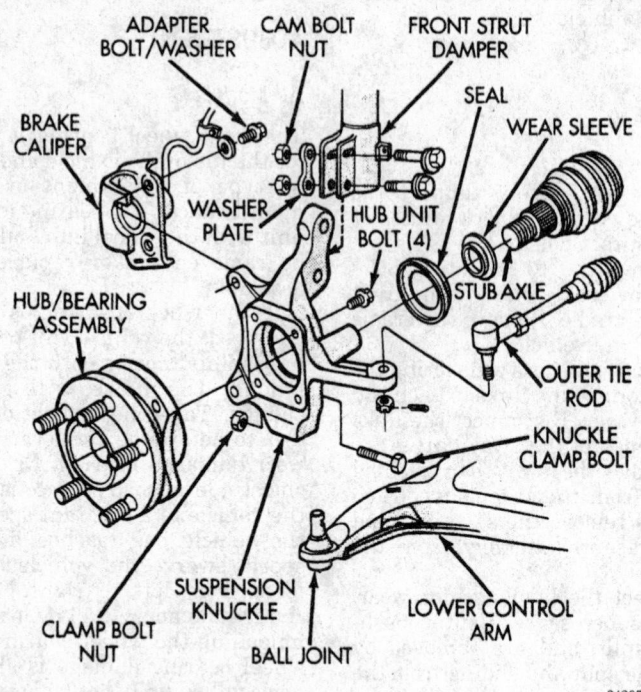

Bolt-on hub and bearing assembly mounting—1993–95

4. Remove the front wheels. Remove the center hub nut and washer.

5. Remove the brake caliper assembly and support it with a piece of wire. Do not permit the caliper to hang from the brake hose. Remove the brake rotor, inner brake pad and caliper mounting adapter.

6. Disconnect the tie rod end from the steering knuckle arm with a tie rod puller tool. Disconnect the front brake hose bracket from the strut.

7. Remove the clamp bolt that secures the ball joint to the steering knuckle.

——— WARNING ———
Be careful not to allow the CV-Joint to separate when pulling away on the steering knuckle. Do not allow the halfshaft to hang by the inner CV-Joint. Support the halfshaft assembly.

8. Ensure that the splined halfshaft is loose in the hub by tapping lightly with a brass drift and hammer. Separate the ball joint and steering knuckle by prying down on the lower control arm. Pull the knuckle assembly out and away from the halfshaft.

9. Remove the steering knuckle from the strut assembly. Note the lo-

cation of the cam bolt, it must be installed in the same location upon installation.

10. Remove the 4 hub and bearing retaining bolts. Remove the wheel hub and bearing assembly.

To install:

11. Install the new hub and bearing assembly and tighten the mounting bolts in a crisscross manner to 45 ft. lbs. (65 Nm).

12. Install a new hub and bearing seal in the recess of the steering knuckle. Use a seal installer tool for this and be sure to seat the seal fully into the steering knuckle recess.

13. Install a new wear sleeve seal. Lubricate the sealing surfaces with multi-purpose grease.

14. Install the steering knuckle to the strut assembly. Install the mounting bolt, cam bolt, washer plate, nuts and torque the bolts to 75 ft. lbs. (100 Nm) plus ¼ turn.

15. Install the halfshaft through the hub and steering knuckle. Fully lubricate the circumference of the seal and wear sleeve with Mopar® multi-purpose lubricant. Connect the ball joint to the knuckle and tighten the clamp bolt to 100 ft. lbs (136 Nm).

16. Install the tie rod end and tighten the retaining nut to 35 ft. lbs. (47 Nm). Install a new cotter pin.

17. Install the brake adapter, pads, rotor and caliper. Reconnect the brake hose bracket to the strut.

18. Install the center hub washer and retaining nut. Apply the brakes and tighten the nut to 180 ft. lbs. (244 Nm). Install the spring washer, nut and new cotter pin. Install the wheel and tire assembly. Tighten the lug nuts, in a star pattern sequence, to 95 ft. lbs. (129 Nm). Lower the vehicle.

19. Reconnect the negative battery cable.

20. Check the vehicle wheel alignment and adjust, if necessary.

1996–97

1. Disconnect the negative battery cable.

2. Remove the cotter pin from the end of the stub axle. Remove the nut lock and spring washer. With the brakes applied, loosen, but do NOT remove the axle nut and washer with the vehicle still on the ground or damage to the wheel bearing will result.

3. Raise and safely support the vehicle. Remove the wheel.

4. Remove the front brake caliper assembly from the steering knuckle assembly and support from the strut assembly using a strong piece of wire.

5. Remove the front brake rotor from the hub/bearing assembly.

6. Remove the retaining nut and washer from the halfshaft stub axle.

7. Separate the outer tie rod from the steering knuckle.

8. Remove the ABS wheel speed sensor from the steering knuckle.

9. Remove the wheel stop from the steering knuckle, if equipped.

10. Remove the nut and bolt that clamps the steering knuckle to the ball joint stud. Using a prybar, pry the control arm down to release the ball stud from the steering knuckle. Be careful not to tear the ball joint grease seal when prying down from the steering knuckle.

11. Separate the outer CV-joint splined shaft from the hub and bearing assembly by holding the CV-joint housing and pulling the steering knuckle away. Be careful not to damage the outer CV-joint wear sleeve or separate the inner CV-joint.

12. Support the halfshaft assembly at the CV-joint housing. Pull the knuckle assembly away from the halfshaft. Take care not to separate the halfshaft inner CV-Joint, or damage to the CV-Joint will occur. Support the halfshaft.

WARNING

The steering knuckle-to-strut assembly attaching bolts are of the serrated type and must not be turned during the removal procedure. Be sure to hold the bolts stationary in the steering knuckle while removing the nuts

---

13. Remove the 2 steering knuckle-to-strut clevis bracket mounting bolts and remove the steering knuckle from the vehicle.

14. Remove the 4 hub and bearing assembly mounting bolts from behind the steering knuckle.

15. Remove the hub and bearing assembly from the steering knuckle.

To install:

16. Thoroughly clean the mating surfaces of the steering knuckle and the hub and bearing assembly of any foreign material or nicks so the surfaces are clean and smooth.

17. Install the new hub and bearing assembly and tighten the mounting bolts in a crisscross pattern to 45 ft. lbs. (65 Nm). Be sure the hub and bearing assembly is seated squarely against the front steering knuckle.

NOTE: If equipped with eccentric strut assembly attaching bolts, it must be installed in the bottom (slotted) hole of the strut clevis bracket.

18. Install the steering knuckle into the strut clevis bracket of the strut assembly. Install the strut-to-steering knuckle attaching bolts and torque both attaching bolts to 65 ft. lbs. (90 Nm) plus an additional ¼ turn.

19. Thoroughly clean the steering knuckle and hub and bearing area of all debris and moisture, where the CV-joint will be installed into the steering knuckle. Also thoroughly clean the bearing shield of the outer CV-joint.

20. Pull the front strut out and insert the splined outer CV-joint into the front hub.

21. Insert the ball joint stud into the steering knuckle clamp. Install a **new** steering knuckle-to-ball joint stud clamping nut and bolt. Be sure to use an exact replacement nut and bolt during installation. Torque the bolt to 105 ft. lbs. (145 Nm).

22. Install the tie rod end into the steering knuckle. Torque the tie rod end-to-steering knuckle nut to 45 ft. lbs. (61 Nm).

23. Install the disc brake rotor.

24. Install the brake caliper assembly onto the steering knuckle.

25. Install the axle washer and nut. Tighten but do not torque.

26. Install the ABS wheel speed sensor.

27. Install the wheel and lug nuts. Torque the lug nuts, in sequence, to 95 ft. lbs. (129 Nm).

28. Lower the vehicle. Do NOT roll the vehicle until the axle nut has been properly torqued or damage to the front wheel bearings will result.

29. With the vehicle's brakes applied, torque the axle nut to 180 ft. lbs. (244 Nm). Install the spring washer, nut lock and a new cotter pin. Wrap the cotter pin prongs tightly around the axle nut lock.

30. Reconnect the negative battery cable. Check the wheel alignment.

REAR SUSPENSION

Shock Absorber

REMOVAL AND INSTALLATION

1. Raise and safely support the vehicle with safety stands.

2. Support the rear axle with a floor jack.

3. Remove the top and bottom shock absorber bolts.

4. Remove the shock absorbers.

To install:

5. Place the new shock in position and install the mounting bolts. For 1993–95, tighten to 80 ft. lbs. (108 Nm) for the lower bolts and 85 ft. lbs (115 Nm) for the upper bolts. For 1996–97, torque the shock absorber mounting bolts to 75 ft. lbs. (101 Nm).

6. Remove the floor jack supporting the rear axle and lower the vehicle to the ground.

Leaf Spring

REMOVAL AND INSTALLATION

1993–95

Front Wheel Drive Vehicles

1. Raise and safely support the rear of the vehicle on safety stands. Locate the safety stands under the frame contact points just ahead of the rear spring fixed ends.

2. Raise the rear axle just enough to relieve the weight on the springs and support on safety stands.

3. Disconnect the actuator assembly of the rear brake proportioning valve. Disconnect the lower ends of

the shock absorbers at the rear axle bracket.

4. Loosen and remove the nuts from the U-bolts. Remove the washer and U-bolts.

5. Lower the rear axle assembly to permit the rear springs to hang free. Support the spring and remove the 4 bolts that mount the fixed end spring bracket. Remove the rear spring shackle nuts and plate. Remove the shackle from the spring.

6. Remove the spring. Remove the fixed end mounting bolts from the bracket and remove the bracket. Remove the front pivot bolt from the front spring hanger.

To install:

7. Install the spring on the rear shackle and hanger. Start the shackle nuts but do not tighten completely.

8. Assemble the front spring hanger on the spring. Raise the front of the spring and install the 4 mounting bolts. Tighten the mounting bolts to 45 ft. lbs. (61 Nm).

9. Reconnect the actuator assembly for the rear brake proportioning valve.

10. Raise the axle assembly and align the spring center bolts in correct position. Install the mounting U-bolts. Tighten the nuts to 60 ft. lbs. (81 Nm).

11. Install the rear shock absorber to the lower brackets.

12. Lower the vehicle to the ground so the full weight is on the springs. Tighten the mounting components as follows:

    a. Front pivot bolt: 105 ft. lbs. (142 Nm)

    b. Shackle nuts: 35 ft. lbs. (47 Nm)

    c. Shock absorber bolts: 80 ft. lbs. (108 Nm)

13. Raise and safely support the vehicle. Adjust the rear brake proportioning valve.

14. Lower the vehicle.

All Wheel Drive Vehicles

1. Raise and safely support the rear of the vehicle on safety stands. Locate the safety stands under the chassis, ahead of the springs.

2. Raise the rear axle just enough to relieve the weight on the springs and support on safety stands.

3. Disconnect the rear brake proportioning valve spring. Disconnect the lower ends of the shock absorbers at the rear axle bracket.

4. Loosen and remove the nuts from the U-bolts. Remove the washer and U-bolts.

5. Lower the rear axle assembly to permit the rear springs to hang free.

Support the spring and remove the 4 bolts that mount the fixed end spring bracket. Remove the rear spring shackle nuts and plate. Remove the shackle from the spring.

6. Remove the spring. Remove the fixed end mounting bolts from the bracket and remove the bracket. Remove the front pivot bolt from the front spring hanger.

7. Separate the rear shackle plate from the shackle and pin assembly. Remove the shackle and pin assembly from the spring.

To install:

8. Assemble the shackle and pin assembly, bushing and shackle plate on rear of spring and spring hanger. Start the shackle and pin assembly through bolts, do not tighten.

9. Assemble the front spring hanger to the front of the spring eye and install pivot bolt and nut. Do not tighten.

NOTE: Pivot bolt must inboard to prevent structural damage during spring installation.

10. Raise the front of the spring into position and install the 4 hanger bolts; tighten them to 45 ft. lbs. (61 Nm). Connect the actuator assembly for the proportioning valve.

11. Raise the axle assembly into position, centered under the spring center bolt.

12. Install the U-bolts, nuts and washers. Tighten the U-bolt nuts to 65 ft. lbs. (88 Nm).

13. Install the shock absorbers and start the bolts.

14. Lower the vehicle to the ground, with the full weight of the vehicle on the wheels. Tighten all of the fasteners in the following sequence and to the listed torques:

    a. Front pivot bolts — 105 ft. lbs. (142 Nm)

    b. Shackle and pin assembly through bolt nuts — 45 ft. lbs. (61 Nm)

    c. Shock absorber bolts — 80 ft. lbs. (108 Nm)

15. Raise and safely support the vehicle and connect the proportioning valve.

16. Lower the vehicle.

1996–97

1. Raise and safely support the rear of the vehicle on safety stands. Locate the safety stands under the frame contact points to a comfortable working position.

2. Raise the rear axle just enough to relieve the weight on the springs and support on safety stands.

3. Disconnect the lower ends of the shock absorbers at the rear axle bracket.

4. Loosen and remove the axle plate bolts located at the leaf springs.

5. Lower the rear axle assembly to permit the rear springs to hang free. Support the spring and remove the 4 bolts that mount the front spring hanger. Remove the rear spring shackle nuts and plate. Remove the leaf spring from the shackle.

6. Remove the rear leaf spring assembly from the vehicle. Remove the front pivot bolt from the front leaf spring mount.

To install:

7. Install the front spring mount onto the front of the leaf spring eye and install pivot bolt and nut. Start the pivot nuts but do not tighten completely. Be sure to install the the pivot bolt so it is facing inboard to prevent structural damage.

8. Raise the front of the spring and install the 4 hanger mounting bolts. Tighten the mounting bolts to 45 ft. lbs. (61 Nm).

9. Install the rear of the leaf spring onto the rear spring shackle and install the shackle plate, but do not tighten.

10. Check to be sure the lower leaf spring isolator is in proper position.

11. Raise the axle assembly and align under the leaf spring locator post. Install the rear axle plate bolts and torque to 80 ft. lbs. (108 Nm).

12. Install the rear shock absorber to the lower brackets.

13. Lower the vehicle to the ground so the full weight is on the springs. Tighten the mounting components as follows:

    a. Front pivot bolt: 115 ft. lbs. (156 Nm)

    b. Shackle nuts: 45 ft. lbs. (61 Nm)

    c. Shock absorber bolts: 75 ft. lbs. (101 Nm)

14. Road test the vehicle.

Sway Bar

REMOVAL AND INSTALLATION

All Models

1. Raise and safely support the vehicle.

2. Remove the 2 lower bolts which hold the sway bar to the link arm on each side of the vehicle.

3. Loosen, but do not remove, the bolts that attach the sway bar bushing retainers to the rear axle housing.

4. While holding the sway bar in place, remove the 4 bushing retaining bolts and remove the sway bar from the axle.

5. If the sway bar links need to be replaced, remove the upper link arm-to-bracket bolt, then remove the link arm from the frame rail mounting bracket.

To install:

6. Inspect the clamps, retainers and bushings, and replace any that are damaged or severely worn.

7. Install the sway bar to the rear axle. The slits in the bushing should face up in the installed position. Do not tighten the bolts.

8. Install the 2 lower link bolts, do not tighten these.

9. Lower the vehicle so all the weight is on the wheels. With the vehicle at its curb height, torque all of the bolts as follows:

    a. Bushing-to-axle bracket — 45 ft. lbs. (61 Nm)

    b. Link arm-to-frame rail bracket — 45 ft. lbs. (61 Nm)

    c. Sway bar-to-link arm — 45 ft. lbs. (61 Nm)

    d. Link arm bracket-to-frame rail — 290 inch lbs. (33 Nm)

Wheel Bearings

ADJUSTMENT

1993–95

Front Wheel Drive Vehicles

1. Raise and safely support the vehicle.

2. Remove the rear wheels.

3. Rotate the rear brake drum assembly carefully. Excessive roughness, lateral play or resistance to rotation may indicate dirt intrusion or bearing failure.

4. If the rear wheel bearings exhibit these conditions during inspection, the inner/outer wheel bearings should be replaced.

5. Damaged bearing seals and resulting excessive grease loss may also require bearing replacement. Moderate grease loss from the bearing is considered normal and should not require replacement of the hub and bearing assembly.

6. If the wheel bearings are in good condition but the bearing grease is dirty, the wheel bearings may only need to be removed and cleaned with a good grease solvent. Then repack the wheel bearings using clean, fresh wheel bearing grease.

7. Adjust the wheel bearings as follows:

a. Remove the grease cap, cotter pin and nut-lock. Loosen the adjusting nut.

b. Tighten the adjusting nut to 20–25 ft. lbs. (27–34 Nm) while rotating the wheel.

c. Back off the adjusting nut ¼ turn (90 degrees).

d. Tighten the adjusting nut finger-tight.

e. Position the nut-lock with one pair of slots in line with the cotter pin hole on the stub axle. Install a new cotter pin.

f. Clean and install the grease cap and wheel. Torque the wheel lug nuts, in a star sequence, to 95 ft. lbs. (129 Nm).

8. Lower the vehicle.

All Wheel Drive Vehicles

The rear wheel bearing is designed for the life of the vehicle and requires no type of adjustment or periodic maintenance. The bearing is a sealed unit with the wheel hub and can only be removed and/or replaced as an assembly.

If the wheel bearing is worn or damaged, the vehicle will produce vibration or a growl noise. To inspect the wheel bearings for wear or damage, proceed as follows:

1. Shift the transaxle into the **N** position. Allow the vehicle to coast. A growl noise should continue.

2. Road test the vehicle on a smooth, level road surface to determine the location of the worn/damaged bearing. The road test will also determine if the noise is the rear wheel bearing or differential gear noise.

3. Accelerate the vehicle to a constant speed.

4. Once the vehicle has reached a constant speed, swerve the vehicle back and forth from left to right. This will change the noise level by loading and unloading the wheel bearings.

1996–97

The rear wheel bearing is designed for the life of the vehicle and requires no type of adjustment or periodic maintenance. The bearing is a sealed unit with the wheel hub and can only be removed and/or replaced as an assembly.

If the wheel bearing is worn or damaged, the vehicle will produce vibration or growl noise. With the transaxle shifted into **N** and the vehicle coasting, the noise should continue. Road test the vehicle on a smooth,

level road surface to determine the location of the worn/damaged bearing. The road test will also determine if the noise is the rear wheel bearing or differential gear noise. Accelerate the vehicle to a constant speed. Once the vehicle has reached a constant speed, swerve the vehicle back and forth from left to right. This will change the noise level by loading and unloading the wheel bearings.

REMOVAL AND INSTALLATION

1993–95

Front Wheel Drive Vehicles

1. Raise and safely support the vehicle. Remove the rear wheel.

2. Remove the wheel grease cap, cotter pin, nut-lock and bearing adjusting nut.

3. Remove the thrust washer and outer bearing.

4. Remove the drum from the spindle. Using a seal puller tool, remove the grease seal and inner bearing from the drum or hub. Discard the old grease seal.

5. Thoroughly clean the old lubricant from the bearings and hub cavity. Inspect the bearing rollers for pitting or other signs of wear that would indicate that replacement is necessary. Light discoloration is normal.

6. If the bearing cup requires replacement, remove the cup from the drum using a brass drift or suitable removal tool.

NOTE: Wheel bearings must be replaced as a set, both the cup and the bearing need to be replaced at the same time.

To install:

7. Install the new bearing cup with a bearing cup installation tool, if necessary. Make sure the cup mounting area in the hub is clean and free from nicks and burrs that would keep the cup from completely seating in the hub.

8. Repack the bearings with high temperature multi-purpose EP grease and add a small amount of new grease to the hub cavity. Be sure to force the lubricant between all rollers in the bearing.

9. Install the new greased inner bearing into the grease coated hub and bearing cup. Install a new grease seal using an appropriate seal installation tool.

10. Install the drum on the spindle after coating the polished spindle surfaces with wheel bearing lubricant.

11. Install the outer wheel bearing, thrust washer and adjusting nut.

12. Tighten the adjusting nut to 20–25 ft. lbs. (27–34 Nm) while rotating the wheel.

13. Back off the adjusting nut ¼ turn (90 degrees).

14. Tighten the adjusting nut finger-tight.

15. Position the nut-lock with one pair of slots in line with the cotter pin hole on the stub axle. Install a new cotter pin.

16. Clean and install the grease cap and wheel. Torque the wheel lug nuts, in a star sequence, to 95 ft. lbs. (129 Nm).

17. Lower the vehicle.

All Wheel Drive Vehicles

The rear wheel bearings are serviced with the hub as a unit. The hub and bearing assembly bolts to the knuckle.

1. Raise and safely support the vehicle.

2. Remove the rear wheel. Remove the brake drum.

3. Remove the cotter pin, nut lock, spring washer and hub nut.

4. Remove the halfshaft flange retaining bolts and remove the halfshaft assembly.

5. Remove the wheel bearing mounting bolts and remove the wheel bearing and hub assembly. Replace the grease seal.

To install:

6. Install the hub and bearing assembly. Tighten the bolts to 96 ft. lbs (130 Nm) in a criss-cross pattern.

NOTE: Thoroughly clean the seal and wear sleeve. Lubricate both before installation.

7. Fully lubricate the seal and wear sleeve with a multi-purpose lubricant. Install the halfshaft assembly.

8. Install the washer and hub nut. With the brakes applied torque the nut to 180 ft. lbs. (244 Nm).

9. Install the spring washer, nut lock and new cotter pin. Wrap the cotter pin prongs tightly around the nut lock.

10. Install the brake drum.

11. Install the rear wheel and lug nuts. Torque the lug nuts, in sequence, to 95 ft. lbs. (129 Nm).

12. Lower the vehicle.

1996–97

1. Raise and safely support the vehicle. Remove the rear wheel.

2. Remove the rear brake drum.

3. Remove the brake shoe assemblies.

4. Disconnect the parking brake cable from the parking brake cable actuating lever.

5. Remove the rear wheel speed sensor from the rear wheel hub/bearing flange.

WARNING

During removal of the hub/bearing assembly, be sure not to damage any of the teeth on the ABS tone wheel. Damage to the tone wheel teeth will result in false ABS cycling and corrosion of the ABS tone wheel.

6. Remove the 4 wheel hub and bearing assembly mounting bolts.

7. Remove the wheel hub and bearing assembly from the vehicle.

To install:

8. Install the 4 hub and bearing assembly mounting bolts into the 4 mounting holes in the rear axle flange.

9. Align the rear wheel hub and bearing assembly to the 4 mounting bolts. Start the mounting bolts into the bearing assembly and tighten the

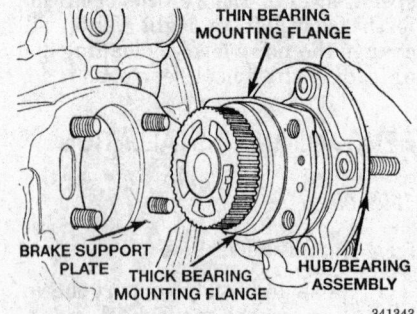

THIN BEARING MOUNTING FLANGE

BRAKE SUPPORT PLATE

THICK BEARING MOUNTING FLANGE

HUB/BEARING ASSEMBLY

341343

Removal of hub and bearing assembly — 1996–97

mounting bolts in a criss-cross pattern until the brake support plate and hub and bearing assembly are both squarely seated onto the rear axle flange.

10. Torque the wheel hub and bearing assembly mounting bolts to 95 ft. lbs. (129 Nm).

11. Install the rear wheel ABS speed sensor into the hub and bearing flange. Torque the speed sensor attaching bolt to 105 inch lbs. (12 Nm).

12. Reconnect the parking brake cable to the cable actuating lever.

13. Install the brake shoe assemblies.

14. Install the brake drum.

15. Install the rear wheels and lug nuts. Torque the wheel lug nuts, in a star sequence, to 95 ft. lbs. (129 Nm).

16. Lower the vehicle. Road test the vehicle.

FORD and MAZDA 4

FORD-Aerostar • Explorer • Mountaineer • Ranger • Splash

MAZDA-B2300 • B3000 • B4000 • Navajo

FIRING ORDERS

NOTE: To avoid confusion, always replace spark plug wires one at a time.

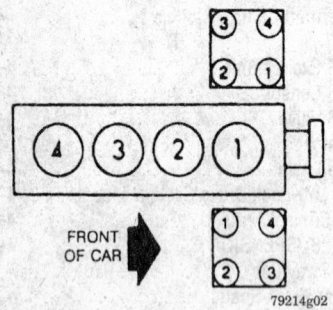

2.3L Engine
Engine Firing Order: 1–3–4–2
Distributorless Ignition System

79214g02

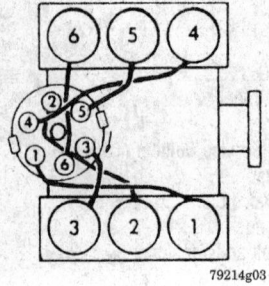

3.0L Engine
Engine Firing Order:
1–4–2–5–3–6
Distributor Rotation: Clockwise

79214g03

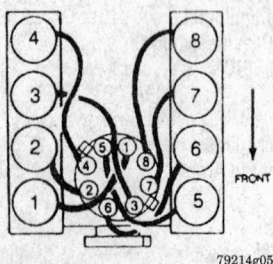

5.0L Engine
Engine Firing Order:
1–5–4–2–6–3–7–8
Distributor Rotation:
Counterclockwise

79214g05

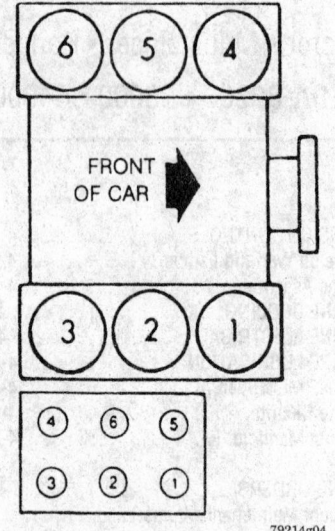

79214g04

4.0L Engine
Engine Firing Order: 1–4–2–5–3–6
Distributor Rotation: Clockwise

ENGINE ELECTRICAL

NOTE: Disconnecting the negative battery cable on some vehicles may interfere with the functions of the on board computer systems and may require the computer to undergo a relearning process, once the negative battery cable is reconnected.

Distributor

REMOVAL AND INSTALLATION

1. Disconnect the negative battery cable.
2. Disconnect the primary wiring connector from the distributor.
3. Mark the position of the cap's No. 1 terminal on the distributor base.
4. Remove the distributor cap and position aside. If it is necessary to remove the spark plug wires, tag each wire and mark it's position on the distributor cap.

5. Disconnect the electrical harness from the distributor.
6. Mark the position of the rotor in relation to the distributor housing and mark the position of the distributor housing in relation to the cylinder block.
7. Remove the distributor holddown bolt and remove the distributor.

To install:
8. If the timing WAS NOT disturbed while the distributor was removed, proceed as follows;
  a. Install the distributor assembly, aligning the marks that were made during the removal procedure.
  b. Install the distributor holddown bolt and leave it snug.
9. If the timing WAS disturbed while the distributor was removed, proceed as follows;
  a. Disconnect the No. 1 spark plug wire and remove the No. 1 spark plug.
  b. Place a finger over the spark plug hole and crank the engine slowly until compression is felt.
  c. Align the timing marks so the engine is set at the initial timing shown on the underhood sticker.
  d. Install the rotor on the shaft and rotate the shaft so the rotor tip points to the No. 1 mark made on the distributor base.
  e. Continue rotating the shaft so the leading edge of the vane is centered on the vane switch assembly.
  f. Position the distributor in the block and rotate the distributor body to align the leading edge of the vane and vane switch. Verify that the rotor tip points to the No. 1 mark on the body.

NOTE: If the vane and vane switch cannot be aligned by rotating the distributor body in the engine, pull the distributor out just far enough to disengage the gears and rotate the shaft to engage a different gear tooth.

  g. Install and finger tighten the hold-down bolt. Snug the bolt so the distributor housing can be moved for timing purposes.
10. Connect the electrical harness to the distributor.
11. Install the distributor cap. If the spark plug wires were removed, install them in their proper position.
12. Connect the negative battery cable.
13. Adjust the timing, as necessary and tighten the distributor holddown bolt.
14. When complete, tighten the hold-down bolt to 25 ft. lbs. (34 Nm).

Ignition Timing

ADJUSTMENT

NOTE: Always refer to the Vehicle Emission Information Label in the engine compartment to verify the timing adjustment procedure.

Distributor Ignition System

1. Place automatic transmission in **PARK** or manual transmission in **NEUTRAL**. The air conditioning and heater controls and all accessories should be in the **OFF** position.
2. Connect an inductive timing light and tachometer to the engine, according to the manufacturer's instructions.
3. Disconnect the single wire inline spout connector or remove the shorting bar from the double wire spout connector.
4. Start the engine and bring to normal operating temperature.

NOTE: To set timing correctly, a remote starter should not be used. Use the ignition key only to start the vehicle. Disconnecting the start wire at the starter relay will cause the TFI module to revert to start mode timing after the vehicle is started. Reconnecting the start wire after the vehicle is running will not correct the timing.

5. Check the idle speed, and adjust if needed.
6. Check the initial timing by aiming the timing light at the timing marks and pointer. Refer to the underhood emission label for specifications.
7. If the marks do not align, shut off the engine and loosen the distributor hold-down bolt. Start the engine, aim the timing light and turn the distributor until the timing marks align. Shut off the engine and tighten the distributor hold-down bolt.
8. Reconnect the single wire inline spout connector or reinstall the shorting bar on the double wire spout connector. Check the timing advance to verify the distributor is advancing beyond the initial setting.
9. Remove the timing light and tachometer.

Distributorless Ignition System

Base timing for distributorless engines is set from the factory at 10 degrees BTDC and is not adjustable.

Alternator

PRECAUTIONS

Several precautions must be observed with alternator equipped vehicles to avoid damage to the unit.
- If the battery is removed for any reason, make sure it is reconnected with the correct polarity. Reversing the battery connections may result in damage to the 1-way rectifiers.
- When utilizing a booster battery as a starting aid, always connect the positive to positive terminals and the negative terminal from the booster battery to a good engine ground on the vehicle being started.
- Never use a fast charger as a booster to start vehicles.
- Disconnect the battery cables when charging the battery with a fast charger.
- Never attempt to polarize the alternator.
- Do not use test lights of more than 12V when checking diode continuity.
- Do not short across or ground any of the alternator terminals.
- The polarity of the battery, alternator and regulator must be matched and considered before making any electrical connections within the system.
- Never separate the alternator on an open circuit. Make sure all connections within the circuit are clean and tight.
- Disconnect the battery ground terminal when performing any service on electrical components.
- Disconnect the battery if arc welding is to be done on the vehicle.

REMOVAL AND INSTALLATION

NOTE: When the battery has been disconnected and reconnected, some abnormal drive symptoms may occur while the Powertrain Control Module (PCM) relearns its adaptive strategy. The vehicle may need to be driven 10 miles (16 km) or more to relearn the strategy.

1. Disconnect the negative battery cable.
2. Disconnect the electrical harness(s) from the rear of the alternator.
3. On 1993–94 2.3L engines, loosen the alternator pivot and adjustment bolts.
4. Remove the adjustment arm bolt. Move the alternator inward and

remove the drive belt from the alternator pulley.
5. On all 1995–97 models, loosen the drive belt tensioner and remove the drive belt from the alternator pulley.
6. Remove the pivot bolt or mounting bolts, then remove the alternator from the bracket. Remove the alternator fan shield, if equipped.

To install:
7. Install the alternator fan shield, if equipped.
8. Position the alternator on the mounting bracket and secure in place with either the pivot bolt or the mounting bolts.
9. On 1995–97 models, tighten the mounting bolts to 30–40 ft. lbs. (40–54 Nm).
10. Install the drive belt on 1995–97 models using the tensioner.
11. On 1993–94 2.3L engines, install the adjustment pivot arm and bolt. Hand-tighten the bolt at this time.
12. Install the drive belt on the pulleys.
13. Adjust the drive belt tension.
14. Connect the wire harness(s) to the alternator.
15. Connect the negative battery cable. Start the vehicle and make sure the electrical system functions correctly.

Drive Belt

REMOVAL AND INSTALLATION

Except 1993–94 2.3L Engine

1. Disconnect the negative battery cable.
2. Using a suitable rachet installed into the belt tensioner. Rotate the tensioner and spring enough to slacken the belt and remove.

To install:
3. Install the belt on to all the accessary pulleys, following the drive belt routing sticker found in most engine compartments.
4. Use a rachet to rotate the belt tensioner enough to install the drive belt around the tensioner. Once the belt is installed, allow the tensioner to slowly spring back and tension the belt.
5. Connect the negative battery cable.

1993–94 2.3L Engine

1. Disconnect the negative battery cable.

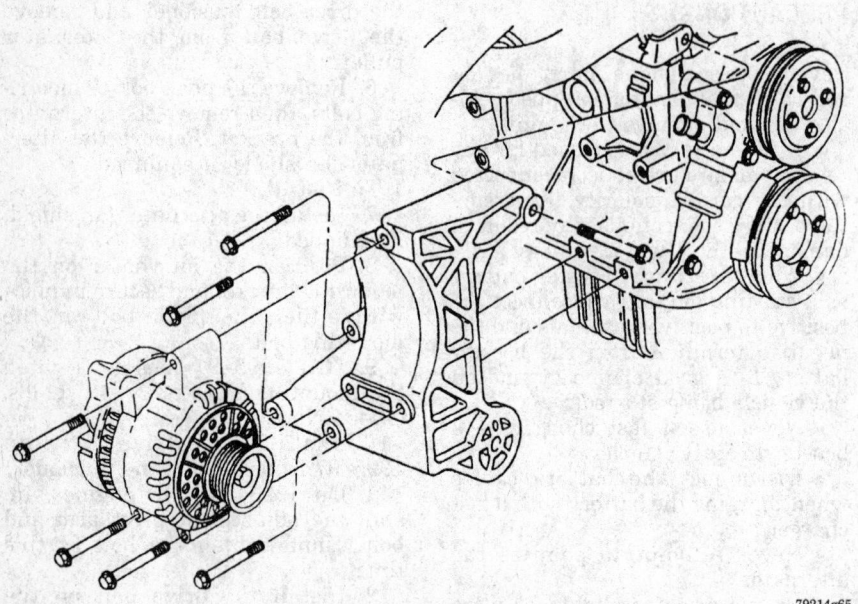

Alternator removal — 2.3L engine

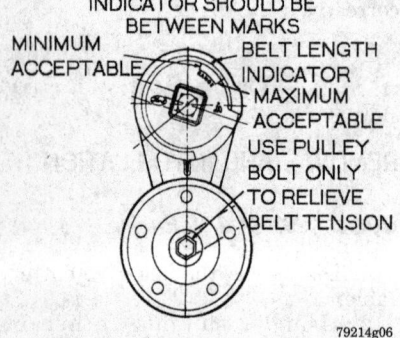

Automatic belt tensioner — 5.0L engine

2. Loosen the adjustment arm and pivot nut and bolt. Do not remove the nut or bolt.

3. Rotate the accessory toward the engine block to slacken the belt tension. Remove the drive belt.

To install:

4. Install the drive belt around the pulleys and accessory. Make sure the belt is properly seated on all the pulleys.

5. Rotate the accessory away from the engine block and apply tension to the drive belt.

6. Adjust the belt tension and tighten the adjustment arm and pivot bolt hardware.

7. Connect the negative battery cable.

ADJUSTMENT

Except 1993–94 2.3L Engine

All 1993–97 engines, with the exception of the 1993–94 2.3L engine are equipped with a spring-loaded drive belt tensioner which maintains the proper belt tension. As a result, no drive belt adjustment is necessary.

1993–94 2.3L Engine

1. Disconnect the negative battery cable. Loosen the alternator adjustment and pivot bolts.

2. Position a suitable belt tension gauge on the belt mid-way between 2 pulleys.

3. Rotate or move the accessory housing using a suitable tool to attain the correct belt tension. Be careful not to damage the accessory housing.

4. The correct belt tension for a new belt should be 150–190 ft. lbs. (195–247 Nm). The correct belt tension for a used belt should be 140–160 ft. lbs. (182–208 Nm).

NOTE: A used belt has more than 10 minutes of operation.

5. Tighten the adjustment bolt and release the pressure. Tighten the pivot bolt.

6. Check the belt tension and re-set, if necessary.

Starter

REMOVAL AND INSTALLATION

1. Disconnect the negative battery cable.

2. Raise the vehicle and support it safely.

3. Disconnect the starter cable and relay harness from the starter.

4. Remove the starter upper and lower mounting bolts, then remove the starter.

To install:

5. Position the starter motor to the engine and install the upper and lower mounting bolts finger-tight.

6. Tighten the mounting bolts to 16–21 ft. lbs. (22–28 Nm).

7. Install the relay harness to the starter. Attach the starter cable and tighten the nut to 91–122 inch lbs. (11–13 Nm).

8. Lower the vehicle. Connect the negative battery cable.

CHASSIS ELECTRICAL

Blower Motor

REMOVAL AND INSTALLATION

NOTE: When the battery has been disconnected and reconnected, some abnormal drive symptoms may occur while the Powertrain Control Module (PCM) relearns its adaptive strategy. The vehicle may need to be driven 10 miles (16 km) or more to relearn the strategy.

Without Air Conditioning

1. Disconnect the negative battery cable.

2. Remove the air cleaner or air inlet duct, as necessary.

3. On Aerostar models, remove the screws attaching the vacuum reservoir to the blower assembly and remove the reservoir.

4. Disconnect the wire harness from the blower motor by pushing

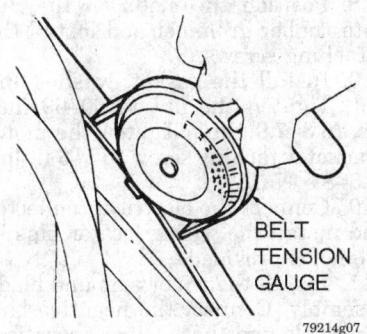

Checking belt tension with a gauge

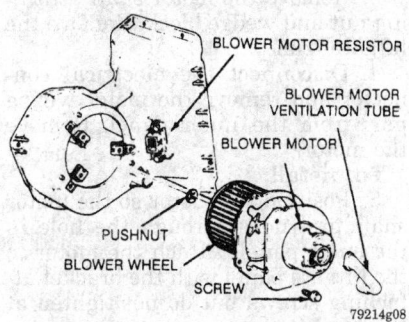

Blower motor assembly — except Aerostar

down on the connector tabs and pulling the connector off of the motor.

5. Disconnect the blower motor cooling tube at the blower motor base.

6. Remove the screws attaching the blower motor and wheel to the heater blower assembly.

7. Holding the cooling tube aside, pull the blower motor and wheel from the blower assembly and remove it from the vehicle.

8. Remove the blower wheel pushnut or clamp from the motor shaft and pull the blower wheel off the motor shaft.

To install:

9. Install the blower wheel on the blower motor shaft.

10. Install the hub clamp or pushnut over the wheel.

11. Holding the cooling tube aside, position the blower motor and wheel on the blower assembly and install the attaching screws.

12. Connect the blower motor cooling tube and the wire harness connector.

13. On Aerostar, install the vacuum reservoir on the hoses with the 2 screws.

14. Install the air cleaner or air inlet duct, as necessary.

15. Connect the negative battery cable and check the system for proper operation.

With Air Conditioning

1. Disconnect the negative battery cable.

2. In the engine compartment, disconnect the wire harness from the motor by pushing down on the tab while pulling the connection off at the motor.

3. Remove the air cleaner or air inlet duct, as necessary.

4. On vehicles equipped with the 5.0L engine, remove the A/C evaporator housing.

5. On all models except Aerostar, remove the solenoid box cover retaining bolts and the solenoid box cover, if equipped.

6. Disconnect the blower motor cooling tube from the blower motor.

7. Remove the 3 blower motor mounting plate attaching screws and remove the motor and wheel assembly from the blower motor housing.

8. Remove the blower motor hub clamp from the motor shaft and pull the blower wheel from the shaft.

To install:

9. Install the blower motor wheel on the blower motor shaft and install a new hub clamp.

10. Install a new motor mounting seal on the blower housing before installing the blower motor.

11. Position the blower motor and wheel assembly in the blower housing and install the attaching screws.

12. Connect the blower motor cooling tube.

13. Connect the electrical wire harness hard shell connector to the blower motor by pushing into place.

14. On all models except Aerostar, position the solenoid box cover, if equipped, into place and install the 3 retaining screws.

15. On vehicles with the 5.0L engine, install the A/C evaporator housing.

16. Install the air cleaner or air inlet duct, as necessary.

17. Connect the negative battery cable and check the blower motor in all speeds for proper operation.

Window Wiper Motor

REMOVAL AND INSTALLATION

NOTE: When the battery has been disconnected and reconnected, some abnormal drive symptoms may occur while the Powertrain Control Module (PCM) relearns its adaptive strategy. The vehicle may need to be driven 10 miles (16 km) or more to relearn the strategy.

Front

Aerostar

1. Turn the wiper switch **ON**. Turn the ignition switch to **RUN** until the wiper blades are in the middle of the windshield, then turn the ignition switch to **OFF** to keep the blades in this position.

2. Disconnect the negative battery cable, then unfasten the electrical connector from the wiper motor.

3. Use tape to mark the relative position of the wiper arms on the windshield.

4. Remove both wiper arms and remove the cowl grille.

5. Remove the linkage retaining clip and disconnect the linkage from the motor crank arm.

6. Remove the motor retaining nuts while holding the motor to keep it from falling.

To install:

7. Position the wiper motor in the cowl and secure in place with the retainer nuts. tighten the nuts to 60–85 inch lbs. (7–9 Nm).

8. Install the linkage and retainer clip to the motor shaft.

9. Install the cowl grille using screws or clips.

10. Install the wiper arms. Use the tape as a marker for wiper arm placement.

11. Connect the negative battery cable. Test the wiper system.

Except Aerostar

1. Turn the wiper switch **ON**. Turn the ignition switch **ON** until the blades are straight up, then turn the ignition **OFF** to keep them there.

2. Disconnect the negative battery cable, then unfasten the harness from the wiper motor.

3. Remove the right wiper arm and blade assembly. Remove the right pivot nut and allow the linkage to drop into the cowl.

4. Remove the linkage access cover, located on the right side of the dash panel, near the wiper motor.

5. Reach through the access cover opening and unsnap the wiper motor clip. Push the clip away from the linkage until it clears the nib on the crank pin, then push the clip off the linkage. Remove the linkage from the crank pin.

6. Remove the 3 attaching screws and remove the wiper motor.

To install:

7. Install the motor using the attaching screws. Tighten to 60–85 inch

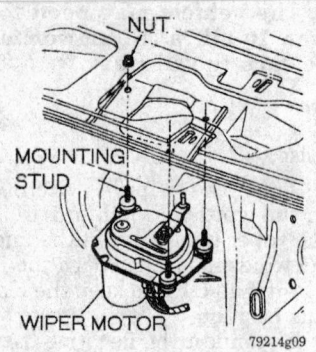

Front wiper motor — Aerostar

lbs. (7–10 Nm). Connect the motor harness.

8. Install the clip completely onto the right linkage, making sure it is fully seated. Do not put the linkage on the motor crank pin and then try to install the clip.

9. Install the left and right linkage arms onto the wiper motor crank pin. Pull the linkage onto the crank pin until it snaps into place. The clip is properly installed if the nib is protruding through the center of the clip.

10. Install the right wiper pivot shaft and nut. Tighten the nut to 84–110 inch lbs. (9–12 Nm).

11. Connect the negative battery cable and turn the ignition **ON**. Turn the wiper switch **OFF** so the wiper motor will park, then turn the ignition **OFF**. Install the right linkage access cover.

12. Install the right wiper blade and arm assembly and test the system.

Rear

Aerostar

1. Disconnect the negative battery cable.
2. Remove the wiper arm.

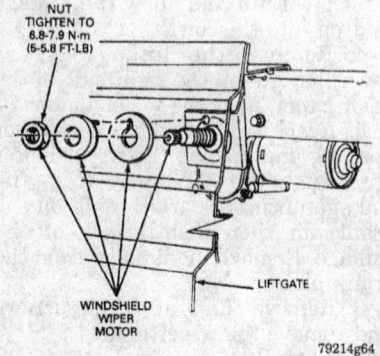

Rear wiper motor and components — Aerostar

3. Remove the motor shaft attaching nut and wedge block. Remove the liftgate trim panel.

4. Disconnect the electrical connector and remove the motor wiring pins from the inner panel. Remove the motor.

To install:

5. Position the motor so the motor shaft protrudes through the hole in the outer panel. Attach the motor to the liftgate panel with the bracket attaching screws but do not tighten at this time.

6. Install the rubber seal and wedge block over the shaft and install the pivot attaching nut. Tighten to 5–6 ft. lbs. (7–9 Nm). Tighten the motor bracket attaching screw to 5–6 ft. lbs. (7–8 Nm).

7. Attach the motor wiring to the liftgate inner panel by installing the wiring pushpins in the holes.

8. Install the articulating arm onto the drive pivot shaft.

9. Connect the negative battery cable and turn the ignition switch **ON**. Operate the wiper switch to cycle and park the wiper system in order to ensure the system linkage is in the **PARK** position before the wiper arm is installed.

10. Locate the blade in the proper position and install the arm onto the shaft with the slide latch in the unlocked position.

11. While applying a downward pressure on the arm head to ensure full seating, raise the other end of the arm sufficiently to allow the latch to slide under the pivot to the locked position. Use finger pressure only to slide the latch, then release the arm and blade against the rear window.

Except Aerostar

1. Disconnect the negative battery cable.
2. Remove the wiper arm and blade assembly.
3. Remove the pivot shaft attaching nut washer and gasket.
4. Remove the liftgate inner trim panel.
5. Remove the motor bracket attaching screw and rectangular plate. On Explorer, remove the 3 motor bracket attaching screws and pull the motor and bracket assembly out of the rubber grommet.
6. Disconnect the electrical connector and disengage the wiring locator pins. Remove the motor.

To install:

7. Position the motor in the liftgate and loosely install the rectangular plate and attaching screw, but do not tighten at this time.

8. Position the motor in the liftgate rubber grommet and install the attaching screws.

9. Install the gasket, washer and nut. Tighten the nut to 60–69 inch lbs. (6.8–7.9 Nm). Tighten the motor bracket attaching screw to 5–6 ft. lbs. (7.5–8.5 Nm).

10. Connect the electrical connector and install the wiring locator pins in the holes provided.

11. Install the wiper arm and blade assembly. Connect the negative battery cable and check wiper operation.

12. Install the liftgate inner trim panel.

Window Wiper Switch

REMOVAL AND INSTALLATION

NOTE: When the battery has been disconnected and reconnected, some abnormal drive symptoms may occur while the Powertrain Control Module (PCM) relearns its adaptive strategy. The vehicle may need to be driven 10 miles (16 km) or more to relearn the strategy.

Front

NOTE: The switch handle is an integral part of the switch and cannot be removed separately.

1. Disconnect the negative battery cable.
2. Remove the steering column cluster finish panel retaining screws.
3. Remove the 3 left control pod retaining screws.
4. On Aerostar models, remove the 2 wiring connectors from the switch without damaging the locking tabs.
5. On all models except Aerostar, remove the 3 wiring connectors from the switch without damaging the locking tabs.
6. Remove the switch handle.

To install:

7. On all models except Aerostar, install the 3 wiring connectors to the rear of the switch handle. Make sure the locking tabs engages fully.

8. On Aerostar models, install the 2 wiring connectors to the rear of the switch handle. Make sure the locking tab engages fully.

9. Align the holes in the switch handle with the mounting holes in the steering column and install the self tapping screws. Tighten the screws to 18–27 inch lbs. (2–3 Nm).

10. Install the steering column finish panel.

11. Connect the negative battery cable. Test the wiper system.

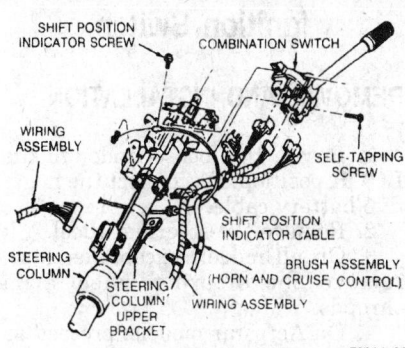

Wiper switch assembly — except Aerostar

Rear

Aerostar

1. Disconnect the negative battery cable.

2. Remove the center trim panel retainer screws and lift off the panel to access the connector at the rear of the switch.

3. Disconnect the harness from the switch, being careful not to break the locking tabs.

4. Remove the from the trim panel.

To install:

5. Install the switch into place in the trim panel.

6. Attach the harness to the rear of the switch making sure the locking tabs engage.

7. Install the center panel to the dash assembly. Secure in place with the retainer screws.

8. Connect the negative battery cable.

9. Test the switch to make sure the rear wipers work correctly.

Except Aerostar

1. Disconnect the negative battery cable.

2. Remove the ashtray retaining screws and lift out the ashtray.

3. Remove the instrument cluster trim panel. Carefully pull the panel out to disengage the clips.

4. Remove the snap-in switch mounting bezel. Disconnect the electrical harness.

5. Remove the switch from the mounting bezel by pushing on the switch from the connector side until the snap-in mounting clips release.

To install:

6. Install the switch into the mounting bezel. Install the mounting bezel into the instrument cluster trim panel Connect the wiring harness to the rear of the switch.

7. Attach the instrument cluster trim panel to the dash panel making sure all the clips engage completely.

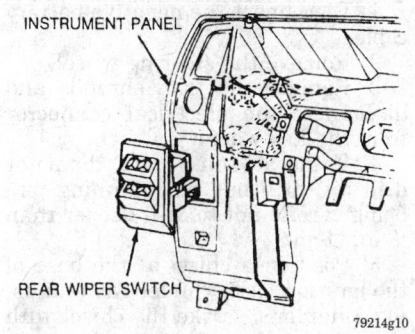

Rear Wiper Switch — Explorer, Mountaineer and Navajo

8. Install the ashtray and secure in place with the retainer screws.

9. Connect the negative battery cable. Test the rear wiper system.

Headlight Switch

REMOVAL AND INSTALLATION

NOTE: When the battery has been disconnected and reconnected, some abnormal drive symptoms may occur while the Powertrain Control Module (PCM) relearns its adaptive strategy. The vehicle may need to be driven 10 miles (16 km) or more to relearn the strategy.

Aerostar

1. Disconnect the negative battery cable.

2. Remove the headlamp switch knob from the shaft by inserting a screwdriver into the knob slot and depressing the clip on the knob. Slide the knob off and place aside.

3. Remove the screws securing the panel below the steering column. remove the instrument cluster trim ring by removing the retainer screws and carefully prying out the trim panel to release the clips.

4. Remove the headlamp switch retainer screws and lift out the switch.

5. Disconnect the electrical connector from the switch.

To install:

6. Attach the harness to the headlamp switch. Position the switch to the dash panel and secure in place using the retainer screws.

7. Install the instrument cluster trim panel to the dash panel. Make sure the clips fasten into place and all the retainer screws are installed.

8. Install the headlamp knob on to the switch.

9. Connect the negative battery cable. Test the headlamp switch to make sure it functions correctly.

Except Aerostar

The headlight switch and dimmer control switch are separate pieces. On these models, the headlight control wires go through the dimmer switch on their way to the headlight switch.

1. Disconnect the negative battery cable.

2. Remove the instrument cluster trim panel, by removing the retainer screws and unfastening the clips.

3. Remove the wiper-washer and fog lamp switch knob if they will interfere with the headlight switch and/or knob removal.

4. Check the switch body for a release button. Press in on the button and remove the knob assembly. If not equipped with a release button, a hook tool may be necessary for knob removal.

5. Unscrew the switch mounting screws from the back of the trim panel. Remove the switch from the trim panel and disconnect the wiring harness.

To install:

6. Install the switch into the trim panel, and install the switch mounting screws.

7. Install the wiring to the switch, then install the switch knob.

8. Install the trim panel and retainer screws.

9. Connect the negative battery cable. Test the headlamp system to make sure it works.

Combination Switch

REMOVAL AND INSTALLATION

The combination switch incorporates the turn signal and wiper control functions. Refer to the wiper switch removal and installation procedures.

Ignition Lock

REMOVAL AND INSTALLATION

Functioning Lock

NOTE: The following procedure should be used on vehicles with functional lock cylinders. Ignition keys are available for these vehicles or the ignition key numbers are known and the proper key can be made.

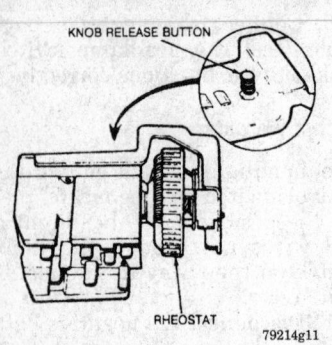

Headlight switch — Explorer, Mountaineer and Navajo

1. Disconnect the negative battery cable.

2. Remove the steering wheel trim shroud, then disconnect the harness from the key warning switch.

3. Turn the lock cylinder to the **RUN** position.

4. Place a 1/8 in. (3mm) diameter pin or small drift punch in the hole located at 4 o'clock and 1 1/4 in. (3cm) from the outer edge of the lock cylinder housing. Depress the retaining pin and pull out the lock cylinder.

To install:

5. Before installing the lock cylinder, lubricate the cylinder cavity, including the drive gear, with a suitable lock lubricant.

6. Turn the lock cylinder to the **RUN** position, depress the retaining pin and insert it into the lock cylinder housing. Make sure the cylinder is fully seated and aligned into the inter-locking washer before turning the key to the **OFF** position. This will permit the cylinder retaining pin to extend into the hole in the lock cylinder housing.

7. Using the ignition key, rotate the lock cylinder to ensure correct mechanical operation in all positions. Connect the electrical connector to the key warning switch.

8. Connect the negative battery cable. Check for proper ignition functions and make sure the column is locked in the **LOCK** position.

9. Install the steering wheel trim shrouds.

Non-Functioning Lock

NOTE: The following procedure should be used on vehicles where the ignition lock is inoperative and the lock cylinder cannot be rotated due to a lost or broken ignition key, the key number is not known or the lock cylinder cap is damaged and/or broken.

1. Disconnect the negative battery cable.

2. Remove the steering wheel.

3. Remove the trim shrouds and disconnect the electrical connector from the key warning switch.

4. Using a 1/8 in. (3mm) diameter drill bit, drill out the retaining pin, being careful not to drill deeper than 1/2 in. (1cm).

5. Position a chisel at the base of the ignition lock cylinder cap and using a hammer, strike the chisel with sharp blows to break the cap away from the lock cylinder.

6. Using a 3/8 in. (9mm) diameter drill bit, drill down the middle of the ignition lock key slot approximately 1 3/4 in. (4cm) until the lock cylinder breaks loose from the breakaway base of the lock cylinder. Remove the lock cylinder and drill shavings from the lock cylinder housing.

7. Remove the snapring, washer and ignition lock drive gear. Thoroughly clean all drill shavings and other foreign material from the casting.

8. Carefully inspect the lock cylinder housing for damage from the removal operation. If the housing is damaged, it must be replaced.

To install:

9. Install the ignition lock drive gear, washer and snapring.

10. Before installing the lock cylinder, lubricate the cylinder cavity, including the drive gear, with a suitable lock lubricant.

11. Turn the lock cylinder to the **RUN** position, depress the retaining pin and insert it into the lock cylinder housing. Make sure the cylinder is fully seated and aligned into the inter-locking washer before turning the key to the **OFF** position. This will permit the cylinder retaining pin to extend into the hole in the lock cylinder housing.

12. Using the ignition key, rotate the lock cylinder to ensure correct mechanical operation in all positions. Connect the electrical connector to the key warning switch.

13. Connect the negative battery cable. Check for proper ignition functions and make sure the column is locked in the **LOCK** position.

14. Install the steering wheel trim shrouds.

Ignition Switch

REMOVAL AND INSTALLATION

1. Rotate the lock cylinder to the **LOCK** position. Disconnect the negative battery cable.

2. Remove the steering wheel.

3. On all models except Aerostar, remove the steering wheel trim shrouds.

4. On Aerostar models, proceed as follows:

   a. Remove the panel to the right of the steering column.

   b. Remove the trim shroud by removing the 4 screws at the bottom of the shroud. Swing the bottom panel of the shroud open and remove the 2 screws attaching the shroud to the retaining plate.

   c. Remove the lock cylinder.

   d. Remove the shroud by first raising the left side so the window for the combination switch lever is up past the lever receptacle, then pull up on the right side of the shroud until the lock cylinder embossment clears the lock cylinder housing.

   e. Work the shroud past the instrument panel.

5. Disconnect the switch harness.

6. Remove the retaining nuts and disengage the ignition switch from the actuator pin.

7. If equipped with break-off head bolts, remove them using an 1/8 in. (3mm) drill bit, then remove the bolts using an "easy-out" tool or equivalent. Disengage the ignition switch from the actuator pin.

NOTE: An alternate method of removing the break-off head bolts is to use a hammer and chisel to turn the bolts 1 revolution counterclockwise. Then, using a pair of adjustable pliers, grab the bolt head and continue to rotate the bolt until removed.

To install:

8. Rotate the ignition key to the **RUN** position and align the holes in the switch casting base with the holes in the lock cylinder housing.

9. Install the nuts or if equipped, new break-off head bolts. Tighten the break-off head bolts until the heads break off.

10. Connect the electrical connector to the ignition switch and install the steering column trim shrouds.

11. On Aerostar, install the lock cylinder.

12. Install the steering wheel. Connect the negative battery cable and

check the ignition switch for proper operation.

Park/Neutral Safety Switch

REMOVAL AND INSTALLATION

1. Disconnect the negative battery cable.
2. Disconnect the harness from the switch.
3. Remove the switch and O-ring using socket tool T74P-77247-A or equivalent.

NOTE: The use of other tools could crush or puncture the walls of the switch.

To install:

4. Install the switch and O-ring using socket T74P-77247 or equivalent. Tighten the switch to 7–10 ft. lbs. (9–14 Nm).
5. Connect the switch wire harness. Connect the negative battery cable.
6. Check the operation of the switch with the parking brake applied. The engine should start only with the transmission selector lever in **N** or **P**. The back-up lights should illuminate only with the selector lever in **R**.

ADJUSTMENT

1. Hold the steering column transmission selector lever against the Neutral stop.
2. Move the sliding block assembly on the neutral switch to the neutral position and insert a 0.091 in. (2.3mm) gauge pin in the alignment hole on the terminal side of the switch.
3. Move the switch assembly housing so the sliding block contacts the actuating pin lever. Secure the switch to the outer tube of the steering column and remove the gauge pin.
4. Check the operation of the switch. The engine should only start in Neutral and Park.

Powertrain Control Module

REMOVAL AND INSTALLATION

The Powertrain Control Module (PCM) is located in the engine compartment, on the left side of the firewall on Aerostar and sport utility

models. It is at the right side cowl region on pick-up models.

NOTE: When the battery has been disconnected and reconnected, some abnormal drive symptoms may occur while the Powertrain Control Module (PCM) relearns its adaptive strategy. The vehicle may need to be driven 10 miles (16 km) or more to relearn the strategy.

1. Disconnect the negative battery cable.
2. From inside the engine compartment, loosen the PCM harness-to-module retainer screw and unplug the connector.
3. Remove the screws securing the PCM and protective cover to the firewall.
4. Remove the PCM from the bracket.

To install:

5. Slide the PCM into the bracket. Install the protective cover over the end of the module, and secure the PCM and cover with the retainer screws. Tighten the screws to 44–62 inch lbs. (5–7 Nm).
6. Install the PCM harness and tighten the module retainer screw.
7. Connect the negative battery cable.

ENGINE COOLING

Radiator

REMOVAL AND INSTALLATION

1. Disconnect the negative battery cable and remove the radiator cap and if equipped reservoir tank cap.

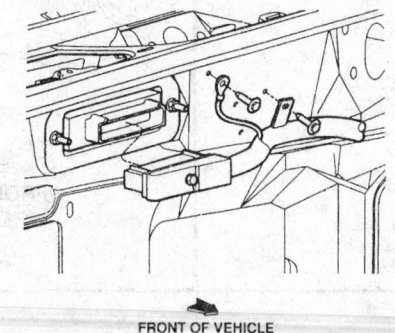

FRONT OF VEHICLE

79214g66

Powertrain control module mounted in firewall

2. Position a drain pan under the radiator and open the draincock and allow the coolant to drain out.
3. Disconnect the overflow hose from the radiator and the fan shroud, if necessary.
4. Remove the shroud/finger guard upper retainer screws. Lift the shroud out of the lower retaining clips and drape it on the fan blade.
5. Disconnect the upper and lower radiator hoses from the radiator.
6. Disconnect and plug the automatic transmission oil cooling lines, if equipped.
7. Remove the radiator upper attaching screws. Tilt the radiator back and lift directly upward, clear of the radiator support and cooling fan.

To install:

8. Make sure the radiator lower support rubber insulators are in place on the lower support.
9. Install the radiator, being careful to clear the fan blade. Make sure the mounting pins on the bottom of the radiator are inserted into the holes in the lower support, and the radiator is firmly seated.
10. Install the radiator upper attaching screws. If equipped with automatic transmission, connect the transmission cooling lines.
11. Connect the radiator upper and lower hoses. Position the shroud in the retainer clips and install the attaching screws.
12. Connect the overflow hose and close the draincock. Connect the negative battery cable. Fill and bleed the cooling system.

Water Pump

REMOVAL AND INSTALLATION

2.3L Engine

1. Disconnect the negative battery cable and drain the cooling system.

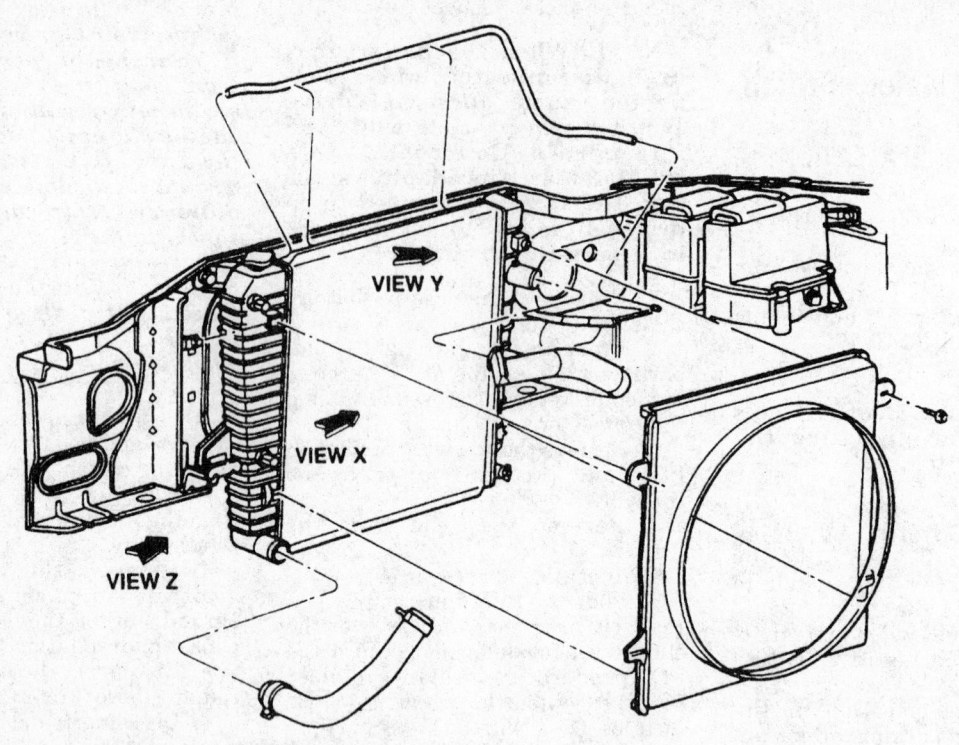

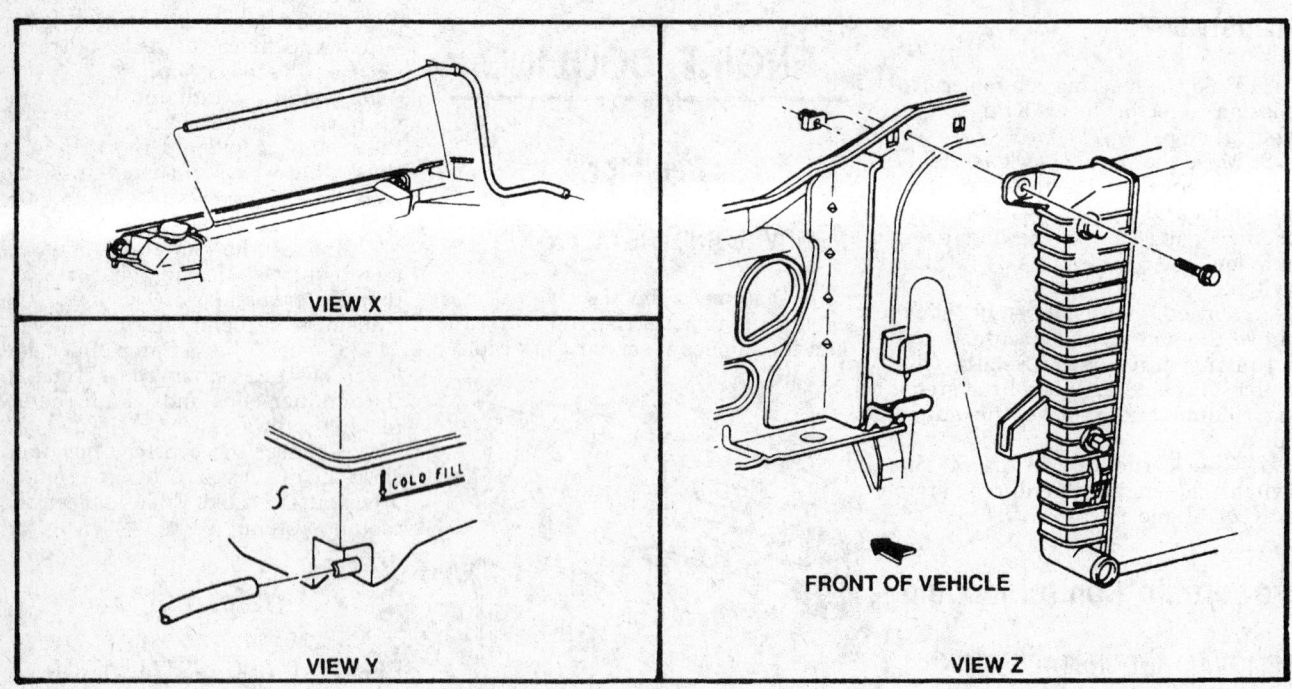

79214g67

Radiator and components — 4.0L shown, others similar

2. Remove the bolts that retain the fan shroud and position the shroud back over the fan.

3. Remove the bolts that retain the cooling fan, then remove the fan and shroud from the vehicle.

4. Remove the accessory drive belts.

5. Remove the water pump pulley and the vent hose to the carbon canister.

6. Remove the heater hose attached to the water pump.

7. Remove the timing belt cover.

8. Remove the lower radiator hose from the water pump.

9. Remove the water pump mounting bolts and the water pump. Clean all gasket mating surfaces.

To install:

10. Position the water pump and new gasket to the engine. Coat the threads of the mounting bolts with sealer and install. Tighten the bolts to 14–21 ft. lbs. (20–30 Nm).

11. Install the timing belt cover.

12. Connect the radiator and heater hoses to the water pump.

13. Install the vent hose to the carbon canister.

14. Install the cooling fan and shroud.

15. Install the drive belts.

16. Connect the negative battery cable.

17. Fill and bleed the cooling system.

3.0L and 4.0L Engines

1. Disconnect the negative battery cable and drain the cooling system.

2. Loosen the nut that attaches the fan clutch to the water pump shaft using tools T84T-6312-C to hold the clutch and T84T6312-d to loosen the nut. A 22mm wrench can also be used. The nut has left-hand thread and must be turned clockwise to remove. Remove the fan and clutch assembly.

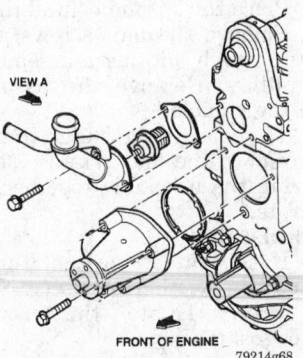

Water pump — 2.3L engine

3. Remove the accessory drive belts.

4. On 4.0L models, remove the belt tensioner and bracket.

5. Loosen the water pump pulley bolts, then the water pump pulley.

6. Remove the alternator harnesses. Remove the alternator from the vehicle.

7. On 1993–94 models, remove the alternator adjusting arm and brace.

8. On Aerostar models, remove the oil filler tube retainer nut and tube.

9. Disconnect the heater hose and lower radiator hose from the pump.

10. On Aerostar models, if the vehicle is equipped with an auxiliary heater, remove the screw retaining the auxiliary heater tube bracket at the power steering pump.

11. On pick-up models, remove the following;

 a. Crank position sensor

 b. Bypass hose and fitting at the intake manifold

 c. Remove the A/C compressor from the bracket and place aside. Do not disconnect the lines.

12. Remove the power steering pump from the bracket and place aside. Do not disconnect the hydraulic lines.

13. Remove the water pump attaching bolts, noting their positions for reinstallation.

14. Remove the water pump. Clean all gasket mating surfaces.

To install:

15. Position the water pump and new gasket in place.

16. On 3.0L engines, apply sealer to the 8mm mounting bolt at the extreme passenger side, prior to installation. Tighten the 6mm mounting bolts to 7 ft. lbs. (10 Nm) and the 8mm mounting bolts to 19 ft. lbs. (25 Nm).

17. On 4.0L engines, tighten all the retainer bolts to 6–9 ft. lbs. (8–12 Nm).

18. Attach the lower radiator and heater hoses to the water pump.

19. On 4.0L engines, install the belt tensioner and bracket.

20. Install the power steering pump, and tighten the retainer bolts to 35–45 ft. lbs. (48–61 Nm) for pick-up models, and 30–40 ft. lbs. (40–54 Nm) for Aerostar models.

21. On pick-up models, install the following;

 a. Position the A/C compressor and tighten the retainer bolts to 17–21 ft. lbs. (22–28 Nm)

 b. Install and tighten the fitting to the intake manifold

 c. Connect the crankshaft position sensor

22. On Aerostar models equipped with the auxiliary heater, install the auxiliary heater tube bracket and tighten the bolt to 6–8 ft. lbs. (8–12 Nm).

23. Install the alternator.

24. Install the water pump pulley using the retainer bolts. Tighten the bolts finger-tight.

25. Install the drive belts. Tighten the water pump pulley bolts to 18–24 ft. lbs. (23–32 Nm).

26. Install the fan, fan clutch and shroud. Tighten the fan clutch nut to 30–100 ft. lbs. (40–135 Nm), and top shroud screws to 53–71 inch lbs. (6–8 Nm).

27. Connect the negative battery cable. Fill and bleed the cooling system.

5.0L Engine

1. Disconnect the negative battery cable.

2. Drain the cooling system.

3. Remove the bolts securing the fan shroud to the radiator, if equipped, and position the shroud over the fan.

4. Disconnect the lower radiator hose, heater hose and by-pass hose at the water pump. Remove the drive belts, fan, fan spacer and clutch. Remove the fan shroud, if equipped.

5. Position the engine control sensor wire harness out of the way.

6. Loosen and remove the water pump pulley bolts and pulley

7. Remove the bolts securing the water pump to the timing chain cover and remove the water pump.

NOTE: The water pump inlet hose clamp at the water pump end of the hose is glued in place and is not removable.

8. Clean all gasket material from the mating surfaces.

To install:

9. Coat a new gasket with sealer and install the water pump and bolts. Tighten the bolts to 15–21 ft. lbs. (20–28 Nm).

10. Reposition the engine control sensor wire harness back in place if moved.

11. Install the water pump pulley a and bolts. Tighten the bolts to 12–18 ft. lbs (16–25 Nm).

12. Connect the lower radiator hose, heater hose and by-pass hose at the water pump.

13. Install the fan shroud. Tighten the bolts to 12–18 ft. lbs (16–25 Nm).

14. Install the fan clutch and fan. Tighten the fan clutch to 35–46 ft. lbs. (47–63 Nm).

15. Install the drive belt.

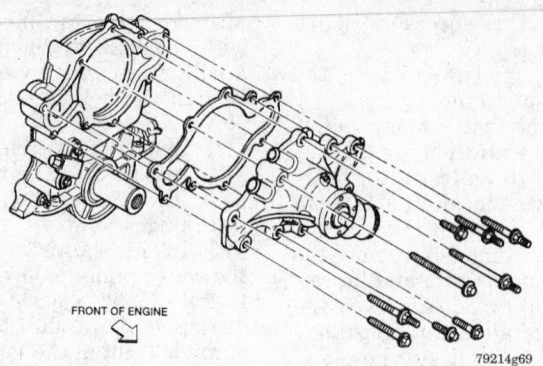

FRONT OF ENGINE

79214g69

Water pump and components — 5.0L engine

16. Connect the negative battery cable.

17. Fill and bleed the cooling system.

Thermostat

NOTE: It is a good practice to check the operation of a new thermostat before it is installed in an engine. Place the thermostat in a pan of boiling water. If it does not open more than ¼ in. (6mm), do not install it in the engine.

REMOVAL AND INSTALLATION

2.3L, 3.0L and 4.0L Engines

1. Disconnect the negative battery cable and drain the cooling system below the level of the thermostat.

2. Disconnect the upper radiator hose and, if equipped, the heater hose from the outlet.

3. Unfasten the water outlet retaining bolts and remove the outlet tube noting the orientation of the thermostat. Remove the thermostat from the outlet.

To install:

4. Clean all gasket mating surfaces.

5. Install the thermostat in the water outlet with the bridge section toward the radiator hose. Turn the thermostat clockwise to lock it in position on the flats cast into the water outlet. Install the outlet tube and new gasket on the engine using the retainer bolts.

NOTE: If the water outlet is equipped with a heater outlet tube opening, check that the full width of the opening is visible within the thermostat port in the assembly. The correct port alignment is required to provide maximum coolant flow to the heater.

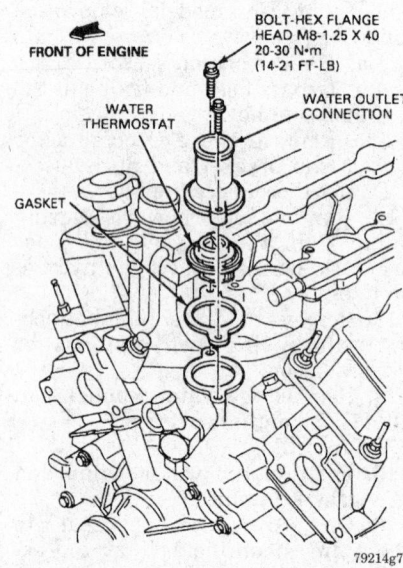

FRONT OF ENGINE

BOLT-HEX FLANGE HEAD M8-1.25 X 40 20-30 N·m (14-21 FT-LB)

WATER THERMOSTAT

WATER OUTLET CONNECTION

GASKET

79214g70

Thermostat and components — 3.0L shown, other similar

6. Install the water outlet with the mounting bolts. Tighten the bolts to 14–21 ft. lbs. (20–30 Nm).

7. Connect the radiator hose and, if equipped, the heater hose to the water outlet.

8. Connect the negative battery cable. Fill and bleed the cooling system.

5.0L Engine

1. Disconnect the negative battery cable.

2. Drain the cooling system below the level of the coolant outlet housing. Use the petcock valve at the bottom of the radiator to drain the system.

3. Disconnect the heater return hose at the thermostat housing.

4. Remove the upper radiator hose.

5. Remove the thermostat housing retainer bolts. Bend the hose and lift the housing with the hose attached to one side.

6. Remove the thermostat and gasket from the intake manifold and clean both mating surfaces.

To install:

7. Install the thermostat in the outlet tube and rotate to the left until it locks into position. Install the thermostat with the bridge (opposite end of the spring) inside the elbow connection and the thermostat flange positioned in the recess in the manifold.

8. Coat a new gasket with water resistant sealer and position it on the outlet of the engine. The gasket must be in place before the thermostat is installed.

9. Position the elbow connection onto the mounting surface of the outlet. Install and tighten the bolts to 12–18 ft. lbs. (16–24 Nm).

10. Install the heater and and upper radiator hose.

11. Connect the negative battery cable.

12. Fill the radiator and operate the engine until it reaches operating temperature. Check the coolant level and adjust if necessary.

Cooling Fan

REMOVAL AND INSTALLATION

2.3L Engines

1. Remove the screws securing the fan shroud to the radiator.

2. Disconnect the overflow tube and lift the shroud assembly off the lower retainer clips.

3. Place the shroud behind the fan.

4. Remove the four screws securing the clutch and fan a assembly to the pulley. Remove the assembly from the vehicle.

5. To separate the clutch from the fan, remove the four retainer screws securing the assembly together, and separate.

To install:

6. If separated, position the fan and clutch together and secure with four screws. Tighten the screws to 4–6 ft. lbs. (6–8 Nm).

7. Position the fan and clutch assembly to the pulley and secure in

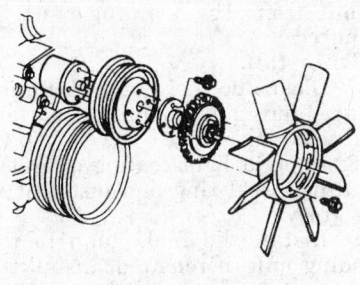

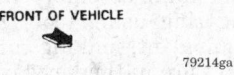

Cooling fan and clutch — 2.3L engine

place with four screws tightened to 12–18 ft. lbs. (16–24 Nm).

8. Install the fan shroud. Tighten the upper retainer screws to 4–6 ft. lbs. (5–6 Nm).

3.0L, 4.0L and 5.0L Engines

1. Remove the screws securing the fan shroud to the radiator.

2. Disconnect the overflow tube and lift the shroud assembly off the lower retainer clips.

3. Place the shroud behind the fan

4. Loosen the nut that attaches the fan clutch to the water pump shaft using tools T84T-6312-C to hold the clutch and T84T6312-d to loosen the nut. A 22mm wrench can also be used. The nut has left-hand thread and must be turned clockwise to remove. Remove the fan and clutch assembly.

5. To separate the clutch from the fan, remove the four retainer screws securing the assembly together, and separate.

To install:

6. If separated, position the fan and clutch together and secure with four screws. Tighten the screws to 50–60 inch lbs. (6–8 Nm).

7. Install the fan assembly on to the water pump shaft. Install the nut

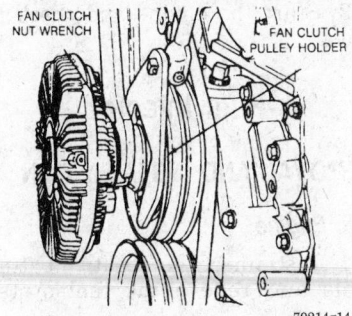

Cooling fan — 3.0L and 4.0L engines

and tighten to 30–100 ft. lbs. (40–135 Nm).

8. Install the fan shroud. Tighten the upper retainer screws to 4–6 ft. lbs. (5–6 Nm).

Cooling System Bleeding

PROCEDURE

When the entire cooling system is drained, the following procedure should be used to remove air from the cooling system and ensure a complete fill.

1. Check the radiator draincock and cylinder block drain plug, if removed are all closed.

2. Fill the cooling system with a 50/50 mixture of anti-freeze and water. Allow several minutes for trapped air to escape. When filling a cross flow radiator, allow time for the coolant to flow through the radiator tubes to the other end of the tank to ensure the radiator is full.

3. Install the radiator cap and or reservoir cap to the pressure relief position by installing the cap to the fully closed position, then backing off to the 1st stop. This will allow any air to escape and will minimize spillage.

4. Slide the heater temperature and mode selection levers to the MAX heat position.

5. Start the engine and operate at fast idle for approximately 3–4 minutes.

6. With the engine **OFF**, wrap the radiator cap and/or reservoir cap with a thick cloth, and carefully remove the cap and add coolant to bring the coolant level up to the correct level.

7. Install the cap to the fully installed position. Then, back off to the 1st stop and operate the engine at fast idle until the thermostat opens and the upper radiator hose is warm. To check the coolant level, turn the engine **OFF**, wrap the cap with a thick cloth and cautiously remove the cap. Add additional coolant, if necessary. Install the cap to the fully installed position.

CAUTION

To avoid personal injury from scalding hot coolant or steam blowing out of the radiator, use extreme care when removing the cap from a hot radiator.

---

8. Fill the coolant recovery reservoir, if equipped to the proper level with a 50/50 mix of anti-freeze and water.

GASOLINE FUEL SYSTEM

Fuel System Service Precautions

Safety is the most important factor when performing not only fuel system maintenance but any type of maintenance. Failure to conduct maintenance and repairs in a safe manner may result in serious personal injury or death. Maintenance and testing of the vehicle's fuel system components can be accomplished safely and effectively by adhering to the following rules and guidelines.

• To avoid the possibility of fire and personal injury, always disconnect the negative battery cable unless the repair or test procedure requires that battery voltage be applied.

• Always relieve the fuel system pressure prior to disconnecting any fuel system component (injector, fuel rail, pressure regulator, etc.), fitting or fuel line connection. Exercise extreme caution whenever relieving fuel system pressure to avoid exposing skin, face and eyes to fuel spray. Please be advised that fuel under pressure may penetrate the skin or any part of the body that it contacts.

• Always place a shop towel or cloth around the fitting or connection prior to loosening to absorb any excess fuel due to spillage. Ensure that all fuel spillage (should it occur) is quickly removed from engine surfaces. Ensure that all fuel soaked cloths or towels are deposited into a suitable waste container.

• Always keep a dry chemical (Class B) fire extinguisher near the work area.

• Do not allow fuel spray or fuel vapors to come into contact with a spark or open flame.

• Always use a backup wrench when loosening and tightening fuel line connection fittings. This will prevent unnecessary stress and torsion to fuel line piping. Always follow the proper torque specifications.

• Always replace worn fuel fitting O-rings with new. Do not substitute fuel hose or equivalent where fuel pipe is installed.

Fuel System Pressure

RELIEVING

1. Disconnect the negative battery cable and remove the fuel filler cap.

2. Remove the cap from the pressure relief valve on the fuel supply manifold. Install pressure gauge T80L-9974-B or equivalent, to the pressure relief valve.

3. Direct the gauge drain hose into a suitable container and depress the pressure relief button.

4. Remove the gauge and replace the cap on the pressure relief valve.

NOTE: As an alternate method, disconnect the inertia switch and crank the engine for 15–20 seconds until the pressure is relieved.

Idle Speed

ADJUSTMENT

The idle speed is controlled by the Powertrain Control Module. Routine adjustment of the idle speed is not necessary.

Mixture

ADJUSTMENT

The idle mixture is controlled by the Powertrain Control Module and cannot be adjusted.

Fuel Filter

REMOVAL AND INSTALLATION

1. Disconnect the negative battery cable and relieve the fuel system pressure.

2. Raise and support the vehicle safely.

3. Remove the bracket cover over the filter, if equipped.

4. Remove the push-connect fittings at both ends of the filter.

5. Place a pan under the filter, then disconnect the fuel lines from the fuel filter.

6. Remove the fuel filter from the bracket and retainer, if equipped. Note the direction of the flow arrow so the replacement filter can be installed correctly.

To install:

7. Install new push-connect fitting at each end of the fuel lines.

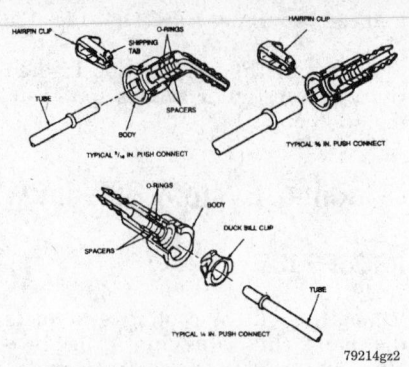

79214gz2

Typical push-connect fittings

8. Install the fuel filter into the bracket making sure the flow direction is correct. Tighten the clamp to 15–25 inch lbs. (2–3 Nm).

9. Install the fuel lines at both ends of the filter. Install the bracket cover, if equipped.

10. Lower the vehicle Start the engine and check for leaks.

Electric Fuel Pump

REMOVAL AND INSTALLATION

1. Disconnect the negative battery cable and relieve the fuel system pressure.

2. Raise and safely support the vehicle.

3. Remove the fuel tank from the vehicle.

4. Remove any dirt that has accumulated around the fuel pump attaching flange so it will not enter the fuel tank during removal and installation.

5. Turn the fuel pump locking ring counterclockwise using tool T90T-9275-A or equivalent. Remove the locking ring using tool T86T-9275-A or equivalent.

6. Remove the fuel pump and discard the seal ring. Separate the fuel

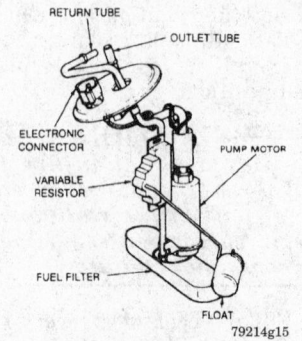

79214g15

Example of typical tank-mounted fuel pump

pump from the sending unit, if required.

To install:

7. Clean the fuel pump mounting flange and tank mounting surface and seal ring groove.

8. Apply a light coating of grease on a new seal ring and install it in the groove.

9. Install the fuel pump to the sending unit, if removed. Install the fuel pump assembly in the tank locking ring using tool T86T-9275-A or equivalent, making sure the locating keys are in the keyways and the seal ring is in place.

10. Hold the fuel pump assembly and the seal ring in place and install the locking ring using tool T90T-9275-A or equivalent. Rotate the ring clockwise using a suitable tool. Find the fuel tank part number on the front bottom of the tank and proceed as follows:

 a. If equipped with part number E59A-9002-CAE tank, tighten the ring to 60–85 ft. lbs. (81–115 Nm), wait 5 minutes and retighten to the same specification. Use only the ring that was removed from the tank, do not replace with a ring from another tank.

 b. If equipped with plastic retaining ring E99A-9A307-D, tighten to 40–55 ft. lbs. (54–74 Nm). Use the same ring that was removed from the tank. If a new tank is installed, use a new ring.

 c. If equipped with part number E69A-9002-PA tank, tighten the locking ring once to 80–113 ft. lbs. (109–153 Nm).

 d. On Ranger and Aerostar, tighten the polyethylene locking ring to 40–45 ft. lbs. (54–61 Nm).

11. Install the fuel tank in the vehicle.

12. Lower the vehicle and fill the fuel tank with at least 10 gallons of fuel. Connect the negative battery cable. Turn the ignition key to **RUN** for 3 seconds repeatedly, 5–10 times, to pressurize the system. Check for leaks.

13. Start the engine and check for leaks.

Fuel Injector

REMOVAL AND INSTALLATION

2.3L Engine

1. Disconnect the negative battery cable and relieve the fuel system pressure.

2. Disconnect the harnesses at the TPS, air bypass valve, water temper-

ature indicator sensor, ACT and ECT sensors and the ignition control assembly. Tag all lines prior to removal to aid reinstallation.

3. Remove the throttle linkage shield and disconnect the throttle linkage and cruise control assembly. Unbolt the accelerator cable from the bracket and position out of the way.

4. Disconnect the vacuum lines at the upper intake manifold vacuum tree, EGR valve, fuel pressure regulator and canister purge line. Tag all lines prior to removal to aid reinstallation.

5. Disconnect the air intake hose and crankcase vent hose.

6. Disconnect the hose for the PCV system from the fitting on the underside of the upper intake manifold.

7. Disconnect the EGR tube from the EGR valve by removing the flange nut.

8. Remove the upper intake manifold retaining bolts and remove the upper intake manifold and throttle body assembly.

9. Disconnect the injector harness.

10. Disconnect the fuel lines from the fuel supply manifold.

11. Remove the 2 fuel supply manifold retaining bolts.

12. Carefully disengage the manifold and fuel injectors from the engine and remove the manifold and injectors.

13. Remove the injectors from the manifold by grasping the injector body and pulling while gently rocking the injector from side-to-side. Remove and discard the injector O-rings.

To install:

14. Lubricate new O-rings with clean light grade oil and install 2 O-rings on each injector.

NOTE: Never use silicone grease at it will clog the injectors.

15. Install the injectors, using a light, twisting, pushing motion.

16. Install the fuel supply manifold, pushing it down to make sure all the O-rings are seated in the fuel rail cups and intake manifold.

17. Install the manifold retaining bolts and tighten to 15–22 ft. lbs. (20–30 Nm) while holding the fuel manifold down.

18. Connect the fuel lines to the supply manifold.

19. After the fuel supply manifold has been installed, and before the fuel injector wires have been connected, connect the negative battery cable and turn the ignition **ON**. This will cause the fuel pump to run for

2–3 seconds and pressurize the fuel system.

20. Check for leaks where the fuel injector is installed into the fuel supply manifold.

21. Disconnect the negative battery cable.

22. Connect the fuel injector harnesses.

23. Make sure the gasket surfaces of the upper and lower intake manifolds are clean.

24. Place a new gasket on the lower intake manifold assembly and place the upper intake manifold and throttle body assembly in position.

25. Install the retaining bolts and tighten in sequence to 15–22 ft. lbs. (20–30 Nm).

26. Connect the EGR tube to the EGR valve and tighten to 18–28 ft. lbs. (25–30 Nm).

27. Connect the PCV system hose to the fitting on the underside of the upper intake manifold.

28. Connect all vacuum lines and harnesses according to the locations that were marked during the removal procedure.

29. Hold the accelerator cable bracket in position on the upper manifold and install the retaining bolts. Tighten to 10–15 ft. lbs. (13.5–20.5 Nm).

30. Install the accelerator cable to the bracket and connect the accelerator cable and cruise control. Install the throttle linkage shield.

31. Connect the air intake hose and crankcase vent hose.

32. Connect the negative battery cable. Start the engine and let it idle for 2 minutes.

33. Turn the engine **OFF** and check for fuel leaks.

3.0L Engine

1. Disconnect the negative battery cable and relieve the fuel system pressure.

2. Remove the air intake throttle body assembly as follows:

   a. Remove the engine air cleaner outlet tube between the air cleaner and throttle body.

   b. Remove the snow shield from the power steering pump bracket and accelerator cable bracket.

   c. Tag and disconnect the vacuum lines at the vacuum fittings on the intake manifold and PCV hose.

   d. Disconnect and remove the accelerator and cruise control cables from the accelerator mounting bracket and throttle lever.

   e. Remove the alternator support brace.

   f. Remove the throttle body-to-lower intake manifold retaining bolts and stud bolts. Remove the throttle body assembly.

3. Disconnect the fuel lines from the fuel supply manifold.

4. Disconnect the wiring harness from the injectors.

5. Disconnect the vacuum line from the fuel pressure regulator.

6. Remove the 4 fuel supply manifold retaining bolts.

7. Carefully disengage the fuel supply manifold from the fuel injectors by lifting and gently rocking the manifold.

8. Remove the injectors by lifting while gently rocking from side-to-side. Remove and discard the O-rings.

To install:

9. Lubricate new O-rings with clean light grade oil and install 2 O-rings on each injector.

NOTE: Never use silicone grease at it will clog the injectors.

10. Install the injectors in the fuel supply manifold, using a light, twisting, pushing motion.

11. Carefully install the fuel supply manifold and injectors into the lower intake manifold, 1 side at a time. Push the fuel supply manifold down to ensure that all injector O-rings are fully seated in the fuel supply manifold cups and intake manifold.

12. While holding the fuel supply manifold in place, install the 2 retaining bolts hand-tight, then tighten to 6–8 ft. lbs. (8–12 Nm).

13. Repeat Steps 11 and 12 for the other side of the fuel supply manifold.

14. Connect the fuel supply and return lines.

15. After the fuel supply manifold has been installed and before the fuel injector wires have been connected, connect the negative battery cable and turn the ignition **ON**. This will cause the fuel pump to run for 2–3 seconds and pressurize the fuel system.

16. Check for leaks where the fuel injector is installed into the fuel supply manifold.

17. Disconnect the negative battery cable.

18. Connect the fuel injector electrical connectors.

19. Install the air intake throttle body. Tighten the bolts, in sequence, to 19 ft. lbs. (25 Nm).

20. Connect the vacuum line to the fuel pressure regulator.

21. Connect the negative battery cable, start the engine and let it idle for 2 minutes.

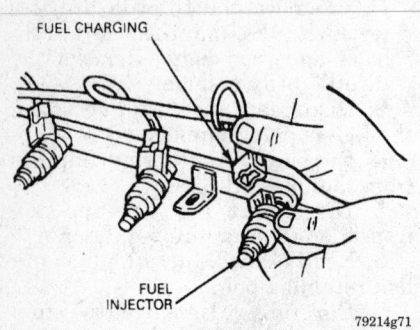

FUEL CHARGING

FUEL INJECTOR

79214g71

Fuel injector and harness — 3.0L engine

22. Turn the engine **OFF** and check for fuel leaks.

4.0L Engine

1. Disconnect the negative battery cable and relieve the fuel system pressure.

2. Disconnect the electrical connectors at the air bypass valve, TPS and ACT sensor.

3. Remove the snow/ice shield to expose the throttle linkage. Remove the throttle cable bracket and disconnect the cable from the ball stud on the throttle body.

4. Remove the air inlet tube from the air cleaner to the throttle body.

5. Disconnect the PCV valve from the valve cover.

6. Disconnect the spark plug wires from the comb at the rear of the manifold.

7. Remove the canister purge line from the fitting in the throttle housing.

8. On Aerostar models, remove the bolt retaining the engine oil dipstick tube.

9. Remove the bolt that retains the A/C line at the upper rear of the upper manifold.

10. Remove the 6 upper intake manifold retaining nuts and remove the upper intake and throttle body assembly.

11. Disconnect the fuel supply line fitting at the fuel manifold.

12. Disconnect the fuel return line from the fuel pressure regulator as follows:

   a. Disengage the locking tabs on the connector retainer and separate the retainer halves.

   b. Inspect the visible internal portion of the fitting for dirt accumulation. Clean the fitting before disassembly.

   c. To disengage the fitting from the regulator, push the fitting toward the regulator, insert the fingers on fuel line coupling key

T90P-9550-A or equivalent, into the slots in the coupling.

   d. Using the tool, pull the fitting from the regulator.

NOTE: If the fitting has been properly disengaged, the fitting should slide off the regulator with minimum effort.

13. Disconnect the harness from the fuel injectors.

14. Remove the 6 bolts retaining the fuel supply manifold, then remove the manifold.

15. Remove the injector retaining clips and remove the injectors from the manifold by grasping the injector body and pulling up while rocking the injector from side-to-side.

16. Remove and discard the injector O-rings.

17. Inspect the injector plastic pintle protection cap and washer for signs of deterioration. Replace the complete injector as required. If plastic pintle protection cap is missing, look for it in the intake manifold.

   To install:

18. Lubricate new O-rings with clean light grade oil and install 2 on each injector.

NOTE: Never use silicone grease at it will clog the injectors.

19. Install the injectors, using a light, twisting, pushing motion.

20. Install the fuel supply manifold, pushing down to make sure all the fuel injector O-rings are fully seated in the fuel supply manifold cups and intake manifold.

21. Install the 6 retaining bolts and tighten to 7–10 ft. lbs. (10–14 Nm). Install the retainer clips.

22. Install the fuel supply line and tighten the fitting to 15–18 ft. lbs. (20–24 Nm).

23. Install the fuel return line to the fuel pressure regulator by pushing it onto the fuel pressure regulator line up to the shoulder on the regulator line.

NOTE: The connector should grip the regulator line securely.

24. Install the connector retainer and snap the 2 halves of the retainer together.

25. Clean and inspect the mounting faces of the fuel manifold and upper intake manifold.

26. Position a new gasket on the mounting studs and install the upper intake manifold on the studs.

27. Install the 6 upper intake manifold retaining nuts and tighten to 15–18 ft. lbs. (20–25 Nm).

28. Attach the A/C line retainer and automatic transmission vacuum line retainer at the upper intake manifold.

29. Install the canister purge line on the throttle body fitting.

30. Connect the vacuum lines to the vacuum tree. Connect the harnesses to the air bypass valve, TPS and ACT sensor.

31. Install the PCV valve in the grommet at the rear of the right valve cover.

32. On Aerostar models, attach the engine oil dipstick tube to the upper intake manifold.

33. Attach the throttle cable bracket to the upper intake manifold, then connect the throttle cable to the ball stud and install the snow/ice shield.

34. After the upper intake manifold has been installed and before the fuel injector wires have been attached, connect the negative battery cable and turn the ignition switch **ON**. This will cause the fuel pump to run for 2–3 seconds and pressurize the system.

35. Check for fuel leaks where the fuel injector is installed into the fuel supply manifold.

36. Turn the ignition switch **OFF** and disconnect the negative battery cable.

37. Connect the injector harnesses and the vacuum line to the regulator.

38. Install the air inlet tube from the throttle body to the air cleaner.

39. Connect the negative battery cable, start the engine and let it idle for 2 minutes.

40. Turn the engine **OFF** and check for fuel leaks.

5.0L Engine

1. Relieve the fuel system pressure.

2. Disconnect the battery ground.

3. Remove the upper intake manifold.

4. Disconnect the wiring at the injectors.

5. Pull upward on the injector body while gently rocking it from side-to-side.

6. Inspect the O-rings on the injector for any sign of leakage or damage. Replace any suspected O-rings.

7. Inspect the plastic cap at the top of each injector and replace it if any sign of deterioration is noticed.

   To install:

8. Lubricate the O-rings with clean engine oil.

9. Install the injectors by pushing them in with a gentle rocking motion.

10. Install the fuel supply manifold.

11. Connect the electrical wiring.

12. Install the upper intake manifold.

EMISSION CONTROLS

Emission Warning Lamps

RESETTING

All vehicles are equipped with a "CHECK ENGINE" or "SERVICE ENGINE SOON" warning lamp located on the instrument cluster. This lamp should come on briefly when the ignition key is turned **ON**, but should turn **OFF** when the engine starts. If the lamp does not come **ON** when the ignition key is turned **ON** or if it comes **ON** and stays **ON** when the engine is running, there is a malfunction in the electronic engine control system. After the malfunction has been remedied, the "CHECK ENGINE" or "SERVICE ENGINE SOON" lamp will go out.

ENGINE MECHANICAL

NOTE: Disconnecting the negative battery cable on some vehicles may interfere with the functions of the on board computer systems and may require the computer to undergo a relearning process, once the negative battery cable is reconnected.

Engine Assembly

REMOVAL AND INSTALLATION

2.3L Engine

1. Disconnect the negative battery cable. Relieve the fuel system pressure.

2. Drain the cooling system. Disconnect the air cleaner tube at the throttle body. Disconnect the idle speed control hose and heat riser tube, if necessary.

3. Mark the location of the hinges on the hood and remove the hood.

4. Disconnect the radiator hoses and, if equipped, disconnect the transmission cooler lines. Remove the fan, shroud and radiator.

5. Remove the oil fill cap. Disconnect the engine wiring harness from the body wiring harness.

6. Disconnect the alternator wiring, starter cable and accelerator cable from the throttle body. If equipped, disconnect the transmission kickdown cable.

7. If equipped, remove the A/C compressor from the mounting bracket and position aside. Leave the refrigerant lines attached.

8. Disconnect the power brake vacuum hose. Disconnect the fuel lines from the fuel supply manifold.

9. Disconnect the heater hoses from the engine.

10. Remove the engine mount nuts. Raise and safely support the vehicle.

11. Drain the engine oil from the crankcase and remove the starter.

12. Disconnect the exhaust pipe at the exhaust manifold. If equipped with manual transmission, remove the dust cover. If equipped with an automatic transmission, remove the converter inspection plate, then remove the converter-to-flywheel bolts.

13. Remove the lower flywheel housing or converter housing attaching bolts and lower the vehicle.

14. Support the transmission and flywheel housing or converter housing with a jack.

15. Remove the flywheel housing or converter housing upper attaching bolts.

16. Attach suitable engine lifting equipment. Carefully lift the engine out of the vehicle and install on a workstand.

To install:

17. Remove the engine from the workstand and carefully lower it into the engine compartment. If equipped with an automatic transmission, insert the converter pilot into the crankshaft.

18. If equipped with manual transmission, insert the transmission input shaft into the clutch disc. It may be necessary to adjust the position of the transmission in relation to the engine, if the input shaft will not insert into the clutch disc. If the engine hangs up after the shaft enters, turn the crankshaft in a clockwise direction slowly until the shaft splines mesh with the clutch disc splines.

19. Install the flywheel or converter housing attaching bolts and remove the engine lifting equipment.

20. Remove the jack from under the vehicle and raise and safely support the vehicle.

21. If equipped with an automatic transmission, install the converter-to-flywheel attaching bolts. Install the lower flywheel housing or converter housing attaching bolts and install the dust plate or converter inspection cover.

22. Connect the exhaust pipe to the exhaust manifold. Install the starter and cable.

23. Lower the vehicle and install the engine mount nuts. Tighten the nuts to 65–85 ft. lbs. (88–115 Nm).

24. Connect the heater hoses to the engine and the fuel lines to the fuel supply manifold. Connect the power brake vacuum hose.

25. Connect the wiring to the alternator and the accelerator cable to the throttle body. If equipped, connect the transmission kickdown rod.

26. If equipped, install the A/C compressor in the mounting brackets.

27. Connect the engine wiring harness to the body harness.

28. Install the fan, shroud and radiator. Connect the radiator hoses and, if equipped, the transmission cooler lines.

29. Install the hood, aligning the hinges with the marks that were made during removal.

30. Connect the air cleaner outlet tube at the throttle body. Connect the idle speed control hose and heat riser tube, if necessary.

31. Fill the crankcase with the proper type and quantity of engine oil. Install the oil cap.

32. Connect the negative battery cable. Fill and bleed the cooling system. Run the engine and check for leaks.

3.0L Engine

Aerostar

1. Disconnect the negative battery cable and relieve the fuel system pressure. Drain the cooling system.

2. Disconnect the upper and lower radiator hoses.

3. Remove the air cleaner hose assembly.

4. Remove the engine fan and shroud.

5. Disconnect the Barometric Manifold Absolute Pressure (BMAP) sensor electrical connector and vacuum line.

6. Remove the shroud covering the throttle linkage and disconnect the linkage at the throttle body.

7. Remove the accessory drive belts.

8. Disconnect the injector harness connector from the main harness. Disconnect the engine coolant temperature sensor, located near the thermostat housing and the engine coolant temperature sender.

9. Disconnect the canister purge solenoid hoses from both sides of the solenoid. If equipped with power steering, disconnect the pump pressure switch.

10. Mark the inlet and outlet heater hoses with chalk prior to removing them from the engine side of the ballast tube.

11. Remove the breather tubes from the air cleaner and rocker arm cover.

12. If equipped with automatic transmission, disconnect the transmission cooler lines from the radiator. Remove the radiator.

13. If equipped with A/C, remove the compressor and retain it to a sidemember with wire.

14. Disconnect the oil fill tube from the alternator bracket. If equipped with an automatic transmission, disconnect the transmission fill tube from the manifold and remove.

15. Disconnect the harnesses from the alternator, and the brake booster vacuum line from the booster.

16. From inside the vehicle, remove the engine cover. Tag and disconnect the electrical connectors for the radio frequency interference suppressor, distributor module and oil pressure sender.

17. Disconnect the fuel lines from the fuel supply manifold.

18. If equipped with manual transmission, place the shift lever in **N** and remove the screws retaining the shift lever boot to the floor. Slide the boot up the shift lever. Remove the bolt retaining the shift lever assembly to the transmission and remove the lever.

19. Raise and safely support the vehicle. Disconnect the oil level sensor connector from the oil pan.

20. Mark the driveshaft to flange position and remove the driveshaft.

21. Remove the bolt retaining the speedometer cable bracket to transmission. Pull the speedometer out of the rear of the transmission.

22. Remove the starter.

23. If equipped with manual transmission, disconnect the coupling at the transmission with tool T88T-70522-A or equivalent, by sliding the white plastic sleeve toward the slave cylinder while applying a slight tug on the tube.

24. If equipped with manual transmission, disconnect the backup lamp switch and neutral sensing switch wires from the transmission.

25. If equipped with automatic transmission, proceed as follows:

   a. Disconnect the electrical connectors for the neutral safety switch and 3–4 shift solenoid connectors. Disconnect the selector and kickdown cable from the transmission lever.

   b. Disconnect the vacuum hose from the transmission vacuum modulator.

   c. Remove the converter access cover and adapter plate bolts from the lower end of the converter housing.

   d. Remove the flywheel-to-converter attaching nuts. Place a 22mm socket and breaker bar on the crankshaft pulley bolt in order to turn the crankshaft and gain access to the flywheel-to-converter nuts.

   e. Disconnect the transmission cooler lines from the transmission.

26. Disconnect the oxygen sensor.

27. Position a jack under the transmission and a safety chain around the transmission. Slightly raise the transmission.

28. Remove the mount-to-crossmember attaching nuts. Remove the nuts and bolts attaching the crossmember to the 2 mounting brackets and remove the crossmember. If required, remove the bolts attaching the mount to the transmission and remove the mount.

29. Remove the converter housing-to-engine fasteners. Move the transmission to the rear so it disengages from the dowel pins and the converter disengages from the flywheel. Lower the transmission from the vehicle.

30. Disconnect and remove the exhaust pipe and catalytic converter. Remove the front wheels.

31. Remove the engine block ground straps, 1 on the cylinder head just behind the power steering pump and the other just above where the exhaust manifold and exhaust pipe connect.

32. Disconnect the stabilizer bar from the lower control arms. Discard the bar nuts.

33. Behind the spindles, disconnect and plug the brake lines at the bracket on the frame.

34. Position a jack under the lower control arm and raise the arm until tension is applied to the coil spring. Remove the bolt and nut retaining the spindle to the upper control arm ball joint. Slowly lower the jack to disconnect the spindle from the ball joint. Install safety chains around the lower control arm and spring seat.

35. Position drive train removal lift 109-00002 or equivalent, under the crossmember and engine assembly.

36. Slowly lower the vehicle until the crossmember rests on the removal lift. Place wood blocks under the front crossmember and rear of the engine block to keep the engine and crossmember assembly level. Install safety chains around the crossmember and lift.

37. With the engine and crossmember securely supported on the lift, remove the 3 nuts that retain the engine crossmember assembly to the frame on each side of the vehicle.

38. Slowly lower the engine assembly out of the vehicle, making sure the A/C compressor and wiring harnesses do not interfere. When the assembly is clear, roll the lift out from under the vehicle.

39. Separate the engine from the crossmember and position on a workstand.

To install:

40. Remove the engine from the workstand. With the front crossmember securely positioned on drive train removal lift 109-00002 or equivalent, slowly lower the engine until the motor mount studs enter the crossmember holes. Install the retaining nuts and tighten to 71–94 ft. lbs. (96–127 Nm).

NOTE: Install wood blocks under the oil pan and crossmember.

41. Roll the removal lift under the vehicle. Align the lift, engine and crossmember assembly so the 3 mounting bolts on each side of the frame are in alignment with the holes in the crossmember.

42. Slowly lower the vehicle so the bolts are piloted in the crossmember holes. Raise the lift or lower the vehicle so the crossmember is against the frame. Install the nuts retaining the crossmember to the frame and tighten to 145–195 ft. lbs. (196–264 Nm). Raise the vehicle and remove the lift.

43. If equipped with an automatic transmission, position the converter to the transmission making sure the converter hub is fully engaged in the pump gear. When the torque converter is fully installed, the distance between the converter pilot and the front of the converter housing should be 7/16–9/16 in. (11–14mm). Make sure the converter rotates freely and is not binding.

44. If equipped with a manual transmission, install the transmis-

sion on a jack and lift into position. Make sure the transmission input shaft engages the pilot bearing in the flywheel and the flywheel housing holes are aligned with the engine block dowel pins. Install the flywheel housing-to-engine block bolts and tighten to 28–38 ft. lbs. (38–51 Nm).

45. If equipped with an automatic transmission, place the transmission on a jack and secure with safety chains. Rotate the converter so the drive studs are aligned with the flywheel holes. Lift the transmission into position and connect the transmission cooler lines to the case. Move the transmission forward into position, making sure the converter housing holes align with the engine block dowel pins. Install the converter housing-to-engine block bolts and tighten to 28–38 ft. lbs. (38–51 Nm).

46. Position the crossmember in the 2 mounting brackets and install the mounting nuts and bolts. Slowly lower the transmission so the mount studs are installed in the proper slots in the crossmember. Install the nuts and tighten to 71–94 ft. lbs. (97–127 Nm). Remove the safety chain and the jack.

47. Install the starter. Connect the starter cable.

48. If equipped with an automatic transmission, connect the modulator vacuum hose. Position the selector cable in the case bracket and press the end of the cable on the ball stud on the lower portion of the selector lever. Install the retainer in the bracket.

49. If equipped with a manual transmission, install the speedometer cable or connect the electronic speedometer wire. Connect the backup lamp and neutral sensor wires.

50. Remove the cap from the hydraulic clutch line. Insert the male coupling into the female coupling on the slave cylinder and make sure the connection is secure.

51. Install the manual transmission shift lever to the shifter. Position the rubber shift boot on the floor and install the screws.

52. If equipped with an automatic transmission, install the kickdown and selector cable. Adjust the cable. Connect the neutral safety switch, converter clutch solenoid and 3–4 shift solenoid connectors. Insert the speedometer driven gear into the transmission and retain with a clamp. Tighten the retaining screw to 20–25 inch lbs. (2.25–2.82).

53. Position a 22mm socket and breaker bar on the crankshaft pulley bolt. Rotate the pulley clockwise, as viewed from the front, to gain access to each torque converter studs. Install the nuts on the studs and tighten to 20–34 ft. lbs. (27–46 Nm).

54. Position the converter access cover and adapter plate on the converter housing. Install the bolts and tighten to 12–16 ft. lbs. (16–22 Nm).

55. Remove the safety chains from around the lower control arms and spring seat. Install a jack under the lower control arms. Slowly raise the control arm until the coil spring is under tension. Continue to raise the arm until the spindle is connected to the upper arm ball joint. Install a new nut and bolt and tighten to 27–37 ft. lbs. (37–50 Nm).

56. Connect the stabilizer bar and new bar nuts.

57. Connect the front brake lines to the caliper hoses at the frame brackets. Install the wheel and tire assemblies.

58. Install the driveshaft, aligning the marks that were made during removal.

59. Install new gaskets on the exhaust manifold and catalytic converter. Install the exhaust pipe and catalytic converter. Tighten the converter-to-muffler nuts and bolts to 18–26 ft. lbs. (25–35 Nm). Tighten the exhaust pipe-to-exhaust manifold nuts to 25–34 ft. lbs. (34–46 Nm).

60. Connect the oxygen sensor connector and install the engine ground straps. Connect the fuel lines and the low-oil level sensor.

61. If equipped with an automatic transmission, connect the transmission oil cooler lines. Install the transmission oil fill tube and lower the vehicle.

62. From inside the vehicle, connect the radio frequency interference suppressor, distributor module connector and oil pressure sender. Install the engine cover.

63. Connect the alternator harness and the brake booster vacuum hose. Connect the breather tube to the oil filler tube and attach the bolt retaining the steering gear to the top of the shaft.

64. If equipped, connect the power steering pressure switch. Connect the heater hoses to the ballast tube by matching the chalk lines to the specific inlet and outlet hoses.

65. Install the radiator. If equipped, connect the transmission oil cooler lines.

66. Connect the canister purge solenoid hoses from both sides of the solenoid. Connect the engine coolant temperature sensor, located near the thermostat housing and connect the engine coolant temperature sender.

67. If equipped with A/C, install the compressor using the attaching bolts.

68. Install the drive belts. Place the injector harness behind the belt tension idler arm and tighten the idler arm.

69. Connect the throttle linkage to the ball stud located on the throttle body. Connect the shroud covering the throttle body.

70. Connect the BMAP sensor harness and vacuum line located on the dash panel.

71. Install the fan and fan shroud. Connect the radiator hoses.

72. Connect the negative battery cable and install the air cleaner and duct assembly.

73. Bleed the brakes and the hydraulic clutch. Fill and bleed the cooling system. Check all fluid levels.

74. Run the engine and check for leaks and proper operation. Check the front end alignment.

Except Aerostar

1. Disconnect the negative battery cable and relieve the fuel system pressure. Drain the cooling system.

2. Mark the position of the hood on the hinges and remove the hood. Remove the air cleaner intake hose.

3. Disconnect the radiator hoses at the radiator. Remove the fan shroud attaching bolts and position the shroud over the fan. Remove the radiator, then remove the shroud.

4. Remove the alternator and bracket and position the alternator aside. Disconnect the alternator ground wire from the cylinder block.

5. Remove the A/C compressor and power steering pump and position aside, if equipped.

6. Disconnect the heater hoses at the intake manifold and water pump. Remove the ground wires from the cylinder block.

7. Disconnect the fuel lines at the chassis to engine connections. Disconnect the throttle cable shield and linkage at the throttle body and intake manifold.

8. Tag and disconnect the vacuum connections at the rear fitting in the upper intake manifold.

9. Tag and disconnect the wires at the ignition coil. Disconnect the 3 body wiring connectors on top of the right rocker arm cover. Disconnect the oil pressure and engine coolant temperature sender harnesses.

10. Disconnect the injector harness, air charge temperature sensor and throttle position sensor. Disconnect the oxygen sensor connector at the

rear of the engine and disconnect the brake booster vacuum hose.

11. Remove the engine front mount-to-crossmember attaching nuts.

12. Raise and safely support the vehicle.

13. Remove the A/C compressor bracket-to-engine bolts and place the compressor aside. Disconnect the wiring from the low oil level sensor and oil pressure sending unit.

14. Remove the retaining bracket holding the transmission cooling lines to the right side of the engine block.

15. Disconnect the exhaust pipes at the manifolds.

16. Remove the starter.

17. If equipped with a manual transmission, disconnect the hydraulic clutch line and remove the flywheel housing-to-engine block bolts.

18. If equipped with automatic transmission, remove the converter inspection cover and disconnect the converter from the flywheel.

19. Disconnect the kickdown and shift cables at the transmission. Remove the converter housing-to-engine block bolts and adapter plate-to-converter housing bolts.

20. Attach suitable engine lifting equipment and position a jack under the transmission.

21. Raise the engine slightly and carefully separate it from the transmission. Carefully lift the engine out of the engine compartment. Install the engine on a workstand.

To install:

22. Remove the engine from the workstand and carefully lower it into the engine compartment. Make sure the exhaust manifolds are aligned with the exhaust pipes.

23. If equipped with manual transmission, insert the transmission input shaft into the clutch disc. It may be necessary to adjust the position of the transmission in relation to the engine, if the input shaft will not enter the clutch disc. If the engine hangs up after the shaft enters, turn the crankshaft in the clockwise direction slowly until the shaft splines mesh with the clutch disc splines.

24. If equipped with an automatic transmission, start the converter pilot into the crankshaft. Make sure the converter rotates freely and is not binding. When the converter is fully installed in the transmission, the distance between the converter pilot and the edge of the converter housing should be $7/16$–$9/16$in. (11–14mm).

25. Install the flywheel housing or converter housing upper bolts, making sure the engine block dowels en-

gage in the housing. If equipped, install the clutch hydraulic line.

26. Remove the jack from under the transmission and remove the engine lifting equipment. If equipped with an automatic transmission, position the kickdown cable on the transmission and engine.

27. Raise and safely support the vehicle. If equipped with automatic transmission, position the transmission linkage bracket and install the remaining converter housing bolts.

28. Install the adapter plate-to-converter housing bolt. Install the converter-to-flywheel nuts and install the inspection cover. Connect the kickdown cable at the transmission.

29. If equipped with a manual transmission, install the flywheel housing attaching bolts.

30. Install the starter and connect the cables. Connect the exhaust pipes at the manifolds.

31. Install the engine front mount nuts and washers or through bolts. Lower the vehicle.

32. Install the ground wires to the engine block. Connect the coil wires, the 3 body wiring connectors and the oxygen sensor harness. Connect the coolant temperature sending unit and oil pressure sending unit. Connect the brake booster vacuum hose.

33. Install the throttle linkage and connect the fuel lines. Connect the heater hoses at the water pump and cylinder block.

34. Install the alternator wiring and drive belt.

35. Install the A/C compressor and power steering pump, if equipped.

36. Install the radiator, hoses, fan and shroud.

37. Connect the negative battery cable. Fill and bleed the cooling system.

38. Bleed the hydraulic clutch, if necessary. Evacuate and charge the air conditioning system, if necessary.

39. Run the engine and check for leaks and proper operation. Install the intake hose. Install the hood, aligning the marks that were made during removal.

4.0L Engine

Aerostar

1. Disconnect the negative battery cable and relieve the fuel system pressure.

2. Remove the front grille. Remove the air cleaner tube and the air cleaner assembly.

3. Discharge the refrigerant from the A/C system, if equipped. Disconnect and remove the A/C compressor.

4. Drain the engine oil. Drain disconnect and remove the power steering oil cooler.

5. Remove the front bumper cover. Drain, disconnect and remove the transmission oil cooler.

6. Drain the radiator and disconnect the radiator hoses. Disconnect the transmission oil cooler lines from the radiator.

7. Disconnect the fan shroud and position it over the fan. Remove the radiator and the shroud.

8. Remove the drive belt and the right front air diverter flap. Remove the center hood latch support and remove the alternator.

9. Remove the engine oil fill tube. From inside the vehicle, remove the engine cover.

10. Remove the power steering pump and bracket.

11. Remove the transmission oil and engine dipsticks and tubes. Disconnect the exhaust system from the exhaust manifolds.

12. Disconnect and remove the starter. Working through the starter opening in the converter housing, remove the converter to flexplate bolts.

13. Remove the transmission oil cooler line bracket retaining bolt and remove the engine mount to frame bolts.

14. Remove the converter housing-to-engine bolts, except the upper bolts.

15. Remove the left motor mount from the engine, then remove the upper converter housing bolts.

16. Disconnect the transmission and transfer case, if equipped harnesses. Disconnect the fuel lines at the fuel supply manifold.

17. Disconnect the throttle linkage and bracket and position aside. Disconnect the heater hoses.

18. Tag and disconnect the vacuum lines from the vapor canister, lower intake manifold and upper intake manifold vacuum tee.

19. Remove the spark plug wires from the coil assembly. Disconnect and remove the coil assembly with mounting bracket.

20. Disconnect the throttle position sensor harness. Remove the throttle body from the upper intake manifold.

21. Tag and disconnect the engine wiring harness main connectors. Install suitable engine lifting equipment and remove the engine from the vehicle. Position the engine on a workstand.

To install:

22. Remove the engine from the workstand and position it in the vehi-

cle. Remove the engine lifting equipment.

23. Connect the engine wiring harness main connectors. Install the throttle body on the upper intake manifold.

24. Install the ignition coil assembly and connect the spark plug wires and coil assembly wiring. Connect the harness for the throttle position sensor.

25. Connect the heater hoses. Connect the vacuum lines to the vapor canister, lower intake manifold and upper intake manifold vacuum tee.

26. Connect the throttle linkage and bracket. Connect the fuel lines to the fuel supply manifold.

27. Install the left motor mount. Install the converter housing-to-engine bolts and connect the transmission and transfer case harnesses.

28. Install the engine mount-to-frame bolts and the transmission oil cooler line bracket retaining bolt.

29. Working through the starter opening in the converter housing, install the converter to flexplate bolts.

30. Install the starter and connect the wires. Connect the exhaust system to the exhaust manifolds.

31. Install the transmission oil and engine oil dipstick tubes and dipsticks.

32. Install the power steering pump and mounting bracket.

33. Install the ice/snow shield and the engine oil fill tube. Install the engine cover.

34. Install and connect the A/C compressor, if equipped. Install and connect the alternator.

35. Install the center hood latch support and the right front air diverter flap. Install the accessory drive belt.

36. Install the radiator and fan shroud. Connect the transmission oil cooler lines and the radiator hoses.

37. Install the transmission oil cooler and the front bumper cover. Install and connect the power steering oil cooler.

38. Fill the engine with the proper type and quantity of engine oil.

39. Install the air cleaner assembly and air tube.

40. Install the front grille and connect the negative battery cable. Fill and bleed the cooling system.

41. If equipped, evacuate and charge the air conditioning system. Check all fluid levels.

42. Run the engine and check for leaks and for proper operation.

Except Aerostar

1. Disconnect the negative battery cable and relieve the fuel system pressure. Drain the cooling system.

2. Mark the position of the hood on the hinges and remove the hood. Remove the air cleaner intake hose.

3. Disconnect the radiator hoses at the radiator. Disconnect the fan shroud and position it over the fan. Remove the radiator, then the shroud.

4. Remove the alternator and bracket and position the alternator aside. Disconnect the alternator ground wire from the cylinder block.

5. Remove the A/C compressor and power steering pump and position aside, if equipped.

6. Disconnect the heater hoses at the intake manifold and water pump. Remove the ground wires from the cylinder block.

7. Disconnect the fuel lines from the fuel supply manifold. Disconnect the throttle cable shield and linkage at the throttle body and intake manifold.

8. Tag and disconnect the vacuum connections at the rear vacuum fitting in the upper intake manifold.

9. Disconnect the wiring from the ignition coil, oil pressure and engine coolant temperature senders. Disconnect the injector harness, air charge temperature sensor and throttle position sensor. Disconnect the brake booster vacuum hose.

10. Raise and safely support the vehicle. Disconnect the exhaust pipes at the manifolds. Remove the starter.

11. Remove the engine front mount-to-crossmember attaching nuts or through-bolts.

12. Remove the converter inspection cover and disconnect the converter from the flywheel. Remove the cable.

13. Remove the converter housing-to-engine block bolts and the adapter plate-to-converter housing bolt. Lower the vehicle.

14. Position a jack under the transmission and install suitable engine lifting equipment.

15. Raise the engine slightly and carefully separate it from the transmission. Carefully lift the engine out of the engine compartment so the rear cover plate is not bent or components damaged. Install the engine on a workstand.

To install:

16. Remove the engine from the workstand and carefully lower it into the engine compartment. Make sure the exhaust manifolds are aligned with the exhaust pipe.

17. At the transmission, start the converter pilot into the crankshaft. Install the converter housing upper bolts, making sure the dowels in the cylinder block engage the flywheel housing. Tighten the bolts to 33–45 ft. lbs. (45–61 Nm).

18. Remove the jack from under the transmission and the engine lifting equipment.

19. Position the kickdown rod on the transmission and engine. Raise and safely support the vehicle.

20. Position the transmission linkage bracket and install the remaining converter housing bolts. Install the adapter plate-to-converter housing bolt. Install the converter-to-flywheel nuts and install the inspection cover. Connect the kickdown rod on the transmission.

21. Install the starter and connect the cable. Connect the exhaust pipes at the manifolds.

22. Install the engine front mount nuts and washers or through bolts. Lower the vehicle.

23. Install the ground wires to the engine block. Connect the ignition coil wiring, then connect the coolant temperature sending unit and oil pressure sending unit. Connect the brake booster vacuum hose.

24. Install the throttle linkage and connect the fuel lines at the fuel supply manifold.

25. Connect the ground cable at the engine block. Connect the heater hoses to the water pump and cylinder block.

26. Install the alternator and bracket. Connect the alternator ground wire to the engine block. Install the drive belt.

27. Install the A/C compressor and power steering pump, if equipped.

28. Position the shroud over the fan. Install the radiator and connect the radiator upper and lower hoses. Install the fan shroud attaching bolts.

29. Connect the negative battery cable. Fill and bleed the cooling system.

30. Run the engine and check for leaks. If equipped, evacuate and charge the air conditioning system.

31. Install the intake hose. Install the hood, aligning the marks that were made during removal.

5.0L Engines

1. Remove the hood.
2. Drain the cooling system and crankcase.
3. Disconnect the battery and alternator cables.

4. Remove the air intake hoses, PCV tube and carbon canister hose.

5. Disconnect the upper and lower radiator hoses.

6. Discharge the A/C system.

7. Disconnect the refrigerant lines at the compressor. Cap all openings immediately.

8. If equipped, disconnect the automatic transmission oil cooler lines.

9. Remove the fan shroud and lay it over the fan.

10. Remove the radiator, fan and shroud.

11. Remove the alternator.

12. Disconnect the oil pressure harness from the sending unit.

13. Disconnect the fuel line and plug.

14. Disconnect the accelerator linkage and speed control linkage at the throttle body.

15. Disconnect the automatic transmission kick-down rod and remove the return spring, if equipped.

16. Disconnect the power brake booster vacuum hose.

17. Disconnect the throttle bracket from the upper intake manifold and swing it out of the way with the cables still attached.

18. Disconnect the heater hoses from the water pump and tee.

19. Remove the upper bellhousing-to-engine attaching bolts.

20. Remove the wiring harness from the left rocker arm cover and position the wires out of the way.

21. Disconnect the ground strap from the cylinder block.

22. Disconnect the A/C compressor clutch wire.

23. Remove the starter.

24. Disconnect the exhaust pipe from the exhaust manifolds.

25. Disconnect the engine mounts from the brackets on the frame.

26. If equipped with an automatic transmissions, remove the converter inspection plate and remove the torque converter-to-flywheel attaching bolts.

27. Remove the remaining bellhousing-to-engine attaching bolts.

28. Lower the vehicle and support the transmission with a jack.

29. Install an engine lifting device.

30. Raise the engine slightly and carefully pull it out of the transmission. Lift the engine out of the engine compartment.

To install:

31. Remove the engine mount brackets from the frame and attach them to the engine mounts. Tighten the mount-to-bracket nuts just enough to hold them securely.

32. Lower the engine carefully into the transmission. Make sure the dowel in the engine block engage the holes in the bellhousing through the rear cover plate. If the engine hangs up after the transmission input shaft enters the clutch disc (manual transmission only), turn the crankshaft with the transmission in gear until the input shaft splines mesh with the clutch disc splines.

33. Install the engine mount nuts and washers. Torque the nuts to 80 ft. lbs. (104 Nm) Tighten the bracket-to-frame bolts to 70 ft. lbs. (91 Nm)

34. Remove the engine lifting device.

35. Install the lower bellhousing-to-engine attaching bolts. Torque the bolts to 50 ft. lbs. (65 Nm)

36. Remove the transmission support jack.

37. If equipped with automatic transmissions, install the torque converter-to-flywheel attaching bolts. Torque the bolts to 30 ft. lbs. (39 Nm)

38. Install the converter inspection plate. Torque the bolts to 60 inch lbs. (7 Nm)

39. Connect the exhaust pipe to the exhaust manifolds. Tighten the exhaust pipe-to-exhaust manifold nuts to 25–35 ft. lbs. (32–45 Nm).

40. Install the starter.

41. Connect the starter cable to the starter.

42. Lower the vehicle.

43. Install the upper bellhousing-to-engine attaching bolts. Torque the bolts to 50 ft. lbs. (65 Nm)

44. Connect the wiring harness at the left rocker arm cover.

45. Connect the ground strap to the cylinder block.

46. Connect the A/C compressor clutch wire.

47. Connect the heater hoses at the water pump and tee.

48. Connect the temperature sending unit wire at the sending unit.

49. Connect the accelerator linkage and speed control linkage at the throttle body.

50. Connect the automatic transmission kick-down rod and install the return spring, if equipped.

51. Connect the power brake booster vacuum hose.

52. Connect the throttle bracket to the upper intake manifold.

53. Connect the fuel tank-to-pump fuel line at the fuel pump. If equipped, connect the chassis fuel line at the fuel rails.

54. Connect the oil pressure sending unit wire.

55. Install the alternator.

56. Connect the refrigerant lines to the compressor.

57. Install the radiator, fan and shroud.

58. Connect the upper and lower radiator hoses, and, if equipped, the automatic transmission oil cooler lines.

59. Install the air intake hoses, PCV tube and carbon canister hose.

60. Connect the battery and alternator cables.

61. Fill the cooling system and crankcase.

62. Charge the air conditioning system.

63. Install the hood.

Engine Mounts

REMOVAL AND INSTALLATION

Front

1. Remove the fan shroud attaching screws.

2. Support the engine using a wood block and a jack placed under the oil pan.

3. Remove the nuts and washers attaching the mounts to the engine bracket. Lift the engine enough to disengage the mount upper stud from the crossmember engine bracket.

4. Remove the bolt attaching the fuel pump shield to the left engine bracket, if necessary.

5. Remove the mount-to-crossmember attaching nut and washer assembly. Remove the engine mount.

To install:

6. Install the engine mount to the crossmember.

7. Install the bolt attaching the fuel pump shield to the left bracket, if necessary.

8. Lower the engine until the mount stud engages in the slot/hole of the engine bracket. Install the attaching nuts.

9. Remove the jack and wood block from the engine oil pan. Install the fan shroud attaching screws.

Rear

1. Place a block of wood and a jack under the transmission.

2. Remove the 2 nuts attaching the mount to the crossmember. Raise the transmission enough to lift the mount from the crossmember.

3. Remove the bolts and nuts attaching the crossmember to the frame side rails and remove the crossmember.

4. If equipped, remove the fasteners attaching the exhaust hanger to the rear engine mount.

5. Remove the 2 bolts attaching the mount to the transmission and remove the mount and retainer assembly.

To install:

6. Position the mount and retainer assembly to the transmission and install the 2 attaching bolts.

7. If equipped, install the fasteners attaching the exhaust hanger to the mount.

8. Install the crossmember to the frame side rails with the attaching nuts and bolts.

9. Lower the transmission and install the mount crossmember attaching nuts. Remove the jack and wood block.

Cylinder Head

REMOVAL AND INSTALLATION

2.3L Engine

1. Disconnect the negative battery cable. Drain the cooling system.

2. Remove the air cleaner assembly. Remove the heater hose retaining screw(s) to the rocker arm cover.

3. If equipped, disconnect the distributor cap and spark plug wires and remove the assembly.

4. Remove the spark plugs. If equipped with distributorless ignition, remove the spark plug wire harnesses.

5. Remove the engine and alternator wiring harnesses. Disconnect the oxygen sensor at the exhaust manifold.

6. Tag and disconnect the required vacuum hoses. Remove the dipstick tube and bracket.

7. Remove the rocker arm cover attaching bolts and remove the cover. Remove the intake manifold attaching bolts.

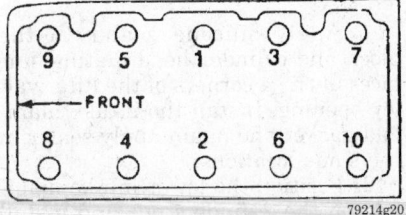

2.3L cylinder head torque sequence

8. Loosen the alternator retaining bolts and remove the belt from the pulley. Remove the mounting bracket-to-head retaining bolts.

9. Remove the upper radiator hose. Remove the timing belt cover. If equipped with power steering, move the power steering pump bracket.

10. Loosen the timing belt idler retaining bolts. Position the idler in the unloaded position and tighten the retaining bolts. Remove the timing belt from the camshaft pulley and auxiliary pulley.

11. Remove the 4 nuts and/or stud bolts retaining the heat stove to the exhaust manifold. Remove the 8 exhaust manifold retaining bolts.

12. Remove the timing belt idler and 2 bracket bolts. Remove the timing belt idler spring stop from the cylinder head.

13. Remove the cylinder head retaining bolts and remove the cylinder head.

14. Clean all gasket mating surfaces. Check the cylinder head for flatness using a straight edge and a feeler gauge. The cylinder head must not be warped more than 0.003 in. in any 6 in. or more than 0.006 in. overall.

To install:

15. Position a new head gasket on the block. Properly position the camshaft in the cylinder head and install the cylinder head on the block.

16. Install the cylinder head bolts and tighten, in sequence, in 2 steps, first to 50–60 ft. lbs. (68–81 Nm) and then to 80–90 ft. lbs. (108–122 Nm).

17. Install a new intake manifold gasket and position the intake manifold to the cylinder head. Install the retaining bolts.

18. Install the timing belt idler spring stop to the cylinder head. Position the timing belt idler to the cylinder head and install the retaining bolts.

19. Install the 8 exhaust manifold retaining bolts and the 4 nuts and/or stud bolts retaining the heat stove to the exhaust manifold.

20. If equipped, align the distributor rotor with the No. 1 spark plug location in the distributor cap.

21. Align the cam gear with the pointer and the crank pulley with the pointer on the timing belt cover.

22. Position the timing belt to the pulleys. Loosen the idler retaining, rotate the engine and check the timing alignment.

23. Adjust the belt tensioner and tighten the retaining bolts. Install the timing belt cover and the retaining bolt(s).

24. Install the upper radiator hose. Position the alternator bracket to the cylinder head and install the retainers. Install the drive belt.

25. Install a new rocker arm cover gasket and rocker cover.

26. Install the spark plugs. Install the spark plug wires and the distributor cap, if equipped.

27. Install the dipstick tube and bracket. Connect the vacuum hoses. Install the retaining heater hose screw(s) to the rocker arm cover.

28. Connect the negative battery cable. Fill and bleed the cooling system.

29. Start the engine and check for leaks. Install the air cleaner hose to the throttle body.

3.0L Engine

1. Disconnect the negative battery cable and relieve the fuel system pressure. Drain the cooling system.

2. Remove the fresh air hose from the throttle body and air cleaner. If equipped, remove the engine oil filler adapter.

3. Disconnect the fuel lines. Tag and disconnect the necessary vacuum lines.

4. Disconnect the upper radiator hose and heater hose and position aside. On Pick-Up models, disconnect the ignition coil electrical harness and remove the coil.

5. Remove the throttle body.

6. Remove the distributor cap, if equipped. Mark the position of the rotor in relation to the distributor housing and the position of the distributor housing in relation to the intake manifold. Remove the distributor hold-down bolt and clamp and remove the distributor. Tag and disconnect the spark plug wires from the spark plugs and remove the distributor cap and wires assembly.

7. If removing the left cylinder head, proceed as follows:

   a. Remove the drive belt(s).

   b. Remove the power steering pump and bracket, leaving the lines connected. Place the assembly aside in a position to prevent fluid leakage.

   c. On Aerostar models, remove the ignition coil and bracket.

   d. Remove the engine oil dipstick tube.

   e. Remove the fuel line retaining bracket bolt from the front of the cylinder head, if equipped.

8. If removing the right cylinder head, proceed as follows:

   a. Remove the drive belt(s).

   b. Remove the accessory drive belt idler or tensioner.

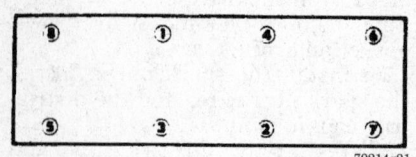

3.0L cylinder head torque sequence

c. Remove the grounding strap throttle cable support bracket, if necessary.

d. Disconnect the alternator harnesses and remove the alternator and bracket assembly.

9. Remove the spark plugs.

10. Disconnect the exhaust pipes and remove the exhaust manifolds.

11. Remove the rocker arm covers. Loosen the rocker arm fulcrum retaining bolts enough to allow the rocker arm to be lifted off the pushrod and rotated to 1 side.

NOTE: Regardless of which head is to be removed, the No. 3 cylinder intake valve pushrod must be removed to allow removal of the intake manifold.

12. Remove the pushrods, marking them so they can be reinstalled in their original positions.

13. Remove the intake manifold.

14. Remove the cylinder head attaching bolts and cylinder head(s).

15. Clean all gasket mating surfaces. Check the cylinder head for flatness using a straight edge and a feeler gauge. The cylinder head must not be warped more than 0.003 in. in any 6 in. or more than 0.006 in. overall.

To install:

16. Position new head gasket(s) on the cylinder block, using the dowels for alignment.

17. Install the cylinder head(s) on the block. Oil the threads of new cylinder head bolts and hand-tighten.

18. Tighten the cylinder head bolts, in sequence, to 59 ft. lbs. (80 Nm). Back off all bolts a minimum of 1 full turn. Retighten the bolts, in sequence, in 2 steps, first to 37 ft. lbs. (50 Nm) and then to 68 ft. lbs. (92 Nm).

19. Install the intake manifold.

20. Install the distributor, if equipped, aligning the marks that were made during removal. Install the hold-down bolt and clamp.

21. Dip each pushrod in heavy engine oil and install them in their original positions.

22. For each valve, rotate the crankshaft until the lifter rests on the base circle of the camshaft lobe, before tightening the rocker arm fulcrum attaching bolts. Position the rocker arms over the valves and pushrods, install the fulcrums and fulcrum bolts and tighten to 24 ft. lbs. (32 Nm).

NOTE: If the original valve train components are being installed, a valve clearance check is not required. If a component has been replaced, perform a valve clearance check.

23. Install the exhaust manifolds and the spark plugs.

24. Install the rocker arm covers. Install the dipstick tube.

25. Install the fuel injector harness to the injectors and inboard rocker arm cover studs. Connect the engine harness to the main harness.

26. Install the distributor cap and connect the spark plug wires to the spark plugs.

27. Install the throttle body. Install the ignition coil and bracket, if necessary and connect the electrical connector.

28. Install the fuel line retaining bracket to the front of the cylinder head, if equipped. Tighten the retaining bolts to 26 ft. lbs. (35 Nm).

29. Install the power steering pump and bracket, if removed. Install the alternator and bracket assembly, if removed and connect the electrical harness.

30. Install the drive belt(s).

31. Connect the fuel lines to the fuel supply manifold. Connect the upper radiator and heater hoses.

32. Connect the vacuum lines. Install the engine oil filler adapter on Aerostar models.

33. Change the engine oil and filter.

NOTE: Engine coolant is corrosive to all engine bearing material. Replacing engine oil after removal of a coolant carrying component helps prevent engine failure later.

34. Install the air cleaner fresh air hose to the throttle body and air cleaner. Connect the negative battery cable.

35. Fill and bleed the cooling system. Run the engine and check for leaks.

36. Check the ignition timing and idle speed and adjust, if necessary.

4.0L Engine

1. Disconnect the negative battery cable and relieve the fuel system pressure. Drain the cooling system.

2. Remove the upper and lower intake manifolds and rocker arm covers.

3. If the left cylinder head is being removed, proceed as follows:

a. Remove the accessory drive belt.

b. Discharge the refrigerant and remove the A/C compressor, if equipped.

c. Remove the power steering pump and bracket and position aside.

d. Remove the spark plugs.

4. If the right cylinder head is being removed, proceed as follows:

a. Remove the accessory drive belt.

b. Remove the alternator and bracket.

c. Remove the ignition coil and bracket assembly.

d. Remove the spark plugs.

5. Disconnect the exhaust pipe and remove the exhaust manifold(s).

6. Remove the rocker arm shaft assembly. Remove the pushrods, marking them so they can be reinstalled in the same positions.

7. Remove and discard the cylinder head attaching bolts and remove the cylinder heads.

8. Clean all gasket mating surfaces. Check the cylinder head for flatness using a straight edge and a feeler gauge. The cylinder head must not be warped more than 0.003 in. in any 6 in. or more than 0.006 in. overall.

To install:

9. Position new cylinder head gasket(s) on the cylinder block. Install cylinder head locating dowels.

NOTE: The cylinder head(s) and intake manifold are torqued alternately and in sequence to insure correct fit and gasket crunch.

10. Install new cylinder head bolts and tighten, in sequence, to 44 ft. lbs. (60 Nm).

11. Apply silicone sealer to the block and cylinder head mating surfaces at the 4 corners of the lifter valley opening. Install the intake manifold gasket and again apply sealer in the same locations.

12. Position the lower intake manifold on the 2 guide studs and install the nuts and bolts hand-tight. Tighten the lower intake manifold bolts, in sequence, to 3–6 ft. lbs. (4–8 Nm).

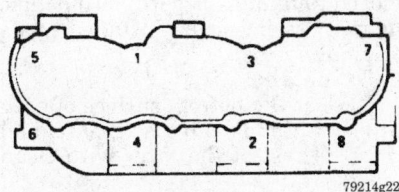

4.0L cylinder head bolt torque sequence

13. Tighten the cylinder head bolts, in sequence, to 59 ft. lbs. (80 Nm).

14. Tighten the intake manifold, in sequence, to 6–11 ft. lbs. (8–15 Nm).

15. Turn the cylinder head bolts 80–85 degrees tighter, in sequence.

16. Tighten the intake manifold, in sequence, to 11–15 ft. lbs. (15–21 Nm) and then to 15–18 ft. lbs. (21–25 Nm), in sequence.

17. Dip both ends of each pushrod in clean heavy engine oil and install. Install the rocker arm and shaft assemblies and tighten the rocker arm shaft support bolts evenly to 46–52 ft. lbs. (62–70 Nm).

18. Apply silicone sealer to the 4 locations at the joint where the intake manifold and cylinder head meet. Install a new rocker arm cover gasket in each cover and install the rocker arm covers. Tighten the rocker arm cover bolts to 3–5 ft. lbs. (4–7 Nm), wait 2 minutes and then retighten to the same specification.

19. Install the upper intake manifold and tighten the nuts to 15–18 ft. lbs. (20–25 Nm).

20. Install the exhaust manifold(s) and connect the exhaust pipe.

21. Install the spark plugs and the ignition coil and bracket assembly.

22. Install the alternator and the accessory drive belt.

23. Install the power steering pump. Install the air conditioning compressor, if equipped.

24. Connect the negative battery cable. Fill and bleed the cooling system. Run the engine and check for leaks.

5.0L Engine

1. Drain the cooling system.

2. Remove the intake manifold and throttle body.

3. Remove the rocker arm cover(s).

4. If the right cylinder head is to be removed, lift the tensioner and remove the drive belt. Loosen the alternator adjusting arm bolt and remove

the alternator mounting bracket bolt and spacer. Swing the alternator down and out of the way. Remove the air cleaner inlet duct.

5. If the left cylinder head is being removed, remove the air conditioning compressor. Persons not familiar with air conditioning systems should exercise extreme caution, perhaps leaving this job to a professional. Remove the oil dipstick and tube. Remove the cruise control bracket.

6. Disconnect the exhaust manifold(s) from the muffler inlet pipe(s).

7. Loosen the rocker arm stud nuts so the rocker arms can be rotated to the side. Remove the pushrods and identify them so they can be reinstalled in their original positions.

8. Remove the cylinder head bolts and lift the cylinder head from the block. Remove the discard the gasket.

To install:

9. Clean the cylinder head, intake manifold, the valve cover and the head gasket surfaces.

10. A specially treated composition head gasket is used. Do not apply sealer to a composition gasket. Position the new gasket over the locating dowels on the cylinder block. Then, position the cylinder head on the block and install the attaching bolts.

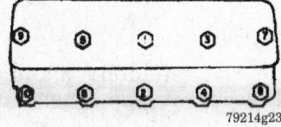

5.0L cylinder head torque sequence

11. The cylinder head bolts are tightened in progressive steps. Tighten all the bolts in the proper sequence to:
    Step 1 — 55–65 ft. lbs.
    Step 2 — 66–72 ft. lbs.

12. Clean the pushrods. Blow out the oil passage in the rods with compressed air. Check the pushrods for straightness by rolling them on a piece of glass. Never try to straighten a pushrod; always replace it.

13. Apply Lubriplate® to the ends of the pushrods and install them in their original positions.

14. Apply Lubriplate® to the rocker arms and their fulcrum seats and install the rocker arms. Adjust the valves.

15. Position a new gasket(s) on the muffler inlet pipe(s) as necessary. Connect the exhaust manifold(s) at the muffler inlet pipe(s).

16. If the right cylinder head was removed, install the alternator, and air cleaner duct. Install the drive belt. If the left cylinder head was removed, install the A/C compressor. Install the dipstick and cruise control bracket.

17. Clean the valve rocker arm cover and the cylinder head gasket surfaces. Place the new gaskets in the covers, making sure the tabs of the gasket engage the notches provided in the cover. Evacuate, charge and leak test the A/C system.

18. Install the intake manifold and related parts. Install the Thermactor® hoses.

19. Fill and bleed the cooling system.

Valve Lifters

REMOVAL AND INSTALLATION

2.3L Engine

1. Disconnect the negative battery cable and remove the air cleaner or air intake duct. Remove the throttle body and EGR supply tube.

2. Remove the rocker arm cover.

3. Rotate the crankshaft so the base circle of the cam is facing the applicable cam follower.

4. Using valve spring compressor lever tool T88T-6565-BH or equivalent, collapse the valve spring and slide the cam follower over the valve lifter and out.

5. Lift out the hydraulic valve lifter.

To install:

6. Rotate the crankshaft so the base circle of the cam is facing the applicable cam follower.

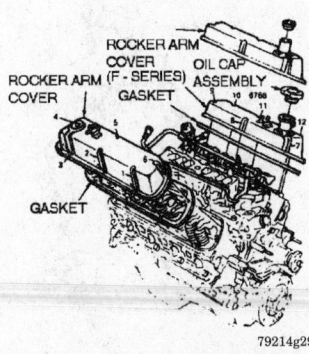

5.0L rocker cover torque sequence

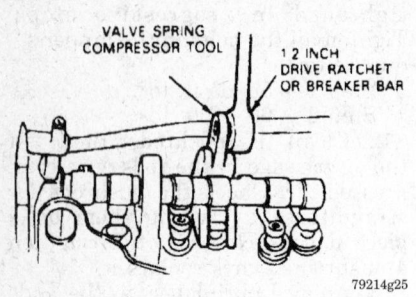

Compressing the valve spring — 2.3L engine

7. Coat the hydraulic lifter with clean engine oil and install it in the bore.

8. Collapse the valve spring using valve spring compressor lever T88T-6565-BH or equivalent. Position the cam follower over the valve lifter and the valve stem.

9. Clean the gasket surfaces of the rocker arm cover and cylinder head.

10. Coat the rocker arm cover and a new gasket with gasket adhesive and install the gasket to the cover.

11. Install the cover and tighten the retaining bolts to 5–8 ft. lbs. (7–11 Nm).

12. Install the throttle body and EGR supply tube, if necessary. Install the air cleaner or air intake duct.

13. Connect the negative battery cable.

3.0L and 4.0L Engines

1. Disconnect the negative battery cable and relieve the fuel system pressure. Drain the cooling system.

2. Remove the intake manifold and rocker arm covers.

3. On 4.0L engines, loosen the rocker arm shaft support bolts 2 turns at a time until the rocker arm and shaft assembly can be removed.

4. On 3.0L engines, loosen the rocker arm fulcrum bolt enough so the rocker arm can be lifted from the pushrod and turned to one side.

5. Remove the pushrods, marking them so they can be reinstalled in their original positions.

6. Remove the lifters. Note the location of each lifter so it can be reinstalled in the same bore. If a lifter is stuck in the bore, use a suitable tool to rotate the lifter back and forth to loosen it from the gum and varnish that may have formed on the lifter.

NOTE: The 4.0L engine is equipped with roller lifters. Roller lifters have an alignment tab which fits into a locating

groove in the lifter bore. Do not attempt to rotate a roller lifter in the bore.

To install:

7. Lubricate the lifters and bores with clean engine oil. Install each lifter in the same bore from which it was removed. On 4.0L engine, install the lifter with the alignment tab in the locating groove of the bore. If a new lifter is being installed, check for free fit in the bore.

8. Check each pushrod for straightness and replace as necessary. Dip each pushrod end in clean engine oil and install in its original position.

9. Lubricate the rocker arm and shaft assembly and install. Draw the shaft support bolts down evenly, 2 turns at a time, until the shafts are fully seated. Tighten the bolts to 46–52 ft. lbs. (62–70 Nm) on 4.0L engines.

10. On 3.0L engines, for each valve, rotate the crankshaft until the lifter rests on the base circle of the camshaft lobe, before tightening the rocker arm fulcrum attaching bolts. Position the rocker arms over the valves and pushrods, install the fulcrums and fulcrum bolts and tighten to 24 ft. lbs. (32 Nm).

11. Install the intake manifold and rocker arm covers.

12. Connect the negative battery cable. Fill and bleed the cooling system.

13. Run the engine and check for leaks.

5.0L Engine

1. Remove the intake manifold.

2. Disconnect the Thermactor® air supply hose at the pump.

3. Remove the rocker arm covers.

4. Loosen the rocker arm fulcrum bolts until the rocker arms can be rotated off the pushrods.

5. Remove the pushrods and keep them in order, for installation.

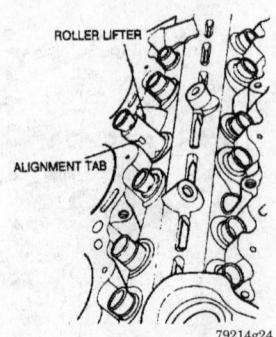

Valve lifter installation — 4.0L engine

6. Using a magnetic lifter removal tool, remove the lifters. Wipe clean the exterior of each lifter as it is removed and mark it with an indelible marker, so it can be installed in its original bore.

To install:

7. Coat the bottom surface of each lifter with multi-purpose grease, and coat the rest of the lifter with clean oil.

8. Install each lifter in the original bore using the magnetic tool.

9. Coat the ends of each pushrod with multi-purpose grease and install each in the original position. Make sure each pushrod is properly seated in the lifter socket.

10. Engage the rocker arms with the pushrods and tighten the rocker arm fulcrum bolts to 18–25 ft. lbs. No valve adjustment should be necessary, however, if there is any question as to post-assembly collapsed lifter clearance, see the hydraulic lifter clearance procedure below.

11. Install the rocker arm covers.

12. Connect the Thermactor® air supply hose at the pump.

13. Install the intake manifold.

Valve Lash

ADJUSTMENT

2.3L Engine

1. Remove the rocker arm cover. Position the camshaft so the base circle of the lobe is facing the cam follower of the valve to be checked.

2. Using tool valve spring compressor lever tool T88T-6565-BH or equivalent, slowly apply pressure to the cam follower until the valve lifter is completely collapsed. Hold the follower in this position and measure the clearance between the base circle of the cam and the follower. The allowable collapsed lifter gap is 0.035–0.055 in. (0.9–1.4mm) at the camshaft.

3. If the clearance is excessive, remove the cam follower and inspect for damage.

4. If the cam follower is not excessively worn, measure the valve spring installed height to make sure the valve is not sticking. The installed height is 1.49–1.55 in. (3.8–3.9cm).

5. If the valve spring installed height is correct, check the camshaft lobe lift. The lobe lift dimension is 0.2381 in. (6mm).

6. If the cam follower, valve spring height and camshaft lobe lift are correct and the base circle-to-follower

clearance is excessive, replace the valve lifter.

3.0L Engine

1. Remove the rocker arm cover.

2. Rotate the crankshaft until the lifter is on the base circle of the cam on the valve to be checked.

3. Using a suitable tool, collapse the lifter fully and measure the clearance between the valve stem tip and rocker arm. The clearance should be 0.085–0.185 in. (2.15–4.69mm).

4. If the clearance is less than specified, install a shorter pushrod. If the clearance is greater than specified, install a longer pushrod.

5.0L Engine

1. Rotate the crankshaft by hand so No. 1 piston is at TDC of the compression stroke. Make a chalk mark on the damper at that point, then, make 2 more chalk marks about 90 degrees apart in a clockwise direction.

2. With No. 1 at TDC, slowly apply pressure, using Lifter Bleed-down wrench T70P-6513-A, or equivalent, to completely bottom the lifter, on the following valves:

No. 1 intake and exhaust
No. 7 intake
No. 5 exhaust
No. 8 intake
No. 4 exhaust

Take care to avoid excessive pressure that might bend the pushrod. Hold the lifter in this position and check the clearance between the rocker arm and the valve stem tip. Allowable clearance is 0.071–0.193 in. (1.8–4.9mm) with a desired clearance of 0.096–0.165 in. (2.4–4.2mm).

3. If the clearance is less than specified, install a shorter pushrod. If the clearance is greater than specified, install a longer pushrod.

4. Rotate the crankshaft clockwise — viewed from the front — 180 degrees, until the next chalk mark is

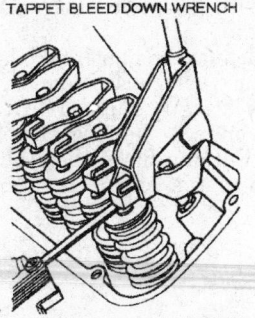

TAPPET BLEED DOWN WRENCH

79214g26

Checking the valve clearance — 5.0L engine

aligned with the timing pointer. Repeat the procedure for:

No. 5 intake
No. 2 exhaust
No. 4 intake
No. 6 exhaust

5. Rotate the crankshaft to the next chalk mark — 90 degrees — and repeat the procedure for:

No. 2 intake
No. 7 exhaust
No. 3 intake and exhaust
No. 6 intake
No. 8 exhaust

Rocker Arms/Shafts

REMOVAL AND INSTALLATION

3.0L Engine

1. Disconnect the negative battery cable. Remove the air cleaner fresh air hose, if necessary.

2. Tag and disconnect the spark plug wires from the spark plugs. Remove the spark plug wire/separator assembly from the rocker arm cover attaching bolt studs and position aside.

3. If the left rocker arm cover is being removed, proceed as follows:

a. Remove the throttle body assembly and the PCV valve.

b. Remove the fuel injector harness stand-offs from the inboard rocker arm cover studs. Move the harness aside.

4. If the right rocker arm cover is being removed, proceed as follows:

a. On Aerostar, remove the oil filler tube assembly and disconnect the closure hose from the oil fill adapter.

b. On Ranger, disconnect the engine harness connectors and remove the air cleaner closure hose from the oil fill adapter.

c. Remove the fuel injector harness stand-offs from the inboard rocker arm cover studs. Move the harness aside.

5. Remove the rocker arm cover attaching bolts and studs, noting their locations. Remove the rocker arm cover.

6. Remove the rocker arm fulcrum bolt and remove the rocker arm and fulcrum.

To install:

7. Lubricate the valve stem tip, pushrod end, fulcrum and rocker arm fulcrum seat with clean engine oil.

8. For each valve, rotate the crankshaft until the lifter rests on the base circle of the camshaft lobe, before tightening the rocker arm fulcrum attaching bolts. Position the

rocker arms over the valves and pushrods, install the fulcrums and fulcrum bolts and tighten to 24 ft. lbs. (32 Nm).

9. Clean the rocker arm cover and cylinder head gasket mating surfaces of all gasket material and/or old silicone sealer.

10. Apply a bead of silicone sealer at the cylinder head to intake manifold rail step and position the rocker arm cover on the cylinder head.

11. Install the bolts/studs in their original locations and tighten to 9 ft. lbs. (12 Nm).

12. Install the remaining components in the reverse order of their removal. Start the engine and check for leaks.

4.0L Engine

1. Disconnect the negative battery cable and relieve the fuel system pressure.

2. Tag and disconnect the spark plug wires. Disconnect the fuel lines.

3. If the left rocker arm cover is being removed, upper intake manifold removal may be required.

4. If the right rocker arm cover is being removed, proceed as follows:

a. Remove the ignition coil and bracket assembly and remove the PCV valve hose and breather.

b. Remove the air inlet duct and the hose attached to the oil fill tube. Remove the accessory drive belt and remove the alternator.

c. Drain the cooling system and remove the upper radiator hose. Remove the low pressure air conditioner hose bracket from the upper intake, if still installed and remove the vacuum hose from the air cleaner.

5. Remove the rocker arm cover attaching screws and load distribution washers. Note the position of the washers so they can be reinstalled in their original positions.

6. Using a light plastic hammer, tap the rocker arm covers to break the seal. Remove the covers.

7. Remove the rocker arm shaft stand attaching bolts by loosening the bolts 2 turns at a time, in sequence. Lift off the rocker arm and shaft assembly.

To install:

8. Apply engine oil to the valve train assembly to provide initial lubrication.

9. Install the rocker arm shaft assembly to the cylinder head, guiding the rocker arms onto the pushrods.

10. Install the rocker arm stand attaching bolts, running them down 2 turns at a time, in sequence, until the

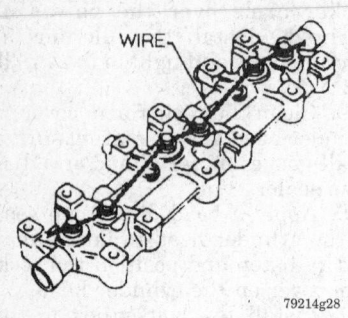

Secure the lifters in place with wire — 3.0L engine

shaft assembly is seated. Tighten the rocker arm stand attaching bolts to 46–52 ft. lbs. (62–70 Nm).

11. Clean all gasket material from the rocker arm cover and cylinder head.

12. Apply silicone sealer to the parting lines where the cylinder head and intake manifold seal. Install the rocker arm cover, using a new gasket.

13. Install the rocker arm cover attaching screws and load distribution washers, making sure the washers are in their original positions. Tighten to 3–5 ft. lbs. (4–7 Nm).

14. Install the remaining components in the reverse order of their removal. Start the engine and check for leaks.

5.0L Engine

1. Remove the intake manifold.
2. Disconnect the Thermactor® air supply hose at the pump.
3. Remove the rocker arm covers.
4. Loosen the rocker arm fulcrum bolts, fulcrum seats and rocker arms; keep all parts in order for installation.

To install:

5. Apply multi-purpose grease to the valve stem tips, the fulcrum seats and sockets.
6. Install the fulcrum guides, rocker arms, seats and bolts. Torque the bolts to 18–25 ft. lbs.
7. Install the rocker arm covers.
8. Connect the Thermactor® air supply hose at the pump.
9. Install the intake manifold.

Intake Manifold

REMOVAL AND INSTALLATION

2.3L Engine

1. Disconnect the negative battery cable and relieve the fuel system pressure. Drain the cooling system.

2. Tag and disconnect the electrical connectors at the throttle position sensor, air charge temperature sensor, engine coolant temperature sensor and air bypass valve, if equipped. Disconnect the knock sensor connector, if equipped.

3. Disconnect the injector wiring harness at the main engine harness and at the water temperature indicator sensor. Disconnect the ignition control assembly connector, if equipped.

4. Tag and disconnect the vacuum lines at the upper intake manifold vacuum tree, EGR valve, fuel pressure regulator and canister purge line.

5. Remove the throttle linkage shield and disconnect the throttle linkage and cruise control. Disconnect the kickdown cable, if equipped. Unbolt the accelerator cable from the bracket and position the cable aside.

6. Disconnect the air intake hose and crankcase vent hose. Disconnect the air bypass hose, if equipped.

7. Disconnect the PCV system by disconnecting the hose from the fitting on the underside of the upper intake. Disconnect the water bypass line at the lower intake manifold.

8. Disconnect the EGR tube from the EGR valve. Remove the attaching

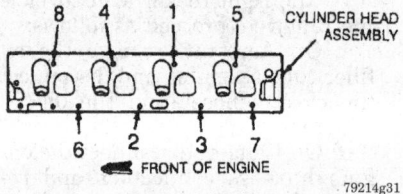

2.3L intake manifold torque sequence

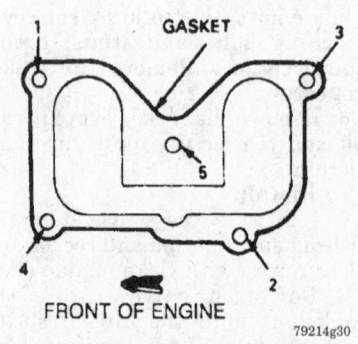

2.3L upper intake manifold bolt torque sequence

bolts and remove the upper intake manifold and throttle body assembly.

9. Remove the engine oil dipstick tube bracket attaching bolt. Disconnect the fuel lines from the fuel supply manifold.

10. Disconnect the electrical connectors from the fuel injectors and position aside. Remove the fuel supply manifold attaching bolts and remove the fuel supply manifold.

11. Remove the attaching bolts and remove the lower intake manifold.

To install:

12. Clean all gasket mating surfaces. Clean and oil the manifold bolt threads.

13. Position a new gasket and install the lower intake manifold. Install the attaching bolts and tighten, in sequence, in 2 steps, first to 5–7 ft. lbs. (7–9 Nm) and then to 15–22 ft. lbs. (20–30 Nm).

14. Install the fuel supply manifold and injectors with the 2 attaching bolts. Tighten to 15–22 ft. lbs. (20–30 Nm). Connect the electrical connectors to the injectors.

15. Position a new gasket on the lower intake manifold and install the upper intake manifold. Install the attaching bolts and tighten, in sequence, to 15–22 ft. lbs. (20–30 Nm).

16. Install the engine oil dipstick tube and retaining bolt. Connect the fuel lines to the fuel supply manifold.

17. Connect the EGR tube to the EGR valve. Tighten to 18–28 ft. lbs. (25–30 Nm).

18. Connect the water bypass line and connect the PCV hose. Connect the vacuum lines to the locations marked during removal.

19. Hold the accelerator cable bracket in position on the upper manifold and install the attaching bolts. Tighten to 10–15 ft. lbs. (13.5–20.5 Nm).

20. Install the accelerator cable to the bracket. Connect the accelerator cable and cruise control. Install the throttle linkage shield.

21. Connect the electrical connectors to the locations marked during removal.

22. Connect the air intake hose and crankcase vent hose. Connect the air bypass hose, if equipped.

23. Connect the negative battery cable. Fill and bleed the cooling system. Run the engine and check for leaks.

3.0L Engine

1. Disconnect the negative battery cable and relieve the fuel system pressure. Drain the cooling system.

2. Remove the air cleaner hoses to the throttle body and rocker arm cover. Disconnect the fuel lines from the fuel supply manifold.

3. Tag and disconnect the necessary vacuum lines.

4. Tag and disconnect the electrical connectors at the air charge temperature sensor, engine coolant temperature sensor, throttle position sensor, air bypass solenoid and coolant temperature sender.

5. Remove the snow shield from the power steering pump bracket and accelerator cable bracket.

6. Disconnect and remove the accelerator and cruise control cables from the accelerator mounting bracket and throttle lever.

7. Remove the alternator support brace.

8. Remove the throttle body-to-lower intake manifold retaining bolts and stud bolts and remove the throttle body assembly.

9. Disconnect the fuel injector harness stand-offs from the inboard rocker arm cover studs and each injector and remove from the engine.

10. Disconnect the upper radiator hose from the thermostat housing and disconnect the heater hoses.

11. Tag and disconnect the spark plug wires.

12. Remove the ignition coil from the rear of the cylinder head, if required.

13. Remove the rocker arm covers. Loosen the No. 3 cylinder intake valve rocker arm fulcrum bolt and rotate the rocker arm away from the valve. Remove the pushrod.

14. Remove the intake manifold bolts. Break the gasket seal by wedging a large prybar between the manifold and the block using the lug on the water pump as a leverage point. Be careful to prevent damage to machines surfaces.

15. Remove the intake manifold.

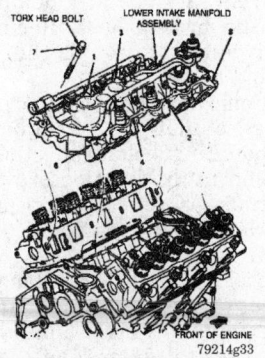

3.0L intake manifold torque sequence

To install:

16. Clean all gasket mating surface.

17. Apply silicone sealer to the intersection of the cylinder block and cylinder head at the 4 corners of the lifter valley opening.

18. Install the front and rear intake manifold seals and secure with the retainers. Position the intake manifold gaskets on the cylinder heads and insert the locking tabs on the cylinder head gaskets.

19. Carefully lower the intake manifold into position being careful not to disturb the silicone sealer. Install the intake manifold bolts and tighten, in sequence, in 2 steps, first to 11 ft. lbs. (15 Nm) and then to 19 ft. lbs. (26 Nm).

20. Install the No. 3 cylinder intake valve pushrod. Apply oil to the pushrod and rocker arm fulcrum and position the rocker arm over the valve and pushrod. Rotate the crankshaft to place the lifter on the base circle of the cam, then tighten the fulcrum bolt to 24 ft. lbs. (32 Nm).

21. Install the rocker arm covers and the fuel injector electrical harness.

22. Install the throttle body using a new gasket. Tighten the throttle body attaching bolts, in sequence, to 19 ft. lbs. (25 Nm).

23. Install the alternator brace. Tighten the nuts to 12 ft. lbs. (16 Nm).

24. Connect the PCV valve hose. Connect the engine coolant temperature sensor, air charge temperature sensor, throttle position sensor, air bypass solenoid and coolant temperature sender connectors.

25. Connect the spark plug wires. Install the ignition coil, if removed.

26. Connect the heater hoses and the upper radiator hose. Connect the vacuum lines to the locations marked during removal.

27. Connect the fuel lines to the fuel supply manifold. Change the engine oil and filter.

NOTE: Engine coolant is corrosive to all engine bearing material. Changing the oil after removal of a coolant carrying component helps prevent engine failure.

28. Connect the negative battery cable. Fill and bleed the cooling system. Install the air cleaner hose.

29. Run the engine and check for leaks. Check the ignition timing, idle speed, throttle linkage and cruise control and adjust, if necessary.

4.0L Engine

1. Disconnect the negative battery cable and relieve the fuel system pressure.

2. Remove the air cleaner air intake duct from the throttle body.

3. Remove the snow/ice shield and disconnect the throttle cable and bracket assembly.

4. Tag and disconnect the vacuum hoses from the fittings on the upper intake manifold.

5. Tag and disconnect the electrical connectors at the throttle body, upper intake manifold, lower intake manifold and injectors.

6. Disconnect the fuel lines from the fuel supply manifold.

7. Remove the ignition coil and bracket assembly.

8. Remove the mounting nuts and remove the upper intake manifold.

9. Remove the rocker arm covers.

10. Remove the intake manifold attaching bolts and nuts. Tap the manifold lightly with a plastic mallet to break the gasket seal and remove the manifold.

To install:

11. Clean all gasket mating surfaces.

12. Apply silicone sealer to the block and cylinder head mating surfaces at the 4 corners of the lifter valley opening. Install the intake manifold gaskets and again apply sealer to the same locations.

13. Position the intake manifold on the 2 guide studs and install the nuts and bolts hand-tight. Tighten the bolts, in sequence, in 4 steps, first to 3–6 ft. lbs. (4–8 Nm), then to 6–11 ft. lbs. (8–15 Nm), then to 11–15 ft. lbs. (15–21 Nm) and finally to 15–18 ft. lbs. (21–25 Nm).

14. Apply silicone sealer to the 4 locations where the intake manifold and the cylinder heads meet. Install the rocker arm covers with new gaskets and tighten evenly to 3–5 ft. lbs. (4–7 Nm). Wait 2 minutes and tighten the bolts again to the same specification.

15. Install the upper intake manifold and tighten the nuts to 15–18 ft. lbs. (20–25 Nm).

16. Install the ignition coil and bracket assembly. Connect the fuel lines to the fuel supply manifold.

17. Connect the electrical connectors at the throttle body, upper intake manifold, lower intake manifold and injectors.

18. Connect the vacuum hoses to the fittings on the upper intake manifold.

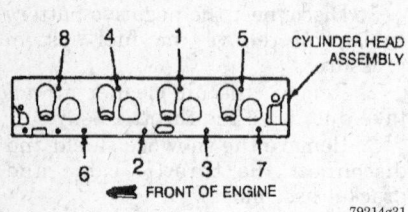

4.0L intake manifold torque sequence

19. Install the throttle cable and bracket assembly and the snow/ice shield to the throttle body.

20. Connect the air cleaner air intake duct to the throttle body.

21. Connect the negative battery cable. Fill and bleed the cooling system. Run the engine and check for leaks.

5.0L Engine

NOTE: Discharge fuel system pressure before starting any work that involves disconnecting fuel system lines.

Upper

1. Remove the air cleaner. Disconnect the electrical connectors at the air bypass valve, throttle position sensor and EGR position sensor.

2. Disconnect the throttle linkage at the throttle ball and the AOD transmission linkage from the throttle body. Remove the bolts that secure the bracket to the intake and position the bracket and cables out of the way.

3. Disconnect the upper manifold vacuum fitting connections by removing all the vacuum lines at the vacuum tree (label lines for position identification). Remove the vacuum lines to the EGR valve and fuel pressure regulator.

4. Disconnect the PCV system by disconnecting the hose from the fitting at the rear of the upper manifold.

5. Remove the 2 canister purge lines from the fittings at the throttle body.

6. Disconnect the EGR tube from the EGR valve by loosening the flange nut.

7. Remove the bolt from the upper intake support bracket to upper manifold. Remove the upper manifold retaining bolts and remove the upper intake manifold and throttle body as an assembly.

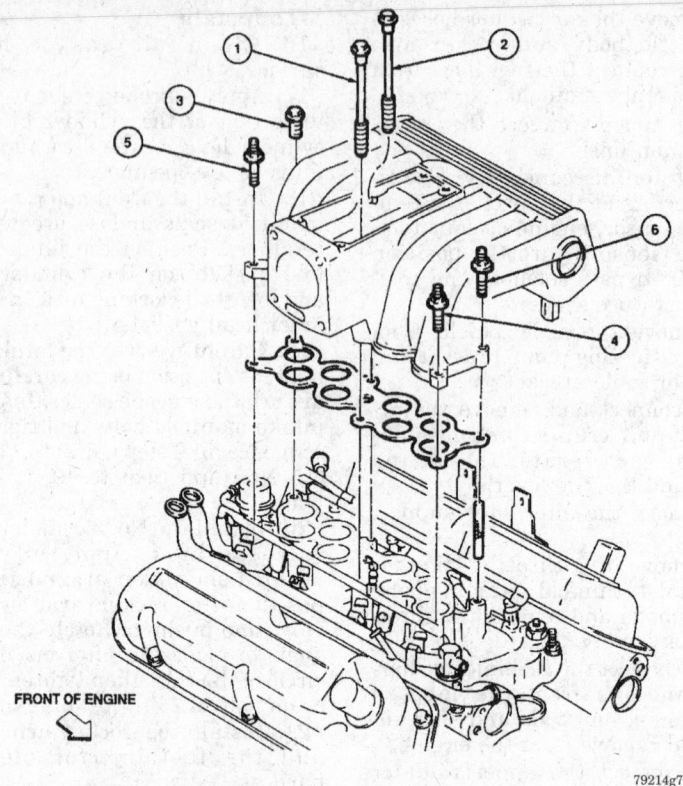

FRONT OF ENGINE

Upper intake manifold bolt torque sequence

8. Clean and inspect all mounting surfaces of the upper and lower intake manifolds.

To install:

9. Position a new mounting gasket on the lower intake manifold.

10. Install the upper intake manifold and throttle body as an assembly. Install the upper manifold retaining bolts and install the bolt at the upper intake support bracket. Mounting bolts are torqued to 12–18 ft. lbs. (16–23 Nm).

11. Connect the EGR tube at the EGR valve.

12. Install the 2 canister purge lines at the fittings at the throttle body.

13. Connect the PCV system hose at the fitting at the rear of the upper manifold.

14. Connect the upper manifold vacuum lines at the vacuum tree. Install the vacuum lines at the EGR valve and fuel pressure regulator.

15. Install the throttle bracket on the intake manifold. Connect the throttle linkage at the throttle ball and the AOD transmission linkage at the throttle body.

16. Connect the electrical connectors at the air bypass valve, throttle position sensor and EGR position sensor.

17. Install the air cleaner.

Lower

1. Upper manifold and throttle body must be removed first.

2. Drain the cooling system.

3. Remove the distributor assembly, cap and wires.

4. Disconnect the electrical connectors at the engine, coolant temperature sensor and sending unit, at the air charge temperature sensor and at the knock sensor.

5. Disconnect the injector wiring harness from the main harness assembly. Remove the ground wire from the intake manifold stud.

6. Disconnect the fuel supply and return lines from the fuel rails.

7. Remove the upper radiator hose from the thermostat housing. Remove the bypass hose. Remove the heater outlet hose at the intake manifold.

8. Remove the air cleaner mounting bracket. Remove the intake manifold mounting bolts and studs. Pay attention to the location of the bolts and studs for reinstallation. Remove the lower intake manifold assembly.

To install:

9. Clean and inspect the mounting surfaces of the heads and manifold.

10. Apply a 1/16 in. (1.5mm) bead of RTV sealer to the ends of the manifold seal (the junction point of the seals and gaskets). Install the end

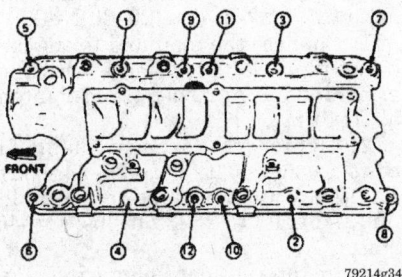

5.0L lower intake manifold torque sequence

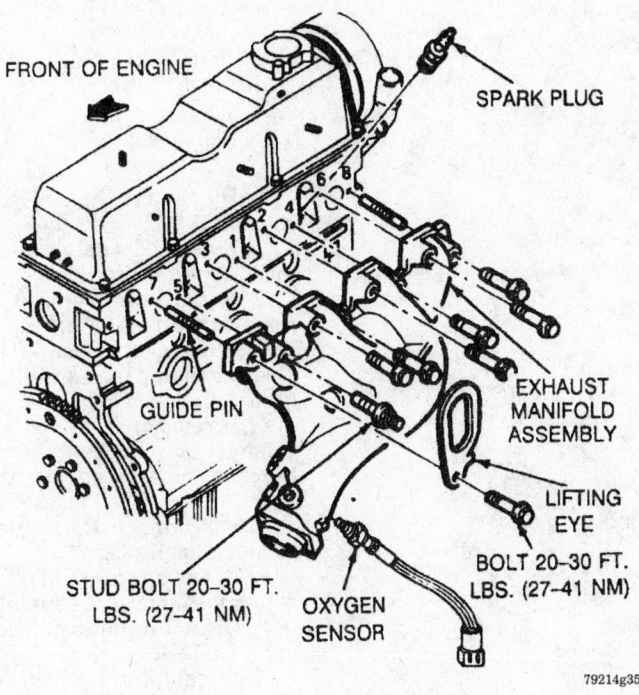

Exploded view of the 2.3L exhaust manifold

seals and intake gaskets on the cylinder heads. The gaskets must interlock with the seal tabs.

11. Install locator bolts at opposite ends of each head and carefully lower the intake manifold into position. Install and tighten the mounting bolts and studs to 23–25 ft. lbs. (30–32 Nm).

12. Install the lower intake manifold assembly. Install the intake manifold mounting bolts and studs. Pay attention to the location of the bolts. Install the air cleaner mounting bracket.

13. Install the heater outlet hose at the intake manifold.

14. Install the bypass hose.

15. Install the upper radiator hose.

16. Connect the fuel supply and return lines at the fuel rails.

17. Connect the injector wiring harness from the main harness assembly. Install the ground wire from the intake manifold stud.

18. Connect the electrical connectors at the engine, coolant temperature sensor, air charge temperature sensor and at the knock sensor.

19. Install the distributor assembly, cap and wires.

20. Fill the cooling system.

Exhaust Manifold

REMOVAL AND INSTALLATION

2.3L Engine

1. Disconnect the negative battery cable. Remove the air cleaner and duct assembly.

2. Remove the EGR tube at the exhaust manifold and loosen at the EGR valve.

3. Remove the check valve at the exhaust manifold and disconnect the hose at the end of the air bypass valve, if equipped.

4. Disconnect the oxygen sensor from the exhaust manifold, if equipped. Remove the sensor, if necessary.

5. Disconnect and remove the cylinder heads coil assembly.

6. Remove the screw attaching the heater hoses to the rocker arm cover. Disconnect the exhaust pipe from the exhaust manifold. Remove the heat shield nuts from the manifold, and place the heat shield aside.

7. Remove the exhaust manifold mounting nuts and bolts and remove the manifold.

8. Clean all gasket material from the mating surfaces.

To install:

9. Position the exhaust manifold to the vehicle and secure in place using the mounting nuts and bolts. Tighten the bolts in two steps, the first to 15–17 ft. lbs. (20–23 Nm), the second step to 20–30 ft. lbs. (27–41 Nm).

10. Connect the exhaust pipe to the manifold, tightening the retainer bolts to 20–30 ft. lbs. (36–46 Nm). install the heat shield and tighten the attaching nuts to 20–30 ft. lbs. (27–40 Nm).

11. Install the coil pack and attaching bolts. Tighten the bolts to 30–40 ft. lbs. (40–55 Nm).

12. Connect the remaining components.

13. Connect the negative battery cable.

3.0L Engine

Left Side

1. Disconnect the negative battery cable.

2. Remove the hot air tube from the exhaust manifold cover. Remove the exhaust manifold cover-to-exhaust manifold bolts and cover.

3. Remove the EGR and the AIR tubes from the exhaust manifold, if equipped.

NOTE: If the alternator is in the way, remove the drive belt and the alternator.

4. Raise and safely support the vehicle.

5. Remove the exhaust pipe-to-exhaust manifold nuts and separate the exhaust pipe from the manifold.

6. Remove the exhaust manifold-to-cylinder head bolts and the manifold from the engine.

7. Clean the gasket mounting surfaces.

To install:

8. Install a new gasket, then position the manifold to the engine. Tighten the exhaust manifold-to-cylinder head nuts to 15–22 ft. lbs. (20–30 Nm) and the exhaust pipe-to-

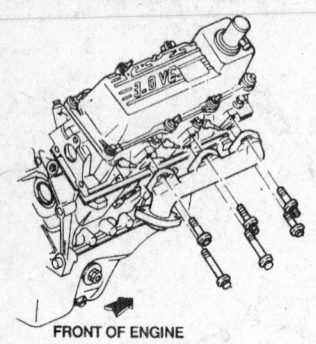

FRONT OF ENGINE

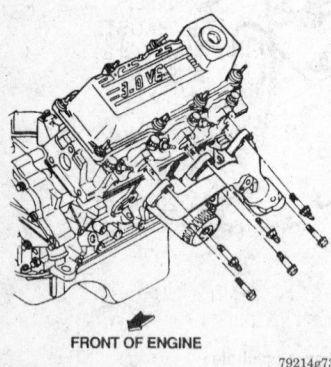

FRONT OF ENGINE

79214g73

Exhaust manifold bolt locations — 3.0L engine

exhaust manifold bolts to 20 ft. lbs. (27 Nm).

9. Install the remaining components.

10. Connect the negative battery cable.

Right Side

1. Disconnect the negative battery cable.

2. Remove the upper/lower exhaust manifold cover-to-exhaust manifold bolts and covers.

3. Remove the AIR tube from the exhaust manifold, if equipped.

4. Raise and safely support the vehicle.

5. Remove the exhaust pipe-to-exhaust manifold bolts and separate the exhaust pipe from the manifold.

6. Remove the exhaust manifold-to-cylinder head bolts and the manifold from the engine.

7. Clean the gasket mounting surfaces.

To install:

8. Install a new gasket, then position the manifold to the engine. Tighten the exhaust manifold-to-cylinder head nuts to 15–22 ft. lbs. (20–30 Nm) and the exhaust pipe-to-

exhaust manifold bolts to 20 ft. lbs. (27 Nm).

9. Install the remaining components.

10. Connect the negative battery cable.

4.0L Engine

1. Disconnect the negative battery cable.

2. Raise and safely support the vehicle, as necessary.

3. If removing the left manifold, remove the engine oil dipstick tube support bracket. Remove the power steering pump pressure and return hoses, if necessary.

4. If removing the right manifold, remove the heater hose support bracket and disconnect the heater hoses.

5. Disconnect the exhaust pipe from the manifold.

6. Remove the manifold attaching bolts and remove the manifold.

7. Clean all mating surfaces.

To install:

8. Install a new gasket, then position the manifold to the engine. Tighten the exhaust manifold-to-cylinder head nuts to 20 ft. lbs. (27 Nm) and the exhaust pipe-to-exhaust manifold bolts to 20 ft. lbs. (27 Nm).

9. Install the remaining components.

10. Connect the negative battery cable.

5.0L Engine

Right Side

1. Disconnect the negative battery cable. Drain the cooling system.

2. Remove the drive belt and belt tensioner.

3. Remove the alternator and electrical harness.

4. Disconnect the exhaust inlet pipe to the exhaust manifold.

5. Access the manifold retainer bolts through the wheel well opening. Remove the fender apron pushpins, then remove the fender apron.

6. Tag and remove the spark plugs where needed.

7. Remove the manifold retainer bolts, then lift off the manifold. Clean all gasket material from the mating surfaces.

To install:

8. Install a new gasket, then position the manifold to the engine. Hand-tighten the exhaust manifold-to-cylinder head bolts at this time.

9. Install the alternator and harness.

10. Install the belt tensioner and drive belt. Tighten the tensioner bolts to 15–22 ft. lbs. (20–30 Nm).

11. Connect the manifold to the exhaust pipe. Secure in place with the attaching nut and bolt. Hand-tighten at this time.

12. Tighten the exhaust manifold bolts to 26–32 ft. lbs. (35–44 Nm). Tighten the inlet pipe-to-manifold nut and bolts to 26–33 ft. lbs. (34–46 Nm).

13. Install the spark plug wires. Position the fender apron in place and secure with the pushpins.

14. Connect the negative battery cable.

Left Side

1. Disconnect the negative battery cable.

2. Remove the drive belt.

3. Remove the A/C compressor and place aside without removing the refrigerant lines.

4. Remove the oil tube and retainer nut.

5. Disconnect the exhaust inlet pipe from the exhaust manifold.

6. Access the manifold retainer bolts through the wheel well opening. Remove the fender apron pushpins, then remove the fender apron.

7. Tag and remove the spark plugs where needed.

8. Remove the manifold retainer bolts, then lift off the manifold. Clean all gasket material from the mating surfaces.

To install:

9. Install a new gasket, then position the manifold to the engine. Hand-tighten the exhaust manifold-to-cylinder head bolts at this time.

10. Install the A/C compressor on the bracket, and tighten the bolts to 16–21 ft. lbs. (21–29 Nm).

11. Install the drive belt.

12. Tighten the exhaust manifold bolts to 26–32 ft. lbs. (35–44 Nm).

13. Connect the spark plug wires.

14. Position the fender apron in place and secure with the pushpins.

15. Connect the manifold to the exhaust pipe. Secure in place with the attaching nut and bolt. Hand-tighten at this time.

16. Tighten the exhaust manifold bolts to 26–32 ft. lbs. (35–44 Nm). Tighten the inlet pipe-to-manifold nut and bolts to 26–33 ft. lbs. (34–46 Nm).

17. Install the spark plug wires. Position the fender apron in place and secure with the pushpins.

18. Connect the negative battery cable.

Timing Chain Front Cover

REMOVAL AND INSTALLATION

3.0L Engine

Aerostar

1. Disconnect the negative battery cable. Remove the air cleaner fresh air hose.

2. Drain the cooling system and the crankcase. Remove the cooling fan.

3. Loosen the water pump hub bolts and remove the accessory drive belts. Remove the water pump pulley.

4. Remove the alternator adjusting arm and brace. Move the alternator aside.

5. Remove the A/C compressor mounting bolts, if equipped. Tie the compressor aside with mechanics wire and remove the bracket.

6. Remove the crankshaft pulley and damper. Remove the water pump, if required.

NOTE: The timing cover can be removed with the water pump installed by not removing the 6mm water pump attaching bolts.

7. Disconnect the lower radiator hose and heater hose.

8. Remove the oil pan assembly. Disconnect the oil level sensor, if equipped, before removal.

9. Remove the front cover attaching bolts and remove the front cover.
   To install:

10. Clean all gasket mating surfaces. Use a seal removal tool to remove the front cover oil seal.

11. Install a new front cover oil seal, using a seal installer. Position a new front cover gasket on the engine block dowel pins.

12. Install the front cover with the attaching bolts. Apply sealer to the 3 attaching bolts on the passenger side of the cover, prior to installation. Tighten the 8mm bolts to 19 ft. lbs. (25 Nm) and the 6mm bolts to 7 ft. lbs. (10 Nm).

13. Install the oil pan. Install the water pump, if removed.

14. Install the crankshaft pulley and damper. Tighten the damper attaching bolt to 107 ft. lbs. (145 Nm).

15. Install the lower radiator hose and heater hose. Install the A/C bracket, compressor and brace, if equipped.

16. Install the alternator assembly, bracket, oil fill tube support and throttle body brace.

17. Install the water pump pulley and accessory drive belts. Install the cooling fan and shroud.

18. Fill the crankcase with the proper type and quantity of engine oil. Connect the negative battery cable.

19. Fill and bleed the cooling system. Run the engine and check for leaks. Install the air cleaner fresh air hose.

Except Aerostar

1. Disconnect the negative battery cable. Drain the cooling system and crankcase.

2. Remove the cooling fan and water pump pulley bolts. Remove the accessory drive belts and the water pump pulley.

3. Remove the alternator adjusting arm and the throttle body brace. Remove the heater air intake duct.

4. Remove the motor mount upper nuts. If equipped with an automatic transmission and A/C, remove the compressor upper bolts, then remove the front cover front nuts.

5. Remove the distributor assembly.

NOTE: Failure to remove the distributor assembly will result in a broken distributor.

6. Raise and safely support the vehicle. Remove the lower A/C compressor bolts and wire the compressor aside. Remove the compressor bracket.

7. Remove the crankshaft pulley and damper. Remove the oil pan. Disconnect the oil level sensor before pan removal.

8. Lower the vehicle and remove the lower radiator hose. Remove the water pump, if required.

NOTE: The timing cover can be removed with the water pump installed by not removing the 6mm water pump attaching bolts.

9. Remove the front cover attaching bolts and remove the front cover.
   To install:

10. Clean all gasket mating surfaces. Use a seal removal tool to remove the front cover oil seal.

11. Install a new front cover oil seal, using a seal installer. Position a new front cover gasket on the engine block dowel pins.

12. Install the front cover with the attaching bolts. Apply sealer to the 3 attaching bolts on the passenger side of the cover, prior to installation. Tighten the 8mm bolts to 19 ft. lbs. (25 Nm) and the 6mm bolts to 7 ft. lbs. (10 Nm).

13. Raise and safely support the vehicle. Install the oil pan and connect the oil level sensor.

14. Install the water pump, if removed.

15. Install the crankshaft pulley and damper. Tighten the damper attaching bolt to 107 ft. lbs. (145 Nm).

16. Install the A/C compressor bracket, if equipped, position the compressor and install the lower bolts. Lower the vehicle.

17. Install the distributor. Install the front cover front nuts and the A/C compressor upper bolts, if equipped.

18. Install the motor mount upper nuts and the heater air intake duct. Install the alternator adjusting arm and brace.

19. Install the water pump pulley and accessory drive belts. Install the cooling fan and coolant hoses.

20. Fill the crankcase with the proper type and quantity of engine oil. Connect the negative battery cable.

21. Fill and bleed the cooling system. Run the engine and check for leaks. Check the ignition timing and adjust, if necessary.

4.0L Engines

1. Disconnect the negative battery cable and drain the cooling system and crankcase.

2. Remove the oil pan and the radiator.

3. Remove the A/C compressor and power steering bracket, if equipped.

4. Remove the alternator and drive belt(s). Remove the fan.

5. Remove the water pump and heater and radiator hoses.

6. Remove the crankshaft pulley/damper assembly. Remove the crankshaft timing sensor.

7. Remove the front cover retaining bolts, noting their positions. If necessary, tap the cover lightly with a plastic hammer to break the gasket seal. Remove the front cover.
   To install:

8. Clean all gasket mating surfaces. Apply sealer to the gasket surfaces on the cylinder block and the back side of the front cover plate. Install the guide sleeves.

9. Apply sealer to the front cover gasket surface and position a new gasket on the front cover.

10. Install the front cover with the retaining screws. Note the different bolt lengths. Tighten the bolts to 13–15 ft. lbs. (17–21 Nm).

11. Install the crankshaft timing sensor.

12. Install the crankshaft pulley/damper assembly. Tighten the at-

taching bolt to 30–37 ft. lbs. (40–50 Nm), then tighten an additional 80–90 degrees.

13. Install the remaining components in the reverse order of their removal. Fill and bleed the cooling system. Run the engine and check for leaks.

5.0L Engine

1. Drain the cooling system and the crankcase.

2. Disconnect the upper and lower radiator hoses from the water pump, transmission oil cooler lines from the radiator, and remove the radiator.

3. Disconnect the heater hose from the water pump. Slide the water pump by-pass hose clamp toward the water pump.

4. Loosen the alternator pivot bolt and the bolt which secures the alternator adjusting arm to the water pump. Position the alternator out of the way.

5. Remove the power steering pump and A/C compressor from their mounting brackets, if equipped.

6. Remove the bolts holding the fan shroud to the radiator, if equipped. Remove the fan, spacer, pulley and drive belts.

7. Remove the crankshaft pulley from the crankshaft damper. Remove the damper attaching bolt and washer and remove the damper with a puller.

8. Disconnect the fuel pump outlet line at the fuel pump. Disconnect the vacuum inlet and outlet lines from the fuel pump. Remove the fuel pump attaching bolts and lay the pump to one side with the fuel inlet line still attached.

9. Remove the oil level dipstick and the bolt holding the dipstick tube to the exhaust manifold.

10. Remove the oil pan-to-cylinder front cover attaching bolts. Use a sharp, thin cutting blade to cut the oil pan gasket flush with the cylinder block. Remove the front cover and water pump as an assembly.

11. Discard the front cover gasket.

To install:

12. Place the front seal removing tool (Ford part no. T70P–6B070–A or equivalent) into the front cover plate and over the front of the seal. Tighten the 2 through bolts to force the seal puller under the seal flange, then alternately tighten the 4 puller bolts a half turn at a time to pull the oil seal from the cover.

13. Coat a new front cover oil seal with Lubriplate® or equivalent and place it onto the front oil seal alignment and installation tool (Ford part

no. T70P–6B070–A or equivalent). Place the tool and the seal onto the end of the crankshaft and push it toward the engine until the seal starts into the front cover.

14. Place the installation screw, washer, and nut onto the end of the crankshaft, then thread the screw into the crankshaft. Tighten the nut against the washer and tool to force the seal into the front cover plate. Remove the tool.

15. Apply Lubriplate® or equivalent to the oil seal rubbing surface of the vibration damper inner hub to prevent damage to the seal. Coat the front of the crankshaft with oil for damper installation.

16. To install the damper, line up the damper keyway with the key on the crankshaft, then install the damper onto the crankshaft. Install the cap screw and washer, and tighten the screw to 80 ft. lbs. (104 Nm). Install the crankshaft pulley.

17. Install the fan, spacer, pulley and drive belts.

18. Install the bolts holding the fan shroud to the radiator, if equipped.

19. Install the power steering pump and A/C compressor.

20. Position and tighten the alternator.

21. Connect the heater hose at the water pump.

22. Install the radiator.

23. Connect the upper and lower radiator hoses, and transmission oil cooler lines.

24. Fill the cooling system and the crankcase.

Front Cover Oil Seal

REPLACEMENT

2.3L, 3.0L and 4.0L Engines

1. Disconnect the negative battery cable.

2. Drain the cooling system and remove the radiator, if necessary to provide access.

3. Remove the drive belts and remove the crankshaft pulley and damper assembly.

4. Remove the seal from the front cover using a seal removal tool. Be careful not to damage the seal housing or crankshaft surfaces.

To install:

5. Coat a new oil seal with clean engine oil and install in the front cover, using a seal installer.

6. Install the crankshaft pulley and damper assembly. Tighten the damper attaching bolt to 85–96 ft.

lbs. (115–130 Nm) on 2.3L engine or 107 ft. lbs. (145 Nm) on 3.0L engine. On 4.0L engine, tighten the attaching bolt to 30–37 ft. lbs. (40–50 Nm), then tighten an additional 80–90 degrees.

7. Install the drive belts. Install the radiator, if removed.

8. Connect the negative battery cable. Fill and bleed the cooling system, if necessary.

5.0L Engine

1. Drain the cooling system and the crankcase.

2. Disconnect the upper and lower radiator hoses from the water pump, transmission oil cooler lines from the radiator, and remove the radiator.

3. Disconnect the heater hose from the water pump. Slide the water pump by-pass hose clamp toward the water pump.

4. Loosen the alternator pivot bolt and the bolt which secures the alternator adjusting arm to the water pump. Position the alternator out of the way.

5. Remove the power steering pump and air conditioning compressor from their mounting brackets, if equipped.

6. Remove the bolts holding the fan shroud to the radiator, if equipped. Remove the fan, spacer, pulley and drive belts.

7. Remove the crankshaft pulley from the crankshaft damper. Remove the damper attaching bolt and washer and remove the damper with a puller.

8. Disconnect the fuel pump outlet line at the fuel pump. Disconnect the vacuum inlet and outlet lines from the fuel pump. Remove the fuel pump attaching bolts and lay the pump to one side with the fuel inlet line still attached.

9. Remove the oil level dipstick and the bolt holding the dipstick tube to the exhaust manifold on the 5.0L engine.

10. Remove the oil pan-to-cylinder front cover attaching bolts. Use a sharp, thin cutting blade to cut the oil pan gasket flush with the cylinder block. Remove the front cover and water pump as an assembly.

11. Discard the front cover gasket.

To install:

12. Place the front seal removing tool (Ford part no. T70P–6B070–A or equivalent) into the front cover plate and over the front of the seal. Tighten the 2 through-bolts to force the seal puller under the seal flange, then alternately tighten the 4 puller bolts a half turn at a time to pull the oil seal from the cover.

13. Coat a new front cover oil seal with Lubriplate® or equivalent and place it onto the front oil seal alignment and installation tool (Ford part no. T70P–6B070–A or equivalent). Place the tool and the seal onto the end of the crankshaft and push it toward the engine until the seal starts into the front cover.

14. Place the installation screw, washer, and nut onto the end of the crankshaft, then thread the screw into the crankshaft. Tighten the nut against the washer and tool to force the seal into the front cover plate. Remove the tool.

15. Apply Lubriplate® or equivalent to the oil seal rubbing surface of the vibration damper inner hub to prevent damage to the seal. Coat the front of the crankshaft with oil for damper installation.

16. To install the damper, line up the damper keyway with the key on the crankshaft, then install the damper onto the crankshaft. Install the cap screw and washer, and tighten the screw to 80 ft. lbs. (104 Nm). Install the crankshaft pulley.

17. Install the fan, spacer, pulley and drive belts.

18. Install the bolts holding the fan shroud to the radiator, if equipped.

19. Install the power steering pump and air conditioning compressor.

20. Position and tighten the alternator.

21. Connect the heater hose at the water pump.

22. Install the radiator.

23. Connect the upper and lower radiator hoses, and transmission oil cooler lines.

24. Fill the cooling system and the crankcase.

Timing Chain and Sprockets

REMOVAL AND INSTALLATION

3.0L Engines

1. Disconnect the negative battery cable and drain the cooling system.

2. Remove the timing chain front cover.

3. Rotate the crankshaft until No. 1 cylinder is at TDC and the crankshaft and camshaft sprocket timing marks are aligned.

4. Remove the camshaft sprocket retaining bolt and remove the sprocket and timing chain.

5. Remove the crankshaft sprocket.

To install:

6. Align the crankshaft sprocket with the key or dowel on the crankshaft and install the sprocket.

7. Make sure the sprocket timing marks are still in alignment.

8. Install the camshaft sprocket and timing chain. Install the camshaft sprocket retaining bolt and tighten to 46 ft. lbs. (63 Nm).

NOTE: The camshaft retaining bolt on the 3.0L engine has a drilled oil passage for timing chain lubrication. If damaged, do not replace with a standard bolt. Clean the oil passage with solvent.

9. Install the timing chain front cover and the remaining components in the reverse order of their removal. Fill the crankcase with the proper type and quantity of engine oil. Fill and bleed the cooling system. Run the engine and check for leaks.

4.0L Engine

1. Disconnect the negative battery cable and drain the cooling system and crankcase.

2. Remove the oil pan and radiator. Remove the accessory drive belt and crankshaft damper.

3. Remove the water pump and timing chain front cover.

4. Remove the camshaft sprocket retaining bolt and the crankshaft sprocket key.

5. Push the timing chain tensioner into the retracted position and install the retaining clip.

6. Remove the crankshaft and camshaft sprockets with the timing chain. Remove the tensioner and guide, as required.

To install:

7. Install the timing chain guide to the cylinder block with the pin of the guide inserted into the oil hole in the block. Install the 2 retaining bolts and tighten to 7–9 ft. lbs. (10–12 Nm).

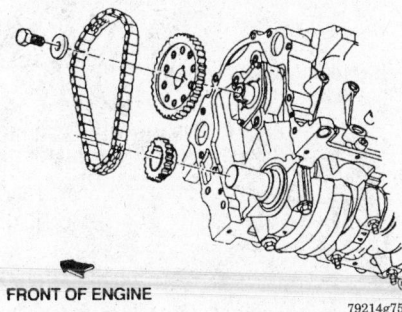

FRONT OF ENGINE

79214g75

Timing chain and sprocket — 3.0L engine

8. Position the camshaft and crankshaft so the sprocket timing marks will align.

9. Install the sprockets and timing chain together. Install the timing chain tensioner with the clip in place to lock the tensioner in the retracted position.

10. Install the crankshaft key and check the timing marks on the sprockets for correct alignment. Make sure the tensioner side of the timing chain is held inward and the guide side of the chain is straight and tight.

11. Install the camshaft sprocket retaining bolt and tighten to 44–50 ft. lbs. (60–68 Nm). Remove the clip from the tensioner assembly.

12. Install the timing chain front cover and the remaining components in the reverse order of their removal. Fill the crankcase with the proper type and quantity of engine oil. Fill and bleed the cooling system. Run the engine and check for leaks.

5.0L Engine

1. Remove the front cover.

2. Rotate the crankshaft counterclockwise to take up the slack on the left side of the chain.

3. Establish a reference point on the cylinder block and measure from this point to the chain.

4. Rotate the crankshaft in the opposite direction to take up the slack on the right side of the chain.

5. Force the left side of the chain out and measure the distance between the reference point and the chain. The timing chain deflection is the difference between the 2 measurements. If the deflection exceeds ½ in. (13mm), replace the timing chain and sprockets.

To install:

6. Turn the crankshaft until the timing marks on the sprockets are aligned vertically.

7. Remove the camshaft sprocket retaining screw and remove the fuel pump eccentric and washers.

8. Alternately slide both of the sprockets and timing chain off the crankshaft and camshaft until free of the engine.

9. Position the timing chain on the sprockets so the timing marks on the sprockets are aligned vertically. Alternately slide the sprockets and chain onto the crankshaft and camshaft sprockets.

10. Install the fuel pump eccentric washers and attaching bolt on the camshaft sprocket. Tighten to 40–45 ft. lbs. (52–58 Nm).

11. Install the front cover.

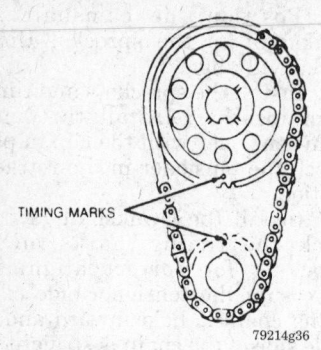

Timing chain alignment marks

Timing Belt Front Cover

REMOVAL AND INSTALLATION

2.3L Engine

1. Disconnect the negative battery cable and drain the cooling system.
2. Loosen the thermactor pump bolts and remove the drive belt, if equipped.
3. Remove the fan blade and 4 water pump pulley bolts.
4. Loosen the alternator retaining bolts and remove the drive belt from the pulleys. Remove the upper radiator hose.
5. Remove the crankshaft pulley bolt and pulley. Remove the thermostat housing.
6. Loosen the power steering pump mounting bracket and position aside.
7. Remove the timing belt front cover retaining bolt(s). Release the cover interlocking tabs, if equipped. Remove the cover.

To install:

8. Install the front cover. If equipped, secure by snapping the interlocking tabs into place. Install the retaining bolt(s).
9. Install the power steering pump mounting bracket.
10. Install the thermostat housing and connect the upper radiator hose.
11. Install the crankshaft pulley and retaining bolt. Tighten to 103–133 ft. lbs. (140–180 Nm).
12. Position the alternator drive belt and adjust the belt tension. Install the water pump pulley and fan.
13. Position the thermactor pump drive belt, if equipped, and adjust the tension.
14. Connect the negative battery cable. Fill and bleed the cooling system. Run the engine and check for leaks.

Timing Belt and Tensioner

ADJUSTMENT

On the 2.3L engine, all timing belt adjustments are accomplished by the timing belt tensioner. No maintenance is needed.

REMOVAL AND INSTALLATION

2.3L Engine

1. Disconnect the negative battery cable.
2. Remove the timing belt front cover.
3. Loosen the belt tensioner adjustment screw. Position belt tension adjusting tool T74P-6254-A or equivalent, on the tension spring rollpin and retract the belt tensioner. Tighten the adjustment screw to hold the tensioner in the retracted position.
4. Remove the bolts holding the timing sensor in place and pull the sensor free of the dowel pin.
5. Remove the crankshaft pulley, hub and belt guide. Remove the timing belt.
6. If the timing belt tensioner is to be removed, remove the adjustment screw and the spring bolt and remove the tensioner.

To install:

7. If removed, install the timing belt tensioner. Install the spring bolt but do not tighten at this time. Position the tensioner in the fully retracted position and tighten the adjustment bolt.
8. Position the crankshaft sprocket to align with the TDC mark and the camshaft sprocket to align with the timing pointer.
9. Install the timing belt over the crankshaft sprocket and then counterclockwise over the auxiliary and camshaft sprockets. Align the belt fore and aft on the sprockets.
10. Loosen the tensioner adjustment bolt to allow the tensioner to move against the belt. If the spring does not have enough tension to move the roller against the belt and the belt hangs loose, it may be necessary to manually push the roller against the belt and tighten the bolt.

NOTE: The spring cannot be used to set belt tension. A wrench must be used on the tensioner assembly.

11. Remove a spark plug from each cylinder to make sure the engine does not jump time during Step 12.

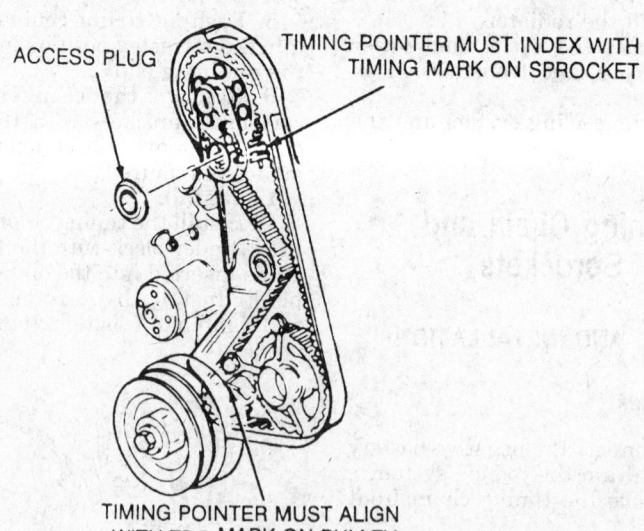

ACCESS PLUG

TIMING POINTER MUST INDEX WITH TIMING MARK ON SPROCKET

TIMING POINTER MUST ALIGN WITH TDC MARK ON PULLEY

ON 1988 VEHICLES DISTRIBUTOR ROTOR MUST ALIGN WITH NO. 1 FIRING POSITION

2.3L timing belt and sprocket alignment

12. Rotate the crankshaft 2 complete turns in the direction of normal rotation to remove the slack from the belt. Tighten the spring bolt to 28–40 ft. lbs. (38–54 Nm) and the adjustment bolt to 14–21 ft. lbs. (19–29 Nm).

13. Install the crankshaft belt guide.

14. Proceed as follows:

a. Install the timing sensor onto the dowel pin and tighten the 2 longer bolts.

b. Rotate the crankshaft 45 degrees counterclockwise and install the crankshaft pulley and hub. Tighten the pulley bolt to 103–133 ft. lbs. (140–180 Nm).

c. Rotate the crankshaft 90 degrees clockwise so the vane of the crankshaft pulley engages with timing sensor positioner tool T89P-6316-A or equivalent. Tighten the 2 shorter sensor bolts.

d. Rotate the crankshaft 90 degrees counterclockwise and remove the sensor positioner tool.

e. Rotate the crankshaft 90 degrees clockwise and measure the outer vane-to-sensor air gap. The air gap must be 0.018–0.039 in. (0.458–0.996mm).

15. Install the remaining components.

16. Connect the negative battery cable.

Timing Sprockets

REMOVAL AND INSTALLATION

2.3L Engine

1. Disconnect the negative battery cable.

2. Remove timing belt front cover and timing belt.

3. Remove timing sprockets retaining bolt(s). Remove the timing sprocket with a suitable puller.

To install:

4. Install the timing sprocket, making sure to align the keyway. Secure the sprocket in place with the retainer bolt.

5. Tighten the camshaft sprocket bolt to 52–70 ft. lbs. (70–95 Nm). Tighten the auxiliary shaft sprocket bolt to 30–40 ft. lbs. (40–54 Nm).

6. Install the timing belt cover.

7. Connect the negative battery cable.

Camshaft

REMOVAL AND INSTALLATION

2.3L Engine

1. Disconnect the negative battery cable and drain the cooling system. Remove the air cleaner assembly.

2. Tag and disconnect the spark plug wires at the plugs and rocker arm cover and position aside. Tag and disconnect the necessary vacuum lines.

3. Remove the rocker arm cover. Remove the alternator mounting bracket-to-cylinder head mounting bolts and position aside.

4. Disconnect and remove the upper radiator hose. Remove the radiator shroud.

5. Remove the timing belt front cover. If equipped with power steering, remove the power steering pump bracket.

6. Remove the timing belt, camshaft followers and camshaft sprocket. Remove the camshaft seal using seal removal tool T74P-6700-B or equivalent.

7. Remove the 2 screws and the camshaft rear retainer.

8. Raise and support the vehicle safely. Remove the front motor mount bolts.

9. Position a jack under the engine and raise the engine carefully as far as it will go. Place blocks of wood between the engine mounts and chassis brackets and remove the jack.

10. Remove the camshaft, being careful to avoid damaging the journals, lobes and bearings.

To install:

11. Make sure the threaded plug is in the rear of the camshaft. If not, remove the threaded plug from the old camshaft and install. Tighten to 12–18 ft. lbs. (16–24 Nm).

12. Coat the camshaft lobes with grease and lubricate the journals

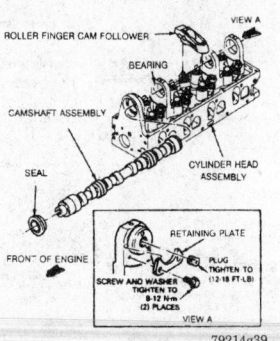

Camshaft installation — 2.3L engine

with heavy engine oil. Carefully slide the camshaft through the bearings.

13. Install the camshaft rear retainer with the 2 screws. Tighten to 6–9 ft. lbs. (8–12 Nm).

14. Install a new camshaft seal using seal installation tool T74P-6150-A or equivalent.

15. Install the remaining components. Fill and bleed the cooling system. Run the engine and check for leaks.

3.0L and 4.0L Engines

1. Disconnect the negative battery cable and relieve the fuel system pressure. Drain the crankcase and the cooling system.

2. Remove the rocker arm covers, rocker arms or rocker arm shaft assemblies and pushrods. Note the position of each component so it can be reinstalled in the same place.

3. Remove the intake manifold.

4. Remove the lifters. Identify each lifter so it can be reinstalled in the original position.

5. Remove the front timing chain cover and the timing chain and sprockets.

6. Remove the thrust plate bolts and remove the thrust plate. Carefully remove the camshaft, being careful not to damage the journals, lobes or bearings.

To install:

7. Coat the camshaft lobes with grease and the journals with heavy engine oil. Carefully install the camshaft, being careful not to damage the journals, lobes or bearings.

8. Install the thrust plate and the thrust plate retaining bolts. Tighten the bolts to 7 ft. lbs. (10 Nm) on 3.0L engine or 7–10 ft. lbs. (10–13 Nm) on 4.0L engine.

9. Check the camshaft end-play using a dial indicator. The end-play should be 0.0008–0.004 in. (0.02–0.10mm) on 4.0L engines or should not exceed 0.007 in. (0.18mm) on 3.0L engine.

10. Install the remaining components. Fill the crankcase with the proper type and quantity of engine oil. Fill and bleed the cooling system. Run the engine and check for leaks.

5.0L Engine

1. Disconnect the negative battery cable.

2. Remove the timing chain and sprocket.

3. Remove the lower intake manifold.

4. Remove the valve tappets and camshaft thrust plate retainer bolts.

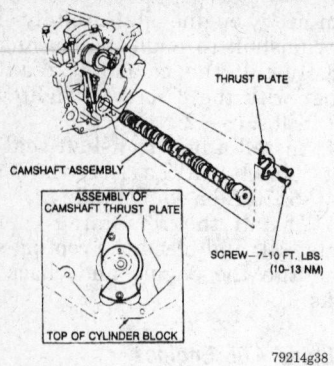

Camshaft installation — 4.0L engine

5. With the thrust plate removed, slowly remove the camshaft from the engine by pulling it toward the front of the engine.

To install:

6. Apply SAE type 50 oil to the camshaft lobes and journals. Slowly install the camshaft into the engine.

7. Apply SAE type 50 oil to the camshaft thrust plate.

8. Position the camshaft thrust plate with the groove toward the cylinder block and install the thrust plate bolts. Tighten the bolts to 9–12 ft. lbs. (13–16 Nm).

9. Install the valve tappets.

10. Install the lower intake manifold.

11. Install the timing chain and sprockets.

12. Start the engine and check for leaks.

Auxiliary Shaft

REMOVAL AND INSTALLATION

1. Disconnect the negative battery cable and drain the cooling system.

2. Remove the timing belt front cover.

3. Remove the timing belt and remove the auxiliary shaft sprocket.

4. Remove the auxiliary shaft cover bolts and the cover.

5. Remove the auxiliary shaft retaining plate screws and remove the retaining plate.

6. Remove the auxiliary shaft, being careful not to damage the journals or bearings.

To install:

7. Coat the auxiliary shaft journals with heavy engine oil. Install the auxiliary shaft, being careful not to damage the journals or bearings.

8. Install the retaining plate. Tighten the retaining plate screws to 6–9 ft. lbs. (8–12 Nm).

9. Install the auxiliary shaft cover and tighten the screws to 6–9 ft. lbs. (8–12 Nm).

10. Install the remaining components. Fill and bleed the cooling system. Run the engine and check for leaks.

Piston and Connecting Rod

POSITIONING

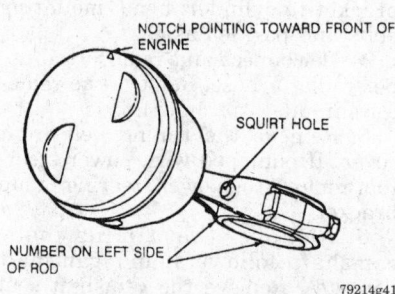

Piston and connecting rod assembly — 2.3L engine

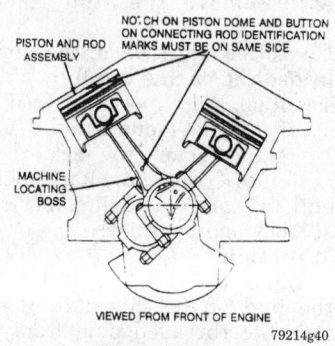

Piston and connecting rod assembly — 3.0L engine

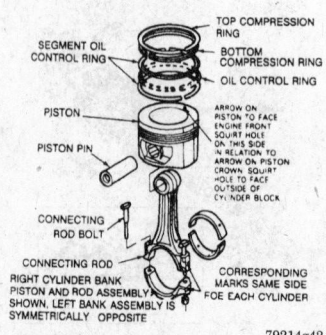

Piston and connecting rod assembly — 4.0L engine

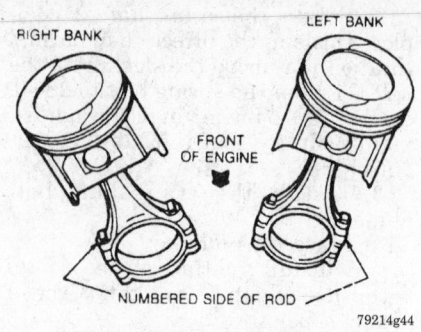

Piston and connecting rod assembly — 2.3L engine

ENGINE LUBRICATION

Oil Pan

REMOVAL AND INSTALLATION

2.3L Engine

1. Disconnect the negative battery cable and remove the air cleaner outlet tube at the throttle body.

2. Remove the engine oil dipstick and remove the engine mount retaining nuts.

3. Disconnect the oil cooler lines at the radiator, if equipped. Remove the fan shroud.

4. If equipped with automatic transmission, remove the radiator retaining bolts and position the radiator upward and wire to the hood.

5. Raise and safely support the vehicle. Drain the engine oil.

6. Remove the starter. Disconnect the exhaust manifold tube to the inlet pipe bracket at the thermactor check valve.

7. Disconnect the catalytic converter at the inlet pipe.

8. Remove the insulator and retainer assembly at the transmission. Remove the transmission mount retaining nuts to the crossmember.

9. If equipped with an automatic transmission, remove the oil cooler lines from the retainer at the block and remove the front crossmember.

10. If equipped with a manual transmission, disconnect the right front lower shock absorber mount.

11. Position a jack under the engine. Raise the engine and position suitable wood blocks between the en-

gine mounts and frame brackets. Remove the jack.

12. If equipped with an automatic transmission, position a jack under the transmission and raise slightly.

13. Remove the oil pan retaining bolts and lower the pan to the chassis. Remove the low oil level sensor assembly and the oil pump drive and pickup tube assembly.

14. If equipped with an automatic transmission, remove the oil pan out the front of the vehicle. If equipped with manual transmission, remove the oil pan out from the rear.

To install:

15. Clean all gasket mating surfaces, the oil pan, oil pump exterior and pickup tube screen.

16. Install the low oil level sensor assembly and tighten to 20–30 ft. lbs. (27–41 Nm).

17. Press a new gasket into the oil pan groove. Retain the gasket in the oil pan by press fitting only.

18. Position the oil pan on the crossmember. Install the oil pump drive and pickup tube assembly.

19. Apply sealer in 6 places on the engine and install the oil pan. Install the oil pan flange bolts tight enough to compress the gasket to the point that the 2 transmission holes are aligned with the 2 tapped holes in the oil pan, but loose enough to allow movement of the pan relative to the block.

20. Install the 2 oil pan-to-transmission bolts and tighten to 30–39 ft. lbs. (40–50 Nm) to align the oil pan with the transmission, then loosen the bolts ½ turn.

21. Tighten all oil pan flange bolts to 90–120 inch lbs. (10–13 Nm), then retighten the 2 oil pan-to-transmission bolts to 30–39 ft. lbs. (40–50 Nm).

22. Install a new oil filter. Position a jack under the engine and raise it enough to remove the wood blocks. Shift the engine/transmission backward to its original position.

23. Install the mount/bracket assembly to the crossmember and lower the engine. Install the front crossmember, if removed.

24. Raise the transmission with the jack and install the mount. Install the stabilizer brackets to the frame, if removed.

25. Connect the automatic transmission oil cooler line retainer clip to the engine, if equipped. Install the transmission mount retaining nuts.

26. Install a new gasket and connect the rear exhaust pipe just behind the catalytic converter.

27. Connect the low oil level sensor wire. Install the starter and connect the starter cable. Lower the vehicle.

28. Connect the vacuum tube to the clip at the front of the automatic transmission, if equipped.

29. Install the radiator and shroud. Connect the oil cooler lines, if equipped.

30. Connect the EGR valve and EGR tube. Install the engine mount retaining nuts.

31. Install the oil dipstick. Fill the crankcase with the proper type and quantity of engine oil.

32. Connect the negative battery cable, start the engine and check for leaks.

3.0L

Aerostar

1. Disconnect the negative battery cable.

2. Remove the engine oil level dipstick. Raise the vehicle and support it safely.

3. If equipped with an oil level sensor, remove the retaining clip at the sensor. Disconnect the harness from the sensor.

4. Drain the crankcase. Remove the starter and transmission inspection cover.

5. Remove the oil pan attaching bolts and remove the pan.

To install:

6. Clean all gasket mating surfaces and the oil pan.

7. Apply a ⅕ in. (5mm) bead of silicone sealer to the junction of the rear main bearing cap and cylinder block and the junction of the front cover assembly and cylinder block.

8. Position the oil pan gasket to the oil pan and secure with sealer. Install the oil pan on the engine block with the attaching bolts and tighten to 9 ft. lbs. (12 Nm).

9. Install the starter and transmission inspection cover. Attach the low oil level sensor harness and install the retainer clip.

10. Lower the vehicle and install the engine oil level dipstick. Fill the crankcase with the proper type and quantity of engine oil.

11. Connect the negative battery cable, start the engine and check for leaks.

Except Aerostar

1. Disconnect the negative battery cable. Remove the engine oil level dipstick.

2. Disconnect the fan shroud and drape it over the fan. Remove the motor mount nuts from the frame.

3. Raise and safely support the vehicle. Remove the low oil level sensor retainer clip at the sensor. Disconnect the electrical connector from the sensor.

4. Drain the crankcase and remove the starter. Remove the transmission inspection cover.

5. Remove the right axle beam on 2WD vehicles.

NOTE: The brake caliper must be removed and wired out of the way.

6. Remove the oil pan bolts. Position a jack under the engine and raise it approximately 2 in. Remove the oil pan.

NOTE: The oil pan fits tightly between the transmission spacer plate and oil pump pickup tube. Use care when removing to avoid damaging the pickup tube.

To install:

7. Clean all gasket mating surfaces and the oil pan.

8. Apply a ⅕ in. (5mm) bead of silicone sealer to the junction of the rear main bearing cap and cylinder block and the junction of the front cover assembly and cylinder block.

9. Position the oil pan gasket to the oil pan and secure with sealer. Install the oil pan on the engine block with the attaching bolts and tighten to 9 ft. lbs. (12 Nm).

10. Install the low oil level sensor harness and the retainer clip. Lower the engine assembly.

11. Install the right axle beam, if removed.

12. Install the transmission inspection cover and the starter. Lower the vehicle.

13. Install the fan shroud and the motor mount nuts. Install the distributor, aligning the marks that were made during removal.

14. Install the engine oil level dipstick. Fill the crankcase with the proper type and quantity of engine oil.

15. Connect the negative battery cable, start the engine and check for leaks.

4.0L Engine

Aerostar

1. Disconnect the negative battery cable.

2. Raise and safely support the vehicle. Remove the starter.

3. On 4WD vehicles, proceed as follows:

   a. Remove the front wheel and tire assemblies.

b. Remove the pivot bolts and nuts from both lower control arms, to allow the control arms to hang.

c. Remove the control arm rear pivot crossmember.

d. Remove the lower nuts from both motor mounts.

e. Remove the front drive axle assembly.

4. Drain the crankcase and remove the oil filter. Disconnect the low oil level sensor from the engine oil pan.

5. Remove the 2 transmission-to-engine oil pan bolts. Remove the oil pan retaining bolts and nuts.

6. On 4WD vehicles, raise the engine approximately 1 in. (3cm)

7. Remove the oil pan.

To install:

8. Clean all gasket mating surfaces and the oil pan.

9. Place a small amount of silicone sealer on the block at the corner where the oil pan, rear seal and block mate.

10. Install a new crankshaft rear main bearing cap wedge seal. The seal should fit snugly into the sides of the rear main bearing cap.

11. Position a new oil pan gasket into the groove in the oil pan and position the 2 oil pan spacers on the oil pan locating pads.

NOTE: If the same oil pan is being reused, the existing spacers may be used. If a new pan is being installed, the pan-to-transmission gap must be measured to find the needed spacer thickness. Failure to use the correct spacer can result in improper clearance between the oil pan and transmission, resulting in oil pan damage and/or an oil leak.

12. If a new oil pan is being installed, find the correct spacer thickness as follows:

a. Position the oil pan on the engine without the spacers and install the retaining nuts on the 4 locating studs.

b. Using a feeler gauge, measure the gap between the locating pads on the pan and the transmission converter housing.

c. If the measured gap is 0.011–0.020 in. (0.27–0.51mm), a 0.010 in. (0.254mm) spacer is required. If the measured gap is 0.021–0.029 in. (0.52–0.76mm), a 0.020 in. (0.508mm) spacer is required. If the measured gap is 0.030–0.039 in. (0.77–1.00mm), a 0.030 in. (0.762mm) spacer is required.

d. Remove the oil pan and position the correct spacers.

13. Install the oil pan with a new gasket on the engine and tighten the retaining bolts and nuts enough to compress the gasket so the transmission bolts align with the holes in the oil pan, but loose enough to allow the pan to move when the transmission bolts are installed.

14. Install the 2 transmission-to-oil pan bolts and tighten to 28–38 ft. lbs. (38–51 Nm) to align the oil pan with the transmission, then loosen the bolts ½ turn.

15. Tighten all the oil pan bolts and nuts evenly to 5–7 ft. lbs. (7–10 Nm), then retighten the 2 transmission-to-oil pan bolts to 28–38 ft. lbs. (38–51 Nm).

16. Connect the low oil level sensor. Install the drain plug and a new oil filter.

17. On 4WD vehicles, proceed as follows:

a. Lower the engine and install the lower motor mount nuts.

b. Install the front drive axle assembly.

c. Install the control arm rear pivot crossmember.

d. Reposition the lower control arms and install the pivot bolts and nuts.

e. Install the front wheel and tire assemblies.

18. Install the starter. Lower the vehicle and fill the crankcase with the proper type and quantity of engine oil.

19. Connect the negative battery cable, start the engine and check for leaks.

Except Aerostar

1. Remove the engine assembly and install on a workstand with the oil pan facing up.

2. Remove the oil pan retaining bolts and remove the pan.

To install:

3. Clean all gasket mating surfaces and the oil pan.

4. Install a new crankshaft rear main bearing cap wedge seal. The seal should fit snugly into the sides of the rear main bearing cap.

5. Position a new oil pan gasket to the engine block and place the oil pan in position on the 4 locating studs. Tighten the retaining nuts and bolts evenly to 5–7 ft. lbs. (7–10 Nm).

6. Measure the gap between the surface of the rear face of the oil pan, at the spacer locations, and the rear face of the engine block as follows:

a. With the oil pan installed on the engine, position a straight edge flat on the rear of the engine block so it extends over 1 of the oil pan/transmission bolt mounting pads.

b. Using a feeler gauge, measure the gap between the mounting pad and the straight edge. Repeat the procedure for the other mounting pad.

c. If the measured gap is 0.011–0.020 in. (0.27–0.51mm), a 0.010 in. (0.254mm) spacer is required. If the measured gap is 0.021–0.029 in. (0.52–0.76mm), a 0.020 in. (0.508mm) spacer is required. If the measured gap is 0.030–0.039 in. (0.77–1.00mm), a 0.030 in. (0.762mm) spacer is required.

d. Install the selected spacers to the mounting pads on the rear of the oil pan before bolting the engine and transmission together.

NOTE: Failure to use the correct spacer can result in improper clearance between the oil pan and transmission, resulting in oil pan damage and/or an oil leak.

7. Remove the engine from the workstand and install in the vehicle.

5.0L Engine

1. Drain the cooling system.

2. Remove the bolts attaching the fan shroud to the radiator and position the shroud over the fan.

3. Remove the upper intake manifold and throttle body.

4. Remove the nuts and lockwashers attaching the engine support insulators to the chassis bracket.

5. If equipped with an automatic transmission, disconnect the oil cooler line at the left side of the radiator.

6. Remove the exhaust system.

7. Raise the engine and place wood blocks under the engine supports.

8. Drain the crankcase.

9. Support the transmission with a floor jack and remove the transmission crossmember.

10. Remove the oil pan attaching bolts and lower the oil pan onto the crossmember.

11. Remove the 2 bolts attaching the oil pump pickup tube to the oil pump. Remove nut attaching oil pump pickup tube to the number 3 main bearing cap stud. Lower the pickup tube and screen into the oil pan.

12. Remove the oil pan from the vehicle.

To install:

13. Clean the oil pan, inlet tube and gasket surfaces. Inspect the gasket sealing surface for damage and distortion due to overtightening of the bolts. Repair and straighten as required.

14. Position a new oil pan gasket and seal to the cylinder block.

15. Position the oil pick-up tube and screen to the oil pump, and install the lower attaching bolt and gasket loosely. Install attaching nut to number 3 main bearing cap stud.

16. Place the oil pan on the crossmember. Install the upper pick-up tube bolt. Tighten the pick-up tube bolts.

17. Position the oil pan to the cylinder block and install the attaching bolts. Tighten to 10–12 ft. lbs. (13–16 Nm)

18. Install the transmission crossmember.

19. Raise the engine and remove the blocks under the engine supports. Bolt the engine to the supports.

20. Install the exhaust system.

21. If equipped with an automatic transmission, connect the oil cooler line at the left side of the radiator.

22. Install the nuts and lockwashers attaching the engine support insulators to the chassis bracket.

23. Install the upper intake manifold and throttle body.

24. Install the fan shroud.

25. Fill the crankcase.

26. Fill and bleed the cooling system.

Oil Pump

REMOVAL AND INSTALLATION

2.3L, 3.0L and 4.0L Engines

1. Disconnect the negative battery cable.

2. Remove the oil pan.

3. Remove the oil pump attaching bolts and, if equipped, remove the oil pump pickup tube retaining nut from the main bearing cap.

4. Remove the oil pump and oil pump driveshaft. Remove and clean the oil pump pickup tube and screen, as necessary.

To install:

5. Install the oil pump pickup tube and screen assembly, if removed.

6. Prime the oil pump by filling either the inlet or outlet port with clean engine oil. Rotate the pump shaft to distribute the oil within the pump body.

7. Insert the oil pump driveshaft into the opening in the block or main bearing cap. On 3.0L engine, assemble the shaft to the oil pump until the retainer clicks into place.

8. Install the oil pump, with a new gasket if equipped, and install the attaching bolts. Tighten the bolts to 14–21 ft. lbs. (19–29 Nm) on 2.3L engine, 35 ft. lbs. (48 Nm) on 3.0L engine or 13–15 ft. lbs. (17–21 Nm) on 4.0L engine.

9. On 2.3L engine, install the pickup tube retaining nut and tighten to 30–41 ft. lbs. (40–55 Nm).

10. Install the oil pan.

11. Fill the crankcase with the proper type and quantity of engine oil. Connect the negative battery cable, start the engine and check for leaks.

5.0L Engine

1. Remove the oil pan.

2. Remove the oil pump inlet tube and screen assembly.

3. Remove the oil pump attaching bolts and remove the oil pump gasket and intermediate driveshaft.

To install:

4. Before installing the oil pump, prime it by filling the inlet and outlet port with oil and rotating the shaft of the pump to distribute it.

5. Position the intermediate driveshaft into the distributor socket.

6. Position the new gasket on the pump body and insert the intermediate driveshaft into the pump body.

7. Install the pump and intermediate driveshaft as an assembly. Do not force the pump if it does not seal readily. The driveshaft may be misaligned with the distributor shaft. To align it, rotate the intermediate driveshaft into a new position.

8. Install the oil pump attaching bolts and tighten them to 20–25 ft. lbs. (26–32 Nm).

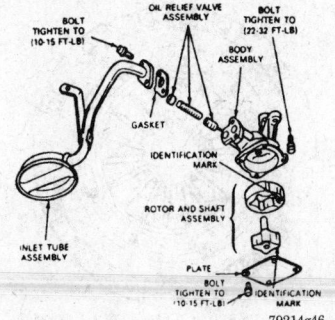

Oil pump assembly — 5.0L engine

TRANSMISSION

Manual Transmission Assembly

REMOVAL AND INSTALLATION

— CAUTION —

The clutch driven disc may contain asbestos, which has been determined to be a cancer causing agent. Never clean clutch surfaces with compressed air! Avoid inhaling any dust from any clutch surface! When cleaning clutch surfaces, use a commercially available brake cleaning fluid.

1. Raise and support the vehicle safely. Prop the clutch pedal in the full up position with a block of wood.

2. Matchmark the driveshaft-to-flange relation.

3. Disconnect the driveshaft at the rear axle and slide it off of the transmission output shaft. Lubricant will leak out of the transmission so be prepared to catch it, or plug the opening with rags or a seal installation tool.

4. Disconnect the speedometer cable at the transmission.

5. Disconnect the shift rods from the shift levers.

6. Remove the shift control from the extension housing and transmission case.

7. On 4-wheel drive models, remove the transfer case.

8. Remove the extension housing-to-rear support bolts.

9. Take up the weight of the transmission with a transmission jack. Chain the transmission to the jack.

10. Raise the transmission just enough to take the weight off of the No. 3 crossmember.

11. Unbolt the crossmember from the frame rails and remove it.

12. Place a jackstand under the rear of the engine at the bellhousing.

13. Lower the jack and allow the jackstand to take the weight of the engine. The engine should be angled slightly downward to allow the transmission to roll backward.

14. Remove the transmission-to-bellhousing bolts.

15. Roll the jack rearward until the input shaft clears the bellhousing. Lower the jack and remove the transmission.

To install:

16. Clean all mating surfaces thoroughly.

17. Install a guide pin in each lower bolt hole. Position the spacer plate on the guide pins.

18. Raise the transmission and start the input shaft through the clutch release bearing.

19. Align the input shaft splines with the clutch disc splines. Roll the transmission forward so the input shaft will enter the clutch disc. If the shaft binds in the release bearing, work the release arm back and forth.

20. Once the transmission is all the way in, install the 2 upper retaining bolts and washers and remove the lower guide pins. Install the lower bolts. Tighten the bolts to 50 ft. lbs. (65 Nm).

21. Raise the transmission just enough to allow installation of the No.3 crossmember.

22. Install the crossmember on the frame rails. Tighten the bolts to 80 ft. lbs. (104 Nm).

23. Lower the transmission onto the crossmember and install the nuts. Tighten the nuts to 70 ft. lbs. (91 Nm).

24. Remove the transmission jack.

25. Install the transfer case.

26. Install the shift control on the extension housing and transmission case.

27. Connect the shift rods at the shift levers.

28. Connect the speedometer cable at the transmission.

29. Slide the driveshaft onto the output shaft and connect the driveshaft at the rear axle, aligning the matchmarks.

Clutch Assembly

REMOVAL AND INSTALLATION

1. Disconnect the negative battery cable.

2. Disconnect the hydraulic clutch master cylinder from the clutch pedal.

3. Raise and safely support the vehicle. Remove the starter.

4. Use coupling disconnect tool T88T-70522-A or equivalent, to slide the white plastic sleeve toward the slave cylinder, then apply a slight tug on the tube to disconnect the hydraulic coupling. Plug the hose.

5. Remove the transmission.

6. Mark the position of the pressure plate on the flywheel so if the pressure plate is reused, it can be reinstalled in the same position.

7. Loosen the pressure plate attaching bolts evenly until the diaphragm spring is expanded. Remove the bolts, pressure plate and clutch disc.

8. Inspect the flywheel for wear, scoring and cracks. Machine or replace, as necessary. Inspect the clutch pilot bearing for wear and free movement. If replacement is necessary, remove using puller tool T58L-101-B or equivalent.

9. Inspect the clutch release bearing for wear and free movement; replace as necessary. Remove the release bearing by twisting it until resistance is felt. Turning further will allow the preload spring to push the bearing assembly off the slave cylinder.

To install:

10. If the pilot bearing was removed, a new one must be installed. Install using replacer tool T71P-7137-C and clutch driver tool T71P-7137-H or equivalent. Install the pilot bearing with the seal facing the transmission so the adapter is not cocked.

11. If the flywheel was removed, make sure the mating surfaces of the crank flange and flywheel are clean, and install the flywheel. Tighten the flywheel bolts to 59 ft. lbs. (80 Nm).

12. Position the clutch disc on the flywheel so alignment tool

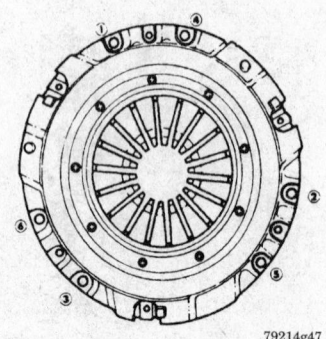

79214g47

Clutch pressure plate bolt torque sequence

T74P-7137-K or equivalent, can enter the pilot bearing and align the disc.

13. Install the pressure plate. If the original pressure plate is being reused, align the marks that were made during the removal procedure. Install the attaching bolts and tighten, in sequence, to 15–24 ft. lbs. (21–32 Nm), then remove the alignment tool.

14. Install the transmission. If equipped, reuse the aluminum washers under the attaching bolts to prevent galvanic corrosion.

15. Connect the hydraulic coupling by pushing the male coupling into the slave cylinder female coupling.

16. Lower the vehicle and connect the clutch master cylinder to the brake pedal. Bleed the clutch system, if necessary.

17. Connect the negative battery cable.

Clutch Master Cylinder

REMOVAL AND INSTALLATION

1. Disconnect the negative battery cable.

2. Disconnect the clutch master cylinder pushrod from the clutch pedal by prying the retainer bushing and pushrod off the pedal pin.

3. Remove the switch from the master cylinder assembly.

4. On sport utility and pick-up models, remove the screw retaining the fluid reservoir to the cowl access cover. On Aerostar models, slide the reservoir out of the relay bracket.

5. Use coupling disconnect tool T88T-70522-A or equivalent, to slide the white plastic sleeve toward the slave cylinder, then apply a slight tug on the tube to disconnect the hydraulic coupling.

6. Remove the retaining bolts and the clutch master cylinder.

To install:

7. Install the pushrod through the hole in the engine compartment. Make sure it is located on the correct side of the clutch pedal. Install the master cylinder and tighten the bolts to 12 ft. lbs. (16 Nm).

8. Insert the coupling end into the slave cylinder and install the tube into the clips.

9. On sport utility and pick-up models, install the fluid reservoir on the cowl access cover with the retaining screw. On Aerostar, slide the reservoir into the relay bracket.

10. Replace the retainer bushing in the clutch master cylinder pushrod if worn or damaged. Install the retainer

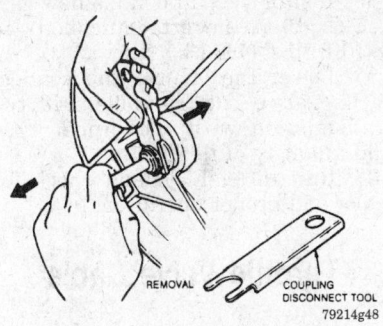

Disconnecting the hydraulic tube

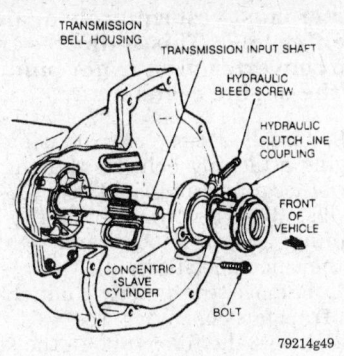

Slave cylinder assembly

and pushrod on the clutch pedal pin. Make sure the flange of the bushing is against the pedal blade. Install the switch.

11. Connect the negative battery cable and bleed the clutch hydraulic system, if necessary.

Clutch Slave Cylinder

REMOVAL AND INSTALLATION

NOTE: Before any vehicle service that requires slave cylinder removal, the clutch master cylinder pushrod must be disconnected from the clutch pedal. If not disconnected, permanent damage to the master cylinder will occur if the clutch pedal is depressed while the slave cylinder is disconnected.

1. Disconnect the negative battery cable.

2. Disconnect the coupling at the transmission using tool T88T-70522-A or equivalent, by sliding the white plastic sleeve toward the slave cylinder while applying a slight tug on the tube.

3. Raise and safely support the vehicle. Remove the transmission and clutch housing.

4. Remove the bolts retaining the slave cylinder to the transmission. Remove the slave cylinder from the transmission input shaft.

5. If necessary, remove the release bearing from the slave cylinder by twisting until resistance is felt, then turning further to allow the preload spring to push the bearing assembly off.

To install:

6. Push the release bearing into place, if removed.

7. Position the slave cylinder over the transmission input shaft with the bleed screw and coupling facing the left side of the transmission.

8. Install the slave cylinder attaching bolts and tighten to 13–19 ft. lbs. (18–26 Nm).

9. Install the transmission.

10. Insert the male coupling into the female coupling on the clutch slave cylinder and make sure the connection is secure.

11. Bleed the clutch hydraulic system. Lower the vehicle and connect the negative battery cable.

Hydraulic Clutch System Bleeding

PROCEDURE

NOTE: Under normal conditions, disconnecting the clutch coupling will not let air into the system. However, if there appears to be air in the system, indicated by a spongy pedal or insufficient bearing travel, the system must be bled.

1. Clean all dirt and grease from around the reservoir cap.

2. Remove the cap and fill the reservoir with DOT 3 heavy duty brake fluid.

3. Raise and safely support the vehicle, as necessary. Loosen the bleed screw, located in the slave cylinder body, next to the inlet connection.

4. Fluid will now begin to flow from the master cylinder, down the tube and into the slave cylinder.

NOTE: Keep the reservoir full at all times to make sure no additional air is drawn into the system.

5. Bubbles should begin to appear at the bleed screw outlet, indicating air is being expelled. When the slave cylinder is full, a steady stream of fluid will come from the slave cylinder outlet. Tighten the bleed screw.

6. Slowly depress the clutch pedal to the floor and hold. Loosen the bleed screw to allow air and excess fluid to be expelled. Retighten the bleed screw when fluid flow stops.

7. Depress and release the clutch pedal slowly, waiting 2 seconds between each cycle. Repeat 5 times.

8. Check the fluid level in the reservoir and add, if necessary. If evidence of air still exists, repeat Steps 6 and 7.

Automatic Transmission Assembly

REMOVAL AND INSTALLATION

1. Disconnect the negative battery cable. Raise and safely support the vehicle.

2. Position a drain pan under the transmission fluid pan. On sport utility models, pry the lower clips of the transmission heat shield back slightly to allow access to the pan bolts.

3. Starting at the rear of the transmission pan and working toward the front, loosen the attaching bolts and allow the fluid to drain. Remove all the bolts except the 2 at the front to allow the fluid to drain. After all fluid has drained, reinstall 2 bolts at the rear of the pan to temporarily hold it in place.

4. Remove the converter access cover from the converter housing. On some applications, it may only be necessary to remove 1 bolt and swing the cover open.

5. Disconnect the starter cable and remove the starter.

6. Place a 22mm socket and breaker bar on the crankshaft pulley attaching bolt. Rotate the pulley clockwise, as viewed from the front, to gain access to each converter attaching nut. Remove the nuts.

NOTE: On 2.3L engines, the converter attaching nuts are accessed through the cover on the engine oil pan. On 3.0L and 4.0L engines, the converter attaching nuts are accessed through the starter mounting hole.

7. Mark the position of the driveshaft on the axle flange and remove the driveshaft. Plug the transmission to prevent fluid leakage.

8. Disconnect the speedometer cable from the transmission.

9. Disconnect the shift rod at the transmission manual lever. Remove the kickdown cable from the ball stud lever. Depress the tab on the cable downshift retainer and remove the cable from the bracket.

10. Disconnect the neutral safety switch wires and converter clutch solenoid connector. Disconnect the vacuum line from the vacuum modulator.

11. Position a transmission jack under the transmission and raise it slightly. Remove the engine rear support-to-crossmember bolts.

12. Remove the crossmember-to-frame side support attaching bolts and remove the crossmember insulator and support and damper.

13. Lower the jack under the transmission and allow the transmission to hang. On sport utility and pick-up models, position a jack to the front of the engine and raise it to gain access to the 2 upper converter housing-to-engine attaching bolts.

14. Disconnect the oil cooler lines at the transmission. Plug the lines and transmission to prevent the entrance of dirt.

15. Remove the lower converter housing-to-engine attaching bolts and remove the transmission filler tube.

16. Secure the transmission to the jack with a safety chain. Remove the 2 upper converter housing-to-engine attaching bolts. Move the transmission to the rear so it disengages from the dowel pins and the converter is disengaged from the flywheel. Lower the transmission from the vehicle.

NOTE: If the transmission is to be removed for an extended period, support the engine with a safety stand and wood block.

To install:

17. Position the converter to the transmission making sure the converter hub is fully engaged in the pump gear. To make sure the converter is fully engaged, push and rotate the converter until 2 "bumps" are felt. Keep pushing and rotating until the distance between the converter pilot and the edge of the converter housing is $7/16$–$9/16$ in. (11–14mm).

18. Place the transmission on a transmission jack and secure with a safety chain. Rotate the converter so the drive studs are in alignment with the holes in the flywheel.

19. Raise the transmission and move it forward into position, being careful not to damage the flywheel and converter pilot.

NOTE: When moving the transmission, do not let the front of the transmission tilt downward. This will cause the converter to move forward and disengage from the pump gear. The con-

verter must rest squarely against the flywheel. This indicates that the converter pilot is not binding in the engine crankshaft.

20. Install 2 converter housing-to-engine attaching bolts at the engine dowel locations and tighten to 28–38 ft. lbs. (38–51 Nm). Install the remaining attaching bolts and tighten to the same specification.

21. Remove the safety chain from the transmission.

22. Insert the filler tube in the stub tube and secure it to the cylinder block with the attaching bolt. Tighten the bolt to 28–38 ft. lbs. (38–51 Nm). If the stub tube is loosened or dislodged, it should be replaced.

23. Install the oil cooler lines in the retaining clip at the cylinder block. Connect the lines to the transmission.

24. Remove the jack supporting the front of the engine.

25. Raise the transmission and position the crossmember, insulator and support and damper to the frame side supports. Install the attaching bolts and tighten to 20–30 ft. lbs. (27–41 Nm).

26. Lower the transmission and install the rear engine support-to-crossmember nut. Tighten the bolt to 60–80 ft. lbs. (82–108 Nm). Remove the transmission jack.

27. Install the vacuum hose on the vacuum modulator and attach the line to the clip. Connect the neutral safety switch plug and the converter clutch solenoid connector.

28. Install the flywheel-to-converter nuts and tighten to 20–34 ft. lbs. (27–46 Nm).

29. Install the converter access cover and adapter plate bolts and tighten to 12–16 ft. lbs. (16–22 Nm). On 2.3L engine, tighten the oil pan access cover bolts to 22–32 inch lbs. (2.5–3.6 Nm).

30. Install the starter and connect the starter cable.

31. Connect the exhaust pipe to the exhaust manifold, if disconnected for removal.

32. Connect the shift rod to the manual lever and the downshift cable to the downshift lever. Connect the speedometer cable.

33. Install the driveshaft, aligning the marks on the axle flange. Adjust the manual and downshift linkage, as required.

34. Remove the bolts temporarily holding the transmission fluid pan and remove the pan. Discard the gasket and clean all old gasket material and dirt from the gasket mating surfaces.

35. Install the pan using a new gasket. Tighten the attaching bolts to 8–10 ft. lbs. (11–13.5 Nm).

36. Lower the vehicle and connect the negative battery cable. Fill the transmission with the proper type and quantity of fluid.

37. Run the vehicle and check for leaks and proper operation.

Throttle Valve Cable

ADJUSTMENT

1. Set the parking brake and put the selector lever in **N**.

2. Remove the protective cover from the cable.

3. Make sure the throttle lever is at the idle stop. If it isn't, check for binding or interference. Never attempt to adjust the idle stop!

4. Make sure the cable is free of sharp bends or is not rubbing on anything throughout its entire length.

5. Lubricate the TV lever ball stud with chassis lube.

6. Unlock the locking tab at the throttle body by prying with a small prybar.

7. Install a spring on the TV control lever, to hold it in the rearmost travel position. The spring must exert at least 10 lbs. of force on the lever.

8. Rotate the transmission outer TV lever 10-30 degrees and slowly allow it to return.

9. Push down on the locking tab until flush.

10. Remove the retaining spring from the lever.

DRIVE AXLE

Driveshaft

REMOVAL AND INSTALLATION

Front

1. Raise and safely support the vehicle.

2. Remove the bolts and straps or the flange bolts retaining the driveshaft to the transfer case. If necessary, remove the boot from the transfer case to gain access to the slip yoke.

3. Remove the bolts and straps retaining the front U-joint to the front axle and remove the front driveshaft.

To install:

4. If equipped, lubricate the slip yoke splines and the edge of the inner diameter of the rubber boot. Slide the driveshaft into the transfer case, making sure the wide-tooth splines are properly indexed. Reposition the boot and install the clamp.

5. If equipped, install the driveshaft to the transfer case flange and install the retaining bolts. Tighten to 12–16 ft. lbs. (17–22 Nm).

6. Install the driveshaft to the front axle flange with the straps and bolts. Tighten to 10–15 ft. lbs. (14–20 Nm).

7. Lower the vehicle.

Rear

1. Raise and safely support the vehicle.

2. Mark the driveshaft in relation to the rear axle flange. If necessary, mark the relation of the driveshaft to the transfer case flange.

3. If equipped with a center bearing assembly, remove the retaining bolts and the spacers under the center bearing bracket, if installed.

4. Remove the attaching bolts and disconnect the driveshaft from the rear axle flange.

5. On 2WD models, slide the driveshaft rearward until the slip yoke clears the transmission extension housing and remove the driveshaft. Plug the extension housing to prevent fluid leakage.

6. On 4WD models, remove the bolts attaching the driveshaft to the transfer case flange and remove the driveshaft.

To install:

7. On 2WD models, lubricate the splines of the slip yoke. Remove the plug from the extension housing and install the driveshaft assembly. Do not allow the slip yoke to bottom on the output shaft with excessive force.

8. On 4WD models, install the driveshaft to the transfer case flange, aligning the marks that were made during removal. Install the attaching bolts and tighten to 61–87 ft. lbs. (83–118 Nm) if equipped with constant velocity U-joints or 12–16 ft. lbs. (17–22 Nm) if equipped with double cardan U-joints.

9. Connect the driveshaft to the rear axle flange, aligning the marks that were made during removal. Install the retaining bolts and tighten to 70–95 ft. lbs. (95–129 Nm).

10. If equipped, install the center bearing attachment bolts and tighten to 27–37 ft. lbs. (37–50 Nm). Make sure the center bearing bracket is installed "square" to the vehicle. If spacers were installed under the center bearing bracket, make sure they are reinstalled.

11. Lower the vehicle.

U-Joints

REMOVAL AND INSTALLATION

1. Remove the driveshaft from the vehicle and place on a clean surface.

—— **WARNING** ——
Do not clamp any driveshaft into a vise. In addition to driveshaft tube failure, the balance of the driveshaft could be effected causing damage to the vehicle under driving conditions.

2. Mark the position of the U-joint components in relation to the driveshaft. To maintain proper driveshaft balance, these components must be install in the same location.

3. Clamp U-joint tool T74P-4635-C or equivalent into a vise.

4. Remove the snaprings from the U-joint bearing caps.

5. Position the slip yoke of the U-joint in the clamp tool and press out the bearing.

6. Remove the slip yoke from the clamp tool and rotate it 180° and position in the vise. Press the bearing to remove the spider. Remove the yoke from the spider.

7. Remove the remaining bearing in the same manner.

To install:

NOTE: Universal joint kits are to be installed as a complete kit. Never mix components from other U-Joints

8. Start a new bearing cup into the yoke of the driveshaft.

9. Position the spider in the driveshaft yoke and press the bearing ¼ in. (6.3mm) below the yoke surface using the clamp.

10. Remove the driveshaft from the clamp, and install a snapring over the new bearing.

11. Start a new bearing cup in the opposite yoke. Place the assembly in the clamp and press the bearing in place. When complete, install another snapring over this bearing to lock in place.

12. Install the remaining bearing in the same manner. Make sure to install a snapring after each bearing has been pressed in place.

13. Before installing the driveshaft into the vehicle, make sure the U-joint rotates in all direction easily. If the joint binds, use a plastic-faced hammer and tap on the driveshaft

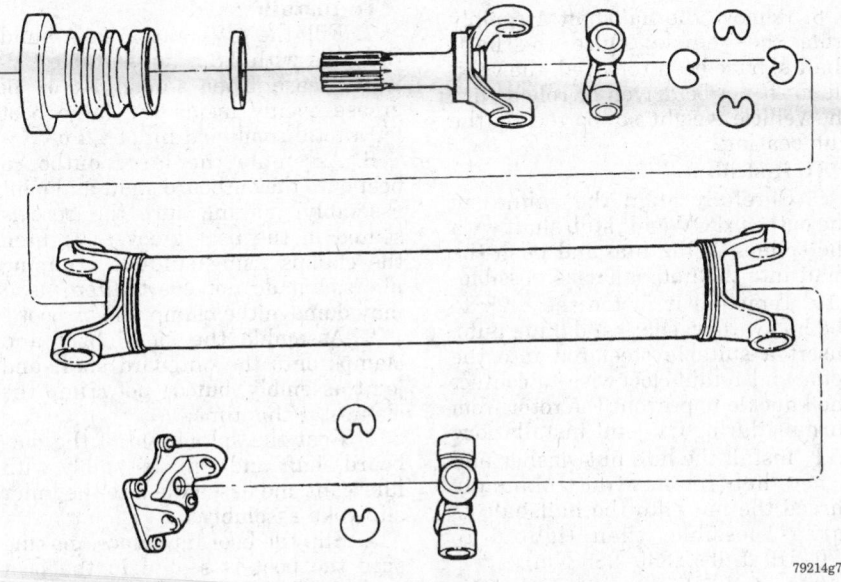

Front driveshaft — Aerostar, rear is similar

79214g77

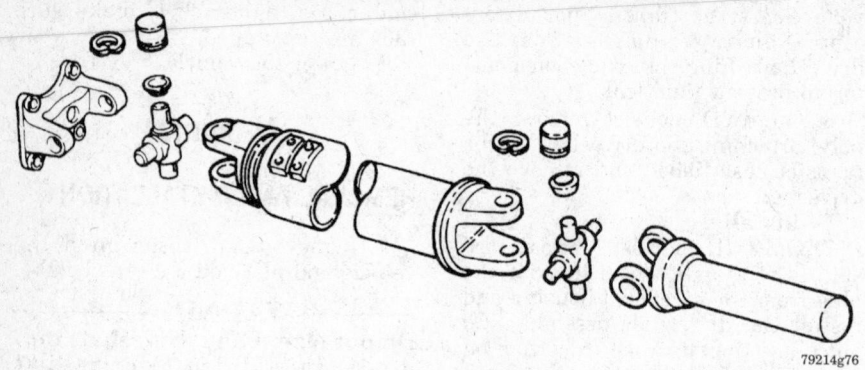

Rear driveshaft — 2WD models

79214g76

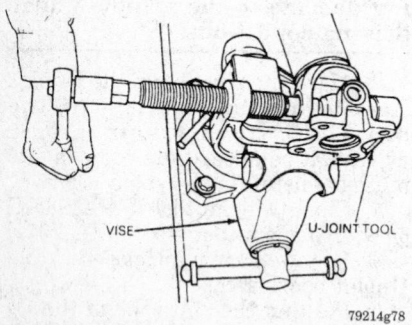

Using a U-joint clamp to remove and install U-joints

79214g78

yoke-side of the joint to seat the bearing.

14. If the driveshaft is equipped with grease fitting, apply a premium long life grease to the joint assembly.

15. Install the driveshaft into the vehicle. Refer to the procedure in this section.

Halfshaft

REMOVAL AND INSTALLATION

1. Raise and safely support the vehicle.

2. Remove the wheel and tire assembly and the hub retainer nut and washer. Discard the nut.

3. Mark the differential shaft flange in relation to the halfshaft so they can be reinstalled in their original position.

4. Remove the bolts and disconnect the halfshaft inboard flanges from the differential axle shaft flanges.

5. Support the end of the shaft by suspending it from an underbody component with mechanics wire. Do not allow the shaft to hang unsupported as damage to the outboard CV-joint may result.

6. Loosen the shock absorber on the lower control arm and move it to the side. Remove the rubber jounce bumper.

7. Separate the outboard CV-joint from the hub using front hub tools T81P-1104-C with T81P-1104-A and adapters T83P-1104-BH1 and T86P-1104-A1 or equivalents, and free the hub, bearing and knuckle assembly from the halfshaft by pushing in the CV-joint outer shaft until it is loose in the assembly.

NOTE: Never use a hammer to separate the outboard CV-joint stub shaft from the hub. Damage to the CV-joint threads and internal components may result.

8. Remove the halfshaft assembly from the vehicle. Once the halfshaft(s) have been removed, the vehicle must not be driven or rolled with the vehicle weight supported by the hub bearing.

To install:

9. Carefully align the splines of the outboard CV-joint stub shaft with the splines in the hub and push the shaft into the hub as far as possible.

10. Temporarily fasten the rotor to the hub with washers and 2 lug nuts. Insert a suitable steel rod into the rotor and rotate clockwise to contact the knuckle to prevent the rotor from turning during CV-joint installation.

11. Install the hub nut washer and a new hub retainer nut. Manually thread the nut onto the halfshaft as far as possible, then tighten to 170–210 ft. lbs. (230–283 Nm).

12. Install the inboard flange of the halfshaft to the differential output flange, aligning the marks that were made during removal. Install the flange bolts and tighten to 22–29 ft. lbs. (30–40 Nm).

13. Install the wheel and tire assembly and lower the vehicle.

CV-Joint Boot

REPLACEMENT

1. Remove the halfshaft assembly from the vehicle.

2. Clamp the halfshaft in a vise equipped with jaw caps to prevent damage to machined surfaces. Do not allow the vise jaws to contact the boot or its clamp.

3. Remove the slip yoke boot clamps and separate the outboard shaft and joint assembly from the inboard slip yoke assembly. Remove and discard the slip yoke boot.

4. Cut the large boot clamp using suitable cutters and peel away from the boot. After removing the clamp, roll the boot back over the shaft and remove and discard the boot.

5. Clean all parts in suitable parts cleaning solvent.

NOTE: Do not submerge the slip yoke assembly U-joint in cleaning solvent.

6. Inspect the CV-joint and slip yoke assembly for excessive wear, pitting, rust and broken parts. If the CV-joint is no longer usable, the complete outer shaft assembly must be replaced.

To install:

7. Fill the CV-joint area around the balls with 2.8 oz. of suitable CV-joint grease. Then spread 1.4 oz. of grease evenly inside the large boot for a total combined fill of 4.2 oz.

8. Assemble the large outboard boot onto the outboard shaft and joint assembly, making sure the boot is seated in the boot grooves. Tighten the clamps using suitable crimping pliers, but do not overtighten, as it may damage the clamp and/or boot.

9. Assemble the small boot and clamps onto the outboard shaft and joint assembly, but do not crimp the clamps at this time.

10. Coat the spline end of the outboard shaft and joint assembly with lubricant and assemble into the inner slip yoke assembly.

11. Slip the boot into place, making sure the boot is seated in the boot grooves. Tighten the clamps using a suitable tool, but do not overtighten, as it may damage the clamp and/or boot.

12. Install the halfshaft in the vehicle.

Automatic Locking Hubs

REMOVAL AND INSTALLATION

1. Raise and support the vehicle safely. Remove the wheel lug nuts and remove the wheel and tire assembly.

2. Remove the retainer washers from the lug nut studs and remove the automatic locking hub assembly from the spindle.

3. Remove the snapring from the end of the spindle shaft.

4. On Dana model 28 axles, remove the axle shaft spacer, needle thrust bearing and the bearing spacer. On Dana model 35 axles, remove the axle shaft spacer.

5. Being careful not to damage the plastic moving cam, pull the cam assembly off the wheel bearing adjusting nut. On Dana model 28 axles, remove the thrust washer and needle thrust bearing from the adjusting nut. On Dana model 35 axles, remove the 2 plastic thrust spacers from the adjusting nut.

6. Using a magnet, remove the locking key. It may be necessary to rotate the adjusting nut slightly to relieve the pressure against the locking key, before the key can be removed.

NOTE: To prevent damage to the spindle threads, look into the spindle keyway under the adjusting nut and remove the separate locking key before removing the adjusting nut.

To install:

7. Loosen the wheel bearing adjusting nut from the spindle using a 2⅜ in. hex socket tool.

8. While rotating the hub and rotor assembly, tighten the wheel bearing adjusting nut to 35 ft. lbs. (47 Nm) to seat the bearings. Spin the rotor and back off the nut ¼ turn.

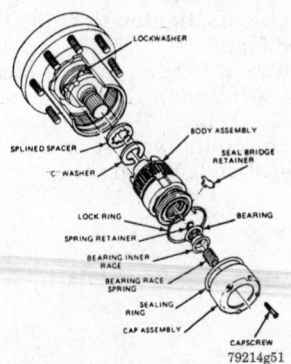

LOCKWASHER
BODY ASSEMBLY
SPLINED SPACER
SEAL BRIDGE RETAINER
"C" WASHER
LOCK RING
BEARING
SPRING RETAINER
BEARING INNER RACE
BEARING RACE SPRING
SEALING RING
CAP ASSEMBLY
CAPSCREW
79214g51

Automatic locking hub assembly

9. Retighten the adjusting nut to 16 inch lbs. (2 Nm) using a torque wrench. Align the closest hole in the wheel bearing adjusting nut with the center of the spindle keyway slot. Advance the nut to the next lug if required. Install the separate locking key in the spindle keyway under the adjusting nut.

NOTE: Extreme care must be taken when aligning the spindle nut adjustment lug with the center of the spindle keyway slot to prevent damage to the separate locking key.

10. On Dana model 28 axles, install the locknut needle bearing and thrust washer in the reverse order of removal. On Dana model 35 axles, install the 2 thrust spacers. Push or press the cam assembly onto the locknut by lining up the key in the fixed cam with the spindle keyway.

NOTE: Extreme care must be taken when aligning the fixed cam key with the spindle keyway to prevent damage to the fixed cam.

11. On Dana model 28 axles, install the bearing thrust washer, needle thrust bearing and axle shaft spacer. On Dana model 35 axles, install the axle shaft spacer.

12. Clip the snapring onto the end of the spindle.

13. Install the automatic locking hub assembly over the spindle by lining up the 3 legs in the hub assembly with the 3 pockets in the cam assembly. Install the retainer washers.

14. Install the wheel and tire assembly. Check the end-play of the wheel and tire assembly on the spindle. Final end-play should be 0–0.003 in. (0–0.08mm). The maximum torque to rotate the hub should be 25 inch lbs. (3 Nm).

15. Lower the vehicle.

Transfer Case Assembly

REMOVAL AND INSTALLATION

———— CAUTION ————
The catalytic converter is located beside the transfer case. Due to the extreme high temperatures generated by the converter, be careful when removing the transfer case or personal injury may result.

Mechanical Shift Type

1. Disconnect the negative battery cable. Raise and safely support the vehicle.

2. If equipped, remove the skid plate from the frame. Remove the damper from the transfer case, if equipped.

3. Place a drain pan under the transfer case, remove the drain plug and drain the fluid. Disconnect the 4WD indicator switch wire connector at the transfer case.

4. If equipped with Borg Warner model 13-50 transfer case, disconnect the front driveshaft from the axle input yoke and pull the driveshaft and front boot assembly out of the transfer case front output shaft.

5. If equipped with Borg Warner model 13-54 transfer case, disconnect the front driveshaft from the transfer case output shaft yoke and wire the driveshaft out of the way.

6. Disconnect the rear driveshaft from the transfer case output shaft flange and wire the driveshaft out of the way.

7. Disconnect the speedometer driven gear from the transfer case rear cover. Disconnect the vent hose from the control lever.

8. Disconnect the nut from the shift lever and remove the shift lever, if necessary.

9. Remove the large and small bolts retaining the shifter to the extension housing. Remove the lever assembly and bushing.

10. Support the transfer case with a transmission jack. Remove the 5 bolts retaining the transfer case to the transmission and extension housing.

11. Slide the transfer case rearward off the transmission output shaft and lower the transfer case from the vehicle. Remove the gasket from between the transfer case and extension housing.

To install:

12. Install a new gasket on the front mounting face of the transfer case assembly.

13. Raise the transfer case with the transmission jack so the transmission output shaft aligns with the transfer case input shaft. Slide the transfer case forward onto the transmission output shaft and onto the dowel pin. Install the 5 retaining bolts and tighten, in sequence, to 25–35 ft. lbs. (34–48 Nm).

14. Remove the transmission jack.

15. Install and adjust the shifter. Always tighten the large bolt retaining the shifter to the extension housing before tightening the small bolt.

16. Install the vent assembly so the white marking on the hose is in position in the notch in the shifter. The upper end of the vent hose should be ³/₄ in. (19mm) above the top of the shifter and positioned just below the floor pan.

17. Connect the speedometer driven gear to the transfer case rear cover. Tighten the screw to 20–25 inch lbs. (2–3 Nm).

18. Connect the rear driveshaft to the transfer case output shaft flange. Tighten the bolts to 61–87 ft. lbs. (83–118 Nm).

19. If equipped with Borg Warner model 13-50 transfer case, clean the transfer case front output shaft female splines. Apply suitable lubricant to the splines and insert the front driveshaft male spline. Connect the front driveshaft to the axle input yoke and tighten the bolts to 12–16 ft. lbs. (16–22 Nm). Push the driveshaft boot to engage the external groove on the transfer case front output shaft.

20. If equipped with Borg Warner model 13-54 transfer case, connect the front driveshaft to the transfer case output shaft yoke. Tighten the bolts to 12–16 ft. lbs. (16–22 Nm).

21. Connect the 4WD indicator switch wire harness at the transfer case.

22. Install the drain plug and tighten to 14–22 ft. lbs. (19–30 Nm). Remove the fill plug and fill the transfer case with the proper type of fluid to the bottom of the fill hole. Install the fill plug and tighten to 14–22 ft. lbs. (19–30 Nm).

23. Install the damper to the transfer case, if equipped. Using new damper bolts, tighten to 25–35 ft. lbs. (34–48 Nm).

24. Install the skid plate, if equipped. Tighten the nuts and bolts to 15–20 ft. lbs. (20–27 Nm).

25. Lower the vehicle and connect the negative battery cable.

Electronic Shift Type

1. Disconnect the negative battery cable. Raise and safely support the vehicle.

2. If equipped, remove the nuts, bolts and skid plate from the frame. Remove the damper from the transfer case, if equipped.

3. Place a drain pan under the transfer case, remove the drain plug and drain the fluid.

4. Remove the wire connector from the feed wire harness at the rear of the transfer case. First squeeze the locking tabs, then pull the connectors apart.

NOTE: Do not pull directly on the wires or pull outwardly on the locking tabs.

5. Remove the connector for the transfer case motor from the mounting bracket, if necessary.

6. If equipped with Borg Warner model 13-50 transfer case, disconnect the front driveshaft from the axle input yoke and pull the driveshaft and front boot assembly out of the transfer case front output shaft.

7. If equipped with Borg Warner model 13-54 transfer case, disconnect the front driveshaft from the transfer case output shaft yoke and wire the driveshaft out of the way.

8. Disconnect the rear driveshaft from the transfer case output shaft flange and wire the driveshaft out of the way.

9. Disconnect the speedometer driven gear from the transfer case rear cover. Disconnect the vent hose from the mounting bracket.

10. Support the transfer case with a transmission jack. Remove the 5 bolts retaining the transfer case to the transmission and extension housing.

11. Slide the transfer case rearward off the transmission output shaft and lower the transfer case from the vehicle. Remove the gasket from between the transfer case and extension housing.

To install:

12. Install a new gasket on the front mounting face of the transfer case assembly.

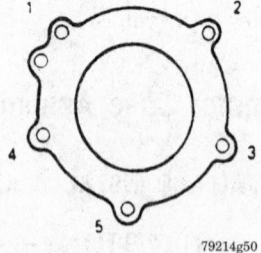

79214g50

Transfer case-to-extension housing bolt torque sequence

13. Raise the transfer case with the transmission jack so the transmission output shaft aligns with the transfer case input shaft. Slide the transfer case forward onto the transmission output shaft and onto the dowel pin. Install the 5 retaining bolts and tighten, in sequence, to 25–35 ft. lbs. (34–48 Nm).

14. Remove the transmission jack.

15. Install the vent hose so the white marking on the hose aligns with the notch in the mounting bracket.

16. Connect the speedometer driven gear to the transfer case rear cover. Tighten the screw to 20–25 inch lbs. (2–3 Nm).

17. Connect the rear driveshaft to the transfer case output shaft flange. Tighten the bolts to 61–87 ft. lbs. (83–118 Nm).

18. If equipped with Borg Warner model 13-50 transfer case, clean the transfer case front output shaft female splines. Apply suitable lubricant to the splines and insert the front driveshaft male spline. Connect the front driveshaft to the axle input yoke and tighten the bolts to 12–16 ft. lbs. (16–22 Nm). Push the driveshaft boot to engage the external groove on the transfer case front output shaft.

19. If equipped with Borg Warner model 13-54 transfer case, connect the front driveshaft to the transfer case output shaft yoke. Tighten the bolts to 12–16 ft. lbs. (16–22 Nm).

20. Attach the connector for the transfer case motor to the mounting bracket, if necessary.

21. Connect the wire connectors on the rear of the transfer case, making sure the retaining tabs lock.

22. Install the drain plug and tighten to 14–22 ft. lbs. (19–30 Nm). Remove the fill plug and fill the transfer case with the proper type of fluid to the bottom of the fill hole. Install the fill plug and tighten to 14–22 ft. lbs. (19–30 Nm).

23. Install the damper to the transfer case, if equipped. Using new damper bolts, tighten to 25–35 ft. lbs. (34–48 Nm).

24. Install the skid plate, if equipped. Tighten the nuts and bolts to 15–20 ft. lbs. (20–27 Nm).

25. Lower the vehicle and connect the negative battery cable.

STEERING

Air Bag

CAUTION

Some vehicles are equipped with an air bag system, also known as the Supplemental Inflatable Restraint (SIR) or Supplemental Restraint System (SRS). The system must be disabled before performing service on or around system components, steering column, instrument panel components, wiring and sensors. Failure to follow safety and disabling procedures could result in accidental air bag deployment, possible personal injury and unnecessary system repairs.

PRECAUTIONS

Several precautions must be observed when handling the inflator module to avoid accidental deployment and possible personal injury.

• Never carry the inflator module by the wires or connector on the underside of the module.

• When carrying a live inflator module, hold securely with both hands, and ensure that the bag and trim cover are pointed away.

• Place the inflator module on a bench or other surface with the bag and trim cover facing up.

• With the inflator module on the bench, never place anything on or close to the module which may be thrown in the event of an accidental deployment.

DISARMING

Driver's Side

CAUTION

The Air Bag system must be disarmed before performing service procedures around the Air Bag or wiring. Failure to do so may cause accidental deployment, resulting in unnecessary repairs and/or personal injury.

1. Disconnect the negative battery cable.
2. Disconnect the positive battery cable.
3. Wait one minute. This is the time required for the backup power supply in the Air Bag diagnostic monitor to deplete the stored energy.

4. Remove the two black cover plugs from the back of the steering wheel and remove the screws and washers securing the module to the steering wheel.
5. Disconnect the Air Bag connector. Remove the module and place aside with the trim facing up.
6. Connect Rotunda Air Bag Simulator 105-R0010 or 105-R0011 or equivalent to the vehicle harness at the top of the steering wheel.

Passenger's Side

1. Disconnect the negative battery cable.
2. Disconnect the positive battery cable.
3. Wait one minute. This is the time required for the backup power supply in the Air Bag diagnostic monitor to deplete the stored energy.
4. Open the glove compartment and push in the side tabs on the compartment to allow it to drop down.
5. Remove the two screws that attach the A/C duct. Remove the duct.
6. Remove the two Air Bag attaching bolts that secure the module in place.
7. Unfasten the harness on the left rear corner of the module.
8. Remove the Air Bag connector from the steel reinforcement by prying the tree out of the hole in the steel reinforcement.
9. Remove the Air Bag and place aside with the cover facing up.
10. Connect Rotunda Air Bag Simulator 105-R000-12 or equivalent to the harness.

ARMING

1. Disconnect the battery cable, id connected. Unfasten the negative cable first, followed by the positive cable. Wait at least one minute.
2. Remove the Air Bag simulator.
3. Connect the Air Bag module. Install the module and components.
4. Connect the positive battery cable.
5. Connect the negative battery cable.
6. Turn ignition switch to **RUN** and verify the air bag indicator is functioning properly.

Steering Wheel

REMOVAL AND INSTALLATION

Without Air Bag

1. Place the steering wheel with the wheel in the straight ahead position.
2. Disconnect the negative battery cable.
3. Remove the retainer nuts from the rear of the steering wheel. Lift off the center pad and place aside.
4. Disconnect the horn and speed control harness, if equipped.
5. Scribe an alignment mark on the steering wheel and column shaft.
6. Remove the steering wheel retainer bolt.
7. Install steering wheel puller T67L-3600-A or equivalent and remove the steering wheel. Route the harness assembly through the steering wheel as the steering wheel is lifted off the shaft.

To install:

NOTE: Make sure the wheels are in a straight ahead position.

8. Route the the harness assembly through the steering wheel opening at the 3 O'clock position, then install the steering wheel on the column shaft. Make sure the shaft and steering wheel alignment marks align correctly.
9. Install a new steering wheel bolt and tighten to 23–33 ft. lbs. (31–45 Nm).
10. Connect the horn and speed control harness, if equipped.
11. Position the center pad and secure with the retainer nuts. Tighten the nuts to 3–4 ft. lbs. (4–6 Nm).
12. Connect the negative battery cable.

With Air Bag

1. Place the steering wheel with the wheel in the straight ahead position.
2. Disconnect the negative, then positive battery cable.
3. Wait one minute, then disarm the air bag module.
4. Disconnect the horn and speed control wire harness, if equipped.
5. Scribe an alignment mark on the steering wheel and column shaft.
6. Remove the steering wheel retainer bolt.
7. Install steering wheel puller T67L-3600-A or equivalent and remove the steering wheel. Route the contact assembly through the steering wheel as the steering wheel is lifted off the shaft.

To install:

NOTE: Make sure the wheels are in a straight ahead position.

8. Route the the contact assembly through the steering wheel opening at the 3 O'clock position, then install the steering wheel on the column shaft. Make sure the shaft and steering wheel alignment marks align correctly.

9. Install a new steering wheel bolt and tighten to 23–33 ft. lbs. (31–45 Nm).

10. Connect the horn and speed control harness, if equipped.

11. Connect the Air Bag module and secure to the steering wheel.

12. Connect the battery cables, positive first. Verify that the Air Bag system functions correctly.

Tie Rod Ends

REMOVAL & INSTALLATION

1. Place the front wheels in the straight ahead position. Raise and safely support the vehicle.

2. Remove the cotter pin and nut from the tie rod end ball stud. Discard the cotter pin.

3. Separate the tie rod end from the spindle or drag link using puller tool T64P-3590-F or equivalent.

4. On Aerostar models, hold the tie rod with a wrench and loosen the tie rod jam nut. Mark the position of the tie rod end on the tie rod threads, grip the tie rod with suitable pliers and remove the tie rod end from the tie rod.

5. sport utility and pick-up models, loosen the bolts on the tie rod adjusting sleeve. Count the number of turns required to remove the tie rod from the tie rod adjusting sleeve and remove the tie rod.

To install:

6. On Aerostar models, thread the replacement tie rod end onto the tie rod to the same location as the 1 that was removed. Hold the tie rod end with a wrench and tighten the jam nut to 35–50 ft. lbs. (48–68 Nm).

7. On sport utility and pick-up models, install the tie rod into the adjusting sleeve the same number of turns required to remove it. With the adjusting sleeve clamps pointed down, tighten the adjusting sleeve nuts to 30–42 ft. lbs. (40–57 Nm).

8. Install the tie rod ball stud into the spindle or drag link. Install the nut and tighten to 50–75 ft. lbs.

(70–100 Nm). Install a new cotter pin.

NOTE: If the cotter pin cannot be installed because the hole in the ball stud does not align with a castellation on the nut, continue to tighten the nut to align them. Never loosen the nut to align the hole and castellation.

9. Lower the vehicle. Check the toe-in setting and adjust, if necessary.

Power Rack and Pinion

REMOVAL AND INSTALLATION

Except 4WD Aerostar

1. Place the steering system in the on center position as follows:

a. Start the engine.

b. Rotate the steering wheel from lock-to-lock and record the number of steering wheel rotations.

c. Divide the number of steering wheel rotations by 2 to give the required number of turns to place the system in the on center position.

d. From the lock position, rotate the steering wheel the number of turns determined in Step c to place the gear in the on center position. Make sure the wheels are in the straight ahead position.

e. Stop the engine.

2. Disconnect the negative battery cable and turn the ignition key to the **ON** position. Raise and safely support the vehicle.

3. Disconnect the pressure and return lines from the steering gear valve housing. Plug the lines and ports in the gear valve housing to prevent the entry of dirt.

4. Remove the bolt retaining the lower intermediate steering column shaft to the steering gear. Disconnect the shaft from the gear.

5. Remove the cotter pins and nuts from the tie rod ends. Separate the tie rod ends from the spindle arms.

6. Remove the stabilizer bar, if equipped for added clearance to remove the assembly.

7. Remove the nuts for the power steering cooler and remove the cooler assembly, if equipped.

8. Support the steering gear and remove the 2 nuts, bolts and washers retaining the gear to the crossmember. Remove the gear from the vehicle.

To install:

9. Position the steering gear on the crossmember. Install the retaining nuts, bolts and washers and tighten to 94–127 ft. lbs. (128–172 Nm).

10. Install the power steering cooler, and tighten the nuts to 50–68 ft. lbs. (68–92 Nm).

11. Connect the pressure and return lines to the gear valve housing ports. Tighten the fittings to 20–25 ft. lbs. (27–34 Nm).

NOTE: The fitting design allows the hoses to swivel when properly tightened. Do not attempt to eliminate looseness by overtightening as this can damage the fittings.

12. With the steering gear, steering wheel and front wheels in the on center position, attach the tie rod ends to the spindle arms.

13. Connect the steering column lower intermediate shaft to the gear. Install the bolt and tighten to 30–42 ft. lbs. (41–56 Nm).

14. Install the stabilizer bar, if equipped.

15. Lower the vehicle and turn the ignition key to the **OFF** position. Connect the negative battery cable.

16. Fill and bleed the steering system. Check the toe setting and adjust, if necessary.

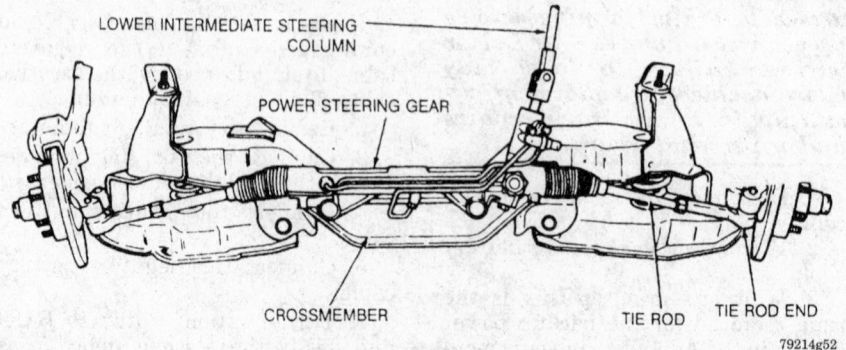

LOWER INTERMEDIATE STEERING COLUMN

POWER STEERING GEAR

CROSSMEMBER

TIE ROD

TIE ROD END

79214g52

Power steering rack assembly

4WD Aerostar

1. Place the steering system in the on center position as follows:

   a. Start the engine.

   b. Rotate the steering wheel from lock-to-lock and record the number of steering wheel rotations.

   c. Divide the number of steering wheel rotations by 2 to give the required number of turns to place the system in the on center position.

   d. From the lock position, rotate the steering wheel the number of turns determined in Step c to place the gear in the on center position. Make sure the wheels are in the straight ahead position.

   e. Stop the engine.

2. Disconnect the negative battery cable and turn the ignition key to the **ON** position. Raise and safely support the vehicle and remove the wheel and tire assemblies. Put transmission in **PARK** position.

3. Raise the vehicle and support on jackstands. Remove the wheels.

4. Disconnect the pressure and return lines from the steering gear valve housing. Plug the lines and ports in the gear valve housing to prevent the entry of dirt.

5. Remove the bolt retaining the lower intermediate steering column shaft to the steering gear. Disconnect the shaft from the gear. Retain any dust seals which may be present.

6. Remove the cotter pins and nuts from the tie rod ends. Separate the tie rod ends from the spindle arms.

7. Remove the top nut and washer with the insulator retaining the shock absorbers to the front crossmember brackets. Remove the bolts retaining the shock absorbers to the bottom of the lower control arms and remove the shock absorbers.

NOTE: Be careful during shock absorber removal as the shock absorber may extend somewhat due to gas pressure.

8. Remove the front stabilizer bar.

9. Install a suitable spring compressor through each coil spring to secure the spring and prevent extension when the control arm pivot bolts are removed.

10. Remove the lower control arm pivot bolts. Reposition and secure the lower control arms away from the brackets. Be careful not to deform the control arm pivot brackets.

11. Remove the 5 nuts from the forward edge of the crossmember lower plate assembly. Remove the nut from the left rear and the bolts from the center and right rear positions at the rear of the assembly. Remove the crossmember lower plate.

12. Support the steering gear and remove the 2 bolts, spacers and bushings retaining the steering gear. Remove the gear from the vehicle.

To install:

13. Position the steering gear on the crossmember with the attaching bolts and tighten to 61–82 ft. lbs. (83–111 Nm). The large end of the spacer is installed facing the steering rack.

14. Install the crossmember lower plate with the nuts and bolt in the proper locations.

15. Position the lower control arms within the brackets and install the pivot bolts. Properly position the retaining flags.

16. Tighten all lower control arm pivot bolt nuts to 100–140 ft. lbs. (136–190 Nm).

17. Install the coil springs and remove the spring compressors.

18. Install the shock absorbers through the lower control arms and install the insulators, nuts and washers. Tighten the nuts to 2-\35 ft. lbs. (34–47 Nm). Install the lower shock absorber retaining bolts and tighten to 16–25 ft. lbs. (22–33 Nm).

19. Install the front stabilizer bar.

20. Connect the pressure and return lines to the gear valve housing ports. Tighten the fittings to 10–15 ft. lbs. (15–20 Nm).

NOTE: The fitting design allows the hoses to swivel when properly tightened. Do not attempt to eliminate looseness by overtightening as this can damage the fittings.

21. With the steering gear, steering wheel and front wheels in the on center position, attach the tie rod ends to the spindle arms. Install the nuts and tighten to 52–73 ft. lbs. (70–100 Nm). If required, advance the nuts to the next castellation and install new cotter pins.

22. Position the intermediate shaft and any dust seals over the steering rack input shaft spline and make sure no rotation from the on center position has occurred. Install the bolt and tighten to 30–42 ft. lbs. (41–56 Nm).

23. Install the wheel and tire assemblies. Lower the vehicle and turn the ignition key to the **OFF** position. Connect the negative battery cable.

24. Fill and bleed the steering system. Check the toe setting and adjust, if necessary.

Power Steering Pump

SYSTEM BLEEDING

1. Fill the power steering fluid reservoir.

2. Disconnect the coil(s) wire.

3. Crank the engine with the starter and continue adding fluid until the level remains constant. Do not prolong cranking as the battery may be drained and the starter damaged.

4. Rotate the steering wheel approximately 30 degrees each side of center while continuing to crank the engine.

5. Recheck the fluid level and fill, as required.

6. Reconnect the coil wire.

7. Start the engine and allow it to run for several minutes.

8. Rotate the steering wheel from stop to stop.

9. Shut off the engine and recheck the fluid level. Add fluid, as required.

10. If air is still trapped in the system, proceed as follows:

    a. Fabricate a purging tool.

    b. Make sure the reservoir fluid level is correct.

    c. Insert the rubber stopper end of the fabricated purging tool tightly into the filler tube.

    d. Connect a suitable length of hose to the purging tool. Connect the other end of the hose to an air conditioner vacuum pump or distributor machine. Do not use engine vacuum.

    e. Start the engine and let it idle for approximately 15 minutes. Turn the steering wheel 1 full cycle every 5 minutes but do not hit the stops. This will assist in removing trapped air.

    f. Stop the engine and disconnect the vacuum source. Remove the purging tool.

    g. Check the fluid level and install the filler tube dipstick.

REMOVAL AND INSTALLATION

1. Remove the power steering fluid from the pump reservoir by disconnecting the fluid return hose at the reservoir and draining the fluid into a container.

2. Remove the pressure hose from the pump. If equipped, disconnect the power steering pump pressure switch.

3. If equipped with a 1993–94 2.3L or 3.0L engine, loosen the alternator or idler pulley assembly pivot and adjustment bolts to slacken belt tension. Remove the drive belt.

4. If equipped with a 1995–97 3.0L or 4.0L engine, slacken the belt tension by lifting the tensioner pulley in a counterclockwise direction. Remove the drive belt from under the tensioner pulley and slowly lower the pulley to stop. Remove the drive belt.

5. Remove the engine oil dipstick tube, if necessary. Remove the power steering pump bracket support brace, if equipped.

6. Install steering pump pulley removal tool T69L-10300-B on the pulley. Hold the pump and rotate the tool nut counterclockwise to remove the pulley. Do not apply in and out pressure on the pump shaft as pressure will damage the internal thrust areas.

7. Remove the bolts attaching the pump to the bracket and remove the pump.

To install:

8. Install the pump on the bracket. Install and tighten the attaching bolts to 30–45 ft. lbs. (41–61 Nm). If equipped with 4.0L engine, tighten to 35–47 ft. lbs. (47–64 Nm), position the support on the bracket and install and tighten the mounting bolts to 35–47 ft. lbs. (47–64 Nm).

9. Install steering pump pulley replacement tool T65P-3A733-C and install the pulley. Remove the tool.

NOTE: Fore and aft location of the pulley on the pump shaft is critical for correct belt alignment. Make sure the pull-off groove on the pulley is facing front and flush with the end of the shaft ± 0.010 in. (0.254mm).

10. Install the drive belt.

11. Install the power steering pump bracket support brace and/or engine dipstick tube, if removed.

12. Install the pressure hose to the pump fitting. Connect the return hose to the pump and tighten the clamp. If equipped, connect the power steering pump pressure switch.

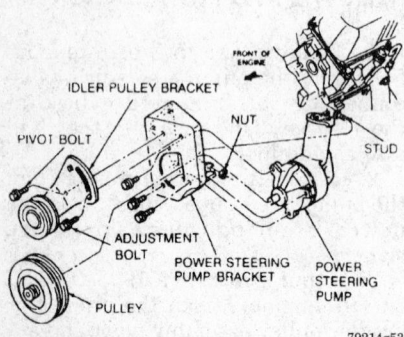

Power steering pump assembly — 3.0L engine

13. Fill and bleed the power steering system. Check for leaks.

BRAKES

Anti-Lock Brake System Service

PRECAUTIONS

• Certain components within the Anti-Lock Brake System (ABS) are not intended to be serviced or repaired individually. Only those components with removal and installation procedures should be serviced.

• Do not use rubber hoses or other parts not specifically specified for and ABS system. When using repair kits, replace all parts included in the kit. Partial or incorrect repair may lead to functional problems and require the replacement of components.

• Lubricate rubber parts with clean, fresh brake fluid to ease assembly. Do not use lubricated shop air to clean parts; damage to rubber components may result.

• Use only specified brake fluid from an unopened container.

• If any hydraulic component or line is removed or replaced, it may be necessary to bleed the entire system.

• A clean repair area is essential. Always clean the reservoir and cap thoroughly before removing the cap. The slightest amount of dirt in the fluid may plug an orifice and impair the system function. Perform repairs after components have been thoroughly cleaned; use only denatured alcohol to clean components. Do not allow ABS components to come into contact with any substance containing mineral oil; this includes used shop rags.

• The Anti-Lock control unit is a microprocessor similar to other computer units in the vehicle. Ensure that the ignition switch is **OFF** before removing or installing controller harnesses. Avoid static electricity discharge at or near the controller.

• If any arc welding is to be done on the vehicle, the control unit should be unplugged before welding operations begin.

Master Cylinder

REMOVAL AND INSTALLATION

1. Disconnect the negative battery cable. If equipped, push the brake pedal down to expel vacuum from the brake booster system.

2. Disconnect the fluid level indicator switch connector from the master cylinder.

3. If equipped with non-power brakes, disconnect the wires from the stop light switch inside the cab below the instrument panel. Remove the lock pin and spacers securing the master cylinder pushrod to the brake pedal assembly. Remove the stop light switch from the pedal.

4. Disconnect and plug the hydraulic lines at the master cylinder. Plug the master cylinder ports.

5. Remove the master cylinder-to-booster or master cylinder-to-dash panel retaining nuts and remove the master cylinder.

NOTE: On Aerostar models, use care when removing the cartridge master cylinder so as not to scratch the exposed primary piston.

To install:

6. If equipped with power brakes, before installing the master cylinder on all except Aerostar, check the distance from the outer end of the vacuum booster assembly pushrod to the front face of the vacuum brake booster assembly. The distance should be 0.980–0.995 in. (24.89–25.27mm). Turn the pushrod adjusting screw in or out, as required, to obtain the proper length.

7. On Aerostar, make sure the interface seal is located in the annular groove in the cartridge master cylinder body where the cartridge master cylinder seals with the booster. Do not assemble the cartridge master cylinder to the booster without this square section interface seal.

8. If equipped with non-power brakes and the dash spacer was removed, coat the spacer with sealer and install on the dash panel.

9. Install the master cylinder with the attaching nuts. On Aerostar, be careful not to scratch the exposed primary piston and make sure the primary piston socket engages the booster pushrod before tightening. Tighten the nuts to 20 ft. lbs. (27 Nm). Connect the fluid level indicator switch.

10. If equipped with non-power brakes, secure the pushrod to the brake pedal assembly with the pin or

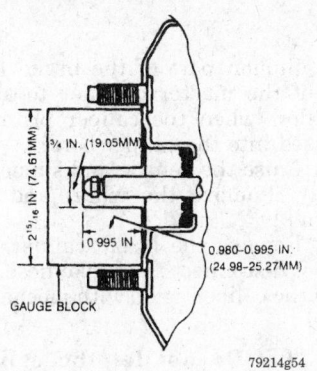

Power brake booster dimensions

shoulder bolt. Make sure the bushings and spacers are installed properly. Install the lockpin and connect the wires to the stop light switch.

11. Connect the brake lines to the master cylinder and fill the master cylinder fluid reservoir. Wrap a shop cloth around the tubing below the fitting to be bled to absorb escaping brake fluid.

12. Have an assistant push the brake pedal to the floor. Crack open the brake line fitting to expel air trapped in the master cylinder. Tighten the fitting, then let the brake pedal return. Repeat this procedure until all air is expelled.

13. Repeat Step 12 on the remaining brake line fitting(s).

14. Final tighten the brake line fittings. Bleed the brake system, as required. Check the master cylinder fluid level.

Power Brake Booster

REMOVAL AND INSTALLATION

1. Disconnect the negative battery cable. Support the master cylinder from the underside with a prop.

2. Disconnect the vacuum hose from the booster check valve and remove the check valve.

3. Remove the master cylinder-to-booster retaining nuts. Pull the master cylinder off the booster and leave it supported by the prop, out of the way enough to allow booster removal.

4. Working inside the cab, remove the hairpin retainer and slide the stoplight switch, valve rod, spacers and bushing off the brake pedal arm. Remove the nuts retaining the booster and remove the booster.

To install:

5. Mount the booster assembly on the engine side of the dash panel by sliding the bracket mounting bolts

and valve operating rod in through the holes in the dash panel.

6. Working inside the cab, install the booster mounting nuts and tighten to 13–25 ft. lbs. (18–33 Nm).

7. Before installing the master cylinder on all except Aerostar models, check the distance from the outer end of the vacuum booster assembly pushrod to the front face of the vacuum brake booster assembly. The distance should be 0.980–0.995 in. (24.89–25.27mm). Turn the pushrod adjusting screw in or out, as required, to obtain the proper length.

8. Install the master cylinder and tighten the retaining nuts to 20 ft. lbs. (27 Nm). Remove the prop from under the master cylinder.

9. Install the booster check valve and connect the vacuum hose. Check the hose routing to make sure the hose is not crimped.

10. Working inside the cab, install the bushing and position the switch on the end of the valve rod, then install the switch and rod on the pedal arm along with the spacers and hairpin retainer.

NOTE: Use only the factory supplied hairpin retainer. Do not substitute other types of retainers.

11. Connect the negative battery cable, start the engine and check brake operation.

Brake Caliper

REMOVAL AND INSTALLATION

Front

1. Siphon part of the brake fluid out of the master cylinder to avoid overflow when the caliper piston is pressed into the caliper bore.

2. Raise the vehicle and support it safely. Remove the wheel and tire assembly.

3. Position an 8 in. C-clamp on the caliper and tighten the clamp to move the caliper piston into the bore approximately ⅛ in. (3mm). Avoid clamp contact with the outer shoe spring clip. Remove the clamp.

NOTE: Do not pry the piston away from the rotor.

4. Clean excess dirt from the pin tab area.

5. Using a ¼ in. drive socket and a light hammer, tap the upper caliper pin towards the outboard side until the pin tabs pass the spindle face.

6. Compress the inboard pin tab, if equipped, with pliers and, with a hammer, drive the pin out until the tab slips into the spindle groove.

7. Place an end of a ⁷⁄₁₆ in. (11mm) diameter punch against the end of the caliper pin and tap the pin out of the caliper slide groove.

8. Repeat Steps 5, 6 and 7 to remove the lower pin.

9. Disconnect and plug the brake hose at the caliper. Remove the caliper from the rotor.

To install:

10. Make sure the caliper mounting surfaces are free of dirt. Lubricate the caliper grooves with disc brake caliper grease and install the caliper.

11. From the caliper outboard side, position the pin between the caliper and spindle grooves. The pin must be positioned so the tabs will be installed against the spindle outer face.

12. Tap the pin on the outboard end with a hammer until the retention tabs on the sides of the pin contact the spindle face.

13. Repeat Steps 11 and 12 for the lower pin.

NOTE: During installation, do not allow the tabs of the caliper pin to be tapped too far into the spindle groove. If this happens, it will be necessary to tap the other end of the caliper pin until the tabs snap in place. The tabs on each end of the pin must be free to catch on the spindle face.

14. Connect the brake hose to the caliper. Bleed the brake system.

15. Install the wheel and tire assembly and lower the vehicle. Check the brake fluid level and check the brakes for proper operation.

Rear

1. Siphon part of the brake fluid out of the master cylinder to avoid overflow when the caliper piston is pressed into the caliper bore.

2. Raise the vehicle and support it safely. Remove the wheel and tire assembly.

3. Position an 8 in. C-clamp on the caliper and tighten the clamp to move the caliper piston into the bore approximately ⅛ in. (3mm). Remove the clamp.

NOTE: Do not pry the piston away from the rotor.

4. Clean excess dirt from the retainer bolt area.

5. Using a Torx® socket, remove the two retainer bolts securing the caliper to the bracket and adaptor plate.

6. Disconnect and plug the brake hose at the caliper. Remove the caliper from the rotor.

To install:

7. Make sure the caliper mounting surfaces are free of dirt. Lubricate the caliper grooves with disc brake caliper grease and install the caliper.

8. Position the caliper to the bracket and secure in place with the retainer bolts. Tighten the bolts to 20 ft. lbs. (27 Nm)

9. Install the caliper brake hose using new washers. Tighten the bolt to 29 ft. lbs. (40 Nm)

10. Fill and bleed the brake system.

11. Install the wheel and tire assembly and lower the vehicle. Check the brake fluid level and check the brakes for proper operation.

Disc Brake Pads

REMOVAL AND INSTALLATION

Front

1. Siphon part of the brake fluid out of the master cylinder to avoid overflow when the caliper piston is pressed into the caliper bore.

2. Raise the vehicle and support it safely. Remove the wheel and tire assembly.

3. Remove the brake caliper, but do not disconnect the brake hose. Secure the caliper aside with mechanics wire.

NOTE: Do not let the caliper hang by the brake hose.

4. Compress the anti-rattle clip and remove the inner brake pad from the caliper.

5. Press each ear of the outer brake pad away from the caliper and slide the torque buttons out of the retention notches.

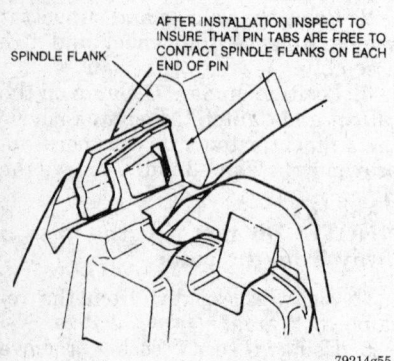

AFTER INSTALLATION INSPECT TO INSURE THAT PIN TABS ARE FREE TO CONTACT SPINDLE FLANKS ON EACH END OF PIN

SPINDLE FLANK

79214g55

Brake caliper pin position

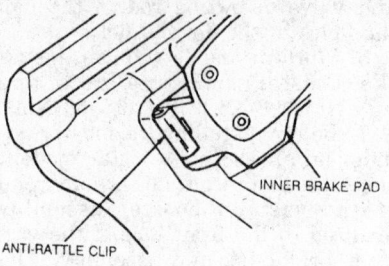

INNER BRAKE PAD

ANTI-RATTLE CLIP

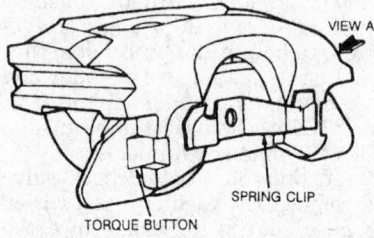

VIEW A

SPRING CLIP

TORQUE BUTTON

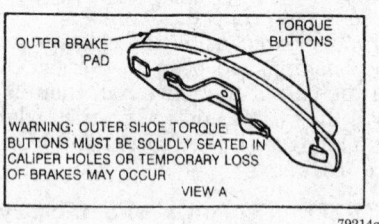

OUTER BRAKE PAD

TORQUE BUTTONS

WARNING: OUTER SHOE TORQUE BUTTONS MUST BE SOLIDLY SEATED IN CALIPER HOLES OR TEMPORARY LOSS OF BRAKES MAY OCCUR

VIEW A

79214g56

Correct brake pad installation

To install:

6. Bottom out the caliper piston in the caliper bore using an 8 in. C-clamp or equivalent and a worn out inner brake pad or block of wood to push against the piston. Do not attempt to bottom out the piston with the outer brake pad installed.

7. Place a new anti-rattle clip on the lower end of the inner brake pad. Make sure the tabs on the clip are properly positioned and the clip is fully seated.

8. Position the inner brake pad and anti-rattle clip in the pad abutment with the ant-rattle clip tab against the pad abutment and the loop-type spring away from the rotor. Compress the anti-rattle clip and slide the upper end of the pad in position.

9. Install the outer pad, making sure the torque buttons on the pad are seated solidly in the matching holes in the caliper.

10. Install the caliper on the spindle.

11. Install the wheel and tire assembly and lower the vehicle. Apply the brakes several times before moving the vehicle to seat the pads.

12. Check the brake fluid level. Check the brakes for proper operation.

Rear

1. Siphon part of the brake fluid out of the master cylinder to avoid overflow when the caliper piston is pressed into the caliper bore.

2. Raise the vehicle and support it safely. Remove the wheel and tire assembly.

3. Remove the brake caliper, but do not disconnect the brake hose. Secure the caliper aside with mechanics wire.

NOTE: Do not let the caliper hang by the brake hose.

4. Remove the inner and outer brake pad from the caliper.

To install:

5. Bottom out the caliper piston in the caliper bore using an 8 in. C-clamp or equivalent and a worn out inner brake pad or block of wood to push against the piston. Do not attempt to bottom out the piston with the outer brake pad installed.

6. Position the inboard brake pad in the caliper and press the retainer spring fully into the caliper piston.

7. Start one end of the outboard brake shoe and lining on the caliper and rotate it down until the locating lugs and the retainer spring are fully seated.

8. Install new shoe slippers on the rear wheel disc brake adaptor.

9. Install the caliper on the spindle.

10. Install the wheel and tire assembly and lower the vehicle. Apply the brakes several times before moving the vehicle to seat the pads.

11. Check the brake fluid level. Check the brakes for proper operation.

Brake Rotor

REMOVAL AND INSTALLATION

Front

1. Raise and safely support the vehicle. Remove the wheel and tire assembly.

2. Remove the caliper and support it aside with wire. Do not let the caliper hang by the brake hose.

3. On 2WD vehicles, remove the dust cap, cotter pin, nut, washer and outer bearing and remove the rotor from the spindle.

4. On 4WD Aerostar models, remove the retainers and brake rotor. On all other 4WD vehicles, remove the locking hub, then remove the brake rotor.

5. Inspect the rotor for scoring, wear and runout; machine or replace as necessary.

To install:

6. Install in the rotor on the spindle, and secure in place with the bearing and retainers. Adjust the wheel bearing.

7. Install the locking hub or dust cap.

8. Install the caliper.

9. Install the wheel and tire assembly and lower the vehicle. Apply the brakes several times before moving the vehicle to seat the pads.

10. Check the brake fluid level. Check the brakes for proper operation.

Rear

1. Raise and safely support the vehicle. Remove the wheel and tire assembly.

2. Remove the caliper and support it aside with wire. Do not let the caliper hang by the brake hose.

3. Remove the press-on keeper nuts. Discard the old nuts.

4. Remove the rear disc brake rotor from the axle shaft.

To install:

5. Position the rotor to the axle shaft and secure with new press-on keeper nuts.

6. Install the caliper.

7. Install the wheel and tire assembly and lower the vehicle. Apply the brakes several times before moving the vehicle to seat the pads.

8. Check the brake fluid level. Check the brakes for proper operation.

Brake Drum

REMOVAL AND INSTALLATION

1. Raise and safely support the vehicle. Remove the wheel and tire assembly.

2. Remove the retaining nuts, if equipped, and remove the brake drum.

NOTE: If the brake drum will not come off, insert a narrow prybar through the brake adjusting hole in the backing plate and disengage the adjusting lever from the adjusting screw. While holding the adjusting lever away from the adjusting screw, loosen the adjusting screw with a brake adjusting tool.

3. Inspect the brake drum surface for wear, scoring and runout. Machine or replace, as necessary.

To install:

4. Install the brake drum and secure in place with the retainer nuts, if equipped.

5. Adjust the rear brakes.

6. Install the wheel. Lower the vehicle.

ADJUSTMENT

Drum Installed

1. Raise and safely support the vehicle.

2. Remove the rubber plug from the adjusting slot on the backing plate.

3. Insert a brake adjusting spoon into the slot and engage the lowest possible tooth on the starwheel. Move the end of the brake spoon downward to move the starwheel upward and expand the adjusting screw. Repeat this operation until the brakes lock the wheels.

4. Insert a small prybar or piece of firm wire (coat hanger wire) into the adjusting slot and push the automatic adjusting lever out and free of the starwheel on the adjusting screw and hold it there.

5. Engage the topmost tooth possible on the starwheel with the brake adjusting spoon. Move the end of the adjusting spoon upward to move the

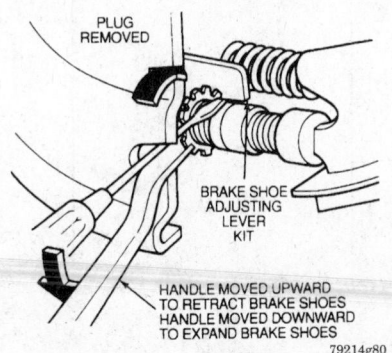

Ranger

TO EXPAND
BRAKE BACKING PLATE
TO CONTRACT
ADJUSTING LEVER MOVED AWAY FROM ADJUSTER
TOOL HANDLE MOVED UPWARD TO EXPAND BRAKES AND DOWN TO CONTRACT

Aerostar

PLUG REMOVED
BRAKE SHOE ADJUSTING LEVER KIT
HANDLE MOVED UPWARD TO RETRACT BRAKE SHOES
HANDLE MOVED DOWNWARD TO EXPAND BRAKE SHOES
79214g80

Using a screwdriver to adjust the drum brakes

adjusting screw starwheel downward and contract the adjusting screw. Back off the adjusting screw starwheel until the wheel spins freely with a minimum of drag. Keep track of the number of turns that the starwheel is backed off, or the number of strokes taken with the brake adjusting spoon.

6. Repeat this operation for the other side. When backing off the brakes on the other side, the starwheel adjuster must be backed off the same number of turns to prevent side-to-side brake pull.

7. When the brakes are adjusted make several stops while backing the vehicle, to equalize the brakes at both of the wheels.

8. Remove the safety stands and lower the vehicle. Road test the vehicle.

Drum Removed

1. Make sure the shoe-to-contact pad areas are clean and properly lubricated.

2. Using an inside caliper check the inside diameter of the drum. Measure across the diameter of the assembled brake shoes, at their widest point.

3. Turn the adjusting screw so the diameter of the shoes is 0.030 in. (0.76mm) less than the brake drum inner diameter.

4. Install the drum.

Brake Shoes

REMOVAL AND INSTALLATION

1. Raise and safely support the vehicle. Remove the wheel and tire assembly and the brake drum.

2. Pull backward on the adjusting lever cable to disengage the adjusting lever from the adjusting screw. Move the outboard side of the adjusting screw upward and back off the pivot nut as far as it will go.

3. Pull the adjusting lever, cable and automatic adjuster spring down and toward the rear to unhook the pivot hook from the large hole in the secondary shoe web. Do not pry the pivot hook from the hole.

4. Remove the automatic adjuster spring and adjusting lever.

5. Remove the secondary shoe-to-anchor spring using a suitable brake spring removal/installation tool. Using the tool, remove the primary shoe-to-anchor spring and unhook the cable anchor. Remove the anchor pin plate, if equipped.

6. Remove the cable guide from the secondary shoe.

7. Remove the shoe hold-down springs, shoes, adjusting screw, pivot nut and socket. Note the color and position of each hold-down spring so they can be reassembled in the same position.

8. Remove the parking brake link and spring. Disconnect the parking brake cable from the parking brake lever.

9. Remove the secondary brake shoe. On 9 in. rear brakes, remove the parking brake lever from the shoe. On 10 in. rear brakes, remove the retainer clip and spring washer and remove the parking brake lever.

To install:

10. Clean the backing plate ledge pads and sand lightly. Apply a light coating of high temperature lithium grease to the points where the brake shoes touch the backing plate. Lubricate the adjusting cable eye and the anchor pin area.

11. Install the parking brake lever on the secondary shoe. On 10 in. brakes, secure with the spring washer and retaining clip.

12. Position the brake shoes on the backing plate and install the hold-down spring pins, springs and cups. Install the parking brake link, spring and washer. Connect the parking

brake cable to the parking brake lever.

13. Install the anchor pin plate, if equipped, and place the cable anchor over the anchor pin with the crimped side toward the backing plate.

14. Install the primary shoe-to-anchor spring using the brake spring removal/installation tool.

15. Install the cable guide on the secondary shoe with the flanged hole fitted into the hole in the secondary shoe. Thread the cable around the cable guide groove.

NOTE: Make sure the cable is positioned in the groove and not between the guide and shoe web.

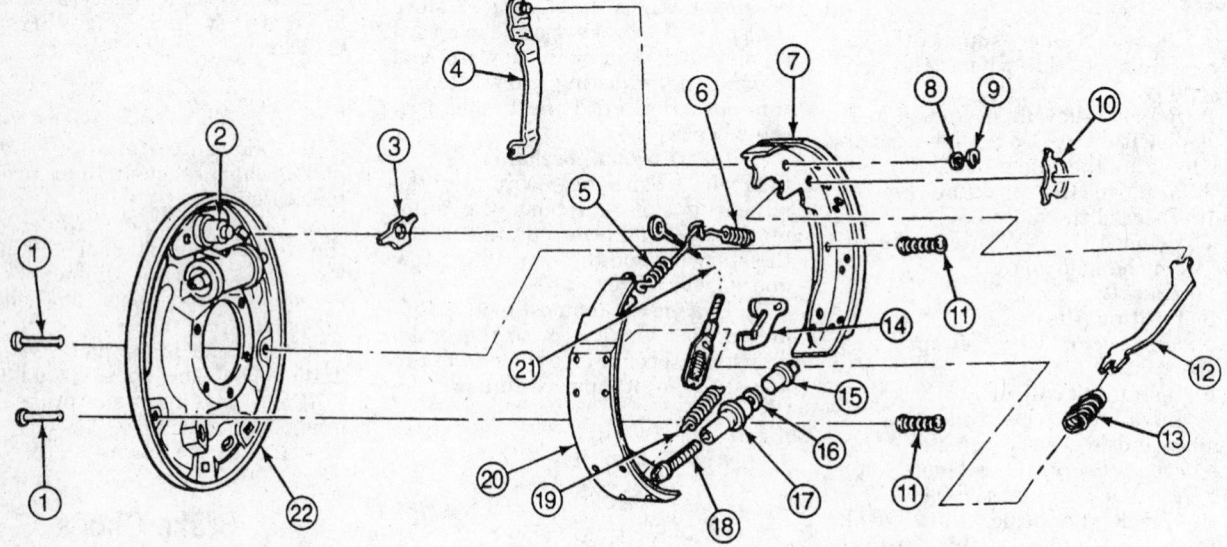

Item	Description
1	Brake Shoe Hold-Down Spring Pin
2	Anchor Pin (Part of 2211)
3	Brake Shoe Anchor Pin Guide Plate
4	Parking Brake Lever
5	Brake Shoe Retracting Spring (Short)
6	Brake Shoe Retracting Spring (Long)
7	Rear Brake Shoe and Lining (Secondary)
8	Washer
9	Parking Brake Lever Pin Retainer
10	Cable Guide
11	Brake Shoe Hold-Down Spring

Item	Description
12	Primary Brake Shoe Parking Brake Lever Link
13	Parking Brake Link Spring
14	Brake Shoe Adjusting Lever
15	Brake Shoe Adjusting Screw Stud
16	Thrust Washer
17	Brake Shoe Adjusting Screw Nut
18	Brake Adjuster Screw
19	Brake Shoe Adjusting Screw Spring
20	Rear Brake Shoe and Lining (Primary)
21	Brake Shoe Adjusting Lever Cable
22	Brake Backing Plate

(Continued)

79214g81

Rear brake shoes and components

16. Install the secondary shoe-to-anchor (long) spring.

NOTE: Make sure the cable end is not cocked or binding on the anchor pin when installed. All parts should be flat on the anchor pin.

17. Apply high temperature lithium grease to the threads and the socket end of the adjusting screw. Turn the adjusting screw into the adjusting pivot nut to the end of the threads and then loosen, ½ turn.

18. Place the adjusting socket on the screw and install the assembly between the shoe ends with the adjusting screw nearest the secondary shoe.

NOTE: Be sure to install the adjusting screw on the same side of the vehicle from which it came. To prevent incorrect installation, the socket end of each adjusting screw is stamped with R or L, to indicate installation on the right or left side of the vehicle. The adjusting pivot nuts have lines machined around the body of the nut, 2 lines indicating the right side nut and 1 line indicating the left side nut.

19. Hook the cable hook into the hole in the adjusting lever from the outboard plate side. The adjusting levers are also stamped with an **R** or **L** to indicate right or left side installation.

20. Place the hooked end of the adjuster spring in the large hole in the primary shoe web and connect the loop end of the spring to the adjuster lever hole.

21. Pull the adjuster lever, cable and automatic adjuster spring down toward the rear to engage the pivot hook in the large hole in the secondary shoe web.

22. After installation, check the action of the adjuster by pulling the section of the cable between the cable guide and the adjusting lever toward the secondary shoe web far enough to lift the lever past a tooth on the adjusting screw wheel. The lever should snap into position behind the next tooth and releasing the cable should cause the adjuster spring to return the lever to its original position. This return action will turn the adjusting screw 1 tooth.

23. If pulling the cable does not produce the action described in Step 22 or if lever action is sluggish instead of positive and sharp, check the position of the lever on the adjusting screw toothed wheel. With the brake in a

vertical position, anchor at the top, the lever should contact the adjusting wheel 1 tooth above the center line of the adjusting screw. If the contact point is below the center line, the lever will not lock on the adjusting screw wheel teeth and the screw will not turn as the lever is actuated by the cable.

24. Adjust the brake shoes using either a brake adjustment gauge or manually with the drums installed.

25. Install the wheels, and lower the vehicle.

Wheel Cylinder

REMOVAL AND INSTALLATION

1. Raise and safely support the vehicle. Remove the wheel and tire assembly, brake drum and brake shoes.
2. Remove the cylinder to shoe connecting pins.
3. Disconnect the brake line from the wheel cylinder.
4. Remove the wheel cylinder retaining bolts and remove the cylinder from the brake backing plate.

To install:

5. Apply silicone rubber D6AZ-19562-AA or equivalent to the wheel cylinder mounting are to seal the backing plate mounting area.
6. Place the wheel cylinder on the backing plate and install the retaining bolts. Tighten the bolts to 106–159 inch lbs. (12–18 Nm).
7. Connect the brake line to the wheel cylinder.
8. Install the brake shoes, internal components and drum.
9. Install the wheel. Bleed the brake system.
10. Lower the vehicle.

Parking Brake Cable

ADJUSTMENT

Aerostar

The parking brake system is self adjusting and requires no adjustment.

Except Aerostar

NOTE: Adjust the drum brakes before adjusting the parking brake. The brake drums must be cold for correct adjustment.

Initial Adjustment

Use this procedure when a new tension limiter is install

1. Apply the parking brake pedal to the fully engaged position.

2. Raise and safely support the vehicle, as necessary. Hold the threaded rod end of the right brake cable to keep it from spinning and thread the equalizer nut 2½ in. up the rod.
3. Check to make sure the cinch strap has slipped and there are less than 1⅜ in. remaining.
4. Release the parking brake and check for proper operation.

Field Adjustment

Use this procedure to correct a slack system if a new tension limiter is not installed.

1. Apply the parking brake pedal to the fully engaged position.
2. Raise and safely support the vehicle, as necessary. Grip the threaded rod to keep it from spinning and tighten the equalizer nut 6 full turns past its original position on the threaded rod.
3. Attach a suitable cable tension gauge in front of the equalizer assembly on the front cable and measure the cable tension. The cable tension should be 400–600 lbs. with the parking brake pedal in the last detent position. If tension is low, repeat Steps 2 and 3.
4. Release parking brake and check for rear wheel drag. There should be no brake drag.

REMOVAL AND INSTALLATION

Front Cable

Aerostar

1. Place the parking brake control in the released position. Release the parking brake cable tension as follows:

   a. Remove the boot cover from the parking brake control assembly.

   b. Insert a steel pin through the pawl lockout pin hole. The pin must be inserted from the inboard side of the control (larger hole) at a slightly upward and forward angle then moved downward and rearward to displace the self adjusting pawl to be inserted through the other hole. This locks out the self adjusting pawl.

   c. Raise and safely support the vehicle with an assistant inside the vehicle. Pull rearward on the equalizer 1–2½ in. (25–38mm) to rotate the self-adjuster reel backward.

   d. Have the assistant insert a steel pin through the self-adjusting spring lock-out holes in the lever and control assembly. This locks

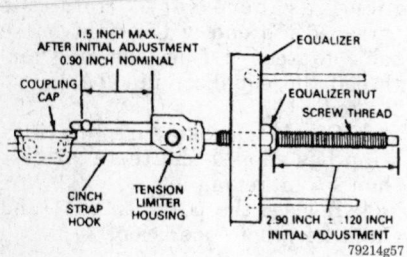

Parking brake cable tension limiter — all models except Aerostar

the ratchet wheel in the cable released position.

NOTE: Do not remove the steel lock pin until the cables are connected to the equalizer. Pin removal releases the tension in the ratchet wheel causing the spring to unwind and release tension, requiring assembly removal to reset spring tension.

2. Raise and safely support the vehicle. Disconnect the rear parking brake cables from the equalizer. Remove the equalizer from the front cable.

3. Remove the bolts retaining the cover to the underbody reinforcement

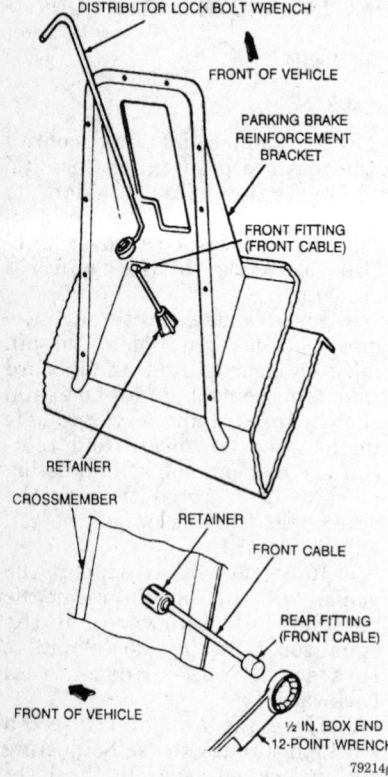

Front parking brake cable — Aerostar

bracket and remove the cover. It may be necessary to loosen fuel tank straps and partially lower the fuel tank to gain access to the cover.

4. Remove the cable anchor pin from the pivot hole in the control assembly ratchet plate. Guide the front cable from the control assembly.

5. Insert a ½ in. box end 12-point distributor lock bolt wrench over the front fitting of the front cable. Push the wrench onto the cable retainer fitting in the crossmember. Compress the retainer fingers and push the retainer rearward through the hole.

6. Insert a ½ in. box end 12-point wrench over the rear fitting of front cable. Push the wrench onto the cable retainer fitting in the crossmember. Compress the retainer fingers and push the retainer forward through the hole.

7. Pull the cable ends through the crossmembers and remove the cable.

To install:

8. Feed the front cable through the holes in both crossmembers. Push the retainers through the holes so the fingers expand over each hole.

9. Route the front cable around the control assembly pulley and insert the cable anchor pin in the pivot hole in the ratchet plate.

10. Slide the return spring over the rear end of the front cable. Connect the equalizer to the front and rear cables.

11. Remove the lock pins from the control assembly to apply cable tension. Position the cover on the reinforcement bracket. Install and tighten the bolts after visually checking to be sure the front cable is attached to the control.

12. Position the boot over the control. Install and tighten the screws.

13. Apply and release the control several times. Make sure the parking brakes are applied, and released and not dragging. Both pins must be removed for proper adjustment.

Except Aerostar

1. Raise and safely support the vehicle.

2. Back off the equalizer nut and remove slug of front cable on Pick-Up models or intermediate cable on sport utility models from the tension limiter.

3. On Pick-Up models, remove the parking brake cable from the cable bracket. On sport utility models, remove the intermediate cable from the bracket and disconnect the intermediate cable from the front cable.

4. Lower the vehicle. Remove the forward ball end of the parking brake

cable from the control assembly clevis.

5. Remove the cable from the control assembly.

6. Using a cord attached to the control lever end of the cable, remove the cable from the vehicle pulling it up into the passenger compartment.

To install:

7. Transfer the cord to the new cable. Position the cable in the vehicle, routing the cable through the dash panel. Remove the cord and secure the cable to the control.

8. Connect the forward ball end of the brake cable to the clevis of the control assembly. Raise and safely support the vehicle.

9. Route the cable through the bracket. On Sport utility models, connect the front cable to the intermediate cable.

10. Connect the slug of the front or intermediate cable cable to the tension limiter connector. Adjust the parking brake cable at the equalizer using initial adjustment or field adjustment, as necessary.

11. Rotate both wheels to make sure the parking brakes are not dragging.

Rear Cable

1. Release parking brake control.

2. On Aerostar models, to release tension on the rear cables, pull rearward on the equalizer assembly about 1–2 in. (25–50mm) and place a clamp on the front cable behind the crossmember.

3. Raise and safely support the vehicle. Remove the wheel and tire assembly and the brake drum.

4. On sport utility and pick-up models, remove the locknut on the threaded rod at the equalizer. Disconnect the rear parking brake cable from the equalizer.

5. Compress the prongs that retain the cable housing to the frame bracket or crossmember and pull out the cable and housing.

6. Working on the wheel side of the backing plate, compress the prongs on the cable retainer so they can pass through the hole in the brake backing plate.

7. Lift the cable out of the slot in the parking brake lever, attached to the secondary brake shoe, and remove the cable through the brake backing plate hole.

To install:

8. Route the cable through the hole in the backing plate. Insert the cable anchor behind the slot in the parking brake lever. Make sure the cable is securely engaged in the park-

ing brake lever so the cable return spring is holding the cable in the parking brake lever.

9. Push the retainer through the hole in the backing plate so the retainer prongs engage the backing plate.

10. Properly route the cable and insert the front of the cable through the frame bracket or crossmember until the prongs expand. Connect the rear cables to the equalizer.

11. On sport utility and pick-up models, rotate the equalizer 90 degrees and reattach the threaded rod to the equalizer.

12. On Aerostar models, remove the clamping device the reapply tension.

13. Install the brake drum and wheel and tire assembly. Adjust the rear brakes.

14. On sport utility and pick-up models, adjust the parking brake tension using the initial adjustment or the field adjustment procedure, as necessary.

15. Apply and release the parking brake control several times. Rotate both wheels to make sure the parking brakes are applied and released and not dragging.

Brake System Bleeding

PROCEDURE

If the master cylinder is known or suspected to have air in the bore, it must be bled before any of the wheel cylinders or calipers.

Master Cylinder

Except RABS and Cartridge Master Cylinders

1. Clean all dirt from the master cylinder filler cap.

2. Loosen the brake line fitting approximately ¾ turn. Wrap a shop cloth around the tubing below the fitting to absorb escaping brake fluid.

3. Have an assistant depress the brake pedal slowly through it's full travel to force air trapped in the master cylinder to escape at the fitting.

4. Tighten the fitting and let the pedal return slowly to the fully released position. Do not release the pedal until the fitting is tightened or air will re-enter the master cylinder.

5. Wait 5 seconds and then repeat the operation until all air bubbles disappear.

6. Repeat Steps a–d on the remaining master cylinder brake line fitting(s).

RABS Models

1. Clean all dirt from the master cylinder filler cap.

2. Place a box wrench on the bleeder fitting On the RABS valve. Attach a rubber drain hose to the bleeder fitting, making sure the end of the hose fits snugly around the bleeder fitting.

3. Submerge the other end of the hose in a container partially filled with clean brake fluid.

4. Loosen the bleeder fitting approximately ¾ turn. Have an assistant slowly press the brake pedal all the way down. Close the bleeder fitting and let the pedal return to the fully released position.

5. Repeat this procedure until no more air bubbles come from the submerged end of the tube. Close the fitting and remove the hose.

Cartridge Master Cylinder

1. Clean all dirt from the master cylinder filler cap.

2. Place a box wrench on the bleeder fitting located on the front of the cartridge master cylinder. Attach a rubber drain hose to the bleeder fitting, making sure the end of the hose fits snugly around the bleeder fitting.

3. Submerge the other end of the hose in a container partially filled with clean brake fluid.

4. Loosen the bleeder fitting approximately ¾ turn. Have an assistant slowly press the brake pedal all the way down. Close the bleeder fitting and let the pedal return to the fully released position.

5. Repeat this procedure until no more air bubbles come from the submerged end of the tube. Close the fitting and remove the hose.

System Bleeding

1. Bleed the brake system by removing the rubber dust cap from the wheel cylinder bleeder fitting at the right-hand rear of the vehicle. Place a box wrench on the bleeder fitting and attach a rubber drain hose to the fitting. The end of the tube should fit snugly around the bleeder fitting. Submerge the other end of the tube in a container partially filled with clean brake fluid and loosen the fitting ¾ turn.

2. Have an assistant push the brake pedal down slowly through it's full travel. Close the bleeder fitting and allow the pedal to slowly return to it's full release position. Wait 5 seconds and repeat the procedure until no bubbles appear at the submerged end of the bleeder tube. Secure the bleeder fitting and remove

the bleeder hose. Install the rubber dust cap on the bleeder fitting.

3. Repeat the procedure in Steps 5 and 6 and bleed the rest of the system in the following sequence: left rear, right front and left front.

4. Refill the master cylinder reservoir after each wheel cylinder or caliper has been bled and install the master cylinder cover and gasket. When brake bleeding is completed, the fluid level should be filled to the maximum level indicated on the reservoir.

5. Always make sure the disc brake pistons are returned to their normal positions by depressing the brake pedal several times until normal pedal travel is established. If the pedal feels spongy, repeat the bleeding procedure.

Rear Anti-lock Brake System (RABS) Module

REMOVAL AND INSTALLATION

1. Disconnect the negative battery cable.

2. Disconnect the wiring harness from the RABS connector by depressing the plastic tab on the connector and pulling the connector off.

3. Remove the 2 nuts that retain the RABS module to the instrument panel anti-shake brace and remove the module.

To install:

4. Install the module to the instrument panel anti-shake brace and secure in place with the nuts.

5. Attach the wire harness to the module.

6. Connect the negative battery cable. Check the system for proper operation.

RABS Valve

REMOVAL AND INSTALLATION

1. Disconnect the negative battery cable.

2. Disconnect and plug the 2 brake lines connected to the RABS valve.

3. Disconnect the wiring harness from the valve harness.

4. Remove the screw retaining the valve and remove the valve.

To install:

5. Position the RABS valve and install the retaining screw. Tighten the retaining screw on Bronco II, Explorer and Ranger to 11–14 ft. lbs. (15–20 Nm) or on Aerostar to 30–40 inch lbs. (3.3–4.5 Nm).

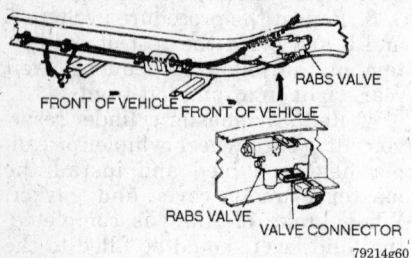

RABS valve location — all models except Aerostar

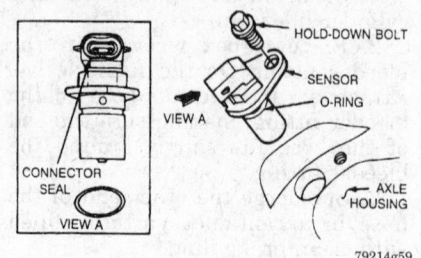

RABS sensor installation

6. Connect the brake valve wiring harness connector.

7. Connect the brake lines to the valve and tighten as follows:

   a. $1/2$–20 threaded fittings — 10–17 ft. lbs. (14–23 Nm).

   b. $7/16$–24 threaded fittings — 10–15 ft. lbs. (14–20 Nm).

   c. $3/8$–24 threaded fittings — 10–15 ft. lbs. (14–20 Nm).

NOTE: Do not overtighten the fittings.

8. Bleed the brake system. It is not necessary to energize the valve electrically to bleed the rear brakes.

9. Connect the negative battery cable.

RABS Sensor

REMOVAL AND INSTALLATION

1. Disconnect the negative battery cable.

2. Pull the wiring harness connector off.

3. Remove the sensor hold-down bolt and remove the sensor from the axle housing.

To install:

4. Clean the axle mounting surface. Use care to prevent dirt from entering the axle housing.

5. Inspect and clean the magnetized sensor pole piece to ensure that it is free from loose metal particles which could cause erratic system operation. Inspect the sensor O-ring for damage and replace, if necessary.

6. Lightly lubricate the sensor O-ring with motor oil, align the sensor bolt hole and install. Do not apply force to the plastic sensor connector. The sensor flange should slide to the mounting surface. This will insure the air gap setting is between 0.005–0.045 in. (0.127–1.14mm).

7. Install the hold down bolt and tighten to 25–30 ft. lbs. (34–40 Nm).

8. Inspect the blue sensor connector seal and replace if missing or damaged. Push the connector on the sensor.

9. Connect the negative battery cable.

RABS Excitor Ring

INSPECTION

1. Remove the RABS sensor.

2. View the excitor ring teeth through the sensor hole. Rotate the rear axle and check the excitor ring teeth for damage or breakage. Dented or broken teeth could cause the RABS system to function when not required.

REMOVAL AND INSTALLATION

To service the excitor ring, the differential case must be removed from the axle housing and the excitor ring pressed off the case.

NOTE: Upon removal, the excitor ring is to be discarded. It is not to be reused.

4-Wheel Anti-lock Brake System Speed Sensor

REMOVAL AND INSTALLATION

Front

1. Inside the engine compartment, disconnect the sensor from the harness.

2. Unclip the sensor cable from the brake hose clips.

3. Remove the retaining bolt from the spindle and slide the sensor from its hole.

4. Installation is the reverse of the removal. Torque the retaining bolt to 40–60 inch lbs.

Rear

1. Disconnect the wiring from the harness.

2. Remove the sensor hold-down bolt and remove the sensor from the axle.

To install:

3. Thoroughly clean the mounting surfaces. Make sure no dirt falls into the axle. Clean the magnetized sensor pole piece. Metal particles can cause sensor problems. Replace the O-ring.

4. Coat the new O-ring with clean engine oil.

5. Position the new sensor on the axle. It should slide into place easily. Correct installation will allow a gap of 0.005–0.045 in.

6. Torque the hold-down bolt to 25–30 ft. lbs.

7. Connect the wiring.

Front Speed Sensor Ring

REMOVAL AND INSTALLATION

1. Raise and safely support the vehicle.

2. Remove the wheels.

3. Remove the caliper, rotor and hub.

4. Using a 3-jawed puller, remove the ring from the hub. The ring cannot be reused; it must be replaced.

To install:

5. Support the hub in a press so the lug studs do not rest on the work surface.

6. Position the new sensor ring on the hub. Using a cylindrical adapter 98mm x 106mm OD, press the ring into place. The ring must be fully seated!

7. The remainder of installation is the reverse of removal.

FRONT SUSPENSION

Shock Absorbers

REMOVAL AND INSTALLATION

1. Raise and safely support the vehicle. Remove the nut and washer attaching the shock absorber to the spring seat or coil spring upper bracket.

2. On Sport Utility and Pick-Up models, remove the nut or nut and

bolt that retains the shock absorber to the radius arm.

3. On Aerostar models, remove the bolts that retain the shock absorber to the bottom of the lower control arm.

4. Slightly compress the shock absorber, as necessary to remove it from the vehicle.

To install:

5. If needed, compress the spring slightly, and install.

6. On Aerostar models attach the bolt that secure the shock to the lower control arm. Tighten the bolt to 16–24 ft. lbs. (22–33 Nm).

7. On Sport Utility and Pick-Up models, install the nut or nut and bolt that retains the shock to the radius arm. Tighten to 42–53 ft. lbs. (57–72 Nm).

Coil Springs

REMOVAL AND INSTALLATION

Aerostar

1. Place the steering wheel to the on center position such that the wheels are straight ahead.

2. Raise and safely support the vehicle. Remove the wheel and tire assembly.

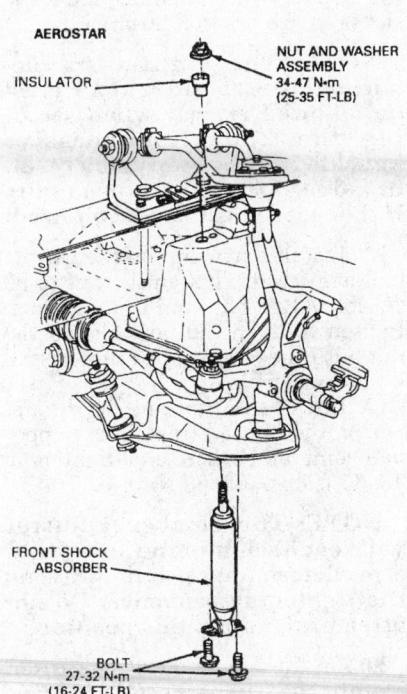

Front shock absorber — Aerostar

3. Disconnect the stabilizer bar link bolt from the lower arm.

4. Remove the nuts or bolts attaching the shock absorber to the lower arm. Remove the upper nut and washer and remove the shock absorber.

5. Using spring compressor tool D78P-5310-A or equivalent, install 1 plate with the pivot ball seat facing downward into the coils of the spring. Rotate the plate so it is flush with the upper surface of the lower arm.

6. Install the other plate with the pivot ball seat facing upward into the coils of the spring. Insert the upper ball nut through the coils of the spring, so the nut rests in the upper plate.

7. Insert the compression rod into the opening in the lower arm, through the upper and lower plate and upper ball nut. Insert the securing pin through the upper ball nut and the compression rod.

8. With the upper ball nut secured, turn the upper plate so it walks up the coil until it contacts the upper spring seat, then back off ½ turn.

9. Install the lower ball nut and thrust washer on the compression rod and screw on the forcing nut. Tighten the forcing nut until the spring is compressed enough so it is free in the seat.

10. Loosen the lower arm pivot bolts. Remove the cotter pin, then loosen but do not remove, the nut attaching the lower ball joint to the spindle. Using puller tool T64P-3590-F or equivalent, loosen the lower ball joint. Remove the puller tool. Support the lower control arm with a jack and remove the ball joint nut. Lower the control arm and remove the spring.

11. If a new spring is to be installed, mark the position of the plates on the spring with chalk. Com-

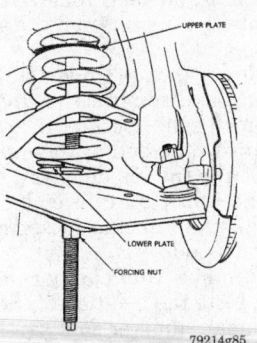

Using spring compressor tool to remove the coil spring

press a new spring for installation and measure the compressed length of the old spring.

12. Loosen the forcing nut to relieve spring tension and remove the tools from the spring.

To install:

13. Assemble the spring compressor and locate in the same positions indicated in Step 11.

14. Before compressing the spring, make sure the upper ball nut securing pin is inserted properly.

15. Compress the coil spring until the spring height reaches the dimension obtained in Step 11.

16. Position the coil spring in the lower control arm.

17. Attach the lower ball joint, tightening the nut to 80–120 ft. lbs. (108–163 Nm). Tighten the lower arm pivot bolts and install a new cotter pin.

18. Loosen and remove spring compressor tool D78P-5310-A. Attach the shock absorber, sway bar link and any other component remaining.

19. Lower the vehicle.

Except Aerostar

1. Raise and safely support the vehicle. Place a support under the rear axle.

2. Remove the nut or nut and bolt attaching the shock absorber to the radius arm.

3. Remove the nut securing the spring to the axle and remove the retainer.

4. Slowly lower the axle to relieve the spring tension. Remove the spring by rotating the upper coil out of the tabs in the upper spring seat.

To install:

5. Install the top of the spring in the upper seat, rotating into position.

6. Raise the axle until the spring is seated in the lower spring seat. Install the lower retainer and tighten the nut to 70–100 ft. lbs. (95–136 Nm).

7. Connect the shock absorber to the radius arm and lower the vehicle.

Upper Ball Joints

INSPECTION

NOTE: Always check and adjust the wheel bearings before ball joint inspection.

1. Raise and safely support the vehicle.

2. On Aerostar models, place a jack under the lower control arm and raise to slightly compress the spring. On Sport Utility and Pick-Up models,

place a jack under the axle beneath the coil spring.

3. Grasp the upper edge of the tire and move the wheel in and out. A $1/32$ in. (0.8mm) or greater movement between the upper spindle arm and the upper control arm or upper part of the axle jaw indicates that the upper ball joint must be replaced.

REMOVAL AND INSTALLATION

Aerostar

If upper ball joint replacement is necessary, the entire upper control arm must be replaced.

2WD Models

1. Raise and safely support the vehicle. Remove the wheel and tire assembly.

2. Remove the brake caliper and support it aside with wire. Do not let the caliper hang by the brake hose.

3. Remove the brake rotor from the spindle. Remove the brake dust shield.

4. Disconnect the steering linkage from the spindle and arm by removing the cotter pin and nut. Remove the tie rod end from the spindle arm.

5. Remove the cotter pin and nut from the lower ball joint stud. Remove the axle clamp bolt from the axle.

6. Remove the camber adjuster from the upper ball joint stud and axle beam.

7. Remove the spindle and ball joint assembly from the axle.

NOTE: Do not use a pickle fork to separate the ball joint from the axle as this will damage the seal and ball joint socket.

To install:

8. Install the spindle assembly in a vise and remove the snapring from the lower ball joint. Remove the lower ball joint from the spindle using C-frame T74P-4635-C or equivalent and a suitable receiver cup to press the ball joint from the spindle.

NOTE: The lower ball joint must be removed first.

9. Repeat Step 8 if removing the upper ball joint.

NOTE: Do not heat the ball joints or the spindle to aid removal.

10. Assemble the C-frame and receiver cup and press in the upper ball joint.

11. Repeat Step 10 if installing the lower ball joint.

NOTE: Do not heat the ball joints or axle to aid installation.

12. Install the snapring onto the ball joint.

13. Place the spindle and ball joints on to the axle. Install the camber adjuster in the upper spindle over the ball joint stud making sure it is properly aligned.

14. Tighten the lower ball joint stud nut to 104–146 ft. lbs. (141–198 Nm). Continue tightening the castellated nut until it lines up with the hole in the stud, then install the cotter pin.

15. Install the clamp bolt into the axle boss and tighten to 48–65 ft. lbs. (65–88 Nm).

16. Install the remaining components.

4WD Models

1. Raise and safely support the vehicle. Remove the front wheel and tire assemblies.

2. Remove the disc brake caliper and wire it to the frame. Do not let the caliper hang by the brake hose.

3. Remove the hub locks, wheel bearings and locknuts.

4. Remove the hub, rotor and outer wheel bearing.

5. Remove the nuts retaining the spindle to the steering knuckle. Tap the spindle with a plastic hammer to loosen the spindle from the knuckle. Remove the splash shield.

6. On the left side of the vehicle, remove the shaft and joint assembly by pulling the assembly out of the carrier. On the right side of the carrier, remove and discard the clamp from the shaft and joint assembly and the stub shaft. Pull the shaft and joint assembly from the splines of the stub shaft.

7. Remove the cotter pin from the tie rod nut and then remove the nut. Separate the tie rod from the steering arm.

8. Remove the upper ball joint snapring and remove the upper ball joint pinch bolt. Loosen the lower ball joint nut to the end of the stud.

9. Strike the inside of the knuckle near the upper and lower ball joints to break the knuckle loose from the ball joint studs.

10. Remove the camber adjuster sleeve. Note the position of the slot in the camber adjuster so it can be reinstalled in the same position during assembly.

11. Remove the lower ball joint nut. Place the knuckle in a vise and re-

move the snapring from the bottom ball joint socket, if equipped.

12. Assemble C-frame T74P-4635-C and ball joint remover T83T-3050-A or equivalents on the lower ball joint. Turn the forcing screw clockwise until the lower ball joint is removed from the steering knuckle.

13. Assemble the C-frame and ball joint remover on the upper ball joint and remove in the same manner.

NOTE: Always remove the lower ball joint first.

To install:

14. Clean the steering knuckle bore and insert the lower ball joint in the knuckle as straight as possible.

15. Assemble C-frame T74P-4635-C, ball joint installer T83T-3050-A and receiver cup T80T-3010-A3 or equivalents to install the lower ball joint. Turn the forcing screw clockwise until the lower ball joint is firmly seated. Install the snapring on the lower ball joint.

NOTE: The lower ball joint must always be installed first.

16. Assemble the C-frame, ball joint installer and receiver cup to install the upper ball joint. Turn the forcing screw clockwise until the ball joint is firmly seated.

17. Install the camber adjuster into the support arm, making sure the slot is in the original position.

NOTE: The tightening sequence in Steps 18 and 19 must be followed exactly when securing the knuckle. Excessive knuckle turning effort may result in reduced steering returnability if this procedure is not followed.

18. Install a new nut on the bottom ball joint stud. Tighten the nut to 90 ft. lbs. (122 Nm) minimum, then tighten to align the next slot in the nut with the hole in the stud. Install a new cotter pin.

19. Install the snapring on the upper ball joint stud. Install the upper ball joint pinch bolt and tighten to 48–65 ft. lbs. (65–88 Nm).

NOTE: The camber adjuster will seat itself into the knuckle at a predetermined position during the tightening sequence. Do not attempt to adjust this position.

20. On the right side of the carrier, install the rubber boot and new clamps on the stub shaft slip yoke. Slide the right shaft and joint assembly into the slip yoke making sure the splines are fully engaged. Slide the

boot over the assembly and crimp the clamp using suitable pliers.

NOTE: The Dana model 28 axle has phased splines; there is only 1 way to assemble the right shaft and joint assembly into the slip yoke. The Dana model 35 axle does not have a blind spline, therefore pay special attention to make sure the yoke ears are in phase (in line) during assembly.

21. On the left side of the carrier, slide the shaft and joint assembly through the knuckle and engage the splines on the shaft in the carrier.
22. Install the splash shield and spindle onto the steering knuckle. Install and tighten the spindle nuts to 45 ft. lbs. (61 Nm).
23. Install the rotor on the spindle and install the outer wheel bearing.
24. Install the wheel bearing, locknut, thrust bearing, snapring and locking hubs.
25. Install the caliper and the wheel and tire assemblies. Lower the vehicle.

Lower Ball Joints

INSPECTION

NOTE: Always check and adjust the wheel bearings before ball joint inspection.

1. Raise and safely support the vehicle.
2. On Aerostar, place a jack under the lower control arm and raise to slightly compress the spring. On sport utility and pick-up models, place a jack under the axle beneath the coil spring.
3. Grasp the lower edge of the tire and move the wheel in and out. A ¹/₃₂ in. (0.8mm) or greater movement between the lower control arm or lower axle and the spindle indicates that the lower ball joint must be replaced.

REMOVAL AND INSTALLATION

Aerostar

If lower ball joint replacement is necessary, the entire lower control arm must be replaced.

2WD Models

1. Raise and safely support the vehicle. Remove the wheel and tire assembly.

2. Remove the brake caliper and support it with wire. Do not let the caliper hang by the brake hose.
3. Remove the dust cap, cotter pin, nut retainer, washer and outer bearing. Remove the brake rotor from the spindle. Remove the brake dust shield.
4. Disconnect the steering linkage from the spindle arm by removing the cotter pin and nut. Remove the tie rod end from the spindle arm.
5. Remove the cotter pin and nut from the lower ball joint stud. Remove the axle clamp bolt from the axle.
6. Remove the camber adjuster from the upper ball joint stud and axle beam.
7. Strike the inside area of the axle to pop the lower ball joint loose from the axle beam. Remove the spindle and ball joint assembly from the axle.

NOTE: Do not use a pickle fork to separate the ball joint from the axle as this will damage the seal and ball joint socket.

To install:
8. Install the spindle assembly in a vise and remove the snapring from the lower ball joint. Remove the lower ball joint from the spindle using C-frame T74P-4635-C or equivalent and a suitable receiver cup to press the ball joint from the spindle.

NOTE: Do not heat the ball joint or the spindle to aid removal.

9. Assemble the C-frame and receiver cup and press in the lower ball joint.

NOTE: Do not heat the ball joint or axle to aid installation.

10. Install the snapring onto the ball joint.
11. Place the spindle and ball joints into the axle. Install the camber adjuster in the upper spindle over the ball joint stud making sure it is properly aligned.
12. Tighten the lower ball joint stud nut to 104–146 ft. lbs. (141–198 Nm). Continue tightening the castellated nut until it lines up with the hole in the stud, then install the cotter pin.
13. Install the clamp bolt into the axle boss and tighten to 48–65 ft. lbs. (65–88 Nm).
14. Install the remaining components.

Except Aerostar

1. Raise and safely support the vehicle. Remove the front wheel and tire assemblies.
2. Remove the disc brake caliper and wire to the frame. Do not let the caliper hang by the brake hose.
3. Remove the hub locks, wheel bearings and locknuts.
4. Remove the hub, rotor and outer wheel bearing.
5. Remove the nuts retaining the spindle to the steering knuckle. Tap the spindle with a plastic hammer to jar it from the knuckle. Remove the splash shield.
6. On the left side of the vehicle, remove the shaft and joint assembly by pulling the assembly out of the carrier. On the right side of the carrier, remove and discard the clamp from the shaft and joint assembly and the stub shaft. Pull the shaft and joint assembly from the splines of the stub shaft.
7. Remove the tie rod from the steering arm.
8. Remove the upper ball joint snapring and remove the upper ball joint pinch bolt. Loosen the lower ball joint nut to the end of the stud.
9. Strike the inside of the knuckle near the upper and lower ball joints to break the knuckle loose from the ball joint studs.
10. Remove the camber adjuster sleeve. Note the position of the slot in the camber adjuster so it can be reinstalled in the same position during assembly.
11. Remove the lower ball joint nut. Place the knuckle in a vise and remove the snapring from the bottom ball joint socket, if equipped.
12. Assemble C-frame T74P-4635-C and ball joint remover T83T-3050-A or equivalents on the lower ball joint. Turn the forcing screw clockwise until the lower ball joint is removed from the steering knuckle.

To install:
13. Clean the steering knuckle bore and insert the lower ball joint in the knuckle as straight as possible.
14. Assemble C-frame T74P-4635-C, ball joint installer T83T-3050-A and receiver cup T80T-3010-A3 or equivalents to install the lower ball joint. Turn the forcing screw clockwise until the lower ball joint is firmly seated. Install the snapring on the lower ball joint.

15. Install the camber adjuster into the support arm, making sure the slot is in the original position.

NOTE: The tightening sequence in Steps 16 and 17 must be followed exactly when securing the knuckle. Excessive knuckle turning effort may result in reduced steering returnability if this procedure is not followed.

16. Install a new nut on the bottom ball joint stud. Tighten the nut to 90 ft. lbs. (122 Nm) minimum, then tighten to align the next slot in the nut with the hole in the stud. Install a new cotter pin.

17. Install the snapring on the upper ball joint stud. Install the upper ball joint pinch bolt and tighten to 48–65 ft. lbs. (65–88 Nm).

NOTE: The camber adjuster will seat itself into the knuckle at a predetermined position during the tightening sequence. Do not attempt to adjust this position.

18. On the right side of the carrier, install the rubber boot and new clamps on the stub shaft slip yoke. Slide the right shaft and joint assembly into the slip yoke making sure the splines are fully engaged. Slide the boot over the assembly and crimp the clamp using suitable pliers.

NOTE: The Dana model 28 axle has phased splines; there is only 1 way to assemble the right shaft and joint assembly into the slip yoke. The Dana model 35 axle does not have a blind spline, therefore pay special attention to make sure the yoke ears are in phase (in line) during assembly.

19. On the left side of the carrier, slide the shaft and joint assembly through the knuckle and engage the splines on the shaft in the carrier.

20. Install the splash shield and spindle onto the steering knuckle. Install and tighten the spindle nuts to 45 ft. lbs. (61 Nm).

21. Install the rotor on the spindle and install the outer wheel bearing in the race.

22. Install the wheel bearing, locknut, thrust bearing, snapring and locking hubs.

23. Install the caliper and the wheel and tire assemblies. Lower the vehicle.

Upper Control Arms

REMOVAL AND INSTALLATION

1. Place the steering wheel in the straight ahead position.

2. Raise the vehicle and support it safely under the front body rails.

NOTE: Only remove and install 1 upper control arm at a time. Never service both sides at the same time.

3. Separate the spindle or steering knuckle, as required.

4. Remove the bolt securing the retainer plate and remove.

5. Mark the position of the control arm mounting brackets.

6. Remove the bolt and washer retaining the front mounting bracket to the flat plate.

7. From under the frame rail, remove the 3 nuts securing the upper control arm mounting brackets to the body.

8. Remove the 3 long bolts retaining the mounting brackets to the body. Rotate the upper control arm out of position in order to access and remove the control arm bolts.

9. Remove the upper control arm, upper ball joint and mounting bracket assembly from the vehicle. If required, remove the damper assembly from the upper control arm.

To install:

10. If removed, install the damper assembly and tighten the retaining bolts to 22–29 ft. lbs. (30–39 Nm).

11. Place the flat plate for the mounting brackets in position on the body rail. Install the bolt and tighten to 10–14 ft. lbs. (14–18 Nm).

12. Place the mounting brackets and upper control arm assembly in position on the flat plate.

13. Install the 3 long bolts and washers retaining the mounting brackets to the body rail. Rotate or rock the upper control arm and bracket assembly until the bolt heads rest against the mounting bracket and the studs extend through the body rail.

14. Install and tighten the nuts retaining the mounting bracket bolts to the body rail to 145–195 ft. lbs. (196–264 Nm). Make sure the mounting brackets do not move from the marked position on the flat plate.

NOTE: The torque figure for the mounting bracket-to-body rail nuts and bolts is critical. They must be tightened to the specified torque.

15. Install and tighten the bolt retaining the front mounting bracket to the flat plate to 35–47 ft. lbs. (47–64 Nm).

16. Place the bolt retaining the plate in position on the mounting bracket and flat plate assembly. Install and tighten the bolt to 10–14 ft. lbs. (14–18 Nm).

17. Install the spindle or steering knuckle, as required.

18. Lower the vehicle and check the front end alignment.

Lower Control Arms

REMOVAL AND INSTALLATION

1. Place the steering wheel in a straight ahead position.

2. Raise the vehicle and support it safely under the frame.

3. Remove the coil spring.

4. Remove the bolts and nuts retaining the control arm to the No. 1 crossmember. Remove the lower control arm.

To install:

5. Position the control arm in the No. 1 crossmember and install the mounting bolts. Install the nuts and tighten until snug. Do not tighten to the specified torque at this time.

6. Install the coil spring.

7. With the vehicle in the normal ride position, tighten the lower control arm retaining nuts and bolts to 100–140 ft. lbs. (136–190 Nm).

Stabilizer Bar

REMOVAL AND INSTALLATION

Aerostar

1. Raise and safely support the vehicle. Remove the nuts retaining the stabilizer bar to the lower control arm link.

2. Remove the insulators and disconnect the bar from the links. If required, remove the nuts retaining the links to the lower control arm. Remove the insulators and remove the links.

3. Remove the bolts retaining the bar mounting bracket to the frame and remove the stabilizer bar. If required, remove the insulators from the stabilizer bar.

To install:

4. If removed, install the insulators to the stabilizer bar.

5. Position the stabilizer bar in place and install the mounting brackets to the frame and secure in place.

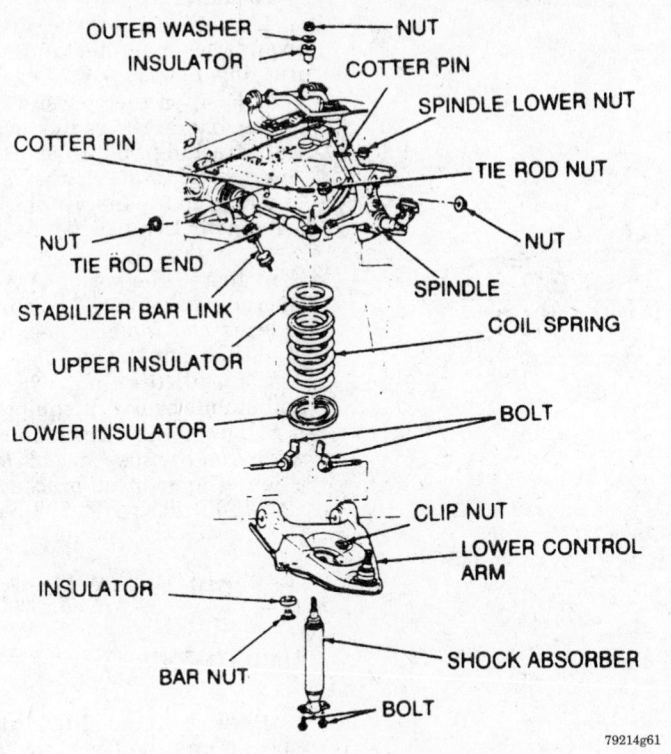

OUTER WASHER — NUT
INSULATOR — COTTER PIN
SPINDLE LOWER NUT
COTTER PIN
TIE ROD NUT
NUT — NUT
TIE ROD END
STABILIZER BAR LINK — SPINDLE
COIL SPRING
UPPER INSULATOR
LOWER INSULATOR — BOLT
CLIP NUT
LOWER CONTROL ARM
INSULATOR
SHOCK ABSORBER
BAR NUT
BOLT

79214g61

Lower control arm — Aerostar

6. Attach the stabilizer bar to the links and secure in place with the nuts.

7. Connect the stabilizer bar to the control arm.

8. Tighten the mounting bracket bolts to 16–24 ft. lbs. (22–33 Nm). Tighten the link nuts to 9–12 ft. lbs. (12–16 Nm).

9. Lower the vehicle.

2WD Explorer and Navajo

1. Raise and safely support the vehicle.

2. Remove the nut and washer and disconnect the stabilizer link assembly from the front I-beam axle.

3. Remove the mounting bolts and remove the stabilizer bar retainers from the stabilizer bar assembly. Remove the stabilizer bar.

To install:

4. Position the stabilizer bar to the vehicle and attach the mounting bolts and retainers.

5. Attach the bar to the front I-beam using the nuts and washers.

6. Tighten the retainer bolts to 35–50 ft. lbs. (47–68 Nm). Tighten the stabilizer bar link nuts to 30–44 ft. lbs. (40–60 Nm).

7. Lower the vehicle.

2WD Ranger, Splash, Mazda B2300, B3000 and B4000

1. Raise and safely support vehicle.

2. Remove the nuts and bolts retaining the stabilizer bar to the end links.

3. Remove the retainers and remove the stabilizer bar and bushings.

To install:

4. Position the stabilizer bar to the vehicle and secure in place with the retainers and bushings.

5. Attach the bar to the end links using the nuts and bolts.

6. Tighten the retainer bolts to 35–50 ft. lbs. (47–68 Nm). Tighten the end link nuts to 30–44 ft. lbs. (40–60 Nm).

7. Lower the vehicle.

4WD Explorer, Mountaineer, Navajo, Ranger, Splash, Mazda B2300, B3000 and B4000

1. Raise and safely support vehicle.

2. Remove the bolts and retainers from the center and right end of the stabilizer bar.

3. Remove the nut, bolt and washer retaining the stabilizer bar to the stabilizer link.

4. Remove the stabilizer bar and bushings.

To install:

5. Position the stabilizer bar and bushings to the vehicle and install the retainers. Connect the nuts, bolts and washers to secure in place.

6. Attach the bolts and retainers to the center and right end of the stabilizer bar.

7. Tighten the retainer bolts to 35–50 ft. lbs. (48–68 Nm). Tighten the stabilizer bar link nut to 30–44 ft. lbs. (40–60 Nm).

Radius Arm

REMOVAL AND INSTALLATION

Explorer, Navajo and Ranger, Splash, Mazda B2300, B3000 and B4000 Models

NOTE: A torque wrench with a capacity of at least 250 ft. lbs. is necessary, along with other special tools, for this procedure.

1. Raise the front of the vehicle and place safety stands under the frame and a jack under the wheel or axle. Remove the wheels.

2. Disconnect the shock absorber from the radius arm bracket.

3. Remove the two spring upper retainer attaching bolts from the top of the spring upper seat and remove the retainer.

4. Remove the nut which attached the spring lower retainer to the lower seat and axle and remove the retainer.

5. Lower the axle and remove the spring.

6. Remove the spring lower seat and shim from the radius arm. The, remove the bolt and nut which attach the radius arm to the axle.

7. Remove the front wheel spindle and stabilizer bar, if equipped.

8. Remove the cotter pin, nut and washer from the radius arm rear attachment.

9. Remove the bushing from the radius arm and remove the radius arm from the vehicle.

10. Inspect and remove if needed the inner bushing from the radius arm.

To install:

11. Position the radius arm to the axle and install the bolt and nut finger-tight.

12. Install the inner bushing, if needed on the radius arm and position the arm to the frame bracket.

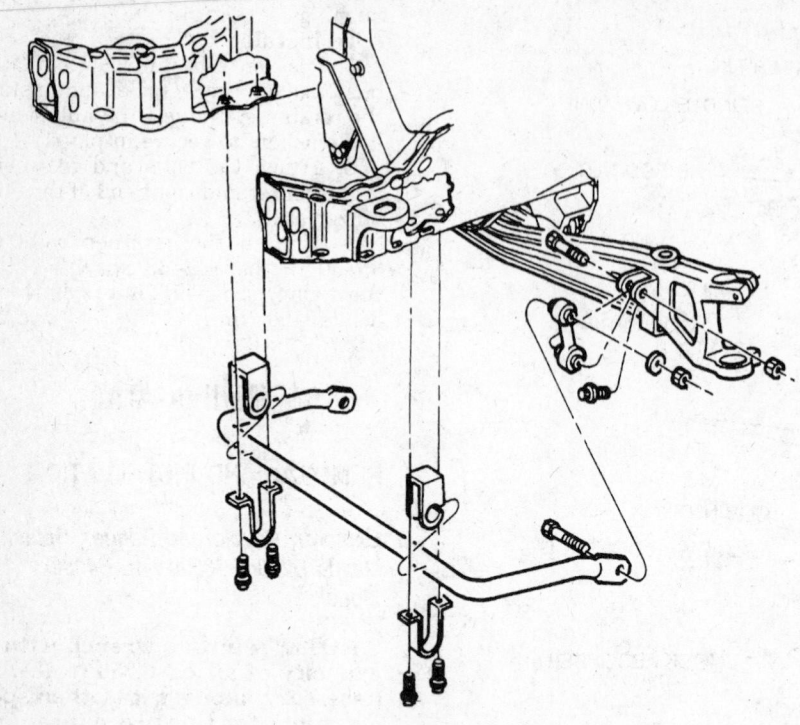

Front stabilizer bar and components — Ford and Mazda pick-up

79214g86

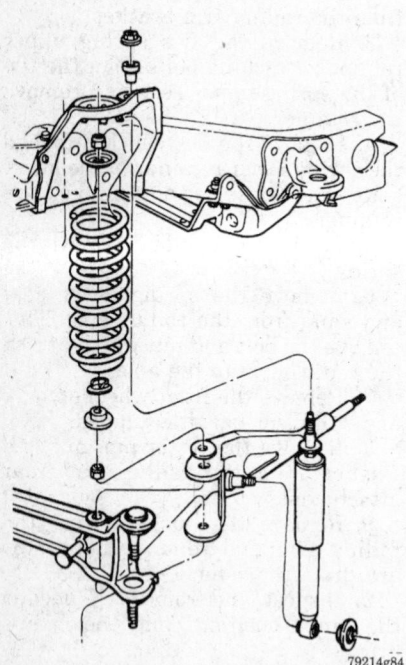

79214g84

Radius arm and components

To install:

4. Position the axle to the frame pivot bracket and install the bolt and nut finger tight.

5. Position the opposite end of the of the axle to the radius arm, install the attaching bolt from underneath through the bracket, the radius arm and the axle. Install the nut and tighten to 191–220 ft. lbs. (258–298 Nm).

6. Install the spring lower seat on the radius arm so the hole in the seat indexes over the arm-to-axle bolt. Install the front spring.

7. Install the front wheel spindle and stabilizer bar, if equipped.

8. Lower the vehicle and with the weight on the suspension, tighten the axle-to-frame pivot bracket bolts to 120–150 ft. lbs. (163–203 Nm).

Front Wheel Bearings

ADJUSTMENT

1. Raise the vehicle until the wheel clears the floor.

2. Remove the hub or wheel cover.

3. Remove the grease cap.

4. Wipe any excess grease from the end of the spindle. Remove the cotter pin and retainer. Discard the cotter pin.

5. Loosen the adjusting nut 3 turns. Obtain a running clearance between the rotor and brake surface by rocking the assembly back and forth. If running clearance cannot be maintained, then remove the caliper and hand aside with wire. Never allow a caliper to hang free.

6. While rotating the assembly, tighten the adjusting nut to 17–25 ft. lbs. (23–24 Nm) to seat the bearing.

7. Loosen the nut ½ turn. Retighten the nut to 18–20 ft. lbs. (23–26 Nm).

8. Place the retainer on the adjusting nut. The castellations on the retainer must be aligned with the cotter pin hole in the spindle. Do not turn the adjusting nut to make the castellation nut line up. Remove the retainer from the nut and re-index the retainer without moving the nut.

9. Install a new cotter pin, and bend the ends in opposite directions.

10. Check the front wheel rotation. If the wheels rotate properly, install the grease cap and hub or cover. If the wheel rotates with a noise or rough motion, then remove the the bearing cones and cups. Inspect the components and lubricate and or replace.

13. Install the bushing, washer, and attaching nut. Tighten the nut to 120 ft. lbs. and install the cotter pin.

14. Tighten the radius arm-to-axle bolt to 269–329 ft. lbs.

15. Install the wheel spindle and stabilizer bar, if equipped.

16. Install the spring seat and insulator on the radius arm so the hole in the seat fits over the arm-to-axle nut.

17. Install the spring.

18. Connect the shock absorber. Torque the nut and bolt to 40–60 ft. lbs.

19. Install the wheels.

20. Lower the vehicle.

I-Beam Axle

REMOVAL AND INSTALLATION

1. Raise and safely support the vehicle. Remove the front wheel spindle, the front spring and the stabilizer bar, if equipped.

2. Remove the spring lower seat from the radius arm and then remove the bolt and nut that attaches the stabilizer bar bracket, if equipped, and radius arm to the front axle.

3. Remove the axle-to-frame pivot bracket bolt and nut.

11. Lower the vehicle, and pump the brake before driving.

REMOVAL AND INSTALLATION

1. Raise the vehicle off the ground until the tires just clear the ground.

2. Remove the wheel cover or hub.

3. Remove the brake caliper from the spindle and use a piece of wire to support it to the vehicle underbody.

4. Remove the grease cap.

5. Wipe any excess grease from the end of the spindle. Remove the cotter pin and retainer. Discard the cotter pin.

6. Remove the outer bearing cone and roller assembly.

7. Pull the hub and rotor assembly off the spindle.

8. Use grease seal removal tool 1175-AC or equivalent, remove and discard the grease retainer. Remove the inner bearing cone and roller assembly.

9. Clean the inner and outer bearing cup assemblies. Inspect the cups for deep scratches or pitting, and replace if needed.

To install:

10. If the bearing cups were removed, install replacement units with a bearing cup replace or equivalent tool.

11. Using your hand or a bearing packer, pack the bearing with a suitable long life bearing grease. Grease the cone surfaces.

12. Place the inner bearing cone and roller assembly in the inner cup. A light film of grease should be included between the lips of the new grease retainer. Install the new retainer using an appropriate driver tool.

13. I install the hub and rotor assembly on the spindle.

14. Install the outer bearing cone and roller. Install the adjusting nut and adjust the wheel bearing.

15. Install a new cotter pin. Attach the grease cap.

16. Install the caliper. Install the wheel, and lower the car.

17. Before driving the vehicle, pump the brake pedal several times to restore brake pedal travel.

REAR SUSPENSION

Shock Absorber

REMOVAL AND INSTALLATION

1. Raise and safely support the vehicle.

2. Place a jack under the rear axle and raise slightly to take the load off the shock absorbers.

3. Remove the shock absorber lower attaching nut and bolt and swing the lower end free of the mounting bracket on the axle housing.

4. Remove the upper attaching bolt or nut(s) and remove the shock absorber.

To install:

5. Install the shock absorber into the upper bracket and secure with attaching bolt.

6. Swing the shock absorber down and position in the lower mounting bracket. Attach the lower mounting nut and bolt.

7. Tighten the lower attaching nut and bolt to 41–53 ft. lbs. (55–72 Nm). On Aerostar, tighten the upper bolt to 41–63 ft. lbs. (55–85 Nm). On pickup models, tighten the upper attaching nut to 41–53 ft. lbs. (55–72 Nm). On Explorer and Navajo, tighten the upper mounting nuts to 15–21 ft. lbs. (21–29 Nm).

Coil Springs

REMOVAL AND INSTALLATION

Aerostar

1. Raise and safely support the vehicle. Place jackstands on the frame rear lift points.

2. Remove the nut and bolt retaining the shock absorber to the axle mount on the lower control arm. Disconnect the shock absorber from the axle bracket.

3. Lower the rear axle until the coil springs are no longer under compression.

4. Remove the nut retaining the lower retainer and spring to the control arm. Remove the bolt retaining the upper retainer and spring to the frame.

5. Remove the spring and retainers. Remove the upper and lower insulators.

To install:

6. Make sure the axle is in the spring unloaded position. Place the lower insulator on the control arm and the upper insulator on top of the spring.

7. Install the coil spring. The small diameter, white colored, tapered coils must face upward with the pigtail resting against the upper insulator rubber stop.

8. With the upper pigtail resting against the rubber stop, rotate the spring with the upper insulator until the lower pigtail points in the 3 o'clock position. Install the upper retainer and bolt and tighten to 30–40 ft. lbs. (40–55 Nm).

9. Install the lower retainer and nut and tighten to 41–65 ft. lbs. (55–88 Nm).

10. Raise the axle to the normal ride position and install the shock absorber. Lower the vehicle.

Leaf Springs

REMOVAL AND INSTALLATION

Except Aerostar

1. Raise the vehicle and safely support with jack stands on the frame until the weight is off the rear spring, with the tires still touching the floor.

2. Remove the nuts from the spring U-bolts and drive the U-bolts from the U-bolt plate.

3. Remove the spring-to-bracket nut and bolt at the front of the spring.

4. Remove the shackle upper and lower nuts and bolts at the rear of the spring. Remove the spring and shackle assembly from the rear shackle bracket.

To install:

5. Position the spring in the shackle and install the upper shackle-to-spring bolt and nut with the bolt head facing outboard.

6. Position the front end of the spring in the bracket and install the bolt and nut. Position the shackle in the rear bracket and install the bolt and nut.

7. On pick-up models, position the spring on top of the axle with the spring tie bolt centered in the hole provided in the seat. On sport utility models, position the spring on the bottom of the axle with the spring tie bolt centered in the hole provided in the seat.

8. Install the spring U-bolts, U-bolt plate and nuts and lower the vehicle. On pick-up models, tighten the

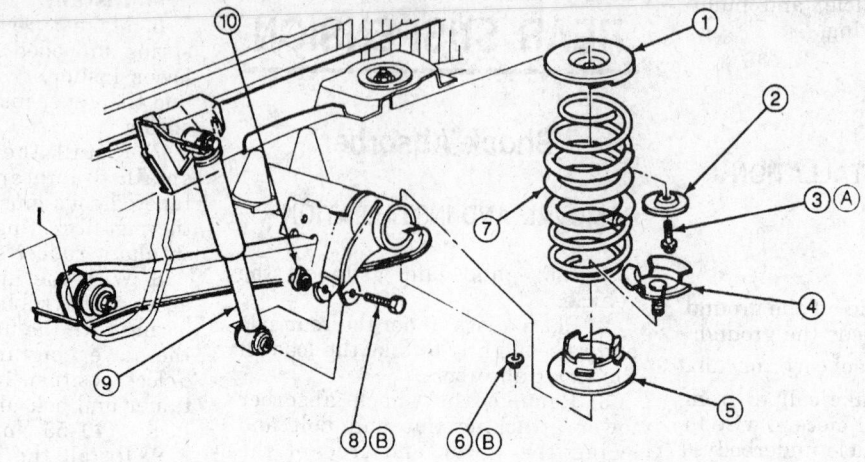

Item	Description
1	Rear Spring Insulator
2	Rear Spring Upper Retainer
3	Bolt
4	Rear Spring Lower Retainer
5	Rear Spring Center Mounting Insulator
6	Nut

Item	Description
7	Rear Spring
8	Bolt
9	Shock Absorber
10	Nut and Retainer
A	Tighten to 40-55 N·m (30-41 Lb-Ft)
B	Tighten to 60-80 N·m (45-60 Lb-Ft)

(Continued)

79214g90

Rear coil spring — Aerostar

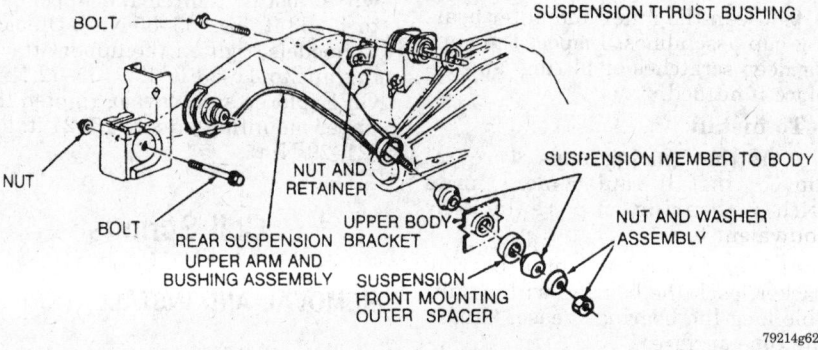

Rear upper control arm — Aerostar

79214g62

spring U-bolt nuts to 65–75 ft. lbs. (88–102 Nm) and the front spring bolt and the rear shackle bolts and nuts to 75–115 ft. lbs. (100–155 Nm). On sport utility models, tighten the spring U-bolt nuts to 88–108 ft. lbs. (119–146 Nm), the front spring bolt and nut to 64–91 ft. lbs. (87–123 Nm) and the rear shackle bolts and nuts to 75–115 ft. lbs. (100–155 Nm).

Rear Control Arms

REMOVAL AND INSTALLATION

Aerostar

Upper Arm

1. Raise the vehicle and support it safely. Place safety stands or equivalent on the rear frame lift points.
2. Remove the nut and bolt retaining the shock absorber to the lower axle bracket. Swing the lower end of the shock absorber free of the axle bracket.
3. Lower the rear axle assembly until the coil springs are no longer under compression.
4. Remove the bolt and nut retaining the upper control arm to the rear axle. Disconnect the upper control arm from the axle. Scribe a mark aligning the position of the cam adjuster in the axle bushing. The cam adjuster controls the rear axle pinion angle for driveline angularity.
5. Remove the bolt and nut retaining the upper control arm to the right frame bracket. Rotate the arm to disengage from the body bracket.
6. Remove the nut and washer retaining the upper control arm to the left frame bracket. Remove the outer insulator and spacer. Remove the control arm from the bracket. Remove the inner insulator from the control arm stud.

NOTE: If the left bracket attachments are loosened prior to disengaging the arm from the right bracket, the uncompressed left bushing will force the arm against the right bracket and make removal difficult.

To install:
7. Position the inner insulator on the control arm stud. Install the control arm so the stud extends through the left frame bracket. Install the spacer and outer insulator over the stud. Install nut and washer assem-

bly and tighten until snug. Do not torque at this time.

8. Position the upper control arm in the right frame bracket. Install the bolt and nut, and tighten until snug. Do not tighten at this time.

9. Align the marks on the cam adjuster and axle bushing. Install the upper control arm to the axle housing. Install the nut and bolt and tighten until snug. Do not tighten at this time.

10. Raise the rear axle to the normal ride position and install the shock absorber. With the axle in the normal ride position, tighten the control arm-to-left frame bracket nut to 60–100 ft. lbs. (81–135 Nm), control arm-to-right frame bracket nut and bolt to 100–133 ft. lbs. (135–170 Nm) and control arm-to-axle housing to 155–210 ft. lbs. (210–284 Nm).

11. Remove the safety stands and lower the vehicle.

Lower Arm

1. Raise and safely support the vehicle. Place safety stands under the frame rear lift points.

2. Remove the nut and bolt retaining the shock absorber to the lower axle bracket. Swing the lower end of the shock absorber free of the axle bracket.

3. Lower the rear axle until the coil springs are no longer under compression.

4. Remove the nut attaching the lower retainer and coil spring to the lower control arm. Remove the insulator from the arm.

5. Remove the bolt and nut retaining the lower control arm to the axle housing.

6. Remove the nut and bolt retaining the lower control arm to the frame bracket and remove the lower control arm.

To install:

7. Position the lower control arm in the bracket on the axle housing. Install the bolt so the head is inboard on the axle bracket. Install the nut but do not tighten at this time.

8. Install the insulator on the lower control arm. With the axle in the spring unloaded position, install the coil spring and lower retainer on the lower control arm.

9. Install the nut attaching the retainer and spring to the lower control arm. Tighten the nut to 41–65 ft. lbs. (55–88 Nm).

10. Raise the axle to the normal ride position. Tighten the lower control arm-to-axle housing nut and bolt and lower control arm-to-frame bracket nut and bolt to 95–130 ft. lbs. (129–177 Nm).

11. Install the shock absorber and lower the vehicle.

Stabilizer Bar

REMOVAL AND INSTALLATION

1. Remove the nuts from the lower ends of the stabilizer bar link.

2. Remove the outer washers and insulators.

3. Disconnect the bar from the links.

4. Remove the inner insulators and washers.

5. Unbolt the link from the frame.

6. Remove the U-bolts, brackets and retainers.

To install:

7. Inspect and replace all worn or cracked rubber parts. Coat all new rubber parts with silicone grease. Assemble all parts loosely and make sure the bar assembly is centered before tightening the fasteners. Observe the following torque figures:

• Stabilizer bar-to-axle nut — 30–42 ft. lbs.

• Link bracket-to-frame nut — 30–42 ft. lbs.

• Link-to-bracket nut — 60 ft. lbs.

• Stabilizer bar-to-link — 15–25 ft. lbs.

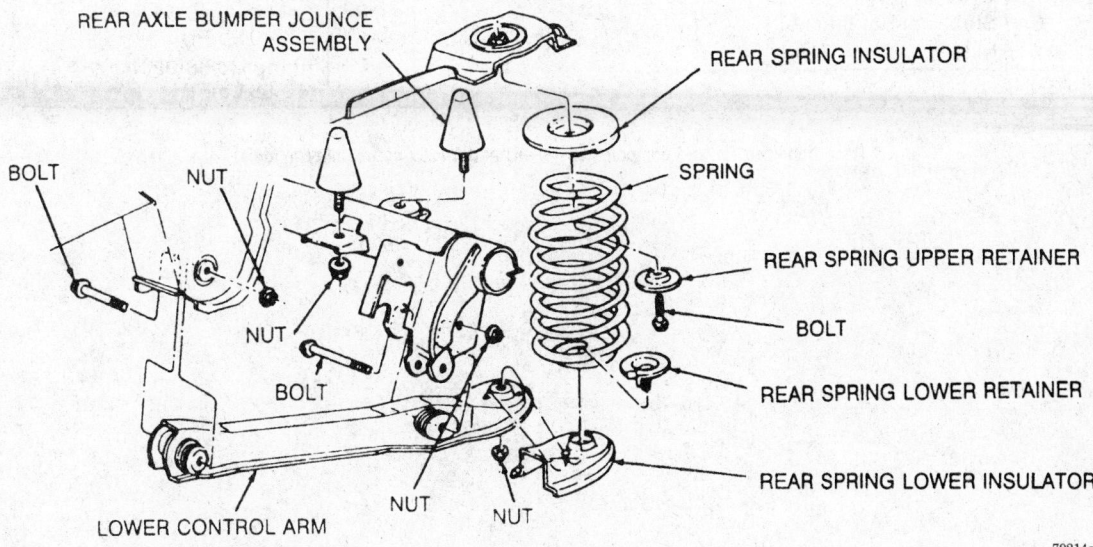

REAR AXLE BUMPER JOUNCE ASSEMBLY
REAR SPRING INSULATOR
SPRING
REAR SPRING UPPER RETAINER
BOLT
REAR SPRING LOWER RETAINER
REAR SPRING LOWER INSULATOR
BOLT
NUT
NUT
BOLT
NUT
NUT
LOWER CONTROL ARM

79214g63

Rear lower control arm — Aerostar

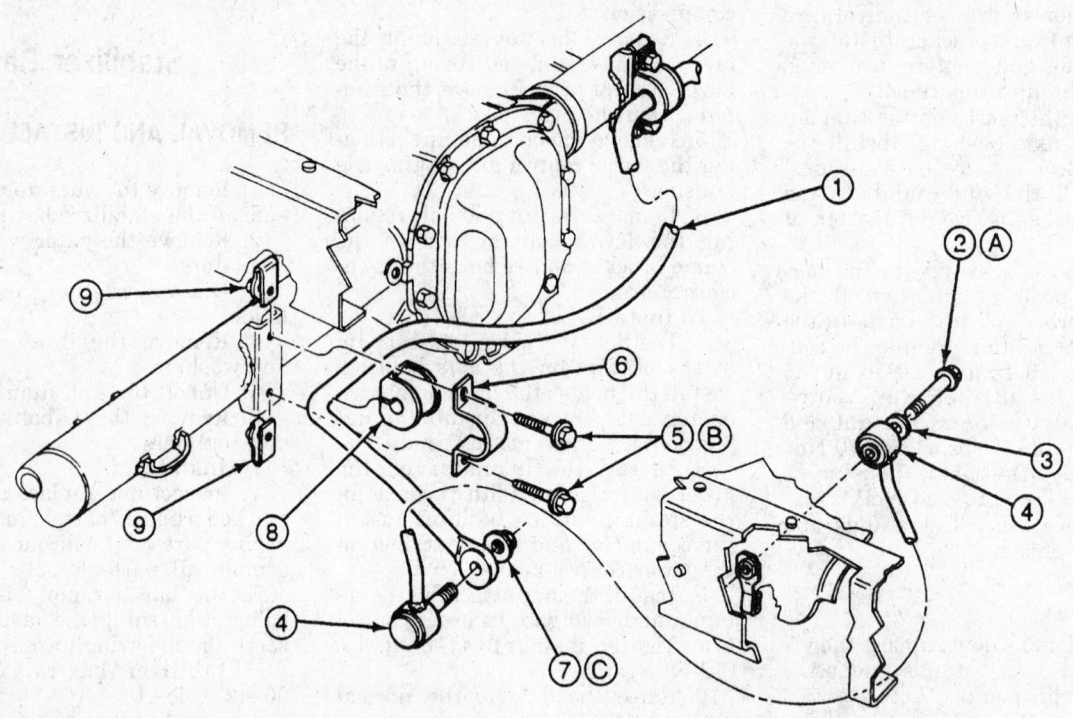

Item	Description
1	Rear Stabilizer Bar
2	Bolt
3	Washer
4	Stabilizer Bar Link
5	Bolt
6	Stabilizer Bar Bracket
7	Nut

(Continued)

Item	Description
8	Lower Suspension Arm Stabilizer Bar Insulator
9	Nut
A	Tighten to 60-80 N·m (44-59 Lb-Ft)
B	Tighten to 40-55 N·m (30-40 Lb-Ft)
C	Tighten to 68-92 N·m (50-68 Lb-Ft)

79214g91

Rear stabilizer bar and components — Ford and Mazda sport utility models

FIRING ORDERS

NOTE: To avoid confusion, always replace spark plug wires one at a time.

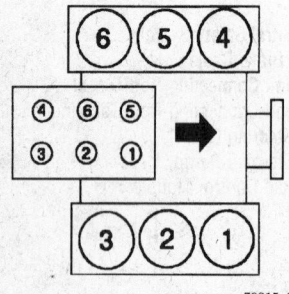

4.2L (VIN 2) Engine
Engine Firing Order: 1–4–2–5–3–6
Distributorless Ignition System (DIS)

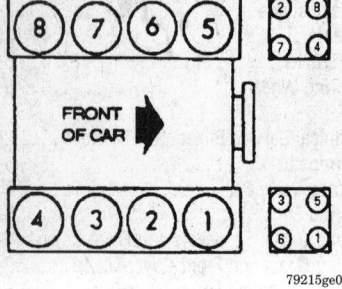

4.6L Engines (VIN W and VIN 6)
Engine Firing Order: 1–3–7–2–6–5–4–8
Distributorless Ignition System (DIS)

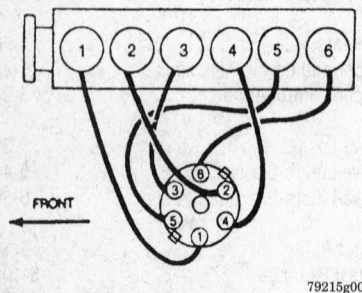

4.9L (VIN Y and Z) Engine
Engine Firing Order: 1–5–3–6–2–4
Distributor Rotation: Clockwise

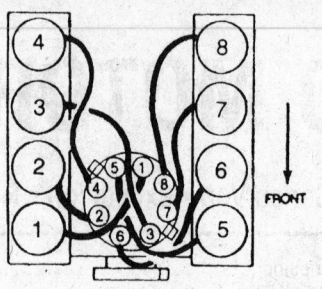

5.0L (VIN N) and 7.5L (VIN G) Engines
Engine Firing Order: 1–5–4–2–6–3–7–8
Distributor Rotation: Counterclockwise

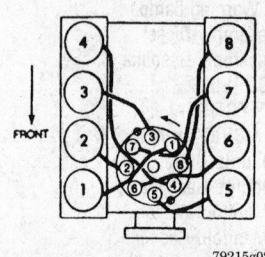

5.8L (VIN H and R) Engines
Engine Firing Order:
1–3–7–2–6–5–4–8
Distributor Rotation:
Counterclockwise

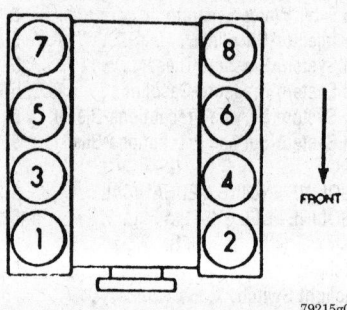

7.3L (VIN C, F, K, and M) Diesel Engines
Engine Firing Order: 1–2–7–3–4–5–6–8

ENGINE ELECTRICAL

NOTE: Disconnecting the negative battery cable on some vehicles may interfere with the functions of the on board computer systems and may require the computer to undergo a relearning process, once the negative battery cable is reconnected.

Distributor

REMOVAL

1. Disconnect the primary wiring connector from the distributor.
2. Mark the position of the cap's No. 1 terminal on the distributor base.
3. Unclip and remove the cap. Remove the adapter.
4. Remove the rotor.
5. Remove the TFI connector.
6. Matchmark the distributor base and engine for installation reference.
7. Remove the hold-down bolt and lift out the distributor.

INSTALLATION

Timing Not Disturbed

1. Visually inspect the distributor. The O-ring should fit tightly onto the housing and be free of cuts. The drive gear should be free of nicks, cracks or excessive wear. The distributor shaft should rotate freely, without any binding.
2. Lubricate the distributor gear teeth with a coating of engine assembly lubricant, such as Ford D9AZ-19579-D, or with fresh motor oil meeting Ford specification ESR-M99C80-A.
3. Align the locating boss and fully seat the distributor rotor on the distributor shaft, if removed.
4. Rotate the distributor shaft so that he distributor rotor blade points toward the marked position on the distributor base adapter.
5. Install the distributor assembly into the engine block with a slight side-to-side twist.

NOTE: If the vane and vane switch assembly cannot be kept on the leading edge after installation, remove the distributor from the cylinder block by pulling upward enough for the distributor gear to disengage the distributor gear from the camshaft gear. Rotate the distributor rotor enough so that the gear will align on the next tooth of the camshaft gear.

6. Install the distributor hold-down clamp and bolt; leave it snug.
7. On V8 engines, position the adapter base in place, then install the attaching bolts.
8. Attach the electrical connector to the distributor.
9. Install the distributor cap. On 4.9L (VIN Y and Z) engines, tighten the distributor cap hold-down screws to 18–23 inch lbs. (2.0–2.6 Nm). On

V8 engines, secure the distributor cap using the spring clips. If the spark plug wires were removed from the distributor cap, install them in their proper position, as marked during the removal procedure.

10. Connect the negative battery cable. Check the initial timing according to the proper procedure.

11. Adjust the timing, as necessary, then tighten the distributor hold-down bolt to 17–25 ft. lbs. (23–34 Nm).

Timing Disturbed

1. Disconnect the No. 1 spark plug wire and remove the No. 1 spark plug.

2. Place a finger over the spark plug hole and crank the engine slowly until compression is felt.

3. Align the TDC mark on the crankshaft pulley with the pointer on the timing cover. This places the No. 1 cylinder at TDC on the compression stroke.

4. Turn the distributor shaft until the rotor points to the No. 1 spark plug tower on the cap.

5. Install the distributor assembly into the engine block with a slight side-to-side twist.

NOTE: If the vane and vane switch assembly cannot be kept on the leading edge after installation, remove the distributor from the cylinder block by pulling upward enough for the distributor gear to disengage the distributor gear from the camshaft gear. Rotate the distributor rotor enough so that the gear will align on the next tooth of the camshaft gear.

6. Install the distributor hold-down clamp and bolt; leave it snug.

7. On V8 engines, position the adapter base in place, then install the attaching bolts.

8. Attach the electrical connector to the distributor.

9. Install the distributor cap. On 4.9L (VIN Y and Z) engines, tighten the distributor cap hold-down screws to 18–23 inch lbs. (2.0–2.6 Nm). On V8 engines, secure the distributor cap using the spring clips. If the spark plug wires were removed from the distributor cap, install them in their proper position, as marked during the removal procedure.

10. Connect the negative battery cable. Check the initial timing according to the proper procedure.

11. Adjust the timing, as necessary, then tighten the distributor hold-down bolt to 17–25 ft. lbs. (23–34 Nm).

Ignition Timing

ADJUSTMENT

NOTE: Always refer to the Vehicle Emission Information Label to verify the timing adjustment procedure and ignition specifics, which may have changed during the manufacture year.

Distributorless Ignition System

Base timing for distributorless engines is set from the factory at 10 degrees BTDC and is not adjustable.

Distributor Ignition System

1. Place automatic transmissions in **P** or manual transmissions in Neutral. The air conditioning and heater controls should be in the **OFF** position.

2. Connect a suitable inductive timing light and a tachometer according to the manufacturer's instructions.

3. Disconnect the single wire in-line spout connector or remove the shorting bar from the double wire spout connector.

4. Start the engine and bring it up to normal operating temperature.

NOTE: To set timing correctly, a remote starter should not be used. Use the ignition key only to start the vehicle. Disconnecting the start wire at the starter relay will cause the TFI module to revert to "start mode timing" after the vehicle is started. Reconnecting the start wire after the vehicle is running will not correct the timing.

5. With the engine running at the timing rpm specified, check the initial timing by aiming the timing light at the timing marks and pointer. Refer to the underhood Vehicle Emission Information Label for specific specifications.

6. If the marks do not align, shut the engine off and loosen the distributor hold-down clamp bolt. Start the engine, and while watching the timing marks with the timing light, turn the distributor until the marks are correctly aligned. Shut the engine off, then tighten the distributor hold-down clamp bolt to 17–25 ft. lbs. (23–34 Nm).

7. Reconnect the single wire in-line spout connector or reinstall the shorting bar on the double wire spout connector. Check the timing advance to verify the distributor is advancing beyond the initial setting.

8. Remove the timing light and tachometer.

Alternator

PRECAUTIONS

Several precautions must be observed with alternator equipped vehicles to avoid damage to the unit.

• If the battery is removed for any reason, make sure it is reconnected with the correct polarity. Reversing the battery connections may result in damage to the one-way rectifiers.

• When utilizing a booster battery as a starting aid, always connect the positive to positive terminals and the negative terminal from the booster battery to a good engine ground on the vehicle being started.

• Never use a fast charger as a booster to start vehicles.

• Disconnect the battery cables when charging the battery with a fast charger.

• Never attempt to polarize the alternator.

• Do not use test lights of more than 12V when checking diode continuity.

• Do not short across or ground any of the alternator terminals.

• The polarity of the battery, alternator and regulator must be matched and considered before making any electrical connections within the system.

• Never separate the alternator on an open circuit. Make sure all connections within the circuit are clean and tight.

• Disconnect the battery ground terminal when performing any service on electrical components.

• Disconnect the battery if arc welding is to be done on the vehicle.

REMOVAL AND INSTALLATION

1. Disconnect the negative battery cable.

2. Disconnect the electrical connectors from the alternator.

3. Remove the alternator drive belt.

4. Remove the pivot bolt and adjusting bolt, or the mounting bolts, then remove the alternator from the engine. Remove the alternator fan shield, if equipped.

To install:

5. Install the fan shield, if removed.

6. Situate the alternator in position, then install the applicable

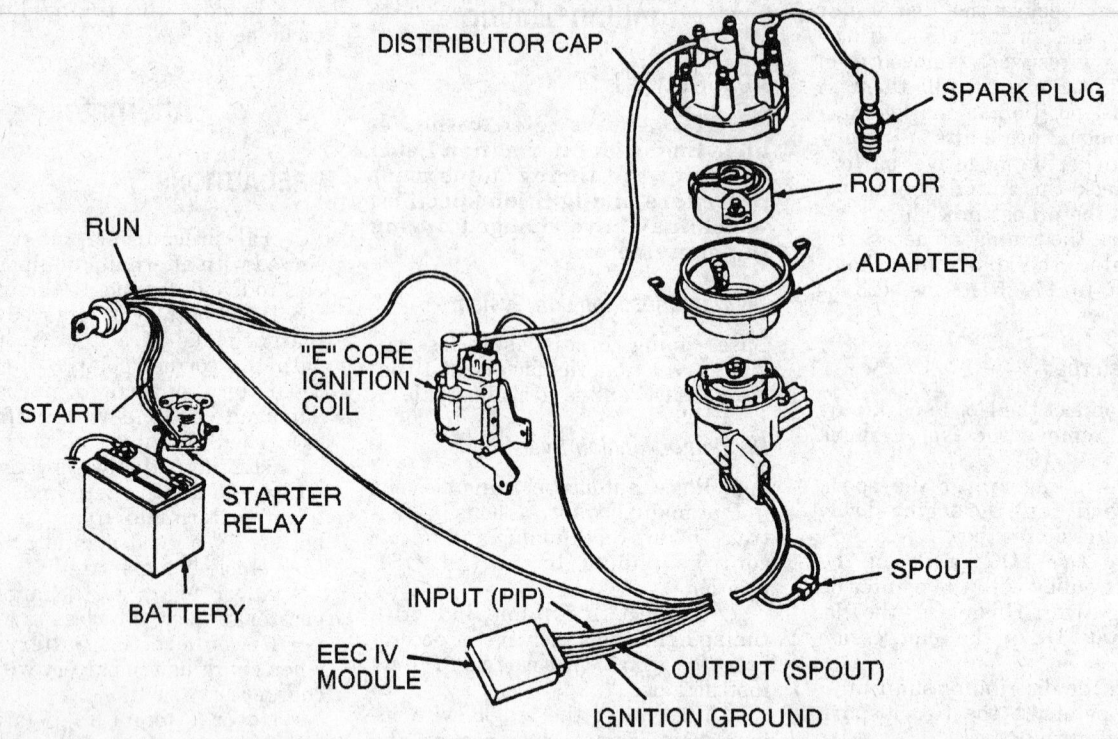

Thick Film integrated (TFI) ignition system with universal distributor — 4.9L (VIN Y and Z), 5.0L (VIN N), 5.8L (VIN H and R) and 7.5L (VIN G) engines

mounting bolts. Tighten the bolts to the following values:

- 4.2L (VIN 2) Engine — 18–22 ft. lbs. (20–30 Nm)
- 4.6L Engines — 18–22 ft. lbs. (20–30 Nm)
- 4.9L (VIN Y and Z) and 7.5L (VIN G) Engines — 16–21 ft. lbs. (21–29 Nm)
- 5.0L (VIN N), 5.8L (VIN H and R) and 7.3L (VIN C, F, K, and M) (Except F-Super Duty) Engines — 16–21 ft. lbs. (21–29 Nm)
- 7.3L (VIN C, F, K, and M) (F-Super Duty) Engine — 30–41 ft. lbs. (40–55 Nm)

7. Install the alternator drive belt. On 7.3L (VIN C, F, K, and M) engines, properly tension the drive belt.

8. Attach all wiring harness electrical connections to the alternator.

NOTE: When the battery has been disconnected and reconnected, some abnormal drive symptoms may occur while the Powertrain Control Module (PCM) relearns its adaptive strategy. The vehicle may need to be driven 10 miles (16 km) or more to relearn its strategy.

9. Connect the negative battery cable.

Drive Belt

REMOVAL AND INSTALLATION

Except 7.5L (VIN G) Alternator Belt

NOTE: All accessory drive belts, except for the alternator drive belt on 7.5L (VIN G) engines, are automatically tensioned. These belts do not require manual adjustment.

1. Using a box end wrench on the tensioner pulley bolt, lift the tensioner arm away from the belt.

2. Remove the old belt. Release the tensioner slowly. Do not allow the tensioner to snap back after the belt is removed because this may damage the tensioner.

To install:

———— WARNING ————
Make sure that the belt is properly seated on all pulleys. One revolution of the engine with an incorrectly seated belt may snap tensile members in the belt, thereby greatly weakening the drive belt.

3. Lift the tensioner arm, then install a new belt over the pulleys making sure that all belt ribs are correctly seated in the pulley.

4. Slowly allow the tensioner to tack up all slack of the drive belt.

7.5L (VIN G) Alternator Belt

1. Loosen the alternator adjustment bolts and pivot bolt.

2. Remove the old belt from the pulleys.

To install:

3. Install the new belt onto the drive pulleys. Make certain that the belt is correctly seated on the pulleys.

4. Tension the drive belt as follows:

a. Use Rotunda Belt Tension Gauge 021–00019, or equivalent, for checking the tension.

b. Loosen the alternator adjustment bolts.

c. Tighten the pivot bolt enough to remove all free-play from the pivot area, but not enough to prevent the alternator from moving.

d. Place an adjustable wrench on the top lug of the alternator and rotate the alternator until the belt is tightened to the proper tension, which is 160–200 lbs. (711–889 N) for new belts or 110–130 lbs. (489–578 N) for used belts. Be sure to use the new belt tension for new belts and the used belt tension for used belts. A used belt is a belt

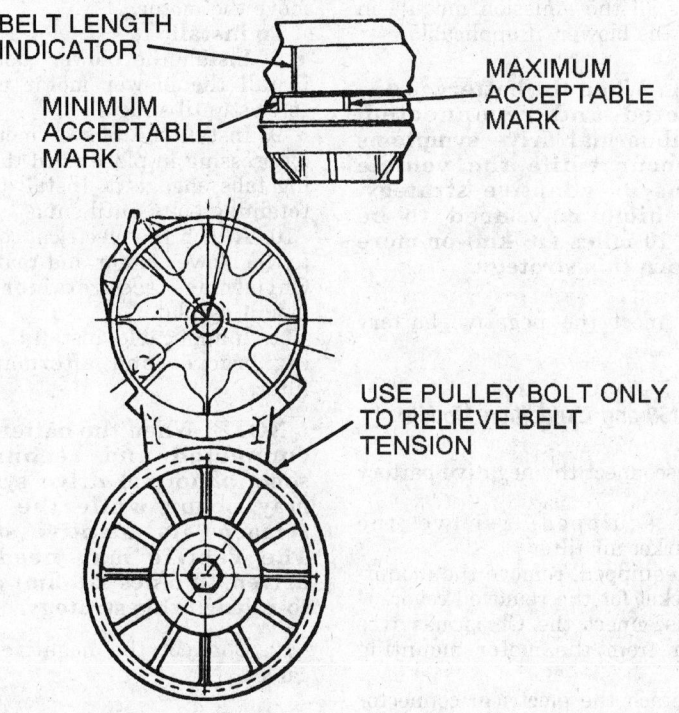

INDICATOR SHOULD BE BETWEEN MARKS

BELT LENGTH INDICATOR

MINIMUM ACCEPTABLE MARK

MAXIMUM ACCEPTABLE MARK

USE PULLEY BOLT ONLY TO RELIEVE BELT TENSION

79215g05

Automatic belt tensioner — 4.9L (VIN Y and Z) engine

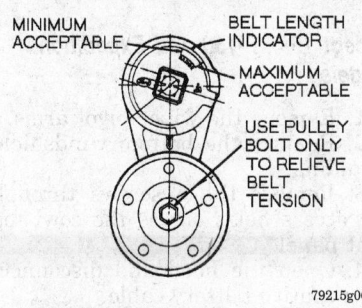

INDICATOR SHOULD BE BETWEEN MARKS

MINIMUM ACCEPTABLE

BELT LENGTH INDICATOR

MAXIMUM ACCEPTABLE

USE PULLEY BOLT ONLY TO RELIEVE BELT TENSION

79215g06

Automatic belt tensioner — 5.0L (VIN N) and 5.8L (VIN H and R) engines

with more than 5 minutes of engine operation.

5. While maintaining the belt tension, tighten the adjustment bolt to 22–29 ft. lbs. (30–40 Nm) for E Series vehicles, and to 30–40 ft. lbs. (41–54 Nm) for F Series and Bronco vehicles.

6. Tighten the pivot bolt to 40–53 ft. lbs. (54–71 Nm).

7. Check and readjust the belt tension, if necessary.

8. Start the engine and allow it to idle for 5 minutes.

9. Recheck the belt tension.

10. If drive belt tension is less than the minimum after 5 minutes operation, then reset belt tension to the used belt tension value.

11. Start the engine again and allow it to idle to 5 minutes, then recheck the belt tension.

12. If the belt will not hold the correct tension, replace it with another one.

Starter

REMOVAL AND INSTALLATION

1. Disconnect the negative battery cable.

2. Raise the vehicle and support it safely.

3. Remove the starter motor solenoid terminal cover, if equipped.

4. Disconnect all wiring harness electrical connections from the starter motor.

5. Remove the starter mounting bolts and remove the starter.

To install:

6. Position the starter motor in place on the engine, then install the mounting bolts to 16–20 ft. lbs. (22–28 Nm).

7. Attach all electrical wiring harness connections to the starter motor.

8. Install the starter motor solenoid terminal cover, if applicable.

9. Lower the vehicle to the ground.

NOTE: When the battery is disconnected and reconnected, some abnormal drive symptoms may occur while the vehicle relearns its adaptive strategy. The vehicle may need to be driven 10 miles (16 km) or more to relearn this strategy.

10. Connect the negative battery.

Diesel Glow Plugs

TESTING

—— **CAUTION** ——
The red-striped wiring harness carries 115v direct current. Severe electrical shock may be received. DO NOT pierce.

1. Turn the ignition switch **OFF**, then detach the electrical connectors from the glow plugs.

2. Using a test light, check for continuity between the glow plug terminal and a power source, with the glow plugs installed in the engine.

3. If there is no continuity, the glow plug must be replaced.

REMOVAL AND INSTALLATION

—— **CAUTION** ——
The red-striped wiring harness carries 115v direct current. Severe electrical shock may be received. DO NOT pierce.

1. Disconnect the negative battery cable.

2. Remove the rocker arm cover.

3. Disconnect the glow plug electrical leads using a pair of pliers.

4. Remove the glow plugs by unscrewing them from the cylinder head with a 10mm socket and wrench.

5. Inspect the tips of the plugs for any evidence of distortion or missing tip ends; replace them if necessary.

To install:

6. Install the glow plug into the cylinder head. Tighten the glow plugs to 14 ft. lbs. (19 Nm).

7. Attach the glow plug electrical connector. Make sure that the glow plug wiring is routed to avoid moving components in the engine bay.

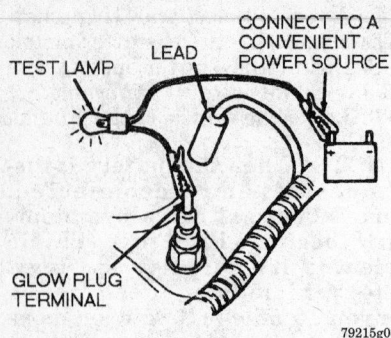

Diesel glow plug testing — 7.3L (VIN C, F, K, and M) engine

8. Install the rocker arm cover.

NOTE: When the battery is disconnected and reconnected, some abnormal drive symptoms may occur while the vehicle relearns its adaptive strategy. The vehicle may need to be driven 10 miles (16 km) or more to relearn this strategy.

9. Connect the negative battery cable.

CHASSIS ELECTRICAL

Heater Blower Motor

REMOVAL AND INSTALLATION

Except 1997 F-150 and Expedition Models

1. Disconnect the battery ground.
2. On vehicles built for sale in California, remove the emission module located in front of the blower.
3. Disconnect the wiring harness at the blower.
4. Disconnect the blower motor cooling tube at the blower.
5. Remove the 3 blower motor mounting bolts.
6. Hold the cooling tube to one side and pull the blower motor from the housing.
   To install:
7. Position the blower motor in the housing, then install the motor retaining bolts.
8. Install the blower motor cooling tube at the blower. Cement the cooling tube on the nipple at the housing using Liquid Butyl Sealer D9AZ–19554–A, or equivalent.

9. Attach the electrical wiring harness connector to the motor.
10. Install the emission module in front of the blower, if applicable.

NOTE: When the battery is disconnected and reconnected, some abnormal drive symptoms may occur while the vehicle relearns its adaptive strategy. The vehicle may need to be driven 10 miles (16 km) or more to relearn this strategy.

11. Connect the negative battery cable.

1997 F-150 and Expedition Models

1. Disconnect the negative battery cable.
2. If equipped, remove the aftermarket air filter.
3. If equipped, remove the mounting bracket for the removed cover.
4. Disconnect the Christmas tree retainer from the motor mounting plate.
5. Detach the electrical connector from the motor.
6. Remove the 3 blower motor cover mounting bolts, release the 3 retaining tabs, then remove the motor cover.

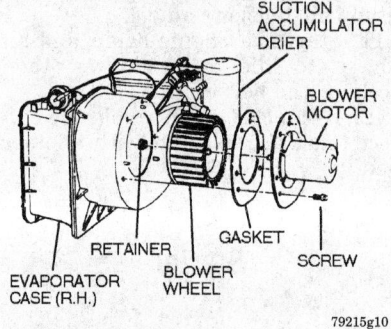

Blower motor and wheel installation — Bronco and F Series

7. Remove the 3 blower motor mounting screws, then carefully remove the motor.
   To install:
8. Install the blower motor, then install the blower motor mounting screws until snug.
9. Install the blower motor cover by pressing in place until the retaining tabs engage it. Install the cover retaining bolts until snug.
10. Attach the electrical connection to the blower motor and reattach the Christmas tree retainer to the mounting plate.
11. If applicable, install the mounting bracket and aftermarket air filter.

NOTE: When the battery is disconnected and reconnected, some abnormal drive symptoms may occur while the vehicle relearns its adaptive strategy. The vehicle may need to be driven 10 miles (16 km) or more to relearn this strategy.

12. Connect the negative battery cable.

Windshield Wiper Motor

REMOVAL AND INSTALLATION

Except 1997 F-150 and Expedition Models

1. Remove the wiper pivot arms.
2. Remove the bottom windshield moulding.
3. Remove the 5 screws through the access holes across the cowl top vent panel.
4. Open the hood and disconnect the negative battery cable.
5. Remove the 10 screws across the vertical surface attaching the cowl top vent panel to the cowl top.
6. Close the hood part way and carefully lift the cowl top vent panel.

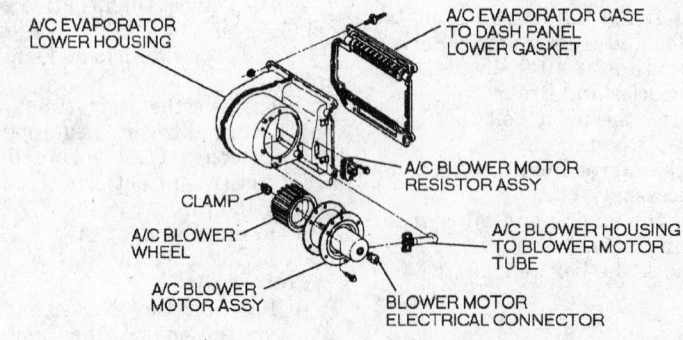

Blower motor and wheel installation — E Series

7. Disconnect or remove the radio antenna or lead wire, as required.

8. Disconnect the washer hose. Remove the cowl top vent panel.

9. Remove the metal shield above the windshield wiper motor.

10. Remove the windshield wire adapter and connecting the arm clip connecting the windshield wiper mounting arm and pivot shaft to the windshield wiper motor drive arm.

11. Detach the electrical connector from the windshield wiper motor.

12. Remove the bolts attaching the windshield wiper motor to the cowl top.

13. Remove the windshield wiper motor from the vehicle.

To install:

14. Install the wiper motor in place, then install the retaining bolts to 60–85 inch lbs. (7–10 Nm).

15. Attach the electrical connector to the wiper motor.

16. Attach the windshield wiper mounting arm and pivot shaft to the windshield wiper motor drive arm.

17. Install the metal shield above the wiper motor.

18. Install the cowl top vent panel, then attach the washer hose.

19. Attach the radio antenna, if necessary.

20. Install the cowl top vent panel retaining fasteners until snug.

NOTE: When the battery has been disconnected and reconnected, some abnormal drive symptoms may occur while the Powertrain Control Module (PCM) relearns its adaptive strategy. The vehicle may need to be driven 10 miles (16 km) or more to relearn the strategy.

21. Connect the negative battery cable.

22. Install the bottom windshield moulding.

23. Before installing the wiper arms, cycle the windshield wiper motor using the wiper switch to make sure that the wiper linkage is in the park position.

24. Install the wiper pivot arms.

1997 F-150 and Expedition Models

1. Disconnect the negative battery cable.

NOTE: The windshield wiper mounting arms and pivot shafts are serviced as an assembly.

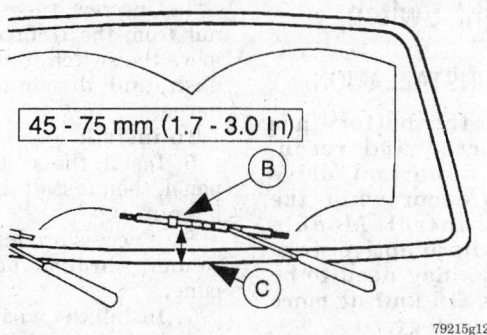

If the left-hand wiper blade is not positioned within specification, reposition as shown — 1997 F-150 and Expedition models

2. Turn the ignition switch to the **ON** position.

3. Turn the windshield wiper switch on, then stop the windshield wiper pivot arms at the highest position by turning the ignition switch to the **OFF** position.

4. Remove the windshield wiper pivot arms as follows:

   a. Remove the two pivot arm nut covers.

   b. Remove the windshield wiper pivot arm retaining nuts.

   c. Remove the windshield wiper pivot arms.

5. Remove the cowl top vent panels.

6. Remove the windshield wiper motor/linkage assembly as a unit as follows:

   a. Detach the electrical connector from the wiper motor.

   b. Remove the 3 assembly hold-down bolts.

   c. Lift the windshield wiper motor/linkage assembly out of the cowl recess.

7. Remove the wiper linkage from the wiper motor by removing the linkage-to-motor shaft retaining bolt,

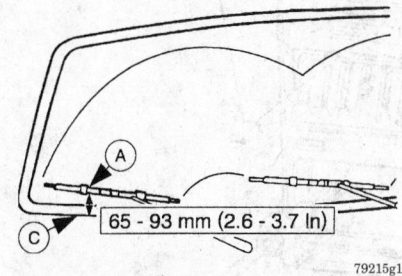

Make certain to position the left-hand wiper blade as shown — 1997 F-150 and Expedition models

then detach the linkage arm from the motor shaft.

8. Remove the 3 wiper motor-to-motor/linkage assembly bolts, then pull the motor out of the assembly.

To install:

9. Situate the wiper motor in position in the motor/linkage assembly, then install the hold-down bolts to 10–12 ft. lbs. (13–17 Nm).

10. Attach the wiper linkage to the motor shaft and install the wiper linkage-to-motor shaft retaining nut to 11–14 ft. lbs. (15–20 Nm).

11. Set the wiper motor/linkage assembly into the windshield cowl recess, then secure with the 3 hold-down bolts to 62–80 inch lbs. (7–9 Nm). Attach the wiper motor electrical connector.

12. Install the cowl top vent panels.

13. Install the wiper arms in the same position as when removed. Install the wiper arm-to-wiper motor shaft nut to 23–29 ft. lbs. (30–40 Nm).

14. Turn the ignition switch **ON** to cycle the wiper arms to the Park position to assure that the wiper arms are installed in the correct position. Turn the wiper switch Off once the wipers are in the Park position.

15. Measure from the center of the left-hand wiper blade to the bottom edge of the windshield moulding. The distance should be 2.6–3.7 in. (65–93mm). Measure the same distance on the right-hand wiper blade. The right-hand wiper blade distance should be 1.7–3.0 in. (45–75mm) above the bottom edge of the windshield moulding.

16. If the distances are not within specifications, reposition the wiper arms on the wiper motor shafts.

17. Connect the negative battery cable.

Headlight Switch

REMOVAL AND INSTALLATION

NOTE: When the battery has been disconnected and reconnected, some abnormal drive symptoms may occur while the Powertrain Control Module (PCM) relearns its adaptive strategy. The vehicle may need to be driven 10 miles (16 km) or more to relearn the strategy.

Except 1997 F-150 and Expedition Models

1. Disconnect the battery ground cable.

2. Depending on the year and model remove the wiper-washer and fog lamp switch knob if they will interfere with the headlight switch knob removal. Check the switch body (behind dash, see Step 3) for a release button. Press in on the button and remove the knob and shaft assembly. If not equipped with a release button, a hook tool may be necessary for knob removal.

3. Remove the steering column shrouds and cluster panel finish panel if they interfere with the required clearance for working behind the dash.

4. Unscrew the switch mounting nut from the front of the dash. Remove the switch from the back of the dash and disconnect the wiring harness.

To install:

5. Install the switch into the dash panel, then install the switch mounting nut.

6. If necessary, install the steering column shrouds or cluster finish panel.

7. Install the wiring to the switch, then install the switch knob.

8. Connect the negative battery cable.

1997 F-150 and Expedition Models

1. Disconnect the negative battery cable.

2. Turn the head lamp switch knob to the head lamp position and pull the knob.

3. Insert a thin tool under the backside off the switch knob (opposite end of switch knob from the indicator mark) to release the head lamp switch knob and remove.

4. Turn the head lamp switch knob 180 degrees and reinstall onto the switch.

5. Turn the knob, now reinstalled, counterclockwise until the back of the head lamp switch knob is in the OFF position.

6. Turn the head lamp switch knob clockwise 360 degrees (one full turn).

7. Pull the head lamp from the instrument panel.

8. Detach the electrical connectors from the switch.

To install:

9. Attach the electrical connectors to the head lamp switch, then press it into the dash panel.

10. Turn the switch knob counterclockwise 360 degrees (one full turn).

11. Turn the switch knob clockwise until the back of the knob is facing the ON position.

12. Once again remove the switch knob from the switch, rotate the knob 180 degrees and reinstall onto the switch (the indicator mark should now be facing the on position).

13. Turn the head lamp switch to the off position.

14. Connect the negative battery cable.

Combination Switch

REMOVAL AND INSTALLATION

NOTE: When the battery has been disconnected and reconnected, some abnormal drive symptoms may occur while the Powertrain Control Module (PCM) relearns its adaptive strategy. The vehicle may need to be driven 10 miles (16 km) or more to relearn the strategy.

1997 F-150 and Expedition Models

1. Disconnect the negative battery cable.

2. Insert the ignition key into the ignition switch lock cylinder and turn to the **RUN** position.

3. Push the ignition switch lock cylinder release tab with a punch while pulling out the ignition switch lock cylinder.

4. If equipped, twist the tilt wheel handle and shank, then remove it.

5. Remove the upper and lower steering column shrouds by removing the 3 retaining screws.

6. Remove the 2 multi-function switch retaining screws (1), then detach the electrical connectors (2) from the switch (3).

7. Remove the switch from the steering column.

To install:

8. Attach the electrical connectors to the switch, position the switch on the steering column, then install the

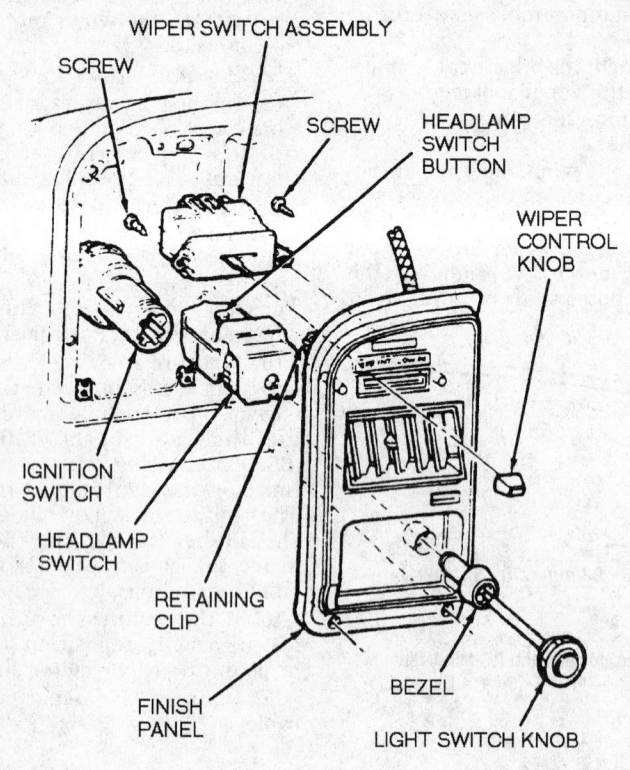

WIPER SWITCH ASSEMBLY
SCREW
SCREW
HEADLAMP SWITCH BUTTON
WIPER CONTROL KNOB
IGNITION SWITCH
HEADLAMP SWITCH
RETAINING CLIP
FINISH PANEL
BEZEL
LIGHT SWITCH KNOB

79215g13

Headlight switch — 1993–96 E Series models

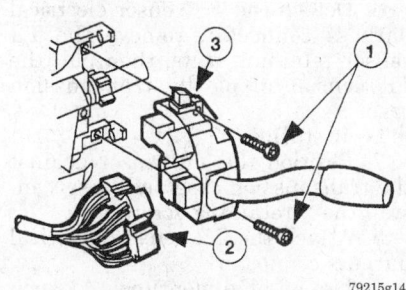

79215g14

The combination switch is mounted to the side of the steering column by 2 retaining screws — 1997 F-150 and Expedition models

retaining screws to 19–25 inch lbs. (2.1–2.9 Nm).

9. Install the upper and lower steering column shrouds.

10. Install the tilt wheel handle and shank, if removed.

11. With the ignition key in the lock cylinder, and in the **RUN** position, push the ignition lock cylinder back into the cylinder mounting recess.

12. Turn the ignition switch **OFF** and remove the ignition key.

13. Connect the negative battery cable.

Except 1997 F-150 and Expedition Models

1. Disconnect the battery ground cable.

2. Remove the steering column shroud.

3. Remove the 2 self-tapping screws that attach the combination switch to the steering column tube casting. Disengage the combination switch from the casting.

4. Detach the 2 electrical connectors from the switch, using caution not to damage the locking tabs. Do not damage the PRNDL cable.

To install:

5. Install the 2 switch electrical connectors to full engagement. The wiring for the switch is to be routed under the PRNDL cable.

6. Align the mounting holes on the combination switch with the holes in the steering column tube casting. Install 2 self-tapping screws making sure to start the screws in the previously tapped holes.

7. For vehicles equipped with automatic transmissions, verify that the PRNDL adjustment is correct.

8. Connect the battery cable.

9. Check the steering column tube for proper operation.

10. Install the steering column shroud.

Ignition Lock Cylinder

REMOVAL AND INSTALLATION

NOTE: When the battery has been disconnected and reconnected, some abnormal drive symptoms may occur while the Powertrain Control Module (PCM) relearns its adaptive strategy. The vehicle may need to be driven 10 miles (16 km) or more to relearn the strategy.

All Models

Functional Cylinder

1. Disconnect the negative battery cable.

2. Insert the ignition key into the ignition switch lock cylinder and turn to the **RUN** position.

3. Push the ignition switch lock cylinder release tab/pin with a punch while pulling out the ignition switch lock cylinder.

To install:

4. While depressing the release tab/pin and with the ignition key in the lock cylinder in the **RUN** position, push the ignition lock cylinder back into the cylinder mounting recess. Make sure that the ignition switch lock cylinder is fully seated and aligned in the interlocking washer before turning the ignition key **OFF**. This will permit the retaining pin of the ignition switch lock cylinder to extend into the hole in the steering column.

5. Turn the ignition switch **OFF**. Make sure that the ignition lock moves to all positions and functions properly, then remove the ignition key.

6. Connect the negative battery cable.

Non-Functional Cylinder

NOTE: The following procedure applies to vehicles in which the ignition lock is inoperative and the lock cylinder cannot be rotated due to a lost or broken key, unknown key number, or a cap that has been damaged or broken to the extent that the lock cylinder cannot be rotated.

1. Disconnect the negative battery cable.

2. Remove the steering wheel.

3. Using a pair of pliers, or similar tool, twist (clockwise direction) the cap off of the ignition switch lock cylinder.

4. Use a 3/8 in. diameter drill bit to drill down the middle of the ignition

lock key slot until the ignition switch lock cylinder breaks loose.

5. Remove and discard the old ignition lock cylinder. Clean all drill shavings from the steering column.

6. Remove the bearing retainer.

7. Remove the bearing and gear from the steering column. Thoroughly clean all drill shavings from the steering column and inspect the bore for any damage.

To install:

8. Install the gear and bearing into their original positions. Install the bearing retainer.

9. Install a new ignition lock cylinder by inserting it into the bore until engaged fully.

10. Install the steering wheel.

11. Connect the negative battery cable.

Ignition Switch

REMOVAL AND INSTALLATION

NOTE: When the battery has been disconnected and reconnected, some abnormal drive symptoms may occur while the Powertrain Control Module (PCM) relearns its adaptive strategy. The vehicle may need to be driven 10 miles (16 km) or more to relearn the strategy.

1997 F-150 and Expedition Models

1. Disconnect the negative battery cable.

2. Release the 4 clips, then remove the steering column opening cover.

3. Release the 3 clips, then remove the lower instrument panel steering column.

4. Remove the hood latch mounting screws, then position the hood latch handle aside.

5. Remove the parking brake release handle mounting screws. Position the parking brake release handle aside.

6. Remove the 2 push clips and release the one expander clip to remove the instrument panel floor duct panel.

7. Remove the 6 upper instrument panel steering column cover retaining bolts, then remove the panel.

8. Remove the ignition switch electrical harness connector retaining bolt, then pull the connector from the switch.

9. Remove the 2 switch retaining screws (A) and pull the switch (B) from the steering column.

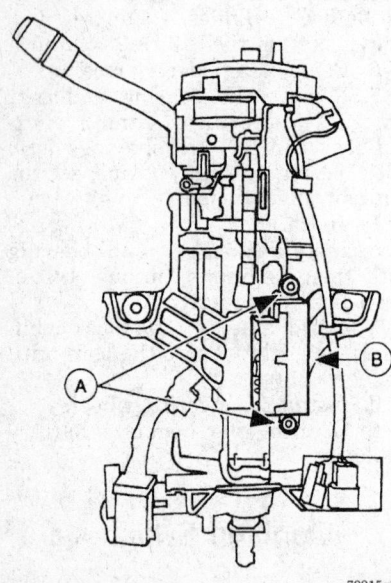

79215g15

The ignition switch is mounted to the steering column assembly by 2 retaining screws — 1997 F-150 and Expedition models

To install:

10. Position the ignition switch in place, then install the 2 retaining screws until snug.

11. Attach the switch electrical harness connector and install the connector-to-switch retaining bolt until secure.

12. Install the instrument panel steering column cover and the instrument panel floor duct panel.

13. Position the parking brake release handle in its original position, then secure in place with the retaining screws. Perform the same with the hood latch handle.

14. Install the lower instrument panel steering column until the 3 clips engage fully. Install the steering column opening cover in the same manner.

15. Connect the negative battery cable.

Except 1997 F-150 and Expedition Models

1. Disconnect the battery ground cable.

2. Remove the steering column shroud and lower the steering column.

3. Disconnect the switch wiring at the multiple plug.

4. Remove the two nuts that retain the switch to the steering column.

5. Lift the switch vertically upward to disengage the actuator rod from the switch and remove the switch.

6. When installing the ignition switch, both the locking mechanism at the top of the column and the switch itself must be in the **LOCK** position for correct adjustment. To hold the mechanical parts of the column in the **LOCK** position, move the shift lever into **P** (with automatic transmissions) or **REVERSE** (with manual transmissions), turn the key to the **LOCK** position, and remove the key. New replacement switches, when received, are already pinned in the **LOCK** position by a metal shipping pin inserted in a locking hole on the side of the switch.

7. Engage the actuator rod in the switch.

8. Position the switch on the column and install the retaining nuts, but do not tighten them.

9. Move the switch up and down along the column to locate the mid-position of rod lash, and then tighten the retaining nuts.

10. Remove the locking pin, connect the battery cable, and check for proper start in **P** or **NEUTRAL**. Also check to make certain that the start circuit cannot be actuated in the **D** and **R** position.

11. Raise the steering column into position at instrument panel. Install steering column shroud.

Park/Neutral Safety Switch

REMOVAL AND INSTALLATION

NOTE: On Ford vehicles, the Park/Neutral Safety Switch is referred to as the Transmission Range (TR) sensor. Although given a different designation, the TR performs the same job as the Park/Neutral Safety Switch.

1997 F-150 and Expedition Models

1. Disconnect the negative battery cable.

2. Raise and support the vehicle safely on jackstands.

3. Disconnect the shift cable from the transmission.

4. Place the transmission in Neutral.

5. Remove and discard the manual control lever nut, then remove the lever from the transmission.

6. Detach the TR sensor electrical harness connector, remove the TR sensor retaining bolts, then pull the TR sensor off of the transmission case.

To install:

7. Position the TR sensor against the transmission case and loosely install the 2 retaining screws.

8. Attach the TR sensor electrical harness connector.

9. Use an alignment tool (A), such as the TR Sensor Alignment Tool T93P-70010-A, to align the TR sensor slots (B).

10. Tighten the TR sensor bolts to 62–89 inch lbs. (7–10 Nm).

11. Position the manual control lever onto the TR sensor, then install the NEW lever nut to 22–26 ft. lbs. (30–35 Nm).

12. Connect the shift cable.

13. Lower the vehicle.

14. Connect the negative battery cable.

Except 1997 F-150 and Expedition Models

C6 Transmission

1. Disconnect the negative battery cable.

2. Raise and support the vehicle safely on jackstands.

3. Remove the downshift linkage rod return spring at the low-reverse servo cover.

4. Coat the outer lever attaching nut with penetrating oil. Remove the nut and lever.

5. Note or matchmark the position of the switch.

6. Remove the 2 switch attaching bolts, disconnect the wiring at the connectors and remove the switch.

To install:

7. Attach the switch wiring connector to the switch, then install the switch itself in the original position. Tighten the switch bolts to 55–75 inch lbs. (6.1–8.4 Nm).

8. Install the outer lever and attaching nut.

9. Install the downshift linkage return spring.

10. Lower the vehicle.

11. Connect the negative battery cable.

AOD Transmission

1. Disconnect the negative battery cable.

2. Raise and support the vehicle safely on jackstands.

3. Disconnect the wiring from the switch.

4. Using a deep socket, unscrew the switch.

To install:

5. Thread the switch into the transmission by hand until the threads are started. Tighten the switch to 10 ft. lbs. (14 Nm).

6. Attach the switch wiring connector, then lower the vehicle.

7. Connect the negative battery cable.

Powertrain Control Module (PCM)

REMOVAL AND INSTALLATION

1997 F-150 and Expedition Models

─────── **WARNING** ───────
Always disconnect the battery ground cable prior to work on the electronic engine control components.

1. Disconnect the negative battery cable.

2. Loosen the PCM electrical connector retaining bolt and separate the connector from the PCM.

3. Remove the right-hand front door scuff plate.

4. Remove the right-hand cowl side trim panel.

5. Remove the PCM bracket clip.

6. Remove the PCM from the vehicle.

To install:

7. Situate the PCM in position in the vehicle, then secure in the PCM bracket clip.

8. Install the cowl side trim panel and the front door scuff plate.

9. Attach the electrical connector to the PCM, then install and tighten the connector-to-PCM retainer bolt to 46–64 inch lbs. (5.2–7.2 Nm).

10. Connect the negative battery cable.

Except 1997 F-150 and Expedition Models

1. Disconnect the negative battery cable.

2. Loosen the PCM electrical connector retaining bolt and separate the connector from the PCM.

3. Remove the 2 PCM seal nuts and loosen the left fender liner screws. Remove the fender liner or bend it down out of the way to clear the PCM.

4. Remove the PCM from the vehicle.

To install:

5. Situate the PCM in position in the vehicle, then install the fender liner.

6. Install the 2 PCM seal nuts.

7. Attach the electrical connector to the PCM, then install and tighten the connector-to-PCM retainer bolt to 35 inch lbs. (4 Nm).

8. Connect the negative battery cable.

ENGINE COOLING

Radiator

REMOVAL AND INSTALLATION

NOTE: When the battery has been disconnected and reconnected, some abnormal drive symptoms may occur while the Powertrain Control Module (PCM) relearns its adaptive strategy. The vehicle may need to be driven 10 miles (16 km) or more to relearn the strategy.

NOTE: Two special tools will be needed for the following procedure. These tools are the Fan Clutch Holding Tool T84T-6312-C and Fan Clutch Wrench T93T-6312-B, or their equivalents.

1. Disconnect the negative battery cable.

2. Drain the cooling system.

3. Remove the front air deflector.

4. Remove the upper radiator hose from the radiator.

5. Remove the radiator coolant recovery reservoir hose from the radiator.

NOTE: The fan clutch has left-hand (reverse) threads.

6. Use the fan clutch holding tool to hold the water pump pulley steady, then remove the fan blade and fan clutch with the fan clutch wrench.

7. Separate the fan clutch from the fan blade assembly.

8. Remove the 2 fan shroud screws, then lift the fan shroud out of the engine compartment.

9. If necessary, disconnect the upper and lower transmission cooler lines from the radiator. Plug the open ends of the lines.

10. Remove the lower radiator hose from the radiator.

11. Remove the 2 upper radiator support bracket bolts. Remove the 2 upper radiator support brackets and the 2 jack handle and wheel nut wrench retainers.

12. Lift the radiator off of the mounting insulators, then remove the insulators themselves.

To install:

13. Install the 2 radiator mounting insulators, then position the radiator on the insulators. Position the 2 upper radiator support brackets and the 2 jack handle and wheel nut wrench retainers. Install the 2 upper radiator support bracket bolts to 18–26 ft. lbs. (25–35 Nm).

14. Connect the lower radiator hose to the radiator.

15. If applicable, connect the lower and upper transmission cooler lines to the radiator. Tighten the fittings to 12–18 ft. lbs. (16–24 Nm).

16. Lower the fan shroud onto the radiator clips, then install the two fan shroud retainer screws until snug.

17. Position the fan clutch on the fan blade assembly, then install the 4 fan clutch bolts to 11–15 ft. lbs. (15–20 Nm).

NOTE: The fan clutch has left-hand threads.

18. Position the fan clutch/blade assembly on the water pump pulley. Using the two special tools, install the fan clutch/blade assembly to the water pump pulley.

19. Push the radiator coolant recovery reservoir hose onto the radiator fitting, then position and tighten the hose clamp.

20. Connect the upper radiator hose to the radiator.

21. Install the front air deflector. Fill the cooling system.

22. Connect the negative battery cable, then start the engine and check for coolant leaks.

Water Pump

REMOVAL AND INSTALLATION

NOTE: When the battery has been disconnected and reconnected, some abnormal drive symptoms may occur while the Powertrain Control Module (PCM) relearns its adaptive strategy. The vehicle may need to be driven 10 miles (16 km) or more to relearn the strategy.

4.2L (VIN 2) Engine

1. Disconnect the negative battery cable.

2. Remove the radiator, fan blade assembly and fan shroud.

3. Remove the accessory drive belt.

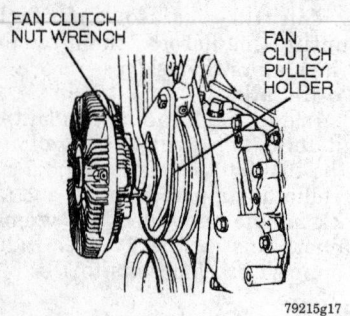

Fan and clutch assembly removal — 7.3L (VIN C, F, K, and M) Diesel engine shown, other engines similar

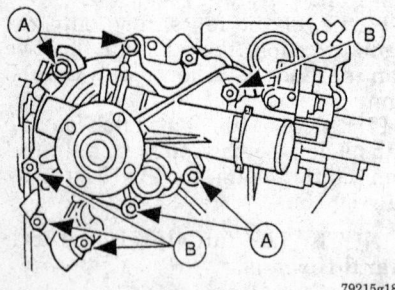

Note the locations of the water pump mounting bolts (A) and nuts (B) — 4.2L (VIN 2) engines

4. Remove the water pump pulley.

5. Remove the heater water outlet tube by removing the wiring harness, then the hold-down bolt.

6. Remove the 4 water pump bolts and the 4 water pump nuts.

7. Remove the water pump stud bolt, the water pump and the water pump housing gasket. Discard the water pump housing gasket.

To install:

8. Before water pump installation, make sure to aptly clean the water pump mounting surfaces of all dirt, grime and old gasket material.

NOTE: All water pump housing bolts, nuts and studs are tightened to 15–22 ft. lbs. (20–30 Nm).

9. Install the water pump onto the engine with a new gasket. Install the water pump stud bolt temporarily finger-tight.

10. Install the water pump mounting nuts and bolts temporarily finger-tight, then tighten all water pump housing fasteners to the correct torque specification in a crisscross fashion.

11. Install the heater water outlet tube, then install and tighten the hold-down bolt to 71–97 inch lbs. (8–11 Nm). Install the wiring harness.

12. Install the water pump pulley and tighten the 4 mounting bolts to 16–21 ft. lbs. (21–29 Nm).

13. Install the accessory drive belt.

14. Install the fan shroud, fan blade assembly and the radiator. Fill the cooling system, then connect the negative battery cable.

4.6L Engines (VIN W and VIN 6)

1. Disconnect the negative battery cable.

2. Remove the fan blade assembly and fan shroud.

3. Remove the accessory drive belt.

4. Remove the water pump pulley.

5. Remove the 4 water pump bolts.

6. Remove the water pump from the engine block. Clean and inspect the mating surfaces.

To install:

7. Before water pump installation, make sure to aptly lubricate and install the O-ring seal.

NOTE: All water pump housing and water pump pulley bolts are tightened to 15–22 ft. lbs. (20–30 Nm).

8. Position the water pump on the engine and loosely install the 4 mounting bolts.

9. Tighten the water pump housing fasteners to the specified value in a clockwise fashion, starting with the uppermost bolt.

10. Install the water pump pulley and also tighten the 4 pulley mounting bolts to the specified value in a clockwise fashion, starting with the right-hand bolt.

11. Install the accessory drive belt.

12. Install the fan shroud and fan blade assembly. Fill the cooling system, then connect the negative battery cable.

4.9L (VIN Y and Z) Engine

1. Drain the cooling system.

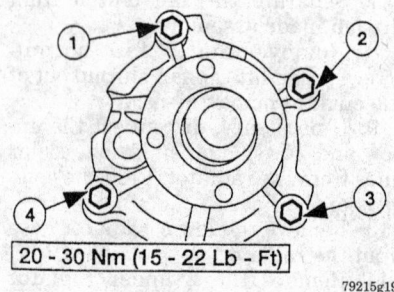

20 - 30 Nm (15 - 22 Lb - Ft)

Make sure to tighten the water pump bolts in the sequence shown — 4.6L engines (VIN W and VIN 6)

2. Disconnect the lower radiator hose from the water pump.

3. Remove the drive belt, fan, fan spacer, fan shroud, if equipped, and water pump pulley.

4. Remove the alternator pivot arm from the pump.

5. Disconnect the heater hose at the water pump.

6. Remove the water pump.

To install:

7. Before installing the old water pump, clean the gasket mounting surfaces on the pump and on the cylinder block. If a new water pump is being installed, remove the heater hose fitting from the old pump and install it on the new one.

8. Coat the new gaskets with sealer on both sides and install the water pump. Torque the mounting bolts to 18 ft. lbs. (25 Nm).

9. Connect the heater hose at the water pump.

10. Install the alternator pivot arm on the pump.

11. Install the water pump pulley fan shroud, fan spacer, fan, and drive belt.

12. Connect the lower radiator hose at the water pump.

13. Fill the cooling system.

5.0L (VIN N), 5.8L (VIN H and R) and 7.5L (VIN G) Engines

1. Drain the cooling system.

2. Remove the bolts securing the fan shroud to the radiator, if equipped, and position the shroud over the fan.

3. Disconnect the lower radiator hose, heater hose and by-pass hose at the water pump. Remove the drive belts, fan, fan spacer and pulley. Remove the fan shroud, if equipped.

4. Loosen the alternator pivot bolt and the bolt attaching the alternator adjusting arm to the water pump. Remove the power steering pump bracket from the water pump and position it out of the way.

5. Remove the bolts securing the water pump to the timing chain cover and remove the water pump.

To install:

6. Coat a new gasket with sealer and install the water pump. Torque the bolts to 18 ft. lbs. (25 Nm).

7. Install the power steering pump bracket.

8. Connect the lower radiator hose, heater hose and by-pass hose at the water pump.

9. Install the fan shroud, if equipped.

10. Install the pulley, fan spacer, fan, and drive belts.

11. Fill the cooling system.

7.3L (VIN C, F, K, and M) Diesel Engine

1. Disconnect both battery ground cables.
2. Drain the cooling system.
3. Remove the radiator shroud halves.
4. Remove the fan clutch and fan.

NOTE: The fan clutch bolts are left hand thread. Remove them by turning them clockwise.

5. Remove the power steering pump belt.
6. Remove the air conditioning compressor belt.
7. Remove the vacuum pump drive belt.
8. Remove the alternator drive belt.
9. Remove the water pump pulley.
10. Disconnect the heater hose at the water pump.
11. If installing a new pump, remove the heater hose fitting from the old pump.
12. Remove the alternator adjusting arm and bracket.
13. Unbolt the air conditioning compressor and position it out of the way; do not disconnect the refrigerant lines!
14. Remove the air conditioning compressor brackets.
15. Unbolt the power steering pump and bracket and position it out of the way; do not disconnect the power steering fluid lines.
16. Remove the bolts attaching the water pump to the front cover and lift off the pump.

To install:

17. Thoroughly clean the mating surfaces of the pump and front cover.
18. Obtain 2 dowel pins, anything that will fit into 2 mounting bolt holes in the front cover, when installing the water pump.
19. Using a new gasket, position the water pump over the dowel pins and into place on the front cover.

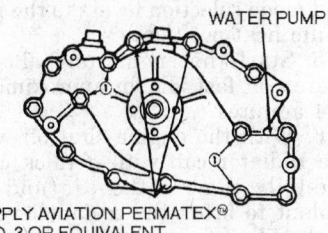

WATER PUMP

APPLY AVIATION PERMATEX®
NO. 3 OR EQUIVALENT
TO THESE BOLTS

① THESE BOLTS 2 3/4 IN. LONG
ALL OTHERS ARE 1 1/2 IN. LONG

79215g20

Water pump — 7.3L (VIN C, F, K, and M) Diesel engine

20. Install the attaching bolts. The 2 top center and 2 bottom center bolts must be coated with RTV silicone sealant prior to installation. Also, the 4 bolts marked No. 1 are a different length than the other bolts. Torque the bolts to 14 ft. lbs. (19 Nm).
21. Install the water pump pulley.
22. Wrap the heater hose fitting threads with Teflon® tape and screw it into the water pump. Torque it to 18 ft. lbs. (25 Nm).
23. Connect the heater hose to the pump.
24. Install the power steering pump and bracket. Install the belt.
25. Install the air conditioning compressor bracket.
26. Install the air conditioning compressor. Install the belt.
27. Install the alternator adjusting arm and install the belt.
28. Install the vacuum pump drive belt.
29. Adjust all the drive belts.
30. Install the fan and clutch. Remember that the bolts are left hand thread. Turn them counterclockwise to tighten them. Torque them to 45 ft. lbs. (61 Nm).
31. Install the fan shroud halves.
32. Fill and bleed the cooling system.
33. Connect the battery ground cables.
34. Start the engine and check for leaks.

Thermostat

NOTE: It is a good practice to check the operation of a new thermostat before it is installed in an engine. Place the thermostat in a pan of boiling water. If it does not open more than ¼ in. (6mm), do not install it in the engine.

REMOVAL AND INSTALLATION

NOTE: When the battery has been disconnected and reconnected, some abnormal drive symptoms may occur while the Powertrain Control Module (PCM) relearns its adaptive strategy. The vehicle may need to be driven 10 miles (16 km) or more to relearn the strategy.

4.2L (VIN 2) Engine

1. Partially drain the cooling system.
2. Remove the upper radiator hose.

3. Remove the water thermostat and gasket assembly as follows:
   a. Remove the 2 water outlet connection bolts.
   b. Remove the water outlet connection.
   c. Remove the water thermostat and gasket assembly.

To install:

4. Clean all gasket surfaces of dirt, grime or old gasket material.
5. Install the water thermostat as follows:
   a. Align the notch in the water thermostat and gasket assembly with the recess in the water outlet connection.
   b. Position the water outlet connection on the lower intake manifold.
   c. Install the 2 water outlet connection bolts to 71–102 inch lbs. (8.0–11.5 Nm).
6. Connect the upper radiator hose.
7. Fill the cooling system, start the engine and check for leaks.

4.6L Engines (VIN W and VIN 6)

1. Partially drain the cooling system.
2. Remove the upper radiator hose.
3. Remove the water outlet connection as follows:
   a. Remove the 2 water outlet connection bolts.
   b. Remove the water outlet connection.
   c. Remove the water thermostat and O-ring. Discard the O-ring.

To install:

4. Clean all gasket surfaces of dirt or grime.
5. Using a new O-ring (A), position the water thermostat (B) in the lower intake manifold (C).
6. Position the water outlet connection on the lower intake manifold. Install the 2 water outlet connection bolts to 15–22 ft. lbs. (20–30 Nm).
7. Connect the upper radiator hose.
8. Fill the cooling system, start the engine and check for leaks.

4.9L (VIN Y and Z) Engine

1. Drain the cooling system below the level of the coolant outlet housing. Use the petcock valve at the bottom of the radiator to drain the system. It is not necessary to remove any of the hoses.
2. Remove the coolant outlet housing retaining bolts and slide the housing with the hose attached to one side.

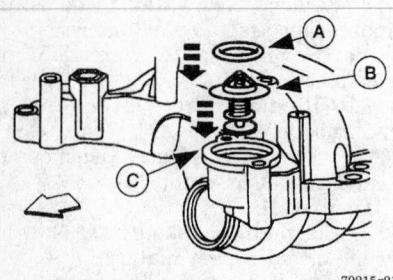

Install a new thermostat (B) in the lower intake manifold (C), then install a new O-ring (A) — 4.6L engines (VIN W and VIN 6)

3. Remove the thermostat and gasket from the cylinder head and clean both mating surfaces.

To install:

4. Coat a new gasket with water resistant sealer and position it on the outlet of the engine. The gasket must be in place before the thermostat is installed.

5. Install the thermostat with the bridge (opposite end of the spring) inside the elbow connection.

6. Position the elbow connection onto the mounting surface of the outlet, so the thermostat flange is resting on the gasket and install the retaining bolts. Torque the bolts to 15 ft. lbs. (20 Nm).

7. Fill the radiator and operate the engine until it reaches operating temperature. Check the coolant level and adjust if necessary.

5.0L (VIN N), 5.8L (VIN H and R) and 7.5L (VIN G) Engines

1. Drain the cooling system below the level of the coolant outlet housing. Use the petcock valve at the bottom of the radiator to drain the system. It is not necessary to remove any of the hoses.

2. Disconnect the bypass hoses at the water pump and intake manifold.

3. Remove the bypass tube.

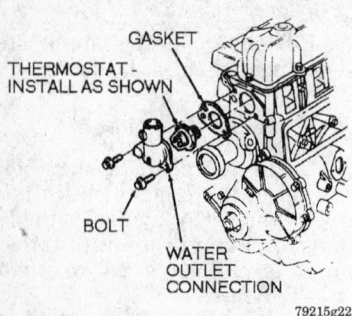

Exploded view of the thermostat and related components found on the 4.9L (VIN Y and Z) engine

4. Remove the coolant outlet housing retaining bolts, bend the hose and lift the housing with the hose attached to one side.

5. Remove the thermostat and gasket from the intake manifold and clean both mating surfaces.

To install:

6. To install the thermostat, coat a new gasket with water resistant sealer and position it on the outlet of the engine. The gasket must be in place before the thermostat is installed.

7. Install the thermostat with the bridge (opposite end of the spring) inside the elbow connection and the thermostat flange positioned in the recess in the manifold.

8. Position the elbow connection onto the mounting surface of the outlet. Torque the bolts to 18 ft. lbs. (25 Nm) on the 5.0L (VIN N) and 5.8L (VIN H and R) engines, or to 28 ft. lbs. (38 Nm) on the 7.5L (VIN G) engine.

9. Install the bypass tube and hoses.

10. Fill the radiator and operate the engine until it reaches operating temperature. Check the coolant level and adjust if necessary.

7.3L (VIN C, F, K, and M) Diesel Engine

—————— **WARNING** ——————
The factory specified thermostat does not contain an internal bypass. On these engines, an internal bypass is located in the block. The use of any replacement thermostat other than that meeting the manufacturer's specifications will result in engine overheating! Use only thermostats meeting the specifications of Ford part number E5TZ-8575-C or Navistar International part number 1807945-C1.

1. Disconnect both battery ground cables.

2. Drain the coolant to a point below the thermostat housing.

3. Remove the alternator and vacuum pump belts

4. Remove the alternator.

5. Remove the vacuum pump and bracket.

6. Remove all but the lowest vacuum pump/alternator mounting casting bolt.

7. Loosen that lowest bolt and pivot the casting outboard of the engine.

8. Remove the thermostat housing attaching bolts, bend the hose and lift the housing up and to one side.

9. Remove the thermostat and gasket.

To install:

10. Clean the thermostat housing and block surfaces thoroughly.

11. Coat a new gasket with waterproof sealer and position the gasket on the manifold outlet opening.

12. Install the thermostat in the manifold opening with the spring element end downward and the flange positioned in the recess in the manifold.

13. Place the outlet housing into position and install the bolts. Torque the bolts to 20 ft. lbs. (27 Nm).

14. Reposition the casting.

15. Install the vacuum pump and bracket.

16. Install the alternator.

17. Adjust the drive belts.

18. Fill and bleed the cooling system.

19. Connect both battery cables.

20. Run the engine and check for leaks.

Cooling System Bleeding

When the entire cooling system is drained, the following procedure should be used to remove air from the cooling system and ensure a complete fill.

1. Close the radiator drain cock and install the cylinder block drain plug, if removed.

2. Fill the cooling system with a 50/50 mixture of anti-freeze and water. Allow several minutes for trapped air to escape. When filling a cross flow radiator, allow time for the coolant to flow through the radiator tubes to the other end of the tank to ensure the radiator is full.

3. Install the radiator cap to the pressure relief position by installing the cap to the fully installed position and then backing off to the 1st stop. This will allow any air to escape and will minimize spillage.

4. Slide the heater temperature and mode selection levers to the maximum heat position.

5. Start the engine and allow to operate at fast idle for approximately 3–4 minutes.

6. With the engine shut off, wrap the radiator cap with a thick cloth, carefully remove the cap and add coolant to bring the coolant level up to the filler neck seat.

7. Install the cap to the fully installed position. Then, back off to the 1st stop and operate the engine at fast idle until the thermostat opens and the upper radiator hose is warm. To check the coolant level, shut the

engine off, wrap the cap with a thick cloth and cautiously remove the cap. Add additional coolant, if necessary. Install the cap to the fully installed position.

— **CAUTION** —

To avoid personal injury from scalding hot coolant or steam blowing out of the radiator, use extreme care when removing the cap from a hot radiator.

8. Fill the coolant recovery reservoir to the proper level with a 50/50 mix of anti-freeze and water.

GASOLINE FUEL SYSTEM

Fuel System Service Precautions

Safety is the most important factor when performing not only fuel system maintenance but any type of maintenance. Failure to conduct maintenance and repairs in a safe manner may result in serious personal injury or death. Maintenance and testing of the vehicle's fuel system components can be accomplished safely and effectively by adhering to the following rules and guidelines.

• To avoid the possibility of fire and personal injury, always disconnect the negative battery cable unless the repair or test procedure requires that battery voltage be applied.

• Always relieve the fuel system pressure prior to disconnecting any fuel system component (injector, fuel rail, pressure regulator, etc.), fitting or fuel line connection. Exercise extreme caution whenever relieving fuel system pressure to avoid exposing skin, face and eyes to fuel spray. Please be advised that fuel under pressure may penetrate the skin or any part of the body that it contacts.

• Always place a shop towel or cloth around the fitting or connection prior to loosening to absorb any excess fuel due to spillage. Ensure that all fuel spillage (should it occur) is quickly removed from engine surfaces. Ensure that all fuel soaked cloths or towels are deposited into a suitable waste container.

• Always keep a dry chemical (Class B) fire extinguisher near the work area.

• Do not allow fuel spray or fuel vapors to come into contact with a spark or open flame.

• Always use a backup wrench when loosening and tightening fuel line connection fittings. This will prevent unnecessary stress and torsion to fuel line piping. Always follow the proper torque specifications.

• Always replace worn fuel fitting O-rings with new. Do not substitute fuel hose or equivalent where fuel pipe is installed.

RELIEVING FUEL SYSTEM PRESSURE

NOTE: A fuel pressure gauge, such as Ford Tool T80L-9974-B, is needed to correctly perform this procedure.

1. Disconnect the negative battery cable and remove the fuel filler cap.
2. Remove the cap from the pressure relief valve on the fuel supply manifold. Install pressure gauge T80L-9974-B or equivalent, to the pressure relief valve.
3. Direct the gauge drain hose into a suitable container and depress the pressure relief button.
4. Remove the gauge and replace the cap on the pressure relief valve.

NOTE: As an alternate method on models except 1997 F-150 and Expedition, disconnect the inertia switch and crank the engine for 15–20 seconds until the pressure is relieved.

Idle Speed and Mixture

ADJUSTMENT

The curb idle and fast idle speeds are controlled by the Powertrain Control

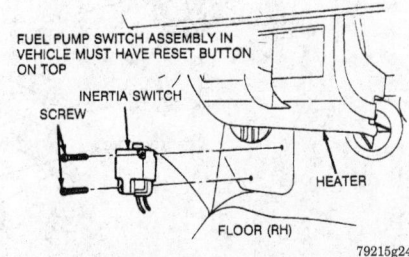

FUEL PUMP SWITCH ASSEMBLY IN VEHICLE MUST HAVE RESET BUTTON ON TOP
INERTIA SWITCH SCREW
HEATER
FLOOR (RH)
79215g24

Typical inertia switch location — this method of fuel pressure relief is not recommended by Ford for the 1997 F-150 and Expedition models

Module (PCM) and the Idle Air Control (IAC) valve. The idle speed is not adjustable.

Fuel Filter

REMOVAL AND INSTALLATION

NOTE: On newer vehicles, especially the 1997 F-150 and Expedition vehicles, a fuel line disconnect tool such as Ford tool T93T-9550-AH is needed for this procedure.

1. Disconnect the negative battery cable and relieve the fuel system pressure.
2. Raise and support the vehicle safely.
3. Disconnect the fuel lines from the fuel filter. Have a drain pan handy to catch any residual fuel once the lines are separated. On newer models, disconnect the fuel lines from the filter as follows:

   a. Disconnect the safety clip from the male hose.

   b. Install and push the fuel line disconnect tool into the female fitting.

   c. Separate the male and female fittings.

   d. Inspect the fuel lines for any damage after the fuel is finished draining.

4. Remove the fuel filter from the bracket and the retainer, if equipped. Note the direction of the flow arrow so the replacement filter can be installed correctly.

To install:

5. Position the fuel filter into the mounting bracket with the flow arrow pointing in the correct direction.
6. Install the fuel lines to the fuel filter. On newer models, align and push the male tube into the female fitting until a click is heard. Pull on the fitting to ensure that it is fully engaged, then install the safety clip.
7. Lower the vehicle to the ground.

NOTE: When the battery has been disconnected and reconnected, some abnormal drive symptoms may occur while the Powertrain Control Module (PCM) relearns its adaptive strategy. The vehicle may need to be driven 10 miles (16 km) or more to relearn the strategy.

8. Connect the negative battery cable.

Electric Fuel Pump

REMOVAL AND INSTALLATION

NOTE: When the battery is disconnected and reconnected, some abnormal drive symptoms may occur while the vehicle relearns its adaptive strategy. The vehicle may need to be driven 10 miles (16 km) or more to relearn the strategy.

1997 F-150 and Expedition Models

1. Relieve the fuel system pressure.
2. Disconnect the negative battery cable.
3. Raise and support the vehicle safely on jackstands.
4. Remove the 6 fuel tank skid plate bolts, then lower the fuel tank skid plate.
5. Drain the fuel from the fuel tank.
6. Disconnect the fuel tank filler pipe hose from the fuel tank filler pipe.
7. Disconnect the fuel tank filler pipe vent hose from the fuel tank filler pipe.
8. Use Fuel Line Removal Tool T93T-9550-AH, or the equivalent, to disconnect the fuel lines from the fuel pump.
9. Detach the fuel tank front connections (EVAP hose and transducer electrical connector).
10. Detach the rear EVAP emissions hose.
11. Separate the fuel pump electrical connector from the fuel pump.
12. Drain the fuel from the fuel tank.
13. Position a hydraulic floor jack, such as the Hi-Lift Jack 014–00210, under the fuel tank.
14. Remove the 2 fuel tank support straps, then lower the fuel tank.
15. Remove the 6 fuel pump retaining bolts, then pull the fuel pump out of the fuel tank.
   To install:
16. Clean the fuel pump mounting flange, then install the new fuel pump into the fuel tank. Tighten the fuel pump-to-fuel tank bolts to 67–92 inch lbs. (7.6–10.4 Nm).
17. Install the fuel tank. Make sure to tighten the fuel tank support strap bolts to 22–30 ft. lbs. (29–40 Nm).
18. Attach all hoses and electrical wiring harness connectors to the fuel tank and fuel pump.
19. Install the fuel tank skid plate and tighten the mounting bolts to 9–13 ft. lbs. (13–17 Nm).

20. Lower the vehicle and fill the fuel tank with fuel. Connect the negative battery cable.
21. Start the vehicle and check for fuel leaks.

Except 1997 F-150 and Expedition Models

1. Release the fuel system pressure. Disconnect the negative battery cable.
2. Raise and support the vehicle safely on jackstands.
3. Remove the fuel tank.
4. Remove the fuel pump in steel fuel tanks as follows:
   a. Disconnect the wiring at the connector.
   b. Remove all dirt from the area of the sender.
   c. Disconnect the fuel lines.
   d. Turn the locking ring counterclockwise to remove it.
5. To remove the fuel pump from a plastic fuel tank, perform the following:
   a. Disconnect the wiring at the connector.
   b. Remove all dirt from the area of the sender.
   c. Disconnect the fuel lines.
   d. Turn the locking ring counterclockwise to remove it.

6. Lift out the fuel pump and sending unit. Discard the gasket.
   To install:
7. Place a new gasket in position in the groove in the tank.
8. Place the sending unit/fuel pump assembly in the tank, indexing the tabs with the slots in the tank. Make sure the gasket stays in place.
9. Hold the assembly in place and position the locking ring.
   a. On steel tanks, turn the locking ring clockwise until the stop is against the retainer ring tab.
   b. On plastic tanks, turn the retaining ring clockwise until handtight. Torque the ring to 40–55 ft. lbs. (54–75 Nm).
10. Make sure the gasket is still in place.
11. Connect the fuel lines and wiring.
12. Install the tank.

Fuel Injector

REMOVAL AND INSTALLATION

4.2L (VIN 2) Engine

1. Disconnect the negative battery cable.
2. Relieve the fuel system pressure.

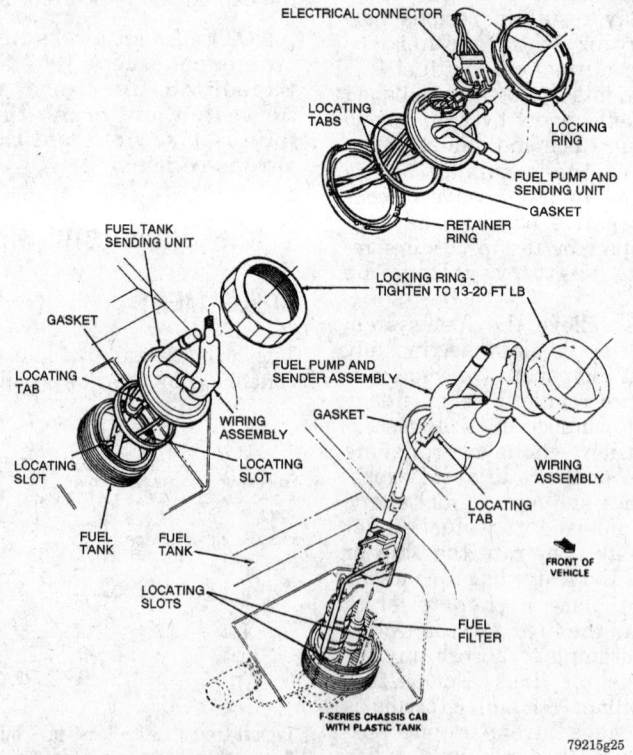

Exploded view of the fuel tank mounted fuel pump — except 1997 F-150 and Expedition models

3. Remove the upper intake manifold.

4. Remove the fuel pressure regulator valve vacuum hose.

------ CAUTION ------
After disconnecting, plug the fuel lines to prevent leakage.

5. Remove the 2 fuel line connector retaining clips, then using Spring Lock Coupling Disconnect Tool D87L-9280-A or equivalent, disconnect the upper fuel line. Use Tool D87L-9280-B to disconnect the lower fuel line.

6. Remove and discard the 4 fuel line O-rings.

7. Detach the 6 fuel injector electrical wiring harness connectors.

8. Remove the 4 fuel injection supply manifold bolts, then lift the manifold off of the engine.

------ WARNING ------
The fuel injectors are deposit resistant. DO NOT clean the fuel injectors.

9. Remove the 6 fuel injectors from the supply manifold, then remove and inspect the 2 O-rings from each fuel injector. Replace the O-rings if any damage is evident.

To install:

10. Lubricate the new O-rings with engine oil before installation. Install the O-rings onto the fuel injectors, make sure that they are correctly seated in their grooves.

11. Install the fuel injectors onto the fuel injection supply manifold, then install the manifold and injectors onto the engine. Prior to installation, make certain that the fuel injector mounting bosses are clean. All of the fuel injectors should be corrected seated into their respective bosses before tightening the manifold retaining fasteners. Tighten the retaining fasteners to 71–102 inch lbs. (8–11 Nm).

12. Attach the electrical wiring harness connectors to the 6 fuel injectors.

13. Attach the 2 fuel lines and reconnect the vacuum hose to the pressure regulator valve.

14. Install the upper intake manifold, then connect the negative battery cable.

4.6L Engines (VIN W and VIN 6)

1. Disconnect the negative battery cable.

2. Relieve the fuel system pressure.

3. Partially drain the cooling system.

4. Remove the power steering reservoir bracket.

5. Remove the accelerator control splash shield.

6. Detach the fuel pressure regulator valve vacuum hose.

------ CAUTION ------
After disconnecting, plug the fuel lines to prevent leakage.

7. Remove the 2 fuel line connector retaining clips.

8. Remove the 2 fuel line connector retaining clips, then using Spring Lock Coupling Disconnect Tool D87L-9280-A or equivalent, disconnect the upper fuel line. Use Tool D87L-9280-B to disconnect the lower fuel line.

9. Remove and discard the 4 fuel line O-rings.

10. Detach the 8 fuel injector electrical wiring harness connectors.

11. Disconnect the heater water inlet tube hose, then remove the brake booster bracket and tube.

12. Remove the PCV hose.

13. Disconnect the Exhaust Gas Recirculation (EGR) components as follows:

a. Remove the EGR valve-to-exhaust manifold tube lower fitting.

b. Remove the EGR valve-to-exhaust manifold tube upper fitting and the EGR valve-to-exhaust manifold tube.

c. Detach the 2 Differential Pressure Feedback EGR (DPFE) transducer hoses.

d. Disconnect the EGR valve vacuum hose.

14. Disconnect the Vapor Management Valve (VMV) hose.

15. Remove the 4 fuel injection supply manifold bolts, then remove the fuel injectors from the supply manifold. Remove and discard the 2 O-rings from each fuel injector.

16. Remove the fuel injection supply manifold from the engine, if necessary.

To install:

17. Lubricate the new O-rings with engine oil before installation. Install the O-rings onto the fuel injectors, make sure that they are correctly seated in their grooves.

18. Position the fuel injection manifold on the engine, then install the fuel injectors onto the fuel injection supply manifold. Prior to installation, make certain that the fuel injector mounting bosses are clean. All of the fuel injectors should be corrected seated into their respective bosses before tightening the manifold re-

taining fasteners. Tighten the retaining fasteners to 71–102 inch lbs. (8–11 Nm).

19. Attach the EGR components as follows:

a. Attach the EGR valve vacuum hose.

b. Attach the 2 DPFE transducer hoses, then install the exhaust manifold tube to the EGR valve and the exhaust manifold tube upper fitting to the EGR valve.

c. Install the exhaust manifold tube lower fitting to the EGR valve.

20. Install the PCV hose.

21. Install the brake booster bracket and tube, then install the heater water inlet tube hose.

22. Attach the electrical wiring harness connectors to the 8 fuel injectors.

23. Attach the fuel lines. Make sure to install 4 new O-rings on the fuel lines and in the fuel line fittings.

24. Attach the fuel pressure regulator valve vacuum hose.

25. Install the accelerator control splash shield, then install the power steering reservoir bracket. Tighten the reservoir bracket mounting bolts to 49–67 ft. lbs. (68–92 Nm).

26. Fill the cooling system, then connect the negative battery cable.

4.9L (VIN Y and Z) Engine

1. Relieve the fuel system pressure.

2. Remove the upper intake manifold assembly.

3. Remove the fuel supply manifold.

4. Disconnect the wiring at each injector.

5. Pull upward on the injector body while gently rocking it from side-to-side.

6. Inspect the O-rings on the injector for any sign of leakage or damage. Replace any suspected O-rings.

7. Inspect the plastic cap at the top of each injector and replace it if any sign of deterioration is noticed.

To install:

8. Lubricate the O-rings with clean engine oil only!

9. Install the injectors by pushing them in with a gentle rocking motion.

10. Install the fuel supply manifold.

11. Connect the electrical wiring.

12. Install the upper intake manifold.

5.0L (VIN N) and 5.8L (VIN H and R) Engines

1. Relieve the fuel system pressure.

2. Disconnect the battery ground.

3. Remove the upper intake manifold.

4. Disconnect the wiring at the injectors.

5. Pull upward on the injector body while gently rocking it from side-to-side.

6. Inspect the O-rings on the injector for any sign of leakage or damage. Replace any suspected O-rings.

7. Inspect the plastic cap at the top of each injector and replace it if any sign of deterioration is noticed.

To install:

8. Lubricate the O-rings with clean engine oil !

9. Install the injectors by pushing them in with a gentle rocking motion.

10. Install the fuel supply manifold.

11. Connect the electrical wiring.

12. Install the upper intake manifold.

7.5L (VIN G) Engine

1. Relieve the fuel system pressure.

2. Disconnect the battery ground.

3. Remove the fuel supply manifold.

4. Disconnect the wiring at the injectors.

5. Pull upward on the injector body while gently rocking it from side-to-side.

6. Inspect the O-rings (2 per injector) on the injector for any sign of leakage or damage. Replace any suspected O-rings.

7. Inspect the plastic cap at the top of each injector and replace it if any sign of deterioration is noticed.

To install:

8. Lubricate the O-rings with clean engine oil only!

9. Install the injectors by pushing them in with a gentle rocking motion.

10. Install the fuel supply manifold.

11. Connect the electrical wiring.

DIESEL FUEL SYSTEM

Fuel System Service Precautions

Safety is the most important factor when performing not only fuel system maintenance but any type of maintenance. Failure to conduct maintenance and repairs in a safe manner may result in serious per-

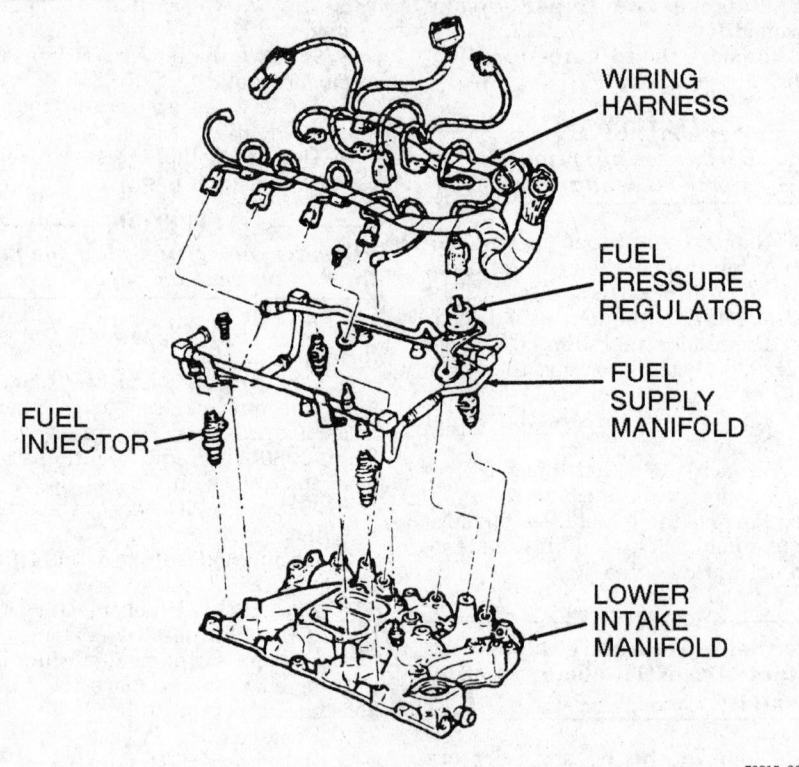

Fuel injector removal — 7.5L (VIN G) engine

sonal injury or death. Maintenance and testing of the vehicle's fuel system components can be accomplished safely and effectively by adhering to the following rules and guidelines.

• To avoid the possibility of fire and personal injury, always disconnect the negative battery cable unless the repair or test procedure requires that battery voltage be applied.

• Always relieve the fuel system pressure prior to disconnecting any fuel system component (injector, fuel rail, pressure regulator, etc.), fitting or fuel line connection. Exercise extreme caution whenever relieving fuel system pressure to avoid exposing skin, face and eyes to fuel spray. Please be advised that fuel under pressure may penetrate the skin or any part of the body that it contacts.

• Always place a shop towel or cloth around the fitting or connection prior to loosening to absorb any excess fuel due to spillage. Ensure that all fuel spillage (should it occur) is quickly removed from engine surfaces. Ensure that all fuel soaked cloths or towels are deposited into a suitable waste container.

• Always keep a dry chemical (Class B) fire extinguisher near the work area.

• Do not allow fuel spray or fuel vapors to come into contact with a spark or open flame.

• Always use a backup wrench when loosening and tightening fuel line connection fittings. This will prevent unnecessary stress and torsion to fuel line piping. Always follow the proper torque specifications.

• Always replace worn fuel fitting O-rings with new. Do not substitute fuel hose or equivalent where fuel pipe is installed.

Fuel System

RELIEVING PRESSURE

———— CAUTION ————
Before removing the fuel tank filler cap, turn the fuel tank filler cap ¼ to ¾ turn counterclockwise and wait for the tank pressure to be relieved. Personal injury may result if the fuel tank filler cap is removed without the pressure fully relieved.

1. Remove the fuel tank filler cap to relieve any pressure in the fuel tank.

2. When servicing the fuel lines, loosen the fuel fitting to allow any

residual fuel line pressure to be relieved.

Idle Speed

ADJUSTMENT

1. Place the transmission in Neutral (manual transmissions) or **P** (automatic transmissions).

2. Bring the engine up to normal operating temperature.

3. Idle speed is measured with the manual transmission in Neutral or the automatic transmission in **D**.

4. Ensure that the curb idle adjusting screw is against the stop. If not, correct the vehicle linkage.

5. Check curb idle speed, using either Rotunda 055–00108 or an equivalent tachometer. Curb idle speed is specified on the Vehicle Emissions Control Information (VECI) decal on the underside of the vehicle's hood. Adjust the idle speed to specification using the idle speed adjusting screw.

6. Place the transmission in Neutral (manual) or **P**. Rev the engine momentarily, then place the transmission in the specified gear and recheck the idle speed. Adjust again if necessary.

7. Remove the tachometer and close the hood.

Fuel Filter/Water Separator

DRAIN WATER

NOTE: Drain water from the water separator manual drain valve whenever the warning light comes ON or every 5000 miles. The WATER IN FUEL light will glow when approximately 3.5 oz. of water accumulates in the separator.

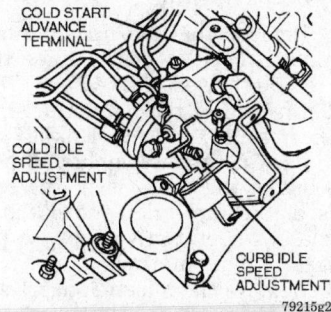

Raise or lower the curb idle speed by turning the curb idle speed adjusting screw — 7.3 Diesel engine

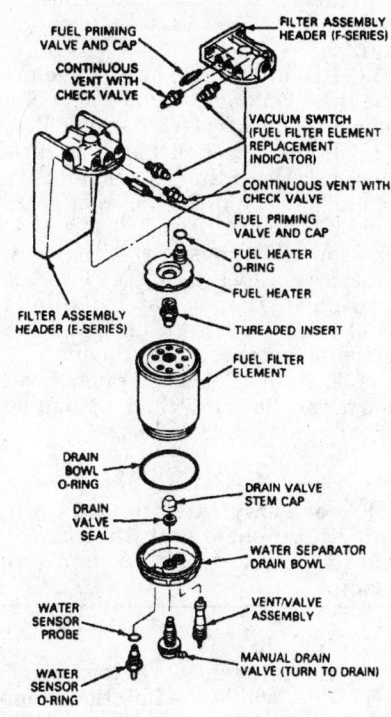

Fuel filter/water separator assembly — 7.3L (VIN C, F, K, and M) Diesel engine.

The Diesel engines are equipped with a fuel/water separator in the fuel supply line. A "Water in Fuel" indicator light is provided on the instrument panel to alert the driver. The light should glow when the ignition switch is in the **start** position to indicate proper light and water sensor function. If the light glows continuously while the engine is running, the water must be drained from the separator as soon as possible to prevent damage to the fuel injection system.

1. Shut off the engine. Failure to shut the engine **OFF** before draining the separator will cause air to enter the system.

2. Unscrew the vent on the top center of the separator unit 2½ to 3 turns.

3. Unscrew the drain screw on the bottom of the separator 1½ to 2 turns and drain the water into an appropriate container.

4. After the water is completely drained, close the water drain finger tight.

5. Tighten the vent until snug, then turn it an additional ¼ turn.

6. Start the engine and check the "Water in Fuel" indicator light; it should not be lit. If it is lit and continues to stay so, there is a problem somewhere else in the fuel system.

REMOVAL AND INSTALLATION

The Diesel engine uses a one-piece spin-on fuel filter. Do not add fuel to the new fuel filter. Allow the engine to draw fuel through the filter.

1. Remove the spin-on filter by unscrewing it counterclockwise.

2. Clean the filter mounting surface.

3. Coat the gasket or the replacement filter with clean Diesel fuel. This helps ensure a good seal.

To install:

4. Tighten the filter by hand until the gasket touches the filter mounting surface.

5. Tighten the filter an additional ½ turn.

NOTE: After changing the fuel filter, the engine will purge the trapped air as it runs. The engine may run roughly and smoke excessively until the air is cleared from the system.

Diesel Injection Pump

REMOVAL AND INSTALLATION

——— **WARNING** ———

Before removing the fuel lines, clean the exterior with clean fuel oil or solvent to prevent entry of dirt into the engine when the fuel lines are removed. Do not wash or steam clean engine while engine is running. Serious damage to injection pump could occur.

1. Disconnect battery ground cables from both batteries.

2. Remove the engine oil filler neck.

3. Remove the bolts attaching injection pump to drive gear.

4. Disconnect the electrical connectors to injection pump.

5. Disconnect the accelerator cable and speed control cable from throttle lever, if equipped.

6. Remove the air cleaner and install clean rags to prevent dirt from entering the intake manifold.

7. Remove the accelerator cable bracket, with cables attached, from the intake manifold and position out of the way.

NOTE: All fuel lines and fittings must be capped using Fuel System Protective Cap Set T83T-9395-A or equivalent, to prevent fuel contamination.

8. Remove the fuel filter-to-injection pump fuel line and cap fittings.

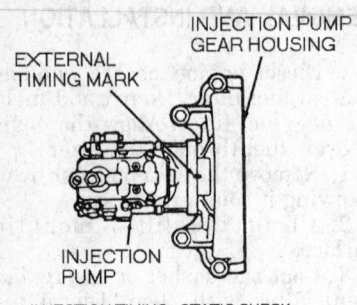

Injection pump timing marks — 7.3L (VIN C, F, K, and M) Diesel engine

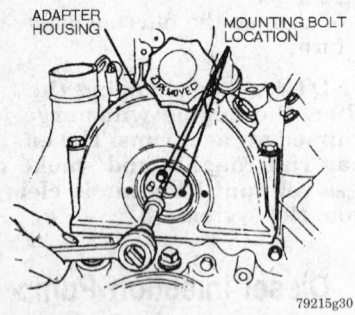

Injection pump drive gear attaching bolt removal — 7.3L (VIN C, F, K, and M) Diesel engine

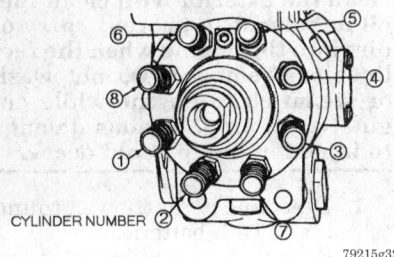

Injection pump cylinder numbering sequence — 7.3L (VIN C, F, K, and M) Diesel engine

9. Remove and cap the injection pump inlet elbow and the injection pump fitting adapter.

10. Remove the fuel return line on injection pump, rotate out of the way, And cap all fittings.

NOTE: It is not necessary to remove injection lines from injection pump. If lines are to be removed, loosen injection line fittings at injection pump before removing it from engine.

11. Remove the fuel injection lines from the nozzles and cap lines and nozzles.

12. Remove the 3 nuts attaching the Injection pump to injection pump adapter using Tool T83T–9000–B.

13. If the injection pump is to be replaced, loosen the injection line retaining clips and the injection nozzle fuel lines with Tool T83T–9396–A and cap all fittings at this time with protective cap set T83T–9395–A or equivalent. Do not install the injection nozzle fuel lines until the new pump is installed in the engine.

14. Lift the Injection pump, with the nozzle lines attached, up and out of the engine compartment.

WARNING

Do not carry injection pump by injection nozzle fuel lines as this could cause lines to bend or crimp.

To install:

15. Install a new O-ring on the drive gear end of the injection pump.

16. Move the injection pump down and into position.

17. Position the alignment dowel on injection pump into the alignment hole on drive gear.

18. Install the bolts attaching the injection pump to drive gear and tighten.

19. Install the nuts attaching injection pump to adapter. Align scribe lines on the injection pump flange and the injection pump adapter and tighten to 14 ft. lbs. (20 Nm).

20. If the injection nozzle fuel lines were removed from the injection pump install at this time.

21. Remove the caps from nozzles and the fuel lines and install the fuel line nuts on the nozzles and tighten to 22 ft. lbs. (30 Nm).

22. Connect the fuel return line to injection pump and tighten the nuts.

23. Install the injection pump fitting adapter with a new O-ring.

24. Clean the old sealant from the injection pump elbow threads, using clean solvent, and dry thoroughly. Apply a light coating of pipe sealant to the elbow threads.

25. Install the elbow in the injection pump adapter and tighten to a minimum of 6 ft. lbs. (8 Nm). Then tighten further, if necessary, to align the elbow with the injection pump fuel inlet line, but do not exceed 360 degrees of rotation or 10 ft. lbs. (14 Nm).

26. Remove the caps and connect the fuel filter-to-injection pump fuel line.

27. Install the accelerator cable bracket to the intake manifold.

28. Remove the rags from the intake manifold and install the air cleaner.

29. Connect the accelerator and speed control cable, if equipped, to throttle lever.

30. Install the electrical connectors on injection pump.

31. Clean the injection pump adapter and oil filler neck sealing surfaces.

32. Apply a 1/8 in. (3mm) bead of RTV sealant on the adapter housing.

33. Install the oil filler neck and tighten the bolts.

34. Connect the battery ground cables to both batteries.

35. Run the engine and check for fuel leaks.

36. If necessary, purge high pressure fuel lines of air by loosening connector 1/2 to 1 turn and cranking engine until solid fuel, free from bubbles flows from connection.

CAUTION

Keep eyes and hands away from nozzle spray. Fuel spraying from the nozzle under high pressure can penetrate the skin.

---

37. Check and adjust injection pump timing.

Diesel Fuel Injector

REMOVAL AND INSTALLATION

NOTE: Before removing the nozzle assemblies, clean the exterior of each nozzle assembly and the surrounding area with clean fuel oil or solvent to prevent entry of dirt into the engine when nozzle assemblies are removed. Also, clean the fuel inlet and fuel leak-off piping connections. Blow dry with compressed air.

1. Remove the fuel line retaining clamp(s) from the nozzle lines that are to be removed.

2. Disconnect the nozzle fuel inlet (high pressure) and fuel leak-off tees from each nozzle assembly and position out of the way. Cover the open ends of the fuel inlet and outlet or nozzles with protective caps, to prevent dirt from entering.

3. Remove the injection nozzles by turning them counterclockwise. Pull the nozzle assembly with the copper washer attached from the engine. Cover the nozzle fuel opening and

spray tip, with plastic caps, to prevent the entry of dirt.

NOTE: Remove the copper injector nozzle gasket from the nozzle bore with special tool, T71P–19703–C, or equivalent, whenever the gasket does not come out with the nozzle.

4. Place the nozzle assemblies in a fabricated holder as they are removed from the heads. The holder should be marked with numbers corresponding to the cylinder numbering of the engine. This will allow for reinstallation of the nozzle in the same ports from which they were removed.
To install:
5. Thoroughly clean the nozzle bore in cylinder head before reinserting the nozzle assembly with nozzle seat cleaner, special tool T83T–9527–A or equivalent. Make certain that no small particles of metal or carbon remain on the seating surface. Blow out the particles with compressed air.
6. Remove the protective cap and install a new copper gasket on the nozzle assembly, with a small dab of grease.

NOTE: Anti-seize compound or equivalent should be used on nozzle threads to aid in installation and future removal.

7. Install the nozzle assembly into the cylinder head nozzle bore.
8. Tighten the nozzle assembly to 33 ft. lbs. (45 Nm).
9. Remove the protective caps from nozzle assemblies and fuel lines.
10. Install the leak-off tees to the nozzle assemblies.

NOTE: Install 2 new O-ring seals for each fuel return tee.

11. Connect the high pressure fuel line and tighten, using a flare nut wrench.
12. Install the fuel line retainer clamps.
13. Start the engine and check for leaks.

EMISSION CONTROLS

Emission Warning Lamps

RESETTING

All gasoline engine equipped light vehicles built for sale outside of California employ this device.

The EMW consists of an instrument panel mounted amber light imprinted with the word EGR, EMISS, or EMISSIONS. The light is connected to a sensor module located under the instrument panel. The purpose is the warn the driver that the 60,000 mile emission system maintenance is required on the vehicle. Specific emission system maintenance requirements are listed in the vehicle's owner's manual maintenance schedule.

1. Turn the key to the **OFF** position.
2. Lightly push a Phillips screwdriver through the 0.2 in. (5mm) diameter hole labeled RESET, and lightly press down and hold it.
3. While maintaining pressure with the screwdriver, turn the key to the **RUN** position. The EMW lamp will light and stay lit as long as pressure is kept on the screwdriver. Hold the screwdriver down for about 5 seconds.
4. Remove the screwdriver. The lamp should go out within 2–5 seconds. If not, repeat Steps 1–3.
5. Turn the key **OFF**.
6. Turn the key to the **RUN** position. The lamp will light for 2–5 seconds and then go out. If not, repeat the rest procedure.

NOTE: If the light comes ON between 15,000 and 45,000 miles or between 75,000 and 105,000 miles, replace the 1,000 hour pretimed module.

GASOLINE ENGINE MECHANICAL

NOTE: Disconnecting the negative battery cable on some vehicles may interfere with the functions of the on board computer systems and may require the computer to undergo a relearning process, once the negative battery cable is reconnected.

Engine Assembly

REMOVAL AND INSTALLATION

NOTE: When the battery is disconnected and reconnected, some abnormal drive symptoms may occur while the vehicle relearns its adaptive strategy. The vehicle may need to be driven 10 miles (16 km) or more to relearn the strategy.

4.2L (VIN 2) Engine

NOTE: To perform the following procedure an engine lifting bracket, such as Ford Tool 014–00730, and an engine hoist or lift are needed.

1. On vehicles equipped with Air Conditioning (A/C), have the A/C system discharged by a qualified mechanic using approved recovery/recycling equipment.
2. Disconnect the negative battery cable.
3. Remove the hood from the vehicle.
4. Drain the cooling system.
5. Relieve the fuel system pressure.
6. Remove the font upper air deflector.
7. Remove the engine air cleaner outlet tube.
8. Remove the radiator, fan blade assembly and fan shroud.
9. Remove the accelerator control splash shield.
10. Disconnect the accelerator cable and, if equipped, the speed control actuator cable. Position both cables aside.
11. Disconnect the Vapor Management Valve (VMV) hose.
12. Detach the manifold vacuum connection.
13. Detach the following vacuum connections from the following intake manifold components:
    a. Intake Manifold Runner Control (IMRC).
    b. Fuel pressure regulator.
    c. IMRC solenoid.
    d. Engine Vacuum Regulator (EVR).
    e. Vacuum reservoir.
14. Detach the EGR valve vacuum connection.
15. Position the power steering reservoir and, on vehicles equipped with A/C, the A/C lines aside.

16. Position the power steering pump aside.

17. Remove the alternator electrical wiring harness connectors.

18. Remove the fuel injector and related wiring.

19. Label, then remove the spark plug wires.

20. Disconnect the 2 heater hoses.

21. Detach the brake booster vacuum hose.

22. Disconnect the Differential Pressure Feedback EGR (DPFE) transducer hose.

23. Remove the breather from the rocker arm cover.

24. Remove the upper intake manifold. Remove and discard the intake manifold upper gasket. Plug the open intake manifold holes with clean shop rags so that nothing (i.e. loose nuts or bolts) can fall down them.

25. Disconnect the fuel lines from the fuel injection supply manifold, then remove the fuel injection supply manifold and injectors.

26. Raise and safely support the vehicle on jackstands.

27. If equipped, detach the block heater cable.

28. Remove the 4 catalytic converter-to-exhaust manifold nuts (2 on each side).

29. Remove the starter motor.

30. Remove the transmission assembly.

31. If equipped, remove the clutch assembly.

32. Remove the right-hand and left-hand engine mount through-bolts .

33. Lower the vehicle.

34. Install the engine lifting bracket 014–00073, or equivalent, as follows:

   a. Install a link in each engine lifting eye.

   b. Connect each end of the lifting bracket to a link.

   c. Connect the chain to an engine lift/hoist. Use the lift/hoist to raise the engine out of the vehicle.

35. Mount the engine on an engine work stand for disassembly.

To install:

36. Install the engine lifting bracket 014–00073, or equivalent, as follows:

   a. Install a link in each engine lifting eye.

   b. Connect each end of the lifting bracket to a link.

   c. Connect the chain to an engine lift/hoist. Use the lift/hoist to raise the engine and position it in the vehicle.

37. remove the lifting bracket and engine hoist.

38. Raise and support the vehicle safely on jackstands.

39. Install the right-hand and left-hand engine mount through-bolts to 51–67 ft. lbs. (68–92 Nm).

40. If equipped, install the clutch assembly.

41. Install the transmission assembly into the vehicle.

42. Install the starter motor.

43. Install the catalytic converter to the exhaust manifolds. Tighten the attaching nuts to 26–34 ft. lbs. (34–46 Nm).

44. Connect the block heater cable, if equipped.

45. Attach the alternator electrical wiring harness connectors.

46. Install the fuel injectors and fuel injection supply manifold onto the engine.

47. Install the upper intake manifold. Make sure to use a new gasket.

48. Install the fuel injector and related wiring.

49. Install the power steering pump in place. Tighten the mounting bolts to 17–20 ft. lbs. (22–28 Nm).

NOTE: When installing the A/C lines, make sure that the O-rings are properly installed.

50. If equipped, install the A/C compressor manifold. Tighten the retaining bolt to 14–18 ft. lbs. (18–24 Nm).

51. Install the power steering reservoir. Tighten the 3 mounting bolts to 80–107 inch lbs. (9–12 Nm).

52. Connect the VMV hose. Also attach the vacuum connectors to the following components:

   a. Intake Manifold Runner Control (IMRC).

   b. Fuel pressure regulator.

   c. IMRC solenoid.

   d. Engine Vacuum Regulator (EVR).

   e. Vacuum reservoir.

53. Attach the EGR valve vacuum connector.

54. Connect the brake booster vacuum hose.

55. Connect the 2 DPFE transducer hoses.

56. Attach the intake manifold vacuum hose.

57. Position the accelerator cable and, if equipped, the speed control actuator cable, then install the bracket bolts. Tighten the accelerator cable bracket bolt to 19–25 inch lbs. (2.1–2.9 Nm) and the speed control actuator cable bracket bolt to 53–72 inch lbs. (5.9–8.1 Nm). Connect the cable ends to the throttle body.

58. Install the accelerator control splash shield.

59. Install the radiator, fan blade assembly and fan shroud.

60. Install the engine air cleaner outlet tube.

61. Install the front air deflector.

62. Fill the cooling system.

63. Install the hood, then connect the negative battery cable.

64. If applicable, have the A/C system evacuated and recharged by a qualified mechanic utilizing a refrigerant recovery/recycling machine.

4.6L Engines (VIN W and VIN 6)

NOTE: To perform the following procedure an engine lifting bracket, such as Ford Tool 014–00730, and an engine hoist or lift are needed.

1. On vehicles equipped with Air Conditioning (A/C), have the A/C system discharged by a qualified mechanic using approved recovery/recycling equipment.

2. Disconnect the negative battery cable.

3. Remove the hood from the vehicle.

4. Remove the radiator.

5. Remove the upper intake manifold.

6. Remove the bulkhead connector cover.

7. Separate the engine bulkhead connector.

8. Disconnect the power steering oil reservoir and bracket.

9. Detach the EGR connections:

   • 2 DPFE transducer hoses (B)

   • EGR valve-to-exhaust manifold tube upper fitting (A).

   • EGR valve-to-exhaust manifold tube lower fitting (C).

10. Slide the hose clamp back, then disconnect the heater water hose.

11. Detach the following electrical wiring harness connectors from each coil assembly:

   • Ignition coil connector

   • Camshaft Position (CMP) sensor connector

   • Radio capacitor connector

12. Remove the ignition coils and brackets.

13. Raise and safely support the vehicle on jackstands.

14. Remove the starter motor.

15. Remove the accessory drive belt, if not already performed.

16. Remove the A/C compressor.

17. On vehicles equipped with automatic transmissions, remove the transmission cooler hoses from the block mounted clip. Also remove the transmission inspection cover, torque converter bolts and transmission-to-engine block bolts.

18. On vehicles equipped with manual transmissions, remove the clutch assembly.

19. Remove the 2 top power steering pump bolts. Remove the 2 lower power steering pump bolts, then position the pump aside.

20. Remove the right-hand exhaust manifold-to-3 way catalytic converter nuts. Remove the right-hand exhaust manifold studs.

21. Remove the left-hand exhaust manifold-to-catalytic converter nuts and position the Y-pipe aside.

22. Lower the vehicle and support the transmission with a hydraulic floor jack.

23. Install the engine lift bracket, or equivalent and the engine floor crane/hoist to the engine block.

24. Lift the engine out of the vehicle and mount on an engine work stand.

To install:

25. On vehicles equipped with manual transmissions, install the clutch assembly.

26. Use the floor crane and lifting bracket to install the engine in the vehicle.

NOTE: Align the engine-to-transmission dowel pins before installing the engine-to-transmission bolts.

27. On vehicles equipped with automatic transmissions, install the transmission-to-engine block bolts, the converter bolts and the transmission inspection cover.

28. Raise and support the vehicle safely on jackstands.

29. Install the right-hand exhaust manifold studs and exhaust manifold-to-catalytic converter nuts. Tighten the nuts to 25–34 ft. lbs. (34–46 Nm).

30. Install the left-hand manifold-to-catalytic converter nuts. Tighten the nuts also to 25–34 ft. lbs. (34–46 Nm).

31. Install the 2 lower, then the 2 upper power steering bolts. Tighten all 4 bolts to 15–20 ft. lbs. (20–30 Nm).

32. Install the Crankshaft Position (CKP) sensor. Tighten the retaining bolts to 72–107 ft. lbs. (8–12 Nm). Attach the CKP sensor electrical wiring harness connector.

33. Install the transmission fluid cooler hoses into the cylinder block mounted clip.

34. Install the A/C compressor.

35. Install the accessory drive belt.

36. Install the starter motor.

37. Install the ignition coils and mounting brackets. Tighten the bracket nuts to 15–23 ft. lbs. (20–30 Nm).

38. Attach the following electrical connections:
- Right-hand and left-hand ignition coil connectors
- Radio capacitor connector
- Camshaft Position (CMP) sensor connector

39. Position the rear heater water hose and compress, then slide the hose clamp into position.

40. Install the following EGR connections:
- EGR valve-to-exhaust manifold tube upper fitting — 26–33 ft. lbs. (35–45 Nm)
- EGR valve-to-exhaust manifold tube lower fitting — 26–33 ft. lbs. (35–45 Nm)
- The 2 DPFE transducer hoses

41. Install the power steering oil reservoir bracket. Tighten the lower 2 bolts to 26–33 ft. lbs. (35–45 Nm) and the upper bolts to 71–107 inch lbs. (8–12 Nm).

42. Attach the engine bulkhead connector.

43. Install the bulkhead connector cover.

44. Install the intake manifold.

45. Install the radiator.

46. Install the hood.

47. Fill the cooling system and fill the engine and transmission with the proper amount and type of fluids.

48. Connect the negative battery cable, start the engine and check for leaks and proper operation.

4.9L (VIN Y and Z) Engine

1. Disconnect the negative battery cable(s) before beginning any work. Label all disconnected hoses, vacuum lines and wires.

2. Drain the cooling system and the crankcase.

3. Remove the hood.

4. Remove the throttle body inlet tubes.

5. Disconnect the positive battery cable.

6. Discharge the air conditioning system.

7. Disconnect the refrigerant lines at the compressor. Cap all openings at once.

8. Remove the compressor.

9. Disconnect the refrigerant lines at the condenser. Cap all openings at once.

10. Remove the condenser.

11. Disconnect the heater hose from the water pump and coolant outlet housing.

12. Disconnect the flexible fuel line from the fuel pump.

13. Remove the radiator.

14. Remove the fan, water pump pulley, and fan belt.

15. Disconnect the accelerator cable.

16. Disconnect the brake booster vacuum hose at the intake manifold.

17. If equipped with automatic transmission, disconnect the transmission kickdown rod at the bell crank assembly.

18. Disconnect the exhaust pipe from the exhaust manifold.

19. Disconnect the Electronic Engine Control (EEC) harness from all the sensors.

20. Disconnect the body ground strap and the battery ground cable from the engine.

21. Disconnect the engine wiring harness at the ignition coil, the coolant temperature sending unit, and the oil pressure sensing unit. Position the wiring harness out of the way.

22. Remove the alternator mounting bolts and position the alternator out of the way.

23. Remove the power steering pump from the mounting brackets and move it to one side, leaving the lines attached.

24. Raise and support the vehicle on jackstands.

25. Remove the starter.

26. Remove the automatic transmission filler tube bracket, if equipped.

27. Remove the rear engine plate upper right bolt.

28. On manual transmission equipped vehicles:
    a. Remove the flywheel housing lower attaching bolts.
    b. Remove the clutch return spring.

29. On automatic transmission equipped vehicles:
    a. Remove the converter housing access cover assembly.
    b. Remove the flywheel-to-converter attaching nuts.
    c. Secure the converter in the housing.
    d. Remove the transmission oil cooler lines from the retaining clip at the engine.
    e. Remove the lower converter housing-to-engine attaching bolts.

30. Remove the nut from each of the 2 front engine mounts.

31. Lower the vehicle and position a jack under the transmission and support it.

32. Remove the remaining bell housing-to-engine attaching bolts.

33. Attach an engine lifting device and raise the engine slightly and carefully pull it from the transmission. Lift the engine out of the vehicle.

To install:

34. Remove the engine mount brackets from the frame. Attach them to the engine mounts, making the nuts just tight enough to hold the brackets securely to the mounts.

35. Place a new gasket on the muffler inlet pipe.

36. Carefully lower the engine into the vehicle. Make sure the dowels in the engine block engage the holes in the bell housing and the mount bracket holes align with the frame holes.

37. On manual transmission equipped vehicles, start the transmission input shaft into the clutch disc. It may be necessary to adjust the position of the engine or transmission in order for the input shaft to enter the clutch disc. If necessary, turn the crankshaft until the input shaft splines mesh with the clutch disc splines.

38. On automatic transmission equipped vehicles, start the converter pilot into the crankshaft. Secure the converter in the housing.

39. Install the bolts securing the mount brackets to the frame.

40. Install the bell housing upper attaching bolts. Torque the bolts to 50 ft. lbs. (68 Nm).

41. Remove the jack supporting the transmission.

42. Remove the lifting device.

43. Install the engine mount nuts and tighten them to 70 ft. lbs. (95 Nm). Tighten the bracket-to-frame bolts to 70 ft. lbs. (95 Nm).

44. Install the automatic transmission coil cooler lines bracket, if equipped.

45. Install the remaining bell housing attaching bolts. Torque them to 50 ft. lbs. (68 Nm).

46. Connect the clutch return spring, if equipped.

47. Install the starter and connect the starter cable.

48. Attach the automatic transmission fluid filler tube bracket, if equipped.

49. If equipped with automatic transmissions, install the transmission oil cooler lines in the bracket at the cylinder block.

50. Connect the exhaust pipe to the exhaust manifold. Tighten the nuts to 25–35 ft. lbs. (34–48 Nm).

51. Connect the engine ground strap and negative battery cable.

52. On a vehicle with an automatic transmission, connect the kickdown rod to the bell crank assembly on the intake manifold.

53. Connect the accelerator linkage.

54. Connect the brake booster vacuum line to the intake manifold.

55. Connect the coil primary wire, oil pressure and coolant temperature sending unit wires, fuel line, heater hoses, and the battery positive cable.

56. Connect the EEC sensors.

57. Install the alternator on its mounting bracket.

58. Install the power steering pump on its bracket.

59. Install the water pump pulley, spacer, fan, and fan belt. Adjust the belt tension.

60. Install the air conditioning compressor. Connect the refrigerant lines.

61. Install the radiator.

62. Install the condenser and connect the refrigerant lines.

63. Charge the refrigerant system.

64. Connect the upper and lower radiator hoses to the radiator and engine.

65. Connect the automatic transmission oil cooler lines, if equipped.

66. Install and adjust the hood.

67. Fill the cooling system.

68. Fill the crankcase.

69. Start the engine and check for leaks.

70. Bleed the cooling system.

71. Adjust the clutch pedal free-play or the automatic transmission control linkage.

72. Install the air cleaner.

5.0L (VIN N) and 5.8L (VIN H and R) Engines

1. Remove the hood.

2. Drain the cooling system and crankcase.

3. Disconnect the battery and alternator cables.

4. Remove the air intake hoses, PCV tube and carbon canister hose.

5. Disconnect the upper and lower radiator hoses.

6. Discharge the air conditioning system.

7. Disconnect the refrigerant lines at the compressor. Cap all openings immediately.

8. If equipped, disconnect the automatic transmission oil cooler lines.

9. Remove the fan shroud and lay it over the fan.

10. Remove the radiator and fan, shroud, fan, spacer, pulley and belt.

11. Remove the alternator pivot and adjusting bolts. Remove the alternator.

12. Disconnect the oil pressure sending unit lead from the sending unit.

13. Disconnect the fuel tank-to-pump fuel line at the fuel pump and plug the line.

14. If equipped, disconnect the chassis fuel line at the fuel rails.

15. Disconnect the accelerator linkage and speed control linkage at the throttle body.

16. Disconnect the automatic transmission kick-down rod and remove the return spring, if equipped.

17. Disconnect the power brake booster vacuum hose.

18. Disconnect the throttle bracket from the upper intake manifold and swing it out of the way with the cables still attached.

19. Disconnect the heater hoses from the water pump and tee.

20. Disconnect the temperature sending unit wire from the sending unit.

21. Remove the upper bell housing-to-engine attaching bolts.

22. Remove the wiring harness from the left rocker arm cover and position the wires out of the way.

23. Disconnect the ground strap from the cylinder block.

24. Disconnect the air conditioning compressor clutch wire.

25. Raise the front of the vehicle and disconnect the starter cable from the starter.

26. Remove the starter.

27. Disconnect the exhaust pipe from the exhaust manifolds.

28. Disconnect the engine mounts from the brackets on the frame.

29. If equipped with automatic transmissions, remove the converter inspection plate and remove the torque converter-to-flywheel attaching bolts.

30. Remove the remaining bell housing-to-engine attaching bolts.

31. Lower the vehicle and support the transmission with a jack.

32. Install an engine lifting device.

33. Raise the engine slightly and carefully pull it out of the transmission. Lift the engine out of the engine compartment.

To install:

34. Remove the engine mount brackets from the frame and attach them to the engine mounts. Tighten the mount-to-bracket nuts just enough to hold them securely.

35. Lower the engine carefully into the transmission. Make sure the dowel in the engine block engage the holes in the bell housing through the rear cover plate. If the engine hangs up after the transmission input shaft enters the clutch disc (manual transmission only), turn the crankshaft with the transmission in gear until the input shaft splines mesh with the clutch disc splines.

36. Install the engine mount nuts and washers. Torque the nuts to 80 ft. lbs. (109 Nm). Tighten the bracket-to-frame bolts to 70 ft. lbs. (95 Nm).

37. Remove the engine lifting device.

38. Install the lower bell housing-to-engine attaching bolts. Torque the bolts to 50 ft. lbs. (68 Nm).

39. Remove the transmission support jack.

40. If equipped with automatic transmissions, install the torque converter-to-flywheel attaching bolts. Torque the bolts to 30 ft. lbs. (41 Nm).

41. Install the converter inspection plate. Torque the bolts to 60 inch lbs. (6.7 Nm).

42. Connect the exhaust pipe to the exhaust manifolds. Tighten the exhaust pipe-to-exhaust manifold nuts to 25–35 ft. lbs. (34–48 Nm).

43. Install the starter. Torque the mounting bolts to 20 ft. lbs. (27 Nm).

44. Connect the starter cable to the starter.

45. Lower the vehicle.

46. Install the upper bell housing-to-engine attaching bolts. Torque the bolts to 50 ft. lbs. (68 Nm).

47. Connect the wiring harness at the left rocker arm cover.

48. Connect the ground strap to the cylinder block.

49. Connect the air conditioning compressor clutch wire.

50. Connect the heater hoses at the water pump and tee.

51. Connect the temperature sending unit wire at the sending unit.

52. Connect the accelerator linkage and speed control linkage at the throttle body.

53. Connect the automatic transmission kick-down rod and install the return spring, if equipped.

54. Connect the power brake booster vacuum hose.

55. Connect the throttle bracket to the upper intake manifold.

56. Connect the fuel tank-to-pump fuel line at the fuel pump. If equipped, connect the chassis fuel line at the fuel rails.

57. Connect the oil pressure sending unit lead to the sending unit.

58. Install the alternator.

59. Connect the refrigerant lines to the compressor.

60. Install the radiator and fan, shroud, fan, spacer, pulley and belt.

61. Connect the upper and lower radiator hoses, and, if equipped, the automatic transmission oil cooler lines.

62. Install the air intake hoses, PCV tube and carbon canister hose.

63. Connect the battery and alternator cables.

64. Fill the cooling system and crankcase.

65. Charge the air conditioning system.

66. Install the hood.

If the torque for a particular fastener was not mentioned above, use the following torque values as a guide:

- ¼ in.-20: 6–9 ft. lbs. (8–12 Nm)
- ⁵/₁₆ in.-18: 12–18 ft. lbs. (16–25 Nm)
- ⅜ in.-16: 22–32 ft. lbs. (30–44 Nm)
- ⁷/₁₆ in.-14: 45–57 ft. lbs. (61–78 Nm)
- ½ in.-13: 55–80 ft. lbs. (75–109 Nm)
- ⁹/₁₆ in.: 85–120 ft. lbs. (116–163 Nm)

7.5L (VIN G) Engine

1. Remove the hood.
2. Drain the cooling system.
3. Disconnect the negative battery cable from the block.
4. Remove the air cleaner assembly.
5. Remove the crankcase ventilation hose.
6. Remove the canister hose.
7. Disconnect the upper and lower radiator hoses.
8. Disconnect the transmission oil cooler lines from the radiator.
9. Disconnect the oil cooler lines at the oil filter adapter.

——— **WARNING** ———
Don't disconnect the lines at the quick-connect fittings behind or at the oil cooler. Disconnecting them may permanently damage them.

10. Discharge the air conditioning system.
11. Disconnect the refrigerant lines at the compressor. Cap the openings at once!
12. Disconnect the refrigerant lines at the condenser. Cap the openings at once!
13. Remove the condenser.
14. Remove the fan shroud from the radiator and position it up, over the fan.
15. Remove the radiator.
16. Remove the fan shroud.
17. Remove the fan, belts and pulley from the water pump.
18. Remove the compressor.

19. Remove the power steering pump from the engine, if equipped, and position it to one side. Do not disconnect the fluid lines.

20. Disconnect the fuel pump inlet line from the pump and plug the line.

21. Disconnect the oil pressure sending unit wire at the sending unit.

22. Remove the alternator drive belts and disconnect the alternator from the engine, positioning it aside.

23. Disconnect the ground cable from the right front corner of the engine.

24. Disconnect the heater hoses.

25. Remove the transmission fluid filler tube attaching bolt from the right side valve cover and position the tube out of the way.

26. Disconnect all vacuum lines at the rear of the intake manifold.

27. Disconnect the accelerator rod and the transmission kickdown rod and secure them out of the way.

28. Disconnect the engine wiring harness at the connector on the fire wall. Disconnect the primary wire at the coil.

29. Remove the upper flywheel housing-to-engine bolts.

30. Raise the vehicle and disconnect the exhaust pipes at the exhaust manifolds.

31. Disconnect the starter cable and remove the starter. Bring the starter forward and rotate the solenoid outward to remove the assembly.

32. Remove the access cover from the converter housing and remove the flywheel-to-converter attaching nuts.

33. Remove the lower the converter housing-to-engine attaching bolts.

34. Remove the engine mount through bolts attaching the rubber insulator to the frame brackets.

35. Lower the vehicle and place a jack under the transmission to support it.

36. Remove the converter housing-to-engine block attaching bolts (left side).

37. Remove the coil and bracket assembly from the intake manifold.

38. Attach an engine lifting device and carefully take up the weight of the engine.

39. Move the engine forward to disengage it from the transmission and slowly lift it from the vehicle.

To install:
40. Remove the engine mount brackets from the frame and attach them to the mounts. Tighten the nuts just enough to hold them securely.

41. Lower the engine slowly into the vehicle.

42. Slide the engine rearward to engage it with the transmission and slowly lower it onto the supports.

43. Tighten the engine support nuts to 74 ft. lbs. (100 Nm). Tighten the bracket bolts to 70 ft. lbs. (95 Nm).

44. Remove the engine lifting device.

45. Install the converter housing-to-engine block upper and left side attaching bolts. Torque the bolts to 50 ft. lbs. (68 Nm).

46. Install the coil and bracket assembly on the intake manifold.

47. Remove the jack from under the transmission.

48. Lower the vehicle.

49. Install the upper converter housing-to-engine attaching bolts. Torque the bolts to 50 ft. lbs. (68 Nm).

50. Install the flywheel-to-converter attaching nuts. Torque the nuts to 34 ft. lbs. (47 Nm).

51. Install the access cover on the converter housing. Torque the bolts to 60–90 inch lbs. (6.7–10 Nm).

52. Install the starter.

53. Connect the starter cable.

54. Raise the vehicle and connect the exhaust pipes at the exhaust manifolds.

55. Connect the engine wiring harness at the connector on the fire wall.

56. Connect the primary wire at the coil.

57. Connect the accelerator rod and the transmission kickdown rod.

58. Connect the speed control cable.

59. Connect all vacuum lines at the rear of the intake manifold.

60. Install the transmission fluid filler tube attaching bolt from the right side valve cover and position the tube out of the way.

61. Connect the heater hoses.

62. Connect the ground cable at the right front corner of the engine.

63. Install the alternator and drive belts.

64. Connect the oil pressure sending unit wire at the sending unit.

65. Connect the fuel pump inlet line at the pump and plug the line.

66. Install the power steering pump and belt.

67. Install air conditioning compressor. Connect the refrigerant lines.

68. Install the fan, belts and pulley on the water pump.

69. Position the fan shroud over the fan.

70. Install the radiator.

71. Attach the fan shroud.

72. Install the condenser.

73. Connect the refrigerant lines at the condenser.

74. Charge the air conditioning system.

75. Connect the oil cooler lines at the oil filter adapter.

76. Connect the transmission oil cooler lines at the radiator.

77. Connect the upper and lower radiator hoses.

78. Connect the canister hose.

79. Connect the crankcase ventilation hose.

80. Connect the negative battery cable from the block.

81. Fill the cooling system.

82. Install the air cleaner assembly.

83. Install the hood.

If the torque for a particular fastener was not mentioned above, use the following torque values as a guide:

- $\frac{1}{4}$ in.-20: 6–9 ft. lbs. (8–12 Nm)
- $\frac{5}{16}$ in.-18: 12–18 ft. lbs. (16–25 Nm)
- $\frac{3}{8}$ in.-16: 22–32 ft. lbs. (30–44 Nm)
- $\frac{7}{16}$ in.-14: 45–57 ft. lbs. (61–78 Nm)
- $\frac{1}{2}$ in.-13: 55–80 ft. lbs. (75–109 Nm)
- $\frac{9}{16}$ in.: 85–120 ft. lbs. (116–163 Nm)

Engine Mounts

REMOVAL AND INSTALLATION

Front Mounts

1997 F-150 and Expedition Models

NOTE: If both engine mounts are to be replaced, remove and install one engine mount at a time. This will allow the engine to remain correctly positioned for easier installation.

1. Remove the engine cooling fan and fan shroud.

2. Raise and safely support the vehicle on jackstands.

3. Remove the mount-to-frame mount bolt on the first engine mount to be removed.

   a. Loosen the other engine mount's frame mount bolt.

   b. Lift the engine slightly with a hydraulic floor crane/hoist, or equivalent.

4. Remove the 3 engine mount-to-cylinder block bolts, then remove the engine mount.

   To install:

5. Situate the new engine mount in position on the engine block, then install the engine mount-to-engine block bolts. Tighten the bolts to 39–53 ft. lbs. (53–72 Nm).

6. Slowly lower the engine onto the engine mounts.

7. If the other engine mount is also going to be removed, leave that mount's through-bolt loose for now, otherwise tighten it to 50–68 ft. lbs. (68–92 Nm).

8. Install the through-bolt in the new engine mount. If the other engine mount is going to be removed, leave the new mount's through-bolt also loose for now, otherwise tighten it to 50–68 ft. lbs. (68–92 Nm).

9. If necessary, repeat the process for the other engine mount. Once both mounts have been installed, make sure that the engine mount through-bolts are tightened to 50–68 ft. lbs. (68–92 Nm).

10. Lower the vehicle to the ground.

11. Install the fan shroud and engine cooling fan.

Except 1997 F-150 and Expedition Models

1. Remove the fan shroud attaching screws.

2. Support the engine using a wood block and a jack placed under the oil pan.

3. Remove the nuts and washers attaching the mounts to the engine bracket. Lift the engine enough to disengage the mount upper stud from the crossmember engine bracket.

4. Remove the bolt attaching the fuel pump shield to the left engine bracket, if necessary.

5. Remove the mount-to-crossmember attaching nut and washer assembly. Remove the engine mount.

   To install:

6. Install the engine mount to the crossmember.

7. Install the bolt attaching the fuel pump shield to the left bracket, if necessary.

8. Lower the engine until the mount stud engages in the slot/hole of the engine bracket. Install the attaching nuts.

9. Remove the jack and wood block from the engine oil pan. Install the fan shroud attaching screws.

Rear Mount

1. Place a block of wood and a jack under the transmission.

2. Remove the 2 nuts attaching the mount to the crossmember. Raise the transmission enough to lift the mount from the crossmember.

3. Remove the bolts and nuts attaching the crossmember to the frame side rails and remove the crossmember.

4. If equipped, remove the fasteners attaching the exhaust hanger to the rear engine mount.

5. Remove the 2 bolts attaching the mount to the transmission and remove the mount and retainer assembly.

To install:

6. Position the mount and retainer assembly to the transmission and install the 2 attaching bolts.

7. If equipped, install the fasteners attaching the exhaust hanger to the mount.

8. Install the crossmember to the frame side rails with the attaching nuts and bolts.

9. Lower the transmission and install the mount crossmember attaching nuts. Remove the jack and wood block.

Cylinder Head

REMOVAL AND INSTALLATION

4.2L (VIN 2) Engine

1. On vehicles equipped with Air Conditioning (A/C), have the A/C system discharged by a qualified mechanic using an approved refrigerant recovery/recycling machine.

2. Disconnect the negative battery cable.

3. Drain the cooling system.

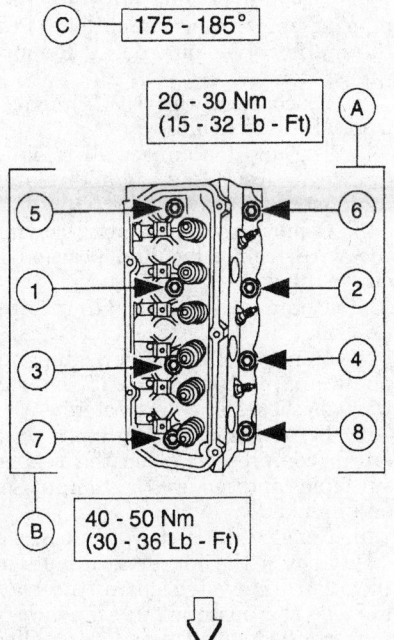

Final-tighten the cylinder head bolts one at a time in the sequence shown here, then tighten an additional 175–185 degrees — 4.2L (VIN 2) engines

4. Remove the upper intake manifold and related components.

5. Remove the lower intake manifold and related components.

6. Remove the rocker arm covers.

7. Remove the exhaust manifold.

8. If removing the left-hand cylinder head, perform the following:

a. Position the power steering pump reservoir aside and remove the A/C compressor.

b. Remove and support the A/C compressor bracket and power steering pump aside.

9. If removing the right-hand cylinder head, perform the following:

a. Remove the alternator.

b. Remove the idler pulley.

c. Remove the alternator bracket.

NOTE: If the cylinder head components, such as rocker arms, valve springs, etc., are to be reinstalled, they must be installed in the same position. Mark the components for original location.

10. Remove the 6 rocker arms by removing the retaining bolts.

11. Pull the pushrods out of the engine. Once again, make sure to label or mark the components removed for reinstallation in their original location.

12. Remove and discard the 8 cylinder head mounting bolts. New bolts are a must for installation.

13. Lift the cylinder head off of the engine block. Remove the cylinder head gasket and discard.

To install:

14. Clean and inspect the cylinder head for flatness.

15. Install a new cylinder head gasket on the cylinder block with the small hole to the front of the engine, then install the cylinder head.

── **WARNING** ──
Always use new cylinder head bolts for installation.

16. Lubricate the cylinder head bolts with clean engine oil prior to installation.

NOTE: Make sure to tighten the cylinder head bolts in three (3) steps.

17. Install the new cylinder head bolts. Tighten the cylinder head bolts in the sequence shown and in three steps to the following values:
- Step 1 — 14 ft. lbs. (20 Nm)
- Step 2 — 29 ft. lbs. (40 Nm)
- Step 3 — 36 ft. lbs. (50 Nm)

── **WARNING** ──
Do not loosen all of the cylinder head bolts at one time. Each cylinder head bolt must be loosened and the final tightening performed prior to loosening the next bolt in the sequence.

18. In the same sequence as used previously, loosen the cylinder head bolt three turns, then tighten the cylinder head bolt to the specific value according to its length. The short bolts should be tightened to 15–32 ft. lbs. (20–30 Nm) and the long bolts to 30–36 ft. lbs. (40–50 Nm). Finally tighten the cylinder head bolt an additional 175–185 degrees. It would be very helpful to utilize a degree socket wrench for this last step.

19. Lubricate the pushrods with clean engine oil prior to installation, then install them into their original positions.

20. Install the rocker arms. Tighten the rocker arm mounting bolts to 23–29 ft. lbs. (30–40 Nm).

21. If the valve train components were replaced with new components, inspect the valve clearance.

22. If installing the right-hand cylinder head, perform the following:

a. Position the alternator bracket in place, then install the 2 long bolts to 31–39 ft. lbs. (41–54 Nm). Install the short bolt and tighten to 18–22 ft. lbs. (24–31 Nm).

b. Install the idler pulley. Tighten the center retaining bolt to 35–46 ft. lbs. (47–63 Nm).

c. Install the alternator.

23. If installing the left-hand cylinder head, complete the following steps:

a. Position the A/C compressor bracket and power steering pump in place, then start the A/C compressor bracket bolt. Install the 3 compressor bracket bolts to 30–40 ft. lbs. (40–55 Nm). Then install the 2 compressor bracket nuts to 16–21 ft. lbs. (21–29 Nm).

b. Install the A/C compressor.

c. Install the power steering pump reservoir. Tighten the hold-down bolts to 80–107 inch lbs. (9–12 Nm).

24. Install the exhaust manifold.

25. Install the 2 rocker arm covers. Inspect the rocker arm cover gaskets for damage prior to installation; replace them if necessary.

26. Install the lower intake manifold and related components.

27. Install the upper intake manifold and related components.

28. Fill the cooling system and connect the negative battery cable.

29. Start the engine and check for any fuel, coolant and vacuum leaks.

4.6L Engines (VIN W and VIN 6)

NOTE: To correctly tighten the cylinder head bolts an angle torque wrench is needed.

1. If the vehicle is equipped with Air Conditioning (A/C), have the A/C system discharged by a qualified mechanic utilizing the appropriate refrigerant recovery/recycling machine.

2. Disconnect the negative battery cable.

3. Remove the cylinder head covers.

4. Remove the intake manifold.

5. Remove the timing chains from the engine.

6. Remove the exhaust manifolds.

7. Remove the 2 heater hose retaining bolts, then compress and slide the hose clamp back to remove the heater water hose.

8. Remove the 10 cylinder head bolts, then lift the cylinder head from the engine block. Discard the cylinder head gasket and clean the engine block surface.

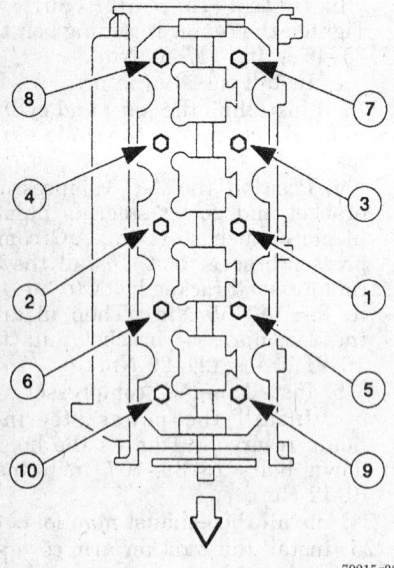

Tighten the left-hand cylinder head bolts in all three steps using the sequence shown — 4.6L (VIN W and VIN 6) engines

79215g38

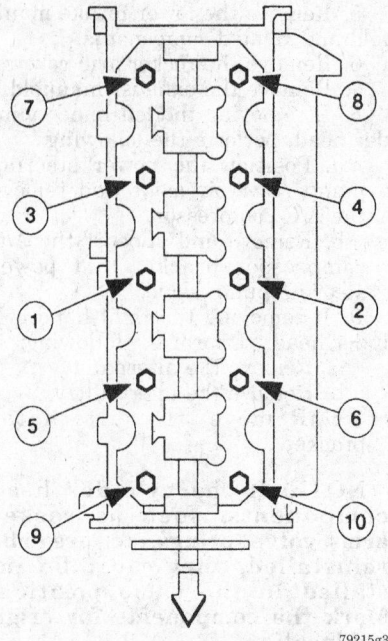

79215g39

Tighten the right-hand cylinder head bolts in all three steps using the sequence shown — 4.6L (VIN W and VIN 6) engines

To install:

——— WARNING ———
Cylinder head bolts must be replaced with new ones. They are tighten-to-yield designed and cannot be reused.

9. Clean and inspect the cylinder head for damage or warpage. Install the cylinder head gasket over the dowel pins. Then install the cylinder head onto the engine block. Loosely install NEW cylinder head bolts.

NOTE: Make sure to tighten the head bolts in 3 steps.

10. Tighten the 10 cylinder head bolts in the sequence shown in 3 steps, as follows:

    a. Step 1 — 27–31 ft. lbs. (37–43 Nm).

    b. Step 2 — tighten an additional 85–95 degrees.

    c. Step 3 — tighten another 85–95 degrees.

11. Install the heater water hose and slide the hose clamp back into position. Install the 2 heater water hose bolts.

12. Install the exhaust manifolds.

13. Install the timing chains.

14. Install the intake manifold.

15. Install the cylinder head covers.

16. Connect the negative battery cable, then start the engine and check for leaks.

17. If the vehicle is equipped with A/C, have the system evacuated and recharged by a qualified mechanic utilizing the appropriate refrigerant recovery/recycling machine.

4.9L (VIN Y and Z) Engine

1. Drain the cooling system. Remove the hood.

2. Remove the throttle body inlet tubes.

3. Remove the air conditioning compressor.

4. Remove the condenser.

5. Disconnect the battery ground cable.

6. Disconnect the heater hoses from the water pump and coolant outlet housing.

7. Disconnect the fuel line at the fuel pump.

8. Remove the radiator.

9. Remove the engine fan and fan drive, the water pump pulley and the drive belt.

10. Disconnect the accelerator cable and retracting spring.

11. Disconnect the power brake hose at the manifold.

12. Disconnect the transmission kickdown rod on vehicles with automatic transmission.

13. Disconnect the muffler inlet pipe at the exhaust manifold. Pull the muffler inlet pipe down. Remove the gasket.

14. Disconnect the EEC harness from all the sensors.

15. Tag and disconnect all remaining wiring from the head and related components.

16. Remove the alternator, leaving the wires connected and position it out of the way.

17. Remove the air pump and bracket.

18. Remove the power steering pump and position it out of the way with the hoses still connected.

19. If the vehicle is equipped with an air compressor, bleed the 2 pressure lines and remove the compressor and bracket.

20. Remove the rocker arm cover.

21. Loosen the rocker arm bolts so they can be pivoted out of the way. Remove the pushrods in sequence so they can be identified and reinstalled in their original positions.

22. Disconnect the spark plug wires at the spark plugs.

23. Remove the cylinder head bolts and remove the cylinder head. Do not pry between the cylinder head and

the block as the gasket surfaces maybe damaged.

To install:

24. Clean the head and block gasket surfaces. If the cylinder head was removed for a gasket change, check the flatness of the cylinder head and block.

25. Position the gasket on the cylinder block.

26. Install a new gasket on the flange of the muffler inlet pipe.

27. Lift the cylinder head above the cylinder block and lower it into position using 2 head bolts installed through the head as guides.

28. Coat the threads of the Nos. 1 and 6 bolts for the right side of the cylinder head with a small mount of water-resistant sealer. Oil the threads of the remaining bolts. Install, but do not tighten, 2 bolts at the opposite ends of the head to hold the head and gasket in position.

29. The cylinder head bolts are tightened in 3 progressive steps. Torque them (in the proper sequence):

   a. First tighten the bolts to 50–55 ft. lbs. (68–75 Nm).

   b. Then tighten the cylinder head bolts to 60–65 ft. lbs. (82–88 Nm)

   c. Finally, tighten the head bolts to 70–85 ft. lbs. (95–116 Nm).

30. Apply Lubriplate® to both ends of the pushrods and install them in their original positions.

31. Apply Lubriplate® to both the fulcrum and seat and position the rocker arms on the valves and pushrods.

32. Adjust the valves, as outlined below.

33. Install the rocker arm cover.

34. Install the air compressor and bracket.

35. Install the power steering pump.

36. Install the air pump and bracket.

37. Install the alternator.

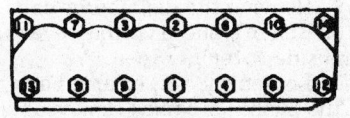

79215g40

Cylinder head bolt torque sequence — 4.9L (VIN Y and Z) engine

38. Connect all wiring at the head and related components.

39. Connect the EEC harness to all the sensors.

40. Connect the muffler inlet pipe at the exhaust manifold.

41. Connect the transmission kickdown rod on vehicles with automatic transmission.

42. Connect the power brake hose at the manifold.

43. Connect the accelerator cable and retracting spring.

44. Install the water pump pulley, the engine fan and fan drive, and the drive belt.

45. Install the radiator.

46. Connect the fuel line at the fuel pump.

47. Connect the heater hoses at the water pump and coolant outlet housing.

48. Connect the battery ground cable.

49. Install the condenser.

50. Install the air conditioning compressor.

51. Install the throttle body inlet tubes.

52. Fill and bleed the cooling system.

53. Install the hood.

5.0L (VIN N) and 5.8L (VIN H and R) Engines

1. Drain the cooling system.

2. Remove the intake manifold and throttle body.

3. Remove the rocker arm cover(s).

4. If the right cylinder head is to be removed, lift the tensioner and remove the drive belt. Loosen the alternator adjusting arm bolt and remove the alternator mounting bracket bolt and spacer. Swing the alternator down and out of the way. Remove the air cleaner inlet duct.

If the left cylinder head is being removed, remove the air conditioning compressor. Persons not familiar with air conditioning systems should exercise extreme caution, perhaps leaving this job to a professional. Remove the oil dipstick and tube. Remove the cruise control bracket.

5. Disconnect the exhaust manifold(s) from the muffler inlet pipe(s).

6. Loosen the rocker arm stud nuts so the rocker arms can be rotated to the side. Remove the pushrods and identify them so they can be reinstalled in their original positions.

7. Disconnect the Thermactor® air supply hoses at the check valves. Cover the check valve openings.

8. Remove the cylinder head bolts and lift the cylinder head from the block. Remove the discard the gasket.

To install:

9. Clean the cylinder head, intake manifold, the valve cover and the head gasket surfaces.

10. A specially treated composition head gasket is used. Do not apply sealer to a composition gasket. Position the new gasket over the locating dowels on the cylinder block. Then, position the cylinder head on the block and install the attaching bolts.

11. The cylinder head bolts are tightened in progressive steps. Tighten all the bolts in the proper sequence to:

5.0L (VIN N) Engine

• Step 1 — 55–65 ft. lbs. (75–89 Nm)

• Step 2 — 66–72 ft. lbs. (90–98 Nm)

5.8L (VIN H and R) Engine

• Step 1 — 85 ft. lbs.(115 Nm)

• Step 2 — 95 ft. lbs. (129 Nm)

• Step 3 — 105–112 ft. lbs. (143–152 Nm)

12. Clean the pushrods. Blow out the oil passage in the rods with compressed air. Check the pushrods for straightness by rolling them on a piece of glass. Never try to straighten a pushrod; always replace it.

13. Apply Lubriplate® to the ends of the pushrods and install them in their original positions.

14. Apply Lubriplate® to the rocker arms and their fulcrum seats and install the rocker arms. Adjust the valves.

15. Position a new gasket(s) on the muffler inlet pipe(s) as necessary. Connect the exhaust manifold(s) at the muffler inlet pipe(s).

16. If the right cylinder head was removed, install the alternator, and air cleaner duct. Install the drive belt. If the left cylinder head was removed, install the compressor. Install the dipstick and cruise control bracket.

17. Clean the valve rocker arm cover and the cylinder head gasket surfaces. Place the new gaskets in the covers, making sure the tabs of the gasket engage the notches provided in the cover. Evacuate, charge and leak test the air conditioning system.

18. Install the intake manifold and related parts. Install the Thermactor® hoses.

19. Fill and bleed the cooling system.

7.5L (VIN G) Engine

1. Drain the cooling system.

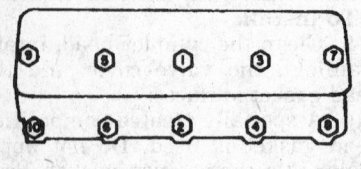

Cylinder head bolt torque sequence — 5.0L (VIN N), 5.8L (VIN H and R) and 7.5L (VIN G) engines

2. Remove the upper and lower intake manifolds..

3. Disconnect the exhaust pipe from the exhaust manifold.

4. Loosen the air conditioning compressor drive belt, if equipped.

5. Loosen the alternator attaching bolts and remove the bolt attaching the alternator bracket to the right cylinder head.

6. Disconnect the air conditioning compressor from the engine and move it aside, out of the way. Do not discharge the air conditioning system.

7. Remove the bolts securing the power steering reservoir bracket to the left cylinder head. Position the reservoir and bracket out of the way. On motor home chassis, remove the oil filler tube.

8. Remove the valve rocker arm covers. Remove the rocker arm bolts, rocker arms, oil deflectors, fulcrums and pushrods in sequence so they can be reinstalled in their original positions.

9. Remove the cylinder head bolts and lift the head and exhaust manifold off the engine. If necessary, pry at the forward corners of the cylinder head against the casting bosses provided on the cylinder block. Do not damage the gasket mating surfaces of the cylinder head and block by prying against them.

To install:

10. Remove all gasket material from the cylinder head and block. Clean all gasket material from the mating surfaces of the intake manifold. If the exhaust manifold was removed, clean the mating surfaces of the cylinder head and exhaust manifold. Apply a thin coat of graphite grease to the cylinder head exhaust port areas and install the exhaust manifold.

11. Position 2 long cylinder head bolts in the 2 rear lower bolt holes of the left cylinder head. Place a long cylinder head bolt in the rear lower

bolt hole of the right cylinder head. Use rubber bands to keep the bolts in position until the cylinder heads are installed on the cylinder block.

12. Position new cylinder head gaskets on the cylinder block dowels. Do not apply sealer to the gaskets, heads, or block.

13. Place the cylinder heads on the block, guiding the exhaust manifold studs into the exhaust pipe connections. Install the remaining cylinder head bolts. The longer bolts go in the lower row of holes.

14. Tighten all the cylinder head attaching bolts in the proper sequence in 3 stages: 80–90 ft. lbs. (109–122 Nm), 100–110 ft. lbs. (136–149 Nm), and finally to 130–140 ft. lbs. (176–190 Nm). When this procedure is used, it is not necessary to retighten the heads after extended use.

15. Make sure the oil holes in the pushrods are open and install the pushrods in their original positions. Place a dab of Lubriplate® to the ends of the pushrods before installing them.

16. Lubricate and install the valve rockers. Make sure the pushrods remain seated in their lifters.

17. Connect the exhaust pipes to the exhaust manifolds.

18. Install the upper and lower intake manifolds.

19. Install the air conditioning compressor.

20. Install the power steering reservoir.

21. Apply oil-resistant sealer to one side of the new valve cover gaskets and lay the cemented side in place in the valve cover. Install the covers.

22. Install the alternator and adjust the drive belt.

23. Adjust the air conditioning compressor drive belt tension.

24. On motor home chassis, install the oil filler tube.

25. Fill and bleed the cooling system.

26. Start the engine and check for leaks.

Valve Lifters

REMOVAL AND INSTALLATION

4.2L (VIN 2) Engine

NOTE: If removing more than one valve lifter, mark the components for proper location.

1. Disconnect the negative battery cable.

2. Remove the lower intake manifold.

3. Remove the rocker arm cover.

4. Remove the rocker arm hold-down bolt, then remove the rocker arm from the cylinder head.

5. Remove the pushrod.

6. Remove the 2 lifter guide plate bolts, then remove the lifter guide plate and retainer.

7. Remove the valve lifter from its bore.

To install:

NOTE: Make certain to install the components in their original locations.

8. Lubricate the valve lifters and pushrods with clean engine oil before installation.

9. Insert the valve lifter into its respective bore, then install the retainer and guide plate. Tighten the guide plate mounting fasteners to 8–10 ft. lbs. (10–14 Nm).

10. Install the pushrods making sure that they are seated into the valve lifter recess fully.

11. Position the rocker arms in place, then install the hold-down bolts. Tighten the bolts to 23–29 ft. lbs. (30–40 Nm).

12. Install the rocker arm cover and the lower intake manifold.

13. Connect the negative battery cable.

4.9L (VIN Y and Z) Engine

1. Disconnect the inlet hose at the crankcase filler cap.

2. Remove the throttle body inlet tubes.

3. Disconnect the accelerator cable at the throttle body. Remove the cable retracting spring. Remove the accelerator cable bracket from the upper intake manifold and position the cable and bracket out of the way.

4. Remove the fuel line from the fuel rail. Be careful not to kink the line.

5. Remove the upper intake manifold and throttle body assembly.

6. Remove the ignition coil and wires.

7. Remove the rocker arm cover.

8. Remove the spark plug wires.

9. Remove the distributor cap.

10. Remove the pushrod cover (engine side cover).

11. Loosen the rocker arm bolts until the pushrods can be removed; keep them in order, for installation.

12. Using a magnetic lifter removal tool, remove the lifters. Wipe clean the exterior of each lifter as it's removed and mark it with an indelible marker, so it can be installed in its original bore.

To install:

13. Coat the bottom surface of each lifter with multi-purpose grease, and coat the rest of the lifter with clean oil.

14. Install each lifter in it original bore using the magnetic tool.

15. Coat each end of each pushrod with multi-purpose grease and install each in its original position. Make sure each pushrod is properly seated in the lifter socket.

16. Engage the rocker arms with the pushrods and tighten the rocker arm bolts enough to hold the pushrods in place.

17. Adjust the valve clearance.

18. Install the pushrod cover (engine side cover).

19. Install the distributor cap.

20. Install the spark plug wires.

21. Install the rocker arm cover.

22. Install the ignition coil and wires.

23. Install the upper intake manifold and throttle body assembly.

24. Install the fuel line at the fuel rail.

25. Install the accelerator cable bracket at the upper intake manifold. Install the cable retracting spring. Connect the accelerator cable at the throttle body.

26. Install the throttle body inlet tubes.

27. Connect the inlet hose at the crankcase filler cap.

5.0L (VIN N) and 5.8L (VIN H and R) Engines

NOTE: The 1993 5.0L (VIN N) engine uses roller lifters.

1. Remove the intake manifold.

2. Disconnect the Thermactor® air supply hose at the pump.

3. Remove the rocker arm covers.

4. Loosen the rocker arm fulcrum bolts until the rocker arms can be rotated off the pushrods.

5. Remove the pushrods and keep them in order, for installation.

6. Using a magnetic lifter removal tool, remove the lifters. Wipe clean the exterior of each lifter as it's removed and mark it with an indelible marker, so it can be installed in its original bore.

To install:

7. Coat the bottom surface of each lifter with multi-purpose grease, and coat the rest of the lifter with clean oil.

8. Install each lifter in it original bore using the magnetic tool.

9. Coat each end of each pushrod with multi-purpose grease and install each in its original position. Make sure each pushrod is properly seated in the lifter socket.

10. Engage the rocker arms with the pushrods and tighten the rocker arm fulcrum bolts to 18–25 ft. lbs. (25–34 Nm). No valve adjustment should be necessary, however, if there is any question as to post-assembly collapsed lifter clearance, see the Hydraulic Lifter Clearance procedure below.

11. Install the rocker arm covers.

12. Connect the Thermactor® air supply hose at the pump.

13. Install the intake manifold.

7.5L (VIN G) Engine

1. Remove the intake manifold.

2. Remove the rocker arm covers.

3. Loosen the rocker arm fulcrum bolts until the rocker arms can be rotated off the pushrods.

4. Remove the pushrods and keep them in order, for installation.

5. Using a magnetic lifter removal tool, remove the lifters. Wipe clean the exterior of each lifter as it's removed and mark it with an indelible marker, so it can be installed in its original bore.

To install:

6. Coat the bottom surface of each lifter with multi-purpose grease, and coat the rest of the lifter with clean oil.

7. Install each lifter in it original bore using the magnetic tool.

8. Coat each end of each pushrod with multi-purpose grease and install each in its original position. Make sure each pushrod is properly seated in the lifter socket.

9. Rotate the crankshaft by hand until No. 1 piston is at TDC of compression. The firing order marks on the damper will be aligned at TDC with the timing pointer.

10. Engage the rocker arms with the pushrods and tighten the rocker arm fulcrum bolts to 18–25 ft. lbs. (25–34 Nm) in the following sequence:

- No. 1 intake and exhaust
- No. 3 intake
- No. 8 exhaust
- No. 7 intake
- No. 5 exhaust
- No. 8 intake
- No. 4 exhaust

11. Rotate the crankshaft 1 full turn — 360 degrees — and re-align the TDC mark and pointer. Tighten

the fulcrum bolt on the following valves:

- No. 2 intake and exhaust
- No. 4 intake
- No. 3 exhaust
- No. 5 intake
- No. 6 exhaust
- No. 6 intake
- No. 7 exhaust

12. Check the valve clearance as described under Hydraulic Valve Clearance, below.

13. Install the intake manifold.

14. Install the rocker arm covers.

Valve Lash

ADJUSTMENT

4.2L (VIN 2) and 4.6L Engines

The 4.2L (VIN 2) and 4.6L engine do not require valve lash adjusting, because they utilize hydraulic lash components in their valve actuation systems; the 4.2L (VIN 2) engine uses hydraulic valve lifters and the 4.6L engines utilize hydraulic lash adjusters, all of which automatically adjust the valve lash. No valve lash adjustment is necessary.

4.9L (VIN Y and Z) Engine

1. Rotate the crankshaft by hand so No. 1 piston is at TDC of the compression stroke. Make a chalk mark on the damper at that point, then, make 2 more chalk marks about 120 degrees apart, dividing the damper into 3 equal parts.

2. With No. 1 at TDC, tighten the rocker arm bolts on No. 1 cylinder intake and exhaust to 17–23 ft. lbs. (23–31 Nm). Then, slowly apply pressure, using Lifter Bleed-down wrench T70P-6513-A, or equivalent, to completely bottom the lifter. Take care to avoid excessive pressure that might bend the pushrod. Hold the lifter in this position and check the clearance between the rocker arm and the valve stem tip. Allowable clearance is 0.10–0.20 in. (2.5–5.0mm) with a desired clearance of 0.125–0.175 in. (3.0–4.5mm)

3. If the clearance is less than specified, install a shorter pushrod. If the clearance is greater than specified, install a longer pushrod.

4. Rotate the crankshaft clockwise — viewed from the front — until the next chalk mark is aligned with the timing pointer. Repeat the procedure for No. 5 intake and exhaust.

5. Rotate the crankshaft to the next chalk mark and repeat the procedure for No. 3 intake and exhaust.

6. Repeat the rotation/checking procedure for the remaining valves in firing order, that is: 6–2–4.

5.0L (VIN N) Engine

1. Rotate the crankshaft by hand so No. 1 piston is at TDC of the compression stroke. Make a chalk mark on the damper at that point, then, make 2 more chalk marks about 90 degrees apart in a clockwise direction.

2. With No. 1 at TDC, slowly apply pressure, using Lifter Bleed-down wrench T70P-6513-A, or equivalent, to completely bottom the lifter, on the following valves:

- No. 1 intake and exhaust
- No. 7 intake
- No. 5 exhaust
- No. 8 intake
- No. 4 exhaust

Take care to avoid excessive pressure that might bend the pushrod. Hold the lifter in this position and check the clearance between the rocker arm and the valve stem tip. Allowable clearance is 0.071–0.193 in. (1.8–4.9mm) with a desired clearance of 0.096–0.165 in. (2.4–4.2mm).

3. If the clearance is less than specified, install a shorter pushrod. If the clearance is greater than specified, install a longer pushrod.

4. Rotate the crankshaft clockwise — viewed from the front — 180 degrees, until the next chalk mark is aligned with the timing pointer. Repeat the procedure for:

- No. 5 intake
- No. 2 exhaust
- No. 4 intake
- No. 6 exhaust

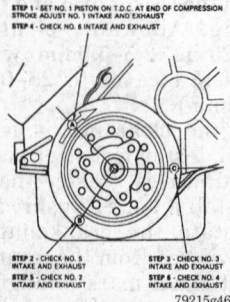

Valve clearance adjustment positions on the crankshaft damper — 4.9L (VIN Y and Z) engine

5. Rotate the crankshaft to the next chalk mark — 90 degrees — and repeat the procedure for:

- No. 2 intake
- No. 7 exhaust
- No. 3 intake and exhaust
- No. 6 intake
- No. 8 exhaust

5.8L (VIN H and R) Engine

1. Rotate the crankshaft by hand so No. 1 piston is at TDC of the compression stroke. Make a chalk mark on the damper at that point, then, make 2 more chalk marks about 90 degrees apart in a clockwise direction.

2. With No. 1 at TDC, slowly apply pressure, using Lifter Bleed-down wrench T70P-6513-A, or equivalent, to completely bottom the lifter, on the following valves:

- No. 1 intake and exhaust
- No. 4 intake
- No. 3 exhaust
- No. 8 intake
- No. 7 exhaust

Take care to avoid excessive pressure that might bend the pushrod. Hold the lifter in this position and check the clearance between the

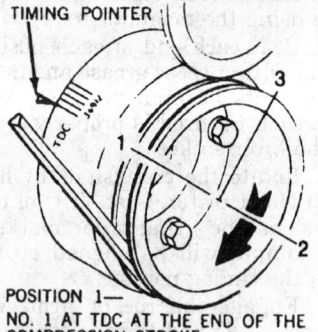

WITH NO. 1 AT TDC AT THE END OF THE COMPRESSION STROKE MAKE A CHALK MARK AT POINTS 2 AND 3 APPROXIMATELY 90 DEGREES APART.

TIMING POINTER

POSITION 1—
NO. 1 AT TDC AT THE END OF THE COMPRESSION STROKE

POSITION 2—
ROTATE THE CRANKSHAFT 180 DEGREES (ONE HALF REVOLUTION) CLOCKWISE FROM POSITION 1

POSITION 3—
ROTATE THE CRANKSHAFT 270 DEGREES (THREE QUARTER REVOLUTION CLOCKWISE FROM POSITION 2

Valve clearance adjustment positions on the crankshaft damper/pulley — 5.0L (VIN N), 5.8L (VIN H and R) and 7.5L (VIN G) engines

rocker arm and the valve stem tip. Allowable clearance is 0.098–0.198 in. (2.5–5.0mm) with a desired clearance of 0.123–0.173 in. (3.1–4.4mm).

3. If the clearance is less than specified, install a shorter pushrod. If the clearance is greater than specified, install a longer pushrod.

4. Rotate the crankshaft clockwise — viewed from the front — 180 degrees, until the next chalk mark is aligned with the timing pointer. Repeat the procedure for:

- No. 3 intake
- No. 2 exhaust
- No. 7 intake
- No. 6 exhaust

5. Rotate the crankshaft to the next chalk mark — 90 degrees — and repeat the procedure for:

- No. 2 intake
- No. 4 exhaust
- No. 5 intake and exhaust
- No. 6 intake
- No. 8 exhaust

7.5L (VIN G) Engine

1. Rotate the crankshaft by hand so No. 1 piston is at TDC of the compression stroke. Make a chalk mark on the damper at that point.

2. With No. 1 at TDC, slowly apply pressure, using Lifter Bleed-down wrench T70P-6513-A, or equivalent, to completely bottom the lifter, on the following valves:

- No. 1 intake and exhaust
- No. 3 intake
- No. 4 exhaust
- No. 7 intake
- No. 5 exhaust
- No. 8 intake and exhaust

Take care to avoid excessive pressure that might bend the pushrod. Hold the lifter in this position and check the clearance between the rocker arm and the valve stem tip. Allowable clearance is 0.075–0.175 in. (1.9–4.4mm) with a desired clearance of 0.100–0.150 in. (2.5–3.8mm).

3. If the clearance is less than specified, install a shorter pushrod. If the clearance is greater than specified, install a longer pushrod.

4. Rotate the crankshaft clockwise — viewed from the front — 360 degrees, until the chalk mark is once again aligned with the timing pointer. Repeat the procedure for:

- No. 2 intake and exhaust
- No. 4 intake
- No. 3 exhaust
- No. 5 intake
- No. 7 exhaust
- No. 6 intake and exhaust

Rocker Arms

REMOVAL AND INSTALLATION

4.9L (VIN Y and Z) Engine

1. Disconnect the inlet hose at the crankcase filler cap.
2. Remove the throttle body inlet tubes.
3. Disconnect the accelerator cable at the throttle body. Remove the cable retracting spring. Remove the accelerator cable bracket from the upper intake manifold and position the cable and bracket out of the way.
4. Remove the fuel line from the fuel rail. Be careful not to kink the line.
5. Remove the upper intake manifold and throttle body assembly.
6. Remove the ignition coil and wires.
7. Remove the rocker arm cover.
8. Remove the spark plug wires.
9. Remove the distributor cap.
10. Remove the pushrod cover (engine side cover).
11. Remove the rocker arm bolts, then lift the rocker arms off of the cylinder head; keep the rocker arms in order for installation.

To install:

12. Engage the rocker arms with the pushrods and tighten the rocker arm bolts enough to hold the pushrods in place.
13. Adjust the valve clearance.
14. Install the pushrod cover (engine side cover).
15. Install the distributor cap.
16. Install the spark plug wires.
17. Install the rocker arm cover.
18. Install the ignition coil and wires.
19. Install the upper intake manifold and throttle body assembly.
20. Install the fuel line at the fuel rail.

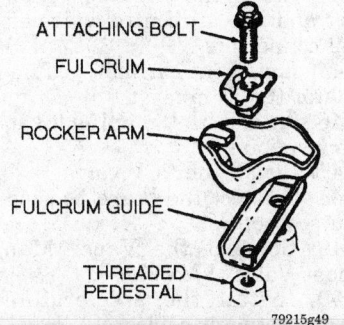

ATTACHING BOLT
FULCRUM
ROCKER ARM
FULCRUM GUIDE
THREADED PEDESTAL

79215g49

Exploded view of the rocker arm components — 4.9L (VIN Y and Z) engine

21. Install the accelerator cable bracket on the upper intake manifold. Install the cable retracting spring. Connect the accelerator cable at the throttle body.
22. Install the throttle body inlet tubes.
23. Connect the inlet hose at the crankcase filler cap.

5.0L (VIN N) and 5.8L (VIN H and R) Engines

1. Remove the intake manifold.
2. Disconnect the Thermactor® air supply hose at the pump.
3. Remove the rocker arm covers.
4. Loosen the rocker arm fulcrum bolts, fulcrum seats and rocker arms; keep all parts in order for installation.

To install:

5. Apply multi-purpose grease to the valve stem tips, the fulcrum seats and sockets.
6. Install the fulcrum guides, rocker arms, seats and bolts. Torque the bolts to 18–25 ft. lbs. (25–34 Nm).
7. Install the rocker arm covers.
8. Connect the Thermactor® air supply hose at the pump.
9. Install the intake manifold.

7.5L (VIN G) Engine

1. Remove the intake manifold.
2. Remove the rocker arm covers.
3. Loosen the rocker arm fulcrum bolts, fulcrum, oil deflector, seat and rocker arms; keep everything in order for installation.

To install:

4. Coat each end of each pushrod with multi-purpose grease.
5. Coat the top of the valve stems, the rocker arms and the fulcrum seats with multi-purpose grease.
6. Rotate the crankshaft by hand until No. 1 piston is at TDC of compression. The firing order marks on the damper will be aligned at TDC with the timing pointer.
7. Install the rocker arms, seats, deflectors and bolts on the following valves:
   - No. 1 intake and exhaust
   - No. 3 intake
   - No. 8 exhaust
   - No. 7 intake
   - No. 5 exhaust
   - No. 8 intake
   - No. 4 exhaust
8. Engage the rocker arms with the pushrods and tighten the rocker arm fulcrum bolts to 18–25 ft. lbs. (25–34 Nm).
9. Rotate the crankshaft on full turn — 360 degrees — and re-align the TDC mark and pointer. Install

the parts and tighten the bolts on the following valves:
   - No. 2 intake and exhaust
   - No. 4 intake
   - No. 3 exhaust
   - No. 5 intake
   - No. 6 exhaust
   - No. 6 intake
   - No. 7 exhaust
10. Install the rocker arm covers.
11. Install the intake manifold.
12. Check the valve clearance as described under Hydraulic Valve Clearance, below.

Intake Manifold

REMOVAL AND INSTALLATION

NOTE: When the battery is disconnected and reconnected, some abnormal drive symptoms may occur while the vehicle relearns its adaptive strategy. The vehicle may need to be driven 10 miles (16 km) or more to relearn the strategy.

4.2L (VIN 2) Engine

1. Remove the engine air cleaner outlet tube.
2. Detach the following ignition coil electrical connections:
   a. Ignition coil electrical connector.
   b. Radio ignition interference capacitor electrical connector.
3. Label, then detach the 6 spark plug wires.
4. Remove the accelerator control splash shield.
5. Disconnect the accelerator cable end and, if equipped, the speed control actuator cable end.
6. Remove the accelerator cable and actuator cable aside, after removing the hold-down bolts.
7. Disconnect the Vapor Management Valve (VMV) hose.
8. Disconnect the brake booster vacuum hose.
9. Disconnect the manifold vacuum connection.
10. Remove the PCV valve from the rocker arm cover.
11. Position the Engine Vacuum Regulator (EVR) bracket aside.
12. Detach the throttle position sensor and idle air control valve electrical connectors.
13. Remove the breather from the rocker arm cover.
14. Remove the 12 upper intake manifold retaining bolts, then lift the manifold off of the engine. Discard the intake manifold upper gasket.

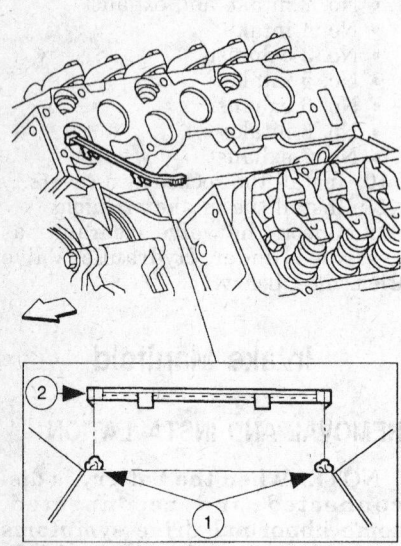

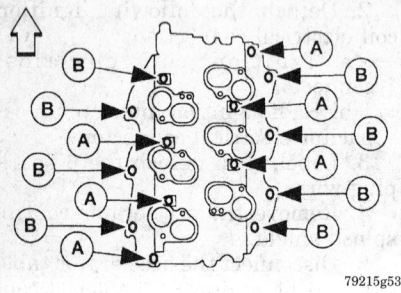

Apply the sealant to the intake manifold/cylinder head seams for the intake manifold end seals as shown — 4.2L (VIN 2) engine

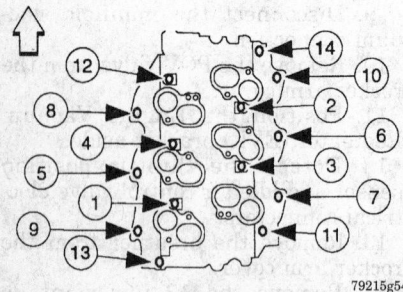

Make certain to install the long bolts (A) and the short bolts (B) in the correct lower intake manifold holes — 4.2L (VIN 2) engine

Tighten the lower manifold bolts in the sequence shown — 4.2L (VIN 2) engine

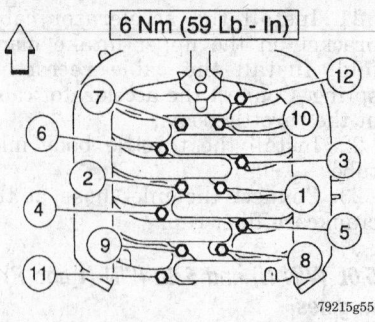

6 Nm (59 Lb - In)

Tighten the upper intake manifold mounting bolts in the sequence shown — 4.2L (VIN 2) engine

15. Detach the 6 fuel injector electrical connectors.
16. Detach the engine coolant temperature sensor and the water temperature indicator sending unit electrical connectors.
17. Disconnect the EGR valve vacuum hose.
18. Remove the EGR valve tube upper fitting.
19. Remove the radiator hose from the lower intake manifold.
20. Position the Intake Manifold Runner Control (IMRC) actuator brackets aside.
21. Disconnect the fuel pressure regulator vacuum line.
22. Disconnect the fuel lines.
23. Disconnect the water pump bypass hose.

NOTE: Remove the lower intake manifold with the fuel injection supply manifold and fuel injectors as one unit.

24. Remove the 6 long bolts and the 8 short bolts, then lift the lower intake manifold off of the engine.
25. Remove and discard the lower intake manifold sealing components.
To install:
26. Clean all components of dirt, grease and old gasket material.
27. Install the lower intake manifold front and rear end seals as follows:
  a. Apply a bead of sealant (Silicone Gasket and Sealant F6AZ-19562-A or equivalent) to the intake manifold front and rear end seal mounting points as illustrated.
  b. Install the lower intake manifold front and rear end seals.
28. Install new lower intake manifold gaskets onto the cylinder heads.

NOTE: The lower intake manifold must be installed within 15 minutes of applying sealant.

29. Apply a bead of the same sealant to the end of the lower intake manifold end seals, where they stop on the cylinder head surface. Position the intake manifold onto the engine block and cylinder heads.
30. Install the lower intake manifold mounting bolts in the correct positions. Refer to the illustration for the correct placement of the long (A) and the short (B) mounting bolts.

NOTE: Make sure to tighten the intake manifold bolts in 2 steps.

31. Tighten the lower intake manifold mounting bolts in the sequence shown, first to 44 inch lbs. (5 Nm), then to 71–101 inch lbs. (8.0–11.5 Nm).
32. Connect the water bypass hose.
33. Connect the fuel lines.
34. Connect the fuel pressure regulator vacuum line.
35. Install the IMRC actuators. Tighten the bolts on the brackets to 71–102 inch lbs. (8.0–11.5 Nm).
36. Install the upper radiator hose to the lower intake manifold.
37. Connect the EGR valve vacuum hose.
38. Install the EGR tube upper fitting to 25–34 ft. lbs. (37–47 Nm).
39. Attach the engine coolant temperature sensor and water temperature indicator sending unit electrical connectors.
40. Install the 6 fuel injector electrical connectors.
41. Install a new intake manifold upper gasket.

NOTE: Make sure to tighten the upper intake manifold bolts in the sequence shown.

42. Position the upper intake manifold onto the lower intake manifold, then tighten the upper intake manifold bolts in the sequence shown to 59 inch lbs. (6 Nm). Tighten the upper intake manifold bolts to 6–8 ft. lbs. (8.0–11.5 Nm) in the sequence shown.
43. Install the breather into the rocker arm cover.
44. Attach the throttle position sensor and idle air control valve electrical connectors.
45. Install the Engine Vacuum Regulator (EVR) bracket.
46. Attach the manifold vacuum connection.
47. Install the PCV valve.
48. Connect the brake booster vacuum hose.
49. Connect the Vapor Management Valve (VMV) hose.
50. Install the accelerator and speed actuator cables.
51. Install the accelerator control splash shield.
52. Install the spark plug wires.

53. Attach the ignition coil and the radio ignition interference capacitor electrical connectors.

54. Install the engine air cleaner outlet tube.

4.6L Engine (VIN W and VIN 6)

1. Disconnect the negative battery cable.

2. Relieve the fuel system pressure.

3. Drain the cooling system.

4. Disconnect the upper radiator hose from the intake manifold.

5. Remove the engine air cleaner outlet tube.

6. Disconnect the accelerator cable from the bracket and the throttle body cam.

7. If equipped, remove the speed control actuator cable from the throttle body.

8. Disconnect all vacuum hoses, fuel lines and electrical wires from the throttle body and intake manifold.

9. Remove the brake booster vacuum hose bracket.

10. Remove the EGR valve-to-exhaust manifold tube.

11. Detach the fuel injector electrical connectors.

12. Label, then remove the spark plug wires, if necessary.

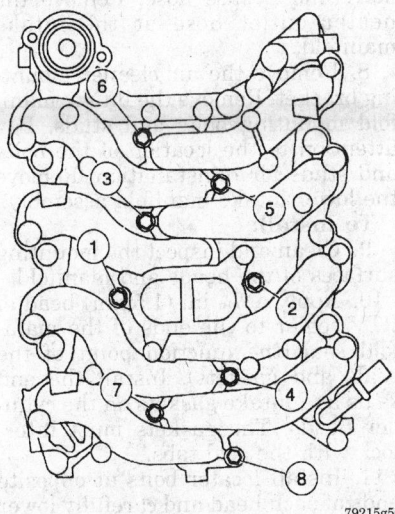

Tighten the lower-to-upper intake manifold bolts in 2 steps following the sequence shown — 4.6L engines (VIN W and VIN 6)

79215g56

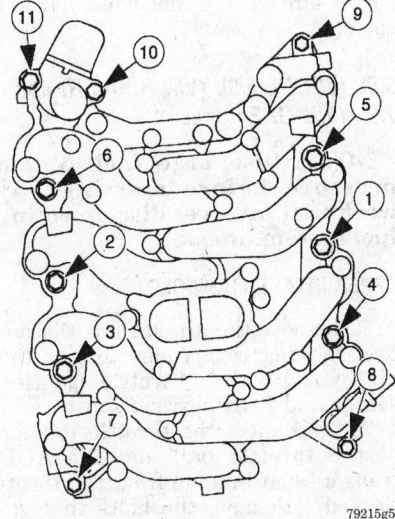

Tighten the upper intake manifold-to-cylinder head mounting bolts in the sequence shown — 4.6L engines (VIN W and VIN 6)

79215g57

13. Remove the accessory drive belt.

14. Remove the alternator.

15. Remove the power steering oil reservoir bracket and set aside.

16. Disconnect the heater water hose from the intake manifold.

17. Remove the 11 intake manifold bolts.

18. Lift the intake manifold off of the engine, then detach the Intake Manifold Tuning Valve (IMTV) electrical connector. Remove and discard the upper intake manifold gaskets.

19. Remove the 8 lower intake manifold bolts, then separate the upper intake manifold from the lower intake manifold. Discard the old gasket.

20. Position the lower intake manifold gasket and the upper intake manifold onto the lower intake manifold, then loosely install the 8 lower intake-to-upper intake manifold bolts.

NOTE: Make sure to tighten the lower-to-upper manifold bolts in 2 steps.

21. Tighten the 8 lower-to-upper intake manifold bolts in 2 steps following the tightening sequence shown. The first step should be to 18 inch lbs. (2 Nm) and the second step to 6–8 inch lbs. (8–12 Nm).

22. Position the 2 upper intake manifold gaskets on the cylinder heads. Set the upper intake manifold in place on the engine, then loosely install the 9 intake manifold-to-cylinder head bolts.

23. Attach the IMTV electrical connector.

NOTE: Check that the thermostat housing is in the correct position before the thermostat housing is installed.

24. Install the thermostat housing and start the 2 housing bolts.

NOTE: Make certain to tighten the intake manifold in 2 steps.

25. Tighten the 11 intake manifold bolts in the sequence shown, first to 18 inch lbs. (2 Nm) then to 15–22 ft. lbs. (20–30 Nm).

26. Install the heater water hose.

27. Position the power steering bracket and install the power steering pump bracket bolts to 71–107 inch lbs. (8–12 Nm).

28. Attach all electrical connections, fuel lines, vacuum tubes and coolant hoses to the intake manifold, fuel injectors and throttle body assembly.

29. Install the alternator and the accessory drive belt.

30. Install the spark plug wires.

31. Install the EGR valve-to-exhaust manifold tube. The tube fittings should be tightened to 26–33 ft. lbs. (35–45 Nm).

32. Install the speed actuator cable, if equipped, and the accelerator cable to the throttle body.

33. Install the engine air cleaner outlet tube.

34. Install the heater water hose.

35. Fill the cooling system with the correct amount and type of coolant.

36. Connect the negative battery cable. Start the engine and check for fuel, vacuum or coolant leaks.

4.9L (VIN Y and Z) Engine

NOTE: The lower intake manifold and the exhaust manifold on the 4.9L (VIN Y and Z) engines is considered a combination manifold. Only the upper intake manifold is covered here.

1. Disconnect the negative battery cable.

2. Label, then detach all electrical connectors from the intake manifold, throttle body and EGR valve.

3. Disconnect all vacuum lines attached to the intake manifold, the throttle body and the EGR valve.

4. Label the 2 EGR transducer hoses, then remove the transducer

hose bracket. Disconnect the transducer hoses.

5. Disconnect the PCV hose from the fitting, located on the underside of the upper intake manifold.

WARNING

When disconnecting the throttle cable from the ball stud, use a pry tool or similar tool close to the ball stud to pry it off. Removing it by hand may damage the cable.

6. Remove the accelerator control splash shield, then disconnect the accelerator cable and speed actuator cable from the throttle body. Position the cables away from the engine.

7. Disconnect the air cleaner outlet tube from the throttle body.

8. Remove the EGR valve-to-exhaust manifold tube.

9. Remove the bolt and washer attaching the intake manifold support to the upper intake manifold.

10. Remove the 7 studs that retain the upper intake manifold.

11. Remove the upper intake manifold and throttle body from the lower intake manifold.

To install:

12. Position a new intake manifold upper gasket on the lower intake manifold, using the lower manifold dowels to position the intake manifold upper gasket.

13. Position the upper intake manifold onto the lower intake manifold. Install the 7 studs to attach the upper manifold to the lower manifold hand-tight.

14. Tighten the 7 upper manifold-to-lower manifold studs to 12–18 ft. lbs. (16–24 Nm).

15. Install the upper manifold support and tighten the mounting bolt to 22–32 ft. lbs. (30–43 Nm).

16. Install the EGR valve-to-exhaust manifold tube. The tube is routed between lower intake manifold runners No. 5 and No. 6. Tighten both of the fittings to 25–35 ft. lbs. (34–47 Nm).

17. Attach the PCV hose to the fitting, located on the valve cover under the upper intake manifold.

18. Connect the accelerator cable, transmission kickdown cable, and speed control actuator cable. Install the accelerator control splash shield.

19. Connect the air cleaner outlet tube to the throttle body.

20. Install the EGR transducer and bracket. Tighten the bolts to 12–18 ft. lbs. (17–24 Nm). Connect the EGR transducer hoses.

21. Reattach all of the vacuum lines removed from the throttle body, the

intake manifold and the EGR valve assemblies. Also install all of the electrical connections to the same components.

22. Connect the negative battery cable.

5.0L (VIN N), 5.8L (VIN H and R) and 7.5L (VIN G) Engines

NOTE: Discharge fuel system pressure before starting any work that involves disconnecting fuel system lines.

UPPER INTAKE MANIFOLD

1. Remove the air cleaner. Disconnect the electrical connectors at the air bypass valve, throttle position sensor and EGR position sensor.

2. Disconnect the throttle linkage at the throttle ball and the AOD transmission linkage from the throttle body. Remove the bolts that secure the bracket to the intake and position the bracket and cables out of the way.

3. Disconnect the upper manifold vacuum fitting connections by removing all the vacuum lines at the vacuum tree (label lines for position identification). Remove the vacuum lines to the EGR valve and fuel pressure regulator.

4. Disconnect the PCV system by disconnecting the hose from the fitting at the rear of the upper manifold.

5. Remove the 2 canister purge lines from the fittings at the throttle body.

6. Disconnect the EGR tube from the EGR valve by loosening the flange nut.

7. Remove the bolt from the upper intake support bracket to upper manifold. Remove the upper manifold retaining bolts and remove the upper intake manifold and throttle body as an assembly.

8. Clean and inspect all mounting surfaces of the upper and lower intake manifolds.

To install:

9. Position a new mounting gasket on the lower intake manifold.

10. Install the upper intake manifold and throttle body as an assembly. Install the upper manifold retaining bolts and install the bolt at the upper intake support bracket. Mounting bolts are torqued to 12–18 ft. lbs. (16–25 Nm).

11. Connect the EGR tube at the EGR valve.

12. Install the 2 canister purge lines at the fittings at the throttle body.

13. Connect the PCV system hose at the fitting at the rear of the upper manifold.

14. Connect the upper manifold vacuum lines at the vacuum tree. Install the vacuum lines at the EGR valve and fuel pressure regulator.

15. Install the throttle bracket on the intake manifold. Connect the throttle linkage at the throttle ball and the AOD transmission linkage at the throttle body.

16. Connect the electrical connectors at the air bypass valve, throttle position sensor and EGR position sensor.

17. Install the air cleaner.

UPPER INTAKE MANIFOLD

1. Upper manifold and throttle body must be removed first.

2. Drain the cooling system.

3. Remove the distributor assembly, cap and wires.

4. Disconnect the electrical connectors at the engine, coolant temperature sensor and sending unit, at the air charge temperature sensor and at the knock sensor.

5. Disconnect the injector wiring harness from the main harness assembly. Remove the ground wire from the intake manifold stud. The ground wire must be installed at the same position it was removed from.

6. Disconnect the fuel supply and return lines from the fuel rails.

7. Remove the upper radiator hose from the thermostat housing. Remove the bypass hose. Remove the heater outlet hose at the intake manifold.

8. Remove the air cleaner mounting bracket. Remove the intake manifold mounting bolts and studs. Pay attention to the location of the bolts and studs for reinstallation. Remove the lower intake manifold assembly.

To install:

9. Clean and inspect the mounting surfaces of the heads and manifold.

10. Apply a 1/16 in. (1.5mm) bead of RTV sealer to the ends of the manifold seal (the junction point of the seals and gaskets). Install the end seals and intake gaskets on the cylinder heads. The gaskets must interlock with the seal tabs.

11. Install locator bolts at opposite ends of each head and carefully lower the intake manifold into position. Install and tighten the mounting bolts and studs to 23–25 ft. lbs. (31–34 Nm).

12. Install the lower intake manifold assembly. Install the intake manifold mounting bolts and studs. Pay attention to the location of the

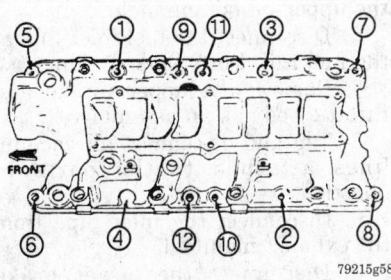

Lower intake manifold bolt sequence — 5.0L (VIN N) and 5.8L (VIN H and R) engines

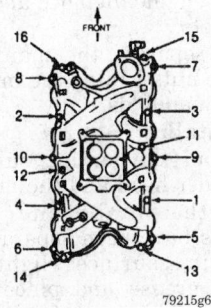

Intake manifold bolt torque sequence — 7.5L (VIN G) engine

bolts. Install the air cleaner mounting bracket.

13. Install the heater outlet hose at the intake manifold.

14. Install the bypass hose.

15. Install the upper radiator hose.

16. Connect the fuel supply and return lines at the fuel rails.

17. Connect the injector wiring harness from the main harness assembly. Install the ground wire from the intake manifold stud.

18. Connect the electrical connectors at the engine, coolant temperature sensor and sending unit, at the air charge temperature sensor and at the knock sensor.

19. Install the distributor assembly, cap and wires.

20. Fill the cooling system.

Exhaust Manifold

REMOVAL AND INSTALLATION

NOTE: When the battery is disconnected and reconnected, some abnormal drive symptoms may occur while the vehicle relearns its adaptive strategy. The vehicle may need to be driven 10 miles (16 km) or more to relearn the strategy.

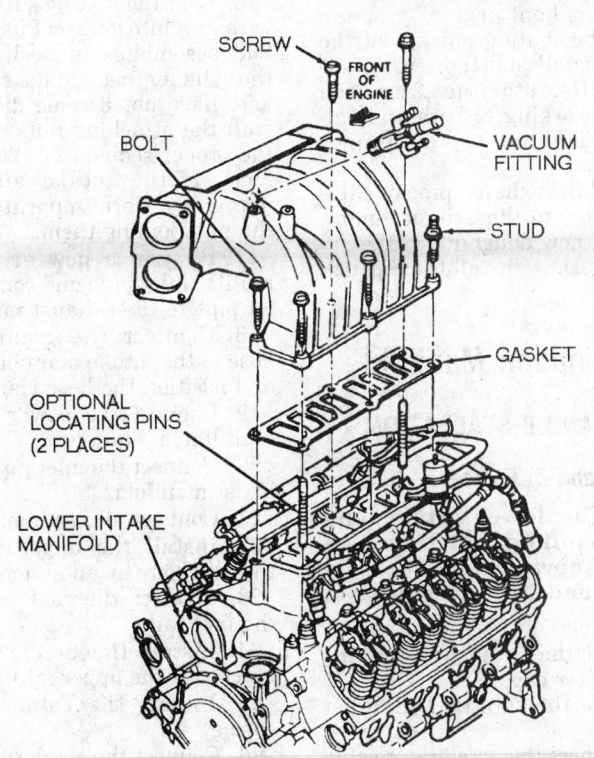

Upper intake manifold installation — 5.0L (VIN N) and 5.8L (VIN H and R) engines

4.2L (VIN 2) Engine

1. Disconnect the negative battery cable.

2. For the right-hand manifold, remove the EGR valve-to-exhaust manifold tube.

3. For the left-hand manifold, remove the oil level indicator tube bracket nut, then remove the oil level indicator tube. Remove and discard the oil level indicator tube O-ring.

4. Raise and safely support the front of the vehicle on jackstands.

5. Detach the heated oxygen sensor electrical connector.

6. Remove the 2 catalytic converter-to-exhaust manifold nuts, then disconnect the Y-pipe from the left-hand exhaust manifold.

7. Remove the exhaust manifold stud bolts, then remove the manifold mounting bolts. Remove the exhaust manifold. Remove and discard the exhaust manifold gasket.

To install:

8. Position the new exhaust manifold gasket onto the engine, then install the exhaust manifold. Tighten the bolts and stud bolts in the sequence shown to 15–22 ft. lbs. (20–30 Nm).

9. Connect the Y-pipe to the exhaust manifold, then install and tighten the catalytic converter nuts to 25–34 ft. lbs. (34–46 Nm).

10. Attach the oxygen sensor connector, then lower the vehicle.

11. For the left-hand exhaust manifold, install a new oil level indicator tube O-ring onto the tube. Insert the tube into the engine block and tighten the bracket retaining nut to 15–22 ft. lbs. (20–30 Nm).

12. For the right-hand exhaust manifold, Install the EGR valve-to-exhaust manifold tube. Tighten the upper and lower fittings to 25–34 ft. lbs. (34–47 Nm).

13. Connect the negative battery cable.

4.6L Engines (VIN W and VIN 6)

1. Raise and safely support the vehicle on jackstands.

2. Remove the front fender splash shield.

3. For the left-hand exhaust manifold, remove the EGR valve-to-exhaust manifold tube and the DPFE transducer hoses.

4. Remove the catalytic converter-to-exhaust manifold bolts.

5. Remove the exhaust manifold mounting bolts, then remove the exhaust manifold itself.

6. Remove and discard the old gasket.

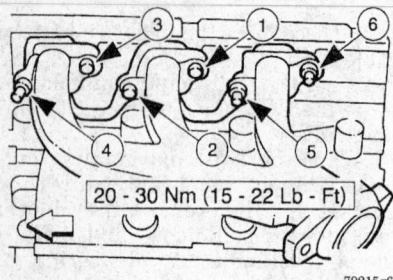

Tighten the left-hand exhaust manifold bolts in the order shown — 4.2L (VIN 2) engines

20 - 30 Nm (15 - 22 Lb - Ft)

79215g62

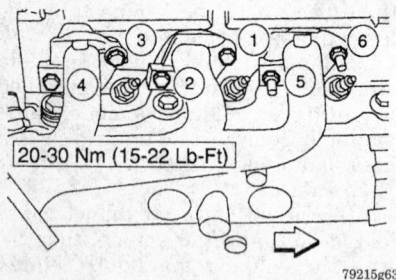

Tighten the right-hand exhaust manifold bolts in the order shown — 4.2L (VIN 2) engines

20-30 Nm (15-22 Lb-Ft)

79215g63

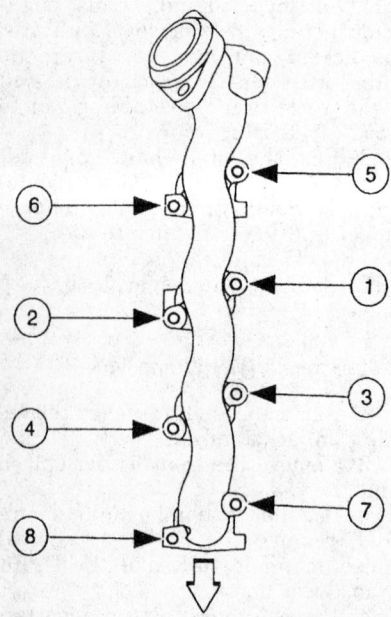

Tighten the exhaust manifold bolts in the sequence shown — 4.6L engines (VIN W and VIN 6)

79215g64

7. Clean and inspect the exhaust manifold for damage.

To install:

8. Position a new gasket and the exhaust manifold onto the engine block. Install the mounting bolts and tighten, in the sequence shown, to 13–16 ft. lbs. (18–22 Nm).

9. Attach the catalytic converter to the exhaust manifold, then install the catalytic converter-to-exhaust manifold bolts and tighten to 25–34 ft. lbs. (34–46 Nm).

10. For the left-hand exhaust manifold, install the DPFE transducer hoses and the EGR valve-to-exhaust manifold tube. Tighten the upper and lower fittings to 26–33 ft. lbs. (35–45 Nm).

11. Install the front fender splash shield.

12. Lower the vehicle to the ground.

5.0L (VIN N), 5.8L (VIN H and R) and 7.5L (VIN G) Engines

1. On the 5.0L (VIN N) engine, remove the dipstick bracket.

2. Disconnect the exhaust pipe or catalytic converter from the exhaust manifold. Remove and discard the doughnut gasket.

3. Remove the exhaust manifold attaching screws and remove the manifold from the cylinder head.

To install:

4. Apply a light coat of graphite grease to the mating surface of the manifold. Install and tighten the attaching bolts, starting from the center and working to both ends alternately. Tighten to the proper specifications.

5. Install the exhaust pipe or catalytic converter to the exhaust manifold using a new doughnut gasket.

6. If necessary, install the dipstick bracket.

Combination Manifold

REMOVAL AND INSTALLATION

4.9L (VIN Y and Z) Engine

NOTE: The lower intake and exhaust manifolds on these engines are known as combination manifolds and are serviced as a unit.

1. Remove the air inlet hose at the crankcase filter cap.

2. Remove the throttle body inlet hoses.

3. Disconnect the accelerator cable at the throttle body.

4. Remove the cable retracting spring.

5. Remove the cable bracket from the upper intake manifold.

6. Disconnect the fuel inlet line at the fuel rail. Don't kink the line!

7. Remove the upper intake and throttle body as an assembly.

8. Tag and disconnect all vacuum lines attached to the parts in question.

9. Disconnect the inlet pipe from the exhaust manifold.

10. Disconnect the power brake vacuum line, if equipped.

11. Remove the bolts and nuts attaching the manifolds to the cylinder head. Lift the manifold assemblies from the engine. Remove and discard the gaskets.

12. To separate the manifold, remove the nuts joining the intake and exhaust manifolds.

To install:

13. Clean the mating surfaces of the cylinder head and the manifolds.

14. If the intake and exhaust manifolds have been separated, coat the mating surfaces lightly with graphite grease and place the exhaust manifold over the studs on the intake manifold. Install the lockwashers and nuts. Tighten them finger tight.

15. Install a new intake manifold gasket.

16. Coat the mating surfaces lightly with graphite grease. Place the manifold assemblies in position against the cylinder head. Make sure the gaskets have not become dislodged. Install the attaching nuts and bolts in the proper sequence to 26 ft. lbs. (35 Nm). If the intake and exhaust manifolds were separated, tighten the nuts joining them.

17. Position a new gasket on the muffler inlet pipe and connect the inlet pipe to the exhaust manifold.

18. Connect the crankcase vent hose to the intake manifold inlet tube and position the hose clamp.

19. Connect the power brake vacuum line, if equipped.

20. Connect the inlet pipe at the exhaust manifold.

21. Connect all vacuum lines.

22. Install the upper intake and throttle body as an assembly.

23. Connect the fuel inlet line at the fuel rail.

24. Install the accelerator cable bracket at the upper intake manifold.

25. Install the cable retracting spring.

26. Connect the accelerator cable at the throttle body.

27. Install the throttle body inlet hoses.

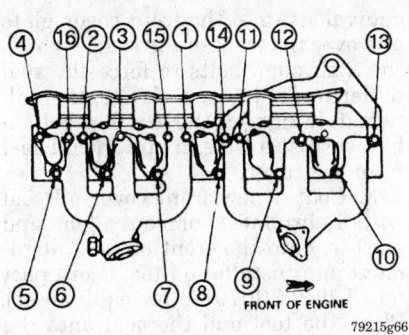

Combination manifold bolt torque sequence — 4.9L (VIN Y and Z) engine

28. Install the air inlet hose at the crankcase filter cap.

Front Cover Seal

REMOVAL AND INSTALLATION

4.2L (VIN 2) Engine

NOTE: To correctly perform the following procedure special tools will be needed. These tools are the Crankshaft Damper Remover T58P-6316-D, the Front Seal Installer/Cover Aligner T88T-6701-A, the Vibration Damper Remover Adapter T82L-6316-B, Seal Remover T92C-6700-CH and the Front Seal Replacer T94P-6701-AH (or their equivalents).

1. Remove the accessory drive belt.
2. Remove the fan blade assembly.
3. Raise and safely support the vehicle on jackstands.

NOTE: Matchmark the crankshaft pulley and damper with each other for installation.

4. Loosen the crankshaft damper bolt, then remove the four crankshaft pulley bolts. Remove the pulley.
5. Remove the crankshaft damper bolt, then attach he crankshaft damper remover tool, T58P-6316-D or equivalent, with 2 crankshaft pulley bolts. Turn the vibration damper remover adapter, T82L-6316-B or equivalent, to pull the damper off of the crankshaft.
6. Use seal remover tool, T92C-6700-CH or equivalent, to remove the oil seal from the front cover. Discard the old seal.

To install:
7. Inspect the crankshaft damper and engine front cover for damage that may cause the front oil seal to fail.

NOTE: Lubricate the parts with clean engine oil before assembly.

8. Use the front seal replacer T94P-6701-AH (spacer), the front seal installer/cover aligner T88T-6701-A and the vibration damper remover adapter T82L-6316-B, or their equivalents, to press the new oil seal into the recess in the front cover.
9. Apply a bead of sealant to the keyway in the crankshaft damper and use the vibration damper remover adapter to install the damper onto the crankshaft. Make sure to utilize Silicone Gasket and Sealant F6AZ-19562-A or equivalent.
10. Install the crankshaft damper bolt and tighten to 103–117 ft. lbs. (140–160 Nm).

NOTE: The crankshaft pulley position on the crankshaft damper was marked before removal, return it to the same position.

11. Install the crankshaft pulley and tighten the 4 pulley mounting bolts to 20–28 ft. lbs. (26–38 Nm).
12. Lower the vehicle.
13. Install the fan blade assembly.
14. Install the accessory drive belt.

4.6L Engines (VIN W and VIN 6)

NOTE: To correctly perform the following procedure special tools will be needed. These tools are the Crankshaft Damper Remover T58P-6316-D, the Front Seal Installer/Cover Aligner T88T-6701-A, the Vibration Damper Replacer T74P-6316-B, and the Seal Remover T74P-6700-A (or their equivalents).

1. Remove the accessory drive belt.
2. Remove the fan blade assembly and fan shroud.
3. Raise and safely support the vehicle on jackstands.
4. Remove the crankshaft pulley bolt.
5. Use the crankshaft damper remover tool, T58P-6316-D or equivalent, to remove the crankshaft pulley.
6. Use seal remover tool, T74P-6700-A or equivalent, to remove the oil seal from the front cover. Discard the old seal.

To install:
7. Inspect the crankshaft damper and engine front cover for damage that may cause the front oil seal to fail.

NOTE: Lubricate the parts with clean engine oil before assembly.

8. Use the front seal installer/cover aligner T88T-6701-A, or the equivalent, to press the new oil seal into the recess in the front cover.
9. Apply a bead of sealant to the keyway in the crankshaft damper and use the vibration damper remover adapter to install the damper onto the crankshaft. Make sure to utilize Silicone Gasket and Sealant F6AZ-19562-A or equivalent.
10. Use the crankshaft damper replacer tool, T74P-6316-B or equivalent, to install the crankshaft pulley.
11. Loosely install the crankshaft pulley bolt.

NOTE: Make certain to tighten the crankshaft pulley bolt in 4 steps.

12. Install the pulley bolt in 4 steps. Some of the steps require tightening, some require loosening of the bolt. Pay close attention, as follows:
   a. Step 1: tighten to 88 ft. lbs. (120 Nm).
   b. Step 2: loosen the bolt one full revolution (360 degrees).
   c. Step 3: tighten to 34–39 ft. lbs. (47–53 Nm).
   d. Step 4: tighten an additional 85–90 degrees (approximately ¼ turn).
13. Lower the vehicle.
14. Install the fan blade assembly and fan shroud.
15. Install the accessory drive belt.

4.9L (VIN Y and Z) Engine

1. Drain the cooling system and disconnect the radiator upper hose at the coolant outlet elbow and remove the 2 upper radiator retaining bolts.
2. Raise the vehicle and drain the crankcase.
3. Remove the splash shield and the automatic transmission oil cooling lines, if equipped, then remove the radiator.
4. Loosen and remove the fan belt, fan and pulley.
5. Use a gear puller to remove the crankshaft pulley damper.
6. Remove the cylinder front cover retaining bolts and gently pry the cover away from the block. Remove the gasket.
7. Drive out the old seal with a pin punch from the rear of the cover. Clean out the recess in the cover.

To install:

8. Coat the new seal with grease and drive it into the cover until it is fully seated. Check the seal to make sure the spring around the seal is in the proper position.

9. Clean the cylinder front cover and the gasket surface of the cylinder block. Apply an oil-resistant sealer to the new front cover gasket and install the gasket onto the cover.

10. Position the front cover assembly over the end of the crankshaft and against the cylinder block. Start, but do not tighten, the cover and pan attaching screws. Slide a front cover alignment tool (Ford part no. T68P–6019–A or equivalent) over the crank stub and into the seal bore of the cover. Tighten all front cover attaching screws to 12–18 ft. lbs. (16–24 Nm) and all oil pan attaching screws to 10–15 ft. lbs. (14–20 Nm). Tighten the oil pan screws first.

NOTE: Trim away the exposed portion of the old oil pan gasket flush with the front of the engine block. Cut and position the required portion of a new gasket to the oil pan and apply sealer to both sides.

11. Lubricate the hub of the crankshaft damper pulley with Lubriplate® to prevent damage to the seal during installation or on initial starting of the engine.

12. Install the fan belt, fan and pulley.

13. Install the radiator.

14. Install the splash shield and the automatic transmission oil cooling lines, if equipped.

15. Fill the crankcase.

16. Connect the radiator upper hose at the coolant outlet elbow and install the 2 upper radiator retaining bolts.

17. Drain the cooling system.

18. Start the engine and check for leaks.

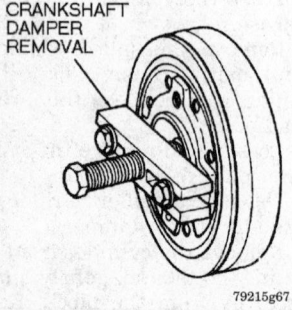

CRANKSHAFT DAMPER REMOVAL

79215g67

Crankshaft damper removal — 4.9L (VIN Y and Z), 5.0L (VIN N), 5.8L (VIN H and R) and 7.5L (VIN G) engines

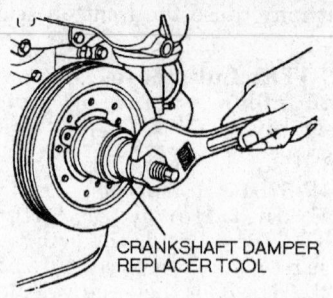

CRANKSHAFT DAMPER REPLACER TOOL

79215g68

Crankshaft damper installation — 4.9L (VIN Y and Z), 5.0L (VIN N), 5.8L (VIN H and R) and 7.5L (VIN G) engines

5.0L (VIN N) and 5.8L (VIN H and R) Engines

1. Drain the cooling system and the crankcase.

2. Disconnect the upper and lower radiator hoses from the water pump, transmission oil cooler lines from the radiator, and remove the radiator.

3. Disconnect the heater hose from the water pump. Slide the water pump by-pass hose clamp toward the water pump.

4. Loosen the alternator pivot bolt and the bolt which secures the alternator adjusting arm to the water pump. Position the alternator out of the way.

5. Remove the power steering pump and air conditioning compressor from their mounting brackets, if equipped.

6. Remove the bolts holding the fan shroud to the radiator, if equipped. Remove the fan, spacer, pulley and drive belts.

7. Remove the crankshaft pulley from the crankshaft damper. Remove the damper attaching bolt and washer and remove the damper with a puller.

8. Disconnect the fuel pump outlet line at the fuel pump. Disconnect the vacuum inlet and outlet lines from the fuel pump. Remove the fuel pump attaching bolts and lay the pump to one side with the fuel inlet line still attached.

9. Remove the oil level dipstick and the bolt holding the dipstick tube to the exhaust manifold on the 5.0L (VIN N) engine.

10. Remove the oil pan-to-cylinder front cover attaching bolts. Use a sharp, thin cutting blade to cut the oil pan gasket flush with the cylinder block. Remove the front cover and water pump as an assembly.

11. Discard the front cover gasket.

To install:

12. Place the front seal removing tool (Ford part no. T70P–6B070–A or equivalent) into the front cover plate and over the front of the seal. Tighten the 2 through bolts to force the seal puller under the seal flange, then alternately tighten the 4 puller bolts a half turn at a time to pull the oil seal from the cover.

13. Coat a new front cover oil seal with Lubriplate® or equivalent and place it onto the front oil seal alignment and installation tool (Ford part no. T70P–6B070–A or equivalent). Place the tool and the seal onto the end of the crankshaft and push it toward the engine until the seal starts into the front cover.

14. Place the installation screw, washer, and nut onto the end of the crankshaft, then thread the screw into the crankshaft. Tighten the nut against the washer and tool to force the seal into the front cover plate. Remove the tool.

15. Apply Lubriplate® or equivalent to the oil seal rubbing surface of the vibration damper inner hub to prevent damage to the seal. Coat the front of the crankshaft with oil for damper installation.

16. To install the damper, line up the damper keyway with the key on the crankshaft, then install the damper onto the crankshaft. Install the cap screw and washer, and tighten the screw to 80 ft. lbs. (109 Nm). Install the crankshaft pulley.

17. Install the fan, spacer, pulley and drive belts.

18. Install the bolts holding the fan shroud to the radiator, if equipped.

19. Install the power steering pump and air conditioning compressor.

20. Position and tighten the alternator.

21. Connect the heater hose at the water pump.

22. Install the radiator.

23. Connect the upper and lower radiator hoses, and transmission oil cooler lines.

24. Fill the cooling system and the crankcase.

7.5L (VIN G) Engine

1. Drain the cooling system and crankcase.

2. Remove the radiator shroud and fan.

3. Disconnect the upper and lower radiator hoses, and the automatic transmission oil cooler lines from the radiator.

4. Remove the radiator upper support and remove the radiator.

5. Loosen the alternator attaching bolts and air conditioning compressor idler pulley and remove the drive belts with the water pump pulley. Re-

move the bolts attaching the compressor support to the water pump and remove the bracket (support), if equipped.

6. Remove the crankshaft pulley from the vibration damper. Remove the bolt and washer attaching the crankshaft damper and remove the damper with a puller. Remove the woodruff key from the crankshaft.

7. Loosen the by-pass hose at the water pump, and disconnect the heater return tube at the water pump.

8. Disconnect and plug the fuel inlet and outlet lines at the fuel pump, and remove the fuel pump.

9. Remove the bolts attaching the front cover to the cylinder block. Cut the oil pan seal flush with the cylinder block face with a thin knife blade prior to separating the cover from the cylinder block. Remove the cover and water pump as an assembly. Discard the front cover gasket and oil pan seal.

To install:

10. Transfer the water pump if a new cover is going to be installed. Clean all of the gasket sealing surfaces on both the front cover and the cylinder block.

11. Coat the gasket surface of the oil pan with sealer. Cut and position the required sections of a new seal on the oil pan. Apply sealer to the corners.

12. Drive out the old front cover oil seal with a pin punch. Clean out the seal recess in the cover. coat a new seal with Lubriplate® or equivalent grease. Install the seal, making sure the seal spring remains in the proper position. A front cover seal tool, Ford part no. T72J-117 or equivalent, makes installation easier.

13. Coat the gasket surfaces of the cylinder block and cover with sealer and position the new gasket on the block.

14. Position the front cover on the cylinder block. Use care not to damage the seal and gasket or misplace them.

15. Coat the front cover attaching screws with sealer and install them.

NOTE: It may be necessary to force the front cover downward to compress the oil pan seal in order to install the front cover attaching bolts. Use a prybar or drift to engage the cover screw holes through the cover and pry downward.

16. Install the fuel pump.
17. Connect the fuel inlet and outlet lines at the fuel pump.

18. Tighten the by-pass hose at the water pump.
19. Connect the heater return tube at the water pump.
20. Install the woodruff key from the crankshaft.
21. Install the damper.
22. Install the crankshaft pulley on the vibration damper.
23. Install the compressor support on the water pump and install the bracket (support), if equipped.
24. Install the drive belts with the water pump pulley.
25. Install the radiator and upper support.
26. Connect the upper and lower radiator hoses, and the automatic transmission oil cooler lines.
27. Install the radiator shroud and fan.
28. Fill the cooling system and crankcase.
Observe the following torques:
• Front cover bolts — 15–20 ft. lbs. (20–27 Nm)
• Water pump attaching screws — 12–15 ft. lbs. (16–20 Nm)
• Crankshaft damper — 70–90 ft. lbs. (95–122 Nm)
• Crankshaft pulley — 35–50 ft. lbs. (48–68 Nm)
• Fuel pump — 19–27 ft. lbs. (26–37 Nm)
• Oil pan bolts — 9–11 ft. lbs. (12.2–15.0 Nm) for the $\frac{5}{16}$ in. (7.9mm) screws and to 7–9 ft. lbs. (9.5–12.2 Nm) for the $\frac{1}{4}$ in. (6.3mm) screws
• Alternator pivot bolt — 45–57 ft. lbs. (61–77 Nm)

Timing Chain, Sprockets and Front Cover

REMOVAL AND INSTALLATION

4.2L (VIN 2) Engine

NOTE: Along with the special tools required to remove the crankshaft pulley and damper, a Synchro Positioning Tool, such as Ford Tool T89p-12200-A, must be obtained prior to installation of the replacement synchronizer assembly. Failure to follow this procedure will result in the fuel system being out of time with the engine, possible causing engine damage.

1. Disconnect the negative battery cable.
2. Remove the radiator, fan blade assembly and fan shroud.
3. Remove the water pump.

4. Remove the camshaft synchronizer as follows:
   a. Remove the EGR valve vacuum hose.
   b. Remove the EGR tube upper fitting, then remove the EGR valve and adapter.
   c. Disconnect the heater water outlet tube.
   d. Remove the Camshaft Position (CMP) sensor connector and mark the orientation of the CMP connector (the direction it is pointing).

─────── **WARNING** ───────
Unless explicitly told to do so in the instructions, do not turn the crankshaft or camshaft after removing the CMP and camshaft synchronizer or the fuel system timing will be out of time with the engine and possible cause engine damage.

   e. Rotate the crankshaft until the TDC mark lines up with the timing mark.
   f. Remove the 2 hold-down bolts, then remove the CMP from the camshaft synchronizer.
   g. Remove the camshaft synchronizer adjustment bolt, then remove the camshaft synchronizer. The oil pump driveshaft may come out with the synchronizer.
5. Remove the crankshaft pulley and damper.
6. Remove the oil pan.

─────── **WARNING** ───────
The cap screw is hidden, make sure to remove or the front cover will be damaged.

7. Remove the all of the front cover retainers (2 stud bolts, 1 bolt, 1 cap screw). The cap screw is located on the bottom edge of the cover, near the oil filter mounting flange.
8. Slide the front cover and gasket off of the dowels. Discard the gasket.
9. Remove the camshaft position sensor drive gear.
10. Rotate the crankshaft until the timing marks (A) and keyways (B) are aligned.
11. Compress the timing chain tensioner, then install a retaining pin to hold in place.
12. Remove the camshaft sprocket, the crankshaft sprocket and the timing chain as an assembly.
13. Remove the timing chain tensioner.

To install:

14. Position the timing chain tensioner in place on the engine, then

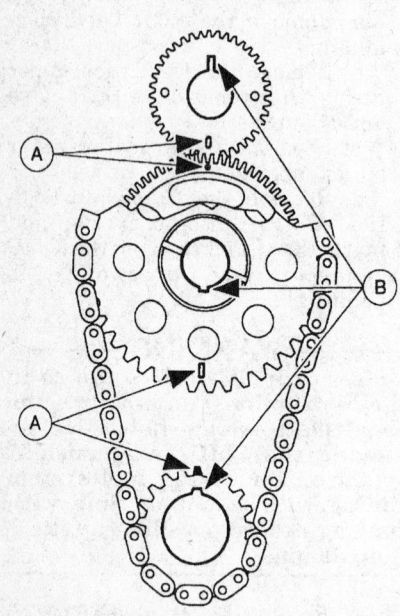

After removing the CMP drive gear, align the timing marks (A) and keyways (B) — 4.2L (VIN 2) engines

Install the camshaft synchronizer in the orientation shown (arrow points to front of engine) — 4.2L (VIN 2) engines

install the 3 retaining bolts to 6–10 ft. lbs. (8–14 Nm).

15. Rotate the crankshaft so that No. 1 piston is at Top Dead Center (TDC) and the crankshaft key is in the 12 o'clock position; this should have already been done during removal.

16. If necessary, retract the tensioner pad mechanism. Use a retaining pin to hold the retracted pad in position.

17. Turn the inner, smaller camshaft sprocket so that the timing mark is on the bottom.

18. Install the timing chain, camshaft outer sprocket and the crankshaft sprocket as an assembly.

19. Make sure that the timing marks (A) and the keyways (B) are aligned as during removal.

20. Position the CMP drive gear onto the camshaft stub, then install the retaining bolt to 30–36 ft. lbs. (40–50 Nm).

21. Remove the timing chain tensioner retaining pin.

22. Clean the front cover and engine block mounting surfaces of dirt, grease or old gasket material. Install the front cover onto the engine dowel pins along with a new gasket. Tighten the 4 front cover fasteners to 15–22 ft. lbs. (20–30 Nm).

23. Install the oil pan.

24. Install the crankshaft pulley and damper.

25. Install the synchro positioning tool on the camshaft synchronizer by rotating the synchro p

NOTE: During installation, the arrow on the synchronizer alignment tool will rotate clockwise as the gears engage.

26. Install the camshaft synchronizer housing assembly so that the arrow on the synchronizer alignment tool points forward and up 54 degrees from the centerline of the engine block.

27. Install the camshaft synchronizer adjustment bolt and tighten to 15–22 ft. lbs. (20–30 Nm).

28. Remove the camshaft synchronizer positioning tool, then set the CMP sensor in position. Install the 2 hold-down bolts to 40–70 inch lbs. (4.5–7.8 Nm).

29. Attach the CMP sensor electrical connector.

NOTE: Reuse the EGR adapter gasket.

30. Install the EGR valve and adapter assembly. Tighten the mounting bolts to 71–102 inch lbs. (8.0–11.5 Nm).

31. Install the heater water outlet tube.

32. Install the upper EGR tube fitting to 25–34 ft. lbs. (34–47 Nm).

33. Attach the EGR valve vacuum hose.

34. Install the water pump.

35. Install the radiator, fan blade assembly and the fan shroud.

36. Fill the cooling system with the correct amount and type of coolant.

37. Connect the negative battery cable.

4.6L Engines (VIN W and VIN 6)

1. Connect the negative battery cable.

2. Remove both of the cylinder head covers.

3. Remove the radiator.

4. Remove the water pump.

5. Disconnect the following electrical connections:
- Both ignition coil connectors
- Camshaft Position (CMP) sensor electrical connector
- Both radio capacitor electrical connectors

6. Remove the ignition coils and brackets.

7. Raise and support the vehicle safely on jackstands.

8. Remove the 2 top power steering pump bolts.

9. Position and support the power steering pump aside. Leave the fluid lines connected.

10. Detach the Crankshaft Position (CKP) sensor electrical connector.

11. Remove the CKP sensor.

12. Drain the engine oil into an adequately-sized drain pan.

13. Remove the front 4 oil pan bolts.

14. Lower the vehicle.

15. Remove the front oil seal.

16. Remove the CMP sensor.

17. Remove the accessory drive belt idler pulley.

18. Remove the accessory drive belt tensioner.

19. Remove the 8 engine front cover bolts and the 7 front cover nuts.

20. Remove the front cover from the engine block.

21. Remove the crankshaft sensor ring from the crankshaft.

22. On VIN 6 engines, use the Camshaft Positioning Tool T96T-6256-A, or equivalent, to position the camshaft.

23. Rotate the crankshaft until both camshaft keyways are 90 degrees from the cylinder head cover surface (refer to the illustration). Make sure that the copper links in the chain line up with the dots on the camshaft sprockets.

24. On VIN W engines, install the Camshaft Positioning Tool T91P-6256-A and the Camshaft Positioning Tool Adapters T92P-6256-A, or their equivalents, on the camshaft.

25. On VIN 6 engines, install the Camshaft Holding Tool T96T-6256-B, or equivalent, on the camshaft.

26. Remove the 2 left-hand timing chain tensioner bolts, then remove the tensioner.

27. Remove the 2 right-hand timing chain tensioner bolts, then remove the tensioner.

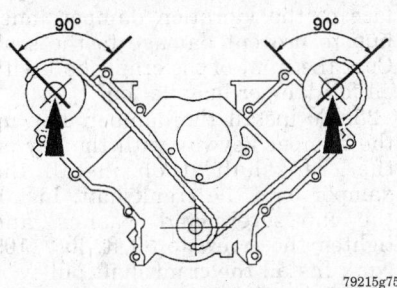

When removing the timing chains, rotate the crankshaft so that the camshafts are positioned as shown — 4.6L engines (VIN W and VIN 6)

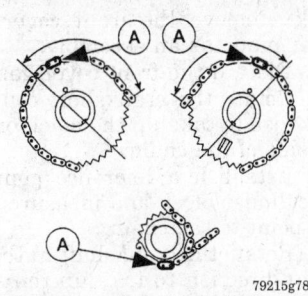

When installing the timing chains, make certain that the copper colored links are aligned with the timing marks — 4.6L engine (VIN W and VIN 6)

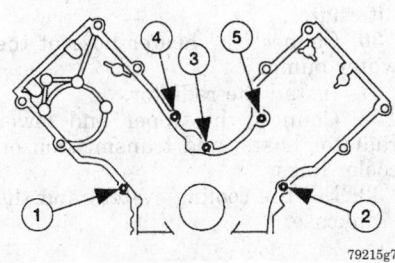

Tighten the first 5 front cover fasteners in the sequence shown —

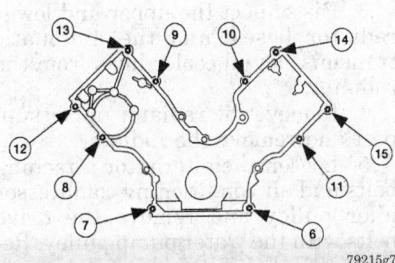

. . . continue tightening the other fasteners as shown here — 4.6L engines (VIN W and VIN 6)

28. Remove the right-hand and left-hand timing chain tensioner guides from the dowel pins.

---- **WARNING** ----

At no time, when the timing chains are removed and the cylinder heads are installed may the crankshaft or camshaft be rotated. Severe piston and valve damage will occur.

29. Remove the right-hand timing chain from the camshaft sprocket, then remove the right-hand chain and outer crankshaft sprocket from

the crankshaft. Perform the same for the left-hand timing chain.

30. Remove the chain guides.

NOTE: The manufacturer does not give procedures for the removal of the camshaft sprockets on the VIN 6 engines. The VIN 6 camshaft sprockets are not attached to the camshafts with a bolt.

31. On VIN W engines, remove the camshaft sprocket bolt, then remove the sprocket from the camshaft.

▶ **To install:**

---- **WARNING** ----

The timing chain procedures must be followed exactly or damage to the valves and pistons will result.

32. On VIN W engines, position the camshaft sprocket on the camshaft. Install the retaining bolt to 81–95 ft. lbs. (110–130 Nm).

33. Release the lock assembly on the timing chain tensioners, then compress them. Install retaining pins to hold them in the retracted position.

34. Install the timing chain guides. Tighten the guide mounting bolts to 71–107 inch lbs. (8–12 Nm).

35. Install the crankshaft sprocket onto the crankshaft, aligning the copper link on the timing chain with the dot on the crankshaft. Position the timing chain over the camshaft sprocket, aligning the copper link with the dot on the camshaft sprocket.

36. Make sure that the copper marks on the timing chain are lined up with the corresponding dots on the crankshaft sprockets.

37. Make sure that the camshaft sprocket keyways are still in the same position as before: 90 degrees from the cylinder head cover mounting surface.

38. Make certain that the right-hand timing chain gear is on the outside of the left-hand timing chain gear with the hubs facing each other.

39. Position the timing chain guides on the dowel pins.

40. Install the timing chain tensioners. Install the mounting bolts until tight.

41. Remove the retaining pins from the tensioners.

42. On VIN W engines, remove the camshaft positioning tool and the camshaft positioning tool adapters from the camshafts.

43. On VIN 6 engines, remove the camshaft holding tool from the camshafts.

44. Install the crankshaft sensor ring on the crankshaft.

45. Apply silicone along the head-to-block surface and the oil pan-to-block surface. Use Silicone Gasket and Sealant F6AZ-19562-A or equivalent.

46. Install the engine front cover on the engine block dowel.

47. Loosely install the front cover bolts and nuts.

NOTE: Make certain to tighten the front cover bolts and nuts in 2 steps.

48. Tighten the front cover bolts and nuts in the sequence shown. In the first step, fasteners numbered 1–5 should be tightened to 15–22 ft. lbs. (20–30 Nm). In the second step, only fasteners numbered 6–15 should be tightened to 29–40 ft. lbs. (40–55 Nm) for VIN 6 engines and to 15–22 ft. lbs. (20–30 Nm) on VIN W engines. In the second step, fasteners numbered 1–5 are not retightened.

49. Install the belt idler pulley and install the belt idler pulley bolt to 15–22 ft. lbs. (20–30 Nm).

50. Install the accessory drive belt tensioner in place, then install and tighten the 3 mounting bolts to 15–22 ft. lbs. (20–30 Nm).

51. Install the CMP sensor. Tighten the CMP sensor mounting bolt to 70–106 inch lbs. (8–12 Nm).

52. Install a new front oil seal.

53. Raise and safely support the front of the vehicle on jackstands.

54. Loosely install the 4 front oil pan bolts.

NOTE: Make sure to tighten the 4 front oil pan bolts in 2 steps.

55. Tighten the 4 front bolts from the left-hand side of the engine to the right-hand side of the engine to 15 ft. lbs. (20 Nm). Then tighten the 4 bolts an additional 60 degrees (one bolt head flat).

56. Position the CKP sensor and install the hold-down bolt. Tighten the hold-down bolt to 72–107 ft. lbs. (8–12 Nm).

57. Attach the CKP sensor electrical connector.

58. Install the power steering pump. Tighten the mounting bolts to 15–20 ft. lbs. (20–30 Nm).

59. Lower the vehicle.

60. Install the ignition coils and brackets.

61. Attach the ignition coil, the radio capacitor and the CMP electrical connectors.

62. Install the cylinder head covers.

63. Install the radiator.

64. Connect the negative battery cable.

65. Fill the engine with the correct amount and type of engine oil, if necessary.

5.0L (VIN N) and 5.8L (VIN H and R) Engines

1. Drain the cooling system and the crankcase.

2. Disconnect the upper and lower radiator hoses from the water pump, transmission oil cooler lines from the radiator, and remove the radiator.

3. Disconnect the heater hose from the water pump. Slide the water pump by-pass hose clamp toward the water pump.

4. Loosen the alternator pivot bolt and the bolt which secures the alternator adjusting arm to the water pump. Position the alternator out of the way.

5. Remove the power steering pump and air conditioning compressor from their mounting brackets, if equipped.

6. Remove the bolts holding the fan shroud to the radiator, if equipped. Remove the fan, spacer, pulley and drive belts.

7. Remove the crankshaft pulley from the crankshaft damper. Remove the damper attaching bolt and washer and remove the damper with a puller.

8. Disconnect the fuel pump outlet line at the fuel pump. Disconnect the vacuum inlet and outlet lines from the fuel pump. Remove the fuel pump attaching bolts and lay the pump to one side with the fuel inlet line still attached.

9. Remove the oil level dipstick and the bolt holding the dipstick tube to the exhaust manifold on the 5.0L (VIN N) engine.

10. Remove the oil pan-to-cylinder front cover attaching bolts. Use a sharp, thin cutting blade to cut the oil pan gasket flush with the cylinder block. Remove the front cover and water pump as an assembly.

11. Discard the front cover gasket.

12. Rotate the crankshaft counter-clockwise to take up the slack on the left side of the chain.

13. Establish a reference point on the cylinder block and measure from this point to the chain.

14. Rotate the crankshaft in the opposite direction to take up the slack on the right side of the chain.

15. Force the left side of the chain out and measure the distance between the reference point and the chain. The timing chain deflection is the difference between the 2 measurements. If the deflection exceeds 1/2 in. (13mm), replace the timing chain and sprockets.

16. Turn the crankshaft until the timing marks on the sprockets are aligned vertically.

17. Remove the camshaft sprocket retaining screw and remove the fuel pump eccentric and washers.

18. Alternately slide both of the sprockets and timing chain off the crankshaft and camshaft until free of the engine.

To install:

19. Position the timing chain on the sprockets so the timing marks on the sprockets are aligned vertically. Alternately slide the sprockets and chain onto the crankshaft and camshaft sprockets.

20. Install the fuel pump eccentric washers and attaching bolt on the camshaft sprocket. Tighten to 40–45 ft. lbs. (54–61 Nm).

21. Place the front seal removing tool (Ford part no. T70P–6B070–A or equivalent) into the front cover plate and over the front of the seal. Tighten the 2 through bolts to force the seal puller under the seal flange, then alternately tighten the 4 puller bolts a half turn at a time to pull the oil seal from the cover.

22. Coat a new front cover oil seal with Lubriplate® or equivalent and place it onto the front oil seal alignment and installation tool (Ford part no. T70P–6B070–A or equivalent). Place the tool and the seal onto the end of the crankshaft and push it toward the engine until the seal starts into the front cover.

23. Place the installation screw, washer, and nut onto the end of the crankshaft, then thread the screw into the crankshaft. Tighten the nut against the washer and tool to force the seal into the front cover plate. Remove the tool.

24. Apply Lubriplate® or equivalent to the oil seal rubbing surface of the vibration damper inner hub to prevent damage to the seal. Coat the front of the crankshaft with oil for damper installation.

25. To install the damper, line up the damper keyway with the key on the crankshaft, then install the damper onto the crankshaft. Install the cap screw and washer, and tighten the screw to 80 ft. lbs. (109 Nm). Install the crankshaft pulley.

26. Install the fan, spacer, pulley and drive belts.

27. Install the bolts holding the fan shroud to the radiator, if equipped.

28. Install the power steering pump and air conditioning compressor.

29. Position and tighten the alternator.

30. Connect the heater hose at the water pump.

31. Install the radiator.

32. Connect the upper and lower radiator hoses, and transmission oil cooler lines.

33. Fill the cooling system and the crankcase.

7.5L (VIN G) Engine

1. Drain the cooling system and crankcase.

2. Remove the radiator shroud and fan.

3. Disconnect the upper and lower radiator hoses, and the automatic transmission oil cooler lines from the radiator.

4. Remove the radiator upper support and remove the radiator.

5. Loosen the alternator attaching bolts and air conditioning compressor idler pulley and remove the drive belts with the water pump pulley. Remove the bolts attaching the compressor support to the water pump and remove the bracket (support), if equipped.

6. Remove the crankshaft pulley from the vibration damper. Remove the bolt and washer attaching the crankshaft damper and remove the damper with a puller. Remove the woodruff key from the crankshaft.

7. Loosen the by-pass hose at the water pump, and disconnect the heater return tube at the water pump.

8. Disconnect and plug the fuel inlet and outlet lines at the fuel pump, and remove the fuel pump.

9. Remove the bolts attaching the front cover to the cylinder block. Cut the oil pan seal flush with the cylinder block face with a thin knife blade prior to separating the cover from the cylinder block. Remove the cover and water pump as an assembly. Discard

the front cover gasket and oil pan seal.

10. Rotate the crankshaft counterclockwise to take up the slack on the left side of the chain.

11. Establish a reference point on the cylinder block and measure from this point to the chain.

12. Rotate the crankshaft in the opposite direction to take up the slack on the right side of the chain.

13. Force the left side of the chain out and measure the distance between the reference point and the chain. The timing chain deflection is the difference between the 2 measurements. If the deflection exceeds $\frac{1}{2}$ in. (13mm), replace the timing chain and sprockets.

14. Turn the crankshaft until the timing marks on the sprockets are aligned vertically.

15. Remove the camshaft sprocket retaining screw and remove the fuel pump eccentric and washers.

16. Alternately slide both of the sprockets and timing chain off the crankshaft and camshaft until free of the engine.

To install:

17. Position the timing chain on the sprockets so the timing marks on the sprockets are aligned vertically. Alternately slide the sprockets and chain onto the crankshaft and camshaft sprockets.

18. Install the fuel pump eccentric washers and attaching bolt on the camshaft sprocket. Tighten to 40–45 ft. lbs. (54–61 Nm).

19. Transfer the water pump if a new cover is going to be installed. Clean all of the gasket sealing surfaces on both the front cover and the cylinder block.

20. Coat the gasket surface of the oil pan with sealer. Cut and position the required sections of a new seal on the oil pan. Apply sealer to the corners.

21. Drive out the old front cover oil seal with a pin punch. Clean out the

seal recess in the cover. coat a new seal with Lubriplate® or equivalent grease. Install the seal, making sure the seal spring remains in the proper position. A front cover seal tool, Ford part no. T72J–117 or equivalent, makes installation easier.

22. Coat the gasket surfaces of the cylinder block and cover with sealer and position the new gasket on the block.

23. Position the front cover on the cylinder block. Use care not to damage the seal and gasket or misplace them.

24. Coat the front cover attaching screws with sealer and install them.

NOTE: It may be necessary to force the front cover downward to compress the oil pan seal in order to install the front cover attaching bolts. Use a prybar or drift to engage the cover screw holes through the cover and pry downward.

25. Install the fuel pump.

26. Connect the fuel inlet and outlet lines at the fuel pump.

27. Tighten the by-pass hose at the water pump.

28. Connect the heater return tube at the water pump.

29. Install the woodruff key from the crankshaft.

30. Install the damper.

31. Install the crankshaft pulley on the vibration damper.

32. Install the compressor support on the water pump and install the bracket (support), if equipped.

33. Install the drive belts with the water pump pulley.

34. Install the radiator and upper support.

35. Connect the upper and lower radiator hoses, and the automatic transmission oil cooler lines.

36. Install the radiator shroud and fan.

37. Fill the cooling system and crankcase.

Observe the following torques:

• Front cover bolts: 15–20 ft. lbs. (20–27 Nm)

• Water pump attaching screws: 12–15 ft. lbs. (16–20 Nm)

• Crankshaft damper: 70–90 ft. lbs. (95–122 Nm)

• Crankshaft pulley: 35–50 ft. lbs. (47–68 Nm)

• Fuel pump: 19–27 ft. lbs. (26–37 Nm)

• Oil pan bolts: 9–11 ft. lbs. (12.2–15.0 Nm) for the $\frac{5}{16}$ in. (7.9mm) screws and to 7–9 ft. lbs. (9.5–12.2 Nm) for the $\frac{1}{4}$ in. (6.3mm) screws

• Alternator pivot bolt: 45–57 ft. lbs. (61–77 Nm)

Timing Gears and Front Cover

REMOVAL AND INSTALLATION

4.9L (VIN Y and Z) Engine

1. Drain the cooling system and disconnect the radiator upper hose at the coolant outlet elbow and remove the 2 upper radiator retaining bolts.

2. Raise the vehicle and drain the crankcase.

3. Remove the splash shield and the automatic transmission oil cooling lines, if equipped, then remove the radiator.

4. Loosen and remove the fan belt, fan and pulley.

5. Use a gear puller to remove the crankshaft pulley damper.

6. Remove the front cover retaining bolts and gently pry the cover away from the block. Remove the gasket.

7. Turn the crankshaft until the timing marks on the camshaft and crankshaft gears are aligned.

8. Use a gear puller to removal both of the timing gears.

To install:

9. Before installing the timing gears, be sure the key and spacer are properly installed. Align the gear key way with the key and install the gear on the camshaft. Be sure the timing marks line up on the camshaft and the crankshaft gears and install the crankshaft gear.

10. Clean the front cover and the gasket surface of the cylinder block. Apply an oil-resistant sealer to the new front cover gasket and install the gasket onto the cover.

11. Position the front cover assembly over the end of the crankshaft and against the cylinder block. Start, but do not tighten, the cover and pan attaching screws. Slide a front cover alignment tool (Ford part no. T68P–6019–A or equivalent) over the crank stub and into the seal bore of the cover. Tighten all front cover and oil pan attaching screws to 12–18 ft. lbs. (16–25 Nm) front cover; 10–15 ft. lbs. (14–20 Nm) oil pan, tightening the oil pan screws first.

NOTE: Trim away the exposed portion of the old oil pan gasket flush with the front of the engine block. Cut and position the required portion of a new gasket to the oil pan and apply sealer to both sides.

12. Lubricate the hub of the crankshaft damper pulley with Lubriplate® to prevent damage to

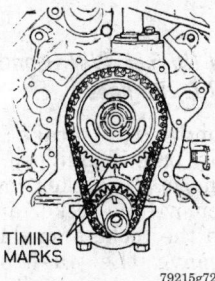

TIMING MARKS

79215g72

Timing chain and sprocket alignment — 5.0L (VINN), 5.8L (VINH and R) and 7.5L (VIN G) engine

the seal during installation or on initial starting of the engine.

13. Install the fan belt, fan and pulley.

14. Install the radiator.

15. Install the splash shield and the automatic transmission oil cooling lines, if equipped.

16. Fill the crankcase.

17. Connect the radiator upper hose at the coolant outlet elbow and install the 2 upper radiator retaining bolts.

18. Drain the cooling system.

19. Start the engine and check for leaks.

Camshaft

REMOVAL AND INSTALLATION

4.2L (VIN 2) Engine

1. Remove the valve lifters.

2. Remove the timing chain and sprockets.

3. Remove the camshaft key from the end of the camshaft, then slide the engine dynamic balance shaft drive gear off of the camshaft.

4. Remove the 2 camshaft thrust plate retaining bolts (1), then remove the thrust plate (2). Remove the camshaft spacer (3), then slide the camshaft (4) out of the front of the engine block. Be cautious not to gouge or scratch the camshaft bearing journals.

To install:

5. Lubricate the camshaft with engine oil prior to installation.

6. Carefully slide the camshaft into the camshaft bore. Do not scratch the bearing surfaces.

7. Install the camshaft thrust plat with the spacer. Tighten the thrust plate mounting bolts to 6–10 ft. lbs. (8–14 Nm).

8. Slide the engine dynamic balance shaft drive gear onto the camshaft. Install the camshaft key to the camshaft groove.

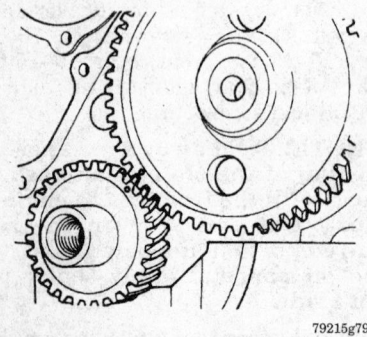

Timing gear alignment — 4.9L (VIN Y and Z) engine

79215g79

9. Install the timing chain and sprockets.

10. Install the valve lifters.

4.6L Engines

1. Remove the cylinder head covers from the engine.

2. Remove the timing chain.

—— **CAUTION** ——

At no time, when the timing chains are removed and the cylinder heads are installed may the crankshaft or camshaft be rotated. Severe piston and valve damage will occur.

3. Remove the camshaft roller lifters.

4. On VIN W engines, remove the timing chain camshaft gear by removing the gear retaining bolt.

5. Remove the camshaft bearing cap bolts, then lift the camshaft bearing caps off of the cylinder head.

6. Lift the camshaft from the cylinder head.

To install:

7. Lubricate the camshaft journals and bearing caps with engine oil which meets Ford specifications WSE-M2C908-A1.

8. Lower the camshaft onto the camshaft bearing journals.

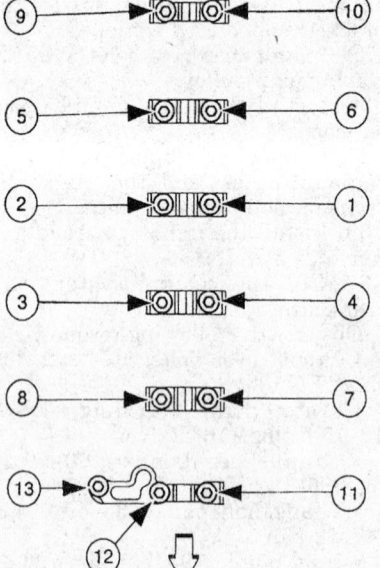

For VIN 6 engines, tighten the camshaft bearing caps in the sequence shown — 4.6L VIN 6 engines

79215g84

9. Install the camshaft bearing caps, then loosely install the bearing cap bolts.

10. Tighten the camshaft bearing cap mounting bolts, in the sequence shown for the particular engine, to 71–107 inch lbs. (8–12 Nm).

11. On VIN W engines, install the camshaft timing chain gear by tightening the retaining bolt to 81–95 ft. lbs. (110–130 Nm).

12. Install the valve lifters.

13. Install the timing chain and sprockets, if applicable.

14. Install the cylinder head covers.

4.9L (VIN Y and Z) Engine

1. Remove the grille, radiator, air conditioner condenser, and timing cover.

2. Remove the distributor, fuel pump, oil pan and oil pump.

3. Align the timing marks. Unbolt the camshaft thrust plate, working through the holes in the camshaft gear.

4. Loosen the rocker arms, remove the pushrods, take off the side cover and remove the valve lifter with a magnet.

5. Remove the camshaft very carefully to prevent nicking the bearings.

To install:

6. Oil the camshaft bearing journals and use Lubriplate® or something similar on the lobes. Install the camshaft, gear, and thrust plate, aligning the gear marks. Tighten down the thrust plate. Make sure the camshaft end-play is not excessive.

7. Install the rocker arms, pushrods and valve lifters.

8. Install the camshaft timing gear.

9. Install the oil pump, oil pan and fuel pump.

10. Install the distributor. The rotor should be at the firing position for no. 1 cylinder, with the timing gear marks aligned.

11. Install the front engine cover, A/C condenser, the radiator and the grille.

5.0L (VIN N), 5.8L (VIN H and R) and 7.5L (VIN G) Engines

1. Remove the intake manifold and valley pan, if equipped.

2. Remove the rocker covers, and either remove the rocker arm shafts or loosen the rockers on their pivots and remove the pushrods. The pushrods must be reinstalled in their original positions.

3. Remove the valve lifters in sequence with a magnet. They must be replaced in their original positions.

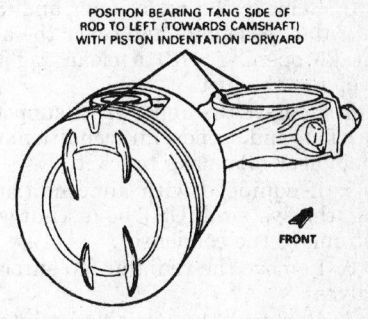

POSITION BEARING TANG SIDE OF ROD TO LEFT (TOWARDS CAMSHAFT) WITH PISTON INDENTATION FORWARD

FRONT

Piston and connecting rod assembly — 4.9L (VIN Y and Z) engine

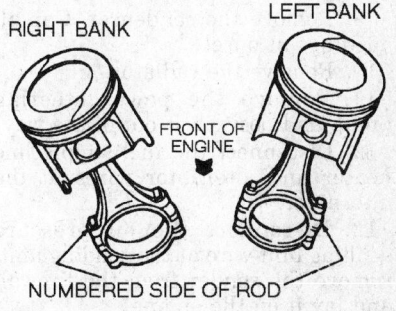

RIGHT BANK — LEFT BANK
FRONT OF ENGINE
NUMBERED SIDE OF ROD

Piston and connecting rod assembly — 5.0L (VIN N) and 5.8L (VIN H and R) engines

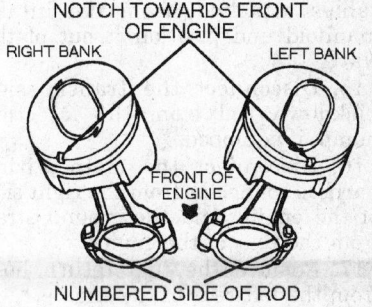

NOTCH TOWARDS FRONT OF ENGINE
RIGHT BANK — LEFT BANK
FRONT OF ENGINE
NUMBERED SIDE OF ROD

Piston and connecting rod assembly — 7.5L (VIN G) engine

9. Install the valve lifters. They must be replaced in their original positions.
10. Install the pushrods, the rocker arms and the rocker arm covers.
11. Install the intake manifold and valley pan, if equipped.

Balance Shaft

REMOVAL AND INSTALLATION

4.2L (VIN 2) Engine

1. Remove the timing chain.
2. Remove the 2 engine dynamic balance shaft thrust plate bolts, then the engine dynamic balance shaft gear, thrust plate and engine dynamic balance shaft.

To install:
3. Turn the camshaft so that the timing mark is at 12 o'clock and install the engine dynamic balance shaft assembly into the cylinder block so that the balance shaft gear timing mark lines up with the camshaft timing mark.

NOTE: If properly aligned, the engine dynamic balance shaft keyway will be at 12 o'clock and the camshaft keyway will be at 6 o'clock on the camshaft.

4. Install the 2 engine dynamic balance shaft thrust plate bolts to 6–10 ft. lbs. (8–14 Nm).
5. Install the timing chain.

Piston and Connecting Rod

POSITIONING

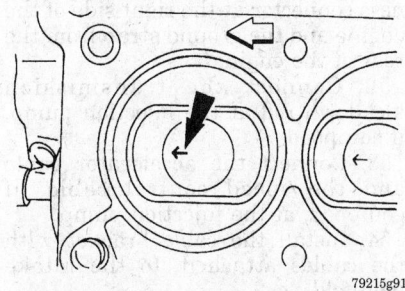

Position the pistons in the engine so that the arrow on the piston top is facing toward the front of the engine block — 4.2L (VIN 2) and 4.6L (VIN W and VIN 6) engines

DIESEL ENGINE MECHANICAL

Engine Assembly

REMOVAL AND INSTALLATION

7.3L (VIN C, F, K, and M) Diesel Engine

1. Remove the hood.
2. Drain the coolant.

3 ... 2
6 ... 5
4 ... 1
3 ... 2
8 -12 Nm (71.4 - 107.1 Lb-In)
6 ... 5
7
8 ... 1
4

For VIN W engines, tighten the camshaft bearing caps in the sequence shown — 4.6L VIN W engines

4. Remove the timing gear cover, timing chain and sprockets.
5. In addition to the radiator and air conditioning condenser, if equipped, it may be necessary to remove the front grille assembly and the hook lock assembly to gain the necessary clearance to code the camshaft out of the front of the engine.

NOTE: A camshaft removal tool, Ford part no. T65L–6250–A and adaptor 14–0314 are needed to remove the Diesel camshaft.

To install:
6. Coat the camshaft with oil liberally before installing it. Slide the camshaft into the engine very carefully so as not to scratch the bearing bores with the camshaft lobes. Install the camshaft thrust plate and tighten the attaching screws to 9–12 ft. lbs. (12–16 Nm). Measure the camshaft end-play. If the end-play is more than 0.009 in. (0.228mm), replace the thrust plate. Assemble the remaining components in the reverse order of removal.
7. Install the radiator, front grille, hood lock assembly and air conditioning condenser, if removed.
8. Install the timing chain and and front cover.

3. Remove the air cleaner and intake duct assembly and cover the air intake opening with a clean rag to keep out the dirt.

4. Remove the upper grille support bracket and upper air conditioning condenser mounting bracket.

5. If equipped with air conditioning, the system MUST be discharged to remove the condenser.

6. Remove the radiator fan shroud halves.

7. Remove the fan and clutch assembly.

8. Detach the radiator hoses and the transmission cooler lines, if equipped.

9. Remove the condenser. Cap all openings at once!

10. Remove the radiator.

11. Remove the power steering pump and position it out of the way.

12. Disconnect the fuel supply line heater and alternator wires at the alternator.

13. Disconnect the oil pressure sending unit wire at the sending unit, remove the sender from the firewall and lay it on the engine.

14. Disconnect the accelerator cable and the speed control cable, if equipped, from the injection pump. Remove the cable bracket with the cables attached, from the intake manifold and position it out of the way.

15. Disconnect the transmission kickdown rod from the injection pump, if equipped.

16. Disconnect the main wiring harness connector from the right side of the engine and the ground strap from the rear of the engine.

17. Remove the fuel return hose from the left rear of the engine.

18. Remove the 2 upper transmission-to-engine attaching bolts.

19. Disconnect the heater hoses.

20. Disconnect the water temperature sender wire.

21. Disconnect the overheat light switch wire and position the wire out of the way.

22. Raise the vehicle and support on it on jackstands.

23. Disconnect the battery ground cables from the front of the engine and the starter cables from the starter.

24. Remove the fuel inlet line and plug the fuel line at the fuel pump.

25. Detach the exhaust pipe at the exhaust manifold.

26. Disconnect the engine insulators from the no. 1 crossmember.

27. Remove the flywheel inspection plate and the 4 converter-to-flywheel

attaching nuts, if equipped with automatic transmission.

28. Remove the jackstands and lower the vehicle.

29. Supporting the transmission on a jack.

30. Remove the 4 lower transmission attaching bolts.

31. Attach an engine lifting sling and remove the engine from the vehicle.

To install:

32. Lower the engine into vehicle.

33. Align the converter to the flexplate and the engine dowels to the transmission.

34. Install the engine mount bolts and torque them to 80 ft. lbs. (109 Nm).

35. Remove the engine lifting sling.

36. Install the 4 lower transmission attaching bolts. Torque the bolts to 65 ft. lbs. (88 Nm).

37. Remove transmission jack.

38. Raise and support the front end on jackstands.

39. If equipped with automatic transmission, install the 4 converter-to-flywheel attaching nuts. Torque the nuts to 34 ft. lbs. (47 Nm).

40. Install the flywheel inspection plate. Torque the bolts to 60–90 inch lbs. (6.7–10.0 Nm).

41. Attach the exhaust pipe at the exhaust manifold.

42. Connect the fuel inlet line.

43. Connect the battery ground cables to the front of the engine.

44. Connect the starter cables at the starter.

45. Lower the vehicle.

46. Connect the overheat light switch wire.

47. Connect the water temperature sender wire.

48. Connect the heater hoses.

49. Install the 2 upper transmission-to-engine attaching bolts. Torque the bolts to 65 ft. lbs. (88 Nm).

50. Connect the fuel return hose at the left rear of the engine.

51. Connect the main wiring harness connector at the right side of the engine and the ground strap from the rear of the engine.

52. Connect the transmission kickdown rod at the injection pump, if equipped.

53. Connect the accelerator cable and the speed control cable, if equipped, at the injection pump.

54. Install the cable bracket with the cables attached, to the intake manifold.

55. Install the oil pressure sending unit.

56. Connect the oil pressure sending unit wire at the sending unit.

57. Connect the fuel supply line heater and alternator wires at the alternator.

58. Install the power steering pump.

59. Install the radiator.

60. Install the condenser.

61. Connect the radiator hoses and the transmission cooler lines, if equipped.

62. Install the fan and clutch assembly.

63. Install the radiator fan shroud halves.

64. If equipped with air conditioning, charge the system.

65. Install the upper grille support bracket and upper air conditioning condenser mounting bracket.

66. Install the air cleaner and intake duct assembly.

67. Fill the cooling system.

68. Install the hood.

Engine Mounts

REMOVAL AND INSTALLATION

Front Mounts

1. Remove the fan shroud attaching screws.

2. Support the engine using a wood block and a jack placed under the oil pan.

3. Remove the nuts and washers attaching the mounts to the engine bracket. Lift the engine enough to disengage the mount upper stud from the crossmember engine bracket.

4. Remove the bolt attaching the fuel pump shield to the left engine bracket, if necessary.

5. Remove the mount-to-crossmember attaching nut and washer assembly. Remove the engine mount.

To install:

6. Install the engine mount to the crossmember.

7. Install the bolt attaching the fuel pump shield to the left bracket, if necessary.

8. Lower the engine until the mount stud engages in the slot/hole of the engine bracket. Install the attaching nuts.

9. Remove the jack and wood block from the engine oil pan. Install the fan shroud attaching screws.

Rear Mount

1. Place a block of wood and a jack under the transmission.

2. Remove the 2 nuts attaching the mount to the crossmember. Raise the transmission enough to lift the mount from the crossmember.

3. Remove the bolts and nuts attaching the crossmember to the frame side rails and remove the crossmember.

4. If equipped, remove the fasteners attaching the exhaust hanger to the rear engine mount.

5. Remove the 2 bolts attaching the mount to the transmission and remove the mount and retainer assembly.

To install:

6. Position the mount and retainer assembly to the transmission and install the 2 attaching bolts.

7. If equipped, install the fasteners attaching the exhaust hanger to the mount.

8. Install the crossmember to the frame side rails with the attaching nuts and bolts.

9. Lower the transmission and install the mount crossmember attaching nuts. Remove the jack and wood block.

Cylinder Head

REMOVAL AND INSTALLATION

7.3L (VIN C, F, K, and M) Diesel Engine

1. Open the hood and disconnect the negative cables from both batteries.

2. Drain the cooling system and remove the radiator fan shroud halves.

3. Remove the radiator fan and clutch assembly using special tool T83T-6312-A and B. This tool is available through the Owatonna Tool Co. whose address is listed in the front of this boot, or through Ford Dealers. It is also available through many tool rental shops.

NOTE: The fan clutch uses a left hand thread and must be removed by turning the nut clockwise.

4. Label and disconnect the wiring from the alternator.

5. Remove the adjusting bolts and pivot bolts from the alternator and the vacuum pump and remove both units.

6. Remove the fuel filter lines and cap to prevent fuel leakage.

7. Remove the alternator, vacuum pump, and fuel filter brackets with the fuel filter attached.

8. Remove the heater hose from the cylinder head.

9. Remove the fuel injection pump.

10. Remove the intake manifold and valley cover.

11. Raise the vehicle and safely support it with jackstands.

12. Disconnect the exhaust pipes from the exhaust manifolds.

13. Remove the clamp holding the oil dipstick tube in place and the bolt attaching the transmission oil dipstick to the cylinder head.

14. Lower the vehicle.

15. Remove the oil dipstick tube.

16. Remove the valve covers, rocker arms and pushrods. Keep the pushrods in order so they can be returned to their original positions.

17. Remove the nozzles and glow plugs.

18. Remove the cylinder head bolts and attach lifting eyes, using special tool T70P-6000 or equivalent, to each end of the cylinder heads.

19. Carefully lift the cylinder heads out of the engine compartment and remove the head gaskets.

NOTE: The cylinder head prechambers may fall out of the heads upon removal.

To install:

20. Position the cylinder head gasket on the engine block and carefully lower the cylinder head in place.

NOTE: Use care in installing the cylinder heads to prevent the prechambers from falling out into the cylinder bores.

21. Install the cylinder head bolt and torque in 4 steps using the sequence.

NOTE: Lubricate the threads and the mating surfaces of the bolt heads and washers with oil.

22. Dip the pushrod ends in clean oil and install the pushrods with the copper colored ends toward the rocker arms, making sure the pushrods are fully seated in the tappet pushrod seats.

23. Install the rocker arms and posts in their original positions. Apply Lubriplate® grease to the valve stem tips. Turn the engine over by hand until the timing mark is at the 11 o'clock position as viewed from the front. Install the rocker arm posts, bolts, and torque to 27 ft. lbs. (37 Nm). Install the valve covers.

24. Install the valley pan and the intake manifold.

25. Install the fuel injection pump.

26. Connect the heater hose to the cylinder head.

27. Install the fuel filter, alternator, vacuum pump, and their drive belts.

28. Install the oil and transmission dip stick.

29. Connect the exhaust pipe to the exhaust manifolds.

30. Reconnect the alternator wiring harness and replace the air cleaner. Connect both battery ground cables.

31. Refill and bleed the cooling system.

32. Run the engine and check for fuel, coolant and exhaust leaks.

NOTE: If necessary, purge the high pressure fuel lines of air by loosening the connector one half to one turn and cranking the engine until a solid stream of fuel, free from any bubbles, flows from the connections.

33. Check the injection pump timing.

34. Install the radiator fan and clutch assembly using special tools T83T-6312A and B or equivalent.

NOTE: The fan clutch uses a left hand thread. Tighten by turning the nut counterclockwise. Install the radiator fan shroud halves.

Valve Lifters

REMOVAL AND INSTALLATION

7.3L (VIN C, F, K, and M) Diesel Engine

1. Remove the intake manifold.

2. Remove the CDR tube and grommet from the valley pan.

3. Remove the valley pan strap from the front of the block.

4. Remove the valley pan drain plug and lift out the valley pan.

5. Remove the rocker arm covers.

6. Remove the rocker arms; keep them in order for installation.

7. Remove the pushrods; keep them in order for installation.

8. Remove the lifter guide retainer.

9. Using a magnetic lifter removal tool, remove the lifters. Wipe clean the exterior of each lifter as it's removed and mark it with an indelible marker, so it can be installed in its original bore.

To install:

10. Coat the bottom surface of each lifter with multi-purpose grease, and coat the rest of the lifter with clean oil.

11. Install each lifter in it original bore using the magnetic tool.

12. Install the lifter guide retainer.

13. Install the pushrods, copper colored end up, into their original locations, making sure they are firmly seated in the lifters.

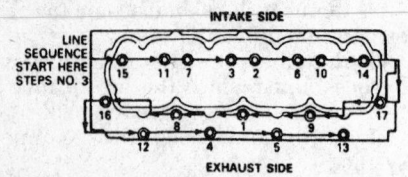

Cylinder head bolt torque sequence — 7.3L (VIN C, F, K, and M) Diesel engine

14. Coat the valve stem tips with multi-purpose grease and install the rocker arms and posts in their original positions.

15. Turn the crankshaft by hand, until the timing mark is at the 11 o'clock position — viewed from the front.

16. Install all the rocker arm post bolts and torque them to 20 ft. lbs. (27 Nm).

17. Install the rocker arm covers.

18. Clean all old RTV gasket material from the block and run a 1/8 in. (3mm) bead of new RTV gasket material at each end of the block. Within 15 minutes, install the valley pan. Install the pan drain plug.

19. Install the CDR tube, new grommet and new O-ring.

20. Install the intake manifold and related parts.

Rocker Arms

REMOVAL AND INSTALLATION

7.3L (VIN C, F, K, and M) Diesel Engine

1. Disconnect the ground cables from both batteries.

2. Remove the valve cover attaching screws and remove both valve cover.

3. Remove the valve rocker arm post mounting bolts. Remove the rocker arms and posts in order and mark them with tape so they can be installed in their original positions.

4. If the cylinder heads are to be removed, then the pushrods can now be removed. Make a holder for the pushrods out of a piece of wood or cardboard, and remove the pushrods in order. It is very important that the pushrods be re-installed in their original order. The pushrods can remain in position if no further disassembly is required.

5. If the pushrods were removed, install them in their original locations. make sure they are fully seated in the tappet seats.

NOTE: The copper colored end of the pushrod goes toward the rocker arm.

6. Apply a polyethylene grease to the valve stem tips. Install the rocker arms and posts in their original positions.

7. Turn the engine over by hand until the valve timing mark is at the 11:00 o'clock position, as viewed from the front of the engine. Install all of the rocker arm post attaching bolts and torque to 20 ft. lbs. (27 Nm).

8. Install new valve cover gaskets and install the valve cover. Install the battery cables, start the engine and check for leaks.

Intake Manifold

REMOVAL AND INSTALLATION

7.3L (VIN C, F, K, and M) Diesel Engine

1. Open the hood and remove both battery ground cables.

2. Remove the air cleaner and install clean rags into the air intake of the intake manifold. It is important that no dirt or foreign objects get into the Diesel intake.

3. Remove the injection pump.

4. Remove the fuel return hose from No. 7 and No. 8 rear nozzles and remove the return hose to the fuel tank.

5. Label the positions of the wires and remove the engine wiring harness from the engine.

NOTE: The engine harness ground cables must be removed from the back of the left cylinder head.

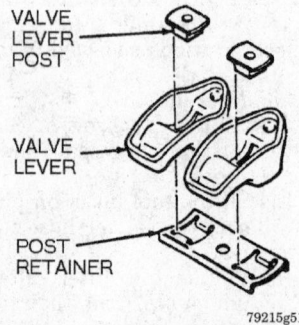

Rocker arm assembly — 7.3L (VIN C, F, K, and M) Diesel engine

6. Remove the bolts attaching the intake manifold to the cylinder heads and remove the manifold.

7. Remove the CDR tube grommet from the valley pan.

8. Remove the bolts attaching the valley pan strap to the front of the engine block, and remove the strap.

9. Remove the valley pan drain plug and remove the valley pan.

To install:

10. Apply a 1/8 in. (3mm) bead of RTV sealer to each end of the cylinder block.

NOTE: The RTV sealer should be applied immediately prior to the valley pan installation.

11. Install the valley pan drain plug, CDR tube and new grommet into the valley pan.

12. Install a new O-ring and new back-up ring on the CDR valve.

13. Install the valley pan strap on the front of the valley pan.

14. Install the intake manifold and torque the bolts to 24 ft. lbs. (33 Nm) using the sequence.

15. Reconnect the engine wiring harness and the engine ground wire located to the rear of the left cylinder head.

16. Install the injection pump.

17. Install the No. 7 and No. 8 fuel return hoses and the fuel tank return hose.

18. Remove the rag from the intake manifold and replace the air cleaner. Reconnect the battery ground cables to both batteries.

19. Run the engine and check for oil and fuel leaks.

NOTE: If necessary, purge the nozzle high pressure lines of air by loosening the connector one half to one turn and cranking the engine until solid stream of fuel, devoid of any bubbles, flows from the connection.

——— CAUTION ———
Keep eyes and hands away from the nozzle spray. Fuel spraying from the nozzle under high pressure can penetrate the skin.

20. Check and adjust the injection pump timing.

Exhaust Manifold

REMOVAL AND INSTALLATION

7.3L (VIN C, F, K, and M) Diesel Engine

1. Disconnect the ground cables from both batteries.

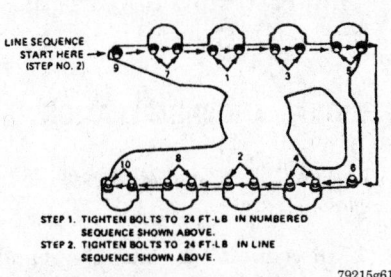

Intake manifold bolt torque sequence — 7.3L (VIN C, F, K, and M) Diesel engine

2. Raise the vehicle and safely support it with jackstands.

3. Disconnect the muffler inlet pipe from the exhaust manifolds.

4. Lower the vehicle to remove the right manifold. When removing the left manifold, jack the tuck up. Bend the tabs on the manifold attaching bolts, then remove the bolts and manifold.

To install:

5. Before installing, clean all mounting surfaces on the cylinder heads and the manifold. Apply an anti-seize compound on the manifold both threads and install the left manifold, using a new gasket and new locking tabs.

6. Torque the bolts to specifications and bend the tabs over the flats on the bolt heads to prevent the bolts from loosening.

7. Raise the vehicle to install the right manifold. Install the right manifold following procedures 5 and 6 above.

8. Connect the inlet pipes to the manifold and tighten. Lower the vehicle, connect the batteries and run the engine to check for exhaust leaks.

Exhaust manifold bolt torque sequence — 7.3L (VIN C, F, K, and M) Diesel engine

Turbocharger

REMOVAL AND INSTALLATION

1. Disconnect the negative battery cable.

2. Remove the two air intake tube assembly bolts, clamps at the turbocharger, crankcase breather assembly, engine air cleaner and air intake tube and hoses.

3. Remove the exhaust outlet clamp from the turbocharger.

4. Raise and safely support the vehicle on jackstands.

5. Remove the engine charge exhaust pipe bolt from the transmission, if so equipped.

6. Remove the bolts and nuts from the catalytic converter-to-engine charge exhaust pipe, if so equipped.

7. Loosen two bolts retaining the turbocharger exhaust inlet pipe to the left exhaust manifold.

8. For automatic transmissions, remove the bolts retaining the left turbocharger exhaust inlet pipe to the turbocharger exhaust inlet adapter.

9. Loosen the two bolts retaining the turbocharger exhaust inlet pipe to the right exhaust manifold.

10. Remove the lower bolt retaining the right turbocharger exhaust inlet pipe to the turbocharger exhaust inlet adapter.

11. Lower the vehicle.

12. For automatic transmissions, remove the upper bolts retaining the right and left turbocharger exhaust inlet pipes to the turbocharger exhaust inlet adapter.

13. Remove the right engine lift hook and bolt.

14. Loosen the air inlet hose clamp at the turbocharger. Disconnect the hose and lay aside.

15. For automatic transmissions, loosen the 4 intake manifold hose clamps, and one clamp retaining the compressor manifold to the turbocharger. Remove the compressor manifold.

16. Remove the four bolts retaining the turbocharger pedestal assembly to the cylinder block.

17. Remove the turbocharger assembly and detach all electrical connectors from it.

NOTE: If the turbocharger is not being removed for service, install the Fuel/Oil turbo Protector Cap Set T94T-9395-AH or equivalent.

18. Remove the oil gallery O-rings.

To install:

19. Install new oil gallery O-rings.

20. Attach the turbocharger electrical connectors and install the turbocharger assembly.

21. Install the 4 bolts retaining the turbocharger pedestal assembly to the engine block. Tighten the bolts to 18 ft. lbs. (25 Nm).

22. Loosely install the 4 bolts retaining the right and left turbocharger exhaust inlet pipes to the turbocharger exhaust inlet adapter.

23. Install the compressor manifold, intake manifold hoses and clamps. Make sure the compressor outlet seal is in position.

24. Install the right engine lift hook and bolt.

25. Raise and safely support the front of the vehicle on jackstands.

26. Tighten the 2 right and left lower bolts (retaining the turbocharger exhaust inlet pipes to the turbocharger exhaust inlet adapter) to 36 ft. lbs. (49 Nm).

27. Tighten the 4 right and left bolts and nuts (retaining the turbocharger exhaust inlet pipes to the exhaust manifolds) to 36 ft. lbs. (49 Nm).

28. Install the catalytic converter to the engine charge exhaust pipe bolts and nuts.

29. Lower the vehicle.

30. Tighten the 2 right and left upper bolts (retaining the turbocharger exhaust inlet pipes to the turbocharger exhaust inlet adapter) to 36 ft. lbs. (49 Nm).

31. Install the exhaust outlet clamp to the turbocharger.

32. Install the air intake tube and hose assembly.

33. Connect the negative battery cable.

Front Cover Seal

REMOVAL AND INSTALLATION

7.3L (VIN C, F, K, and M) Diesel Engine

1. Disconnect both battery ground cables. Drain the cooling system.

2. Remove the air cleaner and cover the air intake on the manifold with clean rags. Do not allow any foreign material to enter the intake.

3. Remove the radiator fan shroud halves.

4. Remove the fan and fan clutch assembly using a puller or ford tool No. T83T-6312-A.

NOTE: The nut is a left hand thread; remove by turning the nut clockwise.

5. Remove the injection pump.

6. Remove the water pump.

7. Raise the vehicle and safely support it with jackstands.

8. Remove the crankshaft pulley and vibration damper.

9. Remove the engine ground cables at the front of the engine.

10. Remove the 5 bolts attaching the engine front cover to the engine block and oil pan.

11. Lower the vehicle.

12. Remove the front cover.

NOTE: The front cover oil seal on the Diesel must be driven out with an arbor press and a 3¼ in. (82.5mm) spacer.

To install:

13. Remove all old gasket material from the front cover, engine block, oil pan sealing surfaces and water pump surfaces.

14. Coat the new front oil seal with Lubriplate® or equivalent grease.

15. The new seal must be installed using a seal installation tool, Ford part no. T83T–6700–A or an arbor press. When the seal bottoms out on the front cover surface, it is installed at the proper depth.

16. Install alignment dowels into the engine block to align the front cover and gaskets. These can be made out of round stock. Apply a gasket sealer to the engine block sealing surfaces, then install the gaskets on the block.

17. Apply a ⅛ in. (3mm) bead of RTV sealer on the front of the engine block. Apply a ¼ in. (6mm) bead of RTV sealer on the oil pan.

18. Install the front cover immediately after applying RTV sealer. The sealer will begin to cure and lose its effectiveness unless the cover is installed quickly.

19. Install the water pump gasket on the engine front cover. Apply RTV sealer to the 4 water pump bolts il-

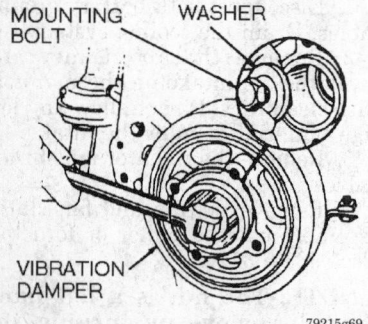

Vibration damper removal — 7.3L (VIN C, F, K, and M) Diesel engine

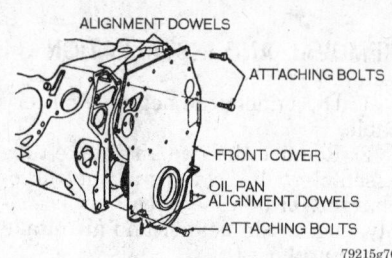

Cylinder block front cover installation — 7.3L (VIN C, F, K, and M) engine

lustrated. Install the water pump and hand tighten all bolts.

—— **WARNING** ——
The 2 top water pump bolts must be no more than 1¼ in. (31.75mm) long. Bolts any longer will interfere with (hit) the engine drive gears.

20. Torque the water pump bolts to 19 ft. lbs. (26 Nm). Torque the front cover bolts to specifications according to bolt size (see Torque Specifications chart).

21. Install the injection pump adaptor and injection pump.

22. Install the heater hose fitting in the pump using pipe sealant, and connect the heater hose to the water pump.

23. Raise the vehicle and safely support it with jackstands.

24. Lubricate the front of the crankshaft with clean oil. Apply RTV sealant to the engine side of the retaining bolt washer to prevent oil seepage past the keyway. Install the crankshaft vibration damper using Ford Special tool T83T–6316B, or equivalent. Tighten the damper-to-crankshaft bolt to 90 ft. lbs. (122 Nm).

25. Install the fan and fan clutch assembly.

NOTE: The nut is a left hand thread; Install by turning the nut clockwise.

26. Install the radiator fan shroud halves.

27. Install the air cleaner.

28. Connect both battery ground cables.

29. Fill the cooling system.

Timing Gears and Front Cover

REMOVAL AND INSTALLATION

7.3L (VIN C, F, K, and M) Diesel Engine

1. Disconnect both battery ground cables. Drain the cooling system.

2. Remove the air cleaner and cover the air intake on the manifold with clean rags. Do not allow any foreign material to enter the intake.

3. Remove the radiator fan shroud halves.

4. Remove the fan and fan clutch assembly using a puller or ford tool No. T83T–6312–A.

NOTE: The nut is a left hand thread; remove by turning the nut clockwise.

5. Remove the injection pump.

6. Remove the water pump.

7. Raise the vehicle and safely support it with jackstands.

8. Remove the crankshaft pulley and vibration damper.

9. Remove the engine ground cables at the front of the engine.

10. Remove the 5 bolts attaching the engine front cover to the engine block and oil pan.

11. Lower the vehicle.

12. Remove the front cover.

NOTE: The front cover oil seal on the Diesel must be driven out with an arbor press and a 3¼ in. (82.5mm) spacer.

13. To remove the crankshaft gear, install the crankshaft drive gear remover Tool T83T–6316–A, and using a breaker bar to prevent crankshaft rotation, or flywheel holding Tool T74R–6375–A, remove the crankshaft gear.

14. Remove the injection pump drive gear using a gear puller.

15. To remove the camshaft gear, perform the following:

   a. Remove the camshaft Allen screw.

   b. Install a gear puller, Tool T83T–6316–A and remove the gear. Remove the fuel supply pump, if necessary.

   c. Install a gear puller, Tool T77E–4220–B and shaft protector T83T–6316–A and remove the fuel pump cam and spacer, if necessary.

   d. Remove the bolts attaching the thrust plate, and remove the thrust plate, if necessary.

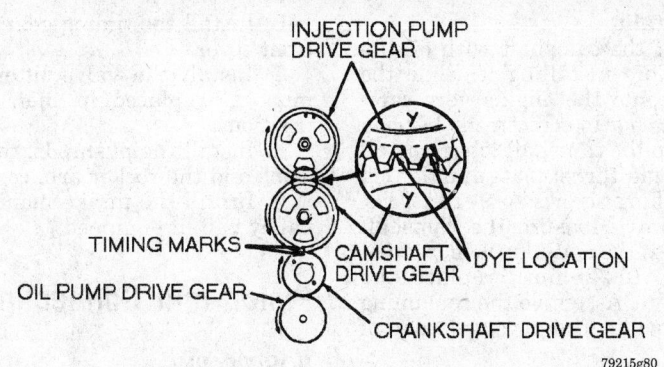

Timing sprocket alignment — 7.3L (VIN C, F, K, and M) Diesel engine

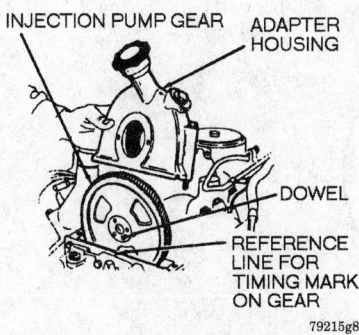

Injection pump drive sprocket cover — 7.3L (VIN C, F, K, and M) Diesel engine

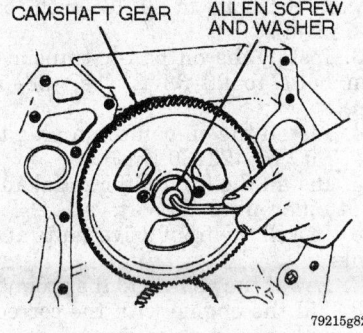

Camshaft drive sprocket removal — 7.3L (VIN C, F, K, and M) Diesel engine

To install:

16. Install the camshaft timing gear as follows:

a. Install a new thrust plate, if removed.

b. Install the spacer and fuel pump cam against the camshaft thrust flange, using installation sleeve and replacer Tool T83T–6316–B, if removed.

c. Install the camshaft drive gear against the fuel pump cam, aligning the timing mark with the timing mark on the crankshaft drive gear, using installation sleeve and replacer Tool T83T–6316–B.

d. Install the camshaft Allen screw and tighten to 18 ft. lbs. (25 Nm).

17. Install the injection pump drive gear in position, aligning all the drive gear timing marks.

NOTE: To determine that the No. 1 piston is at TDC of the compression stroke, position the injection pump drive gear dowel at the 4 o'clock position. The scribe line on the vibration damper should be at TDC. Use extreme care to avoid disturbing the injection pump drive gear, once it is in position.

18. Install the crankshaft gear using Tool T83T–6316–B aligning the crankshaft drive gear timing mark with the camshaft drive gear timing mark.

NOTE: The gear may be heated to 300–350°F (149–260°C) for ease of installation. Heat it in an oven. Do not use a torch.

19. Remove all old gasket material from the front cover, engine block, oil pan sealing surfaces and water pump surfaces.

20. Coat the new front oil seal with Lubriplate® or equivalent grease.

21. The new seal must be installed using a seal installation tool, Ford part no. T83T–6700–A or an arbor press. When the seal bottoms out on the front cover surface, it is installed at the proper depth.

22. Install alignment dowels into the engine block to align the front cover and gaskets. These can be made out of round stock. Apply a gasket sealer to the engine block sealing surfaces, then install the gaskets on the block.

23. Apply a ⅛ in. (3mm) bead of RTV sealer on the front of the engine block. Apply a ¼ in. (6mm) bead of RTV sealer on the oil pan.

24. Install the front cover immediately after applying RTV sealer. The sealer will begin to cure and lose its effectiveness unless the cover is installed quickly.

25. Install the water pump gasket on the engine front cover. Apply RTV sealer to the 4 water pump bolts illustrated. Install the water pump and hand tighten all bolts.

WARNING

The 2 top water pump bolts must be no more than 1¼ in. (31.75mm) long bolts any longer will interfere with (hit) the engine drive gears.

26. Torque the water pump bolts to 19 ft. lbs. (26 Nm). Torque the front cover bolts to specifications according to bolt size (see Torque Specifications chart).

27. Install the injection pump adaptor and injection pump.

28. Install the heater hose fitting in the pump using pipe sealant, and connect the heater hose to the water pump.

29. Raise the vehicle and safely support it with jackstands.

30. Lubricate the front of the crankshaft with clean oil. Apply RTV sealant to the engine side of the retaining bolt washer to prevent oil seepage past the keyway. Install the crankshaft vibration damper using Ford Special tools T83T–6316B. Torque the damper-to-crankshaft bolt to 90 ft. lbs. (122 Nm).

31. Install the fan and fan clutch assembly.

NOTE: The nut is a left hand thread; Install by turning the nut clockwise.

32. Install the radiator fan shroud halves.

33. Install the air cleaner.

34. Connect both battery ground cables.

35. Fill the cooling system.

Camshaft

REMOVAL AND INSTALLATION

7.3L (VIN C, F, K, and M) Diesel Engine

NOTE: Ford recommends removing the Diesel engine from the vehicle for camshaft removal.

1. Remove the intake manifold and valley pan, if equipped.

2. Remove the rocker covers, and either remove the rocker arm shafts or loosen the rockers on their pivots

and remove the pushrods. The pushrods must be reinstalled in their original positions.

3. Remove the valve lifters in sequence with a magnet. They must be replaced in their original positions.

4. Remove the timing gear cover, timing gear and sprockets.

NOTE: A camshaft removal tool, Ford part no. T65L–6250–A and adaptor 14–0314, or equivalents, are needed to remove the Diesel camshaft.

To install:

5. Coat the camshaft with oil liberally before installing it. Slide the camshaft into the engine very carefully so as not to scratch the bearing bores with the camshaft lobes. Install the camshaft thrust plate and tighten the attaching screws to 9–12 ft. lbs. (12–16 Nm). Measure the camshaft end-play. If the end-play is more than 0.009 in. (0.228mm), replace the thrust plate. Assemble the remaining components in the reverse order of removal.

6. Install the timing gear and and front cover.

7. Install the valve lifters. They must be replaced in their original positions.

8. Install the pushrods, the rocker arms and the rocker arm covers.

9. Install the intake manifold and valley pan, if equipped.

Piston and Connecting Rod

POSITIONING

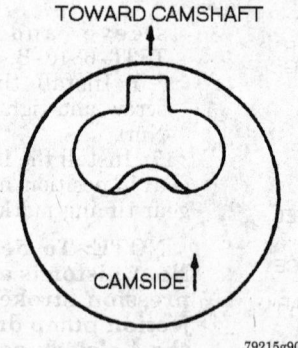

Piston orientation — 7.3L (VIN C, F, K, and M) Diesel engine

ENGINE LUBRICATION

Oil Pan

REMOVAL AND INSTALLATION

4.2L (VIN 2) Engine

1. Raise and safely support the front of the vehicle on jackstands.

2. Remove the oil pan plug and drain the engine oil.

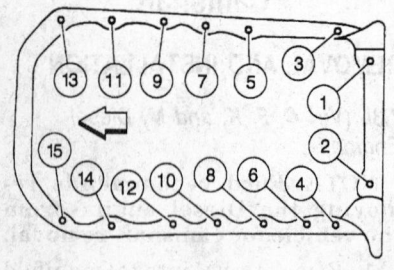

Tighten the oil pan-to-engine block mounting bolts in the sequence shown — 4.2L (VIN 2) engines

3. Remove the 2 front wheel driveshafts and joints, if so equipped.

4. Remove the front differential from the vehicle.

5. Remove the front differential support.

6. Remove the 3 oil pan-to-transmission bolts.

7. Remove the 15 oil pan-to-cylinder block mounting bolts, then lower the oil pan.

To install:

NOTE: If the oil pan is not installed within 15 minutes, remove the sealer and reapply.

8. Temporarily install 2 locator dowels in 2 of the oil pan-to-engine block corner mounting bolt holes.

9. Clean and apply sealant to the rear main bearing cap, the oil pan mating surface, the front cover mounting area, the front cover-to-engine block joints, then install the oil pan rear seal. Make certain to use Silicone Gasket and Sealant F6AZ-19562-A or equivalent for the sealant.

10. Position the oil pan, then install 13 of the oil pan mounting bolts loosely.

11. Remove the 2 locator dowels and install the remaining 2 oil pan-to-engine block bolts.

12. Tighten the 15 oil pan mounting bolts, in the sequence shown, to

36–44 inch lbs. (4–5 Nm), then retighten the bolts to 80–106 inch lbs. (9–12 Nm).

13. Install the oil pan-to-transmission bolts to 28–38 ft. lbs. (38–51 Nm).

14. Install the oil pan drain plug to 16–22 ft. lbs. (22–30 Nm).

15. Install the front differential and front differential support.

16. Install the front driveshafts and joints.

17. Lower the vehicle to the ground.

18. Fill the engine with the correct type and amount of engine oil.

4.6L Engines (VIN W and VIN 6)

1. Raise and safely support the vehicle on jackstands.

2. Remove the front axle housing from the vehicle.

3. Drain the engine oil.

4. Remove the 16 oil pan-to-engine block bolts.

5. Remove the oil pan and old oil pan gasket.

To install:

6. Clean the oil pan and engine block mating surfaces of oil and old gasket material.

7. Install the new oil pan gasket and the oil pan, then install the 16 oil pan-to-engine block bolts loosely.

NOTE: Make sure to tighten the oil pan bolts in 3 steps.

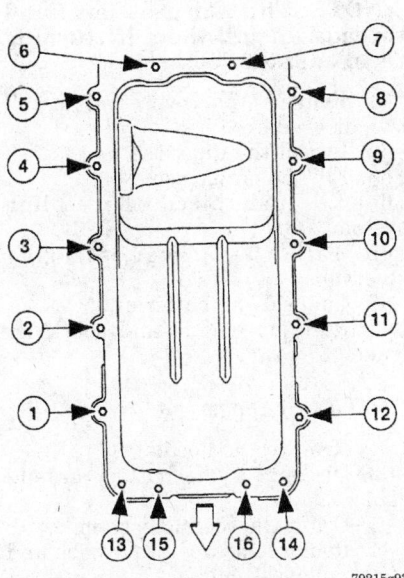

79215g93

Tighten the oil pan-to-engine block bolts in three steps following the sequence shown — 4.6L engines (VIN W and VIN 6)

8. Tighten the oil pan-to-engine bolts in the sequence shown, in the following 3 steps:

   a. Step 1 — 18 inch lbs. (2 Nm).

   b. Step 2 — 15 ft. lbs. (20 Nm).

   c. Step 3 — tighten an additional 60 degrees (one hex head flat of a bolt).

9. Install the oil drain plug.

10. Install the front axle housing.

11. Lower the vehicle.

12. Fill the engine with the correct amount and type of engine oil.

4.9L (VIN Y and Z) Engine

1. Drain the crankcase.

2. Drain the cooling system.

3. Remove the upper intake manifold and throttle body.

4. Remove the starter.

5. Remove the engine front support insulator to support bracket nuts and washers on both supports. Raise the front of the engine with a transmission jack and wood block and place 1 in. (25mm) thick wood blocks between the front support insulators and support brackets. Lower the engine and remove the transmission jack.

6. Remove the oil pan attaching bolts and lower the pan to the crossmember. Remove the 2 oil pump inlet tube and screw assembly bolts and

drop the assembly in the pan. Remove the oil pan. Remove the oil pump inlet tube attaching bolts. Remove the inlet tube and screen assembly from the oil pump and leave it in the bottom of the oil pan. Remove the oil pan gaskets. Remove the inlet tube and screen from the oil pan.

To install:

7. Clean the gasket surfaces of the oil pump, oil pan and cylinder block. Remove the rear main bearing cap to oil pan seal and cylinder front cover to oil pan seal. Clean the seal grooves.

8. Apply oil-resistant sealer in the cavities between the bearing cap and cylinder block. Install a new seal in the rear main bearing cap and apply a bead of oil-resistant sealer to the tapered ends of the seal.

9. Install new side gaskets on the oil pan with oil-resistant sealer. Position a new oil pan to cylinder front cover seal on the oil pan.

10. Clean the inlet tube and screen assembly and place it in the oil pan.

11. Position the oil pan under the engine. Install the inlet tube and screen assembly on the oil pump with a new gasket. Tighten the screws to 5–7 ft. lbs. (6.8–9.5 Nm). Position the oil pan against the cylinder block and install the attaching bolts. Tighten the bolts in sequence to 10–12 ft. lbs. (14–16 Nm).

12. Raise the engine with a transmission jack and remove the wood blocks from the engine front supports. Lower the engine until the front support insulators are positioned on the support brackets. Install the washers and nuts on the insulator studs and tighten the nuts.

13. Install the starter and connect the starter cable.

14. Install the manifold and throttle body.

15. Fill the crankcase and cooling system.

16. Start the engine and check for coolant and oil leaks.

5.0L (VIN N) and 5.8L (VIN H and R) Engines

1. Drain the cooling system.

2. Remove the bolts attaching the fan shroud to the radiator and position the shroud over the fan.

3. Remove the upper intake manifold and throttle body.

4. Remove the nuts and lockwashers attaching the engine support insulators to the chassis bracket.

5. If equipped with an automatic transmission, disconnect the oil

cooler line at the left side of the radiator.

6. Remove the exhaust system.

7. Raise the engine and place wood blocks under the engine supports.

8. Drain the crankcase.

9. Support the transmission with a floor jack and remove the transmission crossmember.

10. Remove the oil pan attaching bolts and lower the oil pan onto the crossmember.

11. Remove the 2 bolts attaching the oil pump pick-up tube to the oil pump. Remove nut attaching oil pump pick-up tube to the number 3 main bearing cap stud. Lower the pick-up tube and screen into the oil pan.

12. Remove the oil pan from the vehicle.

To install:

13. Clean the oil pan, inlet tube and gasket surfaces. Inspect the gasket sealing surface for damages and distortion due to overtightening of the bolts. Repair and straighten as required.

14. Position a new oil pan gasket and seal to the cylinder block.

15. Position the oil pick-up tube and screen to the oil pump, and install the lower attaching bolt and gasket loosely. Install nut attaching to number 3 main bearing cap stud.

16. Place the oil pan on the crossmember. Install the upper pick-up tube bolt. Tighten the pick-up tube bolts.

17. Position the oil pan to the cylinder block and install the attaching bolts. Tighten to 10–12 ft. lbs. (14–16 Nm).

18. Install the transmission crossmember.

19. Raise the engine and remove the blocks under the engine supports. Bolt the engine to the supports.

20. Install the exhaust system.

21. If equipped with an automatic transmission, connect the oil cooler line at the left side of the radiator.

22. Install the nuts and lockwashers attaching the engine support insulators to the chassis bracket.

23. Install the upper intake manifold and throttle body.

24. Install the fan shroud.

25. Fill the crankcase.

26. Fill and bleed the cooling system.

7.3L (VIN C, F, K, and M) Diesel Engine

1. Disconnect both battery ground cables.

2. Remove the oil dipstick.

3. Remove the transmission oil dipstick.

4. Remove the air cleaner and cover the intake opening.

5. Remove the fan and fan clutch.

NOTE: The fan uses left hand threads. Remove them by turning them clockwise.

6. Drain the cooling system.

7. Disconnect the lower radiator hose.

8. Disconnect the power steering return hose and plug the line and pump.

9. Disconnect the alternator wiring harness.

10. Disconnect the fuel line heater connector from the alternator.

11. Raise and support the front end on jackstands.

12. If equipped with automatic transmission, disconnect the transmission cooler lines at the radiator and plug them.

13. Disconnect and plug the fuel pump inlet line.

14. Drain the crankcase and remove the oil filter.

15. Remove the oil filler tube.

16. Disconnect the exhaust pipes at the manifolds.

17. Disconnect the muffler inlet pipe from the muffler and remove the pipe.

18. Remove the upper inlet mounting stud from the right exhaust manifold.

19. Unbolt the engine from the No. 1 crossmember.

20. Lower the vehicle.

21. Install lifting brackets on the front of the engine.

22. Raise the engine until the transmission contact the body.

23. Install wood blocks — 2³/₄ in. (70mm) on the left side; 2 in. (50mm) on the right side — between the engine insulators and crossmember.

24. Lower the engine onto the blocks.

25. Raise and support the front end on jackstands.

26. Remove the flywheel inspection plate.

27. Position fuel pump inlet line No. 1 rearward of the crossmember and position the oil cooler lines out of the way.

28. Remove the oil pan bolts.

29. Lower the oil pan.

NOTE: The oil pan is sealed to the crankcase with RTV silicone sealant in place of a gasket. It may be necessary to separate the pan from the crankcase with a utility knife.

NOTE: The crankshaft may have to be turned to allow the pan to clear the crankshaft throws.

30. Clean the pan and crankcase mating surfaces thoroughly.

To install:

31. Apply a ¹/₈ in. (3mm) bead of RTV silicone sealant to the pan mating surfaces, and a ¹/₄ in. (6mm) bead on the front and rear covers and in the corners; you have 15 minutes within which to install the pan!

32. Install locating dowels into position.

33. Position the pan on the engine and install the pan bolts loosely.

34. Remove the dowels.

35. Torque the pan bolts to 7 ft. lbs. (9.5 Nm) for ¹/₄ in.-20 bolts; 14 ft. lbs. (19 Nm) for ⁵/₁₆ in.-18 bolts; 24 ft. lbs. (33 Nm) for ³/₈ in.-16 bolts.

36. Install the flywheel inspection cover.

37. Lower the truss.

38. Raise the engine and remove the wood blocks.

39. Lower the engine onto the crossmember and remove the lifting brackets.

40. Raise and support the front end on jackstands.

41. Torque the engine-to-crossmember nuts to 70 ft. lbs. (95 Nm).

42. Install the upper inlet pipe mounting stud.

43. Install the inlet pipe, using a new gasket.

44. Install the transmission oil filler tube, using a new gasket.

45. Install the oil pan drain plug.

46. Install a new oil filter.

47. Connect the fuel pump inlet line. Make sure the clip is installed on the crossmember.

48. Connect the transmission cooler lines.

49. Lower the vehicle.

50. Connect all wiring.

51. Connect the power steering return line.

52. Connect the lower radiator hose.

53. Install the fan and fan clutch.

NOTE: The fan uses left hand threads. Install them by turning them counterclockwise.

54. Remove the cover and install the air cleaner.

55. Install the dipsticks.

56. Fill the crankcase.

57. Fill and bleed the cooling system.

58. Fill the power steering reservoir.

59. Connect the batteries.

60. Run the engine and check for leaks.

NOTE: The fan uses left hand threads. Install them by turning them counterclockwise.

61. Remove the cover and install the air cleaner.

62. Install the dipsticks.

63. Fill the crankcase.

64. Fill and bleed the cooling system.

65. Fill the power steering reservoir.

66. Connect the batteries.

67. Run the engine and check for leaks.

7.5L (VIN G) Engine

1. Remove the hood.

2. Disconnect the battery ground cable.

3. Drain the cooling system.

4. Remove the air intake tube and air cleaner assembly.

5. Disconnect the throttle linkage at the throttle body.

6. Disconnect the power brake vacuum line at the manifold.

7. Disconnect the fuel lines at the fuel rail.

8. Disconnect the air tubes at the throttle body.

9. Remove the radiator.

10. Remove the power steering pump and position it out of the way without disconnecting the lines.

11. Remove the oil dipstick tube. On motor home chassis, remove the oil filler tube.

12. Remove the front engine mount through-bolts.

13. Position the air conditioner refrigerant hoses so they are clear of the firewall. If necessary, discharge the system and remove the compressor.

14. Remove the upper intake manifold and throttle body.

15. Drain the crankcase. Remove the oil filter.

16. Disconnect the exhaust pipe at the manifolds.

17. Disconnect the transmission linkage at the transmission.

18. Remove the driveshaft(s).

19. Remove the transmission fill tube.

20. Raise the engine with a jack placed under the crankshaft damper and a block of wood to act as a cushion. Raise the engine until the transmission contacts the underside of the floor. Place wood blocks under the engine supports. The engine **must** remain centralized at a point at least 4 in. (102mm) above the mounts, to remove the oil pan!

21. Remove the oil pan attaching screws and lower the oil pan onto the crossmember. Remove the 2 bolts attaching the oil pump pick-up tube to the oil pump. Lower the assembly from the oil pump. Leave it on the bottom of the oil pan. Remove the oil pan and gaskets. Remove the inlet tube and screen from the oil pan.

To install:

22. Clean the gasket surfaces of the oil pan and cylinder block.

23. Apply a coating of gasket adhesive on the block mating surface and stick the 1-piece silicone gasket on the block.

24. Clean the inlet tube and screen assembly and place on the pump.

25. Position the oil pan against the cylinder block and install the retaining bolts. Torque all bolts to 10 ft. lbs. (14 Nm).

26. Lower the engine and bolt it in place.

27. Install the transmission fill tube.

28. Install the driveshaft(s).

29. Connect the transmission linkage at the transmission.

30. Connect the exhaust pipe at the manifolds.

31. Install the oil filter.

32. Install the upper intake manifold and throttle body.

33. Install the compressor or reposition the hoses.

34. Install the oil dipstick tube. On motor home chassis, install the oil filler tube.

35. Install the power steering pump.

36. Install the radiator.

37. Connect the air tubes at the throttle body.

38. Connect the fuel lines at the fuel rail.

39. Connect the power brake vacuum line at the manifold.

40. Connect the throttle linkage at the throttle body.

41. Install the air intake tube and air cleaner assembly.

42. Fill and bleed the cooling system.

43. Fill the crankcase.

44. Connect the battery ground cable.

45. Install the hood.

Oil Pump

REMOVAL AND INSTALLATION

4.2L (VIN 2) Engine

1. Raise and support the front of the vehicle safely on jackstands.

2. Drain the engine oil and dispose.

3. Remove the oil filter.

4. Remove the 6 oil pump bolts, then remove the oil pump drive gear, the oil pump driven gear, the oil pump O-ring and the oil pump itself. Discard the used oil pump O-ring.

5. Inspect the oil pump components for damage or excessive wear.

6. Check the oil pump face for warpage with a flat edge ruler. The face cannot exhibit more than 0.00157 in. (0.04mm) of distortion.

7. Remove the plug over the oil pressure relief; valve.

8. Remove the oil pressure relief valve ball and spring, then clean the parts.

To install:

NOTE: Lubricate the parts with clean engine oil before assembly.

9. Assemble the oil pressure relief valve ball and spring with a new plug.

10. Install the oil pump, along with a new O-ring, the oil pump driven gear and the drive gear. Install and tighten the 6 oil pump mounting bolts to the torque value specifications indicated in the illustration.

11. Apply a film of clean engine oil to the rubber O-ring on the new filter, then install the filter onto the filter mount.

12. Install the oil pan drain plug.

13. Lower the vehicle.

14. Fill the engine with the correct amount and type of new engine oil.

15. Start the engine and make certain that the oil light on the instrument panel extinguishes within 6–8 seconds after the engine starts.

4.6L Engines (VIN W and VIN 6)

1. Remove the timing chain.

2. Remove the oil pan.

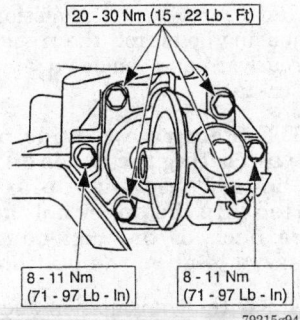

Tighten the oil pump mounting bolts to the specifications shown — 4.2L (VIN 2) engines

3. Remove the 3 oil pump screen and cover bolts, then remove the screen and cover.

4. Remove the oil pump screen and cover spacer.

5. Remove the 4 oil pump mounting bolts, then remove the oil pump from the engine.

To install:

6. Clean and inspect the mating surfaces.

7. Install the oil pump and loosely install the 4 oil pump mounting bolts. Tighten the 4 oil pump bolts in the sequence shown to 71–107 inch lbs. (8–12 Nm).

8. Install the oil pump screen and cover spacer to 15–22 ft. lbs. (20–30 Nm).

9. Install the oil pump screen and cover, then install the 3 oil pump screen and cover bolts. Tighten the bolts near the oil pick-up screen to 15–22 ft. lbs. (20–30 Nm) and the bolts at the opposite end of the pick-up to 70–106 inch lbs. (8–12 Nm).

10. Install the timing chains, then install the oil pan.

7.3L (VIN C, F, K, and M) Diesel Engine

1. Remove the oil pan.

2. Remove the oil pick-up tube from the pump.

3. Unbolt and remove the oil pump.

To install:

4. Assemble the pick-up tube and pump. Use a new gasket.

5. Install the oil pump and torque the bolts to 14 ft. lbs. (19 Nm).

4.9L (VIN Y and Z), 5.0L (VIN N), 5.8L (VIN H and R) and 7.5L (VIN G) Engines

1. Remove the oil pan.

2. Remove the oil pump inlet tube and screen assembly.

3. Remove the oil pump attaching bolts and remove the oil pump gasket and intermediate driveshaft.

To install:

4. Before installing the oil pump, prime it by filling the inlet and outlet port with oil and rotating the shaft of the pump to distribute it.

5. Position the intermediate driveshaft into the distributor socket.

6. Position the new gasket on the pump body and insert the intermediate driveshaft into the pump body.

7. Install the pump and intermediate driveshaft as an assembly. Do not force the pump if it does not seal readily. The driveshaft may be misaligned with the distributor shaft. To

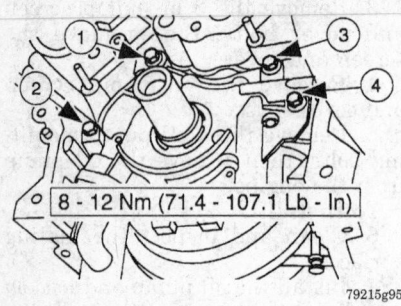

8 - 12 Nm (71.4 - 107.1 Lb - In)

79215g95

Tighten the mounting bolts to 71–107 inch lbs. (8–12 Nm) in the sequence shown — 4.6L engines (VIN W and VIN 6)

align it, rotate the intermediate driveshaft into a new position.

8. Install the oil pump attaching bolts and torque them to 12–15 ft. lbs. (16–20 Nm) on the inline sixes and to 20–25 ft. lbs. (27–34 Nm) on the V8s.

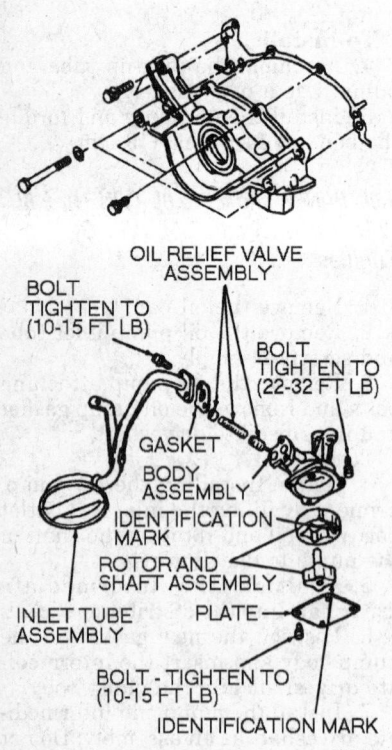

OIL RELIEF VALVE ASSEMBLY

BOLT TIGHTEN TO (10-15 FT LB)

BOLT TIGHTEN TO (22-32 FT LB)

GASKET

BODY ASSEMBLY

IDENTIFICATION MARK

ROTOR AND SHAFT ASSEMBLY

INLET TUBE ASSEMBLY

PLATE

BOLT - TIGHTEN TO (10-15 FT LB)

IDENTIFICATION MARK

79215g96

Oil pump assembly — 5.0L (VIN N) and 5.8L (VIN H and R) engines

TRANSMISSION

Manual Transmission Assembly

REMOVAL AND INSTALLATION

—— CAUTION ——

The clutch driven disc contains asbestos, which has been determined to be a cancer causing agent. Never clean clutch surfaces with compressed air! Avoid inhaling any dust from any clutch surface! When cleaning clutch surfaces, use a commercially available brake cleaning fluid.

NOTE: When the battery is disconnected and reconnected, some abnormal drive symptoms may occur while the vehicle relearns its adaptive strategy. The vehicle may need to be driven 10 miles (16 km) or more to relearn the strategy.

Borg-Warner T-18

2-Wheel Drive Models

1. Remove the rubber boot, floor mat, and the body floor pan cover.
2. Remove the gearshift lever, shift ball and boot as an assembly.
3. Raise the vehicle and support it with jackstands.
4. Drain the transmission.
5. Remove the driveshaft.
6. Disconnect the speedometer cable.
7. Remove the crossmember-to-transmission bolts.
8. Secure the transmission to a transmission jack.
9. Remove the crossmember.
10. Remove the 4 transmission-to-bell housing bolts, roll the transmission rearward and remove it.
   To install:

NOTE: Fabricate 2 guide pins, made by cutting the heads off of 2 long bolts. These guide pins are inserted into the upper bolt holes in the back of the bell housing and serve to align the bolt holes.

11. Raise the transmission and roll it forward using the guide pins to align the bolt holes. Turn the output shaft by hand to align the input shaft splines with the clutch.

12. Install the 2 lower bolts and snug them down.
13. Remove the guide pins and install the 2 remaining bolts. Torque all 4 bolts to 50 ft. lbs. (68 Nm).
14. Install the crossmember. Crossmember-to-frame bolt torque is 55 ft. lbs. (75 Nm).
15. Remove the transmission jack.
16. Install the Crossmember-to-transmission bolts. Torque the bolts to 50 ft. lbs. (68 Nm).
17. Connect the speedometer cable.
18. Install the driveshaft.
19. Install the drain plug and fill the transmission. Drain plug torque is 50 ft. lbs. (68 Nm).
20. Lower the vehicle.
21. Install the gearshift lever, shift ball and boot as an assembly.
22. Install the body floor pan cover, floor mat, and the rubber boot.

4-Wheel Drive Models

1. Remove the rubber boot, floor mat, and the body floor pan cover. Remove the gearshift lever. Remove the weather pad.
2. Remove the transfer case shift lever, shift ball and boot as an assembly.
3. Disconnect the back-up light switch at the rear of the gearshift housing cover.
4. Raise the vehicle and support it with jackstands. Remove the drain and fill plugs and drain the lubricant.
5. Position a transmission jack under the transfer case and disconnect the speedometer cable.
6. Matchmark the flanges and disconnect the rear driveshaft from the transfer case. Wire it up and out of the way.
7. Matchmark and disconnect the front driveshaft at the transfer case. Wire it up and out of the way.
8. Remove the shift link from the transfer case.
9. Unbolt the transfer case from the transmission (6 bolts) and lower the transfer case from the vehicle.
10. Position the transmission jack under the transmission.
11. Remove the 8 rear transmission support-to-transmission bolts.
12. Remove the rear transmission support.
13. Remove the 4 transmission-to-clutch housing attaching bolts.
14. Move the transmission to the rear until the input shaft clears the flywheel housing and lower the transmission.

To install:

NOTE: Fabricate 2 guide pins, made by cutting the heads off of 2 long bolts. These guide pins are inserted into the upper bolt holes in the back of the bell housing and serve to align the bolt holes.

15. Before installing the transmission, apply a light film of grease to the inner hub surface of the clutch release bearing, the release lever fulcrum and the front bearing retainer of the transmission. Do not apply excessive grease because it will fly off onto the clutch disc.

16. Install the transmission in the reverse order of removal. It may be necessary to turn the output shaft with the transmission in gear to align the input shaft splines with the splines in the clutch disc. Fill the transmission with SAE 80W/90 lubricant if it was drained. The transfer case is filled with Dexron®II ATF. Observe the following torque specifications:

- Back-up light switch — 25 ft. lbs. (34 Nm)
- Transmission-to-clutch housing bolts — 65 ft. lbs. (88 Nm)
- Transfer case-to-transmission — 40 ft. lbs. (54 Nm)
- Drain plug — 50 ft. lbs. (68 Nm)
- Fill plug — 50 ft. lbs. (68 Nm)
- Transmission to rear support — 80 ft. lbs. (109 Nm)
- Rear support-to-frame — 55 ft. lbs. (75 Nm)

M5OD 5-Speed

Except 1997 F-150 and Expedition Models

1. Raise and support the vehicle safely. Prop the clutch pedal in the full up position with a block of wood.
2. Matchmark the driveshaft-to-flange relation.
3. Disconnect the driveshaft at the rear axle and slide it off of the transmission output shaft. Lubricant will leak out of the transmission so be prepared to catch it, or plug the opening with rags or a seal installation tool.
4. Disconnect the speedometer cable at the transmission.
5. Disconnect the shift rods from the shift levers.
6. Remove the shift control from the extension housing and transmission case.
7. On 4-wheel drive models, remove the transfer case.
8. Remove the extension housing-to-rear support bolts.
9. Take up the weight of the transmission with a transmission jack. Chain the transmission to the jack.

10. Raise the transmission just enough to take the weight off of the No. 3 crossmember.
11. Unbolt the crossmember from the frame rails and remove it.
12. Place a jackstand under the rear of the engine at the bell housing.
13. Lower the jack and allow the jackstand to take the weight of the engine. The engine should be angled slightly downward to allow the transmission to roll backward.
14. Remove the transmission-to bell housing bolts.
15. Roll the jack rearward until the input shaft clears the bell housing. Lower the jack and remove the transmission.

——————— WARNING ———————
Do not depress the clutch pedal with the transmission removed.

To install:
16. Clean all machined mating surfaces thoroughly.
17. Install a guide pin in each lower bolt hole. Position the spacer plate on the guide pins.
18. Raise the transmission and start the input shaft through the clutch release bearing.
19. Align the input shaft splines with the clutch disc splines. Roll the transmission forward so the input shaft will enter the clutch disc. If the shaft binds in the release bearing, work the release arm back and forth.
20. Once the transmission is all the way in, install the 2 upper retaining bolts and washers and remove the lower guide pins. Install the lower bolts. Torque the bolts to 50 ft. lbs. (68 Nm).
21. Raise the transmission just enough to allow installation of the No.3 crossmember.
22. Install the crossmember on the frame rails. Torque the bolts to 80 ft. lbs. (109 Nm).
23. Lower the transmission onto the crossmember and install the nuts. Torque the nuts to 70 ft. lbs. (95 Nm).
24. Remove the transmission jack.
25. Install the transfer case.
26. Install the shift control on the extension housing and transmission case.
27. Connect the shift rods at the shift levers.
28. Connect the speedometer cable at the transmission.
29. Slide the driveshaft onto the output shaft and connect the driveshaft at the rear axle, aligning the matchmarks.

1997 F-150 and Expedition Models

1. Disconnect the negative battery cable.
2. Place the transmission in Neutral.
3. Remove the gearshift lever, the outer gearshift lever boot and the inner gearshift lever boot.
4. Detach the transmission harness electrical connector by removing the retaining bolt and separating the two halves.
5. Raise and safely support the vehicle on jackstands.
6. Drain the transmission fluid.
7. Matchmark the rear driveshaft to the differential mounting flange and, on 4-wheel drive versions, to the transmission housing, then remove the 4 attaching bolts. Remove the driveshaft from the vehicle.
8. On 4-wheel drive models, remove the transfer case.
9. Use Clutch Coupling Tool T88T-70522-A, or equivalent, to disconnect the clutch hydraulic line.
10. Remove the starter motor.
11. Detach the 2 heated oxygen sensor electrical connectors.
12. Position a transmission jack, or equivalent, under the transmission. Secure the transmission to the jack with a strap.
13. Remove the 2 transmission mount nuts.
14. Remove the 2 heat shield mounting bolts (one on each side).
15. Remove the 6 rear crossmember mounting bolts, then remove the crossmember from the vehicle.
16. Remove the rear transmission mount from the vehicle frame.
17. Remove the exhaust pipe bracket.
18. Remove the fuel line bracket.
19. On 4.2L (VIN 2) engines, remove the 3 oil pan-to-transmission bolts.
20. Remove the 6 engine-to-transmission bolts.
21. Remove the 4 exhaust pipe-to-exhaust manifold nuts (two on each side) and position the exhaust pipe aside.
22. Slowly and carefully lower the transmission from the vehicle.

To install:
23. Raise the transmission slowly and carefully up into the vehicle.
24. Position the exhaust pipe onto the exhaust manifold, then install the 4 mounting nuts to 25–34 ft. lbs. (34–46 Nm).
25. Install the 6 engine-to-transmission bolts and tighten to 30–41 ft. lbs. (40–55 Nm).

26. On 4.2L (VIN 2) engines, install the 3 oil pan-to-transmission bolts to 28–38 ft. lbs. (38–51 Nm).

27. Install the fuel line bracket and tighten the mounting nut to 13–17 ft. lbs. (17–23 Nm).

28. Install the exhaust bracket and tighten the 2 mounting nuts to 25–33 ft. lbs. (34–46 Nm).

29. Install the transmission mount and tighten the 2 mounting bolts to 64–81 ft. lbs. (87–110 Nm).

30. Install the rear crossmember into the vehicle. Tighten the 6 crossmember mounting bolts to 39–53 ft. lbs. (53–72 Nm).

31. Install the 2 heat shield bolts, then install the 2 transmission-to-transmission mount nuts to 64–81 ft. lbs. (87–110 Nm).

32. Lower the transmission jack slowly.

33. Attach the 2 heated oxygen sensor electrical connectors, then install the starter motor.

34. For vehicles equipped with 4-wheel drive, install the transfer case.

35. Install the rear driveshaft. Mount the driveshaft to the differential flange so that the matchmarks align, then tighten the 4 mounting bolts to 65–87 ft. lbs. (88–119 Nm).

36. Connect the hydraulic clutch line. The clutch system must be bled upon completion of transmission installation.

37. Fill the transmission with transmission fluid until it reaches the bottom of the fill port and install the case plug. Make sure to use MERCON Multi-Purpose Automatic Transmission Fluid XT-2-QDX or DDX, or equivalent.

38. Lower the vehicle.

39. Install the shift lever.

40. Attach the transmission harness electrical connector.

41. Attach the negative battery cable.

ZF S5-42 5-Speed

1. Place the transmission in neutral.

2. Remove the carpet or floor mat.

3. Remove the ball from the shift lever.

4. Remove the boot and bezel assembly from the floor.

5. Remove the 2 bolts and disengage the upper shift lever from the lower shift lever.

6. Raise and support the vehicle safely.

7. Disconnect the speedometer cable.

8. Disconnect the back-up switch wire.

9. Place a drain pan under the case and drain the case through the drain plug.

10. Position a transmission jack under the case and safety-chain the case to the jack.

11. Remove the driveshaft.

12. Disconnect the clutch linkage.

13. On 4-wheel drive models, remove the transfer case.

14. Remove the transmission rear insulator and lower retainer.

15. Unbolt and remove the crossmember.

16. Remove the transmission-to-engine block bolts.

17. Roll the transmission rearward until the input shaft clears, lower the jack and remove the transmission.

To install:

18. Install 2 guide studs into the lower bolt holes.

19. Raise the transmission until the input shaft splines are aligned with the clutch disc splines. The clutch release bearing and hub must be properly positioned in the release lever fork.

20. Roll the transmission forward and into position on the front case.

21. Install the bolts and torque them to 50 ft. lbs. (68 Nm). Remove the guide studs and install and tighten the 2 remaining bolts.

22. Install the crossmember and torque the bolts to 55 ft. lbs. (75 Nm).

23. Install the transmission rear insulator and lower retainer. Torque the bolts to 60 ft. lbs. (81 Nm).

24. On 4-wheel drive models, install the transfer case.

25. Connect the clutch linkage.

26. Install the driveshaft.

27. Remove the transmission jack.

28. Fill the transmission.

29. Connect the back-up switch wire.

30. Connect the speedometer cable.

31. Lower the van.

32. Install the boot and bezel assembly.

33. Connect the upper shift to from the lower shift lever. Tighten the bolts to 20 ft. lbs. (27 Nm).

34. Install the carpet or floor mat.

35. Install the ball from the shift lever.

Clutch Assembly

REMOVAL AND INSTALLATION

—————— CAUTION ——————

The clutch driven disc contains asbestos, which has been determined to be a cancer causing agent. Never clean clutch surfaces with compressed air! Avoid inhaling any dust from any clutch surface! When cleaning clutch surfaces, use a commercially available brake cleaning fluid.

Except 1997 F-150 and Expedition Models

1. Raise and support the vehicle safely.

2. If equipped with an externally mounted slave cylinder, remove the clutch slave cylinder. If equipped with an internally mounted slave cylinder, disconnect the quick-disconnect coupling with a spring coupling tool such as T88T-70522-A.

3. Remove the transmission.

4. If equipped with an internally mounted slave cylinder, remove the starter. Remove the flywheel housing attaching bolts and remove the housing. If equipped with tan externally mounted slave cylinder, remove the cover and then remove the release lever and bearing from the clutch housing. To remove the release lever:

  a. Remove the dust boot.

  b. Push the release lever forward to compress the slave cylinder.

  c. Remove the slave cylinder by prying on the steel clip to free the tangs while pulling the cylinder clear.

  d. Remove the release lever by pulling it outward.

5. Mark the pressure plate and cover assembly and the flywheel so they can be reinstalled in the same relative position.

6. Loosen the pressure plate and cover attaching bolts evenly in a staggered sequence a turn at time until the pressure plate springs are relieved of their tension. Remove the attaching bolts.

7. Remove the pressure plate and cover assembly and the clutch disc from the flywheel.

To install:

8. Position the clutch disc on the flywheel so an aligning tool can enter the clutch pilot bearing and align the disc.

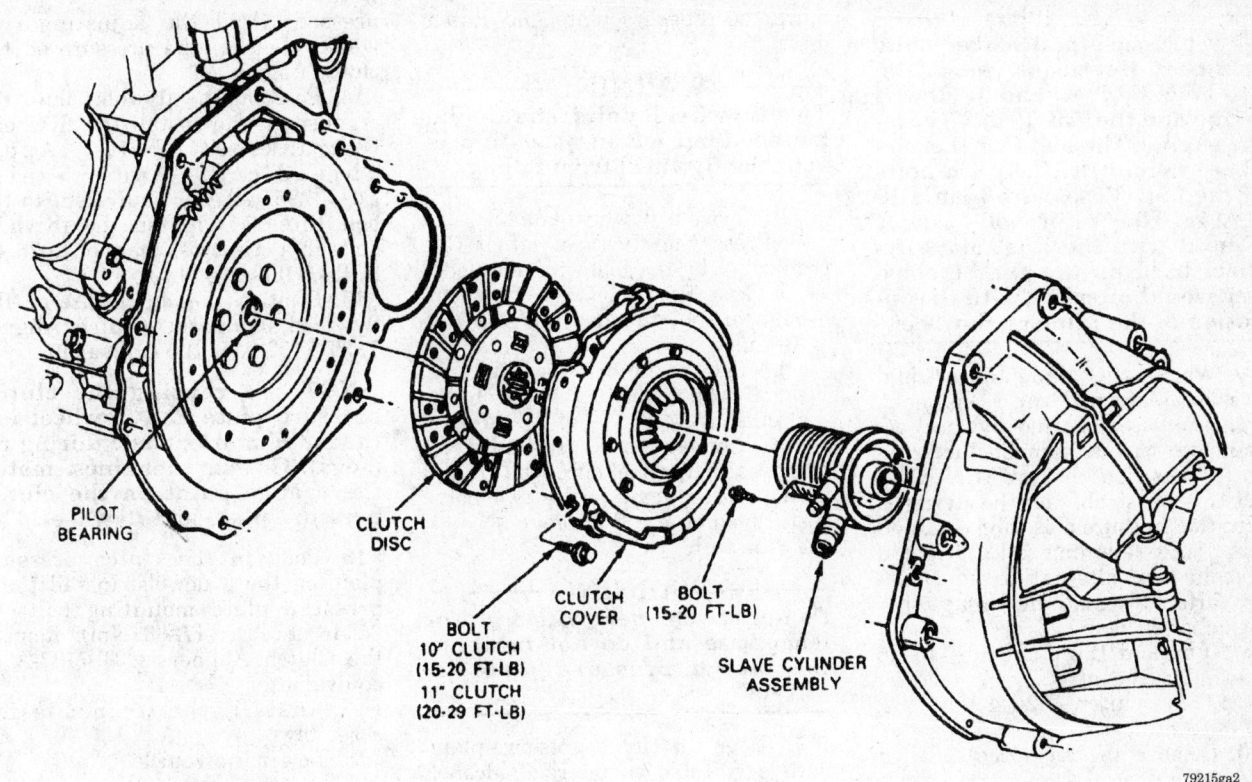

PILOT BEARING

CLUTCH DISC

CLUTCH COVER

**BOLT
10" CLUTCH
(15-20 FT-LB)
11" CLUTCH
(20-29 FT-LB)**

**BOLT
(15-20 FT-LB)**

**SLAVE CYLINDER
ASSEMBLY**

79215ga2

Clutch installation — Bronco, F Series and E Series with 4.9L (VIN Y and Z), 5.0L (VIN N) and 5.8L (VIN H and R) engines

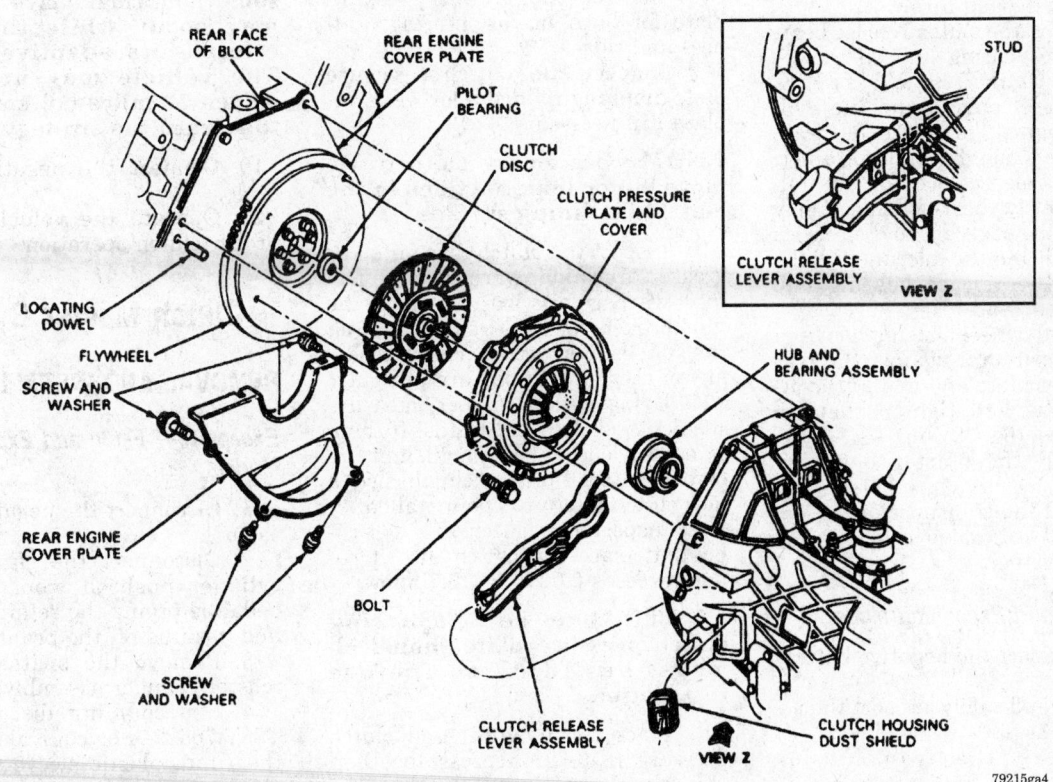

**REAR FACE
OF BLOCK**

**REAR ENGINE
COVER PLATE**

**PILOT
BEARING**

**CLUTCH
DISC**

**CLUTCH PRESSURE
PLATE AND
COVER**

STUD

**CLUTCH RELEASE
LEVER ASSEMBLY**

VIEW Z

**LOCATING
DOWEL**

FLYWHEEL

**SCREW AND
WASHER**

**REAR ENGINE
COVER PLATE**

**SCREW
AND WASHER**

BOLT

**HUB AND
BEARING ASSEMBLY**

**CLUTCH RELEASE
LEVER ASSEMBLY**

VIEW Z

**CLUTCH HOUSING
DUST SHIELD**

79215ga4

Clutch installation — F Series with 7.3L (VIN C, F, K, and M) Diesel and 7.5L (VIN G) engines

——————— **WARNING** ———————

New pressure plate/cover bolts have been issued for use on the 7.3L (VIN C, F, K, and M) Diesel engine and the 7.5L (VIN G) gasoline engine. The bolts for the Diesel are $5/16$ in. x 18 x $3/4$ in. The bolts for the 7.5L (VIN G) are $5/16$ in. x 18 x $59/64$ in. The $59/64$ in. bolts cannot be used with the dual mass flywheel used on the Diesel, since they would interfere with the operation of the primary flywheel.

9. When reinstalling the original pressure plate and cover assembly, align the assembly and flywheel according to the marks made during removal. Position the pressure plate and cover assembly on the flywheel, align the pressure plate and disc, and install the retaining bolts. Tighten the bolts in an alternating sequence a few turns at a time until the proper torque is reached:
- 10 in. and 12 in. clutch — 15–20 ft. lbs. (20–27 Nm)
- 11 in. clutch — 20–29 ft. lbs. (27–39 Nm)

10. Remove the tool used to align the clutch disc.
11. With the clutch fully released, apply a light coat of grease on the sides of the driving lugs.
12. Position the clutch release bearing and the bearing hub on the release lever. On the 7.3L (VIN C, F, K, and M) Diesel engine and the 7.5L (VIN G) engine, clean and lubricate the transmission bearing retainer. Install the release lever on the fulcrum in the flywheel housing. Apply a light coating of grease to the release lever fingers and the fulcrum. Fill the groove of the release bearing hub with grease.
13. If the flywheel housing has been removed, position it against the rear engine cover plate and install the attaching bolts and tighten them to 40–50 ft. lbs. (54–68 Nm).
14. Install the starter motor, if removed.
15. Install the transmission.
16. Install the salve cylinder and bleed the system.

1997 F-150 and Expedition Models

1. Disconnect the negative battery cable.
2. Raise and safely support the vehicle on jackstands.
3. Remove the transmission assembly.
4. If the clutch parts are to be reused, matchmark the clutch pressure plate and flywheel. Remove the 6 clutch pressure plate bolts, then re-

move the pressure plate and clutch disc.

——————— **WARNING** ———————

Two flywheel bolts should be loosened but left in place to prevent the flywheel from falling.

5. Remove 4 flywheel bolts.
6. Press the flywheel off of crankshaft by the following method:
   a. Install two of the flywheel bolts in the threaded holes in the flywheel.
   b. Tighten the bolts evenly until the flywheel is pressed off of the crankshaft.
   c. Remove the bolts used to press the flywheel off of the crankshaft, then remove the two bolts still holding the flywheel on.

To install:

——————— **WARNING** ———————

Do not use cleaners with a petroleum base and do not immerse the clutch pressure plate in solvent.

7. Clean the clutch pressure plate with a suitable commercial alcohol based solvent so that the surface is free from oil film.
8. Inspect the clutch pressure plate for burn marks, scoring, flatness and ridges.
9. Inspect the clutch pressure plate diaphragm fingers for wear; replace it if necessary.

NOTE: Use emery cloth to remove minor imperfections in the clutch disc lining surface.

10. Inspect the clutch disc for oil or grease saturation, worn or loose facings, warpage and loose rivets at the hub, broken springs, wear or rust on the splines. Replace the clutch disc if any of these conditions are present.
11. Using a slide caliper, measure the depth of the rivet heads. If there is not at least 0.012 in. (0.3mm) of clutch material on the clutch disc, a new clutch disc must be installed.
12. Inspect the clutch disc for runout. Replace the clutch disc if runout exceeds 0.0276 in. (0.70mm).

NOTE: The self-adjusting clutch pressure plate should always be adjusted before installation.

13. Place the flywheel and clutch pressure plate in a press together. Use an adapter and depress the clutch diaphragm fingers until the adjusting ring moves freely. Rotate the adjusting ring counterclockwise until the tension springs are com-

pressed. Hold the adjusting ring while releasing the pressure on the clutch fingers.
14. Position the flywheel onto the crankshaft and install the flywheel-to-crankshaft bolts. Apply Threadlock and Sealer E0AZ-19554-AA or equivalent to the bolt threads. Tighten the flywheel bolts in a crisscross manner to 75–85 ft. lbs. (102–115 Nm).
15. Position the clutch disc on the flywheel using the Clutch Aligner T83T-7137-A or the equivalent.

NOTE: If reusing the clutch pressure plate and flywheel use the matchmarks made during removal. On 4.2 L engines, match the orange paint on the clutch pressure plate and flywheel.

16. Position the clutch pressure plate on the 3 dowels. Install the 6 pressure plate mounting bolts to 35–46 ft. lbs. (47–63 Nm). Remove the Clutch Aligner T83T-7137-A or equivalent.
17. Install the transmission assembly.
18. Lower the vehicle.

NOTE: When the battery is disconnected and reconnected, some abnormal drive symptoms may occur while the vehicle relearns its adaptive strategy. The vehicle may need to be driven 10 miles (16 km) or more to relearn the strategy.

19. Connect the negative battery cable.
20. Operate the vehicle to check proper clutch operation.

Clutch Master Cylinder

REMOVAL AND INSTALLATION

Except 1997 F-150 and Expedition Models

1. Disconnect the negative battery cable.
2. Disconnect the clutch master cylinder pushrod from the clutch pedal by prying the retainer bushing and pushrod off the pedal pin.
3. Remove the switch from the master cylinder assembly.
4. Use coupling disconnect tool T88T-70522-A or equivalent, to slide the white plastic sleeve toward the slave cylinder, then apply a slight tug on the tube to disconnect the hydraulic coupling.
5. Remove the retaining bolts and the clutch master cylinder.

To install:

6. Install the pushrod through the hole in the engine compartment. Make sure it is located on the correct side of the clutch pedal. Install the master cylinder and tighten the bolts to 12 ft. lbs. (16 Nm).

7. Insert the coupling end into the slave cylinder and install the tube into the clips.

8. Replace the retainer bushing in the clutch master cylinder pushrod if worn or damaged. Install the retainer and pushrod on the clutch pedal pin. Make sure the flange of the bushing is against the pedal blade. Install the switch.

9. Connect the negative battery cable and bleed the clutch hydraulic system, if necessary.

1997 F-150 and Expedition Models

NOTE: When replacing the clutch master cylinder, remove the tubing from the vehicle for the bench bleeding procedure.

1. Disconnect the negative battery cable.

2. Disconnect the clutch master cylinder pushrod from the clutch pedal by prying the retainer bushing and pushrod off the pedal pin.

3. Remove the switch from the master cylinder assembly.

4. Use coupling disconnect tool T88T-70522-A or equivalent, to slide the white plastic sleeve toward the slave cylinder, then apply a slight tug on the tube to disconnect the hydraulic coupling.

5. Remove the retaining bolts and the clutch master cylinder.

To install:

6. Lightly clamp the clutch master cylinder reservoir in a vise.

7. Fill the clutch master cylinder reservoir to the full line with Ford High Performance DOT 3 Brake Fluid C6AZ-19542-AA or equivalent.

8. Place the clutch slave cylinder end of the hydraulic system tubing in a drain pan or waste container.

NOTE: Make sure that the clutch master cylinder reservoir remains full during the bleeding process to prevent air from entering the clutch master cylinder. Bleed the clutch master cylinder until a solid stream of brake fluid exits the quick connect coupling end of the tube.

9. Open the internal mechanism of the male quick connect coupling and fully depress and hold the clutch master cylinder pushrod.

10. Release the internal mechanism of the quick connect end and release the clutch master cylinder pushrod.

11. Install the pushrod through the hole in the engine compartment. Make sure it is located on the correct side of the clutch pedal. Install the master cylinder and tighten the bolts to 12 ft. lbs. (16 Nm).

12. Insert the coupling end into the slave cylinder and install the tube into the clips.

13. Replace the retainer bushing in the clutch master cylinder pushrod if worn or damaged. Install the retainer and pushrod on the clutch pedal pin. Make sure the flange of the bushing is against the pedal blade. Install the switch.

14. Connect the negative battery cable and bleed the clutch hydraulic system, if necessary.

Clutch Slave Cylinder

REMOVAL AND INSTALLATION

There are 2 types of slave cylinders used: an internally mounted (in the bell housing) and an externally mounted type. The 5.0L (VIN N), 5.8L (VIN H and R) and 7.5L (VIN G) gasoline engines and 7.3L (VIN C, F, K, and M) Diesel engines, equipped with the M50DHD transmission use the externally mounted type; the 4.2L (VIN 2), 4.6L and 4.9L (VIN Y and Z) engine uses the internally mounted type.

Except 1997 F-150 and Expedition Models

——— **WARNING** ———

Prior to any service on models with the externally mounted slave cylinder, that requires removal of the slave cylinder, such as transmission and/or clutch housing removal, the clutch master cylinder pushrod must be disconnected from the clutch pedal. Failure to do this may damage the slave cylinder if the clutch pedal is depressed while the slave cylinder is disconnected.

1. From inside the cab, remove the cotter pin retaining the clutch master cylinder pushrod to the clutch pedal lever. Disconnect the pushrod and remove the bushing.

2. Remove the 2 nuts retaining the clutch reservoir and master cylinder assembly to the firewall.

3. From the engine compartment, remove the clutch reservoir and master cylinder assembly from the firewall. Note here how the clutch tubing routes to the slave cylinder.

4. Push the release lever forward to compress the slave cylinder.

5. On all models with the internally mounted slave cylinder, remove the plastic clip that retains the slave cylinder to the bracket. Remove the slave cylinder.

6. On models with the externally mounted slave cylinder, the steel retaining clip is permanently attached to the slave cylinder. Remove the slave cylinder by prying on the clip to free the tangs while pulling the cylinder clear.

7. Remove the release lever by pulling it outward.

8. Remove the clutch hydraulic system from the vehicle.

To install:

9. Position the clutch pedal reservoir and master cylinder assembly into the firewall from inside the cab, and install the 2 nuts and tighten.

10. Route the clutch tubing and slave cylinder to the bell housing, taking care that the nylon lines are kept away from any hot exhaust system components.

11. Install the slave cylinder by pushing the slave cylinder pushrod into the cylinder. Engage the pushrod into the release lever and slide the slave cylinder into the bell housing lugs. Seat the cylinder into the recess in the lugs.

NOTE: When installing a new hydraulic system, notice that the slave cylinder contains a shipping strap that propositions the pushrod for installation, and also provides a bearing insert. Following installation of the new slave cylinder, the first actuation of the clutch pedal will break the shipping strap and give normal clutch action.

12. Clean the master cylinder pushrod bearing and apply a light film of SAE 30 engine oil.

13. From inside the cab, install the bushing on the clutch pedal lever. Connect the clutch master cylinder pushrod to the clutch pedal lever and install the cotter pin.

14. Check the clutch reservoir and add fluid if required. Depress the clutch pedal at least 10 times to verify smooth operation and proper clutch release.

1997 F-150 and Expedition Models

1. Remove the transmission.

NOTE: Inspect the clutch housing for traces of fluid. If fluid is found, the clutch slave cylinder should be replaced with a new one.

2. Remove the clutch slave cylinder mounting bolts, then slide the slave cylinder off of the transmission input shaft.
3. Remove the clutch release hub and bearing, if necessary, as follows:
   a. Push the clutch release hub and bearing against the spring.
   b. Remove the retainer ring.
   c. Remove the clutch release hub, bearing and spring.

To install:

4. Clean the slave cylinder.
5. Inspect the slave cylinder for the following:
   • Weak spring
   • Worn or damaged piston
   • Worn or damaged boot
   • Leaking fluid

─── **WARNING** ───

The clutch release hub and bearing is pre-lubricated and should not be cleaned with solvent. The clutch release hub and bearing are replaced as an assembly; do not disassemble for inspection or replacement.

6. Wipe all oil and dirt off of the clutch release hub and bearing.
7. Inspect the clutch release hub and bearing as follows:
   a. Rotate the outer race while applying pressure. If the bearing rotation is rough, replace the clutch release hub and bearing.
   b. Inspect for any surface scoring or burrs that may impede the sliding motion of the clutch release hub and bearing. Any scoring or burrs should be polished off with a fine grade of emery paper.
8. Install the clutch release hub and bearing as follows:
   a. Lubricate the clutch release hub and bearing at the sliding points with Premium Long-Life Grease XG-1-C or -K, or equivalent.
   b. Position the spring, release hub and bearing onto the clutch slave cylinder.
   c. Push the clutch release hub and the bearing against the spring.
   d. Install the retainer ring.
9. Position the slave cylinder onto the transmission input shaft, then

tighten the mounting bolts to 14–19 ft. lbs. (19–26 Nm).
10. Install the transmission.

Hydraulic Clutch System

Bleeding

Externally Mounted Slave Cylinder

1. Clean the reservoir cap and the slave cylinder connection.
2. Remove the slave cylinder from the housing.
3. Using a 3/32 in. punch, drive out the pin that holds the tube in place.
4. Remove the tube from the slave cylinder and place the end of the tube in a container.
5. Hold the slave cylinder so the connector port is at the highest point, by tipping it about 30 degrees from horizontal. Fill the cylinder with DOT 3 brake fluid through the port. It may be necessary to rock the cylinder or slightly depress the pushrod to expel all the air.

─── **WARNING** ───

Pushing too hard on the pushrod will spurt fluid from the port!

6. When all air is expelled — no more bubble are seen — install the slave cylinder.

NOTE: Some fluid will be expelled during installation as the pushrod is depressed.

7. Remove the reservoir cap. Some fluid will run out of the tube end into the container. Pour fluid into the reservoir until a steady stream of fluid runs out of the tube and the reservoir is filled. Quickly install the diaphragm and cap. The flow should stop.
8. Connect the tube and install the pin. Check the fluid level.
9. Check the clutch operation.

Internally Mounted Slave Cylinder

Except 1997 F-150 and Expedition Models

NOTE: With the quick-disconnect coupling, no air should enter the system when the coupling is disconnected. However, if air should somehow enter the system, it must be bled.

1. Remove the reservoir cap and diaphragm. Fill the reservoir with DOT 3 brake fluid.
2. Connect a piece of rubber tubing to the slave cylinder bleed screw. Place the other end in a container.

3. Loosen the bleed screw. Gravity will force fluid from the master cylinder to flow down to the slave cylinder, forcing air out of the bleed screw. When a steady stream — no bubbles — flows out, the system is bled. Close the bleed screw.

NOTE: Check periodically to make sure the master cylinder reservoir doesn't run dry.

4. Add fluid to fill the master cylinder reservoir.
5. Fully depress the clutch pedal. Release it as quickly as possible. Pause for 2 seconds. Repeat this procedure 10 times.
6. Check the fluid level. Refill it if necessary. It should be kept full.
7. Repeat Steps 5 and 6 five more times.
8. Install the diaphragm and cap.
9. Have an assistant hold the pedal to the floor while you crack the bleed screw — not too far — just far enough to expel any trapped air. Close the bleed screw, then, release the pedal.
10. Check, and if necessary, fill the reservoir.

1997 F-150 and Expedition Models

NOTE: Be sure to keep the clutch master cylinder reservoir full of brake fluid during the bleeding process to prevent air from entering the clutch master cylinder.

1. Raise and safely support the front of the vehicle on jackstands.

NOTE: It is necessary to have the assistance of a helper to bleed this system.

2. Fill the clutch system reservoir with Ford High Performance DOT 3 Brake Fluid C6AZ-19542-AA or equivalent.
3. Have your assistant depress the clutch pedal rapidly for 5–10 strokes.
4. Wait 1–3 minutes.
5. Repeat Steps 3 and 4 three more times.
6. Loosen the bleeder screw on the transmission for the slave cylinder.
7. Have the helper fully depress the clutch pedal and hold down.
8. Tighten the bleeder screw.
9. The helper should now release the clutch pedal.
10. Apply pressure to the clutch pedal. If the clutch pedal travels more or less than 6–7 in. (15.3–17.7 cm), repeat the bleeding process.

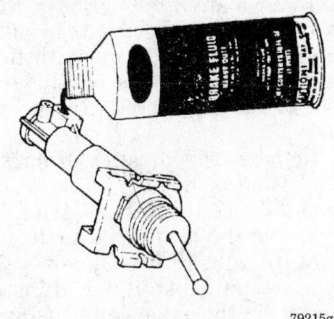

79215ga5

External slave cylinder bleeding —
Bronco, F Series (Pick-Up) and E Series
(Van)

Automatic Transmission Assembly

REMOVAL AND INSTALLATION

C6 Transmission

1. From in the engine compartment, remove the 2 upper converter housing-to-engine bolts.

2. Disconnect the neutral switch wire at the inline connector.

3. Remove the bolt securing the fluid filler tube to the engine cylinder head.

4. Raise and support the vehicle safely.

5. Place the drain pan under the transmission fluid pan. Starting at the rear of the pan and working toward the front, loosen the attaching bolts and allow the fluid to drain. Finally remove all of the pan attaching bolts except 2 at the front, to allow the fluid to further drain. With fluid drained, install 2 bolts on the rear side of the pan to temporarily hold it in place.

6. Remove the converter drain plug access cover from the lower end of the converter housing.

7. Remove the converter-to-flywheel attaching nuts. Place a wrench on the crankshaft pulley attaching bolt to turn the converter to gain access to the nuts.

8. With the wrench on the crankshaft pulley attaching bolt, turn the converter to gain access to the converter drain plug. Place a drain pan under the converter to catch the fluid and remove the plug. After the fluid has been drained, reinstall the plug.

9. On 2WD models, disconnect the driveshaft from the rear axle and slide shaft rearward from the transmission. Install a seal installation tool in the extension housing to prevent fluid leakage.

10. Disconnect the speedometer cable from the extension housing.

11. Disconnect the downshift and manual linkage rods from the levers at the transmission.

12. Disconnect the oil cooler lines from the transmission.

13. Remove the vacuum hose from the vacuum diaphragm unit. Remove the vacuum line retaining clip.

14. Disconnect the cable from the terminal on the starter motor. Remove the 3 attaching bolts and remove the starter motor.

15. On 4WD models remove the transfer case.

16. Remove the 2 engine rear support and insulator assembly-to-attaching bolts.

17. Remove the 2 engine rear support and insulator assembly-to-extension housing attaching bolts.

18. Remove the 6 bolts securing the No. 2 crossmember to the frame side rails.

19. Raise the transmission with a transmission jack and remove both crossmembers.

20. Secure the transmission to the jack with the safety chain.

21. Remove the remaining converter housing-to-engine attaching bolts.

22. Move the transmission away from the engine. Lower the jack and remove the converter and transmission assembly from under the vehicle.

To install:

23. Tighten the converter drain plug.

24. Position the converter on the transmission making sure the converter drive flats are fully engaged in the pump gear.

25. With the converter properly installed, place the transmission on the jack. Secure the transmission on the jack with the chain.

26. Rotate the converter until the studs and drain plug are in alignment with their holes in the flywheel.

27. Move the converter and transmission assembly forward into position, using care not to damage the flywheel and the converter pilot. The converter must rest squarely against the flywheel. This indicates that the converter pilot is not binding in the engine crankshaft.

28. Install the converter housing-to-engine attaching bolts and torque them to 65 ft. lbs. (88 Nm) for the Diesel; 50 ft. lbs. (68 Nm) for gasoline engines.

29. Remove the transmission jack safety chain from around the transmission.

30. Position the No. 2 crossmember to the frame side rails. Install and tighten the attaching bolts.

31. Install transfer case on 4WD models.

32. Position the engine rear support and insulator assembly above the crossmember. Install the rear support and insulator assembly-to-extension housing mounting bolts and tighten the bolts to 45 ft. lbs. (61 Nm).

33. Lower the transmission and remove the jack.

34. Secure the engine rear support and insulator assembly to the crossmember with the attaching bolts and tighten them to 80 ft. lbs. (109 Nm).

35. Connect the vacuum line to the vacuum diaphragm making sure the line is in the retaining clip.

36. Connect the oil cooler lines to the transmission.

37. Connect the downshift and manual linkage rods to their respective levers on the transmission.

38. Connect the speedometer cable to the extension housing.

39. Secure the starter motor in place with the attaching bolts. Connect the cable to the terminal on the starter.

40. Install a new O-ring on the lower end of the transmission filler tube and insert the tube in the case.

41. Secure the converter-to-flywheel attaching nuts and tighten them to 30 ft. lbs. (41 Nm).

42. Install the converter housing access cover and secure it with the attaching bolts.

43. Connect the driveshaft.

44. Adjust the shift linkage as required.

45. Lower the vehicle. Then install the 2 upper converter housing-to-engine bolts and tighten them.

46. Position the transmission fluid filler tube to the cylinder head and secure with the attaching bolts.

47. Make sure the drain pan is securely attached, and fill the transmission to the correct level with the Dexron®II fluid.

AOD Transmission

1. Raise the vehicle on hoist or stands.

2. Place the drain pan under the transmission fluid pan. Starting at the rear of the pan and working toward the front, loosen the attaching bolts and allow the fluid to drain. Finally remove all of the pan attaching bolts except 2 at the front, to allow the fluid to further drain. With fluid drained, install 2 bolts on the rear

side of the pan to temporarily hold it in place.

3. Remove the converter drain plug access cover from the lower end of the converter.

4. Remove the converter-to-flywheel attaching nuts. Place a wrench on the crankshaft pulley attaching bolt to turn the converter to gain access to the nuts.

5. Place a drain pan under the converter to catch the fluid. With the wrench on the crankshaft pulley attaching bolt, turn the converter to gain access to the converter drain plug and remove the plug. After the fluid has been drained, reinstall the plug.

6. On 2WD models, matchmark and disconnect the driveshaft from the rear axle and slide shaft rearward from the transmission. Install a seal installation tool in the extension housing to prevent fluid leakage.

7. Disconnect the cable from the terminal on the starter motor. Remove the 3 attaching bolts and remove the starter motor. Disconnect the neutral start switch wires at the plug connector.

8. Remove the rear mount-to-crossmember attaching bolts and the 2 crossmember-to-frame attaching bolts.

9. Remove the 2 engine rear support-to-extension housing attaching bolts.

10. Disconnect the TV linkage rod from the transmission TV lever. Disconnect the manual rod from the transmission manual lever at the transmission.

11. Remove the 2 bolts securing the bell crank bracket to the converter housing.

12. On 4WD models, remove the transfer case.

13. Raise the transmission with a transmission jack to provide clearance to remove the crossmember. Remove the rear mount from the crossmember and remove the crossmember from the side supports.

14. Lower the transmission to gain access to the oil cooler lines.

15. Disconnect each oil line from the fittings on the transmission.

16. Disconnect the speedometer cable from the extension housing.

17. Remove the bolt that secures the transmission fluid filler tube to the cylinder block. Lift the filler tube and the dipstick from the transmission.

18. Secure the transmission to the jack with the chain.

19. Remove the converter housing-to-cylinder block attaching bolts.

20. Carefully move the transmission and converter assembly away from the engine and, at the same time, lower the jack to clear the underside of the vehicle.

21. Remove the converter and mount the transmission in a holding fixture.

22. Tighten the converter drain plug.

To install:

23. Position the converter on the transmission, making sure the converter drive flats are fully engaged in the pump gear by rotating the converter.

24. With the converter properly installed, place the transmission on the jack. Secure the transmission to the jack with a chain.

25. Rotate the converter until the studs and drain plug are in alignment with the holes in the flywheel.

26. Move the converter and transmission assembly forward into position, using care not to damage the flywheel and the converter pilot. The converter must rest squarely against the flywheel. This indicates that the converter pilot is not binding in the engine crankshaft.

27. Install and tighten the converter housing-to-engine attaching bolts to 40–50 ft. lbs. (54–68 Nm).

28. Remove the safety chain from around the transmission.

29. Install a new O-ring on the lower end of the transmission filler tube. Insert the tube in the transmission case and secure the tube to the engine with the attaching bolt.

30. Connect the speedometer cable to the extension housing.

31. Connect the oil cooler lines to the right side of transmission case.

32. Position the crossmember on the side supports. Torque the bolts to 55 ft. lbs. (75 Nm). Position the rear mount on the crossmember and install the attaching nuts to 90 ft. lbs. (122 Nm).

33. On 4WD models, install the transfer case.

34. Secure the rear support to the extension housing and tighten the bolts to 80 ft. lbs. (109 Nm).

35. Lower the transmission and remove the jack.

E4OD Transmission

1. Raise and support the vehicle safely.

2. Place the drain pan under the transmission fluid pan. Starting at the rear of the pan and working toward the front, loosen the attaching bolts and allow the fluid to drain. Fi-

nally remove all of the pan attaching bolts except 2 at the front, to allow the fluid to further drain. With fluid drained, install 2 bolts on the rear side of the pan to temporarily hold it in place.

3. Remove the dipstick from the transmission.

4. On 4WD models, matchmark and remove the front driveshaft.

5. Matchmark and remove the rear driveshaft. Install a seal installation tool in the extension housing to prevent fluid leakage.

6. Disconnect the linkage from the transmission.

7. On 4WD models, disconnect the transfer case linkage.

8. Remove the heat shield and remove the manual lever position sensor connector by squeezing the tabs and pulling on the connector; never attempt to pry the connector apart!

9. Remove the solenoid body heat shield.

10. Remove the solenoid body connector by pushing on the center tab and pulling on the wiring harness; never attempt to pry apart the connector!

11. On 4WD models, remove the 4x4 switch connector from the transfer case. Be careful not to over-extend the tabs.

12. Pry the harness connector from the extension housing wire bracket.

13. On 4WD models, remove the wiring harness locators from the left side of the connector.

14. Disconnect the speedometer cable.

15. On 4WD models, remove the transfer case.

16. Remove the converter cover bolts.

17. Remove the rear engine cover plate bolts.

18. Disconnect the cable from the terminal on the starter motor. Remove the 3 attaching bolts and remove the starter motor. Disconnect the neutral start switch wires at the plug connector.

19. Remove the converter-to-flywheel attaching nuts. Place a wrench on the crankshaft pulley attaching bolt to turn the converter to gain access to the nuts.

20. Secure the transmission to a transmission jack. Use a safety chain.

21. Remove the rear mount-to-crossmember attaching nuts and the 2 crossmember-to-frame attaching bolts.

22. Disconnect each oil line from the fittings on the transmission. Cap the lines.

23. Remove the 6 converter housing-to-cylinder block attaching bolts.

24. Carefully move the transmission and converter assembly away from the engine and, at the same time, lower the jack to clear the underside of the vehicle.

25. Remove the transmission filler tube.

26. Install Torque Converter Handles T81P-7902-C, or equivalent, at the 12 o'clock and 6 o'clock positions.

To install:

27. Install the converter with the handles at the 12 o'clock and 6 o'clock positions. Push and rotate the converter until it bottoms out. Check the seating of the converter by placing a straightedge across the converter and bell housing. There must be a gap between the converter and straightedge. Remove the handles.

28. Install the transmission filler tube.

29. Rotate the converter to align the studs with the flywheel mounting holes.

30. Carefully raise the transmission into position at the engine. The converter must rest squarely against the flywheel.

31. Install the 6 converter housing-to-cylinder block attaching bolts. Snug them alternately and evenly,

79215ga7

Torque converter installation — E4OD automatic transmission — Bronco, F Series (Pick-Up) and E Series (Van)

then, tighten them alternately and evenly to 40–50 ft. lbs. (54–68 Nm).

32. Install the converter drain plug cover.

33. Connect each oil line at the fittings on the transmission.

34. Install the rear mount-to-crossmember attaching nuts and the 2 crossmember-to-frame attaching bolts. Torque the nuts and bolts to 50 ft. lbs. (68 Nm).

35. Remove the transmission jack.

36. Install the converter-to-flywheel attaching nuts. Place a wrench on the crankshaft pulley attaching bolt to turn the converter to gain access to the nuts. Torque the nuts to 20–30 ft. lbs. (27–41 Nm).

37. Install the starter motor. Connect the cable at the terminal on the starter motor. Connect the neutral start switch wires at the plug connector.

38. Install the rear engine cover plate bolts.

39. Install the converter cover bolts.

40. On 4WD models, install the transfer case.

41. Connect the speedometer cable.

42. On 4WD models, install the wiring harness locators at the left side of the connector.

43. Connect the harness connector at the extension housing wire bracket.

44. On 4WD models, install the 4x4 switch connector at the transfer case.

45. Install the solenoid body connector. An audible click indicates connection.

46. Install the solenoid body heat shield.

47. Install the manual lever position sensor connector and the heat shield.

48. On 4WD models, connect the transfer case linkage.

49. Connect the linkage at the transmission.

50. Install the rear driveshaft.

51. On 4WD models, install the front driveshaft.

52. Install the dipstick.

53. Install the drain pan using a new gasket and sealer.

54. Lower the vehicle.

55. Refill the transmission and check for leaks.

4R70W Transmission

1. Disconnect the negative battery cable.

2. Place the transmission in Neutral.

3. Detach the transmission harness electrical connector by removing

the retaining bolt and separating the two halves.

4. Raise and safely support the vehicle on jackstands.

5. Drain the transmission fluid.

6. Matchmark the rear driveshaft to the differential mounting flange and, on 4-wheel drive versions, to the transmission housing, then remove the 4 attaching bolts. Remove the driveshaft from the vehicle.

7. On 4-wheel drive models, remove the transfer case.

8. Remove the transmission inspection cover.

9. Remove the 4 torque converter nuts by rotating the crankshaft to gain access to all of the nuts.

10. Remove the starter motor.

11. Detach the transmission shift linkage.

12. Disconnect the transmission fluid cooler lines.

13. Detach the 2 heated oxygen sensor electrical connectors.

14. Position a transmission jack, or equivalent, under the transmission. Secure the transmission to the jack with a strap.

15. Remove the 2 transmission mount nuts.

16. Remove the 2 heat shield mounting bolts (one on each side)>

17. Remove the 6 rear crossmember mounting bolts, then remove the crossmember from the vehicle.

18. Remove the rear transmission mount from the vehicle frame.

19. Remove the fluid fill tube.

20. Remove the exhaust pipe bracket.

21. Remove the fuel line bracket.

22. On 4.2L (VIN 2) engines, remove the 9 engine-to-transmission bolts.

23. Remove the 6 engine-to-transmission bolts on 4.6L engines.

24. Remove the 4 exhaust pipe-to-exhaust manifold nuts (two on each side) and position the exhaust pipe aside.

--- **CAUTION** ---

The torque converter is heavy and may result in injury if allowed to fall out of the transmission. Secure the torque converter in the transmission.

25. Slowly and carefully lower the transmission from the vehicle.

To install:

NOTE: Make sure that the torque converter is fully seated in the transmission before positioning the transmission to the engine.

26. Raise the transmission, with the torque converter installed, slowly and carefully up into the vehicle. When mating the transmission to the engine, align the orange balancing marks on the converter stud and flywheel bolt hole.

27. Position the exhaust pipe onto the exhaust manifold, then install the 4 mounting nuts to 25–34 ft. lbs. (34–46 Nm).

28. On 4.6L engines, install the 6 engine-to-transmission bolts and tighten to 30–41 ft. lbs. (40–55 Nm).

29. On 4.2L (VIN 2) engines, install the 9 engine-to-transmission bolts to 28–38 ft. lbs. (38–51 Nm).

30. Install the fuel line bracket and tighten the mounting nut to 13–17 ft. lbs. (17–23 Nm).

31. Install the exhaust bracket and tighten the 2 mounting nuts to 25–33 ft. lbs. (34–46 Nm).

32. Install the transmission mount and tighten the 2 mounting bolts to 64–81 ft. lbs. (87–110 Nm).

33. Install the rear crossmember into the vehicle. Tighten the 6 crossmember mounting bolts to 39–53 ft. lbs. (53–72 Nm).

34. Install the 2 heat shield bolts, then install the 2 transmission-to-transmission mount nuts to 64–81 ft. lbs. (87–110 Nm).

35. Lower the transmission jack slowly.

36. Attach the 2 heated oxygen sensor electrical connectors, then install the starter motor.

37. Connect the transmission fluid cooler lines.

38. Slide the transmission shift linkage into the cable bracket until the tabs are fully seated, then connect the shift cable to the transmission selector lever.

39. Install the starter motor.

40. Rotate the crankshaft to gain access to the torque converter-to-flywheel studs and install the mounting nuts finger-tight. Install all of the nuts in this manner, then rotate the crankshaft a second time and tighten the mounting nuts to 10 ft. lbs. (14 Nm). Rotate the crankshaft a third time and finally tighten the nuts to 20–34 ft. lbs. (27–46 Nm).

41. Install the transmission inspection cover. Tighten the mounting bolts to 22–30 ft. lbs. (30–40 Nm).

42. Install the fluid fill tube.

43. For vehicles equipped with 4-wheel drive, install the transfer case.

44. Install the rear driveshaft. Mount the driveshaft to the differential flange so that the matchmarks

align, then tighten the 4 mounting bolts to 65–87 ft. lbs. (88–119 Nm).

45. Lower the vehicle.

46. Attach the transmission electrical wiring harness connector.

47. Connect the negative battery cable.

48. Fill the transmission with transmission fluid. Make sure to use MERCON Multi-Purpose Automatic Transmission Fluid XT-2-QDX or DDX, or equivalent.

Throttle Valve Cable

ADJUSTMENT

AOD Transmission with EFI Fuel System

Adjustment with Engine OFF

1. Set the parking brake and put the selector lever in **N**.

2. Remove the protective cover from the cable.

3. Make sure the throttle lever is at the idle stop. If it isn't, check for binding or interference. Never attempt to adjust the idle stop!

4. Make sure the cable is free of sharp bends or is not rubbing on anything throughout its entire length.

5. Lubricate the TV lever ball stud with chassis grease.

6. Unlock the locking tab at the throttle body by prying with a small prybar.

7. Install a spring on the TV control lever, to hold it in the rearmost travel position. The spring must exert at least 10 lbs. (14 Nm) of force on the lever.

8. Rotate the transmission outer TV lever 10–30 degrees and slowly allow it to return.

9. Push down on the locking tab until flush.

10. Remove the retaining spring from the lever.

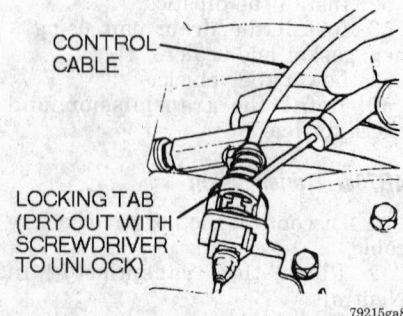

CONTROL CABLE

LOCKING TAB (PRY OUT WITH SCREWDRIVER TO UNLOCK)

79215ga8

Unlocking TV cable locking tab — 5.0L (VIN N) engine

DRIVE LINE

Driveshaft

REMOVAL AND INSTALLATION

One Piece Rear Driveshaft

Except Bronco Models

1. With the vehicle in Neutral, block the front wheels, then raise and safely support the rear of the vehicle with jackstands.

2. Matchmark the driveshaft yoke and axle pinion flange.

3. Remove the driveshaft flange or Universal joint-to-differential flange nuts and bolts.

4. Separate the yoke from the differential flange. It may be necessary to pry it free with a small prybar. Immediately after separation, wrap tape around the U-joint caps to keep them from falling off.

5. Slip the driveshaft off the transmission splines.

To install:

6. Slide the front driveshaft yoke into the rear transmission extension housing so that the splines are aligned.

7. Situate the rear end of the driveshaft in position on the differential flange, making sure that the matchmarks made earlier line up.

8. Install and tighten the driveshaft-to-differential mounting nuts and bolts. On 1997 F-150 and Expedition models, tighten the bolts to 70–95 ft. lbs. (95–129 Nm). On all other models, tighten the U-bolts to 15 ft. lbs. (20 Nm).

9. Lower the vehicle, apply the parking brake and remove the wheel chocks.

Bronco Models

1. With the vehicle in Neutral, block the front wheels, then raise and safely support the rear of the vehicle with jackstands.

2. Matchmark the rear driveshaft yoke and axle flange.

3. Matchmark the front cardan joint and the transfer case yoke.

4. Remove the U-bolt nuts and U-bolts attaching the yoke to the axle flange.

5. Separate the yoke from the flange. It may be necessary to pry it free with a small prybar. Immediately after separation, wrap tape around the U-joint caps to keep them from falling off.

6. Remove the cardan joint-to-transfer case yoke bolts and separate the cardan joint from the yoke.

To install:

7. Position the front driveshaft flange against the transfer case or transmission flange. Make sure that the matchmarks align, then install the mounting bolts to 25 ft. lbs. (34 Nm).

8. Situate the rear end of the driveshaft in position on the differential flange, making sure that the matchmarks made earlier line up.

9. Install and tighten the driveshaft-to-differential mounting nuts and bolts. Tighten the U-bolts to 15 ft. lbs. (20 Nm).

10. Lower the vehicle, apply the parking brake and remove the wheel chocks.

Two Piece Rear Driveshaft

1. With the vehicle in Neutral, block the front wheels, then raise and safely support the rear of the vehicle with jackstands.

2. Matchmark the driveshaft yoke and axle pinion flange.

3. Remove the nuts and U-bolts, if applicable, attaching the yoke to the axle flange.

4. Separate the yoke from the flange. It may be necessary to pry it free with a small prybar. Immediately after separation, wrap tape around the U-joint caps to keep them from falling off.

5. Slip the driveshaft off the coupling shaft splines.

6. Remove the center bearing.

7. Slide the coupling shaft from the transmission shaft splines.

To install:

8. Clean all parts and check for damage. Do not remove the blue plastic coating from the male splines.

9. Slide the coupling driveshaft into the transmission output shaft, then position it and the center bearing support onto the vehicle. Tighten the center bearing support mounting bolts to 39–53 ft. lbs. (53–72 Nm).

10. Slide the rear driveshaft into the mating splines of the coupling shaft, then situate the rear driveshaft onto the rear differential input flange making sure to align the matchmarks made earlier. Tighten the attaching bolts to 70–95 ft. lbs. (95–129 Nm) for 1997 F-150 and Expedition models. For vehicles with conventional U-bolts and nuts,

tighten them to the following specifications dependant on size:

- $5/16$ in.-18: 15 ft. lbs. (20 Nm)
- $3/8$ in.-18: 17–26 ft. lbs. (23–35 Nm)
- $7/16$ in.-20: 30–40 ft. lbs. (41–54 Nm)

11. Lower the vehicle to the ground and remove the wheel chocks.

Front Driveshaft

1. Matchmark the driveshaft yoke and axle pinion flange.

2. Matchmark the driveshaft yoke and transfer case flange.

3. Remove the U-bolt nuts and U-bolts attaching the yoke to the axle flange.

4. Separate the yoke from the flange. It may be necessary to pry it free with a small prybar. Immediately after separation, wrap tape around the U-joint caps to keep them from falling off.

5. Remove the U-bolts and nuts (bolts for the F-350, 1997 F-150 and Expedition) and disconnect the driveshaft from the transfer case. It may be necessary to pry it free with a small prybar. Immediately after separation, wrap tape around the U-joint caps to keep them from falling off.

NOTE: Avoid separating the driveshaft parts at the slip joint. If the driveshaft should become separated, follow the procedure, below.

To install:

6. Situate the rear end of the driveshaft in position on the differential flange, making sure that the matchmarks made earlier line up.

7. Install and tighten the driveshaft-to-differential mounting nuts and bolts. Tighten the U-bolts (except F-350 and 1997 F-150 and Expedition models) to 15 ft. lbs. (20 Nm). Tighten the F-350 bolts to 20–28 ft. lbs. (27–38 Nm). Tighten the 1997 F-150 and Expedition driveshaft-to-front differential bolts to 65–88 ft. lbs. (88–119 Nm).

8. Position the front driveshaft flange against the transfer case or transmission flange. Make sure that the matchmarks align, then install the mounting bolts to 20–28 ft. lbs. (27–38 Nm) for F-350, to 65–88 ft. lbs. (88–119 Nm) for 1997 F-150 and Expedition models, and to 15 ft. lbs. (20 Nm) for all others.

9. Lower the vehicle, apply the parking brake and remove the wheel chocks.

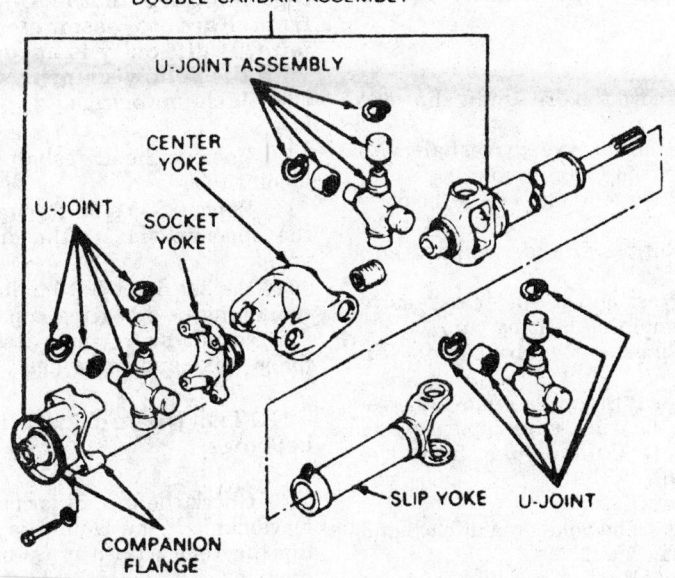

DOUBLE CARDAN ASSEMBLY

U-JOINT ASSEMBLY

CENTER YOKE

U-JOINT

SOCKET YOKE

COMPANION FLANGE

SLIP YOKE   U-JOINT

79215gb0

Double Cardan U-joint assembly — F-350 and Bronco

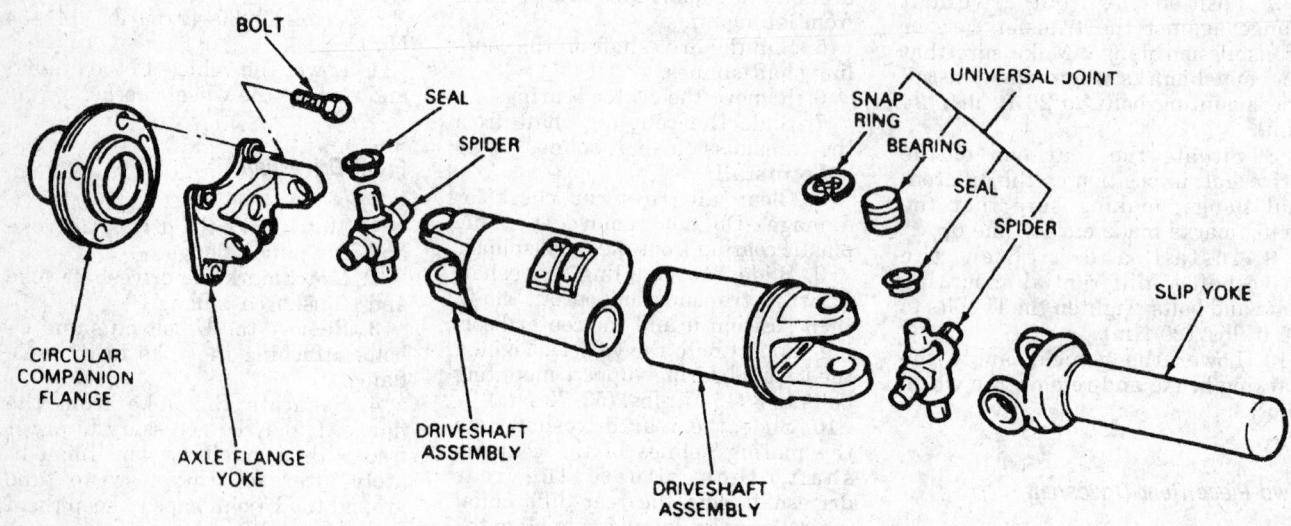

One-piece driveshaft — Bronco, F Series (Pick-Up) and E Series (Van)

79215gb1

U-Joints

REMOVAL AND INSTALLATION

Single Cardan-Type U-Joint

NOTE: To correctly disassembly and reassemble the Universal joints (U-joints), Ford Tool T74P-4635-C, or its equivalent, is necessary.

1. Remove the driveshaft, from which the U-joint is to be removed, from the vehicle.

———— WARNING ————
Do not clamp the driveshaft in the jaws of a vise or similar holding fixture.

2. Place the driveshaft on a suitable work table, being careful not to damage the driveshaft tube.

NOTE: Make sure to mark all components, otherwise the driveshaft assembly may be assembled so that it is out of balance.

3. Matchmark the positions of the driveshaft components.
4. Clamp U-joint Tool T74P-4635-C in a vise.

5. Remove the snaprings from the U-joint ends.

NOTE: If necessary, use a pair of pliers to remove a bearing cup if it cannot be pressed out all the way.

6. Remove the driveshaft slip yoke as follows:
   a. Position the driveshaft slip yoke in the U-joint tool.
   b. Press out one of the bearing cups.
   c. Rotate the driveshaft slip yoke.
   d. Press on the spider to remove the remaining bearing cup.
   e. Remove the driveshaft slip yoke.
7. Repeat Step 6 to remove the remaining bearing cups, spiders and driveshaft flange yoke from the driveshaft.
To install:
8. Clean the yoke area at each end of the driveshaft.
9. Install the bearing cup as follows:
   a. Start a new bearing cup into the driveshaft yoke.
   b. Position the new spider in the driveshaft yoke.

   c. Position the driveshaft in the U-joint tool.
   d. Press the bearing cup ¼ in. (6.3mm) below the yoke surface.

NOTE: Use the yellow snaprings supplied in the U-joint kit from Ford to assemble the U-joint. If difficulty is encountered with the yellow snaprings, install the black snaprings.

10. Remove the driveshaft from the U-joint tool.
11. Repeat Steps 9a through 9d for the opposite side of the driveshaft yoke.
12. Repeat Steps 8–10 to install the remaining new bearing cups, spider, driveshaft slip yoke, driveshaft flange yoke and snaprings.

NOTE: Do not strike the bearings.

13. Check the U-joint for freedom of movement. If the U-joint is binding, tap the yoke with a brass or plastic hammer.
14. If necessary, use Premium Long-Life Grease XG-1-C or -K, or equivalent, to lubricate the U-joints.
15. Install the driveshaft in the vehicle.

Double Cardan-Type U-Joints

NOTE: To correctly disassembly and reassemble the Universal joints (U-joints), Ford Tool T74P-4635-C, or its equivalent, is necessary.

1. Remove the driveshaft, from which the U-joint is to be removed, from the vehicle.

——————— **WARNING** ———————

Do not clamp the driveshaft in the jaws of a vise or similar holding fixture.

2. Place the driveshaft on a suitable work table, being careful not to damage the driveshaft tube.

NOTE: Make sure to mark all components, otherwise the driveshaft assembly may be assembled so that it is out of balance.

3. Matchmark the positions of the spiders, the center yoke, and the centering socket yoke as related to the stud yoke, which is welded to the front of the driveshaft tube.

4. Remove the snaprings that secure the bearings in the front of the center yoke.

5. Using the U-joint tool, press out one of the bearings until it protrudes approximately ⅜ in. (9mm) out of the yoke.

6. Clamp the bearing in a vise and tap on the center yoke to free it from the bearing.

7. Lift the two bearing cups from the spider.

8. Reposition the tool on the yoke and move the remaining bearing in the opposite direction so that it protrudes approximately ⅜ in. (9mm) out of the yoke.

9. Clamp the bearing in a vise, then tap on the center yoke to free it from the bearing.

10. Remove the spider from the center yoke.

11. Pull the centering socket yoke off the center stud. Remove the rubber seal from the centering ball stud.

12. Remove the snap rings from the center yoke and the driveshaft yoke.

13. Position the U-joint tool on the driveshaft yoke and press the bearing outward until the inside of the center yoke almost contacts the slinger ring at the front of the driveshaft yoke.

NOTE: Pressing beyond this point can distort the slinger ring.

14. Clamp the exposed end of the bearing in a vise and drive on the center yoke with a soft-faced hammer to free it from the bearing.

15. Reposition the tool and press on the spider to remove the opposite bearing.

16. Remove the center yoke from the spider.

17. Remove the spider from the driveshaft yoke in the same manner.

18. Clean all serviceable parts in cleaning solvent. If using a repair kit, install all of the parts supplied in the kit. If the driveshaft is damaged, replace the complete shaft to be assured of a balanced assembly.

To install:

NOTE: Universal joints are to be installed as complete units only. Do not mix any components from other universal joints.

19. To assemble the double Cardan joints, position the spider in the driveshaft yoke. Make sure the spider bosses (or lubrication plugs on kits) will be in the same position as originally installed. Press in the bearing using the U-joint tool. Install the snaprings.

20. Pack the socket relief and the ball with Premium Long-Life Grease XG-1-C or -K or equivalent, then position the center yoke over the spider ends and press in the bearing. In the snaprings.

21. Install a new seal on the centering ball stud. Position the centering socket yoke on the stud.

22. Place the front spider in the center yoke. Make sure the spider bosses are properly positioned.

23. With the spider loosely positioned on the center stop, proceed to seat the first pair of bearings into the centering socket yoke, then press the second pair into the centering yoke. Install the snaprings.

24. Apply pressure on the centering socket yoke and install the remaining bearing cup.

25. Lubricate the U-joint assemblies, if equipped with grease fittings, with Premium Long-Life Grease XG-1-C or -K, or equivalent grease.

26. Install the driveshaft into the vehicle.

Halfshaft

REMOVAL AND INSTALLATION

1997 F-150 and Expedition Models

1. Break the front wheel lug nuts loose while the weight of the vehicle is resting on the front wheels.

2. Raise and safely support the front of the vehicle on jackstands.

3. Remove the front wheels.

4. Remove the front hub cotter pin, retainer and nut.

5. Using a floor hydraulic jack, support the lower suspension arm.

6. Remove the upper ball joint cotter pin and castle nut.

7. Use a Pitman arm puller, such as Ford Tool T64P-3590-F, to separate the front wheel knuckle from the front suspension upper arm.

8. Lower the lower suspension arm and steering knuckle slightly to facilitate easier halfshaft removal.

9. Remove the 2 disc caliper mounting bolts, then lift the front disc caliper off of the front disc brake caliper anchor plate and position aside. Do not allow the caliper to hang by the brake hose; suspend it from the vehicle's frame with strong cord or wire.

10. Remove the 6 front halfshaft-to-differential bolts.

——————— **WARNING** ———————

Use care to avoid damaging the hub seal when removing the front halfshaft.

11. Remove the inboard end of the halfshaft from the differential case or extension axle case. Separate the front halfshaft and joints from the hub, then remove the halfshaft and joints from the vehicle.

To install:

12. Slide the halfshaft outboard end into the hub, making sure that the splines engage.

13. Situate the inboard end of the halfshaft against the front differential flange and install the 6 halfshaft-to-differential bolts. Tighten the halfshaft bolts to 51–67 ft. lbs. (68–92 Nm).

14. Install the front disc brake caliper onto the rotor and anchor plate, then install and tighten the 2 caliper mounting bolts to 21–26 ft. lbs. (28–36 Nm).

15. Lift the lower suspension arm and steering knuckle up until the upper ball joint stud is inserted into the steering knuckle. Install the upper ball joint castle nut and tighten to 57–76 ft. lbs. (77–104 Nm). Install a new cotter pin.

16. Install the hub nut onto the halfshaft and tighten the hub nut to 188–254 ft. lbs. (255–345 Nm).

17. Install the hub nut retainer and a new cotter pin.

18. Install the front wheels and tighten the lug nuts in a star-shaped sequence to 83–112 ft. lbs. (113–153 Nm).

19. Lower the vehicle to the ground.

CV-Joint Boot

REMOVAL AND INSTALLATION

1997 F-150 and Expedition Models

NOTE: To adequately perform this procedure CV Boot Clamp Installer T95P-3514-A, or its equivalent, is necessary.

1. Remove the front halfshaft from the vehicle.
2. Slide the 2 inboard clamp protectors off of the boot clamps.

——— WARNING ———
Be careful not to damage the halfshaft boot.

3. Remove the 2 inboard boot clamps.
4. Slide the inboard boot off of the inboard CV-joint housing.
5. Remove the inboard CV-joint retaining ring.
6. Remove the inboard CV-joint housing from the front differential housing.
7. Matchmark the inner race and the ball cage for reassembly.
8. Remove the 6 balls from the ball cage.
9. Remove the snapring retaining the inner race onto the halfshaft.
10. Slide the inner race and ball cage off of the halfshaft.
11. Slide the inboard CV-joint boot off of the halfshaft.
12. Remove the 2 outboard clamp protectors, then grab the boot clamps and peel them away from the halfshaft boot.
13. Remove the outboard CV-joint boot from the halfshaft.
14. Clean and inspect the joint for wear. Replace the joint if worn or damaged.

To install:
15. Pack the outboard CV-joint with Ford High Temp Constant Velocity Joint Grease, or equivalent. Spread any remaining grease from the service kit evenly inside of the outboard halfshaft boot.
16. Clean the CV-joint boot mounting surface, then position the outboard CV-joint boot on the shaft and in the CV-joint grooves. Position the boot clamps in place.

NOTE: Tighten the through-bolt until the installer is in the closed position.

17. Using CV Boot Clamp Installer T95P-3514-A, tighten the boot clamps.
18. Install the 2 outboard clamp protectors.

19. Position the inboard clamp protector and the boot clamp on the halfshaft and CV-joint.
20. Position the inboard boot on the halfshaft.
21. Position the ball cage on the CV-joint section of the halfshaft with the tapered end installed first.

NOTE: Line up the marks made during disassembly.

22. Position the inner race on the halfshaft with the counterbored end installed on the shaft first.
23. Install a new snapring in the groove at the inner end of the halfshaft.
24. Lubricate the 6 balls with Ford High Temp Constant Velocity Joint Grease, or equivalent, then position them in the ball cage.
25. Position the clamp protector and the inboard boot clamp on the inboard CV-joint housing (front differential).
26. Fill the inboard CV-joint housing with 8.29 oz. (235 g) of CV-joint high temp grease.
27. Position the inboard CV-joint housing into the differential housing flange, then install the retaining ring to secure in place.
28. Remove any excess grease from the mating surface, then position the inboard CV-joint boot over the CV-joint and differential housing flange.
29. Adjust the CV-joint-to-boot spacing and overall halfshaft assembled length. The assembled length should be 16.43 in. (417.25mm).

NOTE: The air should be released only after adjusting the CV-joint-to-boot spacing.

30. Insert a flat pry tool under the edge of the CV-joint boot to release built up air pressure in the boot.
31. Use the CV-joint boot installer tool, then tighten the 2 inboard boot clamps.
32. Position the clamp protectors over the inboard boot clamps.
33. Install the front halfshaft into the vehicle.

Transfer Case Assembly

REMOVAL AND INSTALLATION

——— CAUTION ———
The catalytic converter is located beside the transfer case. Due to the extreme high temperatures generated by the converter, be careful when removing the transfer case or personal injury may result.

Borg-Warner Model 13–45

1. Raise and support the vehicle safely.
2. Drain the fluid from the transfer case.
3. Disconnect the 4WD indicator switch wire connector at the transfer case.
4. Remove the skid plate from the frame, if equipped.
5. Matchmark and disconnect the front driveshaft from the front output yoke.
6. Matchmark and disconnect the rear driveshaft from the rear output shaft yoke.
7. Disconnect the speedometer driven gear from the transfer case rear bearing retainer.
8. Remove the retaining rings and shift rod from the transfer case shift lever.
9. Disconnect the vent hose from the transfer case.
10. Remove the heat shield from the frame.
11. Support the transfer case with a transmission jack.
12. Remove the bolts retaining the transfer case to the transmission adapter.
13. Lower the transfer case from the vehicle.

To install:
14. When installing place a new gasket between the transfer case and the adapter.
15. Raise the transfer case with the transmission jack so the transmission output shaft aligns with the splined transfer case input shaft. Install the bolts retaining the transfer case to the adapter.
16. Remove the transmission jack from the transfer case.
17. Connect the rear driveshaft to the rear output shaft yoke. Torque the bolts to 15 ft. lbs. (20 Nm).
18. Install the shift lever to the transfer case and install the retaining nut.
19. Connect the speedometer driven gear to the transfer case.
20. Connect the 4WD indicator switch wire connector at the transfer case.
21. Connect the front driveshaft to the front output yoke. Torque the bolts to 15 ft. lbs. (20 Nm).
22. Position the heat shield to the frame crossmember and the mounting lug on the transfer case. Install and tighten the retaining bolts.
23. Install the skid plate to the frame.
24. Install the drain plug. Remove the filler plug and install 6 pints of

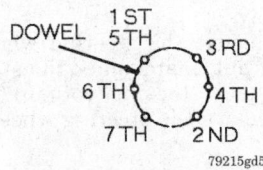

Transfer case-to-adapter bolt torque sequence — Borg-Warner 13–45

Dexron®II type transmission fluid or equivalent.

25. Lower the vehicle.

Borg-Warner Model 13–56

Manual Shift Version

1. Raise and support the vehicle safely.
2. Drain the fluid from the transfer case.
3. Disconnect the 4WD indicator switch wire connector at the transfer case.
4. Remove the skid plate from the frame, if equipped.
5. Matchmark and disconnect the front driveshaft from the front output yoke.
6. Matchmark and disconnect the rear driveshaft from the rear output shaft yoke.
7. Disconnect the speedometer driven gear from the transfer case rear bearing retainer.
8. Remove the retaining rings and shift rod from the transfer case shift lever.
9. Disconnect the vent hose from the transfer case.
10. Remove the heat shield from the frame.
11. Support the transfer case with a transmission jack.
12. Remove the bolts retaining the transfer case to the transmission adapter.
13. Lower the transfer case from the vehicle.
   To install:
14. When installing place a new gasket between the transfer case and the adapter.
15. Raise the transfer case with the transmission jack so the transmission output shaft aligns with the splined transfer case input shaft. Install the bolts retaining the transfer case to the adapter. Torque the bolts to 40 ft. lbs. (54 Nm).
16. Remove the transmission jack from the transfer case.

17. Connect the rear driveshaft to the rear output shaft yoke. Torque the bolts to 15 ft. lbs. (20 Nm).
18. Install the shift lever to the transfer case and install the retaining nut.
19. Connect the speedometer driven gear to the transfer case.
20. Connect the 4WD indicator switch wire connector at the transfer case.
21. Connect the front driveshaft to the front output yoke. Torque the bolts to 15 ft. lbs. (20 Nm).
22. Position the heat shield to the frame crossmember and the mounting lug on the transfer case. Install and tighten the retaining bolts.
23. Install the skid plate to the frame.
24. Install the drain plug. Remove the filler plug and install 6 pints of Dexron®II type transmission fluid or equivalent.
25. Lower the vehicle.

Electronic Shift Version

1. Raise and support the vehicle safely.
2. Drain the fluid from the transfer case.
3. Disconnect the wire connector at the transfer case.
4. Remove the skid plate from the frame, if equipped.
5. Matchmark and disconnect the front driveshaft from the front output yoke.
6. Matchmark and disconnect the rear driveshaft from the rear output shaft yoke.
7. Disconnect the speedometer driven gear from the transfer case rear bearing retainer.
8. Disconnect the vent hose from the transfer case.
9. Remove the heat shield from the frame.
10. Support the transfer case with a transmission jack.

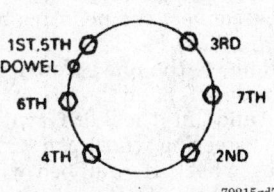

Transfer case-to-adapter bolt torque sequence — Borg-Warner 13–56 electronic and manual shift transfer case — Bronco, F Series (Pick-Up) and E Series (Van)

11. Remove the bolts retaining the transfer case to the transmission adapter.
12. Lower the transfer case from the vehicle.
13. When installing place a new gasket between the transfer case and the adapter.
   To install:
14. Raise the transfer case with the transmission jack so the transmission output shaft aligns with the splined transfer case input shaft. Install the bolts retaining the transfer case to the adapter. Torque the bolts to 40 ft. lbs. (54 Nm).
15. Remove the transmission jack from the transfer case.
16. Connect the rear driveshaft to the rear output shaft yoke. Torque the bolts to 28 ft. lbs. (38 Nm).
17. Install the shift lever to the transfer case and install the retaining nut.
18. Connect the speedometer driven gear to the transfer case. Tighten the bolt to 25 inch lbs. (2.8 Nm).
19. Connect the wire connector at the transfer case.
20. Connect the front driveshaft to the front output yoke. Torque the bolts to 15 ft. lbs. (20 Nm).
21. Position the heat shield to the frame crossmember and the mounting lug on the transfer case. Install and tighten the retaining bolts.
22. Install the skid plate to the frame.
23. Install the drain plug. Remove the filler plug and install 6 pints of Dexron®II type transmission fluid or equivalent.
24. Lower the vehicle.

Borg-Warner 44–06

1. Disconnect the negative battery cable.
2. Raise and safely support the vehicle on jackstands.
3. Drain the transfer case fluid into a large pan or container; dispose of the used fluid.
4. Remove the front and rear driveshafts from the transfer case.
5. Remove the torsion bars and the torsion bar rear support.
6. Remove the transfer case skid plate by removing the 4 mounting bolts, then lowering the slid plate from the vehicle.
7. On manual shift vehicles, use Shift Linkage Insulator Tool T67P-7341-A, or equivalent, to disconnect the shift rod.
8. Detach the Vehicle Speed Sensor (VSS) electrical connector.

9. On manual shift vehicles, detach the 4WD indicator switch and the transfer case coil electrical connectors.

10. On electrical shift vehicles, detach the electric shift motor wiring harness connector.

11. Support the transfer case with an hydraulic floor transmission jack, or equivalent. Secure the transfer case to the jack with a safety strap.

12. Remove the 6 transfer case-to-transmission bolts (3 on each side).

13. Lower the transfer case out of the vehicle.

To install:

14. Raise the transfer case into position with the transmission jack.

15. Install the 6 transfer case-to-transmission bolts to 30–40 ft. lbs. (40–54 Nm). remove the transfer case-to-transmission jack securing strap.

16. On manual shift vehicles, attach the transfer case coil wiring harness connector. Also attach the 4WD indicator switch electrical connector.

17. Attach the VSS wiring harness connector.

18. On manual shift transfer cases, use the shift linkage tool to connect the shift rod.

19. On electrical shift models, attach the electric shift motor wiring harness connector.

20. Situate the transfer case skid plate into position, then install the 4 mounting bolts to 115–150 inch lbs. (13–17 Nm).

21. Install the torsion bars and the torsion bar rear support.

22. Install the driveshafts.

23. Fill the transfer case using Motorcraft MERCON Multi-Purpose Automatic Transmission Fluid XT-2-QDX or DDX, or equivalent fluid.

24. Lower the vehicle.

25. Connect the negative battery cable.

STEERING

Air Bag

— **CAUTION** —
Some vehicles are equipped with an air bag system, also known as the Supplemental Inflatable Restraint (SIR) or Supplemental Restraint System (SRS). The system must be disabled before performing service on or around system components, steering column, instrument panel components, wiring and sensors. Failure to follow safety and disabling procedures could result in accidental air bag deployment, possible personal injury and unnecessary system repairs.

PRECAUTIONS

Several precautions must be observed when handling the inflator module to avoid accidental deployment and possible personal injury.

• Never carry the inflator module by the wires or connector on the underside of the module.

• When carrying a live inflator module, hold securely with both hands, and ensure that the bag and trim cover are pointed away.

• Place the inflator module on a bench or other surface with the bag and trim cover facing up.

• With the inflator module on the bench, never place anything on or close to the module which may be thrown in the event of an accidental deployment.

DISARMING

For the Air Bag system on the Ford full-sized trucks, the positive battery cable must be disconnected for a minimum of one minute before beginning any air bag work to de-energize the backup power supply. It is a good idea to disconnect both the positive and negative battery cables to ensure that the Air Bag system is definitely discharged.

Steering Wheel

REMOVAL AND INSTALLATION

Without Air Bag

1. Set the front wheel in the straight ahead position.
2. Disconnect the negative battery cable.
3. Remove the one screw from the underside of each steering wheel spoke, and lift the horn switch assembly (steering wheel pad) from the steering wheel. If equipped with the sport steering wheel option, pry the button cover off with a prybar.
4. Disconnect the horn switch wires at the connector and remove the switch assembly. If equipped with speed control, squeeze the J-clip ground wire terminal firmly and pull it out of the hole in the steering wheel. Don't pull the wire out without squeezing the clip.
5. Remove the horn switch assembly.
6. Remove the steering wheel retaining nut, matchmark the steering wheel and steering column shaft, then remove the steering wheel with a puller.

— **WARNING** —
Never hammer on the wheel or shaft to remove it! Never use a knock-off type puller.

To install:

7. Slide the steering wheel onto the steering column shaft so that the matchmarks align, then draw the steering wheel onto the column fully by tighten the steering wheel retaining nut to 40 ft. lbs. (54 Nm).
8. Install the horn switch assembly to the steering wheel, then attach the speed control wiring harness connector, if applicable, and the horn switch connector.
9. Position the horn pad onto the steering wheel, then install the retaining screws until snug.
10. Connect the negative battery cable.

With Air Bag

1. Set the front wheel in the straight ahead position.

— **CAUTION** —
The residual power supply must be discharged before any air bag component is serviced.

2. Disconnect the battery-to-starter relay cable for at least one minute to allow the Air Bag system backup power supply to discharge.
3. Remove the 4 air bag module retaining nuts (2 screws for 1997 F-150 and Expedition models) from the air bag module on the back side of the steering wheel, then lift the module away from the steering wheel.
4. Detach the air bag wiring harness connector, and remove the module from the steering wheel.
5. Detach the horn/speed control wiring harness connector.
6. Remove the steering wheel retaining bolt.
7. Using a 2-jawed puller, such as Steering Wheel Puller T77L-4220-B1, remove the steering wheel from the steering column shaft.

— **WARNING** —
Never hammer on the wheel or shaft to remove it! Never use a knock-off type puller.

To install:

8. Make sure that the vehicle's wheels are pointing straight ahead.

9. Route the air bag sliding contact wire harness through the steering wheel opening at the 3 o'clock position. Situate the steering wheel on the steering column shaft so that the matchmarks are aligned. Be sure that the air bag wire is not pinched.

------- **WARNING** -------

Be sure that the wiring does not get trapped between the steering wheel and the air bag sliding contact.

10. Install a new steering wheel retaining bolt and tighten it to 23–32 ft. lbs. (31–44 Nm).

11. Attach the horn/speed control wiring harness to the contact harness, and snap the connector onto the steering wheel clip.

12. Attach the air bag wiring harness from the sliding contact to the air bag module, then install the module to the steering wheel. Tighten the module retaining nuts to 35–53 inch lbs. (4–6 Nm) and the screws until secure.

NOTE: When the battery is disconnected and reconnected, some abnormal driving symptoms may occur while the Powertrain Control Module (PCM) relearns its adaptive strategy. The vehicle may need to be driven 10 miles (16 km) or more for the PCM to relearn the strategy.

13. Connect both battery cables to the battery. Verify that the air bag warning indicator is functioning properly.

Tie Rod Ends

REMOVAL AND INSTALLATION

Except 1997 F-150 and Expedition Models

Inner and Outer Tie Rod Ends

1. Raise and support the front end on jackstands.

2. Place the wheels in a straight-ahead position.

3. Remove the ball stud from the pitman arm using a tie rod end remover.

4. Loosen the nuts on the adjusting sleeve clamp. Remove the ball stud from the adjuster, or the adjuster from the tie rod. Count the number of turns it takes to remove the sleeve from the tie rod or ball stud from the sleeve.

To install:

5. Install the sleeve on the tie rod, or the ball in the sleeve the same number of turns noted during removal. Make sure the adjuster clamps are in the correct position, illustrated, and torque the clamp bolts to 40 ft. lbs. (54 Nm).

6. Keep the wheels straight ahead and install the ball studs. Torque the nuts to 75 ft. lbs. (102 Nm). Use new cotter pins.

7. Install the drag link and connecting rod.

8. Have the front end alignment checked.

1997 F-150 and Expedition Models

Outer Tie Rod End

1. Break the front wheel lug nuts loose while the vehicle is resting on the ground.

2. Raise and safely support the front of the vehicle on jackstands.

3. Remove the tire and wheel assemblies.

4. Matchmark the outer tie rod end to the tie rod sleeve, and the tie rod sleeve to the inner tie rod end.

5. Remove the outer tie rod-to-steering knuckle cotter pin, then remove the castle nut from the outer tie rod ball joint stud.

6. Separate the outer tie rod end from the steering knuckle by using a Pitman arm puller, such as Ford Tool T64P-3590-F.

7. Loosen the outer tie rod sleeve jam nut, then, while counting the revolutions needed to remove, turn the outer tie rod end out of the sleeve.

8. Then, if necessary, loosen the inner tie rod sleeve jam nut, then, also while counting the revolutions needed to remove, turn the tie rod sleeve off of the inner tie rod.

To install:

9. Thread the tie rod sleeve onto the inner tie rod end threads. Turn the sleeve onto the inner tie rod the same number of turns as was needed during removal, then fine adjust the sleeve by aligning the matchmarks.

10. Tighten the inner tie rod sleeve jam nut to 57–77 ft. lbs. (77–104 Nm).

11. Thread the outer tie rod end into the sleeve threads. Turn the outer tie rod into the sleeve the same number of turns as was needed during removal, then fine adjust the outer tie rod end by aligning the matchmarks.

12. Tighten the outer tie rod sleeve jam nut to 57–77 ft. lbs. (77–104 Nm).

13. Install the outer tie rod ball joint stud into the steering knuckle, then install and tighten the ball joint stud castle nut to 57–77 ft. lbs. (77–104 Nm). Install a new cotter pin through the castle nut's castellations and the ball joint studs hole; if the hole and any of the castellations do not line up, tighten the nut until they do.

14. Install the wheel and tire assembly. Install and tighten the lug nuts until snug.

15. Lower the vehicle until the front tires are touching the ground enough to hold them stationary, then tighten them in a star-shaped sequence to 83–112 ft. lbs. (113–153 Nm).

16. Lower the vehicle the rest of the way.

Inner Tie Rod End

1. Break the front wheel lug nuts loose while the vehicle is resting on the ground.

2. Raise and safely support the front of the vehicle on jackstands.

3. Remove the tire and wheel assemblies.

4. Matchmark the inner tie rod end to the tie rod sleeve.

5. Remove the inner tie rod-to-center link (drag link) cotter pin, then remove the castle nut from the inner tie rod ball joint stud.

6. Separate the inner tie rod end from the center link by using a Pitman arm puller, such as Ford Tool T64P-3590-F.

7. Loosen the inner tie rod sleeve jam nut, then, while counting the revolutions needed to remove, turn the inner tie rod end out of the sleeve.

To install:

8. Thread the inner tie rod end into the sleeve threads. Turn the inner tie rod into the sleeve the same number of turns as was needed during removal, then fine adjust the inner tie rod end by aligning the matchmarks.

9. Tighten the inner tie rod sleeve jam nut to 57–77 ft. lbs. (77–104 Nm).

10. Install the inner tie rod ball joint stud into the center link, then install and tighten the ball joint stud castle nut to 57–77 ft. lbs. (77–104 Nm). Install a new cotter pin through the castle nut's castellations and the ball joint stud hole; if the hole and any of the castellations do not line up, tighten the nut until they do.

11. Install the wheel and tire assembly. Install and tighten the lug nuts until snug.

12. Lower the vehicle until the front tires are touching the ground enough to hold them stationary, then tighten them in a star-shaped se-

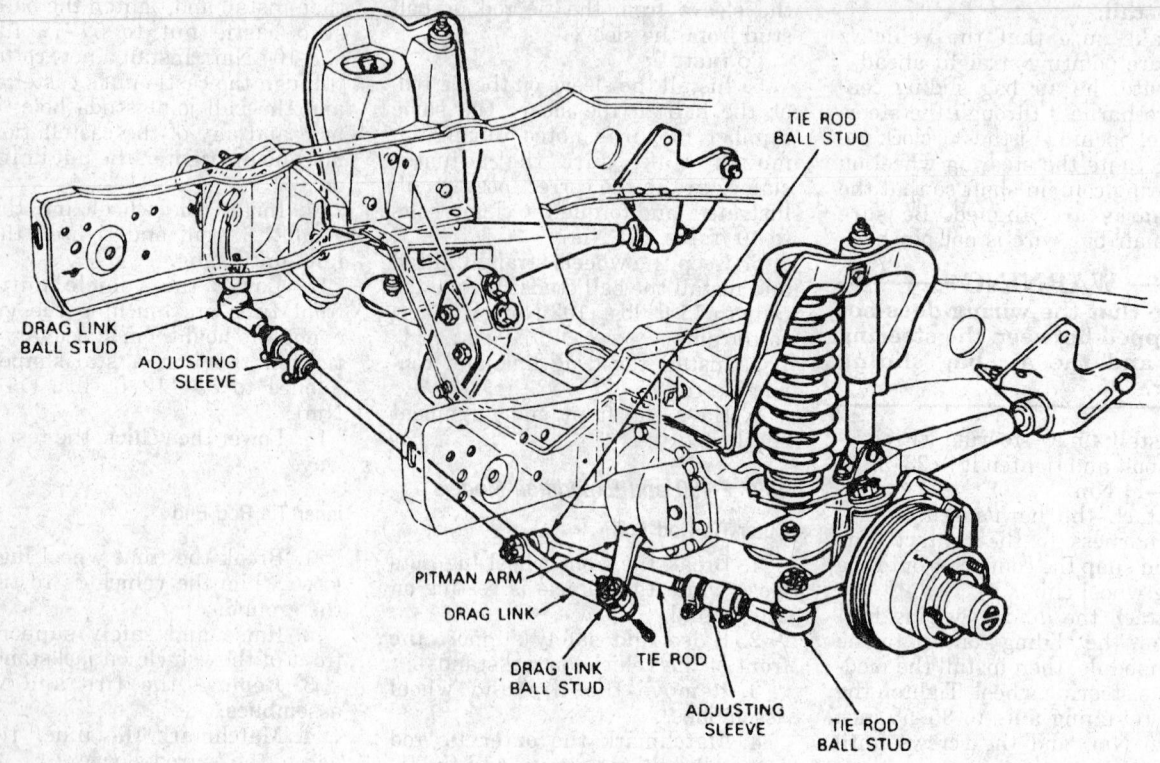

TIE ROD
BALL STUD

DRAG LINK
BALL STUD

ADJUSTING
SLEEVE

PITMAN ARM

DRAG LINK

DRAG LINK
BALL STUD

TIE ROD

ADJUSTING
SLEEVE

TIE ROD
BALL STUD

79215gb7

Steering linkage — Bronco and 4WD 1993–96 F-150 models

quence to 83–112 ft. lbs. (113–153 Nm).

13. Lower the vehicle the rest of the way.

Power Steering Gear

ADJUSTMENT

Except F-Super Duty, Motor Home and 1997 F-150 and Expedition Models

The XR-50 Power Steering Gear is used on all 1993–97 models except for the 1993–97 F-Super Duty stripped chassis and motor home chassis, and the 1997 F-150 and Expedition models.

1. Raise and support the front end on jackstands.
2. Matchmark the pitman arm and gear housing.
3. Set the wheels in a straight-ahead position.
4. Disconnect the pitman arm from the sector shaft.
5. Disconnect the fluid RETURN line at the pump reservoir and cap the reservoir nipple.
6. Place the end of the return line in a clean container and turn the steering wheel lock-to-lock a few times to expel the fluid from the gear.

7. Turn the steering wheel all the way to the right stop. Place a small piece of masking tape on the steering wheel rim as a reference and rotate the steering wheel 45° from the right stop.
8. Disconnect the battery ground.
9. Remove the horn pad.
10. Using an inch-pound torque wrench on the steering wheel nut, record the amount of torque needed to turn the steering wheel 1/8 turn counterclockwise. The preload reading should be 4–9 inch lbs. (0.45–1.00 Nm).
11. Center the steering wheel (1/2 the total lock-to-lock turns) and record the torque needed to turn the steering wheel 90° to either side of center. On a vehicle with fewer than 5000 miles, the meshload should be 15–25 inch lbs. (1.68–2.80 Nm). On a vehicle with 5000 or more miles (8000 km), the meshload should be 7 inch lbs. (0.78 Nm) more than the preload torque. On vehicles with fewer than 5000 miles (8000 km), if the meshload is not within specifications, it should be reset to a figure 14–18 inch lbs. (1.57–2.02 Nm) greater than the recorded preload torque. On vehicles with 5000 or more miles (8000 km), if the meshload is not within specifications, it should be reset to a figure 10–14

inch lbs. (1.12–1.57 Nm) greater than the recorded preload torque

12. If an adjustment is required, loosen the adjuster locknut and turn the sector shaft adjuster screw until the necessary torque is achieved.
13. Once adjustment is completed. hold the adjuster screw and tighten the locknut to 45 ft. lbs. (61 Nm).
14. Recheck the adjustment readings and reset if necessary
15. Connect the return line and refill the reservoir.
16. Install the pitman arm.
17. Install the horn pad.

1997 F-150 and Expedition Models

NOTE: The engine should NOT be running for this test.

1. Turn the steering wheel from the right lock to the left lock at least once.

— **CAUTION** —
The residual power supply must be discharged before any air bag component is serviced.

2. Remove the driver side air bag, as follows:
   a. Disconnect the battery-to-starter relay cable for at least one minute to allow the Air Bag system backup power supply to discharge.

b. Remove the 4 air bag module retaining nuts (2 screws for 1997 F-150 and Expedition models) from the air bag module on the back side of the steering wheel, then lift the module away from the steering wheel.

c. Detach the air bag wiring harness connector, and remove the module from the steering wheel.

3. Raise and support the front of the vehicle safely on jackstands.

4. Remove the steering sector shaft arm drag link castle nut and cotter pin.

5. Using a Pitman arm puller, such as Ford Tool T65P-3590-F, separate the steering sector shaft arm from the drag link.

6. From inside the vehicle, attach an inch pound torque wrench to the steering wheel retaining bolt and measure the rotating torque required to turn the steering wheel 90 degrees from the center position. If the steering wheel requires more force than 13 inch lbs. (1.5 Nm), adjust the steering gear meshload as follows:

a. Hold the steering gear sector shaft and loosen the locknut.

b. Turn the sector shaft until the correct value is measured with the torque wrench at the steering wheel.

c. Hold the sector shaft and tighten the locknut to 19–26 ft. lbs. (26–35 Nm).

7. Install the steering sector shaft arm drag link, castle nut and cotter pin. Tighten the castle nut to 57–76 ft. lbs. (77–104 Nm).

8. Lower the vehicle.

9. Install the driver's side air bag as follows:

a. Route the air bag sliding contact wire harness through the steering wheel opening at the 3 o'clock position. Situate the steering wheel on the steering column shaft so that the matchmarks are aligned. Be sure that the air bag wire is not pinched.

─── **WARNING** ───

Be sure that the wiring does not get trapped between the steering wheel and the air bag sliding contact.

F-Super Duty and Motor Home Chassis

Adjustments must be made with the steering gear removed and mounted in a vise.

NOTE: Backlash is correct when a 4–18 inch lb. increase in rotational torque is noted at the

input shaft as it is rotated and the piston passes the mid-point of its total travel in the housing. The torque increase should occur only at mid-point and should disappear after mid-point.

1. Loosen the locknut and turn the adjusting screw counterclockwise as far as it will go.

2. Using an inch-pound torque wrench, rotate the input shaft as far as it will go in one direction, then, counting the number of full turns and noting the rotational torque, rotate it to the opposite stop.

3. Turn the shaft back ½ the total number of turns to the mid-point.

4. Rotate the shaft 180° to both sides of the mid-point, noting the change in rotational torque. Turn the adjusting screw ⅛–¼ turn at a time until the proper reading of 4–18 inch lb. increase in torque is noted over the mid-point. This increase in torque must be plus the total rotational torque.

5. When the adjustment is correct, hold the adjusting screw and, using a crow's foot adapter, torque the locknut to 74–88 ft. lbs. (100–120 Nm).

6. Check the adjustment to make sure it hasn't changed. Rotate the shaft through its entire travel. It must rotate smoothly.

REMOVAL AND INSTALLATION

Except F-Super Duty, Motor Home and 1997 F-150 and Expedition Models

The XR-50 Power Steering Gear is used on all 1993–;97 models except for the 1993–97 F-Super Duty stripped chassis and motor home chassis, and the 1997 F-150 and Expedition models.

1. Raise and support the front end on jackstands.

2. Place the wheels in the straight-ahead position.

3. Place a drain pan under the gear and disconnect the pressure and return lines. Cap the openings.

4. Remove the splash shield from the flex coupling.

5. Disconnect the flex coupling at the gear.

6. Matchmark and remove the pitman arm from the sector shaft.

7. Support the steering gear and remove the mounting bolts.

8. Remove the steering gear. It may be necessary to work it free of the flex coupling.

To install:

9. Place the splash shield on the steering gear lugs.

10. Slide the flex coupling into place on the steering shaft. Make sure the steering wheel spokes are still horizontal.

11. Center the steering gear input shaft with the indexing flat facing downward.

12. Slide the steering gear input shaft into the flex coupling and into place on the frame side rail. Install the flex coupling bolt and torque it to 30–42 ft. lbs. (41–57 Nm).

13. Install the gear mounting bolts and torque them to 65 ft. lbs. (88 Nm).

14. Make sure the wheels are still straight ahead and install the pitman arm. Torque the nut to 170–288 ft. lbs. (231–392 Nm).

15. Connect the pressure, then, the return lines. Torque the pressure line to 25 ft. lbs. (34 Nm).

16. Snap the flex coupling shield into place.

17. Fill the steering reservoir.

18. Run the engine and turn the steering wheel lock-to-lock several times to expel air. Check for leaks.

1997 F-150 and Expedition Models

1. Raise and safely support the front of the vehicle with jackstands.

2. If equipped remove the undercarriage skid plate.

3. Remove the lower radiator air deflector by removing the 3 screws and 5 push clips.

4. Remove the steering sector shaft arm drag link cotter pin, then remove the castle nut.

5. Use a Pitman arm puller, such as T65P-3590-F, to separate the steering sector shaft arm drag link.

6. Remove the dust shield from the valve housing.

7. Remove the intermediate shaft pinch bolt.

8. Slide the intermediate shaft off the steering gear input shaft.

9. Disconnect the power steering fluid lines from the gear.

10. Remove the 3 steering gear bolts and the steering gear.

11. If the sector shaft arm must be removed to install on a new gear, mount the gear into a vise, then remove the sector shaft nut and lockwasher. Use the Pitman arm puller to draw the sector shaft arm off of the steering gear shaft.

To install:

12. If necessary, install the sector shaft arm on the steering gear shaft, then install the retaining nut and new lockwasher to 170–228 ft. lbs. (234–316 Nm).

13. Situate the power steering gear in position, then install the 3 mounting bolts to 50–68 ft. lbs. (68–92 Nm).

14. Attach the power steering fluid lines to the steering gear, then tighten the fluid line fittings to 20–30 ft. lbs. (27–41 Nm).

15. Position the intermediate shaft onto the steering gear input shaft, then install the pinch bolt to 30–42 ft. lbs. (41–57 Nm).

16. Install the dust shield onto the valve housing.

17. Attach the drag link to the sector shaft arm, then install the castle nut to 57–76 ft. lbs. (77–104 Nm). Tighten the castle nut further until a castellation is aligned with the hole in drag link, then install a new cotter pin.

18. Install the lower radiator air deflector.

19. Install the undercarriage skid plate, if applicable.

20. Lower the vehicle.

F-Super Duty and Motor Home Chassis

1. Raise and support the front end on jackstands.

2. Thoroughly clean all connections.

3. Place a drain pan under the area.

4. Disconnect the hydraulic lines at the gear. Cap all openings at once.

5. Remove the retaining bolt and nut and disconnect the pitman arm from the sector shaft.

6. Remove the bolt and nut securing the input shaft and U-joint.

7. Support the gear and remove the gear-to-frame bolts and nuts.

To install:

8. Position the gear on the frame and install all the bolts and nuts. Torque the nuts to 150–200 ft. lbs. (203–272 Nm).

9. Install the U-joint bolt and nut. Torque the nut to 50–70 ft. lbs. (68–95 Nm).

10. Install the pitman arm.

— **CAUTION** —
Never hammer the pitman shaft onto the sector shaft! Hammering will damage the gear. Use a cold chisel to separate the pitman arm opening.

11. Install the bolt and nut. Torque the nut to 220–300 ft. lbs. (299–408 Nm).

12. Connect the hydraulic lines, fill the reservoir, run the engine and check for leaks.

Power Steering Pump

SYSTEM BLEEDING

Except 1997 F-150 and Expedition Models

1. Fill the power steering fluid reservoir.

2. Disconnect the coil wire.

3. Crank the engine with the starter and continue adding fluid until the level remains constant. Do not prolong cranking as the battery may be drained and the starter damaged.

4. Rotate the steering wheel approximately 30 degrees each side of center while continuing to crank the engine.

5. Recheck the fluid level and fill, as required.

6. Reconnect the coil wire.

7. Start the engine and allow it to run for several minutes.

8. Rotate the steering wheel from stop to stop.

9. Shut off the engine and recheck the fluid level. Add fluid, as required.

10. If air is still trapped in the system, proceed as follows:

   a. Fabricate a purging tool.

   b. Make sure the reservoir fluid level is correct.

   c. Insert the rubber stopper end of the fabricated purging tool tightly into the filler tube.

   d. Connect a suitable length of hose to the purging tool. Connect the other end of the hose to an air conditioner vacuum pump or distributor machine. Do not use engine vacuum.

   e. Start the engine and let it idle for approximately 15 minutes. Turn the steering wheel 1 full cycle every 5 minutes but do not hit the stops. This will assist in removing trapped air.

   f. Stop the engine and disconnect the vacuum source. Remove the purging tool.

   g. Check the fluid level and install the filler tube dipstick.

1997 F-150 and Expedition Models

1. Remove the ignition coil fuse (No. 30) to disable the vehicle.

2. Fill the power steering pump reservoir with MERCON approved fluid (XT-2-QDX) or equivalent.

3. Raise and support the front of the vehicle safely on jackstands.

4. Get an assistant to crank the engine with the starter motor while you add fluid until the level remains constant.

— **WARNING** —
Do not hold the steering wheel at either lock position for more than 5 seconds; damage to the steering pump may occur.

5. With the front wheels off of the ground and while cranking the engine, rotate the steering wheel from lock-to-lock.

6. Check the fluid level and add fluid if necessary.

7. Install the ignition fuse (No. 30) into the fuse block.

8. Start the engine and allow it to run for several minutes.

9. Lower the vehicle.

10. Rotate the steering wheel from lock-to-lock.

11. Turn the engine **OFF** and check the fluid level. Add fluid if necessary.

REMOVAL AND INSTALLATION

C-II Power Steering Pump

This pump is used by all models except the F-Super Duty stripped chassis and motor home models, and the 1997 F-150 and Expedition models.

1. Disconnect the return line at the pump and drain the fluid into a container.

2. Disconnect the pressure line from the pump.

3. Loosen the pump bracket nuts and remove the drive belt. On the 4.9L (VIN Y and Z) and 5.0L (VIN N) with a serpentine drive belt, remove belt tension by lifting the tensioner out of position.

4. Remove the nuts and lift out the pump/bracket assembly.

5. If a new pump or bracket is being installed, remove the pulley from the pump; use a press and adapters.

To install:

6. Position the steering pump on the engine and install the mounting fasteners. Tighten the fasteners to the following specifications:

   • Pivot bolt (4.9L (VIN Y and Z) and 5.0L (VIN N)) — 45 ft. lbs. (61 Nm)

   • Pump-to-adjustment bracket — 45 ft. lbs. (61 Nm)

   • Support bracket-to-engine (5.8L (VIN H and R)) — 65 ft. lbs. (88 Nm)

   • Support bracket-to-water pump housing (4.9L (VIN Y and Z)) — 17 ft. lbs. (23 Nm)

   • Support bracket-to-water pump housing (5.0L (VIN N), 5.8L (VIN H and R)) — 45 ft. lbs. (61 Nm)

   • Pressure line-to-fitting — 29 ft. lbs. (39 Nm)

   • Adjustment bracket-to-support bracket:

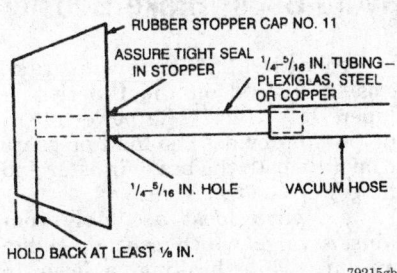

Fabricated power steering system air purging tool — except 1997 F-150 and Expedition models

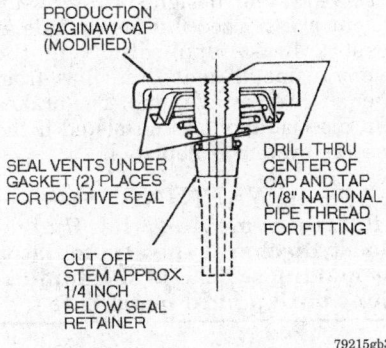

Modified pump reservoir cap — E Series

- 4.9L (VIN Y and Z), 5.0L (VIN N), 5.8L (VIN H and R) — 45 ft. lbs. (61 Nm)
- 7.5L (VIN G) and Diesel — Long bolt — 65 ft. lbs. (88 Nm)
- Short bolt — 45 ft. lbs. (61 Nm)
7. Install the accessory drive belt.
8. Attach the fluid lines to the pump, then bleed the hydraulic steering system.

C-III Power Steering Pump

The C-III steering pump is used on the 1997 F-150 and Expedition models.
1. Remove the engine air cleaner assembly.
2. Remove the accessory drive belt.
3. Disconnect the power steering reservoir pump hose from the power steering pump and drain the fluid into a large catch pan.
4. Disconnect the power steering pressure hose from the steering gear.
5. Remove the 2 top power steering pump bolts.
6. Raise and safely support the front of the vehicle on jackstands.
7. Remove the 2 lower pump bolts and remove the power steering pump from the engine.

8. Lower the vehicle.
9. If necessary, remove the power steering pressure hose fitting from the pump by holding the pump in a vise.
10. If necessary, remove the power steering pump pulley. Hold the Steering Pump Pulley Remover T69L-10300-B, or equivalent, and rotate the forcing bolt to remove the pulley from the pump.

To install:
11. If the pulley was removed, thread the forcing bolt into the power steering pump and turn the steering pump pulley replacer tool, or equivalent, to install the pump pulley.
12. Install a new power steering Teflon® seal onto the pressure hose fitting by stretching the Teflon® seal over the Teflon® seal replacer set D90P-3517-A, or equivalent, until it is large enough to slip over the tube nut.
13. Install the power steering pressure hose, then tighten the fitting to 20–30 ft. lbs. (27–41 Nm).
14. Position the steering pump and loosely install the 4 mounting bolts.
15. Tighten the 2 upper mounting bolts to 15–20 ft. lbs. (20–30 Nm).
16. Raise and support the front of the vehicle safely on jackstands.
17. Tighten the 2 lower mounting bolts to 15–20 ft. lbs. (20–30 Nm).
18. Lower the vehicle.
19. Install the other power steering pressure hose to the steering gear, then tighten to 13–17 ft. lbs. (17–23 Nm).
20. Connect the power steering reservoir pump hose to the power steering pump.
21. Fill and bleed the hydraulic power steering system.

ZF Power Steering Pump

The ZF power steering pump is used on the F-Super Duty stripped chassis and Motor Home models.
1. Disconnect the pressure line from the pump and tie up the ends of both hoses in a raised position. Cap the openings.
2. Loosen the pump pivot and adjusting bolts and remove the drive belt.
3. Remove the bolts and lift out the pump.

To install:
4. Install the pump in the vehicle.
5. Install the pivot bolt and nut finger-tight.
6. Install the accessory drive belt, then adjust the belt tension.

7. Tighten the mounting (pivot and adjusting) bolts to 30–45 ft. lbs. (41–61 Nm).
8. Connect the hoses, then refill the reservoir.
9. Bleed the system, then run the engine and check for leaks.

BRAKES

Anti-Lock Brake System Service

PRECAUTIONS

- Certain components within the Anti-Lock Brake System (ABS) are not intended to be serviced or repaired individually. Only those components with removal and installation procedures should be serviced.
- Do not use rubber hoses or other parts not specifically specified for and ABS system. When using repair kits, replace all parts included in the kit. Partial or incorrect repair may lead to functional problems and require the replacement of components.
- Lubricate rubber parts with clean, fresh brake fluid to ease assembly. Do not use lubricated shop air to clean parts; damage to rubber components may result.
- Use only specified brake fluid from an unopened container.
- If any hydraulic component or line is removed or replaced, it may be necessary to bleed the entire system.
- A clean repair area is essential. Always clean the reservoir and cap thoroughly before removing the cap. The slightest amount of dirt in the fluid may plug an orifice and impair the system function. Perform repairs after components have been thoroughly cleaned; use only denatured alcohol to clean components. Do not allow ABS components to come into contact with any substance containing mineral oil; this includes used shop rags.
- The Anti-Lock control unit is a microprocessor similar to other computer units in the vehicle. Ensure that the ignition switch is **OFF** before removing or installing controller harnesses. Avoid static electricity discharge at or near the controller.
- If any arc welding is to be done on the vehicle, the control unit should be unplugged before welding operations begin.

Master Cylinder

REMOVAL AND INSTALLATION

1. With the engine **OFF**, depress the brake pedal several times to expel any vacuum.

2. Disconnect the fluid level warning switch wire.

3. Disconnect the hydraulic system brake lines at the master cylinder.

4. Remove the master cylinder retaining nuts and remove the master cylinder.

To install:

5. Position the master cylinder assembly on the booster and install the retaining nuts. Tighten the nuts to 18–25 ft. lbs. (25–34 Nm).

6. Connect the hydraulic brake system lines to the master cylinder.

7. Connect the wiring.

8. Bleed the master cylinder.

BRAKE BOOSTER PUSHROD ADJUSTMENT

The pushrod has an adjustment screw to maintain the correct relationship between the booster control valve plunger and the master cylinder piston. If the plunger is too long it will prevent the master cylinder piston from completely releasing hydraulic pressure, causing the brakes to drag. If the plunger is too short it will cause excessive pedal travel and an undesirable clunk in the booster area. Remove the master cylinder for access to the booster pushrod.

To check the adjustment of the screw, place a gauge against the master cylinder mounting surface of the booster body. Adjust the pushrod screw by turning it until the end of the screw just touches the inner edge of the slot in the gauge. Install master cylinder and bleed the system.

Diesel Brake Booster Vacuum Pump

The Diesel is equipped with a vacuum pump, which is driven by a single belt off of the alternator. This pump is located on the top right side of the engine.

Diesel pick-ups are also equipped with a low vacuum indicator switch which actuates the BRAKE warning lamp when available vacuum is below a certain level. The switch senses vacuum through a fitting in the vac-

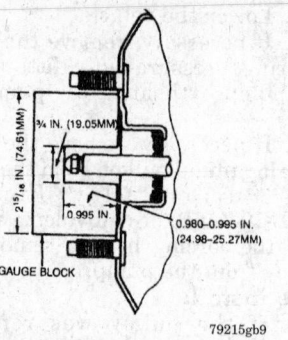

Fabricate a template to adjust the power booster pushrod as shown

uum manifold that intercepts the vacuum flow from the pump. The low vacuum switch is mounted on the right side of the engine compartment, adjacent to the vacuum pump on F-250 and F-350 models.

NOTE: The vacuum pump cannot be disassembled. It is only serviced as a unit (the pulley is separate).

REMOVAL AND INSTALLATION

1. Remove the hose clamp and disconnect the pump from the hose on the manifold vacuum outlet fitting.

2. Loosen the vacuum pump adjustment bolt and the pivot bolt. Slide the pump downward and remove the drive belt from the pulley.

3. Remove the pivot and adjustment bolts and the bolts retaining the pump to the adjustment plate. Remove the vacuum pump and adjustment plate.

To install:

4. Install the pump-to-adjustment plate bolts and tighten to 11–18 ft. lbs. (15–25 Nm). Position the pump and plate on the vacuum pump bracket and loosely install the pivot and adjustment bolts.

5. Connect the hose from the manifold vacuum outlet fitting to the pump and install the hose clamp.

6. Install the drive belt on the pulley. Place a 3/8 in. drive breaker bar or ratchet into the slot on the vacuum pump adjustment plate. Lift up on the assembly until the proper belt tension is obtained. Tighten the pivot and adjustment bolts to 11–18 ft. lbs. (15–25 Nm).

7. Start the engine and make sure the brake system functions properly.

NOTE: The BRAKE light will glow until brake vacuum builds up to the normal level.

Hydro-Boost Brake Booster

A hydraulically powered brake booster is used on the F-Series F-Super Duty truck. The power steering pump provides the fluid pressure to operate both the brake booster and the power steering gear.

The Hydro-Boost assembly contains a valve which controls pump pressure while braking, a lever to control the position of the valve and a boost piston to provide the force to operate a conventional master cylinder attached to the front of the booster. The Hydro-Boost also has a reserve system, designed to store sufficient pressurized fluid to provide at least 2 brake applications in the event of insufficient fluid flow from the power steering pump. The brakes can also be applied unassisted if the reserve system is depleted.

———— **WARNING** ————

Before removing the Hydro-Boost, discharge the accumulator by making several brake applications until a hard pedal is felt.

REMOVAL AND INSTALLATION

———— **CAUTION** ————

Do not depress the brake pedal with the master cylinder removed!

1. Remove the master cylinder from the Hydro-Boost unit. Do not disconnect the brake lines from the master cylinder! Position the master cylinder out of the way.

2. Disconnect the 3 hydraulic lines from the Hydro-Boost unit.

3. Disconnect the pushrod from the brake pedal.

4. Remove the booster mounting nuts and lift the booster from the firewall.

———— **CAUTION** ————

The booster should never be carried by the accumulator. The accumulator contains high pressure nitrogen and can be dangerous if mishandled! If the accumulator is to be disposed of, do not expose it to fire or other forms of incineration! Gas pressure can be relieved by drilling a 1/16 in. (1.5mm) hole in the end of the accumulator can. Always wear safety goggles during the drilling!

5. Installation is the reverse of removal. Tighten the booster mounting nuts to 25 ft. lbs. (34 Nm); the master cylinder nuts to 25 ft. lbs. (34 Nm);

connect the hydraulic lines, refill and bleed the booster as follows:

  a. Fill the pump reservoir with Dexron®II ATF.

  b. Disconnect the coil wires and crank the engine for several seconds.

  c. Check the fluid level and refill, if necessary.

  d. Connect the coil wires and start the engine.

  e. With the engine running, turn the steering wheel lock-to-lock twice. Shut off the engine.

  f. Depress the brake pedal several times to discharge the accumulator.

  g. Start the engine and repeat Step 5e.

  h. If foam appears in the reservoir, allow the foam to dissipate.

  i. Repeat Step 5e as often as necessary to expel all air from the system.

NOTE: The system is, in effect, self-bleeding and normal vehicle operation will expel any further trapped air.

Brake Caliper

REMOVAL AND INSTALLATION

Front Brake Calipers

Rail Slider Type

This brake system utilizes two press-in rails to retain the caliper, and on which the calipers slide to apply brake pressure to the rotors. This system is used by the following vehicles:

• 1993 models — Bronco, F-150, F-250, F-350, F-Super Duty, and E-150

• 1994 models — F-250, F-350 and F-Super Duty

• 1995–97 models — F-Super Duty

1. Raise and safely support the vehicle and remove the wheels.

2. Place an 8 in. (203mm) C-clamp on the caliper and tighten the clamp to bottom the caliper piston in the cylinder bore. Bear the clamp on the outer pad; never press directly on the piston! Remove the C-clamp.

3. Clean the excess dirt from around the caliper pin tabs.

4. Drive the upper caliper pin inward until the tabs on the pin touch the spindle.

5. Insert a small prybar into the slot provided behind the pin tabs on the inboard side of the pin.

6. Using needle nose pliers, compress the outboard end of the pin

while, at the same time, prying with the prybar until the tabs slip into the groove in the spindle.

7. Place the end of a 7/16 in. (11mm) punch against the end of the caliper pin and drive the pin out of the caliper slide groove.

8. Repeat this procedure for the lower pin.

9. Lift the caliper off of the rotor.

10. Remove the brake pads from the caliper.

11. Disconnect the brake hose from the caliper, then plug the brake line to prohibit contamination of the brake fluid by water or dirt.

To install:

12. Connect the brake hose to the caliper. When connecting the brake fluid hose to the caliper, it is recommended that a new copper washer be used at the connection of the brake hose and caliper.

13. Thoroughly clean the areas of the caliper and spindle assembly which contact each other during the sliding action of the caliper.

14. Install the brake pads onto the caliper.

15. Position the caliper on the spindle assembly. Lightly lubricate the caliper sliding grooves with caliper pin grease.

16. Position the a new upper pin with the retention tabs next to the spindle groove.

NOTE: Don't use the bolt and nut with the new pin.

17. Carefully drive the pin, at the outboard end, inward until the tabs contact the spindle face.

18. Repeat the procedure for the lower pin.

─── **WARNING** ───

Don't drive the pins in too far, or it will be necessary to drive them back out until the tabs snap into place. The tabs on each end of the pin must be free to catch on the spindle sides!

19. Install the wheels.

Pin Slider Type

The calipers used on this brake system utilize two bolts, or pins, to retain the calipers to the anchor plates. The calipers slide on these bolts to apply force against the brake rotors. The following vehicles utilize this system of caliper retention:

• 1993 models — E-250 and E-350

• 1994 models — Bronco, F-150, E-150, E-250 and E-350

• 1995–97 models — Bronco, Expedition, F-150, F-250, F-350, E-150, E-250 and E-350

1. Break the front wheel lug nuts loose, then raise and support the front of the vehicle safely on jackstands.

2. Remove the front wheels.

3. Remove the front brake hose bolt, then remove the copper washers and plug the front brake hose.

4. Remove the 2 front disc brake caliper slide pins, then lift the caliper off of the front caliper anchor plate.

To install:

5. Install the front disc brake caliper onto the caliper anchor plate, then install the 2 slide pins. Tighten the slider pins/bolts to the following values:

• 1994–97 Bronco, F-150 and E-150 Models — 21–26 ft. lbs. (28–36 Nm)

• 1996–97 F-250, F-350, E-250 and E-350 Models — 141–190 ft. lbs. (191–259 Nm)

• 1995 F-250, F-350, E-250 and E-350 Models — 85–100 ft. lbs. (115–135 Nm)

• 1993–94 E-250 and E-350 Models — 85–100 ft. lbs. (115–135 Nm)

6. Using new copper washers, attach the front brake hose to the brake caliper, then install and tighten the retaining bolt to 23–29 ft. lbs. (30–40 Nm) for 1997 models, to 10–15 ft. lbs. (14–20 Nm) for 1993–96 models.

7. Bleed the brake system.

8. Clean the wheel hub mounting surface.

9. Install the front wheels and snug the lug nuts to fully seat the wheel against the hub.

10. Lower the vehicle until some of the vehicle's weight rests on the front tires, then tighten the lug nuts to 83–112 ft. lbs. (113–153 Nm).

11. Lower the vehicle completely.

12. Make sure that the brakes are operating correctly.

Rear Brake Caliper

The only Ford Trucks covered by this manual equipped with rear disc brakes are the 1995–97 F-Super Duty models (stripped chassis and motor home).

1. Remove sufficient brake fluid from the brake master cylinder reservoir to allow for pressing the caliper pistons into the bores. Discard the used brake fluid.

2. Raise and safely support the vehicle.

3. Remove the wheels.

4. Place an 8 in. (203mm) C-clamp on the caliper, with he clamp frame on the disc brake caliper and the clamp screw on the outboard brake hoes and lining backing plate, then tighten the clamp to press the caliper

pistons in the cylinder bores only enough to give removal clearance. Remove the C-clamp.

5. Remove the rear brake hose-to-caliper flow bolt and discard the used copper washers. Plug the brake hose so that contamination of the brake fluid does not occur.

6. Using the Hydraulic Caliper Pin Remover D89T-2196-A, or equivalent, drive the upper and lower caliper locking pins out.

7. Remove the disc brake caliper from the vehicle.

To install:

8. If necessary, install the brake shoe and anti-rattle clip.

9. Position the disc brake caliper in the support bracket.

10. Lubricate the caliper locking pins with Ford Silicone Dielectric Compound D7AZ-19A331-A or equivalent.

11. Drive the locking pins into the caliper/support bracket assembly until the tabs at each end of the pin snap into place.

12. Using new copper washers, attach the rear brake hose to the disc brake caliper. Tighten the rear brake hose-to-caliper flow bolt to 22–29 ft. lbs. (30–39 Nm).

13. Install the wheel, then lower the vehicle.

NOTE: Check the brake fluid level in the brake master cylinder reservoir and maintain sufficient level to prevent air from entering the system.

14. Pump the brake pedal several times to drive the caliper pistons into contact with the brake shoe and lining.

15. Fill the brake master cylinder reservoir with new DOT 3 brake fluid.

16. Bleed the brake system and test the brakes for proper operation.

Disc Brake Pads

REMOVAL AND INSTALLATION

Front Brake Pads

NOTE: Never replace the pads on one side only! Always replace pads on both wheels as a set!

1. Break the front wheel lug nuts loose, then raise and support the front of the vehicle safely on jackstands.

2. Remove the front wheels.

3. Remove the front disc brake calipers.

4. Note the position and orientation of the brake pads and anti-rattle clip. Remove the brake shoes and lings from the front disc brake caliper anchor plate, then remove the anti-rattle clips.

To install:

5. Thoroughly clean the areas of the caliper and spindle assembly which contact each other during the sliding action of the caliper.

6. Place a new anti-rattle clip on the lower end of the inboard shoe. Make sure the tabs on the clip are positioned correctly and the loop-type spring is away from the rotor.

7. Place the lower end of the inner brake pad in the spindle assembly pad abutment, against the anti-rattle clip, and slide the upper end of the pad into position. Be sure the clip is still in position.

8. Check and make sure the caliper piston is fully bottomed in the cylinder bore. Use a large C-clamp, bearing on a piece of wood, to bottom the piston, if necessary.

9. Position the outer brake pad on the caliper, and press the pad tabs into place with your fingers. If the pad cannot be pressed into place by hand, use a C-clamp. Be careful not to damage the lining with the clamp. Bend the tabs to prevent rattling.

10. Lightly lubricate the caliper sliding grooves with caliper pin grease.

11. Install the brake caliper onto the anchor plate.

12. Bleed the brake system.

13. Clean the wheel hub mounting surface.

14. Install the front wheels and snug the lug nuts to fully seat the wheel against the hub.

15. Lower the vehicle until some of the vehicle's weight rests on the front tires, then tighten the lug nuts to 83–112 ft. lbs. (113–153 Nm).

16. Lower the vehicle completely.

17. Make sure that the brakes are operating correctly.

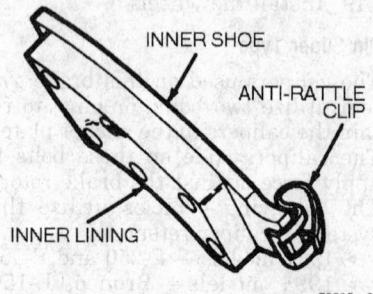

Installing anti-rattle clip on inner pad — single piston caliper — Bronco, F Series (Pick-Up) and E Series (Van)

Rear Brake Pads

NOTE: Never replace the pads on one side only! Always replace pads on both wheels as a set!

1. Remove the rear brake caliper from the rear hub without disconnecting the brake hose.

——— WARNING ———
Do not allow the caliper to hand by the brake hose.

2. Remove the brake pads and anti-rattle spring.

3. Thoroughly clean the areas of the caliper and caliper support assembly which contact each other during the sliding action of the caliper.

To install:

4. Position the brake shoes and anti-rattle clips on the disc brake caliper support bracket.

5. Install the rear disc brake caliper onto the rear support bracket.

6. Bleed the brake system.

7. Clean the wheel hub mounting surface.

8. Install the front wheels and snug the lug nuts to fully seat the wheel against the hub.

9. Lower the vehicle until some of the vehicle's weight rests on the front tires, then tighten the lug nuts to 83–112 ft. lbs. (113–153 Nm).

10. Lower the vehicle completely.

11. Make sure that the brakes are operating correctly.

Brake Rotor

REMOVAL AND INSTALLATION

Front Brake Rotor

Except 1997 F-150 and Expedition Models

1. Remove the front disc brake caliper and brake pads.

2. Remove the 2 front disc brake caliper anchor plate bolts, then lift the anchor plate off of the rotor.

3. Remove the hub grease cap.

4. Remove the cotter pin, then remove the locknut nut.

5. Remove the bearing retaining nut.

6. Remove the front wheel outer bearing retainer washer.

7. Remove the outer front wheel bearing.

8. Slide the front disc brake rotor off of the spindle.

9. Remove the bearing cone and roller. Discard the wheel hub grease seal.

10. On F-Super Duty models, the front disc brake rotor can be removed

from the wheel hub by removing the 10 Torx® machine screws.

To install:

11. On 4-wheel drive vehicles, position the front disc brake rotor on the hub, after cleaning the rotor with Ford Metal Brake Parts Cleaner F3AZ-19579-SA or equivalent. The machine screw threads must be coated with a suitable adhesive before attaching the front disc brake rotor to the wheel hub, then tighten the Torx® screws to 74–89 ft. lbs. (100–120 Nm).

12. Thoroughly clean and inspect the front wheel bearings and the front disc brake hub and rotor.

13. Lubricate the front wheel bearings with Premium Long-Life Grease Xg-1-C or -K, or equivalent.

14. Install the front inner wheel bearing, then press in a new inner wheel hub grease seal.

15. Position the front disc rotor and hub on the spindle.

16. Install the outer front wheel bearing into the rotor and hub.

17. Install the outer bearing retainer washer, then install the wheel bearing adjusting nut finger-tight.

NOTE: The front disc brake rotor should be rotated while adjusting the front wheel bearing end-play.

18. Tighten the wheel bearing adjusting nut to 30 ft. lbs. (40 Nm).

19. Loosen the spindle nut 2 complete revolutions (720 degrees).

20. Spin the rotor counterclockwise and tighten the spindle nut to 17–24 ft. lbs. (23–34 Nm).

21. Loosen the spindle nut just shy of 1/2 revolution (175 degrees).

22. Tighten the spindle nut 17 inch lbs. (2 Nm).

23. Install the retainer nut, a new cotter pin and the hub grease cap.

24. Position the front disc brake caliper anchor plate, then install the mounting bolts to 125–170 ft. lbs. (170–230 Nm).

25. Install the brake pads and brake caliper.

26. Bleed the brake system.

27. Clean the wheel hub mounting surface.

28. Install the front wheels and snug the lug nuts to fully seat the wheel against the hub.

29. Lower the vehicle until some of the vehicle's weight rests on the front tires, then tighten the lug nuts to 83–112 ft. lbs. (113–153 Nm).

30. Lower the vehicle completely.

31. Make sure that the brakes are operating correctly.

1997 F-150 and Expedition Models

1. Remove the front disc brake caliper and brake pads.

2. Remove the 2 front disc brake caliper anchor plate bolts, then lift the anchor plate off of the rotor.

3. On 4-wheel drive vehicles, slide the front disc brake rotor off of the front hub assembly.

4. On 2-wheel drive vehicles, remove the front disc rotor as follows:

a. Remove the hub grease cap.

b. Remove the cotter pin, then remove the retainer nut.

c. Remove the spindle nut.

d. Remove the front wheel outer bearing retainer washer.

e. Remove the outer front wheel bearing.

f. Slide the front disc brake rotor off of the spindle.

5. Remove the wheel hub grease seal and inner front wheel bearing.

To install:

6. On 4-wheel drive vehicles, position the front disc brake rotor on the hub, after cleaning the rotor with Ford Metal Brake Parts Cleaner F3AZ-19579-SA or equivalent.

7. On 2-wheel drive vehicles, install the rotor as follows:

a. Thoroughly clean and inspect the front wheel bearings and the front disc brake hub and rotor.

b. Lubricate the front wheel bearings with Premium Long-Life Grease Xg-1-C or -K, or equivalent.

c. Install the front inner wheel bearing, then press in a new inner wheel hub grease seal.

d. Position the front disc rotor and hub on the spindle.

e. Install the outer front wheel bearing into the rotor and hub.

f. Install the outer bearing retainer washer, then install the spindle nut finger-tight.

NOTE: The front disc brake rotor should be rotated while adjusting the front wheel bearing end-play.

g. Tighten the spindle nut to 30 ft. lbs. (40 Nm).

h. Loosen the spindle nut 2 complete revolutions (720 degrees).

i. Spin the rotor counterclockwise and tighten the spindle nut to 17–24 ft. lbs. (23–34 Nm).

j. Loosen the spindle nut just shy of 1/2 revolution (175 degrees).

k. Tighten the spindle nut 17 inch lbs. (2 Nm).

l. Install the retainer nut, a new cotter pin and the hub grease cap.

8. Position the front disc brake caliper anchor plate, then install the

mounting bolts to 125–170 ft. lbs. (170–230 Nm).

9. Install the brake pads and brake caliper.

10. Bleed the brake system.

11. Clean the wheel hub mounting surface.

12. Install the front wheels and snug the lug nuts to fully seat the wheel against the hub.

13. Lower the vehicle until some of the vehicle's weight rests on the front tires, then tighten the lug nuts to 83–112 ft. lbs. (113–153 Nm).

14. Lower the vehicle completely.

15. Make sure that the brakes are operating correctly.

Rear Brake Rotor

1. Raise and safely support the vehicle. Remove the wheel.

2. Remove the caliper assembly and support it to the frame with a piece of wire without disconnecting the brake fluid hose.

3. Remove the axle shaft.

NOTE: Press in on the tool to disengage the nut ratchet.

4. Remove the wheel bearing hub nut using the Hub Locknut Wrench T88T-4252-A, or equivalent, so that the drive tangs engage the four slots of the hub nut.

5. Remove the outer wheel bearing cone and roller.

6. Protect the rear axle spindle threads with cardboard tubing, tape or equivalent.

7. Remove the rear hub and rotor assembly using care not to damage the rear axle spindle threads.

8. With the rotor facing upward, secure the rear hub and rotor assembly in a vise with protective jaw caps.

9. Remove the 10 rotor-to-rear hub bolts and remove the rotor from the rear hub.

To install:

10. Position the rotor on the rear hub, then install the 10 rotor-to-rear hub bolts to 84–112 ft. lbs. (113–153 Nm).

11. Protect the rear axle spindle threads with cardboard tubing or tape.

12. Carefully position the rear hub and rotor assembly on the rear axle spindle, then remove the thread protection.

13. Install the outer wheel bearing cone and roller, then start the bearing hub nut making sure that the hub nut tab is properly located in the keyway prior to thread engagement.

14. Rotate the hub occasionally while tightening to seat the bearings. Position the Hub Locknut Wrench T88T-4252-A, or equivalent, so that the drive tangs of the tool engage the four slots of the nut. Apply inward pressure to the socket to separate the ratcheting components of the nut, then tighten the nut to 65–75 ft. lbs. (88–102 Nm). Back the nut off 90 degrees (¼ turn), then retighten to 15–20 ft. lbs. (20–27 Nm).

15. Prior to installation of the axle shaft, clean and remove any debris in the hub bolt holes.

16. Inspect the axle shaft for cracked material around the holes and for oversized holes. Replace the axle shaft if these conditions are found.

17. Install the axle shaft and rear wheel gasket.

WARNING

Do not tighten the axle shaft-to-rear hub bolts until the lug nuts are tightened to specification.

18. Using new axle shaft-to-rear hub bolts, coat the bolt threads with a suitable thread adhesive and install the bolts only until they seat.

19. Install the disc brake caliper.

20. Install the wheels. Tighten the lug nuts until snug.

21. Lower the vehicle, then tighten the lug nuts to 140 ft. lbs. (190 Nm).

22. Tighten the axle shaft-to-rear hub bolts to 84–113 ft.lbs. (113–153 Nm).

Brake Drums

REMOVAL AND INSTALLATION

1. Relieve tension on the parking brake system by pulling on the front parking brake cable and conduit, then inserting a thin pry tool into the bottom hole of the parking brake pedal assembly to hold the pedal in place.

2. Raise and safely support the vehicle. Remove the wheel and tire assembly.

3. Remove the spring retaining nuts, if equipped, and remove the brake drum.

4. If the brake drum will not come off, insert a narrow prybar through the brake adjusting hole in the backing plate and disengage the adjusting lever from the adjusting screw. While holding the adjusting lever away from the adjusting screw, loosen the adjusting screw with a brake adjusting tool.

5. Inspect the brake drum surface for wear, scoring and runout. Machine or replace, as necessary.

To install:

6. Before installing a new brake drum, be sure to remove any protective coating with carburetor degreaser.

7. Slide the brake drum onto the wheel lug studs until completely seated against the brake assembly.

8. If applicable, install new push retaining nuts on the lug studs to secure the drum in position.

9. Adjust the rear brake drums so that a slight drag can be felt against the drum.

10. Install the rear wheels, then lower the vehicle.

11. Remove the thin pry tool from the parking brake pedal assembly.

12. Test the brakes to ensure they are functioning correctly.

ADJUSTMENT

Drum Installed

1. Raise and safely support the vehicle.

2. Remove the rubber plug from the adjusting slot on the backing plate.

3. Insert a brake adjusting spoon into the slot and engage the lowest possible tooth on the starwheel. Move the end of the brake spoon downward to move the starwheel upward and expand the adjusting screw. Repeat this operation until the brakes lock the wheels.

4. Insert a small prybar or piece of firm wire (coat hanger wire) into the adjusting slot and push the automatic adjusting lever out and free of the starwheel on the adjusting screw and hold it there.

5. Engage the topmost tooth possible on the starwheel with the brake adjusting spoon. Move the end of the adjusting spoon upward to move the adjusting screw starwheel downward

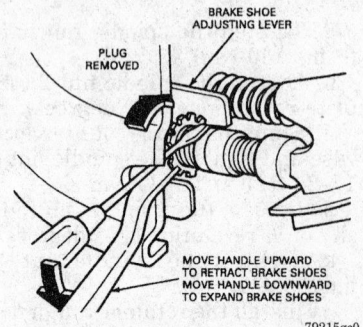

Adjust the brake shoes in, if the rear brake drum will not slide off the brake assembly

and contract the adjusting screw. Back off the adjusting screw starwheel until the wheel spins freely with a minimum of drag. Keep track of the number of turns that the starwheel is backed off, or the number of strokes taken with the brake adjusting spoon.

6. Repeat this operation for the other side. When backing off the brakes on the other side, the starwheel adjuster must be backed off the same number of turns to prevent side-to-side brake pull.

7. When the brakes are adjusted make several stops while backing the vehicle, to equalize the brakes at both of the wheels.

8. Remove the safety stands and lower the vehicle. Road test the vehicle.

Drum Removed

1. Make sure the shoe-to-contact pad areas are clean and properly lubricated.

2. Using an inside caliper check the inside diameter of the drum. Measure across the diameter of the assembled brake shoes, at their widest point.

3. Turn the adjusting screw so the diameter of the shoes is 0.030 in. (0.76mm) less than the brake drum inner diameter.

4. Install the drum.

Brake Shoes

REMOVAL AND INSTALLATION

Bronco, F-150 and E-150 Models

1. Raise and safely support the vehicle. Remove the wheel and tire assembly and the brake drum.

2. Pull backward on the adjusting lever cable to disengage the adjusting lever from the adjusting screw. Move the outboard side of the adjusting screw upward and back off the pivot nut as far as it will go.

3. Pull the adjusting lever, cable and automatic adjuster spring down and toward the rear to unhook the pivot hook from the large hole in the secondary shoe web. Do not pry the pivot hook from the hole.

4. Remove the automatic adjuster spring and adjusting lever.

5. Remove the secondary shoe-to-anchor spring using a suitable brake spring removal/installation tool. Using the tool, remove the primary shoe-to-anchor spring and unhook the cable anchor. Remove the anchor pin plate, if equipped.

6. Remove the cable guide from the secondary shoe.

7. Remove the shoe hold-down springs, shoes, adjusting screw, pivot nut and socket. Note the color and position of each hold-down spring so they can be reassembled in the same position.

8. Remove the parking brake link and spring. Disconnect the parking brake cable from the parking brake lever.

9. Remove the secondary brake shoe. On 9 in. rear brakes, remove the parking brake lever from the shoe. On 10 in. rear brakes, remove the retainer clip and spring washer and remove the parking brake lever.

To install:

10. Clean the backing plate ledge pads and sand lightly. Apply a light coating of high temperature lithium grease to the points where the brake shoes touch the backing plate. Lubricate the adjusting cable eye and the anchor pin area.

11. Install the parking brake lever on the secondary shoe. On 10 in. brakes, secure with the spring washer and retaining clip.

12. Position the brake shoes on the backing plate and install the hold-down spring pins, springs and cups. Install the parking brake link, spring and washer. Connect the parking brake cable to the parking brake lever.

13. Install the anchor pin plate, if equipped, and place the cable anchor over the anchor pin with the crimped side toward the backing plate.

14. Install the primary shoe-to-anchor spring using the brake spring removal/installation tool.

15. Install the cable guide on the secondary shoe with the flanged hole fitted into the hole in the secondary shoe. Thread the cable around the cable guide groove.

NOTE: Make sure the cable is positioned in the groove and not between the guide and shoe web.

16. Install the secondary shoe-to-anchor (long) spring.

NOTE: Make sure the cable end is not cocked or binding on the anchor pin when installed. All parts should be flat on the anchor pin.

17. Apply high temperature lithium grease to the threads and the socket end of the adjusting screw. Turn the adjusting screw into the adjusting pivot nut to the end of the threads and then loosen, ½ turn.

18. Place the adjusting socket on the screw and install the assembly between the shoe ends with the adjusting screw nearest the secondary shoe.

NOTE: Be sure to install the adjusting screw on the same side of the vehicle from which it came. To prevent incorrect installation, the socket end of each adjusting screw is stamped with R or L, to indicate installation on the right or left side of the vehicle. The adjusting pivot nuts have lines machined around the body of the nut, 2 lines indicating the right side nut and 1 line indicating the left side nut.

19. Hook the cable hook into the hole in the adjusting lever from the outboard plate side. The adjusting levers are also stamped with an **R** or **L** to indicate right or left side installation.

20. Place the hooked end of the adjuster spring in the large hole in the primary shoe web and connect the loop end of the spring to the adjuster lever hole.

21. Pull the adjuster lever, cable and automatic adjuster spring down toward the rear to engage the pivot hook in the large hole in the secondary shoe web.

22. After installation, check the action of the adjuster by pulling the section of the cable between the cable guide and the adjusting lever toward the secondary shoe web far enough to lift the lever past a tooth on the adjusting screw wheel. The lever should snap into position behind the next tooth and releasing the cable should cause the adjuster spring to return the lever to its original position. This return action will turn the adjusting screw 1 tooth.

23. If pulling the cable does not produce the action described in Step 22 or if lever action is sluggish instead of positive and sharp, check the position of the lever on the adjusting screw toothed wheel. With the brake in a vertical position, anchor at the top, the lever should contact the adjusting wheel 1 tooth above the center line of the adjusting screw. If the contact point is below the center line, the lever will not lock on the adjusting screw wheel teeth and the screw will not turn as the lever is actuated by the cable.

24. To find the cause of the condition described in Step 23, proceed as follows:

  a. Check the cable and fittings. The cable should completely fill or extend slightly beyond the crimped section of the fittings. If this does not happen, the cable assembly may be damaged and should be replaced.

  b. Check the cable guide for damage. The cable groove should be parallel to the shoe web and the body of the guide should lie flat against the web. Replace the guide if it shows damage.

  c. Check the pivot hook on the lever. The hook surfaces should be square with the body on the lever for proper pivoting. Repair the hook or replace the lever if the hook shows damage.

  d. Be sure the adjusting screw socket is properly seated in the notch in the shoe web.

25. Adjust the brake shoes using either a brake adjustment gauge or manually with the drums installed.

26. If using a brake adjustment gauge, proceed as follows:

  a. Measure the inside diameter of the brake drum with the gauge.

  b. Reverse the tool and adjust the brake shoes until they touch the gauge. The gauge contact points on the shoes must be parallel to the vehicle with the center line through the center of the axle.

  c. Install the drum and wheel and tire assembly. Lower the vehicle.

  d. Apply the brakes sharply several times while driving the vehicle in reverse. Check brake operation by making several stops while driving forward.

27. If manually adjusting the brakes, proceed as follows:

  a. a. Install the brake drum and wheel and tire assembly.

  b. Remove the cover from the adjusting hole at the bottom of the backing plate and turn the adjust-

Rear brake shoe assembly — Bronco, F-150 and E-150

79215gc1

ing screw, using a suitable brake adjusting tool, to expand the brake shoes until they drag against the brake drum.

c. When the shoes are against the drum, insert a narrow prybar through the brake adjusting hole and disengage the adjusting lever from the adjusting screw. While holding the adjusting lever away from the adjusting screw, loosen the adjusting screw with the brake adjusting tool, until the drum rotates freely without drag.

d. Install the adjusting hole cover and lower the vehicle.

e. Apply the brakes. If the pedal travels more than halfway to the floor, there is too much clearance between the brake shoes and drums. Repeat the adjustment procedure.

F-250, F-350, E-250 and E-350 Models

1. Raise and support the vehicle.
2. Remove the wheel and drum.
3. Remove the parking brake lever assembly retaining nut from behind the backing plate and remove the parking brake lever assembly.
4. Remove the adjusting cable assembly from the anchor pin, cable guide, and adjusting lever.
5. Remove the brake shoe retracting springs.
6. Remove the brake shoe hold-down spring from each shoe.
7. Remove the brake shoes and adjusting screw assembly.
8. Disassemble the adjusting screw assembly.
9. Clean the ledge pads on the backing plate. Apply a light coat of Lubriplate® to the ledge pads (where the brake shoes rub the backing plate).

To install:

10. Apply Lubriplate® to the adjusting screw assembly and the hold-down and retracting spring contacts on the brake shoes.

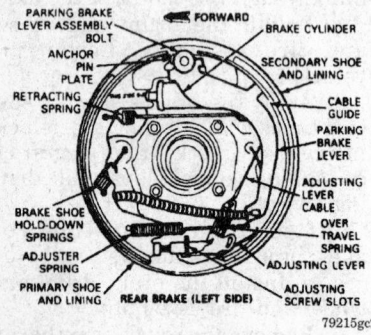

Rear brake shoe assembly — F-250, F-350, E-250 and E-350

79215gc2

11. Install the upper retracting spring on the primary and secondary shoes and position the shoe assembly on the backing plate with the wheel cylinder pushrods in the shoe slots.

12. Install the brake shoe hold-down springs.

13. Install the brake shoe adjustment screw assembly with the slot in the head of the adjusting screw toward the primary shoe, lower retracting spring, adjusting lever spring, adjusting lever assembly, and connect the adjusting cable to the adjusting lever. Position the cable in the cable guide and install the cable anchor fitting on the anchor pin.

14. Install the adjusting screw assemblies in the same locations from which they were removed. Interchanging the brake shoe adjusting screws from one side of the vehicle to the other will cause the brake shoes to retract rather than expand each time the automatic adjusting mechanism is operated. To prevent incorrect installation, the socket end of each adjusting screw is stamped with an **R** or an **L** to indicate their installation on the right or left side of the vehicle. The adjusting pivot nuts can be distinguished by the number of lines machined around the body of the nut. Two lines indicate a right hand nut; one line indicates a left hand nut.

15. Install the parking brake assembly in the anchor pin and secure with the retaining nut behind the backing plate.

16. Adjust the brakes before installing the brake drums and wheels. Install the brake drums and wheels.

17. Lower the vehicle and road test the brakes. New brakes may pull to one side or the other before they are seated. Continued pulling or erratic braking should not occur.

Wheel Cylinder

REMOVAL AND INSTALLATION

1. Raise and safely support the vehicle. Remove the wheel and tire assembly, brake drum and brake shoes.
2. Remove the cylinder-to-shoe connecting pins.
3. Disconnect the brake line from the wheel cylinder. Plug the brake lines immediately after removal so that contamination of the brake fluid does not occur.
4. Remove the wheel cylinder retaining bolts and remove the cylinder from the brake backing plate.

To install:

5. Position the wheel cylinder against the backing plate, then install the mounting bolts. Tighten the mounting bolts to 11–19 ft. lbs. (15–26 Nm).
6. Attach the brake line to the wheel cylinder. Tighten the fitting to 10–14 ft. lbs. (15–20 Nm).
7. Install the wheel cylinder-to-brake shoe connecting pins, then install the brake shoes and other components.
8. Install the brake drum, then adjust the brake shoes.
9. Install the wheels and lower the vehicle.
10. Bleed the brake system.

Parking Brake Cable

ADJUSTMENT

System Adjustment

Except F-Super Duty, and 1997 F-150 and Expedition Models

NOTE: Adjust the drum brakes before adjusting the parking brake.

1. Raise and safely support the vehicle.
2. The brake drums should be cold.
3. Make sure the parking brake pedal is fully released.
4. While holding the tension equalizer, tighten the equalizer nut 6 full turns past its original position.
5. Fully depress the parking brake pedal. Using a cable tension gauge, check rear cable tension. Cable tension should be at least 350 lbs. (158 kg).
6. Fully release the parking brake. No drag should be noted at the wheels.
7. If drag is noted on F-250 and F-350 models, remove the drums and adjust the clearance between the parking brake lever and cam plate. Clearance should be 0.015 in. (0.38mm). Clearance is adjusted at the parking brake equalizer adjusting nut. If the tension limiter on the F-150 and Bronco doesn't release the drag, the tension limiter will have to be replaced.

F-Super Duty

1. Fully release the brake pedal.
2. Spray penetrating oil on the adjusting clevis, jam nut and threaded end of the cable.
3. Loosen the jam nut and remove the locking pin from the clevis.

4. Back off on the clevis until there is slack in the cable.

5. Screw on the clevis until the pin can be inserted while the lever and cable are held tightly in the applied position, Then, remove the pin, let go of the cable and lever, and turn the clevis 10 full turns counterclockwise (loosen).

6. Install the pin.

1997 F-150 and Expedition Models

The 1997 F-150 and Expedition models utilize fully self-adjusting parking brake assemblies and do not require manual adjustment.

Initial Adjustment

NOTE: Use this procedure when a new tension limiter is installed.

Except F-Super Duty, and 1997 F-150 and Expedition Models

1. Raise and safely support the vehicle.

2. Depress the parking brake pedal fully.

3. Hold the tension limiter, install the equalizer nut and tighten it to a point 2½ in. ± ⅛ in. (63.5mm ± 3mm) up the rod.

4. Check to make sure the cinch strap has 1⅜ in. (35mm) remaining.

1997 F-150 and Expedition Models

The 1997 F-150 and Expedition models utilize fully self-adjusting parking brake assemblies and do not require manual adjustment.

Field Adjustment

Except 1997 F-150 and Expedition Models

NOTE: Use this procedure to correct a slack system if a new tension limiter is not installed.

1. Apply the parking brake pedal to the fully engaged position.

2. Raise and safely support the vehicle, as necessary. Grip the threaded

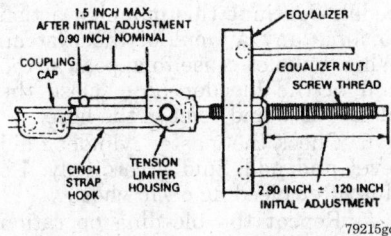

Parking brake cable tension limiter assembly — except F-Super Duty, and 1997 F-150 and Expedition models

rod to keep it from spinning and tighten the equalizer nut 6 full turns past its original position on the threaded rod.

3. Attach a suitable cable tension gauge in front of the equalizer assembly on the front cable and measure the cable tension. The cable tension should be 400–600 lbs. (181–272 kg) with the parking brake pedal in the last detent position. If tension is low, repeat Steps 2 and 3.

4. Release parking brake and check for rear wheel drag. There should be no brake drag.

1997 F-150 and Expedition Models

The 1997 F-150 and Expedition models utilize fully self-adjusting parking brake assemblies and do not require manual adjustment.

REMOVAL AND INSTALLATION

Except F-Super Duty, and 1997 F-150 and Expedition Models

Parking Brake Control

1. Raise and safely support the vehicle.

2. Loosen the adjusting nut at the equalizer.

3. Working in the engine compartment, remove the nuts attaching the parking brake control to the firewall.

4. Remove the cable from the control assembly clevis by compressing the conduit end prongs.

5. Installation is the reverse of removal. Torque the attaching nuts to 15 ft. lbs. (20 Nm).

Equalizer-To-Control Assembly Cable

1. Raise and safely support the vehicle.

2. Back off the equalizer nut and disconnect the cable from the tension limiter.

3. Remove the parking brake cable from the mount.

4. Disconnect the forward end of the cable from the control assembly.

5. Using a cord attached to the upper end of the cable, pull the cable from the vehicle.

6. Installation is the reverse of removal. Adjust the parking brake.

Equalizer-To-Rear Wheel Cable

1. Raise and safely support the vehicle.

2. Remove the wheels and brake drums.

3. Remove the tension limiter.

4. Remove the locknut from the threaded rod and disconnect the cable from the equalizer.

5. Disconnect the cable housing from the frame bracket and pull the cable and housing out of the bracket.

6. Disconnect the cables from the brake backing plates.

7. With the spring tension removed from the lever, lift the cable out of the slot in the lever and remove the cable through the backing plate hole.

8. Installation is the reverse of removal. On the F-250 and F-350, check the clearance between the parking brake operating lever and the cam plate. Clearance should be 0.015 in. (0.38mm) with the brakes fully released.

9. Adjust the brakes.

F-Super Duty

NOTE: To replace the brake shoes, or any other component, the parking brake cable must be disconnected.

1. Place the transmission in gear.

2. Fully release the parking brake pedal.

3. Raise and safely support the vehicle.

4. Disconnect the speedometer cable.

5. Spray penetrating oil on the adjusting clevis, jam nut and threaded end of the cable.

6. Loosen the jam nut and remove the locking pin from the clevis pin.

7. Remove the clevis pin, clevis and jam nut from the cable.

8. Remove the cable from the bracket on the case.

9. Matchmark the driveshaft and disconnect it from the flange.

10. Remove the 6 hex-head bolts securing the parking brake unit to the transmission extension housing and lift off the unit.

NOTE: The unit is filled with Ford Type H automatic transmission fluid.

To install:

11. Refill the unit through the filler plug to the bottom of the plug hole. Install the plug and tighten it to 45 ft. lbs. (61 Nm).

12. Position the unit on the extension housing using 2 guide pins.

13. Using 6 NEW hex-bolts, attach the unit and torque the bolts to 40 ft. lbs. (54 Nm).

14. Connect the driveshaft and torque the bolts to 20 ft. lbs. (27 Nm).

15. Assemble the cable components. Screw on the clevis until the pin can be inserted while the lever and cable are held tightly in the applied position. Then, remove the pin, let go of the cable and lever, and turn the

clevis 10 full turns counterclockwise (loosen).

16. Install the pin.

1997 F-150 and Expedition Models

Front Cable and Pedal Assembly

NOTE: Make certain that the parking brake control is fully released.

1. Pull the front parking brake cable and conduit, then insert a thin pry tool into the hole on the bottom of the parking brake pedal assembly to hold it in the released position.

2. Disconnect the front cable from the cable equalizer, located under the left-hand side of the truck.

3. Remove the front door scuff plate, then remove the left-hand cowl side trim panel.

4. Remove the lower instrument panel steering column cover.

5. Remove the parking brake remote release screw and route the release handle through the opening in the instrument panel.

6. Remove the parking brake pedal assembly as follows:

   a. Detach the parking brake signal switch connector.

   b. Remove the 3 parking brake control nuts.

   c. Separate the parking brake control from the parking brake control mounting bracket.

7. Disconnect the front cable and conduit from the parking brake pedal assembly.

8. Pry the parking brake cable rubber seal from the front floor pan, then compress the retainer and release the front cable and conduit from the bracket.

9. Remove the front cable and conduit.

To install:

10. Route the front cable and conduit into the vehicle, then push the cable/conduit into the bracket until it snaps in place.

11. Push the rubber seal back into place in the floor pan.

12. Attach the parking brake cable and conduit to the pedal assembly.

13. Position the parking brake pedal assembly in place, then install the retaining nuts to 11–15 ft. lbs. (15–20 Nm).

14. Attach the parking brake signal switch connector to the pedal assembly.

15. Install the parking brake remote release screw and route the release handle through the opening in the instrument panel.

16. Install the lower instrument panel steering column cover.

17. Install the left-hand cowl side trim panel.

18. Install the front door scuff plate.

19. Attach the front parking brake cable to the cable equalizer.

20. Release the thin pry tool from the parking brake pedal assembly. Apply and release the parking brake a few times to allow it to self-adjust.

Rear Cable

NOTE: Make certain that the parking brake control is fully released.

1. Pull the front parking brake cable and conduit, then insert a thin pry tool into the hole on the bottom of the parking brake pedal assembly to hold it in the released position.

2. Pull the front parking brake rear cable and conduit, then disconnect the cable equalizer.

3. Compress the retainer and release the rear parking brake rear cable and conduit from the bracket.

4. Raise and support the rear of the vehicle safely on jackstands.

5. Unclip the rear parking brake rear cable and conduit.

6. Remove the rear brake shoes and related components, including the parking brake lever.

7. Remove the rear cable from the backing plate by removing the cable end spring, compressing the retainers (using an aptly-sized box end wrench for this works extremely well), then pulling the rear cable out of the backing plate hole.

To install:

8. Route the rear cable through the backing plate hole and push it in until the retainers engage fully. Install the cable end spring, then install the rear brake components making sure to install the cable end to the brake lever prior to assembly.

9. Route the cable and conduit along the original path, clipping it to the various retainers along the way.

10. Attach the rear cable to the equalizer.

11. Lower the vehicle, then remove the thin pry tool from the brake pedal assembly.

12. Apply and release the parking brake a few times to ensure that it is functioning correctly.

Brake Hydraulic System

BLEEDING

When any part of the hydraulic system has been disconnected for repair or replacement, air may get into the lines and cause spongy pedal action (because air can be compressed and brake fluid cannot). To correct this condition, it is necessary to bleed the hydraulic system after it has been properly connected to be sure all air is expelled from the brake cylinders and lines.

When bleeding the brake system, bleed one brake cylinder at a time, beginning at the cylinder with the longest hydraulic line (farthest from the master cylinder) first. Keep the master cylinder reservoir filled with brake fluid during bleeding operation. Never use brake fluid that has been drained from the hydraulic system, no matter how clean it is.

It will be necessary to centralize the pressure differential valve after a brake system failure has been corrected and the hydraulic system has been bled.

The primary and secondary hydraulic brake systems are individual systems and are bled separately. During the entire bleeding operation, do not allow the reservoir to run dry. Keep the master cylinder reservoirs filled with brake fluid.

Wheel Cylinders and Calipers

1. Clean all dirt from around the master cylinder fill cap, remove the cap and fill the master cylinder with brake fluid until the level is within 1/4 in. (6mm) of the top of the edge of the reservoir.

2. Clean off the bleeder screws at the wheel cylinders and calipers.

3. Attach the length of rubber hose over the nozzle of the bleeder screw at the wheel to be done first. Place the other end of the hose in a glass jar, submerged in brake fluid.

4. Open the bleed screw valve 1/2–3/4 turn.

5. Have an assistant slowly depress the brake pedal. Close the bleeder screw valve and have your assistant slowly return the brake pedal. Continue this pumping action to force any air out of the system. When bubbles cease to appear at the end of the bleeder hose, close the bleed valve and remove the hose.

6. Check the master cylinder fluid level and add fluid accordingly. Do this after bleeding each wheel.

7. Repeat the bleeding operation at the remaining 3 wheels in the following order: right-hand rear, left-hand rear, right-hand front, left-hand front. Fill the master cylinder reservoir.

Master Cylinder

1. Fill the master cylinder reservoirs.
2. Place absorbent rags under the fluid lines at the master cylinder.
3. Have an assistant depress and hold the brake pedal.
4. With the pedal held down, slowly crack open the hydraulic line fitting, allowing the air to escape. Close the fitting and have the pedal released.
5. Repeat Steps 3 and 4 for each fitting until all the air is released.

Wheel Speed Sensor

REMOVAL AND INSTALLATION

Rear Anti-lock Brake System (RABS)

Rear Axle Sensor

1. Disconnect the negative battery cable.
2. Detach the wiring harness connector from the anti-lock sensor.
3. Remove the sensor hold-down bolt and remove the sensor from the axle housing.
   To install:
4. Clean the axle mounting surface. Use care to prevent dirt from entering the axle housing.
5. Inspect and clean the magnetized sensor pole piece to ensure that it is free from loose metal particles which could cause erratic system operation. Inspect the sensor O-ring for damage and replace, if necessary.
6. Lightly lubricate the sensor O-ring with motor oil, align the sensor bolt hole and install. Do not apply force to the plastic sensor connector. The sensor flange should slide to the mounting surface. This will insure the air gap setting is between 0.005–0.045 in. (0.127–1.14mm).
7. Install the hold down bolt and tighten to 25–30 ft. lbs. (34–40 Nm).
8. Inspect the blue sensor connector seal and replace if missing or damaged. Push the connector on the sensor.
9. Connect the negative battery cable.

4-Wheel Anti-lock Brake System

Front Wheel Speed Sensor

1. Inside the engine compartment, disconnect the sensor from the harness.
2. Unclip the sensor cable from the brake hose clips.
3. Remove the retaining bolt from the spindle and slide the sensor from its hole.

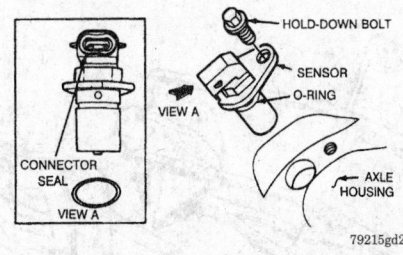

RABS sensor installation

4. Installation is the reverse of the removal. Torque the retaining bolt to 40–60 inch lbs. (4.5–6.7 Nm)

Rear Speed Sensor

1. Disconnect the wiring from the harness.
2. Remove the sensor hold-down bolt and remove the sensor from the axle.
   To install:
3. Thoroughly clean the mounting surfaces. Make sure no dirt falls into the axle. Clean the magnetized sensor pole piece. Metal particles can cause sensor problems. Replace the O-ring.
4. Coat the new O-ring with clean engine oil.
5. Position the new sensor on the axle. It should slide into place easily. Correct installation will allow a gap of 0.005–0.045 in. (0.127–1.143mm).
6. Tighten the hold-down bolt to 25–30 ft. lbs. (34–40 Nm).
7. Connect the wiring.

FRONT SUSPENSION

Shock Absorbers

REMOVAL AND INSTALLATION

F-150 and Bronco Models

Except 4WD Models With Quad Shocks, and 1997 F-150 and Expedition Models

1. Remove the upper nut while holding the shock absorber stem.
2. Remove the lower mounting bolt/nut from the bracket.
3. Compress the shock and remove it.
4. Installation is the reverse of removal. Hold the stud while tightening the upper nut to 30 ft. lbs. (41 Nm). Torque the lower bolt/nut to 60 ft. lbs. (81 Nm).

4WD Models with Quad Shocks

1. Remove the upper nut while holding the shock absorber stem on both forward and rearward shocks.
2. Remove the lower mounting bolt/nut from the rearward shock bracket; the nut and washer from the forward shock bracket.
3. Compress the shocks and remove them.
4. Cut the insulators from the upper spring seat.
5. Install new one piece insulators into the top surface of the upper spring seat. Coat them with a soap solution to aid in installation.
6. Installation of the shocks is the reverse of removal. Use a new steel washer under the upper nut. Hold the stud while tightening the upper nut to 30 ft. lbs. (41 Nm). Torque the lower bolt/nut to 60 ft. lbs. (81 Nm).

1997 F-150 and Expedition Models With 2-Wheel Drive

1. Remove the upper shock absorber nut and washer by holding the shock stem from rotating while loosening the nut and washer.
2. Raise and support the front of the vehicle safely on jackstands.
3. On 2-wheel drive models, remove the 2 lower shock absorber mounting nuts, then lower the shock from inside of the coil spring.
4. On 4-wheel drive models, remove the lower shock absorber mounting bolt and nut. Remove the shock absorber from the vehicle.
   To install:
5. Position the shock absorber in the vehicle so that the top post is inserted through the top mounting hole of the vehicle's upper spring pad.
6. On 4-wheel drive models, install the lower mounting bolt and nut to 56–76 ft. lbs. (76–103 Nm).
7. On 2-wheel drive models, install the 2 lower shock absorber mounting bolts and tighten to 19–25 ft. lbs. (26–34 Nm).
8. Install the upper shock absorber nut and washer. While holding the shock stem, tighten the nut to 34–46 ft. lbs. (47–63 Nm) for 2WD models, and to 57–77 ft. lbs. (77–104 Nm).
9. Lower the vehicle.

F-250 and F-350 Models

1. Remove the nut/bolt retaining the shock to the upper bracket.
2. Remove the lower mounting bolt/nut from the bracket.

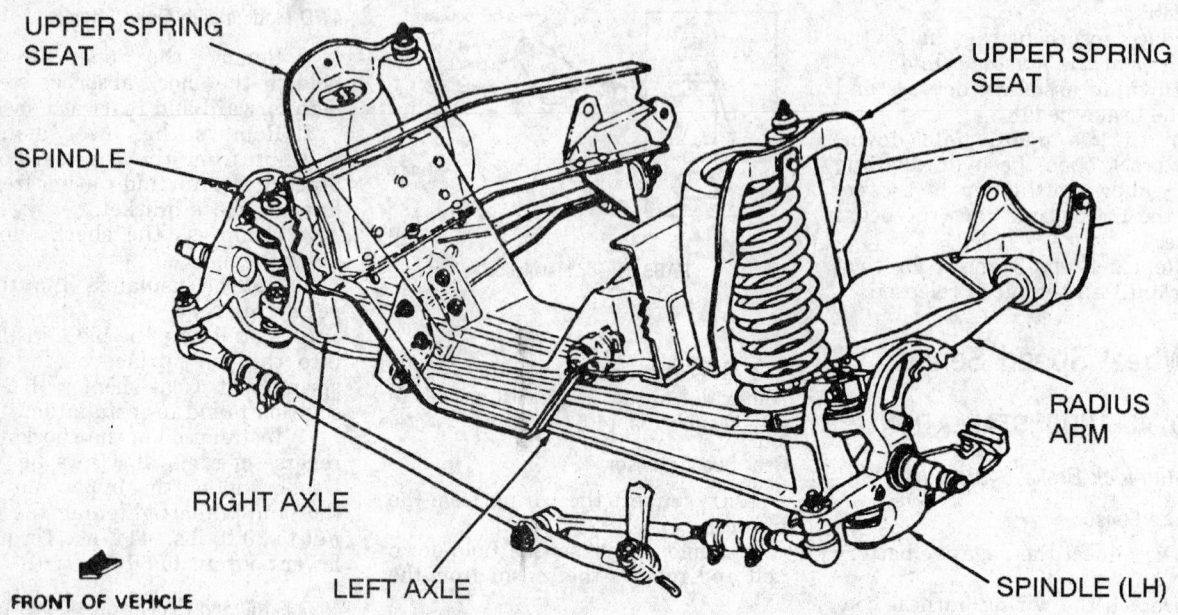

UPPER SPRING SEAT

SPINDLE

UPPER SPRING SEAT

RADIUS ARM

RIGHT AXLE

LEFT AXLE

SPINDLE (LH)

FRONT OF VEHICLE

79215gd0

Front suspension assembly — 2WD 1993–96 F Series models

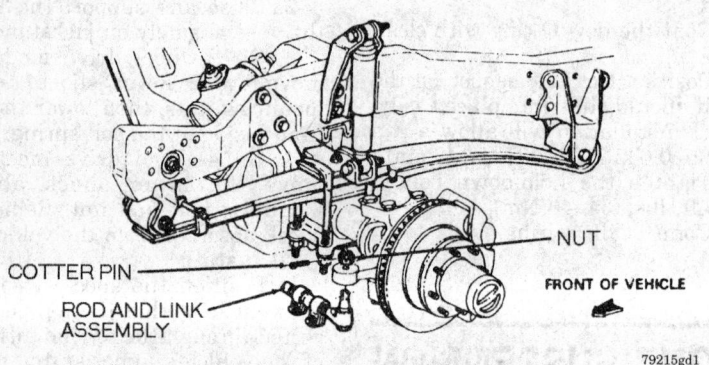

COTTER PIN

ROD AND LINK ASSEMBLY

NUT

FRONT OF VEHICLE

79215gd1

Front suspension assembly — F-350 models

3. Compress the shock and remove it.

4. Installation is the reverse of removal. Tighten the upper and lower nut/bolt to 70 ft. lbs. (95 Nm).

Coil Springs

REMOVAL AND INSTALLATION

Except 1997 F-150 and Expedition Models

1. Raise the front of the vehicle and place jackstands under the frame and a jack under the axle.

2. Remove the wheels.

3. Disconnect the shock absorber from the lower bracket.

4. Remove one bolt and nut and remove the rebound bracket.

5. Remove the two spring upper retainer attaching bolts from the top of the spring upper seat and remove the retainer.

6. Remove the nut attaching the spring lower retainer to the lower seat and axle and remove the retainer.

7. Place a safety chain through the spring to prevent it from suddenly coming loose. Slowly lower the axle and remove the spring.

To install:

8. Place the spring in position and raise the front axle.

9. Position the spring lower retainer over the stud and lower seat, and install the two attaching bolts.

10. Position the upper retainer over the spring coil and against the spring upper seat, and install the two attaching bolts.

11. Tighten the upper retaining bolts to 13–18 ft. lbs. (18–25 Nm), the lower retainer attaching nuts to 70–100 ft. lbs. (95–135 Nm).

12. Connect the shock absorber to the lower bracket. Torque the bolt and nut to 40–60 ft. lbs. (54–81 Nm). Install the rebound bracket.

13. Remove the jack and safety stands.

1997 F-150 and Expedition Models

1. Raise and support the front of the vehicle safely on jackstands.

2. Remove the wheels.

3. Remove the disc brake rotor splash shield.

4. Remove the shock absorber.

5. Remove the brake hose bracket from the front suspension lower control arm.

6. Remove the front stabilizer bar end-link nut.

7. Use Coil Spring Compressor D78P-5310-A, or its equivalent, to compress the coil spring.

8. Remove the lower ball joint cotter pin and castle nut.

9. Use a Pitman arm puller, such as Ford Tool T64P-3590-F, to separate the lower ball joint stud from the front wheel spindle.

10. Matchmark the position of the lower control arm left-hand cam (offset washer under the control arm-to-frame bolt).

11. Remove the 2 control arm lower nuts, then remove the lower bolts.

12. Remove the control arm and coil spring from the vehicle.

To install:

NOTE: The end of the coil spring must cover the first hole and be visible in the second hole — the holes are located in the coil spring mounting groove of the lower control arm.

13. Install the front coil spring, already compressed, into the lower control arm.

NOTE: The lower control arm forward nut must first be tightened while the control arms are held at the curb position ride height. Also, when installing the left-hand control arm bolt, align the matchmarks of the bolt cam.

14. Position the lower control arm and coil spring in the vehicle, then install the control arm-to-frame bolts and nuts. Hold the lower control arm at ride height, then tighten the nuts and bolts to 197–241 ft. lbs. (270–330 Nm). Make sure that the matchmarks are correctly aligned.

15. Insert the lower ball joint stud, along with the ball joint shield, into the steering knuckle, then install the castle nut. Tighten the castle nut to 83–113 ft. lbs. (113–153 Nm). Install a new cotter pin. If the cotter pin hole and castle nut grooves are not aligned, tighten, never loosen, the nut until they are aligned.

16. Install and tighten the stabilizer bar end-link nut to 15–21 ft. lbs. (21–29 Nm).

17. Slowly remove the coil spring compressor, while making sure that the spring is correctly seated in the lower control arm and the upper spring pad.

18. Install the shock absorber.

19. Install the disc brake rotor splash shield and the front wheels, then lower the vehicle.

20. Have the front end alignment inspected.

Leaf Springs

REMOVAL AND INSTALLATION

2-Wheel Drive Models

1. Raise and support the front end on jackstands with the tires still touching the ground.

2. Using jacks, take up the weight of the axle, off the U-bolts.

3. Disconnect the lower end of each shock absorber.

4. Disconnect the spring from the front bracket or shackle.

5. Disconnect the spring from the rear bracket or shackle.

6. Remove the U-bolt nuts.

7. Remove the U-bolts.

8. Disconnect the jack bracket or stabilizer bar as necessary.

9. Lower the axle slightly and remove the spring. Take note of the position of the spring spacer.

To install

10. Position the spring on its seat on the axle and raise it to align the front of the spring with the bracket or shackle.

11. Coat the bushing with silicone grease.

12. Carefully guide the attaching bolt through the bracket or shackle, and the bushing.

13. Install the nut and torque it to:
- chassis/cab spring-to-shackle — 120–150 ft. lbs. (163–203 Nm)
- chassis/cab shackle-to-frame — 150–210 ft. lbs. (203–286 Nm)
- stripped chassis or motor home chassis spring-to-bracket — 148–207 ft. lbs. (201–282 Nm)

14. In a similar fashion, attach the rear of the spring. The torques are:
- chassis/cab spring-to-bracket — 150–210 ft. lbs. (203–286 Nm)
- stripped chassis or motor home chassis spring-to-shackle or shackle-to-bracket — 74–110 ft. lbs. (101–150 Nm)

15. Position the spacer on the spring.

16. Install the U-bolts. Install the jack bracket or stabilizer bar bracket on the forward U-bolt. Install the U-bolt nuts. Torque the nuts, evenly and in gradual increments, in a criss-cross fashion, to:
- chassis/cab — 150–210 ft. lbs. (203–286 Nm)
- stripped chassis and motor home chassis — 220–300 ft. lbs. (299–408 Nm)

17. Connect the shock absorbers. Torque them to:
- chassis/cab models, shock absorber-to-bracket nuts to 52–74 ft. lbs. (71–101 Nm)
- stripped chassis or motor home chassis models, lower attaching bolt to 220–300 ft. lbs. (299–408 Nm)

4-Wheel Drive Models

1. Raise the vehicle frame until the weight is off the front spring with the wheels still touching the floor. Support the axle to prevent rotation.

2. Disconnect the lower end of the shock absorber from the U-bolt spacer. Remove the U-bolts, U-bolt cap and spacer. On F-350 models, remove the 2 bolts retaining the track bar to the spring cap and the track bar bracket.

3. Remove the nut from the hanger bolt retaining the spring at the rear and drive out the hanger bolt.

4. Remove the nut connecting the front shackle and spring eye and drive out the shackle bolt and remove the spring.

To install:

5. Position the spring on the spring seat. Install the shackle bolt through the shackle and spring. Torque the nuts to 150 ft. lbs.

6. Position the rear of the spring and install the hanger bolt. Torque the nut to 150 ft. lbs. (203 Nm).

7. Position the U-bolt spacer and place the U-bolts in position through the holes in the spring seat cap. Install but do not tighten the U-bolt nut. On the F-350, install the track bar. Torque the track bar-to-bracket bolts to 200 ft. lbs. (272 Nm).

8. Connect the lower end of the shock absorber to the U-bolt spacer. Torque the fasteners to 60 ft. lbs. (81 Nm) on the F-250; 70 ft. lbs. (95 Nm) on the F-350.

9. Lower the vehicle and tighten the U-bolt nuts to 120 ft. lbs. (163 Nm).

Torsion Bar

REMOVAL AND INSTALLATION

1997 F-150 and Expedition Models

1. Raise and support the front of the vehicle safely on jackstands.

2. Make an alignment mark on the torsion bar and the torsion bar cross-member support.

3. Measure and record the length from the torsion bar adjuster bolt head to the adjuster bracket.

4. Relieve the torsion bar tension as follows:

a. Install the Torsion Bar Tool T95T-5310-A with the Adapter Plates T96T-5310-A, or their equivalents, onto the frame rail with the tool shaft pointing toward the torsion bar adjuster.

b. Tighten the tool until it touches the torsion bar adjuster.

c. Remove the torsion bar adjuster bolt and nut.

d. Slowly remove the torsion bar tool, then remove the torsion bar adjuster.

e. Repeat Steps 4a through 4d for the other side of the vehicle.

5. Remove the 6 torsion bar crossmember support bolts.

6. Remove the torsion bar crossmember support, then remove the torsion bars from the vehicle.

To install:

7. Position the torsion bars in the lower control arms, then install the torsion bar crossmember support. Tighten the 6 support bolts; the center bolts on each side should be tightened to 30–40 ft. lbs. (40–55 Nm), whereas the outer 4 bolts and accompanying nuts should be tightened to 39–53 ft. lbs. (53–72 Nm).

8. Install the torsion bar adjuster, then install the torsion bar tool and adapter plates.

9. Tighten the torsion bar tool to load the torsion bar.

10. Install the torsion bar adjuster nut and bolt.

11. Remove the torsion bar tool.

12. Turn the torsion bar adjuster until the reference marks align, then repeat the procedure for the other side of the vehicle.

13. Install the front brake anti-lock sensor wire bracket bolt, if applicable.

14. Have the front end alignment and the ride height checked.

Upper and Lower Ball Joints

REMOVAL AND INSTALLATION

Except F-150, Bronco and Expedition Models

1. Raise and support the vehicle safely.

2. Remove the spindles and left and right shafts and joint.

3. Remove the tie rod nut and disconnect the tie rod from the steering arm.

4. Remove the cotter pin from the top ball joint stud. Remove the nut from the top stud and loosen the nut on the lower stud inside the knuckle.

5. Hit the top stud sharply with a plastic mallet to free the knuckle from the axle arm. Remove and discard the bottom nut. New nuts should be used at assembly.

6. Note the positioning of the camber adjuster carefully for reassembly. Remove the camber adjuster. If it's hard to remove, use a puller.

7. Place the knuckle in a vise and remove the snapring from the bottom ball joint. Not all ball joints will have this snapring.

8. Remove the plug from C-frame tool T74P-4635-C and replace it with plug T80T-3010-A. Assemble C-frame tool T74P-4635-C and receiving cup D79T-3010-G (Bronco, F-150 and 250) or T80T-3010-A2 (F-250 Heavy Duty and F-350). on the knuckle.

9. Turn the forcing screw inward until the ball joint is separated from the knuckle.

10. Assemble the C-frame tool with receiving cup D79P-3010-BG on the upper ball joint and force it out of the knuckle.

NOTE: Always force out the bottom ball joint first.

11. Clean the ball joint bores thoroughly.

12. Insert the lower joint into its bore as straight as possible.

13. On the Bronco, F-150 and F-250, assemble the C-frame tool, receiving cup T80T-3010-A3 and installing cup D79T-3010-BF onto the lower ball joint. On the F-250 Heavy Duty and F-350, assemble the C-frame tool, receiving cup T80T-3010-A3 and receiving cup D79T-3010-BG on the lower ball joint.

14. Turn the screw clockwise until the ball joint is firmly seated.

NOTE: If the ball joint cannot be installed to the correct depth, realign the receiving cup on the tool.

15. On all models, assemble the C-frame, receiving cup T80T-3010-A3 and replacer T80T-3010-A1 on the upper ball joint.

16. Turn the screw clockwise until the ball joint is firmly seated.

17. Place the knuckle into position on the axle arm. Install the camber adjuster on the upper ball joint stud with the arrow point to positive or negative as noted before disassembly.

18. Install a new nut on the bottom stud, finger tight. Install a new nut on the top stud finger tight.

19. Tighten the bottom nut to 80 ft. lbs. (109 Nm).

20. Tighten the top nut to 100 ft. lbs. (136 Nm), then, (tighten) advance the nut until the cotter pin hole align with the castellations. Install a new cotter pin.

21. Again tighten the bottom nut, this time to 110 ft. lbs. (149 Nm).

22. Install all other parts.

F-150, Bronco and Expedition Models

1993–96 Vehicles

1. Remove the spindle.

2. Remove the snapring from the ball joints. Assemble the C-frame assembly T74P-4635-C and receiver cup D81T-3010-A, or equivalents, on the upper ball joint. Turn the forcing screw clockwise until the ball joint is removed from the axle.

3. Repeat Step 2 on the lower ball joint.

NOTE: The upper ball joint must always be removed first. DO NOT heat the ball joint or spindle!

To install:

NOTE: The lower ball joint must be installed first.

4. To install the lower ball joint, assemble the C-frame with ball joint receiver cup D81T-3010-A5 and installation cup D81T-3010-A1, and turn the forcing screw clockwise until the ball joint is seated. DO NOT heat the ball joint to aid in installation!

5. Install the snapring onto the ball joint.

6. Install the upper ball joint in the same manner as the lower ball joint.

7. Install the spindle assembly.

1997 Vehicles

The upper and lower ball joints on the 1997 F-150 and Expedition models are integral to the control arms. If the ball joints are deemed faulty, the control arm and ball joint must be replaced as an assembly.

INSPECTION

1. Before an inspection of the ball joints, make sure the front wheel bearings are properly packed and adjusted.

2. Jack up the front of the vehicle and safely support it with jackstands, placing the stands under the I-beam axle beneath the spring as shown in the accompanying illustration.

3. Have a helper grab the lower edge of the tire and move the wheel assembly in and out.

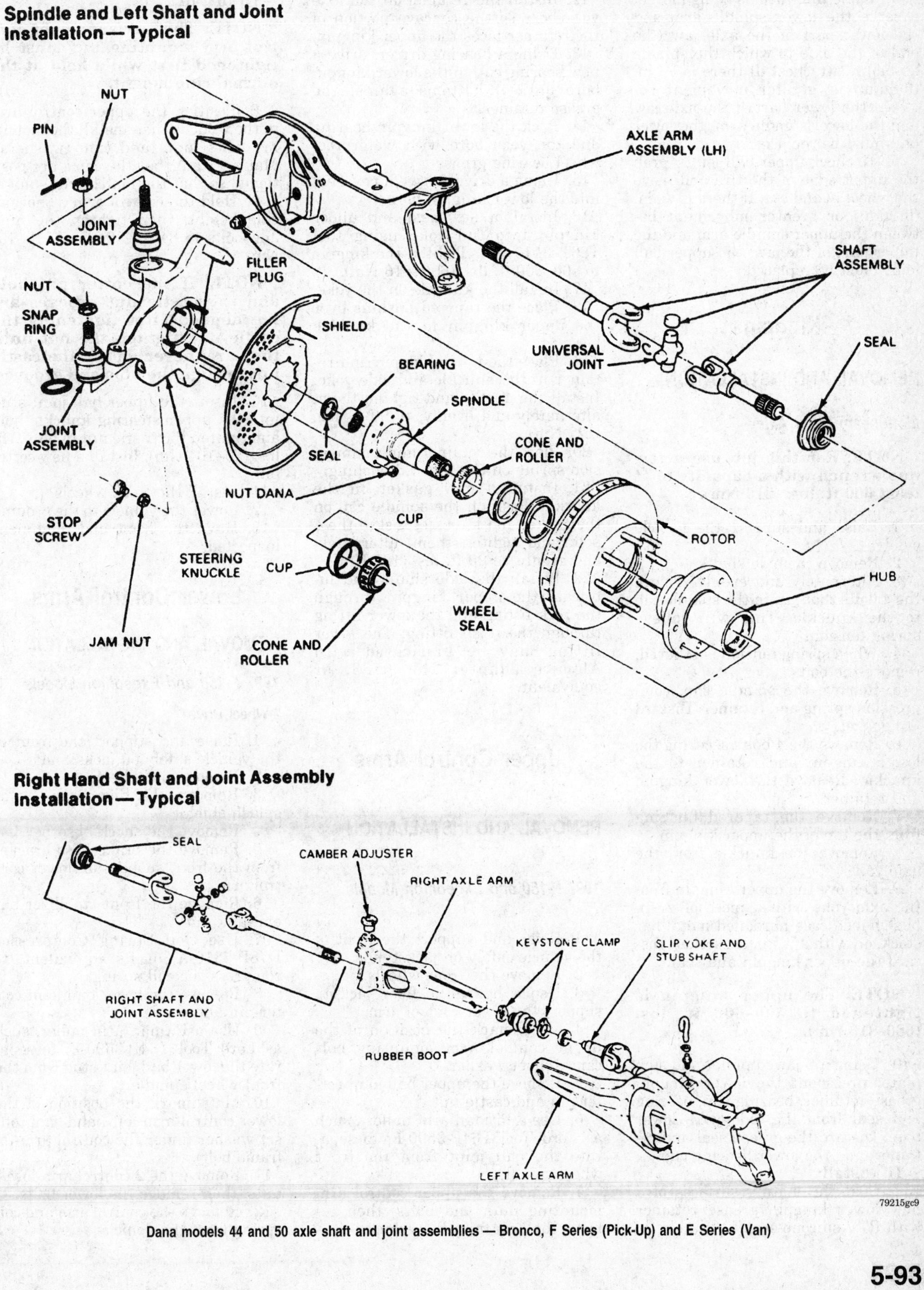

**Spindle and Left Shaft and Joint
Installation — Typical**

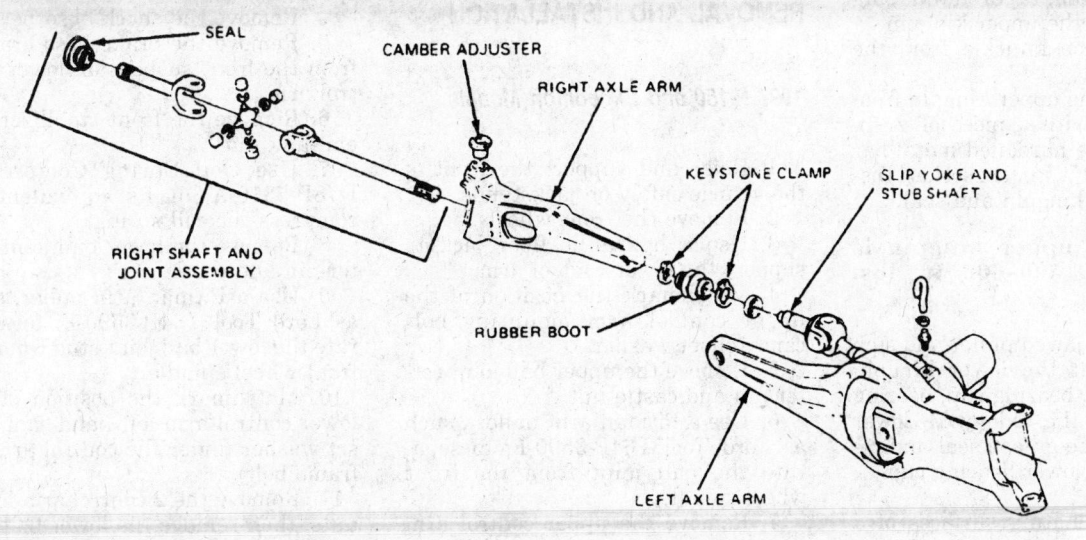

**Right Hand Shaft and Joint Assembly
Installation — Typical**

Dana models 44 and 50 axle shaft and joint assemblies — Bronco, F Series (Pick-Up) and E Series (Van)

79215gc9

4. While the wheel is being moved, observe the lower spindle arm and the lower part of the axle jaw (the end of the axle to which the spindle assembly attaches). If there is $1/32$ in. (0.8mm) or greater movement between the lower part of the axle jaw and the lower spindle arm, the lower ball must be replaced.

5. To check upper ball joints, grab the upper edge of the tire and move the wheel in and out. If there is $1/32$ in. (0.8mm) or greater movement between the upper spindle arm and the upper part of the jaw, the upper ball joint must be replaced.

Kingpins

REMOVAL AND INSTALLATION

Monobeam Front Axle

NOTE: For this job use a torque wrench with a capacity of at least 600 ft. lbs. (816 Nm).

1. Raise and support the vehicle safely.
2. Remove the axle shafts.
3. Alternately and evenly remove the 4 bolts that retain the spindle cap to the knuckle. This will relieve spring tension.
4. When spring tension is relieved, remove the bolts.
5. Remove the spindle cap, compression spring and retainer. Discard the gasket.
6. Remove the 4 bolts securing the lower kingpin and retainer to the knuckle. Remove the lower kingpin and retainer.
7. Remove the tapered bushing from the top of the upper kingpin.
8. Remove the knuckle from the axle yoke.
9. Remove the upper kingpin from the axle yoke with a piece of $7/8$ in. hex-shaped case hardened metal bar stock, or, with a $7/8$ in. hex socket. Discard the upper kingpin and seal.

NOTE: The upper kingpin is tightened to 500–600 ft. lbs. (680–816 Nm).

10. Using a 2-jawed puller and step plate, press out the lower kingpin grease retainer, bearing cup, bearing and seal from the axle yoke lower bore. Discard the grease seal and retainer, and the lower bearing cup.

To install:
11. Coat the mating surfaces of a new lower kingpin grease retainer with RTV silicone sealer.

12. Install the retainer in the axle yoke bore so the concave portion of the retainer faces the upper kingpin.
13. Using a bearing driver, drive a new bearing cup in the lower kingpin bore until it bottoms against the grease retainer.
14. Pack the lower kingpin bearing and the yoke bore with waterproof wheel bearing grease.
15. Using a driver, drive a new seal into the lower kingpin bore.
16. Install a new seal and upper kingpin into the yoke using tool T86T-3110-AH. Tighten the kingpin to 500–600 ft. lbs. (680–816 Nm).
17. Install the knuckle on the yoke.
18. Place the tapered bushing over the upper kingpin in the knuckle bore.
19. Place the lower kingpin and retainer in the knuckle and axle yoke. Install the 4 bolts and tighten them, alternately and evenly, to 90 ft. lbs. (122 Nm).
20. Place the retainer and compression spring on the tapered bushing.
21. Install a new gasket on the knuckle. Position the spindle cap on the gasket and knuckle. Install the 4 bolts and tighten them, alternately and evenly, to 90 ft. lbs. (122 Nm).
22. Install the axle shafts and lubricate the upper kingpin through the zerk fitting and the lower fitting through the flush fitting. The lower fitting may be lubricated with Alemite adapter No. 6783, or equivalent.

Upper Control Arms

REMOVAL AND INSTALLATION

1997 F-150 and Expedition Models

1. Raise and support the front of the vehicle safely on jackstands.
2. Remove the front wheels.
3. Use a hydraulic floor jack to support the lower control arm.
4. Matchmark the position of the upper control arm mounting bolt cams (offset washers).
5. Remove the upper ball joint cotter pin and castle nut.
6. Use a Pitman arm puller , such as Ford Tool T64P-3590-F, to separate the ball joint from the front wheel spindle.
7. Remove the upper control arm mounting nuts and bolts, then remove the arm from the vehicle.

To install:

NOTE: The forward upper control arm mounting nut must be tightened first while held at the normal ride height.

8. Position the upper control arm in the vehicle, then install the mounting bolts, nuts and cam (washers). Make certain that the cams are positioned so that the matchmarks align.
9. Hold the control arm at normal ride height, then tighten the nuts and bolts to 84–112 ft. lbs. (113–153 Nm).

NOTE: If the cotter pin hole and the castle nut grooves are not aligned after tightening the castle nut, tighten the nut until they are; never loosen the castle nut to align the hole and grooves.

10. Insert the upper ball joint stud into the upper steering knuckle hole and tighten the castle nut to 56–77 ft. lbs. (76–104 Nm). Install a new cotter pin.
11. Install the front wheels.
12. Lower the vehicle to the ground.
13. Have the front end alignment inspected.

Lower Control Arms

REMOVAL AND INSTALLATION

1997 F-150 and Expedition Models

2-Wheel Drive

1. Raise and support the front of the vehicle safely on jackstands.
2. Remove the wheels.
3. Remove the disc brake rotor splash shield.
4. Remove the shock absorber.
5. Remove the brake hose bracket from the front suspension lower control arm.
6. Remove the front stabilizer bar end-link nut.
7. Use Coil Spring Compressor D78P-5310-A, or its equivalent, to compress the coil spring.
8. Remove the lower ball joint cotter pin and castle nut.
9. Use a Pitman arm puller, such as Ford Tool T64P-3590-F, to separate the lower ball joint stud from the front wheel spindle.
10. Matchmark the position of the lower control arm left-hand cam (offset washer under the control arm-to-frame bolt).
11. Remove the 2 control arm lower nuts, then remove the lower bolts.
12. Remove the control arm and coil spring from the vehicle.

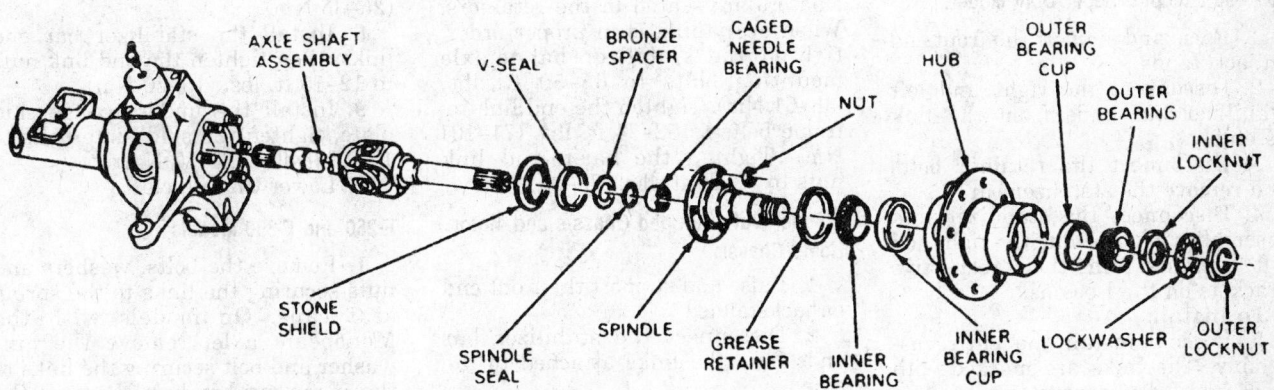

Dana model 60 Monobeam front drive axle assembly

To install:

NOTE: The end of the coil spring must cover the first hole and be visible in the second hole — the holes are located in the coil spring mounting groove of the lower control arm.

13. Install the front coil spring, already compressed, into the lower control arm.

NOTE: The lower control arm forward nut must first be tightened while the control arms are held at the curb position ride height. Also, when installing the left-hand control arm bolt, align the matchmarks of the bolt cam.

14. Position the lower control arm and coil spring in the vehicle, then install the control arm-to-frame bolts and nuts. Hold the lower control arm at ride height, then tighten the nuts and bolts to 197–241 ft. lbs. (270–330 Nm). Make sure that the matchmarks are correctly aligned.

15. Insert the lower ball joint stud, along with the ball joint shield, into the steering knuckle, then install the castle nut. Tighten the castle nut to 83–113 ft. lbs. (113–153 Nm). Install a new cotter pin. If the cotter pin hole and castle nut grooves are not al-

igned, tighten, never loosen, the nut until they are aligned.

16. Install and tighten the stabilizer bar end-link nut to 15–21 ft. lbs. (21–29 Nm).

17. Slowly remove the coil spring compressor, while making sure that the spring is correctly seated in the lower control arm and the upper spring pad.

18. Install the shock absorber.

19. Install the disc brake rotor splash shield and the front wheels, then lower the vehicle.

20. Have the front end alignment inspected.

4-Wheel Drive

1. Raise and safely support the front of the vehicle on jackstands.

2. Remove the front wheel hub.

3. Remove the torsion bars.

4. Use strong wire or cord to suspend the upper control arm from the vehicle body or frame.

5. If necessary, suspend the half-shaft with wire or cord also.

6. Remove the stabilizer bar link nut from the front suspension lower arm.

7. Remove the shock absorber lower bolt and nut from the control arm.

8. Remove the lower ball joint cotter pin and castle nut, then separate the lower ball joint from the steering knuckle using a Pitman arm puller, such as Ford Tool T64P-3590-F.

9. Remove the 2 lower control arm nuts and bolts, then remove the control arm from the vehicle's frame.

To install:

10. Install the lower control arm to the vehicle's frame, then install the mounting bolts and nuts. Tighten the lower control arm bolts and nuts to 121–148 ft. lbs. (164–200 Nm).

11. Install the front stabilizer bar link to the control arm. Tighten the nut to 16–21 ft. lbs. (22–28 Nm).

12. Install the shock absorber lower bolt and nut to 56–76 ft. lbs. (76–103 Nm).

13. Install the lower ball joint stud into the lower steering arm hole, then tighten the castle nut to 83–208 (113–153 Nm). Then install a new cotter pin.

14. Install the torsion bars.

15. Install the wheel hub and bearing unit.

16. Install the front wheels and lower the vehicle.

17. Have the front end alignment and ride height inspected.

Stabilizer Bar

REMOVAL AND INSTALLATION

2-Wheel Drive Models

1993–96 Except F-Super Duty Models

1. Raise and support the front end on jackstands.
2. Disconnect the right and left stabilizer bar ends from the link assembly.
3. Disconnect the retainer bolts and remove the stabilizer bar.
4. Disconnect the stabilizer link assemblies by loosening the right and left locknuts from their respective brackets on the I-beams.

To install:

5. Loosely install the entire assembly. The links are marked with an **R** and **L** for identification.
6. Tighten the link-to-stabilizer bar and axle bracket fasteners to 70 ft. lbs. (95 Nm).
7. Check to make sure the insulators are properly seated and the stabilizer bar is centered.
8. Tighten the stabilizer bar to crossmember attaching bolts to 35 ft. lbs. (48 Nm). On the F-250, F-350, E-250 and E-350 models, tighten the stabilizer bar-to-frame retainer bolts to 35 ft. lbs. (48 Nm). Torque the frame mounting bracket nuts/bolts to 65 ft. lbs. (88 Nm).

1997 F-150 and Expedition Models

1. Raise and safely support the front of the vehicle on jackstands.
2. Remove the stabilizer bar nut from the lower control arms.
3. Remove the left-hand and right-hand stabilizer bar links.
4. Remove the stabilizer bar retaining bolts, then lower the bar off of the front frame rails.
5. Remove the stabilizer bar insulators.

To install:

6. Install the stabilizer bar insulators and position the stabilizer bar onto the frame rails. Tighten the 4 mounting bolts to 19–26 ft. lbs. (26–35 Nm).
7. Install the stabilizer bar end links, then tighten the end link nuts to 12–15 ft. lbs. (15–20 Nm).
8. Lower the vehicle.

F-Super Duty Chassis and Cab Models

1. Raise and support the front end on jackstands.
2. Disconnect each end of the bar from the links.
3. Disconnect the bar from the axle.

4. Unbolt and remove the links from the frame.
5. Installation is the reverse of removal. Replace any worn or cracked rubber parts. Install the bar loosely and make sure it is centered between the leaf springs. Make sure the insulators are seated in the retainers. When everything is in proper order, tighten the stabilizer bar-to-axle mounting bolts to 35–50 ft. lbs. (48–61 Nm). Tighten the end link-to-frame bolts to 52–74 ft. lbs. (71–101 Nm). Tighten the bar-to-end link nuts to 15–25 ft. lbs. (20–34 Nm).

F-Super Duty Stripped Chassis and Motor Home Chassis

1. Raise and support the front end on jackstands.
2. Disconnect the stabilizer bar ends from the links attached to the axle.
3. Remove the bar-to-frame bolts and remove the bar.
4. Remove the links from the axle brackets.
5. Installation is the reverse of removal. Replace any worn or cracked rubber parts. Assemble all parts loosely and make sure the assembly is centered on the frame. make sure the insulators are seated in the retainers. When everything is in proper order, tighten the bar-to-frame brackets bolts to 30–47 ft. lbs. (41–64 Nm). Tighten the link-to-axle bracket bolts to 57–81 ft. lbs. (78–110 Nm). Tighten the bar-to-link nuts to 15–25 ft. lbs. (20–34 Nm).

4-Wheel Drive Models

1993–96 F-150 and Bronco

1. Unbolt the stabilizer bar from the connecting links.
2. Unbolt the stabilizer bar retainers.
3. If necessary to remove the stabilizer bar mounting bracket, remove the coil springs as described above.
4. Installation is the reverse of removal. Torque the retainer nuts to 35 ft. lbs. (48 Nm), then torque all other nuts at the links to 70 ft. lbs. (95 Nm).

1997 F-150 and Expedition Models

1. Raise and safely support the front of the vehicle on jackstands.
2. Remove the undercarriage skid plate.
3. Remove the stabilizer bar nut from the lower control arms.
4. Remove the left-hand and right-hand stabilizer bar links.
5. Remove the stabilizer bar retaining bolts, then lower the bar off of the front frame rails.

6. Remove the stabilizer bar insulators.

To install:

7. Install the stabilizer bar insulators and position the stabilizer bar onto the frame rails. Tighten the 4 mounting bolts to 19–26 ft. lbs. (26–35 Nm).
8. Install the stabilizer bar end links, then tighten the end link nuts to 12–15 ft. lbs. (15–20 Nm).
9. Install the undercarriage skid plate; tighten the skid plate bolts to 10–13 ft. lbs. (14–17 Nm).
10. Lower the vehicle.

F-250 and F-350 Models

1. Remove the bolts, washers and nuts securing the links to the spring seat caps. On models with the Monobeam axle, remove the nut, washer and bolt securing the links to the mounting brackets. Remove the nuts, washers and insulators connecting the links to the stabilizer bar. Remove the links.
2. Unbolt and remove the retainers from the mounting brackets.
3. Remove the stabilizer bar.
4. Installation is the reverse of removal. Torque the connecting links-to-spring seat caps to 70 ft. lbs. (95 Nm). Torque the nuts securing the connecting links to the stabilizer bar to 25 ft. lbs. (34 Nm). Torque the retainer-to-mounting bracket nuts to 35 ft. lbs. (48 Nm).

Radius Arm

REMOVAL AND INSTALLATION

2-Wheel Drive Models

NOTE: A torque wrench with a capacity of at least 350 ft. lbs. (476 Nm) is necessary, along with other special tools, for this procedure.

1. Raise the front of the vehicle and place safety stands under the frame and a jack under the wheel or axle. Remove the wheels.
2. Disconnect the shock absorber from the radius arm bracket.
3. Remove the two spring upper retainer attaching bolts from the top of the spring upper seat and remove the retainer.
4. Remove the nut which attached the spring lower retainer to the lower seat and axle and remove the retainer.
5. Lower the axle and remove the spring.
6. Remove the spring lower seat and shim from the radius arm. The,

remove the bolt and nut which attach the radius arm to the axle.

7. Remove the cotter pin, nut and washer from the radius arm rear attachment.

8. Remove the bushing from the radius arm and remove the radius arm from the vehicle.

9. Remove the inner bushing from the radius arm.

10. Position the radius arm to the axle and install the bolt and nut finger-tight.

11. Install the inner bushing on the radius arm and position the arm to the frame bracket.

12. Install the bushing, washer, and attaching nut. Tighten the nut to 120 ft. lbs. (162 Nm) and install the cotter pin.

13. Tighten the radius arm-to-axle bolt to 269–329 ft. lbs. (366–447 Nm).

14. Install the spring seat and insulator on the radius arm so the hole in the seat fits over the arm-to-axle nut.

15. Install the spring.

16. Connect the shock absorber. Torque the nut and bolt to 40–60 ft. lbs. (54–82 Nm).

17. Install the wheels.

4-Wheel Drive Models

1. Raise the vehicle and position safety stands under the frame side rails.

2. Remove the shock absorber lower attaching bolt and nut and pull the shock absorber free of the radius arm.

3. Remove the lower spring retaining bolt from the inside of the spring coil.

4. Loosen the axle pivot bolt.

5. Remove the nut attaching the radius arm to the frame bracket and remove the radius arm rear insulator. Lower the axle and allow the axle to move forward.

NOTE: The axle must be supported on a floor jack throughout this procedure, and must not be permitted to hang from the brake hose. If the length of the brake hose does not provide sufficient clearance it may be necessary to remove and support the brake caliper.

6. Remove the spring as described above.

7. Remove the bolt and stud attaching the radius arm and bracket to the axle.

8. Move the axle forward and remove the radius arm from the axle. Then, pull the radius arm from the frame bracket.

9. Install the components in the reverse order of removal. Install new bolts and stud type bolts which attach the radius arm and bracket to the axle. Torque the bracket-to-axle bolts to 25 ft. lbs. (34 Nm). Torque the lower radius arm-to-axle bolt to 330 ft. lbs. (449 Nm). Tighten the upper stud-type radius arm-to-axle bolt to 250 ft. lbs. (340 Nm). Torque the radius arm rear attaching nut to 120 ft. lbs. (163 Nm). Torque the lower spring retainer nut to 100 ft. lbs. (136 Nm). Torque the upper spring retainer bolts to 15 ft. lbs. (20 Nm). Torque the axle pivot bolt to 150 ft. lbs. (204 Nm). Torque the lower shock absorber bolt to 60 ft. lbs. (82 Nm).

Twin I-Beam Axles

REMOVAL AND INSTALLATION

NOTE: A torque wrench with a capacity of at least 350 ft. lbs. (476 Nm) is necessary, along with other special tools, for this procedure.

1. Raise and support the front end on jackstands.

2. Remove the spindles.

3. Remove the springs.

4. Remove the stabilizer bar.

5. Remove the lower spring seats from the radius arms.

6. Remove the radius arm-to-axle bolts.

7. Remove the axle-to-frame pivot bolts and remove the axles.

To install:

8. Position the axle on the pivot bracket and loosely install the bolt/nut.

9. Position the other end on the radius arm and install the bolt. Torque the bolt to 269–329 ft. lbs. (366–447 Nm).

10. Install the spring seats.

11. Install the springs.

12. Torque the axle pivot bolts to 120–150 ft. lbs. (162–203 Nm).

13. Install the spindles.

14. Install the stabilizer bar.

Solid I-Beam Axle

REMOVAL AND INSTALLATION

1. Raise and safely support the vehicle. Remove the front wheel spindle, the front spring and the stabilizer bar, if equipped.

2. Remove the spring lower seat from the radius arm and then remove the bolt and nut that attaches the stabilizer bar bracket, if equipped, and radius arm to the front axle.

3. Remove the axle-to-frame pivot bracket bolt and nut.

To install:

4. Position the axle to the frame pivot bracket and install the bolt and nut finger tight.

5. Position the opposite end of the of the axle to the radius arm, install the attaching bolt from underneath through the bracket, the radius arm and the axle. Install the nut and tighten to 191–220 ft. lbs. (258–298 Nm).

6. Install the spring lower seat on the radius arm so the hole in the seat indexes over the arm-to-axle bolt. Install the front spring.

7. Install the front wheel spindle and stabilizer bar, if equipped.

8. Lower the vehicle and with the weight on the suspension, tighten the axle-to-frame pivot bracket bolts to 120–150 ft. lbs. (163–203 Nm).

Track Bar

REMOVAL AND INSTALLATION

F-Super Duty Chassis and Cab Models

1. Raise and support the vehicle safely.

2. Remove the wheel cover and the grease cap from the hub. Remove the cotter pin and the retainer. Discard the cotter pin.

3. Loosen the adjusting nut 3 turns.

4. Obtain running clearance between the brake rotor and disc brake pads by rocking the entire wheel and tire assembly in and out several times to push the caliper and brake pads away from the rotor.

NOTE: Do not pry on the caliper piston to obtain clearance.

5. While rotating the wheel, tighten the adjusting nut to 17–25 ft. lbs. (23–34 Nm) to seat the bearings.

6. Back off the adjusting nut ½ turn. Retighten the nut to 18–20 inch lbs. (2.0–2.3 Nm).

7. Install the retainer on the adjusting nut so the castellations line up with the hole in the spindle without moving the nut. Install a new cotter pin.

8. Check the front wheel rotation. If the wheel rotates properly, reinstall the grease cap and the wheel cover. If rotation is noisy or rough, remove, inspect and lubricate the bearings and bearing races.

9. Before driving the vehicle, pump the brake pedal several times to restore normal brake travel.

2-Wheel Drive Front Wheel Bearings

ADJUSTMENT

1993–96 Bronco, F-150, F-250 and F-350 Models

1. Raise and support the front end on jackstands.
2. Remove the grease cap and remove excess grease from the end of the spindle.
3. Remove the cotter pin and nut lock shown in the illustration.
4. Loosen the adjusting nut 3 full turns. Obtain a clearance between the brake rotor and brake pads by rocking the wheel in and out several times to push the pads away from the rotor. If that doesn't work, remove the caliper. The rotor must turn freely.
5. Tighten the adjusting nut to 17–25 ft. lbs. (23–34 Nm) while rotating the rotor in opposite directions.
6. Back off the adjusting nut 120–180 degrees (1/3–1/2 turn).
7. Install the retainer and cotter pin without additional movement of the locknut.
8. If a dial indicator is available, check the end-play at the hub. End-play should be 0.00024–0.0050 in. (0.006–0.127mm).
9. Install the grease cap.
10. If removed, install the caliper.

1997 F-150 and Expedition Models

1. Raise and support the front end on jackstands.
2. Remove the grease cap and remove excess grease from the end of the spindle.
3. Remove the cotter pin and locknut.
4. Loosen the adjusting nut 3 full turns. Obtain a clearance between the brake rotor and brake pads by rocking the wheel in and out several times to push the pads away from the rotor. If that doesn't work, remove the caliper. The rotor must turn freely.
5. Spin the rotor counterclockwise and tighten the spindle nut to 17–24 ft. lbs. (23–34 Nm).
6. Loosen the spindle nut just shy of 1/2 revolution (175 degrees).
7. Tighten the spindle nut 17 inch lbs. (1.9 Nm).
8. Install the retainer nut, a new cotter pin and the hub grease cap.

9. Lower the vehicle completely.
10. Make sure that the brakes are operating correctly.

REMOVAL AND INSTALLATION

1993–96 F-150, F-250 and F-350

1. Raise and support the front end on jackstands.
2. Remove the wheel cover. Remove the wheel.
3. Remove the caliper from the disc and wire it to the underbody to prevent damage to the brake hose.
4. Remove the grease cap from the hub. Then, remove the cotter pin, nut lock, adjusting nut and flat washer from the spindle. Remove the outer bearing assembly from the hub.
5. Pull the hub and disc assembly off the wheel spindle.
6. Remove and discard the old grease retainer. Remove the inner bearing cone and roller assembly from the hub.
7. Clean all grease from the inner and outer bearing cups with solvent. Inspect the cups for pits, scratches, or excessive wear. If the cups are damaged, remove them with a drift.
8. Clean the inner and outer cone and roller assemblies with solvent and shake them dry. If the cone and roller assemblies show excessive wear or damage, replace them with the bearing cups as a unit.
9. Clean the spindle and the inside of the hub with solvent to thoroughly remove all old grease.
   To install:
10. Covering the spindle with a clean cloth, brush all loose dirt and dust from the brake assembly. Remove the cloth carefully so as to not get dirt on the spindle.
11. If the inner and/or outer bearing cups were removed, install the replacement cups on the hub. Be sure the cups seat properly in the hub.
12. It is imperative that all old grease be removed from the bearings and surrounding surfaces before repacking. The new lithium-based grease is not compatible with the sodium base grease used in the past.
13. Install the hub and disc on the wheel spindle. To prevent damage to the grease retainer and spindle threads, keep the hub centered on the spindle.
14. Install the outer bearing cone and roller assembly and the flat washer on the spindle. Install the adjusting nut.
15. Adjust the wheel bearings by torquing the adjusting nut to 17–25 ft. lbs. (23–34 Nm) with the wheel rotating to seat the bearing. Then back

off the adjusting nut 1/2 turn. Retighten the adjusting nut to 10–15 inch lbs. (1.12–1.68 Nm). Install the locknut so the castellations are aligned with the cotter pin hole. Install the cotter pin. Bend the ends of the cotter pin around the castellations of the locknut to prevent interference with the radio static collector in the grease cap. Install the grease cap.

WARNING

New bolts must be used when servicing floating caliper units. The upper bolt must be tightened first.

---

16. Install the wheels.
17. Install the wheel cover.

1997 F-150 and Expedition Models

1. Remove the front disc brake caliper and brake pads.
2. Remove the 2 front disc brake caliper anchor plate bolts, then lift the anchor plate off of the rotor.
3. Remove the hub grease cap.
4. Remove the cotter pin, then remove the retainer nut.
5. Remove the spindle nut.
6. Remove the front wheel outer bearing retainer washer.
7. Remove the outer front wheel bearing.
8. Slide the front disc brake rotor off of the spindle.
9. Remove the wheel hub grease seal and inner front wheel bearing.
   To install:
10. If the inner and/or outer bearing cups were removed, install the replacement cups on the hub. Be sure the cups seat properly in the hub.
11. Thoroughly clean and inspect the front wheel bearings and the front disc brake hub and rotor.
12. Lubricate the front wheel bearings with Premium Long-Life Grease Xg-1-C or -K, or equivalent.
13. Install the front inner wheel bearing, then press in a new inner wheel hub grease seal.
14. Position the front disc rotor and hub on the spindle.
15. Install the outer front wheel bearing into the rotor and hub.
16. Install the outer bearing retainer washer, then install the spindle nut finger-tight.

NOTE: The front disc brake rotor should be rotated while adjusting the front wheel bearing end-play.

17. Tighten the spindle nut to 30 ft. lbs. (40 Nm).
18. Loosen the spindle nut 2 complete revolutions (720 degrees).

19. Spin the rotor counterclockwise and tighten the spindle nut to 17–24 ft. lbs. (23–34 Nm).

20. Loosen the spindle nut just shy of ½ revolution (175 degrees).

21. Tighten the spindle nut 17 inch lbs. (2 Nm).

22. Install the retainer nut, a new cotter pin and the hub grease cap.

23. Position the front disc brake caliper anchor plate, then install the mounting bolts to 125–170 ft. lbs. (170–230 Nm).

24. Install the brake pads and brake caliper.

25. Bleed the brake system.

26. Clean the wheel hub mounting surface.

27. Install the front wheels and snug the lug nuts to fully seat the wheel against the hub.

28. Lower the vehicle until some of the vehicle's weight rests on the front tires, then tighten the lug nuts to 83–112 ft. lbs. (113–153 Nm).

29. Lower the vehicle completely.

30. Make sure that the brakes are operating correctly.

4-Wheel Drive Front Wheel Bearings

REMOVAL AND INSTALLATION

Except 1997 F-150 and Expedition Models

WITH MANUAL LOCKING HUBS

1. Raise the vehicle and install safety stands.

2. Refer to Manual Locking Hub Removal and Installation and remove the hub assemblies.

3. On Bronco, F-150 and F-250 LD with the Dana 44 axle: apply inward pressure on the bearing adjusting nut, using a socket made for that purpose, available at most auto parts stores, to disengage the adjusting nut locking splines, while turning it counterclockwise to remove it. On F-250 HD (Dana 50 axle) and F-350, use the hub nut tool to unscrew the outer locking nut. Then, remove the lock ring from the bearing adjusting nut. Use the locknut socket to remove the bearing adjusting nut.

4. Remove the caliper and suspend it out of the way.

5. Slide the hub and disc assembly off of the spindle. The outer wheel bearing will slide out as the hub is removed, so be prepared to catch it.

6. Lay the hub on a clean work surface. Carefully drive the inner bearing cone and grease seal out of the hub using Tool T69L–1102–A, or equivalent.

7. Inspect the bearing cups for pits or cracks. If necessary, remove them with a drift. If new cups are installed, install new bearings.

To install:

8. Lubricate the bearings with Multi-Purpose Lubricant Ford Specification, ESA-MIC7-B or equivalent. Clean all old grease from the hub. Pack the cones and rollers. If a bearing packer is not available, work as much lubricant as possible between the rollers and the cages.

9. Drive new cups into place with a driver, making sure they are fully seated.

10. Position the inner bearing cone and roller in the inner cup and install the grease retainer.

11. Carefully position the hub and disc assembly on the spindle.

12. Install the outer bearing cone and roller, and the adjusting nut.

13. On Bronco, F-150 and F-250 LD with the Dana 44 axle:

a. Make sure the metal stamping on the adjusting nut faces inboard and the inner diameter key on the nut enters the spindle keyway.

b. Apply inward pressure on the hub nut wrench and tighten the adjusting nut to 70 ft. lbs. (95 Nm) while rotating the hub back and forth to seat the bearings.

c. Apply inward pressure on the wrench and back off the nut about 90 degrees (¼ turn) then, retighten the nut to 15–20 ft. lbs. (20–27 Nm).

d. Remove the wrench. End-play of the hub/rotor assembly should be 0 (zero) and the torque required to rotate the hub assembly should not exceed 20 inch lbs. (2.24 Nm).

14. Install the outer bearing cone and roller, and the adjusting nut.

15. On the F-250 HD (Dana 50 axle) and F-350:

NOTE: The adjusting nut has a small dowel on one side. This dowel faces outward to engage the locking ring.

a. Using the hub nut socket and a torque wrench, tighten the bearing adjusting nut to 50 ft. lbs. (68 Nm), while rotating the wheel back and forth to seat the bearings.

b. Back off the adjusting nut approximately 90 degrees (¼ turn).

c. Install the lock ring by turning the nut to the nearest hole and inserting the dowel pin.

> ——— **WARNING** ———
> **The dowel pin must seat in a lock ring hole for proper bearing adjustment and wheel retention!**

d. Install the outer lock nut and tighten to 160–205 ft. lbs. (218–279 Nm). Final end-play of the wheel on the spindle should be 0–0.004 in. (0–0.15mm).

16. Assemble the hub parts.

17. Install the caliper.

18. Remove the safety stands and lower the vehicle.

WITH AUTOMATIC LOCKING HUBS

1. Raise the vehicle and install safety stands.

2. Refer to Automatic Locking Hub Removal and Installation and remove the hub assemblies.

3. Using a socket made for that purpose, available at most auto parts stores, use the hub nut tool to unscrew the outer locking nut.

4. Remove the lock ring from the bearing adjusting nut.

5. Use the locknut socket to remove the bearing adjusting nut.

6. Remove the caliper and suspend it out of the way.

7. Slide the hub and disc assembly off of the spindle. The outer wheel bearing will slide out as the hub is removed, so be prepared to catch it.

8. Lay the hub on a clean work surface. Carefully drive the inner bearing cone and grease seal out of the hub using Tool T69L–1102–A, or equivalent.

9. Inspect the bearing cups for pits or cracks. If necessary, remove them with a drift. If new cups are installed, install new bearings.

To install:

10. Lubricate the bearings with Multi-Purpose Lubricant Ford Specification, ESA-MIC7-B or equivalent. Clean all old grease from the hub. Pack the cones and rollers. If a bearing packer is not available, work as much lubricant as possible between the rollers and the cages.

11. Drive new cups into place with a driver, making sure they are fully seated.

12. Position the inner bearing cone and roller in the inner cup and install the grease retainer.

13. Carefully position the hub and disc assembly on the spindle.

14. Install the outer bearing cone and roller, and the adjusting nut.

NOTE: The adjusting nut has a small dowel on one side. This dowel faces outward to engage the locking ring.

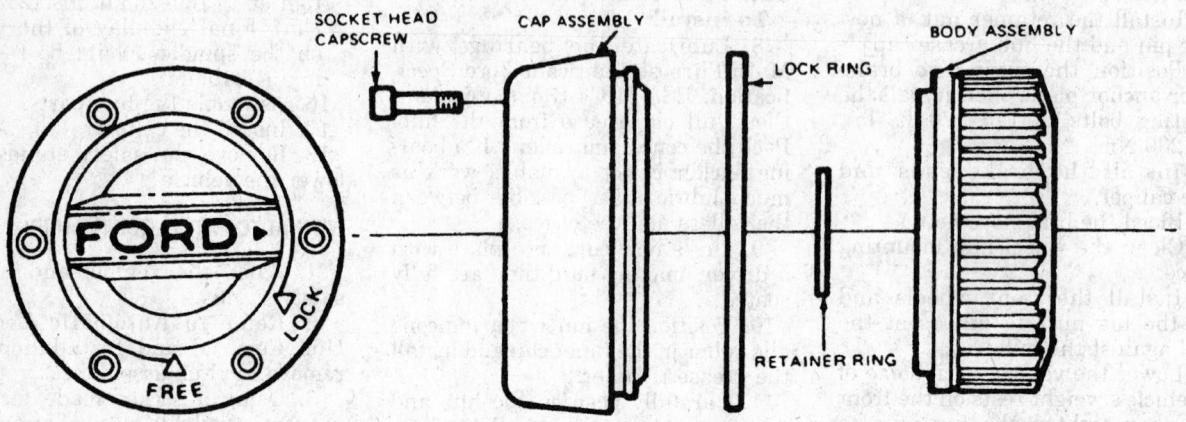

Manual locking hub assembly — 1993–96 Bronco, F Series (Pick-Up) and E Series (Van)

79215gc7

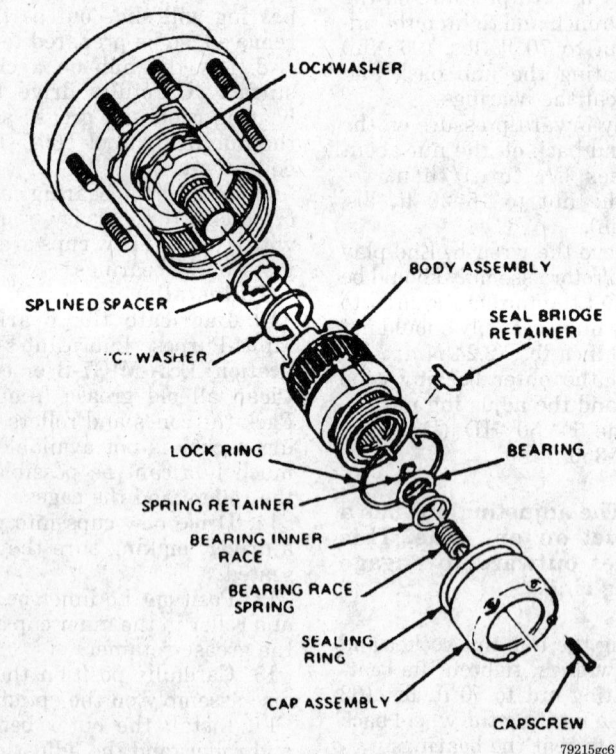

Automatic locking hub assembly — 1993–96 Bronco, F Series (Pick-Up) and E Series (Van)

79215gc6

15. Using the hub nut socket and a torque wrench, tighten the bearing adjusting nut to 50 ft. lbs. (68 Nm), while rotating the wheel back and forth to seat the bearings.

16. Back off the adjusting nut approximately 90 degrees (¼ turn).

17. Install the lock ring by turning the nut to the nearest hole and inserting the dowel pin.

NOTE: The dowel pin must seat in a lock ring hole for proper bearing adjustment and wheel retention.

18. Install the outer lock nut and tighten to 160–205 ft. lbs. (218–279 Nm). Final end-play of the wheel on the spindle should be 0–0.004 in. (0–0.15mm).

19. Assemble the hub parts.

20. Install the caliper.

21. Remove the safety stands and lower the vehicle.

1997 F-150 and Expedition Models

NOTE: The front wheel bearings are an integral, sealed part of the front wheel hub; they must be replaced as a single unit.

1. Raise and safely support the front of the vehicle on jackstands.

2. Remove the front wheels.

3. Remove the front disc brake rotor.

4. If equipped with the 4-wheel anti-lock brake system, remove the 3 front disc brake rotor shield bolts, then remove the rotor shield. Also remove the front brake anti-lock sensor.

5. Remove the front axle hub cotter pin, retainer and hub nut.

────── WARNING ──────
Do not over-extend the CV-joint and boots when removing the hub and bearing assembly.

6. Remove the 3 wheel hub bolts.

NOTE: The CV-joint is a slip fit into the wheel hub and bearing; a puller is not normally required.

7. Push the CV-joint inward, while removing the hub and bearing unit.

8. Remove the hub and bearing seal from the steering knuckle.

To install:

9. Using a seal replacer tool (T96T-1175-A), a threaded drawbar (T77F-1176-A) and a bearing cup replacer (T80T-4000-P), or their equivalents, install a new seal into the steering knuckle.

10. Position the wheel hub on the front wheel halfshaft and into the steering knuckle.

11. Install the 3 hub and bearing unit retaining bolts to 110–148 ft. lbs. (149–201 Nm).

12. If equipped with 4-wheel ABS, install the anti-lock sensor and the disc brake rotor shield.

13. Install the front axle hub nut to 188–254 ft. lbs. (255–345 Nm).

14. Install the nut retainer and a new cotter pin.

15. Install the front disc rotor.

16. Install the wheels, then lower the vehicle to the ground.

REAR SUSPENSION

Shock Absorber

REMOVAL AND INSTALLATION

1. Raise and support the rear end on jackstands.

2. Remove the self-locking nut, steel washer and bolt from the lower end of the shock absorber. Swing the lower end away from the bracket.

3. Remove the upper mounting nut and washer.

To install:

4. Attach the upper end first, then the lower end; don't tighten the nuts yet. If installing new gas shocks, attach the upper end loosely, aim the lower end at its bracket and cut the strap holding the shock compressed. Once extended, these shocks are very difficult to compress by hand!

5. Once the upper and lower ends are attached, tighten the nuts, for all models, as follows:

• 1993–96 Lower nut (except F-Super Duty stripped chassis and motor home chassis) — 52–74 ft. lbs. (71–101 Nm)

• 1993–96 Upper nut (except F-Super Duty stripped chassis and motor home chassis) — 40–60 ft. lbs. (54–81 Nm)

• 1997 Lower nut (F-150 and Expedition models) — 44–60 ft. lbs. (60–81 Nm)

• 1997 Upper nut (F-150 and Expedition models) — 22–29 ft. lbs. (30–40 Nm)

• Upper and lower nuts (F-Super Duty stripped chassis and motor home chassis) — 220–300 ft. lbs. (299–408 Nm)

Leaf Springs

REMOVAL AND INSTALLATION

Except 1997 F-150 and Expedition Models

1. Raise the vehicle by the frame until the weight is off the rear spring with the tires still on the floor.

2. Remove the nuts from the spring U-bolts and drive the U-bolts from the U-bolt plate. Remove the auxiliary spring and spacer, if equipped.

3. Remove the spring-to-bracket nut and bolt at the front of the spring.

4. Remove the upper and lower shackle nuts and bolts at the rear of the spring and remove the spring and shackle assembly from the rear shackle bracket.

5. Remove the bushings in the spring or shackle, if they are worn or damaged, and install new ones.

NOTE: When installing the components, snug down the fasteners. Don't apply final torque to the fasteners until the vehicle is back on the ground.

6. Position the spring in the shackle and install the upper shackle-to-spring nut and bolt with the bolt head facing outward.

7. Position the front end of the spring in the bracket and install the nut and bolt.

8. Position the shackle in the rear bracket and install the nut and bolt.

9. Position the spring on top of the axle with the spring center bolts centered in the hole provided in the seat. Install the auxiliary spring and spacer, if equipped.

10. Install the spring U-bolts, plate and nuts.

11. Lower the vehicle to the floor and tighten the attaching hardware as follows:

U-bolt nuts:

• Bronco, F-150 and F-250 under 8,500 lb. GVW — 75–115 ft. lbs. (102–156 Nm)

• F-250 HD and F-350 — 150–210 ft. lbs. (203–286 Nm)

• F-Super Duty chassis/cab — 200–270 ft. lbs. (272–367 Nm)

• F-Super Duty stripped chassis and motor home chassis — 220–300 ft. lbs. (299–408 Nm)

Spring-to-front spring hanger fasteners:

• F-150 2WD — 75–115 ft. lbs. (102–156 Nm)

• F-250 2WD, F-350 2WD and Bronco — 150–210 ft. lbs. (203–286 Nm)

• F-150, 250, 350 4WD — 150–175 ft. lbs. (203–238 Nm)

• F-Super Duty — 255–345 ft. lbs. (347–469 Nm)

Spring-to-rear spring hanger fasteners:

• All except F-250 and F-350 2WD Chassis Cab — 75–115 ft. lbs. (102–156 Nm)

• F-250 and F-350 2WD Chassis Cab; F-Super Duty — 150–210 ft. lbs. (203–286 Nm)

1997 F-150 and Expedition Models

1. Raise and safely support the rear of the vehicle on jackstands. Make sure to block the front wheels for safety.

2. Remove the rear wheels.

3. Support the rear axle under the differential case with an hydraulic floor jack.

4. Remove the 4 leaf spring U-bolt nuts, then remove the U-bolts.

NOTE: It is easier to remove the spring from one side of the vehicle at a time, since the spring which is still attached will help position the axle during installation.

5. Remove the leaf spring plate and spacer (4x4 models), then slowly lower the rear axle.

6. Remove the leaf spring-to-front frame bracket bolt and nut.

7. Remove the spring-to-shackle upper bolt and nut.

8. Remove the leaf spring from under the vehicle.

To install:

9. Position the rear leaf spring under the truck.

10. Install the spring-to-shackle upper bolt and nut to 72–97 ft. lbs. (98–132 Nm).

11. Install the spring-to-front frame bracket bolt and nut to 72–97 ft. lbs. (98–132 Nm).

12. Raise the rear axle with the floor jack until it rests against the leaf spring, then install the spacer (4x4 models) and the plate.

13. Install the U-bolts and nuts to 72–97 ft. lbs. (98–132 Nm).

14. Repeat the procedure for the other leaf spring, if necessary.

15. Install the rear wheels and lower the vehicle to the ground.

ADJUSTMENTS

Except 1997 F-150 and Expedition Models

Side-to-side lean can be adjusted by about ³⁄₈ in. (10mm) by installing a shim between the spring and axle on the low side. A vehicle that is low in the rear on both sides can be similarly raised by the insertion of 1 shim on each side.

If the side-to-side lean is greater than ½ in., try switch the springs from one side to the other.

Stabilizer Bar

REMOVAL AND INSTALLATION

1. Remove the nuts from the lower ends of the stabilizer bar link.

2. Remove the outer washers and insulators.

3. Disconnect the bar from the links.

4. Remove the inner insulators and washers.

5. Unbolt the link from the frame.

6. Remove the U-bolts, brackets and retainers.

7. Installation is the reverse of removal. Replace all worn or cracked rubber parts. Coat all new rubber parts with silicone grease. Assemble all parts loosely and make sure the bar assembly is centered before tight-

ening the fasteners. Observe the following torque figures:

• Stabilizer bar-to-axle nut, except F-Super Duty — 30–42 ft. lbs. (41–57 Nm)

• Stabilizer bar-to-axle bolt, F-Super Duty chassis/cab — 27–37 ft. lbs. (37–50 Nm)

• Stabilizer bar-to-axle bolt, F-Super Duty stripped chassis and motor home chassis — 30–47 ft. lbs. (41–64 Nm)

• Link bracket-to-frame nut, 4WD — 30–42 ft. lbs. (41–57 Nm)

• Link-to-bracket nut, 4WD — 60 ft. lbs. (81 Nm)

• Link-to-frame nut, 2WD — 60 ft. lbs. (81 Nm)

• Stabilizer bar-to-link — 15–25 ft. lbs. (20–34 Nm)

Wheel Bearings

REMOVAL AND INSTALLATION

NOTE: The specific axle designation is found on the rear axle identification tag, located on the differential cover.

Ford 7.5 and 8.8 Inch Rear Axles

1. Raise and safely support the vehicle.

2. Remove the rear wheel and tire assemblies and the brake drums.

3. Clean all dirt from the carrier cover area. Position a drain pan under the carrier, remove the cover and drain the rear axle.

NOTE: Whenever a plastic rear axle cover is removed, it must be replaced with a new cover and bolts. Steel rear axle covers may be reused.

4. For all 8.8 in. axles and all 7.5 in. axles except 3.73:1 and 4.10:1 ratio axles, proceed as follows:

a. Remove the differential pinion shaft lock bolt and pinion shaft.

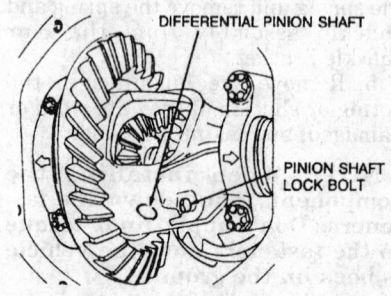

DIFFERENTIAL PINION SHAFT

PINION SHAFT LOCK BOLT

79215gd3

Differential pinion shaft and pinion shaft lockbolt location — Ford 7.5 and 8.8 inch rear differentials

b. Push the flanged end of the axle shafts toward the center of the vehicle and remove the C-lock from the button end of the axle shaft.

c. Remove the axle shaft from the housing.

d. Reinstall the pinion shaft and lock bolt to ensure the pinion gears remain in place.

5. On 7.5 in. axles equipped with 3.73:1 and 4.10:1 axle ratios, proceed as follows:

a. Remove the pinion shaft lock bolt.

b. Push out the pinion shaft until the step on the shaft contacts the ring gear.

c. Remove the C-lock from the axle shaft.

d. Remove the axle shaft from the housing.

e. Reinstall the pinion shaft and lock bolt to ensure the pinion gears remain in place.

6. Using bearing remover T83T-1225-A, or equivalent, and a suitable slide hammer, remove the axle bearing and seal as a unit.

To install:

7. Lubricate a new bearing with rear axle lubricant and install in the housing bore using a suitable driver.

8. Apply grease to the lips of a new axle seal and install, using a seal installer.

9. Remove the pinion shaft lock bolt and pinion shaft. On 7.5 in. axles equipped with 3.73:1 and 4.10:1 axle ratios, push out the pinion shaft until the step contacts the ring gear.

10. Slide the axle shaft into the axle housing, being careful not to damage the seal or axle bearing. Start the splines into the side gear and push firmly until the button end of the axle shaft can be seen in the differential case.

11. Install the C-lock on the button end of the axle shaft, then pull the shaft outboard until the shaft splines engage and the C-lock seats in the counterbore of the differential side gear.

NOTE: On 8.8 in. axles, a rubber O-ring is used to hold the C-lock in position on the axle shaft. Make sure the O-ring is in the groove at the button end of the axle shaft before installing the C-lock.

12. Slide the pinion shaft through the case and pinion gears, aligning the hole in the shaft with the lock bolt hole. Install the lock bolt and tighten to 15–30 ft. lbs. (20–40 Nm).

13. Clean all old sealer from the carrier and cover surfaces. Apply a bead of RTV sealer ⅛–¼ in. wide. The

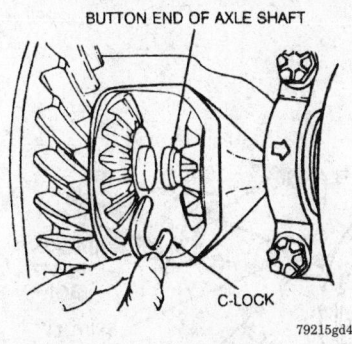

BUTTON END OF AXLE SHAFT

C-LOCK

79215gd4

Axle shaft C-lock removal and installation — Ford 7.5 and 8.8 inch rear differentials

bead should be continuous and should not pass through or outside the holes.

14. Install the cover and tighten the bolts to 15–20 ft. lbs. (21–27 Nm) if equipped with a plastic cover or 25–35 ft. lbs. (34–37 Nm) if equipped with a steel cover. Fill the carrier with the proper type and quantity of fluid.

15. Install the brake drums and the wheel and tire assemblies. Lower the vehicle.

Dana 30 and 35–1A Rear Axles

1. Raise and safely support the vehicle.

2. Remove the rear wheel and tire assemblies and the brake drums.

3. Working through the hole provided in the axle shaft flange, remove the nuts that secure the wheel bearing retainer plate. These are torque prevailing nuts and must not be reused.

4. Pull the axle shaft assembly out of the axle housing using puller adapter tool T66L-4234-A or equivalent, and a suitable slide hammer.

5. Remove the axle shaft carefully so as not to damage the outer seal. Remove the bearing race from the housing using a slide hammer type puller and bearing race puller tool T77F-1102-AA or equivalent. Remove the brake backing plate and wire it to the chassis.

6. Mount the axle shaft in a suitable fixture. Drill a ¼–½ in. hole in the outside diameter of the inner retainer ring to a depth approximately ⅜ in. the thickness of the retainer ring. Do not drill all the way through the retainer ring as this will damage the axle shaft.

7. After drilling the retainer ring, use a chisel positioned across the drilled hole and strike sharply to split the retainer ring.

8. Put the outer bearing race on the axle shaft assembly and place the axle shaft in tool T75L-1165-A, B, C or equivalent. Assemble the 2 halves of the remover collet and tighten the bolts.

9. Press the bearing and seal assembly off of the shaft. Never use heat as this would damage the axle shaft. Inspect the retainer plate for possible distortion and replace if damaged.

To install:

10. Install the outer retainer plate, if removed. Make sure it is not installed backwards.

11. Place a new lubricated seal and bearing on the axle shaft, making sure the race rib ring is facing the axle flange.

12. Press the tapered-bearing and seal assembly onto the axle shaft using tools T75L-1165-B, service plate and adapter tool T75L-1165-DA or equivalents. Apply enough pressure to seat the bearing against the axle shaft shoulder. Do not attempt to press on the bearing retainer at the same time.

13. Position a new bearing retainer on the shaft, then press it into position firmly against the bearing.

14. Apply lubricant to the outer diameter of the race and seal. Install the brake backing plate and attaching bolts.

15. Before sliding the shaft assembly into the axle housing, make sure the outer seal is fully mounted on the bearing.

16. Carefully slide the axle shaft into the housing, start the axle splines into the side gear and push the shaft in until the bearing bottoms in the housing.

17. Install the bearing retainer plate and nuts. Tighten the nuts to 25–35 ft. lbs. (34–47 Nm).

18. Install the brake drum and the wheel and tire assembly.

19. Add lubricant through the filler hole until the level reaches ⅛–⅝ in. (3–16mm) below the bottom of the filler hole. The axle must be in the running position and the vehicle level.

20. Lower the vehicle.

Full Floating Rear Axle

F-250 HD and F-350

The wheel bearings on the Ford 10½ in. full floating rear axle are packed with wheel bearing grease. Axle lubricant can also flow into the wheel hubs and bearings, however, wheel bearing grease is the primary lubricant. The wheel bearing grease provides lubrication until the axle lubri-

cant reaches the bearings during normal operation.

The wheel bearings on the full floating rear axle are packed with wheel bearing grease. Axle lubricant can also flow into the wheel hubs and bearings, however, wheel bearing grease is the primary lubricant. The wheel bearing grease provides lubrication until the axle lubricant reaches the bearings during normal operation.

1. Set the parking brake and loosen the axle shaft bolts.

2. Raise the rear wheels off the floor and place jackstands under the rear axle housing so the axle is parallel with the floor.

3. Remove the wheels.

4. Remove the brake drums.

5. Remove the axle shaft bolts.

6. Remove the axle shaft and discard the gaskets.

7. With the axle shaft removed, remove the gasket from the axle shaft flange studs.

8. Install Hub Wrench T85T-4252-AH, or equivalent, so the drive tangs on the tool engage the slots in the hub nut.

NOTE: The hub nuts are right hand thread on the right hub and left hand thread on the left hub. The hub nuts should be stamped RH and LH. Never use power or impact tools on these nuts! The nuts will ratchet during removal.

9. Remove the hub nut.

10. Install step plate adapter tool D80L-630-7 or equivalent, in the hub.

11. Install puller D80L-1002-L or equivalent, and loosen the hub to the point of removal. Remove the puller and step plate.

12. Remove the hub, taking care to catch the outer bearing as the hub comes off.

13. Install the hub in a soft-jawed vise and pry out the hub seal.

14. Lift out the inner bearing.

15. Drive out the inner and outer bearing races with a drift.

16. Wash all the old grease or axle lubricant out of the wheel hub, using a suitable solvent.

17. Wash the bearing races and rollers and inspect them for pitting, galling, and uneven wear patterns. Inspect the roller for end wear. Replace any bearing and race that appears in any way damaged. Always replace the bearings and races as a set.

To install:

18. Coat the race bores with a light coat of clean, waterproof wheel bearing grease and drive the races

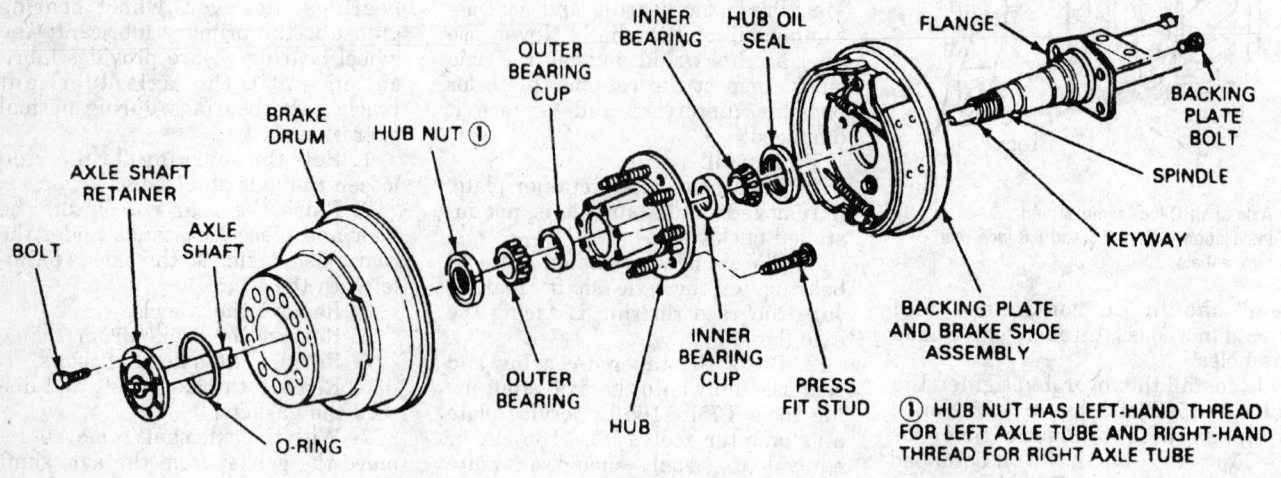

INNER
BEARING
HUB OIL
SEAL
OUTER
BEARING
CUP
FLANGE
BRAKE
DRUM
HUB NUT ①
BACKING
PLATE
BOLT
AXLE SHAFT
RETAINER
SPINDLE
AXLE
SHAFT
KEYWAY
BOLT
BACKING PLATE
AND BRAKE SHOE
ASSEMBLY
O-RING
OUTER
BEARING
INNER
BEARING
CUP
HUB
PRESS
FIT STUD

① HUB NUT HAS LEFT-HAND THREAD
FOR LEFT AXLE TUBE AND RIGHT-HAND
THREAD FOR RIGHT AXLE TUBE

79215gc5

Full-floating axle shaft, bearing and hub assembly — F-250 Heavy Duty and F-350 models

squarely into the bores until they are fully seated.

19. Pack each bearing cone and roller with a bearing packer.

20. Place the inner bearing cone and roller assembly in the wheel hub.

NOTE: When installing the new seal, the words OIL SIDE must go inwards towards the bearing!

21. Place the seal squarely in the hub and drive it into place. The best tool for the job is a seal driver such as T85T–1175–AH, which will stop when the seal is at the proper depth.

NOTE: If the seal is misaligned or damaged during installation, a new seal must be installed.

22. Clean the spindle thoroughly. If the spindle is excessively pitted, damaged or has a predominately bluish tint (from overheating), it must be replaced.

23. Coat the spindle with 80W/90 oil.

24. Pack the hub with clean, waterproof wheel bearing grease.

25. Pack the outer bearing with clean, waterproof wheel bearing grease.

26. Place the outer bearing in the hub and install the hub and bearing together on the spindle.

27. Install the hub nut on the spindle. Make sure the nut tab is located in the keyway prior to thread engagement. Turn the hub nut onto the threads as far as possible by hand, noting the thread direction.

28. Install the hub wrench tool and tighten the nut to 55–65 ft. lbs. (75–88 Nm). Rotate the hub occasionally during nut tightening.

29. Ratchet the nut back 5 teeth; make sure you hear 5 clicks!

30. Inspect the axle shaft O-ring seal and replace it.

31. Install the axle shaft.

32. Coat the axle shaft bolt threads with waterproof seal and install them by hand until they seat; do not tighten them with a wrench at this time!

33. Check the diameter across the center of the brake shoes. Check the diameter of the brake drum. Adjust the brake shoes so their diameter is 0.030 in. (0.76mm) less than the drum diameter.

34. Install the brake drum

35. Install the wheel.

36. Loosen the differential filler plug. If lubricant starts to run out, retighten the plug. If not, remove the plug and fill the housing with 80W/90 gear oil.

37. Lower the vehicle to the floor.

38. Tighten the wheel lugs to 140 ft. lbs. (190 Nm).

39. Now tighten the axle shaft bolts. Torque them to 60–80 ft. lbs. (81–109 Nm).

FIRING ORDERS

NOTE: To avoid confusion, always replace spark plug wires one at a time.

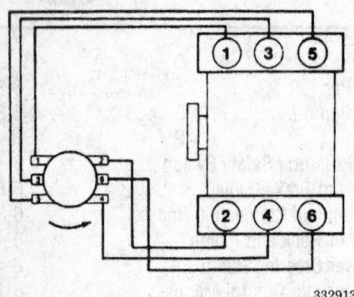

332913

3.0L Engine
Engine Firing Order: 1-2-3-4-5-6
Distributor Rotation : Counterclockwise

ENGINE ELECTRICAL

NOTE: Disconnecting the negative battery cable on some vehicles may interfere with the functions of the on board computer systems and may require the computer to undergo a relearning process, once the negative battery cable is reconnected.

Distributor

REMOVAL AND INSTALLATION

These vehicles use a camshaft-driven distributor. The Camshaft Position Sensor (CMP) is built into the distributor. The CMP sensor monitors engine speed (rpm) and piston position and sends signals to the Powertrain Control Module (PCM). The 3.0L engine uses a power transistor, resistor and condenser and ignition coil mounted separately from the distributor. As the power transistor grounds the primary circuit, the inductive charge built up in the secondary circuit sends a spark from the ignition coil to the distributor. The distributor rotor and cap then send a spark to each spark plug. The spark advance and retard functions are controlled by the PCM. If engine knocking occurs, the Knock Sensor detects the condition and a signal is set to the

PCM. The PCM retards the ignition timing to prevent engine knocking. The base ignition timing is programmed in the anti-knocking zone, if the recommended fuel is used. Therefore, the knock sensor system does not operate under normal driving conditions.

Engine firing order is 1-2-3-4-5-6. Number one cylinder is on the rear bank, forward cylinder.

1. Disconnect the negative battery cable.

2. Remove the distributor cover.

3. Loosen the three distributor cap screws and set aside the distributor cap with all the wires still attached.

4. Disconnect the distributor ground connector from the tab on the distributor housing.

5. Disconnect the distributor harness electrical connector and remove it from its bracket.

6. Rotate the crankshaft until the No. 1 piston is at TDC on the compression stroke. Check to be sure that the timing mark (may be yellow) on the crankshaft pulley and the timing pointer on the front cover are aligned.

7. Note the relation of the distributor rotor to the engine. Make a mark on a nearby engine component to assist with installation.

8. Remove the distributor hold-down bolt and lift up the distributor with the base gasket.

9. Note that if the rotor is being removed, a small retainer bolt on the side must first be removed.

To install:

10. Verify that the crankshaft is still at TDC No. 1 cylinder, compression stroke (firing position).

11. Install a new distributor base gasket. Lower the distributor into the engine. Note that during installation, the distributor rotor will tend to turn as the gears engage and mesh. Make sure that the distributor rotor aligns with the mark made on the engine component during removal. Install the hold-down bolt only finger-tight.

12. Install the distributor electrical connector and the ground connector.

13. Install the distributor cap and wires. Make sure the cap is seated properly. Install the cover.

14. Connect the negative battery cable.

15. Start the engine and allow to warm to operating temperature. Adjust the ignition timing using the recommended procedure. After all adjustments are made, tighten the distributor hold-down bolt to 10–12 ft. lbs. (14–17 Nm).

Ignition Timing

ADJUSTMENT

1. Apply the parking brake and make sure that the vehicle is in PARK.

2. Start and run the engine until it reaches normal operating temperature.

3. Rev the engine to 2500–3000 rpm 2 or 3 times and allow the engine to return to idle.

4. Turn off all electrical loads.

5. Disconnect the Idle Air Control (IAC) solenoid.

6. Check the idle speed. The idle speed should be 750 ± 50 rpm.

7. Connect a timing light to the No. 1 cylinder spark plug wire at the distributor end and check the ignition timing. Make sure that the timing pointer is pointing to the 15 degrees BTDC mark on the crankshaft pulley.

NOTE: Each notch on the crankshaft pulley represents 5 degrees.

8. If the timing is not within the specification, loosen the distributor mounting bolt and adjust the distributor until the timing is at the proper specification.

9. Tighten the distributor mounting bolt to 10–12 ft. lbs, (14–17 Nm).

Alternator

PRECAUTIONS

Several precautions must be observed with alternator equipped vehicles to avoid damage to the unit.

• If the battery is removed for any reason, make sure it is reconnected with the correct polarity. Reversing the battery connections may result in damage to the 1-way rectifiers.

• When utilizing a booster battery as a starting aid, always connect the positive to positive terminals and the negative terminal from the booster battery to a good engine ground on the vehicle being started.

• Never use a fast charger as a booster to start vehicles.

• Disconnect the battery cables when charging the battery with a fast charger.

• Never attempt to polarize the alternator.

• Do not use test lights of more than 12 volts when checking diode continuity.

• Do not short across or ground any of the alternator terminals.

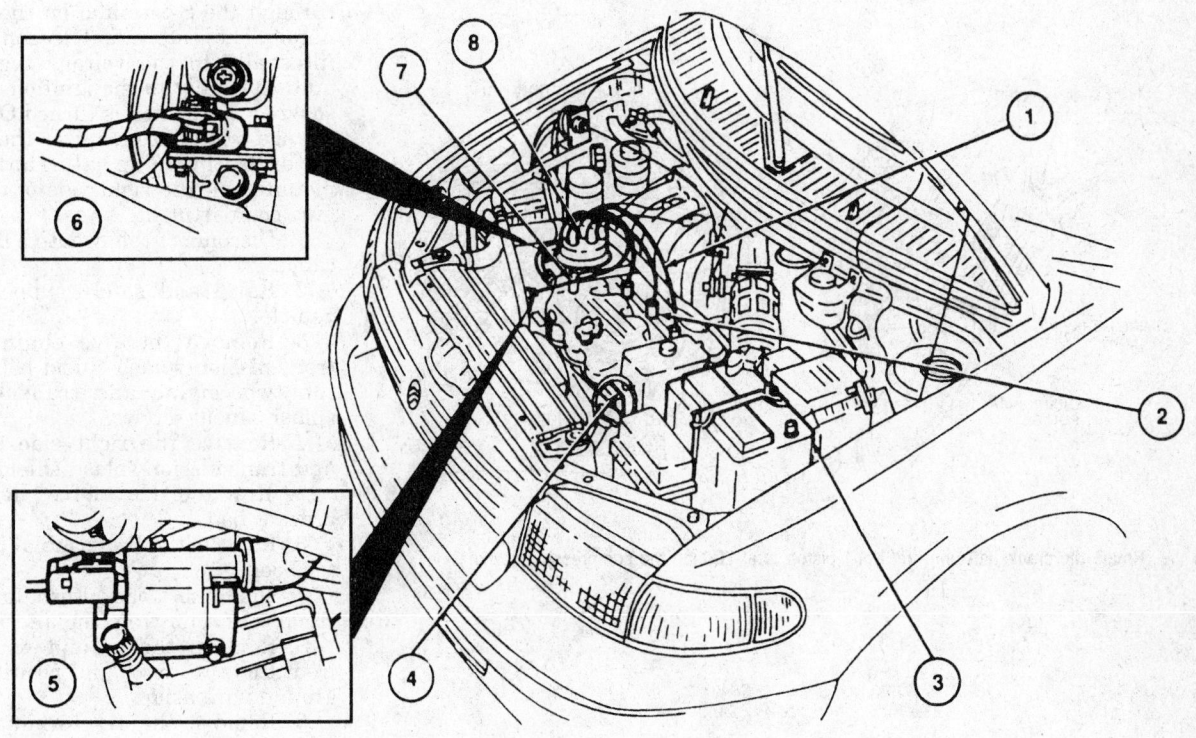

1. Distributor to spark plug wire
2. Spark plug
3. Battery
4. Main fuse
5. Ignition coil
6. Power transistor
7. Ignition coil to distributor high tension wiring
8. Distributor

354325

Ignition system components and locations

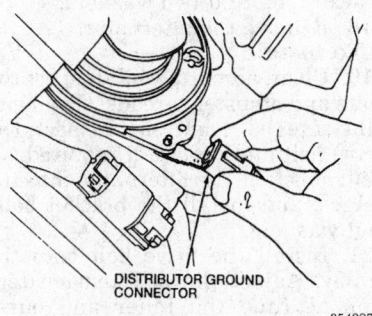

DISTRIBUTOR GROUND
CONNECTOR

354327

**Disconnect the distributor ground connector
from the tab on the distributor housing**

• The polarity of the battery, alternator and regulator must be matched and considered before making any electrical connections within the system.
• Never separate the alternator on an open circuit. Make sure all connections within the circuit are clean and tight.
• Disconnect the battery ground terminal when performing any service on electrical components.
• Disconnect the battery if arc welding is to be done on the vehicle.

REMOVAL AND INSTALLATION

These vehicles use a conventional 12-volt negative ground electrical system. The alternator (also called a generator) has a built-in non-adjustable electronic voltage regulator designed to limit the charging system voltage to an operating range of 14.1 to 14.7 volts. When the engine is started, the alternator begins to generate Alternating Current (AC) which is converted to Direct Current (DC) by the alternator's internal rectifier. This current is then supplied to the vehicle's electrical system through the B+ terminal located on the rear of the alternator. The inter-

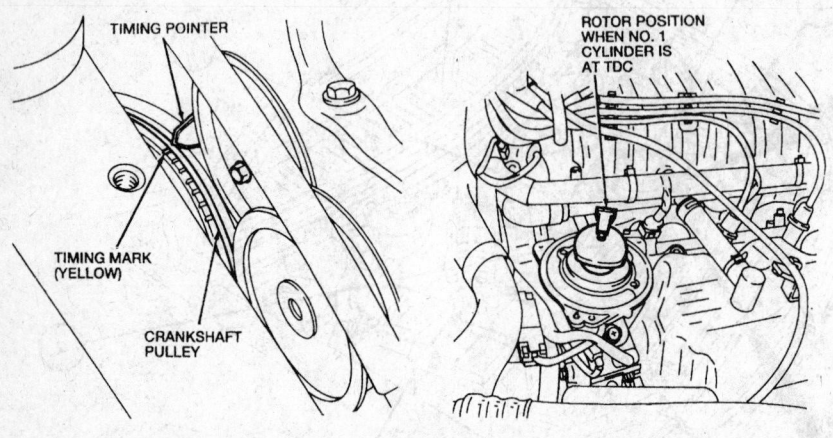

Rotate the crankshaft until the No. 1 piston is at TDC on the compression stroke

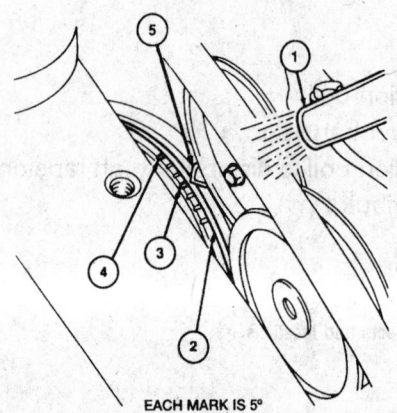

1. Inductive timing light
2. Crankshaft pulley
3. 15 degree BTDC
4. Yellow timing mark (TDC)
5. Timing pointer

IGNITION SYSTEM SPECIFICATIONS

Description	Specification
Plug Type	AGSP-32C
Spark Plug Gap	0.8-0.9mm (0.031-0.035 inch)
Firing Order	1-2-3-4-5-6
Ignition Timing	15° ± 2° BTDC

354388

Timing should be 15 degrees Before Top Center (three notches from TDC)

nal voltage regulator varies the alternator's rotor field excitation according to the electrical current demand of the vehicle which is sampled through the S-terminal on the alternator. The L-terminal is connected internally to the voltage regulator and externally to the ignition switch so when the switch is turned ON, the voltage regulator energizes the alternator rotor. The alternator is mounted on the right side of the engine compartment.

1. Disconnect the negative battery cable.
2. Raise and safely support the vehicle.
3. Remove the five engine and transmission splash shield bolts and the two engine and transmission splash shield screws.
4. Remove the right side engine and transmission splash shield.
5. Remove the wire harness bracket bolt.
6. Remove the B+ terminal insulator boot.
7. Disconnect the alternator electrical connector from the alternator.
8. Remove the ground wire nut from the alternator and position the ground wire aside.
9. Remove the B+ terminal nut from the alternator and position the B+ terminal wire aside.
10. Loosen the inner and outer alternator-to-bracket bolts.
11. Loosen the alternator locking bolt.
12. Loosen the alternator adjustment bolt.
13. Remove the A/C compressor belt, if required.
14. Remove the drive belt from the alternator pulley.
15. Remove the alternator locking bolt and washer.
16. Remove the inner alternator-to-bracket bolt and two washers.
17. Remove the outer alternator-to-bracket bolt and two washers.
18. Remove the alternator.

To install:

19. Clean all parts well. Inspect the bolts and adjuster threads. Clean and lubricate as required. Inspect the drive belt and replace if required.
20. Position the alternator into the vehicle and install the bracket bolts and washers.
21. Install the drive belt onto the pulley. Adjust the belt tension tension. Torque the inner and outer bracket bolts to 17-19 ft. lbs. (23-26 Nm).
22. Install the A/C compressor belt, if required.

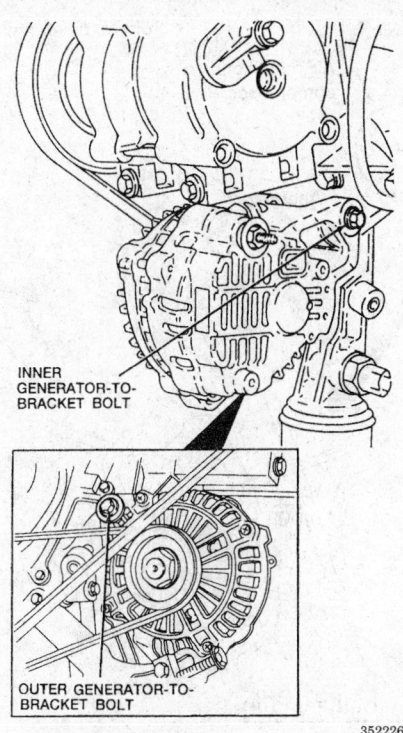

INNER
GENERATOR-TO-
BRACKET BOLT

OUTER GENERATOR-TO-
BRACKET BOLT

352226

Inner and outer alternator-to-bracket bolt locations

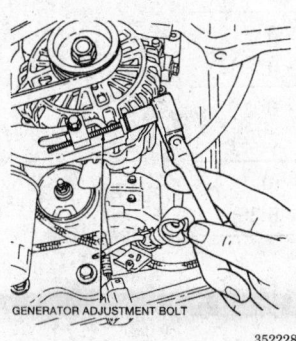

GENERATOR ADJUSTMENT BOLT

352228

Adjust drive belt tension using the alternator adjustment bolt

23. Connect the B+ wire, the ground wire and the alternator harness connector. Make sure the B+ protective boot is properly positioned.
24. Install the under-vehicle splash panels.
25. Connect the negative battery cable.
26. Start vehicle and verify proper alternator operation.

Drive Belt

REMOVAL AND INSTALLATION

NOTE: When replacing more than one belt it is a good idea, to make note or mark what belt goes around what pulley. This will make the installation fast and easy. Also, when a new belt is installed, the manufacturer recommends rechecking the drive belt tension after the vehicle has been driven 1,000 miles.

Alternator

1. Disconnect the negative battery cable.
2. Loosen the pivot and mounting bolts of the alternator.
3. On models with the locking bolt, loosen the locking bolt on the alternator adjusting bolt. The alternator adjusting bolt will loosen and tighten the alternator drive belt tension.
4. On models without the locking bolt, pry the component inward to relieve the tension on the drive belt. Always be careful where using the pry bar not to damage the alternator or surrounding components.
5. When there is enough slack in the belt, remove the belt from the alternator pulley.
   To Install:
6. Verify that the new belt and the old belt have the same length and width. These measurements must be the same or problems will occur when the new belt is adjusted.
7. Correctly route the belt around the pulleys.
8. After new belt is installed correctly, adjust the tension of the new belt.
9. Tighten the mounting, pivot, and lock bolts.
10. Reconnect the negative battery cable.

Air Conditioning Compressor

1. Disconnect the negative battery cable.
2. Loosen the lock bolt for the idler pulley.
3. Loosen the idler pulley adjusting bolt. When the idler pulley adjustment bolt is loosened, the drive belt tension will slowly be released.
4. When there is enough slack in the belt, remove the belt from the pulleys.
   To Install:
5. Verify that the new belt and the old belt have the same length and width. These measurements must be

the same or problems will occur when the new belt is adjusted.
6. Correctly route the belt around the pulleys.
7. After new belt is installed correctly, adjust the tension of the new belt.
8. Tighten the mounting and pivot bolts.
9. Reconnect the negative battery cable.

Power Steering Oil Pump

With Adjustable Idler Pulley

1. Disconnect the negative battery cable.
2. Loosen the lock bolt for the idler pulley.
3. Loosen the idler pulley adjusting bolt. When the idler pulley adjustment bolt is loosened, the drive belt tension will slowly be released.
4. When there is enough slack in the belt, remove the belt from the pulleys.
   To Install:
5. Verify that the new belt and the old belt have the same length and width. These measurements must be the same or problems will occur when the new belt is adjusted.
6. Correctly route the belt around the pulleys.
7. After new belt is installed correctly, adjust the tension of the new belt.
8. Tighten the mounting, pivot, and lock bolts.
9. Reconnect the negative battery cable.

With Non–adjustable Idler Pulley

1. Disconnect the negative battery cable.
2. Loosen the power steering oil pump mounting and pivot bolts.
3. Loosen the drive belt adjustment locking bolt. The drive belt adjustment bolt is located on the power steering oil pump. The drive belt adjustment bolt will move the power steering oil pump and increase or decrease belt tension.
4. Turn the power steering oil pump adjusting bolt until there is enough slack in the drive belt to remove it.
5. Remove the drive belt from around the pulleys.
   To Install:
6. Verify that the new belt and the old belt have the same length and width. These measurements must be the same or problems will occur when the new belt is adjusted.
7. Correctly route the belt around the pulleys.

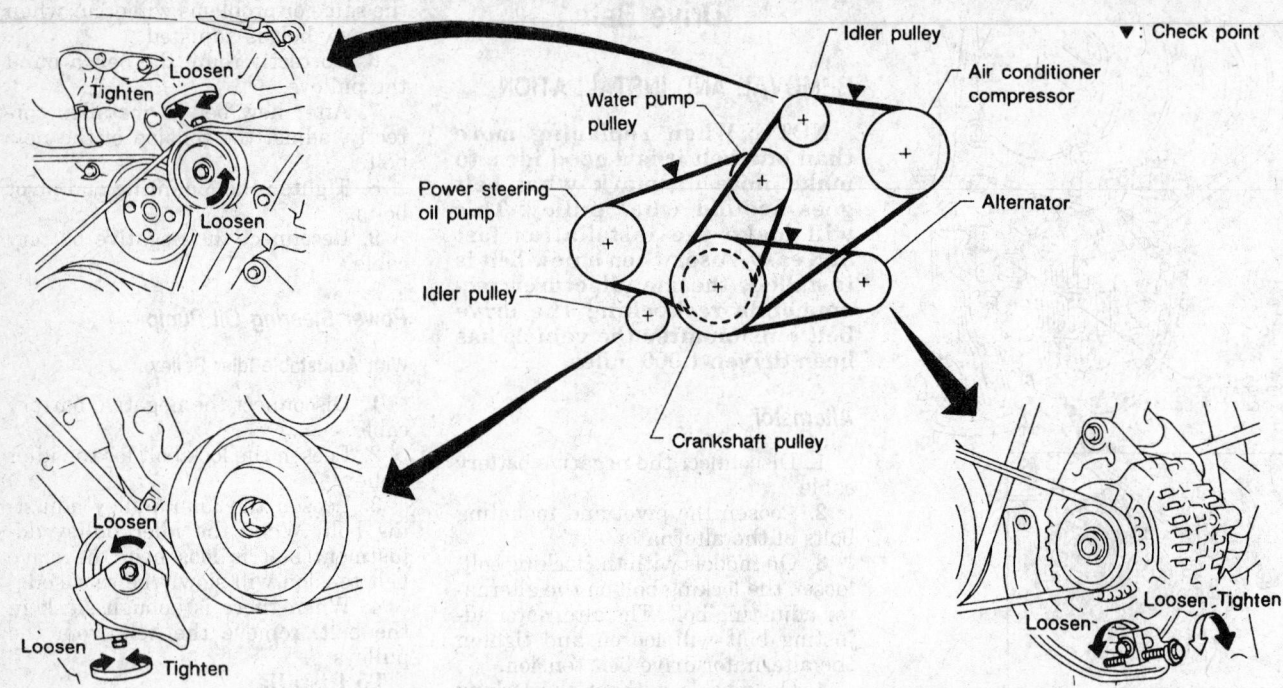

▼ : Check point

Idler pulley

Air conditioner compressor

Water pump pulley

Power steering oil pump

Alternator

Idler pulley

Crankshaft pulley

Loosen Tighten Loosen Loosen Tighten Loosen Loosen Tighten

Unit: mm (in)

	Used belt deflection		Deflection of new belt
	Limit	Deflection after adjustment	
Alternator	12 (0.47)	7 - 9 (0.28 - 0.35)	6 - 8 (0.24 - 0.31)
Air conditioner compressor	10 (0.39)	5 - 7 (0.20 - 0.28)	4 - 6 (0.16 - 0.24)
Power steering oil pump	16 (0.63)	10 - 12 (0.39 - 0.47)	8 - 10 (0.31 - 0.39)
Applied pushing force	98 N (10 kg, 22 lb)		

Inspect drive belt deflection when engine is cold.

255708

Engine drive belts and deflection chart-VG30E

8. After the drive belt is installed correctly, adjust the drive belt tension.

9. Tighten the mounting, pivot, and lock bolts.

10. Reconnect the negative battery cable.

Starter

REMOVAL AND INSTALLATION

These vehicles are equipped with a starter motor that uses a planetary gear reduction system to reduce internal torque. The starter motor is a 12-volt unit that has a solenoid mounted on the gear end housing. The amount of resistance in any vehicle's starting circuit must be kept to a minimum to provide the maximum current for starter operation. A discharged or damaged battery, loose or corroded connections or partially broken cables will result in slower than normal cranking speeds or even a No-Start condition.

A Park/Neutral Position (PNP) switch circuit within the Transmission Range (TR) switch senses when the gearshift lever is in the Neutral or Park position and completes the starter motor circuit.

CAUTION

When servicing the starter motor or performing other underhood work in the vicinity of the starter motor, be aware that the heavy gauge battery input lead at the starter motor is "electrically hot" at all times. A protective boot should be in place over this terminal and must be replaced after servicing. Be sure to disconnect the battery negative (ground) cable before servicing the starter motor.

1. Disconnect the negative battery cable.

2. Remove the engine air cleaner assembly.

3. Slide back the protective boot and remove the nut securing the battery positive cable on the B+ terminal of the starter solenoid. Pull the positive cable from the solenoid.

4. Disconnect the S-terminal electrical connector.

5. Remove the two starter retainer bolts and remove the starter from the vehicle.

To install:

6. Clean all parts well. If reusing the original starter motor, inspect the unit for damage. The starter motor can be disassembled for overhaul and brush replacement. Check the pinion gear and the flywheel teeth for damage. If the flywheel is damaged, it can quickly fail a replacement starter.

7. The starter drive pinion gear depth adjustment can be checked using the following procedure.

   a. Locate and disconnect the M-terminal wire.

   b. Connect the positive (+) battery lead to the starter motor S-terminal.

CAUTION

Do not apply battery voltage for more than 10 seconds.

---

   c. Connect the battery ground cable to a clean, rust-free portion of the starter motor frame.

   d. Measure the clearance between the starter drive and the stop ring. The gap should measure 0.020–0.078 inch (0.5–2.0 mm).

   e. If the pinion gap is not within specifications, increase or decrease the number of pinion thrust washers between the magnetic switch (solenoid) and the starter frame. The gap becomes smaller and the number of washers increases.

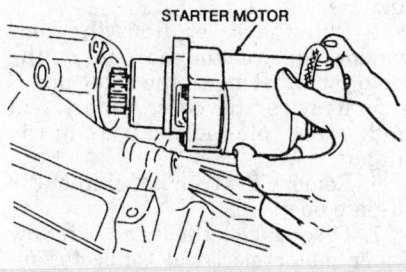

355552

Removing the starter motor

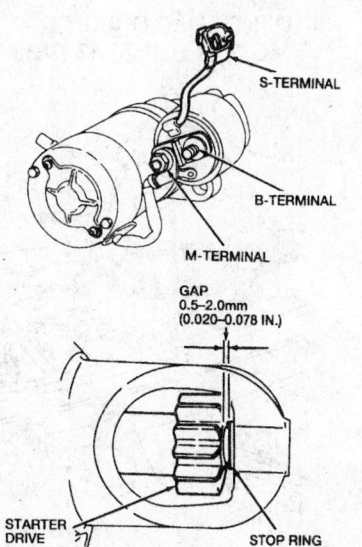

S-TERMINAL

B-TERMINAL

M-TERMINAL

GAP
0.5–2.0mm
(0.020–0.078 IN.)

STARTER
DRIVE

STOP RING

355555

The starter drive pinion gear depth should be checked and adjusted if necessary

8. Install the starter motor. Torque the retainer bolts to 17–19 ft. lbs. (23–26 Nm).

9. Connect the electrical connector and the battery B+ cable. Torque the cable retainer nut to 87–104 inch lbs. (10–12 Nm). Install the protective boot.

10. Install the engine air cleaner assembly.

11. Connect the negative battery cable. Test starter operation to verify repair.

CHASSIS ELECTRICAL

Blower Motor

REMOVAL AND INSTALLATION

The blower motor circuits include two fuses identified as 20A FRONT BLOWER MOTOR fuses located in the interior fuse junction panel. In addition, the 30A IGN SW fuse should also be checked when checking a blower motor inoperative complaint.

1. Disconnect the negative battery cable.

2. Remove 1 plastic rivet and 4 screws holding the right side lower finish panel in place.

3. Disconnect the wiring connector to the blower motor.

4. Remove the blower motor housing tube.

5. Remove the 3 screws holding the blower motor to the housing. Remove the blower motor.

To install:

6. If necessary, the blower motor wheel (also called the fan, or the squirrel cage) can be replaced. Pry off the center retainer and pull the blower motor wheel from the motor shaft. Press the replacement blower motor wheel onto the motor shaft and secure the a new clip.

7. Position the blower motor and secure the 3 retaining screws.

8. Install the blower motor housing tube.

9. Connect the wiring connector to the blower motor.

10. Install the right side lower finish panel.

11. Connect the negative battery cable.

12. Check blower motor operation.

Windshield Wiper Motor

REMOVAL AND INSTALLATION

The front windshield wiper motor is located in the engine compartment and is directly connected to the wiper linkage. The windshield wiper motor is a two-speed unit. The windshield wiper/washer control knob on the multi-function switch controls the interval wiper function. The wiper governor is wired in series between the multi-function switch and the wiper motor. It operates as a timer mechanism for the interval wiper function. The governor is located on the left side strut tower directly below the hood support rod pivot.

The rear window wiper is driven by a single-speed motor mounted in the liftgate. The rear wiper/washer switch is located on the upper right side of the instrument panel finish panel.

Front

1. Disconnect the negative battery cable.

2. Lift the wiper arm and slide out the lock tab located at the bottom of wiper arm. Pull upward and remove

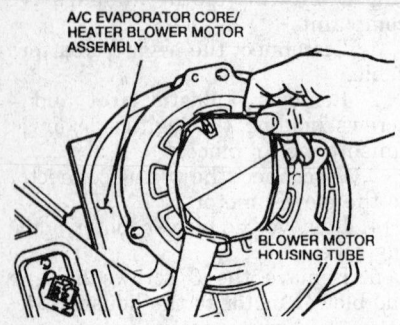

Blower motor and hose assembly

the wiper arm. Repeat for the other wiper arm. Note that the left side and right side wiper arms are not interchangeable. The passenger side wiper arm is longer than the driver's side arm. The wiper arms are marked on the middle of the back side with a "D" for the driver's side and an "A" for the passenger side.

3. Remove the cowl top vent panel by removing the center screw and unlocking the ten locking screws. Raise or lower the hood as necessary to gain clearance for removal of the cowl vent.

4. Disconnect the washer hoses.

5. Remove the wiper motor from the vehicle.

6. Disconnect the wiper motor electrical connection.

7. Remove the six wiper motor bolts and remove the motor and bracket assembly from the vehicle.

8. Remove the four bracket-to-wiper motor locknuts and bolts to separate the motor from the bracket and linkage assembly.

To install:

9. If removed, assemble the motor to the bracket and linkage assembly and install the four bolts and locknuts. Torque the bolts to 54–91 inch lbs. (6–7 Nm).

10. Install the wiper motor and bracket assembly to the vehicle and install the six motor bolts. Torque the bolts to 54–91 inch lbs. (6–7 Nm).

11. Connect the wiper motor electrical connection.

12. Connect the washer hose to the cowl vent.

13. Install the cowl vent, lock the ten locking screws and install the center cowl screw.

14. Install the wiper arms.

15. Reconnect the negative battery cable.

16. Check for proper operation.

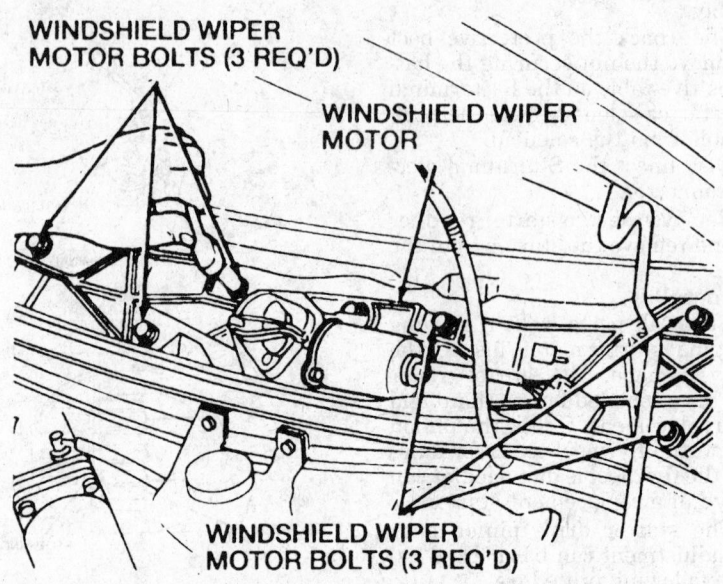

Remove the six bolts and remove the front wiper and linkage assembly from the vehicle

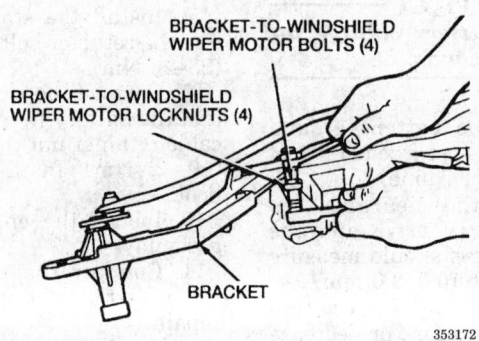

Remove the four bracket-motor locknuts and bolts to separate the front motor from the linkage

Rear

The rear window wiper/washer switch is located on the right side of the instrument panel. The rear window wiper operates at one speed. It completes a cycle every 10 to 15 seconds. If equipped, the opening back window glass must be closed for the rear window washer system to operate.

1. Disconnect the negative battery cable.

2. Depress the locking clip on the inside base of the rear window wiper pivot arm and remove the rear window wiper pivot arm and blade assembly.

3. Disconnect the rear window washer hose from the washer nozzle and bracket.

4. Pull the cover from the rear window wiper motor shaft nut on the motor shaft. Remove the shaft nut.

5. Remove the outer collar, seal and white plastic seal from the liftgate.

6. Remove the trim panel from the liftgate door.

7. Disconnect the rear window wiper motor electrical connector inside the liftgate.

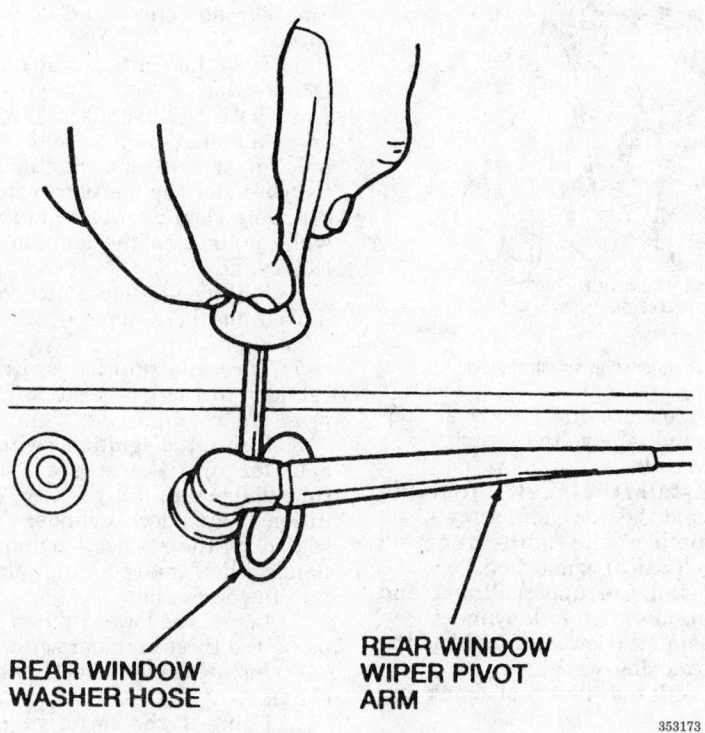

REAR WINDOW WASHER HOSE

REAR WINDOW WIPER PIVOT ARM

353173

Rear wiper arm removal

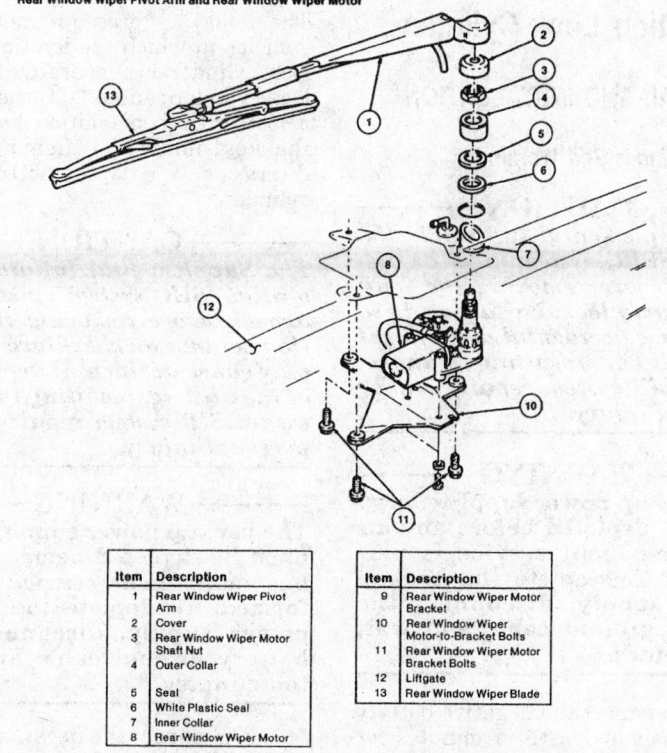

Rear Window Wiper Pivot Arm and Rear Window Wiper Motor

Item	Description
1	Rear Window Wiper Pivot Arm
2	Cover
3	Rear Window Wiper Motor Shaft Nut
4	Outer Collar
5	Seal
6	White Plastic Seal
7	Inner Collar
8	Rear Window Wiper Motor

Item	Description
9	Rear Window Wiper Motor Bracket
10	Rear Window Wiper Motor-to-Bracket Bolts
11	Rear Window Wiper Motor Bracket Bolts
12	Liftgate
13	Rear Window Wiper Blade

353174

Rear window wiper pivot and motor arrangement

8. Remove the three wiper motor bracket bolts. Remove the inner collar from the motor shaft. Remove the two wiper motor-to-bracket bolts to separate the motor from the bracket. Slide the wiper motor electrical connector from the bracket.

To install:

9. Assemble the motor to the bracket. Tighten the two wiper motor-to-bracket bolts to 44 inch lbs. (5 Nm).

10. Install the motor and bracket assembly to the liftgate and tighten the three motor bracket bolts to 54–61 inch lbs. (6–7 Nm). Make sure the electrical connector is secure.

11. Assemble the motor shaft components and tighten the shaft nut to 54–70 inch lbs. (6–8 Nm).

12. Install the liftgate trim panel and luggage compartment lamps, if equipped.

13. Connect the negative battery cable. Cycle the wiper motor at least once. When satisfied that the rear window wiper motor is in its PARK position, install the rear window wiper arm.

Headlight Switch

REMOVAL AND INSTALLATION

The three-position headlamp switch is located on the left side of the instrument panel on the finish panel. The three functions of the three-position switch are:

FIRST POSITION: No lamps are illuminated. The headlamp switch must be in this position for the autolamp system to operate.

SECOND POSITION: All lamps operate except the headlamps.

THIRD POSITION: All lamps operate including the headlamps.

The high beam and flash-to-pass features are incorporated in the multi-function switch.

1. Disconnect the negative battery cable.

2. Pull the headlamp switch/autolamp (if equipped)/ light switch rheostat resistor assembly from the instrument panel.

3. Disconnect the electrical connector from the switch.

4. Disconnect the autolamp/light switch rheostat resistor assembly electrical connector.

5. Pull the headlamp switch knob off of the headlamp switch.

6. Remove the two headlamp switch screws and the headlamp switch.

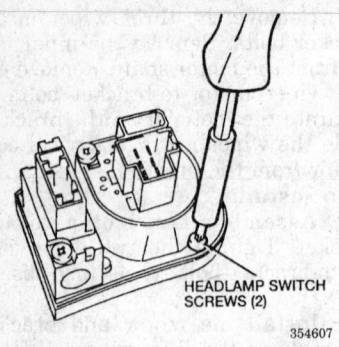

HEADLAMP SWITCH
SCREWS (2)

354607

**The switch is retained to the bezel by
two screws**

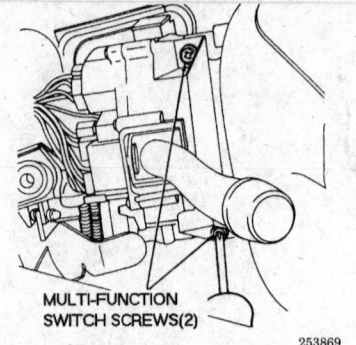

MULTI-FUNCTION
SWITCH SCREWS(2)

253869

Multi function switch removal

7. If the miniature bulb that illuminates the headlamp switch is defective, turn the bulb holder counterclockwise to remove the bulb.

To install:

8. If the miniature bulb that illuminates the headlamp switch was
being replaced, turn the new bulb
holder clockwise to install and seat
the bulb.

9. Fit the headlamp switch to the
bezel and install the two screws.

10. Fit the knob back onto the
switch.

11. Connect the electrical connectors. Verify that they fit securely.

12. Carefully push the headlamp
switch back into the instrument
panel trim panel.

13. Connect the negative battery
cable.

14. Test all headlamp functions and
verify correct operation.

Combination Switch

REMOVAL AND INSTALLATION

1. Disconnect the negative battery
cable and wait ten minutes to deplete
the back up power supply for the air
bag.

─── **CAUTION** ───

*The air bag system (SRS or SIR)
must be disarmed before removing the multi function switch.
Failure to do so may cause accidental deployment, property damage or personal injury.*

2. Unscrew the tilt wheel handle
and shank and remove.

3. Remove the three steering column shroud screws and shroud.

4. Remove the ignition switch lock
cylinder.

5. Remove the upper shroud.

6. Remove the two multi function
switch attaching screws.

7. Disconnect the electrical connectors and remove the switch.

To install:

8. Install the multi function
switch and the attaching screws.

9. Connect the multi function
switch electrical connection.

10. Install the upper shroud and
the ignition switch lock cylinder.

11. Install the lower shroud and the
three attaching screws.

12. Install the tilt wheel shank and
handle.

13. Connect the negative battery
cable.

14. Check for proper operation.

Ignition Lock Cylinder

REMOVAL AND INSTALLATION

Lock Cylinder Still Working

─── **CAUTION** ───

*The Supplemental Inflatable Re-
straint (SIR) system must be disarmed before removing the ignition column lock. Failure to do so
may cause accidental deployment
of the air bag, resulting in unnecessary SIR system repairs and/or
personal injury.*

─── **WARNING** ───

**The backup power supply energy
must be depleted before any air
bag component service is performed. To deplete the backup
power supply, disconnect the
battery ground cable and wait
ten minutes.**

1. Disconnect the negative battery
cable. Wait at least ten minutes for
the air bag backup power supply to
deplete before beginning work.

2. If equipped, unscrew and remove the tilt wheel handle and
shank.

3. Remove the three steering column shroud screws and remove the
shroud.

4. Turn the ignition switch the the
ON position.

5. Place a ⅛-inch (3 mm) wire pin
or small drift punch in the access
hole in the steering column tube
flange under the ignition switch lock
cylinder. Depress the retaining pin
while pulling on the ignition switch
lock cylinder.

6. Pull the ignition switch lock cylinder from the column.

To install:

7. Turn the ignition switch lock
cylinder to the ON position and depress the retaining pin.

8. Insert the ignition switch lock
cylinder into the steering column
tube flange until fully seated. The ignition switch lock cylinder is fully
seated in the steering column tube
flange when the retaining pin snaps
into the access hole.

9. Install the lower column shroud
using the three retainer screws.

10. Install the tilt wheel handle and
shank.

11. Connect the negative battery
cable. Test the ignition switch
operation.

Lock Cylinder Not Working

The following procedure applies to
vehicles in which the ignition switch
lock cylinder is inoperative and the
lock cylinder cannot be turned due to
a lost or broken ignition key where
the key number is unknown, or a
damaged ignition switch lock
cylinder.

─── **CAUTION** ───

*The Supplemental Inflatable Re-
straint (SIR) system must be disarmed before removing the ignition column lock. Failure to do so
may cause accidental deployment
of the air bag, resulting in unnecessary SIR system repairs and/or
personal injury.*

─── **WARNING** ───

**The backup power supply energy
must be depleted before any air
bag component service is performed. To deplete the backup
power supply, disconnect the
battery ground cable and wait
ten minutes.**

1. Disconnect the negative battery
cable. Wait at least ten minutes for
the air bag backup power supply to
deplete before beginning work.

2. Remove the steering wheel using the recommended procedure.

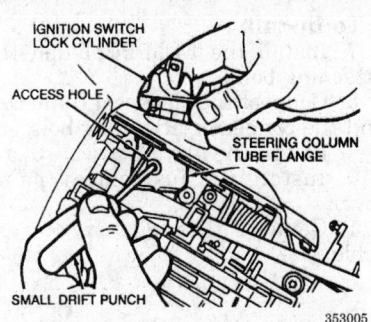

IGNITION SWITCH
LOCK CYLINDER

ACCESS HOLE

STEERING COLUMN
TUBE FLANGE

SMALL DRIFT PUNCH

353005

Use a ⅛ inch punch or wire to depress the retaining pin while pulling out the lock cylinder

3. Using channel-lock type pliers or vise-grip type pliers, twist the ignition switch lock cylinder cap until it separates from the ignition switch lock cylinder.

4. Using a ⅜-inch drill bit, drill down the middle of the key slot approximately 1.75 inches until the ignition switch lock cylinder breaks loose from the lock housing. Remove the ignition switch lock cylinder and the drill shavings from the lock housing.

> **— CAUTION —**
> *Carefully note the position of the steering column upper bearing retainer prior to removal.*

5. Remove the steering column upper bearing retainer by inserting a pick or other suitable tool with a 90 degree bend on its tip, between the steering column upper bearing retainer and the steering column lock housing bearing and prying up.

6. Insert the tip of a screwdriver into the double-D slot of the steering column lock housing bearing then rotate 90 degrees. Remove the steering column lock housing bearing.

> **— CAUTION —**
> *Carefully note the relationship of the steering column lock gear to the position of the steering column lock actuator.*

7. Remove the steering column lock gear.

8. Thoroughly clean all drill shavings and other foreign materials from the lock housing. Carefully inspect the lock housing for damage from the above procedure. If the lock housing is damaged, the steering column tube flange must be replaced.

To install:

9. Replace the steering column tube flange, if damaged.

NOTE: The position of the steering column lock gear is correct if the last tooth of the steering column lock gear is meshed with the last tooth on the steering column lock actuator.

10. Position the steering column lock gear in the base of the lock housing to the position noted during removal (Step 18).

11. Position the steering column lock housing bearing in the lock housing. Insert the tip of a small screwdriver into the double-D slot of the steering column lock housing bearing, then rotate 90 degrees.

12. Press the steering column upper bearing retainer into the lock housing. Make sure that the steering column upper bearing retainer is in its original position.

13. Line up the flats if the steering column lock gear and the steering column lock housing bearing by pulling down on the steering column lock cam.

14. With the flats of the steering column lock gear and the steering column lock housing bearing aligned, install the ignition switch lock cylinder.

15. Install the steering wheel using the recommended procedure.

16. Connect the negative battery cable.

17. Confirm the proper operation of the ignition switch lock cylinder and the ignition switch.

Ignition Switch

REMOVAL AND INSTALLATION

> **— CAUTION —**
> *The Supplemental Inflatable Restraint (SIR) system must be disarmed before removing the ignition switch. Failure to do so may cause accidental deployment of the air bag, resulting in unnecessary SIR system repairs and/or personal injury.*

> **— WARNING —**
> **The backup power supply energy must be depleted before any air bag component service is performed. To deplete the backup power supply, disconnect the battery ground cable and wait ten minutes.**

1. Disconnect the negative battery cable. Wait at least ten minutes for the air bag backup power supply to deplete before beginning work.

2. Remove the five lower instrument panel steering column cover screws, then unsnap and remove the panel.

3. Remove the four left side instrument panel lower reinforcement bolts and remove lower reinforcement.

4. Remove the ignition switch harness bolt. Disconnect the ignition switch harness from the ignition switch.

5. Remove the two ignition switch screws.

6. Remove the ignition from the steering column.

To install:

7. Install the ignition switch on to the steering column. Install the two screws to secure the switch.

8. Install the ignition switch harness bolt. Make sure the electrical connection is secure.

9. Install the lower instrument panel reinforcement. Do not overtighten the bolts. Torque to just 29-39 inch lbs. (3.2-4.4 Nm)

10. Install the instrument panel steering column cover and screws.

11. Connect the negative battery cable.

12. Verify correct function of the ignition switch.

Park/Neutral Safety Switch

REMOVAL AND INSTALLATION

The switch unit is bolted to the left side of the transmission shift lever. The switch prevents the engine from being started in any position except **P** or **N**. It also controls the backup lights.

1. Raise and support the vehicle safely.

2. Remove the transaxle control linkage connection.

3. Disconnect electrical harness, remove switch retaining screws. Remove the switch from the vehicle.

To install:

4. Position switch on transaxle.

5. Install retaining screws and connect wiring.

6. Disconnect the manual control linkage from the manual shaft.

7. Set the manual shaft to the **N** position.

8. Loosen the inhibitor switch mounting screws enough to allow for movement of the switch.

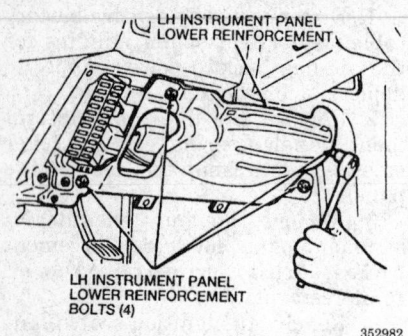

LH INSTRUMENT PANEL LOWER REINFORCEMENT

LH INSTRUMENT PANEL LOWER REINFORCEMENT BOLTS (4)

352982

Remove the left side instrument panel lower reinforcement

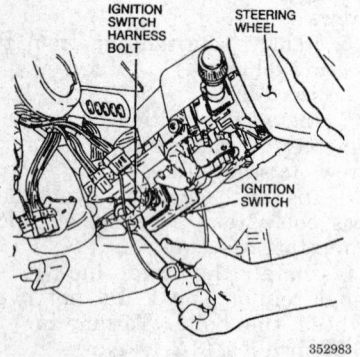

IGNITION SWITCH HARNESS BOLT

STEERING WHEEL

IGNITION SWITCH

352983

Remove the ignition switch harness bolt

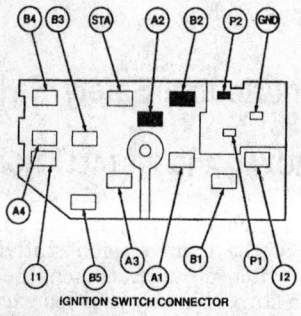

IGNITION SWITCH CONNECTOR

Pin Number	Circuit	Circuit Function
A1	DA20 (W/GN)	Interior Fuse Junction Panel
A2	—	NOT USED
A3	LI70 (BK/PK)	Rear Dome Lamp
A4	DA21 (BR)	Front Heater Blower Motor Relay
B1	DA16 (W/P)	Power Supply
B2	—	NOT USED
B3	LIE4 (BK)	Ground
B4	DA14 (W/P)	Power Supply
B5	DA15 (W/P)	Power Supply
I1	DA19 (W/R)	Interior Fuse Junction Panel
I2	DA22 (R/GN)	Interior Fuse Junction Panel
P1	NP03 (LG/BK)	Low Oil Level Relay
P2	—	NOT USED
STA	ST01 (R)	Inhibit Relay
GND	NPE2 (BK)	Ground

352985

Ignition switch terminal identification — 1995 shown, others similar

9. Insert a 0.16 in. (4 mm) diameter pin and move the switch until the pin falls through the locating holes in the inhibitor switch and manual shaft. Tighten the switch screws equally.

10. Remove the pin and connect the manual control linkage to the shaft.

11. Make sure while holding the brakes on, that the engine will start only in **P** or **N** and that the backup lights only illuminate in reverse.

12. Connect the transaxle control linkage.

Powertrain Control Module

REMOVAL AND INSTALLATION

The Electronic Control Module (ECM) is also called the Powertrain Control Module (PCM) which more accurately reflects its expanded role, doing more than just engine management. The ECM/PCM is a complex and sensitive electronic component and is easily damaged. Always make sure the ignition switch is turned to the **OFF** position and the battery negative (ground) cable is disconnected and isolated before attempting to work on any part of the system. Never touch any terminals with fingers since static electricity will damage the unit. Use care doing tests. Never use a conventional test light or attempt to check resistance when troubleshooting any electronic system. On these vehicles, the ECM/PCM is located behind the glove compartment.

1. Disconnect the negative battery cable.

2. Remove the lower instrument panel cover.

3. Remove the glove box interior.

4. Remove the bolt attaching the electrical connector to the ECM/PCM.

5. Remove the two bolts attaching the ECM/PCM to the bracket.

6. Remove the ECM/PCM.

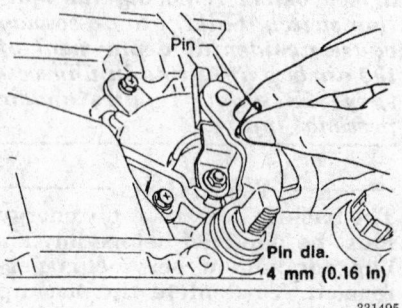

Pin

Pin dia. 4 mm (0.16 in)

331495

Installation of park / neutral switch adjustment pin

To install:

7. Install the ECM/PCM and the attaching bolts.

8. Connect the electrical connector and the connector attaching bolt.

9. Install the glove box.

10. Install the instrument panel cover.

11. Connect the negative battery cable.

ENGINE COOLING

Radiator

REMOVAL AND INSTALLATION

These vehicle use an aluminum core, crossflow-type radiator with plastic tanks. The radiator has an expansion joint that can be mistaken for a separation. The joint appears as a gap or saw cut in the aluminum rails that run across the top and bottom of the radiator core. The joint is needed to allow for expansion and contraction of the radiator and should not be eliminated. Do not mistake the function of the joint in the radiator.

1. Drain the coolant from the radiator and store it properly.

2. Disconnect the negative battery cable.

3. Disconnect the cooling fan wiring harness clamp and disconnect the wiring connector.

4. Remove the overflow tube from the neck of the radiator and then remove the overflow reservoir.

5. Remove the center engine compartment fuse junction panel retaining bolt and position the fuse panel out of the way.

6. Remove the clamps from the upper and lower hoses and remove the hoses.

7. Remove the cooling fan and shroud assembly bolts.

NOTE: The cooling fan and shroud assembly bolts may fall out of the fan and shroud assembly when the nuts are removed. Be careful not to lose them.

8. Remove the fan and shroud assembly.

9. Disconnect and cap the transmission cooler lines.

10. Remove the 2 radiator support brackets and remove the radiator.

To install:

11. Install the radiator on the isolation mounts.

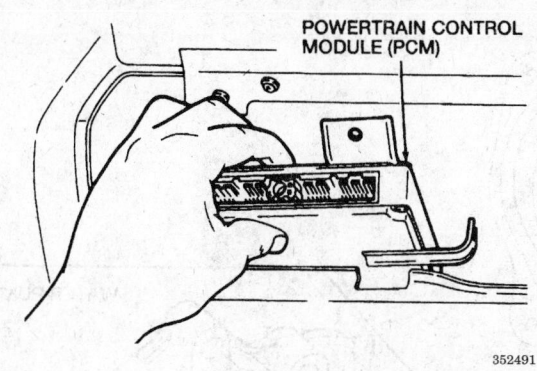

POWERTRAIN CONTROL MODULE (PCM)

352491

Remove the Powertrain Control Module

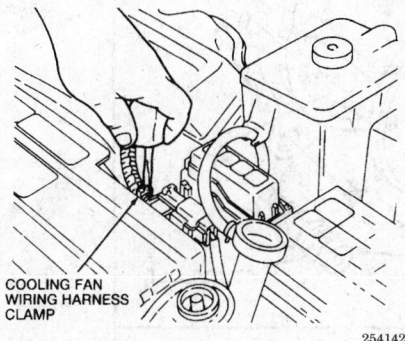

COOLING FAN WIRING HARNESS CLAMP

254142

Cooling fan connectors and fuse block

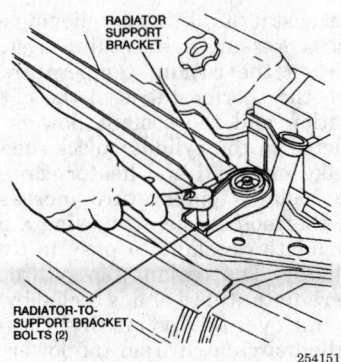

RADIATOR SUPPORT BRACKET

RADIATOR-TO-SUPPORT BRACKET BOLTS (2)

254151

Radiator support bracket

12. Connect the transmission cooler lines.

13. Connect the lower radiator hose and tighten the clamp. The factory recommends that new hose clamps be used. Do not reuse the original hose clamps. It is also recommended that coolant NOT BE USED as a lubricant when installing the radiator hoses. Although this may have been common practice in the past, the factory feels that coolant will prevent the hoses from properly sealing to the fittings.

14. Install the fan and shroud assembly and reconnect the cooling fan wiring. Be sure to install the assembly into the lower mounting slots.

NOTE: Use only specified bolts when installing the radiator support brackets. If a bolt longer than 0.470 inch (12mm) is used, damage may occur to the wiring harness.

15. Secure the radiator support brackets.

16. Install the upper radiator hose and tighten the clamp.

17. Install the fuse panel retaining bolt.

18. Install the overflow bottle and reconnect the overflow tube to the radiator.

19. Reconnect the negative battery cable.

20. Refill the cooling system.

21. Start the engine and check for leaks. Bleed the cooling system using the recommended procedure, if necessary. Verify that the cooling fan is operating correctly.

Water Pump

REMOVAL AND INSTALLATION

1. Drain the cooling system.
2. Disconnect the negative battery cable.
3. Remove the alternator drive belt, the water pump and power steering pump drive belt and the A/C compressor drive belt (if equipped).
4. Use a strap wrench to hold the water pump pulley while removing the four water pump pulley bolts.
5. Remove the water pump pulley from the water pump.
6. Remove the crankshaft pulley using the following procedure.
   a. Raise and safely support the vehicle.
   b. Remove the five right side inner engine and transmission splash shield bolts and two screws

and remove the inner engine and transmission shield.
   c. Remove the four right side outer engine and transmission splash shield bolts and two screws and remove the right side outer engine and transmission splash shields.
   d. Use a strap wrench to hold the crankshaft pulley while removing the crankshaft pulley bolt.
   e. Use a crankshaft damper remover to draw the crankshaft pulley off the front of the crankshaft.
7. Remove the five lower engine front cover bolts and take of the front cover.
8. Remove the six water pump bolts. Make note of the locations of the bolts since one should be a stud/bolt and must be returned to its original location. Remove the water pump.

To install:
9. Clean all parts well. The bolt threads should be cleaned of any old sealer or corrosion. Make sure the mating surfaces between the water pump and the engine block are cleaned of any old sealant. Apply a continuous bead of gasket maker type sealer approximately $1/8$-inch wide onto the water pump and position the water pump on the engine block.
10. Install the six water pump bolts. Refer to any notes made at removal so the bolts can be returned to their original locations. Do not overtighten the water pump bolts. Torque the water pump bolts evenly to 12–15 ft. lbs. (16–21 Nm).
11. Position the water pump pulley on the water pump and install the four pulley bolts. Use a strap wrench to hold the pulley as the bolts are torqued to 12–15 ft. lbs. (16–21 Nm).
12. Install the front engine cover and the five lower front cover bolts. Torque to 27–44 inch lbs. (3–5 Nm).
13. Install the crankshaft pulley using the following procedure.
   a. Install the crankshaft pulley and pulley bolt.
   b. Hold the pulley with a strap wrench. Tighten the crankshaft pulley bolt to 90–98 ft. lbs. (123–132 Nm).
   c. Install the inner and outer engine and transmission splash shields.
14. Install and adjust the drive belts.
15. Connect the negative battery cable.
16. Refill the cooling system.
17. Start the engine, bleed the cooling system using the recommended procedure and verify no leaks.

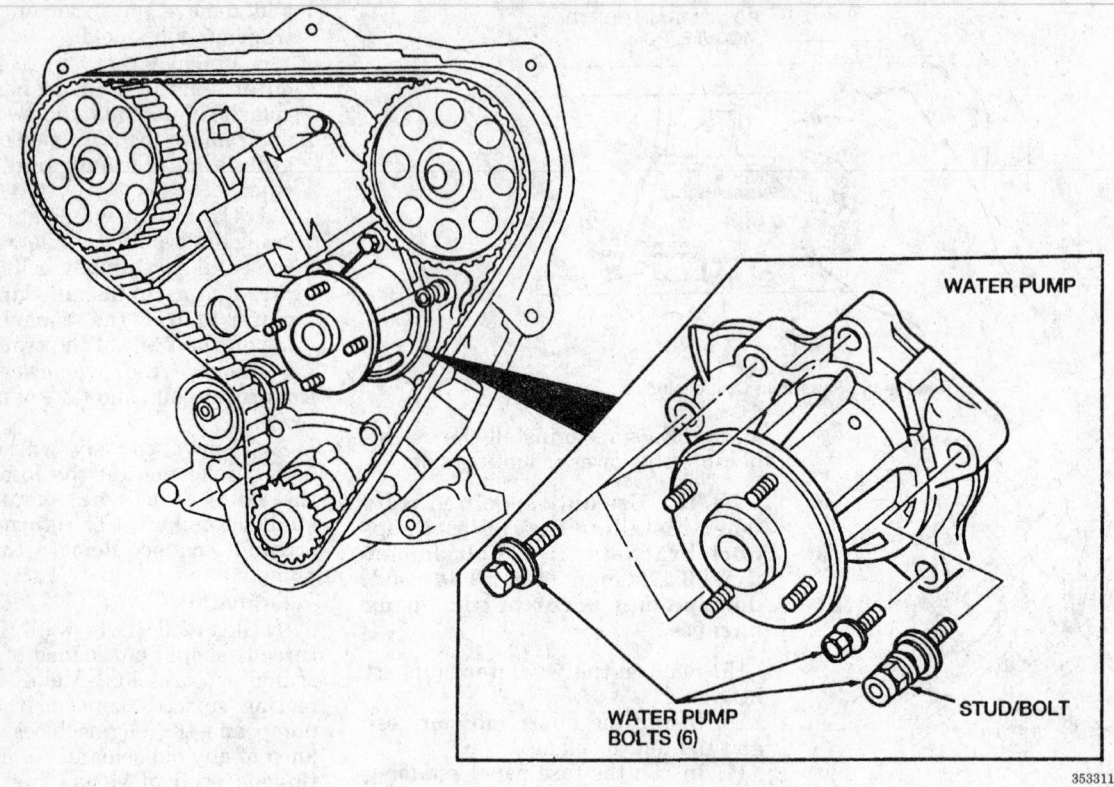

Water pump mounting arrangement. Note the location of the stud/bolt

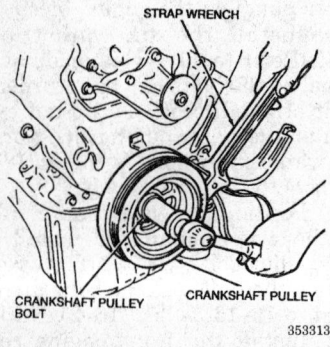

Use a strap wrench to hold the crankshaft pulley for removal and installation

Thermostat

REMOVAL AND INSTALLATION

The water thermostat is located inside the thermostat housing above the drive belts. The thermostat ensures rapid engine warm-up by restricting coolant flow at lower operating temperatures. It also assists in keeping the engine operating temperature within its predetermined limits. The thermostat is equipped with a jiggle valve that allows a small amount of coolant to flow through the

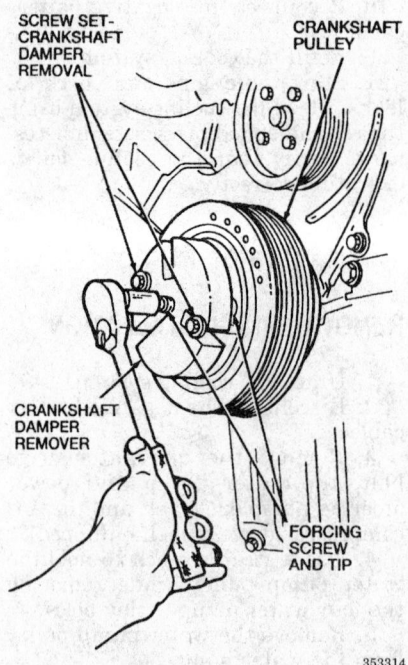

Use a puller to draw the crankshaft pulley off the front of the crankshaft

thermostat at all times, relieving any excess pressure in the cooling system.

When the coolant temperature is cold, the thermostat is in the closed position and the coolant flow is restricted to the cylinder block, heads, intake manifold and heater core. As the coolant temperature increases, the thermostat opens allowing a portion of the coolant to pass into the radiator. The coolant flows through the radiator tubes and is cooled by air passing over the cooling fins. Coolant is then circulated from the lower radiator outlet through the water pump and into the cylinder block to complete the circuit.

1. Disconnect the negative battery cable.

2. Drain the coolant.

3. Remove the upper radiator hose bracket.

4. Reposition the water bypass hose clamp and remove the hose from the water hose connection.

5. If equipped with A/C, unbolt the drier to evaporator liquid line bracket to gain access to the thermostat housing.

6. If equipped with A/C, remove the suction accumulator/drier bolts and bracket. Position the suction/accumulator/drier and liquid line out of the way.

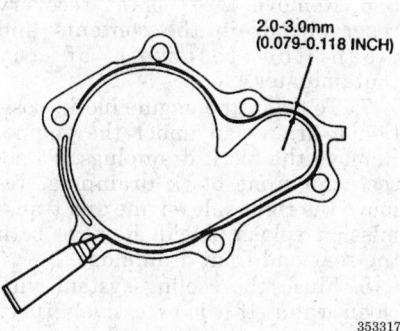

2.0-3.0mm
(0.079-0.118 INCH)

353317

Apply Gasket Maker type sealer as indicated

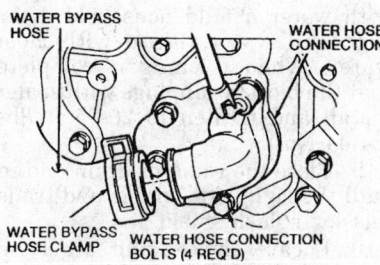

WATER BYPASS HOSE

WATER HOSE CONNECTION

WATER BYPASS HOSE CLAMP

WATER HOSE CONNECTION BOLTS (4 REQ'D)

355636

Remove the four bolts and water hose connection from the thermostat housing

7. Remove the four water hose connection (thermostat housing) and remove the assembly. Pull the thermostat out of the housing.

To install:

8. Clean all parts well. Clean the thermostat housing cover mating surface and groove of any old sealant. Dry off any coolant from these areas. Apply a continuous bead of silicone sealer to the inner surface and the outside edge of the thermostat housing. The sealant should be 0.080–0.120 inch (2.0–3.0 mm) wide on all sealing areas. Work quickly. The thermostat and housing assem-

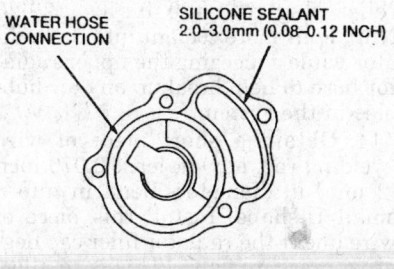

WATER HOSE CONNECTION

SILICONE SEALANT 2.0–3.0mm (0.08–0.12 INCH)

355637

Apply silicone sealer to the thermostat housing as shown

bly should be installed within five minutes of applying the sealer.

9. Position the thermostat in the housing so that the jiggle valve is facing up when installed. Install the housing/thermostat assembly to the engine and install the four bolts. Torque to 12–15 ft. lbs. (16–21 Nm).

10. Connect the water bypass hose to the thermostat housing and clamp in place.

11. Install the upper radiator hose bracket and tighten the bolt securely.

12. Install the A/C line bracket removed earlier, if equipped with A/C.

13. Wait at least 30 minutes for the sealant to completely dry before refilling with coolant. Then refill the cooling system. Connect the negative battery cable, start the engine and check for leaks. Bleed the cooling system if required.

Electric Cooling Fan

REMOVAL AND INSTALLATION

These vehicles use an electric cooling fan. The fan is installed behind the radiator and uses a two-speed engine cooling fan motor mounted within a shroud. Three cooling fan relays (LO, HI 1 and HI 2) govern the operation of the cooling fan motor. The cooling fan relays are located within the left side engine compartment relay block. The Engine Coolant Temperature (ECT) sensor has a thermistor which senses temperature changes. Resistance decreases as the temperature rises. The ECT sensor detects the coolant temperature, modifies the voltage being sent from the PCM and provides this information to the PCM and an input signal. If the ECT sensor malfunctions, the fan motor will operate at high speed as soon as the engine is started. The ECT sensor is located on the top of the water hose connection.

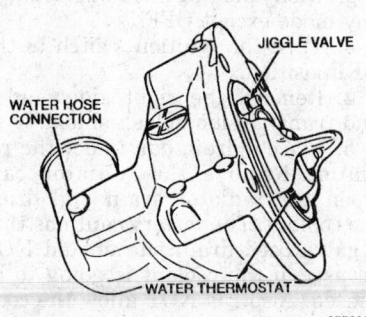

JIGGLE VALVE

WATER HOSE CONNECTION

WATER THERMOSTAT

355638

Position the thermostat so that the jiggle valve will be pointing upward when installed

The engine cooling fan motor, fan blades and fan shroud assembly are removed as an assembly.

1. Disconnect the negative battery cable.

2. Disconnect the two engine cooling fan motor electrical connectors from the fan shroud.

3. Remove the two engine cooling fan motor, fan blade and fan shroud assembly nuts and bolts.

4. Pull the engine cooling fan motor, fan blades and fan shroud assembly from the mounting slots at the bottom of the radiator.

5. To changeout the electric fan motor:

   a. Lay the assembly down and pry the retaining clip from the back of the fan blade.

   b. Remove the three fan motor nuts and remove the heat shield.

   c. With the aid of an assistant to hold the fan blade steady, use a drift punch to tap the fan motor from the fan blade. Remove the motor from the shroud.

To install:

6. If the motor, shroud and fan were disassembled, use the following procedure:

   a. Place the cooling fan motor in the fan shroud.

   b. Install the fan motor, the head shield and the three motor nuts.

   c. Align the fan grooves on the fan blades with the roll pin in the motor shaft and install the retaining clip on the back of the fan blade.

7. Install the fan motor, blade and shroud assembly into the vehicle using care to align the shroud assembly with the mounting slots at the bottom of the radiator.

8. Install the two nut and bolt assemblies that retain the shroud.

9. Connect the two electrical connectors.

10. Connect the negative battery cable.

11. Warm up the vehicle to normal operating temperature and verify that the cooling fan operates correctly.

Cooling System Bleeding

The coolant level should be checked at least once a month. After the engine has cooled down, verify that the coolant level is between the MAX and MIN marks on the radiator coolant recovery reservoir. If needed, fill the coolant system.

If the coolant has not been changed in the last 36 months or 30,000 miles or if an inspection reveals hoses, gas-

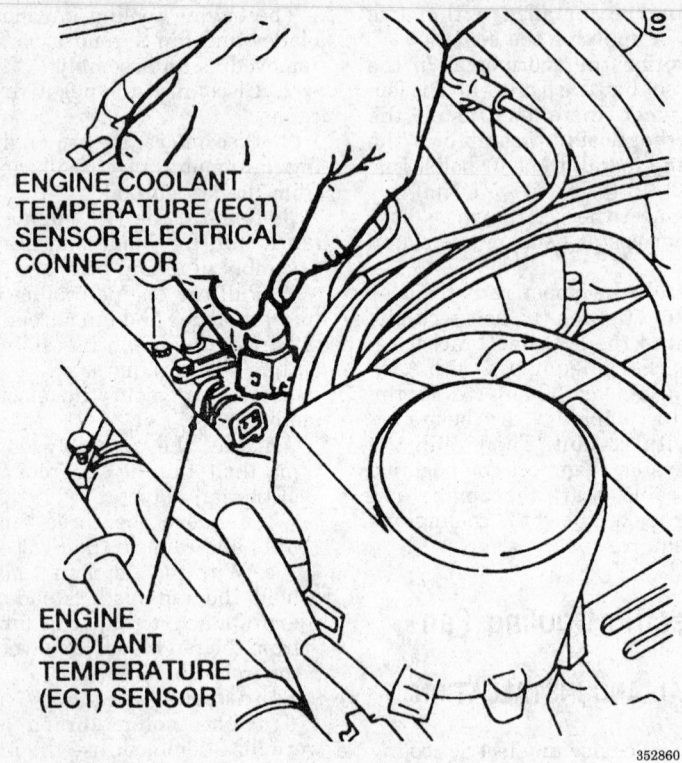

The EngineCoolant Tempreture (ECT) sensor provides the PCM with a tempreture signal. The PCM controls the cooling fan

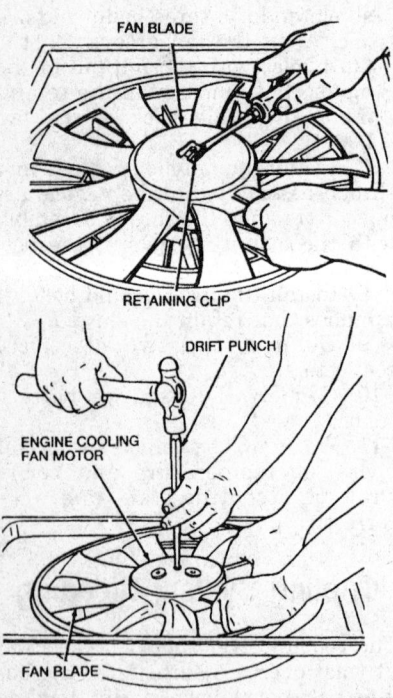

To changeout the cooling fan motor, remove the clip and three mounting nuts. With an assistant, tap the motor form the blade

kets or other components requiring service or replacement, the coolant must be drained and discharged or saved for reuse, depending on its condition. Use care working around the cooling system. The electric cooling fan may come on at any time when the ignition switch is in the **ON** position.

1. Allow the vehicle to become cool (room temperature) before starting this procedure.

2. Turn the ignition switch to the **ON** position and set the front temperature control panel selectors in the following positions: turn the front blower switch to 4, the front temperature control knob to the full WARM position the rear control switch to the 4 position and the mode selector to any mode except OFF.

3. Turn the ignition switch to the **OFF** position.

4. Remove the right side engine and transmission splash shield.

5. Place a drain pan under the radiator. Remove the radiator cap. Open the radiator drain and drain the coolant. The factory cautions that a galvanized drain pan should NOT be used if the coolant is going to be reused. Also DO NOT allow the coolant to contact the drive belts or they will be damaged.

6. Remove the coolant recovery reservoir, drain the contents and clean the reservoir of any contaminates.

7. To drain the engine block, position a drain pan under the engine. Remove the block drainplugs. To access the front block drainplug, remove the right side engine and transmission splash shield. Remove both the rear and front drainplugs.

8. Flush the cooling system with clean water. If it is excessively rusty or dirty, reverse-pressure flush or use an approved flushing compound. Do not use caustic cleaning solutions or copper/brass radiator cleaning agents on this aluminum radiator. Flush with water a mild household detergent and water. Rinse with clean water. When flushing is complete, coat the block drainplugs with sealer, install and tighten to 25-35 ft. lbs. (34-44 Nm).

9. Close the radiator drain and install the right side engine and transmission splash shield.

10. Locate the engine air relief plug. It located near the Idle Air Control-Fast Idle Control (IAC-FIC) assembly. Remove the engine air relief plug.

11. Locate and with the appropriate size (should be 8 mm) Allen wrench, loosen BUT DO NOT REMOVE the radiator air relief plug. Loosen the plug three turns only.

12. Support the coolant recovery reservoir at a height above the engine. Verify that the hose between the coolant recovery reservoir and the radiator filler neck is still attached.

13. Add coolant to the reservoir until coolant just starts to drip from the radiator air relief plug. Close the plug. Slowly pour a 50/50 water-antifreeze mix into the radiator filler neck. Allow several minutes for the air to escape. Continue adding coolant until coolant no longer lowers in the filler neck. If coolant comes out of the radiator air relief plug hole, close the plug and tighten the radiator air relief plug to 14-20 ft. lbs. (20-26 Nm). Pour more coolant into the radiator while squeezing the upper radiator hose to help breakup any air bubbles in the system.

14. Obtain a short piece of wire (welding rod, etc.) at least 0.079 inch (2 mm) in diameter. Bend in into a small U-shape. Install this piece of wire under the radiator filler cap negative pressure valve. This is to allow a flow of air and coolant regardless of pressure. Install the radiator cap

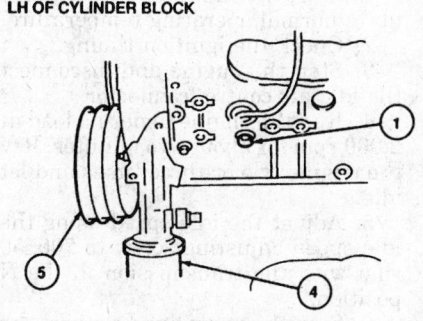

LH OF CYLINDER BLOCK **RH OF CYLINDER BLOCK**

Item	Description
1	Front Cylinder Block Drainplug
2	Rear Cylinder Block Drainplug

Item	Description
3	Front Wheel Driveshaft and Joint
4	Oil Filter
5	Crankshaft Pulley

352936

Location of the engine block drains

IDLE AIR CONTROL-
FAST IDLE CONTROL
(IAC-FIC) ASSEMBLY

INTAKE
PLENUM

ENGINE AIR
RELIEF PLUG

332671

Engine air relief plug

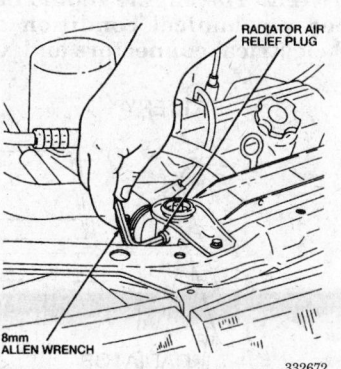

RADIATOR AIR
RELIEF PLUG

8mm
ALLEN WRENCH

332672

Radiator air relief plug

ing temperature. Watch the temperature gauge closely. If the gauge begins to rise above normal, stop the engine. Allow to cool completely and refill the radiator and reservoir as necessary.

18. With the engine idling and cooling system "HOT" squeeze the upper radiator hose by hand. This will help "burp" the system of air by increasing system pressure.

19. Stop the engine and allow to cool completely. During cool-down, make sure the coolant in the reservoir does not drop below the MIN mark.

20. Return the reservoir to its proper location and install.

21. With the engine cool, remove the radiator cap and extract the U-shaped piece of wire installed earlier. Failure to remove the wire will reduce cooling system performance.

22. Refill the radiator and coolant recovery reservoir as necessary.

FUEL SYSTEM

Fuel System Service Precautions

Safety is the most important factor when performing not only fuel system maintenance but any type of maintenance. Failure to conduct maintenance and repairs in a safe manner may result in serious personal injury or death. Maintenance and testing of the vehicle's fuel system components can be accomplished safely and effectively by adhering to the following rules and guidelines.

• To avoid the possibility of fire and personal injury, always disconnect the negative battery cable unless the repair or test procedure requires that battery voltage be applied.

• Always relieve the fuel system pressure prior to disconnecting any fuel system component (injector, fuel rail, pressure regulator, etc.), fitting or fuel line connection. Exercise extreme caution whenever relieving fuel system pressure to avoid exposing skin, face and eyes to fuel spray. Please be advised that fuel under pressure may penetrate the skin or any part of the body that it contacts.

• Always place a shop towel or cloth around the fitting or connection prior to loosening to absorb any excess fuel due to spillage. Ensure that

onto the radiator with the wire in place.

15. Add a 50/50 water-antifreeze mix into the coolant recovery reservoir up to the MAX mark.

16. Close the engine air relief plug, if not done so earlier.

17. With the gearshift lever in PARK, start and run the engine at 2,000 rpm for 15-30 minutes allowing the cooling system to reach full operating temperature. Make sure the radiator coolant reservoir tank fluid level does not drop below the MIN level. The engine cooling fan motor will turn ON and OFF indicating that the system has reached operat-

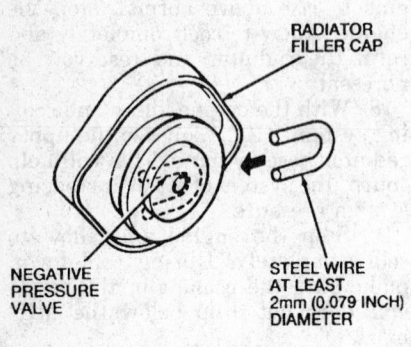

RADIATOR
FILLER CAP

NEGATIVE
PRESSURE
VALVE

STEEL WIRE
AT LEAST
2mm (0.079 INCH)
DIAMETER

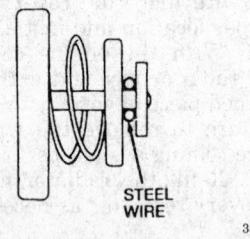

STEEL
WIRE

352935

Make U-shape piece of welding wire to temporarily wedge the cap's pressure relief open so air and coolant will circulate regardless of pressure

all fuel spillage (should it occur) is quickly removed from engine surfaces. Ensure that all fuel soaked cloths or towels are deposited into a suitable waste container.

• Always keep a dry chemical (Class B) fire extinguisher near the work area.

• Do not allow fuel spray or fuel vapors to come into contact with a spark or open flame.

• Always use a backup wrench when loosening and tightening fuel line connection fittings. This will prevent unnecessary stress and torsion to fuel line piping. Always follow the proper torque specifications.

• Always replace worn fuel fitting O-rings with new. Do not substitute fuel hose or equivalent, where fuel pipe is installed.

Fuel System Pressure

RELIEVING

────── CAUTION ──────
Fuel injection systems remain under pressure, even after the engine has been turned OFF. The fuel system pressure must be relieved before disconnecting any

fuel lines. Failure to do so may result in fire and/or personal injury.

Relieve the fuel system pressure using the following procedure.

1. Remove the left side engine compartment relay panel cover.

2. Locate and remove the fuel pump relay from the relay panel.

3. Start the engine.

4. Allow the engine to run until it stalls from fuel starvation. After the engine stalls, crank the engine over two more times to ensure all pressure has been released.

5. Turn the ignition switch to the **OFF** position and install the fuel pump relay.

6. Most service work that follows fuel pressure relief also requires that the negative battery cable (ground) be disconnected before service work begins. This also prevents accidental fuel pump energizing that could repressurize the system.

Idle Speed

ADJUSTMENT

NOTE: The engine should be in good mechanical condition and all electrical connectors and vac-

uum hoses connected before making this adjustment. Block the engine drive wheels and apply the parking brake.

1. Start the engine and let it warm up to normal operating temperature.

2. Check the ignition timing.

3. Stop the engine and disconnect the idle air control connector.

4. Run the engine under no load at 2,000 rpm for about two minutes. Rev the engine two or three times and let idle

5. Adjust the idle speed using the idle speed adjusting screw to 700±50 rpm with the transmission in the N position.

6. Stop the engine and connect the throttle position sensor and the idle air control connector.

7. Run the engine under no load at 2,000 rpm for about two minutes. Rev the engine two or three times and let it idle for one minute.

8. Check that the the idle speed is 750±50 rpm.

9. If the idle speed is not correct, check the idle air control circuit and repair as necessary.

10. If the idle speed is still not correct, substitute a known good ECM.

NOTE: The ECM may be the cause of the problem but this is rarely the case.

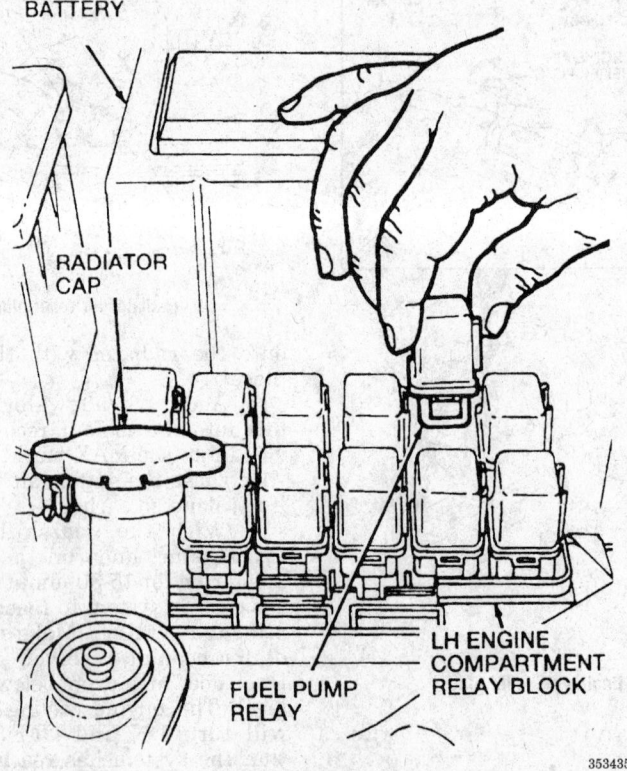

BATTERY

RADIATOR
CAP

FUEL PUMP
RELAY

LH ENGINE
COMPARTMENT
RELAY BLOCK

353435

Fuel pump relay removal — 1995 shown, others similar

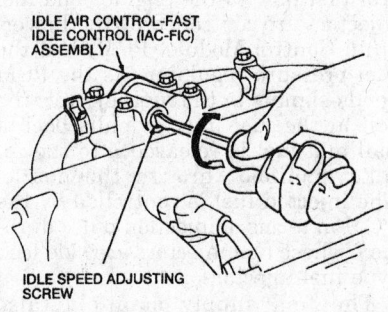

IDLE AIR CONTROL-FAST
IDLE CONTROL (IAC-FIC)
ASSEMBLY

IDLE SPEED ADJUSTING
SCREW

333336

Idle adjusting screw

Mixture

ADJUSTMENT

The air/fuel mixture is automatically adjusted by the engine control system. If the mixture is too rich or too lean, use the appropriate diagnostic procedure to locate the problem.

Fuel Filter

REMOVAL AND INSTALLATION

Clean fuel is key to a trouble-free fuel injection system. These vehicle use several filters. A replaceable in-line filter is the primary fuel filter. This is a high-pressure in-line type that provides extremely fine filtration to protect the small metering orifices of the fuel injection nozzles. The filter is of one-piece construction which cannot be cleaned. If the fuel filter becomes clogged or restricted, it should be replaced.

A low-pressure nylon filter is inside the fuel tank attached to the fuel pump/level sensor. It protects the fuel pump from larger contaminates in the tank. This strainer also allows the passage of small quantities of water which may accumulate in the fuel tank. The design and placement of this filter is such that if this filter becomes clogged, the fuel tank must be removed to service the filter. This provides the opportunity to clean the inside of the fuel tank.

A fuel injector screen is located at the top of each fuel injector on the engine and is not serviceable. If the fuel injector screen becomes clogged, the entire fuel injector must be replaced.

To replace the in-line fuel filter, use the following procedure.

—————— **CAUTION** ——————
Fuel injection systems remain under pressure after the engine has been turned OFF. The fuel system pressure must be relieved before disconnecting any fuel lines. Failure to do so may result in fire and/or personal injury.

1. Relieve the fuel system pressure using the recommended procedure.
2. Disconnect the negative battery cable.
3. Raise and safely support the vehicle.
4. Remove the fuel hose clamps.
5. Disconnect and plug the hoses to prevent leakage.
6. Remove the fuel filter from the bracket.
To install:
7. Install the fuel filter into the bracket with the arrow facing up, in the direction of the fuel travel to the engine.
8. Reconnect the fuel hoses.
9. Install and tighten the hose clamps. Verify that the clamps are properly tightened. System operating pressure is approximately 36 psi and fuel will leak is connections are not properly made.
10. Lower the vehicle.
11. Reconnect the negative battery cable.
12. Check for leaks.

Fuel Pump

REMOVAL AND INSTALLATION

These vehicles are equipped with Sequential Multiport Fuel Injection (SFI) and an electric fuel pump. The fuel pump system consists of: the fuel pump and bracket assembly, an Inertia Fuel Shutoff (IFS) switch, a fuel

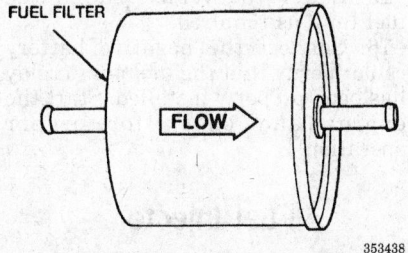

FUEL FILTER

FLOW ▷

353438

When replacing the filter, use only the specified part. Pay attention to the fuel flow direction markings

level sensor and sensor filter, an in-line fuel filter, an evaporative emission valve, a fuel pump relay, a fuel pressure regulator, the fuel supply manifold (fuel rail) and the fuel injectors. Note that the fuel pressure regulator maintains the fuel system pressure at 36 to 38 psi with the engine running.

The SFI system has a fuel pump relay located in the left side engine compartment relay panel. The fuel pump relay is controlled by the Powertrain Control Module (PCM). The fuel pump is, in turn, controlled by the fuel pump relay. When the ignition switch is turned to the **ON** position, the PCM sends a signal to the fuel pump relay to close its contacts. The fuel pump is energized for five seconds prior to vehicle start-up. When the ignition switch is turned to the **OFF** position, the PCM sends a signal to the fuel pump relay to open its contacts to shut the fuel pump off.

This system uses an Inertia Fuel Shutoff (IFS) switch. Be aware of this when troubleshooting a No-Start or No-Fuel Pressure/Fuel Pump Inoperative complaint. The IFS switch is located behind the left side cowl trim panel below the hood release handle. In the event of a collision, the electric contacts in the IFS switch open. When the contacts open, the fuel delivery circuit is open, disabling the fuel system. The engine may continue to run for several seconds after the IFS switch is opened, due to fuel pressure remaining in the fuel supply line. Once the fuel pressure has been released, the engine will stall. To reset the IFS switch, depress the reset button on the IFS switch.

The fuel tank must be removed from the vehicle to service the fuel pump.

—————— **CAUTION** ——————
Fuel injection systems remain under pressure, even after the engine has been turned OFF. The fuel system pressure must be relieved before disconnecting any fuel lines. Failure to do so may result in fire and/or personal injury.

1. Relieve the fuel system pressure using the following procedure.
   a. Remove the left side engine compartment relay panel cover.
   b. Locate and remove the fuel pump relay from the relay panel.
   c. Start the engine.
   d. Allow the engine to run until it stalls from fuel starvation. After the engine stalls, crank the engine

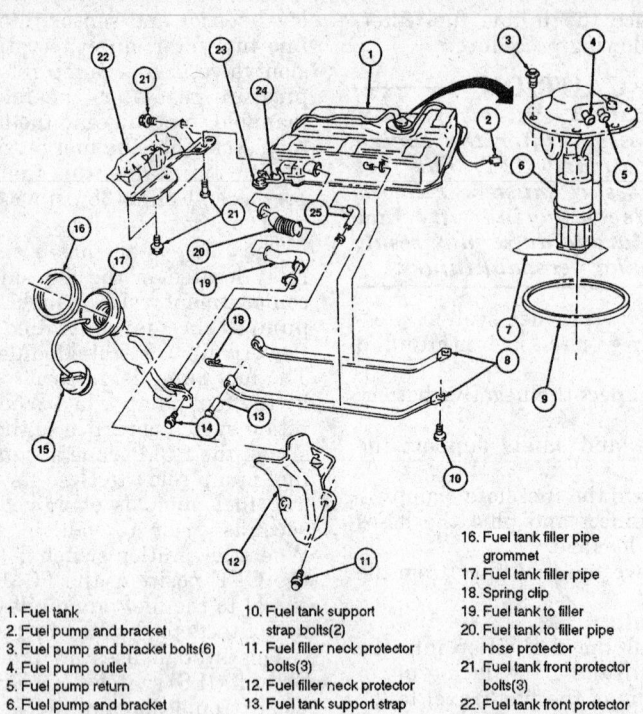

1. Fuel tank
2. Fuel pump and bracket
3. Fuel pump and bracket bolts(6)
4. Fuel pump outlet
5. Fuel pump return
6. Fuel pump and bracket
7. Fuel level sensor filter
8. Fuel tank support straps
9. Fuel level sensor gasket
10. Fuel tank support strap bolts(2)
11. Fuel filler neck protector bolts(3)
12. Fuel filler neck protector
13. Fuel tank support strap hinge
14. Fuel tank filler pipe bolts(2)
15. Fuel tank filler cap
16. Fuel tank filler pipe grommet
17. Fuel tank filler pipe
18. Spring clip
19. Fuel tank to filler
20. Fuel tank to filler pipe hose protector
21. Fuel tank front protector bolts(3)
22. Fuel tank front protector
23. Evaporator emission valve
24. Evaporator emission hose
25. Relief discharge vent hose

353334

Fuel supply system arrangement

over two more times to ensure all pressure has been released.

e. Turn the ignition switch to the **OFF** position and install the fuel pump relay.

2. After relieving the fuel system pressure, disconnect the negative battery cable.

3. Raise and safely support the vehicle.

4. Remove the fuel tank using the recommended draining and tank removal procedure.

CAUTION

Observe all applicable safety precautions when working around fuel. Do not allow fuel spray or fuel vapors to come in contact with a spark or open flame. Keep a dry chemical (Class B) fire extinguisher near the work area. Never drain store fuel in an open container due to the possibility of fire or explosion.

5. Remove the six fuel pump bolts.

6. Lift the fuel pump out of the fuel tank. Use care. The fuel level sensor and fuel pump and bracket must be tipped to remove it from the fuel tank. Do not lift the fuel sensor and pump assembly straight out of the fuel tank or damage to the level sensor may occur.

7. Remove the two bolts attaching the level sensor to the fuel pump.

8. Remove the fuel pump level sensor and the gasket.

9. Discard the gasket.

10. Remove the fuel pump from the bracket.

To install:

11. Position the fuel level sensor on the fuel pump and bracket and install the two bolts.

12. Install a new level sensor gasket. Carefully install the level sensor and pump assembly.

13. Install the six fuel pump bolts. Do not over-tighten the bolts. Torque the bolts to just 18 to 21 inch lbs. (2 to 3 Nm).

14. Install the fuel tank using the recommended procedure.

15. Lower the vehicle. Refill the fuel tank as required.

16. Connect the negative battery cable. Verify that the fuel pump relay has been properly installed. Start the engine and check for proper operation.

Fuel Injector

REMOVAL AND INSTALLATION

The fuel injectors are electronically controlled solenoid valves that con-trol fuel flow to the engine. The fuel injectors are controlled by the Powertrain Control Module (PCM) and the fuel pressure regulator. As the PCM sends signals to the fuel injector, the coil in the fuel injector pulls back a ball and fuel is released into the intake manifold through the nozzle. The injected fuel is controlled by the PCM in terms of injection pulse duration. These fuel injectors are side feed type fuel injectors.

The fuel supply manifold (also called the fuel rail) includes a tubular rail connected to a fuel tube hose. The fuel injection supply manifold delivers high-pressure (approximately 36–38 psi) fuel from the fuel supply line to the six fuel injectors.

CAUTION

Fuel injection systems remain under pressure, even after the engine has been turned OFF. The fuel system pressure must be relieved before disconnecting any fuel lines. Failure to do so may result in fire and/or personal injury.

1. Relieve the fuel system pressure using the following procedure.

a. Remove the left side engine compartment relay panel cover.

b. Locate and remove the fuel pump relay from the relay panel.

c. Start the engine.

d. Allow the engine to run until it stalls from fuel starvation. After the engine stalls, crank the engine over two more times to ensure all pressure has been released.

e. Turn the ignition switch to the **OFF** position and install the fuel pump relay.

2. After relieving the fuel system pressure, disconnect the negative battery cable.

3. If removing a rear fuel injector, remove the upper intake manifold (also called a plenum). This is a lengthy procedure. Use the recommended intake manifold removal procedure.

4. Disconnect the fuel injector electrical connector.

5. If removing a rear fuel injector, use a screwdriver bit socket to remove the fuel injector cap screws. If removing a front fuel injector, use a screwdriver bit socket on an extension to reach the fuel injector cap screws.

6. With cap screws removed, remove the fuel injector cap then pull the fuel injector from the fuel supply manifold (fuel rail).

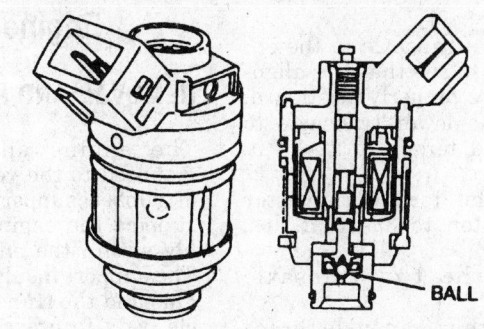

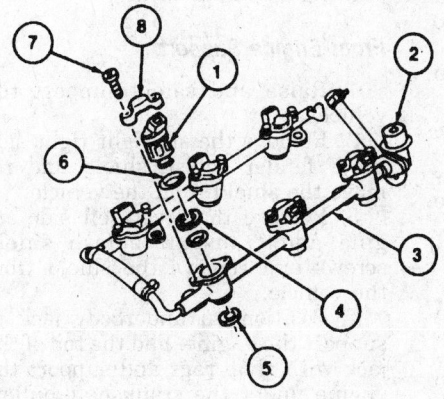

Item	Description
1	Fuel Injector
2	Fuel Pressure Regulator
3	Fuel Injection Supply Manifold
4	Fuel Injector Grommet
5	Fuel Injector Insulator
6	O-Ring
7	Fuel Injector Cap Screw (2 Req'd)
8	Fuel Injector Cap

353459

Fuel injector and the fuel supply manifold (fuel rail)

To install:

7. Install a new fuel injector grommet, fuel injector insulator and O-rings on the fuel injector.

8. Install the fuel injector into the fuel supply manifold (fuel rail). Install the injector cap and the two cap screws.

9. Install the electrical connector.

10. If a rear fuel injector was being serviced, install the upper intake manifold (also called a plenum) using the recommended intake manifold installation procedure.

11. Connect the negative battery cable. Do not start the engine yet.

12. Cycle the ignition switch several times from **OFF** to **RUN** without starting the engine.

13. Check the system for fuel leaks.

ENGINE MECHANICAL

Engine Assembly

REMOVAL AND INSTALLATION

The engine is removed with the transaxle attached. The engine and transaxle are lowered from the vehicle as an assembly.

— CAUTION —
Fuel injection systems remain under pressure even after the engine has been turned OFF. The fuel system pressure must be relieved before disconnecting any fuel lines. Failure to do so may result in fire and/or personal injury.

1. Relieve the fuel system pressure using the recommended procedure.

2. Disconnect the negative battery cable.

3. Drain and properly contain the coolant from the cooling system.

4. Remove the air intake tube and resonator.

5. Drain the engine oil.

6. Disconnect the radiator overflow hose from the radiator filler neck and remove the reservoir.

7. Disconnect the Vehicle Speed Sensor (VSS).

8. Disconnect the valve body wiring harness.

9. Disconnect the starter, EGR solenoid and the Throttle Position (TP) sensor.

10. Disconnect the Transmission Range (TR) switch, idle switch and the Mass Air Flow (MAF) sensor.

11. Disconnect and plug the fuel tubes.

12. Disconnect the vacuum hoses from the evaporative emission canister.

13. Remove the ignition coil and the distributor cap. Use care. Disconnect the high tension wires with a twisting motion to avoid damage.

14. Disconnect the main engine wiring harness from the crankcase vent tube brackets.

15. Disconnect the fuel injector electrical connectors.

16. Remove the two ground connections from the upper intake manifold.

17. Disconnect the power transistor, Engine Coolant Temperature (ECT) sensor, water temperature indicator sender and the two main harness connectors from the upper intake manifold.

18. Disconnect the A/C compressor clutch and the pressure cut-off switch.

19. Disconnect the speed control actuator, if equipped.

20. Disconnect the throttle cable.

21. Remove the upper and lower radiator hoses.

22. If equipped with A/C, remove the two upper A/C compressor bolts.

23. Disconnect the heater hoses.

24. Disconnect the brake booster hose.

25. Remove the ground cable from the oil filler tube.

26. If equipped with A/C, remove the A/C drive belt.

27. Raise and safely support the vehicle.

28. Remove the inner and outer engine and transmission splash shields.

29. Remove the shift cable nut from the Transmission Range (TR) switch.

30. Using a screwdriver release the shift cable locking pin from the shift cable bracket.

31. Disconnect the oil pressure sensor.

32. Disconnect the low oil level sensor.

33. Disconnect the alternator electrical connectors.

34. Remove the alternator drive belt.

35. Remove the power steering drive belt.

36. Remove the power steering pump. Access the front power steering pump-to-bracket bolts by inserting a socket through the pulley holes.

37. If equipped with A/C, remove the two lower compressor bolts and position the compressor aside without disconnecting the refrigerant lines.

38. Remove the exhaust inlet pipe.

39. Remove the front halfshafts and CV-Joints using the recommended procedure.

40. Disconnect and label the oil cooler tubes.

41. Disconnect the transaxle cooler lines.

42. Disconnect the transaxle ground strap.

43. Place a suitable jack or lift under the engine transaxle assembly.

44. Remove the three front transaxle mount bolts.

45. Remove the three transaxle rear mount nuts.

46. Remove the two rear refrigerant/heater pipes hold-down bracket bolts.

47. Remove the four transverse member (crossmember) bolts and the transverse member.

48. Lower the assembly from the vehicle.

49. Remove the upper transaxle to engine bolts.

50. Remove the bolts from both transaxle braces.

51. Remove the lower transaxle bolt.

52. Remove the torque converter bolts.

53. Separate the transaxle from the engine.

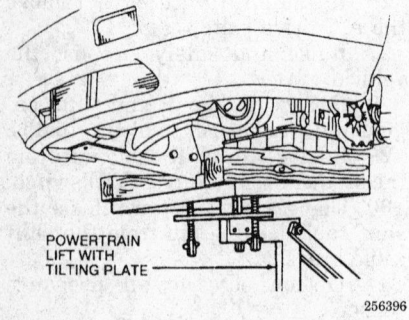

POWERTRAIN LIFT WITH TILTING PLATE →

256396

Lowering engine

To install:

54. Install the transaxle to the engine. Use care to see that the alignment dowels are properly positioned.

55. Install the lower transaxle to engine bolt and torque to 22–30 ft. lbs. (30–40 Nm).

56. Install the torque converter bolts and tighten to 33–43 ft. lbs. (44–59 Nm).

57. Install the two transaxle braces.

58. Torque the transaxle brace bolts to 22–30 ft. lbs. (30–40 Nm.

59. Install the exhaust bracket on to the transaxle and torque to 22–30 ft. lbs. (30–40 Nm.

60. Install the upper transaxle to engine bolts.

61. Torque the bolts to 29–36 ft. lbs. (39–49 Nm).

62. Raise the assembly into the vehicle and install the four transverse bolts.

63. Torque the transverse bolts to 58–65 ft. lbs. (78–88 Nm).

64. Install the two rear A/C heater brackets to the transaxle.

65. Install the three rear transaxle support bracket nuts and torque to 32–42 ft.lbs. (43–55 Nm).

66. Install the three front transaxle mount bolts and torque to 30–38 ft. lbs (41–52 Nm).

67. Lower the engine lift and remove.

68. Reconnect the transaxle ground strap.

69. Reconnect the transaxle cooler lines.

70. Connect the oil cooler lines to the proper connections.

71. Install both halfshafts. Install all of the remaining components.

72. Install the exhaust inlet pipe.

73. Install the A/C compressor, power steering pump and the alternator.

74. Install the drive belts.

75. Connect the alternator.

76. Reconnect the low oil level sensor and the oil pressure sensor.

77. Attach the shift cable and locking pin to the bracket and install the bracket to the Transaxle range switch.

78. Install the splash shields.

79. Lower the vehicle. Install the remaining components.

80. Fill the crankcase with the correct amount of oil.

81. Fill the cooling system.

82. Reconnect the negative battery cable.

83. Bleed the cooling system.

84. Start the engine.

85. Check for leaks and proper operation.

Engine Mounts

REMOVAL AND INSTALLATION

The engine and transaxle are mounted to the vehicle with a series of rubber support insulators which support the engine/transaxle assembly within the engine compartment. The support insulators isolate the engine and the transaxle from the vehicle body to prevent transferring engine and transaxle vibration to the passenger compartment.

Front Engine Support

1. Raise and safely support the vehicle.

2. Remove the six right right side inner fender splash shield and remove the shield from the vehicle.

3. Remove the seven left side engine and transaxle splash shield screws and remove the shield from the vehicle.

4. Position an underbody jack to support the engine. Pad the top of the jack with shop rags and support the engine under the crankshaft pulley. Do not try to support the engine by the oil pan.

5. Remove the engine oil filter to allow removal of the two front engine insulator mounting bolts. Remove the bolts.

6. Remove the two rear engine insulator mounting bolts.

7. If equipped, remove the auxiliary A/C tubing bracket bolts to allow the repositioning of the auxiliary A/C tubing.

8. If equipped, remove the auxiliary A/C tubing shield bolts.

9. Remove the four transverse member (crossmember) bolts and remove the transverse member.

10. Remove the front engine support insulator through bolt and nut and remove the insulator assembly.

To install:

11. Install the front engine support insulator to the vehicle and position its so that the through bolt and nut can be installed. Torque the nut to 58–65 ft. lbs. (78–88 Nm).

12. Install the transverse member. Torque the four bolts to 58–65 ft. lbs. (78–88 Nm).

13. If equipped, install the auxiliary A/C tubing shield bolts. Correctly position the A/C tubing and install the retainer bracket bolts.

14. Install the two rear engine insulator mounting bolts and torque the bolts to 58–65 ft. lbs. (78–88 Nm).

15. Install the two front engine insulator mounting bolts and torque

1   Rear Transaxle Support Insulator Through Bolt
2   Rear Transaxle Support Bracket Brace
3   Rear Transaxle Support Bracket Bolt (3 Req'd)
4   Rear Transaxle Support Bracket
5   Rear Transaxle Support Insulator Through Bolt Nut
6   Rear Transaxle Support Insulator
7   Rear Transaxle Support Insulator Bracket Bolts (4 Req'd)
8   Rear Transaxle Support Insulator Bracket
9   Rear Transaxle Support Insulator Nut (3 Req'd)
10 Front Transaxle Support Insulator Through Bolt Nut
11 Front Transaxle Support Bracket
12 Front Transaxle Support Insulator Through Bolt
13 Front Transaxle Support Insulator Bolt (3 Req'd)
14 Front Transaxle Support Insulator
15 Front Engine Support Insulator Through Bolt
16 Front Engine Support Bracket Bolt
17 Front Engine Support Bracket
18 Front Engine Support Insulator
19 Engine Insulator Mounting Bolt Nut (4 Req'd)
20 Transverse Member
21 Transverse Member Nuts (4 Req'd)
22 Transverse Member Bolts (4 Req'd)
23 Engine Insulator Mounting Bolt, Front (2 Req'd)
24 Engine Insulator Mounting Bolt, Rear (2 Req'd)
25 Rear Engine Support Insulator Through Bolt
26 Rear Engine Support Bracket Bolt (2 Req'd)
27 Rear Engine Support Bracket
28 Rear Engine Support Insulator Through Bolt Nut
29 Rear Transaxle Support Insulator
30 Front Engine Support Insulator Through Bolt Nut
A   Tighten to 43-55 N·m (32-41 Lb-Ft)
B   Tighten to 41-52 N·m (30-38 Lb-Ft)
C   Tighten to 64-74 N·m (47-54 Lb-Ft)
D   Tighten to 78-88 N·m (58-65 Lb-Ft)

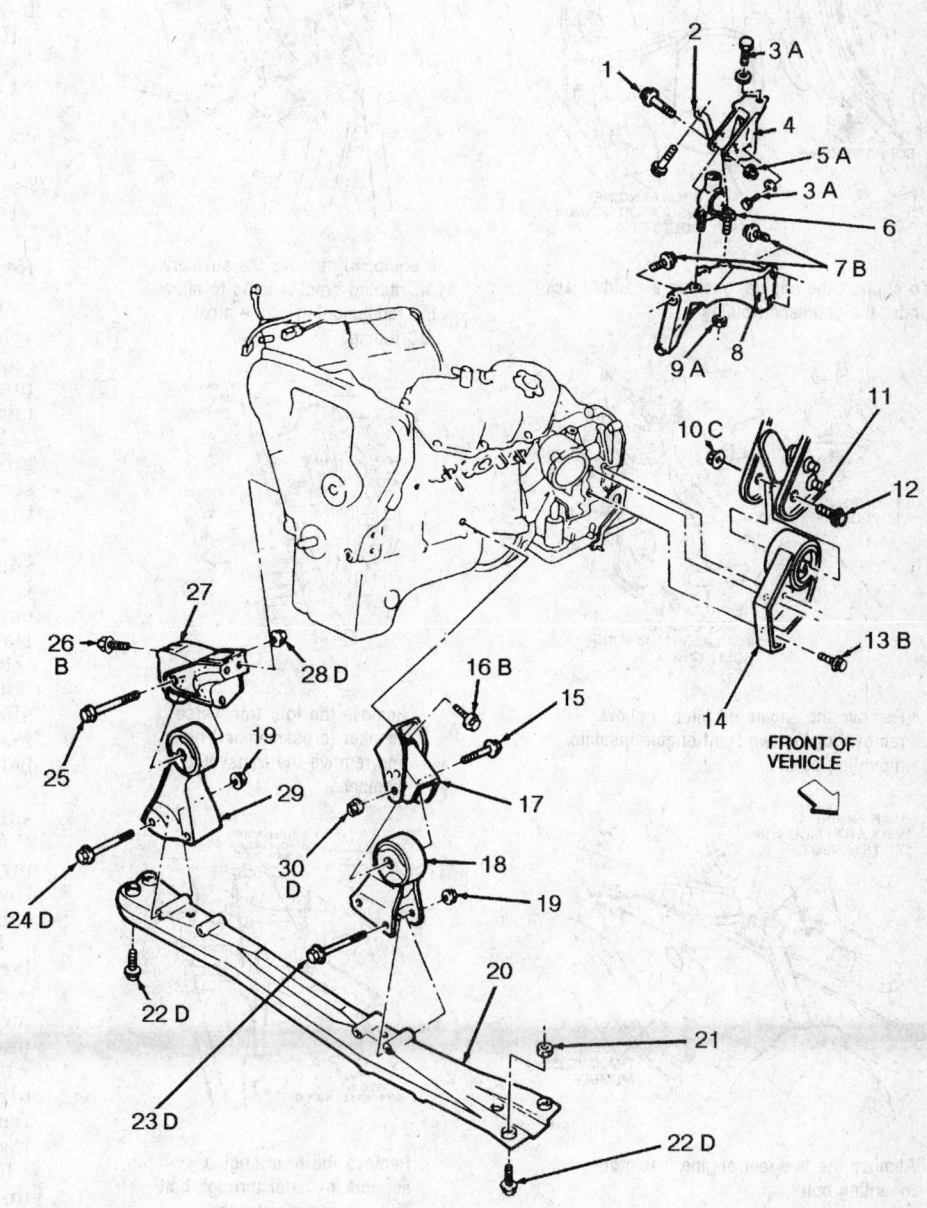

Engine and transaxle mounting

356563

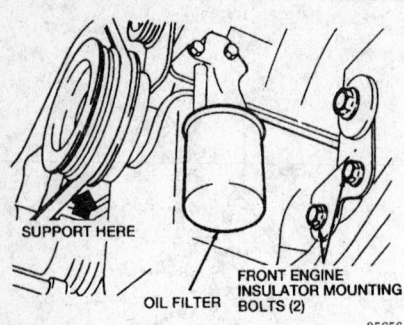

SUPPORT HERE

OIL FILTER

FRONT ENGINE INSULATOR MOUNTING BOLTS (2)

356564

To support the engine, position a padded jack under the crankshaft pulley

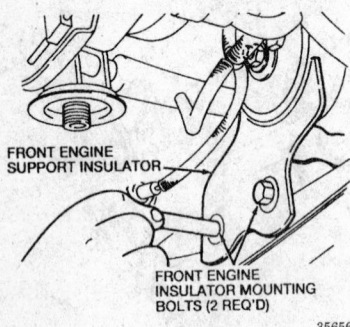

FRONT ENGINE SUPPORT INSULATOR

FRONT ENGINE INSULATOR MOUNTING BOLTS (2 REQ'D)

356565

Remove the engine oil filter to allow removal of the two front engine insulator mounting bolts

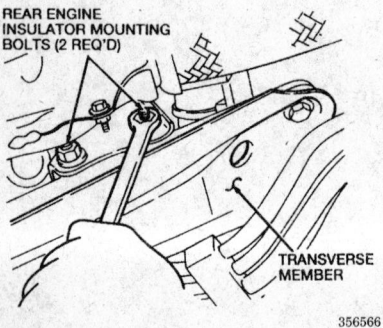

REAR ENGINE INSULATOR MOUNTING BOLTS (2 REQ'D)

TRANSVERSE MEMBER

356566

Remove the two rear engine insulator mounting bolts

the bolts to 58–65 ft. lbs. (78–88 Nm). Install a new oil filter.

16. Remove the support from under the crankshaft pulley.

17. Install the splash shields.

18. Lower the vehicle.

Front Transaxle Support

1. Remove the battery.

2. Remove the engine air cleaner assembly.

3. Remove the radiator coolant recovery reservoir.

4. Remove the battery tray.

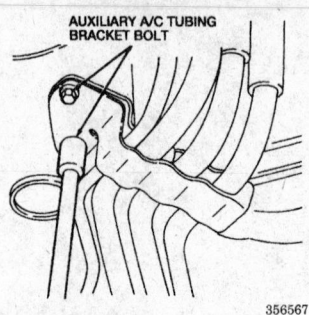

AUXILIARY A/C TUBING BRACKET BOLT

356567

If equipped, remove the auxiliary A/C tubing bracket bolts to allow the repositioning of the auxiliary A/C tubing

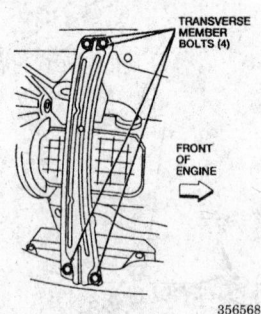

TRANSVERSE MEMBER BOLTS (4)

FRONT OF ENGINE

356568

Remove the four transverse member (crossmember) bolts and remove the transverse member

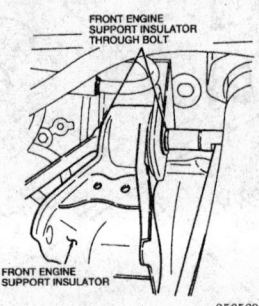

FRONT ENGINE SUPPORT INSULATOR THROUGH BOLT

FRONT ENGINE SUPPORT INSULATOR

356569

Remove the front engine support insulator through bolt and nut and remove the insulator assembly

5. Support the transaxle assembly with a floor jack.

6. Remove the front transaxle support insulator through bolt and nut.

7. Remove the three front transaxle support insulator bolts and remove the front transaxle support insulator from the vehicle.

To install:

8. Install the front transaxle support insulator to the vehicle. Install the bolts and torque the bolts to 30–38 ft. lbs. (41–52 Nm).

9. Align the bolt holes and install the front transaxle support insulator through bolt and nut. Torque the nut to 47–54 ft. lbs. (64–74 Nm).

10. Remove the transaxle assembly support.

11. Install the battery tray.

12. Install the radiator coolant recovery reservoir.

13. Install the engine air cleaner assembly.

14. Install the battery.

Rear Engine Support

1. Raise and safely support the vehicle.

2. Remove the six right right side inner fender splash shield and remove the shield from the vehicle.

3. Remove the seven left side engine and transaxle splash shield screws and remove the shield from the vehicle.

4. Position an underbody jack to support the engine. Pad the top of the jack with shop rags and support the engine under the crankshaft pulley. Do not try to support the engine by the oil pan.

5. Remove the engine oil filter to allow removal of the two front engine insulator mounting bolts. Remove the bolts.

6. Remove the two rear engine insulator mounting bolts.

7. If equipped, remove the auxiliary A/C tubing bracket bolts to allow the repositioning of the auxiliary A/C tubing.

8. If equipped, remove the auxiliary A/C tubing shield bolts.

9. Remove the four transverse member (crossmember) bolts and remove the transverse member.

10. Remove the right side halfshaft (driveshaft) and joint stone guard from the rear engine support insulator.

11. Remove the rear engine support insulator through bolt and nut and remove the insulator from the vehicle.

To install:

12. Install the rear engine support insulator to the vehicle. Align the bolt holes and install the through bolt and nut. Torque the nut to 47–54 ft. lbs. (64–74 Nm).

13. Install the right side halfshaft (driveshaft) and joint stone guard to the rear engine support insulator.

14. Install the transverse member. Torque the four bolts to 58–65 ft. lbs. (78–88 Nm).

15. If equipped, install the auxiliary A/C tubing shield bolts. Correctly position the A/C tubing and install the retainer bracket bolts.

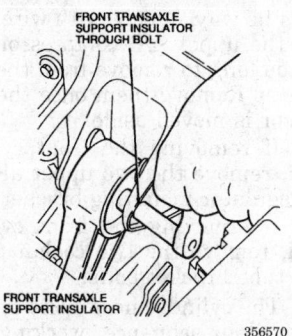

FRONT TRANSAXLE
SUPPORT INSULATOR
THROUGH BOLT

FRONT TRANSAXLE
SUPPORT INSULATOR

356570

**Remove the front transaxle
support insulator through bolt and
nut**

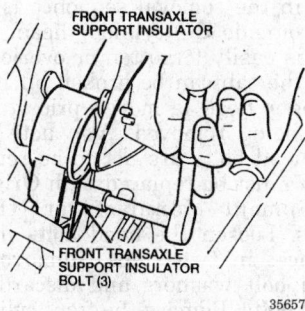

FRONT TRANSAXLE
SUPPORT INSULATOR

FRONT TRANSAXLE
SUPPORT INSULATOR
BOLT (3)

356571

**Remove the three front transaxle
support insulator bolts and remove
the front transaxle support insulator
from the vehicle**

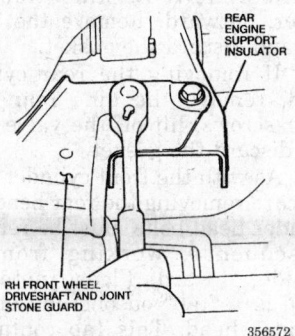

REAR
ENGINE
SUPPORT
INSULATOR

RH FRONT WHEEL
DRIVESHAFT AND JOINT
STONE GUARD

356572

**Remove the right side halfshaft
and stone guard from the rear
engine support insulator**

16. Install the two rear engine insulator mounting bolts and torque the bolts to 58–65 ft. lbs. (78–88 Nm).

17. Install the two front engine insulator mounting bolts and torque the bolts to 58–65 ft. lbs. (78–88 Nm). Install a new oil filter.

18. Remove the support from under the crankshaft pulley.

19. Install the splash shields.

20. Lower the vehicle.

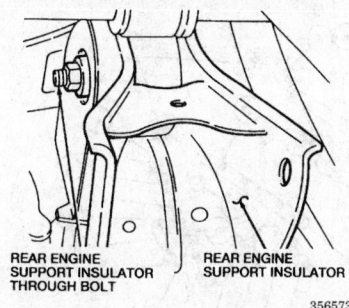

REAR ENGINE
SUPPORT INSULATOR
THROUGH BOLT

REAR ENGINE
SUPPORT INSULATOR

356573

**Remove the rear engine support insulator
through bolt and nut and remove the
insulator from the vehicle**

Cylinder Head

REMOVAL AND INSTALLATION

The 3.0L SOHC V-6 engine is a freewheeling engine design with an inhead camshaft with 12 valves, one each for intake and exhaust per cylinder.

The factory specifies that the cylinder head bolts ARE NOT to be reused. Obtain the proper replacement parts before beginning this procedure. Check carefully that all bolts are removed before attempting to remove a cylinder. A tab, part of the head, contains one lightly torqued head bolt that is external to the valve cover. Do not overlook this "hidden" bolt or the head will be damaged.

—————— **CAUTION** ——————
Fuel injection systems remain under pressure, even after the engine has been turned OFF. The fuel system pressure must be relieved before disconnecting any fuel lines. Failure to do so may result in fire and/or personal injury.

1. Properly relieve the fuel system pressure using the recommended procedure.

2. Drain the cooling system.

3. Disconnect the negative battery cable.

4. Remove the air intake tube.

5. Remove the timing belt using the recommended procedure.

6. Remove the upper intake manifold (plenum) using the recommended intake manifold procedure.

7. Tag each spark plug wire for identification. This saves time at assembly. Disconnect the spark plug wires and ignition coil to distributor high tension wires.

8. Remove the distributor using the recommended procedure.

9. Tag for identification and remove the following electrical connectors:
   a. Exhaust Gas Recirculation Control (EGRC) solenoid
   b. Mass Air Flow (MAF) sensor
   c. Vehicle Speed Sensor (VSS)
   d. Valve body wiring harness
   e. Transmission Range (TR) switch
   f. A/C clutch pulley
   g. Power transistor
   h. A/C cut-off switch
   i. Fuel injectors
   j. Ignition coil
   k. Water temperature indicator sender unit.
   l. Engine Coolant Temperature (ECT) sensor.

10. Remove the main wiring harness bracket from the water hose connection.

11. Remove the fuel tube bracket bolt from the EGR valve bracket.

12. Remove the four Allen-head fuel injection supply manifold (fuel rail) and position the fuel rail and injectors aside.

13. Disconnect the upper heater water hose.

14. Using two steps, remove the four Allen-head intake manifold to cylinder head bolts and the four nuts. Work from the outer fasteners, inward. Remove the lower intake manifold from the vehicle. Discard the gaskets.

15. If removing the front cylinder head, remove the nine valve cover screws, take off the cover and discard the gasket.

16. Use Camshaft Pulley Holding Tool T92P-6312-AH or equivalent to hold the camshaft sprocket while removing the sprocket retaining bolt. Remove the camshaft sprocket and the four seal plate bolts and seal plate.

17. If removing the front cylinder head, remove the oil level indicator (dipstick) bolts and take off the dipstick tube bracket.

18. Raise and safely support the vehicle.

19. If removing the front cylinder head, remove the three exhaust manifold to inlet pipe nuts. Remove and discard the gasket.

20. Remove the front exhaust manifold to mounting bracket bolt.

21. If removing the front cylinder head, remove the two lower A/C compressor bolts, if equipped.

22. Lower the vehicle.

23. If removing the front cylinder head, and if equipped with A/C, loosen the two upper A/C compressor bolts and remove the A/C compressor

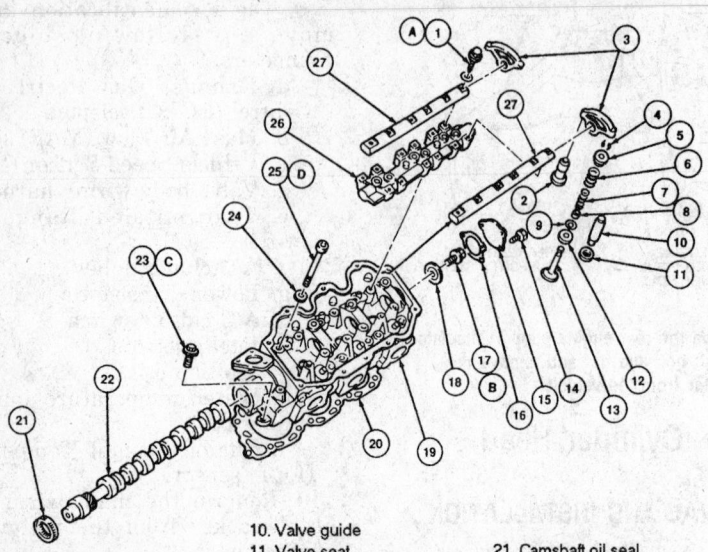

1. Rocker arm shaft bolt(2)
2. Valve tappet
3. Rocker arms
4. Valve spring retainer keys
5. Valve spring retainer, upper
6. Valve spring, outer
7. Valve spring, inner
8. Valve stem seal
9. Valve spring seat, inner

10. Valve guide
11. Valve seat
12. Valve spring seat, outer
13. Exhaust valve
14. Camshaft end plate bolt(3)
15. Camshaft end plate
16. Camshaft end plate gasket
17. Camshaft thrust plate bolt
18. Camshaft thrust plate
19. Cylinder head
20. Cylinder head gasket

21. Camshaft oil seal
22. Camshaft
23. Cylinder head bolt (A)
24. Cylinder head bolt washer
25. Cylinder head bolt(14)
26. Rocker arm shaft support
27. Rocker arm shaft
A. 13-16 ft. lb.(18-22 Nm)
B. 58-65 ft. lb.(78-80 Nm)
C. 70 in. lb.(8 Nm)

354263

3.0L SOHC engine cylinder head and related component arrangement

LOOSENING SEQUENCE

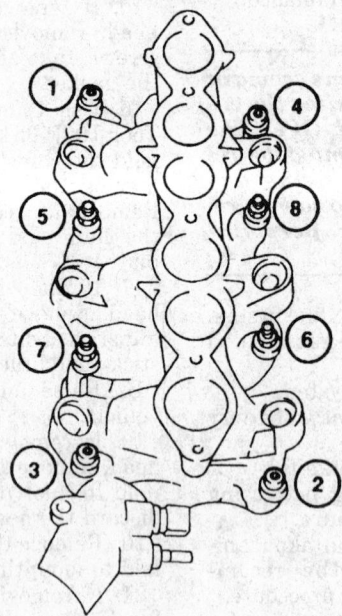

354265

Using two steps, remove the four Allen-head intake manifold to cylinder head bolts and the four nuts

out of he way. Secure with wire. Note that the upper A/C compressor bolts are too long to remove from the compressor. Remove them once the compressor is moved aside.

24. If removing the front cylinder head, remove the two upper alternator regulator mounting bracket bolts.

25. If removing the front cylinder head, remove the two coolant crossover tube bracket bolts.

26. The cylinder head bolts must be removed in sequence, working from the outside, inward. Please note that there is a "tab" on one end of each cylinder head. This tab contains a head bolt. Make sure that the first bolt in the removal sequence is this bolt outside the cylinder head. This bolt is easily forgotten or overlooked and the tab can be broken off if the cylinder head is moved prior to the bolt being removed. Also note that the head bolts are not to be reused. They must be replaced with Original Equipment Manufacturer (OEM) parts. Loosen the head bolts in sequence, in two steps. Remove the head bolt washers and discard the head bolts. Remove the front cylinder head along with the front exhaust manifold and discard the gasket.

27. If removing the rear cylinder head, remove the six rear exhaust manifold nuts working from the center, outward. Remove the manifold and discard the gasket.

28. If removing the rear cylinder head, remove the nine rear valve cover screws, lift off the valve cover and discard the gasket.

29. As with the front cylinder head, use care removing the rear head. The cylinder head bolts must be removed in sequence, working from the outside, inward. Please note that there is a "tab" on one end of each cylinder head. This tab contains a head bolt. Make sure that the first bolt in the removal sequence is this bolt outside the cylinder head. This bolt is easily forgotten or overlooked and the tab can be broken off if the cylinder head is moved prior to the bolt being removed. Also note that the head bolts are not to be reused. They must be replaced with Original Equipment Manufacturer (OEM) parts. Loosen the head bolts in sequence, in two steps. Remove the head bolt washers and discard the head bolts. Remove the rear cylinder head and discard the gasket.

To install:

30. Clean all parts well. With the intake and exhaust valves in place to protect the valve seats, remove deposits from the combustion chambers

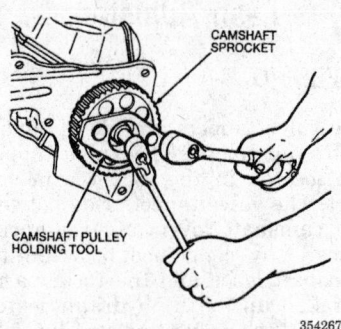

Hold the camshaft sprocket while removing the sprocket retaining bolt

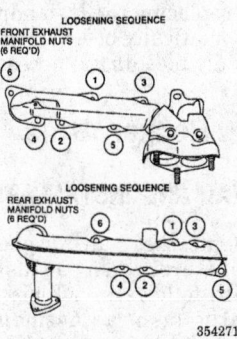

To avoid warping the exhaust manifold, use this sequence

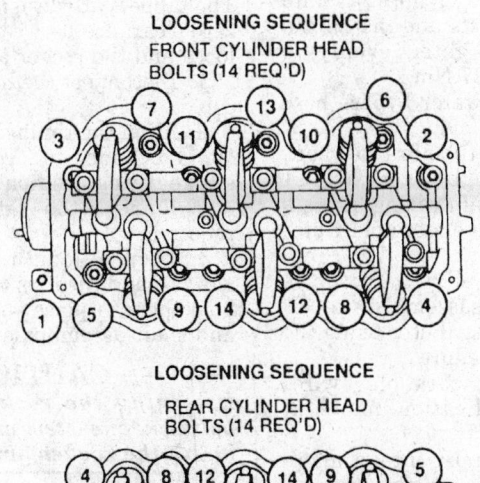

LOOSENING SEQUENCE
FRONT CYLINDER HEAD BOLTS (14 REQ'D)

LOOSENING SEQUENCE
REAR CYLINDER HEAD BOLTS (14 REQ'D)

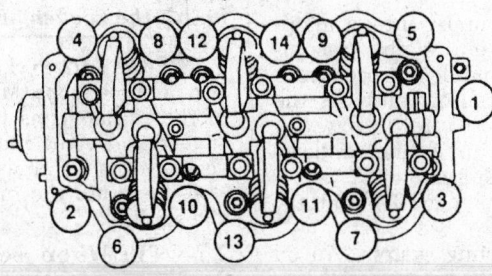

The cylinder head bolts must be removed in sequence, working from the outside, inward. Note the "hidden" bolt in position No. 1

To install:

30. Clean all parts well. With the intake and exhaust valves in place to protect the valve seats, remove deposits from the combustion chambers and valve heads with a scraper and wire brush. Be careful not to damage the head gasket surface. After the valves are removed, clean the guide bores. Use a suitable solvent to remove dirt, grease and other deposits. Clean all head bolt holes. Remove all deposits from the valves with a fine wire brush.

31. Inspect the cylinder head for damage, cracks and leakage of water and oil. If necessary, replace the head. Check the head gasket surface for burrs and nicks. If the head is cracked, it must be replaced.

32. Damaged spark plug threads can be repaired with commercially-available thread inserts. A properly installed insert will be flush to 0.03937 inch (1 mm) below the spark plug counterbore seat.

33. Using a straightedge and a feeler gauge, check the head for flatness. Measure lengthwise and across the head. Maximum distortion is 0.004 inch (0.10mm). If the head distortion exceeds this specification, the head should be resurfaced. Use care. The overall height of the cylinder head must not be reduced too much.

The cylinder head must be replaced if the head height is not within 4.205–4.220 inches (106.8–107.2mm) tall.

34. Position a new head gasket and either the front or rear cylinder head on the block. Examine the head bolt washers. Note that the washers have a chamfer or bevel on one side. The beveled side should face "up" when installed. Examine the new replacement head bolts. There are different lengths. The head bolts in positions 4, 5, 12 and 13 are 5.00 inches (127 mm) long and the rest are 4.17 inches (106 mm) long. Make sure the new cylinder head bolts are installed in the correct positions. Tighten the new head bolts in the following sequence:

   a. Torque all head bolts to 22 ft. lbs. (29 Nm).

   b. Torque all head bolts to 43 ft. lbs. (59 Nm).

   c. Loosen all of the bolts completely using the loosening sequence (loosening from the center outwards).

   d. Torque the head bolts again to 22 ft. lbs. (29 Nm).

   e. Turn all cylinder head bolts 60–65 degrees clockwise. If a torque angle wrench is not available, tighten the head bolts between 40–47 ft. lbs. (54–64 Nm).

   f. Finally install the one head bolt located outside the head, through the tab and tighten to 6 ft. lbs. (8 Nm).

35. If removed, install the exhaust manifold using a new gasket. Torque the nuts from the center, outward to 13–16 ft. lbs. (18–22 Nm).

36. If installing the front cylinder head, position the coolant crossover tube on the bracket on the cylinder head and install the bolt.

37. If installing the front cylinder head, install the two upper alternator regulator mounting bracket bolts and tighten securely.

38. If installing the front cylinder head and if equipped with A/C, position the A/C compressor on the alternator regulator mounting bracket and install the upper compressor bolts. Torque to 33–44 ft. lbs. (45–60 Nm).

39. Raise and safely support the vehicle.

40. If installing the front cylinder head and if equipped with A/C, install the two lower compressor bolts. Torque to 33–44 ft. lbs. (45–60 Nm).

41. Install a new gasket between the front exhaust manifold and rear manifold crossover tube, install the two nuts and one bolt. Install the

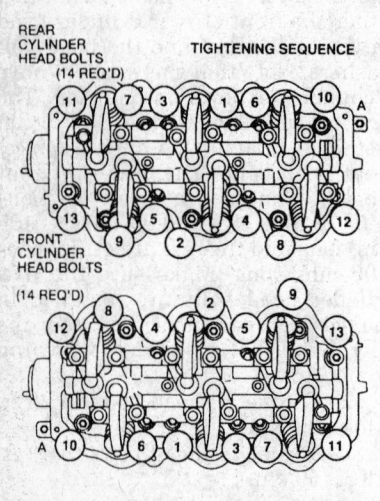

354274

Follow the torque sequence and the steps carefully to achieve a good head gasket seal. The "hidden" bolt is indicated as "A"

- Step 1: 3-5 N·m (27-44 lb-in)
- Step 2: 16-20 N·m (12-14 lb-ft)
- Step 3: 16-20 N·m (12-14 lb-ft)

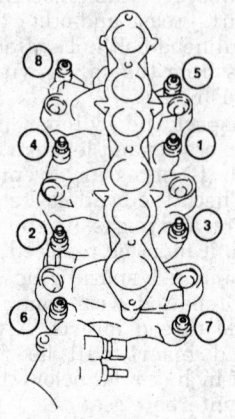

354276

Use this torque sequence on the lower intake manifold

front exhaust manifold mounting bracket bolt and tighten securely. Install a new gasket and position the exhaust inlet pipe onto the manifold. Torque the nuts to 32–40 ft. lbs. (44–54 Nm).

42. Lower the vehicle.

43. If installing the front cylinder head, install the dipstick tube and bracket and secure.

44. Position the rear engine front cover install the four seal plate bolts. Do not over-torque. Tighten to 27–44 inch lbs. (3–5 Nm).

45. Using the Camshaft Pulley Holding Tool or equivalent, install the camshaft sprocket bolt and torque to 58–65 ft. lbs. (78–88 Nm).

46. Using new gasket(s), install the valve cover(s), front or rear, as required. Install the nine bolts. Do not over-torque. Tighten to just 9–26 inch lbs. (1–3 Nm).

47. Install new lower intake manifold gaskets on the cylinder heads and lay the manifold in lace. Install the four Allen-head bolts and four manifold nuts. Torque in sequence, working from the center, outward as follows:

  a. Step 1: Torque to 27–44 inch lbs. (3–5 Nm).

  b. Step 2: Torque to 12–14 ft. lbs. (16–20 Nm).

  c. Step 3: Torque again to 12–14 ft. lbs. (16–20 Nm).

48. Connect the upper water hose.

49. Position the fuel injectors into their respective ports and install the fuel rail bolts. Tighten evenly to 17–20 ft. lbs. (24–27 Nm).

50. Connect the water bypass hose to the intake manifold.

51. Bolt the fuel tube bracket back onto the EGR valve bracket.

52. Attach the main wiring harness bracket to the intake manifold water hose connection.

53. Connect the electrical connectors removed at disassembly.

54. Install the distributor using the recommended procedure.

55. Connect the spark plug wires using the identification made at removal.

56. Install the upper intake manifold (plenum) using the recommended intake manifold procedure.

57. Install the timing belt using the recommended procedure.

58. Install the air cleaner and intake tubes as required.

59. Connect the negative battery cable.

60. Fill the cooling system. An oil and filter change is recommended.

61. Start the vehicle and check for leaks. Check the ignition timing and adjust as required.

Lash Adjusters

BLEEDING

The valve tappets provide automatic lash adjustment. The valve tappets are located in the rocker arm support. The valve tappets ride between the camshaft loves and the rocker arms. Any clearance between the camshaft lobes and the rocker arms is taken up by the hydraulic extension of each valve tappet. The valve tappets are designed to maintain 0.0 inch (0.0mm) clearance between the camshaft lobes and the rocker arms.

After replacing the lash adjuster, it will automatically bleed itself. No additional air bleeding is necessary.

Rocker Arms

REMOVAL AND INSTALLATION

This engine uses valve tappets (also called lifters, lash adjusters or camshaft followers) which provide automatic lash adjustment. The valve tappets are located in the rocker arm shaft support. The valve tappets ride between the camshaft lobes and the rocker arms. Any clearance between the camshaft lobes and the rocker arms is taken up by the hydraulic extension of each tappet. The tappets are designed to maintain zero clearance between the camshaft lobes and the rocker arms.

1. Disconnect the negative battery cable.

2. Align the timing marks to bring No. 1 cylinder to TDC compression stroke (firing position).

3. Remove the valve cover(s) as required.

4. Loop a length of mechanic's wire around the tops of the tappets to hold them in place when rocker arm and shaft assembly is removed.

— CAUTION —
Loosening the rocker arm shaft bolts in one step may distort or break the rocker arm shaft.

5. Loosen the rocker shaft bolts in two or three steps. Mark the location of the rocker arms to ease installation. Remove the rocker shafts with rocker arms, as an assembly.

6. Separate the rocker arms from the shaft.

NOTE: When separating the rocker arms from the rocker arm shafts, be sure to keep the parts in order for reinstallation, if any of the parts are to be reused.

7. Check the rocker arms and the shafts for damage. If necessary, replace the damaged components.

To install:

8. Clean all parts well. Any work parts should be replaced.

9. Attach a loop of mechanic's wire to the top of the lifters so that they will not drop from the lifter guide during assembly. Carefully remove the lifter guide and lifters from the cylinder head. Put an identification mark on the lifters to avoid mixing them up if they are removed from the guide and to be reused. If the lifters are damaged replace them as necessary.

10. Install new lifters if replacing them or install the old lifters to their original locations. Coat all parts with clean engine oil as they are assembled.

---- **CAUTION** ----

When installing the rocker arm shafts, be certain that they are installed in their original positions.

11. Slide the rocker arms onto the shafts in their proper positions.

12. Make sure that cylinder No. 1 is still at TDC.

13. Install the rocker arm/shaft and lifter guide assembly. When seated, remove the wire that held the lifters in place. Coat the bolt threads and seat surfaces with clean engine oil before installing them. Tighten the bolts gradually in three steps to 13–16 ft. lbs. (18–22 Nm).

14. Rotate the crankshaft clockwise 180 degrees ½-turn to bring cylinder No. 4 to TDC. Install the right cylinder head rocker arm/shaft and lifter guide assembly. Coat the bolt threads and seat surfaces with clean engine oil before installing them. Tighten the bolts gradually in three steps to 13–16 ft. lbs. (18–22 Nm).

15. Install the rocker covers with new gaskets. Torque the rocker cover bolts to 12–24 inch lbs. (1–3 Nm).

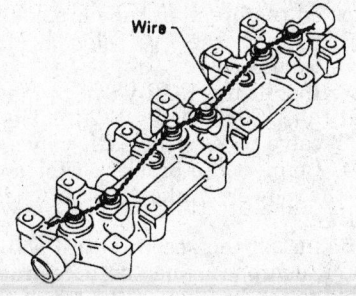

255178

Proper method of secure lifters to rocker assembly

16. Connect the negative battery cable.

17. An oil and filter change is recommended.

18. Test run the engine and verify no valve noise.

Intake Manifold

REMOVAL AND INSTALLATION

This Multiport Fuel Injection (MFI) system is classified as a multi-point, pulse time, mass airflow fuel injection system. This system supplies the engine with the air/fuel mixture necessary for combustion. An air induction system and fuel injection system work in conjunction with the Electronic Engine Control (EEC) system which consists of various sensors, switches and a Powertrain Control Module (PCM). The PCM uses the signals that it receives from the various sensors and switches to compute fuel injector timing and pulse width.

The intake manifold is a two-piece design. The upper half, usually called the plenum, mounts the throttle body, control cables, fast idle solenoid, idle air control solenoid and emission connections. The lower manifold bolts directly to the engine and carries the fuel injectors.

Air enters the system through the air cleaner intake tube. The air then flows through the dry element air cleaner and is metered by the Mass Air Flow (MAF) sensor. The metered air passes through the air cleaner to intake manifold tube and enters the throttle body. From the throttle body, the air passes through the upper intake manifold to the lower intake manifold where it is mixed with fuel for combustion. To reduce intake noise, three engine air intake resonators are part of the system. These three components absorb air flow pulsations as air is drawn into the system.

Upper Intake Manifold (Plenum)

---- **CAUTION** ----

Fuel injection systems remain under pressure, even after the engine has been turned OFF. The fuel system pressure must be relieved before disconnecting any fuel lines. Failure to do so may result in fire and/or personal injury.

1. Relieve the fuel system pressure.

2. Disconnect the negative battery cable.

3. Drain the cooling system.

4. Remove the air cleaner intake tube and resonator. Use the following procedure.

a. Remove the battery.

b. Remove the air cleaner intake tube bracket bolt.

c. Separate the air cleaner intake tube from the engine air cleaner.

d. Remove and separate the air cleaner intake tube from the engine air intake resonator No. 1.

5. Disconnect the idle switch electrical connector and the Throttle Position (TP) sensor. Both connectors are located near the throttle body opening. Slide the TP sensor electrical wiring out of the bracket.

6. Remove the coolant hose from the throttle body.

7. Disconnect the Exhaust Gas Recirculation Control (EGRC) solenoid electrical connector.

8. Disconnect the Evaporative Emission canister vacuum hose from the throttle body.

9. Remove the fuel pressure regulator-to-upper intake manifold vacuum hose from the upper intake manifold.

10. Disconnect the No. 1, 3 and 5 cylinder distributor to spark plug wires from the spark plugs, then remove the spark plug wire boots from between the upper intake manifold runners. Remove the two spark plug wire bracket bolts and spark plug wire brackets from the upper intake manifold.

11. Remove the distributor dust cover. Loosen the three distributor cap screws. Note that the screws are not removable from the distributor cap. Remove the cap from the distributor. Position the cap and plug wires aside to gain access to the main engine wiring harness.

12. Disconnect the rear heater valve vacuum hose from the side of the upper intake manifold, if equipped.

13. Remove the two wire harness bracket bolts and remove the wire harness bracket from the upper intake manifold and set aside.

14. Disconnect the speed control actuator from the throttle lever, if equipped. Disconnect the accelerator cable from the throttle lever.

15. Remove the speed control actuator from the hanging bracket, if equipped.

16. Remove the accelerator cable from the bracket attached to the top of the upper intake manifold.

17. Loosen the accelerator cable locknut and remove the accelerator

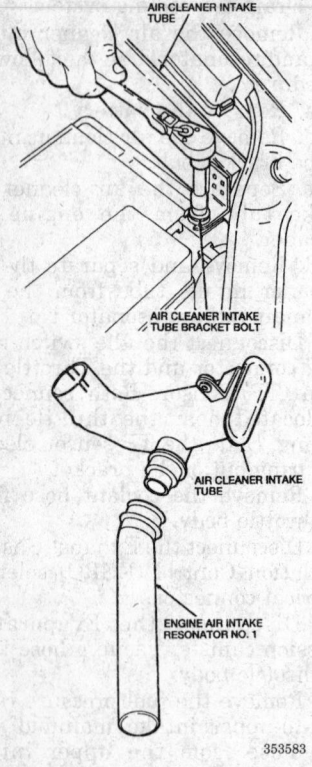

Removing the air cleaner intake tube bracket and No. 1 resonator

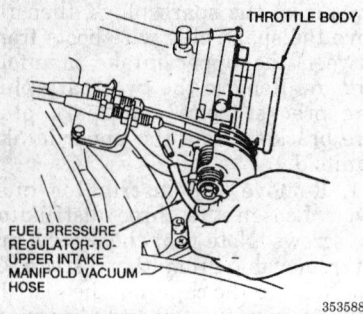

Remove the fuel pressure regulator-to-upper intake manifold vacuum hose from the upper intake manifold

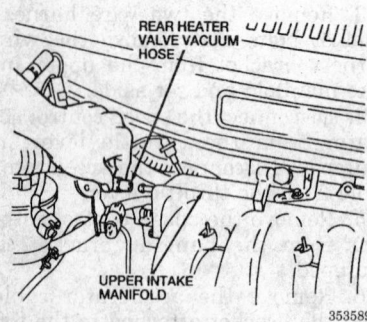

Disconnect the rear heater valve vacuum hose from the side of the upper intake manifold, if equipped

cable from the bracket on top of the upper intake manifold.

18. Loosen the speed control actuator locknut, if equipped, and remove the speed control actuator from the bracket on the top of the upper intake manifold and position aside.

19. Disconnect the brake booster hose from the power brake booster check valve.

20. Remove the Exhaust Gas Recirculation (EGR) valve-to-Back Pressure Transducer (BPT) valve tube from the EGR valve. Remove the EGR valve to manifold tube from the EGR valve.

21. Disconnect the Fast Idle Control solenoid electrical connector.

22. Disconnect the main engine wiring harness clips from the breather tube brackets. Remove the two ground wire bolts from the upper intake manifold and position aside the ground wires and the harness.

23. Remove the two front breather tube bracket-to-upper intake manifold bolts. Disconnect the front breather hose from the front valve cover.

24. Remove the fuel tube mounting bracket bolt.

25. Disconnect the EGR temperature sensor electrical connector.

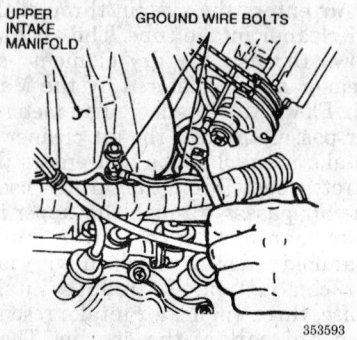

Remove the two ground wire bolts from the upper intake manifold

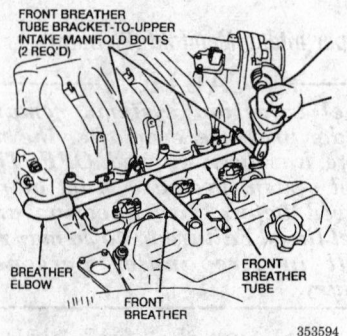

Remove the two front breather tube bracket bolts and disconnect the front breather hose

26. Disconnect the Bypass Air (BPA) valve electrical connector.

27. Disconnect the Idle Air Control (IAC) solenoid electrical connector.

28. Disconnect the Positive Crankcase Ventilation (PCV) hose from the upper intake manifold.

29. Remove the five Allen-head upper intake manifold to lower intake manifold bolts. Separate the upper intake manifold from the lower intake manifold and discard the gasket.

To install:

30. Clean all parts well. Use care working with light alloy parts, not to scratch or damage the gasket sealing surface or any threaded openings. Make sure all traces of old gasket and or sealer are removed.

31. Lay a new gasket on the lower intake manifold carefully aligning all openings. Set the upper intake manifold into place and install the five Allen-head bolts. Tighten the upper manifold bolts evenly, working from the center, outward, to 13–16 ft. lbs. (18–22 Nm). Do not over-tighten.

32. Connect the PCV hose to the upper intake manifold.

33. Connect the IAC solenoid electrical connector.

34. Connect the BPA valve electrical connector.

35. Connect the EGR temperature sensor electrical connector.

36. Connect the coolant hose to the top heater core pipe.

37. Install the fuel tube mounting bracket.

38. Connect the front breather hose to the front valve cover.

39. Position the front breather tube onto the upper intake manifold and install the two front breather tube bracket-to-upper intake manifold bolts. Tighten the front breather tube bracket-to-upper manifold bolts.

40. Position the ground wires and main wiring harness into place. Install and tighten the ground wire bolts. Make sure the connection is clean and tight.

41. Connect the main engine wiring clips on the breather tube brackets.

42. Connect the FIC solenoid electrical connector.

43. Install the EGR valve to manifold tube. Connect the EGR valve to BPT valve tube to the EGR valve.

44. Connect the brake booster hose to the power brake booster check valve.

45. Install the speed control actuator, if equipped, and the accelerator cable to the bracket on the top of the upper intake manifold.

46. Connect the speed control actuator on the throttle lever, if equipped.

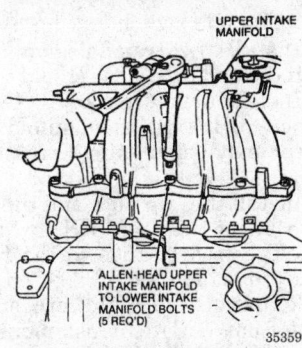

Remove the five Allen-head upper intake manifold to lower intake manifold bolts

47. Connect the accelerator cable on the throttle lever.

48. Tighten the nut on the speed control actuator, if equipped, and the accelerator cable nut.

49. Install the wire harness bracket on the rear of the upper intake manifold. Install and tighten the two wire harness bracket bolts.

50. Connect the rear heater valve vacuum hose to the upper intake manifold, if equipped.

51. Install the distributor cap and tighten the three distributor cap screws. Install the distributor cap dust cover.

52. Install the spark plug wire bracket on the upper intake manifold and install the bracket bolts. Install the spark plug wire boots in between the upper intake manifold runners and connect the No. 1, 3 and 5 cylinder spark plug wires onto the spark plugs.

53. Connect the fuel pressure regulator vacuum hose to the intake manifold. Connect the EVAP canister hose to the throttle body.

54. Connect the EGRC solenoid electrical connector.

55. Connect the coolant hose to the throttle body.

56. Slide the TP sensor electrical wiring into the bracket and connect the TP sensor connector. Connect the idle switch connector.

57. Install the air cleaner intake tube and engine air intake resonator.

58. Fill the cooling system.

59. Connect the negative battery cable.

60. Start the engine and check for leaks and for proper engine operation.

Lower Intake Manifold

── CAUTION ──

Fuel injection systems remain under pressure, even after the engine has been turned OFF. The fuel system pressure must be relieved before disconnecting any fuel lines. Failure to do so may result in fire and/or personal injury.

1. Relieve the fuel system pressure.

2. Disconnect the negative battery cable.

3. Remove the upper intake manifold (plenum) using the above procedure.

4. Disconnect the six fuel injector electrical connectors.

5. Disconnect the rear breather hose from the rear valve cover.

6. Remove the crossover breather tube bracket bolt and the crossover breather tube from the upper engine front cover.

7. Remove the four fuel injection supply manifold (fuel rail) bolts and position the supply manifold and the insulators out of the way.

8. Disconnect the Engine Coolant Temperature (ECT) sensor electrical connector.

9. Disconnect the water temperature indicator sender unit electrical connector.

10. Remove the upper radiator hose from the water hose connection. Remove the upper radiator hose bracket bolt.

11. Remove the water bypass hose from the water hose connection.

12. Remove the bolt securing the water hose connection to the upper engine front cover.

13. Remove the heater hose from the top heater core pipe.

14. Remove the four Allen-head lower intake manifold bolts and the

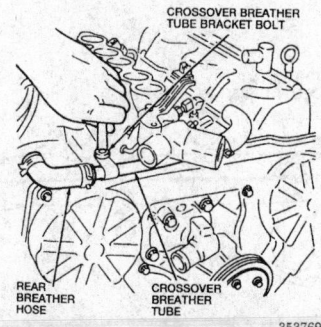

Remove the crossover breather tube bracket bolt and the crossover breather tube

four manifold nuts, in two steps in the reverse order of the tightening sequence indicated, working from the outside fasteners, inward.

15. Remove the lower intake manifold washers. Lift the lower intake manifold from the engine block. Discard the gaskets.

To install:

16. Clean all parts well. Use care working with light alloy parts, not to scratch or damage the gasket sealing surface or any threaded openings. Make sure all traces of old gasket and or sealer are removed.

17. Lay a new lower intake manifold gaskets on the cylinder heads and position the lower intake manifold on the engine block. Install the lower intake manifold washers and the Allen-head bolts and the manifold to cylinder head nuts.

18. Working from the center outward, tighten the lower intake manifold bolts and nuts in three steps using following sequence:

   a. Tighten the lower intake manifold to engine block bolts to 26 to 44 inch lbs. (3 to 5 Nm).

   b. Tighten the lower intake manifold to engine block bolts to 12 to 14 ft. lbs. (16 to 20 Nm) and the nuts to 17 to 20 ft. lbs. (24 to 27 Nm).

   c. Again, tighten the lower intake manifold to engine block bolts to 12 to 14 ft. lbs. (16 to 20 Nm) and the nuts to 17 to 20 ft. lbs. (24 to 27 Nm).

19. Install the heater hose to the top heater core pipe.

20. Install the bolt securing the water hose connection to the upper engine front cover.

21. Install the water bypass tube or hose onto the water hose connection.

22. Install the upper radiator hose bracket and bolt. Connect the upper radiator hose onto the water hose connection.

23. Connect the water temperature indicator sender unit connector. Connect the Engine Coolant Temperature sensor wire.

24. Position the fuel rail with the insulators in place and install the four bolts.

25. Install the breather tube assembly and connect the rear breather to the rear valve cover. Install the crossover breather tube and the crossover tube bracket bolt.

26. Connect the six fuel injector electrical connectors.

27. Install the upper intake manifold (plenum assembly) using the recommended procedure given above.

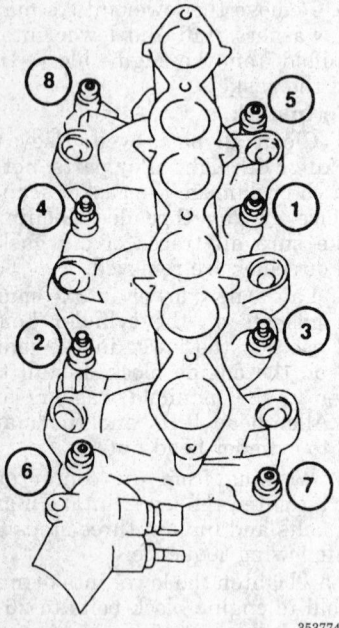

TIGHTENING
SEQUENCE

353774

Tighten the bolts and nuts in three
steps in this sequence

Exhaust Manifold

REMOVAL AND INSTALLATION

Rear (RH) Exhaust Manifold

1. Disconnect the negative battery
cable and wait at least 90 seconds
before performing any work. This allows time for the SRS or air bag system to deplete its back-up energy
supply.
2. Disconnect the radiator overflow hose from the radiator.
3. Slide the radiator coolant-recov-
ery reservoir off of the bracket and
remove the reservoir.
4. Remove the air cleaner intake
tube and the engine air intake
resonator.
5. Remove the six rear (RH) exhaust manifold crossover tube heat-
shield bolts and remove the heat
shields.
6. Remove the two nuts and the
one bolt securing the rear (RH) exhaust manifold tube to the front (LH)
exhaust manifold. Discard the
gasket.
7. Remove the transmission fluid
level indicator tube heat shield.
8. Disconnect the following electrical connectors:
    a. The idle switch.

b. The throttle position sensor.
c. The exhaust gas recirculation
control solenoid.
9. Raise and safely support the
vehicle.
10. Remove the exhaust gas recirculation valve to back-pressure
transducer valve tube nut and position it out of the way.
11. Remove the two EGR valve to
exhaust manifold tube nuts and remove the EGR valve to exhaust manifold tube.
12. Remove the six rear exhaust
manifold nuts in the reverse order of
the tightening sequence.
13. Safely lower the vehicle, remove
the exhaust manifold and discard the
exhaust manifold gasket.
    To Install:
14. Raise and safely support the
vehicle.
15. Make sure that both the exhaust manifold and the cylinder head
mating surfaces are clean of any old
gasket material.
16. Position the rear (RH) exhaust
manifold gasket onto the exhaust
manifold mounting studs.
17. Lower the vehicle safely.
18. Place the rear (RH) exhaust
manifold onto the studs.
19. Safely raise the vehicle and install the six rear (RH) exhaust manifold nuts. Tighten the nuts in sequence to 13–16 ft. lbs. (18–22 Nm).
20. Install the EGR valve to exhaust manifold tube and install the
two EGR valve to exhaust manifold
tube nuts. Tighten the EGR valve to
exhaust manifold tube nuts.
21. Position the EGR valve to the
back-pressure transducer valve tube
nut into place. Tighten the EGR
valve to the BPT valve tube nut.
22. Lower the vehicle carefully.
23. Reconnect the following electrical connectors:
    a. The exhaust gas recirculation
solenoid.
    b. The throttle position sensor.

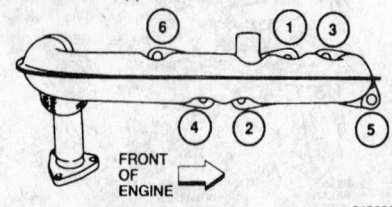

REAR (RH) EXHAUST
MANIFOLD NUTS (6)

FRONT
OF
ENGINE

343820

Rear (RH) exhaust manifold mounting bolts
tightening sequence

c. The idle switch.
24. Install the transmission fluid
level indicator tube heat shield.
25. Install a new gasket between
the front (LH) exhaust manifold and
the rear exhaust manifold crossover
tube.
26. Install the two nuts and the one
bolt securing the rear (RH) exhaust
manifold crossover tube to the front
(LH) exhaust manifold. Tighten the
rear exhaust manifold crossover
tube-to-front (LH) exhaust manifold
nuts and bolt.
27. Reinstall the rear(RH) exhaust
manifold crossover tube heat shield
with the six mounting bolts.
28. Tighten the rear (RH) exhaust
manifold crossover tube bolts.
29. Install the air cleaner intake
tube and the engine air intake
resonator.
30. Install the radiator coolant recovery reservoir and reconnect the
radiator overflow hose to the
radiator.
31. Reconnect the negative battery
cable, start the engine and check for
leaks and proper operation.

Front (LH) Exhaust Manifold

1. Disconnect the negative battery
cable and wait at least 90 seconds
before performing any work. This allows time for the SRS or air bag system to deplete its back up energy
supply.
2. Remove the two nuts and the
one bolt securing the front (LH) exhaust manifold to the rear (RH) exhaust manifold crossover tube. Discard the gasket.
3. Remove the transmission fluid
level indicator tube heat shield.
4. Loosen the six front (LH) exhaust manifold nuts in two steps in
the reverse order of the tightening sequence. Do not remove the three
lower front (LH) exhaust manifold
nuts.
5. Remove the front (LH) exhaust
manifold-to-mounting bracket bolt.
6. Raise and safely support the
vehicle.
7. Disconnect the heated oxygen
sensor electrical connector.
8. Remove the three front (LH) exhaust manifold-to-inlet pipe nuts.
9. Remove the exhaust system flex
tube bracket bolt.
10. Remove the LH inner engine
and transmission splash shield bolts
and screws and remove the LH inner
engine and transmission splash
shield.
11. Remove the three lower exhaust
manifold nuts.

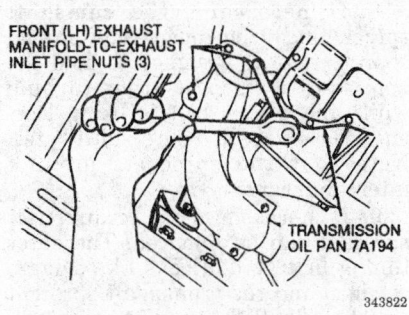

FRONT (LH) EXHAUST
MANIFOLD-TO-EXHAUST
INLET PIPE NUTS (3)

TRANSMISSION
OIL PAN 7A194

343822

Removing the front exhaust pipe from the exhaust manifold

12. Remove the front (LH) exhaust manifold and discard the exhaust manifold gasket.

To Install:

13. Make sure that both the exhaust manifold and the cylinder head mating surfaces are clean of any old gasket material.

14. Position a new front exhaust manifold gasket in place and install the front (LH) exhaust manifold. Install the three lower exhaust manifold mounting nuts. Do not tighten the nuts at this time.

15. Install the LH inner engine and transmission splash shield with their mounting bolts and screws.

16. Reinstall the exhaust system flex tube bracket bolt.

17. Install the three exhaust manifold-to-exhaust inlet pipe nuts.

18. Reconnect the heated oxygen sensor electrical connector.

19. Lower the vehicle.

20. Install the front (LH) exhaust manifold-to-mounting bracket bolt.

21. Install the three upper exhaust manifold mounting bolts and tighten all six exhaust manifold mounting bolts in sequence. Torque the bolts to 13–16 ft. lbs. (18–22 Nm).

22. Install the transmission fluid level indicator tube heat shield.

23. Install the two nuts and the one bolt securing the front (LH) exhaust

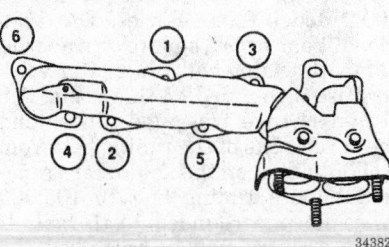

FRONT (LH) EXHAUST
MANIFOLD NUTS (6)

343824

Front (LH) exhaust manifold mounting bolts tightening sequence

manifold to the rear (RH) exhaust manifold crossover tube.

24. Reconnect the negative battery cable, start the engine, check for leaks and road test for proper operation.

Front Crankshaft Seal

REMOVAL AND INSTALLATION

NOTE: The front oil seal is a part of the oil pump body.

1. Disconnect the negative battery cable.

2. Remove the timing belt and the crankshaft sprocket.

3. Remove the oil pump assembly.

4. Remove the oil seal from the oil pump body using a pry tool. Be careful not to damage the oil pump body during seal removal.

To install:

5. Apply clean engine oil to the new oil seal. Install the seal using proper size driver.

6. Install the oil pump assembly to the engine.

7. Install the remaining components in reverse order of removal.

8. Connect the negative battery cable.

Timing Belt, Sprockets, Tensioner and Front Cover

REMOVAL AND INSTALLATION

On this vehicle, right side refers to the "rear" components and left side refers to the "front" components.

1. If the timing belt is to be removed, it is good practice to turn the crankshaft until the engine is at Top Dead Center (TDC) of the number one cylinder, compression stroke (firing position) before beginning work. This should align all timing marks and serve as a reference for all work

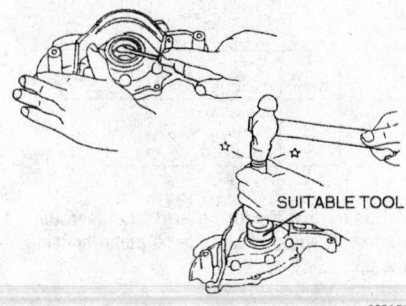

SUITABLE TOOL

339151

Removing and installing the front crankshaft oil seal

that follows. After verifying that the engine is at TDC for number one cylinder, do not crank the engine or allow the crankshaft or camshaft sprockets to be turned or engine timing will be lost.

2. Drain the cooing system.

3. Disconnect the negative battery cable.

4. Remove the alternator drive belt, water pump and power steering pump belt and the A/C compressor belt (if equipped), using the recommended drive belt removal procedure.

5. If equipped with A/C, remove the three A/C compressor drive belt idler pulley bolts and remove the idle pulley.

6. Remove the upper radiator hose bracket bolt. Remove the upper hose with the bracket from the vehicle.

7. Remove the water bypass hose from between the thermostat housing and the lower water hose connection.

8. Remove the main wiring harness from the upper engine front cover.

9. Remove the eight upper engine front cover bolts and remove the upper cover.

10. Raise and safely support the vehicle.

11. Remove the right side front wheel and tire assembly.

12. Remove the four right side engine and transmission splash shield bolts and two screws and remove the right side outer engine and transaxle splash shield.

13. Use a strap wrench to hold the water pump pulley. Remove the four pulley bolts and remove the water pump pulley.

14. Use a strap wrench to hold the crankshaft pulley. Remove the center pulley bolt and remove the crankshaft pulley using a harmonic balancer (damper) puller to draw the pulley from the front of the crankshaft.

15. Remove the five lower engine front cover bolts and remove the lower engine front cover.

16. Make sure that the timing marks between the crankshaft sprocket and the oil pump housing line up.

17. If the timing belt is to be reused, mark and arrow on the belt indicating the direction of rotation. The directional arrow is necessary to ensure that the timing belt, if it to be reused, can reinstalled in the same direction.

18. Loosen the timing belt tensioner nut and slip the timing belt off of the sprockets.

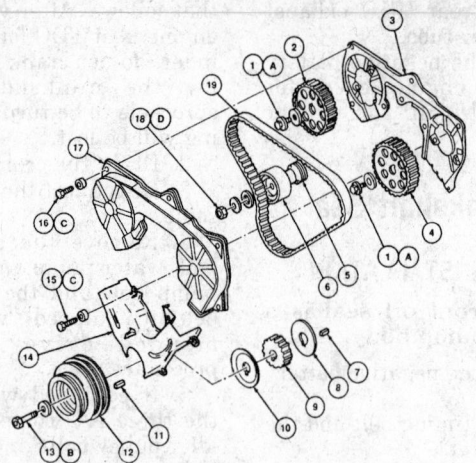

Item	Description	Item	Description
1	Camshaft Sprocket Retaining Bolt	12	Crankshaft Pulley
2	Camshaft Sprocket, Rear	13	Crankshaft Pulley Bolt
3	Seal Plate	14	Engine Front Cover, Lower
4	Camshaft Sprocket, Front	15	Lower Engine Front Cover Bolt
5	Timing Chain / Belt Tensioner Spring	16	Upper Engine Front Cover Bolt
6	Timing Chain / Belt Tensioner	17	Engine Front Cover, Upper
7	Crankshaft Key	18	Timing Chain / Belt Tensioner Nut
8	Inner Timing Chain / Belt Guide	19	Timing Chain / Belt
9	Crankshaft Sprocket	A	Tighten to 78-88 N·m (58-65 Lb-Ft)
10	Outer Timing Chain / Belt Guide	B	Tighten to 123-132 N·m (90-98 Lb-Ft)
11	Crankshaft Key	C	Tighten to 3-5 N·m (27-44 Lb-In)
		D	Tighten to 43-58 N·m (32-43 Lb-Ft)

357156

Camshaft sprockets, timing belt and engine front covers

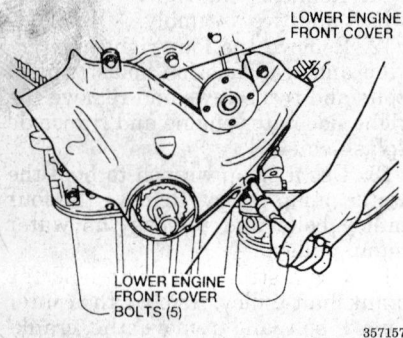

357157

Removing the lower engine timing belt cover

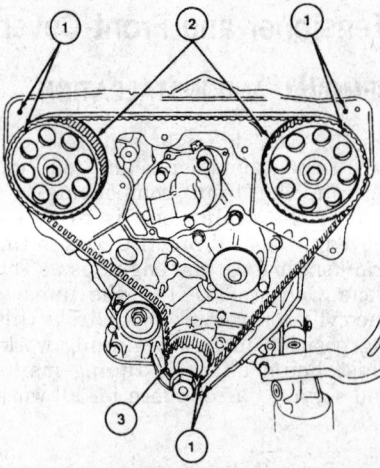

Item	Description
1	Timing Marks
2	Camshaft Sprockets
3	Crankshaft Sprocket

357158

Make sure that the timing marks between the crankshaft sprocket and the oil pump housing line up

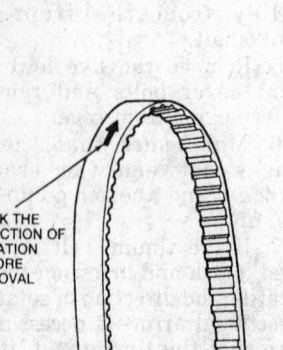

357159

If the timing belt is to be reused, mark and arrow on the belt indicating the direction of rotation

19. If necessary, the camshaft sprockets can be removed. A special spanner tool is designed to hold the sprocket to keep it from turning while the center bolt is being loosened. Use care if using substitutes. Note that the sprockets are not interchangeable.

20. If necessary, the crankshaft sprocket can be removed. The outer timing belt guide (looks like a large washer) and the crankshaft sprocket simply pull off the front of the crankshaft. Note that there are two crankshaft keys. Use care not to loose them.

To install:

21. Clean all parts well. If removed, inspect the crankshaft sprocket for warping or abnormal wear. Check the sprocket teeth for wear, deformation, chipping or other damage. Replace as necessary. Clean the sprocket mounting surface to ease installation. Install the key. Slip the sprocket onto the crankshaft. Tap in place with a suitably sized socket.

22. If removed, inspect the camshaft sprockets for damage and wear. Replace as required. The sprockets should be marked **L3** to designate the front, or left side camshaft and **R3** to designate the rear, or right side camshaft. Use care to install the sprockets properly. A special spanner tool is designed to hold the sprocket to keep it from turning while the center bolt is being tightened. Use care if using substitutes. Torque the camshaft sprocket center bolts to 58 to 65 ft. lbs. (78 to 88 Nm). Verify that the timing marks on the camshaft sprockets and the timing marks on the rear cover (called the seal plate) are aligned.

23. Use an Allen wrench to turn the timing belt tensioner clockwise until the belt tensioner spring is fully extended. Temporarily tighten the tensioner nut to 32 to 43 ft. lbs. (43 to 58 Nm).

24. **IMPORTANT:** Pay special attention to the tooth shape of the timing belt. There was a major change in belt design between the 1993 and 1994 Model Year vehicles. The 1993 Model Year uses a square or trapezoidal tooth timing belt while later vehicles use a rounded tooth design. The design change was made to extend the service life of the timing belt from 60,000 miles for the square or trapezoidal tooth timing belt to 105,000 miles for the rounded tooth belt. If the proper timing belt is not used, the vehicle operator may complain of a "whining" noise coming from the engine compartment. If the camshaft

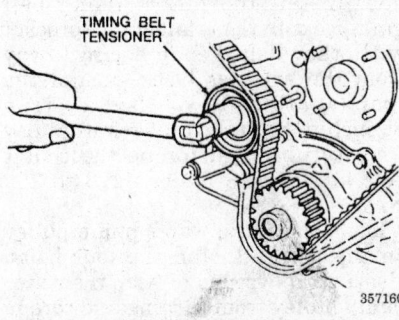

Loosen the timing belt tensioner nut and slip the timing belt off of the sprockets

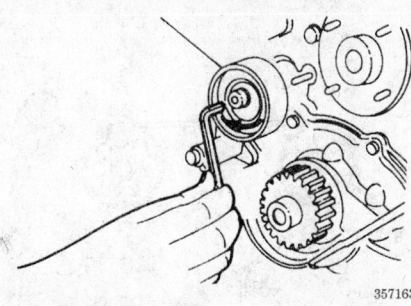

Use an Allen wrench to turn the timing belt tensioner clockwise until the belt tensioner spring is fully extended

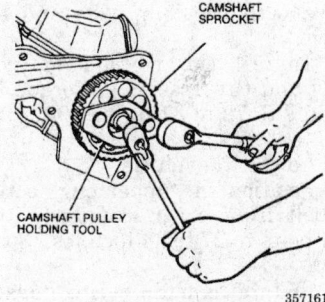

A special spanner tool is designed to hold the sprocket to keep it from turning while the center bolt is being loosened

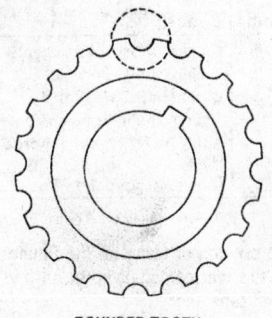

ROUNDED TOOTH

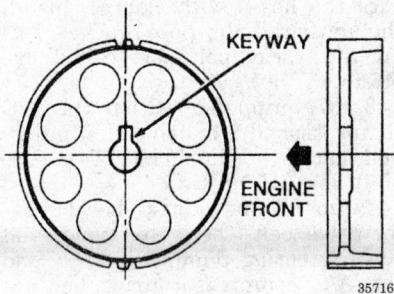

Inspect the camshaft sprockets for damage and wear. Replace as required

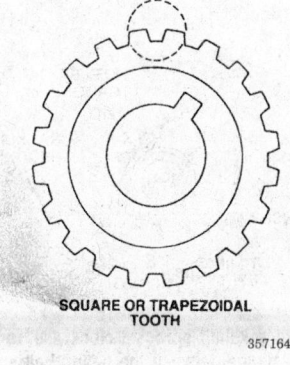

SQUARE OR TRAPEZOIDAL TOOTH

There was a major change in belt design between the 1993 and 1994 Model Year vehicles

sprocket has a square cut in the valley of the gear tooth, it will require a square cut trapezoidal tooth timing belt. If the camshaft sprocket has a rounded cut in the valley of the gear tooth, it will require a rounded tooth timing belt. Use care when procuring replacement parts.

25. If a new timing belt is to be installed, look for a printed arrow on the belt. Make sure the arrow is pointing away from the engine. If the original timing belt is to be reused, make sure that the directional arrow

that was marked at disassembly is facing the correct direction.

26. A new Original Equipment Manufacture (OEM) timing belt should have three white timing marks on it that indicate the correct timing positions of the camshafts and the crankshaft. These marks are to help ensure that the engine is properly timed. When the engine is properly timed, each white timing mark on the timing belt will be aligned with the corresponding camshaft and crankshaft timing mark on the sprocket. Because the white timing marks are not evenly spaced, the technician needs to use care in in-

stalling the belt. There should be 40 timing belt teeth between the timing marks on the front and rear camshaft sprockets and 43 teeth between the timing mark on the front camshaft sprocket and the timing mark on the crankshaft sprocket.

27. Verify that the camshaft timing marks are aligned with the timing marks on the rear cover (seal plate) and that the crankshaft sprocket timing mark is aligned with the timing mark on the oil pump housing.

28. Install the timing belt starting at the crankshaft sprocket and moving around to the camshaft sprockets following a counterclockwise path. Do not allow any slack in the timing belt between the sprockets. After all of the timing marks are matched up with the timing belt installed, slip the timing belt onto the belt tensioner.

29. While holding the timing belt tensioner with an Allen wrench, loosen the tensioner nut. Allow the tensioner to put pressure on the timing belt. Use an Allen wrench to tun the timing belt tensioner 70 to 80 degrees clockwise and tighten the timing belt tensioner nut to 32 to 43 ft. lbs. (43 to 58 Nm).

30. Rotate the crankshaft clockwise twice and align the number one piston to TDC on the compression stroke (firing position).

31. Apply 22 pounds of force on the timing belt between the rear camshaft sprocket and the timing belt tensioner. An assistant may be needed. While holding the timing belt tensioner steady with an Allen wrench, loosen the timing belt tensioner nut. Remove the Allen wrench and adjust the timing belt tensioner using the following procedure:

   a. Install a 0.0138 inch (0.35mm) thick and 0.500 inch (12.7mm) wide feeler gauge where the timing belt just starts to go around the tensioner (approximately the 4 o'clock position, looking at the tensioner.

   b. Turn the crankshaft sprocket clockwise which should force the feeler gauge between the timing belt and the tensioner, up to a position on the tensioner of about 1 o'clock.

   c. Tighten the timing belt tensioner nut to 32 to 43 ft. lbs. (43 to 58 Nm).

   d. Turn the crankshaft clockwise to rotate the feeler gauge out from between the timing belt tensioner and the timing belt.

32. Rotate the crankshaft clockwise twice and once again align the num-

Timing Belt Tooth Spacing

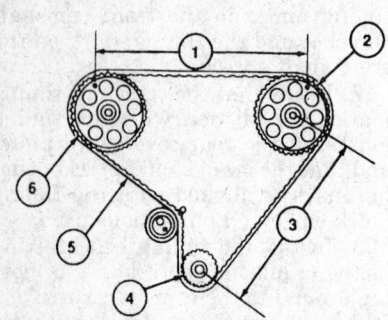

Item	Description
1	40 Timing Chain/Belt Teeth
2	Camshaft Sprocket, Front
3	43 Timing Chain/Belt Teeth
4	Crankshaft Sprocket
5	Timing Chain/Belt
6	Camshaft Sprocket, Rear

357165

Verify 40 belt teeth between the timing marks on the front and rear camshaft sprockets and 43 teeth between the timing mark on the front camshaft sprocket and the mark on the crankshaft sprocket

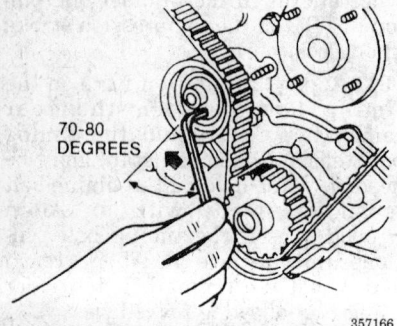

357166

Use an Allen wrench to tun the timing belt tensioner 70 to 80 degrees clockwise

357168

Install a 0.0138 inch thick by 0.50 inch (12.7mm) wide feeler gauge where the timing belt just starts to go around the tensioner, approximately the 4 o'clock position, looking at the tensioner

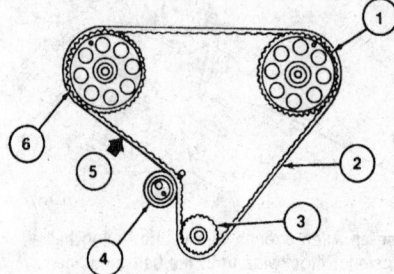

Item	Description
1	Camshaft Sprocket, Front
2	Timing Chain/Belt
3	Crankshaft Sprocket
4	Timing Chain/Belt Tensioner
5	Apply 98 N (22 lbs) of Force Here
6	Camshaft Sprocket, Rear

357167

Apply 22 pounds of force on the timing belt between the rear camshaft sprocket and the timing belt tensioner

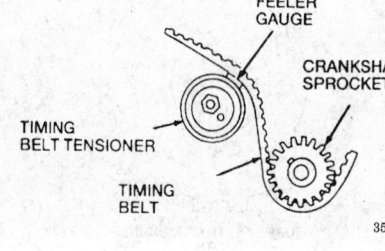

357169

Turn the crankshaft sprocket clockwise, forcing the feeler gauge between the timing belt and the tensioner, up to a position on the tensioner of about 1 o'clock

ber one piston to TDC on the compression stroke (firing position).

33. Apply 22 pounds of force on the timing belt between the front and rear camshaft sprockets. Measure the amount of belt deflection. Belt deflection should be between 0.51 to 0.59 inch (13 to 15mm). If belt deflection is out of specification, repeat Steps 29 through 33. If the timing belt deflection cannot be bought into specification, the timing belt will have to be replaced.

34. Position the lower engine front cover and install the five lower cover bolts. Do not over-tighten. Torque to 27 to 44 inch lbs. (3 to 5 Nm).

35. Install the outer timing belt guide next to the crankshaft sprocket with the dished side facing away from the cylinder block. Install the crankshaft pulley. Use a strap wrench to keep the crankshaft pulley from turning and torque the center bolt to 90 to 98 ft. lbs. (123 to 132 Nm).

36. Position the water pump pulley on the pump. Install the four bolts. Use a strap wrench to keep the water pump pulley from turning and torque the four water pump pulley bolts to 12 to 15 ft. lbs. (16 to 21 Nm).

37. Position the right side outer engine and transaxle splash shield and secure with the four bolts and two screws.

38. Install the right side front wheel and tire assembly. Torque the lug nuts to 72 to 87 ft. lbs. (98 to 118 Nm).

39. Lower the vehicle.

40. Position the upper engine timing belt front cover and torque the eight bolts to 27 to 44 inch lbs. (3 to 5 Nm).

41. Install the main wiring harness on the upper engine front cover.

42. Position the water bypass hose between the thermostat housing and water connection. Install the upper radiator hose between the radiator and the water hose connection. Secure the hoses with clamps. Install the upper radiator hose bracket. Torque the bracket bolt to 34 to 58 ft. lbs. (46 to 65 Nm).

43. If equipped, position the A/C compressor drive belt idler pulley and install the three bolts. Torque to 15 ft. lbs. (21 Nm).

44. Install and adjust the alternator drive belt, the water pump and power steering pump drive belt and the A/C compressor drive belt (of equipped).

45. Connect the battery cable.

46. Fill the cooling system.

47. Start the engine and allow to warm to operating temperature. Check and adjust the ignition timing. Road test to verify correct engine operation.

Camshaft

REMOVAL AND INSTALLATION

1. Disconnect the negative battery cable.

2. Drain the coolant from the cooling system and the engine. Coolant from the engine can be drained by removing the drain plug on the cylinder block.

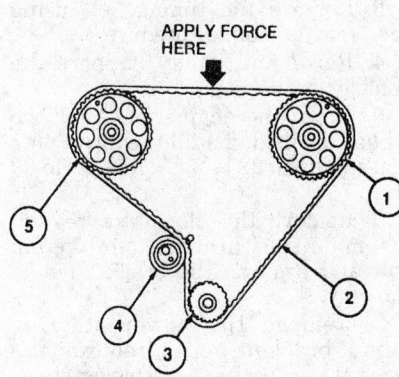

APPLY FORCE HERE

Item	Description
1	Camshaft Sprocket, LH
2	Timing Chain / Belt
3	Crankshaft Sprocket
4	Timing Chain / Belt Tensioner
5	Camshaft Sprocket, RH

357170

Apply 22 pounds of force on the belt between the front and rear camshaft sprockets. Measure the belt deflection

3. Remove the timing belt assembly.

4. Remove the collector assembly and intake manifold.

5. Remove the cylinder head from the engine.

6. With cylinder head mounted on a suitable workbench, remove the rocker shafts with rocker arms. Bolts should be loosened in 2–3 steps.

7. Remove hydraulic valve lifters and lifter guide.

8. Hold hydraulic valve lifters with wire so they will not drop from lifter guide.

9. Remove the camshaft front oil seal and slide camshaft out the front of the cylinder head assembly.

To install:

10. Install camshaft, locate plate, cylinder head rear cover and front oil seal. Set camshaft knock pin at 12:00 o'clock position. Install cylinder head with new gasket to engine.

11. Install valve lifter guide assembly. Assemble valve lifters in their original position. After installing them in the correct location remove the wire holding them in lifter guide.

12. Install rocker shafts in correct position with rocker arms. Tighten bolts in 2–3 stages to 13–16 ft. lbs. Before tightening, be sure to set camshaft lobe at the position where

lobe is not lifted or the valve closed. You can set each cylinder one at a time or follow the procedure below (timing belt must be installed in the correct position):

a. Set No. 1 piston at TDC on its compression stroke and tighten rocker shaft bolts for No. 2, No. 4 and No. 6 cylinders.

b. Set No. 4 piston at TDC on its compression stroke and tighten rocker shaft bolts for No. 1, No. 3 and No. 5 cylinders.

c. Torque specification for the rocker shaft retaining bolts is 13–16 ft. lbs.

13. Install the intake manifold and collector assembly.

14. Install the timing belt cover and camshaft sprocket. The left and right camshaft sprockets are different parts. Install the correct sprocket in the correct position.

15. Install the timing belt assembly, fill coolant and set engine timing to specifications.

Piston and Connecting Rod

POSITIONING

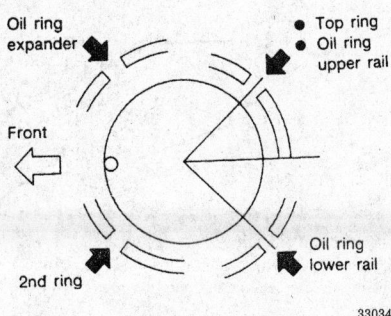

Oil ring expander
Top ring
Oil ring upper rail
Front
Oil ring lower rail
2nd ring

330342

Piston ring positioning

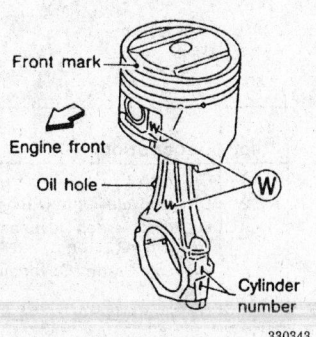

Front mark
Engine front
Oil hole
W
Cylinder number

330343

Piston and connecting rod — front marks

ENGINE LUBRICATION

Oil Pan

REMOVAL AND INSTALLATION

1. Disconnect the negative battery cable.

2. Raise and safely support the vehicle.

3. Drain the engine oil. When all the oil has been drained, install the drain plug and torque to 22–25 ft. lbs. (29–39 Nm).

4. The front and rear engine mounts must be disconnected. This means the engine must be safely supported. Place an underbody type screw-jack or equivalent under the crankshaft pulley. Cover the end of the jack with a pad made of shop rags to prevent damage to the crankshaft pulley. Take up just enough of the engine's weight to allow the engine mounts to be disconnected.

5. Remove the front engine mount (support) insulator through bolt. Remove the rear engine mount (support) through bolt.

6. Remove the two rear refrigerant/heater pipe hold down bracket bolts.

7. Remove the four crossmember (also called transverse member) bolts and remove the crossmember.

8. Remove the exhaust inlet pipe.

9. Remove the four rear transaxle-to-engine brace bolts and the five front transaxle-to-engine brace bolts.

10. Remove the front transaxle-to-engine brace.

11. Disconnect the low oil level sensor electrical connector.

12. Remove the 18 oil pan bolts in the reverse order of the tightening sequence, working from the outside, towards the center bolts.

13. Remove the pan and discard the seals.

To install:

14. Clean all parts well. Make sure that all old sealing material is removed from the oil pan and engine mating surfaces.

Apply Loctite® Ultra Gray 599 Silicone Sealer or equivalent to the seal ends and to the oil pan gasket rail

355955

1. Install new oil pan seals. Apply Loctite® Ultra Gray 599 Silicone Sealer or equivalent to the ends of the oil pan seals.

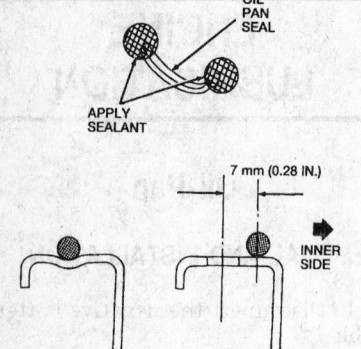

2. Apply a bead of Loctite® Ultra Gray 599 Silicone Sealer or equivalent to the oil pan gasket rail inboard of the bolt holes.

3. Install the oil pan on the engine block. Tighten the 18 oil pan bolts in sequence, working from the inside, towards the outer bolts. Do not over-tighten. Torque to 62 to 70 inch lbs. (7 to 8 Nm).

4. Connect the low oil level sensor electrical connector.

5. Install the front and rear tran-saxle braces. Torque all bolts to 22 to 30 ft. lbs. (30 to 40 Nm).

6. Install the exhaust inlet pipe.

7. Install the crossmember (also called transverse member) and tor-que the bolts 58 to 65 ft. lbs. (78 to 88 Nm).

8. Install both engine support through bolts and torque to 58 to 65 ft. lbs. (78 to 88 Nm).

9. Remove the support jack from under the crankshaft pulley.

10. Lower the vehicle.

11. Fill the engine with the speci-fied engine oil to the required level.

12. Connect the negative battery cable. Start the engine and check for leaks.

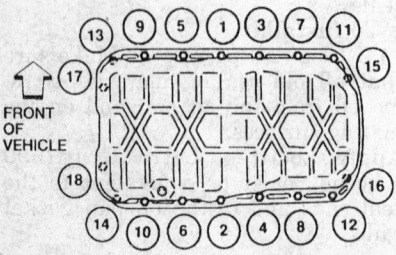

Tighten the 18 oil pan bolts in sequence, working from the inside, towards the outer bolts

Oil Pump

REMOVAL AND INSTALLATION

The engine lubrication system on the 3.0L engine uses a positive-displace-ment, crankshaft driven oil pump that draws oil from the oil pan, through the oil pump screen cover and tube. The oil pump then sends the oil to the oil filter where it is di-rected to the engine block and the cylinder heads. The oil then drains back into the pan. The oil filter adapter is mounted on the oil pump. A full-flow oil filter is mounted to the oil filter adapter. The oil filter adapter uses an oil pressure relief valve that allows engine oil pressure to bypass a clogged or blocked filter and continue to lubricate the engine main oil gallery. This condition is not allowable for long periods of time be-cause the engine oil is not being filtered. Under the recommended normal engine operating conditions, all of the engine oil will pass through the oil filter.

1. Disconnect the negative battery cable.

2. Remove the alternator belt, water pump and power steering belt and the A/C compressor belt, if equipped, using the recommended procedure.

3. Remove the timing belt using the recommended procedure.

4. Raise and safely support the vehicle.

5. Drain the engine oil. After the oil has drained, install the drain plug and torque to 22 to 29 ft. lbs. (29 to 39 Nm).

6. Remove the alternator regula-tor mounting bracket-to-oil pump bolt and position the bracket out of the way.

7. Remove the power steering pump bracket bolts, remove the bracket and position the power steer-ing pump aside.

8. Remove the crankshaft pulley.

9. Remove the outer timing belt guide, the crankshaft drive sprocket and the inner timing belt guide.

10. Disconnect the oil pressure sen-sor electrical connector.

11. Remove the oil filter adapter bolts and remove the oil fil-ter/adapter assembly. Discard the two oil filter adapter O-rings.

12. Remove the oil pan using the recommended procedure.

13. Remove the two oil pump screen and pickup tube assembly bolts and remove the tube assembly. Discard the oil pump pickup tube O-rings.

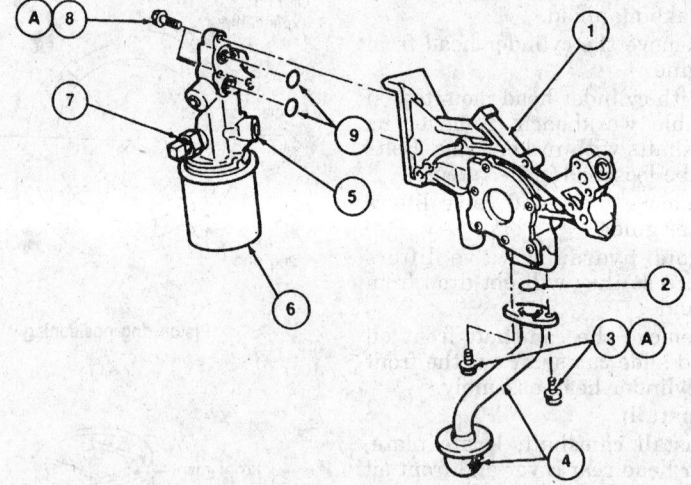

Item	Description
1	Oil Pump
2	Oil Pump Inlet Tube O-Ring
3	Oil Pump Screen Cover and Tube Bolts (2 Req'd)
4	Oil Pump Screen Cover and Tube
5	Oil Filter Adapter

Item	Description
6	Oil Filter
7	Oil Pressure Sensor
8	Oil Filter Adapter Bolt (3 Req'd)
9	Oil Filter Adapter O-Rings (2 Req'd)
A	Tighten to 16-21 N·m (12-15 Lb-Ft)

Engine oil pump, screen cover, pickup tube and oil filter adapter arrangement

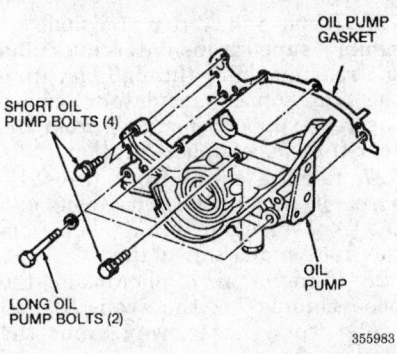

OIL PUMP GASKET

SHORT OIL PUMP BOLTS (4)

OIL PUMP

LONG OIL PUMP BOLTS (2)

355983

Remove the oil filter adapter bolts and remove the oil filter/adapter assembly

14. Remove the six oil pump bolts and remove the pump assembly from the engine. Discard the oil pump gasket.

To install:

15. Clean all parts well. If the pump is to be disassembled for checking, wash all parts in solvent dry thoroughly. Use a brush to clean the inside of the oil pump body and pressure relief chamber. Make sure all dirt and metal particles are removed.

16. Use a feeler gauge to inspect tooth tip clearance. Maximum clearance is 0.007 inch (0.18 mm). If measurement is above the maximum clearance specification, replace the oil pump.

17. Use a feeler gauge to inspect outer rotor-to-oil pump body clearance. Outer rotor-to-oil pump body clearance should be 0.004 to 0.008 inch (0.114 to 0.200 mm). If measurement is above the maximum clearance specification, replace the oil pump.

18. Use a straightedge and a feeler gauge to inspect the oil pump inner rotor side clearance. The inner rotor side clearance should be 0.0020 to 0.0035 inch (0.05 to 0.09 mm). If measurement is above the maximum

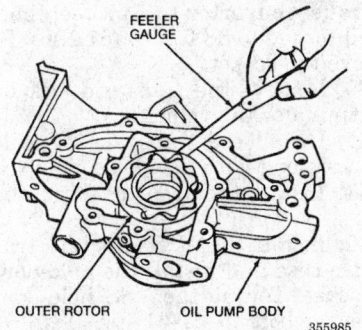

FEELER GAUGE

OUTER ROTOR

OIL PUMP BODY

355985

Use a feeler gauge to inspect outer rotor-to-oil pump body clearance

clearance specification, replace the oil pump.

19. Use a straightedge and a feeler gauge to inspect the oil pump outer rotor side clearance. The outer rotor side clearance should be 0.0020 to 0.0043 inch (0.05 to 0.11 mm). If measurement is above the maximum clearance specification, replace the oil pump.

20. With a micrometer, measure the inner rotor ridge-to-oil pump body clearance. Measure the ridge, then the pump body. Subtract the inner rotor ridge measurement from the oil pump body measurement to get the clearance. The clearance should be 0.0018 to 0.0036 inch (0.045 to 0.091 mm). If measurement is above the maximum clearance specification, replace the oil pump.

21. Measure the pressure regulator valve-to-valve housing diameter. Measure the regulator valve diameter. Subtract to get the valve-to-housing clearance. The clearance should be 0.0016 to 0.0036 inch (0.040 to 0.097 mm). If measurement is above the maximum clearance specification, replace the oil pump.

22. Scrape any dirt or metal particles from the inside of the oil pan.

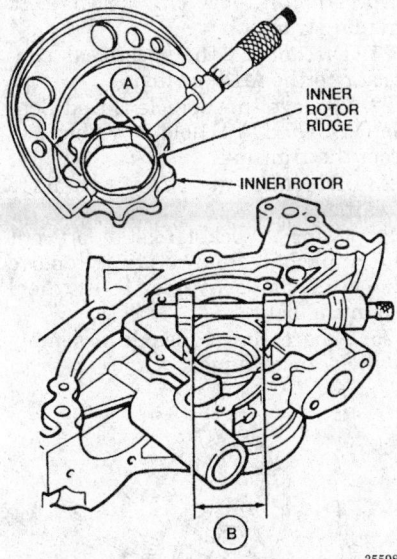

INNER ROTOR RIDGE

INNER ROTOR

355988

Measure the rotor ridge (A) and the pump body (B). Subtract A from B to get inner rotor ridge-to-pump clearance

Scrape all old gasket material from the gasket surface. Wash the oil pan in solvent and dry thoroughly. Inspect the pan for cracks, holes or damaged oil pan and drain plug threads. Check the gasket surface for damage caused by over-tightened bolts. Replace with a new oil pan if repairs cannot be made.

23. With all pump and engine block sealing surfaces clean, install the oil pump using a new gasket. Torque the long bolts to 108 to 144 inch lbs. (12 to 16 Nm) and the short bolts to 38 to 45 inch lbs. (6 to 7 Nm).

24. Install a new O-ring on the oil pump pickup tube and install the tube/screen assembly. Torque the two bolts to 144 to 180 inch lbs. (16 to 21 Nm).

25. Install the oil pan using the recommended procedure.

26. Install two new O-rings to the oil filter adapter and fit to the oil pump. Tighten the three oil filter adapter bolts to 144 to 180 inch lbs. (16 to 21 Nm).

27. Connect the oil pressure sensor electrical connector.

28. Install the inner timing belt guide, the crankshaft drive sprocket and the outer timing belt guide.

29. Install the crankshaft pulley. Use a strap wrench to hold the pulley and torque the center bolt to 90 to 98 ft. lbs. (123 to 132 Nm).

30. Position the power steering pump and bracket on the engine. Torque the bolts to 22 to 27 ft. lbs. (30 to 36 Nm).

31. Position the alternator regulator bracket on the oil pump, install the bolt and torque to 15 ft. lbs. (21 Nm).

32. Lower the vehicle.

33. Install the timing belt using the recommended procedure. Great care must be exercised to properly align all timing marks.

34. Install and adjust all drive belts using the recommended adjustment procedure.

35. Fill the engine with the specified engine oil to the required level.

36. Connect the battery ground cable. Start the engine and check for leaks. Oil pressure can be verified by installing a mechanical oil pressure gauge in the pressure sensor outlet, just above the oil filter rim. Oil pressure with the engine running at no load should be approximately 17 psi at idle and 57 to 70 psi at 3200 rpm. Install the oil pressure sensor with sealant.

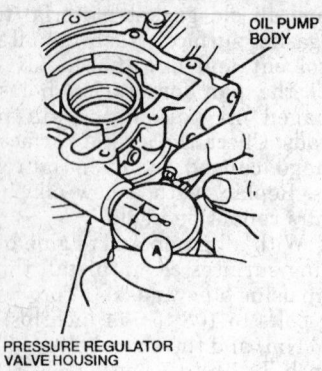

OIL PUMP BODY

PRESSURE REGULATOR VALVE HOUSING

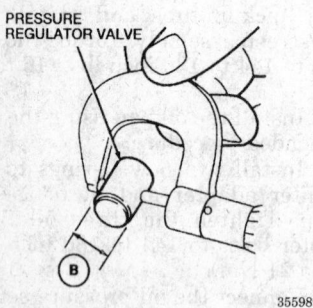

PRESSURE REGULATOR VALVE

355989

Measure the pressure regulator valve-to-valve housing diameter. Measure the regulator valve diameter. Subtract to get the valve-to-housing clearance

TRANSAXLE

Automatic Transaxle Assembly

REMOVAL AND INSTALLATION

1. Disconnect the negative battery cable.

2. Remove the starter motor from the vehicle using the recommended procedure.

3. Disconnect the Park/Neutral Position (PNP, also called the Transmission Range or TR switch) switch electrical connector from the transaxle and disconnect the 2 electrical connectors secured on the electrical connector retaining brackets.

4. Remove the 2 retaining brackets for the electrical connectors.

5. Remove the upper and lower mounting bolts for the oil level indicator and remove the oil level indicator.

6. Remove the upper transaxle-to-engine mounting bolts.

7. Raise and safely support the vehicle.

8. Remove the left side inner engine and transaxle splash shield.

9. Remove the right and left front wheel halfshaft assemblies using the recommended procedure.

10. Place an oil pan under the transaxle and drain the transaxle fluid from the unit.

11. Remove the catalytic converter inlet pipe-to-exhaust bracket retaining bolt.

12. Place a suitable jack or lift under the transaxle assembly.

13. Remove the 2 exhaust bracket-to-transaxle assembly nuts. Remove the exhaust bracket.

14. Remove the 3 rear transaxle-to-engine brace mounting bolts. Remove the rear transaxle-to-engine brace.

15. Remove the 4 front transaxle-to-engine brace mounting bolts. Remove the front transaxle-to-engine brace.

16. Remove the transaxle-to-engine mounting bolt. Remove the 2 exhaust system bracket-to-transaxle studs and torque converter inspection plate.

17. Remove the shift cable nut and shift cable from the Manual Lever Position (MLP) switch.

18. Using a screwdriver release the shift cable locking pin from the shift cable bracket and remove the shift cable from the shift cable bracket.

19. Disconnect the wiring harness electrical connector to the valve body.

20. Disconnect the transaxle ground strap.

21. Disconnect the electrical connector to the MLP switch.

22. Remove the Vehicle Speed Sensor (VSS) and VSS hold-down bracket from the transaxle.

23. Disconnect the transaxle breather tube.

24. Using a socket tool to prevent the crankshaft from rotating, remove the 4 torque converter-to-flywheel mounting bolts.

25. Secure the transaxle assembly to the transaxle jack.

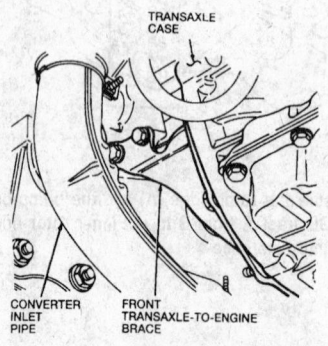

TRANSAXLE CASE

CONVERTER INLET PIPE

FRONT TRANSAXLE-TO-ENGINE BRACE

257028

Removal of the front transaxle-to-engine brace

26. Remove the 3 rear transaxle assembly support insulator nuts and the rear insulator through bolt from the rear support insulator. Remove the rear support insulator from the rear transaxle support bracket.

27. Remove the 3 front transaxle assembly support insulator bolts and the front insulator through bolt from the front support insulator.

28. Using a pair of pliers, slide the hose clamps for the transaxle oil cooler tube hose away from the transaxle.

29. Disconnect and plug the transaxle oil cooler tubes from the transaxle assembly.

30. Carefully separate the transaxle assembly from the engine assembly. Lower the assembly from the vehicle.

To install:

31. Be sure that the transaxle is secured firmly to the transaxle jack.

32. Carefully raise the transaxle into the vehicle and align the transaxle to the engine assembly, making sure that the alignment dowels are positioned properly.

33. Install the lower transaxle-to-engine bolt and torque to 22–30 ft. lbs. (30–40 Nm).

34. Install the rear transaxle support insulator and torque the 3 insulator mounting nuts to 32–41 ft. lbs. (43–55 Nm).

35. Install and torque the rear transaxle support insulator through bolt to 32–41 ft. lbs. (43–55 Nm).

36. Install the front transaxle support insulator and torque the 3 insulator mounting nuts to 30–38 ft. lbs. (41–52 Nm).

37. Install and torque the front transaxle support insulator through bolt to 47–54 ft. lbs. (64–74 Nm).

38. Using a socket on the crankshaft pulley bolt to prevent the crankshaft from turning, install the 4 torque converter-to-flywheel bolts and torque to 38 ft. lbs. (51 Nm). Remove the socket.

39. Remove the transaxle jack out from under the vehicle.

40. Install the transaxle oil cooler tubes along with new hose clamps.

41. Install the transaxle breather tube.

42. Install the VSS into the transaxle case and install the hold-down bracket. Torque the VSS hold-down bracket bolt to 44–61 inch lbs. (5–7 Nm).

43. Install the shift cable into the shift cable bracket and snap the cable locking pin into position. Align the cable and bracket, then tighten the cable nut securely.

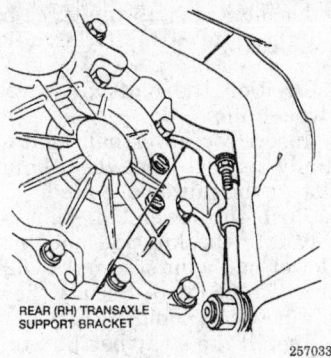

REAR (RH) TRANSAXLE
SUPPORT BRACKET

257033

Removal of the rear transaxle support bracket

44. Reconnect the transaxle ground strap.

45. Install the torque converter inspection plate and exhaust bracket-to-transaxle studs.

46. Install the front and rear transaxle-to-engine braces. Torque the brace bolts to 22–30 ft. lbs. (30–40 Nm).

47. Install the exhaust bracket on to the transaxle and torque to 27 ft. lbs. (35 Nm).

48. Install and torque the catalytic converter inlet pipe-to-exhaust bracket bolt to 32 ft. lbs. (43 Nm).

49. Reconnect the electrical connectors to the MLP switch and the valve body wiring harness.

50. Install the right and left halfshaft assemblies.

51. Install the inner engine and transaxle splash shield.

52. Lower the vehicle.

53. Install the upper transaxle to engine bolts.

54. Torque the bolts to 33 ft. lbs. (44 Nm).

55. Install the transaxle oil level indicator and tighten the upper and lower retaining bolts.

56. Install the 2 electrical connector retaining brackets and tighten the retaining bolts. Reconnect the electrical connectors on the retaining brackets.

57. Reconnect the electrical connector to the PNP switch.

58. Install the starter motor. Reconnect the starter motor wiring.

59. Reconnect the negative battery cable. Refill the transaxle with the correct amount and type of clean transaxle fluid.

60. Start the engine.

61. Check for leaks and proper operation.

DRIVE AXLE

Halfshaft

REMOVAL AND INSTALLATION

The front wheel driveshaft and joints, also known as halfshafts, are the mechanical links that transfer engine power from the transaxle and differential to the front wheels. At the transaxle end, the halfshaft joints are splined to the differential side gears. Disengagement of the left side halfshaft from the differential side gears is prevented by an expanding spring steel circlip. During installation, the circlip compresses around the shaft as it enters the gear. Once through the differential side gear, the circlip expands into a counterbore machined in the back of the differential side gear. The right side halfshaft and joint is secured by a front axle bearing.

The wheel ends of the halfshafts and joints are splined to the wheel hubs which are supported on one-piece wheel bearings. Disengagement of the halfshaft from the wheel hub is prevented by the front axle hub retainers, nut and washers, secured with a cotter pin.

Backlash between the wheel hub and halfshaft is eliminated by the splines. The wheel hub splines are machined straight while the halfshaft joint splines are machined with a slight helical cut. The difference in splines provides a tight backlash-free coupling without the removal and installation problems associated with an interference fit.

Halfshafts and their Constant Velocity Joints (CV-Joints) are precision made and should be handled with care. Do not angle a halfshaft CV-Joint more than about 20 degrees or the joint can be damaged. Never allow a halfshaft to hang if it is disconnected from the transaxle or the hub. Wire the halfshaft to a convenient underbody component. Never strike a halfshaft or CV-Joint component with a metal hammer. Do not drop an assembled halfshaft since internal components may be damaged or the protective boot can be cut allowing the special grease to run out of the joint and dirt and water into the joint, shortening the joint's service life.

1. Raise and safely support the vehicle.

2. Remove the wheel and tire assembly.

3. If necessary, remove the six right side inner fender splash shield screws and remove the right side inner splash shield. Remove the seven right side or seven left side engine and transmission splash shield screws. Remove the splash shields.

4. Remove and discard the cotter pin.

5. Remove the nut retainer, the nut and the hub retainer washers from the front hub assembly.

6. Remove and discard the cotter pin from the lower ball joint nut. Loosen the ball joint nut until it contacts the halfshaft joint. Strike the front wheel knuckle with a soft-faced hammer while pulling down on the lower control arm until the lower ball joint separates from the knuckle. There should now be enough clearance to remove the ball joint nut.

7. Disconnect the sway bar (also called the stabilizer bar) from the lower control arm at the sway bar link nut.

8. Carefully pry down on the lower control arm to separate the ball joint stud completely from the knuckle.

9. Use a prybar to separate the sway bar link from the lower control arm.

10. Separate the halfshaft and CV-Joint from the wheel hub.

11. Position a drain pan under the transaxle since some fluid may run out when the inner joint is disengaged from the transaxle.

12. A pry bar is used to separate the inner CV-Joint from the transaxle. Use great care that the prybar does not damage the transaxle case, differential oil seal, outer race or boot. If removing the left side halfshaft, position prybars on both sides of the outer race, between the outer race and the transaxle case. Gently pry outward to unseat the circlip.

13. When removing the right side halfshaft, it is not be necessary to remove the halfshaft bearing retainer bracket from the cylinder block. Remove the three bearing retainer bolts and pull the right side halfshaft CV-Joint with the bearing retainer from the differential side gear.

14. Support the halfshafts and remove them from the vehicle. Use care not to damage the boots. Place the halfshafts on a flat, protected work area.

To install:

— **CAUTION** —
Do not reuse the circlip used on the left side halfshaft.

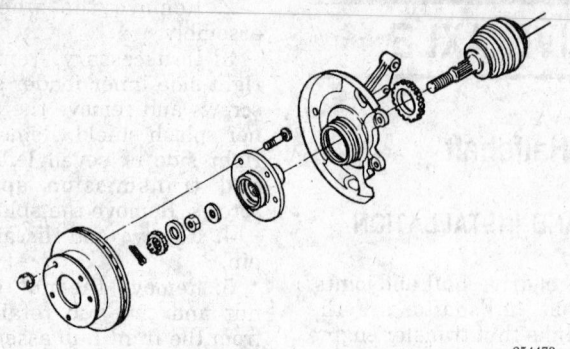

254473

Front halfshaft, hub/bearing assembly, disc brake rotor and ABS sensor arrangement

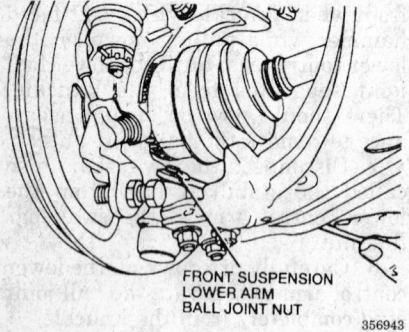

FRONT SUSPENSION LOWER ARM BALL JOINT NUT

356943

Loosen the ball joint nut until it contacts the CV-Joint

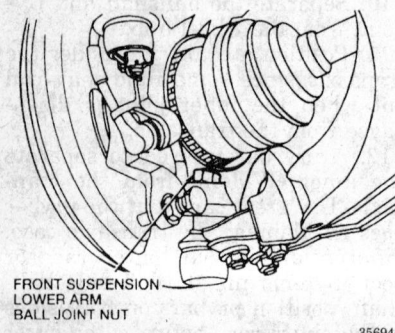

FRONT SUSPENSION LOWER ARM BALL JOINT NUT

356945

With the ball joint loosened from the knuckle, remove the ball joint stud nut

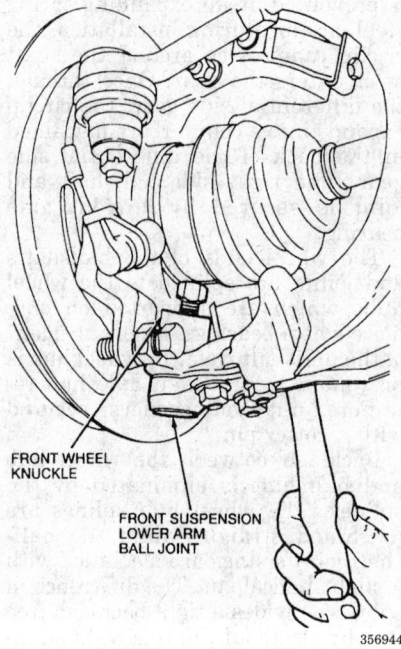

FRONT WHEEL KNUCKLE

FRONT SUSPENSION LOWER ARM BALL JOINT

356944

Strike the knuckle with a soft-faced hammer while pulling down on the lower control arm

15. To prevent over-expanding the circlip, install the circlip carefully, starting one end in the shaft groove and then working the circlip over the CV-Joint splined end. Always use a new circlip. No circlip is used on the right side halfshaft.

16. Inspect the CV-Joint boots. If service is required, replace the CV-Joint boots using the recommended procedure.

17. Inspect the differential oil seals. If damaged, the factory recommends using a hook-type puller and slide hammer arrangement to remove the seals. A seal driver is used to install the replacement differential oil seals.

18. If installing the left side halfshaft and CV-Joint assembly, position the CV-Joint so the splines are aligned with the differential side gear splines, then push the halfshaft joint into the differential case. As the circlip locks into the differential side gear groove, a click will be felt.

19. If installing the right side halfshaft and CV-Joint assembly, simply push the CV-Joint into the differential side gear. Position the bearing retainer onto the bearing retainer bracket which should still be on the

cylinder block. Install the three bolts and tighten to 8–14 ft. lbs. (13–19 Nm).

20. Position the halfshaft through the wheel hub.

21. Insert the lower ball joint stud partially through the wheel knuckle and start the nut on the stud.

22. Push the lower ball joint completely into the knuckle and torque the lower ball joint stud nut to 52–63 ft. lbs. (71–86 Nm). Secure the nut with a new cotter pin.

23. Install the sway bar link to the lower control arm and torque the link nut to 12–16 ft. lbs. (16–22 Nm).

24. Install the front wheel outer bearing retainer, washer and axle nut. Torque the hub nut to 174–231 ft. lbs. (235–314 Nm). Install the nut retainer and secure with a new cotter pin.

25. If removed, install the right side inner fender splash shield and tighten the screws securely. Position the engine and transmission splash shields and tighten the screws securely.

26. Install the wheel and tire assembly. Torque the lug nuts to 72–87 ft. lbs. (98–118 Nm).

27. Lower the vehicle.

28. Check the transaxle fluid level.

29. Road test the vehicle to verify correct operation and no noise or vibration.

CV-Joint Boot

REPLACEMENT

Regularly inspect the front wheel driveshafts (also called halfshafts). Inspect the CV-Joint boots for cuts, undercoating and loose CV-Joint boot clamps. With a damaged CV-Joint boot, the special lubricant will be thrown out, water and dirt will get in and the joint will wear prematurely. Look for any unusual shiny spots on any suspension components. A shiny spot indicates a component is bent, worn or broken and is causing mechanical interference. Protect the CV-Joint boots during undercoating and rustproofing procedures. Compounds that stick to the boot plates will cause rapid deterioration of the boots.

The halfshafts require some disassembly to changeout the CV-Joint boots. Use the following procedure.

1. Remove the front wheel driveshaft(s), also called halfshaft(s) using the recommended procedure.

2. Place the removed halfshaft n a flat, protected work area for inspec-

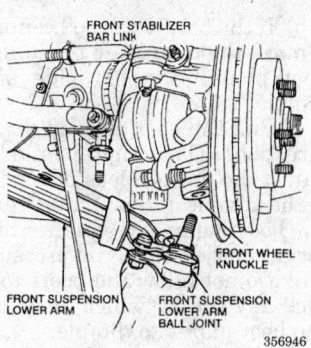

Use a prybar to separate the stabilizer bar link from the lower control arm

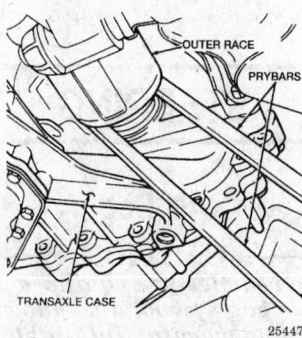

Removing the left side halfshaft by gently prying with two prybars to unseat the circlip

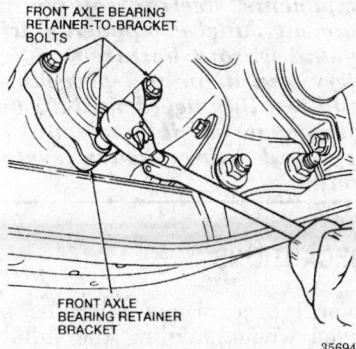

Right side halfshaft bearing retainer bracket

tion. Use care when working with the halfshafts. Support them and do not angle the joint beyond its capacity. Twenty degrees is all the joint should be tilted. Beyond that, the joint can be damaged. Do not drop the halfshaft and joint assembly. Use care not to damage any machined surfaces during disassembly and assembly. Cleanliness is very important when servicing any front wheel drive CV-Joint. A reassembled halfshaft may

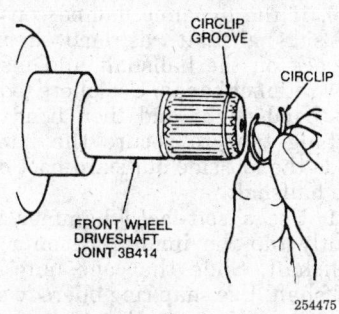

Always use a new circlip. Start one end in the groove then work the circlip over the splined shaft

be damaged if it is over-plunged outward from the outer race.

3. To begin boot removal, place the halfshaft and CV-Joint assembly in a vise equipped with protective jaw caps to prevent damage to any machined parts.

4. Use boot clamp pliers or similar to remove both clamps. Discard the clamps.

5. If the CV-Joint boots are being removed because they are torn or damaged, inspect the grease for contamination by rubbing the grease between the fingers.

a. If there is a gritty feeling, disassemble, clean and inspect the

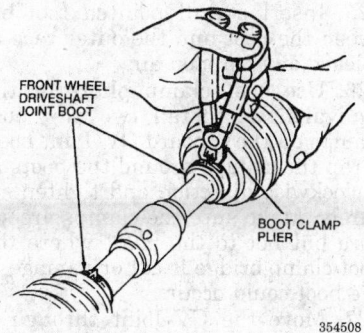

Remove and discard the small and large boot clamps

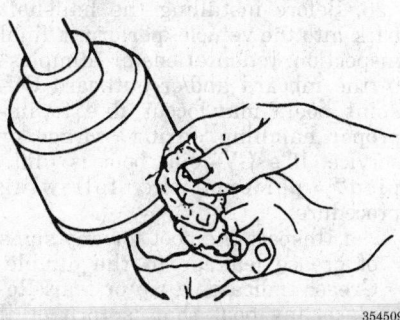

If the boot was damaged, check the grease for grit, disassembly and joint cleaning may be required

CV-Joint assembly. Service the joint as necessary.

b. If the grease is not contaminated and the CV-Joint has been operating satisfactorily, replace only the joint boot and add the required amount of specified lubricant.

6. Slide the inboard CV-Joint boot back and clean out the grease from the inboard joint and boot.

7. Slide off the outer race from the end of the halfshaft.

8. Remove the six ball bearings from the cage and clean them in a suitable solvent.

9. Use snap ring pliers to remove the small snap ring from the end of the halfshaft.

10. Use a 2-Jaw Puller to remove the inner race. Slide off the cage.

11. If the CV-Joint boot is to be reused, wrap the halfshaft splines with tape before removing the boot. The splines are sharp and can damage the boot. Slide the boot off the shaft and set aside.

12. If it is necessary to replace the outboard CV-Joint boot on the right side halfshaft, slide the boot from the halfshaft.

13. If it is necessary to replace the outboard CV-Joint boot on the left side halfshaft, remove the dynamic damper using the following procedure.

a. Use a small prybar to pry up the damper retaining bands.

b. Use pliers to remove the damper retaining bands.

c. Pull the dynamic damper from the halfshaft.

14. Slide the outboard CV-Joint boot from the halfshaft.

To install:

15. Use care when assembling the halfshafts and CV-Joint boots. The outboard and inboard CV-Joint boots are different. Failure to correctly install the proper boot on the proper CV-Joint could lead to premature boot and/or CV-Joint wear.

16. Wrap the halfshaft splines with tape before installing the outboard CV-Joint boot. Slide the outboard boot onto the halfshaft.

17. Fill the outboard joint boot with approximately 6 to 7 ounces of a quality Constant Velocity Joint Grease such as Ford part number E2FZ-19590-B grease which should be readily available. New replacement boot kits usually come with containers of suitable grease. Position the boot on the outboard CV-Joint. Make sure the boot is fully seated in the halfshaft grooves and the outer race.

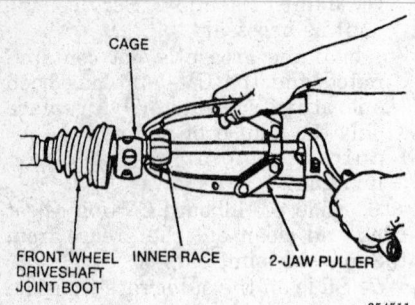

A puller is required to remove the joint inner race

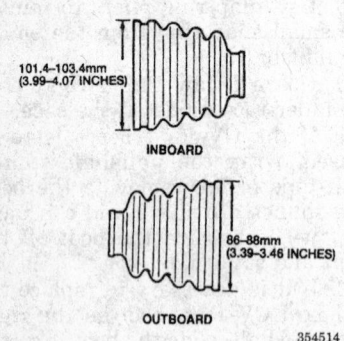

The outboad and inboard boots are different. Use care to correctly identify the proper boot

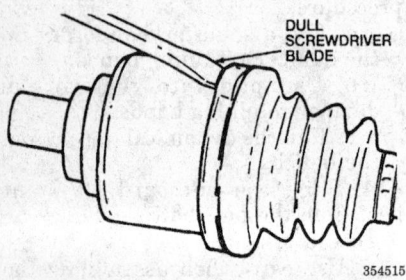

Air can be trapped inside the boot and should be released by venting with a tool as shown

18. Insert a dull pointed tool between the boot and the outer race to release any trapped air.

19. Use boot clamp pliers or the equivalent to install two new boot clamps on the outboard CV-Joint boot. Wrap the clamps around the boots in a clockwise direction and tighten securely. Make sure the clamps are secure but not to the point where the boot clamp bridge is cut or damage to the boot could occur.

20. If the dynamic damper on the left side halfshaft was removed, slide it back on the halfshaft and install new retainer bands. Use pliers to pull the bands tight and then bend the locking tabs to secure the bands. Slide the left side outboard boot onto the halfshaft.

21. Use a soft-faced hammer and gently tap the inner race onto the halfshaft. Slide the cage onto the halfshaft. Use snap ring pliers to seat the snap ring on the end of the halfshaft.

22. Fill the inboard CV-Joint boot with approximately 7½ to 8¼ ounces of a quality High-Temp Constant Velocity Joint Grease such as Ford part number E43Z-19590-A grease which should be readily available. Note that this specification grease is different from that specified for the outboard joint. Inboard joints require a high temperature grease. New replacement boot kits usually come with containers of suitable grease.

23. Install the six ball bearings in the cage. Lubricate with a small amount of the CV-Joint grease to keep them in place in the cage. Slide the outer race onto the end of the halfshaft, over the ball bearings.

24. Position the inboard CV-Joint boot onto the joint. Make sure the boot is fully seated in the halfshaft grooves and the outer race.

25. Insert a dull pointed tool between the boot and the outer race to release any trapped air.

26. Use boot clamp pliers or the equivalent to install two new boot clamps on the inboard CV-Joint boot. Wrap the clamps around the boots in a clockwise direction and tighten securely. Make sure the clamps are secure but not to the point where the boot clamp bridge is cut or damage to the boot could occur.

27. Move the CV-Joint through it full range of travel at various angles. The CV-Joint should flex, extend and compress smoothly.

28. Before installing the halfshaft back into the vehicle, perform a final inspection. Indentations or "dimples" in the inboard and/or outboard CV-Joint boots may occur due to improper handling during storage or service. If a CV-Joint boot is "dimpled", perform the following procedure.

 a. Inspect the boot for any signs of grease leakage in the dimple. Grease indicates a rip or tear. Replace any boot that is torn or if there is other damage.

 b. If the boot is in good condition, remove the dimple by grasping the boot pleat on either side of the dimple.

 c. Pull the boot pleat in opposite directions. The dimple should "pop" out. If the dimple does not "pop" or if the dimple forms again, remove one boot clamp and equalize the internal and external air pressure.

 d. Do not allow the boots to contact any object which may cause the boot plates to dimple.

29. When satisfied with the boot installation and their condition, install the halfshafts using the recommended procedure.

STEERING

Air Bag

CAUTION
Some vehicles are equipped with an air bag system, also known as the Supplemental Inflatable Restraint (SIR) or Supplemental Restraint System (SRS). The system must be disabled before performing service on or around system components, steering column, instrument panel components, wiring and sensors. Failure to follow safety and disabling procedures could result in accidental air bag deployment, possible personal injury and unnecessary system repairs.

PRECAUTIONS

Several precautions must be observed when handling the inflator module to avoid accidental deployment and possible personal injury.

• Never carry the inflator module by the wires or connector on the underside of the module.

• When carrying a live inflator module, hold securely with both hands, and ensure that the bag and trim cover are pointed away.

• Place the inflator module on a bench or other surface with the bag and trim cover facing up.

• With the inflator module on the bench, never place anything on or close to the module which may be thrown in the event of an accidental deployment.

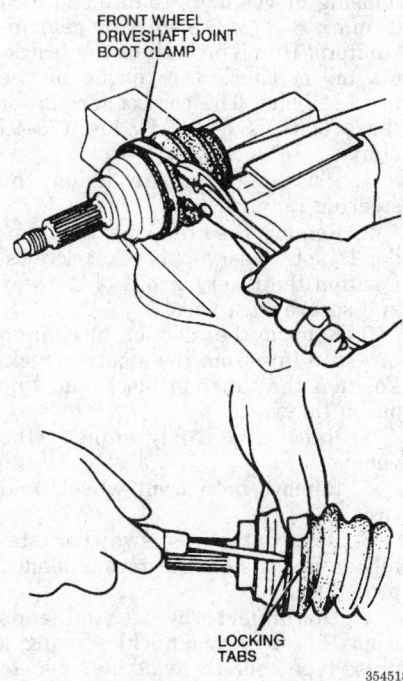

FRONT WHEEL
DRIVESHAFT JOINT
BOOT CLAMP

LOCKING
TABS

354518

Use boot clamp pliers or similar to install the inner and outer clamps

DISARMING

——— CAUTION ———
To avoid rendering the SRS (Supplemental Restraint System) inoperative, which could lead to personal injury or death in the event of a severe frontal collision, extreme caution must be taken when servicing the electrical related systems.

NOTE: All SRS electrical wiring harnesses and connectors are covered with **YELLOW** outer insulation. Do not use electrical test equipment on any circuit related to the SRS (air bag) sensors. When installing SRS components, always install with the arrow marks facing the front of the vehicle.

Disarming

To disarm the SRS system turn the ignition switch to **OFF** position. Then disconnect the both battery cables starting with the negative cable first and wait at least 10 minutes after the cables are disconnected. Be sure to insulate the battery terminal ends.

Arming

To arm the SRS system turn the ignition switch to **OFF** position. Connect the both battery cables starting with the positive cable first.

NOTE: The SRS or air bag system is equipped with a self-diagnostic operation. After turning the ignition key to the **ON** or **START** position, the AIR BAG warning lamp will illuminate for 7 seconds. After 7 seconds, the AIR BAG lamp will extinguish if no malfunction is detected. If the AIR BAG lamp does not extinguish after 7 seconds, check the SRS self diagnostic system for a malfunction.

Steering Wheel

REMOVAL AND INSTALLATION

——— CAUTION ———
The Supplemental Inflatable Restraint (SIR) system must be disarmed before removing the steering wheel. Failure to do so may cause accidental deployment of the air bag, resulting in unnecessary SIR system repairs and/or personal injury.

——— WARNING ———
The backup power supply energy must be depleted before any air bag component service is performed. To deplete the backup power supply, disconnect the battery ground cable and wait ten minutes.

1. Disconnect the negative battery cable. Wait at least ten minutes for the air bag backup power supply to deplete before beginning work.
2. There are two plastic covers (called steering wheel bottom access covers) on the sides of the steering wheel hub that cover the air bag bolts. Remove these left and right side air bag bolt covers.
3. Remove the steering wheel bottom access cover and disconnect the air bag electrical connector.
4. Remove and **discard** the air bag attaching bolts. The bolts must be replaced with new at assembly.

——— CAUTION ———
When carrying a live air bag, make sure the bag and trim cover are pointed away from the body. In the unlikely event of an accidental deployment, the bag will ten deploy with minimal chance of injury. When placing a live air bag on a bench or other surface, always face the bag and trim cover up, away from the surface. This will reduce the motion of the module if it is accidentally deployed.

5. Remove the air bag module.
6. Remove the bolt attaching the steering wheel to the column shaft.
7. Disconnect the speed control (if equipped) and horn electrical connector.
8. Using a suitable puller remove the steering wheel.
   To install:
9. Install the steering wheel onto the steering column. Use care to align the splines.
10. Install the steering wheel attaching bolt and torque to 22 to 29 ft. lbs. (29 to 39 Nm).
11. Reconnect the speed control (if equipped) and horn electrical connector.
12. Install the air bag module using new bolts and torque to 14 ft. lbs. (20Nm).
13. Reconnect the air bag electrical connection.
14. Install the steering wheel covers.
15. Reconnect the negative battery cable.

Tie Rod Ends

REMOVAL AND INSTALLATION

The steering gear (rack and pinion assembly) converts the turning motion of the steering wheel into side-to-side motion. The front wheel spindle tie rods and tie rod ends connect the steering gear to the front wheel spindles. The front wheel spindles convert the side-to-side motion into the turning motion of the wheels.

This procedure may be performed with the steering gear installed in the vehicle.

1. Raise and safely support the vehicle.
2. Remove the front wheel and tire assembly.
3. Remove and discard the cotter pin from the tie rod end retaining nut.
4. Remove the tie rod end retaining nut.
5. Use a Tie Rod Separator tool to press the ball stud from the spindle. Do not hammer on the tie rod end with a wedge-type tool.
6. Underbody components are subject to rust and corrosion. Wire brush the tie rod end and jam nut as well as

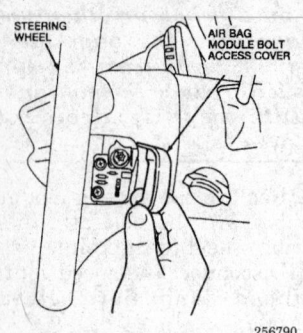

256790

Remove the LH and RH air bag module bolts access covers

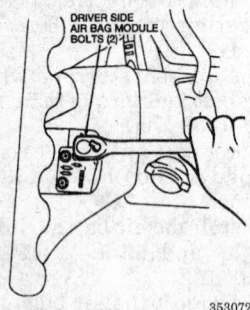

353072

Remove and discard the air bag module bolts. These bolts must be replaced with new parts at assembly

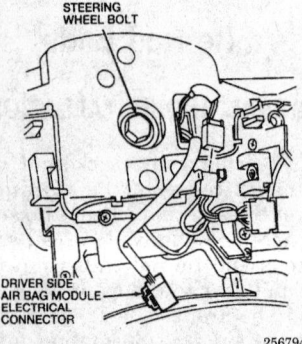

256794

Steering wheel attaching bolt

the exposed threads on the spindle tie rod to make removal easier. A liberal coating of penetrating oil may be helpful.

7. If the tie rod end is to be reused, paint or mark an alignment mark on the outer tie rod end, the tie rod end nut and the spindle tie rod.

8. Loosen the tie rod end jam nut on the spindle rod.

9. Unscrew the tie rod end from the spindle tie rod in a counterclockwise direction. Remove the tie rod end.

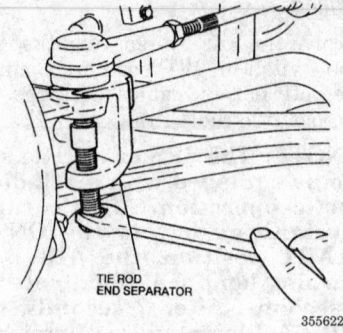

355622

Use a Tie Rod Separator tool to press the ball stud from the spindle

To install:

10. Inspect the rubber bellows for signs of tears or damage. If necessary, the rubber bellows can now be removed and replaced. Remove the outboard clamp and cut the inner clamp to remove. Slide the old bellows off and new bellows into place. Replacement bellows should come with new clamps.

11. The inner spindle tie rod seldom needs service, except accident damage or damaged threads on the outboard ends. If the spindle tie rod is to be replaced, remove the dust seal bellows and unscrew the spindle rod fitting. Install the replacement service part and tighten the fitting securely.

12. Screw the tie rod end onto the front wheel spindle tie rod in a clockwise direction.

13. If the original outer tie rod end is being reused, align the alignment matchmarks on the tie rod end, jam nut and spindle tie rod. Tighten the tie rod end jam nut to 58 to 72 ft. lbs. (78 to 98 Nm).

14. Install the tie rod end in the front wheel spindle. Tighten the retainer nut to 22 to 29 ft. lbs. (29 to 39 Nm). If the cotter pin holes do not align, tighten the nut slowly until they do. DO NOT loosen the nut to align the cotter pin holes. Secure the nut with a new cotter pin.

15. Install the front wheel and tire assembly. Torque the lug nuts evenly to 72 to 87 ft. lbs. (98 to 118 Nm).

16. Lower the vehicle.

17. Check the front toe and if necessary, adjust the tie rod ends.

Power Rack and Pinion

REMOVAL AND INSTALLATION

The power steering gear (also called the Rack and Pinion Assembly) is held in position by two steering gear brackets and insulators. Note that the housing may move slightly when

the steering wheel is turned. If the housing moves more than 0.080 inch (2 mm), replace the steering gear insulators. If one or both of the brackets move, check the torque of the bracket bolts. The correct torque for these bolts is 54–72 ft. lbs. (73–97 Nm).

1. Place a drain pan under the steering rack.

2. Remove the brake master cylinder remote reservoir bracket screws. Position the reservoir out of the way and secure with wire.

3. Remove the junction block/high pressure line from the steering rack. Position the junction block and line out of the way.

4. Raise and safely support the vehicle.

5. Remove both front wheels and tires.

6. Remove the front sway bar (stabilizer bar) using the recommended procedure.

7. Disconnect the tie rod ends from the steering knuckles using a press-type tool to avoid damage to the tie rod ends.

8. Pull back the steering ball stud dust seal and have an assistant turn the steering rack tie rod until the clamp bolt on the lower steering column is accessible.

9. Remove the lower steering column shaft clamp bolt.

10. Remove the power steering fluid return hose and position out of the way.

11. Remove the five steering rack clamp bracket bolts.

12. Lower the steering rack from the vehicle.

To install:

13. Carefully slide the steering gear rack and pinion assembly in place from the left side of the vehicle. Position the input shaft so it is just below the lower steering column shaft clamp.

14. Raise the steering gear until the plastic aligning tab on the input shaft enters the clamp bolt gap on the lower column shaft. Do not install the clamp bolt yet.

15. Examine the steering gear brackets. They should be marked UP with arrows pointing to one end of the bracket. Make sure the brackets are installed correctly. Torque the five steering gear bracket bolts in sequence, working counterclockwise from the number one bolt (upper right side).

16. Connect the fluid return line to the steering gear.

17. Install the steering column shaft clamp bolt. Torque to 17 to 22 ft

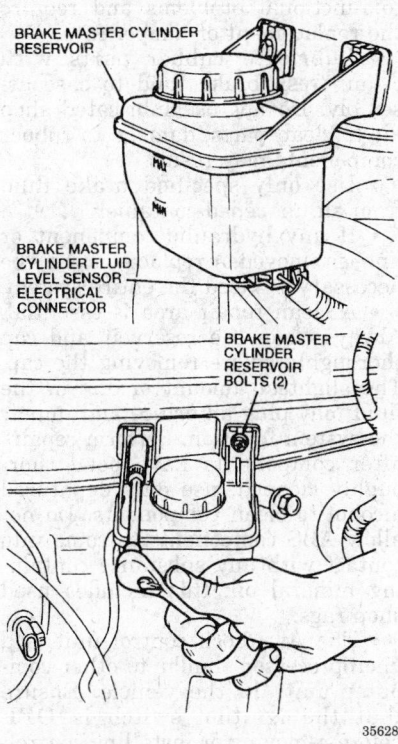

Unbolt the brake master cylinder reservoir and wire it out of the way

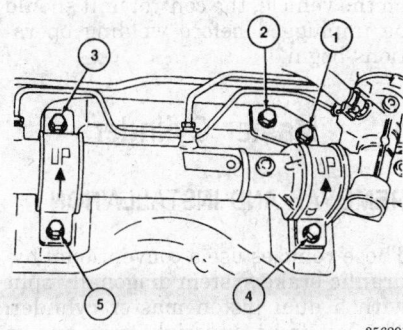

Torque the power steering gear retaining in the sequence shown

lbs. (24 to 29 Nm). Install the dust cover.

18. Connect the tie rod ends using the recommended procedure.

19. Install the stabilizer bar using the recommended procedure.

20. Install the front tire and wheel assemblies. Torque the lug nuts to 72 to 87 ft. lbs. (98 to 118 Nm).

21. Lower the vehicle.

22. Install the junction block. Torque the high pressure line to 11 to 18 ft. lbs. (15 to 25 Nm).

23. Install the brake master cylinder reservoir.

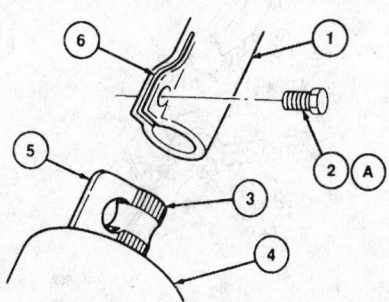

Item	Description
1	Lower Steering Column Shaft
2	Lower Steering Column Shaft Clamp Bolt
3	Power Steering Gear Input Shaft
4	Steering Gear
5	Plastic Aligning Tab
6	Clamp Bolt Gap
A	Tighten to 24-29 N·m (17-22 Lb-Ft)

At installation, align the tab on input shaft with the clamp bolt gap on the lower column shaft

24. Bleed the power steering system using the recommended procedure.

25. Check for leaks and proper operation.

Power Steering Pump

BLEEDING

When replacing or removing the power steering pump, perform one of the following procedures. If no jack or hoist is available to raise the front wheels, use Procedure 1. If the front wheels can be safely raised off the ground, use Procedure No. 2.

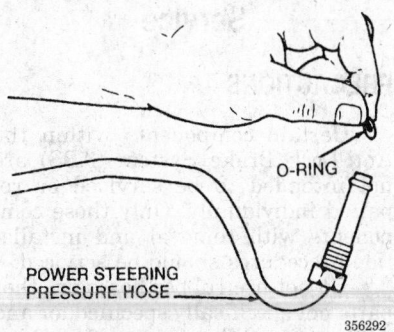

Replace the O-rings, where used, whenever a pressure hose is disconnected

Procedure 1 — Hoist Not Available

The replacement power steering pump must be primed with the specified power steering fluid prior to installation in the vehicle.

1. Remove the fitting from the pressure port and pour power steering fluid into the pressure port with the port facing up to maximize the amount of fluid added to the power steering pump.

2. Slowly turn the pump driveshaft and continue to add fluid.

3. Install the power steering pump in the vehicle. Use care not to let fluid spill out of the pump.

4. Attach the power steering pressure hose and return hose to the power steering pump and oil reservoir.

5. Start the engine and allow it to run for several minutes.

— **CAUTION** —

Do not hold the steering wheel against a stop for more than 15 seconds at a time.

6. Turn the steering wheel fully left and right from stop to stop several times.

7. Turn off the engine and check the fluid level in the power steering fluid remote reservoir. Add fluid if necessary.

Procedure 2 — Hoist Available

1. Install the unprimed replacement power steering pump in the vehicle.

2. Attach the power steering pressure hose and return hose to the power steering pump and oil reservoir.

3. Fill the power steering remote reservoir to the MAX mark.

4. Disconnect the ignition coil-to-distributor high tension wire.

5. Raise and safely support the vehicle so the front wheels are off the ground.

6. Crank the engine with the starter motor. Add fluid to the power steering remote reservoir until the level remains constant.

— **CAUTION** —

Do not hold the steering wheel against a stop for more than 15 seconds at a time.

7. While cranking the engine, turn the steering wheel from stop to stop. Check the fluid level and add fluid as necessary.

8. Connect the ignition coil-to-distributor high tension wiring.

9. Start the engine and allow to idle for several minutes.

CAUTION

Do not hold the steering wheel against a stop for more than 15 seconds at a time.

10. Turn the steering wheel from stop to stop several times.
11. Turn the engine off and check the fluid level. Add fluid as necessary.
12. Lower the vehicle. Road test to very proper power steering operation and no excess system noise.

REMOVAL AND INSTALLATION

Power steering is a hydraulic assisted system that reduces the amount of effort required to turn the steering wheel. The power steering pump draws fluid from a remote reservoir through a hose and compresses the fluid and pumps it to the steering gear (rack and pinion assembly) through a pressure hose. Power steering fluid does not require periodic changing although it should be checked regularly. Fill the power steering reservoir to the MIN mark with the recommended power steering fluid. The power steering pump shares a drive belt with the water pump.

1. Disconnect the negative battery cable.
2. The power steering pump pulley must be removed to access the pump bolts. Use the following procedure.
   a. Raise and safely support the vehicle.
   b. Remove the water pump and power steering pump drive belt using the recommended procedure.
   c. Use a strap wrench to hold the power steering pump pulley while removing the pulley nut.
   d. Remove the pump pulley from the pump shaft.
3. Place a drain pan under the power steering pump.
4. Remove the pressure hose bolt from the pump and secure the pressure hose out of the way.
5. Remove the two reservoir to pump hose bolts and secure the reservoir hose out of the way.
6. Remove the three front power steering pump bolts and the rear pump bolt. Remove the pump from the vehicle.
To install:
7. Clean all parts well. Install the pump, without the pulley, to the vehicle and install the three front bolts and the one rear bolt. Torque the bolts to 11 to 15 ft. lbs. (15 to 20 Nm).

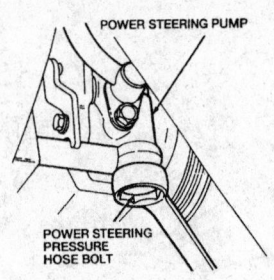

POWER STEERING PUMP

POWER STEERING PRESSURE HOSE BOLT

355652

Remove the pressure hose bolt from the pump and secure the pressure out of the way

8. Position the reservoir hose connection and install the two reservoir to pump hose bolts. Torque the bolts to 10 to 13 ft. lbs. (14 to 18 Nm).
9. Position the high pressure hose connection and install the pressure connection bolt. Torque the bolts to 51 to 58 ft. lbs. (69 to 78 Nm).
10. Install the pump pulley. Use a strap wrench to hold the power steering pump pulley while tightening the pulley nut. Torque the nut to 40 to 50 ft. lbs. (54 to 68 Nm).
11. Install and adjust the water pump and power steering pump drive belt using the recommended procedure.
12. Lower the vehicle and refill the remote reservoir with the recommended power steering fluid. Turning the steering wheel from side to side while the engine is idling will help bleed air from the system.
13. Road test the vehicle to verify correct steering system operation and no leaks.

BRAKES

Anti-Lock Brake System Service

PRECAUTIONS

• Certain components within the Anti-Lock Brake System (ABS) are not intended to be serviced or repaired individually. Only those components with removal and installation procedures should be serviced.
• Do not use rubber hoses or other parts not specifically specified for and ABS system. When using repair kits, replace all parts included in the kit. Partial or incorrect repair may lead to functional problems and require the replacement of components.
• Lubricate rubber parts with clean, fresh brake fluid to ease assembly. Do not use lubricated shop air to clean parts; damage to rubber components may result.
• Use only specified brake fluid from an unopened container.
• If any hydraulic component or line is removed or replaced, it may be necessary to bleed the entire system.
• A clean repair area is essential. Always clean the reservoir and cap thoroughly before removing the cap. The slightest amount of dirt in the fluid may plug an orifice and impair the system function. Perform repairs after components have been thoroughly cleaned; use only denatured alcohol to clean components. Do not allow ABS components to come into contact with any substance containing mineral oil; this includes used shop rags.
• The Anti-Lock control unit is a microprocessor similar to other computer units in the vehicle. Ensure that the ignition switch is **OFF** before removing or installing controller harnesses. Avoid static electricity discharge at or near the controller.
• If any arc welding is to be done on the vehicle, the control unit should be unplugged before welding operations begin.

Master Cylinder

REMOVAL AND INSTALLATION

These vehicles use a conventional hydraulic brake system diagonally split with a dual piston master cylinder. The left front and right rear are on one circuit while the right front and left rear are on another brake circuit. A vacuum booster reduces required pedal pressure. The master cylinder fluid reservoir is mounted to the bulkhead, just above the master cylinder. The reservoir has a sensor which activates the brake system warning indicator in the instrument cluster when the fluid level is too low. Service this vehicle with DOT 3 brake fluid.

1. Disconnect the negative battery cable.
2. Clean the area around the master cylinder and tube connections to keep dirt from getting into the system.
3. Remove and plug the brake master cylinder connector tube (front brake line) and the master cylinder outlet rear tube from the master cyl-

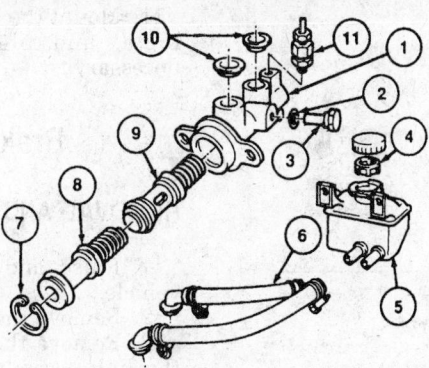

Item	Description
1	Brake Master Cylinder
2	O-Ring
3	Valve Stopper Bolt
4	Float
5	Brake Master Cylinder Reservoir
6	Master Cylinder Reservoir Hoses
7	Snap Ring
8	Primary Piston
9	Secondary Piston
10	Rubber Bushings
11	Dampener Valve

353318

Master cylinder and components

inder. Use care to keep brake fluid from any painted surfaces.

4. Remove and plug the two master cylinder reservoir hoses from the master cylinder. Use care. Brake fluid will tend to drain as soon as the hoses are loosened.

5. Remove the two master cylinder nuts and remove the master cylinder from the vehicle.

To install:

6. Position the master cylinder on the brake booster and install the nuts. Torque to 71-97 inch lbs. (8-11 Nm).

7. Unplug and connect the master cylinder connector tube and outlet rear tube to the master cylinder. Torque the fittings to 11-13 ft. lbs. (15-18 Nm).

8. Unplug and install the two master cylinder reservoir hoses to the master cylinder.

9. Full the fluid reservoir with fresh DOT 3 brake fluid.

10. Bleed the brake system using the recommended procedure.

11. Road test the vehicle to verify correct braking operation.

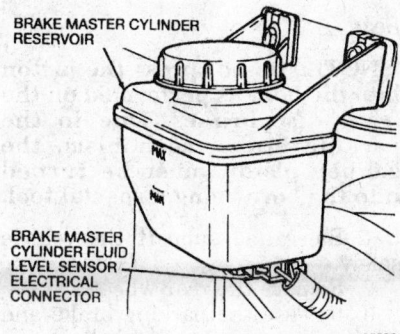

353321

Brake fluid reservoir. Note the level marks

Brake Caliper

REMOVAL AND INSTALLATION

The front disc brake caliper slides on two stainless steel locating pins. The front disc brakes use a conventional pin slider-type front disc brake caliper with a 10.875 inch front disc rotor. The front disc brake caliper is attached to the front suspension with two Torx® head brake caliper bolts. Rubber insulators isolate the stainless steel locating pins from direct contact with the front disc brake cali-

per. The front disc brake calipers must be removed to replace the front brake pads (shoes and linings). Service this vehicle with DOT 3 brake fluid.

1. Raise and safely support the vehicle.

2. Remove the wheel and tire.

3. If the brake caliper is being removed for just brake pad replacement. DO NOT disconnect the brake hose. If the caliper is to be completely removed from the vehicle for overhaul, disconnect the brake fluid line by removing the banjo bolt. Discard the copper washer.

4. Remove the two caliper pin bolts. Most applications will require a Torx® T-40 bit to remove the two brake caliper bolts.

5. If the brake caliper is being removed just for brake service, with the brake hose still attached to the caliper, use a length of wire to support the caliper from the front shock absorber. Do not let the caliper hang by the brake hose. If the caliper is being completely removed from the vehicle for overhaul, use care not to drip brake fluid on the paintwork.

NOTE: If both calipers are being completely removed from the vehicle at the same time, mark them Left and Right so the calipers can be reinstalled to their original locations. The reason for this is that the bleeder screws must be positioned on the top of the front disc brake caliper when installed on the vehicle.

To install:

6. Clean all parts well. Use a C-clamp and a used brake pad to push the caliper piston fully in the piston bore. Inspect the caliper pins and clean any dirt and debris.

7. Install the caliper onto the rotor. Make sure the inboard and outboard brake pads are properly positioned.

8. Lubricate the stainless steel locating pins with a Silicone Dielectric Compound such as Ford DZAZ-19A331-A or equivalent silicone grease. Install the two caliper pin bolts and torque to 18–25 ft. lbs. (24–34 Nm).

9. If disconnected, install the brake hose using a new replacement copper washer, install the banjo bolt and torque to 12–14 ft. lbs. (17–20 Nm).

10. If the brake hose had been disconnected, bleed the brake system.

11. Install the wheel and tire.

12. Torque the lug nuts to 72–87 ft. lbs. (98–118 Nm).

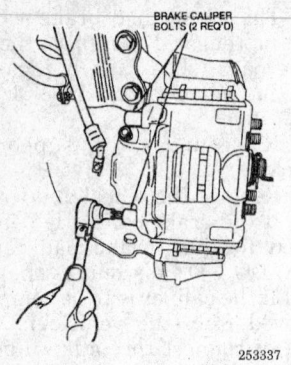

253337

Caliper pin bolt removal

13. Check the master cylinder reservoir and add fresh DOT 3 brake fluid as required.

14. Lower the vehicle. Pump the brake pedal slowly until a firm brake pedal is obtained, indicating that the brake pads are properly seated, before attempting to move the vehicle. Road test and check for proper brake operation.

Disc Brake Pads

REMOVAL AND INSTALLATION

——————— CAUTION ———————

Brake shoes contain asbestos, which has been determined to be a cancer causing agent. Never clean the brake surfaces with compressed air. Avoid inhaling any dust from any brake surface. When cleaning brake surfaces, use a commercially available brake cleaning fluid.

Front

1. Raise and support the front of the vehicle, then remove the wheels.
2. Remove the bottom guide pin from the caliper and swing the caliper cylinder body upward; support the caliper with a wire.
3. Remove the brake pad retainers and the pads.
To install:
4. Compress the piston of the disc brake caliper.
5. Install the brake pads and caliper assembly. Torque the guide pin to 23–30 ft. lbs. (31–41 Nm).
6. Install the wheels.
7. Apply the brakes a few times to seat the pads. Check the master cylinder and add fluid if necessary. Bleed the brakes, if necessary.

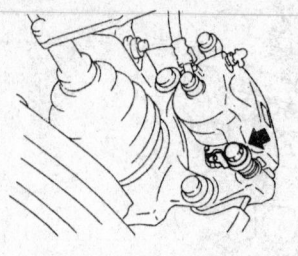

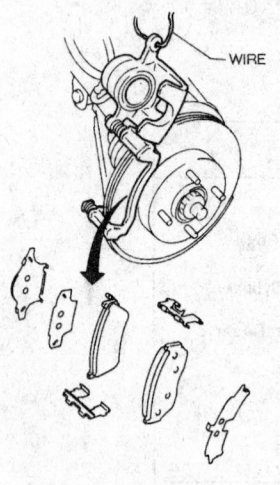

268902

Removing caliper for disc brake replacement

Rear

NOTE: Do not press the piston into the bore as performed on the front disc brakes. Due to the parking brake mechanism, the caliper piston must be turned into the bore using a special tool.

1. Raise and support the vehicle safely.
2. Remove the rear wheels.
3. Release the parking brake and remove the cable bracket bolt.
4. Remove the pin bolts and lift off the caliper body.
5. Pull out the pad springs and then remove the pads and shims.
To install:
6. Clean the piston end of the caliper body and the area around the pin holes. Be careful not to get oil on the rotor.
7. Using the proper tool, carefully turn the piston clockwise back into the caliper body. Take care not to damage the piston boot.
8. Coat the pad contact area on the mounting support with a silicone based grease.
9. Install the pads, shims, and the pad springs. Always use new shims.
10. Position the caliper body in the mounting support and tighten the pin bolts to 28–38 ft. lbs. (38–52 Nm).

11. Mount the wheels, lower the vehicle, and bleed the system if necessary.

Brake Rotor

REMOVAL AND INSTALLATION

1. Raise and safely support the vehicle.
2. Remove the tire and wheel.
3. Remove the brake caliper from the hub assembly using the recommended procedure. Do not disconnect the brake hose. Position the caliper out of the way and support it with a length of wire tied to the shock absorber housing. Do not let the caliper hang by the brake hose.
4. Remove the rotor by pulling it off of the wheel hub bolts. If the rotor is difficult to remove, note the following.
   a. If excessive force must be used to remove the front rotor, the rotor should be checked for excessive runout (warpage) prior to installation.
   b. If additional force is required to remove the rotor, apply penetrating oil on the rotor and wheel hub mating surfaces. Strike the rotor between the wheel studs with a plastic hammer.
   c. Pull the rotor from the wheel hub.
To install:
5. Clean all parts well. If the rotor is to be reused, it should be carefully checked. Use the following procedure.
   a. The front rotor minimum thickness is shown on each rotor. The thickness at which the rotor becomes unsafe is called the Discard Thickness. The discard thickness is 0.945 inch (24 mm). To find the minimum thickness to which the rotor can be machined, add 0.030 inch (0.762 mm) to the discard limit marked on the rotor. This 0.974 inch (24.762 mm) machining limit allows for rotor wear AFTER the rotor has been resurfaced and returned to service.
   b. A special disc brake rotor micrometer is used to measure the front rotor thickness. This type of micrometer has a pointed tip on one or both anvils to get an accurate reading of the rotor's thickness even if deeply grooved. If the thickness of the rotor is less than the machining limit allowed, discard the rotor and install a new one.
   c. Use care when measuring the rotor. Machining a front rotor thin-

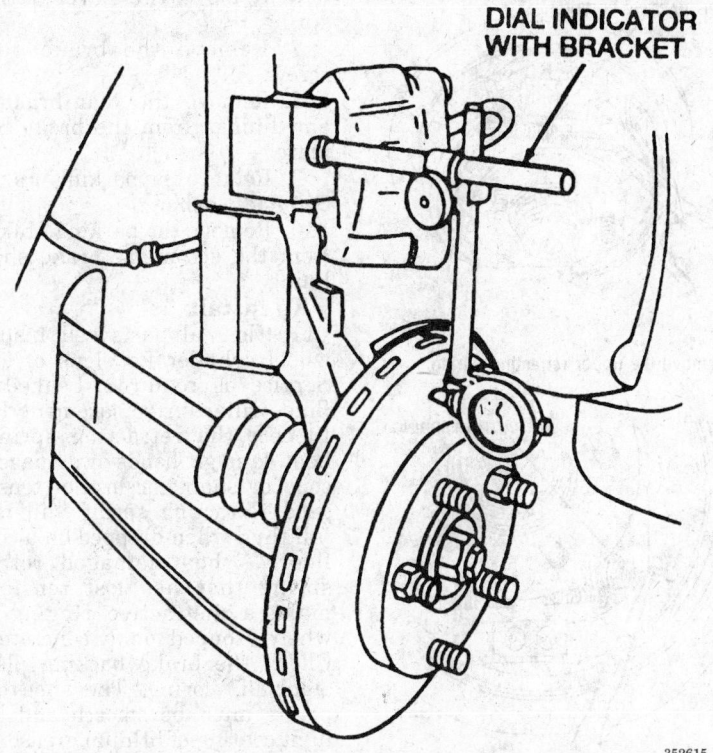

DIAL INDICATOR WITH BRACKET

352615

Using a dial indicator to check front rotor runout

ner than the machining limit could permit the rotor to wear past the safe discard limit before the brake pads wear out. It could also result in severe overheating and fade because the rotor may be unable to absorb the heat generated during braking.

d. If troubleshooting a complaint of a pulsating brake pedal, check the rotor's runout. To check rotor runout (warped rotor), install the rotor onto the wheel hub and secure it with at least two lug nuts. Make sure the front wheel bearing end play is within specification. Jig up a dial indicator (possibly with magnetic base to attach to the front strut). The indicator probe should be centered on the rotor's braking surface. Slowly rotate the rotor and check for runout. Maximum allowed runout is 0.0028 inch (0.07 mm).

e. If the rotor is not within specification, reposition the rotor on the hub to obtain the lowest possible runout. Shift the rotor one hole and secure to the hub. Measure the runout again. Repeat this until the minimum runout is found. If runout is now within specification paint a matchmarks on the rotor and a wheel wheel stud so the rotor can always be returned to this low-

runout position after service. If the rotor cannot be bought into specification, refinish or replace the rotor.

f. To check the rotor for parallelism (thickness variation) use a rotor micrometer and measure in several places around the rotor. The rotor must not vary by more than 0.0004 inch (0.01 mm). The rotor can be resurfaced if it does not fall below the minimum thickness specification (0.974 inch) after machining. If rotor thickness variation is below specification after resurfacing, replace the rotor.

g. If there are light surface irregularities on the rotor, resurface the rotor by lightly sanding the rotor face with fine emery cloth. Lightly sand both sides of the rotor. The back side of the rotor can be sanded at the front wheel knuckle where the caliper normally rides. If scratches or scoring exceeds 0.009 inch (0.22 mm), the rotor should be refinished.

6. If the rotor is being replaced with a new part, check the braking surfaces. Many service replacement brake parts come with a protective coating to avoid rusting while in storage. Clean any protective coating with a suitable solvent. If the original rotor is being installed, clean and in-

spect as listed above. Make sure the rotor is fully seated on the wheel hub.

7. Install the brake caliper and pads using the recommended procedure.

8. If the caliper was completely removed from the vehicle for overhaul and the brake hose has been disconnected, attach the hose with a new copper washer and bleed the brake system.

9. Install the tire and wheel. Torque the lug nuts to 72–87 ft. lbs. (98–118 Nm).

10. Lower the vehicle.

11. Slowly pump the brake pedal several times to seat the brake pads against the rotor. Do not attempt to move the vehicle until a firm pedal has been obtained. Road test the vehicle and check for proper operation.

Brake Drums

REMOVAL AND INSTALLATION

The rear drum brakes used on these vehicles are conventional expanding shoe-type with the brake shoe lining applied to the inside of the rotating drum. An incremental brake adjuster screw is designed to actuate whenever sufficient wear occurs.

1. Raise and safely support the vehicle.

2. Remove the wheel and tire.

3. Remove the brake drum by pulling it from the wheel studs.

4. If necessary for brake drum removal, pry off the access hole plug from the access hole. Insert a screwdriver and a brake adjustment tool. Press the screwdriver against the adjusting lever to disengage it from the adjuster. Loosen the adjuster using the brake adjusting tool.

To install:

5. Clean all parts well. It is good practice to inspect the wheel cylinder for leaks anytime the brake drum is removed. If a new replacement brake drum is being installed, inspect it for a protective coating on the machined inside braking surface. Remove any coating with suitable solvent.

6. Install the brake drum onto the wheel studs.

7. In most all cases, manual brake adjustment IS NOT recommended. Adjustment is performed by driving the vehicle and applying the brakes.

8. Install the tire and wheel and torque the fasteners to 72–87 ft. lbs. (98–118 Nm).

9. Lower the vehicle.

10. Adjust the rear brake shoes by sharply applying the brakes several

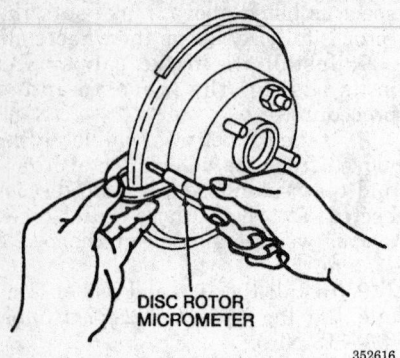

DISC ROTOR MICROMETER

352616

Measuring rotor thickness

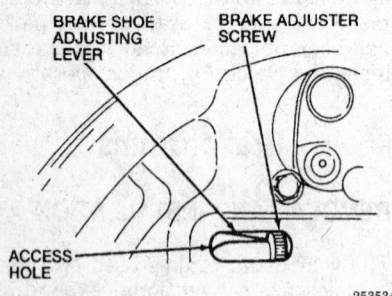

BRAKE SHOE ADJUSTING LEVER

BRAKE ADJUSTER SCREW

ACCESS HOLE

253534

Brake shoe adjustment may need to be loosened to remove the brake drum

times while driving the vehicle alternately forwards and backwards. Check the brake operation by making several stops while driving forward.

Brake Shoes

REMOVAL AND INSTALLATION

The rear drum brakes use an internal rear wheel cylinder with expanding shoes and lining that are applied against a rotating brake drum. An incremental brake adjuster screw is actuated whenever sufficient wear occurs. Brake adjustment takes place in forward or reverse braking but not with parking brake application.

1. Raise and safely support the vehicle.

2. Remove the wheel and tire assembly. Remove the brake drum using the recommended procedure.

3. Disconnect the parking brake rear cable and conduit from the parking brake lever.

4. Remove the two brake shoe hold-down springs and the two brake shoe hold-down pins.

5. Remove the upper retracting spring.

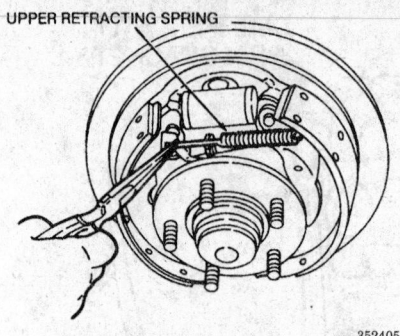

UPPER RETRACTING SPRING

352405

Remove the upper retracting spring

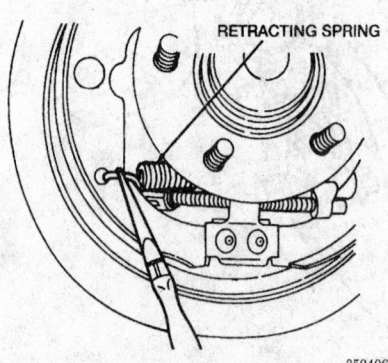

RETRACTING SPRING

352406

Remove the lower retracting spring

6. Remove the lower retracting spring.

7. Remove the brake adjuster screw.

8. Remove the rear brake shoes and linings from the brake backing plate.

9. Remove the parking brake lever clip and washer.

10. Remove the parking brake lever from the secondary brake shoe and lining.

To install:

11. Clean all parts well. Inspect the wheel cylinder for signs of leaking. Service as required. Leaked brake fluid will ruin replacement linings. Inspect the retracting springs for heat damage, bends or damage to the coils or shank or loss of tension. A good retracting spring will make a full thud when dropped on a concrete floor. A heat-damaged retracting spring that has lost tension will make a distinctive ringing sound when dropped on a concrete floor. Check the brake backing plate for signs of scoring. The shoe contact points must be smooth and have a light coating of lithium grease. Verify that the brake lining thickness is between 0.059 to 0.232 inch (1.5 to 5.9 mm). Failure to replace worn rear brake shoes will result in a scored drum.

Brake Drum — Exploded View

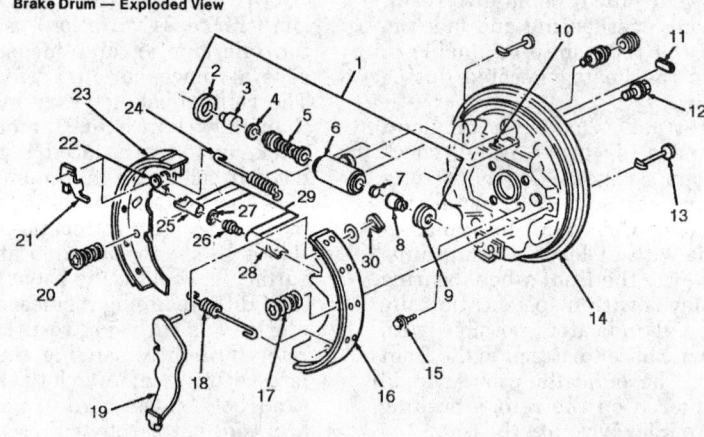

1. Rear wheel cylinder	16. Secondary brake shoe and lining
2. Dust boot	17. Brake shoe hold-down spring
3. Piston	18. Lower retracting spring
4. Cup	19. Parking brake lever
5. Wheel cylinder piston cup spring	20. Brake shoe hold-down spring
6. Wheel cylinder housing	21. Brake shoe adjusting lever
7. Cup	22. Adjuster lever pin
8. Piston	23. Primary brake shoe and lining
9. Dust boot	24. Upper retracting spring
10. Access hole	25. Adjuster socket (Part of 2041)
11. Access hole plug	26. Brake adjuster stud
12. Rear wheel cylinder bolt	27. Washer
13. Brake shoe hold-down pin	28. Adjuster nut
14. Rear brake backing plate	29. Brake adjuster screw
15. Rear brake backing plate bolts	30. Parking brake lever clip

352402

Rear drum brake assembly and related components

12. Inspect the brake drum for scratches, scoring, bell mouth and out-of-round conditions. Remove minor scores on a brake drum with sandpaper. Do not refinish brake drums to remove score marks. A brake drum surface which is highly polished can cause the brakes to lock up. Remove polished surfaces with sandpaper or refinish the brake drum. Refinish a brake drum that is out-of-round enough to cause vehicle vibration or noise when braking. Remove only enough surface metal to true-up the brake drum. Brake drum maximum inside diameter is shown on each drum. If the maximum inside diameter shown on the brake drum is exceeded through wear or refinishing, replace the brake drum. After a brake drum is refinished, wipe the refinished surface with a cloth soaked in clean denatured alcohol. If one brake drum is refinished, the brake drum on the opposite side of the vehicle should also be refinished to the same diameter. The standard inner brake drum diameter is 9.840 inches (250.0 mm). Replace the brake drum if worn beyond 9.900 inches (251.5 mm).

13. Install the parking brake lever to the secondary brake shoe and lining with a new parking brake lever clip.

14. Position the secondary (rear) shoe on the backing plate and install the brake shoe hold-down spring and pin.

15. Position the primary (front) shoe on the backing plate and install the brake shoe hold-down spring and pin.

16. Attach the parking brake rear cable and conduit to the parking brake lever.

17. Attach the lower retracting spring to the rear brake shoes.

18. Apply a light coat of high-quality grease to the threaded areas of the adjuster nut and adjuster socket. Turn the adjuster nut all the way down on the brake adjuster screw, then loosen the adjuster one-half turn. Install the adjuster screw in the slots on the rear brake shoes. The wider slot on the socket must fit in the slot on the primary (front) brake shoe. The slot on the adjuster nut end must fit into the slots in the secondary (rear) brake shoe and parking brake lever.

19. Install the brake shoe adjusting lever on the adjuster lever pin.

20. Install the upper retracting spring in the slot on the secondary shoe and in the slot on the brake shoe adjusting lever. The brake shoe adjusting lever should contact the brake adjuster screw.

21. Install the brake drum onto the wheel studs.

22. Install the tire and wheel and torque the fasteners to 72–87 ft. lbs. (98–118 Nm).

23. Lower the vehicle.

NOTE: In most all cases, manual brake adjustment IS NOT recommended. Adjustment is performed by driving the vehicle and applying the brakes.

24. The rear brakes do not require adjustment when being serviced to obtain a firm brake pedal feel. To achieve a firm brake pedal after servicing the rear brakes, sharply apply the brake pedal several times while driving the vehicle alternately forwards and backwards. Check the brake operation by making several stops while driving forward. The self-adjusting mechanism will sufficiently adjust the rear brake shoes without any manual tightening at the brake shoe adjuster. If the rear brake shoes are manually adjusted, the additional action of the brake shoe adjuster can cause the brakes to become over-tightened and result in binding or overheated rear brakes.

Wheel Cylinder

REMOVAL AND INSTALLATION

When the brake pedal is pressed, hydraulic fluid pressure in the rear wheel cylinders forces the pistons outward, moving the rear brake shoes and linings toward the brake drums. The wheels cylinders should be inspected anytime the rear brake drums are removed. Do not disassemble the rear wheel cylinders unless they are leaking. Pull back each dust boot and look for leakage. A slight amount of brake fluid is almost always present and acts as a lubricant for the pistons. Excessive brake fluid indicates leakage past the piston seals and a need for wheel cylinder overhaul or replacement. If excessive leakage is noted, disassemble, inspect and overhaul the rear wheel cylinder. Inspect the rear wheel cylinder bore for scratches or pitting. If damaged, replace the wheel cylinder. Service this vehicle with DOT 3 brake fluid.

1. Raise and safely support the vehicle.

2. Remove the wheel and tire assembly.

3. Remove the brake drum and brake shoes using the recommended procedures.

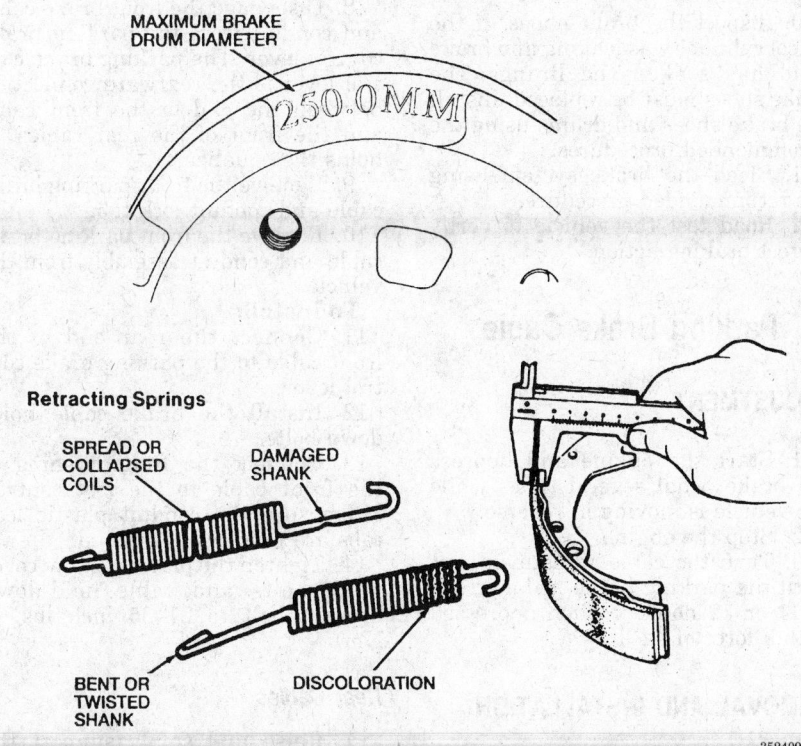

MAXIMUM BRAKE DRUM DIAMETER

250.0MM

Retracting Springs

SPREAD OR COLLAPSED COILS

DAMAGED SHANK

BENT OR TWISTED SHANK

DISCOLORATION

These size checks should be made and the retracting springs' condition checked. Replace questionable parts

2. Remove the wheel and tire assembly.

3. Remove the brake drum and brake shoes using the recommended procedures.

4. Remove the cylinder to brake shoe connecting pins.

5. Disconnect the brake line from the wheel cylinder. Underbody components are subject to rust and corrosion. It is good practice to wire brush the area around the brake line fitting to remove rust. A liberal application of penetrating oil also helps. If the brake line is rusted to the brake line fitting, it may twist off requiring replacement of the brake line.

6. Remove the wheel cylinder retaining bolts and remove the cylinder from the brake backing plate.

7. Inspect the wheel cylinder. If the cylinder was leaking, the following procedure may be used to overhaul the cylinder.

a. Disengage the dust boots from the grooves in the wheel cylinder housing. Remove each dust boot and piston from the wheel cylinder.

b. Remove the cups and wheel cylinder piston cup spring from the wheel cylinder bore.

c. Remove the wheel cylinder bleeder screw from the wheel cylinder housing.

d. Discard all rubber parts.

e. Wash all parts in clean, denatured alcohol. Do not use petroleum based solvents.

f. Inspect the pistons for scratches, scoring or other visible damage. Replace if necessary.

g. Inspect the wheel cylinder bore for scratches or other signs of visible damage. If necessary, hone the cylinder bore. Do not hone more than 0.003 inch beyond the wheel cylinder original diameter of 1.00 inch (25.40 mm). If honing cannot cleanup the bore, replace the wheel cylinder assembly.

h. Wash the wheel cylinder housing in clean, denatured alcohol after honing. Inspect the bleeder screw hole to make sure it is free of dirt and rust.

i. If using a wheel cylinder overhaul kit, use all the parts supplied. Assemble the wheel cylinder using the new parts. Apply a light coat of DOT 3 brake fluid to all parts to aid assembly.

To install:

8. Install the assembled wheel cylinder to the backing plate. Install the bolts and torque to 9 to 13 ft. lbs. (12 to 18 Nm).

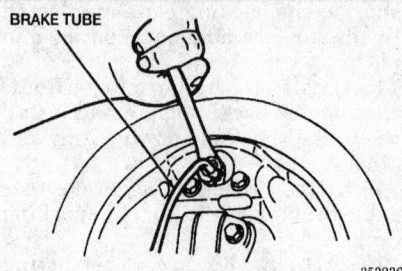

Disconnect the brake line from the wheel cylinder

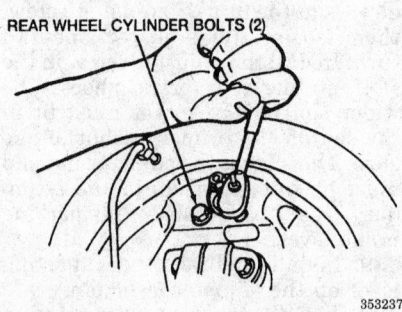

Remove the wheel cylinder bolts from the backing plate

9. Connect the brake line and tighten to 10 to 17 ft. lbs. (14 to 24 Nm).

10. Inspect the brake shoes. If the wheel cylinder was leaking and brake fluid has soaked the linings, the brake shoes must be replaced. Install the brake shoes and drums using the recommended procedures.

11. Bleed the brake system using the recommended procedure.

12. Road test the vehicle to verify correct braking action.

Parking Brake Cable

ADJUSTMENT

1. Start the engine and depress the brake pedal several times while the vehicle is moving in reverse.

2. Stop the engine.

3. Turn the cable adjustment nut until the parking brake pedal stroke is 11 or 12 notches when depressed with a force of 44 lbs.

REMOVAL AND INSTALLATION

The parking brake control system is independent from the hydraulic system and is operated by the parking brake control. The parking brake control is mounted under the left side of the instrument as a foot operated control.

When the parking brake control is applied, the front parking brake cable pulls the cable equalizer forward. The rear cables which are attached to the equalizer, pull the parking brake levers that are connected to the primary brake shoes. This action forces the shoes and linings against against the brake drum. The parking brake control is held in position by a spring-loaded pawl and ratchet mechanism.

Front Cable

1. Remove the left and right side inner kick panels.

2. Remove the left and right side cowl side trim panel.

3. Remove the accelerator stop screw and the stop (rubber block under the accelerator pedal).

4. Fold back the front carpet to the front seats.

5. Remove the 4 parking brake opening cover plate bolts. Remove the plate.

6. Disengage the brake cable bushing from the lever.

7. Remove the brake cable hold-down bolts from the front hold-down bracket.

8. Disengage the front brake cable and conduit from the parking brake control lever. The parking brake control lever is the rearward connection between the end of the front cable and the front of the rear cables. It holds the equalizer.

9. Remove the front parking brake cable and conduit lock tab.

10. Remove the front parking brake cable and conduit assembly from the vehicle.

To install:

11. Connect the rear end of the front cable to the parking brake control lever.

12. Install the brake cable hold-down bolts.

13. Connect the front bushing of the front cable to the foot control. Make sure the conduit plastic lock tabs are properly positioned.

14. Tighten the parking brake cover plate bolts and cable hold-down bracket bolt to 27–35 inch lbs. (4 Nm).

Rear Cables

1. Raise and safely support the vehicle.

2. Remove the cable adjustment nut.

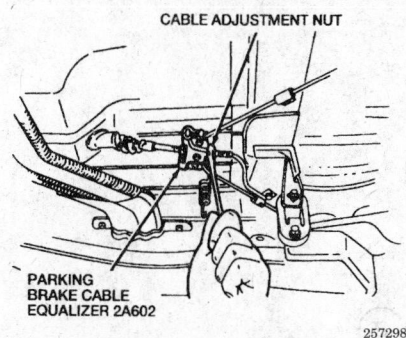

CABLE ADJUSTMENT NUT

PARKING
BRAKE CABLE
EQUALIZER 2A602

257298

Parking brake adjustment

3. Disconnect the parking brake equalizer cable spring from the bracket and remove the rear cable and conduit.

4. Remove the 8 cable hold-down bracket bolts and the 6 cable hold-down brackets on the right side parking brake cable.

5. Remove the 7 cable hold-down bracket bolts and the 5 cable hold-down brackets on the left side parking brake cable.

6. Compress the lock tabs and remove the cables from the brackets.

7. Remove the rear wheels and brake drums.

8. Disconnect the parking brake brake cable from the parking brake levers on both sides.

9. Compress the locking tabs at the backing plate and remove the cable.

To install:

10. Install the replacement rear cables through the openings in the brake backing plates. Make sure locking tabs are properly positioned. Connect the cable ends to the parking brake levers on the brake shoes.

11. Install the rear wheels and brake drums.

12. Thread the front ends of the rear cables through the hold-down brackets and secure the brackets with the bolts. Tighten the hold-down bracket bolts to 27 to 35 inch lbs. (4 Nm).

13. Connect the front ends of the rear cables to the equalizer.

14. Position the parking brake equalizer, install the cable adjustment nut and the spring.

15. Adjust the parking brake using the recommended adjustment procedure.

Brake System

BLEEDING

When any part of the hydraulic system is disconnected for service, air can enter the system and cause spongy brake pedal action. Bleeding is a procedure used to remove air from the hydraulic circuits. Use only clean, fresh DOT 3 brake fluid from a sealed container to service this vehicle. Never use brake fluid that has been drained from the hydraulic system or that has been standing in an open container for an extended period of time.

System Priming

When a new brake master cylinder has been installed or the brake system emptied or partially emptied, fluid may not flow to the bleeder screws during normal bleeding. It may be necessary to prime the master cylinder (also called bleeding the master cylinder) using the following procedure.

1. Use a tubing wrench to remove the brake master cylinder connector tube and the master cylinder outlet rear tube from the master cylinder.

2. Install shop-made short pieces of brake line in the master cylinder and position them in the remote fluid reservoir. Make sure the ends are submerged in the brake fluid in the reservoir.

3. Fill the remote fluid reservoir with fresh DOT brake fluid.

4. Cover the master cylinder remote reservoir with a clean shop towel.

--- **CAUTION** ---

Brake fluid is harmful to painted and plastic surfaces. If brake fluid is spilled on a painted or plastic surface, immediately wash it with water.

---

5. Pump the brakes until clear, bubble-free fluid flows from both brake tubes.

6. Remove the short brake tubes and reinstall the brake master cylinder connector tube and brake master cylinder outlet tube on the master cylinder.

7. Bleed each brake tube at the master cylinder using the following procedure:

a. Have an assistant pump the brake pedal 10 times and then hold firm pressure on the brake pedal.

b. Loosen the rear brake tube flare nut with a tubing wrench until a stream of brake fluid comes

out. Have an assistant maintain pressure on the brake pedal until the brake tube flare nut is tightened again.

c. Repeat this operation until clear, bubble-free fluid comes out from around the tube flare nut. Refill the master cylinder remote reservoir as necessary.

d. Repeat this bleeding operation at the front brake tube flare nut.

8. If any of the brake tubes, front disc brake calipers or rear wheel cylinders have been removed, it may be helpful to prime the system by gravity bleeding. This should be done after the master cylinder is primed and bled. To gravity bleed the brake system:

a. Fill the remote fluid reservoir with fresh DOT brake fluid.

b. Loosen both of the rear wheel cylinder bleeder screws (one at each rear wheel brake backing plate) and leave them open until clear brake fluid flows. Be sure to check the reservoir fluid level often and do not let it run dry.

c. Tighten the bleeder screws. Do not over-torque. Specification is only 61-87 inch lbs. (7-9 Nm).

d. One at a time, loosen the caliper bleeder screws. Leave the bleeder screws open until clear fluid flows. Check the reservoir fluid level often and do not let it run dry.

e. Tighten the caliper bleeder screws. Torque to 12-17 ft. lbs. (17-24 Nm).

9. After the brake system has been primed, use the following procedures as necessary and bleed the brake system at each wheel.

Manual Brake System Bleeding

This procedure requires two technicians.

1. Clean all dirt from the master cylinder remote fluid reservoir filler cap. Important: The reservoir must be at least 3/4-full during the bleeding procedure. Fill the reservoir as necessary. Use only clean, fresh DOT 3 brake fluid from a sealed container. Fill to the MAX level line on the reservoir.

2. If the master cylinder is known or suspected to contain air, it must be bled before the wheel cylinders or calipers. To bleed the master cylinder, loosen one brake line flare nut and have an assistant push the brake pedal slowly its full travel. While the assistant holds down the brake pedal, tighten the flare nut. Repeat this procedure for the other brake line flare

Parking Brake System — Exploded View

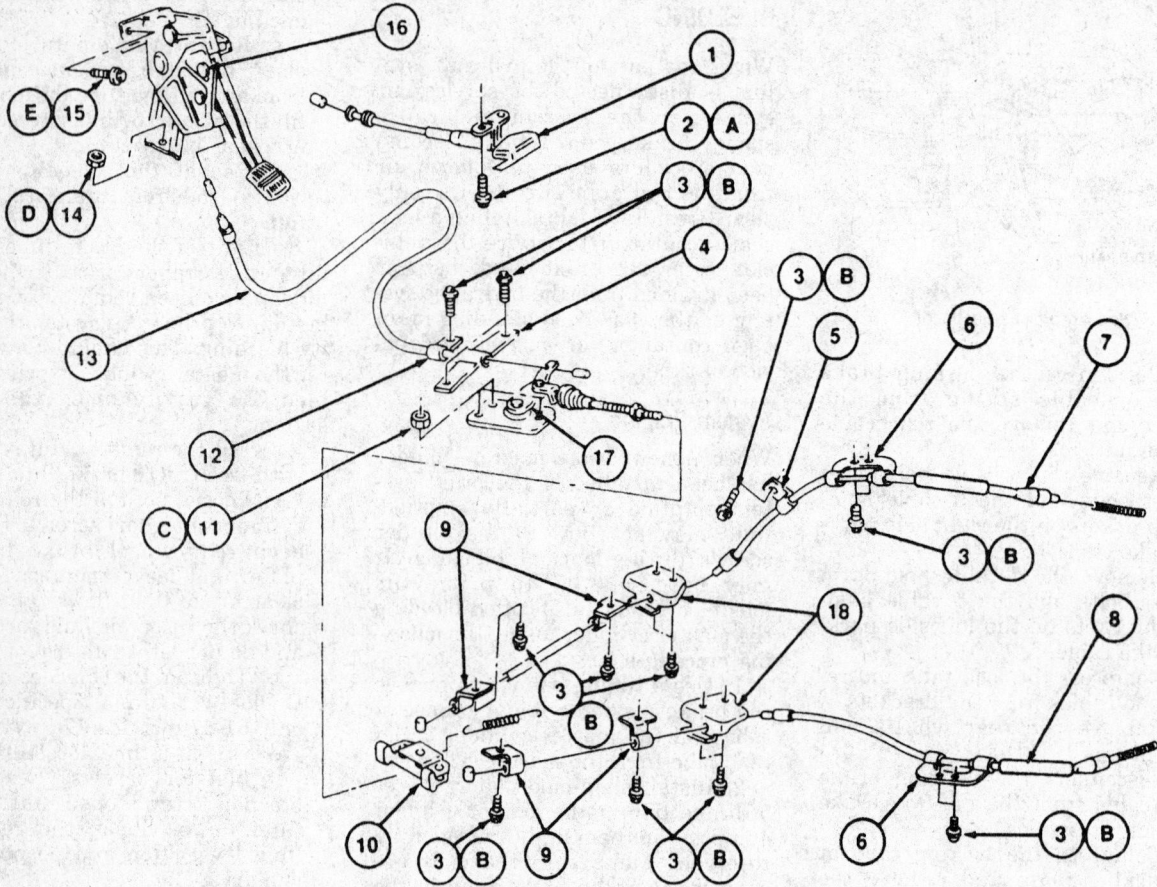

Item	Description	Item	Description
1	Parking Brake Release Handle and Cable	12	Cable Hold-Down Bracket
2	Parking Brake Release Lever Screws (2 Req'd)	13	Front Parking Brake Cable and Conduit
3	Cable Hold-Down Bracket Bolts	14	Parking Brake Control Nut (2 Req'd)
4	Cable Hold-Down Bracket	15	Parking Brake Control Bolt (2 Req'd)
5	Cable Hold-Down Bracket	16	Parking Brake Control
6	Cable Hold-Down Bracket	17	Parking Brake Control Lever
7	Parking Brake Rear Cable and Conduit, RH	18	Cable Hold-Down Bracket
8	Parking Brake Rear Cable and Conduit, LH	A	Tighten to 2-3 N·m (18-27 Lb-In)
9	Cable Hold-Down Bracket	B	Tighten to 3-4 N·m (27-35 Lb-In)
10	Parking Brake Cable Equalizer	C	Tighten to 8-10 N·m (71-88 Lb-In)
11	Lever Mechanism Nuts (3 Req'd)	D	Tighten to 8-11 N·m (71-97 Lb-In)
		E	Tighten to 17-23 N·m (12-17 Lb-Ft)

Parking brake system - exploded view

356165

nut. Repeat this entire procedure several times to ensure all air has been removed from the master cylinder.

3. Manual system bleeding must be performed in the following order: Left Front, Right Front, Left Rear, Right Rear.

4. Remove the protective bleeder screw cap from the appropriate brake.

5. Position a wrench on the bleeder screw.

6. Attach a hose to the bleeder screw. The hose must fit tightly around the bleeder screw.

7. Submerge the free end of the hose in a container partially filled with brake fluid.

8. Loosen the bleeder screw approximately ¾ of a turn.

9. Have the assistant push the brake pedal slowly through its full travel and hold down. The brake pedal must remain depressed until the bleeder screw is tightened.

10. Tighten the bleeder screw.

11. Have the assistant release the brake pedal.

12. Repeat Steps 8 through 11 until there are no more air bubbles at the submerged end of the bleeder hose.

13. Tighten the front disc brake caliper bleeder screws to 12-17 ft. lbs. (17-24 Nm) and the rear wheel cylinder bleeder screws to 61-87 inch lbs. (7-9 Nm).

14. Remove the bleeder hose and attach the protective screw caps.

15. Repeat these steps at each brake.

16. Check the fluid level in the remote reservoir, refilling with DOT 3 brake fluid as necessary.

17. Check the brake pedal feel. If spongy, repeat the bleeding process and/or look for defective system components.

Pressure Brake System Bleeding

When using pressure bleeding equipment, use a bladder-type bleeder tank. In this type of bleeder, the brake fluid is separated from the air by a rubber diaphragm. The bleeder tank must contain enough brake fluid to complete the bleeding operation and should be charged with only 10-30 psi. Never exceed 50 psi.

1. Clean all dirt from the master cylinder remote fluid reservoir filler cap. Important: The reservoir must be at least ¾-full during the bleeding procedure. Fill the reservoir as necessary. Use only clean, fresh DOT 3 brake fluid from a sealed container.

Fill to the MAX level line on the reservoir.

2. Install the bleeder adapter tool on the master cylinder and attach the hose from the bleeder tank to the fitting on the adapter. Follow the manufacturer's instructions when installing and connecting the master cylinder adapter.

3. Open the valve on the bleeder tank.

4. If the master cylinder is known or suspected to contain air, it must be bled before the wheel cylinders or calipers. To bleed the master cylinder, loosen and tighten the brake line flare nuts. Beginning at the front, alternately loosen and tighten the brake line flare nuts. Allow the fluid to flow for several seconds before tighten the flare nut. Repeat this operation several times to make sure all air has been removed from the master cylinder.

5. Pressure system bleeding must be performed in the following order: Left Front, Right Front, Left Rear, Right Rear.

6. Remove the protective bleeder screw cap from the appropriate brake.

7. Position a wrench on the bleeder screw.

8. Attach a hose to the bleeder screw. The hose must fit tightly around the bleeder screw.

9. Submerge the free end of the hose in a container partially filled with brake fluid.

10. Loosen the bleeder screw approximately ¾ of a turn. When the fluid entering the catch container is completely free of bubbles, tighten the bleeder screw. Tighten the front disc brake caliper bleeder screws to 12-17 ft. lbs. (17-24 Nm) and the rear wheel cylinder bleeder screws to 61-87 inch lbs. (7-9 Nm).

11. Remove the bleeder hose and attach the protective screw caps.

12. Repeat these steps at each brake.

13. Close the valve at the bleeder tank, disconnect the hose from the master cylinder adapter and remove the master cylinder adapter.

14. Check the fluid level in the remote reservoir, refilling with DOT 3 brake fluid as necessary.

15. Check the brake pedal feel. If spongy, repeat the bleeding process and/or look for defective system components.

FRONT SUSPENSION

Strut

REMOVAL AND INSTALLATION

1. Disconnect the negative battery cable.

2. Match mark the front strut upper mounting bracket and the chassis strut tower.

3. Raise and safely support the vehicle.

4. Remove the wheel.

5. If equipped, remove the two front brake anti-lock sensor cable bracket bolts and position the anti-lock sensor cable out of the way.

6. Detach the brake tube from the strut.

7. Support the control arm.

8. Remove the strut-to-steering knuckle bolts.

9. Support the strut and remove the 3 upper strut-to-chassis nuts. Remove the strut from the vehicle.

———— WARNING ————
Never loosen the strut center nut until the spring is compressed or serious injury or vehicle damage may occur.

10. Place the strut and coil spring assembly in a suitable vise and remove the strut nut cover.

11. Loosen, but do not remove, the front strut nut.

12. Using an approved coil spring compressor, compress the coil spring.

13. Remove the strut assembly top nut.

14. Remove the following components from the strut assembly:
   a. The upper mounting bracket.
   b. The strut bearing.
   c. The bearing seat.
   d. The upper coil spring seat and dust boot.
   e. The coil spring.

15. Slowly release the tension of the coil spring compressor and remove the coil spring from the compressor tool.

16. Remove the coil spring insulator and slide the jounce bumper off of the strut assembly.

To install:

17. Slide the jounce bumper onto the strut assembly and install the coil spring insulator.

18. Carefully compress the coil spring with an approved coil spring compressor.

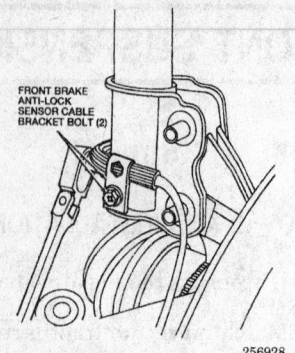

Brake hose attaching bracket

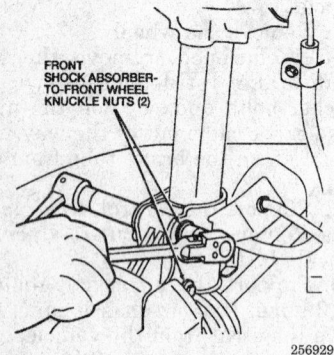

Strut to steering knuckle bolts

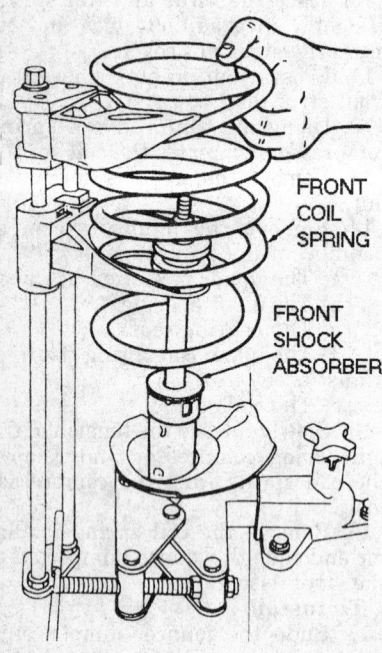

Compressing the coil spring

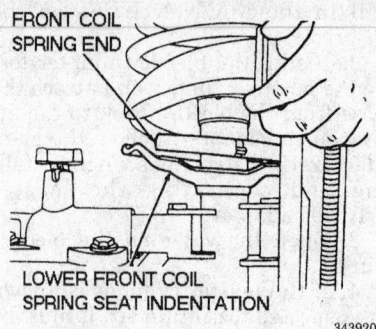

Correctly seating the coil spring to the strut assembly

19. Reinstall the following components to the strut assembly.
   a. The coil spring.

NOTE: Install the coil spring to the strut assembly with the end of the spring in the lower coil spring seat indentation.

   b. The upper coil spring seat and dust boot.
   c. The bearing seat and the bearing.
   d. The upper mounting bracket.
20. Install and tighten the strut assembly nut and torque the nut to 43–58 ft. lbs. (59–78 Nm).
21. Install the strut assembly onto the vehicle and torque the following:
   • Strut-to-body nuts: 29–40 ft. lbs. (39–54 Nm)
   • Strut-to-knuckle bolts: 89–91 ft. lbs. (113–123 Nm)
22. Reattach the brake tube to the strut assembly.
23. Install and tighten the two front brake anti-lock sensor cable bracket bolts.
24. Reinstall the tire and wheel assembly.
25. Connect the negative battery cable and the adjustable strut electrical connectors, if equipped.
26. Check and/or adjust the wheel alignment.

Lower Ball Joints

REMOVAL AND INSTALLATION

To check if ball joint replacement is required, raise and safely support the vehicle clear of the floor and try to rock the wheel up and down. If any play is felt, have an assistant rock the wheel while observing the front suspension lower arm ball joint at the bottom of the steering knuckle. If any movement is seen, the ball joint should be replaced. If not, any wheel play indicates wheel bearing wear.

1. Raise and safely support the vehicle.
2. Remove the tire and wheel.
3. Remove and discard the ball joint cotter pin. Loosen the ball joint attaching nut from the steering knuckle. Because of tight clearance, the nut likely cannot be removed until the ball joint stud is loosened and lower slightly.
4. Strike the front knuckle with a hammer while pulling down on the lower control arm. There should now be enough clearance to allow removal of the ball joint stud nut. Separate the ball joint from the steering knuckle.
5. Remove the three bolts attaching the ball joint to the control arm.
6. Remove the ball joint from the control arm.
   To install:
7. Install the ball joint to the control arm and install the attaching bolts.
8. Torque the bolts to 56–80 ft. lbs. (76–109 Nm).
9. Install the ball joint into the steering knuckle, just enough to get the nut started on the stud. Then push the ball joint stud fully in place. Torque the nut to 52–63 ft. lbs. (71–86 Nm). Secure the nut with a new cotter pin.
10. Install the tire and wheel.
11. Lower the vehicle.
12. A front end alignment check is recommended.

Lower Control Arms

REMOVAL AND INSTALLATION

The front suspension lower control arms control lateral (side-to-side) movement of each front wheel. The inner pivot attachment is the pivot point for the suspension. The lower control arms mount the ball joints that connect to the front wheel steering knuckles. The front suspension lower control arm ball joints and the control arm mounting bolt bushings can be replaced individually.

1. Raise and safely support the vehicle.
2. Remove the wheel and tire assembly.
3. Disconnect the lower ball joint using the recommended procedure.
4. Disconnect the stabilizer bar link nut, washer and insulator and separate the stabilizer bar link shaft from the lower control arm

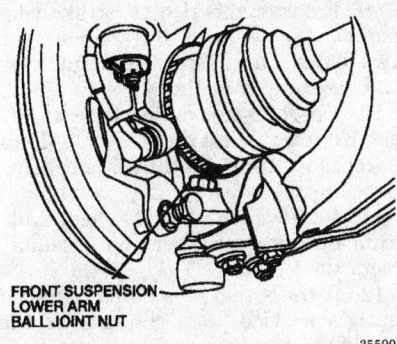

FRONT SUSPENSION
LOWER ARM
BALL JOINT NUT

355901

Loosen the lower ball joint nut

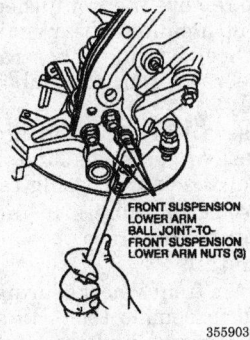

FRONT SUSPENSION
LOWER ARM
BALL JOINT-TO-
FRONT SUSPENSION
LOWER ARM NUTS (3)

355903

Remove the three bolts attaching the ball joint to the control arm

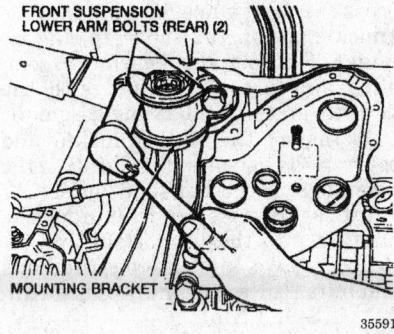

FRONT SUSPENSION
LOWER ARM BOLTS (REAR) (2)

MOUNTING BRACKET

355917

Remove the two lower control arm nut rear bolts

5. Remove the two bolts attaching the control arm to the frame.
6. Remove the lower arm nut.
7. Pull the rear of the lower control arm down and gently pry the control arm forward and off of the lower arm gusset.

To install:
8. Install the lower control to its mounting on the gusset. Install the lower control arm nut and torque to 94–115 ft. lbs. (128–156 Nm).
9. Install the two rear control arm bolts to the frame and torque to 87–108 ft. lbs. (118–147 Nm).

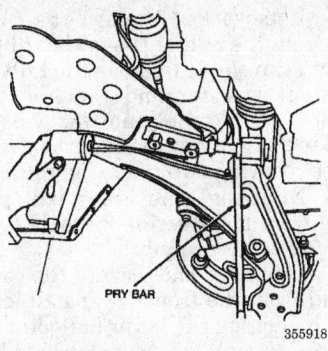

PRY BAR

355918

Pull the control arm down and pry forward and off of the arm gusset

10. Assemble the stabilizer bar link nut, washer and insulator and install on the lower control arm. Torque the nut to 34–38 ft. lbs. (46–51 Nm).
11. Install the lower ball joint using the recommended procedure.
12. Install the wheel and tire assembly.
13. Lower the vehicle.
14. A front end alignment check is recommended.

Sway Bar

REMOVAL AND INSTALLATION

The front sway bar (also called a stabilizer bar) redirects forces to control vehicle roll during cornering. All suspension mounting points are rubber insulated to minimize transfer of road noise and vibration to the body and interior. All front stabilizer bar components can be replaced individually.
1. Raise and safely support the vehicle.
2. Remove the left side and right side front stabilizer bar-to-stabilizer link attaching nuts.
3. Remove the left side and right side stabilizer bar mounting bracket nuts.

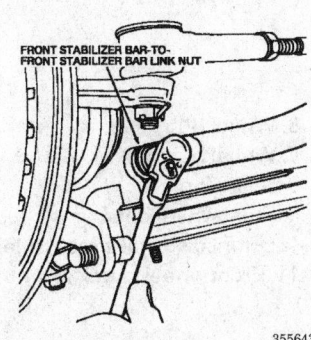

FRONT STABILIZER BAR-TO-
FRONT STABILIZER BAR LINK NUT

355643

Remove the left side and right side front stabilizer bar-to-stabilizer link attaching nuts

4. Remove the left side and right side stabilizer bar-to-lower arm gusset bolts.
5. Gently pry the ends of the stabilizer bar off the front stabilizer bar links and remove the stabilizer bar from the vehicle.
6. If the front stabilizer bar links must be removed, hold the link shaft with a wrench and remove the link lower nut. Remove the link and the two insulators and washers.

To install:
7. If the front stabilizer bar links were be removed, assemble the insulator and washers to the link then install and hold the link shaft with a wrench while tightening the link lower nut to 12–16 ft. lbs. (16–22 Nm).
8. Align the ends of the stabilizer bar with the links and gently push to bar into place.
9. Position the mounting brackets and torque the nuts to 23–31 ft. lbs. (31–42 Nm).
10. Install and tighten the gusset bolts to 23–31 ft. lbs. (31–42 Nm).
11. Install the stabilizer bar to link nuts and torque the nuts to 34–38 ft. lbs. (46–51 Nm).
12. Lower the vehicle. Road test to verify no suspension noise.

Front Wheel Bearings

REMOVAL AND INSTALLATION

The front wheel knuckles transmit steering input, pivoting on the lower control arm ball joints and upper front strut bearing, house driveline components and support the disc brake calipers.

The front wheel hubs and the front steering knuckles can be replaced independently. The front hub/knuckle assemblies can also be disassembled in order to replace the knuckles, hubs, wheel hub bolts, front wheel bearings and front disc brake rotor shield. The front wheels attach to the wheel hub and the brake rotor. The wheel hub and front disc brake rotor are supported by a one-piece front wheel bearing pressed into the knuckle.
1. Raise and safely support the vehicle.
2. Remove the wheel and tire.
3. Remove the brake caliper assembly. DO NOT disconnect the brake hose. Hang the caliper on a piece of wire from a near by support such as the strut.
4. Remove the brake rotor.

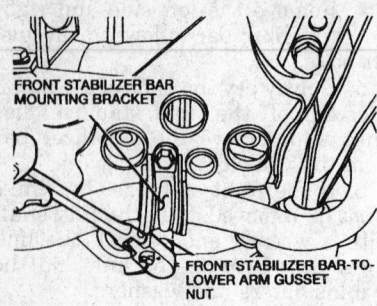

Remove the left side and right side stabilizer bar mounting bracket nuts

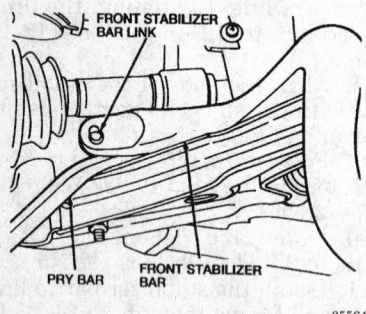

Gently pry the ends of the stabilizer bar off the front stabilizer bar links

5. Remove and discard the cotter pin from the end of the outboard CV-Joint stub shaft. Remove the hub nut retainer, washer and the hub nut. There should be another washer under the hub nut that acts as a front wheel bearing outer bearing retainer.

6. Disengage the lower ball joint stud from the steering knuckle using the following procedure.

　a. Remove and discard the cotter pin from the front lower ball joint.

　b. Loosen the lower ball joint nut until it contacts the front halfshaft joint.

　c. Strike the front knuckle with a hammer while pulling down on the lower control arm until the ball joint stud separates from the knuckle.

　d. Remove the ball joint nut.

　e. Disengage the lower ball joint stud from the steering knuckle.

7. Disengage the outer tie rod end stud from the steering knuckle using the following procedure.

　a. Remove and discard the cotter pin from the outer tie rod end stud.

　b. Remove the outer tie rod end retaining nut.

　c. Use a tie rod end puller to carefully press the tie rod end from the steering knuckle.

8. Remove the front ABS sensor bolt.

9. Remove the two front strut-to-front knuckle nuts and remove the two bolts. Disengage the strut from the steering knuckle.

10. Use a 2-jaw puller to separate the front halfshaft outboard CV-Joint stub shaft from the knuckle/bearing assembly.

11. Remove the front wheel hub, knuckle and wheel bearing assembly from the vehicle.

12. If the knuckle is being replaced with a service part, changeover the steering stop bolt and jam nut from the old knuckle to the replacement part.

13. To remove the front wheel bearing, jig up a puller to bear against the front wheel bearing inner race and pull the race from the hub/knuckle assembly.

14. Use a shop press to press out damaged wheel studs and also to press out the outer bearing race.

15. Use a shop press to press out the inner bearing race.

　To install:

16. If the front wheel bearings were removed, assemble the ABS sensing ring, if removed and the disc brake dust shield under the steering knuckle. Use a shop press to push in new front wheel bearing inner and outer races. Support the knuckle and press the front wheel bearing into the knuckle and install the snap ring retainer. Support the bearing assemblies and press the hub onto the knuckle and wheel bearing assembly.

17. Install the hub, knuckle and bearings as an assembly. Position the assembly on the halfshaft outer CV-Joint stub axle end. Guide the knuckle into the front strut and install the two knuckle-to-strut bolts and nuts. Torque the nuts to 83–91 ft. lbs. (113–123 Nm).

18. Install the ABS sensor bolt. Do not overtighten. Torque to just 16–21 inch lbs. (1.8– 2.4 Nm).

19. Install the outer tie rod end to the steering knuckle. Torque the nut to 22–29 ft. lbs. (29–39 Nm). If the cotter pin holes do not align, tighten the nut slightly until they do. Never loosen the nut to align the holes. Secure the nut with a new cotter pin.

20. Start the lower ball joint stud to the steering knuckle and partially install the nut, then push the ball joint stud fully in place. Tighten the ball joint stud nut to 52– 63 ft. lbs. (71– 86 Nm). Secure the nut with a new cotter pin.

21. Install the front wheel outer bearing retaining washer and the hub retainer nut. Torque to 174–231 ft. lbs. (235–314 Nm). Install the nut

Wheel Hub, Wheel Knuckle and Wheel Bearing Assembly - Exploded View

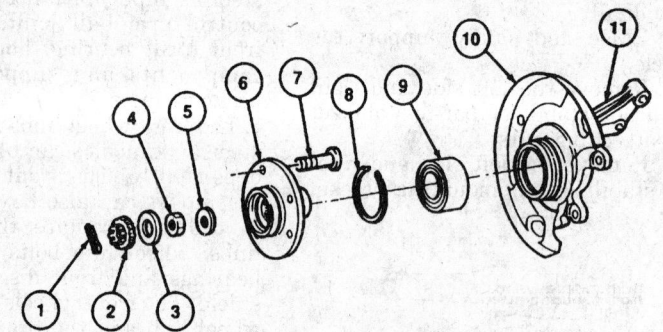

1. Cotter pin
2. Nut retainer
3. Insulator
4. Front axle wheel hub retainer
5. Front wheel outer bearing retainer washer

6. Wheel hub
7. Wheel hub bolt
8. Snap ring
9. Front wheel bearing
10. Front disc brake rotor shield
11. Front wheel knuckle

Front hub, bearing and knuckle arrangement

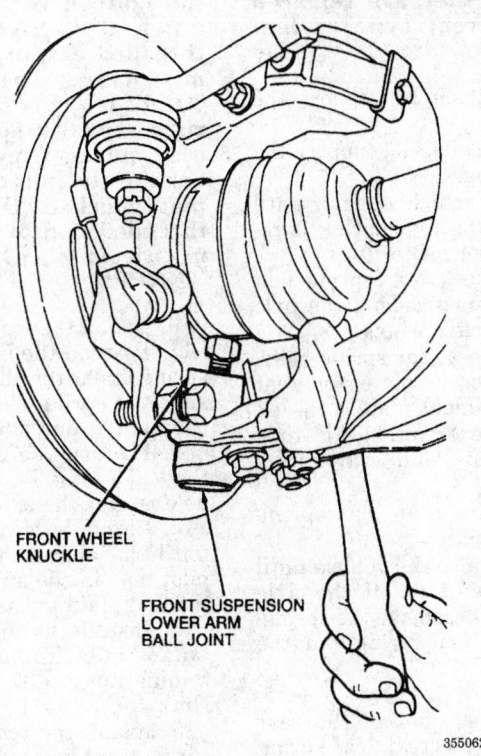

355062

Strike the front knuckle with a hammer while pulling down on the lower control arm

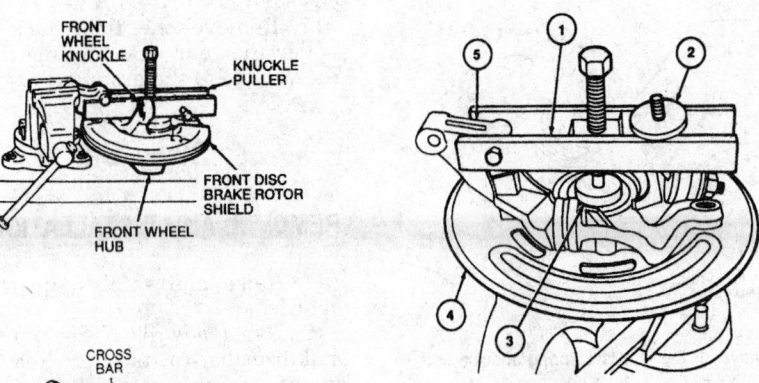

1. Knuckle puller
2. Knuckle puller adapter
3. Step plate adapter
4. Front disc brake rotor shield
5. Front wheel knuckle

355067

Example of a puller set up to bear against the front wheel bearing inner race

retainer, insulator and a new cotter pin.

22. Install the front brake rotor and install the disc brake caliper using the recommended procedure.

23. If removed, install the steering stop bolt.

24. Install the tire and wheel assembly. Tighten the lug nuts to 72 to 87 ft. lbs. (98 to 118 Nm).

25. Lower the vehicle. Pump the brake pedal slowly to seat the front brake pads. Do not move the vehicle until a firm pedal is obtained.

26. A front end alignment is recommended.

REAR SUSPENSION

Shock Absorber

REMOVAL AND INSTALLATION

1. Raise and safely support the vehicle.

2. Support the rear axle and slightly lower the vehicle enough to lessen tension on the shock absorber.

3. Remove the lower shock absorber retaining nut and washer.

4. Disconnect the lower end of the shock absorber from the mounting stud.

5. Remove the shock absorber upper end retaining nut and washer.

6. Remove the shock absorber from the vehicle.

To install:

7. Install the shock absorber onto the upper and lower mounting studs of the vehicle.

8. Install the washers and retaining nuts. Torque the upper and lower retaining nuts to 22–30 ft. lbs. (30–41 Nm).

9. Lower the vehicle.

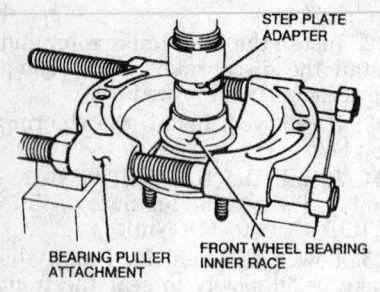

Use a shop press to press out the inner bearing race

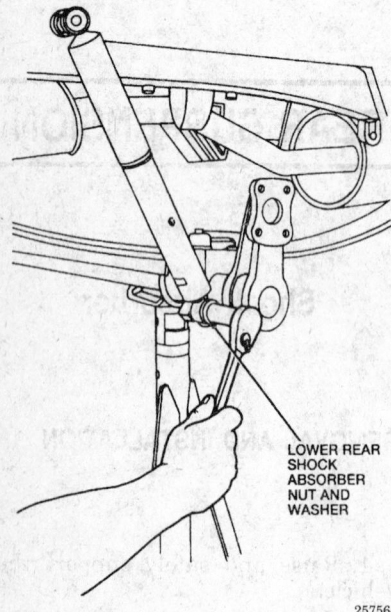

Removal of lower shock absorber retaining nut and washer

Leaf Spring

REMOVAL AND INSTALLATION

NOTE: There are 2 different leaf spring assemblies which appear almost identical in appearance. The optional Handling Package includes a modified spring rate leaf spring and stabilizer bar. If it does NOT have the Handling Package, it will NOT have a rear stabilizer bar or modified spring rate leaf spring. Be sure that when the leaf springs are replaced, they are replaced with the correct type of leaf springs.

1. Raise and safely support the vehicle.
2. Using a floor jack, support the rear axle housing.
3. Lower the vehicle slightly until the tension on the rear spring is relieved by the floor jack.
4. Remove the 4 leaf spring bolt nuts and washers securing the axle housing to the leaf springs.
5. Remove the 2 rear spring bolts, alignment bolt cover plate and rear spring bolt alignment plate.
6. Remove the 2 rear shackle nuts, shackle end plate and rear spring shackle.
7. Remove the rear leaf spring front nut and bolt.
8. Carefully raise the vehicle until the vehicle weight is off the axle housing enough to remove the spring. Remove the rear leaf spring from the vehicle.

To install:

9. Position the front of the rear leaf spring into the rear leaf spring front mounting plates.

NOTE: When installing the left rear leaf spring, install the rear spring front bolt with the head of

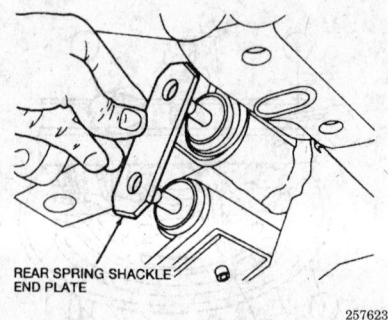

Removal of the rear leaf spring shackle end plate

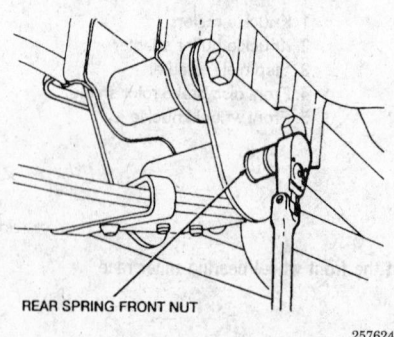

Removing the rear spring front nut

the bolt on the mounting plate marked "INNER", and the threaded end of the bolt on the mounting plate marked "OUTER". When installing the right rear leaf spring, install the rear spring front bolt with the head of the bolt on the mounting plate marked "OUTER", and the threaded end of the bolt on the mounting plate marked "INNER".

10. Install the leaf spring front bolt.
11. Position the rear end of the leaf spring under the chassis bushing and slide the spring shackle through the chassis bushing and rear spring.
12. Position the alignment bolt on the leaf spring. Be sure that the pilot hole is over the axle pad.
13. Carefully lower the the vehicle until the alignment bolt enters the pilot hole in the axle pad and install the end plate to the spring shackle.
14. Install the spring shackle retaining nuts. Torque the shackle retaining nuts to 37–50 ft. lbs. (50–68 Nm).
15. Install the retaining nut to the spring front bolt and torque to 37–50 ft. lbs. (50–68 Nm).
16. Install the 2 rear spring bolts, alignment plate and alignment bolt cover plate. Install the 4 spring bolt washers and nuts. Torque the nuts to 53–72 ft. lbs. (71–98 Nm).
17. Remove the floor jack from under the rear axle housing. Lower the vehicle to the ground.

Sway Bar

REMOVAL AND INSTALLATION

1. Raise and safely support the vehicle.
2. Using a wrench to secure the stabilizer bar-to-stabilizer link bushing studs, remove the stabilizer bar-to-stabilizer link nuts. Disconnect the stabilizer bar from the stabilizer link.
3. Loosen the stabilizer bar upper mounting bracket bolts but do not remove. Remove the lower stabilizer bar mounting bracket bolts.
4. Slide the stabilizer bar down until the mounting brackets clear the loosened upper mounting bolts and remove the stabilizer bar from the vehicle.

To install:

5. Position the stabilizer bar and mounting brackets under the loosened upper mounting bracket bolts. Torque the upper mounting bracket bolts to 23–31 ft. lbs. (31–42 Nm).

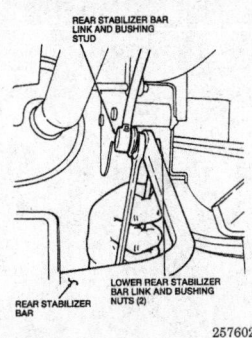

Removal of the stabilizer bar-to-stabilizer link and bushing retaining nuts

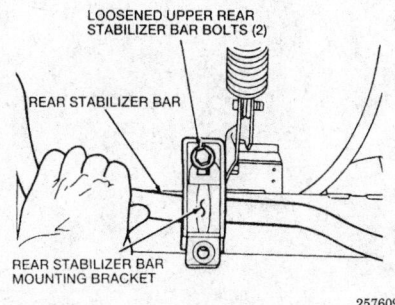

Removing stabilizer bar from the from the vehicle

6. Install the lower stabilizer bar mounting bracket bolts and torque to 23–31 ft. lbs. (31–42 Nm).

7. Connect the stabilizer bar ends to the stabilizer links. Install the stabilizer bar link and bushing nuts and torque to 29–33 ft. lbs. (39–45 Nm).

8. Lower the vehicle.

Wheel Bearings

REMOVAL AND INSTALLATION

1. Raise and safely support the vehicle.

2. Remove the rear wheel(s).

3. Remove the brake drum.

4. Remove the grease cap for the hub.

5. Remove and discard the cotter pin.

6. Remove the wheel bearing nut and washer.

7. Remove the rear wheel hub and bearing assembly.

8. Using a pair of snap ring pliers, remove the bearing snapring from the hub unit.

9. Using bearing cup replacer tool T77F-1202-A or equivalent driver and a press, press out the rear wheel bearing from the hub.

To install:

10. Supporting the hub with dust shield replacer tool T87C-1175-B or equivalent, press the wheel bearing into the hub using pinion bearing cup replacer tool T80T-4000-E or equivalent driver.

11. Install the wheel bearing snapring into the rear of the hub unit.

12. Install the rear wheel hub and bearing assembly onto the vehicle.

13. Install the rear wheel bearing washer and nut and torque the bearing nut to 145–210 ft. lbs. (196–284 Nm). Install a new cotter pin.

14. Install the wheel hub grease cap. Install the brake drum.

15. Install the rear wheel(s) and lug nuts. Torque the lug nuts, in a star sequence, to 72–87 ft. lbs. (98–118 Nm).

16. Lower the vehicle.

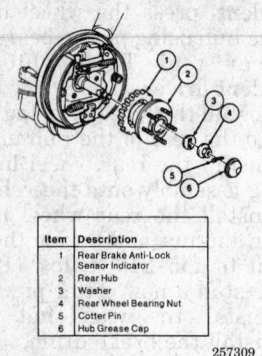

Item	Description
1	Rear Brake Anti-Lock Sensor Indicator
2	Rear Hub
3	Washer
4	Rear Wheel Bearing Nut
5	Cotter Pin
6	Hub Grease Cap

257309

Rear wheel hub and bearing assembly

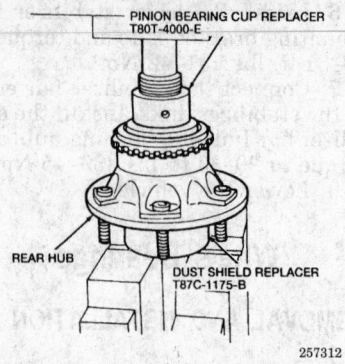

257312

Installing the wheel bearing into the hub unit

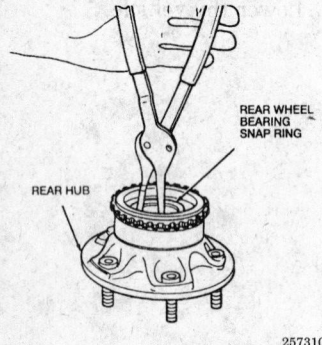

257310

Removing the wheel bearing snapring

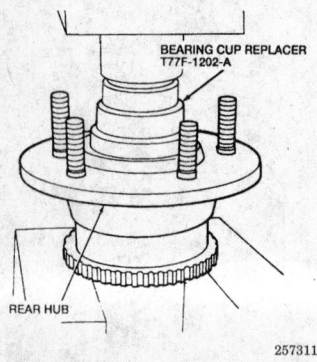

257311

Removing the wheel bearing from the hub unit

FIRING ORDERS

NOTE: To avoid confusion, always replace spark plug wires one at a time.

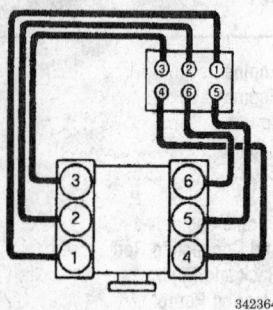

3.0L (VIN U) Engine
Engine Firing Order: 1-4-2-5-3-6
Distributorless Ignition System

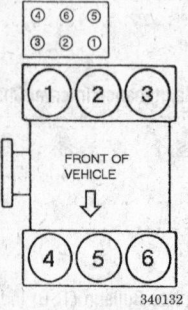

3.8L (VIN 4) Engine
Engine Firing Order:
1-4-2-5-3-6
Distributorless
Ignition System

ENGINE ELECTRICAL

NOTE: Disconnecting the negative battery cable on some vehicles may interfere with the functions of the on board computer systems and may require the computer to undergo a relearning process, once the negative battery cable is reconnected.

Ignition Timing

ADJUSTMENT

The ignition timing is preset at 10 degrees BTDC and is not adjustable.

Alternator

PRECAUTIONS

Several precautions must be observed with alternator equipped vehicles to avoid damage to the unit.

• If the battery is removed for any reason, make sure it is reconnected with the correct polarity. Reversing the battery connections may result in damage to the 1-way rectifiers.

• When utilizing a booster battery as a starting aid, always connect the positive to positive terminals and the negative terminal from the booster battery to a good engine ground on the vehicle being started.

• Never use a fast charger as a booster to start vehicles.

• Disconnect the battery cables when charging the battery with a fast charger.

• Never attempt to polarize the alternator.

• Do not use test lights of more than 12 volts when checking diode continuity.

• Do not short across or ground any of the alternator terminals.

• The polarity of the battery, alternator and regulator must be matched and considered before making any electrical connections within the system.

• Never separate the alternator on an open circuit. Make sure all connections within the circuit are clean and tight.

• Disconnect the battery ground terminal when performing any service on electrical components.

• Disconnect the battery if arc welding is to be done on the vehicle.

REMOVAL AND INSTALLATION

3.0L (VIN U) Engine

1. Disconnect the negative battery cable.
2. Tag and disconnect the electrical harness connectors from the alternator (generator) and voltage regulator.
3. Rotate the accessory drive belt tensioner counterclockwise and remove the drive belt from the alternator pulley.
4. Remove 2 alternator brace retaining bolts and 1 nut and remove the alternator brace.
5. Remove 2 alternator retaining bolts and remove the alternator.
To install:
6. Place the alternator in position and install 2 alternator retaining bolts. Do not tighten at this time.

7. Install the alternator brace and 2 retaining bolts and 1 nut. Tighten the alternator brace retaining nut to 15–22 ft. lbs. (20–30 Nm).
8. Rotate the accessory drive belt tensioner counterclockwise and install the drive belt on the alternator pulley.
9. Tighten 2 alternator brace bolts to 15–22 ft. lbs. (20–30 Nm) and 2 alternator retaining bolts to 30–41 ft. lbs. (40–55 Nm).
10. Connect the electrical harness connectors to the alternator and voltage regulator. Tighten the output terminal nut to 80–97 inch lbs. (9–11 Nm).
11. Connect the negative battery cable.
12. Start the engine and check for proper charging system operation.

3.8L (VIN 4) engine

1995 Models

1. Disconnect the negative battery cable.
2. Tag and disconnect the electrical harness connectors from the alternator (generator) and voltage regulator.
3. Loosen the alternator pivot bolt and remove the mounting brace bolt from the alternator.
4. Rotate the belt tensioner counterclockwise and remove the drive belt from the alternator pulley.
5. Remove the alternator brace.
6. Remove the alternator pivot bolt and remove the alternator/regulator assembly from the vehicle.
To install:
7. Place the alternator/regulator assembly in position on the engine.
8. Install the alternator pivot and mounting brace bolts. Do not tighten the bolts until the drive belt is installed and tensioned.
9. Install the alternator brace and retaining nut. Tighten the retaining nut to 15–22 ft. lbs. (20–30 Nm).
10. Rotate the belt tensioner counterclockwise and install the drive belt to the alternator pulley.
11. Tighten the mounting brace bolt to 15–22 ft. lbs. (20–30 Nm) and the pivot bolt to 30–41 ft. lbs. (80–97 Nm).
12. Connect the alternator and voltage regulator electrical harness connectors. Tighten the output terminal retaining nut to 80–97 inch lbs. (9–11 Nm).
13. Connect the negative battery cable.
14. Start the engine and check the charging system for proper operation.

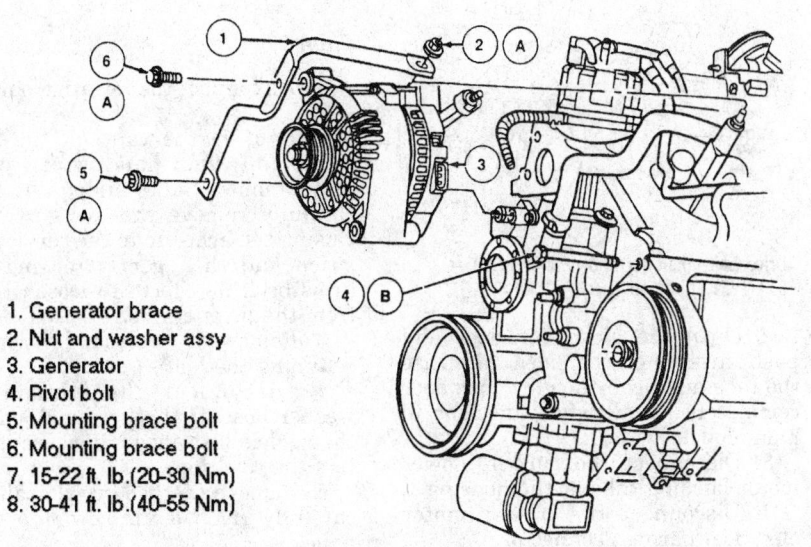

1. Generator brace
2. Nut and washer assy
3. Generator
4. Pivot bolt
5. Mounting brace bolt
6. Mounting brace bolt
7. 15-22 ft. lb.(20-30 Nm)
8. 30-41 ft. lb.(40-55 Nm)

327444

Alternator removal — 1995 3.8L (VIN 4) engine

1996–97 Models

1. Disconnect the negative battery cable.

2. Tag and disconnect the electrical harness connectors from the alternator (generator) and voltage regulator.

3. Rotate the belt tensioner counterclockwise and remove the drive belt from the alternator pulley.

4. Remove 3 bolts retaining the alternator to the engine.

5. Remove the alternator/voltage regulator assembly from the vehicle.

To install:

6. Place the alternator/regulator assembly in position on the engine.

7. Install 3 alternator retaining bolts and tighten to 30–41 ft. lbs. (40–55 Nm).

8. Rotate the belt tensioner counterclockwise and install the accessory drive belt to the alternator pulley.

9. Connect the alternator and voltage regulator electrical harness connectors. Tighten the output terminal retaining nut to 80–97 inch lbs. (9–11 Nm).

10. Connect the negative battery cable.

11. Start the engine and check the charging system for proper operation.

Drive Belt

REMOVAL AND INSTALLATION

————— **CAUTION** —————
Use care when removing or installing the accessory drive belt to ensure that the tool does not slip, possibly causing damage to the vehicle or bodily harm.

1. Disconnect the negative battery cable.

2. Place a 15mm socket and handle or a box wrench on the bolt head retaining the drive belt tensioner pulley. Rotate the drive belt tensioner clockwise to remove tension from the drive belt.

3. Remove the drive belt from the pulleys.

4. To remove the drive belt tensioner, remove the drive belt tensioner retaining bolt and the tensioner. On the 3.8L (VIN 4) engine, remove the tensioner from the right side cylinder head.

NOTE: If the drive belt is to be reinstalled, mark the belt's direction of rotation to prevent belt noise.

To install:

5. If removed, install the drive belt tensioner and retaining bolt. Tighten

the retaining bolt on the 3.0L (VIN U) engine to 30–41 ft. lbs. (40–55 Nm) and the 3.8L (VIN 4) engine to 52–70 ft. lbs. (70–95 Nm).

6. Inspect the pulleys for wear or damage. Clean the pulley grooves of any belt residue.

7. Install the drive belt over all of the pulleys except the drive belt tensioner. If the original belt is being installed, make sure that it is installed to run in its original direction.

8. Rotate the drive belt tensioner carefully and install the drive belt over the tensioner pulley. Ensure that all V-grooves make proper contact with the pulleys.

9. Relax the tensioner and remove the tool.

10. Connect the negative battery cable.

11. Start the engine and check the accessory drive belt for proper operation.

Starter

REMOVAL AND INSTALLATION

1. Disconnect the negative battery cable.

2. Raise and safely support the vehicle.

3. Disconnect the starter cable and the push-on connector from the starter solenoid.

NOTE: To disconnect the hardshell connector from the solenoid S terminal, grasp the plastic shell and pull off; do not pull on the wire. Pull straight off to prevent damage to the connector and S terminal.

4. Remove the upper and lower starter motor retaining bolts and remove the starter from the vehicle.

To install:

5. Place the starter motor in position and install the upper and lower starter motor retaining bolts. Tighten the bolts to 15–20 ft. lbs. (20–27 Nm).

6. Connect the starter solenoid connector making sure to push the connector straight on while listening for it to lock into position.

7. With the starter cable on the **B** terminal, install the retaining nut. Tighten the nut to 80–124 inch lbs. (9–14 Nm).

8. Install the red starter solenoid safety cap.

9. Lower the vehicle.

10. Connect the negative battery cable.

11. Check the starter motor for proper operation.

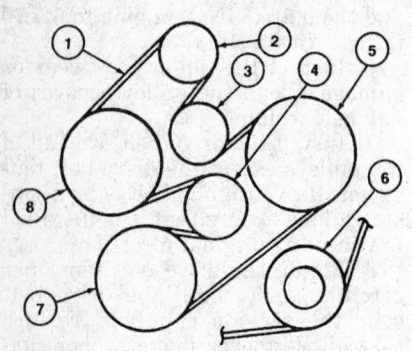

Item	Description
1	Drive Belt
2	Generator
3	Drive Belt Tensioner Pulley
4	Drive Belt Tensioner
5	Power Steering Pump Pulley
6	A/C Compressor
7	Crankshaft Pulley
8	Water Pump Pulley

352884

Accessory drive belt routing — 3.0L (VIN U) engine

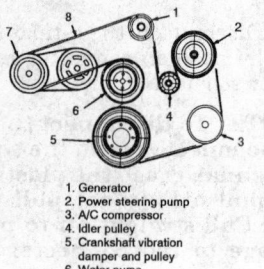

1. Generator
2. Power steering pump
3. A/C compressor
4. Idler pulley
5. Crankshaft vibration damper and pulley
6. Water pump
7. Drive belt tensioner
8. Drive belt

337502

Drive belt routing and pulley locations — 3.8L (VIN 4) engine

CHASSIS ELECTRICAL

Blower Motor

REMOVAL AND INSTALLATION

1. Disconnect the negative battery cable.

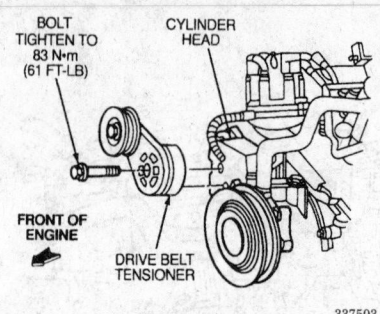

Drive belt tensioner and retaining bolt — 3.8L (VIN 4) engine

2. Open the glove box door and push inwards on the sides to release the tabs and open the glove box door completely by allowing the door to hang down.

3. Disconnect the rubber blower motor housing tube at the housing.

4. Disconnect the blower motor electrical harness connector.

5. Remove 4 heater blower motor retaining screws and remove the heater blower motor and wheel assembly from the housing.

6. To separate the blower wheel from the blower motor, pry off the pushnut from the blower motor shaft.

7. Slide the blower wheel off of the shaft.

To install:

8. If removed, place the blower wheel onto the blower motor shaft and install a new pushnut.

9. Place the blower motor and wheel assembly to the housing and install 4 blower motor retaining screws. Install the remaining components.

10. Connect the negative battery cable.

11. Check the blower motor for proper operation.

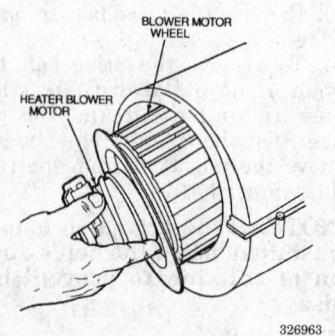

326963

Removing the blower motor and wheel assembly

Windshield Wiper Motor

REMOVAL AND INSTALLATION

Front

1. Disconnect the negative battery cable.

2. Remove the caps on the left-hand and right-hand wiper pivot arms, remove the retaining nuts and carefully remove the arms by first raising the arms up to the service position and then gently rocking the arms back and forth to loosen them from the pivot shafts.

3. Remove 4 rivets and 7 screws retaining the leaf screen.

4. Disconnect the windshield washer hose at the left-hand side of the vehicle just ahead of where it enters the cowl top.

5. Remove the hood cowl seal and carefully remove the cowl top vent panel.

6. Remove 3 retaining screws to separate the cowl drain tube from the top cowl.

7. Disconnect the wiring harness leading to the windshield wiper motor.

8. Remove 5 screws retaining the cowl top to the vehicle and remove the cowl top and wiper linkage assembly from the vehicle.

9. Remove the screw retaining the linkage drive arm to the wiper motor shaft. The linkage can be moved from side to side to aid removal of the screw.

10. Remove 3 screws retaining the windshield wiper motor to the linkage assembly and separate the parts.

11. Disconnect the electrical harness connector from the windshield wiper motor.

12. Remove the windshield wiper motor from the vehicle.

To install:

13. Position the windshield wiper motor to the linkage and install 3 retaining screws. Tighten the screws to 10–12 ft. lbs. (13–17 Nm).

14. Install the electrical harness connector to the windshield wiper motor.

15. Place the linkage drive arm to the wiper motor shaft assembly and install the retaining screw. Tighten the screw to 11–15 ft. lbs. (15–20 Nm). Install the remaining components.

16. Temporarily connect the negative battery cable and cycle the windshield wiper motor to park the linkage in the correct position. Disconnect the negative battery cable.

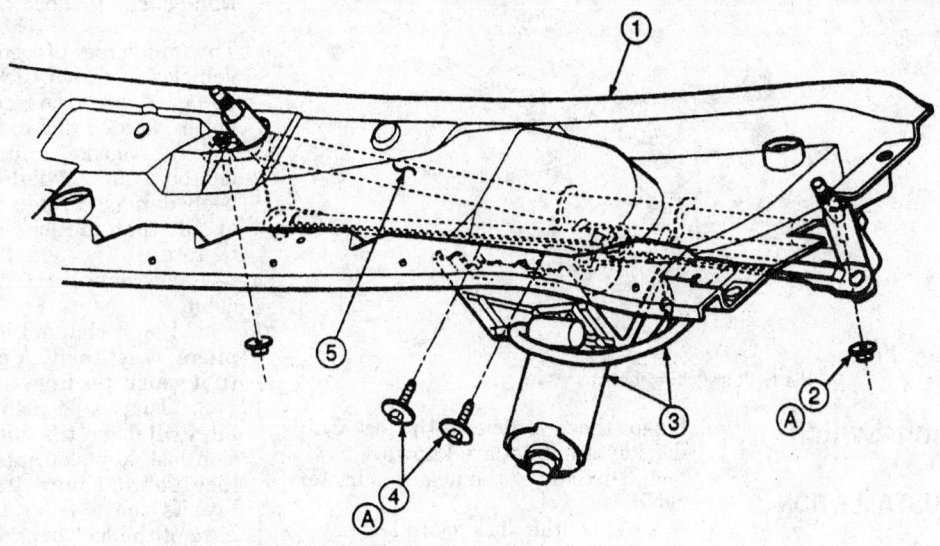

1. Cowl top panel
2. Nut (2)
3. Windshield wiper motor
4. Bolt (2)
5. Windshield wiper mounting arm & pivot shaft
A. 80-106 in. lb.(9-12 Nm)

327885

Front wiper motor assembly

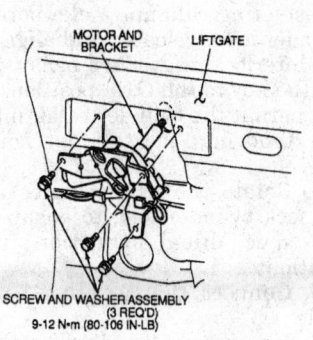

MOTOR AND BRACKET   LIFTGATE

SCREW AND WASHER ASSEMBLY
(3 REQ'D)
9-12 N·m (80-106 IN-LB)

327887

Rear wiper motor assembly

17. Install the wiper arm and blade pivot arms to the pivot shafts and position the pivot arms so that they do not contact the cowl top vent panel or the A-pillar when cycled. Tighten the retaining nuts to 22–29 ft. lbs. (30–40 Nm). Install the caps.

18. Connect the negative battery cable.

19. Check for proper windshield wiper motor operation.

Rear

1. Disconnect the negative battery cable.

2. Remove the wiper pivot arm and blade cap and nut. Carefully rock the arm to loosen the arm from the pivot shaft and lift the arm off of the shaft.

3. Disconnect the electrical harness connector to the wiper motor. Pull on the shell connector not the wiring.

4. Remove 3 screws retaining the wiper motor to the liftgate and remove the rear wiper motor.

To install:

5. Place the rear wiper motor to the liftgate and install 3 retaining screws. Tighten the screws to 80–106 inch lbs. (9–12 Nm).

6. Install the electrical harness connector to the wiper motor.

7. Temporarily connect the negative battery cable and cycle the rear wiper motor to park in the correct position. Disconnect the negative battery cable.

8. Install the wiper pivot arm and blade to the pivot shaft and position the pivot arm so that it has a 51 ± 5mm gap between the blade and the lower glass moulding. Tighten the retaining nut to 10–13 ft. lbs. (14–18 Nm). Install the cap.

9. Connect the negative battery cable.

10. Check the rear wiper motor for proper operation.

Headlight Switch

REMOVAL AND INSTALLATION

1. Disconnect the negative battery cable.

2. Pull off the headlight switch knob and remove the retaining nut.

3. Carefully pry off the left-hand instrument panel finish panel.

4. Remove 2 screws retaining the headlight switch to the instrument panel.

5. Pull the headlight switch from the panel, disconnect the wiring connector and remove the switch.

To install:

6. Connect the electrical harness connector to the headlight switch and position the switch to the instrument panel. Install 2 retaining screws and tighten to 18–27 inch lbs. (2–3 Nm).

7. Install the instrument panel finish panel.

8. Install the headlight switch retaining nut and switch knob.

9. Connect the negative battery cable.

10. Check for proper headlight switch operation.

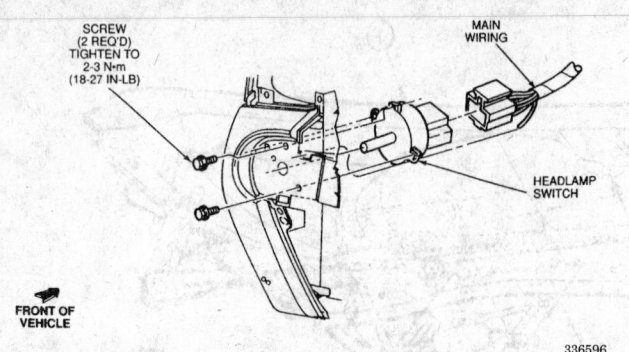

SCREW
(2 REQ'D)
TIGHTEN TO
2-3 N·m
(18-27 IN-LB)

MAIN
WIRING

HEADLAMP
SWITCH

FRONT OF
VEHICLE

336596

Headlight switch removal

Combination Switch

REMOVAL AND INSTALLATION

1. Disconnect the negative battery cable.

2. Remove the upper and lower steering column shrouds.

3. Remove 2 self-tapping screws that secure the combination switch (turn signal and windshield wiper switch) to the steering column.

4. Move the combination switch away from the steering column and disconnect 2 electrical harness connectors. Be careful not to damage the electrical connector locking tabs and the shift indicator cable.

5. Remove the combination switch from the vehicle.

   To install:

6. Connect 2 electrical harness connectors to the combination switch making sure not to damage the shift indicator cable.

7. Place the combination switch in position on the steering column housing and install 2 self-tapping screws into the previously tapped holes. Tighten the screws to 18–27 inch lbs. (2–3 Nm).

8. Verify that the shift indicator adjustment for PRNDL is correct.

9. Install the upper and lower steering column shrouds.

10. Connect the negative battery cable.

11. Check all combination switch functions for proper operation.

Ignition Lock Cylinder

REMOVAL AND INSTALLATION

Functional Lock Cylinder

The following procedure applies to vehicles that have functional lock cylinders. Replacement keys are available for these vehicles if the lock cylinder key numbers are known.

1. Disconnect the negative battery cable.

2. Turn the lock cylinder key to the **RUN** position.

3. Using a ⅛ inch (3.17mm) diameter wire pin or a small drift, depress the lock cylinder retaining pin through the access hole, while pulling out on the lock cylinder to remove it from the steering column housing.

   To install:

4. Install the lock cylinder by turning it to the **RUN** position and depressing the retaining pin. Insert the lock cylinder into its housing in the steering column. Make sure the cylinder is fully seated and aligned in the interlocking washer before turning the key to the **OFF** position. This will permit the cylinder retaining pin to extend into the cylinder housing hole.

5. Rotate the lock cylinder using the lock cylinder key, to ensure correct mechanical operation in all positions.

6. Connect the negative battery cable.

7. Verify proper lock cylinder operation.

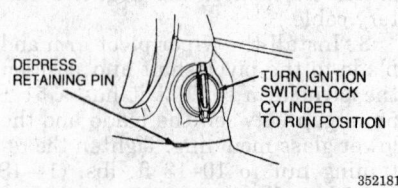

DEPRESS
RETAINING PIN

TURN IGNITION
SWITCH LOCK
CYLINDER
TO RUN POSITION

352181

Depressing the lock cylinder retaining pin

Non-Functional Lock Cylinder

The following procedure applies to vehicles in which the ignition lock is inoperative and the lock cylinder cannot be rotated due to a lost or broken lock cylinder key, unknown key number or a lock cylinder cap that has been damaged and/or broken to the extent that the lock cylinder cannot be rotated.

1. Disconnect the negative battery cable.

2. Using channel lock or vise grip pliers, twist the lock cylinder cap until it separates from the lock cylinder.

3. Using a ⅜ inch diameter drill bit, drill down the middle of the ignition lock key slot approximately 1 ¾ inch (44mm) until the lock cylinder breaks loose from the breakaway base of the lock cylinder. Remove the lock cylinder and drill shavings from the lock cylinder housing in the steering column.

   NOTE: It may be helpful to drill a small pilot hole before using a ⅜ inch drill.

4. Remove the retainer, washer, ignition switch and actuator. Thoroughly clean all drill shavings and other foreign materials from the housing.

5. Inspect the lock cylinder housing for damage from the removal operation. If the housing is damaged, it must be replaced.

   To install:

6. Position the lock drive gear in the same position as that noted during the removal procedure. The position of the lock drive gear is correct if the last tooth on the drive gear is meshed with the last tooth on the rack.

7. Position the bearing retainer in the ignition switch lock cylinder housing. Insert the tip of a screwdriver or similar tool into the double-D slot of bearing and then rotate 90 degrees.

8. Press the blue plastic bearing retainer into the lock cylinder housing. Make sure the retainer is in its original position.

9. Line up the flats of the drive gear with the flats of the washer by pulling down on the column lock actuator.

10. Install the lock cylinder by turning it to the **RUN** position and depressing the retaining pin. Insert the lock cylinder into its housing. Make sure the cylinder is fully seated and aligned in the interlocking washer before turning the key to the **OFF** position. This will permit the

cylinder retaining pin to extend into the cylinder housing hole.

11. Connect the negative battery cable.

12. Check that the vehicle will start in **PARK** and **NEUTRAL** only. Also, check to ensure that the steering column is locked in the **LOCK** position.

Ignition Switch

REMOVAL AND INSTALLATION

1. Disconnect the negative battery cable.

2. Remove the upper steering column shroud and the key release lever, if equipped.

3. Remove the lower steering column shroud.

4. Disconnect the ignition switch electrical harness connector.

5. Turn the ignition switch lock cylinder to the **RUN** position.

6. Remove 2 screws retaining the ignition switch and disengage the ignition switch from the actuator.

To install:

7. Adjust the ignition switch by sliding the carrier to the ignition switch **RUN** position. A new replacement switch assembly will already be set in the **RUN** position.

8. Ensure that the ignition switch lock cylinder is in the **RUN** position. The **RUN** position is achieved by rotating the key lock cylinder approximately 90 degrees from the lock position.

9. Install the ignition switch into the actuator pin. It may be necessary to move the switch slightly back or forth to align the switch mounting holes with the column lock housing threaded holes.

10. Install the ignition switch retaining screws and tighten to 50–69 inch lbs. (6–8 Nm).

11. Connect the electrical harness connector to the ignition switch.

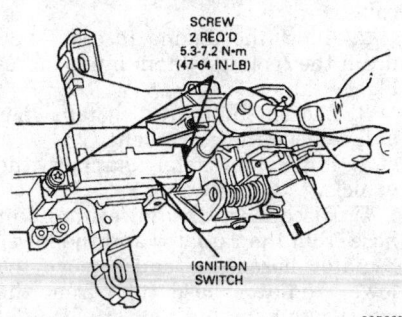

SCREW
2 REQ'D
5.3-7.2 N•m
(47-64 IN-LB)

IGNITION
SWITCH

335669

Ignition switch and retaining screw location

12. Connect the negative battery cable.

13. Check the ignition switch for proper operation, including the **START** and **ACC** positions. Make sure the steering column is locked with the switch in the **LOCK** position.

14. Install the lower and upper steering column shrouds and the key release lever, if equipped.

Manual Lever Position Sensor

REMOVAL AND INSTALLATION

NOTE: The Transmission Range (TR) sensor or manual lever position sensor, only allows the engine to start in Neutral or Park positions as well as being responsible for supplying information to the PCM to help determine gear selection and electronic pressure controls within the transaxle.

1. Disconnect the negative battery cable.

2. Raise and safely support the vehicle.

3. Disconnect the electrical harness connector at the TR sensor.

4. Remove the nut securing the manual control lever to the TR sensor and move the manual control lever and cable aside.

5. Remove 2 TR sensor retaining bolts.

6. Remove the TR sensor from the transaxle.

To install:

7. Place the TR sensor onto the transaxle and loosely install 2 retaining bolts.

8. Install the manual control lever onto the TR sensor shaft. Do not install the retaining nut at this time.

9. Place the shift control selector lever in the **NEUTRAL** position.

10. Align the TR sensor slots using Alignment Tool T92P-70010-AH, or equivalent.

11. Tighten the TR sensor retaining bolts to 7–9 ft. lbs. (9–12 Nm).

12. Install the retaining nut on the manual control lever and tighten to 12–16 ft. lbs. (16–22 Nm).

13. Connect the TR sensor electrical harness connector.

14. Connect the negative battery cable.

15. Check the TR sensor for proper operation. Set the parking brake, the engine should only start in **PARK** or **NEUTRAL**.

16. Road test the vehicle and check the transmission for proper operation.

Powertrain Control Module

REMOVAL AND INSTALLATION

——— WARNING ———
Never disconnect a Powertrain Control Module (PCM) with the battery connected. Be sure to wear a grounding device when removing or installing a PCM to prevent damage to the unit due to static electricity.

1. Disconnect the negative battery cable.

2. Loosen the retaining bolt on the engine control wiring connector and remove the connector from the PCM.

3. Remove 2 PCM cover retaining nuts and remove the PCM cover.

4. Pull the PCM out of its bracket and remove from the vehicle.

To install:

5. Slide the PCM into its bracket and install the PCM cover.

6. Install 2 PCM cover retaining nuts and tighten to 35–49 inch lbs. (4–6 Nm).

7. Carefully install the engine control wiring connector to the PCM. Tighten the retaining bolt to 32 inch lbs. (4 Nm).

8. Connect the negative battery cable.

9. Run the engine and check for proper operation.

ENGINE COOLING

Radiator

REMOVAL AND INSTALLATION

1. Disconnect the negative battery cable.

2. Drain the engine cooling system into a suitable container.

3. Remove the fasteners securing the grille opening panel to the radiator support and front fascia and remove the grille opening panel.

4. Remove the left-hand and right-hand headlamp assemblies and the cornering lamps.

5. Remove the bolts securing the radiator grille opening panel brackets to the radiator support upper and lower reinforcements and remove the

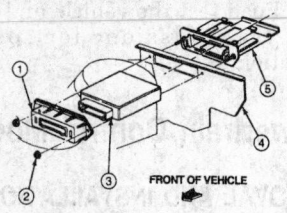

1. PCM cover
2. Nut and washer assy
3. Powertrain control module
4. Dash panel
5. Powertrain control module bracket

338505

Powertrain control module and bracket

radiator grille opening panel brackets.

6. Remove the bolts securing the hood latch support brace to the radiator support upper and lower reinforcements and move the hood latch away from the radiator and A/C condenser assembly.

7. Remove the fasteners securing the harness cover to the radiator support upper reinforcement.

8. Disconnect the underhood lamp assembly from the radiator support upper reinforcement.

9. Disconnect the wiring harness from the engine cooling fan and fan shroud and position the wiring harness out of the way.

10. Remove the bolts securing the radiator support upper reinforcement to the radiator support. Remove the bolts securing the radiator to the radiator support upper reinforcement and remove the radiator support upper reinforcement.

11. Remove the radiator overflow hose from the radiator and plug the radiator overflow hose to prevent coolant loss.

12. Remove the upper and lower radiator hoses from the radiator.

13. Remove the transmission oil cooler line clips and then disconnect 2 oil cooler lines from the transmission using Oil Cooler Line Disconnect T86P-77265-AH, or equivalent.

NOTE: Do not turn the fittings on the radiator when removing the transmission oil cooler lines or damage to the internal O-ring fittings may result.

14. Remove the bolts securing the A/C condenser to the radiator and separate the radiator from the condenser.

15. Remove the radiator and the engine cooling fan and shroud from the vehicle as an assembly.

16. Remove the screws securing the engine cooling fan and shroud to the

radiator and separate the radiator from the shroud.

17. If the radiator is being replaced or sent out for repairs, remove the radiator mounting insulators and install on the new or repaired radiator.

To install:

18. If the upper and/or lower radiator hoses are to be replaced, do so now before installing the radiator.

19. Place the mounting insulators to the radiator support lower reinforcement.

20. Place the engine cooling fan and shroud to the radiator and install the retaining screws. Tighten the screws to 89–106 inch lbs. (10–12 Nm).

21. Carefully lower the radiator and engine cooling fan and shroud assembly into position. Make sure that the moulded pins at the bottom on each end of the radiator are properly fitted to the mounting insulators.

22. Place the radiator to the A/C condenser and install the retaining bolts.

23. Snap the quick-connect fittings of the transmission oil cooler lines to the radiator fittings and install the transmission oil cooler line clips.

24. Connect the upper and lower radiator hoses, making sure to correctly position the hoses and clamps. Tighten the hose clamps to 20–30 inch lbs. (2–3 Nm). Install the remaining components.

25. Fill and bleed the engine cooling system.

26. Connect the negative battery cable.

27. Start the engine and allow to reach normal operating temperature while checking for leaks.

28. Road test the vehicle and check for proper engine and cooling system operation.

Water Pump

REMOVAL AND INSTALLATION

3.0L (VIN U) Engine

1. Disconnect the negative battery cable.

2. Allow the engine to cool. Remove the radiator cap and drain the cooling system.

3. Loosen 4 retaining bolts securing the water pump pulley to the water pump hub. Do not remove.

4. Remove the accessory drive belt.

5. Remove 2 nuts and 1 bolt from the automatic tensioner and remove the tensioner.

6. Disconnect and remove the heater hose and lower radiator hose from the water pump.

7. Remove 11 water pump-to-engine front cover retaining bolts.

8. Lift the water pump and pulley up and out of the vehicle.

To install:

9. Clean the gasket surfaces on the water pump and engine front cover.

10. Install a new gasket on the water pump using a suitable gasket adhesive.

11. Place the water pump in position on the engine with the pulley and 4 water pump pulley retaining bolts loosely installed on the hub.

NOTE: The water pump retaining bolts are of different lengths and must be installed in their correct locations.

12. Lightly oil the retaining bolts except those requiring sealant and install the water pump retaining bolts. Tighten the bolts designated by reference No.1 to 15–22 ft. lbs. (20–30 Nm) and the bolts designated by reference No. 2 to 71–102 inch lbs. (8–12 Nm).

13. Hand tighten the water pump pulley retaining bolts.

14. Install the drive belt tensioner. Tighten the retaining nuts and bolt to 35 ft. lbs. (48 Nm).

15. Install the accessory drive belt.

16. Tighten the water pump pulley retaining bolts to 15–22 ft. lbs. (20–30 Nm).

17. Install the heater hose and lower radiator hose and clamp securely.

18. Fill the engine cooling system.

19. Connect the negative battery cable.

20. Start the engine and allow it to reach normal operating temperature.

21. Check for leaks and proper operation.

3.8L (VIN 4) Engine

1. Disconnect the negative battery cable.

2. Allow the engine to cool, then drain the cooling system into a suitable container.

3. Loosen the drive belt tensioner and remove the drive belt.

4. Raise and safely support the vehicle.

5. Disconnect the lower radiator hose from the radiator and the lower radiator hose tube and remove the lower radiator hose tube from the water pump.

6. Remove the lower nut on both engine support insulators (mounts).

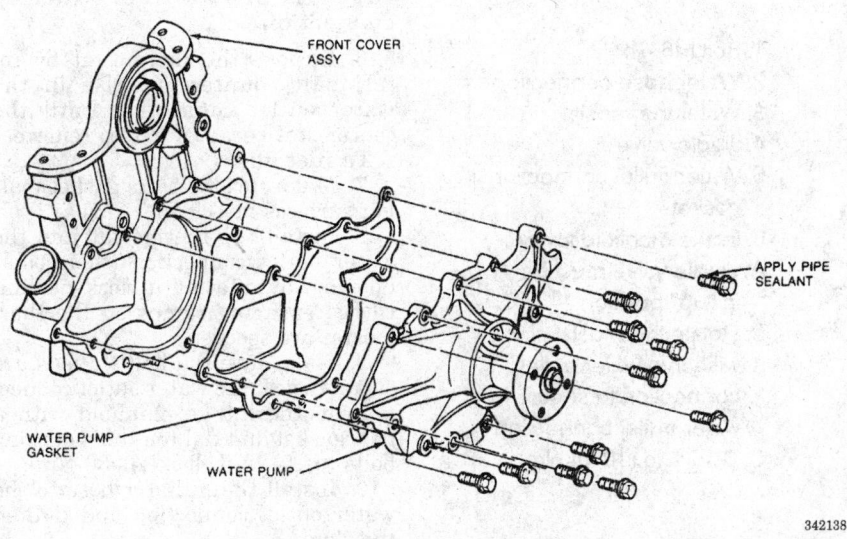

FRONT COVER ASSY

APPLY PIPE SEALANT

WATER PUMP GASKET

WATER PUMP

342138

Water pump removal — 3.0L (VIN U) engine

7. Lower the vehicle.

8. Remove the alternator from the power steering pump support.

9. Remove the power steering reservoir filler cap. Position a drain pan under the power steering pressure line and disconnect the line from the power steering pump using Fuel Line Disconnect T90T-9550-S, or equivalent.

10. Disconnect the bypass hose and the oil cooler hose from the heater water outlet tube.

11. Remove the retaining bolts and disconnect the heater water outlet tube from the water pump.

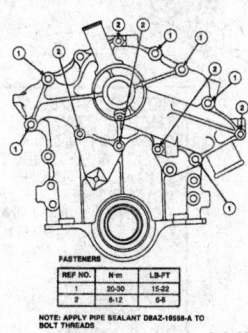

FASTENERS

REF NO.	N·m	LB-FT
1	20-30	15-22
2	8-12	6-8

NOTE: APPLY PIPE SEALANT D8AZ-19558-A TO BOLT THREADS

342139

Water pump retaining bolt location and torque specifications — 3.8L (VIN 4) engine

12. Raise the front of the engine approximately 2 inches (51mm) using appropriate lifting equipment and support the engine with blocks of wood or equivalent.

13. Remove 4 water pump pulley retaining bolts and the water pump pulley.

14. Remove the drive belt tensioner pulley retaining bolt and the drive belt tensioner from the power steering pump support.

15. Remove the power steering pump support retaining bolts and nut. Move the power steering pump and support assembly aside. Place in a position to prevent fluid leakage.

16. Remove the water pump retaining bolts and nuts and remove the water pump from the vehicle.

17. Discard the water pump gasket.

To install:

18. Lightly oil all bolt and stud threads before installation except those that require sealant. Thoroughly clean the water pump and front cover gasket contact surfaces.

19. Place a new water pump gasket on the water pump sealing surface using a coating of contact adhesive to hold the gasket in position.

20. Place the water pump on the engine front cover and install the retaining bolts. Be sure to install the correct length bolt in the correct position. Tighten the retaining bolts to 15–22 ft. lbs. (20–30 Nm).

21. Position the power steering pump support and install the retaining nuts and bolts. Tighten the nuts and bolts to 30–45 ft. lbs. (40–62 Nm).

22. Install the drive belt tensioner pulley and retaining bolt. Tighten the bolt to 34–47 ft. lbs. (47–63 Nm).

23. Position the water pump pulley on the water pump hub and install the retaining bolts. Tighten the bolts to 30–40 ft. lbs. (40–55 Nm).

24. Fill the engine cooling system with the specified type of coolant.

25. Fill the power steering reservoir with the correct fluid.

26. Connect the negative battery cable.

27. Run the engine and allow to reach normal operating temperature while checking for leaks.

28. Road test the vehicle and check for proper engine operation.

29. Recheck the coolant level and fill as necessary.

Thermostat

REMOVAL AND INSTALLATION

3.0L (VIN U) Engine

1. Disconnect the negative battery cable.

2. Drain the engine cooling system.

3. Remove the upper radiator hose from the thermostat housing.

4. Remove 3 retaining bolts from the thermostat housing.

5. Remove the housing and the thermostat as an assembly.

To install:

6. Ensure that all sealing surfaces are free of old gasket material.

7. Install the thermostat into the housing.

8. Make sure the bolt threads are clean. Position a new gasket onto the housing.

9. Install the thermostat housing and thermostat assembly and 3 retaining bolts. Tighten the bolts to 8–10 ft. lbs. (10–14 Nm).

10. Install the upper radiator hose and tighten the clamp.

NOTE: Ensure that the hose clamps are beyond the bead and placed in the center of the clamping surface of the connection. Any used hose clamps must be replaced with a new clamp to ensure proper sealing at the connection. Tighten the hose clamps to 20–30 inch lbs. (2–3 Nm).

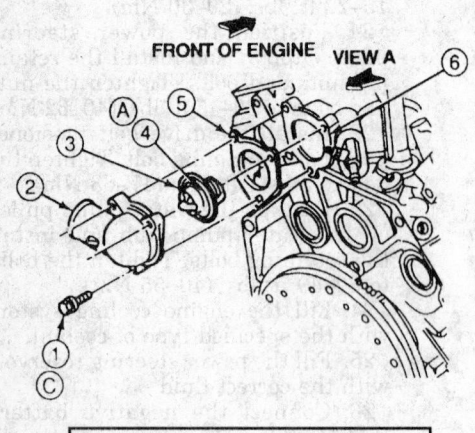

1. Bolt, M6 - 1x32
2. Water hose connection
3. Water thermostat
4. Jiggle valve
5. Water outlet connection gasket
6. Intake manifold, lower
A. Jiggle valve must be in "up" position
B. Rotate thermostat clockwise into water outlet connection to secure in water outlet connection
C. 8-10 ft. lb.(10-14 Nm)

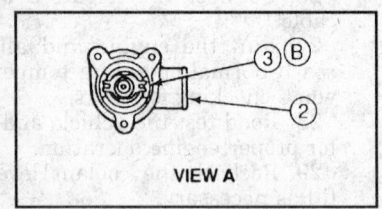

341911

Thermostat and related components — 3.0L (VIN U) engine

11. Fill and bleed the engine cooling system.
12. Connect the negative battery cable.
13. Start the engine and allow it to reach normal operating temperature.
14. Check for coolant leaks and proper cooling system operation.
15. Check the coolant level and add as required.

3.8L (VIN 4) Engine

1. Disconnect the negative battery cable.
2. Place a suitable drain pan below the radiator. Attach a short length of tubing to the drain tube to direct coolant into the drain pan.
3. Allow the engine to cool, then remove the radiator cap and open the draincock. Allow the coolant in the radiator to fall to a level below the water outlet connection (thermostat housing) and then close the draincock.
4. Loosen the top radiator hose clamp at the water outlet connection, remove the hose end and set it out of the way.

5. Remove 2 water outlet connection retaining bolts and lift the water outlet clear of the engine.

NOTE: Do not pry on the housing of the water outlet connection.

6. Remove the thermostat by rotating it counterclockwise in the water outlet connection until the thermostat becomes free to remove.

To install:

7. Make sure that the gasket mating surfaces are clean.
8. Install the thermostat into the water outlet connection (thermostat housing) by rotating it clockwise until the engaging ramps on the thermostat are secure.
9. Make sure the bolt threads are clean. Install the water outlet connection on the intake manifold with a new gasket and tighten the mounting bolts to 15-22 ft. lbs. (20-30 Nm).
10. Install the radiator hose to the water outlet connection and tighten the clamp.

NOTE: Make sure the hose clamp is placed in the center of the clamping surface. A used hose clamp must be replaced with a new clamp to ensure proper sealing at the connection. Tighten the hose clamp to 20-30 inch lbs. (2-3 Nm).

1. Bolt (2)
2. Water outlet connection
3. Lower intake manifold
4. Water outlet connection gasket
5. Water thermostat
A. 15-22 ft. lb.(20-30 Nm)

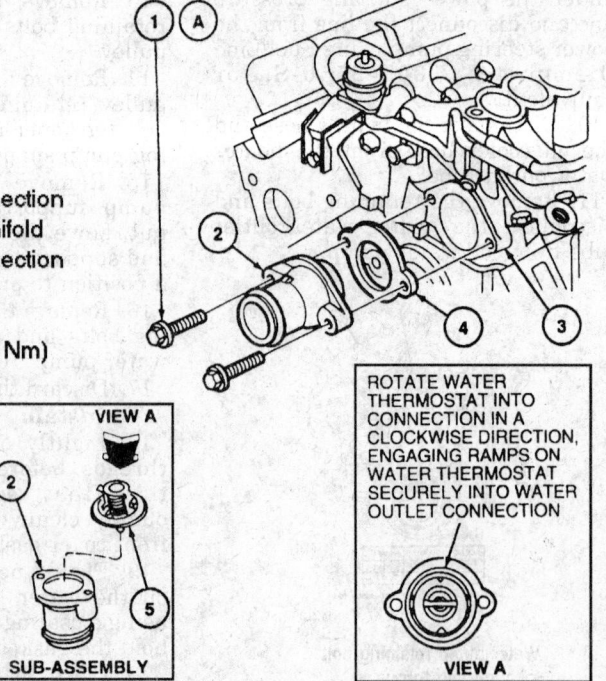

338585

Thermostat and related components — 3.8L (VIN 4) engine

11. Fill and bleed the engine cooling system.

12. Connect the negative battery cable.

13. Run the engine and allow to reach normal operating temperature while checking for coolant leaks.

14. Road test the vehicle and check for proper cooling system operation.

15. Allow the engine to cool, then recheck the coolant level and add as required.

Electric Cooling Fan

REMOVAL AND INSTALLATION

1. Disconnect both battery cables, negative cable first.

2. Raise and safely support the vehicle.

3. Loosen the screw securing the electrical harness connector to the Constant Control Relay Module (CCRM) and remove the electrical connector.

4. Lower the vehicle.

5. Remove the battery and the battery tray.

6. Remove the upper screw securing the CCRM.

7. Rotate the CCRM rearward to gain access to the lower retaining screw.

8. Remove the lower retaining screw and the CCRM.

9. Disconnect the engine cooling fan electrical harness connectors.

10. Remove the fan shroud-to-radiator retaining nuts and bolts.

11. Slide the fan shroud and cooling fan assembly clear of the radiator tank hose connector and lift the assembly up and out of the engine compartment.

NOTE: The fan shroud contains 2 engine cooling fans. The removal procedure is the same for both cooling fan blades and motors.

12. To remove one of the cooling fan motors, remove the fan blade retaining clip and fan blade from the motor shaft.

13. Remove the cooling fan motor retaining bolts and remove the cooling fan from the shroud.

To install:

14. Install the cooling fan motor(s) to the shroud and secure with the retaining bolts. tighten the retaining bolts to 23–33 inch lbs. (3–4 Nm).

15. Position the fan blade onto the motor shaft and install the fan blade retaining clip.

16. Install the fan shroud and cooling fan assembly into the engine com-

partment and slide it in position to the radiator.

17. Install the fan shroud-to-radiator retaining nuts and bolts. Tighten the retaining nuts and bolts to 71–106 inch lbs. (8–12 Nm).

18. Connect the engine cooling fan electrical harness electrical connectors.

19. Position the CCRM and install the lower retaining screw. Do not tighten the screw at this time.

20. Rotate the CCRM forward and install the upper retaining screw. Tighten both screws to 19–26 inch lbs. (2–3 Nm).

21. Install the battery tray and the battery.

22. Raise and safely support the vehicle.

23. Connect the CCRM electrical harness connector.

24. Lower the vehicle.

25. Connect both battery cables, negative cable last.

26. Run the engine and check the cooling fans for proper operation.

Cooling System Bleeding

—— CAUTION ——

Never remove the coolant pressure relief cap or draincock while the engine is running. Allow a hot engine to cool down before cap or draincock removal. When removing the pressure relief cap, always place a heavy shop towel around the cap and slowly turn the cap until coolant pressure begins to release. Once the pressure has been released, push down on the cap and finish removal.

1. Close the radiator draincock if open.

2. With the engine **OFF**, add a ⁵⁰/₅₀ mixture of anti-freeze and water to the radiator up to the bottom of the radiator fill neck.

3. Wait approximately 5 minutes. As the level of the coolant in the radiator drops, slowly add more coolant until the level remains at the filler neck.

4. Install the radiator cap to the first notch to keep spillage to a minimum but still allowing air to escape from the cooling system.

5. Fill the radiator coolant recovery reservoir to the **FULL COLD** mark, then add an additional 1–1.5 quarts of coolant to the radiator coolant recovery reservoir.

6. Start the engine and let idle until the upper radiator hose becomes warm indicating that the thermostat has opened and that coolant is flow-

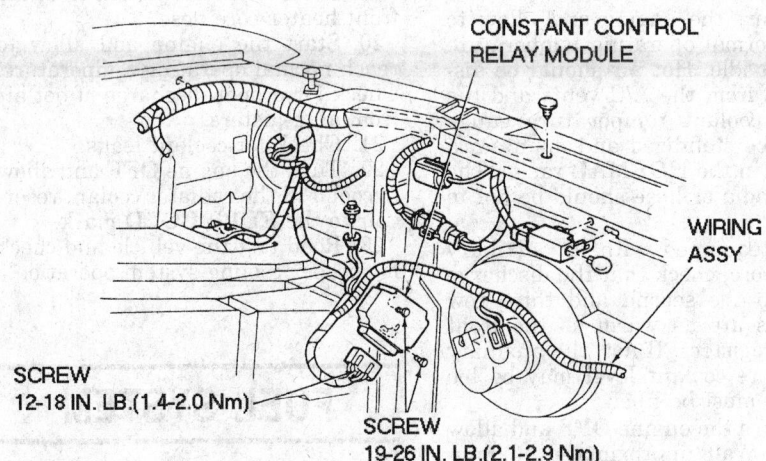

CONSTANT CONTROL RELAY MODULE

WIRING ASSY

SCREW
12–18 IN. LB. (1.4–2.0 Nm)

SCREW
19–26 IN. LB. (2.1–2.9 Nm)

337117

Constant control relay module location

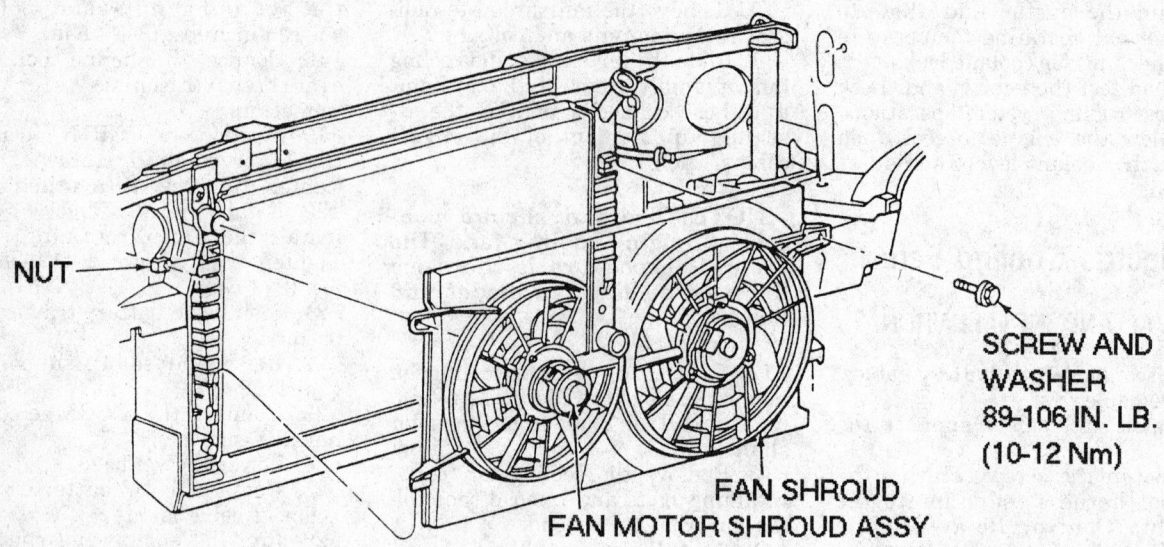

Fan motor and shroud assembly

NUT

SCREW AND
WASHER
89-106 IN. LB.
(10-12 Nm)

FAN SHROUD
FAN MOTOR SHROUD ASSY

337118

ing through the engine cooling system.

7. Turn the engine **OFF**, then using a heavy cloth, carefully remove the radiator cap.

8. Wait approximately 5 minutes. As the level of the coolant in the radiator drops, slowly add more coolant until the level remains at the bottom of the filler neck.

—————— CAUTION ——————

Because the radiator cap is only at the first notch, there should be little or no pressure in the cooling system but the steam and possible hot coolant splashed up by trapped air being released can cause severe burns.

9. Install the radiator cap securely to fully seat the cap.

10. Set the temperature selector to maximum heat and the blower motor to the high speed setting.

11. Adjust the vents to blow through the A/C ducts.

12. Start the engine and allow it to idle. With the engine idling, feel for hot air from the A/C vents. If the discharge air remains cool and the engine coolant temperature gauge does not move, the coolant level is too low and must be filled.

13. Turn the engine **OFF** and allow to cool. Once the engine has cooled sufficiently, add coolant to the system to raise the level to the bottom of the filler cap. Wait approximately 5 minutes. As the level of the coolant in the radiator drops, slowly add more coolant until the level remains at the filler neck.

14. Start the engine and allow to reach normal operating temperature while at idle. Hot air should be discharged from the A/C vents and the engine coolant temperature gauge should be stabilized and maintain a reading in the **NORMAL** range. The upper radiator hose should be hot to the touch.

15. If equipped with an auxiliary heater core, check that the discharge air from the second and third row registers are as warm as the front row of registers. If not, the auxiliary hater core coolant level may be too low and must be filled.

16. Turn the engine **OFF** and allow to cool. Wait approximately 5 minutes. As the level of the coolant in the radiator drops, slowly add more coolant until the level remains at the filler neck.

17. Force coolant into the auxiliary heater core by pinching the front (main) heater core hose at the T in the heater water hose. This will force

all available coolant to the auxiliary heater core.

18. Turn the engine **OFF** and allow to cool. Wait approximately 5 minutes. As the level of the coolant in the radiator drops, slowly add more coolant until the level remains at the filler neck.

19. Remove the restriction from the front heater core hose.

20. Start the engine and allow to reach normal operating temperature. Check for proper discharge of hot air from all registers.

21. Check for coolant leaks.

22. Turn the engine **OFF** and allow to cool. Fill the radiator coolant reservoir to the **FULL COLD** mark.

23. Road test the vehicle and check for proper cooling system operation.

FUEL SYSTEM

Fuel System Service Precautions

Safety is the most important factor when performing not only fuel system maintenance but any type of maintenance. Failure to conduct maintenance and repairs in a safe manner may result in serious personal injury or death. Maintenance and testing of the vehicle's fuel system components can be accomplished safely and effectively by adhering to the following rules and guidelines.

• To avoid the possibility of fire and personal injury, always disconnect the negative battery cable unless the repair or test procedure requires that battery voltage be applied.

• Always relieve the fuel system pressure prior to disconnecting any fuel system component (injector, fuel rail, pressure regulator, etc.), fitting or fuel line connection. Exercise extreme caution whenever relieving fuel system pressure to avoid exposing skin, face and eyes to fuel spray. Please be advised that fuel under pressure may penetrate the skin or any part of the body that it contacts.

• Always place a shop towel or cloth around the fitting or connection prior to loosening to absorb any excess fuel due to spillage. Ensure that all fuel spillage (should it occur) is quickly removed from engine surfaces. Ensure that all fuel soaked cloths or towels are deposited into a suitable waste container.

• Always keep a dry chemical (Class B) fire extinguisher near the work area.

• Do not allow fuel spray or fuel vapors to come into contact with a spark or open flame.

• Always use a backup wrench when loosening and tightening fuel line connection fittings. This will prevent unnecessary stress and torsion to fuel line piping. Always follow the proper torque specifications.

• Always replace worn fuel fitting O-rings with new. Do not substitute fuel hose or equivalent, where fuel pipe is installed.

Fuel System Pressure

RELIEVING

—————— CAUTION ——————
Fuel injection systems remain under pressure, even after the engine has been turned OFF. The fuel system pressure must be relieved before disconnecting any fuel lines. Failure to do so may result in fire and/or personal injury.

1. Relieve the fuel system pressure as follows:
   a. Disconnect the negative battery cable.
   b. Open the fuel tank fill cap to relieve tank pressure.
   c. From the engine compartment, connect Fuel Pressure Gauge T80L-9974-B or equivalent, to the fitting on the fuel injection supply manifold.
   d. Direct the drain hose into a suitable container and depress the pressure relief button until all fuel pressure is relieved.
   e. Close the fuel tank fill cap.
2. An alternate method of relieving fuel pressure is as follows:
   a. Open the fuel tank fill cap to relieve tank pressure.
   b. Disconnect the fuel pump electrical connector located approximately 10 inches (24mm) behind the fuel filter on the backside of the underbody crossmember.
   c. Crank the engine for 15–20 seconds to relieve fuel system pressure.
   d. Turn the ignition switch to the **OFF** position.
   e. Disconnect the negative battery cable.
   f. Reconnect the fuel pump electrical connector.
   g. Close the fuel tank fill cap.

Idle Speed

ADJUSTMENT

This vehicle is equipped with On Board Diagnostics II (OBD II). Engine idle speed is controlled by the Powertrain Control Module (PCM) and the Idle Air Control (IAC) valve. The throttle body does not allow for adjustments and must not be cleaned or damage to the special "sludge tolerant" coating in the throttle body bore will be harmed. If a problem is encountered with the engine idle speed, the OBD II system must be thoroughly inspected before a proper repair can be performed.

Mixture

ADJUSTMENT

The air/fuel mixture is electronically controlled by the Powertrain Control Module (PCM) and the Idle Air Control (IAC) valve, and cannot be adjusted. The PCM continuously adjust the air/fuel ratio in response to signals received from operator controls,

sensor and switch signals monitoring the engines running condition.

Fuel Filter

REMOVAL AND INSTALLATION

—————— CAUTION ——————
Fuel injection systems remain under pressure, even after the engine has been turned OFF. The fuel system pressure must be relieved before disconnecting any fuel lines. Failure to do so may result in fire and/or personal injury.

1. Relieve the fuel system pressure.
2. Raise and safely support the vehicle.
3. Remove the fuel tube safety clip from the stainless steel connector at the forward (outlet) end of the fuel filter.
4. Remove the steel tube and push-connect fitting using Disconnect Tool T90T-9550-S, or equivalent. Be sure to use the correct tool to prevent damage to the push-connect fitting.
5. Locate the push-connect fitting at the rear (inlet) end of the fuel filter.
6. Prepare the plastic hair-pin type clip for removal by bending the shipping tab downward so that it will clear the fitting.
7. Spread the clip legs apart to disengage them from the fitting and push the legs up into the fitting.
8. Lightly pull on the triangular end of the clip and work the clip out of the fitting.
9. Remove the fitting and hose assembly from the fuel filter.
10. Remove the fuel filter by either pulling the fuel filter towards the front of the vehicle or expanding the fuel filter bracket slightly and sliding the fuel filter forward and out of the bracket.
11. Properly dispose of the fuel filter.
   To install:
12. Install the new fuel filter into the bracket making sure that the direction of flow (arrow) is towards the front of the vehicle.
13. Align the seam in the fuel filter with the groove in the fuel filter bracket.
14. Lubricate the O-rings in the fuel line fittings with clean engine oil.
15. Insert the forward (outlet) fitting to the fuel filter and push while listening for the fitting to produce an audible click sound indicating that it

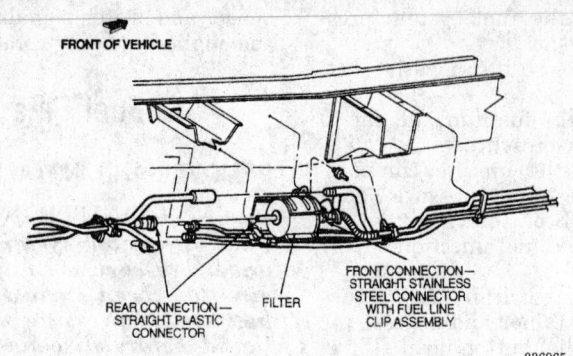

FRONT OF VEHICLE

REAR CONNECTION—
STRAIGHT PLASTIC
CONNECTOR

FILTER

FRONT CONNECTION—
STRAIGHT STAINLESS
STEEL CONNECTOR
WITH FUEL LINE
CLIP ASSEMBLY

336965

Fuel filter location and mounting

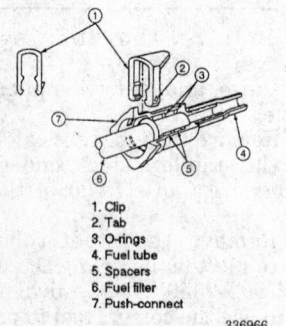

1. Clip
2. Tab
3. O-rings
4. Fuel tube
5. Spacers
6. Fuel filter
7. Push-connect

336966

Clip and components used on the inlet side of the fuel filter

is fully seated. Pull on the fitting to ensure that it is engaged.

16. Install the fuel tube safely clip.

17. Install a new hair-pin type clip into the rear (inlet) fitting. Install the fitting to the fuel filter and push while listening for the fitting to produce an audible click sound indicating that it is fully seated. Pull on the fitting to ensure that it is engaged.

18. Lower the vehicle.

19. Connect the negative battery cable.

20. Run the engine and check for leaks.

Fuel Pump

REMOVAL AND INSTALLATION

——— CAUTION ———

Fuel injection systems remain under pressure, even after the engine has been turned OFF. The fuel system pressure must be relieved before disconnecting any fuel lines. Failure to do so may result in fire and/or personal injury.

1. Relieve the fuel system pressure.

2. Raise and safely support the vehicle.

3. Remove the fuel tank and set on a suitable workbench.

——— CAUTION ———

Always wear eye protection when working around fuel. Clean the area around the fuel pump attaching flange to prevent dirt from entering the fuel tank.

4. Turn the fuel pump module locking ring counterclockwise to loosen. Remove the locking ring.

5. Carefully remove the fuel pump module from the fuel tank.

6. Remove the fuel pump module mounting gasket and discard.

7. If required, separate the fuel pump from the sending unit by disconnecting the fuel tank sending unit electrical connector and removing 2 screws retaining the sending unit to the fuel pump.

To install:

8. Clean the fuel pump module mounting flange, the fuel tank mounting surface groove and the locking ring.

9. If removed, position the fuel sending unit to the fuel pump and install 2 retaining screws. Tighten the screws securely and connect the fuel tank sending unit electrical connector.

10. Apply a suitable grease to a new mounting gasket and install it in the groove in the fuel tank.

11. Carefully install the fuel pump module into the fuel tank. Do not damage the fuel sock (filter) during installation.

12. Make sure that the fuel pump module locking keys are properly fitted to the keyways in the fuel tank and that the mounting gasket has not moved out of position.

13. While holding the fuel pump module and mounting gasket in place, install the locking ring and ro-

tate the locking ring clockwise until it contacts the stop.

14. Install the fuel tank.

15. Lower the vehicle.

16. Connect the negative battery cable.

17. Start the engine and allow to idle for several minutes while checking for fuel leaks.

18. Road test the vehicle and check for proper engine operation.

Fuel Injector

REMOVAL AND INSTALLATION

3.0L (VIN U) Engine

1. Disconnect the negative battery cable.

2. Relieve the fuel system pressure.

3. Remove the throttle body/upper intake manifold assembly.

4. Disconnect the fuel supply and return lines using tools D87L-9280-A (³⁄₈ inch) and D87L-9280-B (¹⁄₂ inch), or equivalents.

5. Disconnect the electrical harness connectors at the fuel injectors.

6. Disconnect the vacuum line from the fuel pressure regulator.

7. Remove 4 fuel injection supply manifold retaining bolts.

8. Carefully disengage the fuel injection supply manifold from the fuel injectors by lifting and gently rocking the fuel injection supply manifold.

9. Remove the fuel injectors by lifting while gently rocking from side to side.

To install:

10. Remove the O-rings on each fuel injector. Inspect the fuel injector end caps, body and washers for wear or deterioration. Replace fuel injectors as needed.

11. Lubricate new O-rings with clean engine oil and install 2 on each fuel injector.

12. Install the fuel injectors on the fuel injection supply manifold using a light twisting-pushing motion.

13. Carefully install the fuel injection supply manifold and fuel injectors into the lower intake manifold, one side at a time. Make sure the O-rings are seated by pushing down on the fuel injection supply manifold.

14. While holding the fuel injection supply manifold in place, install 4 retaining bolts and tighten to 71–106 inch lbs. (8–12 Nm).

15. Connect the fuel supply and return lines and the fuel pressure regulator vacuum line.

16. Before connecting the electrical harness connectors to each fuel injec-

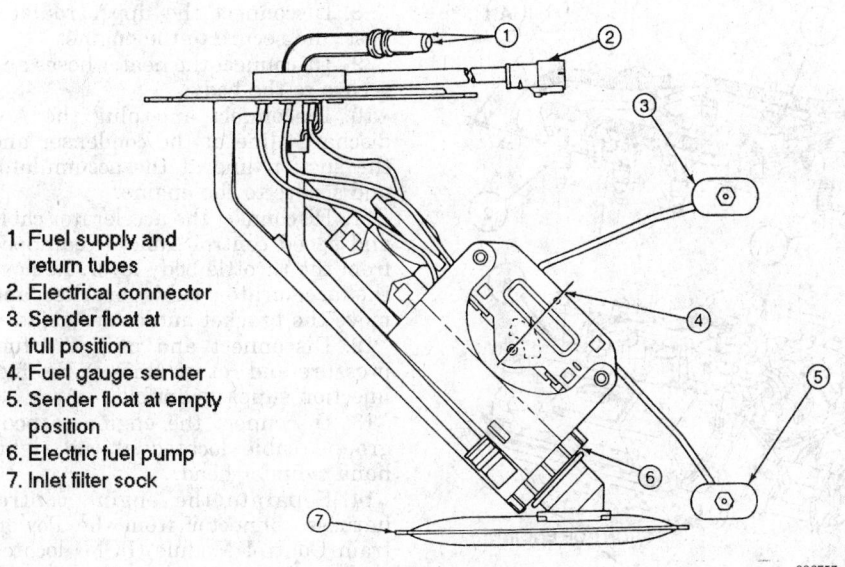

1. Fuel supply and return tubes
2. Electrical connector
3. Sender float at full position
4. Fuel gauge sender
5. Sender float at empty position
6. Electric fuel pump
7. Inlet filter sock

Fuel pump and fuel tank sending unit module

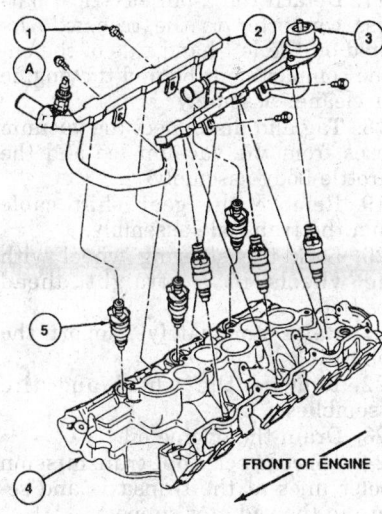

1 Bolt (4 Req'd)
2 Fuel Injection Supply Manifold
3 Fuel Pressure Regulator
4 Intake Manifold
5 Fuel Injector (6 Req'd)
A Tighten to 8-12 N·m (71-106 Lb-In)

Fuel injection supply manifold and fuel injector removal — 3.0L (VIN U) engine

tor, temporarily connect the negative battery cable and turn the ignition switch to the **ON** position, pressurizing the fuel system.

17. Using a clean paper towel. check for leaks where the fuel injectors are fitted to the fuel injection supply manifold. Turn the ignition switch **OFF** and disconnect the negative battery cable.

18. Connect the electrical harness connectors to each fuel injector.

19. Install the throttle body/upper intake manifold assembly.

20. Connect the negative battery cable.

21. Start the engine and allow it to idle for several minutes while checking for fuel leaks.

22. Road test the vehicle and check for proper engine operation.

3.8L (VIN 4) Engine

1. Disconnect the negative battery cable.

2. Relieve the fuel system pressure.

3. Remove the upper intake manifold.

4. Remove the safety clips from the fuel supply and return lines and disconnect the fuel lines from the fuel injection supply manifold using tools

D87L-9280-A and D87L-9280-B, or equivalents.

5. Disconnect and remove the fuel injector electrical harness connectors, if not already done.

6. Remove 4 fuel injection supply manifold retaining bolts (2 on each side).

7. Carefully disengage the fuel injection supply manifold from the intake manifold. The fuel injectors may come out with the fuel injection supply manifold. If so, separate the fuel injectors once the fuel supply manifold is removed.

8. Remove the fuel injectors from the fuel injection supply manifold or intake manifold using a rocking, side-to-side motion.

To install:

9. Install new O-rings (2 each) on each fuel injector and lubricate the O-rings with clean engine oil. Do not use silicone grease as it will clog the injectors.

10. Install each fuel injector using a light, twisting motion while pushing the fuel injector into the cylinder head pockets.

11. Position the fuel supply manifold over the fuel injectors and carefully install.

12. While pushing down on the fuel injection supply manifold, install 4 retaining bolts. Tighten the bolts to 71–97 inch lbs. (8–11 Nm).

13. Install the fuel supply and return lines and their safety clips.

14. Temporarily connect the negative battery cable and turn the ignition switch to the **RUN** position. This will pressurize the fuel system and any leaks will be detected before installing the upper intake manifold.

15. If no fuel leaks are found, turn the ignition switch to the **OFF** position and disconnect the negative battery cable.

16. Connect the fuel injector wiring harness connectors.

17. Install the upper intake manifold.

18. Connect the negative battery cable.

19. Run the engine at idle for at least 2 minutes. Turn the engine **OFF** and check for fuel leaks.

20. Road test the vehicle and check for proper engine operation.

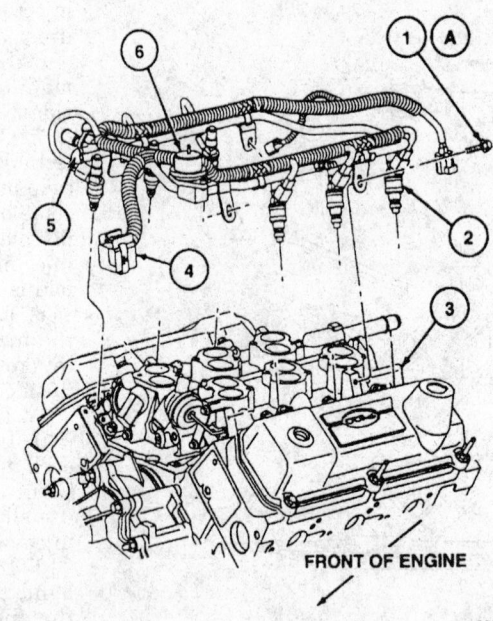

1. Bolt(4)
2. Fuel injector(6)
3. Lower intake manifold
4. Fuel charging wiring
5. Fuel injection supply manifold
6. Fuel pressure regulator
A. 71-97 in. lb.(8-11 Nm)

Fuel injectors and supply manifold assembly — 3.8L (VIN 4) engine

336864

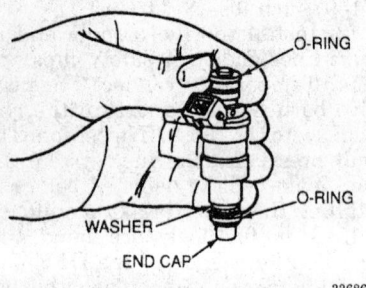

Fuel injector and O-rings

336865

ENGINE MECHANICAL

Engine Assembly

REMOVAL AND INSTALLATION

3.0L (VIN U) Engine

— CAUTION —

Fuel injection systems remain under pressure, even after the engine has been turned OFF. The fuel system pressure must be relieved before disconnecting any fuel lines. Failure to do so may result in fire and/or personal injury.

1. Disconnect the negative battery cable.
2. Drain the engine cooling system.
3. Properly recover the refrigerant from the air conditioning system, if equipped.
4. Remove the cowl top vent panel.
5. Relieve the fuel system pressure.
6. Disconnect the alternator electrical harness connectors.
7. Remove the engine air cleaner outlet tube.
8. Disconnect the upper radiator hose and secure to the engine.
9. Disconnect the heater hoses and secure to the body.
10. Disconnect and plug the A/C discharge line at the condenser and the suction line at the accumulator and secure to the engine.
11. Disconnect the accelerator cable and speed control cable, if equipped from the throttle body lever. Remove the accelerator cable bracket and move the bracket and cables aside.
12. Disconnect and plug the fuel pressure and return lines at the fuel injection supply manifold.
13. Disconnect the engine to body ground cable located at the right-hand cylinder head.
14. Separate the engine control harness connector from the Powertrain Control Module (PCM) located at the upper rear corner of the engine compartment.
15. Disconnect the A/C high and low pressure sensor connections.
16. Disconnect the 42-pin electrical harness connector attached to the power brake booster.
17. Detach the 2-pin electrical harness connector to the transaxle located on the left-hand side of the engine compartment behind the engine air cleaner assembly.
18. Tag and disconnect the vacuum hoses from the vacuum tee and the throttle body assembly.
19. Remove the gear shift cable from the transaxle assembly.
20. Lock the steering wheel with the wheels in a straight ahead position.
21. Raise and safely support the vehicle.
22. Remove the wheel and tire assemblies.
23. Drain the engine oil.
24. Disconnect the transmission cooler lines at the transaxle and secure to the radiator support.
25. Disconnect the electrical harness connectors from the heated oxygen sensors.
26. Remove the dual converter Y-pipe retaining bolts at the pipe outlet and the flex joint and install Exhaust Connector Holding Tool T94T-6000-AH, or equivalent.
27. Remove the nuts retaining the dual converter Y-pipe to the exhaust manifolds and remove the dual converter Y-pipe assembly.
28. Note the routing of the battery ground cable and remove the cable.

The cable must be installed in the exact same routing as removed.

29. Disconnect the starter motor wiring and remove the starter motor.

30. Remove the engine rear plate and 4 flywheel-to-torque converter retaining nuts.

31. Remove the nut retaining the battery cable support bracket.

32. Disconnect the power steering cooler lines.

33. Disconnect the lower radiator hose and heater hoses.

34. Remove the upper bolts from the stabilizer bar links.

35. Remove the dust boot from the steering rack and pinion support by gently spreading the integral tension ring and pushing upward.

36. Remove the steering column pinch bolt from the intermediate shaft at the steering gear.

37. Remove the right-hand and left-hand stabilizer bar links from the stabilizer bar.

38. Separate both lower ball joints from the steering knuckles.

39. Separate both tie rods at the steering knuckles.

40. Remove both half shafts.

41. Support the sub-frame, engine and transaxle assembly with Powertrain Lift 014-00765 or equivalent.

42. Remove 4 sub-frame retaining bolts.

43. Lower the engine, transaxle and sub-frame from the vehicle as an assembly.

44. Separate the engine and transaxle from the sub-frame using suitable lifting equipment.

45. Separate the transaxle from the engine.

To install:

46. Install the engine to the transaxle.

47. Torque the engine to transaxle retaining bolts to 30–44 ft. lbs. (40–60 Nm).

48. Install 4 torque converter retaining nuts and tighten to 20–34 ft. lbs. (27–46 Nm).

49. Install the engine and transaxle assembly to the sub-frame using suitable lifting equipment.

50. Install the left-hand and right-hand engine support insulators.

51. Connect the power steering pressure hose.

52. Raise the engine, transaxle and sub-frame assembly into the engine compartment.

53. Align the sub-frame to the body and install 4 sub-frame retaining bolts. Tighten the bolts to 57–76 ft. lbs. (77–103 Nm).

54. Remove the powertrain lift or equivalent.

55. Partially lower the vehicle.

56. Install both halfshafts. Install the remaining components.

57. Fill the crank case with the correct type and quantity of engine oil.

58. Fill and bleed the power steering system.

59. Fill and bleed the engine cooling system.

60. Start the engine and check for leaks.

61. Evacuate and recharge the A/C system.

62. Check and refill all fluids.

63. Install the cowl top vent panel.

NOTE: Whenever the vehicles sub-frame is removed or lowered, the wheel alignment should be checked.

64. Check the alignment and adjust if necessary.

3.8L (VIN 4) Engine

1. Lock the steering column with the front wheels pointed in a straight-ahead position.

2. Relieve the fuel system pressure.

3. Disconnect the wiring connector to the engine compartment lamp.

4. Mark the positions of the hood hinges and with the help of an assistant, remove the hood.

5. Drain the engine cooling system into a suitable container.

6. Remove the cowl top vent panel.

7. Disconnect the wiring to the alternator and voltage regulator.

8. Remove the engine air cleaner and outlet tube.

9. Disconnect the upper radiator hose from the engine.

10. Disconnect the heater hoses and the heater hose bracket bolt from the left-hand side of the engine.

11. Disconnect the oil cooler inlet tube and the oil cooler tube using Disconnect Tool T86P-77265-AH, or equivalent.

12. Properly recover the refrigerant from the A/C system using proper refrigerant recovery equipment.

13. Disconnect the manifold and tube assembly from the A/C compressor.

14. Remove the accelerator and speed control cables from the throttle body and remove the cable bracket.

15. Disconnect and plug the fuel supply and return lines from the fuel injection supply manifold.

16. Tag and disconnect the ground wire, Powertrain Control Module (PCM) and engine control wiring electrical connectors and the bolt retaining the electrical harness bracket.

17. Tag and disconnect the necessary vacuum hoses, including the brake vacuum hose.

18. Disconnect the gear shift cable from the transaxle.

19. Disconnect the electrical harness connector at the bulkhead.

20. Disconnect the transaxle pressure switches and the throttle valve control actuating cable.

21. Raise and safely support the vehicle.

22. Drain the engine oil into a suitable container. Reinstall the drain plug and move the drain pan aside.

23. Remove the engine oil filter element.

24. Remove the front wheel and tire assemblies.

25. Remove the disc brake calipers and position out of the way. Do not allow the brake calipers to hang by the hydraulic brake hoses.

26. Disconnect the electrical harness connectors at the wheel speed sensors.

27. Disconnect the Heated Oxygen Sensor (HO2S) electrical harness connectors.

28. Remove the dual converter Y-pipe to flex pipe retaining bolts.

29. Install Exhaust Connector Holder T94T-6000-AH or equivalent. This will prevent damage to the flex pipe while the dual converter Y-pipe is removed.

30. Remove the dual converter Y-pipe retaining nuts at the exhaust manifolds and remove the dual converter Y-pipe.

31. Remove the battery cable form the starter motor and the nut securing the ground and positive battery cable bracket.

32. Disconnect the HO2S electrical harness connector from the chassis.

33. Disconnect and plug the power steering fluid cooler lines.

34. Remove the lower radiator hose.

35. Remove the pinch bolt from the rack and pinion (steering gear) and separate the steering shaft from the rack and pinion assembly.

36. Partially lower the vehicle.

37. Remove the strut upper retaining bolts.

38. Raise and safely support the vehicle.

39. Place a suitable engine table under the sub-frame assembly. Raise

the table to support the sub-frame assembly.

NOTE: The use of a proper engine table is highly recommended. An alternate method would be to operate the hoist to raise and lower the vehicle instead of the engine table if one is not available. A stable rest must still be used for the engine/transaxle and sub-frame assembly to sit on that is capable of handling the load of the assembly.

40. Remove 4 sub-frame retaining bolts and carefully lower the engine/transaxle and sub-frame assembly from the vehicle.
41. Install lifting eyes to the engine/transaxle and sub-frame assembly and connect a suitable hoist to lift the assembly from the engine table.
42. Place the assembly on the ground.
43. Separate the sub-frame from the engine/transaxle assembly as follows:
  a. Remove the stabilizer links from the struts.
  b. Disconnect the halfshafts from the transaxle.
  c. Disconnect the power steering lines from the steering gear.
  d. Remove the engine support insulator (mount) nuts.
44. Remove the engine/transaxle assembly from the sub-frame using the engine lifting eyes and suitable lifting equipment.
45. Remove the starter motor.
46. Remove the transaxle inspection cover and 4 torque converter retaining nuts.
47. Disconnect the remaining electrical harness connectors from the transaxle.
48. Remove the bolts securing the right-hand engine support insulator to the transaxle.
49. Remove the engine-to-transaxle bolts and separate the engine from the transaxle.
50. Remove 6 flywheel retaining bolts and the flywheel.
51. Using the engine lifting eyes and suitable lifting equipment, mount the engine to a suitable engine stand.
  To install:
52. Remove the engine from the engine stand using the lifting eyes and suitable lifting equipment.
53. Install the flywheel. Coat the threads of 6 flywheel bolts with pipe sealant and install the bolts. Tighten the flywheel bolts, in an alternating pattern, to 54–64 ft. lbs. (73–87 Nm).

54. Place the engine to the transaxle using care to properly fit the torque converter to the flywheel.
55. Install the engine-to-transaxle retaining bolts and tighten to 41–50 ft. lbs. (56–68 Nm).
56. Install the bolts securing the right-hand engine support insulator bracket to the transaxle. Tighten the bolts to 30–40 ft. lbs. (40–55 Nm).
57. Connect the electrical harness connectors to the transaxle.
58. Install 4 torque converter retaining nuts and the transmission inspection cover.
59. Install the starter motor.
60. Lift the engine/transaxle assembly and position the assembly on the sub-frame.
61. Install the engine to sub-frame mounting nuts. Tighten the mounting nuts to 40–52 ft. lbs. (54–71 Nm).
62. Connect the power steering lines to the power steering gear.
63. Install the halfshafts to the transaxle.
64. Connect the stabilizer links to the struts and install the retaining bolts.
65. Lift the engine/transaxle and sub-frame assembly using the engine lifting eyes and suitable lifting equipment and place onto the engine table or similar device.
66. Remove the engine lifting equipment.
67. Carefully install the engine/transaxle and sub-frame assembly into the vehicle.
68. Install 4 sub-frame retaining bolts and tighten to 57–80 ft. lbs. (77–103 Nm).
69. Remove the engine table or similar device.
70. Partially lower the vehicle.
71. Install the upper strut retaining nuts.
72. Raise and safely support the vehicle. Install the remaining components.
73. Install a new oil filter, then fill the crankcase with oil.
74. Fill and bleed the engine cooling system.
75. With the help of an assistant, install the hood and match the alignment marks on the hood and hinges. Tighten the retaining bolts and adjust the hood alignment if necessary.
76. Connect the engine compartment lamp electrical connector.
77. Fill the engine with the correct grade and quantity of engine oil.
78. Connect the negative battery cable.
79. Run the engine and check for leaks.

80. Bleed the power steering system, if needed.
81. Check and add transmission fluid, if needed.
82. Evacuate and charge the A/C system.
83. Road test the vehicle and check for proper operation.

Engine Mounts

REMOVAL AND INSTALLATION

3.0L (VIN U) Engine

Front and Rear

1. Disconnect the negative battery cable.
2. Install Three Bar Engine Support 014-00750 and Adapter 014-00792 or equivalents and a suitable engine lifting equipment to support the engine.
3. Remove the bolts retaining the front/rear engine support insulator (engine mount) to the mounting bracket.
4. Raise and safely support the vehicle.
5. Remove the front/rear engine support insulator lower retaining nut.
6. Lower the vehicle.
7. Raise the engine approximately 1 inch (25mm) using the engine support equipment, or equivalent.
8. Raise and safely support the vehicle.
9. Remove the front/rear engine support insulator.
  To install:
10. Install the front/rear engine support insulator.
11. Lower the vehicle.
12. Lower the engine using the engine support equipment, or equivalent.
13. Raise and safely support the vehicle.
14. Install the front/rear engine support insulator lower retaining nut. Tighten the nut to 65–87 ft. lbs. (88–119 Nm).
15. Lower the vehicle.
16. Install the front/rear engine support insulator bolts to the mounting bracket. Tighten the bolts to 30–40 ft. lbs. (40–55 Nm).
17. Relax the engine support and remove the engine support and engine lifting equipment from the vehicle.
18. Connect the negative battery cable.
19. Road test the vehicle and check for proper operation.

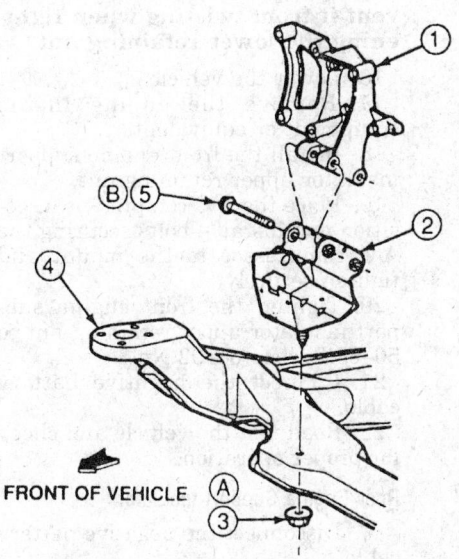

1. A/C compressor mounting bracket
2. Front engine support insulator
3. Nut M 12-1.75
4. Front sub-frame
5. Bolt M 10 x 1.5 x 103.3(2)
A. 65-87 ft. lb.(88-119 Nm)
B. 30-40 ft. lb.(40-55 Nm)

FRONT OF VEHICLE

353359

Front engine support insulator removal — 3.0L (VIN U) engine

1. Case rear bracket
2. Bolt M 10 x 1.5 x 103.3(2)
3. Transmission support insulator
4. Front sub-frame
5. Nut M 12-1.75
A. 30-40 ft. lb.(40-55 Nm)
B. 65-87 ft. lb.(88-119 Nm)

FRONT OF VEHICLE

353360

Rear engine support insulator removal — 3.0L (VIN U) engine

Transaxle Support Insulator

1. Disconnect the negative battery cable.
2. Raise and safely support the vehicle.
3. Remove the left-hand wheel and tire assembly.
4. Place a suitable transmission jack under the transaxle and support the transaxle.
5. Remove 2 nuts retaining the transaxle support insulator to the transaxle support.
6. Remove 2 through-bolts retaining the transaxle support insulator to the sub-frame.
7. Raise the transaxle with the transmission jack enough to unload the transaxle support insulator.
8. Remove 4 bolts retaining the transaxle support to the transaxle.
9. Rotate the transaxle support counterclockwise to disengage the upper stud on the transaxle support insulator and remove the transmission support insulator.

To install:

10. Loosely install the transaxle support insulator to the transaxle support.
11. Place the transaxle support in position and install 4 transaxle support-to-transaxle retaining bolts. Tighten the retaining bolts to 56–75 ft. lbs. (76–103 Nm).
12. Secure the transaxle support insulator to the sub-frame using 2 through-bolts. Tighten the through-bolts to 56–75 ft. lbs. (76–103 Nm).
13. Lower the transaxle enough to load the transaxle support insulator.
14. Install 2 transaxle support insulator-to-transaxle support retaining nuts. Tighten to 65–87 ft. lbs. (88–119 Nm).
15. Remove the transmission jack or equivalent.
16. Install the wheel and tire assembly. Torque the lug nuts to 85–105 ft. lbs. (115–142 Nm).
17. Lower the vehicle.
18. Connect the negative battery cable.
19. Road test the vehicle and check for proper operation.

3.8L (VIN 4) Engine

Front Engine Support Insulator

1. Disconnect the negative battery cable.
2. Remove 4 bolts securing the A/C compressor to its mount. Do not discharge the A/C system.
3. Remove the front engine support insulator (engine mount) upper retaining nut.

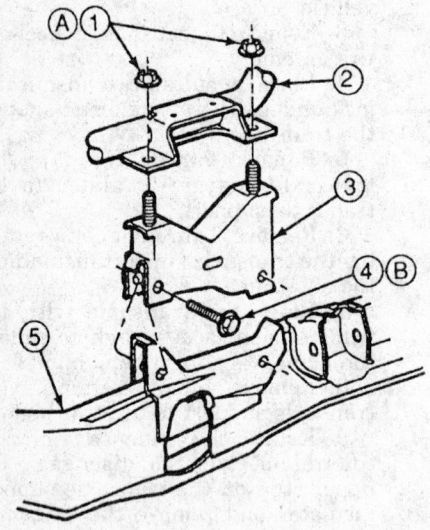

1. Nut M 12-1.75(2)
2. Engine and transmission support
3. Engine and transmission support insulator
4. Bolt M 12-1.75 × 57.5(2)
5. Front sub-frame
A. 65-87 ft. lb.(88-119 Nm)
B. 56-75 ft. lb.(76-103 Nm)

353361

Transaxle support insulator removal — 3.0L (VIN U) engine

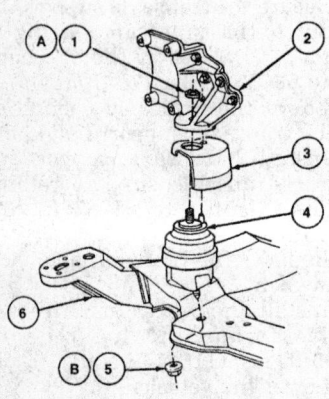

1. Nut
2. A/C compressor mounting bracket
3. Engine mount heat shield
4. Front engine support insulator
5. Nut
6. Front sub-frame
A. 50-68 ft. lb.(68-92 Nm)
B. 83-113 ft. lb.(113-153 Nm)

337610

Front engine support insulator removal — 3.8L (VIN 4) engine

4. Install Engine Support 014-00750 or an equivalent tool capable of raising and supporting the engine to allow clearance for removal of the front engine support insulator. Do not raise the engine at this time.

5. Raise and safely support the vehicle.

6. Loosen and remove the front engine support insulator lower retaining nut.

7. Lower the vehicle.

8. Raise the engine approximately 1 inch (25mm) using the engine lifting equipment or an equivalent method.

9. Raise and safely support the vehicle.

10. Remove the front engine support insulator.

To install:

11. Install the front engine support insulator.

12. Lower the vehicle.

13. Lower the engine using the engine lifting equipment or an equivalent method.

14. Raise and safely support the vehicle.

15. Install the front engine support insulator lower retaining nut and

tighten to 83–113 ft. lbs. (113–153 Nm).

NOTE: Hold the lower portion of the support insulator to prevent it from twisting when tightening the lower retaining nut.

16. Lower the vehicle.

17. Remove the engine lifting equipment or equivalent.

18. Install the front engine support insulator upper retaining nut.

19. Place the A/C compressor in position and install 4 bolts securing the A/C compressor to the mount and tighten securely.

20. Tighten the front engine support insulator upper retaining nut to 50–68 ft. lbs. (68–92 Nm).

21. Connect the negative battery cable.

22. Road test the vehicle and check for proper operation.

Rear Engine Support Insulator

1. Disconnect the negative battery cable.

2. Raise and safely support the vehicle.

3. Remove the rear engine support insulator (engine mount) lower retaining nut at the sub-frame.

4. Lower the vehicle.

5. Install Engine Support 014-00750 or an equivalent tool capable of raising and supporting the engine to allow clearance for removal of the rear engine support insulator. Use the alternator bracket for lifting.

6. Raise the engine approximately 1 inch (25mm).

7. Remove the rear engine support insulator upper retaining nut.

8. Raise and safely support the vehicle.

9. Remove the rear engine support insulator and heat shield from the vehicle.

To install:

10. Place the rear engine support insulator and heat shield in position and loosely install the upper retaining nut. Ensure that the anti-rotation pin on the rear engine support insulator is properly located in the transaxle bracket.

11. Lower the vehicle.

12. Tighten the rear engine support insulator upper retaining nut to 50–68 ft. lbs. (68–92 Nm).

13. Lower the engine using the engine lifting equipment or an equivalent method.

14. Raise and safely support the vehicle.

15. Install and tighten the rear engine support insulator lower retaining nut to 55–75 ft. lbs. (78–103 Nm).

16. Lower the vehicle.

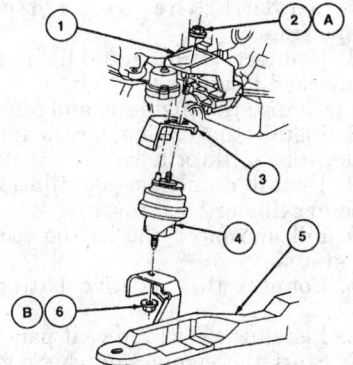

1. Transmission/transaxle bracket
2. Nut
3. Engine mount heat shield
4. Front engine support insulator
5. Front sub-frame
6. Nut
A. 50-68 ft. lb.(68-92 Nm)
B. 57-76 ft. lb.(78-103 Nm)

337612

Rear engine support insulator removal — 3.8L (VIN 4) engine

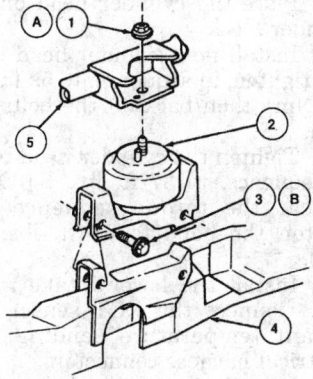

1. Nut
2. Engine & transmission support insulator
3. Bolt (2)
4. Front sub-frame
5. Engine & transmission support
A. 57-76 ft. lb.(77-103 Nm)
B. 56-76 ft. lb.(76-103 Nm)

337613

Transaxle support insulator removal — 3.8L (VIN 4) engine

Cylinder Head

REMOVAL AND INSTALLATION

3.0L (VIN U) Engine

> ——— **CAUTION** ———
> *Fuel injection systems remain under pressure, even after the engine has been turned OFF. The fuel system pressure must be relieved before disconnecting any fuel lines. Failure to do so may result in fire and/or personal injury.*

17. Remove the engine lifting equipment or equivalent.
18. Connect the negative battery cable.
19. Road test the vehicle and check for proper operation.

Transaxle Support Insulator

1. Disconnect the negative battery cable.
2. Raise and safely support the vehicle.
3. Remove the left-hand front wheel and tire assembly.
4. Place a floor jack and a wood block or equivalent, in a suitable location underneath the transaxle and support the transaxle.
5. Remove the nut securing the transaxle support insulator (mount) to the transaxle support.
6. Remove 2 through-bolts securing the transaxle support insulator to the sub-frame.
7. Raise the transaxle with the floor jack or equivalent, enough to remove the load on the transaxle support insulator.
8. Remove 4 bolts securing the transaxle support to the transaxle.
9. Remove the transaxle support by rotating the assembly counterclockwise to disengage the upper stud on the transaxle support insulator.

10. Remove the transaxle support insulator from the vehicle.

To install:

11. Loosely install the transaxle support insulator.
12. Place the transaxle support to the transaxle and install 4 retaining bolts. Tighten the bolts to 39–53 ft. lbs. (53–72 Nm).
13. Place the transaxle support insulator to the sub-frame and install 2 through-bolts. Tighten the bolts to 56–76 ft. lbs. (76–103 Nm).
14. Lower the transaxle using the floor jack or equivalent, just enough to load to the transaxle support insulator. Ensure that when lowering the transaxle to align the transaxle support insulator upper stud and anti-rotation pin to the transaxle support.
15. Install the transaxle support insulator upper retaining nut and tighten to 58–76 ft. lbs. (78–103 Nm).
16. Remove the floor jack and the block of wood, or equivalent.
17. Install the wheel and tire assembly. Torque the lug nuts to 83–114 ft. lbs. (113–153 Nm).
18. Lower the vehicle.
19. Connect the negative battery cable.
20. Road test the vehicle and check for proper operation.

1. Relieve the fuel system pressure.
2. Rotate the crankshaft until the piston for No.1 cylinder is at TDC on its compression stroke.
3. Drain the engine cooling system into a suitable container.
4. Remove the cowl top vent panel.
5. Remove the engine air cleaner outlet tube at the throttle body.
6. Label and disconnect the vacuum lines from the throttle body.
7. Disconnect the hoses from the EGR valve. Loosen the lower EGR tube nut and rotate the tube away from the valve.
8. Label and disconnect the wiring from the Intake Air Temperature (IAT) sensor, Throttle Position Sensor (TPS), Idle Air Control (IAC) valve and the EGR backpressue transducer.
9. Remove the fuel line safety clips and disconnect the fuel supply and return lines from the fuel injection supply manifold.
10. Remove the throttle body/upper intake manifold assembly and discard the gasket.
11. Label and disconnect the engine control wiring harness from the cylinder head cover studs and fuel injectors.
12. Label and disconnect the ignition wires from the spark plugs and the cylinder head cover studs.
13. Remove the ignition coil and bracket from the left-hand (front) cylinder head and set aside.
14. Disconnect the upper radiator and heater hoses.
15. Remove 2 retaining bolts and the Camshaft Position (CMP) sensor and housing.
16. Disconnect the Engine Coolant Temperature (ECT) sensor and temperature sending unit electrical harness connectors.

17. If the left-hand (front) cylinder head is being removed, perform the following:

a. Disconnect the alternator electrical harness connectors.

b. Rotate the tensioner clockwise and remove the accessory drive belt.

c. Remove the automatic belt tensioner assembly.

d. Remove the alternator.

e. Remove the power steering mounting bracket retaining bolts. Leave the hoses connected and place the pump aside in a position to prevent fluid from leaking out.

f. Remove the engine oil dipstick tube from the exhaust manifold.

18. If the right-hand (rear) cylinder head is being removed, perform the following:

a. Remove the alternator belt tensioner bracket.

b. Remove the heater supply tube retaining brackets from the exhaust manifold.

c. Remove the Vehicle Speed Sensor (VSS) cable retaining bolt.

d. Remove the EGR vacuum regulator solenoid and bracket.

19. Remove the cylinder head covers.

20. Loosen the rocker arm seat retaining bolts enough to allow the rocker arm to be rotated away from the pushrods.

NOTE: Regardless of the cylinder head being removed, the No. 3 cylinder intake valve pushrod must be removed to allow removal of the intake manifold.

21. Remove the pushrods and label their positions. The pushrods must be installed in their original position during reassembly.

22. Remove the lower intake manifold.

23. Remove the spark plugs.

24. Remove the exhaust manifolds.

25. Remove and discard the cylinder head bolts and remove the cylinder heads from the engine. Remove and discard the old cylinder head gaskets.

To install:

26. Clean the cylinder head bolts holes in the cylinder block with a tap. Clean the cylinder head, intake manifold, cylinder head cover and cylinder head gasket contact surfaces. Check the flatness of the cylinder head and block gasket surfaces. If the cylinder head is warped more than 0.004 inch, it must be machined flat. Do not remove more than 0.010 inch of material.

27. Place new cylinder head gaskets on the cylinder block using the

dowels in the block for alignment. If the dowels are damaged, they must be replaced.

28. Place the cylinder head on the cylinder block.

29. Install new cylinder head bolts and tighten, in sequence, to 59 ft. lbs. (80 Nm), then back off the bolts one turn.

30. Tighten the cylinder head bolts, in sequence, to 37 ft. lbs. (50 Nm). Repeat the torque sequence and tighten the bolts to 68 ft. lbs. (92 Nm).

31. Install the lower intake manifold. Connect the ECT sensor and coolant temperature sending unit electrical harness connectors.

32. Install the pushrods, making sure they are seated in the valve lifters.

33. Lubricate the pushrods and rocker arms with clean engine oil. Move the rocker arms into position with the pushrods and snug the rocker arm seat retaining bolts.

34. For each valve, rotate the crankshaft to position the camshaft lobe of the rocker arm being tightened, on its heel or base circle before tightening the rocker arm seat retaining bolt. Tighten each rocker arm seat retaining bolt to 5–11 ft. lbs. (7–15 Nm) each time having the camshaft lobe on its base circle. Then

Cylinder head gasket installation — 3.0L (VIN U) engine

tighten the retaining bolts again to 19–28 ft. lbs. (26–38 Nm) with the camshaft lobes in any position.

35. Install the remaining components.

36. Connect all vacuum lines to premarked locations.

37. Change the engine oil and filter.

38. Install the engine air cleaner outlet tube to throttle body.

39. Install crankcase ventilation tube to cylinder head cover.

40. Fill and bleed the engine cooling system.

41. Connect the negative battery cable.

42. Install the cowl top vent panel.

43. Start the engine and check for coolant, fuel, oil, vacuum and exhaust leaks.

44. Road test the vehicle and check for proper engine operation.

3.8L (VIN 4) Engine

NOTE: Before beginning this procedure, make sure to have available 4 long and 4 short cylinder head bolts, per side. Once removed, these bolts lose their torque holding ability or retention capability and must not be reused.

1. Disconnect the negative battery cable.

2. Relieve the fuel system pressure.

3. Remove the cowl top vent panel, if needed.

4. Drain the engine cooling system into a suitable container.

5. Remove the engine air cleaner including the air cleaner outlet tube.

6. Rotate the drive belt tensioner and remove the accessory drive belt.

7. If the left-hand cylinder head is being removed, perform the following:

a. Remove the engine oil filler cap.

b. Remove the power steering pump and bracket leaving the power steering lines attached. Leaving the hoses connected, place the power steering pump/alternator bracket assembly aside in a position to prevent the fluid from leaking out.

c. Remove the alternator and the alternator mounting bracket.

8. If the right-hand cylinder head is being removed, perform the following:

a. Remove the drive belt tensioner.

b. Remove the PCV valve.

9. Remove the upper intake manifold.

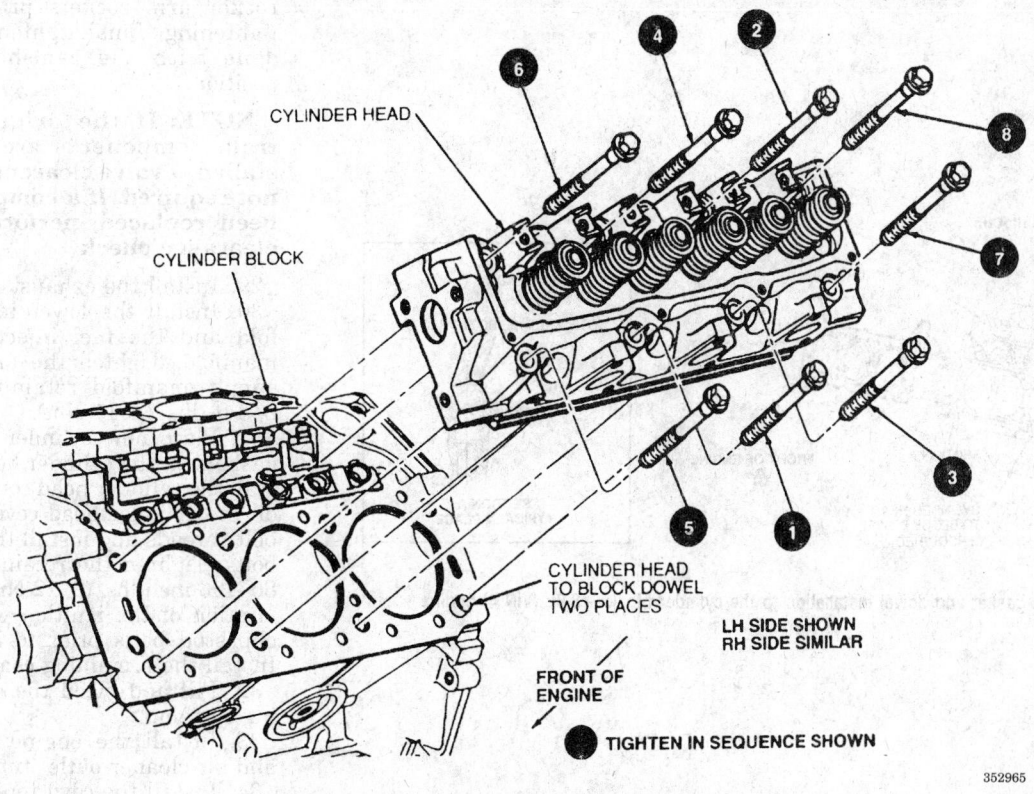

CYLINDER HEAD

CYLINDER BLOCK

CYLINDER HEAD
TO BLOCK DOWEL
TWO PLACES

LH SIDE SHOWN
RH SIDE SIMILAR

FRONT
OF
ENGINE

● TIGHTEN IN SEQUENCE SHOWN

352965

Cylinder head bolt tightening sequence — 3.0L (VIN U) engine

10. Disconnect the fuel supply and return lines and fuel injector wiring harness. Remove the fuel injection supply manifold.

11. Remove the lower intake manifold.

12. Tag to identify, then disconnect the ignition wires from the spark plugs. Remove the ignition wire routing clips from the cylinder head cover retaining bolt studs.

13. Remove the cylinder head covers

14. Remove the exhaust manifolds.

15. Loosen the rocker arm seat (fulcrum) retaining bolts enough to allow the rocker arm to be lifted off the pushrod and rotated aside.

NOTE: If the rocker arms are to be removed from the cylinder head, keep them in the order that they were removed for reassembly.

16. Remove the pushrods. Keep the pushrods in order so they can be installed in their original position during assembly.

17. Remove the cylinder head bolts and discard.

18. Remove the cylinder heads.

19. Clean all gasket mating surfaces.

20. Check the flatness of the cylinder head gasket surface using a straight-edge and a feeler gauge. The allowable warpage is 0.003 inch for every 6.0 inches. Do not machine more than 0.010 inch.

To install:

21. Using new cylinder head bolts, lightly oil the threads of the 4 long cylinder head bolts and apply pipe sealant with Teflon® to the threads of the 4 short cylinder head bolts for each head being installed.

22. Place new cylinder head gaskets on the cylinder block using the dowels for alignment.

23. Carefully position the cylinder heads on the cylinder block and install the new cylinder head bolts hand-tight.

24. For 1995 vehicles, torque the cylinder head bolts in the proper sequence as follows:
- Step 1: 15 ft. lbs. (20 Nm)
- Step 2: 29 ft. lbs. (40 Nm)
- Step 3: 37 ft. lbs. (50 Nm)
- Step 4: Loosen each bolt one at a time in sequence 2–3 turns.
- Step 5: Torque the long cylinder head bolts to 11–19 ft. lbs. (15–25 Nm), then rotate the bolt an additional 85–95° (¼ turn). Go to the next bolt in sequence. Torque the short cylinder head bolts to 7–15 ft. lbs. (10–20 Nm), then rotate the bolt an additional 85–95° (¼ turn). Go to the next bolt in sequence.

25. For 1996–97 vehicles, torque the cylinder head bolts in the proper sequence as follows:
- Step 1: 15 ft. lbs. (20 Nm)
- Step 2: 29 ft. lbs. (40 Nm)
- Step 3: 37 ft. lbs. (50 Nm)
- Step 4: Loosen each bolt one at a time in sequence 2–3 turns.
- Step 5: Torque the long cylinder head bolts to 29–37 ft. lbs. (40–50 Nm), then rotate the bolt an additional 175–185°. Go to the next bolt in sequence. Torque the short cylinder head bolts to 15–22 ft. lbs. (20–30 Nm), then rotate the bolt an additional 175–185°. Go to the next bolt in sequence.

26. Lubricate each pushrod with engine assembly lubricant or heavy engine oil, and install in their original positions.

27. For each valve, rotate the crankshaft until the lifter rests on the heel or base circle of the camshaft lobe (pushrod all the way down). Make sure the rocker arm seat is seated properly, then tighten the rocker arm seat retaining bolt to 44 inch lbs. (5 Nm).

28. Lubricate the rocker arm assemblies with heavy engine oil and final tighten the fulcrum bolts to 19–25 ft. lbs. (25–35 Nm). Fulcrums must be fully seated in cylinder head and pushrods must be seated in

rocker arm sockets prior to final tightening. Final tightening can be done with the camshaft in any position.

NOTE: If the original valve train components are being installed, a valve clearance check is not required. If a component has been replaced, perform a valve clearance check.

29. Install the exhaust manifolds.
30. Install the lower intake manifold and the fuel injection supply manifold. Tighten the fuel injection supply manifold retaining bolts to 6–8 ft. lbs. (8–11 Nm).
31. Place new cylinder head cover gaskets on the cylinder heads and install the cylinder head covers. Lightly oil the cylinder head cover retaining bolt threads and install the retaining bolts. Tighten the retaining bolts to 80–106 inch lbs. (9–12 Nm). Note the location of the ignition wire routing clip stud bolts prior to installation. Install the remaining components.
32. Fill and bleed the engine cooling system.
33. Install the engine air cleaner and air cleaner outlet tube.
34. Install the cowl top vent panel, if removed.
35. Connect the negative battery cable.
36. Start the engine and check for coolant, fuel and oil leaks.
37. Road test the vehicle and check for proper engine operation.

Valve Lifters

REMOVAL AND INSTALLATION

NOTE: Before replacing a valve lifter for noisy operation make sure the noise is not caused by improper valve to rocker arm clearance or by worn rocker arms or pushrods.

3.0L (VIN U) Engine

— **CAUTION** —
Fuel injection systems remain under pressure, even after the engine has been turned OFF. The fuel system pressure must be relieved before disconnecting any fuel lines. Failure to do so may result in fire and/or personal injury.

1. Rotate the crankshaft until the piston for No.1 cylinder is at TDC on its compression stroke.
2. Disconnect the negative battery cable.

Head Gasket Installation

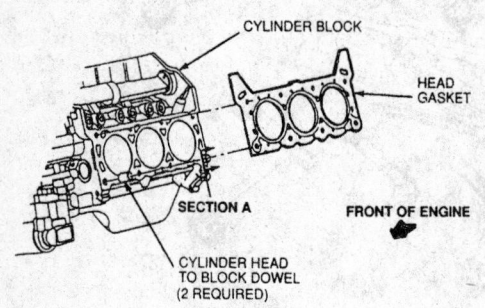

Head gasket and dowel installation to the cylinder block — 3.8L (VIN 4) engine

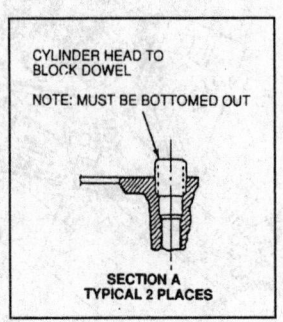

Cylinder head assembly and tightening sequence left side shown — 3.8L (VIN 4) engine

3. Drain the engine coolant.

4. Remove the throttle body/upper intake manifold assembly.

5. Remove both cylinder head covers.

6. Loosen cylinder No. 3 intake valve rocker arm seat retaining bolt and rotate the rocker arm off the pushrod and away from the top of the valve stem. Remove the pushrod.

7. Remove the lower intake manifold.

8. Loosen the rocker arm seat retaining bolts enough to allow the rocker arms to be lifted off the pushrods and rotated to one side.

9. Remove the pushrods. Identify each pushrod for installation to its original location.

10. Loosen 2 valve lifter guide plate and retainer bolts. Remove the guide plate and retainer from the lifter valley.

11. Remove 6 valve lifter guide plates from between each pair of valve lifters by lifting straight up.

NOTE: If the valve lifters are stuck in their bore, it may be necessary to use a claw-type tool to aid removal. Rotate the valve lifter back and forth to loosen it from the deposits.

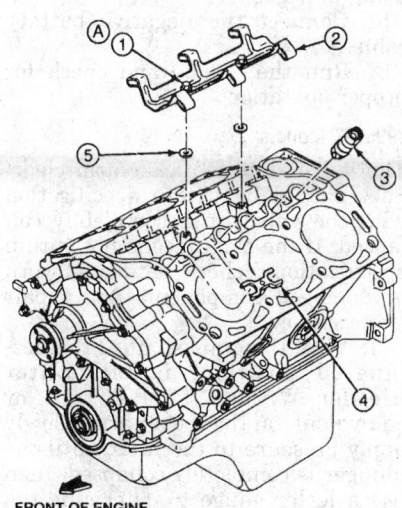

FRONT OF ENGINE

1. Bolt (2 req'd)
2. Tappet guide plate and retainer
3. Valve tappet (12 req'd)
4. Valve tappet guide plate (6 req'd)
5. Washer (2 req'd)
A. Tighten to 10-14 Nm (8-10 lb-ft)

340886

Guide plates and valve lifter removal — 3.0L (VIN U) engine

12. To remove the valve lifter, grasp the valve lifter and pull straight in line with the bore.

To install:

13. Clean the gasket sealing surfaces of the intake manifold and cylinder head. Lightly oil all retaining bolt and stud bolt threads before installation.

14. Lubricate the valve lifters and their bores.

15. Install the valve lifters into their bores. Ensure that the valve lifters move freely in the bores.

16. Aligning the valve lifters flats, install 6 valve lifter guide plates. Install the plates with the word **UP** and/or a button mark visible.

17. Install the valve lifter guide plate and retainer over the valve lifter guide plates. Install the 2 retaining bolts, then tighten to 8–10 ft. lbs. (10–14 Nm).

18. Install the lower intake manifold.

19. Install the pushrods, making sure they are seated in the valve lifters.

20. Lubricate the pushrods and rocker arms with clean engine oil. Move the rocker arms into position with the pushrods and snug the rocker arm seat retaining bolts.

21. For each valve, rotate the crankshaft to position the camshaft lobe of the rocker arm being tightened, on its heel or base circle before tightening the rocker arm seat retaining bolt. Tighten each rocker arm seat retaining bolt to 5–11 ft. lbs. (7–15 Nm) each time having the camshaft lobe on its base circle. Then tighten the retaining bolts again to 19–28 ft. lbs. (26–38 Nm) with the camshaft lobes in any position.

22. Install the cylinder head covers.

23. Install the throttle body/upper intake manifold assembly.

24. Fill and bleed the engine cooling system.

25. Connect the negative battery cable.

26. Start the engine and allow to reach normal operating temperature while checking for leaks.

27. Road test the vehicle and check for proper engine operation.

3.8L (VIN 4) Engine

1. Disconnect the negative battery cable.

2. Tag and disconnect the ignition wires at the spark plugs.

3. Remove the ignition wire separators from the studs on the cylinder head cover retaining bolts. Lay the ignition wires with the wire

separators toward the front of the engine.

4. Remove the cowl top vent panel, if needed.

5. Remove the upper intake manifold.

6. Remove the left and right-hand cylinder head covers.

7. Remove the lower intake manifold.

8. Sufficiently loosen each rocker arm seat retaining bolt to allow the rocker arm to be lifted off the pushrod and rotated aside.

9. Remove the pushrods. The location of each pushrod must be identified so they can be installed in their original location.

10. Remove the 4 bolts holding the guide plate retainers in place; the bolts are held in the retainers. Remove the 6 guide plates from the adjacent valve lifters.

NOTE: If the valve lifters are stuck in the bores due to excessive varnish or gum deposits, it may be necessary to use a claw-type tool to aid removal. When using a remover tool, rotate the valve lifter back and forth to loosen it from gum or varnish that may have formed in the bore or on the valve lifter.

11. Remove the valve lifters using a magnet. The location of each valve lifter must be identified so they can be installed in their original position.

To install:

12. Clean the cylinder head cover and cylinder head gasket mating surfaces.

13. Install each valve lifter in the bore from which it was removed. If new valve lifters are being installed, soak the lifters in engine oil. Check the new lifters for free fit in the valve lifter bores in the cylinder block.

14. Align the flats on the side of the valve lifters and install 6 guide plates between the adjacent valve lifters. Ensure that the word **UP** and/or a dimple is showing. Install 2 guide plate retainers and tighten 4 retaining bolts to 89–124 inch lbs. (8–14 Nm).

15. Dip each pushrod in engine assembly lubricant and install in its original position.

NOTE: The rocker arm seats must be fully seated in the cylinder head and the pushrods must be seated in the rocker arm sockets prior to final tightening.

16. For each valve, rotate the crankshaft until the valve lifter rests on the heel or base circle of the camshaft lobe (pushrod all the way

down). Position the rocker arms over the pushrods, install the rocker arm seats and tighten the rocker arm seat retaining bolts to a maximum of 44 inch lbs. (5 Nm).

17. Once all rocker arms are initially tightened, torque the rocker arm seat retaining bolts again to 23–30 ft. lbs. (30–40 Nm). For final tightening, the camshaft may be in any position.

18. Install the lower intake manifold.

19. Place new gaskets on the cylinder heads and install the cylinder head covers. Lightly oil the cylinder head cover retaining bolt threads and install the retaining bolts. Tighten the retaining bolts to 71–97 inch lbs. (8–11 Nm). Note the location of the ignition wire separator stud bolts prior to installation.

20. If the left-hand cylinder head cover was removed, install the oil filler cap.

21. If the right-hand cylinder head cover was removed, install the PCV valve.

22. Install the upper intake manifold.

23. Install the engine air cleaner assembly.

24. Install the ignition wire separators and connect the ignition wires to the spark plugs.

25. Install the cowl top vent panel, if removed.

26. Connect the negative battery cable.

27. Start the engine and allow to idle for several minutes while checking for oil leaks.

28. Road test the vehicle and check for proper engine operation.

Valve Lash

ADJUSTMENT

3.0L (VIN U) Engine

Hydraulic valve lifters allowing no valve clearance adjustment. A clearance check of the rocker arm-to-valve stem gap is required when machining has been done to the cylinder heads, valves, valve seats or cylinder block head gasket surfaces or when new valve train components have been installed. The clearance check is also useful in determining loose, worn or damaged parts when there is a concern with the valve train. Clearance must be checked when the valve lifter is completely collapsed.

1. To check valve clearance, use Lifter Bleed Down Wrench T71P-6513-B or equivalent, to slowly push down on the pushrod end of the rocker arm and bleed the oil from the valve lifter.

2. Once the valve lifter is totally collapsed, insert the appropriate thickness feeler gauge between the rocker arm and valve stem to check the clearance. There should be 0.085–0.185 inch (2.15–4.69mm) clearance between the valve stem and the rocker arm.

3. Rotate the crankshaft until the piston for No.1 cylinder is at TDC on its compression stroke. With the engine at this position, the following valve clearances can be checked:

- No. 1 intake
- No. 1 exhaust
- No. 3 exhaust
- No. 2 intake
- No. 6 exhaust
- No. 4 intake

4. Rotate the crankshaft 360 degrees and check the following valves in the same manner:

- No. 3 intake
- No. 2 exhaust
- No. 6 intake
- No. 5 intake
- No. 4 exhaust
- No. 5 exhaust

5. If the clearance is not correct, check for loose, worn or damaged components. If no problems are found, the clearance may be adjusted by the use of shorter or longer pushrods.

3.8L (VIN 4) Engine

1995 Models

NOTE: The valve lifters are hydraulic and require no adjustment. A valve clearance check is available to determine if an engine noise concern is valvetrain related by measuring the gap between the valve stem and the rocker arm.

1. Disconnect the negative battery cable.

2. Remove the rocker arm covers.

3. Rotate the crankshaft using a socket and bar on the crankshaft pulley, to place No.1 piston on TDC at the end of the compression stroke.

NOTE: The engine will turn over easier if the spark plugs are removed.

4. For the following valves, collapse the valve lifters using bleed down wrench T82L-6500-A or equivalent, by placing the tool on the rocker arm and applying a downward pressure.

Number 1 piston at TDC on the compression stroke:
- No. 1 intake and No. 1 exhaust
- No. 3 intake and No. 2 exhaust
- No. 6 intake and No. 4 exhaust

5. With each valve lifter collapsed, measure and record the clearance between the valve stem and the rocker arm. The clearance should be 0.09–0.19 inch (2.25–4.79 mm).

6. Rotate the crankshaft 360 degrees, placing No. 1 piston on TDC at the end of its exhaust stroke.

7. For the following valves, collapse the valve lifters using bleed down wrench T82L-6500-A or equivalent, by placing the tool on the rocker arm and applying a downward pressure.

Number 1 piston at TDC on the exhaust stroke:
- No. 2 intake and No. 3 exhaust
- No. 4 intake and No. 6 exhaust
- No. 5 intake and No. 5 exhaust

8. With each valve lifter collapsed, measure and record the clearance between the valve stem and the rocker arm. The clearance should be 0.09–0.19 inch (2.25–4.79 mm).

9. If any of the valve clearances are not within specifications, inspect the valvetrain components for wear or damage and replace parts as required.

10. Install the spark plugs if removed.

11. Install the rocker arm covers using new gaskets.

12. Connect the negative battery cable.

13. Run the engine and check for proper operation.

1995–97 Models

The valve stem-to-rocker arm clearance should be within specification with the valve lifter completely collapsed. If the clearance is not within specifications, check for loose, worn or damaged components and repair as necessary.

1. With the crankshaft in the designated position, install Lifter Bleeder Wrench T71P-6513-B or equivalent, on the rocker arm. Slowly apply pressure to the lifter until the plunger is completely collapsed, then use a feeler gauge to determine the valve stem-to-rocker arm clearance.

2. Rotate the engine until the piston for No.1 cylinder is at TDC on its compression stroke. Check the valve stem-to-rocker arm clearance for the following valves:

- No. 1 intake and No. 1 exhaust
- No. 3 intake and No. 2 exhaust
- No. 6 intake and No. 4 exhaust

3. Rotate the crankshaft 360 degrees and check the valve stem-to-

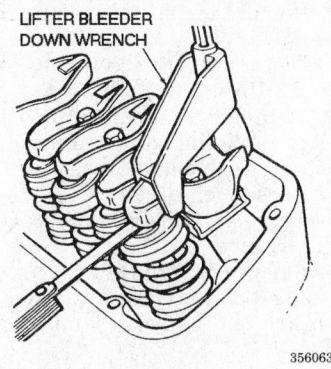

Checking valve stem-to-rocker arm clearance

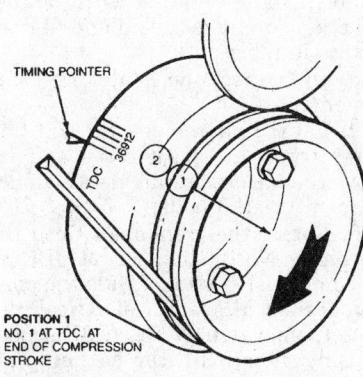

CYL. NO.	CRANKSHAFT POSITION	
	1	2
	SET GAP OF VALVES NOTED	
1	INT — EXH	NONE
2	EXH	INT
3	INT	EXH
4	EXH	INT
5	NONE	INT — EXH
6	INT	EXH

356062

Crankshaft positioning — 1995–97 3.8L models

rocker arm clearance for the following valves.

- No. 2 intake and No. 3 exhaust
- No. 4 intake and No. 6 exhaust
- No. 5 intake and No. 5 exhaust

4. The valve stem-to-rocker arm clearance should be 0.08465–0.18465 inch (2.15–4.69mm) for all valves with the valve lifters fully collapsed.

Rocker Arms

REMOVAL AND INSTALLATION

3.0L (VIN U) Engine

1. Disconnect the negative battery cable.
2. Label and disconnect the ignition wires from the spark plugs.
3. Remove the ignition wire/separator assembly from the cylinder head cover retaining stud bolts.
4. If the left-hand cylinder head cover is being removed, disconnect the air cleaner closure system hose and remove the fuel injector harness from the inboard cylinder head cover studs.

NOTE: Some cylinder head covers (composite) are designed with integral gaskets which should last the life of the vehicle. Use care when removing theses cylinder head covers to prevent damage to the gaskets.

5. If the right-hand cylinder head cover is being removed, remove the throttle body/upper intake manifold.
6. Remove the cylinder head cover retaining bolts, covers and gaskets from the engine.
7. Remove the rocker arm seat retaining bolts, seats (fulcrums) and the rocker arms. Keep all parts in order so they can be reinstalled to their original positions.
8. Remove the pushrods, if necessary. Keep them in order so they can be reinstalled in their original positions.
9. Inspect the rocker arms, seats and pushrods for wear and/or damage. Replace as necessary.

To install:

10. Clean the cylinder head and cylinder head cover gasket sealing surfaces.
11. Install the pushrods, if removed, making sure they seat in the valve lifter sockets.
12. Coat the valve and pushrod tips, rocker arm and seat contact areas with engine assembly lubricant. Lightly oil all bolt and stud bolt threads before installation.
13. Rotate the crankshaft until the valve lifter is on the heel or base circle of the camshaft lobe for each rocker arm to be tightened.
14. Install the rocker arm and components making sure that the pushrods are positioned in the valve lifter and rocker arm sockets. Torque the rocker arm seat retaining bolts in 2 steps, first tighten to 5–11 ft. lbs. (7–15 Nm) and final tighten to 20–28

ft. lbs. (26–38 Nm). Ensure that the valve lifter is on the base circle of the camshaft for each rocker arm as it is installed.

15. Apply a bead of silicone sealant at 4 joints where the cylinder heads and intake manifold meet.
16. Install new cylinder head cover gaskets on the cylinder head covers, if needed and place on the engine. Install the cylinder head cover retaining bolts and stud bolts. Tighten to 8–10 ft. lbs. (10–14 Nm), in the proper sequence.
17. If the left-hand cylinder head cover was removed, connect the crankcase ventilation tube and install the fuel injector wiring to the inboard cylinder head cover stud bolts.
18. If the right-hand cylinder head cover was removed, install the fuel injector wiring to the inboard cylinder head cover stud bolts, install the throttle body/upper intake manifold using a new gasket, install the PCV valve and hose and connect the EGR valve tube. Tighten the EGR valve tube retaining nuts to 26–48 ft. lbs. (35–65 Nm).
19. Connect the negative battery cable.
20. Start the engine and check for oil leaks.
21. Road test the vehicle and check for proper engine operation.

3.8L (VIN 4) Engine

1. Disconnect the negative battery cable.
2. Tag to identify, then disconnect the ignition wires from the spark plugs. Remove the ignition wire separators from the cylinder head cover retaining bolt studs.
3. Remove the upper intake manifold.
4. Remove the left and right-hand cylinder head covers.
5. Remove the rocker arm, seat (fulcrum) and bolt assemblies. Keep each assembly together and identify the assemblies so they may be reinstalled in their original positions.
6. If necessary, remove the pushrods. Identify each pushrod as it is removed so it can be reinstalled in its original location.
7. Inspect the rocker arm assemblies and pushrods for wear and/or damage. Replace parts, as necessary.

To install:

8. Clean the gasket mating surfaces on the cylinder head covers and cylinder heads.
9. If removed, install the pushrods, making sure they are seated in the valve lifters.

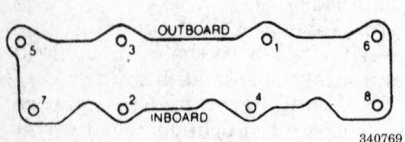

Cylinder head cover bolt torque sequence — 3.0L (VIN U) engine

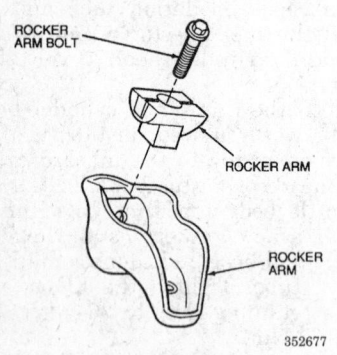

Rocker arm, seat and bolt assembly — 3.8L (VIN 4) engine

10. Apply engine assembly lubricant to the pushrod tips and valve stem tips. Lubricate the rocker arm seats and rocker arms with engine assembly lubricant and install them over the pushrods and valve stems.

NOTE: If new rocker arms and fulcrums are being used, soak the new components in a break-in additive for at least 3 minutes before installation.

11. For each valve, rotate the crankshaft until the valve lifter is on the heel or base circle of the camshaft lobe. Install the rocker arm seat bolt and tighten to 44 inch lbs. (5 Nm) maximum. Make sure the pushrod and rocker arm seats are fully seated prior to tightening.

12. Lubricate all rocker arm assemblies with clean engine oil. Final tighten the rocker arm seat bolts to 23–30 ft. lbs. (30–40 Nm). When final tightening, the camshaft may be in any position.

13. Place new gaskets on the cylinder heads and install the cylinder head covers. Lightly oil the cylinder head cover retaining bolt threads and install the retaining bolts. Tighten the retaining bolts to 71–97 inch lbs. (8–11 Nm). Note the location of the

ignition wire separator stud bolts prior to installation.

14. Install the remaining components. Connect the negative battery cable.

15. Start the engine and check for leaks.

16. Road test the vehicle and check for proper engine operation.

Intake Manifold

REMOVAL AND INSTALLATION

3.0L (VIN U) Engine

— **CAUTION** —

Fuel injection systems remain under pressure, even after the engine has been turned OFF. The fuel system pressure must be relieved before disconnecting any fuel lines. Failure to do so may result in fire and/or personal injury.

1. Disconnect the negative battery cable.
2. Drain the engine cooling system.
3. Relieve the fuel system pressure.
4. Remove the crankcase ventilation tube from the cylinder head cover and engine air cleaner outlet tube.
5. Remove the engine air cleaner outlet tube at the throttle body.
6. Remove the fuel line safety clips and disconnect the fuel supply and return lines at the fuel injection supply manifold.
7. Label and disconnect the vacuum hoses from the throttle body/upper intake manifold assembly.

NOTE: The throttle body and upper intake manifold are manufactured as one assembly and will be referred to as the throttle body assembly.

8. Disconnect the Throttle Position (TP) sensor, Intake Air Temperature (IAT) sensor, Camshaft Position (CMP) sensor, Engine Coolant Temperature (ECT) sensor and Idle Air Control (IAC) valve electrical harness connectors.
9. Disconnect the wiring to the ignition coil, water temperature sender and the EGR backpressure transducer.
10. Disconnect the upper radiator hose from the engine.
11. Remove the alternator-to-throttle body support brace stud bolt.

12. Remove the shield from the IAC valve.
13. Disconnect the accelerator cable from the throttle body lever.
14. Remove 2 accelerator cable bracket retaining bolts and move the accelerator cable and bracket aside.
15. Loosen the EGR valve-to-exhaust manifold tube nuts and remove the EGR tube.
16. Remove 5 throttle body retaining bolts and 1 stud bolt and remove the throttle body assembly and gasket.
17. Label and disconnect the fuel injector electrical harness connectors from each fuel injector. Separate the wiring harness from the cylinder head covers and move the wiring aside. The lower intake manifold assembly can be removed with the fuel injection supply manifold and fuel injectors in place.
18. Disconnect the heater hoses at the engine.
19. Label and remove the ignition wires from the spark plugs and remove the harness retainer-to cylinder head cover stud bolts.
20. Rotate the crankshaft until the piston for No.1 cylinder is at TDC on its compression stroke. Note the position of the CMP sensor electrical connector, then remove the CMP sensor housing along with the oil pump intermediate shaft.
21. Remove the ignition coil from the rear of the left-hand cylinder head.
22. Remove both cylinder head covers.
23. Loosen the intake valve rocker arm retaining bolt from No. 3 cylinder and rotate the rocker arm away from the valve stem and pushrod. Remove the pushrod.
24. Remove the lower intake manifold retaining bolts. Use a suitable prybar to loosen the intake manifold. Pry upward using the area between the thermostat and transaxle as a leverage point. Remove the manifold, gaskets and seals.

To install:
25. Clean the gasket mating surfaces of the throttle body assembly, lower intake manifold and cylinder heads. Lay a shop rag in the lifter valley to catch any gasket material. After scraping, carefully lift the cloth from the lifter valley, being careful not to let any particles enter the oil drain holes or cylinder heads. If necessary, use a suitable solvent to remove old rubber sealant.
26. Clean and lightly oil all retaining bolts and stud threads before installation.

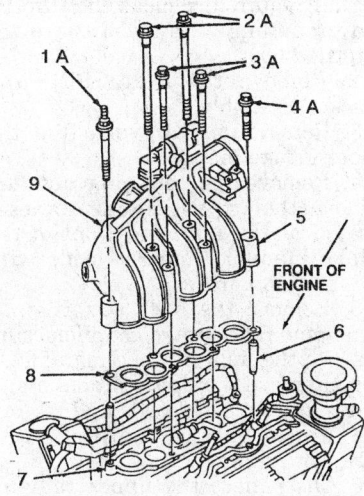

1. Stud bolt
2. Bolt (2 req'd)
3. Bolt (2 req'd)
4. Bolt
5. Throttle body
6. Guide pin (2 req'd)
7. Intake manifold
8. Intake manifold upper gasket
9. Intake manifold vacuum outlet
   fitting and cap
A. Tighten to 20-30 Nm (15-22 lb. ft.)

343159

Throttle body (upper intake manifold) and related components — 3.0L (VIN U) engine

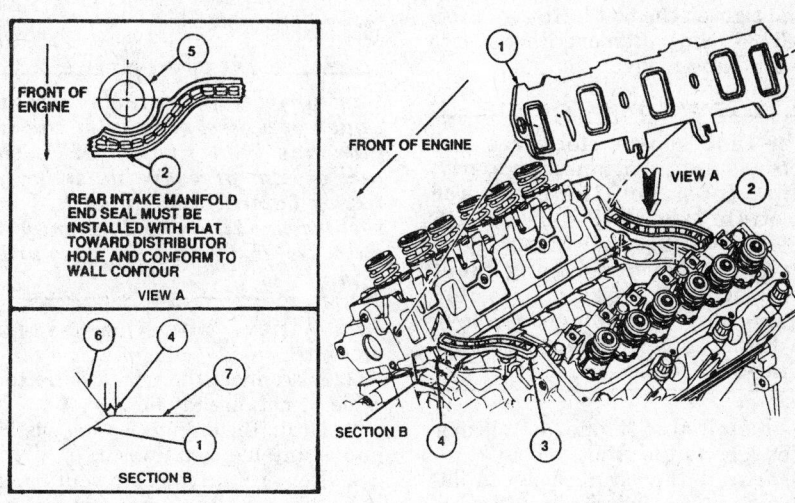

1	Intake Manifold Gasket (2 Req'd)		5	Distributor Hole
2	End Seal		6	Cylinder Head (2 Req'd)
3	End Seal		7	Cylinder Block
4	Silicone Rubber (4 Places)			

343160

Lower intake manifold gasket installation — 3.0L (VIN U) engine

27. Apply a 0.25 inch (5–6mm) bead of a suitable silicone rubber sealer to the 4 intersections of the cylinder block end rails and cylinder heads. Be careful not to let sealer that may block oil passages fall into the engine.

NOTE: When using a silicone sealer, assembly must occur within 15 minutes after the sealer has been applied. After this time, the sealer may start to set-up and its sealing quality may be reduced. In high temperature/humidity conditions, the sealant will start to set up in approximately 5 minutes.

28. Install the front and rear intake manifold end seals in place and secure. Install the intake manifold gaskets, aligning the locking tabs to the provisions on the cylinder head gaskets.

29. Carefully lower the intake manifold into position on the cylinder block and cylinder heads to prevent smearing the silicone sealer and possibly causing gasket voids.

30. Install the lower intake manifold retaining bolts hand tight. Torque the bolts in sequence, in 2 steps to 15–22 lbs. (20–30 Nm) and finally to 20–23 ft. lbs. (26–32 Nm).

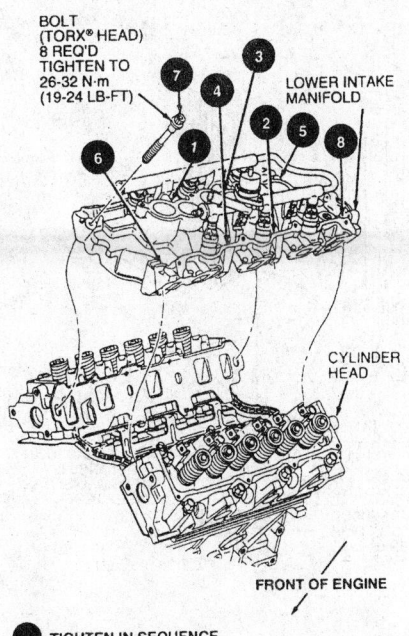

BOLT (TORX® HEAD) 8 REQ'D TIGHTEN TO 26-32 N·m (19-24 LB-FT)

LOWER INTAKE MANIFOLD

CYLINDER HEAD

FRONT OF ENGINE

● TIGHTEN IN SEQUENCE

354189

Lower intake manifold torque sequence — 3.0L (VIN U) engine

31. Install the remaining components.

— **WARNING** —
Syncro Positioning Tool T93P-12200-A or equivalent, must be obtained prior to installing the CMP sensor housing. Failure to follow this procedure will result in improper CMP sensor alignment. This will result in the ignition and fuel systems being out of time with the engine, possibly causing engine damage.

32. Install the CMP sensor housing as follows:
   a. Engage the CMP sensor housing vane into the radial slot of Syncro Positioning Tool T93P-12200-A, or equivalent. Rotate the tool on the CMP sensor housing until the tool boss engages the notch in the CMP sensor.
   b. Install the CMP sensor housing along with the oil pump intermediate shaft. Install the CMP sensor housing so that drive gear engagement occurs when the arrow on the locator tool is pointed approximately 30 degrees counterclockwise from the rear face of the cylinder block. This step will locate the CMP sensor electrical connector in the pre-removal position.

c. Install the hold-down clamp and tighten the bolt to 15–22 ft. lbs. (20–30 Nm). Remove the syncro positioning tool.

— WARNING —

If the CMP sensor electrical connector is not positioned properly, do not reposition the connector by rotating the CMP sensor housing. This will result in the ignition and fuel systems being out of time with the engine, possibly causing engine damage. Remove the housing and repeat the installation procedure.

33. Install the engine air cleaner outlet tube to the throttle body.
34. Install the crankcase ventilation tube to the cylinder head cover.
35. Fill and bleed the engine cooling system.
36. Connect the negative battery cable.
37. Turn the ignition switch to the **RUN** position several times without starting the engine to pressurize the fuel system and to check for fuel leaks.
38. Start the engine and check for fuel and coolant leaks.
39. Install the IAC valve shield.
40. Road test the vehicle and check for proper engine operation.

3.8L (VIN 4) Engine
1995 Models

— CAUTION —

Fuel injection systems remain under pressure, even after the engine has been turned OFF. The fuel system pressure must be relieved before disconnecting any fuel lines. Failure to do so may result in fire and/or personal injury.

1. Relieve the fuel system pressure.
2. Disconnect the negative battery cable, if not already done.
3. Drain the engine cooling system into a suitable container.
4. Remove the cowl top vent panel.
5. Remove the engine air cleaner and the air cleaner outlet tube.
6. Disconnect the accelerator cable at the throttle body. Disconnect the speed control cable, if equipped.
7. Remove the retaining bolts from the accelerator cable mounting bracket and position the cables aside.
8. Tag and disconnect the necessary vacuum lines and electrical connectors.
9. Disconnect the crankcase ventilation tube between the PCV valve and the upper intake manifold.
10. Remove the alternator brace.

11. If required, remove the throttle body retaining nuts and remove the throttle body.
12. Remove the Delta PFE and bracket assembly.
13. Remove the EGR valve from the upper intake manifold.
14. Remove the retaining nut and remove the wiring retainer bracket located at the left-hand front of the intake manifold and set aside with the ignition wires.
15. Remove the bolts securing the front and rear intake manifold supports to the upper intake manifold.
16. Remove the upper intake manifold retaining bolts/studs and remove the upper intake manifold and gasket.
17. Disconnect the upper radiator hose at the thermostat housing and the coolant bypass hose at the lower intake manifold.
18. Remove the heater hose.
19. Remove the bolt securing the power steering line bracket to the ignition coil bracket.
20. Remove the bolt securing the fuel injection supply manifold bracket and the rear intake manifold support to the cylinder head. Remove the rear intake manifold support.
21. Remove the bolt and nut securing the ignition coil bracket to the intake manifold.
22. Remove the fuel injection supply manifold and the fuel injectors.
23. Remove the nut securing the front intake manifold support to the intake manifold and remove the intake manifold support.
24. Tag and disconnect the electrical connectors for the water temperature sender and the engine coolant temperature sensor.

NOTE: The intake manifold is sealed at each corner with silicone sealer. To break the seal, it may be necessary to pry on the front of the manifold with a small prybar. If it is necessary to pry on the manifold, use care to prevent damage to the machined surfaces.

25. Remove the intake manifold retaining bolts and remove the intake manifold, manifold gaskets and the end seals.
   To install:
26. Clean all gasket mating surfaces. Lightly oil all retaining bolt and stud threads.

NOTE: If the lower manifold is being replaced, coat the threads of the sensors and fittings with pipe sealant with Teflon®during installation.

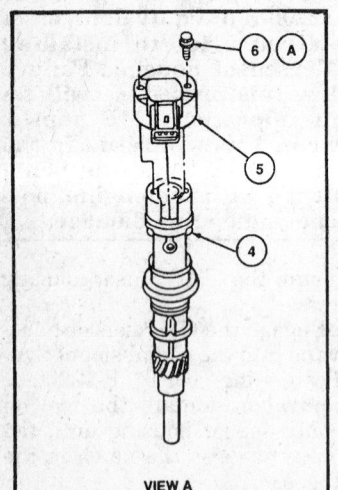
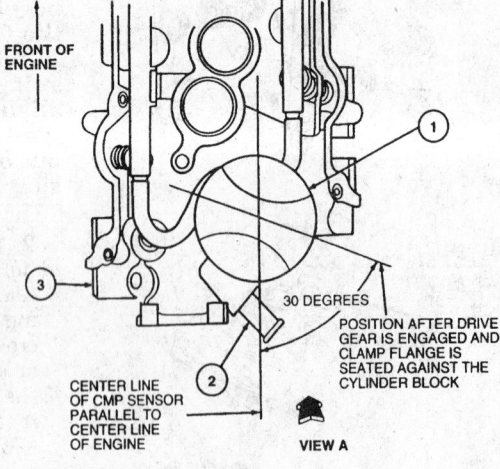

1. Syncro positioning tool
2. Fitting
3. Intake manifold
4. Camshaft position sensor housing
5. Camshaft position sensor
6. Screw (2 req'd)
A. Tighten to 2.5-3.5 Nm (22-31 lb-in)

354190

Camshaft position sensor — 3.0L (VIN U) engine

27. Apply gasket adhesive to each cylinder head mating surface. Press new intake manifold gaskets in place, using the locating pins as necessary to aid in the installation.

28. Apply a ⅛ inch (3–4mm) bead of silicone sealer at each corner where the cylinder head joins the cylinder block. Install the front and rear intake manifold end seals.

29. Carefully lower the intake manifold into place on the cylinder heads and cylinder block. Use locating pins as necessary to guide the manifold.

30. Install 12 bolts and 2 stud bolts in their original locations and tighten the bolts, in sequence, in 2 steps to 13 ft. lbs. (18 Nm) and then 16 ft. lbs. (22 Nm).

31. Connect the electrical harness connectors for the water temperature sender and the engine coolant temperature sensor.

32. Position the front intake manifold support to the intake manifold and install the retaining nut. Tighten the nut to 15–22 ft. lbs. (20–30 Nm).

33. Install new O-rings on the fuel injectors and install the fuel injectors and the fuel injection supply manifold.

34. Install the bolt and nut securing the ignition coil bracket to the intake manifold. Tighten the bolt and nut to 15–22 ft. lbs. (20–30 Nm).

35. Position the rear intake manifold support bracket from the fuel injection supply manifold to the cylinder head. Tighten the bolt to 15–22 ft. lbs. (20–30 Nm).

36. Position the power steering line bracket to the ignition coil bracket and install the retaining bolt. Tighten the bolt securely.

37. Install the heater hose.

38. Connect the coolant bypass hose to the intake manifold.

39. Connect the upper radiator hose to the thermostat housing.

40. Place a new upper intake manifold gasket onto the intake manifold. Use locating pins to temporarily hold the gasket in place and to aid in aligning the upper intake manifold during assembly.

41. Carefully position the upper intake manifold onto the new gasket and install 4 bolts and 2 stud bolts in their original locations. Tighten the bolts in sequence and in 3 steps, first to 8 ft. lbs. (10 Nm), then to 15 ft. lbs. (20 Nm), and finally to 24 ft. lbs. (32 Nm).

42. Install the bolts securing the front and rear intake manifold supports to the upper intake manifold. Tighten the bolts to 15–22 ft. lbs. (20–30 Nm).

43. Position the wiring retainer bracket and install the retaining nut located at the left-hand front of the intake manifold.

44. Install the EGR valve. Tighten the retaining bolts to 15–22 ft. lbs. (20–30 Nm).

45. Install the Delta PFE and bracket assembly.

46. If removed, install the throttle body on the upper intake manifold using a new gasket. Alternately tighten the retaining nuts to 15–22 ft. lbs. (20–30 Nm).

47. Connect the crankcase ventilation tube to the PCV valve and the upper intake manifold.

48. Connect the necessary electrical harness connectors and vacuum lines.

49. Position the accelerator cable bracket and install the retaining bolts. Tighten the bolts to 15–22 ft. lbs. (20–30 Nm).

50. Connect the speed control cable, if equipped.

51. Install the engine air cleaner and air cleaner outlet tube.

52. Install the cowl top vent panel.

53. Fill and bleed the engine cooling system.

54. Connect the negative battery cable.

55. Run the engine and check for leaks.

56. Road test the vehicle and check for proper operation.

1996–97 Models — Upper Manifold

> **CAUTION**
>
> *Fuel injection systems remain under pressure, even after the engine has been turned OFF. The fuel system pressure must be relieved before disconnecting any fuel lines. Failure to do so may result in fire and/or personal injury.*

1. Disconnect the negative battery cable.

2. Remove the engine air cleaner outlet tube.

3. Disconnect the accelerator cable and speed control cable if equipped, at the throttle body lever.

4. Remove 2 retaining bolts securing the accelerator cable bracket and move the bracket and cables aside.

5. Tag and disconnect the necessary electrical and vacuum lines and connectors.

6. Disconnect the PCV hose and the crankcase vent hose.

7. If required, remove the throttle body and the Idle Air Control (IAC) valve from the upper intake manifold.

8. Remove 12 upper intake manifold retaining bolts and remove the upper intake manifold.

9. Inspect the integral upper intake manifold-to-lower intake manifold gaskets.

To install

10. Clean the upper and lower intake manifold gasket sealing surfaces.

11. Ensure that the upper intake manifold seals are completely installed in the manifold grooves and that the seal show no signs of cuts or tears. Replace as necessary.

12. Place the upper intake manifold in position and install 12 retaining bolts to their original locations. Tighten the bolts in sequence to 71–106 inch lbs. (8–12 Nm).

13. If removed, install the IAC valve and the throttle body assembly.

14. Connect the crankcase vent hose and the PCV hose.

15. Connect the electrical and vacuum lines and connectors.

16. Install the accelerator cable bracket with 2 retaining bolts. Torque the bolts to 71–106 inch lbs. (8–12 Nm).

17. Connect the speed control and the accelerator cables to the throttle body lever.

18. Install the engine air cleaner outlet tube.

338305

Intake manifold and retaining bolt torque sequence — 3.8L (VIN 4) engine

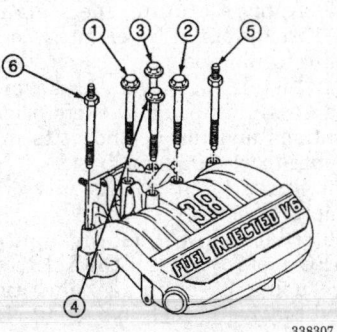

338307

Intake plenum and retaining bolt torque sequence — 1995 3.8L (VIN 4) engine

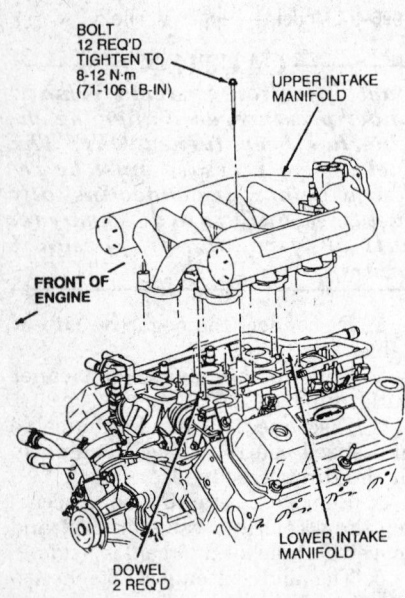

BOLT
12 REQ'D
TIGHTEN TO
8-12 N·m
(71-106 LB-IN)

UPPER INTAKE
MANIFOLD

FRONT OF
ENGINE

LOWER INTAKE
MANIFOLD

DOWEL
2 REQ'D

353864

Upper intake manifold removal — 1996–97 3.8L (VIN 4) engine

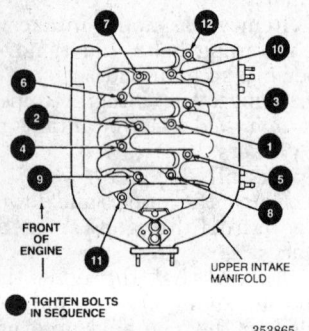

FRONT
OF
ENGINE

UPPER INTAKE
MANIFOLD

TIGHTEN BOLTS
IN SEQUENCE

353865

Upper intake manifold torque sequence — 1996–97 3.8L (VIN 4) engine

19. Connect the negative battery cable.
20. Start engine and check for vacuum leaks.
21. Road test the vehicle and check for proper engine operation.

Lower Intake Manifold

1. Disconnect the negative battery cable.
2. Remove the upper intake manifold.
3. Drain the engine cooling system.
4. Relieve the fuel system pressure.

5. Disconnect the upper radiator hose at the lower intake manifold.
6. Disconnect the water bypass hose and tube.
7. Disconnect the water temperature sender and ECT sensor electrical harness connectors.
8. Remove the vacuum lines from the IMRC motors.
9. Disconnect the engine control wiring to the fuel injectors.
10. Disconnect the fuel supply and return lines at the fuel injection supply manifold.
11. Remove the fuel injection supply manifold and the fuel injectors.
12. Disconnect the vacuum motor and the bracket assemblies.
13. Using a prying tool, separate the valve assembly linkage from the IMRC lever and bushing.
14. Remove the tube retaining bolts and remove the old bushing from the lever.
15. Remove the EGR valve and the adapter.
16. Remove the lower intake manifold retaining bolts.

NOTE: The lower intake manifold is sealed at each corner with sealer. It may be necessary to pry the front of the lower intake manifold to break the seal for removal.

17. Remove the lower intake manifold gaskets and end seals.
To install:
18. Remove any remaining sealer and gasket material from the sealing surfaces.
19. Install 2 new lower intake manifold-to-cylinder head gaskets.
20. Apply a 0.125 inch (3mm) bead of silicone sealer to the 4 corners where the cylinder heads join the cylinder block.
21. Install the front and rear end seals to the cylinder block.
22. Carefully place the lower intake manifold in position using locating pins if necessary to ensure gasket alignment.
23. Apply a coat of pipe sealant to the threads of the lower intake manifold retaining bolts.
24. Install the lower intake manifold retaining bolts to their original locations and torque the bolts in sequence to 71–106 inch lbs. (8–12 Nm).
25. Install the EGR valve and adapter assembly.
26. Place new bushings on the IMRC motors. Install the IMRC motors using 2 retaining bolts, tightened to 71–106 inch lbs. (8–12 Nm).
27. Install the fuel injection supply manifold and the fuel injectors as an assembly.

28. Connect the fuel supply and return lines and the engine control wiring to the fuel injectors.
29. Install the water bypass tube to the lower intake manifold and torque the retaining bolts to 71–97 inch lbs. (8–11 Nm).
30. Connect the water bypass tube hose.
31. Connect the electrical harness connectors to the water temperature sender and the ECT sensor.
32. Connect the vacuum lines to the IMRC motors.
33. Connect the upper radiator hose.
34. Install the upper intake manifold.
35. Fill and bleed the engine cooling system.
36. Connect the negative battery cable.
37. Start the engine and check for coolant, fuel and vacuum leaks.
38. Road test the vehicle and check for proper engine operation.

Exhaust Manifold

REMOVAL AND INSTALLATION

3.0L (VIN U) Engine

Left-Hand Exhaust Manifold

1. Disconnect the negative battery cable.
2. Remove the oil level indicator tube support bracket retaining nut, dipstick and tube.
3. Raise and safely support the vehicle.
4. Remove the exhaust manifold-to-dual converter Y-pipe retaining nuts.
5. Remove the exhaust manifold retaining bolts and stud bolts.
6. Remove the exhaust manifold.
To install:
7. Clean the mating surfaces of the exhaust manifold, cylinder head and dual converter Y-pipe.
8. Lightly oil all bolt and stud bolt threads prior to installation.
9. Place the exhaust manifold to the cylinder head. Install and tighten the exhaust manifold retaining bolts and stud bolts to 15–22 ft. lbs. (20–30 Nm).
10. Connect the dual converter Y-pipe to the exhaust manifold and install the retaining nuts. Tighten the nuts to 25–34 ft. lbs. (34–47 Nm).
11. Install the oil level indicator tube, retaining nut and the oil level dipstick. Tighten the retaining nut to 11–14 ft. lbs. (15–20 Nm).

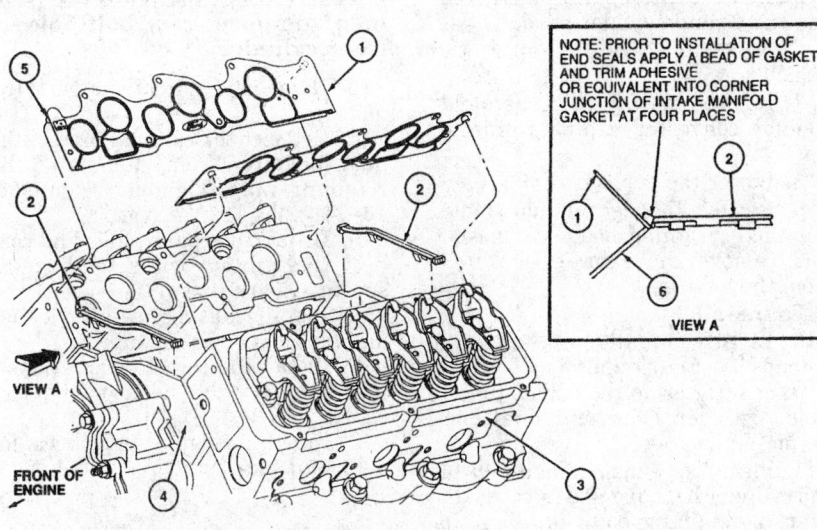

NOTE: PRIOR TO INSTALLATION OF END SEALS APPLY A BEAD OF GASKET AND TRIM ADHESIVE OR EQUIVALENT INTO CORNER JUNCTION OF INTAKE MANIFOLD GASKET AT FOUR PLACES

VIEW A

1	Intake Manifold Gasket (2 Req'd)
2	End Seal (2 Req'd)
3	Cylinder Head (2 Req'd)
4	Cylinder Block
5	Locating Pin
6	Head Gasket (2 Req'd)

353867

Lower intake manifold gasket installation — 1996–97 3.8L (VIN 4) engine

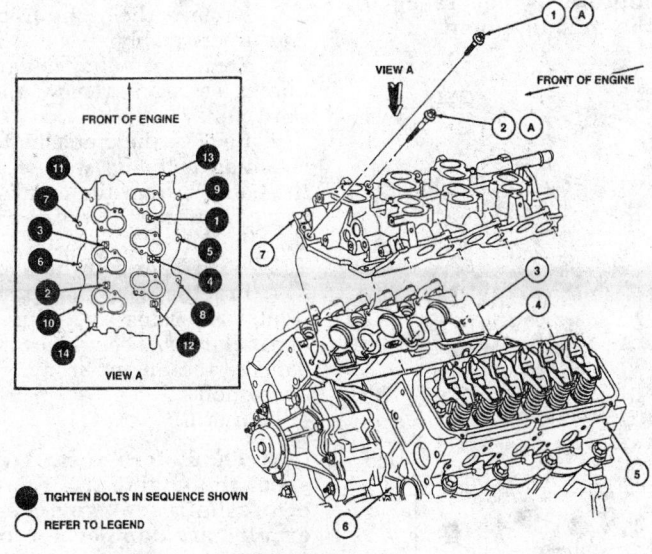

● TIGHTEN BOLTS IN SEQUENCE SHOWN
○ REFER TO LEGEND

1. Bolt (8),lower holes
2. Bolt (6),upper holes
3. Intake manifold gasket
4. End seal(2)
5. Cylinder head(2)
6. Cylinder block
7. Intake manifold
A. 71–106 in. lb.(8–12 Nm)

353868

Lower intake manifold installation and torque sequence — 1996–97 3.8L (VIN 4) engine

12. Connect the negative battery cable.

13. Start the engine and check for exhaust leaks.

14. Road test the vehicle and check for proper engine operation.

Right-Hand Exhaust Manifold

1. Disconnect the negative battery cable.

2. Remove the cowl top vent panel.

3. Disconnect the EGR backpressure transducer hoses from the EGR tube fittings. Remove the EGR tube from the exhaust manifold. Use a back-up wrench on the lower adapter.

4. Raise and safely support the vehicle.

5. Remove the exhaust manifold-to-dual converter Y-pipe retaining nuts.

6. Lower the vehicle.

7. Remove the exhaust manifold retaining bolts and stud bolts.

8. Remove the exhaust manifold from the vehicle.

To install:

9. Clean the mating surfaces of the exhaust manifold, cylinder head and dual converter Y-pipe.

10. Lightly oil all bolt and stud bolt threads prior to installation.

11. If replacing the exhaust manifold, remove the EGR tube adapter and install on the new exhaust manifold.

12. Place the exhaust manifold to the cylinder head. Install and tighten the exhaust manifold retaining bolts and stud bolts to 15–22 ft. lbs. (20–30 Nm).

13. Raise and safely support the vehicle.

14. Connect the dual converter Y-pipe to the exhaust manifold and install the retaining nuts. Tighten the nuts to 25–34 ft. lbs. (34–47 Nm).

15. Lower the vehicle.

16. Connect the EGR tube to the exhaust manifold. Tighten the EGR tube fitting to 26–48 ft. lbs. (35–65 Nm).

17. Connect the EGR backpressure transducer hoses.

18. Install the cowl top vent panel.

19. Connect the negative battery cable.

20. Start the engine and check for exhaust leaks.

21. Road test the vehicle and check for proper engine operation.

3.8L (VIN 4) Engine

Left Exhaust Manifold

1. Disconnect the negative battery cable.

2. Remove the oil level dipstick tube support bracket.

3. Tag and disconnect the ignition wires from the spark plugs and move aside.

4. If necessary, remove the EGR valve-to-exhaust manifold tube.

5. Raise and safely support the vehicle.

6. Remove the left-hand exhaust manifold-to-dual converter Y-pipe retaining nuts.

7. Lower the vehicle.

8. Remove the exhaust manifold retaining bolts and remove the manifold and gasket from the vehicle.

To install:

9. Lightly oil all bolt and stud threads before installation. Clean the mating surfaces on the exhaust manifold, cylinder head and dual converter Y-pipe.

10. Place the left-hand exhaust manifold on the cylinder head using a new gasket. Install the lower front bolt hole on No. 5 cylinder as a pilot bolt.

11. Install the remaining left-hand exhaust manifold retaining bolts. Tighten the bolts to 15–22 ft. lbs. (20–30 Nm) following the correct sequence.

NOTE: A slight warpage in the exhaust manifold may cause a misalignment between the bolt holes in the head and the manifold. Elongate the holes in the exhaust manifold as necessary to correct the misalignment, if apparent. Do not elongate the pilot hole, the lower front bolt on No. 5 cylinder.

12. Raise and safely support the vehicle.

13. Connect the dual converter Y-pipe to the exhaust manifold. Tighten the retaining nuts to 16–24 ft. lbs. (21–32 Nm).

14. Lower the vehicle.

15. If removed, install the EGR valve-to-exhaust manifold tube.

16. Connect the ignition wires to the spark plugs.

17. Install the dipstick tube support bracket and retaining nut. Tighten the nut to 15–22 ft. lbs. (20–30 Nm).

18. Connect the negative battery cable.

19. Run the engine and check for exhaust leaks.

Right Exhaust Manifold

1. Disconnect the negative battery cable.

2. Remove the cowl top vent panel.

3. Remove the engine air cleaner and air cleaner outlet tube.

4. Tag and disconnect the ignition wires from the spark plugs.

5. Remove the spark plugs from the right-hand cylinder head.

6. Raise and safely support the vehicle.

7. Remove the exhaust manifold-to-dual converter Y-pipe retaining nuts.

8. Lower the vehicle.

9. Remove the right-hand exhaust manifold retaining bolts and gasket and remove the exhaust manifold from the vehicle.

To install:

10. Lightly oil all bolt and stud threads before installation. Clean the mating surfaces on the exhaust manifold, cylinder head and dual converter Y-pipe.

11. Place the exhaust manifold on the cylinder head using a new gasket. Start 2 retaining bolts to align the manifold with the cylinder head.

12. Install the remaining bolts and tighten to 15–22 ft. lbs. (20–30 Nm) following the correct sequence.

NOTE: A slight warpage in the exhaust manifold may cause a misalignment between the bolt holes in the head and the manifold. Elongate the holes in the exhaust manifold as necessary to correct the misalignment, if ap-

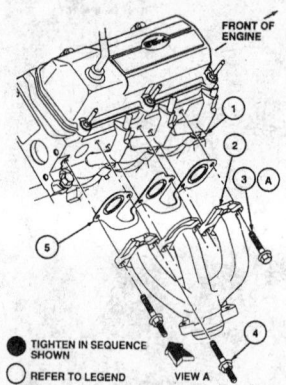

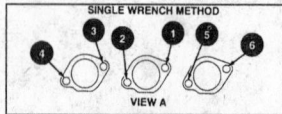

1. Cylinder head,RH
2. Exhaust manifold,RH
3. Bolt(4)
4. Stud bolt(2)
5. Exhaust manifold gasket
A. 15-22 ft. lb.(20–30 Nm)

337663

Right-hand exhaust manifold installation and torque sequence — 1996–97 3.8L (VIN 4) engine, left side similar

parent. Do not elongate the pilot hole, the lower rear bolt hole on No. 2 cylinder.

13. Raise and safely support the vehicle.

14. Connect the exhaust pipe to the dual converter Y-pipe and install the retaining nuts. Tighten the nuts to 16–24 ft. lbs. (21–32 Nm).

15. Install the spark plugs and connect the ignition wires to their respective spark plugs.

16. Install the engine air cleaner and air cleaner outlet tube.

17. Install the cowl top vent panel.

18. Connect the negative battery cable.

19. Run the engine and check for exhaust leaks.

Front Cover Seal

REMOVAL AND INSTALLATION

3.0L (VIN U) Engine

1. Disconnect the negative battery cable.

2. Remove the accessory drive belt.

3. Raise and safely support the vehicle.

4. Remove the right front wheel and tire assembly.

5. Remove 4 pulley-to-damper retaining bolts and remove the crankshaft pulley.

6. Remove the crankshaft damper retaining bolt and washer. Remove the damper from the crankshaft using Damper Removal Tool T58P-6316-D, and Adapter T82L-6316-B, or equivalents.

7. Pry the seal from the engine front cover using a jet plug remover or similar tool, being careful not to damage the engine front cover and crankshaft.

To install:

NOTE: Before installation, inspect the engine front cover and crankshaft seal surface of the crankshaft damper for damage, nicks, burrs or other roughness which may cause the new seal to fail. Service or replace components as necessary.

8. Lubricate the front cover oil seal lip with clean engine oil and install the seal using Damper/Seal Replacer T82L-6316-A and T70P-6B070-A, or equivalents.

9. Coat the crankshaft damper sealing surface with clean engine oil. Apply a suitable silicone sealer to the keyway of the crankshaft damper

just prior to installation. Install the damper using Damper/Seal Replacer T82L-6316-A, or equivalent.

10. Install the damper retaining bolt and washer. Tighten to 93–121 ft. lbs. (125–165 Nm).

11. Place the crankshaft pulley to the damper and install 4 retaining bolts. Tighten the bolts to 30–44 ft. lbs. (40–60 Nm).

12. Install the right front wheel and tire assembly. Torque the lug nuts to 85–105 ft. lbs. (115–142 Nm).

13. Lower the vehicle.

14. Install the accessory drive belt. Check the drive belt for proper routing and engagement in the pulleys.

15. Connect the negative battery cable.

16. Start the engine and check for oil leaks and proper engine operation.

3.8L (VIN 4) Engine

1. Disconnect the negative battery cable.

2. Rotate the drive belt tensioner and remove the accessory drive belt from the crankshaft pulley.

3. Raise and safely support the vehicle.

4. Remove the wheel and tire assembly.

5. Remove the crankshaft pulley retaining bolts and remove the crankshaft pulley from the damper.

6. Remove the crankshaft damper retaining bolt and washer.

7. Remove the damper from the crankshaft using Damper Remover T58P-6316-D and Adapter T82L-6316-B, or equivalents.

8. Pry the front cover oil seal from the front cover with a suitable tool being careful not to damage the front cover and crankshaft.

To install:

NOTE: **Before installation, inspect the front cover and shaft seal surface of the crankshaft damper for damage, nicks, burrs or other roughness which may cause the new seal to fail. Service or replace components as necessary.**

9. Lubricate the seal lip with clean engine oil and install the seal using Seal Replacer T94P-6701-AH and Seal Aligner T88T-6701-A1, or equivalents.

10. Coat the crankshaft damper sealing surface with clean engine oil.

11. Apply silicone sealer to the keyway of the damper prior to installation.

12. Install the damper using Damper Replacer T82L-6316-A, or equivalent.

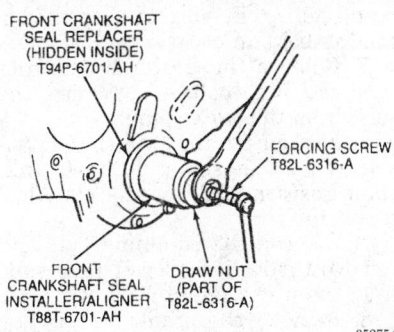

Installing the front cover oil seal — 3.8L (VIN 4) engine

13. Install the damper retaining bolt and washer. Tighten the retaining bolt to 103–132 ft. lbs. (140–180 Nm).

14. Position the crankshaft pulley on the damper and install the crankshaft pulley retaining bolts. Tighten the bolts to 19–28 ft. lbs. (26–38 Nm).

15. Rotate the drive belt tensioner and position the accessory drive belt over the crankshaft pulley. Check the drive belt for proper routing and engagement in the pulleys.

16. Install the wheel and tire assembly. Torque the lug nuts to 83–114 ft. lbs. (113–153 Nm).

17. Lower the vehicle.

18. Connect the negative battery cable.

19. Start the engine and check for oil leaks.

20. Road test the vehicle and check for proper engine operation.

Timing Chain, Sprockets and Front Cover

REMOVAL AND INSTALLATION

3.0L (VIN U) Engine

1. Disconnect the negative battery cable.

2. Drain the engine cooling system.

3. Loosen 4 water pump pulley bolts while the accessory drive belt is in place.

4. Remove the accessory drive belt.

5. Remove the drive belt tensioner retaining bolt.

6. Raise and safely support the vehicle.

7. Remove the right-hand front wheel and tire assembly.

8. Remove the lower radiator hose and the heater hose from the water pump and engine front cover.

9. Remove the crankshaft pulley and damper.

10. Drain the engine oil and remove the oil pan.

11. Remove the screw retaining the power steering line bracket to the sub-frame located at the right-hand front side of the engine.

12. Disconnect the engine control wiring from the Crankshaft Position (CKP) sensor and locating stud.

13. Remove the bolts retaining the water pump and engine front cover to the cylinder block.

14. Remove the water pump and engine front cover as an assembly.

15. Rotate the crankshaft until the piston for No.1 cylinder is at TDC on its compression stroke and the timing marks are aligned.

16. Check the timing chain and sprockets for excessive wear. If timing chain deflection exceeds 6 degrees, the timing chain and sprockets need to be replaced.

17. Remove the camshaft sprocket retaining bolt and washer. Slide the camshaft and crankshaft sprockets and timing chain forward and remove from the engine as an assembly.

To install:

18. Clean and inspect all parts. Clean the gasket sealing surfaces of the engine oil pan, cylinder block and engine front cover.

19. Slide both sprockets and the timing chain onto the camshaft and crankshaft with the timing marks aligned. Install the camshaft retaining bolt and washer and torque to 37–51 ft. lbs. (50–70 Nm). Apply clean engine oil to the timing chain and sprockets after installation.

NOTE: **The camshaft retaining bolt has a drilled oil passage in it for timing chain lubrication. Prior to installation, clean the passage and make sure it is clear. Never replace the camshaft bolt with a standard bolt.**

20. Lightly oil all bolt and stud threads except bolts 1, 2 and 3 that require a suitable pipe sealant.

21. Inspect the front cover oil seal for wear or damage and replace, if necessary.

22. Install a new engine front cover gasket over the cylinder block dowels.

23. Install the engine front cover/water pump assembly onto the cylinder block with the water pump pulley loosely installed to the water pump hub with 4 retaining bolts.

24. Apply a suitable pipe sealant to the threads of bolt numbers 1, 2 and 3 and hand start with the rest of the engine front cover retaining bolts. Tighten bolts 1-10 to 19 ft. lbs. (25

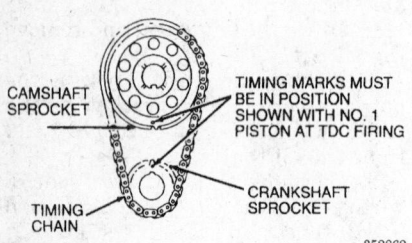

CAMSHAFT SPROCKET

TIMING MARKS MUST BE IN POSITION SHOWN WITH NO. 1 PISTON AT TDC FIRING

TIMING CHAIN

CRANKSHAFT SPROCKET

352862

Camshaft and crankshaft sprocket alignment — 3.0L (VIN U) engine

Nm) and bolts 11-15 to 84 inch lbs. (10 Nm).

25. Install the engine oil pan. Tighten the retaining bolts to 108 inch lbs. (12 Nm).

26. Hand tighten the water pump pulley retaining bolts.

27. Install the crankshaft damper and pulley. Torque the damper retaining bolt to 93–121 ft. lbs. (125–165 Nm) and 4 pulley retaining bolts to 30–44 ft. lbs. (40–60 Nm).

28. Install the drive belt tensioner and retaining bolt. Tighten the bolt to 30–41 ft. lbs. (40–55 Nm).

29. Install the accessory drive belt and tighten 4 water pump pulley retaining bolts to 15–22 ft. lbs. (20–30 Nm).

30. Install the lower radiator hose and the heater hose and tighten the clamps.

31. Install the power steering line bracket to the sub-frame. Install the retaining screw and tighten securely.

32. Install the wheel and tire assembly. Tighten the lug nuts to 85–105 ft. lbs. (115–142 Nm).

33. Lower the vehicle.

34. Fill the crankcase with the correct type and quantity of engine oil.

35. Fill and bleed the engine cooling system.

36. Connect the negative battery cable.

37. Start the engine and check for coolant and oil leaks.

38. Road test the vehicle and check for proper engine operation.

3.8L (VIN 4) Engine

1. Disconnect the negative battery cable.

2. Remove the engine from the vehicle and position on a suitable workstand.

3. Remove the accessory drive belt and the drive belt tensioner, if not already removed.

4. Remove the water pump pulley retaining bolts and remove the water pump pulley. Be sure to keep the retaining bolts in order.

5. Remove the heater hose outlet tube retaining bolts and remove the tube from the water pump.

6. Disconnect the Camshaft Position (CMP) sensor and the Crankshaft Position (CKP) sensor electrical connectors.

7. Remove the retaining bolt and remove the radiator lower hose tube from the water pump.

8. Remove the crankshaft pulley retaining bolt and washer.

9. Remove the crankshaft damper and pulley using a suitable puller.

NOTE: If the crankshaft pulley and vibration damper have to be separated, mark the damper and pulley so they may be reassembled in the same relative position. This is important as the damper and pulley are initially balanced as a unit. If the crankshaft damper is being replaced, check if the original damper has balance pins installed. If so, new balance pins must be installed on the new damper in the same position as the original damper. The crankshaft pulley, new or original, must also be installed in the same relative position as originally installed.

10. Remove the engine oil filter element.

11. Remove the engine oil pan.

NOTE: The engine front cover cannot be removed without removing the oil pan.

12. Remove the engine front cover retaining bolts. It is not necessary to remove the water pump.

——— WARNING ———
Do not overlook the cover retaining bolt located behind the oil pump and filter adapter. The front cover will break if pried on and all retaining bolts are not removed.

13. Remove the engine front cover and water pump as an assembly. Remove and discard the cover gasket.

NOTE: The engine front cover contains the oil pump and water pump. If a new front cover is to be installed, remove the water pump and oil pump from the old front cover.

14. Loosen the camshaft sprocket retaining bolt. Remove the camshaft sprocket retaining bolt and washer and slide off the camshaft position sensor drive gear.

15. Remove the crankshaft sprocket, camshaft sprocket and timing chain all at once by working the assembly off the crankshaft and the camshaft. If the crankshaft sprocket is difficult to remove, use 2 small prybars underneath the crankshaft sprocket to force it up and off of the crankshaft.

16. If required to remove the timing chain tensioner, first pull back on the ratcheting mechanism and then place a pin through the hole in the bracket to relieve the tension on the spring.

17. Remove 3 timing chain tensioner retaining bolts from the front of the cylinder block and remove the tensioner.

To install:

18. Clean all gasket mating surfaces. If reusing the engine front cover, replace the front cover oil seal.

19. Rotate the crankshaft to position the piston for No.1 cylinder at TDC on its compression stroke. The crankshaft keyway should be at the 12 o'clock position.

20. Install the timing chain tensioner using 3 retaining bolts. Tighten the bolts to 70–124 inch lbs. (8–14 Nm). Make sure that the ratcheting mechanism is in the retracted position with the pin sticking out of the bracket hole.

21. Lubricate the timing chain with clean engine oil.

22. Install the camshaft sprocket, crankshaft sprocket and timing chain as an assembly. Make sure the timing marks align. The timing marks must be positioned across from each other. Ensure that the engine balance shaft drive gear and the engine balance shaft driven gears are properly aligned.

23. Remove the pin from the timing chain tensioner to load the timing chain tensioner against the timing chain.

24. Install the camshaft position sensor drive gear and the camshaft sprocket bolt and washer. Tighten the camshaft sprocket bolt to 30–37 ft. lbs. (40–50 Nm).

25. If a new engine front cover is being installed, install the oil pump and water pump from the engine front cover that is being replaced. Install the oil pump and water pump using new gaskets. Tighten 4 larger oil pump retaining bolts to 17–23 ft. lbs. (23–32 Nm), and 2 smaller retaining bolts to 6–8 ft. lbs. (8–11 Nm). Tighten the water pump retaining bolts to 15–22 ft. lbs. (20–30 Nm).

26. Install a new front crankshaft seal into the engine front cover and

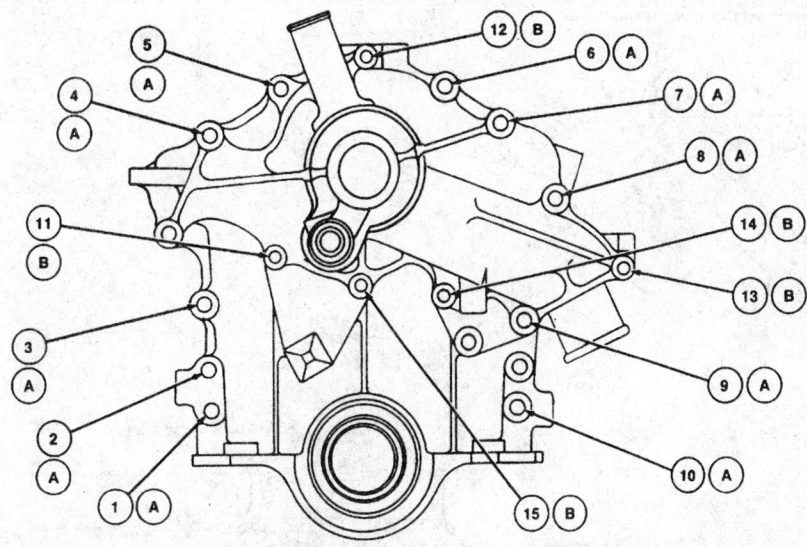

Fastener And Hole No.	Fasteners		Torque Specifications	
	Size	Fastener Application	N·m	LB-FT
1A	M8 x 1.25 x 43.5	F/C TO BLOCK	20-30	15-22
2A	M8 x 1.25 x 43.5	F/C TO BLOCK	20-30	15-22
3A	M8 x 1.25 x 73	W/P & F/C TO BLOCK	20-30	15-22
4A	M8 x 1.25 x 104.3	W/P & F/C TO BLOCK	20-30	15-22
5A	M8 x 1.25 x 73	F/C TO BLOCK	20-30	15-22
6A	M8 x 1.25 x 73	W/P & F/C TO BLOCK	20-30	15-22
7A	M8 x 1.25 x 73	W/P & F/C TO BLOCK	20-30	15-22
8A	M8 x 1.25 x 104.3	W/P & F/C TO BLOCK	20-30	15-22
9A	M8 x 1.25 x 104.3	W/P & F/C TO BLOCK	20-30	15-22
10A	M8 x 1.25 x 52	F/C TO BLOCK	20-30	15-22
11B	M6 x 1 x 28.5	W/P TO F/C	8-12	71-106 (lb-in)
12B	M6 x 1 x 28.5	W/P TO F/C	8-12	71-106 (lb-in)
13B	M6 x 1 x 28.5	W/P TO F/C	8-12	71-106 (lb-in)
14B	M6 x 1 x 28.5	W/P TO F/C	8-12	71-106 (lb-in)
15B	M6 x 1 x 28.5	W/P TO F/C	8-12	71-106 (lb-in)

W/P—Water Pump

F/C—Engine Front Cover

352865

Engine front cover bolt location and identification — 3.0L (VIN U) engine

lubricate the seal lip with clean engine oil.

27. Place a new engine front cover gasket on the cylinder block and install the front cover using the dowel pins for proper alignment. Lightly oil the retaining bolt threads and install the engine front cover retaining bolts. Tighten the retaining bolts to 15–22 ft. lbs. (20–30 Nm). Install and tighten the bolt, located closest to the oil pump filter adapter flange last, after applying a suitable thread locking compound to the bolt threads prior to installation.

28. Install the engine oil pan and a new engine oil filter element.

29. Coat the crankshaft damper sealing surface with clean engine oil. Apply a small amount of silicone sealer to the crankshaft keyway.

30. Position the crankshaft pulley key in the crankshaft keyway and install the damper, using Damper Replacer T82L-6316-A, or equivalent.

31. Install the damper washer and retaining bolt and tighten to 103–132 ft. lbs. (140–180 Nm). Install the crankshaft pulley to the damper and tighten the retaining bolts to 20–28

ft. lbs. (26–38 Nm). Install the remaining componants.

32. Fill the engine with the proper grade and quantity of engine oil.

33. Fill and bleed the engine cooling system.

34. Connect the negative battery cable.

35. Start the engine and check for leaks.

36. Road test the vehicle and check for proper engine operation.

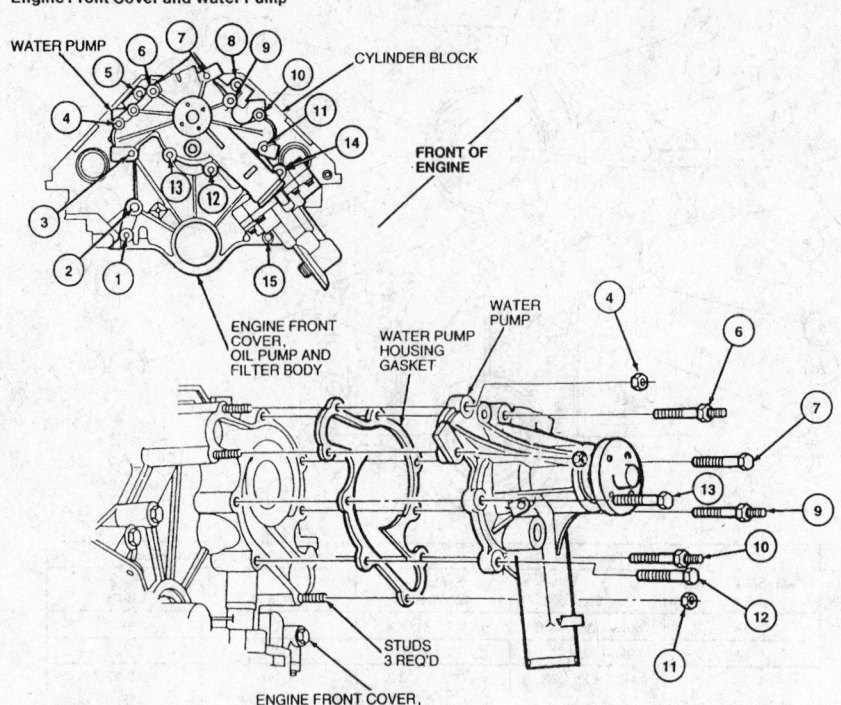

Front cover and water pump assembly — 3.8L (VIN 4) engine

Fastener and Hole No.	Hole No. Water Pump	Hole No. Front Cover	Fasteners Part Name
1		4	Stud
2		2	Stud
3	2	9	Stud
4	1	8	Stud
5		10	Bolt
6	9	15	Stud Bolt
7	8	16	Bolt
8		11	Bolt
9	7	17	Stud Bolt
10	6	1	Stud Bolt
11	5	7	Stud

Fastener and Hole No.	Hole No. Water Pump	Hole No. Front Cover	Fasteners Part Name
12	4	13	Bolt
13	3	14	Bolt
14		6	Bolt
15		5	Cap Screw
3, 4, 11	2, 1, 5	9, 8, 7	Nut

352674

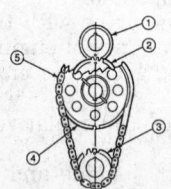

1. Engine balance shaft driven gear
2. Engine balance shaft drive gear
3. Crankshaft sprocket
4. Camshaft sprocket
5. Timing chain

338614

Timing chain sprocket and balance shaft gear timing marks — 3.8L (VIN 4) engine

Camshaft

REMOVAL AND INSTALLATION

3.0L (VIN U) Engine

— CAUTION —

Fuel injection systems remain under pressure, even after the engine has been turned OFF. The fuel system pressure must be relieved before disconnecting any fuel lines. Failure to do so may result in fire and/or personal injury.

1. Disconnect the negative battery cable.

2. Drain the engine cooling system.

3. Relieve the fuel system pressure.

4. Drain the engine oil.

5. Remove the engine from the vehicle and position in a suitable holding fixture.

6. Remove the engine oil pan and engine front cover.

7. Remove the throttle body/upper intake manifold assembly.

8. Remove both cylinder head covers.

9. Remove the lower intake manifold leaving the fuel injection supply manifold and fuel injectors in place.

10. Loosen the rocker arm seat retaining bolts and rotate each rocker arm off the pushrods. Remove the pushrods and reference so they can be installed in their original positions.

11. Remove 2 bolts and the valve lifter guide plate and retainer from the valve lifter valley.

12. Remove 6 valve lifter guide plates from between each pair of valve lifters by lifting straight up.

13. Using a magnet or valve lifter removal tool, remove the valve lifters, keeping them in order so they can be installed in their original positions. If the valve lifters are stuck in the bores by excessive varnish, use a valve lifter puller to remove the valve lifters.

14. Align the timing marks on the camshaft and crankshaft sprockets. Check the camshaft end-play as follows:

   a. Push the camshaft toward the rear of the engine and install a dial indicator tool, so the indicator point is on the camshaft sprocket attaching screw.

   b. Zero the dial indicator. Position a small prybar or equivalent, between the camshaft sprocket and block.

   c. Pull the camshaft forward and release it. Compare the dial indicator reading with the camshaft end-play service limit specification of 0.005 inch (0.13mm).

   d. If the camshaft end-play is over the amount specified, replace the thrust plate.

15. Remove the timing chain and sprockets.

16. Remove 2 retaining bolts and the camshaft thrust plate. Carefully remove the camshaft by pulling it toward the front of the engine. Remove it slowly to avoid damaging the bearings, journals and lobes.

NOTE: If the camshaft is replaced, new valve lifters should also be installed.

17. Inspect the camshaft journals and lobes for wear and/or damage. Replace as necessary.

To install:
18. Clean all gasket mating surfaces.

19. Lubricate the camshaft lobes and journals with engine assembly lubricant. Carefully insert the camshaft through the bearings into the cylinder block.

20. Install the camshaft thrust plate and 2 retaining bolts. Tighten the retaining bolts to 84 inch lbs. (10 Nm).

21. Install the timing chain and sprockets. Tighten the camshaft sprocket retaining bolt to 46 ft. lbs. (63 Nm).

NOTE: The camshaft bolt has a drilled oil passage it it for timing chain lubrication. Make sure the passage is clean prior to bolt installation. If the bolt is damaged, do not replace the camshaft bolt with a standard bolt or engine damage may result.

22. Lubricate the valve lifters and their bores.

23. Install the valve lifters into their bores. Ensure that the valve lifters move freely in the bores.

24. Aligning the valve lifters flats, install 6 valve lifter guide plates. Install the plates with the word **UP** and/or a button mark visible.

25. Install the valve lifter guide plate and retainer over the valve lifter guide plates. Install 2 retaining bolts and tighten to 8–10 ft. lbs. (10–14 Nm).

26. Install the lower intake manifold.

27. Install the pushrods, making sure they are seated in the valve lifters.

28. Lubricate the pushrods and rocker arms with clean engine oil. Move the rocker arms into position with the pushrods and snug the rocker arm seat retaining bolts.

29. For each valve, rotate the crankshaft to position the camshaft lobe of the rocker arm being tightened, on its heel or base circle before tightening the rocker arm seat retaining bolt. Tighten each rocker arm seat retaining bolt to 5–11 ft. lbs. (7–15 Nm) each time having the camshaft lobe on its base circle. Tighten the retaining bolts again, to 19–28 ft. lbs. (26–38 Nm) with the camshaft lobes in any position.

30. Install the oil pan and the engine front cover.

31. Install the cylinder head covers.

32. Install the throttle body/upper intake manifold assembly.

33. Install the engine in the vehicle.

34. Fill and bleed the engine cooling system.

35. Fill the crankcase with the correct quantity and type of engine oil.

36. Connect the negative battery cable.

37. Start the engine and allow to reach normal operating temperature while checking for leaks.

38. Road test the vehicle and check for proper engine operation.

3.8L (VIN 4) Engine

1. Disconnect the negative battery cable.

2. Remove the engine from the vehicle and place on a suitable work stand.

3. Remove the upper intake manifold.

4. Remove the cylinder head covers.

5. Remove the lower intake manifold.

6. Loosen the rocker arms and remove the pushrods keeping them in the order of removal for installation reference.

7. Remove the guide plates and valve lifters keeping the valve lifters in the order of removal for installation reference.

8. Remove the engine oil pan.

9. Remove the engine front cover. Remove the timing chain and sprockets.

10. Remove 2 retaining bolts and the camshaft thrust plate.

11. Remove the camshaft through the front of the engine, being careful not to damage the camshaft bearing surfaces.

To install:

NOTE: Inspect the camshaft rear bearing cover. If damaged or leaking, replace the camshaft rear bearing cover. Inspect the camshaft and camshaft bearings for signs of wear or damage and replace as necessary. If the camshaft is replaced, new valve lifters should be installed.

12. Lubricate the camshaft lobes and journals with engine assembly lubricant.

13. Install the camshaft, being careful not to damage the lobes or bearing surfaces while sliding it into position.

14. Install the camshaft thrust plate. Install and tighten the 2 retaining bolts to 72–120 inch lbs. (8–14 Nm).

15. Check the camshaft end-play as follows:

   a. Temporarily install the camshaft sprocket retaining bolt.

   b. Push the camshaft toward the rear of the engine.

   c. Install a dial indicator to the front of the cylinder block and position the dial indicator to rest on the face of the camshaft sprocket retaining bolt.

   d. Zero the dial indicator.

   e. Pull the camshaft forward.

   f. Note the reading on the dial indicator. The end-play should

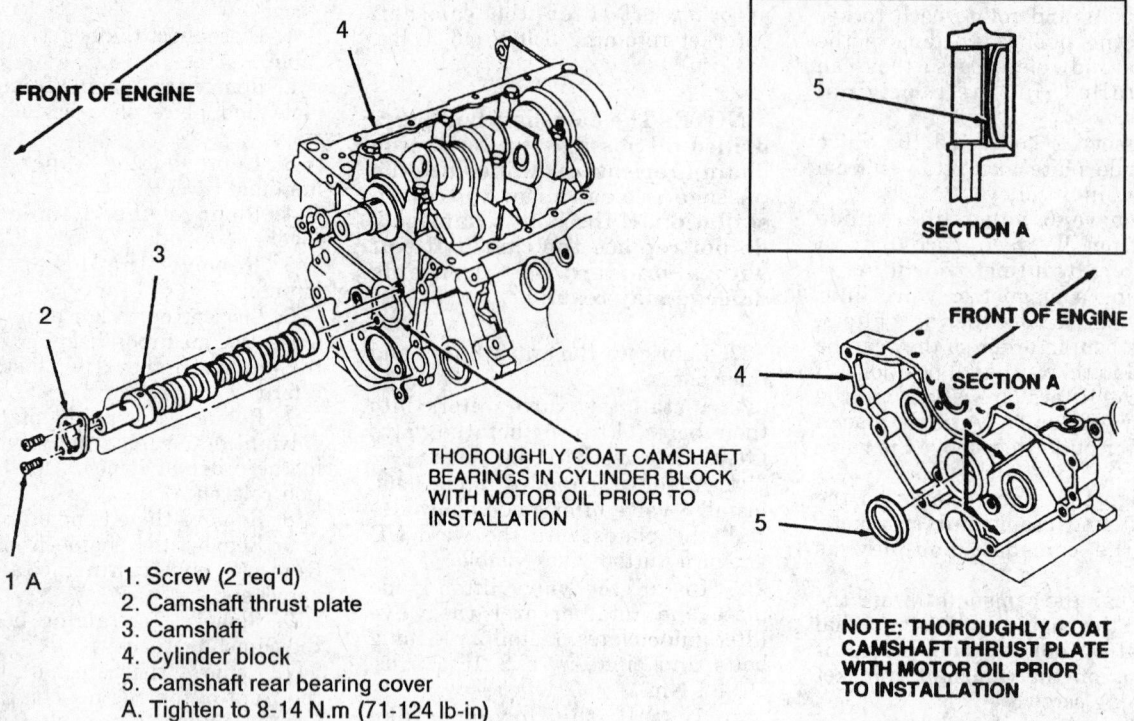

FRONT OF ENGINE

SECTION A

FRONT OF ENGINE

SECTION A

THOROUGHLY COAT CAMSHAFT BEARINGS IN CYLINDER BLOCK WITH MOTOR OIL PRIOR TO INSTALLATION

NOTE: THOROUGHLY COAT CAMSHAFT THRUST PLATE WITH MOTOR OIL PRIOR TO INSTALLATION

1 A

1. Screw (2 req'd)
2. Camshaft thrust plate
3. Camshaft
4. Cylinder block
5. Camshaft rear bearing cover
A. Tighten to 8-14 N.m (71-124 lb-in)

337101

Camshaft, thrust plate and rear bearing cover — 3.8L (VIN 4) engine

measure between 0.001–0.006 inch (0.025–0.150mm).

g. If the reading is excessive, replace the camshaft thrust plate and recheck. If the camshaft end-play is still excessive, check for a worn camshaft or cylinder block.

h. Remove the camshaft sprocket retaining bolt.

16. Install the engine oil pan.

17. Install the timing chain and sprockets making sure that the camshaft, crankshaft and balance shaft are properly timed.

18. Install the engine front cover.

19. Install the valve lifters, guide plates and pushrods.

20. Rotate the rocker arms into position over the pushrods and valve stems.

21. For each valve, rotate the crankshaft until the valve lifter is on the heel or base circle of the camshaft lobe. Install the rocker arm seat bolt and tighten to 44 inch lbs. (5 Nm) maximum. Make sure the pushrod and rocker arm seats are fully seated prior to tightening.

22. Lubricate all rocker arm assemblies with clean engine oil. Final tighten the rocker arm seat bolts to 23–30 ft. lbs. (30–40 Nm). When final tightening, the camshaft may be in any position.

23. Install the lower intake manifold.

24. Install the cylinder head covers. Tighten the cylinder head cover retaining bolts to 71–97 inch lbs. (8–11 Nm).

25. Install the upper intake manifold.

26. Remove the engine from the work stand and install in the vehicle.

27. Restore all fluid levels.

28. Connect the negative battery cable.

29. Start the engine and check for leaks.

30. Road test the vehicle and check for proper engine operation.

Balance Shaft

REMOVAL AND INSTALLATION

3.8L (VIN 4) Engine

1. Disconnect the negative battery cable.

2. Remove the engine from the vehicle and place on a suitable work stand.

3. Remove the engine oil pan.

4. Remove the engine front cover.

5. Remove the timing chain and sprockets.

6. Remove the balance shaft drive gear and spacer.

7. Remove 2 bolts retaining the balance shaft thrust plate.

NOTE: Use care not to damage the balance shaft bearing surfaces.

8. Carefully slide the balance shaft, driven gear and thrust plate out the front of the cylinder block.

To install:

NOTE: Inspect the balance shaft rear bearing cover. If damaged or leaking, replace the balance shaft rear bearing cover. Inspect the balance shaft and bearings for signs of wear or damage and replace as necessary.

9. Lubricate the balance shaft journals with engine assembly lubricant.

10. Install the balance shaft, driven gear and thrust plate, being careful not to damage the bearing surfaces while sliding it into position. Install 2 thrust plate retaining bolts and tighten to 70–124 inch lbs. (8–14 Nm).

11. Install the balance shaft drive gear and spacer.

12. Align the balance shaft timing marks.

13. Install the remaining components. Fill the crankcase with the proper grade and quantity of engine oil.

14. Fill and bleed the engine cooling system.

15. Connect the negative battery cable.

16. Start the engine and check for leaks.

17. Road test the vehicle and check for proper engine operation.

Piston and Connecting Rod

POSITIONING

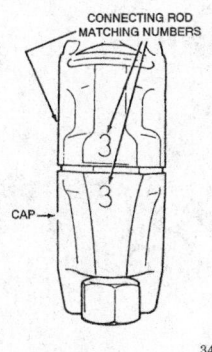

Connecting rod and cap positioning

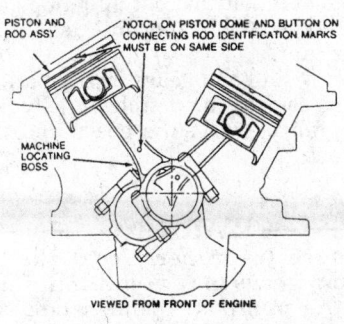

Piston to cylinder block positioning (notch on piston must face front of engine)

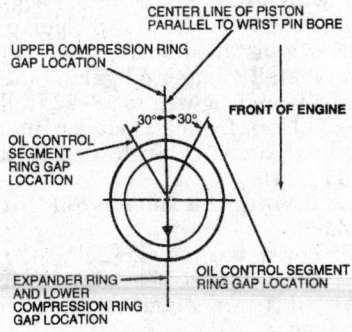

Piston ring positioning

ENGINE LUBRICATION

Oil Pan

REMOVAL AND INSTALLATION

3.0L (VIN U) Engine

1. Disconnect the negative battery cable.

2. Remove the engine oil level dipstick.

3. Raise and safely support the vehicle.

4. If equipped with a low oil level sensor, remove the retainer clip at the sensor. Disconnect the electrical harness connector from the sensor.

5. Drain the engine oil.

6. Disconnect the electrical harness connectors from the heated oxygen sensors.

7. Remove the dual converter Y-pipe.

8. Remove the starter motor and brace.

9. Remove the engine rear plate from the torque converter housing.

10. Remove 16 oil pan retaining bolts and carefully remove the engine oil pan from the cylinder block. Remove the oil pan gasket.

To install:

11. Clean the gasket sealing surfaces on the cylinder block and the engine oil pan. Apply a ¼ inch (6mm) bead of silicone sealer to the junction of the engine front cover and cylinder block and to the junction of the rear main bearing cap and cylinder block.

NOTE: When using a silicone sealer, the assembly process should occur within 5 minutes after the sealer has been applied. Make sure the sealer does not fall into the engine and form plugs that could block oil passages.

12. Place the oil pan gasket on the engine oil pan and secure with a suitable contact adhesive.

13. Place the engine oil pan into position on the cylinder block. Install 16 engine oil pan retaining bolts. Tighten the retaining bolts to 8–10 ft. lbs. (10–14 Nm). Back off all bolts and retighten. Install the remaining components.

14. Lower the vehicle.

15. Install the engine oil level dipstick.

16. Connect the negative battery cable.

17. Fill the crankcase with the proper type and quantity of engine oil.

18. Start the engine and check for oil and exhaust leaks.

19. Road test the vehicle and check for proper engine operation.

3.8L (VIN 4) Engine

1. Disconnect the negative battery cable.

2. Raise and safely support the vehicle.

3. Drain the engine oil into a suitable container and remove the oil filter element. Install the drain plug and tighten it to 15–25 ft. lbs. (20–34 Nm). Move the drain pan aside.

4. Remove the dual converter Y-pipe assembly.

5. Remove the starter motor and the rear engine plate.

6. Remove 18 oil pan retaining bolts and remove the oil pan.

7. To remove the oil pump screen and tube, remove 2 retaining bolts and the support bracket nut and remove the oil pump screen and tube and gasket.

To install:

8. Clean the gasket sealing surfaces on the cylinder block and oil pan.

9. If removed, clean the oil pump screen and tube mounting gasket surfaces and install a new oil pump screen and tube gasket.

10. Place the oil pump screen and tube in position using a new gasket, and install 2 retaining bolts and the support bracket nut. Tighten the retaining bolts to 15–22 ft. lbs. (20–30 Nm) and the nut to 30–40 ft. lbs. (40–55 Nm).

11. Trial fit the oil pan to the cylinder block. Ensure that enough clearance has been provided to allow the oil pan to be installed without sealant being accidentally scraped off when the oil pan is positioned under the engine.

12. Ensure that the gasket mating surfaces are clean of engine oil. Apply a bead of silicone sealer to the oil pan flange. Also apply a bead of sealer to the engine front cover/cylinder block joint and fill the grooves on both sides of the rear main seal cap.

NOTE: When using silicone rubber sealer, assembly must occur within 15 minutes after sealer application. After this time, the sealer may start to harden and its sealing effectiveness may be reduced.

13. Install the oil pan and secure to the cylinder block with 18 retaining

bolts. Tighten the retaining bolts to 84–108 inch lbs. (9–12 Nm).

14. Install a new oil filter element.

15. Install the rear engine cover and the starter motor.

16. Install the dual converter Y-pipe assembly.

17. Lower the vehicle.

18. Fill the crankcase with the correct grade and amount of engine oil.

19. Connect the negative battery cable.

20. Start the engine and check for oil leaks.

Oil Pump

REMOVAL AND INSTALLATION

3.0L (VIN U) Engine

1. Disconnect the negative battery cable.

2. Raise and safely support the vehicle.

3. Drain the engine oil.

4. Remove the engine oil pan.

5. Remove the oil pump retaining bolt and remove the oil pump and the oil pump intermediate shaft from the engine.

6. If replacing the engine oil pump, separate the intermediate shaft from the oil pump.

To install:

7. If replacing the engine oil pump, insert the oil pump intermediate shaft into the new oil pump assembly until the intermediate shaft retaining ring clicks into place.

8. Prime the new oil pump by filling either the inlet or the outlet port with engine oil. Rotate the pump shaft to distribute the oil within the oil pump body cavity.

9. Insert the oil pump intermediate shaft assembly through the hole in the rear main bearing cap and place the oil pump onto the locating pins.

10. Install the oil pump retaining bolt and torque the bolt to 30–40 ft. lbs. (40–55 Nm).

11. Install the engine oil pan.

12. Lower the vehicle.

13. Fill the crankcase with the proper type and quantity of engine oil.

14. Connect the negative battery cable.

15. Start engine and check for leaks and proper oil pressure.

16. Road test the vehicle and check for proper engine operation.

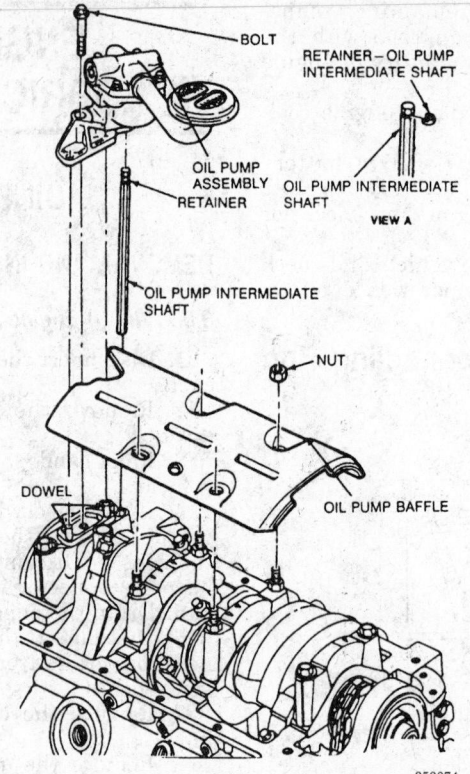

Oil pump installation — 3.0L (VIN U) engine

3.8L (VIN 4) Engine

NOTE: The oil pump is mounted on the engine front cover assembly. Oil pan removal is necessary only for pick-up tube/screen replacement or service.

1. Disconnect the negative battery cable.

2. Raise and safely support the vehicle.

3. Remove the engine oil filter element. Allow the engine oil to drain into a suitable container.

4. Remove 6 oil pump body-to-engine front cover retaining bolts and remove the oil pump body.

5. If required, remove the oil pump gears from the pocket in the engine front cover.

6. Inspect the oil pump body, gears and engine front cover pocket for wear and/or damage. Replace parts as necessary.

To install:

7. Clean the oil pump body-to-engine front cover gasket mating surface. Place a straight-edge across the oil pump mounting surface and check for wear or warpage using a feeler gauge. If the surface is out of flat by more than 0.0016 inch (0.04mm), replace the oil pump body.

8. Lightly pack the oil pump gear pocket in the engine front cover with petroleum jelly or coat all pump gear surfaces with engine assembly lubricant.

9. Install the gears in the pocket. Make certain the petroleum jelly fills the gap between the gears and the pocket.

——— WARNING ———
Failure to properly coat the oil pump gears may result in failure of the pump to prime when the engine is started.

10. Place the oil pump body in position with a new gasket on the engine front cover. Ensure the oil pump is correctly installed on the dowel pins.

11. Install 4 larger oil pump retaining bolts and tighten to 18–22 ft. lbs. (25–30 Nm). Install 2 smaller retaining bolts and tighten to 6–8 ft. lbs. (8–11 Nm).

12. Install the engine oil filter element.

13. Lower the vehicle.

14. Add 1 quart of engine oil if a new oil filter element has been installed.

15. Connect the negative battery cable.

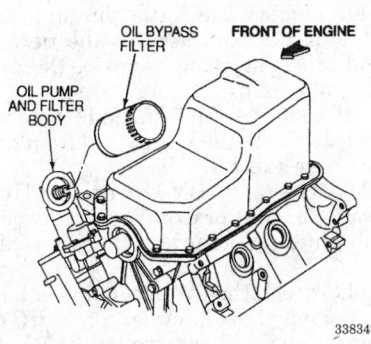

Oil pump location on the engine front cover — 3.8L (VIN 4) engine

16. Start the engine and check for oil leaks and proper engine oil pressure.

17. Road test the vehicle and check for proper engine operation.

18. Recheck the engine oil level.

TRANSAXLE

Automatic Transaxle Assembly

REMOVAL AND INSTALLATION

AX4S

1. Disconnect the negative battery cable.

2. Remove the air cleaner assembly, hoses and tubes.

3. Remove the cowl top vent panel.

4. Disconnect the transaxle electrical harness connectors from the engine and remove the bolt retaining the main wiring harness bracket. Position the wiring harness out of the way.

5. Disconnect the shift cable and bracket from the transaxle.

6. Install engine lifting eyes to the left-hand front exhaust manifold stud and the right-hand rear exhaust manifold stud.

7. Place Engine Support D88L-6000-A or equivalent, in position and attach to the engine lifting eyes using suitable lifting cables. Remove any slack in the cables.

8. Remove the transaxle dipstick and tube.

9. Remove 4 transaxle housing-to-cylinder block retaining bolts from the top of the transaxle.

10. Raise and safely support the vehicle.

11. Place a drain pan under the transaxle. Loosen the pan bolts and allow the transmission fluid to drain. Once the fluid has drained to the level of the pan, finish removing the pan bolts and the pan and drain the remainder of transmission fluid.

12. Reinstall the transaxle oil pan.

13. Remove the front wheel and tire assemblies.

14. Disconnect the Heated Oxygen Sensor (HO2S) electrical connectors.

15. Install Flex Tube Support T94T-6000-AH or equivalent, to prevent damage to the exhaust flex tube. Make sure that the steering is in a straight-ahead position.

16. Remove the dual converter Y-pipe.

17. Disconnect the steering shaft from the steering gear rack and pinion assembly.

18. Place a drain pan under the rack and pinion assembly.

19. Disconnect the power steering fluid lines from the rack and pinion assembly and allow the fluid to drain into the pan.

20. Remove and discard the left-hand and right-hand nuts retaining the stabilizer bar link to the stabilizer bar. Use care not to damage the stabilizer bar link ball stud seal by holding the stud with a 6mm Allen wrench while turning the retaining nut.

21. Remove the cotter pin and the castellated nut from the left-hand and right-hand outer tie rod ends and discard the cotter pins and nuts.

22. Separate the outer tie rod ends from the steering knuckles using Tie Rod Remover T81P-3504-W, or equivalent.

23. Remove the left-hand and right-hand steering knuckle-to-ball joint pinch bolts and nuts and discard.

24. Spread the pinch joint, if required, on each steering knuckle and separate the ball joints from the steering knuckles using a suitable prybar or equivalent.

25. Remove 2 retaining bolts from the engine/transaxle support insulator and 4 retaining bolts from the engine/transaxle support and remove the support.

26. Place Sub-frame Removal Brace 014-00751 or equivalent and a suitable transmission jack to the sub-frame.

27. Remove 4 sub-frame to body retaining bolts. Lower and remove the sub-frame.

28. Disconnect the wiring harness for the starter, low oil sensor and the Vehicle Speed Sensor (VSS). Disconnect the speedometer cable, if equipped.

29. Remove and plug both transaxle oil cooler lines from the transaxle using Disconnect Tool T90T-9550-B, or equivalent.

30. Remove 2 starter motor retaining bolts and place the starter motor aside.

31. Remove the bolt from the transaxle housing cover and remove the transaxle housing cover.

32. Rotate the crankshaft using a ½ inch drive ratchet and a ⅞ inch socket on the crankshaft pulley retaining bolt to access the torque converter-to-flywheel retaining nuts.

33. Remove 4 torque converter-to-flywheel retaining nuts.

34. Remove the engine-to-transaxle retaining bolts.

35. Remove both halfshafts from the transaxle and position out of the way.

36. Install transmission plugs into the differential case to prevent the gears from becoming misaligned.

37. Install Transaxle Brace Adapter 014-00763 or equivalent, by removing 3 transaxle oil pan bolts for mounting the adapter to the transaxle.

38. Place a suitable transmission jack to the transaxle brace adapter to support the transaxle.

39. Remove 2 remaining lower transaxle housing-to-engine retaining bolts.

40. Separate the transaxle from the engine and carefully lower the transaxle out of the vehicle.

To install:

NOTE: Clean the transaxle cooler lines thoroughly before installing the transaxle assembly.

41. Place the transaxle assembly onto a suitable transmission jack and transaxle brace adapter.

42. Carefully raise the transaxle assembly into place and align the torque converter bolts to the flywheel.

43. Install 2 lower transaxle housing-to-engine retaining bolts. Tighten to 41–50 ft. lbs. (55–68 Nm).

44. Install 4 torque converter-to-flywheel nuts. Use a ½ inch drive ratchet and a ⅞ inch socket on the crankshaft pulley retaining bolt, to rotate the crankshaft and gain access to the torque converter studs. Tighten the torque converter-to-flywheel nuts to 23–39 ft. lbs. (31–53 Nm).

45. Remove the transmission jack and transaxle brace adapter.

46. Install the halfshaft assemblies.

47. Install the transaxle housing cover and bolt.

48. Install the starter motor and 2 retaining bolts. Tighten the bolts to 30–40 ft. lbs. (41–54 Nm).

49. Unplug and install 2 transaxle oil cooler lines to the transaxle.

50. Position the sub-frame on Sub-frame Removal Brace 014-00751 or equivalent and a suitable transmission jack.

51. Raise the sub-frame into position and install 4 sub-frame to body retaining bolts. Tighten the bolts to 55–75 ft. lbs. (75–102 N.m).

52. Remove the sub-frame removal brace and transmission jack.

53. Install the engine/transaxle support. Install and tighten 4 retaining bolts to 40–55 ft. lbs. (54–75 Nm).

54. Install the engine/transaxle support insulator and tighten 2 bolts to 60–85 ft. lbs (81–116 Nm).

55. Replace the transaxle oil pan gasket if needed, and properly torque the transaxle oil pan retaining bolts. Install the remaining components.

56. Lower the vehicle.

57. Install the remaining transaxle housing-to-engine retaining bolts. Tighten to 41–50 ft. lbs. (55–68 Nm).

58. Remove the engine support, cables and engine lifting eyes.

59. Install the transaxle oil filler tube and dipstick.

60. Install the shift cable and bracket.

61. Position the main wiring harness and bracket.

62. Install the electrical harness connectors to the engine.

63. Install the engine air cleaner assembly, hoses and tubes.

64. Install the cowl top vent seal.

65. Connect the negative battery cable.

66. Fill the transaxle with the correct amount and type of transmission fluid.

67. Make sure that the parking brake is engaged.

68. Run the engine through all gear ranges and check for leaks.

69. Adjust the shift cable as required.

NOTE: Whenever the vehicles sub-frame is removed or lowered, the wheel alignment should be checked.

70. Check the alignment and adjust if necessary.

71. Road test the vehicle and check for proper transaxle operation.

DRIVE AXLE

Halfshaft

REMOVAL AND INSTALLATION

NOTE: Do not begin this procedure without having a new wheel hub retainer nut, retainer circlip and a new lower ball joint-to-wheel knuckle retaining bolt and nut, per side. Once removed, these parts lose their torque holding ability or retention capability and must not be reused. This procedure is the same for the left-hand or right-hand halfshafts.

1. Disconnect the negative battery cable.

2. Raise and safely support the vehicle.

3. Remove the front wheel and tire assembly.

4. Snug 2 of the lug nuts back onto the rotor.

5. Insert the tapered end of a pry bar or steel rod into one of the cooling slots of the disc brake rotor and place the bar against the disc brake anchor plate, to keep the rotor from turning.

6. Loosen and remove the wheel hub retaining nut. Discard the nut.

7. Remove the ball joint-to-wheel knuckle pinch bolt and nut. Discard the bolt and nut.

8. Remove the anti-lock brake sensor and move aside, if equipped.

9. Using a pry bar or similar tool, separate the ball joint from the wheel knuckle. Use care not to damage the bushing or ball joint boot.

10. Remove the nut from the stabilizer bar link and separate the stabilizer bar from the strut (shock absorber) using a tie rod end removal tool.

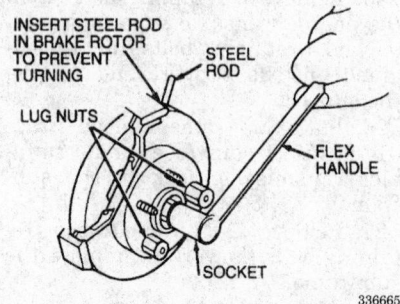

336665

Loosening the wheel hub retaining nut

11. Remove the cotter pin and castellated nut that secures the tie rod end to the steering knuckle. Discard the cotter pin.

12. Using a tie rod end removal tool, separate the tie rod end from the steering knuckle.

13. Install CV-Joint Puller T86P-3514-A1 or equivalent, between the inner CV-joint and the transaxle case.

14. Attach the corresponding extension and slide hammer to the CV-joint puller and remove the halfshaft and CV-joint from the transaxle case.

NOTE: Make sure that the CV-joint puller does not contact the transmission speed sensor or damage may result.

15. Support the transaxle end of the halfshaft with a length of wire to a convenient underbody component to prevent damage to the CV-joint boots.

16. Separate the outer CV-joint and halfshaft from the wheel hub using Hub Remover/Replacer T81P-1104-C or equivalent, and its associated components, or equivalent.

—————— **WARNING** ——————
Never use a hammer to separate the halfshaft from the front wheel hub as damage to the threads or internal components may result.

17. Remove the wire supporting the halfshaft and remove the halfshaft assembly from the vehicle.

18. Remove the circlip on the stub shaft end of the inboard CV-joint and discard.

To install:

19. Install a new circlip on the inboard CV-joint stub shaft. Start one end of the circlip into the groove and work the circlip over the housing end and into the groove. This will avoid over-expanding the circlip.

20. Carefully align the splines of the inboard CV-joint (install the halfshaft and both CV-joints as an assembly) with the splines in the transaxle case and push it into the differential side gear until the circlip is felt to seat.

21. Position the outer CV-joint with halfshaft and carefully align the splines of the outer CV-joint with the splines of the wheel hub.

22. Push the CV-joint shaft into the wheel hub as far as possible.

23. Install a new wheel hub retaining nut onto the exposed threads of the outer CV-joint and manually thread the nut on as far as possible.

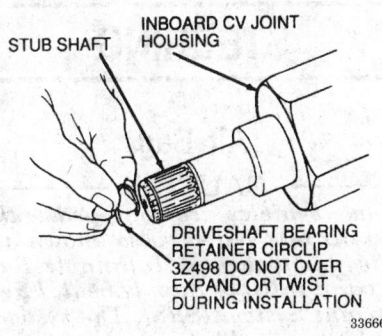

Installing a new circlip on the inboard CV-joint stub shaft

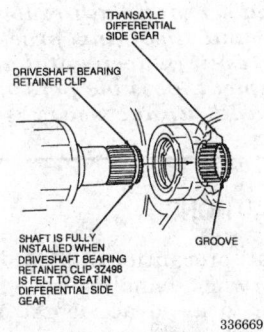

Aligning the inboard CV-joint stub shaft to the differential side gear

24. Install the front suspension lower control arm ball joint into the wheel knuckle.

25. Install a new lower control arm-to-wheel knuckle pinch bolt and nut. Hold the bolt and torque the nut to 40–55 ft. lbs. (54–74 Nm).

26. Install the anti-lock brake sensor, if equipped.

27. Connect the stabilizer bar link to the stabilizer bar and install the nut. Torque the nut to 35–48 ft. lbs. (47–65 Nm).

28. Insert the tapered end of a pry bar or similar tool, into one of the cooling slots in the disc brake rotor and jam to prevent the rotor from turning.

29. Finish running the wheel hub retaining nut up and torque to 157–212 ft. lbs. (213–287 Nm). Do not use power or impact tools to tighten the hub nut.

30. Remove the pry bar or similar tool.

31. Install the tie rod end on the wheel knuckle. Install the castellated nut and torque to 35–47 ft. lbs. (48–63 Nm). Install a new cotter pin.

32. Remove the lug nuts temporarily installed to stabilize the brake rotor.

33. Install the wheel and tire assembly. Torque the lug nuts to 85–105 ft. lbs. (115–142 Nm).

34. Lower the vehicle.

35. Connect the negative battery cable.

36. Check the transmission fluid level and replace any fluid lost during the service procedure.

37. Road test the vehicle and check for proper operation.

CV-Joint Boot

REPLACEMENT

NOTE: The inboard and outboard CV-joint boot removal is the same for the left-hand and right-hand sides.

Inboard

1. Remove the halfshaft from the vehicle.

2. Place the halfshaft on a bench or in a soft jaw vise.

3. Cut off both CV-joint boot clamps.

4. Cut the CV-joint boot off or slide the boot back onto the halfshaft.

5. Slide the CV-joint stub shaft housing off to gain access to the tripod assembly.

6. Using snapring pliers, move the stop ring back onto the halfshaft.

7. Slide the tripod assembly back onto the halfshaft to allow access to the driveshaft bearing retainer circlip.

8. Remove the driveshaft bearing retainer circlip from the halfshaft.

9. Slide off the tripod assembly.

10. Slide off the CV-joint boot if not already cut off.

TRIPOD ASSY
337043

Moving the stop ring on the halfshaft

To install:

11. Slide the small boot clamp onto the halfshaft.

12. Slide the CV-joint boot onto the halfshaft, small end first.

13. Position and tighten the small boot clamp.

NOTE: Inspect the constant velocity joint grease for contamination. If contaminated, the CV-joint assembly should be thoroughly cleaned and inspected for wear.

14. Slide the tripod assembly (chamfered side first) onto the halfshaft.

15. Install a new driveshaft bearing retainer circlip into the halfshaft groove.

16. Slide the tripod assembly over the bearing retainer circlip compressing the clip to expose the stop ring groove at the other end.

17. Using snapring pliers, move the stop ring into its groove and make sure it is fully seated.

18. Install the Trilobe insert onto the CV-joint housing (if removed) and fill the housing with the proper grease. Place the remaining grease evenly into the boot area.

19. Install the CV-joint stub shaft housing onto the tripod assembly. Make sure it is fully seated.

20. Place a screwdriver under the lip of the boot, being careful not to damage the boot, to allow trapped air to escape.

21. Place the large clamp into position and tighten the clamp.

22. Wipe off any remaining grease.

23. Install the halfshaft into the vehicle and check for proper operation.

Outboard

1. Cut off both CV-joint boot clamps.

2. Cut off the CV-joint boot or push back onto the halfshaft.

3. Angle the outer CV-joint away from the halfshaft to expose the inner bearing race.

4. Using a brass drift and hammer, apply a sharp tap to the inner bearing race to dislodge the internal driveshaft bearing retainer circlip.

5. Remove the outer CV-joint from the halfshaft.

6. Slide off the CV-joint boot if not already cut off.

To install:

7. Install the small boot clamp.

8. Install the CV-joint boot, small end first. Tape wrapped around the splines will aid in installing the new boot.

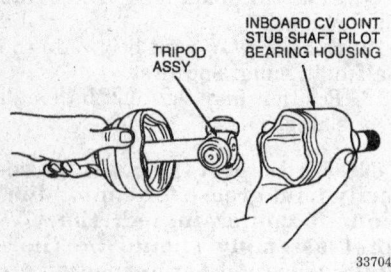

Removing the CV-joint stub shaft housing

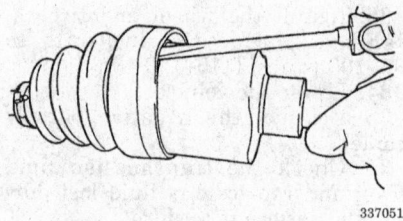

Method of removing excess air pressure

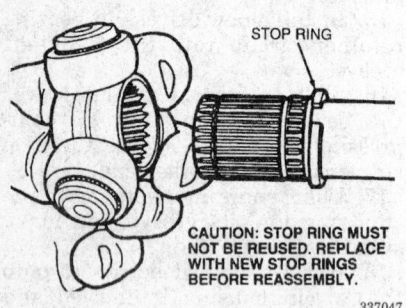

CAUTION: STOP RING MUST NOT BE REUSED. REPLACE WITH NEW STOP RINGS BEFORE REASSEMBLY.

Installing the tripod onto the halfshaft

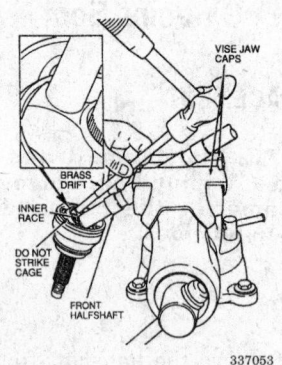

Removing the outer CV-joint

13. Using a plastic faced hammer, tap the CV-joint onto the halfshaft until it is fully seated.

14. Wipe off any remaining grease.

15. Install the large boot clamp and tighten the clamp.

16. Install the halfshaft into the vehicle and check for proper operation.

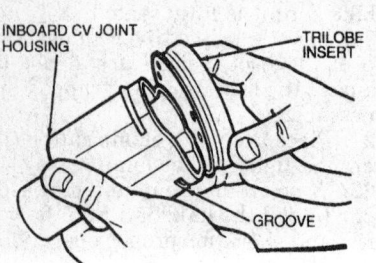

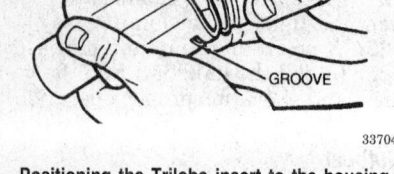

Positioning the Trilobe insert to the housing

9. Position and tighten the small boot clamp.

NOTE: Inspect the CV-joint grease for contamination. If contaminated, the CV-joint assembly should be thoroughly cleaned and inspected for wear.

10. Install a new driveshaft bearing retainer circlip onto the halfshaft.

11. Fill the CV-joint with the proper grease.

12. Place the CV-joint onto the halfshaft and align the splines.

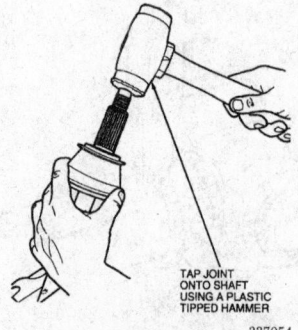

Installing the outer CV-joint onto the halfshaft

STEERING

Air Bag

CAUTION

Some vehicles are equipped with an air bag system, also known as the Supplemental Inflatable Restraint (SIR) or Supplemental Restraint System (SRS). The system must be disabled before performing service on or around system components, steering column, instrument panel components, wiring and sensors. Failure to follow safety and disabling procedures could result in accidental air bag deployment, possible personal injury and unnecessary system repairs.

PRECAUTIONS

Several precautions must be observed when handling the inflator module to avoid accidental deployment and possible personal injury.

- Never carry the inflator module by the wires or connector on the underside of the module.
- When carrying a live inflator module, hold securely with both hands, and ensure that the bag and trim cover are pointed away.
- Place the inflator module on a bench or other surface with the bag and trim cover facing up.
- With the inflator module on the bench, never place anything on or close to the module which may be thrown in the event of an accidental deployment.

DISARMING

CAUTION

The Supplemental Restraint System (SRS) must be disarmed before performing service around SRS components or SRS wiring. Failure to do so may cause accidental deployment of the air bag, resulting in unnecessary SRS repairs and/or personal injury.

1. Place the vehicles front wheels in a straight ahead position.

2. Disconnect both battery cables, negative cable first.

3. Wait at least 1 minute for the air bag backup power supply to drain before continuing.

4. Proceed with the repair.

5. Once complete, connect the battery cables, negative cable last.

6. Prove out the air bag system by turning the ignition key to the **RUN** position and visually monitoring the air bag indicator lamp in the instrument cluster. The indicator lamp should illuminate for approximately 6 seconds and then turn **OFF**. If the indicator lamp does not illuminate, stays on, or flashes at any time, a fault has been detected by the air bag diagnostic monitor.

Steering Wheel

REMOVAL AND INSTALLATION

—— CAUTION ——

The Supplemental Inflatable Restraint (SIR) system must be disarmed before performing service around SIR system components or SIR system wiring. Failure to do so may cause accidental deployment of the air bag, resulting in unnecessary SIR system repairs and/or personal injury.

1. Place the vehicle with the front wheels in a straight ahead position.
2. Disconnect both battery cables, negative cable first.

—— CAUTION ——

WAIT at least 1 minute for the air bag backup power supply to drain before continuing.

3. Pry off 2 plugs covering the air bag module retaining screws; 1 on each side of the steering wheel.
4. Remove 2 air bag module retaining screws. Lift the module away from the steering wheel and disconnect the air bag electrical harness connector from the air bag module. Remove the module from the steering wheel.

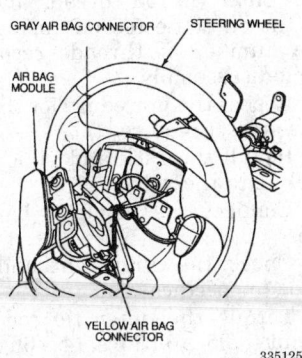

GRAY AIR BAG CONNECTOR   STEERING WHEEL

AIR BAG MODULE

YELLOW AIR BAG CONNECTOR

335125

Air bag module removal

—— CAUTION ——

When carrying a live air bag, make sure the air bag and trim cover are pointed away from the body. In the unlikely event of an accidental deployment, the bag will then deploy with minimal chance if injury. When placing a live air bag on a bench or other surface, always face the bag and trim cover up, away from the surface. This will reduce the motion of the module if it is accidently deployed.

5. Disconnect the horn wire connector and the cruise control wire harness connector (if equipped) from the steering wheel.
6. Remove and discard the steering wheel retaining bolt.
7. Install a suitable steering wheel puller and remove the steering wheel. Route the contact assembly wire harness through the steering wheel as the wheel is lifted off the shaft.
8. Rotate the air bag sliding contact to the right or left to engage the service lock. Make sure that the air bag sliding contact internal service lock is engaged. The air bag sliding contact should not be turned more than 45 degrees to the right or left.

To install:

NOTE: If a new air bag sliding contact is being installed, make sure to remove the plastic lock tool after the intermediate shaft is connected to the steering gear, if removed.

9. Make sure the vehicle's front wheels are in the straight-ahead position.
10. Route the contact assembly wire harness through the steering wheel opening at the 3 o'clock position and install the steering wheel on the shaft. The steering wheel and shaft alignment marks should be aligned. Make sure the air bag contact wire is not pinched.
11. Install a new steering wheel retaining bolt and tighten to 25–34 ft. lbs. (34–46 Nm).
12. Connect the horn and cruise control wire harness (if equipped) to the steering wheel. Make sure the wiring does not get trapped between the steering wheel and contact assembly.
13. Connect the air bag wire harness to the air bag module and install the module to the steering wheel. Make sure that all air bag-to-steering wheel seams are flush and no gaps are visible. Tighten the air bag mod-

ule retaining screws to 106 inch lbs. (12 Nm).
14. Install 2 plugs covering the air bag module retaining screws.
15. Connect the positive, then negative battery cables.
16. Prove out the air bag system by turning the ignition key to the **RUN** position and visually monitoring the air bag indicator lamp in the instrument cluster. The indicator lamp should illuminate for approximately 6 seconds and then turn off. If the indicator lamp does not illuminate, stays on, or flashes at any time, a fault has been detected by the air bag diagnostic monitor.
17. Check the steering wheel, horn and steering column for proper operation.
18. Road test the vehicle and check the speed control for proper operation.

Tie Rod Ends

REMOVAL AND INSTALLATION

NOTE: Before beginning this procedure, make sure that a new outer tie rod end castellated nut is available, per side. The outer tie rod ends may be serviced with the rack and pinion assembly in the vehicle, however the vehicle manufacturer recommends removing the rack and pinion assembly from the vehicle when servicing the inner tie rod ends. It is recommended to replace both inner tie rod ends while the rack and pinion assembly is out of the vehicle.

Outer Tie Rod End

1. Disconnect the negative battery cable.
2. Remove the wheel and tire assembly.
3. Remove the cotter pin and the castellated nut from the outer tie rod end and discard the cotter pin and nut.
4. Separate the outer tie rod end from the steering knuckle using Tie Rod End Remover Tool 3290-D, or equivalent.
5. Hold the outer tie rod end with a wrench and loosen the tie rod end jam nut.
6. Note the depth that the outer tie rod end jam nut is located.
7. Remove the outer tie rod end from the inner tie rod spindle. Count and record the number of turns required to remove the outer tie rod end.

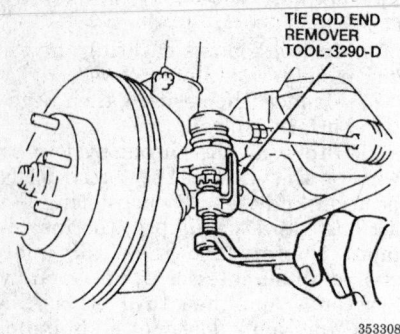

TIE ROD END
REMOVER
TOOL-3290-D

353308

Outer tie rod end removal

8. Remove the outer tie rod end from the vehicle.

To install:

9. Clean the threads on the inner tie rod spindle (front wheel spindle connecting rod).

10. Thread the new outer tie rod end onto the inner tie rod. Use the same number of turns recorded during disassembly. The jam nut should also indicate that the new outer tie rod end is positioned properly.

11. Place the outer tie rod end stud into the steering knuckle. Set the front wheels in a straight ahead position.

12. Install a new castellated nut onto the outer tie rod end stud. Torque the nut to 41 ft. lbs. (55 Nm). Continue to tighten the castellated nut until a new cotter pin can be inserted through the hole in the stud. Install a new cotter pin.

13. Install the wheel and tire assembly. Torque the lug nuts to 85–105 ft. lbs. (115–142 Nm).

14. Connect the negative battery cable.

15. Check the alignment and set the toe to specification.

16. Torque the outer tie rod end jam nut to 35–46 ft. lbs. (47–63 Nm).

17. Road test the vehicle and check for proper operation.

Inner Tie Rod End

1. Remove the rack and pinion assembly from the vehicle and secure to a bench mounted holding fixture.

2. Working on one side, remove the outer tie rod end making sure to record the number of turn required for removal.

3. Remove the outer tie rod end jam nut.

4. Remove the outer clamp securing the inner tie rod bellows to the tie rod spindle.

5. Loosen the larger inner clamp using a wrench or screwdriver and remove the bellows.

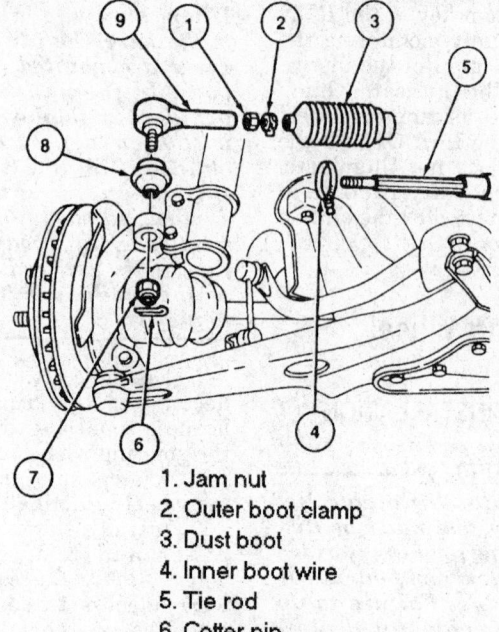

1. Jam nut
2. Outer boot clamp
3. Dust boot
4. Inner boot wire
5. Tie rod
6. Cotter pin
7. Tie rod end nut
8. Tie rod end boot
9. Tie rod end

353310

Tie rod assembly components

6. Gently tap around the rivet head that is securing the inner tie rod to the rack and pinion assembly with a sharp chisel. Raise the rivet enough to be removed with side cutters.

NOTE: Use caution not to shear the rivet head off.

7. Adjust the rack and pinion assembly so that several teeth of the rack are exposed.

8. Hold the rack with an adjustable wrench on the end teeth only, while loosening the inner tie rod nut (ball joint nut) with Nut Wrench T74P-3504-U, or equivalent.

9. Once the inner tie rod is loose, remove by hand.

10. Remove the inner tie rod from the vehicle and inspect the rack and pinion assembly for seal leakage. Replace the rack and pinion assembly if there is excessive leakage from the rack seals.

To install:

11. Install a new inner tie rod assembly.

12. Adjust the rack and pinion assembly so that several teeth of the rack are exposed.

13. While holding the rack with an adjustable wrench nearest the rack end, tighten the inner tie rod nut (ball joint nut) using Nut Wrench

T74P-3504-U or equivalent, to 55–65 ft. lbs. (75–88 Nm).

14. Install a new roll pin (coiled pin) in the inner tie rod nut and tap into position lightly with a hammer.

15. Apply a small amount of grease to the lip of the bellows where it clamps to the inner tie rod spindle to allow the shaft to turn without twisting the bellows.

16. Install the bellows with the larger inner clamp and tighten with a wrench or screwdriver.

17. Install a new outer clamp using needlenose pliers.

18. Install the jam nut onto the tie rod spindle.

19. Apply a small amount of grease to the outer tie rod threads and install the outer tie rod end using the same number of threads recorded during disassembly.

20. Repeat the procedure for the opposite side, if required.

21. Install the rack and pinion assembly into the vehicle.

22. Connect the negative battery cable.

23. Check the alignment and set the toe to specification.

24. Torque the outer tie rod end jam nut to 35–46 ft. lbs. (47–63 Nm).

25. Road test the vehicle and check for proper operation.

Power Rack and Pinion

REMOVAL AND INSTALLATION

NOTE: This procedure requires the use of a twin post hoist. Before continuing with this procedure, make sure to have available, new pinch bolts for the steering gear shaft, castellated nuts for the outer tie-rod ends and TFE O-rings for the fluid line fittings. Once removed, these parts lose their torque holding ability or retention capability and must not be reused.

1. Disconnect the negative battery cable.
2. Working from inside the vehicle, remove the nuts retaining the steering column weather seal to the dash panel.
3. Remove the pinch bolt retaining the steering column input shaft coupling to the steering gear (rack and pinion assembly) input shaft.
4. Set the steering column weather seal aside. Remove the pinch bolt at the lower steering column shaft and remove the steering column input shaft and coupling (intermediate shaft).
5. Raise and safely support the vehicle on a twin post lift.
6. Remove the front wheel and tire assemblies.
7. Support the vehicle with jackstands under the front jack pads.
8. Remove the cotter pin and the castellated nut from the left-hand and right-hand outer tie rod ends and discard the cotter pins and nuts.
9. Separate the outer tie rod ends from the steering knuckles using Tie Rod End Remover Tool 3290-D, or equivalent.
10. Hold each outer tie rod end with a wrench and loosen the tie rod end jam nut.
11. Note the depth that the outer tie rod end jam nut is located.
12. Remove the outer tie rod end from the inner tie rod spindle. Count and record the number of turns required to remove the outer tie rod end.
13. Remove the outer tie rod end from the vehicle.
14. Remove the front stabilizer bar (sway bar) from the vehicle.
15. Remove 2 nuts from the steering gear-to-sub frame retaining bolts.

16. Make sure that the jackstands are in position. Remove 2 rear bolts securing the sub-frame to the body.

NOTE: Support the flex tube before disconnecting the exhaust system to prevent premature failure.

17. Install Flex Tube Holding Tool T94T-6000-AH, or equivalent and disconnect the exhaust system flex tube at the dual converter Y-pipe.
18. Slowly lower the twin post hoist carefully until the rear of the sub-frame separates from the body approximately 4 inches.
19. Remove the heat shield band and fold the heat shield down.
20. Rotate the steering gear to clear the bolts from the sub-frame while pulling the steering gear to the lest side of the vehicle to allow access for removing the power steering fluid lines.
21. Place a drain pan under the steering gear and remove the pressure and return lines. Allow the steering gear and lines to drain into the drain pan.
22. Remove the steering gear assembly through the left-hand side of the vehicle.

To install:
23. Install new plastic seals (TFE O-rings) on the line fittings.
24. If removed from the steering gear or the steering gear is being replaced, be sure to install 2 steering gear retaining bolts.
25. Install the steering gear through the left-hand wheel well.
26. Install the pressure and return lines to the steering gear. Tighten the lines to 15–25 ft. lbs. (20–35 Nm). Swivel movement of the lines is normal when the fittings are properly tightened.
27. Place the steering gear in position on the sub-frame.
28. Thread the new outer tie rod end onto the inner tie rod. Use the same number of turns recorded during disassembly. The jam nut should also indicate that the new outer tie rod end is positioned properly.
29. Place the outer tie rod end stud into the steering knuckle. Set the front wheels in a straight-ahead position.
30. Install a new castellated nut onto the outer tie rod end stud. Torque the nut to 41 ft. lbs. (55 Nm). Continue to tighten the castellated nut until a new cotter pin can be inserted through the hole in the stud. Install a new cotter pin.
31. Install the band on the heat shield.
32. Install the stabilizer bar.

33. Install 2 steering gear retaining nuts. Tighten the nuts to 83–113 ft. lbs. (113–153 Nm).
34. Slowly raise the twin post hoist until the rear of the sub-frame contacts the body.
35. Install 2 sub-frame retaining bolts. Tighten the bolts to 57–80 ft. lbs. (77–103 Nm).
36. Install the exhaust system flex tube to the dual converter Y-pipe and remove the flex tube support, or similar device.
37. Install the wheel and tire assemblies. Torque the lug nuts to 85–105 ft. lbs. (115–142 Nm).
38. Remove the jackstands.
39. Lower the vehicle.
40. Place the steering column weather seal on the steering column gear input shaft coupling.
41. Using new pinch bolts, install the steering column gear input shaft coupling on the steering gear input shaft and the lower steering column shaft. Tighten the upper pinch bolt to 30–40 ft. lbs. (40–54 Nm) and the lower pinch bolt to 25–34 ft. lbs. (34–46 Nm).
42. Install the steering column weather seal to the dash panel. Tighten the nuts to 40–55 inch lbs. (4.5–6.3 Nm).
43. Connect the negative battery cable.
44. Fill the power steering pump reservoir and bleed the power steering system.
45. Check for power steering leaks.

NOTE: Whenever the vehicles sub-frame is removed or lowered, the wheel alignment should be checked.

46. Check the alignment and adjust if necessary.
47. Road test the vehicle and check for proper operation.

Power Steering Pump

BLEEDING

1. Disable the ignition system.
2. Raise the vehicle until the front tires are just off of the ground and safely support. Make sure that the transaxle is not in gear.
3. Fill the power steering pump reservoir.
4. Crank the engine for 30 seconds without turning the steering wheel and recheck the fluid level. Add fluid as the level drops in the power steering reservoir as needed.
5. Crank the engine for 30 seconds while turning the steering wheel lock

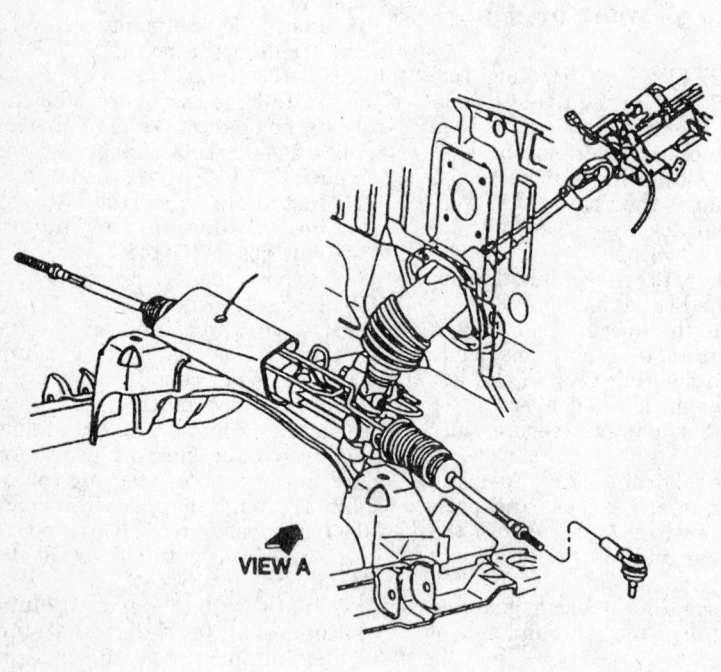

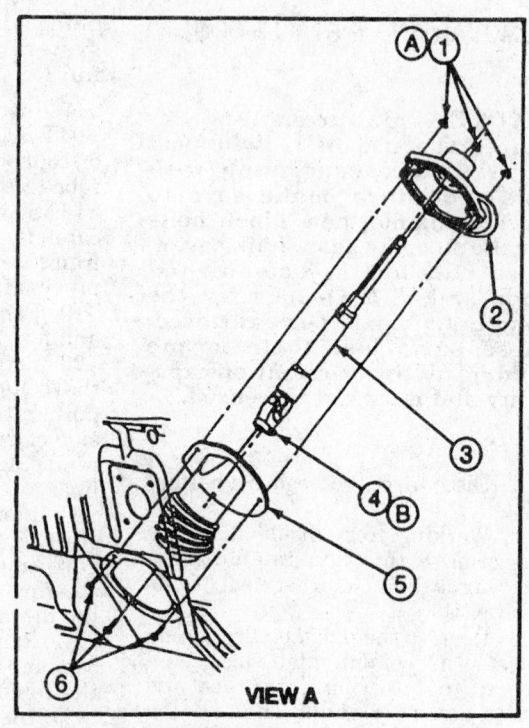

1	Nut
2	Steering Column Opening Weather Seal (Primary)
3	Steering Column Intermediate Shaft Coupling
4	Pinch Bolt

5	Steering Column Opening Weather Seal (Secondary)
6	Stud
A	Tighten to 4.5-6.3 N·m (40-55 Lb-In)
B	Tighten to 34-46 N·m (25-33 Lb-Ft)

335392

Steering column shaft and coupling assembly

to lock. Add fluid as the level drops in the power steering reservoir as needed.

— **WARNING** —
Do not hold the steering wheel against a stop for more than 5 seconds as damage to the steering pump could result.

6. Lower the vehicle.
7. Restore the ignition system.
8. Start the engine and check that the power steering system is free of air. If not, repeat the procedure. If required, purge the system of air using an external vacuum source fol-lowing the equipment manufacturers instructions.

REMOVAL AND INSTALLATION

3.0L (VIN U) Engine

1. Disconnect the negative battery cable.
2. Remove the accessory drive belt.
3. Remove the alternator.
4. Remove the drive belt tensioner.
5. Remove the power steering pump pulley using Puller T69L-10300-B, or equivalent.

6. Place a drain pan under the power steering pump and disconnect the power steering pressure and return hoses from the power steering pump. Allow the power steering fluid to drain.
7. Remove 3 power steering pump retaining bolts from the front of the pump and 1 retaining bolt from the rear of the pump.
8. Remove the power steering pump from the vehicle.
To install:
9. Place the power steering pump in position and install 4 retaining bolts. Tighten the retaining bolts to 30-40 ft. lbs. (40-55 Nm).

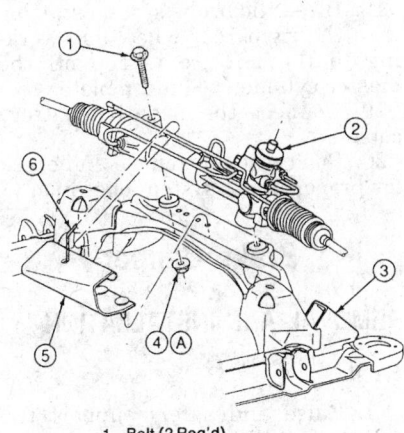

1 Bolt (2 Req'd)
2 Steering Gear
3 Front Sub-Frame
4 Nut (2 Req'd)
5 Steering Shaft U-Joint Shield
6 Strap
A Tighten to 113-153 N·m
  (83-113 Ft-Lb)

335395

Rack and pinion mounting to the sub-frame

10. Install the power steering pressure and return hoses to the power steering pump.

11. Install the drive belt tensioner.

12. Install the alternator assembly.

13. Place the power steering pump pulley on the power steering pump shaft and install Pulley Replacer T65P-3A733-C, or equivalent. When using this tool, the small diameter threads must be fully engaged in the pump shaft before pressing on the pulley. Hold the head screw and turn the nut to install the pulley. Install the pulley face flush with the pump shaft to within 0.10 inch (0.25mm) of the end of the pump shaft.

14. Install the accessory drive belt.

15. Fill the power steering pump reservoir to the proper level.

16. Connect the negative battery cable.

17. Start the engine and bleed air from the power steering system, as necessary while checking for power steering leaks.

18. Road test the vehicle and check for steering system proper operation.

3.8L (VIN 4) Engine

1. Disconnect the negative battery cable.

2. Remove the accessory drive belt.

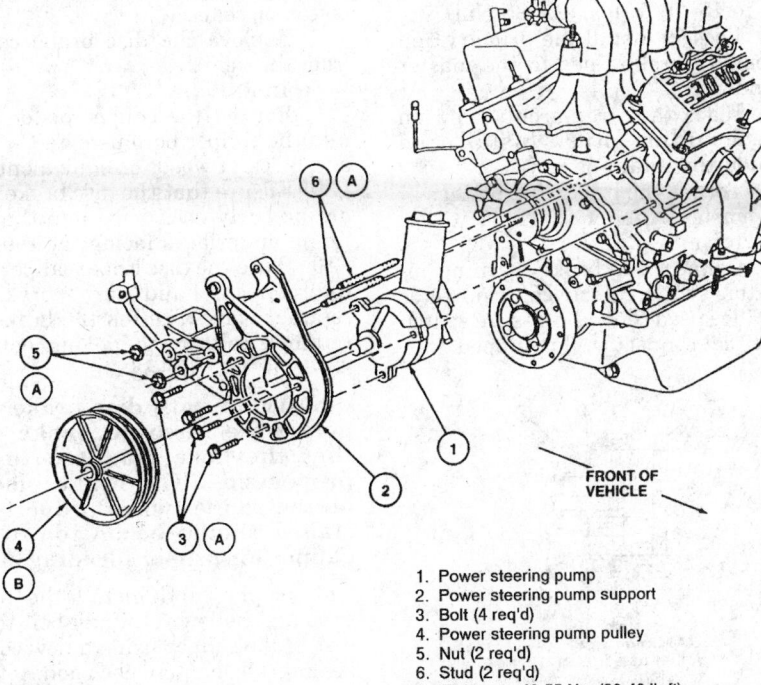

1. Power steering pump
2. Power steering pump support
3. Bolt (4 req'd)
4. Power steering pump pulley
5. Nut (2 req'd)
6. Stud (2 req'd)
A. Tighten to 40-55 Nm (30-40 lb-ft)
B. Press flush with end of power
   steering pump shaft

353345

Power steering pump and related components — 3.0L (VIN U) engine

3. Remove the alternator.

4. Remove the power steering pump pulley using Puller T69L-10300-B, or equivalent.

5. Place a drain pan under the power steering pump and disconnect the power steering pressure and return hoses from the power steering pump. Allow the power steering fluid to drain.

6. Remove the power steering pump retaining bolts and remove the power steering pump from the vehicle.

To install:

7. Place the power steering pump in position and install the retaining bolts. Tighten the retaining bolts to 30–40 ft. lbs. (40–55 Nm).

8. Install the power steering pressure and return hoses to the power steering pump.

9. Place the power steering pump pulley on the power steering pump shaft and install Pulley Replacer T65P-3A733-C, or equivalent. When using this tool, the small diameter threads must be fully engaged in the pump shaft before pressing on the pulley. Hold the head screw and turn the nut to install the pulley. Install the pulley face flush with the pump shaft to within 0.10 inch (0.25mm) of the end of the pump shaft.

10. Install the alternator.

11. Install the accessory drive belt.

12. Fill the power steering pump reservoir to the proper level.

13. Connect the negative battery cable.

14. Start the engine and bleed air from the power steering system, as necessary while checking for power steering leaks.

15. Road test the vehicle and check for steering system proper operation.

BRAKES

Anti-Lock Brake System Service

PRECAUTIONS

• Certain components within the Anti-Lock Brake System (ABS) are not intended to be serviced or repaired individually. Only those components with removal and installation procedures should be serviced.

• Do not use rubber hoses or other parts not specifically specified for and

ABS system. When using repair kits, replace all parts included in the kit. Partial or incorrect repair may lead to functional problems and require the replacement of components.

• Lubricate rubber parts with clean, fresh brake fluid to ease assembly. Do not use lubricated shop air to clean parts; damage to rubber components may result.

• Use only specified brake fluid from an unopened container.

• If any hydraulic component or line is removed or replaced, it may be necessary to bleed the entire system.

• A clean repair area is essential. Always clean the reservoir and cap thoroughly before removing the cap. The slightest amount of dirt in the fluid may plug an orifice and impair the system function. Perform repairs after components have been thoroughly cleaned; use only denatured alcohol to clean components. Do not allow ABS components to come into contact with any substance containing mineral oil; this includes used shop rags.

• The Anti-Lock control unit is a microprocessor similar to other computer units in the vehicle. Ensure that the ignition switch is **OFF** before removing or installing controller harnesses. Avoid static electricity discharge at or near the controller.

• If any arc welding is to be done on the vehicle, the control unit should be unplugged before welding operations begin.

Master Cylinder

REMOVAL AND INSTALLATION

1. Disconnect the negative battery cable.
2. Apply the brake pedal several times to exhaust all vacuum in the system.
3. Disconnect the fluid level indicator electrical connector and the speed control pressure switch electrical connector, if equipped.
4. Remove the brake lines from the primary and secondary outlet ports on the brake master cylinder.
5. Remove the wrap-around clip and the retaining nut from the mounting studs.
6. Remove 2 nuts retaining the master cylinder to the power brake booster mounting studs.
7. Slide the master cylinder forward and remove it from the vehicle.
To install:
8. Install the speed control pressure switch from the old master cyl-

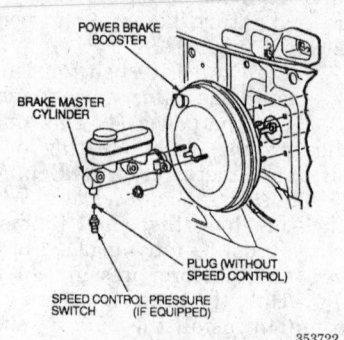

Master cylinder removal

inder, if equipped or the plug if not included in the new master cylinder.
9. Bench bleed the new master cylinder.
10. Before installing the master cylinder onto the power brake booster, check the brake booster pushrod length by fabricating a gauge block that when set against the brake booster will determine if the pushrod needs to be lengthened or shortened. The pushrod should protrude out of the brake booster 0.980–0.995 inch (24.89–25.27 mm). The pushrod is adjusted by turning the adjustment screw on the end of the pushrod.
11. Place the brake master cylinder onto the mounting studs of the power brake booster.
12. Install 2 retaining nuts and torque to 18–26 ft. lbs. (25–35 Nm).
13. Loosely install the primary and secondary brake lines to the master cylinder.
14. Place the wrap-around clip on the mounting stud. Install and tighten the retaining nut.
15. Torque the brake line fittings at the master cylinder to 11–15 ft. lbs. (15–20 Nm).
16. Connect the brake warning indicator switch electrical connector and the speed control pressure switch electrical connector, if equipped.

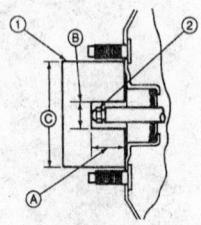

1. Gauge block
2. Adjustment screw
A. 0.980-0.995 in.(24.89-25.27 mm)
B. 3/4 in.(19-05 mm)
C. 2-15/16 in.(74.61 mm)

353723

Dimensions of tool to measure pushrod length

17. Fill the master cylinder with the proper brake fluid to just below the full line.
18. Bleed the brake system starting with the right rear wheel and working to the left front. Top off the master cylinder when complete.
19. Connect the negative battery cable.
20. Road test the vehicle and check for proper brake system operation.

Brake Caliper

REMOVAL AND INSTALLATION

Front

1. Raise and safely support the vehicle.
2. Remove the wheel and tire assembly.
3. Mark the disc brake caliper to avoid mixing the left-hand and right-hand components.
4. Disconnect the brake hose from the disc brake caliper by loosening and removing the hollow retaining bolt. Discard the 2 copper sealing washers and plug the brake hose.
5. Remove the 2 brake pin retainer bolts.
6. Lift the disc brake caliper off of the disc brake rotor using a rotating motion. Do not pry against the caliper piston which may damage the piston or seals.
7. Remove the disc brake caliper from the vehicle.
To install:
8. Retract the caliper piston fully into the caliper bore using a C-clamp and block of wood, or equivalent.
9. Ensure that the disc brake pads are properly positioned and that the lining material is facing the rotor.
10. Place the disc brake caliper over the rotor and hand start 2 brake pin retainer bolts. Tighten the brake pin retainer bolts to guide pin bolts to 23–28 ft. lbs. (31–38 Nm).

NOTE: If both disc brake calipers were removed, make sure that they are mounted to the proper side. The brake bleeder on the caliper when properly installed should be on top of the caliper for proper bleeding of air.

11. Unplug and install the brake hose and hollow retaining bolt to the disc brake caliper using a new copper sealing washer on each side of the hose fitting. Tighten the retaining bolt to 35–46 ft. lbs. (47–63 Nm).
12. Bleed the brake system and install the rubber bleeder screw caps when complete.

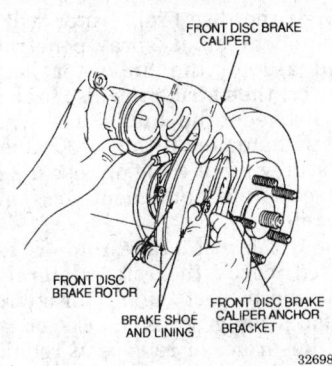

FRONT DISC BRAKE CALIPER

FRONT DISC BRAKE ROTOR

BRAKE SHOE AND LINING

FRONT DISC BRAKE CALIPER ANCHOR BRACKET

326983

Removing the front disc brake caliper

13. Install the wheel and tire assembly. Torque the lug nuts to 85–105 ft. lbs. (115–142 Nm).

14. Lower the vehicle.

15. Pump the brake pedal several times to position the brake pads before attempting to move the vehicle.

16. Check and fill the brake master cylinder as required.

17. Road test the vehicle and check for proper brake operation.

Rear

1. Remove and discard ½ of the brake fluid from the brake master cylinder reservoir.

2. Raise and safely support the vehicle.

3. Remove the wheel and tire assembly.

4. Disconnect the brake hose from the disc brake caliper by loosening and removing the hollow retaining bolt. Discard 2 copper sealing washers and plug the brake hose.

5. Using a C-clamp or equivalent, position the clamp frame on the inboard side of the disc brake caliper housing. Place the clamp screw on the outboard disc brake pad and tighten the clamp enough to press the caliper piston into the caliper housing releasing pressure on the disc brake pads.

6. Remove 2 disc brake caliper retaining bolts.

7. Remove the caliper by swinging out the bottom of the caliper first.

8. Remove the disc brake pads, if necessary.

To install:

9. Retract the disc brake caliper piston fully into the caliper bore using a C-clamp and block of wood, or equivalent.

10. Ensure that the disc brake pads are properly positioned and that the lining material is facing the rotor.

11. Install the caliper over the disc brake rotor and position on the brake

adapter. Install 2 disc brake caliper retaining bolts and tighten to 11–14 ft. lbs. (15–20 Nm).

12. Unplug and install the brake hose and hollow retaining bolt to the disc brake caliper using a new copper sealing washer on each side of the hose fitting. Tighten the retaining bolt to 35–46 ft. lbs. (47–63 Nm).

13. Bleed the brake system and install the rubber bleeder screw caps when complete.

14. Install the wheel and tire assembly. Torque the lug nuts to 85–105 ft. lbs. (115–142 Nm).

15. Lower the vehicle.

16. Pump the brake pedal several times to position the brake pads before attempting to move the vehicle.

17. Check and fill the brake master cylinder as required.

18. Road test the vehicle and check for proper brake operation.

Disc Brake Pads

REMOVAL AND INSTALLATION

Front

1. Remove ½ of the brake fluid from the brake master cylinder reservoir. Properly dispose of the brake fluid.

2. Raise and safely support the vehicle.

3. Remove the wheel and tire assembly.

4. Remove 2 disc brake caliper brake pin retainers. Do not remove the brake hose from the caliper.

5. Lift the disc brake caliper off of the disc brake rotor using a rotating motion. Do not pry against the caliper piston which may damage the piston or seals.

6. Hang the disc brake caliper with a length of wire or equivalent to prevent damage to the brake hose.

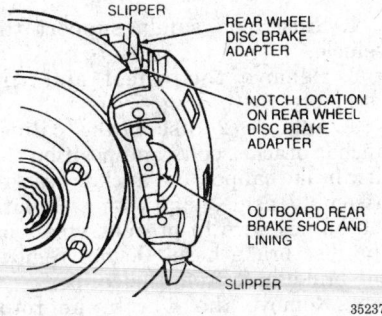

SLIPPER

REAR WHEEL DISC BRAKE ADAPTER

NOTCH LOCATION ON REAR WHEEL DISC BRAKE ADAPTER

OUTBOARD REAR BRAKE SHOE AND LINING

SLIPPER

352377

Rear disc brake caliper mounting

7. Remove the inner and outer disc brake pads and the anti-rattle clip.

8. Inspect the disc brake rotor surfaces for grooves, cracks or glazing. Resurface or replace as required. If resurfacing, observe the minimum thickness specification.

To install:

9. Retract the caliper piston fully into the caliper bore using a C-clamp and wood block or equivalent. This will allow room for the new disc brake pads.

10. Install new inner and outer disc brake pads and the anti-rattle clip. Ensure that the disc brake pads are properly positioned and that the lining material is facing the rotor.

11. Place the disc brake caliper over the rotor and install 2 disc brake caliper brake pin retainers. Tighten the brake pin retainers to 23–28 ft. lbs. (31–38 Nm).

12. Install the wheel and tire assembly. Torque the lug nuts to 85–105 ft. lbs. (115–142 Nm).

13. Lower the vehicle.

14. Pump the brake pedal to position the brake pads before attempting to move the vehicle.

15. Check and fill the brake master cylinder reservoir, as required.

16. Road test the vehicle and check for proper brake system operation.

Rear

1. Remove ½ of the brake fluid from the brake master cylinder reservoir. Properly dispose of the brake fluid.

2. Raise and safely support the vehicle.

3. Remove the wheel and tire assembly.

4. Using a C-clamp or equivalent, position the clamp frame on the inboard side of the disc brake caliper housing. Place the clamp screw on the outboard disc brake pad and tighten the clamp enough to press the caliper piston into the caliper housing releasing pressure on the disc brake pads.

5. Remove 2 disc brake caliper retaining bolts. Do not remove the disc brake caliper brake hose from the caliper.

6. Work the disc brake caliper off the brake rotor and disc brake adapter. Move the disc brake caliper aside and secure with wire or equivalent to prevent damage to the brake hose.

7. Remove the slippers from the anchor plate abutments by gently prying them off the rails and discard the slippers.

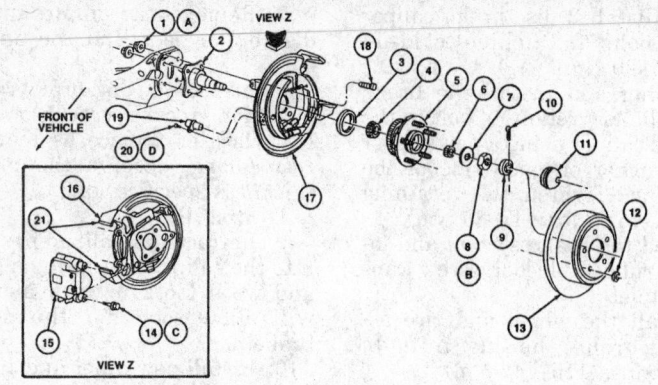

VIEW Z

FRONT OF VEHICLE

VIEW Z

1	Nut, Rear Wheel Spindle-to-Rear Axle
2	Rear Wheel Spindle
3	Wheel Hub Grease Seal
4	Bearing Cone and Roller, Rear
5	Rear Hub
6	Rear Wheel Bearing
7	Washer
8	Rear Hub Nut
9	Retainer
10	Cotter Pin
11	Hub Grease Cap
12	Keeper Nut
13	Rear Disc Brake Rotor
14	Caliper Bolt
15	Rear Disc Brake Caliper
16	Rear Wheel Disc Brake Adapter

17	Rear Wheel Disc Brake Shield
18	Bolt, Rear Wheel Spindle-to-Rear Axle
19	Rear Brake Anti-Lock Sensor
20	Bolt, Rear Brake Anti-Lock Sensor
21	Slippers
A	Tighten to 65-80 N·m (48-59 Lb-Ft)
B	To set bearing preload, tighten to 24-31 N·m (18-22 Lb-Ft) while rotating the rear hub. Back off and retighten to 2 N·m (17 Lb-In)
C	Tighten to 15-20 N·m (11-14 Lb-Ft)
D	Tighten to 8-10 N·m (71-88 Lb-In)

(Continued)

352498

Rear disc brake components

8. Remove the inner and outer disc brake pads.

9. Inspect the disc brake rotor surfaces for grooves, cracks or glazing. Resurface or replace as required. If resurfacing, observe the minimum thickness specification.

To install:

10. Retract the disc brake caliper piston fully into the caliper bore using a C-clamp and block of wood, or equivalent. This will make room for the new disc brake pads.

11. Install new anti-wear slippers on the rail abutments by snapping them in place.

12. Install new inner and outer disc brake pads. Ensure that the disc brake pads are properly positioned and that the lining material is facing the rotor.

13. Install the disc brake caliper over the brake rotor and place on the brake adapter. Ensure that the notches on the upper ends of the brake pads are seated over the upper ledge of the disc brake adapter and the lower tabs are placed on the lower ledge of the disc brake adapter.

14. Lubricate 2 disc brake caliper retaining bolts with a suitable grease and install. Tighten the retaining bolts to 11–14 ft. lbs. (15–20 Nm).

15. Install the wheel and tire assembly. Torque the lug nuts to 85–105 ft. lbs. (115–142 Nm).

16. Lower the vehicle.

17. Pump the brake pedal several times to position the brake pads before attempting to move the vehicle.

18. Check and fill the brake master cylinder reservoir, as required.

19. Road test the vehicle and check for proper brake operation.

Brake Rotor

REMOVAL AND INSTALLATION

Front

1. Raise and safely support the vehicle.

2. Remove the wheel and tire assembly.

3. Remove 2 disc brake caliper anchor bracket bolts and position the disc brake caliper and anchor bracket assembly aside, supporting it with mechanics wire to prevent stress on the disc brake hose. Do not remove the brake hose from the caliper.

4. Remove the disc brake rotor from the hub by pulling it off of the hub studs.

5. If the disc brake rotor will not come off by hand, spray penetrating fluid around the hub/rotor mating surface, then strike the disc brake rotor between the hub studs with a plastic hammer. If the disc brake rotor still will not come off, use a 2 or 3 jaw puller or equivalent, to remove the rotor.

6. If the disc brake rotor is to be reused, check the rotors lateral runout and inspect the disc brake rotor surfaces for grooves, cracks or glazing. Resurface or replace as required. If resurfacing, observe the minimum thickness specification.

To install:

7. Clean the hub and rotor mating surfaces.

8. If the brake rotor is being replaced, clean off the protective coating on the rotor surfaces.

9. Apply a small amount of suitable grease to the pilot opening of the disc brake rotor where it contacts the hub and install the disc brake rotor onto the hub.

10. Install the disc brake caliper and anchor bracket assembly with 2 anchor bracket retaining bolts. Tighten the bolts to 65–87 ft. lbs. (88–118 Nm).

11. Install the wheel and tire assembly. Tighten the lug nuts to 85–105 ft. lbs. (115–142 Nm).

12. Lower the vehicle.

13. Pump the brake pedal several times to position the brake pads before attempting to move the vehicle.

14. Check and fill the brake master cylinder, as required.

15. Road test the vehicle and check for proper brake operation.

Rear

1. Raise and safely support the vehicle.

2. Remove the wheel and tire assembly.

3. Using a C-clamp or equivalent, position the clamp frame on the inboard side of the disc brake caliper housing. Place the clamp screw on the outboard disc brake pad and tighten the clamp enough to press the caliper piston into the caliper housing releasing pressure on the disc brake pads.

4. Remove 2 disc brake caliper retaining bolts.

5. Remove the caliper by swinging out the bottom of the caliper first and support the caliper assembly with mechanics wire to prevent stress on the disc brake hose. Do not remove the brake hose from the caliper.

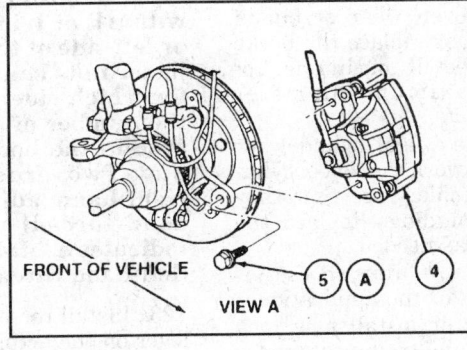

1. Front wheel driveshaft joint
2. Front wheel hub and spindle
3. Front disc brake rotor
4. Disc brake caliper
5. Bolt
A. 65–87 ft. lb.(88–118 Nm)

327036

Front disc brake rotor and related components

6. Remove the keeper nuts from the disc brake rotor, if equipped.

7. Remove the plug from the adjustment hole for the parking brake shoes from the rear of the brake shield.

8. Using a suitable tool, turn the parking brake adjuster clockwise for the drivers side or counterclockwise for the passenger side to back off the parking brake shoes to allow rotor removal.

9. Remove the disc brake rotor from the hub.

10. If the disc brake rotor will not come off by hand, spray penetrating fluid around the hub/rotor mating surface, then strike the disc brake rotor between the hub studs with a plastic hammer. If the disc brake rotor still will not come off, use a 2 or 3 jaw puller or equivalent, to remove the rotor.

11. If the disc brake rotor is to be reused, check the rotors lateral runout and inspect the disc brake rotor surfaces for grooves, cracks or glazing. Resurface or replace as required. If resurfacing, observe the minimum thickness specification.

To install:

12. Clean the hub and rotor mating surfaces.

13. If the brake rotor is being replaced, clean off the protective coating on the rotor surfaces.

14. Apply a small amount of suitable grease to the pilot opening of the disc brake rotor where it contacts the hub and install the disc brake rotor onto the hub.

NOTE: It is not necessary to install the keeper nuts on the brake rotor.

15. Install the disc brake caliper and 2 caliper retaining bolts. Tighten the bolts to 11–14 ft. lbs. (15–20 Nm).

16. Adjust the parking brake shoes and install the access plug to the brake shield.

17. Install the wheel and tire assembly. Torque the lug nuts to 85–105 ft. lbs. (115–142 Nm).

18. Lower the vehicle.

19. Pump the brake pedal several times to position the brake pads before attempting to move the vehicle.

20. Check and fill the brake master cylinder, as required.

21. Road test the vehicle and check for proper brake operation.

Brake Drums

REMOVAL AND INSTALLATION

1. Raise and safely support the vehicle.

2. Remove the wheel and tire assembly.

3. Remove the retainers holding the drum to the hub, if installed and discard.

4. Grasp the drum and remove.

5. If the drum will not slide off with light force, the brake shoes will need to be backed off. Remove the rubber plug on the backing plate and insert a screwdriver and a brake adjusting tool into the slot. Hold the adjuster lever away from the adjuster wheel with the screwdriver and back of the adjuster wheel with the brake adjusting tool.

6. Remove the brake drum. Inspect the drum for wear and/or damage. Machine or replace as necessary. If machining, observe the maximum diameter specification.

To install:

7. If a new brake drum is being installed, remove the protective coating from the inner brake surface.

8. Use a suitable brake adjustment gauge to measure the inside diameter of the brake drum.

9. Adjust the brake shoes to match the inside diameter of the brake drum.

10. Slide the brake drum onto the hub. Make sure that the brake shoes are not tight to the brake drum.

11. Install the rubber plug in the access hole. Retainers do not need to be reused to hold the drum.

12. Install the wheel and tire assembly. Torque the lug nuts to 85–105 ft. lbs. (115–142 Nm).

13. Lower the vehicle.

14. Check and fill the brake master cylinder as required.

15. Road test the vehicle and check for proper brake operation.

Brake Shoes

REMOVAL AND INSTALLATION

1. Raise and safely support the vehicle.

2. Remove the wheel and tire assembly.

3. Remove the brake drum.

4. Disconnect the parking brake cable from the trailing brake shoe parking brake lever.

5. Remove 2 brake shoe hold-down springs and pins.

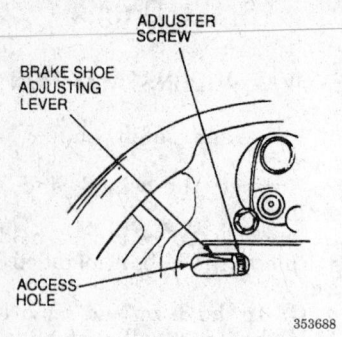

Location of the access hole on the brake backing plate

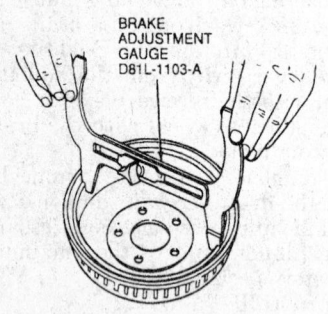

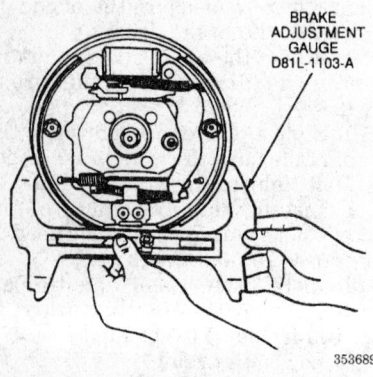

Measuring the brake shoes and drum

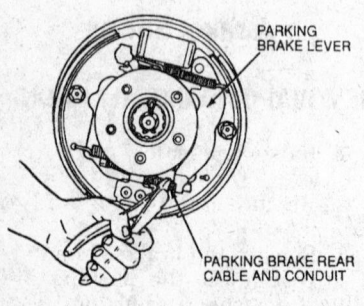

Removing the parking brake cable from the lever

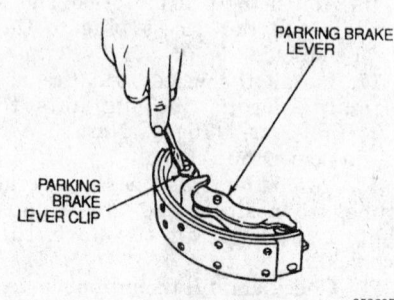

Parking brake lever and clip (pin not shown)

6. Remove the brake shoe adjusting screw spring (upper spring) and the brake shoe adjusting lever.

7. Remove the brake shoe adjuster assembly.

8. Remove the brake shoe retracting spring (lower spring).

9. Remove the brake shoes from the backing plate.

10. On the trailing shoe equipped with the parking brake lever, remove and discard the parking brake lever clip. Remove the washer and the parking brake lever.

To install:

11. Inspect the brake drum for glazing, cracks, uneven wear or out of round. Machine or replace the brake drum as required. If machining, observe the maximum diameter specification.

12. Inspect the wheel cylinder for leakage. Gently work the wheel cylinder pistons to make sure that they move without binding. Replace the wheel cylinder, as needed.

13. Check the new brake shoes that they are correct for the application.

14. Swap over or install new pins into the leading and trailing shoes for the parking brake lever and the brake shoe adjusting lever using the old shoes as a guide.

15. Apply a suitable grease to the backing plate/brake shoe contact areas.

16. Apply a light coat of a suitable grease on the threads of the adjuster nut and socket of the brake shoe adjuster assembly. Screw the adjuster nut and socket to its shortest position and then lengthen it ½ turn.

17. Install the parking brake lever to the trailing shoe with the washer and a new parking brake lever clip.

18. Position the trailing shoe onto the backing plate and install the brake shoe hold-down spring and pin.

19. Position the leading shoe to the backing plate and install the brake shoe hold-down spring and pin.

20. Attach the parking brake cable to the parking brake lever on the trailing brake shoe.

21. Attach the brake shoe retracting spring (lower spring) between the brake shoes.

22. Install the brake adjuster assembly in the slots between the brake shoes. The wider slot on the socket end must fit in the slot on the leading brake shoe. The slot on the adjuster nut end must fit in the slots on the trailing shoe and the parking brake lever.

NOTE: The socket end of each brake adjuster screw is marked with a R or L indicating the right or left side of the vehicle. The adjuster nuts can also be identified for which side they belong to by the number of grooves machined around the body of the adjuster nut. Two grooves indicates a right-hand adjuster nut (right-hand thread) while one groove indicates a left-hand adjuster nut (left-hand thread).

23. Install the brake shoe adjusting lever on the adjusting lever pin.

24. Install the brake shoe adjusting screw spring (upper spring) in the slot on the trailing shoe and in the slot on the brake shoe adjusting lever. Make sure that the brake shoe adjusting lever contacts the brake adjuster screw.

25. Use a suitable brake adjustment gauge to measure the inside diameter of the brake drum.

26. Adjust the brake shoes to match the inside diameter of the brake drum.

27. Slide the brake drum onto the hub. Make sure that the brake shoes are not tight to the brake drum.

28. Install the wheel and tire assembly. Torque the lug nuts to 85–105 ft. lbs. (115–142 Nm).

29. Lower the vehicle.

30. Check and fill the brake master cylinder, as required.

31. Road test the vehicle and check for proper brake operation.

Wheel Cylinder

REMOVAL AND INSTALLATION

1. Raise and safely support the vehicle.

2. Remove the wheel and tire assembly.

3. Remove the retainers holding the drum to the hub, if installed and discard.

4. Remove the brake drum.

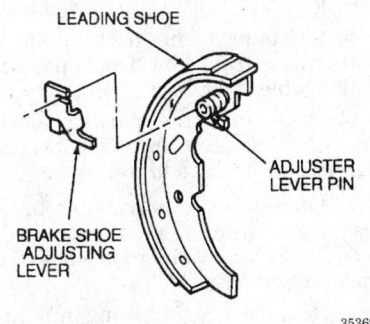

Brake shoe adjusting lever and pin

LEADING SHOE

ADJUSTER LEVER PIN

BRAKE SHOE ADJUSTING LEVER

353699

5. Remove the brake shoes.

6. Disconnect the wheel cylinder brake line from the wheel cylinder. Avoid twisting the brake line.

7. Remove 2 wheel cylinder retaining bolts.

8. Remove the wheel cylinder from the backing plate.

To install:

9. Place the new wheel cylinder to the backing plate and install 2 retaining bolts. Tighten the bolts to 9–13 ft. lbs. (12–18 Nm).

10. Install the brake line to the wheel cylinder and tighten the fitting to 11–15 ft. lbs. (15–20 Nm).

11. Install the brake shoes. Adjust the brakes to fit the brake drum and install the brake drum.

NOTE: The retainers holding the brake drum to the hub, if originally equipped, do not need to be replaced.

12. Install the wheel and tire assembly. Torque the lug nuts to 85–105 ft. lbs. (115–142 Nm).

13. Lower the vehicle.

14. Bleed the brake system using only clean DOT 3 or equivalent brake fluid from a closed container.

15. Pump the brake pedal several times to position the brake shoes before attempting to move the vehicle.

16. Check and fill the brake master cylinder as required.

17. Road test the vehicle and check for proper brake operation.

Parking Brake Cable

ADJUSTMENT

1. On vehicles equipped with rear drum brakes, the parking brake is self-adjusting and should maintain correct parking brake adjustment as long as all components are in good working order and the brake shoes are properly adjusted. Cable tension is adjusted whenever the parking brake is applied and released.

2. On vehicles equipped with rear disc brakes, proceed as follows:

a. Raise and safely support the vehicle.

b. Remove the plug from the parking brake adjuster access hole.

c. Insert a suitable tool and expand the parking brake adjuster while turning the wheel and tire assembly. Extend the parking brake shoes until a drag is felt on the wheel and tire assembly.

d. Back off the parking brake adjuster 1 full turn.

e. Repeat the adjustment procedure to the opposite side, if needed.

3. Lower the vehicle.

4. Check the parking brake system for proper operation.

5. Pull up on the parking brake control lever while listening to the number of clicks from the parking pawl mechanism required to apply resistance to the parking brake system. If properly adjusted, resistance should be felt after 5 clicks from the parking pawl mechanism.

NOTE: If servicing the parking brake system, the tension on the parking brake cables must first be released.

6. To release parking brake cable tension, proceed as follows:

a. Place the parking brake control in the released position.

b. Remove the parking brake lever boot.

c. Using a screwdriver, push down on the tension arm until it is fully depressed and lock the tension arm with a second screwdriver or similar tool, through the hole in the parking brake control.

7. To reset the parking brake cable tension, proceed as follows:

a. Remove the screwdriver or similar tool to allow the parking brake control to set the cable tension. Keep fingers clear during this step.

b. Install the parking brake lever boot.

8. Apply and release the parking brake control lever several times. Make sure that the rear brakes are not dragging when the parking brake control lever is released.

REMOVAL AND INSTALLATION

Front Cable

1. Place the parking brake control in the released position.

2. Remove the parking brake lever boot.

3. Using a screwdriver, push down on the tension arm until it is fully depressed and lock the tension arm with a second screwdriver or similar tool, through the hole in the parking brake control.

4. Disconnect the parking brake rear cables from the front parking brake cable equalizer.

5. Guide the front parking brake cable from the parking brake control.

6. Slightly lift the carpet and remove the front cable grommet out of the floor.

7. Pull the cable with the equalizer through the opening in the floor and remove from the vehicle.

To install:

8. Install the front parking brake cable around the control assembly pulley and insert the cable in the hole in the tensioner arm.

9. Run the front cable equalizer bracket through the hole in the floor and snap the grommet in place.

10. Connect the equalizer to the rear cables by bending and inserting the rear cables into the equalizer and then straightening the cables.

11. Remove the screwdriver or similar tool to allow the parking brake control to set the cable tension. Keep fingers clear during this step.

12. Install the parking brake lever boot.

13. Apply and release the parking brake control lever several times. Make sure that the rear brakes are not dragging when the parking brake control lever is released.

Rear Cables With Drum Brakes

NOTE: The removal and installation procedure given is for the removal and installation of one cable but the process is the same for the left-hand or right-hand sides.

1. Place the parking brake control in the released position.

2. Remove the parking brake lever boot.

3. Using a screwdriver, push down on the tension arm until it is fully depressed and lock the tension arm with a second screwdriver or similar tool, through the hole in the parking brake control.

4. Raise and safely support the vehicle.

5. Remove the rear wheel and tire assembly.

6. Remove the rear brake drum.

7. Disconnect the parking brake rear cable from the front parking brake cable equalizer. Remove the brake shoes if needed, to allow the

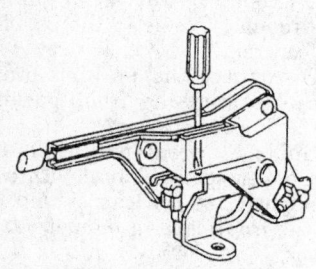

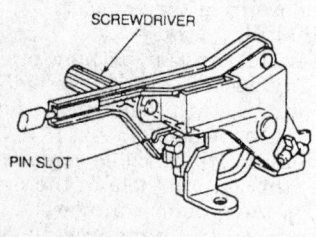

SCREWDRIVER

PIN SLOT

354457

Releasing and locking the tension arm in the parking brake control

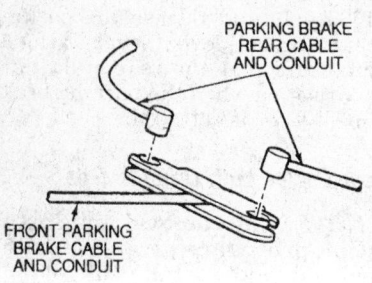

PARKING BRAKE REAR CABLE AND CONDUIT

FRONT PARKING BRAKE CABLE AND CONDUIT

335465

Removing the rear cables from the equalizer

removal of the rear parking brake cable end.

8. Disconnect the parking brake cable from the parking brake lever attached to the trailing shoe by lifting the cable end out of the slot of the parking brake lever.

9. Compress the retaining fingers on the parking brake cable conduit at the backing plate using a ⁹⁄₁₆ flare wrench and pull the cable and con-

duit through the opening in the brake backing plate.

10. Remove the rear parking brake cable and conduit through the rail brackets and from the vehicle.

To install:

11. Run the rear parking brake cable through the opening in the brake backing plate and insert the cable anchor behind the slot in the parking brake lever which is attached to the trailing shoe.

12. Push the rear parking brake cable conduit into the opening in the brake backing plate so that the retaining fingers engage the brake backing plate.

13. Install the brakes if removed for access. Install the brake drum.

14. Install the wheel and tire assembly. Torque the lug nuts to 85–105 ft. lbs. (115–142 Nm).

15. Route the rear parking brake cable and install the front of the rear parking brake cable through the rail brackets until the prongs expand, securing the cables position.

16. Connect the end of the cable to the parking brake cable equalizer.

17. Partially lower the vehicle.

18. Remove the screwdriver or similar holding tool from the parking brake control to set the tension on the parking brake cables. Keep fingers clear of the parking brake control.

19. Apply and release the parking brake control several times.

20. Pump the brake pedal several times to position the brake shoes.

21. Adjust the rear brakes.

22. Rotate the rear wheel to ensure that the brakes are not dragging.

23. Lower the vehicle.

Rear Cables With Disc Brakes

NOTE: The removal and installation process given is for the removal and installation of one cable but the procedure is the same for the left-hand or right-hand sides.

1. Place the parking brake control in the released position.

2. Remove the parking brake lever boot.

3. Using a screwdriver, push down on the tension arm until it is fully depressed and lock the tension arm with a second screwdriver or similar tool, through the hole in the parking brake control.

4. Raise and safely support the vehicle.

5. Disconnect the parking brake rear cable from the front parking brake cable equalizer.

6. Disconnect the parking brake rear cable conduit from the bracket on the disc brake adapter.

7. Disconnect the parking brake rear cable from the parking brake lever at the top of the rear wheel disc brake adapter.

8. Remove the retaining nut and the parking brake cable bracket from the rear axle.

9. Guide the rear parking brake cable and conduit out of the frame and remove from the vehicle.

To install:

10. Guide the rear parking brake cable and conduit through the frame and position properly.

11. Install the parking brake cable bracket to the rear axle and install the retaining nut. Tighten the nut to 14–19 ft. lbs. (20–25 Nm).

12. Connect the parking brake rear cable to the parking brake lever at the top of the rear wheel disc brake adapter.

13. Connect the parking brake rear cable conduit to the bracket on the disc brake adapter.

14. Connect the parking brake rear cable to the front parking brake cable equalizer.

15. Lower the vehicle.

16. Remove the screwdriver or similar holding tool from the parking brake control to set the tension on the parking brake cables. Keep fingers clear of the parking brake control.

17. Apply and release the parking brake control several times.

18. Pump the brake pedal several times to position the parking brake shoes.

19. If needed, adjust the parking brake brakes.

20. Lower the vehicle.

21. Check the parking brake system for proper operation.

22. Pull up on the parking brake control lever while listening to the number of clicks from the parking pawl mechanism required to apply resistance to the parking brake system. If properly adjusted, resistance should be felt after 5 clicks from the parking pawl mechanism.

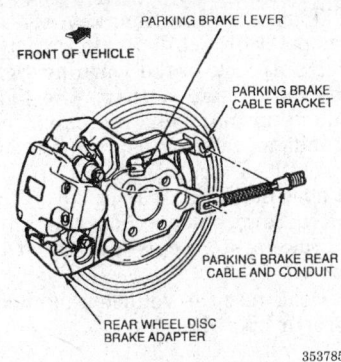

PARKING BRAKE LEVER

FRONT OF VEHICLE

PARKING BRAKE CABLE BRACKET

PARKING BRAKE REAR CABLE AND CONDUIT

REAR WHEEL DISC BRAKE ADAPTER

353785

Rear disc brake parking brake cable removal

Brake System

BLEEDING

Conventional System

NOTE: If only bleeding air trapped after the anti-lock brake system Hydraulic Control Unit (HCU) such as a caliper or wheel cylinder, the conventional brake bleeding system should be sufficient to remove any air in that portion of the brake system. Whenever the brake master cylinder is removed or replaced, follow the brake master cylinder bleeding procedure first and then if necessary, bleed any trapped air in the HCU using the New Generation Star Tester or equivalent, following the correct procedure and then conventionally bleed the system. If the HCU is replaced, bleed any trapped air in the HCU using the New Generation Star Tester or equivalent, following the correct procedure and then conventionally bleed the system.

1. Clean the area around the brake master cylinder fill cap and remove the cap.
2. Fill the reservoir with clean DOT 3 or equivalent brake fluid from a closed container and install the cap.
3. If the master cylinder is known or suspected to have air in the bore, it must be bled before any of the wheel cylinders or calipers. To bleed the brake master cylinder, place a shop towel under the front master cylinder outlet fitting and loosen the fitting approximately ¾ turn. Have an assistant depress the brake pedal slowly through its full travel. Close the outlet fitting and let the pedal return slowly to the fully released position.

Wait 5 seconds, then repeat the operation until all air bubbles disappear.

WARNING

Be careful not to spill brake fluid on painted surfaces, as it can destroy the finish. If any brake fluid is spilled, rinse the area immediately with water.

4. Repeat the brake master cylinder bleeding procedure at the rear master cylinder outlet fitting.
5. Check the brake master cylinder reservoir and fill as necessary. Continue with priming (gravity bleeding) the brake system.
6. Ensure that the brake master cylinder reservoir is full. Open the bleeder screws on both rear wheels and leave open until clear brake fluid flows from both bleeder screws, then close the rear bleeder screws. Do not allow the brake master cylinder reservoir to run dry during the priming procedure.
7. One at a time, open the front bleeder screws and leave open until clear brake fluid flows from the bleeder screws. Close the bleeder screws and check the brake master cylinder reservoir fluid level. Continue with bleeding the brake system.
8. Place a box–end wrench or equivalent, on the brake bleeder at the right-rear wheel.

NOTE: Always bleed the longest line first.

9. Place a drain tube to the brake bleeder screw and the free end into a container half full of clean brake fluid.
10. Loosen the bleeder screw while an assistant applies the brake pedal slowly through its entire travel.
11. Close the brake bleeder and then have the assistant release the brake pedal.
12. Continue bleeding the right-rear wheel until the fluid is free of air bubbles.

NOTE: DO NOT allow the master cylinder to run out of brake fluid. Only use fresh fluid from a closed container.

13. Repeat the procedure on the left-front wheel, then the left-rear and finally the right-front wheel.
14. Throughout the brake bleeding process, continually check the brake master cylinder reservoir and top off the fluid as required.
15. If the brake pedal is spongy, repeat the brake bleeding procedure. If available, bleed the brake system using a pressure bleeder following the

manufacturers recommendations. Do not exceed 50 psi (345 kPa).
16. When the bleeding procedure is complete, top off the brake master cylinder reservoir and install the fill cap.
17. Road test the vehicle and check for proper brake system operation.

Using the NGS Tester

NOTE: This anti-lock brake system bleeding procedure requires the use of Rotunda Tool 007-00500 (New Generation Star Tester) or similar scan tool with the anti-lock brake system bleeding programming, and should be used whenever the Hydraulic Control Unit (HCU) is replaced.

The ABS diagonal braking system requires that the front brakes are bled first allowing for a shorter flow path of trapped air in the system. Once the front brakes have been thoroughly bled, the only air remaining in the system to be removed is in the upper section of the HCU and the rear brake lines.

1. Clean the area around the master cylinder fill cap and remove the cap.
2. Fill the brake master cylinder reservoir with clean DOT 3 or equivalent, brake fluid from a closed container and maintain throughout the procedure.
3. If the brake master cylinder is known or suspected to have air in the bore, it must be bled before any of the wheel cylinders or calipers.

WARNING

Be careful not to spill brake fluid on painted surfaces, as it can destroy the finish. If any brake fluid is spilled, rinse the area immediately with water.

4. Connect a clear waste line to the left-front brake bleeder screw.
5. Have an assistant cycle the brake pedal slowly through its entire travel 25–30 times or until no more air bubbles can be seen in the clear waste line.

NOTE: One cycle of the brake pedal is movement from the full upright position to the full extension of the brake pedal back to the full upright position again.

6. Close the left-front bleeder screw and then have the assistant release the brake pedal.
7. Repeat the bleeding procedure at the right-front bleeder screw followed by the left-rear and finally the right-rear bleeder screw. Make sure

to bleed each until the clear waste line is free of air bubbles.

NOTE: DO NOT allow the master cylinder to run out of brake fluid. Only use fresh fluid from a closed container.

8. Repeat the procedure on the left-front wheel, then the left-rear and finally the right-front wheel. Top off the brake fluid in the brake master cylinder reservoir between each bleeding.

9. Connect the New Generation Star Tester 007-00500, or equivalent.

10. Start the engine and allow it to idle. The engine must be running to supply enough electrical current to the ABS module.

11. Depress the brake pedal to half of its full extension.

12. Begin the NGS tester or similar scan tool brake bleeding program routine following the tool manufacturers directions.

NOTE: The NGS program removes trapped air from the lower sections of the HCU into the upper sections of the HCU which can then be bled at the individual brake bleeders.

13. Preform a minimum of 7 conventional pressure bleed cycles on the left-front and right-front brakes.

14. Repeat the conventional pressure bleed cycles on the left-front and right-front brakes 2 more times and on the last bleed cycle make sure to bleed the rear brakes to ensure that all air is removed. Top off the brake fluid in the master cylinder reservoir between each bleeding.

NOTE: One conventional brake bleed cycle consists of advancing the brake pedal to its extended position, opening the bleeder screw, releasing the brake fluid, closing the bleeder screw and releasing the brake pedal to its full upright position.

15. Turn **OFF** the engine.
16. Top off the brake master cylinder reservoir.
17. Road test the vehicle and check for proper brake system operation.

Wheel Speed Sensor

REMOVAL AND INSTALLATION

Front

1. Disconnect the negative battery cable.

2. Raise and safely support the vehicle.

3. Disconnect the speed sensor (anti-lock brake sensor) electrical harness connector.

4. Remove the speed sensor wiring harness from the brake hose clips.

5. Remove the speed sensor cable clip retaining bolt from the steering knuckle.

6. Remove the speed sensor retaining bolt from the steering knuckle.

7. Slide the speed sensor out of the mounting hole in the steering knuckle and remove from the vehicle.

To install:

8. Slide the speed sensor into the mounting hole in the steering knuckle.

9. Install the speed sensor retaining bolt and tighten to 71–88 inch lbs. (8–10 Nm).

10. Install the speed sensor wiring harness to the brake hose clips.

11. Connect the speed sensor (anti-lock brake sensor) electrical harness connector.

12. Lower the vehicle.

13. Connect the negative battery cable.

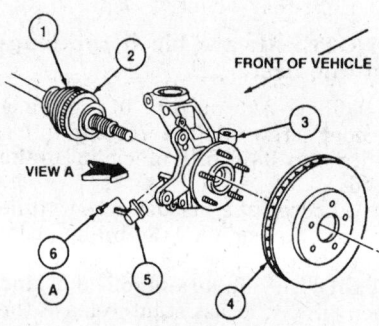

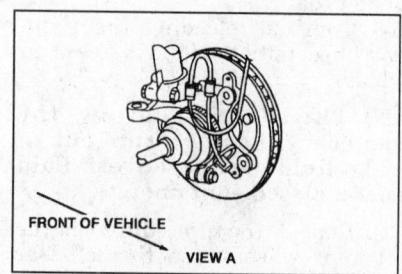

1. Front wheel driveshaft and joint
2. Speed sensor indicator
3. Front wheel knuckle
4. Front disc brake rotor
5. Front brake anti-lock sensor
6. Bolt
A. Tighten to 8-10 Nm (71-88 lb-in)

341823

Front speed sensor removal

14. Turn the ignition key to the **RUN** position while watching the amber anti-lock brake warning light in the instrument cluster. The light should illuminate for approximately 2 seconds and extinguish proving out the anti-lock brake system. If the light does not illuminate or stays illuminated, a problem with the anti-lock brake system exist and must be corrected.

15. Road test the vehicle and check for proper operation.

Rear

1. Disconnect the negative battery cable.

2. Raise and safely support the vehicle.

3. Disconnect the rear speed sensor (anti-lock brake sensor) electrical harness connector.

4. Remove the sensor wiring harness from the brake hose clips and feed the speed sensor wiring harness through the frame.

5. Remove the speed sensor retaining bolt from the disc brake adapter if equipped with rear disc brakes, or the mounting bracket if equipped with rear drum brakes.

6. Remove the rear speed sensor from the vehicle.

To install:

7. Place the speed sensor in position.

8. Install the speed sensor retaining bolt and tighten to 71–88 inch lbs. (8–10 Nm).

9. Feed the speed sensor wiring harness through the frame and install the sensor wiring harness to the brake hose clips.

10. Connect the speed sensor (anti-lock brake sensor) electrical harness connector.

11. Lower the vehicle.

12. Connect the negative battery cable.

13. Turn the ignition key to the **RUN** position while watching the amber anti-lock brake warning light in the instrument cluster. The light should illuminate for approximately 2 seconds and extinguish proving out the anti-lock brake system. If the light does not illuminate or stays illuminated, a problem with the anti-lock brake system exist and must be corrected.

14. Road test the vehicle and check for proper operation.

FRONT SUSPENSION

Strut

REMOVAL AND INSTALLATION

NOTE: Before beginning this procedure, make sure that the following new parts are available: hub retaining nut, steering knuckle-to-ball joint pinch bolt and nut, steering knuckle-to-strut pinch bolt, outer tie rod-to-steering knuckle castellated nut and cotter pin, stabilizer bar link-to-strut retaining nut, per side. Once removed, these parts lose their torque holding ability or retention capability and must not be reused.

1. Disconnect the negative battery cable.
2. Leave the steering column in the unlocked position.
3. Raise and safely support the vehicle.
4. Remove the wheel and tire assembly.
5. Remove the disc brake caliper and the disc brake rotor.
6. Remove the outer tie-rod end cotter pin and castellated nut. Discard the cotter pin and nut.
7. Separate the outer tie-rod end from the steering knuckle using Tie Rod End Remover T81P-3504-W, or equivalent.
8. Remove and discard the nut retaining the stabilizer bar link to the strut. Use care not to damage the stabilizer bar link ball stud seal by holding the stud with a 6mm Allen wrench while turning the retaining nut.
9. Remove the steering knuckle-to-ball joint pinch bolt and nut and discard.

10. Spread the pinch joint on the steering knuckle and separate the ball joint from the steering knuckle using a suitable prybar or equivalent.
11. Remove the hub retaining nut and washer and discard.
12. Separate the outer CV-joint and halfshaft from the hub using Front Hub Remover/Replacer T81P-1104-C or equivalent and the related adapters.

NOTE: To prevent internal damage, do not allow the outer CV-joint to over-extend.

13. Once removed, support the end of the halfshaft with a length of safety wire.
14. Remove and discard the steering knuckle to strut pinch bolt. Spread the pinch joint slightly if required for removal.
15. Separate the strut from the steering knuckle by using a large prybar or similar tool.
16. Lower the vehicle enough to gain access to the strut upper mounting bracket retaining nuts.
17. Remove 3 strut upper mounting bracket retaining nuts and remove the strut and coil spring assembly from the vehicle.

— **CAUTION** —

Do not attempt to remove the coil spring from the strut without first compressing the coil spring with the appropriate tool.

18. Install Spring Compressor 086–00036 or equivalent, to the coil spring and compress the spring until the spring tension is relieved from the spring seat.
19. Place a 10mm box wrench on the top of the strut shaft while removing the strut retaining nut with a 21mm crow foot wrench and ratchet, or equivalent.
20. Remove the strut retaining nut, washer, mounting bracket, bearing

and seal, washers, spring insulator and the coil spring.
21. Replace parts as necessary. If coil spring is being replaced, relax the tension on the coil spring and remove the compressor.

To install:
22. Compress the coil spring if removed from the spring compressor.
23. Install the spring insulator, coil spring, lower washer, bearing and seal and the upper mounting bracket to the strut body.

NOTE: Both upper and lower strut washers should have the cupped side toward the strut tower.

24. Install the upper washer and strut retaining nut onto the strut, hand tight.
25. Place a 10mm box wrench on the top of the strut shaft while tightening the strut retaining nut with a 21mm crow foot wrench and ratchet, or equivalent. Tighten the nut to 40–46 ft. lbs. (55–63 Nm).
26. Relax the tension on the coil spring and remove the coil spring compressor.
27. With the vehicle partially raised and safely supported, position the strut/coil spring assembly up into the strut tower and install but do not tighten 3 strut upper mounting bolts.
28. Place the lower portion of the strut tube into the steering knuckle.
29. Raise and safely support the vehicle.
30. Install a new steering knuckle-to-strut pinch bolt. Tighten the bolt to 85–97 ft. lbs. (115–132 Nm).
31. Lubricate the splines of the outer CV-joint with clean engine oil and insert the stub shaft of the outer CV-joint into the hub splines as far as possible using hand pressure only making sure that the splines are properly aligned.
32. Install the hub retaining washer and a new wheel hub retaining nut.
33. Install the ball joint to the steering knuckle making sure that the stud groove is properly aligned.
34. Install a new steering knuckle-to-ball joint pinch bolt and nut. Tighten the nut to 46–52 ft. lbs. (62–71 Nm).
35. Install the stabilizer bar link to the strut and install a new retaining nut. Tighten the retaining nut to 66–77 ft. lbs. (90–104 Nm). Use care not to damage the stabilizer bar link ball stud seal by holding the stud with a 6mm Allen wrench while turning the retaining nut.
36. Install the outer tie-rod end to the steering knuckle and install a

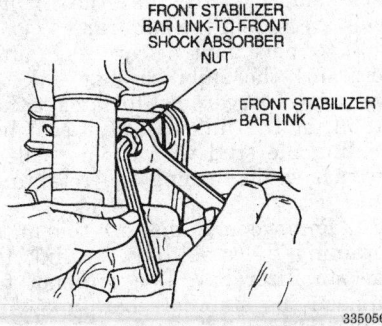

FRONT STABILIZER
BAR LINK-TO-FRONT
SHOCK ABSORBER
NUT

FRONT STABILIZER
BAR LINK

335056

Proper method of removing the stabilizer bar link nut

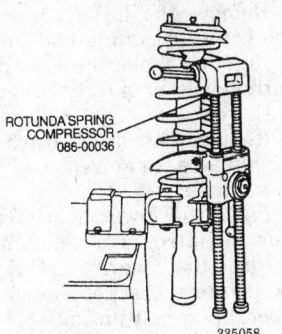

ROTUNDA SPRING
COMPRESSOR
086-00036

335058

Strut and coil spring assembly in spring compressor

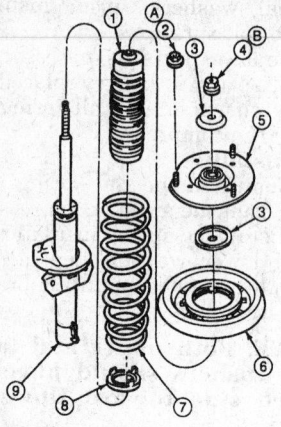

1 Dust Boot
2 Front Shock Absorber Mounting Bracket Nut (3 Req'd)
3 Washer
4 Front Shock Absorber Nut
5 Front Shock Absorber Mounting Bracket
6 Front Suspension Bearing and Seal
7 Front Coil Spring
8 Front Spring Insulator
9 Front Shock Absorber
A Tighten to 35-40 N·m (25-30 Ft-Lb)
B Tighten to 55-63 N·m (40-46 Ft-Lb)

335059

Strut and coil spring components

new castellated nut. Tighten the nut to 46–52 ft. lbs. (62–71 Nm). Install a new cotter pin.

37. Install the disc brake rotor and disc brake caliper.

38. Install the wheel and tire assembly. Torque the lug nuts to 85–105 ft. lbs. (115–142 Nm).

39. Lower the vehicle enough to gain access to the strut upper mounting bracket retaining nuts.

40. Tighten 3 strut upper mounting bracket nuts to 25–30 ft. lbs. (35–40 Nm).

41. Lower the vehicle.

42. Tighten the wheel hub retaining nut to 170–202 ft. lbs. (230–275 Nm) using hand tools only.

NOTE: Do not use power or impact tools for tightening the hub retaining nut.

43. Connect the negative battery cable.

44. Apply the brake pedal several times to position the disc brake pads before moving the vehicle.

45. Check the alignment and adjust if necessary.

46. Road test the vehicle and check for proper operation.

Lower Ball Joints

REMOVAL AND INSTALLATION

The lower ball joint is an integral component of the lower control arm. If the ball joint is defective, the lower control arm must be replaced.

Lower Control Arms

REMOVAL AND INSTALLATION

NOTE: Before beginning this procedure, make sure to have available a new lower control arm-to-sub frame retaining bolt and nut, lower arm strut-to-lower control arm retaining nut and a steering knuckle-to-ball joint pinch bolt and nut, per side. Once removed, these parts lose their torque holding ability or retention capability and must not be reused.

1. Disconnect the negative battery cable.

2. Leave the steering column in the unlocked position.

3. Raise and safely support the vehicle.

4. Remove the wheel and tire assembly.

5. Remove and discard the lower arm strut retaining nut from the lower control arm. Remove the dished washer.

NOTE: Do not allow the half-shaft to move outward or the tripod CV-joint internal parts could separate, causing failure of the joint.

6. Remove the steering knuckle-to-ball joint pinch bolt and nut and discard.

7. Spread the pinch joint on the steering knuckle and separate the ball joint from the steering knuckle using a suitable prybar or equivalent.

8. Remove the lower control arm-to-sub frame retaining bolt and nut.

9. Remove the lower control arm from the lower arm strut and remove from the vehicle.

10. Replace the lower control arm or bushings, as required.

To install:

11. Place the lower arm strut into the lower control arm inner bushing.

12. Place the lower control arm into the sub frame bracket and install a new retaining bolt and nut. Hold the lower control arm in a horizontal position and tighten the lower control arm-to-sub frame nut to 85–97 ft. lbs. (115–132 Nm).

13. Assemble the lower control arm ball joint stud into the steering knuckle. Make sure that the ball joint stud groove is properly positioned.

14. Install a new steering knuckle-to-ball joint pinch bolt and nut. Tighten the nut to 46–52 ft. lbs. (62–71 Nm).

15. Clean the lower arm strut threads. Make sure that the inboard washer is installed.

16. Install the lower arm strut into the lower control arm and install the dished washer and a new strut-to-lower control arm retaining nut. Make sure that the dished side of the washer is facing away from the lower control arm strut bushing.

17. Tighten the strut-to-lower control arm retaining nut to 85–97 ft. lbs. (115–132 Nm).

18. Install the wheel and tire assembly. Torque the lug nuts to 85–105 ft. lbs. (115–142 Nm).

19. Lower the vehicle.

20. Connect the negative battery cable.

21. Check the alignment and adjust if necessary.

22. Road test the vehicle and check for proper operation.

Stabilizer Bar

REMOVAL AND INSTALLATION

NOTE: Before starting this procedure, make sure to have available, 4 new stabilizer bar (sway bar) link retaining nuts, 2 new stabilizer bar insulators and 2 new sub-frame to body retaining bolts. Once removed, these parts lose their torque holding ability or retention capability and must not be reused.

1. Disconnect the negative battery cable.

2. Raise and safely support the vehicle making sure that the lifting pads are not on the sub-frame.

3. Remove and discard the nuts retaining the stabilizer bar link to the strut. Use care not to damage the stabilizer bar link ball stud seal by holding the stud with a 6mm Allen wrench while turning the retaining nut.

4. Remove and discard the nuts retaining the stabilizer bar link to the stabilizer bar. Use care not to damage the stabilizer bar link ball stud seal by holding the stud with a 6mm Allen wrench while turning the retaining nut.

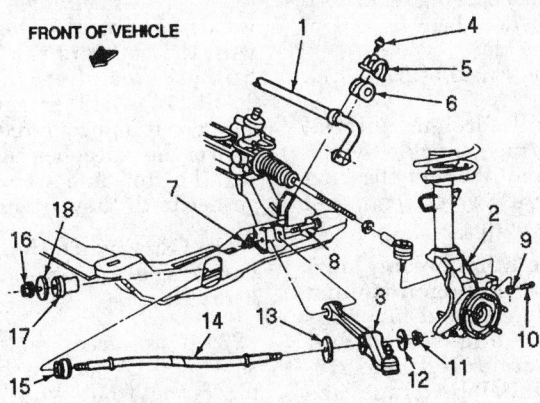

FRONT OF VEHICLE

1. Front stabilizer bar
2. Front wheel hub and spindle
3. Front suspension lower arm
4. Stabilizer bar bracket bolt
5. Stabilizer bar bracket
6. Stabilizer bar bracket insulator
7. Front suspension lower arm-to-front sub-frame bolt nut
8. Front suspension lower arm-to-front sub-frame bolt
9. Tie rod end-to-front wheel hub and spindle nut
10. Cotter pin
11. Front suspension lower arm strut-to-front suspension lower arm nut
12. Washer
13. Washer
14. Front suspension lower arm strut
15. Front suspension strut insulator
16. Front suspension lower arm strut-to-front sub-frame nut
17. Front suspension lower arm strut inner bushing
18. Washer

335490

Lower control arm, strut and related components

5. Position a set of screwjacks or jackstands under the sub-frame and remove the rear sub-frame to frame bolts. Lower the sub-frame at the rear to gain access to the stabilizer bar brackets.

6. Remove 4 stabilizer bar U-bracket bolts and remove the stabilizer bar from the vehicle.

NOTE: When removing the stabilizer bar, replace the insulators and the U-bracket bolts with new ones.

To install:

7. Clean the stabilizer bar and the stabilizer bar mounting area.

8. Lubricate the inside diameter of the new stabilizer bar insulator with a rubber lubricant. Do not use petroleum or mineral based lubricants, as they will deteriorate the rubber.

9. Install the stabilizer bar insulators onto the stabilizer bar and position the insulators in their approximate locations.

10. Place the stabilizer bar in the vehicle and install the insulators, U-brackets and new retaining bolts. Tighten 4 retaining bolts to 40–46 ft. lbs. (55–63 Nm).

11. Raise the sub-frame using the screwjacks or jackstands and install 2 new sub-frame to body retaining bolts. Tighten the bolts to 85–99 ft. lbs. (115–135 Nm).

12. Place the stabilizer bar links to the struts and install new retaining nuts. Tighten the retaining nuts to 66–77 ft. lbs. (90–104 Nm). Use care not to damage the stabilizer bar link ball stud seal by holding the stud with a 6mm Allen wrench while turning the retaining nut.

NOTE: Check the stabilizer bar links for markings such as "TOP LH" or "TOP RH" indicating location.

13. Place the stabilizer bar links to the stabilizer bar and install new retaining nuts. Tighten the retaining nuts to 66–77 ft. lbs. (90–104 Nm). Use care not to damage the stabilizer bar link ball stud seal by holding the stud with a 6mm Allen wrench while turning the retaining nut.

14. Remove the screwjacks or jackstands.

15. Lower the vehicle.

16. Connect the negative battery cable.

NOTE: Whenever the vehicles sub-frame is removed or lowered, the wheel alignment should be checked.

17. Check the alignment and adjust if necessary.

18. Road test the vehicle and check for proper operation.

Front Wheel Bearings

REMOVAL AND INSTALLATION

NOTE: Before beginning this procedure, make sure to have available, 1 new wheel hub retaining nut, 1 steering knuckle-to-ball joint pinch bolt and nut, 1 steering knuckle-to-strut pinch bolt, 1 outer tie rod-to-steering knuckle castellated nut and cotter pin, 1 stabilizer bar link-to-strut retaining nut, per side. Once removed, these parts lose their torque holding ability or retention capability and must not be reused.

1. Disconnect the negative battery cable.

2. Raise and safely support the vehicle.

3. Remove the wheel and tire assembly.

4. Remove the disc brake caliper and rotor.

5. Support the disc brake caliper with wire, or equivalent. Do not let the caliper hang by the brake hose.

6. Remove 3 disc brake rotor shield rivets and remove the disc brake rotor dust shield.

7. Remove the anti–lock brake speed sensor retaining bolt and place the anti-lock brake sensor out of the way.

8. Remove the outer tie-rod end cotter pin and castellated nut. Discard the cotter pin and nut.

9. Separate the outer tie-rod end from the steering knuckle using Tie Rod End Remover T81P-3504-W, or equivalent.

10. Remove the hub retaining nut and washer and discard.

11. Remove the steering knuckle-to-ball joint pinch bolt and discard the nut and bolt.

12. Spread the pinch joint on the steering knuckle and separate the ball joint from the steering knuckle using a suitable prybar or equivalent.

13. Remove and discard the nut retaining the stabilizer bar link to the strut. Use care not to damage the stabilizer bar link ball stud seal by holding the stud with a 6mm Allen wrench while turning the retaining nut.

14. Loosen but do not remove 3 strut upper mounting bracket nuts.

15. Separate the outer CV-joint and halfshaft from the hub using Front Hub Remover/Replacer T81P-1104-C

or equivalent and the related adapters.

NOTE: To prevent internal damage, do not allow the outer CV-joint to overextend.

16. Once removed, support the end of the halfshaft with a length of safety wire.

17. Support the steering knuckle with a length of safety wire and remove the steering knuckle to strut pinch bolt.

18. Work the steering knuckle off of the strut. Carefully remove the support wire and remove the steering knuckle from the vehicle.

19. Place the steering knuckle assembly onto a suitable workbench.

20. Install a 2-jaw puller and a suitable shaft protector and separate the hub from the steering knuckle.

21. Remove and discard the snapring securing the wheel bearing in the steering knuckle.

22. Using a suitable hydraulic press, place bearing spacer T86P-1104-A2 or equivalent, on the press plate with the step side facing up and position the steering knuckle with the outboard side up on the spacer. Install bearing remover T83P-1104-AH2 or equivalent, cen-

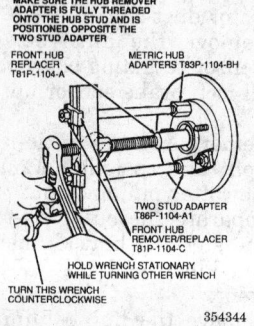

MAKE SURE THE HUB REMOVER ADAPTER IS FULLY THREADED ONTO THE HUB STUD AND IS POSITIONED OPPOSITE THE TWO STUD ADAPTER

FRONT HUB REPLACER T81P-1104-A

METRIC HUB ADAPTERS T83P-1104-BH

TWO STUD ADAPTER T86P-1104-A1

FRONT HUB REMOVER/REPLACER T81P-1104-C

HOLD WRENCH STATIONARY WHILE TURNING OTHER WRENCH

TURN THIS WRENCH COUNTERCLOCKWISE

354344

Separating the halfshaft from the hub (shown with rotor installed)

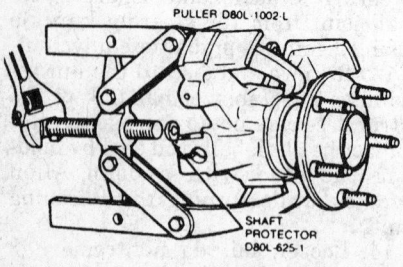

PULLER D80L-1002-L

SHAFT PROTECTOR D80L-625-1

354345

Removing the hub from the steering knuckle

tered on the wheel bearing inner race and press the wheel bearing out of the steering knuckle.

23. Discard the wheel bearing.

To install:

24. Remove all foreign material from the steering knuckle wheel bearing bore and the hub bearing journal to ensure correct seating of the new wheel bearing.

NOTE: If the hub bearing journal is scored or damaged it must be replaced. The wheel bearings are pregreased and sealed and require no scheduled maintenance. The wheel bearings are preset and cannot be adjusted. If a wheel bearing is disassembled for any reason, it must be replaced as a unit, as individual service seals, rollers and races are not available.

25. Place Bearing Spacer T86P-1104-A2 or equivalent, with the step side down on the hydraulic press plate and position the steering knuckle with the outboard side down on the spacer. Position a new wheel bearing in the inboard side of the steering knuckle. Install Bearing Installer T86P-1104-A3 or equivalent, with the undercut side facing the wheel bearing, on the wheel bearing outer race and press the wheel bearing into the steering knuckle. Make sure the wheel bearing seats completely against the shoulder of the steering knuckle bore.

NOTE: Wheel Bearing Installer T86P-1104-A3 or equivalent, must be positioned as indicated above to prevent wheel bearing damage during installation.

26. Install a new snapring (part of the bearing kit) in the steering knuckle groove.

27. Place Bearing Spacer T86P-1104-A2 or equivalent. on the arbor press plate and position the hub on the on the bearing spacer

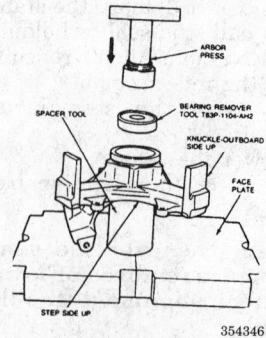

ARBOR PRESS

SPACER TOOL

BEARING REMOVER TOOL T83P-1104-AH2

KNUCKLE-OUTBOARD SIDE UP

FACE PLATE

STEP SIDE UP

354346

Pressing the wheel bearing from the steering knuckle

with the wheel lugs facing downward. Position the steering knuckle with the outboard side down, on the hub. Place Bearing Remover T83P-1104-AH2 or equivalent, flat side down and centered on the inner race of the wheel bearing while pressing the hub and wheel bearing together until they are fully seated.

NOTE: Ensure that the hub rotates freely on the steering knuckle.

28. If required, remove the halfshaft and CV-joint assembly and replace the front wheel dust shield prior to installing the steering knuckle into the vehicle. Replace the front wheel dust shield on the outer CV-joint with a new dust shield from the wheel bearing kit. Make sure the dust shield flange faces outward toward the wheel bearing. Use Drive Tube T83T-3132-A1 and Front Bearing Dust Seal Installer T86P-1104-A4, or equivalents.

29. Install the halfshaft, if removed.

30. Place the steering knuckle in position and secure with safety wire to hold the assembly during the installation procedure.

31. Place the steering knuckle to the lower strut tube.

32. Lubricate the splines of the outer CV-joint with clean engine oil and insert the stub shaft of the outer CV-joint into the hub splines as far as possible using hand pressure only making sure that the splines are properly aligned.

33. Temporarily install the disc brake rotor and secure with washers and 2 lug nuts. Insert a steel rod or similar tool into the cooling fins and rotate the rotor until the steel rod contacts the steering knuckle to prevent it from turning.

34. Install the hub retaining washer and a new hub retaining nut. Tighten the hub retaining nut to 170–202 ft. lbs. (230–275 Nm) using hand tools only.

NOTE: Do not use power or impact tools for tightening the hub retaining nut.

35. Place the stabilizer bar link to the strut and install a new retaining nut. Tighten the retaining nut to 66–77 ft. lbs. (90–104 Nm). Use care not to damage the stabilizer bar link ball stud seal by holding the stud with a 6mm Allen wrench while turning the retaining nut.

36. Install a new steering knuckle-to-strut pinch bolt. Tighten the pinch bolt to 85–97 ft. lbs. (115–132 Nm).

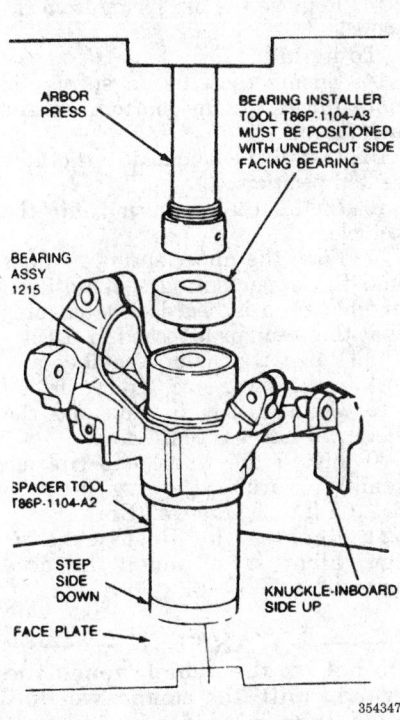

Pressing the wheel bearing into the steering knuckle

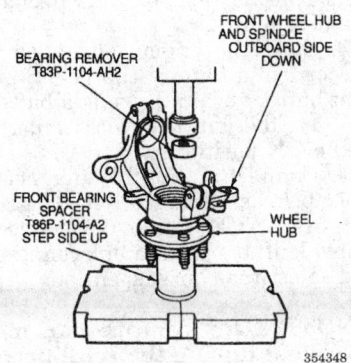

Pressing the hub into the steering knuckle

37. Install a new steering knuckle-to-ball joint pinch bolt and nut. Tighten the nut to 46–52 ft. lbs. (62–71 Nm).

38. Place the outer tie-rod end to the steering knuckle and install a new castellated nut. Tighten the nut to 40–46 ft. lbs. (55–63 Nm). Install a new cotter pin.

39. Install the anti-lock brake speed sensor and retaining bolt. Tighten the retaining bolt to 40–60 inch lbs. (5–7 Nm).

40. Install the disc brake rotor dust shield to the steering knuckle using 3 new rivets.

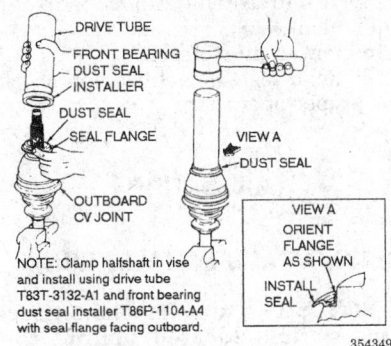

Installing the front wheel dust shield

41. Tighten 3 strut upper mounting bracket nuts to 22–30 ft. lbs. (35–40 Nm).

42. Apply a suitable grease to the pilot diameter of the disc brake rotor and install the disc brake rotor to the hub.

43. Install the disc brake caliper assembly.

44. Install the wheel and tire assembly. Torque the lug nuts to 85–105 ft. lbs. (115–142 Nm).

NOTE: Do not use an impact gun to tighten the lug nuts or damage to the wheel bearing may result.

45. Lower the vehicle.

46. Connect the negative battery cable.

47. Pump the brake pedal to position the brake pads before attempting to move the vehicle.

48. Road test the vehicle and check the steering, brakes and suspension for proper operation.

REAR SUSPENSION

Shock Absorber

REMOVAL AND INSTALLATION

—————— WARNING ——————
The lower control arm or axle assembly must be supported before removal of upper or lower shock absorber retaining bolts to prevent damage to related components.

NOTE: If equipped with air suspension, the air suspension service switch, located behind the jack storage area, must be deactivated before the vehicle can be raised.

1. Disconnect the negative battery cable.

2. Raise and safely support the vehicle.

3. Remove the wheel and tire assembly.

4. Position a jack stand under the axle assembly. Raise the axle slightly to set the suspension in its normal position.

5. Remove both shock absorber mounting bolts and nuts.

6. Lower the axle slightly to help aid in the removal of the shock absorber bolts and nuts.

7. Remove the shock absorber from the vehicle.

To install:

8. Place the shock absorber to the upper mounting bracket and install the retaining bolt and nut.

9. Slightly raise the axle assembly using a jackstand until the shock absorber lower bolt hole aligns with the bolt hole in the axle mounting bracket.

10. Install the lower shock absorber retaining bolt and nut.

11. Tighten both shock absorber retaining bolts to 50–68 ft. lbs. (68–92 Nm.).

12. Install the wheel and tire assembly. Torque the lug nuts to 85–105 ft. lbs. (115–142 Nm).

13. Remove the jackstand from under the axle assembly.

14. If equipped with air suspension, partially lower the vehicle but do not put any load on the wheel and tire assemblies. Activate the air suspension switch and allow the air springs to fill for approximately 90 seconds.

15. Lower the vehicle.

16. Connect the negative battery cable.

17. Road test the vehicle and check for proper operation.

Coil Spring

REMOVAL AND INSTALLATION

NOTE: If a twin post hoist is to be used, the vehicle must be supported on jackstands placed under the pads of the underbody forward of the axle trailing arm bracket.

1. Raise and safely support the vehicle.

2. Place a floor jack under the lower axle assembly and raise the suspension to normal curb height.

3. Remove the wheel and tire assembly.

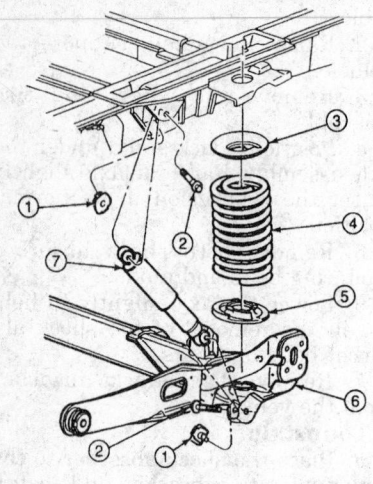

1 J-Nut
2 Bolt M 12-1.75 x 66 Hex
  Flanged Head
3 Rear Spring Insulator
  (Upper)
4 Rear Spring
5 Rear Spring Insulator
  (Lower)
6 Axle Assembly
7 Shock Absorber

335340

Rear shock absorber installation

4. Remove the retaining bolt securing the shock absorber to the axle assembly.

5. Slowly lower the axle assembly with the floor jack until the tension on the coil spring has relaxed.

6. Remove the coil spring from the vehicle.

To install:

7. If removed, position the lower spring insulator on the axle assembly and press the insulator downward into place. Ensure that the insulator is properly seated.

8. Place the upper spring insulator on top of the coil spring.

9. Install the coil spring on the axle assembly. Ensure that the spring is properly seated.

10. With the floor jack or equivalent, slowly raise the axle assembly. Guide the upper spring insulator onto the upper spring underbody seat.

11. Place the shock absorber to the bracket on the axle assembly and install a new nut and bolt. Tighten to 15–19 ft. lbs. (19–26 Nm).

12. Using the floor jack, raise the suspension to normal curb height.

13. Install the wheel and tire assembly. Torque the lug nuts to 85–105 ft. lbs. (115–142 Nm).

14. Remove the floor jack or equivalent.

15. Lower the vehicle.

16. Road test the vehicle and check for proper operation.

Air Spring

REMOVAL AND INSTALLATION

NOTE: The air suspension service switch, located behind the jack storage area, must be deactivated before the vehicle can be raised.

1. Turn the air suspension service switch to the **OFF** position.

2. Raise and safely support the vehicle.

3. Remove the wheel and tire assembly.

4. Raise the axle assembly to normal ride height using floor jacks or equivalent.

5. Disconnect the air spring solenoid valve electrical harness connector.

6. Deflate the air spring and disconnect the air line by pushing the air line into the valve and holding the plastic ring on the solenoid, then while holding the plastic ring pull the air line out of the solenoid.

7. Remove the metal solenoid retaining clip.

8. Turn the solenoid counterclockwise to the stop then pull outward to release the air pressure.

9. Once the air pressure has been released rotate the solenoid counterclockwise the the next stop and remove the solenoid.

10. Disconnect the air spring at the lower spring seat by depressing the tabs under the lower spring seat.

11. Lower the axle assembly slowly to gain access to the air spring.

12. Disconnect the retaining tabs at the top of the air spring.

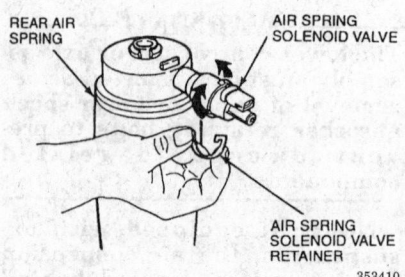

REAR AIR
SPRING

AIR SPRING
SOLENOID VALVE

AIR SPRING
SOLENOID VALVE
RETAINER

353410

Removing the solenoid retainer clip

13. Remove the air spring from the vehicle.

To install:

14. Ensure that the air spring did not unroll at the bottom of the membrane.

15. Install the solenoid to the fully seated position.

16. Install the air spring into the vehicle.

17. Push the upper spring retainer into the upper spring seat until an audible snap is heard, This ensures that the retainer is locked in place.

18. Raise the axle assembly and lock in the lower spring retainer.

19. Connect the air line and the electrical harness connectors.

20. Install the wheel and tire assembly. Torque the lug nuts to 85–105 ft. lbs. (115–142 Nm).

21. Remove the floor jacks or equivalent from under the axle assembly.

——— WARNING ———
Do not let the vehicle touch the ground until the compressor has been turned on for at least 90 seconds.

22. Partially lower the vehicle but do not put any load on the wheel and tire assemblies.

23. Start the engine and allow to idle or use a battery charger to prevent battery drain. If using a battery charger, the ignition switch must be in the **ON** position.

24. Connect a suitable diagnostic tool such as Rotunda New Generation Star (NGS) Tester 007-00500 or equivalent, to the data link connector and select rear air suspension.

NOTE: The compressor may over heat during the refill procedure, If this happens a circuit breaker will turn the compressor OFF for approximately 15 minutes before resetting.

25. Turn **ON** the air spring solenoid valves and compressor until the air springs are full.

26. Turn OFF the air spring solenoid valves and compressor, exit the diagnostic mode and remove the diagnostic tool.

27. Activate the air suspension service switch if not already done.

28. Lower the vehicle.

29. Close all doors on the vehicle. Open the drivers door momentarily to activate the ride height adjustment mode.

30. Road test the vehicle and check for proper operation.

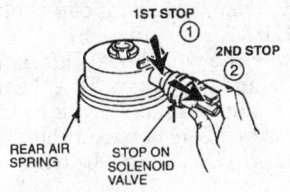

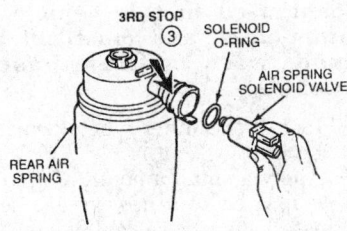

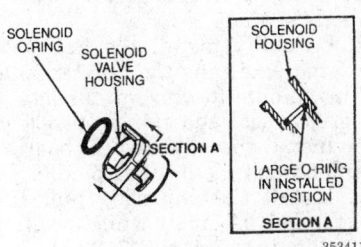

353411

Removing the air spring solenoid

Transverse Arms

REMOVAL AND INSTALLATION

NOTE: If equipped with air suspension, the air suspension service switch, located behind the jack storage area, must be deactivated before the vehicle can be raised.

1. Disconnect the negative battery cable.
2. Raise and safely support the vehicle.
3. Remove the retaining bolt securing the transverse arm (track bar) to the axle assembly.
4. Remove the retaining bolt securing the transverse arm to the transverse arm mounting bracket at the frame.
5. Remove the transverse arm from the vehicle.
   To install:
6. Place the transverse arm to the transverse arm mounting bracket at the frame and install the retaining bolt. Do not tighten the bolt at this time.
7. Place the opposite end of the transverse arm to the axle assembly and install the retaining bolt.

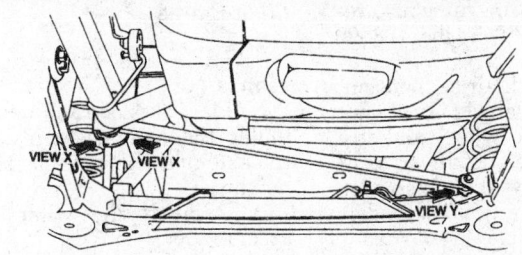

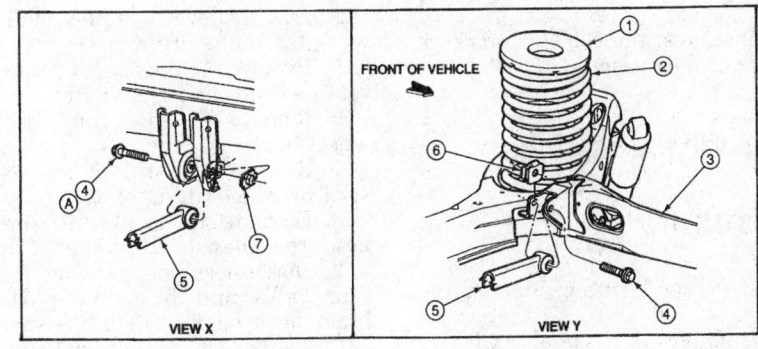

1	Spring Insulator (Upper)	5	Track Bar
2	Rear Spring	6	Nut
3	Axle Assembly	7	Clip
4	Bolt	A	Tighten to 68-92 N·m (50-68 Ft-Lb)

354235

Transverse arm (track bar) with coil spring suspension

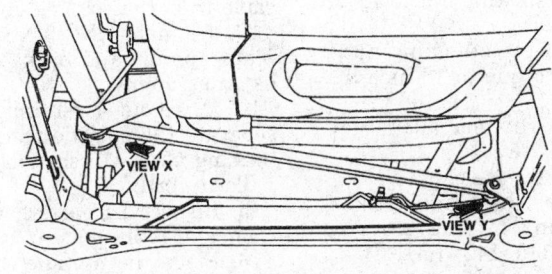

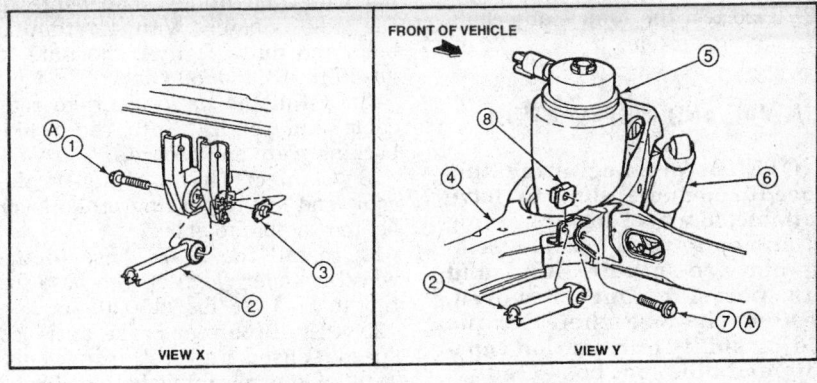

1	Track bar bolt	6	Shock absorber
2	Track bar	7	Track bar bolt
3	Clip nut	8	Clip nut
4	Rear axle assembly	A	Tighten to 68-92 Nm (50-68 lb-ft)
5	Rear air spring		

354236

Transverse arm (track bar) with air suspension

8. Tighten both transverse arm retaining bolts to 50–68 ft. lbs. (68–92 Nm).

9. If equipped with air suspension, partially lower the vehicle but do not put any load on the wheel and tire assemblies. Activate the air suspension switch and allow the air springs to fill for approximately 90 seconds.

10. Lower the vehicle.

11. Connect the negative battery cable.

12. Road test the vehicle and check for proper operation.

Wheel Bearings

ADJUSTMENT

1. Raise and safely support the vehicle.

2. Remove the wheel and tire assembly.

3. Remove the grease cap from the bearing and hub assembly.

4. Remove the cotter pin and retainer. Discard the cotter pin.

5. Back the adjusting nut off several turns.

6. Rotate the hub assembly while tightening the adjusting nut to 18–23 ft. lbs. (24–31 Nm).

7. Back off of the adjusting nut 2 or 3 turns and torque to 18 inch lbs. (2 Nm).

8. Install the retainer and a new cotter pin.

9. Install the grease cap.

10. Install the wheel and tire assembly. tighten the lug nuts to 85–105 ft. lbs. (115–142 Nm).

11. Lower the vehicle.

12. Road test the vehicle and check for proper operation.

REMOVAL AND INSTALLATION

NOTE: Before beginning this procedure, make sure to have available, 1 wheel hub retaining nut and cotter pin, 4 brake backing plate-to-spindle bolts and nuts, per side. Once removed, these parts lose their torque holding ability or retention capability and must not be reused.

NOTE: If equipped with air suspension, the air suspension service switch, located behind the jack storage area, must be deactivated before the vehicle can be raised and reactivated before the vehicle is completely lowered.

Drum Brakes

1. Raise and safely support the vehicle.

2. Place a jackstand or screw jack under the axle assembly and raise the axle to its normal position at curb height.

3. Remove the wheel and tire assembly.

4. Remove the pushnuts retaining the brake drum, if equipped and remove the brake drum.

5. Remove the dust cap, cotter pin, retainer and the adjusting nut.

6. Remove the wheel hub with the wheel bearings.

7. Remove the anti-lock brake sensor from the backing plate.

8. Disconnect and plug the brake hose from the wheel cylinder.

9. Disconnect the parking brake rear cable end from the parking brake lever behind the brake shoe.

10. Using a 9/16 inch box end wrench or equivalent, compress the parking brake rear cable prongs at the brake backing plate and pull the parking brake cable and conduit out of the backing plate.

11. Remove 4 retaining nuts securing the spindle to the axle assembly.

12. Remove the emergency brake cable from the routing clip.

13. Remove the brake backing plate, spindle and brake components as an assembly.

14. Drive out 4 retaining bolts and separate the spindle from the brake backing plate, if necessary.

To install:

15. Inspect all parts for damage or wear and replace as necessary.

16. Place the spindle, brake backing plate and brake components to the axle assembly with 4 retaining bolts and nuts. Tighten the nuts to 48–59 ft. lbs. (60–80 Nm).

17. Install the parking brake rear cable and conduit into the brake backing plate and secure.

18. Connect the parking brake cable end to the parking brake lever behind the brake shoe.

19. Install the brake hose to the wheel cylinder. Tighten the hose fitting to 11–14 ft. lbs. (15–20 Nm).

20. Install the rear brake anti-lock speed sensor and retaining bolt. Tighten the retaining bolt to 68–92 inch lbs. (8–10 Nm).

NOTE: If a new hub is being used, remove the protective coating with carburetor degreaser before assembly.

21. If the inner or outer bearing cups (races) were removed, install new ones using bearing cup replacer T73T-1202-A or equivalent.

22. Pack the wheel bearings using a wheel bearing packer. If a bearing packer is not available, work as much grease as possible between the rollers and cages. Grease the bearing cone surface.

NOTE: The wheel bearing lubricant used in this vehicle is lithium-based and must not be mixed with sodium-based lubricants.

23. Install the inner wheel bearing into the hub.

24. Apply a small amount of grease to the lip area of a new grease seal and install the grease seal using seal replacer T83T-1175-B or equivalent. Make sure that the grease seal is properly seated.

25. Carefully install the hub onto the spindle keeping the hub centered on the spindle to prevent damage to the grease seal and spindle threads.

26. Install the outer wheel bearing and flat washer onto the spindle.

27. Install the adjusting nut. Rotate the hub assembly while torquing the adjusting nut to 17–23 ft. lbs. (23–31 Nm). Back off of the adjusting nut 2 or 3 turns and retorque to 18 inch lbs. (2 Nm).

28. Install the retainer and a new cotter pin.

29. Install the grease cap and brake drum.

30. Bleed the brake system.

31. Install the wheel and tighten the lug nuts to 85–105 ft. lbs. (115–142 Nm).

32. If equipped with air suspension, partially lower the vehicle but do not put any load on the wheel and tire assemblies. Activate the air suspension switch and allow the air springs to fill for approximately 90 seconds.

33. Lower the vehicle.

34. Before moving the vehicle, pump the brake pedal several times to position the brake shoes.

35. Road test the vehicle and check for proper operation.

Disc Brakes

1. Raise and safely support the vehicle.

2. Place a jackstand or screw jack under the axle assembly and raise the axle to its normal position at curb height.

3. Remove the wheel and tire assembly.

4. Remove the disc brake caliper and rotor.

5. Remove the dust cap, cotter pin, retainer and the adjusting nut.

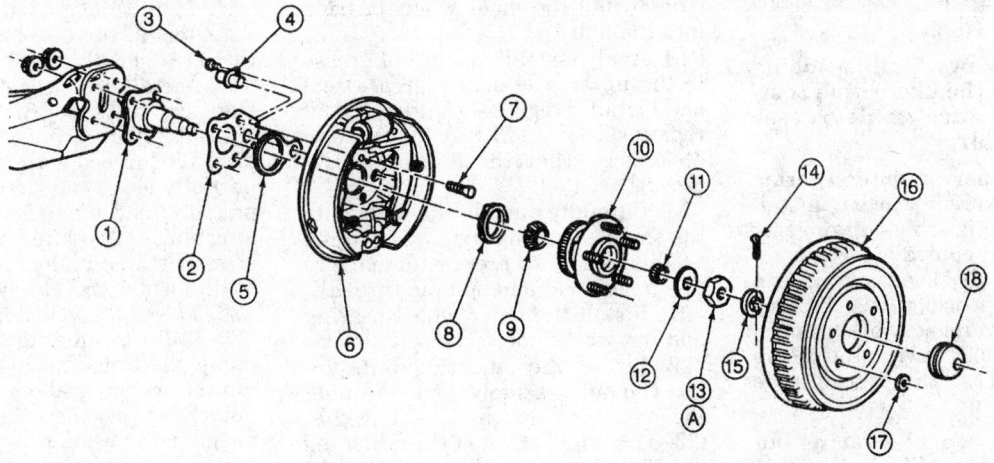

1	Rear Wheel Spindle	11	Bearing Outer and Roller
2	Anti-Lock Brake Control Module Bracket	12	Washer
3	Bolt	13	Nut
4	Rear Brake Anti-Lock Sensor	14	Pin
5	Rear Wheel Gasket	15	Retainer
6	Rear Brake Backing Plate	16	Brake Drum
7	Bolt	17	Nut
8	Oil Seal	18	Hub Grease Cap
9	Bearing Cone and Roller	A	Tighten to 23-38 N·m (17-28 Ft-Lb) to Set End Play. Back Off Nut and Retighten to 2.0 N·m (18 In-Lb).
10	Rear Hub		

335284

Hub, spindle and bearings with drum brakes

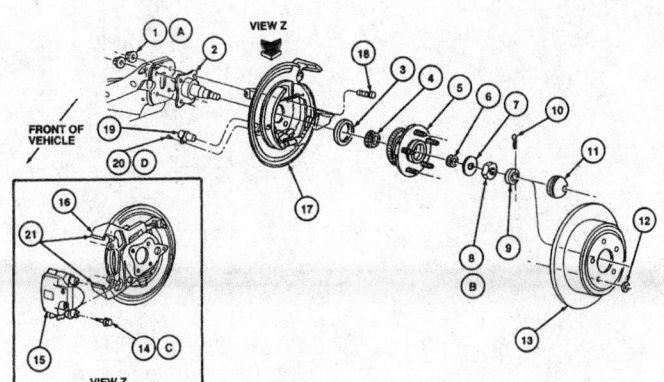

1	Nut, Rear Wheel Spindle-to-Rear Axle	17	Rear Wheel Disc Brake Shield
2	Rear Wheel Spindle	18	Bolt, Rear Wheel Spindle-to-Rear Axle
3	Wheel Hub Grease Seal	19	Rear Brake Anti-Lock Sensor
4	Bearing Cone and Roller, Rear	20	Bolt, Rear Brake Anti-Lock Sensor
5	Rear Hub	21	Slippers
6	Rear Wheel Bearing	A	Tighten to 65-80 N·m (48-59 Lb-Ft)
7	Washer	B	To set bearing preload, tighten to 24-31 N·m (18-22 Lb-Ft) while rotating the rear hub. Back off and retighten to 2 N·m (17 Lb-In).
8	Rear Hub Nut		
9	Retainer		
10	Cotter Pin		
11	Hub Grease Cap	C	Tighten to 15-20 N·m (11-14 Lb-Ft)
12	Keeper Nut		
13	Rear Disc Brake Rotor	D	Tighten to 8-10 N·m (71-88 Lb-In)
14	Caliper Bolt		
15	Rear Disc Brake Caliper		
16	Rear Wheel Disc Brake Adapter		

354221

Hub, spindle and bearings with disc brakes

6. Remove the wheel hub with the wheel bearings.

7. Remove the anti-lock brake sensor from the disc brake adapter.

8. Disconnect the parking brake rear cable end from the bracket and the parking brake lever at the top of the disc brake adapter.

9. Remove 4 retaining nuts securing the spindle to the axle assembly.

10. Remove the disc brake adapter, spindle and parking brake components as an assembly.

11. Drive out 4 retaining bolts and separate the spindle from the disc brake adapter, if necessary.

To install:

12. Inspect all parts for damage or wear and replace as necessary.

13. Place the spindle, disc brake adapter and parking brake components to the axle assembly with 4 retaining bolts and nuts. Tighten the nuts to 48–59 ft. lbs. (60–80 Nm).

14. Install the parking brake rear cable and conduit into the parking brake bracket and onto the parking brake lever.

15. Install the rear brake anti-lock speed sensor and retaining bolt.

Tighten the retaining bolt to 68–92 inch lbs. (8–10 Nm).

NOTE: If a new hub is being used, remove the protective coating with carburetor degreaser before assembly.

16. If the inner or outer bearing cups (races) were removed, install new ones using bearing cup replacer T73T-1202-A or equivalent.

17. Pack the wheel bearings using a wheel bearing packer. If a bearing packer is not available, work as much grease as possible between the rollers and cages. Grease the bearing cone surface.

NOTE: The wheel bearing lubricant used in this vehicle is lithium-based and must not be mixed with sodium-based lubricants.

18. Install the inner wheel bearing into the hub.

19. Apply a small amount of grease to the lip area of a new grease seal and install the grease seal using seal replacer T83T-1175-B or equivalent. Make sure that the grease seal is properly seated.

20. Carefully install the hub onto the spindle keeping the hub centered on the spindle to prevent damage to the grease seal and spindle threads.

21. Install the outer wheel bearing and flat washer onto the spindle.

22. Install the adjusting nut. Rotate the hub assembly while torquing the adjusting nut to 17–23 ft. lbs. (23–31 Nm). Back off of the adjusting nut 2 or 3 turns and retorque to 18 inch lbs. (2 Nm).

23. Install the retainer and a new cotter pin.

24. Install the grease cap.

25. Install the disc brake rotor and caliper assembly.

26. Install the wheel and tighten the lug nuts to 85–105 ft. lbs. (115–142 Nm).

27. If equipped with air suspension, partially lower the vehicle but do not put any load on the wheel and tire assemblies. Activate the air suspension switch and allow the air springs to fill for approximately 90 seconds.

28. Lower the vehicle.

29. Before moving the vehicle, pump the brake pedal several times to position the brake pads.

30. Road test the vehicle and check for proper operation.

GEO and SUZUKI

8

GEO-Tracker **SUZUKI**-Samurai • Sidekick • X-90

FIRING ORDERS

NOTE: To avoid confusion, always replace spark plug wires one at a time.

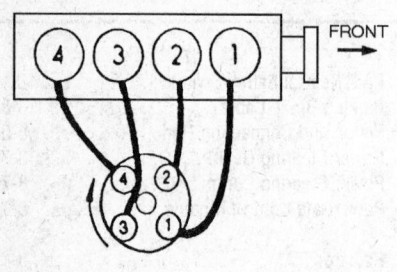

342882

1.3L (VIN 3) and 1.6L (VIN 0), (VIN U) Engines
Engine Firing Order: 1–3–4–2
Distributor Rotation: Clockwise

343974

1.8L (VIN 2) Engine
Engine Firing Order: 1–3–4–2
Distributorless Ignition

ENGINE ELECTRICAL

NOTE: Disconnecting the negative battery cable on some vehicles may interfere with the functions of the on board computer systems and may require the computer to undergo a relearning process, once the negative battery cable is reconnected.

Distributor

REMOVAL AND INSTALLATION

1.3L (VIN 3) and 1.6L (VIN 0), (VIN U) Engines

1. Disconnect the negative battery cable.

2. Disconnect the distributor CAS coupler.

NOTE: Do not bend or twist the spark plug wires to avoid internal damage. Grip the the wire boot when removing or installing the wires.

3. Remove the distributor cap and note which tower the rotor is pointing towards.
4. Mark the rotor position on the distributor housing and mark the distributor housing position on the engine.
5. Remove the distributor flange bolt and remove the distributor.

NOTE: Do not crank the engine with the distributor removed. Distributor assembly exploded view — 1.3L (VIN 3) engine

181538

To install:

1. Engine not disturbed (Engine was not cranked with the distributor removed):
   a. Install the distributor; align the reference marks made during the removal procedure.
   b. Verify that the distributor rotor is pointing towards the same tower on the distributor cap as it was prior to removal.
   c. Tighten the flange bolt to 11 ft. lbs. and install the distributor cap.
   d. Connect the distributor CAS coupler.
   e. Connect the negative battery cable, start the engine, and set the ignition timing to specification.
2. If the crankshaft was rotated while the distributor was removed from the engine, the piston in No. 1 cylinder must be brought to Top Dead Center (TDC) on the compression stroke. Proceed as follows:
   a. Remove the spark plug from No. 1 cylinder.
   b. Place a finger over the spark plug hole and rotate the crankshaft until compression is felt.
   c. Once compression is felt, continue rotating the crankshaft until the timing mark on the crankshaft pulley is aligned with the **0** mark on the timing indicator.

NOTE: After aligning the 2 marks, remove the cylinder head cover to visually check that the rocker arms are not riding on the camshaft cams at No. 1 cylinder. If the arms are found to be riding

on the cams, turn the crankshaft 360 degrees to realign the 2 marks.

   d. Reinstall the rocker arm cover.
   e. Reinstall the spark plug in No.1 cylinder.
   f. Turn the distributor shaft until the rotor is aligned with the No. 1 cylinder spark plug tower on the distributor cap. Install the distributor with the rotor in this position.
   g. Tighten the flange bolt to 11 ft. lbs. and install the distributor cap.
   h. Connect the distributor (CAS) coupler.
   i. Reconnect the negative battery cable, start the engine, and set the ignition timing to specification.

Ignition Timing

ADJUSTMENT

1.3L (VIN 3) and 1.6L (VIN 0), (VIN U) Engines

1. Start the engine and warm to normal operating temperature. Prior to any adjustment, be sure all the electrical accessories are **OFF**.
2. After warming the engine, make sure the idle speed is 800 rpm.
3. Make sure that A/C is OFF, manual transmission is in NEUTRAL; automatic transmission in P range, and parking brake lever is pulled fully.
4. Remove cap from the monitor coupler next to battery and connect terminals **C** and **D** (1993–95) or **E** and **D** (1996) with a jumper wire.
5. Connect the timing light according to manufacturer's instructions and use the to the No. 1 cylinder spark plug wire as an ignition pickup.

NOTE: When terminals C and D are connected, observe if ignition timing is varying. If ignition timing is varying, that indicates ungrounded D terminal which prevents accurate inspection and adjustment. Make sure to ground the D terminal securely.

6. With the engine running at the specified idle speed, direct the timing light to the crankshaft pulley. If the specified timing mark on the timing tab is aligned with the timing notch on the crankshaft pulley, the ignition is properly timed.
7. Initial ignition timing should be 8 degrees (1.3L) or 5 degrees (1.6L)

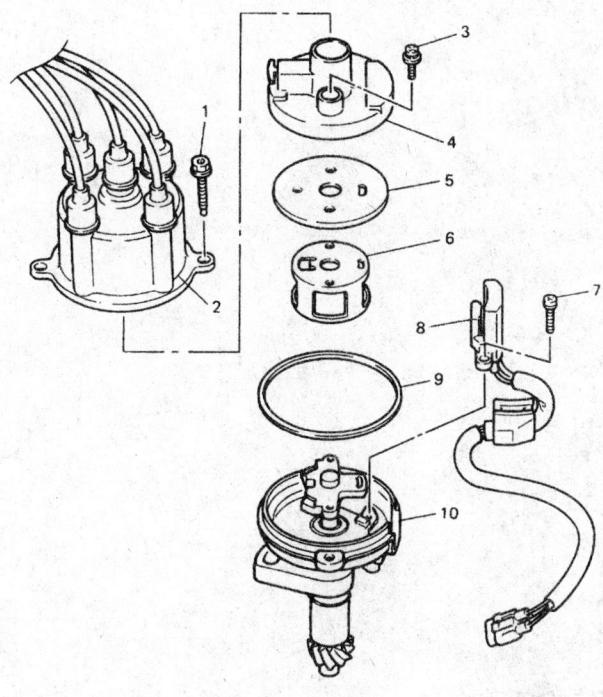

1. Cap bolt
2. Cap
3. Rotor screw
4. Rotor
5. Shield cover
6. Sigal rotor
7. CAS screw
8. CAS
9. Cap seal
10. Housing assembly

181538

BTDC at 800 rpm. The firing order is 1–3–4–2.

8. If the timing is out of adjustment, loosen the distributor flange bolt and turn the distributor housing to advance or retard the timing.

9. After the adjustment, tighten the flange bolt and recheck the timing.

10. After checking and/or adjusting ignition timing, disconnect service wire from monitor coupler.

1.8L (VIN 2) Engines

The ignition timing is computer controlled and is not adjustable.

Alternator

PRECAUTIONS

Several precautions must be observed with alternator equipped vehicles to avoid damage to the unit.

• If the battery is removed for any reason, make sure it is reconnected with the correct polarity. Reversing the battery connections may result in damage to the 1–way rectifiers.

• When utilizing a booster battery as a starting aid, always connect the positive to positive terminals and the negative terminal from the booster battery to a good engine ground on the vehicle being started.

• Never use a fast charger as a booster to start vehicles.

• Disconnect the battery cables when charging the battery with a fast charger.

• Never attempt to polarize the alternator.

• Do not use test lights of more than 12 volts when checking diode continuity.

• Do not short across or ground any of the alternator terminals.

• The polarity of the battery, alternator and regulator must be matched and considered before making any electrical connections within the system.

• Never separate the alternator on an open circuit. Make sure all connections within the circuit are clean and tight.

• Disconnect the battery ground terminal when performing any service on electrical components.

• Disconnect the battery if arc welding is to be done on the vehicle.

REMOVAL AND INSTALLATION

1. Disconnect the negative battery cable.

2. On the 1.8L (VIN 2), remove the air inlet hose.

3. Disconnect the wire connector and white lead wire from the alternator.

4. On the 1.3L (VIN 3), uncouple the brake pipe from the pipe clamp on the radiator under cover. Remove the under cover.

5. On the 1.8L (VIN 2), remove the clamps from the alternator bracket and engine mounting bracket.

6. Loosen and remove the alternator drive belt.

7. If necessary, remove the charcoal canister and mounting bracket.

8. Remove the alternator mounting bolt and alternator drive belt adjusting bolt.

9. Remove the alternator from the vehicle.

To install:

10. On the 1.8L, mount the alternator to the alternator bracket with the mounting bolts. Torque the alternator bolts to 17 ft. lbs. 923 Nm).

11. On the 1.3L and 1.6L, install the alternator and loosely install the mounting bolt and drive belt adjusting bolt.

12. On the 1.8L (VIN 2), install the clamps on the alternator bracket and engine mounting bracket.

13. Install the drive belt and adjust the belt tension. Tighten the alternator mounting bolts to 20 ft. lbs. (27 Nm).

14. Install the under cover, and reclamp the brake pipe.

15. If removed, install the charcoal canister and mounting bracket.

16. Reconnect the wiring to the alternator.

17. On the 1.8L (VIN 2), install the air inlet hose.

18. Reconnect the negative battery cable.

Drive Belt

REMOVAL AND INSTALLATION

1.3L (VIN 3) and 1.6L (VIN 0), (VIN U) Engines

1. Loosen the upper and lower alternator mounting bolts and rotate the alternator towards the center of the engine.

2. If equipped with power steering only (without A/C), loosen the power steering pump mounting bolt and adjusting bolt at the power steering pump. Rotate the pump inward to remove the belt.

3. If equipped with either A/C and power steering or A/C only, loosen the adjusting bolt, pivot bolt and pivot nut at the A/C compressor. Remove the A/C belt from the engine.

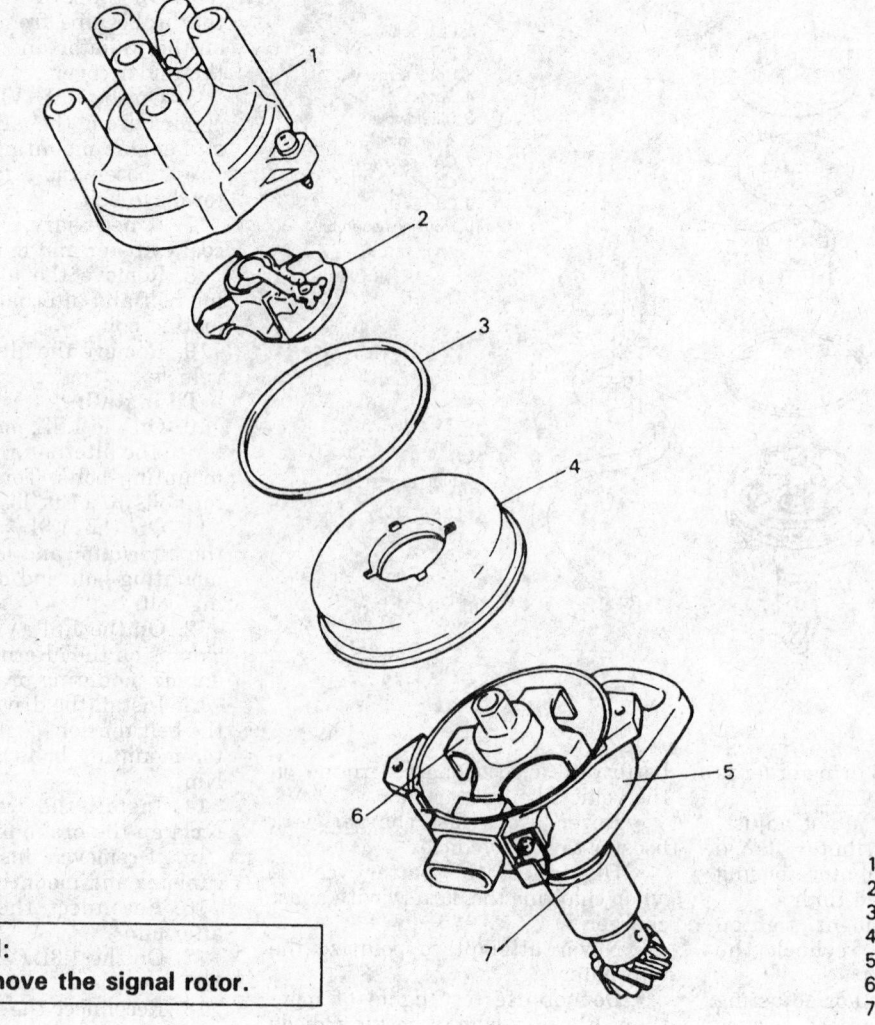

1. Cap
2. Rotor
3. Cap seal
4. Cover
5. Housing assembly
6. Signal rotor
7. O ring

324123

CAUTION:
Don't remove the signal rotor.

Exploded view of the distributor components — 1.6L (VIN 0), (VIN U) TFI or MFI engines

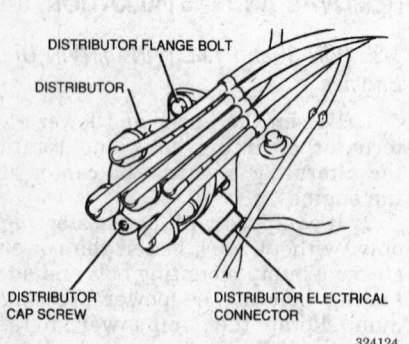

DISTRIBUTOR FLANGE BOLT

DISTRIBUTOR

DISTRIBUTOR CAP SCREW

DISTRIBUTOR ELECTRICAL CONNECTOR

324124

Distributor mounted on engine — 1.6L (VIN 0), (VIN U) TFI or MFI engines

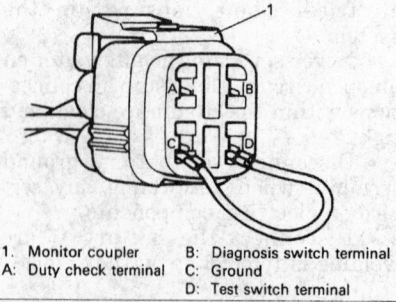

1. Monitor coupler
A: Duty check terminal
B: Diagnosis switch terminal
C: Ground
D: Test switch terminal

181639

Coupler exploded view (C to D terminals) — 1.3L (VIN 3) and 1.6L (VIN 0), (VIN U) engines

To install:

4. Install the belts on the vehicle making sure the belts are in the proper grooves and not crossed.

5. If equipped with A/C and power steering or A/C only, pull back on the A/C compressor to tighten the drive belt. The belt tension should be between 0.24–0.35 inch (6–9 mm) of deflection at 22 lbs. of pressure. Tighten the adjusting bolt, pivot bolt and pivot nut.

6. If equipped with power steering only, pull back on the power steering pump to tighten the drive belt. The belt tension should be between 0.24–0.35 inch (6–9 mm) of deflection

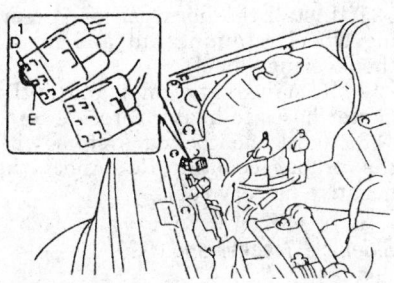

1. Monitor coupler

D: Ground
E: Test switch terminal

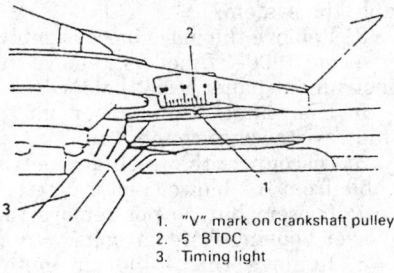

1. "V" mark on crankshaft pulley
2. 5 BTDC
3. Timing light

332403

Coupler E to D terminals

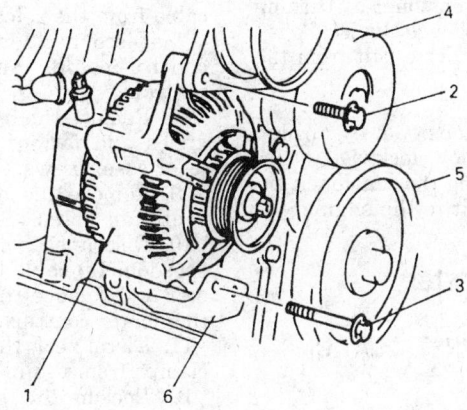

1. Generator
2. Upper generator bolt (Short)
3. Lower generator bolt (Long)
4. Generator belt tensioner
5. Crankshaft pulley
6. Generator bracket

331499

Alternator bolts — 1.8L (VIN 2) engine

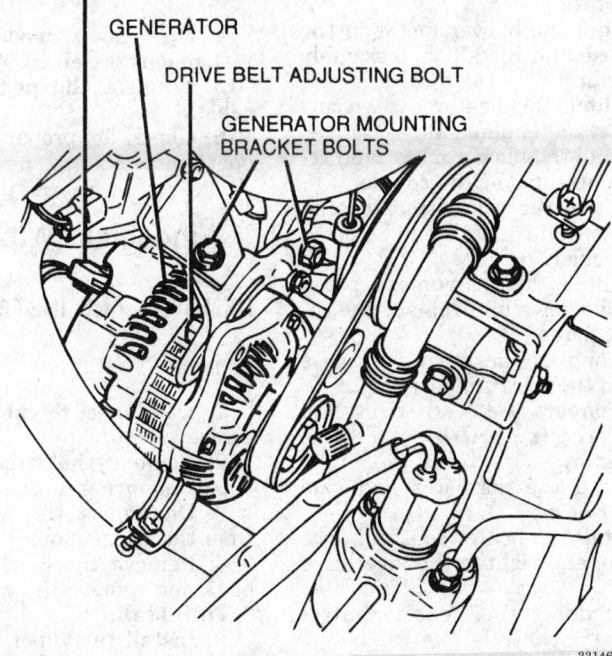

GENERATOR ELECTRICAL CONNECTOR

GENERATOR

DRIVE BELT ADJUSTING BOLT

GENERATOR MOUNTING
BRACKET BOLTS

331462

Alternator mounting components — 1.6L (VIN 0), (VIN U) engine; 1.3L (VIN 3) is similar

at 22 lbs. of pressure. Tighten the pivot bolt and adjusting bolt.

7. For the alternator belt, pull back on the alternator to tighten the belt. The belt tension should be between 0.24–0.32 inch (6–8 mm) of deflection at 22 lbs. of pressure. Tighten the adjusting bolt and pivot bolt.

1.8L (VIN 2) Engine

Cooling Fan Belt

1. Loosen the adjusting bolt and pivot bolt to the cooling fan pulley.
2. Slacken the belt by turning the cooling fan pulley.
3. Remove the cooling fan/clutch from the cooling fan pulley.
4. Remove the cooling fan belt from the engine.
   To install:
5. Install the cooling fan belt to the engine.
6. Install the cooling fan to the cooling fan pulley.
7. Adjust the cooling fan tension by turning the cooling fan pulley. Cooling fan tension should be between 0.20–0.27 inch (5–7 mm) at 22 lbs. (10 kg) of force.
8. Once the cooling fan is tightened, tighten the adjusting bolt and pivot bolt. Torque the two bolts to 37 ft. lbs. (50 Nm).

Alternator, Power Steering, and A/C Drive Belt

1. Loosen the tensioner by turning the tensioner pulley clockwise.
2. Remove the drive belt from the engine.

To install:

3. Loosen the tensioner by turning the tensioner pulley clockwise.
4. While holding the tensioner, install the drive belt to the engine.

Starter

REMOVAL AND INSTALLATION

All Engines

1. Disconnect the negative battery cable.
2. Raise and support the vehicle safely.
3. Disconnect the lead wires and battery cable from the starter motor.
4. Support the starter and remove the 2 mounting bolts.
5. Remove the starter.

To install:

6. Install the starter and tighten the 2 mounting bolts to 22 ft. lbs. (30 Nm).
7. Connect the lead wires to the starter, and the battery cable to the starter.
8. Lower the vehicle and connect the negative battery cable.
9. Check the starter for proper operation.

CHASSIS ELECTRICAL

Blower Motor

REMOVAL AND INSTALLATION

Samurai

1. Disconnect the negative battery cable and drain the cooling system.
2. Disconnect the inlet and outlet heater hoses from the heater core.
3. Remove the horn pad and the steering wheel retaining nut and remove the steering wheel by using the special tool (09944–36010) or equivalent.
4. Disconnect and tag the radio and cigar lighter wires. Remove the radio from the vehicle.

5. Remove the ash tray and mounting plate.
6. Disconnect the hood release cable from the release lever.
7. Disconnect and tag the heater control cables and wires at the controls.
8. Remove the heater control lever knobs and facing plate. Loosen the lever case screws.
9. Remove the defroster and side ventilator hoses.
10. Disconnect the lead wires and speedometer cable from the speedometer and remove the lead wires from the heater controls.
11. Disconnect the wiring harness clamps from the instrument panel.
12. Loosen the instrument panel mounting screws and remove the instrument panel.

NOTE: When removing the heater lever case which is fitted in the steering column holder, be very careful not to damage it.

13. Loosen the front door opening stop screws and remove the steering column bracket.
14. Disconnect and tag the blower motor and resistor connections at the coupler.
15. Loosen the heater case securing nut on the engine side.
16. Remove the heater assembly from the vehicle.
17. Remove the blower motor from the case.

To install:

18. Install the blower motor in the heater case and install the assembly in the vehicle.
19. Tighten the heater case securing nut on the engine side.
20. Install the blower motor and resistor connections at the coupler.
21. Tighten the front door opening stop screws and install the steering column holder.
22. Tighten the instrument panel mounting screws and replace the instrument panel.
23. Reconnect the wiring harness clamps to the instrument panel.
24. Reconnect the lead wires and speedometer cable to the speedometer.
25. Install the defroster and side ventilator hoses.
26. Install the heater control knobs and plate, and tighten the lever case screws.
27. Reconnect the heater control cables at the controls.
28. Reconnect the hood release cable to the release lever.
29. Install the ash tray and mounting plate.

30. Reconnect the radio and cigar lighter wires. Install the radio in the vehicle.
31. Install the horn pad and steering wheel retaining nut and install the steering wheel.
32. Reconnect the inlet and outlet heater hoses to heater core.
33. Refill the cooling system with the proper coolant. Reconnect the negative battery cable.

Sidekick, Tracker and X-90

1. Disconnect the negative battery cable.
2. On models with air bags, disable the system.
3. Remove the glove box assembly.
4. On 1996–97 models, remove the instrument panel assist holder.
5. Disconnect the blower motor and resistor wire connectors.
6. Disconnect the fresh air control cable from the blower motor case.
7. Loosen, but do not remove the blower housing fastener bolts.
8. Remove the 3 blower motor mounting screws.
9. Remove the blower motor.

To install:

10. Install the blower motor and secure with the 3 screws.
11. Tighten the blower housing fastener bolts.
12. Connect the fresh air control cable to the blower motor case.
13. Connect the electrical connectors and install the glove box assembly.
14. On 1996–97 models, install the instrument panel assist holder.
15. Connect the negative battery cable.
16. Check for proper blower motor operation.

Windshield Wiper Motor

REMOVAL AND INSTALLATION

Samurai

1. Disconnect the negative battery cable.
2. Remove the wiper linkage to wiper mounting nut.
3. Disconnect the wire connector from the wiper motor.
4. Remove the 3 wiper mounting bolts and remove the wiper motor.

To install:

5. Install the wiper motor and install the 3 mounting bolts.
6. Connect the wire connector and linkage to the wiper motor. Tighten the linkage mounting nut.

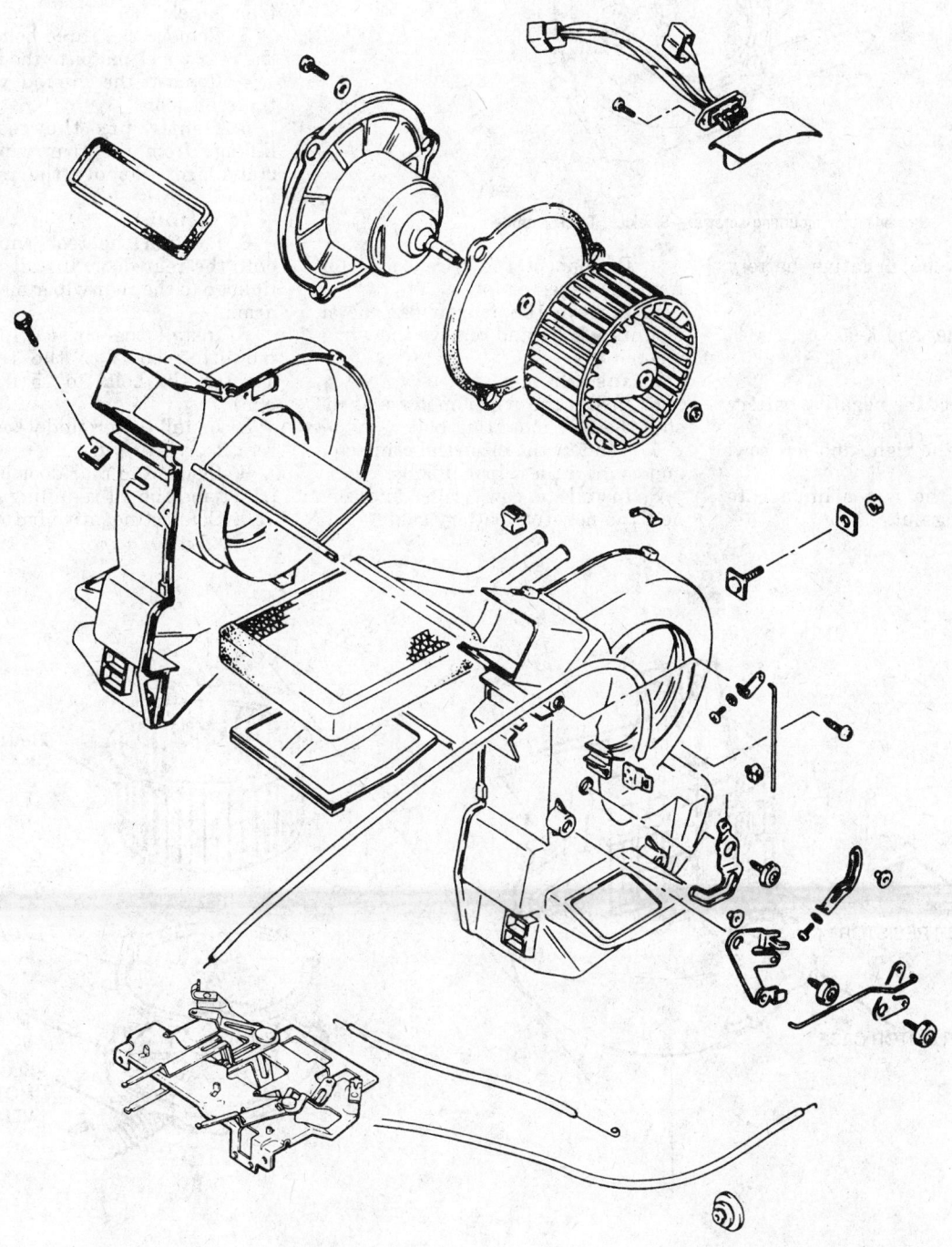

178940

Exploded view of the blower motor components — Samurai models

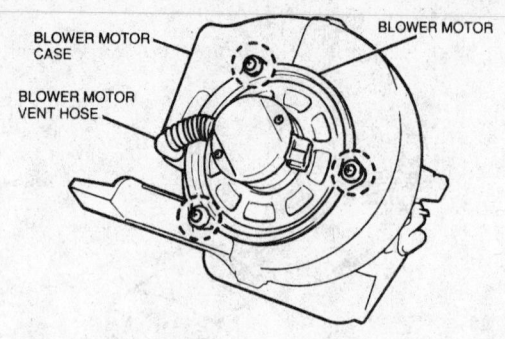

Blower motor mounting screws — Sidekick, Tracker models

7. Connect the negative battery cable.

Sidekick, Tracker and X-90

Front Wiper Motor

1. Disconnect the negative battery cable.

2. Remove the right and left cowl grilles.

3. Remove the wiper linkage to wiper mounting nut.

4. Disconnect the wire connector from the wiper motor.

5. Remove the four wiper motor mounting bolts and remove the wiper motor.

To install:

6. Install the wiper motor and install the four mounting bolts.

7. Connect the electrical connector and connect the wiper linkage.

8. Install the cowl grilles and connect the negative battery cable.

Rear Wiper Motor

1. Disconnect negative battery cable.

2. Remove the 14 plastic retaining clips, and then the rear door inner trim panel.

3. Remove the three bolts holding the rear wiper motor to the rear door.

4. Remove the ground wire from the rear door.

5. Gently pry the rear wiper linkage from the rear wiper motor crank arm. Remove the rear wiper motor.

To install:

6. Position the rear wiper motor onto the rear door. Install the wiper linkage to the rear wiper motor crank arm.

7. Install the three wiper motor mounting bolts to the rear door. Tighten the bolts to 15 ft. lbs. (20 Nm).

8. Install the ground wire from the rear door.

9. Install the rear door inner trim panel and the 14 retaining clips.

10. Connect negative battery cable.

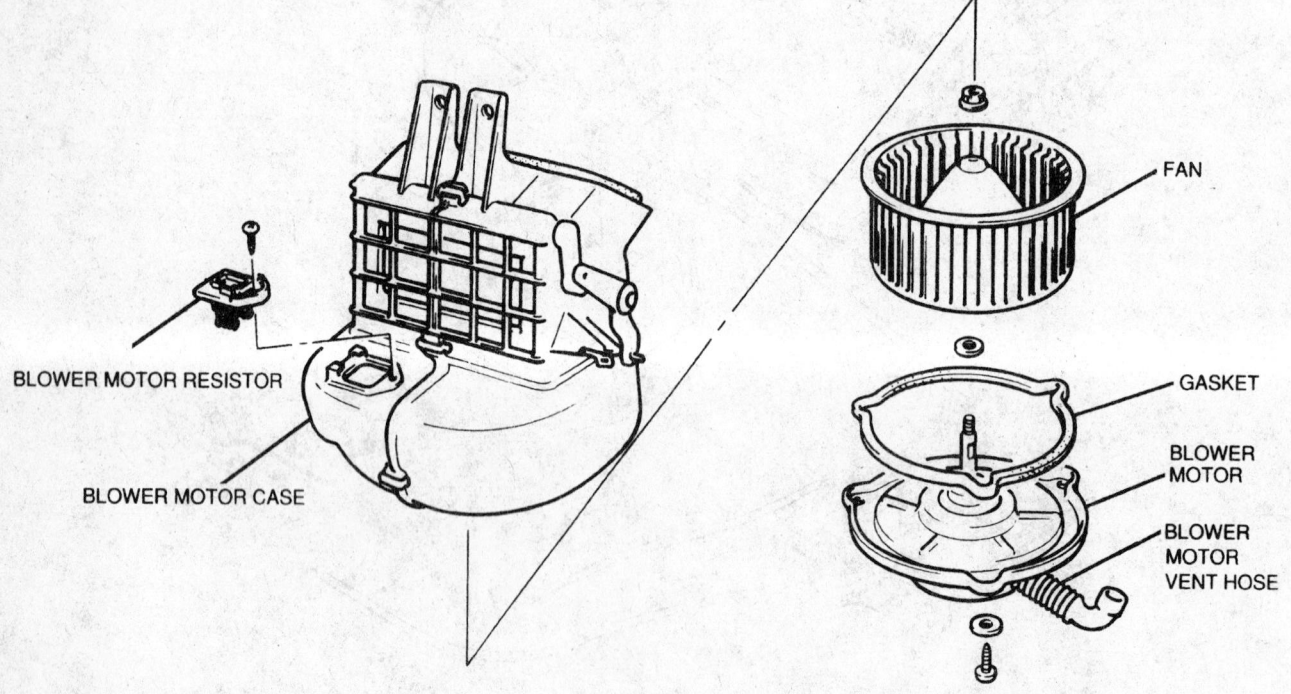

Exploded view of the blower motor components — Sidekick, Tracker models

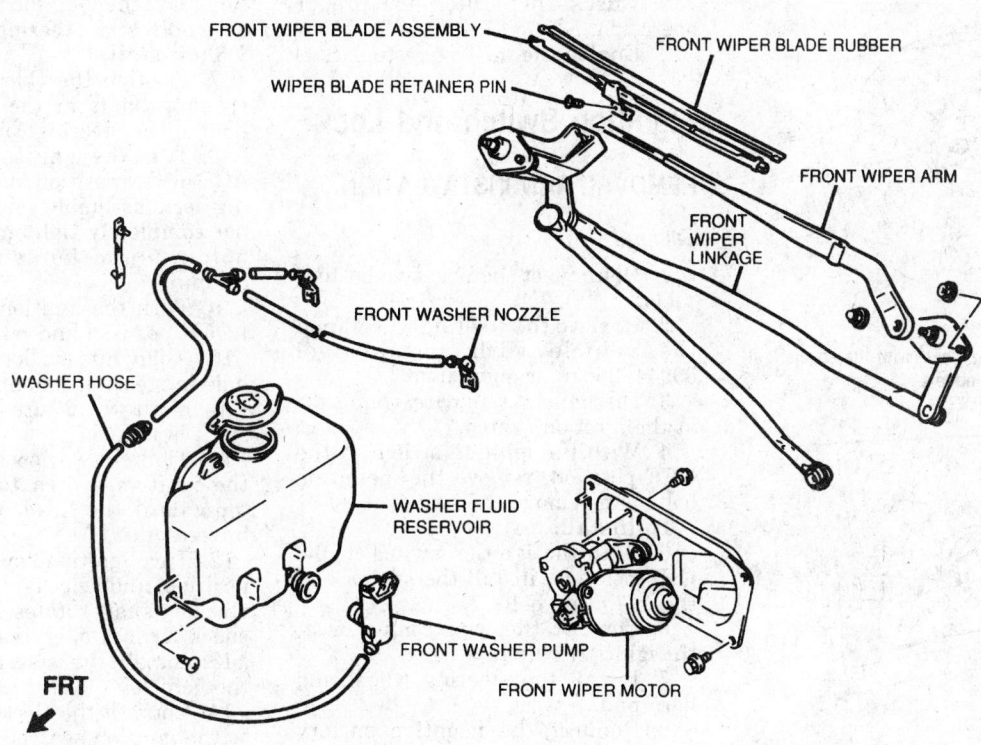

FRT

324051

Exploded view of the windshield wiper system components — Sidekick, Tracker and X-90 models

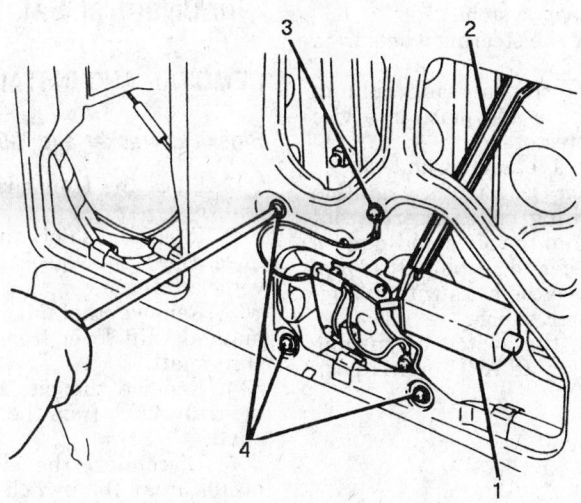

1 Rear wiper motor
2 Rear wiper linkage
3 Rear wiper motor ground wire
4 Rear wiper motor retaining screws

324054

Rear wiper motor and linkage — Sidekick, Tracker and X-90 models

Combination Switch

REMOVAL AND INSTALLATION

> **CAUTION**
>
> *The 1996–97 Sidekick, Tracker and X-90 are equipped with the Supplemental Inflatable Restraint (SIR) system. The Supplemental Inflatable Restraint (SIR) system must be disarmed before removing the steering wheel. Failure to do so may cause accidental deployment of the air bag, resulting in unnecessary SIR system repairs and/or personal injury.*

1. See the procedures under Steering Wheel Removal and Installation, and remove the steering wheel.
2. On 1996–97 Sidekick, Tracker and X-90, remove the hole cover and knee bolster panel.
3. Remove the upper and lower column cover screws and remove the covers.
4. Disconnect the lead wires from the combination switch at the connector.
5. Remove the combination switch assembly screws.
6. Remove the combination switch from the steering column.

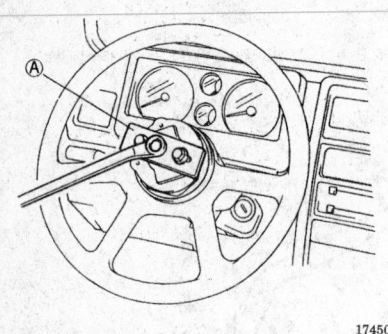

Remove the steering wheel with the special tool (A) — Samurai models

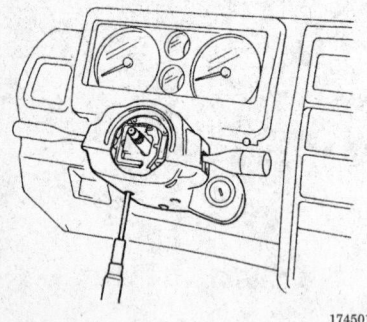

Remove the steering column cover screws — Samurai models

To install:

7. Install the combination switch to the steering column and install the assembly screws.

8. Reconnect the lead wires at the connector to the combination switch.

9. Install the lower and upper column covers, and reinstall the steering wheel. Torque the shaft nut to 19–28 ft. lbs.

10. On 1996–97 Sidekick, Tracker and X-90, install the knee bolster and hole cover.

11. Install the horn button, and connect the negative battery cable.

12. Check the switch for proper operation.

13. Enable the air bag system.

Ignition Switch and Lock

REMOVAL AND INSTALLATION

Samurai

1. Disconnect the negative battery cable.

2. Remove the steering wheel from the vehicle with special tool 09944–36010 or equivalent.

3. Disconnect the wire connector at the ignition switch.

4. With the ignition switch in the **OFF** position, remove the mounting bolts and remove the switch.

To install:

5. With the ignition switch in the **OFF** position, install the switch and the mounting bolts.

6. Connect the wire connector at the ignition switch.

7. Install the steering wheel and horn pad.

8. Connect the negative battery cable.

Sidekick, Tracker and X-90

1. Disconnect the negative battery cable.

2. If equipped with an air bag, disable the air bag system.

3. Remove the steering wheel from the vehicle.

4. Remove the steering column from the vehicle and place in a vice with jaw protectors.

5. Using a hammer and chisel, create slots on the top of the ignition switch mounting bolts. Insert a flat bladed tool into the slots and remove the bolts. (Refer to graphic) It is possible to remove the bolts with a hammer and center punch.

6. Remove the switch by turning the ignition key to **ACC** or **ON** posi-

Remove the combination switch retaining screws — Samurai models

tion and removing the ignition switch assembly from steering column.

To install:

7. Position the oblong hole in the steering shaft in the center of the hole in the steering column.

8. Turn the ignition switch key to ACC or ON position and install steering lock assembly onto column. Do not completely tighten the two bolts holding ignition switch to the column.

9. Turn the ignition switch to the LOCK position and pull out key.

10. Align hub on lock with oblong hole in steering column shaft and rotate shaft to assure that steering shaft is locked.

11. Tighten two new bolts attaching the ignition switch to steering column until the head of each bolt is broken off.

12. Turn ignition key to ACC or ON position and check to be sure that steering shaft rotates smoothly. Also check for lock operation.

13. Install the steering column in the vehicle.

14. Connect the electrical connector to the ignition switch.

15. If equipped with an air bag, enable the air bag.

16. Connect the negative battery cable.

Park/Neutral Safety Switch

REMOVAL AND INSTALLATION

Sidekick, Tracker and X-90

1. Place the transmission in neutral (**N**).

2. Raise and safely support the vehicle. Block the wheel to prevent the vehicle from moving.

3. Remove the nut, washer and manual shift lever from the manual shift shaft.

4. Remove the one bolt and the neutral switch from the manual shift shaft.

5. Disconnect the electrical connector from the switch and remove the switch.

To install:

6. Connect the electrical connector to the neutral safety switch.

7. Install the switch onto the manual shaft and install the washer and nut.

8. Install the manual shift lever on the shaft and install the washer and nut.

9. Lower the vehicle.

10. Make sure that the transmission operates properly in all postions.

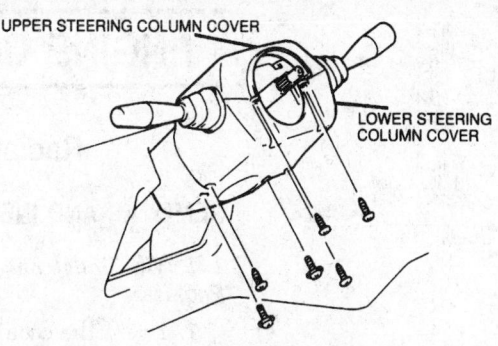

Combination switch and mounting screws — Sidekick, Tracker models

176482

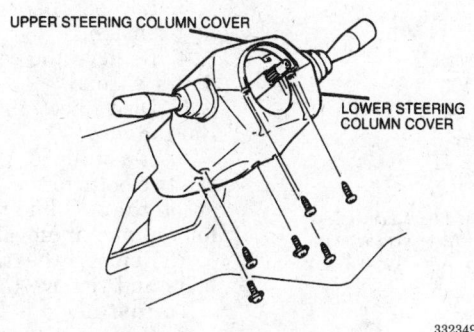

Steering column covers and mounting screws — Sidekick, Tracker and X-90 models

332349

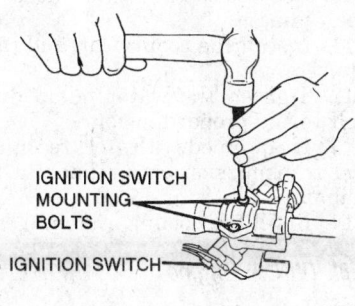

Removing the ignition switch shear bolts — Sidekick, Tracker and X-90 models

331256

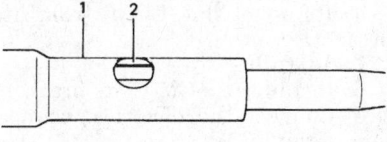

1. Steering column
2. Steering shaft

331258

Aligning oblong hole of steering shaft with hole in column — Sidekick, Tracker and X-90 models

Powertrain Control Module

REMOVAL AND INSTALLATION

Samurai

1. Disconnect the negative battery cable.
2. Remove the fuel pump relay, and the main relay from the Electronic Control Module (ECM).
3. Disconnect the couplers from the ECM while releasing the coupler lock..

4. Remove the three mounting screws, and remove the ECM.
   To install:
5. Position the ECM, and install the mounting screws.
6. Connect the couplers to the ECM.
7. Install the fuel pump relay and the main relay.
8. Connect the negative battery cable.

1993–95 Sidekick, Tracker

1. Disconnect the negative battery cable.
2. Remove the two screws from the left front speaker cover on the instrument panel.
3. Remove the speaker mounting screws and remove the speaker.
4. Remove the two bolts and the Electronic Control Module (ECM) bracket from the instrument panel support brace.
5. Remove the two screws and separate the fuse box from the ECM bracket.
6. Remove the two screws and separate the ECM from the ECM bracket.
7. Disconnect the ECM electrical connectors.
   To install:
8. Connect the electrical connectors to the ECM.
9. Install the ECM in the mounting bracket and install the mounting screws.
10. Connect the fuse box to the ECM bracket and install the two screws.
11. Position the ECM bracket on the instrument panel support and install the mounting bolts.
12. Install the speaker and speaker mounting screws.
13. Install the speaker cover and speaker cover screws.
14. Connect the negative battery cable.

1996–97 Sidekick, Tracker and X-90

With 1.6L (VIN 0), (VIN U) Engine

1. Disconnect the negative battery cable.

NOTE: It is recommended that a grounding strap be worn when handling the ECM. The grounding strap will prevent a static electricity discharge from damaging the ECM.

2. Remove the two bolts and remove the Electronic Control Module (ECM) bracket from the instrument panel support brace. If equipped with Automatic Transmission, remove the TCM (Transmission Control Module).
3. Disconnect the ECM electrical connectors and remove the ECM from the vehicle.
   To install:
4. Connect the electrical connectors to the ECM.
5. Position the ECM bracket on the instrument panel support and install the mounting bolts. If equipped with A/T, connect the TCM with the ECM.

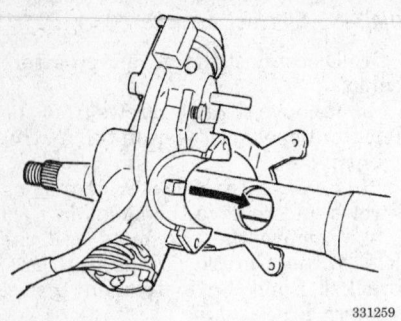

Aligning hob on lock with oblong hole of steering shaft — Sidekick, Tracker and X-90 models

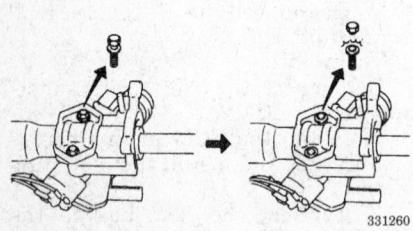

Tightening the new bolts — Sidekick, Tracker and X-90 models

6. Connect the negative battery cable.

With 1.8L (VIN 2) Engine

1. Disconnect the negative battery cable.

NOTE: It is recommended that a grounding strap be worn when handling the ECM. The grounding strap will prevent a static electricity discharge from damaging the ECM.

2. Remove the steering column undercover by removing the four screws.

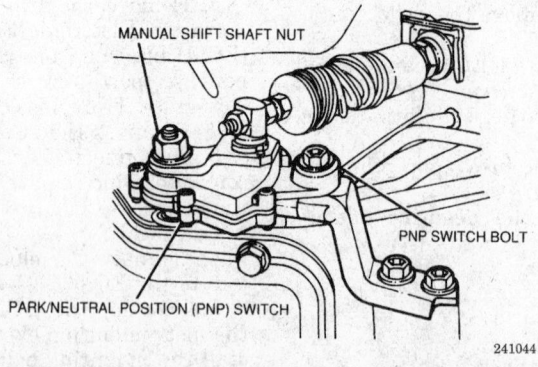

Neutral safety switch components — Sidekick, Tracker and X-90 models

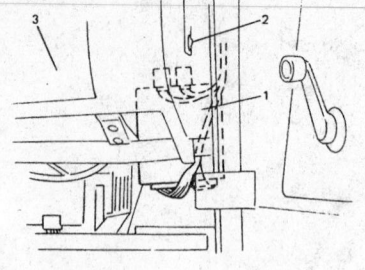

1. ECM
2. Instrument main panel
3. Glove box

ECM location — Samurai models

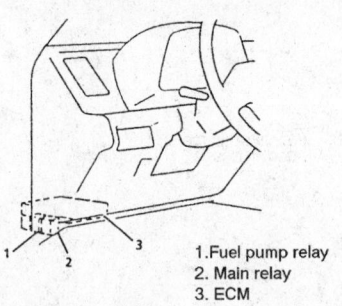

1. Fuel pump relay
2. Main relay
3. ECM

ECM location — 1993–95 Sidekick, Tracker models

3. Disconnect the electrical connectors from the ECM.
4. Remove the ECM from its bracket.
To install:
5. Install the ECM to its bracket.
6. Connect the electrical connectors to the ECM.
7. Install the steering column undercover by installing the four screws.
8. Connect the negative battery cable.

ENGINE COOLING

Radiator

REMOVAL AND INSTALLATION

1.3L (VIN 3) and 1.6L (VIN 0), (VIN U) Engines

1. Drain the cooling system.
2. If the vehicle has an automatic transmission, place an oil pan under the radiator and disconnect the A/T fluid hoses from radiator.
3. Loosen the water pump drive belt tension.
4. Remove the cooling fan and radiator shroud.
5. Disconnect the water hoses to the radiator.
6. If equipped with A/C, remove the two bolts to the A/C line and reposition the A/C line to allow clearance for radiator removal.
7. Remove the radiator retaining bolts and remove the radiator.
To install:
8. Replace the radiator mounting bolts and install the radiator.
9. Connect the A/C line to the mounting bracket with the two bolts.
10. Reconnect the water hoses to the radiator.
11. Install the cooling fan and radiator shroud.
12. Tighten the water pump drive belt to the proper tension.
13. If equipped with A/T, reconnect the transmission cooler lines to the radiator.
14. Refill the cooling system.

1.8L (VIN 2) Engine

1. Drain the coolant from the cooling system by loosening the drain plug to the radiator.
2. If equipped with A/T, place a oil pan under the radiator and disconnect the A/T fluid hoses from the radiator.
3. Loosen the nuts to the cooling fan/clutch nuts.
4. Loosen the cooling fan belt tension by loosening the adjusting bolt and pivot bolt to the fan pulley.
5. Remove the nuts to the cooling fan/clutch.
6. Disconnect the radiator inlet hose from the radiator.
7. Remove the power steering oil tank.
8. Remove the radiator shroud bolts.
9. Disconnect the outlet hose from the radiator.

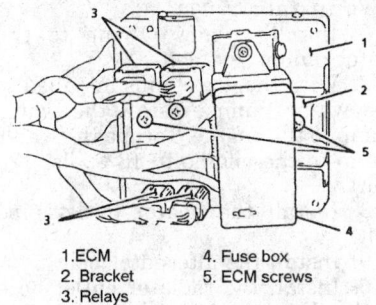

1. ECM
2. Bracket
3. Relays
4. Fuse box
5. ECM screws

182098

Removing the ECM — 1993–95 Sidekick, Tracker models

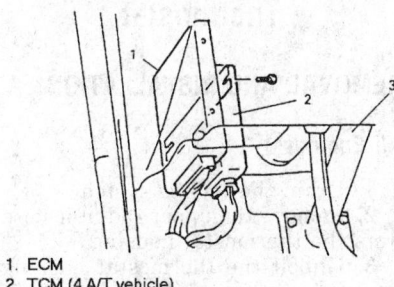

1. ECM
2. TCM (4 A/T vehicle)
3. Steering column

323894

ECM location — 1996–97 Sidekick, Tracker and X-90

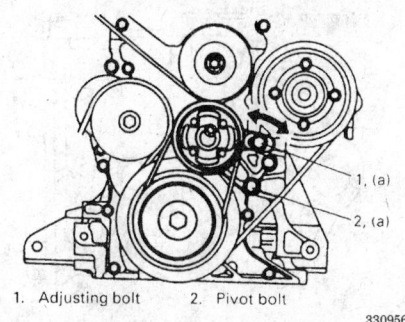

1. Adjusting bolt
2. Pivot bolt

330956

Fan pulley adjusting bolt and pivot bolt — Sidekick models with 1.8L (VIN 2) engines

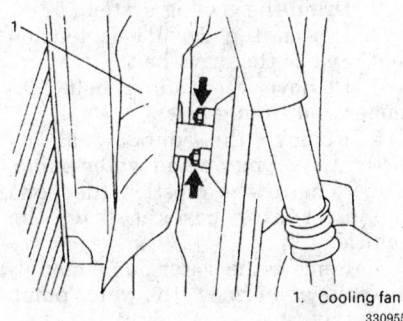

1. Cooling fan

330955

Fan/clutch nuts — Sidekick models with 1.8L (VIN 2) engines

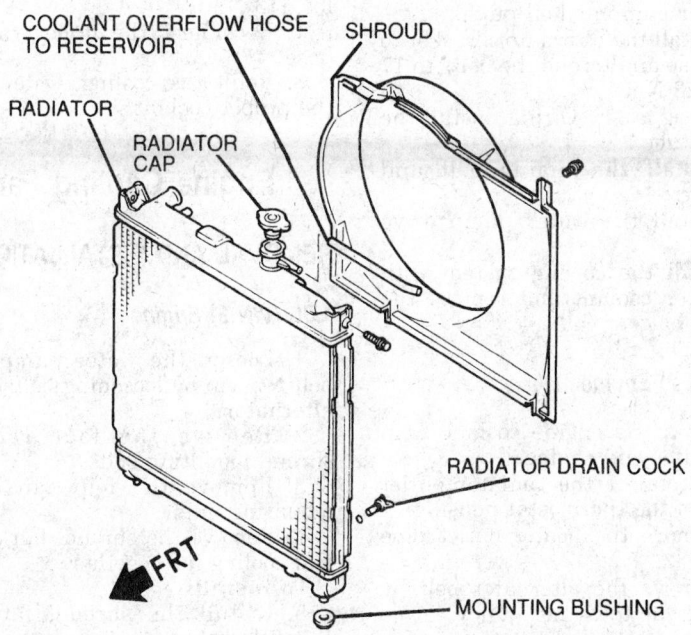

COOLANT OVERFLOW HOSE TO RESERVOIR

SHROUD

RADIATOR

RADIATOR CAP

RADIATOR DRAIN COCK

FRT

MOUNTING BUSHING

241065

Radiator and fan shroud — Sidekick, Tracker and X-90 models with 1.6L (VIN 0), (VIN U) engines

10. Disconnect the reservoir hose from the radiator.
11. Remove the radiator from the vehicle.
 To install:
12. Install the radiator to the vehicle.
13. Connect the reservoir hose to the radiator.
14. Connect the outlet hose to the radiator.
15. Install the cooling fan/clutch and radiator shroud.
16. Install the radiator shroud bolts.
17. Install the cooling fan/clutch nuts to hold the fan to the fan pulley.
18. Install the power steering tank.
19. Connect the radiator inlet hose to the radiator.
20. Tighten the cooling fan belt to 0.20–0.27 inch (5–7 Nm) at 22 lbs. Tighten the adjusting bolt and pivot bolt.
21. If equipped with A/T, connect the A/T fluid hoses to the radiator.
22. Fill the engine and radiator with coolant.
23. Start the engine and check for leaks.

Water Pump

REMOVAL AND INSTALLATION

1.3L (VIN 3) and 1993–95 1.6L (VIN 0), (VIN U) Engines

1. Drain the cooling system.
2. Loosen the drive belt tension and remove the drive belt. If equipped, loosen the air conditioning belt tension and remove the air conditioning belt. If not equipped with A/C, loosen the power steering pump and remove belt.
3. Remove the radiator fan shroud and radiator fan mounting bolts. Remove the radiator shroud, fan, and water pump pulley from the vehicle.
4. Rotate the engine to TDC.
5. Remove the crankshaft pulley bolts and remove the crankshaft pulley.

NOTE: The crankshaft pulley bolt can be removed without removing the center crankshaft bolt.

6. Remove the timing belt cover mounting bolts and remove the cover.

— **WARNING** —
Place the engine at TDC prior to removing the timing belt.

7. Loosen the timing belt tensioner adjusting bolt and pivot nut.

Hold the tensioner to loosen the timing belt and remove the belt from the camshaft pulley.

8. Remove the timing belt tensioner mounting bolts and remove the tensioner plate and spring.

9. If necessary, remove the dipstick tube and the alternator bracket from the vehicle.

10. Remove the water pump mounting bolts and remove the water pump assembly.

To install:

11. Clean and inspect the surface of the engine before installation.

12. Using a new gasket, install the new water pump on the engine. Torque the mounting bolts to 7.5–9.0 ft. lbs. (10–12 Nm).

13. Install the rubber seals between the water pump and cylinder head and water pump and oil pump.

14. Install the timing belt tensioner plate, tensioner, and spring.

15. Align the marks on the timing belt and the camshaft sprocket. Install the timing belt in the same position on the camshaft sprocket as when removed.

16. Adjust the timing belt to be free of any slack. Torque the tensioner bolts to 7.5–9.0 ft. lbs. (10–12 Nm). Install the timing belt cover.

17. Install the crankshaft and water pump pulleys. Torque the crankshaft and water pump pulley bolts to 7.5–9.0 ft. lbs. (10–12 Nm).

18. Install the cooling fan/clutch, shroud, and drive belts.

19. Adjust the valve lash and the drive belt tension.

20. Refill the cooling system with the proper coolant.

21. Start the engine and check for leaks. Check and/or adjust the ignition timing as necessary.

1996–97 1.6L (VIN 0), (VIN U) Engine

1. Disconnect the negative battery cable.

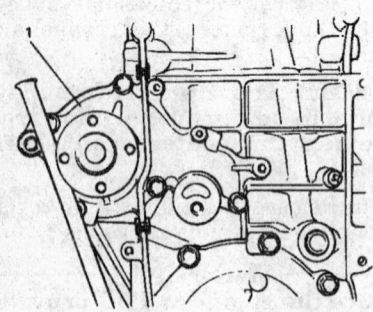

1. Water pump
181935

Water pump location — 1.3L (VIN 3) engine

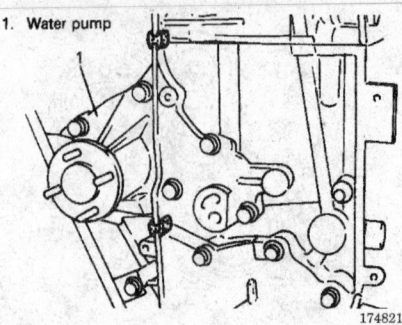

1. Water pump
174821

Water pump assembly — 1993–95 1.6L (VIN 0), (VIN U) engines

2. Drain the cooling system.

3. Loosen the drive belt tension and remove the drive belt.

4. Remove the timing belt tensioner and timing belt.

5. Remove the rubber seal between the oil pump and water pump.

6. Remove the dipstick tube and the alternator bracket from the vehicle.

7. Remove the water pump mounting bolts and remove the water pump assembly.

To install:

8. Clean and inspect the surface of the engine before installation.

9. Using a new gasket, install the new water pump on the engine. Torque the mounting bolts to 9 ft. lbs. (12 Nm).

10. Install the rubber seals between the water pump and oil pump.

11. Install the alternator bracket to the vehicle and torque the bolts to 17 ft. lbs. (23 Nm).

12. Using a new O-ring, install the dipstick tube.

13. Install the timing belt and tensioner.

14. Install the water pump drive belt.

15. Refill the cooling system with the proper coolant and replace the negative battery cable.

1.8L (VIN 2) Engine

1. Drain the engine coolant from the radiator and engine.

2. Disconnect the radiator outlet hose from the thermostat housing.

3. Remove the heater outlet pipe bolt.

4. Remove the alternator belt by loosening the tensioner pulley.

5. Remove the water pump assembly by removing the four bolts.

NOTE: Do not lose the dowel pins when removing the water pump.

To install:

6. Install a new O-ring to the water pump.

7. With the dowel pins installed to the water pump, install the water pump to the engine. Torque the water pump to the engine to 18 ft. lbs. (25 Nm).

8. Install the heater outlet pipe bolt.

9. Install the alternator belt.

10. Install the radiator outlet hose to the thermostat housing.

11. Fill the engine and radiator with coolant.

Thermostat

REMOVAL AND INSTALLATION

All Engines

1. Drain the cooling system.

2. Remove the upper radiator hose from the thermostat housing.

3. Unbolt the thermostat housing from the intake manifold and remove the thermostat.

To install:

4. Clean and inspect the surfaces of the housing and the engine.

5. Install the new thermostat with the spring facing towards the engine.

6. Install a new gasket and the thermostat housing to the intake manifold. Torque the bolts to 10 ft. lbs. (13 Nm).

7. Reconnect the upper radiator hose.

8. Refill the cooling system with the proper coolant.

Engine Cooling Fan

REMOVAL AND INSTALLATION

1.3L (VIN 3) Engine

1. Loosen the water pump drive belt tension by loosening bolts to the alternator.

2. Remove the four radiator shroud mounting bolts.

3. Remove the four fan clutch mounting nuts.

4. Remove the shroud, fan clutch, and cooling fan together.

To install:

5. Install the shroud, fan, and clutch together.

6. Install the fan clutch mounting nuts. Torque the nuts to 7–8.5 ft. lbs. (9–12 Nm).

7. Install the fan shroud mounting bolts.

1. Heater outlet pipe bolt
2. Thermostat cap
3. Exhaust manifold cover

330973

Water pump outlet pipe bolt and thermostat housing — 1.8L (VIN 2) engine

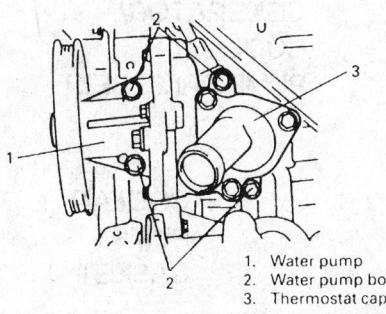

1. Water pump
2. Water pump bolt
3. Thermostat cap

330974

Water pump bolts — 1.8L (VIN 2) engine

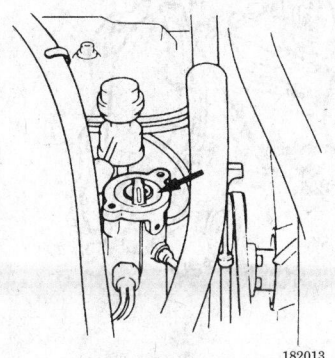

182013

Thermostat installation — 1.3L (VIN 3) engine

8. Tighten the water pump belt. Belt tension should be 0.24–0.32 in. (6 to 8 mm) at 22 lbs (10kg).

1.6L (VIN 0), (VIN U) Engine

1. If equipped with A/C or power steering, drain the cooling system by loosening the drain plug of the radiator.

NOTE: The cooling system does not have to be completely drained. It only has to be drained below the level of the upper radiator hose.

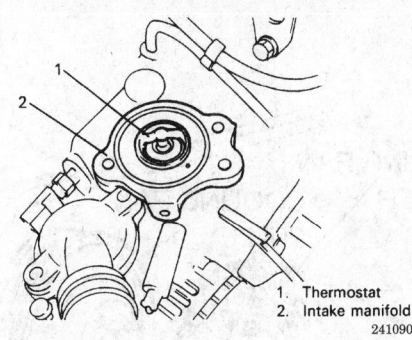

1. Thermostat
2. Intake manifold

241090

Thermostat installation — 1.6L (VIN 0), (VIN U) engine

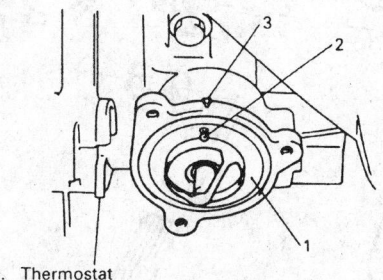

1. Thermostat
2. Air bleed valve
3. Match mark

330958

Thermostat installation — 1.8L (VIN 2) engine

2. Disconnect the upper radiator hose from the radiator.
3. If equipped with A/C, remove the A/C pipe mounting bolts.
4. Remove the four radiator shroud mounting bolts.
5. Loosen water pump drive belt tension by loosening the bolts to the alternator.
6. Remove the four fan clutch mounting nuts.
7. Remove the shroud, fan clutch, and cooling fan together.
To install:
8. Install the shroud, fan and clutch together.
9. Install the fan clutch mounting nuts. Torque the nuts to 8 ft. lbs. (11 Nm).
10. Install the fan shroud mounting bolts.
11. If equipped with A/C, install the A/C pipe mounting bolts.
12. If equipped with A/C or power steering, connect the upper radiator hose to the radiator.
13. If equipped with A/C or power steering, tighten the drain plug.
14. Tighten the water pump belt. Belt tension should be 0.24–0.32 in. (6 to 8 mm) at 22 lbs (10kg).
15. Refill the cooling system.

1.8L (VIN 2) Engine

1. Drain the coolant from the radiator and engine.
2. Disconnect the radiator inlet hose from the radiator.
3. Loosen the cooling fan/clutch nuts.
4. Loosen the cooling fan belt by loosening the adjusting bolt and pivot bolt.
5. Remove the cooling fan/clutch nuts.
6. Remove the radiator shroud bolts.
7. Remove the cooling fan/clutch and the radiator shroud from the engine compartment.
To install:
8. Install the cooling fan/clutch and the radiator shroud.
9. Install the radiator shroud bolts.
10. Connect the cooling fan/clutch to the engine with the four nuts.
11. Tighten the cooling fan belt to 0.20–0.27 inch (5–7 Nm) at 22 lbs.
12. Connect the radiator inlet hose to the radiator.
13. Fill the cooling system with coolant.
14. Start the vehicle and check for leaks.

Cooling System

BLEEDING

All Engines

1. With the engine **OFF**, fill the radiator to the bottom of the filler neck with a 50/50 mix of antifreeze and water.
2. Start the engine.
3. Allow the engine to reach normal operating temperature at idle.
4. When the thermostat opens (the upper radiator hose gets hot), turn the engine **OFF**. Top off the system with coolant.
5. Install the radiator cap.
6. Fill the coolant recovery reservoir to the full hot mark.

NOTE: When installing the reservoir tank cap, align the arrow marks on the tank and cap.

7. Shut off the engine and allow to cool.
8. With the engine cool, check the level of coolant in the recovery bottle and top off as necessary.

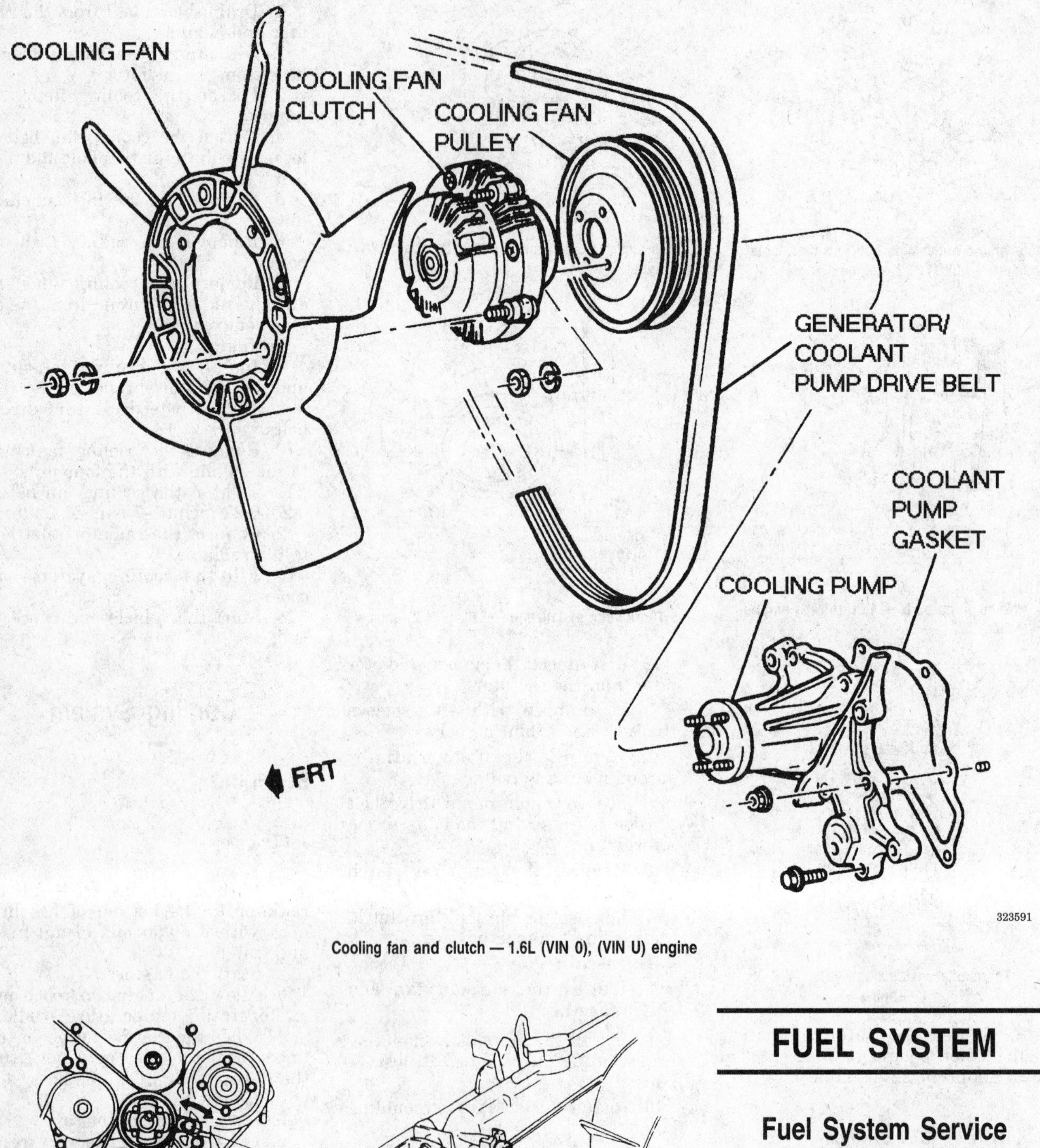

COOLING FAN

COOLING FAN CLUTCH

COOLING FAN PULLEY

GENERATOR/ COOLANT PUMP DRIVE BELT

COOLANT PUMP GASKET

COOLING PUMP

◀ FRT

323591

Cooling fan and clutch — 1.6L (VIN 0), (VIN U) engine

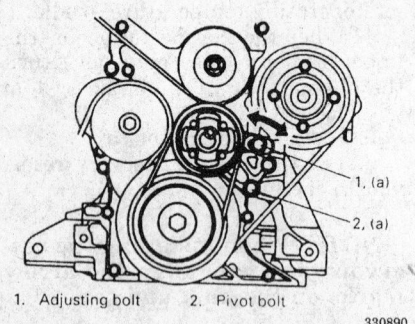

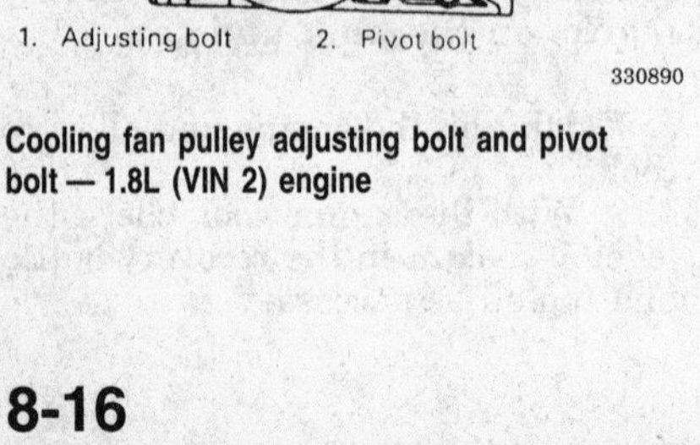

1. Adjusting bolt   2. Pivot bolt

330890

Cooling fan pulley adjusting bolt and pivot bolt — 1.8L (VIN 2) engine

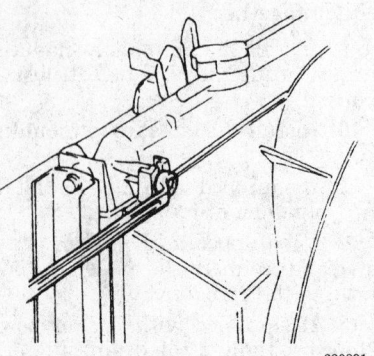

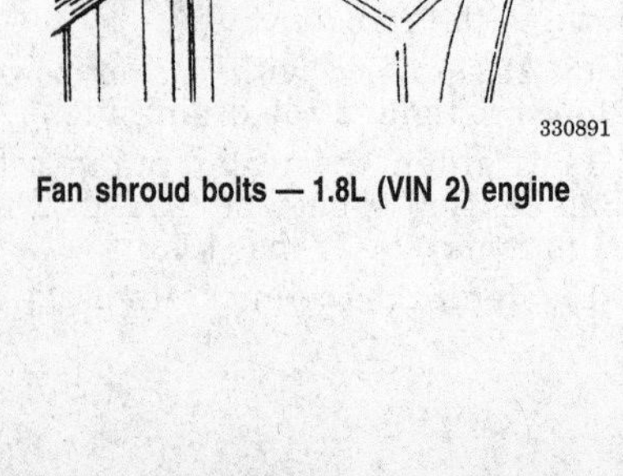

330891

Fan shroud bolts — 1.8L (VIN 2) engine

FUEL SYSTEM

Fuel System Service Precautions

Safety is the most important factor when performing not only fuel system maintenance but any type of maintenance. Failure to conduct maintenance and repairs in a safe manner may result in serious personal injury or death. Maintenance and testing of the vehicle's fuel system components can be accomplished

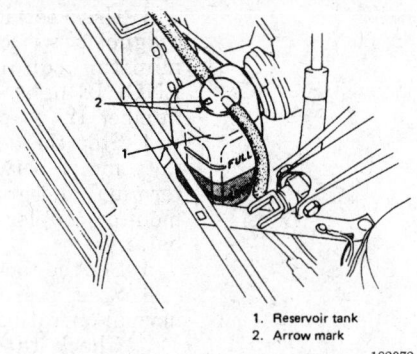

1. Reservoir tank
2. Arrow mark

182072

Coolant reservoir tank — 1.3L (VIN 3) engine

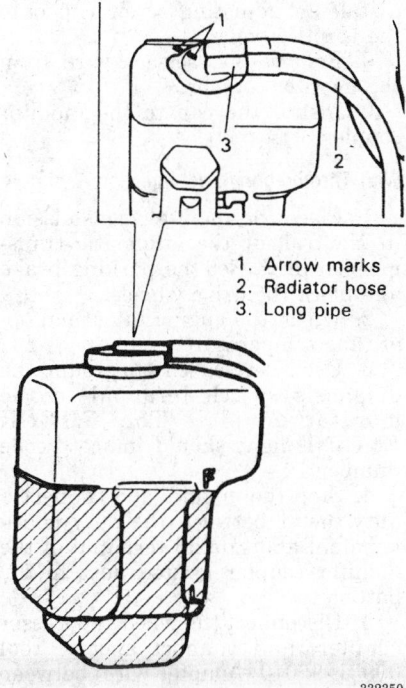

1. Arrow marks
2. Radiator hose
3. Long pipe

332350

**Coolant reservoir tank — 1.6L (VIN 0), (VIN U)
MFI engine**

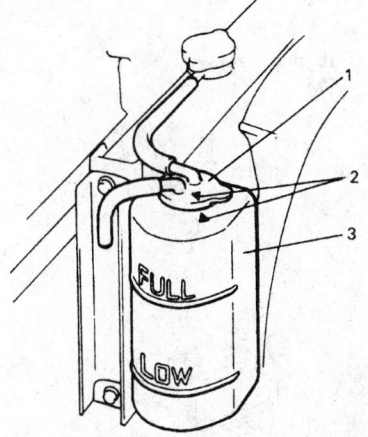

1. Reservoir tank cap
2. Arrow marks
3. Reservoir tank

332338

**Coolant reservoir tank — 1.6L (VIN 0), (VIN U)
TFI engine**

• Always use a backup wrench when loosening and tightening fuel line connection fittings. This will prevent unnecessary stress and torsion to fuel line piping. Always follow the proper torque specifications.

• Always replace worn fuel fitting O-rings with new. Do not substitute fuel hose or equivalent, where fuel pipe is installed.

Fuel System Pressure

RELIEVING

——————— **CAUTION** ———————

Care should be used when working around the fuel system. DO NOT smoke or expose the fuel system to any open flames or sparks. When the fuel tank is removed from the vehicle DO NOT allow it to sit in direct sunlight. Even if the tank appears empty the fuel vapors in the tank can explode if subjected to extreme heat. Always keep a suitable fire extinguisher handy when servicing the fuel system.

————————————————————

1. Remove the fuel filler cap from the fuel filler neck to release the fuel vapor pressure in the fuel tank.

2. Disconnect the electrical connector from the fuel pump relay.

3. Start the vehicle and allow the engine to run until it stalls.

4. Crank the engine for three more seconds to eliminate any remaining pressure in the fuel lines.

5. Disconnect the negative battery cable.

6. Connect the electrical connector to the fuel pump relay.

Idle Speed

ADJUSTMENT

1.3L (VIN 3) Engine

Before performing the idle speed adjustment, make sure of the following:

• All emissions and vacuum hoses are connected.

• The accelerator cable has some play, that is, it is not to tight.

• All accessories are OFF.

• The ignition timing is set correctly.

1. Place the transmission in Neutral, set the parking brake and block the drive wheels.

2. Warm the engine to normal operating temperature.

3. Connect a spare fuse to the diagnostic switch terminal in the fuse

safely and effectively by adhering to the following rules and guidelines.

• To avoid the possibility of fire and personal injury, always disconnect the negative battery cable unless the repair or test procedure requires that battery voltage be applied.

• Always relieve the fuel system pressure prior to disconnecting any fuel system component (injector, fuel rail, pressure regulator, etc.), fitting or fuel line connection. Exercise extreme caution whenever relieving fuel system pressure to avoid exposing skin, face and eyes to fuel spray. Please be advised that fuel under

pressure may penetrate the skin or any part of the body that it contacts.

• Always place a shop towel or cloth around the fitting or connection prior to loosening to absorb any excess fuel due to spillage. Ensure that all fuel spillage (should it occur) is quickly removed from engine surfaces. Ensure that all fuel soaked cloths or towels are deposited into a suitable waste container.

• Always keep a dry chemical (Class B) fire extinguisher near the work area.

• Do not allow fuel spray or fuel vapors to come into contact with a spark or open flame.

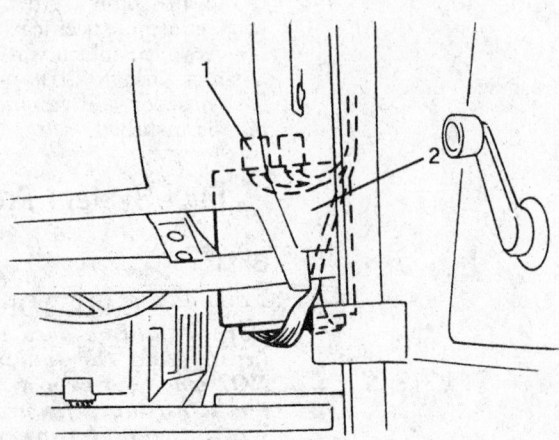

1 Fuel pump relay
2 ECM

182743

Fuel pump relay location — 1.3L (VIN 3) engine

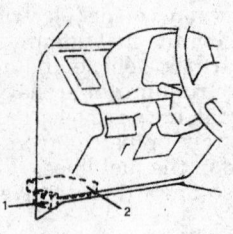

1 Fuel pump relay (Relay with Pink wire)
2 ECM

182736

Fuel pump relay location — 1993–95 1.6L (VIN 0), (VIN U) engines

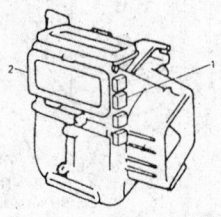

1. Fuel pump relay (Relay with Pink wire)
2. Heater unit

332393

Fuel pump relay location — 1996–97 1.6L (VIN 0), (VIN U) and 1.8L (VIN 2) engines

box. The **CHECK ENGINE** light should indicate code number 12.

4. Stop the engine, and connect a duty meter between the duty check terminal and ground terminal of the monitor coupler located next to the battery.

5. Set the tachometer.

6. Turn **ON** the ignition switch and wait for five seconds. Restart the engine, and run it at about 2000 rpm for five minutes to warm up the engine, then let it slow down to idle speed.

7. Check the idle speed. If the speed is not between 750–850 rpm, adjust by turning the idle speed adjusting screw on the throttle body.

8. When adjustment is completed, install an adjusting screw cap onto the throttle body and remove the spare fuse.

1.6L (VIN 0), (VIN U) Engine

With Multi-port Fuel Injection

1. Place the manual transmission in Neutral, or the automatic transmission in **P**. Set the parking brake and block the drive wheels.

2. Warm the engine to normal operating temperature.

3. Use a service wire to ground the diagnosis switch terminal in the monitor coupler. The **CHECK ENGINE** light should indicate code number **12**.

4. Stop the engine, and connect a duty meter between the duty check terminal and ground terminal of the monitor coupler located next to the battery.

5. Set the tachometer.

6. Start and warm the engine to normal operating temperature.

7. Check the idle speed. If the speed is not between 750–850 rpm, adjust by turning the idle speed adjusting screw on the throttle body.

8. When adjustment is completed, install an adjusting screw cap onto the throttle body.

9. Remove the service wire from the monitor coupler.

10. Install the cap to the monitor coupler.

With Throttle-body Fuel Injection

1. Place the manual transmission in Neutral, or the automatic transmission in **P**. Set the parking brake and block the drive wheels.

2. Warm the engine to normal operating temperature.

3. Use a service wire to ground the diagnosis switch terminal in the monitor coupler. The **CHECK ENGINE** light should indicate code number 12.

4. Stop the engine, and connect a duty meter between the duty check terminal and ground terminal of the monitor coupler located next to the battery.

5. Disconnect the noise suppressor coupler, and connect special tool 09931–96010 (Adapter wire) between the suppressor and coupler. Set the tachometer.

6. Turn **ON** the ignition switch and wait for five seconds. Restart the engine, and run it at about 2000 rpm for five minutes to warm up the engine, then let it slow down to idle speed.

7. Check the idle speed. If the speed is not between 750–850 rpm, adjust by turning the idle speed adjusting screw on the throttle body.

8. When adjustment is completed, install an adjusting screw cap onto the throttle body.

9. Remove the service wire from the monitor coupler.

10. Install the cap to the monitor coupler.

1.8L (VIN 2) Engine

1. Place the manual transmission in Neutral, or the automatic trans-

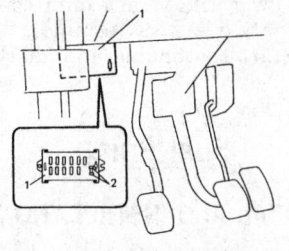

1 Fuse box
2 Diagnosis switch terminal

194283

Grounding the diagnostic switch terminal — 1.3L (VIN 3) engine

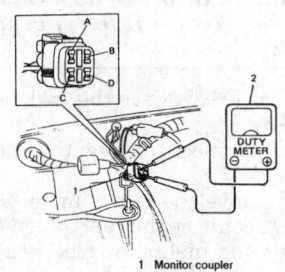

1 Monitor coupler
2 Duty meter
A: Duty check terminal
C: Ground terminal

194284

Connecting the duty meter — 1.3L (VIN 3) engine

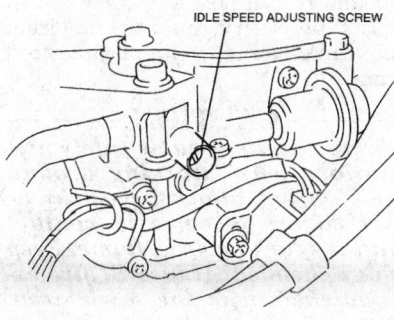

IDLE SPEED ADJUSTING SCREW

194278

Idle speed adjusting screw — 1.3L (VIN 3) engine

mission in **P**. Set the parking brake and block the drive wheels.

2. Warm the engine to normal operating temperature.

3. Use a service wire to ground the diagnosis switch terminal in the monitor coupler.

4. Stop the engine, and connect a duty meter between the duty output terminal and ground terminal of the monitor coupler located next to the battery.

5. Set the tachometer.

6. Start and warm the engine to normal operating temperature.

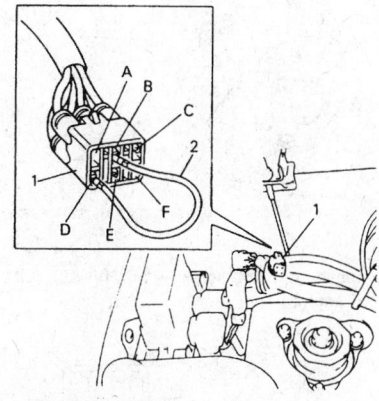

1 Monitor coupler
2 Service wire

A: Blank
B: Diagnosis switch terminal
C: Diagnosis output terminal
D: Ground terminal
E: Test switch terminal
F: Duty output terminal

332366

Grounding the diagnostic switch terminal — 1.6L (VIN 0), (VIN U) MFI engine

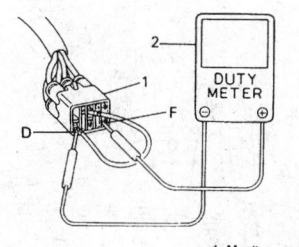

1 Monitor coupler
2 Duty meter
D: Ground terminal
F: Duty output terminal

332367

Connecting the duty meter — 1.6L (VIN 0), (VIN U) MFI and 1.8L (VIN 2) engines

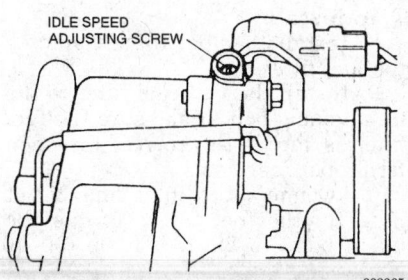

IDLE SPEED ADJUSTING SCREW

332365

Idle speed adjusting screw — 1.6L (VIN 0), (VIN U) MFI engine

7. Check the idle speed. If the speed is not between 750–850 rpm, adjust by turning the idle speed adjusting screw on the throttle body.

8. When adjustment is completed, install an adjusting screw cap onto the throttle body.

9. Remove the service wire from the monitor coupler.

10. Install the cap to the monitor coupler.

Mixture

ADJUSTMENT

All Engines

The air/fuel mixture is controlled by the engine control system and is not adjustable. The engine control system consists of sensors which detect engine conditions and the Powertrain Control Module (PCM) which controls the system based on sensor signals and actuators which operate under the control of the PCM. The injector drive times and injector timing are PCM controlled for optimum air/fuel mixture. The PCM provides a richer air/fuel mixture in "open-loop" operation when the engine is cold or under extremely high load. When warm or operating normally, the PCM controls the air/fuel mixture by using the heated oxygen sensor signal to carry out "closed-loop" control for best emission controlling mixture. Since the PCM regulates the air/fuel mixture under all conditions, the mixture is not adjustable.

Fuel Filter

REMOVAL AND INSTALLATION

— **CAUTION** —
The fuel system pressure must be relieved before disconnecting any fuel lines. Failure to do so may result in personal injury.

1. Disconnect the negative battery cable.

2. Remove the fuel filler cap to release the fuel vapor pressure in the fuel tank. After releasing pressure, reinstall filler cap.

3. Raise and support the vehicle safely.

4. Properly release the fuel pressure.

5. Place a fuel container under fuel filter.

6. Disconnect the inlet and outlet hoses from the fuel filter.

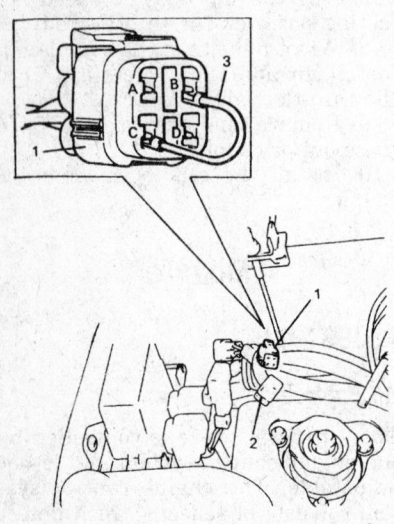

A: Duty check terminal   1 Monitor coupler
B: Diagnosis switch terminal   2 Cap
C: Ground terminal   3 Service wire
D: Test switch terminal

194351

Grounding the diagnostic switch terminal — 1.6L (VIN 0), (VIN U) TFI engine

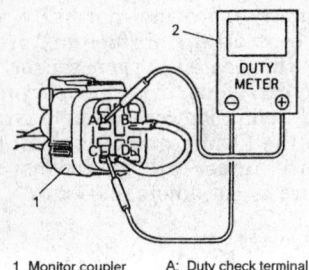

1 Monitor coupler   A: Duty check terminal
2 Duty meter   C: Ground terminal

194352

Connecting the duty meter — 1.6L (VIN 0), (VIN U) TFI engine

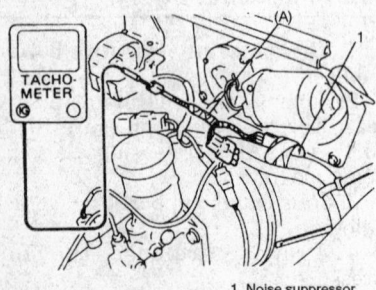

1 Noise suppressor

194364

Special tool 09931–96010 — 1.6L (VIN 0), (VIN U) TFI engine

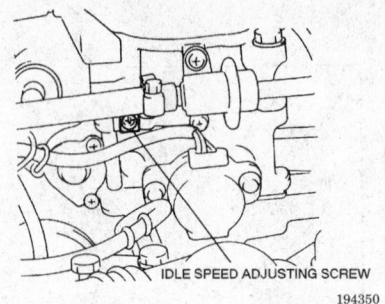

IDLE SPEED ADJUSTING SCREW

194350

Idle speed adjusting screw — 1.6L (VIN 0), (VIN U) TFI engine

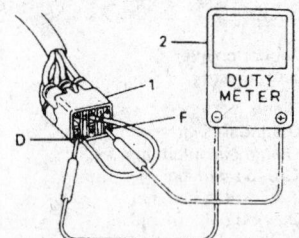

1. Monitor coupler   D: Ground terminal
2. Duty meter   F: Duty output terminal

341248

Connecting the duty meter — 1.8L (VIN 2) Engine

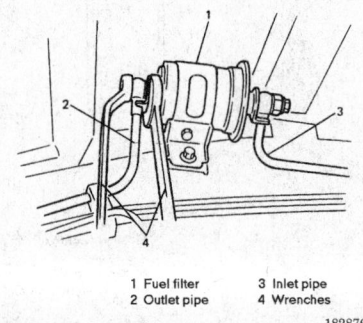

1 Fuel filter   3 Inlet pipe
2 Outlet pipe   4 Wrenches

182876

Fuel filter mounting location and fuel line connections — 1.3L (VIN 3) engine

7. Remove the fuel filter from the chassis frame by removing bolt and removing fuel filter from fuel filter clamp.

To install:

8. Install the fuel filter clamp to the fuel filter.

9. Install the new fuel filter to the frame using bolt. Make sure the fuel filter is facing the correct direction for proper fuel flow.

10. Connect the inlet and outlet hoses to the fuel filter. Torque the union bolts to 22–28.5 ft. lb (30–40 Nm).

11. Lower the vehicle and connect the negative battery terminal.

12. Start the engine and check for leaks.

Fuel Pump

REMOVAL AND INSTALLATION

─── **CAUTION** ───

Fuel injection systems remain under pressure, even after the engine has been turned OFF. The fuel system pressure must be relieved before disconnecting any fuel lines. Failure to do so may result in fire and/or personal injury.

1. Properly relieve the fuel system pressure.

2. Disconnect the negative battery cable.

3. Remove the rear bumper cover.

4. Disconnect the fuel gauge sending unit and fuel pump electrical connectors from the fuel tank.

5. Disconnect the fuel tank filler hose cover, filler hose and fuel tank inlet valve (breather hose).

6. Disconnect the inlet pipe from the fuel filter and remove the fuel vapor and return hoses.

7. If necessary, drain the fuel from the tank using a hand operated pump.

─── **CAUTION** ───

Observe all applicable safety precautions when working around fuel. Do not allow fuel spray or fuel vapors to come in contact with a spark or open flame. Keep a dry chemical (Class B) fire extinguisher near the work area. Never drain or store fuel in an open container due to the possibility of fire or explosion.

8. Disconnect the fuel tank protector and remove the fuel tank and cover from the vehicle.

9. Remove the six fuel pump mounting screws and remove the fuel pump from the fuel tank.

To install:

10. Install the fuel pump to the fuel tank using a new gasket. Install the six mounting screws.

11. Install the fuel tank and cover to the vehicle and connect the fuel tank protector.

12. Replace the fuel vapor and return hoses and connect the inlet pipe to the fuel filter.

13. Reconnect the fuel tank inlet valve (breather hose), filler hose and fuel tank filler hose cover.

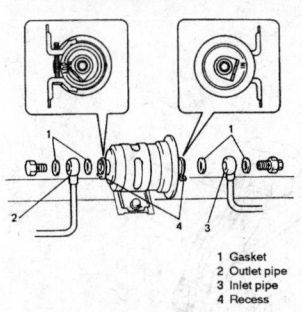

1 Gasket
2 Outlet pipe
3 Inlet pipe
4 Recess

182877

Exploded view of the fuel filter components — 1.3L (VIN 3) engine

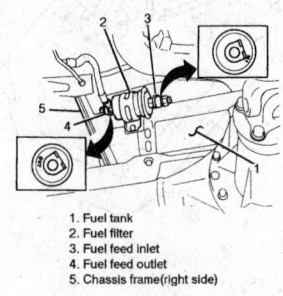

1. Fuel tank
2. Fuel filter
3. Fuel feed inlet
4. Fuel feed outlet
5. Chassis frame(right side)

241124

Fuel filter mounting location and fuel line connections — 1.6L (VIN 0), (VIN U) and 1.8L (VIN 2) engines

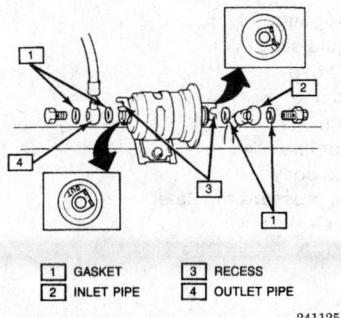

| 1 | GASKET | 3 | RECESS |
| 2 | INLET PIPE | 4 | OUTLET PIPE |

241125

Exploded view of the fuel filter components — 1.6L (VIN 0), (VIN U) and 1.8L (VIN 2) engines

14. Connect the fuel pump and fuel gauge sending sending unit electrical connectors to the fuel tank.
15. Replace the bumper cover, and reconnect the negative battery cable.
16. If necessary, refill the fuel tank with any fuel that was drained.
17. Start the engine and check for leaks.

Fuel Injector

REMOVAL AND INSTALLATION

1.3L (VIN 3) and 1.6L (VIN 0), (VIN U) TFI Engines

1. Disconnect the negative battery cable.
2. Release the fuel pressure in the fuel feed line.
3. Remove the the air intake case from the throttle body.
4. Remove the fuel feed pipe clamp from the intake manifold and disconnect the fuel feed pipe from the throttle body.
5. Remove the injector cover.
6. Disconnect the injector coupler, release its wire harness from the clamp and remove its grommet from the throttle body.
7. Place a cloth over the injector and a hand on top of it. Using an air nozzle, blow low pressure compressed air into the fuel inlet port of the throttle body, and the injector can be removed.

NOTE: Be precise about the pressure of the compressed air. Using excessively high pressure may force the injector to jump out and may cause damage, not only to the injector itself but also to other parts.

To install:
8. Apply thin coating of gasoline to O-rings and then install the fuel injector to the throttle body.
9. Connect fuel injector wire connector and install injector cover. Torque the injector cover screw to 1.4 ft. lbs. (2 Nm).
10. Connect fuel feed pipe to throttle body.
11. Connect the negative battery cable and turn the ignition **ON** for 3 seconds and then **OFF** until fuel pressure is felt at the return hose.
12. Install air intake case assembly.
13. Secure all wires and check for any leaks in the system.

1993–95 1.6L (VIN 0), (VIN U) MFI Engines

1. Release the fuel pressure in the fuel feed line. Disconnect the negative battery cable.
2. Remove the throttle cover.
3. Remove the air intake pipe with the hose as follows:
   a. Disconnect the Idle Air Control (IAC) valve hose, and the PCV hose from the air intake pipe.

   b. Carefully remove the radiator cap to relieve the engine coolant pressure, and then reinstall the cap.
   c. Disconnect the water hoses from the air intake pipe. Cover the hose-to-pipe joint with a shop rag to catch any escaping coolant.
   d. Unclamp the automatic transmission throttle cable, if equipped.
   e. Remove the air intake pipe.
4. Remove the Exhaust Gas Recirculation (EGR) modulator from the EGR modulator bracket.
5. Remove the 2 bolts and the EGR modulator bracket from the intake manifold.
6. Disconnect the 4 fuel injector connectors. Remove the 2 bolts and fuel pressure regulator from the fuel rail.
7. Remove the fuel pulsation damper and fuel feed pipe from the fuel rail.
8. Remove the 3 bolts and the fuel rail from the intake manifold. Remove the 4 fuel injectors from the intake manifold.

To install:
9. Install new O-rings onto the fuel injectors; be careful not to damage the O-rings. Install the grommets onto the fuel injectors.

NOTE: Inspect the insulators for breakage or scoring. If necessary, replace them.

10. Install the insulators and cushions to the intake manifold.
11. Lubricate the O-rings with fuel and install the injectors onto the fuel rail and then the whole assembly to the intake manifold.

NOTE: Make sure the fuel injectors rotate smoothly. If not, an O-ring may be installed incorrectly; replace it.

12. Install the 3 fuel rail bolts and torque to 13–20 ft. lbs. (18–28 Nm).
13. Lubricate the new gaskets with fuel. Install the fuel feed pipe, fuel pulsation damper and fuel rail. Torque the fuel pulsation damper bolts to 18.5–25 ft. lbs. (25–35 Nm).
14. Install the fuel pressure regulator and torque the bolts to 6–8.5 ft. lbs. (8–12 Nm).
15. Connect the 4 fuel injectors. Install the EGR modulator bracket and torque the bracket-to-intake manifold bolts to 17 ft. lbs. (23 Nm).
16. Install the EGR modulator to the EGR bracket.
17. Install the air intake pipe with the hose.
18. Install the throttle cover.
19. Connect negative battery cable.

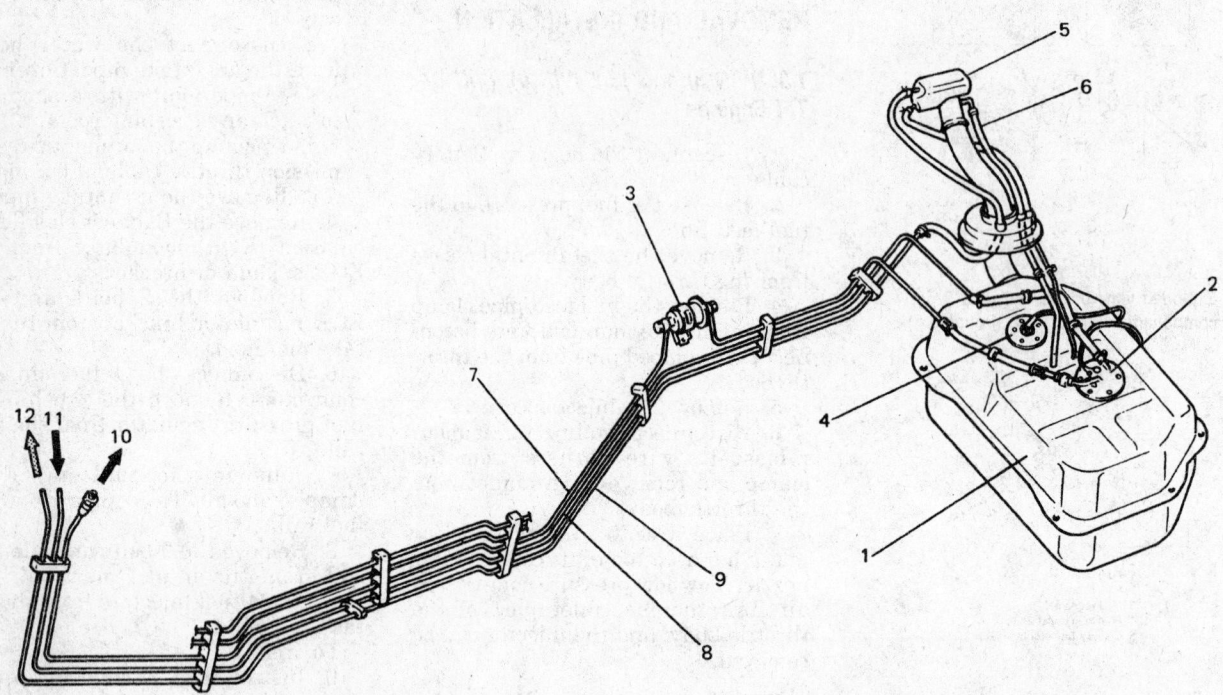

1 Fuel tank
2 Fuel pump
3 Fuel filter
4 Fuel level gauge
5 Vapor liquid seperator
6 Breather hose
7 Fuel feed line
8 Fuel return line
9 Fuel vapor line
10 To throttle body
11 From fuel pressure regulator
12 To canister

183107

Fuel tank components — Samurai

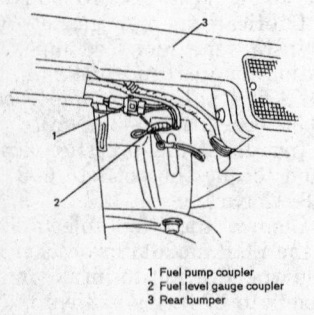

1 Fuel pump coupler
2 Fuel level gauge coupler
3 Rear bumper

183116

**Fuel pump and gauge electrical
connector locations — Samurai**

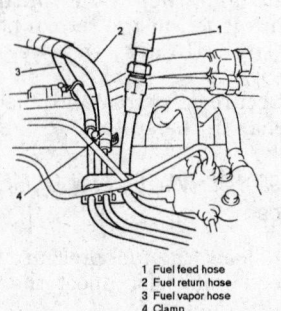

1 Fuel feed hose
2 Fuel return hose
3 Fuel vapor hose
4 Clamp

183108

**Fuel tank hose and pipe
connections — Samurai**

1996–97 1.6L (VIN 0), (VIN U) MFI Engines

1. Release the fuel pressure in the fuel feed line. Disconnect the negative battery cable.

2. Remove the throttle cover.

3. Remove the air intake pipe with the hose as follows:

a. Disconnect the Idle Air Control (IAC) valve hose, and the PCV hose from the air intake pipe.

b. Carefully remove the radiator cap to relieve the engine coolant pressure, and then reinstall the cap.

c. Disconnect the water hoses from the air intake pipe. Cover the

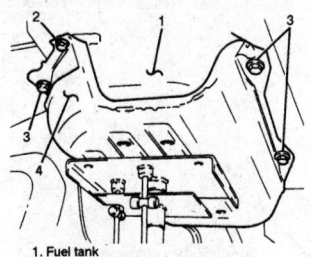

1. Fuel tank
2. Fuel tank protector-top-fuel tank bolt
3. Fuel tank protector and fuel tank mounting bolts
4. Fuel tank protector

331048

Removing the fuel tank — Sidekick, Tracker and X-90

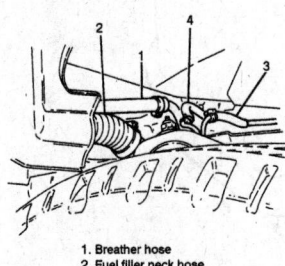

1. Breather hose
2. Fuel filler neck hose
3. Fuel return hose
4. Fuel vapor hose

331050

Fuel tank hose and pipe connections — Sidekick, Tracker and X-90

hose-to-pipe joint with a shop rag to catch any escaping coolant.

d. Unclamp the automatic transmission throttle cable, if equipped.

e. Remove the air intake pipe.

4. Remove the intake manifold No. 2 stiffener from the intake manifold and surge tank.

5. Disconnect the four fuel injector connectors.

6. Remove the bolts and fuel pressure regulator from the fuel rail.

7. Remove the clamp bolts for the fuel feed pipe and return pipe.

8. Remove the three bolts and the fuel rail from the intake manifold. Remove the 4 fuel injectors from the intake manifold.

To install:

9. Install new O-rings onto the fuel injectors; be careful not to damage the O-rings. Install the grommets onto the fuel injectors.

NOTE: Inspect the insulators for breakage or scoring. If necessary, replace them.

10. Install the insulators and cushions to the intake manifold.

11. Lubricate the O-rings with fuel and install the injectors onto the fuel

rail and then the whole assembly to the intake manifold.

NOTE: Make sure the fuel injectors rotate smoothly. If not, an O-ring may be installed incorrectly; replace it.

12. Install the three fuel rail bolts and torque to 17 ft. lbs. (23 Nm). Make sure the fuel injectors rotate freely.

13. Install the fuel pressure regulator and torque the bolts to 8 ft. lbs. (10 Nm).

14. Lubricate the new gaskets with fuel. Install the fuel feed pipe and return pipe.

15. Connect the four fuel injector connectors.

16. Install the air intake pipe with the hose.

17. Install the intake manifold No. 2 stiffener to the intake manifold and surge tank.

18. Install the throttle cover.

19. Connect negative battery cable.

20. Start the engine and check for leaks.

1.8L (VIN 2) Engine

1. Release the fuel pressure in the fuel feed line.

2. Disconnect the negative battery cable from the battery.

3. Disconnect the four fuel injector connectors.

4. Remove the three bolts and the fuel rail from the intake manifold. Remove the 4 fuel injectors from the intake manifold.

5. If necessary, disconnect the supply line and return line for the fuel rail from the fuel pipes.

To install:

6. If removed, connect the supply line and return line to the fuel pipes.

7. Install new O-rings onto the fuel injectors; be careful not to damage the O-rings. Install the grommets onto the fuel injectors.

NOTE: Inspect the insulators for breakage or scoring. If necessary, replace them.

8. Install the insulators and cushions to the intake manifold.

9. Lubricate the O-rings with fuel and install the injectors onto the fuel rail and then the whole assembly to the intake manifold.

NOTE: Make sure the fuel injectors rotate smoothly. If not, an O-ring may be installed incorrectly; replace it.

10. Install the three fuel rail bolts and torque to 22 ft. lbs. (30 Nm).

Make sure the fuel injectors rotate freely.

11. Connect the four fuel injector connectors.

12. Connect negative battery cable to the battery.

13. Start the engine and check for leaks.

ENGINE MECHANICAL

Engine Assembly

REMOVAL AND INSTALLATION

1.3L (VIN 3) and 1993–95 1.6L (VIN 0), (VIN U) Engines

——— **CAUTION** ———
The fuel system pressure must be relieved before disconnecting any fuel lines. Failure to do so may result in personal injury.

1. Disconnect the battery cables from the battery, negative cable first.

2. Mark the position of the hood on the hinges for installation reference, then remove the hood with the aid of an assistant.

3. Drain the cooling system.

4. Remove the radiator reservoir tank, fan shroud, cooling fan and radiator.

5. Properly discharge the air conditioning system and remove the air conditioning condenser, if equipped.

6. Remove the air cleaner outlet hose.

7. Disconnect the accelerator and the automatic transmission kickdown cable from the throttle body, if equipped.

8. Disconnect and tag the throttle opener VSV and EGR VSV wires at the coupler.

9. Disconnect and tag the water temperature, oil pressure, air temperature and ground cable wires at the intake manifold.

10. Disconnect and tag the injector, throttle position sensor and idle speed control solenoid valve wires at their couplers, if equipped.

11. Disconnect the PTC heater wires at the coupler for automatic transmission vehicle, if equipped.

12. Disconnect and tag the wires at the starter and alternator.

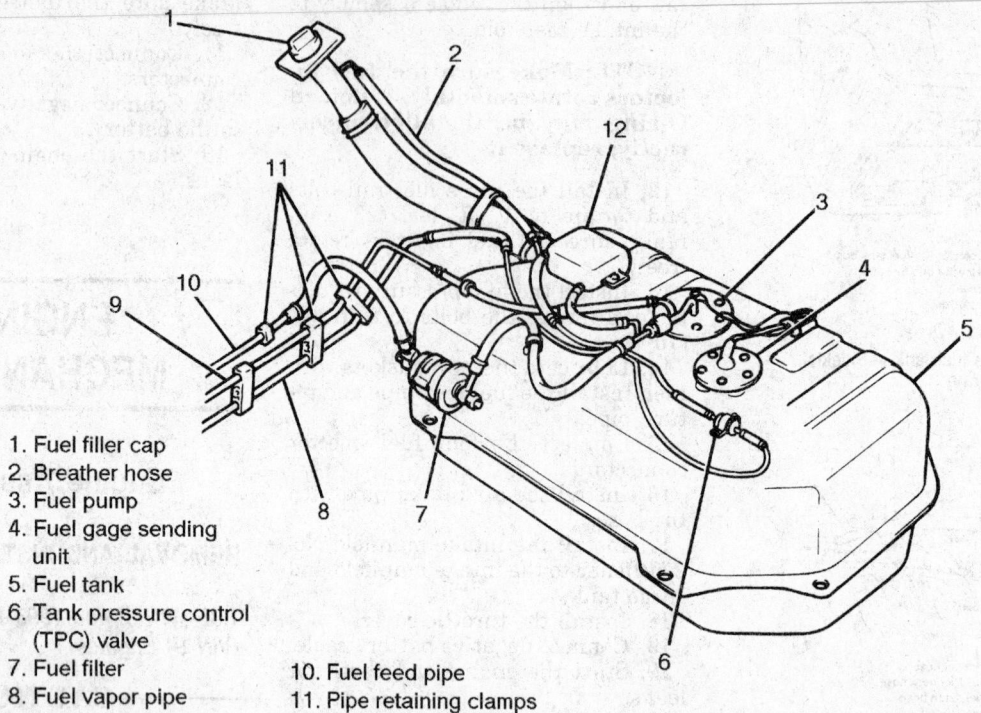

1. Fuel filler cap
2. Breather hose
3. Fuel pump
4. Fuel gage sending unit
5. Fuel tank
6. Tank pressure control (TPC) valve
7. Fuel filter
8. Fuel vapor pipe
9. Fuel return pipe
10. Fuel feed pipe
11. Pipe retaining clamps
12. Fuel vapor separator

331052

Fuel tank components — Sidekick, Tracker and X-90

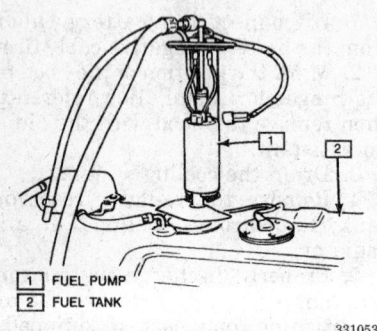

| 1 | FUEL PUMP |
| 2 | FUEL TANK |

331053

Removing the pump assembly from the fuel tank

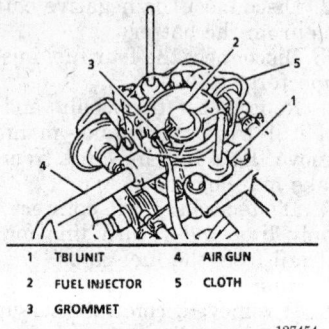

1	TBI UNIT	4	AIR GUN
2	FUEL INJECTOR	5	CLOTH
3	GROMMET		

127454

Fuel injector components — Throttle-body Fuel Injection (TFI) system

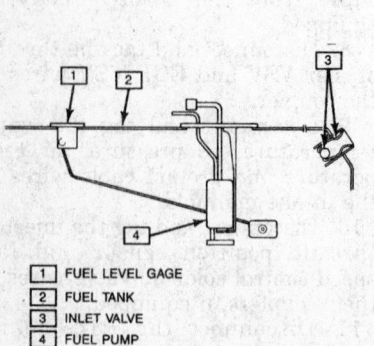

1	FUEL LEVEL GAGE
2	FUEL TANK
3	INLET VALVE
4	FUEL PUMP

331055

Fuel pump and gauge assembly — Sidekick, Tracker and X-90

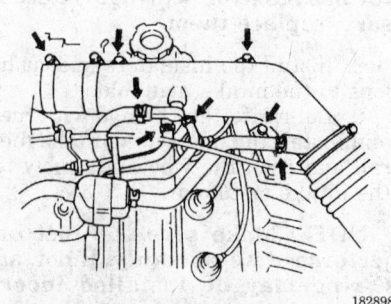

182898

Air intake pipe removal — Multi-port Fuel Injected (MFI) engines

13. Disconnect and tag the oxygen sensor and distributor wires at their couplers. Remove the coil wire.

14. Remove the starter motor and disconnect the ground wires from the distributor assembly.

15. Remove the fuel tank filler cap to relieve the pressure. Reinstall the cap.

16. Relieve the fuel pressure in the fuel feed line. Disconnect and tag the fuel feed and return hoses.

17. Remove the gear shift lever mounting bolts and remove the shifter.

18. Disconnect the canister purge hose and remove the pressure sensor hose from the fuel filter, if equipped.

19. Disconnect the brake booster hose from the intake manifold.

20. Remove the vacuum hose for the automatic transmission from the intake manifold, if equipped.

21. Disconnect the heater hoses from the heater core outlet pipe and the intake manifold. Disconnect the water inlet hose from the water inlet pipe.

22. Raise and safely support the vehicle.

23. Drain the engine oil and remove the exhaust pipe from the exhaust manifold and muffler.

24. Disconnect the clutch cable, if equipped with manual transmission.

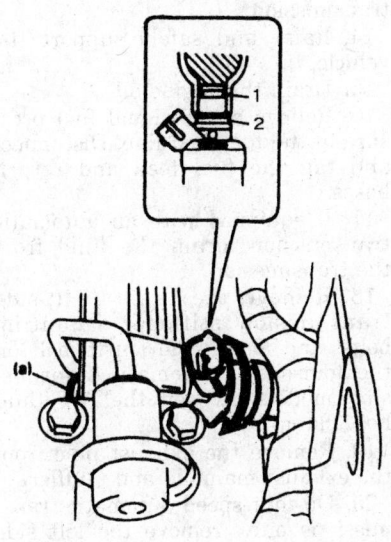

1 O-ring
2 Grommet

182900

Rotate injector — Multi-port Fuel Injected (MFI) engines

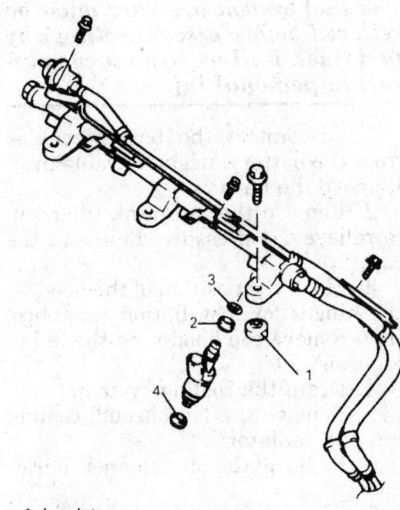

1. Insulator
2. Grommet
3. O ring
4. Cushion

331074

Fuel injector and rail assembly — 1.8L (VIN 2) engine

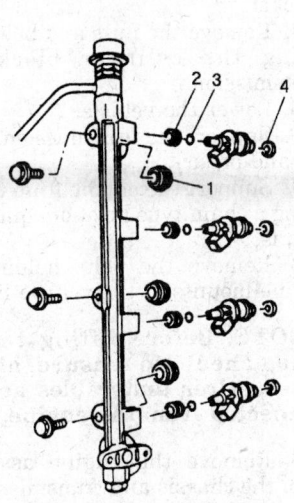

1 Insulator
2 Grommet
3 O-ring
4 Cushion

329031

Exploded view of the fuel delivery rail and injectors — 1.6L (VIN 0), (VIN U) engine

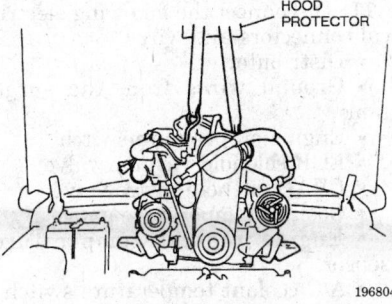

HOOD PROTECTOR

196804

Hoisting the engine — 1.3L (VIN 3) engine

25. Drain the automatic transmission fluid, if equipped.

26. Disconnect the clutch (torque converter) housing lower plate.

27. Remove the lock drive plate, using a suitable tool, if equipped with automatic transmission.

28. Lower the vehicle.

29. Remove the nuts and bolts fastening the cylinder block and transmission.

30. Support the transmission, using a stand or jack.

31. Support the engine from the top using a chain type hoist or equivalent means.

32. Remove the engine mounts with the chassis side mounting brackets.

NOTE: Before lifting the engine, check to ensure all the hoses, wires and cables are disconnected from the engine.

33. Remove the engine assembly from the chassis and transmission by sliding towards the front side and carefully hoisting the engine.

To install:

34. Install the engine into the engine compartment and connect to the transmission.

35. Install the engine mounts with the chassis side mounting brackets. Tighten nuts **A** to 29–36 ft. lbs. (40–50 Nm), and bolts **B** to 37–43 ft. lbs. (50–60 Nm).

36. Remove the lifting device.

37. Install the nuts and bolts fastening the cylinder block and transmission. Tighten to 37–43 ft. lbs. (50–60 Nm).

38. Raise and safely support the vehicle.

39. Install the lock driveplate, if equipped.

40. Connect the clutch (torque converter) housing lower plate.

41. Connect the automatic transmission lines, if equipped.

42. Reconnect the clutch cable to the bracket, if equipped with manual transmission.

43. Install the exhaust pipe to the exhaust manifold and muffler.

44. Lower the vehicle.

45. Reconnect the heater hoses to the heater core outlet pipe and the intake manifold. Reconnect the water inlet hose.

46. Install the vacuum hose for the automatic transmission to the intake manifold, if equipped.

47. Reconnect the brake booster hose to the intake manifold.

48. Install the pressure sensor hose to the fuel filter and connect the canister purge hose, if equipped.

49. Connect the fuel return and feed hoses.

50. Install the starter motor and connect the ground wire to the distributor.

51. Connect the coil wire and connect the oxygen sensor and distributor wires to their couplers.

52. Reconnect the wires at the alternator and starter motor.

53. Connect the PTC heater wires at the coupler, if equipped with automatic transmission.

54. Connect the idle speed control solenoid valve, throttle position sensor and injector wires at their couplers, if equipped.

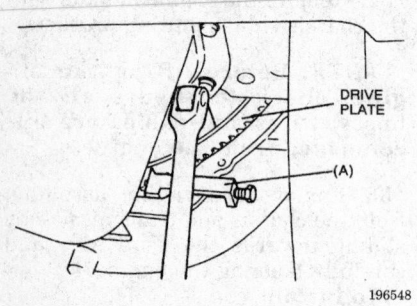

Locking the drive plate (automatic transmission only) — 1.6L (VIN 0), (VIN U) engine

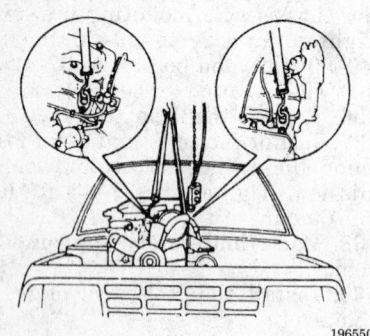

Hoisting the engine — 1.6L (VIN 0), (VIN U) engine

55. Connect the ground cable, air temperature, oil pressure and water temperature wires to the intake manifold.

56. Reconnect the throttle opener VSR and EGR VSR wires to their couplers, if equipped.

57. Connect the accelerator cable and the automatic transmission kickdown cable, if equipped, to the throttle body.

58. Install the air cleaner outlet hose. Refill the engine oil and the transmission oil to the proper lever.

59. Install the air conditioning condenser and properly recharge the air conditioning system, if equipped.

60. Install the radiator, cooling fan, fan shroud and radiator reservoir tank.

61. Install the hood, aligning the marks made during removal.

62. Fill the cooling system with the proper coolant.

63. Reconnect the positive and negative battery cables.

64. Adjust the accelerator, clutch and the automatic transmission kickdown cable, if equipped.

65. Before starting the engine, check to see if all the parts disassembled are back in place securely.

66. Start the engine and check the timing. Check for any oil leaks.

1996–97 1.6L (VIN 0), (VIN U) Engines

— CAUTION —

The fuel system pressure must be relieved before disconnecting any fuel lines. Failure to do so may result in personal injury.

1. Disconnect the battery cables from the battery, negative cable first. Remove the battery.

2. Remove the fuel tank filler cap to relieve the pressure. Reinstall the cap.

3. Mark the position of the hood on the hinges for installation reference, then remove the hood with the aid of an assistant.

4. Drain the cooling system.

5. Remove the fan shroud, cooling fan and radiator.

6. Remove the air cleaner outlet hose.

7. Disconnect the accelerator and if equipped with an automatic transmission, the A/T kickdown cable from the throttle body.

8. Remove the intake manifold bracket, and then release the wiring harnesses from the clamps.

9. Disconnect and tag the wires at the starter and alternator.

10. Remove the starter motor and disconnect the ground wires from the distributor assembly.

11. Disconnect the following electrical connectors and wires:
- Distributor
- Ground wires from the surge tank
- Engine oil pressure switch
- EGR solenoid vacuum valve
- EVAP solenoid purge valve
- Coolant temperature gauge
- Engine coolant temperature sensor
- A/C coolant temperature switch
- Injector connectors
- Throttle position sensor
- IAC valve
- Heated oxygen sensor
- Alternator
- EGR solenoid vacuum valve

12. Disconnect the following hoses:
- Canister purge hose from the EVAP canister purge valve
- Brake booster hose from the surge tank
- Radiator outlet hose from the inlet pipe
- Heater inlet and outlet hose from the heater unit
- Vacuum hose for the A/T from the intake manifold
- Tank pressure control solenoid vacuum valve hose from the solenoid vacuum valve
- Manifold differential pressure sensor hose from the pressure sensor

13. Loosen the top nuts and bolts fastening the cylinder block to the transmission.

14. Raise and safely support the vehicle.

15. Drain the engine oil.

16. Relieve any residual fuel pressure in the fuel feed line. Disconnect and tag the fuel feed and return hoses.

17. If equipped with an automatic transmission, drain the fluid from the transmission.

18. Remove the three right-side transmission stiffener mounting bolts, and loosen the fourth bolt on the four-speed automatic transmission models. Loosen the A/T fluid hose clamp bolt.

19. Remove the exhaust pipe from the exhaust manifold and muffler.

20. On four-speed automatic transmissions only, remove the left side stiffener.

21. If equipped with an automatic transmission, disconnect the clutch (torque converter) housing lower plate.

22. If equipped with an automatic transmission, remove the torque converter bolts.

23. Leaving the hoses connected, remove from the cylinder block the power steering pump and A/C compressor (as equipped) with the bracket.

24. Remove the nuts and bolts fastening the cylinder block and transmission.

25. Lower the vehicle.

26. Support the transmission, using a stand or jack.

27. Support the engine from the top using a chain type hoist or equivalent means.

28. Remove the bolts holding the engine mounts to the vehicle frame.

NOTE: Before lifting the engine, check to ensure all the hoses, wires and cables are disconnected from the engine.

29. Remove the engine assembly from the chassis and transmission by sliding towards the front side and carefully hoisting the engine.

To install:

30. Install the engine into the engine compartment and connect to the transmission.

31. Install the engine to transmission bolts. Do not torque the bolts at this time.

32. Install and torque the engine mounting bolts to the vehicle frame. Torque the bolts to 37 ft. lbs. (50 Nm).

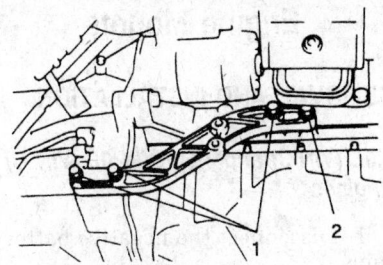

1. Right side transmission stiffener bolts (remove)
2. Right side transmission stiffener bolt (loosen)

328616

Right side transmission stiffener — 1996–97 1.6L (VIN 0), (VIN U) engines

33. Tighten the bolts and nuts holding the engine to the transmission to 62 ft. lbs. (85 Nm).

34. Remove the lifting device.

35. Raise and safely support the vehicle.

36. Install the A/C compressor and bracket, and the power steering pump.

37. If equipped with a 4 speed automatic transmission, install the left-side stiffener and torque the mounting bolts to 37 ft. lbs. (50 Nm).

38. If equipped with an automatic transmission, install the torque converter bolts. Tighten the torque converter bolts to 47 ft. lbs. (65 Nm).

39. If equipped with an automatic transmission, connect the clutch (torque converter) housing lower plate.

40. Install the exhaust pipe to the exhaust manifold and muffler.

41. Install the right-side stiffener. On 4A/T models, tighten the loosened bolt to 37 ft. lbs. (50 Nm); on all automatic transmissions install and tighten the three removed bolts to 21 ft. lbs. (29 Nm).

42. Connect the fuel return and feed hoses. Tighten the flare nut on the fuel feed hose to 29–36 ft. lbs. (40–50 Nm).

43. Lower the vehicle.

44. Connect the following hoses:
- Canister purge hose to the EVAP canister purge valve
- Brake booster hose to the surge tank
- Radiator outlet hose to the inlet pipe
- Heater inlet and outlet hose to the heater unit
- Vacuum hose for the A/T to the intake manifold
- Tank pressure control solenoid vacuum valve hose to the solenoid vacuum valve
- Manifold differential pressure sensor hose to the pressure sensor

45. Connect the following electrical connectors and wires:
- Distributor
- Ground wires to the surge tank
- Engine oil pressure switch
- EGR solenoid vacuum valve
- EVAP solenoid purge valve
- Coolant temperature gauge
- Engine coolant temperature sensor
- A/C coolant temperature switch
- Injector connectors
- Throttle position sensor
- IAC valve
- Heated oxygen sensor
- Alternator
- EGR solenoid vacuum valve

46. Install the starter motor and connect the electrical connectors.

47. Clamp the wiring harnesses, and install the intake manifold bracket.

48. Connect the accelerator cable and the automatic transmission kickdown cable, if equipped, to the throttle body.

49. Install the air cleaner outlet hose. Refill the engine oil and the transmission oil to the proper lever.

50. Install the radiator, cooling fan and fan shroud.

51. Install the hood, aligning the marks made during removal.

52. Fill the cooling system with the proper coolant.

53. Install the battery. Reconnect the positive and negative battery cables.

54. Adjust the accelerator, clutch and the automatic transmission kickdown cable, if equipped.

55. Adjust the belt tensions.

56. Before starting the engine, check to see if all the parts disassembled are back in place securely.

57. Start the engine and check the timing. Check for any leaks.

1.8L (VIN 2) Engine

—————— CAUTION ——————

The fuel system pressure must be relieved before disconnecting any fuel lines. Failure to do so may result in personal injury.

1. Disconnect the negative battery cable from the battery.

2. Remove the fuel tank filler cap to relieve the pressure. Reinstall the cap.

3. Mark the position of the hood on the hinges for installation reference, then remove the hood with the aid of an assistant.

4. Drain the cooling system.

5. Remove the fan shroud, cooling fan and radiator.

6. Disconnect the accelerator and if equipped with an automatic transmission, disconnect the A/T kickdown cable from the throttle body.

7. Remove the strut tower bar by removing the two bolts and two nuts.

8. Remove the air cleaner outlet hose and air cleaner case.

9. Remove the engine oil level gauge and if equipped with an A/T, remove the A/T fluid level gauge guide.

10. Disconnect the electrical connectors from the following:
- Injector wire
- Camshaft position sensor
- Ignition coil wire
- Throttle position sensor
- Mass air flow sensor
- Idle Air Control valve
- Crankshaft position sensor
- Ground wire from the intake manifold
- EVAP canister purge valve
- EGR valve
- Oxygen sensor
- Engine Control Temperature sensor
- Coolant temperature switch
- Alternator
- Starter
- Oil pressure
- Power steering pump
- Ground wire from the starter
- Wire harness clamps

11. Disconnect the following hoses:
- Fuel feed hose from the fuel delivery pipe
- Fuel return hose from the fuel return pipe
- Heater hoses from the heater core
- EVAP canister hose from the canister pipe
- Brake booster vacuum hose
- MAP sensor hose from the intake manifold

12. With the power steering hoses connected, remove the power steering pump from the engine. Support the pump to the side of the engine. Do not allow the pump to hang from the power steering hoses.

13. With the A/C hoses still connected, remove the A/C compressor from the engine. Support the compressor to the side of the engine. Do not allow the pump to hang from the compressor hoses.

14. Raise and safely support the front of the vehicle.

15. If equipped with four wheel drive, remove the front differential from the vehicle.

16. Remove the front exhaust pipe from the vehicle.

17. If equipped with an automatic transmission, remove the A/T fluid

hose clamps from the right side of the transmission stiffener.

18. Remove the right side transmission stiffener.

19. Remove the clutch housing lower plate.

20. If equipped with an automatic transmission, remove the torque converter bolts.

21. Lower the vehicle.

22. Remove the starter motor.

23. Support the transmission.

24. Remove the bolts and nuts fastening the engine to the transmission.

25. Install a lifting device to the engine.

26. Disconnect the engine mounts from the engine by removing the nuts.

27. Check that all hoses, wires and cables are disconnected from the engine.

28. Remove the engine assembly from the vehicle.

To install:

29. Install the engine to the vehicle.

30. Connect the engine mounts to the engine and torque the nuts and bolts to 37 ft. lbs. (50 Nm).

31. Install the bolt and nuts to hold the engine to the transmission. Torque the bolt and nuts to 58 ft. lbs. (80 Nm).

32. Remove the lifting device from the engine.

33. Install the starter.

34. Raise the vehicle.

35. If equipped with an automatic transmission, install the torque converter bolts and torque the bolts to 47 ft. lbs. (65 Nm).

36. Install the clutch housing lower plate.

37. Install the transmission stiffener and bolts. Torque the stiffener bolts to 37 ft. lbs. (50 Nm).

38. If equipped with an automatic transmission, install the automatic

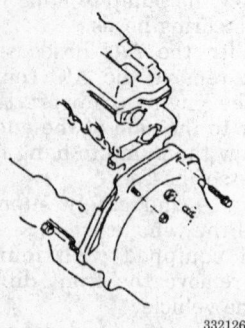

332126

Transmission-to-engine bolts and nuts — 1.8L (VIN 2) engine

transmission fluid hose clamps to the right side transmission stiffener.

39. Install the front exhaust pipe. Torque the bolts and nuts to 37 ft. lbs. (50 Nm).

40. Install the front differential housing.

41. Lower the vehicle.

42. Install the A/C compressor to the engine.

43. Install the power steering pump to the engine.

44. Connect the following hoses:
• Fuel feed hose to the fuel delivery pipe
• Fuel return hose to the fuel return pipe
• Heater hoses to the heater core
• EVAP canister hose to the canister pipe
• Brake booster vacuum hose
• MAP sensor hose to the intake manifold

45. Connect the electrical connectors from the following:
• Injector wire
• Camshaft position sensor
• Ignition coil wire
• Throttle position sensor
• Mass air flow sensor
• Idle Air Control valve
• Crankshaft position sensor
• Ground wire to the intake manifold
• EVAP canister purge valve
• EGR valve
• Oxygen sensor
• Engine Control Temperature sensor
• Coolant temperature switch
• Alternator
• Starter
• Oil pressure
• Power steering pump
• Ground wire from the starter
• Wire harness clamps

46. Install the engine oil level gauge.

47. If equipped with A/T, install the fluid level gauge guide.

48. Install the air cleaner upper case and outlet hose.

49. Install the strut tower bar.

50. Connect and adjust the accelerator cable and A/T throttle cable to the throttle body.

51. Install the radiator, radiator fan shroud and cooling fan.

52. Fill the engine with oil and the radiator with engine coolant.

53. Install the hood.

54. Connect the negative battery cable to the battery.

55. Start the vehicle and check for leaks.

Engine Mounts

REMOVAL AND INSTALLATION

1.3L (VIN 3) and 1.6L (VIN 0), (VIN U) Engines

1. Disconnect the negative battery cable.

2. Remove the four bolts and reposition the fan shroud.

3. Raise and safely support the vehicle.

4. Remove the four bolts and the front skid plate.

5. Support the engine using a suitable jack.

6. On the right side, disconnect the starter wiring.

7. On the left side, remove the exhaust pipe-to-exhaust bracket nut.

8. Remove the 3 (right side) or 2 (left side) engine mount-to-frame bracket bolts.

9. On the left side, remove the 2 engine mount-to-exhaust bracket bolts and remove the exhaust bracket.

10. Remove the 3 engine mount-to-engine bracket bolts.

11. Raise the engine slightly with the jack.

12. Remove the engine mount assembly.

13. Remove the frame and engine brackets from the engine mount.

To install:

14. Install the engine and frame brackets on the engine mount.

15. Install the engine mount and lower the engine into position.

16. Install the engine mount-to-frame bracket bolts. DO NOT tighten at this time.

17. On the left side, install the exhaust bracket and install the two exhaust bracket mounting bolts.

18. Install the engine mount-to-engine bracket bolts. Tighten the brackets mounting bolts to: 29–36 ft. lbs. (40–50 Nm) for the nuts, and 36–43 ft. lbs (50–60 Nm) for the bolts.

19. Connect the lead wires to the starter.

20. Remove the engine support.

21. On the left side, connect the exhaust pipe to the exhaust bracket and install the mounting bolt. Tighten all fasteners to 40 ft. lbs. (55 Nm).

22. Torque the three engine mount-to-frame bracket bolts to 40 ft. lbs. (54 Nm).

23. Install the front skid plate.

24. Lower the vehicle.

25. Install the radiator fan shroud and mounting bolts.

26. Connect the negative battery cable.

1.8L (VIN 2) Engine

1. Disconnect the negative battery cable.
2. Raise and safely support the vehicle.
3. Remove the four bolts and the front skid plate.
4. Support the engine using a suitable jack.
5. Remove the nut holding the engine mount to the mounting bracket.
6. Remove the nut holding the engine mount to the vehicle frame.
7. Raise the engine slightly and remove the engine mount from the vehicle.
8. Remove the engine mount bracket from the engine by removing the four bolts.
   To install:
9. Install the engine bracket on the engine by installing the four bolts. Torque the bolts to 37 ft. lbs. (50 Nm).
10. Install the engine mount to the vehicle.
11. Install the nut to hold the engine mount to the engine bracket. Torque the nut to 37 ft. lbs. (50 Nm).
12. Lower the vehicle.

13. Install the nut to hold the engine mount to the vehicle frame. Torque the nut to 37 ft. lbs. (50 Nm).
14. Remove the engine support.
15. Install the front skid plate.
16. Lower the vehicle.
17. Connect the negative battery cable.

Cylinder Head

REMOVAL AND INSTALLATION

1.3L (VIN 3) and 1.6L (VIN 0), (VIN U) TFI Engines

———— CAUTION ————
The fuel system pressure must be relieved before disconnecting any fuel lines. Failure to do so may result in personal injury.

1. Disconnect the negative battery cable and drain the cooling system.
2. Disconnect the air cleaner assembly.
3. Disconnect the brake booster hose from the intake manifold and the air valve water hose from the throttle body.
4. Disconnect the accelerator cable and if equipped, the automatic transmission kickdown cable, from the throttle body.

5. Disconnect and tag the throttle opener VSV and EGR VSV wires at the coupler.
6. Disconnect and tag the water temperature sensor, oil pressure sender, air temperature sensor, and ground cable wires at the intake manifold.
7. Disconnect and tag the fuel injector, throttle position sensor, and idle speed control solenoid valve wires at their couplers, if equipped.
8. Disconnect the wire harness from its holding clamps.
9. Disconnect the MAP sensor vacuum hose and water bypass hose from the intake manifold.
10. Disconnect the radiator inlet hose from the thermostat housing. Disconnect the heater inlet hose from the intake manifold.
11. If equipped with automatic transmission, disconnect the PTC heater wires at the coupler.
12. Disconnect and tag the oxygen sensor and distributor wires at their couplers. Remove the coil wire.
13. Disconnect the ground wires from the distributor assembly.
14. Remove the fuel tank filler cap to relieve the pressure and reinstall the cap.
15. Properly release the fuel pressure in the fuel feed line.
16. Disconnect the fuel feed and return lines.
17. Disconnect the canister purge hose and if equipped, remove the pressure sensor hose from the intake manifold.
18. Remove the vacuum hose for the automatic transmission from the intake manifold, if equipped.
19. Disconnect the radiator cooling fan, fan shroud, water pump drive belt, and water pump pulley.
20. Disconnect the crankshaft pulley, timing belt cover, and the timing belt.
21. Raise and safely support the vehicle.
22. Disconnect the exhaust pipe from the exhaust manifold and lower the vehicle.
23. If equipped, remove the air conditioner compressor adjusting arm from the cylinder head.
24. Remove the cylinder head cover mounting bolts and remove the cylinder head cover.
25. Loosen all the valve adjusting screw locknuts and turn the adjusting screws back all the way to allow all the valves to close.
26. Disconnect any remaining connections and remove the intake manifold from the cylinder head.

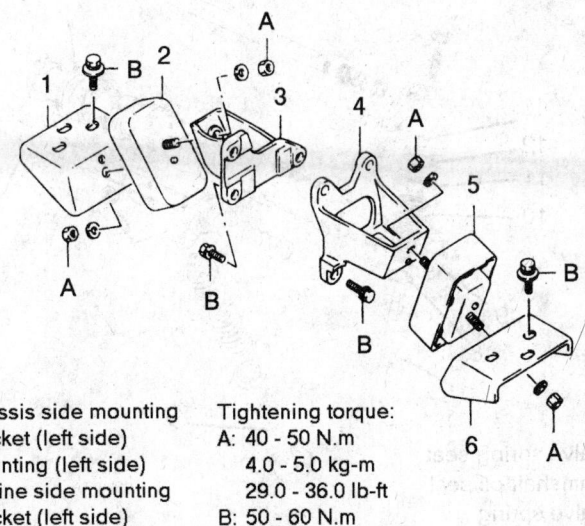

1. Chassis side mounting bracket (left side)	Tightening torque:
2. Mounting (left side)	A: 40 - 50 N.m
3. Engine side mounting bracket (left side)	4.0 - 5.0 kg-m
	29.0 - 36.0 lb-ft
4. Engine side mounting bracket (right side)	B: 50 - 60 N.m
	5.0 - 6.0 kg-m
5. Mounting (right side)	36.5 - 43.0 lb-ft
6. Chassis side mounting bracket (right side)	

241747

Exploded view of the engine mount components — 1.3L (VIN 3) and 1.6L (VIN 0), (VIN U) engines

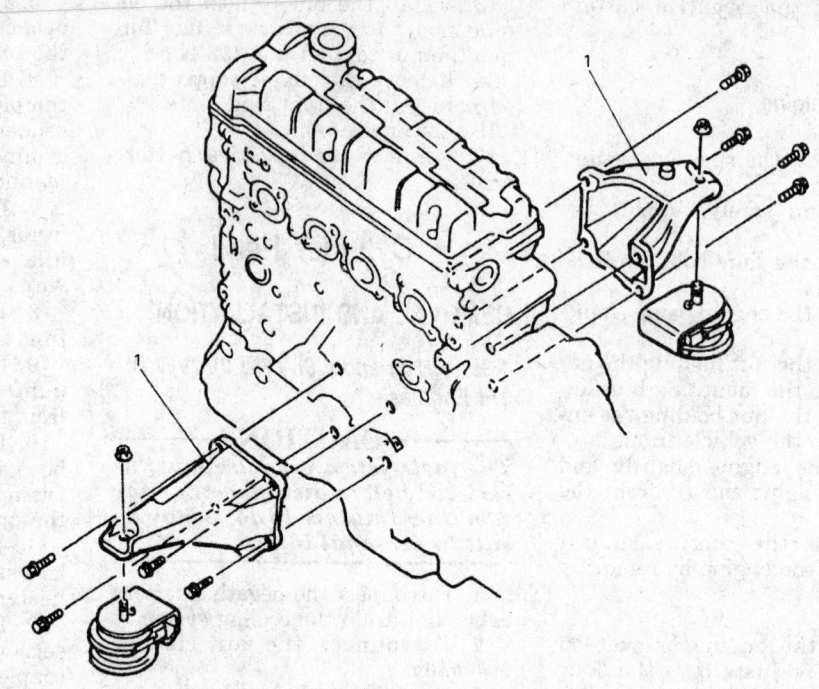

1. Engine side mounting bracket

332132

Engine mount components — 1.8L (VIN 2) engine

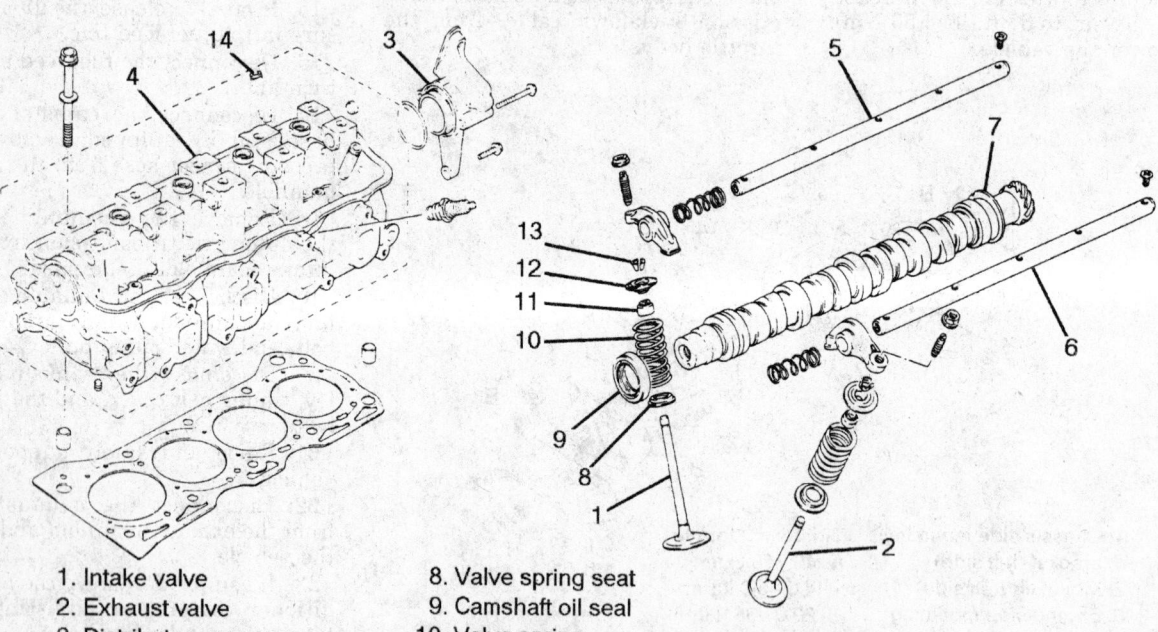

1. Intake valve
2. Exhaust valve
3. Distributor gear case
4. Cylinder head
5. Intake rocker arm shaft
6. Exhaust rocker arm shaft
7. Camshaft

8. Valve spring seat
9. Camshaft oil seal
10. Valve spring
11. Valve stem seal
12. Valve spring retainer
13. Valve keeper
14. Valve guide

196317

Cylinder head components — 1.3L (VIN 3) and 1.6L (VIN 0), (VIN U) SOHC TFI engines

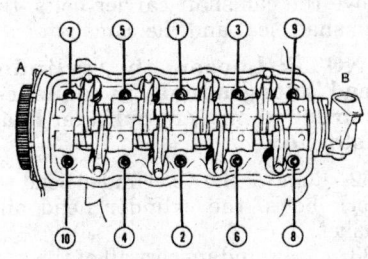

A CAMSHAFT PULLEY SIDE
B DISTRIBUTOR SIDE

196316

Cylinder head bolt torque sequence — 1.3L (VIN 3) and 1.6L (VIN 0), (VIN U) engines

27. Remove the distributor and distributor housing from the cylinder head.

28. Remove the cylinder head mounting bolts and remove the cylinder head from the engine. Remove the bolts in reverse order of the tightening sequence.

29. Remove any oil and water in the cylinder bores and on top of the pistons.

30. Clean and inspect the sealing surfaces of the cylinder head and the engine block.

To install:

31. Install the cylinder head using a new gasket.

32. Lubricate the cylinder head mounting bolts with engine oil and install the cylinder head mounting bolts. Tighten the bolts in 3 even and sequential steps to 51–54 ft. lbs. (70–75 Nm). Tightening should begin in the center and work towards both ends simultaneously.

33. Install the intake manifold to the cylinder head.

34. Raise and safely support the vehicle.

35. Connect the exhaust pipe to the exhaust manifold and safely lower the vehicle.

36. If removed, install the A/C compressor adjusting arm.

37. Reconnect the timing belt, timing belt cover, and the crankshaft pulley.

38. Install the distributor and distributor housing to the cylinder head. Make sure the distributor is installed in the correct firing position.

39. Adjust the valve lash.

40. Tighten all the valve adjusting screw nuts.

41. Install the cylinder head cover and tighten the mounting bolts 4 ft. lbs. (5 Nm).

42. Install the water pump pulley, water pump drive belt, fan shroud, and the radiator cooling fan.

43. Install the vacuum hose for the automatic transmission to the intake manifold, if equipped.

44. If removed, install the pressure sensor hose to the intake manifold and connect the canister purge hose.

45. Install the fuel return and feed lines.

46. Connect the ground wires to the distributor assembly.

47. Install the coil wire. Connect the oxygen sensor and distributor wires to their couplers.

48. Connect the radiator inlet hose to the thermostat housing. Connect the heater inlet hose to the intake manifold.

49. If removed, connect the PTC wires to the coupler.

50. Connect the MAP sensor vacuum hose and the water bypass hose to the intake manifold.

51. Reconnect the wiring harness to its holding clamps.

52. Connect the idle speed control valve, throttle position sensor, and the fuel injector wires to their couplers, if equipped.

53. Reconnect the ground cable, air temperature sensor, oil pressure sender, and water temperature sensor wires to the intake manifold.

54. Install the throttle opener VSV and EGR VSV wires to their couplers.

55. Connect the accelerator cable and if equipped, the automatic transmission kickdown cable, to the throttle body.

56. Install the air valve water hose to the throttle body and connect the brake booster hose to intake manifold.

57. Reconnect the air cleaner assembly.

58. Refill the cooling system with the proper coolant and connect the negative battery.

59. Start the engine and check for any water, fuel, or oil leaks when finished.

60. Check and/or adjust the ignition timing as necessary.

1993–95 1.6L (VIN 0), (VIN U) MFI Engines

— CAUTION —

The fuel system pressure must be relieved before disconnecting any fuel lines. Failure to do so may result in personal injury.

1. Disconnect the negative battery cable.

2. Remove the fuel filler cap to release the pressure in the tank, then replace the fuel filler cap. Properly relieve the fuel system pressure.

3. Drain the cooling system. Disconnect the air intake pipe from the ACL intake hose and throttle hose.

4. Remove the throttle cover bolts from the intake manifold, and remove the cover.

5. If equipped with an automatic transmission, disconnect the accelerator and kickdown cables from the throttle body bell housing.

6. Remove 1 bolt, nut, screw and accelerator cable bracket from the throttle body.

7. Disconnect the electrical connectors from the following components:
- Throttle Position (TP) sensor
- Idle Air Control (IAC) valve
- Engine Coolant Temperature (ECT) sensor
- ECT sending unit
- A/C ECT switch (if equipped with A/C)
- Evaporative Emissions Solenoid Purge (EVAP SP) valve
- Exhaust Gas Recirculation (EGR) temperature sensor (Calif. vehicles)
- EGR solenoid vacuum valve
- Fuel injectors
- Engine ground wire from intake surge tank

8. Label and disconnect the vacuum hoses from the following:
- EVAP SP valve
- Vacuum modulator supply hose (if equipped with automatic transmission)
- Brake booster supply hose
- EGR valve
- EGR valve modulator

9. Disconnect the coolant hose from the IAC valve and the IAC hose from the IAC valve. Disconnect the coolant hoses from the fast idle air valve below the throttle body.

10. Disconnect the fuel feed hose at the fuel feed hose union and the fuel feed hose from the fuel return line. Disconnect the PCV hose from the PCV valve.

11. Loosen the upper radiator hose clamp at the thermostat housing. Disconnect the upper radiator hose from the thermostat housing. Disconnect the coolant bypass hose from the intake manifold.

12. Remove the generator adjusting arm bracket from the intake manifold. Remove the front and rear intake manifold reinforcement brackets from the intake manifold.

13. Remove the lower intake manifold support bracket from the intake manifold. Remove the intake manifold nuts/bolts from the cylinder head. Remove the intake manifold,

the intake surge tank and the throttle body from the cylinder head.

14. Remove the upper exhaust manifold heat shield from the exhaust manifold.

15. Raise and safely support the vehicle.

16. Remove the exhaust manifold reinforcement bracket from the exhaust manifold and engine mount.

17. Remove the lower exhaust manifold heat shield from the exhaust manifold. Disconnect the Three-Way Catalytic converter (TWC) assembly from the exhaust manifold. Lower the vehicle.

18. Remove the exhaust manifold with gasket and engine hanger from the cylinder head.

19. If equipped, loosen the upper and lower A/C compressor mounting bolts. Loosen the upper and lower power steering pump mounting bolts. Remove the A/C compressor, If equipped, and/or the power steering drive belt.

20. Loosen the upper and lower generator mounting bolts. Remove the generator/water pump drive belt from the pulleys. Remove the cooling fan and the water pump pulley from the water pump.

21. Raise and safely support the vehicle. If equipped, remove the skid plate from the undercarriage. Remove the 2 lower radiator shroud bolts and lower the vehicle.

22. If equipped, remove the 2 A/C suction line bracket bolts at the right side of the radiator core support and carefully reposition the suction line for radiator shroud removal access.

23. Remove the 2 upper radiator shroud bolts and the shroud from the radiator.

NOTE: If the shroud is difficult to remove, drain the cooling system and disconnect the upper radiator hose from the radiator to gain access.

24. Remove the 5 crankshaft pulley bolts and the pulley. Remove the oil pressure sending unit wire conduit from the timing belt cover.

NOTE: It is not necessary to remove the crankshaft center bolt when removing the crankshaft pulley.

25. To remove the timing belt, perform the following procedure:

　a. Remove the timing belt cover from the engine.

NOTE: There are 2 sets of timing marks which must be aligned to ensure correct engine timing upon installation. A notch in the

camshaft timing belt gear designated as E must be aligned with the notch in the cylinder head. A punch mark on the crankshaft timing belt gear should align with the arrow in the oil pump casting. Make sure to align both sets of marks prior to timing belt removal.

　b. Rotate the engine to align the timing marks on the cylinder head cover and camshaft timing belt gear; this should align the timing marks on the oil pump casting and crankshaft timing belt gear.

　c. Loosen the timing belt tensioner bolt. Remove the timing belt tensioner spring from the timing belt tensioner plate.

　d. Remove the timing belt from the camshaft and crankshaft timing belt gears.

26. Secure the camshaft sprocket, using a special tool (Geo: J-41840; Suzuki: 09917-68220) or equivalent, then remove the camshaft sprocket bolt and the camshaft sprocket.

27. Remove the 2 inner timing belt cover bolts from the cylinder head, and remove the cover. Remove the 2 A/C compressor mounting bracket bolts from the cylinder head, and remove the bracket.

28. Remove the distributor from the distributor case. Remove the 6 cylinder head cover mounting bolts and remove the cover. Remove the cylinder head cover gasket and O-rings from the cylinder head cover.

NOTE: A small amount of oil may drain from the distributor case upon removal from the cylinder head. Place a suitable container under the distributor case or use a shop towel to catch and absorb the oil.

29. Remove the 3 distributor case bolts from the cylinder head, and remove the distributor case.

30. Loosen all the valve adjusting screw locknuts. Loosen all the valve adjusting screws until all rocker arms move freely.

NOTE: Always loosen the camshaft carrier bolts gradually in sequence, in order to relieve tension on the camshaft; if the camshaft carrier bolts are removed at random, damage to the camshaft may occur.

31. Remove the 12 camshaft carrier cap bolts from the cylinder head. Re-

move the camshaft carrier bolts, the camshaft seal and the camshaft.

NOTE: Loosen the cylinder head bolts gradually in sequence, in order to prevent cylinder head distortion.

32. Remove the 10 cylinder head-to-block bolts, the cylinder head and gasket.

33. Clean and inspect all of the gasket mating surfaces. Inspect and/or replace any damaged parts.

To install:

34. Install a new cylinder head gasket and the cylinder head with the distributor case onto the cylinder block. Torque the cylinder head bolts, in sequence, in 3 steps:

　a. Step 1: 26 ft. lbs. (35 Nm)

　b. Step 2: 41 ft. lbs. (55 Nm)

　c. Step 3: 52 ft. lbs. (70 Nm)

35. Lubricate the camshaft with clean oil. Apply RTV silicone rubber sealant to the bottom of the No. 6 camshaft carrier cap.

36. Install the camshaft and camshaft carrier caps onto the cylinder head, then, torque the camshaft carrier cap-to-cylinder head bolts to 89 inch lbs. (10 Nm).

NOTE: Always tighten the camshaft carrier cap bolts gradually in sequence; if the camshaft carrier cap bolts are tightened at random. damage to the camshaft may occur.

37. Lubricate the new camshaft seal lip with clean engine oil. Install the new camshaft seal into the cylinder head until it is flush with the camshaft carrier surface.

38. Apply RTV silicone rubber sealant to the surface of the distributor case that mates with the rear of the rocker arm shaft.

39. Install the distributor case onto the cylinder head and torque the distributor case mounting bolts to 89 inch lbs. (10 Nm).

40. Install the A/C mounting bracket with compressor and torque the 2 mounting bracket-to-cylinder head bolts to 89 inch lbs. (10 Nm).

41. Install the inner timing belt cover and torque the 2 inner timing belt cover mounting bolts to 89 inch lbs. (10 Nm).

NOTE: During camshaft timing belt gear installation, align the camshaft dwell pin with the slot in the camshaft timing belt gear designated as E.

42. Install the camshaft sprocket, using holding tool #: 09917-68220, or equivalent, onto the camshaft and torque the camshaft sprocket bolt,

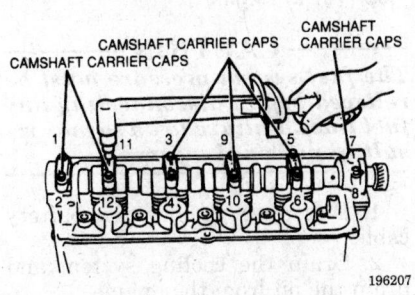

CAMSHAFT CARRIER CAPS
CAMSHAFT CARRIER CAPS
CAMSHAFT CARRIER CAPS

196207

Camshaft carrier cap bolt torque sequence — 1993–95 1.6L (VIN 0), (VIN U) MFI engines

onto the camshaft, to 44 ft. lbs. (60 Nm).

43. Install the timing belt tensioner plate and tensioner; secure with timing belt tensioner stud and bolt only finger-tight. Push the timing belt tensioner plate up when installing the timing belt.

NOTE: When installing the timing belt, the directional arrows on the timing belt must be matched with the rotation of the crankshaft; if not, excessive wear and timing belt failure may occur.

44. Install the timing belt by performing the following procedures:
   a. Align the timing marks of the camshaft sprocket with the cylinder head cover and the oil pump housing with the crankshaft sprocket.
   b. Install the timing belt onto the camshaft and crankshaft sprockets.
   c. Rotate the crankshaft through 2 complete revolutions, to remove any slack from the timing belt and to properly seat the timing belt.
   d. Inspect both sets of timing marks to ensure that they are aligned. Install the belt tensioner spring to the timing belt tensioner plate.
   e. Torque the timing belt tensioner stud to 89 inch lbs. (10 Nm) and the bolt to 18 ft. lbs. (25 Nm).
45. Install the distributor into the distributor case. Make sure the distributor rotor is facing the correct firing position.
46. Install the timing belt cover and torque the nut/bolts to 89 inch lbs. (10 Nm). Connect the oil pressure sending unit wire to the timing belt cover.
47. Install the crankshaft pulley to the crankshaft and torque the bolts

to 12 ft. lbs. (16 Nm). Install the radiator shroud and torque both bolts to 89 inch lbs. (10 Nm). Reconnect the oil pressure sending unit wire conduit to the timing belt cover.

48. If equipped, carefully, position the A/C suction line and brackets to the right side of the radiator core support and torque both bolts to 89 inch lbs. (10 Nm).

49. Raise and safely support the vehicle. Install the 2 lower radiator shroud bolts and torque to 89 inch lbs. (10 Nm). If equipped, install the front skid plate and torque the 4 bolts to 40 ft. lbs. (54 Nm). Lower the vehicle.

50. Install the water pump pulley and fan; torque the 4 nuts to 97 inch lbs. (11 Nm). Install the generator/water pump drive belt onto the pulleys. Install the A/C compressor, if equipped, and the power steering drive belt.

51. Using new gaskets, install the exhaust manifold and torque the exhaust manifold mounting nuts to 17 ft. lbs. (23 Nm). Raise and safely support the vehicle.

52. Using new gaskets, connect the front pipe/TWC assembly to the exhaust manifold and torque the 3 bolts to 37 ft. lbs. (50 Nm). Install the lower exhaust manifold heat shield and torque both bolts to 89 inch lbs. (10 Nm). Install the exhaust manifold bracket; torque the engine mount-to-exhaust bracket bolts to 40 ft. lbs. (54 Nm) and the exhaust bracket-to-exhaust manifold nut to 37 ft. lbs. (50 Nm). Lower the vehicle.

53. Install the upper exhaust manifold heat shield and torque the nuts/bolts to 89 inch lbs. (10 Nm). Install the air intake pipe bracket and torque the bracket-to-cylinder head bolts to 89 inch (10 Nm).

54. Using a new gasket, install the intake manifold, with intake surge tank and throttle body, and torque the intake manifold mounting bolts onto the cylinder head, in sequence, to 17 ft. lbs. (23 Nm).

55. Install the lower intake manifold support bracket, the rear and the front intake manifold reinforcement bracket; torque the bracket mounting bolts to 37 ft. lbs. (50 Nm).

56. Install the generator adjusting arm bracket and torque the bracket mounting nut/bolt to 37 ft. lbs. (50 Nm).

57. Adjust the valve lash.

58. Install a new cylinder head cover gasket and 4 O-rings to the cyl-

inder head cover. Install the cylinder head cover and torque the 6 bolts to 89 inch lbs. (10 Nm).

59. Install the following hoses:
   • Coolant bypass hose to the intake manifold.
   • Upper radiator hose to the thermostat housing; secure with a clamp.
   • PCV hose to the PCV valve.
   • Fuel return hose to the fuel return line.
   • Fuel feed hose at the fuel feed hose union.
   • Coolant hoses to the fast idle air valve below the throttle body.
   • IAC air hose to the IAC valve.
   • Coolant hose to the IAC valve.

60. Install the following vacuum hoses:
   • EGR valve modulator.
   • EGR valve.
   • Brake booster supply hose.
   • Vacuum modulator supply hose, if equipped with an automatic transmission.
   • EVAP SP valve.

61. Connect the following electrical connectors:
   • Engine ground wire to intake surge tank.
   • Fuel injectors.
   • EGR solenoid vacuum valve.
   • EGR temperature sensor, California vehicles.
   • Evaporative Emission Solenoid Purge (EVAP SP).
   • A/C ECT switch, if equipped with A/C.
   • ECT sensor sending unit.
   • Engine Coolant Temperature (ECT) sensor.
   • Idle Air Control (IAC).
   • Throttle Position (TP) sensor.

62. Install the accelerator cable bracket to the throttle body and torque the nut/bolt to 17 ft. lbs. (23 Nm).

63. If equipped with an automatic transmission, connect the accelerator cable and kickdown cable to the throttle body bell housing; adjust the cables.

64. Install the throttle cover to the intake manifold and torque the bolts to 11 ft. lbs. (15 Nm). Refill the cooling system, as necessary.

65. Install the air intake pipe to the throttle body hose and the ACL intake hose. Connect the negative battery cable.

66. Start the engine and check oil, fuel, or coolant leaks.

67. Check and/or adjust the ignition timing as necessary.

1996–97 1.6L (VIN 0), (VIN U) MFI Engines

— CAUTION —

The fuel system pressure must be relieved before disconnecting any fuel lines. Failure to do so may result in personal injury.

1. Disconnect the negative battery cable.
2. Drain the cooling system.
3. Remove the intake manifold stiffener from the intake manifold and engine block.
4. Disconnect the electrical connectors from the following components:
   - Distributor
   - Engine ground wire from intake surge tank
   - EGR solenoid vacuum valve
   - Evaporative Emissions Solenoid Purge (EVAP SP) valve
   - Throttle Position (TP) sensor
   - Idle Air Control (IAC) valve
   - Engine Coolant Temperature (ECT) sensor and gauge
   - Exhaust Gas Recirculation (EGR) temperature sensor
   - EGR solenoid vacuum valve
   - Fuel injectors
   - Heated oxygen sensor
5. Label and disconnect the vacuum hoses from the following:
   - EVAP canister purge hose
   - Brake booster supply hose
   - Manifold differential pressure sensor
   - Tank pressure control solenoid vacuum valve hose from the solenoid vacuum valve
6. Disconnect the fuel feed and return hoses from each pipe.
7. Remove the cylinder head cover.
8. Fully loosen all the valve lash adjusting screws.
9. Disconnect the following engine cooling water hoses:
   - Radiator inlet hose
   - Heater inlet hose
   - IAC valve outlet
   - Bypass hose
10. Remove the timing belt from the camshaft and crankshaft pulleys.
11. Disconnect the exhaust pipe from the exhaust manifold and remove the exhaust manifold stiffener.
12. Loosen the cylinder head bolts in order. Once each bolt is loose, remove the bolts from the cylinder head.
13. Check to make sure all components are removed or disconnected before removing the cylinder head.

14. Remove the cylinder head with the intake manifold, exhaust manifold and distributor as an assembly.

To install:

15. Install a new cylinder head gasket and the cylinder head with the distributor case onto the cylinder block. Torque the cylinder head bolts, in sequence, in 3 Steps:
   a. Step 1: 26 ft. lbs. (35 Nm)
   b. Step 2: 41 ft. lbs. (55 Nm)
   c. Step 3: 49.5 ft. lbs. (68 Nm)
16. Connect the exhaust pipe to the exhaust manifold and install the exhaust manifold stiffener.
17. Install the timing belt to the camshaft and crankshaft pulleys.
18. Connect the following engine cooling water hoses:
   - Radiator inlet hose
   - Heater inlet hose
   - IAC valve outlet
   - Bypass hose
19. Adjust the valve lash with the adjusting screws.
20. Install the cylinder head cover.
21. Connect the fuel feed and return hoses to each pipe.
22. Connect the vacuum hoses:
   - EVAP canister purge hose
   - Brake booster supply hose
   - Manifold differential pressure sensor
   - Tank pressure control solenoid vacuum valve hose to the solenoid vacuum valve
23. Connect the electrical connectors to the following components:
   - Distributor
   - Engine ground wire to the intake surge tank
   - EGR solenoid vacuum valve
   - Evaporative Emissions Solenoid Purge (EVAP SP) valve
   - Throttle Position (TP) sensor
   - Idle Air Control (IAC) valve
   - Engine Coolant Temperature (ECT) sensor and gauge
   - Exhaust Gas Recirculation (EGR) temperature sensor
   - EGR solenoid vacuum valve
   - Fuel injectors
   - Heated oxygen sensor
24. Install the intake manifold stiffener to the intake manifold and engine block.
25. Adjust all cables and belts. Check that all electrical and vacuum lines are connected correctly.
26. Fill the engine coolant and check all fluids.
27. Connect the negative battery cable to the battery.
28. Start the engine and check for leaks.

1.8L (VIN 2) Engine

— CAUTION —

The fuel system pressure must be relieved before disconnecting any fuel lines. Failure to do so may result in personal injury.

1. Disconnect the negative battery cable.
2. Drain the cooling system and drain the oil from the engine.
3. Remove the strut tower bar by removing the bolts and nuts.
4. Remove the air outlet hose.
5. Disconnect the accelerator cable and, if equipped with A/T, A/T throttle cable from the throttle body.
6. Remove the timing chains from the engine.
7. Remove the camshafts and valve lash adjuster from the engine.
8. Disconnect the following electrical connectors from the engine:
   - EGR valve connector
   - IAC valve connector
   - Throttle position sensor connector
   - MAF sensor connector
   - EVAP solenoid purge valve connector
   - Ground solenoid from the intake manifold
   - Heated oxygen sensor connector
   - Coolant temperature switch connector
   - ECT sensor connector
   - Injector wire harness connector
   - Wire harness clamps
9. Disconnect the following hoses from the engine:
   - Brake booster hose from the intake manifold
   - MAP sensor hose from the intake manifold
   - Canister purge hose from the EVAP canister
   - Coolant hose from the throttle body
   - Fuel feed hose from the delivery pipe
   - Fuel return hose from the fuel pressure regulator
   - Heater hose from the heater outlet pipe
   - Radiator inlet hose from the water outlet pipe
10. Remove the intake manifold stiffener.
11. Disconnect the coolant pipe from the intake manifold by removing the bolts.
12. Raise and safely support the vehicle.
13. Disconnect the front exhaust pipe from the exhaust manifold.
14. Remove the exhaust manifold stiffener.

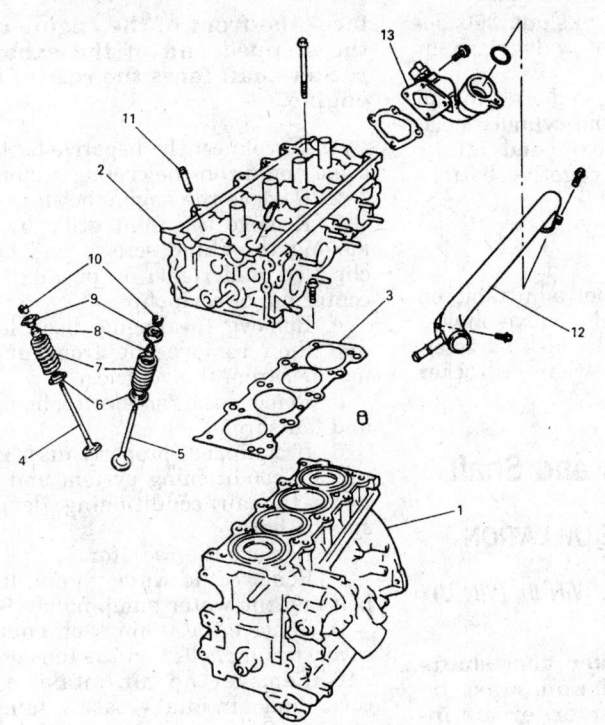

1. Cylinder block
2. Cylinder head
3. Cylinder head gasket
4. Intake valve
5. Exhaust valve
6. Valve spring seat
7. Valve spring
8. Valve stem oil seal
9. Valve spring retainer
10. Valve cotter
11. Valve guide
12. Water outlet pipe
13. Water outlet cap

333337

Cylinder head component assembly — 1.8L (VIN 2) engine

15. Loosen and remove the cylinder head bolts in the reverse order of the torque sequence.

16. Check to make sure all components are removed or disconnected from the cylinder head.

17. Remove the cylinder head with the intake manifold, exhaust manifold and water outlet pipe.

To install:

18. Clean the mating surfaces of the cylinder head and engine block.

19. Make sure the cylinder head guide pins are in place.

20. Using a new gasket, install the cylinder head to the engine.

21. Apply engine oil to the cylinder head bolts and tighten the bolts in order as follows:

   a. Tighten all bolts to 39 ft. lbs. (53 Nm) in the order shown.

   b. In the same order, further tighten each bolt to 61 ft. lbs. (84 Nm).

   c. Fully loosen each bolt in order shown.

   d. Tightening all bolts in tightening sequence to 27 ft. lbs. (37 Nm).

   e. In the same order, torque each bolt to 76 ft. lbs. (105 Nm). Tighten the outside head bolt to 8 ft. lbs. (11 Nm).

1. Crankshaft pulley side
2. Flywheel side

333344

Cylinder head bolts tightening sequence — 1.8L (VIN 2) engine

22. Check that the crankshaft key aligns with the timing mark on the engine block.

23. Install the exhaust manifold stiffener.

24. Install the exhaust pipe to the exhaust manifold.

25. Connect the coolant pipe to the intake manifold.

26. Install the intake manifold front stiffener.

27. Connect the following hoses:
- Brake booster hose to the intake manifold
- MAP sensor hose to the intake manifold

- Canister purge hose to the EVAP canister
- Coolant hose to the throttle body
- Fuel feed hose to the delivery pipe
- Fuel return hose to the fuel pressure regulator
- Heater hose to the heater outlet pipe
- Radiator inlet hose to the water outlet pipe

28. Connect the following electrical connectors as follows:
- EGR valve connector
- IAC valve connector
- Throttle position sensor connector
- MAF sensor connector
- EVAP solenoid purge valve connector
- Ground solenoid from the intake manifold
- Heated oxygen sensor connector
- Coolant temperature switch connector
- ECT sensor connector
- Injector wire harness connector
- Wire harness clamps

29. Install the camshaft and valve lash adjuster.

30. Install the timing chains to the engine.

31. Connect and adjust the A/T throttle cable and accelerator cable.

32. Install the air cleaner outlet cable.

33. Install the strut tower bar.

34. Fill the radiator and engine with coolant.

35. Fill the engine with oil.

36. Connect the negative battery cable to the battery.

37. Run the engine and check for leaks and proper timing.

Valve Lash

ADJUSTMENT

1.3L (VIN 3) and 1.6L (VIN 0), (VIN U) Engines

1993 Models

Valve lash refers to the gap between the rocker arm adjusting screw and valve stem. Valve lash can be adjusted with the engine hot or cold. Specifications are provided for both adjustments.

Valve lash — 1.3L (VIN 3) and 1.6L (VIN 0), (VIN U) engines 198721

1. Disconnect the negative battery cable and remove the air intake case.

2. Remove the cylinder head cover assembly.

3. Turn the crankshaft clockwise so the **V** mark on the crankshaft pul-

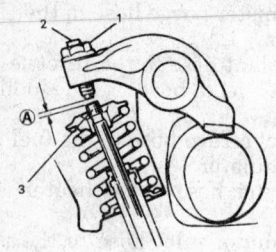

1. Adjusting screw lock nut
2. Adjusting screw
3. Valve stem

198721

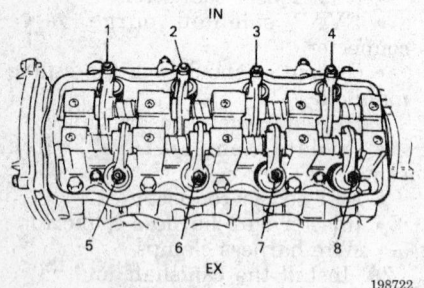

Valve numbered locations — 1.3L (VIN 3) and 1.6L (VIN 0), (VIN U) engines

ley is aligned with the **0** mark on the timing belt cover.

4. Remove the distributor cap and confirm that the rotor is facing the No. 1 firing position. If the rotor is out of place, turn the crankshaft 360 degrees.

5. With the engine in this position, check the valve lash at valves 1, 2, 5 and 7. Clearance should be:
- Cold
- Intake 0.005–0.007 inch (0.13–0.17mm)
- Exhaust 0.006–0.008 inch (0.15–0.19mm)
- Hot
- Intake 0.009–0.011 inch (0.23–0.27mm)
- Exhaust 0.010–0.011 inch (0.25–0.29mm)

NOTE: The valves are adjusted by loosening the locknut on the valve adjuster and turning the adjusting screw to obtain the proper clearance. Once the proper clearance is obtained, the locknut must be torqued to 11–13 ft. lbs. (15–19 Nm) while holding the adjusting screw. Recheck the clearance after the locknut is torqued.

6. Rotate the crankshaft 360 degrees and check the valve lash at valves 3, 4, 6 and 8.

7. After adjusting and checking all the valves, install the cylinder head cover, distributor cap, and intake hose. Connect the negative battery cable.

1.8L (VIN 2) Engine

Valve clearance is not adjustable on these vehicles. If the valve makes noise, look for a leaky lash adjuster, excessive camshaft wear or rocker arm wear.

Rocker Arms and Shaft

REMOVAL AND INSTALLATION

1.3L (VIN 3) and 1.6L (VIN 0), (VIN U) TFI Engines

NOTE: The rocker arm shafts are not identical and must be kept in the proper order for installation. If the shafts get mixed up before installation, the intake rocker shaft has a 0.55 inch (14mm) stepped end and the exhaust rocker shaft has a 0.59 inch (15mm) stepped end. The stepped end of the intake rocker shaft

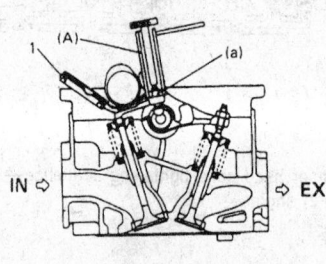

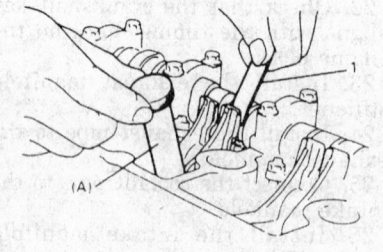

Adjusting the valves with the special tool (A) — 1.6L (VIN 0), (VIN U) MFI engines

faces the front of the engine and the stepped end of the exhaust rocker shaft faces the rear of the engine.

1. Disconnect the negative battery cable and drain the cooling system.
2. Remove the engine hood.
3. Remove the front grille by removing the three screws and each clip (left and right) by pushing the center pin of each clip.
4. Remove the engine hood lock, and then remove the front upper member from the vehicle.
5. Remove the radiator cooling fan and fan shroud.
6. If equipped, properly discharge the air conditioning system and remove the air conditioning flexible suction hose.
7. Remove the radiator.
8. Remove the water pump drive belt and the water pump pulley.
9. Remove the timing belt outside cover, timing belt, and the tensioner.
10. Remove the air intake case cover and air intake case retaining screws. Disconnect the PCV hose, and remove the air intake case from the throttle body.
11. Remove the PCV hose from the rocker arm cover. Disconnect the spark plug wires and the accelerator cable from the retaining clamps. Remove the rocker arm cover.
12. Insert a 0.35 in. (9mm) rod into the hole in the camshaft to lock the camshaft in place. Loosen the camshaft timing sprocket retaining bolt and remove the sprocket and timing belt inside cover.
13. Loosen all the valve lash adjusting screw locknuts and turn the adjusting screws out all the way.
14. Remove the rocker arm shaft screws.
15. Remove the intake and exhaust rocker arm shafts, rocker arms and springs. Keep all parts in order so they can be reinstalled in their original locations.
16. Inspect the rocker arms and shafts for wear and/or damage and replace parts as necessary.
To install:
17. Apply engine oil to the rocker arms, springs, and shafts. Install the shafts in the correct direction, into the cylinder head by placing the rocker arms and springs on the shafts as they are installed. With the rocker arms, springs and shafts installed, torque the rocker shaft mounting screws to 80–106 inch lbs. (9–12 Nm).
18. Lock the camshaft in place with the rod as described in the removal procedure, then install the inside

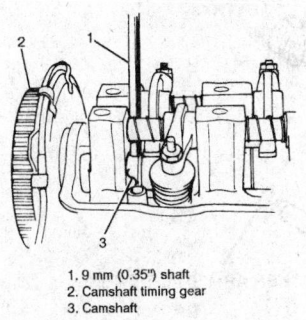

1. 9 mm (0.35") shaft
2. Camshaft timing gear
3. Camshaft

203638

**Locking the camshaft in position —
1.3L (VIN 3) and 1.6L (VIN 0), (VIN
U) TFI engines**

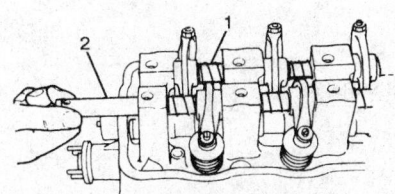

1. Intake rocker arm shaft
2. Exhaust rocker arm shaft

203639

**Removing the rocker arm shaft — 1.3L (VIN 3)
and 1.6L (VIN 0), (VIN U) TFI engines**

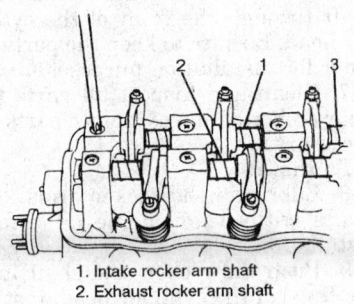

1. Intake rocker arm shaft
2. Exhaust rocker arm shaft
3. Rocker arm spring

203640

**Removing the rocker arm shaft screws —
1.3L (VIN 3) and 1.6L (VIN 0), (VIN U) TFI
engines**

timing belt cover and camshaft
sprocket.

19. With the camshaft locked,
tighten the camshaft sprocket bolt to
41–46 ft. lbs (56–64 Nm).

20. Install the tensioner, timing
belt, and the timing belt cover.

21. Adjust the valve lash.

22. Install the water pump pulley
and the water pump drive belt.

23. Install the radiator.

24. Install the radiator cooling fan
and fan shroud. If equipped, connect

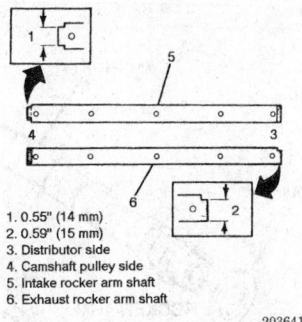

1. 0.55" (14 mm)
2. 0.59" (15 mm)
3. Distributor side
4. Camshaft pulley side
5. Intake rocker arm shaft
6. Exhaust rocker arm shaft

203641

**Identification of the rocker shafts —
1.3L (VIN 3) and 1.6L (VIN 0), (VIN U)
TFI engines**

the air conditioning flexible suction
hose.

25. Clean the rocker arm cover and
cylinder head mating surfaces.

26. Install the rocker arm cover, using a new gasket. Tighten the bolts to
34–44 inch lbs. (4–5 Nm). Connect
the spark plug wires and accelerator
cable to the retaining clamps and
connect the PCV hose.

27. Install the air intake case to the
throttle body, making sure the air intake seal is properly seated.

28. Attach the PCV hose to the air
intake case and install the air intake
case retaining screws. Install the air
intake case cover.

29. Install the front upper member
and the engine hood lock.

30. Install the front grille.

31. Install the engine hood.

32. Connect the negative battery
cable.

33. Refill and bleed the cooling system. If equipped, evacuate and
charge the air conditioning system.

34. Start the engine and check for
leaks. Check and/or adjust the ignition timing as necessary.

1.6L (VIN 0), (VIN U) MFI Engine

1. Disconnect the negative battery
cable and drain the cooling system.

2. Remove the 3 screws and 2 clips
and remove the grille from the radiator core support. Remove the hood
latch-to-header panel bolts and the
latch.

3. Disconnect the 2 electrical connectors from the horn. Remove the 12
header panel bolts and the panel
from the vehicle.

4. Remove the radiator.

5. Align the timing marks and remove the timing belt and tensioner
from the engine.

— **WARNING** —

**After the timing belt is removed,
never turn the camshaft and
crankshaft independently more**

than ±90°. If turned, interference
may occur among the piston and
valves causing possible damage
to the effected parts.

6. Using camshaft sprocket holding tool (Geo: J-41840; Suzuki:
09917–68220) or equivalent, hold the
sprocket stationary and remove the
camshaft sprocket bolt.

7. Remove the cylinder head cover
as follows:

a. Label and disconnect the idle
air control valve and PCV valve
hoses from the air intake pipe.

b. Disconnect the coolant hoses
from the air intake pipe.

c. Loosen the air cleaner-to-air
intake pipe and air intake pipe-to-
throttle body hose clamps. Remove
the 3 bolts and the air intake pipe.

d. Remove the air intake pipe
bracket from the cylinder head and
the throttle cover from the intake
manifold.

e. Disconnect the accelerator
cable and, if equipped, kickdown
cable from the throttle body. Remove the accelerator cable bracket
from the throttle body.

f. Remove the PCV valve from
the cylinder head cover. Label and
disconnect the spark plug wires
from the spark plugs and cylinder
head cover.

g. Remove the 6 bolts and the
cylinder head cover. Remove the
cover gasket and 4 O-rings from
the cylinder head.

8. Remove the distributor cap and
distributor assembly.

9. Remove the distributor case
from the cylinder head. Place a suitable container under the distributor
case prior to removal, as a small
quantity of oil may drain from the
case upon removal.

10. Loosen all of the valve adjusting
screw locknuts. Loosen all valve adjusting screws until the rocker arms
move freely.

— **WARNING** —

**Always remove the camshaft car-
rier bolts gradually, in sequence,
in order to relieve tension on the
camshaft; if the bolts are removed at random, damage may
occur to the camshaft.**

11. Gradually, remove the 12
camshaft carrier cap bolts, in sequence. Remove the camshaft carrier,
the camshaft seal and the camshaft
from the cylinder head.

12. Remove the rocker arm shaft
plug from the cylinder head.

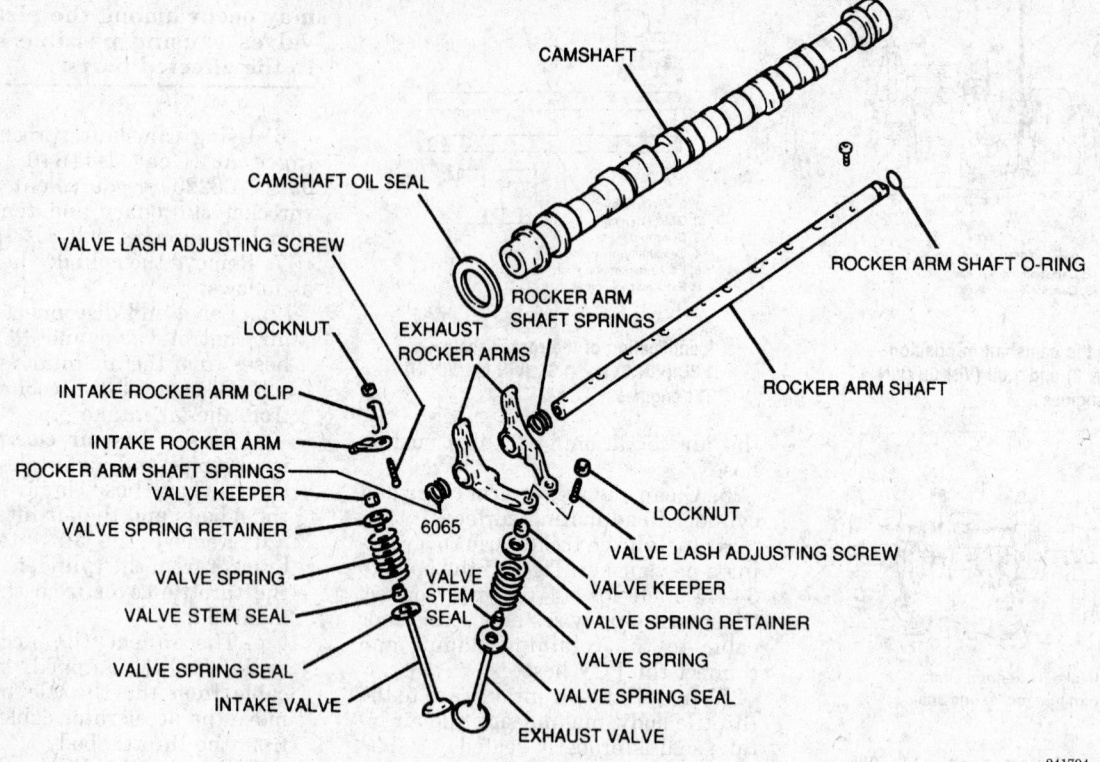

Exploded view of the valve train components — 1.6L (VIN 0), (VIN U) MFI engines

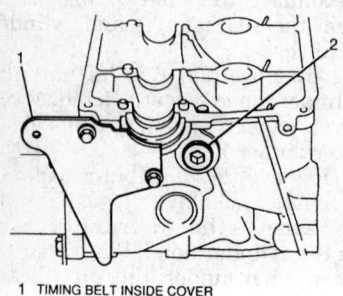

1 TIMING BELT INSIDE COVER
2 ROCKER ARM SHAFT PLUG

Rocker arm shaft plug and timing belt inside cover — 1.6L (VIN 0), (VIN U) MFI engines

13. Remove the timing belt inner cover-to-cylinder head bolts and the cover.

14. Remove all intake rocker arms with clips from the rocker arm shaft. Keep all parts in order so they can be reinstalled in their original locations.

15. Remove the 6 rocker arm shaft-to-cylinder head bolts. Push the rocker arm shaft through the rear of the cylinder head until the end of the rocker arm shaft appears. Remove the O-ring from the rear of the rocker arm shaft.

16. Remove the exhaust rocker arms, rocker arm springs and rocker

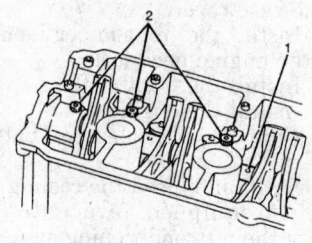

1 ROCKER ARM SHAFT
2 ROCKER ARM SHAFT BOLTS

Rocker arm shaft bolts — 1.6L (VIN 0), (VIN U) MFI engines

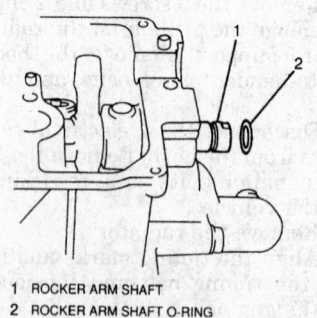

1 ROCKER ARM SHAFT
2 ROCKER ARM SHAFT O-RING

Rocker arm shaft O-ring — 1.6L (VIN 0), (VIN U) MFI engines

arm shaft by pulling the rocker arm shaft through the front of the cylinder head. Be sure to keep the parts in order for installation purposes.

17. Clean and inspect all parts for wear and/or damage; replace parts as necessary.

To install:

18. Lubricate the rocker arms and shafts with clean engine oil before installation.

19. Push the rocker arm shaft into the front of the cylinder head; install the exhaust rocker arms and springs as the rocker arm shaft is being installed into the cylinder head.

20. Push the rocker arm shaft through the rear of the cylinder head. Install a new O-ring onto the rocker arm shaft.

21. Rotate the rocker arm shaft so the flat machined surface is horizontal and facing downward, parallel with the cylinder head mating surface and slide the shaft back into the cylinder head.

22. Install the 6 rocker arm shaft bolts and torque the rocker arm-to-cylinder head bolts to 89 inch lbs. (10 Nm). Fill the rocker arm shaft bolt holes with clean engine oil.

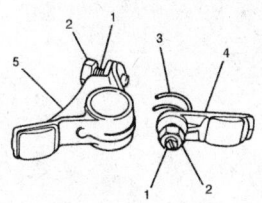

1 VALVE LASH ADJUSTING SCREW
2 LOCKNUT
3 INTAKE ROCKER ARM CLIP
4 INTAKE ROCKER ARM
5 EXHAUST ROCKER ARM

241799

Exhaust and intake rocker arms — 1.6L (VIN 0), (VIN U) MFI engines

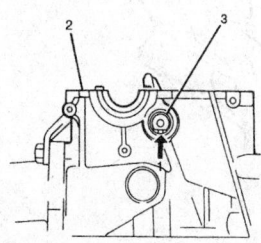

1 ROCKER ARM SHAFT FLAT
2 CYLINDER HEAD
3 ROCKER ARM SHAFT

241801

Rocker arm shaft installation position — 1.6L (VIN 0), (VIN U) MFI engines

23. Install the intake rocker arms with clips onto the rocker arm shaft.

NOTE: The camshaft carrier caps are embossed with numbers and arrows to ensure correct assembly. The No. 1 camshaft carrier cap must be installed at the front of the cylinder head with the remaining carrier caps following in numerical order. The directional arrows must always point toward the front of the cylinder head.

24. Lubricate the camshaft with clean engine oil. Apply RTV silicone sealant to the bottom of the No. 6 camshaft carrier cap. Install the camshaft and camshaft carrier caps onto the cylinder head and gradually, torque the 12 bolts, in sequence, to 89 inch lbs. (10 Nm).

— **WARNING** —

If the camshaft carrier cap bolts are tightened at random, damage to the camshaft may occur.

25. Lubricate the new camshaft seal lip with clean engine oil and install it into the cylinder head until it

is flush with the camshaft carrier surface.

26. Install the timing belt inner cover and torque the cover-to-cylinder head bolts to 89 inch lbs. (10 Nm). Install the rocker arm shaft plug into the cylinder head and torque to 24 ft. lbs. (33 Nm).

NOTE: During camshaft timing belt sprocket installation, align the camshaft dowel pin with the slot in the camshaft timing belt gear designated as E.

27. Install the camshaft sprocket. Using holding tool (Geo: J-41840; Suzuki: 09917–68220) or equivalent, to hold the sprocket in place, torque the camshaft sprocket bolt to 44 ft. lbs. (60 Nm).

28. Push the timing belt tensioner plate up when installing the timing belt.

NOTE: When installing the timing belt, the directional arrows on the timing belt must be matched with the rotation of the crankshaft; if not, excessive wear and timing belt failure may occur.

— **WARNING** —

After the timing belt is removed, never turn the camshaft and crankshaft independently more than ±90°. If turned, interference may occur among the piston and valves causing possible damage to the effected parts.

29. Install the timing belt by performing the following procedures:
   a. Align the timing marks of the camshaft sprocket with the cylinder head cover and the oil pump housing with the crankshaft sprocket.
   b. Install the timing belt onto the camshaft and crankshaft sprockets.
   c. Rotate the crankshaft through 2 complete revolutions, to remove any slack from the timing belt and to properly seat the timing belt.
   d. Inspect both sets of timing marks to ensure that they are aligned. Install the belt tensioner spring to the timing belt tensioner plate.
   e. Torque the timing belt tensioner stud to 89 inch lbs. (10 Nm) and the bolt to 18 ft. lbs. (25 Nm).

30. Install the timing belt cover and torque the nut/bolts to 89 inch lbs. (10 Nm). Connect the oil pressure sending unit wire to the timing belt cover.

31. Apply RTV silicone rubber sealant to the surface of the distributor

case that mates with the rear of the rocker arm shaft. Install the distributor case and torque the 3 case-to-cylinder head bolts to 89 inch lbs. (10 Nm).

NOTE: With the timing marks aligned on the sprockets and the timing belt installed, the number four piston is at TDC of the compression stroke.

32. Install the distributor into the distributor case. Make sure the rotor is aligned with the No. 4 tower on the distributor cap. Install the distributor cap.

33. Adjust the valve lash.

34. Install the cylinder head cover onto the cylinder head, in the reverse order of removal. Clean all sealing surfaces and use a new gasket and O-rings. Tighten the cylinder head cover bolts to 89 inch lbs. (10 Nm).

35. Install the radiator into the radiator core support. Reconnect the radiator hoses.

36. Install the header panel and torque the 12 bolts to 89 inch lbs. (10 Nm). Connect the 2 electrical connectors to the horn.

37. Install the hood latch to the header panel and torque both bolts to 89 inch lbs. (10 Nm).

38. Install the front grille to the radiator core support.

39. Connect the negative battery cable and refill the cooling system.

40. Start the engine; allow it to reach normal operating temperature and check for leaks.

41. Check and/or adjust the ignition timing as necessary.

Intake Manifold

REMOVAL AND INSTALLATION

1.3L (VIN 3) and 1.6L (VIN 0), (VIN U) TFI Engines

— **CAUTION** —

Fuel injection systems remain under pressure, even after the engine has been turned OFF. The fuel system pressure must be relieved before disconnecting any fuel lines. Failure to do so may result in fire and/or personal injury.

1. Properly relieve the fuel system pressure.

2. Disconnect the negative battery cable. Drain the cooling system.

3. Remove the air intake case from the manifold.

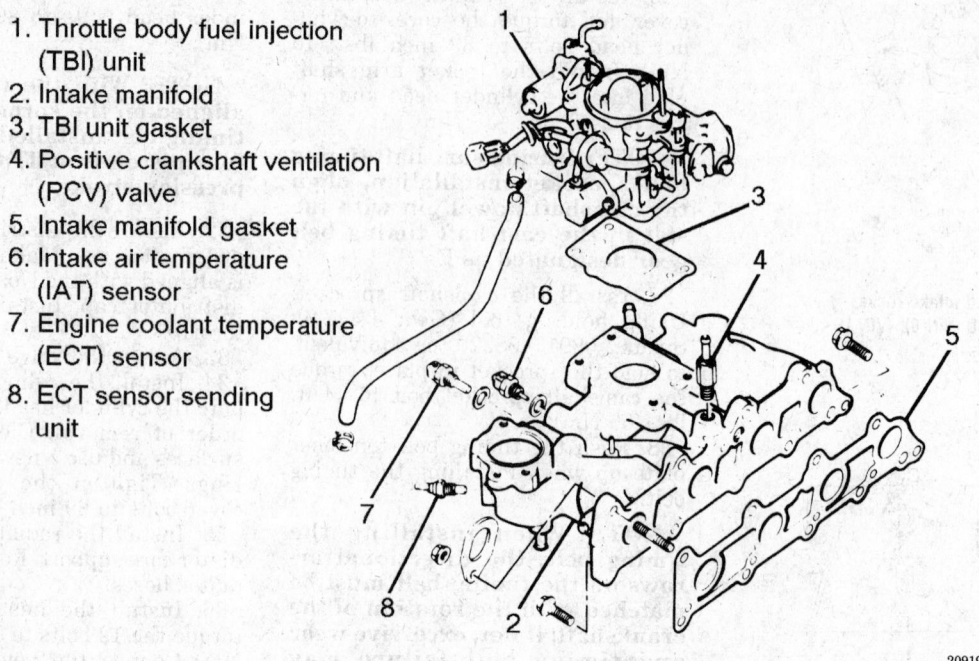

1. Throttle body fuel injection (TBI) unit
2. Intake manifold
3. TBI unit gasket
4. Positive crankshaft ventilation (PCV) valve
5. Intake manifold gasket
6. Intake air temperature (IAT) sensor
7. Engine coolant temperature (ECT) sensor
8. ECT sensor sending unit

Intake manifold components — 1.6L (VIN 0), (VIN U) TFI engines; 1.3L (VIN 3) engines are similar

4. Disconnect the accelerator cable and the automatic transmission kickdown cable, if equipped.

5. Disconnect the injector, throttle position sensor and idle speed control solenoid valve wires at their couplers, if equipped.

6. Disconnect the vacuum hoses from the throttle body and throttle opener. Remove the water hose from the air valve, if equipped.

7. Disconnect the fuel line from the throttle body and intake manifold.

8. Remove the fuel return hose.

9. Disconnect the throttle body from the intake manifold and remove the PCV hose from the cylinder head cover.

10. Disconnect the pressure sensor hose from the fuel filter, if equipped, and remove the brake booster hose from the intake manifold.

11. Disconnect the vacuum hose for the automatic transmission from the intake manifold, if equipped.

12. Remove the Solenoid Vacuum Valve (for throttle opener) hose from the intake manifold.

13. Disconnect the water hose from the thermostat cap. Remove the heater inlet and water bypass hose from the intake manifold.

14. Disconnect the hoses at the EGR valve and remove the ground wire from the intake manifold.

15. Disconnect the wire couplers from the air temperature sensor, water temperature sensor, water temperature gauge and PTC heater, if equipped with automatic transmission.

16. Disconnect the wire harnesses from their clamps.

17. Remove the intake manifold mounting bolts and remove the intake manifold from the cylinder head.

18. Remove the PCV valve, EGR valve, fuel filter, thermostat from the intake manifold.

19. Clean and inspect the sealing surfaces of the intake manifold and the cylinder head.

To install:

20. Install the thermostat, fuel filter, EGR valve and PCV valve to the intake manifold.

21. Using a new gasket, install the intake manifold and tighten the mounting bolts to 13–20 ft. lbs. (17–27 Nm).

22. Reconnect the wiring harnesses to their clamps and fasten the wire couplers to the air temperature sensor, water temperature sensor, water temperature gauge and PTC heater, if equipped with automatic transmission.

23. Reconnect the ground wire to the intake manifold and replace the hoses at the EGR valve.

24. Install the water heater inlet and bypass hose to the intake manifold. Replace the water hose to the thermostat cap.

25. Replace the SVV (for throttle opener) hose to the intake manifold.

26. Connect the automatic transmission vacuum hose to the intake manifold, if equipped.

27. Replace the brake booster hose to the intake manifold and connect the pressure sensor to the fuel filter, if equipped.

28. Reconnect the fuel line to the throttle body.

29. Replace the water hose to the air valve. Connect the vacuum hoses to the throttle body and throttle opener, if equipped.

30. Reconnect the injector, throttle position sensor and idle speed control wires to their couplers, if equipped.

31. Connect the accelerator cable and the automatic transmission kickdown cable, if equipped, to the throttle body.

32. Install the air intake case to the manifold.

33. Refill the cooling system with the proper coolant. Connect the negative battery cable.

34. Check for vacuum, water and oil leaks when finished.

1.6L (VIN 0), (VIN U) MFI Engines

CAUTION

Fuel injection systems remain under pressure, even after the ignition is switched OFF. The fuel system pressure must be relieved before disconnecting any fuel lines. Failure to do so may result in fire and/or personal injury.

1. Disconnect the negative battery cable and drain the cooling system.
2. Remove the air intake pipe from the manifold.
3. Disconnect the accelerator cable from the throttle body.
4. If equipped with an automatic transmission, disconnect the automatic transmission kickdown/throttle cable.
5. Disconnect the fuel injector, throttle position sensor, and idle speed control solenoid valve wires at their couplers.
6. Disconnect the electrical wires for EGR solenoid vacuum valve, ground wires for the intake surge tank, and the EVAP solenoid purge valve.
7. Disconnect the wire couplers from the air temperature sensor, water temperature sensor, water temperature gauge, and the A/C coolant temperature switch.
8. Remove the brake booster hose from the intake manifold and the canister purge vacuum hose from the canister.
9. If equipped with an automatic transmission (3 model), disconnect the vacuum hose for the automatic transmission from the intake manifold.
10. Disconnect the water hose from the thermostat cap. Remove the heater inlet and water bypass hose from the intake manifold. Remove the water hose from the idle air control valve.
11. Disconnect the hoses at the EGR valve.
12. Remove the fuel filler cap to release the fuel pressure in the fuel tank. Reinstall the cap.
13. Release the fuel pressure in the fuel line.
14. Disconnect the fuel feed line and remove the fuel return hose.
15. Remove the PCV hose from the cylinder head cover.
16. Disconnect the wire harnesses from their clamps.
17. Remove the alternator adjust arm bracket (stiffener).
18. Remove the intake manifold bracket (stiffener), No. 1 stiffener and No. 2 stiffener with the EGR modulator.

NOTE: Verify that all electrical and vacuum connections have been disconnected prior to removing the intake manifold.

19. Remove the intake manifold mounting bolts and remove the intake manifold, along with the surge tank and throttle body, from the cylinder head.
20. Remove the surge tank and throttle body from the intake manifold.
21. Clean and inspect the sealing surfaces of the intake manifold and the cylinder head.

To install:

22. Install the throttle body and surge tank to the intake manifold.
23. Using a new gasket, install the intake manifold and tighten the mounting bolts to 13–20 ft. lbs. (17–27 Nm).
24. Install the three stiffener brackets removed to gain access to the intake manifold. Tighten the **A** bolts to 29–43 Nm (40–60 Nm).
25. Reconnect the wiring harnesses to their clamps.
26. Reconnect the fuel feed and return lines.
27. Reconnect the EGR hoses.

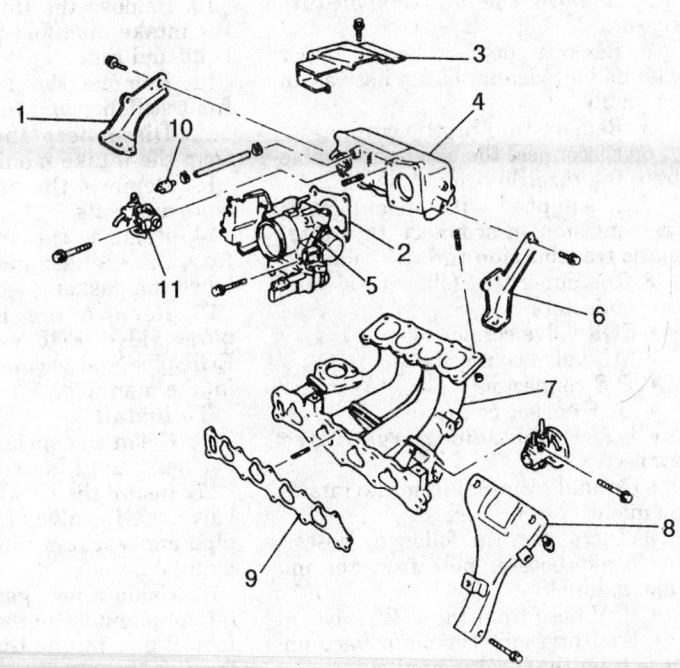

1 Intake manifold No.1 stiffener
2 Gasket
3 Throttle cover
4 Intake surge tank
5 Throttle body
6 No.2 stiffener
7 Intake manifold
8 Intake manifold stiffener
9 Gasket
10 PCV valve
11 IAC valve

328944

Exploded view of the intake manifold components — 1.6L (VIN 0), (VIN U) MFI engines

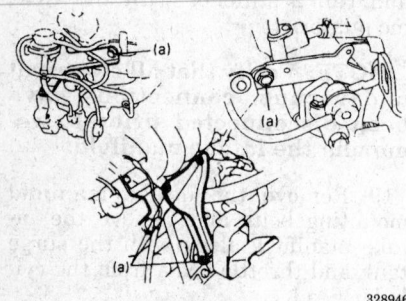

Intake manifold brackets (A= bolt locations) — 1.6L (VIN 0), (VIN U) MFI engines

328946

28. Reconnect the water hose to the idle air control valve.

29. Reconnect the heater inlet and water bypass hoses to the intake manifold.

30. Reconnect the radiator hose to the thermostat cap and the PCV hose to the PCV valve.

31. If equipped, reconnect the vacuum hose to the intake for the automatic transmission.

32. Reconnect the brake booster and canister purge vacuum hoses to the intake manifold and canister.

33. Reconnect the wire couplers to the air temperature sensor, water temperature sensor, water temperature gauge, and the A/C coolant temperature switch.

34. Reconnect the wire couplers to the EGR solenoid vacuum valve and the EVAP solenoid purge valve. Reconnect the ground wires for the intake surge tank to the intake manifold.

35. Reconnect the fuel injector, throttle position sensor, and idle speed control wires to their couplers.

36. Reconnect the accelerator cable and the automatic transmission kickdown/throttle cable to the throttle body.

37. Install the air intake pipe to the manifold.

38. Refill the cooling system with the proper coolant. Reconnect the negative battery cable.

39. Start the engine and check for vacuum, water, or fuel leaks when finished.

1.8L (VIN 2) Engine

——— CAUTION ———
Fuel injection systems remain under pressure, even after the ignition is switched OFF. The fuel system pressure must be relieved before disconnecting any fuel lines. Failure to do so may result in fire and/or personal injury.

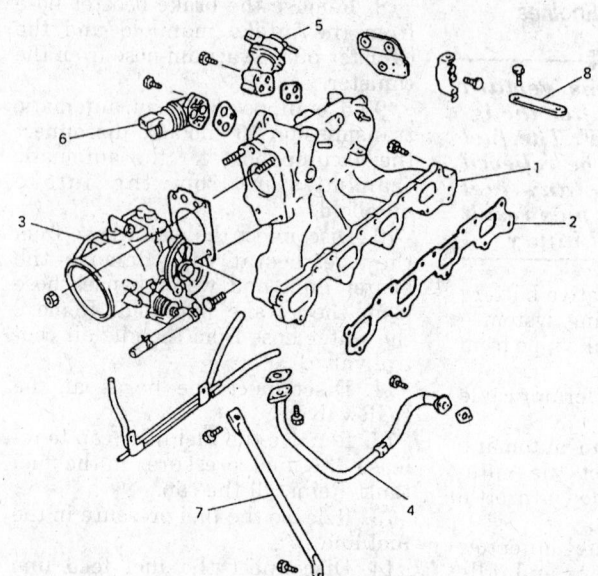

1. Intake manifold
2. Intake manifold gasket
3. Throttle body
4. EGR pipe
5. EGR valve
6. IAC valve
7. Intake manifold front stiffener
8. Intake manifold rear stiffener

331450

Intake manifold assembly — 1.8L (VIN 2) engines

1. Disconnect the negative battery cable and drain the cooling system.

2. Remove the strut tower bar by removing the two bolts and two nuts.

3. Remove the air temperature sensor.

4. Remove the air cleaner upper case and air cleaner outlet hose as an assembly.

5. Remove the throttle cover.

6. Disconnect the accelerator cable from the throttle body.

7. If equipped with an automatic transmission, disconnect the automatic transmission throttle cable.

8. Disconnect the following electrical connectors:
- EGR valve connector
- IAC valve connector
- TPS connector
- MAF sensor connector
- EVAP solenoid purge valve connector
- Ground terminal from the intake manifold

9. Disconnect the following hoses:
- Brake booster hose from the intake manifold
- PCV hose from the PCV valve
- Fuel pressure regulator vacuum hose from the intake manifold
- MAP sensor hose from the intake manifold
- Canister purge hose from the intake manifold

- Coolant hose from the throttle body and coolant bypass pipe
- Breather hose from the throttle body

10. Remove the throttle body from the intake manifold by removing the bolts and nuts.

11. Remove the intake manifold front stiffener and rear stiffener.

12. Disconnect the coolant pipe from the intake manifold.

13. Remove the intake manifold bolts and nuts.

14. Remove the intake manifold from the cylinder head and then remove the gasket.

15. Remove the EVAP solenoid purge valve, EGR valve, IAC valve, EGR pipe and vacuum pipe from the intake manifold.

To install:

16. Clean the surfaces to the cylinder head and intake manifold.

17. Install the EVAP solenoid purge valve, EGR valve, IAC valve, EGR pipe and vacuum pipe to the intake manifold.

18. Using a new gasket, install the intake manifold to the cylinder head. Install and torque the bolts to 17 ft. lbs. (23 Nm).

19. Connect the coolant pipe to the intake manifold.

20. Install the intake manifold front stiffener and rear stiffener. Tor-

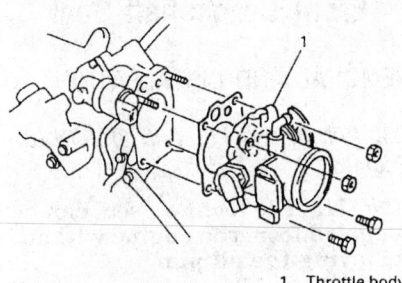

1. Throttle body
331454

Throttle body bolts and nuts — 1.8L (VIN 2) engines

que the bolts to the front stiffener to 37 ft. lbs. (50 Nm) and the rear stiffener to 18 ft. lbs. (25 Nm).

21. Install the throttle body to the intake manifold with a new gasket. Torque the bolts and nuts to 10 ft. lbs. (13 Nm).

22. Connect the following hoses:
• Brake booster hose to the intake manifold
• PCV hose to the PCV valve
• Fuel pressure regulator vacuum hose to the intake manifold
• MAP sensor hose to the intake manifold
• Canister purge hose to the intake manifold
• Coolant hose to the throttle body and coolant bypass pipe
• Breather hose to the throttle body

23. Connect the following electrical connectors:
• EGR valve connector
• IAC valve connector
• TPS connector
• MAF sensor connector
• EVAP solenoid purge valve connector
• Ground terminal to the intake manifold

24. Connect and adjust the accelerator cable and the automatic transmission throttle cable to the throttle body.

25. Install the throttle cover.

26. Install the air cleaner upper case and air cleaner outlet hose.

27. Install the air temperature sensor to the air cleaner box.

28. Install the strut tower bar and torque the bolts and nuts to 66 ft. lbs. (90 Nm).

29. Check that all removed parts are back in place.

30. Fill the engine and radiator with coolant.

31. Connect the negative battery cable to the battery.

32. Start the engine and check for leaks.

Exhaust Manifold

REMOVAL AND INSTALLATION

1.3L (VIN 3) and 1.6L (VIN 0), (VIN U) Engines

1. Raise and safely support the vehicle.

2. On 1996–97 engines, remove the air intake pipe and bracket.

3. Disconnect the three exhaust pipe bolts connecting the exhaust pipe to the exhaust manifold. Lower the vehicle after exhaust manifold is disconnected from the exhaust pipe.

4. Disconnect the oxygen sensor lead wire at the coupler.

5. Disconnect the exhaust manifold upper and lower covers or heat shields from the exhaust manifold.

6. Remove the exhaust manifold mounting bolts, and remove the exhaust manifold from the cylinder head.

To install:

7. Clean and inspect the sealing surfaces of the exhaust manifold and the cylinder head.

8. Using new gaskets, install the exhaust manifold to the cylinder head and tighten the mounting bolts to 13–20 ft. lbs. (17–27 Nm).

9. Connect the exhaust manifold upper and lower covers or heat shields from the exhaust manifold.

10. Connect the oxygen sensor lead wire at the coupler.

11. Raise the vehicle. Install the three exhaust pipe bolts connecting the exhaust pipe to the exhaust manifold and torque to 29–43 ft. lbs. (40–60 Nm).

12. On 1996–97 engines, install the air intake pipe and bracket.

13. Lower the vehicle.

14. Check for exhaust leaks when finished.

1.8L (VIN 2) Engine

1. Remove the strut tower bar.

2. Disconnect the oxygen sensor connector.

3. Remove the exhaust manifold cover from the exhaust manifold by removing the bolts.

4. Remove the exhaust stiffener bar by removing the bolt and nut.

5. Raise and safely support the front of the vehicle.

6. Disconnect the front exhaust pipe from the exhaust manifold by removing the two bolts.

7. Remove the exhaust manifold bolt and nuts. Remove the exhaust manifold and gasket from the engine.

To install:

8. Install a new gasket to the engine cylinder head.

9. Install the exhaust manifold to the engine with the nuts and bolt. Torque the nuts to 17 ft. lbs. (23 Nm).

10. Install the front exhaust pipe gasket and pipe to the exhaust manifold. Install the two bolts and torque the bolts to 37 ft. lbs. (50 Nm).

11. Install the exhaust manifold stiffener with the nut and bolt. Torque the nut to 40 ft. lbs. (55 Nm) and the bolt to 37 ft. lbs. (50 Nm).

12. Install the upper cover to the exhaust manifold.

13. Connect the oxygen sensor electrical connector.

14. Install the strut tower bar and torque the bolts and nuts to 66 ft. lbs. (90 Nm).

15. Start the engine and check for exhaust leaks.

Front Cover Seal

REMOVAL AND INSTALLATION

1.8L (VIN 2) Engine

1. Raise and safely support the front of the vehicle.

2. Remove the crankshaft pulley bolt.

3. Using a crankshaft puller, remove the crankshaft pulley from the engine.

———— **CAUTION** ————
When removing the oil seal, make sure not to damage the crankshaft or the timing cover.

4. Using a knife, cut off the oil seal lip.

5. Tape the end of a flat bladed tool to avoid damaging crankshaft. Pry out the oil seal using the taped end of the tool.

6. Inspect the oil seal riding surface on the crankshaft for signs of wear or damage.

To install:

7. Wipe the seal bore with a clean rag.

8. Apply multipurpose grease the lip of a new oil seal.

9. Drive the oil seal into place using a seal installer tool. Make sure the seal surface is flush with the timing cover. Be extremely careful not to damage the seal.

10. Install the crankshaft pulley to the engine. Install the bolt and torque the bolt to 109 ft. lbs. (150 Nm).

11. Lower the vehicle.

12. Start the engine and check for leaks.

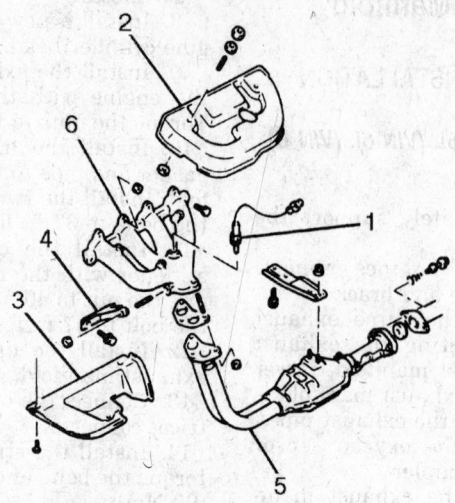

1. Oxygen (O2S) sensor
2. Upper exhaust manifold cover
3. Lower exhaust manifold cover
4. Exhaust bracket
5. Front pipe/3-way catalytic converter (TWC) assembly
6. Exhaust manifold

174495

Exhaust manifold and related components — 1993–95 1.6L (VIN 0), (VIN U) engines

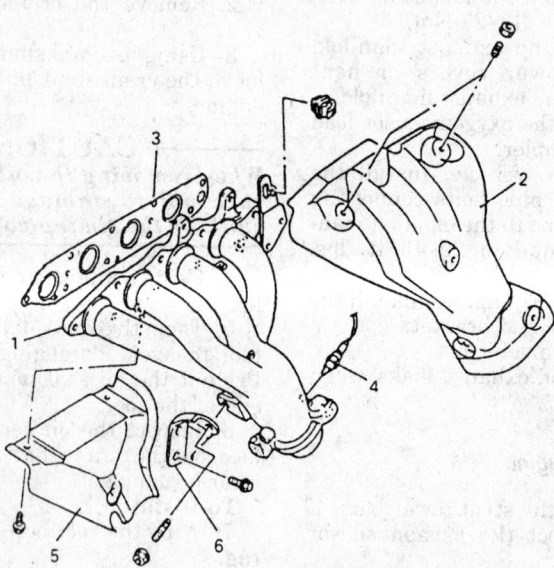

1. Exhaust manifold
2. Upper cover
3. Gasket
4. Heated oxygen sensor
5. Lower cover
6. Exhaust manifold stiffener

324069

Exhaust manifold component assembly — 1996–97 1.6L (VIN 0), (VIN U) engines

Front Crankshaft Seal

REMOVAL AND INSTALLATION

1.3L (VIN 3) and 1.6L (VIN 0), (VIN U) Engines

NOTE: The front oil seal can be removed from the engine without removing the oil pump.

1. Disconnect the negative battery cable from the battery.
2. Remove the front covers and the timing belt.
3. Remove the crankshaft timing belt sprocket.

—— **WARNING** ——
When removing the front seal, be extremely careful not to damage the crankshaft.

4. Using a knife, cut off the oil seal lip.
5. Tape the end of a flat bladed tool to avoid damaging crankshaft. Pry out the oil seal using the taped end of the tool.
6. Inspect the oil seal riding surface on the crankshaft for signs of wear or damage.
To install:
7. Wipe the seal bore with a clean rag.
8. Apply multipurpose grease the lip of a new oil seal.
9. Drive the oil seal into place using a seal installer tool. Make sure the seal surface is flush with the oil pump case edge. Work from the front of the cover. Be extremely careful not to damage the seal.
10. Install the crankshaft sprocket.
11. Install the timing belt and front covers.
12. Connect the negative battery cable to the battery.
13. Start the engine and check for leaks.

Timing Chain, Sprockets and Front Cover

REMOVAL AND INSTALLATION

1.8L (VIN 2) Engine

1. Disconnect the negative battery cable from the battery.
2. Raise and safely support the front of the vehicle.
3. Drain the engine oil from the engine.
4. Remove the oil pan and oil pump strainer from the engine.

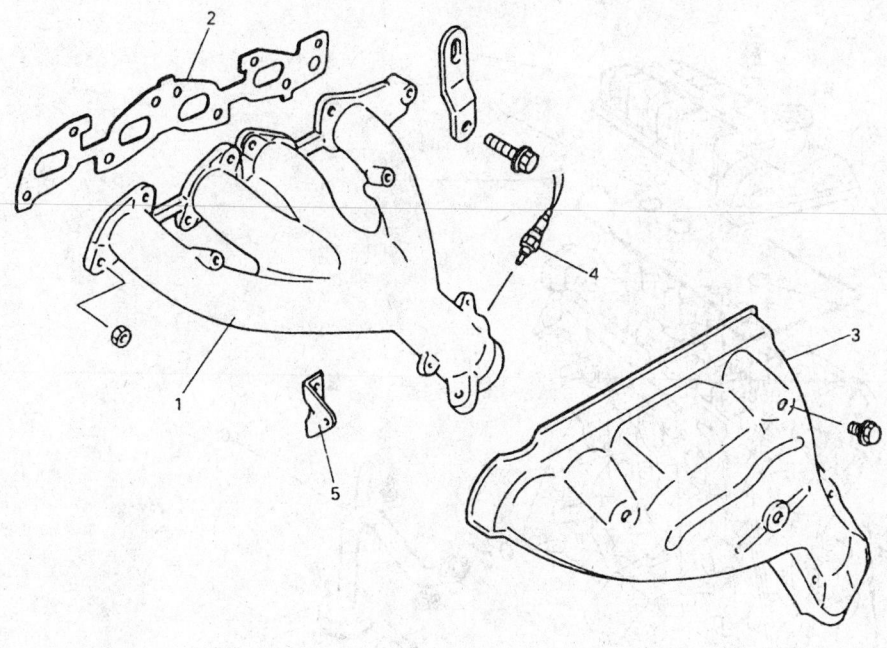

1. Exhaust manifold
2. Exhaust manifold gasket
3. Heated oxygen sensor
4. Exhaust manifold upper cover
5. Exhaust manifold stiffener

331013

Exhaust manifold assembly — 1.8L (VIN 2) engines

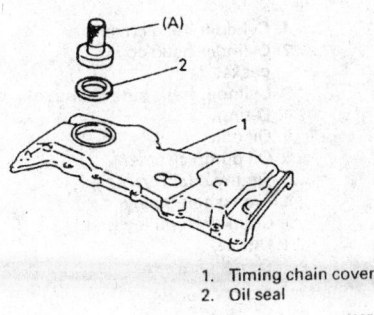

1. Timing chain cover
2. Oil seal

332500

Front cover oil seal location — 1.8L (VIN 2) engines

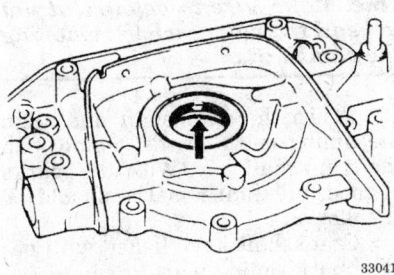

330410

Front crankshaft oil seal location — 1.3L (VIN 3) and 1.6L (VIN 0), (VIN U) engines

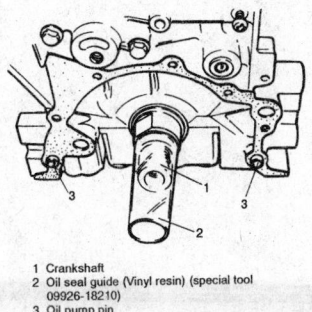

1 Crankshaft
2 Oil seal guide (Vinyl resin) (special tool 09926-18210)
3 Oil pump pin

325892

Oil seal guide tool — 1.3L (VIN 3) and 1.6L (VIN 0), (VIN U) engine

5. Remove the cylinder head cover as follows:

a. Remove the ignition coil cover.

b. Disconnect ignition coil connectors and remove the ignition coils from the engine.

c. Disconnect the accelerator cable from the clamp on the cylinder head cover.

d. Remove the oil level gauge.

e. Disconnect the breather hose and PCV hose from the cylinder head cover.

f. Remove the cylinder head cover by removing the nuts.

6. Remove the water bypass pipe and bypass hose No. 2.

7. Remove the cooling fan from the cooling fan pulley.

8. Remove the cooling fan belt by loosening the fan pulley bolts and turning the fan pulley. Remove the cooling fan pulley from the engine.

9. Remove the alternator fan belt by turning the belt tensioner clockwise.

10. Remove the water pump pulley.

11. Remove the drive belt tensioner.

12. Remove the drive belt idler pulley.

13. Disconnect the radiator outlet hose from the thermostat housing.

14. Without removing the A/C lines, disconnect the A/C compressor from the compressor bracket.

15. Remove the A/C compressor bracket.

16. Remove the crankshaft pulley bolt.

17. Using a crankshaft puller, remove the crankshaft pulley from the engine.

18. Remove the bolts and remove the timing chain cover from the vehicle.

19. Check the oil seal lip for damage. Replace as necessary.

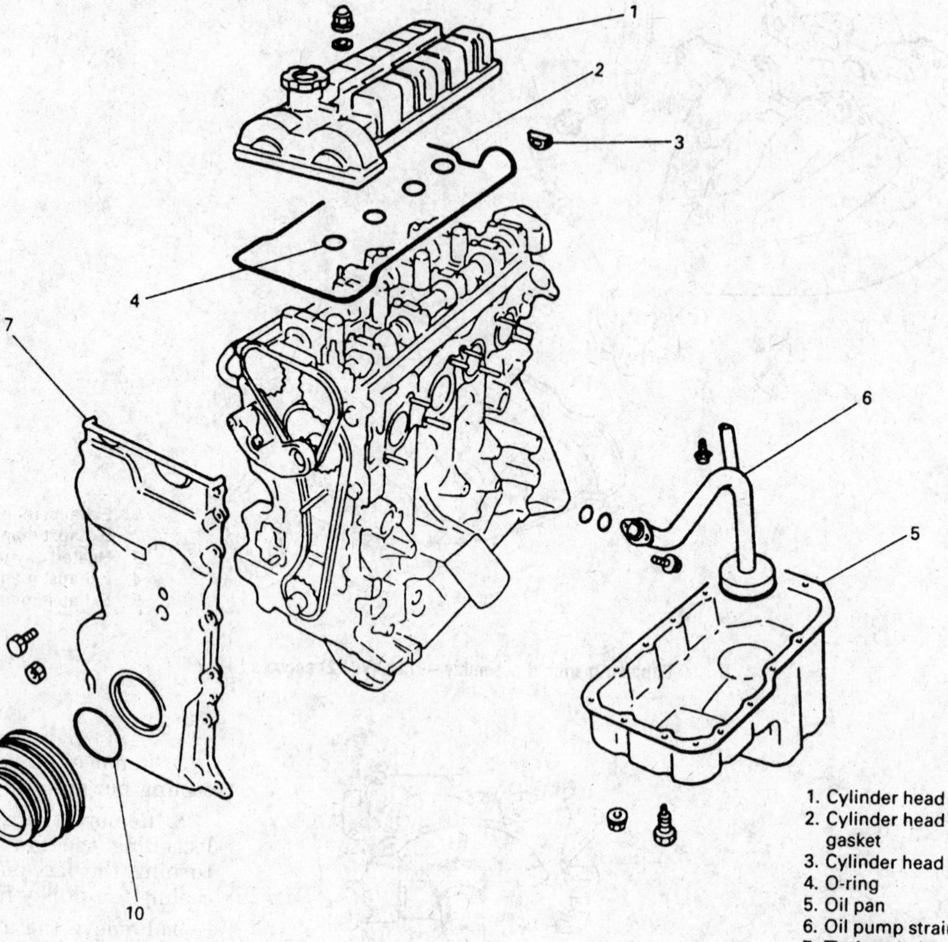

1. Cylinder head cover
2. Cylinder head cover gasket
3. Cylinder head side seal
4. O-ring
5. Oil pan
6. Oil pump strainer
7. Timing chain cover
8. Crankshaft pulley
9. Crankshaft pulley bolt
10. Oil seal

332536

Timing chain cover assembly overview — 1.8L (VIN 2) engines

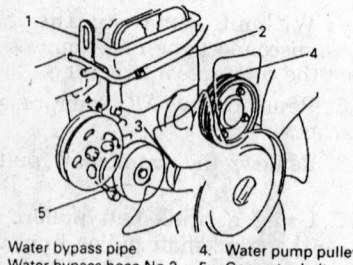

1. Water bypass pipe
2. Water bypass hose No.2
3. Cooling fan pulley
4. Water pump pulley
5. Generator belt tensioner

332537

Timing chain cover components for removal and installation — 1.8L (VIN 2) engines

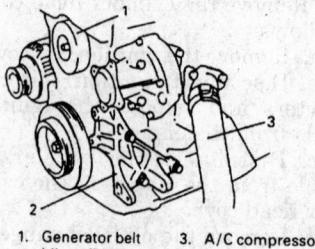

1. Generator belt idle pulley
2. Radiator outlet hose
3. A/C compressor bracket

333078

Timing chain cover components for removal and installation continued — 1.8L (VIN 2) engines

—— **CAUTION** ——
This engine is a interference engine. Make sure to be careful not to bend the valves when removing the camshafts.

20. Turn the crankshaft and align the timing mark on the engine with the crankshaft key. With the marks aligned, all timing marks should be as follows:

• Crankshaft key aligned with engine mark timing mark
• Arrow mark on idler sprocket points upward, vertically.
• The marks on the camshaft sprockets align with the marks on the cylinder head.

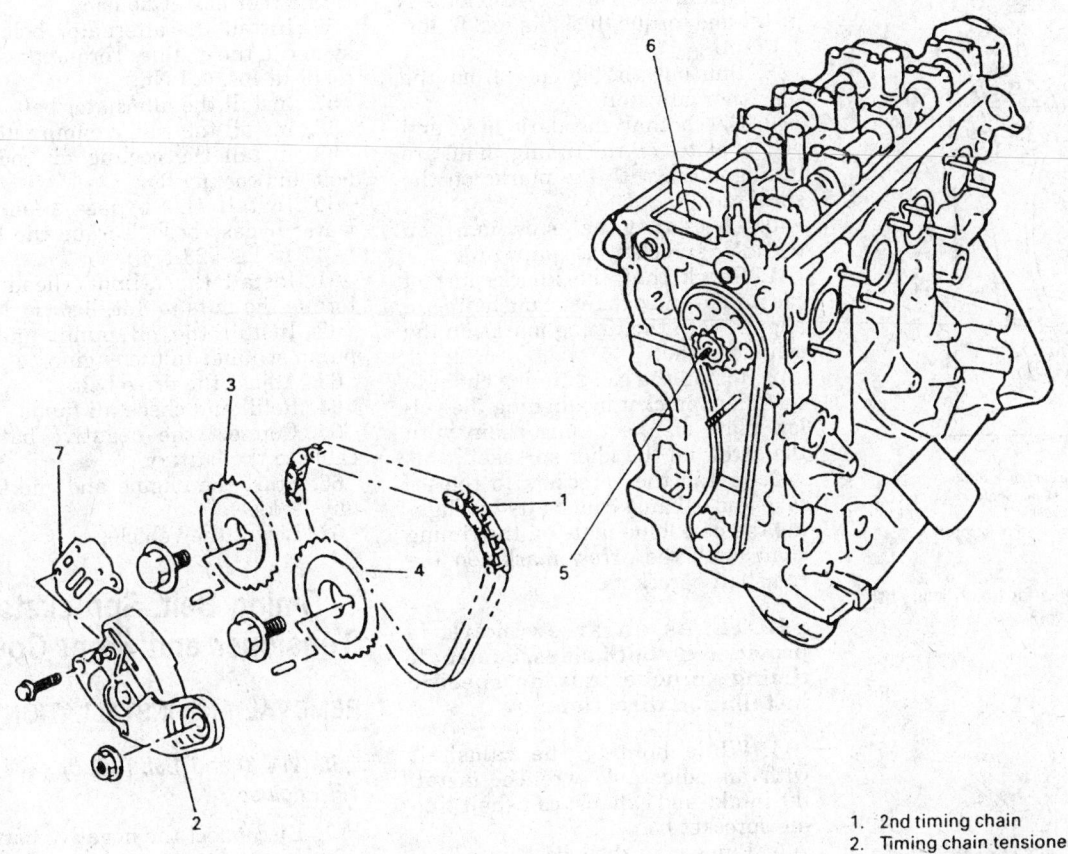

1. 2nd timing chain
2. Timing chain tensioner adjuster No.2
3. Intake camshaft timing sprocket
4. Exhaust camshaft timing sprocket
5. Idler sprocket
6. Timing chain guide No.2
7. Tensioner adjuster No.2 gasket

333047

No. 2 timing chain component assembly — 1.8L (VIN 2) engines

21. Remove the timing chain tensioner adjuster. To remove the timing chain tensioner adjuster, slacken the No. 2 timing chain by turning the intake camshaft slightly counterclockwise while pushing back on pad. Do not turn the intake camshaft too far.

22. Using an adjustable wrench connected to the hexagon part of the camshaft, remove the intake and exhaust camshaft timing sprocket bolts.

23. Remove the camshaft timing sprockets and No. 2 timing chain.

— CAUTION —

Do not turn the camshaft more than necessary. If turned excessively, valve damage may occur.

24. Remove the No. 1 timing chain guide by removing the bolts.

25. Remove the No. 1 timing chain tensioner adjuster by removing the nut and spacer.

26. Remove the timing chain tensioner by removing the nut.

27. Remove the idler sprocket and No. 1 timing chain.

28. Remove the crankshaft timing sprocket.

To install:

29. Check that the crankshaft timing sprocket is aligned with the timing mark on the cylinder block.

30. Install the crankshaft timing sprocket.

31. Apply oil to bush of idler sprocket.

32. Install the idler sprocket and sprocket shaft to the engine. The idler sprocket mark should be straight up.

33. Install the No. 1 timing chain by aligning the dark blue plate on the timing chain with the mark on the idler sprocket.

34. Bring yellow plate of the No. 1 timing chain into alignment with the mark on the crankshaft timing sprocket.

35. Install the spacer, nut, and timing chain tensioner.

36. With the latch of the No. 1 tensioner adjuster returned and plunger

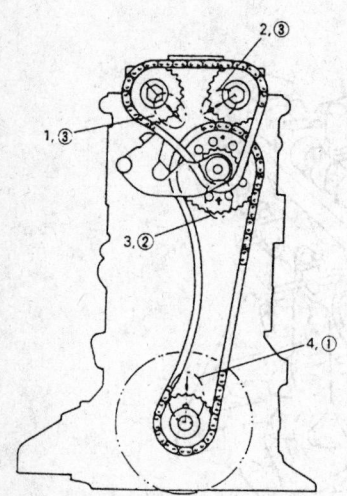

1. Timing marks of intake camshaft timing sprocket
2. Timing marks of exhaust camshaft timing sprocket
3. Arrow mark on idler sprocket
4. Timing marks of crankshaft timing sprocket

333053

Aligning the timing marks for No. 2 timing belt removal — 1.8L (VIN 2) engines

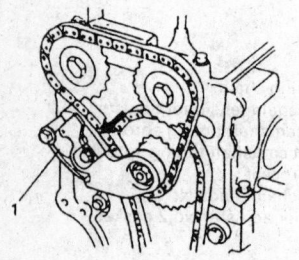

1. Timing chain tensioner adjuster No.2

333059

Removing the timing chain tensioner adjuster — 1.8L (VIN 2) engines

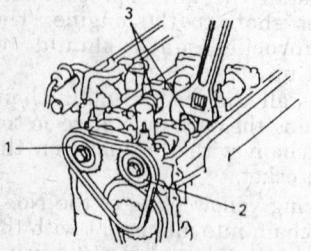

1. Intake camshaft timing sprocket
2. Exhaust camshaft timing sprocket
3. Hexagonal part of camshaft

333060

Removing the intake and exhaust timing sprockets — 1.8L (VIN 2) engines

pushed back into body, insert stopper into latch and body. After inserting stopper, check to make sure that the plunger will not come out.

37. Install the No. 1 tensioner adjuster and torque the bolts to 8 ft. lbs. (11 Nm).

38. Pull out the stopper from the tensioner adjuster.

39. Check that the dark blue and yellow plates on the timing chain are in alignment with the marks on the sprockets.

40. Check that the arrow mark on the idler sprocket faces upward.

41. Check that the knock pins of the intake and exhaust camshafts are aligned with the timing marks on the cylinder head.

42. Install the No. 2 timing chain to the idler sprocket by aligning the yellow plate on the timing chain with the arrow on the idler sprocket.

43. Install the sprockets to the intake and exhaust camshafts by aligning the dark blue plate on the timing chain with the arrow marks on the camshaft sprockets.

NOTE: As an arrow mark is provided on both sides, camshaft timing sprocket has no specific installation direction.

44. While holding the camshaft with an adjustable wrench, install the intake and exhaust camshaft timing sprocket bolts.

45. Plunger pushed back into timing chain tensioner, insert stopper into timing chain tensioner body. After inserting it, check to make sure that the plunger will not come out.

46. Install the timing chain tensioner adjuster with the gasket. Torque the nut to the idler sprocket to 33 ft. lbs. (45 Nm) and the bolts for the timing chain tensioner to 8 ft. lbs. (11 Nm).

47. Pull out the stopper from the timing chain tensioner adjuster.

48. Turn the crankshaft two rotations clockwise. Align the timing mark on the crankshaft to the timing mark on the cylinder block.

49. Apply oil to timing chains, tensioner, tensioner adjusters, sprockets, sprockets and guides.

50. Apply engine oil to the oil seal lip.

51. Install the timing chain cover to the engine and install the bolts. Torque the bolts to 8 ft. lbs. (11 Nm).

52. Install the crankshaft pulley with the bolt. Torque the bolt to 109 ft. lbs. (150 Nm).

53. Install the A/C bracket to the engine and install the bolts. Torque the bolts to 40 ft. lbs. (55 Nm).

54. Install the alternator belt idler pulley and torque the nut to 33 ft. lbs. (45 Nm).

55. Connect the radiator outlet hose to the thermostat housing.

56. Install the alternator belt tensioner to the engine. Torque the bolts to 19 ft. lbs. (25 Nm).

57. Install the alternator belt.

58. Install the water pump pulley.

59. Install the cooling fan pulley, belt and cooling fan.

60. Install the bypass pipe and water bypass hose. Torque the bolts to 17 ft. lbs. (23 Nm).

61. Install the cylinder head and torque the nuts to 8 ft. lbs. (11 Nm).

62. Install the oil pump and oil pump strainer to the engine.

63. Adjust the drive belt.

64. Refill and check all fluids.

65. Connect the negative battery cable to the battery.

66. Start the engine and check for any leaks.

67. Lower the vehicle.

Timing Belt, Sprockets, Tensioner and Front Cover

REMOVAL AND INSTALLATION

1.3L (VIN 3) and 1.6L (VIN 0), (VIN U) TFI Engines

1. Disconnect the negative battery cable.

2. Drain the engine coolant and remove the radiator cooling fan and fan shroud.

3. If equipped, disconnect the air conditioning compressor drive belt, properly discharge the air conditioning system and remove the air conditioning compressor flexible suction hose.

4. Loosen the alternator mounting bolts and remove the water pump drive belt and pulley.

5. Remove the crankshaft pulley mounting bolts and remove the crankshaft pulley.

NOTE: The crankshaft drive belt pulley can be removed without loosening the center crankshaft bolt.

6. Disconnect the timing belt cover mounting bolts and remove the timing belt cover. Loosen but do not remove the tensioner bolt.

7. Disconnect the air intake case from the intake manifold.

8. Loosen the timing belt tensioner adjusting bolt and pivot nut. Hold pressure on the tensioner to loosen the timing belt and remove the

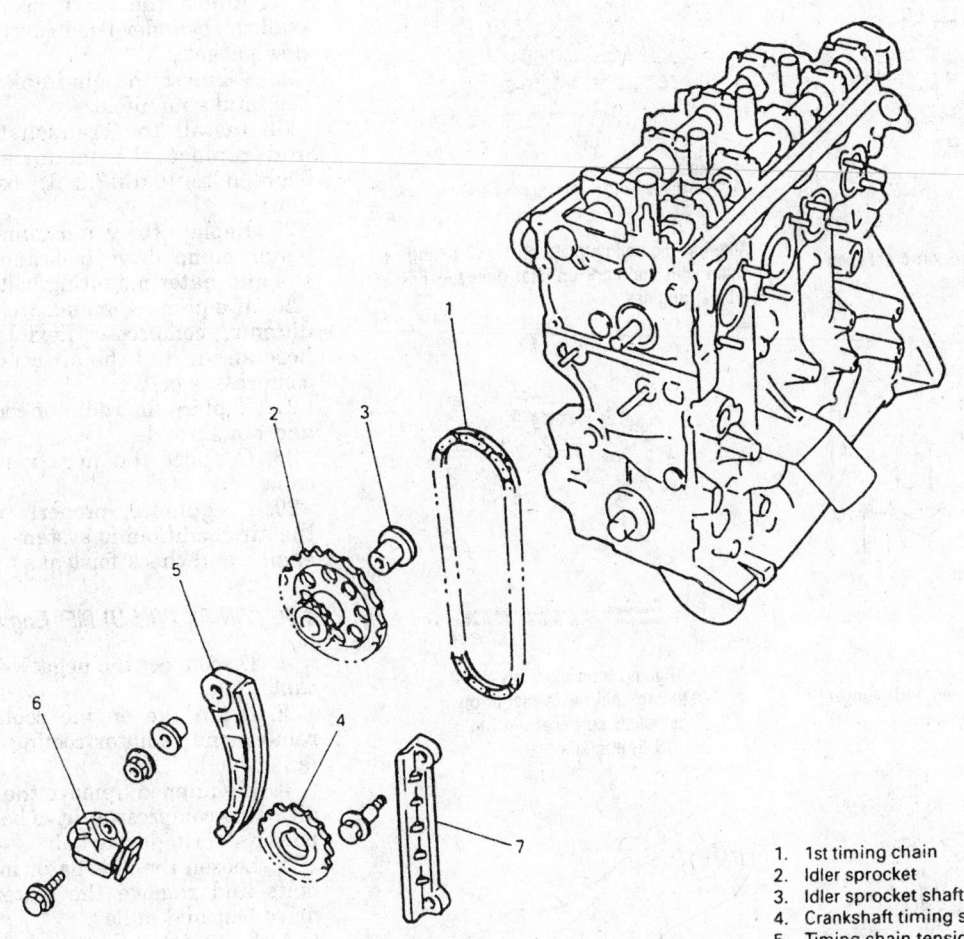

1. 1st timing chain
2. Idler sprocket
3. Idler sprocket shaft
4. Crankshaft timing sprocket
5. Timing chain tensioner
6. Timing chian tensioner adjuster No.1
7. Timing chain guide No.1

332741

No. 1 timing chain component assembly — 1.8L (VIN 2) engines

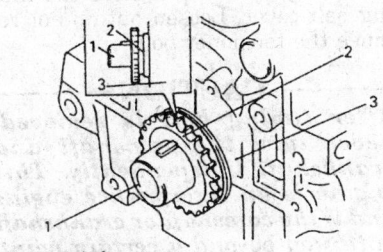

1. Crankshaft
2. Crankshaft timing sprocket
3. Cylinder block

332743

Installing the crankshaft sprocket — 1.8L (VIN 2) engines

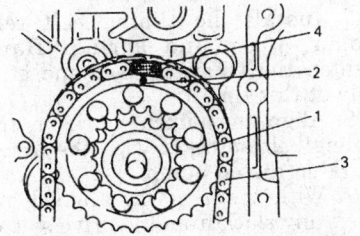

1. Idler sprocket
2. Match mark on idler sprocket
3. 1st timing chain
4. Dark blue plate

332744

Aligning the dark blue plate with mark on idler sprocket — 1.8L (VIN 2) engines

timing belt from the camshaft and crankshaft sprockets.

9. Remove the timing belt tensioner, tensioner plate and tensioner spring.

10. Remove the cylinder head cover and loosen all the valve adjusting screws to permit rotation of the camshaft.

11. Remove the camshaft timing belt sprocket bolt by inserting a 0.35 in. (9mm) rod into the hole in the camshaft to lock the camshaft.

12. Remove the camshaft sprocket mounting bolt and sprocket.

13. Remove the crankshaft sprocket bolt by using a suitable gear stopper to hold the flywheel. Remove the

8-49

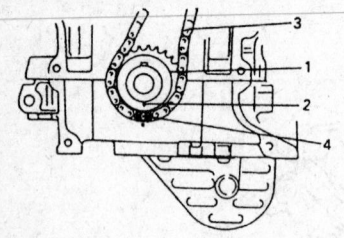

1. Crankshaft timing sprocket   3. 1st timing chain
2. Match mark   4. Yellow plate

332745

Aligning yellow plate with mark on crankshaft sprocket — 1.8L (VIN 2) engines

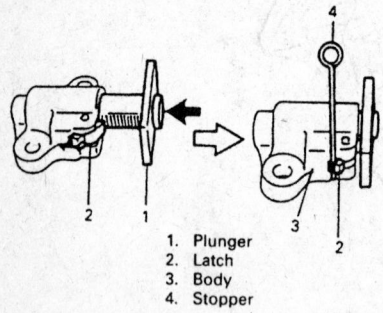

1. Plunger
2. Latch
3. Body
4. Stopper

332746

No. 1 tensioner adjuster returned and plunger pushed back — 1.8L (VIN 2) engines

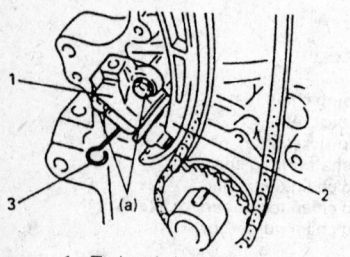

1. Timing chain tensioner adjuster No.1
2. Timing chain tensioner
3. Stopper

332747

No. 1 timing chain tensioner installed on engine — 1.8L (VIN 2) engines

crankshaft timing belt sprocket bolt, sprocket and key.

To install:

14. Install the camshaft sprocket and bolt on the camshaft.

15. Lock the camshaft using the rod and tighten the sprocket bolt to 41–46 ft. lbs. (55–62 Nm). Remove the rod.

16. With the crankshaft locked, install the crankshaft timing belt sprocket and key. Install the crankshaft sprocket bolt and tighten to 47–54 ft. lbs. (64–73 Nm) for the 1.6L or 76–83 ft. lbs. (105–115 Nm) for the 1.3L.

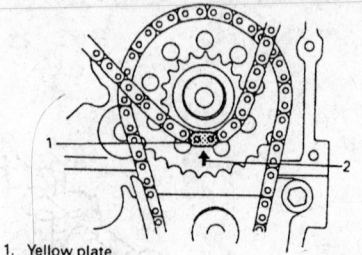

1. Yellow plate
2. Match mark of 2nd timing chain (Arrow mark)

333065

Aligning the yellow plate on No. 2 timing chain with the mark on idler pulley — 1.8L (VIN 2) engines

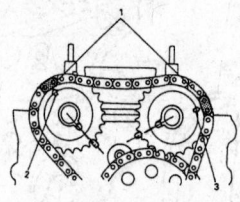

1. Dark blue
2. Arrow mark on intake camshaft timing sprocket
3. Arrow mark on exhaust camshaft timing sprocket

333066

Aligning blue mark on No. 2 timing chain with marks on camshaft sprockets — 1.8L (VIN 2) engines

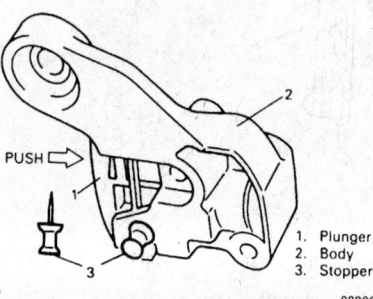

1. Plunger
2. Body
3. Stopper

333067

Installing the timing chain tensioner with stopper installed — 1.8L (VIN 2) engines

17. Install the timing belt tensioner, plate and spring. Hand tighten the tensioner bolt and stud only at this time.

18. Align the punch mark on the timing belt sprocket with the arrow mark on the oil pump.

19. With the 4 marks aligned, remove any slack from the drive side of the belt. Tighten the tensioner bolt to 17.5–21.5 ft. lbs. (24–30 Nm).

20. To allow the belt to be free of any slack, turn the crankshaft clockwise 2 full rotations. Confirm that the 4 marks are aligned.

21. Install the timing cover and tighten the bolts to 7.0–8.5 ft. lbs. (9–12 Nm).

22. Adjust the valve lash and install the cylinder head cover, using a new gasket.

23. Connect the air intake case to the intake manifold.

24. Install the crankshaft pulley and replace the mounting bolts. Tighten to 10.5–13.0 ft. lbs. (14–18 Nm).

25. Replace the water pump pulley, water pump drive belt and tighten the alternator mounting bolts.

26. If equipped, install the air conditioning compressor flexible suction hose and install the air conditioning compressor belt.

27. Replace the radiator cooling fan and fan shroud.

28. Connect the negative battery cable.

29. If equipped, properly recharge the air conditioning system. Run the engine and check for leaks.

1.6L (VIN 0), (VIN U) MFI Engine

1. Disconnect the negative battery cable.

2. Drain the engine coolant and remove the radiator cooling fan and fan shroud.

3. If equipped, remove the air conditioning compressor drive belt or the power steering drive belt.

4. Loosen the alternator mounting bolts and remove the water pump drive belt and pulley.

5. Remove the five crankshaft pulley mounting bolts and remove the crankshaft pulley.

NOTE: The crankshaft drive belt pulley can be removed without loosening the center crankshaft bolt.

6. Disconnect the timing belt cover mounting bolts and remove the timing belt cover. Loosen but do not remove the tensioner bolt.

— **CAUTION** —
After timing belt is removed, never turn the camshaft and crankshaft independently. This engine is an interference engine and if the camshaft or crankshaft is turned beyond a certain point, damage to the valves could occur.

7. Loosen the timing belt tensioner adjusting bolt and pivot nut. Hold pressure on the tensioner to loosen the timing belt and remove the timing belt from the camshaft and crankshaft sprockets.

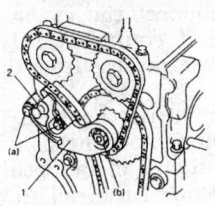

1. Timing chain tensioner adjuster No.2
2. Stopper

333068

No. 2 timing chain tensioner installed to engine — 1.8L (VIN 2) engines

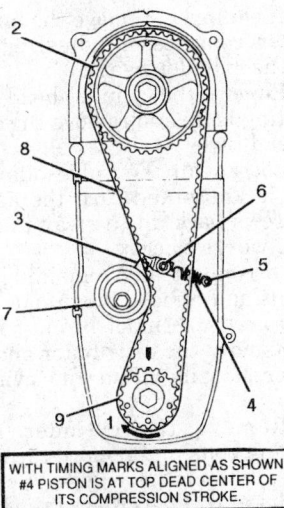

WITH TIMING MARKS ALIGNED AS SHOWN, #4 PISTON IS AT TOP DEAD CENTER OF ITS COMPRESSION STROKE.

1. Crankshaft timing belt gear rotation direction
2. Camshaft timing gear
3. Tensioner plate
4. Tensioner spring
5. Tensioner spring screw
6. Tensioner stud
7. Tensioner
8. Timing belt
9. Crankshaft timing belt gear

174537

Timing belt overview — 1993–95 1.6L (VIN 0), (VIN U) engines

8. Remove the timing belt tensioner, tensioner plate and tensioner spring.

9. Remove the camshaft timing belt sprocket bolt by inserting a 0.35 in. (9mm) rod into the hole in the camshaft to lock the camshaft.

10. Remove the camshaft sprocket mounting bolt and sprocket.

11. Remove the crankshaft sprocket bolt by using a suitable gear stopper to hold the flywheel. Remove the crankshaft timing belt sprocket bolt, sprocket and key.

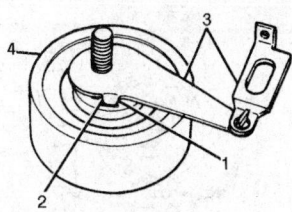

1. Tensioner slot
2. Tensioner plate lock tab
3. Tensioner plate
4. Tensioner

174541

Tensioner assembly lock tab and slot — 1993–95 1.6L (VIN 0), (VIN U) engines

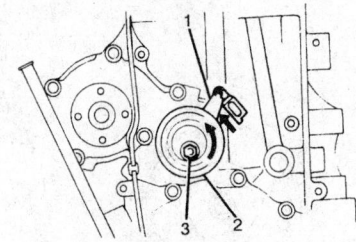

1. Tensioner plate
2. Tensioner
3. Tensioner retaining bolt

174540

Tensioner plate adjustment — 1993–95 1.6L (VIN 0), (VIN U) engines

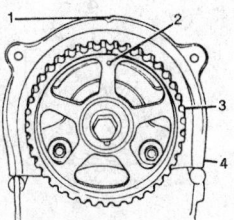

1. "V" timing mark
2. Camshaft timing gear timing mark
3. Camshaft timing gear
4. Timing belt inside cover

174539

Camshaft timing mark alignment — 1993–95 1.6L (VIN 0), (VIN U) engines

To install:

12. Install the camshaft sprocket and bolt on the camshaft.

13. Lock the camshaft using the rod and tighten the sprocket bolt to 41–46 ft. lbs. (55–62 Nm). Remove the rod.

14. With the crankshaft locked, install the crankshaft timing belt sprocket and key. Install the crankshaft sprocket bolt and tighten to 47–54 ft. lbs. (64–73 Nm).

15. Install the timing belt tensioner, plate and spring. Hand tighten the tensioner bolt and stud only at this time.

16. Turn the camshaft sprocket clockwise and align the timing marks.

17. Turn the crankshaft clockwise, using a 17mm wrench to crank the timing belt sprocket bolt.

18. Align the punch mark on the timing belt sprocket with the arrow mark on the oil pump.

19. With the 4 marks aligned, remove any slack from the drive side of the belt. Tighten the tensioner bolt to 16–20 ft. lbs. (22–28 Nm).

20. To allow the belt to be free of any slack, turn the crankshaft clockwise 2 full rotations. Confirm that the 4 marks are aligned.

21. Install the timing cover and tighten the bolts to 7.0–8.5 ft. lbs. (9–12 Nm).

22. Install the crankshaft pulley and replace the mounting bolts. Tighten to 10.5–13.0 ft. lbs. (14–18 Nm).

23. Replace the water pump pulley, water pump drive belt and tighten the alternator mounting bolts.

24. If equipped, install the air conditioning compressor belt or power steering belt.

25. Replace the radiator cooling fan and fan shroud.

26. Connect the negative battery cable.

27. Run the engine and check for leaks.

Camshaft

REMOVAL AND INSTALLATION

1.3L (VIN 3) and 1.6L (VIN 0), (VIN U) TFI Engines

— **CAUTION** —
The fuel system pressure must be released before disconnecting any fuel lines. Failure to do so may result in personal injury.

1. Disconnect the negative battery cable and drain the cooling system.

2. Disconnect the air cleaner assembly.

3. Disconnect the brake booster hose from the intake manifold and the air valve water hose from the throttle body.

4. Disconnect the accelerator cable and if equipped, the automatic transmission kickdown cable, from the throttle body.

5. Disconnect and tag the throttle opener VSV and EGR VSV wires at the coupler.

6. Disconnect and tag the water temperature sensor, oil pressure

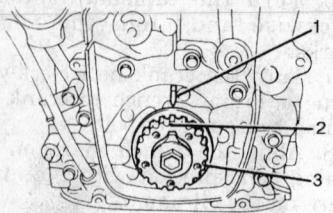

1. Arrow timing mark
2. Punch mark
3. Crankshaft timing belt gear

174538

Crankshaft timing mark alignment — 1993–95 1.6L (VIN 0), (VIN U) engines

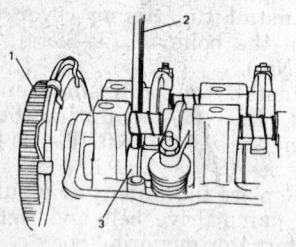

1 Camshaft timing belt pulley
2 Proper size rod
3 Camshaft

194749

Insert a rod to lock the camshaft — 1.3L (VIN 3) and 1.6L (VIN 0), (VIN U) TFI engine

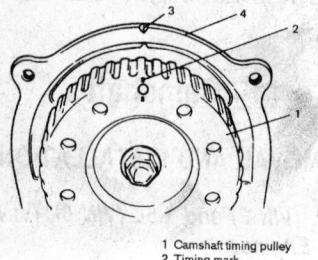

1 Camshaft timing pulley
2 Timing mark
3 "V" mark
4 Belt inside cover

194746

Camshaft timing mark alignment — 1.3L (VIN 3) engine

sender, air temperature sensor, and ground cable wires at the intake manifold.

7. Disconnect and tag the fuel injector, throttle position sensor, and idle speed control solenoid valve wires at their couplers, if equipped.

8. Disconnect the wire harness from its holding clamps.

9. Disconnect the MAP sensor vacuum hose and water bypass hose from the intake manifold.

10. Disconnect the radiator inlet hose from the thermostat housing.

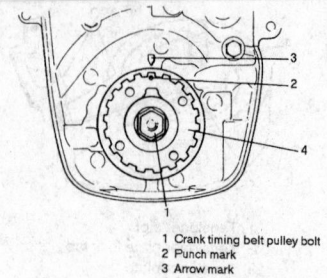

1 Crank timing belt pulley bolt
2 Punch mark
3 Arrow mark
4 Crank timing belt pulley

194747

Crankshaft timing mark alignment — 1.3L (VIN 3) engine

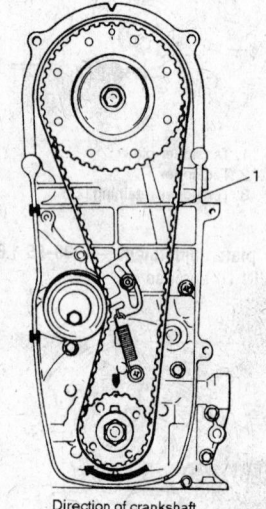

Direction of crankshaft

1 Drive side of belt

194748

Timing belt installation — 1.3L (VIN 3) engine

Disconnect the heater inlet hose from the intake manifold.

11. If equipped with automatic transmission, disconnect the PTC heater wires at the coupler.

12. Disconnect and tag the oxygen sensor and distributor wires at their couplers. Remove the coil wire.

13. Disconnect the ground wires from the distributor assembly.

14. Remove the fuel tank filler cap to relieve the pressure and reinstall the cap.

15. Properly release the fuel pressure in the fuel feed line.

16. Disconnect the fuel feed and return lines.

17. Disconnect the canister purge hose and if equipped, remove the pressure sensor hose from the intake manifold.

18. Remove the vacuum hose for the automatic transmission from the intake manifold, if equipped.

19. Disconnect the radiator cooling fan, fan shroud, water pump drive belt, and water pump pulley.

20. Disconnect the crankshaft pulley, timing belt cover, and the timing belt.

21. Raise and safely support the vehicle.

22. Disconnect the exhaust pipe from the exhaust manifold and lower the vehicle.

23. If equipped, remove the air conditioner compressor adjusting arm from the cylinder head.

24. Remove the cylinder head cover mounting bolts and remove the cylinder head cover.

25. Loosen all the valve adjusting screw locknuts and turn the adjusting screws back all the way to allow all the valves to close.

26. Disconnect any remaining connections and remove the intake manifold from the cylinder head.

27. Remove the distributor and distributor housing from the cylinder head.

28. Remove the cylinder head mounting bolts and remove the cylinder head from the engine. Remove the bolts in reverse order of the tightening sequence.

29. Remove any oil and water in the cylinder bores and on top of the pistons.

30. Clean and inspect the sealing surfaces of the cylinder head and the engine block.

31. Remove the timing belt sprocket from the camshaft and remove the rocker arms, springs and rocker arm shafts.

32. Keep all the valve train parts in the order that they were removed. The valve train parts must be reinstalled in their original position.

33. Carefully remove the camshaft from the rear of the cylinder head.

To install:

34. Lubricate the lobes and journals of the camshaft and the oil seal on the cylinder head with clean engine oil.

35. Install the camshaft to the cylinder head.

36. Install the rocker arm shafts, rocker arms, and springs in their original position. Make sure the rocker shafts are installed on the correct side and facing the right direction.

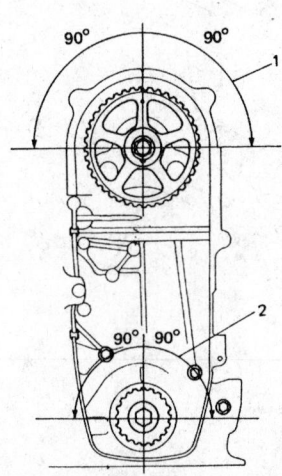

1. Camshaft allowable turning range - - - By timing mark, within 90° from "V" mark on head cover on both right and left.
2. Crankshaft allowable turning range - - - by punch mark, within 90° from arrow mark on oil pump case on both right and left.

328723

Allowable sprocket movement — 1.6L (VIN 0), (VIN U) MFI Engine

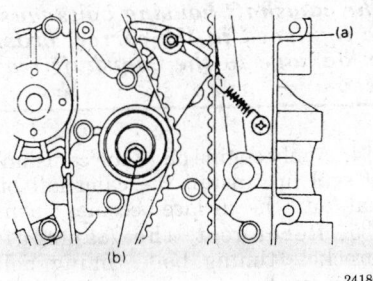

241863

Tensioner assembly on engine — 1.6L (VIN 0), (VIN U) MFI Engine

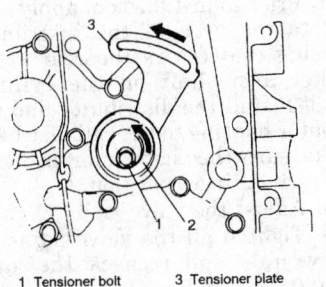

1 Tensioner bolt   3 Tensioner plate
2 Tensioner

241864

Tensioner plate adjustment — 1.6L (VIN 0), (VIN U) MFI Engine

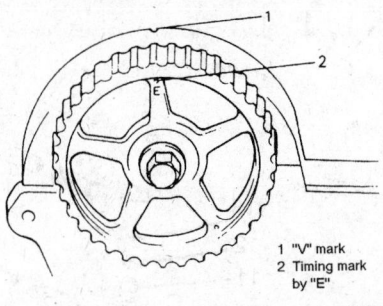

1 "V" mark
2 Timing mark by "E"

241865

Camshaft timing mark alignment — 1.6L (VIN 0), (VIN U) MFI Engine

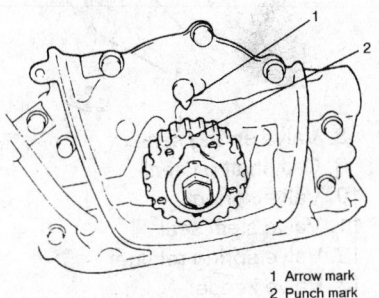

1 Arrow mark
2 Punch mark

241866

Crankshaft timing mark alignment — 1.6L (VIN 0), (VIN U) MFI Engine

37. Install the timing belt inside cover and camshaft timing belt pulley. Tighten the pulley mounting bolt to 41–46 ft. lbs. (56–64 Nm).
38. Install the cylinder head using a new gasket.
39. Lubricate the cylinder head mounting bolts with engine oil and install the cylinder head mounting bolts. Tighten the bolts evenly and in sequence to 46–54 ft. lbs. (62–73 Nm). Tightening should begin in the center and work towards both ends simultaneously.
40. Install the intake manifold to the cylinder head.
41. Raise and safely support the vehicle.
42. Connect the exhaust pipe to the exhaust manifold and safely lower the vehicle.
43. If removed, install the A/C compressor adjusting arm.
44. Reconnect the timing belt, timing belt cover, and crankshaft pulley.
45. Install the distributor and distributor housing to the cylinder head. Make sure the distributor is installed in the correct firing position.
46. Adjust the valve lash.
47. Tighten all the valve adjusting screw nuts.

48. Install the cylinder head cover and tighten the mounting bolts 4 ft. lbs. (5 Nm).
49. Install the water pump pulley, water pump drive belt, fan shroud, and the radiator cooling fan.
50. Install the vacuum hose for the automatic transmission to the intake manifold, if equipped.
51. If removed, install the pressure sensor hose to the intake manifold and connect the canister purge hose.
52. Install the fuel return and feed lines.
53. Connect the ground wires to the distributor assembly.
54. Install the coil wire. Connect the oxygen sensor and distributor wires to their couplers.
55. Connect the radiator inlet hose to the thermostat housing. Connect the heater inlet hose to the intake manifold.
56. If removed, connect the PTC wires to the coupler.
57. Connect the MAP sensor vacuum hose and the water bypass hose to the intake manifold.
58. Reconnect the wiring harness to its holding clamps.
59. Connect the idle speed control valve, throttle position sensor, and fuel injector wires to their couplers, if equipped.
60. Reconnect the ground cable, air temperature sensor, oil pressure sender, and water temperature sensor wires to the intake manifold.
61. Install the throttle opener VSV and EGR VSV wires to their couplers.
62. Connect the accelerator cable and if equipped, the automatic transmission kickdown cable, to the throttle body.
63. Install the air valve water hose to the throttle body and connect the brake booster hose to intake manifold.
64. Reconnect the air cleaner assembly.
65. Refill the cooling system with the proper coolant and connect the negative battery.
66. Start the engine and check for any water, fuel, or oil leaks when finished.
67. Check and/or adjust the ignition timing as necessary.

1.6L (VIN 0), (VIN U) MFI Engine

1. Disconnect the negative battery cable.
2. Remove the front grille and engine hood lock. Disconnect the lead wire from the horn and remove the front upper member from the body.
3. Drain the cooling system and remove the radiator.

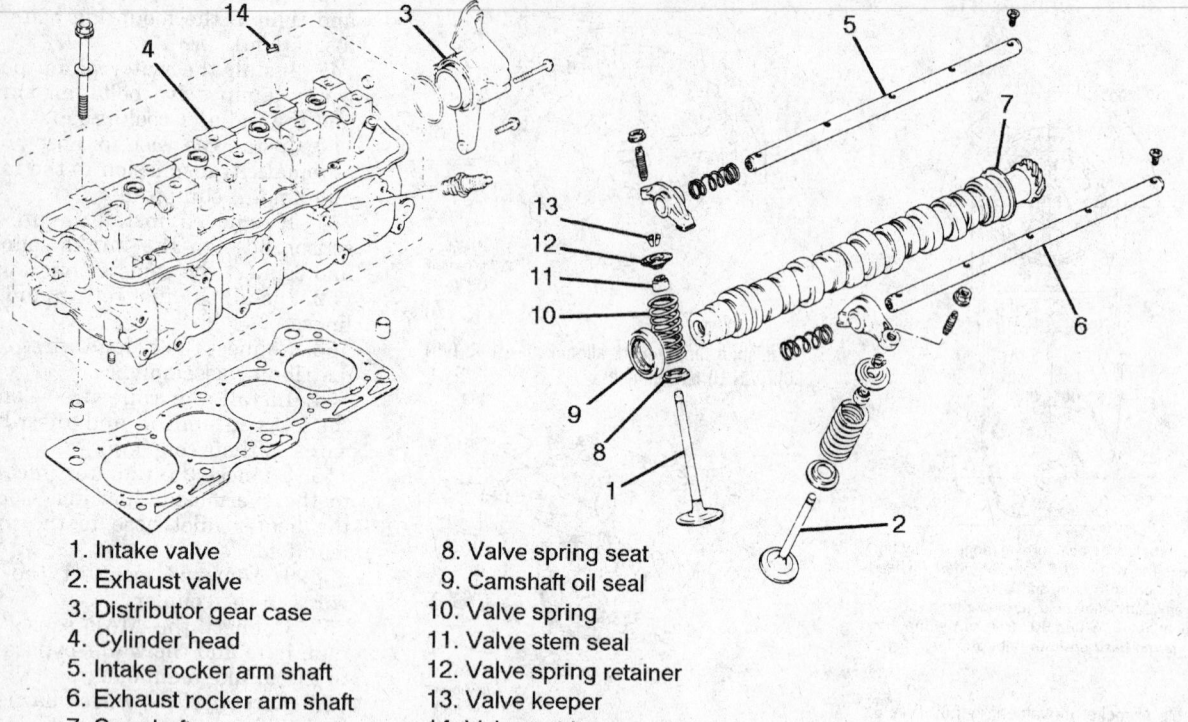

1. Intake valve
2. Exhaust valve
3. Distributor gear case
4. Cylinder head
5. Intake rocker arm shaft
6. Exhaust rocker arm shaft
7. Camshaft
8. Valve spring seat
9. Camshaft oil seal
10. Valve spring
11. Valve stem seal
12. Valve spring retainer
13. Valve keeper
14. Valve guide

195828

Camshaft components — 1.3L (VIN 3) and 1.6L (VIN 0), (VIN U) TFI engines

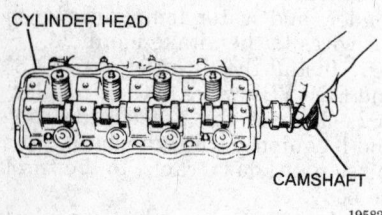

CYLINDER HEAD

CAMSHAFT

195829

Removing the camshaft — 1.3L and 1.6L (VIN 0), (VIN U) TFI engines

4. Remove the water pump belt, water pump pulley, crankshaft pulley, timing belt cover, and the timing belt.

5. Remove the camshaft sprocket.

6. Remove the air intake pipe. Remove the cylinder head cover mounting bolts and remove the cylinder head cover.

7. Remove the distributor and distributor case. Use a shop rag to catch the oil that will leak out of the case as it is removed.

8. Loosen all the valve adjusting screw locknuts and turn the adjusting screws back all the way to allow all the valves to close.

9. Remove the camshaft housing bolts, housings, and the camshaft.

——— CAUTION ———
The camshaft housing bolts must be removed in the correct order or damage to the camshaft may occur.

To install:

10. Lubricate the lobes and journals of the camshaft with clean engine oil.

11. Install the camshaft on the cylinder head. Install the camshaft housing to the camshaft and cylinder head, starting with the number one housing.

NOTE: Embossed marks are provided on each camshaft housing, indicating position and direction for installation.

12. Apply engine oil to sliding surface of each housing against camshaft journal. Apply sealant to the mating surface of the number six housing which will mate with the cylinder head.

13. Apply engine oil to the housing bolts, and hand tighten the bolts into the housing. Follow the tightening sequence in three to four even stages, finishing with a final torque of 7.0–8.5 ft. lbs. (9–12 Nm).

——— CAUTION ———
The camshaft housing bolts must be tightened in the correct order or damage to the camshaft may occur.

14. Apply engine oil to the camshaft oil seal lip. Install the camshaft oil seal until the surface becomes flush.

15. Reconnect the camshaft sprocket, timing belt, timing belt cover, crankshaft pulley, water pump pulley, and the water pump belt. Make sure the pin on the camshaft fits into the slot at the **E** mark on the camshaft sprocket. Tighten the sprocket bolt to 41–46 ft. lbs. (56–64 Nm).

16. Prior to installation, apply sealant to the area of the distributor housing that covers the rear of the rocker arm shaft on the cylinder head. Install the distributor and distributor housing to the cylinder head. Make sure the distributor is facing the correct firing position.

17. Adjust the valve lash.

18. Tighten all the valve adjusting screw nuts and recheck the valve lash.

19. Using a new gasket, install the cylinder head cover and connect the air intake pipe.

20. Install the radiator.

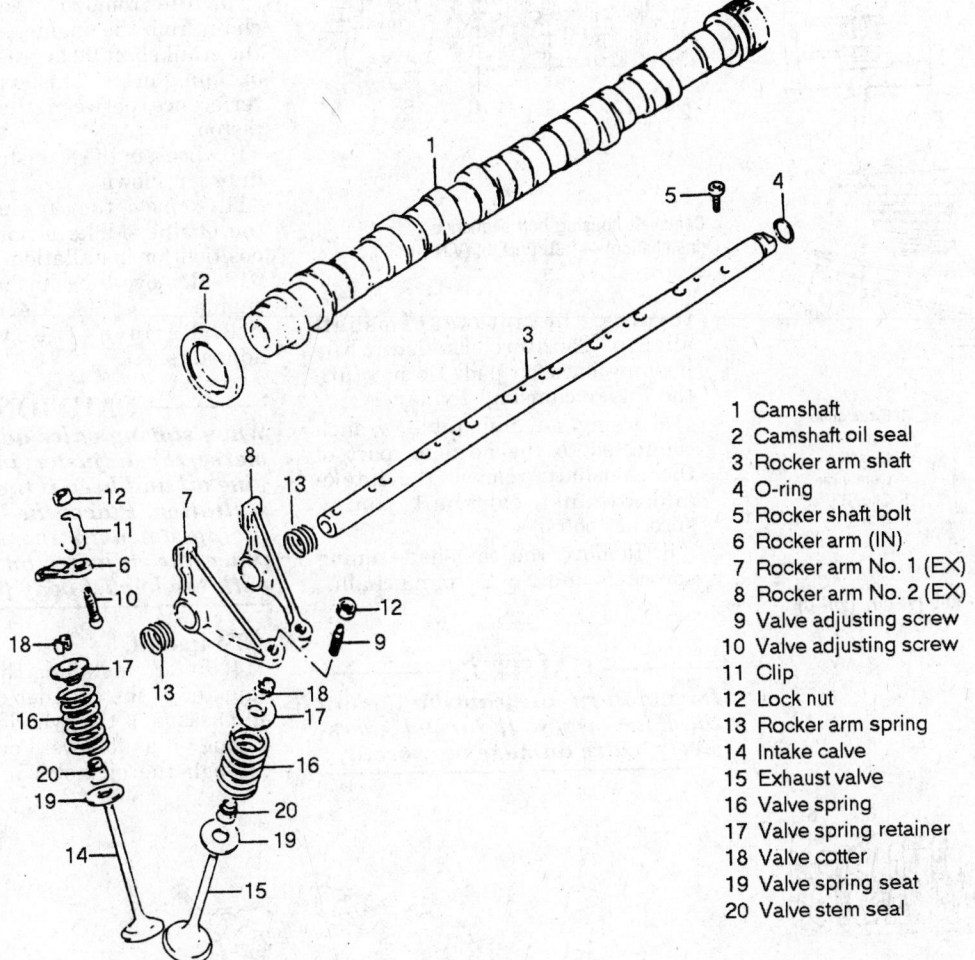

1 Camshaft
2 Camshaft oil seal
3 Rocker arm shaft
4 O-ring
5 Rocker shaft bolt
6 Rocker arm (IN)
7 Rocker arm No. 1 (EX)
8 Rocker arm No. 2 (EX)
9 Valve adjusting screw
10 Valve adjusting screw
11 Clip
12 Lock nut
13 Rocker arm spring
14 Intake calve
15 Exhaust valve
16 Valve spring
17 Valve spring retainer
18 Valve cotter
19 Valve spring seat
20 Valve stem seal

328902

Exploded view of the camshaft components — 1.6L (VIN 0), (VIN U) MFI engines

21. Install the hood lock, connect the lead wire to the horn, and install the front upper member.

22. Install the front grille.

23. Refill the cooling system with the proper coolant and connect the negative battery.

24. If equipped with an automatic transmission, check and top off the A/T fluid level as necessary.

25. Start the engine and check for any water or oil leaks when finished.

26. Check and/or adjust the ignition timing as necessary.

1.8L (VIN 2) Engine

— **CAUTION** —
This engine is a interference engine. Make sure to be careful not to bend the valves when removing the camshafts.

1. Disconnect the negative battery cable from the battery.

2. Drain the engine oil from the engine.

3. Drain the coolant from the cooling system and engine.

4. Remove the oil pan and oil pump strainer.

5. Remove the cylinder head cover.

6. Remove the timing chain cover.

7. Remove the No. 2 timing chain from the engine as follows:

a. Turn the crankshaft and align the timing mark on the engine with the crankshaft key. With the marks aligned, all timing marks should be as follows:

• Crankshaft key aligned with cylinder block timing mark

• Arrow mark on idler sprocket points upward, vertically.

• The marks on the camshaft sprockets align with the marks on the cylinder head.

b. Remove the timing chain tensioner adjuster. To remove the timing chain tensioner adjuster, slacken the No. 2 timing chain by

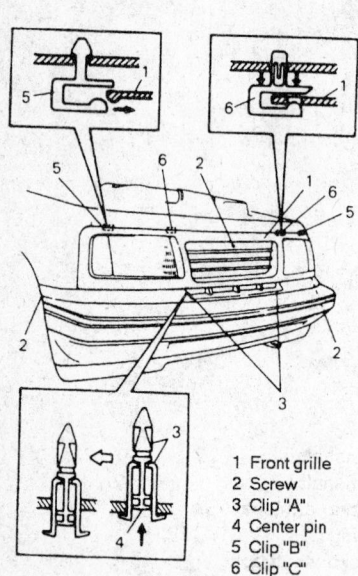

Front grille assembly — 1.6L (VIN 0), (VIN U) MFI engines

1 Front grille
2 Screw
3 Clip "A"
4 Center pin
5 Clip "B"
6 Clip "C"

328904

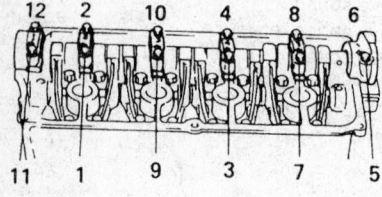

Camshaft housing bolt sequence (removal) — 1.6L (VIN 0), (VIN U) MFI engines

328903

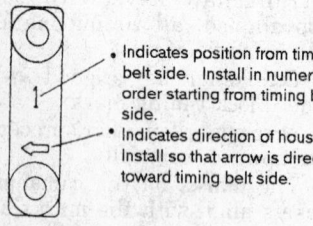

• Indicates position from timing belt side. Install in numerical order starting from timing belt side.
• Indicates direction of housing. Install so that arrow is directed toward timing belt side.

328905

Camshaft housing identification — 1.6L (VIN 0), (VIN U) MFI engines

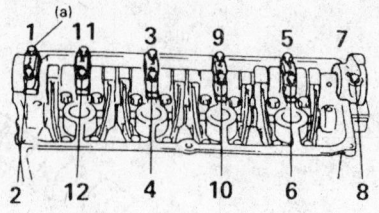

328906

Camshaft housing bolt sequence (installation) — 1.6L (VIN 0), (VIN U) MFI engines

turning the intake camshaft slightly counterclockwise while pushing back on pad. Do not turn the intake camshaft too far.

c. Using an adjustable wrench connected to the hexagon part of the camshaft, remove the intake and exhaust camshaft timing sprocket bolts.

d. Remove the camshaft timing sprockets and No. 2 timing chain.

— CAUTION —
Do not turn the camshaft more than necessary. If turned excessively, valve damage may occur.

8. Remove the camshaft position sensor from the engine by disconnecting the electrical connector and removing the bolt.

9. After removing the No. 2 timing chain from the engine, set the key on the crankshaft 90 degrees to the right of timing mark. This is to prevent interference between the valves and piston.

10. Loosen the camshaft cap bolts in order shown.

11. Remove the camshaft caps from the engine. Make a note of the cap position for installation.

12. Remove the camshafts from the engine.

13. Remove the valve lash adjusters.

— CAUTION —
When storing valve adjusters, immerse the adjuster in clean engine oil and keep it there until installation. Placet the bucket body facing down in the oil. Do not place the adjuster on its side or with the bucket body facing up.

To install:

14. Before installing the valve lash adjuster to the cylinder head, fill the oil passage of the cylinder head with engine oil as follows: Pour engine oil through the oil holes in the sliding

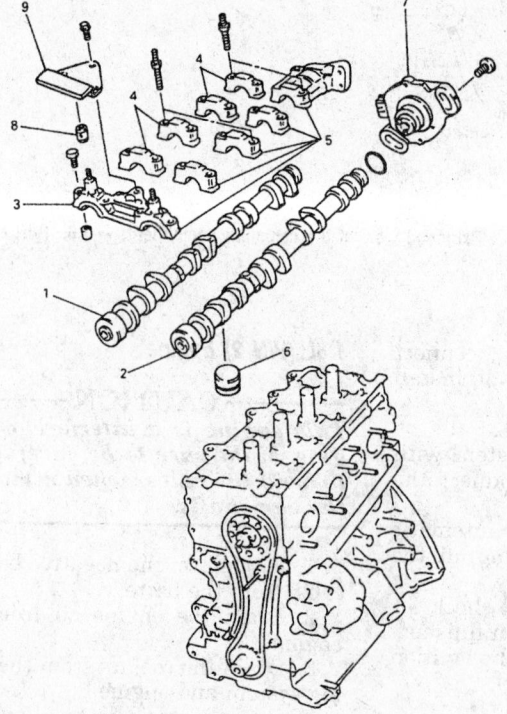

1. Intake camshaft
2. Exhaust camshaft
3. Camshaft housing
4. Intake camshaft housing
5. Exhaust camshaft housing
6. Valve lash adjuster
7. CMP sensor
8. Oil relief valve
9. Timing chain guide No.2

332652

Camshaft and valve lash adjusters removal and installation — 1.8L (VIN 2) engines

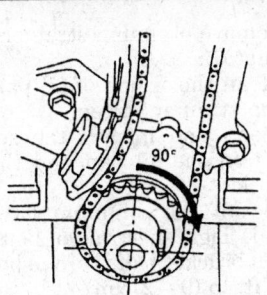

Turning crankshaft 90 degrees to prevent interference between valves and piston — 1.8L (VIN 2) engines

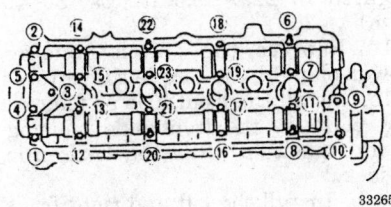

Camshaft cap bolts loosening sequence — 1.8L (VIN 2) engines

part of the valve lash adjuster. Perform this check on both the intake and exhaust sides.

15. Install the valve lash adjusters to the cylinder head. Apply engine oil around the valve lash adjuster and then install it to the cylinder head.

16. Apply oil to sliding surface of each camshaft and camshaft journal. Install the camshafts to the engine. Make sure the timing marks are aligned with the marks on the camshafts.

17. Install the camshaft housing pins.

18. Apply sealant to the exhaust camshaft end housing sealing surface area as shown.

19. Install the camshaft caps to their original positions and direction. Each cap is marks indicating its posi-

tion and direction. Install the housings as indicated by these marks.

20. Apply engine oil to the cap bolts and install the bolts. Uniformly tighten each bolt in sequence shown. Repeat the sequence two or three times before each bolt is torqued to 8 ft. lbs. (11 Nm).

21. Turn the crankshaft back 90 degrees and align the timing mark on the engine with the crankshaft key.

22. Install the camshaft sensor to the engine.

23. Install the No. 2 timing chain as follows:

a. Check that the crankshaft timing sprocket is aligned with the timing mark on the cylinder block.

b. Check that the arrow mark on the idler sprocket faces upward.

c. Check that the knock pins of the intake and exhaust camshafts

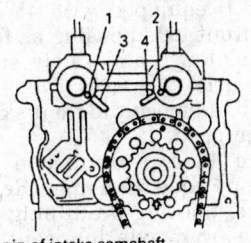

1. Knock pin of intake camshaft
2. Knock pin of exhaust camshaft
3. Match mark of intake camshaft
4. Match mark of exhaust camshaft

Installing camshaft to cylinder head-Camshaft positioning — 1.8L (VIN 2) engines

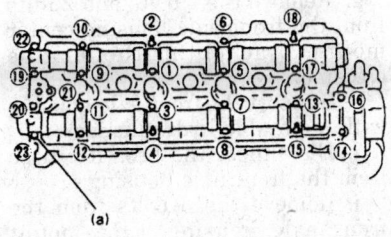

Camshaft cap bolts tightening sequence — 1.8L (VIN 2) engines

are aligned with the timing marks on the cylinder head.

d. Install the No. 2 timing chain to the idler sprocket by aligning the yellow plate on the timing chain with the arrow mark on the idler sprocket.

e. Install the sprockets to the intake and exhaust camshafts by aligning the dark blue plate on the timing chain with the arrow marks on the camshaft sprockets.

NOTE: As an arrow mark is provided on both sides, camshaft timing sprocket has no specific installation direction.

f. While holding the camshaft with an adjustable wrench, install the intake and exhaust camshaft timing sprocket bolts.

g. Plunger pushed back into timing chain tensioner, insert stopper into timing chain tensioner body. After inserting it, check to make sure that the plunger will not come out.

h. Install the timing chain tensioner adjuster with the gasket. Torque the nut to the idler sprocket to 33 ft. lbs. (45 Nm) and the bolts for the timing chain tensioner to 8 ft. lbs. (11 Nm).

i. Pull out the stopper from the timing chain tensioner adjuster.

j. Turn the crankshaft two rotations clockwise. Align the timing mark on the crankshaft to the timing mark on the cylinder block.

k. Apply oil to timing chains, tensioner, tensioner adjusters, sprockets, sprockets and guides.

24. Install the timing chain cover.

25. Install the cylinder head cover.

26. Install the oil pan and oil pump strainer.

27. Fill the cooling system with coolant.

28. Fill the front differential with gear oil.

29. Fill the engine with engine oil.

30. Connect the negative battery cable to the battery.

31. Start the engine and check the engine timing.

32. Check for leaks.

Piston and Connecting Rod

POSITIONING

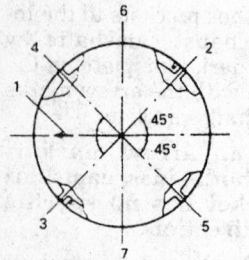

1. Arrow mark
2. 1st ring end gap
3. 2nd ring end gap and oil ring spacer gap
4. Oil ring upper rail gap
5. Oil ring lower rail gap
6. Intake side
7. Exhaust side

333514

Piston ring positioning — all engines

ENGINE LUBRICATION

Oil Pan

REMOVAL AND INSTALLATION

1.3L (VIN 3) Engine

1. Raise and safely support the vehicle.
2. Drain the engine oil.
3. Remove the oil pan mounting bolts and separate the pan from the engine block.
4. Remove oil pump strainer mounting bolts and remove the strainer and oil pan together.

NOTE: It may be necessary to loosen the engine mounts and raise the engine slightly to gain clearance for oil pan removal. Be careful when raising the engine as not to damage any wires or components.

To install:
5. Clean the inside of oil pan and oil pump strainer screen.
6. Clean and inspect the sealing surfaces on the oil pan and the engine block.
7. Install the oil pump strainer and seal into the oil pan. Using a silicone type sealant, install the oil pan and strainer in the vehicle.
8. Install the strainer into the oil pump housing and tighten the strainer bolt first and then the

bracket bolt. Torque the strainer bolts to 7.0–8.5 ft. lbs. (9–12 Nm).
9. Install the oil pan mounting bolts and torque them to 7.0–8.5 ft. lbs. (9–12 Nm).

NOTE: Tightening should begin at the center moving outward on both sides.

10. Install the oil drain plug and tighten to 22.0–28.5 ft. lbs (30–40 Nm).
11. If the engine was raised, lower it and secure the engine mounts.
12. Lower the vehicle.
13. Refill the engine with engine oil. Start the engine and check for leaks.

1.6L (VIN 0), (VIN U) Engine

1. Raise and safely support the vehicle.
2. If equipped with 4WD, remove the front axle housing as follows:
   a. Raise and safely support the vehicle.
   b. Remove the front skid plate, if equipped.
   c. Place a drain pan under the front axle housing. Remove the axle housing drain plug and drain the front axle lubricant. Reinstall the drain plug.
   d. Remove the left and right halfshafts from the vehicle.
   e. Remove the left inner axle shaft.
   f. Mark the position of the driveshaft on the front axle flange for installation reference, then remove the driveshaft.
   g. Remove the 2 bolts and 2 nuts from the front axle housing center mount bracket at the engine crossmember.
   h. Support the front axle housing with a suitable hydraulic jack.
   i. Disconnect the breather hose from the front axle housing.
   j. Remove the 3 bolts from the front axle housing right mount bracket and the 4 bolts from the left mount bracket.
   k. Lower the hydraulic jack and remove the front axle housing from the vehicle.
3. On 1996–97 engines, disconnect the crankshaft position sensor electrical connector.
4. On 1996–97 engines, remove the crankshaft position sensor from the oil pan by removing the bolt.
5. Drain the engine oil.
6. Remove the clutch housing lower plate or the torque converter housing lower plate.
7. Remove the oil pan mounting bolts and remove the pan.

8. Remove oil pump strainer.
To install:
9. Clean the inside of oil pan and oil pump strainer screen.
10. Clean and inspect the sealing surfaces on the oil pan and the engine block.
11. Install the oil pump strainer and seal. Tighten strainer bolt first and then bracket bolt. Torque bolts to 7.0–8.5 ft. lb (9–12 Nm).
12. Using new gaskets, install the oil pan and tighten the oil pan bolts to 7.0–8.5 ft. lbs. (9–12 Nm).

NOTE: Tightening should begin at the center moving outward on both sides.

13. Install the oil drain plug and tighten to 22.0–28.5 ft. lbs (30–40 Nm).
14. Replace the clutch housing lower plate or the torque converter.
15. Install the front axle as follows:
   a. Raise the front axle housing into position with the hydraulic jack.
   b. Install the left and right front axle housing mount bracket bolts and tighten to 43 ft. lbs. (60 Nm).
   c. Connect the breather hose.
   d. Remove the hydraulic jack from under the front axle housing.
   e. Install the 2 bolts and 2 nuts into the front axle housing center mount bracket at the engine crossmember and tighten to 43 ft. lbs. (60 Nm).
   f. Align the driveshaft with the reference marks made during removal and install the driveshaft. Tighten the nuts and bolts to 43 ft. lbs. (60 Nm).
   g. Install the left inner axle shaft.
   h. Install the left and right halfshafts.
   i. Tighten the front axle housing drain plug to 33 ft. lbs. (45 Nm).
   j. Remove the oil level/filler plug from the front axle housing and fill the front axle housing with the proper type and quantity of lubricant. Install the oil level/filler plug and tighten to 33 ft. lbs. (45 Nm).
   k. Install the skid plate, if equipped, and tighten the bolts to 40 ft. lbs. (54 Nm).
16. On 1996–97 engines, install the crankshaft position sensor with the bolts.
17. On 1996–97 engines, connect the crankshaft position sensor electrical connector.
18. Lower the vehicle.
19. Refill the engine with engine oil. Run the engine and check for leaks.

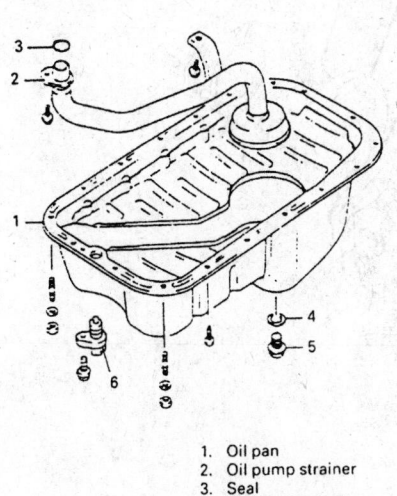

1. Oil pan
2. Oil pump strainer
3. Seal
4. Drain plug gasket
5. Drain plug
6. CKP sensor

328015

Exploded view of the oil pan components — 1.6L (VIN 0), (VIN U) engines

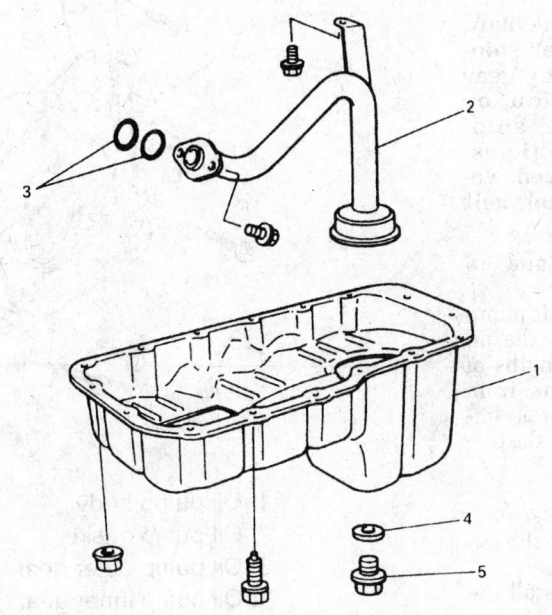

1. Oil pan
2. Strainer
3. O-ring
4. Gasket
5. Drain plug

331545

Oil pan components — 1.8L (VIN 2) engines

1.8L (VIN 2) Engine

1. Raise and safely support the vehicle.
2. Remove the front wheels.
3. Drain and remove the front differential assembly from the chassis, if equipped with 4WD.
4. Remove the tie rod, center link and idler arm.
5. Drain the engine oil by removing the drain plug.
6. Remove the clutch housing or torque converter housing lower plate.
7. Remove the oil pan mounting bolts and nuts. Lower the oil pan until it stops. The cross member and oil pump strainer will keep the oil pan from being removed from the engine.
8. With a wrench inserted between oil pan and crank case, remove the oil pump strainer mounting bolts. Remove the oil pan and strainer as an assembly.

To install:

9. Clean the inside of oil pan and oil pump strainer screen.
10. Clean and inspect the sealing surfaces on the oil pan and the engine block.
11. Apply sealant to the oil pan mating surface and place the oil pan

and strainer in position for installation.

12. Using new O-rings, install the oil pump strainer to the engine. The pan must be in the lowered position when installing the oil pump strainer. Torque the oil pump strainer bolts to 8 ft. lbs. (11 Nm).
13. Install the oil pan and tighten the oil pan bolts to 8 ft. lbs. (11 Nm).

NOTE: Tightening should begin at the center moving outward on both sides.

14. Install the oil drain plug and tighten to 22.0–28.5 ft. lbs (30–40 Nm).
15. Replace the clutch housing lower plate or the torque converter.
16. Install the tie rod, center arm and idler arm.
17. Install the front differential housing.
18. Fill the front differential housing with gear oil.
19. Install the wheels.
20. Lower the vehicle.
21. Refill the engine with engine oil.
22. Connect the negative battery cable to the battery.
23. Run the engine and check for leaks. Check the wheel alignment.

Oil Pump

REMOVAL AND INSTALLATION

1.3L (VIN 3), 1.6L (VIN 0), (VIN U)

1. Disconnect the negative battery cable.
2. Remove the radiator cooling fan, fan shroud, water pump pulley and water pump drive belt.
3. Remove the crankshaft pulley by removing the five bolts.
4. Remove the timing belt outside cover, timing belt and timing belt tensioner.
5. Disconnect the alternator and remove the bracket.
6. If equipped with A/C or power steering, disconnect the A/C compressor and the power steering unit from the engine. Do not disconnect the lines from either unit.
7. If equipped with A/C or power steering, remove the power steering bracket and A/C compressor bracket.
8. Raise and safely support the vehicle. Drain the engine oil.
9. Remove the clutch (torque converter) housing lower plate.

10. With the crankshaft locked in position, remove the crankshaft gear bolt, guide and gear.

NOTE: To lock the crankshaft, engage a special tool (gear stopper) with the flywheel ring gear for manual transmission or driveplate ring gear for automatic transmission vehicles. With the crankshaft locked, remove the crankshaft timing belt pulley bolt.

11. Remove the oil pan and oil pump strainer.

12. Remove the seven oil pump mounting bolts and remove the oil pump. Take notice to the lengths of the bolts and the positions from which they were removed. It is important to replace the bolts in the correct positions.

To install:

13. Clean and inspect the sealing surfaces of the oil pump and the engine block.

14. Using a new gasket, install the oil pump to the engine.

15. Install the seven oil pump mounting bolts and tighten to 7.0–8.5 ft. lbs. (9–12 Nm). There are two different sized bolts, so be sure to install the bolts in the correct positions.

NOTE: To prevent the oil seal lip from being damaged when installing the oil pump to the crankshaft, use a proper seal guide tool when installing. After installing the oil pump, check to be sure the oil lip is not upturned, then remove the special tool. The edge of the oil pump gasket might bulge out. If it does, cut off bulge with knife.

16. Install the oil strainer and pan.

17. With the crankshaft locked, install the crankshaft crankshaft gear, guide and bolt. Torque the bolt to 80 ft. lbs. (110 Nm).

18. If equipped, install the air conditioning bracket and power steering bracket.

19. If equipped, install the air conditioning compressor and power steering pump.

20. Replace the alternator and the alternator bracket.

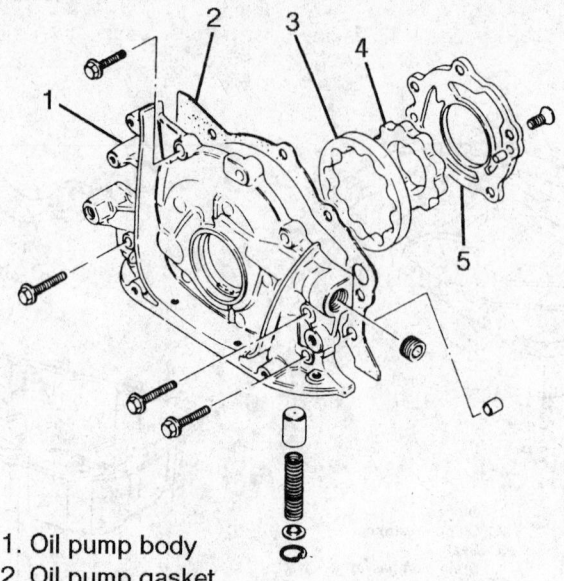

1. Oil pump body
2. Oil pump gasket
3. Oil pump outer gear
4. Oil pump inner gear
5. Gear plate

328952

Oil pump components — 1.6L (VIN 0), (VIN U) engine; 1.3L (VIN 3) is similar

21. Install the timing belt tensioner, timing belt and timing belt outside cover.

22. Install the water pump drive belt, water pump pulley, radiator cooling fan and fan shroud.

23. Refill the engine with engine oil and connect the negative battery cable.

24. Lower the vehicle.

25. Run the engine and check for any leaks.

1.8L (VIN 2) Engine

1. Raise and safely support the vehicle. Drain the engine oil.

2. Remove the oil pan and oil pump strainer.

3. Remove the oil pump sprocket cover by removing the bolts.

4. Remove the oil pump from the lower crankcase and chain by removing the bolts.

——— WARNING ———
Do not remove the sprocket from the oil pump. Damage to the oil pump center shaft and/or abnormal operation of the oil pump could result if the sprocket is removed.

To install:

5. Install the oil pump to the chain and lower crank case and torque the bolts to 15 ft. lbs. (20 Nm).

NOTE: When installing the oil pump, be careful not to allow the pins to fall off.

6. Install the oil pump sprocket cover and torque the bolt to 8 ft. lbs. (11 Nm).

7. Install the oil pan and strainer.

8. Fill the engine with engine oil.

9. Lower the vehicle.

10. Run the engine and check for oil pressure and any leaks.

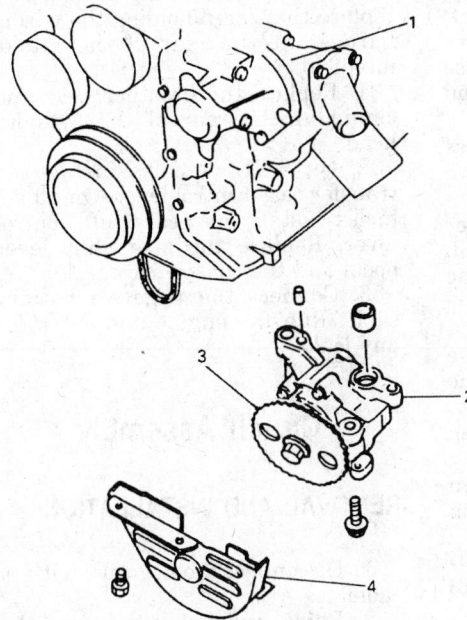

1. Cylinder block
2. Oil pump
3. Oil pump sprocket
4. Oil pump sprocket cover

331558

Oil pump assembly overview — 1.8L (VIN 2) engines

TRANSMISSION

Manual Transmission Assembly

REMOVAL AND INSTALLATION

Samurai

1. Disconnect the negative battery cable.

2. Remove the console cover, remove the 4 gear shift boot mounting bolts and slide the boot upwards on the gear shifter.

3. Remove the boot clamp and remove the second boot from the shift lever case.

4. Loosen the gear shift lever case cover bolts and remove the gear shift lever out of the case.

5. Disconnect the back-up light and fifth switch lead wires at the electrical connector.

6. Raise and safely support the vehicle. Drain the oil from the transmission and the transfer case.

7. Disconnect the starter lead wires and mounting bolts and remove the starter. Remove the fuel hose clamps from the transmission case.

8. Remove the flange bolts from the driveshaft between the transmission to transfer case. Remove and mark the shaft.

9. Remove the flange bolts from the driveshaft between the transfer case and the front differential. Remove and mark the shaft.

10. Disconnect the clutch cable and remove the clutch housing lower plate.

11. Disconnect the bolts fastening the cylinder block to the transmission case.

12. Remove the transmission crossover protection pipe located under the transmission case. The mounting bolts are located on each side of the frame.

13. Disconnect the center exhaust pipe.

14. Position a transmission jack, support the transmission, and remove the transmission rear mounting member from the vehicle.

15. Remove the transmission from the vehicle.

To install:

16. Position and raise the transmission on the transmission jack; install the transmission assembly into the clutch disc.

17. Install the engine rear mounting member and tighten the bolts to 29–43 ft. lbs. (40–60 Nm).

18. Remove the transmission jack and replace the center exhaust pipe.

19. Install the transmission crossover protection pipe.

20. Connect the bolts attaching the engine to the transmission case and tighten to 14–20 ft. lbs. (18–28 Nm).

21. Replace the clutch housing lower plate and connect the clutch cable. Adjust the clutch cable.

22. Install the two driveshafts, and replace the flange bolts. Tighten to 17–21 ft. lbs. (23–30 Nm).

23. Install the starter and replace the mounting bolts and the lead wires to the starter.

24. Reconnect the fuel hose clamps to the transmission case.

25. Refill the transmission and if equipped, the transfer case with the recommended gear oil.

26. Lower the vehicle.

27. Connect the back-up light and fifth switch lead wires at the electrical connector.

28. Install the gear shift control lever and tighten the shift lever bolts to 5 ft. lbs. (7 Nm). Replace the gear shift lever boots and the console cover.

29. Connect the negative battery cable. Start the engine; check for any leaks and proper clutch operation.

Sidekick, Tracker and X-90

1. Disconnect the negative battery cable.

2. Remove the console cover, remove the 4 gear shift boot mounting bolts and slide the boot upwards on the gear shifter.

3. Remove the boot clamp and remove the second boot from the shift lever case.

4. Push the gear shift lever control case down with fingers, turn it counterclockwise and remove the shift control lever.

5. Remove the transfer case shift control lever the same way, if equipped with 4WD.

6. Disconnect the breather hose and clamp at the rear of the cylinder head.

7. Remove the clamp at the rear of the intake manifold to free up the wiring harness. Disconnect the harness coupler.

8. Raise and safely support the vehicle. Drain the oil from the transmission and the transfer case, if equipped with 4WD.

9. Disconnect the starter lead wires and mounting bolts and remove the starter.

10. Remove the fuel line clamp on the transmission. Disconnect the

bolts fastening the engine to the transmission.

11. Remove the flange bolts from the front driveshaft. Remove and mark the shaft, if equipped with 4WD.

12. Remove the flange bolts from the rear driveshaft. Remove and mark the shaft.

13. Disconnect the clutch cable and remove the clutch housing lower plate.

14. Disconnect the center exhaust pipe and remove the nuts from the joint with the engine.

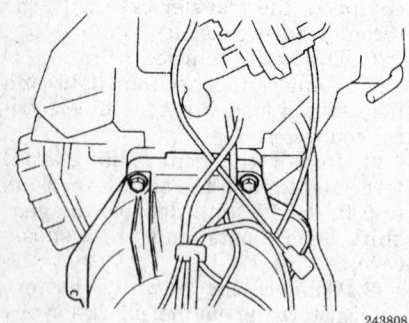

Transmission mounting bolt locations — Sidekick, Tracker and X-90 models

243808

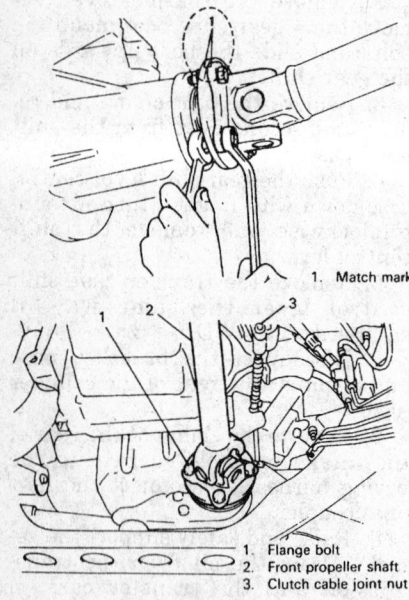

1. Match mark
1. Flange bolt
2. Front propeller shaft
3. Clutch cable joint nut

243809

Driveshaft flanges — Sidekick, Tracker and X-90 models

15. Disconnect the speedometer cable from the transfer case.

16. Position a transmission jack and remove the engine rear mounting member from the vehicle. Move the transmission and transfer case toward the rear of the vehicle and lower.

17. Disconnect the wiring harness and breather hose at the transmission.

18. Separate the gear shift lever case and transfer case, if equipped, from the transmission. Remove the transmission from the vehicle.

To install:

19. Install the gear shift lever case and transfer case, if equipped, to the transmission.

20. Install the transmission wiring harness and breather hose.

21. Raise the transmission and transfer case on the transmission jack and place under the vehicle.

22. Install the engine rear mounting member and tighten the bolts to 29–43 ft. lbs. (40–60 Nm).

23. Connect the speedometer cable to the transfer case and join with the engine.

24. Remove the transmission jack and replace the center exhaust pipe. Torque the mounting bolts to 29–43 ft. lbs. (40–60 Nm).

25. Replace the clutch housing lower plate and connect the clutch cable. Adjust the clutch cable.

26. Install the front and rear driveshaft and replace the flange bolts and tighten to 37–43 ft. lbs. (50–60 Nm).

27. Connect the bolts attaching the engine to the transmission and tighten to 51–72 ft. lbs. (70–100 Nm), and replace the fuel line clamp on the transmission.

28. Install the starter and replace the mounting bolts and the lead wires to the starter.

29. Lower the vehicle. Refill the transmission and transfer case, if

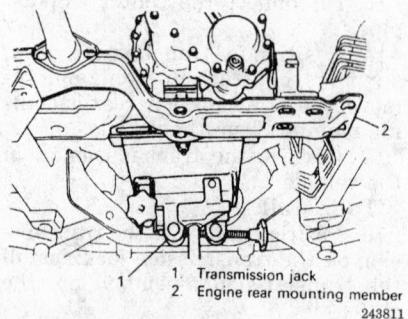

1. Transmission jack
2. Engine rear mounting member

243811

Rear engine crossmember — Sidekick, Tracker and X-90 models

equipped, with the recommended gear oil.

30. Connect the wiring coupler and replace the clamp holding the wiring harness at the rear of the intake manifold.

31. Replace the breather hose and clamp at the rear of the cylinder head.

32. On 4WD vehicles install the transfer case control lever. On all vehicles install the gear shift control lever. Replace the gear shift lever boots and the console cover.

33. Connect the negative battery cable. Run the engine and check for any leaks.

Clutch Assembly

REMOVAL AND INSTALLATION

1. Disconnect the negative battery cable.

2. Raise and safely support the vehicle.

3. Support the engine, and remove the transmission from the vehicle.

4. Support the pressure plate, and remove the 6 pressure plate to flywheel mounting bolts.

5. Remove the pressure plate and the clutch disc from the flywheel.

6. Inspect the flywheel for wear and/or scoring; machine or replace, as necessary. If necessary, remove the flywheel mounting bolts and remove the flywheel.

To install:

7. Install the flywheel and tighten the mounting bolts to 54–57 ft. lbs. (75–80 Nm).

8. Inspect the condition of the clutch release bearing and the input shaft bearing and replace as necessary.

9. Align the clutch disc to the flywheel using a clutch disc alignment tool.

10. Install the pressure plate and evenly tighten the pressure plate bolts to 13–20 ft. lbs. (18–28 Nm).

NOTE: Before assembling, make sure the clutch disc and the pressure plate are clean and dry.

11. Remove the clutch disc alignment tool.

12. With the engine supported, install the transmission in the vehicle.

13. Lower the vehicle and connect the negative battery cable.

14. Check and adjust the clutch pedal height. Check the clutch operation when finished.

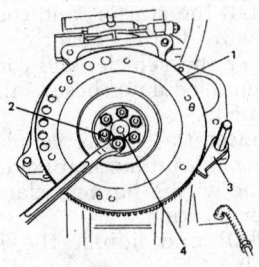

1. Flywheel
2. Flywheel bolt
3. Special tool (Flywheel holder 09924-17810)
4. Input shaft bearing

197734

Flywheel mounting bolts

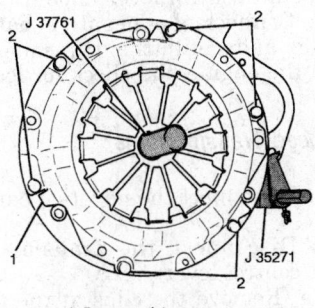

J 37761

J 35271

1. Pressure plate cover
2. Pressure plate cover bolts

243668

Pressure plate mounting bolts

Clutch Cable

ADJUSTMENT

1. Depress the clutch pedal until resistance is felt. Measure the distance to determine the pedal free play. The travel distance should be between 0.6–1.1 inch (15–30 mm).
2. Make sure the outer cable nuts are tightened around the center of the outer cable thread area. If the pedal is out of specification, loosen the clutch cable joint nut, and adjust the travel.

1. Flywheel
2. Clutch disc
3. Pressure plate
4. Pressure plate cover
5. Clutch pilot bearing

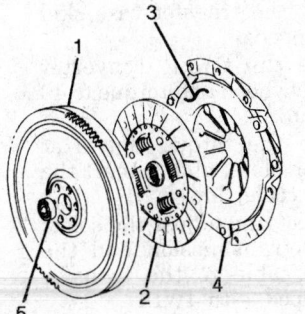

197601

Exploded view of the clutch components

b. Pedal free travel

"b"

333203

Clutch pedal free travel — Sidekick, Tracker and X-90 models

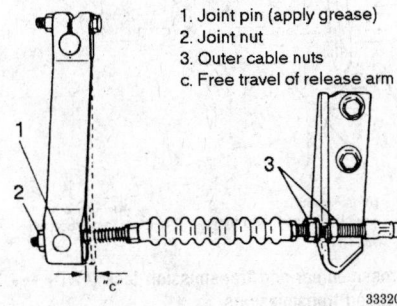

1. Joint pin (apply grease)
2. Joint nut
3. Outer cable nuts
c. Free travel of release arm

"c"

333204

Clutch cable outer nut location — Sidekick, Tracker and X-90 models

3. Tighten the clutch cable joint nut, and test the operation of the pedal.

Automatic Transmission Assembly

REMOVAL AND INSTALLATION

3-Speed Transmissions

1. Disconnect the negative battery cable.
2. Disconnect the transmission shift control lever. Remove the trans-

fer case shift control lever knob, if equipped with 4WD.
3. Disconnect the breather hose from the clamp at the rear of the cylinder head.
4. Disconnect the wiring harness clamp at the rear end of the intake manifold to free up the harness.
5. Disconnect the wiring harness coupler and remove the kick-down cable at the throttle body.
6. Remove the vacuum modulator hose at the intake manifold.
7. Raise and safely support the vehicle.
8. Disconnect the starter lead wires and mounting bolts, remove the starter motor.
9. Drain the oil from the transmission and transfer case, if equipped with 4WD.
10. Disconnect the flange bolts from the front driveshaft, remove and mark the shaft.
11. Disconnect the flange bolts from the rear driveshaft, remove and mark the shaft.
12. Disconnect the select cable from the transmission and the speedometer cable from the transfer case.
13. Remove the torque converter housing lower plate and disconnect and plug the oil cooler lines.
14. Hold the flywheel in place with Special tool 09927–56010, or equivalent, and remove the three torque converter mounting bolts at the flywheel.
15. If equipped with 4WD, disconnect the transfer case skid plate.
16. Remove the exhaust bracket at the catalytic converter and at the transmission. Disconnect the center exhaust pipe, if necessary.
17. Remove the transmission to engine retaining bolts and nuts.

— CAUTION —
The transmission assembly, and transfer case (if equipped), may tilt rearward on the jack. Use an auxiliary arm on the jack to secure and stabilize the assembly.

18. Support the transmission using a transmission jack or equivalent. On 4WD vehicles, disconnect the transmission crossmember mounting bolts and remove the crossmember
19. Disconnect the lead wires and breather hose at the transmission and lower the transmission with the transfer case from the vehicle.
20. Remove the 12 transmission-to-transfer case bolts and separate the transmission from the transfer case, if equipped with 4WD.

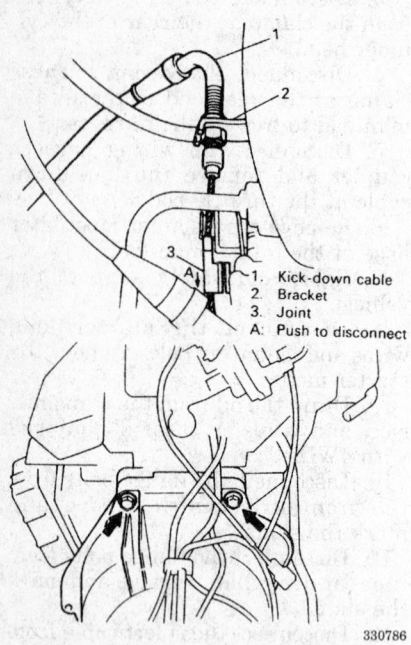

1. Kick-down cable
2. Bracket
3. Joint
A: Push to disconnect

330786

Kick-down cable — 3-speed transmissions

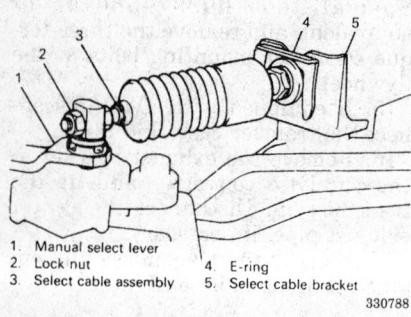

1. Manual select lever
2. Lock nut
3. Select cable assembly
4. E-ring
5. Select cable bracket

330788

Select cable — 3-speed transmissions

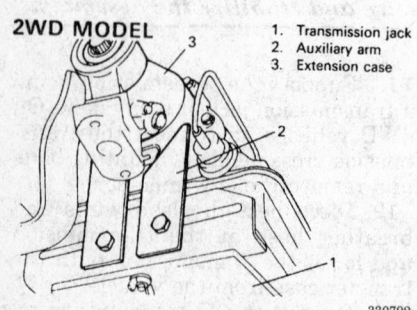

2WD MODEL

1. Transmission jack
2. Auxiliary arm
3. Extension case

330792

Transmission jack (2WD) — 3-speed transmissions

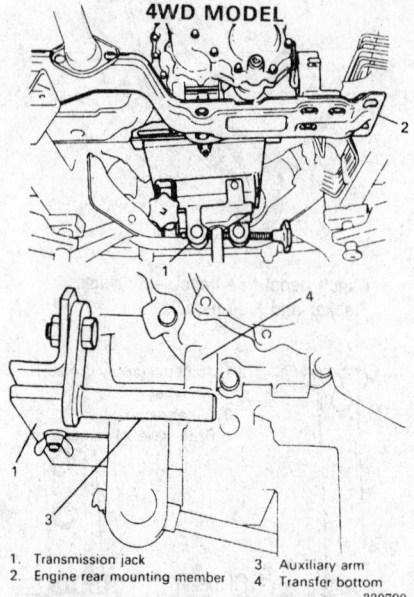

4WD MODEL

1. Transmission jack
2. Engine rear mounting member
3. Auxiliary arm
4. Transfer bottom

330790

Crossmember and transmission jack (4WD) — 3-speed transmissions

To install:

21. Connect the transfer case to the transmission and tighten the bolts to 17 ft. lbs. (23 Nm), if equipped with 4WD.

22. Raise the transmission, connect the lead wires and breather to the transmission.

23. Replace the transmission to engine retaining bolts and tighten to 62 ft. lbs. (85 Nm).

24. Install the transmission cross member and tighten the bolts to 36 ft. lbs. (50 Nm).

25. Connect the exhaust bracket to the catalytic converter and the transmission.

26. Reconnect the center exhaust pipe, if necessary.

27. Connect the transfer case skid plate, if equipped.

28. Replace the torque converter bolts at the flywheel, and torque to 40 ft. lbs. (55 Nm).

29. Replace the torque converter housing lower plate, and connect the transmission cooler lines.

30. Reconnect and adjust the select cable to the transmission and the speedometer cable to the transfer case, if equipped with 4WD.

31. Replace the front and rear driveshafts and tighten the flange bolts to 37 ft. lbs.

32. Install the starter, and connect the lead wires to the starter.

33. Lower the vehicle, and connect the vacuum modulator hose to the intake manifold.

34. Connect the wiring coupler at the rear of the intake manifold, and replace the wiring harness clamp in the proper position.

35. Install and adjust the kick-down cable.

36. Connect the breather hose to the clamp at the rear of the cylinder head.

37. Install the transmission shift control lever and replace the transfer case shift control lever knob.

38. Connect the negative battery cable, and refill the transmission. Run the engine and check for leaks.

4-Speed Transmissions

1. Disconnect the negative battery cable.

2. Disconnect the transmission shift control lever:

   a. Remove the clips at the front of the rear console box by pushing in on the center pin first.

   b. Remove the screws at the front and clips at the rear, and remove the front console box.

3. Remove the transfer case shift control lever, if equipped with 4WD:

   a. Remove the boot cover and boot number two.

   b. Remove the boot clamp, and boot number one from the lever.

   c. Push the transfer case cover down, turn counterclockwise and remove the lever.

4. Remove the battery, dipstick and oil filler tube.

5. Raise and safely support the vehicle.

6. Disconnect the throttle cable from the throttle cam and bracket.

7. Disconnect and tag the wiring harness couplers.

8. Remove the starter mounting bolts and remove the starter motor.

9. Remove the transmission-to-engine bolts. The right side bolt is longer.

10. Drain the oil from the transmission and transfer case, if equipped with 4WD.

11. Disconnect the flange bolts from the front driveshaft, matchmark and remove the shaft.

12. Disconnect the flange bolts from the rear driveshaft, matchmark and remove the shaft.

13. Disconnect the gear select cable from the transmission by removing the nut from the end of the cable, and the E-ring from the bracket. Remove

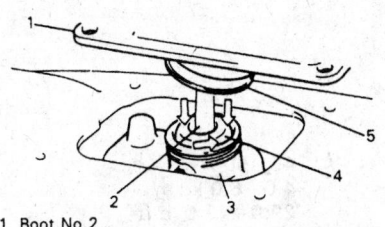

1. Boot No. 2
2. Transfer shift control lever
3. Transfer shift lever case
4. Transfer shift control case cover
5. Boot No. 1

330797

Transfer shift control lever — 4-speed transmissions

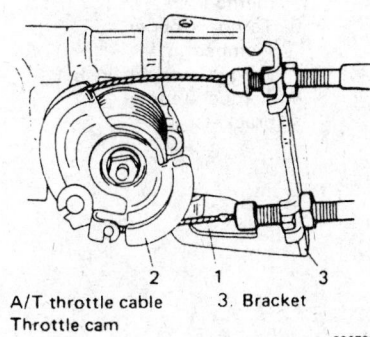

1. A/T throttle cable
2. Throttle cam
3. Bracket

330798

Throttle cable — 4-speed transmissions

the two bracket bolts and the bracket.

14. Remove the two exhaust pipes blocking the removal of the transmission.

15. Remove the left-side transmission case stiffener. Remove the four **A** bolts from the right-side stiffener, and loosen the one **B** bolt, as shown in the graphic.

16. Remove the torque converter housing lower plate and disconnect and plug the oil cooler lines.

17. Hold the flywheel in place with Special tool 09927–56010, or equivalent, and remove the three tor-

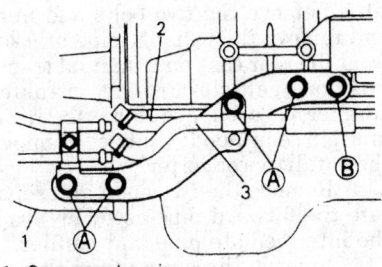

1. Converter housing
2. Outlet hose
3. Inlet hose

330800

Right stiffener bolt locations — 4-speed transmissions

que converter mounting bolts at the flywheel.

18. Remove the transmission-to-engine retaining nuts.

19. Remove the speedometer end nut, and disconnect the cable.

─────── **CAUTION** ───────

The transmission assembly, and transfer case (if equipped), may tilt rearward on the jack. Use an auxiliary arm on the jack to secure and stabilize the assembly.

20. Support the transmission using a transmission jack or equivalent. Disconnect the transmission crossmember mounting bolts and remove the crossmember. Remove the torque stopper member and the torque stopper bushing after removing their mounting bolts.

21. Remove the wiring harness and breather hose.

22. Lower the transmission with the transfer case from the vehicle.

23. Remove the 12 transmission-to-transfer case bolts and separate the transmission from the transfer case, if equipped with 4WD.

To install:

24. Connect the transfer case to the transmission and tighten the bolts to 14–20 ft. lbs. (18–28 Nm), if equipped with 4WD.

25. Raise the transmission, connect the lead wires and breather to the transmission.

26. Install the transmission-to-engine retaining nuts and tighten to 51–72 ft. lbs. (70–100 Nm).

27. Install the transmission crossmember, torque stopper member and bushing; tighten the bolts to 29–43 ft. lbs. (40–60 Nm).

28. Install the speedometer cable.

29. Install the torque converter bolts at the flywheel, and torque to 44–50 ft. lbs. (60–70 Nm). Remove the special tool 09927–56010, or equivalent.

30. Install the torque converter housing lower plate and connect the transmission cooler lines.

31. Position and install the left and right side stiffeners. Tighten the mounting bolts to 29–43 ft. lbs. (40–60 Nm).

32. Install the removed exhaust pipes.

33. Install the select bracket and tighten the bolts to 14–20 ft. lbs. (18–28 Nm). Install and adjust the select cable.

34. Align the matchmarks and install the front and rear driveshafts. Tighten the flange bolts to 36–43 ft. lbs. (50–60 Nm).

35. Install the transmission-to-engine bolts and tighten to 51–72 ft. lbs. (70–100 Nm).

36. Install the starter.

37. Connect the wiring couplers.

38. Install and adjust the throttle cable.

39. Install the oil filler tube, dipstick, and battery.

40. Install the front console box.

41. Install the transmission shift control lever and the transfer case shift control lever, if equipped.

42. Connect the negative battery cable and refill the transmission. Start the engine and check for leaks.

TRANSFER CASE

Transfer Case Assembly

REMOVAL AND INSTALLATION

Samurai

NOTE: Even the 2WD Samurai uses a transfer case. Rather than use a straight line driveshaft from the transmission to the differential, the Samurai leaves the transfer case in place and replaces its internal working with a simple offset gear mechanism, thus keeping 2 of the 3 driveshafts used in the 4WD vehicle.

1. Raise and safely support the vehicle.

2. Support the transfer case using a suitable transmission jack.

3. Remove the mounting bolts from each universal-joint flange connection to disconnect the driveshafts from the transfer case assembly.

4. On the 4WD model, remove the clamp and boot from the shift lever.

5. On the 4WD model, twist the control lever guide counterclockwise while pushing down on it to remove the lever.

6. Drain the oil from the transfer case.

7. Disconnect the speedometer cable from the transfer case.

8. On the 4WD model, disconnect the 4WD switch lead wire at the coupler.

9. Remove the three mounting bolts securing the transfer case to the chassis.

10. Remove the transfer case from the vehicle.

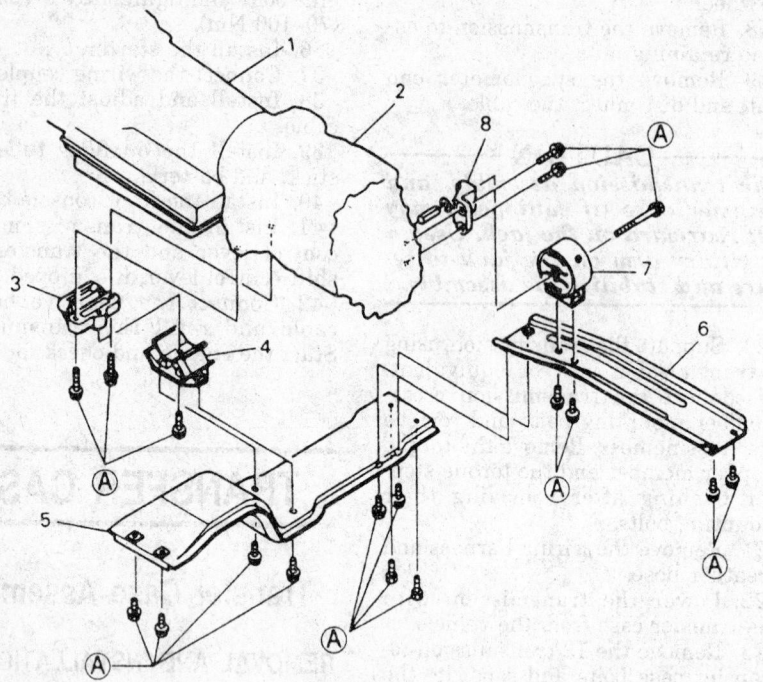

Ⓐ : 40—60 N·m
4.0—6.0 kg-m
29.0—43.0 lb-ft

1. Automatic transmission
2. Transfer
3. Engine rear mounting bracket
4. Engine rear mounting
5. Engine rear mounting member
6. Torque stopper member
7. Torque stopper bush
8. Torque stopper bracket

330803

Crossmember exploded view and torque values — 4-speed transmissions

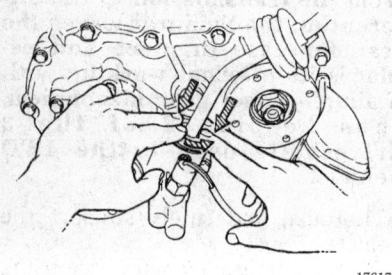

176172

Push down and twist to remove the lever — 4WD Samurai models

To install:

11. Position and install the transfer case. Tighten the mounting bolts to 18.5–25 ft. lbs. (25–34 Nm).

12. On the 4WD model, connect the 4WD switch lead wire at the coupler.

13. Connect the speedometer cable to the transfer case.

14. Replace the oil in the transfer case with approved fluid.

15. On the 4WD model, install the shift lever making sure to push down and twist it clockwise into position.

16. On the 4WD model, install the lever boot and clamp.

17. Install the universal-joint flange connections to the transfer

case. Torque the bolts to 36–43 ft. lbs. (50–60 Nm). Remove the support.

18. Lower the vehicle.

Sidekick, Tracker and X-90

1. Remove the two screws and the plastic retainers and console box from the floor.

2. Remove the six screws and boot cover and gearshift control lever boot from the control lever.

3. Remove the one clamp and gearshift lever case boot from the gearshift lever case.

4. Remove the gearshift control lever from the gearshift lever case by pushing down on the gearshift control lever pivot and turning counterclockwise 90° and lifting up.

5. Disconnect the four wheel drive electrical connector from the four wheel drive switch.

6. Remove the four bolts from the fan shroud at the radiator and safely raise and support the vehicle.

7. Remove the two bolts and separate the skid plate from the transfer case crossmember, if equipped.

8. Place a drain pan beneath the transfer case and remove the transfer case plug and drain the oil.

NOTE: An index mark should be made on the front and rear driveshaft pinion flange yokes

and the front and rear differential pinion flanges to ensure the driveshaft is reinstalled in the proper position to prevent any vibration or driveline imbalance.

9. Remove the four bolts and nuts to the front driveshaft and remove the front driveshaft from the vehicle.

10. Remove the four bolts and nuts to the rear driveshaft and remove the rear driveshaft from the vehicle.

11. Disconnect the speedometer cable from the speedometer driven gear case.

12. Remove one bolt on the speedometer cable clip and ground wire from the torque stopper housing.

13. Remove the two bolts and nuts and remove the exhaust pipe bracket from the rear case on a manual transmission vehicle or the transfer adapter case on a automatic transmission equipped vehicle. Remove the catalytic converter.

14. Remove the two bolts and separate the forward pipe assembly from the intermediate pipe and muffler.

15. Remove three nuts from the exhaust manifold and remove the forward pipe/three way catalytic converter assembly from the vehicle.

16. Support the transfer case with a suitable hydraulic jack.

17. Remove the two bolts from the torque stopper bracket.

18. Remove the six bolts and transfer case crossmember from the undercarriage.

19. Place a block of wood between the distributor gear housing and the bulkhead to prevent the distributor from being damaged when the transfer case is lowered.

20. Safely lower the transfer case slowly until the engine contacts its support point on the block of wood.

21. Remove the one clamp and breather hose from the gearshift lever case.

22. Remove the twelve transfer case to the transmission bolts.

23. Slide the transfer case of the transmission output shaft and slowly lower the transfer case making sure there are no obstructions.

To install:

24. Raise the transfer case into position and slide it onto the transmission output shaft.

25. Install and tighten the twelve transfer case to transmission bolts.Tighten to 21 ft. lbs. (21 Nm).

26. Install the breather hose to gearshift lever case; secure it with one clamp.

27. Raise the transfer case slowly until the engine is no longer in contact with the wood block between the bulkhead and the distributor gear housing.

28. Remove the wood block.

29. Install the transfer case crossmember to the undercarriage; secure with six bolts and tighten to 37 ft. lbs. (50 Nm).

30. Install two bolts to the torque stopper bracket. Tighten the bolts to 37 ft. lbs. (50 Nm).

31. Remove the jack from under the transfer case.

32. Install the forward pipe/three way catalytic convertor to the exhaust manifold; secure with three bolts and tighten to 37 ft. lbs. (50 Nm).

33. Install the exhaust pipe bracket to the rear case on manual transmissions or the transfer adapter case on vehicles with automatic transmissions and three way catalytic convertors; secure with two bolts and tighten to 44 ft. lbs. (60 Nm).

34. Install the speedometer cable clip and ground wire to the torque stopper housing; secure with one bolt and tighten to 89 inch lbs. (10 Nm).

35. Install the speedometer cable to the speedometer driven gear case.

36. Align the reference marks and install the front drive shaft; secure with four bolts and four nuts, tighten to 37 ft. lbs. (50 Nm).

37. Apply pipe sealant to the transfer case drain plug, install the plug into the transfer case and tighten to 21 ft. lbs. (28 Nm).

38. Remove the transfer case filler plug and add approximately 1.8 quarts (1.7 liters) of synthetic 75W-90 GL4 lubricant GM part number 1052080 or equivalent, into the transfer case. The oil level should be even with the bottom of the filler plug hole Apply sealant to the filler plug and tighten to 21 ft. lbs. (28 Nm).

39. If equipped, install the transfer case skid plate plate, secure the plate to the crossmember with two bolts and tighten to 40 ft. lbs. 54 Nm).

40. Lower the vehicle.

41. Install the four bolts to the fan shroud at the radiator and tighten the bolts to 89 inch lbs. (10 Nm).

42. Install the four wheel drive switch electrical connector to the four wheel drive switch.

43. Install the gearshift control lever into the gearshift lever case and pushing down on the shift control lever pivot and turning clockwise 90 degrees (1/4 turn) and releasing.

44. Install the gearshift lever case boot to the gearshift lever case; secure with one clamp.

45. Install the gearshift control lever boot and boot cover onto the gearshift control lever; secure with one screw.

46. Position the console box to the floor; secure with two plastic retainers and two screws.

47. Road test the vehicle.Check for proper operation of the transfer case and make sure there are no leaks, vibrations or abnormal noises.

DRIVE AXLE

Driveshaft

REMOVAL AND INSTALLATION

Samurai

The 4WD Samurai uses three separate driveshafts, the shafts are designated as No. 1, No. 2, and No. 3. The No. 1 shaft connects the transmission to the transfer case. The No. 2 shaft connects the transfer case to the front differential. The No. 3 shaft connects the transfer case to the rear differential.

The 2WD vehicle uses two separate driveshafts the No. 1 and No. 3 driveshafts. The driveshafts are used in the 2WD vehicles the same way as they are used in the 4WD vehicles.

1. Raise and safely support the vehicle.

2. Matchmark the driveshafts to the yokes on the transfer case or transmission and the differential.

3. Support the driveshaft and remove the attaching bolts.

4. Remove the driveshaft from the vehicle.

NOTE: The transmission side end of the No. 1 shaft does not have a flange piece; the end is splined and slides onto the output shaft inside the extension case. To remove the shaft, just pull the shaft off of the extension case. When removing the No. 1 shaft, the transmission fluid will not leak if the front and the rear of the vehicle are raised evenly and the transmission fluid is filled to specification. If the vehicle is raised unevenly or the fluid is above specification, drain the transmission fluid prior to removing the No. 1 shaft.

To install:

5. The No. 2 and No. 3 shafts are equipped with plunge joints on the shafts where the driveshaft can expand (lengthen) or contract (shorten) itself. Pull the driveshaft apart and liberally fill the driveshaft splines with chassis grease. When reassembling the driveshaft, align the matchmarks on the splines and the shaft to prevent noise or vibration. Verify that the rubber boot is pulled over and is protecting the driveshaft splines.

6. Install the driveshaft, by aligning the matchmarks made during disassembly.

7. Install and tighten the mounting bolts to 37–43 ft. lbs. (50–60 Nm).

8. Lubricate the driveshaft U-joints at the grease fittings and lower the vehicle.

9. If fluid was drained or leaked, refill the transmission to the proper level with the proper lubricant.

Sidekick, Tracker and X-90

1. Raise and safely support the vehicle.

2. Matchmark the driveshaft(s) to the yokes on the transfer case or transmission and the differential.

3. Drain the transfer case oil when servicing the 4WD vehicles.

4. Support the driveshaft and remove the attaching nuts and bolts.

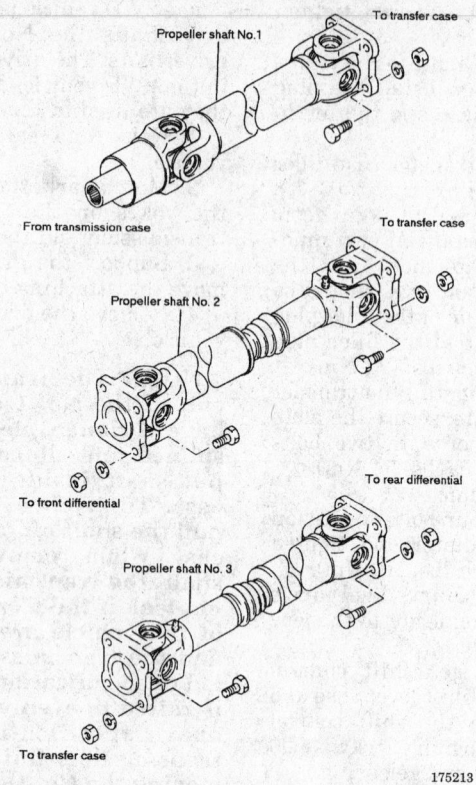

Driveshafts No. 1, No. 2, and No. 3 — 4WD Samurai models

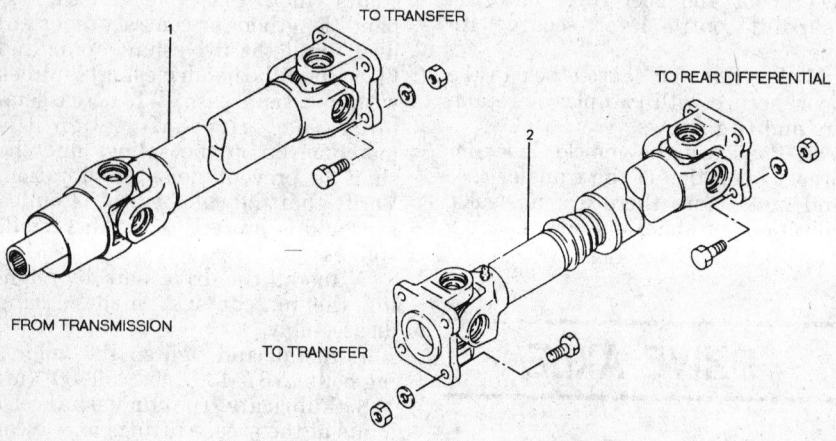

1. Propeller shaft No. 1
2. Propeller shaft No. 3

Tightening torque for propeller shaft (Universal joint flange) bolts and nuts	N·m	kg-m	lb-ft
	50 — 60	5.0 — 6.0	36.5 — 43.0

196017

Driveshaft assemblies No. 1 and No. 3 — 2WD Samurai models

5. Remove the driveshaft from the vehicle.

To install:

6. Install the driveshaft, by aligning the matchmarks to the vehicle.

7. Tighten the mounting bolts and nuts to 40 ft. lbs. (55 Nm).

8. Refill the transfer case and apply Loctite™ pipe sealant to the threads of the oil lever filler plug.

9. Safely lower the vehicle.

U-Joints

REMOVAL AND INSTALLATION

1. Remove the driveshaft or axle shaft (Samurai front axle) from the vehicle.

2. Remove the two snaprings from the pinion flange.

3. Use a 18mm socket as a driver and a 24mm socket as a cup. Place the driveshaft in a soft jaw vice with the 18mm socket against one of the bearing caps and the 24mm positioned so the bearing cap will slide into the socket. Compress the vice until the bearing cap has moved out 3–4mm.

4. Remove the driveshaft and sockets from the vice. Reposition the driveshaft so the vice jaws can be clamped down on the exposed portion of the bearing cap. With the cap clamped snugly, tap upward on the driveshaft until the cap comes free.

5. Using the 18mm socket and a hammer remove the other bearing cap by driving it out of the pinion flange.

6. Remove the pinion flange.

7. Remove the two snaprings from the driveshaft yoke.

8. Use a 18mm socket as a driver and a 24mm socket as a cup. Place the driveshaft in a soft jaw vice with the 18mm socket against one of the bearing caps and the 24mm positioned so the bearing cap will slide into the socket. Compress the vice until the bearing cap has moved out 3–4mm.

9. Remove the driveshaft and sockets from the vice. Reposition the driveshaft so the vice jaws can be clamped down on the exposed portion of the bearing cap. With the cap clamped snugly, tap upward on the driveshaft until the cap comes free.

10. Using the 18mm socket and a hammer remove the other bearing cap by driving it out of the driveshaft.

11. Remove the U-joint.

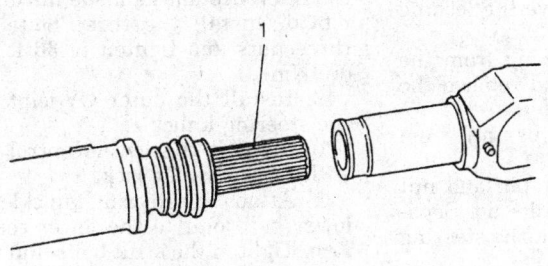

● GREASE SPLINES LIBERALLY,
   FILLING GROOVES WITH GREASE.

1 GREASE (CHASSIS GREASE)

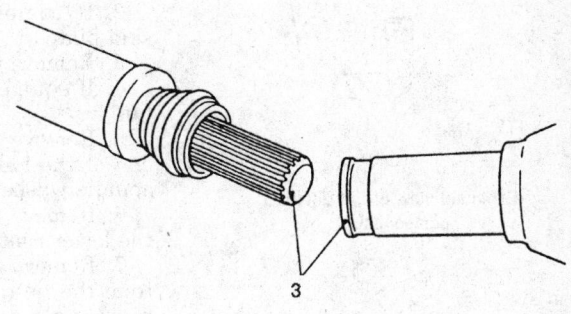

3 MATCH MARKS

195983

Greasing No. 2 and No. 3 driveshaft splines and aligning the matchmarks — Samurai models

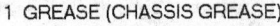

1. Pinion flange yoke
2. Snap ring

240856

Removing the snapring from the driveshaft yoke (pinion flange yoke similar)

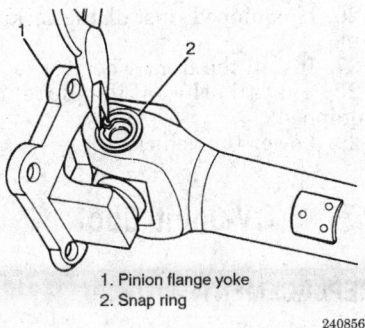

1. Spider
2. Bearing cap

240858

Removing the bearing cap from the driveshaft (pinion flange similar)

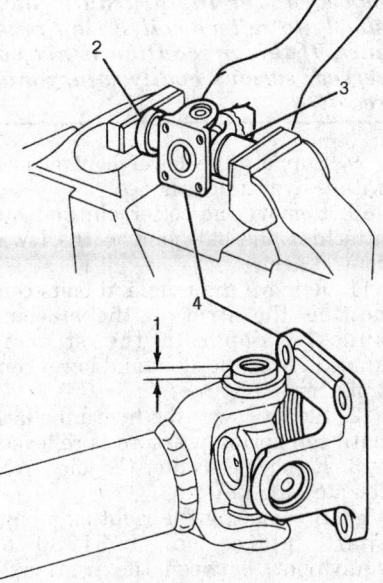

1. 3-4 mm (0.12-0.16")
2. 18 mm socket
3. 24 mm socket
4. pinion flange yoke

240857

Pressing out the driveshaft yoke bearing caps (pinion flange yoke similar)

To install:

12. Install the U-joint into the driveshaft yoke.

NOTE: DO NOT force the bearing caps into place. If the U-joint will not move freely, one of the bearing cap needle bearings may have become unseated. Remove the bearing caps and check needle bearing position.

13. Install one bearing cap into the driveshaft yoke. Fit the end of the U-joint into the bearing cap and using a hammer lightly tap the cap until it is flush with the yoke. Using an 18mm socket, tap the bearing cap down until the snapring groove is visible.

14. Install the second bearing cap and position the U-joint so the it is part way into each bearing cap. Lightly tap the second bearing cap into place. Using an 18mm socket tap the U-joint down until the snapring groove is visible.

15. Install the two snaprings.

16. Verify the U-joint moves freely.

17. Install the pinion flange over the U-joint.

18. Install one bearing cap into the pinion flange yoke. Fit the end of the U-joint into the bearing cap and using a hammer lightly tap the cap until it is flush with the yoke. Using an 18mm socket, tap the bearing cap

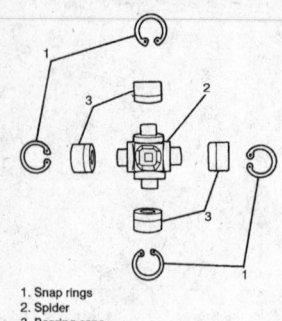

1. Snap rings
2. Spider
3. Bearing caps

240859

Exploded view of the U-joint components

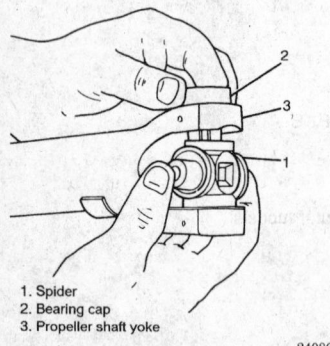

1. Spider
2. Bearing cap
3. Propeller shaft yoke

240860

Install the bearing cap onto the U-joint

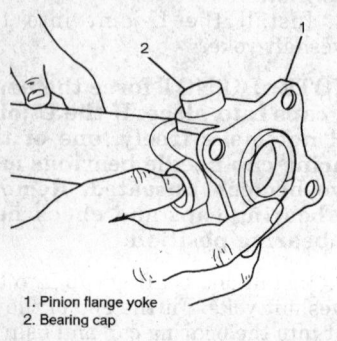

1. Pinion flange yoke
2. Bearing cap

240861

Installing the pinion flange

down until the snapring groove is visible.

19. Install the second bearing cap and position the U-joint so the it is part way into each bearing cap. Lightly tap the second bearing cap into place. Using an 18mm socket tap the U-joint down until the snapring groove is visible.

20. Install the two snaprings.

21. Verify the U-joint moves freely.

22. Lubricate the U-joint at the grease fitting and install the driveshaft in the vehicle.

Halfshaft

REMOVAL AND INSTALLATION

1. Raise and safely support the vehicle.

2. If equipped, remove the front skid plate.

3. Remove the front wheel.

4. If equipped, remove the locking hub.

5. Remove the snapring from the end of the halfshaft and remove the spindle washer.

6. Remove the sway bar nut from the lower control arm.

7. Remove the cotter pin and nut from the tie rod end ball stud. Separate the tie rod end from the steering knuckle.

8. Remove the brake caliper from the knuckle and suspend with wire, without disconnecting the brake line. Do not let the caliper hang from the brake hose.

CAUTION
The coil spring is under extreme pressure. Make sure the control arm is firmly supported with a hydraulic jack before removing the lower ball joint nut. After the lower ball joint nut has been removed, lower the hydraulic jack slowly to relieve coil spring pressure. If this precaution is not observed, serious bodily injury may result.
---

9. Support the lower control arm with a hydraulic jack.

10. Remove the cotter pin and nut attaching the ball joint to the lower control arm.

11. Remove the nuts and bolts connecting the strut to the steering knuckle. Separate the steering knuckle from the strut and lower control arm.

12. Slowly lower the hydraulic jack until coil spring pressure is relieved.

13. Remove the outer CV-joint from the steering knuckle.

14. If removing the right side halfshaft, place tool J 37780 or equivalent, between the front axle housing and the inner CV-joint. Gently tap the inner CV-joint away and out of the front axle housing.

15. If removing the left side halfshaft, scribe a reference mark on the left inner axle shaft flange and the inner CV-joint flange to ensure correct installation. Remove the three bolts and three nuts and separate the inner CV-joint from the left inner axle shaft.

16. Remove the halfshaft from the vehicle.

To install:

17. If installing the right halfshaft, install the inner CV-joint into the axle housing, making sure the snapring seats in the differential side gear.

18. If installing the left halfshaft, install the left inner axle shaft flange to the inner CV-joint flange, aligning the reference marks made during removal. Install the three bolts and three nuts and tighten to 36 ft. lbs. (50 Nm).

19. Install the outer CV-joint into the steering knuckle.

20. Support the lower control arm with the hydraulic jack.

21. Attach the steering knuckle and lower ball joint to the lower control arm. Tighten the strut bolts and nuts to 65 ft. lbs. (90 Nm). Tighten the ball joint nut to 42 ft. lbs. (58 Nm) and install a new cotter pin.

22. Remove the hydraulic jack from the lower control arm.

23. Install the brake caliper to the knuckle.

24. Install the tie rod end to the steering knuckle and tighten the nut to 30 ft. lbs. (40 Nm). Install a new cotter pin.

25. Install the spindle washer and snapring to the end of the halfshaft.

26. If equipped, install the locking hub.

27. Install the front wheel.

28. Install the skid plate, if equipped.

29. Lower the vehicle.

CV-Joint Boot

REPLACEMENT

Sidekick, Tracker and X-90

NOTE: During differential-side and wheel-side joint service, index marks (reference marks) should be placed on the differential-side joint and wheel-side housing. A corresponding matching mark should be placed on the drive axle shaft as well. This will establish the joint to axle position and ensure that all components are installed in the same position from which they were removed. If this precaution is not observed and the components are installed in a different position, uneven or premature component wear may result.

1. Raise and safely support the vehicle.

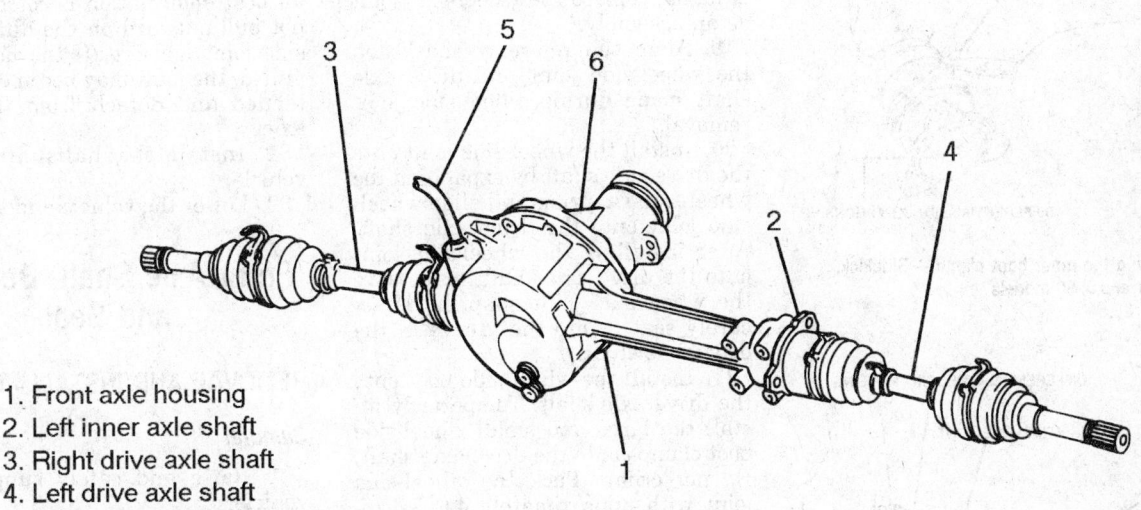

1. Front axle housing
2. Left inner axle shaft
3. Right drive axle shaft
4. Left drive axle shaft
5. Front axle housing breather hose
6. Differential carrier

127767

Halfshafts and front drive unit — Sidekick, Tracker models

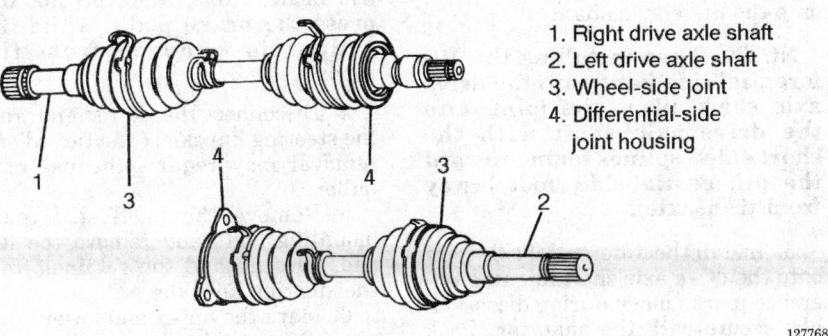

1. Right drive axle shaft
2. Left drive axle shaft
3. Wheel-side joint
4. Differential-side joint housing

127768

Left and right side halfshafts — Sidekick, Tracker models

2. Remove the halfshaft from the vehicle.

3. Secure the shaft in a vise, placing wood or soft metal between the jaws to protect the shaft.

4. Remove the large differential-side boot clamp from the differential-side boot by drawing the clamp hooks together.

5. Remove the small differential-side boot clamp from the differential-side boot. Temporarily slide the differential-side boot toward the center of the drive axle shaft. Place an index mark (reference mark) on the differential-side joint housing and drive axle shaft to ensure correct assembly.

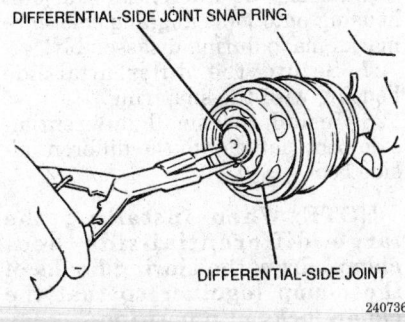

DIFFERENTIAL-SIDE JOINT SNAP RING

DIFFERENTIAL-SIDE JOINT

240736

Removing the differential side joint — Sidekick, Tracker and X-90 models

6. Remove the snapring securing the differential-side joint housing.

7. Remove the differential-side joint housing from the joint. Place an index mark (reference mark) on the joint and drive axle shaft to ensure correct reassembly.

8. Remove the snapring and joint from the drive axle shaft.

9. Remove the differential-side boot from the drive axle shaft.

10. Remove the large wheel-side boot clamp from the wheel-side boot.

11. Remove the small wheel-side boot clamp from the wheel-side boot.

12. Remove the wheel-side boot from the drive axle shaft. Place an index mark (reference mark) on the wheel-side joint and the drive axle shaft to ensure correct assembly.

13. Remove the wheel-side joint from the drive axle shaft by ex-

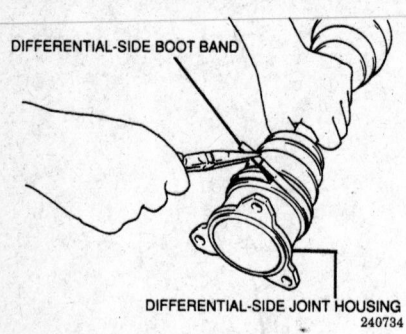

Removing the inner boot clamp — Sidekick, Tracker and X-90 models

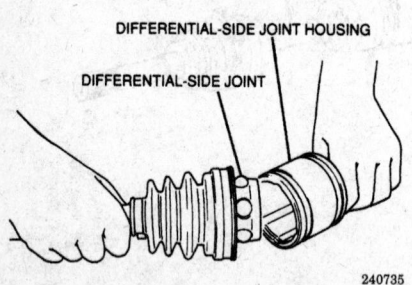

Removing the differential side joint housing — Sidekick, Tracker and X-90 models

panding the wheel-side joint snapring.

NOTE: Do not disassemble the joints. If any abnormality is found in the joint, replace it as an assembly. Do not wash drive axle boots in solvent. Washing drive axle boots in degreaser or other solvents causes deterioration of boots. DO NOT wash the differential side joint in degreaser. Washing the joint in degreaser will remove all lubrication in the joint needle bearings. Clean the joint assembly and drive axle boots with a clean, dry, solvent-free rag.

To install:
14. Clean the differential-side and the wheel-side boots with a clean, dry, solvent-free cloth.
15. Clean the differential side joint and the wheel-side joint with a clean, dry, solvent-free cloth.
16. Inspect the differential-side and the wheel-side boots for tears, damage or fatigue. Replace as necessary.
17. Inspect the differential side joint for excessive wear or damage. If any excessive wear, damage or abnormality is found, replace the joint as an assembly.

18. Inspect the wheel-side joint for excessive wear or damage. If any excessive wear, damage or abnormality is found, replace the wheel-side joint as an assembly.
19. Align the reference marks on the wheel-side joint and drive axle shaft made during wheel-side joint removal.
20. Install the wheel-side joint onto the drive axle shaft by expanding the wheel-side snapring and slide wheel-side joint onto the drive axle shaft. After installing the wheel-side joint onto the drive axle shaft, make sure the wheel-side joint snapring is securely seated into the groove in the drive axle shaft.
21. Install the wheel-side boot onto the drive axle shaft. Temporarily install the large and small wheel-side boot clamps onto the drive axle shaft. Do not crimp. Pack the wheel-side joint with approximately 4.6–5.3 oz. (130–150 g) of black grease provided in the wheel-side boot kit.
22. Install the small wheel-side boot clamp onto the wheel-side boot.
23. Install the large wheel-side boot clamp onto the wheel-side boot. Temporarily install the large and small differential-side boot clamps onto the drive axle shaft. Do not crimp.
24. Install the differential-side boot onto the drive axle shaft.

NOTE: When installing the differential side joint onto the drive axle shaft, place the joint onto the drive axle shaft with the short sided splines facing toward the differential-side boot (away from transaxle).

25. Install the differential side joint onto the drive axle shaft aligning reference marks made during disassembly; secure with the snapring. Pack the differential-side joint housing with approximately 8.1–8.8 oz. (230–250 g) of the lubricant provided in the differential-side boot kit.
26. Install the differential-side joint housing onto joint aligning reference marks made during disassembly.
27. Secure the differential-side housing with the snap ring.
28. Install the small differential-side boot clamp onto the differential-side boot.

NOTE: When installing the large differential-side boot clamp, draw the closing hooks of the clamp together so that the clamp locks into position.

29. Install the large differential-side boot clamp onto the differential-side boot.

30. Inspect both boots for distortion or dents. Correct by pulling outward on the boot in the desired areas until all boot deformation is corrected. Do not pull outward on the differential-side joint housing. If the housing is pulled, the joint may become over-extended and detach from the drive axle.
31. Install the halfshaft in the vehicle.
32. Lower the vehicle and road test.

Front Axle Shaft, Bearing, and Seal

REMOVAL AND INSTALLATION

Samurai

1. Raise and safely support the vehicle.
2. Drain the oil in the front differential.
3. Remove the front wheels and disconnect the brake caliper. There is no need to disconnect the brake hose. Support the brake caliper with a piece of wire; do not allow the caliper to hang from the brake hose.

NOTE: Be careful not to twist the brake hose. Also, do not depress the brake pedal while the caliper is removed from the rotor.

4. Disconnect the tie rod end from the steering knuckle. (The tie rod end removal may require the use of a puller.)
5. Remove the 8 oil seal cover mounting bolts and remove the felt pad, oil seal, and the retainer from the steering knuckle.
6. Mark the upper and lower kingpins. Remove the 4 mounting bolts and disconnect the kingpins from steering knuckle. The kingpins must be kept separated so as to prevent an error during reassembly.
7. Remove the axle shaft from the housing with the steering knuckle attached.

NOTE: At this time, the lower kingpin bearing sometimes falls out. So remove the bearing while pulling off the knuckle gradually.

8. If necessary, remove the oil seal from the axle shaft taking note of the original positioning of the seal on the shaft.
To install:
9. If the front axle shaft seal was removed, apply grease to the lip portion of the seal, and slide the seal into place on the axle shaft.

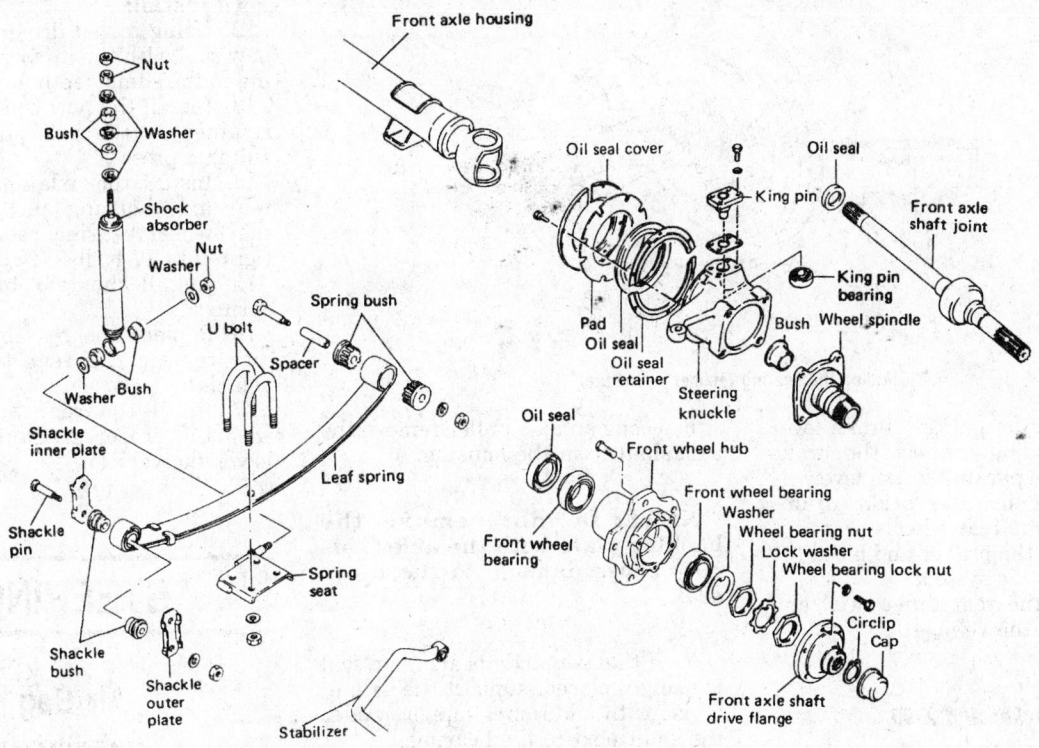

Front suspension components — Samurai

Rear Axle Shaft, Bearing, and Seal

REMOVAL AND INSTALLATION

Samurai

1. Raise and safely support the vehicle. Drain the rear differential assembly.
2. Make sure the rear parking brake is released.
3. Remove the rear wheels and remove the rear brake drums from the vehicle.
4. Disconnect the parking brake cables from the levers. Remove the parking brake lever stop plates.
5. Disconnect and plug the brake lines to the wheel cylinders.
6. Remove the backing plate mounting bolts.
7. Using a slide hammer, remove the rear axles with the backing plates attached.
8. Using a suitable prying tool, remove the axle seal from the housing.
9. If the axle, axle bearing, or backing plate is being replaced, support the axle in a vise with additional support under the shaft next to the bearing.

— CAUTION —

Eye protection must be worn during the following 3 steps. Failure to do so could cause injury.

10. With the axle shaft supported properly, grind the top and bottom of the axle bearing retainer. This will enable it to be removed without damaging the axle shaft.
11. Using a chisel, break the retainer and remove the retainer from the axle shaft.
12. Using a press or suitable bearing puller, remove the axle shaft bearing from the axle shaft.
13. Remove the backing plate from the axle shaft.

To install:

14. Using a seal driver, install the new seal with the lip facing the housing to the same depth as the old seal.
15. Install the backing plate on the axle shaft and using a press, install the bearing and the retainer on the axle shaft.
16. Install the axle shaft in the housing.
17. Install the backing plate mounting bolts and torque them to 14–20 ft. lbs. (19–27 Nm). Connect the brake lines to the wheel cylinders.

10. If a new axle shaft is being installed, transfer the steering knuckle to the new axle shaft. Install the axle shaft into the differential housing.
11. Install the lower kingpin bearing and the upper and lower kingpins in their correct location. Tighten the mounting bolts to 15–22 ft. lbs (20–30 Nm).
12. Install the steering knuckle oil seal, felt pad, retainer, and the 8 mounting bolts. Tighten the mounting bolts to 6–8 ft. lbs (8–12 Nm).
13. Connect the tie rod end to the steering knuckle and tighten the nut to 22–39 ft. lbs. (30–55 Nm). Install a new cotter pin.
14. Install the front brake caliper assembly. If the brake line was disconnected, bleed the brake system.
15. Install the front wheels and refill the differential with the proper fluid.
16. Lower the vehicle.

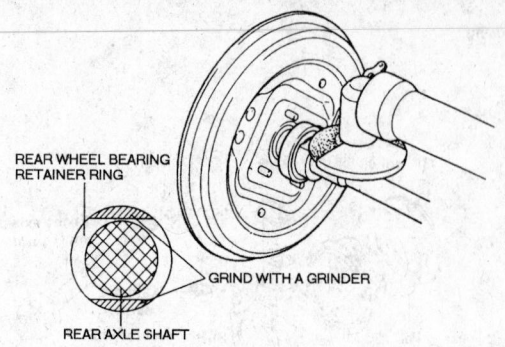

Grinding the bearing retainer — Samurai

175784

18. Install the parking brake lever stop plates and connect the brake cables to the parking brake lever.

19. Install the rear brake drums and install the rear wheels.

20. Adjust the brakes and bleed the brake hydraulic system.

21. Refill the rear differential and safely lower the vehicle.

Sidekick, Tracker and X-90

1. Raise and safely support the vehicle.

2. Remove the rear wheels and remove the rear brake drums from the vehicle.

3. Drain the gear oil from the rear axle housing.

4. Remove the rear brake return springs.

NOTE: If both axles are being removed mark the axles left and right. The axles are different lengths and must be installed in the correct position.

5. Remove the rear wheel bearing retainer nuts from the rear axle housing.

6. Using an axle puller remove the axle shaft from the housing.

NOTE: Do not remove the backing plate with the axle. This may cause damage to the inner seal.

7. If the axle, axle bearing or seal is being replaced, support the axle in a vise with additional support under the shaft next to the bearing.

——— CAUTION ———
Eye protection must be worn during rear axle bearing and seal removal. Failure to do so could cause injury.

8. With the axle shaft supported properly, grind the top and bottom of the axle bearing retainer until they are flat. DO NOT grind the axle, component failure could result.

9. Using a chisel and hammer, finish removing the retainer from the axle shaft.

10. Using a press or suitable bearing puller, remove the axle shaft bearing from the axle shaft.

11. Using a prying tool, remove the seal from the axle housing.

To install:

12. Using a seal driver, install the new seal with the lip facing the housing to the same depth as the old seal.

13. Install the new bearing and the retainer on the axle shaft using a suitable press.

14. Install the axle shaft into the rear axle housing and replace the rear wheel bearing retaining nuts, tighten to 17 ft. lbs. (23 Nm).

15. Install the rear brake return springs.

16. Replace the rear brake drums and replace the rear tires on the vehicle.

17. Refill the rear axle housing with the proper gear oil and safely lower the vehicle.

STEERING

Air Bag

——— CAUTION ———
Some vehicles are equipped with an air bag system, also known as the Supplemental Inflatable Restraint (SIR) or Supplemental Restraint System (SRS). The system must be disabled before performing service on or around system components, steering column, instrument panel components, wiring and sensors. Failure to follow safety and disabling procedures could result in accidental air bag deployment, possible personal injury and unnecessary system repairs.

PRECAUTIONS

Several precautions must be observed when handling the inflator module to avoid accidental deployment and possible personal injury.

• Never carry the inflator module by the wires or connector on the underside of the module.

• When carrying a live inflator module, hold securely with both hands, and ensure that the bag and trim cover are pointed away.

• Place the inflator module on a bench or other surface with the bag and trim cover facing up.

• With the inflator module on the bench, never place anything on or close to the module which may be thrown in the event of an accidental deployment.

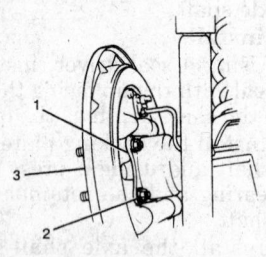

1. Rear axle housing
2. Rear axle shaft retaining nuts
3. Brake backing plate

240822

Rear axle shaft retaining bolts — Sidekick, Tracker and X-90

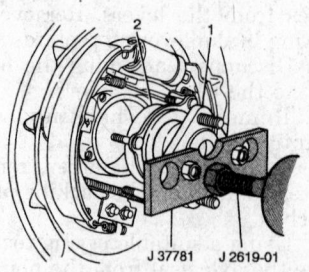

J 37781   J 2619-01

1. Rear axle shaft
2. Rear axle shaft bearing retainer

240823

Removing the rear axle shaft — Sidekick, Tracker and X-90

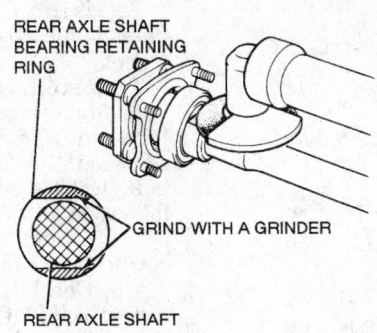

240824

Grinding the bearing retainer — Sidekick, Tracker and X-90

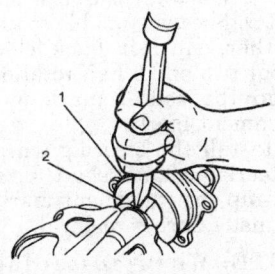

1. Rear axle shaft bearing
2. Rear axle shaft bearing retaining ring

240825

Removing the bearing retainer — Sidekick, Tracker and X-90

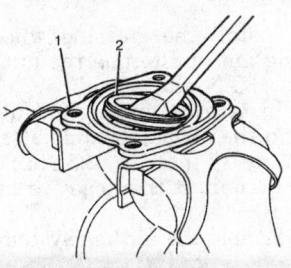

1. Rear axle shaft bearing retainer
2. Rear axle shaft outer oil seal

240826

Removing the rear axle seal — Sidekick, Tracker and X-90

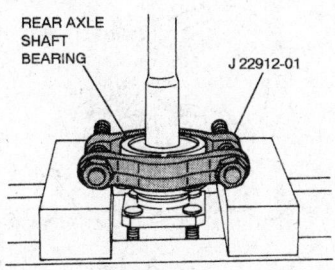

240827

Removing the rear axle bearing — Sidekick, Tracker and X-90

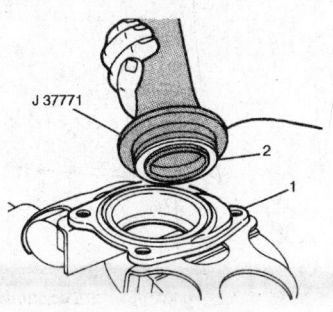

1. Rear axle shaft bearing retainer
2. Rear axle shaft outer oil seal

240828

Installing the seal — Sidekick, Tracker and X-90

DISARMING

——— CAUTION ———

The Supplemental Inflatable Restraint (SIR) system must be disarmed before performing service around SIR components or SIR wiring. Failure to do so may cause accidental deployment of the air bag, resulting in unnecessary SIR system repairs and/or personal injury.

Disarming

1. Turn the steering wheel so the front wheels are in the straight ahead position.
2. Turn the ignition switch to the **LOCK** position and remove the key.
3. Remove the AIR BAG fuse from the air bag fuse box.
4. Remove the steering wheel side cap and disconnect the yellow connector inside the inflator module housing.

5. Remove the glove box and disconnect the yellow passenger air bag inflator module connector.

Enabling

1. Turn the steering wheel so the front wheels are in the straight ahead position.
2. Turn the ignition switch to the **LOCK** position.
3. Reconnect the passenger air bag inflator module yellow connector. Install glove box.
4. Reconnect the yellow connector inside the inflator module housing on the driver's air bag. Install the inflator module onto the connector stay.
5. Install the plastic access cover.
6. Turn the ignition switch to the **ON** position. Verify that the air bag indicator lamp flashes 7 times and then turns OFF. If the lamp does not function as specified, there is a malfunction in the SIR system.

Steering Wheel

REMOVAL AND INSTALLATION

1993–95 Models

1. Disconnect the negative battery cable.
2. Disconnect the horn button and remove the steering wheel shaft nut.
3. Make matchmarks on the steering wheel and the shaft to use as a guide during reinstallation.
4. Remove the steering wheel, using a steering wheel puller.
   To install:
5. Install the steering wheel onto the shaft, aligning the matchmarks.
6. Install and tighten the shaft nut to 24 ft. lbs. (32 Nm).
7. Install the horn button and connect the negative battery cable.

1996–97 Models

——— CAUTION ———

The Supplemental Inflatable Restraint (SIR) system must be disarmed before removing the steering wheel. Failure to do so may cause accidental deployment of the air bag, resulting in unnecessary SIR system repairs and/or personal injury.

1. Disable the air bag system.
2. Disconnect the negative battery cable.
3. Remove the steering wheel side cap from the right side and disconnect the horn connectors.

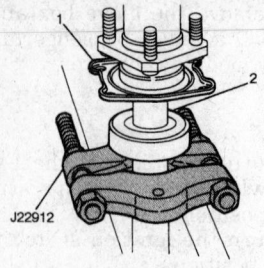

1. Rear axle shaft bearing retainer
2. Rear axle shaft bearing

J22912

240830

Installing the bearing and retainer — Sidekick, Tracker and X-90

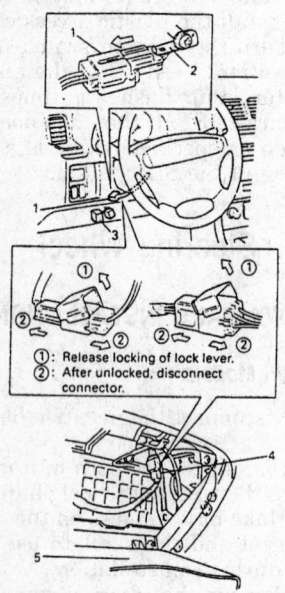

①: Release locking of lock lever.
②: After unlocked, disconnect connector.

1. Yellow connector of driver air bag (inflator) module
2. Connector stay
3. Air bag fuse box
4. Yellow connector of passenger air bag (inflator) module
5. Glove box

322403

Location of driver and passenger air bag connectors — 1996–97 Sidekick, Tracker and X-90 models

4. Remove the air bag module attaching bolts and the air bag assembly from the vehicle.

CAUTION

When carrying a live air bag, make sure the bag and trim cover are pointed away from the body. In the unlikely event of an accidental deployment, the bag will then deploy with minimal chance of injury. When placing a live inflator module on a bench or other surface, always place the bag and trim cover up, away from the sur-

---

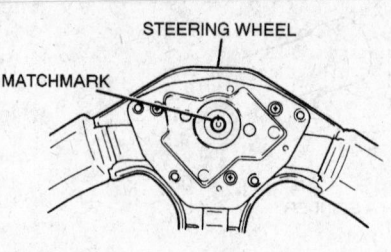

STEERING WHEEL

MATCHMARK

128197

Steering wheel horn contact plate — 1993–95 models

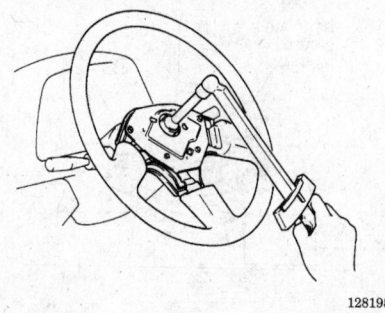

128198

Tightening the steering wheel nut — 1993–95 models

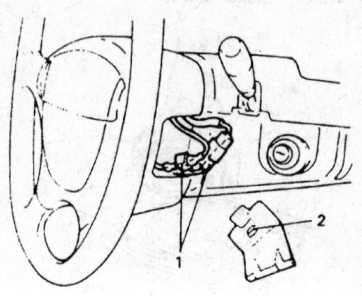

1. Horn connectors
2. Projection of cap

322455

Steering wheel side cap — 1996–97 models

face. This will reduce the motion of the module if it is accidently deployed.

WARNING

The air bag system coil assembly is easily damaged if the correct steering wheel puller tools are not used.

---

5. Remove the steering wheel nut.
6. Scribe alignment marks across the steering wheel and shaft to ease installation.

---

7. Using a suitable puller (09944–36010), remove the steering wheel.

To install:

8. Make sure the wheels are facing forward and the contact coil is centered.

9. Center the contact coil as follows:

  a. Check that the vehicle's wheels are facing straight ahead.

  b. Check that the ignition switch is at the LOCK position.

  c. Turn the contact coil counterclockwise slowly with a light force. Turn the coil until the contact coil will not turn any further.

  d. From the position where contact coil became unable to turn any further, turn it back clockwise about two and a half rotations and align the center mark with the alignment mark.

10. Install the steering wheel onto the steering column shaft. Be sure to match up the alignment marks that were made during removal.

NOTE: When installing the steering wheel, make sure to install the steering wheel to the steering shaft with the two lugs contact coil fitted in the two grooves in the back of the steering wheel.

11. Install the steering wheel retaining nut and torque the nut to 24 ft. lbs. (33 Nm).

12. Install the air bag module and tighten the air bag module attaching screws to 17 inch lbs. (23 Nm).

13. Reconnect the negative battery cable.

14. Enable the air bag system.

Manual Steering Gear

REMOVAL AND INSTALLATION

Samurai

1. Remove the steering shaft coupler bolt and disconnect the coupler from the gear. Raise and support the vehicle safely.

2. Remove the radiator under cover and disconnect the pitman arm from the drag link.

3. Disconnect the steering damper from the pitman arm.

4. Support the steering gear and remove the mounting bolts.

5. Remove the steering gear from the vehicle.

To install:

6. Install the steering gear in the vehicle and tighten the mounting nuts to 55 ft. lbs. (75 Nm).

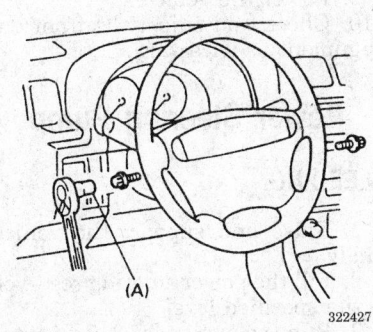

Air bag retaining bolts — 1996–97 models

322427

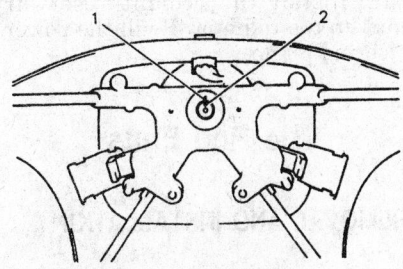

1 STEERING SHAFT ALIGNMENT MARK
2 STEERING SHAFT ALIGNMENT MARK

322425

Aligning the steering wheel to the steering column shaft — 1996–97 models

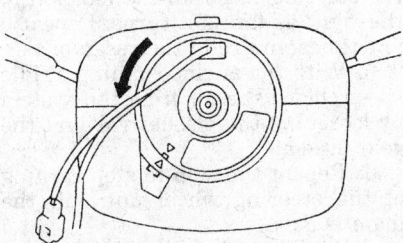

Turn slowly till coil stops.

322456

Turning the contact counterclockwise — 1996–97 models

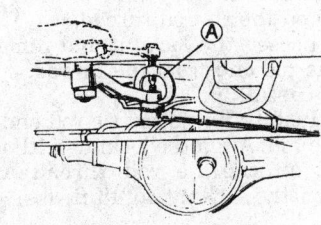

179081

Separating the pitman arm from the drag link — Samurai

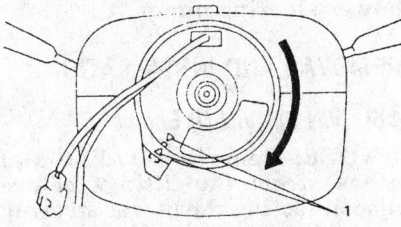

Turn contact coil back about 2 and a half turns.

1. Align marks

322458

Aligning the contact coil marks — 1996–97 models

7. Install the steering damper to the pitman arm tighten the nut to 30 ft. lbs. (40 Nm).

8. Connect the drag link to the pitman arm. Tighten the nut to 50 ft. lbs. (70 Nm).

9. Install the radiator under cover.

10. Connect the steering coupler to the steering box and tighten the bolt to 18 ft. lbs. (24 Nm). Lower the vehicle.

Sidekick, Tracker

1. Raise and support the vehicle safely.

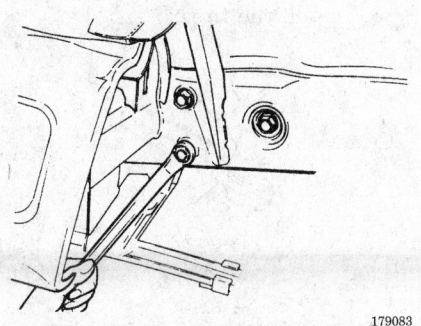

179083

Steering gear mounting bolts — Samurai

2. Remove the lower skid plate.

3. Disconnect the steering lower shaft mounting bolts.

4. Using special tool J29107, disconnect the center link end from the pitman arm.

5. Remove the 3 steering gear box mounting bolts.

6. Disconnect the steering lower shaft joint and remove the steering gear.

7. If replacing the steering gear, remove the nut connecting the pitman arm to the steering gear. Use a pitman arm puller to remove the pitman arm from the steering gear.

To install:

8. Connect the pitman arm to the steering gear and install the nut. Torque the nut to 116 ft. lbs. (160 Nm).

9. Install the steering gear box by connecting to the lower shaft joint.

NOTE: Align the flat part of the steering gear worm shaft with the bolt hole of the lower shaft joint.

10. Replace the steering gear box mounting bolts and tighten to 61 ft. lbs. (85 Nm).

11. Attach the center link to the pitman arm and tighten the nut to 37 ft. lbs. (30–68 Nm).

12. Connect the lower shaft mounting bolts and tighten to 18 ft. lbs. (25 Nm).

13. Install the lower skid plate.

14. Lower the vehicle.

Power Steering Gear

REMOVAL AND INSTALLATION

Sidekick, Tracker and X-90

1. Remove the coolant reservoir tank from the radiator.

2. Disconnect the steering column lower shaft from the gear box by removing the bolt.

3. Raise and safely support the vehicle.

4. Remove the center link nut and lock washer holding the pitman arm to the center link. Using a pitman arm puller, disconnect the center link from the pitman arm.

5. Lower the vehicle and place a fluid catch pan under the power steering gear box. Remove the pressure hose from the power steering gear assembly and plug the line.

6. Disconnect the return hose and plug the line.

7. Remove the three power steering gear mounting bolts.

8. Remove the power steering gear.

9. Remove the nut holding the pitman arm to the power steering box. Place alignment marks on the pitman arm and steering box. Using a pitman arm puller, remove the pitman arm from the steering box.

To install:

10. Align the matchmarks on the pitman arm and the power steering gear sector shaft. Install the pitman arm to the gear assembly and tighten the nut to 101 ft. lbs. (140 Nm).

11. Install the power steering gear assembly on the vehicle and tighten the mounting bolts to 66 ft. lbs. (85 Nm).

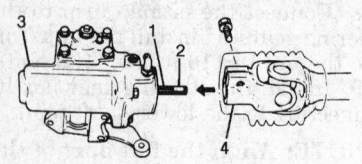

1. Input shaft
2. Alignment
3. Power steering gear
4. Lower steering column shaft

341296

Lower steering column shaft and power steering gear connection — Sidekick, Tracker and X-90

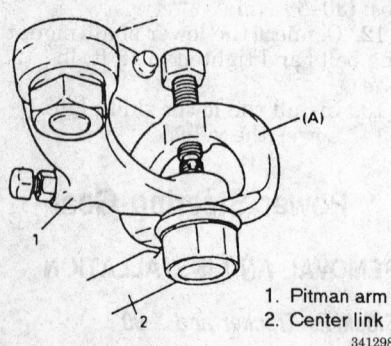

1. Pitman arm
2. Center link

341298

Disconnecting the Pitman arm from the center link — Sidekick, Tracker and X-90

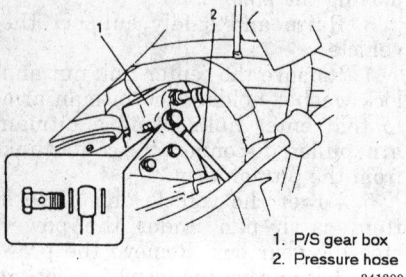

1. P/S gear box
2. Pressure hose

341300

Pressure hose union bolt — Sidekick, Tracker and X-90

12. Connect the power steering pressure and return hoses. Using new gaskets, torque the union bolt for the pressure line to 26 ft. lbs. (35 Nm).

13. Raise and safely support the vehicle.

14. Install the center link to the pitman arm and tighten the nut to 37 ft. lbs. (50 Nm). Lower the vehicle.

15. Connect the steering column lower shaft to the gear assembly and tighten the bolts to 18 ft. lbs. (25 Nm).

---

16. Install the coolant reservoir tank to the radiator. Refill the power steering pump.

Tie Rod Ends

REMOVAL AND INSTALLATION

1. Raise and safely support the vehicle and remove the wheels.

2. Remove the cotter pin and castle nut from the outer tie rod end.

3. Separate the tie rod end from the steering knuckle, using a tie rod end remover tool.

4. Mark the tie rod end locknut position on the tie rod thread.

5. Loosen the locknut and remove the tie rod end from the tie rod.

To install:

6. Install the tie rod tie rod end to the tie rod. Align the locknut with the mark on the tie rod thread and tighten the locknut to 48 ft. lbs. (65 Nm).

7. Connect the tie rod end to the steering knuckle. Tighten the castle nut until the holes of the split pin are aligned but only within the specified torque 33 ft. lbs. (44 Nm). Install a new cotter pin

8. Install the tires.

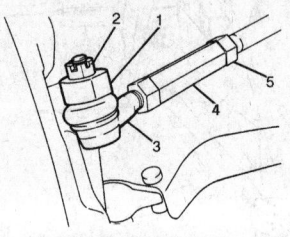

1. Steering knuckle
2. Castle nut
3. Tie rod end
4. Tie rod connector sleeve
5. Tie rod locknut

328787

Tie rod components

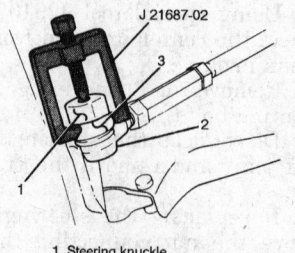

J 21687-02

1. Steering knuckle
2. Tie rod end
3. Tie rod end dust seal

328788

Separating the tie rod end from the steering knuckle

---

9. Lower the vehicle.

10. Check and adjust the front end alignment as necessary.

Power Steering Pump

BLEEDING

1. Raise and support the vehicle safely.

2. Fill the power steering reservoir to the specified level.

3. Run the engine for 3–5 minutes, stop the engine and add fluid if necessary to reach specified level.

4. With the engine stopped, turn the steering wheel to the left and to the right as far as it turns. Repeat a few times and refill the reservoir.

5. With the engine running at idle speed, bleed the air from the system by loosening the bleeder valve at the gear assembly.

6. Repeat the stop to stop turning of the steering wheel until all the foam is gone.

7. Tighten the bleed valve securely. Recheck the fluid level in the reservoir.

NOTE: When air bleeding is not complete, it is indicated by a foaming fluid on the level indicator or a humming noise from the power steering pump.

REMOVAL AND INSTALLATION

1.6L (VIN 0), (VIN U) Engine

NOTE: Remove all dirt and grease from the fittings before disconnecting the power steering pressure and return lines.

1. Remove the power steering belt.

2. Remove the coolant reservoir tank from the radiator.

3. If equipped with A/C, loosen the air conditioning compressor adjusting and pivot bolts.

4. If not equipped with A/C, loosen the power steering pump adjusting and mounting bolts.

5. Remove the power steering belt.

6. Disconnect the power steering pressure and return hose and plug.

7. Disconnect the power steering pressure switch lead wire at the switch terminal.

8. Remove the engine oil filter.

9. Remove the power steering pump mounting and adjusting bolts.

10. Remove the power steering pump.

To install:

11. Install the power steering pump and replace the pump mounting bolts. Do not tighten.

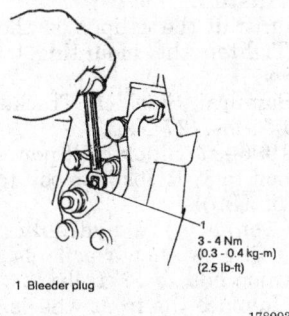

1 Bleeder plug

3 - 4 Nm
(0.3 - 0.4 kg-m)
(2.5 lb-ft)

178993

**Bleeder plug location — 1993–95
Sidekick, Tracker**

12. Install the power steering pump pressure switch lead wire to the switch terminal.

13. Replace the power steering pressure and return hoses. Torque the pressure hose union bolt to 44 ft. lbs. (60 Nm).

14. Install the power steering belt and tighten the power steering pump mounting bolts to 21 ft. lbs. (28 Nm).

15. If equipped with A/C, tighten the air conditioning mounting bolts to 21 ft. lbs. (28 Nm).

16. Replace the coolant reservoir tank to the radiator. Refill the power steering pump.

17. Replace the oil filter and fill the crankcase to the proper level.

18. Run the engine and operate the power steering. Check for any leaks.

1.8L (VIN 2) Engine

NOTE: Remove all dirt and grease from the fittings before disconnecting the power steering pressure and return lines.

1. Remove the coolant reservoir tank from the radiator.

2. Remove the air cleaner outlet hose.

3. Remove the alternator drive belt.

4. Using a syringe, remove the fluid from the power steering pump.

5. Disconnect the power steering pressure switch lead wire at the switch terminal.

6. Disconnect the power steering pump discharge hose from the power steering pump. Plug the line to prevent fluid from spilling from the line.

7. Remove the union bolt and disconnect the suction line from the power steering pump. Hold the discharge connector with a wrench to prevent it from getting loose and draining power steering fluid.

8. Remove the three mounting bolts from the power steering pump.

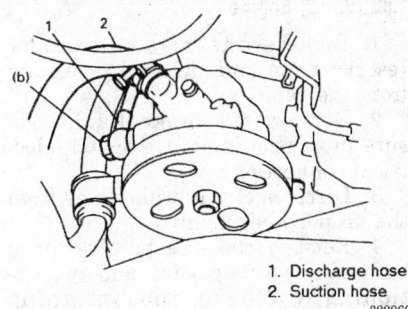

1. Discharge hose
2. Suction hose

322960

Power steering pressure and discharge hoses — 1.8L (VIN 2) engines

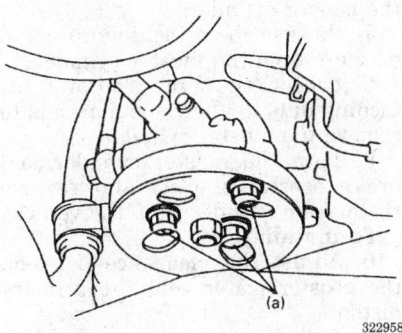

322958

Power steering pump bolts — 1.8L (VIN 2) engines

9. Remove the power steering pump from the engine.

To install:

10. Install the power steering pump with the three mounting bolts. Torque the bolts to 37 ft. lbs. (50 Nm).

11. Connect the pressure hose and suction hose to the power steering pump. Torque the union bolt to 44 ft. lbs. (60 Nm). Use new gaskets for the union bolt.

12. Connect the power steering pressure switch lead wire to the switch terminal.

13. Install the alternator belt.

14. Install the air cleaner outlet hose.

15. Fill the power steering pump with Dexron®III A/T fluid.

16. Bleed the air in the power steering pump.

17. Start the engine and check for leaks.

BRAKES

Anti-Lock Brake System Service

PRECAUTIONS

• Certain components within the Anti-Lock Brake System (ABS) are not intended to be serviced or repaired individually. Only those components with removal and installation procedures should be serviced.

• Do not use rubber hoses or other parts not specifically specified for and ABS system. When using repair kits, replace all parts included in the kit. Partial or incorrect repair may lead to functional problems and require the replacement of components.

• Lubricate rubber parts with clean, fresh brake fluid to ease assembly. Do not use lubricated shop air to clean parts; damage to rubber components may result.

• Use only specified brake fluid from an unopened container.

• If any hydraulic component or line is removed or replaced, it may be necessary to bleed the entire system.

• A clean repair area is essential. Always clean the reservoir and cap thoroughly before removing the cap. The slightest amount of dirt in the fluid may plug an orifice and impair the system function. Perform repairs after components have been thoroughly cleaned; use only denatured alcohol to clean components. Do not allow ABS components to come into contact with any substance containing mineral oil; this includes used shop rags.

• The Anti-Lock control unit is a microprocessor similar to other computer units in the vehicle. Ensure that the ignition switch is **OFF** before removing or installing controller harnesses. Avoid static electricity discharge at or near the controller.

• If any arc welding is to be done on the vehicle, the control unit should be unplugged before welding operations begin.

Master Cylinder

REMOVAL AND INSTALLATION

Samurai

1. Remove the air cleaner case, if necessary.

2. Disconnect the reservoir lead wire.

3. Clean the outside of the reservoir, and drain the fluid from the reservoir.

4. Remove the reservoir connector screw, and remove the reservoir.

5. Disconnect and plug the two brake fluid lines at the master cylinder.

6. Remove the two master cylinder to booster mounting bolts and remove the master cylinder.

To install:

7. Install the new master cylinder and tighten the mounting bolts to 7.5–11.5 ft. lbs. (10–15 Nm).

8. Install the hydraulic brake lines to the master cylinder and tighten and tighten the flare nuts to 13 ft. lbs. (17 Nm).

9. Position the reservoir, and install the connector screw.

10. Fill the reservoir with the specified brake fluid.

11. Connect the reservoir lead wire. Install the air cleaner case, if removed.

12. Bleed the air from the brake hydraulic system and check the brake pedal play.

Sidekick, Tracker and X-90

1.6L (VIN 0), (VIN U) Engine

1. Disconnect the reservoir lead wire.

2. Clean the outside of the reservoir and remove the fluid from the reservoir.

3. For vehicles without ABS, disconnect and plug the brake fluid lines at the master cylinder.

4. For vehicles with ABS, disconnect pipes 1 and 2 from the master cylinder and ABS actuator, and pipe 3 from the ABS actuator.

5. Remove the master cylinder mounting nuts and remove the master cylinder.

To install:

6. Install the new master cylinder. Tighten the mounting nuts to 9.5 ft. lbs. (13 Nm).

7. Install the hydraulic brake lines to the master cylinder and tighten and tighten the flare nuts to 11 ft. lbs. (15 Nm).

8. Reconnect the reservoir lead wire.

9. Fill the reservoir with the specified brake fluid.

10. Bleed the air from the brake hydraulic system and check the brake pedal play.

11. Road test the vehicle and verify proper brake system operation.

1.8L (VIN 2) Engine

1. Remove the master cylinder reservoir cap and siphon brake fluid from the vehicle.

2. Remove the brake fluid pressure proportioning valve switch electrical connection.

3. Disconnect the brake lines from the proportioning valve.

4. Remove the one bolt securing both the two way joint and proportioning valve to the retaining bracket.

5. Remove the master cylinder level switch electrical connector.

6. Disconnect the brake lines from the master cylinder.

7. Remove the attaching nuts and washers from the master cylinder.

8. Loosen the brake booster attaching nuts to allow enough room to remove the master cylinder.

9. Turn the master cylinder and brake booster slightly and remove the master cylinder from the vehicle.

To install:

10. Adjust the clearance between the booster piston and the primary piston.

11. Turn the brake booster slightly and install the master cylinder to the brake booster.

12. Torque the brake booster nuts to 10 ft. lbs. (13 Nm).

13. Install and torque the master cylinder attaching bolts to 10 ft. lbs. (13 Nm).

14. Install the brake lines to the master cylinder. Torque the flare nuts to 12 ft. lbs. (16 Nm).

15. Connect the master cylinder electrical connector.

16. Install the proportioning valve to the master cylinder and install the bolt. Torque the bolt to 8 ft. lbs. (10 Nm).

17. Connect and tighten the four brake lines to the proportioning valve to 12 ft. lbs. (16 Nm).

18. Connect the proportioning valve electrical connector.

19. Refill the brake fluid reservoir with DOT approved brake fluid.

20. Bleed the brake system.

Brake Caliper

REMOVAL AND INSTALLATION

1. Raise and safely support the vehicle.

2. Remove the wheels.

3. Disconnect and plug the brake line.

4. Remove the caliper mounting bolts (guide pins) and remove the caliper from the vehicle.

To install:

5. Install the caliper on the vehicle. Tighten the mounting bolts as follows:

• Samurai, Sidekick, Tracker/X-90 to 20 ft. lbs. (27 Nm).

• 1996–97 Sidekick Sport — bottom bolt to 37 ft. lbs.; top bolt to 42 ft. lbs. (58 Nm)

6. Connect the hydraulic brake line, using two new washers. Torque the union bolt to 17 ft. lbs. (23 Nm).

7. Replace the front wheels.

8. Lower the vehicle.

9. Fill the brake reservoir and bleed the hydraulic brake system.

Disc Brake Pads

REMOVAL AND INSTALLATION

Samurai

1. Raise and safely support the vehicle.

2. Remove the wheels.

3. Remove the caliper anti-rattle clip. Remove the two guide pin caps and remove the two guide pins with a 6mm hex wrench.

4. Lift off the brake caliper but, do not remove the brake hose.

NOTE: Do not allow the caliper to hang from the brake hose; support it with a wire or by the mounting bracket. Also, do not twist the brake hose or depress the brake pedal with the caliper removed from the rotor.

5. Remove the disc pads from the caliper.

To install:

6. Use a large pair of channel locks or a C-clamp to press the caliper piston back into its bore. Make sure the piston is not cocked and is pressed in straight.

7. Install the brake pads onto caliper assembly.

8. Install the brake caliper. Install the guide pins and caps. Replace the anti-rattle clip.

9. Install the front wheels.

10. Lower the vehicle.

NOTE: Be sure to pump the brake pedal several times to seat the new disc brake pads before attempting to move the vehicle.

Sidekick, Tracker and X-90

1. Siphon about ⅔ of the fluid out of the master cylinder.

2. Raise and safely support the vehicle.

3. Remove the wheels.

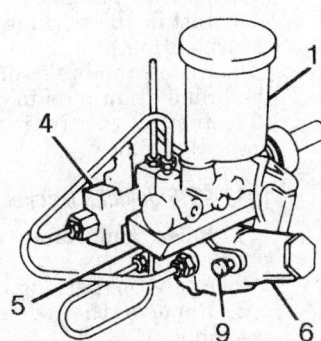

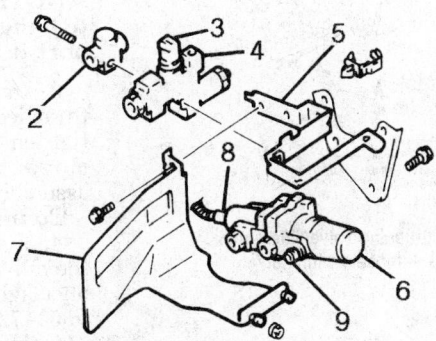

1. Master cylinder fluid reservoir
2. Two-way joint
3. Brake fluid pressure differential switch
4. Proportioning/differential valve
5. Retaining bracket

6. Pressure limiting valve
7. Heat protector shield
8. Brake fluid pressure limiting valve electrical connector
9. Bleeder valve

128291

Master cylinder mounting components — 1993 Sidekick, Tracker models

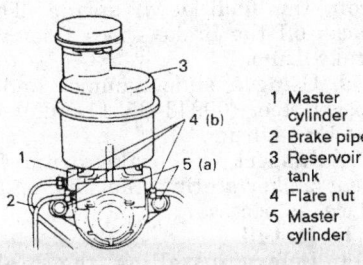

1 Master cylinder
2 Brake pipe
3 Reservoir tank
4 Flare nut
5 Master cylinder

177140

Master cylinder brake line connections —
1994–95 Sidekick, Tracker models

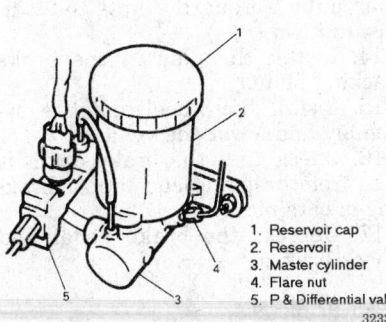

1. Reservoir cap
2. Reservoir
3. Master cylinder
4. Flare nut
5. P & Differential valve

323376

Master cylinder — 1996–97 Sidekick and X-90
models with 1.8L engines

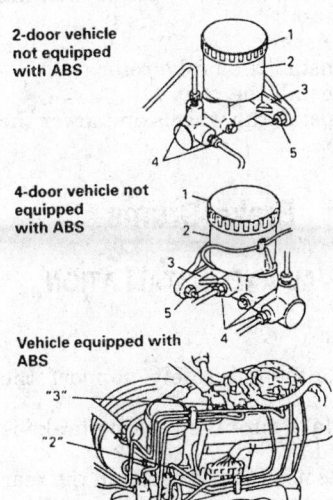

2-door vehicle not equipped with ABS

4-door vehicle not equipped with ABS

Vehicle equipped with ABS

"3"

"2"

"1"

1. Reservoir cap
2. Reservoir
3. Master cylinder
4. Flare nuts
5. Attaching nut

323325

Master cylinder and surrounding components —
1996–97 Sidekick, Tracker and X-90 models
with 1.6L (VIN 0), (VIN U) engines

4. Remove the brake caliper mounting bolts and remove the caliper from the mounting bracket.

—— WARNING ——
Do not allow the caliper to hang from the brake hose.

5. Support the caliper with a wire.
6. Remove the disc brake pads and any shims from the caliper mounting bracket.
To install:
7. Install the brake pads and any shims removed into the caliper mounting bracket.
8. Using a large pair of plies or a C-clamp compress the caliper piston back into the bore.
9. Install the caliper on the mounting bracket and install the mounting bolts. Tighten the mounting bolts to 20 ft. lbs. (27 Nm).
10. Install the front wheels and lower the vehicle.

—— CAUTION ——
Do not attempt to drive the vehicle until after the following step is performed.

11. Depress the brake pedal repeatedly until a firm pedal is obtained. Do not attempt to drive the vehicle unless a firm pedal is obtained.

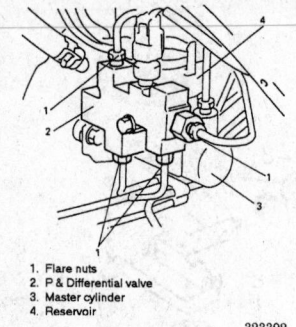

1. Flare nuts
2. P & Differential valve
3. Master cylinder
4. Reservoir

323392

Proportioning valve/differential valve — models with 1.8L engines

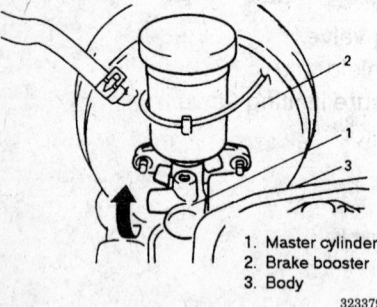

1. Master cylinder
2. Brake booster
3. Body

323379

Turning the brake booster and master cylinder — 1996–97 Sidekick and X-90 models with 1.8L engines

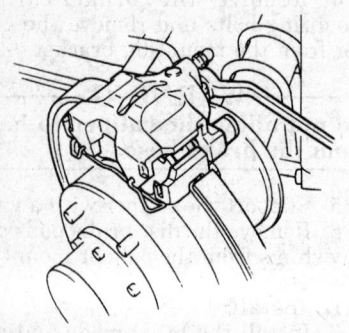

174764

Remove the brake line from the caliper — Samurai models

12. Check the fluid level in the master cylinder. Add fresh brake fluid, as necessary.
13. Road test the vehicle.

Brake Rotor

REMOVAL AND INSTALLATION

Samurai

1. Raise and safely support the vehicle. Remove the front wheels.

2. Remove the brake caliper mounting bracket, with the caliper and the brake line attached, move it out of the way and support it.

NOTE: Do not allow the caliper to hang from the brake hose. Support it by the mounting bracket.

3. Install two 8 mm bolts into the threaded holes in the brake rotor and tighten them evenly. This will remove the rotor from the hub assembly.
To install:
4. Install the new brake rotor on the hub and install the caliper assembly. Tighten the caliper bracket bolts to 51–72 ft. lbs.
5. Install the front wheels and tighten to 65 ft. lbs.
6. Lower the vehicle.

Sidekick, Tracker and X-90

1. Raise and safely support the vehicle.
2. Remove the wheel and tire assembly.
3. Remove the caliper mounting bracket and position the caliper out of the way. Support the caliper aside; do not let the caliper hang by the brake hose.
4. Remove the brake rotor from the wheel hub by installing two 8mm bolts into the rotor and tightening them evenly.
To install:
5. Install the brake rotor.
6. Install the caliper.
7. Install the wheels and lower the vehicle.

Brake Drums

REMOVAL AND INSTALLATION

Samurai

1. Raise and safely support the vehicle.
2. Make sure the parking brake is released.
3. Remove the wheels and the rear drum nuts from the vehicle.
4. To increase the clearance between the brake shoe and drum, remove the parking brake shoe lever return spring, and disconnect the parking brake cable joint from the shoe lever.
5. Remove the parking brake shoe lever stopper plate.
6. Remove the brake drum, using a slide hammer.
To install:
7. Install the parking brake shoe lever stopper plate.

8. Connect the brake cable joint to the parking brake shoe lever by using the joint pin. Insert the pin down from the top, and then install the clip into the joint pin hole.
9. Install the parking brake lever return spring.
10. Install the brake drum, tighten the brake drum nuts to 58 ft. lbs.
11. Install the rear wheels and lower the vehicle.

1993–95 Sidekick, Tracker

1. Raise and safely support the vehicle.
2. Apply the parking brake lever.
3. Remove the wheel and tire assembly.
4. Release the parking brake.
5. Remove the four brake drum nuts from the brake drum.
6. Release the parking brake cable at the brake lever.
7. Remove the parking brake lever cover screws and pull up on the brake lever cover. Loosen the parking brake cable locking nut.
8. Remove the backing plate plug to gain access to the hold down spring. Once the plug is removed, insert a screwdriver into the plug hose till its tip contacts the shoe hold down spring. Push the spring in to release the parking brake shoe lever from the hold down spring. This backs off the brake shoes from the brake drum.
9. Using a slide hammer and a special tool (09943–35511), pull off the brake drum.
10. Inspect the brake drum for wear and/or scoring. Machine or replace, as necessary.
To install:
11. Before installing the brake drum, maximize the brake shoe-to-drum clearance by putting a screwdriver between the rod and racket and pushing down the ratchet.
12. Put the brake shoe hod down spring back to its original position.
13. Install the brake drum with the four nuts. Torque the nuts to 58 ft. lbs. (80 Nm).
14. Install the plug in the brake backing plate.
15. Install the wheel and tire assembly and lower the vehicle.
16. Check that the brake drum is free from dragging and proper braking is obtained.
17. Depress the brake pedal and check brakes.

1996–97 Models

1. Raise and safely support the vehicle.
2. Remove the rear wheel(s).

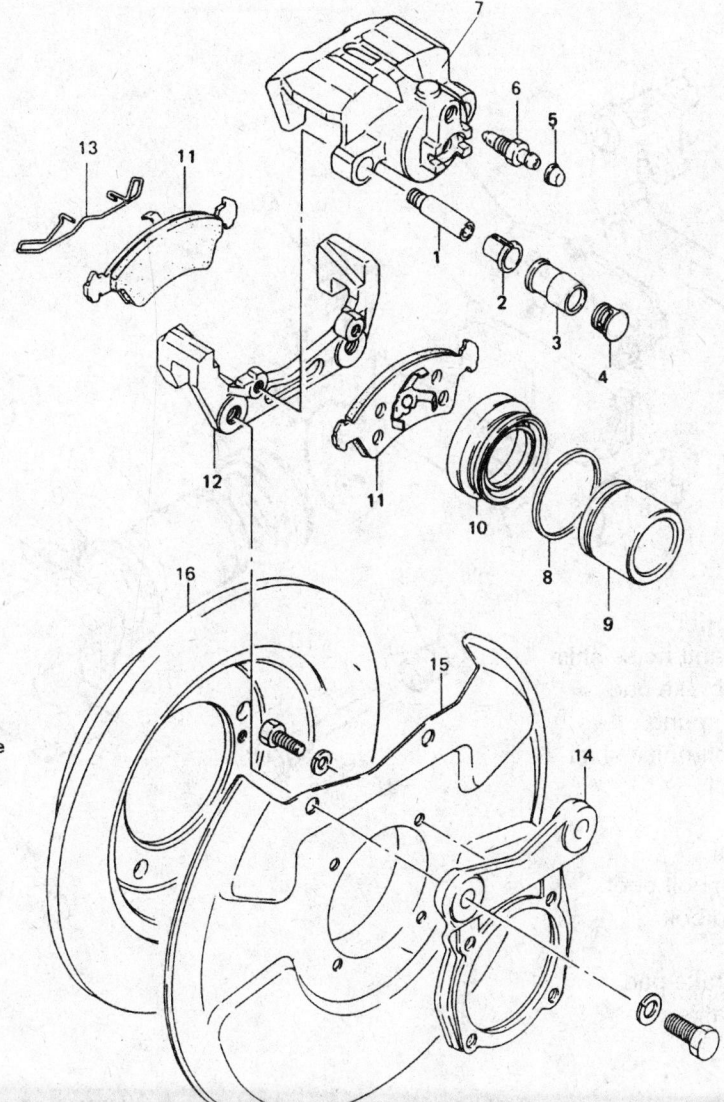

1 Caliper guide pin
2 Caliper guide pin sleeve
3 Guide pin boot
4 Guide pin cap
5 Bleeder plug cap
6 Bleeder plug
7 Disc brake caliper
    (Disc brake cylinder)
8 Piston seal
9 Disc brake piston
10 Cylinder boot
11 Disc brake pad
12 Disc brake carrier
13 Caliper antirattle clip
14 Caliper holder
15 Dust cover
16 Brake disc

174763

Front disc brake assembly — Samurai models

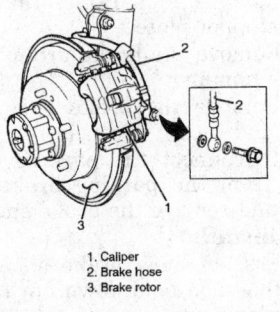

1. Caliper
2. Brake hose
3. Brake rotor

239922

**Brake line components —
Sidekick, Tracker models**

3. Release the parking brake.

4. Remove the parking brake lever cover screws and loosen the brake cable locking nut.

5. Install two 8mm bolts into the brake drum holes and uniformly tighten each bolt. Tighten each bolt until the brake drum is removed from the vehicle. If there is difficulty in removing the drum, insert a small tool through the hole in the rear of the backing plate, and hold the automatic adjusting lever away from the adjuster. Using another narrow, flat tool at the same time, reduce the brake shoe adjuster by turning the adjusting wheel.

To install:

6. Install the brake drum and pull the parking brake lever all the way up until a clicking sound can no longer be heard.

7. Verify that the rear wheels will not turn. If the rear wheels turn, adjust the parking brake cable as necessary.

8. Release the parking brake and remove the brake drum. Measure the diameter of the brake shoes. Outer diameter should be as follows:

• For 2 door models: 8.638±0.0012 inches (219±0.3mm)

• For 4 door models:9.980±0.0079 inches (253.5±0.2mm)

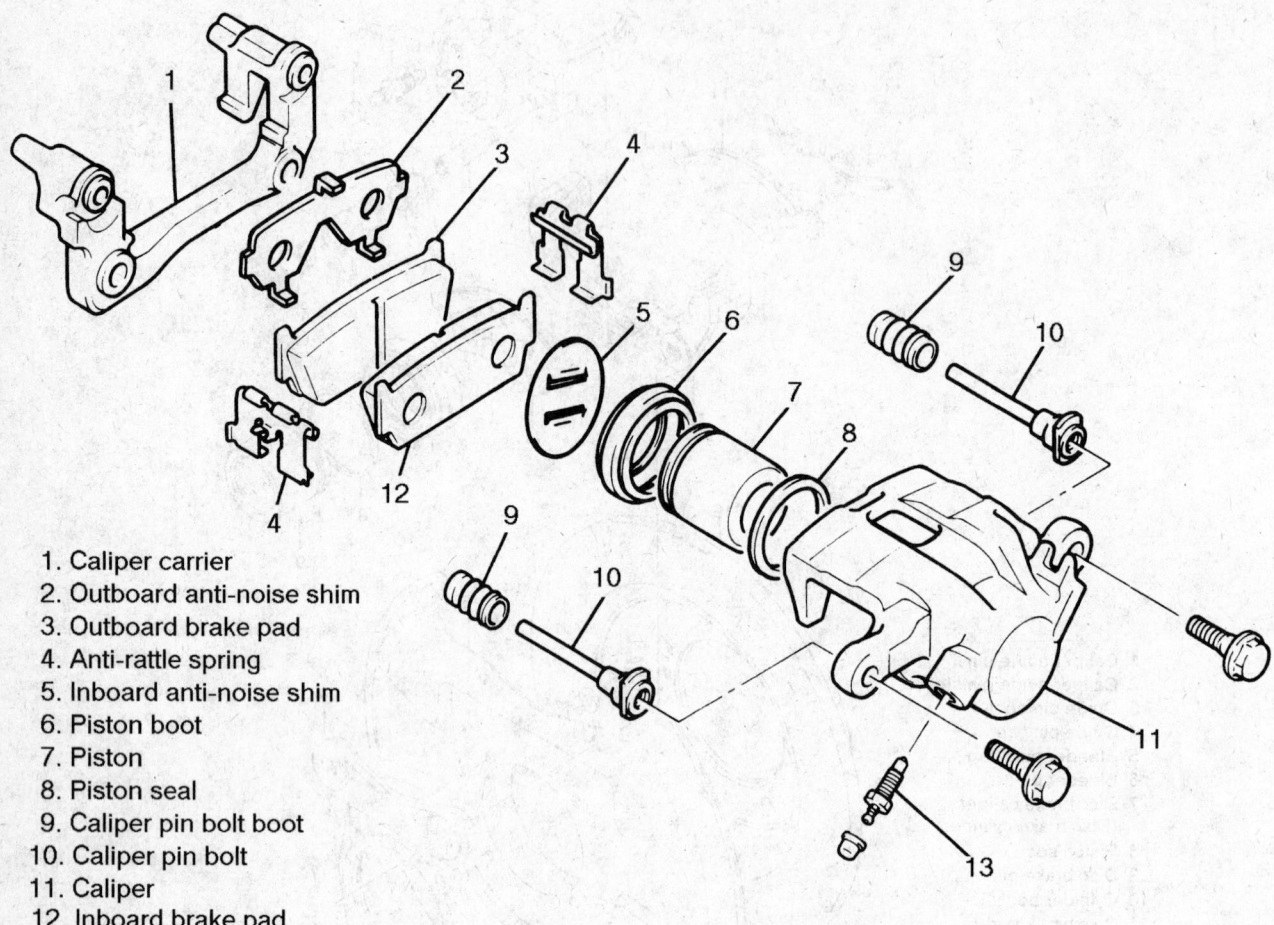

1. Caliper carrier
2. Outboard anti-noise shim
3. Outboard brake pad
4. Anti-rattle spring
5. Inboard anti-noise shim
6. Piston boot
7. Piston
8. Piston seal
9. Caliper pin bolt boot
10. Caliper pin bolt
11. Caliper
12. Inboard brake pad
13. Bleeder valve

329175

Front brake components — Sidekick, Tracker and X-90 models

9. If the brake shoe clearance is not correct, adjust the brake shoes until the clearance is correct.

10. Reinstall the brake drum, replace the wheel(s), and safely lower the vehicle.

11. Adjust the parking brake and install the cover with the two screws.

12. Road test the vehicle for proper brake operation.

Brake Shoes

REMOVAL AND INSTALLATION

Samurai

1. Raise and safely support the vehicle.

2. Make sure the parking brake is released.

3. Remove the wheels and the rear drum nuts from the vehicle.

4. To increase the clearance between the brake shoe and drum, remove the parking brake shoe lever return spring, and disconnect the parking brake cable joint from the shoe lever.

5. Remove the parking brake shoe lever stopper plate.

6. Remove the brake drum, using a slide hammer.

7. Remove the brake shoe hold-down and return springs.

8. Disconnect the parking brake cable from the parking brake shoe lever and remove the brake shoes.

To install:

9. Position new brake shoes. Install the shoe hold down springs by pushing them down in place, and turning the hold down pins.

10. Install the parking brake shoe lever stopper plate.

11. Connect the brake cable joint to the parking brake shoe lever by using

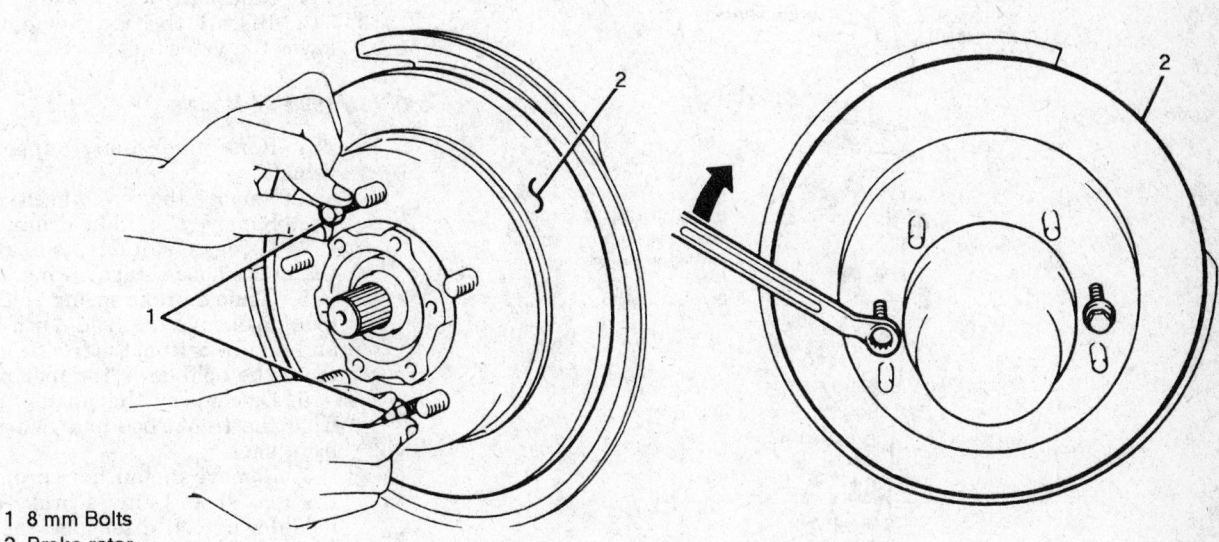

1 8 mm Bolts
2 Brake rotor

329171

Insert 8mm bolts into rotor — Sidekick, Tracker and X-90 models

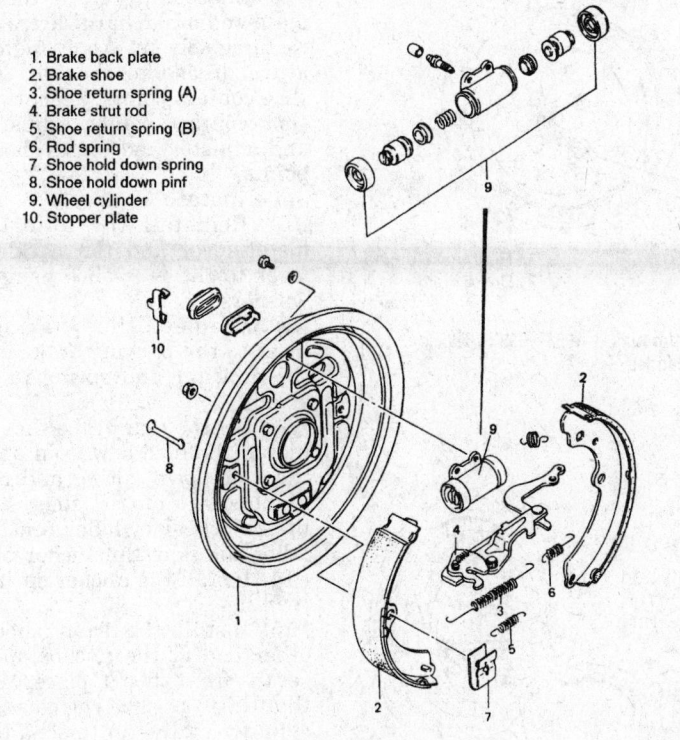

1. Brake back plate
2. Brake shoe
3. Shoe return spring (A)
4. Brake strut rod
5. Shoe return spring (B)
6. Rod spring
7. Shoe hold down spring
8. Shoe hold down pinf
9. Wheel cylinder
10. Stopper plate

174847

Rear drum brake assembly — Samurai models

the joint pin. Insert the pin down from the top, and then install the clip into the joint pin hole.

12. Install the parking brake lever return spring.

13. Install the brake drum, tighten the brake drum nuts to 58 ft. lbs.

14. Install the rear wheels and lower the vehicle.

1993–95 Sidekick, Tracker

1. Raise and safely support the vehicle.

2. Remove the rear wheels from the vehicle. Remove the rear brake drums.

3. Remove the brake shoe hold-down springs by rotating the pin.

4. Remove the return springs and anti-rattle springs.

5. Remove the brake lever and ratchet. Remove the parking brake rod and upper return springs.

6. Remove the brake shoes.

To install:

7. Install the brake shoes and install the parking brake lever and cable to the assembly.

8. Install the upper return springs and parking brake rod. Install the brake lever and ratchet.

9. Install the return springs and anti-rattle springs.

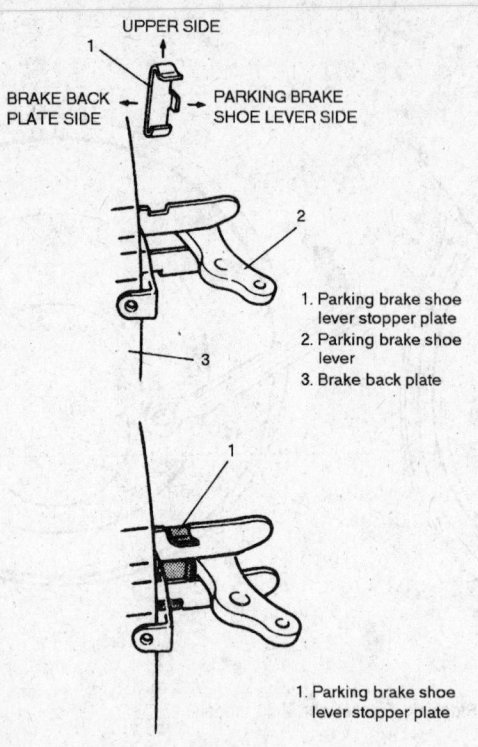

UPPER SIDE

BRAKE BACK PLATE SIDE ← → PARKING BRAKE SHOE LEVER SIDE

1. Parking brake shoe lever stopper plate
2. Parking brake shoe lever
3. Brake back plate

1. Parking brake shoe lever stopper plate

174848

Parking brake shoe lever assembly — Samurai models

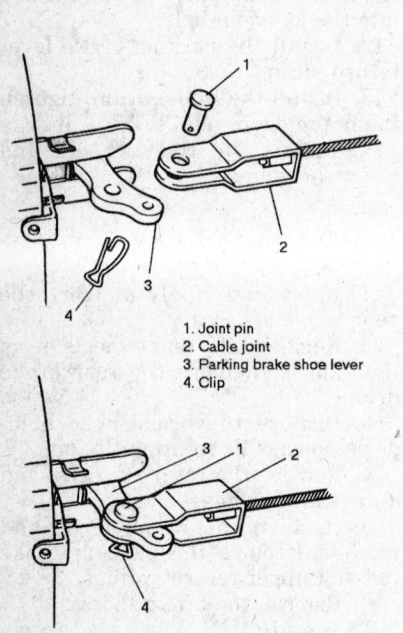

1. Joint pin
2. Cable joint
3. Parking brake shoe lever
4. Clip

174870

Brake cable joint assembly — Samurai models

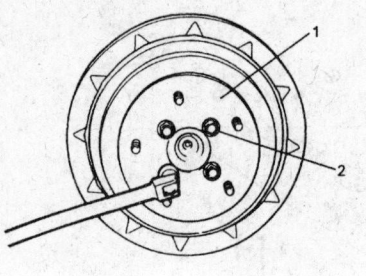

1. Rear brake drum
2. Brake drum nuts

323246

Brake drum nuts — 1993–1995 Sidekick, Tracker models

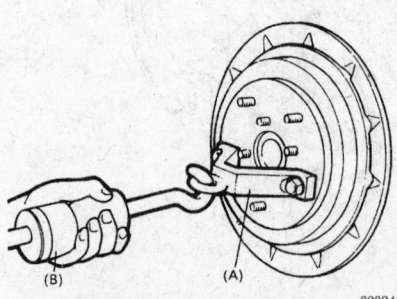

323248

Removing the brake drum with a puller — 1993–1995 Sidekick, Tracker models

10. Install the brake shoe hold-down springs and install the rear brake drums.

11. Adjust the rear brakes.

12. Install the rear wheels and lower the vehicle.

1996–97 Models

1. Raise and safely support the vehicle.

2. Remove the rear wheel(s).

3. Remove the brake drum.

4. Using a suitable tool, remove the brake shoe return spring.

5. Using a brake spring hold-down tool, disengage the hold-down spring and retainers from the front shoe. Remove the hold-down retainer pin.

6. Disconnect the anchor spring from the front shoe and remove the front shoe.

7. Remove the anchor spring from the rear shoe. Using a brake spring hold-down tool, disengage the hold-down spring and retainers from the rear shoe. Remove the hold-down pin.

8. Disengage the parking brake lever from the parking brake cable and remove the rear shoe.

9. Remove the C-washer, the automatic adjuster lever and spring, the C-washer, and the parking brake lever from the rear shoe.

10. Thoroughly clean the backing plate and brake hardware with brake cleaning solvent. Apply high temperature grease to the backing plate shoe contact points, anchor plate and shoe contact points, adjusting bolt, and adjuster and brake shoe contact points.

To install:

11. Reinstall the automatic adjuster lever and the parking brake lever to the rear shoe using new C-washers.

12. Connect the parking brake lever to the parking brake cable. Set the adjuster and spring to the rear shoe.

13. Set the rear brake shoe in place, install the hold-down pin and install the hold-down spring and retainers. Make sure that the shoe is inserted in the wheel cylinder and that the other end is in the anchor plate.

14. Install the anchor spring to the rear shoe.

15. Install the front shoe to the other end of the anchor spring and set the front shoe in place. Make sure that the front shoe engages the wheel cylinder, adjuster mechanism and spring, and the anchor plate.

16. Reinstall the front brake shoe hold-down pin and secure with the hold-down spring and retainers using a suitable tool.

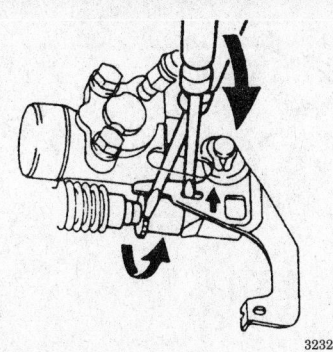

Reducing the adjuster to remove the brake drum — 1996–97 models

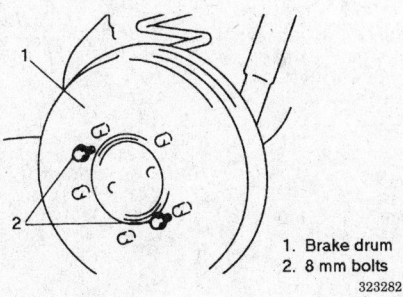

1. Brake drum
2. 8 mm bolts

Removing the brake drum with the two 8mm bolts — 1996–97 models

17. Install the return spring.

18. Install the brake drum and pull the parking brake lever all the way up until a clicking sound can no longer be heard.

19. Verify that the rear wheels will not turn. If the rear wheels turn, adjust the parking brake cable as necessary.

20. Release the parking brake and remove the brake drum. Measure the diameter of the brake shoes. Brake diameter should be as follows:

• For 2 door models: 8.638±0.0012 inches (219±0.3 mm)
• For 4 door models: 9.980±0.0079 inches (253.5±0.2 mm)

21. If the brake shoe clearance is not correct, adjust the brake shoes until the clearance is correct.

22. Reinstall the brake drum, replace the wheel(s), and safely lower the vehicle.

23. Road test the vehicle for proper brake operation.

Wheel Cylinder

REMOVAL AND INSTALLATION

1. Raise and safely support the vehicle.

2. Remove the rear wheels and brake drums from the vehicle.

3. Remove the rear brake shoes and disconnect the brake line from the rear of the wheel cylinder. Plug the brake lines.

4. Remove the 2 rear wheel cylinder mounting bolts. Remove the rear wheel cylinder.

To install:

5. Install the wheel cylinder and tighten the mounting bolts to 7 ft. lbs. (84 inch lbs.) (9.5 Nm).

6. Connect the brake line to the wheel cylinder and torque the flare nut to 10–13 ft. lbs. (14–17 Nm).

7. Install the rear brake shoes.

8. Install the rear brake drums and the rear wheels.

9. Lower the vehicle.

10. Bleed the brake system and check for any leaks when finished.

Parking Brake Cable

ADJUSTMENT

Samurai

1. Loosen the stopper nut and adjust the parking brake lever by loosening or tightening the self locking nut at the park brake lever. While turning the adjusting nut, hold the holding nut with a wrench to prevent the inner cable from twisting. After the adjustment, tighten the stopper nut against the pin.

2. The proper adjustment is when the park brake lever is within 3–8 notches, when the lever is pulled up at 44 lbs.

3. When the lever is applied by the aforementioned notches, the rear wheels should not move. The rear wheels should turn after the parking brake is released. Check the rear drum for dragging after adjustment.

Sidekick, Tracker and X-90

1. Ensure the following conditions are met before adjusting the parking brake cable:

a. No air is trapped in the brake system.

b. Brake pedal travel in within specifications.

c. The brake pedal has been depressed a few times with about 44 lbs. of force applied and released.

d. The parking brake lever has been applied up a few times with about 44 lbs. of force applied and released.

e. Rear brakes are not worn beyond the limits and the self adjusting mechanism operates properly.

2. Adjust the parking brake cable by loosening or tightening the self locking nut at the park brake lever.

3. The proper adjustment is when the parking brake lever is within 7–9 notches, when the lever is pulled up at 44 lbs.

4. Road test the vehicle and check the rear drum for dragging after adjustment.

REMOVAL AND INSTALLATION

Samurai

1. Disconnect the parking brake cable from the parking lever.

2. Raise and safely support the vehicle. Remove the rear wheels and brake drums from the vehicle.

3. Remove the rear brake shoes and disconnect the parking brake cable from the parking brake shoe lever.

4. Remove the cable from the brake backing plate by squeezing the parking brake cable stop ring.

5. Remove the parking brake cable from its chassis holding clamps.

To install:

6. Install the cable to the backing plate and to the brake shoe lever. Make sure to route the cable through its chassis holding clamps.

7. Install the brake shoes and install the rear brake drums.

8. Install the rear wheels and connect the cable to the parking brake lever. Now, lower the vehicle.

9. Make sure the parking brake is functioning properly and adjust the cable as necessary.

Sidekick, Tracker and X-90

1. Block the vehicle wheels and release the parking brake lever.

2. Remove the two screws and one clip securing the parking brake lever cover in place and remove the cover.

3. Remove the parking brake cable locknut and washer.

4. Disconnect the connecting rod from the equalizer.

5. Raise and safely support the vehicle. Remove the rear wheels and brake drums from the vehicle.

6. Remove the rear brake shoes.

7. Disconnect the parking brake cable from the parking brake shoe lever.

8. Remove the cable from the brake backing plate by squeezing the parking brake cable stop ring.

9. Remove the four bolts securing the cable housing to the underbody.

10. Remove the cable from the vehicle.

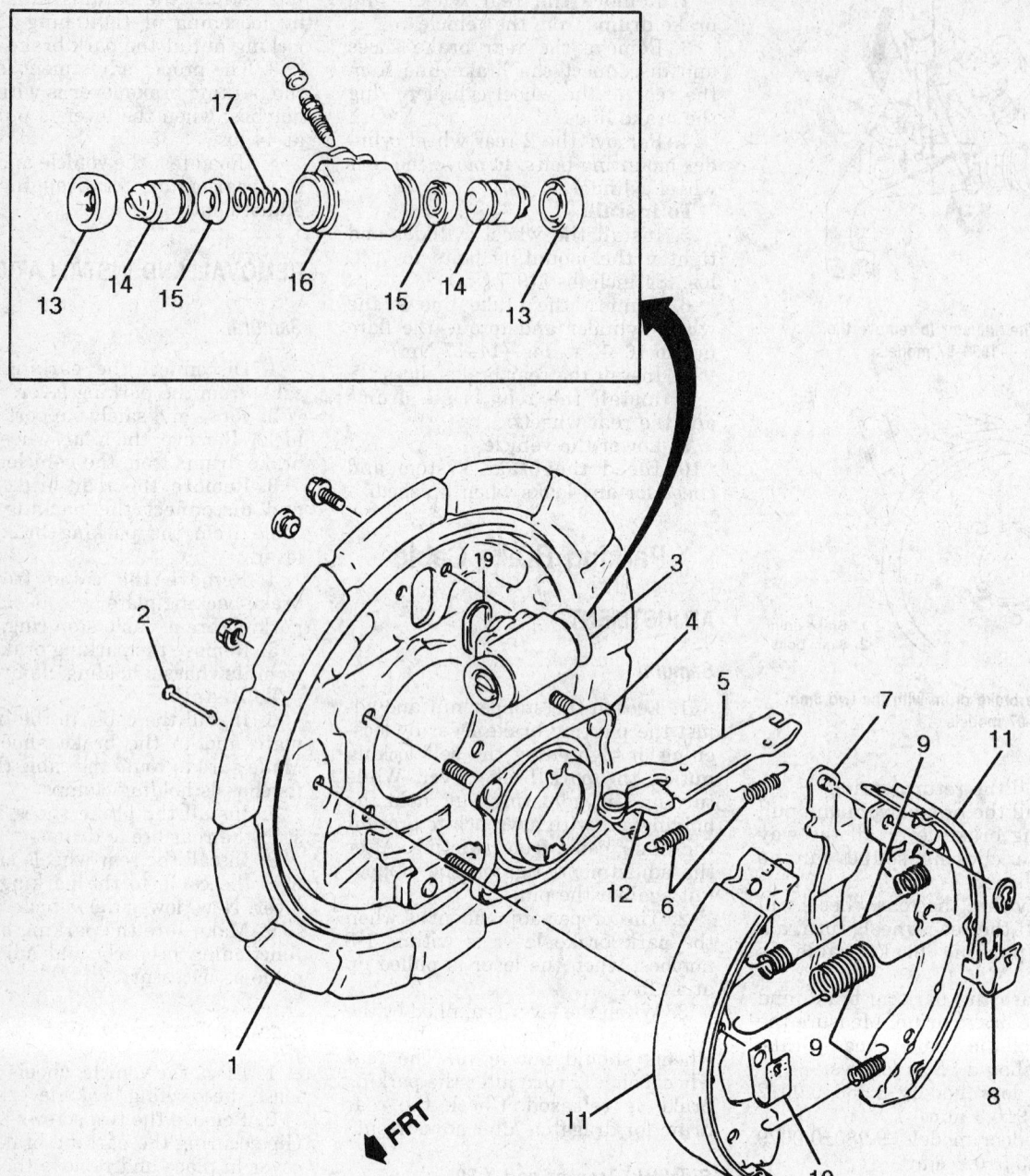

1. Backing plate
2. Hold down pin
3. Wheel cylinder
4. Wheel bearing retainer
5. Parking brake rod
6. Upper return springs
7. Parking brake lever
8. Hold down spring
9. Return springs
10. Retaining spring
11. Brake shoes
12. Ratchet
13. Wheel cylinder boot
14. Wheel cylinder piston
15. Wheel cylinder cup seal
16. Wheel cylinder body
17. Wheel cylinder spring

239956

Brake shoe components — 1993–95 Sidekick, Tracker models

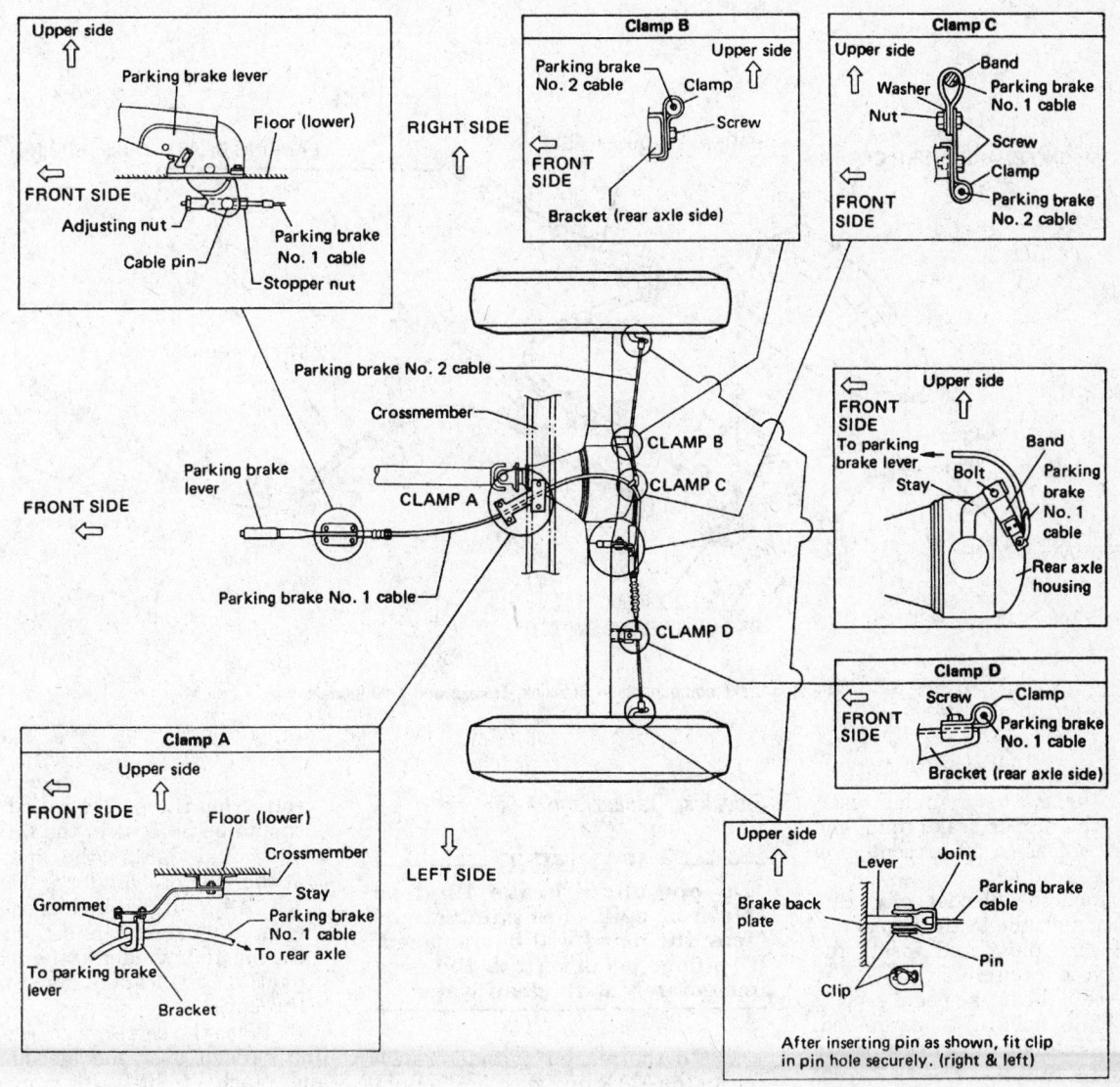

174933

Parking brake cable assembly — Samurai models

To install:

11. Position the cable in the vehicle and install the four underbody mounting bolts. Feed the cable up through the hole in the floor and seat the grommet in place.

12. Connect the cable housing to the backing plate and install the cable stopper ring. Connect the cable end to the brake shoe lever.

13. Install the brake shoes and install the rear brake drums.

14. Install the rear wheels.

15. Lower the vehicle.

16. Connect the parking brake cable to the equalizer.

17. Connect the equalizer to the connecting rod and install the washer and nut. DO NOT tighten the nut at this time.

18. Adjust the parking brake, and tighten the nut.

19. Install the brake lever cover and install the clip and two mounting bolts.

20. Apply the parking brake and remove the wheel blocks.

Brake Hydraulic System

BLEEDING

Samurai

NOTE: Air bleeding is required at four places: right and left front wheels, proportioning valve (P+B valve) and the left rear wheel cylinder.

──── **WARNING** ────

Make sure to tighten each bleeder plug to 7 ft. lbs. after completion. Check the entire brake circuit to make sure that no fluid leakage exists.

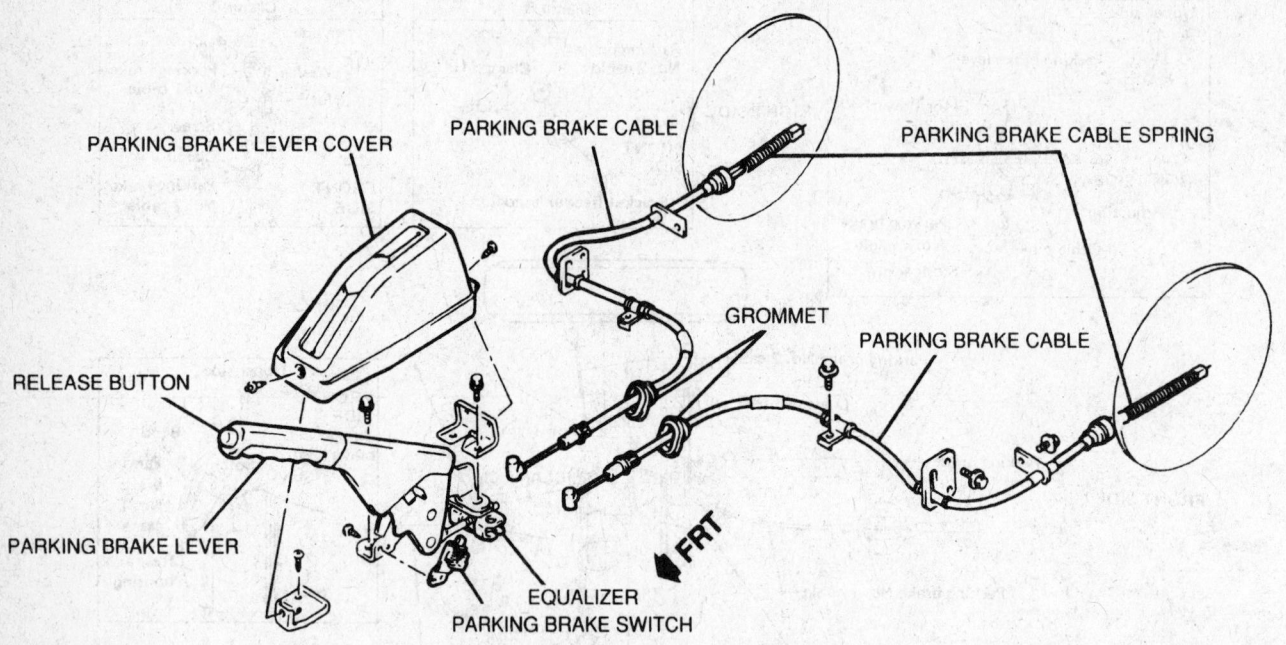

PARKING BRAKE LEVER COVER

PARKING BRAKE CABLE

PARKING BRAKE CABLE SPRING

RELEASE BUTTON

GROMMET

PARKING BRAKE CABLE

PARKING BRAKE LEVER

FRT

EQUALIZER
PARKING BRAKE SWITCH

333762

Parking brake components — Sidekick, Tracker and X-90 models

1. Fill the master cylinder reservoir with the proper brake fluid, and maintain at least a 50 percent level during the bleeding.

2. Remove the bleeder plug cap. Attach a vinyl tube to the wheel cylinder bleeder plug, and insert the open end into a container.

3. Depress the brake pedal several times. While holding it depressed, loosen the bleeder plug one-third to one-half of a turn.

4. When the fluid pressure is almost gone, retighten the bleeder plug.

5. Repeat the operation until all air bubbles are gone from the hydraulic line. Depress and hold the brake pedal, and tighten the bleeder plug.

6. Replace the bleeder plug cap.

7. After completing the bleeding procedure, apply fluid pressure to the pipe line while checking for leaks.

8. Fill the brake fluid reservoir to the specified full level.

9. Check brake pedal for a spongy feeling; if any evidence exists, repeat entire procedure.

Sidekick, Tracker and X-90

----------- WARNING -----------
Do not allow brake fluid to splash or spill onto painted surfaces; the paint will be damaged. If spillage occurs, flush the area immediately with clean water.
----

1. Fill the master cylinder reservoir to the MAX line with brake fluid and keep it at least half full throughout the bleeding procedure.

2. If the master cylinder has been removed or disconnected, it must be bled before any brake unit is bled. To bleed the master cylinder:
  a. Disconnect the front brake line from the master cylinder and allow fluid to flow from the front connector port.
  b. Reconnect the line to the master cylinder and tighten it until it is fluid tight.
  c. Have a helper press the brake pedal down one time and hold it down.
  d. Loosen the front brake line connection at the master cylinder. This will allow trapped air to escape, along with some fluid.
  e. Again tighten the line, release the pedal slowly and repeat the sequence (steps c-d-e) until only fluid

runs from the port. No air bubbles should be present in the fluid.
  f. Final tighten the line fitting at the master cylinder to 11 ft. lbs.
  g. After all the air has been bled from the front connection, bleed the master cylinder at the rear connection by repeating Steps a through e.

3. Place the correct size box-end or line wrench over the bleeder valve and attach a tight-fitting transparent hose over the bleeder. Allow the tube to hang submerged in a transparent container of clean brake fluid. The fluid must remain above the end of the hose at all times, otherwise the system will ingest air instead of fluid.

4. Have an assistant pump the brake pedal several times slowly and hold it down.

5. Slowly unscrew the bleeder valve (1/4-1/2 turn is usually enough). After the initial rush of air and fluid, have the assistant slowly release the brake pedal. When the pedal is released, tighten the bleeder.

6. Repeat Steps 4 and 5 until no air bubbles are seen in the hose or container. If air is constantly appearing after repeated bleeding, the system must be examined for the source of the leak or loose fitting.

7. If the entire system must be bled, begin with the right rear, then

the left front, left rear and right front brake in that order. After each brake is bled, check and top off the fluid level in the reservoir.

——————— **WARNING** ———————
Do not reuse brake fluid which has been bled from the brake system.

Rear Wheel Speed Sensor

REMOVAL AND INSTALLATION

All with 1.6L (VIN 0), (VIN U) Engine

NOTE: 1993–95 models are equipped with rear ABS only.

1. Disconnect the negative battery cable from the battery.
2. Raise and safely support the rear of the vehicle.
3. Disconnect the speed sensor cover.
4. Disconnect the electrical connector to the speed sensor.
5. Remove the speed sensor from the differential carrier by removing the bolt.
To install:
6. Check that no foreign material is attached to the sensor and/or rotor.
7. Install the speed sensor to the differential carrier and install the bolt. Torque the bolt to 16 ft. lbs. (21 Nm).
8. Connect the electrical connector to the speed sensor.
9. Install the speed sensor cover.

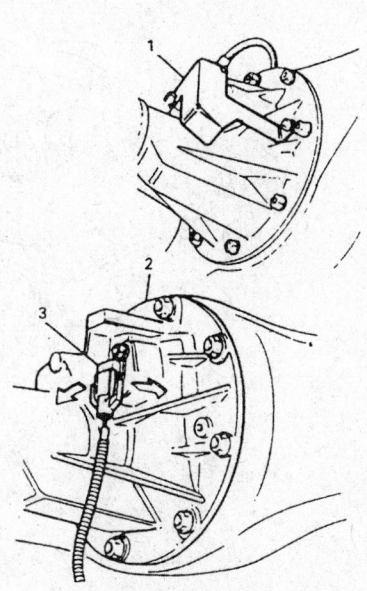

1. Rear wheel speed sensor cover
2. Rear differential carrier
3. Rear wheel speed sensor coupler

333391

Rear speed sensor — with 1.6L engines

10. Connect the negative battery cable to the battery.
11. Check the speed sensor signal.

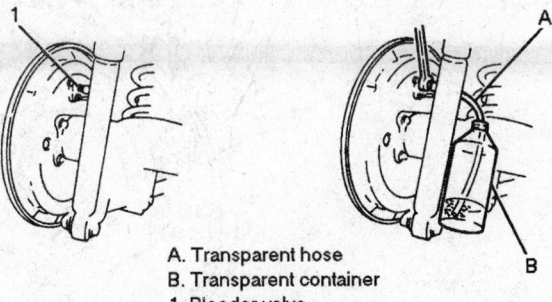

A. Transparent hose
B. Transparent container
1. Bleeder valve

245340

Bleeding brake wheel cylinders — Sidekick, Tracker and X-90 models

All with 1.8L (VIN 2) Engine

1. Disconnect the negative battery cable from the battery.
2. Raise and safely support the rear of the vehicle.
3. Disconnect the rear wheel sensor connector and detach the wire harness from the vehicle body and rear axle.
4. Remove the harness clamp bolts and remove the rear wheel sensor from the rear axle.
5. Remove the rear speed sensor bolt and remove the speed sensor from the knuckle.
To install:
6. Check that no foreign material is attached to the sensor and/or rotor.
7. Install the rear speed sensor with the bolt to the knuckle. Torque the bolt to 8 ft. lbs. (10 Nm).
8. Install the speed sensor harness to the rear axle with the bolts. Torque the bolts to 15 ft. lbs. (21 Nm).
9. Connect the speed sensor connector.
10. Lower the vehicle.
11. Connect the negative battery cable to the battery.
12. Check the speed sensor signal.

Front Wheel Speed Sensor

REMOVAL AND INSTALLATION

1. Disconnect the negative battery cable from the battery.
2. Raise and safely support the front of the vehicle.
3. Disconnect the speed sensor connector(s).
4. Remove the front speed sensor from the steering knuckle by removing the bolt.
To install:
5. Check that no foreign material is attached to the sensor and/or rotor.
6. Install the front speed sensor to the steering knuckle and install the bolt. Torque the bolt to 8 ft. lbs. (10 Nm).
7. Install the front speed sensor connector to the speed sensor.
8. Connect the negative battery cable to the battery.
9. Check the speed sensor signal.

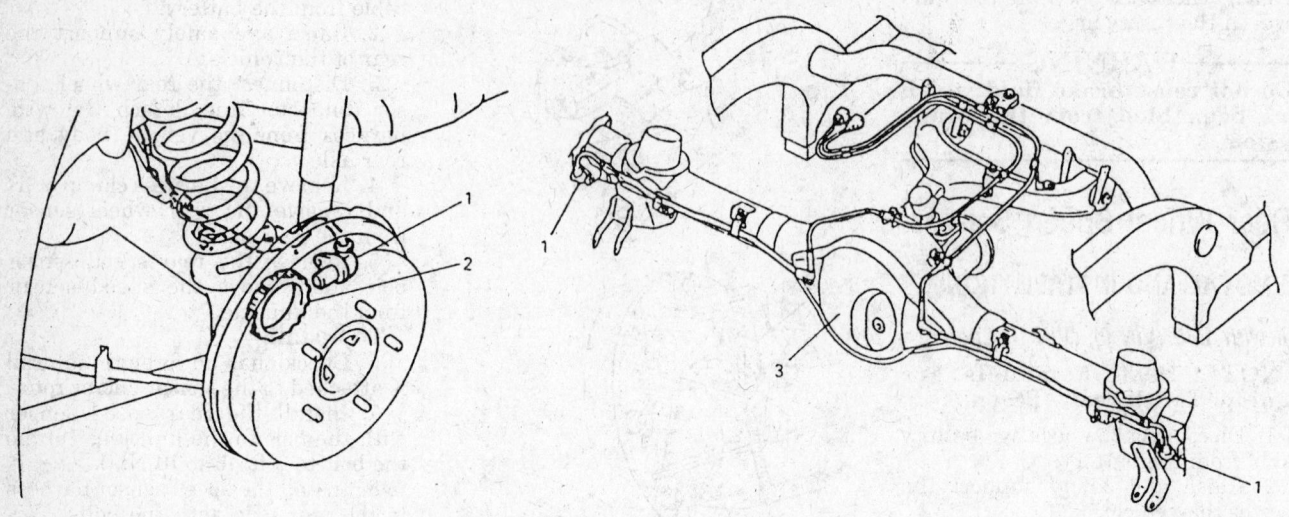

1. Rear wheel speed sensor
2. Sensor rotor
3. Rear axle housing

333370

Rear speed sensor — with 1.8L (VIN 2) engines

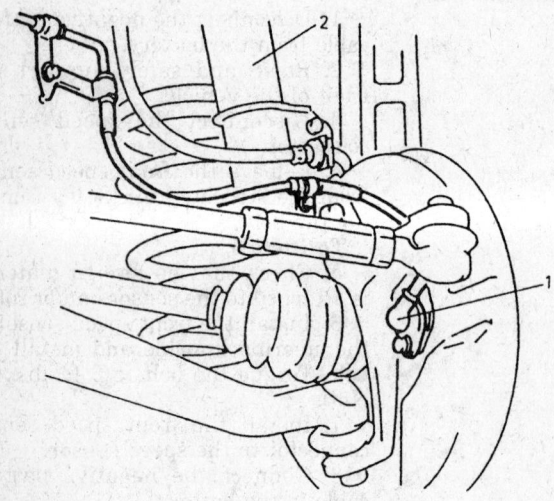

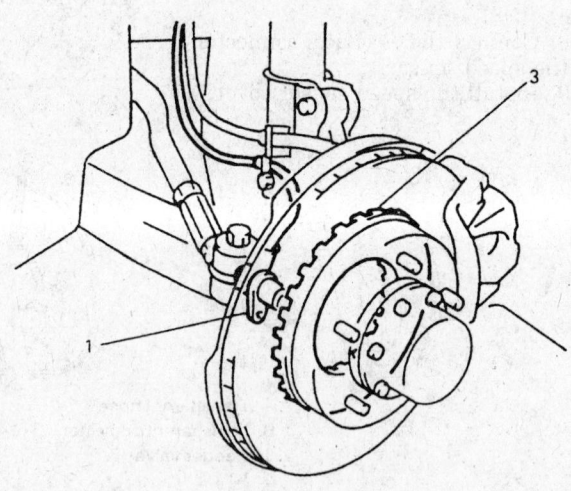

1. Left front wheel speed sensor
2. Left front strut
3. Sensor rotor

333371

Front speed sensor

FRONT SUSPENSION

Strut

REMOVAL AND INSTALLATION

Sidekick, Tracker and X-90

1. Raise and safely support the vehicle. Allow the front suspension to hang free.
2. Remove the front wheel.
3. Disconnect the E-clip mounting the brake hose and remove the brake hose from the strut bracket.
4. Support the control arm with a stand or floor jack.

─────── CAUTION ───────

The coil spring is under extreme pressure. Make sure control arm is firmly supported with a hydraulic jack before continuing with procedure. If this precaution is not observed, serious bodily injury may result.

5. Matchmark the strut lower bracket to the steering knuckle. Remove the strut bracket to steering knuckle bolts.

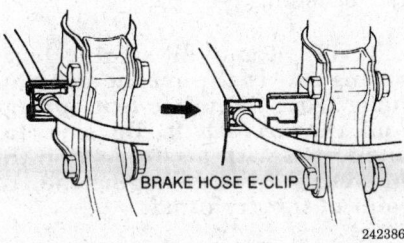

Removing the brake line E-clip — Sidekick, Tracker and X-90 models

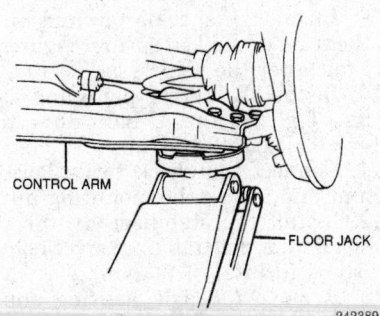

Positioning of the floor jack when supporting the lower control arm — Sidekick, Tracker and X-90 models

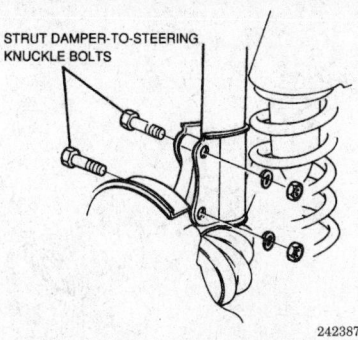

Strut lower mounting nuts and bolts — Sidekick, Tracker and X-90 models

6. Remove the upper strut mounting nuts. Hold the strut to prevent it from falling. Remove the strut assembly from the vehicle.
To install:
7. Install the strut assembly and tighten the upper strut mounting nuts to 18 ft. lbs. (24 Nm).
8. Install the strut bracket to steering knuckle bolts and line up the matchmarks and tighten the nuts and bolts to 66 ft. lbs. (90 Nm).
9. Remove the stand or jack.
10. Connect the brake hose to the strut bracket using the E-clip.
11. Install the front wheels and lower the vehicle.
12. Check the vehicle's alignment.

Shock Absorber

REMOVAL AND INSTALLATION

Samurai

1. Raise and safely support the vehicle.
2. Support the axle assembly and remove the upper shock absorber mounting nut.
3. Remove the lower shock absorber mounting nut and remove the shock absorber.

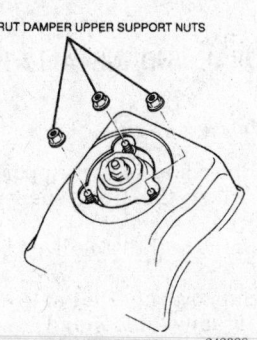

Upper strut mounting nuts — Sidekick, Tracker and X-90 models

To install:
4. Install the shock absorber and install the lower mounting nut to the bolt.
5. Torque the upper mounting nut to 20 ft. lbs. (27 Nm), and the lower nut to 33 ft. lbs. (45 Nm).
6. Remove the axle support and lower the vehicle.

Coil Spring

REMOVAL AND INSTALLATION

Sidekick, Tracker and X-90

1. Raise and safely support the vehicle with jackstands under the frame.
2. Remove the front wheel.
3. Remove the engine skid plate, if equipped.
4. Support the lower control arm with a floor jack.
5. Remove the 3 nuts and bolts from the control arm, separating the control arm from the ball joint.
6. Disconnect the stabilizer bar from the control arm.

─────── CAUTION ───────

The coil spring is under pressure. Make sure control arm is firmly supported with a hydraulic jack before continuing with procedure. If this precaution is not observed, serious bodily injury may result.

7. Remove the lower strut mounting bracket bolts and disconnect the strut bracket from the steering knuckle.
8. Lower the jack until all tension is removed from the coil spring. Remove the coil spring from the vehicle.
To install:
9. Install the coil spring onto the control arm and slowly raise the jack.

NOTE: The bottom of the spring has a larger diameter than the top. Make sure that the spring is installed correctly.

10. Install the strut bracket to steering knuckle mounting nuts and bolts and tighten to 66 ft. lbs.
11. Connect the stabilizer link to the control arm. Torque the nut to 21 ft. lbs. (28 Nm).
12. Install the 3 nuts and bolts connecting the control arm and ball joint. Torque the nuts to 63 ft. lbs. (85 Nm).
13. Install the engine skid plate. Torque the bolts to 40 ft. lbs. (54 Nm).
14. Install the front wheels, lower the vehicle.

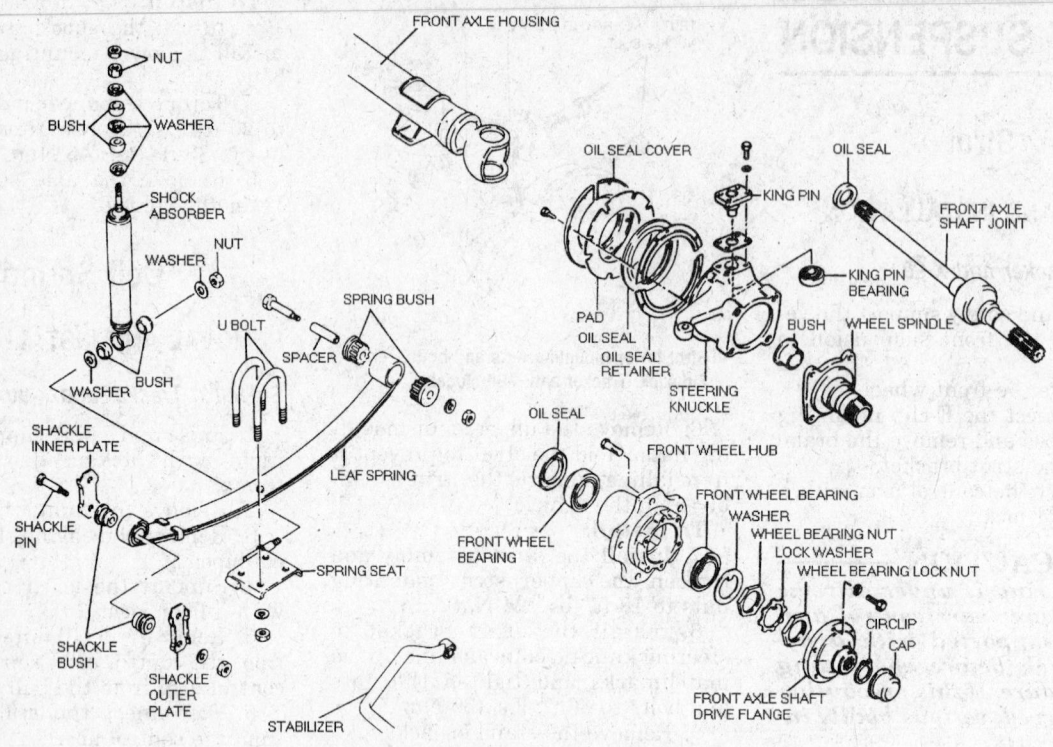

Front suspension assembly exploded view — Samurai models

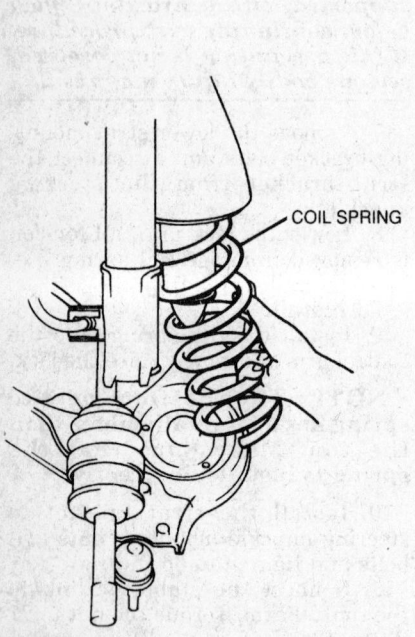

Removing the coil spring — Sidekick, Tracker and X-90 models

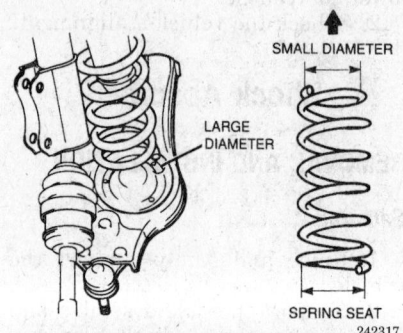

Proper positioning of the coil spring — Sidekick, Tracker and X-90 models

Leaf Spring

REMOVAL AND INSTALLATION

Samurai

1. Raise and safely support the vehicle. Allow the front suspension to hang free.
2. Remove the stabilizer bar pivot bolt.
3. Support the front axle assembly with an adjustable stand.
4. Remove the leaf spring to spring plate mounting U-bolts.
5. Remove the lower shock mounting bolt and remove the spring plate.

6. Remove the shackle pin and the nut from the front of the leaf spring.
7. Disconnect the leaf spring bolt at the rear of the spring and remove the leaf spring.

NOTE: Removal of the leaf spring causes the axle housing to hang from the other leaf spring and the driveshaft. Be sure to support it with a safety stand to prevent it from damaging the U-joint of the driveshaft.

To install:
8. Install the leaf spring and leaf spring rear bolt. Tighten the rear mounting bolt after the vehicle is lowered.
9. Install the shackle pin and nut to the front of the leaf spring. Tighten the nut after the vehicle is lowered.
10. Install the spring plate and U-bolts. Tighten the U-bolt nuts to 44–58 ft. lbs. (60–80 Nm).
11. Connect the shock to its lower mount and install the mounting nut.
12. Install the stabilizer bar pivot bolt. Tighten the stabilizer pivot bolts to 51–65 ft. lbs. (70–90 Nm).
13. Remove the axle housing support, lower the vehicle.
14. After the vehicle is on the ground, tighten the rear mounting bolt to 44–61 ft. lbs. (60–85 Nm).

Tighten the front shackle nut to 22–40 ft. lbs. (30–55 Nm).

Lower Ball Joints

REMOVAL AND INSTALLATION

Sidekick, Tracker and X-90

1. Raise and safely support the vehicle with jackstands under the frame.
2. Remove the front wheel.
3. Remove the engine skid plate, if equipped.
4. Support the lower control arm with a floor jack.
5. Remove the 3 nuts and bolts from the control arm, separating the control arm from the ball joint.
6. Disconnect the stabilizer bar from the control arm.

—— CAUTION ——

The coil spring is under pressure. Make sure control arm is firmly supported with a hydraulic jack before continuing with procedure. If this precaution is not observed, serious bodily injury may result.

7. Remove the lower strut mounting bracket bolts and disconnect the strut bracket from the steering knuckle.
8. Lower the jack until all tension is removed from the coil spring. Remove the coil spring from the vehicle.
9. Unbolt the ball joint from the control arm.
To install:
10. Position the new ball joint on the control arm. Install the bolts and tighten the nuts to 63 ft. lbs. (85 Nm).
11. Install the coil spring onto the control arm and slowly raise the jack.

NOTE: The bottom of the spring has a larger diameter than the top. Make sure that the spring is installed correctly.

12. Install the strut bracket to steering knuckle mounting nuts and bolts and tighten to 66 ft. lbs.
13. Connect the stabilizer link to the control arm. Torque the nut to 21 ft. lbs. (28 Nm).
14. Install the 3 nuts and bolts connecting the control arm and ball joint. Torque the nuts to 63 ft. lbs. (85 Nm).
15. Install the engine skid plate. Torque the bolts to 40 ft. lbs. (54 Nm).

16. Install the front wheels, lower the vehicle.

Lower Control Arms

REMOVAL AND INSTALLATION

Sidekick, Tracker and X-90

1. Raise and safely support the vehicle with jackstands under the frame.
2. Remove the front wheel.
3. Remove the engine skid plate, if equipped.
4. Support the lower control arm with a floor jack.
5. Remove the 3 nuts and bolts from the control arm, separating the control arm from the ball joint.
6. Disconnect the stabilizer bar from the control arm.

—— CAUTION ——

The coil spring is under pressure. Make sure control arm is firmly supported with a hydraulic jack before continuing with procedure. If this precaution is not observed, serious bodily injury may result.

7. Remove the lower strut mounting bracket bolts and disconnect the strut bracket from the steering knuckle.
8. Lower the jack until all tension is removed from the coil spring. Remove the coil spring from the vehicle.
9. Remove the front and rear through-bolts from the control arm and remove the control arm from the vehicle.
To install:
10. Position the new control arm. Install the bolts and tighten the nuts to 74 ft. lbs. (100 Nm).
11. Install the coil spring onto the control arm and slowly raise the jack.

NOTE: The bottom of the spring has a larger diameter than the top. Make sure that the spring is installed correctly.

12. Install the strut bracket to steering knuckle mounting nuts and bolts and tighten to 66 ft. lbs.
13. Connect the stabilizer link to the control arm. Torque the nut to 21 ft. lbs. (28 Nm).
14. Install the 3 nuts and bolts connecting the control arm and ball joint. Torque the nuts to 63 ft. lbs. (85 Nm).

15. Install the engine skid plate. Torque the bolts to 40 ft. lbs. (54 Nm).
16. Install the front wheels, lower the vehicle.

Sway Bar

REMOVAL AND INSTALLATION

Samurai

1. Raise and safely support the vehicle.
2. Disconnect the stabilizer bar pivot bolts.
3. Remove the stabilizer bar mount bushing bracket bolts and nuts.
4. Remove the stabilizer bar.
To install:
5. Install the new stabilizer bar, using new bushings.
6. Torque the stabilizer bar pivot bolts to 51–65 ft. lbs. (69–88 Nm), and the mounting bracket nuts to 13–20 ft. lbs. (17–27 Nm).
7. Lower the vehicle.

Sidekick, Tracker and X-90

1. Raise and safely support the vehicle.
2. Remove the engine skid plate, if equipped.
3. Disconnect the left and the right sway links from the lower control arm.
4. Remove the sway bar mount bushing bracket bolts and nuts.
5. Remove the sway bar.
6. Remove the sway bar link nuts from the sway bar and remove the sway bar links.
To install:
7. Install the sway bar links to the sway bar and install the nuts. Torque the nuts to 37 ft. lbs. (50 Nm).
8. Install the sway bar, using new bushings. Install the sway bar bracket bushing bolts and nuts and connect the sway bar link kits. DO NOT tighten any mounting hardware until all nuts and bolts are in place and the sway bar is centered in the vehicle.
9. Torque the sway bar link nuts to the lower control arm to 21 ft. lbs. (28 Nm).
10. Torque the sway bar bracket bolts to 18 ft. lbs. (25 Nm) and the sway bar bracket nuts to 62 ft. lbs. (85 Nm).
11. Install the engine skid plate, if removed.
12. Lower the vehicle.

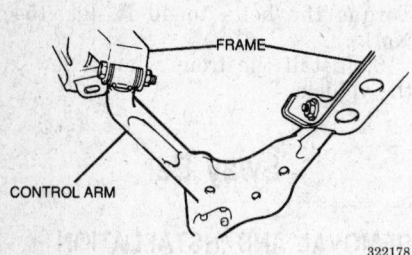

Lower control arm mounting bolts — Sidekick, Tracker and X-90 models

Front Wheel Bearings and Locking Hubs

ADJUSTMENT

Samurai

NOTE: Wheel bearing starting preload should be within 2.2–6.6 lb. (13 kg).

1. Raise the front end of the vehicle, and remove the front wheels.
2. Remove the wheel bearing locknut and lock washer.

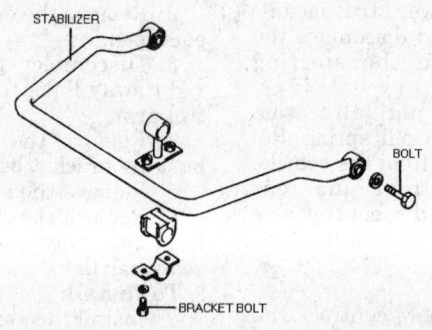

Sway bar (exploded view) — Samurai models

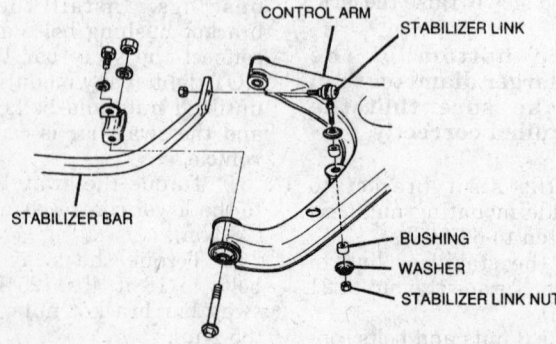

Exploded view of the sway bar mounting components — Sidekick, Tracker and X-90 models

3. Tighten the bearing nut to 57 ft. lbs. (80 Nm) while spinning the hub by hand.
4. Loosen the nut to completely, and then tighten it to the proper torque of 7–10 ft. lbs. (10–15 Nm). This will give the proper bearing preload.
5. Insert the lock washer after adjustment, and tighten the locknut to 44–65 ft. lbs. (60–90 Nm). Bend a part of the lock washer toward the bearing nut (on body side), and another toward the locknut (outside). this will lock the two nuts.
6. Recheck the bearing starting preload, making sure it is within specifications.

Sidekick, Tracker and X-90

The front wheel bearings are a cartridge type design and cannot be adjusted. To check for a loose wheel bearing, proceed as follows:
1. Raise and safely support the vehicle.
2. Remove the front wheel.
3. Compress the caliper piston to free the caliper assembly.
4. Using a suitable dial indicator, measure the thrust play.
5. Push and pull the brake rotor by hand. If rotor movement exceeds

0.002 in. (0.05mm), replace the wheel bearings.
6. Install the wheel and lower the vehicle.
7. Apply the brakes several times before moving the vehicle, to seat the caliper piston.

REMOVAL AND INSTALLATION

Samurai

1. Raise and safely support the vehicle. Remove the front wheels.
2. Remove the caliper mounting bolts and move the caliper out of position with the brake line attached.

NOTE: Do not allow the caliper to hang on the brake hose. Support it by the mounting bracket.

3. Install 2 (8mm) bolts into the threaded holes and tighten evenly. This will remove the rotor from the hub assembly.
4. If equipped with a locking hub:
   a. Thread a bolt into the axle shaft and pull the axle shaft out towards you. Remove the snapring.
   b. Remove the locking hub body assembly.
5. If not equipped with a locking hub, remove the front axle shaft cap and the circlip. Remove the drive flange from the steering knuckle.
6. Straighten the bent lock washer and remove the hub nut and washer.
7. Remove the front wheel hub and bearing from the spindle.
8. Remove the oil seal and race from the wheel hub.
9. Clean and inspect the hub and bearing seats. Repack the wheel bearings. Install the new bearing, race and grease seal in the same position.

To install:
10. Install the front wheel hub and bearing to the knuckle.
11. Tighten the bearing nut to 57 ft. lbs. (80 Nm) while spinning the hub by hand.
12. Loosen the nut to 0 ft. lbs., and then tighten it to the proper torque of 7–10 ft. lbs. (10–15 Nm). This will give the proper bearing preload.
13. Insert the lock washer after adjustment, and tighten the locknut to 44–65 ft. lbs. (60–90 Nm). Bend a part of the lock washer toward the bearing nut (on body side), and another toward the locknut (outside). this will lock the two nuts.
14. Install the drive flange to the steering knuckle.
15. If not equipped with free wheeling hub, install the front axle cap and circlip. Connect the rotor to the hub assembly.

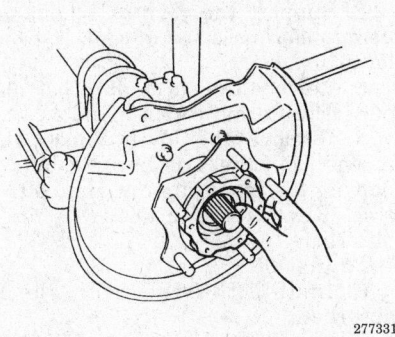

Removing the front wheel hub — Samurai models

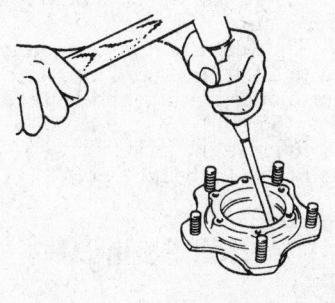

Removing the oil seal and outer race of wheel bearing — Samurai models

16. If equipped with a locking hub:

a. Install a new gasket onto the locking hub assembly.

b. Install the locking hub body assembly to the wheel hub flange and tighten the hub body bolts to 18 ft. lbs. (24 Nm).

c. Thread a bolt into the axle shaft and pull the axle shaft out towards you. Install the snapring and remove the bolt from the axle shaft.

d. Install a new gasket in the manual locking hub cover.

e. Before installing the locking hub cover, make sure of the following:

• The selector knob is in the FREE position.

• The clutch should be lifted (retracted) towards the cover. The clutch must be positioned properly to ensure proper hub operations.

• The gasket is centered and installed correctly.

f. Install the locking hub cover and tighten the bolts to 9 ft. lbs. (13 Nm).

NOTE: The mark on the hub knob must be facing the FREE position.

g. Check that the hub assembly is working correctly. If there are

problems with the operation, Remove the hub cover and repeat steps 9–12.

17. Install the brake rotor.

18. Place the brake caliper into position and install the caliper mounting bolts.

19. If removed, reconnect the locking hub assembly and install the front wheels.

20. Lower the vehicle.

21. Pump the brake pedal several times to seat the front brake pads. Road test the vehicle and verify proper operation.

Sidekick, Tracker and X-90

1. Raise and safely support the vehicle.

2. Remove the wheels.

3. If equipped with 4WD and automatic locking hubs:

a. Unscrew the automatic hub cover and remove the cover and O-ring.

b. Remove the hub assembly mounting bolts and remove the hub assembly.

4. If equipped with 4WD and manual locking hubs:

a. Remove the six manual hub cover mounting bolts and remove he manual hub cover and gasket.

b. Remove the six manual hub mounting bolts.

c. Remove the manual hub and O-ring.

5. If equipped with 2WD, remove the hub cap from the hub.

6. Remove the caliper mounting bracket and position the caliper out of the way. Support the caliper.

7. Remove the brake rotor from the wheel hub. Remove the front wheel bearing lock plate screws, lock plate and washer.

8. Remove the wheel bearing lock nut and washer and remove the

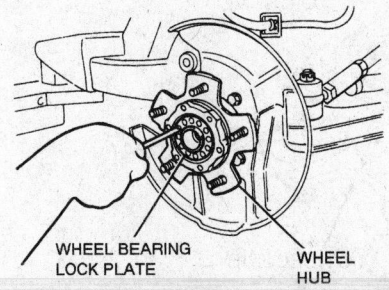

Removing the lockplate screws — Sidekick, Tracker and X-90 models

wheel hub complete with bearings and seals.

NOTE: If wheel hub can not be removed by hand, use special tool J37781 with a J2619–01 for Geo and 09943–35511 with 09930–30102 for Suzuki. These tools are a hub remover and sliding hammer.

9. If equipped with ABS brakes, remove the sensor rotor from the wheel hub.

NOTE: Pull out the sensor rotor from the wheel hub gradually and evenly. Pulling it out partially may cause it to deform.

10. Remove the inner bearing grease seal.

11. Remove the snapring and remove the inner bearing.

12. Clean and inspect the hub and bearing seats.

To install:

13. Install the inner race, wheel bearing snapring and seal in the hub. Apply lithium grease to the lip portion of the oil seal.

14. If equipped with ABS brakes, install the sensor rotor.

15. Install the hub on the spindle and install the outer bearing, lock nut, washer, and lock plate. Torque the lock nut to 152 ft. lbs. (210 Nm).

16. Install the four lock plate mounting screws. Torque the screws to 12 inch lbs. (1.5 Nm).

17. Install the brake rotor.

18. Install the caliper and torque the bolts to 61 ft. lbs. (85 Nm)..

19. If equipped with 2WD, install the locking hub cap.

20. If equipped with 4WD and automatic locking hubs:

a. Install a new O-ring on the hub assembly.

b. Install the hub assembly onto the wheel flange. Make sure the tab on the hub fits into the notch on the spindle.

c. Install the six mounting bolts and tighten to 18 ft. lbs. (25 Nm).

d. Install a new O-ring on the hub cover.

e. Install the hub cover.

21. If equipped with 4WD and manual locking hubs:

a. Install a new O-ring on the manual hub body.

b. Install the manual hub onto the wheel flange and install the six mounting bolts. Tighten the mounting bolts to 18 ft. lbs. (25 Nm).

c. Install a new gasket on the manual hub cover.

d. Install the hub cover on the hub. The lever must be in the

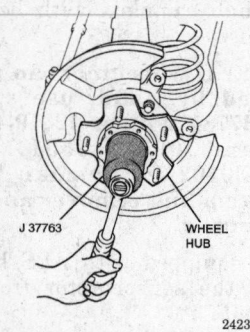

Removing the wheel bearing locknut — Sidekick, Tracker and X-90 models

242337

WHEEL HUB

242339

Removing the inner wheel bearing snapring — Sidekick, Tracker and X-90 models

FREE position with the clutch pulled out toward the cover.

e. Install the six mounting bolts. Tighten the bolts to 106 inch lbs. (12 Nm).

22. Install the wheels and lower the vehicle.

REAR SUSPENSION

Shock Absorber

REMOVAL AND INSTALLATION

Samurai

1. Raise and safely support the vehicle. Support the rear axle housing.
2. Remove the upper and lower shock absorber mounting bolts.
3. Remove the rear shock absorber.
   To install:
4. Install the the rear shock absorber.
5. Tighten the upper mounting bolt to 25 ft. lbs (34 Nm), and the lower mounting bolt to 23–39 ft. lbs. (31–53 Nm).
6. Remove the rear axle support and lower the vehicle.

Sidekick, Tracker and X-90

1. Raise and safely support the vehicle.

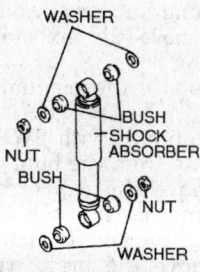

WASHER

BUSH
SHOCK
ABSORBER
NUT
BUSH

NUT

WASHER

179717

Rear shock absorber (exploded view) — Samurai models

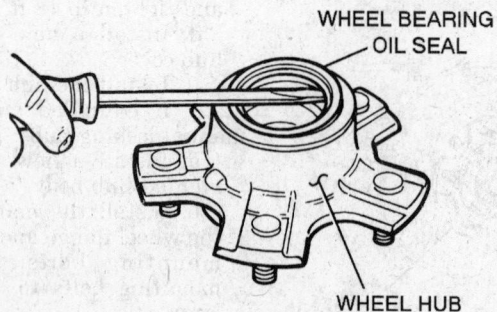

WHEEL BEARING
OIL SEAL

WHEEL HUB

242338

Removing the inner grease seal — Sidekick, Tracker and X-90 models

2. Support the rear axle housing using floor jack to prevent it from lowering.
3. Remove the shock absorber upper lock nut and retaining nut.
4. Remove the lower shock absorber from the axle housing by removing the mounting nut and bolt.
5. Remove the rear shock absorber.
   To install:
6. Install the the rear shock absorber.
7. Install the lower mounting nut and bolt. The lower bolt should head should point in toward the center of the vehicle.
8. Install the upper retaining nut and locknut.
9. Tighten the upper mounting nuts to 21 ft. lbs. (29 Nm), and the lower mounting nuts and bolts to 63 ft. lbs. (85 Nm).
10. Remove the jack.
11. Lower the vehicle.

Coil Spring

REMOVAL AND INSTALLATION

Sidekick, Tracker and X-90

1. Raise and safely support the vehicle. Remove the rear wheels.
2. Support the rear axle housing using a floor jack.
3. Remove the shock absorber lower mounting nut and bolt. Remove the bolts and nuts from both shock absorbers.
4. Disconnect the parking brake cables from the hangers on the lower control arms.
5. Lower the rear axle housing so the coil spring can be removed.

——————— **CAUTION** ———————
Take care to avoid stretching the brake hose!

6. Remove the coil spring from the vehicle.
   To install:
7. Install the coil spring to the spring seat and raise the axle housing. Make sure the spring is seated correctly.
8. Install the lower shock absorber mounting bolts and nuts but do not tighten.
9. Connect the parking brake cable hangers and install the rear wheels.
10. Lower the vehicle and tighten the lower shock absorber nuts to 64 ft. lbs. (86 Nm)

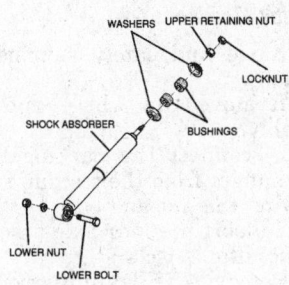

Rear shock absorber upper mounting components — Sidekick, Tracker and X-90 models

243630

Rear shock absorber components — Sidekick, Tracker and X-90 models

243629

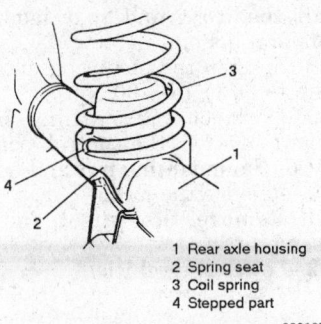

1 Rear axle housing
2 Spring seat
3 Coil spring
4 Stepped part

329187

Rear coil spring placement — Sidekick, Tracker and X-90 models

Leaf Spring

REMOVAL AND INSTALLATION

Samurai

1. Raise and safely support the vehicle. Remove the rear wheels.
2. Safely support the axle housing and disconnect the shocks.

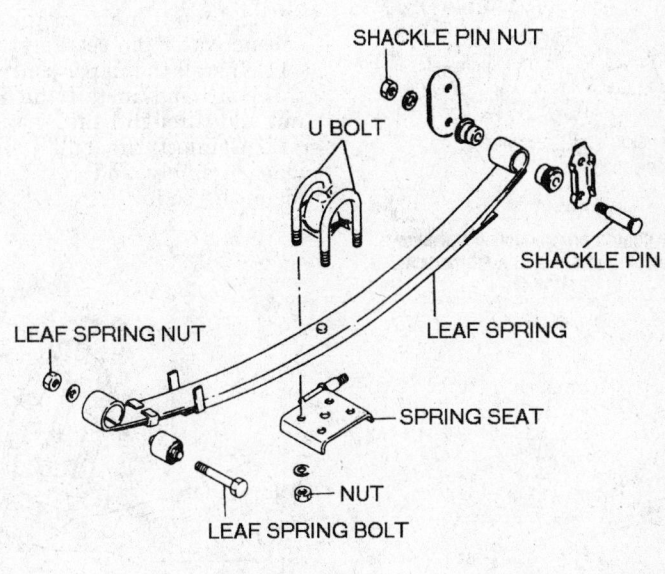

Rear leaf spring (exploded view) — Samurai models

179713

3. Disconnect the stabilizer bar from the shackle plate under the leaf spring.

NOTE: Do not let the axle housing hang on the brake hoses or lines.

4. Remove the rear axle housing U-bolt nuts and remove the bolts.
5. Raise the rear axle housing to release spring tension and remove the shackle plate.
6. Support the leaf spring and disconnect the rear leaf spring mounting bolts.
7. Remove the leaf spring assembly.
   To install:
8. Install the leaf spring and shackle assembly and tighten the mounting bolts after the vehicle is lowered.
9. Install the U-bolt nuts and tighten to 50 ft. lbs. (68 Nm).
10. Connect the stabilizer bar under the shackle plate. Install the shocks and tighten the mounting bolts to 30 ft. lbs. (40 Nm).
11. Install the rear wheels, remove the axle housing support, and lower the vehicle.
12. With the vehicle on the ground, tighten the rear leaf spring mounting bolts to 44–61 ft. lbs. (60–85 Nm).

Tighten the rear leaf spring shackle pin nuts to 22–40 ft. lbs. (30–55 Nm).

Upper Control Arms

REMOVAL AND INSTALLATION

Sidekick, Tracker and X-90

1. Raise and safely support the vehicle.
2. Remove the tire and wheel assembly.
3. If necessary, remove bracket from upper control arm.
4. Position a jack under the axle.
5. Remove the four bolts securing the ball joint to the differential carrier.
6. Remove the upper control arm-to-body mounting nut and through bolt.
7. Remove the control arm.
8. If the control arm is being replaced, remove the ball joint assembly from the control arm.
9. To remove the ball joint bracket:
   a. Remove the cotter pin from ball joint.
   b. Remove ball joint castle nut.
   c. Using a bearing puller, remove bracket from ball joint stud bolt.

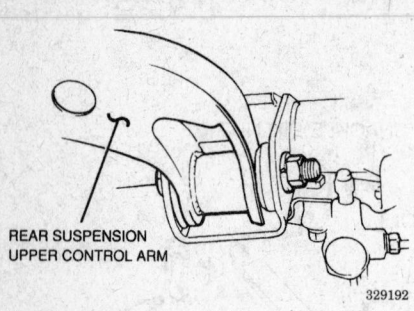

REAR SUSPENSION
UPPER CONTROL ARM

329192

Rear upper control arm mounting nut and bolt — Sidekick, Tracker and X-90 models

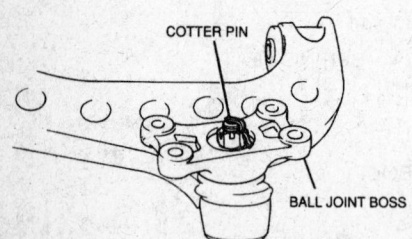

COTTER PIN

BALL JOINT BOSS

329193

Rear upper control arm — Sidekick, Tracker and X-90 models

d. Remove ball joint boot set ring and ball joint boot.
To install:
10. To install ball joint:
a. When installing ball joint boot, be sure to fit boot set wire into ring groove in boot.
b. After installing ball joint bracket to ball joint stud bolt, tighten castle nut till split pin hole

in stud bolt aligns with slot in nut but within range of specified torque. Specified torque is 42 ft. lbs. (58 Nm).
c. Install new cotter pin and bend cotter pin securely.
11. Install the upper control arm to the body and install the mounting nut and through bolts.
12. Connect the ball joint to the axle assembly and install the four mounting bolts.

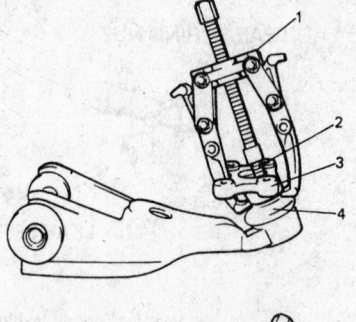

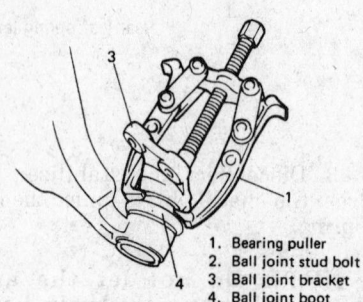

1. Bearing puller
2. Ball joint stud bolt
3. Ball joint bracket
4. Ball joint boot

329194

Bearing puller used to remove the bracket from the ball stud — Sidekick, Tracker and X-90 models

13. Tighten the ball joint-to-axle bolts to 37 ft. lbs. (50 Nm) and the control arm mounting bolts to 66 ft. lbs. (90 Nm).
14. If necessary, connect the upper bracket to upper control arm. Torque bolts to 13–20 ft. lb (18–28 Nm).
15. Remove the jack.
16. Install the tire and wheel assembly.
17. Lower the vehicle.

Trailing Arms

REMOVAL AND INSTALLATION

Sidekick, Tracker, X-90

1. Raise and safely support the vehicle.
2. Remove the wheel and tire assembly.
3. Disconnect the parking brake cable hanger from the trailing rod by removing the nut and bolt.
4. Support the rear axle assembly with a suitable jack.
5. Remove the trailing rod rear nut, bolt and washer.
6. Remove the trailing rod front nut, bolt and washer and remove the trailing rod.
To Install:
7. Install the control arm and install the front and rear mounting nuts and bolts.
8. Tighten the mounting nuts and bolts to 66 ft. lbs. (90 Nm).
9. Connect the parking brake cable hanger to the control arm and install the mounting nut and bolt.
10. Remove the jack.
11. Remove the wheel and tire assembly.
12. Lower the vehicle.

GENERAL MOTORS and ISUZU

GM-Astro • Blazer • Bravada • Jimmy • Safari • S10/15 Pick-Up • Sonoma • Typhoon

Isuzu-Hombre

FIRING ORDERS

NOTE: To avoid confusion, always replace spark plug wires one at a time.

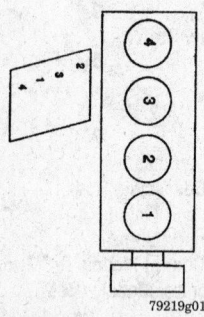

2.2L (VIN W and 4)
Engine
Engine Firing Order:
1–3–4–2
Distributorless Ignition
System

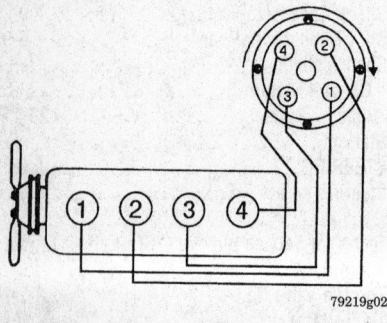

2.5L (VIN A) Engine
Engine Firing Order: 1–3–4–2
Distributor Rotation: Clockwise

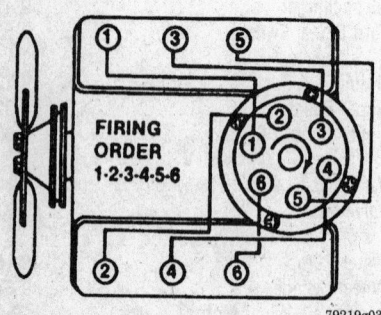

2.8L (VIN R) Engine
Engine Firing Order: 1–2–3–4–5–6
Distributor Rotation: Clockwise

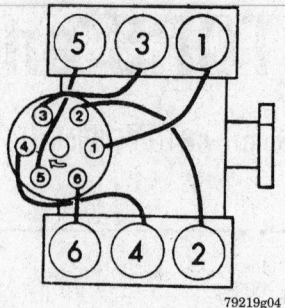

4.3L (VIN W, X, and Z) Engine
Engine Firing Order: 1–6–5–4–3–2
Distributor Rotation: Clockwise

ENGINE ELECTRICAL

NOTE: Disconnecting the negative battery cable on some vehicles may interfere with the functions of the on board computer systems and may require the computer to undergo a relearning process, once the negative battery cable is reconnected.

—— WARNING ——
NEVER disconnect the negative battery cable with the ignition ON. Removing power from the computer control module with the ignition ON may destroy the module.

Distributor

Most distributor equipped vehicles covered by this manual utilize a commonly recognizable distributor ignition system. However, some 1995–97 4.3L (VIN W) engines utilize a distributor-like ignition system in which the commonly recognized distributor is replaced by a High Voltage Switch (HVS). The HVS is mounted in a fixed position so the Powertrain Control Module (PCM) or Vehicle Control Module (VCM) may control ignition timing using input from a camshaft position sensor (usually integral to the HVS). Because of this, proper rotor alignment when removing and installing the HVS is even more critical

than with the conventional distributor systems.

REMOVAL

1. Disconnect the negative battery cable.
2. Remove all necessary components in order to gain access to the distributor assembly. On most V-type engines it will be necessary to remove the air cleaner assembly for access.
3. Tag and disengage the distributor electrical connectors. If equipped, disconnect the vacuum line.
4. Remove the distributor cap. If necessary, tag and disconnect the spark plug wires.
5. Matchmark the rotor and the distributor body. Matchmark the distributor assembly and the engine block.

NOTE: Although it is only necessary to matchmark the distributor position for ease of installation, it may be advisable to first position the engine at TDC. If the distributor is being removed for further engine repair which may involve rotating the crankshaft (such as to align timing marks), setting the engine to TDC at this time will keep the distributor matchmarks valid during installation.

6. Remove the distributor holddown bolt. Carefully remove the distributor.

NOTE: As the distributor is removed from the engine, the rotor will turn counterclockwise. Observe and mark the finish position of the rotor. When reinstalling, position the rotor at the last mark and set the distributor into the engine. As the distributor drops into place, the rotor should turn to its original position, providing the engine crankshaft has not been rotated with the distributor out.

INSTALLATION

Timing Not Disturbed

NOTE: To ensure correct ignition timing if the engine has not been disturbed, the distributor must be installed with the rotor in the same position as when removed.

1. Align the rotor to the last mark made and install the distributor in the engine.

NOTE: On vehicles not equipped with the HVS system, if the distributor shaft cannot drop into the engine, remove the distributor and use a small prytool through the mounting hole to rotate the oil pump driveshaft until it can align with the distributor gear.

2. As the distributor is fully seated, the rotor should turn and end up at the first mark made. Ensure the distributor and oil pump rod are fully engaged.

NOTE: When the distributor is fully seated on the engine, make sure the rotor is properly aligned with the first timing mark. If equipped with the HVS system, if the rotor dos not align with the mark, the gear teeth of the HVS and camshaft have meshed 1 or more teeth out of time and the HVS procedure for timing disturbed should be used to assure proper installation.

3. Reconnect the distributor cap and wires. If applicable, connect the vacuum line to the distributor.

4. Tighten the distributor holddown bolt, then install the air cleaner or duct, if removed for access.

NOTE: If a check engine light is illuminated after HVS installation (Astro/Safari) and a Diagnostic Trouble Code (DTC) 1345 is found, the HVS has been installed incorrectly. Proceed to the timing disturbed installation procedure for the 1995–97 Astro/Safari Van.

5. Check and adjust the ignition timing.

Timing Disturbed

Except 1995–97 HVS Ignition Systems

1. Set the engine to TDC: Remove the No. 1 spark plug. Place a finger over the spark plug hole and rotate the engine in the normal direction of rotation slowly, until compression is felt.

2. Align the timing mark on the crankshaft pulley to the **0** on the engine timing indicator by rotating the engine in the same direction slowly. The engine is now set on No. 1 TDC.

NOTE: An alternate method may be used to assure the engine is at TDC if the valve cover is removed. Watch the rocker arms for the No. 1 cylinder as the en-

gine is turned. If the valves move as the crankshaft timing marker approaches the scale, the No. 1 cylinder is on its exhaust stroke. If the valves remain closed as the timing mark approaches the scale, then the No. 1 cylinder is approaching TDC of the compression stroke.**

3. Install the distributor to the engine so the rotor is pointing to the No. 1 spark plug tower on the distributor cap once the distributor is fully seated in the engine.

4. Install the distributor cap, spark plug, wiring and connectors. If applicable, connect the vacuum line to the distributor.

5. If removed for access, install the air duct and/or air cleaner assembly, as applicable.

6. Check and adjust ignition timing.

1995–97 HVS Ignition Systems

1. Set the engine to TDC: Remove the No. 1 spark plug. Place a finger over the spark plug hole and rotate the engine in the normal direction of rotation slowly, until compression is felt.

2. Align the timing mark on the crankshaft pulley to the **0** on the engine timing indicator by rotating the engine in the same direction slowly. The engine is now set on No. 1 TDC.

NOTE: An alternate method may be used to assure the engine is at TDC if the valve cover is removed. Watch the rocker arms for the No. 1 cylinder as the engine is turned. If the valves move as the crankshaft timing marker approaches the scale, the No. 1 cylinder is on its exhaust stroke. If the valves remain closed as the timing mark approaches the scale, then the No. 1 cylinder is approaching TDC of the compression stroke.

3. Remove the HVS cap screws and cap to expose the rotor.

4. Align the pre-drilled indent hole in the HVS driven gear with the arrow cast into the upper portion of the shaft housing. The rotor should point to the cap hold-down mount nearest the flat side of the housing.

5. Using a long prytool, align the oil pump driveshaft in the engine to the mating drive tab in the HVS.

6. Guide the HVS into place, making sure the locating slot in the HVS base fits over the dowel pin in the intake manifold.

7. Once the HVS is FULLY SEATED, the rotor tip should be aligned with the pointer cast into the HVS base. this pointer will have a "6" cast into it, indicating the HVS component is designed for use in a 6-cylinder engine. If the rotor tip does not align within a few degrees of the pointer, the gear mesh between the HVS and camshaft is likely off by a tooth or more. If so, repeat the procedure again to achieve proper alignment.

8. Install the cap and mounting screws.

9. Install the HVS mounting clamp and tighten to 20 ft. lbs. (27 Nm).

10. Engage the 3-wire camshaft position sensor connector to the base of the HVS assembly.

11. Connect the spark plug and coil leads to the HVS cap. If a check engine light is illuminated and a Diagnostic Trouble Code (DTC) 1345 is found, either the HVS has been installed incorrectly or an incorrect HVS assembly has been installed.

Ignition Timing

The timing may be adjusted only on those that are equipped with a conventional distributor ignition. If equipped with the distributorless Electronic Ignition (EI) system, the control module sets timing and makes all necessary spark changes. On these systems, the crankshaft position sensor is mounted in a fixed position, therefore not allowing for adjustment. The HVS system found on some 1995–97 vehicles is also not adjustable.

NOTE: The 1995–97 4.3L (VIN W) engine is equipped with a new distributor-like ignition component called the High Voltage Switch (HVS). On the HVS ignition system the Powertrain Control Module (PCM) utilizes a camshaft position sensor to determine spark timing, dwell and firing of the ignition coil. Because of this, positioning of the HVS (and ignition timing) is fixed and non-adjustable. Do not attempt to adjust the ignition timing by rotating the HVS or cross/mis-firing may occur.

Connect the timing light and tachometer to the engine according to the tool manufacturers' instructions. Make sure the timing light is connected to the No. 1 spark plug wire. digital tachometer must be used.

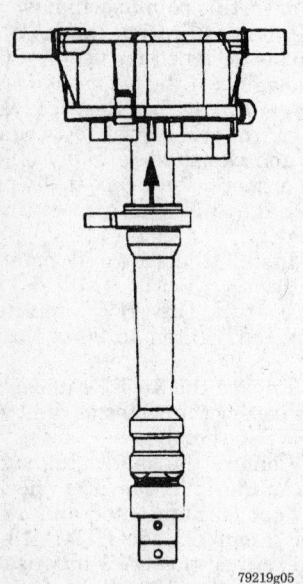

Proper HVS alignment — 1995–97 vehicles with HVS ignition

ADJUSTMENT

1. Locate and clean the timing marks on the crankshaft pulley and the front of the timing case cover.

2. Use chalk or white paint to color the mark on the scale that will indicate the correct timing, when aligned with the mark on the pulley or the pointer.

3. Attach a tachometer and a timing light to the engine.

4. On early model vehicles that are not equipped with Electronic Spark Control (ESC), disconnect and plug the vacuum lines to the distributor.

5. If equipped with ESC, the electronic spark timing must be disabled or bypassed to prevent the control module from advancing timing while attempting to set it. This would obviously lead to an incorrect base timing setting. There are 2 possible methods of disabling the EST system, depending on the type of engine:

• On 2.5L engines, ground the **A** and **B** terminals on the ALDL connector under the dash before adjusting the timing.

• On all other engines using the EST distributor, disengage the timing connector wire. Refer to the Vehicle Emission Control Information (VECI) label for details on the partic-

ular engine. Most vehicles are equipped with a single wire timing bypass connector. The bypass wire is normally a tan wire with a black stripe, that breaks out of the wiring harness conduit adjacent to the distributor, but on some later vehicles (1994–97) it may break out of a taped section just below the heater case in the passenger compartment.

6. Start the engine, then check and adjust the idle speed, as necessary.

7. Loosen the distributor lock bolt slightly to permit the distributor to be turned.

8. Adjust the idle to the correct specification.

9. With the timing light aimed at the pulley and the marks on the engine, turn the distributor in the direction of rotor rotation to retard the spark or in the opposite direction of rotor rotation to advance the spark. Align the marks on the pulley and the engine with the flashes of the timing light.

10. Turn the engine **OFF**, tighten the distributor hold-down bolt and recheck the timing.

Alternator

PRECAUTIONS

Several precautions must be observed with alternator equipped vehicles in order to avoid damage to the unit.

• If the battery is removed for any reason, make sure it is reconnected with the correct polarity. Reversing the battery connections may result in damage to the 1-way rectifiers.

• When utilizing a booster battery as a starting aid, always connect the positive to positive terminals and the negative terminal from the booster battery to a good engine ground on the vehicle being started.

• Never use a fast charger as a booster to start vehicles.

• Disconnect the battery cables when charging the battery with a fast charger.

• Never attempt to polarize the alternator.

• Do not use test lamps of more than 12 volts when checking diode continuity.

• Do not short across or ground any of the alternator terminals.

• The polarity of the battery, alternator and regulator must be matched and considered before making any electrical connections within the system.

• Never separate the alternator on an open circuit. Make sure all connections within the circuit are clean and tight.

• Disconnect the battery ground terminal when performing any service on electrical components.

• Disconnect the battery if arc welding is to be done on the vehicle.

REMOVAL AND INSTALLATION

1. Disconnect the negative battery cable.

2. Remove the necessary components in order to gain access to the alternator assembly. On early models of the Astro/Safari vans, remove the upper radiator fan shroud. On most Astro/Safari vans, remove the engine cover from inside the vehicle.

3. If necessary, remove the air cleaner and/or duct work.

4. Tag and disengage the electrical connectors at the alternator.

5. Remove the alternator belt. If equipped with a serpentine belt, relieve the belt tension, the carefully remove the belt from the alternator pulley.

6. If equipped, remove the alternator brace. On 1994–97 vehicles equipped with a 2.2L engine, this may be accomplished working through the wheel well.

7. Remove the alternator retaining bolts and remove the alternator.

8. Installation is the reverse of the removal procedure. Check and adjust the belt tension, as required.

Drive Belt

REMOVAL AND INSTALLATION

NOTE: Do not allow the drive belt tensioner to snap into the free position. This may result in damage to the tensioner.

Except Astro and Safari

1. Disconnect the negative battery cable.

2. Use a ½ in. breaker bar with a socket placed on the tensioner pulley axis bolt and rotate the tensioner to the left (counterclockwise).

3. Remove the belt.

To install:

4. Route the belt over all the pulleys, except the belt tensioner.

5. Use a ½ in. breaker bar with a socket places on the tensioner pulley axis bolt and rotate the tensioner to the left (counter clockwise).

6. Route the belt over the tensioner pulley.

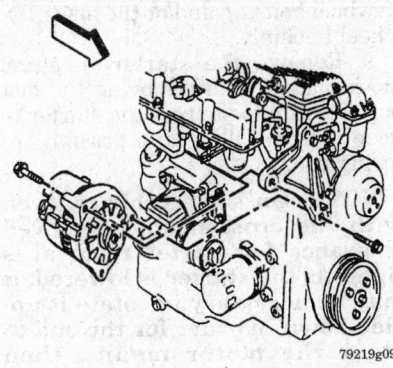

Alternator mounting — 2.2L engine

7. Check the belt for correct groove tracking around each pulley.

8. Connect the negative battery cable.

Astro and Safari

1. Disconnect the negative battery cable.

2. Install a ³/₈ in. drive on the tensioner arm and rotate the arm counterclockwise.

3. Remove the drive belt.

To install:

4. Route the belt over all the pulleys except the alternator.

5. Place a ³/₈ in. drive on the tensioner arm and rotate the arm counterclockwise.

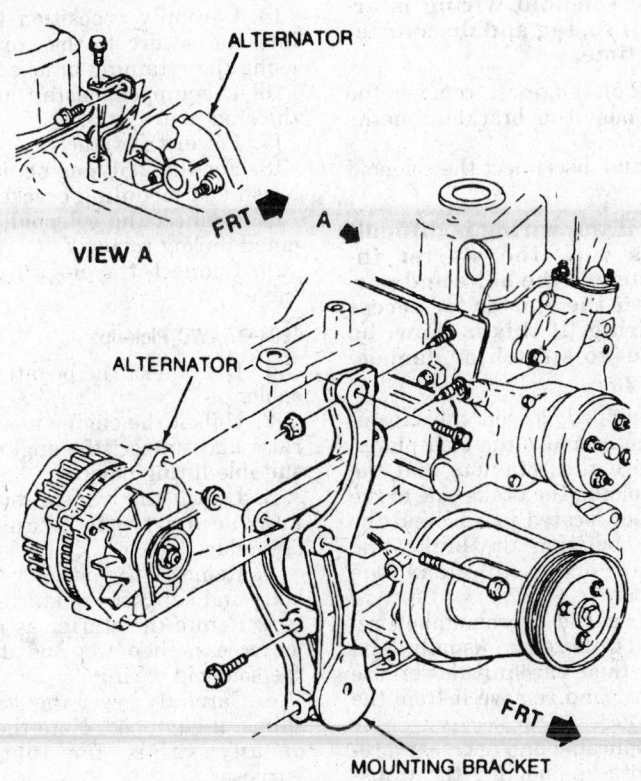

Alternator mounting — 2.5L engine

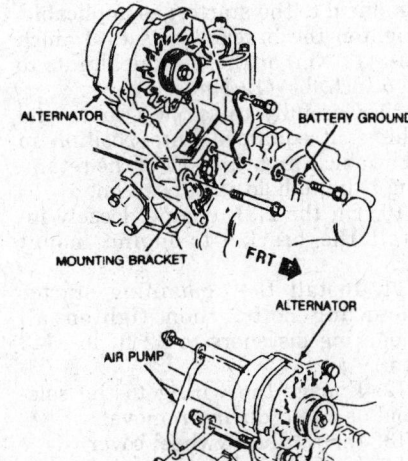

Alternator mounting — 2.8L engine

6. Route the belt over the alternator pulley.

7. Check the belt for correct tracking around each pulley.

8. Connect the negative battery cable.

Starter

REMOVAL AND INSTALLATION

Except Astro and Safari

1993 2WD Pick-ups

1. Disconnect the negative battery cable.

2. Raise and support the vehicle safely.

3. On 2.5L engines, remove the brush end mounting bracket from the starter motor.

4. Tag and disconnect the solenoid wiring.

NOTE: If the wiring is difficult to access with the starter installed, remove the bolts and partially lower the starter for access to the wiring. If this is done, be careful not to stretch or damage the wiring.

5. Remove the 2 bolts and washers (noting the location of any shims); then remove the starter and the shim.

To install:

6. Position the starter in the vehicle (along with any shims which were removed), then support while threading the starter mounting bolts.

7. Tighten the starter mounting bolts to 30–33 ft. lbs. (40–45 Nm).

8. Engage the starter solenoid wiring as noted during removal.

9. On 2.5L engines, install the brush end mounting bracket.

10. Lower the vehicle and connect the negative battery cable.

1994–97 2WD Pick-ups

1. Disconnect the negative battery cable.

2. Raise and support the vehicle safely.

3. Remove the cover in order to provide access to the flywheel.

4. Tag and disconnect the solenoid wiring.

5. For the 2.2L engine, remove the attaching bracket-to-engine mount bolt.

6. Remove the starter-to-engine block bolts, then carefully remove the starter noting the position of any shims (if used).

7. If necessary, remove the bracket (2.2L engine) or the shield

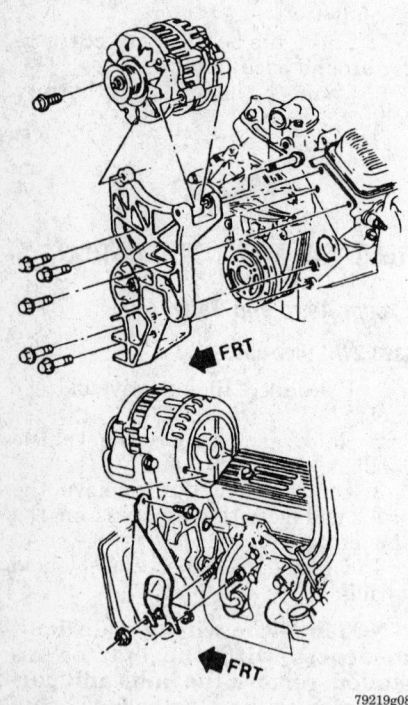

Common alternator mounting for the 4.3L engine

79219g08

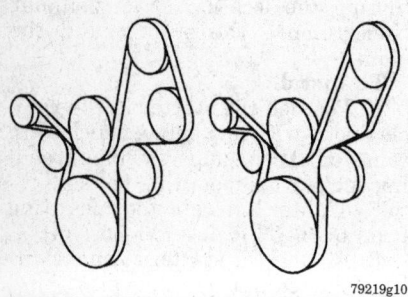

Drive belt routing — 2.2L with A/C (power steering on left and manual steering on right)

79219g10

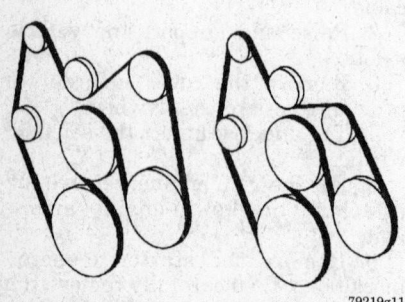

Drive belt routing — 2.2L without A/C (power steering on left and manual steering on right)

79219g11

(4.3L engine) from the starter assembly.

To install:

8. If removed, install the bracket or shield to the starter, as applicable. Tighten the bracket nuts to 97 inch lbs. (11 Nm) or the the shield nuts to 106 inch lbs. (12 Nm).

9. Carefully raise the starter and shims, if equipped, into position in the vehicle and thread 1 of the retaining bolts to hold it in position.

10. On the 2.2L engine, loosely install the bracket-to-engine mount bolt.

11. Install the remaining starter mounting bolts, then tighten all mounting fasteners to 32 ft. lbs. (43 Nm).

12. Engage the wiring to the solenoid as noted during removal.

13. Install the flywheel cover.

14. Lower the vehicle and connect the negative battery cable.

1993 4WD Pick-ups and 1994–97 Blazer, Bravada and Jimmy

1. Disconnect the negative battery cable.

2. Raise and support the vehicle safely.

NOTE: On some vehicles access to the wiring may be easier from above. Before raising and supporting the vehicle, check to see if the solenoid wiring is accessible. If so, tag and disconnect it at this time.

3. On 2.5L engines, remove the brush end mounting bracket from the starter motor.

4. Tag and disconnect the solenoid wiring.

NOTE: If the wiring is difficult to access with the starter installed, remove the bolts and partially lower the starter for access to the wiring. If this is done, be careful not to stretch or damage the wiring.

5. If equipped, loosen the retaining bolts and remove the skid plate.

6. Remove the retainers and the brackets holding the brake line to the crossmember located just behind the oil pan. Reposition the brake line slightly in order to clear the crossmember.

7. Remove the crossmember retaining bolts, there are usually 3 on each side, then carefully lower the crossmember and remove it from the vehicle for access.

8. As applicable and necessary, remove the bracket holding the transmission fluid cooler lines to the flywheel housing, brace rod to the

flywheel housing and/or the lower flywheel housing.

9. Remove the starter-to-engine block bolts. When removing the last bolt, be sure to support the starter to keep it from falling and possibly injuring you.

NOTE: On some vehicles, even with the crossmember removed clearance for starter removal is tight. As the starter is lowered, it may be necessary to rotate it upside down in order for the end to clear the motor mount, then lower the nose behind the bellhousing and rotate it back so the solenoid is on top and the starter may be removed.

10. Carefully lower the starter and shims, if equipped.

To install:

11. Position the starter in the vehicle (along with any shims which were removed) and support while threading the starter mounting bolts. Tighten the starter mounting bolts to 30–33 ft. lbs. (40–45 Nm).

12. If removed, install the lower flywheel housing.

13. If equipped, install the transmission cooler line bracket and/or the brace rod to the housing.

14. Install the crossmember to the frame and secure using the retaining bolts.

15. Carefully reposition the brake line and secure to the crossmember using the retaining brackets.

16. If equipped, install and secure the skid plate.

17. Lower the vehicle.

18. For the 2.5L engine, install the brush end mounting bracket.

19. Connect the solenoid wiring as noted during removal.

20. Connect the negative battery cable.

1994–97 4WD Pick-ups

1. Disconnect the negative battery cable.

2. Unbolt the engine mounts, then raise and support the engine using a suitable lifting device.

3. Unbolt the transmission mount and support the transmission assembly.

4. Remove the starter-to-engine bolts and support the starter.

5. Rotate the starter as necessary for access, then tag and disconnect the solenoid wiring.

6. Carefully lower the starter and shims, if equipped. Note the location of any shims for installation purposes.

7. If necessary, remove the shield from the starter assembly.

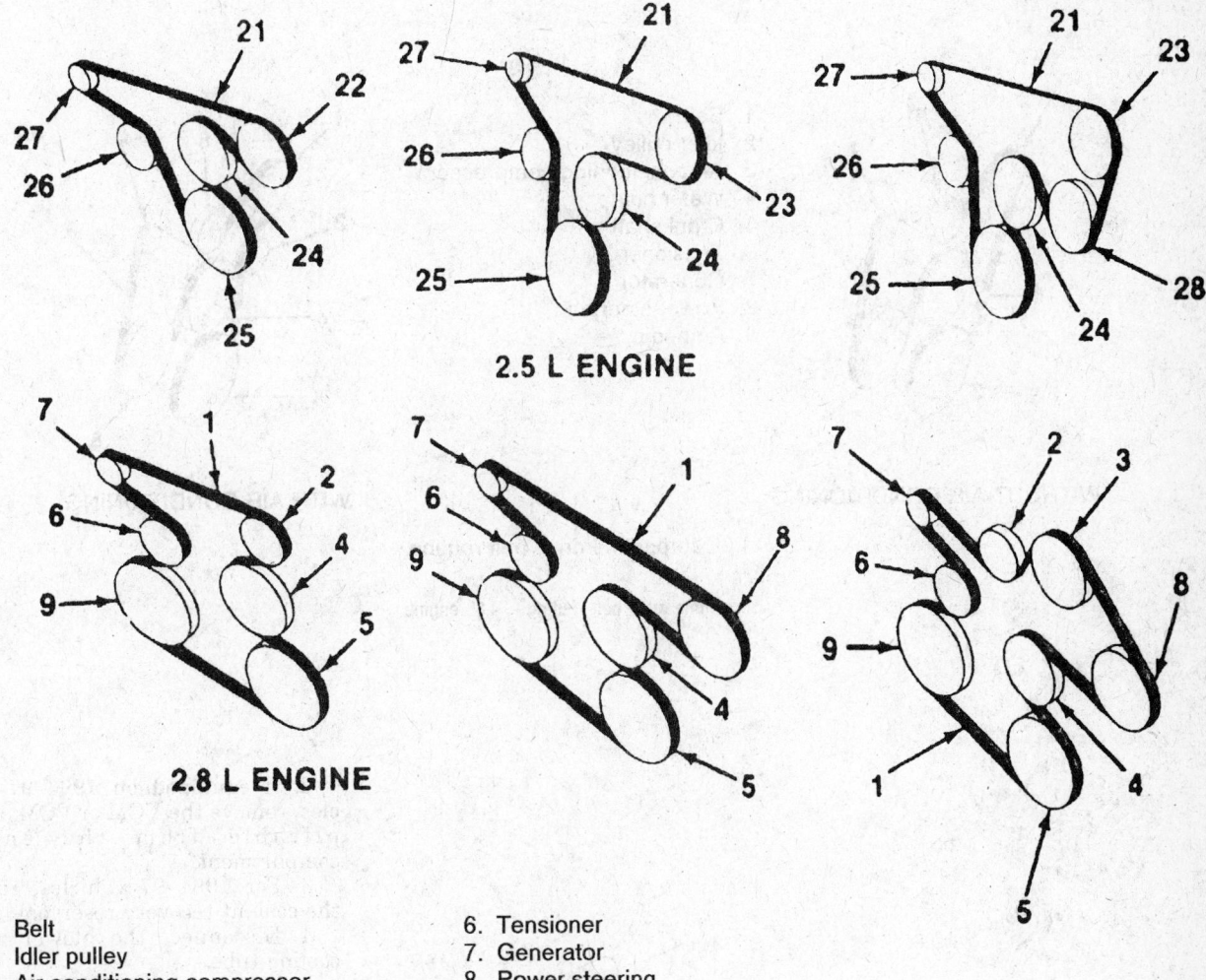

2.5 L ENGINE

2.8 L ENGINE

1. Belt
2. Idler pulley
3. Air conditioning compressor
4. Water pump
5. Crankshaft

6. Tensioner
7. Generator
8. Power steering
9. Air pump

79219g12

Serpentine drive belt routing — 2.5L and 2.8L engines

To install:

8. Raise the starter into position in the vehicle along with any shims (making sure they are in their original positions), then tighten the mounting bolts to 32 ft. lbs. (43 Nm).

9. If removed, install the shield to the starter assembly and tighten the retaining nuts to 106 inch lbs. (12 Nm).

10. Engage the wiring to the solenoid as noted during removal.

11. Install the transmission mount and remove the supports.

12. Lower the engine and secure the engine mounts, then remove the lifting device.

13. Connect the negative battery cable.

Astro and Safari

1. Disconnect the negative battery cable.

2. Raise and support the vehicle safely.

3. If access is possible with the starter installed, tag and disconnect the solenoid wiring.

4. Remove the 2 bolts and washers (along with the shims, if equipped), then carefully the starter. If the wiring was not disconnected previously, support the starter, then tag and disconnect the wiring at this time.

5. Installation is the reverse of removal. Install the shim, if used, and torque the mounting bolts to 32 ft. lbs. (43 Nm).

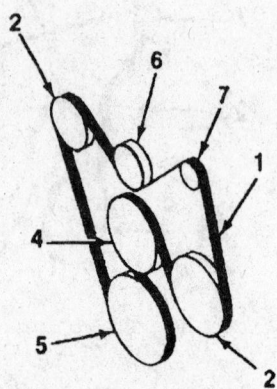

1. Belt
2. Idler pulley
3. Air conditioning compressor
4. Water pump
5. Crankshaft
6. Tensioner
7. Generator
8. Power steering
9. Air pump

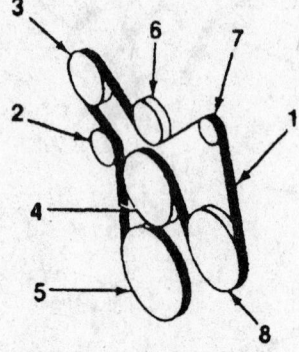

WITHOUT AIR CONDITIONING

WITH AIR CONDITIONING

4.3L serpentine drive belt routing

Serpentine drive belt routing — 4.3L engine

79219g13

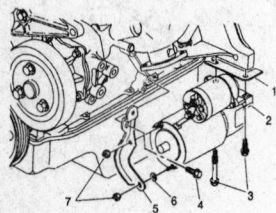

1. SHIM
2. STARTER ASSEMBLY
3. BOLT, 43 N·m (32 LBS. FT.)
4. BOLT, 43 N·m (32 LBS. FT.)
5. BRACKET, STARTER MOTOR
6. WASHER
7. NUT, 11 N·m (97 LBS. IN.)

79219g14

Starter mounting — 2.2L engine shown

CHASSIS ELECTRICAL

Blower Motor

REMOVAL AND INSTALLATION

Except Astro and Safari

1. Disconnect the negative battery cable.

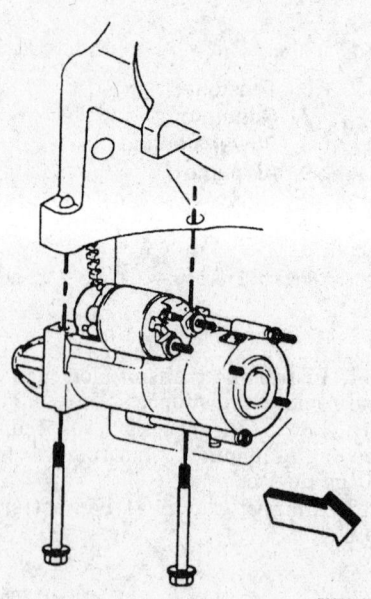

79219g15

Starter mounting — 1996 4.3L exc. Astro and Safari

2. If equipped on 1994–97 vehicles, remove the VCM or PCM, as applicable from the engine compartment.

3. For 1994–97 vehicles, remove the coolant recovery reservoir.

4. Disconnect the blower motor cooling tube.

5. Disengage the electrical connector(s) from the blower motor, as necessary.

6. For 1996–97 vehicles, remove the screws, the lower screw is on the bottom front of the cover. Detach the harness from the blower motor.

7. For vehicles through 1995, remove the blower motor-to-case screws, then carefully withdraw the blower motor from the case.

8. For 1996–97 vehicles, use a razor blade or utility knife to cut through the cover on the cut line, then proceed as follows:

a. Starting with the upper half of the cover, tear the remaining part of the access cover from the remaining portion.

b. Remove the lower cover using the same procedure from the upper half, then remove the blower motor.

To install:

9. For 1996–97 vehicles, install the blower motor, then proceed as follows:

    a. Fasten the upper and lower halves of the access covers together using the 3 flange clips.

    b. Using black duct tape, place a piece of duct tape along the bottom lower edge of the lower half only. The tape must be the full width of the cover.

    c. Install the access cover onto the case and secure with the retaining screws.

    d. Align the cut areas, then seal the cut line areas using black weatherstrip adhesive.

10. For 1995 vehicles install the blower motor to the case and secure using the retaining screws.

11. Engage the electrical connector(s) to the motor, as necessary.

12. Connect the blower motor cooling tube.

13. For 1994–97 vehicles, install the coolant recovery reservoir, then if equipped, install the VCM or PCM.

14. Connect the negative battery cable.

Astro and Safari

1. Disconnect the negative battery cable.

2. Tag and disengage the necessary electrical connections.

3. Remove the engine coolant recovery bottle.

4. Disconnect and reposition the windshield washer bottle.

5. For 1996–97 vehicles, remove the two screws retaining the cover in the blower motor area. Cut the acoustic cover between the 5 clip lands. Keep this piece for installation purposes.

6. Disengage the cooling tube from the heater blower assembly and remove the blower motor retaining screws. If necessary, remove the blower motor relay bracket.

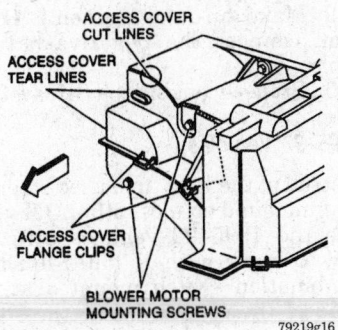

Access cover cut and tear lines — 1996–97 Except Astro and Safari

79219g16

7. Remove the blower motor.

8. Installation is the reverse of the removal procedure. Transfer the blower motor cage, as required.

Windshield Wiper Motor

REMOVAL AND INSTALLATION

Blazer, Bravada, Hombre, Jimmy, Sonoma, Typhoon and 1995–97 Astro/Safari

1. Disconnect the negative battery cable.

2. Remove the wiper arms from the linkage so the cowl may be removed.

3. Remove the cowl vent grille and screen.

4. Disengage the wiring from the wiper motor.

5. Except for 1994–97 vehicles, loosen but do not remove the nuts which hold the drive link to the motor crank arm. Detach the drive link from the crank arm.

6. For 1994–97 vehicles, remove the wiper transmission from the wiper motor drive link using J-39232 or equivalent tool.

7. Remove the wiper motor-to-cowl screws. Carefully rotate the motor and guide the drive link from the hole in the cowl, then remove the motor.

To install:

8. Guide the motor drive link through the hole in the cowl, then install the motor and secure using the retainers.

9. For 1994–97 vehicles, install the drive link socket onto the crank arm ball using J-39529 or equivalent tool. The wiper transmission assembly must be installed to the crank arm PAST the 2nd detent so the seal is compressed to a maximum height of 1 in. (25.5mm)

NOTE: Before assembling the transmission drive link socket to the crank arm ball, lubricate the inside of the socket using white lithium grease.

10. Except for 1994–97 vehicles, position the drive link socket onto the crank arm ball of the wiper motor assembly. Evenly tighten the socket nuts to secure the connection.

11. Engage the wiring to the wiper motor.

12. Install the cowl vent grille and screen.

13. Install the wiper arm and blade assemblies to the transmission linkage.

14. Connect the negative battery cable and verify proper operation.

1993–94 Astro and Safari

1. Disconnect the negative battery cable.

2. Remove the wiper arms from the linkage so the cowl may be removed.

3. Remove the cowl vent grille.

4. Disengage the wiring from the wiper motor.

5. Disconnect the transmission link from the crank arm on the motor by prying it toward the rear of the vehicle.

6. Remove the motor mounting bolts and remove the motor.

7. Installation is the reverse of removal.

Headlight Switch

REMOVAL AND INSTALLATION

Except Astro and Safari

1993 Vehicles

1. Disconnect the negative battery cable. Make sure the headlight switch if off.

2. From under the headlight switch assembly, remove the trim plate/switch assembly retaining screws.

NOTE: For some vehicles equipped with a fog lamps, the fog lamp switch may have to be removed in order to access the headlight trim plate/switch screw.

3. Pull the switch trim plate from the instrument panel. Pivot the trim plate outward at the bottom, then pull the plate downward to release it.

4. Disengage the electrical wiring connectors from the rear of the switch/trim plate assembly.

5. Remove the headlight switch from the trim plate assembly; if necessary, replace the switch.

6. To install, reverse the removal procedures.

1994–97 VEHICLES

1. Disconnect the negative battery cable. Make sure the headlight switch is off.

2. Remove the instrument cluster trim bezel.

3. Disengage the electrical connector.

4. Remove the switch-to-bezel retaining screws.

5. Remove the switch from the bezel.

6. Installation is the reverse of removal. To make sure the switch is working properly, temporarily con-

nect the negative battery cable and check switch operation BEFORE the bezel is fully installed.

Astro and Safari

1. Disconnect the negative battery cable.

2. Remove the screws (usually 5) holding the instrument panel cluster trim plate to the panel. There may also be another screw behind the heater control panel.

3. Pull the trim panel forward for access, then disengage the 2 electrical connectors from the back of the headlight switch.

4. Either remove the screws (usually 4) securing the headlight switch to the back of the trim plate or remove the switch from the plate by releasing the retaining tabs on the sides of the switch and pulling the switch out of the plate.

5. Remove the switch.

6. Installation is the reverse of the removal.

Turn Signal Switch

REMOVAL AND INSTALLATION

NOTE: When servicing any components on the steering column, should any fasteners require replacement, be sure to use only nuts and bolts of the same size and grade as the original fasteners. Using screws that are too long could prevent the column from collapsing during a collision.

NOTE: The following procedures require the use of a lock plate compressor tool such as J-23653 or equivalent.

Except 1995–97 Vehicles

1. If equipped, make sure the wheels are locked in the straight-ahead position, then properly disable the SIR (air bag) system.

2. Disconnect the negative battery cable.

3. Matchmark and remove the steering wheel.

4. If equipped with an SIR system, remove the SIR coil assembly retaining ring, then remove the coil assembly and allow it to hang freely from the wiring. Remove the wave washer.

5. On non-SIR equipped vehicles, remove the shaft lock cover.

6. Push downward on the shaft lock assembly until the snapring is exposed using the shaft lock compressor tool.

7. Remove the shaft lock retaining snapring, then carefully release the tool and remove the shaft lock from the column.

8. Remove the turn signal cancelling cam assembly.

9. For standard columns not equipped with SIR, remove the upper bearing spring and thrust washer.

10. For tilt columns or SIR equipped standard columns, remove the upper bearing spring, inner race seat and inner race.

11. Move the turn signal lever upward to the "Right Turn" position.

12. Remove the access cap and disengage the multi-function lever harness connector, then grasp the lever and pull it from the column.

13. Loosen and remove the hazard knob retaining screw, then remove the screw, button, spring and knob.

14. Remove the screw and the switch actuator arm.

15. Remove the turn signal switch retaining screws, then pull the switch forward and allow it to hang from the wires. If the switch is only being removed for access to other components, this may be sufficient.

16. If the switch is to be replaced, cut the wires near the top of the switch and discard the switch. Before cutting the wires, verify that the wire color codes are the same. Secure the connector of the new switch to the old wires, and pull the new harness down through the steering column while removing the old switch.

17. If the original switch is to be reused, attach a piece of wire or string around the connector and pull the harness up through the column, while pulling the string up through the column and leaving the string or wire in position to help with reinstallation later.

18. After freeing the switch wiring protector from its mounting, pull the turn signal switch straight up and remove the switch, switch harness, and the connector from the column.

NOTE: On some vehicles access to the connector may be difficult. If necessary, remove the column support bracket assembly and properly support the column, and/or remove the wiring protectors.

To install:

19. Install the switch and wiring harness to the vehicle. If the switch was completely removed, use the length of mechanic's wire or string to pull the switch harness through the column, then engage the connector.

NOTE: If the column support bracket or wiring protectors were removed, install them before proceeding.

20. Position the switch on the column and secure using the retaining screws.

21. Install the switch actuator arm and retaining screw.

22. Install the hazard knob assembly, then install the multi-function lever.

23. Install the thrust washer and upper bearing spring (standard columns without SIR) or the inner race, upper bearing race seat and upper bearing spring (tilt columns or standard with SIR), as applicable.

24. Lubricate the turn signal cancelling cam using a suitable synthetic grease (usually included in the service kit), then install the cam assembly.

25. Position the shaft lock and a new snapring, then use the lock compressor to hold the lock down while seating the new snapring. Make sure the ring is firmly seated in the groove, then carefully release the tool.

NOTE: The coil assembly will become uncentered if the steering column is separated from the steering gear and allowed to rotate or if the centering spring is pushed down, letting the hub rotate while the coils assembly is removed from the steering column.

26. If equipped with SIR, make sure the coil is centered, then install the wave washer, followed by the coil and the retaining ring. The coil ring must be firmly seated in the shaft groove.

27. On non SIR columns, install the shaft lock cover.

28. Align and install the steering wheel.

29. Make sure the ignition is **OFF**, then connect the negative battery cable.

30. Properly enable the SIR system.

1995–97 Vehicles

Instead of the long time used steering column found on most other GM vehicles the 1995–97 vehicles utilize a new column with a multi-function combination switch mounted at the head of the column below the steering wheel and an upper/lower shroud assembly. The combination switch performs such functions as the wiper

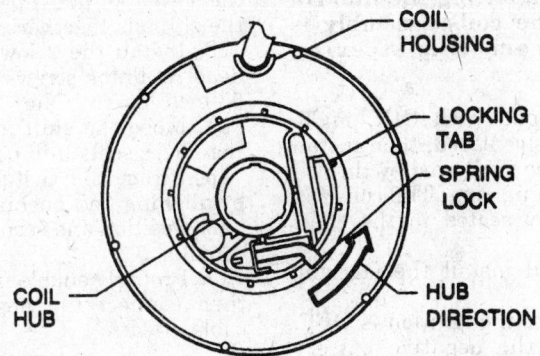

PERFORM THE FOLLOWING STEPS TO CENTER COIL ASSEMBLY

A. WHEELS STRAIGHT AHEAD.
B. REMOVE COIL ASSEMBLY.
C. HOLD COIL ASSEMBLY WITH BOTTOM UP.
D. WHILE HOLDING COIL ASSEMBLY, DEPRESS SPRING LOCK TO ROTATE HUB IN DIRECTION OF ARROW UNTIL IT STOPS.
E. THE COIL RIBBON SHOULD BE WOUND UP SNUG AGAINST CENTER HUB.
F. ROTATE COIL HUB IN OPPOSITE DIRECTION APPROXIMATELY TWO AND A HALF (2-1/2) TURNS. RELEASE SPRING LOCK BETWEEN LOCKING TABS.

79219gf1

Centering the SIR coil

switch and the turn signal switch along with any other duties of the multi-function lever.

Combination Switch

REMOVAL AND INSTALLATION

1995–97 Vehicles

NOTE: Removal of the SIR coil is not necessary during this procedure. avoid removing the coil and make sure the steering column, if disconnected from the gear, is not allowed to rotate excessively. This is to prevent uncentering and damaging the coil. Should the coil become uncentered, it must be removed, centered and repositioned on the steering column.

1. Disconnect the negative battery cable.
2. Properly disable the SIR (air bag) system.
3. Either lower the steering column from the instrument panel for access or unbolt and remove the column. If the column is removed, prevent it from rotating so the SIR coil does not become uncentered.

4. If applicable, remove the tilt lever by pulling outward.
5. Remove the 2 pan head tapping screws from the lower column shroud, then tilt the shroud down and slide it back to disengage the locking tabs. Remove the lower shroud.
6. Remove the 2 Torx head screws from the upper shroud.
7. Lift the upper shroud for access to the lock cylinder hole. Hold the key in the **START** position and use a 1/16 in. Allen wrench to push on the lock cylinder retaining pin.
8. Release the key to the **RUN** position and pull the steering column lock cylinder set from the lock module assembly. Remove the upper shroud.
9. If necessary, remove the shift lever clevis, then remove the lever.
10. Remove the wiring harness straps (noting the positioning for installation purposes), then disengage the steering column bulkhead connector from the vehicle wiring harness.
11. Disengage the grey and black connectors for the turn signal and multi-function switch from the column bulkhead connector.
12. Remove the 2 pan head switch retaining screws, then remove the switch from the steering column.

To install:
13. Position the multi-function switch assembly, then use a suitable small bladed tool to compress the electrical contact while moving the switch into position. Make sure the electrical contact rests on the cancelling cam assembly.
14. Install the switch retaining screws and tighten to 53 inch lbs. (6.0 Nm).
15. Engage the grey and black multi-function switch connectors to the column bulkhead connector.
16. Engage the steering column bulkhead-to-vehicle connector.
17. If removed, install the shift lever and secure the clevis.
18. Install the wiring harness straps as noted during removal.
19. Position the shift lever and multi-function lever seals to ease installation of the upper and lower shrouds.
20. Install the upper shroud and lock cylinder. With the key installed to the lock cylinder and turned to the **RUN** position, make sure the sector in the lock module is also in this position.
21. Install the lock cylinder to the upper shroud, then align the locking tab and positioning tab with the slots in the lock module assembly. With the tabs aligned, carefully push the cylinder into position.
22. Install the upper shroud Torx head retaining screws and tighten to 12 inch lbs. (1.4 Nm).
23. Install the lower shroud, making sure the slots on the shroud engage with the upper shroud tabs. Tilt the lower shroud upward and snap the shrouds together.
24. Install the 2 lower shroud pan head retaining screws and tighten to 53 inch lbs. (6.0 Nm).
25. Move the shift and multi-function lever seals into position, then
26. If removed, install the tilt lever by aligning and pushing inward.
27. Position and secure the steering column.
28. Properly enable the SIR system, then connect the negative battery cable.

Ignition Lock Cylinder

REMOVAL AND INSTALLATION

1993–94 Vehicles

1. If equipped, make sure the wheels are locked in the straight-ahead position, then properly disable the SIR (air bag) system.

2. Disconnect the negative battery cable.

3. Matchmark and remove the steering wheel.

4. If equipped with an SIR system, remove the SIR coil assembly retaining ring, then remove the coil assembly and allow it to hang freely from the wiring. Remove the wave washer.

NOTE: On some SIR equipped vehicles, it may be necessary to completely remove the coil and wiring from the steering column before removing the lock cylinder assembly. If so, attach a length of mechanic's wire to the coil connector at the base of the column, then carefully pull the harness and wire through the steering column towards the top. Leave the wire in position inside the column in order to pull the harness back down into position during installation.

5. Remove the turn signal switch from the column and allow it to hang from the wires (leaving them connected).

6. Remove the buzzer switch assembly. On some vehicles it may be necessary to temporarily remove the key from the lock cylinder in order to remove the buzzer. If so, the key should be reinserted before the next step.

7. Carefully remove the lock cylinder screw and the lock cylinder. If possible, use a magnetic tipped screwdriver on the screw in order to help prevent the possibility of dropping it.

─────── CAUTION ───────
If the screw is dropped upon removal, it could fall into the steering column, requiring complete disassembly in order to retrieve the screw and prevent damage.

To install:

8. Align and install the lock cylinder set.

9. Push the lock cylinder all the way in, then carefully install the retaining screw. Tighten the screw to 22 inch lbs. (2.5 Nm) on tilt columns or to 40 inch lbs. (4.5 Nm) on standard non-tilt columns.

10. If necessary, install the buzzer switch assembly.

11. Reposition and secure the turn signal switch assembly

NOTE: The coil assembly will become uncentered if the steering column is separated from the steering gear and allowed to rotate or if the centering spring is

pushed down, letting the hub rotate while the coils assembly is removed from the steering column.

12. If equipped with SIR, make sure the coil is centered, then install the wave washer, followed by the coil and the retaining ring. The coil ring must be firmly seated in the shaft groove.

13. Align and install the steering wheel.

14. Make sure the ignition is **OFF**, then connect the negative battery cable.

15. Properly enable the SIR system.

1995–97 Vehicles

1. Disconnect the negative battery cable.

2. Properly disable the SIR (air bag) system.

3. Either lower the steering column from the instrument panel for access or unbolt and remove the column. If the column is removed, prevent it from rotating so the SIR coil does not become uncentered.

4. If applicable, remove the tilt lever by pulling outward.

5. Remove the 2 pan head tapping screws from the lower column shroud, then tilt the shroud down and slide it back to disengage the locking tabs. Remove the lower shroud.

6. Remove the 2 Torx head screws from the upper shroud.

7. Lift the upper shroud for access to the lock cylinder hole. Hold the key in the **START** position and use a $1/16$ in. Allen wrench to push on the lock cylinder retaining pin.

8. Release the key to the **RUN** position and pull the steering column lock cylinder set from the lock module assembly. Remove the upper shroud.

To install:

9. Install the upper shroud and lock cylinder. With the key installed to the lock cylinder and turned to the **RUN** position, make sure the sector in the lock module is also in this position.

10. Install the lock cylinder to the upper shroud, then align the locking tab and positioning tab with the slots in the lock module assembly. With the tabs aligned, carefully push the cylinder into position.

11. Install the upper shroud Torx head retaining screws and tighten to 12 inch lbs. (1.4 Nm).

12. Install the lower shroud, making sure the slots on the shroud engage with the upper shroud tabs. Tilt

the lower shroud upward and snap the shrouds together.

13. Install the 2 lower shroud pan head retaining screws and tighten to 53 inch lbs. (6.0 Nm).

14. Move the shift and multi-function lever seals into position, then

15. If removed, install the tilt lever by aligning and pushing inward.

16. Position and secure the steering column.

17. Properly enable the SIR system, then connect the negative battery cable.

Ignition Switch

REMOVAL AND INSTALLATION

1993–94 Vehicles

1. If equipped, make sure the wheels are locked in the straight-ahead position, then properly disable the SIR (air bag) system.

2. Disconnect the negative battery cable.

3. Remove the lower column trim panel, then remove the steering column-to-instrument panel fasteners and carefully lower the column for access to the switch.

4. On some vehicles, the dimmer switch must be removed in order to remove the ignition switch. If necessary, remove the dimmer switch.

5. Place the ignition switch in the **OFF-LOCK** position.

NOTE: If the lock cylinder was removed, the switch slider should be moved to the extreme right position, then 1 detent to the left.

6. Remove the ignition switch-to-steering column retainers and disengage the switch wiring, then remove the assembly.

To install:

7. Before installing the ignition switch, place it in the **OFF-LOCK** position, then make sure the lock cylinder and actuating rod are in the Locked position (1st detent from the top or 1st detent to the right of far left detent travel).

NOTE: Most replacement switches are pinned in the OFF-LOCK position for installation purposes. If so, the pins must be removed after installation or damage may occur.

8. Install the activating rod into the ignition switch and assemble the switch onto the steering column. Once the switch is properly positioned, tighten the ignition switch-to-

steering column retainers to 35 inch lbs. (4.0 Nm).

NOTE: When installing the ignition switch, use only the specified screws since over length screws could impair the collapsibility of the column.

9. If removed, install the dimmer switch.

10. Raise the column into position and secure, then install any necessary trim plates.

11. Make sure the ignition is **OFF**, then connect the negative battery cable.

12. Properly enable the SIR system.

1995–97 Vehicles

1. Disconnect the negative battery cable.

2. Properly disable the SIR (air bag) system.

3. Either lower the steering column from the instrument panel for access or unbolt and remove the column. If the column is removed, prevent it from rotating so the SIR coil does not become uncentered.

4. Remove the combination switch from the steering column.

5. If equipped, remove the alarm switch from the lock module assembly by gently prying the retaining clip on the alarm switch using a small blade prytool. Then, rotate the alarm switch ¼ turn and remove.

6. Remove the 2 ignition switch self-tapping retaining screws.

7. Disengage the connector, then remove the wiring harness from the slot in the steering column. Remove the ignition and key alarm switch.

To install:

8. Position the switch to the column. Route the wire harness through the slot in the column housing assembly. Secure the harness using a wire strap through the hole located in the bottom of the housing assembly.

9. Install the switch retaining screws and tighten to 12 inch lbs. (1.4 Nm) in order to secure the switch.

10. If applicable, install the alarm switch to the lock module assembly by aligning the switch (with the retaining clip) parallel to the lock cylinder, then rotating the switch ¼ turn until locked in place.

11. Install the combination switch to the steering column.

12. Position and secure the steering column.

13. Properly enable the SIR system, then connect the negative battery cable.

Neutral Safety Switch

ADJUSTMENT

Column Mounted Switch

Most vehicles covered by this manual utilize a column mounted, ratcheting, self-adjusting switch.

1. Move the switch all the way toward the **L** gear position.

2. Move the selector to the **P** position.

3. The main housing and housing back should ratchet, providing proper switch adjustment.

Floor Console Mounted Switch

1. Remove the center console for access to the switch.

2. Place the shift control lever in Neutral.

3. Align the carrier tang on the switch with the slot on the shifter.

4. Loosen the retaining switch nuts.

5. Rotate the switch to align the service adjustment hole with the carrier tang hold, then use a 0.09 in. (2.34mm) gauge pin to complete adjustment. Insert the pin in the service adjustment hole and rotate the switch until it drops to a depth of 0.59 in. (15mm). Hold the switch in this position and tighten the retaining nuts to 30 inch lbs. (3.4 Nm).

REMOVAL AND INSTALLATION

Column Mounted Switch

Most vehicles covered by this manual should be equipped with ratcheting self-adjusting switches.

1. Disconnect the negative battery cable.

2. If necessary for access, remove the steering column insulator/filler panel for access to the switch.

3. Disengage the electrical harness connector from the switch.

4. Remove the switch by grasping and pulling it straight out of the steering column jacket.

To install:

5. Align the actuator on the switch with the holes in the shift tube.

6. Set the parking brake and place the gear selector in Neutral.

7. Press down on the front of the switch until the tangs snap into the rectangular holes in the steering column jacket.

8. Adjust the switch by moving the gear selector to **P**. The main housing and the housing back should ratchet, providing the proper switch adjustment.

9. Engage the harness connector to the switch.

10. Connect the negative battery cable, then verify proper switch operation. Make sure the reverse lights work and that the ignition will only work in the Neutral or **P** positions. If necessary, readjust the switch. For ratcheting type switches, move the gear selector all the way to the Low position, then repeat the adjustment.

11. If applicable, install the steering column insulator/filler panel.

Floor Console Mounted Switch

1. Disconnect the negative battery cable.

2. Remove the center console for access to the switch assembly.

3. Disengage the switch electrical connector.

4. Remove the retaining nuts, then remove the switch.

5. If necessary, remove the gauge pin from the switch.

6. If installing a new switch:

   a. Place the shift control lever in Neutral.

   b. Align the carrier tang on the back-up lamp/neutral safety switch with the slot on the shifter.

NOTE: Replacement switches are pinned in the Neutral position to ease installation. If the switch has been rotated or the switch is broken, install the switch using the "old switch" installation and adjustment procedure.

   c. Tighten the switch retaining nuts to 30 inch lbs. (3.4 Nm), then engage the switch connector.

   d. Move the shift control lever out of Neutral in order to shear the plastic pin, then remove the accessible piece(s) of the gauge pin.

7. If installing an old switch, install and adjust the switch to assure proper operation:

   a. Place the shift control lever in Neutral.

   b. Align the carrier tang on the switch with the slot on the shifter.

   c. Loosely install the retaining switch nuts and engage the wiring connector.

   d. Rotate the switch to align the service adjustment hole with the carrier tang hold, then use a 0.09 in. (2.34mm) gauge pin to complete adjustment. Insert the pin in the service adjustment hole and rotate the switch until it drops to a depth of 0.59 in. (15mm). Hold the switch in this position and tighten the retaining nuts to 30 inch lbs. (3.4 Nm).

8. Install the center console.
9. Connect the negative battery cable.
10. Verify proper switch operation.

Powertrain Control Module (PCM) or Vehicle Control Module (VCM)

For most applications, the control module is located inside the passenger compartment, on the right side of the vehicle. For these vehicles it is usually found either under/behind the glove box or behind the kick panel. Some late model applications equipped with a PCM or VCM have the control module located in the engine compartment mounted on the right side, on/near the fender and wheel well. But not all VCMs or late model PCMs are mounted in the engine compartment.

REMOVAL AND INSTALLATION

1. Make sure the ignition is turned **OFF**, then disconnect the negative battery cable.
2. Locate the computer control module. If not readily visible on the right side of the engine compartment, it is probably mounted under the dash.
3. If equipped with a passenger compartment mounted module, remove the dash panel located below the glove box and/or passenger side kick panel, as necessary.
4. Carefully disengage the connectors from the control module.

NOTE: On some modules, it may be easier to unfasten and rotate the module for easier access to the wiring. If necessary, do this before attempting to disengage the connectors.

5. Loosen and remove the mounting hardware, then remove the module from the vehicle.
To install:
6. If the module is being replaced, remove the access covers and CAREFULLY replace the PROM chip or CALPAK. Refer to the procedure later in this section for details and cautions.
7. Make sure the access cover is properly installed.

———— CAUTION ————
Remember, the ignition MUST be OFF when connecting the control module wiring.

8. Position the computer control module to the vehicle and secure using the hardware. If it was necessary to pivot the module for wire removal, then make sure the connectors are engaged before positioning the module.
9. Engage the wiring connectors to the module.
10. If removed for access, install the dash and/or kick panel.
11. Make sure the ignition is still **OFF**, then connect the negative battery cable.
12. If equipped with a VCM-A, the EEPROM must be reprogrammed using a suitable service system and the latest available software. In all likelihood, the vehicle must be towed to a dealer or repair shop containing the suitable equipment for this service.

ENGINE COOLING

Radiator

REMOVAL AND INSTALLATION

1. Disconnect the negative battery cable.
2. Drain the engine cooling system.
3. If equipped with the MFI-Turbo engine:
    a. Remove the air cleaner and duct, then remove the turbocharger inlet elbow.
    b. Disconnect the positive battery cable, then remove the battery and the battery tray.
    c. Disconnect the heater hose and clamp from the radiator.
4. Disconnect the overflow hose from the radiator, then disconnect the upper and lower radiator hoses.
5. If equipped with A/C, it may be necessary to remove the A/C hose retaining clip and reposition the hose for shroud and/or radiator removal. Do not disconnect any refrigerant fittings.

NOTE: If equipped with a 1 piece shroud, the shroud may be unbolted from the radiator support and pushed back over the cooling fan instead of removing it completely.

6. Most vehicles should be equipped with a 2 piece radiator/fan shroud. Remove the upper fan shroud-to-radiator support bolts and the upper fan shroud-to-lower fan shroud retainers. Remove the upper shroud.
7. If equipped with an A/T, disconnect and plug the fluid cooler lines at the radiator. Plug all openings to prevent system contamination or excessive fluid loss.

NOTE: On some late model Astro and Safari vehicles, the automatic transmission fluid cooler lines utilize a quick-connect fitting. Use J-37088-2A or equivalent quick-disconnect tool when servicing these vehicles.

8. If equipped with a factory engine oil cooler which is integral to the radiator, disconnect and plug the oil cooler lines at the radiator. Plug all openings to prevent system contamination or excessive fluid loss. Plug all openings to prevent system contamination or excessive fluid loss.
9. Lift the radiator straight upward from the supports. Be careful to lift the radiator straight upward and not to tilt it excessively as the radiator will still contain a significant amount of coolant and, if applicable, transmission fluid/engine oil.
To install:
10. Lower the radiator into position on the supports.
11. If equipped, remove the plugs, then connect the engine oil cooler lines and tighten the fittings.
12. If equipped with an A/T, remove the plugs, then connect the transmission fluid cooler lines and tighten the fittings.
13. Install the upper fan shroud and secure using the support and lower shroud retainers.
14. If applicable, reposition the A/C refrigerant hose and secure using the retaining clip.
15. Connect the overflow hose, upper and lower radiator hoses.
16. If equipped with the MFI-Turbo engine:
    a. Connect the heater hose and clamp to the radiator.
    b. Install the battery tray, then install the battery and connect the positive battery cable
    c. Install the turbocharger inlet elbow, then install the air cleaner and duct.

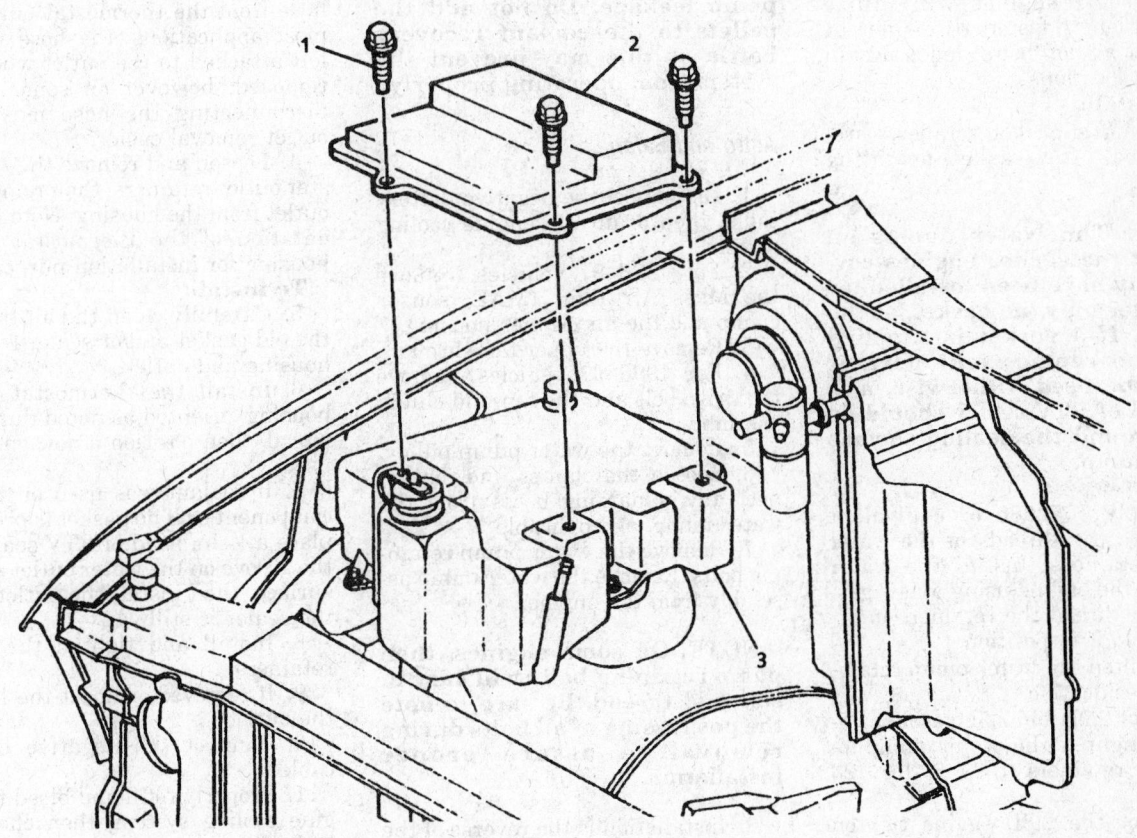

1 **VCM ATTACHING BOLT - TIGHTEN TO 9 TO 11 N·m (7 TO 8 lb. ft.)**

2 **VEHICLE CONTROL MODULE (VCM)**

3 **COOLANT RESEVOIR**

79219gf2

Vehicle Control Module (VCM) location — 1995 exc. Astro and Safari

17. Connect the negative battery cable.

NOTE: Whenever the cooling system in a 1995–97 vehicle (except diesel engines or U-Series vehicles) is completely drained and refilled with fresh coolant, 2 sealant pellets GMSPO part 3634621 must be added to the radiator. Failure to use the correct sealant pellets may result in premature coolant pump leakage. Do not add the pellets to the coolant recovery bottle as this may prevent the system from operating properly.

18. Properly refill the engine cooling system.

Water Pump

REMOVAL AND INSTALLATION

Except Astro and Safari

1. Disconnect the negative battery cable, then drain the engine cooling system.

2. Relieve the belt tension, then remove the accessory drive belts or the serpentine drive belt, as applicable.

3. Remove the upper fan shroud, then remove the fan or fan and clutch assembly, as applicable.

4. Remove the water pump pulley.

5. Loosen the clamp and disconnect the coolant hose(s) from the water pump.

NOTE: For the hoses on some engines, such as the heater hose on the early 2.8L, removal may be easier if the hose is left attached until the pump is free from the block. Once the pump is removed from the engine, the pump may be pulled (giving a better grip and greater leverage) from the tight hose connection.

6. Remove the retainers, then remove the water pump from the engine. Note the positions of all retainers as some engines will utilize different length fasteners in different locations and/or bolts and studs in different locations.

To install:

7. Using a gasket scraper, carefully clean the gasket mounting surfaces.

NOTE: The water pumps on some of the earlier engines covered may have been installed using sealer only, no gasket, at the factory. If a gasket is supplied with the replacement part, it should be used. Otherwise, a ⅛ in. bead of RTV sealer should be used around the sealing surface of the pump.

8. Apply 1052080 or equivalent sealant to the threads of the water pump retainers. Install the water pump to the engine using a new gasket, then thread the retainers in order to hold it in position.

9. Tighten the water pump retainers to specification:

   a. For 2.2L and 2.5L gasoline engines tighten the water pump-to-engine retainers to 17 ft. lbs. (23 Nm).

   b. For the 2.8L engine tighten the retainers to 15 ft. lbs. (21 Nm) and 89 inch lbs. (10 Nm) depending on fastener location. The central 3 lower bolts are tightened to the inch lbs. specification.

   c. For the 4.3L engine, tighten the bolts and studs to 30 ft. lbs. (41 Nm).

10. Connect the coolant hose(s) and secure using the retaining clamp(s).

11. Install the water pump pulley, then install the fan or fan and clutch assembly.

12. If equipped with a serpentine drive belt, position the belt over the pulleys, then carefully allow the tensioner back into contact with the belt.

13. If equipped with V-belts, install the accessory drive belts and adjust the tension.

14. Install the upper fan shroud, then connect the negative battery cable.

15. Properly refill the engine cooling system, then run the engine and check for leaks.

NOTE: Whenever the cooling system in a 1995–97 vehicle is completely drained and refilled with fresh coolant, 2 sealant pellets GMSPO part 3634621 must be added to the radiator. Failure to use the correct sealant pellets may result in premature coolant pump leakage. Do not add the pellets to the coolant recovery bottle as this may prevent the system from operating properly.

Astro and Safari

1. Disconnect the negative battery cable, then drain the engine cooling system.

2. For 1996–97 vehicles, remove the Mass Air Flow (MAF) sensor clamp and the air cleaner housing.

3. Remove the upper fan shroud.

4. For 1996–97 vehicles, remove the drive belt and the fan and clutch assembly.

5. Remove the water pump pulley.

6. Loosen the clamps and disconnect any remaining hoses from the water pump, as applicable.

7. Remove the water pump retaining bolts. Remove the water pump assembly from the engine.

NOTE: On some engines, then pump retaining bolts will vary in size and thread. Be sure to note the positioning of all bolts during removal to assure proper installation.

8. Installation is the reverse of the removal procedure. Use a new gasket and carefully tighten the bolts.

Thermostat

REMOVAL AND INSTALLATION

1. Disconnect the negative battery cable.

2. Drain the radiator until the level is below the thermostat (below the level of the intake manifold, cyl-

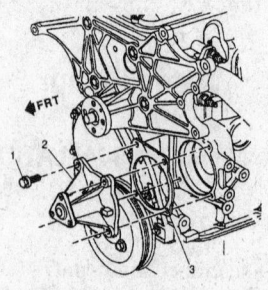

1. BOLT
2. PUMP, COOLANT
3. GASKET

79219g17

Water pump mounting — 2.2L engine

inder head or other thermostat housing, as applicable).

3. If necessary, disconnect the hose from the thermostat outlet. On most applications, the hose may be left attached to the outlet when it is removed, however on some models disconnecting the hose may make outlet removal easier.

4. Loosen and remove the thermostat outlet retainers, then remove the outlet from the housing. Note the orientation of the thermostat in the housing for installation purposes.

To install:

5. Carefully clean the all traces of the old gasket and/or sealer from the housing and outlet.

6. Install the thermostat to the housing, oriented as noted during removal, then position a new gasket (if used).

7. If sealant was used on the old component or if no gasket is provided, place a ⅛ in. bead of RTV sealant in the groove on the water outlet sealing surface, then install the outlet while the sealer is still wet.

8. Install and tighten the outlet retainers.

9. If removed, connect the hose to the outlet.

10. Connect the negative battery cable.

11. Properly refill and bleed the engine cooling system, then check for leaks.

NOTE: Whenever the cooling system in a 1995–97 vehicle is completely drained and refilled with fresh coolant, 2 sealant pellets GMSPO part 3634621 must be added to the radiator. Failure to use the correct sealant pellets may result in premature coolant pump leakage. Do not add the pellets to the coolant recovery bottle as this may prevent the system from operating properly.

Cooling Fan

REMOVAL AND INSTALLATION

NOTE: DO NOT use or repair a damaged fan assembly. An unbalanced fan assembly could fly apart and cause personal injury or property damage. Replace damaged assemblies with new ones.

1. Disconnect the negative battery cable.

2. Remove the upper radiator shroud and, if desired for additional

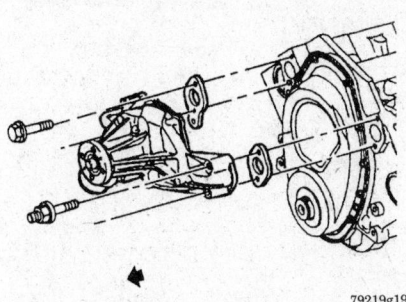

Water pump assembly mounting — 4.3L engine

clearance, remove the radiator from the vehicle.

NOTE: Although it is not necessary in most cases, the radiator may be removed from the vehicle for easier access to the fan retainers. If the radiator is left in place, use extra caution to prevent damage to the fragile radiator fins.

3. Remove the fan assembly attaching nuts (clutch type) or bolts (standard type), then remove the fan assembly from the engine.

NOTE: Some vehicles use a spacer between the fan and the water pump pulley. If used, be sure to retain the spacer for installation.

4. If necessary, the clutch may be removed from the fan by removing the attaching nuts or bolts (as applicable).
To install:
5. If removed, install the fan to the clutch and secure using the fasteners.
6. Position the spacer (if used) and fan assembly to the water pump pulley and secure using the fasteners.
7. If removed for clearance, install the radiator.
8. Install the upper fan shroud.
9. Connect the negative battery cable.
10. If the radiator was removed, properly refill the engine cooling system.

Cooling System Bleeding

NOTE: Whenever the cooling system in a 1995–97 vehicles is completely drained and refilled with fresh coolant, 2 sealant pellets GMSPO part 3634621 must be added to the radiator. Failure to use the correct sealant pellets may result in premature coolant pump leakage. Do not add the pellets to the coolant recovery bottle as this may prevent the system from operating properly.

1. To bleed the system, start with the system cool, the radiator cap off and the radiator filled to about an inch below the filler neck.
2. Start the engine and run it at slightly above normal idle speed. If air bubbles appear and the coolant level drops, fill the system with a 50/50 antifreeze/water mixture to bring the level back to the proper level.
3. Run the engine until the thermostat opens and coolant flow is visible.
4. At this point, air is often expelled and the level may drop again. Keep refilling the system until the level is near the top of the radiator and remains constant.
5. Fill the radiator to the filler neck, then install the radiator filler cap.
6. Check and make sure the coolant reservoir is filled to the correct level.

FUEL SYSTEM

Fuel System Service Precaution

When working with the fuel system certain precautions should be taken; always work in a well ventilated area, keep a dry chemical (Class B) fire extinguisher near the work area. Always disconnect the negative battery cable and do not make any repairs to the fuel system until all the necessary steps for repair have been reviewed.

Fuel System Pressure

Before loosening or disconnecting any fuel fitting or system component, always relieve the fuel system pressure in order to help prevent the danger of fire or injury.

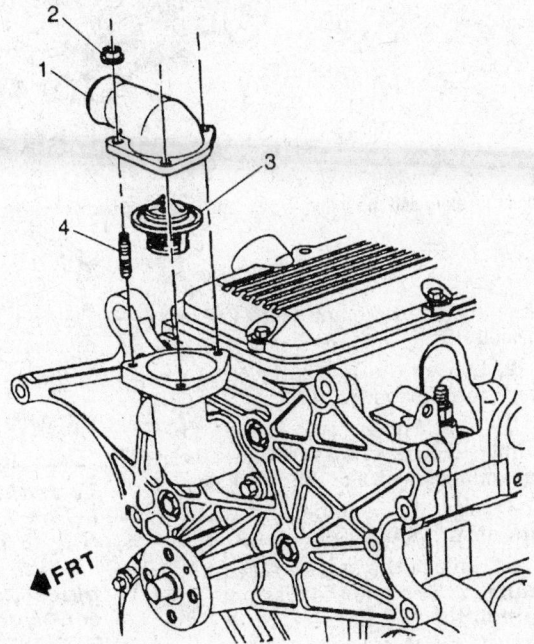

1. OUTLET, COOLANT
2. NUT, 10 N·m (89 LBS. IN.)
3. THERMOSTAT
4. STUD, 10 N·m (89 LBS. IN.)

Thermostat mounting — 2.2L engine

1. Stud
2. Bolt
3. Bolt
4. Outlet
5. Gasket
6. Thermostat
7. Housing
8. Bolt
9. Gasket
10. Bolt
11. Gasket
12. Thermostat
13. Outlet

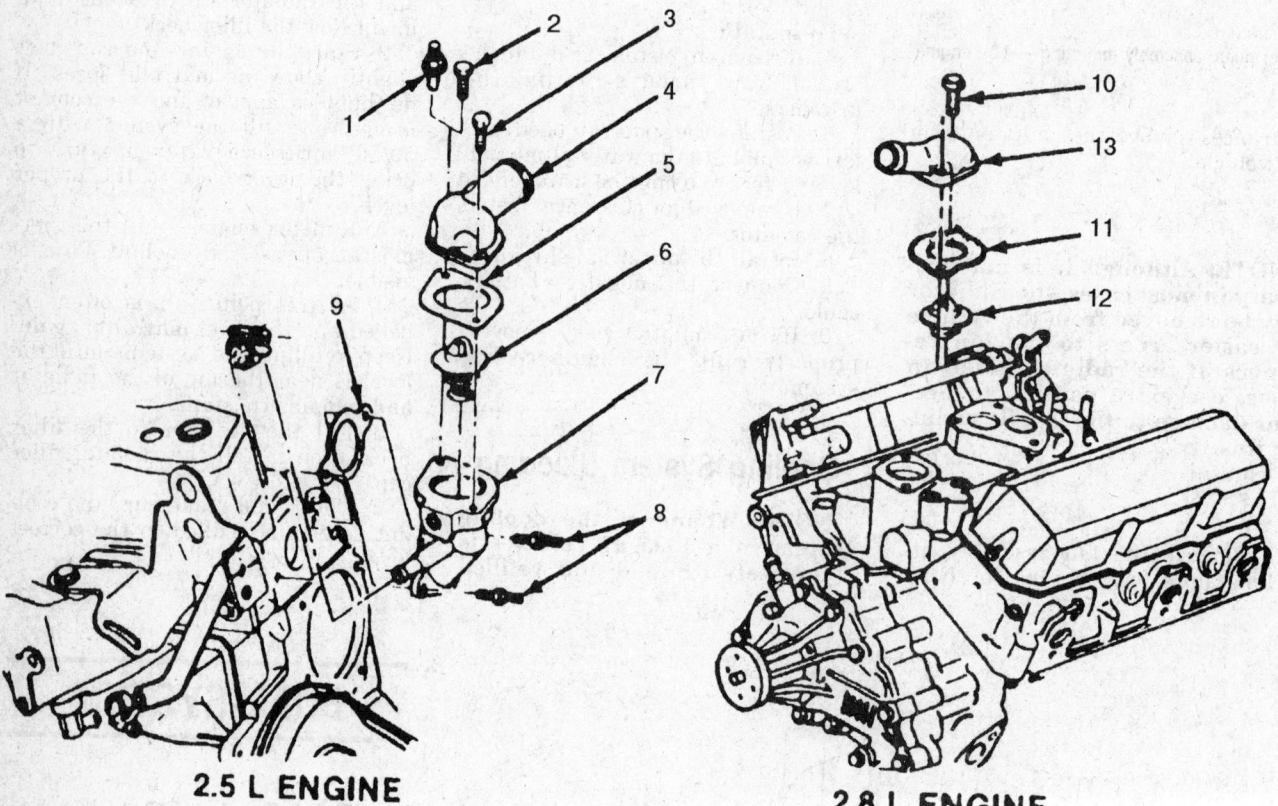

2.5 L ENGINE

2.8 L ENGINE

79219g21

Thermostat locating and mounting — 2.5L and 2.8L engines

RELIEVING

Throttle Body Fuel Injection System

2.5L (VIN A) Engine

Unlike most GM vehicle Throttle Body Fuel Injection (TBI) engines, then Throttle Body (TBI) unit used on the 2.5L engine does not contain an automatic pressure bleed down feature. Because of this, and the lack of service fittings, the pressure is relieved by disabling the fuel pump and allowing the engine to run until fuel pressure drops.

1. Place the transmission selector in **P** for automatic transmissions or **N** for manual transmissions, then set the parking brake and block the drive wheels.

2. Loosen the fuel filler cap to relieve tank pressure.

3. Disengage the fuel pump/sending unit 3 terminal electrical connector at the fuel tank.

4. Start the engine and allow to run until it stops due to lack of fuel.

5. Engage the starter (turn key to start) for 3 seconds to dissipate pressure in the fuel lines.

6. Turn the ignition **OFF**, then reengage the connector at the fuel tank.

7. Disconnect the negative battery cable to prevent fuel spillage should the ignition key accidentally be turned **ON** with a fuel fitting disconnected.

8. When service is finished, tighten the fuel filler cap and connect the negative battery cable.

— **CAUTION** —
To reduce the chance of personal injury when disconnecting a fuel line, always cover the line with cloth to collect escaping fuel, then place the cloth in an approved container.

EXCEPT 2.5L (VIN A) Engine

Unlike most 2.5L TBI engine, the remaining GM vehicle TBI engines utilize an automatic pressure bleed down

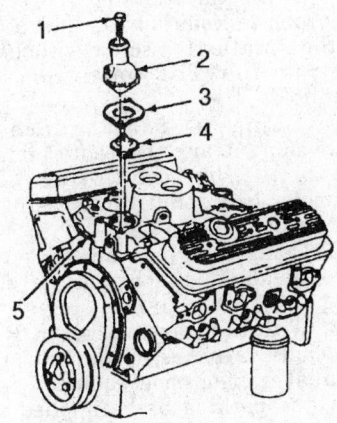

1. Bolt
2. Water outlet
3. Gasket
4. Thermostat
5. Inlet manifold

79219g22

Thermostat mounting — 4.3L (VIN Z) engine

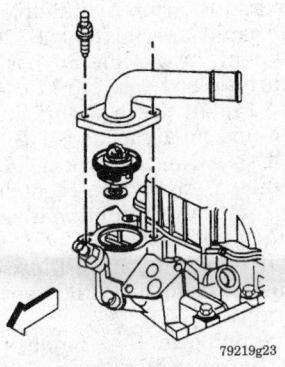

79219g23

Thermostat mounting — 4.3L (VIN W) engine

feature. But, some fuel pressure related steps should still be taken to assure safer working conditions.

1. Disconnect the negative battery cable to prevent fuel spillage should the ignition key accidentally be turned **ON** with a fuel fitting disconnected.

2. Loosen fuel filler cap to relieve fuel tank pressure.

3. The internal constant bleed feature of the Model 220 TBI unit relieves fuel pump system pressure when the engine is turned **OFF**.

Therefore, no further action is required.

NOTE: Allow the engine to set for 5–10 minutes; this will allow the orifice (in the fuel system) to bleed off the pressure.

4. When fuel service is finished, tighten the fuel filler cap and connect the negative battery cable.

Multi-Port Fuel Injection Systems

The MFI, MFI-Turbo, CMFI and Central SFI fuel systems used on GM vehicles all operate under high fuel pressures. It is very important that the pressure be properly relieved prior to servicing the system or any of its components.

A schrader valve is provided on these fuel systems in order to conveniently test or release the system pressure. A fuel pressure gauge and adapter will be necessary to connect the gauge to the fitting. Most of the MFI and MFI-Turbo systems utilize a service valve on 1 end of the fuel rail assembly. The CMFI and CSFI systems covered here uses a valve located on the inlet pipe fitting, immediately before it enters the CMFI/CSFI assembly (towards the rear of the engine).

1. Disconnect the negative battery cable to assure the prevention of fuel spillage if the ignition switch is accidentally turned **ON** while a fitting is still disconnected.

2. Loosen the fuel filler cap to release the fuel tank pressure.

3. Make sure the release valve on the fuel gauge is closed, then connect the fuel gauge to the pressure fitting located on the inlet fuel pipe fitting.

NOTE: When connecting the gauge to the fitting, be sure to wrap a rag around the fitting to avoid spillage. After repairs, place the rag in an approved container.

4. Install the bleed hose portion of the fuel gauge assembly into an approved container, then open the gauge release valve and bleed the fuel pressure from the system.

5. When the gauge is removed, be sure to open the bleed valve and drain all fuel from the gauge assembly.

6. When fuel service is finished, tighten the fuel filler cap and connect the negative battery cable.

Fuel Filter

REMOVAL AND INSTALLATION

2.5L (VIN A) Engine

1. Properly relieve the fuel system pressure.

2. Disconnect the negative battery cable.

3. Locate the fuel filter by tracing a fuel line either from the tank forward or from the engine rearward. The filter is normally found along the frame rail or in the engine compartment.

4. Disengage the fuel line connections from the filter.

5. Remove the filter mounting clamp bolt and remove the filter.

6. Installation is the reverse of the removal procedure.

2.8L (VIN R) Engine

1. Disconnect the negative battery cable.

2. Remove the fuel line connections from the filter. If equipped with quick connect fittings, use tool J–37088A or equivalent to separate the lines.

NOTE: Before disengaging quick-connect fittings, always twist each side of the connection ¼ turn (in opposite directions) in order to loosen any dirt, then use compressed air (while wearing safety glasses) to blow the dirt free of the fittings.

3. Remove the filter mounting clamp bolt and remove the filter.

4. Installation is the reverse of the removal procedure. Before installing any quick-connect fittings, apply a few drops of clean engine oil to the male connector to assure proper seal and prevent a possible leak.

Except 2.5L and 2.8L (VIN R) Engines

The fuel filter is normally located along the frame rail of the vehicle. On some vehicles however, it may have been relocated to the engine compartment. When in doubt, trace a fuel line from the engine backwards or from the tank forward in order to locate the filter.

1. Properly relieve the fuel system pressure.

2. Disconnect the negative battery cable.

3. Raise and support the vehicle safely.

4. Disengage the fuel line connections from the filter.

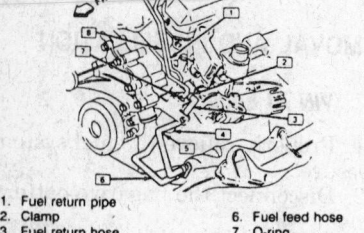

1. Fuel return pipe
2. Clamp
3. Fuel return hose
4. In-line fuel filter
5. Clamp
6. Fuel feed hose
7. O-ring
8. Tighten nut to 20 ft. lbs. (26 Nm)

79219g24

Fuel filter mounting — 2.8L engine

5. Remove the bolt from the filter mounting clamp, then remove the clamp and filter assembly. Separate the filter from the clamp.

6. To install, reverse the removal procedures ensuring the filter flow is in the correct direction. Start the engine and check for leaks.

NOTE: The filter has an arrow (fuel flow direction) on the side of the case, be sure to install it correctly in the system, the with arrow facing away from the fuel tank.

Fuel Pump

REMOVAL AND INSTALLATION

1. Properly relieve the fuel system pressure.
2. Disconnect the negative battery cable.
3. Drain and remove the fuel tank from the vehicle
4. Using a suitable spanner wrench, turn the fuel pump/sending unit assembly locking ring (located on top of the fuel tank) counterclockwise, then carefully lift the assembly from the tank and remove the pump from the fuel lever sending device.

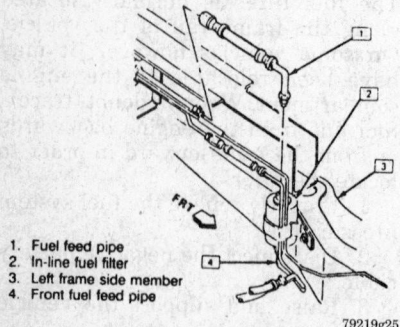

1. Fuel feed pipe
2. In-line fuel filter
3. Left frame side member
4. Front fuel feed pipe

79219g25

Common fuel filter mounting for the 4.3L engine

5. Installation is the reverse of the removal procedure. The fuel pump/sending unit assembly O-ring should be replaced whenever the tank is removed.

Idle Speed

ADJUSTMENT

Idle speed and mixture adjustments on these engines are controlled by the computer modules, regulated through the IAC valve and fuel injector(s), respectively. No periodic check or adjustments are necessary for these systems.

However, the minimum idle speed on some 1993 systems may be adjusted if the throttle body has been replaced AND an incorrect idle speed cannot be obtained. These adjustments should not be performed unless all other possible causes (vacuum leaks, bad fuel pressure, faulty wiring or components) have already been eliminated.

2.5L (VIN A) Engine

The throttle stop screw that is used to adjust the idle speed of the vehicle is preset at the factory. The throttle stop screw is then covered with a

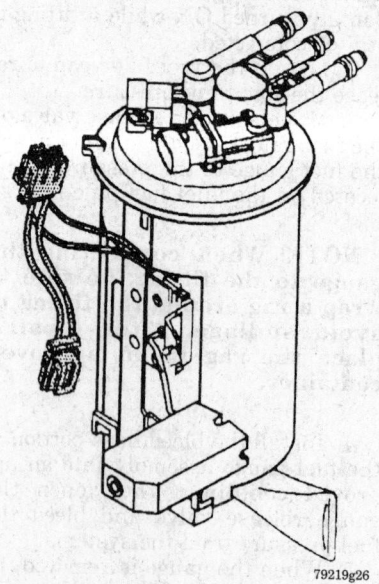

79219g26

View of the fuel pump assembly — 1996 2.2L engine shown

steel plug to prevent adjustment in the field. If it is necessary to gain access to the throttle stop screw, the following procedure will allow access to the throttle stop screw without removing the TBI unit from the manifold

1. Using a small punch or equivalent, mark the center line of the throttle stop screw. Drill a $5/32$ in. diameter hole through the casting of the hardened steel plug.

2. Using a $5/16$ in. diameter punch or equivalent, punch out the steel plug.

3. With the transmission in **P** for automatic transmission or neutral for manual transmission equipped vehicles, the parking brake applied and the drive wheels blocked, remove the air cleaner and plug the thermac vacuum port, as required.

4. If equipped with automatic transmission, remove the transmission detent cable from the throttle control bracket in order to gain access to the minimum air adjustment screw.

5. Connect a tachometer to the engine and disconnect the idle air control motor connector.

6. Start the engine and let the engine reach normal operating temperature and the rpm to stabilize.

7. Install special tool J-33047 or equivalent, in the idle air passage of the throttle body. Be sure to seat the tool in the air passage until it is bottomed and no air leaks exist.

8. Use a No. 20 Torx head bit or equivalent, turn the throttle stop screws until the minimum idle rpm is within specification.

9. If removed install the isolator. Install the transmission detent cable, as required.

10. Shut down the engine and remove the special tool or equivalent from the throttle body.

11. Reconnect the idle air control motor connector and seal the hole drilled through the throttle body housing with silicone sealant or equivalent.

12. Check the throttle position sensor voltage as required. Install the air cleaner and thermac vacuum line.

2.8L and 4.3L (VIN Z) TBI Engines

1. Remove the air cleaner, adapter and gaskets. Discard the gaskets. Plug any vacuum line ports, as necessary.

2. Leave the Idle Air Control (IAC) valve connected and with the engine **OFF**, ground the diagnostic terminal (ALDL connector).

3. Turn the ignition switch to the **ON** position, do not start the engine. Wait for at least 30 seconds; this allows the IAC valve pintle to extend and seat in the throttle body.

4. With the ignition switch still in the **ON** position, disconnect IAC electrical connector.

5. Remove the ground from the diagnostic terminal and start the engine. Let the engine reach normal operating temperature.

6. Apply the parking brake and block the drive wheels. Remove the plug from the idle stop screw by piercing it with a suitable tool and then applying leverage to the tool to lift the plug out.

7. With the engine in the proper shift selector range, adjust the idle stop screw to set the minimum idle to specification.

8. Turn the ignition **OFF** and reconnect the IAC valve connector. Unplug any plugged vacuum line ports and install the air cleaner, adapter and new gaskets.

CONTROLLED IDLE SPEED CHECK

Idle speed adjustments are not provided for most 1994–97 vehicles. However, a controlled idle speed

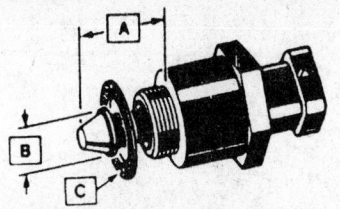

A	DISTANCE OF PINTLE EXTENSION
B	DIAMETER AND SHAPE OF PINTLE
C	IAC VALVE GASKET

79219g28

A common threaded IAC valve used on GM vehicles

check may be made on most vehicles to determine if the computer module is successfully regulating the idle speed. If an improper idle speed is indicated by the check, inspect all components of the engine control systems.

1. Using a scan tool, check to be sure no codes are stored in the computer self-diagnosis memory. If codes are stored, diagnose and correct the problems before proceeding.

2. Verify that the idle air control system is working correctly, then check to make sure the ignition timing is properly set.

3. Connect the Tech 1® or equivalent scan tool to the Diagnostic

Link Connector (DLC). Make sure the tool is in the OPEN mode.

4. Start and run the engine at normal operating temperature.

5. If equipped with an automatic transmission, use the scan tool to check for a proper state of the PRNDL position (R-D-L) switch.

6. Using the scan tool, compare readings of the controlled idle speed and IAC valve pintle position to the specifications for that engine.

7. If the readings agree with specifications, the system is operating properly. If not, the system must be thoroughly checked for faulty connections or components.

Mixture

ADJUSTMENT

All fuel/air mixture adjustments are regulated by the computer control module (PCM or VCM, as applicable). No adjustments are necessary or possible.

Fuel Injector

REMOVAL AND INSTALLATION

TBI Engines

——————— **CAUTION** ———————
When removing the injector(s), be careful not to damage the electrical connector pins (on top of the injector), the injector fuel filter and the nozzle. The fuel injector is serviced as a complete assembly ONLY, it is an electrical component and should not be immersed in any kind of cleaner.

1. Properly relieve the fuel system pressure, then disconnect the negative battery cable.

2. Remove the air cleaner assembly.

3. At the injector connector(s), squeeze the 2 tabs together and pull straight up to disengage connector from the injector.

4. Except for the 2.5L engine (which is equipped with the model 700 TBI unit), loosen the fuel meter cover retaining screws, then remove the cover from the fuel meter body, but leave the cover gasket in place.

5. For the 2.5L engine (model 700 TBI unit), loosen the injector retainer screw, then remove the screw and retainer from the top of the throttle body.

| 1 | IDLE STOP SCREW ASSEMBLY |
| 2 | IDLE STOP SCREW PLUG |

79219g27

Removing idle stop screw plug for adjustment — TBI engines

CONTROLLED IDLE SPEED
NOTE: Engine at operating temperature

Engine	Transmission	Gear (Drive/Neutral)	Idle Speed (RPM)	IAC Counts *	Open/Closed Loop
4.3L (Central MFI)					
M-VAN	AUTO	DRIVE	550	5-40	CL
L-VAN	AUTO	DRIVE	625	15-50	CL

* Add 2 counts for engines with less than 500 miles. Add 2 counts for every 1000 ft. above sea level.

79219g29

Controlled idle speed check specifications — Astro and Safari

CONTROLLED IDLE SPEED
NOTE: Engine at operating temperature

Engine	Transmission	Gear (Drive/Neutral)	Idle Speed (RPM)	IAC Counts *	Open/Closed Loop **
2.2L S-TRK	MAN	NEUTRAL	950	5-20	CL
	AUTO	DRIVE	800	15-40	CL
4.3L (TBI)	AUTO	DRIVE	650	5-30	CL
S/T-PICKUP	MANUAL	NEUTRAL	725	3-30	
4.3L (TBI)					
S/T UTILITY	MANUAL	NEUTRAL	700	5-30	CL
4.3L (CMFI)					
S/T UTILITY	AUTO	DRIVE	600	5-40	CL
PICKUP	AUTO	DRIVE	625	15-50	CL
	MANUAL	NEUTRAL	650	15-50	CL

* On a manual transmission vehicle the Tech 1 will display RDL in neutral.

* Add 2 counts for engines with less than 500 miles. Add 2 counts for every 1000 ft. above sea level.

** Let engine idle until proper fuel control status ("Open/Closed Loop") is reached.

79219g30

Controlled idle speed check specifications — Except Astro and Safari

6. Using a small pry bar and a round fulcrum, carefully pry the injector until it is free, then remove the injector from the fuel meter body.

7. Remove the small O-ring from the nozzle end of the injector. If equipped and removal is necessary, carefully rotate the injector's fuel filter back and forth to remove it from the base of the injector.

8. If applicable (except for the 2.5L engine), remove and discard the fuel meter cover gasket.

9. Remove the large O-ring and back-up washer, if equipped, from the top of the counterbore of the fuel meter body injector cavity.

To install:

NOTE: Be sure the replacement injectors have an identical part number. For some applications, the 4.3L (VIN Z) uses 2 different injectors that have different flow rates. Injectors with part no. 5235134 (color coded orange and green) are located on the throttle lever side, and those with part no. 5235342 (color coded pink and brown) are located on the TPS side.

10. If removed, with the larger end of the filter facing the injector (so the filter covers the raised rib of the injector base) install the filter by twisting it into position on the injector.

11. Lubricate the new O-rings with clean automatic transmission fluid, then install the small O-ring on the nozzle end of the injector. Be sure the O-ring is pressed up against the injector or injector filter (as applicable).

12. Install the steel backup washer, if equipped, in the top counterbore of the fuel meter body's injector cavity, then install the new large O-ring directly over the backup washer. Make sure the O-ring is properly seated in the cavity and is flush with the top of the fuel meter body casting surface.

NOTE: If the backup washer and large O-ring are not properly installed before the fuel injector, a fuel leak will likely result.

13. For the model 220 TBI unit (all engines except the 2.5L), install the fuel injector into the cavity by aligning the raised lug on the injector base with the cast notch in the fuel meter body cavity. Once the injector is aligned, carefully push down on the injector by hand until it is fully seated in the cavity. When properly aligned and installed, the injector terminals will be approximately parallel to the throttle shaft.

14. For the 2.5L engine (model 700 TBI unit), carefully align and install the injector to the cavity. Push straight down until the injector is properly seated. The injector connector should be installed parallel to the casting support rib and facing in the general direction of the cut-out in the fuel meter body provided for the wire grommet.

15. Except for the 2.5L engine, position a new fuel meter cover gasket, then install the cover to the body, making sure the gasket remains in position. Using a suitable threadlocking compound, install and tighten the cover retainers to 30 inch lbs. (4 Nm).

16. For the model 2.5L engine, install the injector retainer and coat the retainer screw threads with a suitable threadlocking compound, then install and tighten the screw to secure the injector.

17. Engage the injector electrical connector(s).

18. Connect the negative battery cable, then turn the **ON** to pressurize the fuel system and check for leaks.

19. Install the air cleaner assembly, then start the engine and check for leaks.

2.2L MFI Engine

The bottom feed fuel injectors used on the 2.2L MFI engine are installed to the lower intake manifold assembly. For access, the upper intake must first be removed.

NOTE: Take care when servicing the lower intake and fuel injectors to prevent dirt or contaminants from entering the fuel system. All openings in the fuel lines and passages should be capped or plugged while disconnected.

1. Properly relieve the fuel system pressure, then disconnect the negative battery cable.

2. Remove the accelerator bracket retaining bolts/nuts, then remove or reposition the bracket.

3. Remove the upper intake manifold assembly.

4. Remove the fuel return line bracket nut, then remove the return line retaining bracket and position it away from the pressure regulator.

5. Remove the fuel pressure regulator assembly.

NOTE: Do not attempt to remove the injectors from their bores while lifting upward on the retaining bracket or damage may occur. Do not attempt to remove the bracket without first removing the pressure regulator.

6. Remove the fuel injector retainer bracket attaching screws, then remove the bracket by carefully sliding it off in order to clear the injector slots and regulator.

7. Tag and disengage the injector electrical connectors.

8. Remove the fuel injectors from the lower intake manifold assembly, then remove and discard the old O-rings.

To install:

NOTE: Because each injector is calibrated for a specific flow rate, make sure to only replace fuel injectors using an identical part number to the old injectors.

9. Lubricate the new O-ring seals with clean engine oil, then position them on the injectors.

10. Carefully install the injectors assemblies into the lower manifold sockets, making sure the electrical connectors are facing inward.

11. Position the injector bracket to the retaining slots and regulator are aligned with the bracket slots.

12. Engage the injector electrical connectors as tagged during removal.

13. Install the fuel pressure regulator assembly.

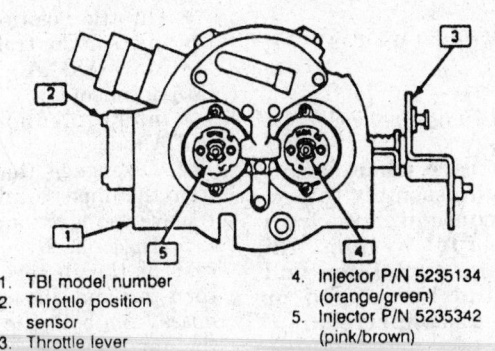

1. TBI model number
2. Throttle position sensor
3. Throttle lever
4. Injector P/N 5235134 (orange/green)
5. Injector P/N 5235342 (pink/brown)

79219g31

Fuel injector location — Model 220 TBI unit (used on engines, except the 2.5L)

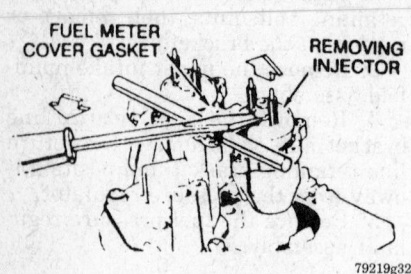

Removing the fuel injector — Model 220 TBI unit

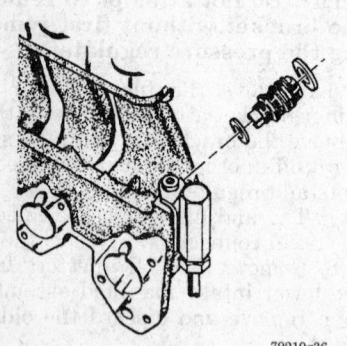

Fuel injector removal — 1996 2.2L engine shown

14. Make sure the threads of the injector retainer bracket screws are coated with a suitable threadlocking compound such as Loctite®262 or equivalent, then install and tighten them to 31 inch lbs. (3.5 Nm).

15. Install the upper intake manifold assembly.

16. If not done already, install the accelerator cable bracket. Tighten the retaining nut to 22 ft. lbs. (30 Nm) and the retaining bolts to 18 ft. lbs. (25 Nm).

17. Connect the negative battery cable.

18. Pressurize the fuel system by cycling the ignition (without attempting to start the engine), then check for leaks.

19. If not done already, install the air inlet duct

4.3L (VIN Z) MFI-Turbo Engines

NOTE: Take care when servicing the fuel rail assembly to prevent dirt or contaminants from entering the fuel system. All openings in the fuel lines and passages should be capped or plugged while disconnected.

1. Before disassembly, clean the fuel rail assembly with a spray type cleaner such as AC Delco® X-30A or equivalent. After removal, do not immerse the fuel rail in liquid solvent.

2. Properly relieve the fuel system pressure and disconnect the negative battery cable.

3. Remove the upper intake plenum, if equipped with turbocharger.

4. Remove the fuel supply and return lines from the fuel rail by squeezing the tabs and pulling the lines apart or by loosening the fittings, depending upon the application. Cap all open lines to prevent dirt and contaminants from entering.

5. Disconnect the vacuum line from the pressure regulator, then tag and disconnect the fuel injector wire connectors.

6. Remove the fuel rail securing bolts and lift the rail up with equal force on both sides.

7. Remove the injector retaining clips and remove the injectors.

8. Remove and discard All old O-rings from the fuel injectors and the fuel lines.

9. Installation is the reverse of the removal procedure. Always use new O-rings on the injectors and fuel line fittings.

4.3L (VIN W) CMFI and CSFI Engines

The non-repairable CMFI/CSFI assembly or injection unit consists of a fuel meter body, gasket seal, fuel pressure regulator, fuel injector and 6 poppet nozzles with fuel tubes. The assembly is housed in the lower intake manifold. Should a failure occur in the CMFI/CSFI assembly, the entire component must be replaced as a unit.

1. Remove the plastic cover and properly relieve the fuel system pressure.

2. Disconnect the negative battery cable, then remove the air cleaner and air inlet duct.

3. Disengage the wiring harness from the necessary upper intake components including:
- Throttle Position (TP) sensor
- Idle Air Control (IAC) motor
- Manifold Absolute Pressure (MAP) sensor
- Intake Manifold Tuning Valve (IMTV)

4. Disengage the throttle linkage from the upper intake manifold, then remove the ignition coil.

5. Disconnect the PCV hose at the rear of the upper intake manifold, then tag and disengage the vacuum hoses from both the front and rear of the upper intake.

6. Remove the upper intake manifold bolts and studs, making sure to note or mark the location of all studs to assure proper installation. Remove the upper intake manifold from the engine.

7. Disengage the injector wiring harness connector at the CMFI/CSFI assembly.

8. Remove and discard the fuel fitting clip.

9. Disconnect the fuel inlet and return tube and fitting assembly. Discard the old O-rings.

10. Squeeze the poppet nozzle locktabs together while lifting each nozzle out of the casting socket. Once all 6 nozzles are released, carefully lift the CMFI/CSFI assembly out of the casting.

11. Disassemble the lower hold-down plate and nuts. While pulling the poppet nozzle tube downward, push with a small prytool down between the injector terminals until the injector is removed. Remove and discard the injector O-rings.

To install:

12. Lubricate new injector O-ring seals with clean engine oil and install on the injector assembly.

13. Assembly the CSFI fuel injector assembly into the fuel meter body injector socket.

14. Install the lower hold-down plate and nuts. Tighten the nuts to 27 inch lbs. (3 Nm).

15. Align the assembly grommet with the casting grommet slots and push downward until it is seated in the bottom guide hole.

CAUTION
To reduce the risk of fire and personal injury, be absolutely sure the poppet nozzles are firmly seated and locked into their casting sockets. An unlocked poppet nozzle could work loose from its socket resulting in a dangerous fuel leak.

---

16. Carefully insert the poppet nozzles into the casting sockets. Make sure they are firmly seated and locked into the casting sockets.

17. Position new O-ring seals (lightly coated with clean engine oil), then connect the fuel inlet and return tube and fitting assembly.

18. Install a new fuel fitting clip.

19. Temporarily connect the negative battery cable, then pressurize the fuel system by cycling the ignition switch **ON** for 2 seconds, then **OFF** for 10 seconds and repeating, as necessary. Once the fuel system is pressurized, check for leaks.

20. Disconnect the negative battery cable.

21. Position a new upper intake manifold gasket on the engine, mak-

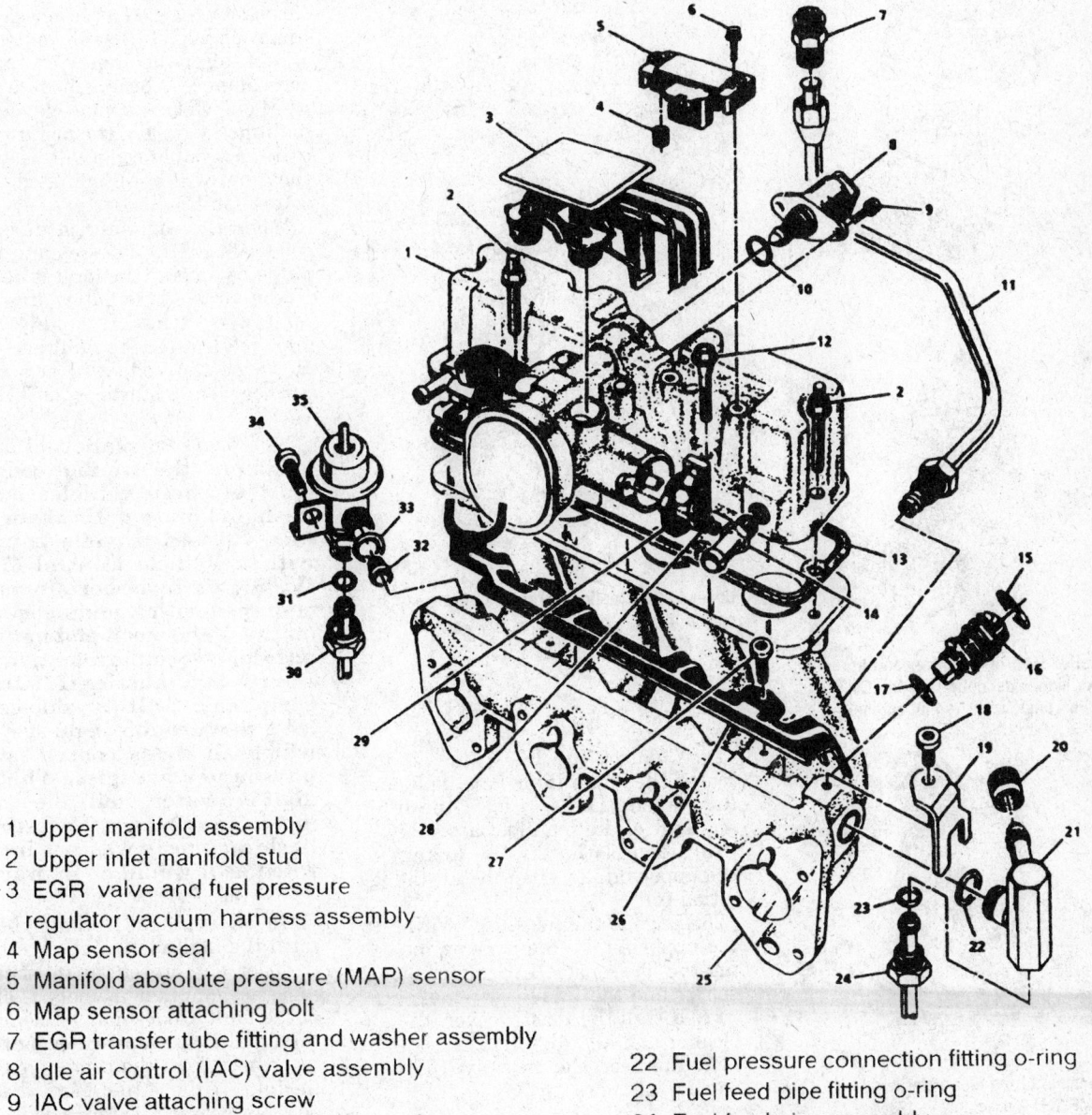

1 Upper manifold assembly
2 Upper inlet manifold stud
3 EGR valve and fuel pressure
   regulator vacuum harness assembly
4 Map sensor seal
5 Manifold absolute pressure (MAP) sensor
6 Map sensor attaching bolt
7 EGR transfer tube fitting and washer assembly
8 Idle air control (IAC) valve assembly
9 IAC valve attaching screw
10 IAC valve o-ring
11 EGR transport tube assembly
12 Upper inlet manifold bolt
13 Upper inlet manifold gasket
14 Power brake fitting
15 Fuel injector upper o-ring
16 MFI fuel injector assembly (bottom feed)
17 Fuel injector lower o-ring
18 Fuel pressure connection fitting screw
19 Fuel pressure connection fitting bracket
20 Fuel pressure connection cap
21 Fuel pressure connection fitting

22 Fuel pressure connection fitting o-ring
23 Fuel feed pipe fitting o-ring
24 Fuel feed pipe assembly
25 Lower manifold assembly
26 Injectors retainer
27 Injectors retainer attaching screw
28 TP sensor attaching screw
29 Throttle position (TP) sensor
30 Fuel return pipe assembly
31 Fuel return pipe fitting o-ring
32 Filter (if so equipped) screen
33 Fuel pressure regulator o-ring
34 Fuel pressure regulator attaching screw
35 Fuel pressure regulator assembly

79219g33

Exploded view of the upper and lower intake manifold assemblies along with related components — 2.2L MFI engine

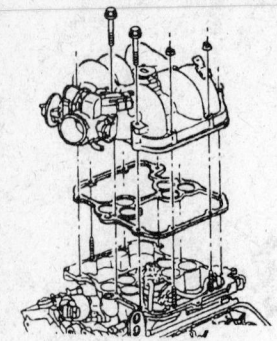

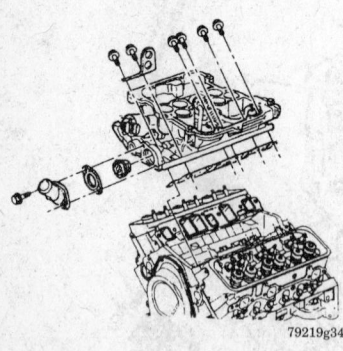

79219g34

Exploded view of the upper and lower intake manifolds along with the CMFI system components — 4.3L engine

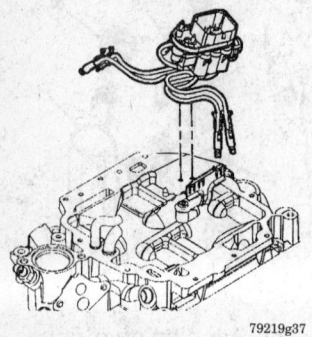

79219g37

Removal of the fuel meter body (CMFI/CSFI) assembly — 1996 vehicle shown

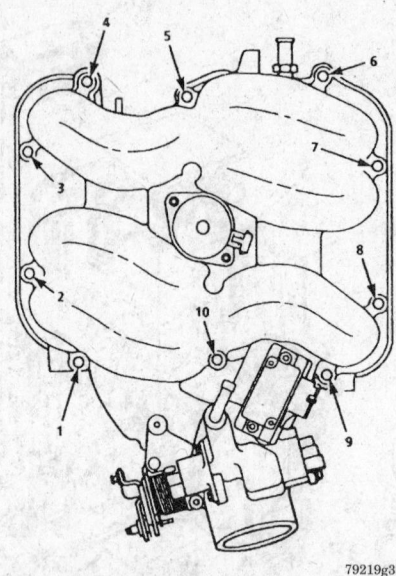

79219g35

Upper intake manifold torque sequence — 4.3L engine through 1995

24. Connect the PCV hose to the rear of the upper intake manifold and the vacuum hoses to both the front and rear of the manifold assembly.

25. Connect the throttle linkage to the upper intake, then install the ignition coil.

26. Engage the necessary wiring to the upper intake components including the TP sensor, IAC motor, MAP sensor and the IMTV.

27. Install the plastic cover, the air cleaner and air inlet duct.

28. Connect the negative battery cable.

ing sure the green sealing lines are facing upward.

22. Install the upper intake manifold being careful not to pinch the fuel injector wires between the manifolds.

23. Install the manifold retainers, making sure the studs are properly positioned, then tighten them using the proper sequence to 124 inch lbs. (14 Nm) for vehicles through 1995. For 1996 vehicles, tighten the retainers 83 inch lbs. (10 Nm).

EMISSION CONTROLS

Emission Warning Lamps

RESETTING

These vehicles are equipped with a SERVICE ENGINE SOON Malfunction Indicator Lamp (MIL). The MIL will illuminate when the key is turned to the **ON** position as a systems check, but will extinguish shortly after the engine is started. If the computer control module (ECM, PCM or VCM, as applicable) detects a malfunction in 1 of the monitored circuits, a trouble code will be set and the lamp will be illuminated to indicate a fault.

When the computer module sets a code, the MIL will remain illuminated as long as the fault is detected. If the problem is intermittent, the light will extinguish approximately 10 seconds after the fault goes away. However the code will stay in the memory for 50 starts or until cleared.

NOTE: If a scan tool is not available, the trouble codes for most of these vehicles may be flash diagnosed. However, certain 1994–97 models equipped with a Vehicle Control Module (VCM), such as certain models equipped with a module mounted on the fender well and not in the passenger compartment, are On-Board Diagnostic II (OBD-II) compliant. OBD-II vehicles utilize a new trouble code system in which all codes consist of 5 alpha-numeric digits. The first digit is a letter, while the remaining 4 are numbers. Nevertheless, flash diagnosis was not incorporated as it would be impractical.

Obviously, the MIL cannot be reset until the malfunction is corrected (or at least not present if intermittent) and the computer memory is cleared of the fault. Codes may be cleared using a suitable scan tool or by removing power from the computer control module. If the latter method is chosen, power must be removed from the module for a minimum of 30 seconds. The lower the ambient temperature, the longer period of time the power must be removed.

Remember, to prevent control module damage, the ignition must always be **OFF** when removing power. Power may be removed through various methods, depending on how the vehicle is equipped. The preferable method is to remove the control module fuse from the fuse box. The control module pigtail may also be unplugged. If necessary, the negative battery cable may be disconnected, but other on-board data such as radio presets will be lost.

ENGINE MECHANICAL

NOTE: Disconnecting the battery cable on most vehicles may interfere with the functions of the on-board computer systems and may require the computer to undergo a relearning process.

Engine Assembly

REMOVAL AND INSTALLATION

Except Astro and Safari

2.5L and 2.8L (VIN R) Engines

1. Properly relieve the fuel system pressure and disconnect the negative battery cable.

2. Matchmark the hood hinges for installation reference and remove the hood.

3. Drain the cooling system and remove the upper and lower radiator hoses. Disconnect the coolant overflow hose. Disconnect the heater hoses at the engine.

4. Remove the upper and lower fan shrouds. On automatic transmission equipped vehicles, disconnect and plug the transmission cooler lines.

5. Remove the air cleaner assembly and cover the carburetor/throttle body with a rag.

6. Label and disconnect all necessary hoses, vacuum lines and wires from the engine, transmission and transfer case, if equipped.

7. Disconnect the throttle cable, transmission TV cable and cruise control cable, if equipped.

8. Raise and support the vehicle safely. Disconnect the exhaust pipes at the manifold.

9. Remove the front driveshaft and skid plates on 4WD vehicles.

10. If applicable, disconnect the strut rods at the bellhousing.

11. For automatic transmission vehicles the body will have to be raised away from the engine in order for the top transmission-to-engine bolt(s) to clear the cowl. Remove the body mounting bolts, then raise the front of the body away from the frame. Support the body using blocks of wood. Remove the top transmission-to-engine mounting bolts, then re-

move the wood and lower the body back into position.

NOTE: 4WD vehicles equipped with automatic transmissions usually do not require transmission/transfer case removal when removing the engine. If equipped with a 2.8L engine, 4WD and a manual transmission, the transfer case and transmission must usually be removed prior to removing the engine.

12. Remove the rear driveshaft.

13. Support the transmission and remove the transmission crossmember.

14. If equipped with an automatic transmission, perform the following:

    a. Remove the torque converter cover.

    b. Remove the torque converter-to-flexplate attaching bolts.

    c. If transmission removal is desired, remove the transmission shift linkage and, if equipped, the transfer case shift linkage. Remove the remaining transmission-to-engine mounting bolts, and remove the transmission and transfer case, if equipped, as an assembly.

15. If equipped with a manual transmission, perform the following:

    a. Remove the clutch slave cylinder and set aside.

    b. If transmission removal is desired or necessary, remove the transmission shift linkage and shifter, and if equipped, remove the 4WD transfer case shift linkage and shifter.

    c. Remove the transmission-to-bellhousing bolts, if the transmission is being removed, carefully lower the transmission and transfer case, if equipped, as an assembly.

NOTE: If the manual transmission is removed, leave the bellhousing in place to protect the clutch during engine removal. Leaving the bellhousing in place will also prevent the necessity of raising the body off the frame to access the top bellhousing-to-engine bolts which would otherwise be blocked by the cowl.

16. Remove the accessory drive belts. Remove the fan.

17. If equipped, remove the power steering pump, A/C compressor and air pump with their brackets and place aside in the engine compartment with the lines intact.

NOTE: Do not disconnect the fluid or refrigerant lines.

18. Verify that nothing else is attached to the engine and, if necessary, disconnect the remaining component(s).

19. Attach a suitable lifting device to the engine and remove. Pause several times while lifting the engine to make sure no hoses or wiring have been snagged by the powerplant.

To install:

20. Using the lifting device, carefully lower the engine into position. Loosely install the engine mount bolts at this time to hold the engine in position.

21. If equipped, reposition and secure the power steering pump and/or A/C compressor to the engine.

22. Install the engine cooling fan and the accessory drive belts.

23. If removed, install the transmission and, if equipped, transfer case.

24. Install the transmission crossmember and remove the support. Install all accessible transmission-to-engine/bellhousing bolts (as applicable) at this time.

25. Tighten the engine and transmission fasteners:

- Engine mount-to-engine: 35 ft. lbs. (47 Nm).
- Engine mount-to-frame mount: 52 ft. lbs. (70 Nm).
- Transmission mount-to-transmission: 45 ft. lbs. (61 Nm).
- Transmission mount-to-crossmember: 24 ft. lbs. (33 Nm).

26. For automatic transmission vehicles, install the torque converter-to-flexplate bolts and install the torque converter cover.

27. If the transmission/transfer case was removed, reconnect the shift linkage.

28. For manual transmission vehicles, position and install the slave cylinder assembly.

29. Install the rear driveshaft.

30. Carefully raise the front of the body away from the frame, then install the remaining transmission-to-engine bolts. Carefully lower the body back into position and secure the body mounts.

31. If applicable, connect the strut rods at the bellhousing.

32. For 4WD vehicles, install the front driveshaft and, if equipped, the skid plates.

33. Connect the throttle cable, transmission TV cable and cruise control cable, if equipped.

34. Reconnect all necessary hoses, vacuum lines and wires from the engine, transmission and transfer case, if equipped, as noted during removal.

35. Remove the cover from the throttle body/carburetor, then install the air cleaner assembly.

36. For automatic transmission equipped vehicles, remove the plugs, then reconnect the transmission cooler lines.

37. Install the upper and lower fan shrouds.

38. Connect the upper and lower radiator hoses, the coolant overflow hose and the heater hoses at the engine.

39. Align and install the hood using the matchmarks made during removal.

40. Connect the negative battery cable, then proper refill the engine cooling system.

41. Check all powertrain fluid levels and add, as necessary.

4.3L (VIN W, X, and Z) Engine — 1993 2WD

1. Disconnect the negative battery cable and properly relieve the fuel system pressure.

2. Scribe matchmarks for installation purposes, then remove the hood.

3. Properly drain the engine cooling system, then disconnect the upper radiator hose from the radiator.

4. Disconnect the overflow hose, then remove the upper fan shroud.

5. If equipped, disconnect the automatic transmission cooler lines from the radiator assembly. Plug the openings to prevent system contamination or excessive fluid loss.

6. Remove the radiator assembly.

7. Remove the engine cooling fan, then disconnect and plug the heater hoses.

8. Remove the air cleaner assembly, then tag and disconnect all necessary vacuum hoses.

9. Tag and disconnect all necessary wires at the bulkhead, ground wires and main feed wires.

10. Disconnect the throttle and cruise control cables, as equipped.

11. Remove the distributor cap.

12. Raise and support the vehicle safely.

13. Remove the catalytic converter-to-exhaust pipe bolts, then disconnect the exhaust pipes at the manifold.

14. Disconnect the strut rods at the bellhousing.

15. If equipped, remove the flywheel cover, then remove the torque converter bolts.

16. Remove the shield at the rear of the catalytic converter.

17. Disconnect the converter hanger at the exhaust pipe.

18. Remove the lower fan shroud.

19. Disconnect the fuel lines and loosen all fuel line clamps from the TBI unit and engine. Move the lines to the rear of the engine compartment and tie them in place, but take care not to stress or damage the lines.

20. Remove the 2 outer air dam bolts.

21. Remove the left body mount bolts, then carefully raise the body from the frame and support.

22. Remove the bellhousing retaining bolts, then remove the supports and carefully lower the body back to the frame.

23. Remove the motor mount through-bolts, then lower the vehicle.

24. Remove the A/C compressor and/or power steering pump from the engine, then position them aside with the lines intact. There is no need to disconnect the lines from either of these components.

25. Support the transmission, then install a suitable lifting device and carefully lift the engine. Pause several times while lifting the engine to make sure no wires or hoses have become snagged.

To install:

26. Carefully lower the engine into the vehicle and engage it with the transmission. Remove the support from the transmission and the lifting device from the engine.

27. Raise and support the vehicle safely.

28. Install the motor mount through-bolts.

29. Raise the body from the frame for access, then install and tighten the bellhousing bolts. Lower the body back into position and install the body mount bolts.

30. If equipped, reposition and secure the power steering pump and/or A/C compressor.

31. Install and tighten the 2 outer air dam bolts.

32. Reposition the fuel lines and connect them to the TBI unit. Be sure to properly secure the lines using the clamps.

33. Install the lower fan shroud, then install the converter hanger to the exhaust pipe.

34. Install the shield at the rear of the converter

35. If applicable, install the torque converter bolts, then install the flywheel cover.

36. Connect the strut rods at the bellhousing.

37. Connect the exhaust pipes to the manifolds, then install the converter-to-exhaust pipe bolts.

38. Lower the vehicle.

39. Install the distributor cap.

40. Connect the throttle and, if equipped, cruise control cables.

41. Engage all wires at the bulkhead, ground wires and main feed wires as tagged during removal.

42. Install the vacuum hoses as noted during removal, then install the air cleaner assembly.

43. Connect the heater hoses, then install the fan.

44. Install the radiator and the upper fan shroud.

45. Connect the overflow hose, the radiator hoses and, if equipped, the transmission cooler lines.

46. Align and install the hood, then check all powertrain fluid levels.

47. Connect the negative battery cable, then properly fill the engine cooling system.

48. Start and run the engine, then check for leaks.

4.3L (VIN W, X, and Z) Engine — 1993 4WD; Except MFI-Turbo

1. Disconnect the negative battery cable and properly relieve the fuel system pressure.

2. If equipped, disconnect the underhood light.

3. Scribe matchmarks for installation purposes, then remove the hood.

4. Properly drain the engine cooling system, then raise and support the the vehicle safely.

5. Loosen the front and remove the 2 body mounts located near or under the cab.

6. Remove the outer bolts from the front air dam.

7. Raise the body away from the frame to gain the necessary clearance for transmission bolt removal. Remove the top transmission-to-engine bolts, then carefully lower the body back into position.

8. Support the transmission, then remove the remaining transmission-to-engine bolts.

9. Remove the 2nd crossmember (located 2nd back from the front of the vehicle).

10. Disconnect the exhaust pipes at the manifolds, then remove the catalytic converter hanger.

11. Remove the torque converter cover bolts and disconnect the front driveshaft from the front differential, then remove the torque converter cover.

12. Release the transmission oil cooler lines at the engine clips.

13. Remove the motor mount bolts.

14. Remove the flexplate-to-torque converter bolts.

15. Remove the front splash shield, then remove the lower fan shroud retaining bolts.

16. Carefully lower the vehicle.

17. Remove the upper fan shroud, then disconnect the hoses from the radiator.

18. Disconnect the oil filter pipe at the remote oil filter.

19. Remove the radiator assembly, then remove the engine cooling fan.

20. Remove the air cleaner assembly.

21. Remove the A/C compressor and/or power steering pump from the engine, then position them aside with the lines intact. There is no need to disconnect the lines from either of these components.

22. Disconnect the fuel lines and loosen all fuel line clamps from the TBI unit and engine. Move the lines to the rear of the engine compartment and tie them in place, but take care not to stress or damage the lines.

23. Tag and disconnect all necessary wires, vacuum lines and emission hoses.

24. Disconnect the accelerator cable and, if equipped, the cruise control cable.

25. Disconnect the engine wiring harness at the bulkhead connector.

26. Disconnect the heater hoses at the engine.

27. Make sure the transmission is supported, then install a suitable lifting device and carefully lift the engine. Pause several times while lifting the engine to make sure no wires or hoses have become snagged.

To install:

28. Carefully lower the engine into the vehicle and engage it with the transmission. Remove the support from the transmission and the lifting device from the engine.

29. Raise and support the vehicle safely.

30. Install the motor mount through-bolts.

31. Raise the body from the frame for access, then install and tighten the bellhousing bolts. Lower the body back into position and install the body mount bolts.

32. Install the flexplate-to-torque converter bolts, then install the torque converter cover.

33. Install the front driveshaft assembly, then install the catalytic converter hanger.

34. Connect the exhaust pipe to the manifolds.

35. Install the crossmember.

36. Install the lower fan shroud retaining bolts, then install the front splash shield.

37. Secure the transmission oil cooler lines at the engine clips.

38. Install the front air dam bolts, then carefully lower the vehicle.

39. Engage the engine wiring harness at the bulkhead connector.

40. Connect the accelerator cable, and if equipped, the cruise control cable.

41. Connect the heater hoses, then connect the wires, vacuum lines and emission hoses as noted during removal.

42. Reposition and connect the fuel lines making sure they are properly routed in their clamps.

43. If equipped, reposition and secure the power steering pump and/or A/C compressor.

44. Install the engine cooling fan, then install the radiator assembly.

45. Connect the oil filter pipe at the remote oil filter using new O-rings.

46. Connect the radiator hoses and install the drive belt(s).

47. Install the upper fan shroud.

48. Install the hood and check all powertrain fluid levels.

49. Connect the negative battery cable and properly refill the engine cooling system.

50. Start and run the engine, then check for leaks.

4.3L MFI-Turbo Engine

1. Disconnect the negative battery cable, followed by the positive cable, then properly relieve the fuel system pressure.

2. Properly drain both the engine cooling system and the turbocharger air cooler system.

3. Scribe matchmarks for installation purposes, then remove the hood.

4. Remove the air cleaner and duct assembly.

5. Remove the turbocharger air inlet elbow.

6. Remove the upper fan shroud, then remove the fan and pulley nuts.

7. Remove the serpentine drive belt, then remove the fan and pulley.

8. Remove the battery tray and vacuum tank.

9. Raise and support the vehicle safely.

10. Remove the front tire and wheel assemblies, then remove the wheel house panels

11. Disconnect the mufflers and tailpipe from the catalytic converter, then remove the catalytic converter support bolts.

12. Remove the turbocharger outlet pipe bracket and outlet pipe nuts.

Move the outlet pipe and catalytic converter away from the turbocharger.

13. Disengage the electrical wiring connector from the charge air cooler radiator temperature sensor, then disconnect the radiator hoses.

14. Remove the charge air cooler radiator.

15. Disconnect the exhaust crossover pipe. Underhood access is necessary, either install the front wheels and lower the vehicle or adjust lift for access.

16. Disconnect the transmission oil cooler lines from the radiator. Plug the openings to prevent system contamination or excessive fluid loss.

17. Remove the upper and lower radiator hoses.

18. Disconnect the engine oil cooler lines from the radiator. Plug the openings to prevent system contamination or excessive fluid loss.

19. Disconnect the heater hoses and the overflow hoses from the radiator.

20. Remove the radiator.

21. Disconnect the oil pipes at the filter adapter.

22. Remove the power steering pump hoses from the steering gear. Plug the openings to prevent system contamination or excessive fluid loss.

23. Remove the engine coolant reservoir.

24. Disconnect the A/C compress from the bracket and position aside with the lines intact.

25. Loosen the charge air cooler clamps, then disconnect the ducts and hoses.

26. Remove the charge air cooler from the supports.

27. Remove the throttle body, gasket and cable bracket from the upper intake manifold and position aside.

28. If not done already, disconnect the heater hose from the lower intake manifold.

29. Tag and disengage the electrical connectors from the upper intake manifold assembly.

30. Disconnect the fuel pipes from the fuel rail assembly.

31. Tag and disengage the remaining electrical connectors and vacuum hoses form the engine assembly.

32. If additional access is needed, further raise and support the vehicle safely.

33. Remove the rear driveshaft.

34. Position a support under the transmission, then remove the transmission crossmember and mount.

35. Remove the front driveshaft.

36. Remove the torque converter cover, then disconnect the shift linkage from the transmission.

37. Remove the torque converter bolts.

38. Disconnect the fuel pipes from the hoses near the transfer case, then remove the fuel line bracket from the transfer case.

39. Disengage the fuel line clip and the electrical clips and connectors from the transmission and transfer case.

40. Remove the transfer case and gasket.

41. Disconnect the TV cable from the transmission.

42. Disconnect the transmission cooler lines from the transmission. Be sure to plug all openings.

43. Remove the torque converter housing bolts, then carefully lower the transmission.

44. Remove the transmission oil cooler lines from the oil pan.

45. Remove the fuel line clip, oil line and electrical harness clips from the cylinder heads.

46. Tag and disengage the starter motor electrical connections, then remove the starter from the engine.

47. Remove the engine mount through-bolts and nuts. Underhood access is necessary, either install the front wheels and lower the vehicle or adjust the lift for access.

48. Disengage the electrical connections form the oil pressure and knock sensors, then remove the ground strap.

49. Install a suitable lifting device and carefully lift the engine. Pause several times while lifting the engine to make sure no wires or hoses have become snagged.

To install:

50. Carefully lower the engine into the vehicle.

51. Engage the electrical connections to the oil pressure and knock sensors, then install the ground strap.

52. Make sure the engine is properly positioned on the engine mounts, then route the wiring harnesses into position.

53. Install the engine mount through-bolts and nuts, then tighten the bolts to 61 ft. lbs. (83 Nm) or the nuts to 52 ft. lbs. (70 Nm).

54. Install the fuel line clip, oil line and electrical harness clips to the cylinder heads.

55. Install the starter motor and engage the wiring as noted during removal.

56. Carefully raise the transmission into position in the vehicle, then install the torque converter housing bolts and tighten to 35 ft. lbs. (41 Nm).

57. Connect the transmission oil cooler lines to the transmission housing and tighten to 21 ft. lbs. (28 Nm).

58. Connect the TV cable to the transmission.

59. Install the transfer case and gasket, then tighten the retaining bolts to 39 ft. lbs. (53 Nm).

60. Install the transmission cooler lines to the oil pan clip.

61. Install the fuel line pipes to the hoses, then connect the fuel line bracket to the transfer case. Tighten the fuel line hose fittings to 19 ft. lbs. (26 Nm) and the bracket bolt to 26 ft. lbs. (35 Nm).

62. Engage the fuel line clip and the electrical wiring clips and connectors to the transmission and transfer case.

63. Install the torque converter bolts and tighten to 46 ft. lbs. (63 Nm), then connect the shift linkage to the transmission.

64. Install the torque converter cover and tighten the retaining bolts to 35 ft. lbs. (47 Nm).

65. Install the front driveshaft and tighten the retaining bolts to 52 ft. lbs. (70 Nm).

66. Install the transmission crossmember and mount, then tighten the crossmember bolts to 35 ft. lbs. (47 Nm).

67. Install the rear driveshaft and tighten the retainers to 15 ft. lbs. (20 Nm).

68. Install the turbocharger outlet pipe and nuts, then install the turbocharger outlet pipe bracket. Tighten the pipe nuts to 41 ft. lbs. (55 Nm) and the pipe support bolt to 22 ft. lbs. (30 Nm).

69. Install the catalytic converter support bolts, then install the mufflers and tailpipe to the converter. Tighten the support bolts to 25 ft. lbs. (34 Nm) and the tailpipe bolts to 24 ft. lbs. (32 Nm).

70. Connect the crossover pipe and tighten the nuts to 12 ft. lbs. (16 Nm).

71. Install the charge air cooler radiator, connect the cooler radiator hoses and engage the electrical connection.

72. Underhood access is necessary, either install the front wheels and lower the vehicle or adjust the lift for access.

73. Engage the electrical connectors and vacuum lines to the engine.

74. Connect the fuel pipes to the fuel rail assembly, then install the upper intake manifold assembly and gasket. Tighten the upper intake manifold bolts (starting at the 2 middle bolts, moving outward to each side) to 18 ft. lbs. (24 Nm).

75. Engage the electrical connectors to the upper intake manifold assembly, then connect the heater hose to the lower intake manifold.

76. Install the throttle body, gasket and cable bracket to the upper intake manifold assembly. Tighten the throttle body bolts to 18 ft. lbs. (24 Nm).

77. Install the charge air cooler to the supports, then install the charge air cooler ducts and hoses (secure using the clamps).

78. Reposition the A/C compressor to the bracket and secure.

79. Install the engine coolant reservoir.

80. Install the power steering pump hoses to the steering gear and tighten the fittings to 21 ft. lbs. (28 Nm).

81. Install the oil pipes to the filter adapter and tighten the retaining bolt to 26 ft. lbs. (35 Nm).

82. Install the radiator assembly.

83. Connect the heater hoses and the overflow hose to the radiator.

84. Install the engine oil and transmission fluid cooler lines to the radiator. Tighten the engine oil line fittings to 26 ft. lbs. (35 Nm) and the transmission fluid lines to 20 ft. lbs. (27 Nm).

85. Connect the upper and lower radiator hoses.

86. Install the battery tray and vacuum tank.

87. Install the fan, pulley and nuts.

88. Install the serpentine drive belt, then tighten the pulley nuts to 18 ft. lbs. (24 Nm).

89. Install the upper radiator shroud.

90. Install the turbocharger air inlet elbow.

91. Install the air cleaner and duct.

92. Align and install the hood using the matchmarks made during removal.

93. Install the front wheel house panels.

94. Install the front tire and wheel assemblies, then carefully lower the vehicle.

95. Check all powertrain fluid levels and add, as necessary.

96. Connect the positive battery cable, followed by the negative battery cable.

97. Properly refill the charge air cooling system and bleed the air.

98. Properly refill the engine cooling system.

99. Start and run the engine, then check for leaks.

1994–97 4.3L (VIN W, X, and Z) Engine

1. Disconnect the negative battery cable and properly relieve the fuel system pressure.

2. Disconnect the vacuum reservoir and/or the underhood light from the hood (as equipped), then remove the outer cowl vent grilles.

3. Matchmark and remove the hood.

4. Raise and support the front of the vehicle safely. It will be most convenient if the vehicle can be supported so underhood access is still possible. Otherwise, the vehicle will have to be raised and lowered multiple times during the procedure for the necessary access.

5. Drain the engine cooling system and the engine oil into separate drain pans.

6. Disconnect the oxygen sensor and/or wiring.

7. Disconnect the exhaust at the manifolds and loosen the hanger at the catalytic converter. This is necessary to remove the rear catalytic converter cushion mounts for removal of the exhaust assembly.

8. If equipped, remove the skid plate.

9. Remove the pencil braces from the engine to the transmission.

10. If equipped, remove the slave cylinder and position aside.

11. Disconnect the line clamp at the bellhousing.

12. Tag and remove the wiring from the starter, remove the flywheel cover and remove the starter.

13. Remove the oil filter.

14. Remove the engine mount through-bolts and remove all of the bellhousing bolts, except the upper left.

15. Disconnect the battery ground (negative) cable from the engine.

16. On 4WD vehicles, remove the front drive axle bolts and roll the axle downward.

17. Remove the air cleaner assembly and duct work.

18. Remove the upper radiator shroud, then remove the fan assembly.

19. Remove the multi-ribbed serpentine drive belt, then remove the water pump pulley.

20. Disconnect the upper radiator hose, then remove the A/C compressor, if equipped, and position aside with the lines intact.

21. Disconnect the lower radiator hose, then disconnect the oil cooler and overflow lines from the radiator. Plug the cooler line openings to prevent system contamination or excessive fluid loss.

22. Remove the radiator from the vehicle, then remove the lower radiator shroud.

23. Disconnect the power steering hoses from the steering gear, then cap the openings to prevent system contamination or excessive fluid loss.

24. Disconnect the heater hoses from the intake manifold and the water pump.

25. Tag, disconnect and remove the wiring harness and vacuum lines from the engine

26. Disconnect the throttle cables, then remove the distributor cap.

27. Remove the remaining bolt from the bellhousing.

28. Disconnect the fuel lines and remove the bracket.

29. Remove the ground strap(s) from the rear of the cylinder head.

30. On 4WD vehicles, loosen the front body mount bolts.

31. Support the transmission.

32. Install a suitable lifting device and carefully lift the engine. Pause several times while lifting the engine to make sure no wires or hoses have become snagged.

To install:

33. Carefully lower the engine into the vehicle.

34. On 4WD vehicles, tighten the front body mount bolts.

35. Install the ground strap(s) to the rear of the cylinder head.

36. Connect the fuel lines and install the bracket.

37. Install the upper left bellhousing bolt.

38. Install the distributor cap and wires.

39. Connect the throttle cables.

40. Connect the vacuum lines and wiring harness connectors as noted during removal.

41. Connect the heater hoses, then uncap and connect the power steering hoses.

42. Install the lower shroud, then install the radiator.

43. Uncap and connect the oil cooler lines to the radiator, then connect the overflow hose.

44. Connect the lower radiator hose, then if equipped, reposition and secure the A/C compressor to the engine.

45. Install the upper radiator hose, then install the water pump pulley.

46. Install the serpentine drive belt, then install the fan assembly.

47. Install the upper radiator shroud, then install the air cleaner and ducts.

48. For 4WD vehicles, roll the front axle up into position, then install and tighten the retaining bolts.

49. Connect the battery ground strap to the engine block.

50. Install the remaining bellhousing bolts.

51. Install the engine mount through-bolts and tighten to 49 ft. lbs. (66 Nm).

52. Install a new oil filter, then install the starter motor.

53. Install the flywheel cover.

54. If equipped, reposition and secure the clutch slave cylinder.

55. Install the pencil brace and the skid plate, as equipped.

56. Install the catalytic converter Y-pipe assembly and hangers.

57. Carefully lower the vehicle.

58. Align the marks made during removal and install the hood.

59. Install the outer cowl vent grilles, then connect the vacuum reservoir and/or the underhood light to the hood (as equipped).

60. Check all powertrain fluid levels and add, as necessary. Be sure to properly fill the engine crankcase with clean engine oil.

61. Connect the negative battery cable and properly fill the engine cooling system.

62. Start and run the engine, then check for leaks.

2.2L (VIN 4) Engine

NOTE: The manufacturer recommends the discharge and recovery of the A/C system R-134a refrigerant for this procedure. Do not attempt this without the proper equipment. R-134a should not be mixed with R-12 refrigerant.

1. Disconnect the negative battery cable and properly relieve the fuel system pressure.

2. Disconnect the vacuum reservoir and/or the underhood light from the hood (as equipped).

3. Disconnect the windshield washer line from the hood, then remove the outer cowl vent grilles.

4. Matchmark and remove the hood.

5. Raise and support the vehicle safely. It will be most convenient if the vehicle can be supported so underhood access is still possible. Otherwise, the vehicle will have to be raised and lowered multiple times during the procedure for the necessary access.

6. Properly recover the R-134a refrigerant from the A/C system.

7. Drain the engine cooling system and the engine oil into separate drain pans.

8. Disconnect the oxygen sensor and/or wiring.

9. Disconnect the exhaust at the manifolds and loosen the hanger at the catalytic converter.

10. Remove the pencil braces from the engine to the transmission.

11. Remove the inspection cover.

12. Tag and remove the wiring from the starter, then remove the starter.

13. Remove the engine mount through-bolts, then remove the bellhousing bolts.

14. Remove the battery from the vehicle, then disconnect the battery ground (negative cable) from the engine.

15. Remove the air cleaner assembly and duct work.

16. Remove the upper fan shroud, then remove the engine cooling fan.

17. Remove the multi-ribbed serpentine drive belt, then remove the water pump assembly.

18. Disconnect the upper radiator hose, then remove the A/C compressor, if equipped, and position aside.

19. Disconnect the lower radiator hose, then remove the radiator and lower fan shroud.

20. Disconnect the power steering hoses from the pump, then cap the openings to prevent system contamination or excessive fluid loss.

21. Disconnect the heater hose from the intake manifold, then disconnect the ground straps at the rear of the engine.

22. Tag, disconnect and remove the wiring harness and vacuum lines from the engine.

23. Disconnect the throttle cable, then disconnect the fuel lines from the fuel rail and engine.

24. Support the transmission.

25. Install a suitable lifting device and carefully lift the engine. Pause several times while lifting the engine to make sure no wires or hoses have become snagged.

To install:

26. Carefully lower the engine into the vehicle, then remove the support and engine lifting device.

27. Connect the fuel lines and install the bracket(s).

28. Connect the throttle cable.

29. Connect the vacuum lines and wiring harness connectors as noted during removal. Engage the wiring harness clips and connect the ground straps to the rear of the engine.

30. Connect the heater hose, then uncap and connect the power steering hoses.

31. Install the lower shroud, then install the radiator.

32. Reposition and secure the A/C compressor to the engine.

33. Install the upper radiator hose, then install the water pump.

34. Install the serpentine drive belt, then install the fan assembly.

35. Install the upper radiator shroud, then install the air cleaner and ducts.

36. Connect the battery ground strap to the engine block.

37. Install the bellhousing bolts.

38. Install the engine mount through-bolts.

39. Install the starter motor.

40. Install the flywheel cover.

41. Install the pencil braces.

42. Install the catalytic converter pipe assembly and hangers.

43. Carefully lower the vehicle.

44. Align the marks made during removal and install the hood.

45. Install the outer cowl vent grilles, then connect the vacuum reservoir and/or the underhood light to the hood (as equipped).

46. Connect the windshield washer hoses.

47. Check all powertrain fluid levels and add, as necessary. Be sure to properly fill the engine crankcase with clean engine oil.

48. Connect the negative battery cable and properly fill the engine cooling system.

49. Start and run the engine, then check for leaks.

Astro and Safari

1. If equipped, properly discharge and recover the refrigerant from the A/C system.

2. Disconnect the negative battery cable.

3. Drain the cooling system.

4. Raise and safely support the vehicle. Disconnect the exhaust pipes at the manifolds.

5. If applicable, disconnect the strut rods at the flywheel housing.

6. Remove the torque converter cover, then remove the torque converter bolts.

7. Remove the starter assembly, then drain the oil and remove the and oil filter. Disconnect the wires at the transmission. Disconnect the fuel lines.

8. Tag and disengage the wires at the engine and frame. Disconnect the fuel lines at the frame.

9. Disconnect the transmission and engine oil cooler lines at the radiator.

10. Remove the lower fan shroud retainers, then remove the motor mount bolts.

11. Lower the vehicle. Remove the headlight bezels and/or grille, as nec-

essary. On 1994–96 vehicles, remove the horns.

12. Remove the radiator close out panel and the radiator support brace.

13. Remove the hood latch mechanism. If necessary, remove the master cylinder.

14. Remove the air cleaner assembly and ducts.

15. Remove the upper fan shroud.

16. For 1993 vehicles:

　　a. Remove the upper radiator core support, then if equipped, remove the A/C condenser.

　　b. Remove the radiator filler panels, then remove the radiator.

　　c. Remove the lower radiator shroud, then remove the engine cover.

　　d. If equipped, disconnect the A/C hose at the accumulator.

　　e. Remove the multi-ribbed accessory drive belt, then remove the fan.

17. For 1994–97 vehicles:

　　a. If equipped, remove the A/C condenser.

　　b. Remove the fan and clutch assembly, then remove the lower fan shroud.

　　c. If not done already, disconnect the remaining oil cooler-to-radiator lines.

　　d. Remove the radiator.

　　e. Remove the engine cover.

　　f. If equipped, remove the A/C accumulator.

　　g. If not done already, remove the multi-ribbed engine accessory drive belt.

18. Disconnect the power steering pump lines at the gearbox (1993) or from the hydro-boost, oil cooler and reservoir (1994–97).

19. If equipped, remove the A/C compressor pencil braces at the engine block.

20. Remove the power steering pump, bracket and A/C compressor as an assembly.

21. Disengage the alternator wiring, then remove the alternator and bracket assembly.

22. Disengage the wiring harness at the bulkhead. Except for 1995–96 vehicles, remove the right kick panel.

23. Disengage the wiring from the knock sensor module.

24. Disconnect the upper and lower radiator hoses, then disconnect the heater hose from the water pump.

25. Remove the oil filler tube, then remove the transmission filler tube (top bolt only).

26. Tag and disconnect the vacuum hoses at the intake manifold.

27. If equipped, remove the cruise control servo and bracket.

28. Matchmark and remove the distributor assembly or the High Voltage Switch (HVS) assembly, as applicable.

29. If equipped with the 4.3L (VIN W) engine, remove the upper intake manifold assembly, then disconnect the fuel lines and remove the lower intake manifold.

30. If equipped with the 4.3L (VIN Z) engine, disconnect the fuel lines from the TBI unit, tag and disengage all cables, wiring and hoses, then remove the TBI unit from the engine. Remove the MAP sensor bracket, then disconnect the heater hose from the engine block with bracket from the exhaust manifold.

31. Raise and support the vehicle safely, then If equipped, remove the transfer case brace.

32. For 1994–97 vehicles, remove the fuel line bracket and ground wire from the back of the left cylinder head.

33. Remove the transmission oil level indicator tube.

34. Disengage the necessary wiring from the transmission.

35. Remove the bellhousing bolts, then lower the vehicle.

36. For 1995–97 vehicles the tie bar must be cut from the vehicle in order

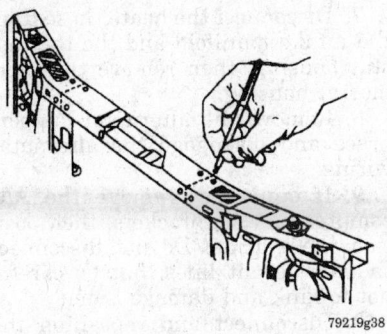

79219g38

Scribing marks for cutting the tie bar — 1995–97 vehicles

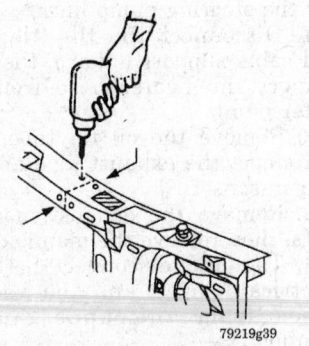

79219g39

Drilling out holes for tie bar replacement brackets — 1995–97 vehicles

to create sufficient clearance for engine removal:

a. Remove the master cylinder retaining nuts, then reposition the cylinder assembly out of the way.

b. Scribe marks for cutting the tie bar assembly. The marks should be made at the centerline between the indentations on the right and left side of the bar assembly.

c. Using the replacement brackets from the service kit as a template over the indentations, center punch the holes for drilling.

d. Drill out 8mm holes for the brace bolts.

e. Carefully cut the tie bar cross section using a reciprocating power saw or hack saw.

NOTE: Extreme care must be taken when cutting out the tie bar cross section. The tie bar will be attached using brackets from the service kit. the cut out portion of the bar and the brackets must be treated with anti-corrosion materials and painted. Care taken during cutting will help save time on surface preparation and installation.

37. Attach a suitable lifting device to the engine and support the transmission, then carefully remove the engine.

To install:

38. Carefully lower the engine into position and engage it to the transmission assembly. If possible, thread the bellhousing bolts to secure the engine to the transmission.

39. Remove the engine lifting device, then raise and support the vehicle safely. Install any remaining bellhousing bolts, then tighten the bolts and remove the transmission support.

40. Engage the wiring to the transmission assembly.

41. Install the transmission oil level indicator tube.

42. On 1994–97 vehicles install the fuel line bracket and ground wire to the back of the left cylinder head.

43. If equipped, install the transfer case brace.

44. Lower the vehicle.

45. If equipped with the 4.3L (VIN Z) engine, connect the heater hose to the engine block with the exhaust manifold bracket, then install the MAP sensor bracket. Instal the TBI unit, connecting all wiring, cables and hoses. Connect the fuel lines.

46. If equipped with the 4.3L (VIN W) engine, instal the lower intake manifold assembly, then connect the

fuel lines and install the upper intake manifold assembly.

47. Align and install the distributor or the HVS assembly, as equipped.

48. If equipped, install the cruise control servo and bracket.

49. Connect the vacuum hoses to the intake manifold.

50. Install the transmission filler tube (upper bolt) and the oil filler tube.

51. If removed and applicable, install the ignition coil.

52. Except for 1996–97 vehicles, the air cleaner and ducts may be installed at this time.

53. Connect the heater and radiator hoses.

54. Connect the wiring harness to the knock sensor module, then for 1993–94 vehicles install the kick panel.

55. Engage the wiring harness at the bulkhead.

56. Install the alternator and bracket as an assembly, then engage the wiring.

57. Install the power steering pump, bracket and A/C compressor assembly. Connect the compressor pencil braces to the block and connect the hoses to the power steering pump, oil cooler and reservoir. Make sure all components are secure.

58. For 1993 vehicles, install the fan.

59. Position the multi-ribbed drive belt.

60. Install the accumulator and/or connect the refrigerant hoses to the accumulator assembly, as applicable.

61. Install the lower radiator shroud, then for 1994–97 vehicles, install the fan and clutch.

62. For 1995–97 vehicles install the tie bar assembly:

a. File the rough edges of the tie bar and removed cross section.

b. Clean the assembly, cross section and brackets using a wax and grease remover.

c. Treat all bare metal surfaces with an anticorrosion primer.

d. Apply primer surfaces to the tie bar assembly, cross section and brackets.

e. Paint the components and allow to dry.

f. Install the front brackets to the tie bar cross section and to the bar assembly using the 2 bolts and nuts facing the front of the vehicle.

g. Install the U-nuts to the rear tie bar cross section and tie bar assembly.

h. Install the rear bracket and remaining nuts and bolts, then tighten to 24 ft. lbs. (31 Nm)

i. Reposition and secure the master cylinder assembly. If necessary, cut out indication hole in the air cleaner snorkel.

63. Install the radiator and connect the hoses.

64. If equipped, install the A/C condenser.

65. Instal the upper fan shroud and, if applicable, the upper radiator core support.

66. Instal the hood latch mechanism, then install the core support brace.

67. On 1993–94 vehicles, install the radiator lower close out panel.

68. On 1994–97 vehicles, install the horns.

69. Install the grille and, if applicable, the headlight bezels.

70. If not done earlier, install the air cleaner and intake ducts at this time.

71. If not done earlier, connect the fuel lines at the engine.

72. Raise and support the vehicle safely, then install the motor mount fasteners.

73. If not done earlier, install the lower fan shroud retainers.

74. Connect the engine and transmission oil cooler lines./

75. Secure the fuel line bracket at the frame.

76. Engage the wiring to the engine and frame, as necessary.

77. Install the oil filter, then install the starter assembly.

78. Install the torque converter bolts, then install the cover.

79. Connect the exhaust pipes, then lower the vehicle.

80. Refill the engine crankcase with engine oil, then connect the negative battery cable.

81. Properly refill the engine cooling system.

82. If equipped, make sure all refrigerant lines are properly connected, then evacuate and charge the A/C system.

Engine Mounts

NOTE: When lifting or raising the engine for any reason, do not support the assembly under the oil pan, any sheetmetal or the crankshaft pulley.

------ CAUTION ------
When working on the engine mounts, make sure the engine is SECURELY supported at all times. If necessary, use blocks of wood to help make sure it cannot shift suddenly.

REMOVAL AND INSTALLATION

2.2L and 2.5L (VIN A) Engines

1. Disconnect the negative battery cable.

2. Support the engine with a suitable lifting fixture. Do not load the engine mounts, just take the engine weight off them.

3. Remove the engine mount through-bolt and nut, then raise the engine just sufficiently to permit mount removal.

4. Remove the engine mount bolts, nuts and washers.

5. Remove the mount assembly attaching bolts, nuts and washers. Remove the mount assembly.

6. Installation is the reverse of the removal procedure.

2.8L (VIN R) Engine

1. Disconnect the negative battery cable.

2. Raise and support the vehicle safely. Support the engine with a suitable lifting fixture.

3. Remove the fan shroud and front wheel.

4. Remove the engine mount through-bolt and nut.

5. Raise the engine enough to permit removal of the engine mounting and block in position.

6. If necessary for access, remove the tie rod at the drag link, then remove the stabilizer link from the control arm. Remove the lower shock absorber bolts. Remove the lower control arm pivot bolts and position control aside.

7. Remove the mount assembly attaching bolts, nuts and washers. Remove the mount assembly.

8. Installation is the reverse of the removal procedure.

4.3L (VIN W, X, and Z) Engine

1. Disconnect the negative battery cable.

2. Except Astro and Safari or if access from underneath is preferable, raise and support the vehicle safely.

3. Except Astro and Safari, remove the cab or body mounting bolts. Raise the body for clearance and block in position.

4. Support the engine with a suitable lifting fixture. Just take the weight off the mounts, do not load them.

NOTE: Do not position the jack under the oil pan, any sheetmetal or the crankshaft pulley, otherwise damage may occur.

5. Remove the engine mount through-bolt and nut.

6. Raise the engine just sufficiently to permit removal of the engine mounting and block in position. Keep a close eye on the engine-to-cowl clearance when raising it.

7. Remove the mount assembly to frame retainers. Remove the mount assembly.

8. Installation is the reverse of the removal procedure.

Cylinder Head

REMOVAL AND INSTALLATION

2.2L (VIN W and 4) Engine

1. Properly relieve the fuel system pressure, then disconnect the negative battery cable.

2. Drain the engine cooling system, then disconnect the air duct from the air inlet.

3. Disconnect the upper radiator hose, then remove the upper fan shroud.

4. Remove the radiator assembly, then remove the lower fan shroud.

5. Remove the fan assembly, then remove the serpentine drive belt.

6. Remove the water pump pulley.

7. Disconnect the heater hose from the intake manifold and the thermostat housing, then remove the thermostat housing.

8. Remove the alternator support brace and disengage the alternator wiring.

9. If equipped, remove the A/C compressor with brackets, then position them aside. Do not disconnect the refrigerant lines, but be careful not to kink and damage them.

10. Disconnect and reposition the accessory bracket along with the alternator and power steering pump still attached. Be careful not to damage the steering pump lines.

11. Disconnect the throttle cable and cable support linkage, then disconnect the heater hose from the water pump.

12. Remove the oil fill tube, then disconnect the exhaust pipe and the oxygen sensor.

13. Remove the exhaust manifold bolts, then remove the manifold.

14. Tag and disconnect both the electrical wiring and the vacuum hoses from the upper intake manifold.

15. Remove the upper intake manifold, then tag and disconnect the wiring from the lower intake manifold.

16. Disconnect the fuel lines, then tag and disconnect the spark plug wires.

17. Remove the lower intake manifold from the engine.

18. Remove the rocker arm cover from the cylinder head.

19. Remove the rocker arms and pushrods.

20. Disconnect the engine lift bracket from the rear of the engine.

21. Remove the cylinder head bolts and studs, then carefully lift the cylinder head from the engine.

To install:

22. Carefully clean and inspect the gasket mounting surfaces.

NOTE: The gasket surfaces on both the head and block must be clean of any foreign matter and free of nicks or heavy scratches. The cylinder bolt threads in the block and thread on the bolts must be cleaned (dirt will affect the bolt torque).

23. Place a new gasket over the dowel pins (do not use any sealer on the gasket), then position the cylinder head over the gasket and dowels.

24. Apply a coating of 1052080 or equivalent sealer to the cylinder head bolt threads. Install the cylinder head bolts (within 15 minutes of sealer application), then tighten them in the proper sequence first to a torque of 46 ft. lbs. (63 Nm) for long bolts or to 43 ft. lbs. (58 Nm) for short bolts and then tighten all bolts an additional 90 degree turn using a torque angle meter.

25. Install the engine lift bracket.

26. Install the rocker arms and pushrods.

27. Install the rocker arm cover.

28. Install the lower intake manifold.

29. Connect the spark plug wires and the fuel lines.

30. Engage the wiring to the lower intake manifold.

31. Install the upper intake manifold.

32. Connect the vacuum hoses and electrical wiring to the upper intake, as tagged during removal.

33. Install the oil fill tube assembly.

34. Install the exhaust manifold, then connect the exhaust pipe and oxygen sensor.

35. Connect the heater hose to the water pump, then connect the throttle cable support and throttle cable.

36. Install the accessory support bracket and components.

37. If equipped, reposition and secure the A/C compressor.

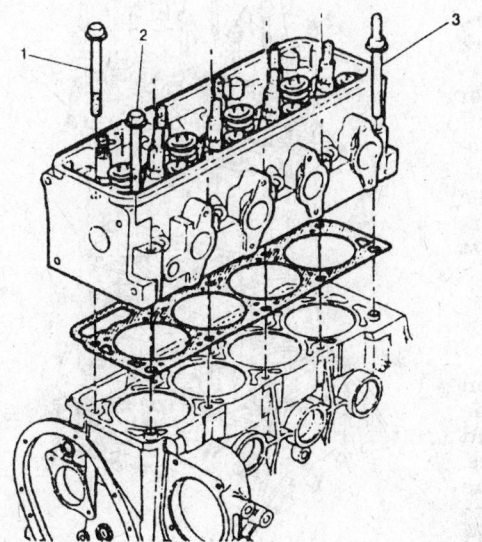

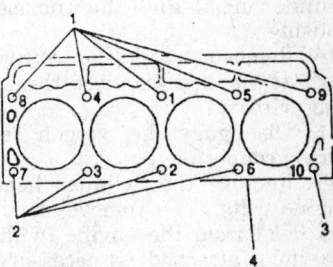

1 LONG BOLTS
2 SHORT BOLTS
3 STUD
4 NUMBERS ON GASKET INDICATE TORQUE SEQUENCE

CYLINDER HEAD BOLT TORQUE PROCEDURE

1 TIGHTEN BOLTS IN SEQUENCE (ITEM 4) TO:
LONG BOLTS: 63 N·m (46 LBS. FT.)
SHORT BOLTS: 58 N·m (43 LBS. FT.)

2 TIGHTEN ALL BOLTS AN ADDITIONAL ANGLE
OF 90° IN SEQUENCE (ITEM 4) USING
J 36660 OR EQUIVALENT

79219g40

Cylinder head bolt torque sequence — 2.2L engine

38. Install the power steering support brace and the alternator support brace. Engage the alternator wiring.

39. Install the thermostat housing, then connect the heater hose to the housing.

40. Install the water pump pulley and the serpentine drive belt.

41. Install the fan assembly, then install the radiator and the lower fan shroud.

42. Install the upper fan shroud, then connect the upper radiator hose.

43. Connect the air inlet duct work, then connect the negative battery cable.

44. Properly refill the engine cooling system and check for leaks.

2.5L (VIN A) Engine

NOTE: Let the vehicle sit overnight before attempting to remove the cylinder head. The engine must be cold.

CAUTION

Relieve the pressure on the fuel system before disconnecting any fuel line connection.

1. Properly relieve the fuel system pressure, then disconnect the negative battery cable.

2. Drain the engine cooling system, then remove the air cleaner assembly.

3. If equipped, remove and reposition the A/C compressor with the lines intact. Make sure the lines are not kinked or otherwise damaged.

4. Remove the rocker arm cover.

5. Loosen the rocker arms and remove the pushrods.

6. Tag and disconnect the vacuum lines, fuel lines and wiring at the TBI unit. Remove the vacuum lines from the studs on the intake manifold.

7. Disconnect the accelerator, the cruise control and the TVS cables, as equipped.

8. Remove the alternator along with the brackets and position aside.

9. Remove the water pump bypass and heater hoses from the intake manifold.

10. Disconnect the exhaust pipe from the exhaust manifold.

11. Disconnect the upper radiator hose, then remove the vacuum tube from the coolant outlet stud.

12. Remove the fuel filter and fuel line bracket at the rear of the cylinder head.

13. Remove the dipstick tube, then disconnect the wiring harness and ground strap at the rear of the cylinder head.

14. Tag and disengage the wiring connectors from the sensors and the cylinder head and the thermostat housing.

15. Remove the ignition coil wires, then tag and disconnect the spark plug wires.

16. Disengage the oxygen sensor wiring connector.

17. Remove the cylinder head-to-engine bolts, then remove the cylinder head from the engine (with the manifolds attached). If necessary, remove the intake and the exhaust manifolds from the cylinder head.

To install:

18. Carefully clean and inspect the gasket mounting surfaces.

NOTE: The gasket surfaces on both the head and block must be clean of any foreign matter and free of nicks or heavy scratches. The cylinder bolt threads in the block and thread on the bolts must be cleaned (dirt will affect the bolt torque).

19. Place a new gasket over the dowel pins, then position the cylinder head on the block over the dowel pins and gasket.

20. Apply a coating of 1052080 or equivalent sealer to the threads of the cylinder head bolts, then thread the bolts into position. Make sure the bolts are threaded within 15 minutes of the sealant application in order to allow the sealer to properly cure.

21. Tighten the cylinder head bolts to specification, using multiple passes of the proper sequence:
• Torque all bolts gradually to 18 ft. lbs. (25 Nm).
• Torque all bolts except the left front bolt (No. 9 in the sequence) to 26 ft. lbs. (35 Nm), then retorque number 9 to 18 ft. lbs. (25 Nm).
• Torque all bolts an additional 1/4 turn (90°).

22. Engage the oxygen sensor connector.

23. Connect the spark plug and ignition coil wiring.

24. Engage the wiring connectors to the sensors on the cylinder head and thermostat housing.

25. Connect the wiring harness bracket and ground strap to the rear of the cylinder head.

26. Install the dipstick tube, then install the fuel filter and fuel line bracket.

27. Secure the vacuum tube at the coolant outlet stud, then connect the upper radiator hose.

28. Connect the exhaust pipe to the manifold.

29. Connect the water pump bypass and heater hoses.

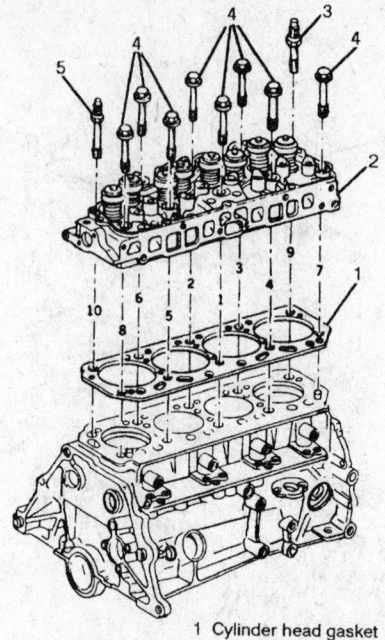

1 Cylinder head gasket
2 Cylinder head
3 Cylinder head bolt
4 Cylinder head bolt
5 Cylinder head bolt

79219g41

Cylinder head bolt torque sequence — 2.5L engine

30. Reposition and secure the alternator with the brackets.

31. Connect the accelerator, cruise control and TVS cables, as equipped.

32. Secure the vacuum hoses to the intake manifold studs, then engage the vacuum hoses, fuel lines and electrical wires to the TBI unit as tagged during removal.

33. Install the pushrods and secure the rocker arms.

34. Install the rocker arm cover.

35. If equipped, reposition and secure the A/C compressor along with the brackets.

36. Install the air cleaner assembly, then connect the negative battery cable.

37. Properly refill the engine cooling system, then check for leaks.

2.8L (VIN R) Engine

—— **CAUTION** ——
Relieve the pressure on the fuel system before disconnecting any fuel line connection.

1. Properly relieve the fuel system pressure, then connect the negative battery cable.

2. Drain the engine cooling system.

3. Remove the intake manifold assembly.

4. If necessary, raise and support the vehicle safely. If it decided to disconnect the exhaust manifold and not the pipe in the next step, it may not be necessary to raise the vehicle.

5. Either disconnect the exhaust pipe from the exhaust manifold (so the manifold may be removed with the cylinder head) or, remove the exhaust manifold-to-cylinder block retainers and support the manifold out of the way.

6. If raised, lower the vehicle.

7. If removing the left side cylinder head, remove the dipstick tube, then disconnect the ground strap and the sensor connector from the cylinder head.

8. If removing the right side cylinder head, remove the drive belt, alternator, and AIR pump with mounting bracket.

NOTE: If valve train components, such as the pushrods, are to be reused, they must be tagged or arranged to insure installation in their original locations.

9. Loosen the rocker arm nuts, turn the rocker arms and remove the pushrods. Keep the pushrods in the same order as removed.

10. Loosen the cylinder head bolts using multiple passes in the reverse order of the tightening sequence.

11. Remove the cylinder head; do not pry on the head to loosen it.

To install:

12. Carefully clean and inspect the gasket mounting surfaces.

NOTE: The gasket surfaces on both the head and block must be clean of any foreign matter and free of nicks or heavy scratches. The cylinder bolt threads in the block and thread on the bolts must be cleaned (dirt will affect the bolt torque).

13. Place a new gasket over the dowel pins with the words "This Side Up" facing upwards, then position the cylinder head on the block over the dowel pins and gasket.

14. Apply a coating of 1052080 or equivalent sealer to the threads of the cylinder head bolts, then thread the bolts into position and tighten them to specification using the proper torque sequence:
• First, torque all bolts to 40 ft. lbs. (55 Nm).
• Then, torque all bolts an additional 1/4 turn (90°).

15. Install the pushrods, then reposition the rocker arms and adjust the valves.

16. For the right side, install the alternator and AIR pump along with

the mounting bracket, then install and tighten the drive belt.

17. For the left side, connect the ground strap and sensor connector, then install the dipstick tube.

18. If necessary, raise and support the vehicle safely.

19. Either connect the exhaust pipe to the manifold or the manifold to the cylinder head, as applicable. If raised, lower the vehicle.

20. Install the intake manifold assembly.

21. Connect the negative battery cable, then properly refill the engine cooling system.

22. Check and/or adjust the ignition timing.

4.3L (VIN W, X, and Z) Engine

CAUTION

Relieve the pressure on the fuel system before disconnecting any fuel line connection.

---

1. Properly relieve the fuel system pressure, then disconnect the negative battery cable.

2. Drain the engine cooling system.

3. Remove the rocker arm cover.

4. Remove the intake manifold.

5. Remove the exhaust manifold.

6. If removing the right cylinder head, remove or disconnect:
- Electrical connector at the sensor.
- Dipstick tube at the cylinder head bracket.
- Air conditioning compressor (position it aside with the refrigerant lines attached), if equipped.
- A/C compressor, if equipped/belt tensioner bracket.

Cylinder head bolt torque sequence — 2.8L engines

79219g42

7. If removing the left cylinder head, remove or disconnect:
- Alternator (position it aside).
- Left side engine accessory bracket with power steering pump (position the pump aside with the lines attached) and brackets, if equipped.

8. Tag and disconnect the wiring from the spark plugs. If necessary, remove the spark plugs from the cylinder head.

9. Loosen the rocker arms and remove the pushrods.

NOTE: If valve train components, such as the rocker arms or pushrods, are to be reused, they must be tagged or arranged to insure installation in their original locations.

10. Remove the cylinder head bolts by loosening them in the reverse of the torque sequence, then carefully remove the cylinder head.

To install:

11. Carefully clean and inspect the gasket mounting surfaces.

NOTE: The gasket surfaces on both the head and block must be clean of any foreign matter and free of nicks or heavy scratches. The cylinder bolt threads in the block and thread on the bolts must be cleaned (dirt will affect the bolt torque).

NOTE: Do not apply sealer to composition steel-asbestos gaskets.

12. If using a steel only gasket, apply a thin and even coat of sealer to both sides of the gaskets.

13. Place a new gasket over the dowel pins with the bead or the words "This Side Up" facing upwards (as applicable), then carefully lower the cylinder head into position over the gasket and dowels.

14. Apply a coating of 1052080 or equivalent sealer to the threads of the cylinder head bolts, then thread the bolts into position until finger-tight. Using the proper torque sequence, tighten the bolts in 3 steps:
- First, tighten the bolts to 25 ft. lbs. (34 Nm).
- Next, tighten the bolts to 45 ft. lbs. (61 Nm).
- Finally, tighten the bolts to 65 ft. lbs. (90 Nm).

15. Install the pushrods, secure the rocker arms and adjust the valves.

16. If removed, install the spark plugs. Engage the spark plug wires.

17. If the left cylinder head was removed, reposition and secure the engine accessory bracket with the

power steering pump and brackets, as equipped. Install the alternator.

18. If the right cylinder head was removed, install the A/C compressor, if equipped, and A/C compressor/belt tensioner bracket, then install the dipstick tube bracket and engage the sensor electrical connector.

19. Install the exhaust manifold.

20. Install the intake manifold.

21. Install the rocker arm cover.

22. Connect the negative battery cable, then properly refill the engine cooling system.

23. Run the engine to check for leaks, then check and/or adjust the ignition timing.

Valve Lifters

REMOVAL AND INSTALLATION

When installing new lifters, a pre-lube should always be applied to the lifter body. It is also a good idea to prime hydraulic lifters by submerging them in clean engine oil and depressing the plunger using an old pushrod. This allows the internal components of the hydraulic lifter to coat with oil before initial operation in the engine. All lifters should be replaced when a new camshaft is installed.

2.2L (VIN W and 4) Engine

1. Remove the rocker arm cover.

2. Remove the cylinder head assembly from the engine.

3. Remove the bolts retaining the lifter anti-rotation brackets, then remove the brackets.

4. Remove the hydraulic roller lifters from the bores.

To install:

NOTE: If installing a new lifter, coat the lifter body with a suitable camshaft pre-lube

5. Install each hydraulic lifter to its bore being careful to align the flat sides (top) of the lifters with the flat sides of the anti-rotation brackets. When properly installed the flat sides of each lifter are aligned parallel to the anti-rotation bracket. The roller at the bottom of the lifter is parallel to the camshaft lobe.

NOTE: Make sure to properly align and install each lifter as improper installation of the lifters or brackets could result in engine damage.

6. Install the anti-rotation bracket retaining bolts and tighten to 97 inch lbs. (11 Nm).

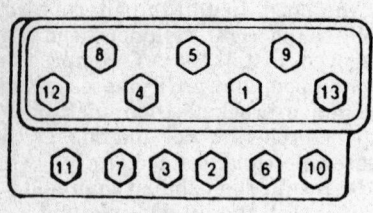

FRONT

79219g43

Cylinder head bolt torque sequence — 4.3L engine

7. Install the cylinder head.
8. Install the rocker arm cover.

2.5L (VIN A) Engine

1. Remove the rocker arm covers.
2. Remove the pushrod cover:
   a. Remove the alternator and bracket.
   b. Disengage the ignition coil wires, then remove the spark plug wires and bracket from the intake manifold.
   c. Remove the fuel pipes and clips from the pushrod cover.
   d. Remove the oil pressure gage sender, then remove the wiring harness brackets from the pushrod cover.
   e. Unscrew the nuts from the cover attaching studs, reverse 2 of the nuts so the washers face outward and screw them back onto the 2 inner studs.
   f. Assemble the 2 remaining nuts to the same 2 inner studs with the washers facing inward, then using a small wrench on the inner nut (on each stud) jam the nuts slightly together.
   g. Again using the wrench on the inner stud, unscrew the studs until the cover breaks loose, then remove the nuts from the studs and remove the cover from the engine. Remove the studs from the cover and reinstall them to the engine. Tighten the studs to 88 inch lbs. (10 Nm).

NOTE: Keep all components in order. If reusing components, install them into their original positions. If a new hydraulic lifter is being installed, all sealer coating inside the lifter must be removed.

3. Remove the pushrods.
4. Remove the lifter studs and retainers.
5. Remove the lifter guides and lifters.

6. Inspect the lifter and lifter bore for wear and scuffing. Examine the roller for freedom of movement and/or flat spots on the roller surface.
7. Installation is the reverse of removal, be sure to lubricate the lifter and lifter body using clean engine oil. If installing new lifters, all sealer coating inside the lifter must first be removed.
8. When installing the pushrod cover, use a thin coating of RTV sealant around the entire cover flange. Install the cover retaining nuts and tighten to 106 inch lbs. (12 Nm) starting at the front of the engine and working toward the nuts in the rear.

2.8L (VIN R) Engine

Some engines have both standard size and 0.25mm (0.010 in.) oversize valve lifters. The cylinder block will be marked with a white paint mark and 0.25mm O.S. stamp where the oversize lifters are used.

If lifter replacement is necessary, use new design lifters with a narrow flat along the lower ¾ of the body length. This provides additional oil to the cam lobe and lifter surfaces.

NOTE: Use of a Hydraulic Lifter Remover tool J-9290-1 (slide hammer type) or J-3049-A (pliers type) will greatly ease the removal of stuck lifters.

1. Remove the rocker arm covers.
2. Remove the intake manifold.
3. Remove the rocker arm nuts and balls.
4. Remove the rocker arms and pushrods.

NOTE: If any valve train components (lifters, pushrods, rocker arms) are to be reused, they must be tagged or arranged during removal to assure installation in their original locations.

5. Remove the lifters from the bores.
To install:
6. For proper rotation during engine operation, the lifter bottom must be convex. Check the lifter bottom for proper shape using a straight edge. If the lifter bottom is not convex, replace the lifter. Chances are if lifters are in need of replacement, so is the camshaft.
7. Lubricate and install the lifters. If installing new lifters, coat the lifter body and foot using Molykote® or equivalent prelube, then add 1051396 or equivalent engine oil supplement to the crankcase.

8. Install the pushrods, rocker arms, rocker arm nuts and balls, then properly adjust the valve lash.
9. Install the intake manifold.
10. Install the rocker arm covers.

4.3L (VIN W, X, and Z) Engine

1. Remove the rocker arm cover.
2. Remove the intake manifold assembly.

NOTE: If any valve train components (lifters, pushrods, rocker arms) are to be reused, they must be tagged or arranged during removal to assure installation in their original locations.

3. Remove the rocker arm and pushrod assemblies.
4. Remove the hydraulic lifter retainer bolts, them remove the retainers or retainers and restrictors, as equipped.
5. Remove the lifters from the engine.
To install:
6. Inspect the lifter and lifter bore for wear and scuffing. Examine the roller for freedom of movement and/or flat spots on the roller surface.
7. If new lifters are being installed, lubricate the lifters using a suitable pre-lube and add GM engine oil supplement or equivalent additive to the crankcase.
8. Install the hydraulic lifters along with the restrictors and/or retainers. Secure the retainers using the bolts and tighten to 12 ft. lbs. (16 Nm).
9. Install the pushrods and rocker arm assemblies, then adjust the valve lash.
10. Install the intake manifold assembly.
11. Install the rocker arm covers.

Valve Lash

ADJUSTMENT

2.2L (VIN W and 4) Engine

Because the rocker arm fasteners are secured and torqued, valve lash is not adjustable on the 2.2L engine. If a valve train problem is suspected, check that the rocker arm nuts are tightened to 22 ft. lbs. (30 Nm). Be very careful not to overtighten the rocker arm nuts. ONLY torque the nuts when the hydraulic lifter is resting on the base circle of the camshaft and not when it is held upward on the lobe. When valve lash falls out of specification (valve tap is heard), replace the rocker arm, pushrod and

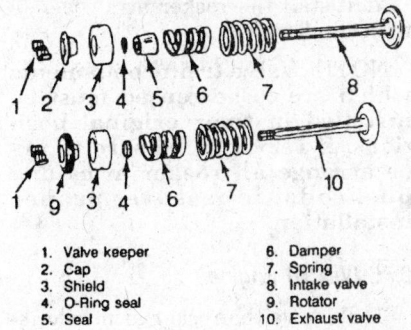

1. Valve keeper
2. Cap
3. Shield
4. O-Ring seal
5. Seal
6. Damper
7. Spring
8. Intake valve
9. Rotator
10. Exhaust valve

79219g44

Valves and related components — 4.3L engine

hydraulic lifter on the offending cylinder.

2.5L (VIN A) Engine

Because the rocker arm fasteners are secured and torqued, valve lash is not adjustable on the 2.5L engine. If a valve train problem is suspected, check that the rocker arm bolts are tightened to 22 ft. lbs. (30 Nm). Be sure to only tighten the rocker arm bolts when the hydraulic lifter for that rocker arm is on the base circle of the camshaft and not when it is held upward on the lobe. When valve lash falls out of specification (valve tap is heard), replace the rocker arm, pushrod and hydraulic lifter on the offending cylinder.

2.8L (VIN R) Engine

NOTE: These engines utilize hydraulic valve lifters which means that a valve adjustment is not a regular maintenance item. The valves must only be adjusted if the rockers arms have been disturbed for any reason such as cylinder head, camshaft, pushrod or lifter removal.

1. Remove the air cleaner and the rocker arm cover(s).

2. Rotate the crankshaft until the mark on the crankshaft pulley aligns with the **0** mark on the timing plate. Make sure the No. 1 cylinder is positioned on the compression stroke. The No. 1 piston is on it's compression stroke when both the intake and exhaust valves remain closed as the crankshaft damper mark approaches the timing scale.

NOTE: Another method to tell when the piston is coming up on the compression stroke is by removing the spark plug and placing a finger over the hole in order to feel air being forced out of the spark plug hole. Stop turning the crankshaft when pressure is

felt and the TDC timing mark on the crankshaft pulley is directly aligned with the timing mark pointer or the 0 mark on the scale.

3. When the engine is on the No. 1 firing position, adjust the following valves:

- Intake — 1, 5 and 6
- Exhaust — 1, 2 and 3

4. To adjust the valves, back-out the adjusting nut until lash can be felt at the pushrod, then turn the nut until all of the lash is removed.

NOTE: To determine is all of the lash is removed, turn the pushrod between 2 fingers until the movement is removed.

5. When all of the lash has been removed, turn the adjusting an additional 1½ turns; this will center the lifter plunger.

6. Crank the engine 1 complete revolution until the timing tab (**0** degree mark) and the crankshaft pulley mark are again in alignment. Now the engine is in the No. 4 firing position. Adjust the following valves:

- Intake — 2, 3 and 4
- Exhaust — 4, 5 and 6

7. Install the rocker arm cover(s).

8. Start and run the engine, then check and adjust the timing, as necessary.

4.3L (VIN Z) Engine

NOTE: This engine utilizes hydraulic valve lifters which means that a valve adjustment is not a regular maintenance item. The valves must only be adjusted if the rockers arms have been disturbed for any reason such as cylinder head, camshaft, pushrod or lifter removal.

For 1993–94, the 4.3L (VIN Z) engine may be equipped with either of 2 rocker arm retaining systems. If the engine utilizes screw-in type rocker arm studs with positive stop shoulders, no valve lash adjustment is necessary or possible. All 1995–97 4.3L engines utilize this system. If so equipped, please refer to the 4.3L (VIN W) valve lash information. If however, the engine utilizes the pressed-in rocker arm studs, use the following procedure to tighten the rocker arm nuts and properly center the pushrod on the hydraulic lifter:

1. To prepare the engine for valve adjustment, rotate the crankshaft until the mark on the damper pulley aligns with the **0** degree mark on the timing plate and the No. 1 cylinder is on the compression stroke. When the

No. 1 piston is on it's compression stroke both the intake and exhaust valves will remain closed as the crankshaft damper mark approaches the timing scale.

NOTE: Another method to tell when the piston is coming up on the compression stroke is by removing the spark plug and placing a finger over the hole in order to feel air being forced out of the spark plug hole. Stop turning the crankshaft air is felt and the TDC timing mark on the crankshaft pulley is directly aligned with the timing mark pointer or the 0 mark on the scale.

2. With the engine at No. 1 TDC, adjust the exhaust valves of cylinders No. 1, 5 and 6 and the intake valves of cylinders No. 1, 2 and 3 by performing the following procedures:

a. Back out the adjusting nut until lash can be felt at the pushrod.

b. While rotating the pushrod, turn the adjusting nut inward until all of the lash is removed.

c. When the play has disappeared, turn the adjusting nut inward 1 additional turn for 1993 engines or 1¾ additional turns for 1994 engines.

3. Rotate the crankshaft 1 complete revolution and align the mark on the damper pulley with the **0** degree mark on the timing plate; the engine is now positioned on the No. 4 firing position. This time the No. 4 cylinder valves remain closed as the timing mark approaches the scale. Adjust the exhaust valves of cylinders No. 2, 3 and 4 and the intake valves of cylinders No. 4, 5 and 6, by performing the following procedures:

a. Back out the adjusting nut until lash can be felt at the pushrod.

b. While rotating the pushrod, turn the adjusting nut inward until all of the lash is removed.

c. When the play has disappeared, turn the adjusting nut inward 1 additional turn for 1993 engines or 1¾ additional turns for 1994 engines.

4. Install the remaining components, then start the engine and check for oil leaks.

4.3L (VIN W) Engine

The 4.3L (VIN W) engine and some later 4.3L (VIN Z) engines (including all 1995 models) are equipped with screw-in type rocker arm studs with positive stop shoulders. Because the shoulders allow the rocker arms to be torqued into proper position, no adjustments are necessary or possible. If a valve train problem is suspected, check that the rocker arm nuts are tightened to 20 ft. lbs. (27 Nm). When valve lash falls out of specification (valve tap is heard), replace the rocker arm, pushrod and hydraulic lifter on the offending cylinder.

Rocker Arms and Shafts

REMOVAL AND INSTALLATION

2.2L (VIN W and 4) Engine

1. Remove the rocker arm cover from the cylinder head.
2. Remove the rocker arm retaining nut, then remove the arm and ball. If necessary, withdraw the pushrod form the cylinder head.

NOTE: Valve train components which are to be reused must be installed in their original positions. If removed, be sure to tag or arrange all rocker arms and pushrods to assure proper installation.

To install:
3. Inspect the rocker arms, balls and pushrods for damage or wear and replace, as necessary:
   a. Check the rocker arms, balls and their mating surfaces. Make sure the surfaces are smooth and free from scoring or other damage.
   b. Check the rocker arm areas that contact the valve stems and the sockets that contact the pushrods, make sure these areas are smooth and free of both damage and wear.
   c. Make sure the pushrods are not bent; this can be determined by rolling them on a flat surface. Check the ends of the pushrods for scoring or roughness
   d. Inspect the rocker arm bolts for thread damage. Check the

rocker arm bolts in the shoulder area for contact damage with the rocker arm.
4. If removed, install the pushrods making sure they are seated within the lifters.
5. If installing new rocker arms and balls, coat the friction surfaces using Dri-Slide Molykote® or equivalent pre-lube.

NOTE: When tightening the rocker arm retainers, make sure the lifter for that valve is resting on the base circle of the camshaft not on the lobe. Do not overtighten the retainers.

6. Install the rocker arms and ball, then tighten the retaining nuts to 22 ft. lbs. (30 Nm).

NOTE: Valve lash is not adjustable on the 2.2L engine.

7. Install the rocker arm cover, then start and run the engine to check for leaks.

2.5L (VIN A) Engine

1. Remove the rocker arm cover from the cylinder head.
2. Remove the rocker arm bolt, the ball washer and the rocker arm.

NOTE: If only the pushrod is to be removed, back off the rocker arm bolt, swing the rocker arm aside and remove the pushrod. When removing more than 1 assembly, at the same time, be sure to keep them in order for reassembly purposes.

3. If necessary, remove the pushrods and guides.
4. Inspect the rocker arms and ball washers for scoring and/or other damage, replace them (if necessary).

NOTE: If replacing worn components with new ones, be sure to coat the new parts with Molykote® or equivalent pre-lube before installation.

To install:
5. If removed, install the pushrods and guides.
6. Install the rocker arms, ball washers and retaining bolts. Tighten the rocker arm-to-cylinder head bolt for each valve to 22 ft. lbs. (30 Nm), but only tighten the bolt with the hydraulic lifter for that valve on the base circle of the camshaft. Do not tighten the bolts while the lifter is resting on the raised portion of the lobe and do not overtighten the bolts.

NOTE: Valve lash is not adjustable on the 2.5L engine.

7. Install the rocker arm cover to the cylinder head.

NOTE: Valve train components which are to be reused must be installed in their original positions. If removed, be sure to tag or arrange all rocker arms and pushrods to assure proper installation.

2.8L (VIN R) Engine

1. Remove the rocker arm cover(s) from the cylinder head.
2. Remove the rocker arm nut, the rocker arm and the ball washer.

NOTE: If only the pushrod is to be removed, loosen the rocker arm nut, swing the rocker arm to the side and remove the pushrod.

3. Withdraw the pushrod from the cylinder head.
To install:
4. Inspect and replace components if worn or damaged.
5. Coat the bearing surfaces of the rocker arms and the rocker arm ball washers with Molykote® or equivalent pre-lube.
6. Install the pushrods making sure they seat properly in the lifter.
7. Install the rocker arms, ball washers and the nuts, then tighten the rocker arm nuts until there is little or no valve lash.

NOTE: Each valve must be adjusted when the lifter is sitting on the base circle of the camshaft, not the raised section of the lobe.

8. Properly adjust the valve lash.
9. Install the rocker arm cover.
10. Start and run the engine, then check for leaks. Check and adjust the timing, as necessary.

NOTE: Valve train components which are to be reused must be installed in their original positions. If removed, be sure to tag or arrange all rocker arms and pushrods to assure proper installation.

4.3L (VIN W, X, and Z) Engine

1. Remove the rocker arm cover(s) from the cylinder head.
2. Remove the rocker arm nut, the rocker arm and the ball washer.

NOTE: If only the pushrod is to be removed, loosen the rocker arm nut, swing the rocker arm to the side and remove the pushrod.

3. Withdraw the pushrod from the cylinder head.

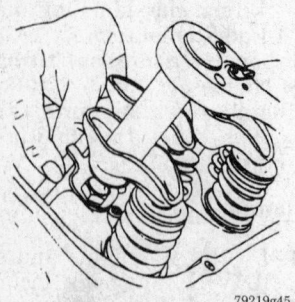

79219g45

When adjusting valve lash, turn the pushrod between 2 fingers while tightening to remove lash

To install:

4. Inspect and replace components if worn or damaged.

5. Coat the bearing surfaces of the rocker arms and the rocker arm ball washers with Molykote® or equivalent pre-lube.

6. Install the pushrods making sure they seat properly in the lifter.

7. Install the rocker arms, ball washers and the nuts.

8. For the 4.3L (VIN W) engine and any 1993–96 4.3L (VIN Z) engines which are equipped with screw-in type rocker arm studs with positive stop shoulders, tighten the rocker arm adjusting nuts against the stop shoulders to 20 ft. lbs. (27 Nm). No further adjustment is necessary or possible.

9. If possible, properly adjust the valve lash.

10. Install the rocker arm cover(s) to the cylinder head.

11. Start and run the engine, then check for leaks and for proper ignition timing adjustment.

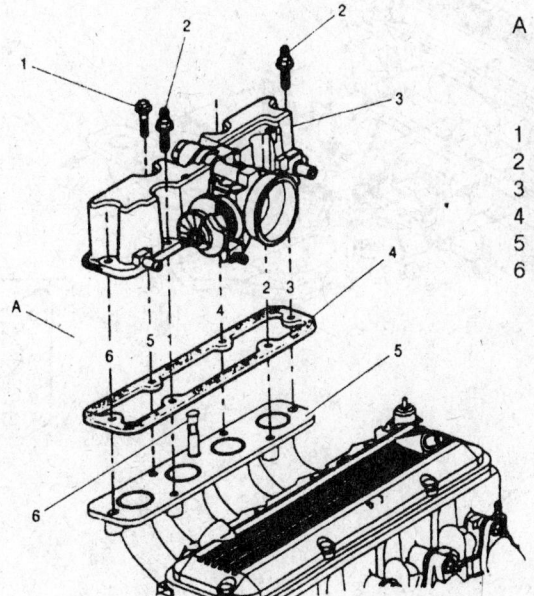

A Upper intake manifold assembly tightening sequence

1 Bolt
2 Stud
3 Upper intake manifold assembly
4 Gasket
5 Lower intake manifold
6 EGR valve injector

79219g46

Upper intake manifold mounting and torque sequence — 2.2L engine

Intake Manifold

REMOVAL AND INSTALLATION

2.2L (VIN W and 4) Engine

The 2.2L (VIN 4) engine was introduced in 1994 and utilizes a Multi-Port Fuel Injection (MFI) system. The intake manifold is an assembly of separate components, an upper and a lower manifold.

1. Properly release the fuel system pressure (if the lower manifold assembly is being removed) and disconnect the negative battery cable.

2. Remove the air cleaner duct work.

3. Disconnect the throttle cable support and cable from the manifold.

4. Remove the MAP sensor and the EGR solenoid valve from the upper intake manifold and engine (if the upper manifold is not being replaced, simply disengage the wiring and hoses).

5. Tag and disengage all wiring and vacuum hoses from the upper intake manifold.

6. Loosen the retainers, then remove the upper intake manifold from the engine and lower manifold assembly.

7. Disconnect the fuel lines.

8. Tag and disconnect the spark plug wires from the Electronic Ignition (EI) coil pack.

9. Remove the lower intake manifold retaining nuts, then remove the lower intake manifold and gasket.

To install:

10. Carefully remove all traces of gasket material from the mating surfaces. Check the EGR passage to be sure it is free of excessive carbon deposits and clean, as necessary.

11. Install the lower intake manifold using a new gasket, then tighten the retaining nuts to 24 ft. lbs. (33 Nm) using the proper sequence.

12. Connect the spark plug wires to the EI coil pack and noted during removal.

13. Connect the fuel lines.

14. Install the upper intake manifold using a new gasket, then tighten the retainers to 22 ft. lbs. (30 Nm) using the proper sequence..

15. Engage the wiring connectors and vacuum hoses to the upper intake manifold assembly.

16. Install the MAP sensor and EGR solenoid valve.

17. Install and secure the throttle cable support and cable. Tighten the cable bracket bolts to 18 ft. lbs. (25 Nm).

18. Install the air cleaner duct work.

19. Connect the negative battery cable, then start and run the engine to check for leaks.

2.5L (VIN A) Engine

1. Properly relieve the fuel system pressure, then disconnect the negative battery cable.

2. Drain the engine cooling system and remove the air cleaner assembly.

3. Label and disconnect the wiring harnesses and connectors at the intake manifold.

4. Disconnect the accelerator, TVS, and cruise control cables (as equipped) with brackets.

5. Disconnect the EGR vacuum line.

6. Remove the emission sensor bracket at the manifold.

7. Tag and disconnect the fuel lines, vacuum lines and wiring from the TBI unit.

8. Disconnect the water pump bypass hose from the intake manifold.

9. Remove the alternator rear bracket.

10. Tag and disconnect the vacuum hoses and pipes from the intake manifold and the vacuum hold-down at the thermostat and manifold.

11. If applicable, disconnect the heater hose at the intake.

12. Disconnect the coil wires.

13. Remove the intake manifold bolts, then remove the manifold from the engine.

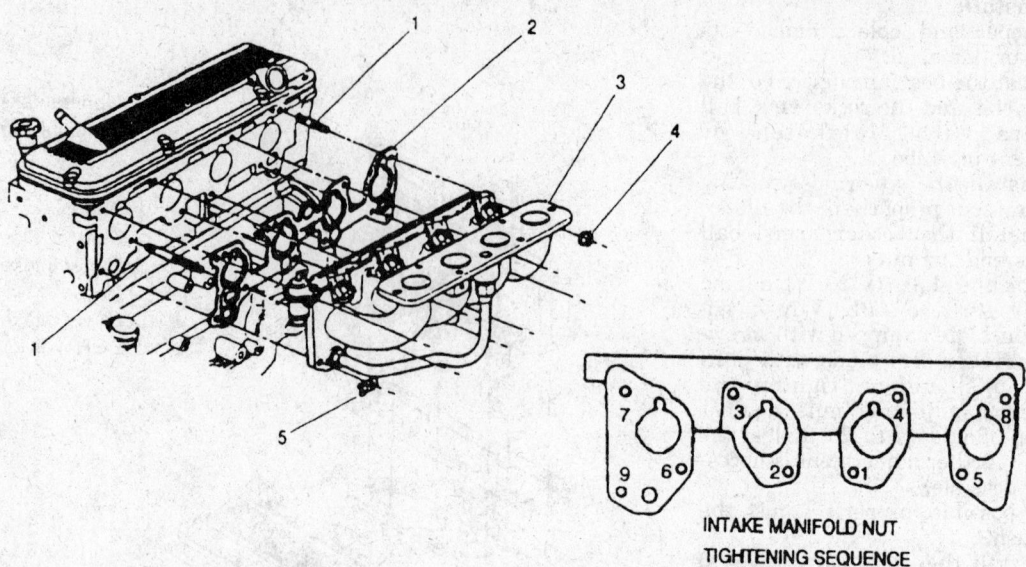

1 STUD
2 GASKET
3 INTAKE MANIFOLD
4 NUT
5 CLIP

INTAKE MANIFOLD NUT
TIGHTENING SEQUENCE

79219g47

Lower intake manifold mounting and installation torque sequence — 2.2L engine

To install:

14. Using a gasket scraper, clean the gasket mounting surfaces.

15. Install the intake manifold to the engine using a new gasket and carefully thread the retainers.

16. Slowly and evenly tighten all of the retainers to 25 ft. lbs. (34 Nm).

17. Engage the coil wires.

18. If applicable, connect the heater hose at the intake.

19. Connect the vacuum hoses and pipes to the manifold and the vacuum hold-down at the thermostat as noted during removal.

20. Install the alternator rear bracket, then connect the coolant by-pass hose.

21. Connect the fuel lines, vacuum lines and the wiring to the TBI unit.

22. Install the emissions sensor bracket, then connect the EGR valve hose.

23. Connect the accelerator, TVS and cruise control cables (as equipped) with brackets.

24. Engage and secure the wiring harness connectors at the intake, as noted during removal.

25. Install the air cleaner assembly, then connect the negative battery cable.

26. Properly refill the engine cooling system, then check for leaks.

2.8L (VIN R) Engine

1. Properly relieve the fuel system pressure, then disconnect the negative battery cable.

2. Drain the engine cooling system and remove the air cleaner.

3. Tag and disengage the electrical connectors, vacuum hoses, fuel lines and accelerator cable(s) from the TBI unit.

4. If equipped with an AIR management system, remove the hose and the mounting bracket.

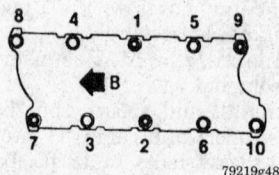

79219g48

Intake manifold torque sequence — 2.8L engine

5. Label and disconnect the spark plug wires from the spark plugs and the electrical connectors from the ignition coil. Disengage the coolant switch electrical connectors on the intake manifold.

6. Remove the distributor cap (with the wires connected). Match-mark and remove the distributor from the intake manifold and engine.

7. Disconnect the EGR vacuum line and the evaporative emission hoses. Remove the pipe brackets from the rocker arm covers.

8. Remove the heater and upper radiator hoses from the intake manifold.

9. If equipped, remove the power brake vacuum hoses from the intake manifold.

10. Remove the rocker arm covers.

11. Remove the intake manifold-to-engine nuts and bolts, then the intake manifold from the engine.

To install:

12. Using a gasket scraper, clean the mating surfaces of remaining gasket and sealer. Since the manifold is made from aluminum, be sure to inspect it for warpage and/or cracks; if necessary, replace it.

13. Position new intake manifold gaskets to the engine and apply a 3/16

in. (5mm) bead of RTV sealant to the front and rear of the engine block.

NOTE: The gaskets are marked "Right Side" and "Left Side"; do not interchange them. The gaskets may have to be cut slightly to fit past the center pushrods; do not cut any more material than necessary. Hold the gaskets in place by extending the ridge bead of sealer 1/4 in. onto the gasket ends.

14. Position the intake manifold taking care not to disturb the gaskets. Make sure the areas between the case ridges and the intake manifold are completely sealed, then carefully thread the retainers. For 1993 engines, before threading the 2 center retainers (No.s 1 & 2 in the torque sequence) apply a coating of 9985427 or equivalent sealer to the threads.

15. Tighten the intake manifold-to-cylinder head fasteners to 23 ft. lbs. (31 Nm) using the proper torque sequence.

16. Install the rocker arm covers.

17. If equipped, connect the power brake vacuum hoses at the intake manifold.

18. Connect the heater and upper radiator hoses from the intake manifold.

19. Install the pipe brackets to the rocker arm covers, then connect the EGR vacuum line and the evaporative emission hoses.

20. Align the marks made during removal and install the distributor assembly, then install the distributor cap and spark plug wires.

21. Engage the electrical connectors to the ignition coil and to the coolant switch on the intake manifold.

22. If equipped with an AIR management system, install the hose and the mounting bracket.

23. Engage the electrical connectors, vacuum hoses, fuel lines and the accelerator cable(s) to the TBI unit.

24. Install the air cleaner assembly, then connect the negative battery cable.

25. Properly refill the engine cooling system, then check for leaks.

26. Adjust the ignition timing and check the coolant level after the engine has warmed up.

4.3L (VIN W and X) Engine

NOTE: If only the upper intake manifold is being removed, the fuel system pressure does not

need to be released. **ALWAYS release the pressure before disconnecting any fuel lines.**

1. Remove the plastic cover, then properly relieve the fuel system pressure and disconnect the negative battery cable.

2. Drain the engine cooling system, then remove the air cleaner and air inlet duct.

3. For vehicles through 1995, disengage the wiring harness from the necessary upper intake components including:
- Throttle Position (TP) sensor
- Idle Air Control (IAC) motor
- Manifold Absolute Pressure (MAP) sensor
- Intake Manifold Tuning Valve (IMTV)

4. For 1996–97 vehicles, wiring harness connectors and brackets and move aside.

5. Disengage the throttle linkage from the upper intake manifold, then remove the ignition coil.

6. Disconnect the fuel lines and bracket from the rear of the lower intake manifold. Disconnect the brake booster vacuum hose at the upper intake manifold.

7. For vehicles through 1995, disconnect the PCV hose at the rear of the upper intake manifold, then tag and disengage the vacuum hoses from both the front and rear of the upper intake.

8. For 1996–97 vehicles, remove the purge solenoid and bracket.

9. Remove the upper intake manifold bolts and studs, making sure to note or mark the location of all studs to assure proper installation. Remove the upper intake manifold from the engine.

10. Disengage the distributor or HVS wiring (as equipped), then matchmark the distributor or HVS and remove the assembly from the engine.

11. Disconnect the upper radiator hose at the thermostat housing and the heater hose at the lower intake manifold.

12. Remove any necessary wiring harnesses and brackets.

13. For 1996–97 vehicles, remove the automatic transmission dipstick tube. Remove the EGR tube, clamp and tube.

14. Remove the pencil brace (A/C compressor bracket-to-lower intake manifold).

15. For 1996–97 vehicles, remove the alternator bracket bolts next to the thermostat housing.

16. For vehicles through 1995, disengage the wiring harness connectors

from the necessary lower intake components including:
- Fuel injector
- Exhaust Gas Recirculation (EGR) valve
- Engine Coolant Temperature (ECT) sensor

17. Remove the lower intake manifold retaining bolts, then remove the manifold from the engine.

18. Using a gasket scraper, carefully clean the gasket mounting surfaces. Be sure to inspect the manifold for warpage and/or cracks; if necessary, replace it.

To install:

19. Position the gaskets to the cylinder head with the port blocking plates to the rear and the "this side up" stamps facing upward, then apply a 3/16 in. (5mm) bead of RTV sealant to the front and rear of the engine block at the block-to-manifold mating surface. Extend the bead 1/2 in. (13mm) up each cylinder head to seal and retain the gaskets.

20. Install the lower intake manifold taking care not to disturb the gaskets, then tighten the manifold retainers to 35 ft. lbs. (48 Nm) using the proper torque sequence for vehicles through 1995.

21. For 1996–97 vehicles, tighten the bolts, in sequence in three steps, as follows:
   a. 1st step to 26 inch lbs. (3 Nm).
   b. 2nd step to 106 inch lbs. (12 Nm).
   c. 3rd step to 11 ft. lbs.)15 Nm).

22. For 1996–97 vehicles, install the alternator bracket bolt next to the thermostat housing. Install the EGR tube, clamp and bolt.

23. Engage the wiring harness to the lower manifold components, including the injector, EGR valve and ECT sensor.

24. Install the pencil brace to the A/C compressor bracket and the lower intake manifold.

25. If necessary, install the transmission oil dipstick tube.

26. Connect the fuel supply and return lines to the rear of the lower intake. Temporarily reconnect the negative battery cable, then pressurize the fuel system (by cycling the ignition without starting the engine) and check for leaks. Disconnect the negative battery cable and continue installation.

27. Connect the heater hose to the lower intake and the upper radiator hose to the thermostat housing.

28. Align the matchmarks and install the distributor assembly, then engage the wiring.

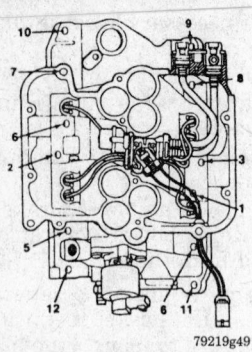

Lower intake manifold torque sequence — 4.3L VIN W engines through 1995

◀ FRT

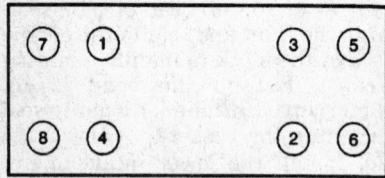

INTAKE SEQUENCE

79219g51

Lower intake manifold tightening sequence — 1996–97 4.3L (VIN W and X) engines

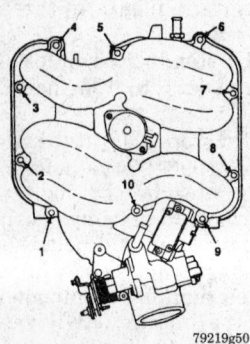

79219g50

Upper intake manifold torque sequence — 4.3L VIN W engines through 1995

29. Connect the vacuum hoses to the upper and lower intake manifold.

30. Position a new upper intake manifold gasket on the engine, making sure the green sealing lines are facing upward.

31. Install the upper intake manifold being careful not to pinch the fuel injector wires between the manifolds.

32. Install the manifold retainers, making sure the studs are properly positioned, then tighten them using the proper sequence to 124 inch lbs.

(14 Nm) for vehicles through 1995. For 1996–97 vehicles, tighten the bolts to 88 inch lbs. (10 Nm).

33. For 1996–97 vehicles, install the purge solenoid and bracket. Connect the brake booster vacuum hose at the upper intake manifold.

34. If removed, connect the PCV hose to the rear of the upper intake manifold and the vacuum hoses to both the front and rear of the manifold assembly.

35. Connect the throttle linkage to the upper intake, then install the ignition coil.

36. If necessary, engage the necessary wiring to the upper intake components including the TP sensor, IAC motor, MAP sensor and the IMTV.

37. Install the plastic cover, the air cleaner and air inlet duct.

38. Connect the negative battery cable, then properly refill the engine cooling system.

4.3L (VIN Z) Non-Turbo Engine

Except Astro and Safari

1. Disconnect the negative battery cable and properly relieve the fuel system pressure.

2. Drain the engine cooling system.

3. Remove the air cleaner and heat stove tube.

4. Remove the 2 braces at the rear of the serpentine drive belt tensioner.

5. Disconnect the upper radiator hose.

6. Remove the emissions relays along with the bracket, then disconnect the wiring harness from the retaining clips and position aside. Disconnect the ground cable from the intake manifold stud.

7. Remove the power brake vacuum pipe, then disconnect the heater hose pipe at the manifold and fuel lines at the TBI unit.

8. Remove the ignition coil, then disengage the electrical connectors at the sensors on the manifold.

9. Matchmark and remove the distributor from the engine. For details, please refer to the procedure earlier in this section.

NOTE: For ease of installation, do not crank the engine with the distributor removed.

10. Tag and disengage the wires and hoses from the TBI unit.

11. Disconnect the the EGR hose, then disconnect the throttle, TVS and cruise control cables (as equipped).

12. Remove the intake manifold retaining studs and/or bolts, then remove the manifold and gaskets.

To install:

13. Using a gasket scraper, carefully clean the gasket mounting surfaces. Be sure to inspect the manifold for warpage and/or cracks; if necessary, replace it.

14. Position the gaskets to the cylinder head with the port blocking plates to the rear, then apply a 3/16 in. (5mm) bead of RTV sealant to the front and rear of the engine block at the block-to-manifold mating surface. Extend the bead 1/2 in. (13mm) up each cylinder head to seal and retain the gaskets.

15. Install the intake manifold taking care not to disturb the gaskets, then tighten the manifold retainers to 35 ft. lbs. (48 Nm) using the proper torque sequence.

16. Engage the TVS, cruise control and/or throttle cables, as equipped.

17. Connect the EGR hose then engage the wires and hoses at the TBI unit as noted during removal

18. Align and install the distributor assembly.

19. Install the ignition coil, then connect the fuel pipes.

20. Connect the heater hose pipe and the power brake vacuum pipe.

21. Connect the ground cable to the intake manifold stud, then position and secure the wiring harness using the clips.

22. Install the emissions relays along with their bracket, then connect the upper radiator hose.

23. Install the brace at the rear of the drive belt tensioner, then install the air cleaner and heat stove tube.

24. Connect the negative battery cable, then properly refill the engine cooling system.

25. Run the engine and check for leaks.

Astro and Safari

1. Disconnect the negative battery cable. Remove the engine cover assembly. Remove the air cleaner assembly. Drain the cooling system.

2. Remove the distributor cap and ignition wires. Disconnect the ESC connector and remove the distributor.

3. Remove the cruise control transducer, if equipped.

4. Remove the detent, cruise and accelerator cables.

5. Remove the transmission and engine oil filler tubes at the alternator brace.

6. If equipped, remove the air conditioning compressor and idler pulley at the alternator brace. Remove the alternator brace.

7. Disconnect the fuel lines. Remove the necessary vacuum hoses and electrical wires.

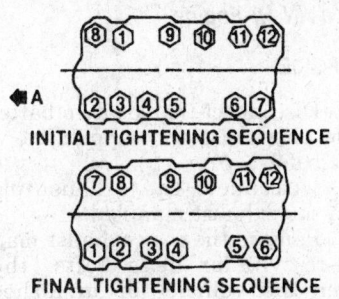

INITIAL TIGHTENING SEQUENCE

FINAL TIGHTENING SEQUENCE

A. Front Of Engine

79219g52

Intake manifold torque sequence — Non-Turbo 4.3L (VIN Z) engine 1993–94

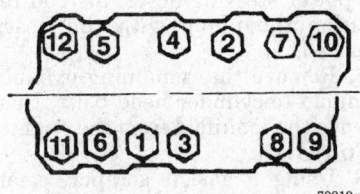

79219g53

Intake manifold torque sequence — 1995 Non-Turbo 4.3L (VIN Z) engine

8. Remove the AIR hoses and brackets, if equipped. Remove the upper radiator hose. Remove the heater hose at the manifold. As required, remove the TBI unit.

9. Remove the intake manifold retaining bolts. Remove the intake manifold from the engine.

To install:

10. The gaskets are marked for right and left side installation. Do not interchange them. Clean the sealing surface of the engine block and apply a 5/16 in. bead of silicone sealer to each ridge.

11. Install the new gaskets onto the heads. The gaskets will have to be cut slightly to fit past the center pushrods. Do not cut any more material than necessary. Hold the gaskets in place by extending the ridge bead of sealer 1/4 in. onto the gasket ends. (When the intake manifold is installed the area between the ridges and the manifold should be completely sealed.)

12. Install the intake manifold and torque the bolts in sequence to 35 ft. lbs. (47 Nm) except position No. 9 which is torqued to 41 ft. lbs. (56 Nm).

13. Install the AIR hoses and brackets, if equipped. Install the up-

per radiator hose. Install the heater hose at the manifold. Install the TBI unit.

14. Connect the fuel lines. Install the necessary vacuum hoses and electrical wires.

15. If equipped, install the air conditioning compressor and idler pulley at the alternator brace. Install the alternator brace.

16. Install the transmission and engine oil filler tubes at the alternator brace.

17. Install the detent, cruise and accelerator cables.

18. Install the cruise control transducer, if equipped.

19. Install the distributor, cap and ignition wires. Connect the ESC connector to the distributor.

20. Install the air cleaner assembly. Fill the cooling system with the proper type and quantity of antifreeze.

21. Connect the negative battery cable. Install the engine cover assembly.

4.3L (VIN Z) MFI-Turbo Engine

1. Properly relieve the fuel system pressure and disconnect the negative battery cable.

2. Drain the engine cooling system and the charge air cooling system.

3. Loosen the wing nut at the air cleaner, then position the air cleaner and duct aside.

4. Loosen the charge air cooler clamps, then remove the ducts and hoses.

5. Remove the charge air cooler from the supports, then remove the air cooler lateral (center) support.

6. Disengage the electrical connectors and hoses from the EVRV solenoid, the ignition coil and the MAP sensor located on the multi-use bracket.

7. Remove the multi-use bracket from the lower intake manifold assembly.

8. Tag and disengage the hoses and electrical connectors from the upper intake manifold assembly.

9. Loosen the throttle body retaining bolts, then remove the throttle body, gasket and cable bracket from the upper intake manifold.

10. Remove the upper intake manifold retaining bolts, then remove the manifold and gasket.

11. Disconnect the heater hose from the lower intake manifold.

12. Remove the charge air cooler coolant inlet pipe from the lower intake manifold, then disconnect it from the radiator by loosening the clamp and disengaging the hose.

13. Tag and disengage the wiring connectors from the fuel injectors, then disconnect the fuel pipes from the fuel rail assembly.

14. Loosen and remove the retaining bolts, then remove the fuel rail assembly (rail and injectors) from the lower manifold.

15. If equipped, remove the rear A/C brace,.

16. Disconnect the turbocharger coolant return line, then remove the upper radiator hose from the lower intake manifold.

17. Remove the ground strap from the lower intake manifold, then disengage the coolant sensor connector.

18. Tag and disconnect the spark plug wires from the distributor, remove the distributor cap and remove the distributor from the engine. Be sure to matchmark the distributor and rotor with the engine before removal.

19. Remove the lower intake manifold bolts and studs, then remove the manifold from the engine.

To install:

20. Using a gasket scraper, carefully clean the gasket mounting surfaces. Be sure to inspect the manifold for warpage and/or cracks; if necessary, replace it.

21. Position the gaskets to the cylinder head with the port blocking plates to the rear, then apply a 3/16 in. (5mm) bead of RTV sealant to the front and rear of the engine block at the block-to-manifold mating surface. Extend the bead 1/2 in. (13mm) up each cylinder head to seal and retain the gaskets.

22. Install the lower intake manifold taking care not to disturb the gaskets, then tighten the manifold retainers to 35 ft. lbs. (48 Nm).

23. Align the marks made during removal and install the distributor to the engine. Install the distributor cap and engage the spark plug wires.

24. Engage the coolant sensor electrical connector, then connect the ground strap.

25. Connect the upper radiator hose and the turbocharger coolant return line, then install the A/C compressor rear brace, if equipped.

26. Install the fuel rail and injectors, then connect the fuel pipes to the rail and engage the wiring connectors to the injectors. Tighten the fuel rail bolts to 58 inch lbs. (6.5 Nm) and the pipe fittings to 16 ft. lbs. (22 Nm).

27. Place the charge air cooler coolant inlet pipe into position on the lower intake manifold, then connect

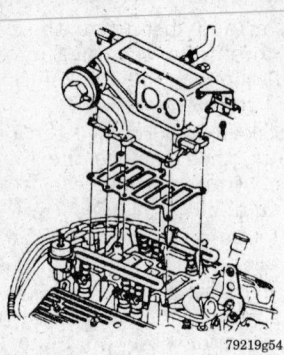

**Upper intake manifold
mounting — 4.3L MFI-turbo engine**

the hose and clamp from the radiator to the charge air cooler inlet pipe.

28. Connect the heater hose to the lower intake manifold.

29. Install the upper intake manifold using a new gasket, then tighten the retaining bolts (starting from the 2 middle bolts working outward to each side) to 18 ft. lbs. (24 Nm).

30. Install the throttle body and cable bracket using a new throttle body gasket, then tighten the retaining bolts to 18 ft. lbs. (24 Nm).

31. Engage the upper intake manifold connectors and hoses, then install the multi-use bracket. Engage the hoses and wiring to the MAP sensor, ignition coil and EVRV solenoid.

32. Install the charge air cooler lateral support and finger-tighten the retaining nut.

33. Install and secure the charge air cooler supports, then tighten the lateral support nut.

34. Install the charge air cooler hoses and ducts, then secure using the clamps.

35. Connect the negative battery cable.

36. Properly refill the charge air cooling system and the engine cooling system.

Exhaust Manifold

REMOVAL AND INSTALLATION

2.2L (VIN W and 4) Engine

1. Disconnect the negative battery cable, then remove the air cleaner and duct work.

2. Either disengage the wiring or remove the oxygen sensor from the manifold. If the manifold or sensor is to be replaced, remove the sensor.

3. Remove the oil fill tube assembly.

4. Loosen and remove the exhaust manifold retaining nuts, first disconnect the pipe from the manifold, then remove the manifold from the engine.

To install:

5. Carefully clean the threads of the exhaust manifold retainers, then remove all remaining traces of gasket from the mating surfaces.

6. Install the manifold to the engine using a new gasket, then tighten the manifold nuts to 115 inch lbs. (13 Nm). Connect the pipe to the manifold and tighten the retainers.

7. Install the oil fill tube assembly.

8. Either install the oxygen sensor or engage the wiring, as applicable. If the sensor or manifold was replaced, tighten the oxygen sensor to 30 ft. lbs. (41 Nm).

9. Install the air cleaner and duct work.

10. Connect the negative battery cable.

2.5L (VIN A) Engine

1. Disconnect the negative battery cable.

2. Remove the air cleaner and heat stove pipe.

3. Remove the A/C compressor, if equipped, drive belt, and the rear adjusting bracket (if used). Lay the compressor aside in the engine compartment.

4. Remove the dipstick tube and bracket.

5. Disconnect the exhaust pipe from the exhaust manifold.

6. Disengage the electrical connector from the oxygen sensor.

7. Remove the exhaust manifold-to-engine bolts/washers and the manifold from the engine.

To install:

8. Using a gasket scraper, clean the gasket mounting surfaces.

9. Install the manifold to the engine using a new gasket, then tighten the center retainers 36 ft. lbs. (49 Nm) and the outer retainers to 32 ft. lbs. (43 Nm) using the proper torque sequence.

10. Engage the oxygen sensor electrical connector, then connect and secure the exhaust pipe to the manifold.

11. Install the dipstick tube and bracket, then reposition and install the A/C compressor, if equipped, rear adjusting bracket (if used) and the drive belt.

12. Install the air cleaner and the heat stove pipe.

13. Connect the negative battery cable.

2.8L (VIN R) Engine

LEFT SIDE

1. Disconnect the negative battery cable, then raise and support the vehicle safely.

2. Disconnect the exhaust pipe from the exhaust manifold.

3. Remove the rear exhaust manifold-to-cylinder head bolts, then lower the vehicle for underhood access.

4. If necessary, disconnect the air management hoses and wiring.

5. If applicable, disconnect the heat stove tube.

6. If equipped, remove the P/S pump and bracket; do not disconnect the power steering hoses, instead reposition the pump with the hoses attached.

7. Remove the remaining exhaust manifold-to-cylinder head bolts, then remove the manifold from the engine.

To install:

8. Using a gasket scraper, clean the gasket mounting surfaces. Inspect the exhaust manifold for distortion, cracks or damage; replace if necessary.

9. Install the exhaust manifold to the cylinder using a new gasket, then tighten the exhaust manifold-to-cylinder head bolts to 25 ft. lbs. (34 Nm) using a circular pattern, working from the center towards the outer ends.

10. If equipped, reposition and secure the P/S pump and bracket

11. If applicable, connect the heat stove tube.

12. If necessary, reconnect the air management hoses and wiring.

13. Raise and support the vehicle safely.

14. Connect the exhaust pipe to the manifold.

15. Carefully lower the vehicle, then connect the negative battery cable.

RIGHT SIDE

1. Disconnect the negative battery cable, then raise and support the vehicle safely.

2. Disconnect the exhaust pipe from the exhaust manifold.

3. Remove the rear exhaust manifold-to-cylinder head bolts, then lower the vehicle for underhood access.

4. Remove the AIR system diverter valve and heat shield, then remove the AIR system pump and alternator brackets.

5. If necessary, disconnect the air management hoses and wiring.

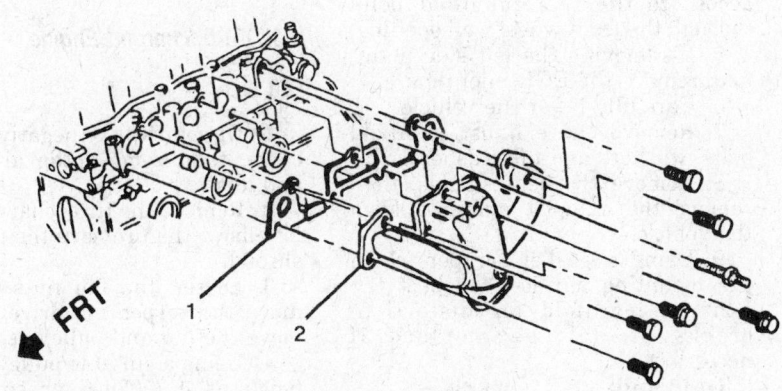

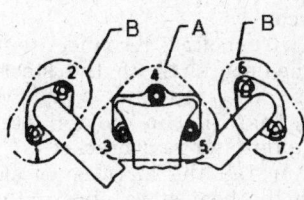

TIGHTENING SEQUENCE

79219g55

1. Exhaust manifold
2. Exhaust manifold gasket

Exhaust manifold installation and torque sequence — 2.5L engine

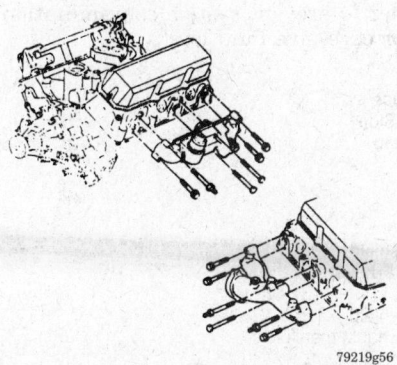

79219g56

Exhaust manifold installation — 2.8L engine

6. Remove the remaining exhaust manifold-to-cylinder head bolts, then remove the manifold from the engine.

To install:

7. Using a gasket scraper, clean the gasket mounting surfaces. Inspect the exhaust manifold for distortion, cracks or damage; replace if necessary.

8. Install the exhaust manifold to the cylinder using a new gasket, then tighten the exhaust manifold-to-cylinder head bolts to 25 ft. lbs. (34 Nm) using a circular pattern, working from the center to the outer ends.

9. If necessary, reconnect the air management hoses and wiring.

10. Install the AIR system pump and alternator brackets, then install the AIR system diverter valve and heat shield.

11. Raise and support the vehicle safely.

12. Connect the exhaust pipe to the manifold.

13. Lower the vehicle, then connect the negative battery cable.

4.3L Non-Turbocharged Engines

1993–95 Vehicles

1. Disconnect the negative battery cable.

2. On van models, remove the engine cover.

3. Raise and support the vehicle safely.

4. Disconnect the exhaust pipe from the exhaust manifold.

5. Lower the vehicle for underhood access.

6. Tag and disconnect the spark plug wires from the plugs and from the retaining clips.

7. If removing the left side manifold:

  a. Remove the air cleaner with heat stove pipe and cold air intake pipe.

  b. Remove the power steering/alternator rear bracket.

  c. Check for sufficient clearance between the manifold and the intermediate steering shaft. On some models it will be necessary to disconnect the intermediate shaft from the steering gear in order to reposition the shaft for clearance.

8. If necessary when removing the right side manifold, unbolt the A/C compressor and bracket, then position the assembly aside. Do not disconnect the lines or allow them to become kinked or otherwise damaged.

9. If necessary for the right side manifold, remove the spark plugs, dipstick tube and wiring.

10. Unbend the locktangs then remove the exhaust manifold retaining bolts, washers and tab washers. Remove the exhaust manifold, then remove and discard the old gaskets.

To install:

11. Using a gasket scraper, clean the gasket mounting surfaces. Inspect the exhaust manifold for distortion, cracks or damage; replace if necessary.

12. Install the exhaust manifold to the cylinder using a new gasket, then tighten the exhaust manifold-to-cylinder head bolts to 26 ft. lbs. (36 Nm) on the center exhaust tube and to 20 ft. lbs. (28 Nm) on the front and rear exhaust tubes. Once the bolts are tightened, bend the tabs on the wash-

ers back over the heads of all bolts in order to lock them in position.

13. If removed on the right side, install the spark plugs, dipstick tube and wiring.

14. If unbolted, reposition and secure the A/C compressor and bracket assembly.

15. If the left manifold was removed:

   a. If unbolted, reconnect the intermediate shaft to the steering gear.

   b. Install the power steering/alternator rear bracket.

   c. Install the air cleaner along with the heat stove pipe and cold air intake pipe.

16. Connect the spark plug wires to the retainer clips and to the plugs as noted during removal.

17. Raise and support the vehicle safely.

18. Connect the exhaust pipe to the manifold.

19. Lower the vehicle.

20. On Van models, instal the engine cover.

21. Connect the negative battery cable.

1996–97 Vehicles

1. Disconnect the negative battery cable.

2. Raise and safely support the vehicle.

3. Disconnect the exhaust pipe from the exhaust manifold.

4. Remove the front tires to gain access to the rear manifold bolts through the front wheel well opening.

5. If removing the left side manifold, remove the EGR inlet pipe.

6. Carefully lower the vehicle.

7. Remove the exhaust manifold bolts, washers and tab washers.

8. Remove the heat shields, then remove the exhaust manifold from the vehicle.

9. Using a gasket scraper, clean the mounting surfaces. Inspect the exhaust manifold for distortion, cracks or damage; replace if necessary.

To install:

10. Position the exhaust manifold and the heat shields.

11. Install the manifold bolts, washers and tab washers. Tighten the bolts to 22 ft. lbs. (30 Nm), then bend the tab washers over the heads of all the bolts.

12. If installing the left side manifold, connect the EGR inlet pipe.

13. Raise and safely support the vehicle.

14. Install the rear manifold bolts through the front wheel opening. Install the front tires.

15. Connect the front exhaust pipe to the manifold.

16. Carefully lower the vehicle, then connect the negative battery cable.

4.3L Turbocharged Engine

Left Side

1. Disconnect the negative battery cable, then remove the air cleaner and duct.

2. Remove the turbocharger air inlet elbow, then remove the upper fan shroud.

3. Loosen the fan nuts, then remove the serpentine drive belt. Remove the fan and pulley assembly.

4. Using a suitable pulley remover (such as J-25034-B or equivalent) pull the power steering pump pulley from the pump.

5. Remove the rear brace from the alternator.

6. Raise and support the vehicle safely.

7. Remove the left tire and wheel, then remove the left wheel house panel.

8. Disconnect the power steering inlet and outlet hoses from the pump. Immediately plug all openings in order to prevent system contamination or excessive fluid loss.

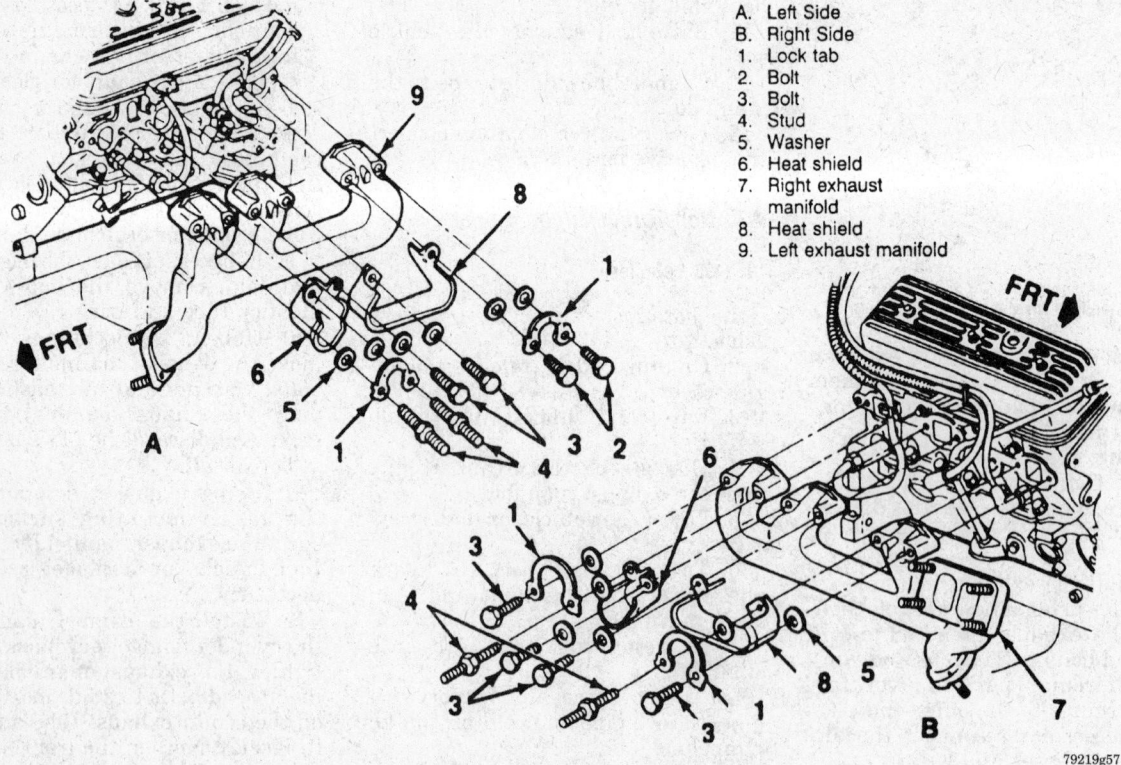

A. Left Side
B. Right Side
1. Lock tab
2. Bolt
3. Bolt
4. Stud
5. Washer
6. Heat shield
7. Right exhaust manifold
8. Heat shield
9. Left exhaust manifold

79219g57

Exhaust manifold mounting — 1993–95 4.3L engines

9. Disconnect the oil filter line bracket from the power steering pump.

10. Disconnect the intermediate shaft.

11. Unbolt the exhaust crossover pipe from the manifold.

12. Remove the power steering pump and rear brace as an assembly from the front bracket.

13. Tag and disconnect the wires from the spark plugs, then remove the plugs from the left cylinder head.

14. Bend back the locktabs, then remove the exhaust manifold bolts, studs, locktabs, washers and heat shields. Remove the exhaust manifold.

To install:

15. Using a gasket scraper, clean the gasket mounting surfaces. Inspect the exhaust manifold for distortion, cracks or damage; replace if necessary.

16. Install the exhaust manifold to the cylinder using a new gasket, then tighten retainers to 33 ft. lbs. (45 Nm). Once the retainers are tightened, bend the tabs on the washers back over the hex-heads of all bolts/studs in order to lock them in position.

17. Install and tighten the spark plugs, then connect the spark plug wires.

18. Install the power steering pump and rear brace to the front bracket. Tighten the bracket bolts to 37 ft. lbs. (50 Nm) and the rear brace nut to 33 ft. lbs. (45 Nm).

19. Connect the crossover pipe and tighten the retaining nuts to 12 ft. lbs. (16 Nm).

20. Connect the intermediate shaft, then secure the oil filter line bracket to the power steering pump.

21. Remove the plugs, then connect the power steering outlet and inlet hoses to the pump. Tighten the hose fittings to 20 ft. lbs. (27 Nm).

22. Install the left wheel housing panel, then install the left front tire and wheel assembly.

23. Carefully lower the vehicle.

24. Install the rear brace to the alternator and tighten the retaining bolt to 18 ft. lbs. (25 Nm).

25. Install the power steering pump pulley using a suitable installer tool.

26. Install the fan and pulley to the water pump, then install the serpentine drive belt. With the belt installed, tighten the fan and pulley nuts to 18 ft. lbs. (24 Nm).

27. Install the upper fan shroud, then install the turbocharger air inlet elbow.

28. Install the air cleaner and duct.

29. Connect the negative battery cable.

30. Refill the power steering pump and bleed air from the system.

Right Side

1. Disconnect the negative battery cable, followed by the positive cable from the battery. Loosen the battery hold-down clamp and remove the battery.

2. Drain the engine cooling system, then remove the battery tray and vacuum tank.

3. Disconnect the turbocharger oil feed hose and the coolant return pipe.

4. Disengage the oxygen sensor connector, then raise and support the vehicle safely.

5. Remove the right front tire and wheel, then remove the right wheel housing panel.

6. Disconnect the mufflers and tailpipe from the catalytic converter.

7. Remove the catalytic converter support bolts, then carefully lower the vehicle for underhood access.

8. Disengage the turbocharger solenoid electrical connector and loosen the turbocharger coolant feed line.

9. Raise and support the vehicle safely, then remove the turbocharger outlet pipe nuts.

10. Remove the turbocharger outlet pipe support bolt. Move the outlet pipe and catalytic converter away from the turbocharger.

11. Remove the turbocharger oil return pipe from the turbocharger, then remove the turbocharger mounting nuts.

12. Disconnect the coolant feed line from the turbocharger, then remove the turbocharger assembly.

13. Disconnect the exhaust crossover pipe from the manifold.

14. Tag and disconnect the wires from the spark plugs, then remove the plugs from the left cylinder head.

15. Remove the charger air cooler lower supports.

16. Bend back the locktabs, then remove the exhaust manifold bolts, studs, locktabs, washers and heat shields. Remove the exhaust manifold.

To install:

17. Using a gasket scraper, clean the gasket mounting surfaces. Inspect the exhaust manifold for distortion, cracks or damage; replace if necessary.

18. Install the exhaust manifold to the cylinder using a new gasket, then tighten retainers to 33 ft. lbs. (45 Nm). Once the retainers are tightened, bend the tabs on the washers back over the hex-heads of all bolts/studs in order to lock them in position.

19. Install and tighten the spark plugs, then connect the spark plug wires.

20. Install the charge air cooler lower supports.

21. Connect the exhaust crossover pipe to the manifold, then tighten the retaining nuts to 12 ft. lbs. (16 Nm).

22. Position the turbocharger, then connect the coolant feed pipe and loosely install the bolt.

23. Loosely install the turbocharger mounting nuts followed by the return pipe and bolts. Tighten the coolant feed pipe bolt to 26 ft. lbs. (35 Nm), then mounting nuts to 33 ft. lbs. (45 Nm) and the oil return pipe bolts to 35 inch lbs. (4 Nm).

24. Install the turbocharger outlet pipe and nuts along with the outlet pipe support bolt. Tighten the pipe nuts to 41 ft. lbs. (55 Nm) followed by the support bolt to 22 ft. lbs. (30 Nm).

25. Engage the turbocharger solenoid electrical connector.

26. Loosely install the catalytic converter support bolts, then engage the mufflers and tailpipe to the catalytic converter. Tighten the support bolts to 25 ft. lbs. (34 Nm), then tighten the tailpipe bolts to 24 ft. lbs. (32 Nm).

27. Install the wheel housing panel.

28. Install the tire and wheel assembly, then carefully lower the vehicle.

29. Engage the oxygen sensor electrical connector.

30. Connect the turbocharger oil feed hose and the coolant return pipe. Tighten the oil hose fitting to 13 ft. lbs. (17 Nm) and the coolant pipe fittings to 16 ft. lbs. (22 Nm).

31. Install the battery tray and vacuum tank.

32. Install the battery and secure using the hold-down clamp.

33. Connect the positive cable, then the negative cable to the battery.

34. Properly refill the engine cooling system.

Turbocharger

The Turbocharger and multi-port fuel injection are special to the Typhoon. They require some service procedures which are unique to the rest of these vehicles.

PRECAUTIONS

Proper maintenance practices must be followed in order to prolong the performance and life of the turbocharger. Failures in the turbocharger system are often caused by oil lag, restriction or lack of oil flow, dirty oil or foreign objects entering the turbocharger.

When working on or around the turbocharger system always follow these precautions:

• Do not allow dust, sand or other foreign material to enter the turbocharger. Dust or sand will erode the compressor wheel blades, while uneven blade wear can produce shaft motion causing bearing failure. Large or heavy objects will completely destroy the turbocharger and could cause severe damage to the engine.

• The air cleaner system must be properly maintained. A plugged or restricted air cleaner will reduce air pressure and volume at the compressor air inlet. The pressure drop will lower turbocharger performance and may cause oil pullover during idle. Oil pullover is a compressor end oil seal leak without seal part failure.

• The oil lube system must be properly maintained. If dirt or foreign material is introduced into the turbocharger bearing system by lube oil, the center housing bearing bore surfaces will wear. Contaminants act as an abrasive cutting tool and will eventually wear through bearing surfaces causing turbocharger noise and poor performance. Excessive noise and/or oil smoke may be noticed. Remember that oil will completely bypass the filter if the element becomes clogged. Adhere closely to oil and filter change intervals on turbocharged vehicles.

• Do not allow sludge to build up in the turbocharger. Sludge may occur if oil oxidation or breakdown occurs. Possible causes would include, engine overheating, excessive combustion products from piston blow-by, noncompatible oils, engine coolant leakage into the oil, incorrect grade/quality of oil or improper oil change intervals. If turbine end oil leakage is noted, check the turbocharger oil drain tube and the crankcase breathers. Remove any restrictions and check again for leaks.

• The charge air cooler system must be properly maintained. Only use the proper type and quantity of coolant in the system. If coolant becomes dirty or contaminated, the system must be flushed and refilled with fresh coolant.

REMOVAL AND INSTALLATION

1. Drain the engine cooling system, then remove the air intake duct and the turbocharger air intake duct.
2. Disconnect the negative battery cable, followed by the positive cable from the battery. Loosen the battery hold-down clamp and remove the battery.
3. Remove the battery tray and vacuum tank.
4. Remove the nut and disconnect the turbocharger oil feed line, then disengage the electrical connector from the solenoid.
5. Loosen the turbocharger coolant return line clamp nut, then disconnect the coolant return line assembly.
6. Raise and support the vehicle safely.
7. Remove the right tire and wheel, then remove the wheel house panel retaining screws and remove the panel for access.
8. If necessary, disengage the oxygen sensor electrical connector, then remove the nut from the turbocharger and the bolt from the outlet pipe support bracket. Remove the clamp from the catalytic converter and bolts from the converter support, then remove the turbocharger outlet pipe.
9. Remove the retaining bolts from the turbocharger oil return pipe, then remove the return pipe and gasket from the turbocharger assembly.
10. Remove the turbocharger mounting nuts.
11. Disconnect the coolant feed line from the turbocharger, then remove the turbocharger assembly.

To install:

12. Inspect and clean the turbocharger and manifold mounting surfaces.
13. Position the turbocharger, then connect the coolant feed pipe with gaskets and loosely install the bolt.

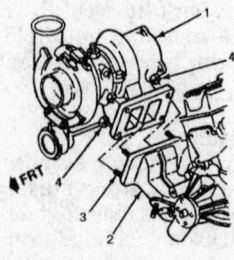

1. Turbocharger assembly
2. Manifold
3. Stud
4. Nut

79219g58

Turbocharger assembly mounting — Typhoon

14. Loosely install the turbocharger mounting nuts followed by the oil feed pipe line fitting. Tighten the coolant feed pipe bolt to 26 ft. lbs. (35 Nm), then the turbocharger mounting nuts to 33 ft. lbs. (45 Nm) and the oil feed line fitting to 13 ft. lbs. (16 Nm).
15. Connect the turbocharger oil return pipe and gaskets, then tighten the line bolt to 35 inch lbs. (4 Nm).
16. Connect the turbocharger coolant return line assembly and tighten the line nut to 16 ft. lbs. (22 Nm). Secure the coolant return line clamp and tighten to 16 inch lbs. (1.8 Nm).
17. Install the turbocharger outlet pipe to the catalytic converter and turbocharger. Install the nuts to the turbocharger, then clamp to the converter, the bolts to the converter support and the the bolt to the outlet pipe bracket. Tighten the turbocharger nuts to 41 ft. lbs. (55 Nm), the converter clamp nuts to 42 ft. lbs. (57 Nm), the support bracket bolt to 22 ft. lbs. (30 Nm) and the converter support bolts to 25 ft. lbs. (34 Nm).
18. If disengaged, connect the oxygen sensor wiring to the harness.
19. Install the wheel housing panel, then install the tire and wheel assembly. Carefully lower the vehicle.
20. Engage the turbocharger solenoid electrical connector.
21. Install the battery tray and vacuum tank.
22. Install the battery and secure using the hold-down clamp.
23. Install the air intake duct and the turbocharger air intake duct.
24. Connect the positive cable, then the negative cable to the battery.
25. Properly refill the engine cooling system.

Front Cover Oil Seal

REPLACEMENT

On most gasoline engines it is possible to replace the timing cover front oil seal without removing the cover. For the 2.2L gasoline engine, the manufacturer does not recommend this procedure and suggests that the front cover be removed if seal replacement is necessary.

Except 2.2L (VIN W and 4) Engine

1. Disconnect the negative battery cable.
2. Remove the accessory drive belt.
3. Remove the crankshaft damper and pulley.

4. Pry the seal out of the front cover using a small prytool. Be very careful not to distort the front cover or to score the end of the crankshaft.

To install:

5. Lightly coat the lips of the replacement crankshaft seal with clean engine oil, then position the seal with the open end facing inward the engine. Use a suitable seal installation driver to position the seal in the front cover.

6. Install the crankshaft damper and pulley.

7. Install the accessory drive belt.

8. Connect the negative battery cable.

2.2L (VIN W and 4) Engine

1. Remove the crankcase (timing) front cover from the engine.

2. Carefully remove the old crankshaft seal from the cover using a suitable prytool.

To install:

3. Lubricate the lips of a new seal with clean engine oil, then use a seal centering tool (such as J-35468 or equivalents) to install the seal to the front cover. Leave the tool in position in the seal until the cover is installed.

4. Install the timing cover to the engine.

Timing Chain, Sprockets and Front Cover

It is recommended by the manufacturer that the front cover oil seal be replaced whenever the cover is removed.

REMOVAL AND INSTALLATION

2.2L (VIN W and 4) Engine

1. Disconnect the negative battery cable.

2. Remove the power steering fluid reservoir from the radiator shroud, then remove the upper fan shroud.

3. Carefully release the belt tension, then remove the serpentine drive belt.

4. Remove the alternator and brackets from the engine, then position them aside.

5. Remove the crankshaft pulley and hub.

6. As necessary, disconnect the lower radiator hose clamp at the

water pump, then loosen and/or remove the oil pan.

NOTE: There may be bolts attaching the front of the oil pan to the timing cover. If so, make sure they are removed before attempting to remove the cover.

7. Remove the crankcase (timing) front cover bolts, then remove cover from the engine. Make sure all bolts are removed and be careful not to force and damage the cover.

8. If necessary, carefully remove the old crankshaft seal from the cover using a suitable prytool.

9. Turn the crankshaft until the timing marks on the sprockets are in alignment. The marks should also be in alignment with the tabs on the tensioner.

10. Remove the tensioner retaining bolts.

11. Remove the camshaft sprocket retaining bolts, then remove the sprocket and timing chain.

12. If necessary, remove the crankshaft sprocket using J-22888-20 or equivalent puller.

To install:

13. If removed, install the crankshaft sprocket using a suitable installer. Make sure the sprocket is fully seated against the crankshaft.

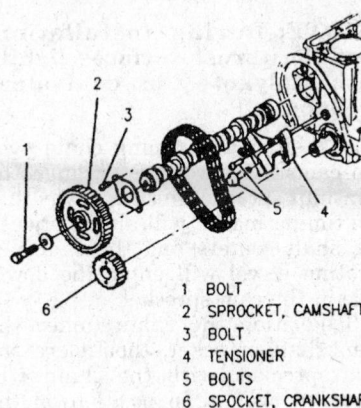

1	BOLT
2	SPROCKET, CAMSHAFT
3	BOLT
4	TENSIONER
5	BOLTS
6	SPOCKET, CRANKSHAFT

A ALIGN TABS ON TENSIONER WITH MARKS ON CAMSHAFT & CRANKSHAFT SPROCKETS.

79219g59

Timing chain, sprocket and camshaft mounting — 2.2L engine

14. Compress the tensioner spring and insert a cotter pin or nail in the hole provided to hold the tensioner in position.

15. Loosely install the tensioner retaining bolts.

16. Position the camshaft sprocket in the timing chain, position the chain under the crankshaft sprocket and the camshaft sprocket to the camshaft.

17. Verify that the timing marks are all properly aligned, then loosely install the camshaft sprocket bolt.

18. Tighten the tensioner bolts to 17 ft. lbs. (23 Nm), then tighten the camshaft sprocket bolt to 77 ft. lbs. (105 Nm).

19. Remove the cotter pin or nail holding the tensioner in position off the chain.

20. Carefully remove all traces of gasket or sealant from the mating surfaces.

21. If removed, lubricate the lips of a new seal with clean engine oil, then use a seal centering tool (such as J-35468 or equivalent) to install the seal to the front cover. Leave the tool in position in the seal until the cover is installed.

22. Apply a $\frac{3}{8}$ in. (10mm) wide by $\frac{5}{16}$ (5mm) thick bead of RTV sealer to the oil pan at the front crankcase cover sealing surface. Then apply a $\frac{1}{4}$ in. (6mm) by $\frac{1}{8}$ in. (3mm) thick bead of RTV to the crankcase front cover at the block sealing surface.

23. Install the crankcase front cover to the engine using the seal tool to assure it is properly centered and prevent damage to the hub. Tighten the cover retaining bolts to 97 inch lbs. (11 Nm), then remove the seal centering tool.

24. Install the lower radiator hose clamp at the water pump, then install and secure the oil pan, as applicable.

25. Install the crankshaft pulley and hub.

26. Reposition and secure the alternator with brackets.

27. Install the serpentine drive belt, then install the upper fan shroud.

28. Reposition and secure the power steering reservoir.

29. Connect the negative battery cable.

2.8L (VIN R) Engine

1. Disconnect the negative battery cable, then drain the engine cooling system.

2. Remove the serpentine drive belt.

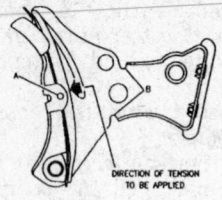

A INSERT PIN AFTER TENSION HAS BEEN APPLIED
B TABS, USED FOR CAMSHAFT AND CRANKSHAFT ALIGNMENT

79219g60

Locking the timing chain tensioner into position for chain installation — 2.2L engine

3. Remove the water pump from the engine.
4. Remove the power steering pump bracket.
5. Remove the crankshaft pulley bolt, crankshaft pulley and hub (torsional damper).

NOTE: The outer ring (weight) of the torsional damper is bonded to the hub with rubber. The damper must be removed with a puller which acts on the inner hub only. Pulling on the outer portion of the damper will break the rubber bond or destroy the tuning of the unit.

6. Disconnect the lower radiator hose at the front cover.
7. Remove the timing cover bolts and cover. Check for bolts threaded from the front of the oil pan to the bottom of the cover. If present, these must be removed before attempting to loosen the cover.
8. If the front seal is to be replaced, it can be pried out of the cover with a small prytool.
9. Rotate the crankshaft until the No. 4 cylinder is on the TDC of it's compression stroke and the camshaft sprocket mark aligns with the mark on the crankshaft sprocket (facing each other at a point closest together in their travel) and in line with the shaft centers.
10. Remove the camshaft sprocket-to-camshaft nut and/or bolts, then remove the camshaft sprocket (along with the timing chain). If the sprocket is difficult to remove, use a plastic mallet to bump the sprocket from the camshaft.

NOTE: The camshaft sprocket (located by a dowel) is lightly pressed onto the camshaft and should come off easily. The chain comes off with the camshaft sprocket.

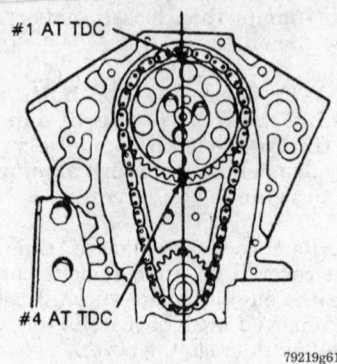

#1 AT TDC

#4 AT TDC

79219g61

Timing mark alignment — 2.8L engine

11. If necessary use J-5825-A or equivalent crankshaft sprocket removal tool to free the timing sprocket from the crankshaft.
To install:
12. Inspect the timing chain and the timing sprockets for wear or damage, replace the damaged parts as necessary.
13. Using a gasket scraper, clean the gasket mounting surfaces of any old gasket material. Using solvent, clean the oil and grease from the gasket mounting surfaces.
14. If removed, use J-5590 or equivalent crankshaft sprocket installation tool and a hammer to drive the crankshaft sprocket onto the crankshaft, without disturbing the position of the engine.

NOTE: During installation, coat the thrust surfaces lightly with Molykote® or equivalent pre-lube.

15. Position the timing chain over the camshaft sprocket. Arrange the camshaft sprocket in such a way that the timing marks will align between the shaft centers and the camshaft locating dowel will enter the dowel hole in the cam sprocket.
16. Position the chain under the crankshaft sprocket, then place the cam sprocket, with the chain still mounted over it, in position on the front of the camshaft. Install and tighten the camshaft sprocket-to-camshaft retainers to 17 ft. lbs. (23 Nm).
17. With the timing chain installed, turn the crankshaft 2 complete revolutions, then check to make certain that the timing marks are in correct alignment between the shaft centers.
18. Clean the gasket mating surfaces of the engine and cover of all remaining gasket or sealer material. Be careful not to score or damage the surfaces.
19. If removed, install a new seal to the cover using a suitable installation

driver such as J-23042 (early model), J-35468 (late model) or equivalent. Be sure to support the back of the seal cover area during installation. Lightly coat the lips of the new seal with clean engine oil.
20. Lightly coat both sides of a new gasket using an anaerobic sealant, then position the gasket and cover to the engine. Install and tighten the retainers. If equipped with an oil pan sealing lip on the cover, apply a bead of sealant to the front cover at the oil pan mating surface.

NOTE: If equipped, do not forget the oil pan-to-front-cover bolts.

21. Install the water pump assembly to the engine.
22. Connect the lower radiator hose to the front cover.
23. Install the crankshaft hub and pulley.
24. Install the power steering pump bracket.
25. Install the serpentine drive belt.
26. Connect the negative battery cable, then properly refill the engine cooling system.
27. Run the engine until normal operating temperature has been reached, then check for leaks.

4.3L (VIN W, X, and Z) Engine

1. Disconnect the negative battery cable and drain the engine cooling system.
2. Remove the crankshaft pulley and damper.

NOTE: The outer ring (weight) of the torsional damper is bonded to the hub with rubber. The damper must be removed with a puller which acts on the inner hub only. Pulling on the outer portion of the damper will break the rubber bond or destroy the tuning of the unit.

3. Remove the water pump assembly.
4. Loosen the oil pan.
5. Remove the front cover bolts and, if equipped, the reinforcements, then remove the front cover from the engine.
6. If the front cover seal is to be replaced, it may be pried front the front cover using a suitable prytool.
7. Rotate the crankshaft until the timing marks on the camshaft and crankshaft sprockets are in proper alignment.
8. Remove the camshaft sprocket-to-camshaft nut and/or bolts, then remove the camshaft sprocket along

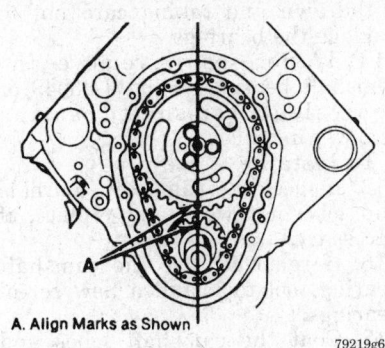

A. Align Marks as Shown

79219g62

Timing mark alignment — 4.3L engines

with the timing chain. If the sprocket is difficult to remove, use a plastic mallet to bump the sprocket from the camshaft.

NOTE: The camshaft sprocket (located by a dowel) is lightly pressed onto the camshaft and should come off easily. The chain comes off with the camshaft sprocket.

9. If necessary use J-5825-A or equivalent crankshaft sprocket removal tool to free the timing sprocket from the crankshaft.

To install:

10. Inspect the timing chain and the timing sprockets for wear or damage, replace the damaged parts as necessary.

11. Using a gasket scraper, clean the gasket mounting surfaces of all remaining traces of old gasket. Using solvent, clean the oil and grease from the gasket mounting surfaces.

12. If removed, use J-5590 or equivalent crankshaft sprocket installation tool and a hammer to drive the crankshaft sprocket onto the crankshaft, without disturbing the position of the engine.

NOTE: During installation, coat the thrust surfaces lightly with Molykote® or equivalent pre-lube.

13. Position the timing chain over the camshaft sprocket. Arrange the camshaft sprocket in such a way that the timing marks will align between the shaft centers and the camshaft locating dowel will enter the dowel hole in the cam sprocket.

14. Position the chain under the crankshaft sprocket, then place the cam sprocket, with the chain still mounted over it, in position on the front of the camshaft. Install and tighten the camshaft sprocket-to-camshaft retainers.

15. With the timing chain installed, turn the crankshaft 2 complete revolutions, then check to make certain that the timing marks are in correct alignment between the shaft centers.

16. Clean the gasket mating surfaces of the engine and cover of all remaining gasket or sealer material. Be careful not to score or damage the surfaces.

NOTE: Beginning in 1992, the manufacturer began recommending to wait until the front cover is mounted to the engine before installing the replacement crankshaft oil seal. This may be to assure the cover is properly supported. On earlier vehicles, the manufacturer allowed for installation with the cover removed or installed, but waiting would be acceptable for all years of the 4.3L engine.

17. If desired on early model engines, install a new seal to the cover using a suitable installation driver, such as J-35468 or equivalent. Be sure to support the back of the seal cover area during installation. Lightly coat the lips of the new seal with clean engine oil.

18. Position a new front cover gasket to the engine or cover using gasket cement to hold it in position. For 1992–96 vehicles, lubricate the front of the oil pan seal with engine oil to aid in reassembly.

19. Install the front cover to the engine. For 1992–96 vehicles, take care while engaging the front of the oil pan seal with the bottom of the cover.

20. Install front cover retaining bolts and tighten to 124 inch lbs. (14 Nm).

21. If removed and not installed earlier, use the seal installation driver to install the new crankshaft seal at this time.

22. Secure the oil pan.

23. Install the water pump.

24. Install the crankshaft damper and pulley.

25. Connect the negative battery cable, then properly refill the engine cooling system.

26. Run the engine until normal operating temperature has been reached, then check for leaks.

Timing Gears

Unlike the rest of the engines utilized by these vehicles, the 2.5L engine does not use a timing chain assembly.

Instead the camshaft timing gear is directly driven by the crankshaft timing gear. The timing gear (camshaft sprocket) is pressed onto the camshaft and requires the use of an arbor press to remove.

REMOVAL AND INSTALLATION

2.5L (VIN A) Engine

1. Disconnect the negative battery cable.

2. Remove the power steering reservoir from the fan shroud, then remove the fan shroud.

3. Remove the serpentine drive belt.

4. Remove the alternator and brackets from the engine (lay them aside).

5. Remove the crankshaft pulley and hub.

NOTE: The outer ring (weight) of the torsional damper is bonded to the hub with rubber. The damper must be removed with a puller which acts on the inner hub only. Pulling on the outer portion of the damper will break the rubber bond or destroy the tuning of the unit.

6. Drain the engine cooling system, then disconnect the lower radiator hose at the water pump.

7. Remove the timing cover bolts and cover. Check for bolts threaded from the front of the oil pan to the bottom of the cover. If present, these must be removed before attempting to loosen the cover.

8. If the front seal is to be replaced, it can be pried out of the cover with a small prytool.

9. Remove the camshaft.

10. Using an arbor press, a press plate and the GM Gear Removal tool J-971 or equivalent, press the timing gear from the camshaft.

NOTE: When pressing the timing gear from the camshaft, be certain that the position of the press plate does not contact the woodruff key.

To install:

11. To assemble, position the press plate to support the camshaft at the back of the front journal. Place the gear spacer ring and the thrust plate over the end of the camshaft, then install the woodruff key. Press the timing gear onto the camshaft, until

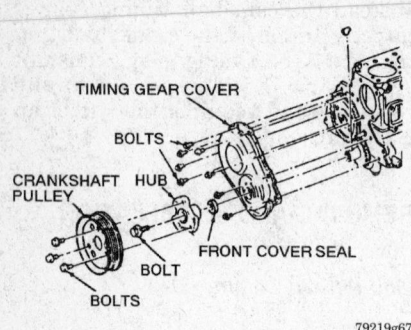

Timing gear front cover — 2.5L engine

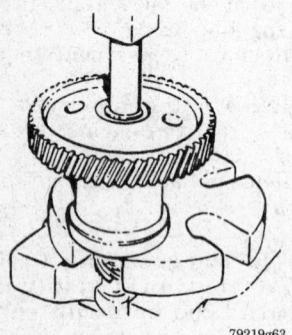

Press the timing gear from the camshaft — 2.5L engine

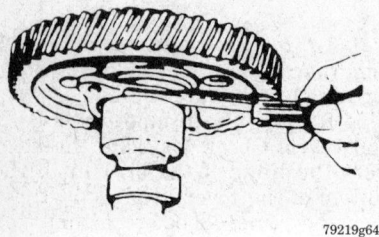

Use a feeler gauge to check thrust plate clearance once the gear and thrust plate are assembled to the camshaft

it bottoms against the gear spacer ring.

NOTE: The end clearance of the thrust plate should be 0.0015–0.005 in. (0.038–0.127mm). If less than 0.0015 in. (0.038mm), replace the spacer ring; if more than 0.005 in. (0.127mm), replace the thrust plate.

12. To complete the installation, align the marks on the timing gears and install the camshaft.

13. If removed, use the GM seal installer/centering tool J-34995 or equivalent to install the replacement cover seal. Be sure to support the seal area of the cover when installing the new seal.

14. Clean all sealing surfaces and apply a ⅜ in. bead of RTV sealant to the oil pan and timing gear cover sealing surfaces.

15. Use the GM Seal Installer/Centering Tool J-34995 or equivalent align the front cover. Install the cover while the RTV sealant is still wet.

16. Install the cover (and pan) retaining bolts, then tighten the bolts to 90 inch lbs. (10 Nm). Remove the centering tool from the timing cover.

17. Connect the lower radiator hose to the water pump.

18. Install the crankshaft hub and pulley.

19. Reposition and secure the alternator and brackets.

20. Install the serpentine drive belt or install and adjust the accessory drive belts, as applicable.

21. Install the upper fan shroud, then reposition and secure the power steering reservoir.

22. Connect the negative battery cable and properly refill the engine cooling system.

23. Run the engine until normal operating temperature is reached, then check for leaks.

Camshaft

REMOVAL AND INSTALLATION

2.2L (VIN W and 4) Engine

1. Properly relieve the fuel system pressure, then disconnect the negative battery cable.

2. Drain the engine cooling system and the engine oil.

3. Remove the radiator.

4. Remove the rocker arm cover.

5. Remove the cylinder head.

6. Remove the anti-rotation bracket bolts and brackets, then remove the valve lifters.

7. Remove the oil pump drive retaining bolt, then remove the drive by lifting and twisting.

8. Remove the crankshaft pulley and hub.

9. Remove the serpentine drive belt idler pulley.

10. Remove the timing cover from the engine.

11. Remove the timing chain and camshaft sprocket.

12. Remove the camshaft thrust plate retaining bolts, then remove the plate from the block.

13. Pull the camshaft straight out of the engine, turning slightly as it is

withdrawn and taking care not to damage the bearings.

14. If necessary, remove the camshaft bearings using J-33049 or equivalent camshaft bearing remover/installer.

To install:

15. Inspect the camshaft, journals and lobes for wear and replace, if necessary.

16. If removed, use the camshaft bearing tool to install a new set of bearings.

17. Coat the camshaft lobes and journals with a high viscosity oil with zinc such as 12345501 or equivalent.

18. Carefully insert the camshaft in the engine, turning it slightly from side to side and it is inserted.

19. Install the thrust plate and tighten the retaining bolts to 106 inch lbs. (12 Nm).

20. Install the timing chain and camshaft sprocket.

21. Install the timing cover to the engine.

22. Install the serpentine drive belt idler pulley.

23. Install the crankshaft pulley and hub.

24. Install the oil pump drive by inserting while twisting, then install the retaining bolt and tighten to 18 ft. lbs. (25 Nm).

25. Install the valve lifters and the anti-rotation brackets.

26. Install the cylinder head.

27. Install the rocker arm cover.

28. Install the radiator.

29. Connect the negative battery cable and properly refill the engine cooling system.

2.5L (VIN A) Engine

1. Disconnect the negative battery cable and drain the engine cooling system.

2. Remove the radiator.

3. If equipped with A/C, disconnect the condenser baffles and the condenser, then raise the condenser and set it aside without disconnecting the refrigerant lines.

4. Remove the grille and filler panel from the front of the vehicle.

5. Remove the timing cover from the engine.

6. Matchmark and remove the distributor.

7. Remove the oil pump drive shaft.

8. Remove the pushrod cover:

   a. Remove the alternator and bracket.

   b. Disengage the ignition coil wires, then remove the spark plug wires and bracket from the intake manifold.

c. Remove the fuel pipes and clips from the pushrod cover.

d. Remove the oil pressure gage sender, then remove the wiring harness brackets from the pushrod cover.

e. Unscrew the nuts from the cover attaching studs, reverse 2 of the nuts so the washers face outward and screw them back onto the 2 inner studs.

f. Assemble the 2 remaining nuts to the same 2 inner studs with the washers facing inward, then using a small wrench on the inner nut (on each stud) jam the nuts slightly together.

g. Again using the wrench on the inner stud, unscrew the studs until the cover breaks loose, then remove the nuts from the studs and remove the cover from the engine. Remove the studs from the cover and reinstall them to the engine. Tighten the studs to 88 inch lbs. (10 Nm).

9. Remove the rocker arm cover, pushrods and valve lifters from the engine.

NOTE: When removing the pushrods and the valve lifters, be sure to keep them in order for reassembly purposes.

10. Position the camshaft gear so access to the thrust plate retainers is possible. Remove the camshaft thrust plate-to-engine bolts.

11. If not done already, align the timing marks, then while supporting the camshaft (to prevent damaging the bearing or lobe surfaces), carefully remove it from the front of the engine.

To install:

12. Inspect the camshaft for scratches, pitting and/or wear on the bearing and lobe surfaces. Check the timing gear teeth for damage.

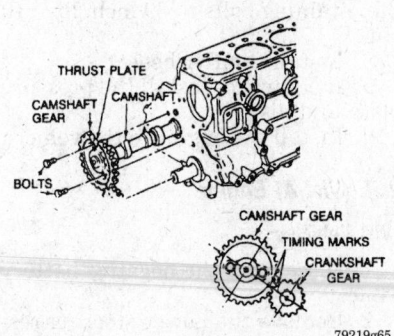

79219g65

Timing gear/camshaft assembly installation and timing mark alignment — 2.5L engine

13. If necessary, replace the camshaft bearings as follows:

a. Install a camshaft bearing removal/installation tool with the shoulder toward the bearing. Ensure that enough threads are engaged.

b. Using 2 wrenches, hold the puller screw while turning the nut. When the bearing has been released from the bore, remove the tool.

c. Assemble the tool on the driver to remove the front and rear bearings.

d. Install the bearings at the correct clocking to ensure the oil holes in the block and bearings align.

e. Install a fresh camshaft bearing rear cover using sealant.

14. Lubricate the camshaft using a high viscosity oil with zinc (such as 12345501 or equivalent), then carefully insert the camshaft into the engine while aligning the timing marks.

15. Turn the camshaft as necessary, then install the camshaft thrust plate-to-engine bolts to and tighten to 90 inch lbs. (10 Nm).

16. Install the valve lifters, then install the pushrods and rocker arms to the engine.

17. Using a thin bead of RTV sealant, install the pushrod cover to the engine, then tighten the stud nuts to 106 inch lbs. (12 Nm) starting at the front nut and working toward the rear of the engine. Reposition and secure the remaining components which were removed to access the pushrod cover.

18. Install the oil pump drive shaft.

19. Align and install the distributor.

20. Install the timing cover to the engine.

21. Install the the grille and filler panel to the front of the vehicle.

22. If equipped with A/C, reposition and secure the condenser and condenser baffles.

23. Install the radiator.

24. Connect the negative battery cable, then properly refill the engine cooling system.

2.8L and 4.3L (VIN W, X, and Z) Engines

1. Properly relieve the fuel system pressure, then disconnect the negative battery cable.

2. Drain the engine cooling system.

3. Remove the radiator.

4. Remove the rocker arm covers from the engine.

5. Remove the intake manifold assembly.

6. Remove the rocker arms, pushrods and lifters.

7. Remove the crankshaft pulley and hub.

8. Remove the engine front (timing) cover.

9. Align the timing marks on the crankshaft and camshaft sprockets.

10. Remove the camshaft sprocket and timing chain.

11. If equipped, remove the balance shaft drive gear.

12. Remove the camshaft thrust plate.

13. Install the sprocket bolts or longer bolts of the same thread into the end of the camshaft as a handle, then remove the camshaft front the front of the engine while turning slightly from side to side, as necessary. Take care not to damage the camshaft bearings when removing the camshaft.

To install:

14. Lubricate the camshaft journals with clean engine oil or a suitable pre-lube, then install the camshaft into the block being extremely careful not to contact the bearings with the cam lobes.

15. Install the camshaft thrust plate.

16. If equipped, install the balance shaft drive gear.

17. Install the timing chain and camshaft sprocket.

18. Install the engine front (timing) cover.

19. Install the crankshaft pulley and hub.

20. Install the valve lifters, then install the pushrods and rocker arms. Properly adjust the valve clearance.

21. Install the intake manifold assembly.

22. Install the rocker arm covers to the engine.

23. Install the radiator to the vehicle.

24. Connect the negative battery cable and properly refill the engine cooling system.

Balance Shaft

REMOVAL AND INSTALLATION

4.3L (VIN W) Engine

1. Disconnect the negative battery cable. Drain the cooling system into a suitable container. Properly relieve the fuel system pressure and discharge the air conditioning system.

2. Remove the air cleaner and air intake duct.

3. Remove the upper radiator shroud.

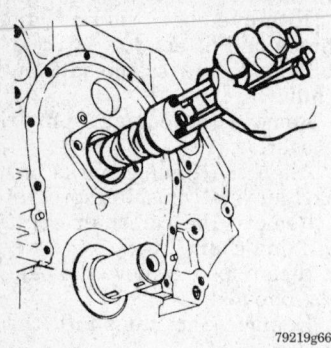

Camshaft removal and installation — 4.3L engine

4. Remove the oil and transmission cooler lines at the radiator.

5. Remove the hoses from the radiator and remove the radiator.

6. Remove the air conditioner condenser and fan assembly.

7. Remove the serpentine belt.

8. Remove the brace at the coolant pump and remove the pump.

9. Remove the torsional damper.

10. Raise and support the vehicle safely.

11. Remove the flywheel cover. Drain the engine oil and loosen the oil pan bolts; remove the 2 nuts and bolts at the front of the oil pan.

12. Remove the front timing cover bolts and cover.

NOTE: Use care when removing the front cover, as not to damage the oil pan gasket.

13. Remove the crankshaft front cover oil seal.

14. Remove the camshaft sprocket, timing chain and balance shaft drive gear.

15. Remove the balance shaft retainer.

16. Remove the intake manifold assembly.

17. Remove the hydraulic lifter retainer.

18. Remove the balance shaft from the bearing using a soft faced mallet.

19. Remove the rear bearing using tool J–38834 and J–36996.

NOTE: The balance shaft and bearings are only to be serviced or replaced as a complete assembly.

To install:

20. Dip the bearings in clean engine oil before installation.

21. Using tool J–38834, install the balance shaft rear bearing with the flat edge and the manufacturer's markings facing the front of engine.

22. Dip the front balance shaft bearing into engine oil. Using tool

J–36996 and J–8092, install the balance shaft into the block.

23. Install the balance shaft retainer and torque the bolts to 120 inch lbs. (14 Nm).

24. Install the balance shaft drive gear. Apply Loctite® to the bolt and torque to 15 ft. lbs. (20 Nm) plus turn the bolt an additional 35 degrees.

25. Install the lifter retainer and turn the balance shaft to ensure proper clearance.

26. Install the balance shaft rear plug.

27. Turn the camshaft so the balance shaft drive gear timing marks are aligned.

28. Install the balance shaft drive gear retaining stud and torque to 12 ft. lbs. (16 Nm).

29. Install the intake manifold assembly.

30. Install the timing chain and gears with the timing marks dot-to-dot; this is the No. 4 cylinder firing position, so set the distributor rotor to the No. cylinder firing position when installing. Torque the camshaft bolts and nut to 21 ft. lbs. (28 Nm).

31. Install the distributor assembly.

32. Install the timing cover oil seal and install the cover.

33. Raise and support the vehicle safely.

34. Install the flywheel cover.

35. Tighten the oil pan bolts; install the 2 nuts and bolts at the front of the oil pan.

36. Install the torsional damper.

37. Install the brace at the coolant pump and install the pump.

38. Install the serpentine belt.

39. Install the air conditioner condenser and fan assembly.

40. Install the radiator and hoses.

41. Install the oil and transmission cooler lines at the radiator.

42. Install the upper radiator shroud.

43. Install the air cleaner and air intake duct.

44. Connect the negative battery cable. Fill the cooling system with proper type and quantity of antifreeze. Properly recharge the air conditioning system. Fill the engine with the correct quantity and grade of engine oil.

Piston and Connecting Rod

POSITIONING

During disassembly, note the location of the connecting rod bearing tang slots, on some engines, the slots must

be on the side opposite of the camshaft. This does not apply to the 3.8L engine.

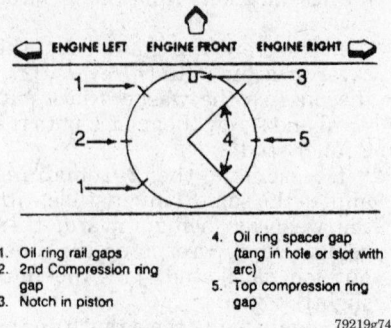

1. Oil ring rail gaps
2. 2nd Compression ring gap
3. Notch in piston
4. Oil ring spacer gap (tang in hole or slot with arc)
5. Top compression ring gap

Piston and ring gap positioning

ENGINE LUBRICATION

Oil Pan

REMOVAL AND INSTALLATION

2.2L (VIN W and 4) Engine

1. Remove the engine assembly.

2. If equipped, remove the clutch pressure plate and disc from the engine.

3. Remove the flywheel.

4. Remove the oil pan nuts and bolts, then remove the pan from the bottom of the block.

To install:

5. Carefully clean the gasket mating surfaces of any remaining old gasket or sealer material.

6. Position a new gasket and seal onto the oil pan. Use a thin bead of sealant at either side of the sealer.

7. Install the oil pan and tighten the retaining bolts to 89 inch. lbs. (10 Nm).

8. Install the flywheel.

9. If equipped, install the pressure plate and disc.

10. Install the engine to the vehicle.

2.5L (VIN A) Engine

2WD Vehicles

1. Disconnect the negative battery cable.

2. Remove the power steering reservoir at the fan shroud.

3. Remove the radiator fan shroud.

1. Camshaft sprocket bolt
2. Camshaft sprocket
3. Balance shaft driven gear bolt
4. Balance shaft driven gear
5. Balance shaft drive gear
6. Nut
7. Stud

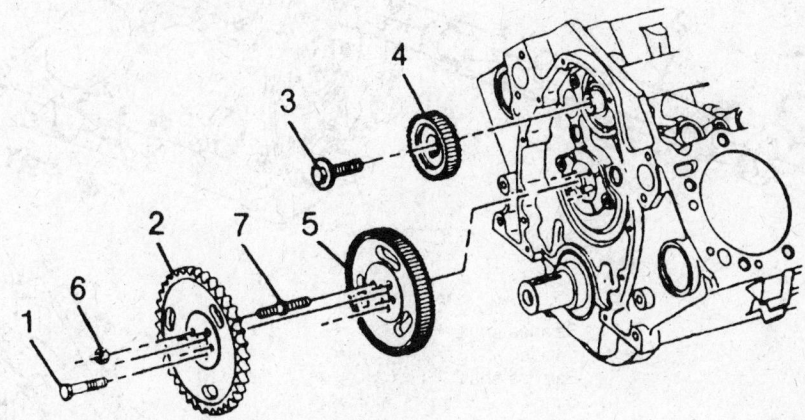

79219g68

Balance shaft drive gears — 4.3L (VIN W) engines

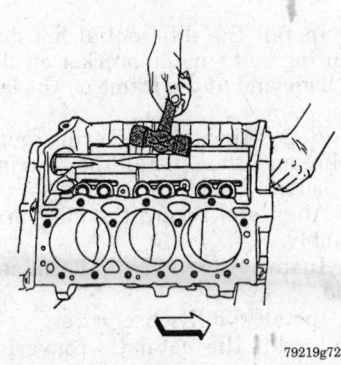

79219g72

Remove the balance shaft using a soft-faced mallet

4. Raise and support the vehicle safely.

5. Drain the engine oil.

6. Remove the strut rods.

7. Disconnect the exhaust pipes at the manifolds.

8. Remove the catalytic converter and exhaust pipe.

9. Remove the flywheel cover.

10. Remove the starter and brace.

11. Remove the brake line at the crossmember. Remove the engine mount through-bolts.

12. Carefully raise the engine, as necessary for clearance, then remove the oil pan bolts and pan.

To install:

13. Using a gasket scraper, clean the gasket mounting surfaces. Make sure all sealing surfaces are clean and free of oil.

14. Apply RTV sealant to the oil pan flange and block.

15. Using care to avoid disturbing the RTV beads, install the oil pan onto the cylinder block. The sealant must be wet during oil pan bolt torquing.

16. Tighten oil pan bolts to 90 inch lbs. (10 Nm).

17. Carefully lower the engine into position, then install and tighten the mount through-bolts.

18. Install the brake line to the crossmember.

19. Install the starter and brace.

20. Install the flywheel cover.

21. Install the converter and exhaust pipe, then connect the exhaust pipe to the manifolds.

22. Install the strut rods, then lower the vehicle.

23. Install the radiator/fan shroud, then install the power steering reservoir.

24. Refill the engine crankcase, then connect the negative battery cable.

25. Start the engine, establish normal operating temperatures and check for leaks.

4WD Vehicles

1. Disconnect the negative battery cable.

2. Remove the power steering reservoir at the fan shroud.

3. Remove the radiator fan shroud.

4. Remove the engine oil dipstick.

5. Raise and support the vehicle safely.

6. Drain the engine oil.

7. Remove the brake line at the crossmember, then remove the crossmember.

8. If equipped, remove the transmission cooler lines.

9. Disconnect the exhaust pipes at the manifolds.

10. Remove the catalytic converter hanger.

11. Remove the flywheel cover.

12. Remove the driveshaft splash shield.

13. Matchmark and remove the idler arm assembly.

14. Remove the steering gear bolts. Pull the steering gear and linkage forward.

15. Remove the differential housing mounting bolts at the bracket on the right side and at the frame on the left side.

16. Remove the starter and brace.

17. Remove the engine mount through-bolts.

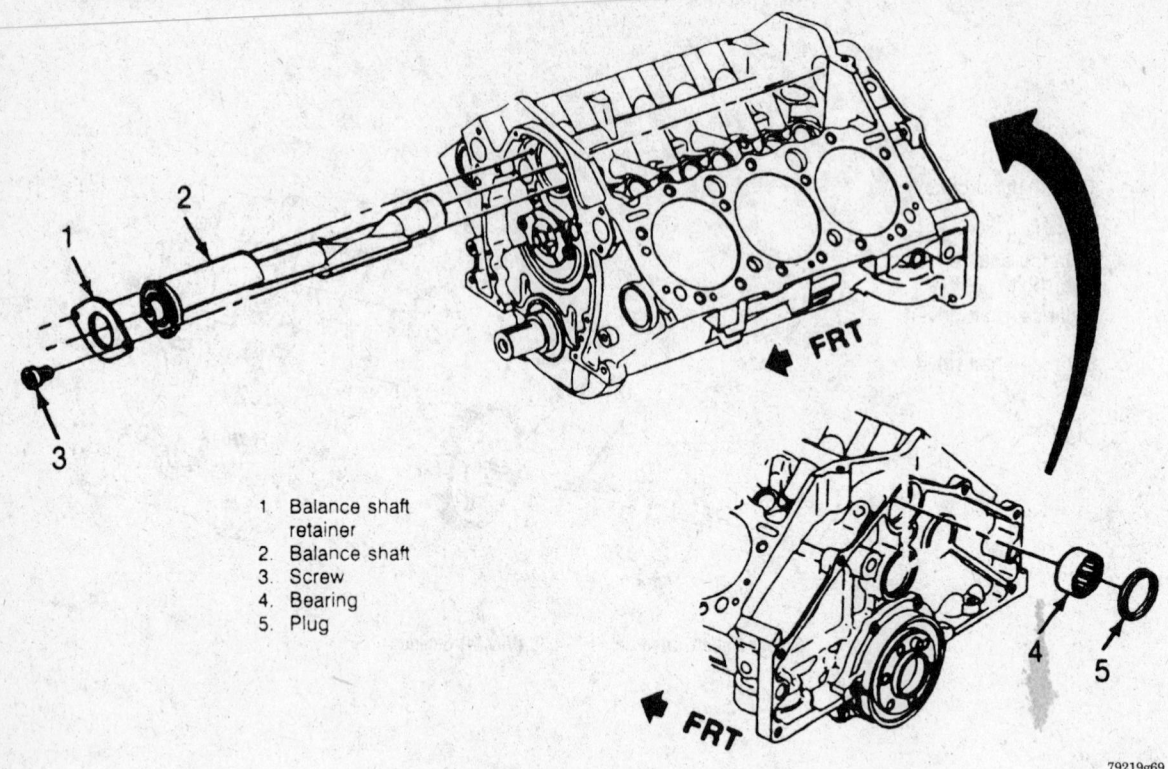

1. Balance shaft retainer
2. Balance shaft
3. Screw
4. Bearing
5. Plug

Balance shaft assembly — 4.3L (VIN W) engines through 1995

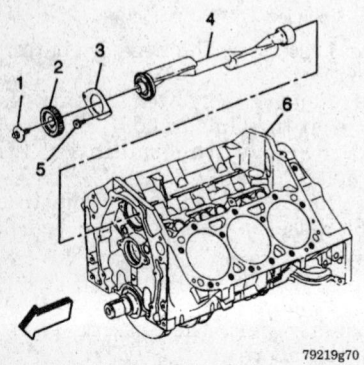

Balance shaft assembly — 1996–97 4.3L engines

18. Remove the oil pan bolts and pan.

To install:

19. Apply RTV sealant to the oil pan flange and block.

20. Using care to avoid disturbing the RTV beads, install the oil pan onto the cylinder block. The sealant must be wet during oil pan bolt torquing.

21. Tighten oil pan bolts to 90 inch lbs. (10 Nm).

22. Install the engine mount through-bolts.

23. Install the starter and brace.

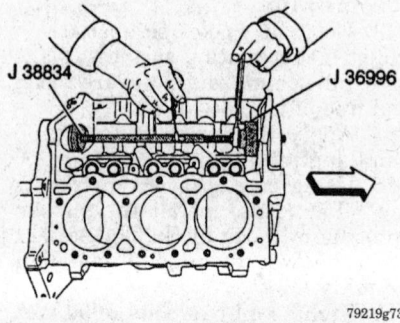

Remove the rear bearing using the special tools

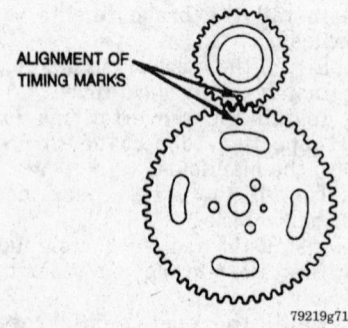

Balance shaft alignment marks — 4.3L (VIN W) engine

24. Install the differential housing mounting bolts at the bracket on the right side and at the frame on the left side.

25. Reposition the steering gear and linkage, then install the steering gear bolts.

26. Align and secure the idler arm assembly.

27. Install the driveshaft splash shield.

28. Install the flywheel cover.

29. Install the catalytic converter hanger.

30. Connect the exhaust pipes at the manifolds.

31. If equipped, connect the transmission cooler lines.

32. Install the crossmember, then secure the brake line at the crossmember.

33. Carefully lower the vehicle.

34. Install the engine oil dipstick.

35. Install the radiator fan shroud.

36. Install the power steering reservoir at the fan shroud.

37. Refill the engine crankcase, then connect the negative battery cable.

38. Start the engine, establish normal operating temperatures and check for leaks.

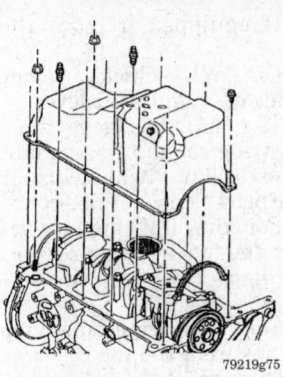

Oil pan removal — 1996 2.2L
engine shown

2.8L and 4.3L (VIN W, X, and Z) Engines

2WD Vehicles

1. Remove the engine.
2. Remove the oil pan retainers (nuts, studs and/or bolts) and rail reinforcements, if equipped. Remove the oil pan from the block.
3. For the 1996–97 4.3L engine, remove the rubber bell housing plugs and gasket.

To install:

4. Using a gasket scraper, clean the gasket mounting surfaces. Make sure all sealing surfaces are clean and free of oil.
5. Apply 1052080 (4.3L), 1052914 (2.8L) or equivalent sealant to the oil pan rail where it contacts the timing cover-to-block joint (front) and the crankshaft rear seal retainer-to-block joint (rear). Continue the bead of sealant about 1 in. (25mm) in both directions from each of the 4 corners.

NOTE: For the 2.8L engine, be sure all RTV is removed from the blind attaching holes to ensure proper fastener tightening can take place.

6. Using a new gasket, install the rubber bell housing plugs (if equipped), the oil pan, reinforcements, if equipped, and retainers. Tighten the retainers to specification:
   2.8L engines:
   Bolts: 18 ft. lbs. (25 Nm)
   1993–95 4.3L engine:
   Bolts: 100 inch lbs. (11 Nm)
   Nuts at corners: 17 ft. lbs. (23 Nm)
   1996–97 4.3L engine:
   Bolts and studs (in sequence): 18 ft. lbs. (25 Nm).

NOTE: The alignments between the rear of the oil pan and the rear of the block is critical. The two surfaces must be flush to allow for proper alignment with the transmission housing.

7. Install the engine into the vehicle. Refill the crankcase with fresh oil. Start the engine, establish normal operating temperatures and check for leaks.

4WD Vehicles

1. Disconnect the negative battery cable.
2. Remove the dipstick.
3. Raise and support the vehicle safely.
4. Remove the drive belt splash shield, the front axle shield, and the transfer case shield.

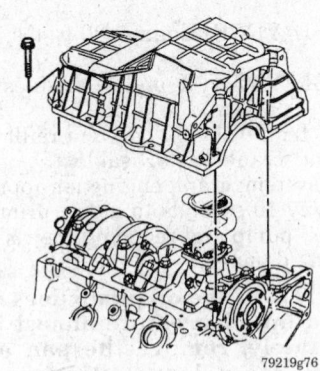

Oil pan removal — 1996 4.3L engine shown

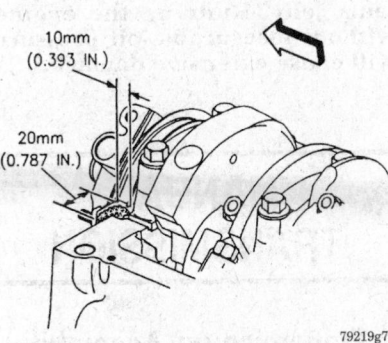

Oil pan sealer area

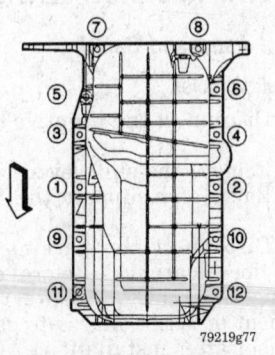

Oil pan tightening sequence for 1996–97 vehicles

5. Remove the front skid plate and drain the engine crankcase oil, then remove the flywheel cover.
6. Remove the left and right motor mount through-bolts.
7. Raise the engine using a suitable lifting device and block in position. This may be accomplished using large wooden blocks between the motor mounts and brackets.

NOTE: Use extreme caution when blocking the engine in position. Get out from under the vehicle and rock the engine slightly once the blocks are in place to be sure the engine is properly supported.

8. Disconnect the oil cooler line, then remove the oil filter adapter.
9. Remove the pitman arm bolt, then disconnect the pitman arm.
10. Remove the idler arm bolts, then disconnect the idler arm.
11. Remove the front differential through-bolts, then disconnect or remove the front driveshaft, if necessary.
12. Roll the differential assembly forward for clearance.
13. Remove the starter motor retaining bolts, then lower the starter and either remove it from the vehicle or suspend it out of the way using mechanic's wire.
14. Remove the oil pan bolts, nuts and reinforcements, then lower the oil pan and gasket.

To install:

15. Using a gasket scraper, clean the gasket mounting surfaces. Make sure all sealing surfaces are clean and free of oil.
16. Apply 1052080 or equivalent sealant to the oil pan rail where it contacts the timing cover-to-block joint (front) and the crankshaft rear seal retainer-to-block joint (rear). Continue the bead of sealant about 1 in. (25mm) in both directions from each of the 4 corners.
17. Using a new gasket, install the oil pan, reinforcements and retainers. Tighten the bolts to 100 inch lbs. (11 Nm) and the nuts at the corners to 17 ft. lbs. (23 Nm) for vehicles through 1995. For 1996–97 vehicles, tighten the retainers, in sequence, to 18 ft. lbs. (25 Nm).
18. Install the starter motor and secure using the mounting bolts.
19. Roll the differential back into position, then install/connect the front driveshaft. Install the front differential through-bolts.
20. Connect the idler arm and secure using the retaining bolts, then connect the pitman arm and secure using the bolts.

21. Install the transfer case shield.
22. Install the flywheel cover, then install the front skid plate.
23. Install the front axle shield, then install the drive belt splash shield.
24. Carefully lower the vehicle.
25. Install the dipstick, then properly refill the engine crankcase.
26. Connect the negative battery cable.
27. Start the engine, establish normal operating temperatures and check for leaks.

Oil Pump

REMOVAL AND INSTALLATION

1. Remove the oil pan.
2. Remove the oil pump attaching bolt and, if equipped, the pickup tube nut/bolt, then remove the pump along with the pickup tube and shaft, as necessary.
3. If necessary for the 2.2L engine, remove the extension shaft and retainer (being careful not the crack the retainer) from the pump.

To install:

4. For the 2.2L engine, if the extension shaft was removed, heat the extension shaft retainer in hot water, then install the shaft and retainer to the oil pump. Make sure the retainer does not crack during installation.
5. Ensure that the pump pickup tube is tight in the pump body. If the tube should come loose, oil pressure will be lost and oil starvation will occur. If the pickup tube is loose it should be replaced.
6. If the pump has been disassembled and is being replaced or for any reason oil has been removed, it must be primed. It can either be filled with oil before installing the cover plate and oil kept within the pump during

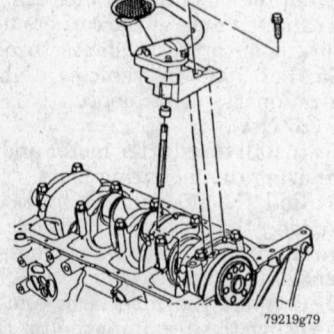

Oil pump removal — 1996 2.2L engine shown, others similar

handling or the entire pump cavity can be filled with petroleum jelly.

NOTE: If the pump is not primed, the engine could be damaged before it receives adequate lubrication when the engine is started.

7. Install the pump aligning the pump shaft with the distributor drive gear as necessary. Tighten oil pump/pickup tube retainer(s) to specification:
   - 2.2L (VIN W and 4) Engine: 32 ft. lbs. (44 Nm)
   - 2.5L (VIN A) Engine:
     Pump bolts 18 ft. lbs. (25 Nm)
     Pickup tube nut to 31 ft. lbs. (42 Nm)
   - 2.8L (VIN R) Engine: 30 ft. lbs. (41 Nm)
   - 4.3L (VIN W, X, and Z) Engines: 65 ft. lbs. (90 Nm)
8. Install the oil pan and refill the engine crankcase. Disable the ignition system; crank engine for approximately 10 seconds to aid in priming the oil pump and reducing the risk of engine damage.

NOTE: If the oil pump does not build up oil pressure almost immediately, remove the pan and check for a loose oil pump-to-pickup tube attachment. If necessary dismantle the pump and pack the pump cavity with petroleum jelly. Running the engine without measurable oil pressure will cause extensive damage.

MANUAL TRANSMISSION

Transmission Assembly

REMOVAL AND INSTALLATION

Except Astro and Safari

1993–95 Vehicles

1. Disconnect the negative battery cable.
2. Remove the shift lever.
3. Raise and support the vehicle safely.
4. Disconnect the parking brake cable for clearance. Before disconnecting the cable, measure the adjustment or scribe marks to ease adjustment after installation.
5. Matchmark and remove the driveshaft/propeller shaft.

6. If equipped, remove the skid plate.
7. On 4WD vehicles, remove the transfer case and shift lever.
8. Tag and disengage the necessary wiring connectors, including the reverse (backup) light switch and Vehicle Speed Sensor (VSS) connectors.
9. For the 4.3L engine, properly relieve the fuel system pressure, then disconnect the fuel lines at the engine.
10. For 2.2L engines, disconnect the fuel lines from the top cover.
11. Disconnect the exhaust pipes.

NOTE: On some 1993 vehicles, access to and removal of the upper bellhousing bolts may be difficult. If necessary, remove the body/cab mounting bolts and carefully raise the front left of the vehicle away from the frame for access.

12. Remove the slave cylinder from the transmission. If the cylinder can be repositioned with the hydraulics intact, simply support it out of the way, but make sure the line is not stretched, kinked or otherwise damaged. If necessary, disconnect the line and remove the cylinder completely.
13. Support the transmission carefully. If possible, secure the transmission to the jack to keep it from shifting suddenly and falling.
14. Remove the transmission mount retainers.
15. If necessary, remove the catalytic converter support hanger.
16. Disconnect or remove the necessary support braces.
17. If necessary, unbolt and remove the transmission crossmember.
18. Make sure the transmission is properly supported, then remove the transmission-to-bellhousing or transmission-to-engine bolts, as necessary.
19. Carefully pull the transmission assembly, straight back to disengage the input shaft splines, Do not allow the transmission to hang on the clutch. For 4.3L engine, it may be necessary to rotate the transmission counterclockwise, then pull back and disengage it from the engine. Once the transmission is disengaged, slowly lower the jack and transmission to remove it.

To install:

20. Place a THIN coat of high-temperature grease on the main drive gear (input shaft) splines, then shift the transmission into height gear.

NOTE: For 4.3L engines, it may be necessary to rotate the transmission clockwise while inserting it to the clutch hub.

79219g79

21. Secure the transmission, then carefully raise it into position, directly behind the engine. Slowly insert the input shaft through the clutch. Rotate the output shaft slowly to engage the splines of the input shaft into the clutch while pushing the transmission forward into place. Do not force the transmission into position, the transmission will easily fall into place when everything is properly aligned.

22. Once the transmission is properly positioned, install the retaining bolts and tighten to 55 ft. lbs. (75 Nm).

23. If removed, install the cross-member to the vehicle.

24. Install or connect the necessary support braces.

25. If removed, install the catalytic converter support hanger.

26. Install the transmission mount bolts and tighten to 37 ft. lbs. (50 Nm).

27. Install the slave cylinder to the transmission. If the hydraulic line was disconnected, the system will have to be bled of air.

28. Connect the exhaust pipes.

29. For 2.2L engines, secure the fuel lines to the top cover.

30. For 4.3L engines, connect the fuel lines at the engine.

31. Engage the necessary wiring harness connectors, including the reverse light switch and the VSS.

32. For 4WD vehicles, install the transfer case.

33. If equipped, install the skid plate.

34. Align and install the driveshaft.

35. Connect and adjust the parking brake cable.

36. Carefully lower the vehicle.

37. Install the shift lever.

38. Connect the negative battery cable.

1996–97 Vehicles

NOTE: Shift the transmission into 3rd or 4th gear position.

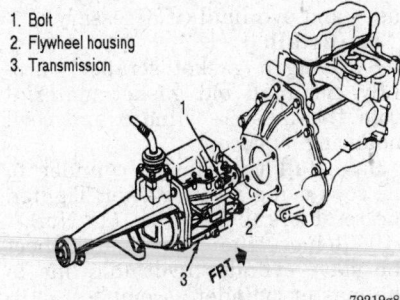

1. Bolt
2. Flywheel housing
3. Transmission

79219g80

New Venture Gear manual transmission — 1993–95, exc. Astro and Safari

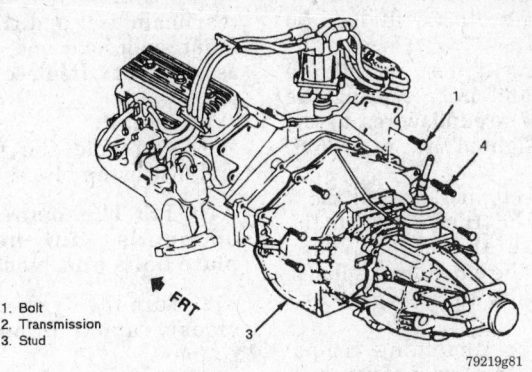

1. Bolt
2. Transmission
3. Stud

79219g81

Borg-Warner T-5 manual transmission — 1993–95, exc. Astro and Safari

1. Disconnect the negative battery cable.

2. Remove the shift lever and the shift housing.

3. Raise and safely support the vehicle.

4. Remove the parking brake cable for clearance.

5. Remove the propeller shaft.

6. Detach the wiring harness connectors from the speed sensor and back-up switch.

7. Remove the wiring harness retainers from the right side of the transmission.

8. Detach the muffler from the catalytic converter.

9. Disconnect the exhaust pipes from the exhaust manifold.

10. Remove the catalytic converter hanger and exhaust section.

11. If equipped with 4WD, perform the following:

  a. Remove the 3 bolts, 2 nuts and washers securing the transfer case shield to the frame rail and transmission.

  b. Unfasten the the 2 bolts and stud securing the left side transfer case brace to the transmission and transfer case.

  c. Remove the 2 bolts securing the right side transfer case brace to the transmission and transfer case, then remove the transfer case.

12. For 2.2L engines, remove the 2 bolts and nut securing the transmission left side brace to the engine and transmission. Remove the 2 bolts securing the transmission right side brace to the engine and transmission.

13. Detach the hydraulic clutch quick-connect from the concentric slave cylinder following one of the two steps:

  a. Use two small prytools at 180 degrees from each other to depress the white plastic sleeve on the quick connect to separate the clutch line from the concentric slave cylinder quick connect.

  b. Use special tool J 36221 to depress the white plastic sleeve on the quick connect to separate the clutch line end from the concentric slave cylinder quick connect.

14. Remove the four bolts securing the clutch housing cover to the transmission.

15. Remove the clutch plate and clutch cover from the flywheel. Support the transmission with a suitable jack.

16. Properly relieve the fuel pressure, then remove the fuel lines and fuel line retainers from the rear crossmember.

17. Remove the rear crossmember from the frame rail.

18. Detach the wiring harness from the front crossmember. Move the wiring harness away from the transmission oil pan. Lower the transmission enough to gain access to the top of the transmission.

19. Remove the fuel line retainers from the top of the transmission.

20. If equipped with a 2.2L engine, remove the bolt, washer, and nut securing the wiring harness ground wires to the engine block. Unfasten the 5 bolts retaining the transmission to the engine. Pull the transmission straight back on the clutch hub splines.

21. If equipped with a 4.3L engine, remove the 7 bolts securing the transmission to the engine. Rotate the transmission counter-clockwise and then pull the transmission back from the clutch hub splines.

22. Lower the transmission using the transmission jack.

To install:

23. Raise the transmission using the jack.

24. Attach the fuel line retainers to the top of the transmission.

25. For 2.2L engines, install the 5 bolts securing the transmission to the engine. Push the transmission straight back onto the hub splines.

Tighten the bolts to 66 ft. lbs. (90 Nm).

26. For 2.2L engines, install the bolt, washer and nut securing the wiring harness ground wires to the engine block. Tighten the nut to 66 ft. lbs. (90 Nm).

27. For 4.3L engines, install the 7 bolts securing the transmission to the engine. Rotate the transmission clockwise onto the clutch hub splines. Tighten the bolts to 35 ft. lbs. (47 Nm).

28. Install the remaining components in the reverse order of removal. Torque the following:

Right and Left side brace bolts to 37 ft. lbs. (50 Nm)

Right and Left side transfer case bolt to 35 ft. lbs. (47 Nm)

Transfer case shield bolts and nuts to 24 ft. lbs. (33 Nm)

29. Connect the negative battery cable.

Clutch Assembly

REMOVAL AND INSTALLATION

1993–95 Vehicles

1. Disconnect the negative battery cable.
2. Remove the transmission assembly. If not done already, remove the flywheel cover.
3. Disconnect the slave cylinder or clutch adjusting rod.
4. Remove the bellhousing retaining bolts. Remove the bellhousing.
5. Remove the throwout spring and fork.
6. Remove the ball stud from the bellhousing.
7. Install the clutch removal tool and support the clutch assembly.

NOTE: Before removing the clutch from the flywheel, matchmark the flywheel, clutch cover and 1 pressure plate lug, so these parts may be assembled in their same relative positions and retain the factory balance.

8. Loosen the clutch plate retaining bolts slowly and evenly 1 at a time until all pressure is released from the pressure plate assembly.
9. Remove the clutch, pressure plate and removal tool. Check the flywheel for damage, repair or replace, as required.
10. Check the clutch assembly, flywheel and pilot bearing for signs of wear, scoring, overheating, etc. If the clutch plate, flywheel or pressure plate is oil-soaked, inspect the engine

rear main seal and the transmission input shaft seal and correct leakage as required. Replace any damaged parts.

To install:

11. Assemble the pressure plate and disc assembly, as required.

NOTE: The manufacturer recommends that new pressure plate bolts and washers be used.

12. Turn the flywheel until the previously applied mark is at the bottom.
13. Install the clutch disc, pressure plate and cover, using a suitable clutch aligning tool.
14. Turn the clutch until the matchmark on the clutch cover aligns with the mark on the flywheel.
15. Install the attaching bolts and tighten them a little at a time in a crossing pattern until the spring pressure is taken up.
16. Remove the aligning tool.
17. Coat the rounded end of the ball stud with high temperature wheel bearing grease.
18. Install the ball stud in the bellhousing. Pack the ball stud from the lubrication fitting.
19. Pack the inside recess and the outside groove of the release bearing with high temperature wheel bearing grease and install the release bearing and fork. The fork must be installed so the fingers and tabs fit into the groove of the release bearing.
20. Install the release bearing seat and spring.
21. Install the bellhousing and retaining bolts, as required.
22. Install the slave cylinder or clutch rod.
23. Install the transmission and the flywheel cover.
24. Connect the negative battery cable.
25. Bleed the hydraulic system or adjust the clutch as required.

1996–97 Vehicles

1. Disconnect the negative battery cable.
2. Remove the transmission.
3. Install J 33169 clutch alignment tool or a used clutch drive gear to support the clutch.
4. Mark the flywheel, clutch cover and a pressure plate lug for alignment when installing.
5. Remove the clutch cover bolts and washers.
6. Remove the clutch cover assembly and the clutch plate. Remove the clutch alignment tool.
7. Clean all parts and inspect for damage.

To install:

8. Install clutch alignment tool J 33169, or equivalent, to support the clutch.
9. Align the marks made during removal or, if new, align the lightest part of the clutch cover, identified by a yellow dot, with the heaviest part identified by an "X".
10. Install the washers and bolts securing the clutch plate and clutch cover assembly to the flywheel.
11. Tighten each screw one turn at a time to avoid warping the clutch cover. Tighten the clutch cover and plate-to-flywheel bolts to 33 ft. lbs. (45 Nm) for 2.2L engines, or to 29 ft. lbs. (40 Nm) for 4.3L engines.
12. Remove the clutch alignment tool, then install the transmission in the vehicle.
13. Connect the negative battery cable.

Clutch Master Cylinder

REMOVAL AND INSTALLATION

1993 Vehicles

1. Disconnect the negative battery cable.
2. Remove hush panel or lower filler panel from under the dash.
3. If necessary, remove the lower left A/C duct.
4. Remove the pushrod retainer, then disconnect pushrod from clutch pedal.
5. Either disconnect the reservoir hose (if the reservoir is not being removed or if there is no clearance for reservoir removal with the hose installed) or remove the reservoir retainers.
6. Disconnect slave cylinder hydraulic line from the clutch master cylinder. Immediately plug all openings to prevent system contamination or excessive fluid loss.
7. Remove the master cylinder-to-cowl brace nuts. Remove master cylinder and overhaul (if necessary).

To install:

8. Using a gasket scraper, carefully clean all old gasket material from the master cylinder and cowl mounting surfaces.
9. Install the master cylinder to the cowl using a new gasket. Tighten the retainers to 13 ft. lbs. (18 Nm).
10. Remove the caps, then connect the slave cylinder hydraulic line to the master cylinder assembly.
11. Either connect the reservoir hose or install the reservoir and secure using the retainers.

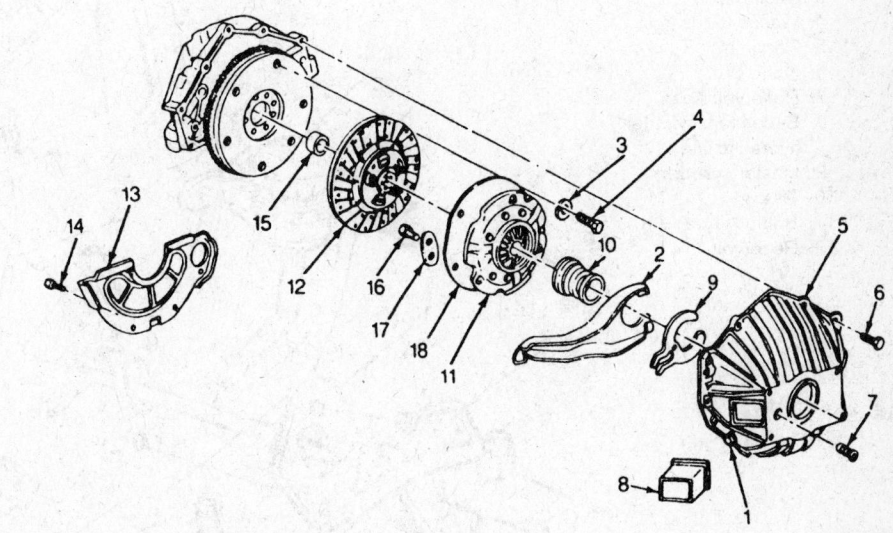

1. Lubrication fitting (R/V and C/K models only)
2. Clutch fork
3. Spring washer
4. Screw
5. Flywheel housing
6. Screw
7. Ball stud
8. Boot
9. Retainer
10. Release bearing
11. Cover assembly
12. Driveplate
13. Cover
14. Screw
15. Pilot bearing
16. Screw
17. Strap
18. Pressure plate

Exploded view of typical clutch assembly — Vehicles through 1995

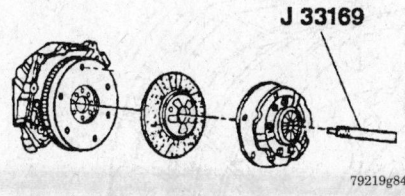

Use the clutch alignment tool to support the clutch — 1996 vehicle shown

12. Connect the pushrod to the pedal and secure using the retainer.

13. If removed, install the lower left A/C duct.

14. Install the hush/filler panel.

15. Connect the negative battery cable.

16. Properly refill the master cylinder reservoir, then bleed the system of air and check for fluid leaks.

1994–97 Vehicles

NOTE: On the 2.2L engine, individual components of the clutch actuating system (master cylinder/slave cylinder) are not be available for service. It is recommended that a complete, prefilled and pre-bled unit should be installed and NO attempts should be made to disconnect the hydraulic lines.

1. Disconnect the negative battery cable.

2. Remove lower filler panel from under the dash.

3. If necessary, remove the lower left A/C duct.

4. Remove the pushrod retainer, then disconnect pushrod from clutch pedal.

NOTE: To expedite the hydraulic clutch bleeding, loose as little fluid as possible when disconnecting hydraulic lines. All openings should be immediately capped or plugged to prevent system contamination or excessive fluid loss.

5. For 1994–95 pick-ups, disconnect the slave cylinder hydraulic line from the master cylinder. Drive out the retaining pin holding the hydraulic tube to the master cylinder using a $\frac{7}{64}$ in. (3mm) punch.

6. For 1996–97 pick-ups, detach the clutch line from the concentric slave cylinder quick connect coupling, using tool J 36221 to depress the while plastic sleeve on the quick connect coupling to separate the line from the cylinder.

7. For Blazer, Bravada and Jimmy, detach the slave cylinder hydraulic line coupling using J 36221, or equivalent. Detach the slave cylinder hydraulic line form the cowl retainer clip.

8. For 1996–97 vehicles, remove the tube clips from the wiring harness bracket and sheet metal.

9. For pick-ups, remove the master cylinder from the cowl panel by grasping the cylinder body and rotating it about 45 degrees clockwise, then withdrawing it from the opening.

10. For Blazer, Bravada and Jimmy, remove the nuts from the master cylinder, then remove the cylinder from the cowl panel.

11. If necessary, remove the bolts from the reservoir, then remove the reservoir from the vehicle.

To install:

12. If removed, install the reservoir to the cowl panel and secure with the retaining bolts. Tighten the bolts to 25 inch lbs. (2.8 Nm).

13. For pick-ups, hold the master cylinder assembly at a 45 degree angle (as during removal), then install the master cylinder to the cowl by inserting and twisting counterclockwise 45 degrees. Be sure not to over

1. Brace
2. Wave washer
3. Push rod
4. Washer
5. Retainer
6. Nut
7. Reservoir hose
8. Secondary cylinder hydraulic line
9. Master cylinder
10. Gasket
11. Bolt
12. Reservoir

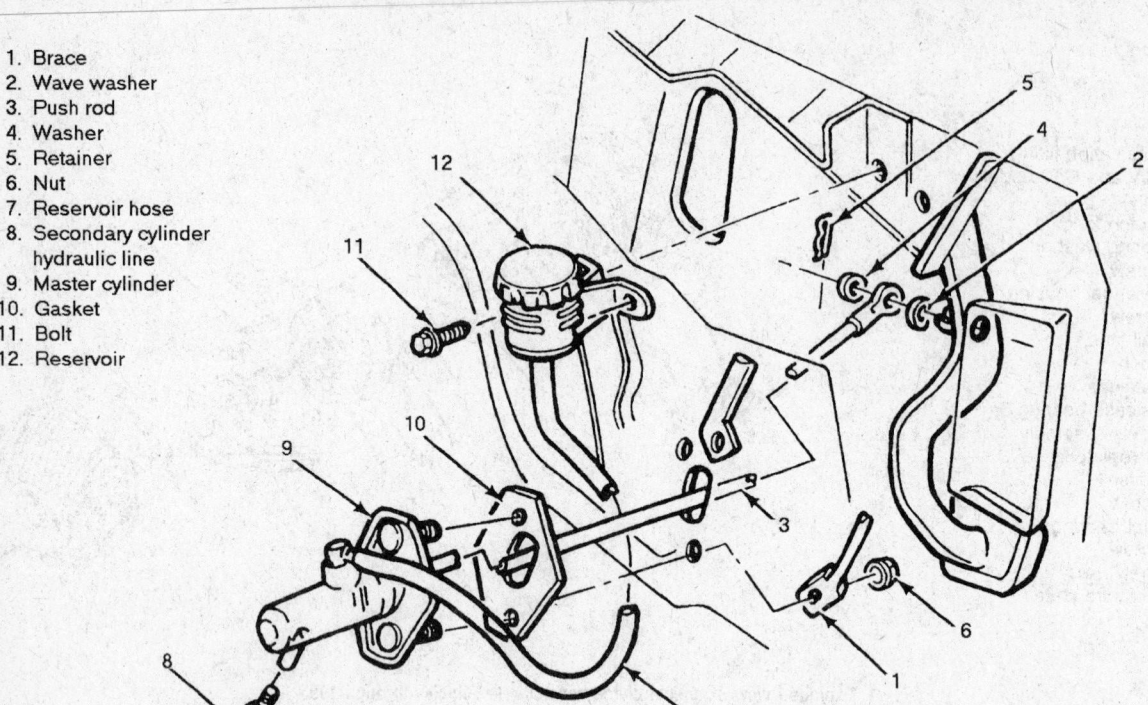

Clutch master cylinder and reservoir — 1993 vehicle shown

79219g85

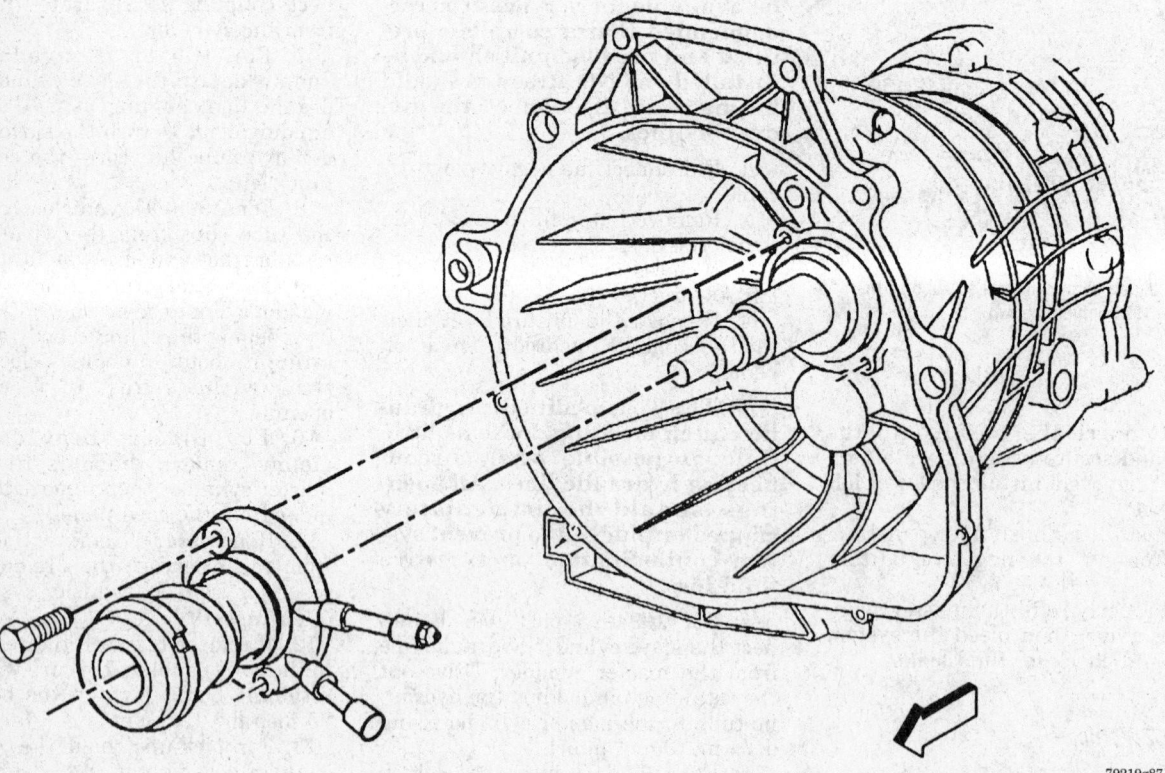

Master cylinder mounting — 1996 vehicle shown

79219g87

rotate the cylinder assembly or damage will occur.

NOTE: Use a new retaining pin and quad seal during hydraulic line installation.

14. For Blazer, Bravada and Jimmy, install the master cylinder to the cowl panel, then install the retaining nuts. Tighten the nuts to 13 ft. lbs. (18 Nm).

15. For Blazer, Bravada and Jimmy, install the hydraulic line coupling, then attach the line to the cowl retaining clip.

16. For pick-ups, remove the caps, then connect the slave cylinder hydraulic line to the master cylinder assembly. On some 4.3L engines, a special band tie is required to secure the tubing to the body.

17. If necessary, attach the tube clips to the wiring harness bracket and the sheet metal.

18. Connect the pushrod to the pedal and secure using the retainer.

19. If removed, install the lower left A/C duct.

20. Install the lower filler panel.

21. Connect the negative battery cable.

22. Properly refill the master cylinder reservoir, then bleed the system of air and check for fluid leaks.

Clutch Slave Cylinder

REMOVAL AND INSTALLATION

1993 Vehicles

1. Disconnect the negative battery cable.

2. Raise and support the front of the vehicle safely.

3. Disconnect the hydraulic line from clutch master cylinder. Immediately plug or cap all openings to prevent system contamination or excessive fluid loss. If equipped, remove the hydraulic line-to-chassis screw and the clip from the chassis.

4. Remove the slave cylinder-to-bellhousing nuts.

5. Remove the pushrod and the slave cylinder assembly from the vehicle, then overhaul it (if necessary).

To install:

6. Remove the plugs or caps, then install and secure the hydraulic line to the slave cylinder.

NOTE: If used, be sure to properly install the clip and hydraulic line-to-chassis screw.

7. Properly bleed the hydraulic system.

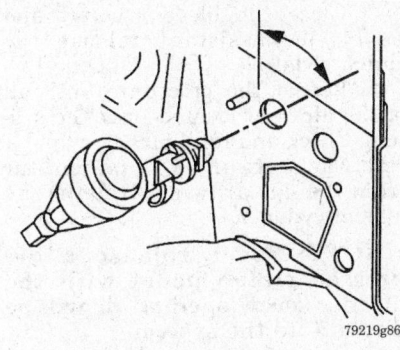

Clutch slave (secondary) cylinder mounting — 1993 vehicle shown

8. Install the slave cylinder to the bellhousing, then secure using the retaining nuts and tighten to 13 ft. lbs. (18 Nm).

9. Carefully lower the vehicle.

10. Connect the negative battery cable.

1994–95 Vehicles

NOTE: On the 2.2L engine, individual components of the clutch actuating system (master cylinder/slave cylinder) may not be available for service. If so it is recommended that a complete, pre-filled and pre-bled unit should be installed and NO attempts should be made to disconnect the hydraulic lines.

1. Disconnect the master cylinder pushrod from the clutch pedal assembly.

NOTE: The master cylinder pushrod should be disconnected to protect the slave cylinder while it is removed from the bellhousing. If not, any attempt to depress the clutch pedal while the slave cylinder is withdrawn from the bellhousing will cause permanent damage to the slave cylinder assembly.

2. Using a 7/64 in. (3mm) punch to drive out the roll pin, disconnect the hydraulic line from the slave cylinder assembly. Immediately cap or plug all openings to prevent system contamination or excessive fluid loss.

NOTE: To expedite the hydraulic clutch bleeding process, loose as little fluid as possible.

3. Remove the slave cylinder-to-bellhousing retaining nuts.

4. Remove the slave cylinder from the bellhousing.

To install:

5. Install the slave cylinder to the bellhousing, then tighten the retaining nuts to 18 ft. lbs. (24 Nm).

6. Uncover the hydraulic line openings, then connect the hydraulic line to the slave cylinder using a new O-ring (lightly lubricated with fluid) and a new roll pin.

7. Connect the master cylinder pushrod to the clutch pedal assembly.

8. Properly bleed the hydraulic system, then check for leaks.

1996–97 Vehicles

1. Remove the transmission and the clutch assembly.

2. Remove the two bolts securing the concentric slave cylinder to the transmission input shaft.

To install:

3. Position the slave cylinder on the transmission input shaft. Make sure that the bleed screw and coupling are positioned with the transmission ports.

4. Install the slave cylinder retaining bolts, then tighten to 18 ft. lbs. (24 Nm).

5. Install the clutch assembly and the transmission.

Hydraulic Clutch System Bleeding

Bleeding air from the hydraulic clutch system is necessary whenever any part of the system has been disconnected or the fluid level (in the reservoir) has been allowed to fall so low, that air has been drawn into the master cylinder.

BLEEDING

1993 Vehicles

1. Fill master cylinder reservoir with new brake fluid conforming to Dot 3 specifications.

——— **CAUTION** ———
Never, under any circumstances, use fluid which has been bled from a system to fill the reservoir as it may be aerated, have too much moisture content and possibly be contaminated.

2. Raise and support vehicle safely.

3. Remove the slave cylinder retainers.

4. Hold slave cylinder at approximately 45 degrees with the bleeder screw located at the highest point. Have an assistant fully depress and hold the clutch pedal, then open the bleeder screw.

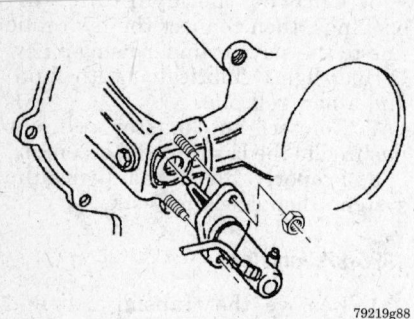

Slave cylinder mounting — 1995 2.2L engine

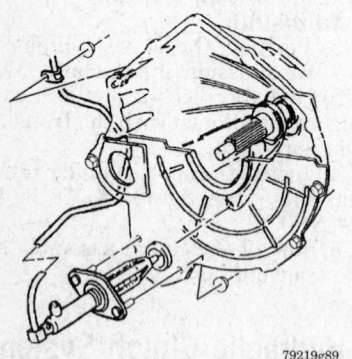

Slave cylinder mounting — 1995 4.3L engine

5. Close the bleeder screw and have your assistant release the clutch pedal.

6. Repeat the procedure until all of the air is evacuated from the system. Check and refill master cylinder reservoir as required to prevent air from being drawn through the master cylinder.

NOTE: Never release a depressed clutch pedal with the bleeder screw open or air will be drawn into the system.

7. Test the clutch for proper operation.

1994–97 Vehicles

1. Fill master cylinder reservoir with new brake fluid conforming to Dot 3 specifications.

────── **CAUTION** ──────

Never, under any circumstances, use fluid which has been bled from a system to fill the reservoir as it may be aerated, have too much moisture content and possibly be contaminated.

2. Have an assistant fully depress and hold the clutch pedal, then open the bleeder screw.

3. Close the bleeder screw and have your assistant release the clutch pedal.

4. Repeat the procedure until all of the air is evacuated from the system. Check and refill master cylinder reservoir as required to prevent air from being drawn through the master cylinder.

NOTE: Never release a depressed clutch pedal with the bleeder screw open or air will be drawn into the system.

5. If the previous steps do not result in satisfactory pedal feel, remove the reservoir cap and pump the clutch pedal very fast for 30 seconds. Stop to let the air escape, then repeat the procedure as necessary to purge all remaining air.

6. Test the clutch for proper operation.

AUTOMATIC TRANSMISSION

Transmission Assembly

REMOVAL AND INSTALLATION

Except Astro and Safari

1993 Vehicles

1. Properly relieve the fuel system pressure, then disconnect the negative battery cable.

2. Remove the air cleaner assembly. If equipped (4L60 model), disconnect the TV cable from the TBI unit.

3. Raise and support the vehicle safely.

4. Drain the transmission fluid.

5. Disconnect the shift linkage from the transmission.

6. Disconnect and remove the fuel lines.

7. Matchmark and remove the front (if used) and rear driveshafts.

8. Remove the support bracket at the catalytic converter. Remove any components necessary for clearance.

9. Support the transmission assembly.

10. Remove the transmission crossmember. Take care not to stretch or damage any cables or wiring when attempting to remove the crossmember.

11. Lower the transmission slightly for access, then remove the dipstick tube and seal. Cover or plug the opening in the transmission housing to prevent system contamination.

12. Disengage the speedometer harness connector and the vacuum modulator line, if used.

13. Disengage the necessary electrical wiring connectors from the transmission assembly.

14. Disconnect the transmission fluid cooler lines. Plug or cap all openings to prevent system contamination or excessive fluid spillage.

15. On 4WD vehicles, disengage the transfer case shifter and position it aside.

16. Remove the transmission support braces. Be sure to tag or note the location of all support braces as they must be installed in their original positions.

17. Remove the torque converter housing cover.

18. Matchmark the flywheel-to-torque converter relationship, then remove the retaining bolts.

19. Support the engine.

20. Remove the transmission-to-engine retaining bolts. Note the positions of any brackets or clips and move them aside.

21. Slide the transmission straight back off the locating pins. Be careful not to drop the torque converter, then as soon as access is possible, install a torque converter retaining strap.

22. Carefully lower the transmission.

To install:

23. Make sure the torque converter is properly seated and a retaining strap is installed.

24. Carefully raise the transmission into position in the vehicle, then remove the converter retaining strap. Slide the transmission straight onto the locating pins while aligning the marks on the flywheel and torque converter.

NOTE: The converter must be flush on the flywheel and rotate freely by hand.

25. Install the transmission-to-engine retainers, taking care to properly reposition all brackets, clips and harness, as noted during removal. The retainers should not be torqued yet. Do not install the dipstick tube or transmission support brace screws at this time.

26. Install and finger-tighten the converter bolts, then tighten them to 50 ft. lbs. (68 Nm).

27. Remove the support from the engine.

28. Install the converter housing cover.

29. Install the transmission support braces as noted during removal.

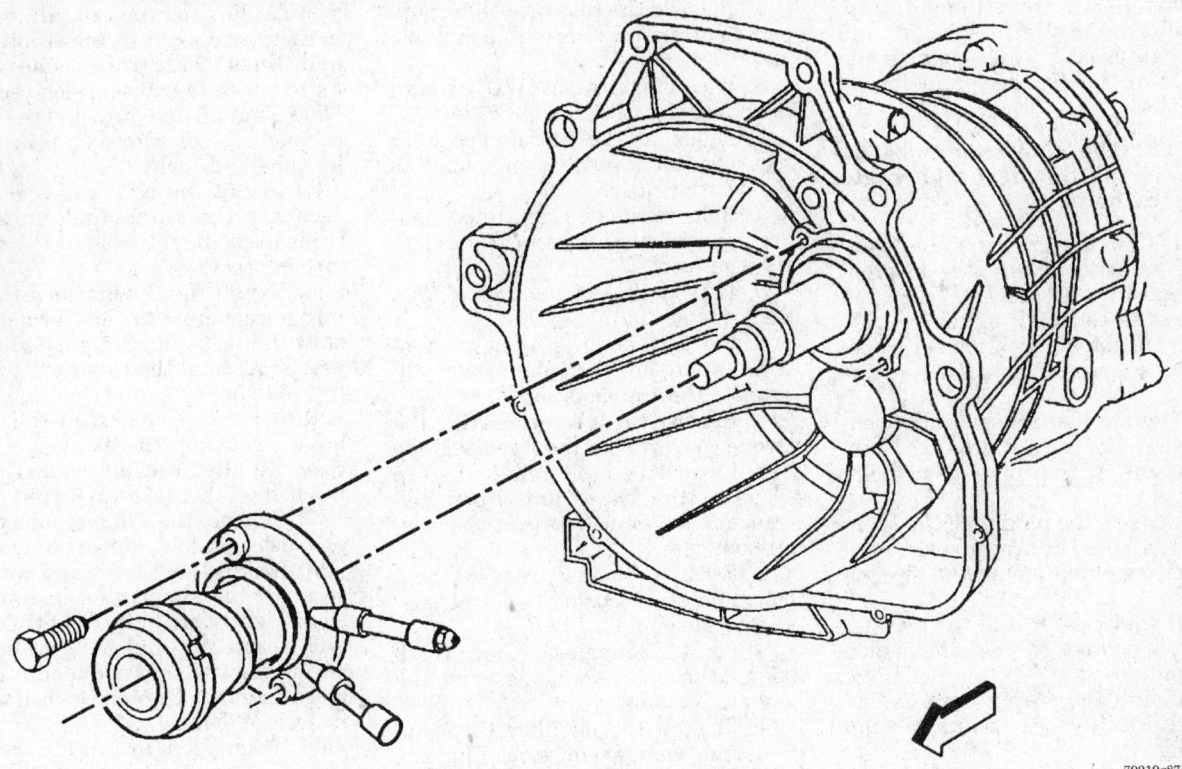

Concentric slave cylinder mounting — 1996 vehicle shown

79219g87

30. On 4WD vehicles, connect the transfer case shifter assembly.

31. Uncap the openings, then connect the transmission cooler lines. Be careful not to twist or bend the lines.

32. If used, connect the vacuum modulator line.

33. Engage the speedometer harness connector and the remaining electrical wiring to the transmission housing.

34. Uncover the opening in the transmission housing and position a new seal, then install the dipstick tube.

35. Tighten the transmission-to-engine retainers and the dipstick tube retainer, then tighten the bolts to 23 ft. lbs. (32 Nm).

36. Raise the rear of the transmission (taking care not to pinch or damage any cables, wires or components), then install the crossmember. Secure the transmission mount and any components which were removed for access.

37. Remove the support from the transmission, then install the catalytic converter bracket.

38. Align and install the front, if equipped, and rear driveshafts.

39. Connect the fuel lines.

40. Connect the shift linkage and adjust, as necessary.

41. Refill the transmission using fresh fluid.

42. Carefully lower the vehicle.

43. If equipped, engage the TV cable, then check and adjust, as necessary.

44. Connect the negative battery cable, then check for proper operation.

1994–95 Vehicles

1. Properly relieve the fuel system pressure, then disconnect the negative battery cable.

2. Raise and support the vehicle safely.

3. Drain the transmission fluid.

4. Disconnect the shift linkage from the transmission.

5. Matchmark and remove the rear driveshaft.

6. For 4WD vehicles, remove the transfer case assembly.

7. Disconnect the fuel lines.

8. Remove the transmission crossmember-to-frame retaining bolts and member-to-transmission retainer(s), then remove the crossmember. Take care not to stretch or damage any cables or wiring when attempting to remove the crossmember.

9. Disconnect the exhaust pipe(s) from the manifold(s).

10. Remove the flywheel inspection cover and matchmark the flywheel-to-torque converter relationship, then remove the flywheel-to-torque converter bolts.

11. Remove the fuel lines and wiring harness from the transmission.

12. Make sure the transmission is properly supported, then remove the transmission-to-engine bolts. Note the positioning of any brackets, clips or harnesses as they are removed.

13. Remove the dipstick tube and seal, then plug or cover the opening in the transmission housing in order to prevent system contamination.

14. Disconnect the cooler lines. Plug or cover all openings in order to prevent system contamination or excessive fluid spillage.

15. Support the engine.

16. Support the transmission, pull it straight back and disengage it from the engine. Carefully lower the transmission and remove it. If possible, install a torque converter retaining strap before attempting to lower the transmission. This will prevent the converter from possibly falling and causing damage and/or personal injury.

To install:

17. Make sure the torque converter is properly seated and a retaining strap is installed.

18. Carefully raise the transmission into position in the vehicle, then

remove the converter retaining strap. Slide the transmission straight onto the locating pins while aligning the marks on the flywheel and torque converter.

NOTE: The converter must be flush on the flywheel and rotate freely by hand.

19. Install the transmission-to-engine retainers, taking care to properly reposition all brackets, clips and harness, as noted during removal. Do not install the dipstick tube or transmission support brace screws at this time.
20. Tighten the transmission-to-engine retainers to 23 ft. lbs. (32 Nm).
21. Connect the transmission cooler lines.
22. Remove the plug from the opening, then position a new seal to the transmission and install the dipstick tube.
23. If equipped, install the transfer case support rods to the transmission housing.
24. Install the wiring harness and fuel lines to the transmission housing.
25. Make sure the matchmarks are aligned, then install and finger-tighten the flywheel-to-converter bolts. Once the converter is properly seated by finger-tightening the bolts, torque them to 46 ft. lbs. (63 Nm).
26. Hook the edge of the flywheel inspection cover under the lip of the engine oil pan, then install and secure the cover.
27. Connect the exhaust pipe(s) to the manifold(s).
28. Install the transmission crossmember to the vehicle. Tighten the member-to-frame bolts to 56 ft. lbs. (77 Nm).
29. On 4WD vehicles, install the transfer case assembly.
30. Align and install the rear driveshaft.
31. Connect the shift linkage then check and adjust, as necessary.
32. Refill the transmission using fresh fluid.
33. Carefully lower the vehicle.
34. Connect the negative battery cable, then check for proper operation.

1996–97 Vehicles

1. Disconnect the negative battery cable. Raise and safely support the vehicle.
2. Drain the transmission fluid from the transmission via the oil pan.
3. Detach the shift cable from the transmission control lever and bracket.

4. Remove the rear propeller. Support the transmission with a suitable jack.
5. If equipped with 4WD, remove the transfer case.
6. Unfasten the nut and washer securing the transmission mount to the crossmember.
7. Remove the two bolts and washers securing the mount to the transmission.
8. Detach the exhaust pipe from the exhaust manifold.
9. For 2.2L engines, remove the 6 bolts securing the converter pan cover to the transmission.
10. Remove the 3 bolts securing the torque converter to the flywheel.
11. Remove the bolt, clip and strap securing the three fuel lines and transmission vent hose to the transmission case.
12. For 2.2L engines, remove the 5 bolts and nut securing the transmission to the engine.
13. For 4.3L engines, remove the 9 bolts securing the transmission to the engine.
14. Remove the oil filler tube and seal from the transmission. Plug the oil filler tube opening.
15. Disconnect and plug the transmission cooler lines from the transmission.
16. Detach the wiring harness connectors from the transmission speed sensor and park/neutral position switch. Remove all vehicle harness wires, clips, tubes, brackets, and lines that may interfere with the removal or the transmission from the vehicle.
17. Support the engine with a jackstand before detaching the transmission from the engine. Remove the transmission from the vehicle.

To install:
18. Make sure the torque converter is seated properly and that J 21366 is in place. Raise the transmission into plate and remove J 21366. Support the transmission with a suitable jack, then slide the transmission straight onto the locating pins while lining up the marks on the flywheel and the torque converter.
19. For 2.2L engines, install the 5 bolts and nut securing the engine to the transmission.
20. For 4.3L engines, install the 9 bolts securing the engine to the transmission.
21. Tighten the bolts to 66 ft. lbs. (90 Nm) for the 2.2L engine or to 34 ft. lbs. (47 Nm) for the 4.3L engine.
22. Attach the wiring harness connectors to the transmission speed sensor and park/neutral position

switch. Position/install all vehicle harness wires, clips, tubes, brackets and lines that were removed or moved prior to transmission removal.
23. Unplug and attach the transmission oil cooler lines and the oil filler tube and seal.
24. Install the bolt, clip and strap securing the three fuel lines and transmission vent hose to the transmission case.
25. Install the 3 bolts securing the torque converter to the flywheel. Install the bolts finger-tight to ensure proper seating, then tighten to 46 ft. lbs. (63 Nm).
26. For 2.2L engines, install the 6 bolts securing the converter pan cover to the transmission. Tighten the bolts to 37 ft. lbs. (50 Nm).
27. Connect the exhaust pipe to the exhaust manifold support brackets.
28. Install the 2 bolts and washers securing the mount to the transmission. Tighten to 35 ft. lbs. (47 Nm).
29. Install the nut and washer securing the transmission mount to the cross member. Tighten the bolts to 38 ft. lbs. (52 Nm).
30. If equipped, install the transfer case.
31. Remove the transmission jack and the engine support stands.
32. Install the rear propeller shaft.
33. Attach the shift cable to the transmission shift lever and bracket.
34. Carefully lower the vehicle. Fill the transmission with new fluid. Connect the negative battery cable.

Astro and Safari

1. Disconnect the negative battery cable. Remove the air cleaner assembly. Disconnect the throttle valve cable and the throttle linkage.
2. Raise and support the vehicle safely.
3. Drain the transmission fluid. Disconnect the shift linkage. Properly relieve the fuel system pressure and remove the fuel lines.
4. Remove the driveshaft. If equipped with a transfer case, also remove the front driveshaft.
5. Disconnect and remove all required exhaust system components. Support the transmission using the proper equipment.
6. Remove the crossmember retaining bolts. Remove the transmission mount retaining bolts. Remove the crossmember from the vehicle and lower the transmission enough to gain access to other components.
7. Remove the transmission dipstick and tube. Disconnect the speedometer cable. If equipped, disconnect the vacuum modulator line.

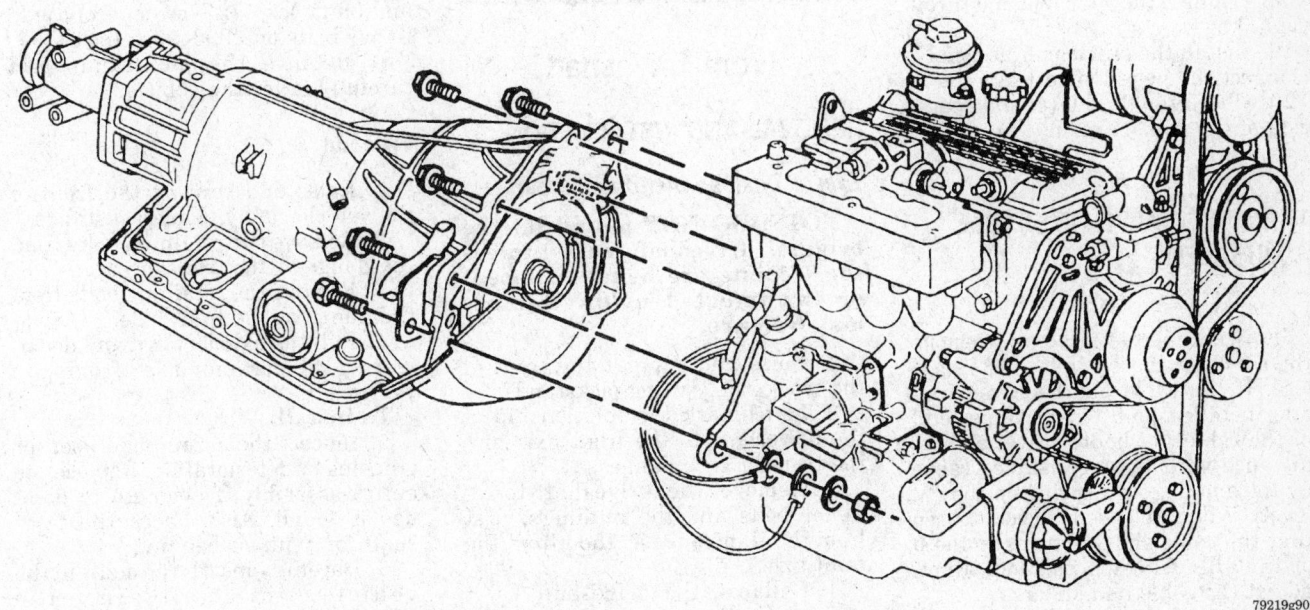

Transmission-to-engine mounting — 1996 2.2L engine shown

79219g90

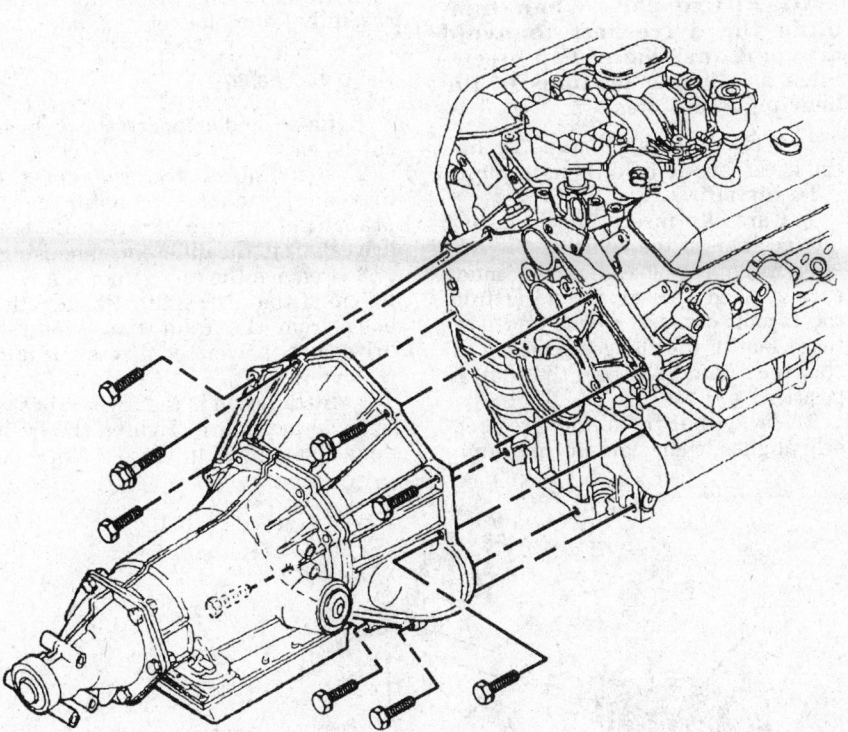

Transmission-to-engine mounting — 1996 4.3L engine shown

79219g91

8. Disconnect all electrical connections from the transmission assembly. Disconnect and plug the fluid cooler lines.

9. Remove the damper and support, as required. Remove the flywheel housing cover and torque converter to flywheel bolts. If equipped, remove the transfer case assembly.

10. Properly support the engine assembly. Remove the transmission to engine retaining bolts. Carefully remove the automatic transmission.

To install:

11. Install the transmission into position, making sure the converter is properly seated.

12. Install the engine-to-transmission and torque converter-to-flywheel bolts. Torque the bolts to specification.

13. Install the converter housing cover. Install the damper and support.

14. Connect all electrical leads to the transmission and connect the speedometer cable.

15. Install the dipstick tube. Install the crossmember and install the transmission mount retaining bolts. Reconnect the fuel lines.

16. Connect all exhaust system components. Install the front driveshaft on 4WD vehicles.

17. Connect the shift linkage and install the skid plate, if equipped. Lower the vehicle.

18. Connect the TV cable and throttle linkage.

19. Install the air cleaner assembly. Connect the negative battery cable.

20. Check the fluid level and operation of the transmission.

THROTTLE VALVE (TV) CABLE ADJUSTMENT

1. If necessary for access, remove the air cleaner assembly.

2. If the cable has been removed and installed, pull on the upper end of the cable. It should travel a short distance with light resistance cause by the small return spring on the TV lever. When released, check to see that the cable slider returns to the 0 or the fully adjusted position; if not, adjust the cable and slider.

 a. Depress and hold the readjust tab on the end of the TV cable.

 b. Move the slider back on the cable (away from the throttle lever) until it stops against the fitting.

 c. Release the readjust tab.

3. Rotate the throttle lever by hand (do not use the accelerator pedal) to the full throttle stop (wide open throttle) position. The slider must move (ratchet) toward the lever in order to adjust the cable when the lever is rotated.

4. Release the throttle lever and check for proper operation.

NOTE: The throttle valve cable adjustment should be rechecked with the engine at normal operating temperature. A cable adjustment may appear to function properly while the engine is cold, but not work properly when the engine is running.

DRIVE AXLE

Front Driveshaft

REMOVAL AND INSTALLATION

Except Typhoon, Astro and Safari

NOTE: DO NOT pound on the original driveshaft ears (unless the U-joints are being replaced) or the injected nylon U-joints may fracture.

1. Raise and support the front of the truck safely using jackstands.

2. Matchmark the relationship of the driveshaft to the front axle and the transfer case flanges.

3. Remove the driveshaft-to-retainer bolts and the retainers, first from the transfer case, then from the front axle.

4. Collapse the driveshaft so it may be disengaged from the transfer case flange.

5. Move the driveshaft rearward (between the transfer case and the chassis) to disengage it from the front axle.

NOTE: Use care when handling the driveshaft to avoid dropping the U-joint cap assemblies and losing portions of the bearing assemblies.

6. Wrap a length of tape around the loose caps to hold them in place.
To install:

7. Carefully insert the driveshaft into position in the vehicle.

8. Align the matchmarks made earlier, then remove the tape from the U-joint cap assemblies and position them to the flanges. Loosely install the retainers to hold the shaft in position.

9. Verify that the marks are properly aligned, then tighten the retainers to 15 ft. lbs. (20 Nm) for vehicles through 1991, or to 92 ft. lbs. (125 Nm) for the transfer case flange bolts and 53 ft. lbs. (72 Nm) for front axle flange bolts on 1993–95 vehicles.

10. Remove the jackstands and carefully lower the vehicle.

Typhoon

1. Raise and support the front of the vehicle safely using jackstands.

2. Remove the retaining bolts from the flange at the transfer case.

3. Remove the retaining bolts from the flange at the front axle.

4. Pull the axle forward and downward, then remove it from the vehicle.

To install:

5. Inspect the shroud and boot for cracking or deterioration. Replace the entire assembly if evidence of damage is found. Also, check the front shaft for dents or bending.

6. Carefully install the shaft to the vehicle while aligning the matchmarks made during removal. Loosely install the retaining bolts to hold the shaft in position.

7. Tighten the transfer case flange bolts, followed by the axle flange bolts to 52 ft. lbs. (70 Nm).

8. Remove the jackstands and carefully lower the vehicle.

Astro and Safari

1. Raise and support the vehicle safely.

2. Matchmark the rear of the driveshaft to the transfer case. Matchmark the front of the driveshaft to the differential housing.

3. Remove the bolts from the rear flange of the driveshaft. Remove the bolts from the front flange of the driveshaft. Lower the driveshaft and remove it.

4. Installation is the reverse of the removal procedure. Tighten the front flange bolts to 53 ft. lbs. (72 Nm) and

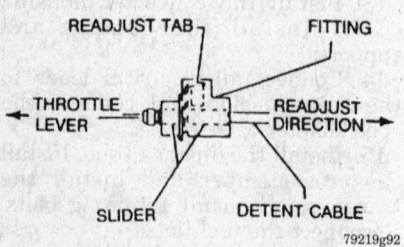

Throttle Valve (TV) cable adjustment

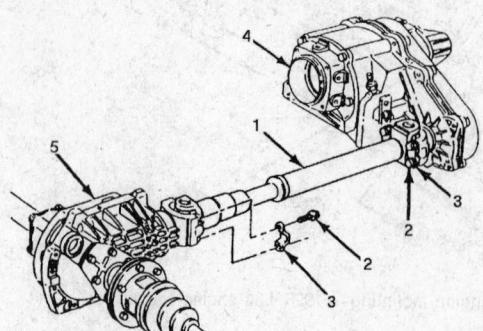

1. Front-Drive propeller shaft
2. Bolt
3. Retainer
4. Transfer case
5. Front axle

79219g93

Front driveline assembly — except Typhoon, Astro and Safari

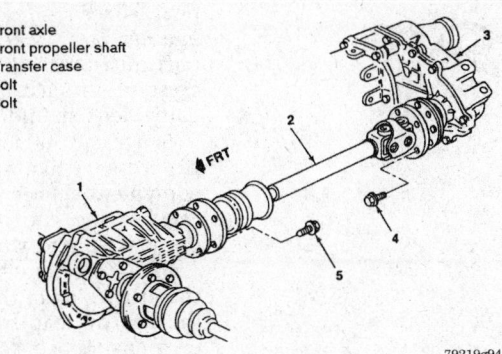

1. Front axle
2. Front propeller shaft
3. Transfer case
4. Bolt
5. Bolt

79219g94

Front axle and driveline assembly — Typhoon

the rear flange bolts to 92 ft. lbs. (125 Nm).

Rear Driveshaft and U-Joints

REMOVAL AND INSTALLATION

NOTE: DO NOT pound on the original propeller shaft yoke ears (unless the U-joints are being replaced) or the injected nylon joints may fracture.

1. Raise and support the rear of the truck safely using jackstands.
2. Matchmark the relationship of the driveshaft-to-pinion flange and the slip yoke-to-front driveshaft or transmission/transfer case housing, as necessary. ALL shaft components which are being removed should be installed in their original positions, including slip yokes and U-joints ears. Disconnect the rear universal joint by removing its retainers. If the bearing caps are loose, tape them together to prevent dropping and loss of bearing rollers.

3. If equipped with a one-piece driveshaft, perform the following procedures:
    a. Slide the driveshaft forward to disengage it from the rear axle flange.
    b. Move the driveshaft rearward to disengage it from the transmission slip-joint, passing it under the axle housing.
4. If equipped with a two-piece driveshaft, perform the following procedures:
    a. Slide the rear driveshaft half forward to disengage it from the rear axle flange.
    b. Slide the driveshaft rearward to disengage it from slip-joint of the front driveshaft half, passing it under the axle housing.
    c. Remove the center bearing support nuts and bolts.
    d. Slide the front driveshaft half rearward to disengage it from the transfer case/transmission slip-joint.

NOTE: DO NOT allow the driveshaft to drop or allow the universal joints to bend to extreme angles, as this might fracture the injected joint internally. Support the propeller shaft during removal.

To install:

5. Inspect the slip-joint splines for damage, burrs or wear, for this will damage the transmission seal. Apply engine oil to all splined propeller shaft yokes.
6. DO NOT use a hammer to force the driveshaft into place. Check for burrs on the transmission output shaft spline, twisted slip yoke splines or possibly the wrong U-joint. Make sure the splines agree in number and fit. To prevent trunnion seal damage, DO NOT place any tool between the yoke and splines.
7. If installing a one-piece driveshaft, perform the following procedures:
    a. Align the matchmarks made during removal, then slide the driveshaft into the transmission/transfer case.
    b. Align the rear universal joint-to-rear axle pinion flange, and make sure the bearings are properly seated in the pinion flange yoke.
    c. Install the rear driveshaft-to-pinion fasteners and verify that all marks are aligned. Torque the fasteners to 15 ft. lbs. (20 Nm).
8. If installing a two-piece driveshaft, perform the following procedures:
    a. Install the front driveshaft half into the transmission/transfer case (aligning the marks made during removal) and bolt the center bearing support in position. Torque the center bearing to support nuts and bolts to 25 ft. lbs. (34 Nm).

NOTE: In most cases, the yoke on the front driveshaft half must be bottomed out in the transmission (fully forward) before installation to the support.

    b. Rotate the shaft as necessary so the front U-joint trunnion is in the correct position to align the matchmarks for the rear driveshaft half.

NOTE: Before installing the rear driveshaft, align the U-joint trunnions using the matchmarks made earlier. In some cases a "key" in the output spline of the front driveshaft half will align with a missing spline in the rear yoke.

    c. Attach the rear U-joint to the axle, then torque the retainers to 15 ft. lbs. (20 Nm).
9. Remove the jackstands and carefully lower the vehicle.
10. Road test the vehicle.

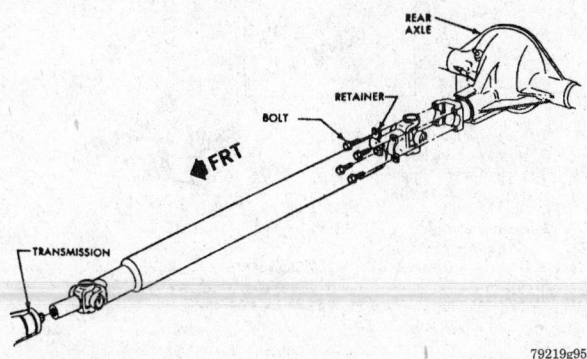

79219g95

Exploded view of a common single piece rear driveshaft mounting

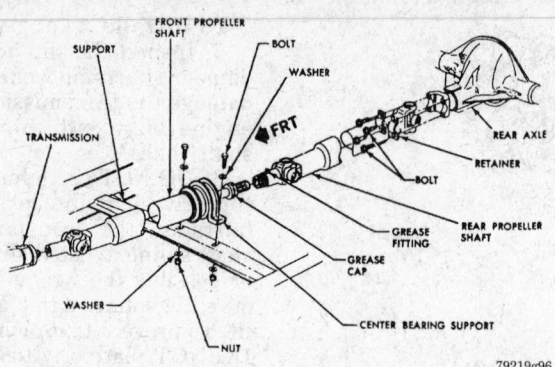

Exploded view of a common two-piece rear driveshaft mounting

U-Joint

With the exception of the particular size socket/pipe which must be used to support the trunnion yoke on some earlier model driveshafts, U-joint replacement is identical between the front and rear driveshaft assemblies. Make sure an appropriately sized socket is used as a support on early model front driveshafts, these will require a 1¼ in. socket as opposed to the 1⅛ in. socket used for later models and most rear driveshafts.

These vehicles utilize 2 different types of U-joint assemblies. At the factory, most vehicles are equipped with a nylon injected assembly. Because these assemblies contain no snapring grooves, and the very action of removing them destroys their retainers, they cannot be reused. Once removed, nylon injected U-joints should be discarded. If your vehicle has already received U-joint service, then it will be equipped with a snapring retained assembly. Once properly removed, these assemblies may be reused if they are in good condition.

REMOVAL AND INSTALLATION

Snapring Type

1. Matchmark and remove the driveshaft assembly from the vehicle. Refer to the procedure earlier in this section.

——— **WARNING** ———
NEVER clamp the driveshaft tube in a vise, for this may dent the tube. Support the driveshaft horizontally and clamp on the yokes of the universal joints using a soft-jawed vise or blocks of wood to protect the yokes.

2. Remove the snaprings from the yoke by pinching the ends together with a pair of pliers. If the snapring is difficult to remove, tap the end of the bearing cap lightly to relieve pressure from the snapring.

3. Support the propeller shaft horizontally in line with the base plate of a bench vise, but never clamp the driveshaft tube.

4. Place the universal joint so the lower ear of the yoke is supported on a 1¼ in. or 1⅛ inch socket (or ID pipe), depending on the application. Most rear driveshafts and late model driveshafts require the 1⅛ in sup-

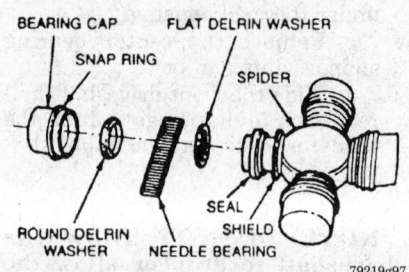

Exploded view of a replacement U-joint assembly — internal snapring type

port. Press the trunnion bearing against the socket/pipe in order to partially press it from the yoke. A cross press such as tool J–9522–3 or equivalent should be used for this.

5. Grasp the cap and work it out. If necessary, use tool J–9522–5 or an equivalent spacer to further push the the bearing cap from the trunnion, then grasp and work it free.

6. Rotate the shaft and support the other side of the yoke, then press the bearing cap from the yoke as in previous steps.

7. Remove the trunnion from the driveshaft yoke.

8. Clean and check the condition of all parts. Use U-joint repair kits to replace all the worn parts or replace the assembly using a new U-joint.

NOTE: If the used universal joints are going to be reinstalled, repack with new grease.

To install:

9. Repack the bearings with chassis grease and replace the trunnion dust seals after any operation that requires disassembly of the U-joint. Be sure the lubricant reservoir at the end of the trunnion is full of lubricant. Fill the reservoirs with lubricant from the bottom.

10. Partially insert the cross into the yoke so one trunnion seats freely in the bearing cap, then rotate the shaft so this trunnion is on the bottom.

11. Install the opposite bearing cap part way. Be sure both trunnions are started straight into the bearing caps.

12. Press against opposite bearing caps, working the cross constantly to be sure the trunnions are free in the bearings. If binding occurs, check the needle rollers to be sure one or more needles have not become lodged under an end of the trunnion.

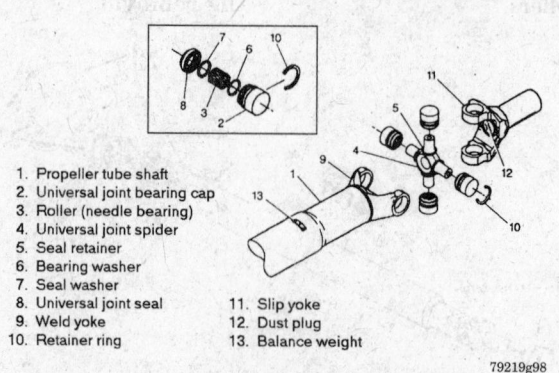

1. Propeller tube shaft
2. Universal joint bearing cap
3. Roller (needle bearing)
4. Universal joint spider
5. Seal retainer
6. Bearing washer
7. Seal washer
8. Universal joint seal
9. Weld yoke
10. Retainer ring
11. Slip yoke
12. Dust plug
13. Balance weight

Exploded view of a replacement U-joint assembly — external snapring type

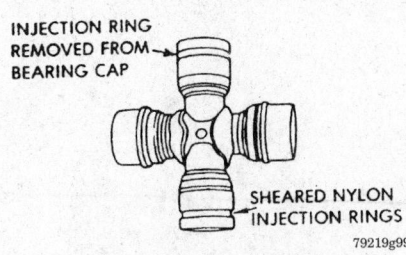

INJECTION RING
REMOVED FROM → BEARING CAP

SHEARED NYLON
INJECTION RINGS

79219g99

Production Nylon type U-joint

13. As soon as one bearing retainer groove is exposed, stop pressing and install the bearing retainer snapring.

NOTE: It may be necessary to strike the yoke with a hammer to align the seating of the bearing retainers.

14. Continue to press until the opposite bearing retainer can be installed. If difficulty installing the snaprings is encountered, tap the yoke with a hammer to spring the yoke ears slightly.

15. Once the driveshaft and U-joints are properly assembled, align the matchmarks and install the driveshaft to the vehicle.

Nylon Injected Type

NOTE: Don't disassemble these joints unless replacing the complete U-joint. These factory installed joints cannot be reused and should instead be replaced by snapring type U-joints.

1. Matchmark and remove the driveshaft assembly from the vehicle. Refer to the procedure earlier in this section.

——— **WARNING** ———
NEVER clamp the driveshaft tube in a vise, for this may dent the tube. Support the driveshaft horizontally and clamp on the yokes of the universal joints using a soft-jawed vise or blocks of wood to protect the yokes.

2. Support the driveshaft horizontally in line with the base plate of a bench vise, but never clamp the driveshaft tube.

3. Place the universal joint so the lower ear of the yoke is supported on a 1¼ in. or 1⅛ inch socket (or ID pipe), depending on the application. Most rear driveshafts and late model driveshafts require the 1⅛ in support.

4. Press the lower bearing cap out of the yoke ear, this will shear the nylon injected ring retaining the lower bearing cap.

5. If the bearing cap is not completely removed, lift the cross (J–9522–3) and insert J–9522–5 or equivalent spacer, then press the cap completely out.

6. Rotate the driveshaft, shear the opposite plastic retainer, and press the other bearing cap out in the same manner.

7. Remove the cross from the yoke.

NOTE: Production U-joints cannot be reassembled. There are no bearing retainer grooves in the caps. Discard all parts that were removed and substitute those in the overhaul kit.

8. If the front U-joint is being removed, separate the bearing caps from the slip yoke in the same manner.

9. Remove the sheared plastic bearing retainer from the yoke. If necessary, drive a small pin or punch through the injection holes to aid removal.

10. Install the new snapring U-joints. Refer to the snapring type installation procedure found earlier in this section.

Halfshaft

REMOVAL AND INSTALLATION

1. Raise and support the vehicle safely. Remove the tire and wheel assemblies.

2. Remove the cotter pin, retainer, nut and washer. Remove the brake line support bracket.

3. Matchmark the axle tube-to-flange location. Loosen but do not remove the axle tube-to-flange bolts. Using the proper tools, remove the tie rods at the steering knuckles.

4. Remove the lower shock absorber retaining bolts and move the shock absorbers aside.

5. Separate the upper ball joint from the steering knuckle. Suspend the steering knuckle from the frame using a piece of wire.

6. As required, remove the skid plate. Remove the halfshaft to axle tube bolts.

7. Move the inner part of the halfshaft forward. Support it away from the frame. Using a suitable tool, remove the shaft from the hub and bearing assembly.

8. Remove the halfshaft.

To install:

9. Install the halfshaft into position. Push it into the hub.

10. Install the shaft retaining nut and washer. Tighten the halfshaft retaining nut to 160–200 ft. lbs. (220–270 Nm).

11. Install the cotter pin in the nut. Install the halfshaft-to-axle tube bolts and tighten to 60 ft. lbs. (80 Nm).

12. Install the lower shock absorber bolts. Connect the tie rods to the steering knuckle.

13. Install the skid plate, if equipped.

14. Install the wheel and tire assemblies. Lower the vehicle

CV-Boot and Joint

REMOVAL AND INSTALLATION

Outer

1. Raise and safely support the vehicle.

2. Remove the tire and wheel assemblies.

3. Remove the halfshaft from the vehicle and place the assembly in a vise with soft jaws.

4. Cut the seal clamps and remove the boot. Slide the boot down away from the joint.

5. Clean the grease around the joint. Spread the snapring and remove the joint.

6. Discard the boot, if damaged.

7. Using a brass drift punch and hammer, tap on the joint cage until it moves enough to remove a ball, then remove the other balls in the same fashion.

8. Remove the cage and inner race by turning the assembly straight up 90 degrees to the center line of the outer housing. Maneuver the cage and inner race out of the outer race housing assembly. Then remove the inner race from the cage assembly.

To install:

9. Thoroughly clean all part in solvent. Apply grease to the ball grooves of the inner and outer race.

10. Install the inner race into the cage.

11. Install the cage assembly into the housing.

12. Tilt the cage up and install the ball into the assembly. Pack the assembly with grease.

13. Install the small boot clamp onto the axle shaft.

14. Install the axle boot onto the shaft and secure the small boot clamp using a suitable tool.

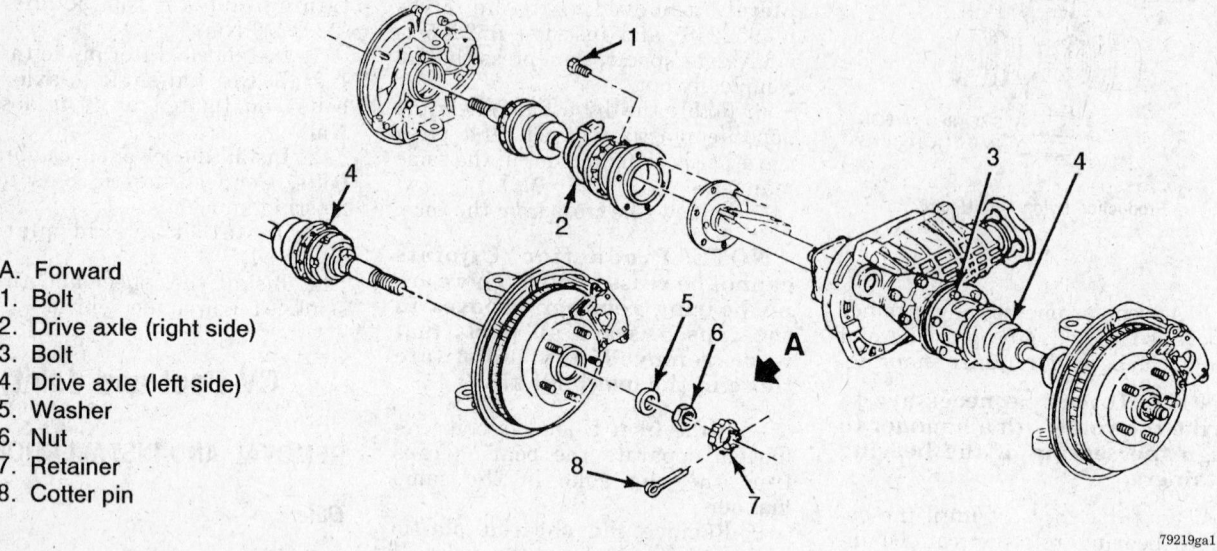

A. Forward
1. Bolt
2. Drive axle (right side)
3. Bolt
4. Drive axle (left side)
5. Washer
6. Nut
7. Retainer
8. Cotter pin

Front drive axle assembly

79219ga1

15. Install the joint assembly onto the axle shaft, the snapring will snap into place automatically. Pack the boot and outer joint with a sufficient quantity of grease.

16. Position the boot over the joint housing. Secure with the large clamp, using a suitable tool.

17. Reinstall the axle shaft assembly into the vehicle.

18. Install the tire and wheel assembly.

19. Lower the vehicle.

Inner

1. Raise and safely support the vehicle.

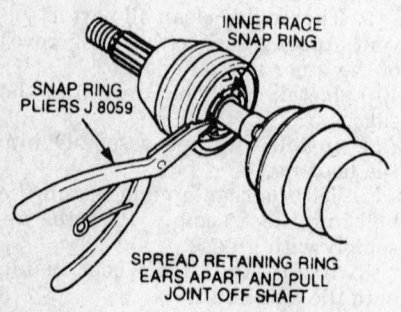

INNER RACE SNAP RING

SNAP RING PLIERS J 8059

SPREAD RETAINING RING EARS APART AND PULL JOINT OFF SHAFT

79219ga2

Separating the outer joint from the axle shaft

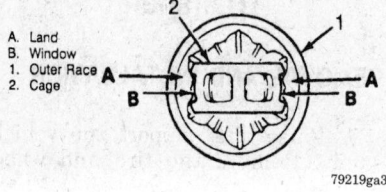

A. Land
B. Window
1. Outer Race
2. Cage

79219ga3

Positioning the outer joint cage and race for removal/installation

2. Remove the tire and wheel assemblies.

3. Remove the halfshaft from the vehicle and place the assembly in a vise with soft jaws.

4. Cut the seal clamps and remove the boot. Slide the joint housing off the joint and shaft assembly.

5. Clean the grease around the joint. Spread the snapring and remove the joint.

6. Discard the boot, if damaged.

To install:

7. Thoroughly clean all part in solvent.

8. Install the small boot clamp onto the axle shaft.

9. Install the axle boot onto the shaft and secure the small boot clamp using a suitable tool.

10. Install the joint assembly with the snapring counterbore facing (out) the axle shaft housing, the snapring will snap into place automatically.

11. Install the spacer ring onto the axle shaft, ensuring the ring is fully seated in the groove.

12. Position the axle housing onto the joint and shaft assembly. Pack the boot and joint with a sufficient quantity of grease.

13. Position the boot over the joint housing. Secure with the large clamp, using a suitable tool.

14. Reinstall the axle shaft assembly into the vehicle.

15. Install the tire and wheel assembly.

16. Lower the vehicle.

Transfer Case Assembly

REMOVAL AND INSTALLATION

Except Astro and Safari

1. Disconnect the negative battery cable.

2. Shift the transfer case into the **4HI** range.

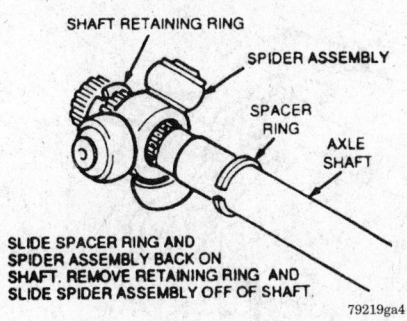

SHAFT RETAINING RING
SPIDER ASSEMBLY
SPACER RING
AXLE SHAFT

SLIDE SPACER RING AND
SPIDER ASSEMBLY BACK ON
SHAFT. REMOVE RETAINING RING AND
SLIDE SPIDER ASSEMBLY OFF OF SHAFT.

79219ga4

Inner joint and snapring removal

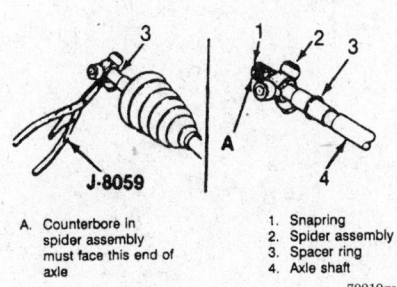

J-8059

A. Counterbore in spider assembly must face this end of axle

1. Snapring
2. Spider assembly
3. Spacer ring
4. Axle shaft

79219ga5

Inner joint installation

3. Raise and support the vehicle safely.

4. If equipped, remove the skid plate/transfer case shield bolts, then remove the skid plate/transfer case from under the transmission/transfer case assembly.

5. Remove the plug and drain the transfer case fluid.

6. Matchmark and remove the front and rear driveshafts/propeller shafts from the transfer case.

7. Tag and disconnect the vacuum lines and/or the electrical connectors, as equipped.

8. If applicable, disconnect the transfer case shift rod/cable from the case.

9. If applicable, remove the support brace-to-transfer case bolts.

10. Support the transfer case, then remove the transfer case-to-transmission retaining bolts

11. Slide the transfer case rearward and off the transmission output shaft, then carefully lower it.

12. Remove all traces of old gasket material from the mating surfaces.

To install:

13. Carefully raise the transfer case into position behind the transmis-

sion. Position a new gasket, using sealer to hold it in position, then slide the transfer case onto the transmission output shaft.

14. Install the transfer case-to-transmission retaining bolts, then tighten to 24 ft. lbs. (33 Nm) for vehicles through 1995 or to 41 ft. lbs. (55 Nm) for 1996–97 vehicles.

15. If equipped, install the support brace bolts and tighten to 35–37 ft. lbs. (47–50 Nm).

16. Remove the support from the transfer case.

17. If equipped, connect the shift rod to the case.

18. Engage the vacuum lines and/or electrical connections, as necessary.

19. Align and install the front and rear driveshafts.

20. Make sure the vehicle is level, then properly refill the transfer case through the filler plug.

21. If equipped, install the skid plate.

22. Carefully lower the vehicle.

23. Connect the negative battery cable.

Astro and Safari

1. Disconnect the negative battery cable. If necessary, shift the transfer case into the **4HI** position to ease linkage removal.

2. Raise and support the vehicle safely.

3. If, equipped, remove the skid plate. Drain the fluid from the transfer case.

4. Drain the fluid from the transfer case.

5. Remove the front and rear propeller shafts.

6. Disconnect the breather hose.

7. Detach the electrical connections.

8. Support the transfer case with a suitable jack.

9. Remove the adapter-to-transfer case bolts.

10. Remove the nuts and washers.

11. Remove the transfer case, then remove the transfer case-to-adapter gasket.

To install:

12. Position a new transfer case-to-adapter gasket. Use sealer to hold the gasket in place.

13. Support the transfer case with a suitable jack.

14. Fasten the transfer case adapter to the transfer case.

15. Install the transfer case retaining bolts and tighten to 38 ft. lbs. (52 Nm).

16. Remove the jack from the transfer case.

17. Install the washers and nuts. Tighten the nuts to 26 ft. lbs. (35 Nm).

18. Install the support brace, washers and bolts. Tighten bolt (53 in figure) to 94 ft. lbs. (128 Nm) and bolts (51 in figure) to 74 ft. lbs. (100 Nm)./

19. Attach the electrical connectors.

20. Connect the breather hose.

21. Install the front and rear propeller shafts.

22. Fill the transfer case with the proper lubricant.

23. Lower the vehicle, then connect the negative battery cable.

STEERING

Air Bag

──── **CAUTION** ────

Some vehicles are equipped with an air bag system, also known as the Supplemental Inflatable Restraint (SIR) system. The system must be disabled before performing service on or around system components, steering column, instrument panel components, wiring and sensors. Failure to follow safety and disabling procedures could result in accidental air bag deployment, possible personal injury and unnecessary system repairs.

PRECAUTIONS

Several precautions must be observed when handling the inflator module to avoid accidental deployment and possible personal injury.

• Never carry the inflator module by the wires or connector on the underside of the module.

• When carrying a live inflator module, hold securely with both hands, and ensure that the bag and trim cover are pointed away.

• Place the inflator module on a bench or other surface with the bag and trim cover facing up.

• With the inflator module on the bench, never place anything on or close to the module which may be thrown in the event of an accidental deployment.

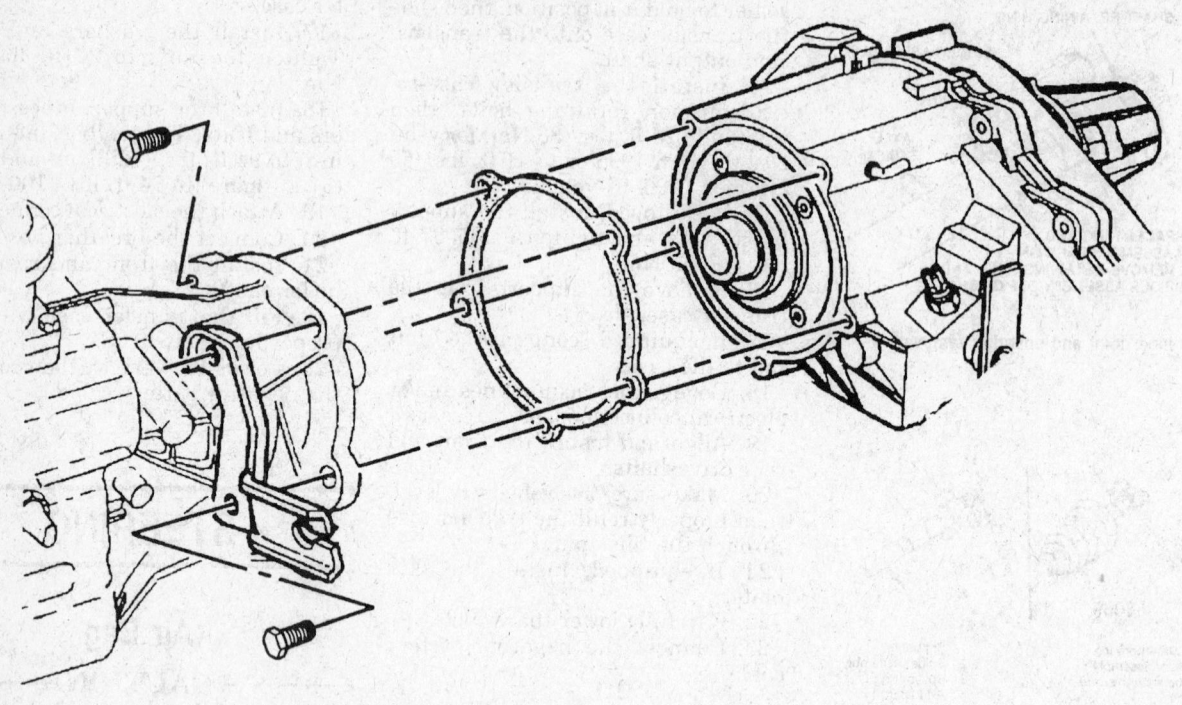

79219ga6

Transfer case-to-manual transmission mounting — exc. Astro and Safari

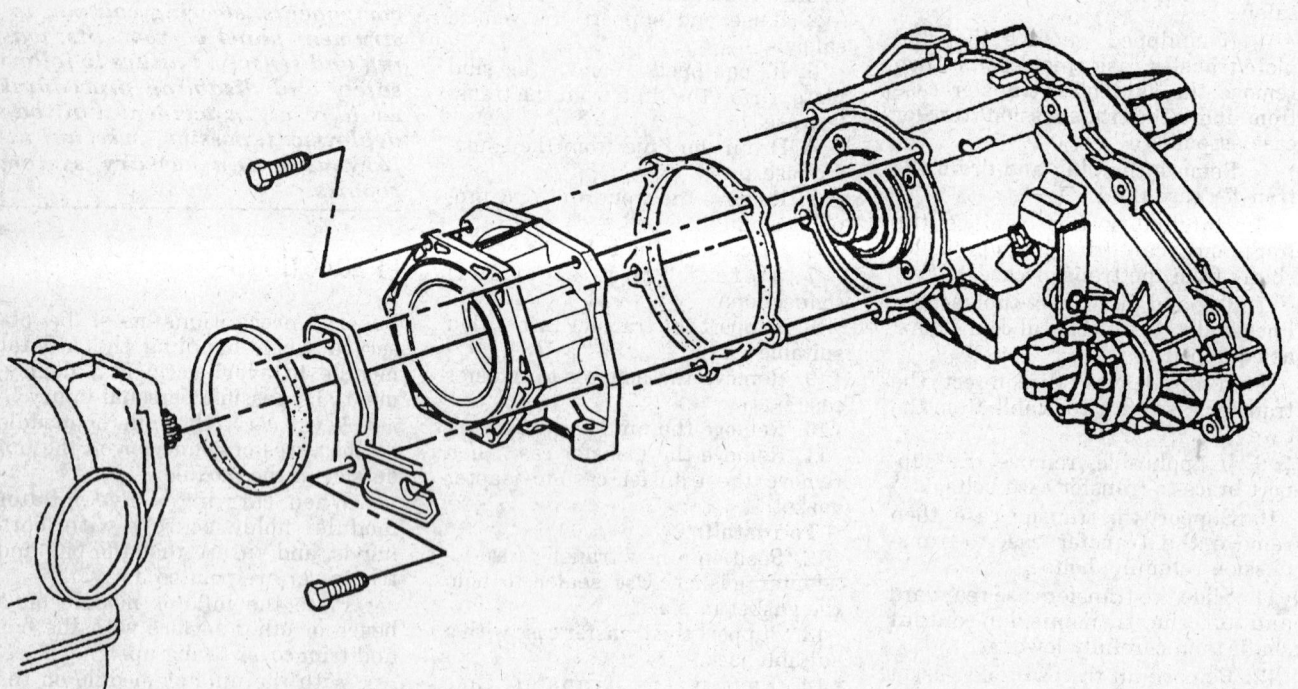

79219ga7

Transfer case-to-automatic transmission mounting — exc. Astro and Safari

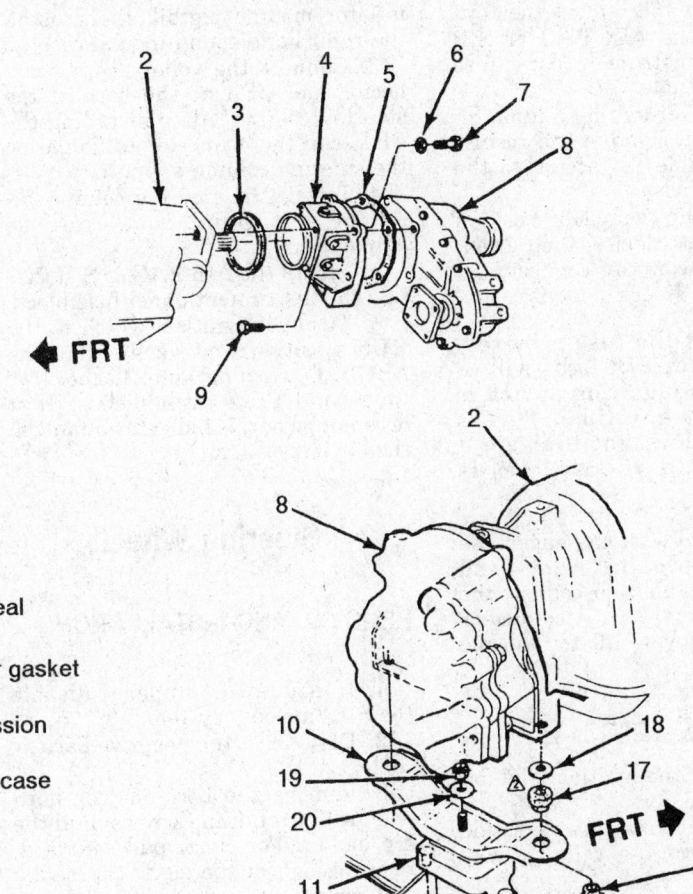

1. Frame
2. Transmission
3. Transfer case adapter seal
4. Transfer case adapter
5. Transfer case to adapter gasket
6. Adapter bolt seal
7. Bolt. adaptor to transmission
8. Transfer case
9. Bolt. adapter to transfer case
10. Transfer case support
11. Transfer case mount
12. Nut
13. Washer
14. Bolt
15. Washer
16. Lower insulator
17. Upper insulator
18. Washer
19. Nut
20. Washer

Transfer case mounting — Astro and Safari

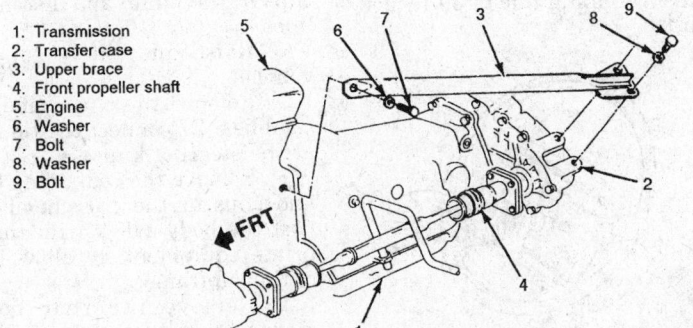

1. Transmission
2. Transfer case
3. Upper brace
4. Front propeller shaft
5. Engine
6. Washer
7. Bolt
8. Washer
9. Bolt

Transfer case support brace — exc. Astro and Safari

DISARMING

—— CAUTION ——
The Supplemental Inflatable Restraint (SIR) system must be disarmed before performing many in-vehicle service procedures. Failure to do so may cause accidental deployment of the air bag, resulting in unnecessary SIR system repairs and/or personal injury.

Disabling the SIR System

1. Turn the steering wheel so the vehicle's wheels are pointing straight ahead.

2. Turn the ignition switch to the **LOCK** position and remove the key.

3. Remove the AIR BAG or SIR fuse from the instrument panel fuse block, as applicable.

4. Remove the steering column filler panel or left hand sound insulator, as applicable for access to the SIR wiring harness.

5. Remove the Connector Position Assurance (CPA) device, then disengage the yellow 2-way connector at the base of the steering column.

NOTE: With the fuse removed, the AIR BAG or SIR light will illuminate if the ignition switch is turned ON at any time. This is normal and does not indicate a problem when the system is disarmed.

6. If equipped with passenger side air bags, remove the right hand sound insulator, then disconnect the CPA and yellow 2-way connector from the passenger inflator module pigtail.

Enabling the SIR System

1. Make sure the ignition is in the **LOCK** position.

2. If equipped with passenger side air bags, connect the yellow 2-way

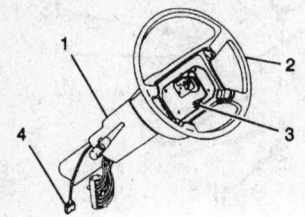

1. STEERING COLUMN
2. STEERING WHEEL
3. INFLATOR MODULE CONNECTOR
4. SIR STEERING COLUMN CONNECTOR

79219gb1

Yellow 2-way SIR connector

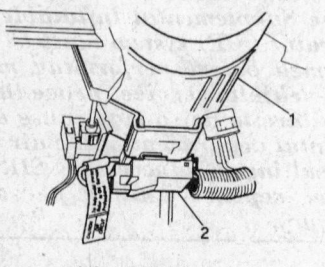

1. Steering column
2. Connector,sir(yellow)

325682

Yellow 2-way air bag CPA connector for steering column

connector and CPA to the passenger inflator module pigtail, then install the right hand sound insulator.

3. Connect the yellow 2-way connector and CPA at the base of the steering column. After installing the CPA, clip the connector to flange on the steering column support.

4. Install the steering column filler or left hand insulator, as applicable.

5. Install the AIR BAG or SIR fuse into the instrument panel fuse block.

6. Turn the ignition switch to the **RUN** position and verify that the AIR BAG warning lamp flashes 7–9 times and then extinguishes. If it does not go out, it indicates a fault in the air bag system.

Steering Wheel

REMOVAL AND INSTALLATION

1. If equipped, properly disable the SIR (air bag) system.

2. Disconnect the negative battery cable.

3. Remove the horn pad or horn pad/air bag retaining screws. Pull the air bag and/or horn pad outward, then disconnect the electrical lead(s) and remove.

4. Matchmark the steering wheel and the shaft.

5. Remove the steering wheel retaining clip, if equipped, and nut.

6. If necessary, remove the horn plunger contact.

7. Matchmark the relationship between the steering wheel splines to the steering shaft. Remove the steering wheel, using a suitable puller.

8. Installation is the reverse of removal. Align the matchmarks made during removal and tighten the steering wheel retaining nut to 30 ft. lbs. (40 Nm).

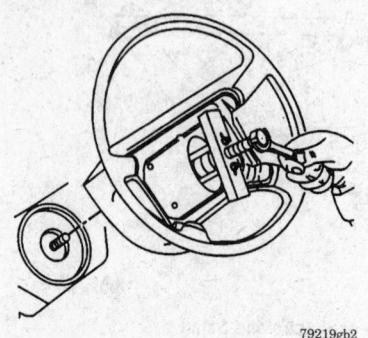

79219gb2

Remove the steering wheel using J 1859-A or equivalent puller

Tie Rod Ends

REMOVAL AND INSTALLATION

1. Raise and support the vehicle safely. Remove the tire and wheel assemblies.

2. Remove the cotter pins and nuts. Using the proper removal tool, separate the outer tie rod from the steering knuckle using Universal Steering Linkage Puller J 24319-B for Astro and Safari or Tie Rod Puller J 6627-A for all others.

3. Disconnect the inner tie rod from the relay rod using J 6627-A or equivalent tie rod puller. Remove the tie rod ends from the adjuster tubes.

To install:

4. Installation is the reverse of the removal procedure. Tighten the inner tie rod ball stud nut to 35 ft. lbs. (47 Nm). Tighten the outer tie rod ball stud to the steering knuckle to 35 ft. lbs. (47 Nm). The number of threads on both the inner and outer tie rod ends must be equal within 3 threads.

5. Install the tire and wheel assemblies, then carefully lower the vehicle.

6. Adjust the front end alignment, as required.

Power Rack and Pinion

REMOVAL AND INSTALLATION

1. Disconnect the negative battery cable.

2. Remove the air cleaner assembly.

3. Remove the dust boot from the steering gear.

4. Remove the intermediate shaft lower pinch bolt and disconnect the intermediate shaft from the lower stub shaft.

5. Remove the fluid line retaining clips at the pump and disconnect the lines.

6. Raise and safely support the vehicle.

7. Remove the wheel and tire assemblies. Disconnect the tie rod ends at the steering knuckle.

8. Remove the remaining brackets and clips at the crossmember. Support the body safely with the appropriate equipment, to allow lowering of the subframe.

9. Remove the rear subframe mounting bolts and carefully lower the rear of the subframe approximately 5 in. (128mm).

10. Remove the rack and pinion mounting bolts and remove the rack through the left wheel opening.

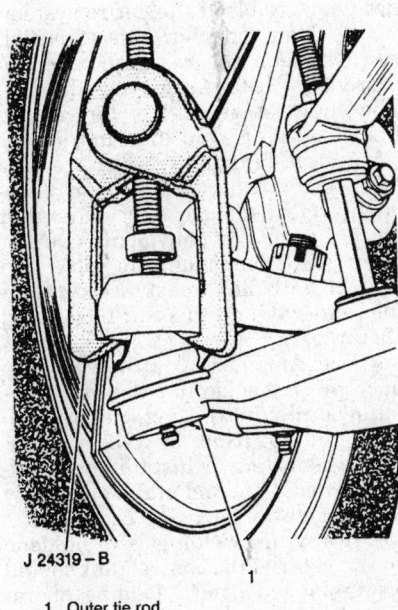

J 24319 – B

1. Outer tie rod

79219gb3

Removing the outer tie rod ball stud — Astro and Safari shown

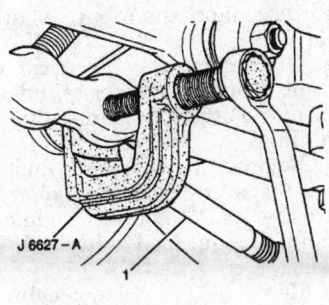

J 6627 – A

1. INNER TIE ROD

79219gb4

Removing the inner tie rod ball stud — Astro and Safari shown

To install:

11. Install the rack and pinion through the left wheel opening.

12. Install the rack and pinion mounting nuts, tighten to 70 ft. lbs. (95 Nm).

13. Raise the subframe assembly and install the rear mounting bolts.

14. Remove any supports and install the brackets and clips to the crossmember.

15. Install the wheel and tire assemblies. Lower the vehicle.

16. Connect the fluid lines at the pump and tighten to 18 ft. lbs. (25 Nm).

17. Install the line retaining clips. Connect the intermediate shaft to the stub shaft.

18. Install the dust boot over the steering gear.

19. Install the air cleaner assembly and connect the negative battery cable.

20. Fill and bleed the steering system.

Power Steering Pump

SYSTEM BLEEDING

1993–95 Vehicles

1. Fill the reservoir to the proper level and let the fluid remain undisturbed for at least 2 minutes.

2. Start the engine and run it for only about 2 seconds.

3. Add fluid as necessary.

4. Repeat Steps 1–3 until the level remains constant.

5. Raise the front of the vehicle so the front wheels are off the ground. Set the parking brake and block both rear wheels front and rear. Manual transmissions should be in neutral; automatic transmissions should be in **P**.

6. Start the engine and run it at approximately 1500 rpm.

7. Turn the wheels (off the ground) to the right and left, lightly contacting the stops.

8. Add fluid as necessary.

9. Lower the vehicle and turn the wheels right and left on the ground.

10. Check the level and refill as necessary.

11. If the fluid is extremely foamy, let the vehicle stand for a few minutes with the engine off and repeat the procedure. Check the belt tension and check for a bent or loose pulley. The pulley should not wobble with the engine running.

12. Check that no hoses are contacting any parts of the vehicle, particularly sheetmetal.

13. Check the fluid level and refill as necessary.

14. Check for air in the fluid. Aerated fluid appears milky. If air is present, repeat the above operation. If it is obvious that the pump will not respond to bleeding after several attempts, a pressure test may be required.

1996–97 Vehicles

1. Make sure the ignition switch is **OFF**.

2. Turn the steering wheel full left.

3. Fill the fluid reservoir to the "FULL COLD" level. Leave the cap off.

4. Raise the front wheels off the ground.

5. With an assistant checking the fluid level and condition, turn the steering wheel lock-to-lock at least 20 times. Make sure the engine remains OFF. Keep note of the following:

• On systems with long return lines or fluid coolers, turn the steering wheel lock-to-lock at least 40 times.

• Trapper air may cause fluid to overflow. Thoroughly clean any spilled fluid to allow for leak check.

• Keep the fluid level at "FULL COLD".

• Be alert to periodic bubbles that could indicate a loose connection or leaky O-ring seal.

6. While turning the wheel, check the fluid constantly. No bubbles are allowed. For any sign of bubbles, recheck the connections, then repeat Step 5.

7. Start the engine. With the engine idling, maintain fluid level, then reinstall the cap.

8. Return the wheels to center. Lower the front wheels to the ground.

9. Keep the engine running for two minutes.

10. Turn the steering wheel in both directions and verify the following: Smooth power assist, noiseless operation, proper fluid level, no system leaks and proper fluid condition.

11. If all proper conditions apply, the procedure is complete.

REMOVAL AND INSTALLATION

Except Turbocharged Engine

1. Disconnect the negative battery cable.

2. Disconnect and cap the power steering pump hoses. Remove the accessory drive belt.

3. As required, remove the power steering pump pulley using a suitable puller tool or equivalent.

4. Remove the pump mounting bolts. Remove the pump.

5. Installation is the reverse of the removal procedure. Bleed the power steering system.

Turbocharged Engine

1. Disconnect the negative battery cable.

2. Remove the air cleaner and duct assembly.

3. Remove the upper fan shroud.

4. Loosen the fan nuts and remove the serpentine belt.

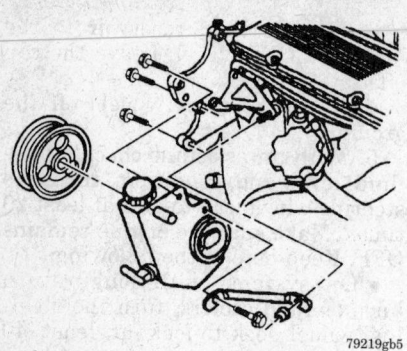

Power steering pump mounting — 1996 2.2L engine shown

79219gb5

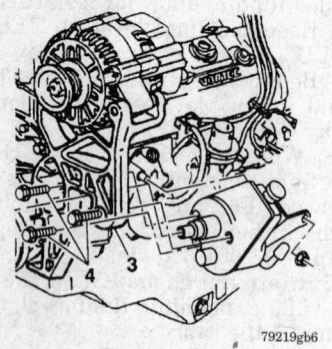

Power steering pump mounting — 1996 4.3L engine shown

79219gb6

5. Remove the fan and pulley assembly.

6. Remove the power steering pump pulley as follows:

   a. Install tool J–25034–B, ensure the pilot bolt bottoms in the pump shaft.

   b. Hold the pilot bolt with a suitable wrench.

   c. Turn the shaft locknut counterclockwise and remove the pulley.

7. Raise and safely support the vehicle.

8. Remove the left tire and wheel assembly.

9. Remove the left wheel house panel.

10. Remove the power steering hose bracket.

11. Place a drain pan below the pump and remove the power steering pressure and return lines, capping the lines to prevent dirt from entering.

12. Remove the bolts from the rear bracket at the alternator.

13. Lower the vehicle and remove the assembly.

14. Remove the bracket from the pump as necessary.

To install:

15. Install the bracket to the pump, if removed.

16. Install the pump assembly and torque the bolts to 37 ft. lbs. (50 Nm).

17. Install the bolts to the rear bracket at the alternator.

18. Install the power steering pressure and return lines.

19. Install the power steering hose bracket.

20. Raise and safely support the vehicle.

21. Install the left wheel house panel.

22. Install the left tire and wheel assembly. Lower the vehicle.

23. Install the power steering pump pulley as follows:

   a. Place the pulley on the shaft.

   b. Install tool J–25033–B, ensure the pilot bolt bottoms in the pump shaft.

   c. Hold the pilot bolt with a suitable wrench.

   d. Turn the shaft locknut clockwise and install the pulley.

24. Install the fan and pulley assembly.

25. Tighten the fan pulley nuts and install the serpentine belt.

26. Install the upper fan shroud.

27. Install the air cleaner and duct assembly.

28. Connect the negative battery cable. Fill and bleed the power steering system. Check system for leaks.

BRAKES

Anti-Lock Brake System Service

PRECAUTIONS

• Certain components within the Anti-Lock Brake System (ABS) are not intended to be serviced or repaired individually. Only those components with removal and installation procedures should be serviced.

• Do not use rubber hoses or other parts not specifically specified for an ABS system. When using repair kits, replace all parts included in the kit. Partial or incorrect repair may lead to functional problems and require the replacement of components.

• Lubricate rubber parts with clean, fresh brake fluid to ease assembly. Do not use lubricated shop air to clean parts; damage to rubber components may result.

• Use only specified brake fluid from an unopened container.

• If any hydraulic component or line is removed or replaced, it may be necessary to bleed the entire system.

• A clean repair area is essential. Always clean the reservoir and cap thoroughly before removing the cap. The slightest amount of dirt in the fluid may plug an orifice and impair the system function. Perform repairs after components have been thoroughly cleaned; use only denatured alcohol to clean components. Do not allow ABS components to come into contact with any substance containing mineral oil; this includes used shop rags.

• The Anti-Lock control unit is a microprocessor similar to other computer units in the vehicle. Ensure that the ignition switch is **OFF** before removing or installing controller harnesses. Avoid static electricity discharge at or near the controller.

• If any arc welding is to be done on the vehicle, the control unit should be unplugged before welding operations begin.

Master Cylinder

REMOVAL AND INSTALLATION

1. Disconnect the negative battery cable.

2. Disconnect the electrical connections from the master cylinder, as required. Disconnect and plug the fluid lines.

3. Remove the master cylinder to power booster retaining bolts. Remove the RWAL control module assembly, if equipped with anti-lock brakes.

4. Remove the master cylinder. Remove the vacuum booster pushrod.

To install:

5. Install the master cylinder in position on the booster. Connect the booster pushrod.

6. Install the master cylinder retaining bolts and tighten to 20 ft. lbs. (27 Nm).

7. Connect the fluid lines to the master cylinder. Connect the RWAL control unit, if equipped with anti-lock brakes, to the bracket.

8. Connect the negative battery cable. Refill the master cylinder and bleed the brake system.

Brake Caliper

REMOVAL AND INSTALLATION

1. Remove ⅔ of the brake fluid from the master cylinder reservoir.

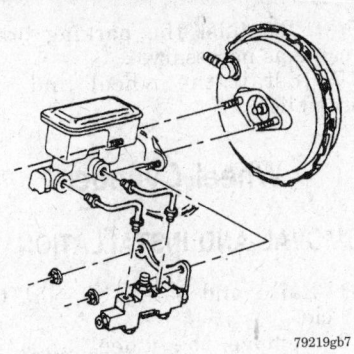

Master cylinder mounting — 1996 exc. Astro and Safari

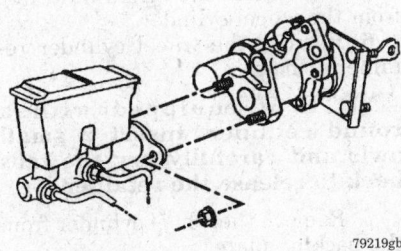

Master cylinder mounting — 1996 Astro and Safari

2. Raise and support the vehicle safely. Remove the tire and wheel assembly.

3. Disconnect and plug the caliper fluid line. Remove the bolts retaining the caliper to the rotor. Remove the caliper from the rotor.

NOTE: If the caliper is being removed for brake pad replacement, the fluid line does not need to be disconnected.

4. Remove the disc brake pads from the caliper. Remove the disc brake pad retaining clips from inside the caliper.

To install:

5. Clean and lubricate the sleeves and bushings with silicon grease. Install the pads in the caliper.

6. Install the caliper in position over the rotor and install the mounting bolts. Tighten the mounting bolts to 38 ft. lbs. (51 Nm).

7. Connect the fluid lines to the caliper, if disconnected, and tighten to 33 ft. lbs. (45 Nm).

8. Install the wheel and tire assembly.

9. Lower the vehicle and refill the master cylinder to the correct level. Bleed the brake system if the fluid lines were disconnected from the caliper.

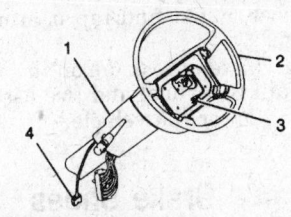

1. STEERING COLUMN
2. STEERING WHEEL
3. INFLATOR MODULE CONNECTOR
4. SIR STEERING COLUMN CONNECTOR

Exploded view of a common caliper assembly — compact vehicle

Disc Brake Pads

REMOVAL AND INSTALLATION

1. Remove ⅔ of the brake fluid from the master cylinder.

2. Raise and safely support the vehicle.

3. Place a C-clamp around the outer pad and caliper; tighten the C-clamp until the piston is fully compressed in the caliper. Remove the brake caliper.

4. Remove the inboard pad and retaining spring from the caliper.

5. Remove the outboard pad from the caliper.

6. Remove the sleeves and bushings.

To install:

7. Clean and lubricate the sleeves and bushing with silicone lubricant and install them in the caliper.

8. Clip the retaining spring onto the inboard pad and install the pad in the caliper.

9. Install the outboard pad into the caliper.

10. Install the caliper in position over the rotor and install the mounting bolts. Bend the tabs, on the outboard brake pad, over the caliper.

11. Install the wheel and tire assemblies.

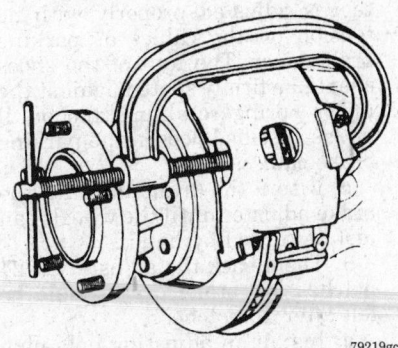

Compressing the caliper piston with a C-clamp

12. Lower the vehicle, refill the master cylinder and pump pedal to attain full brake pedal before road testing the vehicle.

Brake Rotor

REMOVAL AND INSTALLATION

2WD Vehicles

1. Raise and support the front of the vehicle safely using jackstands.

2. Remove the tire and wheel assembly.

3. Remove the brake caliper mounting bolts and carefully remove the caliper (along with the brake pads) from the rotor. Do not disconnect the brake line; instead, wire the caliper out of the way with the line still connected.

NOTE: Once the rotor is removed from the vehicle, the wheel bearings may be cleaned and repacked or the bearings and races may be replaced.

4. Carefully pry out the grease cap, then remove the cotter pin, spindle nut, and washer. Remove the hub, being careful not to drop the outer wheel bearings. As the hub is pulled forward, the outer wheel bearings will often fall forward and may easily be removed at this time.

To install:

5. Carefully install the wheel hub over the spindle.

6. Using your hands, firmly press the outer bearing into the hub.

7. Loosely install the spindle washer and nut, but do not install the cotter pin or dust cap at this time.

8. Install the brake caliper.

9. Install the tire and wheel assembly.

10. Properly adjust the wheel bearings:

a. Spin the wheel forward by hand and tighten the nut to 12 ft. lbs. (16 Nm) in order to fully seat the bearings and remove and burrs from the threads.

b. Back off the nut until it is just loose, then finger-tighten the nut.

c. Loosen the nut ¼–½ turn until either hole in the spindle lines up with a slot in the nut, then install a new cotter pin. This may appear to be too loose, but it is the correct adjustment.

d. Proper adjustment creates 0.001–0.005 in. (0.025–0.127mm) end-play.

11. Install the dust cap.

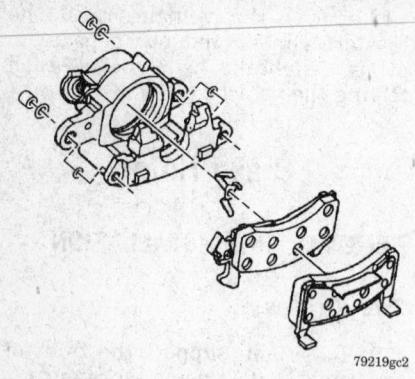

79219gc2

Exploded view of the disc brake assembly

12. Install the wheel/hub cover, then remove the supports and carefully lower the vehicle.

4WD and AWD Vehicles

1. Raise and support the front of the truck safely using jackstands under the frame.
2. Remove the tire and wheel assembly.
3. Remove the brake caliper mounting bolts and carefully remove the caliper (along with the brake pads) from the rotor. Do not disconnect the brake line; instead wire the caliper out of the way with the line still connected.
4. If equipped, remove the lockwashers from the hub studs in order to free the rotor.
5. Remove the brake rotor from the wheel hub.
To install:
6. Inspect the rotor for nicks, scores and/or damage, then replace, if necessary.
7. Install the rotor over the wheel hub studs.
8. If used, install the lockwashers over the studs.
9. Install the brake caliper and pads.
10. Install the tire and wheel assembly.
11. Remove the jackstands and carefully lower the vehicle. DO NOT attempt to move the vehicle unless a firm brake pedal is felt.

Brake Drums

REMOVAL AND INSTALLATION

1. Raise and safely support the vehicle.
2. Remove the wheel and tire assembly.
3. Remove the brake drum. If the drum will not pull of the axle, use a rubber mallet and tap it around the edge.
4. Install the drum on the axle and install the wheel and tire assembly.
5. Lower the vehicle.

Brake Shoes

REMOVAL AND INSTALLATION

1. Raise and safely support the vehicle.
2. Remove the wheel and tire assembly.
3. Remove the brake drum.
4. Remove the return springs from the brake shoes. Remove the shoe guide.
5. Remove the hold-down springs and pins. Remove the actuator lever and pivot.
6. Remove the lever return spring. Remove the actuator link.
7. Remove the parking brake strut and spring. Remove the parking brake lever.
8. Remove the brake shoes and the adjuster assembly.
To install:
9. Lubricate the contact points on the backing plate and the adjuster with lithium grease.
10. Install the parking brake lever, adjusting screw and spring assembly.
11. Install the shoe assembly onto the backing plate.
12. Install the parking brake lever, strut and strut spring.
13. Install the actuator lever and lever pivot. Install the actuator link.
14. Install the lever spring, the hold-down pins and springs.
15. Install the shoe guide. Install the return springs and install the brake drum in position.
16. Adjust the brakes as follows:
    a. Remove the knockout area in the backing plate, behind the adjuster assembly.
    b. Ensure the parking brake system is adjusted properly with no tension on the cables or parking brake lever. The tops of the shoes should be firmly seated against the upper spring retaining anchor, if not as specified, loosen the parking brake cables.
    c. Install the drum and turn the brake adjuster until the wheels can just be turned by hand.
    d. Then, back the adjuster off 24 notches. No brake drag should be felt after 12 notches.
    e. Install an adjusting hole plug in the backing plate to prevent dirt and moisture from entering.
    f. Readjust the parking brake cable as necessary.
17. Install the wheel and tire assemblies.

Wheel Cylinder

REMOVAL AND INSTALLATION

1. Raise and safely support the vehicle.
2. Remove the wheel and tire assemblies.
3. Remove the brake drum.
4. Remove the brake shoes, as necessary.
5. Disconnect the brake fluid line from the wheel cylinder.
6. Remove the wheel cylinder retainer or bolt.

NOTE: If equipped with a round retainer, insert 2 small awls and carefully pry the tabs back to release the retainer.

7. Remove the wheel cylinder from the backing plate.
To install:
8. Install the wheel cylinder in position on the backing plate.
9. Install the retainer or bolt. Connect the brake line to the wheel cylinder.

NOTE: If equipped with a round retainer, a socket may be used as a driver to ease installation.

10. Install the brake linings and the brake drum.
11. Install the wheel and tire assembly. Lower the vehicle.
12. Bleed the brake system.

Parking Brake Cable

ADJUSTMENT

Except Astro and Safari

1993–94 Vehicles

The rear brakes serve a dual purpose. They are used as service brakes and as parking brakes. To obtain proper adjustment of the parking brake, the service brakes must first be properly adjusted. Inspect the cables for binding or sticking.

1. Raise and support the vehicle safely. Loosen the equalizer nut. Some vehicles may require the removal of the cable guide on the equalizer.
2. Set the parking brake pedal 2 clicks for 2WD vehicles or 3 clicks for 4WD vehicles.

1. Hold-down pins
2. Backing plate
3. Parking brake lever
4. Secondary shoe
5. Shoe guide
6. Parking brake strut
7. Actuator lever
8. Actuator link
9. Return spring
10. Return spring
11. Hold-down springs
12. Lever pivot
13. Lever return spring
14. Strut spring
15. Adjusting screw assembly
16. Adjusting screw spring
17. Primary shoe

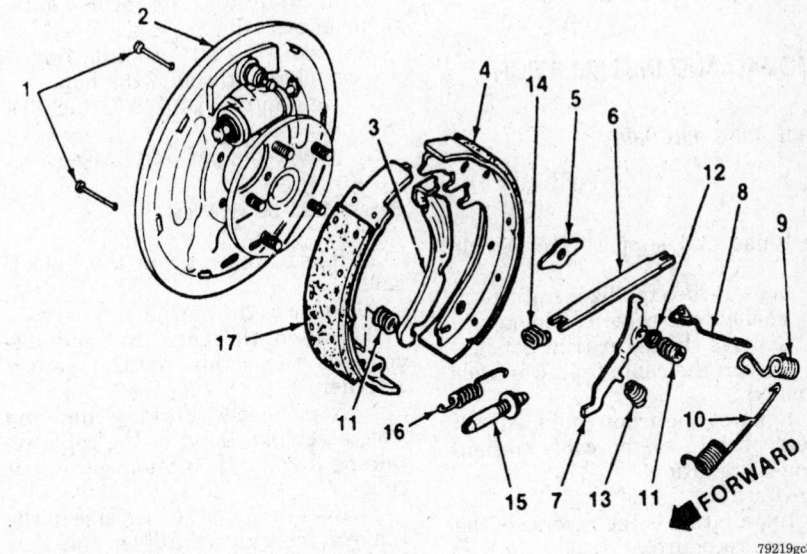

Exploded view of the drum brake components

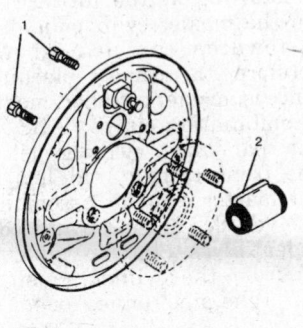

1. Bolts
2. Wheel cylinder

Wheel cylinder mounting — bolt retained type

3. Tighten the equalizer nut until the rear wheels will not rotate without excessive force in the forward motion.

4. Back off the equalizer nut until there is light drag when the wheels are rotated in the forward motion.

5. If removed, install the cable guide. Release the parking brake.

6. Rotate the rear wheels in the forward motion. There should be no brake drag.

1995–97 vehicles

The parking brake must be adjusted any time the parking brake cables

have been replaced or disconnected, or if heavy foot pressure on the pedal travel is less than half the pedal total travel. The rear brakes must be properly adjusted before proceeding with the parking brake adjustment.

1. Block the drive wheels.

2. Raise and safely support the rear axle with jackstands.

3. Loosen the equalize nut.

4. Fully release the parking brake pedal.

5. Tighten the equalizer nut until the rear wheels will not rotate without excessive force in a forward direction.

6. Loosen the equalizer nut until there is little or no drag when the rear wheels are rotated in a forward direction.

7. Carefully lower the vehicle.

8. Remove the blocks from the front wheels.

Astro and Safari

1993–95 Vehicles

The parking brake must be adjusted any time the parking brake cables have been replaced or disconnected, or if under heavy foot pressure the pedal travel is 5 to 8 clicks. The rear brakes must be adjusted properly

before proceeding with the parking brake adjustment.

1. Block the front wheels.

2. Raise and safely support the rear axle with safety stands.

3. Loosen the equalizer. Adjust the equalizer until the brake pedal effort gauge (J 28662) shows 100 lbs. at 10 clicks.

4. Release the parking brake and rotate the rear wheels. There should be no brake drag in either direction.

5. Lower the vehicle, then unblock the front wheels.

1996–97 vehicles

The parking brake must be adjusted any time the parking brake cables have been replaced or disconnected, or if parking brake holding ability is inadequate. The rear brakes must be properly adjusted before proceeding with the parking brake adjustment.

1. Tighten the adjuster nut on the equalizer, periodically turning the rear wheels by hand, until the brake shoes begin to contact the brake drums. A rubbing noise from the brake drums should be heard when this occurs.

2. Fully apply and release the parking brake lever to the floor three times.

3. Repeat Step 1.

4. Loosen the adjuster nut on the equalizer two revolutions. There should be no brake drag in either direction.

5. Carefully lower the vehicle.

REMOVAL AND INSTALLATION

Except Astro and Safari

Front

1. Raise and support the vehicle safely.
2. Loosen the equalizer nut.
3. Remove the cable retainer.
4. Remove the lower trim panel.
5. Detach the cable from the lever assembly.
6. Remove the front cable. Attach a piece of wire to the cable to help during installation.

To install:

7. Installation is the reverse of the removal procedure.
8. Make sure all retaining fingers are completely through the hole.
9. Tighten the bolts to 71 inch lbs. (8 Nm).
10. Adjust the parking brake.

Rear

1. Raise and support the vehicle safely.
2. Loosen the nut at the equalizer. Detach the cable connector.
3. Remove the tire and wheel assembly. Remove the brake drum and lining.
4. Detach the retainer at the backing plate.
5. Remove the retainer.
6. Unfasten the bolt and clip, then remove the rear cable.
7. Installation is the reverse of the removal procedure. Adjust the rear brakes, as required. Adjust the parking brake.

Astro and Safari

Front — 1993–95 Vehicles

1. Raise and support the vehicle safely.
2. Loosen the equalizer nut and detach the front cable from the connector.
3. Remove the retaining bolts and clips.
4. Bend the retaining fingers and remove the grommet.
5. Remove the fuse box cover.
6. Detach the cable from the lever assembly.,

7. Remove the front cable. Attach a piece of wire to the cable to help during installation.

To install:

8. Installation is the reverse of the removal procedure.
9. Make sure all retaining fingers are completely through the hole.
10. Tighten the bolts to 71 inch lbs. (8 Nm).
11. Adjust the parking brake.

Rear — 1993–95 Vehicles

1. Raise and support the vehicle safely.
2. Remove the spring.
3. Loosen the equalizer and disconnect the cable at the center retainer.
4. Remove the brake drum and shoe assembly. Bend in the cable retaining fingers, then remove the rear cable.
5. Installation is the reverse of the removal procedure. Adjust the rear brakes, as required. Adjust the parking brake.

Front — 1996–97 Vehicles

1. Raise and safely support the vehicle with safety stands.
2. Loosen the equalizer.
3. Detach the front cable from the connector at the left rear cable.
4. Disconnect the release cable from the lever assembly.
5. Remove the cowl side panel trim.
6. Remove the lever assembly. Detach the cable from the lever assembly.
7. Unfasten the front four screws on the step trim plate. Lift the rocker panel trim and peel back the carpet to expose the parking brake cable grommet.
8. Unclip the front cable and remove by pulling it through the body.
9. Installation is the reverse of the removal procedure.

Rear — 1996–97 Vehicles

1. Raise and safely support the vehicle.
2. Remove the spring from the equalizer, then loosen the equalizer.
3. Detach the connector at the front cable.
4. Remove the brake drum and shoe assembly.
5. Bend in the cable retaining fingers at the backing plate, then remove the rear cable.
6. Installation is the reverse of the removal procedure.

Bleeding the Hydraulic Brake System

The hydraulic brake system must be bled any time a line is disconnected or any time air enters the system. If a point in the system, such as a wheel cylinder or caliper brake line is the only point which was opened, the bleeder screws down stream in the hydraulic system are the only ones which must be bled. If however, the master cylinder fittings are opened or if the reservoir level drops sufficiently that air is drawn into the system, air must be bled from the entire hydraulic system. If the brake pedal feels spongy upon application, and goes almost to the floor but regains height when pumped, air has entered the system. It must be bled out. If no fittings were recently opened for service, check for leaks that would have allowed the entry of air and repair them before attempting to bleed the system.

As a general rule, once the master cylinder (and the brake pressure modulator valve or combination valve on ABS systems) is bled, the remainder of the hydraulic system should be bled starting at the furthest wheel from the master cylinder and working towards the nearest wheel. Therefore, the correct bleeding sequence is: master cylinder, modulator or combination valve (ABS only), right rear wheel cylinder, left rear, right front caliper and left front. Most master cylinder assemblies on these vehicles are not equipped with bleeder valves, therefore air must be bled from the cylinders using the front brake pipe connections.

MANUAL BLEEDING

1. Clean the top of the master cylinder, remove the cover and fill the reservoirs with clean fluid. To prevent squirting fluid, and possibly damaging painted surfaces, install the cover during the procedure, but be sure to frequently check and top off the reservoirs with fresh fluid.

—————— **CAUTION** ——————
Never reuse brake fluid which has been bled from the system.

2. The master cylinder must be bled first if it is suspected to contain air. If the master cylinder was removed and bench bled before installation it must still be bled, but it

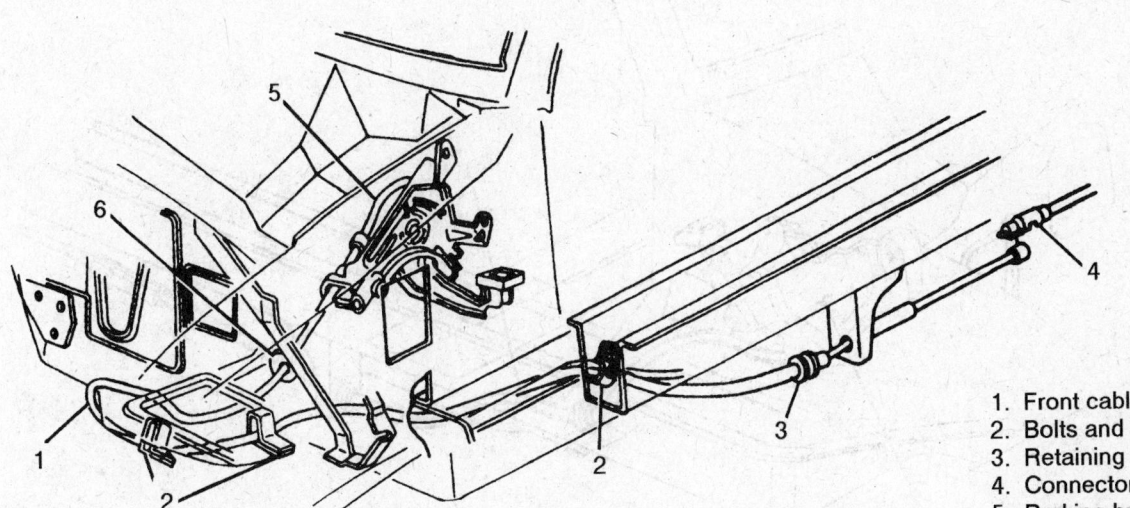

1. Front cable
2. Bolts and clips
3. Retaining fingers
4. Connector
5. Parking brake lever
6. Grommet

79219gc5

Front parking brake cable routing — 1995 Astro and Safari shown

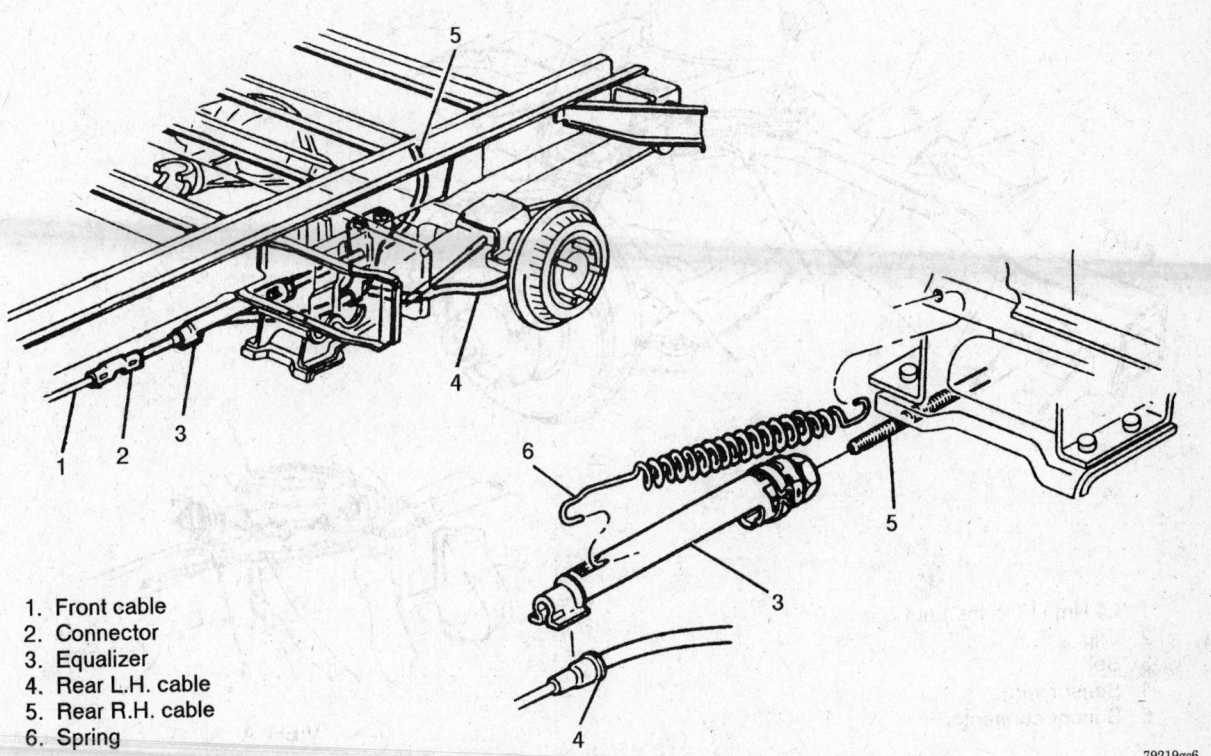

1. Front cable
2. Connector
3. Equalizer
4. Rear L.H. cable
5. Rear R.H. cable
6. Spring

79219gc6

Equalizer and rear cables — 1993–95 Astro and Safari

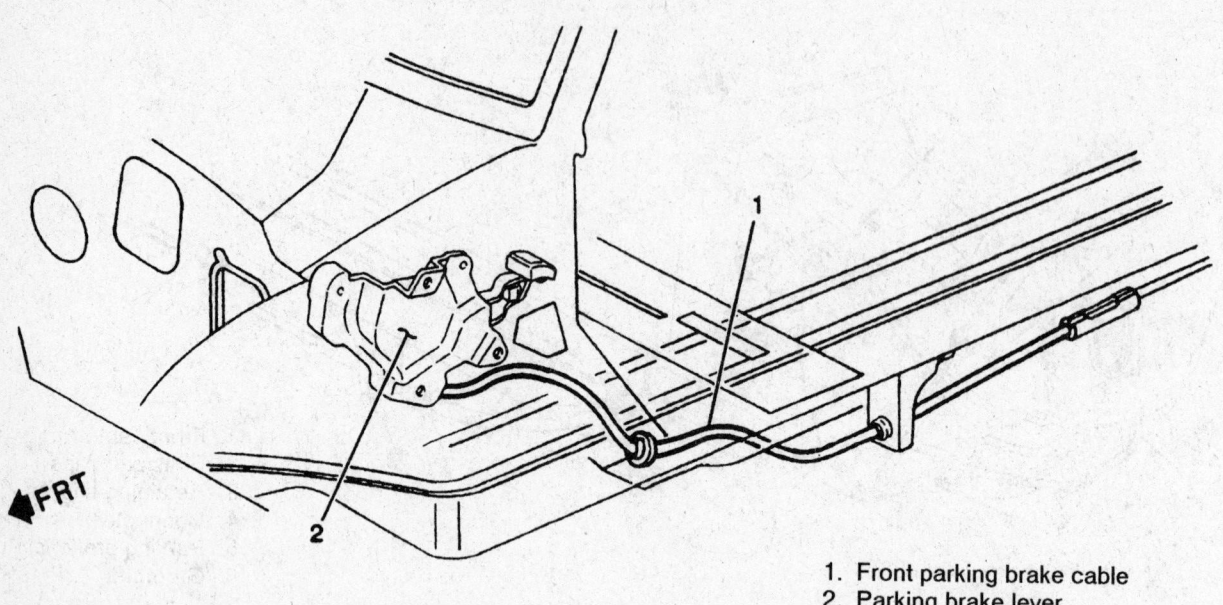

1. Front parking brake cable
2. Parking brake lever

79219gc7

Front parking brake cable routing — 1996 Astro and Safari shown

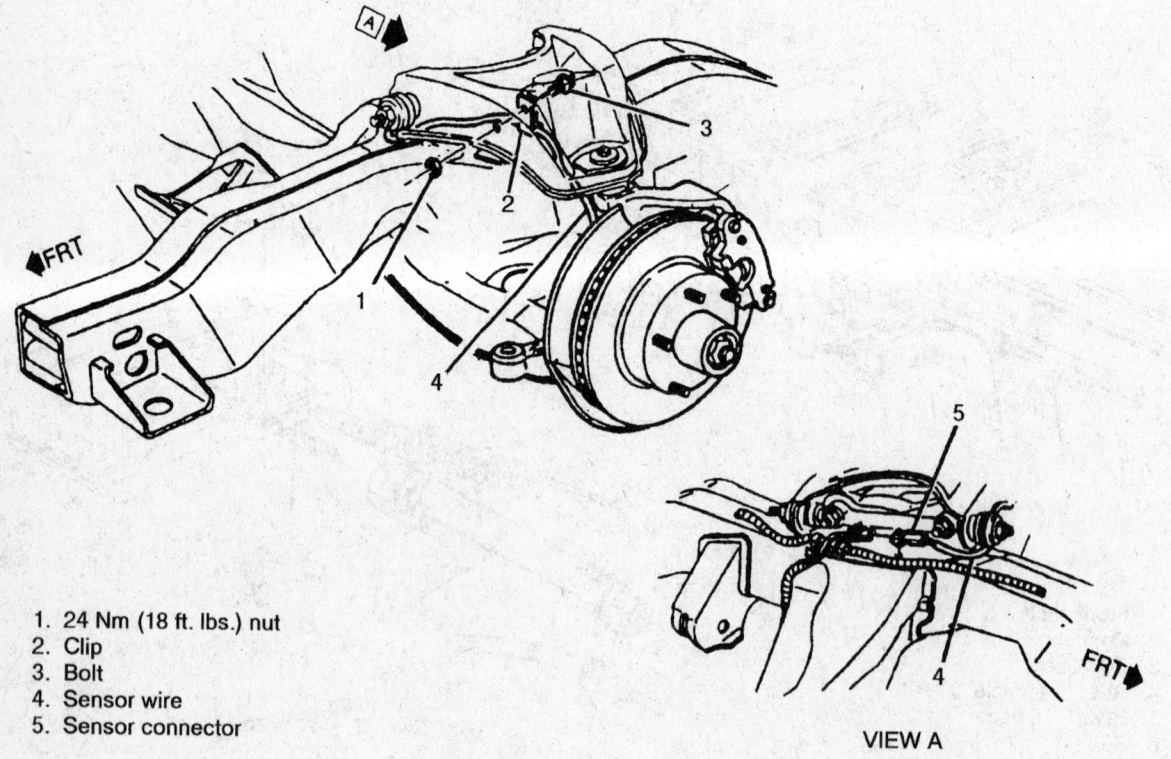

1. 24 Nm (18 ft. lbs.) nut
2. Clip
3. Bolt
4. Sensor wire
5. Sensor connector

VIEW A

79219gc8

Equalizer and rear cables — 1996 Astro and Safari

should take less time and effort. Bleed the master cylinder as follows:

a. Position a container under the master cylinder to catch the brake fluid.

——————— **WARNING** ———————

Do not allow brake fluid to spill on or come in contact with the vehicle's finish as it will remove the paint. In case of a spill, immediately flush the area with water.

b. Loosen the front brake line at the master cylinder and allow the fluid to flow from the front port.

c. Have an assistant depress the brake pedal slowly and hold (air and/or fluid should be expelled from the loose fitting). Tighten the line, then release the brake pedal and wait 15 seconds. Loosen the fitting and repeat until all air is removed from the master cylinder bore.

d. When finished, tighten the line fitting.

e. Repeat the sequence at the master cylinder rear pipe fitting.

NOTE: During the bleeding procedure, make sure your assistant does not release the brake pedal while a fitting is loosened or while a bleeder screw is opening. Air will be drawn back into the system.

3. Check and refill the master cylinder reservoir.

NOTE: Remember, if the reservoir is allowed to empty of fluid during the procedure, air will be drawn into the system and bleeding procedure must be restarted at the master cylinder assembly.

4. On late model ABS equipped vehicles, perform the special ABS procedures. On 4 wheel ABS systems the Brake Pressure Modulator Valve (BPMV) must be bled (if it has been replaced or if it is suspected to contain air) and on most Rear Wheel Anti-Lock (RWAL) systems the combination valve must be held open. In both cases, special combination valve depressor tools should be used during bleeding and a scan tool must be used for ABS function tests.

5. If a single line or fitting was the only hydraulic line disconnected, then only the caliper(s) or wheel cylinder(s) affected by that line must be bled. If the master cylinder required bleeding, then all calipers and wheel cylinders must be bled in the proper sequence:

a. Right rear

b. Left rear
c. Right front
d. Left front

6. Bleed the individual calipers or wheel cylinders as follows:

a. Place a suitable wrench over the bleeder screw and attach a clear plastic hose over the screw end. Be sure the hose is seated snugly on the screw or you may be squirted with brake fluid.

NOTE: Be very careful when bleeding wheel cylinders and brake calipers. The bleeder screws often rust in position and may easily break off if forced. Installing a new bleeder screw will often require removal of the component and may include overhaul or replacement of the wheel cylinder/caliper. To help prevent the possibility of breaking a bleeder screw, spray it with some penetrating oil before attempting to loosen it.

b. Submerge the other end of the tube in a transparent container of clean brake fluid.

c. Loosen the bleed screw, then have a friend apply the brake pedal slowly and hold. Tighten the bleed screw, release the brake pedal and wait 15 seconds. Repeat the sequence (including the 15 second pause) until all air is expelled from the caliper or cylinder.

d. Tighten the bleed screw when finished.

7. Check the pedal for a hard feeling with the engine not running. If the pedal is soft, repeat the bleeding procedure until a firm pedal is obtained.

8. If the brake warning light is on, depress the brake pedal firmly. If there is no air in the system, the light will go out.

9. After bleeding, make sure a firm pedal is achieved before attempting to move the vehicle.

PRESSURE BLEEDING

A proper pressure bleeder tool will utilize a rubber diaphragm between the air source and brake fluid in order to prevent air, moisture oil and other contaminants from entering the hydraulic system.

1. Prepare a pressure bleeder tool such as J-29567 or equivalent by making sure the pressure tank is at least $2/3$ full of fresh, clean brake fluid. In most cases, the bleeder must be bled each time fluid is added. Charge the bleeder tool to 20–25 psi (140–170 kPa).

2. Install a suitable combination valve depressor tool such as J-39177 or equivalent, to the combination valve in order to hold the valve open during the bleeding operation.

3. Install the pressure bleeder tool to the master cylinder reservoir.

4. On 4 wheel ABS systems, bleed the Brake Pressure Modulator Valve (BPMV) of air.

5. Bleed each wheel cylinder or caliper in the proper sequence:

a. Right rear
b. Left rear
c. Right front
d. Left front

6. Connect a hose from the bleeder tank to the adapter at the master cylinder, then open the tank valve.

7. Attach a clear vinyl hose to the brake bleeder screw, then immerse the opposite end into a container partially filled with clean brake fluid.

8. Open the bleeder screw $3/4$ turn and allow the fluid to flow until no air bubbles are seen in the fluid, then close the bleeder screw and tighten.

9. Repeat the bleeding process at each wheel.

10. Inspect the brake pedal for sponginess and if necessary, repeat the entire bleeding procedure.

11. Remove the depressor tool from the combination valve and the bleeder adapter from the master cylinder.

12. Refill the master cylinder to the proper level with brake fluid.

13. Do not attempt to move the vehicle unless a firm brake pedal is obtained.

Bleeding the RWAL Brake System

The use of a power bleeder is recommended, but the system may also be bled manually. If a power bleeder is used, it must be of the diaphragm type and provide isolation of the fluid from air and moisture.

Do not pump the pedal rapidly when bleeding; this can make the circuits very difficult to bleed. Instead, press the brake pedal slowly 1 time and hold it down while bleeding takes place. Tighten the bleeder screw, release the pedal and wait 15 seconds before repeating the sequence. Because of the length of the brake lines and other factors, it may take 10 or more repetitions of the sequence to bleed each line properly. When necessary to bleed all 4 wheels, the correct order is right rear, left rear, right front and left front.

PROCEDURE

─── **CAUTION** ───

Do not move the vehicle until a firm brake pedal is achieved. Failure to properly bleed the system may cause impaired braking and the possibility of injury and/or property damage

1. Make sure the ignition is in the **OFF** position to prevent setting false trouble codes.
2. After properly bleeding the master cylinder, install J-39177 or equivalent combination valve depressor tool to the combination valve. This tool is used to hold the internal valve open allowing the entire system to be completely bled.
3. Recheck the master cylinder fluid level and add, as necessary.
4. Bleed the wheel cylinders as described earlier in this section.
5. Attach the Tech-1 or equivalent scan tool, then perform 3 RWAL function tests.
6. Re-bleed the rear wheel cylinders.
7. Check for a firm brake pedal, if necessary repeat the entire bleeding procedure.

Bleeding the 4WAL Brake System

The EHCU/BPMV module is the 1 component which adds to the complexity of bleeding the 4WAL brake systems. For the most part the system is bled in the same manner as the non-ABS vehicles. But because of the EHCU/BPMV's complex internal valving additional steps are necessary if the unit has been replaced or if it is suspected to contain air. These bleeding steps are not necessary if the only connection/fitting(s) opened were downstream of the unit. These steps may or may not be necessary after master cylinder replacement. If in doubt (or without the necessary special tools) thoroughly bleed the system and see if a firm brake pedal can be obtained, if not, the EHCU/BPMV must be bled as well.

As with the RWAL brake system, the use of a power bleeder is recommended, but the system may also be bled manually. If a power bleeder is used, it must be of the diaphragm type and provide isolation of the fluid from air and moisture.

Do not pump the pedal rapidly when bleeding; this can make the circuits very difficult to bleed. Instead, press the brake pedal slowly 1 time

and hold it down while bleeding takes place. Tighten the bleeder screw, release the pedal and wait 15 seconds before repeating the sequence. Because of the length of the brake lines and other factors, it may take 10 or more repetitions of the sequence to bleed each line properly. When necessary to bleed all 4 wheels, the correct order is right rear, left rear, right front and left front.

─── **CAUTION** ───

Do not move the vehicle until a firm brake pedal is achieved. Failure to properly bleed the system may cause impaired braking and the possibility of injury and/or property damage

If the EHCU/BPMV requires bleeding, the following procedures may be used to free all trapped air from the component. The procedures differ because the component used on the MFI-Turbo is equipped with external bleeders in additional to the internal bleeders found on the 1994 and later units. In either case, 3 combination valve depressor tools and a scan tool are required. The combination valve depressor tools are used to hold the internal passages (combination valve and EHCU/BPMV accumulator bleed stems open allowing the entire system to be completely bled.

Finally, remember to always bleed the 4WAL brake system with the ignition **OFF** to prevent setting false trouble codes.

PROCEDURE

MFI-Turbo

The EHCU used on the MFI-Turbo is equipped with the internal bleeders AND a pair of external bleeder screws. These external bleeders look like normal brake bleeders and are found on top of the unit. Like any bleeder screw, they MUST remain closed when the unit is not pressurized.

The Internal Bleed Valves on either side of the unit must be opened ¼–½ turn before bleeding begins. These valves open internal passages within the unit. The valve located on the left side (nearest the fender) is used for the rear brake section, while the valve on the right (nearest the engine) is used for the front brakes. Actual bleeding is performed at the 2 bleeders on the top of the EHCU module. The bleeders must not be opened when the system is not pressurized. The ignition switch must be

OFF or false trouble codes may be set.

1. Make sure the ignition is in the **OFF** position to prevent setting false trouble codes.
2. Open the internal bleed valves ¼–½ turn each.
3. Install J-35856 or equivalent combination valve depressor tool on the left accumulator bleed stem of the EHCU. Install 1 tool on the right bleed stem and install the 3rd tool on the combination valve (rear).
4. Inspect the fluid level in the master cylinder, filling if needed.
5. Have an assistant slowly depress the brake pedal and hold it down.
6. Open the left bleeder on top of the unit. Allow fluid to flow until no air is seen or until the brake pedal bottoms.
7. Close the left bleeder, then have your assistant release the pedal slowly and wait 15 seconds.
8. Repeat these steps starting with depressing the brake pedal (including the 15 second pause), until no air is seen in the fluid.
9. Tighten the left internal bleed valve to 60 inch lbs. (7 Nm).
10. Bleed air from the right bleeder screw on top of the EHCU in the same manner as the left screw.
11. When bleeding of the right port is complete, tighten the right internal bleed valve to 60 inch lbs. (7 Nm).
12. Remove the 3 special combination valve tools.
13. Check the master cylinder fluid level, refilling as necessary.
14. Bleed the individual brake circuits at each wheel.
15. Switch the ignition **ON** and use the hand scanner tool to perform 3 function tests on the system.
16. Evaluate the brake pedal feel and repeat the bleeding procedure if it is not firm.
17. Carefully test drive the vehicle at moderate speeds; check for proper pedal feel and brake operation. If any problem is noted in feel or function, repeat the entire bleeding procedure.

Except MFI-Turbo

Unlike the The EHCU/BPMV used on the MFI-Turbo, the component used on other 4WAL systems is usually not equipped with external bleeder screws. Therefore, the unit can only be bled through the downstream bleeder screws (wheel cylinders/calipers). To accomplish this the internal bleeder and the accumulator stems/combination valves must be opened to allow air/fluid to pass through the unit. The Internal Bleed

Valves on either side of the unit must be opened $\frac{1}{4}$–$\frac{1}{2}$ turn before bleeding begins. As with most ABS systems found on these vehicles, the ignition switch must be **OFF** or false trouble codes may be set.

1. Make sure the ignition is in the **OFF** position to prevent setting false trouble codes.

2. If necessary, properly bleed the master cylinder assembly. Check and add additional fluid, as necessary.

3. Open the internal bleed valves $\frac{1}{4}$–$\frac{1}{2}$ turn each.

4. Install one J–39177 or equivalent combination valve depressor tool on the left accumulator bleed stem of the EHCU. Install 1 tool on the right accumulator bleed stem and install the 3rd tool on the combination valve.

5. Properly bleed the wheel cylinders and calipers.

6. Remove the 3 special tools.

7. Check the master cylinder fluid level, refilling as necessary.

8. Switch the ignition **ON** (engine not running) and use a hand scanner to perform 6 function tests on the system.

9. Repeat the wheel cylinder and caliper bleeding procedure to remove all air that was purged from the BPMV during the function tests.

10. Check for a firm brake pedal. If necessary, repeat the entire procedure until a firm pedal is obtained.

11. Carefully test drive the vehicle at moderate speeds; check for proper pedal feel and brake operation. If any problem is noted in feel or function, repeat the entire bleeding procedure.

Front Wheel Speed Sensor

REMOVAL AND INSTALLATION

4WAL System

2WD

1. Disconnect the negative battery cable.

2. Raise and support the vehicle safely.

3. Remove the wheel and tire assembly.

4. Remove the brake caliper, hub and rotor assembly.

5. Remove the sensor wire from the clips on the upper control arm and disconnect the connector.

6. Remove the sensor and backing plate attaching bolts and remove the assembly.

7. Installation is the reverse of the removal procedure.

4WD

1. Disconnect the negative battery cable.

2. Raise and support the vehicle safely.

3. Remove the wheel and tire assembly.

4. Remove the hub and rotor assembly.

5. Disconnect the sensor wire connector.

6. Remove the bolts securing the sensor and sensor wire.

7. Remove the sensor from the spindle.

8. Installation is the reverse of the removal procedure.

FRONT SUSPENSION

Shock Absorbers

REMOVAL AND INSTALLATION

Except Astro and Safari

2WD Vehicles

1. Raise and safely support the vehicle with jackstands.

2. Remove the wheel and wire assembly.

3. Remove the nut. Hold the shock absorber stem with a wrench while backing the nut off.

4. Unfasten the retainer and remove the grommet.

5. Remove the bolts. Pull the shock absorber out from below. Lower grommet and retainer are on the stem. Replace the parts, as necessary.

6. Remove the nuts, if damaged or worn.

To install:

7. Install the retainer and grommet on the stem. Fully extend the stem.

8. Maneuver the shock absorber up through the lower control arm and spring. Insert the stem end through the hold in the upper control arm frame bracket.

9. Fasten the grommet and retainer to the stem.

10. Install the nut and tighten to 54 ft. lbs. (73 Nm).

11. If needed, install new nuts.

12. Install the bolt through the pivot holes to the lower control arm holes. Tighten to 54 ft. lbs. (73 Nm).

13. Install the wheel and tire assembly, then carefully lower the vehicle.

4WD Vehicles

1. Raise and safely support the vehicle.

2. Remove the wheel and tire assembly.

3. Remove the lower nut and bolt, then collapse the shock absorber.

4. Unfasten the upper nut and bolt, then remove the shock absorber from the vehicle.

To install:

5. Position the shock absorber to the bracket.

6. Install the bolts, the install the nuts. Tighten the nuts to 54 ft. lbs. (73 Nm).

7. Install the tire and wheel assembly, then carefully lower the vehicle.

Astro and Safari

1. Raise and safely support the vehicle. Support the lower control arm.

2. Remove the tire and wheel assembly and inner wheel well splash shield.

3. Remove the lower nut, washer and bolt.

4. Remove the upper nut, washer and bolt. Collapse the shock absorber.

5. Remove the shock absorber from the vehicle.

6. Installation is the reverse of the removal procedure.

Coil Springs

REMOVAL AND INSTALLATION

2WD Vehicle

1. Raise and support the vehicle safely. Remove the wheel and tire assembly. Remove the shock absorber lower retaining bolts.

2. Push the shock absorber through the control arm and into the spring.

3. With the vehicle supported so the control arms hang free, install tool J–23028 or equivalent, onto a support and into the lower control arm bushings. Remove the stabilizer bar from the control arm.

4. Remove the stabilizer to lower control arm attachment. Raise and remove the tension on the lower control arm bolts.

5. Install a safety chain around the spring and through the lower control arm. Remove the lower control arm rear pivot bolt, than remove the other pivot bolt.

6. Lower and allow the lower control arm to hang free. Remove the spring assembly.

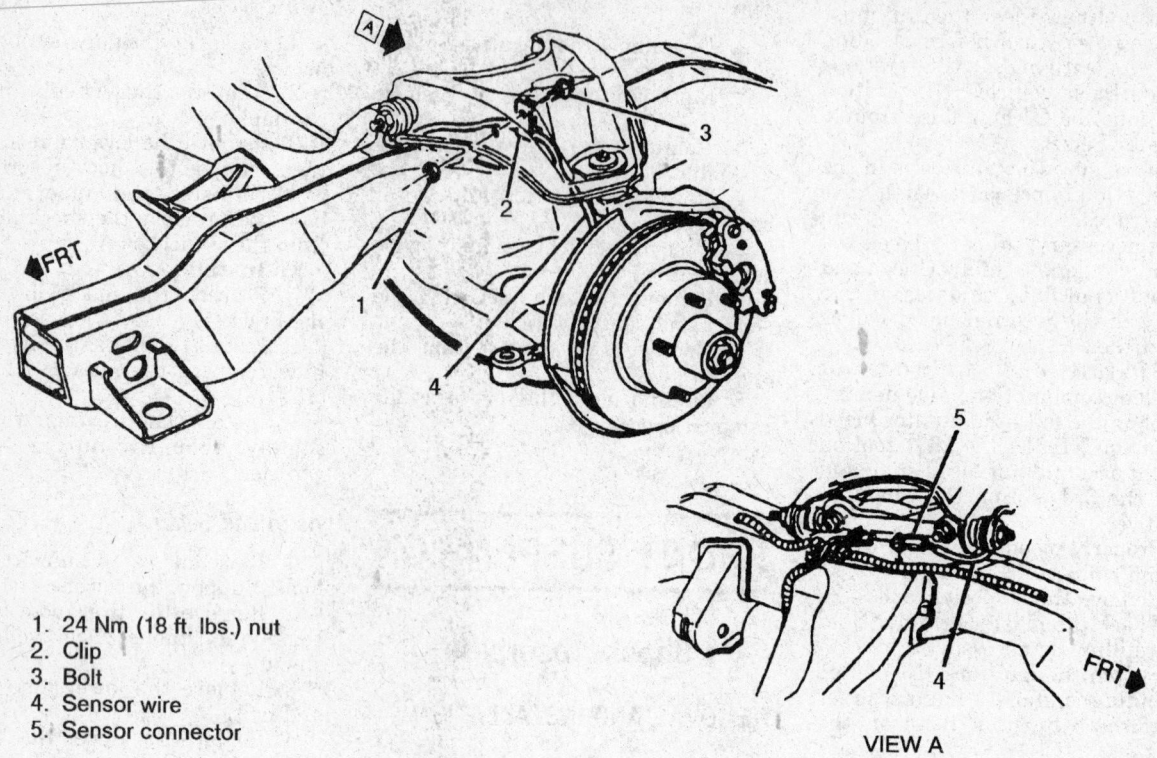

1. 24 Nm (18 ft. lbs.) nut
2. Clip
3. Bolt
4. Sensor wire
5. Sensor connector

VIEW A

79219gc8

Front wheel speed sensor — 1993 2WD shown

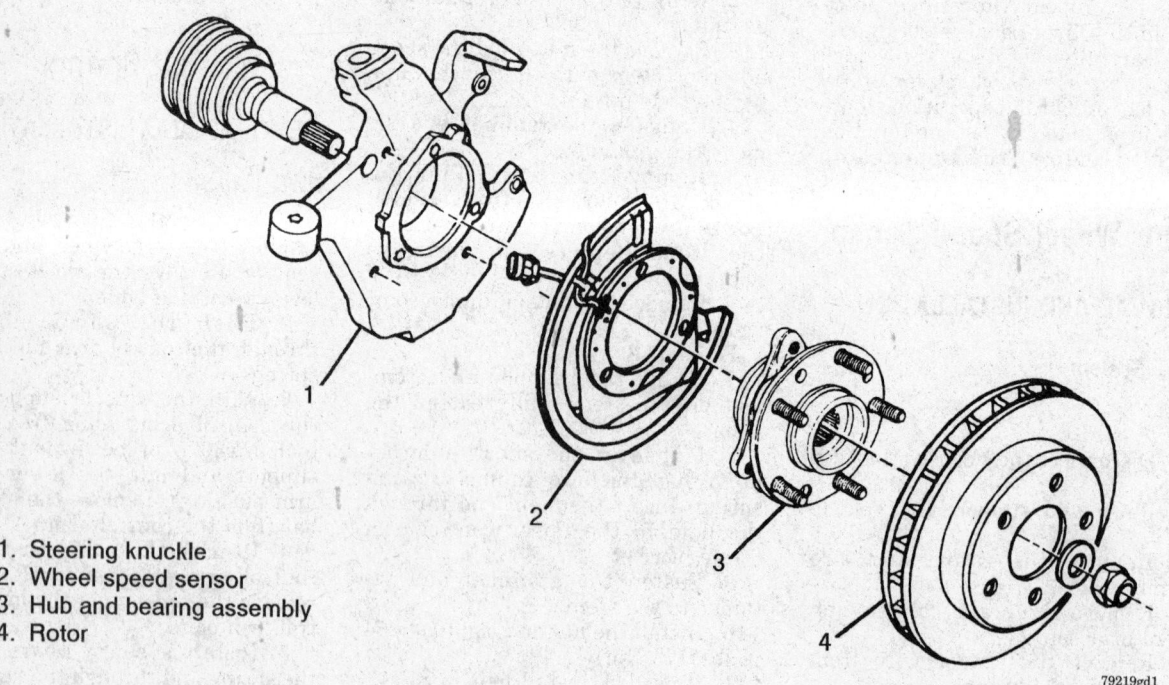

1. Steering knuckle
2. Wheel speed sensor
3. Hub and bearing assembly
4. Rotor

79219gd1

Front wheel speed sensor — 1993 4WD shown

1. 26 Nm (19 ft. lbs.) bolts
2. 16 Nm (12 ft. lbs.) bolt
3. Nut
4. Steering knuckle
5. Gasket
6. Splash shield
7. Wheel speed sensor
8. Rotor

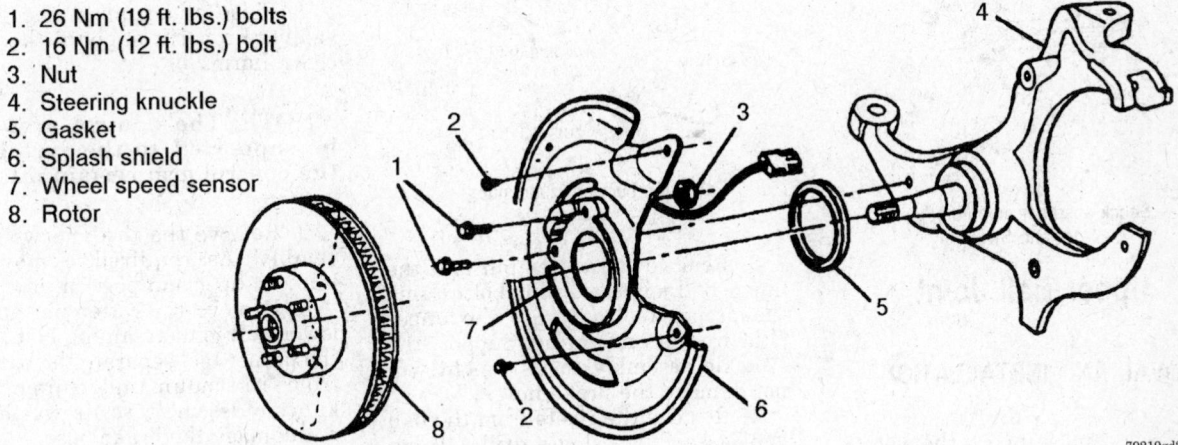

79219gd2

Front wheel speed sensor — 1996 vehicle shown

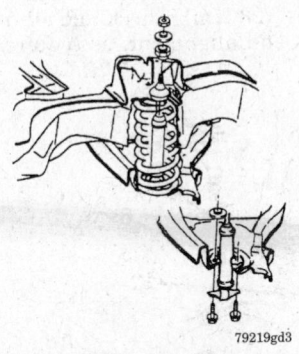

79219gd3

Shock absorber attachment — 1996 2WD Except Astro and Safari

7. Installation is the reverse of the removal procedure. When positioning the spring in the lower control arm, be sure the spring insulator is in the proper position before lifting the control arm in place.

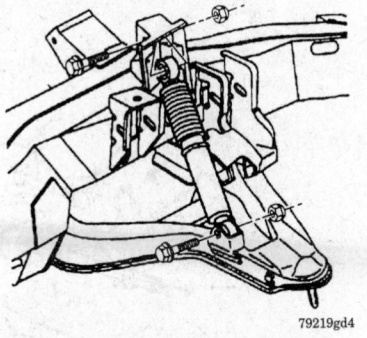

79219gd4

Shock absorber attachment — 1996 4WD Except Astro and Safari

Torsion Bars and Support

REMOVAL AND INSTALLATION

1. Raise and safely support the vehicle. Remove the wheel and tire assemblies.
2. If necessary, remove the transmission shield.
3. Remove the torsion bar adjusting bolt using tool J–36202 or equivalent. Count the number of tool turns required to remove the bolt.
4. Remove the torsion bar support retainer plate and insulator.

5. Remove the torsion bar by sliding it forward into the control arm and lowering it.
6. Remove the torsion bar support. Remove the adjusting arm and the adjusting arm bolt.
To install:
7. Install the adjusting arm to the support and loosely install the adjusting bolt.
8. Install the support to the frame and the insulator to the frame end.
9. Install the retainer to the support. Install the retainer mounting bolts and tighten to 26 ft. lbs. (35 Nm).
10. Tighten the center retainer bolt to 25 ft. lbs. (34 Nm).
11. Install the torsion bar to the lower control arm and raise and slide the torsion bar into the adjusting arm. The torsion bar should have 6mm clearance at the support.
12. Attach tool J–36202 to the support and tighten it against the adjusting arm the recorded number of turns.
13. Install the adjusting bolt and turn it in until it contacts the adjusting arm. Remove the tool.
14. If necessary, install the transmission shield.
15. Install the wheel and tire assemblies. Lower the vehicle.

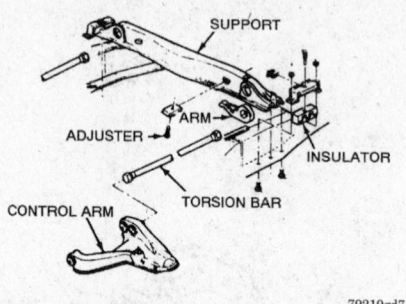

1. Washer
2. Nut
3. Shock absorber
4. Lower arm kit
5. Bolt
6. Bolt

79219gd5

**Shock absorber mounting —
Astro and Safari**

Upper Ball Joint

REMOVAL AND INSTALLATION

1. Raise and support the vehicle safely. Properly support the lower control arm.

NOTE: The control arm must be supported so the spring and the control arm remain intact.

2. Remove the tire and wheel assembly. As required, remove the brake caliper and position it aside.

Torsion bar spring mounting

79219gd7

3. Remove the cotter pin and the upper ball joint retaining bolt. Using the proper tool separate the upper joint from its mounting. Support the knuckle assembly so its weight will not damage the brake hose.

4. Remove the rivets from the ball joint assembly, using a drill with an 1/8 in. and then 1/2 in. bit. Remove the ball joint from the upper control arm.

5. Installation is the reverse of the removal procedure. Tighten the replacement bolts to 17 ft. lbs. (23 Nm). Check and adjust the front end alignment, as required.

Lower Ball Joint

REMOVAL AND INSTALLATION

1. Raise and support the vehicle safely. Properly support the lower control arm.

NOTE: The control arm must be supported so the spring and the control arm remain intact.

2. Remove the tire and wheel assembly. As required, remove the brake caliper and position it aside.

3. Remove the cotter pin and the lower ball joint retaining bolt. Using the proper tool separate the ball joint from its mounting. Support the knuckle assembly so its weight will not damage the brake hose.

4. Remove the rivets from the ball joint assembly, using a drill with a 1/8 in. bit. Remove the ball joint from the control arm.

5. Installation is the reverse of the removal procedure. Be sure to use the nuts and bolts that are supplied with the replacement ball joint assembly. Tighten the replacement bolts to 17 ft. lbs. (23 Nm). Check and adjust the front end alignment, as required.

---

1. Coil spring
2. Lower control arm
3. Upper control arm
4. Upper ball joint
5. Nut 6
6. Cotter pin
7. Lower ball joint
8. Bolt
9. Nut
10. Bolt
11. Insulator
12. Bumper
13. Bushing
14. Bushing
15. Bolt
16. Nut
17. Shaft
18. Nut
19. Shim
20. Nut
21. Retainer
22. Bushing

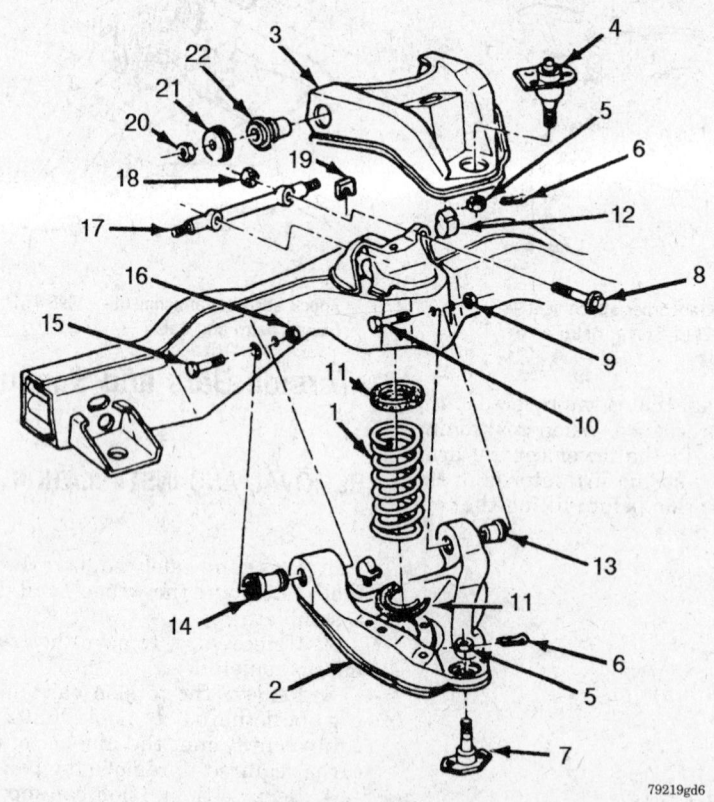

79219gd6

Exploded view of the coil spring and related suspension components

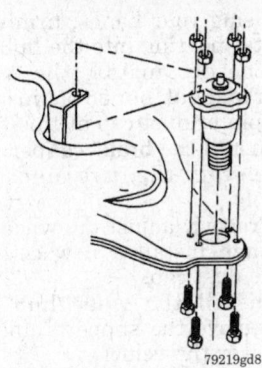

Upper ball joint installation — 1996 2WD shown

Upper Control Arms

REMOVAL AND INSTALLATION

1. Note and record the amount of shims used at the control arm retaining bolts. These shims must be installed in the same location as removed. Remove the nuts and the shims.

2. Raise and support the vehicle safely. Properly support the lower control arm. The control arm must be supported so the spring and the control arm remain intact.

3. Remove the wheel and tire assembly. Separate the upper ball joint from the steering knuckle, using the proper tool. Support the hub assembly.

4. Remove the upper control arm retaining bolts. Remove the upper control arm.

5. Installation is the reverse of the removal procedure. Tighten the upper control arm nuts to 65 ft. lbs. (88 Nm). Check and adjust the front end alignment, as required.

Lower Control Arms

REMOVAL AND INSTALLATION

1. Raise and support the vehicle safely.

2. Remove the wheel and tire assemblies. Properly support the lower control arm assembly.

3. Remove the coil spring.

4. Remove the lower ball joint cotter pin and retaining nut.

5. Using the proper tool, separate the lower ball joint from the steering knuckle.

6. Remove the lower control arm.

7. Installation is the reverse of the removal procedure. Check and adjust front alignment, as required.

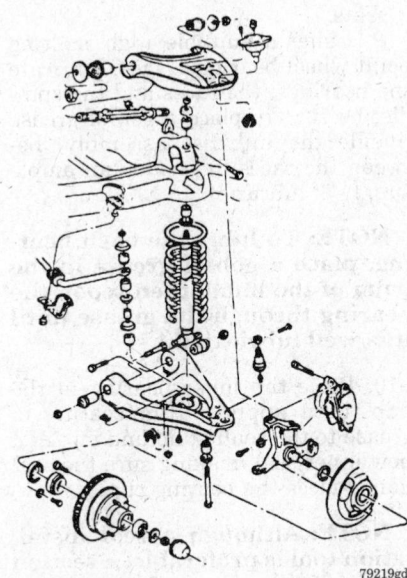

Front suspension components — 2-wheel drive

Stabilizer Shaft/Sway Bar

REMOVAL AND INSTALLATION

1. Raise and safely support the vehicle.

2. Remove the wheel and tire assembly.

3. Remove the left and right side stabilizer mounting bolts. Keep the sides separate for installation.

4. Remove the center stabilizer insulators and lower the stabilizer.

To install:

5. Install the stabilizer in position in the vehicle. Tighten the left and right mounting bolts to 24 ft. lbs. and the center bushing supports to 35 ft. lbs.

6. Install the wheel and tire assemblies and lower the vehicle.

Front Wheel Bearings

ADJUSTMENT

2WD Models

1. Raise and safely support the vehicle.

2. If equipped, remove the wheel/hub cover for access, then remove the dust cap from the hub.

3. Remove the cotter pin and loosen the spindle nut.

4. Spin the wheel forward by hand and tighten the nut to 12 ft. lbs. (16 Nm) in order to fully seat the bearings and remove any burrs from the threads.

5. Back off the nut until it is just loose, then finger-tighten the nut.

6. Loosen the nut $1/4$–$1/2$ turn until either hole in the spindle lines up with a slot in the nut, then install a new cotter pin. This may appear to be too loose, but it is the proper adjustment.

7. Proper adjustment creates 0.001–0.005 in. (0.025–0.127mm) end-play.

REMOVAL AND INSTALLATION

2WD Vehicles

1. Raise and support the front of the vehicle safely using jackstands.

2. Remove the tire and wheel assembly.

3. Remove the brake caliper mounting bolts and carefully remove the caliper (along with the brake pads) from the rotor. Do not disconnect the brake line; instead wire the caliper out of the way with the line still connected.

4. Carefully pry out the grease cap, then remove the cotter pin, spindle nut, and washer. Remove the hub, being careful not to drop the outer wheel bearings. As the hub is pulled forward, the outer wheel bearings will often fall forward and they may easily be removed at this time.

5. If not done already, remove the outer roller bearing assembly from the hub. The inner bearing assembly will remain in the hub and may be removed from the rear of the hub after prying out the inner seal with a small prybar. Discard the seal after removal.

To install:

6. Clean all parts in solvent and allow to air dry, then check for excessive wear or damage. Inspect all of the parts for scoring, pitting or cracking and replace if necessary.

NOTE: DO NOT remove the bearing races from the hub, unless they show signs of damage.

7. If it is necessary to remove the wheel bearing races, use the GM front bearing race removal tool No. J-29117 or equivalent, to drive the races from the hub/disc assembly. A hammer and drift may be used to drive the races from the hub, but the race removal tool is quicker.

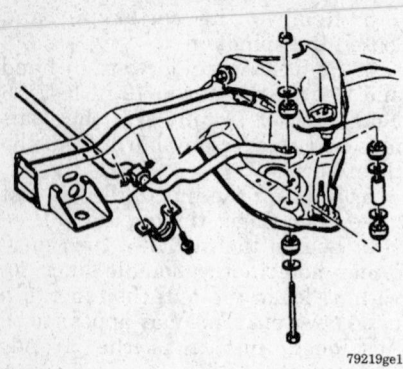

79219ge1

Stabilizer bar mounting — exc. Astro and Safari

8. If the bearing races were removed, position the replacement races in the freezer for a few minutes and then install them to the hub:

a. Lightly lubricate the inside of the hub/disc assembly using wheel bearing grease.

b. Using the GM seal installation tools No. J-8092 and J-8850 or equivalent, drive the inner bearing race into the hub/disc assembly until it seats. Make sure the race is properly seated against the hub shoulder and is not cocked.

NOTE: When installing the bearing races, be sure to support the hub/disc assembly with GM tool No. J-9746-02 or equivalent.

c. Using the GM seal installation tools No. J-8092 and J-8457 or equivalent, drive the outer race into the hub/disc assembly until it seats.

9. Using a suitable high melting point wheel bearing grease, lubricate the bearings, the races and the spindle; be sure to place a gob of grease (inside the hub/disc assembly) between the races to provide an ample supply of lubricant.

NOTE: To lubricate each bearing, place a gob of grease in the palm of the hand, then scoop the bearing through the grease until it is well lubricated.

10. Place the inner bearing in the hub, then apply a thin coating of grease to the sealing lip and install a new inner seal, making sure the seal flange faces the bearing cup.

NOTE: Although a seal installation tool is preferable, a section of pipe with a smooth edge or a suitably sized socket may be used to drive the seal into position. Make sure the seal is flush with the outer surface of the hub assembly.

11. Carefully install the wheel hub over the spindle.

12. Using your hands, firmly press the outer bearing into the hub.

13. Loosely install the spindle washer and nut, but do not install the cotter pin or dust cap at this time.

14. Install the brake caliper.

15. Install the tire and wheel assembly.

16. Properly adjust the wheel bearings, then install a new cotter pin and the dust cap.

17. Install the wheel/hub cover, then remove the supports and carefully lower the vehicle.

4WD Vehicles

1. Raise and support the front of the vehicle safely using jackstands.

2. Properly unload the torsion bar. For details, please refer to the torsion bar procedure located earlier in this section.

3. Remove the tire and wheel assembly.

4. Install an axle shaft boot seal protector to the tri-pot axle joint.

5. At the wheel hub, remove the cotter pin and retainer, then loosen and remove the castle nut and the thrust washer. In order to hold the hub from turning when loosening the nut, insert a drift through the caliper and into the rotor vanes.

A. Hold stabilizer shaft even
   with frame when tightening insulator
1. Insulator
2. Washer
3. Link
4. Bolt
5. Bolt
6. Clamp
7. Insulator
8. Stabilizer shaft
9. Nut

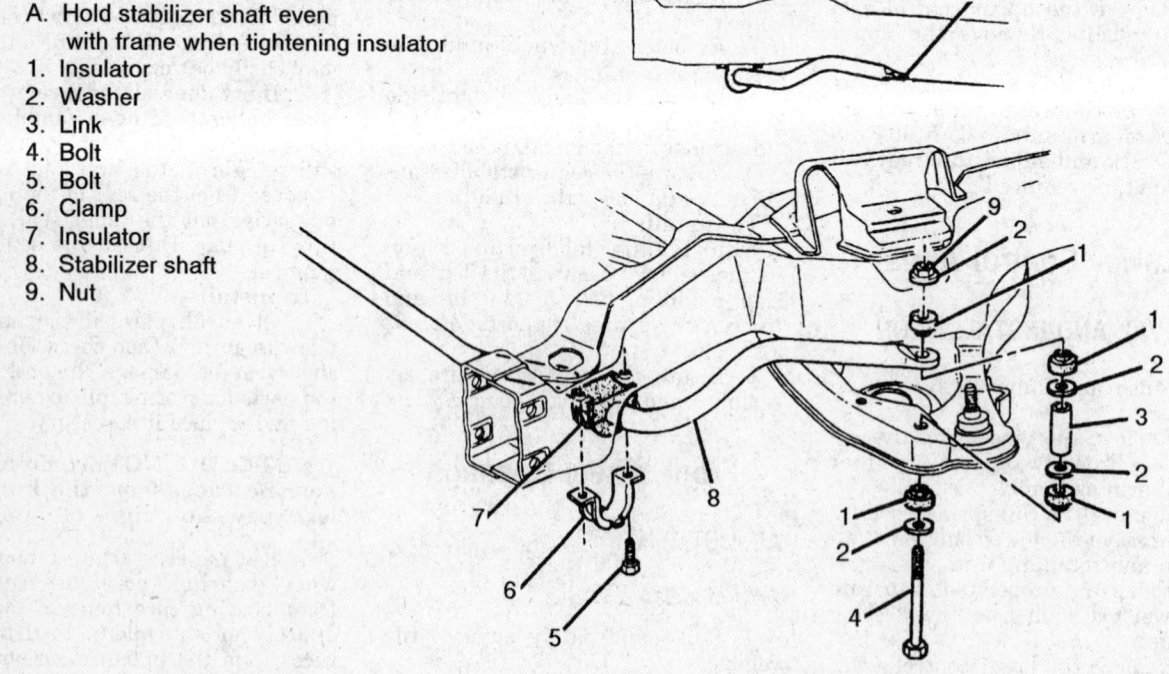

79219ge2

Stabilizer shaft assembly — 2WD Astro and Safari

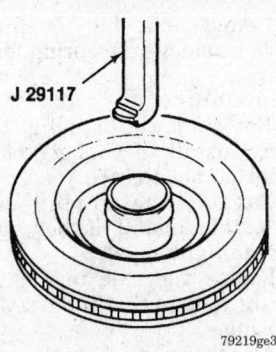

J 29117

79219ge3

Removing the front wheel bearing races — 2WD vehicles

6. Remove the brake caliper and support it aside using wire or a coat hanger. Make sure the brake line is not stretched or damaged.

7. Remove the brake disc from the wheel hub.

8. Remove the bolts retaining the hub/bearing assembly to the knuckle, then carefully pull the assembly from the splined end of the halfshaft. If available, use J-28733-A, or an equivalent spindle remover to prevent damage to the shaft or hub/bearing assembly.

NOTE: When removed, lay the hub and bearing assembly on the hub bolt (outboard) side in order to prevent damage or contamination of the bearing seal.

9. Remove the splash shield.

10. Remove the cotter pin and castle nut from the tie rod end, then separate the end from the knuckle using a suitable steering linkage puller.

11. Remove the cotter pins from the ball joints, then loosen the stud nuts.

12. Use the ball joint separator tool J-36607 or equivalent to loosen the ball joints in the steering knuckle.

13. Remove the ball joint nuts, then separate the ball joints from the knuckle and remove the knuckle from the vehicle.

14. Remove the spacer and the seal from the steering knuckle.

15. Clean and inspect the parts for nicks, scores and/or damage, then replace them as necessary.

To install:

16. Install a new seal into the steering knuckle, using a knuckle seal installation tool such as J-28574 or equivalent.

17. Install the spacer, then position the knuckle and insert the upper and lower ball joints.

18. Install the upper and lower ball joint stud nuts and tighten to specification, then install new cotter pins.

19. Align the splash shield to the knuckle, then install the hub and

bearing assembly, aligning the threaded holes. Install the retaining bolts and tighten to 77 ft. lbs. (105 Nm).

20. Install the tie rod end of the steering knuckle, then secure using the retaining nut and a new cotter pin.

21. Install the brake disc.

22. Reposition and secure the brake caliper.

23. Install the washer and retaining nut to the end of the halfshaft. Insert a brass drift to keep the rotor and hub from turning, then tighten the shaft nut to 180 ft. lbs. (245 Nm).

24. Install the retainer and a new cotter pin, but DO NOT back off specification in order to insert the cotter pin.

25. Remove the torsion bar unloader tool and the drive axle boot protector.

26. Install the tire and wheel assembly.

27. Remove the jackstands and carefully lower the vehicle.

28. Check and/or adjust the vehicle trim height, as necessary.

REAR SUSPENSION

Shock Absorbers

REMOVAL AND INSTALLATION

Except Astro and Safari

1. Raise and support the vehicle safely.

2. Properly support the rear axle assembly.

3. Remove the frame bracket nuts and bolts from the shock absorber.

4. Remove the anchor plate nut and washer from the shock absorber.

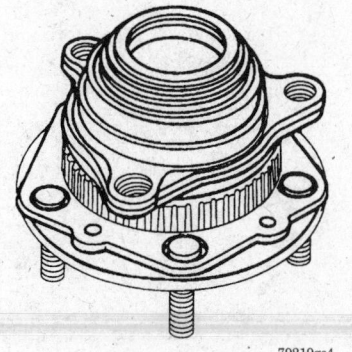

79219ge4

Hub and bearing assembly — 4WD vehicles

5. Remove the shock absorber from the vehicle.

To install:

6. Position the shock absorber to the lower anchor plate. Do not attach the washer or nut.

7. Install the shock absorber-to-frame bracket with the bolts and nuts.

8. Install the lower anchor plate washer and nut and tighten to 62 ft. lbs. (84 Nm).

9. Lower the vehicle.

Astro and Safari

1. Raise and support the vehicle safely.

2. Properly support the rear axle assembly.

3. Remove the top shock absorber nut, washers and bolt.

4. Remove the bottom shock absorber nut and bolt.

5. If necessary, remove the parking brake bracket on the right shock absorber.

6. Remove the shock absorber.

To install:

7. Position the shock to the vehicle.

8. If necessary, install the parking brake bracket on the right shock absorber bracket.

9. Install the bottom and top shock retainers. Tighten the upper and lower nuts to 75 ft. lbs. (100 Nm).

10. Lower the vehicle.

Leaf Springs

REMOVAL AND INSTALLATION

Except Astro and Safari

1. Raise and support the vehicle safely. Properly support the rear axle assembly to relieve tension on the springs.

2. Remove the shock absorbers. Remove the U-bolt nuts, washers, anchor plates and the U-bolts.

3. If equipped, remove the spare tire.

4. Remove the shackle to frame bolt, washers and nut.

5. On pick-ups and 4-door Blazer, Bravada and Jimmy, remove the fuel tank.

6. Remove the spring assembly-to-front bracket nut, washers and bolt.

7. Remove the spring assembly.

8. Remove the shackle-to-spring nut, washers and spring bolt, then remove the shackle.

To install:

9. Position to the shackle to the rear spring eye. Attach the shackle to

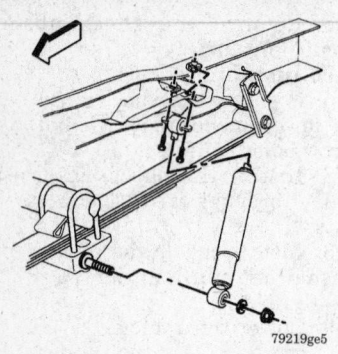

Rear shock absorber mounting — pickup and 2-door Blazer, Bravada and Jimmy

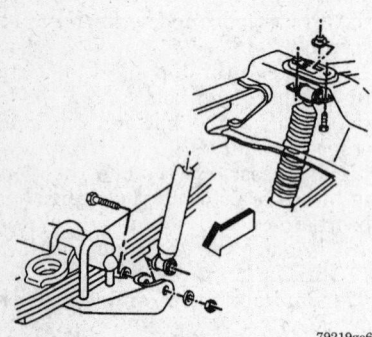

Rear shock absorber mounting — 4-door Blazer, Bravada and Jimmy

the spring bolt with the washers and nut. Do not tighten yet.

10. Place the spring assembly in the vehicle. Attach the spring to the front bracket bolt with the washers and nut. Do not tighten yet.

11. If removed, install the fuel tank.

12. Install the shackle-to-frame bolt, washers and nut. Do not tighten.

13. Remove the spring support. If equipped, install the spare tire.

14. Install the U-bolts, anchor plate, washers and U-bolt nuts.

15. Tighten the U-bolt nuts, in two steps to 18 ft. lbs. (25 Nm), then to 73 ft. lbs. (100 Nm) in the sequence shown in the accompanying figure.

NOTE: Support the axle in such a way that there is a distance of about 6.7 in. (170mm) between the axle tube and the bumper bracket metal surface.

16. Tighten the front bracket nut and the rear shackle nut to 89 ft. lbs. (122 Nm).

17. Install the shock absorber, then carefully lower the vehicle.

Astro and Safari

1. Raise and support the vehicle safely. Properly support the rear axle assembly to relieve tension on the springs.

2. If necessary for access to the lower plate front nut, remove the axle bumper and nut.

3. Remove the U-bolt and lower plate nuts. With a stabilizer shaft, it will be necessary to remove the lower nuts, washers and clamps.

4. Remove the U-bolts, lower plate and anchor plate. Lower the axle away from the spring.

WARNING

Do NOT let the axle hang by the brake hose. Damage to the hose may occur.
---

5. Remove the shackle nut and bolt. Detach the spring from the shackle.

6. Remove the hanger nut and bolt, then remove the spring form the hanger.

To install:

7. Position the spring to the hanger. Install the hanger bolt and nut, but do not tighten yet.

8. Attach the spring to the shackle, then install the bolt and nut, but do no tighten yet.

9. Fasten the axle to the spring. Raise the axle so that it butts against the spring.

10. Attach the anchor plate to the top of the spring. Apply rubber lubricant to the isolator on the spring to aid in installation of the anchor plate to the spring.

11. Install the lower plate and U-bolt around the axle and through the anchor plate. With a stabilizer shaft, it will be necessary to install the clamps, washers and bolts.

12. Attach the nuts to the lower plate and U-bolt. Starting with the inner (lower plate side) nuts, gradually tighten the four nut so that the anchor plate moves uniformly, side to side, over the spring. Tighten the anchor plate nuts to 41 ft. lbs. (56 Nm).

NOTE: After tightening the fasteners to the specified torque, there should be no gap between the anchor plate, axle tube bracket and lower plate. A metal to metal condition should exist.

13. Install the axle bumper and nut. Tighten the axle bumper nut to 33 ft. lbs. (45 Nm). Tighten the hanger nut and bolt and shackle nut and bolt to 74 ft. lbs. (100 Nm).

14. Inspect the rear suspension trim height, it should be 5.3–5.7 in. (135–145mm).

15. Carefully lower the vehicle.

1. Spring bolt
2. Frame bolt
3. Washers
4. Nut
5. Washers
6. Nut
7. Shackle
8. Spring assembly
9. Washer
10. U-bolt nut
11. Anchor plate
12. Nut
13. U-bolts
14. Washer
15. Bolt
16. Bushing

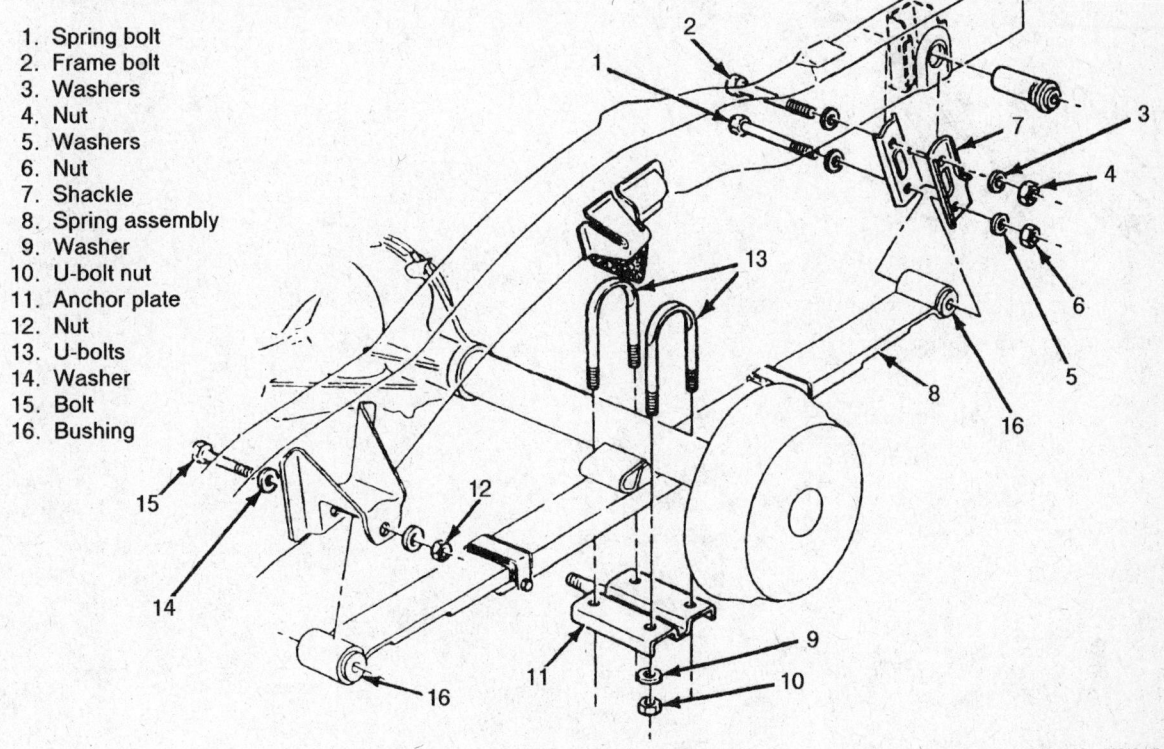

Spring assembly — Blazer, Bravada and Jimmy

79219ge7

1. Spring bolt
2. Frame bolt
3. Washers
4. Nut
5. Washers
6. Nut
7. Shackle
8. Spring assembly
9. Washer
10. U-bolt nut
11. Anchor plate
12. Nut
13. U-bolts
14. Washer
15. Bolt
16. Bushing

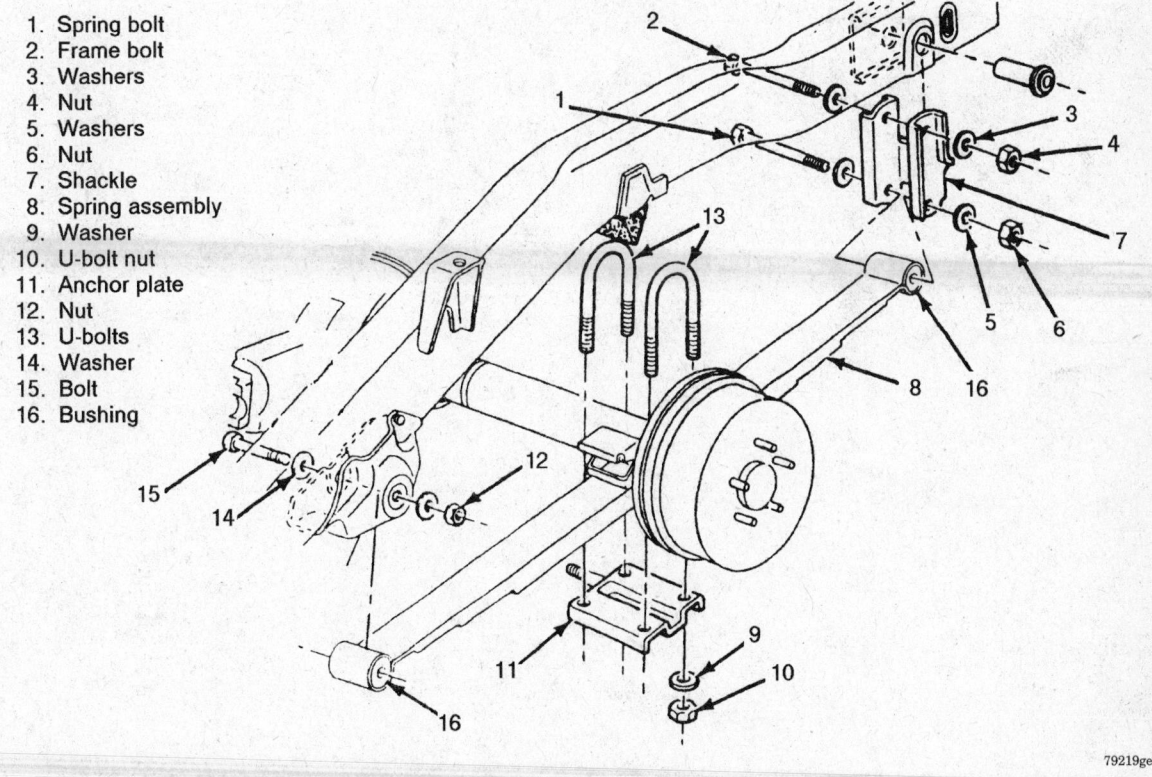

Spring assembly — Pick-up

79219ge8

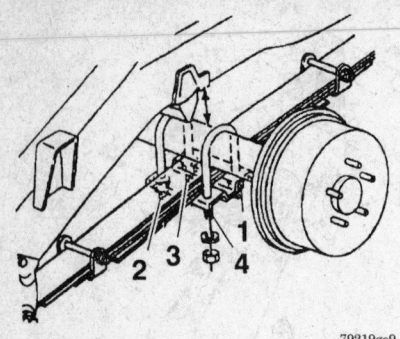

79219ge9

U-bolt tightening sequence — Except Astro and Safari

FIRING ORDERS

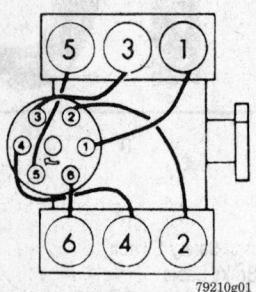

79210g01

4.3L (VIN W and Z) Engines
Engine Firing Order:
1–6–5–4–3–2
Distributor Rotation:
Clockwise

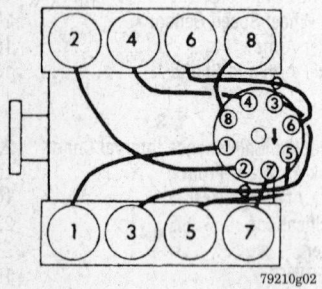

79210g02

5.0L (VIN H and M), 5.7L (VIN K and R), and 7.4L (VIN N and J) Engines
Engine Firing Order:
1–8–4–3–6–5–7–2
Distributor Rotation: Clockwise

ENGINE ELECTRICAL

NOTE: Disconnecting the negative battery cable on some vehicles may interfere with the functions of the on board computer systems and may require the computer to undergo a relearning process, once the negative battery cable is reconnected.

――――― **WARNING** ―――――
NEVER disconnect the negative battery cable with the ignition ON. Removing power from the computer control module with the ignition ON may destroy the module.

Distributor

Most distributor equipped vehicles covered by this manual utilize a commonly recognizable distributor ignition system. However, the 1996–97 4.3L (VIN W), 5.0L (VIN M), 5.7L (VIN R) and 7.4L (VIN J) engines utilize a distributor-like ignition system in which the commonly recognized distributor is replaced by a High Voltage Switch (HVS). The HVS is mounted in a fixed position so the Powertrain Control Module (PCM) may control ignition timing using input from a camshaft position sensor (usually integral to the HVS). Because of this, proper rotor alignment when removing and installing the HVS is even more critical than with the conventional distributor systems.

REMOVAL

1. Disconnect the negative battery cable.
2. Remove all necessary components in order to gain access to the distributor assembly. On most V-type engines it will be necessary to remove the air cleaner assembly for access.
3. Tag and disengage the distributor electrical connectors. If equipped, disconnect the vacuum line.
4. Remove the distributor cap. If necessary, tag and disconnect the spark plug wires.
5. Matchmark the rotor and the distributor body. Matchmark the distributor assembly and the engine block.

NOTE: Although it is only necessary to matchmark the distributor position for ease of installation, it may be advisable to first position the engine at TDC. If the distributor is being removed for further engine repair which may involve rotating the crankshaft (such as to align timing marks), setting the engine to TDC at this time will keep the distributor matchmarks valid during installation.

6. Remove the distributor hold-down bolt. Carefully remove the distributor.

NOTE: As the distributor is removed from the engine, the rotor will turn counterclockwise. Observe and mark the finish position of the rotor. When reinstalling, position the rotor at the last mark and set the distributor into the engine. As the distributor

drops into place, the rotor should turn to its original position, providing the engine crankshaft has not been rotated with the distributor out.

INSTALLATION

Timing Not Disturbed

NOTE: To ensure correct ignition timing if the engine has not been disturbed, the distributor must be installed with the rotor in the same position as when removed.

1. Align the rotor to the last mark made and install the distributor in the engine.

NOTE: On vehicles not equipped with the HVS system, if the distributor shaft cannot drop into the engine, remove the distributor and use a small pry tool through the mounting hole to rotate the oil pump driveshaft until it can align with the distributor gear.

2. As the distributor is fully seated, the rotor should turn and end up at the first mark made. Ensure the distributor and oil pump rod are fully engaged.

NOTE: When the distributor is fully seated on the engine, make sure the rotor is properly aligned with the first timing mark. If equipped with the HVS system, if the rotor dos not align with the mark, the gear teeth of the HVS and camshaft have meshed 1 or more teeth out of time and the HVS procedure for timing disturbed should be used to assure proper installation.

3. Reconnect the distributor cap and wires. If applicable, connect the vacuum line to the distributor.
4. Tighten the distributor hold-down bolt, then install the air cleaner or duct, if removed for access.
5. Check and adjust the ignition timing.

Timing Disturbed

Except 1995–97 HVS Ignition Systems

1. Set the engine to TDC: Remove the No. 1 spark plug. Place a finger over the spark plug hole and rotate the engine in the normal direction of rotation slowly, until compression is felt.
2. Align the timing mark on the crankshaft pulley to the 0 on the engine timing indicator by rotating the

engine in the same direction slowly. The engine is now set on No. 1 TDC.

NOTE: An alternate method may be used to assure the engine is at TDC if the valve cover is removed. Watch the rocker arms for the No. 1 cylinder as the engine is turned. If the valves move as the crankshaft timing marker approaches the scale, the No. 1 cylinder is on its exhaust stroke. If the valves remain closed as the timing mark approaches the scale, then the No. 1 cylinder is approaching TDC of the compression stroke.

3. Install the distributor to the engine so the rotor is pointing to the No. 1 spark plug tower on the distributor cap once the distributor is fully seated in the engine.

4. Install the distributor cap, spark plug, wiring and connectors. If applicable, connect the vacuum line to the distributor.

5. If removed for access, install the air duct and/or air cleaner assembly, as applicable.

6. Check and adjust ignition timing.

1995–97 HVS Ignition Systems

1. Set the engine to TDC: Remove the No. 1 spark plug. Place a finger over the spark plug hole and rotate the engine in the normal direction of rotation slowly, until compression is felt.

2. Align the timing mark on the crankshaft pulley to the **0** on the engine timing indicator by rotating the engine in the same direction slowly. The engine is now set on No. 1 TDC.

NOTE: An alternate method may be used to assure the engine is at TDC if the valve cover is removed. Watch the rocker arms for the No. 1 cylinder as the engine is turned. If the valves move as the crankshaft timing marker approaches the scale, the No. 1 cylinder is on its exhaust stroke. If the valves remain closed as the timing mark approaches the scale, then the No. 1 cylinder is approaching TDC of the compression stroke.

3. Remove the HVS cap screws and cap to expose the rotor.

4. Align the pre-drilled indent hole in the HVS driven gear with the arrow cast into the upper portion of the shaft housing. The rotor should point to the cap hold-down mount nearest the flat side of the housing.

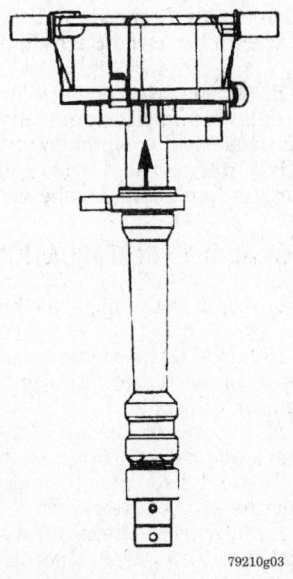

79210g03

Proper HVS alignment — 1995–97 HVS ignition

5. Using a long pry tool, align the oil pump driveshaft in the engine to the mating drive tab in the HVS.

6. Guide the HVS into place, making sure the locating slot in the HVS base fits over the dowel pin in the intake manifold.

7. Once the HVS is FULLY SEATED, the rotor tip should be aligned with the pointer cast into the HVS base. this pointer will have a "6" or an "8" cast into it, indicating the HVS component is designed for use in a 6-cylinder engine. If the rotor tip does not align within a few degrees of the pointer, the gear mesh between the HVS and camshaft is likely off by a tooth or more. If so, repeat the procedure again to achieve proper alignment.

8. Install the cap and mounting screws.

9. Install the HVS mounting clamp and tighten to 20 ft. lbs. (27 Nm).

10. Engage the 3-wire camshaft position sensor connector to the base of the HVS assembly.

11. Connect the spark plug and coil leads to the HVS cap. If a check engine light is illuminated and a Diagnostic Trouble Code (DTC) 1345 is found, either the HVS has been installed incorrectly or an incorrect HVS assembly has been installed.

Ignition Timing

On gasoline engines, the timing may be adjusted only on those that are equipped with a conventional distributor ignition. If equipped with the distributorless Electronic Ignition (EI) system, the control module sets timing and makes all necessary spark changes. On these systems, the crankshaft position sensor is mounted in a fixed position, therefore not allowing for adjustment. The HVS system found on some 1995–96 vehicles is also not adjustable.

NOTE: The 1995 4.3L (VIN Z) and 1996–97 4.3L (VIN W) engines are equipped with a new distributor-like ignition component called the High Voltage Switch (HVS). On the HVS ignition system the Powertrain Control Module (PCM) utilizes a camshaft position sensor to determine spark timing, dwell and firing of the ignition coil. Because of this, positioning of the HVS (and ignition timing) is fixed and non-adjustable. Do not attempt to adjust the ignition timing by rotating the HVS or cross/misfiring may occur.

Connect the timing light and tachometer to the engine according to the tool manufacturers' instructions. Make sure the timing light is connected to the No. 1 spark plug wire. If equipped with a diesel engine, a special timing light and a digital tachometer must be used.

ADJUSTMENT

1. Locate and clean the timing marks on the crankshaft pulley and the front of the timing case cover.

2. Use chalk or white paint to color the mark on the scale that will indicate the correct timing, when aligned with the mark on the pulley or the pointer.

3. Attach a tachometer and a timing light to the engine.

4. On early model vehicles that are not equipped with Electronic Spark Control (EST), disconnect and plug the vacuum lines to the distributor.

5. If equipped with EST, the electronic spark timing must be disabled or bypassed to prevent the control module from advancing timing while attempting to set it. This would obviously lead to an incorrect base timing setting. There are 2 possible methods of disabling the EST system, depending on the type of engine:

6. Using the EST distributor, disengage the timing connector wire. Refer to the Vehicle Emission Control Information (VECI) label for details on the particular engine. Most vehicles are equipped with a single wire timing bypass connector. The bypass wire is normally a tan wire with a black stripe, that breaks out of the wiring harness conduit adjacent to the distributor, but on some later vehicles (1994–97) it may break out of a taped section just below the heater case in the passenger compartment.

7. Start the engine, then check and adjust the idle speed, as necessary.

8. Loosen the distributor lock bolt slightly to permit the distributor to be turned.

9. Adjust the idle to the correct specification.

10. With the timing light aimed at the pulley and the marks on the engine, turn the distributor in the direction of rotor rotation to retard the spark or in the opposite direction of rotor rotation to advance the spark. Align the marks on the pulley and the engine with the flashes of the timing light.

11. Turn the engine **OFF**, tighten the distributor hold-down bolt and recheck the timing.

Alternator

PRECAUTIONS

Several precautions must be observed with alternator equipped vehicles to avoid damage to the unit.

• If the battery is removed for any reason, make sure it is reconnected with the correct polarity. Reversing the battery connections may result in damage to the 1-way rectifiers.

• When utilizing a booster battery as a starting aid, always connect the positive to positive terminals and the negative terminal from the booster battery to a good engine ground on the vehicle being started.

• Never use a fast charger as a booster to start vehicles.

• Disconnect the battery cables when charging the battery with a fast charger.

• Never attempt to polarize the alternator.

• Do not use test lights of more than 12 volts when checking diode continuity.

• Do not short across or ground any of the alternator terminals.

• The polarity of the battery, alternator and regulator must be matched

and considered before making any electrical connections within the system.

• Never separate the alternator on an open circuit. Make sure all connections within the circuit are clean and tight.

• Disconnect the battery ground terminal when performing any service on electrical components.

• Disconnect the battery if arc welding is to be done on the vehicle.

REMOVAL AND INSTALLATION

1. Disconnect the negative battery cable.

2. Remove the necessary components in order to gain access to the alternator assembly.

3. If necessary, remove the air cleaner and/or duct work.

4. Tag and disengage the electrical connectors at the alternator.

5. Remove the alternator belt. If equipped with a serpentine belt, relieve the belt tension, the carefully remove the belt from the alternator pulley.

6. If equipped, remove the alternator brace.

7. Remove the alternator retaining bolts and remove the alternator.

8. Installation is the reverse of the removal procedure. Check and adjust the belt tension, as required.

Drive Belt

REMOVAL AND INSTALLATION

1. Using a ½ in. breaker bar with a socket placed on the tensioner pulley bolt, rotate the tensioner to relieve belt tension.

2. Remove the serpentine belt.

To install:

3. Route the belt over all the pulleys except the tensioner.

4. Place the the breaker bar and socket on the tensioner pulley bolt and rotate the tensioner to the released position.

5. Install the belt and and return the pulley to its original position.

6. Check that the belt is properly seated in each pulley.

Starter

REMOVAL AND INSTALLATION

1. Disconnect the negative battery cable.

2. If equipped, remove any brackets and/or shields.

3. If accessible from above, tag and disconnect the solenoid wiring at this time.

4. Raise and support the vehicle safely.

5. If not done earlier, tag and disconnect the solenoid wiring.

6. As necessary, remove the flywheel cover, the and/or exhaust crossover pipe.

7. Remove the starter mounting bolts and/or retaining nuts, then remove the starter assembly from the vehicle noting the positioning of any shims which may be used.

8. Installation is the reverse of the removal procedure. Install any shims that were removed with the starter and torque the retaining bolts to 35 ft. lbs. (45 Nm). Where equipped, tighten the bracket bolt on diesel engines to 24 ft. lbs. (33 Nm).

9. Once installed, check the flywheel-to-starter pinion clearance with a wire gauge and adjust (using shims), if necessary.

Glow Plugs

TESTING

Inhibit Switch

1. Check the temperature controlled switch to make sure it is closed at low temperatures or open at temperatures above 125°F (52°C).

2. Remove the connector from the inhibit switch when the engine temperature is below 100°F (38°C).

3. Set the ohmmeter on a low range or use a self powered test light.

4. Test across the terminals. The switch should be closed (test light **ON** or a reading of less than 0.1 ohm on the meter).

5. Test terminals to ground with a test light or the ohmmeter on a high range. The light should be **OFF** or the meter show greater than 1.0 mega-ohm.

6. Replace the switch if it tests open across the terminals or if either terminal is closed to the ground.

7. Disconnect the plug from the switch terminals when the engine is above 125°F.

8. Set the ohmmeter on the highest scale or use a self powered test light and test across the terminals. Test across each terminal to ground.

9. The switch should be open (test light **OFF** or high ohm reading of greater than 1 mega-ohm on the meter).

10. Replace the switch if it is closed. Use a socket wrench when installing

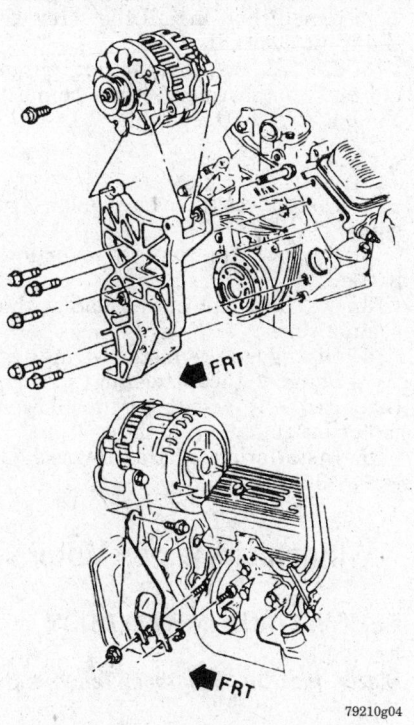

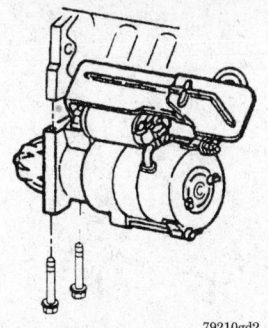

Starter mounting location for the gasoline engine

79210gd2

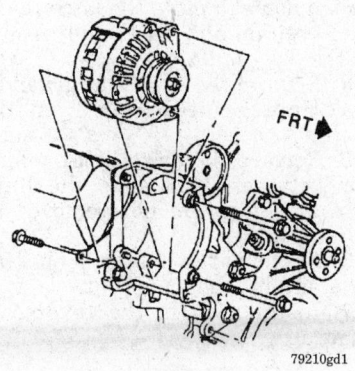

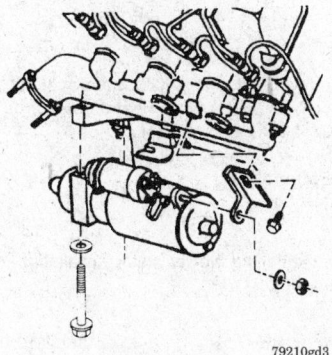

79210gd3

Starter mounting location for the diesel engine

Alternator mounting for the gasoline engine

79210g04

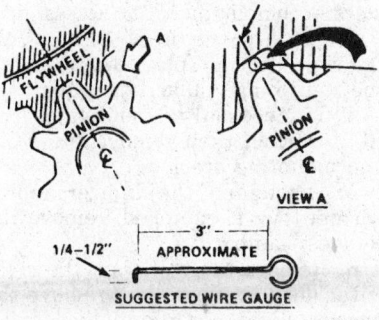

79210g05

Measuring flywheel-to-starter pinion clearance during starter installation

79210gd1

Alternator mounting for the diesel engine

5. If the light works right but the after start glow plug feature does not, replace the controller.

REMOVAL AND INSTALLATION

1. Disconnect the negative battery cables.

2. On the 6.2L engine and the left side only of the 6.5L, disconnect the glow plug lead wires and then remove the plugs. You'll need a $^3/_8$ in. (9.525mm) deep-well socket.

3. On the right side of the 6.5L engine, raise the truck and support it with safety stands. Remove the right front tire.

4. Remove the inner splash shield from the fender well.

5. Remove the lead wire from the plug at the No. 2 cylinder. Remove the lead wires from plugs in the Nos. 4 and 6 cylinders at the harness connectors.

6. Remove the heat shroud for the plug in the No. 4 cylinder. Remove the heat shroud for the plug in the No. 6 cylinder. Slide the shrouds back just far enough to allow access so you can unplug the wires.

7. Remove the plugs in cylinders No. 2, 4 and 6.

8. Reach up under the vehicle and disconnect the lead wire at No. 8. Remove the glow plug. You may find that removing the exhaust down pipe make this a bit easier when working on Nos. 6 and 8.

9. Installation is the reverse of removal. Install all glow plugs and carefully tighten them to 13 ft. lbs. (17 Nm) for the right side on the 6.5L; 17 ft. lbs. (23 Nm) on all others.

---

the switch and torque to 17 ft. lbs. (21 Nm).

Controller

The glow plug controller provides glow plug operation after starting a cold engine.

1. With the engine cold 80°F (27°C), turn the engine control switch to the **RUN** position and let the glow plugs cycle.

2. After 2 minutes of letting the glow plugs cycle, crank the engine for 1 second; it is not important that the engine starts. Return the engine con-

trol switch to **RUN**. The glow plugs should cycle at least once after cranking.

3. If the plugs do not turn on, disconnect the controller connector and check terminal **B** with a grounded 12 volt test light. The light should be **OFF** with the engine control switch in **RUN**, and **ON** when the engine is cranked.

4. If the light does not operate as described, repair a short or open in the engine harness purple wire.

CHASSIS ELECTRICAL

Blower Motor

REMOVAL AND INSTALLATION

Without Air Conditioning

Except G-Series

1. Disconnect the negative battery cable.

2. If necessary, remove the right rear quarter trim panel.

3. Mark the position of the blower motor in relation to its case.

4. Remove the electrical connection at the motor.

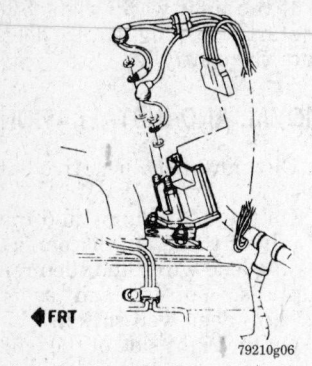

79210g06

Glow plug controller

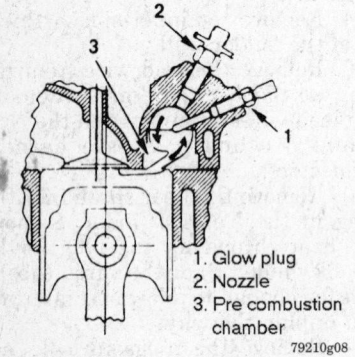

1. Glow plug
2. Nozzle
3. Pre combustion chamber

79210g08

Diesel engine glow plug location

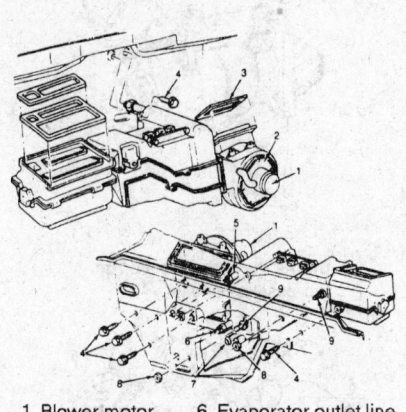

1. Blower motor
2. Screws
3. Gasket
4. Screw
5. Evaporator inlet line
6. Evaporator outlet line
7. Heater core tubes
8. Nut
9. Stud

79210g09

Air conditioning blower motor installation — pick-up

5. Remove the blower attaching screws and remove the assembly.

6. The blower wheel can be removed from the motor shaft by removing the retaining nut.

7. Installation is the reverse of the removal procedure. Apply a bead of sealer to the mounting flange before installation.

G-Series

1. Disconnect the negative battery cable.

2. On 1993 vehicles, remove the retaining screws, then remove the cover for access to the blower motor and wiring.

3. Disengage the necessary wiring.

4. For 1994–97 vehicles, remove the cooling tube.

5. Remove the attaching screws, then carefully withdraw the blower motor assembly.

6. Installation is the reverse of removal procedure.

With Air Conditioning

Blazer, Pick-Up, Suburban, Tahoe and Yukon

1. Disconnect the negative battery cable.

2. Remove the instrument panel storage compartment for access.

3. Remove the front screw from the right door sill plate, then remove the right hinge pillar trim panel.

4. If necessary, disengage the ECM wiring, then remove the ECM and mounting bracket.

5. Disengage the blower motor wiring, then if equipped, remove the courtesy lamp.

6. Remove the bolt from the right lower dash support, then remove the blower motor cover and disconnect the cooling tube.

7. Loosen and remove the motor flange screws, then remove the blower motor pulling forward carefully to prevent damaging the fan. It may be necessary to pry very carefully on the back right side of the instrument panel.

To install:

8. Install the blower motor to the case and secure using the retaining screws.

9. Connect the cooling tube and install the cover.

10. Install the bolt to the right lower dash support, then install the courtesy lamp, if equipped.

11. Engage the blower motor wiring. If applicable, install the ECM and mounting bracket, then engage the wiring.

12. Install the right hinge pillar trim panel, then install the screw to the front door sill plate.

13. Install the instrument panel storage compartment, then connect the negative battery cable.

G-Series

1. Disconnect the negative battery cable.

2. Remove the coolant overflow bottle.

3. On 1993 vehicles, remove the cooling tube.

4. Disengage the motor wiring.

5. Remove the attaching screws, then carefully withdraw the blower motor assembly.

6. Installation is the reverse of removal.

Windshield Wiper Motor

REMOVAL AND INSTALLATION

Blazer, Pick-Up, Suburban, Tahoe and Yukon

1. Disconnect the negative battery cable.

2. Pivot the wiper arm away from the windshield, move the latch to the open position and lift the wiper arm off the driveshaft.

3. Remove the cowl vent grille.

4. Unplug the wiring from the motor.

5. Remove the drive link-to-crank arm retainers and slide the links from the arm. Do not remove the crank arm.

6. Remove the motor mounting bolts and lift the motor out.

7. Installation is the reverse of removal.

G-Series

1. Make sure the wipers are parked. The wiper arms should be in their normal **OFF** position.

2. Disconnect the negative battery cable.

3. Remove the wiper arms.

4. Remove the cowl vent grille.

5. Loosen the nuts holding the transmission linkage, then separate the linkage from the wiper motor crank arm.

6. Disconnect the power feed to the wiper motor at the connector next to the radio. For 1995–96 vehicles, remove the radio.

7. Separate the left defroster outlet from the flex hose.

8. Remove the screw holding the left hand heater duct to the engine

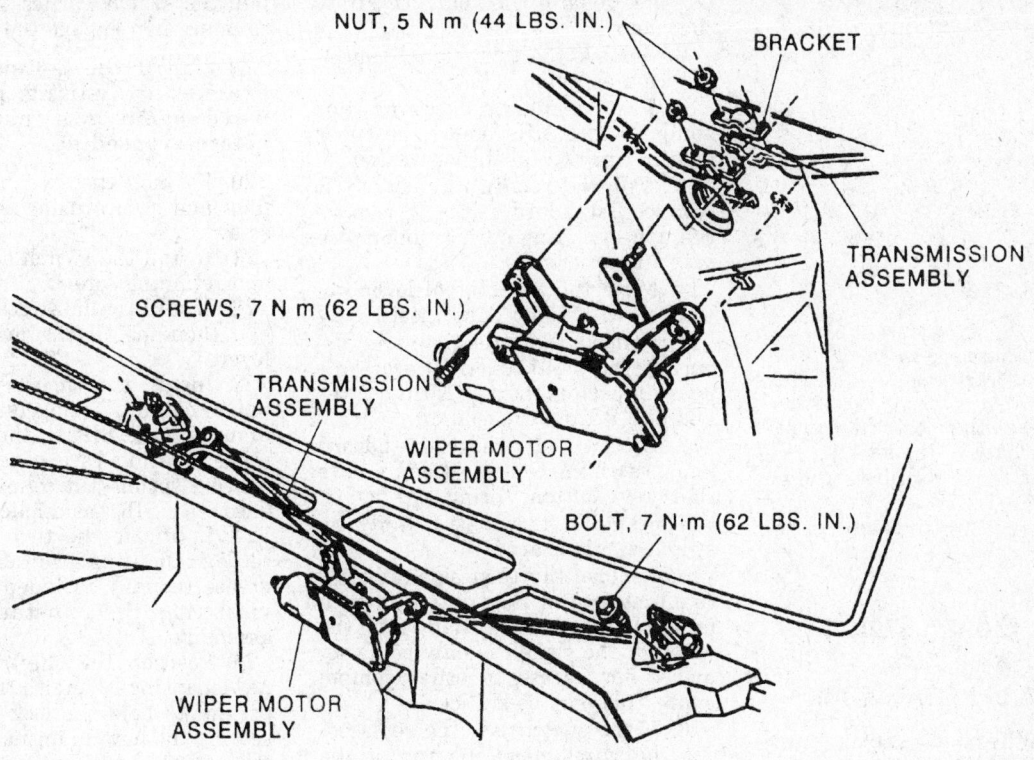

NUT, 5 N m (44 LBS. IN.)

BRACKET

TRANSMISSION ASSEMBLY

SCREWS, 7 N m (62 LBS. IN.)

TRANSMISSION ASSEMBLY

WIPER MOTOR ASSEMBLY

BOLT, 7 N·m (62 LBS. IN.)

WIPER MOTOR ASSEMBLY

79210gd5

Wiper motor mounting — 1996 pick-up shown

shroud, then twist the heater duct down and out.

9. Remove the screws (usually 3) holding the wiper motor to the cowl and remove the motor.

To install:

10. Position the wiper motor to the vehicle, then secure using the retaining screws.

11. If applicable, install the windshield washer hoses at the pump.

12. Position the heater duct (twisting in and up), then install the screw holding the left hand heater duct to the engine cover shroud.

13. Connect the left defroster outlet to the flex hose.

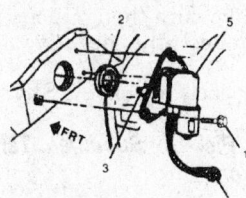

1. Bolt 35 in. lbs. (4 nm)
2. Seal
3. Crank arm
4. Wiper motor harness
5. Wiper motor assembly

79210g11

Wiper motor mounting — G-Series

14. Engage the electrical connector to the wiper motor assembly.

15. Lubricate the inside of the crank arm socket using white lithium grease, then reach in through the access hole and connect the crank arm to the socket. Tighten the nuts holding the transmission linkage to the wiper motor crank arm.

NOTE: For 1994–97 vehicles, use a pair of pliers to carefully squeeze the drive link onto the crank arm. The wiper transmission assembly must be installed to the crank arm PAST the 2nd detent so the seal is compressed to a maximum height of 1 in. (25.5mm)

16. If removed, install the radio assembly.

17. Install the cowl vent grille.

18. Install the wiper arms.

19. Connect the negative battery cable.

Headlight Switch

REMOVAL AND INSTALLATION

Blazer, Pick-Up, Suburban, Tahoe and Yukon

1. Disconnect the negative battery cable.

2. Remove the instrument cluster bezel.

3. Disengage the switch wiring connector.

4. If equipped, remove the headlight switch retaining screws and pull the switch away from the bezel, if not unsnap the switch from the bezel. Remove the switch.

5. Installation is the reverse of removal.

G-Series

1. Disconnect the negative battery cable.

2. Press the switch knob retaining pin and remove the knob. In most cases, the switch must be pulled out to the low beam position in order to depress the pin and remove the knob.

3. Remove the left instrument panel trim or lower trim panel, as applicable.

4. Remove the retaining nut securing the switch.

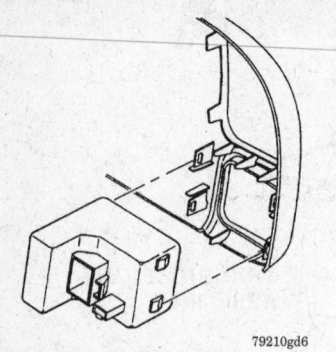

View of the headlamp switch location — 1996 pick-up

5. Disengage the electrical connector from the back of the switch.

6. Remove the switch from the instrument panel.

7. Reverse the procedure for installation.

Turn Signal Switch

REMOVAL AND INSTALLATION

NOTE: When servicing any components on the steering column, should any fasteners require replacement, be sure to use only nuts and bolts of the same size and grade as the original fasteners. Using screws that are too long could prevent the column from collapsing during a collision.

NOTE: The following procedures require the use of a lock plate compressor tool such as J-23653 or equivalent.

Except 1995–97 Pick-Up, Suburban, Tahoe and Yukon

1. If equipped, make sure the wheels are locked in the straight-ahead position, then properly disable the SIR (air bag) system.

2. Disconnect the negative battery cable.

3. Matchmark and remove the steering wheel.

4. If equipped with an SIR system, remove the SIR coil assembly retaining ring, then remove the coil assembly and allow it to hang freely from the wiring. Remove the wave washer.

5. On non-SIR equipped vehicles, remove the shaft lock cover.

6. Push downward on the shaft lock assembly until the snapring is exposed using the shaft lock compressor tool.

7. Remove the shaft lock retaining snapring, then carefully release the

tool and remove the shaft lock from the column.

8. Remove the turn signal cancelling cam assembly.

9. For standard columns not equipped with SIR, remove the upper bearing spring and thrust washer.

10. For tilt columns or SIR equipped standard columns, remove the upper bearing spring, inner race seat and inner race.

11. Move the turn signal lever upward to the "Right Turn" position.

12. Remove the access cap and disengage the multi-function lever harness connector, then grasp the lever and pull it from the column.

13. Loosen and remove the hazard knob retaining screw, then remove the screw, button, spring and knob.

14. Remove the screw and the switch actuator arm.

15. Remove the turn signal switch retaining screws, then pull the switch forward and allow it to hang from the wires. If the switch is only being removed for access to other components, this may be sufficient.

16. If the switch is to be replaced, cut the wires near the top of the switch and discard the switch. Before cutting the wires, verify that the wire color codes are the same. Secure the connector of the new switch to the old wires, and pull the new harness down through the steering column while removing the old switch.

17. If the original switch is to be reused, attach a piece of wire or string around the connector and pull the harness up through the column, while pulling the string up through the column and leaving the string or wire in position to help with reinstallation later.

18. After freeing the switch wiring protector from its mounting, pull the turn signal switch straight up and remove the switch, switch harness, and the connector from the column.

NOTE: On some vehicles access to the connector may be difficult. If necessary, remove the column support bracket assembly and properly support the column, and/or remove the wiring protectors.

To install:

19. Install the switch and wiring harness to the vehicle. If the switch was completely removed, use the length of mechanic's wire or string to

pull the switch harness through the column, then engage the connector.

NOTE: If the column support bracket or wiring protectors were removed, install them before proceeding.

20. Position the switch in the column and secure using the retaining screws.

21. Install the switch actuator arm and retaining screw.

22. Install the hazard knob assembly, then install the multi-function lever.

23. Install the thrust washer and upper bearing spring (standard columns without SIR) or the inner race, upper bearing race seat and upper bearing spring (tilt columns or standard with SIR), as applicable.

24. Lubricate the turn signal cancelling cam using a suitable synthetic grease (usually included in the service kit), then install the cam assembly.

25. Position the shaft lock and a new snapring, then use the lock compressor to hold the lock down while seating the new snapring. Make sure the ring is firmly seated in the groove, then carefully release the tool.

NOTE: The coil assembly will become uncentered if the steering column is separated from the steering gear and allowed to rotate or if the centering spring is pushed down, letting the hub rotate while the coils assembly is removed from the steering column.

26. If equipped with SIR, make sure the coil is centered, then install the wave washer, followed by the coil and the retaining ring. The coil ring must be firmly seated in the shaft groove.

27. On non SIR columns, install the shaft lock cover.

28. Align and install the steering wheel.

29. Make sure the ignition is **OFF**, then connect the negative battery cable.

30. Properly enable the SIR system.

1995–97 Pick-Up, Suburban, Tahoe and Yukon

Instead of the steering column found on most other GM vehicles the 1995–97 pick-up, Suburban, Tahoe and Yukon utilize a new column with a multi-function combination switch mounted at the head of the column below the steering wheel and an upper/lower shroud assembly. The com-

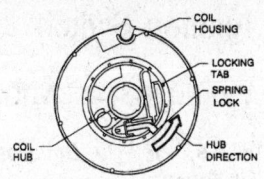

PERFORM THE FOLLOWING STEPS TO CENTER COIL ASSEMBLY

A. WHEELS STRAIGHT AHEAD.
B. REMOVE COIL ASSEMBLY.
C. HOLD COIL ASSEMBLY WITH BOTTOM UP.
D. WHILE HOLDING COIL ASSEMBLY, DEPRESS SPRING LOCK TO ROTATE HUB IN DIRECTION OF ARROW UNTIL IT STOPS.
E. THE COIL RIBBON SHOULD BE WOUND UP SNUG AGAINST CENTER HUB.
F. ROTATE COIL HUB IN OPPOSITE DIRECTION APPROXIMATELY TWO AND A HALF (2-1/2) TURNS. RELEASE SPRING LOCK BETWEEN LOCKING TABS.

79210g12

Centering the SIR coil

bination switch performs such functions as the wiper switch and the turn signal switch along with any other duties of the multi-function lever.

Combination Switch

REMOVAL AND INSTALLATION

NOTE: Removal of the SIR coil is not necessary during this procedure. avoid removing the coil and make sure the steering column, if disconnected from the gear, is not allowed to rotate excessively. This is to prevent uncentering and damaging the coil. Should the coil become uncentered, it must be removed, centered and repositioned on the steering column.

1. Disconnect the negative battery cable.
2. Properly disable the SIR (air bag) system.
3. Either lower the steering column from the instrument panel for access or unbolt and remove the column. If the column is removed, prevent it from rotating so the SIR coil does not become uncentered.
4. If applicable, remove the tilt lever by pulling outward.
5. Remove the 2 pan head tapping screws from the lower column shroud, then tilt the shroud down and slide it back to disengage the locking tabs. Remove the lower shroud.
6. Remove the 2 Torx head screws from the upper shroud.
7. Lift the upper shroud for access to the lock cylinder hole. Hold the key in the **START** position and use a $1/16$ in. Allen wrench to push on the lock cylinder retaining pin.
8. Release the key to the **RUN** position and pull the steering column lock cylinder set from the lock mod-

ule assembly. Remove the upper shroud.
9. If necessary, remove the shift lever clevis, then remove the lever.
10. Remove the wiring harness straps (noting the positioning for installation purposes), then disengage the steering column bulkhead connector from the vehicle wiring harness.
11. On column shift models, remove the axial position assurance connector from the electrical Brake Transmission Shift Interlock (BTSI) actuator. Disengage the wiring connector from the actuator.
12. Disengage the grey and black connectors for the turn signal and multi-function switch from the column bulkhead connector.
13. Remove the 2 pan head switch retaining screws, then remove the switch from the steering column.
To install:
14. Position the multi-function switch assembly, then use a suitable small bladed tool to compress the electrical contact while moving the switch into position. Make sure the electrical contact rests on the cancelling cam assembly.
15. Install the switch retaining screws and tighten to 53 inch lbs. (6.0 Nm).
16. Engage the grey and black multi-function switch connectors to the column bulkhead connector.
17. On models with a column shift engage the wiring connector to the BTSI actuator, then secure using the axial position assurance connector.
18. Engage the steering column bulkhead-to-vehicle connector.
19. If removed, install the shift lever and secure the clevis.
20. Install the wiring harness straps as noted during removal.
21. Position the shift lever and multi-function lever seals to ease installation of the upper and lower shrouds.
22. Install the upper shroud and lock cylinder. With the key installed to the lock cylinder and turned to the **RUN** position, make sure the sector in the lock module is also in this position.
23. Install the lock cylinder to the upper shroud, then align the locking tab and positioning tab with the slots in the lock module assembly. With the tabs aligned, carefully push the cylinder into position.
24. Install the upper shroud Torx head retaining screws and tighten to 12 inch lbs. (1.4 Nm).
25. Install the lower shroud, making sure the slots on the shroud en-

gage with the upper shroud tabs. Tilt the lower shroud upward and snap the shrouds together.
26. Install the 2 lower shroud pan head retaining screws and tighten to 53 inch lbs. (6.0 Nm).
27. Move the shift and multi-function lever seals into position, then
28. If removed, install the tilt lever by aligning and pushing inward.
29. Position and secure the steering column.
30. Properly enable the SIR system, then connect the negative battery cable.

Ignition Lock Cylinder

REMOVAL AND INSTALLATION

Except 1995–97 Pick-Up, Suburban, Tahoe and Yukon

1. If equipped, make sure the wheels are locked in the straight-ahead position, then properly disable the SIR (air bag) system.
2. Disconnect the negative battery cable.
3. Matchmark and remove the steering wheel.
4. If equipped with an SIR system, remove the SIR coil assembly retaining ring, then remove the coil assembly and allow it to hang freely from the wiring. Remove the wave washer.

NOTE: On some SIR equipped vehicles, it may be necessary to completely remove the coil and wiring from the steering column before removing the lock cylinder assembly. If so, attach a length of mechanic's wire to the coil connector at the base of the column, then carefully pull the harness and wire through the steering column towards the top. Leave the wire in position inside the column in order to pull the harness back down into position during installation.

5. Remove the turn signal switch from the column and allow it to hang from the wires (leaving them connected).
6. Remove the buzzer switch assembly. On some vehicles it may be necessary to temporarily remove the key from the lock cylinder in order to remove the buzzer. If so, the key should be reinserted before the next step.
7. Carefully remove the lock cylinder screw and the lock cylinder. If possible, use a magnetic tipped screwdriver on the screw in order to

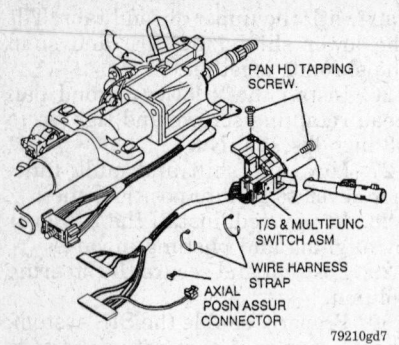

PAN HD TAPPING SCREW.

T/S & MULTIFUNC SWITCH ASM

WIRE HARNESS STRAP

AXIAL POSN ASSUR CONNECTOR

79210gd7

Exploded view of combination switch — 1996 pick-up

help prevent the possibility of dropping it.

--------- CAUTION ---------

If the screw is dropped upon removal, it could fall into the steering column, requiring complete disassembly in order to retrieve the screw and prevent damage.

To install:

8. Align and install the lock cylinder set.

9. Push the lock cylinder all the way in, then carefully install the retaining screw. Tighten the screw to 22 inch lbs. (2.5 Nm) on tilt columns or to 40 inch lbs. (4.5 Nm) on standard non-tilt columns.

10. If necessary, install the buzzer switch assembly.

11. Reposition and secure the turn signal switch assembly

NOTE: The coil assembly will become uncentered if the steering column is separated from the steering gear and allowed to rotate or if the centering spring is pushed down, letting the hub rotate while the coils assembly is removed from the steering column.

12. If equipped with SIR, make sure the coil is centered, then install the wave washer, followed by the coil and the retaining ring. The coil ring must be firmly seated in the shaft groove.

13. Align and install the steering wheel.

14. Make sure the ignition is **OFF**, then connect the negative battery cable.

15. Properly enable the SIR system.

1995–97 Pick-Up, Suburban, Tahoe and Yukon

1. Disconnect the negative battery cable.

2. Properly disable the SIR (air bag) system.

3. Either lower the steering column from the instrument panel for access or unbolt and remove the column. If the column is removed, prevent it from rotating so the SIR coil does not become uncentered.

4. If applicable, remove the tilt lever by pulling outward.

5. Remove the 2 pan head tapping screws from the lower column shroud, then tilt the shroud down and slide it back to disengage the locking tabs. Remove the lower shroud.

6. Remove the 2 Torx head screws from the upper shroud.

7. Lift the upper shroud for access to the lock cylinder hole. Hold the key in the **START** position and use a 1/16 in. Allen wrench to push on the lock cylinder retaining pin.

8. Release the key to the **RUN** position and pull the steering column lock cylinder set from the lock module assembly. Remove the upper shroud.

To install:

9. Install the upper shroud and lock cylinder. With the key installed to the lock cylinder and turned to the **RUN** position, make sure the sector in the lock module is also in this position.

10. Install the lock cylinder to the upper shroud, then align the locking tab and positioning tab with the slots in the lock module assembly. With the tabs aligned, carefully push the cylinder into position.

11. Install the upper shroud Torx head retaining screws and tighten to 12 inch lbs. (1.4 Nm).

12. Install the lower shroud, making sure the slots on the shroud engage with the upper shroud tabs. Tilt the lower shroud upward and snap the shrouds together.

13. Install the 2 lower shroud pan head retaining screws and tighten to 53 inch lbs. (6.0 Nm).

14. Move the shift and multi-function lever seals into position, then

15. If removed, install the tilt lever by aligning and pushing inward.

16. Position and secure the steering column.

17. Properly enable the SIR system, then connect the negative battery cable.

Ignition Switch

REMOVAL AND INSTALLATION

Except 1995–97 Pick-Up, Suburban, Tahoe and Yukon

1. If equipped, make sure the wheels are locked in the straight-ahead position, then properly disable the SIR (air bag) system.

2. Disconnect the negative battery cable.

3. Remove the lower column trim panel, then remove the steering column-to-instrument panel fasteners and carefully lower the column for access to the switch.

4. On some vehicles, the dimmer switch must be removed in order to remove the ignition switch. If necessary, remove the dimmer switch.

5. Place the ignition switch in the **OFF-LOCK** position.

NOTE: If the lock cylinder was removed, the switch slider should be moved to the extreme right position, then 1 detent to the left.

6. Remove the ignition switch-to-steering column retainers and disengage the switch wiring, then remove the assembly.

To install:

7. Before installing the ignition switch, place it in the **OFF-LOCK** position, then make sure the lock cylinder and actuating rod are in the Locked position (1st detent from the top or 1st detent to the right of far left detent travel).

NOTE: Most replacement switches are pinned in the OFF-LOCK position for installation purposes. If so, the pins must be removed after installation or damage may occur.

8. Install the activating rod into the ignition switch and assemble the switch onto the steering column. Once the switch is properly positioned, tighten the ignition switch-to-steering column retainers to 35 inch lbs. (4.0 Nm).

NOTE: When installing the ignition switch, use only the specified screws since over length screws could impair the collapsibility of the column.

9. If removed, install the dimmer switch.

10. Raise the column into position and secure, then install any necessary trim plates.

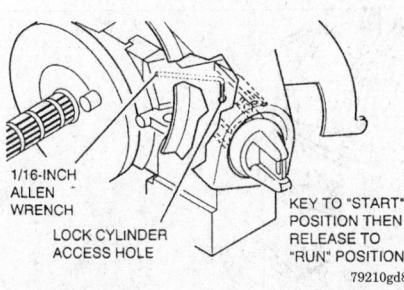

Removing the lock cylinder — 1996 pick-up

11. Make sure the ignition is **OFF**, then connect the negative battery cable.

12. Properly enable the SIR system.

1995–97 Pick-Up, Suburban, Tahoe and Yukon

1. Disconnect the negative battery cable.

2. Properly disable the SIR (air bag) system.

3. Either lower the steering column from the instrument panel for access or unbolt and remove the column. If the column is removed, prevent it from rotating so the SIR coil does not become uncentered.

4. Remove the combination switch from the steering column.

5. If equipped, remove the alarm switch from the lock module assembly by gently prying the retaining clip on the alarm switch using a small blade pry tool. Then, rotate the alarm switch ¼ turn and remove.

6. Remove the 2 ignition switch self-tapping retaining screws.

7. Disengage the connector, then remove the wiring harness from the slot in the steering column. Remove the ignition and key alarm switch.

To install:

8. Position the switch to the column. Route the wire harness through the slot in the column housing assembly. Secure the harness using a wire strap through the hole located in the bottom of the housing assembly.

9. Install the switch retaining screws and tighten to 12 inch lbs. (1.4 Nm) in order to secure the switch.

10. If applicable, install the alarm switch to the lock module assembly by aligning the switch (with the retaining clip) parallel to the lock cylinder, then rotating the switch ¼ turn until locked in place.

11. Install the combination switch to the steering column.

12. Position and secure the steering column.

13. Properly enable the SIR system, then connect the negative battery cable.

Park/Neutral Safety Switch

REMOVAL AND INSTALLATION

Column Mounted Switch

Most vehicles covered by this manual should be equipped with ratcheting self-adjusting switches.

1. Disconnect the negative battery cable.

2. If necessary for access, remove the steering column insulator/filler panel for access to the switch.

3. Disengage the electrical harness connector from the switch.

4. Remove the switch by grasping and pulling it straight out of the steering column jacket.

To install:

5. Align the actuator on the switch with the holes in the shift tube.

6. Set the parking brake and place the gear selector in Neutral.

7. Press down on the front of the switch until the tangs snap into the rectangular holes in the steering column jacket.

8. Adjust the switch by moving the gear selector to **P**. The main housing and the housing back should ratchet, providing the proper switch adjustment.

9. Engage the harness connector to the switch.

10. Connect the negative battery cable, then verify proper switch operation. Make sure the reverse lights work and that the ignition will only work in the Neutral or **P** positions. If necessary, readjust the switch. For ratcheting type switches, move the gear selector all the way to the Low position, then repeat the adjustment.

11. If applicable, install the steering column insulator/filler panel.

Floor Console Mounted Switch

1. Disconnect the negative battery cable.

2. Remove the center console for access to the switch assembly.

3. Disengage the switch electrical connector.

4. Remove the retaining nuts, then remove the switch.

5. If necessary, remove the gauge pin from the switch.

6. If installing a new switch:

   a. Place the shift control lever in Neutral.

   b. Align the carrier tang on the back-up lamp/neutral safety switch with the slot on the shifter.

NOTE: Replacement switches are pinned in the Neutral position to ease installation. If the switch has been rotated or the switch is broken, install the switch using the "old switch" installation and adjustment procedure.

   c. Tighten the switch retaining nuts to 30 inch lbs. (3.4 Nm), then engage the switch connector.

   d. Move the shift control lever out of Neutral in order to shear the

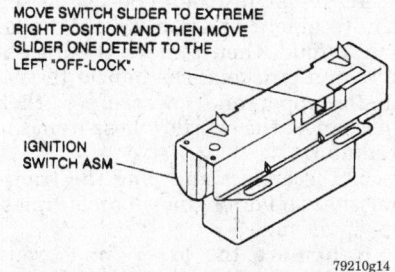

Ignition switch OFF-LOCK positioning for removal — G Van shown

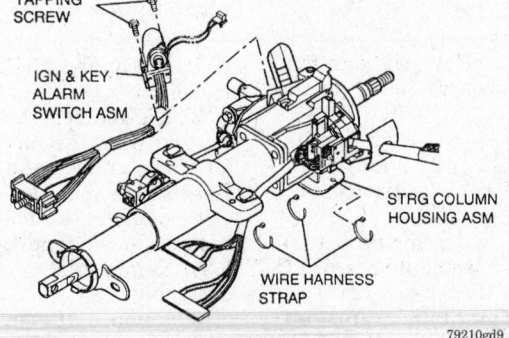

Exploded view of the ignition switch and related components — 1996 pick-up

plastic pin, then remove the accessible piece(s) of the gauge pin.

7. If installing an old switch, install and adjust the switch to assure proper operation:

a. Place the shift control lever in Neutral.

b. Align the carrier tang on the switch with the slot on the shifter.

c. Loosely install the retaining switch nuts and engage the wiring connector.

d. Rotate the switch to align the service adjustment hole with the carrier tang hold, then use a 0.09 in. (2.34mm) gauge pin to complete adjustment. Insert the pin in the service adjustment hole and rotate the switch until it drops to a depth of 0.59 in. (15mm). Hold the switch in this position and tighten the retaining nuts to 30 inch lbs. (3.4 Nm).

8. Install the center console.

9. Connect the negative battery cable.

10. Verify proper switch operation.

Transmission Mounted Switch

1. Place the transmission in park and disconnect the negative battery cable.

2. Raise the vehicle and support it with jackstands.

3. Disconnect the shift cable end from the shift control lever and remove the nut securing the shift control lever to the manual shaft.

4. Disengage the electrical connector from the switch and remove the switch retainers.

5. Remove the switch from the transmission.

To install:

6. Adjust the switch as outlined in this section.

7. Install the switch and tighten the switch retainers. Tighten the switch retainers to 20 ft. lbs. (27 Nm).

8. Engage the electrical connector and install the control lever to the manual shaft.

9. Install the control lever nut and tighten to 20 ft. lbs. (27 Nm).

10. Lower the vehicle and connect the negative battery cable.

11. Check the switch for proper operation. The vehicle should start in **P** or **N** only.

12. If adjustment is required, loosen the switch retaining bolts and rotate the switch slightly, tighten the bolts and check switch operation.

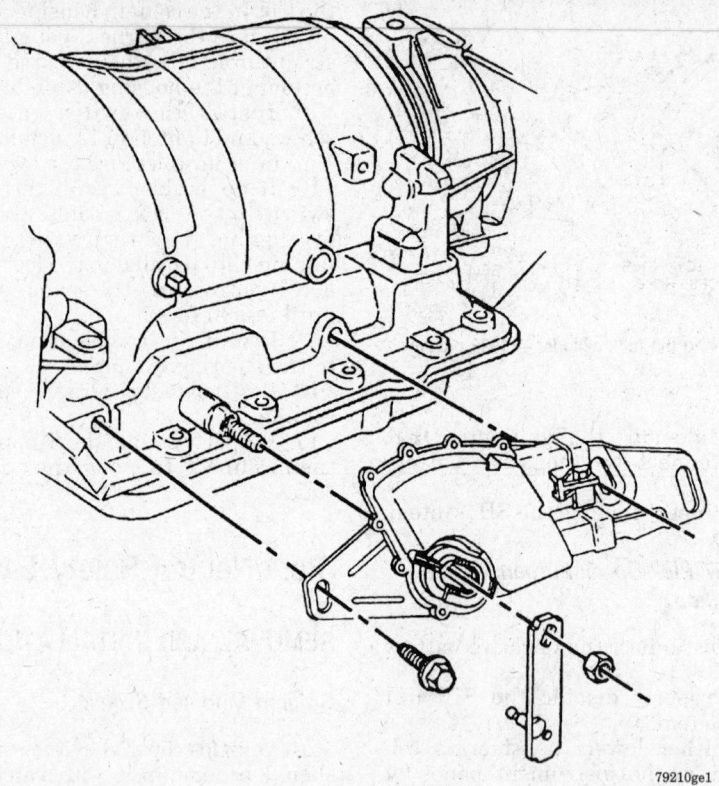

View of the transmission mounted Park/Neutral switch — 1996 pick-up

Powertrain Control Module

REMOVAL AND INSTALLATION

RWAL System

1. Disconnect the negative battery cable.

2. Disconnect the electrical connectors from the module.

3. Remove the module from the master cylinder/proportioning valve bracket.

4. Installation is the reverse of the removal procedure.

4WAL System

1. Disconnect the negative battery cable.

2. Disconnect the electrical connectors from the module.

3. Disconnect and plug the brake lines at the module.

4. Remove the bolts attaching the module to the fenderwell.

5. Remove the module and bracket assembly.

6. Remove the bracket from the module.

7. Installation is the reverse of the removal procedure.

ENGINE COOLING

Radiator

REMOVAL AND INSTALLATION

Except G-Series with Diesel Engines

1. Disconnect the negative battery cable.

2. Drain the cooling system into a suitable container.

3. For the G-Series, disconnect the radiator upper and lower hoses, then disconnect the overflow hose from the radiator. Remove the fan shroud.

4. Except for the G-Series, loosen the retainers and remove the upper fan shroud, then remove the insulators and brackets. Disconnect the radiator upper and lower hoses, then disconnect the overflow hose from the radiator.

5. Disconnect and plug the transmission and/or engine oil cooler lines, as applicable.

6. Remove the lower fan shroud bolts, then remove the lower shroud.

7. If necessary, remove the clutch fan.

8. Remove the radiator.

9. Installation is the reverse of the removal procedure.

NOTE: Whenever the cooling system in a 1995–97 vehicle (except diesel engines or U-Series vehicles) is completely drained and refilled with fresh coolant, 2 sealant pellets GMSPO part 3634621 must be added to the radiator. Failure to use the correct sealant pellets may result in premature coolant pump leakage. Do not add the pellets to the coolant recovery bottle as this may prevent the system from operating properly.

G-Series with Diesel Engines

1. Disconnect the negative battery cables.
2. Drain the engine cooling system.
3. Remove the air intake snorkel.
4. Remove the windshield washer bottle.
5. Remove the hood release cable.
6. Remove the upper fan shroud.
7. Disconnect the upper radiator hose.
8. Disconnect and plug the transmission cooler lines.
9. Disconnect the low coolant sensor wire.
10. Disconnect the overflow hose.
11. Disconnect the engine oil cooler lines.
12. Disconnect the lower radiator hose.
13. Unbolt the brake master cylinder from the booster and reposition aside leaving the lines attached. Make sure the brake lines are not kinked, stretched or otherwise damaged.
14. Unbolt and remove the radiator.
To install:
15. Install the radiator.
16. Secure the brake master cylinder on the booster.
17. Connect the lower radiator hose.
18. Connect the engine oil cooler lines.
19. Connect the overflow hose.
20. Connect the low coolant sensor wire.
21. Connect the transmission cooler lines.
22. Connect the upper radiator hose.
23. Install the upper fan shroud.
24. Install the hood release cable.
25. Install the windshield washer bottom.

26. Install the air intake snorkel.
27. Fill the radiator with the proper type and quantity of coolant and inspect the system for leaks.

Water Pump

REMOVAL AND INSTALLATION

Gasoline Engines

1. Disconnect the negative battery cable.
2. Except for G-Series, remove the upper fan shroud
3. On the G-Series, remove the air intake duct, then remove the upper radiator shroud.
4. Drain the engine coolant into a suitable container.
5. Remove the drive belt(s) from the water pump pulley.
6. Remove fan, clutch and pulley.
7. If necessary, remove any accessory brackets that will interfere with water pump removal.
8. Disconnect the coolant hoses from the water pump assembly.
9. Remove the bolts, pump assembly and old gasket from the engine.
To install:
10. Ensure the gasket surfaces on the pump and engine are clean.
11. Install the pump assembly with a new gasket. Tighten the bolts to 30 ft. lbs. (40 Nm).
12. Connect the hoses to the water pump assembly.
13. Install any accessory brackets which were removed for access.
14. Install the fan, clutch and pulley.
15. Install the accessory drive belt(s), as applicable.
16. Install the upper fan or radiator shroud and the air intake duct, as applicable.
17. Connect the battery, then properly refill the engine cooling system.

NOTE: Whenever the cooling system in a 1995–97 vehicle (except diesel engines) is completely drained and refilled with fresh coolant, 2 sealant pellets GMSPO part 3634621 must be added to the radiator. Failure to use the correct sealant pellets may result in premature coolant pump leakage. Do not add the pellets to the coolant recovery bottle as this may prevent the system from operating properly.

Diesel Engines

1. Disconnect the negative battery cables.
2. Remove the fan and fan shroud.
3. Drain the engine coolant into a suitable container.
4. If necessary, remove the air conditioning hose bracket and/or the oil filler tube, as required.
5. Remove the engine accessory drive belt(s).
6. Raise and support the vehicle safely.
7. Remove the vacuum pump mounting bracket nuts, then remove the bolt holding the pump and alternator. Remove the vacuum pump and bracket.
8. Remove the power steering pump and bracket, then support the assembly aside.
9. Lower the vehicle, then disconnect the coolant hoses from the pump.
10. Remove the water pump plate retaining bolts, then remove the pump and plate assembly from the engine.
11. Remove the bolt on the rear of the water pump plate, then separate the pump and gasket from the plate.
To install:
12. Install the water pump and a new gasket to the plate. Tighten the retaining bolt (at the rear of the plate) to 17 ft. lbs. (23 Nm).
13. Make sure the block mating surface and the plate flanges are free of oil. Apply an anaerobic sealer GM part 1052357 or equivalent.

NOTE: The sealer must be wet to the touch when the bolts are torqued.

14. Attach the water pump and plate assembly, then install and tighten the retainers.
15. Connect the coolant hoses to the pump assembly.
16. Raise and support the vehicle safely, then reposition and secure the power steering pump and bracket.
17. Install the vacuum pump and bracket, along with the bolt holding the pump and alternator.
18. Lower the vehicle, then install the fan and pulley.
19. Install the engine accessory drive belt(s).
20. If removed, install the oil filler tube and/or air conditioning hose bracket nuts.
21. Install the fan shroud.
22. Connect the batteries.
23. Fill the radiator with the proper type and quantity of antifreeze.

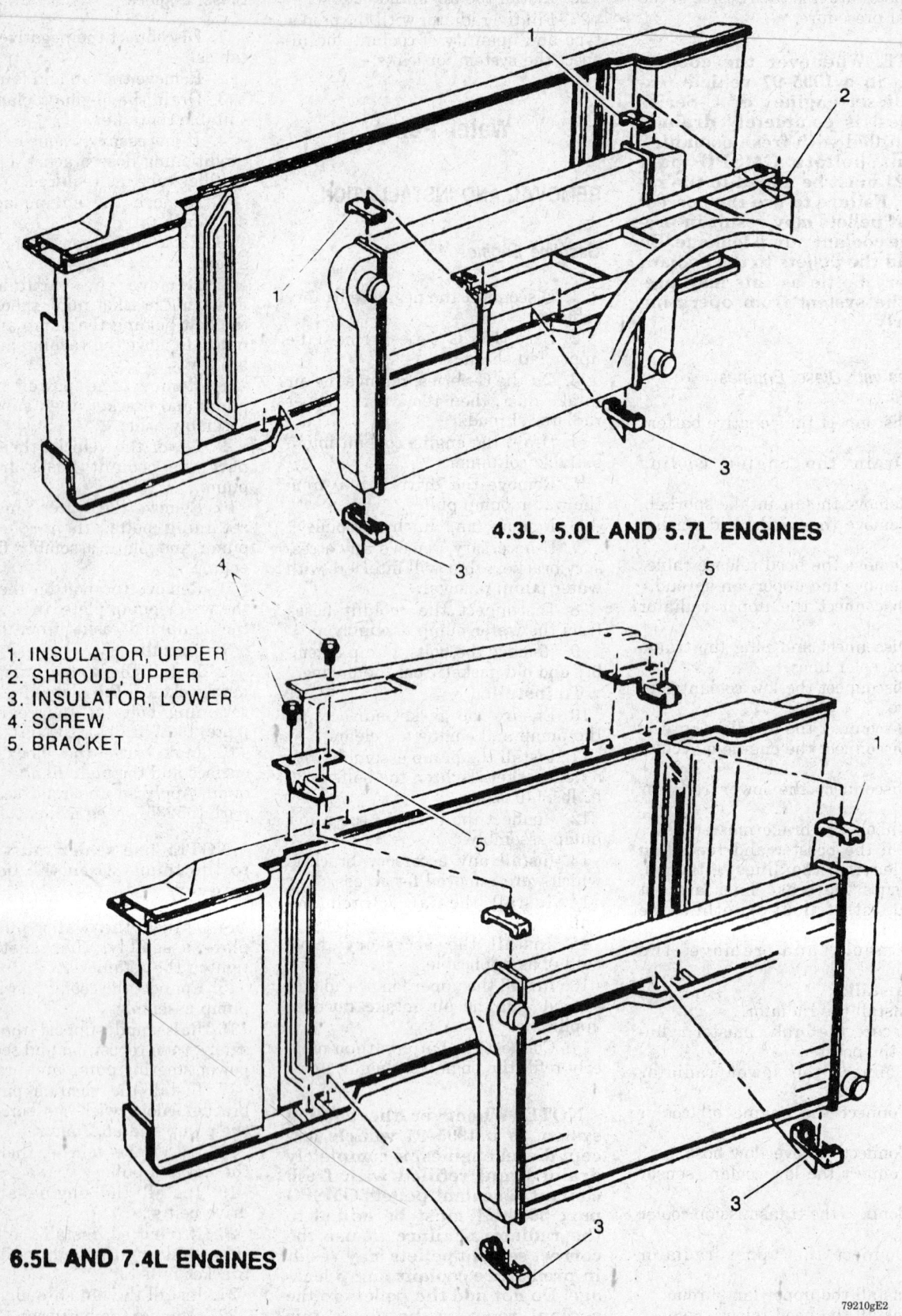

4.3L, 5.0L AND 5.7L ENGINES

1. INSULATOR, UPPER
2. SHROUD, UPPER
3. INSULATOR, LOWER
4. SCREW
5. BRACKET

6.5L AND 7.4L ENGINES

79210gE2

View of the radiator and mounting components — 1996 pick-up

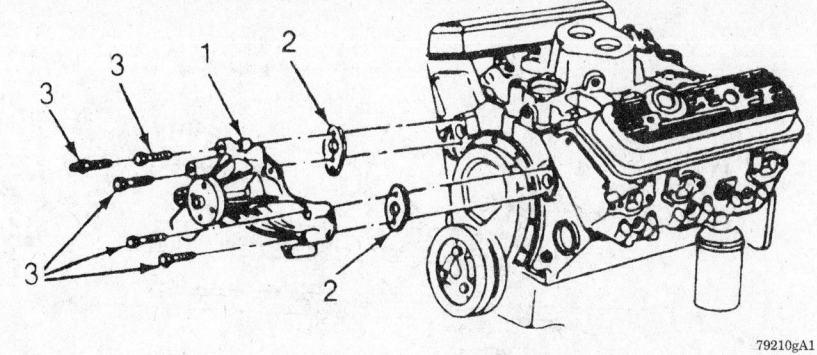

1. Water pump
2. Gasket
3. Bolt

79210gA1

Exploded view of the water pump assembly and related components — 4.3L engine

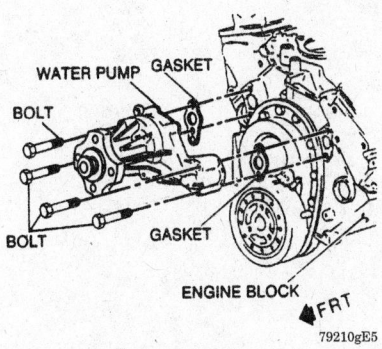

Exploded view of the water pump assembly and related components — 5.0L, and 5.7L engines

Thermostat

REMOVAL AND INSTALLATION

Except Diesel Engines

1. Disconnect the negative battery cable.
2. Drain the radiator until the level is below the thermostat (below the level of the intake manifold, cylinder head or other thermostat housing, as applicable).
3. If necessary, disconnect the hose from the thermostat outlet. On most applications, the hose may be left attached to the outlet when it is removed, however on some models disconnecting the hose may make outlet removal easier.
4. Loosen and remove the thermostat outlet retainers, then remove the outlet from the housing. Note the orientation of the thermostat in the housing for installation purposes.
   To install:
5. Carefully clean the all traces of the old gasket and/or sealer from the housing and outlet.
6. Install the thermostat to the housing, oriented as noted during re-

moval, then position a new gasket (if used).
7. If sealant was used on the old component or if no gasket is provided, place a ⅛ in. bead of RTV sealant in the groove on the water outlet sealing surface, then install the outlet while the sealer is still wet.
8. Install and tighten the outlet retainers.
9. If removed, connect the hose to the outlet.
10. Connect the negative battery cable.
11. Properly refill and bleed the engine cooling system, then check for leaks.

NOTE: Whenever the cooling system in a 1995–97 vehicle (except diesel engines) is completely drained and refilled with fresh coolant, 2 sealant pellets GMSPO part 3634621 must be added to the radiator. Failure to use the correct sealant pellets may result in premature coolant pump leakage. Do not add the pellets to the coolant recovery bottle as this may prevent the system from operating properly.

Diesel Engines

1. Disconnect the negative battery cables.
2. Remove the upper fan shroud.
3. Drain the cooling system to a point below the thermostat.
4. Remove the engine oil dipstick tube brace and the oil fill brace.
5. If necessary, disconnect the radiator inlet hose.
6. Loosen the retainers, then remove the water outlet.
7. Remove the thermostat and gasket.

NOTE: When cleaning the gasket mating surfaces, look for traces of RTV sealant. If no traces

are found, install the replacement gasket dry (without additional sealant).

8. Installation is the reverse of removal. Use a new gasket and torque the outlet retainers to 31 ft. lbs. (42 Nm).

Cooling Fan

REMOVAL AND INSTALLATION

4.3L, 5.0L, and 5.7L Engines

1. Disconnect the negative battery cable.
2. Remove the upper fan shroud.
3. Remove the drive belt(s).
4. Loosen the retainers and remove the fan.
   To install:
5. Install the fan and tighten the retainers.
6. Install the drive belts.
7. Install the fan shroud and connect the negative battery cable.

7.4L and Diesel Engines

1. Disconnect the negative battery cable.
2. Remove the upper fan shroud.
3. Locate the yellow dot on the fan clutch and matchmark the water pump pulley.
4. Unfasten the retainers and remove the fan and fan clutch from the vehicle.
   To install:
5. Install the fan and align the matchmarks on the water pump pulley and fan clutch.
6. Install the retainers and tighten to 18 ft. lbs. (24 Nm).
7. Install the fan shroud and connect the negative battery cable.

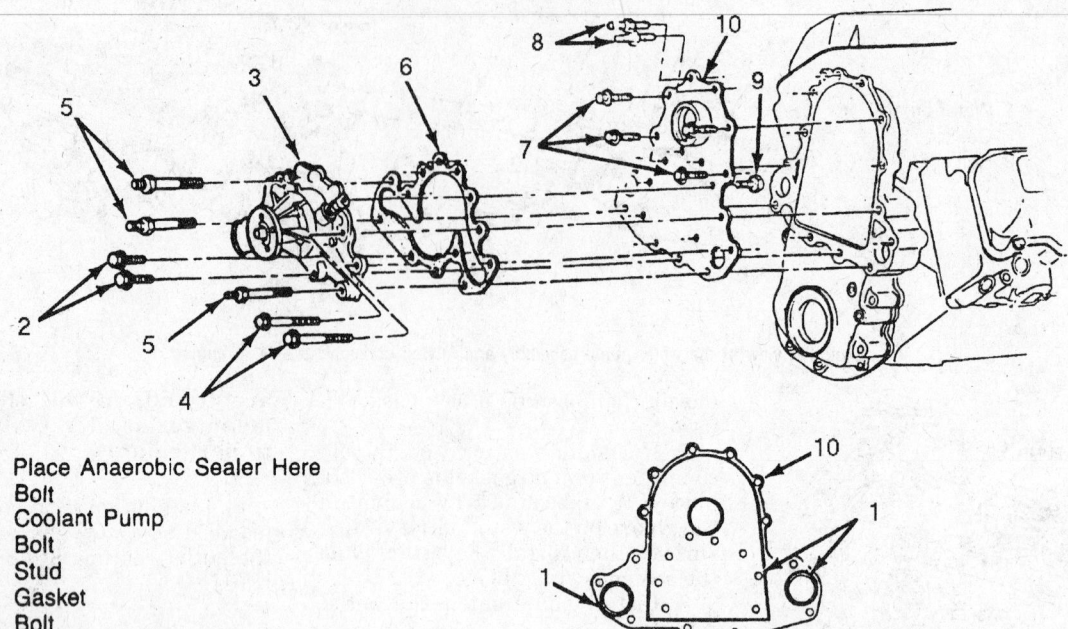

1. Place Anaerobic Sealer Here
2. Bolt
3. Coolant Pump
4. Bolt
5. Stud
6. Gasket
7. Bolt
8. Stud
9. Bolt
10. Coolant Pump Plate

79210gE6

Exploded view of the water pump assembly and related components — diesel engines

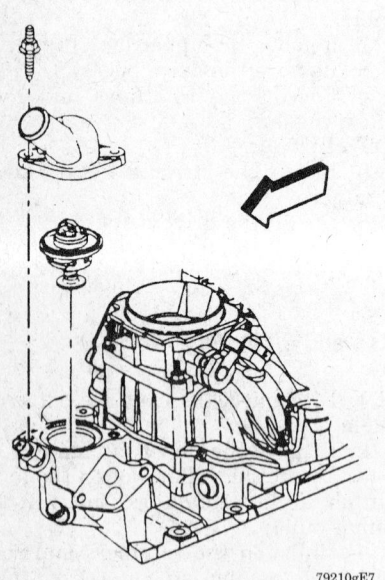

79210gE7

Exploded view of the thermostat and related components — 1996 4.3L

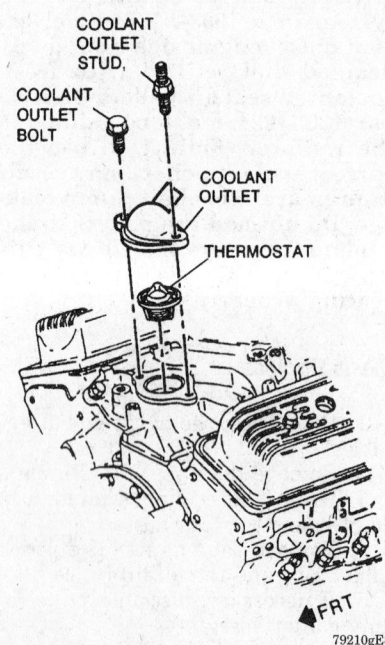

COOLANT OUTLET STUD,
COOLANT OUTLET BOLT
COOLANT OUTLET
THERMOSTAT

FRT
79210gE8

Exploded view of the thermostat and related components — 1996 5.0L and 5.7L engines

Auxiliary Electric Cooling Fan

An electric, auxiliary fan is used on some. The purpose of the fan is to provide additional cooling during extended idle and slow moving vehicle operation. The system consists of an engine coolant temperature sensor, relay and the electric fan motor. When the engine coolant temperature reaches a predetermined point, the sensor will close the circuit to the relay, energizing it, and causing the relay to apply 12 volts to the fan motor. Once temperature falls below that point, the sensor will open the circuit, de-energizing the relay and causing it to cut power from the fan motor.

REMOVAL AND INSTALLATION

1. Disconnect the negative battery cable.
2. On the G-Series and if necessary on the other trucks, remove the grille for access.
3. Unplug the fan harness connector.
4. Remove the fan-to-brace bolts and lift out the fan.
5. Installation is the reverse of removal.

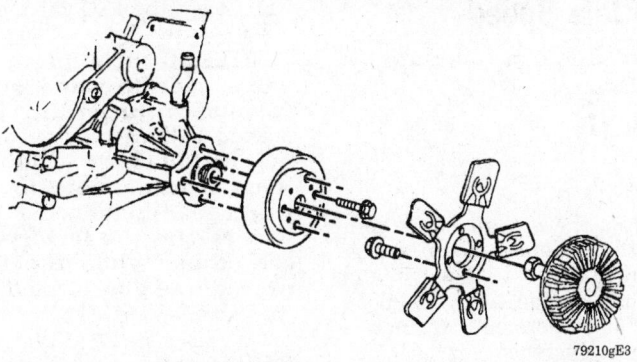

Cooling fan and fan clutch assemblies — 1996 4.3L. 5.0L, 5.7L engines

79210gE3

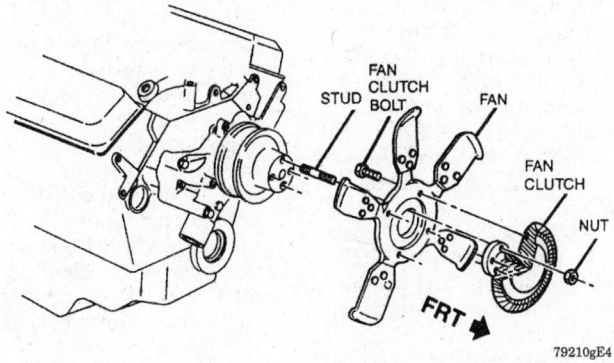

Cooling fan and fan clutch assemblies — 1996 7.4L (VIN J) and diesel engines

79210gE4

Cooling System Bleeding

NOTE: Whenever the cooling system in a 1995–97 vehicle (except diesel engines) is completely drained and refilled with fresh coolant, 2 sealant pellets GMSPO part 3634621 must be added to the radiator. Failure to use the correct sealant pellets may result in premature coolant pump leakage. Do not add the pellets to the coolant recovery bottle as this may prevent the system from operating properly.

1. To bleed the system, start with the system cool, the radiator cap off and the radiator filled to about an inch below the filler neck.
2. Start the engine and run it at slightly above normal idle speed. If air bubbles appear and the coolant level drops, fill the system with a 50/50 antifreeze/water mixture to bring the level back to the proper level.
3. Run the engine until the thermostat opens and coolant flow is visible.
4. At this point, air is often expelled and the level may drop again.

Keep refilling the system until the level is near the top of the radiator and remains constant.
5. Fill the radiator to the filler neck, then install the radiator filler cap.
6. Check and make sure the coolant reservoir is filled to the correct level.

GASOLINE FUEL SYSTEM

Fuel System Service Precautions

Safety is the most important factor when performing not only fuel system maintenance but any type of maintenance. Failure to conduct maintenance and repairs in a safe manner may result in serious personal injury or death. Maintenance and testing of the vehicle's fuel system components can be accomplished safely and effectively by adhering to the following rules and guidelines.

• To avoid the possibility of fire and personal injury, always disconnect the negative battery cable unless the repair or test procedure requires that battery voltage be applied.

• Always relieve the fuel system pressure prior to disconnecting any fuel system component (injector, fuel rail, pressure regulator, etc.), fitting or fuel line connection. Exercise extreme caution whenever relieving fuel system pressure to avoid exposing skin, face and eyes to fuel spray. Please be advised that fuel under pressure may penetrate the skin or any part of the body that it contacts.

• Always place a shop towel or cloth around the fitting or connection prior to loosening to absorb any excess fuel due to spillage. Ensure that all fuel spillage (should it occur) is quickly removed from engine surfaces. Ensure that all fuel soaked cloths or towels are deposited into a suitable waste container.

• Always keep a dry chemical (Class B) fire extinguisher near the work area.

• Do not allow fuel spray or fuel vapors to come into contact with a spark or open flame.

• Always use a backup wrench when loosening and tightening fuel line connection fittings. This will prevent unnecessary stress and torsion to fuel line piping. Always follow the proper torque specifications.

• Always replace worn fuel fitting O-rings with new. Do not substitute fuel hose or equivalent, where fuel pipe is installed.

Fuel System Pressure

RELIEVING

Before loosening or disconnecting any fuel fitting or system component, always relieve the fuel system pressure in order to help prevent the danger of fire or injury.

Throttle Body Fuel Injection Systems

GM vehicles with TBI engines utilize an automatic pressure bleed down feature. But, some fuel pressure related steps should still be taken to assure safer working conditions.

1. Disconnect the negative battery cable to prevent fuel spillage should the ignition key accidentally be turned **ON** with a fuel fitting disconnected.

2. Loosen fuel filler cap to relieve fuel tank pressure.

3. The internal constant bleed feature of the Model 220 TBI unit relieves fuel pump system pressure when the engine is turned **OFF**. Therefore, no further action is required.

NOTE: Allow the engine to set for 5–10 minutes; this will allow the orifice (in the fuel system) to bleed off the pressure.

4. When fuel service is finished, tighten the fuel filler cap and connect the negative battery cable.

Multi-Port Fuel Injection Systems

The MFI and CMFI fuel systems used on GM vehicles all operate under high fuel pressures. It is very important that the pressure be properly relieved prior to servicing the system or any of its components.

A schraeder valve is provided on these fuel systems in order to conveniently test or release the system pressure. A fuel pressure gauge and adapter will be necessary to connect the gauge to the fitting. Most of the MFI systems utilize a service valve on 1 end of the fuel rail assembly. The CMFI system covered here uses a valve located on the inlet pipe fitting, immediately before it enters the CMFI assembly (towards the rear of the engine).

1. Disconnect the negative battery cable to assure the prevention of fuel spillage if the ignition switch is accidentally turned **ON** while a fitting is still disconnected.

2. Loosen the fuel filler cap to release the fuel tank pressure.

3. Make sure the release valve on the fuel gauge is closed, then connect the fuel gauge to the pressure fitting located on the inlet fuel pipe fitting.

NOTE: When connecting the gauge to the fitting, be sure to wrap a rag around the fitting to avoid spillage. After repairs, place the rag in an approved container.

4. Install the bleed hose portion of the fuel gauge assembly into an approved container, then open the gauge release valve and bleed the fuel pressure from the system.

5. When the gauge is removed, be sure to open the bleed valve and drain all fuel from the gauge assembly.

6. When fuel service is finished, tighten the fuel filler cap and connect the negative battery cable.

Idle Speed

ADJUSTMENT

Idle speed on these engines are controlled by the computer modules, regulated through the IAC valve and fuel injector(s), respectively. No periodic check or adjustments are necessary for these systems.

However, the minimum idle speed on some systems on the 7.4L (VIN N) engine TBI system used 1993–97 may be adjusted if the throttle body has been replaced and incorrect idle speed cannot be obtained. These adjustments should not be performed unless all other possible causes (vacuum leaks, bad fuel pressure, faulty wiring or components) have already been eliminated.

This procedure should be performed only if parts of the throttle body have been replaced. The engine should be at normal operating temperature

1. Remove the air cleaner, adapter and gaskets. Discard the gaskets. Plug any vacuum line ports, as necessary.

2. Leave the Idle Air Control (IAC) valve connected and ground the diagnostic terminal (ALDL connector).

3. Turn the ignition switch to the **ON** position; do not start the engine. Wait for at least 10 seconds; this allows the IAC valve pintle to extend fully and seat in the throttle body.

4. With the ignition switch still in the **ON** position, disconnect IAC electrical connector.

5. Remove the ground from the diagnostic terminal and start the engine. Let the engine reach normal operating temperature.

6. Apply the parking brake and block the drive wheels. Remove the plug from the idle stop screw by piercing it first with a suitable tool, then applying leverage to lift the plug out.

7. Connect a suitable tachometer to the engine.

8. Ensure that the transmission is in the specified (N or D) position, with the ECM in "Open or Closed" loop as specified.

9. Adjust the idle stop screw to obtain the specified RPM reading.

10. Turn the ignition **OFF** and reconnect the IAC valve connector. Unplug any plugged vacuum line ports and install the air cleaner, adapter and new gaskets.

11. Reset the IAC valve as follows:

NOTE: If installing a new IAC valve, measure and adjust the valve accordingly. If reinstalling a used IAC valve, do not push or pull on the pintle to adjust pintle length or damage to the IAC worm gear might occur. The valve is preset at the factory and will self-adjust when the following procedure is performed.

a. Set a new IAC valve by measuring the distance between the tip of the pintle and the valve mounting surface.

b. If greater than 1.10 in. (28mm), use light finger pressure to slowly retract the pintle. The force required to retract a new valve will not damage the valve.

c. Install the valve and connect the wire connector.

d. Reset a used IAC valve pintle position by depressing the accelerator pedal slightly, start the engine and run for 5 seconds, turn the key **OFF** for 10 seconds, then restart the vehicle and check for proper idle operation.

CONTROLLED IDLE SPEED CHECK

Idle speed adjustments are not provided for most 1994–97 vehicles (with the exception of the 7.4L (VIN N) engine). However, a controlled idle speed check may be made on most vehicles to determine if the computer module is successfully regulating the idle speed. If an improper idle speed is indicated by the check, inspect all components of the engine control systems.

1. Using a scan tool, check to be sure no codes are stored in the computer self-diagnosis memory. If codes are stored, diagnose and correct the problems before proceeding.

2. Verify that the idle air control system is working correctly, then check to make sure the ignition timing is properly set.

3. Connect the Tech 1® or equivalent scan tool to the Diagnostic Link Connector (DLC). Make sure the tool is in the OPEN mode.

4. Start and run the engine at normal operating temperature.

5. If equipped with an automatic transmission, use the scan tool to check for a proper state of the PRNDL position (R-D-L) switch.

6. Using the scan tool, compare readings of the controlled idle speed

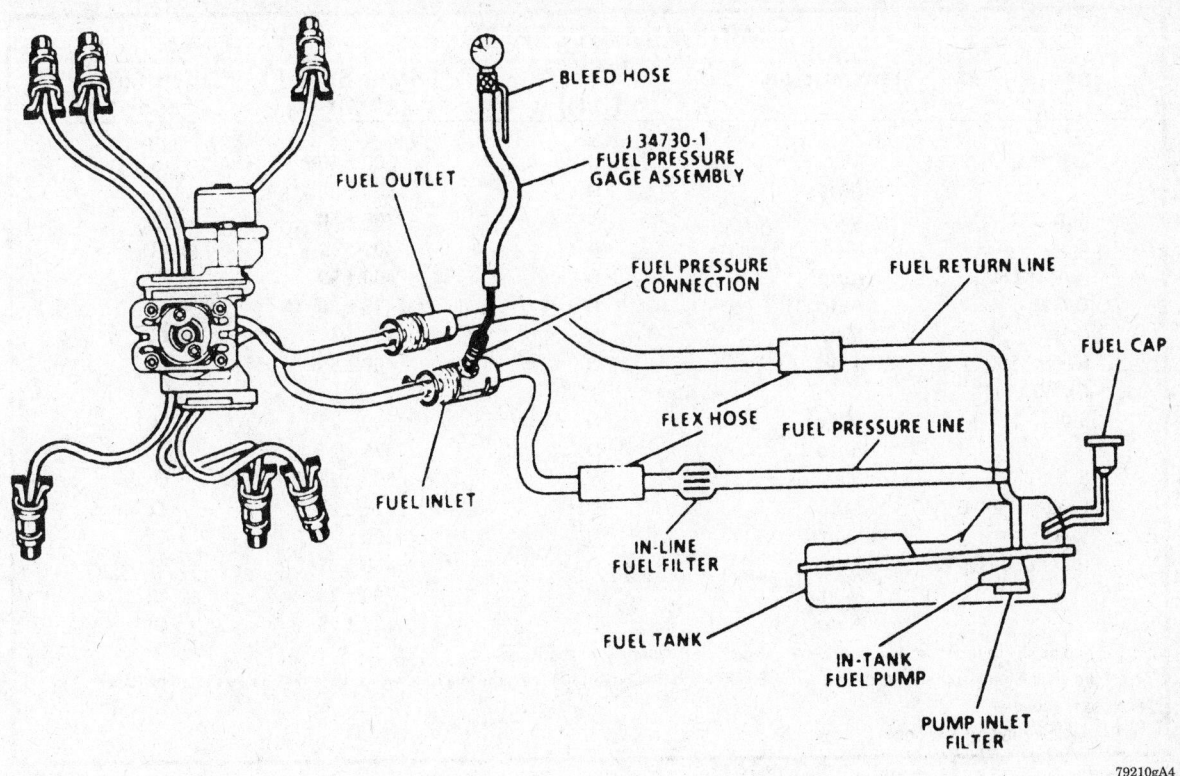

CMFI fuel system pressure testing — 4.3L (VIN W) engine

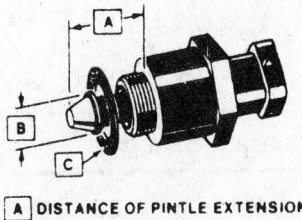

A DISTANCE OF PINTLE EXTENSION
B DIAMETER AND SHAPE OF PINTLE
C IAC VALVE GASKET

79210g17

A threaded IAC valve used on GM vehicles

and IAC valve pintle position to the specifications for that engine.

NOTE: For purposes of the minimum idle speed and IAC valve specifications, the G-Series Van should be treated as the other trucks.

7. If the readings agree with specifications, the system is operating properly. If not, the system must be thoroughly checked for faulty connections or components.

Mixture

ADJUSTMENT

Fuel mixture adjustments on these engines are controlled by the computer modules, regulated through the IAC valve and fuel injector(s), respectively. No periodic check or adjustments are necessary for these systems.

Fuel Filter

REMOVAL AND INSTALLATION

The fuel filter is normally located along the frame rail of the vehicle. On some vehicles however, it may have been relocated to the engine compartment. When in doubt, trace a fuel line from the engine backwards or from the tank forward in order to locate the filter.

1. Properly relieve the fuel system pressure.
2. Disconnect the negative battery cable.
3. Raise and support the vehicle safely.
4. Disengage the fuel line connections from the filter.

5. Remove the bolt from the filter mounting clamp, then remove the clamp and filter assembly. Separate the filter from the clamp.
6. To install, reverse the removal procedures ensuring the filter flow is in the correct direction. Start the engine and check for leaks.

NOTE: The filter has an arrow (fuel flow direction) on the side of the case, be sure to install it correctly in the system, the with arrow facing away from the fuel tank.

Fuel Pump

REMOVAL AND INSTALLATION

1. Properly relieve the fuel system pressure.
2. Disconnect the negative battery cable.
3. Drain and remove the fuel tank from the vehicle
4. Using a suitable spanner wrench, turn the fuel pump/sending unit assembly locking ring (located on top of the fuel tank) counterclockwise, then carefully lift the assembly from the tank and remove the pump from the fuel lever sending device.

MINIMUM IDLE SPEED

Engine	Transmission	Gear (D/N)	Engine Speed (RPM)**	Open/Closed Loop*
2.5L	Man.	N	600 ± 50	CL
	Auto.	N	500 ± 50	CL
2.8L	Man.	N	700 ± 50	OL
4.3L	Man.	N	400-525	CL
(under 8500	Auto.	D	400 ± 50	CL
GVW)	Auto.(1)	D	475 ± 50	CL
4.3L	Man.	N	400-525	CL
(Over 8500	Auto.	D	400 ± 50	CL
GVW)				
5.0L	Man.	N	500 ± 25	OL
	Auto.	D	425 ± 25	CL
5.7L	Man.	N	500 ± 25	OL
(under 8500 GVW)	Auto.	D	425 ± 25	CL
5.7L	Man.	N	550 ± 25	CL
(over 8500 GVW)	Auto.	D	450 ± 25	CL
7.4L	Man.	N	700 ± 25	OL
	Auto.	D	625 ± 25	OL

* Let engine idle until proper fuel control status (open/closed loop) is reached

** If the engine has less than 500 miles or is checked at altitudes above 1500 feet, the idle rpm with a seated IAC valve should be lower than values above.

(1) 4.3L High-Output ML Van Series

79210gA5

Minimum idle air rate — 1993 vehicles

CONTROLLED IDLE SPEED
Note: Engine at operating temperature 92°C to 104°C (196°F to 222°F)

Engine	Transmission	Gear (Drive/Neutral)	Idle Speed (RPM)	IAC Counts *	Open/Closed Loop **
4.3L (TBI) UNDER 8500 GVW C/K-TRK	AUTO MANUAL	DRIVE NEUTRAL	590 ± 25 550 ± 25	5-30 5-30	CL
4.3L (TBI) C/K-TRK (OVER 8500 GVW)	AUTO MANUAL	DRIVE NEUTRAL	650 ± 25 700 ± 25	5-30 5-30	CL CL
5.0L C/K-TRK	MAN AUTO	NEUTRAL DRIVE	650 ± 25 550 ± 25	5-30 5-30	CL CL
5.7L C/K (under 8500 GVW)	MAN AUTO	NEUTRAL DRIVE	660 ± 25 525 ± 25	5-30 5-30	CL CL
5.7L C/K (over 8500 GVW)	MAN AUTO	NEUTRAL DRIVE	590 ± 25 550 ± 25	5-30 5-30	CL CL
7.4L C/K	MAN AUTO	NEUTRAL DRIVE	750 ± 25 675 ± 25	5-30 5-30	CL CL

* On manual transmission vehicles the Tech 1 will display RDL in neutral.
  Add 2 counts for engines with less than 500 miles. Add 2 counts for every 1000 ft. above sea level (4.3L and V8).

** Let engine idle until proper fuel control status ("Open/Closed Loop") is reached.

79210gA6

Controlled idle speed check specifications

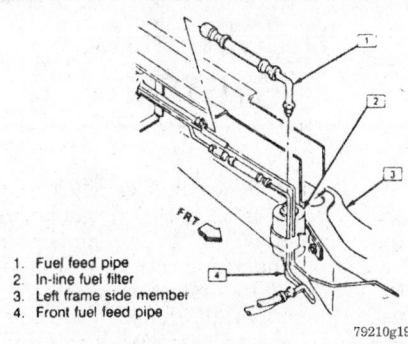

1. Fuel feed pipe
2. In-line fuel filter
3. Left frame side member
4. Front fuel feed pipe

79210g19

Fuel filter mounting for the 4.3L engine

5. Installation is the reverse of the removal procedure. The fuel pump/sending unit assembly O-ring should be replaced whenever the tank is removed.

Fuel Injector

REMOVAL AND INSTALLATION

TBI Engines

------- CAUTION -------

When removing the injector(s), be careful not to damage the electrical connector pins (on top of the injector), the injector fuel filter and the nozzle. The fuel injector is serviced as a complete assembly ONLY, it is an electrical component and should not be immersed in any kind of cleaner.

1. Properly relieve the fuel system pressure, then disconnect the negative battery cable.
2. Remove the air cleaner assembly.
3. At the injector connector(s), squeeze the 2 tabs together and pull straight up to disengage connector from the injector.
4. Losen the fuel meter cover retaining screws, then remove the cover from the fuel meter body, but leave the cover gasket in place.
5. Using a small pry bar and a round fulcrum, carefully pry the injector until it is free, then remove the injector from the fuel meter body.
6. Remove the small O-ring from the nozzle end of the injector. If equipped and removal is necessary, carefully rotate the injector's fuel filter back and forth to remove it from the base of the injector.
7. If applicable, remove and discard the fuel meter cover gasket.
8. Remove the large O-ring and back-up washer, if equipped, from the top of the counterbore of the fuel meter body injector cavity.

To install:

NOTE: **Be sure the replacement injectors have an identical part number. For some applications, the 4.3L (VIN Z) uses 2 different injectors that have different flow rates. Injectors with part no. 5235134 (color coded orange and green) are located on the throttle lever side, and those with part no. 5235342 (color coded pink and brown) are located on the TPS side.**

9. If removed, with the larger end of the filter facing the injector (so the filter covers the raised rib of the injector base) install the filter by twisting it into position on the injector.
10. Lubricate the new O-rings with clean automatic transmission fluid, then install the small O-ring on the nozzle end of the injector. Be sure the O-ring is pressed up against the injector or injector filter (as applicable).
11. Install the steel backup washer, if equipped, in the top counterbore of the fuel meter body's injector cavity, then install the new large O-ring directly over the backup washer. Make sure the O-ring is properly seated in the cavity and is flush with the top of the fuel meter body casting surface.

NOTE: **If the backup washer and large O-ring are not properly installed before the fuel injector, a fuel leak will likely result.**

12. For the model 220 TBI unit, install the fuel injector into the cavity by aligning the raised lug on the injector base with the cast notch in the fuel meter body cavity. Once the injector is aligned, carefully push down on the injector by hand until it is fully seated in the cavity. When properly aligned and installed, the injector terminals will be approximately parallel to the throttle shaft.
13. Position a new fuel meter cover gasket, then install the cover to the body, making sure the gasket remains in position. Using a suitable thread locking compound, install and tighten the cover retainers to 30 inch lbs. (4 Nm).
14. Engage the injector electrical connector(s).
15. Connect the negative battery cable, then turn the **ON** to pressurize the fuel system and check for leaks.
16. Install the air cleaner assembly, then start the engine and check for leaks.

4.3L (VIN W) CMFI Engine

The non-repairable CMFI assembly or injection unit consists of a fuel meter body, gasket seal, fuel pressure regulator, fuel injector and 6 poppet nozzles with fuel tubes. The assembly is housed in the lower intake manifold. Should a failure occur in the CMFI assembly, the entire component must be replaced as a unit.

1. Remove the plastic cover and properly relieve the fuel system pressure.
2. Disconnect the negative battery cable, then remove the air cleaner and air inlet duct.
3. Disengage the wiring harness from the necessary upper intake components including:
• Throttle Position (TP) sensor
• Idle Air Control (IAC) motor
• Manifold Absolute Pressure (MAP) sensor
• Intake Manifold Tuning Valve (IMTV)
4. Disengage the throttle linkage from the upper intake manifold, then remove the ignition coil.
5. Disconnect the PCV hose at the rear of the upper intake manifold, then tag and disengage the vacuum hoses from both the front and rear of the upper intake.
6. Remove the upper intake manifold bolts and studs, making sure to note or mark the location of all studs to assure proper installation. Remove the upper intake manifold from the engine.
7. Disengage the injector wiring harness connector at the CMFI assembly.
8. Remove and discard the fuel fitting clip.
9. Disconnect the fuel inlet and return tube and fitting assembly. Discard the old O-rings.
10. Squeeze the poppet nozzle locktabs together while lifting each nozzle out of the casting socket. Once all 6 nozzles are released, carefully lift the CMFI assembly out of the casting.

To install:

11. Align the CMFI assembly grommet with the casting grommet slots and push downward until it is seated in the bottom guide hole.

------- CAUTION -------

To reduce the risk of fire and personal injury, be absolutely sure the poppet nozzles are firmly seated and locked into their casting sockets. An unlocked poppet nozzle could work loose from its socket resulting in a dangerous fuel leak.

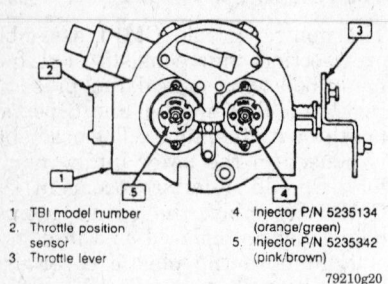

1. TBI model number
2. Throttle position sensor
3. Throttle lever
4. Injector P/N 5235134 (orange/green)
5. Injector P/N 5235342 (pink/brown)

79210g20

Fuel injector location — Model 220 TBI unit (used on engines, except the 2.5L)

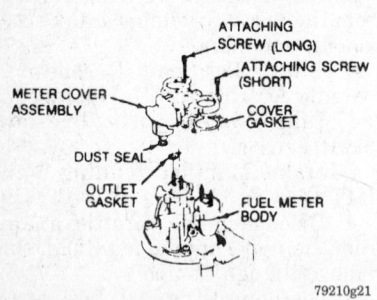

79210g21

Exploded view of the fuel meter cover mounting — Model 220 TBI unit

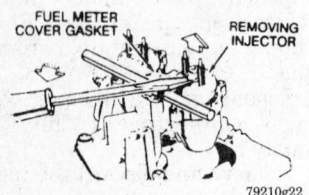

79210g22

Removing the fuel injector — Model 220 TBI unit

12. Carefully insert the poppet nozzles into the casting sockets. Make sure they are firmly seated and locked into the casting sockets.

13. Position new O-ring seals (lightly coated with clean engine oil), then connect the fuel inlet and return tube and fitting assembly.

14. Install a new fuel fitting clip.

15. Temporarily connect the negative battery cable, then pressurize the fuel system by cycling the ignition switch **ON** for 2 seconds, then

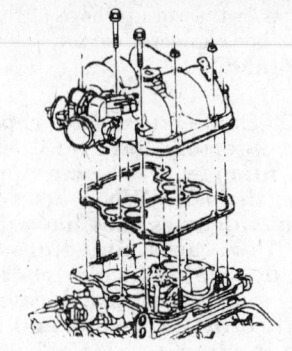

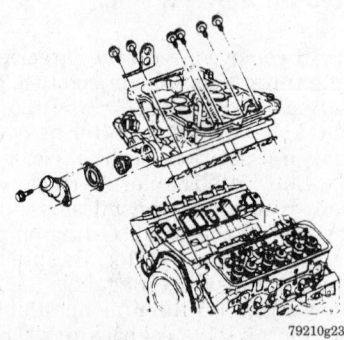

79210g23

Exploded view of the upper and lower intake manifolds along with the CMFI system components — 4.3L engine

OFF for 10 seconds and repeating, as necessary. Once the fuel system is pressurized, check for leaks.

16. Disconnect the negative battery cable.

17. Position a new upper intake manifold gasket on the engine, making sure the green sealing lines are facing upward.

18. Install the upper intake manifold being careful not to pinch the fuel injector wires between the manifolds.

19. Install the manifold retainers, making sure the studs are properly positioned, then tighten them using the proper sequence to 124 inch lbs. (14 Nm).

20. Connect the PCV hose to the rear of the upper intake manifold and the vacuum hoses to both the front and rear of the manifold assembly.

21. Connect the throttle linkage to the upper intake, then install the ignition coil.

22. Engage the necessary wiring to the upper intake components including the TP sensor, IAC motor, MAP sensor and the IMTV.

23. Install the plastic cover, the air cleaner and air inlet duct.

24. Connect the negative battery cable.

DIESEL FUEL SYSTEM

Starting on some 1993 vehicles and on all 1994–97 vehicles (except the non-turbocharged, heavy emission VIN Y engine), an electronically controlled diesel fuel injection pump was used. The major difference in the new electronic system is the use of the Powertrain Control Module (PCM) to control emission output by regulating the emission systems, monitoring engine operation and electronically controlling the diesel injection pump. Most system removal and installation procedures remain similar or the same as the mechanical system with subtle differences for additional or revised components.

Fuel System Service Precautions

Safety is the most important factor when performing not only fuel system maintenance but any type of maintenance. Failure to conduct maintenance and repairs in a safe manner may result in serious personal injury or death. Maintenance and testing of the vehicle's fuel system components can be accomplished safely and effectively by adhering to the following rules and guidelines.

• To avoid the possibility of fire and personal injury, always disconnect the negative battery cable unless the repair or test procedure requires that battery voltage be applied.

• Always relieve the fuel system pressure prior to disconnecting any fuel system component (injector, fuel rail, pressure regulator, etc.), fitting or fuel line connection. Exercise extreme caution whenever relieving fuel system pressure to avoid exposing skin, face and eyes to fuel spray. Please be advised that fuel under pressure may penetrate the skin or any part of the body that it contacts.

• Always place a shop towel or cloth around the fitting or connection prior to loosening to absorb any excess fuel due to spillage. Ensure that all fuel spillage (should it occur) is quickly removed from engine surfaces. Ensure that all fuel soaked cloths or towels are deposited into a suitable waste container.

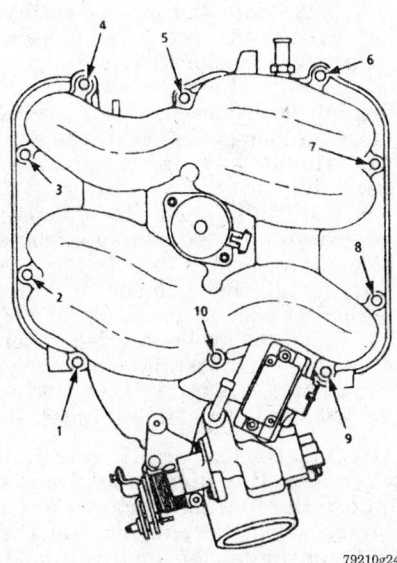

Upper intake manifold torque sequence — 4.3L engine

79210g24

- Always keep a dry chemical (Class B) fire extinguisher near the work area.
- Do not allow fuel spray or fuel vapors to come into contact with a spark or open flame.
- Always use a backup wrench when loosening and tightening fuel line connection fittings. This will prevent unnecessary stress and torsion to fuel line piping. Always follow the proper torque specifications.
- Always replace worn fuel fitting O-rings with new. Do not substitute fuel hose or equivalent, where fuel pipe is installed.

Fuel System

RELIEVING PRESSURE

Fuel system pressure can be released by wrapping a fuel fitting in a heavy shop towel and cracking loose the fitting. NEVER perform this with any source of ignition nearby!

BLEEDING AIR

1. Open the air bleed valve on the fuel manager/filter.
2. Connect a hose to the air bleed valve and place the other of the hose in a suitable container.

---- CAUTION ----

The diesel/water mixture is flammable and may be hot. to avoid personal injury or property damage, do not allow the diesel/water mixture to come in contact with skin, open flame or a hot engine. Do not overfill the container holding the fuel mixture as heat from a warm engine or any another heat sorce may cause the fuel to expand and leak from the container which may lead to a fire.

3. Remove the F/SOL fuse from the fuse panel.
4. Crank the engine in short intervals of 10-to-15 seconds until clear fuel is observed at the air bleed hose (wait for one minute between cranking intervals).
5. Remove the hose and close the air bleed valve.
6. Install the F/SOL fuse and start the vehicle. Allow the vehicle to run at idle for 5 minutes.
7. Check for fuel leaks and clear any DTC's.

Idle Speed

ADJUSTMENT

Idle speed is controlled by the PCM on electronically controlled injection pump engines. On mechanical pump engines, the idle speed may be adjusted:

NOTE: A special tachometer suitable for diesel engines must be used. A gasoline engine type tachometer will not work with the diesel engine.

1. Set the parking brake and block the drive wheels.
2. Run the engine up to normal operating temperature. The air cleaner must be mounted and all accessories turned **OFF**.
3. Install the diesel tachometer as per the manufacturer's instructions.
4. Adjust the low idle speed screw on the fuel injection pump to manu-facturer's specification per the emission control label.

NOTE: All idle speeds are to be set within 25 rpm of the specified values.

5. Adjust the fast idle speed as follows:
   a. Remove the connector from the fast idle solenoid. Connect a jumper wire from the battery positive terminal to the solenoid terminal to energize the solenoid.
   b. Open the throttle momentarily to ensure that the fast idle solenoid plunger is energized and fully extended.
   c. Adjust the fast idle by turning the hex-head screw to manufacturer's specification per the emission control label.
   d. Remove the jumper wire and reinstall the connector to the fast idle solenoid.
6. Disconnect and remove the tachometer.

Fuel Manager/Filter

The fuel manager/filter is an inline filter with several functions. These functions include acting as a fuel filter, water detector, water separator, water drain and a fuel heater.

DRAINING WATER FROM THE SYSTEM

Water is the worst enemy of the diesel fuel injection system. The injection pump and injectors, which are designed and constructed with extremely close tolerances, can be easily damaged if enough water is forced through them in the fuel. Engine performance will also be drastically affected and engine damage can occur. Diesel fuel is much more susceptible than gasoline to water contamination. Diesel engine vehicles are equipped with an indicator light system located in the instrument panel.

On 1993–97 vehicles, the water sensor is located in the fuel filter. Once the water level in the collector at the bottom of the filter assembly reaches a predetermined point (approximately 2.2 fluid ounces on 1993 vehicles) the indicator light will illuminate. The light will come ON for 2–5 seconds each time the ignition is turned ON, assuring the driver the light is working. If there is water in the fuel, the light will come back ON

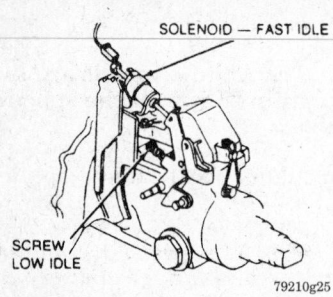

SOLENOID — FAST IDLE

SCREW
LOW IDLE

79210g25

Mechanical diesel injection pump idle adjustment locations

after a 15–20 second off delay, and then remain ON.

NOTE: Water should be drained from the fuel system as soon as possible after the warning light illuminates. It is recommended that the water be drained after no longer than 2 hours of engine operation after the light comes on.

Purging The Fuel Tank

If equipped with the tank based water warning system, the fuel tank may be purged of water using the system components. The fuel tank is equipped with a filter which screens out the water and lets it lay in the bottom of the tank below the fuel pickup. When the water level reaches a point where it could be drawn into the system, a warning light flashes in the cab. A built-in siphoning system starting at the fuel tank and going to the rear spring hanger on some models, and at the midway point of the right frame rail on other models permits attaching a hose at the shut-off and siphon out the water. If it becomes necessary to drain water from the fuel tank, also check the primary fuel filter for water.

Purging The Fuel Filter

1. Make sure the engine is turned **OFF**, then firmly set the parking brake and block the drive wheels.
2. Remove the fuel filler cap to relieve any pressure or vacuum from the tank.
3. Position a suitable container under the filter drain hose, then open the drain valve 2–3 turns.
4. Start the engine and allow it to idle for 1–2 minutes or until clear fuel is observed.
5. Close the drain valve and stop the engine.
6. Install the fuel filler cap.

7. Dispose of the contaminated fuel in a proper manner.

REMOVAL AND INSTALLATION

1993 Vehicles

1. Turn the ignition **OFF**. Remove the fuel tank cap to release any pressure or vacuum in the tank.
2. Drain the fuel from the fuel filter by opening both the air bleed and the water drain valve allowing the fuel to drain into a suitable container.
3. Unstrap both bail wires with a suitable tool and remove the filter by pulling straight away from the filter base.
   To install:
4. Before installing the new filter, ensure that both filter mounting plate fittings are clear of dirt.
5. Position the filter, then snap into place with the bail wires.
6. Close the water drain valve and open the air bleed valve. Connect a 1/8 in. (3mm) I.D. hose to the air bleed port and place the other end into a suitable container.
7. Disconnect the fuel injection pump shutdown solenoid wire.
8. Crank the engine for 10–15 seconds, then wait 1 minute for the starter motor to cool. Repeat until clear fuel is observed coming from the air bleed.

NOTE: If the engine is to be cranked or started with the air cleaner removed, care must be taken to prevent dirt from being pulled into the air inlet manifold which could result in engine damage.

9. Close the air bleed valve, reconnect the injection pump solenoid wire and replace the fuel tank cap.
10. Start the engine, allow it to idle for 5 minutes and check the fuel filter for leaks.

1994–97 Vehicles

1. Turn the ignition **OFF**. Remove the fuel tank cap to release any pressure or vacuum in the tank.

NOTE: It is not necessary to drain all the fuel from the header in order to change the element since the fuel will remain in the header's cavity.

2. Remove the element nut, turning it by hand to the left. If necessary, a strap wrench may be used to loosen the nut.

3. Remove the element by lifting straight up and out of the header assembly.
   To install:
4. Make sure the mating surface between the element assembly and the header assembly is clean.
5. Install the new element by aligning the widest key slot located under the element assembly cap with the widest key in the header assembly.
6. Carefully push the element downward until the mating surfaces make contact.
7. Install the element nut and tighten securely by hand.
8. Open the air bleed valve on top of the fuel manager/filter assembly, then connect a length of hose placing the other end in a suitable container.

NOTE: Be extremely cautious when handling diesel fuel. Do not expose the fuel to sparks or open flames. Also, be cautious as the fuel coming out of the drain hose could be hot.

9. Disconnect the fuel injection pump shutdown solenoid wire.
10. Crank the engine for 10–15 seconds, then wait 1 minute for the starter motor to cool. Repeat until clear fuel is observed coming from the air bleed.
11. Close the air bleed valve, reconnect the injection pump solenoid wire and replace the fuel tank cap.
12. Start the engine, allow it to idle for 5 minutes and check the fuel manager/filter assembly for leaks.

Diesel Injection Pump

Though most earlier vehicles (1993 models) utilize a mechanical injection pump, most later vehicles are equipped with an electronically controlled pump. The electronic pump is still driven by gear and rotates at the same speed as the camshaft. The major difference between the mechanical and electronic pumps is an electronic stepper motor used to control injection timing and a fuel solenoid driver used to control the fuel injection solenoid on the electronic model.

REMOVAL AND INSTALLATION

1. Disconnect both battery negative cables.
2. Remove the intake manifold.
3. Remove the fuel injection lines.
4. If necessary, disconnect the detent and/or accelerator cable at the injection pump.

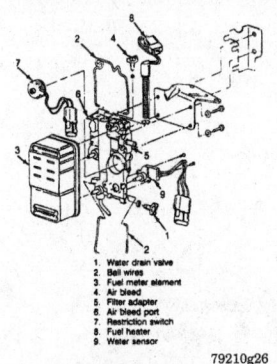

Diesel engine fuel filter — 1993 models

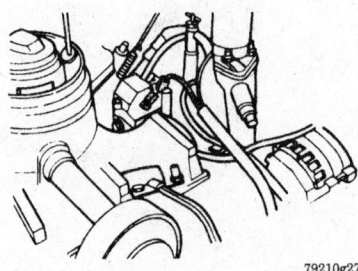

79210g27

Diesel shutdown solenoid location

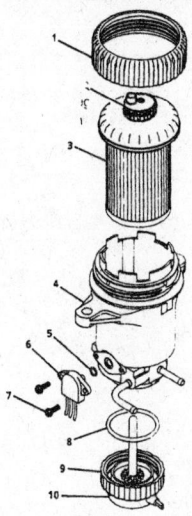

1	ELEMENT NUT
2	AIR BLEED VALVE
3	ELEMENT ASSEMBLY
4	HOUSING
5	SENSOR SEAL
6	WATER IN FUEL SENSOR
7	SENSOR MOUNTING SCREW
8	CAP SEAL
9	CAP NUT
10	FUEL HEATER

79210gF1

Exploded view of the fuel manager/filter assembly

Fuel Injector

REMOVAL AND INSTALLATION

1. Disconnect the negative cable on both batteries.
2. Disconnect the fuel line clip, then remove the fuel return hose.
3. Remove the fuel injection line. Immediately cap the nozzle and lines to prevent system contamination and damage.
4. Using GM special tool J–29873 or equivalent, remove the injector. Always remove the injector by turning the 30mm hex portion of the injector.
5. Install the injector using a new gasket and torque to 50 ft. lbs. (70 Nm).
6. Connect the injection line and torque the nut to 20 ft. lbs. (25 Nm).
7. Install the fuel return hose, then install the fuel line clip and connect the batteries.

GASOLINE ENGINE MECHANICAL

NOTE: Disconnecting the battery cable on most vehicles may interfere with the functions of the on-board computer systems and may require the computer to undergo a relearning process.

Engine Assembly

REMOVAL AND INSTALLATION

Blazer, Pick-Up, Suburban, Tahoe and Yukon

1. Relieve the fuel system pressure, then disconnect the negative battery cable.
2. Remove the hood.
3. Drain the cooling system.
4. Remove the air cleaner.
5. Remove the accessory drive belt, fan and water pump pulley.
6. Remove the radiator and shroud.
7. Disconnect the heater hoses at the engine.
8. Remove the air conditioning condenser (if equipped).
9. Disconnect the accelerator, cruise control and detent linkage, as applicable.

---

5. Tag and disconnect the necessary wires and hoses at the injection pump.
6. Disconnect the fuel return line at the top of the injection pump.
7. If equipped on mechanical pump engines, remove the air conditioning hose retainer bracket.
8. If necessary, disconnect the fuel feed line at the injection pump.
9. Remove the oil fill tube (on the mechanical pump engines this should include the Crankcase Depression Regulator (CDR) valve vent hose assembly). Remove the grommet.

NOTE: Do not engage the starter in order to rotate the engine with the injection pump removed. The pump driven gear could jam in the front housing resulting in a sheared crankshaft or camshaft gear key and possible valve train damage.

10. Scribe or paint a matchmark on the front cover and the injection pump flange.
11. Rotate the crankshaft by hand and remove the injection pump driven gear bolts, accessing the bolts through the oil filler neck hole.
12. Remove the injection pump-to-front cover attaching nuts. Remove

the pump. Be sure to cap all open lines and nozzles in order to prevent system contamination and damage.

To install:

13. Position and new pump gasket.
14. Align the locating pin on the pump hub with the slot in the injection pump driven gear (the SLOT not the hole in the gear). At the same time, align the timing marks.
15. Attach the injection pump to the front cover, checking the timing marks before torquing the nuts to 30 ft. lbs. (40 Nm).
16. Install the driven gear-to-injection pump bolts, torquing the bolts to 20 ft. lbs. (25 Nm).
17. Install the grommet and oil fill tube.
18. If applicable, install the air conditioning bracket.
19. Connect the fuel feed line and torque to 20 ft. lbs. (25 Nm).
20. If removed, connect the fuel return line to the pump.
21. If necessary, connect the detent and/or accelerator cables.
22. Connect all wires and hoses previously removed.
23. Connect the injector lines and install the intake manifold.
24. Connect the negative battery cables. Start the engine and check for leaks.

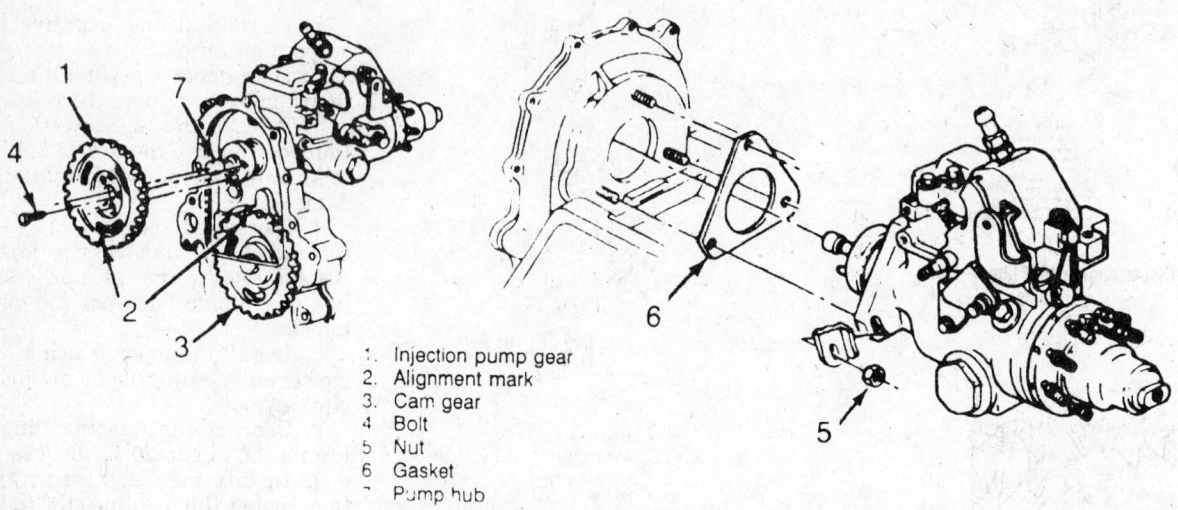

1. Injection pump gear
2. Alignment mark
3. Cam gear
4. Bolt
5. Nut
6. Gasket
7. Pump hub

79210g28

Exploded view of the diesel injection pump

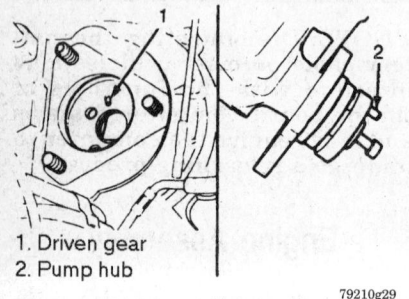

1. Driven gear
2. Pump hub

79210g29

Diesel injection pump locating pin

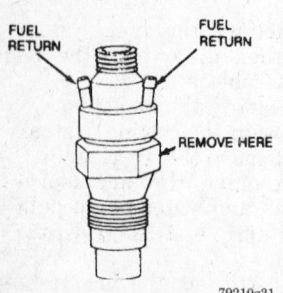

FUEL RETURN   FUEL RETURN

REMOVE HERE

79210g31

Diesel injection nozzles

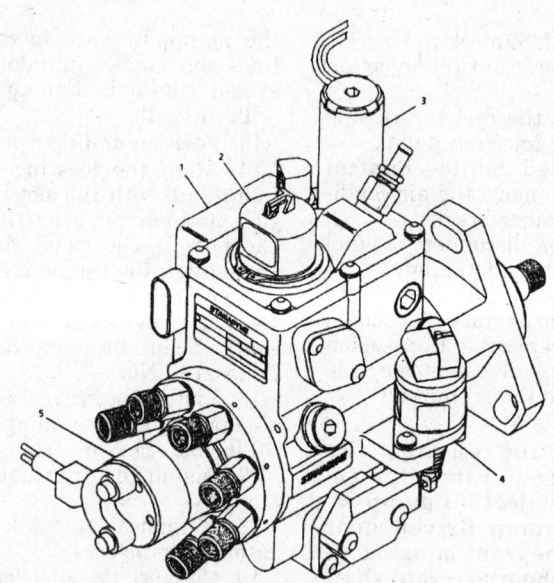

1. Fuel solenoid driver
2. Optical/fuel temperature sensor
3. Engine shut off solenoid
4. Injection timing stepper motor
5. Fuel solenoid

79210g30

Electronic injection pump components

10. Disconnect the air conditioning compressor, if equipped, and lay aside.

11. Remove the power steering pump, if used, and lay aside.

12. Remove the alternator and compressor.

13. Disconnect the engine wiring from the engine.

14. Disconnect the fuel line(s).

15. Disconnect the vacuum lines from the intake manifold.

16. Remove the distributor.

17. Raise the vehicle and support it safely.

18. Drain the engine oil.

19. Disconnect the exhaust pipes from the manifolds.

20. Disconnect the strut rods at the engine mountings, if used.

21. Remove the flywheel or torque converter cover.

22. Disconnect the wiring along the oil pan rail.

23. Remove the starter.

24. If applicable, disconnect the wire for the fuel gauge.

25. If equipped with automatic transmission, remove the converter to flexplate bolts.

26. Lower the vehicle and suitably support the transmission. Attach a suitable lifting fixture to the engine.

27. Remove the bellhousing to engine bolts.

28. Remove the engine mounting to frame bolts and remove the engine.

To install:

29. Raise the vehicle and support it safely.

30. Lower the engine and install the engine mounting bolts.

31. Install the bellhousing to engine bolts and torque to 35 ft. lbs. (44 Nm).

32. Install the converter to flex bolts and torque.

33. Install the fuel gauge wiring (if applicable) and starter.

34. Install the flywheel or torque converter cover.

35. Connect the strut rods at the engine mountings, if used.

36. Install the exhaust pipes at the manifold.

37. Lower the vehicle.

38. Install the distributor.

39. Connect the vacuum lines to the intake manifold.

40. Install the fuel line(s).

41. Connect the engine wiring harness.

42. Install the alternator and bracket.

43. Install the power steering pump, if used.

44. Connect the air conditioning compressor, if used.

45. Connect the accelerator, cruise control and detent linkage.

46. Connect the heater hoses.

47. Install the air conditioning condenser (if equipped).

48. Install the radiator and shroud.

49. Install the accessory drive belts.

50. Install the air cleaner.

51. Install the hood.

52. Install the proper quantity and grade of coolant and engine oil.

53. Connect the negative battery cable.

G-Series

4.3L Engine

1. If equipped, properly recover the refrigerant from the A/C system.

2. Properly relieve the fuel system pressure, then disconnect the negative battery cable.

3. If necessary, remove the glove box.

4. Drain the engine cooling system.

5. Remove the engine cover.

6. Remove the air cleaner assembly and air intake duct.

7. Remove the power steering reservoir.

8. Remove the upper fan shroud, then remove the fan and pulley. Remove the radiator assembly.

9. Remove the air conditioning condenser, then remove the compressor and bracket.

10. Remove the alternator.

11. If equipped, remove the cruise control servo.

12. Tag and disconnect all vacuum hoses.

13. Disconnect the accelerator linkage, cruise control and TVS cables with the mounting brackets from the throttle body, as applicable.

14. Properly relieve the fuel system pressure and remove the TBI unit.

15. Remove the distributor cap with the wires attached. If necessary, remove the coil and mounting bracket.

16. Remove the MAP sensor and mounting bracket.

17. Tag and disengage the engine wiring harness connectors from the necessary sensors and switches. Lay the harness aside.

18. Remove the upper half of the engine dipstick tube.

19. Remove the oil filler tube.

20. If necessary, remove the power steering pump and position aside.

21. Remove the headlight bezels and the grille.

22. Remove the upper radiator support.

23. If necessary, remove the lower fan shroud and filler panel. Remove the hood latch support.

24. Raise and support the vehicle safely.

25. Drain the engine oil.

26. Disconnect the exhaust pipes at the manifolds.

27. Remove the strut rods at the torque converter underpan.

28. Remove the torque converter cover.

29. If equipped, disconnect the oil cooler lines from the engine.

30. Remove the starter.

31. Remove the flexplate-to-torque converter bolts.

32. Remove the engine mounting through bolts.

33. Remove the bellhousing-to-engine bolts.

34. Lower the vehicle and support the transmission using the proper equipment.

35. Attach an engine crane to the engine, pull the engine forward and upward and remove it.

To install:

36. Lower the engine into position and engage the transmission.

37. Install and tighten the engine mount through-bolts.

38. Install and tighten the bellhousing-to-engine bolts.

39. Raise and support the vehicle safely for access.

40. Install the flexplate-to-torque converter bolts.

41. Install the starter.

42. Install the torque converter cover.

43. Install the strut rods at the torque converter underpan.

44. If equipped, connect the oil cooler lines to the engine.

45. Connect the exhaust pipes at the manifolds, then lower the vehicle.

46. Install the radiator support.

47. Install the head light bezel and grill.

48. Install oil fill tube and the upper half of the dipstick tube.

49. If applicable, install the MAP sensor and/or ignition coil and brackets.

50. Install the distributor cap and spark plug wires.

51. Engage the engine wiring harness connectors and harness retaining clamps.

52. Install the TBI unit, then connect the accelerator cables (with cruise control and/or TVS cables as applicable) along with the brackets.

53. Connect the vacuum hoses to the engine.

54. Install the cruise control servo.

55. Install the alternator.

56. If equipped, install the air conditioning compressor and the condenser.

57. Install the radiator and lower shroud.

58. Install the cooling fan and pulley, then install the upper fan shroud.

59. Install the power steering reservoir.

60. Install the hood release cable.

61. Install the air duct and the air cleaner assembly.

62. Refill the engine crankcase.

63. Connect the negative battery cable.

64. Refill the cooling system with the proper type and quantity of antifreeze.

65. Install the engine cover.

66. If removed, install the glove box.

67. If equipped, properly evacuate and charge the air conditioning system.

5.0L, and 5.7L Engines

1. If equipped, properly recover the refrigerant from the A/C system.

2. Properly relieve the fuel system pressure, then disconnect the negative battery cable.

3. Drain the engine cooling system.

4. Remove the radiator coolant reservoir bottle.

5. Remove the grille and the lower grille valance.

6. Remove the upper radiator support.

7. Remove the air conditioning condenser from in front of the radiator.

8. Remove the radiator.

9. Remove the power steering pump and position it aside with the lines intact.

10. Remove the engine cover.

11. Remove the air cleaner.

12. Remove the TBI unit.

13. Disconnect the engine wiring harness from the firewall connection.

14. Tag and disengage all necessary vacuum lines and wiring harness connectors.

15. Disconnect the heater pipe or hoses at the engine, as applicable.

16. Remove the thermostat housing.

17. Remove the oil filler tube.

18. Remove the cruise control servo, bracket and transducer, as necessary.

19. Raise and support the vehicle safely.

20. Drain the engine oil.

21. Disconnect the exhaust pipes at the manifolds.

22. Remove the driveshaft from the transmission and plug the end of the transmission case to prevent fluid leakage.

23. Disconnect the transmission shift linkage and the speed sensor connector.

24. Disconnect the fuel and vapor return lines at the engine.

25. Remove the rear engine/transmission mount bolts.

WARNING
Do not jack under the oil pan, crankshaft pulley or any sheet metal for any reason. Due to the small clearance between the oil pan and pump screen, jacking against the pan could cause it to be bent up against the pump screen resulting in a damaged oil pickup unit.

---

26. Support the transmission and engine.

27. Remove the engine mount bracket-to-frame bolts.

28. Remove the engine mount through bolts.

29. Raise the engine slightly and remove the engine mounts.

30. Support the engine with wood blocks.

31. Lower the vehicle.

32. Install a suitable engine lifting device.

33. Remove the engine and transmission from the vehicle as an assembly.

34. Separate the engine and transmission, as required.

To install:

35. If separated, joint the engine and transmission and support using a suitable lifting device.

36. Install the engine and transmission assembly to the vehicle, then install the engine mounts and secure.

37. Remove the engine lifting device, then raise and support the vehicle safely.

38. Connect the fuel and vapor return lines at the engine.

39. Connect the transmission shift linkage, then engage the speed sensor connector.

40. Install the driveshaft.

41. Connect the exhaust pipes at the manifolds, then lower the vehicle.

42. If equipped, install the cruise control servo, bracket and transducer.

43. Install the oil filler tube.

44. Install the thermostat housing.

45. Connect the heater hoses or pipe at the engine.

46. Connect the vacuum hoses and wiring, as tagged during removal.

47. Connect the engine wiring harness connector.

48. Install the TBI unit.

49. Install the air cleaner.

50. Reposition and secure the power steering pump.

51. Install the radiator.

52. If equipped, install the A/C condenser.

53. Install the upper radiator support, the grille and the lower grille valance.

54. Install the radiator coolant reservoir bottle.

55. Refill the engine crankcase.

56. Install the engine cover.

57. Connect the negative battery cable.

58. Refill the cooling system with the proper type and quantity of antifreeze.

59. If equipped, properly evacuate and charge the air conditioning system.

7.4L Engine

1. If equipped, properly recover the refrigerant from the A/C system.

2. Properly relieve the fuel system pressure, then disconnect the negative battery cable.

3. Remove the engine cover.

4. Drain the engine cooling system, then drain the crankcase of engine oil.

5. Remove the air intake duct and the air cleaner assembly.

6. Disconnect the upper radiator hoses at the engine and the upper heater hose and the rear of the engine.

7. Tag and disconnect all wiring at the TBI unit, then disconnect the accelerator linkage (including the cruise control and/or TV linkage).

8. Tag and disconnect all vacuum hoses.

9. Remove the distributor cap (lay aside) and all spark plug wiring. Matchmark and remove the distributor assembly.

10. Disconnect the fuel lines and bracket. If equipped, remove the cruise control bracket.

11. Remove the ignition coil and bracket.

12. Disconnect the ECS and MAP sensors with bracket(s).

13. Remove the transmission and engine oil dipstick tubes.

14. If applicable, remove the EGR solenoid.

15. Remove the air conditioning compressor brackets (usually 2).

16. Remove the TBI unit.

17. Disconnect the wiring at the rear of the engine.

18. Remove the headlight bezels, grille, lower grille valance or bumper filler panel and all necessary sheet metal to ease removal and installation of the engine and transmission assembly.

19. If equipped, remove the cruise control servo.

20. Remove the hood latch and the radiator coolant reservoir bottle.

21. Remove the upper radiator brackets.

22. If not done earlier, remove the front end sheet metal cross panel, then remove the sheet metal vertical support with electric cooling fan.

23. Remove the air conditioning condenser from in front of the radiator. Cap all lines.

24. Remove the windshield wiper reservoir and washer pump.

25. Remove the upper cooling fan shroud.

26. Disconnect the lower radiator and heater hoses, then disconnect the engine oil and transmission cooler lines.

27. Remove the radiator, then remove the lower shroud.

28. Remove the cooling fan, then remove the multi-ribbed belt and the pump pulley.

29. Remove A/C line from the accumulator and alternator.

30. Remove the air conditioning line assembly from the back of the compressor. Remove the right side engine accessory mounting bracket with the A/C compressor and idler pulley.

31. Remove the power steering pump and position aside.

32. Disconnect the alternator wires, then remove the left engine accessory bracket along with the alternator and belt tensioner.

33. Raise and support the vehicle safely.

34. Remove the exhaust crossover pipe heat shields, then disconnect the pipe from the manifolds.

35. Remove the starter motor heat shield, then disconnect the starter motor and, if applicable, knock sensor wires. Remove the starter motor assembly.

36. Disconnect the transmission wiring and linkage.

37. Remove the driveshaft from the transmission, then plug the end of the transmission case to prevent fluid leakage.

38. Remove the engine mount through-bolts. The bolts may be difficult to remove until the engine weight is taken off the mounts, if necessary, loosen them and wait until the engine is supported.

39. Do not position the jack under the oil pan, crankshaft pulley or any sheet metal!.

40. Disconnect the transmission crossmember.

41. Remove the wiring from the front of the engine.

42. Lower the vehicle for access and engine/transmission assembly removal.

43. Attach a suitable lifting device to the engine.

44. Raise the engine as necessary and maneuver the engine/transmission assembly.

45. If necessary, separate the engine from the transmission, then mount the engine on a work stand.

To install:

46. If removed, connect the transmission to the engine and prepare them for installation.

47. Raise the engine/transmission and guide the assembly into position in the vehicle.

48. With the assembly still supported, carefully install the transmission crossmember and engine mount fasteners.

49. Install the driveshaft.

50. Connect the transmission shift linkage and wiring.

51. Install the starter motor assembly, then engage the starter motor and, if applicable, then knock sensor wiring. Install the starter motor heat shield.

52. Connect the exhaust crossover pipe, then install the crossover pipe heat shields.

53. Lower the vehicle as necessary for access.

54. Install the left engine accessory bracket along with the alternator and belt tensioner. Engage the alternator wiring.

55. Reposition and secure the power steering pump.

56. Install the right accessory accessory mounting bracket with the A/C compressor and idler pulley. Connect the air conditioning line assembly to the back of the compressor.

57. Install the A/C line to the accumulator and alternator. Connect the lower radiator hose to the engine.

58. Install the water pump pulley, then install the belt and the cooling fan.

59. Install the radiator and the lower shroud.

60. Connect the lower radiator and heater hoses, along with the engine oil and transmission cooler lines.

61. Install the upper cooling fan shroud.

62. Install the windshield wiper reservoir and washer pump.

63. Install the air conditioning condenser.

64. Install the sheet metal vertical support with electric cooling fan, then install the front end sheet metal cross panel.

65. Install the upper radiator brackets.

66. Install the radiator coolant reservoir bottle and the hood latch.

67. If equipped, install the cruise control servo.

68. Install the remaining front end sheet metal or trim components including the lower grille valance or bumper filler panel, grille and headlight bezels.

69. Connect the wiring at the rear of the engine.

70. Install the TBI unit.

71. Install the air conditioning compressor brackets (usually 2).

72. If applicable, install the EGR solenoid.

73. Install the transmission and engine oil dipstick tubes.

74. Connect the ECS and MAP sensors with bracket(s).

75. Install the ignition coil and bracket.

76. If equipped, install the cruise control bracket.

77. Connect the fuel lines and bracket.

78. Align and install the distributor assembly, then install the distributor cap and spark plug wiring.

79. Connect all vacuum hoses as tagged during removal.

80. Connect the accelerator linkage (including the cruise control and/or TV linkage), then connect all wiring at the TBI unit.

81. Connect the upper radiator hoses at the engine and the upper heater hose and the rear of the engine.

82. Install the air cleaner assembly and the air intake duct.

83. Refill the engine crankcase.

84. Connect the negative battery cable.

85. Refill the cooling system with the proper type and quantity of antifreeze.

86. If equipped, properly evacuate and charge the air conditioning system.

87. Install the engine cover.

Engine Mounts

NOTE: When lifting or raising the engine for any reason, do not support the assembly under the oil pan, any sheet metal or the crankshaft pulley.

CAUTION

When working on the engine mounts, make sure the engine is SECURELY supported at all times. If necessary, use blocks of wood to help make sure it cannot shift suddenly.

REMOVAL AND INSTALLATION

1. Disconnect the negative battery cable.
2. If access from underneath is preferable, raise and support the vehicle safely.
3. Support the engine with a suitable lifting fixture. Just take the weight off the mounts, do not load them.

NOTE: Do not position the jack under the oil pan, any sheet metal or the crankshaft pulley, otherwise damage may occur.

4. Remove the engine mount through-bolt and nut.
5. Raise the engine just sufficiently to permit removal of the engine mounting and block in position. Keep a close eye on the engine-to-cowl clearance when raising it.
6. Remove the mount assembly to frame retainers. Remove the mount assembly.
7. Installation is the reverse of the removal procedure.

Cylinder Head

REMOVAL AND INSTALLATION

4.3L Engines

1. Properly relieve the fuel system pressure, then disconnect the negative battery cable.
2. On Van models, remove the engine cover for access.
3. Drain the engine cooling system.

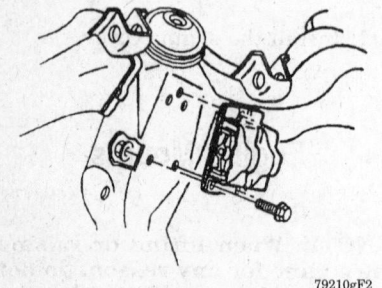

Exploded view of the front engine mount — 1996 2-wheel drive

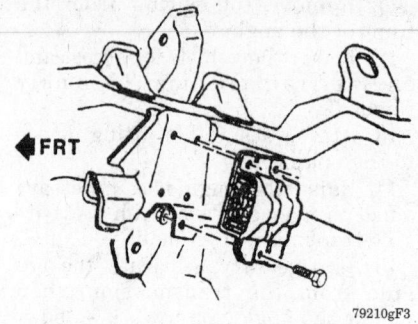

79210gF3

Exploded view of the front engine mount — 1996 4-wheel drive

4. Remove the intake manifold.
5. Remove the exhaust manifold.
6. Except for Van models, if equipped with AIR, remove the air pipe at the rear of the head, then remove the air pump bolt and spacer at the right cylinder head.
7. If necessary, remove the alternator and bracket.
8. If necessary, remove the A/C compressor, if equipped, and position aside with the lines intact, then remove the alternator.
9. Remove the engine accessory bracket bolts and studs at the cylinder head. On some models, left cylinder head removal may require loosening the remaining bracket bolts in order to provide the necessary clearance.
10. Remove the spark plug wires from the brackets
11. Remove the necessary wiring harness connections from the head components and mounting clips. On most models this will include a clip and/or ground strap at the rear of the cylinder head and a connector at the coolant temperature sensor.
12. On Van models, remove the fuel pipes and bracket from the rear of the cylinder head, then remove the cruise control transducer bracket, if equipped.
13. If not done earlier, remove the rocker arm covers, then loosen the rocker arms and remove the pushrods. Tag or arrange all valve train components to assure installation in their original locations.
14. If necessary, remove the spark plugs.
15. Remove the cylinder head bolts by loosening them in the reverse of the torque sequence, then carefully remove the cylinder head.

To install:

16. Carefully clean and inspect the gasket mounting surfaces.

NOTE: The gasket surfaces on both the head and block must be clean of any foreign matter and free of nicks or heavy scratches. The cylinder bolt threads in the block and thread on the bolts must be cleaned (dirt will affect the bolt torque).

NOTE: Do not apply sealer to composition steel-asbestos gaskets.

17. If using a steel only gasket, apply a thin and even coat of sealer to both sides of the gaskets.
18. Place a new gasket over the dowel pins with the bead or the words "This Side Up" facing upwards (as applicable), then carefully lower the cylinder head into position over the gasket and dowels.
19. Apply a coating of 1052080 or equivalent sealer to the threads of the cylinder head bolts, then thread the bolts into position until finger-tight. Using the proper torque sequence, tighten the bolts in 3 steps as follows:

1988–95 models:
- First pass: 25 ft. lbs. (34 Nm)
- Second pass: 45 ft. lbs. (61 Nm)
- Final pass: 65 ft. lbs. (90 Nm)

On 1996–97 models install the bolts in sequence to 22 ft. lbs. (30 Nm). The bolts must then be tightened again in sequence in the following order:
- Short length bolt: (11, 7, 3, 2, 6, 10) 55 degrees
- Medium length bolt: (12, 13) 65 degrees
- Long length bolts: (1, 4, 8, 5, 9) 75 degrees

20. Install the pushrods, secure the rocker arms and adjust the valves. Install the rocker arm covers.
21. If removed, install the spark plugs. Engage the spark plug wires.
22. On the G-Series, install the cruise control transducer bracket to the left cylinder head, if equipped.
23. Position the wiring harness engage the connections and secure in the clips.
24. On Van models, secure the fuel pipes and bracket to the rear of the cylinder head.
25. install the engine accessory bracket bolts and studs.
26. Except for Van models, install the AIR pump mounting bolt and spacer to the right cylinder head, then install the air pipe.
27. Install the exhaust manifold.
28. Install the intake manifold.

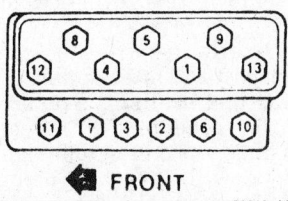

**Cylinder head bolt torque
sequence — 4.3L engine**

79210gA8

29. On the G-Series, reposition and secure the A/C compressor, if equipped, then install the alternator.

30. On Van models, install the engine cover.

31. Connect the negative battery cable, then properly refill the engine cooling system.

5.0L and 5.7L Engines

1. Properly relieve the fuel system pressure, then disconnect the negative battery cable.

2. For the G-Series, remove the engine cover for access.

3. Drain the engine cooling system.

4. Remove the intake manifold.

5. Remove the exhaust manifold.

6. Disconnect the ground strap from the rear of the right cylinder head.

7. If removing the right side cylinder head:

   a. Except for the G-Series, if equipped with AIR, remove the pump bolt and spacer at the cylinder head.

   b. Remove the A/C compressor and position it aside with the lines attached.

   c. Remove the fuel pipe, plug wire and wiring harness brackets at the rear of the cylinder head.

8. If removing the left side cylinder head:

   a. For the G-Series, remove the alternator, then remove the power steering pump and position it aside with the lines attached.

   b. Remove the nut and stud attaching the main accessory bracket to the cylinder head. It may be necessary to loosen the remaining bolts and studs and move the bracket forward slightly in order to gain the necessary clearance to remove the cylinder head.

9. If necessary, remove the spark plugs.

10. Remove the valve covers. Back off the rocker arm nuts and pivot the rocker arms out of the way, then remove the pushrods. Mark the pushrods so they can be installed in their original positions.

11. Remove the cylinder head bolts and remove the heads.

To install:

12. Install the cylinder heads using new gaskets. Install the gaskets with the bead up. Coat a steel gasket on both sides with sealer. If a composition gasket is used, do not use sealer.

13. Clean the cylinder head bolts, apply a coating of 1052080 or equivalent sealer to the threads. Thread the bolts into position until finger-tight. Tighten the bolts using 3 passes of the proper torque sequence:

- First, tighten the bolts to 25 ft. lbs. (34 Nm).
- Next, tighten the bolts to 45 ft. lbs. (61 Nm).
- Finally, tighten the bolts to 65 ft. lbs. (90 Nm).

14. Install the pushrods as noted or arranged during removal, then reposition and secure the rocker arms. Properly adjust the valve lash, then install the valve covers.

15. If removed, install the spark plugs.

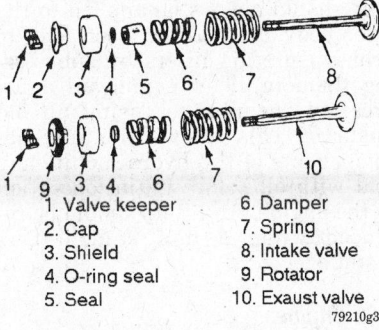

1. Valve keeper
2. Cap
3. Shield
4. O-ring seal
5. Seal
6. Damper
7. Spring
8. Intake valve
9. Rotator
10. Exaust valve

79210g35

Valves and components — 4.3L, 5.0L and 5.7L engines

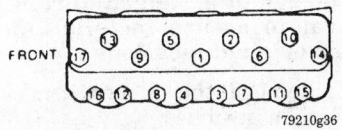

Cylinder head bolt torque sequence — 5.0L and 5.7L engines

79210g36

16. Install the intake manifold.

17. Install the exhaust manifold.

18. If the left cylinder head was removed, install the necessary components in the reverse order of their removal.

19. If the right cylinder head was removed install the necessary components in the reverse order of their remove.

20. Connect the ground strap to the rear of the right cylinder head.

21. On the G-Series, install the engine cover.

22. Connect the negative battery cable, then properly refill the engine cooling system.

7.4L Engine

1. Properly relieve the fuel system pressure, then disconnect the negative battery cable.

2. Remove the intake manifold.

3. If removing the left cylinder head, remove the alternator and power steering pump, if equipped, along with the brackets. Position the power steering pump and bracket aside with the lines attached.

4. If removing the right cylinder head, remove the A/C compressor and/or the AIR pump with brackets, as applicable, then position aside.

5. Remove the exhaust manifolds.

6. Remove the rocker arm covers.

7. If necessary, remove the spark plugs.

8. Except for the G-Series, remove the AIR pipe bolts at the rear of the cylinder head and push the pipe out of the way.

9. Disconnect the ground strap at the rear of the head.

10. Disconnect the sensor wire.

11. Back off the rocker arm nuts and pivot the rocker arms out of the way so the pushrods can be removed. Identify the pushrods so they can be installed in their original positions.

12. Remove the cylinder head bolts and remove the head.

To install:

13. Thoroughly clean the mating surfaces of the head and block. Clean the bolt holes thoroughly.

14. Install the cylinder heads using new gaskets. Install the gaskets with the bead up.

NOTE: Coat a steel gasket on both sides with sealer. If a composition gasket is used, do not use sealer.

15. Clean the cylinder head bolts, apply a coating of 1052080 or equivalent sealer to the threads. Thread the bolts into position until

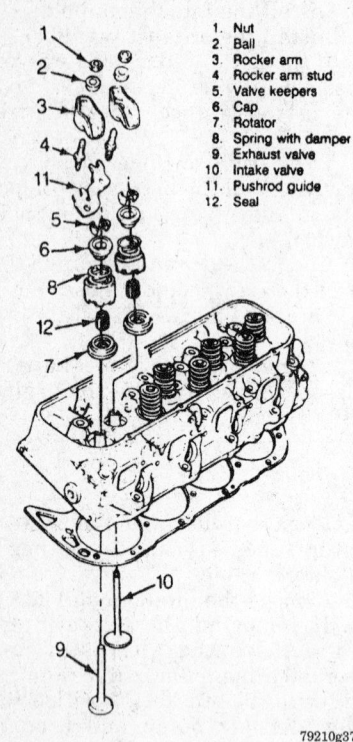

1. Nut
2. Ball
3. Rocker arm
4. Rocker arm stud
5. Valve keepers
6. Cap
7. Rotator
8. Spring with damper
9. Exhaust valve
10. Intake valve
11. Pushrod guide
12. Seal

79210g37

Cylinder head and components — 7.4L engine

finger-tight. Tighten the bolts using 3 passes of the proper torque sequence:
- First, tighten the bolts to 30 ft. lbs. (40 Nm).
- Next, tighten the bolts to 60 ft. lbs. (80 Nm).
- Finally, tighten the bolts to 80 ft. lbs. (110 Nm).

16. Install the pushrods as noted or arranged during removal, then reposition and secure the rocker arms. Properly adjust the valve lash.

17. Connect the sensor wiring and the ground strap.

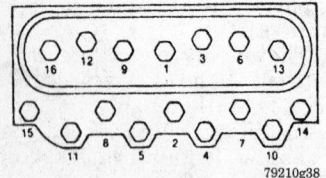

79210g38

Cylinder head bolt torque sequence — 7.4L engine

18. Except for the G-Series, connect the AIR pipe to the rear of the cylinder head.

19. For the G-Series, install the intake manifold at this time.

20. Install the rocker arm covers.

21. If removed, install the spark plugs.

22. Except for the G-Series, install the exhaust manifolds.

23. Install the power steering pump and/or alternator with brackets to the left cylinder head.

24. Install the A/C compressor and/or AIR pump with brackets to the right cylinder head.

25. Install the intake and exhaust manifolds.

26. Install the spark plugs.

27. Install the rocker arm cover.

28. Install the air conditioning compressor and the forward mounting bracket.

29. For the G-Series, install the exhaust manifolds.

30. Except for the G-Series, install the intake manifolds.

31. Connect the negative battery cable.

Valve Lifters

REMOVAL AND INSTALLATION

When installing new lifters, a prelube should always be applied to the lifter body. It is also a good idea to prime hydraulic lifters by submerging them in clean engine oil a depressing the plunger using an old pushrod. This allows the internal components of the hydraulic lifter to coat with oil before initial operation in the engine. All lifters should be replaced when a new camshaft is installed.

4.3L Engines

1. Remove the rocker arm cover.
2. Remove the intake manifold assembly.

NOTE: If any valve train components (lifters, pushrods, rocker arms) are to be reused, they must be tagged or arranged during removal to assure installation in their original locations.

3. Remove the rocker arm and pushrod assemblies.

4. Remove the hydraulic lifter retainer bolts, them remove the retain-

ers or retainers and restrictors, as equipped.

5. Remove the lifters from the engine.

To install:

6. Inspect the lifter and lifter bore for wear and scuffing. Examine the roller for freedom of movement and/or flat spots on the roller surface.

7. If new lifters are being installed, lubricate the lifters using a suitable pre-lube and add GM engine oil supplement or equivalent additive to the crankcase.

8. Install the hydraulic lifters along with the restrictors and/or retainers. Secure the retainers using the bolts and tighten to 12 ft. lbs. (16 Nm).

9. Install the pushrods and rocker arm assemblies, then adjust the valve lash.

10. Install the intake manifold assembly.

11. Install the rocker arm covers.

5.0L, 5.7L, and 7.4L Engines

1. Properly relieve the fuel system pressure, then disconnect the negative battery cable and drain the engine cooling system.

2. Remove the rocker arm covers.

3. Remove the intake manifold.

4. Loosen the rocker arms and remove the pushrods.

NOTE: If any valve train components (lifters, pushrods, rocker arms) are to be reused, they must be tagged or arranged during removal to assure installation in their original locations.

5. Remove the lifters from their bores. A stuck lifter should be removed with a slide hammer type lifter removal tool such as J-9290-01 or equivalent.

To install:

6. Lubricate the lifter body and foot using a high viscosity oil with Zinc such as 12345501 or equivalent. If installing new lifters, change the engine oil and use a high viscosity oil with Zinc when refilling the crankcase.

7. Install the intake manifold.

8. Install the pushrods, then secure the rocker arms and properly adjust the valve lash.

9. Install the rocker arm covers.

10. Connect the negative battery cable, then properly refill the engine cooling system.

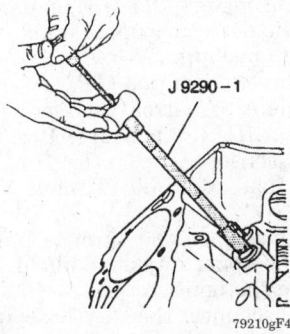

J 9290 – 1

79210gF4

Removing the valve lifters — 1996
5.0L and 5.7L engines shown

Valve Lash

ADJUSTMENT

4.3L (VIN Z) Engine

NOTE: This engine utilizes hydraulic valve lifters which means that a valve adjustment is not a regular maintenance item. The valves must only be adjusted if the rockers arms have been disturbed for any reason such as cylinder head, camshaft, pushrod or lifter removal.

For 1993–94, the 4.3L (VIN Z) engine may be equipped with either of 2 rocker arm retaining systems. If the engine utilizes screw-in type rocker arm studs with positive stop shoulders, no valve lash adjustment is necessary or possible. All 1995–97 4.3L engines utilize this system. If so equipped, please refer to the 4.3L (VIN W) valve lash information. If however, the engine utilizes the pressed-in rocker arm studs, use the following procedure to tighten the rocker arm nuts and properly center the pushrod on the hydraulic lifter:

1. To prepare the engine for valve adjustment, rotate the crankshaft until the mark on the damper pulley aligns with the **0** degree mark on the timing plate and the No. 1 cylinder is on the compression stroke. When the No. 1 piston is on it's compression stroke both the intake and exhaust valves will remain closed as the crankshaft damper mark approaches the timing scale.

NOTE: Another method to tell when the piston is coming up on the compression stroke is by removing the spark plug and placing a finger over the hole in order to feel air being forced out of the spark plug hole. Stop turning the crankshaft air is felt and the TDC timing mark on the crank-

shaft pulley is directly aligned with the timing mark pointer or the 0 mark on the scale.

2. With the engine at No. 1 TDC, adjust the exhaust valves of cylinders No. 1, 5 and 6 and the intake valves of cylinders No. 1, 2 and 3 by performing the following procedures:

a. Back out the adjusting nut until lash can be felt at the pushrod.

b. While rotating the pushrod, turn the adjusting nut inward until all of the lash is removed.

c. When the play has disappeared, turn the adjusting nut inward 1 additional turn for 1992–93 engines or 1¾ additional turns for 1994 engines.

3. Rotate the crankshaft 1 complete revolution and align the mark on the damper pulley with the **0** degree mark on the timing plate; the engine is now positioned on the No. 4 firing position. This time the No. 4 cylinder valves remain closed as the timing mark approaches the scale. Adjust the exhaust valves of cylinders No. 2, 3 and 4 and the intake valves of cylinders No. 4, 5 and 6, by performing the following procedures:

a. Back out the adjusting nut until lash can be felt at the pushrod.

b. While rotating the pushrod, turn the adjusting nut inward until all of the lash is removed.

c. When the play has disappeared, turn the adjusting nut inward 1 additional turn for 1993 engines or 1¾ additional turns for 1994 engines.

4. Install the remaining components, then start the engine and check for oil leaks.

4.3L (VIN W) Engine

The 4.3L (VIN W) engine and some later 4.3L (VIN Z) engines (including all 1995–97 models) are equipped with screw-in type rocker arm studs with positive stop shoulders. Because the shoulders allow the rocker arms to be torqued into proper position, no adjustments are necessary or possible. If a valve train problem is suspected, check that the rocker arm nuts are tightened to 20 ft. lbs. (25 Nm). When valve lash falls out of specification (valve tap is heard), replace the rocker arm, pushrod and hydraulic lifter on the offending cylinder.

5.0L and 5.7L Engines

NOTE: This engine utilizes hydraulic valve lifters which means that a valve adjustment is not a regular maintenance item. The valves must only be adjusted if the rockers arms have been disturbed for any reason such as cylinder head, camshaft, pushrod or lifter removal.

1. Remove the valve covers and gaskets.

2. Crank the engine until the mark on the damper aligns with the **TDC** or **0** mark on the timing tab and the engine is in No. 1 cylinder firing position. This can be determined by placing a finger on the No. 1 cylinder valves as the marks align. If the valves do not move, it is in the No. 1 firing position. If the valves move, it is in No. 6 firing position and the crankshaft should be rotated 1 more complete revolution to the No. 1 firing position.

3. With the engine in No. 1 firing position, adjust the following valves: Exhaust — 1, 3, 4, 8 Intake — 1, 2, 5, 7

4. To adjust the valves:

a. Back off the adjusting nut until lash is felt at the pushrod.

b. Turn in the adjusting nut until all lash is removed. This is determined by rotating the pushrod gently between 2 fingers, the point at which it can no longer be rotated is the point of no lash.

c. Turn the adjusting nut in an 1 additional turn to center the lifter plunger.

5. Crank the engine 1 full revolution until the marks are again in alignment. This is the No. 6 firing position. Adjust the remaining valves in the with the engine in this position: Exhaust — 2, 5, 6, 7 Intake — 3, 4, 6, 8. Adjustment is again performed by bringing the rocker arm nut to the point of no pushrod lash, then tightening 1 additional turn.

6. Reinstall the valve covers using new gaskets.

7.4L Engine

Because the rocker arm fasteners are secured and torqued, valve lash is not adjustable on the 7.4L engine. If a valve train problem is suspected, check that the rocker arm bolts are tightened to 40 ft. lbs. (54 Nm). Be very careful not to overtighten the rocker arm bolts. In most cases it is desirable to install and tighten the bolts when the when the hydraulic lifter is resting on the base circle of the camshaft, not when it is held up-

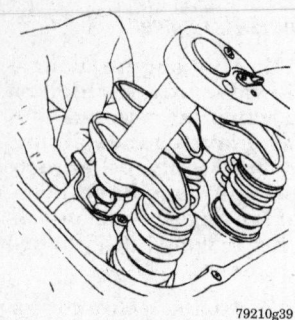

79210g39

When adjusting valve lash, turn the pushrod between 2 fingers while tightening to remove lash

ward on the lobe. This will help assure proper seating of the rocker arm and valve train components. When valve lash falls out of specification (valve tap is heard), replace the rocker arm, pushrod and hydraulic lifter on the offending cylinder.

Rocker Arms

REMOVAL AND INSTALLATION

4.3L, 5.0L and 5.7L Engines

1. On Van models, remove the engine cover for access.
2. Remove the rocker arm cover(s) from the cylinder head.
3. Remove the rocker arm nut, the rocker arm and the ball washer.

NOTE: If only the pushrod is to be removed, loosen the rocker arm nut, swing the rocker arm to the side and remove the pushrod.

4. Withdraw the pushrod from the cylinder head.
To install:
5. Inspect and replace components if worn or damaged.
6. Coat the bearing surfaces of the rocker arms and the rocker arm ball washers with Molykote® or equivalent pre-lube.
7. Install the pushrods making sure they seat properly in the lifter.
8. Install the rocker arms, ball washers and the nuts.
9. For the 4.3L (VIN W) engine and any 1993–96 4.3L (VIN Z) engines which are equipped with screw-in type rocker arm studs with positive stop shoulders, tighten the rocker arm adjusting nuts against the stop shoulders to 20 ft. lbs. (27 Nm). No further adjustment is necessary or possible.
10. For some 4.3L (VIN Z) engines, the 5.7L engine and the 5.7L engine (which are not equipped with screw-

in type rocker arm studs and positive stop shoulders), properly adjust the valve lash.
11. Install the rocker arm cover(s) to the cylinder head.
12. On Van models, install the engine cover.
13. Start and run the engine, then check for leaks and for proper ignition timing adjustment.

7.4L Engine

1. Remove the rocker arm cover from the cylinder head.
2. Remove the rocker arm bolt.

NOTE: If only the pushrod is to be removed, back off the rocker arm bolt, swing the rocker arm aside and remove the pushrod. When removing more than 1 assembly, at the same time, be sure to keep them in order for reassembly purposes.

3. Remove the rocker arm and ball.
To install:
4. Inspect and replace components if worn or damaged.
5. If new rocker arms are being installed, coat the friction surfaces using a high viscosity oil with zinc such as 12345501 or equivalent. If installing old rocker arms, it is still a good idea to coat the surfaces using clean engine oil.
6. Install the rocker arm with ball, then install the retaining bolt and tighten to 40 ft. lbs. (54 Nm).
7. Install the rocker arm cover(s).
8. Start the engine and check for leaks.

Intake Manifold

REMOVAL AND INSTALLATION

4.3L (VIN W) Engine

NOTE: If only the upper intake manifold is being removed, the fuel system pressure does not need to be released. ALWAYS release the pressure before disconnecting any fuel lines.

1. Remove the plastic cover, then properly relieve the fuel system pressure and disconnect the negative battery cable.
2. Drain the engine cooling system, then remove the air cleaner and air inlet duct.

3. Disengage the wiring harness from the necessary upper intake components including:
- Throttle Position (TP) sensor
- Idle Air Control (IAC) motor
- Manifold Absolute Pressure (MAP) sensor
- Intake Manifold Tuning Valve (IMTV)
4. Disengage the throttle linkage from the upper intake manifold, then remove the ignition coil.
5. Disconnect the PCV hose at the rear of the upper intake manifold, then tag and disengage the vacuum hoses from both the front and rear of the upper intake.
6. Remove the upper intake manifold bolts and studs, making sure to note or mark the location of all studs to assure proper installation. Remove the upper intake manifold from the engine.
7. Disengage the distributor or HVS wiring (as equipped), then matchmark the distributor or HVS and remove the assembly from the engine.
8. Disconnect the upper radiator hose at the thermostat housing and the heater hose at the lower intake manifold.
9. Disconnect the fuel supply and return lines at the rear of the lower intake manifold.
10. Remove the pencil brace (A/C compressor bracket-to-lower intake manifold).
11. Disengage the wiring harness connectors from the necessary lower intake components including:
- Fuel injector
- Exhaust Gas Recirculation (EGR) valve
- Engine Coolant Temperature (ECT) sensor
12. Remove the lower intake manifold retaining bolts, then remove the manifold from the engine.
13. Using a gasket scraper, carefully clean the gasket mounting surfaces. Be sure to inspect the manifold for warpage and/or cracks; if necessary, replace it.
To install:
14. Position the gaskets to the cylinder head with the port blocking plates to the rear and the "this side up" stamps facing upward, then apply a 3/16 in. (5mm) bead of RTV sealant to the front and rear of the engine block at the block-to-manifold mating surface. Extend the bead 1/2 in. (13mm) up each cylinder head to seal and retain the gaskets.
15. Install the lower intake manifold taking care not to disturb the gaskets, then tighten the manifold

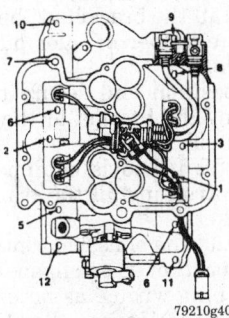

Lower intake manifold
torque sequence — 4.3L
(VIN W) engine

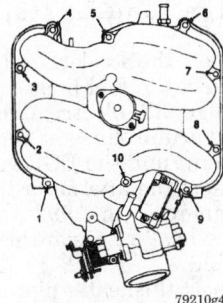

Upper intake manifold
torque sequence — 4.3L
(VIN W) engine

retainers to 35 ft. lbs. (48 Nm) using the proper torque sequence.

16. Engage the wiring harness to the lower manifold components, including the injector, EGR valve and ECT sensor.

17. Install the pencil brace to the A/C compressor bracket and the lower intake manifold.

18. Connect the fuel supply and return lines to the rear of the lower intake. Temporarily reconnect the negative battery cable, then pressurize the fuel system (by cycling the ignition without starting the engine) and check for leaks. Disconnect the negative battery cable and continue installation.

19. Connect the heater hose to the lower intake and the upper radiator hose to the thermostat housing.

20. Align the matchmarks and install the distributor assembly, then engage the wiring.

21. Position a new upper intake manifold gasket on the engine, making sure the green sealing lines are facing upward.

22. Install the upper intake manifold being careful not to pinch the fuel injector wires between the manifolds.

23. Install the manifold retainers, making sure the studs are properly positioned, then tighten them using the proper sequence to 124 inch lbs. (14 Nm).

24. Connect the PCV hose to the rear of the upper intake manifold and the vacuum hoses to both the front and rear of the manifold assembly.

25. Connect the throttle linkage to the upper intake, then install the ignition coil.

26. Engage the necessary wiring to the upper intake components including the TP sensor, IAC motor, MAP sensor and the IMTV.

27. Install the plastic cover, the air cleaner and air inlet duct.

28. Connect the negative battery cable, then properly refill the engine cooling system.

4.3L (VIN Z) Engine

1. Disconnect the negative battery cable and properly relieve the fuel system pressure.

2. Drain the engine cooling system.

3. Remove the air cleaner and heat stove tube.

4. Remove the 2 braces at the rear of the serpentine drive belt tensioner.

5. Disconnect the upper radiator hose.

6. Remove the emissions relays along with the bracket, then disconnect the wiring harness from the retaining clips and position aside. Disconnect the ground cable from the intake manifold stud.

7. Remove the power brake vacuum pipe, then disconnect the heater hose pipe at the manifold and fuel lines at the TBI unit.

8. Remove the ignition coil, then disengage the electrical connectors at the sensors on the manifold.

9. Matchmark and remove the distributor from the engine. For details,

INITIAL TIGHTENING SEQUENCE

FINAL TIGHTENING SEQUENCE

A Front Of Engine

79210g42

Intake manifold torque sequence —
1993–94 4.3L (VIN Z) engine

please refer to the procedure earlier in this section.

NOTE: For ease of installation, do not crank the engine with the distributor removed.

10. Tag and disengage the wires and hoses from the TBI unit.

11. Disconnect the the EGR hose, then disconnect the throttle, TVS and cruise control cables (as equipped).

12. Remove the intake manifold retaining studs and/or bolts, then remove the manifold and gaskets.

To install:

13. Using a gasket scraper, carefully clean the gasket mounting surfaces. Be sure to inspect the manifold for warpage and/or cracks; if necessary, replace it.

14. Position the gaskets to the cylinder head with the port blocking plates to the rear, then apply a 3/16 in. (5mm) bead of RTV sealant to the front and rear of the engine block at the block-to-manifold mating surface. Extend the bead 1/2 in. (13mm) up each cylinder head to seal and retain the gaskets.

15. Install the intake manifold taking care not to disturb the gaskets, then tighten the manifold retainers to 35 ft. lbs. (48 Nm) using the proper torque sequence.

16. Engage the TVS, cruise control and/or throttle cables, as equipped.

17. Connect the EGR hose then engage the wires and hoses at the TBI unit as noted during removal

18. Align and install the distributor assembly.

19. Install the ignition coil, then connect the fuel pipes.

20. Connect the heater hose pipe and the power brake vacuum pipe.

21. Connect the ground cable to the intake manifold stud, then position and secure the wiring harness using the clips.

22. Install the emissions relays along with their bracket, then connect the upper radiator hose.

23. Install the brace at the rear of the drive belt tensioner, then install the air cleaner and heat stove tube.

24. Connect the negative battery cable, then properly refill the engine cooling system.

25. Run the engine and check for leaks.

5.0L and 5.7L Engines

1. Disconnect the negative battery cable and properly relieve the fuel system pressure.

2. On Van models, remove the engine cover.

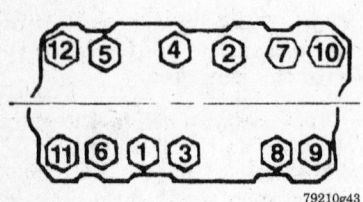

Intake manifold torque sequence — 1995–97 4.3L (VIN Z) engine

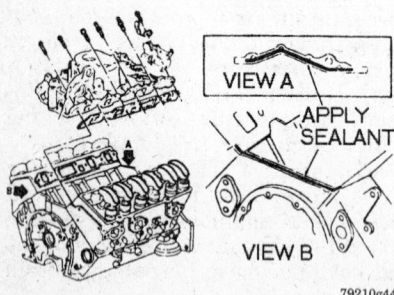

Intake manifold mounting — 4.3L (VIN Z) engine; note the bolt/stud usage may vary

3. Drain the engine cooling system.

4. Remove the air cleaner assembly.

5. Disconnect the heater pipe and upper radiator hose at the intake manifold.

6. If necessary, disconnect the heater hose at the rear of the manifold.

7. Disconnect the alternator rear brace at the manifold.

8. Tag and disconnect the vacuum hoses at the manifold, TBI unit and the EGR valve.

9. Tag and disengage all electrical connections from the manifold and TBI unit.

10. Disconnect the fuel lines from the TBI unit, then disengage the accelerator and cruise control linkage (as equipped).

11. Matchmark and remove the distributor.

12. If equipped, remove the A/C compressor rear bracket (except vans) or the compressor and bracket (van models), then position aside.

13. Remove the power brake vacuum line from the vacuum booster to the manifold.

14. Disconnect the coil or spark plug wiring, as necessary, then re-

move the emission control sensors and bracket on the right side.

15. On Van models equipped with cruise control, remove the transducer and bracket.

16. Remove the fuel line bracket at the rear of the intake manifold and reposition the lines. Remove the bracket at the rear of the belt tensioner or idler pulley, as applicable.

17. If necessary, remove the TBI unit.

18. Remove the intake manifold bolts, studs and engine lift hooks, if necessary. Remove the intake manifold, then remove and discard the old gaskets.

To install:

NOTE: Before installing the intake manifold, ensure the gasket surfaces are thoroughly clean.

19. Apply $3/16$ in. (5mm) bead of sealant to the front and rear sealing surfaces of the engine block manifold. Extend the bead $1/2$ in. (13mm) up each cylinder head to seal and retain the new gaskets.

20. Carefully install the manifold.

21. Install the intake manifold retainers and tighten to 35 ft. lbs. (48 Nm) using multiple passes of the proper sequence.

22. If removed, install the TBI unit.

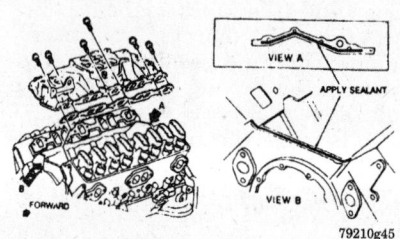

Intake manifold installation — 5.0L and 5.7L engines

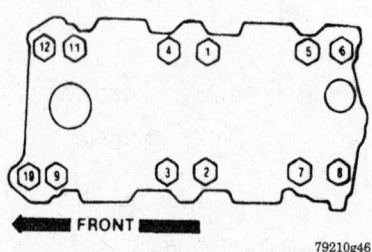

Intake manifold torque sequence — 5.0L and 5.7L engines

23. Install the bracket at the rear of the belt tensioner or idler pulley, as applicable.

24. Reposition and secure the fuel lines using the bracket at the rear of the manifold.

25. On Van models equipped with cruise control, install the transducer and bracket.

26. Install the emission control sensors and bracket, then engage the coil or spark plug wiring, as necessary.

27. Install the vacuum line between the vacuum booster and manifold.

28. If equipped, reposition and secure the A/C compressor and/or bracket.

29. Align and install the distributor.

30. Connect the accelerator linkage.

31. Connect the fuel lines.

32. Engage all electrical connections and vacuum lines at the manifold, TBI unit and the EGR valve.

33. Connect the alternator brace.

34. If removed on 1992 vehicles, connect the heater hose at the rear of the manifold.

35. Connect the heater pipe and the upper radiator hose.

36. Install the air cleaner assembly.

37. On Van models, install the engine cover.

38. Connect the negative battery cable, then properly refill the engine cooling system.

7.4L Engine

1. Properly relieve the fuel system pressure, then disconnect the negative battery cable.

2. On Van models, remove the engine cover.

3. Drain the engine cooling system.

4. Remove the air cleaner assembly.

5. Remove the upper radiator hose and the water pump bypass hose.

6. Disconnect the heater hose and pipe at the TBI unit.

7. Disengage the sensor wire at the front of the manifold.

8. Tag and disengage all electrical connections, along with all vacuum lines from the manifold and TBI unit.

9. Disconnect the accelerator, cruise control and TVS linkage, as equipped.

10. Remove the wiring harness from the retaining clips or bracket and position aside.

11. Either remove the coil with bracket and/or disconnect the coil wiring.

12. Disconnect the fuel lines at the TBI unit and on Van models, at the rear support.

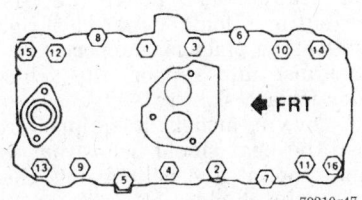

Intake manifold torque sequence — 7.4L engine

79210g47

13. If necessary, remove the TBI unit.

14. Remove the crankcase ventilation hoses.

15. Matchmark and remove the distributor assembly.

16. Remove the sensors and bracket on the right side of the engine (except vans) or remove the sensor and bracket from the front of the intake (van models).

17. On Van models, disconnect the automatic transmission dipstick tube from the manifold and move aside for clearance, then remove the MAP sensor and bracket.

18. If applicable, remove the cruise control transducer.

19. Remove the air conditioning compressor rear bracket, if equipped, then remove the alternator and rear bracket (van models) or just the rear bracket (except vans).

20. Remove the intake manifold retaining bolts.

21. Remove the intake manifold, then remove and discard both the old gaskets and seals.

To install:

22. Install the intake manifold using new gaskets and seals.

NOTE: If RTV sealer was used on the old manifold seals, apply a 3/16 in. (5mm) bead of RTV sealer to the front and rear block where the seals join the cylinder heads.

23. Install the intake manifold bolts, then tighten to 30 ft. lbs. (40 Nm). using the proper torque sequence. After all bolts are torqued, retighten them in sequence to assure proper torquing.

24. Install the alternator and/or rear bracket, as applicable. If equipped, install the the air conditioning compressor rear bracket.

25. If applicable, install the cruise control transducer.

26. On Van models, install the MAP sensor and bracket.

27. Except for Van models, install the sensors and bracket on the right side of the engine.

28. If removed, install the ignition coil and bracket. Connect the coil wiring.

29. Align and install the distributor assembly.

30. If removed, install the TBI unit.

31. Connect the vacuum hoses as tagged during removal, then install the crankcase ventilation hoses.

32. Connect the fuel lines at the TBI unit and on Van models, at the rear support.

33. On Van models, install the sensor and bracket on the front of the intake.

34. Reposition and secure the wiring harness to the retaining clips or bracket.

35. Connect the accelerator, cruise control and TVS linkage, as equipped.

36. Engage all electrical connections as tagged during removal, along with the sensor wire at the front of the manifold.

37. Connect the heater hose and pipe at the TBI unit.

38. Install the upper radiator hose and the water pump bypass hose.

39. On Van models, connect the automatic transmission dipstick tube to the manifold.

40. Install the air cleaner assembly.

41. Install the engine cover.

42. Connect the negative battery cable and properly refill the engine cooling system.

Exhaust Manifold

REMOVAL AND INSTALLATION

4.3L Engines

1. Disconnect the negative battery cable.

2. On Van models, remove the engine cover.

3. Raise and support the vehicle safely.

4. If necessary, remove the EGR inlet pipe (left side manifold).

5. Disconnect the exhaust pipe from the exhaust manifold.

6. Lower the vehicle for underhood access.

7. Tag and disconnect the spark plug wires from the plugs and from the retaining clips.

8. If removing the left side manifold:

   a. Remove the air cleaner with heat stove pipe and cold air intake pipe.

   b. Remove the power steering/alternator rear bracket.

   c. Check for sufficient clearance between the manifold and the intermediate steering shaft. On some models it will be necessary to disconnect the intermediate shaft from the steering gear in order to reposition the shaft for clearance.

9. If necessary when removing the right side manifold, unbolt the A/C compressor and bracket, then position the assembly aside. Do not disconnect the lines or allow them to become kinked or otherwise damaged.

10. If necessary for the right side manifold, remove the spark plugs, dipstick tube and wiring.

11. Unbend the lock tangs then remove the exhaust manifold retaining bolts, washers and tab washers. Remove the exhaust manifold, then remove and discard the old gaskets.

To install:

12. Using a gasket scraper, clean the gasket mounting surfaces. Inspect the exhaust manifold for distortion, cracks or damage; replace if necessary.

13. Install the exhaust manifold to the cylinder using a new gasket, then tighten the exhaust manifold-to-cylinder head bolts, on 1993–95 models to 26 ft. lbs. (36 Nm) on the center exhaust tube and to 20 ft. lbs. (28 Nm) on the front and rear exhaust tubes. On 1996–97 models tighten the bolts in two steps:

   a. First step; 11 ft. lbs. (15 Nm).

   b. Final step; 22 ft. lbs. (30 Nm).

14. Once the bolts are tightened, bend the tabs on the washers back over the heads of all bolts in order to lock them in position.

15. If removed on the right side, install the spark plugs, dipstick tube and wiring.

16. If unbolted, reposition and secure the A/C compressor and bracket assembly.

17. If the left manifold was removed:

   a. If unbolted, reconnect the intermediate shaft to the steering gear.

   b. Install the power steering/alternator rear bracket.

   c. Install the air cleaner along with the heat stove pipe and cold air intake pipe.

18. Connect the spark plug wires to the retainer clips and to the plugs as noted during removal.

19. Raise and support the vehicle safely.

20. Connect the exhaust pipe to the manifold.

21. Lower the vehicle.

22. On Van models, install the engine cover.

23. Connect the negative battery cable.

5.0L and 5.7L Engines

1. Disconnect the negative battery cable.

2. For Van models, remove the engine cover.

3. Remove the air cleaner.

4. Raise and support the vehicle safely.

5. Disconnect the exhaust pipe at the manifold, then lower the vehicle for underhood access.

6. If removing a manifold with the oxygen sensor mounted to it, disengage the sensor wiring.

7. If necessary, disconnect the AIR hose at the check valve.

8. Disconnect the heat stove pipe and remove the dipstick tube bracket, if working on the right side of the engine.

9. If removing the left side manifold, remove the power steering pump rear bracket at the manifold.

10. If necessary on 1992 vehicles, loosen the alternator and remove the lower bracket. Then if necessary, remove the air conditioner compressor

rear bracket and the diverter valve and bracket.

NOTE: On models with air conditioning, it may be necessary to remove the compressor, and tie it out of the way. Do not disconnect the compressor lines.

11. Remove the manifold bolts and remove the manifold(s). Some models have lock tabs on the front and rear manifold bolts which must be removed before removing the bolts.

12. Installation is the reverse of removal. Tighten the 2 center retaining bolts to 26 ft. lbs. (36 Nm) and the outer manifold retaining bolts to 20 ft. lbs. (28 Nm).

7.4L Engine

1. Disconnect the negative battery cable.

2. On Van models, remove the engine cover.

3. If removing the right side manifold, remove the heat stove pipe and the dipstick tube.

4. If removing a manifold with the oxygen sensor mounted to it, disengage the sensor wiring.

5. If applicable, disconnect the AIR hose at the check valve.

6. Remove the spark plugs and wires.

7. On Van models, remove the exhaust manifold bolts and the spark plug heat shields. Leave the front bolt (left manifold) or rear bolt (right manifold) in place for support.

8. Raise and support the vehicle safely for access.

9. On Van models, if equipped, remove the heat shield bolts from the engine mount and bellhousing, then remove the shield.

10. Disconnect the exhaust pipe at the manifold.

11. Remove the manifold bolts (or bolt in the case of Van models). Except for Van models (on which it was done earlier) remove the spark plug heat shields.

NOTE: It may be necessary to raise the engine slightly to gain sufficient clearance for removal of the manifold on some G-Series vehicles.

12. Remove the exhaust manifold.

13. Clean the mating surfaces and the retainer threads.

14. Install the manifold, spark plug heat shields and bolts. Tighten the bolts to 40 ft. lbs. (54 Nm). starting from the center bolts and working towards the outside.

15. Complete the remainder of the installation by reversing the removal procedures.

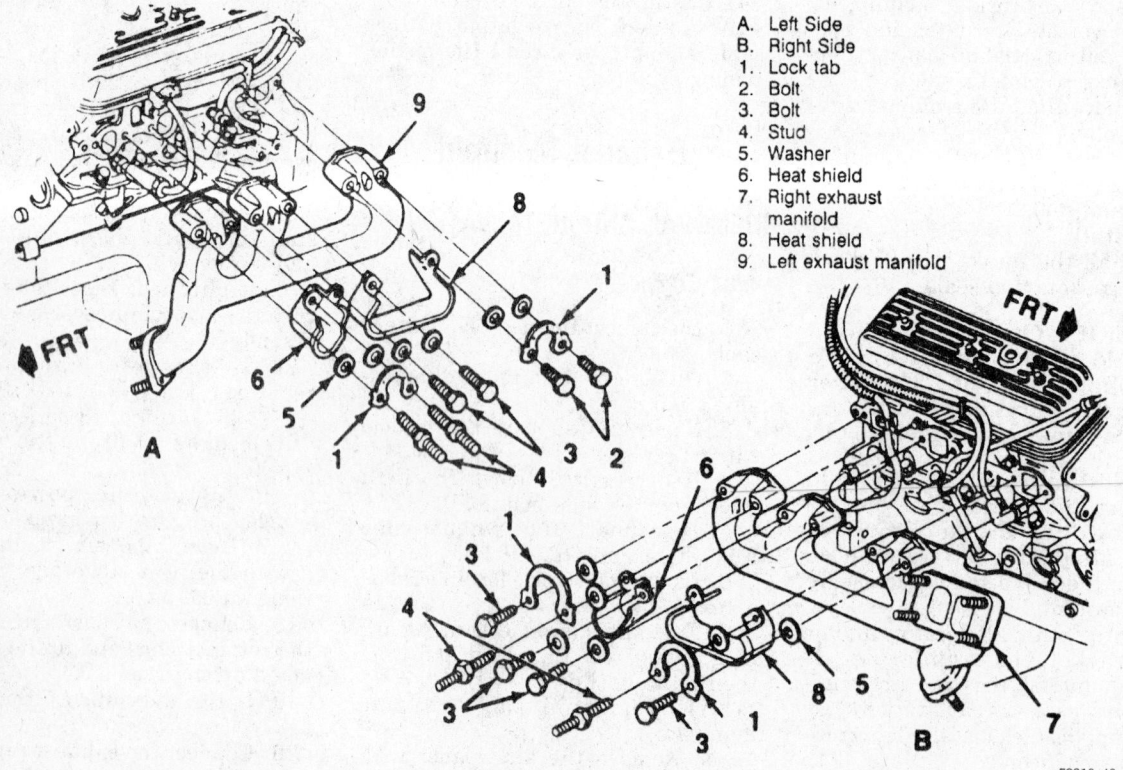

A. Left Side
B. Right Side
1. Lock tab
2. Bolt
3. Bolt
4. Stud
5. Washer
6. Heat shield
7. Right exhaust manifold
8. Heat shield
9. Left exhaust manifold

79210g48

Exhaust manifold mounting — 4.3L engine

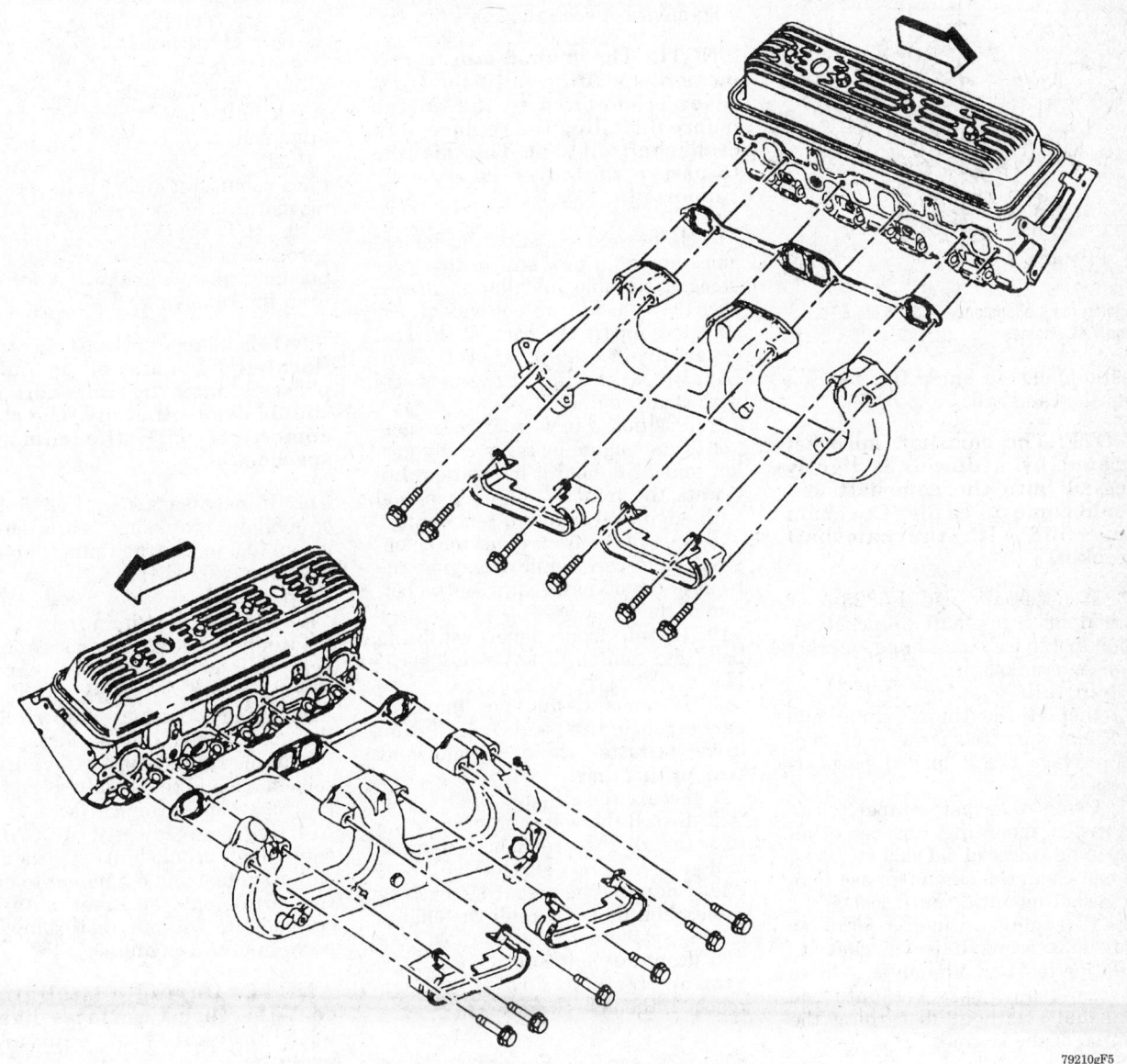

Exploded view of the exhaust manifold — 1996 5.0L and 5.7L engines

79210gF5

Timing Chain, Sprockets and Front Cover

It is recommended by the manufacturer that the front cover oil seal be replaced whenever the cover is removed.

REMOVAL AND INSTALLATION

4.3L Engines

1. Disconnect the negative battery cable and drain the engine cooling system.

2. Remove the crankshaft pulley and damper.

NOTE: The outer ring (weight) of the torsional damper is bonded to the hub with rubber. The damper must be removed with a puller which acts on the inner hub only. Pulling on the outer portion of the damper will break the rubber bond or destroy the tuning of the unit.

3. Remove the water pump assembly.

4. Loosen the oil pan.

5. Remove the front cover bolts and, if equipped, the reinforcements, then remove the front cover from the engine.

6. If the front cover seal is to be replaced, it may be pried front the front cover using a suitable pry tool.

7. Rotate the crankshaft until the timing marks on the camshaft and crankshaft sprockets are in proper alignment.

8. Remove the camshaft sprocket-to-camshaft nut and/or bolts, then remove the camshaft sprocket (along with the timing chain). If the sprocket is difficult to remove, use a

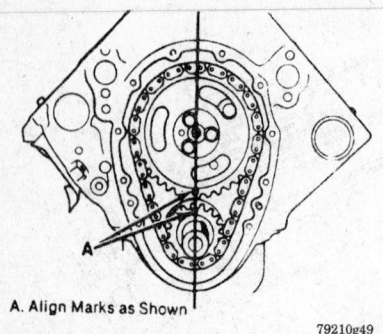

A. Align Marks as Shown

79210g49

Timing mark alignment — 4.3L, 5.0L, 5.7L, and 7.4L engines

plastic mallet to bump the sprocket from the camshaft.

NOTE: The camshaft sprocket (located by a dowel) is lightly pressed onto the camshaft and should come off easily. The chain comes off with the camshaft sprocket.

9. If necessary use J-5825-A or equivalent crankshaft sprocket removal tool to free the timing sprocket from the crankshaft.

To install:

10. Inspect the timing chain and the timing sprockets for wear or damage, replace the damaged parts as necessary.

11. Using a gasket scraper, clean the gasket mounting surfaces of all remaining traces of old gasket. Using solvent, clean the oil and grease from the gasket mounting surfaces.

12. If removed, use J-5590 or equivalent crankshaft sprocket installation tool and a hammer to drive the crankshaft sprocket onto the crankshaft, without disturbing the position of the engine.

NOTE: During installation, coat the thrust surfaces lightly with Molykote® or equivalent pre-lube.

13. Position the timing chain over the camshaft sprocket. Arrange the camshaft sprocket in such a way that the timing marks will align between the shaft centers and the camshaft locating dowel will enter the dowel hole in the cam sprocket.

14. Position the chain under the crankshaft sprocket, then place the cam sprocket, with the chain still mounted over it, in position on the front of the camshaft. Install and tighten the camshaft sprocket-to-camshaft retainers.

15. With the timing chain installed, turn the crankshaft 2 complete revolutions, then check to make certain

that the timing marks are in correct alignment between the shaft centers.

NOTE: The manufacturer recommends waiting until the front cover is mounted to the engine before installing the replacement crankshaft oil seal. This may be to assure the cover is properly supported.

16. If desired on early model engines, install a new seal to the cover using a suitable installation driver, such as J-35468 or equivalent. Be sure to support the back of the seal cover area during installation. Lightly coat the lips of the new seal with clean engine oil.

17. Position a new front cover gasket to the engine or cover using gasket cement to hold it in position. Lubricate the front of the oil pan seal with engine oil to aid in reassembly.

18. Install the front cover to the engine. Take care while engaging the front of the oil pan seal with the bottom of the cover.

19. Install front cover retaining bolts and tighten to 124 inch lbs. (14 Nm).

20. If removed and not installed earlier, use the seal installation driver to install the new crankshaft seal at this time.

21. Secure the oil pan.

22. Install the water pump.

23. Install the crankshaft damper and pulley.

24. Connect the negative battery cable, then properly refill the engine cooling system.

25. Run the engine until normal operating temperature has been reached, then check for leaks.

5.0L, 5.7L, and 7.4L Engines

1. Disconnect the negative battery cable.

2. Drain the cooling system. Remove the fan shroud assembly.

3. Remove the belts, pulleys and water pump assembly.

4. Remove the crankshaft pulley and damper.

5. Remove the oil pan-to-front cover bolts.

6. Remove the screws holding the timing chain cover to the block, pull the cover forward enough to cut the front oil pan seal. Cut the seal flush with the block on both sides.

NOTE: On some models it will be necessary to completely remove the oil pan in order to properly remove and install the front cover, assuring against oil leaks.

7. Pull off the cover and gaskets.

8. Use a suitable tool to pry the old seal out of the front face of the cover.

9. Rotate the crankshaft until the timing marks on the camshaft and crankshaft sprockets are in proper alignment.

10. Remove the camshaft sprocket-to-camshaft nut and/or bolts, then remove the camshaft sprocket (along with the timing chain). If the sprocket is difficult to remove, use a plastic mallet to bump the sprocket from the camshaft.

NOTE: The camshaft sprocket (located by a dowel) is lightly pressed onto the camshaft and should come off easily. The chain comes off with the camshaft sprocket.

11. If necessary use J-5825-A or equivalent crankshaft sprocket removal tool to free the timing sprocket from the crankshaft.

To install:

12. Inspect the timing chain and the timing sprockets for wear or damage, replace the damaged parts as necessary.

13. Using a gasket scraper, clean the gasket mounting surfaces of all remaining traces of old gasket. Using solvent, clean the oil and grease from the gasket mounting surfaces.

14. If removed, use J-5590 or equivalent crankshaft sprocket installation tool and a hammer to drive the crankshaft sprocket onto the crankshaft, without disturbing the position of the engine.

NOTE: During installation, coat the thrust surfaces lightly with Molykote® or equivalent pre-lube.

15. Position the timing chain over the camshaft sprocket. Arrange the camshaft sprocket in such a way that the timing marks will align between the shaft centers and the camshaft locating dowel will enter the dowel hole in the cam sprocket.

16. Position the chain under the crankshaft sprocket, then place the cam sprocket, with the chain still mounted over it, in position on the front of the camshaft. Install and tighten the camshaft sprocket-to-camshaft retainers.

17. With the timing chain installed, turn the crankshaft 2 complete revolutions, then check to make certain that the timing marks are in correct alignment between the shaft centers.

18. Using seal driver J-22102 or equivalent, install the new seal so the

open end is toward the inside of the cover.

NOTE: Coat the lip of the new seal with oil prior to installation.

19. Install a new front pan seal, cutting the tabs off.
20. Coat a new cover gasket with adhesive sealer and position it on the block.
21. Apply a 1/8 in. bead of RTV gasket material to the front cover. Install the cover carefully onto the locating dowels.
22. Tighten the attaching screws.
23. If removed, install the oil pan.
24. Tighten the cover-to-pan bolts.
25. Install the torsional damper.
26. Install the water pump assembly.
27. Connect the negative battery cable.
28. Fill the cooling system with the proper type and quantity of antifreeze.

Camshaft

REMOVAL AND INSTALLATION

4.3L Engines

1. Properly relieve the fuel system pressure, then disconnect the negative battery cable.
2. Drain the engine cooling system.
3. Remove the radiator.
4. Remove the rocker arm covers from the engine.
5. Remove the intake manifold assembly.
6. Remove the rocker arms, pushrods and lifters.
7. Remove the crankshaft pulley and hub.
8. Remove the engine front (timing) cover.
9. Align the timing marks on the crankshaft and camshaft sprockets.
10. Remove the camshaft sprocket and timing chain.
11. If equipped, remove the balance shaft drive gear.
12. Remove the camshaft thrust plate.
13. Install the sprocket bolts or longer bolts of the same thread into the end of the camshaft as a handle, then remove the camshaft front the front of the engine while turning slightly from side to side, as necessary. Take care not to damage the camshaft bearings when removing the camshaft.
To install:
14. Lubricate the camshaft journals with clean engine oil or a suitable

pre-lube, then install the camshaft into the block being extremely careful not to contact the bearings with the cam lobes.
15. Install the camshaft thrust plate.
16. If equipped, install the balance shaft drive gear.
17. Install the timing chain and camshaft sprocket.
18. Install the engine front (timing) cover.
19. Install the crankshaft pulley and hub.
20. Install the valve lifters, then install the pushrods and rocker arms. Properly adjust the valve clearance.
21. Install the intake manifold assembly.
22. Install the rocker arm covers to the engine.
23. Install the radiator to the vehicle.
24. Connect the negative battery cable and properly refill the engine cooling system.

5.0L and 5.7L Engines

1. Disconnect the negative battery cable, Drain the cooling system and properly relieve the fuel system pressure.
2. On the G-Series, remove the engine cover.
3. Remove the air cleaner.
4. On the G-Series, remove the grille and center support.
5. Remove the air conditioning condenser and swing the condenser forward from its mounting, if equipped.
6. Remove the fan, the shroud and the radiator.
7. Remove the valve covers.
8. Remove the water pump assembly.
9. Align the timing marks and remove the torsional damper.
10. Remove the timing chain cover.
11. Disconnect the electrical and vacuum connections at the intake manifold.
12. Mark the distributor rotor-to-housing location. Remove the distributor assembly.
13. Remove the intake manifold, pushrods and hydraulic lifters.
14. Remove the camshaft sprocket bolts, camshaft sprocket and timing chain. Tap the sprocket on its lower edge to loosen it.
15. Remove the crankshaft sprocket, as required.
16. As required, remove the front engine mount through bolts and raise the engine to gain sufficient clearance for camshaft removal.

17. Install two or three 5/16–18 bolts 4–5 in. long into the camshaft threaded holes; carefully pull the camshaft from the block.
18. Inspect the shaft for signs of excessive wear or damage.
To install:
19. Liberally coat camshaft and bearing with heavy engine oil or engine assembly lubricant and insert the cam into the engine.
20. Lower the engine and install the engine mount through bolts.
21. Align the timing marks on the camshaft and crankshaft gears.
22. Install the camshaft sprocket and chain and tighten the bolts to specification.
23. Install the hydraulic lifters and pushrods and adjust the valves.
24. Install the distributor assembly.
25. Install the timing chain cover.
26. Install the torsional damper.
27. Install the water pump.
28. Install the valve covers.
29. Install the fan, the shroud and radiator.
30. Install the air conditioning condenser, if equipped.
31. On the G-Series, install the grille and center support.
32. Install the air cleaner.
33. On the G-Series, install the engine cover.
34. Connect the battery cable and fill the cooling system.

7.4L Engine

1. Disconnect the negative battery cable. Properly relieve the fuel system pressure.
2. Remove the engine cover, G-Series. Remove the air cleaner assembly.
3. Remove the grille and center support section, as required.
4. Properly discharge the air conditioning system. Remove the air conditioning compressor, condenser and auxiliary fan, if equipped.
5. Drain the cooling system.
6. Remove the fan, the shroud and radiator.
7. Remove the alternator belt, remove the alternator assembly, as required.
8. Remove the valve covers.
9. Disconnect the hoses from the water pump.
10. Remove the water pump.
11. Align the timing marks at TDC. Remove the harmonic balancer and pulley.
12. Remove the engine front cover.
13. Mark the distributor rotor-to-housing location. Remove the distributor assembly.

14. Remove the intake manifold assembly.

15. Remove the lifters, pushrods, and rocker arms.

16. Rotate the camshaft so the timing marks align.

17. Remove the camshaft sprocket bolts. Remove the camshaft sprocket and timing.

18. Remove the engine mount through bolts. Raise and support the engine to aid in camshaft removal, as required.

19. Install two or three $^5/_{16}$–18 bolts in the holes in the front of the camshaft and carefully pull the camshaft from the block.

To install:

20. Liberally coat camshaft and bearing with heavy engine oil or engine assembly lubricant and insert the cam into the engine.

21. Align the timing marks on the camshaft sprocket and crankshaft gears.

22. Install the camshaft sprocket and chain and tighten the bolts to specification. Lower the engine and install the engine mount bolts.

23. Install the lifters and pushrods and adjust the valves.

24. Install the intake manifold.

25. Install the distributor using the locating marks made during removal.

26. Install the engine front cover.

27. Install the harmonic balancer and pulley.

28. Install the water pump.

29. Connect the hoses at the water pump.

30. Install the valve covers.

31. Install the alternator.

32. Install the fan shroud and radiator.

33. Fill the cooling system with the proper type and quantity of antifreeze.

34. Install the air conditioning condenser and compressor.

35. Install the grille and center support.

36. Install the air cleaner assembly.

37. Connect the battery.

Balance Shaft

REMOVAL AND INSTALLATION

4.3L (VIN W) Engine

1. Disconnect the negative battery cable. Drain the cooling system into a suitable container. Properly relieve the fuel system pressure and discharge the air conditioning system.

2. Remove the air cleaner and air intake duct.

3. Remove the upper radiator shroud.

4. Remove the oil and transmission cooler lines at the radiator.

5. Remove the hoses from the radiator and remove the radiator.

6. Remove the air conditioner condenser and fan assembly.

7. Remove the serpentine belt.

8. Remove the brace at the coolant pump and remove the pump.

9. Remove the torsional damper.

10. Raise and support the vehicle safely.

11. Remove the flywheel cover. Drain the engine oil and loosen the oil pan bolts; remove the 2 nuts and bolts at the front of the oil pan.

12. Remove the front timing cover bolts and cover.

NOTE: Use care when removing the front cover, as not to damage the oil pan gasket.

13. Remove the crankshaft front cover oil seal.

14. Remove the camshaft sprocket, timing chain and balance shaft drive gear.

15. Remove the balance shaft retainer.

16. Remove the intake manifold assembly.

17. Remove the hydraulic lifter retainer.

18. Remove the balance shaft from the bearing using a soft faced mallet.

19. Remove the rear bearing using tool J–38834 and J–26941.

NOTE: The balance shaft and bearings are only to be serviced or replaced as a complete assembly.

To install:

20. Dip the bearings in clean engine oil before installation.

21. Using tool J–38834, install the balance shaft rear bearing with the flat edge and the manufacturer's markings facing the front of engine.

22. Dip the front balance shaft bearing into engine oil. Using tool J–36996 and J–8092, install the balance shaft into the block.

23. Install the balance shaft retainer and torque the bolts to 124 inch lbs. (14 Nm).

24. Install the balance shaft drive gear. Apply Loctite® to the bolt and torque to 15 ft. lbs. (20 Nm) plus turn the bolt an additional 35 degrees.

25. Install the lifter retainer and turn the balance shaft to ensure proper clearance.

26. Install the balance shaft rear plug.

27. Turn the camshaft so the balance shaft drive gear timing marks are aligned.

28. Install the balance shaft drive gear retaining stud and torque to 15 ft. lbs. (20 Nm).

29. Install the intake manifold assembly.

30. Install the timing chain and gears with the timing marks dot-to-dot; this is the No. 4 cylinder firing position, so set the distributor rotor to the No. cylinder firing position when installing. Torque the camshaft bolts and nut to 21 ft. lbs. (28 Nm).

31. Install the distributor assembly.

32. Install the timing cover oil seal and install the cover.

33. Raise and support the vehicle safely.

34. Install the flywheel cover.

35. Tighten the oil pan bolts; install the 2 nuts and bolts at the front of the oil pan.

36. Install the torsional damper.

37. Install the brace at the coolant pump and install the pump.

38. Install the serpentine belt.

39. Install the air conditioner condenser and fan assembly.

40. Install the radiator and hoses.

41. Install the oil and transmission cooler lines at the radiator.

42. Install the upper radiator shroud.

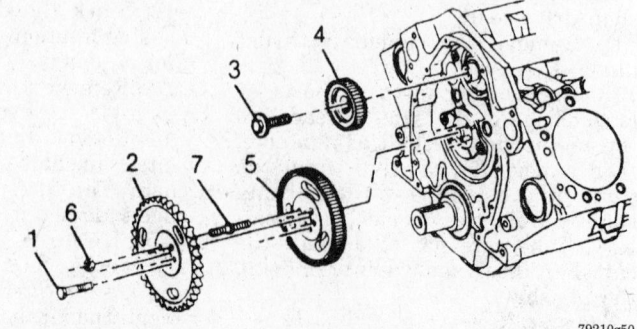

1. Camshaft sprocket bolt
2. Camshaft sprocket
3. Balance shaft driven gear bolt
4. Balance shaft driven gear
5. Balance shaft drive gear
6. Nut
7. Stud

Balance shaft drive gears — 4.3L (VIN W) engine

79210g50

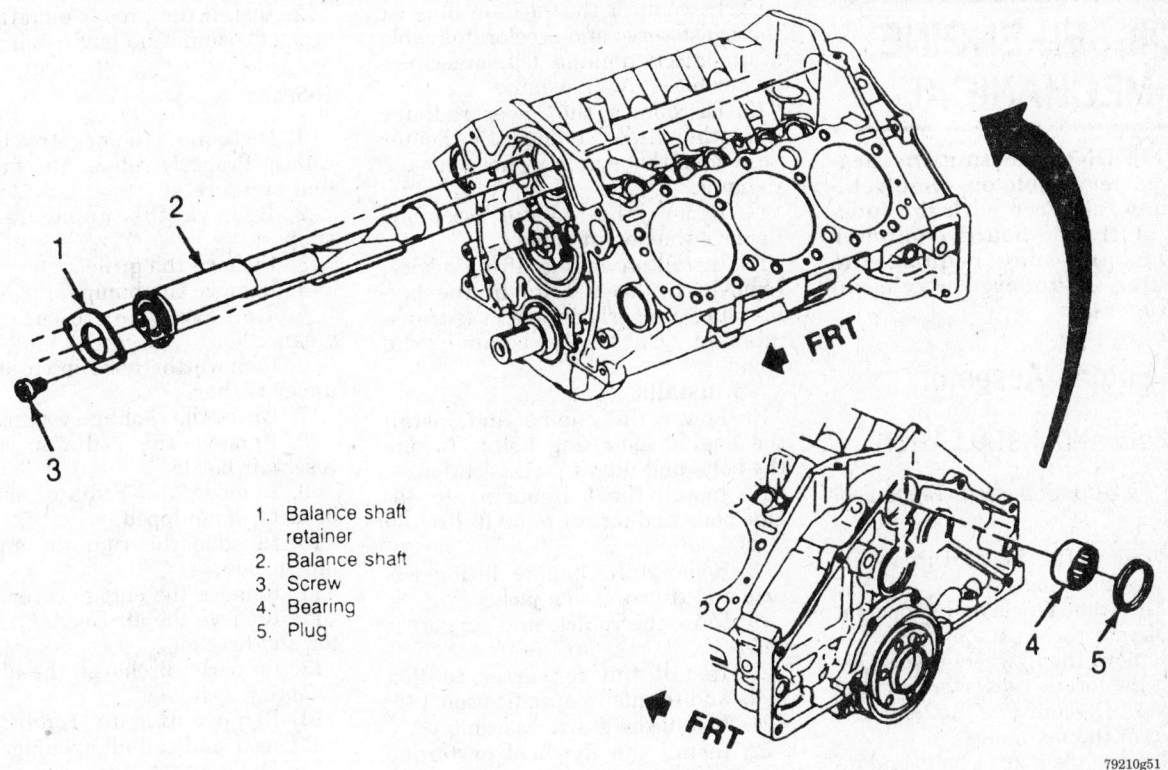

1. Balance shaft retainer
2. Balance shaft
3. Screw
4. Bearing
5. Plug

79210g51

Balance shaft assembly — 4.3L (VIN W) engine

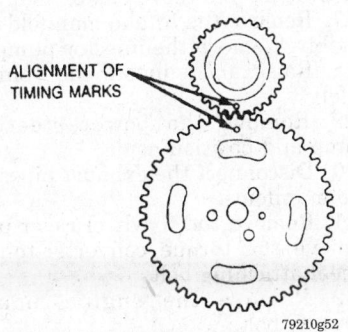

ALIGNMENT OF TIMING MARKS

79210g52

Balance shaft alignment marks — 4.3L (VIN W) engine

43. Install the air cleaner and air intake duct.

44. Connect the negative battery cable. Fill the cooling system with proper type and quantity of antifreeze. Properly recharge the air conditioning system. Fill the engine with the correct quantity and grade of engine oil.

Piston and Connecting Rod

POSITIONING

During disassembly, note the location of the connecting rod bearing tang slots, on some engines, the slots must be on the side opposite of the camshaft.

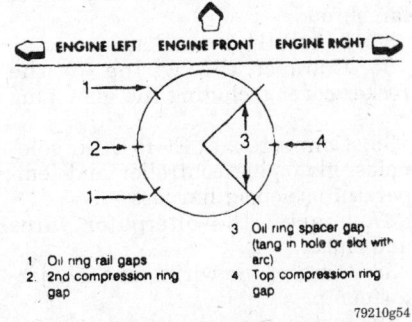

ENGINE LEFT ENGINE FRONT ENGINE RIGHT

1. Oil ring rail gaps
2. 2nd compression ring gap
3. Oil ring spacer gap (tang in hole or slot with arc)
4. Top compression ring gap

79210g54

Piston ring gap locations — all engines

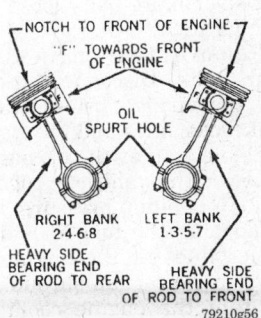

NOTCH TO FRONT OF ENGINE
"F" TOWARDS FRONT OF ENGINE
OIL SPURT HOLE
RIGHT BANK 2-4-6-8
LEFT BANK 1-3-5-7
HEAVY SIDE BEARING END OF ROD TO REAR
HEAVY SIDE BEARING END OF ROD TO FRONT

79210g56

Correct relationship of the piston and rod — 4.3L, 5.0L and 5.7L engines

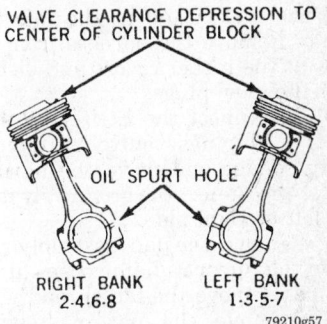

VALVE CLEARANCE DEPRESSION TO CENTER OF CYLINDER BLOCK
OIL SPURT HOLE
RIGHT BANK 2-4-6-8
LEFT BANK 1-3-5-7

79210g57

Correct relationship of the piston and rod — 7.4L engine

DIESEL ENGINE MECHANICAL

NOTE: Disconnecting the negative battery cable on some vehicles may interfere with the functions of the on-board computer systems and may require the computer to undergo a relearning process.

Engine Assembly

REMOVAL AND INSTALLATION

Blazer, Pick-Up, Suburban, Tahoe and Yukon

1. Remove the hood. Disconnect and remove the batteries.
2. Raise the vehicle and support it safely. Drain the engine oil.
3. Remove the flywheel cover. Disconnect the torque converter from the flexplate. Disconnect the exhaust pipes from the manifolds.
4. Remove the starter bolts and remove the starter. Remove the transmission bellhousing to engine bolts, leaving 1 or more bolts loose to prevent separation.
5. Remove the engine mount through bolts. Disconnect the block heater, the wiring harness, transmission oil cooler lines and the front battery cable clamp at the oil pan.
6. Lower the vehicle. Properly relieve the fuel pump pressure. Disconnect and plug the fuel lines and the oil cooler lines at the engine block. Remove the lower fan shroud bolts.
7. Drain the engine coolant. Remove the air cleaner assembly. Disconnect the ground cable from the alternator bracket. Disconnect the alternator wires and clips.
8. Disconnect the TPS, EGR–EPR and the fuel cut-off at the injection pump. Remove the harness from the clips at the rocker covers and disconnect the glow plugs.
9. Disconnect the EGR–EPR solenoids, glow plugs, controller, temperature sender and move the harness aside. Disconnect the ground strap on the left or right side.
10. Remove the fan assembly. Remove the upper radiator hoses at the engine. Remove the fan shroud.
11. Remove the power steering pump and belt. Remove the reservoir and lay the pump and reservoir aside. If equipped with air conditioning, remove the compressor and position aside.

12. Disconnect the vacuum lines at the cruise servo and accelerator cable at the injection pump. Disconnect the heater hoses at the engine.
13. Disconnect the lower radiator hose, the oil cooler lines, the heater hose and the overflow hose at the radiator.
14. Remove the radiator assembly. Remove the detent cable.
15. Install an engine lifting device, remove the loose bolts in the bellhousing. Properly support the transmission. Carefully remove the engine.

To install:

16. Lower the engine and install the engine mounting bolts. Torque the bolts and nuts to specification
17. Install the bellhousing to engine bolts and torque to 30 ft. lbs. (40 Nm).
18. Remove the engine lifting fixture and transmission jack.
19. Raise the vehicle and support it safely.
20. Install the converter to flex bolts and torque to specification.
21. Install the starter assembly.
22. Install the flywheel or torque converter cover.
23. Install the starter.
24. Install the exhaust pipes at the manifold.
25. Connect the wiring harness, transmission cooler lines and a battery cable clamp at the oil pan.
26. Connect the fuel return lines at the engine.
27. Connect the oil cooler lines at the engine.
28. Lower the vehicle.
29. Install the radiator.
30. Install the heater hose to the engine.
31. Connect the accelerator, cruise control and detent cables at the injection pump.
32. Connect the power steering pump and reservoir.
33. Install the fan and the upper fan shroud.
34. Install the ground strap.
35. Connect the wiring to the rocker cover including the glow plug wires.
36. Connect the EGR-EPR solenoids, glow plug controller and temperature solenoid harness.
37. Connect the alternator wires and clips.
38. Connect the wiring at the injector pump.
39. Install the air cleaner and air conditioning compressor.
40. Install the hood.
41. Connect the negative battery cable.

42. Install the proper quantity and type of coolant and engine oil.

G-Series

1. Disconnect the negative battery cables. Properly relieve the fuel system pressure.
2. Remove the upper radiator support.
3. Remove the grille.
4. Remove the bumper.
5. Remove the lower grille valance.
6. Remove the hood latch and the upper tie bar.
7. Drain the cooling system.
8. Remove the radiator coolant reservoir bottle.
9. Remove the radiator support bracket, if equipped.
10. Remove the radiator and the fan shroud.
11. Remove the engine cover.
12. Remove the air cleaner, resonator and bracket.
13. Properly discharge the air conditioning system.
14. Remove the air conditioning condenser and cap all openings.
15. Disconnect the air cleaner bracket at the valve cover.
16. Remove the crankcase ventilator bracket and move aside.
17. Remove the intake manifold assembly. Remove the injector pump.
18. Raise and support the vehicle safely.
19. Remove the power steering pump and position aside.
20. Disconnect the exhaust pipes at the manifolds.
21. Remove the flywheel cover and remove the torque converter-to-flywheel attaching bolts.
22. Remove the engine mount through bolts.
23. Disconnect the blocker heater wires.
24. Remove the starter assembly.
25. Disconnect the fuel line at the fuel pump (lift pump).
26. Lower the vehicle.
27. Remove the bellhousing-to-engine bolts.
28. Remove the cruise control transducer, if equipped.
29. Remove the air conditioning compressor.
30. Remove the oil fill tube upper bracket.
31. Remove the wiring harness and connections from engine.
32. Unbolt the transmission dipstick tube and position aside.
33. Remove the heater hose at the engine.
34. Remove the alternator upper bracket.

35. Disconnect the glow plug temperature inhibit switch connector at the coolant crossover.

36. Remove the coolant crossover/thermostat assembly.

37. Connect a suitable engine lifting device to the center intake manifold bolt holes.

38. Support the transmission and remove the engine.

To install:

39. Install the engine into the vehicle. Raise and safely support the vehicle.

40. Install the engine mount through bolts and transmission bellhousing bolts.

41. Install the flywheel-to-converter attaching bolts.

42. Install the starter and converter housing underpan.

43. Connect the block heater wires.

44. Connect the exhaust pipes to the manifold assembly.

45. Install the power steering pump. Lower the vehicle.

46. Install the coolant crossover/thermostat assembly.

47. Install the alternator upper bracket.

48. Install the heater hoses, transmission dipstick, air cleaner resonator and bracket assembly.

49. Install the wiring harness and connectors to the engine assembly.

50. Install the oil fill tube upper bracket.

51. Install the air conditioning compressor.

52. Install the injection pump and intake manifold.

53. Install the radiator, fan and lower shroud.

54. Install the air conditioning condenser.

55. Install the upper tie bar.

56. Install the fan lower shroud, coolant recovery bottle and battery cables.

57. Install the hood latch, lower valance, bumper, grille, and headlight bezels.

58. Fill the cooling system with the proper type and quantity of antifreeze. Check all fluid levels.

59. Evacuate and recharge the air conditioning system.

60. Install the engine cover. Inspect engine for leaks.

Engine Mounts

REMOVAL AND INSTALLATION

1. Disconnect the negative battery cable.

2. Raise and safely support the vehicle.

3. Properly support the engine.

NOTE: Do not position the jack under the oil pan, any sheet metal or the crankshaft pulley, otherwise damage may occur.

4. Remove the engine mount through-bolt.

5. Raise the engine to gain sufficient clearance for the mount to be removed.

6. Remove the engine mount attaching bolts, nuts and washers.

7. Remove the mount assembly.

8. Installation is the reverse of the removal procedure.

Cylinder Head

REMOVAL AND INSTALLATION

Right Side

1. Disconnect the negative battery cable, relieve the fuel system pressure and drain the coolant system. Properly discharge the air conditioning system, if equipped.

2. Remove the intake manifold. Remove the fan upper shroud. Remove the compressor assembly, if equipped.

3. Raise and support the vehicle safely.

4. Remove the exhaust manifold. Lower the vehicle.

5. Remove the valve cover, rocker arm assemblies and pushrods. Mark all components so they may be returned to their original location.

6. Remove the air cleaner resonator and bracket.

7. Remove the transmission and oil dipstick tube; remove the oil fill tube from the coolant crossover pipe.

8. Remove the heater, radiator and bypass hoses.

9. Remove the alternator upper bracket and alternator.

10. Remove the fuel bleeder valve at the coolant crossover pipe.

11. Remove the fuel return crossover line clamp bolts from both cylinder heads.

12. Disconnect the wire connector from the sensor in the coolant crossover pipe.

13. Remove the coolant crossover pipe/thermostat assembly.

14. Remove the head bolts and the cylinder head.

To install:

15. Clean the mating surfaces of the head and block thoroughly.

16. Install a new head gasket on the engine block. Do not coat the gaskets with any sealer on either engine.

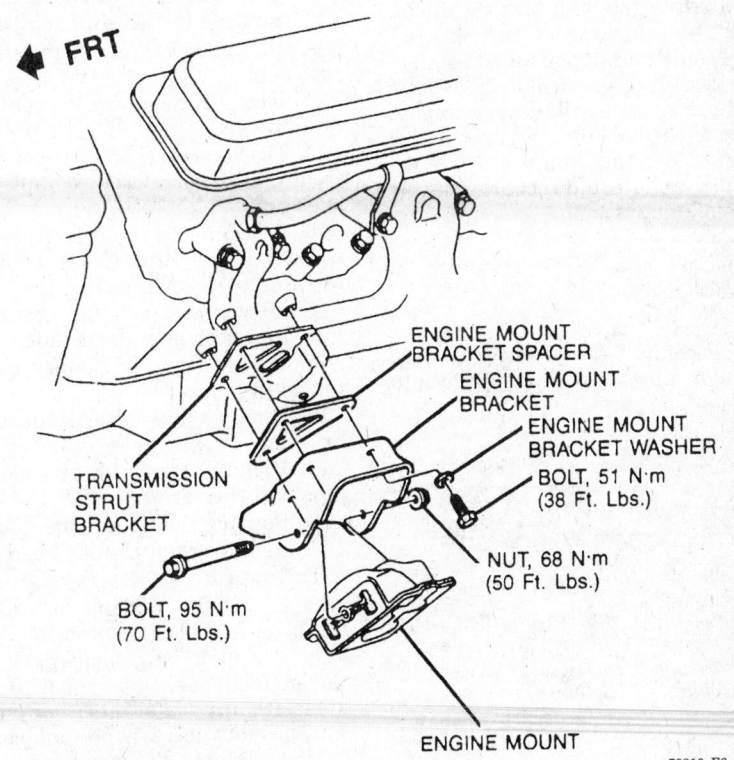

◄ FRT

ENGINE MOUNT BRACKET SPACER

ENGINE MOUNT BRACKET

ENGINE MOUNT BRACKET WASHER

BOLT, 51 N·m (38 Ft. Lbs.)

NUT, 68 N·m (50 Ft. Lbs.)

TRANSMISSION STRUT BRACKET

BOLT, 95 N·m (70 Ft. Lbs.)

ENGINE MOUNT

79210gF6

Exploded view of the front engine mount and brackets — 1996 6.5L diesel engine

The gaskets have a special coating that eliminates the need for sealer. Install the cylinder head onto the block.

17. Clean the head bolts thoroughly. Coat the threads of the head bolts with sealing compound GM part 1052080 or equivalent, before installation. Tighten the head bolts to 20 ft. lbs. (25 Nm), in the proper sequence, next tighten all bolts to 50 ft. lbs. (65 Nm) in the proper sequence, and finally tighten all bolts an additional 90 degree (¼ turn).

18. Install the coolant crossover pipe and thermostat.

19. Install the fuel valve and alternator assembly.

20. Connect the bypass hose.

21. Connect the upper radiator hose.

22. Connect the heater hoses at the head.

23. Install the transmission and oil dipstick tube.

24. Install the air cleaner resonator and bracket.

25. Install the pushrods, hardened ends facing up.

26. Install the rocker arm assemblies.

27. Adjust the valves.

28. Install the valve cover. Install the alternator assembly.

29. Raise and support the vehicle safely.

30. Install the exhaust manifold. Lower the vehicle.

31. Install the upper fan shroud.

32. Install the intake manifold. Connect the negative battery cables.

33. Fill the cooling system with the proper type and quantity of antifreeze. Evacuate and recharge the air conditioning system.

Left Side

1. Disconnect the negative battery cables. Drain the cooling system into

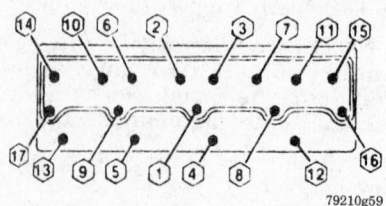

Cylinder head bolt torque sequence — 6.2L and 6.5L diesel engines

a suitable container. Properly discharge the air conditioning system.

2. Remove the intake manifold.

3. Remove the air conditioning compressor belt.

4. Remove the air conditioning compressor.

5. Remove the valve cover.

6. Remove the rocker arm assemblies. Mark the parts to ensure installation in their original location.

7. Remove the pushrods. Keep them in order.

8. Raise and support the vehicle safely.

9. Remove the exhaust manifold.

10. Remove the power steering pump and rear bracket, position the assembly aside. Lower the vehicle.

11. Remove the oil dipstick tube.

12. Disconnect the transmission detent cable.

13. Remove the glow plug controller and bracket.

14. Disconnect the wire connector from the sensor at the coolant crossover pipe.

15. Disconnect the oil fill tube, fuel bleeder valve and hoses from the coolant crossover pipe.

16. Disconnect the fuel crossover line from both cylinder heads.

17. Remove the air cleaner resonator and bracket.

18. Remove the alternator upper bracket.

19. Remove the coolant crossover pipe and thermostat.

20. Remove the head bolts.

21. Remove the cylinder head.

To install:

22. Clean the mating surfaces of the head and block thoroughly.

23. Install a new head gasket on the engine block. Do not coat the gaskets with any sealer on either engine. The gaskets have a special coating that eliminates the need for sealer. Install the cylinder head onto the block.

24. Clean the head bolts thoroughly. Coat the threads of the head bolts with sealing compound GM part 1052080 or equivalent, before installation. Install the rear cylinder head bolt first. Tighten the head bolts to 20 ft. lbs. (25 Nm). in the proper sequence, next tighten all bolts to 50 ft. lbs. (65 Nm) in the proper sequence, and finally tighten all bolts an additional 90 degree (¼ turn).

25. Install the coolant crossover pipe and thermostat.

26. Install the fuel valve and alternator assembly.

27. Connect the bypass hose.

28. Connect the upper radiator hose.

29. Connect the heater hoses at the head.

30. Install the transmission and oil dipstick tube.

31. Install the air cleaner resonator and bracket.

32. Install the pushrods, hardened ends facing up.

33. Install the rocker arm assemblies.

34. Adjust the valves.

35. Install the valve cover. Install the air conditioner compressor and bracket assembly.

36. Raise and support the vehicle safely.

37. Install the exhaust manifold. Lower the vehicle.

38. Install the upper fan shroud.

39. Install the intake manifold. Connect the negative battery cables.

40. Fill the cooling system with the proper type and quantity of antifreeze. Evacuate and recharge the air conditioning system.

Valve Lifters

REMOVAL AND INSTALLATION

1. Disconnect the negative battery cables.

2. Remove the valve covers, rocker arm shafts and pushrods. Position all components in the exact order which they were removed, so they may be reinstalled in their original bore.

3. Remove the cylinder head, as required.

4. Remove the valve lifter clamps and guide plates.

5. Remove the hydraulic lifters. Position the lifters in the exact order which they were removed, so they may be reinstalled in their original bore.

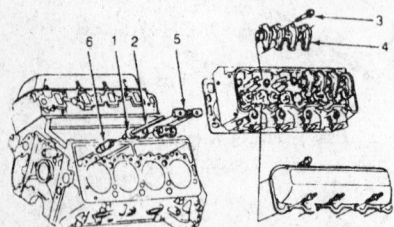

1. Hydraulic lifter
2. Pushrod
3. Bolt
4. Rocker arm assembly
5. Clamp
6. Guide plate

79210g58

Cylinder head and components — 6.2L and 6.5L diesel engines

To install:

6. Coat the roller tips with GM part 1052365 or equivalent.

NOTE: All new lifters must be primed by working the plunger while submerged in kerosene or diesel fuel. Some engines will have 2 sizes of lifters being used, standard and oversized. The oversized lifter will be etched "10" on the side and the block will be stamped O.S. on the cast pad adjacent to the lifter bore.

7. Install the lifters in their original bores.

8. Install the guide plates and clamps. Torque the clamps bolts to 18 ft. lbs. (26 Nm).

9. Turn the engine 2 full turns and verify the lifters move freely in the guide plates. If the engine does not turn freely, 1 or more lifters are binding in the guide plate.

10. Install the cylinder head assembly.

11. Install the rocker arm shafts and pushrods assembly. Position all components in the exact order they were removed.

12. Install the valve covers.

13. Connect the negative battery cables.

Valve Lash

ADJUSTMENT

All engines use hydraulic lifters, which require no periodic adjustment.

Rocker Arm and Shafts

REMOVAL AND INSTALLATION

1. Disconnect the negative battery cables. On the G-Series, remove the engine cover.

2. Remove all the necessary components in order to gain access to the engine valve covers. As required, properly relieve the fuel system pressure before disconnecting any fuel lines.

3. Remove the valve cover retaining bolts. Remove the valve cover from the engine.

4. Remove the rocker arm assemblies. Keep them in order for reinstallation.

5. Installation is the reverse of the removal procedure. Be sure to use new gaskets or RTV sealant, as necessary.

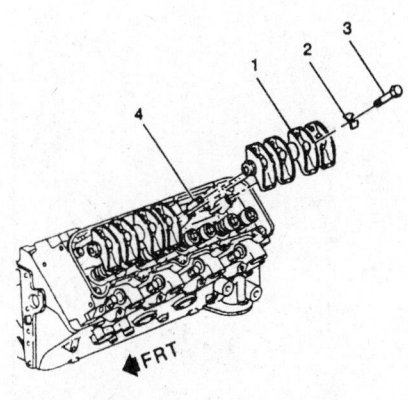

1. ROCKER ARM
2. RETAINER
3. BOLT
4. PUSHROD

79210gF7

Exploded view of the rocker arms and related components — 1996 diesel engines

Intake Manifold

REMOVAL AND INSTALLATION

1. Disconnect both negative battery cables.

2. Drain the cooling system and properly relieve the fuel system pressure.

3. Remove the engine cover, G-Series.

4. Remove the air cleaner assembly.

5. Remove the EPR/EGR valve bracket from the intake manifold.

6. Remove the CDR valve.

7. Remove the crankcase ventilator hose and EGR.

8. Remove the air conditioning rear bracket, if equipped.

9. Remove the fuel line bracket and ground strap.

10. Remove the fuel filter bracket at the intake manifold.

11. Remove the intake manifold bolts. The injection line clips are retained by these bolts.

12. Remove the intake manifold.

NOTE: If the engine is to be further serviced with the manifold removed, install protective covers over the intake ports.

To install:

13. Clean the manifold gasket surfaces on the cylinder heads and install new gaskets before installing the manifold.

NOTE: The gaskets have an opening for the EGR valve on light duty installations. An insert covers this opening on heavy duty installations.

14. Install the intake manifold.

15. Install the intake manifold bolts and fuel injection line clips.

16. Install the fuel filter bracket at the intake manifold.

17. Install the fuel line bracket and ground strap.

18. Install the air conditioning rear bracket, if equipped.

19. Install the crankcase ventilator hose and EGR.

20. Install the CDR valve.

21. Install the EPR/EGR valve bracket from the intake manifold.

22. Install the air cleaner assembly.

23. Install the engine cover, G-Series.

24. Fill the cooling system with the proper type and quantity of antifreeze.

25. Connect both negative battery cables. Inspect for leaks.

Exhaust Manifold

REMOVAL AND INSTALLATION

1. Disconnect the batteries.

2. Raise and support the vehicle safely.

3. Disconnect the exhaust pipe from the manifold flange and lower the vehicle.

4. Remove the engine cover and disconnect the glow plug wires.

5. Remove the air cleaner duct bracket.

6. Remove the glow plugs. Remove the turbocharger assembly, as required.

7. Remove the air conditioner compressor rear bracket, as required.

8. Remove the manifold bolts and remove the manifold.

9. Installation is the reverse of the removal procedure. Torque the manifold bolts to 26 ft. lbs. (35 Nm).

Timing Chain, Sprockets and Front Cover

REMOVAL AND INSTALLATION

1. Disconnect both negative battery cables. Drain the cooling system.

LEFT SIDE

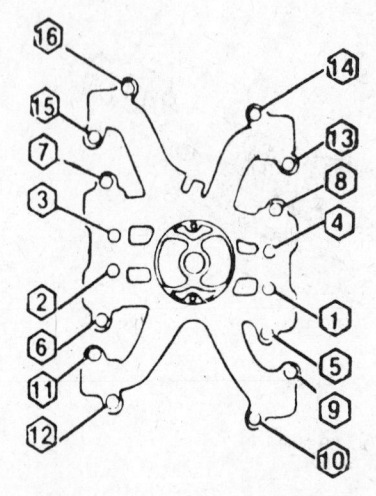

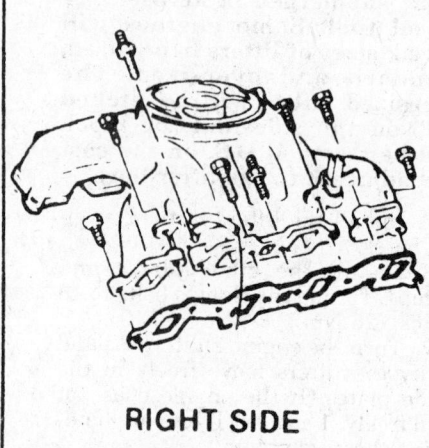

RIGHT SIDE

79210g60

Intake manifold installation and torque sequence — 6.2L and 6.5L diesel engines

2. Remove the water pump and pulleys.

3. Rotate the crankshaft to align the marks on the torsional damper with the **0** mark on the timing tab.

4. Scribe a mark aligning the injection pump flange and the front cover, if not already marked.

5. Remove the crankshaft pulley and torsional damper.

6. Remove the front cover-to-oil pan bolts (4).

7. Remove the 2 fuel return line clips.

8. Remove the injection pump gear.

9. Remove the injection pump retaining nuts from the front cover.

10. Remove the baffle. Remove the remaining cover bolts and remove the front cover.

11. Remove the injection pump gear.

12. Align the camshaft timing gear marks and remove the bolt and washer attaching the camshaft gear.

13. Remove the camshaft sprocket with the timing chain. Remove the crankshaft sprocket.

To install:

14. Install the cam sprocket, timing chain and crankshaft sprocket as a unit, aligning the timing marks on the sprockets.

15. Rotate the crankshaft to align the injection pump and camshaft gears. Install the injection pump gear.

16. If the front cover oil seal is to be replaced, it can now be pried out of the cover with a suitable prying tool. Press the new seal into the cover evenly.

17. Clean both sealing surfaces until all traces of old sealer are gone. Apply a $3/32$ in. (2mm) bead of GM sealant 1052357 or equivalent to the sealing surface. Apply a $3/16$ in. (5mm) bead of RTV type sealer to the bottom portion of the front cover which attaches to the oil pan. Install the front cover.

18. Install the baffle.

19. Install the injection pump, making sure the scribe marks on the pump and front cover are aligned. Tighten the nuts to 31 ft. lbs. (42 Nm).

20. Install the injection pump driven gear, making sure the marks on the cam gear and pump are aligned. Torque the injection pump gear bolts to 17 ft. lbs. (23 Nm).

NOTE: Verify that there is a minimum clearance of 0.040 in. (1.0mm) between the injection pump gear and baffle or noise may be result.

21. Install the fuel line clips, the front cover-to-oil bolts, and the torsional damper and crankshaft pulley. Torque the pan and the damper bolt to specification.

22. Install the water pump and pulley assembly.

23. Fill the cooling system with the proper type and quantity of antifreeze.

24. Connect the negative battery cables. Inspect engine for leaks.

Camshaft

REMOVAL AND INSTALLATION

Blazer, Pick-Up, Suburban, Tahoe and Yukon

1. Disconnect the battery cables and relieve the fuel system pressure.

2. Drain the cooling system.

3. Remove the radiator, the shroud and fan assembly.

4. Remove the grille and parking light assembly.

5. Remove the hood latch and brace assembly.

6. Remove the oil pump drive.

7. Remove the power steering pump, alternator and air conditioner compressor and position aside.

8. Remove the rocker arm covers.

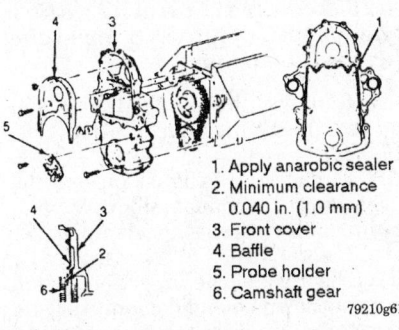

1. Apply anarobic sealer
2. Minimum clearance
   0.040 in. (1.0 mm)
3. Front cover
4. Baffle
5. Probe holder
6. Camshaft gear

79210g61

Front cover and components — 6.2L and 6.5L diesel engines

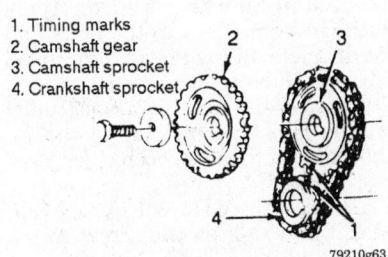

1. Timing marks
2. Camshaft gear
3. Camshaft sprocket
4. Crankshaft sprocket

79210g63

Injection pump gear and timing marks — 6.2L and 6.5L diesel engines

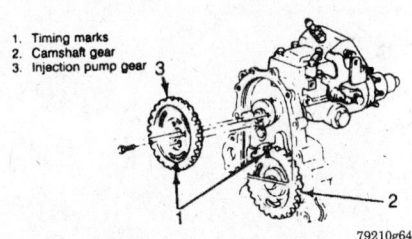

1. Timing marks
2. Camshaft gear
3. Injection pump gear

79210g64

Camshaft and sprockets — 6.2L and 6.5L diesel engines

9. Remove the rocker arm assemblies and pushrods. Mark them so they can be returned to their original position.
10. Remove the hydraulic lifters and keep them in order so they can be returned to their original bore.
11. Remove the front cover.
12. Remove the timing chain and camshaft sprocket.
13. Remove the injector pump.
14. Raise the engine and support it safely.
15. Remove the front engine mounting through bolts.
16. Remove the air conditioner condenser mounting bolts and lift the condenser out.

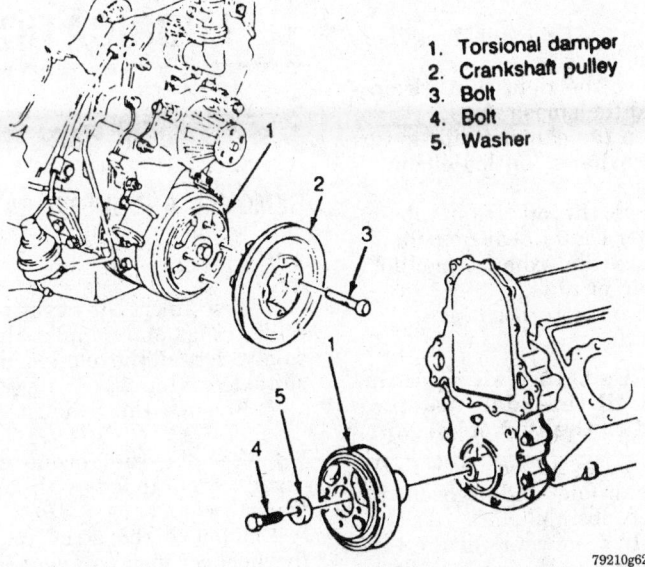

1. **Torsional damper**
2. **Crankshaft pulley**
3. **Bolt**
4. **Bolt**
5. **Washer**

79210g62

Torsional damper and crankcase pulley installation — 6.2L and 6.5L diesel engines

17. Remove the thrust plate bolts and thrust plate.
18. Carefully remove the camshaft from the block.
19. Remove the thrust plate spacer, if necessary.

To install:

20. Install the spacer with the ID chamfer toward the camshaft.

NOTE: It is recommended that the engine oil, oil filter and hydraulic lifters be replaced when installing a new camshaft.

21. Coat the camshaft lobes with Molykote® or equivalent.
22. Lubricate the camshaft journals with engine oil.
23. Insert the camshaft carefully into the block, install the thrust plate and bolts. Torque to 17 ft. lbs. (23 Nm).
24. Lower the engine and install the engine mount through bolts.
25. Align the timing marks and install the timing chain and sprockets.
26. Install the air conditioner condenser, if equipped.
27. Install the injector pump.
28. Install the front cover.
29. Install the hydraulic lifters in the same bore as they were removed.
30. Install the rocker arm assemblies and pushrods in their original locations.
31. Install the rocker arm covers.
32. Install the power steering pump, alternator and air conditioner compressor.
33. Install the oil pump drive.
34. Install the hood latch and brace.
35. Install the grille and parking light assembly.
36. Install the radiator, the shroud and fan assembly.
37. Fill the cooling system with the proper type and quantity of antifreeze.
38. Connect the negative battery cables.

G-Series

1. Disconnect the battery cables and relieve the fuel system pressure.
2. Remove the headlight bezels.
3. Remove the grille, bumper and lower valance panel.
4. Remove the hood latch.
5. Remove the coolant recovery bottle.
6. Remove the upper tie bar.
7. Remove the air conditioner compressor.
8. Drain the cooling system and remove the radiator and fan.
9. Remove the oil pump drive.
10. Remove the cylinder heads to gain clearance for lifter removal.

11. Remove the alternator lower bracket.

12. Remove the water pump.

13. Remove the torsional damper.

14. Remove the front cover.

15. Remove the injection pump.

16. Remove the rocker arm covers.

17. Remove the rocker arm assemblies and pushrods. Mark them so they can be returned to their original position.

18. Remove the hydraulic lifters and keep them in order so they can be returned to their original bore.

19. Remove the timing chain and camshaft sprocket.

20. Remove the thrust plate bolts and thrust plate.

21. Carefully remove the camshaft from the block.

22. Remove the thrust plate spacer, if necessary.

To install:

23. Install the spacer with the ID chamfer toward the camshaft.

NOTE: It is recommended that the engine oil, oil filter and hydraulic lifters be replaced when installing a new camshaft.

24. Coat the camshaft lobes with Molykote or equivalent.

25. Lubricate the camshaft journals with engine oil.

26. Insert the camshaft carefully into the block, install the thrust plate and bolts and torque to 17 ft. lbs.

27. Align the timing marks and install the timing chain and sprockets.

28. Install the hydraulic lifters in the same bore as they were removed.

29. Install the rocker arm assemblies and pushrods in their original locations.

30. Install the rocker arm covers.

31. Install the fuel pump.

32. Install the front cover.

33. Install the torsional damper and water pump.

34. Install the alternator lower bracket.

35. Install the cylinder heads.

36. Install the oil pump drive.

37. Install the radiator and fan.

38. Install the air conditioner compressor.

39. Install the upper tie bar.

40. Install the coolant recovery bottle.

41. Install the hood latch.

42. Install the grille, bumper and lower valance panel.

43. Install the headlight bezels.

44. Install the battery cables.

45. Fill the cooling system.

46. Evacuate and charge the air conditioner system.

Piston and Connecting Rod

POSITIONING

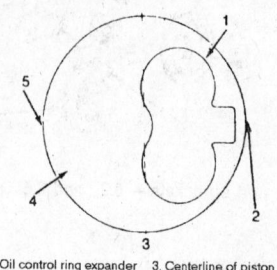

1. Oil control ring expander gap
2. Second compression ring gap
3. Centerline of piston pin
4. Oil control ring gap
5. Top compression ring gap

79210g65

Piston ring gap locations — 6.2L and 6.5L diesel engines

Turbocharger

REMOVAL AND INSTALLATION

1. Disconnect the negative battery cable.

2. Remove the air inlet duct.

3. Disconnect the oil feed line from the top of the turbocharger.

4. Remove the screw retaining the Crankcase Depression Regulator (CDR) valve vent bracket.

5. Remove the CDR valve and vent tube.

6. Loosen the fasteners retaining the air cleaner assembly to the wheel well and remove the air cleaner assembly.

7. Loosen the heat shield retainers and remove the shield.

8. Remove the right front tire assembly and the splash shield.

9. Loosen the exhaust pipe-to-turbocharger exhaust outlet elbow V-band clamp.

10. Remove the oil drain tube-to-turbocharger center bearing bolts.

11. Remove the exhaust manifold-to-turbocharger nuts.

12. Remove the turbocharger.

To install:

NOTE: Use antisieze compound on all threaded fasteners connected to the turbocharger

13. Engage the turbocharger to the exhaust manifold and tighten the nuts to 37 ft. lbs. (50 Nm).

14. Install a new oil drain tube flange gasket and the oil drain tube. Tighten the bolts to 19 ft. lbs. (26 Nm).

NOTE: Use 1-to-2cc of engine oil to feed the oil feed hole at the top of the turbocharger and hand rotate the compressor wheel/shaft. This will pre-lube the shaft bearings

15. Connect the oil feed line and tighten the connection to 13 ft. lbs. (17 Nm).

16. Engage the exhaust pipe to the turbocharger exhaust elbow V-band clamp and tighten the clamp to 71 inch. lbs. (8 Nm).

17. Disengage the injection pump fuel shutdown solenoid connector and crank the engine for no more 15 seconds to prime the oil system. Do not let the engine start.

18. Install the right front splash shield and wheel.

19. Install the heat shield. Apply Loctite or equivalent to the bolts and install the bolts and tighten to 56 inch. lbs. (6 Nm).

20. Install the air cleaner assembly and the rubber connector to the intake duct and the turbocharger compressor outlet.

21. Install the CDR valve, tube and bracket and tighten the screw.

22. Install the air intake duct.

23. Install the CDR vent tubes into the air intake duct.

NOTE: Operate the engine at idle for at least 3 minutes after installing the turbocharger

ENGINE LUBRICATION

Oil Pan

REMOVAL AND INSTALLATION

4.3L Engines

1. Disconnect the negative battery cable. Raise and support the vehicle safely. Drain the engine oil into a suitable container.

2. Remove the exhaust crossover pipe.

3. Remove the torque converter cover, if equipped with automatic transmission.

4. Remove the strut rods at the flywheel cover, if equipped.

5. Remove the strut rod at the front engine mounts, if equipped.

6. Remove the starter assembly.

7. Remove the oil pan bolts, nuts and reinforcements.

8. Remove the oil pan and gaskets.

To install:

9. Thoroughly clean all gasket surfaces and install a new gasket, using only a small amount of sealer at the front and rear corners of the oil pan.

10. Install the oil pan and new gaskets.

11. Install the oil pan bolts, nuts and reinforcements. Torque the pan bolts to specification.

12. Install the starter assembly.

13. Install the strut rod brackets at the front engine mounts.

14. Install the strut rods at the flywheel cover.

15. Install the torque converter cover, if equipped with automatic transmission.

16. Install the exhaust crossover pipe. Lower the vehicle.

17. Connect the negative battery cable.

18. Fill the engine with the proper quantity and type of oil.

5.0L and 5.7L Engines

1. Disconnect the negative battery cable. Raise the vehicle, support it safely, and drain the engine oil.

2. Remove the exhaust crossover pipe.

3. Remove the flywheel or torque converter cover.

4. Remove the strut rods at the front engine mountings, if used.

5. Remove the oil pan bolts, nuts and reinforcements.

6. Remove the oil pan and gaskets.

To install:

7. Thoroughly clean all gasket surfaces and install a new gasket, using a small amount of sealer at the front and rear corners of the oil pan.

8. Install the oil pan and new gaskets.

9. Install the oil pan bolts, nuts and reinforcements. Torque the pan bolts to specification.

10. Install the strut rods at the front engine mountings.

11. Install the torque converter or flywheel cover.

12. Install the exhaust crossover pipe.

13. Connect the negative battery cable.

14. Fill the crankcase with the proper type and quantity of oil.

Diesel Engine

Blazer, Pick-Up, Suburban, Tahoe and Yukon

1. Disconnect the battery cables.

2. Raise the vehicle and support it safely.

3. Drain the engine oil. Remove the oil dipstick.

4. Remove the flywheel cover.

5. Disconnect the exhaust pipes from the manifolds.

6. Remove the front engine mount through bolts and raise the engine.

7. Remove the oil pan bolts and remove the oil pan.

8. Remove the oil pan rear seal.

To install:

9. Clean the old RTV sealant from the oil pan and block.

10. Apply a ³⁄₁₆ in. (5mm) bead of RTV sealant to the oil pan sealing surface, inboard of the bolt holes. The sealant must be wet to the touch when the oil pan is to be installed.

11. Install the oil pan rear seal.

12. Install the oil pan to the engine and install the retaining bolts. Torque all except the rear 2 bolts to 84 inch lbs. Torque the rear 2 bolts to 17 ft. lbs.

13. Lower the engine.

14. Install the engine mounting through bolt and nut.

15. Install the oil dipstick. Install the exhaust pipes to the manifolds.

16. Install the flywheel cover and lower the vehicle.

17. Refill with the proper grade and quantity of oil.

18. Install the battery cables.

G-Series

1. Disconnect the battery cables.

2. Remove the engine cover.

3. Remove the engine oil dipstick.

4. Remove the engine oil dipstick tube at the rocker cover.

5. Raise the vehicle and support it safely.

6. Remove the transmission flywheel cover.

7. Drain the engine oil.

8. Disconnect the oil cooler lines at the block.

9. Remove the starter. Remove the battery cables, transmission cooler lines and attaching clamps from the oil pan.

10. Remove the oil pan bolts and remove the oil pan and oil pan rear seal.

To install:

11. Apply a ³⁄₁₆ in. (5mm) bead of RTV sealant to the oil pan sealing surface, inboard of the bolt holes. The sealant must be wet to the touch when the oil pan is to be installed.

12. Install the oil pan rear seal.

13. Install the oil pan to the engine and install the retaining bolts. Torque all bolts to specifications.

14. Install the starter. Install the transmission, battery cables and attaching clamps to the oil pan.

15. Install the engine oil cooler lines.

16. Install the transmission flywheel cover.

17. Lower the vehicle.

18. Install the engine oil dipstick tube at the rocker cover.

19. Install the engine oil dipstick.

20. Install the engine cover.

21. Refill with the proper grade and quantity of oil.

22. Install the battery cables.

7.4L Engine

NOTE: Removal of the transmission may be necessary on the G-Series vehicles.

1. Disconnect the negative battery cable.

2. Remove the fan shroud.

3. Remove the air cleaner.

4. Remove the distributor cap.

5. Raise and support the vehicle safely.

6. Drain the engine oil. Remove the starter assembly, if equipped with manual transmission.

7. Remove the torque converter or clutch housing cover.

8. Remove the oil filter.

9. Remove the oil pressure line from the side of the block.

10. Support the engine.

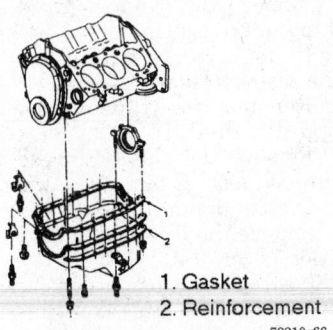

1. Gasket
2. Reinforcement

79210g66

Oil pan mounting — 4.3L engine

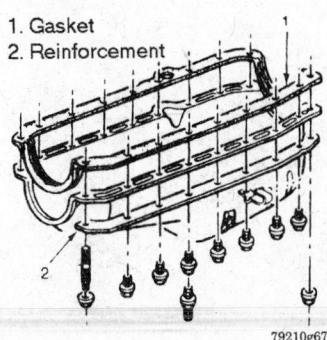

1. Gasket
2. Reinforcement

79210g67

Oil pan mounting — 5.0L and 5.7L engines

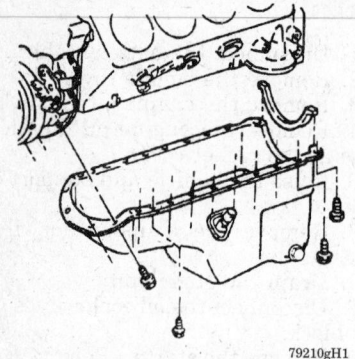

Oil pan mounting — 1996 6.5L diesel engine

11. Remove the engine mount through bolts.

12. Raise the engine just enough to remove the pan.

13. Remove the oil pan bolts, the oil pan and discard the gaskets.

To install:

14. Clean all mating surfaces thoroughly.

15. Apply RTV gasket material to the front and rear corners of the gaskets.

16. Coat the gaskets with adhesive sealer and position them on the block.

17. Install the rear pan seal in the pan with the seal ends mating with the gaskets.

18. Install the front seal on the bottom of the front cover, pressing the locating tabs into the holes in the cover.

19. Install the oil pan.

20. Install the pan bolts, clips and reinforcements. Torque the pan bolts to specification.

21. Lower the engine onto the mounts.

22. Install the engine mount through-bolts.

23. Install the oil pressure line.

24. Install the oil filter. Install the starter assembly, if removed.

25. Install the torque converter or clutch housing cover.

26. Install the distributor cap.

27. Install the air cleaner.

28. Install the fan shroud.

29. Connect the battery.

30. Fill the crankcase with the proper type and quantity of oil.

Oil Pump

REMOVAL AND INSTALLATION

Gasoline Engines

1. Remove the oil pan.

2. Remove the oil pump attaching bolt and, if equipped, the pickup tube nut/bolt, then remove the pump along with the pickup tube and shaft, as necessary.

To install:

3. Ensure that the pump pickup tube is tight in the pump body. If the tube should come loose, oil pressure will be lost and oil starvation will occur. If the pickup tube is loose it should be replaced.

4. If the pump has been disassembled and is being replaced or for any reason oil has been removed, it must be primed. It can either be filled with oil before installing the cover plate and oil kept within the pump during handling or the entire pump cavity can be filled with petroleum jelly.

NOTE: If the pump is not primed, the engine could be damaged before it receives adequate lubrication when the engine is started.

5. Install the pump aligning the pump shaft with the distributor drive gear as necessary. Tighten oil pump/pickup tube retainer(s) to 65 ft. lbs. (90 Nm) on all 4.3L and all V8 Engines.

6. Install the oil pan and refill the engine crankcase. Disable the ignition system; crank engine for approximately 10 seconds to aid in priming the oil pump and reducing the risk of engine damage.

NOTE: If the oil pump does not build up oil pressure almost immediately, remove the pan and check for a loose oil pump-to-pickup tube attachment. If necessary dismantle the pump and pack the pump cavity with petroleum jelly. Running the engine without measurable oil pressure will cause extensive damage.

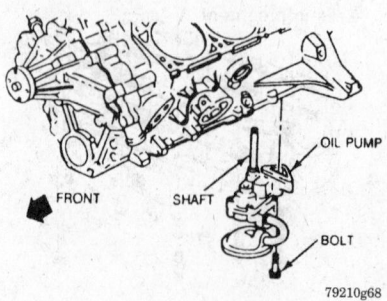

Oil pump mounting — gasoline engines

Diesel Engines

1. Raise the vehicle and support it with jackstands.

2. Drain the oil and remove the oil pan.

3. Remove the oil pump to crankshaft rear main bearing attaching bolt.

4. Remove the oil pump and hex drive.

To install:

5. Inspect the oil pan pick up tube and screen for damage and the hex drive for cracks.

6. Install the oil pump and extension shaft to the engine. Align the extension shaft hex with the drive hex, the oil pump should push easily into place.

7. Install the oil pump bolt and tighten to 65 ft. lbs. (90 Nm).

8. Install the oil pan and lower the vehicle.

9. Fill the crankcase with the proper grade and amount of oil.

MANUAL TRANSMISSION

Manual Transmission Assembly

REMOVAL AND INSTALLATION

Blazer, Pick-Up, Suburban, Tahoe and Yukon

1. Disconnect the negative battery cable. Remove the shifter boot and lever.

2. Raise and support the vehicle safely. Drain the transmission.

3. Remove driveshaft after marking the position of the shaft to the axle flange.

4. Remove the exhaust pipes and parking brake cables.

5. Disconnect the wiring harness at the transmission.

6. Remove the transfer case, if equipped with 4WD.

7. Remove the clutch slave cylinder and support it out of the way.

8. On the 85mm (MG5) transmission, remove the flywheel housing inspection cover.

9. Position a transmission jack or equivalent, under the transmission for support.

10. Remove the crossmember. Visually inspect to see if other equipment,

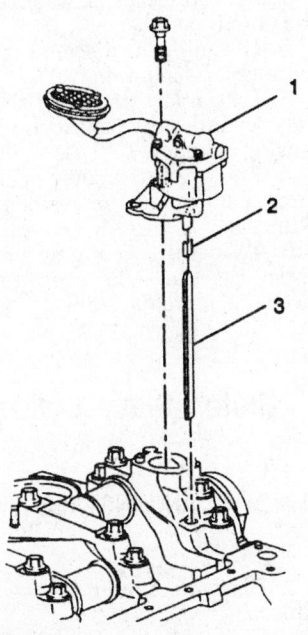

1. OIL PUMP
2. PLASTIC RETAINER
3. HEX DRIVE

79210gH2

Exploded view of the oil pump assembly — diesel engine

brackets or lines, must be removed to permit removal of transmission.

NOTE: Mark position of crossmember when removing to prevent incorrect installation. The tapered surface should face the rear.

11. Remove the top 2 transmission to housing bolts and insert 2 guide pins.

NOTE: The use of guide pins will not only support the transmission but will prevent damage to the clutch disc. Guide pins can be made by taking 2 bolts, the same as those just removed only longer, and cutting off the heads. Make an adjustment slot. Be sure to support the clutch release bearing and support assembly during removal of the transmission. This will prevent the release from falling out of the flywheel housing.

12. Remove the remaining bolts and slide transmission straight back from engine. Use care to keep the transmission drive gear straight in line with clutch disc hub.
13. Remove the transmission.

To install:

14. Place the transmission in high gear. Coat the input shaft splines with high temperature grease.
15. Raise the transmission into position.
16. Install the guide pins in the top 2 bolt holes.
17. Roll the transmission forward and engage the clutch splines. Keep pushing the transmission forward until it mates with the engine.
18. Install the bolts, removing the guide pins. Torque the bolts to specification.
19. Install the transfer case, if equipped.
20. Install the crossmember. Torque the crossmember bolts to specification. Remove the transmission jack.
21. Install the slave cylinder.
22. Install the driveshaft.
23. Install the exhaust system.
24. Install the inspection cover on the 85mm transmission.
25. Install the gearshift lever.
26. Connect the wiring harness at the transmission.
27. Fill the transmission with the proper type and quantity of oil.

NOTE: Do not force the transmission into the clutch disc hub. Do not let the transmission hang unsupported in the splined portion of the clutch disc.

G-Series

1. Raise and support the vehicle safely.
2. Drain the transmission.
3. Disconnect the speedometer cable, back-up light and TCS switch.
4. Remove the shift controls from the transmission.
5. Disconnect the driveshaft and remove it.
6. Support the transmission using the proper equipment.
7. Remove the transfer case, if equipped with 4WD.
8. Inspect the transmission to be sure all necessary components have been removed or disconnected.
9. Mark the front of the crossmember to be sure it is installed correctly.
10. Remove the flywheel housing under pan and transmission mounting bolts.
11. Move the transmission slowly away from the engine, keeping the mainshaft in alignment with the clutch disc hub. Support the clutch release bearing to prevent it from falling out of the flywheel housing when the transmission is removed.

12. Remove the transmission.
To install:
13. Lightly coat the mainshaft with high temperature grease. Do not use much grease, since, under normal operation, the grease will be thrown onto the clutch, causing it to fail.
14. Raise the transmission into position under the vehicle.
15. Roll the unit forward engaging the spline of the mainshaft with the splines in the clutch hub. Continue pushing forward until the transmission mates with the bellhousing.
16. Install and tighten the transmission-to-bellhousing bolts. Install the transfer case, if equipped.
17. Install the flywheel housing under pan.
18. Install the crossmember. Torque the bolts to 50 ft. lbs.
19. Inspect the transmission to be sure all necessary components have been installed or connector.
20. Connect the driveshaft.
21. Install the shift controls.
22. Connect the speedometer cable.
23. Connect the back-up light switch.
24. Connect the TCS switch.
25. Fill the transmission with the proper quantity and type of oil.
26. Road test the vehicle.

Clutch Assembly

REMOVAL AND INSTALLATION

1. Disconnect the negative battery cable.
2. Remove the transmission assembly. If not done already, remove the flywheel cover.
3. Disconnect the slave cylinder or clutch adjusting rod.
4. Remove the bellhousing retaining bolts. Remove the bellhousing.
5. Remove the throw-out spring and fork.
6. Remove the ball stud from the bellhousing.
7. Install the clutch removal tool and support the clutch assembly.

NOTE: Before removing the clutch from the flywheel, match-mark the flywheel, clutch cover and 1 pressure plate lug, so these parts may be assembled in their same relative positions and retain the factory balance.

8. Loosen the clutch plate retaining bolts slowly and evenly 1 at a time until all pressure is released from the pressure plate assembly.
9. Remove the clutch, pressure plate and removal tool. Check the fly-

wheel for damage, repair or replace, as required.

10. Check the clutch assembly, flywheel and pilot bearing for signs of wear, scoring, overheating, etc. If the clutch plate, flywheel or pressure plate is oil-soaked, inspect the engine rear main seal and the transmission input shaft seal and correct leakage as required. Replace any damaged parts.

To install:

11. Assemble the pressure plate and disc assembly, as required.

NOTE: The manufacturer recommends that new pressure plate bolts and washers be used.

12. Turn the flywheel until the previously applied mark is at the bottom.

13. Install the clutch disc, pressure plate and cover, using a suitable clutch aligning tool.

14. Turn the clutch until the matchmark on the clutch cover aligns with the mark on the flywheel.

15. Install the attaching bolts and tighten them a little at a time in a crossing pattern until the spring pressure is taken up.

16. Remove the aligning tool.

17. Coat the rounded end of the ball stud with high temperature wheel bearing grease.

18. Install the ball stud in the bellhousing. Pack the ball stud from the lubrication fitting.

19. Pack the inside recess and the outside groove of the release bearing with high temperature wheel bearing grease and install the release bearing and fork. The fork must be installed so the fingers and tabs fit into the groove of the release bearing.

20. Install the release bearing seat and spring.

21. Install the bellhousing and retaining bolts, as required.

22. Install the slave cylinder or clutch rod.

23. Install the transmission and the flywheel cover.

24. Connect the negative battery cable.

25. Bleed the hydraulic system or adjust the clutch as required.

Clutch Master Cylinder

REMOVAL AND INSTALLATION

1. Disconnect the negative battery cable.

2. Remove the lower steering column cover(s) for access.

3. If necessary, remove the lower left side air conditioning duct.

4. Disconnect the pushrod from the clutch pedal.

5. If equipped, disconnect the remote reservoir hose.

6. Disconnect and plug the secondary cylinder hydraulic line to the master cylinder.

7. Remove the master cylinder retaining nuts and remove the master cylinder.

8. To install, use a new gasket, reverse the removal procedure and bleed the clutch system.

Clutch Slave Cylinder

REMOVAL AND INSTALLATION

1. Disconnect the negative battery cable.

2. Raise the vehicle and support it safely.

3. Disconnect and plug the hydraulic line at the slave cylinder.

4. Remove the slave cylinder retaining bolts. Remove the slave cylinder.

5. Installation is the reverse of the removal procedure. Bleed the system, as required.

1. Lubrication fitting (R/V and C/K models only)
2. Clutch fork
3. Spring washer
4. Screw
5. Flywheel housing
6. Screw
7. Ball stud
8. Boot
9. Retainer
10. Release bearing
11. Cover assembly
12. Driveplate
13. Cover
14. Screw
15. Pilot bearing
16. Screw
17. Strap
18. Pressure plate

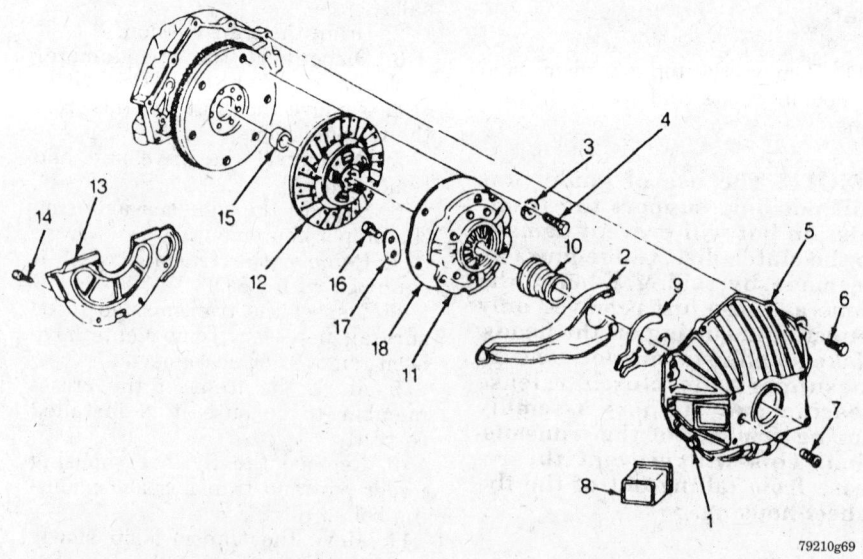

79210g69

Exploded view of typical clutch assembly

Hydraulic Clutch System Bleeding

Bleeding air from the hydraulic clutch system is necessary whenever any part of the system has been disconnected or the fluid level (in the reservoir) has been allowed to fall so low, that air has been drawn into the master cylinder.

1. Fill master cylinder reservoir with new brake fluid conforming to Dot 3 specifications.

——————— CAUTION ———————

Never, under any circumstances, use fluid which has been bled from a system to fill the reservoir as it may be aerated, have too much moisture content and possibly be contaminated.

2. Raise and support vehicle safely.

3. Remove the slave cylinder retainers.

4. Hold slave cylinder at approximately 45 degrees with the bleeder screw located at the highest point. Have an assistant fully depress and hold the clutch pedal, then open the bleeder screw.

5. Close the bleeder screw and have your assistant release the clutch pedal.

6. Repeat the procedure until all of the air is evacuated from the system. Check and refill master cylinder reservoir as required to prevent air from being drawn through the master cylinder.

NOTE: Never release a depressed clutch pedal with the bleeder screw open or air will be drawn into the system.

7. Test the clutch for proper operation.

Automatic Transmission Assembly

REMOVAL AND INSTALLATION

2-Wheel Drive

NOTE: It may be necessary to disconnect and remove the exhaust crossover pipe on 8-cylinder engines, and to disconnect the catalytic converter and remove its support bracket, if equipped.

1. Disconnect the battery ground cable. Remove the air cleaner and disconnect the detent cable.

2. Raise and support the vehicle safely. Drain the transmission.

3. Matchmark the axle-to-driveshaft flanges and remove the driveshaft.

4. Disconnect the speedometer cable, downshift cable, vacuum modulator line, shift linkage, throttle linkage, fuel lines and fluid cooler lines at the transmission, as required. Remove the filler tube.

5. Support the transmission. Unbolt and remove the crossmember.

6. Remove the torque converter underpan, Matchmark the flywheel and converter, and remove the converter bolts.

7. Support the engine and lower the transmission slightly for access to the upper transmission to engine bolts.

8. Remove the transmission to engine bolts and pull the transmission back and out of the vehicle.

NOTE: Keep the front of the transmission up so the converter doesn't fall out.

9. Installation is the reverse of the removal procedure.

NOTE: Lubricate the internal yoke splines at the transmission end of the driveshaft with lithium base grease.

4-Wheel Drive

1. Disconnect the battery ground cable and remove the transmission dipstick. Detach the downshift cable. Remove the transfer case shift lever knob and boot.

2. Raise and support the vehicle safely.

3. Remove the skid plate, if any. Remove the flywheel cover.

4. Matchmark the flywheel and torque converter, remove the bolts and secure the converter so it doesn't fall out of the transmission.

5. Detach the shift linkage, speedometer cable, vacuum modulator line, downshift cable, throttle linkage and cooler times at the transmission. Remove the filler tube.

6. Remove the exhaust crossover pipe to manifold bolts.

7. Unbolt the transfer case adapter from the crossmember. Support the transmission and transfer case. Remove the crossmember.

8. Move the exhaust system aside. Detach the driveshafts after matchmarking their flanges. Disconnect the parking brake cable.

9. Unbolt the transfer case from the frame bracket. Support the engine. Unbolt the transmission from

the engine, pull the assembly back, and remove.

10. Reverse the procedure for installation.

Throttle Valve (TV) Cable

ADJUSTMENT

1. If necessary for access, remove the air cleaner assembly.

2. If the cable has been removed and installed, pull on the upper end of the cable. It should travel a short distance with light resistance caused by the small return spring on the TV lever. When released, check to see that the cable slider returns to the 0 or the fully adjusted position; if not, adjust the cable and slider:

    a. Depress and hold the readjust tab on the end of the TV cable.

    b. Move the slider back on the cable (away from the throttle lever) until it stops against the fitting.

    c. Release the readjust tab.

3. Rotate the throttle lever by hand (do not use the accelerator pedal) to the full throttle stop (wide open throttle) position. The slider must move (ratchet) toward the lever in order to adjust the cable when the lever is rotated.

4. Release the throttle lever and check for proper operation.

NOTE: The throttle valve cable adjustment should be rechecked with the engine at normal operating temperature. A cable may adjustment may appear to function properly while the engine is cold, but no work properly when warmed.

DRIVE AXLE

Driveshaft and U-Joints

REMOVAL AND INSTALLATION

1. Raise and support the vehicle safely.

2. Mark the relationship of the driveshaft to the front axle and transfer case flange.

3. Remove the skid plate, if equipped.

4. Remove the bolts, nuts, washers and U-bolts/retainers from the front axle flange-to-driveshaft.

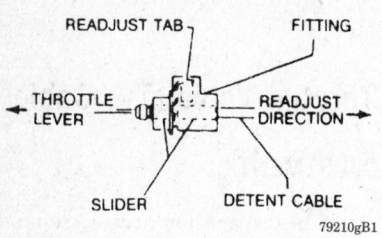

Throttle valve (TV) cable adjustment

5. Remove the slip yoke from the front axle.

NOTE: Do not let the U-joint caps fall off the yoke, tape the bearing caps in place to avoid loss of the bearing rollers.

6. Remove the bolts at the transfer case flange, slide the driveshaft forward and disengage the assembly from the transfer case.

7. Place the driveshaft into a holding fixture.

8. Remove the retaining rings.

9. Using 2 sockets, 1 with a diameter just smaller than the bearing cap; the other with an opening large enough to accept a bearing cap, press out the bearing caps in a vise.

10. Remove the trunnion from the slip yoke.

To install:

11. Place the trunnion in between the yoke ears of the driveshaft and begin installing both bearing caps by hand, finish the installation using a vise

12. Use a socket with a smaller diameter than the bearing cap and press the bearing cap in past the retainer ring groove. Install the retaining ring.

13. Turn the driveshaft over and repeat the procedure.

14. Install the slip yoke over the trunnion and repeat the procedure for the remaining bearing caps.

15. Install the driveshaft, U-bolt/retainers, nuts and bolts, aligning the shaft with the previous made marks on the transfer case and front axle.

16. Lower the vehicle and road test.

Halfshaft

REMOVAL AND INSTALLATION

4-Wheel Drive

1. Raise and support the vehicle safely.

2. Remove the wheel and tire assembly.

3. Remove the skid plate, as required.

4. Remove the drive axle hub nut and washer.

5. Remove the brake line support bracket from the upper control arm to allow extra travel of the control arm.

6. Remove the left outer tie rod attaching nut and cotter pin. Separate the tie rod from the steering knuckle using a suitable tie rod end splitter.

7. Position the tie rod aside and push steering linkage to the opposite side of the vehicle.

8. Remove the lower shock attaching nut and bolt; position the shock aside.

9. Remove the left stabilizer bar bracket and bushing at the frame. Remove the stabilizer bar bolt, spacer and bushings at the lower control arm.

10. Lower the vehicle, taking pressure off the upper control arm by placing a support below the lower control arm between the spring seat and the ball joint.

11. Remove the upper ball joint cotter pin and loosen (do not remove) the upper ball joint attaching nut. Using a suitable ball joint splitter, separate the ball joint stud from the steering knuckle. Remove the attaching nut.

NOTE: Cover the shock mounting bracket and lower ball joint stud with a towel to prevent the axle boot from tearing during Removal and Installation.

12. Separate the axle shaft from the hub and rotor using tool J–28733 or equivalent.

13. Remove the axle shaft inner flange bolts. Remove the shaft.

To install:

14. Lubricate the axle and hub splines with an approved high temperature wheel bearing grease. Position the shaft in the hub and install the inboard CV-joint-to-flange bolts.

15. Install the upper ball joint to steering knuckle and torque the stud nut to 61 ft. lbs. (83 Nm). Install a cotter pin through the upper ball joint stud and nut. Lubricate the ball joint as required.

16. Install the left stabilizer bar bracket and bushing at the frame. Install the stabilizer bar bolt, spacer and bushings at the lower control arm.

17. Position the lower shock in the mount bracket and install the attaching nut and bolt.

18. Connect the left tie rod end at the steering knuckle. Torque the nut to 35 ft. lbs. (47 Nm). Install a cotter pin through the tie rod stud and nut.

19. Connect the brake line bracket to the control arm, ensuring the line and/or hose is not twisted or kinked.

20. Install the skid plate, as required.

21. Install the axle hub washer and nut. Insert a drift through the rotor vanes to keep the axle from turning and torque the hub nut to 180 ft. lbs. (245 Nm) and the inboard CV-joint flange bolts to 60 ft. lbs. (80 Nm).

22. Remove the drift, install the wheel and tire assembly. Remove stands and lower the vehicle.

CV-Joint Boot

REMOVAL AND INSTALLATION

Outer

1. Raise and support the vehicle safely.

2. Remove the wheel and tire assembly.

3. Remove the skid plate, as required.

4. Remove the drive axle and place the assembly in a suitable holding fixture.

5. Remove the large (swage) ring using a chisel, being careful not to damage the housing surface.

6. Cut and remove the small clamp on the CV-boot.

7. Slide the boot back off the joint housing. Clean excess grease from the joint area.

8. Spread the ears of the joint retaining clip and slide the joint off the axle shaft.

9. Remove the boot from the axle shaft.

To install:

10. Place the small clamp on the small neck of the CV-boot. Do not crimp the clamp.

11. Slide the boot onto the axle shaft and into the groove on the axle shaft.

12. Crimp the small clamp using tool J–35910 or equivalent.

13. Disassemble the joint and thoroughly clean out old grease. Place a sufficient quantity of approved CV-joint grease inside the boot and in the joint.

14. Pinch the large CV-boot securing ring in an oval shape and slide the ring onto the boot large end.

15. Push the CV-joint onto the axle shaft until the retaining ring is seated in the groove on the axle shaft.

16. Slide the large diameter of the CV-boot with the ring in place over the outside of the CV-joint housing and locate the seal lip in the housing groove. Release any air trapped inside the boot using a thin flat blunt tool between the boot and housing.

17. Using a suitable ring installer tool, press the ring onto the housing until fully seated.

18. Install the halfshaft assembly, skid plate and the wheel and tire.

19. Lower the vehicle and road test.

Inner

1. Raise and support the vehicle safely.

2. Remove the wheel and tire assembly.

3. Remove the skid plate, as required.

4. Remove the drive axle and place the assembly in a suitable holding fixture.

5. Remove the large (swage) ring using a chisel, being careful not to damage the housing surface.

6. Cut and remove the small clamp on the CV-boot.

7. Slide the boot back off the joint housing. Remove the joint and axle shaft from the joint housing.

8. Spread the ears of the joint retaining clip and slide the joint back on the axle shaft. Remove the forward retaining ring and slide the joint off the shaft. Handle the joint with care to avoid disassembly.

9. Remove the rear retaining ring and remove the boot from the axle shaft.

To install:

10. Place the small clamp on the small neck of the CV-boot. Do not crimp the clamp.

11. Slide the boot onto the axle shaft and into the groove on the axle shaft.

12. Crimp the small clamp using tool J–35910 or equivalent. Install the forward joint retaining ring onto the axle shaft.

13. Thoroughly clean the old grease from the joint, boot and housing assembly. Place a sufficient quantity of approved CV-joint grease inside the boot.

14. Slide the joint assembly with the counterbore facing the axle shaft onto the shaft against the retaining ring.

15. Install the forward retaining ring in the groove on the axle shaft. Slide the joint towards the end of the shaft and move the inside retaining ring into the groove.

16. Pack the joint sufficient quantity of approved CV-joint grease. Pinch the large CV-boot securing ring in an oval shape and slide the ring onto the boot large end.

17. Push the CV-joint onto the axle shaft.

18. Slide the large diameter of the CV-boot with the ring in place over the outside of the CV-joint housing and locate the seal lip in the housing groove. Release any air trapped inside the boot using a thin flat blunt tool between the boot and housing.

19. Using a suitable ring installer tool, press the ring onto the housing until fully seated.

20. Install the halfshaft assembly, skid plate and the wheel and tire.

21. Lower the vehicle and road test.

Transfer Case Assembly

REMOVAL AND INSTALLATION

1. Disconnect the negative battery cable. If necessary, shift the transfer case into the **4HI** position to ease linkage removal.

2. Raise and support the vehicle safely.

3. If, equipped, remove the skid plate. Drain the fluid from the transfer case.

4. Drain the fluid from the transfer case.

5. Matchmark the transfer case front output shaft yoke and driveshaft for reassembly. Disconnect the driveshaft from the transfer case.

6. Matchmark the rear axle yoke and the driveshaft for reassembly. Remove the driveshaft.

7. If equipped, disconnect the speedometer cable.

8. Disconnect any vacuum or electrical connections at the transfer case.

9. If necessary, remove the parking cables.

10. If applicable, remove the catalytic converter hanger bolts at the converter assembly.

11. Support the transmission and transfer case assembly. If necessary, remove the transmission mount retaining bolts and lower the transmission and transfer case assembly to gain additional clearance.

12. Remove the transfer case retaining bolts.

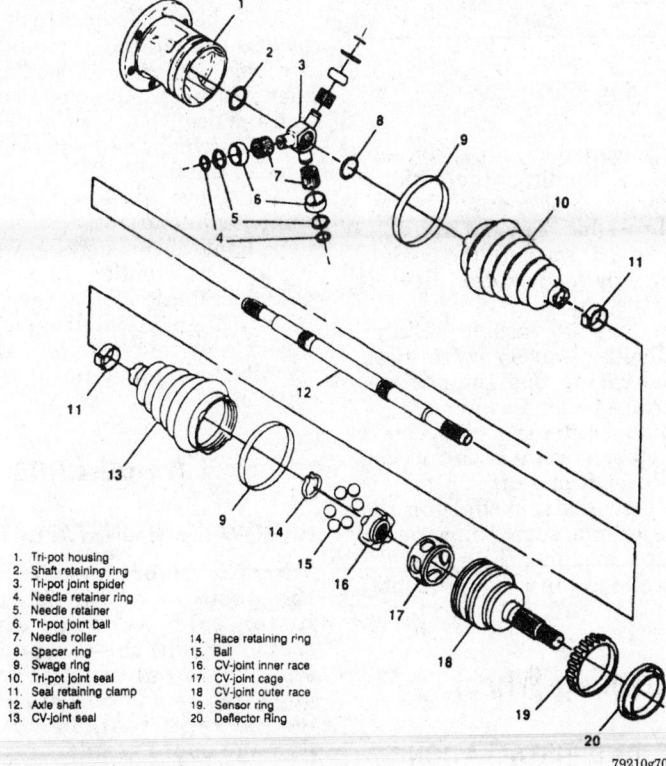

1. Tri-pot housing
2. Shaft retaining ring
3. Tri-pot joint spider
4. Needle retainer ring
5. Needle retainer
6. Tri-pot joint ball
7. Needle roller
8. Spacer ring
9. Swage ring
10. Tri-pot joint seal
11. Seal retaining clamp
12. Axle shaft
13. CV-joint seal
14. Race retaining ring
15. Ball
16. CV-joint inner race
17. CV-joint cage
18. CV-joint outer race
19. Sensor ring
20. Deflector Ring

79210g70

Exploded view of the halfshaft assembly — 4-wheel drive

SWAGE CLAMP SIZE CHART		
TOOL NO.	DESCRIPTION	APPLICATION
J 36652-1	Split Plate Swage Clamp	K 10/20
J 36652-2	Split Plate Swage Clamp	K 30 (Outboard)
J 36652-3	Split Plate Swage Clamp	K 30 (Inboard)

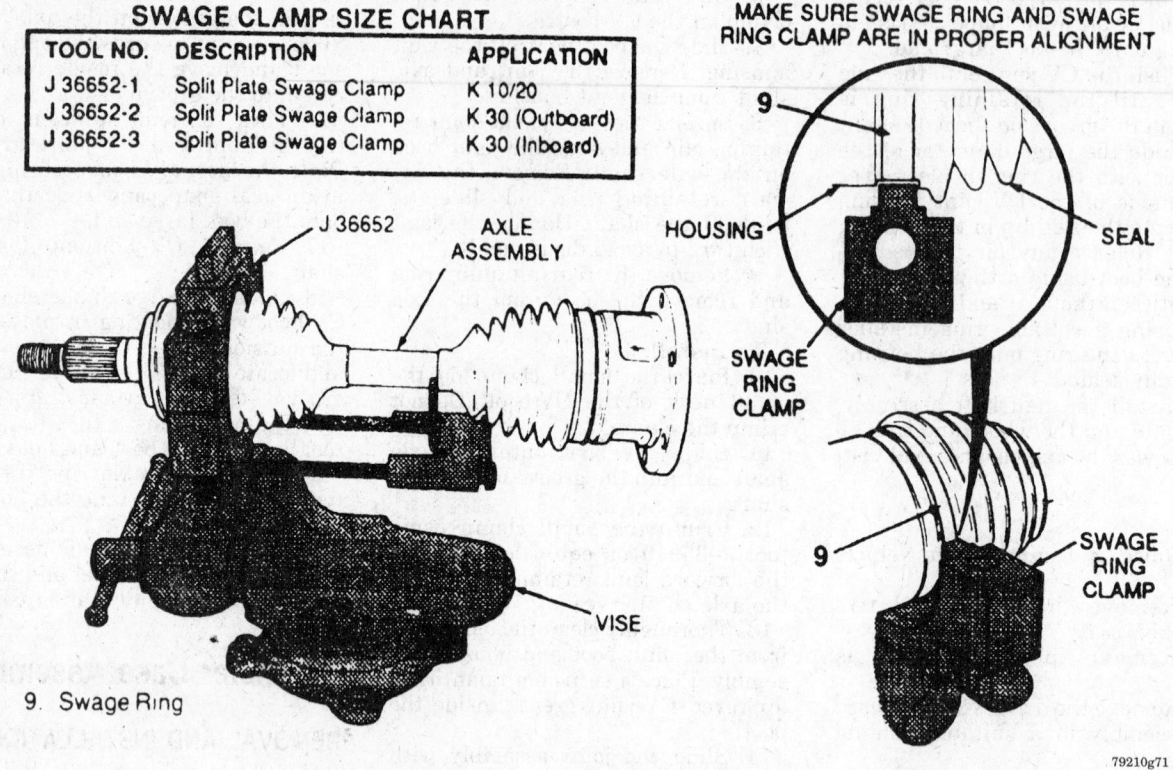

9. Swage Ring

Pressing the axle boot in place — 4-wheel drive

79210g71

13. As required, remove the shift lever bracket mounting bolts from the transfer case adapter in order to remove the upper left transfer case retaining bolt.

14. Separate the transfer case from its mounting and remove it.

15. Installation is the reverse of removal.

STEERING

Air Bag

— CAUTION —

Some vehicles are equipped with an air bag system, also known as the Supplemental Inflatable Restraint (SIR) or Supplemental Restraint System (SRS). The system must be disabled before performing service on or around system components, steering column, instrument panel components, wiring and sensors. Failure to follow safety and disabling procedures could result in accidental air bag deployment, possible personal injury and unnecessary system repairs.

PRECAUTIONS

Several precautions must be observed when handling the inflator module to avoid accidental deployment and possible personal injury.

• Never carry the inflator module by the wires or connector on the underside of the module.

• When carrying a live inflator module, hold securely with both hands, and ensure that the bag and trim cover are pointed away.

• Place the inflator module on a bench or other surface with the bag and trim cover facing up.

• With the inflator module on the bench, never place anything on or close to the module which may be thrown in the event of an accidental deployment.

Steering Wheel

REMOVAL AND INSTALLATION

1. If equipped, properly disable the SIR (air bag) system.

2. Disconnect the negative battery cable.

3. Remove the horn pad or horn pad/air bag retaining screws. Pull the air bag and/or horn pad outward, then disconnect the electrical lead(s) and remove.

4. Matchmark the steering wheel and the shaft.

5. Remove the steering wheel retaining clip and nut.

6. Remove the steering wheel, using a suitable puller.

7. Installation is the reverse of removal. Align the matchmarks made during removal and tighten the steering wheel retaining nut to 30 ft. lbs. (40 Nm).

Tie Rod Ends

REMOVAL AND INSTALLATION

NOTE: Before servicing, note the position of the tie rod adjuster tube and the direction from which the bolts are installed. Do not attempt to disengage the tie rod ball stud using a wedge type tool, because seal damage could result.

1. Raise and support the vehicle safely. As required, remove the tire and wheel assembly.

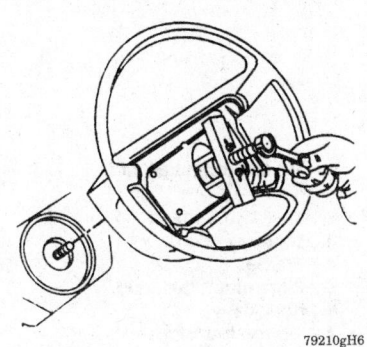

79210gH6

**Removing the steering wheel using a
steering wheel puller**

2. Remove the cotter pins and nuts. Using the proper removal tool J–6627A separate the outer tie rod from the steering knuckle.

3. Disconnect the inner tie rod from the relay rod using tool J–6627A. Remove the tie rod ends from the adjuster tubes, counting the number of turns to aid in installation.

To install:

4. Grease the threads and turn the new tie rod end in as many turns as were needed to remove it. This will give approximately correct toe-in. Tighten the clamp bolts to 14 ft. lbs.

5. Secure the tie rod ends to the relay rod and steering knuckle with a new nut. Tighten the nuts to 40 ft. lbs. (54 Nm) and install new cotter pins. Tighten the nut to align the cotter pin, do not loosen it.

6. Install the tire and wheel assembly.

7. Lower the vehicle and check the front end alignment.

Power Rack and Pinion

REMOVAL AND INSTALLATION

1. Disconnect the negative battery cable.

2. Remove the air cleaner assembly.

3. Remove the dust boot from the steering gear.

4. Remove the intermediate shaft lower pinch bolt and disconnect the intermediate shaft from the lower stub shaft.

5. Remove the fluid line retaining clips at the pump and disconnect the lines.

6. Raise and safely support the vehicle.

7. Remove the wheel and tire assemblies. Disconnect the tie rod ends at the steering knuckle.

8. Remove the remaining brackets and clips at the crossmember. Sup-

port the body safely with the appropriate equipment, to allow lowering of the subframe.

9. Remove the rear subframe mounting bolts and carefully lower the rear of the subframe approximately 5 in. (128mm).

10. Remove the rack and pinion mounting bolts and remove the rack through the left wheel opening.

To install:

11. Install the rack and pinion through the left wheel opening.

12. Install the rack and pinion mounting nuts, tighten to 70 ft. lbs. (95 Nm).

13. Raise the subframe assembly and install the rear mounting bolts.

14. Remove any supports and install the brackets and clips to the crossmember.

15. Install the wheel and tire assemblies. Lower the vehicle.

16. Connect the fluid lines at the pump and tighten to 18 ft. lbs. (25 Nm).

17. Install the line retaining clips. Connect the intermediate shaft to the stub shaft.

18. Install the dust boot over the steering gear.

19. Install the air cleaner assembly and connect the negative battery cable.

20. Fill and bleed the steering system.

Power Steering Pump

SYSTEM BLEEDING

1. Fill the reservoir to the proper level and let the fluid remain undisturbed for at least 2 minutes.

2. Start the engine and run it for only about 2 seconds.

3. Add fluid as necessary.

4. Repeat Steps 1–3 until the level remains constant.

5. Raise the front of the vehicle so the front wheels are off the ground. Set the parking brake and block both rear wheels front and rear. Manual transmissions should be in neutral; automatic transmissions should be in **P**.

6. Start the engine and run it at approximately 1500 rpm.

7. Turn the wheels (off the ground) to the right and left, lightly contacting the stops.

8. Add fluid as necessary.

9. Lower the vehicle and turn the wheels right and left on the ground.

10. Check the level and refill as necessary.

11. If the fluid is extremely foamy, let the vehicle stand for a few minutes with the engine off and repeat the procedure. Check the belt tension and check for a bent or loose pulley. The pulley should not wobble with the engine running.

12. Check that no hoses are contacting any parts of the vehicle, particularly sheet metal.

13. Check the fluid level and refill as necessary.

14. Check for air in the fluid. Aerated fluid appears milky. If air is present, repeat the above operation. If it is obvious that the pump will not respond to bleeding after several attempts, a pressure test may be required.

REMOVAL AND INSTALLATION

1. Place a drain pan under the pump.

2. Disconnect the negative battery cable.

3. Disconnect and cap the hoses at the pump.

NOTE: If equipped with a remote reservoir, disconnect and cap the reservoir hose at the pump.

4. Loosen the pump adjusting bolts and nuts and remove the pump belt.

5. Remove the pulley from the pump as required using a suitable pulley remover/installer tool J–29785–A or equivalent, as required.

6. Remove the adjusting bolts, nuts and brackets and remove the pump assembly.

To install:

7. Connect the brackets to the pump.

8. Place the pulley on the end of the pump shaft and install tool J–25033–B or equivalent.

NOTE: On models with a remote reservoir fill the pump housing with as much fluid as possible before mounting.

9. Install the pump assembly and attaching parts loosely to the engine.

10. Install the hoses to the pump and fill the reservoir. Bleed the pump by turning the pulley backwards (counterclockwise as viewed from the front) until the air bubbles cease to appear.

11. Tighten all retaining bolts and nuts.

12. Install the pump belt over the pulley and adjust.

13. Fill and bleed the system.

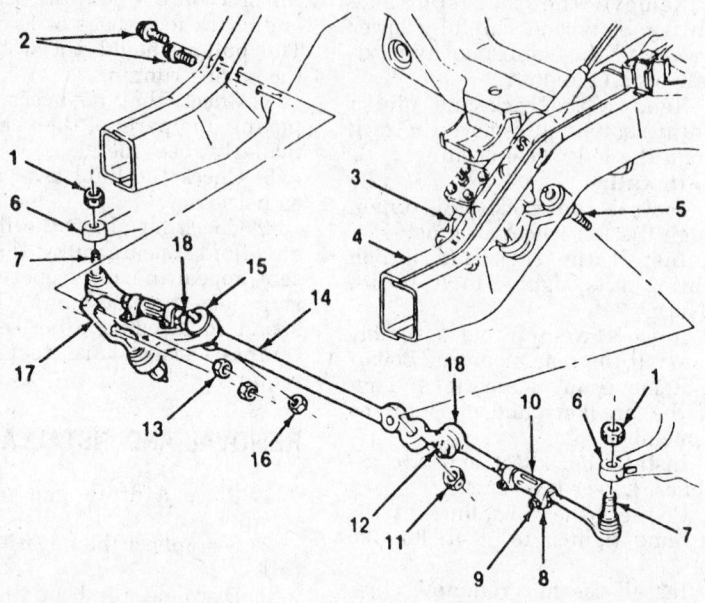

1. Tie rod outer ball joint nut
2. Idler arm frame bolts
3. Steering gear
4. Frame
5. Pitman arm ball stud
6. Knuckle
7. Tie rod ball stud
8. Clamp
9. Clamp nut
10. Adjuster tube
11. Pitman arm nut
12. Tie rod inner ball joint nut
13. Idler arm frame nut
14. Relay rod
15. Idler arm ball joint
16. Idler arm ball joint nut
17. Idler arm mounting bracket
18. Tie rod inner ball joint

79210g72

Steering linkage — 2-wheel drive truck

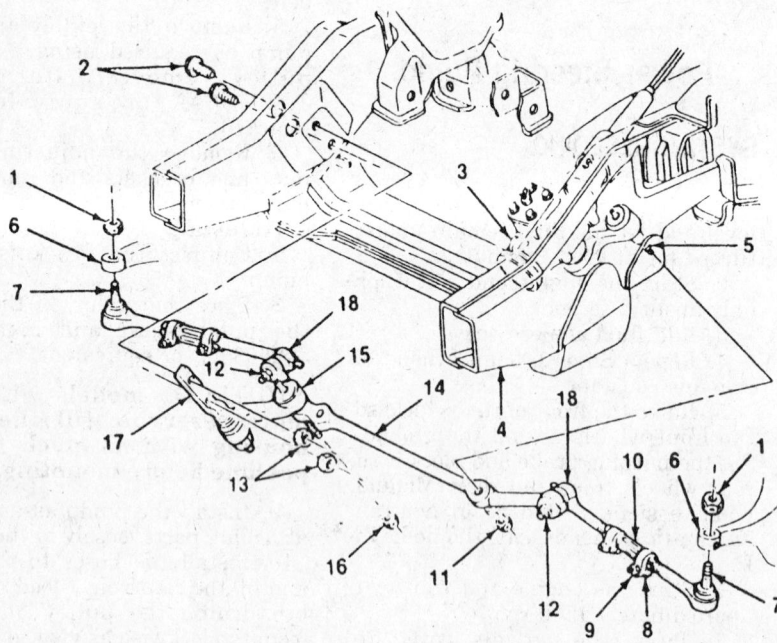

1. Tie rod outer ball joint nut
2. Idler arm frame bolts
3. Steering gear
4. Frame
5. Pitman arm ball stud
6. Knuckle
7. Tie rod ball stud
8. Clamp
9. Clamp nut
10. Adjuster tube
11. Pitman arm nut
12. Tie rod inner ball joint nut
13. Idler arm frame nut
14. Relay rod
15. Idler arm ball joint
16. Idler arm ball joint nut
17. Idler arm mounting bracket
18. Tie rod inner ball joint

79210g73

Steering linkage — 4-wheel drive truck

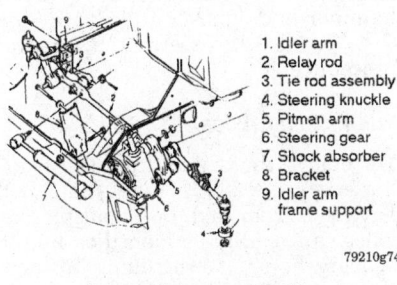

1. Idler arm
2. Relay rod
3. Tie rod assembly
4. Steering knuckle
5. Pitman arm
6. Steering gear
7. Shock absorber
8. Bracket
9. Idler arm frame support

79210g74

Steering linkage — G-Series

BRAKES

For all brake system repair and service procedures not detailed below, please refer to "Brakes" in the Unit Repair section.

Anti-Lock Brake System Service

PRECAUTIONS

• Certain components within the Anti-Lock Brake System (ABS) are not intended to be serviced or repaired individually. Only those components with removal and installation procedures should be serviced.

• Do not use rubber hoses or other parts not specifically specified for and ABS system. When using repair kits, replace all parts included in the kit. Partial or incorrect repair may lead to functional problems and require the replacement of components.

• Lubricate rubber parts with clean, fresh brake fluid to ease assembly. Do not use lubricated shop air to clean parts; damage to rubber components may result.

• Use only specified brake fluid from an unopened container.

• If any hydraulic component or line is removed or replaced, it may be necessary to bleed the entire system.

• A clean repair area is essential. Always clean the reservoir and cap thoroughly before removing the cap. The slightest amount of dirt in the fluid may plug an orifice and impair the system function. Perform repairs after components have been thoroughly cleaned; use only denatured alcohol to clean components. Do not allow ABS components to come into contact with any substance contain-

ing mineral oil; this includes used shop rags.

• The Anti-Lock control unit is a microprocessor similar to other computer units in the vehicle. Ensure that the ignition switch is **OFF** before removing or installing controller harnesses. Avoid static electricity discharge at or near the controller.

• If any arc welding is to be done on the vehicle, the control unit should be unplugged before welding operations begin.

Master Cylinder

REMOVAL AND INSTALLATION

1. Disconnect the negative battery cable. Disconnect any electrical connections from the master cylinder, as required. Disconnect and plug the fluid lines.
2. If equipped with power brakes, remove the master cylinder to power booster retaining bolts. Remove the Rear Wheel Anti-Lock (RWAL) control module assembly, if equipped. Do not allow fluid to leak onto the module.
3. If equipped with manual brakes, remove the master cylinder pushrod from the brake pedal and remove the RWAL control module assembly.
4. Remove the master cylinder. If equipped with power brakes remove the vacuum booster pushrod.
5. Installation is the reverse of the removal procedure. Bench bleed the master cylinder prior to installation. If equipped with power brakes be sure to install the vacuum booster pushrod. Bleed the system, as required.

Brake Caliper

REMOVAL AND INSTALLATION

NOTE: There are 2 caliper designs and they can be identified by the method used to secure the assembly to the spindle bracket. The Delco 3000/3100 caliper is secured by a bolt and sleeve combination. The Bendix caliper assembly is secured by a slider, spring and bolt.

1. Remove the cover on the master cylinder and siphon enough fluid out of the reservoirs to bring the level to ⅓ full. This step prevents spilling fluid when the piston is pushed back.

2. Raise and support the vehicle safely. Remove the front wheels and tires.
3. Position a C-clamp around the outside pad and caliper; tighten the C-clamp until the caliper piston bottoms in its bore.
4. Remove the brake hose from the caliper by removing the inlet fitting.
5. Remove the bolt and sleeve or bolt and slider assemblies which hold the caliper and then lift the caliper off the rotor.
6. Remove the inboard and outboard shoe.
   To install:
7. Install the pads onto the caliper.
8. Position the caliper onto the knuckle/rotor assembly and secure the assembly with the mounting bolts or sliders.
9. Reconnect the brake line to the caliper.
10. Pump the brake pedal and verify there is minimal brake pedal travel.
11. Check the brake fluid level. Install the tire and wheel assembly.
12. Lower the vehicle.

Disc Brake Pads

REMOVAL AND INSTALLATION

Delco Type

1. Remove the cover on the master cylinder and siphon out ⅔ of the fluid. This step prevents spilling fluid when the piston is pushed back into the caliper bore.
2. Raise and support the vehicle safely.
3. Remove the wheels.
4. Compress the brake piston back into its bore using a C-clamp.
5. Remove the 2 bolts which hold the caliper and then lift the caliper off the disc.

NOTE: Do not let the caliper assembly hang by the brake hose.

6. Remove the inboard and outboard shoe.
7. Remove the pad support spring from the piston, if equipped.
   To install:
8. Thoroughly inspect, clean and lubricate all caliper slide points, bolts and hardware.
9. Position the retainer spring on the inner pad and insert the assembly into the center cavity of the piston.
10. Push down on the inner pad until it lays flat against the caliper. It is important to push the piston all the

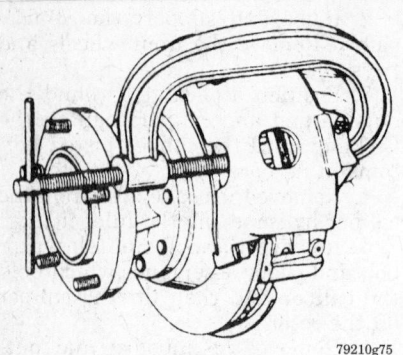

Compressing the caliper piston

way into the caliper if new linings are installed or the caliper will not fit over the rotor.

11. Position the outboard pad with the ears of the pad over the caliper ears and the tab at the bottom engaged in the caliper cutout.

12. With the 2 pads in position, place the caliper over the brake disc and align the holes in the caliper with those of the mounting bracket.

NOTE: Make certain the brake hose is not twisted or kinked.

13. Install the mounting bracket bolts through the sleeves in the inboard caliper ears and through the mounting bracket, making sure the ends of the bolts pass under the retaining ears on the inboard pad.

14. Tighten the mounting bolts to 38 ft. lbs. (51 Nm). After both calipers are mounted pump the brake pedal to seat the pad against the rotor. Use a pair of channel lock pliers to bend over the upper ears of the outer pad so it isn't loose.

15. Install the wheels and lower the vehicle.

16. Add fluid to the master cylinder reservoirs so they are ¼ in. (6.35mm) from the top.

17. Test the brake pedal by pumping it to obtain a hard pedal. Check the fluid level again and add fluid as necessary. Do not move the vehicle until a pedal is obtained.

Bendix Type

1. Remove approximately ⅓ of the brake fluid from the master cylinder. Discard the used brake fluid.

2. Raise and support the vehicle safely and remove the wheel.

3. Push the piston back into its bore. This can be done by using a C-clamp.

4. Remove the bolt at the caliper slider. Use a brass drift pin to remove the slider and spring.

5. Rotate the caliper up and forward from the bottom and lift it off the caliper support.

6. Tie the caliper out of the way with a piece of wire. Be careful not to damage the brake line.

7. Remove the inner shoe from the caliper support. Discard the inner shoe clip.

8. Remove the outer shoe from the caliper.

To install:

9. Thoroughly clean, inspect and lubricate the caliper, slider and spring with silicone.

10. Install a new inboard shoe clip on the shoe.

11. Install the lower end of the inboard shoe into the groove provided in the support. Slide the upper end of the shoe into position. Be sure the clip remains in position.

12. Position the outboard shoe in the caliper with the ears at the top of the shoe over the caliper ears and the tab at the bottom of the shoe engaged in the caliper cutout. If assembly is difficult, a C-clamp may be used. Be careful not to damage the lining.

13. Position the caliper over the brake disc, top edge first. Rotate the caliper downward onto the support.

14. Place the spring over the caliper support key, install the assembly between the support and lower caliper groove. Tap into place until the key retaining screw can be installed.

15. Install the screw and torque to 15 ft. lbs. (20 Nm). The boss must fit fully into the circular cutout in the key.

16. Install the wheel and add brake fluid as necessary.

Brake Drums

REMOVAL AND INSTALLATION

Semi-Floating Axles

1. Raise and support the vehicle safely.

2. Mark the relationship of the wheel to the hub and remove the wheel.

3. Mark the relationship of the drum to the hub and pull the drum from the brake assembly. If the brake drums have been scored from worn linings, the brake adjuster must be backed off so the brake shoes will retract from the drum. The adjuster can be backed off by inserting a brake adjusting tool through the access hole provided. In some cases the access hole is provided in the brake drum. A

metal cover plate is over the hole. This may be removed by using a hammer and chisel.

4. To install, reverse the removal procedure.

Full Floating Axles

To remove the drums from full floating rear axles, the axle shaft will have to be removed. Full floating rear axles can readily be identified by the bearing housing protruding through the center of the wheel.

1. Raise and support the vehicle safely.

2. Remove the wheel.

3. Remove the axle shaft.

4. Remove the retaining ring, key and adjusting nut.

5. Remove the hub and drum.

To install:

6. Install the hub and drum to the tube.

7. Install the adjusting nut and torque to specification.

8. Install the key and retaining ring.

9. Install the axle shaft and wheel.

Brake Shoes

REMOVAL AND INSTALLATION

Leading/Trailing Brakes

NOTE: This brake system is used on the lower GVW rated vehicles.

1. Raise the vehicle and support it safely.

2. Remove the tire and wheel assembly.

3. Remove the brake drums.

NOTE: The brake pedal must not be depressed while the drums are removed.

4. Raise the lever arm of the actuator until the upper end is clear of the slot in the adjuster screw.

5. Slide the actuator off the adjuster pin. Disconnect the actuator spring from the shoe.

6. Remove the hold-down spring assemblies and pins.

7. Pull the bottom ends of the shoes apart and lift the lower return spring over the anchor plate. Allow the shoe ends to come together and remove the spring.

8. Remove the shoe assembly, along with the upper return spring and the adjusting screw assembly.

9. Remove the upper return spring and the adjusting screw assembly from the shoes.

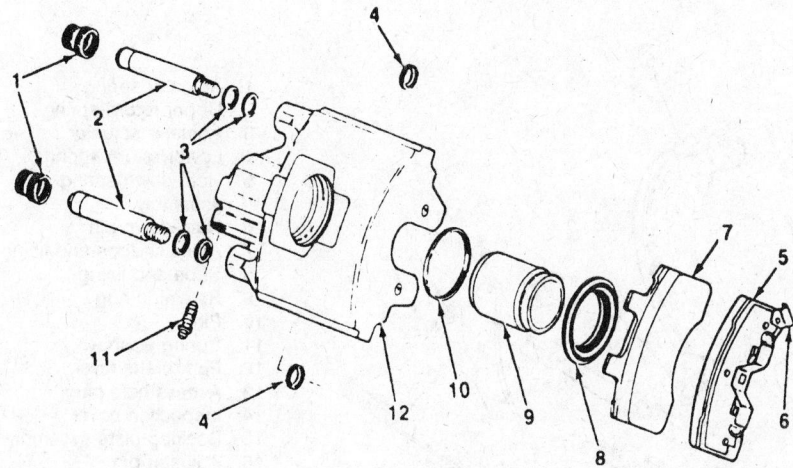

1. Bolt boot
2. Mounting bolt assembly
3. Bushing
4. Mounting bolt seal
5. Outboard shoe and lining
6. Wear sensor
7. Inboard shoe and lining
8. Boot
9. Piston
10. Piston seal
11. Bleeder valve
12. Caliper housing

79210g76

Replacing the disc brake pads — Delco type

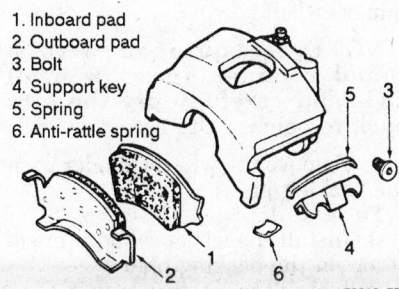

1. Inboard pad
2. Outboard pad
3. Bolt
4. Support key
5. Spring
6. Anti-rattle spring

79210g77

Replacing the disc brake pads — Bendix type

10. Remove the retaining ring, pin, spring washer, and parking brake lever.

To install:

11. Clean adjuster wheel and the backing plates with a suitable cleaner. Lubricate the backing plate contact points, levers and adjuster with a suitable lubricant.

12. Assemble the parking lever, spring washer (concave side facing the brake lever), pin, and retaining ring onto the rearward shoe.

13. Install the adjuster pin in the forward shoe with the pin projecting 0.276 in. (7mm) from the side of the

shoe web where the adjuster actuator is installed.

14. With the brake shoes laying on a flat surface (the shoe with the parking lever to the rear of the vehicle), install the upper return spring.

15. Install the adjuster screw assembly with the spring clip facing the backing plate.

16. Place the shoes in position on the backing plate. Do not place the lower shoe webs under the anchor plate.

17. Install the lower return spring, spread the bottom of the shoes and position the shoe against the backing plate.

18. Install the hold-down pins and spring assemblies.

19. Install the adjuster actuator over the end of the adjuster pin so the top leg engages the notch in the adjuster screw.

20. Install the actuator spring, being careful not to over-stretch it more than 3.27 in. (83mm).

21. Install the parking brake cable to the lever.

22. Adjust the parking brake if the shoes will not totally retract.

23. Install the drum, tire and wheel assembly. Adjust the rear brakes and lower the vehicle.

Duo-Servo Brakes

1. Raise the vehicle and support it safely.

2. Remove the tire and wheel assembly.

3. Remove the brake drums.

NOTE: The brake pedal must not be depressed while the drums are removed.

4. Using a brake tool, remove the shoe return springs.

5. Remove the shoe guide.

6. Remove the hold-down springs and pins.

7. Remove the actuator lever and pivot.

8. Remove the lever return spring.

9. Remove the actuator link, parking brake strut, spring retaining ring.

10. Remove the parking brake lever and washer.

11. Remove the shoe assemblies.

12. Remove the adjuster screw and spring from the shoe assembly.

To install:

13. Use a brake cleaning fluid to remove dirt from the brake drum. Check the drums for scoring, cracks and for out-of-round; service the drums as necessary.

14. Check the wheel cylinders by carefully pulling the lower edges of

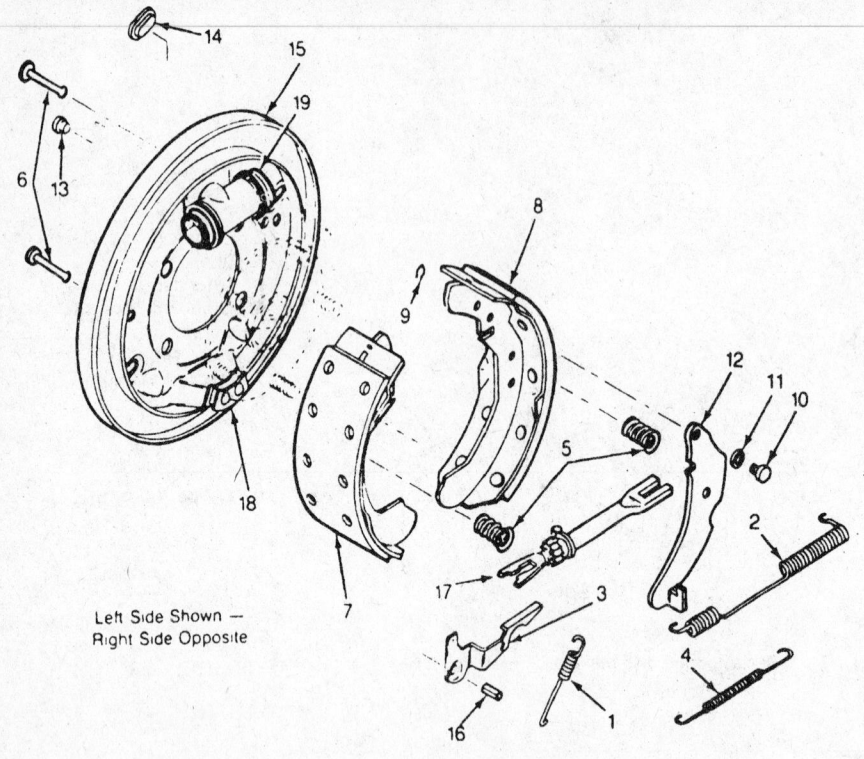

1. Actuator spring
2. Upper return spring
3. Adjuster actuator
4. Lower return spring
5. Hold-down spring assembly
6. Hold-down pin
7. Adjuster shoe and lining
8. Shoe and lining
9. Retaining ring
10. Pin
11. Spring washer
12. Park brake lever
13. Access hole plug
14. Inspection cover
15. Backing plate assembly
16. Adjuster pin
17. Adjusting screw assembly
18. Anchor plate
19. Wheel cylinder assembly

79210g78

Left Side Shown —
Right Side Opposite

Exploded view of the rear drum brake assembly — Leading/Trailing type

the wheel cylinder boots away from the cylinders. If there is excessive leakage, the inside of the cylinder will drip fluid; repair or replace as necessary.

15. Check the flange plate, which is located around the axle, for leakage of differential lubricant.

16. Lightly lubricate the parking brake cable, parking brake lever where it enters the shoe and the backing plate-to-shoe contact points. Use high temperature, waterproof, grease or special brake lube.

17. Install the parking brake lever into the secondary shoe with the attaching bolt, spring washer, lockwasher, and nut. It is important that the lever move freely before the shoe is attached. Move the assembly and check for proper action.

18. Lubricate the adjusting screw and make sure it works freely.

19. Connect the adjuster screw and spring to the bottom portion of both shoes. Ensure the spring does not interfere with the adjuster rotation when installed. The primary (smaller shoe pad area) to the front and secondary shoe (larger shoe pad area) to the rear of the vehicle.

20. Install the shoe assembly. Ensuring the shoe webs are positioned correctly against the wheel cylinder.

21. Install the parking brake cable.
22. Secure the primary shoes with the hold-down pin and spring.
23. Install the parking brake strut and the strut spring.
24. Install the actuator lever and pivot, securing the assembly with the hold-down pin and spring. Install the actuator link and spring.
25. Install the return springs.
26. Check the operation of the self-adjusting mechanism by moving the actuating lever by hand.
27. Adjust the brakes and install the drum.
28. Adjust the parking brake.
29. Install the tire and wheel assembly.
30. Lower the vehicle.

Wheel Cylinder

REMOVAL AND INSTALLATION

1. Raise and safely support the vehicle.
2. Remove the wheel and tire assemblies.
3. Remove the brake drum.
4. Remove the brake shoes, as necessary.
5. Disconnect the brake fluid line from the wheel cylinder.

6. Remove the wheel cylinder retainer or bolt.

NOTE: If equipped with a round retainer, insert 2 small awls and carefully pry the tabs back to release the retainer.

7. Remove the wheel cylinder from the backing plate.
To install:
8. Install the wheel cylinder in position on the backing plate.
9. Install the retainer or bolt. Connect the brake line to the wheel cylinder.

NOTE: If equipped with a round retainer, a socket may be used as a driver to ease installation.

10. Install the brake linings and the brake drum.
11. Install the wheel and tire assembly. Lower the vehicle.
12. Bleed the brake system.

Parking Brake Cable

ADJUSTMENT

The rear brakes serve a dual purpose. They are used as service brakes and as parking brakes. To obtain proper adjustment of the parking brake, the

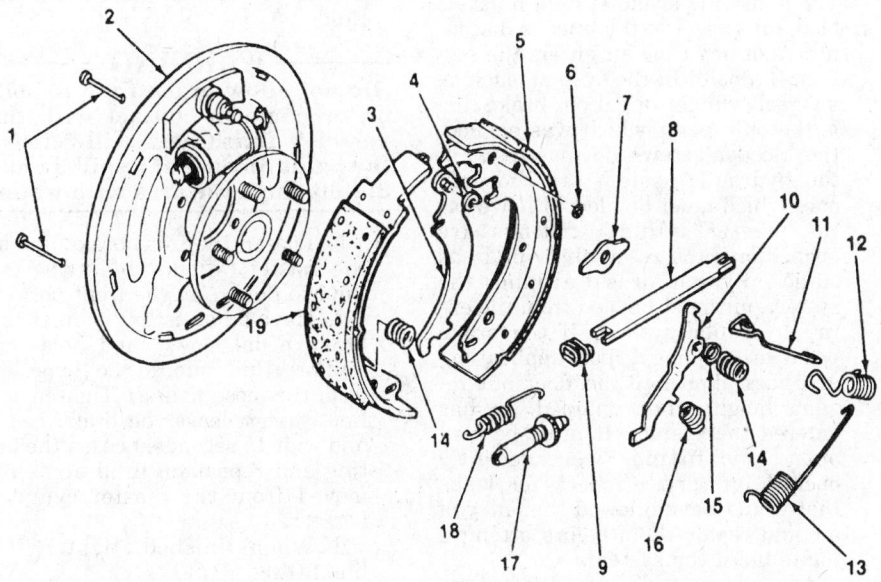

1. Hold-down pins
2. Backing plate
3. Parking brake lever
4. Washer
5. Secondary shoe
6. Retaining ring
7. Shoe guide
8. Parking brake strut
9. Strut spring
10. Actuator lever
11. Actuator link
12. Return spring
13. Return spring
14. Hold down spring
15. Lever pivot
16. Lever return spring
17. Adjusting screw assembly
18. Adjusting screw spring
19. Primary shoe

79210g79

Exploded view of the rear drum brake assembly — Duo-Servo type

service brakes must first be properly adjusted. Inspect the cables for binding or sticking.

G-Series

Foot Pedal Type

1. Apply the parking brake 4 notches from the fully released position.
2. Raise and support the vehicle.
3. Loosen the jam nut at the equalizer.
4. Tighten or loosen the adjusting nut until a light drag is felt when the rear wheels are rotated forward.
5. Tighten the check nut.
6. Release the parking brake and rotate the rear wheels. No drag should be felt. If even a light drag is felt, readjust the parking brake.
7. Lower the vehicle.

NOTE: If a new parking brake cable is being installed, prestretch it by applying the parking brake hard about 3 times before making adjustments.

Lever Type

1. Raise and support the vehicle safely. Turn the adjusting knob on the parking brake lever counterclockwise until it stops.
2. Apply the parking brake. Loosen the equalizer nut.

3. Tighten the equalizer nut until light drag is felt while rotating the rear wheels in the forward motion.
4. Adjust the knob on the parking brake lever until a definite snap over center is felt.
5. Release the parking brake. Rotate the rear wheels in the forward motion. There should be no brake drag.

Driveshaft Type

1. Raise and support the vehicle safely. Remove the clevis pin connecting the pull rod and the relay lever.
2. Rotate the brake drum to align the access hole with the adjusting screw. If equipped with manual transmission the access hole is located at the bottom of the backing plate. If equipped with automatic transmission the access hole is located at the top of the shoe.
3. For first time adjustment it will be necessary to remove the driveshaft and the drum in order to remove the lanced area from the drum and clean out the metal shavings.
4. Adjust the screw until the drum cannot be rotated by hand. Back off the adjusting screw 10 notches, the drum should rotate freely.
5. Position the parking brake lever in the fully released position.

Take up the slack in the cable to overcome spring tension.
6. Adjust the clevis of the pull rod to align with the hole in the relay lever. Install the clevis pin. Install a new cover in the drum access hole.

Blazer, Pick-Up, Suburban, Tahoe and Yukon

1. Raise and support the vehicle safely. Matchmark the wheel to the axle flange. Remove the tire and wheel assembly.
2. Matchmark the drum to the axle flange. Remove the brake drum.
3. Using tool J–21177A or equivalent, measure and record the brake drum inside diameter.
4. Turn the adjuster nut and adjust the shoe and lining to a diameter 0.010–0.020 in. less than the measured inside diameter of the brake drum.
5. Be sure the stops on the parking brake levers are against the edge of the brake shoe web. If not, loosen the parking brake cable adjustment.
6. Tighten the parking brake cable at the adjuster nut until the lever stops begin to move off the shoe webs. Loosen the adjustment nut until the lever stops move back, barely touching the shoe webs. The final clear-

ance between the stops and either web should be 0.5mm.

7. Install the drums and wheels. Align the assemblies with the matchmarks made during removal.

8. Apply and release the service brake pedal 30–35 times using normal pedal force. Pause about 1 second between each pedal application.

9. Depress the parking brake 6 clicks. Check the rear wheels they should not rotate.

10. Release the parking brake lever. Check for free wheel rotation.

REMOVAL AND INSTALLATION

Front Cable

1. Raise and support the vehicle safely.

2. Remove adjusting nut from equalizer.

3. Remove retainer clip from rear portion of front cable at frame and from lever arm.

4. Disconnect front brake cable from parking brake pedal or lever assemblies. Remove front brake cable. On some models, it may assist installation of the new cable if a heavy cord is tied to other end of cable in order to guide new cable through proper routing.

5. Installation is the reverse of the removal procedure.

6. Adjust the parking brake.

Center Cable

1. Raise and support the vehicle safely.

2. Remove adjusting nut from equalizer.

3. Unhook connector at each end and disengage hooks and guides.

4. Install new cable by reversing removal procedure.

5. Adjust parking brake.

6. Apply parking brake 3 times with heavy pressure and repeat adjustment.

Rear Cable

1. Raise and support the vehicle safely.

2. Remove rear wheel and brake drum.

3. Loosen adjusting nut at equalizer.

4. Disengage rear cable at connector.

5. Bend retainer fingers.

6. Disengage cable at brake shoe operating lever.

7. Install new cable by reversing removal procedure.

8. Adjust parking brake.

Brake System Bleeding

The hydraulic brake system must be bled any time 1 of the lines is disconnected or any time air enters the system. If a point in the system, such as a wheel cylinder or caliper brake line is the only point which was opened, the bleeder screws down stream in the hydraulic system are the only ones which must be bled. If however, the master cylinder fittings are opened or if the reservoir level drops sufficiently that air is drawn into the system, air must be bled from the entire hydraulic system. If the brake pedal feels spongy upon application, and goes almost to the floor but regains height when pumped, air has entered the system. It must be bled out. If no fittings were recently opened for service, check for leaks that would have allowed the entry of air and repair them before attempting to bleed the system.

As a general rule, once the master cylinder (and the brake pressure modulator valve or combination valve on ABS systems) is bled, the remainder of the hydraulic system should be bled starting at the furthest wheel from the master cylinder and working towards the nearest wheel. Therefore, the correct bleeding sequence is: master cylinder, modulator or combination valve (ABS only), right rear wheel cylinder, left rear, right front caliper and left front. Most master cylinder assemblies on these vehicles are not equipped with bleeder valves, therefore air must be bled from the cylinders using the front brake pipe connections.

MANUAL BLEEDING

1. Clean the top of the master cylinder, remove the cover and fill the reservoirs with clean fluid. To prevent squirting fluid, and possibly damaging painted surfaces, install the cover during the procedure, but be sure to frequently check and top off the reservoirs with fresh fluid.

--- CAUTION ---
Never reuse brake fluid which has been bled from the system.

2. The master cylinder must be bled first if it is suspected to contain air. If the master cylinder was removed and bench bled before installation it must still be bled, but it

should take less time and effort. Bleed the master cylinder as follows:

a. Position a container under the master cylinder to catch the brake fluid.

--- WARNING ---
Do not allow brake fluid to spill on or come in contact with the vehicle's finish as it will remove the paint. In case of a spill, immediately flush the area with water.

b. Loosen the front brake line at the master cylinder and allow the fluid to flow from the front port.

c. Have an assistant depress the brake pedal slowly and hold (air and/or fluid should be expelled from the loose fitting). Tighten the line, then release the brake pedal and wait 15 seconds. Loosen the fitting and repeat until all air is removed from the master cylinder bore.

d. When finished, tighten the line fitting.

e. Repeat the sequence at the master cylinder rear pipe fitting.

NOTE: During the bleeding procedure, make sure your assistant does not release the brake pedal while a fitting is loosened or while a bleeder screw is opening. Air will be drawn back into the system.

3. Check and refill the master cylinder reservoir.

NOTE: Remember, if the reservoir is allowed to empty of fluid during the procedure, air will be drawn into the system and bleeding procedure must be restarted at the master cylinder assembly.

4. On late model ABS equipped vehicles, perform the special ABS procedures. On 4 wheel ABS systems the Brake Pressure Modulator Valve (BPMV) must be bled (if it has been replaced or if it is suspected to contain air) and on most Rear Wheel Anti-Lock (RWAL) systems the combination valve must be held open. In both cases, special combination valve depressor tools should be used during bleeding and a scan tool must be used for ABS function tests.

5. If a single line or fitting was the only hydraulic line disconnected, then only the caliper(s) or wheel cylinder(s) affected by that line must be bled. If the master cylinder required bleeding, then all calipers and wheel cylinders must be bled in the proper sequence:

a. Right rear

b. Left rear

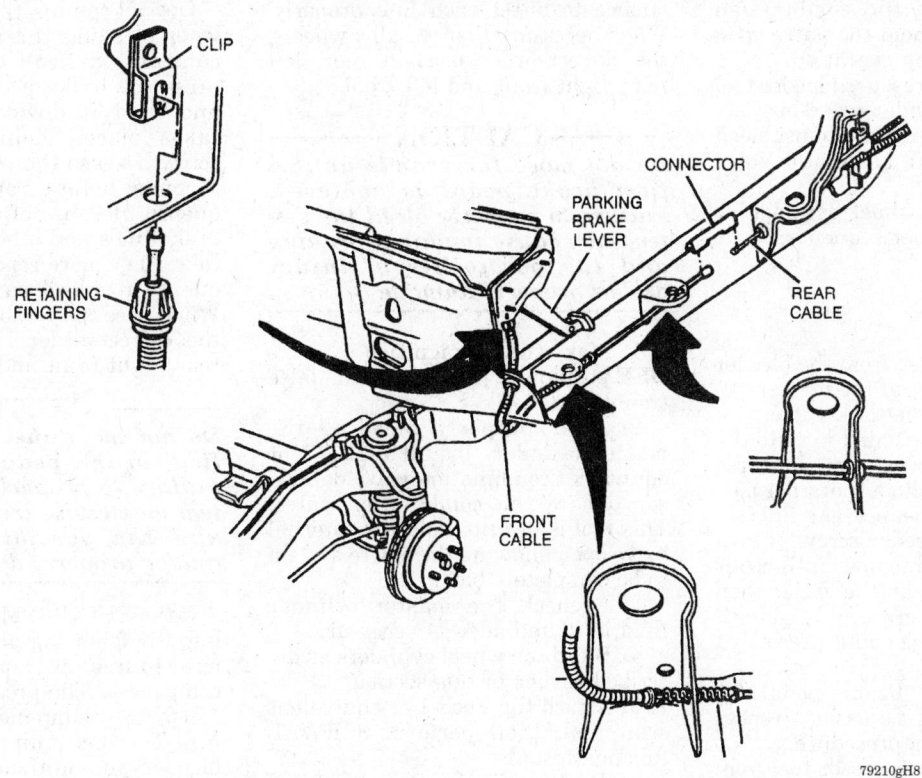

Exploded view of the front parking brake cable routing and its related components — 1996 pick-up shown

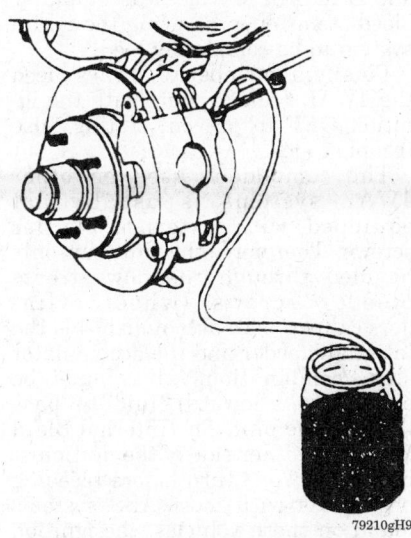

Submerge one end of a clear tube in a transparent container of clean brake fluid and engage the other end to a bleeder screw

c. Right front
d. Left front

6. Bleed the individual calipers or wheel cylinders as follows:

a. Place a suitable wrench over the bleeder screw and attach a clear plastic hose over the screw end. Be sure the hose is seated snugly on the screw or you may be squirted with brake fluid.

NOTE: Be very careful when bleeding wheel cylinders and brake calipers. The bleeder screws often rust in position and may easily break off if forced. Installing a new bleeder screw will often require removal of the component and may include overhaul or replacement of the wheel cylinder/caliper. To help prevent the possibility of breaking a bleeder screw, spray it with some penetrating oil before attempting to loosen it.

b. Submerge the other end of the tube in a transparent container of clean brake fluid.

c. Loosen the bleed screw, then have a friend apply the brake pedal slowly and hold. Tighten the bleed screw, release the brake pedal and wait 15 seconds. Repeat the sequence (including the 15 second

pause) until all air is expelled from the caliper or cylinder.

d. Tighten the bleed screw when finished.

7. Check the pedal for a hard feeling with the engine not running. If the pedal is soft, repeat the bleeding procedure until a firm pedal is obtained.

8. If the brake warning light is on, depress the brake pedal firmly. If there is no air in the system, the light will go out.

9. After bleeding, make sure a firm pedal is achieved before attempting to move the vehicle.

PRESSURE BLEEDING

A proper pressure bleeder tool will utilize a rubber diaphragm between the air source and brake fluid in order to prevent air, moisture oil and other contaminants from entering the hydraulic system.

1. Prepare a pressure bleeder tool such as J-29567 or equivalent by making sure the pressure tank is at least $2/3$ full of fresh, clean brake fluid. In most cases, the bleeder must be bled each time fluid is added. Charge the bleeder tool to 20–25 psi (140–170 kPa).

2. Install a suitable combination valve depressor tool such as J-39177

or equivalent, to the combination valve in order to hold the valve open during the bleeding operation.

3. Install the pressure bleeder tool to the master cylinder reservoir.

4. On 4 wheel ABS systems, bleed the Brake Pressure Modulator Valve (BPMV) of air.

5. Bleed each wheel cylinder or caliper in the proper sequence:
   a. Right rear
   b. Left rear
   c. Right front
   d. Left front

6. Connect a hose from the bleeder tank to the adapter at the master cylinder, then open the tank valve.

7. Attach a clear vinyl hose to the brake bleeder screw, then immerse the opposite end into a container partially filled with clean brake fluid.

8. Open the bleeder screw 3/4 turn and allow the fluid to flow until no air bubbles are seen in the fluid, then close the bleeder screw and tighten.

9. Repeat the bleeding process at each wheel.

10. Inspect the brake pedal for sponginess and if necessary, repeat the entire bleeding procedure.

11. Remove the depressor tool from the combination valve and the bleeder adapter from the master cylinder.

12. Refill the master cylinder to the proper level with brake fluid.

13. Do not attempt to move the vehicle unless a firm brake pedal is obtained.

Bleeding the RWAL Brake System

On RWAL equipped vehicles, a few steps (listed below) should be added to the bleeding sequence in order to ease the procedure and assure all air is removed from the system. These extra steps may be used on all RWAL vehicles to assure proper bleeding.

The use of a power bleeder is recommended, but the system may also be bled manually. If a power bleeder is used, it must be of the diaphragm type and provide isolation of the fluid from air and moisture.

Do not pump the pedal rapidly when bleeding; this can make the circuits very difficult to bleed. Instead, press the brake pedal slowly 1 time and hold it down while bleeding takes place. Tighten the bleeder screw, release the pedal and wait 15 seconds before repeating the sequence. Because of the length of the brake lines and other factors, it may take 10 or more repetitions of the se-

quence to bleed each line properly. When necessary to bleed all 4 wheels, the correct order is right rear, left rear, right front and left front.

— CAUTION —
Do not move the vehicle until a firm brake pedal is achieved. Failure to properly bleed the system may cause impaired braking and the possibility of injury and/or property damage

1. Make sure the ignition is in the **OFF** position to prevent setting false trouble codes.

2. After properly bleeding the master cylinder, install J-39177 or equivalent combination valve depressor tool to the combination valve. This tool is used to hold the internal valve open allowing the entire system to be completely bled.

3. Recheck the master cylinder fluid level and add, as necessary.

4. Bleed the wheel cylinders as described earlier in this section.

5. Attach the Tech-1 or equivalent scan tool, then perform 3 RWAL function tests.

6. Re-bleed the rear wheel cylinders.

7. Check for a firm brake pedal, if necessary repeat the entire bleeding procedure.

Bleeding the 4WAL Brake System

The EHCU/BPMV module is the 1 component which adds to the complexity of bleeding the 4WAL brake systems. For the most part the system is bled in the same manner as the non-ABS vehicles. But because of the EHCU/BPMV's complex internal valving additional steps are necessary if the unit has been replaced or if it is suspected to contain air. These bleeding steps are not necessary if the only connection/fitting(s) opened were downstream of the unit. These steps may or may not be necessary after master cylinder replacement. If in doubt (or without the necessary special tools) thoroughly bleed the system and see if a firm brake pedal can be obtained, if not, the EHCU/BPMV must be bled as well.

As with the RWAL brake system, the use of a power bleeder is recommended, but the system may also be bled manually. If a power bleeder is used, it must be of the diaphragm type and provide isolation of the fluid from air and moisture.

Do not pump the pedal rapidly when bleeding; this can make the circuits very difficult to bleed. Instead, press the brake pedal slowly 1 time and hold it down while bleeding takes place. Tighten the bleeder screw, release the pedal and wait 15 seconds before repeating the sequence. Because of the length of the brake lines and other factors, it may take 10 or more repetitions of the sequence to bleed each line properly. When necessary to bleed all 4 wheels, the correct order is right rear, left rear, right front and left front.

— CAUTION —
Do not move the vehicle until a firm brake pedal is achieved. Failure to properly bleed the system may cause impaired braking and the possibility of injury and/or property damage

If the EHCU/BPMV requires bleeding, the following procedures may be used to free all trapped air from the component. The procedures differ because the component used on the MFI-Turbo is equipped with external bleeders in additional to the internal bleeders found on the 1994 and later units. In either case, 3 combination valve depressor tools and a scan tool are required. The combination valve depressor tools are used to hold the internal passages (combination valve and EHCU/BPMV bleed accumulator bleed stems open allowing the entire system to be completely bled.

Finally, remember to always bleed the 4WAL brake system with the ignition **OFF** to prevent setting false trouble codes.

The component used on other 4WAL systems is usually not equipped with external bleeder screws. Therefore, the unit can only be bled through the downstream bleeder screws (wheel cylinders/calipers). To accomplish this the internal bleeder and the accumulator stems/combination valves must be opened to allow air/fluid to pass through the unit. The Internal Bleed Valves on either side of the unit must be opened 1/4–1/2 turn before bleeding begins. As with most ABS systems found on these vehicles, the ignition switch must be **OFF** or false trouble codes may be set.

1. Make sure the ignition is in the **OFF** position to prevent setting false trouble codes.

2. If necessary, properly bleed the master cylinder assembly. Check and add additional fluid, as necessary.

3. Open the internal bleed valves 1/4–1/2 turn each.

4. Install one J–39177 or equivalent combination valve depressor tool on the left accumulator bleed stem of the EHCU. Install 1 tool on the right accumulator bleed stem and install the 3rd tool on the combination valve.

5. Properly bleed the wheel cylinders and calipers.

6. Remove the 3 special tools.

7. Check the master cylinder fluid level, refilling as necessary.

8. Switch the ignition **ON** (engine not running) and use a hand scanner to perform 6 function tests on the system.

9. Repeat the wheel cylinder and caliper bleeding procedure to remove all air that was purged from the BPMV during the function tests.

10. Check for a firm brake pedal. If necessary, repeat the entire procedure until a firm pedal is obtained.

11. Carefully test drive the vehicle at moderate speeds; check for proper pedal feel and brake operation. If any problem is noted in feel or function, repeat the entire bleeding procedure.

Front Wheel Speed Sensor

REMOVAL AND INSTALLATION

4WAL System

2-Wheel Drive

1. Disconnect the negative battery cable.

2. Raise and support the vehicle safely.

3. Remove the wheel and tire assembly.

4. Remove the brake caliper, hub and rotor assembly.

5. Remove the sensor wire from the clips on the upper control arm and disconnect the connector.

6. Remove the sensor and backing plate attaching bolts and remove the assembly.

7. Installation is the reverse of the removal procedure.

4-Wheel Drive

1. Disconnect the negative battery cable.

2. Raise and support the vehicle safely.

3. Remove the wheel and tire assembly.

4. Remove the hub and rotor assembly.

5. Disconnect the sensor wire connector.

6. Remove the bolts securing the sensor and sensor wire.

7. Remove the sensor from the spindle.

8. Installation is the reverse of the removal procedure.

Rear Wheel Speed Sensor

REMOVAL AND INSTALLATION

4-Wheel Anti-Lock System

1. Disconnect the negative battery cable.

2. Raise and support the vehicle safely.

3. Remove the wheel, tire and drum assembly.

4. Remove the primary (forward) brake shoe.

5. Disconnect the sensor wire connector and remove the sensor wire from the rear axle clips.

6. Remove the bolts securing the sensor and sensor wire.

7. Remove the sensor from the backing plate.

8. Installation is the reverse of the removal procedure.

FRONT SUSPENSION

Strut

All front wheel drive vehicles are equipped with a MacPherson strut front suspension.

REMOVAL AND INSTALLATION

NOTE: Do not remove the top center nut from the strut assembly. This nut should only be removed when the strut assembly is out of the vehicle, mounted in a holding fixture and the coil spring is in a compressed position using the proper strut coil spring compressor.

1. Remove the 3 nuts that retain the top of the strut assembly.

2. Raise and safely support the vehicle.

3. Remove the wheel and tire assembly. Remove the brake line bracket from the strut mount.

4. Remove the lower strut mounting bolts.

5. Remove the strut assembly from the vehicle and place the strut in a suitable holding fixture.

6. Disassemble the strut as follows:

   a. With the strut coil spring in a compressed position approximately ½ its normal length, remove the nut from the top of the strut.

   b. Place tool J–34013–27 or equivalent guide rod on top of the damper shaft. Use the rod to guide the damper shaft straight down through the bearing cap while decompressing the spring.

   c. Remove the coil spring and other components.

7. Installation is the reverse of the removal procedure. Ensure the spring seat flat should face 10 degrees forward of the centerline of the strut assembly spindle.

8. Tighten the strut lower bolts to 140 ft. lbs. (190 Nm) and the upper mounting nuts to 18 ft. lbs. (25 Nm).

Shock Absorbers

REMOVAL AND INSTALLATION

G-Series

1. Raise and support the vehicle safely. Properly support the lower control arm assembly, as required. Remove the tire and wheel assembly.

2. Remove the upper shock absorber retaining bolt. Remove the lower shock absorber retaining bolt. Vehicles equipped with quad shocks have a spacer between them.

3. Remove the shock absorber.

4. Installation is the reverse of the removal procedure.

2-Wheel Drive Truck

1. Remove the upper shock absorber retaining bolt. Raise and support the vehicle safely.

2. Properly support the lower control arm, as required. Remove the lower shock absorber retaining bolt.

3. Remove the shock absorber.

4. Installation is the reverse of removal. Tighten the upper nut to 100 inch lbs. (11 Nm); tighten the lower mounting bolts to 20 ft. lbs. (27 Nm).

4-Wheel Drive Truck

1. Raise and support the vehicle safely.

2. Remove the upper end bolt, nut and washer.

3. Remove the lower end bolt, nut and washer.

4. Remove the shock absorber and inspect the rubber bushings for wear and the shock for leaks, replace the shock absorber assembly necessary.

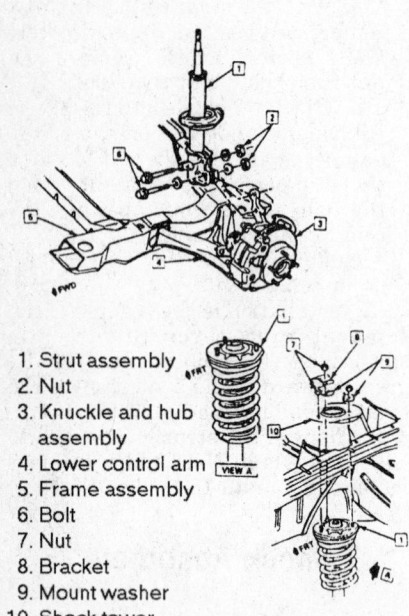

1. Strut assembly
2. Nut
3. Knuckle and hub assembly
4. Lower control arm
5. Frame assembly
6. Bolt
7. Nut
8. Bracket
9. Mount washer
10. Shock tower

79210g80

MacPherson strut assembly/disassembly

5. Installation is the reverse of removal. Torque the both nuts to 66 ft. lbs. (90 Nm). Make sure the bolts are inserted in the proper direction. The bolt head on the upper end should be forward; the bottom end bolt head is rearward.

Coil Springs

REMOVAL AND INSTALLATION

G-Series

1. Raise and support the vehicle safely under the frame rails. The control arms should hang freely.
2. Remove the wheel.
3. Disconnect the shock absorber at the lower end and move it aside.
4. Disconnect the stabilizer bar from the lower control arm.
5. Support the lower control arm and install a spring compressor on the spring or chain the spring to the control arm as a safety precaution.

NOTE: If equipped with an air cylinder inside the spring, remove the valve core from the cylinder and expel the air by compressing the cylinder with a prybar. With the cylinder compressed, replace the valve core so the cylinder will stay in the com-

pressed position. Push the cylinder as far as possible towards the top of the spring.

6. Raise to remove the tension from the lower control arm cross-shaft and remove the 2 U-bolts securing the cross-shaft to the crossmember.

NOTE: The cross-shaft and lower control arm keeps the coil spring compressed. Use care when lowering the assembly.

7. Slowly lower the control arm until the spring can be removed. Be sure all compression is relieved from the spring.
8. If the coil spring was chained, remove the chain and spring. If a compressor was used, remove the spring and slowly release the compressor.
9. Remove the air cylinder, if equipped.
To install:
10. Install the air cylinder so the protector plate is towards the upper control arm. The schraeder valve should protrude through the hole in the lower control arm.
11. Install the chain and spring or compress the spring and install the assembly.
12. Slowly lower control arm. Align the indexing hole in the shaft with the crossmember attaching studs.
13. Install the 2 U-bolts securing the cross-shaft to the crossmember. Torque the nuts to 85 ft. lbs. (115 Nm).
14. Remove the support.
15. Connect the stabilizer bar to the lower control arm. Torque the nuts to 24 ft. lbs.
16. Connect the shock absorber at the lower end. Torque the bolt to specification.
17. If equipped with air cylinders, inflate the cylinder to 60 psi.
18. Install the wheel.
19. Lower the vehicle. Once the weight of the vehicle is on the wheels, reduce the air cylinder pressure to 50 psi.
20. Check the front end alignment.

Blazer, Pick-Up, Suburban, Tahoe, Yukon

1. Raise and support the vehicle safely. Allow the control arms to hang free. Remove the tire and wheel assembly. Remove the shock absorber assembly, as required.
2. Install tool J-23028 under the lower control arm and a jack. Install a safety chain around the spring and through the lower control arm.

3. Remove the stabilizer shaft from the lower control arm. Raise and remove the tension on the lower control arm bolts.
4. Remove the lower control arm rear bolt, than remove the other retaining bolt.
5. Lower allow the lower control arm to hang free. Remove the spring assembly.
To install:
6. Install the chain and spring. If you used spring compressors, install the spring and compressors.
   a. Make sure the insulator is in place.
   b. Make sure the tape is at the lower end. New springs will have an identifying tape.
   c. Make sure the gripper notch on the top coil is in the frame bracket.
   d. Make sure 1 drain hole in the lower arm is covered by the bottom coil and the other is open.
7. Slowly raise the lower control arm. Guide the control arm into place with a prybar.
8. Install the pivot shaft bolts, front 1 first. The bolts must be installed with the heads towards the front of the vehicle. Remove the safety chain or spring compressors.

NOTE: Do not torque the bolts yet. The bolts must be torqued with the vehicle at its proper ride height.

9. Remove the jack.
10. Connect the stabilizer bar to the lower control arm. Torque the nuts to specification.
11. Install the shock absorber.
12. Install the wheel.
13. Lower the vehicle. Once the weight of the vehicle is on the wheels check the "Z" height as follows:
   a. Lift the front bumper about 1½ in. (38mm) and let it drop.
   b. Repeat this procedure 2–3 more times.
   c. Draw a line on the side of the lower control arm from the centerline of the control arm pivot shaft, dead level to the outer end of the control arm.
   d. Measure the distance between the lowest corner of the steering knuckle and the line on the control arm. Record the figure.
   e. Push down about 38mm on the front bumper and let it return. Repeat the procedure 2–3 more times.
   f. Re-measure the distance at the control arm.
   g. Determine the average of the 2 measurements.

h. If the figure is correct, tighten the control arm pivot nuts to 121 ft. lbs. (165 Nm). Align the front end. If the figure is incorrect, replace the coil spring.

Leaf Spring

REMOVAL AND INSTALLATION

1. Raise the vehicle and support it with safety stands.
2. Support the axle separately to eliminate load on the springs.
3. Remove the wheel assembly.
4. Remove the shock absorber.
5. Using tool J 6627-A or its equivalent, separate the stabilizer link from the stabilizer shaft.
6. Remove the stabilizer link from the axle.
7. Remove the spacer and spring spacer.
8. Separate the leaf spring from the rear shackle by removing the nut and bolt.
9. Separate the leaf spring from the front hanger by removing and bolt.
10. Remove the leaf spring by pulling it back and out.

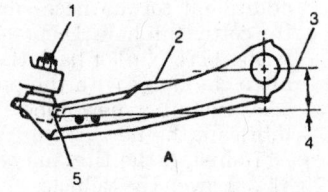

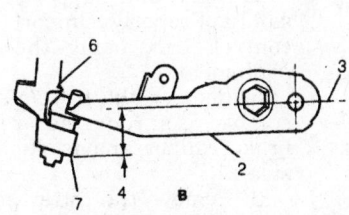

A. C-Series
B. K-Series
C. K-Series torsion bar
D. C/K-Series rear suspension
1. Lower ball joint
2. Lower control arm
3. Pivot bolt center line
4. "Z" Height
   C1, 2, 3–89–101mm
   K1, 2–151-163mm
   K3–139-151mm
5. Lower ball joint extrusion
6. Steering knuckle
7. Steering knuckle lower corner
8. Nut
9. Torsion bar support assembly
10. Torsion bar adjustment arm
11. Bolt–1 turn equals 6mm height change
12. Frame
13. Bottom surface of jounce bracket
14. "D" Height
15. Rear axle
16. Jounce bumper

79210g81

Trim height adjustment — exc. G-series

To install:
11. Install the leaf spring with the double wrap end towards the the front of the vehicle.
12. Engage the spring to the hangers, install the nut and bolt and tighten the nuts to 136 ft. lbs. (185 Nm).
13. Position the spring spacer onto the axle and install the spacer U-bolts, washers and nuts. Tighten the nuts in two steps in a diagonal sequence. First pass 18 ft. lbs. (25 Nm) and second pass 92 ft. lbs.
14. Install the stabilizer retainers and insulators and tighten the nuts until the distance between each retainer is 1.5 in. (38 mm).
15. Engage the stabilizer link to the stabilizer shaft, install the nut and washer and tighten to 50 ft. lbs. (68Nm).
16. Install the shock absorber.
17. Install the wheel assembly and lower the vehicle.

Torsion Bars and Support

REMOVAL AND INSTALLATION

4-Wheel Drive

1. Raise and support the vehicle safely.

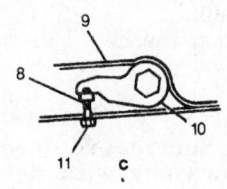

2. Remove the wheels.
3. Support the lower control arm.
4. Matchmark the both torsion bar adjustment bolt positions.
5. Using tool J-36202, increase the tension on the adjusting arm.
6. Remove the adjustment bolt and retaining plate.
7. Move the tool aside.
8. Slide the torsion bars forward.
9. Remove the adjusting arms.
10. Remove the nuts and bolts from the torsion bar support crossmember and slide the support crossmember rearward.
11. Matchmark the position of the torsion bars and note the markings on the front end of each bar. They are not interchangeable. Remove the torsion bars.
12. Remove the support crossmember.
13. Remove the retainer, spacer and bushing from the support crossmember.

To install:
14. Assemble the retainer, spacer and bushing on the support.
15. Position the support assembly on the frame, out of the way.
16. Align the matchmarks and install the torsion bars, sliding them forward until they are supported.
17. Install the adjuster arms on the torsion bars.
18. Bolt the support crossmember into position. Torque the center nut to 18 ft. lbs. (24 Nm); the edge nuts to 46 ft. lbs. (62 Nm).
19. Install the adjuster retaining plate and bolt on each torsion bar.
20. Using tool J-36202, increase tension on both torsion bars.
21. Install the adjustment retainer plate and bolt on both torsion bars.
22. Set the adjustment bolt to the marked position.
23. Release the tension on the torsion bar until the load is take up by the adjustment bolt.
24. Remove the tool.
25. Install the wheels.
26. Check the front end alignment and "Z" height.

Upper Ball Joint

REMOVAL AND INSTALLATION

G-Series

1. Raise and support the vehicle safely. Properly support the lower control arm.
2. Remove the wheel assembly. Remove the brake caliper and position it to the side.

3. Remove the cotter pin and the upper ball joint retaining bolt. Using the proper tool separate the upper joint from its mounting. Support the knuckle assembly so its weight will not damage the brake hose.

4. Remove the rivets from the ball joint assembly, using the proper tools. Remove the ball joint from the upper control arm.

To install:

5. Install the new ball joint into the control arm. Position the bleed vent in the rubber boot facing inward.

6. Secure the ball joint to the control arm using the new nuts and bolts.

7. Lower the upper arm ball joint stud into the steering knuckle.

8. Install the ball stud nut and torque to specification. Tighten the nut to align the cotter pin hole.

9. Install the brake caliper, if removed.

10. Install a new lube fitting and lubricate the new joint.

11. Install the tire and wheel.

12. Lower the vehicle.

2-Wheel Drive

1. Raise and support the vehicle safely. Properly support the lower control arm, using the necessary equipment.

2. Remove the tire and wheel assembly. Remove the brake caliper and position it to the side.

3. Remove the cotter pin and the upper ball joint retaining bolt. Using the proper tool separate the upper joint from its mounting. Support the knuckle assembly so its weight will not damage the brake hose.

4. Remove the rivets from the ball joint assembly, using the proper tools. Remove the ball joint from the upper control arm.

To install:

5. Install the replacement ball joint in the control arm, using the bolts and nuts supplied. Torque the nuts to 18 ft. lbs. (24 Nm).

6. Position the ball stud in the knuckle. Make sure it is squarely seated. Torque the ball stud nut to 74 ft. lbs. (100 Nm).

7. Install a new cotter pin.

8. Install a new lube fitting and lubricate the new joint.

9. If removed, install the brake caliper.

10. Install the wheel and lower the vehicle. Check and align the front end.

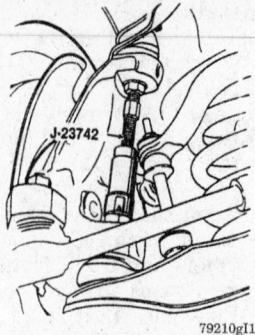

Removing the upper ball joint using a ball joint separator

4-Wheel Drive

1. Raise and support the vehicle safely.

2. Remove the wheel.

3. Unbolt the brake hose bracket from the control arm.

4. Using a ⅛ in. drill bit, drill a pilot hole through each ball joint rivet.

5. Drill out the rivets with a ½ in. drill bit. Punch out any remaining rivet material.

6. Remove the cotter pin and nut from the ball stud.

7. Support the lower control arm.

8. Using a ball joint separator, separate the stud from the knuckle.

To install:

9. Position the new ball joint on the control arm.

NOTE: Service replacement ball joints come with nuts and bolts to replace the rivets.

10. Install the bolts and nuts. Tighten the nuts to 17 ft. lbs. (23 Nm) for 15 and 25-Series; 52 ft. lbs. (70 Nm) for 35-Series.

NOTE: The bolts are inserted from the bottom.

11. Start the ball stud into the knuckle. Make sure it is squarely seated. Install the ball stud nut and pull the ball stud into the knuckle with the nut. Torque the nut after the vehicle wheel are on the ground and the suspension is loaded.

12. Install the wheel.

13. Lower the vehicle. Once the weight of the vehicle is on the wheels tighten the nut to 84 ft. lbs. (115 Nm).

Lower Ball Joint

REMOVAL AND INSTALLATION

G-Series

1. Raise and support the vehicle safely. Properly support the lower control arm, using the necessary equipment.

2. Remove the tire and wheel assembly. As required, remove the brake caliper and position it to the side.

3. Remove the cotter pin and the lower ball joint retaining bolt. Using the proper tool separate the ball joint from its mounting. Support the knuckle assembly so its weight will not damage the brake hose.

4. Press the ball joint out of the lower control arm, using the proper tool.

To install:

5. Start the new ball joint into the control arm. Position the bleed vent in the rubber boot facing inward.

6. Press the ball joint into the control arm until fully seated.

7. Lower the upper arm and insert the lower ball joint stud into the steering knuckle.

8. Install the brake caliper, if removed.

9. Install the ball stud nut and torque to 90 ft. lbs. (122 Nm) plus the additional torque necessary to align the cotter pin hole. Do not exceed 130 ft. lbs. (175 Nm) or back the nut off to align the holes with the pin.

10. Install a new lube fitting and lubricate the new joint.

11. Install the tire and wheel.

12. Lower the vehicle.

2-Wheel Drive

1. Raise and support the vehicle safely. Properly support the lower control arm, using the necessary equipment.

2. Remove the tire and wheel assembly. As required, remove the brake caliper and position it to the side.

3. Remove the cotter pin and the lower ball joint retaining nut. Using the proper tool separate the ball joint from its mounting. Support the knuckle assembly so its weight will not damage the brake hose.

4. Remove the rivets from the ball joint assembly, using the proper tools. Remove the ball joint from the control arm.

To install:

5. Secure the new ball joint to the control arm.

6. Position the ball joint into the knuckle. Install the nut and tighten it to 84 ft. lbs. (115 Nm).

7. Advance the nut to align the cotter pin hole and insert the new cotter pin.

8. Install the brake caliper, if removed.

9. Install a new lube fitting and lubricate the new joint.

10. Install the wheel.

11. Lower the vehicle.

12. Check the front end alignment.

4-WHeel Drive

1. Raise and support the vehicle safely.

2. Remove the wheel.

3. Remove the splash shield from the knuckle.

4. Disconnect the inner tie rod end from the relay rod using a ball joint separator.

5. Remove the hub nut and washer. Insert a long drift or dowel through the vanes in the brake rotor to hold the rotor in place.

6. Remove the axle shaft inner flange bolts.

7. Using a puller, force the outer end of the axle shaft out of the hub. Remove the shaft.

8. Using a 1/8 in. drill bit, drill a pilot hole through each ball joint rivet.

9. Drill out the rivets with a 1/2 in. drill bit. Punch out any remaining rivet material.

10. Remove the cotter pin and nut from the ball stud.

11. Support the lower control arm.

12. Matchmark the both torsion bar adjustment bolt positions.

13. Using tool J-36202, increase the tension on the adjusting arm.

14. Remove the adjustment bolt and retaining plate.

15. Move the tool aside.

16. Slide the torsion bars forward.

17. Using a screw-type forcing tool, separate the ball joint from the knuckle.

To install:

18. Position the new ball joint on the control arm.

NOTE: Service replacement ball joints come with nuts and bolts to replace the rivets.

19. Install the bolts and nuts. Tighten the nuts to 45 ft. lbs.

NOTE: The bolts are inserted from the bottom.

20. Start the ball stud into the knuckle. Make sure it is squarely seated. Install the ball stud nut and pull the ball stud into the knuckle

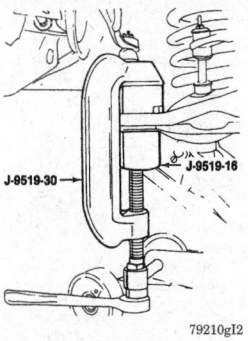

79210gI2

Installing the lower ball joint — 1996 pick-up shown

with the nut. Do not final-torque the nut yet.

21. Using tool J-36202, increase tension on both torsion bars.

22. Install the adjustment retainer plate and bolt on both torsion bars.

23. Set the adjustment bolt to the marked position.

24. Release the tension on the torsion bar until the load is take up by the adjustment bolt.

25. Remove the tool.

26. Position the shaft in the hub and install the washer and hub nut. Leave the drift in the rotor vanes and tighten the hub nut to 175 ft. lbs.

27. Install the flange bolts. Tighten them to 59 ft. lbs. Remove the drift.

28. Connect the inner tie rod end at the steering relay rod. Torque the nut to 35 ft. lbs.

29. Install the splash shield.

30. Install the wheel.

31. Lower the vehicle. Once the weight of the vehicle is on the wheels:

    a. Lift the front bumper about 1 1/2 in. (38mm) and let it drop.

    b. Repeat this procedure 2–3 more times.

    c. Draw a line on the side of the lower control arm from the center-line of the control arm pivot shaft, dead level to the outer end of the control arm.

    d. Measure the distance between the lowest corner of the steering knuckle and the line on the control arm. Record the figure.

    e. Push down about 1 1/2 in. (38mm) on the front bumper and let it return. Repeat the procedure 2–3 more times.

    f. Re-measure the distance at the control arm.

    g. Determine the average of the 2 measurements. The average distance should be as specified.

    h. If the figure is correct, tighten the control arm pivot nuts to 94 ft. lbs. (128 Nm).

    i. If the figure is not correct, tighten the pivot bolts to 94 ft. lbs.

(128 Nm) and have the front end alignment corrected.

Upper Control Arms

REMOVAL AND INSTALLATION

G-Series

1. Note and record the amount of shims. These shims must be installed in the same location as removed. Remove the nuts and the shims.

2. Raise and support the vehicle safely. Properly support the lower control arm, using the necessary equipment. The control arm must be supported so the spring and the control arm remain intact.

3. Remove the tire and wheel assembly. Remove the brake caliper assembly and position it to the side. Loosen the upper ball joint from the steering knuckle, using the proper tool. Support the hub assembly.

4. Remove the upper control arm retaining bolts. Remove the upper control arm.

To install:

5. Place the control arm in position and install the nuts. Before tightening the nuts, insert the caster and camber shims in the same order as when installed.

6. Install the nuts securing the control arm shaft studs to the cross-member bracket. Tighten the nuts to 70 ft. lbs. (95 Nm) for 10/1500-Series; 105 ft. lbs. (142 Nm) for all other Series.

7. Install the ball stud nut. Torque the nut to 50 ft. lbs. (68 Nm) for 10/1500-Series; 90 ft. lbs. (122 Nm) for all other Series.

8. Install the cotter pin. Never back off the nut to install the cotter pin. Always advance it.

9. Install the brake caliper. Remove the spring compressor.

10. Install the wheel.

11. Check the front end alignment.

Blazer, Pick-Up, Suburban, Tahoe and Yukon

1. Raise and support the vehicle safely.

2. Support the lower control arm.

3. Remove the wheel.

4. Unbolt the brake hose bracket from the control arm.

5. Remove the air cleaner extension.

6. Remove the cotter pin from the upper control arm ball stud and loosen the stud nut until the bottom surface of the nut is slightly below the end of the stud.

7. Install a spring compressor on the coil spring for safety.

8. Using a screw-type forcing tool, break loose the ball joint from the knuckle.

9. Remove the nuts and bolts securing the control arm to the frame brackets.

10. The 35-Series bushings are replaceable. The 15/25-Series bushings are welded in place.

To install:

11. Place the control arm in position and install the shims, bolts and new nuts. Both bolt heads must be inboard of the control arm brackets. Tighten the nuts finger-tight for now.

NOTE: Do not torque the bolts yet. The bolts must be torque with the vehicle at its proper ride height.

12. Install the ball stud nut. Torque the nut to 84 ft. lbs. (115 Nm). Install the cotter pin. Never back off the nut to install the cotter pin.

13. Install the brake caliper.

14. Remove the spring compressor or safety chain.

15. Install the wheel.

16. Install the brake hose.

17. Install the air cleaner extension.

18. Install the battery ground cable.

19. Lower the vehicle. Once the weight of the vehicle is on the wheels:

a. Lift the front bumper about 38mm and let it drop.

b. Repeat this procedure 2–3 more times.

c. Draw a line on the side of the lower control arm from the centerline of the control arm pivot shaft, dead level to the outer end of the control arm.

d. Measure the distance between the lowest corner of the steering knuckle and the line on the control arm. Record the figure.

e. Push down about 38mm on the front bumper and let it return. Repeat the procedure 2–3 more times.

f. Re-measure the distance at the control arm.

g. Determine the average of the 2 measurements. The average distance should be as specified.

h. If the figure is correct, tighten the control arm pivot nuts to 139 ft. lbs. (190 Nm).

i. If the figure is not correct, adjust or repair the ride height.

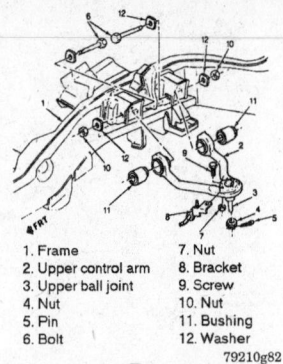

1. Frame
2. Upper control arm
3. Upper ball joint
4. Nut
5. Pin
6. Bolt
7. Nut
8. Bracket
9. Screw
10. Nut
11. Bushing
12. Washer

79210g82

Upper control arm installation — 4-wheel drive

Lower Control Arms

REMOVAL AND INSTALLATION

G-Series

1. Raise and support the vehicle safely. Properly support the lower control arm assembly. Remove the tire and wheel assembly. Remove the brake caliper and position it to the side.

2. Remove the control arm U-bolts and remove the coil spring. Remove the lower ball joint cotter pin and retaining nut. Using the proper tool, separate the lower ball joint from the steering knuckle.

3. Remove the lower control arm.

To install:

4. Install the lower control arm and spring assembly. Torque the U-bolts to 85 ft. lbs. (115 Nm).

5. Install the ball stud nut. Torque the nut to 90 ft. lbs. (122 Nm). Install the cotter pin. Never back off the nut to install the cotter pin.

6. Install the brake caliper.

2-Wheel Drive

1. Raise and support the vehicle safely. Remove the tire and wheel assembly. Properly support the lower control arm assembly. Remove the lower control arm bolts and coil spring.

2. Remove the lower ball joint cotter pin and retaining nut. Using the proper tool, separate the lower ball joint from the steering knuckle.

3. Remove the lower control arm.

To install:

4. Slowly raise the lower control arm with the coil spring. Guide the control arm into place with a prybar.

5. Install the pivot shaft bolts, front 1 first. The bolts must be installed with the heads towards the front of the vehicle. Remove the safety chain or spring compressors.

NOTE: Do not torque the bolts yet. The bolts must be torque with the vehicle at its proper ride height.

6. Remove the jack.

7. Connect the stabilizer bar to the lower control arm. Torque the nuts to specification.

8. Install the shock absorber.

9. Install the wheel.

10. Lower the vehicle. Once the weight of the vehicle is on the wheels proceed as follows:

a. Lift the front bumper about 1½ in. (38mm) and let it drop.

b. Repeat this procedure 2–3 more times.

c. Draw a line on the side of the lower control arm from the centerline of the control arm pivot shaft, dead level to the outer end of the control arm.

d. Measure the distance between the lowest corner of the steering knuckle and the line on the control arm. Record the figure.

e. Push down about 1½ in. (38mm) on the front bumper and let it return. Repeat the procedure 2–3 more times.

f. Re-measure the distance at the control arm.

g. Determine the average of the 2 measurements. The average distance should be 73.6mm ± 6mm.

h. If the figure is correct, tighten the control arm pivot nuts to specification.

i. If the figure is not correct, tighten the pivot bolts and check the front end alignment corrected.

4-Wheel Drive

1. Raise and support the vehicle safely. Remove the wheel. Remove the splash shield from the knuckle.

2. Disconnect the stabilizer bar from the control arm. Remove the shock absorber.

3. Disconnect the tie rod end from the relay rod.

4. Remove the hub nut and washer. Insert a long drift or dowel through the vanes in the brake rotor to hold the rotor in place.

5. Remove the axle shaft inner flange bolts.

6. Using a puller, force the outer end of the axle shaft out of the hub. Remove the shaft.

7. Support the lower control arm. Matchmark the both torsion bar adjustment bolt positions.

8. Using tool J-36202, increase the tension on the adjusting arm.

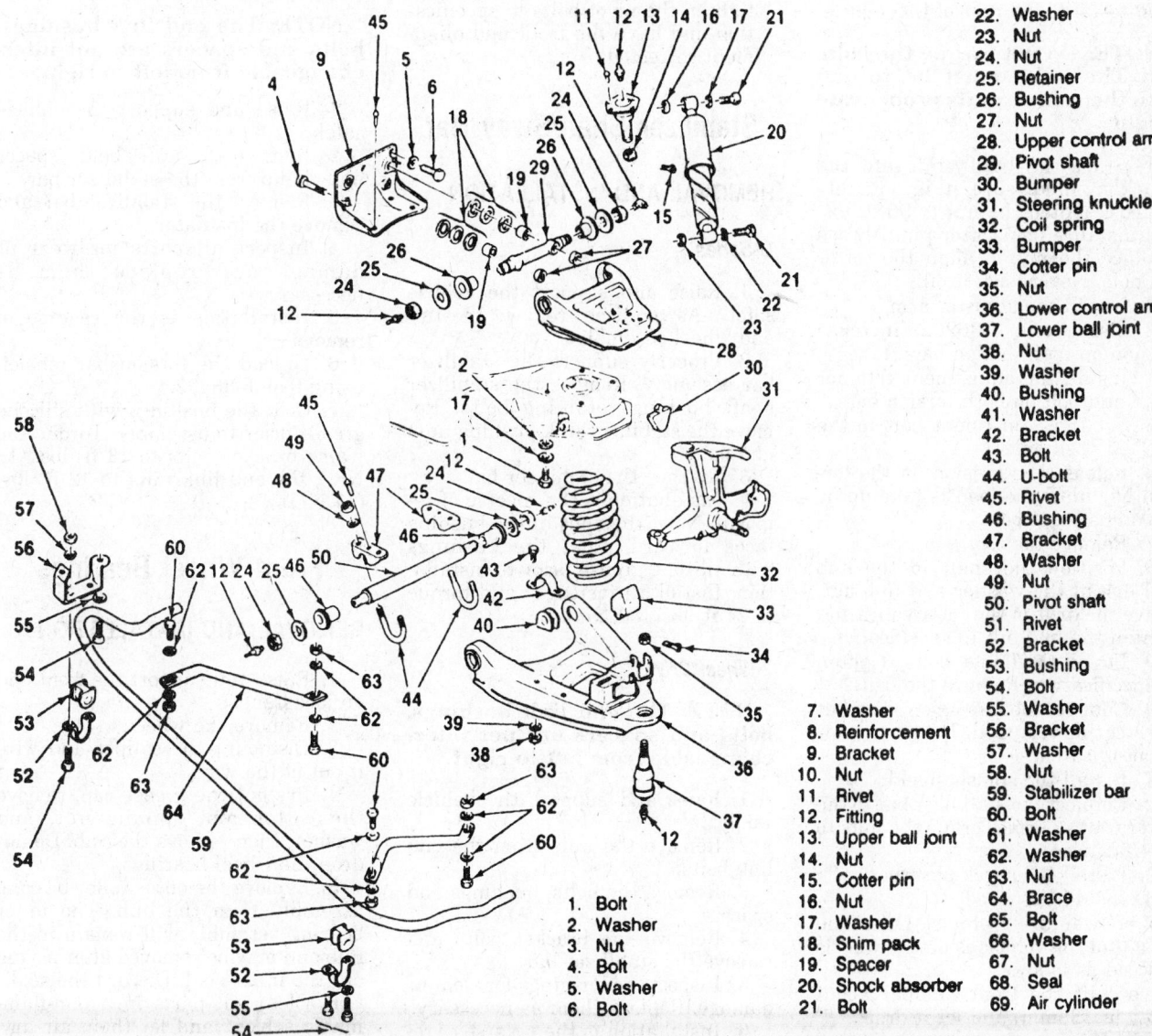

22.	Washer
23.	Nut
24.	Nut
25.	Retainer
26.	Bushing
27.	Nut
28.	Upper control arm
29.	Pivot shaft
30.	Bumper
31.	Steering knuckle
32.	Coil spring
33.	Bumper
34.	Cotter pin
35.	Nut
36.	Lower control arm
37.	Lower ball joint
38.	Nut
39.	Washer
40.	Bushing
41.	Washer
42.	Bracket
43.	Bolt
44.	U-bolt
45.	Rivet
46.	Bushing
47.	Bracket
48.	Washer
49.	Nut
50.	Pivot shaft
51.	Rivet
52.	Bracket
53.	Bushing
54.	Bolt
55.	Washer
56.	Bracket
57.	Washer
58.	Nut
59.	Stabilizer bar
60.	Bolt
61.	Washer
62.	Washer
63.	Nut
64.	Brace
65.	Bolt
66.	Washer
67.	Nut
68.	Seal
69.	Air cylinder

7.	Washer
8.	Reinforcement
9.	Bracket
10.	Nut
11.	Rivet
12.	Fitting
13.	Upper ball joint
14.	Nut
15.	Cotter pin
16.	Nut
17.	Washer
18.	Shim pack
19.	Spacer
20.	Shock absorber
21.	Bolt

1.	Bolt
2.	Washer
3.	Nut
4.	Bolt
5.	Washer
6.	Bolt

79210g83

Front suspension assembly — G-Series

9. Remove the adjustment bolt and retaining plate.

10. Move the tool aside. Slide the torsion bars forward. Remove the adjusting arm.

11. Remove the cotter pin from the lower ball stud and loosen the nut.

12. Loosen the lower ball stud in the steering knuckle using a ball joint stud removal tool. When the stud is loose, remove the nut from the stud. It may be necessary to remove the brake caliper and wire it to the frame to gain clearance.

13. Remove the control arm-to-frame bracket bolts, nuts and washers.

14. Remove the lower control arm and torsion bar as a unit.

15. Separate the control arm and torsion bar.

16. On 1500 and 2500 series, the bushings are not replaceable. If they are damaged, the control arm will have to be replaced. On 3500 series, proceed as follows:

a. Front bushing: Unbend the crimps with a punch. Force out the bushings with tools J-36618-2, J-9519-23, J-36618-4 and 36618-1.

b. Rear bushing: Force out the bushings with tools J-36618-5, J-9519-23, J-36618-3 and J-36618-2. There are no crimps.

To install:

17. On 3500 series, install a new front bushings, then a new rear bushing using the removal tools.

18. Assemble the control arm and torsion bar.

19. Raise the control arm assembly into position. Insert the front leg of the control arm into the crossmember first, then the rear leg into the frame bracket.

20. Install the bolts, front 1 first. The bolts must be installed with the front bolt head heads towards the

front of the vehicle and the rear bolt head towards the rear of the vehicle!

NOTE: Do not torque the bolts yet. The bolts must be torque with the vehicle at its proper ride height.

21. Start the ball joint into the knuckle. Make sure it is squarely seated. Tighten the nut to 96 ft. lbs. and install a new cotter pin. Always advance the nut to align the cotter pin hole. Never back it off.

22. Install the adjuster arm.

23. Using tool J-36202, increase tension on both torsion bars.

24. Install the adjustment retainer plate and bolt on both torsion bars.

25. Set the adjustment bolt to the marked position.

26. Release the tension on the torsion bar until the load is take up by the adjustment bolt.

27. Remove the tool.

28. Position the shaft in the hub and install the washer and hub nut. Leave the drift in the rotor vanes and tighten the hub nut to specification.

29. Install the flange bolts. Tighten to specification. Remove the drift.

30. Connect the inner tie rod end at the steering relay rod. Torque the nut to specification.

31. Install the splash shield.

32. Connect the stabilizer bar to the lower control arm. Torque the nuts to specification.

33. Install the shock absorber.

34. Install the wheel.

35. Lower the vehicle. Once the weight of the vehicle is on the wheels proceed as follows:

   a. Lift the front bumper about 1½ in. (38mm) and let it drop.

   b. Repeat this procedure 2–3 more times.

   c. Draw a line on the side of the lower control arm from the centerline of the control arm pivot shaft, dead level to the outer end of the control arm.

   d. Measure the distance between the lowest corner of the steering knuckle and the line on the control arm. Record the figure.

   e. Push down about 1½ in. (38mm) on the front bumper and let it return. Repeat the procedure 2–3 more times.

   f. Re-measure the distance at the control arm.

   g. Determine the average of the 2 measurements. The average distance should be 73.6mm ± 6mm.

   h. If the figure is correct, tighten the control arm nuts to specification.

   i. If the figure is not correct, tighten the pivot bolts to specification and have the front end alignment corrected.

Stabilizer Shaft/Sway Bar

REMOVAL AND INSTALLATION

G-Series

1. Raise and support the vehicle safely. As required, remove the tire and wheel assemblies.

2. Properly support the stabilizer bar assembly. Remove the stabilizer shaft bushing retaining bolts. Remove the stabilizer link bushing nuts and bolts.

3. Remove the stabilizer bar.

4. Installation is the reverse of removal. Note, the split in the bushing faces forward. Coat the bushings with silicone grease prior to installation. Install all fasteners and torque to 24 ft. lbs. (33 Nm).

2-Wheel Drive

NOTE: The end link bushings, bolts and spacers are not interchangeable from left to right.

1. Raise and support the vehicle safely.

2. Remove the nuts from the end link bolts.

3. Remove the bolts, bushings and spacers.

4. Remove the bracket bolts and remove the stabilizer bar.

5. Inspect the bushings for wear or damage. Replace them as necessary.

6. Installation is the reverse of removal. Coat the bushings with silicone grease prior to assembly. The slit in the bushings faces the front of the vehicle. Torque the frame bracket bolts to 24 ft. lbs. (33 Nm); the end link nuts to 13 ft. lbs. (18 Nm).

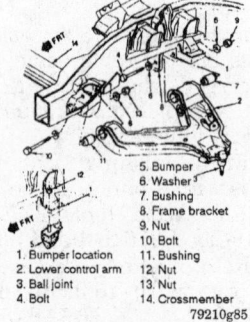

5. Bumper
6. Washer
7. Bushing
8. Frame bracket
9. Nut
10. Bolt
11. Bushing
12. Nut
13. Nut
14. Crossmember

1. Bumper location
2. Lower control arm
3. Ball joint
4. Bolt

79210g85

Lower control arm installation — 4-wheel drive

4-Wheel Drive

NOTE: The end link bushings, bolts and spacers are not interchangeable from left to right.

1. Raise and support the vehicle safely.

2. Remove the nuts, bolts, spacer and clamp from the stabilizer bar.

3. Remove the stabilizer bar and remove the insulator.

4. Inspect all parts for wear or damage and replace them as necessary.

5. Installation is the reverse of removal.

6. Unload the torsion bar tension using tool J–36202.

7. Coat the bushings with silicone grease prior to assembly. Torque the frame bracket bolts to 13 ft. lbs. (18 Nm); the end link nuts to 12 ft. lbs. (17 Nm).

Front Wheel Bearings

REMOVAL AND INSTALLATION

1. Raise and support the front end on jackstands.

2. Remove the wheel.

3. Dismount the caliper and wire it out of the way.

4. Pry out the grease cap, remove the cotter pin, spindle nut, and washer, then remove the hub. Do not drop the wheel bearings.

5. Remove the outer roller bearing assembly from the hub. The inner bearing assembly will remain in the hub and may be removed after prying out the inner seal. Discard the seal.

6. Clean all parts in a non-flammable solvent and let them air dry. Never spin-dry a bearing with compressed air! Check for excessive wear and damage.

7. Using a hammer and drift, remove the bearing races from the hub. They are driven out from the inside out.

To install:

8. Install new bearing races, if required. When installing new races, make sure that they are not cocked and that they are fully seated against the hub shoulder.

9. Pack both wheel bearings using high melting point wheel bearing grease for disc brakes. Ordinary grease will melt and ooze out ruining the pads. Bearings should be packed using a cone-type wheel bearing greaser tool. If one is not available they may be packed by hand. Place a healthy glob of grease in the palm of one hand and force the edge of the

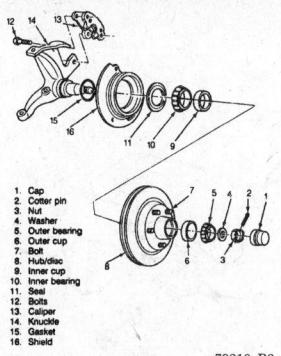

1. Cap
2. Cotter pin
3. Nut
4. Washer
5. Outer bearing
6. Outer cup
7. Bolt
8. Hub/disc
9. Inner cup
10. Inner bearing
11. Seal
12. Bolts
13. Caliper
14. Knuckle
15. Gasket
16. Shield

79210gB3

**Hub and bearing assembly —
2-wheel drive**

bearing into it so that the grease fills the bearing. Do this until the whole bearing is packed.

10. Place the inner bearing in the hub and install a new inner seal, making sure that the seal flange faces the bearing race.

11. Carefully install the wheel hub over the spindle.

12. Using your hands, firmly press the outer bearing into the hub. Install the spindle washer and nut.

13. Spin the wheel hub by hand and tighten the nut until it is just snug — 12 ft. lbs. (16 Nm). Back off the nut until it is loose, then tighten it finger tight. Loosen the nut until either hole in the spindle lines up with a slot in the nut and insert a new cotter pin. There should be 0.001–0.008 in. (0.025–0.200mm) end-play. This can be measured with a dial indicator, if you wish.

14. Replace the dust cap, wheel and tire.

REAR SUSPENSION

Shock Absorbers

REMOVAL AND INSTALLATION

1. Raise and support the vehicle safely. Properly support the rear axle

assembly. Remove the upper and lower shock absorber bolts.

2. Remove the shock absorber.
3. Installation is the reverse of the removal procedure.

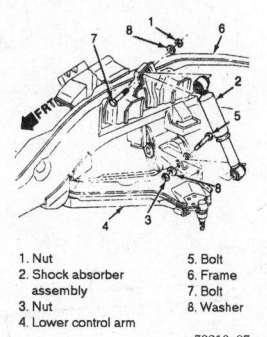

1. Nut
2. Shock absorber assembly
3. Nut
4. Lower control arm
5. Bolt
6. Frame
7. Bolt
8. Washer

79210g87

Rear shock absorber mounting — 4-wheel drive

Leaf Springs

REMOVAL AND INSTALLATION

G-Series

1. Raise and support the vehicle safely. Properly support the rear axle assembly to relieve tension on the springs.

2. If equipped, remove the stabilizer bar. Loosen, but do not remove the spring to shackle nut and bolt.

3. Remove the nut and bolt securing the shackle to the rear hanger.

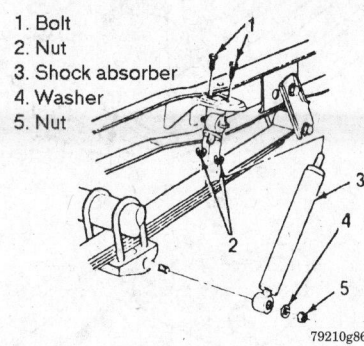

1. Bolt
2. Nut
3. Shock absorber
4. Washer
5. Nut

79210g86

Rear shock absorber mounting — 2-wheel drive

Remove the nut and bolt securing the leaf spring to the front hanger.

4. Remove the leaf spring from the front hanger. Remove the nut and bolt securing the shackle to the leaf spring. Remove the shackle.

5. Remove the nuts and washers holding the spring to the frame. If equipped, remove the rear stabilizer anchor plate, spacers, shims and auxiliary spring.

6. Remove the U-bolts from the assembly. Remove the leaf spring.

7. Installation is the reverse of the removal procedure.

8. Tighten the U-bolt nuts in a diagonal sequence.

Blazer, Pick-Up, Suburban, Tahoe and Yukon

1. Raise and support the vehicle safely. Properly support the rear axle assembly to relieve tension on the springs.

2. Remove the shock absorber. Remove the U-bolt nuts, washers, anchor plate and U-bolt.

3. Remove the shackle to frame bolt, washers and nut. Remove the spring assembly to front bracket nut, washers and bolt.

4. Remove the spring assembly. As required, separate the spring from the shackle.

5. Installation is the reverse of the removal procedure.

6. Tighten the U-bolt nuts in a diagonal sequence to specification. The spring height must be adjusted to obtain a measurement of 7.17 in. (182mm) between the top surface of the axle jounce pad and the bottom surface of the frame jounce pad.

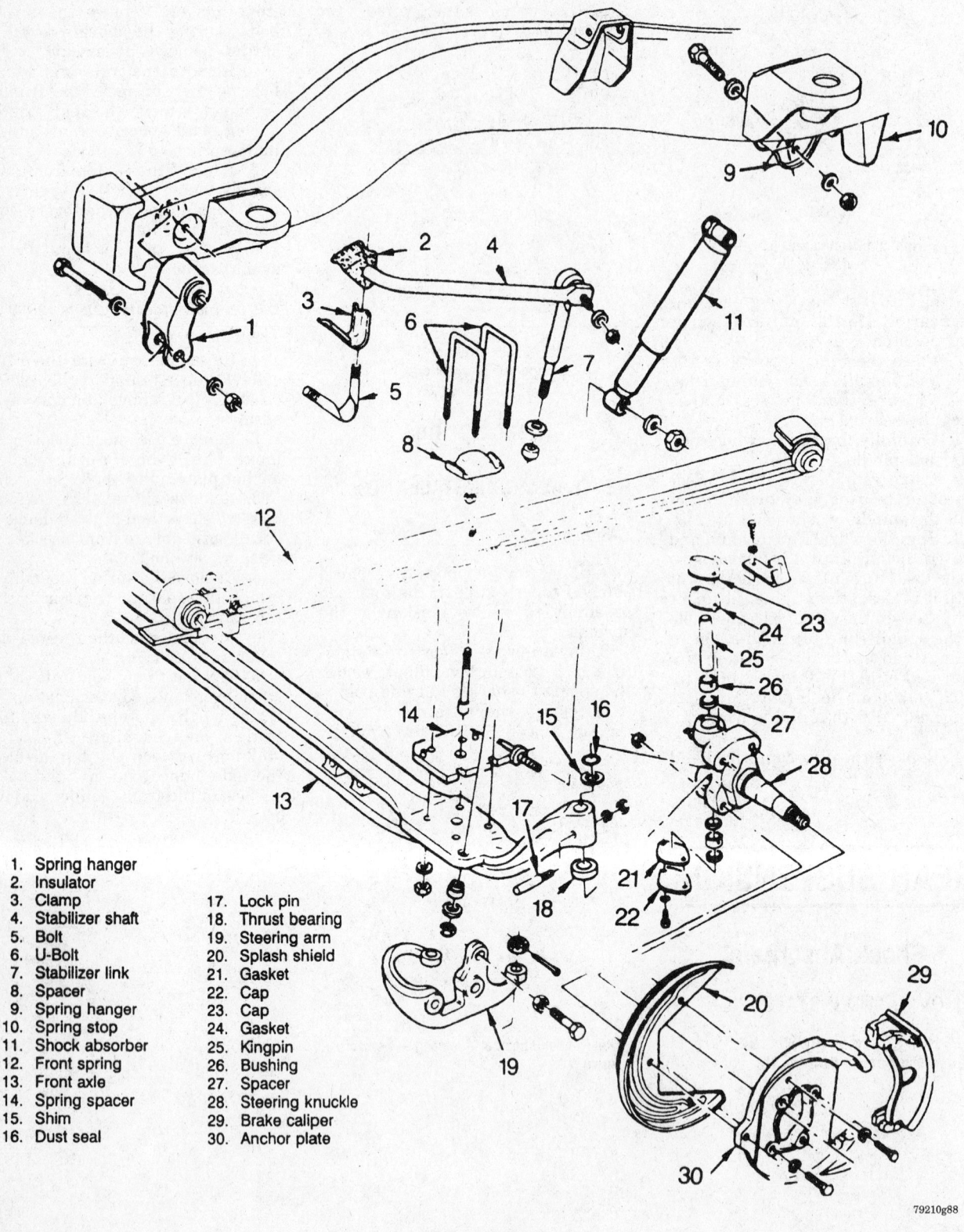

1. Spring hanger
2. Insulator
3. Clamp
4. Stabilizer shaft
5. Bolt
6. U-Bolt
7. Stabilizer link
8. Spacer
9. Spring hanger
10. Spring stop
11. Shock absorber
12. Front spring
13. Front axle
14. Spring spacer
15. Shim
16. Dust seal

17. Lock pin
18. Thrust bearing
19. Steering arm
20. Splash shield
21. Gasket
22. Cap
23. Cap
24. Gasket
25. Kingpin
26. Bushing
27. Spacer
28. Steering knuckle
29. Brake caliper
30. Anchor plate

79210g88

Front I-beam suspension

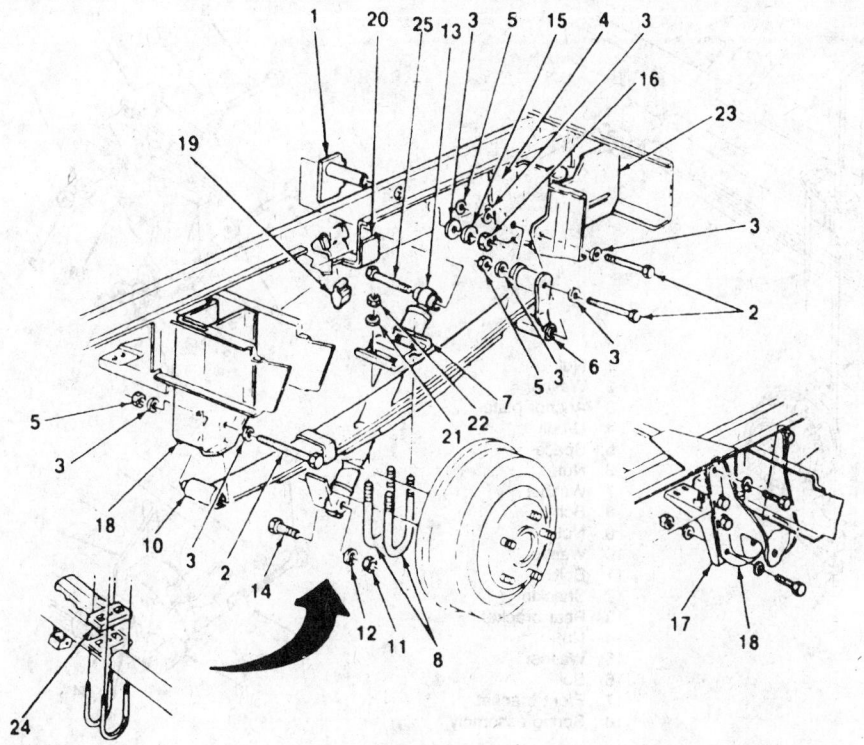

1. Bracket
2. Bolt
3. Washer
4. Rear hanger
5. Nut
6. Rear shackle
7. Anchor plate
8. U-bolt
9. Shim
10. Leaf spring
11. Nut
12. Spring washer
13. Rear shock absorber
14. Bolt
15. Spring washer
16. Nut
17. Front hanger support
18. Front hanger
19. Axle bumper
20. Bumper bracket
21. Washer
22. Nut
23. Rear hanger reinforcement
24. Spacer
25. Bolt

79210g89

Rear suspension assembly — G10/1500

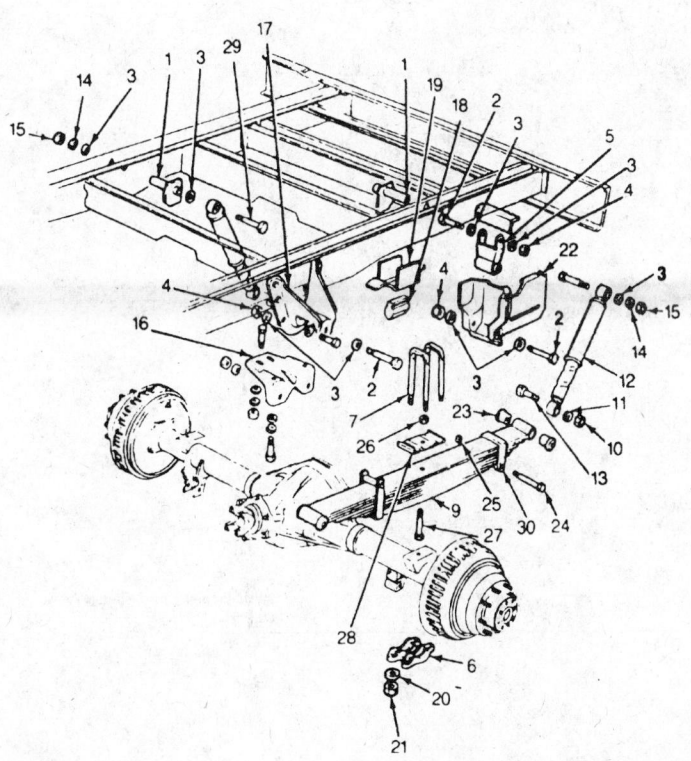

1. Bracket
2. Bolt
3. Washer
4. Nut
5. Rear shackle
6. Anchor plate
7. U-bolt
8. Shim
10. Nut
11. Spring washer
12. Rear shock absorber
13. Bolt
14. Spring washer
15. Nut
16. Front hanger support
17. Front hanger
18. Axle bumper
19. Bumper bracket
20. Washer
21. Nut
22. Rear hanger reinforcement
23. Leaf spring eye bushing
24. Bolt
25. Nut
26. Nut
27. Bolt
28. Spacer
29. Bolt
30. Spring clip

79210g90

Rear suspension assembly — G30/3500

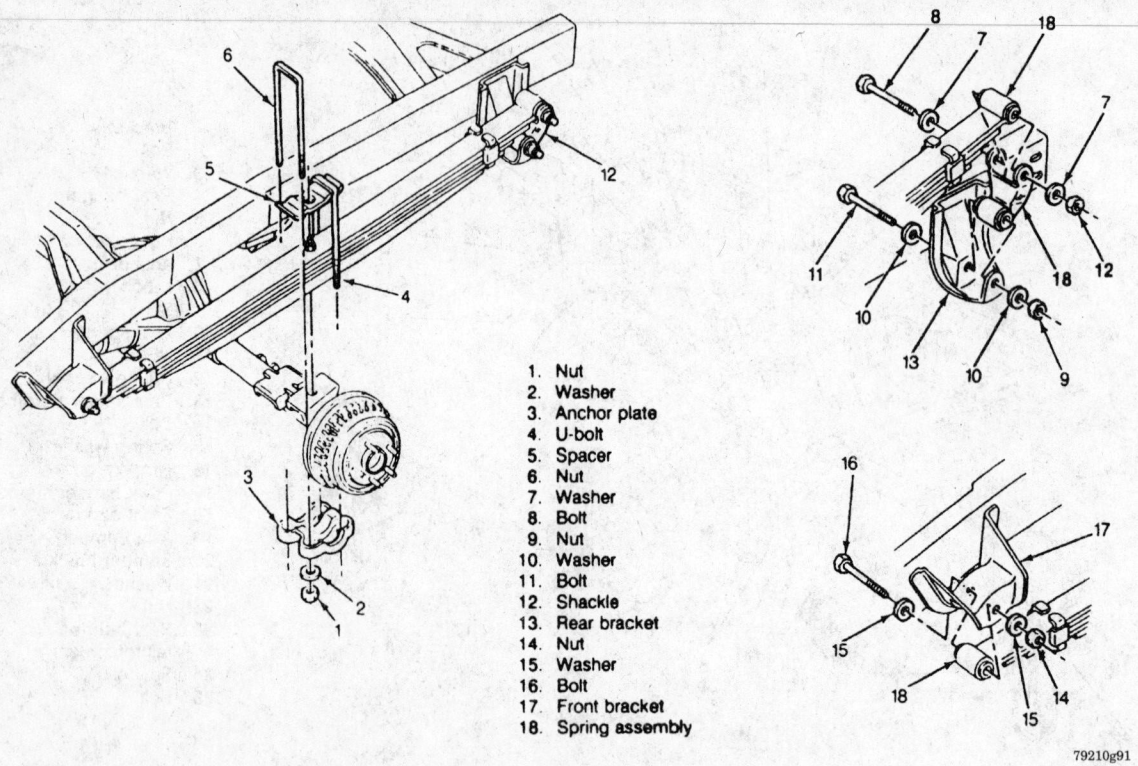

1. Nut
2. Washer
3. Anchor plate
4. U-bolt
5. Spacer
6. Nut
7. Washer
8. Bolt
9. Nut
10. Washer
11. Bolt
12. Shackle
13. Rear bracket
14. Nut
15. Washer
16. Bolt
17. Front bracket
18. Spring assembly

79210g91

Rear leaf spring and components — Blazer, Pick-Up, Suburban, Tahoe and Yukon

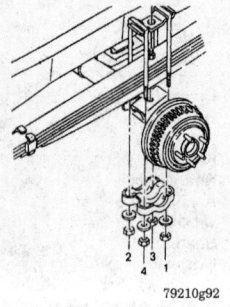

79210g92

U-bolt tightening sequence — Blazer, Pick-Up, Suburban, Tahoe and Yukon

GENERAL MOTORS **11**

CHEVROLET-Lumina APV **OLDSMOBILE**-Silhouette
PONTIAC-Trans Sport

FIRING ORDERS

NOTE: To avoid confusion, always replace spark plug wires one at a time.

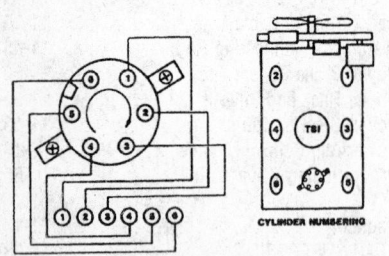

85473004

3.1L (VIN D) Engine
Engine Firing Order: 1–2–3–4–5–6
Distributor Rotation: Clockwise

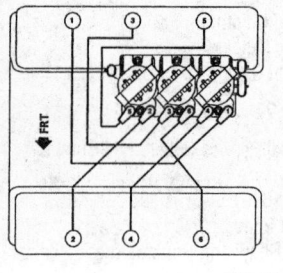

7921a050

3.4L (VIN E) Engine
Engine Firing Order: 1–2–3–4–5–6
Distributorless Ignition System

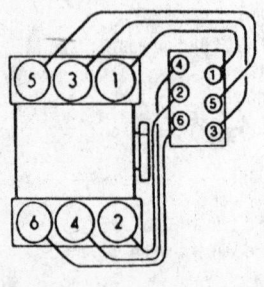

85473005

3.8L (VIN L) Engine
Engine Firing Order: 1–6–5–4–3–2
Distributorless Ignition System

ENGINE ELECTRICAL

NOTE: Disconnecting the negative battery cable on some vehicles may interfere with the functions of the on board computer systems and may require the computer to undergo a relearning process, once the negative battery cable is reconnected.

——— WARNING ———
NEVER disconnect the negative battery cable with the ignition ON. Removing power from the computer control module with the ignition ON may destroy the module.

Distributor

The 3.1L (VIN D) engine is equipped with a distributor ignition system. This distributor is is comprised of an internal magnetic pick-up assembly which contains a permanent magnet, a pole piece with internal teeth and a pick-up coil.

REMOVAL

1. Disconnect the negative battery cable.
2. Remove all necessary components in order to gain access to the distributor assembly. On most V-type engines it will be necessary to remove the air cleaner assembly for access.
3. Tag and disengage the distributor electrical connectors. If equipped, disconnect the vacuum line.
4. Remove the distributor cap. If necessary, tag and disconnect the spark plug wires.
5. Matchmark the rotor and the distributor body. Matchmark the distributor assembly and the engine block.

NOTE: Although it is only necessary to matchmark the distributor position for ease of installation, it may be advisable to first position the engine at TDC. If the distributor is being removed for further engine repair which may involve rotating the crankshaft (such as to align timing marks), setting the engine to TDC at this time will keep the distributor matchmarks valid during installation.

6. Remove the distributor holddown bolt. Carefully remove the distributor.

NOTE: As the distributor is removed from the engine, the rotor will turn counterclockwise. Observe and mark the finish position of the rotor. When reinstalling, position the rotor at the last mark and set the distributor into the engine. As the distributor drops into place, the rotor should turn to its original position, providing the engine crankshaft has not been rotated with the distributor out.

INSTALLATION

Timing Not Disturbed

NOTE: To ensure correct ignition timing if the engine has not been disturbed, the distributor must be installed with the rotor in the same position as when removed.

1. Align the rotor to the last mark made and install the distributor in the engine.

NOTE: On vehicles not equipped with the HVS system, if the distributor shaft cannot drop into the engine, remove the distributor and use a small prytool through the mounting hole to rotate the oil pump driveshaft until it can align with the distributor gear.

2. As the distributor is fully seated, the rotor should turn and end up at the first mark made. Ensure the distributor and oil pump rod are fully engaged.

NOTE: When the distributor is fully seated on the engine, make sure the rotor is properly aligned with the first timing mark. If equipped with the HVS system, if the rotor dos not align with the mark, the gear teeth of the HVS and camshaft have meshed 1 or more teeth out of time and the HVS procedure for timing disturbed should be used to assure proper installation.

3. Reconnect the distributor cap and wires. If applicable, connect the vacuum line to the distributor.
4. Tighten the distributor holddown bolt, then install the air cleaner or duct, if removed for access.
5. Check and adjust the ignition timing.

Timing Disturbed

1. Set the engine to TDC: Remove the No. 1 spark plug. Place a finger over the spark plug hole and rotate the engine in the normal direction of rotation slowly, until compression is felt.

2. Align the timing mark on the crankshaft pulley to the **0** on the engine timing indicator by rotating the engine in the same direction slowly. The engine is now set on No. 1 TDC.

NOTE: An alternate method may be used to assure the engine is at TDC if the valve cover is removed. Watch the rocker arms for the No. 1 cylinder as the engine is turned. If the valves move as the crankshaft timing marker approaches the scale, the No. 1 cylinder is on its exhaust stroke. If the valves remain closed as the timing mark approaches the scale, then the No. 1 cylinder is approaching TDC of the compression stroke.

3. Install the distributor to the engine so the rotor is pointing to the No. 1 spark plug tower on the distributor cap once the distributor is fully seated in the engine.

4. Install the distributor cap, spark plug, wiring and connectors. If applicable, connect the vacuum line to the distributor.

5. If removed for access, install the air duct and/or air cleaner assembly, as applicable.

6. Check and adjust ignition timing.

Ignition Coils

REMOVAL AND INSTALLATION

3.4L (VIN E) AND 3.8L (VIN L) Engines

NOTE: Any 1 of the 3 coils on the 3.4L (VIN E) or 3.8L (VIN L) engine may be removed separately from the other(s).

1. Disconnect the negative battery cable.

2. Tag and disconnect the spark plug wiring from the coil(s).

3. Remove the coil retaining screws (there are usually 2 screws per coil).

4. Separate the coil from the ignition module.

To install:

5. Install the coil to the ignition module assembly.

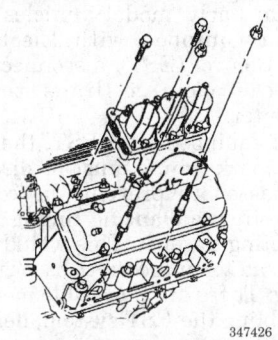

Removing or installing the ignition coils — 3.4L (VIN E) engine

347426

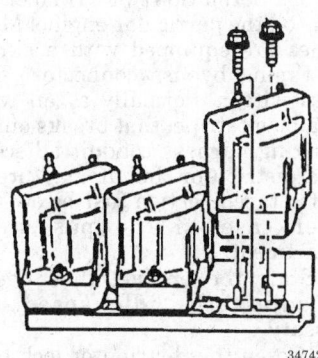

Ignition coils — 3.4L (VIN E) engine

347427

6. Install the coil retaining screws and tighten to 40 inch lbs. (4.5 Nm).

7. Engage the spark plug wire to the coil(s) and noted during removal.

8. Connect the negative battery cable.

Ignition Module

REMOVAL AND INSTALLATION

3.4L (VIN E) AND 3.8L (VIN L) Engines

1. Disconnect the negative battery cable.

2. Disconnect the wire connector from the module.

3. Remove the screws securing the module and coils to the mounting plate.

4. Separate the ignition coils from the module by pulling straight up on the coils.

NOTE: Position the coils and wiring so the assembly will not be hanging from the wires.

5. Installation is the reverse of the removal procedure.

Crankshaft Sensor

REMOVAL AND INSTALLATION

3.4L (VIN E) AND 3.8L (VIN L) Engines

1. Disconnect the negative battery cable.

2. Remove the serpentine belt from the crankshaft pulley.

3. Raise and support the vehicle safely.

4. Remove the right front tire and wheel assembly.

5. Remove the right inner fender access cover.

6. Using a 28mm socket, remove the harmonic balancer retaining bolt.

7. Using a harmonic balancer pulley removal tool such as J-38197 or equivalent, remove the balancer assembly.

8. On 3.4L (VIN E) engine, remove the crankshaft sensor harness clip and electrical connector. Remove the sensor from the engine block.

9. On 3.8L (VIN L) engine, loosen the retaining bolts and remove the sensor shield. Do not use a prybar when removing the shield.

10. Disengage the sensor electrical connector, then remove the sensor from the engine block.

To install:

11. Install the sensor to the engine block.

12. On 3.4L (VIN E) engine, install the sensor securing bolts and torque to 8 ft. lbs. (10 Nm). Install the electrical connector.

13. On 3.8L (VIN L) engine, install the sensor securing bolts and torque to 14–28 ft. lbs. (20–40 Nm).

14. On 3.4L (VIN E) engine, apply sealer to the keyway of the balancer and install the balancer to the crankshaft.

15. On 3.4L (VIN E) engine, apply sealer to the threads of the crankshaft balancer bolt. Torque the bolt to 79 ft. lbs. (107 Nm)

16. On 3.8L (VIN L) engine, install the crankshaft sensor shield.

17. Engage the wire connector.

18. On 3.8L (VIN L) engine, position the balancer onto the crankshaft.

19. On 3.8L (VIN L) engine, apply sealer to the threads of the balancer bolt and torque the bolt to 110 ft. lbs. (150 Nm) plus turn the bolt an additional 76 degrees.

20. Install the fender shield, the install the tire and wheel assembly.

21. Lower the vehicle.

22. Install the serpentine belt, then connect the negative battery cable.

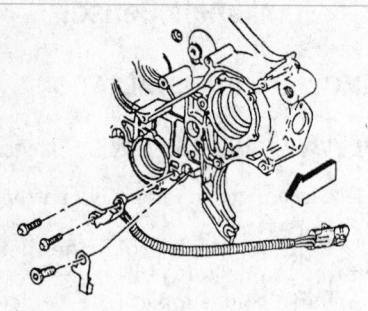

Removing or installing the crankshaft sensor — 3.4L (VIN E) engine

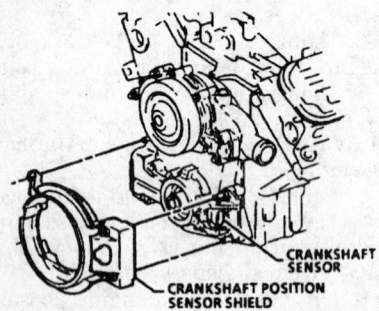

Install the shield over the crankshaft sensor — 3.8L (VIN L) engine

Ignition Timing

The ignition timing may be adjusted only on those engines equipped with a conventional distributor ignition. If equipped with the distributorless Electronic Ignition (EI) system, the control module sets timing and makes all necessary spark changes. On these systems, the crankshaft position sensor is mounted in a fixed position, therefore not allowing for adjustment.

Connect the timing light and tachometer to the engine according to the tool manufacturers' instructions. Make sure the timing light is connected to the No. 1 spark plug wire.

ADJUSTMENT

1. Locate and clean the timing marks on the crankshaft pulley and the front of the timing case cover.

2. Use chalk or white paint to color the mark on the scale that will indicate the correct timing, when aligned with the mark on the pulley or the pointer.

3. Attach a tachometer and a timing light to the engine.

4. On early model vehicles that are not equipped with Electronic Spark Control (EST), disconnect and plug the vacuum lines to the distributor.

5. If equipped with EST, the electronic spark timing must be disabled or bypassed to prevent the control module from advancing timing while attempting to set it. This would obviously lead to an incorrect base timing setting. There are 2 possible methods of disabling the EST system, depending on the type of engine:

• Engines using the EST distributor, disengage the timing connector wire. Refer to the Vehicle Emission Control Information (VECI) label for details on the particular engine. Most vehicles are equipped with a single wire timing bypass connector. The bypass wire is normally a tan wire with a black stripe, that breaks out of the wiring harness conduit adjacent to the distributor, but it may break out of a taped section just below the heater case in the passenger compartment.

6. Start the engine, then check and adjust the idle speed, as necessary.

7. Loosen the distributor lock bolt slightly to permit the distributor to be turned.

8. Adjust the idle to the correct specification.

9. With the timing light aimed at the pulley and the marks on the engine, turn the distributor in the direction of rotor rotation to retard the spark or in the opposite direction of rotor rotation to advance the spark. Align the marks on the pulley and the engine with the flashes of the timing light.

10. Turn the engine **OFF**, tighten the distributor hold-down bolt and recheck the timing.

Alternator

PRECAUTIONS

Several precautions must be observed with alternator equipped vehicles in order to avoid damage to the unit.

• If the battery is removed for any reason, make sure it is reconnected with the correct polarity. Reversing the battery connections may result in damage to the 1-way rectifiers.

• When utilizing a booster battery as a starting aid, always connect the positive to positive terminals and the negative terminal from the booster battery to a good engine ground on the vehicle being started.

• Never use a fast charger as a booster to start vehicles.

• Disconnect the battery cables when charging the battery with a fast charger.

• Never attempt to polarize the alternator.

• Do not use test lamps of more than 12 volts when checking diode continuity.

• Do not short across or ground any of the alternator terminals.

• The polarity of the battery, alternator and regulator must be matched and considered before making any electrical connections within the system.

• Never separate the alternator on an open circuit. Make sure all connections within the circuit are clean and tight.

• Disconnect the battery ground terminal when performing any service on electrical components.

• Disconnect the battery if arc welding is to be done on the vehicle.

REMOVAL AND INSTALLATION

1. Disconnect the negative battery cable.

2. Remove the necessary components in order to gain access to the alternator assembly.

3. If necessary, remove the air cleaner and/or duct work.

4. Tag and disengage the electrical connectors at the alternator.

5. Remove the alternator belt. If equipped with a serpentine belt, relieve the belt tension, the carefully remove the belt from the alternator pulley.

6. If equipped, remove the alternator brace.

7. Remove the alternator retaining bolts and remove the alternator.

8. Installation is the reverse of the removal procedure. Check and adjust the belt tension, as required.

BELT TENSION ADJUSTMENT

Serpentine Belts

Serpentine belts use an automatic tensioner which is spring activated and can be turned to the left or the right to apply or release the pulley tension.

Most tensioners are equipped with a belt wear scale, make sure the belt is not stretched beyond it's serviceable life. Also, the serpentine belt

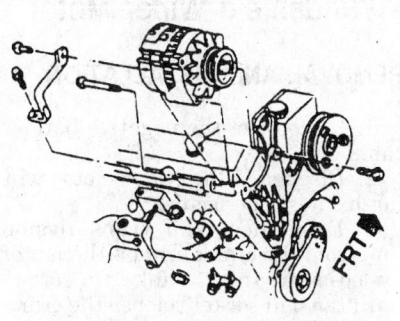

Alternator mounting — 3.1L (VIN D) engine

85473013

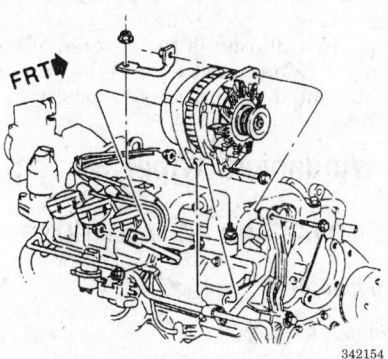

Alternator mounting — 3.4L (VIN E) engine

342154

grooves must match the grooves in the pulleys.

NOTE: Most tensioners are equipped to insert a $1/2$ in. breaker bar into the tensioner pulley in order to apply leverage and release the belt tension.

Starter

REMOVAL AND INSTALLATION

3.1L (VIN D) Engine

1. Disconnect the negative battery cable.
2. Raise and support the vehicle safely.
3. If necessary, remove the air conditioning compressor brace.
4. Remove the solenoid cover, then tag and disconnect the solenoid wiring. Reposition the wiring harness, as necessary.
5. Disengage the oil pressure sensor electrical connector, then remove the sensor.
6. Either remove the 2 bolts or the 1 bolt and 1 nut retaining the starter to the engine block. Carefully remove the starter and the shim(s), noting shim position, if equipped, for installation purposes.

7. Installation is the reverse of removal. Install the shim, if used, and torque the mounting bolts to 32 ft. lbs. (43 Nm).

3.4L (VIN E) AND 3.8L (VIN L) Engines

1. Disconnect the negative battery cable.
2. Raise and support the vehicle safely.
3. Remove the cover from the starter wiring, then tag and disconnect the wiring.
4. Remove the screws and harness clips retaining the wiring, then position the harness out of the way.
5. Remove the retainers and the flywheel cover.
6. Remove the starter bolts, then carefully remove the starter and any shims, if equipped, noting the shim location(s) for installation purposes.
7. Installation is the reverse of removal. Install the shim, if used, and torque the mounting bolts to 32 ft. lbs. (43 Nm).

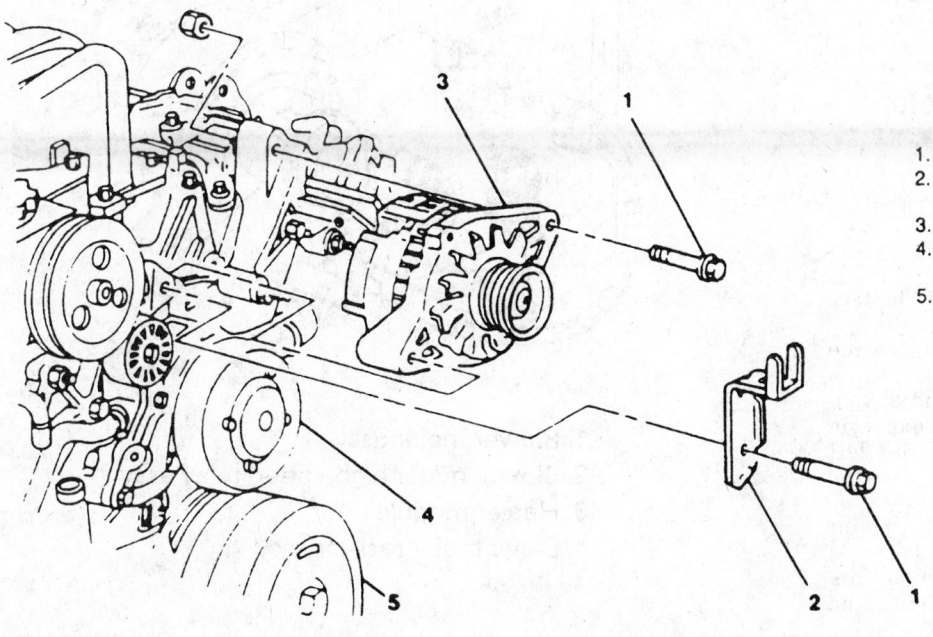

1. Bolt
2. Engine wiring harness bracket
3. Alternator
4. Crankshaft torsional damper
5. Coolant pump pulley

85473014

Alternator mounting — 3.8L (VIN L) engine

1. Engine assembly
2. Starter motor bolt
3. Starter motor assembly

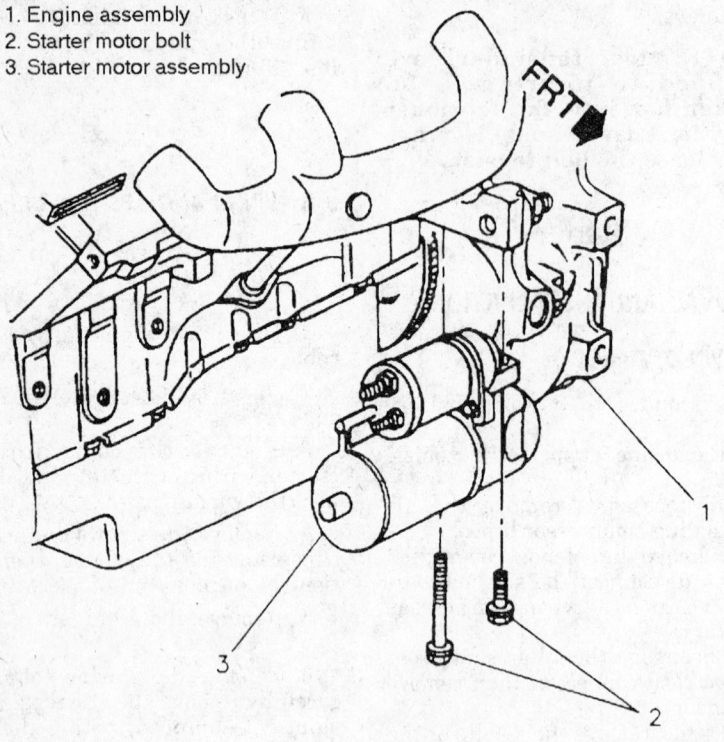

Starter mounting — 3.4L (VIN E) engine

Windshield Wiper Motor

REMOVAL AND INSTALLATION

1. Disconnect the negative battery cable.
2. Disengage the wiper motor wiring harness connector.
3. Disconnect the transmission link from the crank arm on the motor by loosening the 2 crank arm screws until the ball socket release the crank arm ball. The screws are usually 8mm heads.
4. Remove the motor mounting bolts and slide the motor out of its mounting.
5. Installation is the reverse of removal. When installing, the wiper motor must be in the park position.

Windshield Wiper Switch

REMOVAL AND INSTALLATION

Lumina APV

Without Air Bag

The windshield wiper switch in the Lumina is located in the steering col-

CHASSIS ELECTRICAL

Heater Blower Motor

REMOVAL AND INSTALLATION

1. Disconnect the negative battery cable.
2. If necessary, remove the engine air cleaner assembly.
3. Remove the screws and disconnect the left windshield wiper arm linkage. Carefully move the left wiper arm linkage aside for access to the blower motor assembly.
4. Disengage the blower motor electrical harness.
5. Remove the blower motor retaining screws and remove the blower motor assembly.
6. Installation is the reverse of removal.

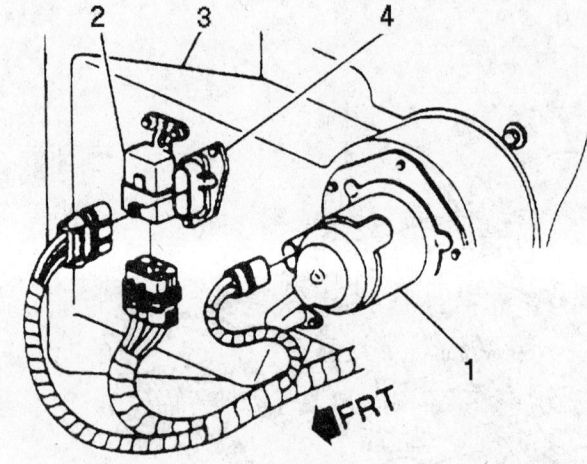

1. Blower motor assy
2. Blower motor high-speed relay assy
3. Heater module
4. Blowr motor resistor assy

Removing or installing the blower motor

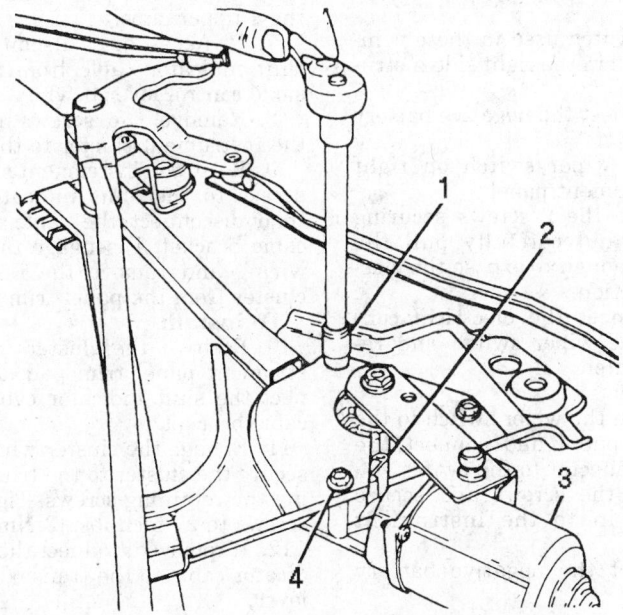

1. Socket,10mm
2. Removal slot
3. Wiper motor assy
4. Screw(3),62 in. lb.(7 Nm)

7921A003

Removing or installing the Windshield wiper motor

umn and is a part of the combination switch or multi-function switch.

1. Disconnect the negative battery cable.

2. Matchmark and remove the steering wheel.

NOTE: Although in some cases the components necessary to remove the wiper switch may be removed with the steering column installed in the vehicle, it is usually necessary to at least unbolt, lower and support the column.

3. Remove the turn signal switch.

4. Remove the lock cylinder assembly.

5. Remove the lock housing cover screws.

6. If applicable, remove the tilt lever.

7. Remove the lock housing cover assembly.

8. If necessary, unbolt the steering column support bracket from the column.

9. Disengage the wiper switch (pivot and pulse switch) connector from the wiring harness, then remove the wiring protector.

10. Attach a length of mechanic's wire to the switch connector, then carefully pull the harness through the column (from the top), leaving the wire in the column for assembly.

11. Remove the switch.

To install:

12. Install the switch to the lock housing cover assembly, then install the pivot pin.

13. Vehicles equipped with floor shift:

a. Carefully pull the switch harness through the steering column using the mechanic's wire, then engage the connector to the harness.

b. If applicable, install the column support bracket.

c. Lubricate the dimmer switch rod actuator using lithium grease, then if removed, install the actuator to the column housing cover end cap.

d. Install the end cap to the lock housing cover assembly. Make sure the bottom edge of the dimmer switch rod actuator is resting on the bend in the dimmer switch rod.

e. Install the lock housing cover assembly, then secure using the retaining screws. Tighten the screw in the 12 o'clock position first, then the 8 o'clock position next and finally the screw in the 3 o'clock position.

14. If removed, install the ignition and dimmer switches.

15. Install the lock cylinder assembly.

16. Install the turn signal switch.

17. If removed or lowered, position and secure the steering column.

18. Align and install the steering wheel.

19. Connect the negative battery cable.

With Air Bag

1. Properly disable the SIR (air bag) system, then disconnect the negative battery cable.

2. Matchmark and remove the steering wheel.

NOTE: Although in some cases the components necessary to remove the wiper switch may be removed with the steering column installed in the vehicle, it is usually necessary to at least unbolt, lower and support the column.

3. Remove the turn signal switch.

4. Remove the lock cylinder assembly.

5. Remove the lock housing cover screws.

6. If applicable, remove the tilt lever.

7. Remove the lock housing cover assembly.

8. Remove the column housing cover end cap (in some cases, this should be done along with the switch rod actuator), then remove the switch actuator pivot pin.

9. If necessary, unbolt the steering column support bracket from the column.

10. Disengage the wiper switch (pivot and pulse switch) connector from the wiring harness, then remove the wiring protector.

11. Attach a length of mechanic's wire to the switch connector, then carefully pull the harness through the column (from the top), leaving the wire in the column for assembly.

12. Remove the switch.

To install:

13. Install the switch to the lock housing cover assembly, then install the pivot pin.

14. Carefully pull the switch harness through the steering column using the mechanic's wire, then engage the connector to the harness.

15. If applicable, install the column support bracket.

16. Lubricate the dimmer switch rod actuator using lithium grease, then if removed, install the actuator to the column housing cover end cap.

17. Install the end cap to the lock housing cover assembly. Make sure the bottom edge of the dimmer switch rod actuator is resting on the bend in the dimmer switch rod.

18. Install the lock housing cover assembly, then secure using the re-

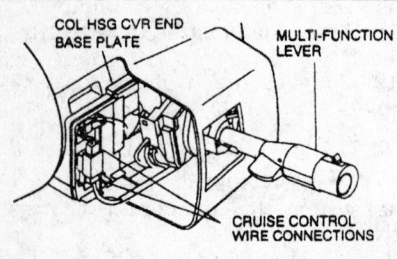

Some columns use a multi-function lever with cruise control wiring

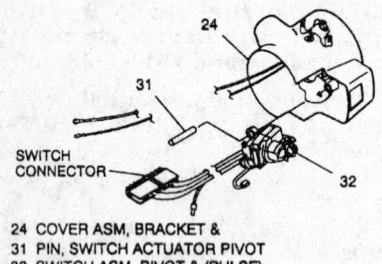

24 COVER ASM, BRACKET &
31 PIN, SWITCH ACTUATOR PIVOT
32 SWITCH ASM, PIVOT & (PULSE)

85473027

Removing the windshield wiper (pivot and pulse) switch

taining screws. Tighten the screw in the 12 o'clock position first, then the 8 o'clock position next and finally the screw in the 3 o'clock position.

19. Install the lock cylinder assembly.

20. Install the turn signal switch.

21. If removed or lowered, position and secure the steering column.

22. Align and install the steering wheel.

23. Connect the negative battery cable, then properly enable the SIR system.

Silhouette and Trans Sport

The wiper switch used in these vehicles is located in the right side instrument panel.

1. Disconnect the negative battery cable.

2. Locate wiper switch on right side of instrument panel.

3. Remove the 2 screws securing the switch and carefully pull the switch out enough to expose the electrical connection.

4. Disconnect the electrical connector at the wiper switch and remove the switch.

To install:

5. Position the wiper switch to the instrument panel and connect the electrical connector to the switch.

6. Install the screws and secure the switch in to the instrument panel.

7. Connect the negative battery cable.

Instrument Cluster

REMOVAL AND INSTALLATION

Lumina APV

1. Disconnect the negative battery cable.

2. Remove the rear window wiper switch from the instrument cluster housing, then disengage the electrical connector.

3. Remove the head light switch from the instrument cluster housing, then disengage the electrical connector.

4. Remove the 4 screws securing the instrument cluster housing to the instrument panel trim pad, then remove the housing from the pad.

5. Remove the left hand sound insulator panel.

6. Remove the steering column opening filler from the instrument

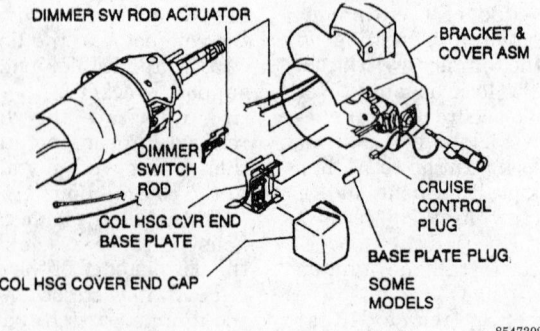

DIMMER SW ROD ACTUATOR
BRACKET & COVER ASM
DIMMER SWITCH ROD
COL HSG CVR END BASE PLATE
CRUISE CONTROL PLUG
COL HSG COVER END CAP
BASE PLATE PLUG.
SOME MODELS

85473026

Exploded view of a common housing cover mounting

panel lower trim pad by disengaging the 4 upper clips.

7. If necessary, disconnect the shift indicator cable from the transaxle control lever bowl.

8. Remove the screws retaining the instrument cluster to the pad.

9. Raise the instrument cluster for access to the shift indicator cable, then disconnect the cable from the cable bracket. Disengage the cluster wiring and remove the instrument cluster from the panel trim pad.

To install:

10. Position the cluster to the instrument panel trim pad, then connect the shift indicator cable to the cable bracket.

11. Engage the cluster wiring, then secure the cluster to the trim pad using the retaining screws. Tighten the screws to 27 inch lbs. (3 Nm).

12. If removed, connect the shift indicator cable to the transaxle control lever:

a. Position the transaxle control lever in the Neutral gate notch.

b. Guide the shift indicator cable around the edge of the transaxle control lever bowl and into a position that locates the cluster pointer to Neutral.

c. Push the cable onto the control lever bowl, taking care to assure the cable rests on the bowl and not the steering column jacket.

d. Check that with the control lever in Neutral, portions of the "N" are visible on either side of the pointer. Make sure each gear is indicated in the same fashion, with the pointer at the center of the indicator.

13. Install the steering column opening filler panel to the instrument panel lower trim pad by engaging the 4 clips.

14. Install the left hand sound insulator panel.

15. Install the 4 screws securing the instrument cluster housing to the instrument panel lower trim pad. Tighten the screws to 18 inch lbs. (2 Nm).

16. Engage the head light switch connector, then install the switch to the cluster housing.

17. Engage the rear wiper switch connector, then install the switch to the cluster housing.

18. Connect the negative battery cable.

Silhouette and Trans Sport

1. Disconnect the negative battery cable.

2. Open the glove box door to access the 2 screws securing the lower

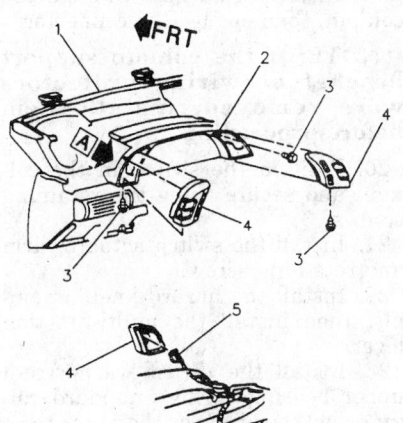

1 PAD-I/P
2 PLATE, I/P CLUSTER TRIM
3 SCREWS
4 SWITCH, WINDSHIELD WIPER
  AND WASHER
5 CONNECTOR, WINDSHIELD WIPER AND
  WASHER

7921A005

Headlight and wiper switch removal and installation — Silhouette and Trans Sport

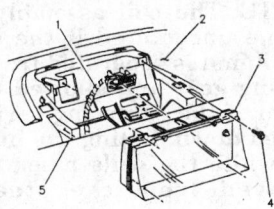

1 CONNECTOR, I/P HARNESS
2 PAD, I/P TRIM
3 CLUSTER, I/P
4 SCREWS
5 PAD, I/P LOWER TRIM

7921A004

Instrument cluster assembly — Lumina APV — 1994–96 Trans Sport

instrument panel trim pad assembly to the instrument panel pad assembly.

3. Remove the 2 screws securing the instrument cluster trim panel to the instrument panel pad.

4. Remove the screw retaining the head light switch pod to the cluster trim panel and remove the head light switch pod. Disconnect the wiring harness to the pod.

5. Remove the screw retaining the windshield wiper switch pod to the cluster trim panel and remove the wiper switch pod. Disconnect the wiring harness to the pod.

6. Remove the 2 screws behind each switch pod securing the cluster trim panel to the left and right instrument cluster mounting brackets.

7. Remove the side window defogger outlet grilles from the instrument panel trim pad, then remove the pad retaining screws from the 2 openings.

8. Carefully unseat the pad. On Canadian vehicles, disengage the daytime running lamp sensor from the pad. Remove the pad.

9. For 1994–96 vehicles, remove the left hand sound insulator panel.

10. Remove the steering column opening filler by disengaging the retaining clips (usually 4).

11. Disconnect the shift indicator "**PRNDL**" cable clip from the control lever bowl.

12. Remove the 2 screws on each side from the left and right cluster retainer brackets and lift up on the cluster for access.

13. If equipped, disconnect the shift indicator cable from the cable bracket.

14. Disengage the instrument panel harness connector from the the instrument cluster.

15. Remove the cluster.

To install:

16. Position the cluster, then connect the instrument panel harness connector.

17. If equipped, connect the shift indicator cable to the cable bracket.

18. Seat the cluster, then secure the cluster retainer brackets using the screws.

19. Connect the shift indicator cable to the transaxle control lever:

   a. Position the transaxle control lever in the Neutral gate notch.

   b. Guide the shift indicator cable around the edge of the transaxle control lever bowl and into a position that locates the cluster pointer to Neutral.

   c. Push the cable onto the control lever bowl, taking care to assure the cable rests on the bowl and not the steering column jacket.

   d. Check that with the control lever in Neutral, portions of the "N" are visible on either side of the pointer. Make sure each gear is indicated in the same fashion, with the pointer at the center of the indicator.

20. Install the steering column opening filler.

21. If equipped, install the left hand sound insulator panel.

22. Loosely position the instrument panel trim pad.

23. If equipped, connect the daytime running lamp sensor to the instrument panel trim pad.

24. Install the 2 screws securing the lower instrument panel trim pad assembly to the instrument panel pad assembly, behind the glove box door.

25. Install the pad retaining screws located in the defogger outlets, then install the defogger grilles.

26. Install the 2 screws behind each switch pod securing the cluster trim panel to the left and right instrument cluster mounting brackets.

27. Connect the wiring harness to the wiper switch pod, then install the pod to the panel.

28. Connect the wiring harness to the head light switch pod, then install the pod to the panel.

29. Connect the negative battery cable.

Headlight Switch

REMOVAL AND INSTALLATION

1. Separate, the head light switch from the instrument cluster housing. Most switches are snapped into position using retaining clips, though some may use a small retaining screw.

2. The headlight switch use in the Lumina is retained by clips. Carefully pry the switch from the instrument panel and disconnect the electrical connector.

3. The headlight switch use in Silhouette and Trans Sport is retained by a small screw. Remove the screw and pull the switch from the instrument panel and disconnect the electrical connector.

4. Installation is the reverse of the removal.

Turn Signal Switch

REMOVAL AND INSTALLATION

NOTE: When servicing any components on the steering column, should any fasteners require replacement, be sure to use only nuts and bolts of the same size and grade as the original fasteners. Using screws that are too long could prevent the column from collapsing during a collision.

NOTE: The following procedures require the use of a lock plate compressor tool such as J-23653 or equivalent.

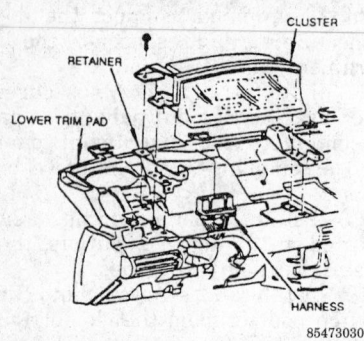

Instrument cluster assembly — Silhouette and 1994–96 Trans Sport

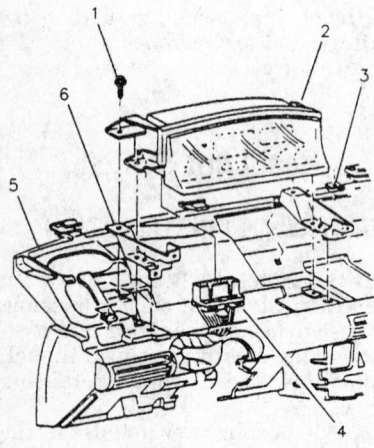

1 BOLTS (4)
2 INSTRUMENT CLUSTER
3 NUT
4 I/P HARNESS
5 I/P LOWER HARNESS
6 RETAINER

7921A006

Removing or installing the headlight switch — Lumina APV

1. If equipped, make sure the wheels are locked in the straight-ahead position, then properly disable the SIR (air bag) system.

2. Disconnect the negative battery cable.

3. Matchmark and remove the steering wheel.

4. If equipped with an SIR system, remove the SIR coil assembly retaining ring, then remove the coil assembly and allow it to hang freely from the wiring. Remove the wave washer.

5. On non-SIR equipped vehicles, remove the shaft lock cover.

6. Push downward on the shaft lock assembly until the snapring is exposed using the shaft lock compressor tool.

7. Remove the shaft lock retaining snapring, then carefully release the tool and remove the shaft lock from the column.

8. Remove the turn signal cancelling cam assembly.

9. For standard columns not equipped with SIR, remove the upper bearing spring and thrust washer.

10. For tilt columns or SIR equipped standard columns, remove the upper bearing spring, inner race seat and inner race.

11. Move the turn signal lever upward to the "Right Turn" position.

12. Remove the access cap and disengage the multi-function lever harness connector, then grasp the lever and pull it from the column.

13. Loosen and remove the hazard knob retaining screw, then remove the screw, button, spring and knob.

14. Remove the screw and the switch actuator arm.

15. Remove the turn signal switch retaining screws, then pull the switch forward and allow it to hang from the wires. If the switch is only being removed for access to other components, this may be sufficient.

16. If the switch is to be replaced, cut the wires near the top of the switch and discard the switch. Before cutting the wires, verify that the wire color codes are the same. Secure the connector of the new switch to the old wires, and pull the new harness down through the steering column while removing the old switch.

17. If the original switch is to be reused, attach a piece of wire or string around the connector and pull the harness up through the column, while pulling the string up through the column and leaving the string or wire in position to help with reinstallation later.

18. After freeing the switch wiring protector from its mounting, pull the turn signal switch straight up and remove the switch, switch harness, and the connector from the column.

NOTE: On some vehicles access to the connector may be difficult. If necessary, remove the column support bracket assembly and properly support the column, and/or remove the wiring protectors.

To install:

19. Install the switch and wiring harness to the vehicle. If the switch was completely removed, use the length of mechanic's wire or string to pull the switch harness through the column, then engage the connector.

NOTE: If the column support bracket or wiring protectors were removed, install them before proceeding.

20. Position the switch in the column and secure using the retaining screws.

21. Install the switch actuator arm and retaining screw.

22. Install the hazard knob assembly, then install the multi-function lever.

23. Install the thrust washer and upper bearing spring (standard columns without SIR) or the inner race, upper bearing race seat and upper bearing spring (tilt columns or standard with SIR), as applicable.

24. Lubricate the turn signal cancelling cam using a suitable synthetic grease (usually included in the service kit), then install the cam assembly.

25. Position the shaft lock and a new snapring, then use the lock compressor to hold the lock down while seating the new snapring. Make sure the ring is firmly seated in the groove, then carefully release the tool.

NOTE: The coil assembly will become uncentered if the steering column is separated from the steering gear and allowed to rotate or if the centering spring is pushed down, letting the hub rotate while the coils assembly is removed from the steering column.

26. If equipped with SIR, make sure the coil is centered, then install the wave washer, followed by the coil and the retaining ring. The coil ring must be firmly seated in the shaft groove.

27. On non SIR columns, install the shaft lock cover.

28. Align and install the steering wheel.

29. Make sure the ignition is **OFF**, then connect the negative battery cable.

30. Properly enable the SIR system.

Ignition Lock Cylinder

REMOVAL AND INSTALLATION

1. If equipped, make sure the wheels are locked in the straight-ahead position, then properly disable the SIR (air bag) system.

2. Disconnect the negative battery cable.

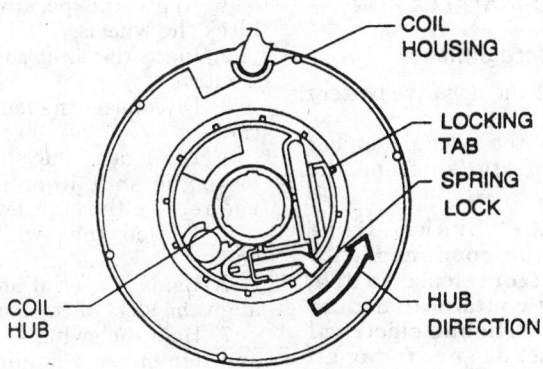

PERFORM THE FOLLOWING STEPS TO CENTER COIL ASSEMBLY

A. WHEELS STRAIGHT AHEAD.
B. REMOVE COIL ASSEMBLY.
C. HOLD COIL ASSEMBLY WITH BOTTOM UP.
D. WHILE HOLDING COIL ASSEMBLY, DEPRESS SPRING LOCK TO ROTATE HUB IN DIRECTION OF ARROW UNTIL IT STOPS.
E. THE COIL RIBBON SHOULD BE WOUND UP SNUG AGAINST CENTER HUB.
F. ROTATE COIL HUB IN OPPOSITE DIRECTION APPROXIMATELY TWO AND A HALF (2-1/2) TURNS. RELEASE SPRING LOCK BETWEEN LOCKING TABS.

85473040

Centering the SIR coil

3. Matchmark and remove the steering wheel.

4. If equipped with an SIR system, remove the SIR coil assembly retaining ring, then remove the coil assembly and allow it to hang freely from the wiring. Remove the wave washer.

NOTE: On some SIR equipped vehicles, it may be necessary to completely remove the coil and wiring from the steering column before removing the lock cylinder assembly. If so, attach a length of mechanic's wire to the coil connector at the base of the column, then carefully pull the harness and wire through the steering column towards the top. Leave the wire in position inside the column in order to pull the harness back down into position during installation.

5. Remove the turn signal switch from the column and allow it to hang from the wires (leaving them connected).

6. Remove the buzzer switch assembly. On some vehicles it may be necessary to temporarily remove the key from the lock cylinder in order to remove the buzzer. If so, the key should be reinserted before the next step.

7. Carefully remove the lock cylinder screw and the lock cylinder. If possible, use a magnetic tipped screwdriver on the screw in order to help prevent the possibility of dropping it.

------- **CAUTION** -------
If the screw is dropped upon removal, it could fall into the steering column, requiring complete disassembly in order to retrieve the screw and prevent damage.

To install:
8. Align and install the lock cylinder set.

9. Push the lock cylinder all the way in, then carefully install the retaining screw. Tighten the screw to 22 inch lbs. (2.5 Nm) on tilt columns or to 40 inch lbs. (4.5 Nm) on standard non-tilt columns.

10. If necessary, install the buzzer switch assembly.

11. Reposition and secure the turn signal switch assembly

NOTE: The coil assembly will become uncentered if the steering column is separated from the steering gear and allowed to rotate or if the centering spring is pushed down, letting the hub rotate while the coils assembly is removed from the steering column.

12. If equipped with SIR, make sure the coil is centered, then install the wave washer, followed by the coil and the retaining ring. The coil ring must be firmly seated in the shaft groove.

13. Align and install the steering wheel.

14. Make sure the ignition is **OFF**, then connect the negative battery cable.

15. Properly enable the SIR system.

Ignition Switch

REMOVAL AND INSTALLATION

1. If equipped, make sure the wheels are locked in the straight-ahead position, then properly disable the SIR (air bag) system.

2. Disconnect the negative battery cable.

3. Remove the lower column trim panel, then remove the steering column-to-instrument panel fasteners and carefully lower the column for access to the switch.

4. On some vehicles, the dimmer switch must be removed in order to remove the ignition switch. If necessary, remove the dimmer switch.

5. On 1992–96 vehicles, place the ignition switch in the **OFF-LOCK** position.

NOTE: If the lock cylinder was removed, the switch slider should be moved to the extreme right position, then 1 detent to the left.

6. Remove the ignition switch-to-steering column retainers and disengage the switch wiring, then remove the assembly.

To install:
7. Before installing the ignition switch, place it in the **OFF-LOCK** position, then make sure the lock cylinder and actuating rod are in the Locked position (1st detent from the top or 1st detent to the right of far left detent travel).

NOTE: Most replacement switches are pinned in the OFF-LOCK position for installation purposes. If so, the pins must be removed after installation or damage may occur.

8. Install the activating rod into the ignition switch and assemble the switch onto the steering column. Once the switch is properly posi-

tioned, tighten the ignition switch-to-steering column retainers to 35 inch lbs. (4.0 Nm).

NOTE: When installing the ignition switch, use only the specified screws since over length screws could impair the collapsibility of the column.

9. If removed, install the dimmer switch.
10. Raise the column into position and secure, then install any necessary trim plates.
11. Make sure the ignition is **OFF**, then connect the negative battery cable.
12. Properly enable the SIR system.

Stoplight Switch

ADJUSTMENT

1994–96 utilize an adjustable switch. All of these vehicles, the stoplight actuator component takes the form of a ribbed tubular switch which fits into an interference clip mounted to the pedal bracket. All adjusting switches should self-adjust during installation or on their first use.

1. Depress the brake pedal and press the brake light switch inward until it seats firmly against the clip.

NOTE: As the switch is being pushed into the clip, audible clicks can be heard.

2. Release the brake pedal, then pull it back against the pedal stop until the audible clicks can no longer be heard. The clicks indicate that the switch is moving into position in the clip for proper adjustment.
3. Verify that the switch operates properly when the pedal is depressed and released.

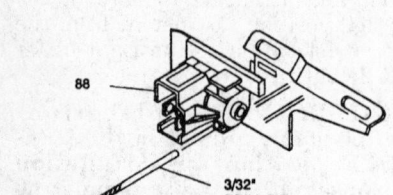

88 SWITCH ASM, DIMMER

85473036

Stoplight switch

REMOVAL AND INSTALLATION

Tubular Interference Switch

1. Disconnect the negative battery cable.
2. Disengage the vacuum and/or electrical connector(s) from the switch.

NOTE: Some brake light switches may be combined with the cruise control release switch, therefore may contain a vacuum hose or an additional electrical connector. If so, be sure to tag all hoses/wiring before disconnecting in order to assure proper installation.

3. Grasp the switch and withdraw it from the clip. If necessary the clip can be removed from the pedal bracket as well. Some clips are designed with lock tangs and may be pivoted in order to pull them from the bracket. On these, you may twist the switch and retainer clip together and withdraw them as an assembly. Other clips are also an interference fit and must be squeezed and withdrawn.

To install:

4. If removed, install the retainer to the pedal bracket.
5. With the pedal depressed so it will not interfere, insert and fully seat the switch into the clip on the pedal bracket.
6. In order to adjust the switch, grasp the pedal and pull it fully back against the stop. While pulling the pedal backwards, a clicking sound should be heard as the switch is pushed backwards in the clip until it is in the proper position. Make sure the switch plunger is seated when the pedal is released and that the plunger fully extends as the pedal is depressed. Check the switch operation and adjust again, if necessary.
7. Engage the vacuum line and/or electrical connector(s) to the switch.
8. Connect the negative battery cable.
9. Verify proper switch operation.

Neutral Safety Switch

ADJUSTMENT

This switch may also be referred to as a Transaxle Range Switch.

NOTE: Replacement switches are pinned in the Neutral position to ease installation. If the switch has been rotated or the switch is broken, adjustment may be performed as follows.

1. Apply the parking brake and block the wheels.
2. Place the shift control lever in Neutral.
3. Disconnect the negative battery cable.
4. Disengage the shift cable, remove the shift lever retaining nut and remove the shift lever.
5. Loosen the switch mounting bolts.
6. Using a special alignment tool, align the slots of the tool and switch.
7. Hold the switch in this position and tighten the retaining nuts to 18 ft. lbs. (25 Nm).
8. Install shift lever in this position. Install the lever retaining nut.
9. Connect the shift cable to the shift lever.
10. Connect the negative battery cable.
11. Verify proper switch operation.

REMOVAL AND INSTALLATION

1. Apply the parking brake and block the wheels.
2. Place the shift control lever in Neutral.
3. Disconnect the negative battery cable.
4. Disengage the switch electrical connector.
5. Remove the shift lever retaining nut and remove the shift lever.
6. Remove the retaining nuts, then remove the switch.
7. If installing a new switch:
   a. Verify that the shift control shaft is in Neutral.
   b. Install the new switch to the shift shaft and loosely install the retaining bolts.
   c. Align the flats of the shift shafts to the flats in the switch. Torque the switch retaining bolts to 18 inch lbs. (25 Nm).

NOTE: Replacement switches are pinned in the Neutral position to ease installation. If the switch has been rotated or the switch is broken, install the switch using the "old switch" installation and adjustment procedure.

8. If installing an old switch, install and adjust the switch to assure proper operation:
   a. Remove the shift lever retaining nut and remove the shift lever.
   b. Loosen the switch mounting bolts.
   c. Using a special alignment tool, align the slots of the tool and switch.

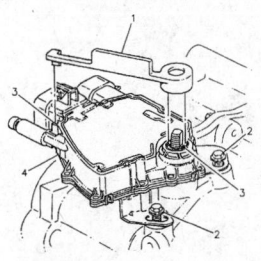

1 ALIGNMENT TOOL
2 BOLT/SCREW
3 ALIGNMENT SLOTS
4 TRANSAXLE RANGE SWITCH

7921A009

Aligning the neutral safety switch

d. Hold the switch in this position and tighten the retaining nuts to 18 ft. lbs. (25 Nm).

9. Install shift lever in this position. Install the lever retaining nut.

10. Connect the electrical connector to the switch and the shift cable.

11. Connect the negative battery cable.

12. Verify proper switch operation.

Powertrain Control Module Electronic Control Module

REMOVAL AND INSTALLATION

The Power Train Control Module (PCM) is located in the passenger compartment behind the right side cowl/kick panel. Earlier units were called an Engine Control Module (ECM).

1. Disconnect the negative battery cable.

2. Place the ignition in the **OFF** position.

3. Remove the right hand hush panel.

4. Remove the connectors from the PCM.

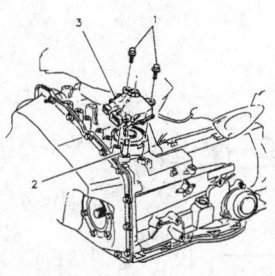

1 BOLT/SCREW
2 SHAFT, TRANSAXLE SHIFT
3 SWITCH, TRANSAXLE RANGE

7921A010

Removing or installing the neutral safety switch

5. Remove the PCM mounting hardware.

6. Remove the PCM from the passenger compartment.

7. Remove the PCM access cover.

8. Remove the PROM from the PCM using two fingers, push both retaining clips back away from the PROM. At the same time, grasp it at both ends and lift it up out of the socket.

NOTE: Replacement PCM is supplied without a PROM, so care should be used when removing it from the defective PCM so it can be reused in the new PCM.

To install:

9. Align the small notches in the PROM socket and gently press down on the ends of the PROM until the clips are against the sides of the PROM. Press inward on the clips until they snap into place.

10. Install the PCM into the passenger compartment.

11. Install the connections to the PCM.

12. Install the right side hush panel.

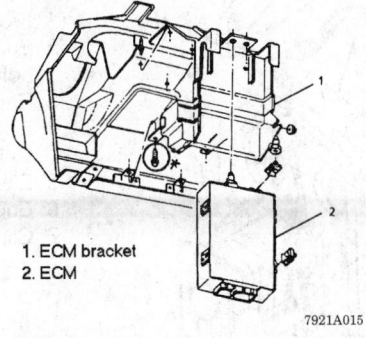

1. ECM bracket
2. ECM

7921A015

Removal and installation of the PCM/ECM — 1994–95

7921A016

Removal and installation of the PCM/ECM — 1996

ENGINE COOLING

Radiator

REMOVAL AND INSTALLATION

1. Disconnect the negative battery cable.

2. Drain the coolant from the radiator.

3. Disconnect the engine forward strut bracket at the radiator, loosen the bolt at the other end and swing the strut rearward.

4. Disconnect the forward lamp harness from the fan frame and unplug the fan connector.

5. Remove the fan attaching bolts, then remove the fan and frame assembly.

6. Scribe the latch location then remove the hood latch from the radiator support.

7. Disconnect the coolant hoses from the radiator and the coolant recovery tank hose from the radiator neck.

8. Disconnect and plug the transaxle oil cooler lines.

9. Remove the radiator-to-support attaching bolts and clamps, then remove the radiator.

10. Installation is the reverse of removal.

Electric Cooling Fan

REMOVAL AND INSTALLATION

1. Disconnect the negative battery cable.

2. If necessary, disconnect the engine forward strut bracket from the radiator frame and swing it rearward.

3. Disconnect the forward lamp harness from the fan frame.

4. Remove the fan attaching bolts.

5. Disconnect the fan wiring.

6. Remove the fan and frame assembly.

7. Installation is the reverse of the removal procedure.

Heater Core

REMOVAL AND INSTALLATION

1. Disconnect the negative battery cable and drain the engine cooling system.

2. Remove the 2 screws attaching the right hand windshield wiper

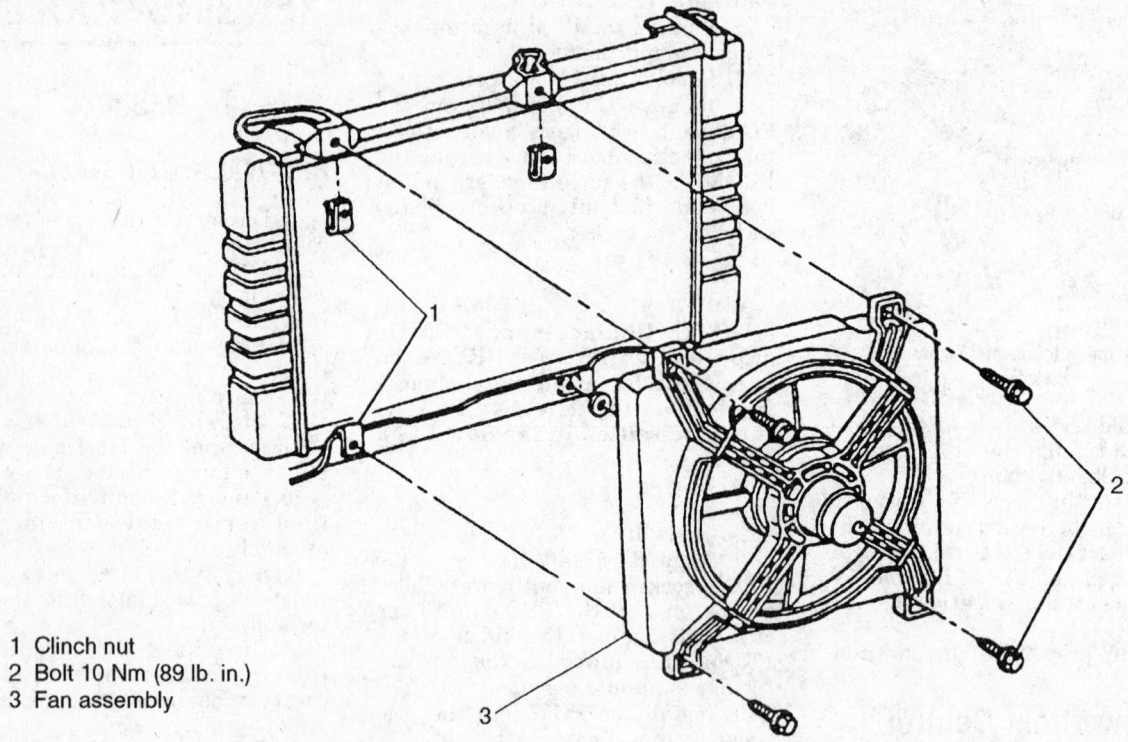

1 Clinch nut
2 Bolt 10 Nm (89 lb. in.)
3 Fan assembly

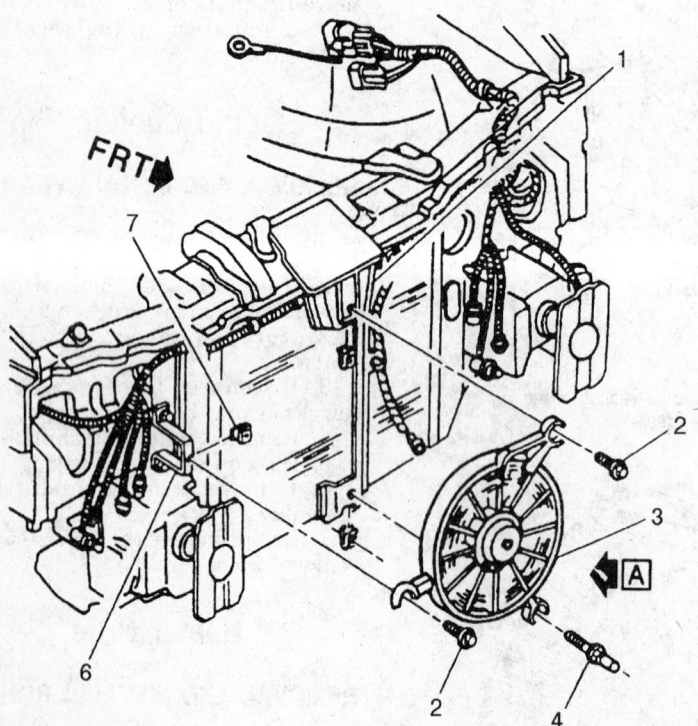

1 Front end lower support
2 Engine auxiliary cooling fan
  bolt, 10 Nm (89 lb. in.)
3 Engine auxiliary cooling fan
4 Engine auxiliary cooling fan stud,
  10 Nm (89 lb. in.)
5 Electrical fan connector
6 Engine auxiliary cooling fan bracket
7 Clinch nut

FRT

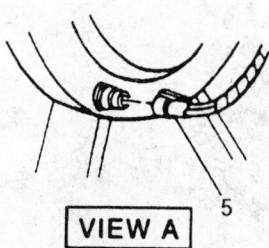

VIEW A

344554

Removing or installing the electric cooling fan

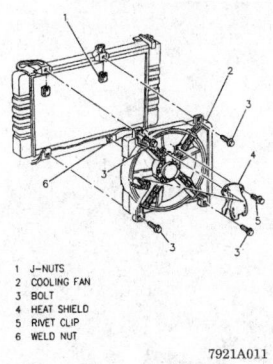

1 J-NUTS
2 COOLING FAN
3 BOLT
4 HEAT SHIELD
5 RIVET CLIP
6 WELD NUT

7921A011

Removing or installing the electric cooling fan

transmission arm to the wiper frame. Move the wiper transmission for access to the heater hoses.

3. Loosen the clamps, then disconnect the outlet and inlet hoses from the heater core tubes.

4. Remove the right hand instrument panel sound insulator panel push-in retainers. Remove the screw and nut from the stud.

5. Carefully pull the sound insulator panel away from the instrument panel, then remove the courtesy lamp from the panel and remove the panel.

6. Open the instrument panel compartment (glove box) door. Remove the 3 screws from the bottom of the box insert, then remove the upper box screws with the 2 bumpers. Disengage the wiring harness connector from the box lamp and switch assembly. Remove the glove box and door from the instrument panel.

7. Remove the 2 screws retaining the instrument panel lower extension housing to the support bracket.

8. Remove the right instrument panel outer support-to-panel lower trim pad attaching bolt, then place a wedge between the outer support and the lower trim pad.

9. Disengage the heater control vacuum harness from the vacuum/electric solenoid assembly. Remove the 2 solenoid retaining screws, then remove the solenoid from the heater core cover.

10. Remove the 6 screws and 2 clips from the heater core cover, then remove the cover and seal from the heater case.

11. Remove the retaining screw and the lower heater core bracket.

12. Remove the left retaining screw from the upper heater core bracket, then loosen the right screw. Reposition the upper bracket to allow for heater core removal.

13. Carefully remove the heater core inlet/outlet tubes from the

tube/case seal, then remove the heater core from the case.

To install:

14. Carefully install the heater core to the case while guiding the tubes through the seal.

15. Position the upper bracket and secure using the retaining bolts.

16. Install and secure the lower bracket.

17. Install the heater core cover and seal and secure using the screws and clips.

18. Install and secure the vacuum/electric solenoid, then connect the vacuum harness.

19. Remove the wedge, then install the outer support to the lower trim pad.

20. Install the screws retaining the instrument panel lower extension housing to the lower extension support brackets.

21. Install the glove box.

22. Install the instrument panel sound insulator.

23. Connect the heater hoses to the core tubes and secure using the retaining clamps.

24. Reposition the windshield wiper transmission and attach it to the wiper frame.

25. Connect the negative battery cable and properly refill the engine cooling system.

Water Pump

REMOVAL AND INSTALLATION

3.1L (VIN D) and 3.8L (VIN L) Engines

1. Disconnect the negative battery cable, then drain the engine cooling system.

2. On the 3.1L (VIN D) engine, disconnect the heater hose, then remove the serpentine drive belt shield.

3. For the 3.8L (VIN L) engine, loosen but do not remove the water pump pulley bolts.

4. Remove the serpentine drive belt. On the 3.1L (VIN D) engine, a ³⁄₈ in. drive breaker bar may be used to pivot the belt tensioner.

5. Remove the water pump pulley.

6. Loosen the clamps and disconnect any remaining hoses from the water pump, as applicable.

7. Remove the water pump retaining bolts. Remove the water pump assembly from the engine.

NOTE: On some engines, then pump retaining bolts will vary in size and thread. Be sure to note

the positioning of all bolts during removal to assure proper installation.

8. Installation is the reverse of the removal procedure. Use a new gasket and carefully tighten the bolts.

Thermostat

REMOVAL AND INSTALLATION

3.1L (VIN D) and 3.8L (VIN L) Engines

1. Disconnect the negative battery cable.

2. Drain the radiator until the level is below the thermostat (below the level of the intake manifold, cylinder head or other thermostat housing, as applicable).

3. If necessary, disconnect the hose from the thermostat outlet. On most applications, the hose may be left attached to the outlet when it is removed, however on some models disconnecting the hose may make outlet removal easier.

4. Loosen and remove the thermostat outlet retainers, then remove the outlet from the housing. Note the orientation of the thermostat in the housing for installation purposes.

To install:

5. Carefully clean the all traces of the old gasket and/or sealer from the housing and outlet.

6. Install the thermostat to the housing, oriented as noted during removal, then position a new gasket (if used).

7. If sealant was used on the old component or if no gasket is provided, place a ⅛ in. bead of RTV sealant in the groove on the water outlet sealing surface, then install the outlet while the sealer is still wet.

8. Install and tighten the outlet retainers.

9. If removed, connect the hose to the outlet.

10. Connect the negative battery cable.

11. Properly refill and bleed the engine cooling system, then check for leaks.

Cooling System Bleeding

1. To bleed the system, start with the system cool, the radiator cap off and the radiator filled to about an inch below the filler neck.

2. Start the engine and run it at slightly above normal idle speed. If air bubbles appear and the coolant level drops, fill the system with a

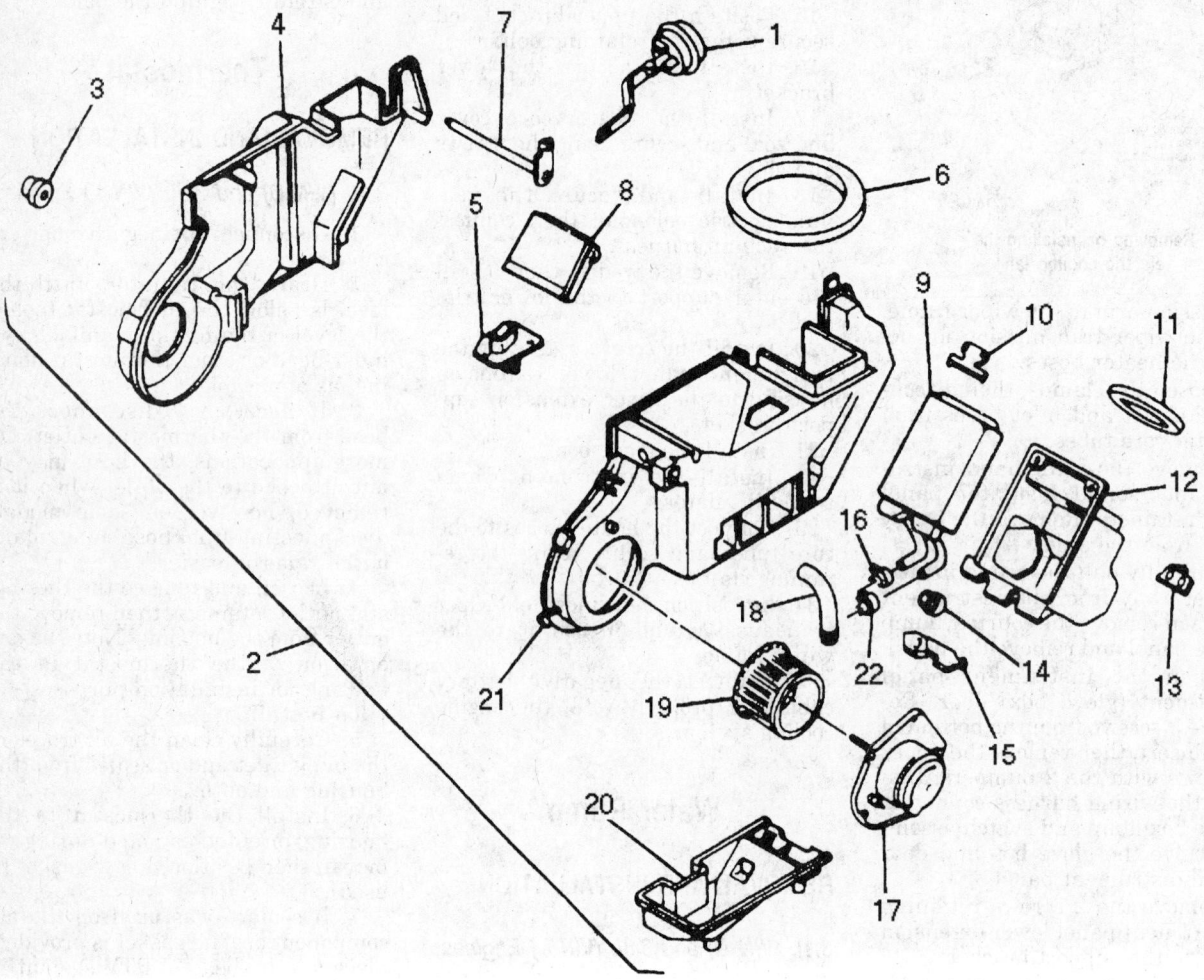

1. Rear temperature valve vacuum actuator
2. Auxiliary heater blower module
3. Tube inlet seal
4. Heater case
5. Auxiliary heater blower motor resistor
6. Outlet gasket
7. Rear temperature valve shaft(w/lever)
8. Rear temperature valve
9. Heater core
10. Heater core clip
11. Defroster outlet gasket

12. Heater outlet case
13. Heater and blower mounting bracket
14. Heater core tube seal
15. Heater core clamp
16. Heater core outlet pipe clamp
17. Auxiliary heater blower motor
18. Blower motor cooling tube
19. Auxiliary heater blower motor impeller
20. Drain case
21. Auxiliary heater blower cover case
22. Heater core inlet pipe clamp

7921A012

Exploded view of the heater control module assembly

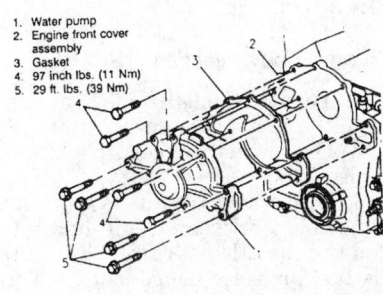

1. Water pump
2. Gasket
3. 89 inch lbs. (10 Nm)
4. Locator – Must be vertical

85473052

Water pump assembly mounting — 3.1L (VIN D) engine, 3.4L (VIN E) is similar

1. Water pump
2. Engine front cover assembly
3. Gasket
4. 97 inch lbs. (11 Nm)
5. 29 ft. lbs. (39 Nm)

85473053

Water pump assembly mounting — 3.8L (VIN L) engine

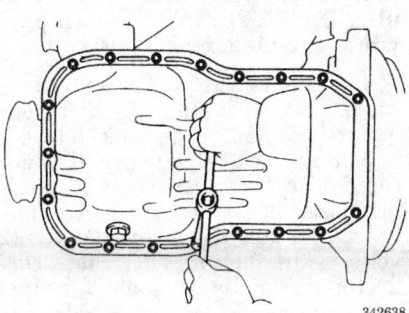

342638

Thermostat mounting — 3.4L (VIN E) engine

50/50 antifreeze/water mixture to bring the level back to the proper level.

3. Run the engine until the thermostat opens and coolant flow is visible.

4. At this point, air is often expelled and the level may drop again. Keep refilling the system until the level is near the top of the radiator and remains constant.

5. Fill the radiator to the filler neck, then install the radiator filler cap.

6. Check and make sure the coolant reservoir is filled to the correct level.

FUEL SYSTEM

Fuel System Service Precaution

When working with the fuel system certain precautions should be taken; always work in a well ventilated area, keep a dry chemical (Class B) fire extinguisher near the work area. Always disconnect the negative battery cable and do not make any repairs to the fuel system until all the necessary steps for repair have been reviewed.

Relieving Fuel System Pressure

Before loosening or disconnecting any fuel fitting or system component, always relieve the fuel system pressure in order to help prevent the danger of fire or injury.

THROTTLE BODY FUEL INJECTION SYSTEMS

3.1L (VIN D) Engine

Unlike most TBI engines, the TBI system used in the 3.1L (VIN D) engine, utilize an automatic pressure bleed down feature. But, some fuel pressure related steps should still be taken to assure safer working conditions.

1. Disconnect the negative battery cable to prevent fuel spillage should the ignition key accidentally be turned **ON** with a fuel fitting disconnected.

2. Loosen fuel filler cap to relieve fuel tank pressure.

3. The internal constant bleed feature of the Model 220 TBI unit relieves fuel pump system pressure when the engine is turned **OFF**. Therefore, no further action is required.

NOTE: Allow the engine to set for 5–10 minutes; this will allow the orifice (in the fuel system) to bleed off the pressure.

4. When fuel service is finished, tighten the fuel filler cap and connect the negative battery cable.

MULTI-PORT FUEL INJECTION SYSTEMS

3.4L (VIN E) and 3.8L (VIN L) Engines

The MFI fuel systems used on GM vehicles all operate under high fuel pressures. It is very important that the pressure be properly relieved prior to servicing the system or any of its components.

A schrader valve is provided on these fuel systems in order to conveniently test or release the system pressure. A fuel pressure gauge and adapter will be necessary to connect the gauge to the fitting. Most of the MFI systems utilize a service valve on 1 end of the fuel rail assembly.

1. Disconnect the negative battery cable to assure the prevention of fuel spillage if the ignition switch is accidentally turned **ON** while a fitting is still disconnected.

2. Loosen the fuel filler cap to release the fuel tank pressure.

3. Make sure the release valve on the fuel gauge is closed, then connect the fuel gauge to the pressure fitting located on the inlet fuel pipe fitting.

NOTE: When connecting the gauge to the fitting, be sure to wrap a rag around the fitting to avoid spillage. After repairs, place the rag in an approved container.

4. Install the bleed hose portion of the fuel gauge assembly into an approved container, then open the gauge release valve and bleed the fuel pressure from the system.

5. When the gauge is removed, be sure to open the bleed valve and drain all fuel from the gauge assembly.

6. When fuel service is finished, tighten the fuel filler cap and connect the negative battery cable.

Fuel Tank

REMOVAL AND INSTALLATION

NOTE: Draining the fuel tank through the filler neck can only be performed after knocking the filler neck check ball into the tank. This will require the fuel pump/sending unit to be removed in order to retrieve and reposition the check ball.

1. Properly relieve the fuel system pressure, then disconnect the negative battery cable.

2. Using an approved pump, drain the fuel from the tank into a suitable container:

   a. Block the front wheels, then raise and support the rear of the vehicle safely so the rear bumper is 28 in. (71cm) above the ground.

   b. Loosen the fuel filler neck tube clamp at the tank, then wrap a shop towel around the neck tube and slowly disconnect the tube from the tank.

   c. Using a socket extension that is approximately 18 in. long, drive the filler neck check ball into the tank.

   d. Use a hand operated pump to drain the fuel from the tank into a suitable container.

3. Raise and support the vehicle safely and level for access to the vehicle underside.

4. For vehicles 1994, unbolt the rear exhaust system hangers from the converter to the tail pipe, then reposition the heat shield for access to the tank straps:

   a. Remove the tail pipe hanger attaching bolt.

   b. Remove the muffler hanger attaching bolts and the muffler hanger.

   c. Loosen the converter hanger attaching nuts.

   d. Remove the heat shield attaching screws. Note the position of the heat shield for installation purposes.

   e. Support the exhaust and move the heat shield to gain access to the right side fuel tank retaining strap attaching bolts.

5. Remove the in-line fuel filter body clips or attaching screws, as applicable.

6. Disengage the quick-connect fuel fittings both at the inlet side of the in-line fuel filter and at the fuel return line fitting located near the filter: Grasp both ends of a fuel line and twist ¼ turn in each direction to loosen any dirt in the quick-connect fitting. Squeeze the plastic tabs of the male ends of the connectors and pull the connections apart.

7. Disengage the fuel meter/sender assembly electrical wiring, then tag and disconnect the fuel vapor/vent hoses at the tank.

8. With the help of an assistant, properly support the fuel tank, then unbolt and remove the retaining strap.

9. Carefully lower the fuel tank. Remove the fuel sender and seal ring using J–35731 or equivalent tool, in order to recover and install the filler neck check ball.

To install:

10. Remove the filler neck check ball from the tank and check for cracks or holes. If damaged, the filler neck extension pipe (tube) must be replaced.

11. Pop the filler neck check ball back into position in the filler neck tube.

12. Install the fuel sender and seal ring using tool J–35731 or equivalent tool.

13. With the aid of an assistant, raise the tank into position and secure using the retaining straps. Tighten the strap bolts to 18 ft. lbs. (24 Nm).

14. Connect the fuel vapor/vent hoses and electrical connections at the sender.

15. Apply a few drops of clean engine oil to the fuel feed/return line quick connect fittings, then push connectors together until a snap is heard. Pull on both lines and verify they are secure.

16. Secure the in-line fuel filter using the attaching screws or 2 new body clips, as applicable.

17. On vehicles 1994, reposition and secure the exhaust heat shield, the tail pipe, muffler and converter hangers. Make sure the heat shield is properly positioned as noted during removal.

18. Lower the vehicle an fill the tank.

19. Connect the negative battery cable.

20. Turn the ignition key **ON** for 2 seconds and then **OFF** for 10 seconds. Cycle the ignition as necessary to pressurize the fuel system; then check the system for leaks.

Fuel Filter

REMOVAL AND INSTALLATION

3.1L (VIN D), 3.4L (VIN E) and 3.8L (VIN L) Engines

1. Disconnect the negative battery cable.

2. Remove the fuel line connections from the filter. If equipped with quick connect fittings, use tool J–37088A or equivalent to separate the lines.

NOTE: Before disengaging quick-connect fittings, always twist each side of the connection ¼ turn (in opposite directions) in order to loosen any dirt, then use compressed air (while wearing safety glasses) to blow the dirt free of the fittings.

3. Remove the filter mounting clamp bolt and remove the filter.

4. Installation is the reverse of the removal procedure. Before installing any quick-connect fittings, apply a few drops of clean engine oil to the male connector to assure proper seal and prevent a possible leak.

Electric Fuel Pump

PRESSURE TESTING

Throttle Body Injection (TBI)

1. Turn the engine **OFF** and relieve the fuel system pressure.

2. Disconnect the negative battery cable.

3. Disconnect the fuel supply line somewhere between the in-line filter and the throttle body. Install a suitable fuel pressure gauge using a T fitting in the fuel line.

NOTE: On some vehicles it may be necessary to remove the air cleaner for easier access to the fuel supply line. If so, be sure to plug the THERMAC vacuum port while the cleaner is removed.

4. Connect the negative battery cable. Verify there is a sufficient quantity of fuel in the tank.

5. Run the fuel pump and check the fuel pressure. There are 3 possible methods to run the fuel pump. Start and run the engine, apply 12 volts to the fuel pressure check connector (normally a single terminal red connector located on the driver's side of the engine compartment) using a fused jumper wire or cycle the ignition without starting the engine. If cycling the ignition is chosen, each time the ignition is turned **ON**, the pump will run for 2 seconds.

6. Observe the fuel pressure with the pump running, it should be 9–13 psi. (62–90 kPa). The system pressure must be checked with the pump running because on most applications the pressure will drop immediately after the pump shuts off.

7. If the fuel pressure reading is not as specified, inspect the fuel pump for proper operation, the lines and filter for kinks or clogging.

8. On testing is complete, depressurize the fuel system, then remove the fuel pressure gauge and adapter. Reconnect the fuel line, start the engine and check for fuel leaks.

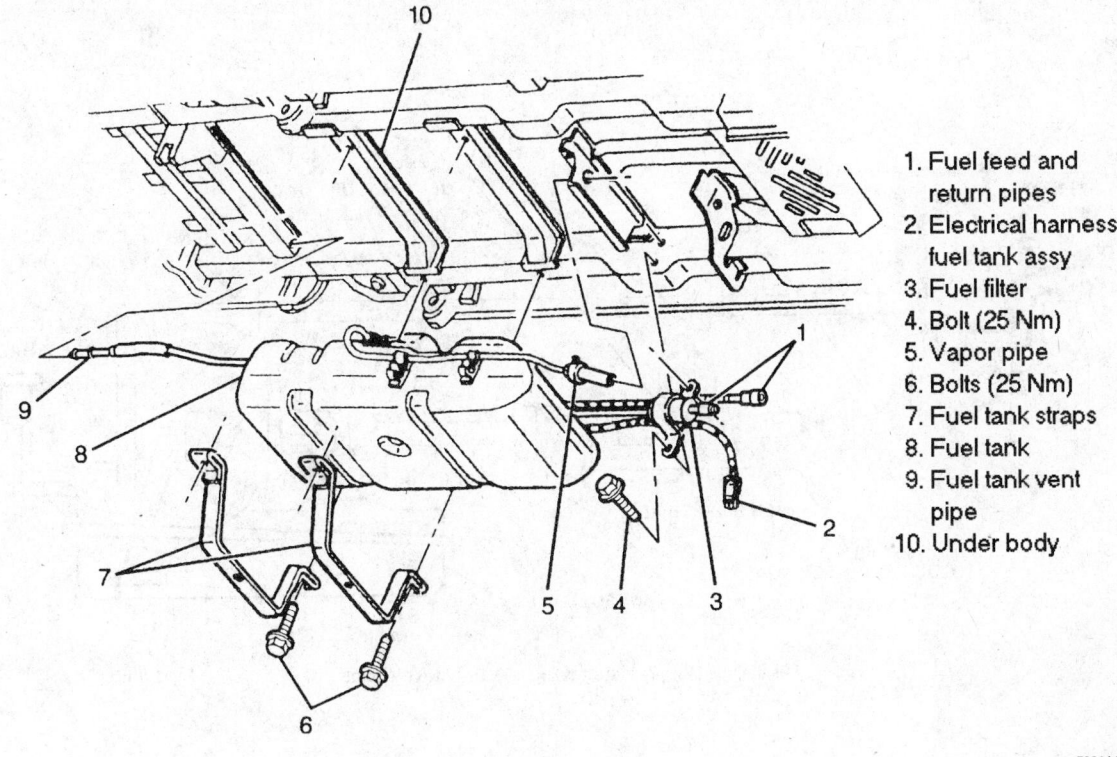

1. Fuel feed and
   return pipes
2. Electrical harness
   fuel tank assy
3. Fuel filter
4. Bolt (25 Nm)
5. Vapor pipe
6. Bolts (25 Nm)
7. Fuel tank straps
8. Fuel tank
9. Fuel tank vent
   pipe
10. Under body

7921A013

Removal or installation of the fuel tank

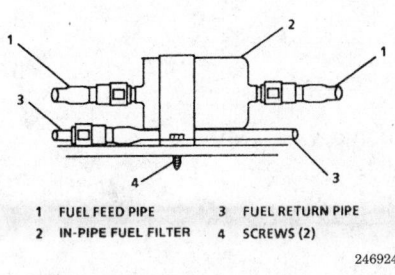

| 1 | FUEL FEED PIPE | 3 | FUEL RETURN PIPE |
| 2 | IN-PIPE FUEL FILTER | 4 | SCREWS (2) |

246924

Removal or installation of the fuel filter

Multi-Port Fuel Injection

3.4L (VIN E) and 3.8L (VIN L) Engines

1. Properly relieve the fuel system pressure.
2. Leave the gauge attached to the pressure fitting on the fuel inlet pipe.
3. If disconnected during the fuel pressure relief procedure, reconnect the negative battery terminal.
4. Verify there is a sufficient quantity of fuel in the tank.
5. Turn the ignition switch **ON**, the fuel pump should run for 2 seconds and turn **OFF**. If necessary, cycle the ignition **OFF** for 10 seconds and then **ON** again in order to build maximum system pressure. The pressure gauge reading should be approximately 41–47 psi. (284–325 kPa).
6. Turn the ignition switch **OFF** and observe the pressure gauge, the reading may vary slightly, then should hold and not leak down.
7. Start the vehicle and observe the gauge, the pressure should be 3–10 psi (21–69 kPa) lower because of vacuum applied to the regulator.
8. If not as specified, inspect the pump, regulator, filter and lines for proper operation, kinks or clogging.

REMOVAL AND INSTALLATION

1. Properly relieve the fuel system pressure.
2. Disconnect the negative battery cable.
3. Drain and remove the fuel tank from the vehicle.
4. Using a suitable spanner wrench, turn the fuel pump/sending unit assembly locking ring (located on top of the fuel tank) counterclockwise, then carefully lift the assembly from the tank and remove the pump from the fuel lever sending device.
5. Installation is the reverse of the removal procedure. The fuel pump/sending unit assembly O-ring should be replaced whenever the tank is removed.

Fuel Injection

IDLE SPEED ADJUSTMENT

Idle speed and mixture adjustments on these engines are controlled by the computer modules, regulated through the IAC valve and fuel injector(s), respectively. No periodic check or adjustments are necessary for these systems.

However, on the 1994 3.1L (VIN D) TBI engine, the minimum idle speed may be adjusted if the throttle body has been replaced AND an incorrect idle speed cannot be obtained. These adjustments should not be performed unless all other possible causes (vacuum leaks, bad fuel pressure, faulty wiring or components) have already been eliminated.

1994 3.1L (VIN D) TBI Engine

1. Remove the air cleaner, adapter and gaskets. Discard the gaskets. Plug any vacuum line ports, as necessary.
2. Leave the Idle Air Control (IAC) valve connected and with the engine **OFF**, ground the diagnostic terminal (ALDL connector).
3. Turn the ignition switch to the **ON** position, do not start the engine. Wait for at least 30 seconds; this al-

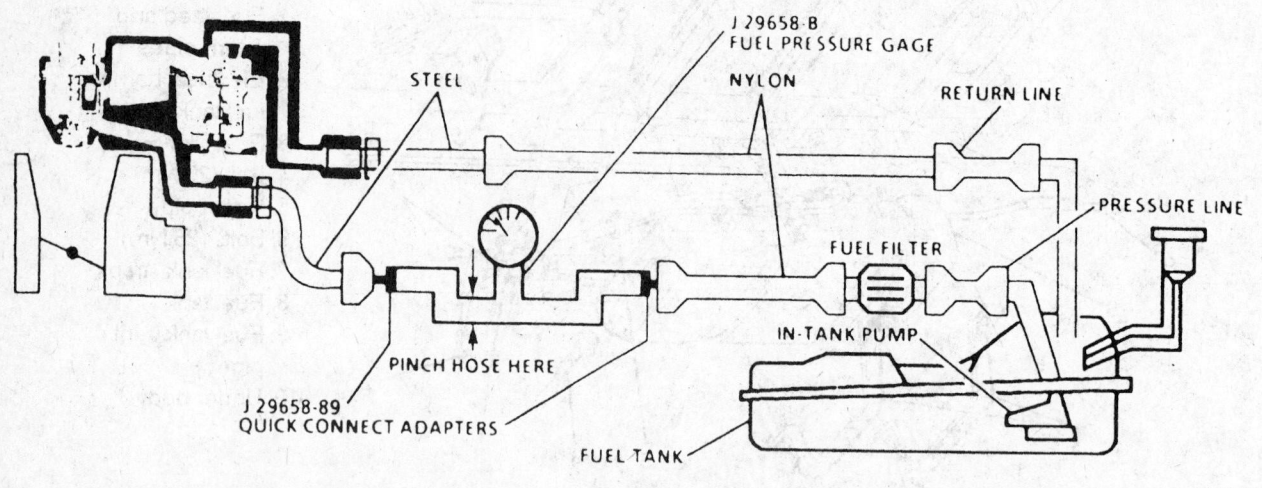

STEEL

J 29658-B
FUEL PRESSURE GAGE

NYLON

RETURN LINE

PRESSURE LINE

FUEL FILTER

PINCH HOSE HERE

IN-TANK PUMP

J 29658-89
QUICK CONNECT ADAPTERS

FUEL TANK

85473058

TBI fuel system pressure testing — 3.1L (VIN D) engine

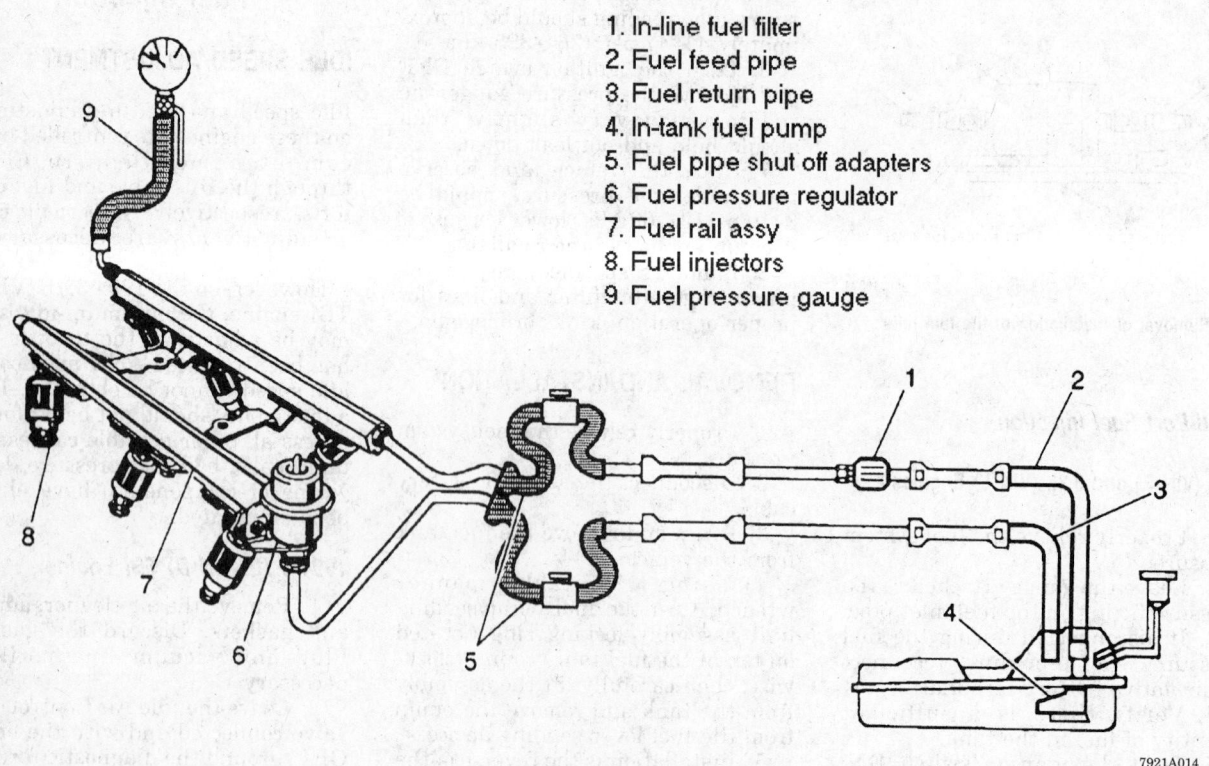

1. In-line fuel filter
2. Fuel feed pipe
3. Fuel return pipe
4. In-tank fuel pump
5. Fuel pipe shut off adapters
6. Fuel pressure regulator
7. Fuel rail assy
8. Fuel injectors
9. Fuel pressure gauge

7921A014

MFI system pressure testing — 3.4L (VIN E) engine

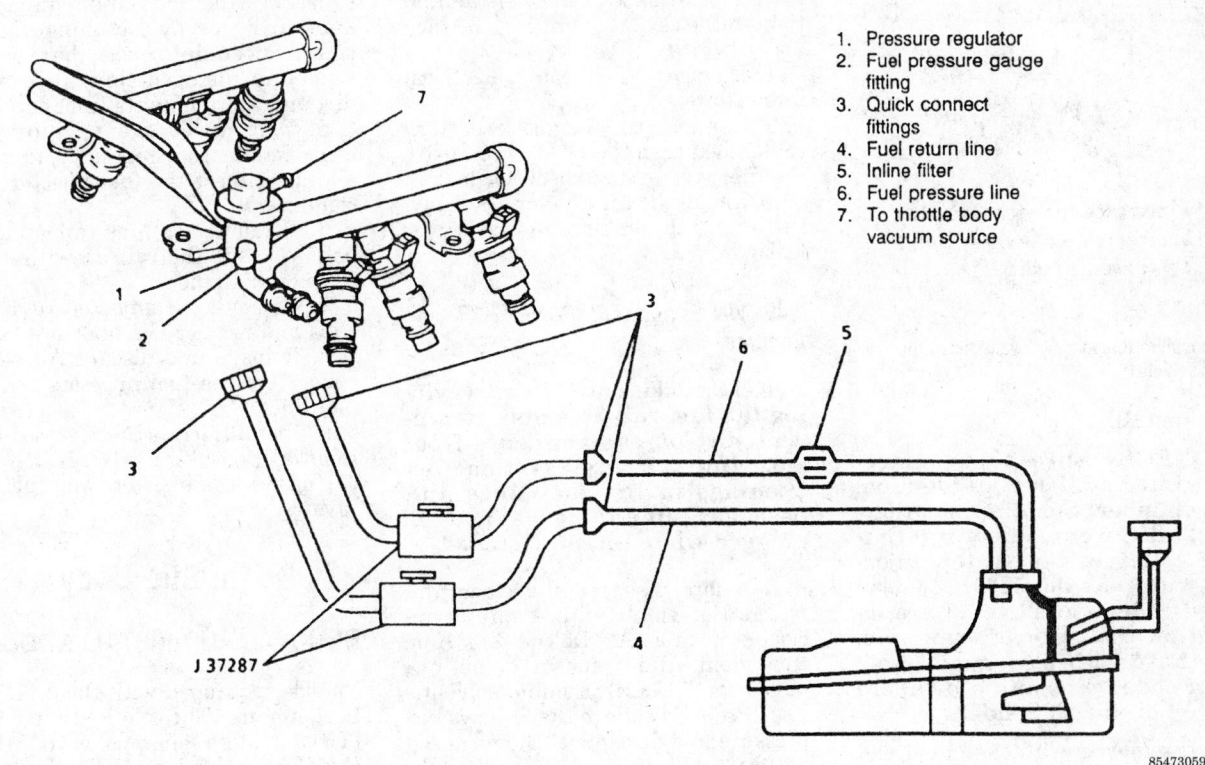

1. Pressure regulator
2. Fuel pressure gauge fitting
3. Quick connect fittings
4. Fuel return line
5. Inline filter
6. Fuel pressure line
7. To throttle body vacuum source

85473059

MFI system pressure testing — 3.8L (VIN L) engine

lows the IAC valve pintle to extend and seat in the throttle body.

4. With the ignition switch still in the **ON** position, disconnect IAC electrical connector.

5. Remove the ground from the diagnostic terminal and start the engine. Let the engine reach normal operating temperature.

6. Apply the parking brake and block the drive wheels. Remove the plug from the idle stop screw by piercing it with a suitable tool and then applying leverage to the tool to lift the plug out.

7. With the engine in the proper shift selector range, adjust the idle stop screw to set the minimum idle to specification.

8. Turn the ignition **OFF** and reconnect the IAC valve connector. Unplug any plugged vacuum line ports and install the air cleaner, adapter and new gaskets.

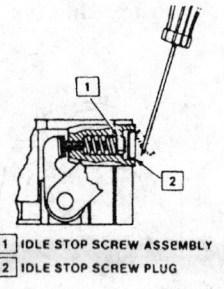

1 IDLE STOP SCREW ASSEMBLY
2 IDLE STOP SCREW PLUG

85473064

Removing idle stop screw plug for adjustment — TBI engines

Fuel Injector

REMOVAL AND INSTALLATION

TBI Engines

— **CAUTION** —
When removing the injector(s), be careful not to damage the electrical connector pins (on top of the injector), the injector fuel filter and the nozzle. The fuel injector is serviced as a complete assembly

ONLY, it is an electrical component and should not be immersed in any kind of cleaner.

1. Properly relieve the fuel system pressure, then disconnect the negative battery cable.

2. Remove the air cleaner assembly.

3. At the injector connector(s), squeeze the 2 tabs together and pull straight up to disengage connector from the injector.

4. Loosen the fuel meter cover retaining screws, then remove the cover from the fuel meter body, but leave the cover gasket in place.

5. Using a small pry bar and a round fulcrum, carefully pry the injector until it is free, then remove the injector from the fuel meter body.

6. Remove the small O-ring from the nozzle end of the injector. If equipped and removal is necessary, carefully rotate the injector's fuel filter back and forth to remove it from the base of the injector.

7. If applicable, remove and discard the fuel meter cover gasket.

8. Remove the large O-ring and back-up washer, if equipped, from the top of the counterbore of the fuel meter body injector cavity.

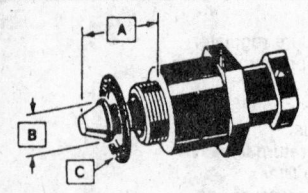

A | DISTANCE OF PINTLE EXTENSION
B | DIAMETER AND SHAPE OF PINTLE
C | IAC VALVE GASKET

85473065

A common threaded IAC valve used on GM vehicles

To install:

NOTE: Be sure the replacement injectors have an identical part number. Some applications use 2 different injectors that have different flow rates. Injectors with part no. 5235134 (color coded orange and green) are located on the throttle lever side, and those with part no. 5235342 (color coded pink and brown) are located on the TPS side.

9. If removed, with the larger end of the filter facing the injector (so the filter covers the raised rib of the injector base) install the filter by twisting it into position on the injector.

10. Lubricate the new O-rings with clean automatic transmission fluid, then install the small O-ring on the nozzle end of the injector. Be sure the O-ring is pressed up against the injector or injector filter (as applicable).

11. Install the steel backup washer, if equipped, in the top counterbore of the fuel meter body's injector cavity, then install the new large O-ring directly over the backup washer. Make sure the O-ring is properly seated in the cavity and is flush with the top of the fuel meter body casting surface.

NOTE: If the backup washer and large O-ring are not properly installed before the fuel injector, a fuel leak will likely result.

12. For the model 220 TBI unit, install the fuel injector into the cavity by aligning the raised lug on the injector base with the cast notch in the fuel meter body cavity. Once the injector is aligned, carefully push down on the injector by hand until it is fully seated in the cavity. When properly aligned and installed, the injector terminals will be approximately parallel to the throttle shaft.

13. Position a new fuel meter cover gasket, then install the cover to the body, making sure the gasket remains in position. Using a suitable

thread locking compound, install and tighten the cover retainers to 30 inch lbs. (4 Nm).

14. Engage the injector electrical connector(s).

15. Connect the negative battery cable, then turn the **ON** to pressurize the fuel system and check for leaks.

16. Install the air cleaner assembly, then start the engine and check for leaks.

3.4L (VIN E) and 3.8L (VIN L) MFI Engines

NOTE: Take care when servicing the fuel rail assembly to prevent dirt or contaminants from entering the fuel system. All openings in the fuel lines and passages should be capped or plugged while disconnected.

1. Before disassembly, clean the fuel rail assembly with a spray type cleaner such as AC Delco® X-30A or equivalent. After removal, do not immerse the fuel rail in liquid solvent.

2. Properly relieve the fuel system pressure and disconnect the negative battery cable.

3. Remove the upper intake plenum, if equipped with turbocharger.

4. Remove the fuel supply and return lines from the fuel rail by

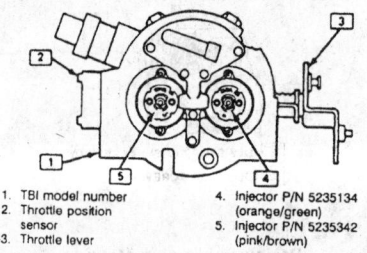

1. TBI model number
2. Throttle position sensor
3. Throttle lever
4. Injector P/N 5235134 (orange/green)
5. Injector P/N 5235342 (pink/brown)

85473069

Fuel injector location — Model 220 TBI unit

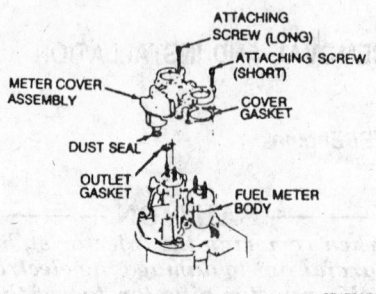

ATTACHING SCREW (LONG)
ATTACHING SCREW (SHORT)
METER COVER ASSEMBLY
COVER GASKET
DUST SEAL
OUTLET GASKET
FUEL METER BODY

85473070

Exploded view of the fuel meter cover mounting — Model 220 TBI unit

squeezing the tabs and pulling the lines apart or by loosening the fittings, depending upon the application. Cap all open lines to prevent dirt and contaminants from entering.

5. Disconnect the vacuum line from the pressure regulator, then tag and disconnect the fuel injector wire connectors.

6. Remove the fuel rail securing bolts and lift the rail up with equal force on both sides.

7. Remove the injector retaining clips and remove the injectors.

8. Remove and discard All old O-rings from the fuel injectors and the fuel lines.

9. Installation is the reverse of the removal procedure. Always use new O-rings on the injectors and fuel line fittings.

Throttle Body

REMOVAL AND INSTALLATION

Vehicles equipped with the 3.1L VIN D engine use Throttle Body Injection (TBI). Unlike engines with Multi-Port Fuel Injection (MFI), the throttle body contains the fuel injectors. Note that the TBI unit does not need to be removed from the intake manifold to service the fuel injectors.

1. Disconnect the negative battery cable.

———— CAUTION ————
Fuel injection systems remain under pressure, even after the engine has been turned OFF. The fuel system pressure must be relieved before disconnecting any fuel lines. Failure to do so may result in fire and/or personal injury.

2. Relieve the fuel system pressure using the following procedure.

 a. Verify that the negative battery cable has been disconnected. This avoids possible fuel discharge if an accidental attempt is made is start the engine.

 b. Loosen the fuel filler cap to relieve tank vapor pressure. Leave the cap loose.

 c. The internal constant bleed feature of the TBI Model 220 relieves fuel pump pressure when the engine is OFF. No further pressure relief is required.

3. Remove the air cleaner assembly.

4. Disconnect the electrical connectors to the Idle Air Control (IAC) valve, Throttle Position (TP) sensor and fuel injectors.

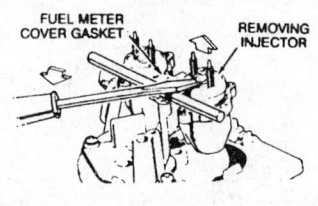

85473071

Removing the fuel injector — Model 220 TBI unit

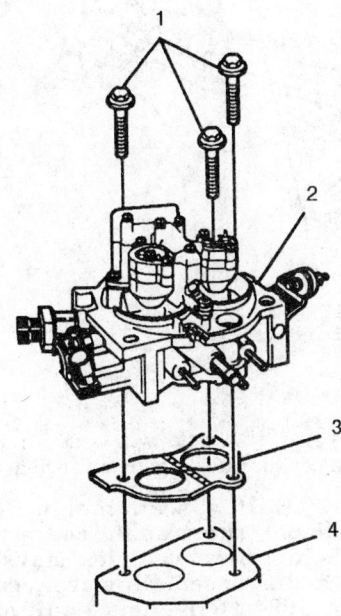

1. Bolt
2. TBI unit
3. Gasket (must be installed with stripe facing up)
4. Engine intake manifold

359227

Removal and installation of the Throttle Body — 1994–95 3.1L (VIN D) engine

5. Disconnect the fuel injector wiring harness.

6. Disconnect the throttle cable, transaxle control cable and cruise control cable, as equipped.

7. Remove the cable support bracket attaching screws and bracket.

8. Disconnect the vacuum hoses and crankcase ventilation valve hose.

9. Disconnect the fuel feed and return pipe. Use a backup wrench on the TBI fuel line nuts to prevent them from turning.

10. Remove the fuel pipe O-rings and discard.

11. Remove the TBI attaching bolts and nuts.

12. Remove the TBI unit and discard the flange gasket.

13. Cover the manifold opening to keep out dirt and debris.

To install:

14. Clean all parts well. DO NOT put the TP sensor, IAC valve, fuel meter cover assembly, fuel injectors and other components containing rubber in a solvent or cleaner bath. A chemical reaction will cause these parts to swell, harden or distort. Do not soak the throttle body with these parts still attached. If the throttle body assembly requires cleaning, keep cold immersion cleaner soaking time to a minimum. Some models have hidden throttle shaft seals that could loose their effectiveness by extended soaking.

15. The hardware used to attach the fuel meter assembly and TP sensor is factory coated with thread-locking compound. If these parts are removed for service, inspect the hardware threads. If the thread-locking compound is no longer coating the threads, clean the threads well and apply Loctite®262 or equivalent thread-locking compound on the threads. Usually, replacement components come with pre-coated hardware.

16. Use care in cleaning old gasket material from machined aluminum surfaces as sharp tools may damage the sealing surfaces.

17. Using a new flange gasket, install the throttle body to the intake manifold. Torque the attaching bolts to 18 ft. lbs. (25 Nm).

18. Install the O-rings on the fuel pipes and connect the fuel feed and return pipes. Torque the feed pipe nut to 25 ft. lbs. (34 Nm). Torque the return pipe nuts to 22 ft. lbs. (30 Nm). Use a backup wrench on the TBI fuel line nuts to keep them from turning as the pipe fittings are tightened.

19. Connect the vacuum and PCV hoses as required.

20. Install the cable support bracket and screws. Tighten the screws to 88 inch lbs. (10 Nm).

21. Connect the throttle cable, transmission control cable and cruise control cable as equipped. Make sure the cables do not hold the throttle open. Adjust as required.

22. Connect all electrical connectors. Make sure the connectors are fully seated and latched.

23. Tighten the fuel filler cap. Connect the negative battery cable.

24. Turn the ignition switch to the **ON** position for two seconds, then turn to the **OFF** position for ten seconds. Again turn the ignition switch to the **ON** position and check for fuel leaks.

25. With the engine OFF, check that the accelerator pedal is free. Depress the pedal to the floor and release.

26. Install the air cleaner assembly.

27. Reset the IAC valve pintle position using the following procedure.

a. Depress the accelerator pedal slightly.

b. Start and run the engine for three seconds.

c. Turn the ignition switch to the **OFF** position for ten seconds.

d. Restart the engine and check for proper idle operation.

28. DO NOT try to remove the plug and readjust the stop screw. Mis-adjustment may result in damage to the IAC valve or throttle body. The throttle stop screw is covered with a plug at the factory following adjustment. The minimum throttle valve position is set at the factory with a stop screw. This setting allows enough air flow by the throttle valve to cause the IAC valve pintle to be positioned a calibrated number of steps (counts) from the seat during controlled idle operation.

EMISSION CONTROLS

Emission Warning Lamps

RESETTING

These vehicles are equipped with a SERVICE ENGINE SOON Malfunction Indicator Lamp (MIL). The MIL will illuminate when the key is turned to the **ON** position as a systems check, but will extinguish shortly after the engine is started. If the computer control module (ECM, PCM or VCM, as applicable) detects a malfunction in 1 of the monitored circuits, a trouble code will be set and the lamp will be illuminated to indicate a fault.

When the computer module sets a code, the MIL will remain illuminated as long as the fault is detected. If the problem is intermittent, the light will extinguish approximately 10 seconds after the fault goes away.

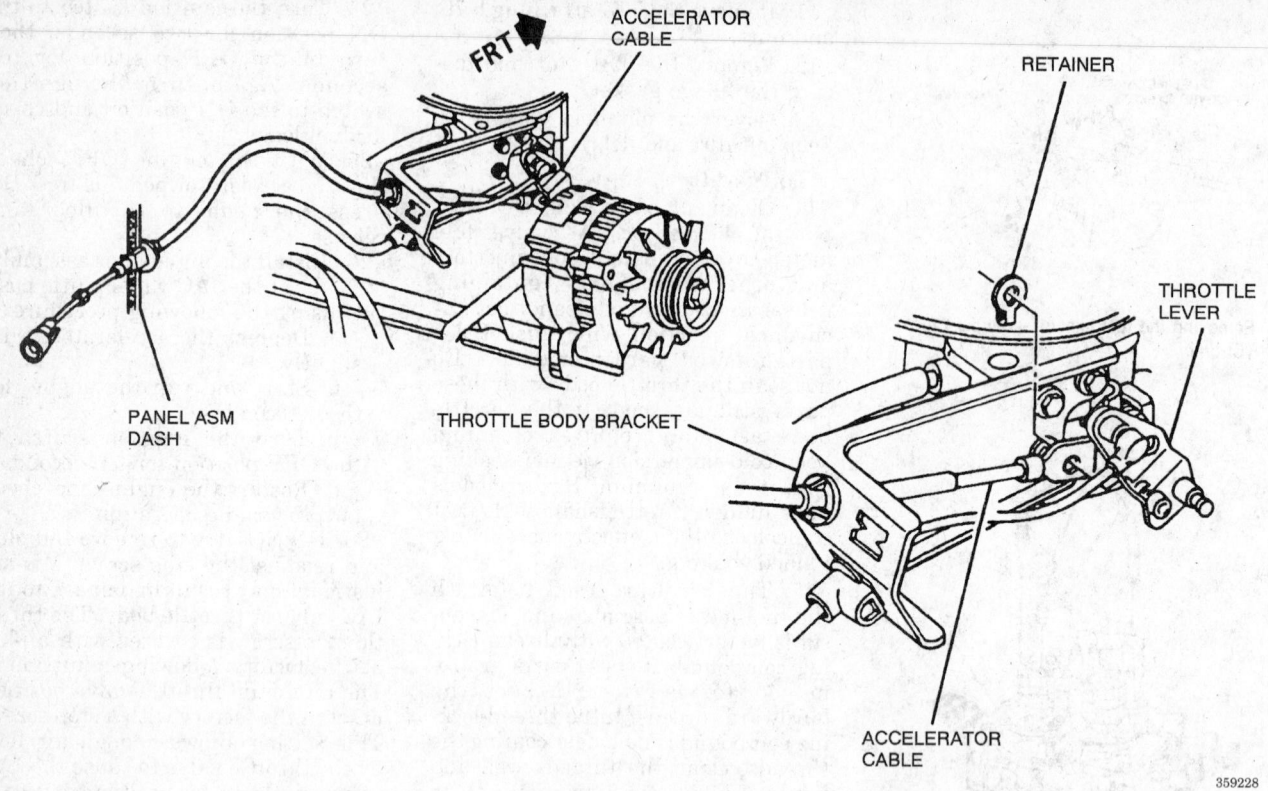

Accelerator cable routing

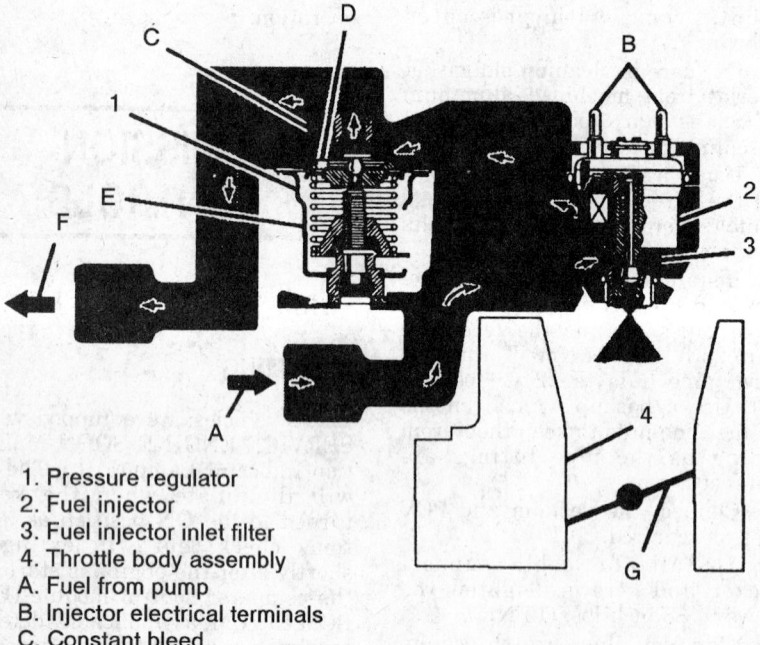

1. Pressure regulator
2. Fuel injector
3. Fuel injector inlet filter
4. Throttle body assembly
A. Fuel from pump
B. Injector electrical terminals
C. Constant bleed
D. Pressure regulator diaphragm assembly
E. Pressure regulator spring
F. Fuel return to tank
G. Throttle valve

Throttle body components

However the code will stay in the memory for 50 starts or until cleared.

NOTE: If a scan tool is not available, the trouble codes for most of these vehicles may be flash diagnosed. However, certain 1994–96 models equipped with a Vehicle Control Module (VCM), such as certain vehicles equipped with a module mounted on the fender well and not in the passenger compartment, are On-Board Diagnostic II (OBD-II) compliant. OBD-II vehicles utilize a new trouble code system in which all codes consist of 5 alpha-numeric digits. The first digit is a letter, while the remaining 4 are numbers. Nevertheless, flash diagnosis was not incorporated as it would be impractical.

Obviously, the MIL cannot be reset until the malfunction is corrected (or at least not present if intermittent) and the computer memory is cleared of the fault. Codes may be cleared using a suitable scan tool or by removing power from the computer control module. If the latter method is chosen, power must be removed from the module for a minimum of 30 seconds. The lower the ambient temperature, the longer period of time the power must be removed.

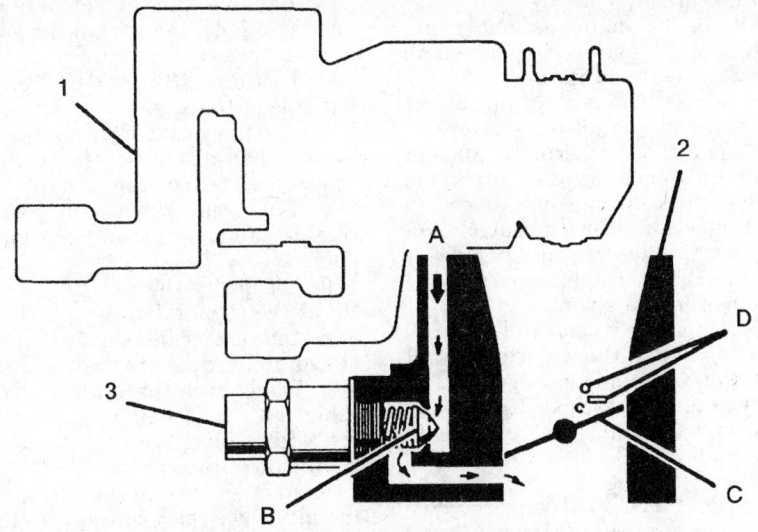

1. Fuel meter cover & body assemblies
2. Throttle body assembly
3. Idle air control (IAC) valve assembly
A. Filtered air inlet
B. Pintle
C. Throttle valve
D. Vacuum ports - for engine or emission controls

359230

IAC valve

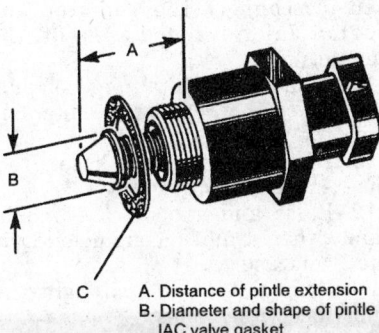

A. Distance of pintle extension
B. Diameter and shape of pintle
IAC valve gasket

359232

IAC valve components

Remember, to prevent control module damage, the ignition must always be **OFF** when removing power. Power may be removed through various methods, depending on how the vehicle is equipped. The preferable method is to remove the control module fuse from the fuse box. The control module pigtail may also be unplugged. If necessary, the negative battery cable may be disconnected, but other on-board data such as radio presets will be lost.

ENGINE MECHANICAL

NOTE: Disconnecting the battery cable on most vehicles may interfere with the functions of the on-board computer systems and may require the computer to undergo a relearning process.

Engine Assembly

REMOVAL AND INSTALLATION

3.1L (VIN D) Engine

1. Disconnect the negative battery cable.
2. Drain the cooling system. Disconnect the air flow tube from the air cleaner.
3. Disconnect the electrical connector from the ECM and push it through to the engine compartment. Disconnect the harness from the clips on the body and lay it across the engine.
4. Disconnect the engine harness at the bulkhead connector. Disconnect the throttle and TV cables.

5. Disconnect the fuel lines. Disconnect the transaxle shift linkage.
6. Disconnect the cooler lines at the radiator. Disconnect the radiator and heater hoses.

NOTE: On some newer vehicles it may be necessary to disconnect the refrigerant lines from the compressor in order to prevent stressing and damaging them. If so, properly recover the refrigerant using a recovery station before loosening ANY fittings. Cap all openings to prevent system contamination or damage.

7. Remove the air conditioning compressor from the bracket and support it aside. Remove the upper engine support strut.
8. Raise and safely support the vehicle. Remove the front wheel and tire assemblies.
9. Remove the stabilizer bar. Disconnect the tie rod ends and the lower control arm ball joints.
10. Disconnect the halfshafts from the transaxle and support them aside. Disconnect the steering shaft pinch bolt.
11. Remove the starter.
12. Remove the flywheel cover, then remove the torque converter bolts.
13. Disconnect the exhaust pipe at the manifold. Support the engine and sub-frame.
14. Remove the sub-frame bolts and lower the engine/transaxle and subframe.
   To install:
15. Raise the engine assembly into position, then install and tighten the subframe bolts.
16. Connect the exhaust pipe at the rear manifold. Install the starter.
17. Connect the steering shaft and install the pinch bolt. Connect the halfshafts to the transaxle.
18. Connect the lower control arm ball joints to the steering knuckles.
19. Install the stabilizer bar. Install the upper engine strut.
20. Install the wheel and tire assemblies. Lower the vehicle. Install the radiator and heater hoses.
21. Install the shift linkage. Connect the fuel lines and the throttle and TV cables.
22. Connect the harness to bulkhead connector. Connect the ECM harness to the ECM.
23. Connect the air cleaner hose and the radiator upper support.
24. Connect the negative battery cable.
25. Fill the cooling system. Install the air conditioning compressor. If lines were disconnected, properly

evacuate and recharge the A/C system.

3.4L (VIN E) Engine

1. Disconnect the negative battery cable.
2. Drain the cooling system. Disconnect the air flow tube from the air cleaner.
3. Disconnect the electrical connector from the ECM and push it through to the engine compartment.
4. Disconnect the harness from the clips on the body and lay it across the engine.
5. Disconnect the engine harness at the bulkhead connector. Disconnect the throttle and TV cables.

— CAUTION —
Fuel injection systems remain under pressure, even after the engine has been turned OFF. The fuel system pressure must be relieved before disconnecting any fuel lines. Failure to do so may result in fire/or personal injury.

NOTE: After relieving system pressure a small amount of fuel may be released when servicing fuel pipes or connections. In order to reduce the chance of personal injury, cover fuel line fittings with a shop towel before disconnecting, to catch any fuel that may leak out. Place the towel in an approved container when disconnect is completed.

6. Relieve fuel system pressure.
7. Disconnect the transaxle shift linkage.
8. Disconnect the cooler lines at the radiator.
9. Disconnect the radiator and heater hoses.
10. Remove the air conditioning compressor from the bracket and support it out of the way. Remove the upper engine support strut.
11. Raise and safely support the vehicle. Remove the front wheel and tire assemblies.
12. Remove the stabilizer bar. Disconnect the tie rod ends and the lower control arm ball joints.
13. Disconnect the halfshafts and support them out of the way. Disconnect the steering shaft pinch bolt.
14. Remove the starter.
15. Disconnect the exhaust pipe at the manifold.
16. Support the engine and subframe with a suitable jack.
17. Remove the sub-frame bolts and lower the engine/transaxle and subframe from the vehicle.

To install:
18. Raise the engine assembly into position and install the subframe bolts. Tighten to 35 ft. lbs.
19. Connect the exhaust pipe at the rear manifold. Install the starter.
20. Connect the steering shaft and install the pinch bolt. Connect the halfshafts to the transaxle.
21. Connect the lower control arm ball joints to the steering knuckles.
22. Install the stabilizer bar. Install the upper engine strut.
23. Install the wheel and tire assemblies. Lower the vehicle. Install the radiator and heater hoses.
24. Connect the fuel lines. For quick-connect fittings, use the following procedure:

— CAUTION —
To reduce the risk of fire and personal injury, before connecting a quick-connect fitting, apply a few drops of clean engine oil to the male pipe end. This will ensure proper reconnection and prevent a possible fuel leak. During normal operation, the O-rings located in the female connector will swell and prevent proper reconnection if not lubricated.

a. Apply a few drops of clean engine oil to the male pipe end.
b. Push both sides of the fitting together to cause the retaining tabs/fingers to snap into place.
c. Once installed, pull on both sides of the fitting to make sure the connection is secure.
25. Install the shift linkage. Connect the throttle and TV cables.
26. Connect the harness to bulkhead connector.
27. Connect the ECM harness to the ECM.
28. Connect the air cleaner hose and the radiator upper support.
29. Check the engine oil level and correct as required. Fill the cooling system. Install the air conditioning compressor and recharge the system, as required.
30. Connect the negative battery cable.

NOTE: Whenever the vehicle sub-frame is removed or lowered, the wheel alignment should be checked.

3.8L (VIN L) Engine

1. If equipped, properly recover the refrigerant from the A/C system.
2. Properly relieve the fuel system pressure, then disconnect the negative battery cable.
3. Remove the air cleaner and duct assembly, then drain the engine cooling system.
4. Disconnect the wiring from the left side of the engine.
5. Disengage the cruise servo connector, then disengage the connector at the emergency jumper box.
6. Disengage the wiring harness retainer from the right side of the engine.
7. Disengage the electrical connector at the cooling fan.
8. Disconnect the shift cable, heat shield and bracket at the transaxle.
9. Disconnect the battery ground cable from the engine, then remove the wiring block connector from the right side of the engine.
10. Remove the screws retaining the multi-use relay bracket to the tie bar.
11. Disengage the A/C compressor and accumulator wiring.
12. Disconnect the engine fuel vapor harness from the engine.
13. Disconnect the accelerator and cruise control cables along with the bracket from the throttle body.
14. Disconnect the brake vacuum hose at the engine.
15. Remove the engine mount strut bolts and nuts from the engine.
16. Disconnect the radiator and heater inlet/outlet hoses at the engine.
17. Remove the retaining bolt and disconnect the refrigerant manifold from the A/C compressor.
18. Disconnect the fuel hoses from the rail assembly.
19. Raise and support the vehicle safely, then drain the engine oil from the crankcase.
20. Remove the left and right tire and wheel assemblies.
21. Disconnect the engine wiring harness at the front of the frame, then disengage the wiring from the starter motor assembly.
22. Remove the flywheel covers from the transaxle, then remove the starter assembly.
23. Remove the torque converter bolts.
24. Remove the steering shaft (intermediate shaft) pinch bolt.
25. Disconnect the tie rod ends from the steering knuckles, then disconnect the right and left lower control arm ball joints.
26. Separate the drive axles from the transaxle, then support them aside.
27. Disconnect the exhaust pipe-to-manifold bolts and springs.
28. Disconnect the transaxle oil cooler line at the transaxle.

29. Remove the engine front mount nuts from the frame.

30. Position and engine/transaxle frame support under the assembly, then lower the vehicle or raise the support so the assembly is secure.

31. Remove the bolts retaining the frame assembly, then raise the vehicle leaving the engine/transaxle frame assembly behind.

32. If necessary, remove the power steering pump from the engine and position aside. Remove the necessary components and separate the engine from the transaxle.

To install:

33. If removed install the engine to the transaxle and frame assembly. Install the engine-to-transaxle retainers and tighten to 55 ft. lbs. (75 Nm). Install the engine-to-transaxle brace and retaining bolts. Tighten the engine side bolts to 70 ft. lbs. (95 Nm) and the transaxle side bolts to 47 ft. lbs. (63 Nm). Install the necessary components, including the power steering pump.

34. Position the engine/transaxle frame assembly under the vehicle, then carefully lower the vehicle until the frame is in position.

35. Install the frame retaining bolts and tighten to 103 ft. lbs. (140 Nm).

36. Raise and support the vehicle safely for access, then connect the transaxle cooler lines.

37. Connect the exhaust pipe to the manifold and tighten the retainers to 18 ft. lbs. (25 Nm).

38. Connect the drive axles to the transaxle.

39. Install the ball joints and tie rods to the knuckles.

40. Connect the intermediate shaft pinch bolt and tighten to 33 ft. lbs. (47 Nm).

41. Install the torque converter bolts and tighten to 46 ft. lbs. (62 Nm).

42. Install the starter assembly to the engine, then install the torque converter cover. Engage the starter wiring.

43. Connect the engine wiring harness and retainers to the frame.

44. Install the front tire and wheel assemblies, then lower the vehicle.

45. Install the A/C manifold to the compressor and secure using the retaining bolt.

46. Connect the fuel lines to the rail.

47. Connect the heater/radiator pipes and hoses to the engine.

48. Install the engine mount strut bolt and nut to the engine, then tighten to 44 ft. lbs. (60 Nm).

49. Connect the vacuum hose from the engine to the brake booster.

50. Install the accelerator and cruise control bracket, heat shield and cables to the throttle body.

51. Connect the engine fuel vapor harness to the engine.

52. Engage all wiring removed earlier in the reverse order of removal.

53. Refill the engine crankcase with oil, then connect the negative battery cable.

54. Properly refill the engine cooling system.

55. If equipped, properly evacuate and recharge the A/C system.

Engine Mounts

NOTE: When lifting or raising the engine for any reason, do not support the assembly under the oil pan, any sheet metal or the crankshaft pulley.

────── **CAUTION** ──────
When working on the engine mounts, make sure the engine is SECURELY supported at all times. If necessary, use blocks of wood to help make sure it cannot shift suddenly.
──────────────

REMOVAL AND INSTALLATION

3.1L (VIN D) Engine

1. Disconnect the negative battery cable.

2. Raise and support the vehicle safely. Support the engine with a suitable lifting fixture just sufficiently to take the weight off the mount.

3. Remove the engine mount nuts from below the engine frame mounting bracket.

4. Raise the engine enough to permit removal of the engine mounting and block the engine in position.

5. Remove the mount assembly to engine bracket nuts and washers. Remove the mount assembly.

6. Installation is the reverse of the removal procedure. Tighten the mount retaining nuts to 35 ft. lbs. (48 Nm), then check the mounts for proper alignment. If the mount is not properly aligned, loosen the nuts and allow the mount to reposition itself before tightening again. If the mount is not properly positioned, drivetrain component failure could occur.

3.4L (VIN E) Engine

These vehicles use nonadjustable powertrain mountings. Broken or de-

teriorated mounts should be replaced immediately. To check the mounts, raise the engine to remove weight from the mount, then place a slight tension on the rubber. Observe the mount while raising the engine. Replace an engine mount if it exhibits a hard rubber surface with heat cracks, if the rubber is separated from the metal plate, if the rubber is split through the center or if fluid is leaking from the hydraulic mount.

Engine Mount

1. Disconnect the negative battery cable.

2. Install engine support fixture J-28467-A, J 28467-90 and J-28467-200 or equivalents. Raise the engine just enough to support the weight.

3. Raise and safely support the vehicle.

4. Remove the mount retaining nuts from the bottom of the frame.

5. Remove the transaxle mount nuts.

6. Remove the mount to engine mount bracket nuts.

7. Remove the mount.

To install:

8. Install the mount to the engine mount bracket.

9. Tighten the nuts to 32 ft. lbs. (44 Nm).

10. Lower the engine.

11. Install the mount to the frame and tighten the bolts to 32 ft. lbs. (44 Nm).

12. Install the transaxle nuts and tighten to 32 ft. lbs. (44 Nm).

13. Lower the vehicle.

14. Remove the engine support fixture.

15. Connect the negative battery cable.

Engine Mount Strut

1. Disconnect the negative battery cable.

2. Remove the engine mount strut to engine mount strut and air conditioning bracket bolt and nut.

3. Remove the engine mount strut to engine mount strut, bracket bolt and nut.

4. Remove the engine mount strut.

To install:

5. Install the engine mount strut and tighten the nuts to 32 ft. lbs. (44 Nm).

6. Connect the negative battery cable.

3.8L (VIN L) Engine

1. Disconnect the negative battery cable, then remove the serpentine drive belt.

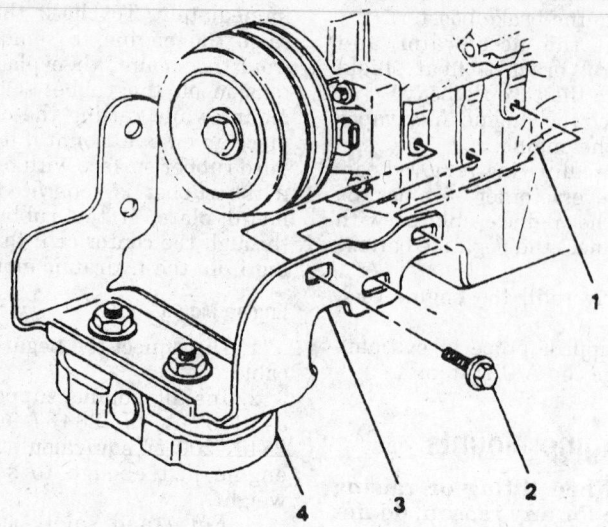

1 **ENGINE ASSEMBLY**
2 **BOLT**
3 **ENGINE MOUNT BRACKET**
4 **ENGINE MOUNT BRACKET TO FRAME**

253445

Engine mount-to-bracket — 3.1L (VIN E) and 3.8L (VIN L)

2. Install an engine support fixture, then raise and support the vehicle safely. Remove the right front tire and wheel assembly, then remove the right splash shield.

3. Remove the flywheel access covers.

4. Remove the crankshaft balancer bolt and balancer using J-39096 and J-38197 or equivalents.

5. Disengage the wiring connector from the crankshaft sensor, then remove the sensor shield from the engine.

6. Disconnect the power steering lines and engine wiring at the front of the frame. Remove the power steering gear heat shield from the gear assembly, then remove the gear assembly from the frame mounts and support aside using wire.

7. Remove the engine front mount-to-frame nuts, then remove the nuts retaining the front and rear transaxle mounts to the frame.

8. Support the frame from below, then remove the frame retaining bolts. Lower the frame.

9. Remove the bolts retaining the front engine mount to the engine, then remove the mount.

10. Installation is the reverse of the removal procedure. Be sure to tighten the front engine mount-to-engine bolts to 66 ft. lbs. (90 Nm), then

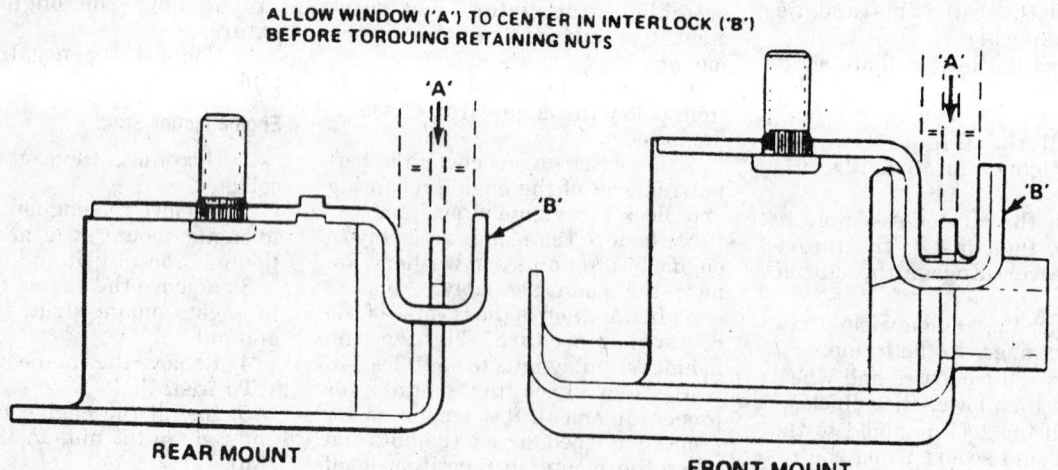

ALLOW WINDOW ('A') TO CENTER IN INTERLOCK ('B')
BEFORE TORQUING RETAINING NUTS

REAR MOUNT   FRONT MOUNT

85473086

Transaxle mount alignment — 3.1L (VIN D) engines

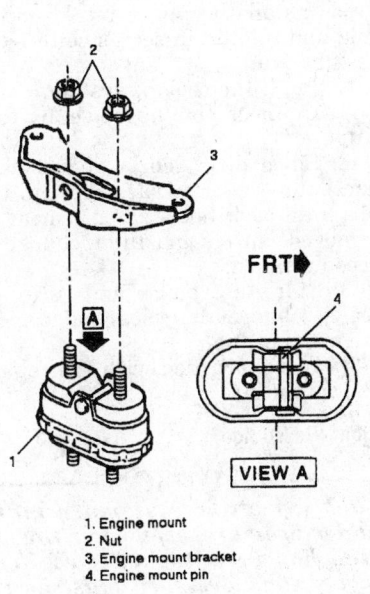

1. Engine mount
2. Nut
3. Engine mount bracket
4. Engine mount pin

347757

Engine mount-to-bracket — 3.4L (VIN E)

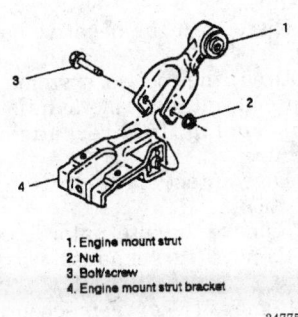

1. Engine mount strut
2. Nut
3. Bolt/screw
4. Engine mount strut bracket

347758

Engine mount strut and bracket — 3.4L (VIN E)

frame retaining bolts to 103 ft. lbs. (140 Nm) and the engine/transaxle mount-to-frame bolts to 33 ft. lbs. (44 Nm).

Cylinder Head

REMOVAL AND INSTALLATION

3.1L (VIN D) Engine

1. Relieve the fuel system pressure, then disconnect the negative battery cable.

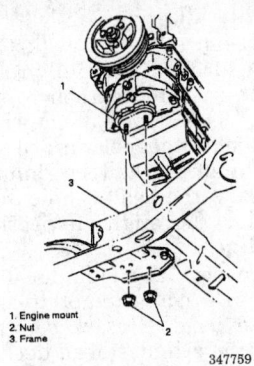

1. Engine mount
2. Nut
3. Frame

347759

Engine mount — 1996 3.4L (VIN E) engine

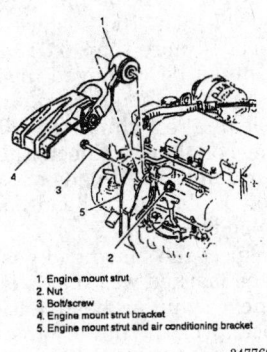

1. Engine mount strut
2. Nut
3. Bolt/screw
4. Engine mount strut bracket
5. Engine mount strut and air conditioning bracket

347760

Engine mount strut — 1996 3.4L (VIN E) engine

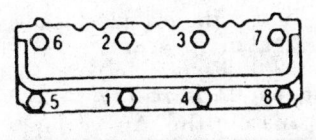

85473089

Cylinder head bolt torque sequence — 3.1L (VIN D) engine

2. Drain the engine cooling system.
3. Remove the valve cover(s).
4. Remove the intake manifold.
5. Raise and support the vehicle safely, then disconnect the exhaust crossover pipe.
6. If necessary, disconnect the dipstick tube attachment and/or the the alternator bracket.
7. Lower the vehicle.

8. Loosen the rocker arms and remove the pushrods. Keep All valve train components in order for reinstallation.
9. Remove the cylinder head retaining bolts. Remove the cylinder head from the engine along with the exhaust manifold.

To install:
10. Ensure that the cylinder bolt threads in the block and threads on the bolts are cleaned, as dirt will affect bolt torque.
11. Position the new gasket over the dowel pins with THIS SIDE UP showing, then instal the cylinder head (along with the exhaust manifold).
12. Coat the threads of the cylinder head bolts with sealing compound 1052080 or equivalent.
13. Install the cylinder head retaining bolts. Torque the cylinder head bolts gradually in the proper sequence to 33 ft. lbs. (45 Nm). Then turn each bolt an additional 90 degrees.
14. Position the intake gasket, then install the pushrods in their original locations. Make sure the lower ends of the pushrods are properly positioned in the lifter seats.
15. Properly adjust the valve lash.
16. Install the intake manifold assembly.
17. Instal the valve cover(s).
18. Raise and support the vehicle safely. If removed, install the dipstick tube and/or the alternator bracket.
19. Connect the exhaust crossover pipe. Lower the vehicle.
20. Connect the negative battery cable and properly refill the engine cooling system.

3.4L (VIN E) Engine

The 3.4L VIN E engine uses aluminum cylinder heads. Use care when working with light alloy parts. Valve guides are pressed in. Roller rocker arms are located on a pedestal in a slot in the cylinder head and are retained on individual threaded bolts.

The cylinder heads are retained by torque-to-yield bolts. A torque angle meter is required for proper torque at assembly. New replacement head bolts are recommended.

Before removing the cylinder head(s) from the engine and before disassembling the valve mechanism, perform a compression test and note the results. During disassembly, be sure that the valve train components are kept together and identified so that they can be installed in their original locations.

—— **CAUTION** ——

Fuel injection systems remain under pressure, even after the engine has been turned OFF. The fuel system pressure must be relieved before disconnecting any fuel lines. Failure to do so may result in fire and/or personal injury.

Left (Front) Side

1. Evacuate the air conditioning system refrigerant and recover, using approved refrigerant recycling equipment.
2. Relieve the fuel system pressure using the recommended procedure.
3. Drain the cooling system.
4. Disconnect the negative battery cable.
5. Remove the upper and lower intake manifold assembly using the recommended intake manifold removal procedure.
6. Remove the exhaust crossover pipe.
7. Disconnect the spark plug wires.
8. Remove the rocker arm covers.

NOTE: Any valve train components that are to be reused must be returned to their original locations. Keep the parts in organized and in order.

9. Loosen the rocker arms until able to remove the push rods using the recommended rocker arm removal procedure. Remove the pushrods.
10. Remove the oil level indicator (dipstick) retainer bolts and remove the dipstick and tube.
11. Remove the A/C compressor using the following procedure.
    a. Remove the top compressor bolts.
    b. Raise and safely support the vehicle.
    c. Remove the bottom A/C compressor bolts.
    d. Disconnect the lines from the back of the A/C compressor.
    e. Remove the A/C compressor from the vehicle.
12. Remove the lower A/C compressor bracket bolts.
13. Lower the vehicle.
14. Remove the top A/C compressor bracket bolts and remove the bracket.
15. Remove the cylinder head bolts.
16. Remove the cylinder head and gasket.

To install:

17. Clean all parts well. Clean all gasket surfaces. Carefully remove all varnish soot and carbon to the bare metal. DO NOT use a motorized wire brush on any gasket surface since the soft aluminum will be damaged. If necessary, the head can be disassembled for thorough inspection and reconditioning.
18. Inspect the cylinder head for cracks. Do not attempt to weld the cylinder head. If cracked, replace it. Check the cylinder head deck, intake and exhaust manifold mating surfaces for flatness. These surfaces may be reconditioned by milling. If the surfaces are "out of flat" by more than 0.005 inch, the surface should be milled. If more than 0.010 inch of metal must be removed from the head, the head should be replaced.
19. Clean the cylinder head bolts and the bolt holes. Check the head bolts for damaged threads or stretching. New replacement head bolts are recommended.
20. Inspect the new head gasket. It should be marked which side is "UP". Place new cylinder head gasket on the block, over the dowel pins.
21. To avoid damage, install the spark plugs after the cylinder head has been installed on the engine block assembly. Install the cylinder head onto block.
22. Coat the cylinder head bolt threads with GM 1052080 sealer or equivalent.
23. Install the cylinder head bolts, and tighten in sequence. Tighten bolts 33 ft. lbs. (45 Nm) and then an additional 90 degrees (¼-turn).
24. Install the A/C compressor bracket and upper bolts.
25. Raise and safely support the vehicle.
26. Install the lower bracket bolts.
27. Install the A/C compressor.
28. Connect the A/C lines to the back of the compressor.
29. Install the lower A/C compressor bolts.
30. Lower the vehicle.
31. Install the top compressor bolts.
32. Install the oil level indicator.
33. Install the push rods and rocker arms.
34. Tighten the rocker arm bolts to 89 inch lbs. (10 Nm) plus an additional 30 degrees. Do not overtighten or the threads in the aluminum head may be damaged.
35. Install the rocker arm covers.
36. Install the exhaust crossover pipe.
37. Install the spark plug wires.

38. Install the lower intake manifold and tighten the bolts to 115 inch lbs. (13 Nm).
39. Install the upper intake manifold and tighten in sequence to 18 ft. lbs. (25 Nm).
40. Refill the coolant system.
41. Connect the negative battery cable.
42. Since dirt, debris and coolant can enter the crankcase through the oil drain-back holes when a head is removed, an oil and filter change is recommended.
43. Start the vehicle and verify no leaks, abnormal noises and correct engine operation.
44. When satisfied with the repair, charge the A/C system.

Right (Rear) Side

—— **CAUTION** ——

Fuel injection systems remain under pressure, even after the engine has been turned OFF. The fuel system pressure must be relieved before disconnecting any fuel lines. Failure to do so may result in fire and/or personal injury.

1. Relieve the fuel system pressure using the recommended procedure.
2. Disconnect the negative battery cable.
3. Drain the coolant system.
4. Remove the intake manifold assembly using the recommended procedure.
5. Disconnect the ignition coil connection.
6. Remove the alternator.
7. Remove the exhaust crossover pipe.
8. Remove the oxygen sensor.
9. Raise and safely support the vehicle.
10. Remove the exhaust pipe from the manifold.
11. Lower the vehicle.
12. Remove the rocker arm cover.

NOTE: Any valve train components that are to be reused must be returned to their original locations. Keep the parts in organized and in order.

13. Loosen the rocker arms until able to remove the push rods using the recommended rocker arm removal procedure. Remove the pushrods.
14. Remove the cylinder head bolts and remove the cylinder head.

To install:

15. Clean all parts well. Clean all gasket surfaces. Carefully remove all

varnish soot and carbon to the bare metal. DO NOT use a motorized wire brush on any gasket surface since the soft aluminum will be damaged. If necessary, the head can be disassembled for thorough inspection and reconditioning.

16. Inspect the cylinder head for cracks. Do not attempt to weld the cylinder head. If cracked, replace it. Check the cylinder head deck, intake and exhaust manifold mating surfaces for flatness. These surfaces may be reconditioned by milling. If the surfaces are "out of flat" by more than 0.005 inch, the surface should be milled. If more than 0.010 inch of metal must be removed from the head, the head should be replaced.

17. Clean the cylinder head bolts and the bolt holes. Check the head bolts for damaged threads or stretching. New replacement head bolts are recommended.

18. Inspect the new head gasket. It should be marked which side is "UP". Place new cylinder head gasket on the block, over the dowel pins.

19. To avoid damage, install the spark plugs after the cylinder head has been installed on the engine block assembly. Install the cylinder head onto block.

20. Coat the cylinder head bolt threads with GM 1052080 sealer or equivalent.

21. Install the cylinder head bolts, and tighten in sequence. Tighten bolts 33 ft. lbs. (45 Nm) and then an additional 90 degrees (¼-turn).

22. Install the push rods and rocker arms.

23. Tighten the rocker arm bolts to 89 inch lbs. (10 Nm) plus an additional 30 degrees. Do not overtighten or the threads in the aluminum head may be damaged.

24. Install the rocker arm cover.

25. Raise and safely support the vehicle.

26. Install the exhaust pipe to the manifold.

27. Lower the vehicle.

28. Install the oxygen sensor.

29. Install the exhaust crossover pipe.

30. Install the alternator.

31. Connect the coil connections.

32. Install the lower intake manifold and tighten to 115 inch lbs. (10 Nm).

33. Install the upper intake manifold and tighten in sequence to 18 ft. lbs. (24 Nm).

34. Refill the coolant system.

35. Connect the negative battery cable.

36. Since dirt, debris and coolant can enter the crankcase through the oil drain-back holes when a head is removed, an oil and filter change is recommended.

37. Start the vehicle and verify no leaks, abnormal noises and correct engine operation.

3.8L (VIN L) Engine

1. Relieve the fuel system pressure, then disconnect the negative battery cable.

2. Drain the engine cooling system.

3. Remove the intake manifold.

4. Remove the exhaust manifold(s).

5. Remove the valve cover(s).

6. Disconnect the electronic ignition and spark plug wires.

7. Remove the alternator bracket and 1 air conditioner bracket bolt. Remove the power steering pump.

8. As necessary, remove the belt tensioner assembly and/or the fuel pipe heat shield.

9. Remove the rocker arms, pushrods and guide plates. Keep them in order for reinstallation.

10. Remove the cylinder head retaining bolts. Remove the cylinder head from the engine.

To install:

11. Ensure that the cylinder bolt threads in the block are cleaned, as dirt will affect bolt torque.

NOTE: This engine uses special torque to yield head bolts. The procedure must be followed carefully and new bolts must be used whenever the head is removed.

12. Position the new gasket with the arrow pointing to the front of the engine.

13. Install the cylinder head onto the engine.

14. Coat the underside of the bolt heads with sealing compound 1052080 or equivalent. Coat the threads of the bolts with a suitable thread locking compound.

15. Install the cylinder head onto the engine. Install the cylinder head retaining bolts and torque as follows:

   a. Torque the cylinder head bolts gradually in the proper sequence to 35 ft. lbs. (47 Nm).

   b. Then turn each bolt an additional 130 degree turn in sequence.

   c. Finally turn the 4 center bolts an additional 30 degrees.

16. Install the rocker arms, pushrods and guide plates in the same position from which they were removed. Apply a suitable thread locking compound to the rocker arm pedestal bolts, then install and tighten to 28 ft. lbs. (38 Nm) for vehicles 1992–93, to 19 ft. lbs. (25 Nm) plus 70 degrees for 1994 vehicles or to 11 ft. lbs. (15 Nm) plus 90 degrees for 1995–96 vehicles.

17. Install the intake manifold and the valve cover(s).

18. Raise and support the vehicle safely. Connect the exhaust manifold(s) to the exhaust pipe(s). Lower the vehicle.

19. Install the air conditioner bracket bolt. Install the alternator and bracket.

20. Install the ignition coil and spark plug wires.

21. Install the belt tensioner, then install power steering pump assembly.

22. Install the fuel pump heat shield.

23. Connect the negative battery cable, then properly refill the engine cooling system.

Valve Lifters

REMOVAL AND INSTALLATION

When installing new lifters, a prelube should always be applied to the lifter body. It is also a good idea to prime hydraulic lifters by submerging them in clean engine oil a depressing the plunger using an old pushrod. This allows the internal components of the hydraulic lifter to coat with oil before initial operation in the engine. All lifters should be replaced when a new camshaft is installed.

3.1L (VIN D) Engine

1. Properly relieve the fuel system pressure, then disconnect the negative battery cable and drain the engine cooling system.

2. Remove the intake manifold.

3. Loosen the rocker arms, then remove the pushrod guides and remove the pushrods.

NOTE: If any valve train components (lifters, pushrods, rocker arms) are to be reused, they must be tagged or arranged during removal to assure installation in their original locations.

4. Remove the lifters.

To install:

5. Coat the foot of each lifter using a suitable prelube.

6. Install the lifters.

7. Install the pushrods and guides.

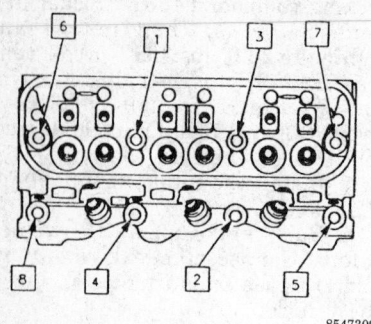

Cylinder head bolt torque sequence — 3.8L (VIN L) engine

85473090

8. Coat the friction surfaces of the rocker arms with prelube, then reposition and secure the rocker arms. Properly adjust the valve lash.

9. Install the intake manifold.

10. Connect the negative battery cable and properly refill the engine cooling system.

3.4L (VIN E) Engine

1. Properly relieve the fuel system pressure, then disconnect the negative battery cable and drain the engine cooling system.

2. Remove the rocker arm covers.

3. Remove the intake manifold.

4. Unbolt the rocker arms and pedestals.

NOTE: If any valve train components (lifters, pushrods, rocker arms) are to be reused, they must be tagged or arranged during removal to assure installation in their original locations.

5. Remove the pushrods.

NOTE: The intake and exhaust pushrods are different lengths. The intake pushrods are 5.68 in. (144.18mm) and the exhaust pushrods are 6.0 in (152.51mm)

6. Remove the lifter guide retainer bolts, then remove the guides.

7. Remove the lifters.

To install:

8. Coat the foot of each lifter using a suitable prelube.

9. Install the lifters.

10. Install the lifter guides, then tighten the retainer bolts to 89 inch lbs. (10 Nm).

11. Install the pushrods.

12. Install the rocker arms and pedestals. Apply a suitable thread locking compound to the rocker arm pedestal bolts, then install and tighten to 89 inch lbs. (10 Nm) plus 30 degrees.

13. Install the intake manifold.

14. Install the rocker arm covers.

15. Connect the negative battery cable and properly refill the engine cooling system.

3.8L (VIN L) Engine

1. Properly relieve the fuel system pressure, then disconnect the negative battery cable and drain the engine cooling system.

2. Remove the rocker arm covers.

3. Remove the intake manifold.

4. Unbolt the rocker arms and pedestals.

NOTE: If any valve train components (lifters, pushrods, rocker arms) are to be reused, they must be tagged or arranged during removal to assure installation in their original locations.

5. Remove the pushrods.

6. Remove the lifter guide retainer bolts, then remove the guides.

7. Remove the lifters.

To install:

8. Coat the foot of each lifter using a suitable prelube.

9. Install the lifters.

10. Install the lifter guides, then tighten the retainer bolts to 22 ft. lbs. (30 Nm).

11. Install the pushrods.

12. Install the rocker arms and pedestals. Apply a suitable thread locking compound to the rocker arm pedestal bolts, then install and tighten to 19 ft. lbs. (25 Nm) plus 70 degrees for 1994 vehicles or to 11 ft. lbs. (15 Nm) plus 90 degrees for 1995–96 vehicles.

13. Install the intake manifold.

14. Install the rocker arm covers.

15. Connect the negative battery cable and properly refill the engine cooling system.

Valve Lash

ADJUSTMENT

3.1L (VIN D) Engine

NOTE: These engines utilize hydraulic valve lifters which means that a valve adjustment is not a regular maintenance item. The valves must only be adjusted if the rockers arms have been disturbed for any reason such as cylinder head, camshaft, pushrod or lifter removal.

1. Remove the air cleaner and the rocker arm cover(s).

2. Rotate the crankshaft until the mark on the crankshaft pulley aligns with the **0** mark on the timing plate.

Make sure the No. 1 cylinder is positioned on the compression stroke. The No. 1 piston is on it's compression stroke when both the intake and exhaust valves remain closed as the crankshaft damper mark approaches the timing scale.

NOTE: Another method to tell when the piston is coming up on the compression stroke is by removing the spark plug and placing a finger over the hole in order to feel air being forced out of the spark plug hole. Stop turning the crankshaft when pressure is felt and the TDC timing mark on the crankshaft pulley is directly aligned with the timing mark pointer or the 0 mark on the scale.

3. When the engine is on the No. 1 firing position, adjust the following valves:

- Intake — 1, 5 and 6
- Exhaust — 1, 2 and 3

4. To adjust the valves, back-out the adjusting nut until lash can be felt at the pushrod, then turn the nut until all of the lash is removed.

NOTE: To determine is all of the lash is removed, turn the pushrod between 2 fingers until the movement is removed.

5. When all of the lash has been removed, turn the adjusting an additional 1½ turns; this will center the lifter plunger.

6. Crank the engine 1 complete revolution until the timing tab (0 degree mark) and the crankshaft pulley mark are again in alignment. Now the engine is in the No. 4 firing position. Adjust the following valves:

- Intake — 2, 3 and 4
- Exhaust — 4, 5 and 6

7. Install the rocker arm cover(s).

8. Start and run the engine, then check and adjust the timing, as necessary.

3.4L (VIN E) and 3.8L (VIN L) Engines

Because the rocker arm fasteners are secured and torqued, valve lash is not adjustable on the 3.4L (VIN E) or 3.8L (VIN L) engines. If a valve train problem is suspected, check that the rocker arm pedestals bolts are tightened to specification. During initial installation the bolts are coated with a suitable thread locking compound. If they are sufficiently loosened to cause valve train noise, they should be removed and thoroughly cleaned. Apply a suitable thread locking compound to the rocker arm pedestal bolts, install and tighten to the following specifications:

- On 3.4L (VIN E) engine, torque to 89 inch lbs. (10 Nm) plus 30 degrees.
- On 3.8L (VIN L) engine, torque to 19 ft. lbs. (25 Nm) plus 70 degrees for 1994 vehicles or to 11 ft. lbs. (15 Nm) plus 90 degrees for 1995–96 vehicles.

When valve lash falls out of specification (valve tap is heard) and tightening the bolts does not solve the problem, replace the rocker arm, pushrod and hydraulic lifter on the offending cylinder.

Rocker Arms

REMOVAL AND INSTALLATION

NOTE: Valve train components which are to be reused must be installed in their original positions. If removed, be sure to tag or arrange all rocker arms and pushrods to assure proper installation.

3.1L (VIN D) Engine

1. Remove the rocker arm cover(s) from the cylinder head.
2. Remove the rocker arm nut, the rocker arm and the ball washer.

NOTE: If only the pushrod is to be removed, loosen the rocker arm nut, swing the rocker arm to the side and remove the pushrod.

3. Withdraw the pushrod from the cylinder head.
To install:
4. Inspect and replace components if worn or damaged.
5. Coat the bearing surfaces of the rocker arms and the rocker arm ball washers with Molykote® or equivalent pre-lube.
6. Install the pushrods making sure they seat properly in the lifter.
7. Install the rocker arms, ball washers and the nuts, then tighten the rocker arm nuts until there is little or no valve lash.

NOTE: Each valve must be adjusted when the lifter is sitting on the base circle of the camshaft, not the raised section of the lobe.

8. Properly adjust the valve lash.
9. Install the rocker arm cover.
10. Start and run the engine, then check for leaks. Check and adjust the timing, as necessary.

3.4L (VIN E) and 3.8L (VIN L) Engine

1. Remove the rocker arm cover from the engine.
2. Remove the rocker arm pedestal retaining bolts.
3. Remove the pedestal and rocker arm assembly.
To install:
4. Inspect and replace components if worn or damaged. Clean all old thread locking material from the pedestal bolts.
5. Install the rocker arms and pedestals. Apply a suitable thread locking compound to the rocker arm pedestal bolts, install and tighten to the following specifications:
- On 3.4L (VIN E) engine, torque to 89 inch lbs. (10 Nm) plus 30 degrees.
- On 3.8L (VIN L) engine, torque to 19 ft. lbs. (25 Nm) plus 70 degrees for 1994 vehicles or to 11 ft. lbs. (15 Nm) plus 90 degrees for 1995–96 vehicles.
6. Install the rocker arm covers.

Intake Manifold

REMOVAL AND INSTALLATION

3.1L (VIN D) Engine

1. Properly relieve the fuel system pressure, then disconnect the negative battery cable.
2. Drain the engine cooling system and remove the air cleaner.
3. Tag and disconnect the necessary wiring and vacuum hoses in order to remove the valve covers.
4. Disconnect the fuel lines from the TBI unit and reposition for access.
5. Remove the valve covers. It will be necessary to remove or reposition the alternator (with brackets) and disconnect some coolant hoses for valve cover removal.
6. Remove the TBI unit from the intake manifold.
7. Remove the power steering pump and carefully position it aside with the lines intact.
8. Matchmark and remove the distributor assembly.
9. Remove the intake manifold bolts, nuts and washers, then remove the intake manifold and discard the old gasket.
To install:
10. Clean the sealing surface of the engine block and apply a 3/16 in. bead of RTV sealer to each ridge.
11. Install the new gaskets onto the heads. Hold the gaskets in place by extending the ridge bead of sealer 1/4 in. onto the gasket ends. (When the intake manifold is installed, the area between the ridges and the manifold should be completely sealed.)
12. Install the intake manifold onto the engine.
13. Coat the threads of the intake manifold studs using a sealer such as 1052080 or equivalent.
14. Install the intake manifold retainers and tighten to 13 ft. lbs. (18 Nm) using the proper sequence, then tighten the retainers (again in sequence) to 19 ft. lbs. (26 Nm).
15. Align and install the distributor assembly.
16. Reposition and secure the power steering pump assembly.
17. Install the TBI unit.
18. Install the valve covers. Reposition and secure the alternator (with brackets) and connect the coolant hoses.
19. Connect the fuel lines to the TBI unit.
20. Connect the wiring and vacuum hoses as tagged during removal.
21. Install the air cleaner.
22. Connect the negative battery cable, then properly refill the engine cooling system.

3.4L (VIN E) Engine

This engine uses a two piece intake manifold. The upper half (often called a plenum) mounts the throttle body. The lower half of the manifold bolts to the engine and contains the fuel injectors. Please note that this engine uses a sequential multiport fuel injection system. Injector connectors must be connected to their appropriate fuel injector assembly or engine emissions and engine performance will be seriously affected. Identify and tag for identification all wiring connectors as well as vacuum and other components as required to assure correct assembly.

This procedure includes both upper and lower intake manifold removal and installation.

——— CAUTION ———
Fuel injection systems remain under pressure, even after the engine has been turned OFF. The fuel system pressure must be relieved before disconnecting any fuel lines. Failure to do so may result in fire and/or personal injury.

1. Disconnect the negative battery cable.
2. Relieve the fuel system pressure using the recommended procedure.

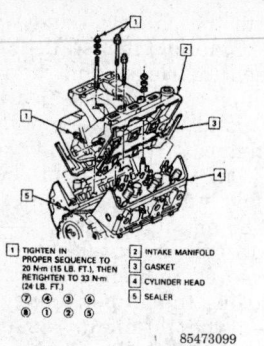

1 TIGHTEN IN PROPER SEQUENCE TO 20 Nm (15 LB. FT.), THEN RETIGHTEN TO 33 Nm (24 LB. FT.) ⑦ ④ ③ ⑥ ⑧ ① ② ⑤	2 INTAKE MANIFOLD 3 GASKET 4 CYLINDER HEAD 5 SEALER

85473099

Intake manifold bolt torque sequence — 3.1L (VIN D) engine

3. Drain the engine coolant. Remove the coolant recovery bottle.

4. Remove the air cleaner and duct assembly.

5. Remove the acoustic (engine sound deadener) cover .

6. Disconnect the throttle and cruise control cables from the throttle body. Remove retaining brackets and set cable assemblies aside.

7. Disconnect the coolant hoses from the manifold.

8. Identify and tag for identification any remaining vacuum lines and disconnect from the intake manifold.

9. Label and remove the front spark plug wires.

10. Move the coil bracket (leaving the coils and solenoids attached) out of the way.

11. Disconnect the electrical connectors from the ignition coil assembly.

12. Remove the manifold air pressure sensor.

13. Remove the brake booster hose.

14. Remove the EGR valve assembly.

15. Remove the thermostat bypass pipe nut from the upper intake manifold.

16. Remove the upper intake manifold studs and bolts then remove the upper intake manifold and gaskets.

17. If the lower intake manifold needs to be removed, remove the fuel injector rail bolts and remove the fuel injector rail assembly.

18. Remove the heater inlet pipe assembly, upper radiator hose and tie straps retaining the heater outlet pipe and ignition wiring assembly. Disconnect the heater pipe from the heater core to the coolant pump.

19. Remove the power steering pump bolts and pump.

20. Remove the rocker arm covers.

21. Remove the lower intake manifold retaining bolts and remove the lower intake manifold and gasket.

To install:

22. Clean all parts well. Use care in cleaning old gasket material from the machined aluminum surfaces on the plenum and manifold as sharp tools may damage sealing surfaces.

23. Clean the mating surfaces to the intake manifold and engine block. Remove any loose pieces of RTV sealer.

24. Install the lower intake manifold to the engine block. Apply sealant GM 12345739 or equivalent at the engine block to manifold mating surface. The bead should be 3.0 mm wide and 5.0 mm thick.

25. Install the lower intake manifold retaining bolts. Apply sealant GM 12345382 or equivalent to the threads of the bolts. Torque bolts in sequence to 115 inch lbs. (13 Nm).

26. Install the valve rocker covers.

27. Connect the heater pipe from the heater core to the coolant pump. Install new tie straps around the heater outlet pipe and ignition harness assembly. Connect the upper radiator hose to the engine and the heater inlet pipe to the manifold assembly.

28. Install the power steering pump and pulley.

29. Remove the injector O-ring seals from both the spray tip ends and the fuel rail end of each injector. Discard the seals. With the spray tip end O-ring removed, the O-ring backup piece may slip off of the injector. Be sure to retain the O-ring backup for reuse. Make sure that the O-ring backup piece is in place on the spray tip end of the injector before installing a new O-ring. Lubricate new injector O-ring seals with clean engine oil and install on the injector assembly.

30. Install the fuel rail assembly to the intake manifold. Tilt the rail assembly to install the injectors. Install the fuel rail attaching bolts and tighten to 89 inch lbs. (10 Nm).

31. Connect the injector electrical connectors.

32. Install new O-rings on the fuel lines and install the fuel feed and return pipes. Tighten the fuel rail nuts to 13 ft. lbs. (17 Nm). Use a backup wrench on the fittings to prevent them from turning.

33. Using new gaskets, install the intake manifold plenum. Be sure to route the MAP sensor electrical connector to the outside of the the plenum gasket. Torque the bolts to 18 ft. lbs. (25 Nm).

34. Install the serpentine drive belt. Install the coolant recovery tank.

35. Install the MAP sensor, braces to the alternator, ignition coil front bolts and the EGR to plenum bolts. Connect the vacuum lines as noted during removal.

36. If the throttle body was removed from the upper intake manifold, inspect the throttle body before installation. Throttle body bore and valve deposits may be cleaned using carburetor cleaner and a parts cleaning brush. DO NOT use a cleaner that contains Methyl Ethyl Ketone (MEK), an extremely strong solvent and not necessary for this type of deposit. The TP sensor and IAC valve should NOT come into contact with solvents or cleaners as they may be damaged. Verify that the gasket surfaces are clean, and, using a new flange gasket, install the throttle body. Torque the fasteners to 18 ft. lbs. (25 Nm).

37. Connect the throttle and cruise control cables.

38. Connect the IAC valve and TP sensor electrical connectors. Connect the air inlet duct. Check that the accelerator pedal is free by depressing the pedal to the floor and releasing.

39. Connect all remaining electrical connections and vacuum lines. Make sure the alternator braces are secure.

40. Refill the cooling system. GM recommends adding two engine coolant sealant pellets GM 3634621 or equivalent. Starting with the 1996 Model Year, these vehicles were filled at the factory with a new type of antifreeze/coolant called GM Goodwrench DEX-COOL™. When adding coolant to 1996 vehicles, it is important that you use GM Goodwrench DEX-COOL™ (orange-colored, silicate-free) coolant. A 50/50 mixture of DEX-COOL™ and clean water will provide all the recommended protection for 1996 vehicles. **DO NOT use DEX-COOL™ in pre-1996 vehicles. DO NOT mix DEX-COOL™ with any other type of antifreeze.**

41. Since coolant can get into the engine's oil system when the intake manifold is removed, change the engine oil and filter.

42. Connect the negative battery cable.

43. Turn the key to the **ON** position several times to pressurize the fuel system and check for fuel leaks.

44. After the engine is running, bleed the cooling system and check for proper idle quality.

3.8L (VIN L) Engine

The 3.8L (VIN L) engine utilizes a 2-piece intake manifold assembly. The entire assembly may be removed

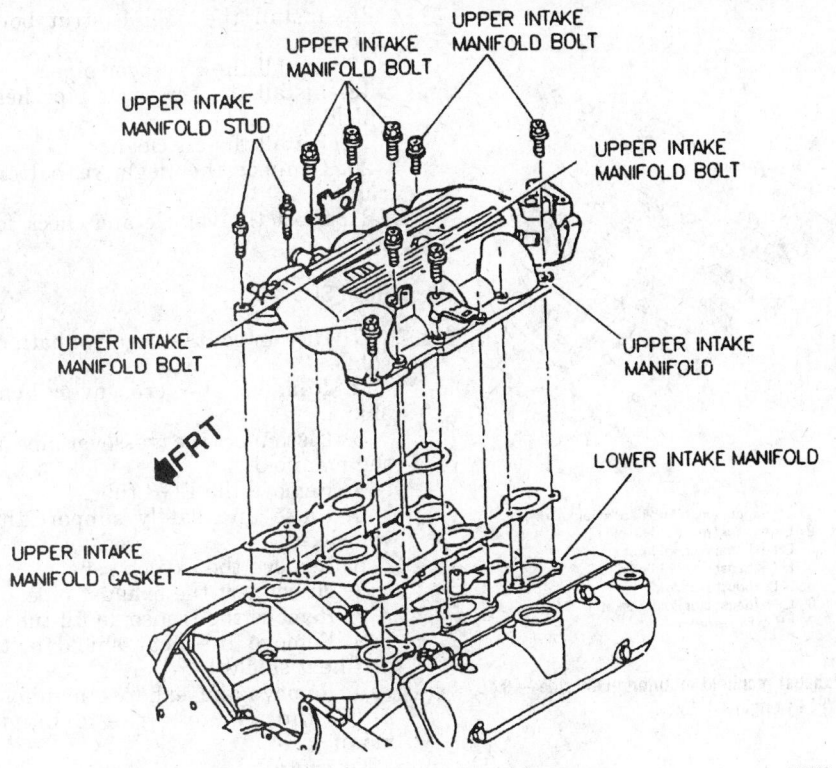

UPPER INTAKE MANIFOLD BOLT

UPPER INTAKE MANIFOLD BOLT

UPPER INTAKE MANIFOLD STUD

UPPER INTAKE MANIFOLD BOLT

UPPER INTAKE MANIFOLD BOLT

UPPER INTAKE MANIFOLD

FRT

UPPER INTAKE MANIFOLD

LOWER INTAKE MANIFOLD

UPPER INTAKE MANIFOLD GASKET

345500

Intake manifold assembly — 3.4L (VIN E) engine

without separating the upper half from the lower half.

1. Properly relieve the fuel system pressure, then disconnect the negative battery cable.

2. Drain the engine cooling system and disconnect the air intake duct.

3. Tag and disconnect the spark plug wires on the right side of the engine, then position the wires aside.

4. Remove the fuel rail.

5. Remove the exhaust crossover heat shield.

6. Remove the cable bracket-to-cylinder head mounting bolt.

NOTE: Do not separate the upper manifold from the lower manifold unless component replacement is necessary.

7. If necessary, loosen and remove the upper intake manifold bolts, then separate the upper manifold from the lower manifold.

8. Remove the power steering pump support bracket.

9. Loosen the alternator and move aside to obtain clearance.

10. Disconnect the heater pipes and bypass hose.

11. Remove the lower intake manifold bolts, then remove the manifold or manifold assembly (as applicable) from the engine.

To install:

12. Thoroughly clean all manifold mating surfaces, bolts and bolt holes. Apply sealant to the ends of the manifold seals and coat the bolt threads with Loctite® or equivalent thread locking compound. Install the lower intake manifold, gasket and bolts. Torque the lower manifold bolts in sequence, twice, to 88 inch lbs. (10 Nm) for vehicles 1994 or to 11 ft. lbs. (15 Nm) for 1995–96 vehicles.

——— **WARNING** ———
The manifold surfaces used on this engine should not be scraped or wire brushed in order to clean the gaskets. The surfaces could be easily damaged if this is ignored. Instead, use a commercially available solvent to clean the mating surfaces.

13. Connect the heater pipes and bypass hose.

14. Reposition and secure the alternator.

15. Install the power steering pump support bracket.

16. If removed, prepare the upper intake manifold mating surface for installation. Apply a 1/16 bead of Loctite Instant Gasket Eliminator GM P/N 1052942 or equivalent to the

mating surface on the lower manifold. Be sure to circle all bolt holes. Install the upper manifold assembly, then apply thread locking compound to the retainers. Install the upper intake manifold retainers and tighten to 11 ft. lbs. (15 Nm) for 1994–96 vehicles.

17. Install the cable bracket-to-cylinder head mounting bolt.

18. Install the exhaust heat shield.

19. Install the fuel rail.

20. Connect the spark plug wires on the right side of the engine.

21. Connect the air intake duct.

22. Connect the negative battery cable, then properly refill the engine cooling system.

Exhaust Manifold

REMOVAL AND INSTALLATION

3.1L (VIN D) Engine

1. Disconnect the negative battery cable.

2. To remove the left (front) exhaust manifold:

   a. Remove the serpentine belt and the air conditioning compressor. Position the compressor aside with the lines intact.

   b. Remove the engine strut and bracket.

   c. Disconnect the crossover pipe.

   d. Remove the exhaust manifold attaching bolts and remove the manifold.

3. To remove the right (rear) exhaust manifold:

   a. Disconnect the oxygen sensor wire.

   b. Remove the crossover pipe.

   c. Raise and support the vehicle safely.

   d. Disconnect the exhaust pipe.

   e. Support the rear center of the frame.

   f. Remove the rear frame mount bolts.

   g. Lower the frame 8–10 in. for access.

   h. Remove the exhaust manifold bolts and remove the assembly.

4. Installation is the reverse of the removal procedure.

3.4L (VIN E) Engine

The exhaust manifolds are conventional nodular iron castings. Left and right manifolds are connected by a crossover pipe. Use care with the exhaust manifold-to-cylinder head fasteners. The cylinder heads are aluminum.

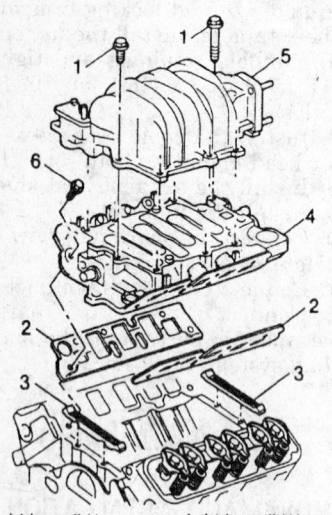

1. Intake manifold upper bolt
2. Intake manifold gasket
3. Intake manifold seal
4. Intake manifold lower
5. Intake manifold upper
6. Intake manifold lower bolt

85473100

Intake manifold assembly — 3.8L (VIN L) engine

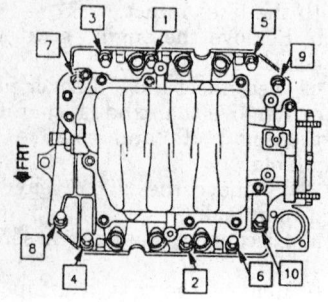

85473101

Intake manifold bolt torque sequence — 3.8L (VIN L) engine

85473113

Exhaust manifold mounting — 3.1L (VIN D) engine

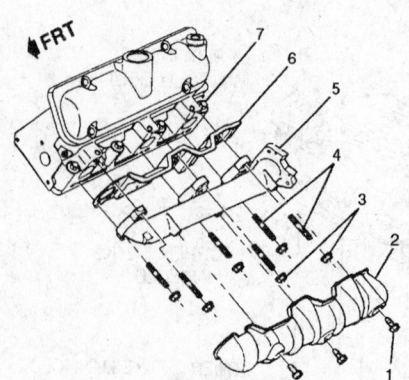

1. LH exhaust manifold heat shield screw
2. LH exhaust manifold shield
3. LH exhaust manifold nut
4. LH exhaust manifold stud
5. LH exhaust manifold
6. LH exhaust manifold gasket
7. LH cylinder head

344375

Exhaust manifold mounting-Left side — 3.4L (VIN E) engine

Left Side

1. Disconnect the negative battery cable.
2. Remove the air cleaner.
3. Remove the crossover heat shield.
4. Remove the crossover pipe nuts and pipe.
5. Remove the engine strut bolts and strut.
6. Remove the serpentine drive belt.
7. Remove the A/C compressor (leaving the hoses attached) and lay it aside.
8. Remove the engine strut and air conditioning bracket.
9. Remove the exhaust manifold heat shield.
10. Remove the exhaust manifold retaining bolts and remove the exhaust manifold.

To install:

11. Clean the mating surfaces, install the exhaust manifold and tighten the retaining bolts to 12 ft. lbs. (16 Nm).
12. Install the exhaust manifold heat shield.
13. Install the engine strut and air conditioning bracket.
14. Install the air conditioning compressor.
15. Install the serpentine belt.

16. Install the engine strut bolts and strut.
17. Install the crossover pipe.
18. Install the crossover pipe heat shield.
19. Install the air cleaner.
20. Connect the negative battery cable.
21. Start the vehicle and check for leaks.

Right Side

1. Disconnect the negative battery cable.
2. Remove the crossover heat shield.
3. Disconnect the crossover pipe at the manifold.
4. Remove the EGR tube.
5. Raise and safely support the vehicle.
6. Remove the oxygen sensor.
7. Disconnect the exhaust pipe.
8. Remove the transaxle fill tube.
9. Remove the heat shield bolts and heat shield.
10. Remove the exhaust manifold nuts and remove the exhaust manifold.

To install:

11. Clean the mating surfaces, install the exhaust manifold and tighten the retaining bolts to 12 ft. lbs. (16 Nm).
12. Install the heat shield.
13. Install the transaxle fill tube.
14. Install the exhaust pipe.
15. Connect the oxygen sensor wire.
16. Lower the vehicle.
17. Install the EGR tube.
18. Install the crossover pipe.
19. Install the crossover heat shield.
20. Connect the negative battery cable.
21. Start the engine and check for leaks.

3.8L (VIN L) Engine

1. Disconnect the negative battery cable.
2. To remove the left (front) exhaust manifold:
   a. Remove the crossover pipe.
   b. Tag and disconnect the spark plug wires from the plugs.
   c. Remove the exhaust manifold bolts/studs and the oil dipstick tube.
   d. Remove the exhaust manifold assembly.
3. To remove the right (rear) exhaust manifolds:
   a. Tag and disconnect the spark plug wires.
   b. Remove the throttle cable bracket.

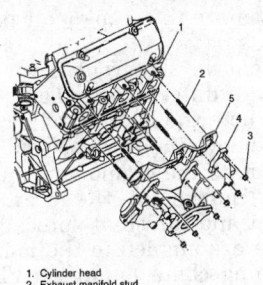

1. Cylinder head
2. Exhaust manifold stud
3. Exhaust manifold nut
4. Manifold right exhaust
5. Right exhaust manifold gasket

344378

Exhaust manifold mounting-Right side — 3.4L (VIN E) engine

c. Remove the crossover pipe heat shield.

d. Remove the transaxle dipstick and tube assembly.

e. Disconnect the oxygen sensor wire.

f. Remove the fasteners connecting the crossover pipe to the manifold.

g. Remove the plastic vacuum tank mounted on the cowl.

h. Raise and support the vehicle safely.

i. Remove the catalytic converter heat shield and hanger.

j. Remove the front exhaust pipe-to-the manifold attaching nuts.

k. Remove the front exhaust pipe from the manifold. Lower the vehicle.

l. Remove the engine lift bracket and remove the manifold attaching nuts.

m. Remove the exhaust manifold assembly.

4. Installation is the reverse of the removal procedure.

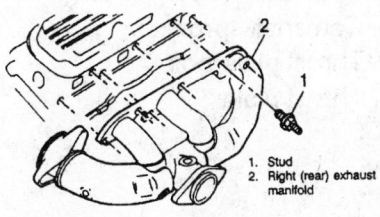

1. Stud
2. Right (rear) exhaust manifold

85473114

Exhaust manifold mounting — 3.8L (VIN L) engine

Front Cover Oil Seal

REPLACEMENT

It is possible to replace the timing cover front oil seal without removing the cover.

1. Disconnect the negative battery cable.

2. Remove the accessory drive belt.

3. Remove the crankshaft damper and pulley.

4. Pry the seal out of the front cover using a small prytool. Be very careful not to distort the front cover or to score the end of the crankshaft.

To install:

5. Lightly coat the lips of the replacement crankshaft seal with clean engine oil, then position the seal with the open end facing inward the engine. Use a suitable seal installation driver to position the seal in the front cover.

6. Install the crankshaft damper and pulley.

7. Install the accessory drive belt.

8. Connect the negative battery cable.

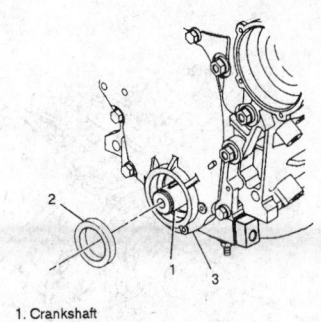

1. Crankshaft
2. Front cover seal
3. Front cover

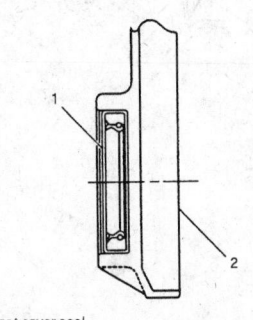

1. Front cover seal
2. Front cover

347859

Removing the front cover seal — 3.4L (VIN E) engine

Timing Chain and Sprockets

REMOVAL AND INSTALLATION

3.1L (VIN D) Engine

1. Disconnect the negative battery cable. Drain the engine cooling system.

2. Remove the right front tire and wheel assembly.

3. Remove the front cover assembly. Ensure the marks on the crankshaft and camshaft gears are aligned using the marks on the damper stamping or cast alignment marks on cylinder and case.

4. Remove the bolts that hold the camshaft sprocket to the camshaft. This sprocket is a light press fit on the camshaft.

5. Remove the timing chain. Using a suitable puller, remove the crankshaft sprocket, as required.

To install:

6. Install the crankshaft sprocket.

7. Lubricate the camshaft thrust plate surface with Molykote® or equivalent. Install the chain onto camshaft sprocket.

8. Holding the sprocket vertically with the chain hanging down, align the marks on the camshaft and crankshaft sprockets and install the assembly onto the camshaft.

9. Install the camshaft to gear attaching bolts and torque to 18 ft. lbs. (24 Nm). After the sprockets are in place, turn the engine 2 full revolutions to make certain the timing marks are in correct alignment between the shaft centers.

10. Lubricate the chain with engine oil and install the front cover.

11. Connect the negative battery cable, then properly refill the engine cooling system.

3.4L (VIN E) Engine

The camshaft drive is a conventional timing chain and sprockets. The front cover (timing chain cover) houses the front crankshaft oil seal and also mounts the water pump.

1. Disconnect the negative battery cable.

2. Raise and safely support the vehicle.

3. Drain the engine oil into a suitable container.

4. Remove the oil pan using the recommended procedure.

5. Lower the vehicle.

6. Drain the coolant system into a suitable container.

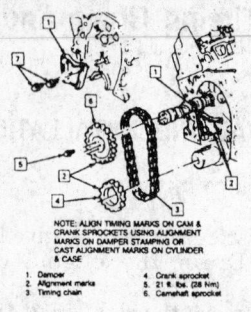

NOTE: ALIGN TIMING MARKS ON CAM &
CRANK SPROCKETS USING ALIGNMENT
MARKS ON DAMPER STAMPING OR
CAST ALIGNMENT MARKS ON CYLINDER
& CASE

1. Damper
2. Alignment marks
3. Timing chain
4. Crank sprocket
5. 21 ft. lbs. (28 N·m)
6. Camshaft sprocket

85473123

Exploded view of timing chain assembly — 3.1L (VIN D) engine

7. Remove the serpentine drive belt.

8. Remove the alternator and brackets.

9. Remove the power steering pump.

10. Remove the serpentine belt belt tensioner.

11. Remove the coolant by-pass adapter.

12. Remove the coolant hose from the water pump.

13. Remove the water pump pulley.

14. Remove the crankshaft balancer using the following procedure.

   a. Raise and safely support the vehicle.

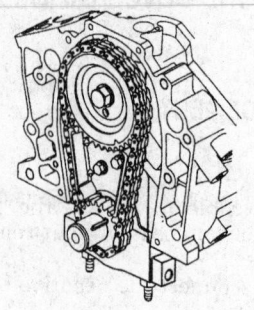

344225

Timing chain alignment — 3.4L (VIN E) engine

   b. Remove the right front tire and wheel.

   c. Remove the inner fender splash shield.

   d. Remove the balancer center bolt. An assistant may be required to keep the flywheel from turning.

   e. The inertia weight section of the crankshaft balancer is assembled to the hub with a rubber sleeve. Use a puller to draw the balancer from the front of the crankshaft. If improperly removed, the inertia weight section of the balancer may shift, destroying the tuning of the balancer.

15. Remove the crankshaft position sensor.

16. Remove the front cover bolts. There are different size bolts so use care to note the location of each bolt.

17. Remove the front cover.

18. Rotate the crankshaft until the timing marks on the crankshaft sprocket and camshaft sprocket locator hole are aligned to the marks on the timing chain dampener. This is the number 1 piston at Top Dead Center.

19. Remove the camshaft sprocket bolt.

20. Remove the camshaft sprocket and timing chain.

21. Remove the crankshaft sprocket using J-5825 -A or equivalent gear puller.

   To install:

22. Clean all parts well. The sealing surfaces of the engine block and the front cover must be clean of old sealer and oil. Use a suitable degreaser solvent. The front seal can be replaced. Pry out the old seal. Drive in the replacement seal until it is flush with the front cover seal bore.

23. Coat all parts with a suitable lubricant. GM recommends their Engine Oil Supplement #1052367. Apply to all parts and especially the thrust face surface of the camshaft sprocket.

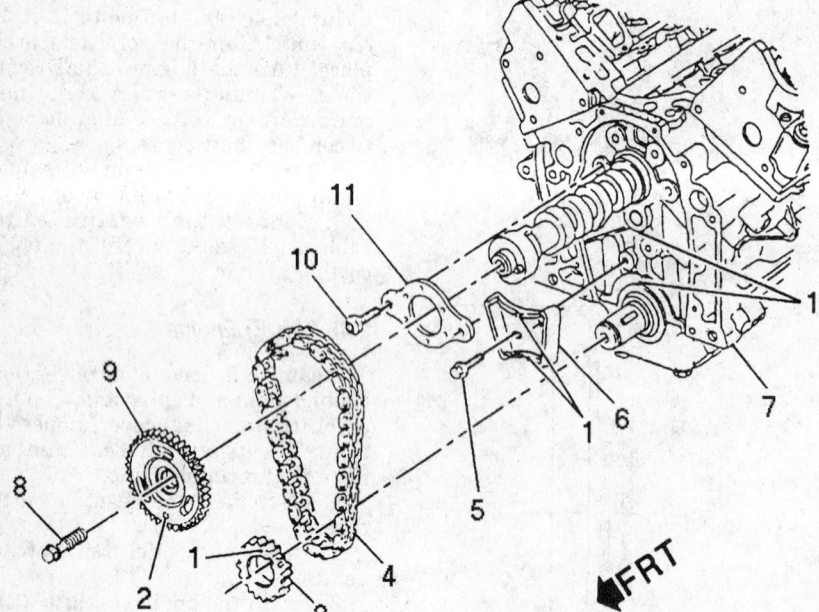

1. Timing alignment marks
2. Locator hole
3. Crankshaft sprocket
4. Timing chain
5. Timing chain dampener bolt
6. Timing chain dampener
7. Engine block
8. Camshaft sprocket bolt
9. Camshaft sprocket
10. Thrust plate bolt
11. Thrust plate

344221

Timing chain components — 3.4L (VIN E) engine

24. Install the crankshaft sprocket using J-38612 or equivalent. This tool threads into the front of the crankshaft to draw the sprocket onto the crankshaft. Use care if using substitutes. Install the sprocket until it is fully seated on the flange of the crankshaft nose.

25. Install the camshaft sprocket and timing chain.

26. Install the chain damper to the block and tighten to 15 ft. lbs. (21 Nm).

27. Align the crankshaft mark to the timing mark on the bottom of the chain damper.

28. Hold camshaft sprocket with the chain hanging down and drape the chain over the crankshaft sprocket.

29. Align the timing mark on the camshaft gear (center line of locator hole) with the timing mark on top of the chain damper.

30. Install the camshaft sprocket bolt and tighten 103 ft. lbs. (140 Nm).

31. Lubricate the timing chain with engine oil.

32. Clean all gasket sealing surfaces.

33. Apply sealer GM #1052080 or equivalent sealer to both sides of the lower edges of the front cover gasket where the gasket contacts the oil pan gasket.

34. Install new front cover gasket.

35. Install the front cover and bolts.

36. Tighten the large bolts to 35 ft. lbs. (47 Nm) and tighten the small bolts to 15 ft. lbs. (21 Nm).

37. Install the crankshaft position sensor.

38. Install the crankshaft balancer and tighten the bolt to 76 ft. lbs. (103 Nm).

39. Install the inner fender splash shield and the right front tire and wheel.

40. Install the water pump pulley.

41. Install the coolant hose to the water pump.

42. Install the coolant by-pass adapter.

43. Install the belt tensioner.

44. Install the power steering pump.

45. Install the alternator brackets.

46. Install the alternator.

47. Install the serpentine drive belt.

48. Refill the cooling system.

49. Install the oil pan.

50. Refill the crankcase with new engine oil. A filter change is recommended.

51. Connect the negative battery cable.

52. Start the engine and verify no leaks.

53. Road test the vehicle.

3.8L (VIN L) Engine

1. Disconnect the negative battery cable, then drain the engine cooling system.

2. Align the marks on the crankshaft damper and timing cover with the engine in the No. 1 firing position. Remove the front cover assembly.

3. Ensure the marks on the crankshaft and camshaft gears are aligned.

4. Remove the timing chain damper and camshaft sprocket.

5. Remove the timing chain. Using a suitable puller, remove the crankshaft sprocket, as required.

To install:

6. Install the crankshaft sprocket.

7. Install the chain onto camshaft sprocket.

8. Holding the sprocket vertically with the chain hanging down, align the marks on the camshaft and crankshaft sprockets and install the assembly onto the camshaft.

9. Install the camshaft to gear attaching bolt and torque to 74 ft. lbs. (100 Nm) and then an additional 105 degree turn. Install the damper and torque to 16 ft. lbs. (22 Nm).

10. After the sprockets are in place, turn the engine 2 full revolutions to make certain the timing marks are in correct alignment between the shaft centers.

11. Lubricate the chain with engine oil and install the front cover.

12. Connect the negative battery cable, then properly refill the engine cooling system.

Timing Chain Front Cover

It is recommended by the manufacturer that the front cover oil seal be

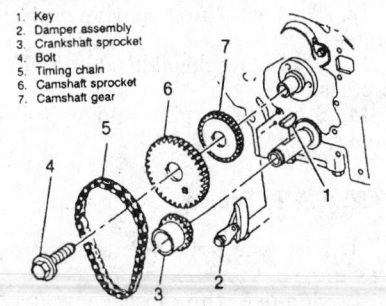

1. Key
2. Damper assembly
3. Crankshaft sprocket
4. Bolt
5. Timing chain
6. Camshaft sprocket
7. Camshaft gear

85473124

Exploded view of timing chain assembly — 3.8L (VIN L) engine

replaced whenever the cover is removed.

REMOVAL AND INSTALLATION

3.1L (VIN D) Engine

1. Disconnect the negative battery cable.

2. Drain the cooling system.

3. Remove the accessory drive belt and tensioner.

4. Remove the power steering pump.

5. Raise and safely support the vehicle. Remove the inner splash shield.

6. Drain the engine oil. Remove the crankshaft pulley and damper. Remove the starter and support it aside.

7. Place a support under the engine-to-transaxle mount.

8. Remove the engine mount bolts and the engine mount. Raise the engine slightly.

9. Remove the lower front cover bolts and lower the oil pan. Remove the radiator hose at the water pump.

10. Remove the heater hose at the cooling system fill pipe. Remove the bypass and overflow hoses.

11. Remove the remaining front cover bolts and remove the front cover.

To install:

12. Clean all gasket mating surfaces. Install a new gasket in position on the engine block.

13. Apply sealer to the lower edges of the front cover. Install the front cover in position on the engine block. Install and tighten the upper retainers.

14. Install the oil pan in position, then secure and tighten the lower front cover bolts.

15. Install the engine mount to the engine and lower the engine into position.

16. Install the crankshaft damper and pulley. Install the flywheel cover and inner splash shield.

17. Install the starter. Connect the heater bypass hose an the slower radiator hose to the water pump. Lower the vehicle.

18. Install the power steering pump bracket. Install the accessory drive belt and tensioner.

19. Refill the cooling system and the crankcase to the correct levels.

20. Connect the negative battery cable. Run the engine to normal operating temperature and check for leaks.

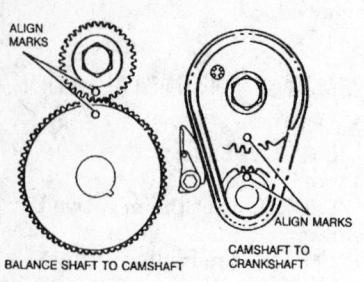

Timing mark alignment — 3.8L (VIN L) engine

85473125

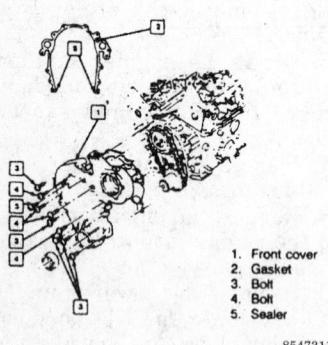

1. Front cover
2. Gasket
3. Bolt
4. Bolt
5. Sealer

85473117

Front cover assembly — 3.1L (VIN D)
engine

3.4L (VIN E) Engine

The front cover (timing chain cover)
houses the front crankshaft oil seal.

1. Disconnect the negative battery
cable.
2. Remove the serpentine belt.
3. Raise and safely support the
vehicle.
4. Remove the right front tire and
wheel assembly.
5. Remove the crankshaft balancer using the following procedure.
   a. Raise and safely support the
vehicle.
   b. Remove the right front tire
and wheel.
   c. Remove the inner fender
splash shield.
   d. Remove the balancer center
bolt. An assistant may be required
to keep the flywheel from turning.
   e. The inertia weight section of
the crankshaft balancer is assembled to the hub with a rubber
sleeve. Use a puller to draw the
balancer from the front of the
crankshaft. If improperly removed,
the inertia weight section of the
balancer may shift, destroying the
tuning of the balancer.
6. Remove the front seal from the
front cover, by prying out on the seal.

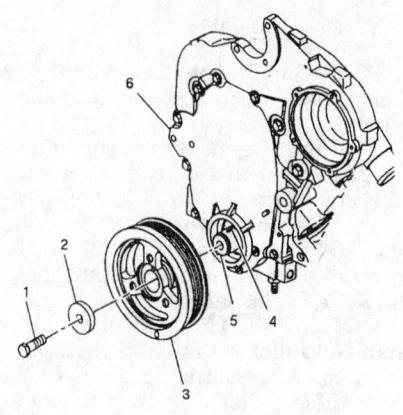

1. Crankshaft balancer bolt
2. Crankshaft balancer washer
3. Crankshaft balancer
4. Crankshaft key
5. Crankshaft
6. Front cover

347858

Front cover assembly — 3.4L (VIN E) engine

Use care not to damage the light alloy front cover or scratch the crankshaft surface.
   To install:
7. Clean all parts well. Lubricate
the new seal with clean engine oil.
8. Install the seal into the front
cover (with the lip facing out) using
tool J-34995 or equivalent. Install the
replacement seal until it is flush with
the front cover seal bore.
9. Install the crankshaft balancer
and tighten the bolt to 76 ft. lbs. (103
Nm). An assistant may be required to
keep the flywheel from turning.
10. Install the inner fender splash
shield and the tire and wheel
assembly.
11. Install the serpentine drive
belt.
12. Connect the negative battery
cable.
13. Check engine oil level and fill as
necessary.
14. Start the engine and verify no
leaks.

3.8L (VIN L) Engine

1. Disconnect the negative battery
cable.
2. Remove the accessory drive
belt.
3. Remove crankshaft damper using GM tool J-38197 or equivalent.

The flywheel must be held from turning using J-37096 or equivalent tool.
4. Remove the crankshaft sensor
shield and sensor.
5. Remove the water pump pulley.
6. Remove the oil pan-to-front
cover bolts and the remaining front
cover attaching bolts.
7. Remove the front cover
assembly.
   To install:
8. Installation is the reverse of the
removal procedure.
9. Coat the threads of the cover retaining bolts with 1052080 or
equivalent sealant, then install the
bolts and tighten to 22 ft. lbs. (30
Nm).
10. Tighten the oil pan retaining
bolts to 125 inch lbs. (14 Nm).
11. Tighten the crankshaft damper
bolt to 111 ft. lbs. (150 Nm), plus an
additional 76 degree turn using a torque angle meter.

Camshaft

REMOVAL AND INSTALLATION

3.1L (VIN D) and 3.8L (VIN L) Engines

1. Disconnect the negative battery
cable.
2. Drain the cooling system.
3. Remove the engine from the vehicle and support it in a suitable
holding fixture.
4. Remove the intake manifold.
Remove the valve cover and the valve
train components.
5. Remove the front cover
assembly.
6. Remove the timing chain and
sprocket. Remove the thrust plate, if
equipped.
7. Remove the camshaft from the
block. Insert 3 bolts approximately 3
in. long into the camshaft gear bolt
holes to supply leverage while removing the camshaft. Use care not to
damage the bearings.
   To install:
8. Lubricate the camshaft with
Molykote or equivalent, before
installation.
9. Install the camshaft into the
cylinder block, use care not to damage the bearings.
10. Install the thrust plate, if
equipped. Install the timing chain
and sprocket. Make sure the timing
marks align correctly.
11. Install the front cover assembly.
12. Install the intake manifold assembly. Install the valve train components and the valve cover.
13. Install the engine into the
vehicle.

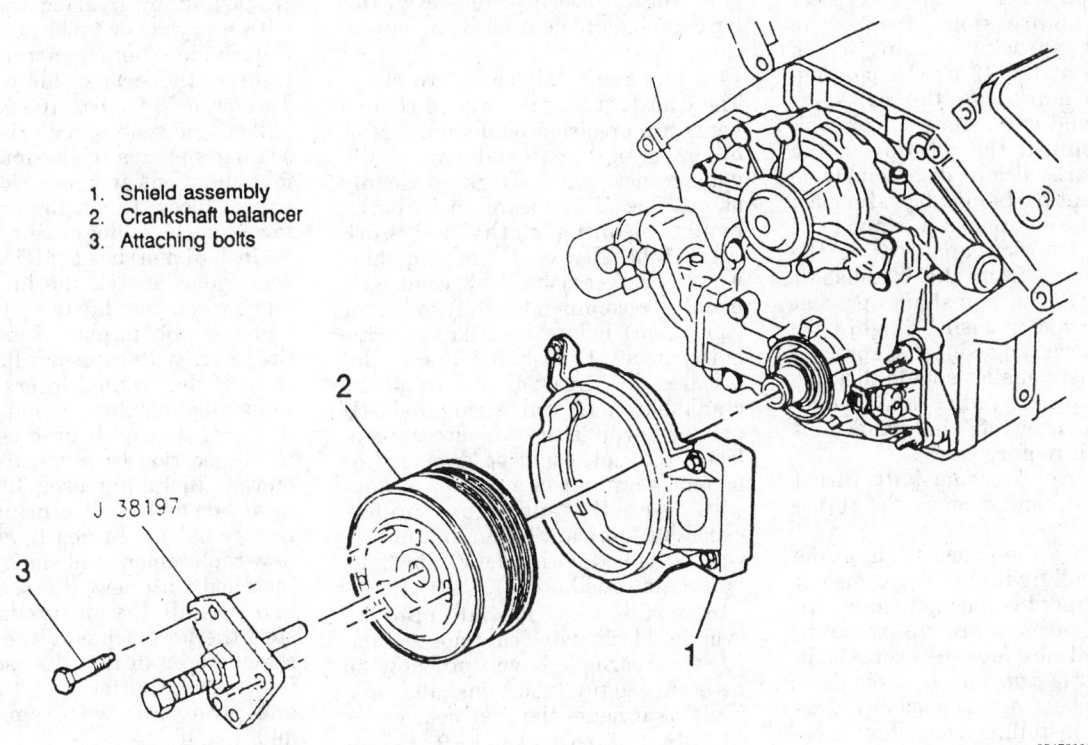

1. Shield assembly
2. Crankshaft balancer
3. Attaching bolts

J 38197

85473118

Removing crankshaft pulley hub assembly — 3.8L (VIN L) engine

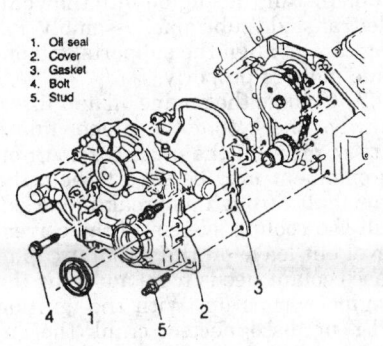

1. Oil seal
2. Cover
3. Gasket
4. Bolt
5. Stud

85473119

Front cover assembly — 3.8L (VIN L) engine

14. Fill the cooling system to the correct level and connect the negative battery cable.

3.4L (VIN E) Engine

Please note that the factory recommends that the engine assembly be removed from the vehicle to remove the camshaft from this engine.

1. Disconnect the negative battery cable.
2. Drain the cooling system.
3. Remove the engine from the vehicle using the proper procedure. This a lengthy process involving the removal of the en-

gine/transaxle/subframe assembly. Safely secure the engine in a suitable holding fixture.

4. Remove the rocker arm covers. If the covers are difficult to remove, shear off by bumping on the end of the rocker arm cover with the palm of the hand or soft rubber mallet. Use care not to scratch the sealing flange.
5. Remove the oil pump drive gear hold down clamp bolt and clamp.
6. Remove the oil pump driven gear assembly.
7. Remove the intake manifold.

NOTE: When removing valve train components, be sure to keep them in order for reassembly purposes. Note that new lifters are a must when replacing a camshaft. Used or worn lifters installed on a new camshaft will rapidly fail the camshaft.

8. Remove the rocker arms and components. Keep the parts in a rack so they may be reinstalled in the same location. Remove the pushrods and the valve lifters. Use care. Valve lifters, if they are to be reused, should be kept in order so they may be reinstalled in their original position. Some engines may have standard and oversize (O.S.) valve lifters. Where O.S. lifters are used, the cylinder block should be marked with a

daub of white paint and "0.25" (mm) O.S. stamped on the lifter boss.

NOTE: The inertia weight section of the torsional damper is assembled to the hub with a rubber sleeve. The removal and installation procedures (with proper tools) must be followed or movement of the inertia weight section on the hub will destroy the tuning of the torsional damper, and the engine timing reference.

9. Remove the dampener retaining bolt. It will probably be necessary to lock the crankshaft by freezing the flywheel with a prybar or other suitable locking device before attempting to remove the crank dampener bolt. Use a puller to draw the dampener off the crankshaft. Use care not to loose the key on the crankshaft.
10. Remove the front cover assembly. Note that the oil pan will need to be removed, or at least loosened.
11. Make sure that the timing marks on the cam and crank sprockets are aligned. Use the alignment marks on the dampener stamping or cast alignment marks on the cylinder block. Piston No. 1 should still be at top dead center of the compression stroke from the distributor removal procedure. Carefully check the camshaft sprocket timing marks.

With piston No. 1 at top dead center of the compression stroke, the camshaft sprocket timing mark should be at the 12 o'clock position, with the marks on the camshaft sprocket and crankshaft sprocket aligned. Confirm the position of the timing marks before disassembly so replacement parts may be assembled in the same relationship. Unbolt the camshaft sprocket and remove the timing chain from the crankshaft sprocket. If the camshaft sprocket does not come off easily, a light blow on the lower edge of the sprocket with a plastic mallet should dislodge the sprocket. A puller may be required to draw off the crankshaft sprocket, if required.

12. Remove the camshaft thrust plate screws and remove the thrust plate.

13. Remove the camshaft from the block by pulling it out. The camshaft is supported by four journals. All camshaft journals are the same diameter and care must be exercised in removing the camshaft to avoid damage to the bearings. A good aid in removing or installing a camshaft is using extra long bolts in the camshaft sprocket bolt holes to act as a handle.

To install:

14. Inspect the camshaft for scratches, pitting and/or wear on the bearing and lobe surfaces. Note the timing marks on the sprockets (cam and crankshaft). It is absolutely necessary that the camshaft sprocket timing mark and the crankshaft timing mark be properly aligned. These marks are often indistinct and hard to see. If new sprockets are to be installed, or as an aid in reinstalling the original sprockets, it may be helpful to place a small dot of light-color paint on each timing mark. This will make aligning the marks easier under the low-light conditions found in many work areas. If the camshaft timing marks are misaligned, the en-

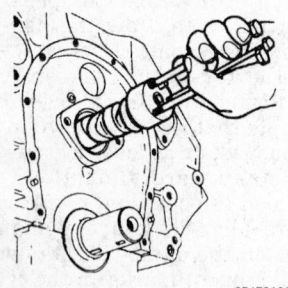

85473128

Camshaft removal and installation — 3.1L (VIN D), 3.4L (VIN E) and 3.8L (VIN L) engines

gine will run poorly or may even suffer engine damage from being out-of-time.

15. Make sure all parts are clean. The camshaft and its related components are precision made and dirt or other foreign material can easily damage new parts. Work as cleanly as possible. This means tools, parts, hands, cleaning cloths and work area. Liberally coat the camshaft with special camshaft lubricant (GM usually recommends Molykote® or equivalent) before installation. This is important. Camshaft lobes are lubricated by throw-off oil from the crankshaft and connecting rods. If not enough oil is thrown onto the cam lobes, the cam could be damaged or become worn out in a matter of minutes. An oil supplement and/or camshaft lubricant (often, but not always supplied with a new camshaft) is recommended.

16. Install the camshaft into the cylinder block, using care not to damage the bearings. It will probably be necessary to turn the camshaft, especially as it nears the rear bearing.

17. Install the camshaft thrust plate and tighten the screws to 89 inch lbs. (10 Nm).

18. Install the crankshaft sprocket, if previously removed.

19. Lubricate the cam sprocket thrust surface with camshaft lube. Hold the cam sprocket with the chain hanging down and slip the chain around the crankshaft sprocket, aligning the marks on the camshaft and crankshaft sprockets. Align the dowel in the camshaft end with the dowel hole in the cam sprocket. Draw the cam sprocket onto the camshaft using the mounting bolts. Torque the bolts to 21 ft. lbs. (28 Nm). Lubricate the timing chain with engine oil. Recheck the timing marks to make sure they are correctly aligned.

20. Inspect the front cover sealing surfaces both on the cover and on the engine block. Clean with a suitable degreaser. Apply sealer to the bottom sealing surface of the front cover. Install the front cover assembly. If the oil pan gasket is damaged, the oil pan should be removed, thoroughly cleaned and reinstalled using a new gasket.

21. Coat the front cover seal contact area (on the dampener) with engine oil. Apply sealant to the key and keyway. Place the dampener in position over the key in the crankshaft and pull the dampener into position. Install the dampener retaining bolt. It will probably be necessary to lock the

crankshaft by freezing the flywheel with a prybar or other suitable locking device before attempting to tighten the crank dampener bolt. Torque to 76 ft. lbs. (103 Nm).

22. Clean sealing material from the sealing surfaces of the intake manifold and front and rear ridges of the engine block. Clean the sealing surfaces with a degreaser. Apply a $\frac{3}{16}$-inch (5 mm) bead of RTV sealer on each ridge. Install the intake manifold gaskets and lay the intake manifold assembly in place. Liberally coat the lifters with camshaft lube and install. If the original lifters are being reinstalled on the original camshaft, use care that each lifter is returned to the location from which it was removed. Installing used lifters on a new camshaft will quickly fail the new camshaft. In nearly all cases, a new replacement camshaft should be installed with new lifters, as a set.

23. Install the pushrods, making sure the lower ends of the pushrods are centered in the lifter seats. Coat the bearing surfaces of the rockers and pivot balls with camshaft lube and install.

24. Tighten the rocker arms to 89 inch lbs. (10 Nm).

25. Install the rocker arm covers.

26. Install the engine into the vehicle/transaxle/subframe assembly using care to align the supports and engine mounts properly.

27. Change the engine oil and filter. Always check that the proper quantity of oil is in the crankcase. An oil supplement may be used to help the camshaft through its break-in period. Fill the cooling system to the correct level but leave off the radiator cap in case coolant needs to be added as the engine warms up. With the ignition OFF or disconnected, crank the engine several times. Listen for any unusual noises or evidence that any parts are binding.

28. Connect the negative battery cable. Start the engine, listening for any unusual noises and checking for leaks. Run the engine at about 1000 rpm until the engine is at operating temperature. Listen for improperly adjusted valves or any unusual noises. Check for oil, fuel, coolant and vacuum leaks. With the engine at operating temperature, set the ignition timing to the underhood tune-up and emission label specification

29. Road test vehicle.

NOTE: Whenever the vehicle sub-frame is removed or lowered, the wheel alignment should be checked.

Balance Shaft

REMOVAL AND INSTALLATION

3.8L (VIN L) Engine

1. Disconnect the negative battery cable.
2. Remove the engine.
3. Remove the flywheel and intake manifold.
4. Remove the lifter guide retainer.
5. Remove the timing chain cover.
6. Remove the balance shaft drive gear bolt, the camshaft sprocket and timing chain.
7. Remove the balance shaft retainer bolts, retainer and gear.
8. Using tool J–6125–B, remove the balance shaft.
9. Remove the balance shaft rear plug. Using tool J–36995–5, remove the balance shaft rear bearing.

NOTE: The balance shaft and bearings are only to be serviced or replaced as a complete assembly.

To install:

10. Dip the bearings in clean engine oil before installation.
11. Using tool J–36995–5, install the balance shaft rear bearing with the rolled edge facing into the engine and the manufacturer's markings facing the flywheel side.
12. Dip the front balance shaft bearing into engine oil. Using tool J–36996, install the balance shaft into the block.
13. Temporarily install the balance shaft retainer and bolts. Install the balance shaft drive gear. Apply Loctite® to the bolt and torque to 14 ft. lbs. (20 Nm) plus turn the bolt an additional 35 degrees.
14. Install the balance shaft rear plug.
15. Measure the balance shaft endplay. Endplay should be 0–0.008 in. (0–0.203mm).
16. Measure the balance shaft radial play at both the front and rear.

The front radial play should be 0–0.0011 in. (0–0.028mm). The rear radial play should be 0.0005–0.0047 in. (0.0127–0.119mm).

17. Temporarily install the camshaft gear and align the timing marks pointed straight down. Remove the camshaft gear and align the balance shaft marks point straight down. Install the camshaft gear and align the marks by turning the balance shaft.
18. Ensure the No. 1 piston is at TDC and install the camshaft sprocket and timing chain.
19. Measure the gear lash at 4 places, every ¼ turn. The lash should be 0.002–0.005 in. (0.050–0.127mm).
20. Tighten the balance shaft front bearing retainer to 22 ft. lbs. (30 Nm).

21. Install the timing chain cover.
22. Install the lifter guide retainer.
23. Install the flywheel and intake manifold.
24. Install the engine into the vehicle.
25. Connect the negative battery cable.

Piston and Connecting Rod

POSITIONING

During disassembly, note the location of the connecting rod bearing tang slots, on some engines, the slots must be on the side opposite of the camshaft. This does not apply to the 3.8L (VIN L) engine.

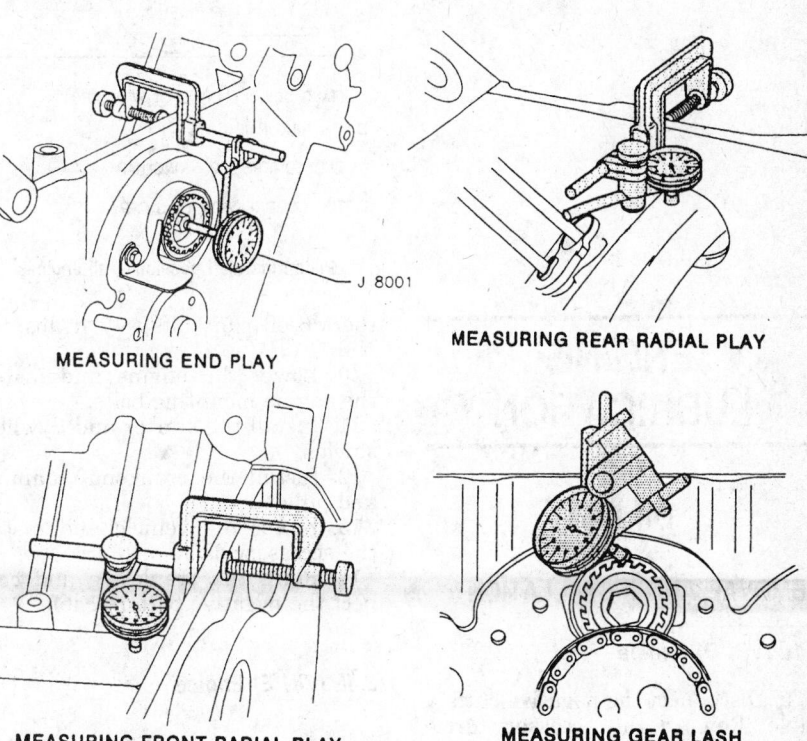

MEASURING END PLAY

MEASURING REAR RADIAL PLAY

MEASURING FRONT RADIAL PLAY

MEASURING GEAR LASH

85473130

Balance shaft clearance measurement — 3.8L (VIN L) engine

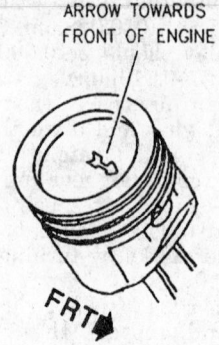

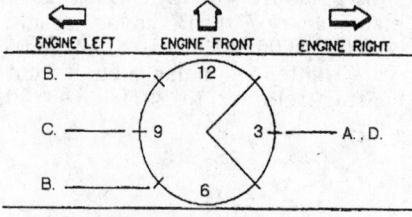

A. OIL RING SPACER GAP
(TANG IN HOLE OR SLOT WITH ARC)

B. OIL RING RAIL GAPS

C. 2ND COMPRESSION RING GAP

D. TOP COMPRESSION RING GAP

330596

Piston installed position — all engines

ENGINE LUBRICATION

Oil Pan

REMOVAL AND INSTALLATION

3.1L (VIN D) Engine

1. Disconnect the negative battery cable. Remove the accessory drive belt.
2. Raise and safely support the vehicle.
3. Remove the crankshaft damper and pulley.
4. Drain the engine oil. Remove the flywheel shields.
5. Remove the starter. Support the engine.
6. Remove the engine mounting bolts.
7. Raise the engine slightly.
8. Remove the oil pan bolts and the oil pan.
   To install:
9. Install a new oil pan gasket and install the oil pan. Tighten M8 oil pan bolts to 19 ft. lbs. (25 Nm) and the M6 oil pan bolts to 7 ft. lbs. (10 Nm).
10. Lower the engine and install the engine mounting bolts.
11. Install the starter and flywheel shields.
12. Install the crankshaft damper and pulley.
13. Lower the vehicle and install the accessory drive belt.
14. Refill the crankcase and connect the negative battery cable.

3.4L (VIN E) Engine

Use care when servicing the oil pan on the 3.4L VIN E engine. The engine main bearing caps are drilled and tapped for structural oil pan side bolts. Do not overlook the side bolts when attempting to remove the oil pan.

1. Disconnect the negative battery cable.
2. The engine mounts will have to be disconnected to perform this service. Install engine support fixtures J-28467-A, J28467-90 and J-28467-200 or their equivalent engine supports to safely suspend the weight of the engine so the engine mounts can be removed later in this procedure.
3. Raise and safely support the vehicle.
4. Drain the engine oil.
5. Unbolt and remove the oil filter drip shield.
6. Disconnect the exhaust pipe from the manifold.
7. Place safety stands under the sub-frame at the front and rear.
8. Remove the engine-to-frame nuts.
9. Remove the transaxle-to-frame nuts.
10. Loosen the rear frame bolts, but do not remove.
11. Remove the front frame bolts and lower the front of the frame.
12. Remove the engine mount.
13. Remove the engine mount bracket.
14. Remove the starter motor.
15. Remove the transaxle brace.
16. Disconnect the oil level wiring harness connector at the oil pan.
17. Remove the oil pan side bolts.
18. Remove the oil pan bottom bolts.
19. Remove the oil pan.
   To install:
20. Clean all parts well. Clean all gasket sealing surfaces on the oil pan flanges, the oil pan rail and the front cover. Clean the threaded holes in

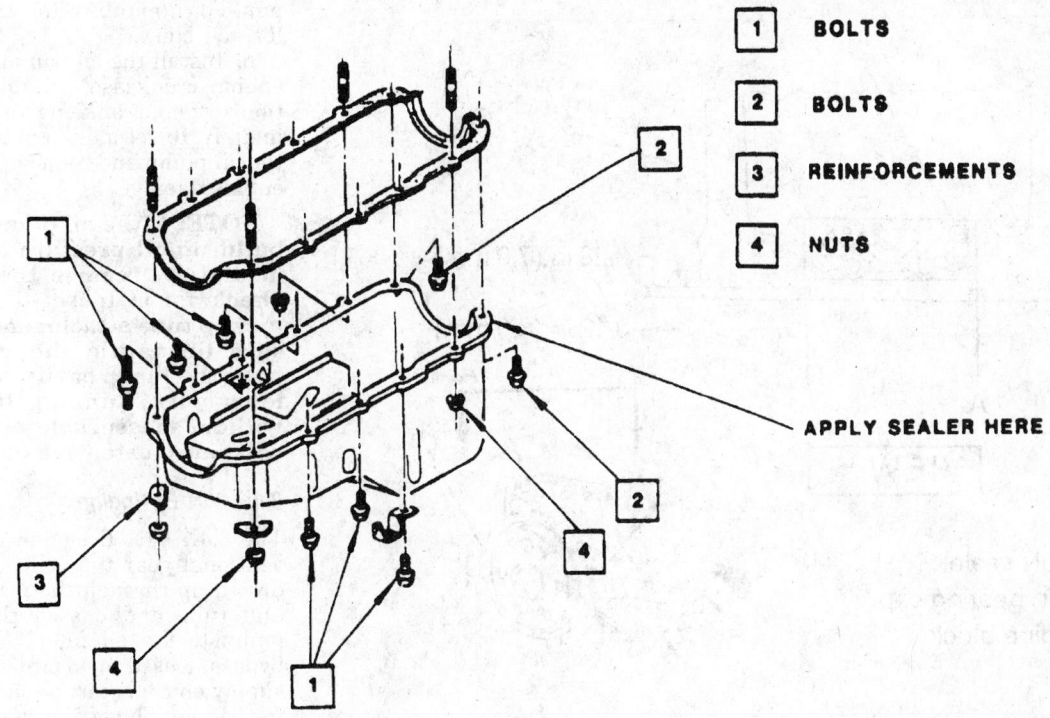

1	BOLTS
2	BOLTS
3	REINFORCEMENTS
4	NUTS

— APPLY SEALER HERE

247651

Oil pan mounting — 3.1L (VIN D) engine

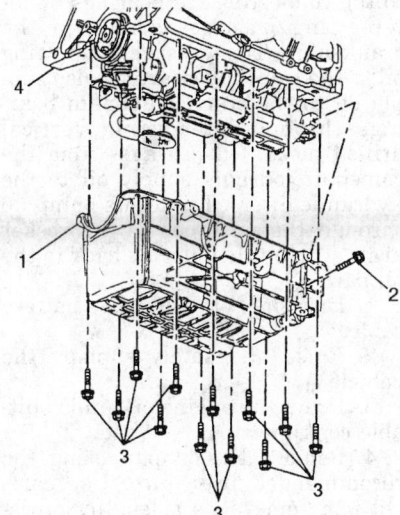

1. Oil pan
2. Oil pan side bolt
3. Oil pan retaining bolt
4. Engine block

343728

Oil pan mounting — 3.4L (VIN E) engine

the main bearing cap and all threaded holes.

NOTE: Apply a small amount of sealer on either side of the rear main bearing cap, where the seal surface on the cap meets the cylinder block. This is important.

21. Install the oil pan and gasket.
22. Install the oil pan bolts and tighten the bottom bolts to 18 ft. lbs. (25 Nm) and tighten the side bolts to 37 ft. lbs. (50 Nm). Tool J 39505 or equivalent is required to properly torque the structural side bolts.
23. Install the oil level wiring harness connector.
24. Install the transaxle brace.
25. Install the starter.
26. Install the engine mount bracket.
27. Install the engine mount.
28. Raise the frame to proper position and install new bolts.
29. Tighten the side-to-crossmember bolts to 40 ft. lbs. (54 Nm) and tighten the left-hand frame insulator bolt to 103 ft. lbs. (140 Nm).
30. Remove the safety stands.
31. Install the exhaust pipe to the manifold.
32. Install the transaxle mount nuts and tighten to 32 ft. lbs. (44 Nm).

33. Install the engine mount nuts and tighten to 32 ft. lbs. (44 Nm).
34. Install the oil filter drip shield.
35. Lower the vehicle.
36. Fill the crankcase with engine oil. A filter change is recommended.
37. Remove the engine support fixtures.
38. Connect the negative battery cable.
39. Start the engine and verify no leaks.

NOTE: Whenever the vehicle sub-frame is removed or lowered, the wheel alignment should be checked.

3.8L (VIN L) Engine

1. Disconnect the negative battery cable. Remove the accessory drive belt.
2. Raise and safely support the vehicle.
3. Remove the crankshaft damper and pulley.
4. Drain the engine oil. Remove the flywheel shields.
5. Remove the starter. Support the engine.
6. Remove the engine mounting bolts.
7. Raise the engine slightly.
8. Remove the oil pan bolts and the oil pan.

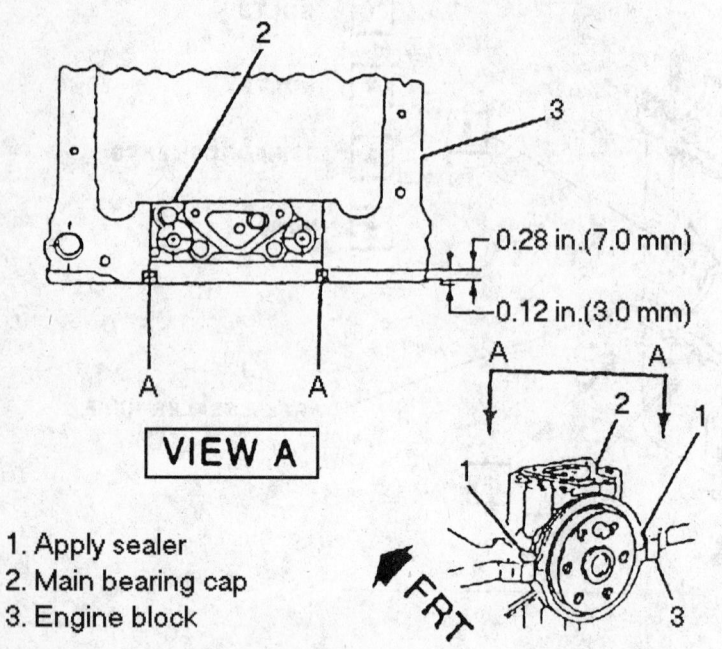

1. Apply sealer
2. Main bearing cap
3. Engine block

Apply sealer to rear main bearing cap — 3.4L (VIN E) engine

To install:

9. Install a new oil pan gasket and install the oil pan. Tighten the oil pan bolts to 124 inch lbs. (14 Nm).

10. Lower the engine and install the engine mounting bolts.

11. Install the starter and flywheel shields.

12. Install the crankshaft damper and pulley.

13. Lower the vehicle and install the accessory drive belt.

14. Refill the crankcase and connect the negative battery cable.

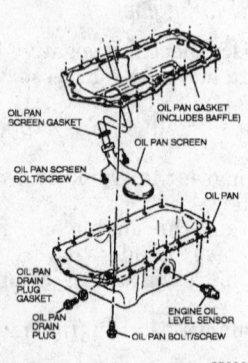

Oil pan mounting — 3.8L (VIN L) engine

Oil Pump

REMOVAL AND INSTALLATION

3.1L (VIN D) Engine

1. Remove the oil pan.

2. Remove the oil pump attaching bolt and, if equipped, the pickup tube nut/bolt, then remove the pump along with the pickup tube and shaft, as necessary.

To install:

3. Ensure that the pump pickup tube is tight in the pump body. If the tube should come loose, oil pressure will be lost and oil starvation will occur. If the pickup tube is loose it should be replaced.

4. If the pump has been disassembled and is being replaced or for any reason oil has been removed, it must be primed. It can either be filled with oil before installing the cover plate and oil kept within the pump during handling or the entire pump cavity can be filled with petroleum jelly.

NOTE: If the pump is not primed, the engine could be damaged before it receives adequate lubrication when the engine is started.

5. Install the pump aligning the pump shaft with the distributor drive

gear as necessary. Tighten oil pump/pickup tube retainer(s) to 40 ft. lbs. (54 Nm).

6. Install the oil pan and refill the engine crankcase. Disable the ignition system; crank engine for approximately 10 seconds to aid in priming the oil pump and reducing the risk of engine damage.

NOTE: If the oil pump does not build up oil pressure almost immediately, remove the pan and check for a loose oil pump-to-pickup tube attachment. If necessary dismantle the pump and pack the pump cavity with petroleum jelly. Running the engine without measurable oil pressure will cause extensive damage.

3.4L (VIN E) Engine

The 3.4L VIN E engine uses a conventional gear type oil pump. Oil is drawn up through the pickup screen and tube and passed through the pump to the oil filter. An oil filter bypass is used to ensure adequate oil supply on cold start or should the filter become plugged or develop excessive pressure drop. The bypass is designed to open at 10 to 12 psi. The engine uses a priority oil delivery system which supplies oil first to the crankshaft journals. Oil from the crankshaft main bearings is supplied to the connecting rod bearings by intersecting passages drilled in the crankshaft. The passages supplying oil to the camshaft bearings also supply oil to the crankshaft main bearings through intersecting vertical drilled holes. Oil passages from the camshaft journals supply oil to the hydraulic lifters. The lifters pump up through the pushrods to the rocker arms. The oil then drains back to the oil pan.

1. Disconnect the negative battery cable.

2. Raise and safely support the vehicle.

3. Drain the engine oil into a suitable container.

4. Remove the oil pan using the recommended procedure. Use care. Oil pan removal is a lengthy procedure and requires loosening and lowering of the subframe. In addition, there are "hidden" bolts that go through the side of the oil pan which must not be overlooked.

5. Remove the oil pump bolt.

6. Remove the oil pump and drive shaft extension.

To install:

7. Clean all parts well. The oil pan may contain sludge which should be removed.

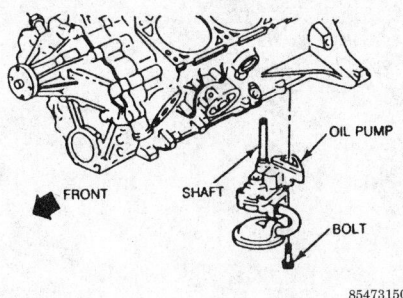

Common oil pump mounting — 3.1L (VIN D) engine

85473150

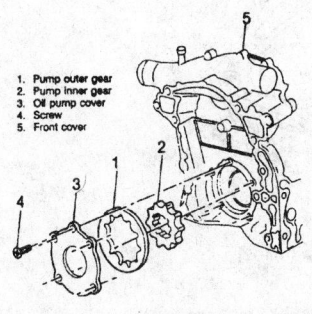

1. Pump outer gear
2. Pump inner gear
3. Oil pump cover
4. Screw
5. Front cover

85473151

Oil pump gears and housing assembly — 3.8L (VIN L) engine

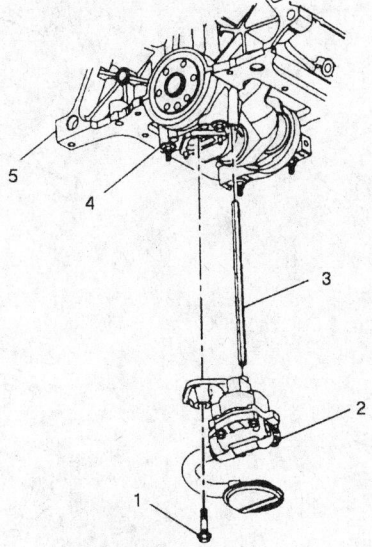

1. Oil pump bolt
2. Oil pump
3. Oil pump drive rod
4. Main bearing cap
5. Engine block

361588

Oil pump removal and installation — 3.4L (VIN E) engine

8. The oil pump can be checked for wear. In actual practice, oil pumps are normally replaced with a new unit. If checking the oil pump is desired, use the following procedure.

a. To check the pump for wear, drain the oil from the pump and remove the driveshaft. DO NOT remove the pickup tube from the oil pump cover unless loose or broken.

b. Unbolt and remove the oil pump cover.

c. Remove the pump gears.

d. The pressure regulator valve can be disassembled by removing the retaining pin. Use caution since the spring and valve are under spring pressure. If the valve is stuck, soak the pump housing in carburetor cleaning solvent.

e. Clean all parts of sludge, oil and varnish. Carburetor cleaning solvent is recommended.

f. Inspect the pump pieces for wear, cracks, chips and scoring. Do not attempt repairs. Replace the pump if any pieces are in doubt.

g. With a feeler gauge, measure the gear lash. Specification is 0.0037 to 0.0077 inch (0.094 to 0.195mm).

h. Check the gear pocket depth. Specification is 1.2020 to 1.2040 inch (30.531 to 30.582mm).

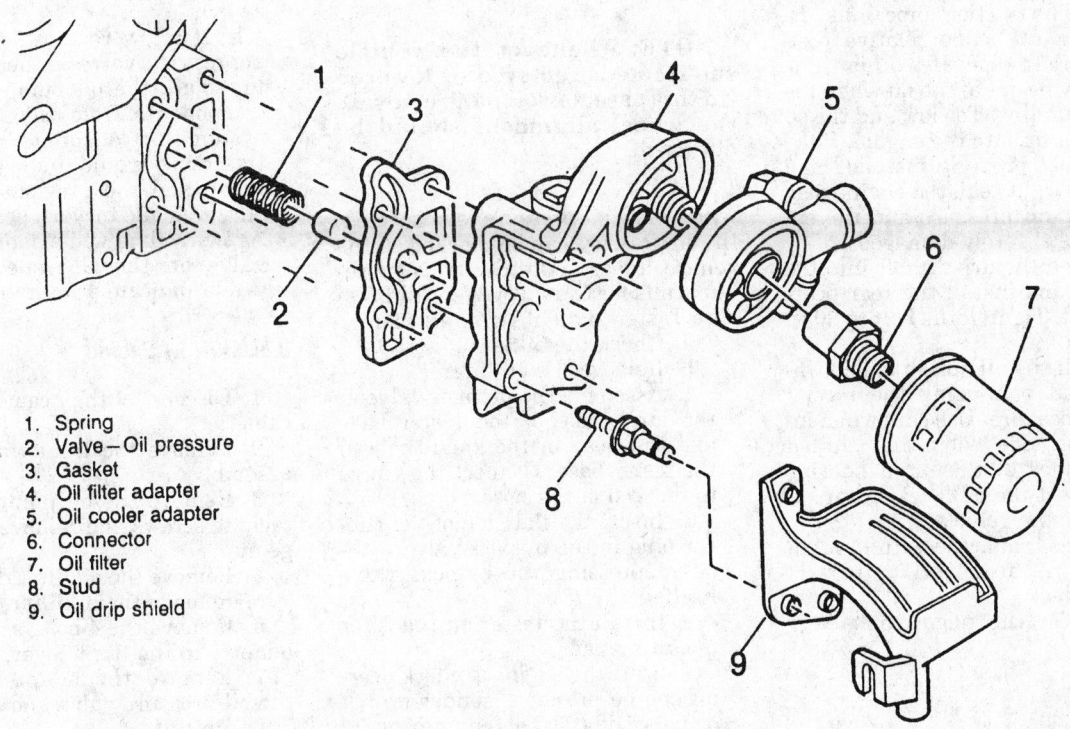

1. Spring
2. Valve—Oil pressure
3. Gasket
4. Oil filter adapter
5. Oil cooler adapter
6. Connector
7. Oil filter
8. Stud
9. Oil drip shield

85473152

Oil filter adapter and oil pressure valve — 3.8L (VIN L) engine

i. Check the gear pocket diameter. Specification is 1.5030 to 1.5050 inch (38.176 to 38.226mm).

j. Check the gear lengths. Specification is 1.1990 to 1.2000 inch (30.455 to 30.480mm).

k. Check the gear diameters. Specification is 1.4980 to 1.5000 inch (38.049 to 38.100mm).

l. Check the gear side clearances. Specification is 0.0010 to 0.0030 inch (0.025 to 0.088mm).

m. Check the gear end clearances. Specification is 0.0020 to 0.0050 inch (0.040 to 0.127mm). Note that when deciding pump serviceability based on end clearance, consider the depth of the wear pattern in the pump cover.

n. If the oil pump body or components are out of specification, replace the pump with a new unit.

9. If the pump was disassembled, lubricate all the internal parts with clean engine oil and assemble the gears to the pump and the pump cover to the pump body. Torque the cover bolts evenly to 89 inch lbs. (10 Nm).

10. If removed, install the pressure regulator valve and the spring. Make sure the retainer pin is secure.

11. If removed, apply sealer GM #1050026 or equivalent and install a new suction pipe. Tap into place with J 21882 or equivalent oil pump tube installer. The suction pipe must be installed in the same relative position as the old pipe. If too high, the pickup may be out of the oil when the engine is running. Too low and the oil pan will not fit onto the engine. If the pickup is not properly installed and with an air tight seal, the engine may not develop oil pressure and the engine will be severely damaged.

12. Engage the drive shaft into the drive gear and install the rear bearing cap bolt. Tighten the bolt to 30 ft. lbs. (41 Nm).

13. Install the oil splash shield. Install the oil pan using the recommended procedure. Use care when installing the side bolts that thread into the sides of the main bearing caps. Connect the oil level sensor.

14. Lower the vehicle.

15. Fill the crankcase with new engine oil. A new oil filter is recommended.

16. Connect the negative battery cable.

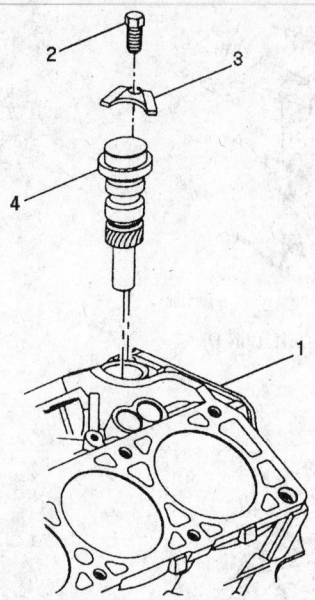

1. Engine block
2. Oil pump drive clamp bolt
3. Oil pump drive clamp
4. Oil pump drive

361597

Oil pump drive arrangement — 3.4L (VIN E) engine

17. Start the engine and verify oil pressure and no leaks.

NOTE: Whenever the vehicle sub-frame is removed or lowered (in this case, for oil pan removal), the wheel alignment should be checked.

18. If low oil pressure is suspected, the oil pressure can be checked on the vehicle. Please note that this procedure calls for GM J 25087-C Oil Pressure Tester and Oil Pump Primer.

a. Check the oil level.

b. Remove the oil filter.

c. Assemble the plunger valve in the large hole of the tester base and the hose in the small hole of the tester base. Connect the gauge to the end of the hose.

d. Insert the flat side of the rubber plug in the by-pass valve without depressing the by-pass valve itself.

e. Install the tester on the filter mounting pad.

f. Start the engine to check overall engine pressure, sender switch or noisy lifters. The engine should be at operating temperature before checking the oil pressure. The pressure should be about 60 psi at 1850 rpm using 10W-30 engine oil.

g. If adequate oil pressure is indicated, check the pressure sending switch.

h. If a low reading is indicated, depress the valve on the tester base to isolate the oil pump and/or its components from the lubricating system. An adequate reading at this time would indicate a good pump and the previous low pressure was due to worn bearings, etc. A low reading while depressing the valve on the GM special tester would indicate a faulty pump.

3.8L (VIN L) Engine

1. Disconnect the negative battery cable.

2. Remove the front timing cover assembly.

3. Remove the oil pump cover attaching screws and remove the pump gears.

4. Remove the oil filter drip shield.

5. Remove the oil filter.

6. Remove the 4 bolts securing the adapter to the front cover.

7. Remove the adapter, gasket, the oil pressure valve and spring.

To install:

8. Clean all part in solvent and remove the old gaskets.

9. Check all parts for scoring, cracks or excessive wear. Check pres-

sure regulator spring for loss of tension and replace as necessary.

10. Install the oil filter adapter, a new gasket, the oil pressure valve and spring.

11. Install the 4 adapter to front cover screws and torque to 22 ft. lbs. (30 Nm).

12. Install the oil filter.

13. Install the oil filter drip shield.

14. Install the pump gears into the cover assembly and pack with petroleum jelly. Install the oil pump cover attaching screws and torque to 97 inch lbs. (11 Nm).

15. Install the front timing cover assembly.

16. Connect the negative battery cable.

NOTE: If the oil pump pickup screen is thought to be possibly clogged or dirty the oil pan should be removed and the screen removed and cleaned with solvent.

CHECKING

1. Disconnect the negative battery cable.

2. Remove the oil pressure sending unit.

3. Connect a suitable oil pressure gauge in place of the sending unit.

4. Reconnect the negative battery cable.

5. Ensure oil level is within specification.

6. Start the engine and set to specified test RPM.

7. Verify oil pressure reading meets specifications.

8. Remove the gauge and reinstall the sending unit.

Rear Main Oil Seal

REMOVAL AND INSTALLATION

1-Piece Seal

Most of these engines will utilize a 1-piece rear main seal assembly. The seal is mounted to a housing on the rear of the block which is normally equipped with removal notches or slots to help free the old seal. A new seal is carefully driven into position

using a threaded seal installation tool.

1. Remove the transmission assembly.

2. If equipped, remove the clutch assembly.

3. Remove the flywheel and verify the rear main seal is leaking.

NOTE: Most seal retainer housings are notched in order to provide a safer prying point for seal removal.

4. Remove the seal by inserting a small prybar into the notch (if provided) or through the dust lip at an angle and prying the seal out. Be careful not to score the crankshaft sealing surface.

To install:

5. Check the crankshaft and seal bore for nicks or damage. Repair as necessary.

6. Lightly coat the inner and outer diameters of the seal with clean engine oil, then position the seal on the installation tool.

7. Position the tool to the crankshaft and thread the tool's screws into the tapped holes. Tighten the screws securely using a screwdriver to attach the tool and assure proper seal installation.

8. Turn the handle of the tool until it bottoms and the seal has been completely seated.

9. Turn the handle of the tool out until it stops, then remove the tool and verify that the is seated squarely in the bore.

10. Install the flywheel to the engine.

11. If equipped, install the clutch assembly.

12. Install the transmission assembly.

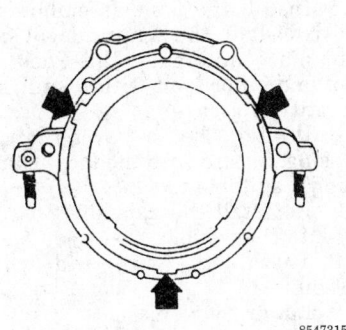

85473153

Most 1-piece rear main seal retainers are equipped with removal notches

AUTOMATIC TRANSAXLE

Transaxle Assembly

REMOVAL AND INSTALLATION

The automatic transaxle can be removed only as an assembly with the engine and subframe. See the procedures under Engine Removal and Installation, earlier in this section.

TV CABLE ADJUSTMENT

1. As required, remove the air cleaner assembly.

2. Depress and hold down the metal readjust tab at the engine end of the throttle valve cable.

3. Move the slider until it stops against the fitting. Release the readjustment tab.

4. Rotate the throttle lever to its full travel position. The slider must move toward the lever when the lever is rotated to its full travel position.

5. Check for proper operation. When the engine is cold, the cable may appear to be functioning properly, check the cable when the engine is hot.

6. Road test the vehicle.

NEUTRAL SAFETY SWITCH ADJUSTMENT

1. Place the transaxle in the **N** position.

2. Loosen the switch attaching screws.

3. Insert a ⅛ in. gauge pin into the service adjustment hole.

4. Rotate the switch on the shifter assembly to align the service hole in the switch with the hole in the carrier. When the gauge pin drops into the service adjustment hole, tighten the bolts and nut.

5. Remove the gauge pin and check the operation of the switch.

6. The vehicle should only start in the **N** or **P** position.

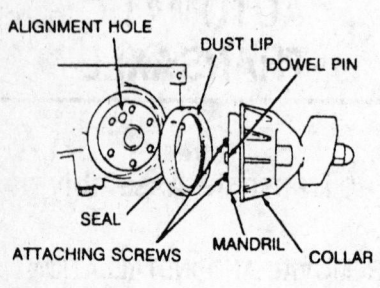

Common 1-piece rear main seal installation

DRIVE AXLE

Halfshaft

REMOVAL AND INSTALLATION

1. Raise and safely support the vehicle.
2. Remove the tire and wheel assemblies.
3. Remove the halfshaft retaining nut and washer.
4. Remove the brake caliper from the rotor and support it aside.
5. Remove the brake rotor from the hub.
6. Disconnect the stabilizer shaft from the control arm and disconnect the ball joint from the steering knuckle.
7. Using a suitable axle seal protector to guard against possible boot damage. Remove the halfshaft from the transaxle using a suitable tool.
8. Remove the halfshaft from the hub and bearing assembly.

To install:
9. Install the halfshaft into the hub and bearing assembly.
10. Connect the lower ball joint to the steering knuckle.
11. Connect the stabilizer shaft to the control arm. Install the rotor.
12. Install the brake caliper. Install a new halfshaft nut and tighten the nut to 185 ft. lbs. (250 Nm).
13. Seat the halfshaft into the transaxle, by pushing it in firmly. Check that the shaft is seated by pulling on it.
14. Install the wheel and tire assemblies.
15. Lower the vehicle.

CV-Boot and Joint

REMOVAL AND INSTALLATION

Outer

1. Raise and safely support the vehicle.
2. Remove the tire and wheel assemblies.
3. Remove the halfshaft from the vehicle and place the assembly in a vise with soft jaws.
4. Cut the seal clamps and remove the boot. Slide the boot down away from the joint.
5. Clean the grease around the joint. Spread the snapring and remove the joint.
6. Discard the boot, if damaged.
7. Using a brass drift punch and hammer, tap on the joint cage until it moves enough to remove a ball, then remove the other balls in the same fashion.
8. Remove the cage and inner race by turning the assembly straight up 90 degrees to the center line of the outer housing. Maneuver the cage and inner race out of the outer race housing assembly. Then remove the inner race from the cage assembly.

To install:
9. Thoroughly clean all part in solvent. Apply grease to the ball grooves of the inner and outer race.
10. Install the inner race into the cage.
11. Install the cage assembly into the housing.
12. Tilt the cage up and install the ball into the assembly. Pack the assembly with grease.
13. Install the small boot clamp onto the axle shaft.
14. Install the axle boot onto the shaft and secure the small boot clamp using a suitable tool.
15. Install the joint assembly onto the axle shaft, the snapring will snap into place automatically. Pack the boot and outer joint with a sufficient quantity of grease.
16. Position the boot over the joint housing. Secure with the large clamp, using a suitable tool.
17. Reinstall the axle shaft assembly into the vehicle.
18. Install the tire and wheel assembly.
19. Lower the vehicle.

Inner

1. Raise and safely support the vehicle.
2. Remove the tire and wheel assemblies.

3. Remove the halfshaft from the vehicle and place the assembly in a vise with soft jaws.
4. Cut the seal clamps and remove the boot. Slide the joint housing off the joint and shaft assembly.
5. Clean the grease around the joint. Spread the snapring and remove the joint.
6. Discard the boot, if damaged.

To install:
7. Thoroughly clean all part in solvent.
8. Install the small boot clamp onto the axle shaft.
9. Install the axle boot onto the shaft and secure the small boot clamp using a suitable tool.
10. Install the joint assembly with the snapring counterbore facing (out) the axle shaft housing, the snapring will snap into place automatically.
11. Install the spacer ring onto the axle shaft, ensuring the ring is fully seated in the groove.
12. Position the axle housing onto the joint and shaft assembly. Pack the boot and joint with a sufficient quantity of grease.
13. Position the boot over the joint housing. Secure with the large clamp, using a suitable tool.
14. Reinstall the axle shaft assembly into the vehicle.
15. Install the tire and wheel assembly.
16. Lower the vehicle.

Front Wheel Hub, Knuckle and Bearing

REMOVAL AND INSTALLATION

1. Raise and safely support the vehicle.
2. Remove the wheel and tire assembly.
3. Remove the brake caliper and support it aside.
4. Remove the brake rotor. Remove the halfshaft retaining nut.
5. Disconnect the tie rod end from the knuckle.
6. Remove the hub and bearing assembly mounting bolts and remove the assembly from the steering knuckle.
7. Matchmark the strut-to-knuckle positioning. Remove the strut mounting bolts from the knuckle. Remove the splash shield from the knuckle. Remove the ball joint from the knuckle.
8. Remove the knuckle from the ball joint and remove the knuckle.

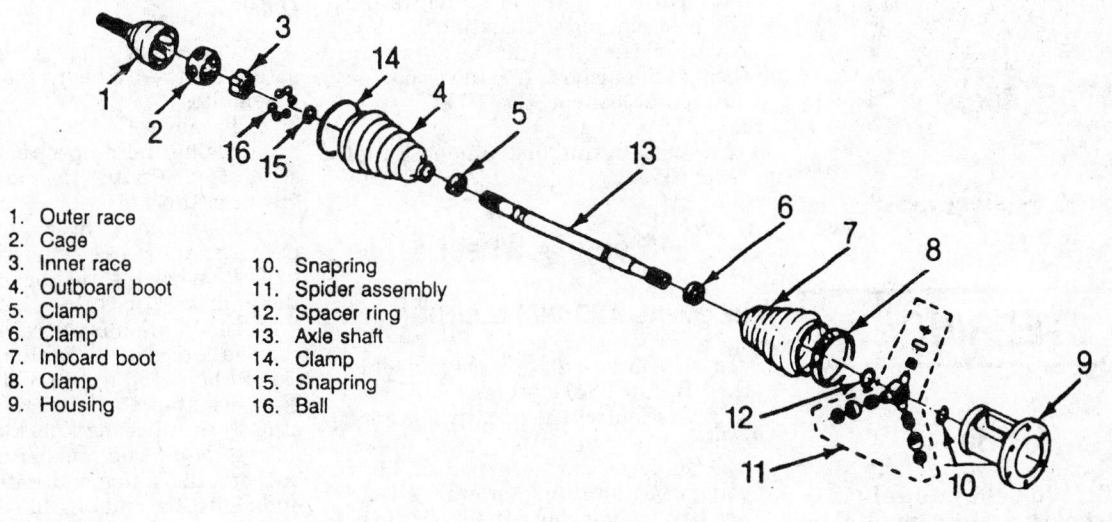

1. Outer race
2. Cage
3. Inner race
4. Outboard boot
5. Clamp
6. Clamp
7. Inboard boot
8. Clamp
9. Housing
10. Snapring
11. Spider assembly
12. Spacer ring
13. Axle shaft
14. Clamp
15. Snapring
16. Ball

85473167

Exploded view of the halfshaft assembly

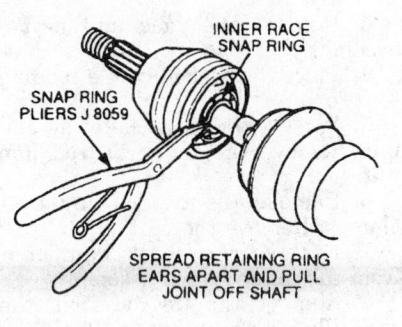

INNER RACE SNAP RING

SNAP RING PLIERS J 8059

SPREAD RETAINING RING EARS APART AND PULL JOINT OFF SHAFT

85473168

Separating the outer joint from the axle shaft

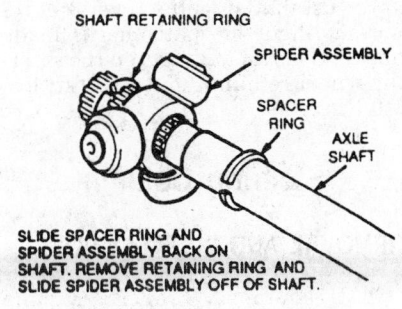

SHAFT RETAINING RING

SPIDER ASSEMBLY

SPACER RING

AXLE SHAFT

SLIDE SPACER RING AND SPIDER ASSEMBLY BACK ON SHAFT. REMOVE RETAINING RING AND SLIDE SPIDER ASSEMBLY OFF OF SHAFT.

85473172

Inner joint and snapring removal

To install:

9. Install the knuckle in position on the ball joint. Slide the knuckle onto the halfshaft.

10. Install the ball joint nut and cotter pin. Install the splash shield to the knuckle.

11. Install the hub and bearing assembly to the knuckle. Tighten the hub mounting bolts to 86 ft. lbs. (116 Nm). Install the halfshaft retaining nut.

12. Connect the tie rod to the knuckle. Install the brake rotor and the brake caliper.

13. Install the wheel and tire assembly. Lower the vehicle.

14. Check the front end alignment.

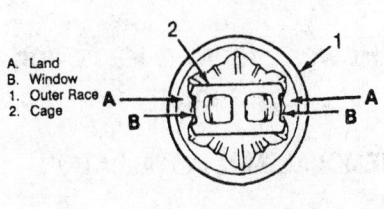

A. Land
B. Window
1. Outer Race
2. Cage

85473171

Positioning the outer joint cage and race for removal/installation

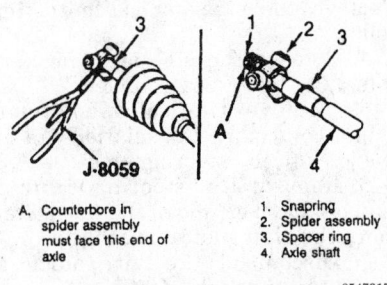

J-8059

A. Counterbore in spider assembly must face this end of axle

1. Snapring
2. Spider assembly
3. Spacer ring
4. Axle shaft

85473173

Inner joint installation

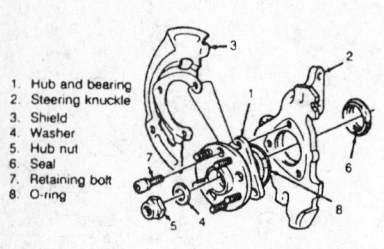

1. Hub and bearing
2. Steering knuckle
3. Shield
4. Washer
5. Hub nut
6. Seal
7. Retaining bolt
8. O-ring

85473182

Front hub and bearing assembly

STEERING

Air Bag

NOTE: 1994–96 models are equipped with a supplemental inflatable restraint (SIR) system or driver's side air bag. Before attempting any work on or near the steering column, ALWAYS disarm the air bag to prevent a costly and possibly dangerous accidental deployment.

DISARMING

1. Turn the wheels to the straight-ahead position, then turn the ignition switch to **LOCK.**
2. Remove the instrument panel lower extension for access to the fuse block.
3. Remove the "AIR BAG" or "SIR" fuse from the block, as applicable.
4. Remove the steering column filler panel or left hand sound insulator, as applicable, for access to the SIR wiring harness.
5. Remove the Connector Position Assurance (CPA) device, then disengage the yellow 2-way connector at the base of the steering column.

NOTE: With the fuse removed, the AIR BAG or SIR light will illuminate if the ignition switch is turned ON at any time. This is normal and does not indicate a problem when the system is disarmed.

To enable:
6. Make sure the ignition is in the **LOCK** position.
7. Engage the yellow SIR connector, then secure using the CPA device.

8. Install the steering column filler or sound insulator panel, as applicable.
9. Install the SIR system fuse to the fuse block.
10. Turn the ignition switch to the **ON** position and verify that the AIR BAG indicator light flashes 7 times, then extinguishes. If it does not go out, troubleshoot the SIR system fault.
11. Install the instrument panel lower extension.

Steering Wheel

REMOVAL AND INSTALLATION

1. If equipped, properly disable the SIR (air bag) system.
2. Disconnect the negative battery cable.
3. Remove the horn pad or horn pad/air bag retaining screws. Pull the air bag and/or horn pad outward, then disconnect the electrical lead(s) and remove.
4. Matchmark the steering wheel and the shaft.
5. Remove the steering wheel retaining clip and nut.
6. Remove the steering wheel, using a suitable puller.
7. Installation is the reverse of removal. Align the matchmarks made during removal and tighten the steering wheel retaining nut to 30 ft. lbs. (40 Nm).

Steering Column

REMOVAL AND INSTALLATION

1. If equipped, properly disable the SIR (air bag) system.
2. Disconnect the negative battery cable.
3. Remove the left instrument panel sound insulator and trim pad. Remove the steering column trim collar.
4. Remove the steering wheel, if column is to be disassembled.
5. Remove the column upper clamp bolt; mark the relationship of the joint to the steering shaft.
6. Remove the steering column support bracket under the dash. Remove the shift indicator cable.
7. Disconnect the wire harness connector under the dash.
8. Remove the shift cable at the actuator and housing holder.
9. Remove the column assembly.
10. Installation is the reverse of the removal procedure.

Tie Rod Ends

REMOVAL AND INSTALLATION

Outer

1. Raise and support the vehicle safely. Remove the tire and wheel assemblies.
2. Remove the cotter pins and nuts. Using the proper removal tool, separate the outer tie rod from the steering knuckle.
3. Disconnect the inner tie rod from the relay rod using the proper tool. Remove the tie rod ends from the adjuster tubes.
4. Installation is the reverse of the removal procedure. Tighten the inner tie rod ball stud nut to 35 ft. lbs. (47 Nm). Tighten the outer tie rod ball stud to the steering knuckle to 35 ft. lbs. (47 Nm). The number of threads on both the inner and outer tie rod ends must be equal within 3 threads.
5. Adjust the front end alignment, as required.

Inner

1. Disconnect the negative battery cable.
2. Raise and support the vehicle safely.
3. Remove the rack and pinion assembly.
4. Place the assembly in a holding fixture.
5. Remove the outer tie rod assembly. Remove the inner tie rod jam nut.
6. Remove the tie rod end boot clamps. Remove the boot.
7. Remove the shock dampener from the inner tie rod assembly.
8. Remove the tie rod from the rack. Place a wrench on the flat of the rack assembly and another wrench on the flats of the inner tie rod housing.
9. Installation is the reverse of the removal procedure.

Power Steering Rack and Pinion

REMOVAL AND INSTALLATION

1. Disconnect the negative battery cable.
2. Remove the air cleaner assembly.
3. Remove the dust boot from the steering gear.
4. Remove the intermediate shaft lower pinch bolt and disconnect the

intermediate shaft from the lower stub shaft.

5. Remove the fluid line retaining clips at the pump and disconnect the lines.

6. Raise and safely support the vehicle.

7. Remove the wheel and tire assemblies. Disconnect the tie rod ends at the steering knuckle.

8. Remove the remaining brackets and clips at the crossmember. Support the body safely with the appropriate equipment, to allow lowering of the subframe.

9. Remove the rear subframe mounting bolts and carefully lower the rear of the subframe approximately 5 in. (128mm).

10. Remove the rack and pinion mounting bolts and remove the rack through the left wheel opening.

To install:

11. Install the rack and pinion through the left wheel opening.

12. Install the rack and pinion mounting nuts, tighten to 70 ft. lbs. (95 Nm).

13. Raise the subframe assembly and install the rear mounting bolts.

14. Remove any supports and install the brackets and clips to the crossmember.

15. Install the wheel and tire assemblies. Lower the vehicle.

16. Connect the fluid lines at the pump and tighten to 18 ft. lbs. (25 Nm).

17. Install the line retaining clips. Connect the intermediate shaft to the stub shaft.

18. Install the dust boot over the steering gear.

19. Install the air cleaner assembly and connect the negative battery cable.

20. Fill and bleed the steering system.

RACK BEARING PRELOAD ADJUSTMENT

1. Raise and support the vehicle safely. Ensure the steering wheel is centered.

2. Loosen the adjuster plug locknut and turn the adjuster plug clockwise until it bottoms in the housing; then back off 50–70 degrees.

3. Check the returnability of the steering wheel after the adjustment.

4. Tighten the locknut to 50 ft. lbs. (70 Nm), while holding the adjuster plug stationary.

5. Lower the vehicle.

Power Steering Pump

REMOVAL AND INSTALLATION

Power steering pump service is different for each engine. Note that the power steering pump, like all belt driven accessories on these engines, is rigidly mounted to the engine. Drive belt tension is maintained by a spring loaded belt tensioner. In addition, use of the proper power steering fluid is important. Use only the fluid specified. For power steering pump removal, use the following procedures.

3.1L (VIN D) Engine

1. Disconnect the negative battery cable.

2. Remove the drive belt guard. Lift or rotate the tensioner using a $^3/_8$-inch breaker bar. Remove the serpentine belt.

3. Remove the fluid lines from the power steering pump.

4. Remove the pump retaining bolts and remove the pump from the engine.

5. Transfer the pulley as necessary.

To install:

6. Install the pump to the engine mounting brackets. Tighten the bolts to 18 ft. lbs. (25 Nm).

7. Connect the fluid lines to the pump.

8. Inspect the serpentine drive belt. Small cracks will not impair belt performance. If the belt is missing sections of the belt ribs or shows signs of fraying or splitting, the belt should be replaced. If the belt is serviceable, or if a new belt is being installed, lift or rotate the tensioner using a $^3/_8$-inch breaker bar and install the serpentine belt. Note that belt tension is maintained by the ten-

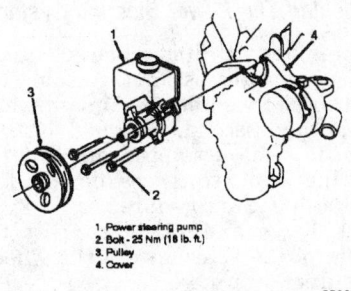

1. Power steering pump
2. Bolt - 25 Nm (18 lb. ft.)
3. Pulley
4. Cover

250651

Power steering pump removal and installation — 3.1L (VIN D) and 3.4L (VIN E) engines

sioner and is not adjustable. Install the belt guard.

NOTE: When adding fluid or making a complete change, always use GM #1050017 or equivalent power steering fluid. Failure to use the proper fluid will cause hose and seal damage and fluid leaks.

9. Fill the system with fluid. Power steering fluid level is indicated either by marks on a see-through fluid reservoir or by marks on a fluid level indicator on the fluid reservoir cap. If the fluid is warmed up (about 170 degrees F., hot to the touch) fluid level should be at the FULL HOT mark. If the fluid is cool, (about 70 degrees F.) the fluid level should be at the FULL COLD mark.

10. After replacing the fluid or servicing the power steering hydraulic system the air must be bled from the system. Air in the system prevents an accurate fluid level reading, causes pump cavitation noise and, over time, could damage the pump. Detailed bleeding procedures are given in the illustration captioned: Fig. 1 GM Recommended Practice For Bleeding The Power Steering System.

3.4L (VIN E) Engine

The power steering pump provides hydraulic pressure for the steering system. The reservoir is separate from the housing and internal parts. A pressure-relief valve, inside the flow control valve (built into the pump body), limits pump pressure. Service this vehicle with GM #1050017 power steering fluid or equivalent. Failure to use the proper fluid will cause hose and seal failure and fluid leaks.

1. Disconnect the negative battery cable.

2. Remove the belt from the pulley.

3. Disconnect the lines at the power steering pump (place a pan under the vehicle to catch the fluid).

4. Remove the power steering pump retaining bolts.

5. Remove the power steering pump.

To install:

6. Transfer the pulley if necessary. A specialized puller and installer tool is required to draw the pulley from the pump shaft and press the pulley back on again. If the pulley is removed or transferred to a replacement pump, the pulley must be pressed on until the face of the pulley hub is flush with the pump drive pul-

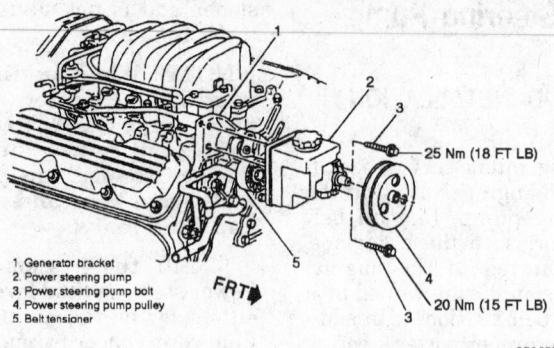

1. Generator bracket
2. Power steering pump
3. Power steering pump bolt
4. Power steering pump pulley
5. Belt tensioner

25 Nm (18 FT LB)

FRT

20 Nm (15 FT LB)

250655

Power steering pump mounting — 3.8L (VIN L) engine

ley. Do not use an arbor or hydraulic shop press to install the pulley or the pump may be ruined.

7. Position the power steering pump into place and install the mounting bolts.

8. Tighten the mounting bolts to 20 ft. lbs. (27 Nm).

9. Connect the lines to the power steering pump and tighten to 20 ft. lbs. (27 Nm).

10. Install the drive belt onto the pulley.

11. Fill the pump with fluid.

12. Connect the negative battery cable.

13. Bleed the air from the system using the recommended procedure.

14. Road test the vehicle and verify proper operation and no leaks.

3.8L (VIN L) Engine

1. Disconnect the negative battery cable.

2. Disconnect the inlet and outlet pipe at the pump.

3. Lift or rotate the tensioner using an 18mm box end wrench on the pulley nut. Remove the serpentine belt.

4. Remove the pump retaining bolts and remove the pump and reservoir assembly from the engine.

5. Transfer the pulley as necessary.

6. Transfer the reservoir as necessary. GM TC-Series pumps have the reservoir retained to the pump by two side clips. An O-ring seals the reservoir to the pump.

To install:

7. If removed, install the reservoir to the pump using a new O-ring.

8. Install the pump to the engine mounting brackets. Tighten the bolts to 20 ft. lbs. (27 Nm).

9. Connect the fluid lines to the pump.

10. Inspect the serpentine drive belt. Small cracks will not impair belt performance. If the belt is missing sections of the belt ribs or shows signs of fraying or splitting, the belt should be replaced. If the belt is serviceable, or if a new belt is being installed, lift or rotate the tensioner using an 18mm box end wrench and install the serpentine belt. Note that belt tension is maintained by the tensioner and is not adjustable.

NOTE: When adding fluid or making a complete change, always use GM #1050017 or equivalent power steering fluid. Failure to use the proper fluid will cause hose and seal damage and fluid leaks.

11. Fill the system with fluid. Power steering fluid level is indicated either by marks on a see-through fluid reservoir or by marks on a fluid level indicator on the fluid reservoir cap. If the fluid is warmed up (about 170 degrees F., hot to the touch) fluid level should be at the FULL HOT mark. If the fluid is cool, (about 70 degrees F.) the fluid level should be at the FULL COLD mark.

12. After replacing the fluid or servicing the power steering hydraulic system the air must be bled from the system. Air in the system prevents an accurate fluid level reading, causes pump cavitation noise and, over time, could damage the pump.

Flushing The Power Steering System

Power steering fluid replacement is a two stage process: first, flushing the old fluid from the system with new fluid; and second, bleeding the system to remove trapped air. The following two sequences outline the steps in each procedure.

1. Raise and safely support the front of the vehicle until the wheels are free to turn.

2. Remove the fluid return line at the pump reservoir inlet connector.

3. Plug the inlet connector port on the pump reservoir.

4. Position the fluid return line toward a large container in order to catch the draining fluid.

5. While an assistant fills the reservoir with the proper power steering fluid, start and run the engine at idle.

6. Turn the steering wheel from stop to stop.

NOTE: Do not hold the wheel against the stops while flushing the system. Holding the steering wheel against the stops will cause high system pressure, overheating and damage to the pump and/or steering gear.

7. Continue draining until all of the old fluid is cleared from the power steering system. Addition of approximately 1 quart of new fluid will be required to flush system.

8. Unplug the pump reservoir inlet and reconnect the return line.

9. Turn the engine off and fill the reservoir to the FULL COLD mark.

10. Continue with the bleeding procedure.

Bleeding The Power Steering System

After replacing the fluid or servicing the power steering hydraulic system, the system must be bled to remove air from the fluid. Air in the system prevents an accurate fluid level reading, causes pump cavitation noise and over time could damage the pump.

1. Fill the reservoir to the proper level and let the fluid remain undisturbed for at least 2 minutes.

2. Start the engine and run it for only about 2 seconds.

3. Add fluid as necessary.

4. Repeat Steps 1–3 until the level remains constant.

5. Raise the front of the vehicle so the front wheels are off the ground. Set the parking brake and block both rear wheels front and rear. Manual transmissions should be in neutral; automatic transmissions should be in P.

6. Start the engine and run it at approximately 1500 rpm.

7. Turn the wheels (off the ground) to the right and left, lightly contacting the stops.

8. Add fluid as necessary.

9. Lower the vehicle and turn the wheels right and left on the ground.

10. Check the level and refill as necessary.

11. If the fluid is extremely foamy, let the vehicle stand for a few minutes with the engine off and repeat the procedure. Check the belt tension and check for a bent or loose pulley.

The pulley should not wobble with the engine running.

12. Check that no hoses are contacting any parts of the vehicle, particularly sheet metal.

13. Check the fluid level and refill as necessary.

14. Check for air in the fluid. Aerated fluid appears milky. If air is present, repeat the above operation. If it is obvious that the pump will not respond to bleeding after several attempts, a pressure test may be required.

BRAKES

For all brake system repair and service procedures not detailed below, please refer to "Brakes" in the Unit Repair section.

Master Cylinder

REMOVAL AND INSTALLATION

Without ABS

The master cylinder is a composite design (plastic reservoir and aluminum body) to be used in a diagonally split system (one front and one diagonally opposite rear brake served by the primary piston, and opposite front and rear brakes served by the secondary piston). It incorporates the function of a standard dual master cylinder, plus it has a fluid level sensor and integral proportioners. The proportioners are designed to provide better front to rear braking balance with heavy brake application.

1. Disconnect the electrical connector from the fluid level sensor.

2. Disconnect brake pipes from master cylinder.

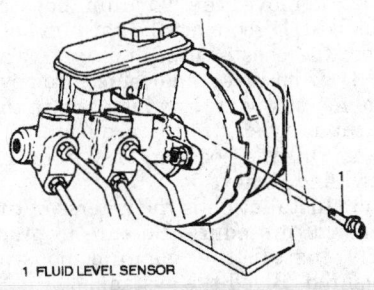

1 FLUID LEVEL SENSOR

246559

Fluid level sensor removal and installation — Without ABS

3. Plug the brake pipes to prevent excess fluid loss and contamination from entering system.

4. Remove the 2 mounting nuts and remove the master cylinder.

To install :

5. Install the master cylinder and mounting nuts.

6. Tighten the mounting nuts to 20 ft. lbs. (27 Nm).

7. Connect the brake pipes to the master cylinder and tighten to 15 ft. lbs. (20 Nm).

8. Connect the electrical connector to the fluid level sensor.

9. Fill master cylinder to proper level with clean brake fluid. Proper level is to the MAX fill indicator level in opening on top reservoir.

10. Bleed the brake system.

11. Recheck fluid level.

12. Road test vehicle and ensure proper operation.

With ABS

When the ABS modulator cylinder pistons are in their uppermost position, each motor has prevailing torque due to the force necessary to ensure each piston is held firmly at the top of its travel. This torque results in "gear tension," or force on each gear that makes motor pack separation difficult. To avoid injury, or damage to the gears, the "Gear Tension Relief Sequence" briefly reverses each motor to eliminate the prevailing torque. This procedure is one of the many functions of GM's Tech 1 scan tool. Use care when using a substitute. In general, make sure the ignition switch is in the **OFF** position. Install the Tech 1 or equivalent with the correct chassis cartridge. Turn the ignition switch to the **ON** position, leaving the engine OFF. Select the proper function. The "Gear Tension Relief Sequence" is F5 on the Tech 1 scan tool. Note that this same scan tool is needed to bleed the system after repairs to the hydraulic system.

Always perform the "Gear Tension Relief Sequence" prior to removing the hydraulic modulator/master cylinder assembly from the vehicle. Each hydraulic modulator gear (large gears) should be able to be turned in one direction and then in the opposite direction when the motor pack is removed. If any gear will not move, replace the hydraulic modulator.

---**CAUTION**---

To perform the Gear Tension Relief Sequence, a Tech 1 scan tool or equivalent scan tool must be

used prior to removal of the ABS modulator/master cylinder assembly.

---**WARNING**---

When servicing the master cylinder on this vehicle the procedure below must be followed in the correct sequence. DO NOT perform any services to the master cylinder without first performing the gear tension release procedure.

1. Perform the gear tension release procedure.

2. Disconnect the two ABS solenoid electrical connectors.

3. Disconnect the electrical connector from the brake fluid level switch.

4. Disconnect the six-way motor pack electrical connectors.

NOTE: When disconnecting the brake lines from the master cylinder and removing the unit, use care not to spill brake fluid on any painted surfaces or electrical connectors.

5. Disconnect and cap the four brake lines from the master cylinder assembly. Plug the master cylinder ports to prevent excessive fluid loss.

6. Remove the master cylinder mounting nuts.

7. Remove the ABS master cylinder and modulator assembly.

To install:

8. Install the ABS master cylinder and modulator assembly on the power booster and tighten the mounting nuts to 20 ft. lbs. (27 Nm).

9. Connect the four brake lines to the master cylinder and tighten the fittings to 18 ft. lbs. (24 Nm).

10. Connect the electrical connector to the brake fluid level sensor.

11. Connect the electrical connectors to the ABS system solenoids.

12. Connect the 6-way motor pack electrical connectors.

13. Fill the master cylinder to the proper level. The proper level is to the **MAX** level indicator on the reservoir.

14. Bleed the brake system following the recommended procedure.

Proportioning Valve

REMOVAL AND INSTALLATION

1. Disconnect the negative battery cable.

2. Remove the electrical connector from the master cylinder.

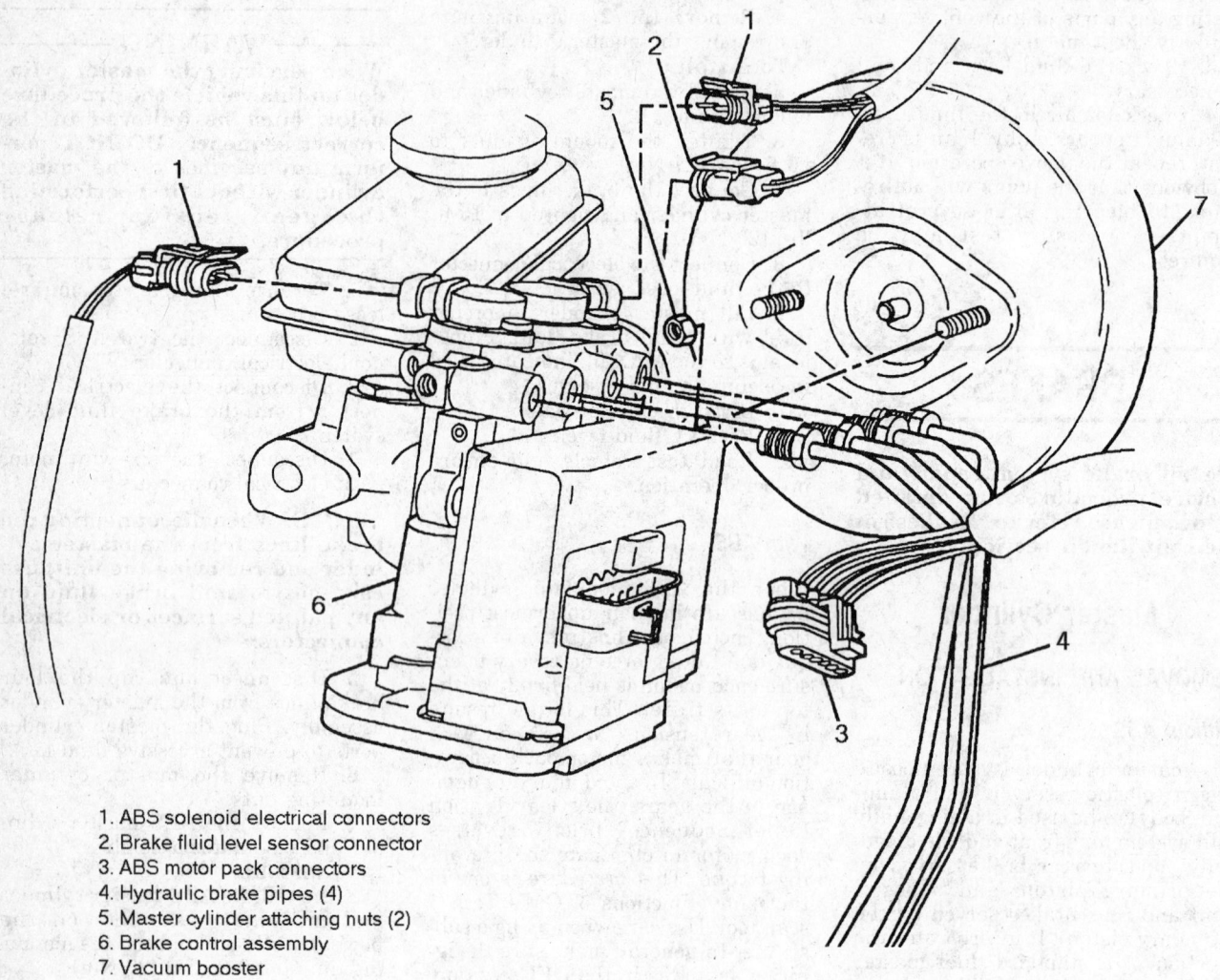

1. ABS solenoid electrical connectors
2. Brake fluid level sensor connector
3. ABS motor pack connectors
4. Hydraulic brake pipes (4)
5. Master cylinder attaching nuts (2)
6. Brake control assembly
7. Vacuum booster

250091

Master cylinder/modulator removal and installation — ABS system

3. Drain and remove the master cylinder reservoir.

4. Remove the proportioning valve caps from the master cylinder.

5. Remove the O-rings, springs and the valve pistons. Use care not to scratch the valves in any way.

6. Remove the valve seals from the valve pistons.

To install:

7. Install new seals on the valve pistons. Lubricate the seals and the pistons with silicon grease.

8. Install the valve pistons and O-rings into the master cylinder.

9. Install the valve cap assemblies and tighten to 20 ft. lbs. (27 Nm).

10. Install the reservoir assembly. Connect the electrical leads.

11. Connect the negative battery cable. Bleed the brake system.

Power Brake Booster

REMOVAL AND INSTALLATION

Vacuum Booster

1. Disconnect the negative battery cable. Do not disconnect the master cylinder fluid lines, unless there is a clearance problem. Remove the master cylinder and position aside.

2. Remove the vacuum booster pushrod. Disconnect the vacuum hose from the booster assembly.

3. From inside the vehicle, remove the mounting studs which secure the vacuum booster to the fire wall.

4. Pull the booster away from the cowl and remove it.

5. Installation is the reverse of the removal procedure. Be sure to properly install the vacuum booster pushrod. Bleed the system.

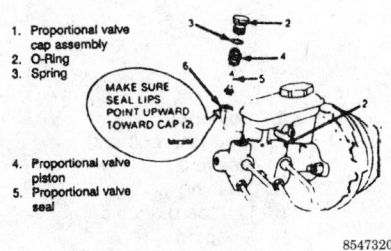

1. Proportional valve cap assembly
2. O-Ring
3. Spring

MAKE SURE SEAL LIPS POINT UPWARD TOWARD CAP (2)

4. Proportional valve piston
5. Proportional valve seal

85473200

Proportioning valve replacement

Brake Caliper

REMOVAL AND INSTALLATION

1. Remove ⅔ of the brake fluid from the master cylinder reservoir.
2. Raise and support the vehicle safely. Remove the tire and wheel assembly.
3. Disconnect and plug the caliper fluid line. Remove the bolts retaining the caliper to the rotor. Remove the caliper from the rotor.

NOTE: If the caliper is being removed for brake pad replacement, the fluid line do not need to be disconnected.

4. Remove the disc brake pads from the caliper. Remove the disc brake pad retaining clips from inside the caliper.

To install:

5. Clean and lubricate the sleeves and bushings with silicon grease. Install the pads in the caliper.
6. Install the caliper in position over the rotor and install the mounting bolts. Tighten the mounting bolts to 38 ft. lbs. (51 Nm).
7. Connect the fluid lines to the caliper, if disconnected, and tighten to 33 ft. lbs. (45 Nm).
8. Install the wheel and tire assembly.
9. Lower the vehicle and refill the master cylinder to the correct level. Bleed the brake system if the fluid lines were disconnected from the caliper.

Disc Brake Pads

REMOVAL AND INSTALLATION

1. Remove ⅔ of the brake fluid from the master cylinder.
2. Raise and safely support the vehicle.
3. Place a C-clamp around the outer pad and caliper; tighten the C-clamp until the piston is fully com-

pressed in the caliper. Remove the brake caliper.
4. Remove the inboard pad and retaining spring from the caliper.
5. Remove the outboard pad from the caliper.
6. Remove the sleeves and bushings.

To install:

7. Clean and lubricate the sleeves and bushing with silicon lubricant and install them in the caliper.
8. Clip the retaining spring onto the inboard pad and install the pad in the caliper.
9. Install the outboard pad into the caliper.
10. Install the caliper in position over the rotor and install the mounting bolts. Bend the tabs, on the outboard brake pad, over the caliper.
11. Install the wheel and tire assemblies.
12. Lower the vehicle, refill the master cylinder and pump pedal to attain full brake pedal before road testing the vehicle.

Brake Rotor

REMOVAL AND INSTALLATION

1. Safely raise and support the vehicle.

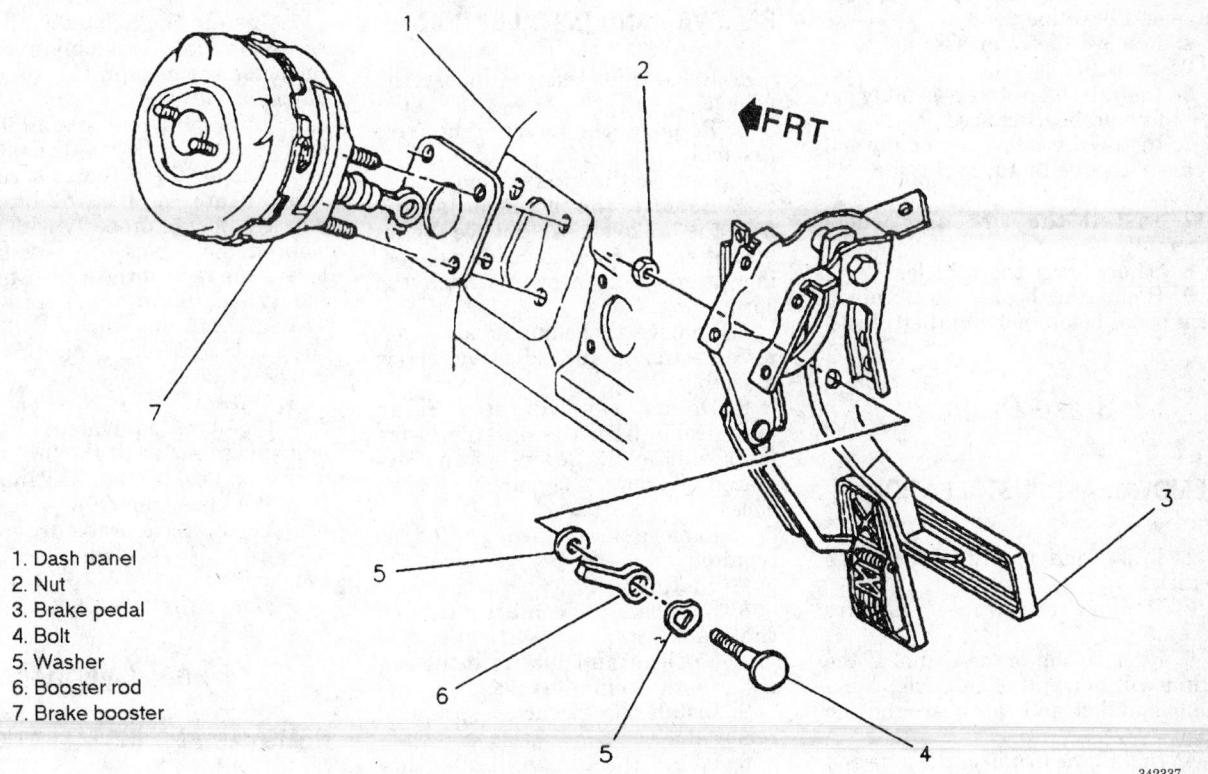

1. Dash panel
2. Nut
3. Brake pedal
4. Bolt
5. Washer
6. Booster rod
7. Brake booster

342337

Power brake booster removal and installation

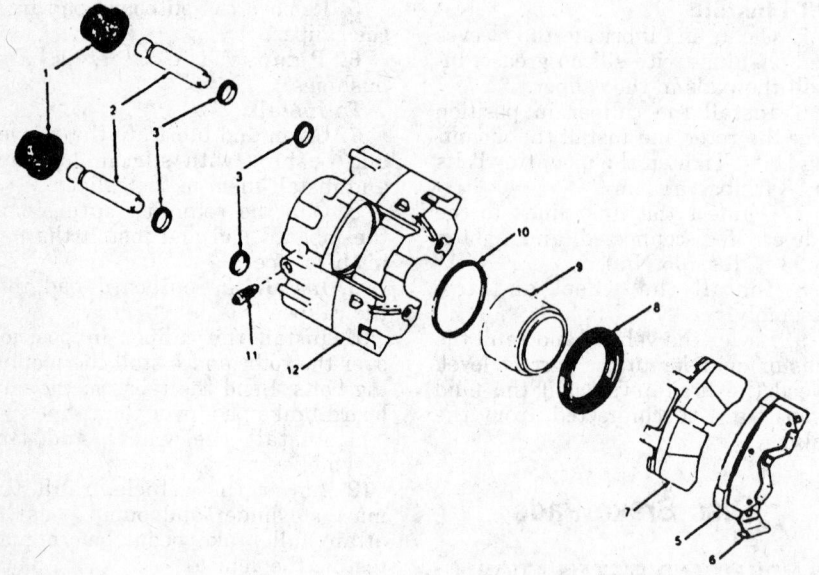

1. Bolt boot
2. Mounting bolt and sleeve
3. Bushing
5. Outboard shoe and lining
6. Wear sensor
7. Inboard shoe and lining
8. Boot
9. Piston
10. Piston seal
11. Bleeder valve
12. Caliper housing

85473201

Exploded view of a common caliper assembly

2. Remove the tire and wheel assembly.

3. Remove the caliper bolts and support the caliper.

4. Remove the rotor assembly.

To install:

5. Install the rotor assembly on the hub and bearing assembly.

6. Install the caliper assembly and tighten the bolts to 38 ft. lbs. (51 Nm).

7. Install the tire and wheel assembly.

8. Safely lower the vehicle.

9. Pump the brakes to obtain a firm pedal before moving the vehicle.

Brake Drums

REMOVAL AND INSTALLATION

1. Raise and safely support the vehicle.

2. Remove the wheel and tire assembly.

3. Remove the brake drum. If the drum will not pull of the axle, use a rubber mallet and tap it around the edge.

4. Install the drum on the axle and install the wheel and tire assembly.

5. Lower the vehicle.

Brake Shoes

REMOVAL AND INSTALLATION

1. Raise and safely support the vehicle.

2. Remove the wheel and tire assembly.

3. Remove the brake drum.

4. Remove the actuator spring from the brake shoes. Remove the retractor spring from the shoe web, being careful not to over stretch the spring.

5. Remove the adjuster shoe, adjuster actuator and adjusting screw assembly.

6. Do not remove the parking brake cable from the parking brake lever, unless the lever is being replaced. Remove the parking brake shoe.

7. Remove the retractor spring, as required.

To install:

8. Lubricate the contact points on the backing plate with lithium grease. Clean and lubricate the adjuster with lithium grease.

9. Install the retractor spring, if removed.

10. Install the parking brake shoe against the backing plate and snap the retractor spring into the slot on

the brake shoe. Install the parking brake lever onto the parking brake shoe.

11. Install the adjuster shoe and adjusting screw assembly. Install the retractor spring into the slot on the adjuster shoe web.

12. Lubricate and install the adjuster actuator onto the adjuster shoe. Install the actuator spring.

13. Ensure the parking brake system is adjusted properly with no tension on the cables or parking brake lever. The tops of the shoes should be firmly seated against the upper spring retaining anchor, if not as specified, loosen the parking brake cables.

14. Adjust the brakes using J–21177–A or equivalent. Turn the adjuster screw until the brake lining diameter is 0.050 in. (1.27mm) less than the inside diameter of the brake drum. Install the brake drum.

15. Install the wheel and tire assembly.

16. Lower the vehicle.

Wheel Cylinder

REMOVAL AND INSTALLATION

1. Raise and safely support the vehicle.

1. Adjuster socket
2. Adjuster screw
3. Pivot nut
4. Retractor spring
5. Adjuster shoe and lining
6. Wheel cylinder
7. Bleeder valve
8. Bolt
9. Access hole plug
10. Backing plate
11. Park brake shoe and lining
12. Park brake lever
13. Actuator spring
14. Adjuster actuator
15. Adjusting screw assembly

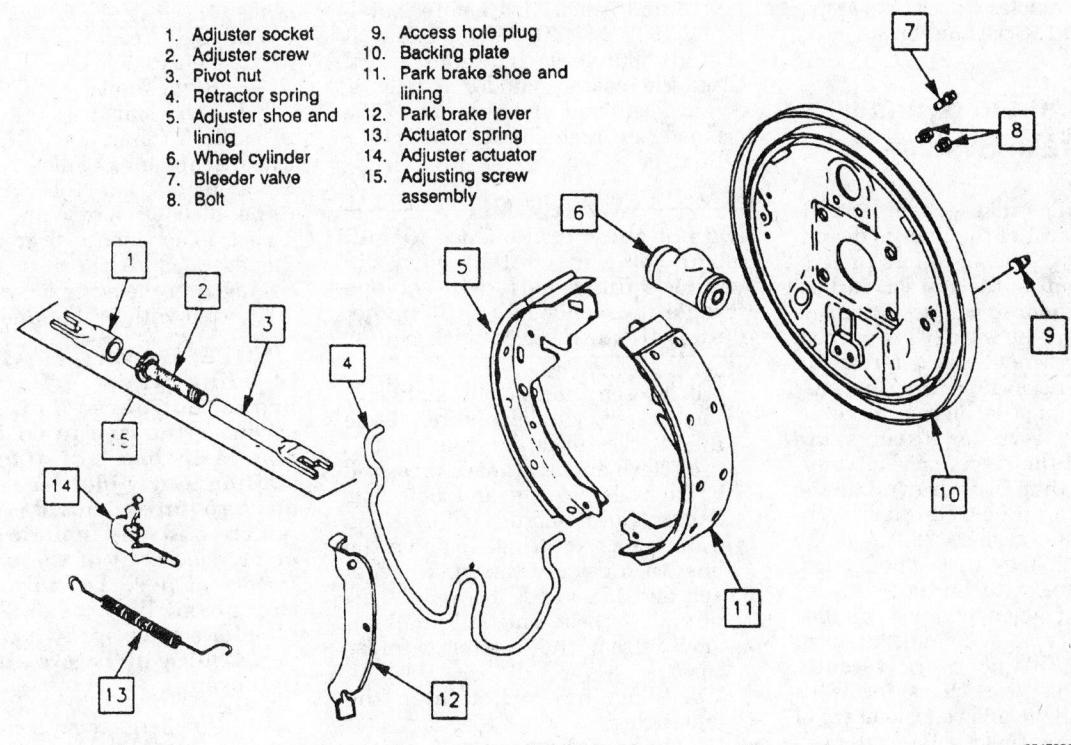

85473206

Exploded view of the drum brake components

2. Remove the wheel and tire assemblies.

3. Remove the brake drum.

4. Remove the brake shoes, as necessary.

5. Disconnect the brake fluid line from the wheel cylinder.

6. Remove the wheel cylinder retainer or bolt.

NOTE: If equipped with a round retainer, insert 2 small awls and carefully pry the tabs back to release the retainer.

7. Remove the wheel cylinder from the backing plate.

To install:

8. Install the wheel cylinder in position on the backing plate.

9. Install the retainer or bolt. Connect the brake line to the wheel cylinder.

NOTE: If equipped with a round retainer, a socket may be used as a driver to ease installation.

10. Install the brake linings and the brake drum.

11. Install the wheel and tire assembly. Lower the vehicle.

12. Bleed the brake system.

Parking Brake Cable

REMOVAL AND INSTALLATION

Rear

1. Raise and support the vehicle safely. Remove the tire and wheel assembly. Remove the brake drum.

2. Loosen the equalizer and disconnect the cable at the center retainer.

3. Compress the plastic retainer fingers and remove the retainer from the frame bracket.

4. Remove the rear brake shoe assembly. Disconnect the parking brake cable. Remove the cable from the frame and from the brake backing plate.

5. Installation is the reverse of the removal procedure. Adjust the rear brakes, as required. Adjust the parking brake.

Front

1. Raise and support the vehicle safely. Loosen the adjuster nut and disconnect the front cable from the connector.

2. Compress the retainer fingers and loosen the assembly at the frame. Remove the supports.

3. Lower the vehicle. As required, remove dash trim panels to gain access to the parking brake pedal assembly.

4. Disconnect the cable from the parking brake pedal, compress the retainer fingers. Remove the cable.

5. Installation is the reverse of the removal procedure. Adjust the parking brake.

ADJUSTMENT

The rear brakes serve a dual purpose. They are used as service brakes and as parking brakes. To obtain proper adjustment of the parking brake, the service brakes must first be properly adjusted. Inspect the cables for binding or sticking.

1. Set the parking brake pedal 4 clicks.

2. Raise and support the vehicle safely. Remove the access plug in the backing plate.

3. Adjust the parking brake until an $1/8$ in. drill can be inserted through the access hole into the space between the shoe web and park brake lever. Satisfactory adjustment will be obtained when an $1/8$ in. drill bit will fit but a $1/4$ in. drill bit will not.

4. Release the parking brake and verify the rear wheels will rotate freely.

5. Replace the access plug and lower the vehicle. Check for proper operation of the parking brake.

Bleeding the Hydraulic Brake System

The hydraulic brake system must be bled any time 1 of the lines is disconnected or any time air enters the system. If a point in the system, such as a wheel cylinder or caliper brake line is the only point which was opened, the bleeder screws down stream in the hydraulic system are the only ones which must be bled. If however, the master cylinder fittings are opened or if the reservoir level drops sufficiently that air is drawn into the system, air must be bled from the entire hydraulic system. If the brake pedal feels spongy upon application, and goes almost to the floor but regains height when pumped, air has entered the system. It must be bled out. If no fittings were recently opened for service, check for leaks that would have allowed the entry of air and repair them before attempting to bleed the system.

As a general rule, once the master cylinder (and the brake pressure modulator valve or combination valve on ABS systems) is bled, the remainder of the hydraulic system should be bled starting at the furthest wheel from the master cylinder and working towards the nearest wheel. Therefore, the correct bleeding sequence is: master cylinder, modulator or combination valve (ABS only), right rear wheel cylinder, left rear, right front caliper and left front. Most master cylinder assemblies on these vehicles are not equipped with bleeder valves, therefore air must be bled from the cylinders using the front brake pipe connections.

MANUAL BLEEDING

1. Clean the top of the master cylinder, remove the cover and fill the reservoirs with clean fluid. To prevent squirting fluid, and possibly damaging painted surfaces, install the cover during the procedure, but be sure to frequently check and top off the reservoirs with fresh fluid.

CAUTION
Never reuse brake fluid which has been bled from the system.

2. The master cylinder must be bled first if it is suspected to contain air. If the master cylinder was removed and bench bled before installation it must still be bled, but it should take less time and effort. Bleed the master cylinder as follows:
a. Position a container under the master cylinder to catch the brake fluid.

WARNING
Do not allow brake fluid to spill on or come in contact with the vehicle's finish as it will remove the paint. In case of a spill, immediately flush the area with water.

b. Loosen the front brake line at the master cylinder and allow the fluid to flow from the front port.
c. Have an assistant depress the brake pedal slowly and hold (air and/or fluid should be expelled from the loose fitting). Tighten the line, then release the brake pedal and wait 15 seconds. Loosen the fitting and repeat until all air is removed from the master cylinder bore.
d. When finished, tighten the line fitting.
e. Repeat the sequence at the master cylinder rear pipe fitting.

NOTE: During the bleeding procedure, make sure your assistant does not release the brake pedal while a fitting is loosened or while a bleeder screw is opening. Air will be drawn back into the system.

3. Check and refill the master cylinder reservoir.

NOTE: Remember, if the reservoir is allowed to empty of fluid during the procedure, air will be drawn into the system and bleeding procedure must be restarted at the master cylinder assembly.

4. On late model ABS equipped vehicles, perform the special ABS procedures. On 4 wheel ABS systems the Brake Pressure Modulator Valve (BPMV) must be bled (if it has been replaced or if it is suspected to contain air) and on most Rear Wheel Anti-Lock (RWAL) systems the combination valve must be held open. In both cases, special combination valve depressor tools should be used during bleeding and a scan tool must be used for ABS function tests.

5. If a single line or fitting was the only hydraulic line disconnected, then only the caliper(s) or wheel cylinder(s) affected by that line must be bled. If the master cylinder required bleeding, then all calipers and wheel cylinders must be bled in the proper sequence:
a. Right rear
b. Left rear
c. Right front
d. Left front

6. Bleed the individual calipers or wheel cylinders as follows:
a. Place a suitable wrench over the bleeder screw and attach a clear plastic hose over the screw end. Be sure the hose is seated snugly on the screw or you may be squirted with brake fluid.

NOTE: Be very careful when bleeding wheel cylinders and brake calipers. The bleeder screws often rust in position and may easily break off if forced. Installing a new bleeder screw will often require removal of the component and may include overhaul or replacement of the wheel cylinder/caliper. To help prevent the possibility of breaking a bleeder screw, spray it with some penetrating oil before attempting to loosen it.

b. Submerge the other end of the tube in a transparent container of clean brake fluid.
c. Loosen the bleed screw, then have a friend apply the brake pedal slowly and hold. Tighten the bleed screw, release the brake pedal and wait 15 seconds. Repeat the sequence (including the 15 second pause) until all air is expelled from the caliper or cylinder.
d. Tighten the bleed screw when finished.

7. Check the pedal for a hard feeling with the engine not running. If the pedal is soft, repeat the bleeding procedure until a firm pedal is obtained.

8. If the brake warning light is on, depress the brake pedal firmly. If there is no air in the system, the light will go out.

9. After bleeding, make sure a firm pedal is achieved before attempting to move the vehicle.

PRESSURE BLEEDING

A proper pressure bleeder tool will utilize a rubber diaphragm between the air source and brake fluid in order to prevent air, moisture oil and other contaminants from entering the hydraulic system.

1. Prepare a pressure bleeder tool such as J-29567 or equivalent by making sure the pressure tank is at least 2/3 full of fresh, clean brake fluid. In most cases, the bleeder must

be bled each time fluid is added. Charge the bleeder tool to 20–25 psi (140–170 kPa).

2. Install a suitable combination valve depressor tool such as J-39177 or equivalent, to the combination valve in order to hold the valve open during the bleeding operation.

3. Install the pressure bleeder tool to the master cylinder reservoir.

4. On 4 wheel ABS systems, bleed the Brake Pressure Modulator Valve (BPMV) of air.

5. Bleed each wheel cylinder or caliper in the proper sequence:
   a. Right rear
   b. Left rear
   c. Right front
   d. Left front

6. Connect a hose from the bleeder tank to the adapter at the master cylinder, then open the tank valve.

7. Attach a clear vinyl hose to the brake bleeder screw, then immerse the opposite end into a container partially filled with clean brake fluid.

8. Open the bleeder screw ¾ turn and allow the fluid to flow until no air bubbles are seen in the fluid, then close the bleeder screw and tighten.

9. Repeat the bleeding process at each wheel.

10. Inspect the brake pedal for sponginess and if necessary, repeat the entire bleeding procedure.

11. Remove the depressor tool from the combination valve and the bleeder adapter from the master cylinder.

12. Refill the master cylinder to the proper level with brake fluid.

13. Do not attempt to move the vehicle unless a firm brake pedal is obtained.

Bleeding the RWAL Brake System

One 1994–96 vehicles a few steps (listed below) should be added to the bleeding sequence in order to ease the procedure and assure all air is removed from the system. These extra steps may be used on all RWAL vehicles to assure proper bleeding.

The use of a power bleeder is recommended, but the system may also be bled manually. If a power bleeder is used, it must be of the diaphragm type and provide isolation of the fluid from air and moisture.

Do not pump the pedal rapidly when bleeding; this can make the circuits very difficult to bleed. Instead, press the brake pedal slowly 1 time and hold it down while bleeding takes place. Tighten the bleeder

screw, release the pedal and wait 15 seconds before repeating the sequence. Because of the length of the brake lines and other factors, it may take 10 or more repetitions of the sequence to bleed each line properly. When necessary to bleed all 4 wheels, the correct order is right rear, left rear, right front and left front.

------ CAUTION ------

Do not move the vehicle until a firm brake pedal is achieved. Failure to properly bleed the system may cause impaired braking and the possibility of injury and/or property damage

1. Make sure the ignition is in the **OFF** position to prevent setting false trouble codes.

2. After properly bleeding the master cylinder, install J-39177 or equivalent combination valve depressor tool to the combination valve. This tool is used to hold the internal valve open allowing the entire system to be completely bled.

3. Recheck the master cylinder fluid level and add, as necessary.

4. Bleed the wheel cylinders as described earlier in this section.

5. Attach the Tech-1 or equivalent scan tool, then perform 3 RWAL function tests.

6. Re-bleed the rear wheel cylinders.

7. Check for a firm brake pedal, if necessary repeat the entire bleeding procedure.

Bleeding the 4WAL Brake System

The EHCU/BPMV module is the 1 component which adds to the complexity of bleeding the 4WAL brake systems. For the most part the system is bled in the same manner as the non-ABS vehicles. But because of the EHCU/BPMV's complex internal valving additional steps are necessary if the unit has been replaced or if it is suspected to contain air. These bleeding steps are not necessary if the only connection/fitting(s) opened were downstream of the unit. These steps may or may not be necessary after master cylinder replacement. If in doubt (or without the necessary special tools) thoroughly bleed the system and see if a firm brake pedal can be obtained, if not, the EHCU/BPMV must be bled as well.

As with the RWAL brake system, the use of a power bleeder is recommended, but the system may also be

bled manually. If a power bleeder is used, it must be of the diaphragm type and provide isolation of the fluid from air and moisture.

Do not pump the pedal rapidly when bleeding; this can make the circuits very difficult to bleed. Instead, press the brake pedal slowly 1 time and hold it down while bleeding takes place. Tighten the bleeder screw, release the pedal and wait 15 seconds before repeating the sequence. Because of the length of the brake lines and other factors, it may take 10 or more repetitions of the sequence to bleed each line properly. When necessary to bleed all 4 wheels, the correct order is right rear, left rear, right front and left front.

------ CAUTION ------

Do not move the vehicle until a firm brake pedal is achieved. Failure to properly bleed the system may cause impaired braking and the possibility of injury and/or property damage

If the EHCU/BPMV requires bleeding, the following procedures may be used to free all trapped air from the component. The procedures differ because the component used on the MFI-Turbo is equipped with external bleeders in additional to the internal bleeders found on the 1994 and later units. In either case, 3 combination valve depressor tools and a scan tool are required. The combination valve depressor tools are used to hold the internal passages (combination valve and EHCU/BPMV bleed accumulator bleed stems open allowing the entire system to be completely bled.

Finally, remember to always bleed the 4WAL brake system with the ignition **OFF** to prevent setting false trouble codes.

The component used on other 4WAL systems is usually not equipped with external bleeder screws. Therefore, the unit can only be bled through the downstream bleeder screws (wheel cylinders/calipers). To accomplish this the internal bleeder and the accumulator stems/combination valves must be opened to allow air/fluid to pass through the unit. The Internal Bleed Valves on either side of the unit must be opened ¼–½ turn before bleeding begins. As with most ABS systems found on these vehicles, the ignition switch must be **OFF** or false trouble codes may be set.

1. Make sure the ignition is in the **OFF** position to prevent setting false trouble codes.

2. If necessary, properly bleed the master cylinder assembly. Check and add additional fluid, as necessary.

3. Open the internal bleed valves $1/4$–$1/2$ turn each.

4. Install one J–39177 or equivalent combination valve depressor tool on the left accumulator bleed stem of the EHCU. Install 1 tool on the right accumulator bleed stem and install the 3rd tool on the combination valve.

5. Properly bleed the wheel cylinders and calipers.

6. Remove the 3 special tools.

7. Check the master cylinder fluid level, refilling as necessary.

8. Switch the ignition **ON** (engine not running) and use a hand scanner to perform 6 function tests on the system.

9. Repeat the wheel cylinder and caliper bleeding procedure to remove all air that was purged from the BPMV during the function tests.

10. Check for a firm brake pedal. If necessary, repeat the entire procedure until a firm pedal is obtained.

11. Carefully test drive the vehicle at moderate speeds; check for proper pedal feel and brake operation. If any problem is noted in feel or function, repeat the entire bleeding procedure.

Anti-Lock Brake System Service

There are various systems used on these vehicles, the Four Wheel Anti-Lock (4WAL) which is available on most vehicles or the Rear Wheel Anti-Lock (RWAL) system.

PRECAUTION

Failure to observe the following precautions may result in system damage.

• Before performing electric arc welding on the vehicle, disconnect the Electronic Brake Control Unit.

• When performing painting work on the vehicle, do not expose the Electronic Brake Control Unit to temperatures in excess of 185°F (85°C) for longer than 2 hours. The system may be exposed to temperatures up to 200°F (95°C) for less than 15 minutes.

• Never disconnect or connect the Electronic Brake Control Unit connector with the ignition switch ON.

• Never disassemble any component of the Anti-Lock Brake System which is designated non-serviceable;

the component must be replaced as an assembly.

• When filling the master cylinder, always use Delco Supreme 11 brake fluid or equivalent, which meets DOT-3 specifications; petroleum base fluid will destroy the rubber parts. Do not allow fluid to be spilled on the Electronic Brake Control Unit.

Control Module

NOTE: Some late model vehicles are equipped with Vehicle Control Modules (VCMs). On these vehicles 1 computer controls both engine/emission operation and the anti-lock brake system (no separate brake control module is used).

REMOVAL AND INSTALLATION

RWAL System

1. Disconnect the negative battery cable.
2. Disconnect the electrical connectors from the module.
3. Remove the module from the master cylinder/proportioning valve bracket.
4. Installation is the reverse of the removal procedure.

4WAL System

1. Disconnect the negative battery cable.
2. Disconnect the electrical connectors from the module.
3. Disconnect and plug the brake lines at the module.
4. Remove the bolts attaching module to the fenderwell.
5. Remove the module and bracket assembly.
6. Remove the bracket from the module.
7. Installation is the reverse of the removal procedure.

Isolation/Dump Valve

REMOVAL AND INSTALLATION

RWAL and 4WAL System

1. Disconnect the negative battery cable.
2. Disconnect and plug the brake line fittings at the isolation/dump valve located under the master cylinder.
3. Remove the master cylinder bracket bolts.
4. Disconnect the Electronic Control Unit connector.

5. Remove the isolation/dump valve.
6. Installation is the reverse of the removal procedure.

Front Wheel Speed Sensor

REMOVAL AND INSTALLATION

4WAL System

1. Disconnect the negative battery cable.
2. Raise and support the vehicle safely.
3. Remove the wheel and tire assembly.
4. Remove the brake caliper, hub and rotor assembly.
5. Remove the sensor wire from the clips on the upper control arm and disconnect the connector.
6. Remove the sensor and backing plate attaching bolts and remove the assembly.
7. Installation is the reverse of the removal procedure.

Rear Wheel Speed Sensor

REMOVAL AND INSTALLATION

4WAL System

1. Disconnect the negative battery cable.
2. Raise and support the vehicle safely.
3. Remove the wheel, tire and drum assembly.
4. Remove the primary (forward) brake shoe.
5. Disconnect the sensor wire connector and remove the sensor wire from the rear axle clips.
6. Remove the bolts securing the sensor and sensor wire.
7. Remove the sensor from the backing plate.
8. Installation is the reverse of the removal procedure.

FRONT SUSPENSION

MacPherson Strut Assembly

All front wheel drive vehicles are equipped with a MacPherson strut front suspension.

REMOVAL AND INSTALLATION

NOTE: Do not remove the top center nut from the strut assembly. This nut should only be removed when the strut assembly is out of the vehicle, mounted in a holding fixture and the coil spring is in a compressed position using the proper strut coil spring compressor.

1. Remove the 3 nuts that retain the top of the strut assembly.
2. Raise and safely support the vehicle.
3. Remove the wheel and tire assembly. Remove the brake line bracket from the strut mount.
4. Remove the lower strut mounting bolts.
5. Remove the strut assembly from the vehicle and place the strut in a suitable holding fixture.
6. Disassemble the strut as follows:

   a. With the strut coil spring in a compressed position approximately ½ its normal length, remove the nut from the top of the strut.

   b. Place tool J–34013–27 or equivalent guide rod on top of the damper shaft. Use the rod to guide the damper shaft straight down through the bearing cap while decompressing the spring.

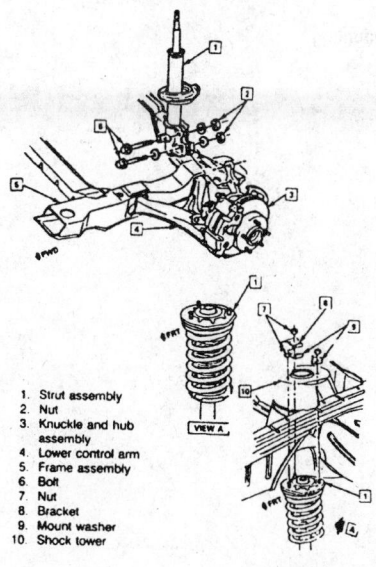

1. Strut assembly
2. Nut
3. Knuckle and hub assembly
4. Lower control arm
5. Frame assembly
6. Bolt
7. Nut
8. Bracket
9. Mount washer
10. Shock tower

85473210

MacPherson strut assembly/disassembly

c. Remove the coil spring and other components.

7. Installation is the reverse of the removal procedure. Ensure the spring seat flat should face 10 degrees forward of the centerline of the strut assembly spindle.
8. Tighten the strut lower bolts to 140 ft. lbs. (190 Nm) and the upper mounting nuts to 18 ft. lbs. (25 Nm).

Lower Ball Joint

INSPECTION

1. Raise and safely support the lower control arm.
2. Wipe the ball joint clean and check the seal for cuts or tears.
3. Check the wheel bearings for proper adjustment.
4. Position a dial indicator against the lowest outside point of the tire. Rock the wheel in and out.
5. Check the reading on the dial indicator. The reading should be no more than 0.125 inch (3.18mm).

REMOVAL AND INSTALLATION

1. Raise and support the vehicle safely. Properly support the lower control arm.

NOTE: The control arm must be supported so the spring and the control arm remain intact.

2. Remove the tire and wheel assembly. As required, remove the brake caliper and position it aside.
3. Remove the pinch bolt from the lower ball joint. Using the proper tool separate the upper joint from the steering knuckle. Support the knuckle assembly so its weight will not damage the brake hose.
4. Remove the rivets from the ball joint assembly, using a drill with an ⅛ in. and then ½ in. bit.
5. Remove the stabilizer shaft bushing assembly nut.
6. Remove the ball joint from the lower control arm.
7. Installation is the reverse of the removal procedure. Tighten the ball joint pinch bolt to 33 ft. lbs. (45 Nm). Check and adjust the front end alignment, as required.

Lower Control Arms

REMOVAL AND INSTALLATION

1. Raise and safely support the vehicle so the suspension hangs freely.

2. Remove the wheel and tire assemblies.
3. Remove the stabilizer shaft-to-control arm mounting bolt. Remove the lower ball joint pinch bolt.
4. Separate the steering knuckle from the lower ball joint.
5. Remove the lower control arm mounting bolts and remove the control arm.

To install:

6. Install the control arm in position on the vehicle frame. Do not tighten the control arm bolts at this time.
7. Install the stabilizer shaft to the control arm, do not tighten the bolts at this time.
8. Connect the steering knuckle to the control arm using a new pinch bolt. Tighten to 33 ft. lbs. (45 Nm). Lower the vehicle so the weight is supported by the control arms.
9. Tighten the control arm bolts to 61 ft. lbs. (83 Nm). Tighten the stabilizer shaft bolts to 32 ft. lbs. (43 Nm).
10. Install the wheel and tire assemblies. Lower the vehicle completely.

Stabilizer Shaft/Sway Bar

REMOVAL AND INSTALLATION

1. Raise and safely support the vehicle.
2. Remove the wheel and tire assembly.
3. Remove the left and right side stabilizer mounting bolts. Keep the sides separate for installation.
4. Remove the center stabilizer insulators and lower the stabilizer.

To install:

5. Install the stabilizer in position in the vehicle. Tighten the left and right mounting bolts to 24 ft. lbs. and the center bushing supports to 35 ft. lbs.
6. Install the wheel and tire assemblies and lower the vehicle.

REAR SUSPENSION

Shock Absorbers

REMOVAL AND INSTALLATION

1. Open the lift gate and open the trim cover.
2. Remove the upper shock mounting nut and grommet.

3. Raise and safely support the vehicle. Properly support the rear axle assembly.

4. If equipped with electronic level control suspension, remove the air line from the shock absorber. Allow the air to bleed off.

5. Remove the lower mounting bolt and remove the shock.

To install:

6. Install the shock in position and install the lower mounting bolt. Tighten to 44 ft. lbs. (59 Nm). Connect the air line to the shock, if equipped.

7. Lower the vehicle and install the upper shock retaining nut, tighten it to 16 ft. lbs. (22 Nm). Install the trim cover.

Track Bar

REMOVAL AND INSTALLATION

1. Raise and safely support the vehicle.

2. Remove the track bar mounting bolts from the body and the axle.

3. Lower the track bar.

To install:

4. Install the track bar at the axle, loosely install the bolt.

5. Connect the other end of the track bar at the frame.

6. Tighten the bolt at the axle to 44 ft. lbs. (60 Nm) and the track bar-to-frame bolt to 35 ft. lbs. (47 Nm).

7. Lower the vehicle.

Coil Springs

REMOVAL AND INSTALLATION

1. Raise and safely support the vehicle.

2. Safely support the rear axle assembly.

3. Remove the right and left brake line-to-axle attaching screws. Allow the brake lines to hang freely.

4. Disconnect the track bar-to-axle attaching bolt.

5. Disconnect the lower shock absorber mounting bolts.

6. Slowly lower the rear axle and remove the springs and insulators.

7. Installation is the reverse of the removal procedure.

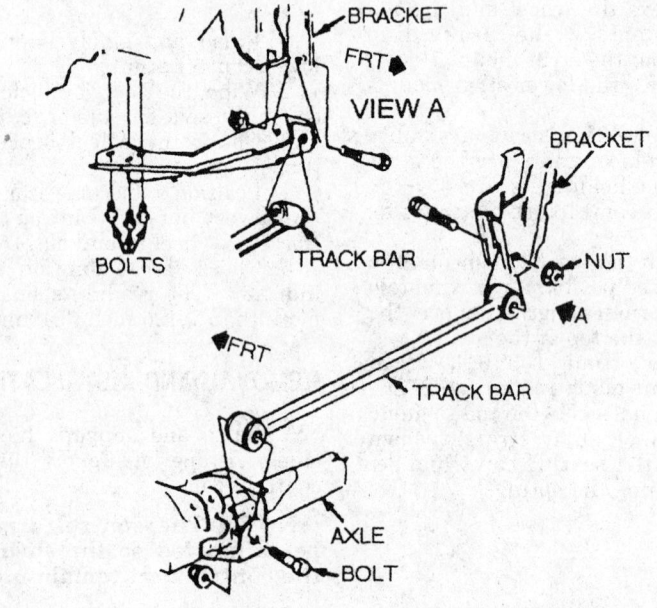

85473232

Rear track bar mounting

FIRING ORDERS

NOTE: To avoid confusion, always replace spark plug wires one at a time.

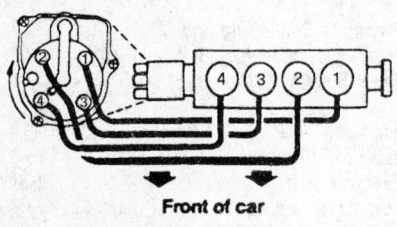

Front of car

302173

2.2L Engine (F22B6)
Engine Firing Order: 1–3–4–2
Distributor Rotation: Clockwise

ENGINE ELECTRICAL

NOTE: Disconnecting the negative battery cable on some vehicles may interfere with the functions of the on-board computer systems and may require the computer to undergo a relearning process, once the negative battery cable is reconnected.

Distributor

REMOVAL AND INSTALLATION

NOTE: The radio may contain a coded theft protection circuit. Always make note of your code before disconnecting the battery.

1. Disconnect the negative battery cable.
2. Unclamp the cruise control cable from the valve cover and carefully move it out of the way.
3. The intake air duct may be removed for more access to the distributor.
4. Set the No. 1 cylinder at TDC for the compression stroke. Once the engine is in this position, it must not be disturbed.
5. Label and disconnect the ignition wires.

DISTRIBUTOR END   CAMSHAFT END

8P CONNECTOR

2P CONNECTOR

NEW O-RING

MOUNTING BOLT
18 N·m (1.8 kgf·m, 13 lbf·ft)

290581

Distributor components

6. Uncouple the 2–P and 8–P connectors and remove them from their clips on the side of the distributor.
7. Remove the three distributor mounting bolts.
8. Remove the distributor from the cylinder head.

To install:

NOTE: If the camshaft or crankshaft has rotated during assembly, rotate the crankshaft to bring the engine to TDC/compression for the No. 1 cylinder. Make sure the UP mark on the camshaft is facing up, and that the crankshaft TDC mark aligns with the pointer on the lower timing cover.

9. Coat a new O-ring with clean engine oil and install it onto the distributor shaft.
10. Install the distributor into the cylinder head. The lugs on the distributor shaft fit into the groove on the end of the camshaft.
11. Install the three mounting bolts, only hand-tighten them at this time.
12. Couple the 2–P and 8–P connectors and install them onto their clips.

13. Connect the ignition wires.
14. Reconnect the negative battery cable.
15. Check and adjust the ignition timing. To adjust the ignition timing: first, loosen the distributor mounting bolts. Then, turn the distributor housing clockwise to retard the timing, or counterclockwise to advance the timing.
16. Tighten the mounting bolts to 13 ft. lbs. (18 Nm).
17. Install the cruise control cable back into its clamp on the valve cover.
18. Install the intake air duct if it was removed.
19. Enter the radio security code.

Ignition Timing

ADJUSTMENT

1. Start the engine and run it at 3000 rpm with the transaxle in the **N** or **P** position. When the radiator fan comes on (normal operating temperature), let the engine run at idle.
2. Remove the connector holder underneath the left edge of the glove box. Remove the 2–P service check connector from the connector holder.
3. Connect a SCS service connector, 07PAZ-0010100 or equivalent, to the service check connector.
4. Check the idle speed:
   a. Connect a test tachometer to the test tachometer connector located on the right side of the firewall near the shock tower.
   b. The idle speed should be 700 ± 50 rpm with the transaxle in **N** or **P**, and all electrical accessories off.
5. Connect a timing light to the No. 1 ignition wire.
6. The timing should be 15° ± 2° BTDC (red timing mark on crankshaft pulley) at 700 ± 50 rpm.
7. If the timing is out of range, adjust it. First, loosen the distributor mounting bolts. Next, turn the distributor housing counterclockwise to advance the timing, or clockwise to retard the timing. Then, tighten the distributor mounting bolts to 13 ft. lbs. (18 Nm).
8. Recheck the ignition timing.
9. Remove the service connector from the check connector. Install the check connector back into its holder. Tuck the connector holder back under the glovebox.

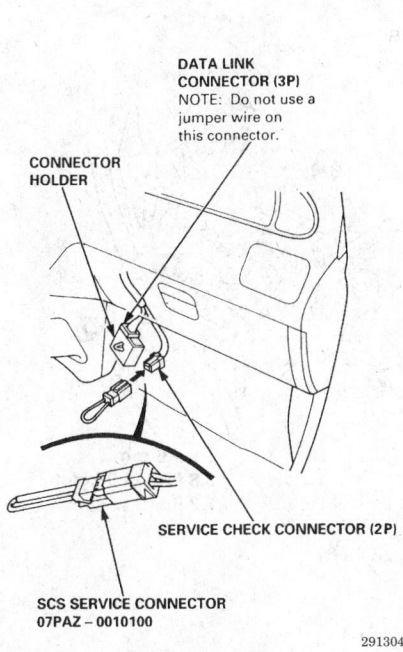

DATA LINK
CONNECTOR (3P)
NOTE: Do not use a
jumper wire on
this connector.

CONNECTOR
HOLDER

SERVICE CHECK CONNECTOR (2P)

SCS SERVICE CONNECTOR
07PAZ – 0010100

291304

Service check connector and special connector

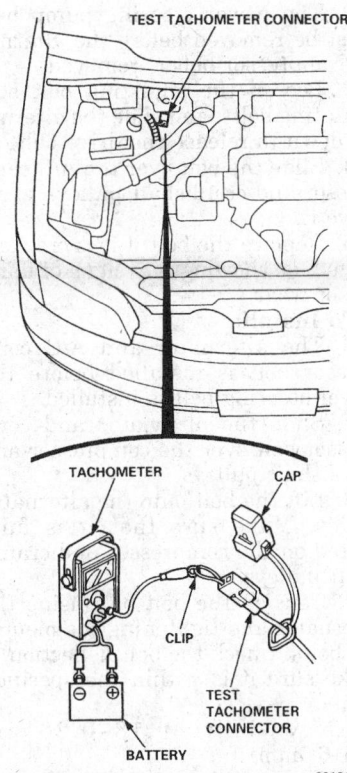

TEST TACHOMETER CONNECTOR

TACHOMETER

CAP

CLIP

TEST
TACHOMETER
CONNECTOR

BATTERY

291310

Test tachometer connector

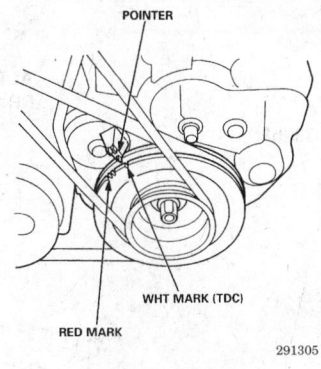

POINTER

WHT MARK (TDC)

RED MARK

291305

Timing marks

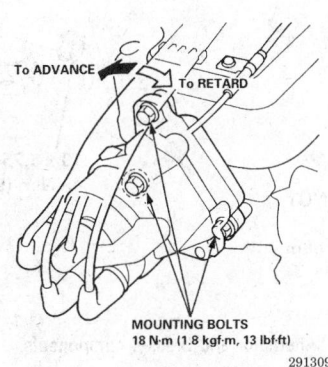

To ADVANCE

To RETARD

MOUNTING BOLTS
18 N·m (1.8 kgf·m, 13 lbf·ft)

291309

Distributor advance and retard

Alternator

PRECAUTIONS

Several precautions must be observed with alternator equipped vehicles to avoid damage to the unit.

• If the battery is removed for any reason, make sure it is reconnected with the correct polarity. Reversing the battery connections may result in damage to the 1–way rectifiers.

• When utilizing a booster battery as a starting aid, always connect the positive to positive terminals and the negative terminal from the booster battery to a good engine ground on the vehicle being started.

• Never use a fast charger as a booster to start vehicles.

• Disconnect the battery cables when charging the battery with a fast charger.

• Never attempt to polarize the alternator.

• Do not use test lights of more than 12 volts when checking diode continuity.

• Do not short across or ground any of the alternator terminals.

• The polarity of the battery, alternator and regulator must be matched and considered before making any electrical connections within the system.

• Never separate the alternator on an open circuit. Make sure all connections within the circuit are clean and tight.

• Disconnect the battery ground terminal when performing any service on electrical components.

• Disconnect the battery if arc welding is to be done on the vehicle.

REMOVAL AND INSTALLATION

NOTE: The radio may contain a coded anti-theft circuit. Always make note of your code number before disconnecting the battery cables.

NOTE: The alternator is equipped with an internal voltage regulator. If the voltage regulator is found to be faulty, the alternator must be rebuilt or replaced.

1. With the ignition **OFF**, disconnect the negative battery cable, and then the positive battery cable.
2. Remove the power steering pump.
3. Disconnect the multi-pin electrical connector from the alternator.
4. Remove the terminal nut and remove the wire from the **B** terminal.
5. Loosen the through bolt, then loosen the adjustment locknut, and then the adjusting bolt.
6. Remove the belt from the alternator pulley.
7. Remove the adjustment bolt and nut.
8. Support the alternator. Remove the through bolt and remove the alternator.
To install:
9. If the alternator brackets were removed, reinstall them. Coat the bolts with a thread sealer or liquid thread lock. Install the bracket bolts and tighten them to 36 ft. lbs. (50 Nm).
10. Fit the alternator into place. Install the through bolt but do not tighten the bolt at this time.
11. Install the adjustment bolt and locknut then the adjusting bolt. Make certain the adjusting bolt is properly installed into the adjustment through bolt.
12. Install the belt and adjust the belt tension. Tighten the alternator adjustment locknut to 16 ft. lbs. (22 Nm).
13. Tighten the through bolt to 33 ft. lbs. (44 Nm).
14. Connect the wire to the **B** terminal and tighten the nut.

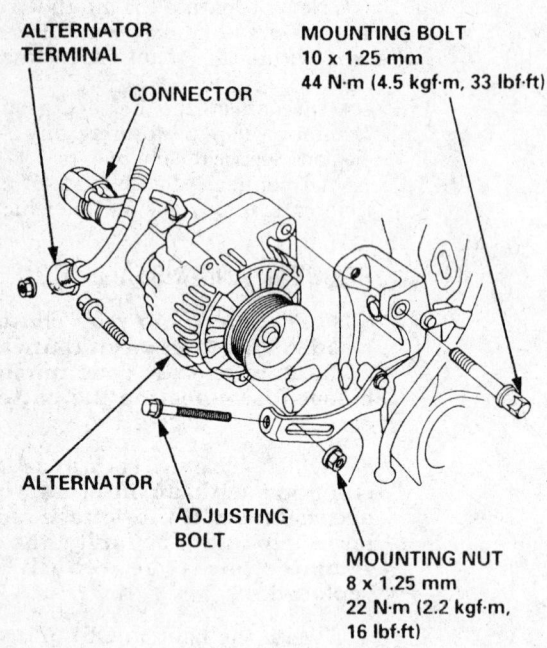

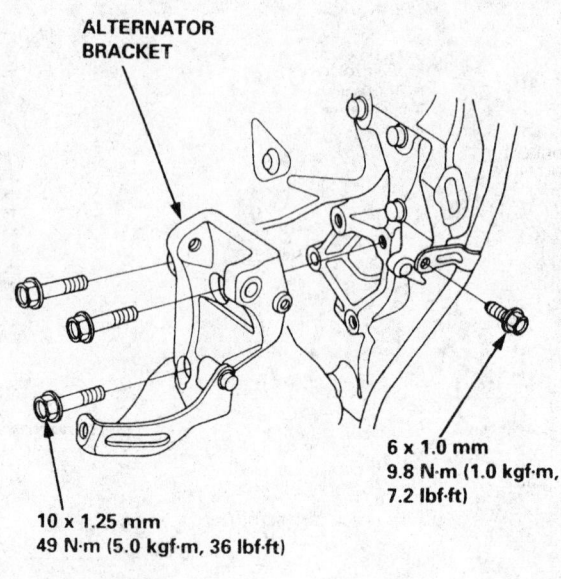

Alternator and bracket components

15. Install the multi-pin connector to the alternator.

16. Install the power steering pump.

17. Adjust the belt tensions.

18. Connect the positive and negative battery cables.

19. Enter the radio security code.

Drive Belt

REMOVAL AND INSTALLATION

NOTE: The radio may contain a coded theft protection circuit. Always make note of your code number before disconnecting the battery.

Power Steering Pump

1. Disconnect the negative battery cable.

NOTE: Due to limited engine clearance, it may be necessary to lift and support the vehicle to remove the left front wheel and the splash shield for extra clearance when removing and installing the drive belts.

2. Loosen the power steering belt tension adjusting bolt.

3. Loosen the power steering pump mounting nuts and pivot the pump downward.

4. Remove the power steering belt.

5. Replace the belt if it is cracked, glazed, or shows signs of damage or wear.

To install:

6. The power steering belt is installed after the alternator/air conditioning compressor belt.

7. Slide the power steering belt onto the crankshaft pulley.

8. Slide the power steering belt onto the pump pulley. Make sure that the belt is fully seated on both the crankshaft and pump pulleys.

9. Tension the power steering belt by raising the pump and tightening the mounting nuts. Check that the belt deflection is within the limits.

• New belt: 0.43–0.49 in. (11.0–12.5mm)

• Used belt: 0.51–0.63 in. (13–16mm)

10. Make final adjustments to the belt deflection by tightening the adjusting bolt.

11. Reconnect the negative battery cable.

Alternator and A/C

1. Disconnect the negative battery cable.

2. The power steering pump belt must be removed before the alternator/compressor belt is removed.

3. Loosen the adjusting bolt and mounting bolts, and slide the alternator down to release the drive belt.

4. Slide the belt over the A/C compressor and crankshaft pulleys to remove it.

5. Replace the belt if it is cracked, glazed, or showing any signs of damage or wear.

To Install:

6. The alternator and A/C compressor belt is installed before the power steering belt is installed.

7. Slide the alternator and compressor belt over the compressor and crankshaft pulleys.

8. Lift the belt onto the alternator pulley. Make sure that it is fully seated on the compressor and crankshaft pulleys.

9. Tension the belt by raising the alternator and tightening the mounting bolts. Check the belt deflection to make sure it is within the specified limits.

• New belt: 0.18–0.26 in. (4.5–6.5mm)

• Used belt: 0.31–0.41 in. (8.0–10.5mm)

10. Make final adjustments to the belt deflection by tightening the adjustment bolt.

11. Reconnect the negative battery cable.

12. Enter the radio security code.

Starter

REMOVAL AND INSTALLATION

NOTE: The radio may contain a coded theft protection circuit. Always make note of your code number before disconnecting the battery.

1. Disconnect the negative and positive battery cables.

2. Certain Honda engines have a bracket on the starter motor housing for the engine wiring harness and upper radiator hose. If equipped with a bracket, unclamp it and move the wiring harness or radiator hose out of the way.

3. Disconnect the starter cable from its terminal.

4. Disconnect the black/white wire from the S terminal on the starter solenoid.

5. Remove the two starter mounting bolts. Remove the starter.

6. Inspect the starter terminals and clean off any corrosion to ensure good electrical contact. Inspect the

MITSUBA:

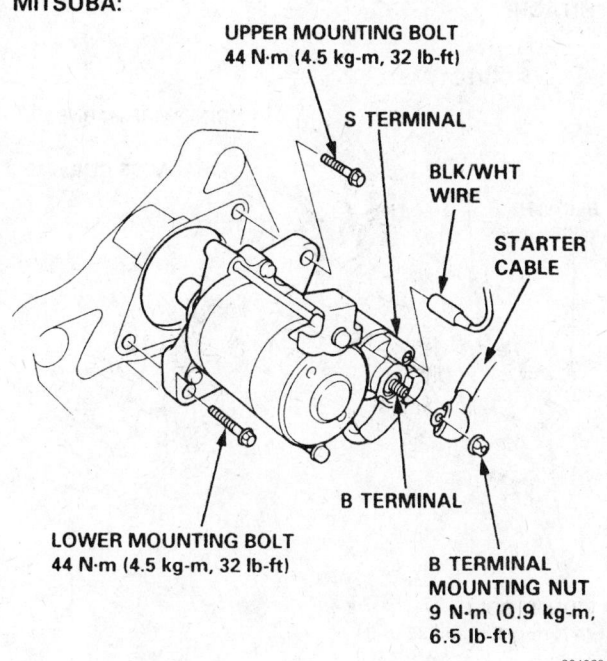

UPPER MOUNTING BOLT
44 N·m (4.5 kg-m, 32 lb-ft)

S TERMINAL

BLK/WHT WIRE

STARTER CABLE

B TERMINAL

LOWER MOUNTING BOLT
44 N·m (4.5 kg-m, 32 lb-ft)

B TERMINAL MOUNTING NUT
9 N·m (0.9 kg-m, 6.5 lb-ft)

304060

Mitsuba starter components and terminals

cables for brittle or cracked insulation and damaged or loose terminals and replace as necessary.

To Install:

7. Install the starter. Install the mounting bolts and tighten them to 32 ft. lbs. (44–45 Nm).

8. Reconnect the starter cable with the crimp side facing up and tighten the nut to 6.5 ft. lbs. (9 Nm). Reconnect the black/white wire.

9. Reconnect the positive and negative battery cables.

10. Test the operation of the starter.

11. If equipped, enter the radio security code.

NIPPONDENSO:

UPPER MOUNTING BOLT
44 N·m (4.5 kg-m, 32 lb-ft)

B TERMINAL MOUNTING NUT
9 N·m (0.9 kg-m, 6.5 lb-ft)

STARTER CABLE

S TERMINAL

BLK/WHT WIRE

LOWER MOUNTING BOLT
44 N·m (4.5 kg-m, 32 lb-ft)

304059

Nippondenso starter components and terminals

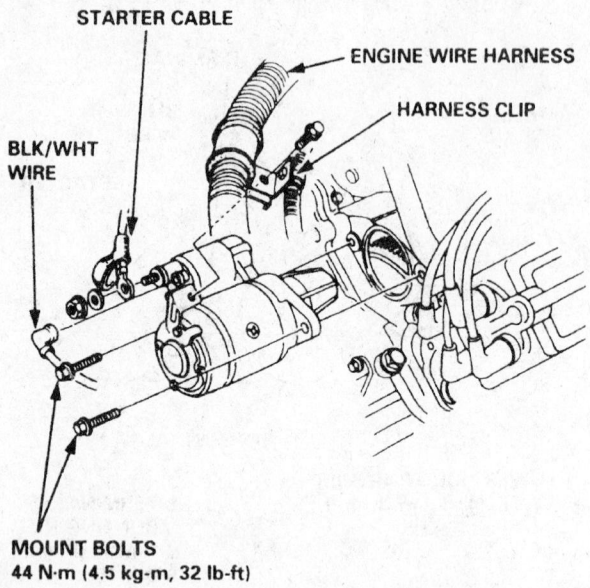

HITACHI:

STARTER CABLE

ENGINE WIRE HARNESS

HARNESS CLIP

BLK/WHT WIRE

MOUNT BOLTS
44 N·m (4.5 kg-m, 32 lb-ft)

304061

Hitachi starter components and terminals

CHASSIS ELECTRICAL

Blower Motor

REMOVAL AND INSTALLATION

Front

———— CAUTION ————
The SRS main wiring harness is routed near the blower motor housing. The SRS wiring harness is encased in yellow insulation. Be careful not to damage the SRS wiring. If any SRS component must be serviced, the SRS system must first be disabled.

NOTE: The radio may contain a coded theft protection circuit. Always make note of your code number before disconnecting the battery.

1. Disconnect the negative and positive battery cables. Wait at least three minutes before working around the air bags.

2. Remove the glove box door and glove box frame.

NOTE: The blower motor can be removed without removing the entire blower housing.

3. Disconnect the blower motor lead.

4. Remove the three self-tapping screws from the bottom of the blower housing. Lower the blower motor out of the housing.

To Install:

5. Fit the blower motor into the housing and install the three self-tapping screws. Make sure the edges of the blower fan rotor do not rub the edge of the housing.

6. Reconnect the blower motor lead.

7. Install the glove box frame and glove box door.

8. Reconnect the negative and positive battery cables. Start the vehicle.

9. Test the operation of the blower motor at all of its speed settings.

10. Enter the radio security code.

Rear

NOTE: The radio may contain a coded theft protection circuit. Always make note of your code number before disconnecting the battery.

1. Disconnect the negative battery cable.

2. Remove the right and left access lids from the rear A/C unit.

3. Carefully pry the switch from the rear A/C unit trim panel. Disconnect the switch.

4. Remove the right and left screw covers from the intake and outlet grilles. Unscrew and remove the intake and outlet grilles.

5. Unscrew and remove the rear A/C housing cover.

6. Remove the four screws securing the A/C blower assembly to the rear A/C housing.

7. Disconnect the blower motor lead and remove the blower assembly.

8. Remove the bracket from the blower housing.

9. Remove the four self-tapping screws from the blower motor, and separate it from the housing.

10. Remove the set screw to separate the blower motor from the tubular air vane.

To Install:

11. Assemble the blower motor and the tubular air vane. Install the set screw.

12. Assemble the blower motor into the housing and install the four screws. Install the bracket onto the housing.

13. Install the blower assembly into the rear A/C housing and reconnect the motor lead.

14. Install the A/C housing cover.

15. Install the intake and outlet grilles and their screw covers.

16. Reconnect and install the A/C switch.

17. Install the left and right access lids.

18. Reconnect the negative battery cable.

19. Test the operation of the rear A/C blower at all of its speed settings. Check for air leaks.

20. Enter the radio security code.

Windshield Wiper Motor

REMOVAL AND INSTALLATION

NOTE: The radio may contain a coded theft protection circuit. Always make note of your code number before disconnecting the battery.

Front

1. Disconnect the negative battery cable.

2. Lift up the nut cover on each wiper arm and remove the 12mm nut

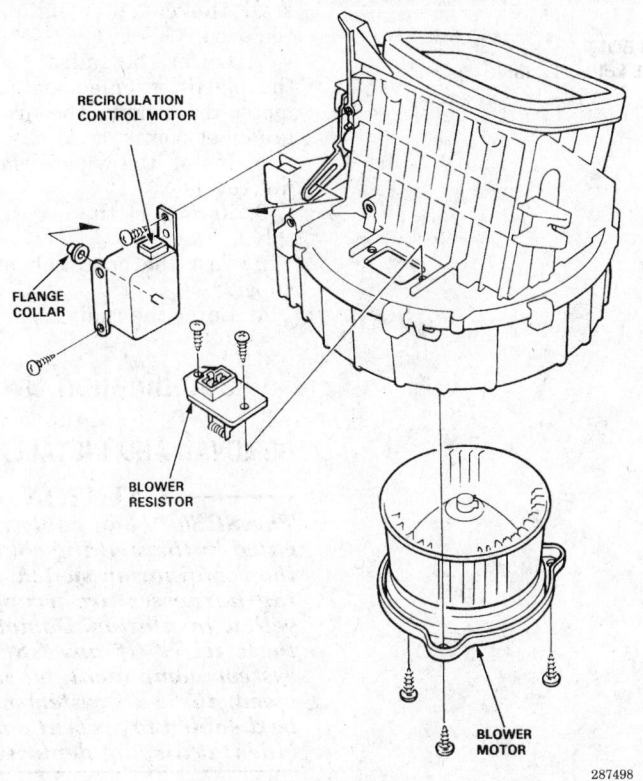

Exploded view of the front blower assembly

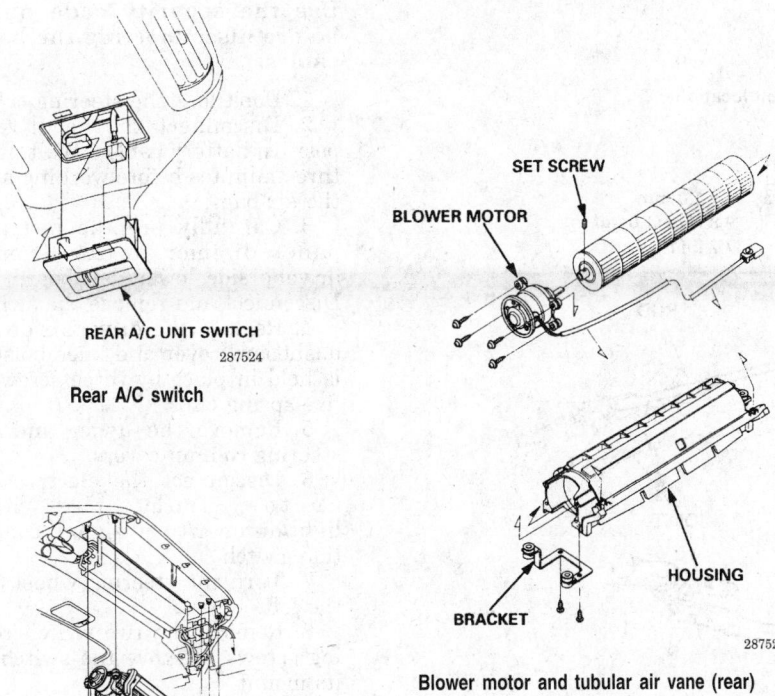

Rear A/C switch

Blower motor and tubular air vane (rear)

Rear blower motor assembly

and retaining clip. Remove the left and right wiper arms.

3. Carefully pry up the clips to remove the air scoop with the rubber hood seal attached.

4. Disconnect the wiper motor 5-P connector.

5. Carefully separate the motor linkage pivot from the wiper motor. Unbolt and remove the wiper motor.

6. If necessary, the wiper linkages can be removed by unbolting them from the joints at either end of the cowl. Then, unbolt the center linkage pivot from the wiper motor bracket. Remove the linkage from the cowl, taking care not to scratch the vehicle's paint.

To Install:

7. Install the wiper linkages if they were removed. Tighten the center pivot and side joint bolts to 7 ft. lbs. (10 Nm).

8. Install the wiper motor. Reconnect the linkage pivot to the wiper motor and tighten the bolt to 13 ft. lbs. (18 Nm).

9. Reconnect the 5-P wiper motor connector.

10. Install the air scoop. Make sure that all the clips are properly seated and not broken. Make sure the hood seal is correctly seated.

11. Install the wiper arms. The wiper arm with the torsion strut belongs on the passenger's side of the vehicle. Install the retaining clips, nuts, and nut covers.

12. Reconnect the negative battery cable.

13. Test the operation of the wipers at all of their speed settings.

14. Enter the radio security code.

Rear

1. Disconnect the negative battery cable.

2. Remove the rear window wiper arm nut cover and nut. Remove the wiper arm and the plastic grommet. Be careful not to scratch the paint.

3. Raise the tailgate. Carefully pry up the clips to remove the side window trim. Unscrew and remove the grab handle. Carefully pry up the clips to remove the lower tailgate trim panel. Disconnect and remove the courtesy light if necessary.

4. Disconnect the wiper motor and remove its three mounting bolts.

To Install:

5. Install the wiper motor into position and install the three mounting bolts. Reconnect the wiper motor connector.

6. Install the tailgate lower trim panel and side window trim. Install the grab handle. Reconnect and in-

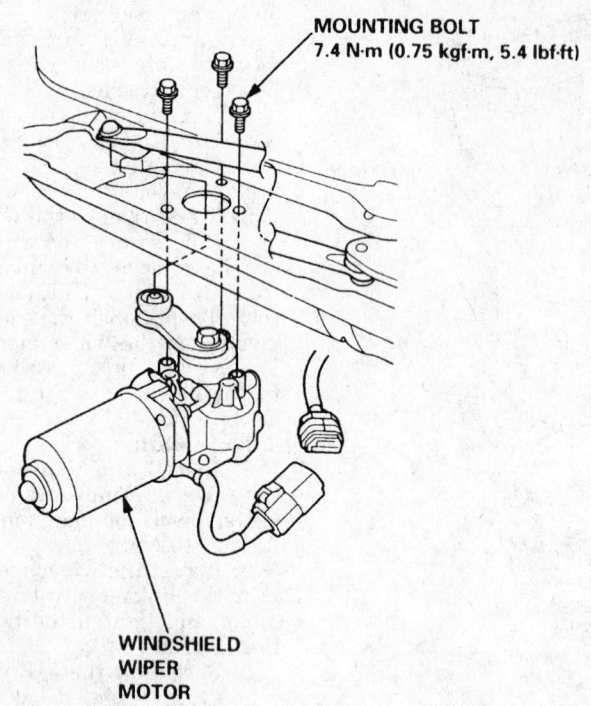

MOUNTING BOLT
7.4 N·m (0.75 kgf·m, 5.4 lbf·ft)

WINDSHIELD
WIPER
MOTOR

287371

Front windshield wiper motor

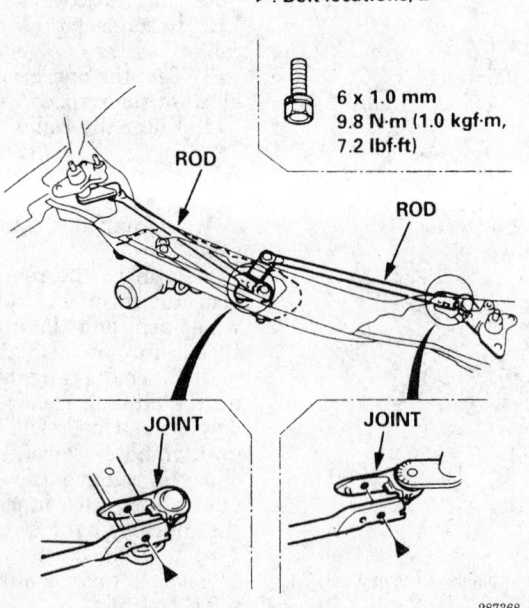

▶: Bolt locations, 2

6 x 1.0 mm
9.8 N·m (1.0 kgf·m, 7.2 lbf·ft)

ROD

ROD

JOINT

JOINT

287368

Front wiper linkage

stall the courtesy light if it was removed.

7. Lower the tailgate and install the plastic grommet onto the wiper motor driveshaft. The arrow on the grommet points down.

8. Install the wiper arm, nut, and nut cover.

9. Reconnect the negative battery cable.

10. Test the operation of the rear wiper.

11. Enter the radio security code.

Combination Switch

REMOVAL AND INSTALLATION

———— CAUTION ————
The SRS/air bag cable reel is located in the steering column near the combination switch. SRS wiring harnesses are wrapped with yellow insulation. Do not damage these wires. If any SRS/air bag system component must be serviced, the SRS system must first be disabled to prevent possible accidental air bag deployment.

NOTE: The original radio contains a coded anti-theft circuit. Use the security code number before disconnecting the battery cables.

1. Don't lock the steering column.

2. Disconnect the negative and positive battery cables. Wait at least three minutes before working around the air bags.

3. Carefully pry the instrument panel dimmer switch from the driver's side lower dashboard cover. Disconnect and remove the switch.

4. Remove the driver's side lower dashboard cover and knee bolster. It is held in place by three screws and five spring clips.

5. Remove the upper and lower steering column covers.

6. Disconnect the electrical connector from the headlight/dimmer/turn signal combination switch.

7. Turn the steering wheel 90° to the left.

8. Remove the two switch retaining screws. Remove the switch from its mount.

To Install:

9. Connect the electrical connector and install the switch into its mount.

10. Install the two retaining screws. Verify that no electrical wires are pinched.

11. Install the steering column covers.

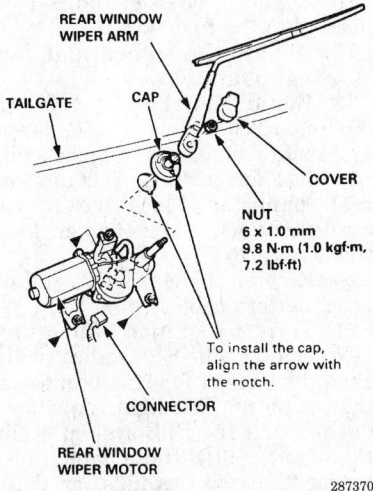

▶: Bolt locations, 3

6 x 1.0 mm
9.8 N·m (1.0 kgf·m,
7.2 lbf·ft)

REAR WINDOW
WIPER ARM

TAILGATE   CAP

COVER

NUT
6 x 1.0 mm
9.8 N·m (1.0 kgf·m,
7.2 lbf·ft)

To install the cap,
align the arrow with
the notch.

CONNECTOR

REAR WINDOW
WIPER MOTOR

287370

Rear wiper motor and wiper arm

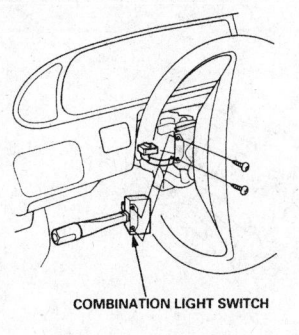

COMBINATION LIGHT SWITCH

287382

Combination switch

12. Install the knee bolster and lower dashboard cover.

13. Reconnect the instrument panel dimmer switch and install it into the driver's side lower dashboard cover.

14. Reconnect the positive and negative battery cables.

15. Turn the ignition switch to the **ON** position, but don't start the engine. The SRS indicator light should turn on for six seconds, and then turn off. This light sequence indicates that the SRS system is functioning normally. If the light stays on longer there is a system malfunction which must be diagnosed.

16. Check the operation of the headlights and turn signals.

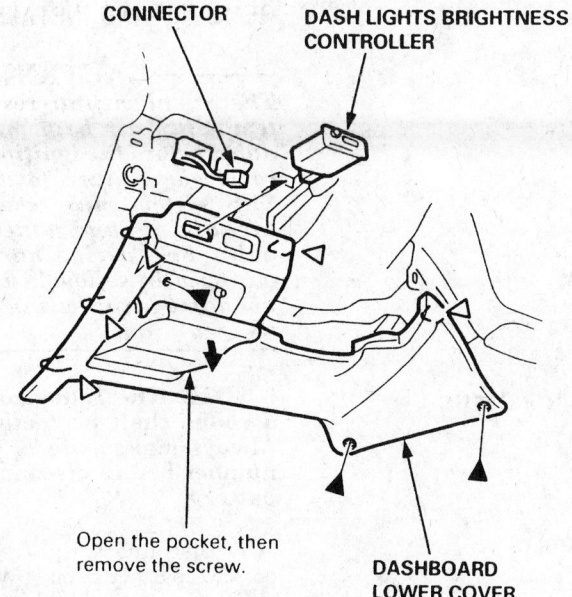

CONNECTOR   DASH LIGHTS BRIGHTNESS
CONTROLLER

Open the pocket, then
remove the screw.

DASHBOARD
LOWER COVER

287379

Instrument panel dimmer switch

Ignition Lock Cylinder

REMOVAL AND INSTALLATION

———— **CAUTION** ————

The Supplemental restraint system (SRS, air bag) must be disabled before the ignition switch is removed. Failure to disable the SRS system may result in personal injury and unnecessary repairs. SRS wiring harnesses are wrapped in yellow insulation. Do not damage the wiring harness or its insulation.

NOTE: The radio may contain a coded theft protection circuit. Always make note of your code number before disconnecting the battery.

1. Disconnect the negative and positive battery cables. Wait at least three minutes before servicing the air bag.

2. Remove the steering wheel access cover. Disconnect the air bag from the cable reel connector. Install the red shorting connector onto the driver's air bag connector.

NOTE: Some vehicles may be equipped with spring-loaded locking air bag connectors. These air bag connectors automatically short themselves when they are disconnected from the SRS harness. When uncoupling spring-loaded connectors, do not pull on the connector body — slide the sleeve toward its stop while securely holding the connector.

3. Carefully pry the panel dimmer switch from the lower dashboard cover. Remove the lower dashboard cover and the knee bolster.

4. Remove the upper and lower steering column covers.

5. Open the fuse box door; and, disconnect the 7–P and 8–P connectors.

6. Remove the steering column mounting nuts and bolts. Lower the steering column. Be careful not to bend or kink the shift cable.

7. Center-punch the shear bolts on the lock cylinder and carefully drill their heads off with a 3/16 in. (5mm) bit. Remove the shear bolts.

———— **WARNING** ————

Be careful not to damage the yellow SRS wiring harness, or the ignition switch body when punching and drilling out the shear bolts.

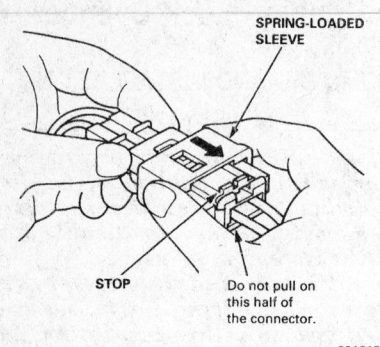

Spring-loaded air bag connectors

291215

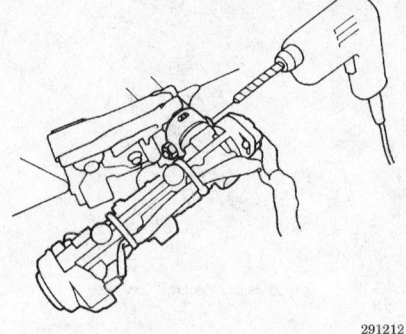

Drilling shear bolts

291212

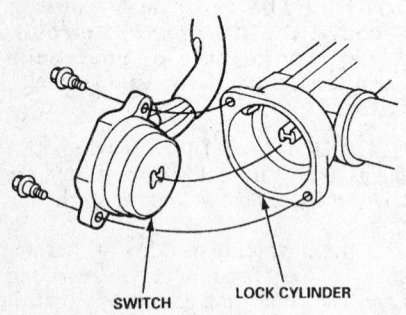

Ignition electrical switch

291211

8. Put the key in the ignition and turn it to the **I** position.

9. Use a suitable tool to push in the lock pin. Then, pull the steering lock out of the switch body.

To Install:

10. Turn the ignition switch to the to the **I** position.

11. Push the lock pin in, and insert the steering lock into the switch body. Make sure the switch clicks securely into place.

12. Install new shear bolts and only hand-tighten them. Make sure the projection on the ignition switch body is aligned with the hole in the steering column.

13. Put the key in the ignition, and make sure the steering lock and ignition switch operate smoothly.

14. Tighten the shear bolts until their heads twist off.

15. Raise the steering column into position. Tighten the mounting nuts to 12 ft. lbs. (16 Nm), and the mounting bolts to 28 ft. lbs. (39 Nm).

16. Connect the 7–P and 8–P connectors to the fuse box.

17. Install the upper and lower steering column covers.

18. Install the knee bolster and dashboard lower cover. Reconnect and install the panel dimmer switch.

19. Reconnect the air bag and cable reel connector. Place the shorting connector back into its holder. Install the access cover.

20. Reconnect the positive and negative battery cables.

21. Turn the ignition switch to the **ON** position. The SRS indicator light should come on for six seconds, and then turn off. This light sequence indicates that the SRS system is functioning normally. If the light stays on longer than six seconds, or doesn't come on, the system must be diagnosed.

22. Test the operation of the ignition switch and steering lock.

23. Enter the radio security code.

Ignition Switch

REMOVAL AND INSTALLATION

———— **CAUTION** ————

The Supplemental restraint system (SRS, air bag) must be disabled before the ignition switch is removed. Failure to disable the SRS system may result in personal injury and unnecessary repairs. SRS wiring harnesses are wrapped in yellow insulation. Do not damage the wiring harness or its insulation.

NOTE: The radio may contain a coded theft protection circuit. Always make note of your code number before disconnecting the battery.

1. Disconnect the negative and positive battery cables. Wait at least three minutes before servicing the air bag.

2. Remove the steering wheel access cover. Disconnect the air bag from the cable reel connector. Install

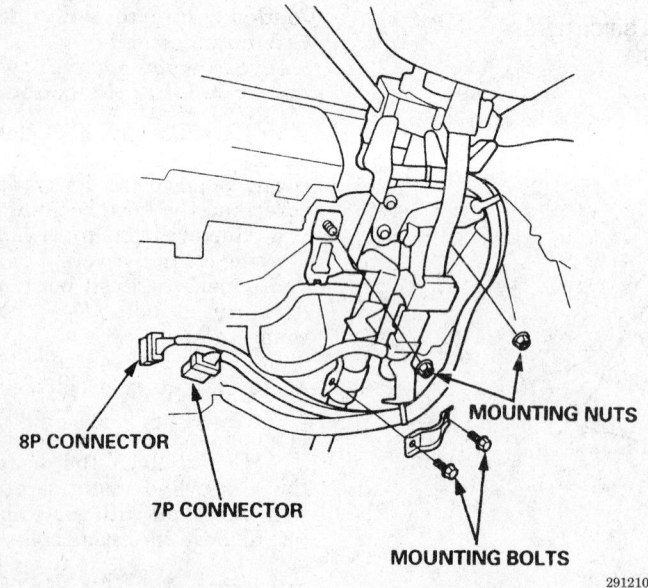

Electrical connections

291210

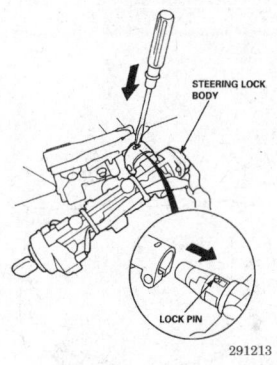

Steering lock pin

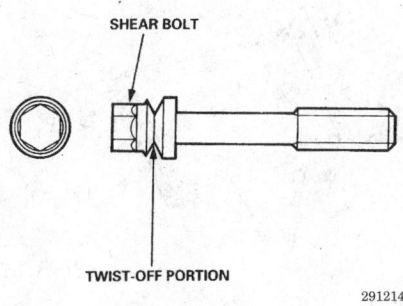

New shear bolt

the red shorting connector onto the driver's air bag connector.

NOTE: Some vehicles may be equipped with spring-loaded locking air bag connectors. These air bag connectors automatically short themselves when they are disconnected from the SRS harness. When uncoupling spring-loaded connectors, do not pull on the connector body — slide the sleeve toward its stop while securely holding the connector.

3. Carefully pry the panel dimmer switch from the lower dashboard cover. Remove the lower dashboard cover and the knee bolster.

4. Remove the upper and lower steering column covers.

5. Open the fuse box door and disconnect the 7–P connector.

6. Put the key in the ignition and turn it to the **O** position.

7. Remove the two screws from the switch body to separate it from the lock cylinder.

To Install:

8. Fit the switch body onto the lock cylinder. Install the two screws.

9. Reconnect the 7–P connector to the fuse box.

10. Install the upper and lower steering column covers.

11. Install the knee bolster and dashboard lower cover. Reconnect and install the panel dimmer switch.

12. Reconnect the air bag and cable reel connectors. Place the shorting connector into its holder. Install the steering wheel access cover.

13. Reconnect the battery cables.

14. Test the operation of the ignition switch.

15. Enter the radio security code.

Park/Neutral Safety Switch

REMOVAL AND INSTALLATION

NOTE: The radio may contain a coded theft protection circuit. Always make note of your code number before disconnecting the battery.

1. Disconnect the negative battery cable.

2. Shift the transaxle into **N** and engage the parking brake.

3. The gear position switch coupling is located to the rear of the starter motor. Remove the coupling from its stay. Uncouple the connectors.

4. Raise and support the vehicle safely.

5. Place a jack under the transaxle. Raise it enough to take up the weight of the transaxle.

6. Unbolt and remove the transaxle mount. Then, lower the jack.

7. Remove the A/T gear position switch from the right side cover of the transaxle. Unbolt the harness clip from the transaxle case.

8. Remove the A/T gear position switch from the right side cover.

To Install:

9. Set the A/T gear position switch into the **N** position. The switch will click into position.

10. Make sure the control shaft is in the **N** position. Install the switch onto the control shaft. Route the wiring over the transaxle case, and in-

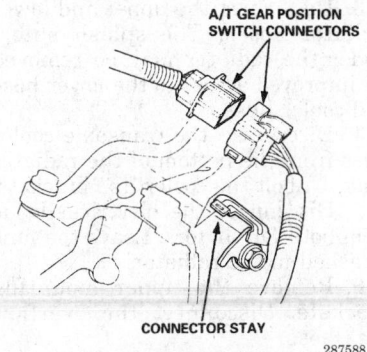

Gear position switch connectors and stay

stall the harness clip bolt. Install the switch cover.

11. Raise the transaxle with a jack.

12. Install the transaxle mount and its through-bolt and nuts. First, tighten the nuts to 28 ft. lbs. (38 Nm). Then, tighten the through-bolt to 47 ft. lbs. (64 Nm).

13. Remove the jack. Lower the vehicle.

14. Couple the A/T gear position switch connectors and place the connection onto the stay.

15. Reconnect the negative battery cable.

16. Turn the ignition switch to the **ON** position, but don't start the engine. Move the shift lever through all the gears, and make sure the gear position switch is synchronized with the gear position indicator.

17. Start the engine. Move the shift lever through all the gears. Make sure the engine only starts in the **P** and **N** positions. Make sure the transaxle cannot be shifted from the **R** position to the **N** positions without first pulling the lever toward you. Make sure the back-up lights come on when the shift lever is in the **R**position.

18. Enter the radio security code.

Engine Control Module

REMOVAL AND INSTALLATION

NOTE: The radio may contain a coded theft protection circuit. Always make note of your code number before disconnecting the battery.

1. Disconnect the negative and positive battery cables. Wait three minutes before working around the airbags.

2. Remove the passenger's side door sill moulding. Remove the right kick panel and right console side panel.

3. Pull back the carpeting far enough to expose the ECM cover.

4. Remove the four nuts and lift up the ECM cover.

5. To remove the ECM, first disconnect the three wiring harnesses. Then, remove the three bolts to separate the ECM from the cover.

To Install:

6. Install the ECM onto the cover. Reconnect the three wiring harnesses.

7. Install the ECM cover onto the four studs on the floor. Install the four nuts.

8. Place the carpet back over the ECM cover.

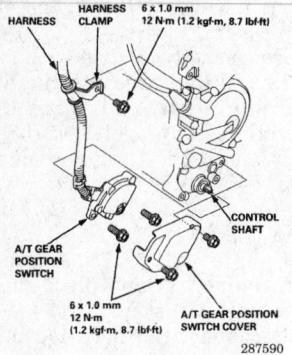

HARNESS
HARNESS CLAMP
6 x 1.0 mm
12 N·m (1.2 kgf·m, 8.7 lbf·ft)
CONTROL SHAFT
A/T GEAR POSITION SWITCH
6 x 1.0 mm
12 N·m (1.2 kgf·m, 8.7 lbf·ft)
A/T GEAR POSITION SWITCH COVER
287590

Gear position switch and cover

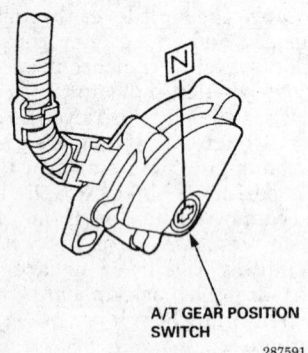

A/T GEAR POSITION SWITCH
287591

Gear position switch neutral position

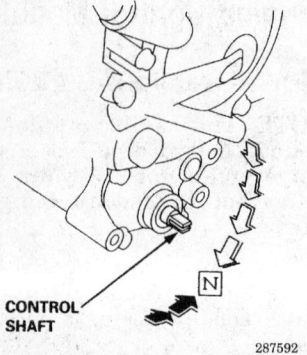

CONTROL SHAFT
287592

Control shaft neutral position

9. Install the right console panel, the right kick panel, and the passenger's side door sill moulding.
10. Reconnect the positive and negative battery cables.
11. Make sure all indicator lights function properly. Make sure that any diagnostic codes have been checked.
12. Road test the vehicle to check engine performance.
13. Enter the radio security code.

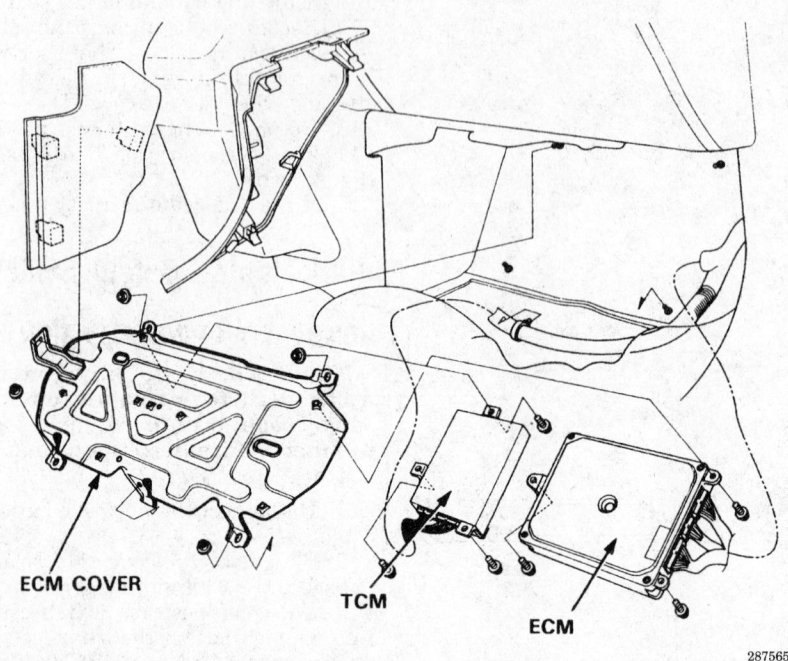

ECM COVER

TCM

ECM

287565

ECM and cover

ENGINE COOLING

Radiator

REMOVAL AND INSTALLATION

NOTE: The radio may contain a coded theft protection circuit. Always make note of your code number before disconnecting the battery.

1. Disconnect the negative battery cable.
2. Drain the coolant from the radiator.
3. Disconnect the upper and lower radiator hoses. The splash shield under the radiator may be removed for improved access to the lower hose and cooler lines.
4. Disconnect the transaxle cooler lines from the bottom of the radiator tank. Unbolt the cooler line brackets.
5. Disconnect the electrical leads from both fan motors. Leave the fans attached to the radiator.
6. Remove the upper mounting brackets. Disconnect the overflow tank hose.
7. Lift the radiator up and out of the vehicle.

8. Remove the cooling fans from the radiator.

To Install:

9. Assemble the cooling fans onto the radiator.
10. Make sure the lower mounting cushions are securely seated.
11. Install the radiator into the vehicle. Install the upper mounting brackets.
12. Reconnect the transaxle cooler lines and install the cooler line brackets onto the bottom of the radiator tank.
13. Reconnect the upper and lower radiator hoses and overflow tank hose.
14. Reconnect the fan motor leads. Make sure the condenser fan resistor is connected.
15. Install the drain plug with a new O-ring.
16. Install the splash shield if it was removed.
17. Open the cooling system bleed bolt. It is located on the thermostat housing.
18. Refill the radiator with a coolant mixture containing 50–60% antifreeze. Use only antifreeze formulated to prevent the corrosion of aluminum parts. Fill the radiator until the coolant draining from the bleed bolt is free of air bubbles. Then,

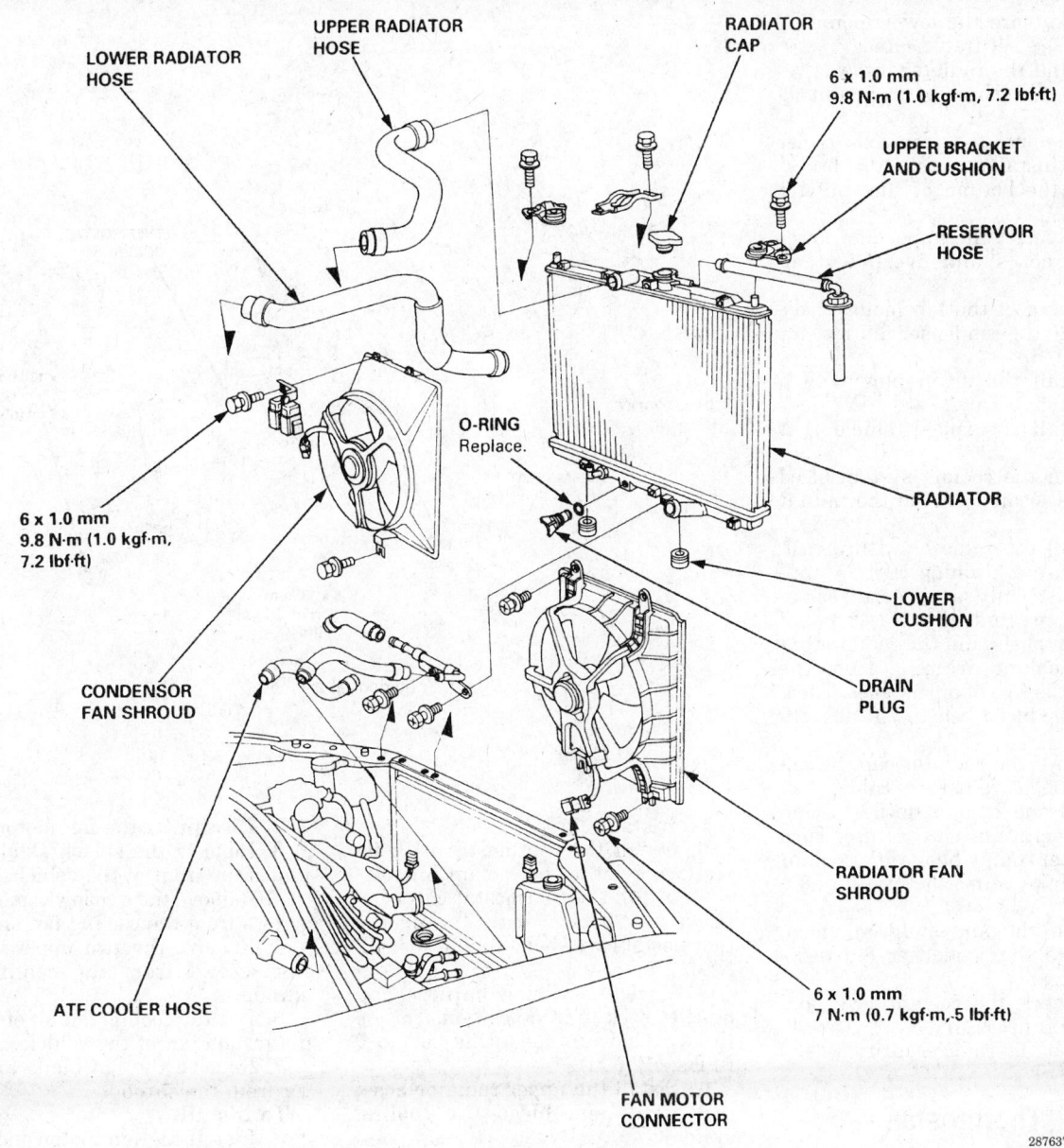

LOWER RADIATOR HOSE

UPPER RADIATOR HOSE

RADIATOR CAP

6 x 1.0 mm
9.8 N·m (1.0 kgf·m, 7.2 lbf·ft)

UPPER BRACKET AND CUSHION

RESERVOIR HOSE

O-RING
Replace.

RADIATOR

6 x 1.0 mm
9.8 N·m (1.0 kgf·m, 7.2 lbf·ft)

LOWER CUSHION

CONDENSOR FAN SHROUD

DRAIN PLUG

RADIATOR FAN SHROUD

ATF COOLER HOSE

6 x 1.0 mm
7 N·m (0.7 kgf·m, 5 lbf·ft)

FAN MOTOR CONNECTOR

287637

Radiator and cooling fan components

tighten the bleed bolt to 7 ft. lbs. (10 Nm).

19. Install the radiator cap. Reconnect the negative battery cable.

20. Run the engine until it is an normal operating temperature. Turn the heater on. Check for coolant leaks. Make sure the cooling fan turns on.

21. Turn the air conditioning on. Make sure the condenser fan turns on.

22. Recheck the coolant level and add more if necessary.

23. Enter the radio security code.

Water Pump

REMOVAL AND INSTALLATION

NOTE: The radio may contain a coded theft protection circuit. Always make note of your code number before disconnecting the battery.

1. Disconnect the negative battery cable.

2. Drain the coolant from the radiator.

3. Disconnect the upper and lower radiator hoses. The splash shield under the radiator may be removed

for improved access to the lower hose and cooler lines.

4. Disconnect the transaxle cooler lines from the bottom of the radiator tank. Unbolt the cooler line brackets.

5. Disconnect the electrical leads from both fan motors. Leave the fans attached to the radiator.

6. Remove the upper mounting brackets. Disconnect the overflow tank hose.

7. Lift the radiator up and out of the vehicle.

8. Remove the cooling fans from the radiator.

To Install:

9. Assemble the cooling fans onto the radiator.

10. Make sure the lower mounting cushions are securely seated.

11. Install the radiator into the vehicle. Install the upper mounting brackets.

12. Reconnect the transaxle cooler lines and install the cooler line brackets onto the bottom of the radiator tank.

13. Reconnect the upper and lower radiator hoses and overflow tank hose.

14. Reconnect the fan motor leads. Make sure the condenser fan resistor is connected.

15. Install the drain plug with a new O-ring.

16. Install the splash shield if it was removed.

17. Open the cooling system bleed bolt. It is located on the thermostat housing.

18. Refill the radiator with a coolant mixture containing 50–60% antifreeze. Use only antifreeze formulated to prevent the corrosion of aluminum parts. Fill the radiator until the coolant draining from the bleed bolt is free of air bubbles. Then, tighten the bleed bolt to 7 ft. lbs. (10 Nm).

19. Install the radiator cap. Reconnect the negative battery cable.

20. Run the engine until it is an normal operating temperature. Turn the heater on. Check for coolant leaks. Make sure the cooling fan turns on.

21. Turn the air conditioning on. Make sure the condenser fan turns on.

22. Recheck the coolant level and add more if necessary.

23. Enter the radio security code.

Thermostat

REMOVAL AND INSTALLATION

NOTE: The radio may contain a coded theft protection circuit. Always make note of your code number before disconnecting the battery.

1. Disconnect the negative battery cable.

2. Drain the coolant into a sealable container.

3. Remove the upper radiator hose.

4. Unbolt the wiring harness bracket from the thermostat cover. Position the wiring harness away from the thermostat.

5. Remove the thermostat cover. Then, remove the thermostat and its rubber seal.

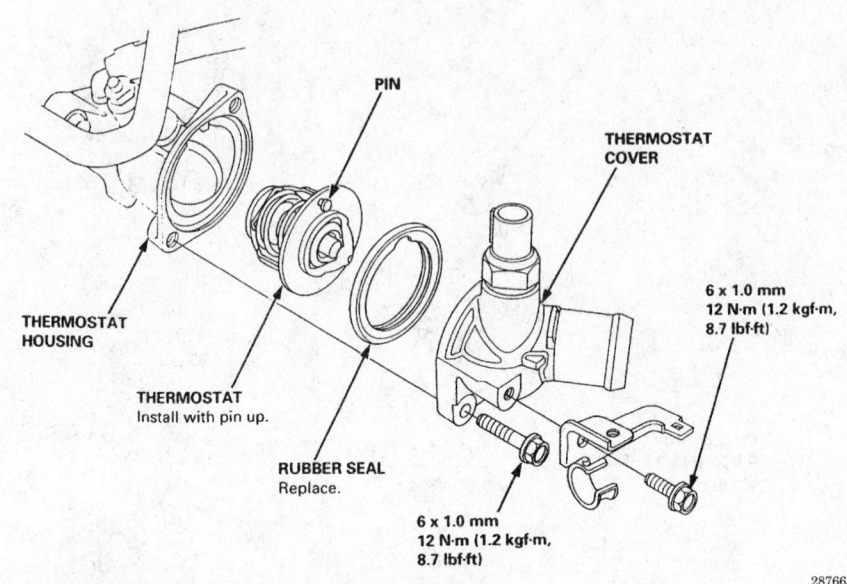

Thermostat and cover

To install:

6. Install the thermostat with its pin facing out of the coolant inlet.

7. Install the thermostat cover using a new rubber seal. Torque the thermostat cover bolts to 9 ft. lbs. (12 Nm).

8. Move the wiring harness and bracket back to into position. Torque the wire harness mounting bolt to 9 ft. lbs. (12 Nm).

9. Install the upper radiator hose.

10. Refill and bleed the cooling system.

11. Connect the negative battery cable and enter the radio security code.

12. Warm up the engine and check for coolant leaks.

Electric Cooling Fan

REMOVAL AND INSTALLATION

NOTE: The radio may contain a coded theft protection circuit. Always make note of your code number before disconnecting the battery.

Cooling fan

1. Disconnect the negative battery cable.

2. Disconnect the fan motor lead.

3. Remove the splash shield from under the front of the vehicle.

4. Remove the two lower mounting screws from the cooling fan shroud.

5. Remove the two upper mounting screws from the cooling fan shroud.

6. Lift the cooling fan off of the radiator and out of the vehicle.

7. Remove the fan blade and motor from the shroud.

To install:

8. Install the fan motor and blade onto the shroud. Use a new self-locking nut on the fan blade.

9. Install the cooling fan onto the radiator and install the mounting screws.

10. Install the splash shield.

11. Reconnect the fan motor lead.

12. Reconnect the negative battery cable.

13. Run the engine until it reaches operating temperature and test the operation of the fan.

14. Enter the radio security code.

Condenser fan

1. Disconnect the negative battery cable.

2. Disconnect the fan motor lead.

3. Remove the splash shield from under the front of the vehicle.

4. Remove the lower mounting screw from the condenser fan shroud.

5. Remove the two upper mounting screws from the condenser fan shroud.

6. Lift the condenser fan off of the radiator and out of the vehicle.

NOTE: If the condenser fan motor is faulty, the entire fan assembly is replaced.

To install:

7. Install the condenser fan onto the radiator and install the mounting screws.

8. Install the splash shield.

9. Reconnect the fan motor lead.

10. Reconnect the negative battery cable.

11. Run the engine until it reaches operating temperature. Turn the air conditioning on, and test the operation of the fan.

12. Enter the radio security code.

Cooling System

BLEEDING

NOTE: The radio may contain a coded theft protection circuit. Always make note of your code number before disconnecting the battery.

1. Mix the recommended antifreeze with an equal amount of water in a clean container.

NOTE: Coolant concentrations greater than 60% will impair cooling efficiency and are not recommended.

2. Loosen the air bleed bolt in the thermostat housing, then fill the radiator to the bottom of the filler neck with the mixed coolant. Tighten the bleed bolt as soon as coolant starts to run out in a steady stream without bubbles. Torque the bleed bolt to 7 ft. lbs. (10 Nm).

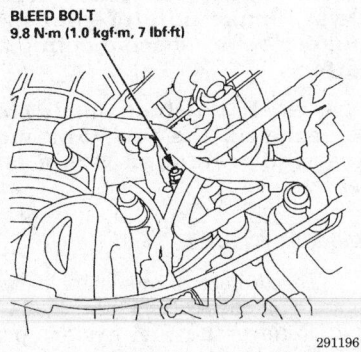

BLEED BOLT
9.8 N·m (1.0 kgf·m, 7 lbf·ft)

291196

Bleed bolt

—— WARNING ——

When pouring engine coolant, be sure to shut the relay box lid and not to let coolant spill on the electrical components or the paint. If any coolant spills, rinse it off immediately.

3. Fill the reservoir to the MAX mark with the coolant mixture.

4. Connect the negative battery cable and enter the radio security code.

5. With the radiator cap removed, start the engine and let it run until warmed up (the radiator fan comes on at least twice). Then, if necessary, top off the radiator with the coolant mixture.

6. Install the radiator cap securely and run the engine to check for coolant leaks.

FUEL SYSTEM

Fuel System Service Precautions

Safety is the most important factor when performing not only fuel system maintenance but any type of maintenance. Failure to conduct maintenance and repairs in a safe manner may result in serious personal injury or death. Maintenance and testing of the vehicle's fuel system components can be accomplished safely and effectively by adhering to the following rules and guidelines.

• To avoid the possibility of fire and personal injury, always disconnect the negative battery cable unless the repair or test procedure requires that battery voltage be applied.

• Always relieve the fuel system pressure prior to disconnecting any fuel system component (injector, fuel rail, pressure regulator, etc.), fitting or fuel line connection. Exercise extreme caution whenever relieving fuel system pressure to avoid exposing skin, face and eyes to fuel spray. Please be advised that fuel under pressure may penetrate the skin or any part of the body that it contacts.

• Always place a shop towel or cloth around the fitting or connection prior to loosening to absorb any excess fuel due to spillage. Ensure that all fuel spillage (should it occur) is quickly removed from engine sur-

---

faces. Ensure that all fuel soaked cloths or towels are deposited into a suitable waste container.

• Always keep a dry chemical (Class B) fire extinguisher near the work area.

• Do not allow fuel spray or fuel vapors to come into contact with a spark or open flame.

• Always use a backup wrench when loosening and tightening fuel line connection fittings. This will prevent unnecessary stress and torsion to fuel line piping. Always follow the proper torque specifications.

• Always replace worn fuel fitting O-rings with new. Do not substitute fuel hose or equivalent, where fuel pipe is installed.

Fuel System Pressure

RELIEVING

—— CAUTION ——

Do not allow fuel spray or fuel vapors to come in contact with a spark or open flame. Keep a dry chemical fire extinguisher nearby. Never store fuel in an open container due to risk of fire or explosion.

NOTE: Always install new sealing washers whenever fuel system service bolts and banjo bolts are loosened or removed. Replace any stripped or damaged banjo bolts.

1. Make sure the ignition is **OFF**.

2. Disconnect the negative battery cable.

3. Remove the fuel filler cap.

4. Hold the fuel rail inlet banjo bolt with a flare wrench. Hold the service bolt with a box wrench.

5. Place a shop towel over the fitting to absorb leakage.

6. Slowly loosen the the service bolt one turn.

7. After service has been completed, always install a new sealing washer on the service bolt.

8. Tighten the service bolt to 8.7 ft. lbs. (12 Nm). If the entire fitting has been removed, torque the service bolt to 16 ft. lbs. (22 Nm).

9. Reconnect the negative battery cable. Install the fuel filler cap.

10. Turn the ignition to the **ON** position, but don't start the engine. Then, turn the ignition **OFF**. Repeat this step two or three times to pressurize the fuel system.

11. Check the fuel fittings for any signs of leaks.

Idle Speed

ADJUSTMENT

NOTE: The radio may contain a coded theft protection circuit. Always make note of your code number before disconnecting the battery or removing the backup/radio fuse.

1. Start engine and warm it up to normal operating temperature at 3000 rpm with the transaxle in the **P** or **N** position and no electrical load. The cooling fan must come on at least once. Switch the ignition off.
2. Connect a test tachometer.
3. Disconnect the connector from the IAC valve.
4. Start the engine with the accelerator pedal slightly depressed. Use the pedal to stabilize the engine speed at 1000 rpm. Slowly release the throttle until the engine idles. Check the idle with air conditioning, cooling fan, and all electrical loads off. On vehicles equipped with daytime running lights, fully engage the parking brake to turn the headlights off.
5. The idle should be 550 ± 50 rpm with the vehicle in **P** or **N**.

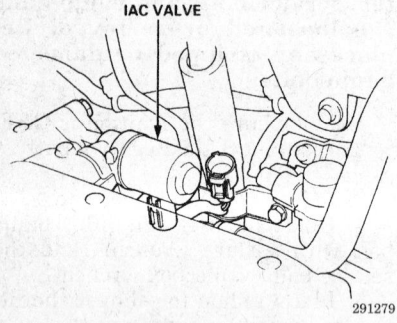

IAC valve connector

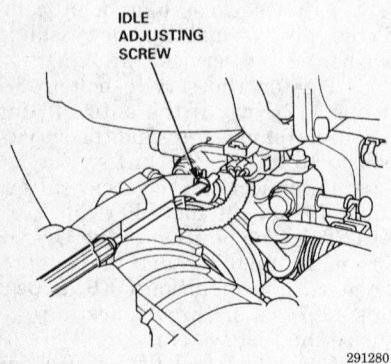

Idle adjusting screw

6. If the rpm is not within specifications, adjust as needed by turning the idle adjusting screw on the throttle body.
7. Switch the ignition off. Reconnect the connector to the IAC valve. Then, reset the ECM by removing the Back-up/Radio 7.5A fuse from the underhood fuse box for at least 10 seconds.
8. Restart the engine, allowing it to idle in **N** or **P** for one minute. With all electrical loads including the air conditioning and cooling fan off, idle speed should be 700 ± 50 rpm.
9. Allow the engine to idle for one minute with the headlights on low beam. The idle speed should be 770 ± 50 rpm.
10. Turn the headlights off. Switch the air conditioner on and the blower fan to high speed. After one minute, the idle should be 770 ± 50 rpm.
11. Enter the radio security code.

Mixture

ADJUSTMENT

The air/fuel mixture is computer controlled according to the needs of the engine and is not adjustable. If the air/fuel mixture is too lean or too rich, other problems with the engine and/or engine control system exist.

Fuel Filter

REMOVAL AND INSTALLATION

———— CAUTION ————
Fuel injection systems remain under pressure after the engine has been turned OFF. Properly relieve fuel pressure before disconnecting any fuel lines. Failure to do so may result in fire or personal injury.

NOTE: The radio may contain a coded theft protection circuit. Always make note of your code number before disconnecting the battery.

1. Disconnect the negative battery cable.
2. Place a shop towel under the fuel filter to absorb leakage.
3. Relieve the fuel pressure:
    a. Remove the fuel filler cap.
    b. Hold the banjo bolt on the fuel rail with a flare wrench.
    c. Wrap a shop towel around the fuel rail fitting.

d. Use a box-end wrench to loosen the service bolt one turn.
4. Remove the service bolt and disconnect the fuel line from the filter. Due to the restricted location of the fuel filter, a flare wrench and socket may be needed to loosen the fuel filter fittings.
5. Use flare wrenches to loosen the fuel inlet line from the bottom of the filter.
6. Unbolt the fuel filter clamp from the vehicle's firewall. Remove the fuel filter.
    To Install:

NOTE: Always use new sealing washers when installing the fuel filter and reconnecting fuel lines. Replace any stripped banjo bolts.

7. Clean the fuel line fittings before installing the fuel filter.
8. Install the fuel filter and bracket. Connect the fuel inlet line to the filter and carefully torque it to 27 ft. lbs. (37 Nm).
9. Connect the fuel line to the top of the filter with new sealing washers. Carefully torque the service bolt to 16 ft. lbs. (22 Nm).
10. Install new sealing washers onto the fuel rail fitting. Then, tighten the fitting to 16 ft. lbs. (22 Nm).
11. Install the fuel filler cap.
12. Reconnect the negative battery cable.
13. Turn the ignition **ON**, but don't start the engine. Then, turn the ignition **OFF**. Repeat this step two or three times to pressurize the fuel system.
14. Check the fuel filter and fuel rail fittings for leakage.
15. Enter the radio security code.

Fuel Pump

REMOVAL AND INSTALLATION

NOTE: The radio may contain a coded theft protection circuit. Always make note of your code number before disconnecting the battery.

NOTE: If the fuel pump must be removed, the fuel tank must first be removed from the vehicle.

Fuel tank removal

1. Disconnect the negative battery cable.
2. Relieve the fuel system pressure by loosening the fuel filler cap and the fuel rail service bolt.

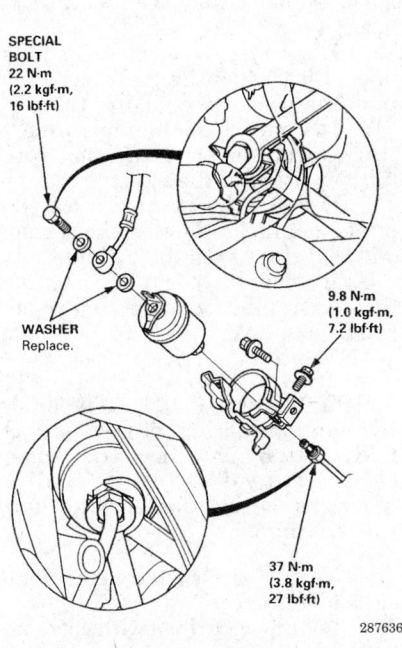

Fuel filter components

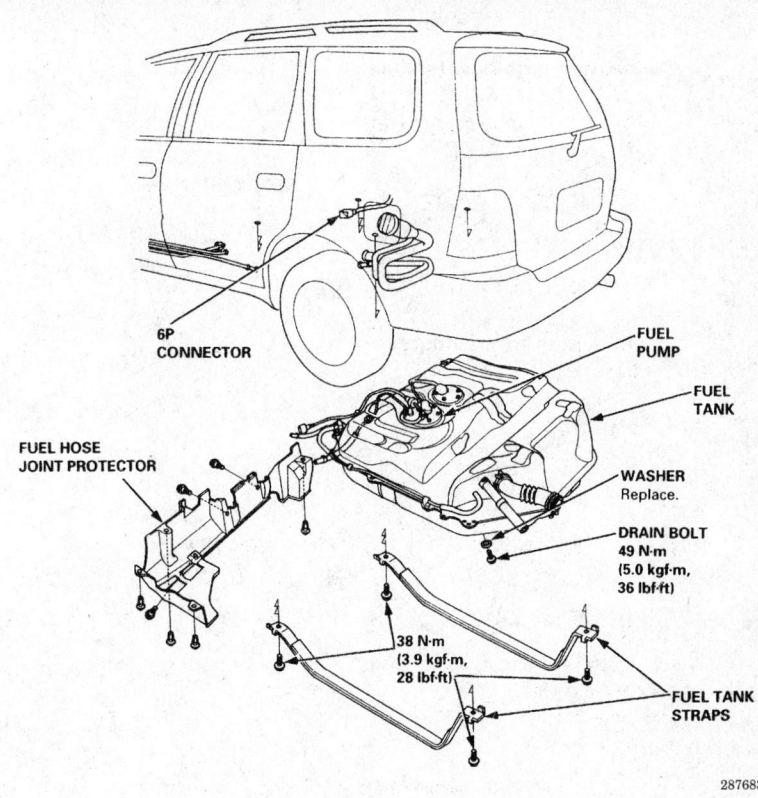

Fuel tank components

CAUTION

Fuel injection systems remain under pressure after the engine has been turned OFF. Properly relieve fuel pressure before disconnecting any fuel lines. Failure to do so may result in fire or personal injury.

3. Raise and safely support the vehicle.
4. Remove the fuel tank drain plug. Drain the fuel into an approved gasoline storage container. Install the drain plug with a new crush washer.

CAUTION

Do not allow fuel spray or fuel vapors to come in contact with a spark or open flame. Keep a dry chemical fire extinguisher nearby. Never store fuel in an open container due to risk of fire or explosion.

5. Remove the fuel hose protector.
6. Disconnect the fuel filler hoses.
7. Disconnect the fuel hose and the quick-connect fitting. Hold the connector with one hand. Press the retainer tabs down and in with your other hand; then, pull the connectors

apart. Plug the fuel lines after removal to keep dirt out.

NOTE: Don't use tools to separate the connectors. If the connection is tight, gently wiggle the connector in and out to free it from retainer. Don't remove the retainer from the fuel pipe.

8. Use a transmission jack and a broad piece of wood under the tank. Adjust the position of the jack as necessary to evenly support the fuel tank.
9. Remove the bolts attaching the tank straps; then, remove the straps.
10. Lower the tank to disconnect the fuel pump connector and the fuel gauge sender connector. If the tank is stuck to the vehicle's undercoating, pry it loose gently with a piece of wood.
11. Remove the fuel tank.

Fuel pump removal and installation

1. Disconnect the 2-P connector from the fuel pump.
2. Unbolt the fuel pump from the fuel tank.
3. Lift the fuel pump up and allow any fuel in it to drain into the tank. Then, remove the fuel pump.
To Install:
4. Fit the fuel pump into the tank.

5. Install the mounting nuts and tighten them to 4–5 ft. lbs. (6 Nm)
6. Reconnect the 2–P connector.

Fuel tank installation

1. Position the fuel tank under the vehicle and raise it enough to connect the fuel pump and fuel gauge sender connectors.
2. Position the tank in the vehicle and install the tank straps. Torque the bolts attaching the straps to 28 ft. lbs. (38 Nm).
3. Remove the transmission jack from under the fuel tank.
4. Clean the quick-connect fuel line fittings and apply a light coat of oil to the contact areas. Install a new retainer on the fuel pump connector.
5. Connect the fuel filler hoses and install new clamps if necessary.
6. Install the fuel hose protector.
7. Make sure the drain plug has been installed. Torque the drain plug to 36 ft. lbs. (49 Nm).
8. Install a new sealing washer on the fuel rail service bolt. Tighten the bolt to 8.7 ft. lbs. (12 Nm).
9. Lower the vehicle.
10. Refill the fuel tank.
11. Connect the negative battery cable and enter the radio security code.

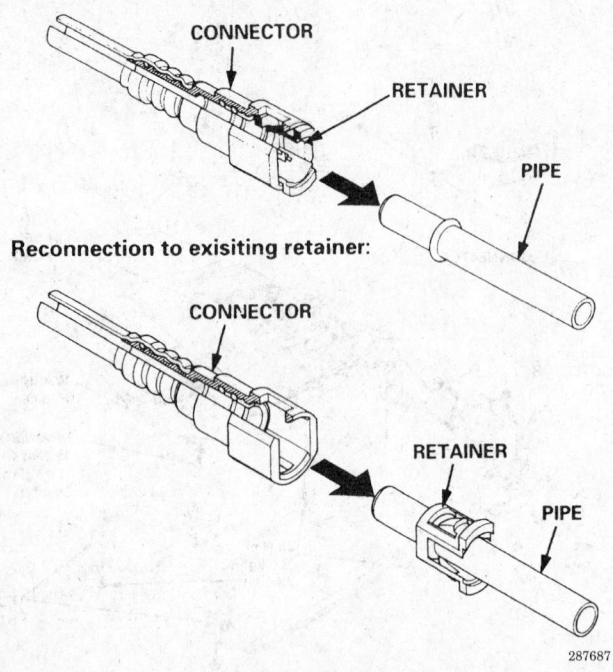

Connection with new retainer:

CONNECTOR

RETAINER

PIPE

Reconnection to exisiting retainer:

CONNECTOR

RETAINER

PIPE

287687

Quick-connect fitting coupling

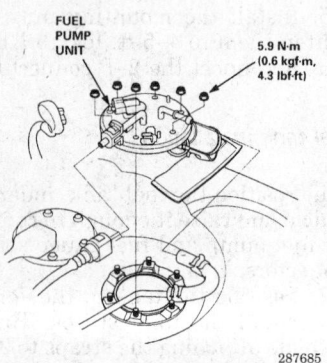

FUEL PUMP UNIT

5.9 N·m (0.6 kgf·m, 4.3 lbf·ft)

287685

Fuel pump components

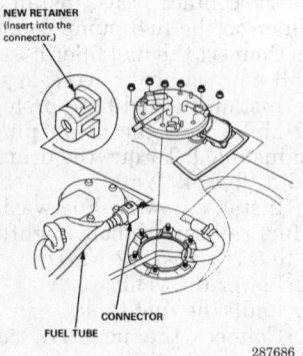

NEW RETAINER (Insert into the connector.)

CONNECTOR

FUEL TUBE

287686

New fitting retainer on fuel pump

12. Switch the ignition **ON** but don't start the engine. The fuel pump should run for approximately two seconds, building pressure within the lines. Switch the ignition **OFF**, then **ON** two or three more times to build full system pressure. Check for fuel leaks.

Fuel Injector

REMOVAL AND INSTALLATION

—— **CAUTION** ——

Fuel injection systems remain under pressure after the engine has been turned OFF. Properly relieve fuel pressure before disconnecting any fuel lines. Failure to do so may result in fire or personal injury.

NOTE: The radio may contain a coded theft protection circuit. Always make note of your code number before disconnecting the battery.

1. Disconnect the negative battery cable.
2. Relieve the fuel pressure by removing the fuel filler cap and then loosening the fuel rail service bolt one turn.

3. Uncouple the fuel injector connectors. Detach the injector wiring harness holder from the fuel rail and move it out of the way.
4. Disconnect the three vacuum lines. Disconnect the fuel return line from the pressure regulator tube.
5. Disconnect the fuel inlet line.
6. Unbolt the fuel rail and carefully lift it off of the injectors.
7. Remove the injectors from the intake manifold. Remove the cushion and seal rings from the fuel injectors.
8. Be sure to clean up any fuel that was spilled on the engine and intake manifold.

To Install:

NOTE: Always use new cushion rings, sealing rings, and O-rings when installing fuel injectors. Always use new sealing washers when connecting fuel line fittings.

9. Install new cushion rings onto each fuel injector.
10. Coat new O-rings with clean engine oil and install them onto each fuel injector.
11. Fit the fuel injectors into the fuel rail.
12. Coat new sealing rings with clean engine oil and install them into the intake manifold orifices.
13. Install the fuel rail and injector assembly onto the intake manifold. Make sure each injector is properly seated.
14. Install the retainer nuts and evenly tighten each to 8.7 ft. lbs. (12 Nm).
15. Install the fuel inlet line onto the rail with new sealing washers. Tighten the fitting to 16 ft. lbs. (22 Nm), and the the service bolt to 8.7 ft. lbs. (12 Nm).
16. Reconnect the three vacuum lines. Reconnect the fuel return hose to the pressure regulator.
17. Couple the wiring harness connectors to the fuel injectors. Fit the harness holder back onto its clips. Make sure the fuel injector wiring harness connectors and grounds are secure.
18. Reconnect the negative battery cable.
19. Turn the ignition to the **ON** position, but don't start the engine. Then, turn the ignition **OFF**. Repeat this step two or three times to pressurize the fuel system.
20. Carefully check the fuel injectors, fuel rail, and fuel line fittings for any signs of leakage.
21. Road test the vehicle and check engine performance.
22. Enter the radio security code.

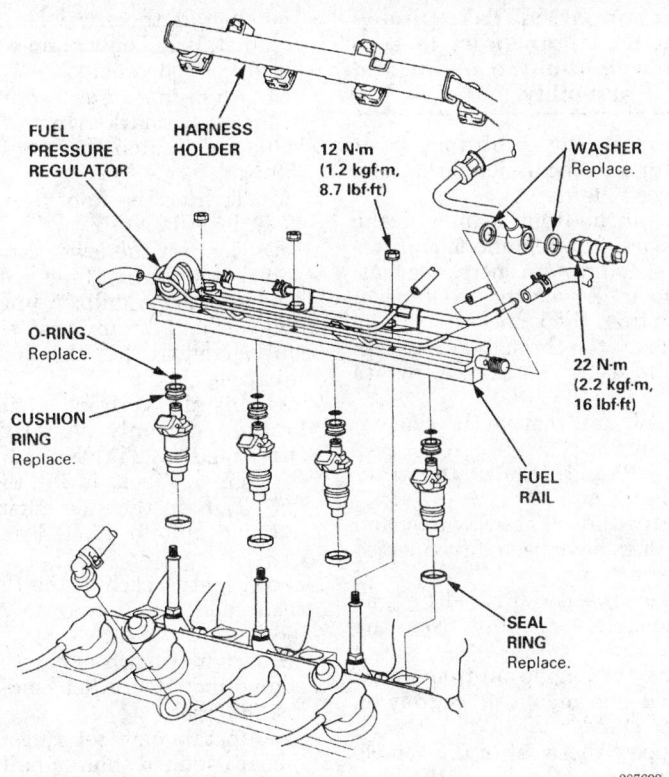

Fuel rail and injector components

FUEL PRESSURE REGULATOR

HARNESS HOLDER

12 N·m (1.2 kgf·m, 8.7 lbf·ft)

WASHER Replace.

O-RING Replace.

CUSHION RING Replace.

22 N·m (2.2 kgf·m, 16 lbf·ft)

FUEL RAIL

SEAL RING Replace.

287692

EMISSION CONTROLS

Service Interval Lamp

RESETTING

Vehicles equipped with a Maintenance Reminder indicator will indicate it is time for scheduled maintenance. When it is near 7,500 miles (12,000 km) since the last maintenance, the indicator will turn yellow. If you exceed 7,500 miles (12,000 km), the indicator will turn red. The indicator can be reset by inserting the ignition key or other similar object into the slot below and to the left of the indicator. This will extinguish the indicator for the next 7500 miles.

INDICATOR

SLOT

276484

Resetting the Maintenance Required indicator

ENGINE MECHANICAL

Engine Assembly

REMOVAL AND INSTALLATION

NOTE: The radio may contain a coded theft protection circuit. Always make note of your code number before disconnecting the battery.

NOTE: The engine and transaxle are removed from the vehicle as a unit. A hydraulic lift is helpful for this procedure since the front subframe must be removed for the engine/transaxle assembly to be lowered from the vehicle.

1. Disconnect negative and positive battery cables.
2. Raise the hood and mark the positions of the hinge plates with a felt-tipped marker. Remove the hood and move it out of the work area to avoid scratching the paint.
3. Remove the battery. Disconnect the ground cable from the battery tray. Unbolt and remove the battery tray and its support bracket.
4. Loosen the throttle cable and cruise control cable locknuts. Remove both cables from their brackets and slip them out of their linkages.
5. Remove the intake air duct.
6. Drain the engine oil, transaxle fluid, and engine coolant into separate, sealable containers.
7. Disconnect the battery cable from the fuse/relay box and ABS relay box.
8. Uncouple the three engine harness connectors located on the right side of the engine compartment.
9. Disconnect the brake booster vacuum hose and EVAP canister hose from the intake manifold plenum.
10. Relieve the fuel pressure by loosening the fuel rail service bolt one turn.

CAUTION

Fuel injection systems remain under pressure after the engine has been turned OFF. Properly relieve fuel pressure before disconnecting any fuel lines. Failure to do so may result in fire or personal injury.

11. Disconnect the fuel line and the return hose from the fuel rail. Clean up any spilled fuel.
12. Disconnect the three vacuum hoses from the left side of the intake manifold. Unclamp the power steering hose.
13. Uncouple the three engine harness connectors located on the left side of the engine compartment. Disconnect the fuel injector resistor connector.
14. Loosen the power steering pump adjusting bolt and mount bolts. Slip the belt off its pulleys.
15. Unbolt and remove the power steering pump, but do not disconnect

its hydraulic lines. Wire the pump away from the engine.

16. Disconnect the alternator. Loosen and remove the alternator belt.

17. Unbolt and remove the alternator and its mounting bracket.

18. Disconnect and remove the upper and lower radiator hoses.

19. Remove the radiator with both cooling fans attached to it.

20. Disconnect the heater hoses from the coolant pipe under the intake manifold.

21. Disconnect the two cooler lines from the front of the transaxle case. Plug the inlet lines on the transaxle case to prevent moisture contamination.

22. Unbolt the A/C compressor from the side of the engine block. Don't disconnect the A/C lines. Wire the compressor to the radiator support so it's out of the work area.

23. Attach a chain hoist to the lifting hooks on either side of the engine block.

24. Raise and safely support the vehicle. Take up the slack of the chain hoist.

25. Disconnect the heated oxygen sensor connector. Leave the sensor installed in the exhaust pipe.

26. Separate exhaust pipe from the catalytic converter and the exhaust manifold and remove it.

27. Matchmark the subframe center beam to the rear beam. Remove the two bolts, but don't remove the center beam at this point.

28. Remove the shift cable cover. Disconnect the shift cable from the control shaft and suspend it out of the way with wire.

29. Remove the splash shield.

30. Remove the front wheels.

31. Remove the damper fork flange bolts from the lower control arms. Unbolt the damper fork from the strut and remove it from the vehicle.

32. Use a ball joint removal tool to separate the lower control arms from the steering knuckles.

33. Use a pry tool to detach the left and right halfshafts from the intermediate shaft and transaxle case. Tap the splined shafts out of the hubs using a plastic mallet. Move the halfshafts out of the way and wire them up. Tie plastic bags over the halfshaft ends to protect the boots and splined shafts.

34. Unbolt and remove the front engine mount bracket.

------------- **WARNING** -------------
The next step involves the removal of the subframe front beam. Make sure the vehicle is securely supported. Take up any slack in the chain hoist to support the weight of the engine and transaxle assembly.

35. Support the subframe front beam with a transmission jack and a sturdy wood plank.

36. Unbolt the front beam with the center beam, front engine mount, radius rods, and lower control arms attached to it. Lower the front beam assembly from the vehicle.

37. Remove the through-bolt to separate the rear mount from its bracket.

38. Unbolt and remove the side engine mount.

39. Unbolt and remove the transaxle side mount.

40. Verify that all hoses, wires, and vacuum lines have been disconnected from the engine.

41. Slowly lower the engine/transaxle assembly from the vehicle.

42. Move the engine out from under the vehicle and mount it securely on an engine stand.

43. Support the front of the vehicle with jackstands and block the rear wheels, as the front suspension has been disassembled. If the vehicle must be moved with the engine out, install the front beam and all the suspension components.

To Install:

NOTE: Use new self-locking nuts and color-coded self-locking bolts when installing the engine mounts, subframe components, and suspension components.

44. Raise and safely support the vehicle.

45. Move the engine/transaxle assembly into position under the vehicle and attach a chain hoist to the engine lifting hooks.

46. Carefully lift the engine/ transaxle assembly into position in the vehicle.

47. Install the side engine mount. Use a 6mm punch or similarly-sized tool to steady the mount. Install the nut and bolts. Tighten the bolt to 47 ft. lbs. (64 Nm), and remove the 6mm punch. Only hand-tighten the nut and bolt on the engine side of the mount.

48. Install the transaxle side mount. Use a 6mm punch or similarly-sized tool to steady the mount. Install and only hand-tighten the nuts. Install the bolt and tighten it to 47 ft. lbs. (64 Nm).

49. Connect the rear mount bracket and tighten its bolt to 47 ft. lbs. (64 Nm).

50. Install the front beam assembly. Install and only hand-tighten the new color-coded bolts.

51. Align the center beam and rear beam matchmarks. Install the two bolts and tighten them to 37 ft. lbs. (50 Nm).

52. Tighten the front beam bolts to 47 ft. lbs. (64 Nm).

53. Connect the lower control arms and stabilizer bar links. Make sure that all the stabilizer link spacers and washers are properly positioned. Only hand-tighten the fasteners at this time.

54. Install the front engine mount bracket and only hand-tighten the three bolts. Tighten the mount through-bolt to 47 ft. lbs. (64 Nm).

55. Tighten the side engine mount nut and bolt to 47 ft. lbs. (64 Nm) each.

56. Tighten each of the three transaxle side mount nuts to 28 ft. lbs. (38 Nm).

57. Tighten each of the three front engine mount bracket bolts to 28 ft. lbs. (38 Nm).

58. Install new set rings onto the inboard splined shaft of halfshaft. Install the halfshafts, making sure that each snaps securely into place.

59. Reconnect the lower control arm to the steering knuckle ball joint. Tighten the castle nut to 36–43 ft. 36–43 ft. lbs. (49–59 Nm). Then, tighten the nut only enough to install a new cotter pin.

60. Connect the shift cable linkage with a new lock washer and tighten the bolt to 7 ft. lbs. (10 Nm). Install the shift cable cover.

61. Install exhaust pipe using new gaskets. Tighten the manifold nuts to 40 ft. lbs. (54 Nm). Tighten the converter flange nuts to 16 ft. lbs. (22 Nm). Tighten the bracket nuts to 13 ft. lbs. (18 Nm). Reconnect the oxygen sensor.

62. Move the A/C compressor back into position and tighten the bolts to 16 ft. lbs. (22 Nm).

63. Reconnect the transaxle cooler line hoses.

64. Install the front wheels and splash shield.

65. Lower the vehicle and remove the chain hoist.

66. With the vehicle on the ground, tighten the damper fork bolt to 47 ft. lbs. (64 Nm). Tighten the pinch bolt to 32 ft. lbs. (44 Nm). Tighten the control arm flange bolt to 40 ft. lbs. (50 Nm). Tighten the stabilizer bar linkage nut to 14 ft. lbs. (19 Nm).

67. Install the radiator and its upper brackets. Reconnect the cooling and condenser fan connectors.

68. Install the upper and lower radiator hoses. Connect the heater hoses to the coolant pipe.

69. Install the alternator bracket and tighten the three bracket bolts to 36 ft. lbs. (49 Nm).

70. Install and connect the alternator. Install the A/C compressor belt and the alternator belt. Tighten the alternator mounting bolt to 33 ft. lbs. (44 Nm). Adjust the tensions of both belts.

71. Move the power steering pump into place and install its mounting nuts. Install the power steering belt. Tighten the mounting nuts to 16 ft. lbs. (22 Nm). Adjust the belt tension.

72. Couple the three engine wiring harness connectors on the left side of the engine compartment. Reconnect the fuel injector resistor connector.

73. Reconnect the vacuum hoses at the intake manifold plenum. Connect the power steering pump clamp.

74. Reconnect the fuel line to the fuel rail using new sealing washers. Tighten the fitting to 16 ft. lbs. (22 Nm), and the service bolt to 8.7 ft. lbs. (12 Nm).

75. Reconnect the EVAP and brake booster hoses to the intake manifold plenum.

76. Couple the three engine wiring harness connectors on the right side of the engine compartment.

77. Refill the engine with fresh oil.

78. Refill the transaxle with fresh ATF.

79. Refill the radiator with fresh coolant and bleed the cooling system.

80. Reconnect the battery cable to the fuse/relay box and the the ABS fuse/relay box.

81. Reconnect the throttle and cruise control cables to their linkages. Replace the cables if they are kinked. Adjust each cable's deflection.

82. Install the intake air duct.

83. Install the battery tray and its support bracket. Connect the ground cable to the bracket.

84. Install the battery and connect the positive and negative cables.

85. Fit the hood into position and loosely install the hinge bolts. Align the hinges with their matchmarks and tighten the bolts to 7 ft. lbs. (10 Nm). Close the hood and check its alignment with the fenders, bumper, and windshield. Make sure the windshield washer fluid tube is connected.

86. Check the shift cable adjustment.

87. Start the engine allow it to reach normal operating temperature.

88. Check all fluid levels and top-up if necessary. Check for signs of fluid or fuel leaks

89. Check the operation of the heater and air conditioner.

90. Check and adjust the front wheel alignment.

91. Enter the radio security code.

92. Road test the vehicle and check the transaxle shift points and the operation of all engine-driven accessories such as power steering and air conditioning.

Engine Mounts

REMOVAL AND INSTALLATION

NOTE: The radio may contain a coded theft protection circuit. Always make note of your code number before disconnecting the battery.

Side Engine Mount

1. Secure the hood as far open as possible.

2. Disconnect the negative and positive battery cables.

3. Attach an engine hoist to the engine lifting points and raise the hoist to remove all slack from the chain.

NOTE: A floor jack with a cushion or wooden plank over its pad may be substituted for the lifting chain.

4. Remove the nut and bolt attaching the side engine mount to the engine.

5. Remove the through bolt from the side engine mount.

To install:

6. Install the side engine mount. Use a 6 x 100 mm bolt to the mount to position the mount in its bracket. Only hand-tighten the nut and bolt attaching the mount to the engine. Torque the mount's through-bolt to 47 ft. lbs. (64 Nm); then, remove the 6 x 100 mm bolt from the mount.

7. Torque the side engine mount nut and bolt to 40 ft. lbs. (54 Nm) each.

8. Remove the hoist equipment from the engine.

9. Install the battery and connect the positive and negative battery cables. Enter the radio security code.

Front Engine Mount

1. Secure the hood as far open as possible.

2. Disconnect the negative and positive battery cables.

3. Attach an engine hoist to the engine lifting points and raise the hoist to remove all slack from the chain.

4. Remove the bolts attaching the front engine mount bracket to the engine.

5. Remove the through-bolt from the front engine mount. Discard the bolt.

6. Remove the bolts attaching the mount to the front beam.

To install:

7. Install the front engine mount to the front beam. Torque the bolts attaching the mount to the beam to 69 ft. lbs. (93 Nm).

8. Install the front mount bracket. Do not tighten the bolts attaching the mount to the engine assembly, only snug the bolts in place. Install a new through-bolt to the front mount. Torque the new through bolt to 47 ft. lbs. (64 Nm).

9. Torque the three bolts attaching the front mount bracket to the engine to 28 ft. lbs. (38 Nm) each.

10. Remove the hoist equipment from the engine.

11. Install the battery and connect the positive and negative battery cables. Enter the radio security code.

Rear Engine Mount

NOTE: The rear engine mount is liquid-filled and equipped with a diaphragm actuator. The actuator is controlled by the ECM and a solenoid valve. As engine rpm changes, intake vacuum is applied to the actuator to adjust the mount's damping capabilities. If the actuator lever doesn't operate correctly, the mount must be replaced.

1. Matchmark the hood hinges with a felt-tipped marker. Remove the hood.

2. Disconnect the negative and positive battery cables.

3. Raise and safely support the vehicle.

4. Attach an engine hoist to the engine lifting points and raise the hoist to remove all slack from the chain.

5. Remove and discard the bolts attaching the rear engine mount bracket to the engine.

6. Remove and discard the through-bolt from the rear engine mount.

7. Disconnect the vacuum hose from the rear engine mount.

8. Remove the bolts attaching the mount to the rear beam.

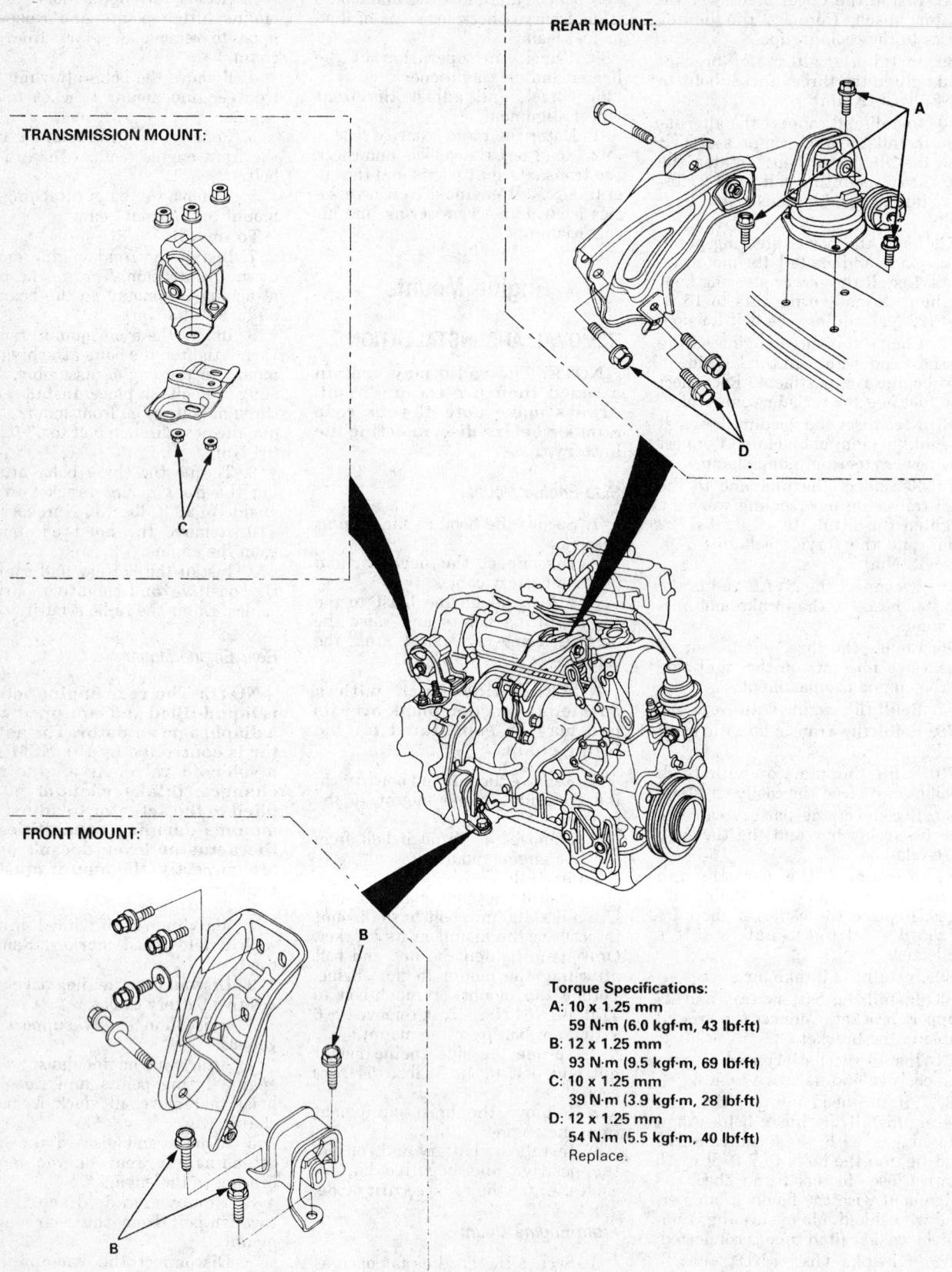

REAR MOUNT:

TRANSMISSION MOUNT:

FRONT MOUNT:

Torque Specifications:
A: 10 x 1.25 mm
   59 N·m (6.0 kgf·m, 43 lbf·ft)
B: 12 x 1.25 mm
   93 N·m (9.5 kgf·m, 69 lbf·ft)
C: 10 x 1.25 mm
   39 N·m (3.9 kgf·m, 28 lbf·ft)
D: 12 x 1.25 mm
   54 N·m (5.5 kgf·m, 40 lbf·ft)
   Replace.

290042

Engine mounts and torque specifications

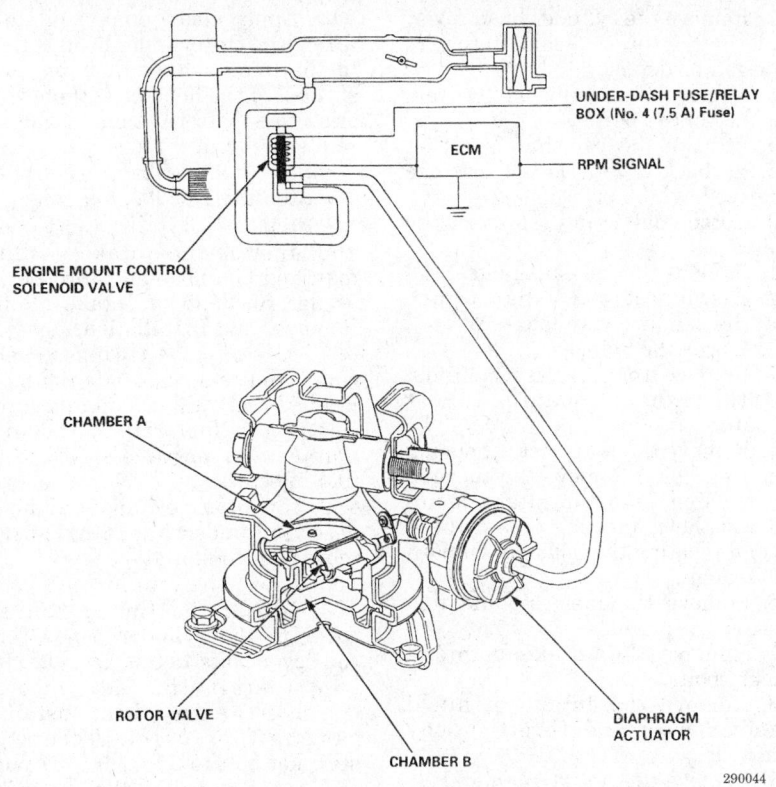

ENGINE MOUNT CONTROL SOLENOID VALVE

UNDER-DASH FUSE/RELAY BOX (No. 4 (7.5 A) Fuse)

ECM

RPM SIGNAL

CHAMBER A

ROTOR VALVE

CHAMBER B

DIAPHRAGM ACTUATOR

290044

Rear mount actuator schematic

To install:

9. Install the rear engine mount to the rear beam. Torque the bolts attaching the mount to the beam to 43 ft. lbs. (59 Nm).

——— WARNING ———
Tighten the bolts attaching the mount to the engine first. Then, tighten the through-bolt. The engine mount may be damaged by excessive engine vibration if this tightening order isn't followed.

10. Install the rear engine mount bracket using new bolts. Torque the new self-locking bolts attaching the mount to the engine to 40 ft. lbs. (54 Nm).
11. Install a new rear engine mount through-bolt. Torque the new through bolt to 47 ft. lbs. (64 Nm).
12. Reconnect the vacuum hose to the rear engine mount.
13. Lower the vehicle.
14. Remove the hoist equipment from the engine.
15. Install the battery and connect the positive and negative battery cables. Enter the radio security code.

Transaxle Mount

1. Matchmark the hood hinges with a felt-tipped marker. Remove the hood.

2. Disconnect the negative and positive battery cables.
3. Remove the battery and the air cleaner assembly to gain better access to the mount.
4. Attach an engine hoist to the engine lifting points and raise the hoist to remove all slack from the chain.
5. Remove the nuts attaching the mount to the transaxle.
6. Remove the through-bolt from the transaxle mount.
7. Remove the transaxle mount from the mounting bracket.
To install:
8. Install the transaxle mount to the mounting bracket. Torque the attaching nuts/bolts to 28 ft. lbs. (39 Nm).
9. Install the transaxle mount. Use a 6 x 100 mm bolt to the mount to position the mount into its bracket. Only hand-tighten the nuts attaching the mount to the transaxle at this time. Torque the through-bolt to 47 ft. lbs. (64 Nm) then remove the 6 x 100 mm bolt from the mount.
10. Torque the nuts attaching the transaxle mount to the transaxle to 28 ft. lbs. (38 Nm).
11. Remove the hoist equipment from the engine.
12. Install the air cleaner assembly.

13. Install the battery and connect the positive and negative battery cables. Enter the radio security code.

Cylinder Head

REMOVAL AND INSTALLATION

NOTE: The radio may contain a coded theft protection circuit. Always make note of your code number before disconnecting the battery.

1. Disconnect the negative and positive battery cables.
2. Remove the valve cover and the upper timing belt cover.
3. Turn the crankshaft to align the TDC marks and set cylinder No.1 to TDC/compression. The white mark on the crankshaft pulley should align with the pointer on the timing belt cover.
4. Drain the engine coolant into a sealable container.
5. Drain the engine oil.
6. Follow the instructions under Intake Manifold Removal and Installation, and remove the intake manifold.
7. Disconnect the throttle cable and throttle control cable from the throttle body. If equipped with cruise control, remove the cruise control cable.

NOTE: Be careful not to bend the cable when removing it. Do not use pliers to remove the cable from the linkage. Always replace a kinked cable with a new one.

8. Remove the intake air duct.
9. Disconnect and label the breather hose, positive crankcase ventilation (PCV) hose, and evaporative emissions (EVAP) control canister hose.
10. Relieve the fuel system pressure by loosening the service bolt on the fuel rail.

——— CAUTION ———
Fuel injection systems remain under pressure after the engine has been turned OFF. Properly relieve fuel pressure before disconnecting any fuel lines. Failure to do so may result in fire or personal injury.

11. Disconnect the fuel feed and return hoses from the fuel rail.
12. Disconnect the vacuum hoses located near the fuel feed and return hoses.
13. Remove the brake booster vacuum hose from the intake manifold.

Label and remove the other vacuum hoses from the intake manifold.

14. Remove the clamp holding the power steering hose to the strut tower.

15. Remove the wire harness clamp and the ground cable from the intake manifold.

16. Remove the connector and the terminal from the alternator. Then, remove the engine wire harness from the valve cover.

17. Remove the mounting bolts and drive belt from the power steering pump. Pull the pump away from the mounting bracket without disconnecting the hoses. Support the pump out of the way.

18. Loosen the adjusting and mounting bolts for the alternator and remove the drive belt.

19. Unclamp the engine wire harness and bypass hose from the lower side of the intake manifold.

20. Disconnect and label the following engine wire harness connectors:

    a. Fuel injector connectors

    b. Intake air temperature (IAT) sensor connector

    c. Idle air control (IAC) valve connector

    d. Throttle position (TPS) sensor connector

    e. Manifold absolute pressure (MAP) sensor connector

    f. Heated oxygen sensor (HO_2S) connector

    g. Engine coolant temperature (ECT) sensor connector

    h. ECT switch connector

    i. ECT gauge sending unit connector

    j. Exhaust gas recirculation (EGR) valve lift sensor

    k. CKP/TDC/CYP sensor connector

    l. Ignition coil connector

21. Label and disconnect the electrical connectors and ignition wires from the distributor.

22. Mark the position of the distributor and remove it from the cylinder head. Disconnect the ignition coil wire from the distributor.

23. Remove the upper radiator hose and the heater inlet hose from the cylinder head.

24. Remove the lower radiator hose from the thermostat housing.

25. Remove the coolant bypass hoses.

26. Use a jack with a cushioned pad to support the weight of the engine. Remove the through-bolt from the side engine mount and remove the mount.

27. Remove the cylinder head cover. Replace the rubber seals if they're damaged or deteriorated.

28. Remove the timing belt covers and the timing belt.

29. Remove the camshaft sprocket and the back cover. Do not lose the sprocket key.

30. Raise and safely support the vehicle.

31. Remove the splash shield.

32. Disconnect the exhaust pipe from the exhaust manifold.

33. Lower the vehicle.

34. Remove the exhaust manifold and the exhaust manifold heat insulator.

35. Remove the thermostat housing mounting bolts. Remove the thermostat housing from the intake manifold and the connecting pipe by pulling and twisting the housing. Discard the O-rings.

36. Remove the fuel rail and fuel injectors.

37. Remove the intake manifold bracket bolts.

38. Remove the intake manifold chamber with the throttle body attached.

39. Remove the intake manifold.

40. Remove the cylinder head bolts in the proper crisscross sequence starting at the outer edges and working inward. Then, remove the cylinder head.

NOTE: To prevent warpage, loosen the bolts in reverse of the removal sequence $1/3$ turn at a time. Repeat the sequence until all bolts are loosened.

To install:

41. Make sure all cylinder head and block gasket surfaces are clean. Check the cylinder head for warpage. If warpage is less than 0.002 in. (0.05 mm), cylinder head resurfacing is not required. The maximum resurface limit is 0.008 in. (0.2 mm) based on a cylinder head total height of 3.94 in. (100 mm).

42. Install a new head gasket.

43. Make sure the No. 1 cylinder is at TDC/compression.

44. Clean the oil control orifice and install a new O-ring. Replace the oil control orifice if necessary.

45. Install the dowel pins to the engine block.

46. Install the bolts that secure the intake manifold to its bracket but do not tighten them.

47. Position the camshaft so that the UP mark is facing upward.

48. Install the cylinder head and make sure it is properly seated onto its dowel pins.

49. Apply clean engine oil to the threads of the cylinder head bolts and to the underside of their heads. Install all of the head bolts. Following a crisscross pattern, tighten the bolts sequentially in three steps:

- Step 1: 29 ft. lbs. (39 Nm)
- Step 2: 51 ft. lbs. (69 Nm)
- Step 3: 72.3 ft. lbs. (98.1 Nm)

50. Install the intake manifold, manifold chamber and throttle body as described under Intake Manifold Removal and Installation.

51. Install new O-rings, cushion rings, and seal rings onto the fuel injectors, fuel rail, and intake orifice. Install the fuel rail to the intake manifold as an assembly with the fuel injectors.

52. Install the exhaust manifold as described under Intake Exhaust Removal and Installation.

53. Lower the vehicle.

54. Install the timing belt back cover to the cylinder head. Torque the cover bolts to 9 ft. lbs. (12 Nm).

55. Install the key into the camshaft groove; then, install the camshaft sprocket. Torque the sprocket bolt to 27 ft. lbs. (37 Nm).

56. Make sure the camshaft sprocket and the crankshaft pulleys are aligned to TDC for the compression stroke. The camshaft sprocket UP mark should face up. The camshaft keyway should also face up.

57. Install the timing belt and the balancer shaft belt.

58. Install the lower timing belt cover and torque the bolts to 9 ft. lbs. (12 Nm).

59. Install a new seal around the adjusting nut. Do not loosen the adjusting nut.

60. Install the crankshaft pulley. Coat the threads and seating face of the pulley bolt with engine oil. Install and torque the bolt to 181 ft. lbs. (245 Nm).

61. Install the side engine mount. Tighten the bolt and nut attaching the mount to the engine to 40 ft. lbs (55 Nm). Torque the through nut and bolt to 47 ft. lbs. (65 Nm). Remove the jack from under the engine.

62. Adjust the valves.

63. Install the upper timing belt cover. Torque the bolt on the intake side of the head to 9 ft. lbs. (12 Nm) and torque the bolt on the exhaust side of the head to 7 ft. lbs. (10 Nm).

64. Torque the crankshaft pulley bolt to 181 ft. lbs. (245 Nm) if it broke loose while adjusting the valves.

65. Install the splash shield.

66. Thoroughly clean the valve cover gasket mating surfaces . Install the valve cover gasket to the groove

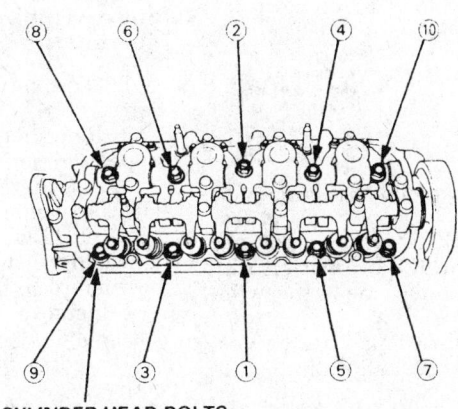

CYLINDER HEAD BOLTS
12 x 1.25 mm
98.1 N·m (10.0 kgf·m, 72.3 lbf·ft)

290759

Cylinder head bolt tightening sequence

of the cylinder head cover. Make sure the gasket is seated securely in the corners of the recesses.

67. Apply sealant to the four corners of the recesses of the valve cover gasket. Do not install the parts if 5 minutes or more have elapsed since applying sealant. After assembly, wait at least 20 minutes before filling the engine with oil.

68. Clean the valve cover contacting surface with a shop towel. Install the valve cover. Tighten the valve cover cap nuts in a clockwise sequence to 7 ft. lbs. (10 Nm).

69. Install a new O-ring to the coolant connecting pipe, and to the thermostat housing. Install the housing to the coolant pipe and the intake manifold. Torque the mounting bolts to 16 ft. lbs. (22 Nm).

70. Install the coolant bypass hoses.

71. Connect the lower radiator hose to the thermostat housing.

72. Connect the upper radiator hose and the heater inlet hose to the cylinder head.

73. Install the distributor to the cylinder head. Only hand-tighten the mounting bolts at this time.

74. Connect the ignition wires to the spark plugs. Connect the distributor electrical connectors. Install the ignition coil wire to the distributor.

75. Connect the following engine wire harness connectors:

  a. Fuel injector connectors

  b. Intake air temperature (IAT) sensor connector

  c. Idle air control (IAC) valve connector

  d. Throttle position (TP) sensor connector

  e. Manifold absolute pressure (MAP) sensor connector

  f. Heated oxygen sensor (HO_2S) connector

  g. Engine coolant temperature (ECT) sensor connector

  h. ECT switch connector

  i. ECT gauge sending unit connector

  j. Exhaust gas recirculation (EGR) valve lift sensor

  k. CKP/TDC/CYP sensor connector

  l. Ignition coil connector

76. Install the engine wire harness and bypass hose to the lower side of the intake manifold.

77. Install and adjust the alternator drive belt.

78. Install the power steering pump to the power steering pump mounting bracket.

79. Install and adjust the power steering belt.

80. Install the alternator wire harness to the valve cover. Connect the

terminal and connector to the alternator.

81. Connect the ground cable and the wire harness clamp to the intake manifold.

82. Connect the power steering hose clamp to the engine block.

83. Connect the brake booster vacuum hose to the intake manifold. Connect the other vacuum hoses to the intake manifold.

84. Connect the vacuum hoses located near the fuel feed and return hoses.

85. Connect the fuel return hose and the fuel feed hose to the fuel rail. Install new washers to the fuel feed hose connection. Torque the fuel feed hose banjo bolt to 16 ft. lbs. (22 Nm). Tighten the service bolt to 8.7 ft. lbs. (12 Nm).

86. Install the breather hose, PCV hose, and the EVAP control canister hose.

87. Install the air intake duct.

88. Connect and adjust the throttle cable. Connect and adjust the throttle control cable and the cruise control cable, if equipped.

89. Drain the engine oil into a sealable container. Install the drain plug with a new crush washer.

90. Refill the engine with clean oil.

91. Fill and bleed the air from the cooling system.

92. Connect the positive and the negative battery cables. Enter the radio security code.

93. Start the engine, checking carefully for any coolant, fuel, oil, or air leaks.

94. Check the ignition timing and adjust it if necessary, then tighten the distributor bolts to 13 ft. lbs. (18 Nm).

Valve Lash

ADJUSTMENT

NOTE: The radio may contain a coded anti-theft circuit. Obtain the customer's security code before disconnecting the battery.

1. Disconnect the negative battery cable.

2. The valves should be adjusted when the engine is cold. If the engine has been run, allow it to cool to below 100° F (38° C) before beginning adjustments.

3. Remove the cylinder head cover and the upper timing belt cover.

4. Rotate the crankshaft to align the white TDC mark on the crankshaft pulley with the pointer on the cover for the No. 1 cylinder compres-

sion stroke. Make sure the **UP** mark on the camshaft sprocket is up and the TDC marks align with the edge of the cylinder head.

5. Hold a No. 1 cylinder rocker arm against the camshaft and use a feeler gauge to check the clearance at the valve stem. Intake valve clearance should be 0.010 in. (026 mm), exhaust valve clearance should be 0.012 in. (0.30 mm). The service limit for both intake and exhaust valves is plus or minus 0.0008 in. (0.02mm). Loosen the locknut and turn the adjusting screw to adjust the clearance. Tighten the locknut and recheck the clearance.

6. The adjustment order is 1–3–4–2. Rotate the crankshaft counterclockwise 180° to bring each cylinder to TDC/compression. Adjust each set of valves.

a. At TDC/compression for the No. 3 cylinder, the UP mark is parallel to the exhaust side of the cylinder head.

b. At TDC/compression for the No. 4 cylinder, the UP mark is pointed straight down, and the TDC marks align with the edge of the cylinder head.

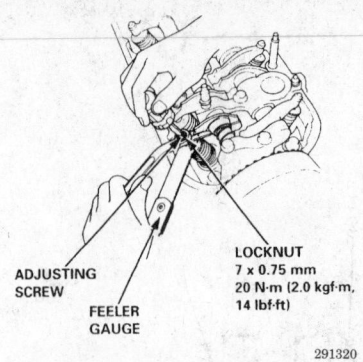

Adjusting screw and feeler gauge

c. At TDC/compression for the No. 2 cylinder, the UP mark is parallel to the intake side of the cylinder head.

7. After adjusting the valves, retorque the crankshaft pulley bolt to 181 ft. lbs. (245 Nm).

8. Install the cylinder head and timing belt covers.

9. Reconnect the negative battery cable. Enter the radio security code.

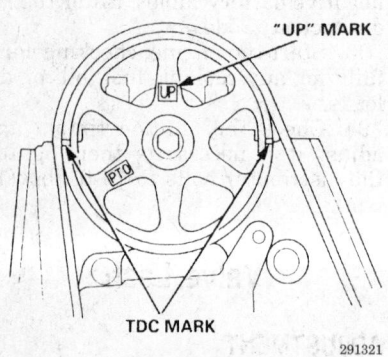

TDC/compression for the No. 1 cylinder

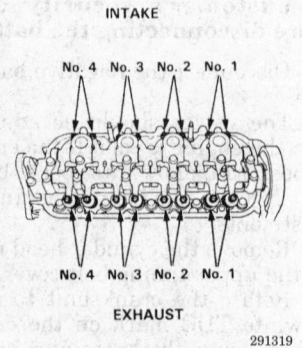

Adjusting screw locations

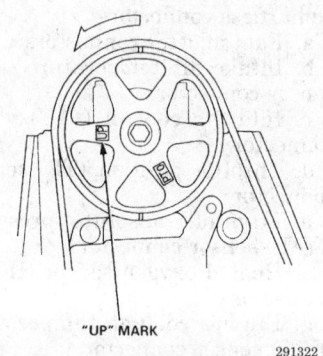

TDC/compression for the No. 3 cylinder

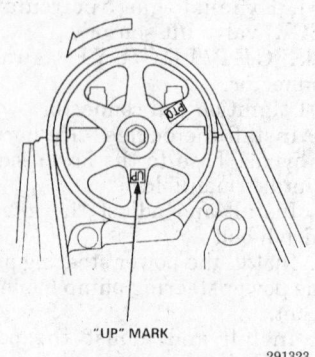

TDC/compression for the No. 4 cylinder

Rocker Arms and Shafts

REMOVAL AND INSTALLATION

NOTE: The original radio contains a coded anti-theft circuit. Obtain the customer's security code number before disconnecting the battery.

1. Disconnect the negative battery cable.

2. Remove the valve cover and the upper timing belt cover.

3. Set the No. 1 cylinder to TDC for the compression stroke. Verify that the TDC marks are correctly aligned. Once the engine is set in this position, it must not be disturbed.

4. Remove the distributor as an assembly.

5. Loosen the valve adjusting screws.

6. Cover the timing belt with a clean shop towel to protect it from engine oil. If the belt is contaminated with oil, it must be replaced.

7. Remove the camshaft holder bolts. Unscrew the bolts two turns at a time in a crisscross pattern to prevent damaging the valves. camshaft, or rocker arm assembly.

NOTE: The rocker arms and shafts are an assembly; they must be removed from the engine as a unit. To prevent warpage, always follow the torque sequence carefully when removing or installing the rocker shaft assembly.

8. Remove the rocker arm and shaft assemblies. Do not remove the camshaft holder bolts. The bolts keep the camshaft bearing caps, springs, and rocker arms in place on the shafts.

9. If the rocker arms or shafts are to be replaced, identify the parts as they are removed from the shafts to ensure reinstallation in the original location.

To install:

10. Verify that the engine is set to TDC/compression for the No. 1 cylinder. The camshaft keyway faces up when the engine is at TDC/compression.

11. Lubricate the camshaft journals and lobes with clean engine oil. Install a new camshaft seal if necessary.

12. Assemble the rocker arms, shafts, and camshaft bearing caps.

13. Apply sealant to the mating surfaces of the No. 1 and No. 6 camshaft bearing caps. Do not allow the sealant to cure before the rocker arm assembly is installed.

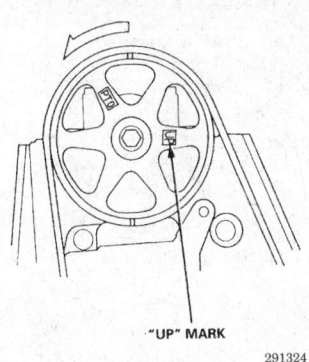

"UP" MARK

291324

TDC/compression for the No. 2 cylinder

14. Set the rocker arm assembly in place and loosely install the bolts. Tighten each bolt two turns at a time in the proper sequence to ensure that the rockers do not bind on the valves. Tighten the 8mm rocker arm bolts to 16 ft. lbs. (22 Nm). Tighten the 6mm bolts to 8.7 ft. lbs. (12 Nm).

15. Verify that the engine is at TDC/compression, and install the distributor.

16. Adjust the valves and tighten the locknuts to 14 ft. lbs. (20 Nm).

17. Install the valve cover and upper timing belt cover.

18. Reconnect the negative battery cable.

19. Check the ignition timing and adjust if necessary. Tighten the distributor mounting bolts to 13 ft. lbs. (18 Nm).

Intake Manifold

REMOVAL AND INSTALLATION

NOTE: The radio may contain a coded theft protection circuit. Always make note of your code number before disconnecting the battery.

1. Disconnect the negative battery cable.

2. Drain the engine coolant into a sealable container.

3. Disconnect the cooling hoses from the intake manifold.

4. Label and disconnect the vacuum hoses and electrical connectors on the manifold and throttle body. Unplug the connector from the Exhaust Gas Recirculation (EGR) valve. Position the wiring harnesses out of the way.

5. Disconnect the throttle cable from the throttle body.

6. Disconnect the cruise control cable from the throttle linkage. Unbolt the cable clamp from the valve cover and move the cable out of the way.

7. Relieve the fuel pressure.

—————— CAUTION ——————

Fuel injection systems remain under pressure after the engine has been turned OFF. Properly relieve fuel pressure before disconnecting any fuel lines. Failure to do so may result in fire or personal injury.

8. Remove the fuel rail and fuel injectors.

9. Remove the thermostat housing mounting bolts. Gently pull and twist the thermostat housing to remove it from the intake manifold and the coolant connecting pipe. Discard the O-rings.

10. It may be necessary to remove the upper intake manifold chamber and throttle body assembly in order to access the nuts securing the manifold to the head.

11. Unbolt and remove the intake manifold support bracket. If necessary, raise and support the vehicle safely to reach the manifold support bracket.

12. Loosen the intake manifold nuts in a crisscross pattern starting at the edges and working toward the center of the manifold. Supporting the intake manifold and remove the nuts, then remove the manifold.

13. Clean any old gasket material from the cylinder head and the intake manifold. Check and clean the chamber and mating surfaces on the cylinder head.

To install:

14. Install a new manifold gasket. Place the manifold into position and support it.

15. Install the support bracket to the manifold. Torque the bolt holding the bracket to the manifold to 16 ft. lbs. (22 Nm).

16. Starting at the center of the manifold, tighten the nuts in a crisscross pattern to 16 ft. lbs. (22 Nm). The tension must be even across the entire face of the manifold to prevent leaks.

17. If the upper intake manifold chamber and throttle body assembly was removed, install it with a new gasket. Tighten the nuts and bolts to 16 ft. lbs. (22 Nm).

18. Install new O-rings onto the coolant connecting pipe and the thermostat housing. Install the housing to the coolant pipe and the intake manifold. Torque the mounting bolts to 16 ft. lbs. (22 Nm).

19. Connect and adjust the throttle cable.

Specified torque:
8 mm bolts: 22 N·m (2.2 kgf·m, 16 lbf·ft)
6 mm bolts: 12 N·m (1.2 kgf·m, 8.7 lbf·ft)

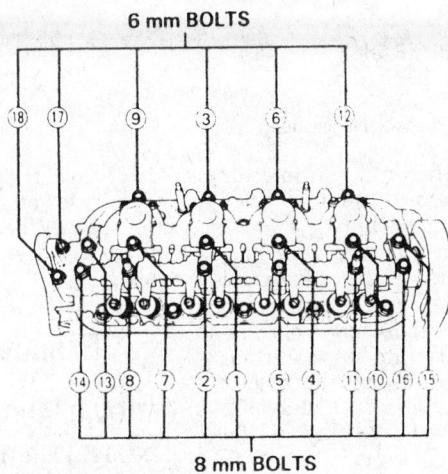

6 mm BOLTS

8 mm BOLTS

290715

Camshaft holder bolt tightening sequence

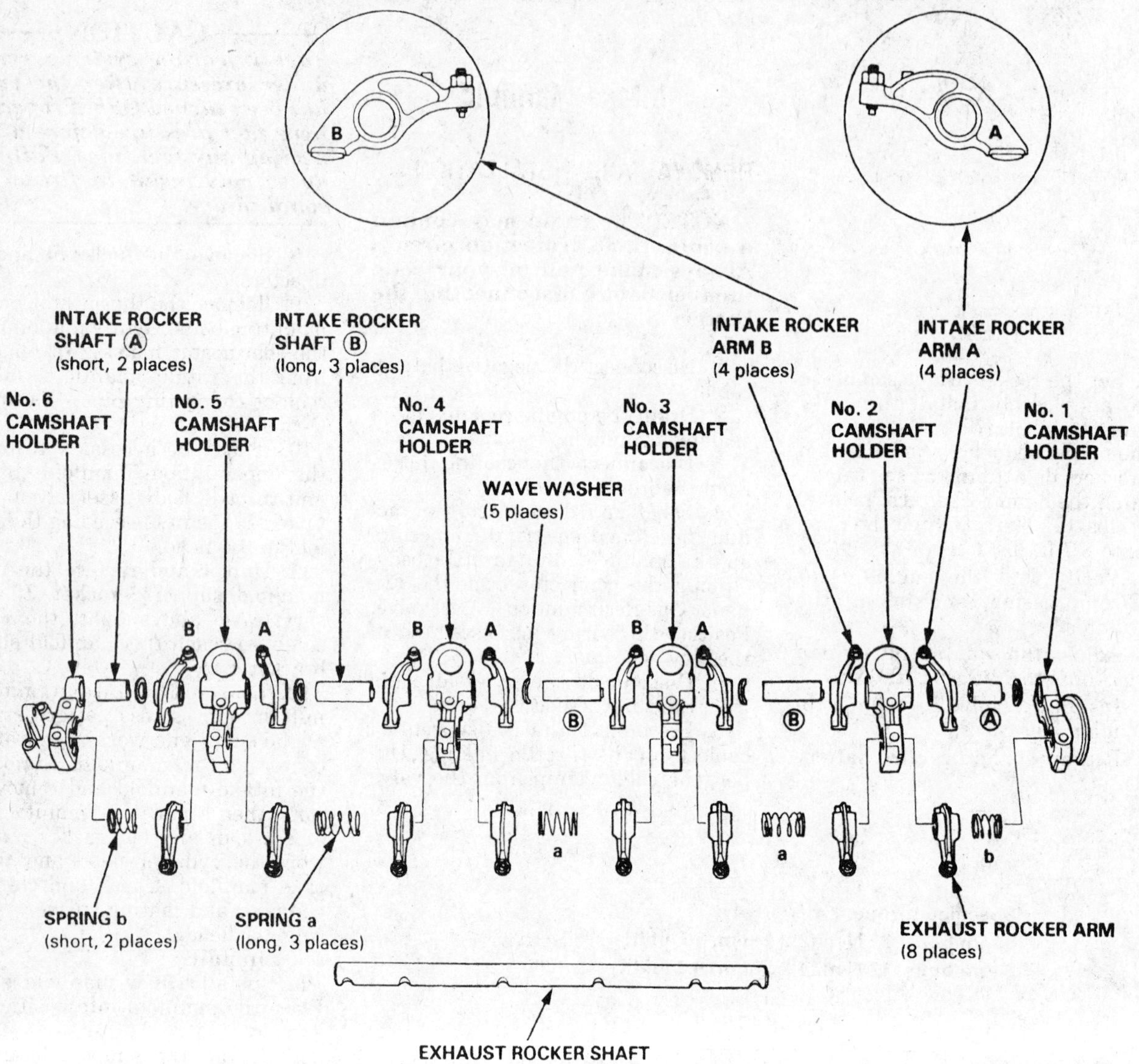

Letter "B" is stamped on rocker arm.

Letter "A" is stamped on rocker arm.

INTAKE ROCKER SHAFT Ⓐ (short, 2 places)

INTAKE ROCKER SHAFT Ⓑ (long, 3 places)

INTAKE ROCKER ARM B (4 places)

INTAKE ROCKER ARM A (4 places)

No. 6 CAMSHAFT HOLDER

No. 5 CAMSHAFT HOLDER

No. 4 CAMSHAFT HOLDER

No. 3 CAMSHAFT HOLDER

No. 2 CAMSHAFT HOLDER

No. 1 CAMSHAFT HOLDER

WAVE WASHER (5 places)

SPRING b (short, 2 places)

SPRING a (long, 3 places)

EXHAUST ROCKER ARM (8 places)

EXHAUST ROCKER SHAFT

290643

Rocker arm and shaft components

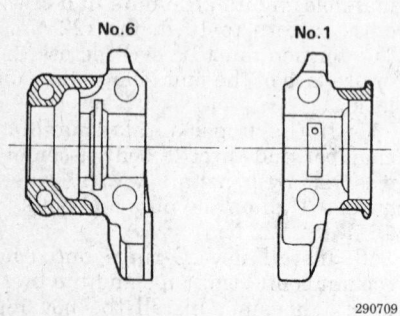

Camshaft bearing cap sealant application points

No.6   No.1

290709

20. Install the fuel rail and injector assembly. Reconnect the fuel lines using new sealing washers.

21. Properly position the wire harnesses and connect the electrical connectors.

22. Connect the cruise control cable and place it back into its clamp on the valve cover. Adjust the cruise control cable so there is 0.20 ± 0.02 in. (5 ± 0.5mm) of free-play at the linkage.

23. Connect the vacuum hoses.

24. Fill and bleed the air from the cooling system.

25. Connect the negative battery cable and enter the radio security code.

26. Start the engine and check carefully for any fuel, coolant, or vacuum leaks. Check the manifold gasket areas carefully for any vacuum leaks.

Exhaust Manifold

REMOVAL AND INSTALLATION

NOTE: The radio may contain a coded theft protection circuit. Always make note of your code number before disconnecting the battery.

1. Disconnect the negative battery cable.

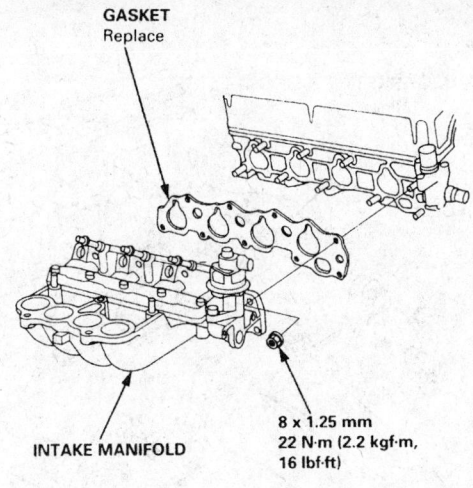

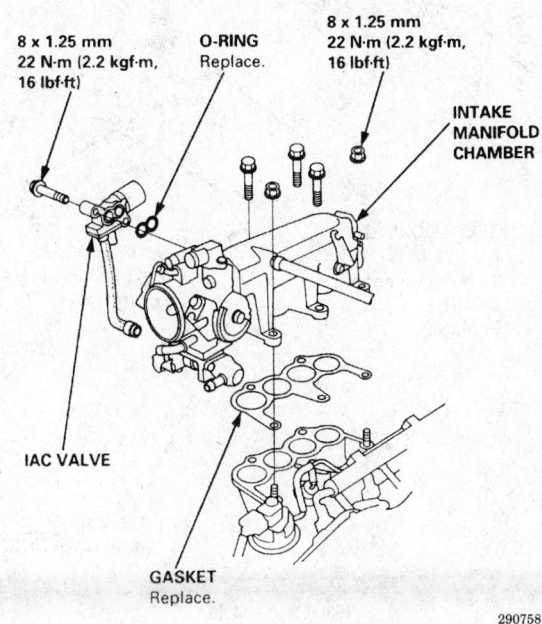

Intake manifold and intake chamber components

2. Safely raise and support the vehicle.

3. Remove the nuts attaching the front exhaust pipe to the exhaust manifold. Separate the pipe from the manifold and discard the gasket. A long extension may be helpful for reaching the nuts.

4. Remove the exhaust manifold heat shield.

5. Remove the exhaust manifold bracket bolts and remove the bracket.

6. Loosen the exhaust manifold nuts in a crisscross pattern starting at the edges of the manifold and working toward its center. Remove the nuts.

7. Remove the manifold and discard the gaskets. Clean the manifold and cylinder head mating surfaces.

To install:

8. Install a new exhaust manifold gasket. Place the manifold into position and support it. Install the nuts snugly onto the studs.

9. Install the support bracket below the manifold. Torque the bracket mounting bolts to 33 ft. lbs. (44 Nm).

10. Starting with the inner or center nuts, tighten the nuts in a crisscross pattern to 23 ft. lbs. (31 Nm). The tension must be even across the entire face of the manifold to prevent leaks.

11. Install the exhaust manifold heat shield and torque its bolts to 16 ft. lbs. (22 Nm).

12. Connect the front exhaust pipe, using new gaskets and self-locking nuts. Tighten the exhaust pipe attaching nuts to 40 ft. lbs. (55 Nm).

13. Connect the negative battery cable and enter the radio security code.

14. Start the engine and check for exhaust leaks.

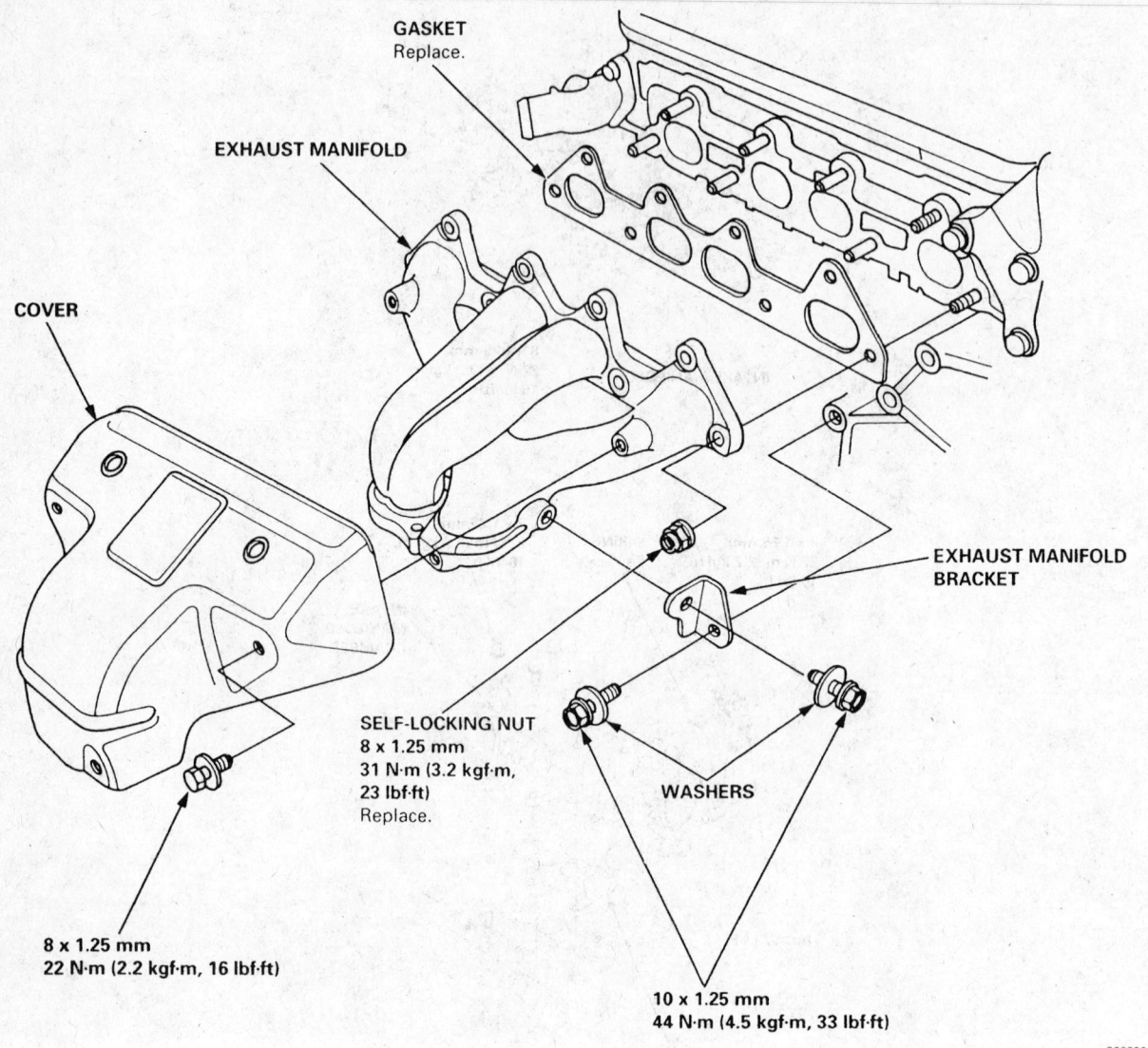

GASKET Replace.

EXHAUST MANIFOLD

COVER

EXHAUST MANIFOLD BRACKET

SELF-LOCKING NUT
8 x 1.25 mm
31 N·m (3.2 kgf·m, 23 lbf·ft)
Replace.

WASHERS

8 x 1.25 mm
22 N·m (2.2 kgf·m, 16 lbf·ft)

10 x 1.25 mm
44 N·m (4.5 kgf·m, 33 lbf·ft)

288293

Exhaust manifold

Front Crankshaft Seal

REMOVAL AND INSTALLATION

NOTE: The original radio may contain a coded anti-theft circuit. Obtain the customer's security code number before disconnecting the battery cables.

1. Disconnect the negative battery cable.
2. Raise and safely support the vehicle.
3. Remove the splash shield.
4. Remove the engine accessory drive belts.

5. Set the engine at TDC for the No. 1 piston on the compression stroke. The crankshaft pulley mark must be aligned with the white mark on the lower timing cover. Once in this position, the engine must not be disturbed.
6. Remove the upper timing belt cover and crankshaft pulley. Remove the lower timing belt cover.

NOTE: Mark the direction of the timing belt's rotation if it is to be reinstalled. If there is any doubt about the condition of the timing belt, or if it has been contaminated by oil, it must be replaced.

7. Remove the timing belt.
8. If equipped with a TDC sensor mounted on the oil pump housing, unbolt the sensor assembly and move it out of the way. Do not get any oil on the sensor assembly.
9. Remove the crankshaft timing sprocket.
10. Drain the engine oil.
11. Use a seal removal tool to remove the seal from the oil pump case.
12. If the balancer shaft oil seal must be replaced, remove the balancer shaft sprocket and use a seal puller to remove the seal.
    To Install:
13. Clean the seal mounting surfaces on the engine block.

14. Apply a thin coat of grease on the crankshaft and seal lips.

15. Install the seal with the part number facing out. Use a seal driver to seat the seal against the oil pump. Clean any excess grease off the crankshaft and make sure the seal lip is not distorted.

16. Install the TDC sensor assembly back into position. Tighten the sensor mounting bolts to 9 ft. lbs. (12 Nm).

17. Install the crankshaft timing sprocket.

18. Refill the engine with oil.

19. Verify that the engine is at TDC for the no. 1 cylinder on the compression stroke. Install and tension the timing belt.

20. Install the timing belt covers and crankshaft pulley. Retighten the crankshaft pulley bolt to 181 ft. lbs. (245 Nm).

21. Install and adjust the accessory drive belts.

22. Verify that all engine components that may have been removed have been reinstalled correctly.

23. Install the splash shield and lower the vehicle.

24. Connect the negative battery cable.

25. Top up the engine oil if necessary.

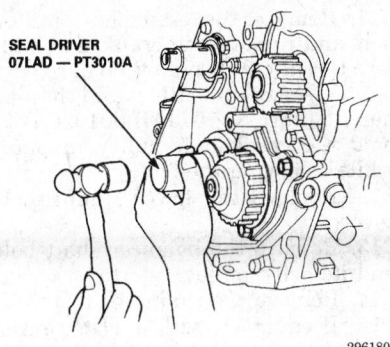

SEAL DRIVER
07LAD — PT3010A

296180

Oil seal driver and oil seal installation

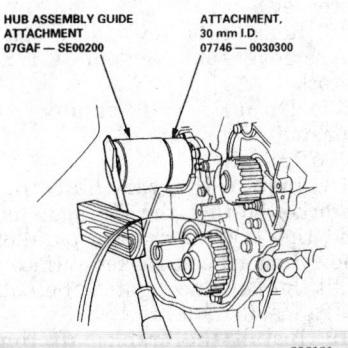

HUB ASSEMBLY GUIDE
ATTACHMENT
07GAF — SE00200

ATTACHMENT,
30 mm I.D.
07746 — 0030300

296181

Balancer shaft oil seal installation

26. Run the engine and check for leaks.

Timing Belt, Sprockets, Tensioner and Front Cover

ADJUSTMENTS

WARNING

It is very highly recommended that a timing belt be replaced any time its tension is released. A timing belt should not be viewed as an adjustable component. The timing belt's tension cannot be increased to compensate for wear. If the engine has been disassembled for mechanical work, a new timing belt should be installed. The small cost of a new timing belt is cheap insurance against expensive engine damage which can be caused by the failure of a re-used timing belt.

Setting the Tension of a New Timing Belt and Balancer Belt

NOTE: After the timing belt is installed, it must be properly tensioned. The timing belt tensioner is spring-loaded to automatically apply the proper tension to the timing belts once its nut is tightened properly. The engine must be cold before belt tension is set.

1. Disconnect the negative and positive battery cables.

2. Remove the valve cover and the upper timing belt cover.

3. Rotate the crankshaft counterclockwise five to six turns to make sure the belt is properly seated.

4. Set the No. 1 piston at TDC for its compression stroke.

5. Rotate the crankshaft counterclockwise so that the camshaft pulley moves only three teeth beyond its TDC mark.

6. Tighten the tensioner adjusting nut to 33 ft. lbs. (45 Nm).

7. Tighten the crankshaft pulley bolt to 181 ft. lbs. (245 Nm).

8. Install the upper timing cover and the valve cover.

9. Reconnect the negative and positive battery cables.

Timing Belt and Balancer Belt Inspection

1. Disconnect the negative and positive battery cables.

2. Label and disconnect the ignition wires and remove the spark plugs. Remove the valve cover and the upper timing belt cover.

3. Remove the crankshaft pulley and the lower timing cover.

4. Install the crankshaft pulley and tighten its bolt to 181 ft. lbs. (245 Nm).

5. Rotate the crankshaft pulley counterclockwise to cycle the belts through their entire rotations.

6. Inspect the entire length of each belt. Look carefully for any signs of the following conditions:

   a. Cracked, chipped, or broken teeth.

   b. Fraying, separation, or heat damage to the belt's rubber and fiber layers.

   c. Oil or coolant leaks which may have contaminated the belt.

   d. Make sure the timing marks align. Misaligned marks may indicate that the belt has jumped one or more teeth, or has been improperly tensioned.

7. Check the camshaft, crankshaft, and balancer shaft oil seals for any signs of leakage. Also check the water pump for leakage. The source of any oil or coolant leaks must be found and corrected before a new belt is installed.

8. Replace the timing and balancer belts if they are damaged in any way, or if you are uncertain as to their condition. Honda's recommended service interval for timing and balancer belt replacement is 90,000 miles (144,000 Km). Additionally, it is recommended that the water pump be replaced at this interval.

9. After inspection, retorque the crankshaft pulley bolt to 181 ft. lbs. (245 Nm). Install the timing belt and valve covers. Install the spark plugs and reconnect the ignition wires. Reconnect the battery cables.

REMOVAL AND INSTALLATION

NOTE: The radio may contain a coded theft protection circuit. Always make note of your code number before disconnecting the battery.

1. Disconnect the negative and positive battery cables.

2. Remove the valve cover.

3. Remove the upper timing belt cover.

4. Turn the engine to align the timing marks and set cylinder No.1 to TDC for the compression stroke. The white mark on the crankshaft pulley should align with the pointer on the timing belt cover. The words UP embossed on the camshaft pulley should be aligned in the upward position. The marks on the edge of the

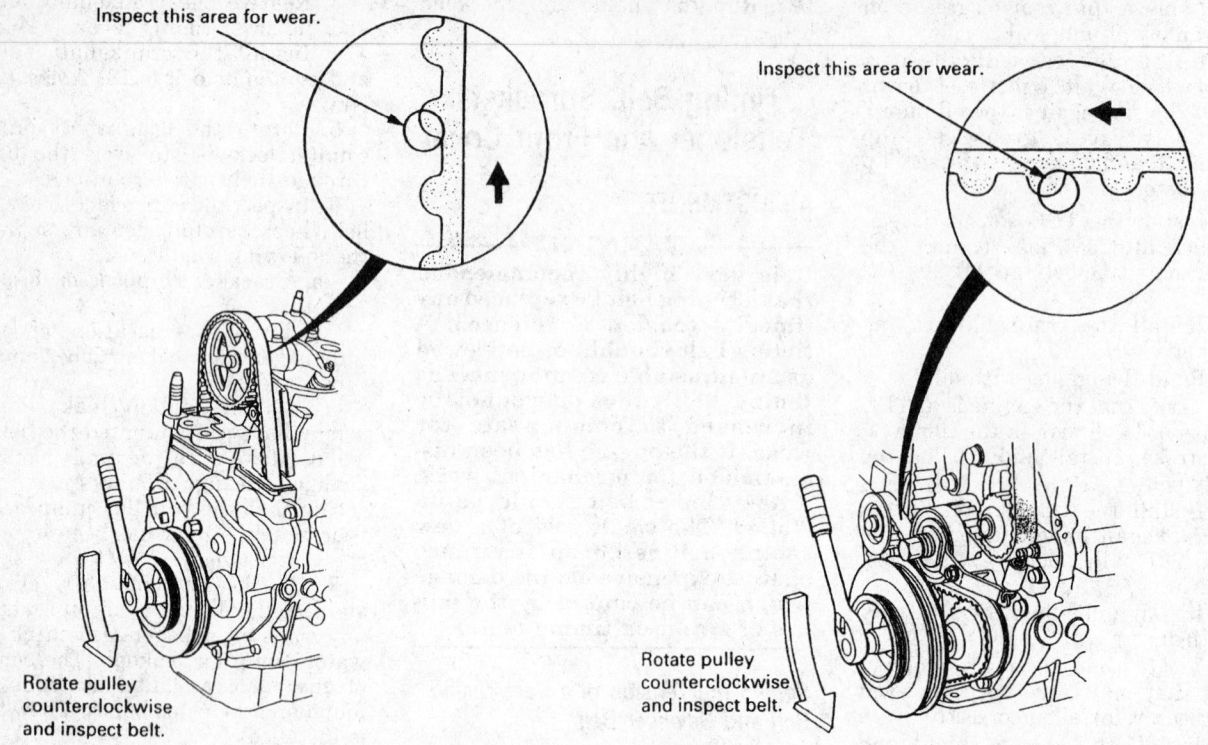

Timing belt and balancer belt inspection points

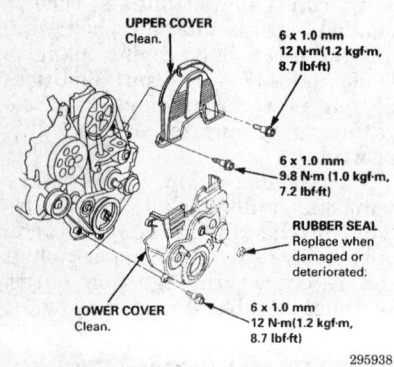

Upper and lower timing covers

pulley should be aligned with the cylinder head or the back cover upper edge. Once in this position, the engine must NOT be turned or disturbed.

5. Remove the splash shield from below the engine.

6. Remove the wheel well splash shield.

7. Loosen and remove the power steering pump belt. Remove the power steering pump.

8. Loosen the adjusting and mounting bolts for the alternator and remove the drive belt.

9. Support the engine with a floor jack cushioned with a piece of wood.

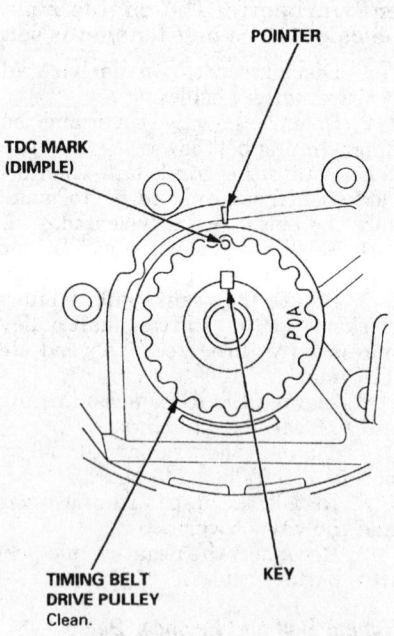

Crankshaft drive sprocket TDC marks

10. Remove the through-bolt for the side engine mount and remove the mount.

11. Remove the crankshaft pulley bolt and remove the crankshaft pulley. Use a crank pulley holder (part No. 07MAB-PY3010A) and holder handle (part No. 07JAB-001020A) to hold the crankshaft pulley in place while removing the bolt.

12. Remove the lower timing belt cover.

13. Remove the balancer shaft belt and its drive pulley.

14. Remove the timing belt.

15. If equipped with a TDC sensor assembly at the crankshaft sprocket, unbolt the assembly and move it to the side before removing the sprocket.

16. Remove the key and the spacers to remove the crankshaft timing sprocket.

17. Unbolt and remove the camshaft timing sprocket.

To Install:

18. Install the camshaft timing sprocket so that the UP mark is up and the TDC marks are parallel to the cylinder head gasket surface. Install the key and tighten the bolt to 27 ft. lbs. (37 Nm)

19. Install the crankshaft timing sprocket so that the TDC mark aligns with the pointer on the oil pump. In-

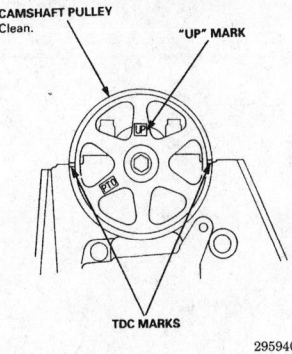

Camshaft sprocket UP mark and TDC marks

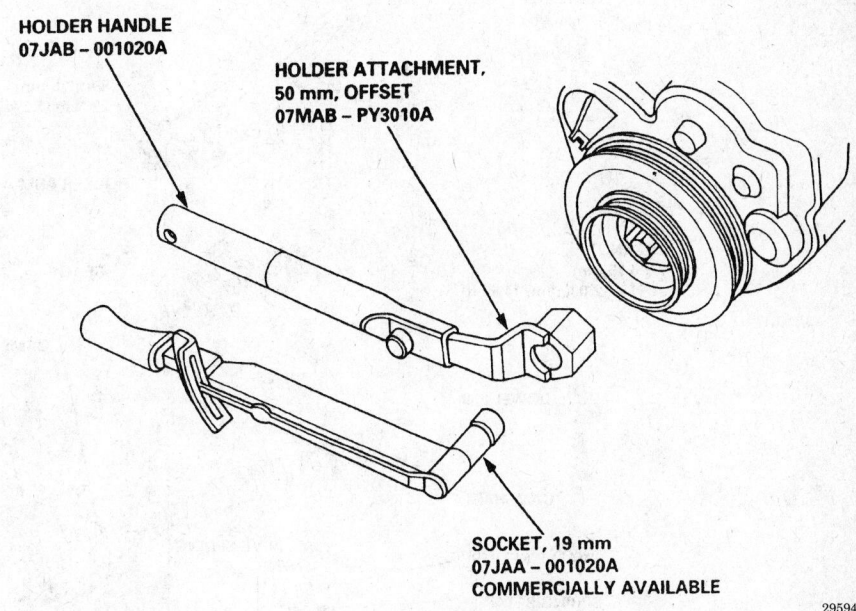

Crankshaft pulley holder tools

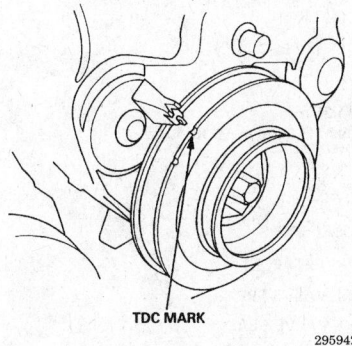

Crankshaft pulley TDC mark

stall the spacers with their concave surfaces facing in. Install the key. Install the TDC sensor assembly back into position before installing the timing belt.

20. Install and tension the timing belt.

21. Install the balancer shaft belt drive pulley.

22. Install and tension the balancer shaft belt.

23. Make sure the timing belts have been tensioned correctly and that all TDC and alignment marks are in their proper positions.

24. Install the lower timing cover and the crankshaft pulley. Apply engine oil to the pulley bolt threads and washer surface. Install the pulley bolt and tighten it to 181 ft. lbs. (245 Nm).

25. Install the upper timing cover and the valve cover. Make sure the seals are properly seated.

26. Install the side engine mount. Tighten the through-bolt to 47 ft. lbs. (64 Nm). Tighten the mount nut and bolt to 40 ft. lbs. (55 Nm) each.

27. Remove the floor jack.

28. Install and tension the alternator belt.

29. Install the power steering pump and tension its belt.

30. Install the splash shields.

31. Reconnect the positive and negative battery cables. Enter the radio security code.

32. Check engine operation.

Camshaft

REMOVAL AND INSTALLATION

1. See the procedures under Rocker Arms and Shaft Removal and Installation and remove the camshaft carrier.

2. Remove the rocker arm and shaft assembly. Leave the camshaft bearing cap bolts in the camshaft holders to hold the rocker arm/shaft assembly together.

3. Remove the camshaft and camshaft seal.

To install:

NOTE: Use new O-rings, seals, and gaskets when installing the camshaft.

4. Clean and inspect the camshaft bearing caps in the cylinder head.

5. Lubricate the lobes and journals of the camshaft prior to installation. Install the camshaft with the keyway facing up so that the engine remains at TDC/compression for the No. 1 piston.

6. Lubricate a new camshaft seal with engine oil. Use camshaft installer shaft, 07NAF-PT0020A; installer cap, 07NAF-PT0010A; and seal guide, 07NAG-PT0010A; or their equivalents to install the camshaft seal.

7. Install the camshaft carrier.

Balance Shafts

REMOVAL AND INSTALLATION

NOTE: The radio may contain a coded theft protection circuit. Always make note of your code number before disconnecting the battery.

1. Disconnect the negative battery cable.

2. Remove the valve cover and the upper timing belt cover.

3. Set the No. 1 cylinder to TDC for the compression stroke.

4. Remove the engine assembly from the vehicle.

5. Remove the rear cover.

6. Mount the engine on an engine stand. Make sure the mounting bolts are tight.

7. Remove the timing belt covers and the timing belts.

8. Drain the engine oil into a sealable container.

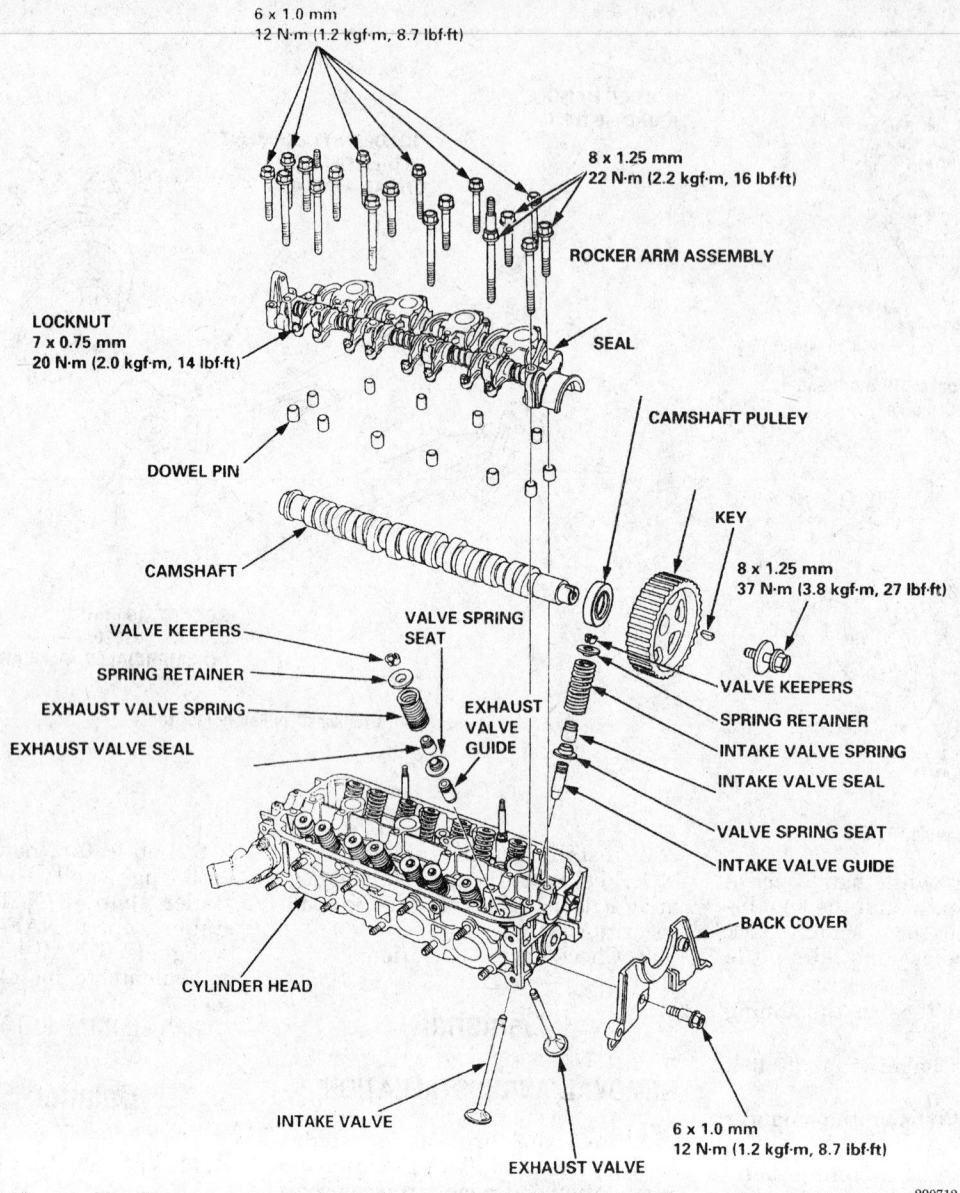

6 x 1.0 mm
12 N·m (1.2 kgf·m, 8.7 lbf·ft)

8 x 1.25 mm
22 N·m (2.2 kgf·m, 16 lbf·ft)

ROCKER ARM ASSEMBLY

LOCKNUT
7 x 0.75 mm
20 N·m (2.0 kgf·m, 14 lbf·ft)

SEAL

CAMSHAFT PULLEY

DOWEL PIN

KEY

CAMSHAFT

8 x 1.25 mm
37 N·m (3.8 kgf·m, 27 lbf·ft)

VALVE KEEPERS

VALVE SPRING SEAT

SPRING RETAINER

EXHAUST VALVE SPRING

EXHAUST VALVE SEAL

EXHAUST VALVE GUIDE

VALVE KEEPERS

SPRING RETAINER

INTAKE VALVE SPRING

INTAKE VALVE SEAL

VALVE SPRING SEAT

INTAKE VALVE GUIDE

BACK COVER

CYLINDER HEAD

INTAKE VALVE

EXHAUST VALVE

6 x 1.0 mm
12 N·m (1.2 kgf·m, 8.7 lbf·ft)

290712

Camshaft and valve components exploded view

CAMSHAFT

SEAL GUIDE
07NAG−PT0010A
Apply oil.

INSTALLER CUP
07NAF−PT0010A

Seal housing surface
should be dry.

CAMSHAFT OIL SEAL
Apply a light coat of oil
to camshaft and inner
lip of seal.

INSTALLER SHAFT
07NAF−PT0020A

290714

Camshaft seal installation tools

9. Remove the oil pan and the oil screen.

10. Remove the timing belt and balancer belt tensioners.

11. Remove the timing belt drive pulley from the crankshaft.

12. Insert a suitable tool into the maintenance hole in the front balancer shaft. Unbolt and remove the balancer driven pulley.

NOTE: For servicing the balance shafts, front refers to the side of the engine facing the radiator. Rear refers to the side of the engine facing the firewall.

13. Align the rear timing balancer pulley using a 6 x 100 mm bolt or rod.

Mark the bolt or rod at a point 2.9 inches (74 mm) from the end. Remove the service bolt from the maintenance hole on the side of the block; insert the bolt/rod into the hole. Align the 74 mm mark with the face of the hole. This pin will hold the shaft in place.

14. Remove the balancer gear case and the dowel pins. Discard the O-ring.

15. Remove the balancer driven gear attaching bolt and the balancer driven gear from the rear balancer shaft.

16. Remove the bolt or rod used to align the rear balancer shaft.

26. Install the oil pump to the engine block. Torque the mounting bolts to 9 ft. lbs. (12 Nm).

27. Install the oil screen. Torque the screen mounting bolts and nuts to 9 ft. lbs. (12 Nm).

28. Install the oil pan.

29. Install the balancer driven pulley to the front balancer belt. Hold the balancer shaft in place with a suitable tool. Torque the attaching bolt to 22 ft. lbs. (29 Nm).

30. Align the rear timing balancer shaft using a 6 x 100 mm bolt or rod inserted into the maintenance hole on the side of the block.

31. Install the balancer driven gear to the rear balancer shaft. Torque the bolt to 18 ft. lbs. (25 Nm).

32. Before installing the balancer driven gear case, apply molybdenum disulfide (lithium grease) to the thrust surfaces of the balancer gears.

33. Align the groove on the pulley edge to the pointer on the balancer gear case.

34. Install the balancer gear case to the engine and install the mounting bolts and nut. The rear balancer shaft should still be held in place with a 6 x 100 mm bolt/rod. Torque the mounting bolts and nut to 18 ft. lbs. (25 Nm).

35. Check the alignment of the pointer on the balancer pulley to the pointer on the oil pump.

36. Install the timing belt tensioners.

37. Install the timing belt drive pulley to the crankshaft.

38. Install the timing belt and the balancer belt.

39. Install the crankshaft pulley and the timing belt covers.

40. Install the engine assembly into the vehicle.

41. Refill the engine with with clean, fresh oil.

42. Connect the negative battery cable and enter the radio security code.

43. Start the engine and check for oil leaks and proper oil pressure.

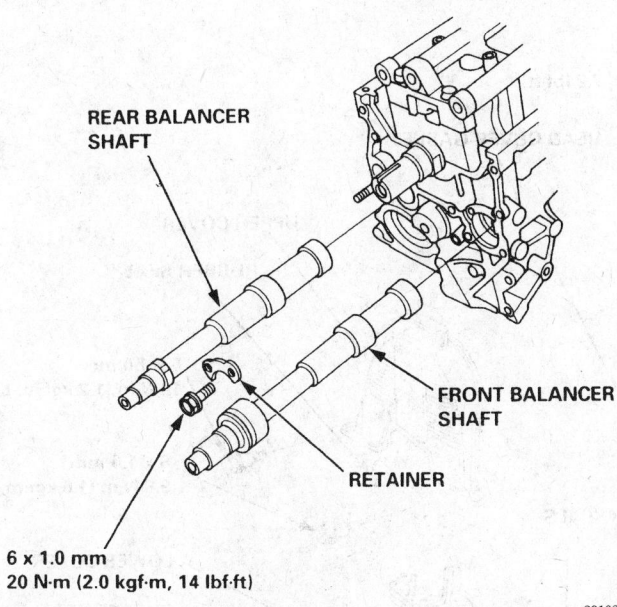

REAR BALANCER SHAFT

FRONT BALANCER SHAFT

RETAINER

6 x 1.0 mm
20 N·m (2.0 kgf·m, 14 lbf·ft)

291083

Balance shafts

17. Remove the oil pump mounting bolts and remove the oil pump assembly. Remove the dowel pins from the engine and clean the oil pump mating surfaces of old gasket material and oil. Discard the O-rings.

18. Carefully pull the rear balancer shaft from the engine block.

19. Unbolt and remove the front balancer shaft retainer.

20. Carefully pull the front balancer shaft from the engine block.

To install:

21. Lubricate the front and rear balancer shafts with clean engine oil.

22. Carefully install the balancer shafts into the engine block. Install the retainer to the front shaft and torque the retainer bolts to 14 ft. lbs. (20 Nm).

23. Install the two oil pump dowel pins and new O-rings to the cylinder block.

24. Install new oil seals in the oil pump housing and rear cover.

25. Clean and dry the oil pump and engine block mating surfaces. Apply a liquid gasket evenly in a narrow bead, centered on the mating surface. Once the sealant is applied, do not wait longer than 20 minutes to install the parts; the sealant will become ineffective. After final assembly, wait at least 30 minutes before adding oil to the engine, giving the sealant time to set. To prevent leakage of oil, apply a suitable thread sealer to the inner threads of the bolt holes.

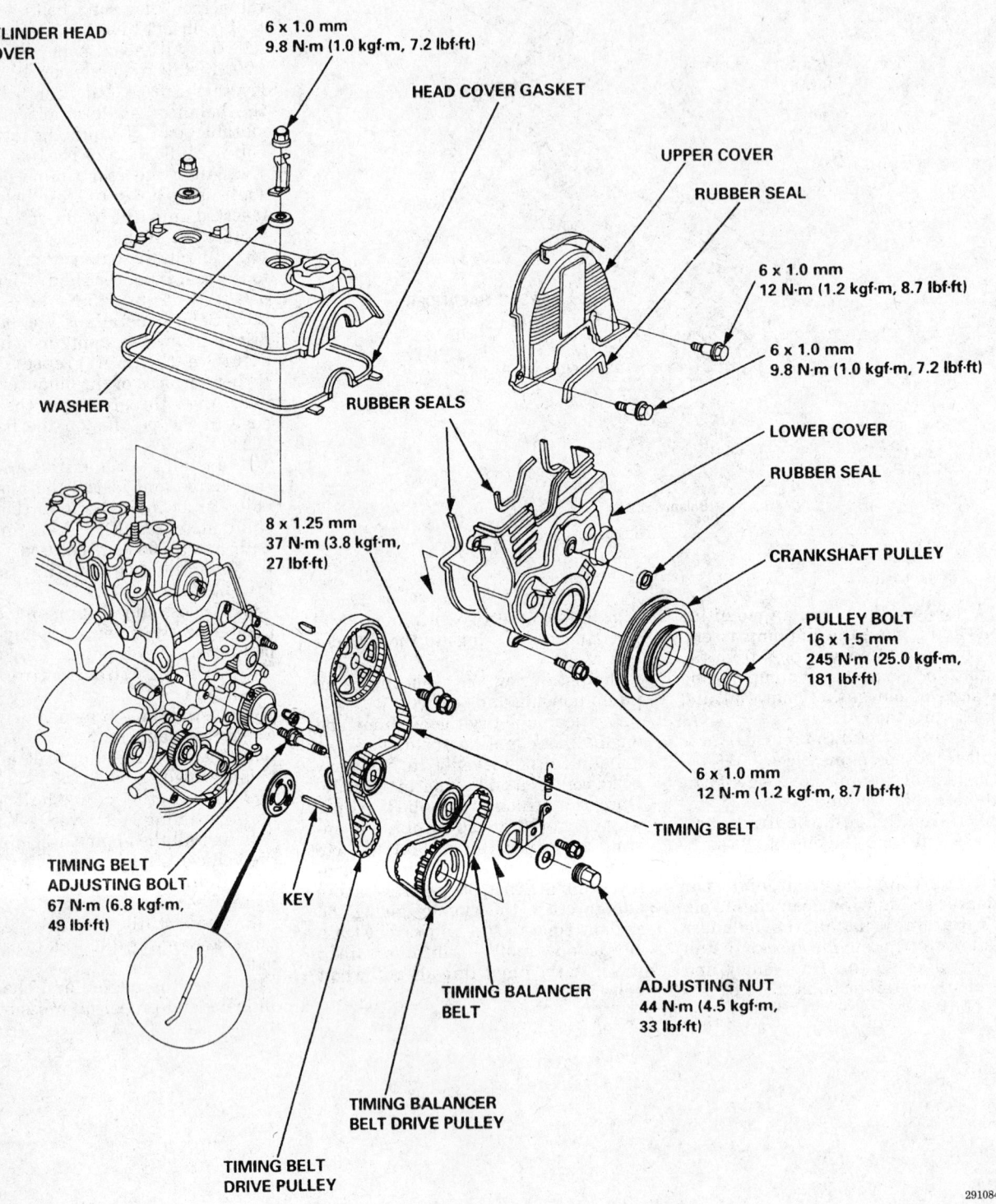

CYLINDER HEAD COVER

6 x 1.0 mm
9.8 N·m (1.0 kgf·m, 7.2 lbf·ft)

HEAD COVER GASKET

UPPER COVER

RUBBER SEAL

6 x 1.0 mm
12 N·m (1.2 kgf·m, 8.7 lbf·ft)

6 x 1.0 mm
9.8 N·m (1.0 kgf·m, 7.2 lbf·ft)

WASHER

RUBBER SEALS

LOWER COVER

RUBBER SEAL

CRANKSHAFT PULLEY

8 x 1.25 mm
37 N·m (3.8 kgf·m, 27 lbf·ft)

PULLEY BOLT
16 x 1.5 mm
245 N·m (25.0 kgf·m, 181 lbf·ft)

6 x 1.0 mm
12 N·m (1.2 kgf·m, 8.7 lbf·ft)

TIMING BELT

TIMING BELT ADJUSTING BOLT
67 N·m (6.8 kgf·m, 49 lbf·ft)

KEY

TIMING BALANCER BELT

ADJUSTING NUT
44 N·m (4.5 kgf·m, 33 lbf·ft)

TIMING BALANCER BELT DRIVE PULLEY

TIMING BELT DRIVE PULLEY

291084

Timing belt components

FRONT BALANCER:

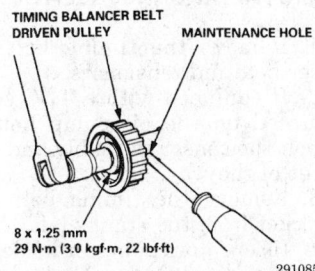

TIMING BALANCER BELT
DRIVEN PULLEY

MAINTENANCE HOLE

8 x 1.25 mm
29 N·m (3.0 kgf·m, 22 lbf·ft)

291085

Front balancer shaft driven pulley

REAR BALANCER:

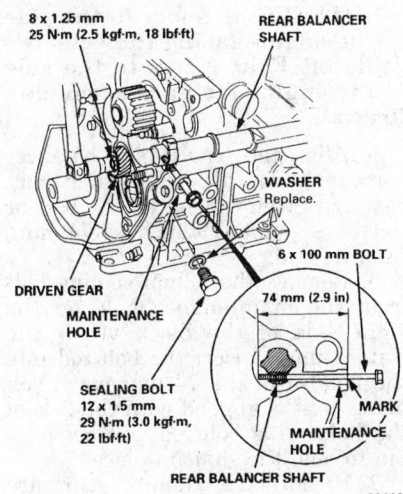

8 x 1.25 mm
25 N·m (2.5 kgf·m, 18 lbf·ft)

REAR BALANCER
SHAFT

WASHER
Replace.

6 x 100 mm BOLT

74 mm (2.9 in)

DRIVEN GEAR

MAINTENANCE
HOLE

SEALING BOLT
12 x 1.5 mm
29 N·m (3.0 kgf·m,
22 lbf·ft)

MARK

MAINTENANCE
HOLE

REAR BALANCER SHAFT

291086

Rear balancer shaft driven gear and
maintenance hole showing 6x100mm holder
bolt

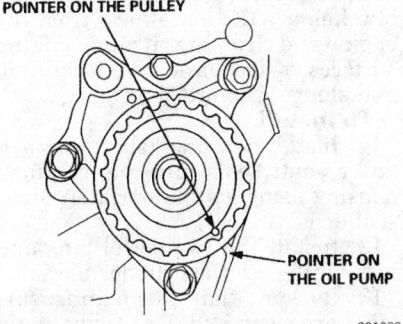

POINTER ON THE PULLEY

POINTER ON
THE OIL PUMP

291088

Gear case pointers after installation

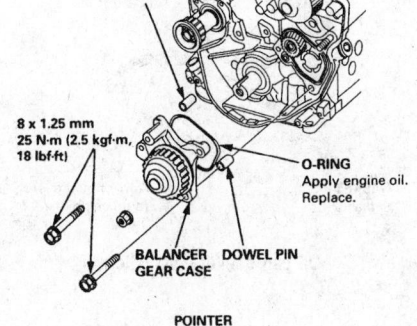

DOWEL PIN

8 x 1.25 mm
25 N·m (2.5 kgf·m,
18 lbf·ft)

O-RING
Apply engine oil.
Replace.

BALANCER   DOWEL PIN
GEAR CASE

POINTER
Align the groove to
pointer.

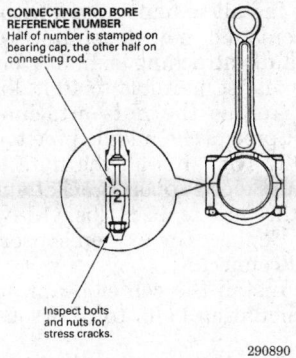

GROOVE

291087

Balancer gear case and alignment pointers

Piston and Connecting Rod

POSITIONING

CONNECTING ROD BORE
REFERENCE NUMBER
Half of number is stamped on
bearing cap, the other half on
connecting rod.

Inspect bolts
and nuts for
stress cracks.

290890

Rod bearing size code number

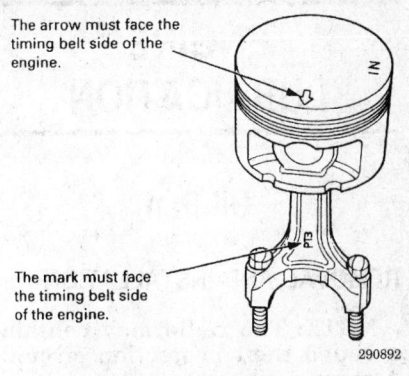

The arrow must face the
timing belt side of the
engine.

The mark must face
the timing belt side
of the engine.

290892

Piston position markings

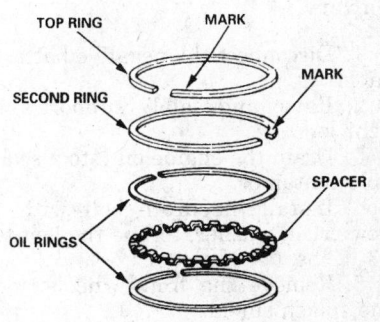

TOP RING   MARK

MARK

SECOND RING

SPACER

OIL RINGS

SECOND RING GAP

DO NOT position any ring gap
at piston thrust surfaces.

Approx. 90°

Approx.
90°

TOP RING GAP

OIL RING
GAP

15°

15°

SPACER GAP

OIL RING GAP

DO NOT position any ring gap
in line with the piston pin hole.

290893

Piston ring end gap positioning

ENGINE LUBRICATION

Oil Pan

REMOVAL AND INSTALLATION

NOTE: The radio may contain a coded theft protection circuit. Always make note of your code number before disconnecting the battery.

1. Disconnect the negative battery cable.
2. Raise and safely support the vehicle.
3. Drain the engine oil into a sealable container.
4. Install the drain bolt with a new crush washer, torque the bolt to 33 ft. lbs. (44 Nm).
5. Remove the front wheels and the splash shield.
6. Remove the subframe center beam.
7. Disconnect the oxygen sensor electrical connector.
8. Remove the nuts attaching exhaust pipe to the exhaust manifold and the mid pipe. Remove exhaust pipe and discard the gaskets.
9. Remove the torque converter cover.
10. Loosen the oil pan nuts and bolts in a crisscross pattern. Remove the oil pan. If necessary, use a seal cutter, or a mallet to tap the corners of the oil pan. Do not pry on the pan to get it loose.
11. Clean the oil pan mounting surface of old gasket material and engine oil.
12. Inspect the oil screen and pick-up tube for blockage, residue, or build-up. Replace the oil screen and pick-up tube if necessary.

To install:

13. Apply sealant where the oil pump and rear oil seal housing attach to the engine block. Work quickly so that the sealant doesn't set before the oil pan is installed.
14. Apply sealant to the corners of the curved section of the oil pan gasket. Install the oil pan and gasket to the engine block.
15. Install the oil pan nuts and bolts. Evenly finger-tighten the nuts and bolts.
16. Torque the nuts and bolts in a three-step, crisscross pattern to 10 ft. lbs. (14 Nm). Do not over-tighten the

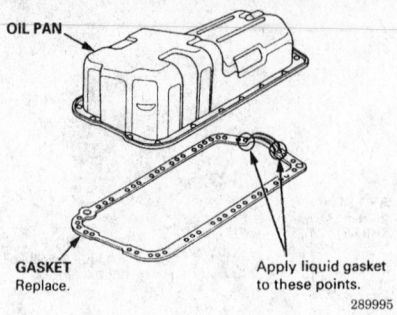

Oil pan gasket installation

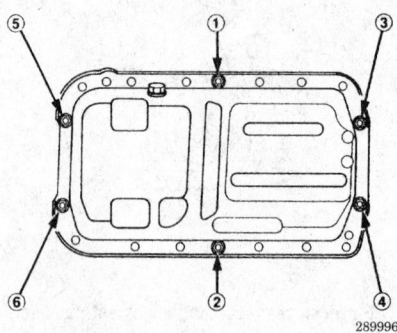

Oil pan torque sequence

bolts, this can distort the gasket and cause oil leakage.
17. Install the torque converter cover and torque the bolts to 9 ft. lbs. (12 Nm).
18. Install exhaust pipe with new gaskets and new locknuts. Torque the nuts attaching exhaust pipe to the exhaust manifold to 40 ft. lbs. (54 Nm), torque the nuts attaching exhaust pipe to the middle pipe to 16 ft. lbs. (22 Nm). Install the nuts to the exhaust pipe support bracket and torque the nuts to 13 ft. lbs. (18 Nm).
19. Connect the oxygen sensor electrical connector.
20. Install the center beam, torque the mounting bolts to 37 ft. lbs. (50 Nm).
21. Install the splash shield.
22. Install the front wheels.
23. Lower the vehicle. Make sure the sealant has cured and fill the engine with oil.
24. Tighten the wheel nuts to 80 ft. lbs. (110 Nm).
25. Connect the negative battery cable and enter the radio security code.
26. Start the engine and check for oil leaks.

Oil Pump

REMOVAL AND INSTALLATION

1. Remove the timing belt, balancer belt and tensioners.
2. If equipped with a TDC sensor mounted on the oil pump housing, unbolt the sensor assembly and move it out of the way.
3. Remove the timing belt drive sprocket from the crankshaft.
4. Insert a pin punch or holder tool into the maintenance hole in the front balancer shaft (located behind the balancer sprocket). Hold the shaft steady with the tool and remove the balancer sprocket. Make sure the tool or bolt used is strong enough to resist bending when torque is applied to the sprocket nut.

NOTE: Front refers to the side of the engine facing the vehicle's radiator. Rear refers to the side of the engine facing the vehicle's firewall.

5. Align the rear timing balancer sprocket using a 6 x 1.00 mm bolt, rod, or pin punch. Mark the bolt or rod at a point 2.9 inches (74 mm) from its end.
6. Remove the 12mm sealing bolt from the maintenance hole on the right side of the block below the water pump. Insert the bolt/rod into the hole until the 74 mm mark you made on it is aligned with the face of the hole. This bolt/rod will act as a pin to hold the shaft in place.
7. Remove the balancer gear case and the dowel pins. Discard the O-ring.
8. Unbolt and remove the balancer driven gear. Leave the holder tool in the maintenance hole.
9. Remove the oil pan and the oil screen. Discard the screen gasket. Replace the oil screen if it shows signs of blockage.
10. Remove the oil pump mounting bolts and remove the oil pump assembly. Remove the dowel pins from the engine and clean the oil pump mating surfaces of old gasket material, oil, and sludge. Discard the O-rings.

To install:

11. Install new crankshaft and balancer shaft seals into the oil pump housing using an appropriately-sized seal driver.
12. Install the two dowel pins and new O-rings to the cylinder block.
13. Be sure that the mating surfaces are clean and dry. Apply liquid gasket evenly in a narrow bead, centered on the mating surface. Once the sealant is applied, do not wait longer

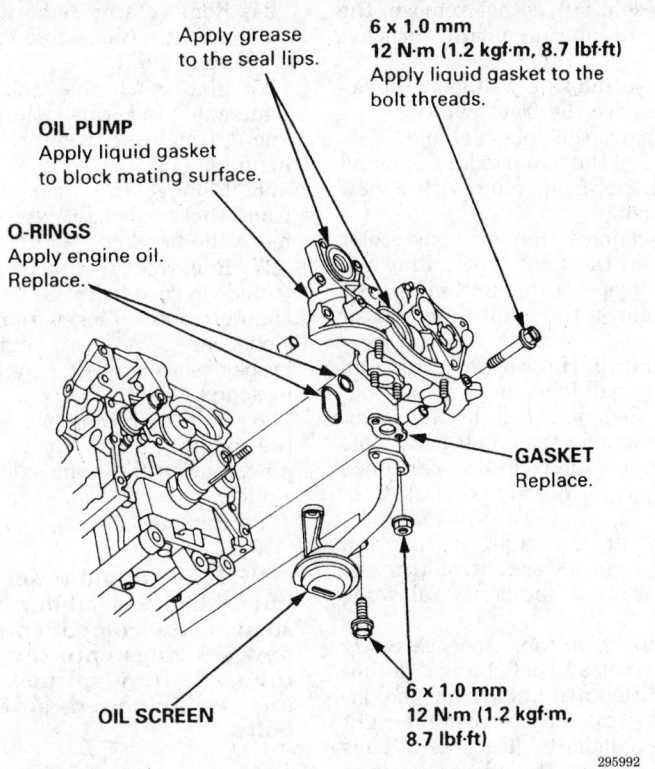

Apply grease to the seal lips.

6 x 1.0 mm
12 N·m (1.2 kgf·m, 8.7 lbf·ft)
Apply liquid gasket to the bolt threads.

OIL PUMP
Apply liquid gasket to block mating surface.

O-RINGS
Apply engine oil. Replace.

GASKET
Replace.

OIL SCREEN

6 x 1.0 mm
12 N·m (1.2 kgf·m, 8.7 lbf·ft)

295992

Oil pump and screen

balancer driven pulley to the front balancer shaft. Torque the attaching bolt to 22 ft. lbs. (29 Nm).

18. Install the balancer driven gear to the rear balancer shaft. Torque the bolt to 18 ft. lbs. (25 Nm).

19. Before installing the balancer driven gear and the gear case, apply molybdenum disulfide (lithium grease) to the thrust surfaces of the balancer gears.

20. Align the groove on the pulley edge to the pointer on the balancer gear case.

21. Install the balancer gear case to the engine with a new O-ring. Install the mounting bolts and nut. The rear balancer shaft should be held in place with a 6 x 100 mm bolt/rod . Torque the mounting bolts and nut to 18 ft. lbs. (25 Nm).

22. Check the alignment of the pointer on the balancer pulley to the pointer on the oil pump.

23. Remove the 6 x 1.00 holder bolt/rod from the maintenance hole. Install the sealing bolt with a new crush washer. Torque it to 22 ft. lbs. (29 Nm).

24. Install the timing belt and tensioners.

than 20 minutes to install the parts; the sealant will become ineffective. After final assembly, wait at least 30 minutes before adding oil to the engine to give the sealant time to set. To prevent leakage of oil, apply a suitable thread sealer to the inner threads of the bolt holes.

14. Install the oil pump to the engine block. Torque the mounting bolts to 9 ft. lbs. (12 Nm).

15. Install the oil screen. Torque the screen mounting bolts and nuts to 9 ft. lbs. (12 Nm).

16. Install the oil pan.

17. Hold the front balancer shaft in place with a suitable tool. Install the

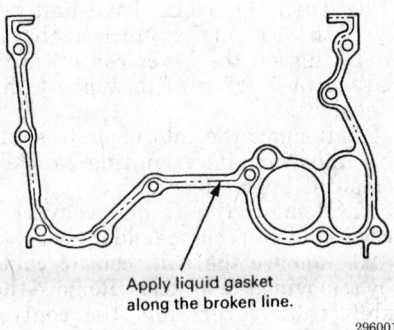

Apply liquid gasket along the broken line.

296001

Oil pump housing sealing surface

TRANSAXLE

Automatic Transaxle Assembly

REMOVAL AND INSTALLATION

NOTE: The radio may contain a code anti-theft circuit. Always obtain the customer's security code number before disconnecting the battery cables.

1. Shift the transaxle into neutral.
2. Disconnect the negative and positive battery cables and remove the battery.
3. Remove the air intake hose, air cleaner case, and battery tray. Disconnect the ground cable and the cable bracket from the battery tray.
4. Disconnect the throttle cable from the throttle control lever.
5. Disconnect the transaxle ground cable and the mainshaft speed sensor connectors. Disconnect the shift and lock-up solenoid valve connectors.
6. Disconnect the starter cables from the starter motor and unbolt the cable bracket from the transaxle case.
7. Disconnect the vehicle speed sensor connector. The VSS is located on the rear of the transaxle case near the cooler line inlet.
8. Disconnect the gear position switch and counter shaft speed sensor connectors.
9. Loosen the four upper transaxle case bolts, but leave them threaded into the engine block. Move the lower radiator hose slightly upward and toward the engine block if more clearance is needed for a socket and extension.

10. Loosen, but do not remove, the three front engine mount bracket bolts.
11. Raise and safely support the vehicle. Remove the front wheels.
12. Remove the splash shield.
13. Drain the transaxle fluid and reinstall the drain plug with a new crush washer.
14. Disconnect the transaxle cooler hoses from the joint pipes. Plug the hoses to keep out dirt and moisture.
15. Remove the subframe center beam.
16. Remove the cotter pins and lower arm ball joint nuts; then, separate the ball joints from the lower arms using a suitable ball joint tool.
17. Remove the right damper pinch bolt. Separate the damper fork from the strut.
18. Unbolt the right radius rod from the right lower control arm and remove it from the front subframe beam.
19. Using a suitable tool, carefully pry the right and left halfshafts out of the differential. Pull on the inboard CV-joints and remove the right and left halfshafts. Tie plastic bags over the halfshaft ends to prevent damage to the CV boots and splines. Don't let the left halfshaft hang by its own weight: use a piece of wire to suspend it out of the way.
20. Turn the right driveshaft toward the front of the vehicle so that it is resting on the lower control arm. Use a piece of wire to support the halfshaft.
21. Remove the intermediate shaft by unbolting its mounting bracket from the engine block.
22. Remove the torque converter cover and shift cable holder.
23. Remove the shift control cable by removing the lockbolt. Remove the shift cable lever from the control shaft. Don't disconnect the control lever from the shift cable. Wire the shift cable out of the work area and be careful not to kink it.

24. Remove the eight drive plate bolts one at a time while rotating the crankshaft pulley.
25. Place a suitable jack under the transaxle and raise the jack just enough to take weight off of the mounts.
26. Remove the transaxle mount from the transaxle case. Don't remove the bracket.
27. Remove the upper and lower transaxle case bolts. Unbolt the rear engine mount bracket from the transaxle case. The rear engine mount bracket through-bolt may have to be loosened first.
28. Pull the transaxle away from the engine until it clears the dowel pins. Lower the transaxle from the vehicle.

To install:

NOTE: Use new self-locking nuts when assembling the front suspension components. Install new set rings onto the halfshaft inboard joint splines. Replace any color-coded self-locking bolts.

29. Flush the transaxle cooler lines:
   a. Use a pressurized flusher (Honda or Kent-Moore part No. J38405–A or equivalent). Use only Honda flushing fluid (Honda part No. J35944–20), other fluids will damage the system.
   b. Fill the flusher with 21 ounces of fluid. Pressurize the flusher to 80–120 PSI, following the procedure on the fluid container and flusher.
   c. Clamp the discharge hose of the flusher to the cooler return line. Clamp the drain hose to the cooler inlet line and route it into a bucket or drain tank. Make sure the drain hose is securely clamped to the drain container.
   d. Connect the flusher to air and water lines. Use hot water if its available.
   e. Open the flusher water valve and flush the cooler for ten seconds.
   f. Depress the flusher trigger to mix flushing fluid with the water. Flush for two minutes, turning the air valve on and off for five seconds every 15–20 seconds. The maximum air pressure for the flushing procedure is 12 psi (845 kpa).
   g. After finishing one flushing cycle, reverse the hoses and flush in the opposite direction.
   h. Dry the cooler lines with compressed air so that NO moisture is left in the cooler system.

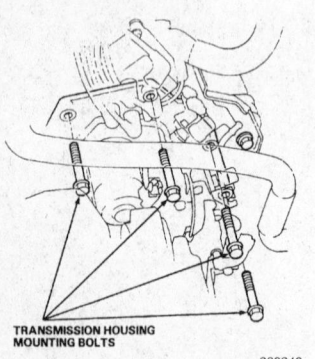

TRANSMISSION HOUSING MOUNTING BOLTS

289249

Upper transaxle case bolts

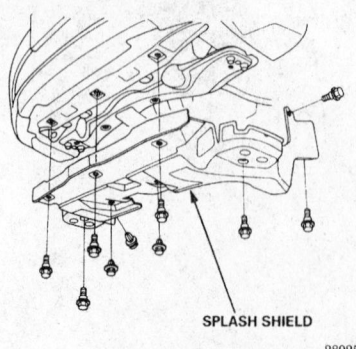

SPLASH SHIELD

289255

Subframe center beam

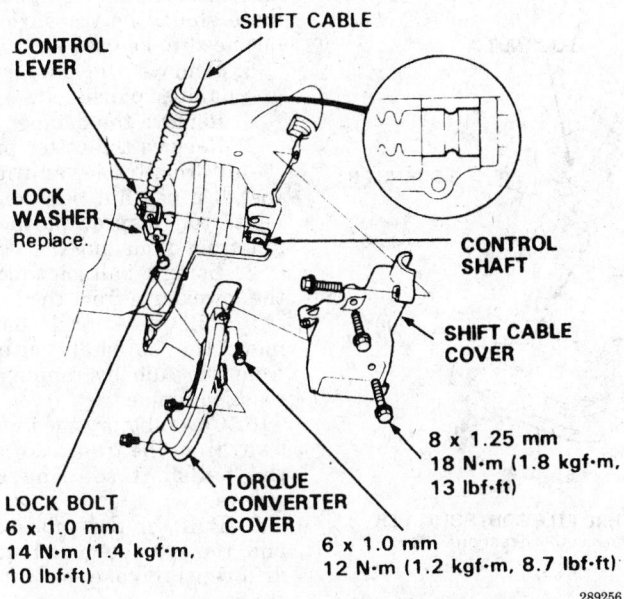

Shift cable linkage components

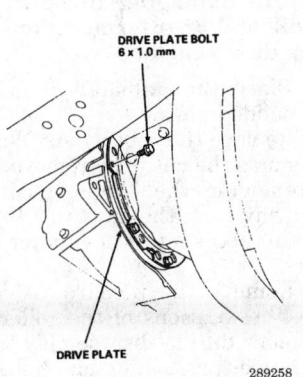

Transaxle mount

30. Make sure the two 14mm dowel pins are installed into the torque converter housing.

31. Install the torque converter onto the transaxle mainshaft with a new hub O-ring. Install the starter motor onto the transaxle case and tighten its mounting bolts to 33 ft. lbs. (44 Nm).

32. Raise the transaxle into position and install the transaxle housing mounting bolts. Evenly tighten the bolts to 47 ft. lbs. (65 Nm).

33. Connect the rear engine mount bracket to the transaxle case and evenly tighten the three new self-locking bolts to 40 ft. lbs. (54 Nm).

Tighten the rear mount through-bolt to 47 ft. lbs. (64 Nm) if it was loosened.

34. Install the intake manifold bracket and tighten the bolts to 16 ft. lbs. (22 Nm).

35. Install the transaxle mount and hand-tighten the through-bolt. Tighten the three nuts to 28 ft. lbs. (38 Nm). Tighten the through-bolt to 47 ft. lbs. (65 Nm).

36. Tighten the three front-engine-mount bracket bolts to 28 ft. lbs. (38 Nm).

37. Remove the transmission jack.

38. Attach the torque converter to the drive plate and hand-tighten the bolts. Tighten the eight bolts in two steps in a crisscross pattern: first to 4.3 ft. lbs. (6 Nm), and finally to 8.7 ft. lbs. (12 Nm). Check for free rotation after tightening the last bolt.

39. Install the shift control cable and control cable holder. Tighten the shift cable lockbolt to 10 ft. lbs. (14 Nm). Tighten the control cable holder bolts to 13 ft. lbs. (18 Nm).

40. Install the torque converter cover and tighten the bolts to 9 ft. lbs. (12 Nm).

41. Install the intermediate shaft and tighten the mounting bolts to 28 ft. lbs. (38 Nm).

42. Install the radius rod and damper fork.

43. Install a new set ring on the end of each halfshaft.

44. Turn the right steering knuckle fully outward and slide the axle into the differential until the set ring snaps into the differential side gear. Repeat the procedure on the left side. Make sure the halfshafts are fully seated in the differential and intermediate shaft.

45. Install the damper fork bolts and ball joint nuts to the lower arms. Tighten the ball joint nut to 40 ft. lbs. (55 Nm) and install a new cotter pin.

46. Install the subframe center beam and tighten its bolts to 37 ft. lbs. (50 Nm). Install the splash shield.

47. Install the front wheels and lower the vehicle.

48. Use a pulley holder tool in conjunction with a torque wrench to tighten the crankshaft pulley bolt to 181 ft. lbs. (245 Nm).

49. Reconnect the speed sensor connector.

50. Raise the right front knuckle with a floor jack until the weight of the vehicle is supported by the jack. Tighten the damper fork pinch bolt to 32 ft. lbs. (44 Nm). Tighten the radius rod bolts to 76 ft. lbs. (103 Nm), and the radius rod nut to 32 ft. lbs. (44 Nm). Hold the damper fork bolt with a wrench and tighten the nut to 40 ft. lbs. (55 Nm).

51. Connect the cables to the starter. Place the radiator hose back into its bracket.

52. Reconnect the throttle control cable.

53. Connect the lock-up control solenoid valve and shift control solenoid valve connectors.

54. Connect the mainshaft and countershaft speed sensor connectors and the transaxle ground cable.

55. Connect the transaxle cooler hoses to the joint pipes.

56. Install the battery base, air cleaner case and air intake hose. Reconnect the ground cable and the cable bracket to the battery base.

57. Install the battery. Connect the positive and negative battery cables.

58. Refill the transaxle with ATF. Use only Honda Premium ATF or an equivalent DEXRON®II or III ATF. Connect the battery cables.

   a. Leave the flusher drain hose attached to the cooler return line.

   b. With the transaxle in park, run the engine for 30 seconds, or until approximately one quart of fluid is discharged. This completes the cooler flushing process.

   c. Remove the drain hose and reconnect the cooler return line.

d. Refill the transaxle fluid to the proper level.

59. Start the engine, set the parking brake and shift the transaxle through all gears three times. Check for proper shift cable and throttle cable adjustment.

60. Let the engine reach operating temperature with the transaxle in **P** or **N**. Then, shut off the engine and check the fluid level.

61. Road test the vehicle. Check for proper shifting.

62. After road testing the vehicle, loosen the front engine mount bracket bolts; then, retighten them to 28 ft. lbs. (39 Nm).

63. Check and adjust the vehicle's front wheel alignment. Tighten the front wheel nuts to 80 ft. lbs. (110 Nm).

64. Enter the radio security code.

Throttle Valve Cable

ADJUSTMENT

NOTE: The radio may contain a coded theft protection circuit. Always make note of your code number before disconnecting the battery.

1. Warm the engine up to operating temperature.
2. Make sure the idle speed is correct.
3. There must be 0.39–0.47 in. (10–12 mm) of play in the throttle cable at the throttle linkage. If the cable deflection is not within these specs, loosen the locknut, and tighten the adjusting nut until the defection is correct. Then, retighten the locknut.
4. Verify that the throttle control cable is securely clamped and that the throttle linkage is in the fully closed position.
5. Disconnect the negative battery cable.

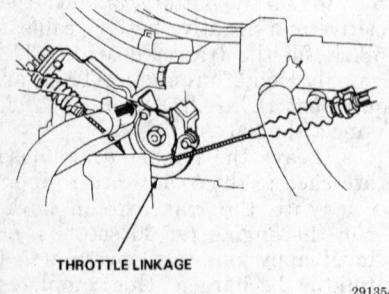

THROTTLE LINKAGE

291354

Throttle linkage

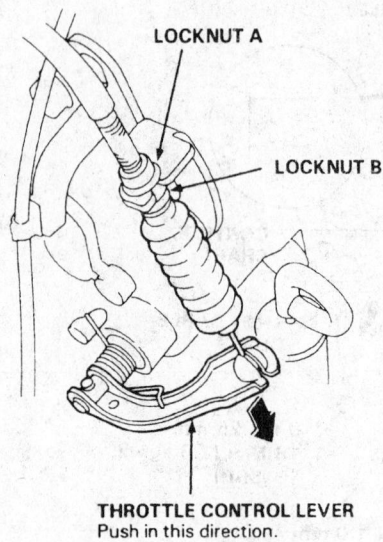

LOCKNUT A

LOCKNUT B

THROTTLE CONTROL LEVER
Push in this direction.

291355

Throttle control lever and locknuts

6. Loosen the locknut on the throttle control cable at the throttle control lever.
7. Turn the locknut to remove the throttle control cable freeplay. Do this while pushing down on the lever to hold it in the fully closed position.
8. Tighten the locknuts.
9. Reconnect the negative battery cable.
10. Inspect the movement of the throttle control lever. Check the cable deflection.
11. Test drive the vehicle and check the transaxle's shift points.

DRIVE AXLE

Halfshaft and CV-Joint Boot

REMOVAL AND INSTALLATION

1. Loosen the front spindle nut.
2. Raise and safely support the vehicle.
3. Remove the front wheels and the spindle nut.

4. Drain the transaxle fluid and install the drain plug with a new crush washer. If the halfshaft to be removed is installed into the intermediate shaft, the transaxle fluid does not need to be drained.

5. Remove the damper fork nut and damper pinch bolt.
6. Remove the damper fork.
7. Remove the cotter pin and castle nut from the lower arm ball joint. Install a hex nut flush onto the ball joint stud to prevent the ball joint tool from damaging the stud threads.
8. Using a ball joint tool, separate the lower arm from the knuckle.
9. Pull the knuckle outward. Remove the halfshaft outboard joint from the hub by tapping it with a plastic hammer.
10. Carefully pry the inner CV-joint away from the transaxle case to force the halfshaft set ring out of the groove.
11. Pull on the inboard CV-joint and remove the halfshaft from the differential case or intermediate shaft.

NOTE: Do not pull on the halfshaft as the CV-joint may come apart. Use care when prying out the assembly and pull it straight to avoid damaging the differential oil seal or intermediate shaft oil or dust seals.

12. Place the halfshaft in a vise with padded jaws.
13. Remove the boot bands. Welded bands must be cut to be removed. After removing the bands, push the boot away from the inboard CV-joint to gain access to the spider and rollers.
14. Remove the inboard CV-joint. Mark the locations of the rollers on the spider during disassembly to ensure proper positioning and halfshaft balance during reassembly.
15. Remove the rollers and the circlip. Mark the position of spider on the shaft.
16. Use a bearing puller to remove the spider from the shaft. Remove the stopper ring.
17. Remove the inboard CV-joint boot.
18. If equipped, remove the dynamic damper.
19. Remove the boot bands and the outboard CV-joint boot.
20. Inspect the outboard CV-joint for excess play and rough movement. The outboard CV-joint can't be disassembled. If it is damaged or worn out, it must be replaced.

To install:

21. Check the CV-joint components and replace any worn parts. Care-

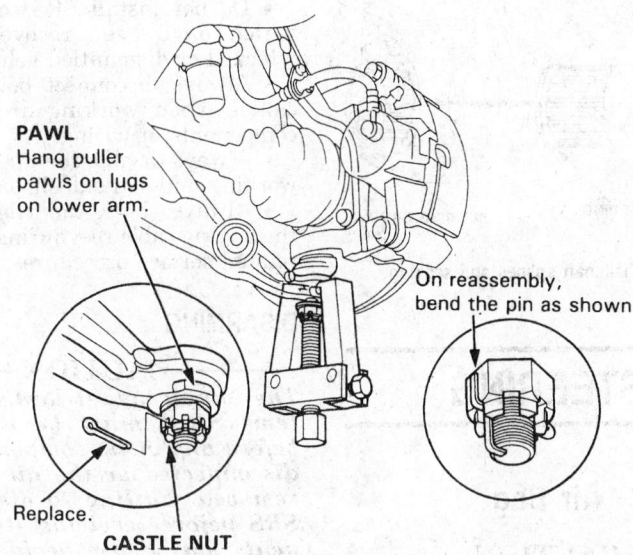

PAWL
Hang puller pawls on lugs on lower arm.

On reassembly, bend the pin as shown.

Replace.

CASTLE NUT

291404

Lower control arm ball joint

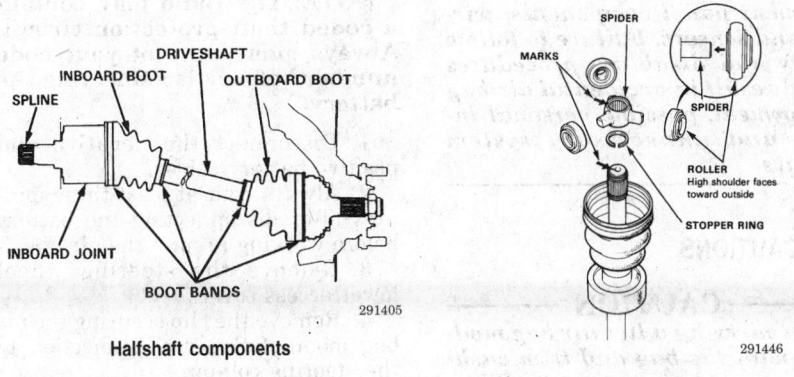

SPLINE

INBOARD BOOT

DRIVESHAFT

OUTBOARD BOOT

INBOARD JOINT

BOOT BANDS

291405

Halfshaft components

SPIDER

MARKS

SPIDER

ROLLER
High shoulder faces toward outside

STOPPER RING

291446

Spider gear

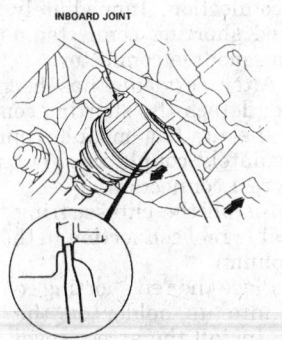

INBOARD JOINT

291406

Inboard joint and prying tool

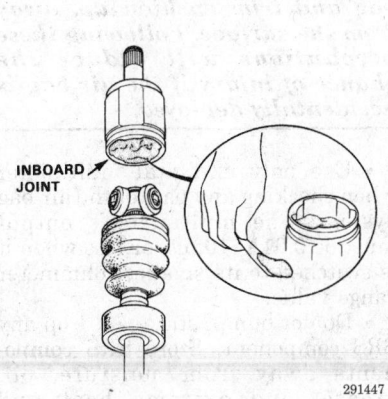

INBOARD JOINT

291447

Inboard CV-joint

fully clean any reusable parts in solvent and dry them with compressed air. Replace the boots if they show any signs of cracks or rips.

22. Wrap the inboard halfshaft splines with tape to prevent the new boots from being ripped.

23. Install the outboard boot onto the shaft and pack it with the grease included in the boot replacement kit. Install the boot onto the outboard joint with a new boot band.

24. Install the dynamic damper with a new band.

25. Install the inboard boot onto the shaft; then, remove the tape from the splines.

26. Install a new stopper ring onto the halfshaft. Align the matchmarks on the spider and the shaft, install the spider onto the halfshaft. Install a new circlip.

27. Install the rollers onto the spider in their original positions. The high shoulders of the rollers must face out.

28. Pack the inboard CV-joint and inboard boot with the grease included in the boot kit. Do not mix different types of grease.

29. Install the inboard joint onto the halfshaft.

30. Adjust the halfshafts so that each boot is partially extended.

31. Install a new boot band on the inboard halfshaft.

32. Install a new set ring onto the inboard shaft groove.

33. Replace the differential oil seal or intermediate shaft seal if either were damaged during removal.

34. Install new set rings on the ends of the halfshafts.

35. Install the halfshafts and make sure the set ring locks in the differential gear groove and the halfshaft bottoms in the differential or intermediate shaft.

36. Install the outboard joint into the hub. Make sure the splines mesh together and the joint is fully seated into the hub.

37. Fit the ball joint stud into the lower control arm. Install the damper fork into position. Torque the upper damper pinch bolt to 32 ft. lbs. (44 Nm) and the fork nut to 47 ft. lbs. (65 Nm).

38. Torque the ball joint castle nut to 40 ft. lbs. (55 Nm); then, tighten the nut just enough to install a new cotter pin.

39. Install the front wheels. Install a new spindle nut, but don't tighten it yet.

40. Lower the vehicle.

41. Torque the spindle nut to 181 ft. lbs. (245 Nm) and stake its tab.

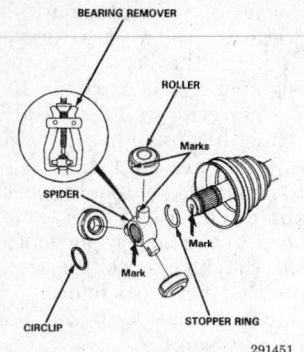

Spider and roller components

291451

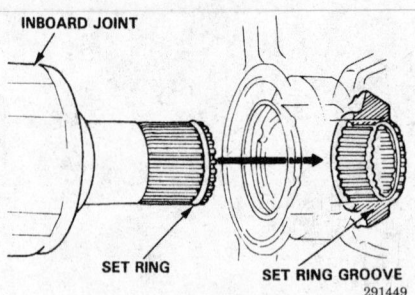

Inboard halfshaft splines and set ring

291449

Halfshaft length

291452

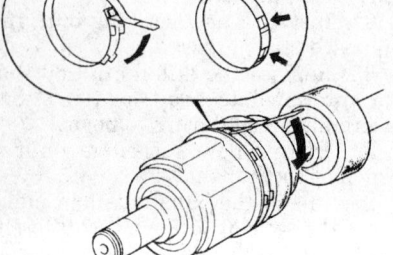

CV-boot bands

291448

Tighten the wheel nuts to 80 ft. lbs. (110 Nm).

42. Fill the transaxle with the proper type and quantity of fluid.

43. Warm the engine up, check the transaxle fluid level, and road test the vehicle.

STEERING

Air Bag

—— CAUTION ——

Some vehicles are equipped with an air bag system, also known as the Supplemental Inflatable Restraint (SIR) or Supplemental Restraint System (SRS). The system must be disabled before performing service on or around system components, steering column, instrument panel components, wiring and sensors. Failure to follow safety and disabling procedures could result in accidental air bag deployment, possible personal injury and unnecessary system repairs.

PRECAUTIONS

—— CAUTION ——

When carrying a live air bag module, point the bag and trim cushion away from your body. When placing a live air bag on a bench or other surface, always face the bag and trim cushion up, away from the surface. Following these precautions will reduce the chance of injury if the air bag is accidentally deployed.

• Use only a digital multimeter when checking any part of the air bag system. The multimeter's output must be 0.01A (10mA) or less when it is switched to its smallest ohmmeter range value.

• Do not bump, strike, or drop any SRS component. Store SRS components away from moisture, oil, grease, and extreme heat and humidity.

• Do not cut, damage, or attempt to alter the SRS wiring harness or its yellow insulation.

• Do not install SRS components which have been recovered from wrecked or dismantled vehicles.

• Always disconnect both battery cables when working around SRS components or wiring.

• Always disable the air bag when working under the dashboard.

• Always check the alignment of the air bag cable reel during steering-related service procedures.

DISARMING

—— CAUTION ——

The Supplemental Restraint System (SRS) must be disarmed before any of its components are disconnected or the air bag are removed. Failing to disarm the SRS before servicing its components may cause accidental deployment of the air bag, resulting in unnecessary SRS repairs and possible personal injury.

Driver's Air Bag

1995 Vehicles

NOTE: The radio may contain a coded theft protection circuit. Always make note of your code number before disconnecting the battery.

1. Disconnect the negative and positive battery cables.

2. Always wait at least three minutes after disconnecting the battery before working around the air bag.

3. Remove the steering wheel lower access cover.

4. Remove the clip securing the air bag module/cable reel connection to the steering column.

5. Remove the red shorting connector from its holder on the access cover.

6. Uncouple the air bag and cable reel connection. Immediately install the red shorting connector onto the air bag module connector.

7. After servicing has been completed, detach the shorting connector from the air bag module connector. Immediately couple the air bag and cable reel connectors.

8. Install the clip securing the air bag/cable reel connection to the steering column.

9. Place the red shorting connector back into its holder on the access cover. Install the access cover.

10. Reconnect the positive and negative battery cables.

11. Turn the ignition switch to the **ON** position, but don't start the engine. The SRS indicator light should turn on for six seconds and then turn off. If the SRS indicator light doesn't come on, or stays on longer than six seconds, the system fault must be diagnosed.

12. Enter the radio security code.

1996–97 Vehicles

NOTE: The radio may contain a coded theft protection circuit. Always make note of your code number before disconnecting the battery.

1. Disconnect the negative and positive battery cables.

2. Always wait at least three minutes after disconnecting the battery before working around the air bag.

3. Remove the steering wheel lower access cover.

4. Remove the clip securing the air bag module/cable reel connection to the steering column. Two types of air bag connections are used: spring-loaded connectors, and connectors requiring a shorting connector upon uncoupling.

5. Uncouple the air bag and cable reel connection. Immediately install

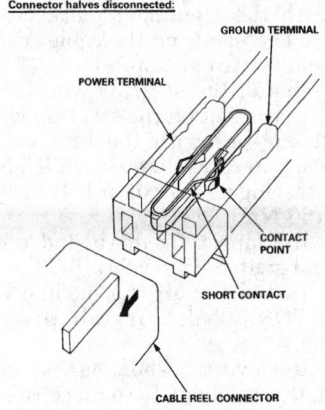

Connector halves disconnected:

GROUND TERMINAL
POWER TERMINAL
CONTACT POINT
SHORT CONTACT
CABLE REEL CONNECTOR

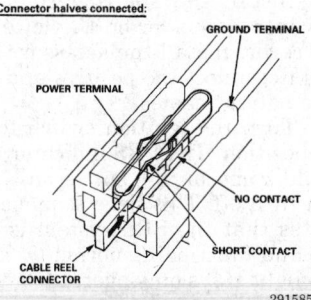

Connector halves connected:

GROUND TERMINAL
POWER TERMINAL
NO CONTACT
SHORT CONTACT
CABLE REEL CONNECTOR

291585

Cut-away view of spring-loaded connectors — 1996–97 Vehicles

the red shorting connector onto the air bag module connector.

NOTE: Spring-loaded air bag connectors contain a spring-contact self-disabling contact. A shorting connector doesn't need to be installed on the driver's air bag connector.

6. If the vehicle is equipped with spring-loaded connectors:
   a. Hold the connector body, not the wiring.
   b. Pull the spring-loaded locking sleeve toward its stop while holding the opposite half of the connector.
   c. After releasing the locking sleeve, uncouple the connectors.

7. After servicing has been completed, couple the air bag and cable reel connectors. For spring-loaded connectors, press the sleeve side of the connector into the pawl side until the sleeve locks the connectors together.

8. Install the clip securing the air bag/cable reel connection to the steering column.

9. Install the access cover.

10. Reconnect the positive and negative battery cables.

11. Turn the ignition switch to the **ON** position, but don't start the engine. The SRS indicator light should turn on for six seconds and then turn off. If the SRS indicator light doesn't come on, or stays on longer than six seconds, the system fault must be diagnosed.

12. Enter the radio security code.

Passenger's Air Bag

1995 Vehicles

NOTE: The radio may contain a coded theft protection circuit. Always make note of your code number before disconnecting the battery.

1. Disconnect the negative and positive battery cables.

2. Always wait at least three minutes after disconnecting the battery before working around the air bag.

3. Remove the glove box door and frame. Remove any lower mounting brackets that may cover the air bag connection.

4. Remove the shorting connector from its holder. On some vehicles, the shorting connector is permanently attached to the passenger's air bag frame: this type of shorting connector stays in place.

5. Uncouple the air bag module connector from the yellow SRS main wiring harness. Immediately connect

the air bag connector to the red shorting connector.

6. After servicing has been completed, detach the shorting connector from the air bag module connector. Immediately couple the air bag and cable reel connectors.

7. Install the clip securing the air bag/SRS harness connection to the passenger's air bag frame.

8. Place the red shorting connector back into its holder.

9. Install any lower mounting brackets that may have been removed. Install the glove box frame and glove box door.

10. Reconnect the positive and negative battery cables.

11. Turn the ignition switch to the **ON** position, but don't start the engine. The SRS indicator light should turn on for six seconds and then turn off. If the SRS indicator light doesn't come on, or stays on longer than six seconds, the system fault must be diagnosed.

12. Enter the radio security code.

1996–97 Vehicles

NOTE: The radio may contain a coded theft protection circuit. Always make note of your code number before disconnecting the battery.

1. Disconnect the negative and positive battery cables.

2. Always wait at least three minutes after disconnecting the battery before working around the air bag.

3. Remove dashboard storage compartment.

4. Uncouple the air bag module connector from the yellow SRS main wiring harness. Immediately connect the air bag connector to the red shorting connector. Two types of air bag connections are used: spring-loaded connectors, and connectors requiring a shorting connector upon uncoupling.

NOTE: Spring-loaded air bag connectors contain a spring-contact self-disabling contact. A shorting connector doesn't need to be installed on the driver's air bag connector.

5. If the vehicle is equipped with spring-loaded connectors:
   a. Hold the connector body, not the wiring.
   b. Pull the spring-loaded locking sleeve toward its stop while holding the opposite half of the connector.
   c. After releasing the locking sleeve, uncouple the connectors.

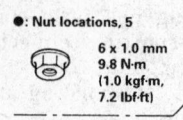

● : Nut locations, 5

6 x 1.0 mm
9.8 N·m
(1.0 kgf·m,
7.2 lbf·ft)

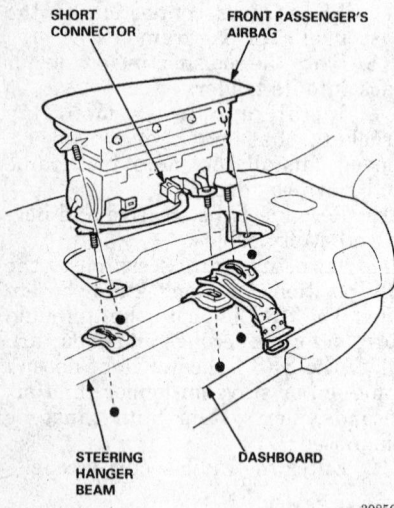

SHORT
CONNECTOR

FRONT PASSENGER'S
AIRBAG

STEERING
HANGER
BEAM

DASHBOARD

308561

Passenger's air bag bracket — 1995 Vehicles

6. After servicing has been completed, detach the shorting connector from the air bag module connector. Immediately couple the air bag and SRS harness connectors. For spring-loaded connectors, press the sleeve side of the connector into the pawl side until the sleeve locks the connectors together.

7. Install the clip securing the air bag/SRS harness connection to the passenger's air bag frame.

8. Install the dashboard storage compartment.

9. Reconnect the positive and negative battery cables.

10. Turn the ignition switch to the **ON** position, but don't start the engine. The SRS indicator light should turn on for six seconds and then turn off. If the SRS indicator light doesn't come on, or stays on longer than six seconds, the system fault must be diagnosed.

11. Enter the radio security code.

Steering Wheel

REMOVAL AND INSTALLATION

── CAUTION ──
The supplemental restraint system (SRS, air bag) must be disabled before removing the steer-

ing wheel. Failure to disarm the SRS system may cause accidental air bag deployment, resulting in unnecessary SRS system repairs and the risk of personal injury.

NOTE: The original radio contains a coded anti-theft circuit. Obtain the customer's security code number before disconnecting the battery cables.

1. Disconnect the negative and positive battery cables. Wait three minutes before servicing the air bag.
2. Remove the lower steering wheel access cover.
3. Uncouple the air bag connector from the cable reel connector.
4. Remove the red shorting connector from the access cover and install it onto the air bag connector.

NOTE: Some vehicles may be equipped with spring-loaded locking air bag connectors. These air bag connectors automatically short themselves when they are disconnected from the SRS harness. When uncoupling spring-loaded connectors, do not pull on the connector body — slide the sleeve toward its stop while securely holding the connector.

5. Remove the covers on either side of the steering wheel.
6. Remove the TORX® bolts from either side of the steering wheel. Carefully lift the air bag module from the steering wheel and remove it from the vehicle.

── CAUTION ──
The air bag module is live. Carry the air bag module with the cushion facing away from your body. Store the air bag module with the cushion facing up on a work bench. Following these precau-

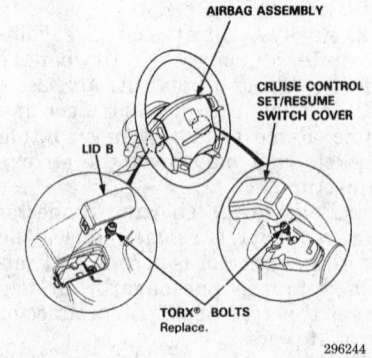

AIRBAG ASSEMBLY

CRUISE CONTROL
SET/RESUME
SWITCH COVER

LID B

TORX® BOLTS
Replace.

296244

Airbag, cable reel, and shorting connectors

tions will lessen the chance of personal injury if the air bag module accidentally deploys.

7. Disconnect the horn and cruise control switch connectors.
8. Remove the steering wheel nut or bolt. Pull the steering wheel off of the shaft by rocking it slightly side-to-side.

── WARNING ──
If a steering wheel puller is used to remove the steering wheel, do not thread the puller bolts more than five threads into the steering wheel. Install a nut five threads up on each puller bolt to act as a stop. If the puller bolts are threaded into the steering wheel more than five threads, the cable reel will be damaged.

To Install:

NOTE: Use new TORX® bolts when reinstalling the air bag module. They are available at a Honda dealer.

9. The cable reel must be centered before installing the steering wheel.
• Rotate the cable reel clockwise until it stops.
• Rotate the cable reel counter-clockwise two turns.
• The yellow gear tooth must line up with the alignment mark on the cover. The arrow on the cable reel label points straight up.
10. Install the steering wheel. Verify that the slot on the steering wheel shaft engages with the tabs on the turn signal cancelling sleeve. Tighten the steering wheel nut or bolt to 36 ft. lbs. (50 Nm).
11. Reconnect the horn and cruise control switch connectors.
12. Install the air bag module with new TORX® bolts. Install the side covers.
13. Remove the shorting connector from the air bag. Reconnect the air bag and cable reel connectors.
14. Install the shorting connector back into its holder on the lower access cover. Install the access cover.
15. Reconnect the positive and negative battery cables.
16. Turn the ignition switch to the **ON** position. The SRS indicator light should come on for six seconds and then turn off. This light sequence indicates that the SRS system is enabled and functioning normally. If the SRS light stays on longer, the system must be diagnosed.
17. Check the operation of the horn and cruise control switches.

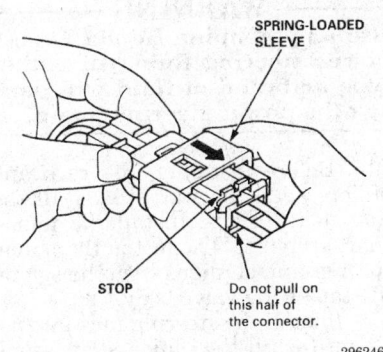

Spring-loaded air bag connectors

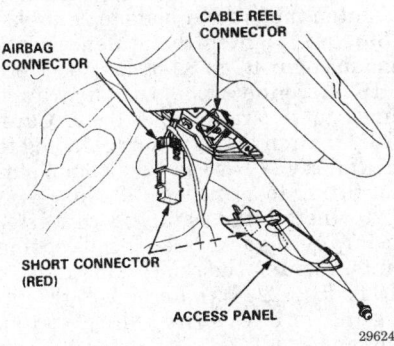

Steering wheel side covers

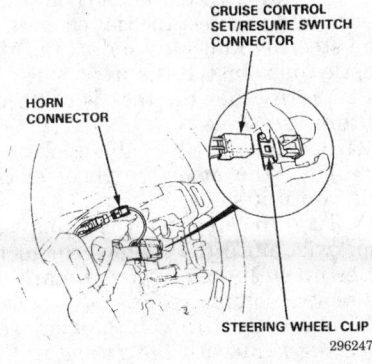

Horn and cruise control switch connectors

18. Turn the steering wheel counter clockwise and verify that the cable reel is aligned.

19. Check the front wheel alignment and the steering wheel spoke angle. Adjust the tie rod ends if the alignment is out of specification.

Tie Rod Ends

REMOVAL AND INSTALLATION

1. Raise and support the vehicle.
2. Remove the front wheels.

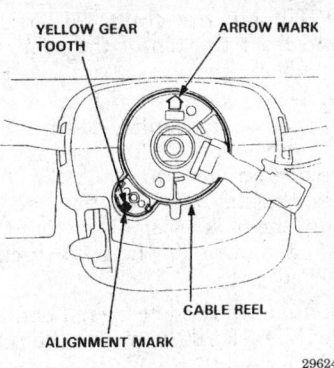

Cable reel alignment marks

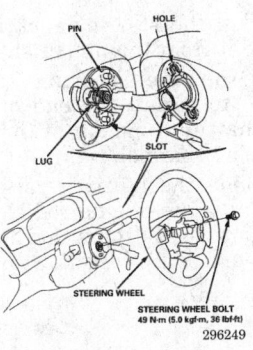

Steering wheel and turn
signal cancelling sleeve tabs

3. Use a ball joint remover to separate the tie rod end from the steering knuckle. Be careful not to tear the tie rod end boots.

4. Loosen the tie rod locknut.

5. Remove the tie rod end. Leave the locknut as a guide for installing and adjusting the new tie rod end.

To Install:

6. Install the tie rod end onto the steering rack ends.

7. Install the tie rod end onto the steering knuckle and tighten the ball joint castle nut to 29–35 ft. lbs. (40–48 Nm). Install a new cotter pin.

8. Install the front wheels and lower the vehicle

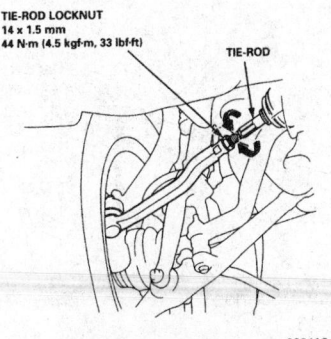

Tie rod end components

9. Check the vehicle's front wheel alignment and adjust the toe by turning the tie rods (shaft extending from the bellows at either end of the rack) equally until the alignment is within the proper specification.

10. After the alignment has been set, tighten the tie rod locknuts to 33 ft. lbs. (44 Nm).

11. Road test the vehicle.

Power Rack and Pinion Steering Gear

REMOVAL AND INSTALLATION

NOTE: The original radio may contain a coded anti-theft circuit. Obtain the customer's security code number before disconnecting the battery cable.

1. Drain the fluid from the power steering system:
   a. Lift the power steering reservoir off of its mount and disconnect the inlet hose.
   b. Insert a length of tubing into the inlet hose and route the tubing into a drain container.
   c. With the engine running at idle, turn the steering wheel lock-to-lock several times until fluid stops running out of the hose. Then, immediately shut off the engine.

2. Position the front wheels straight ahead. Lock the steering column with the ignition key. Reconnect the reservoir inlet hose.

3. Disconnect the negative and positive battery cables. Wait at least three minutes before working around the airbags.

4. Remove the steering joint cover and remove the lower steering joint bolts. Disconnect the steering joint by sliding it up toward the steering column.

5. Raise and safely support the vehicle.

6. Remove the front wheels.

7. Remove the tie-rod-end cotter pins and castle nuts. Install a 10mm nut onto the end of the ball joint stud so the threads won't be damaged by the ball joint remover. Using a ball joint remover tool, disconnect the tie rod ends from the steering knuckles.

8. Remove the self-locking nuts and separate the catalytic converter and the joint pipe from exhaust pipe and the front muffler. Remove the catalytic converter. Be careful not to damage the oxygen sensors: disconnect their electrical leads if necessary.

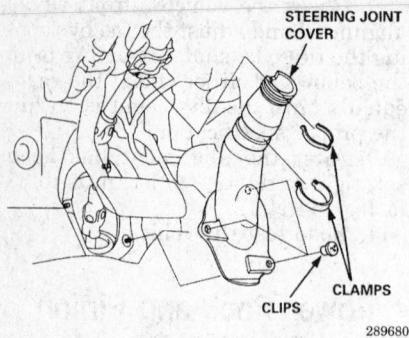

Steering joint cover

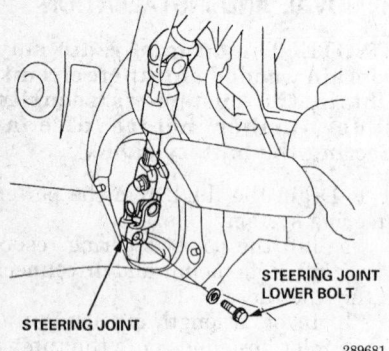

Steering joint and bolts

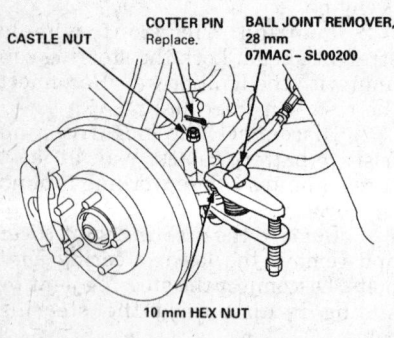

Tie rod ends and ball joint tool

9. Unbolt the fluid return line clamp from the top of the rear subframe beam.

10. Use a flare wrench to disconnect the two hydraulic lines from the rack valve body. Plug the lines to keep dirt and moisture out. Carefully move the disconnected lines to the rear of the rack assembly so that they are not damaged when the rack is removed.

11. Remove the left tie rod end and slide the rack all the way to the right.

12. Remove the rack stiffener plate; then, remove the right steering rack mounting bolts.

13. Pull the steering rack down to release it from the pinion shaft.

14. Drop the steering rack far enough to permit the end of the pinion shaft to come out of the hole in the frame channel.

15. Slide the steering rack to the right until the left tie rod clears the subframe, then drop it down and out of the vehicle to the left.

To Install:

NOTE: Use new gaskets and self-locking nuts when installing the catalytic converter.

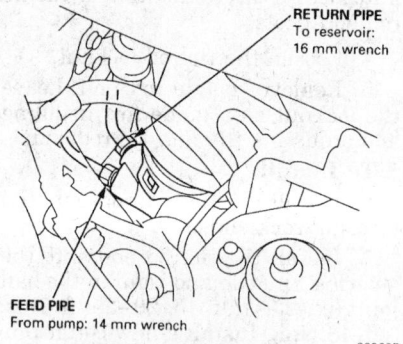

Hydraulic line fittings

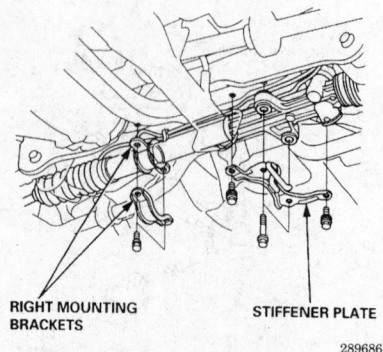

Rack stiffener plate and bracket

16. Before installing the rack and pinion, slide the rack's ends all the way to the right. Install the pinion shaft grommet. The lug on the pinion shaft grommet aligns with the slot on the top of the valve body.

17. Install the steering rack into position. Install the pinion shaft grommet and insert the pinion through the hole in the firewall.

18. Install the rack mounting bolts. Tighten the bracket bolts to 28 ft. lbs. (39 Nm). Tighten the stiffener plate mounting bolts to 32 ft. lbs. (43 Nm).

19. Reconnect the two hydraulic lines to the rack valve body. Carefully tighten the 14mm inlet fitting to 27 ft. lbs. (37 Nm) and the 16mm outlet fitting to 21 ft. lbs. (28 Nm).

20. Install the catalytic converter using new gaskets and self-locking nuts. Tighten the self-locking nuts to 16 ft. lbs. (22 Nm). Reconnect the oxygen sensors if they were disconnected.

21. Center the rack ends within their steering strokes.

22. Install the tie rod ends onto the rack ends. Connect the tie rod ends to the steering knuckles and install the castle nuts. Install the front wheels.

23. Verify that the rack is centered within its strokes.

24. Center the SRS cable reel:
• Turn the steering wheel clockwise until it stops.
• Turn the steering wheel counterclockwise until the yellow gear tooth lines up with the alignment mark on the lower column cover.

25. Line up the bolt hole in the steering joint with the groove in the pinion shaft. Slip the joint onto the pinion shaft. Pull the joint up and down to make sure the splines are fully seated. Tighten the joint bolts to 22 ft. lbs. (30 Nm).

NOTE: Connect the steering joint and pinion shaft with the cable reel and steering rack centered. Verify that the lower joint bolt is securely seated in the pinion shaft groove. If the steering wheel and rack are not centered, reposition the serrations at the lower end of the steering joint.

26. Install the steering joint cover.

27. Torque the ball joint castle nuts to 29–35 ft. lbs. (40–48 Nm). Then, tighten them only enough to install new cotter pins.

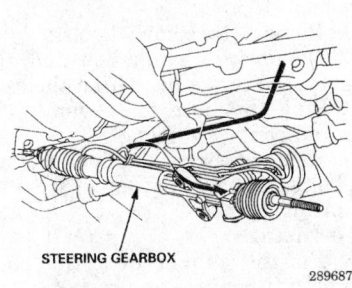

STEERING GEARBOX

289687

Rack removal and installation direction

28. Lower the vehicle. Reconnect the negative battery cable.

29. Make sure the reservoir inlet line has been reconnected. Fill the reservoir to the upper line with Honda power steering fluid. Run the engine at idle and turn the steering wheel lock-to-lock several times to bleed any air from the system and fill the rack valve body with fluid. Recheck the fluid level and add more if necessary.

30. Check the power steering system for leaks.

31. Check the front wheel alignment and steering wheel spoke angle. Make adjustments by turning the left and right tie rod ends equally.

32. Road test the vehicle.

33. Enter the radio security code.

Power Steering Pump

BLEEDING

1. Raise the front of the vehicle and support it with safety stands. Block the rear wheels.

2. Lift the power steering reservoir off of its mount. Disconnect the return hose from the steering rack at the reservoir. Immediately plug the reservoir inlet to prevent fluid loss and contamination. Don't disconnect the hose that connects the pump to the reservoir.

3. Insert a length of rubber tubing into the return hose and route the tubing into a drain container.

4. With the engine running at idle, turn the steering wheel lock-to-lock several times until fluid stops running out of the hose. Immediately shut off the engine.

5. After servicing, reconnect the reservoir return line. Fill the reservoir to the upper line with genuine Honda power steering fluid.

6. Run the engine at idle and turn the steering wheel lock-to-lock sev-

eral times to bleed air from the system and fill the rack valve body.

7. Recheck the fluid level and add more if necessary. Don't overfill the reservoir.

8. Check the power steering system for leaks.

9. Lower the vehicle.

REMOVAL AND INSTALLATION

1. Disconnect the negative battery cable.

2. Drain the power steering fluid from the reservoir.

3. Cover the alternator with some shop rags to protect it from spilled fluid.

4. Loosen the the pump adjusting bolt and mounting bolts. Slip the pump drive belt off of its pulley.

5. Disconnect and plug the outlet and inlet hoses.

6. Remove the pump mounting bolts. Remove the pump.

To Install:

—————— **CAUTION** ——————
Use only genuine Honda power steering fluid. Any other brand and type of fluid will damage the power steering pump.

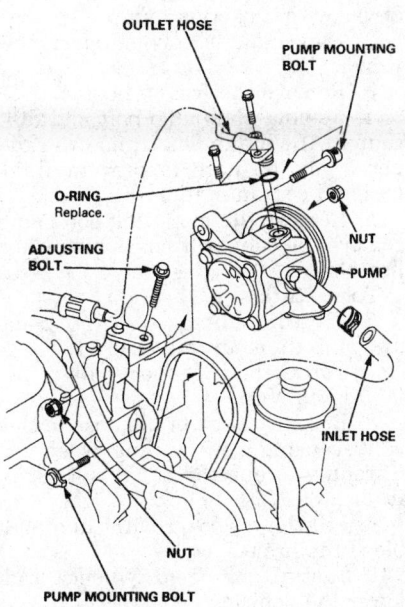

OUTLET HOSE

PUMP MOUNTING BOLT

O-RING
Replace.

ADJUSTING BOLT

NUT

PUMP

NUT

INLET HOSE

PUMP MOUNTING BOLT

289738

Power steering pump components

7. Install the power steering pump into its bracket. Loosely install the mounting and adjusting bolts.

8. Install a new O-ring onto the outlet hose fitting. Connect the outlet and inlet hoses to the power steering pump. Tighten the outlet hose bolts to 8 ft. lbs. (11 Nm).

9. Fit the pump belt onto the pulley.

10. Adjust the pump belt tension to the proper specification for the vehicle.

11. Tighten the mounting nuts and bolts to 17 ft. lbs. (24 Nm).

12. Fill the fluid reservoir to the upper level line with Honda power steering fluid.

13. Connect the negative battery cable.

14. Bleed the power steering system by running the engine at idle and turning the steering wheel lock-to-lock several times. Add more fluid if necessary.

15. Check the hose connections for leaks.

16. Enter the radio security code and reset the clock.

BRAKES

Anti-Lock Brake System Service

PRECAUTIONS

- Certain components within the Anti-Lock Brake System (ABS) are not intended to be serviced or repaired individually. Only those components with removal and installation procedures should be serviced.
- Do not use rubber hoses or other parts not specifically specified for and ABS system. When using repair kits, replace all parts included in the kit. Partial or incorrect repair may lead to functional problems and require the replacement of components.
- Lubricate rubber parts with clean, fresh brake fluid to ease assembly. Do not use lubricated shop air to clean parts; damage to rubber components may result.
- Use only specified brake fluid from an unopened container.
- If any hydraulic component or line is removed or replaced, it may be necessary to bleed the entire system.
- A clean repair area is essential. Always clean the reservoir and cap

thoroughly before removing the cap. The slightest amount of dirt in the fluid may plug an orifice and impair the system function. Perform repairs after components have been thoroughly cleaned; use only denatured alcohol to clean components. Do not allow ABS components to come into contact with any substance containing mineral oil; this includes used shop rags.

• The Anti-Lock control unit is a microprocessor similar to other computer units in the vehicle. Ensure that the ignition switch is **OFF** before removing or installing controller harnesses. Avoid static electricity discharge at or near the controller.

• If any arc welding is to be done on the vehicle, the control unit should be unplugged before welding operations begin.

Master Cylinder

REMOVAL AND INSTALLATION

NOTE: The original radio contains a coded anti-theft circuit. Obtain the customer's security code number before disconnecting the battery.

1. Disconnect the negative battery cable.
2. Disconnect the fluid level sensor wires.
3. Remove the fluid from the reservoir using a suction pump.
4. Use a flare wrench to disconnect the brake lines from the master cylinder. Plug the lines to prevent dirt and dust contamination.
5. Remove the mounting nuts and remove the master cylinder from the power booster.

To Install:

NOTE: The master cylinder piston-to-booster pushrod clearance must be checked and adjusted before the master cylinder is installed.

6. Install a pushrod adjustment gauge, Honda tool No. 07JAG-SD40100, or equivalent, onto the master cylinder body. Turn the tool's adjusting nut until its center shaft touches the master cylinder piston.
7. Reverse the adjustment tool and install it onto the power booster. Install the master cylinder nuts and tighten them to 11 ft. lbs. (15 Nm).
8. Connect a vacuum gauge in-line with the booster vacuum line. Connect the battery cable and run the engine at a steady idle speed to deliver 20 in. (500mm) Hg of vacuum.
9. Measure the clearance with a feeler gauge. The clearance must be within the specification. If the clearance is not correct, it must be adjusted by tightening or loosening the power booster pushrod adjuster (located inside the vehicle). Specifications are 0–0.02 in. (0–0.4mm).
10. Coat a new rod seal with silicon grease and install it onto the master cylinder. Remove the adjustment gauge and install the master cylinder onto the booster. Tighten the nuts to 11 ft. lbs. (15 Nm).
11. Connect the brake lines and carefully tighten the fittings to 14 ft. lbs. (19 Nm).
12. Fill the reservoir with fluid and install the cap. Reconnect the fluid level sensor wire.
13. Bleed the brake system.
14. Pump the pedal several times to build pedal pressure and test the feel of the pedal action.
15. Check brake pedal height and freeplay.
16. Road test the vehicle.

Brake Caliper

REMOVAL AND INSTALLATION

Front Caliper

1. Remove some fluid from the reservoir with a suction pump.
2. Raise and safely support the vehicle.
3. Remove the front wheels.
4. Remove the banjo bolt and disconnect the brake hose from the caliper. Plug the hose to prevent fluid loss and contamination.
5. Remove the mounting bolts and remove the caliper from its mounting bracket.

To Install:

6. Fit the caliper over the pads and onto its mounting bracket.
7. Torque both caliper bolts to 36 ft. lbs. (49 Nm).
8. Reconnect the brake hose to the caliper using new sealing washers. Carefully torque the banjo bolt to 25 ft. lbs. (35 Nm).
9. Fill the reservoir with fluid and bleed the brakes.
10. Install the front wheels and lower the vehicle.

Rear Caliper

1. Remove some fluid from the reservoir with a suction pump.
2. Raise and safely support the vehicle.
3. Remove the rear wheels.
4. Remove the banjo bolt and disconnect the brake hose from the caliper. Plug the hose to prevent fluid loss and contamination.
5. Remove the two caliper mounting bolts. Remove the caliper from its mounting bracket.

To Install:

6. Fit the caliper over the pads and onto its mounting bracket.
7. Tighten the caliper bolts to 17 ft. lbs. (23 Nm) .
8. Reconnect the brake hose with new sealing washers. Tighten the banjo bolt to 17 ft. lbs. (34 Nm).
9. Fill the reservoir with fluid and bleed the brake system. Adjust the parking brake if necessary.
10. Install the rear wheels and lower the vehicle.

Disc Brake Pads

REMOVAL AND INSTALLATION

Front Disc Pads

——————— CAUTION ———————
Brake pads and shoes contain asbestos, which has been determined to be a cancer causing agent. Never clean the brake surfaces with compressed air. Avoid inhaling any dust from brake surfaces. When cleaning brakes, use commercially available brake cleaning fluids.
————————————————————

1. Raise and support the vehicle safely.
2. Remove the front wheels.
3. Remove a small amount of brake fluid from the reservoir using a suction pump.
4. Unbolt the brake hose clamp from the knuckle by removing the retaining bolts.
5. Remove the lower caliper retaining bolt and pivot the caliper upward, off of the pads.
6. Remove the pad shim and pad retainers. Remove the disc brake pads from the caliper.

To install:

7. Clean the caliper thoroughly; remove any rust from the lip of the disc or rotor. Check the brake rotor for grooves or cracks. If any heavy scoring is present, the rotor must be replaced.
8. Install the pad retainers. Apply molybdenum brake grease to both surfaces of the shims and the back of the disc brake pads.

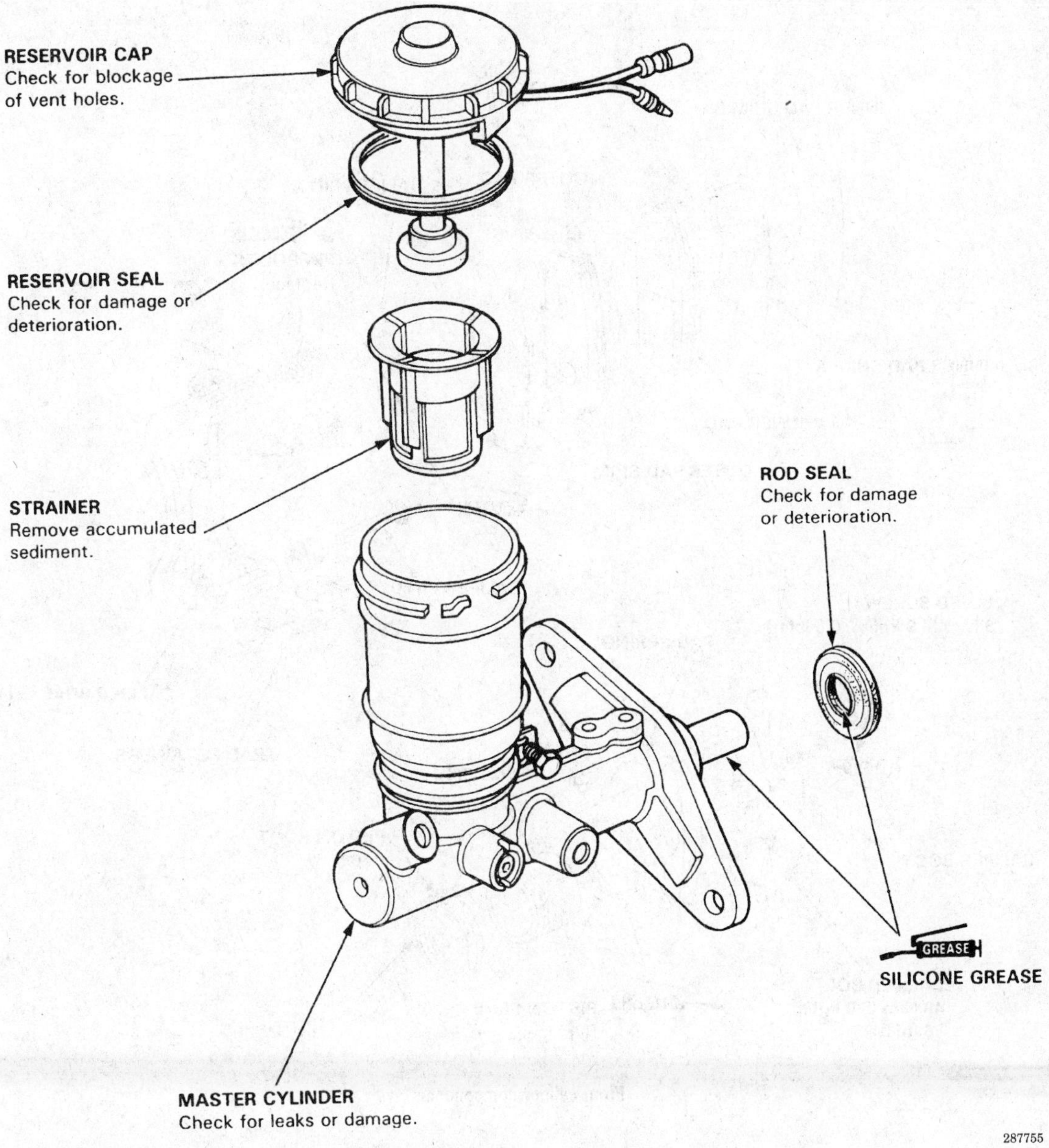

RESERVOIR CAP
Check for blockage of vent holes.

RESERVOIR SEAL
Check for damage or deterioration.

STRAINER
Remove accumulated sediment.

ROD SEAL
Check for damage or deterioration.

SILICONE GREASE

MASTER CYLINDER
Check for leaks or damage.

287755

Master cylinder components

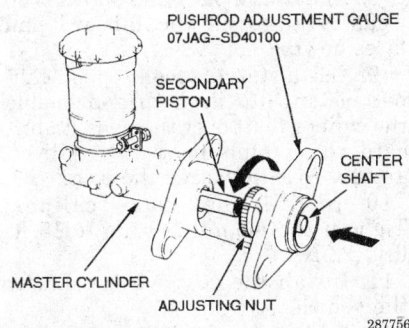

PUSHROD ADJUSTMENT GAUGE
07JAG--SD40100

SECONDARY PISTON

CENTER SHAFT

MASTER CYLINDER

ADJUSTING NUT

287756

Piston and pushrod adjustment gauge

WARNING
DO NOT get any lubricant on the braking surface of the pad.

9. Install the pads and shims. The pad with the wear indicator goes in the inboard position.

10. Push in the caliper piston so the caliper will fit over the pads. This is most easily accomplished with a pad spreader or large C-clamp.

WARNING
As the piston is forced back into the caliper, fluid will be forced back into the master cylinder reservoir. It may be necessary to siphon some fluid out to prevent overflowing.

11. Pivot the caliper down into position and tighten the mounting bolt to 36 ft. lbs. (49 Nm).

12. Connect the brake hose to the knuckle, if removed.

13. Install the wheel and lower the vehicle to the ground.

14. Add brake fluid to the master cylinder reservoir and install the cap.

15. Depress the brake pedal several times and make sure that the movement feels normal. The first brake pedal application may result in a very long pedal action due to the pis-

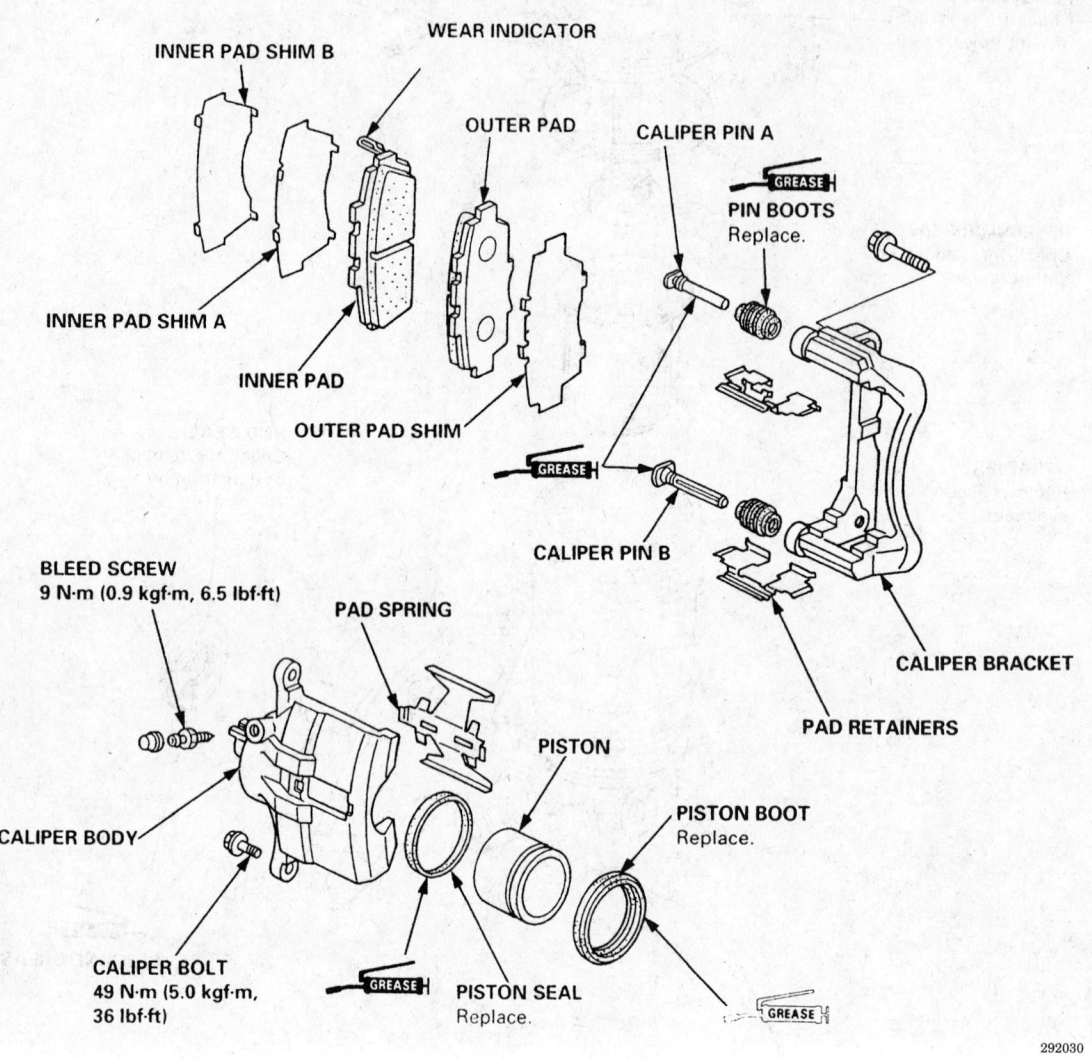

Front caliper components

tons being retracted. Always make several brake applications before starting the vehicle. Bleed the system if necessary.

Rear Disc Pads

1. Raise and safely support the vehicle.

2. Remove a small amount of brake fluid from the reservoir using a suction pump.

3. Remove the rear wheels.

4. Remove the two caliper mounting bolts and remove the caliper from the bracket.

5. Remove the pads, shims, and pad retainers.

To install:

6. Clean the caliper thoroughly; remove any dirt or dust. Check the brake rotor for grooves or cracks and machine or replace, as necessary.

7. Install the pad retainers. Apply molybdenum brake grease to both surfaces of the shims and the back of the disc brake pads.

WARNING

DO NOT get any lubricant on the braking surface of the pad.

---

8. Install the pads and shims. The wear retainer on the inboard pad faces down.

9. Use a suitable tool to push caliper piston into its bore and enable the caliper to fit over the pads. Lubricate the piston boot with silicon grease. Avoid twisting the boot.

10. Install the brake caliper. Tighten the mounting bolts to 17 ft. lbs. (23 Nm).

11. Install the rear wheels. Lower the vehicle.

12. Add brake fluid to the master cylinder reservoir. Depress the brake pedal several times to seat the pads. Bleed the brakes if necessary.

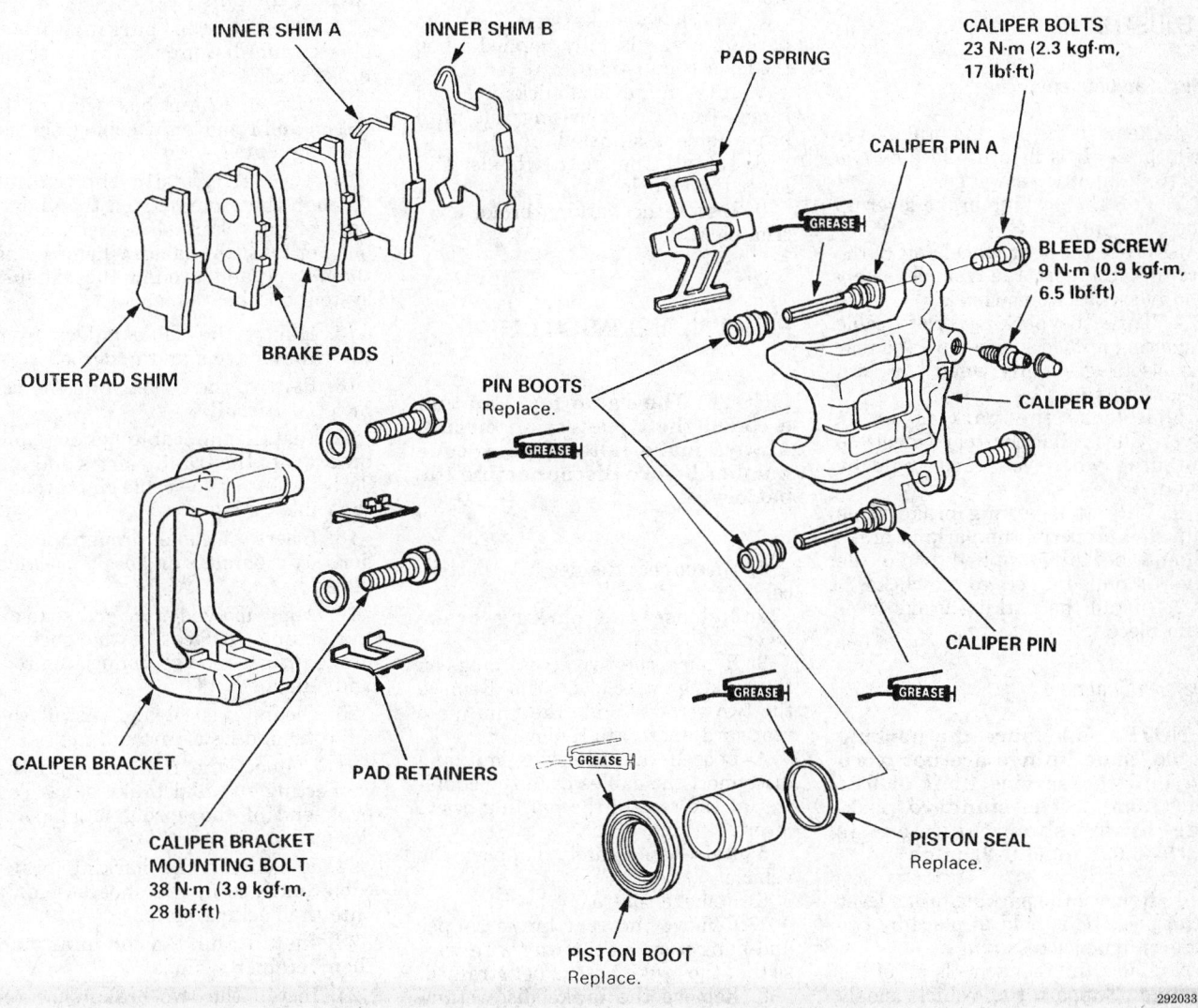

INNER SHIM A

INNER SHIM B

PAD SPRING

CALIPER BOLTS
23 N·m (2.3 kgf·m, 17 lbf·ft)

CALIPER PIN A

BLEED SCREW
9 N·m (0.9 kgf·m, 6.5 lbf·ft)

GREASE

OUTER PAD SHIM

BRAKE PADS

PIN BOOTS
Replace.

GREASE

CALIPER BODY

CALIPER PIN

GREASE

GREASE

CALIPER BRACKET

PAD RETAINERS

GREASE

PISTON SEAL
Replace.

CALIPER BRACKET
MOUNTING BOLT
38 N·m (3.9 kgf·m, 28 lbf·ft)

PISTON BOOT
Replace.

292033

Rear caliper components

Brake Rotor

REMOVAL AND INSTALLATION

1. Raise and safely support the vehicle.
2. Remove the front or rear wheels.
3. Unbolt the brake hose mounting bracket if the front calipers are to be removed.
4. Unbolt the caliper bracket from the knuckle. Hang the caliper out of the way with a length of wire. Do not allow the caliper to hang by the brake hose.

5. Remove the two 6mm retaining screws.
6. Remove the brake rotor. If the rotor is difficult to remove, install two 8mm bolts into the threaded holes and tighten them evenly and alternately to loosen the rotor.
   To Install:
7. Check the rotor for cracks, uneven wear, or other damage. Replace it if necessary.
8. Install the rotor onto the hub.
9. Install the rotor retaining screws and tighten them to 7 ft. lbs. (10 Nm).

10. Install the caliper and bracket onto the rotor. Tighten the front caliper bracket bolts to 80 ft. lbs. (110 Nm). Tighten the rear caliper bracket bolts to 28 ft. lbs. (39 Nm). If necessary, tighten the front caliper body bolts to 36 ft. lbs. (49 Nm); tighten the rear caliper body bolts to 17 ft. lbs. (23 Nm).
11. Connect the brake hose bracket and install its bolts.
12. Install the wheels.
13. Pump the brake pedal several times to check the system. Bleed the brakes if the pedal feels spongy or weak.

Parking Brake Cable

ADJUSTMENT

Minor adjustment

1. Remove the parking brake lever trim piece. It is held in place by two screws under access plugs.
2. Pull the parking brake lever up one click only.
3. Raise the rear wheels off of the ground. Support the vehicle safely and block the front wheels.
4. Turn the parking brake cable adjusting nut clockwise until the rear wheels drag slightly when you turn them by hand.
5. Release the parking brake lever. Check that the rear wheels do not drag when you turn them by hand.
6. With the parking brake cable adjusted properly, the parking brake should be fully applied when the lever is pulled up six to ten clicks.
7. Install the parking brake lever trim piece.

Major adjustment

NOTE: Make sure the parking brake shoe linings are not worn beyond the service limit of 0.04 in. (1.0mm). The standard parking brake shoe thickness is 0.075–0.098 in. (1.9–2.5mm).

1. Remove the parking brake lever trim piece. It is held in place by two screws under access plugs.
2. Raise the rear wheels off of the ground. Support the vehicle safely and block the front wheels.
3. Remove the rear wheels.
4. Release the parking brake lever. Loosen the adjusting nut by turning it counterclockwise.
5. Insert a flat-bladed pry tool into the rear brake shoe adjusting hole.

Turn the adjuster up until the shoes lock; then, back the adjuster off eight stops.
6. Check to make sure that the parking brake is fully applied when the lever is pulled up six to ten clicks.
7. If the number of clicks is out of range, turn the adjusting nut until brake cable is adjusted.
8. Install the rear wheels and lower the vehicle.
9. Install the parking brake lever trim piece.

REMOVAL AND INSTALLATION

NOTE: The radio may contain a coded theft protection circuit. Always make note of your code number before disconnecting the battery.

1. Disconnect the negative battery cable.
2. Release the parking brake lever.
3. Remove the two access plugs on the parking brake lever trim. Remove the two screws. Slide the trim piece rearward and up to remove it.
4. Loosen the adjusting nut and disconnect the cables from the equalizer at the rear of the parking brake lever.
5. Raise and safely support the vehicle.
6. Remove the rear wheels.
7. Remove the rear brake caliper and hang it out of the way with wire so that the brake hose is not strained.
8. Remove the brake disc retaining screws.
9. Remove the brake disc. If the disc has seized onto the hub, evenly screw two 8mm bolts into the threaded holes to pop it loose.
10. Disconnect and remove the two parking brake shoe return springs.

11. Remove the tension pins by pressing the retainer springs in and turning the pins.
12. Remove the parking brake shoes and disconnect the parking brake cable.
13. Use a 12mm box wrench to loosen and remove the cable from the backing plates.
14. Carefully pull the cables through the grommets on the underbody of the vehicle. Be careful not to bend or twist the cables when moving them over and around the exhaust system and heat shields.
15. Unbolt the cable guides from the trailing arms and underbody.
16. Remove the cables from the vehicle.**To install:**
17. Install the cable guides and brackets to the trailing arms and underbody. Insert the cable end through the cable guide.
18. Insert the cable through the underbody grommets and brake backing plates.
19. Apply molybdenum grease to all the sliding surfaces of the parking brake return springs, connecting rod, and adjuster.
20. Clean, grease, and install the adjuster and its return spring.
21. Connect the rod spring to the connecting rod and brake shoe. The hook end of the spring must point downward.
22. Reconnect the parking brake cable and install the shoe assembly onto the backing plate.
23. Install the tension pins and their retainer springs.
24. Install the two brake shoe return springs.
25. Install the brake disc and caliper. Tighten the brake disc screws to 7 ft. lbs. (10 Nm), and the caliper bracket bolts to 28 ft. lbs. (39 Nm).
26. Install the rear wheels.
27. Lower the vehicle.
28. Connect the cable and equalizer assembly to the parking brake lever.
29. Tighten the locknut.
30. Check the operation of the parking brake and adjust it if necessary. The parking brake should be fully applied when the lever is raised six to ten clicks.
31. Verify that the cable guides are evenly tight and that the cable doesn't bind or catch on any of the exhaust components.
32. Reconnect the negative battery cable.
33. Enter the radio security code.

EQUALIZER

ADJUST NUT

287704

Equalizer and adjusting nut

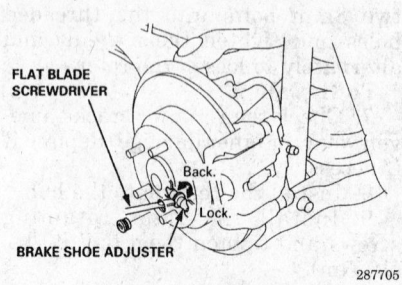

FLAT BLADE SCREWDRIVER

Back.

Lock.

BRAKE SHOE ADJUSTER

287705

Parking brake shoe adjuster

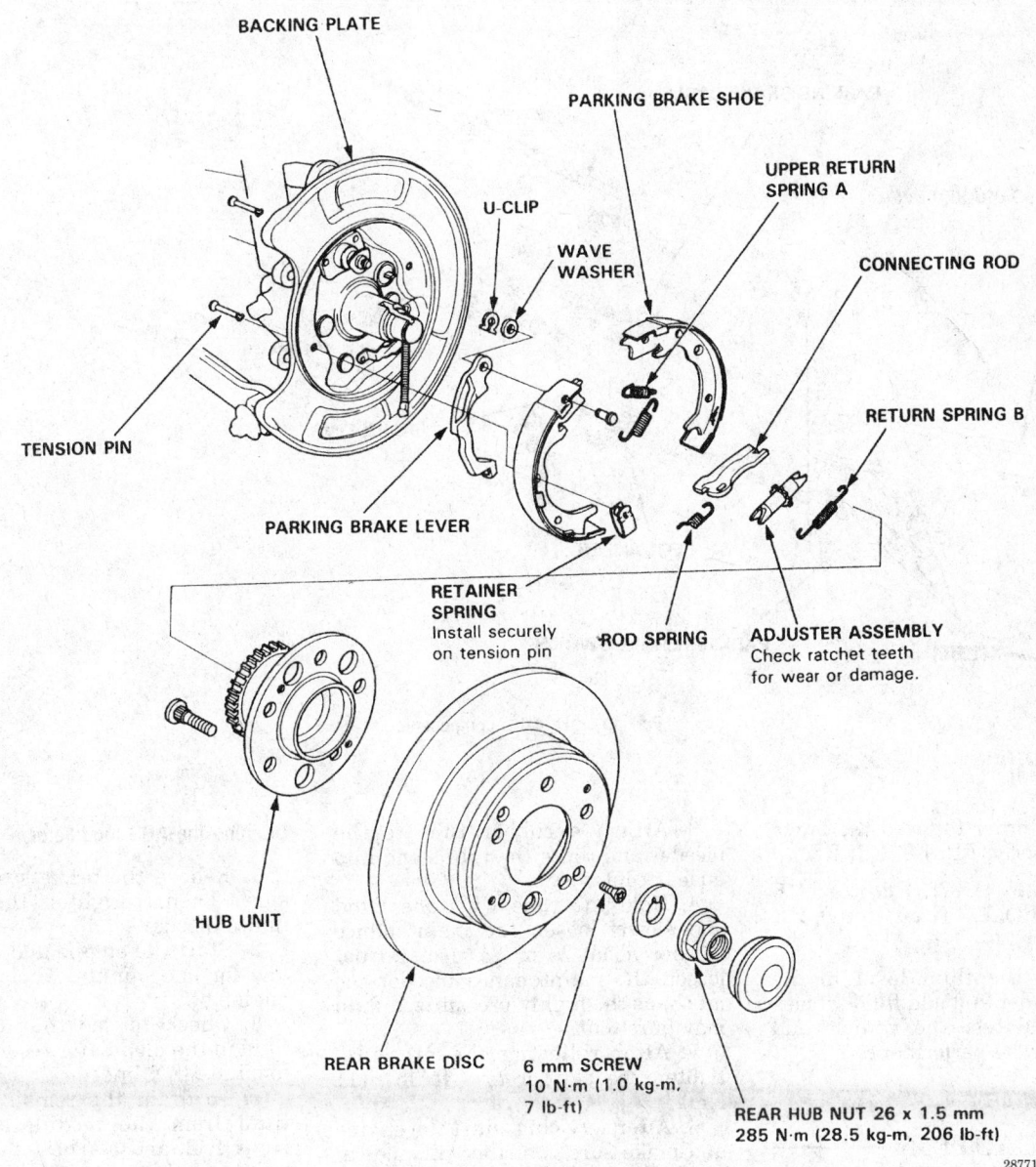

Exploded view of the parking brake shoes

Brake Hydraulic System

BLEEDING

Without ABS or with Rear Wheel ABS

——— CAUTION ———

The Honda anti-lock brake system contains brake fluid under extremely high pressure within the pump, accumulator and modulator assembly. Do not disconnect or loosen any lines, hoses, fittings or components without properly relieving the system pressure. If a tool is required to relieve system pressure, use only a bleeder T-

wrench 07HAA-SG00100 or equivalent to relieve pressure. For modulators with maintenance bleeders, follow the proper depressurizing procedure. Improper procedures or failure to discharge the system pressure may result in severe personal injury and/or property damage.

NOTE: The master cylinder must be full at the start of the bleeding procedure and checked after bleeding the brake at each wheel. Add fluid as required. Use only DOT 3 or 4 brake fluid. If a pressure bleeder is not available,

an assistant will be necessary to perform this brake bleeding operation.

1. Fill the master cylinder.
2. Have an assistant slowly pump the brake pedal several times and then apply a steady pressure to the brake pedal.
3. Attach a bleed hose to the bleed screw and place it into a clear container. Loosen the brake bleed screw at the brake caliper and allow the fluid to flow. Close the bleeder.
4. Repeat this procedure for each wheel, until no air bubbles appear in the brake fluid. Torque the bleeders to 7 ft. lbs. (9 Nm). Use the following

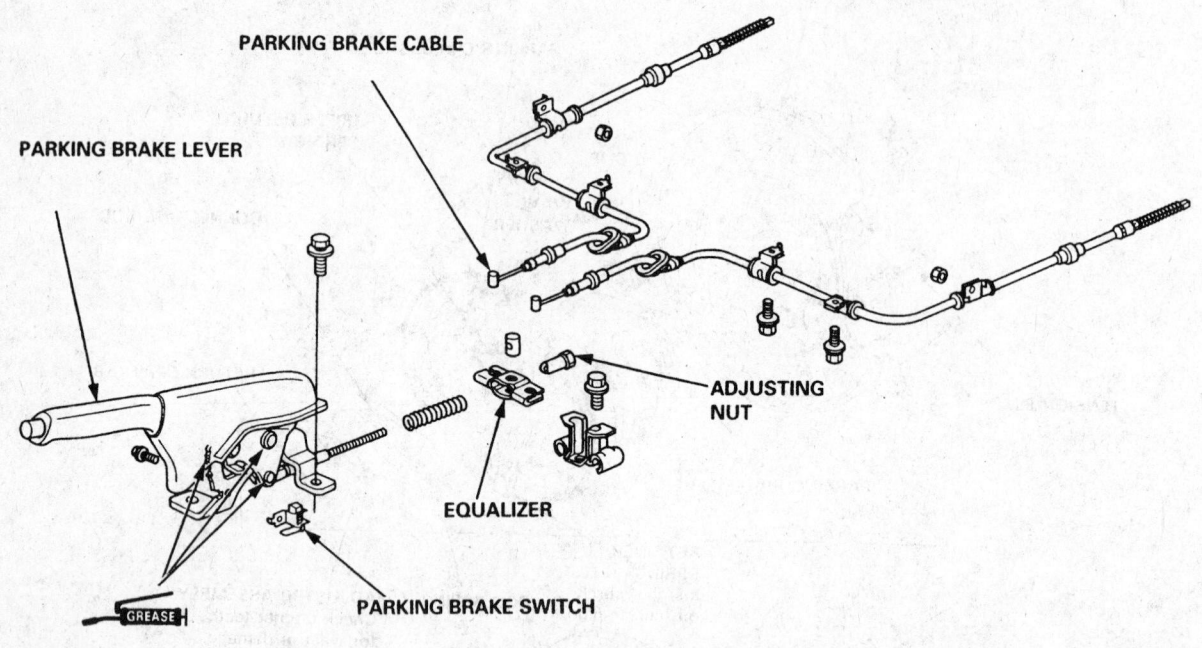

PARKING BRAKE CABLE

PARKING BRAKE LEVER

ADJUSTING NUT

EQUALIZER

GREASE

PARKING BRAKE SWITCH

287714

Parking brake cable components

sequence in order to bleed the brake system properly, RR, LF, LR, RF.

NOTE: LF (Left Front), RF (Right Front), LR (Left Rear), RR (Right Rear).

5. Check the fluid level in the master cylinder and add fluid , if necessary. Road test the vehicle and check the brake performance.

With 4–Wheel ABS

——— **CAUTION** ———
The hydraulic accumulator contains brake fluid and nitrogen gas at extremely high pressures. Certain portions of the hydraulic system also contain brake fluid at high pressure. Do not loosen the relief plug on the accumulator. The system must be depressurized before disconnecting any hoses, lines, or fittings, or personal injury may result.

1. Remove the cap from the modulator maintenance bleeder.
2. Connect a flare or box-end wrench to the maintenance bleeder.

——— **WARNING** ———
Brake fluid will damage the vehicle's paint. Immediately clean up any spills.

3. Attach a rubber tube to the bleeder and route the tube's end into a clear container.
4. Hold the tube with one hand and slowly loosen the maintenance bleeder about ⅛ or ¼ turn. Do not loosen the maintenance bleeder too much, as the highly-pressurized fluid may burst out.
5. After relieving the pressure, tighten the bleeder to 8 ft. lbs. (11 Nm).
6. After servicing, start the engine and make sure that the ABS indicator light turns off. Check the performance of the brake system.

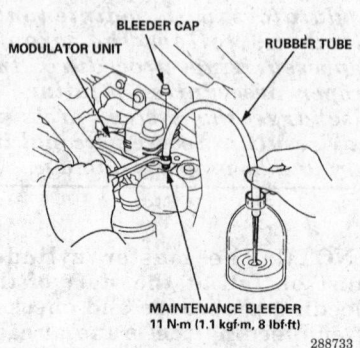

BLEEDER CAP

MODULATOR UNIT

RUBBER TUBE

MAINTENANCE BLEEDER
11 N·m (1.1 kgf·m, 8 lbf·ft)

288733

Modulator maintenance bleeder and bleeding equipment — 4-Wheel ABS

Draining the ABS modulator

1. Relieve the brake system pressure. Then, retighten the maintenance bleeder.
2. Start the engine and allow it to idle for one minute. Then, shut the engine off.
3. Check to see that the brake fluid in the modulator reservoir is below the MAX line.
4. To drain the remaining brake fluid from the modulator, repeat steps 1, 2, and 3. The total fluid capacity of the modulator is 5 fl. oz. (150 ml); about 1.3–1.5 fl. oz. (40–45 ml) of fluid is drained during each cycle.

NOTE: The modulator should be bled if air enters it during the draining and brake fluid replacement processes.

5. Remove the modulator reservoir cap and fill the reservoir to the MAX line with clean brake fluid. Recap the reservoir.
6. Repeat steps 1, 2, and 3 two more times. Then, refill the reservoir with clean brake fluid.
7. Tighten the bleeder to 8 ft. lbs. (11 Nm).
8. After servicing, start the engine and make sure that the ABS indicator light turns off.

Bleeding the ABS modulator

1. Fill the modulator reservoir to the MAX line.

2. Attach a snug-fitting rubber tube to the maintenance bleeder and route the end of the tube into a clean container.

3. Loosen the maintenance bleeder. Then start the engine to active the modulator's pump motor.

4. Tighten the maintenance bleeder when fluid starts to flow out of the bleeder.

5. Shut the engine off after the modulator pump motor stops.

NOTE: If the ABS indicator light turns on and the pump motor stops, restart the engine and repeat steps 3, 4, and 5.

6. Make sure the brake fluid level is at the reservoir's MAX line.

7. After servicing, start the engine and make sure that the ABS indicator light turns off.

Wheel Speed Sensor

REMOVAL AND INSTALLATION

NOTE: The radio may contain a coded theft protection circuit. Always make note of your code number before disconnecting the battery.

Front wheel speed sensor

1. Make sure the ignition switch is turned **OFF**.

2. Disconnect the negative battery cable.

3. Detach the wheel sensor cable from the ABS harness at its connector located on the vehicle's shock tower. Carefully detach the plastic clip from the shock tower.

4. Raise and support the vehicle safely. Remove the front wheels.

5. Unbolt the speed sensor cable brackets from the inside of the shock tower and the steering knuckle.

6. Unbolt the wheel speed sensor assembly from the steering knuckle.

7. Pull the speed sensor cable through the hole in the shock tower. Be careful not to bend or kink the cable if it is to be reused. Remove the speed sensor and cable from the vehicle as an assembly.

To Install:

8. Feed the speed sensor connector through the hole in the shock tower and connect it to the ABS harness.

9. Install the speed sensor cable brackets to the inside of the shock tower and the steering knuckle. Do not bend or twist the speed sensor cable.

10. Install the wheel speed sensor assembly into its cavity on the steering knuckle. Torque the speed sensor cable brackets to 7 ft. lbs. (10 Nm). Torque the wheel speed sensor mounting bolts to 16 ft. lbs. (22 Nm).

11. Check the wheel sensor air gap:

   a. Inspect the pulser wheel for damaged teeth.

   b. Rotate the driveshaft by hand and measure the clearance between the wheel speed sensor and the pulser wheel. Measure the clearance for one full rotation of the driveshaft. The standard air gap clearance is 0.02–0.04 in. (0.4–1.0 mm).

   c. If the air gap exceeds the standard, inspect the steering knuckle and pulser wheel for distortion.

12. Install the front wheels and lower the vehicle.

13. Reconnect the negative battery cable.

14. Turn the ignition to the **ON** position, but don't start the engine. The ABS indicator light should turn on. Start the engine and verify that the ABS indicator light turns off. If the ABS indicator light blinks, doesn't come on, or stays on with the engine running, the system fault must be diagnosed.

Rear wheel speed sensor

1. Make sure the ignition switch is turned **OFF**.

2. Disconnect the negative battery cable.

3. Raise and support the vehicle safely. Remove the rear wheels.

4. Uncouple the wheel speed sensor cable connector from its junction on the rear suspension beam. Detach the plastic cable clips from the rear suspension beam.

5. Unbolt the speed sensor cable brackets from the lower control arm.

6. Unbolt the wheel speed sensor assembly from its cavity on the knuckle. Remove the speed sensor and cable. Be careful not to bend or kink the cable if it is to be reused.

To Install:

7. Install the speed sensor cable brackets to the lower control arm.

8. Install the speed sensor assembly into its cavity on the knuckle. Tighten the speed sensor bolts to 7 ft. lbs. (10 Nm). Tighten the speed sensor cable bracket bolts to 7 ft. lbs. (10 Nm).

9. Reconnect the speed sensor cable connector to its junction on the rear suspension beam. Fasten the plastic cable clips.

10. Check the wheel sensor air gap:

   a. Inspect the pulser wheel for damaged teeth.

   b. Rotate the hub by hand and measure the clearance between the wheel speed sensor and the pulser wheel. Measure the clearance for one full rotation of the hub. The standard air gap clearance is 0.01–0.05 in. (0.3–1.3 mm).

   c. If the air gap exceeds the standard, inspect the knuckle and pulser wheel for distortion.

11. Install the rear wheels and lower the vehicle.

12. Reconnect the negative battery cable.

13. Turn the ignition to the **ON** position, but don't start the engine. The ABS indicator light should turn on. Start the engine and verify that the ABS indicator light turns off. If the ABS indicator light blinks, doesn't come on, or stays on with the engine running, the system fault must be diagnosed.

Front:

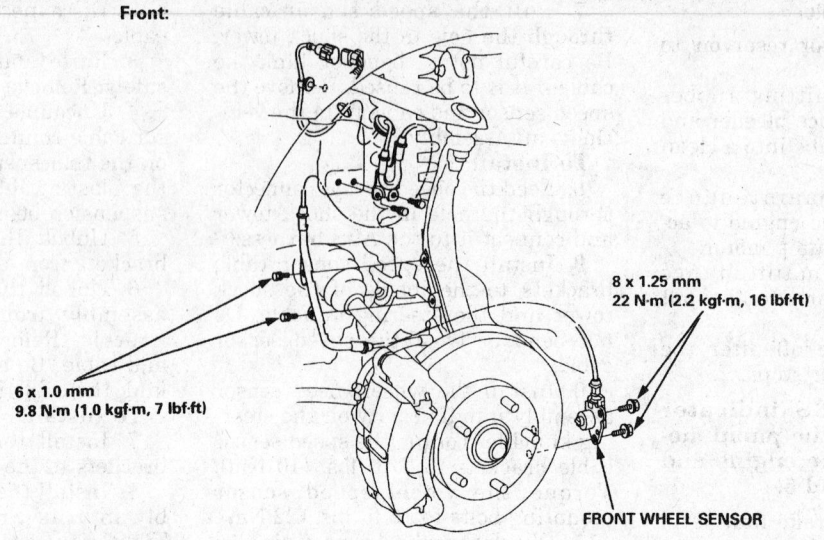

6 x 1.0 mm
9.8 N·m (1.0 kgf·m, 7 lbf·ft)

8 x 1.25 mm
22 N·m (2.2 kgf·m, 16 lbf·ft)

FRONT WHEEL SENSOR

Rear:

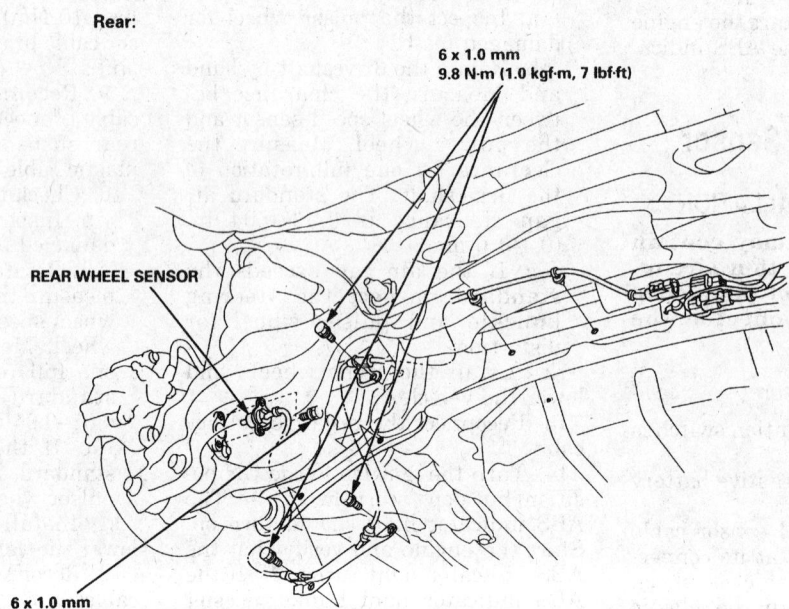

6 x 1.0 mm
9.8 N·m (1.0 kgf·m, 7 lbf·ft)

REAR WHEEL SENSOR

6 x 1.0 mm
9.8 N·m (1.0 kgf·m, 7 lbf·ft)

295846

Front and rear wheel speed sensor components

FRONT SUSPENSION

Strut and Coil Spring

REMOVAL AND INSTALLATION

1. Raise and safely support the vehicle.
2. Remove the front wheels.
3. Remove the brake hose clamp bolts from the strut.
4. Remove the damper fork bolts and remove the damper fork.

5. Remove the three strut mounting nuts. Remove the strut from the vehicle.
6. Place the strut in vice and install a spring compressor onto the coil spring. Follow the spring compressor manufacturer's instructions.
7. Compress the spring and remove the self-locking nut from the top of the strut. Disassemble the strut mounts and remove the coil spring.
8. Inspect the strut mounts for wear and damage. Replace any damaged or worn parts.

To Install:

NOTE: Use new self-locking nuts when assembling and installing the struts.

9. Install the spring compressor onto the coil spring. Set the spring onto the strut cartridge. The flat part of the coil spring is its top.
10. Assemble the strut mount and its washer onto the strut. Tighten the self-locking nut to 22 ft. lbs. (29 Nm). Remove the spring compressor.

NOTE: Use new self-locking bolts when installing the struts and assembling the damper forks.

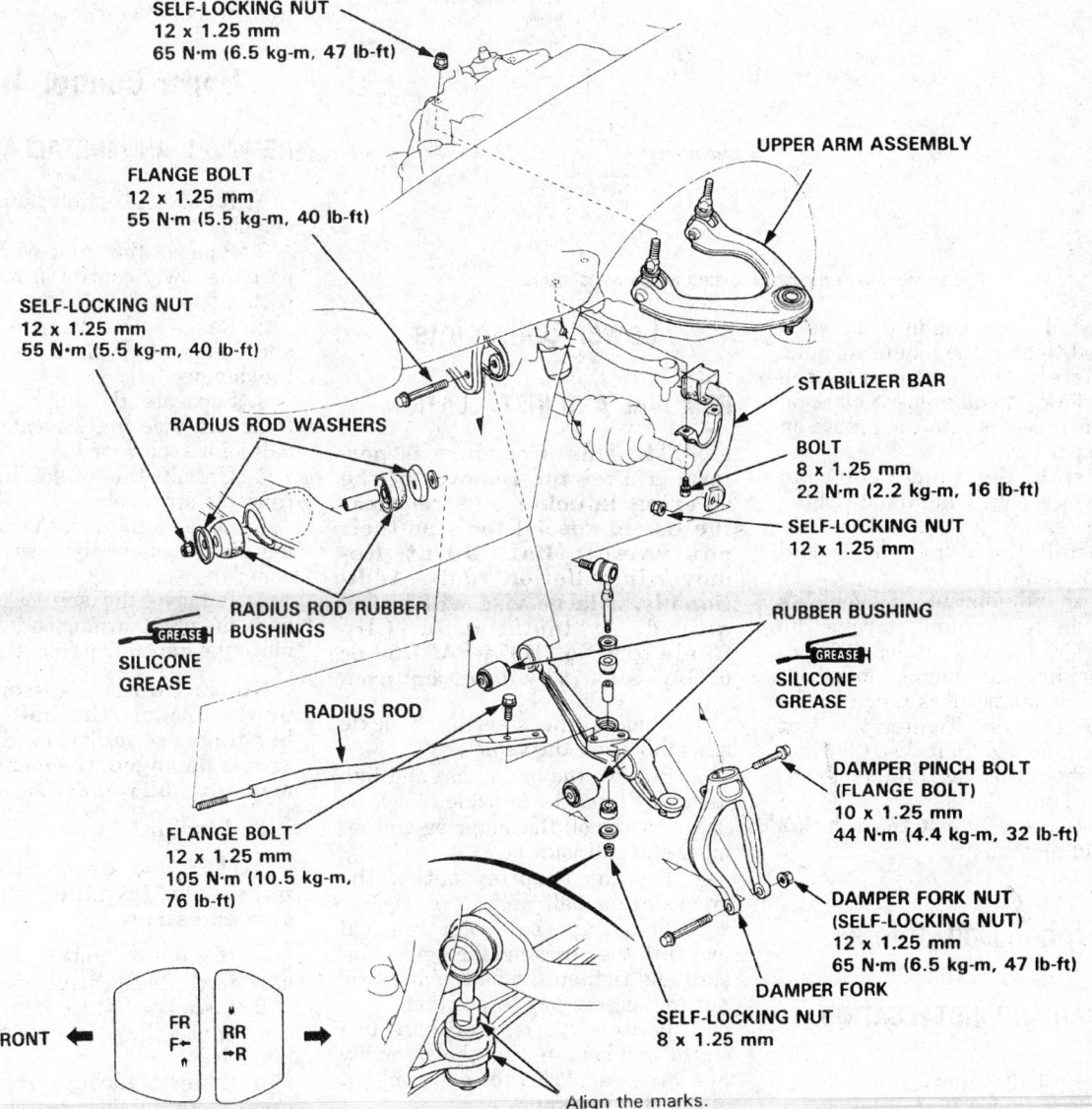

SELF-LOCKING NUT
12 x 1.25 mm
65 N·m (6.5 kg-m, 47 lb-ft)

FLANGE BOLT
12 x 1.25 mm
55 N·m (5.5 kg-m, 40 lb-ft)

SELF-LOCKING NUT
12 x 1.25 mm
55 N·m (5.5 kg-m, 40 lb-ft)

RADIUS ROD WASHERS

RADIUS ROD RUBBER BUSHINGS

GREASE
SILICONE GREASE

RADIUS ROD

FLANGE BOLT
12 x 1.25 mm
105 N·m (10.5 kg-m, 76 lb-ft)

UPPER ARM ASSEMBLY

STABILIZER BAR

BOLT
8 x 1.25 mm
22 N·m (2.2 kg-m, 16 lb-ft)

SELF-LOCKING NUT
12 x 1.25 mm

RUBBER BUSHING

GREASE
SILICONE GREASE

DAMPER PINCH BOLT
(FLANGE BOLT)
10 x 1.25 mm
44 N·m (4.4 kg-m, 32 lb-ft)

DAMPER FORK NUT
(SELF-LOCKING NUT)
12 x 1.25 mm
65 N·m (6.5 kg-m, 47 lb-ft)

DAMPER FORK

SELF-LOCKING NUT
8 x 1.25 mm

FRONT ←

FR F←

RR →R

Align the marks.

289162

Front suspension components

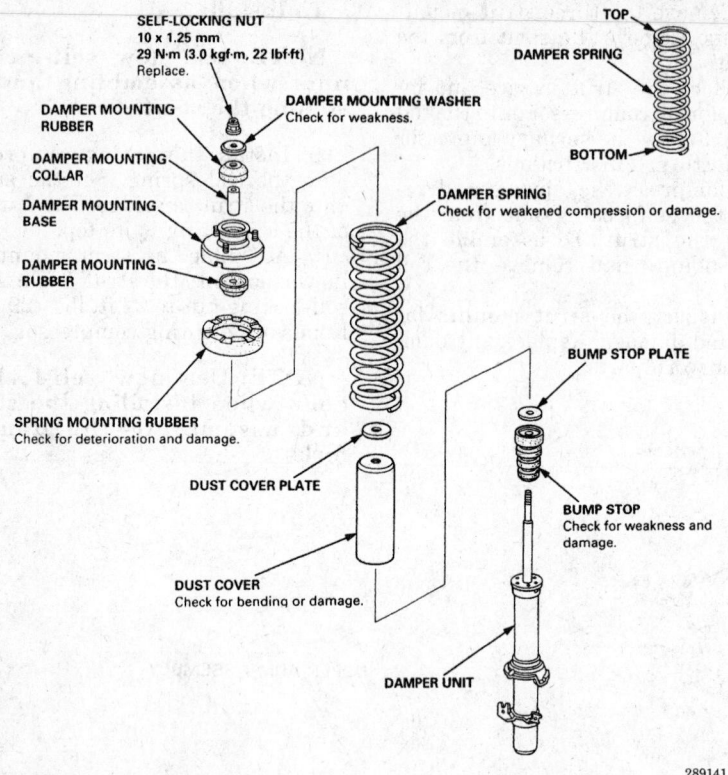

SELF-LOCKING NUT
10 x 1.25 mm
29 N·m (3.0 kgf·m, 22 lbf·ft)
Replace.

DAMPER MOUNTING RUBBER

DAMPER MOUNTING COLLAR

DAMPER MOUNTING BASE

DAMPER MOUNTING RUBBER

DAMPER MOUNTING WASHER
Check for weakness.

TOP

DAMPER SPRING

BOTTOM

DAMPER SPRING
Check for weakened compression or damage.

SPRING MOUNTING RUBBER
Check for deterioration and damage.

DUST COVER PLATE

BUMP STOP PLATE

BUMP STOP
Check for weakness and damage.

DUST COVER
Check for bending or damage.

DAMPER UNIT

289141

Coil spring, strut cartridge, and strut mount components

11. Install the strut into the vehicle. Hand-tighten the mounting nuts.

12. Install the strut into the damper fork. The alignment mark on the strut tube fits into the groove on the damper fork.

13. Install the pinch bolt and damper fork bolt. Only hand-tighten these bolts.

14. Install the front wheels and lower the vehicle.

15. With all four of the vehicle's wheels on the ground, torque the damper fork nut to 47 ft. lbs. (65 Nm) while holding the damper fork bolt. Torque the damper fork pinch bolt to 32 ft. lbs. (44 Nm). Tighten the strut mounting nuts to 28 ft. lbs. (39 Nm).

16. Tighten the wheel nuts to 80 ft. lbs. (110 Nm).

17. Check and adjust the vehicle's front end alignment.

Upper Ball Joints

REMOVAL AND INSTALLATION

The upper ball joints cannot be replaced separately. If the ball joints become worn or damaged, the whole upper arm must be replaced. See Upper Control Arm procedures, later.

Lower Ball Joints

REMOVAL AND INSTALLATION

NOTE: This procedure is performed after the removal of the steering knuckle and requires the use of special tools or their equivalent: Ball Joint Remover/Installation tools. Additionally, a large vise will be required. At installation, Clip Guide tool No. 07974–SA50700 or 07GAG-SD40700 will be required.

1. Remove the steering knuckle assembly from the vehicle.

2. Remove the brake disc and hub assembly from the knuckle.

3. Pry the off the snapring and remove the ball joint boot.

4. Pry the snapring out of the groove in the ball joint.

5. Install the ball joint removal tool with the large end facing out. Install and tighten the ball joint castle nut to hold the tool in position.

6. Position the removal base tool on the ball joint and set the assembly in a large vise. Press the ball joint out of the steering knuckle.

To Install:

7. Position the new ball joint into the hole of the steering knuckle.

8. Install the ball joint installer tool with the small end facing out.

9. Position the installation base tool on the ball joint and set the assembly in a large vise. Press the ball joint into the steering knuckle.

10. Seat the snapring in the groove of the ball joint.

11. Pack the interior of the ball joint boot with grease.

12. Adjust the boot clip guide tool with its adjusting bolt until the end of the tool aligns with the groove on the boot. Slide the clip over the tool and into position on the ball joint boot.

13. Install the hub assembly and brake disc onto the knuckle.

14. Install the steering knuckle assembly into the vehicle.

15. Check and adjust the vehicle's front wheel alignment.

Upper Control Arms

REMOVAL AND INSTALLATION

1. Raise and safely support the vehicle.

2. Remove the front wheels. Support the lower control arm assembly with a floor jack.

3. Remove the damper fork bolt and damper fork pinch bolt. Remove the damper fork.

4. Separate the upper ball joint from the steering knuckle using a ball joint separator tool.

5. Unbolt the brake hose clips from the strut tube.

6. Remove the three strut mounting nuts. Remove the strut from the vehicle.

7. Remove the self-locking nuts from the upper arm anchor bolts. Remove the upper arm from the vehicle.

NOTE: Do not disassemble the upper arm. If the ball joint or bushings are faulty, or the upper arm is damaged, the entire upper arm must be replaced.

To Install:

NOTE: Use new self-locking nuts when installing the upper arm and strut.

8. Install the upper control arm assembly into the strut tower.

9. Install the strut into the vehicle. Connect the damper fork bolt and pinch bolt.

10. Connect the upper ball joint. Connect the brake hose clips to the strut tube.

11. Install the front wheels and lower the vehicle.

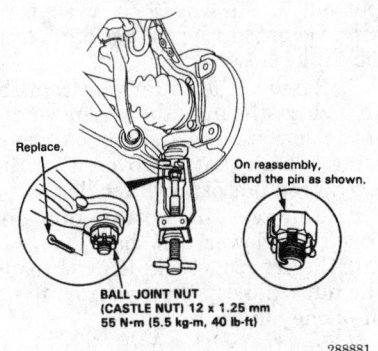

Lower Ball Joint

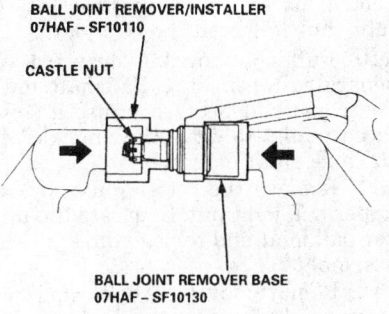

Removing ball joint from steering knuckle

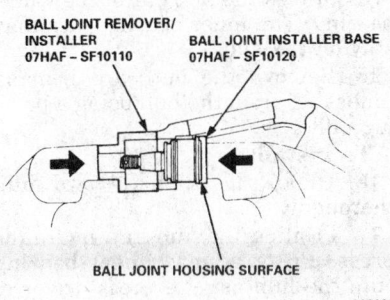

Installing new ball joint into steering knuckle

12. With all four of the vehicle's wheels on the ground, torque the upper control arm nuts to 47 ft. lbs. (65 Nm). Torque the strut mounting nuts to 28 ft. lbs. (39 Nm). Torque the damper fork pinch bolt to 32 ft. lbs. (44 Nm) and the damper fork bolt to 47 ft. lbs. (65 Nm). Torque the castle nut to 32 ft. lbs. (44 Nm); then, tighten it only enough to install a new cotter pin.

13. Tighten the wheel nuts to 80 ft. lbs. (110 Nm).

14. Check and adjust the vehicle's front end alignment.

Ball joint boot clip guide

Lower Control Arms

REMOVAL AND INSTALLATION

1. Raise and safely support the vehicle.

2. Remove the front wheels.

3. Disconnect the sway bar link from the lower control arm.

4. Remove the cotter pin and nut from the lower ball joint. Disconnect the ball joint from the lower control arm a using ball joint separator, tool 07MAC – SL00200, or equivalent.

5. Remove the nut from the radius rod front beam bushing.

6. Unbolt the radius rod from the lower control arm.

7. Support the lower control arm with a floor jack and remove the damper fork bolt.

8. Unbolt the lower control arm from the subframe and remove it from the vehicle.

To install:

NOTE: Use new self-locking nuts when installing the lower control arm.

————— **WARNING** —————

All suspension fasteners must be tightened to torque specifications with all four of the vehicle's wheels on the ground. This step is important for the preloading of the suspension.

9. Install the lower control arm and install the radius rod bolts.

10. Install the lower control subframe flange bolt and tighten to 40 ft. lbs. (55 Nm). Tighten the radius rod bolts to 76 ft. lbs. (105 Nm). Compress the strut with a floor jack and install the damper fork bolt and tighten its flange bolt to 47 ft. lbs. (64 Nm).

11. Connect the lower ball joint and the lower control arm and install the

nut. Tighten the nut to 40 ft. lbs. (55 Nm) and install a new cotter pin.

NOTE: If the hole in the ball joint stud does not align with the nut castellation, further tighten the nut until the cotter pin can be installed; never loosen the nut to allow cotter pin installation.

12. Attach the sway bar link and tighten the nut to 14 ft. lbs. (19 Nm).

13. Install the front wheels and lower the vehicle.

Sway Bar

REMOVAL AND INSTALLATION

1. Raise and safely support the vehicle.

2. Remove the front wheels.

3. Disconnect the stabilizer bar linkages from both lower control arms.

4. Remove the mounting bolts and the stabilizer bar bushing brackets.

5. Remove the stabilizer bar.

NOTE: Examine the rubber bushings very carefully for any splits or deformation. Clean the inner and outer surfaces of the bushings before installation.

To Install:

NOTE: Use new self-locking nuts when installing the stabilizer bar.

6. Make certain the bushings are properly seated in their brackets.

7. Install the stabilizer bar bushing brackets to the subframe.

8. Connect the stabilizer bar linkages to the control arms. Only hand-tighten the nuts and bolts.

9. Install the front wheels.

10. With the vehicle on the ground, tighten the brackets bolts to 16 Nm (12 ft. lbs.) and the linkage bolts to 22 Nm (16 ft. lbs.).

11. Tighten the wheel nuts to 80 ft. lbs. (110 Nm).

Front Wheel Bearings

ADJUSTMENT

1. Raise and support the vehicle safely.

2. Remove the front and/or rear wheels.

3. Install the lug nuts and tighten them to 80 ft. lbs. (110 Nm).

4. Use a dial gauge to measure front bearing end play at the hub flange.

Front/Rear:
Standard: 0 – 0.05 mm (0 – 0.002 in)

Front

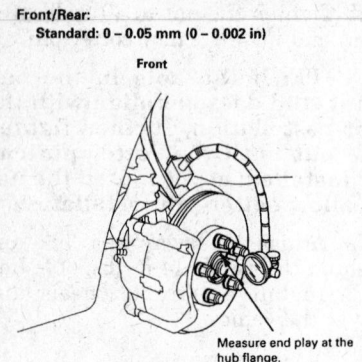

Measure end play at the hub flange.

Rear

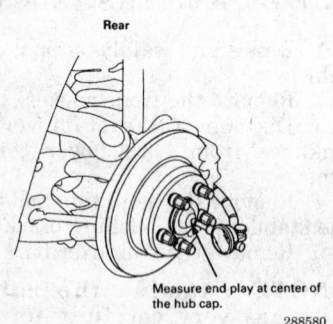

Measure end play at center of the hub cap.

288580

Wheel bearing inspection with a dial gauge

5. Use a dial gauge to measure rear bearing end play at the center of the hub's grease cap.

6. Pull the rotor assembly in and out to measure the bearing play. Compare the dial gauge readings.

7. The standard bearing end play for both front and rear wheels is 0–0.002 in. (0–0.05 mm). If the end play measurement exceeds the standard, the wheel bearings must be replaced. The wheel bearings cannot be adjusted.

REMOVAL AND INSTALLATION

NOTE: Once the wheel bearing is removed, it must be replaced. A hydraulic press and bearing drivers are required to remove and install the wheel bearing. The following Honda tools or their equivalents are needed: hub assembly tool, 07GAF-SE0100; hub bases, 07965–SD90100; bearing driver, 07749-0010000; 52 x 55mm driving attachment, 07746-0010400.

1. Pry the spindle nut stake away from the spindle, then loosen the nut. Do not tighten or loosen a spindle nut unless the vehicle is sitting on all four wheels. The torque is high

Press

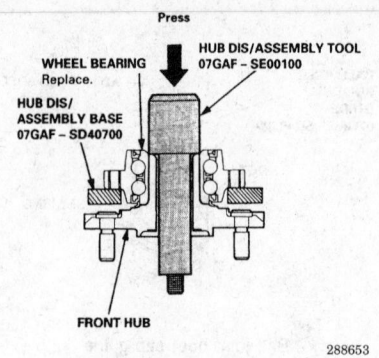

WHEEL BEARING
Replace.

HUB DIS/
ASSEMBLY BASE
07GAF – SD40700

HUB DIS/ASSEMBLY TOOL
07GAF – SE00100

FRONT HUB

288653

Hub disassembly tools

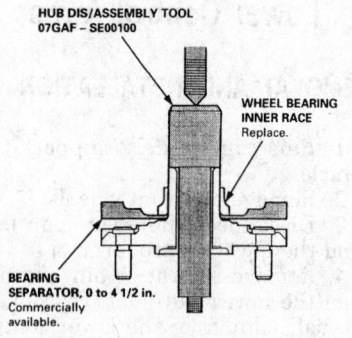

HUB DIS/ASSEMBLY TOOL
07GAF – SE00100

WHEEL BEARING
INNER RACE
Replace.

BEARING
SEPARATOR, 0 to 4 1/2 in.
Commercially
available.

288654

Wheel bearing inner race

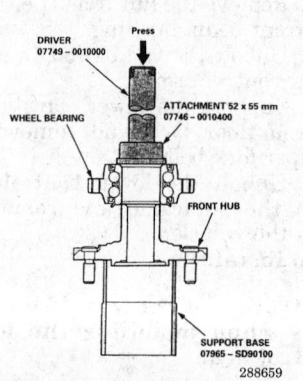

DRIVER
07749 – 0010000

Press

WHEEL BEARING

ATTACHMENT 52 x 55 mm
07746 – 0010400

FRONT HUB

SUPPORT BASE
07965 – SD90100

288659

Hub support base

enough to cause the vehicle to fall even when properly supported.

2. Raise and safely support the vehicle.

3. Remove the wheel and the spindle nut.

4. Unbolt the ABS wheel sensor and its cable from the knuckle. Don't disconnect the cable, wire it out of the way.

5. Remove the brake disc. If the disc if seized, evenly screw two 8mm bolts into the threaded holes to pop it loose.

6. Remove the caliper mounting bolts and the caliper. Support the cal-

iper out of the way with a length of wire. Do not let the caliper hang from the brake hose.

7. Remove the cotter pin from the tie rod castle nut, then remove the nut. Separate the tie rod ball joint using a ball joint remover, then lift the tie rod out of the knuckle.

8. Remove the cotter pin and loosen the lower arm ball joint nut half the length of the joint threads. The nut will keep the arm from flying off of the joint.

9. Separate the ball joint and lower arm using a ball joint puller with the pawls applied to the lower arm. Avoid damaging the ball joint boot. If necessary, apply penetrating lubricant to loosen the ball joint.

10. Pull the knuckle outward to separate it from the halfshaft outboard joint. If necessary, use a soft-faced mallet to drive the knuckle off the axle shaft.

11. Remove the cotter pin and the upper ball joint nut. Separate the upper ball joint and remove the knuckle assembly.

12. Remove the four bolts and remove the hub unit from the knuckle.

13. Remove the splash guard from the knuckle.

14. Position the hub in a hydraulic press. Press the hub out of the wheel bearing. The inner bearing race may stay on the hub.

15. Remove the outboard bearing inner race from the hub using a bearing puller.

To install:

16. Clean the knuckle and hub thoroughly.

17. Position the hub in a hydraulic press. Press a new wheel bearing onto the hub using a press driver of the correct diameter. Make sure the press tool contacts only the inner bearing race.

18. Install the splash shield.

19. Install the hub/bearing assembly onto the knuckle and torque the bolts to 33ft. lbs. (45 Nm).

20. Install the knuckle/hub assembly on the vehicle. Make sure the hub is fully seated onto the axle shaft. Torque the upper ball joint nut and the tie rod nut to 32 ft. lbs. (44 Nm) and install new cotter pins. Tighten the lower ball joint nut to 40 ft. lbs. (55 Nm) and install a new cotter pin.

21. Install the brake disc and caliper. Torque the brake caliper bolts to 80 ft. lbs. (110 Nm). Torque the brake disc retaining screws to 7ft. lbs. (10 Nm).

22. Install the front wheels and lower the vehicle.

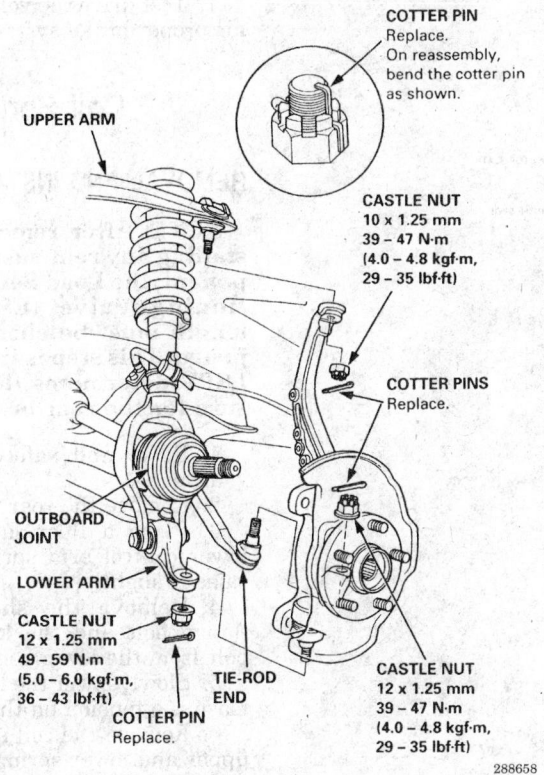

COTTER PIN
Replace.
On reassembly,
bend the cotter pin
as shown.

UPPER ARM

CASTLE NUT
10 x 1.25 mm
39 – 47 N·m
(4.0 – 4.8 kgf·m,
29 – 35 lbf·ft)

COTTER PINS
Replace.

OUTBOARD
JOINT

LOWER ARM

CASTLE NUT
12 x 1.25 mm
49 – 59 N·m
(5.0 – 6.0 kgf·m,
36 – 43 lbf·ft)

COTTER PIN
Replace.

TIE-ROD
END

CASTLE NUT
12 x 1.25 mm
39 – 47 N·m
(4.0 – 4.8 kgf·m,
29 – 35 lbf·ft)

288658

Steering knuckle components

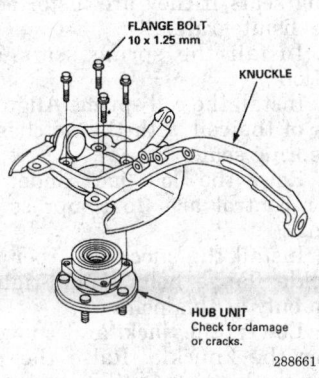

FLANGE BOLT
10 x 1.25 mm

KNUCKLE

HUB UNIT
Check for damage
or cracks.

288661

Hub assembly and knuckle

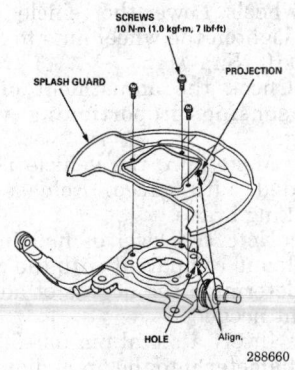

SCREWS
10 N·m (1.0 kgf·m, 7 lbf·ft)

SPLASH GUARD

PROJECTION

HOLE   Align.

288660

Splash guard

23. With all four wheels resting on the ground, install a new spindle nut and torque it to 180 ft. lbs. (245 Nm). After tightening, use a drift to stake the spindle nut shoulder against the spindle.

24. Tighten the wheel nuts to 80 ft. lbs. (110 Nm).

25. Check and adjust the vehicle's front wheel alignment.

REAR SUSPENSION

Strut

REMOVAL AND INSTALLATION

NOTE: After removing and installing any rear suspension component, the Load Sensing Proportioning Valve (LSPV) spring length must be checked and adjusted. This step is important: the LSPV determines the fluid pressure for the rear brakes.

1. Remove the cup holder from the top of the right rear interior trim. Remove the cup holder, storage tray, and jack from the left rear interior trim.

2. Raise and support the vehicle safely.

3. Remove the rear wheels.

4. Place a floor jack under the lower control arm and raise it slightly to compress the spring.

5. Remove the lower shock mount flange bolt and the the knuckle flange bolt.

6. Unbolt the shock mount from inside the vehicle. Remove the shock from the vehicle.

To Install:

7. Check the shock mount and bushings and replace any that are damaged. Assemble the mount, bushing, and stopper on the shock.

8. Install the shock into the vehicle. Install new self-locking upper mounting nuts.

9. Raise the lower control arm with a floor jack. Make sure the coil spring is properly seated.

10. Install the shock absorber and knuckle flange bolts.

11. Raise the jack enough to take up the weight of the vehicle. Tighten both of the flange bolts to 76 ft. lbs. (103 Nm).

12. Lower the floor jack. Install the rear wheels. Lower the vehicle.

13. Tighten the shock mount nuts to 28 ft. lbs. (39 Nm), and the shock piston nut to 22 ft. lbs. (29 Nm). Install the jack, tool tray, and cup holders.

14. Tighten the wheel nuts to 80 ft. lbs. (110 Nm).

15. Check the adjustment of the load sensing proportioning valve spring:

   a. Make sure the vehicle is not loaded with cargo. Release the parking brake.

   b. Note the level of fuel in the tank and compare it with the chart to determine the degree of adjustment needed.

   c. Insert a metal pin 5.0–5.3mm in diameter into the 5mm diameter hole in the LSPV arm.

   d. Use a caliper to measure the distance between the 5mm pin and the 8mm adjusting bolt thread. This is the length of the LSPV spring.

   e. If the measurement is out of specification, loosen the 8mm adjusting bolt and adjust the spring length to specification according to the values on the chart. Tighten the adjusting nut to 9 ft. lbs. (12 Nm).

16. Check and adjust the rear wheel alignment.

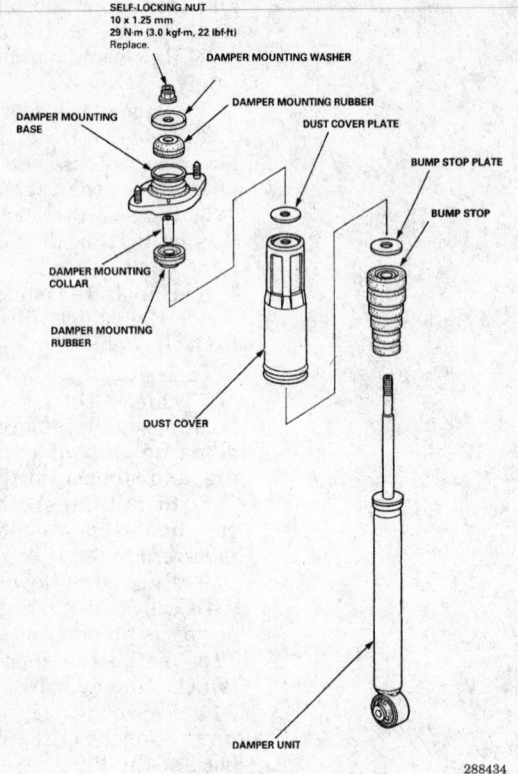

Shock absorber components

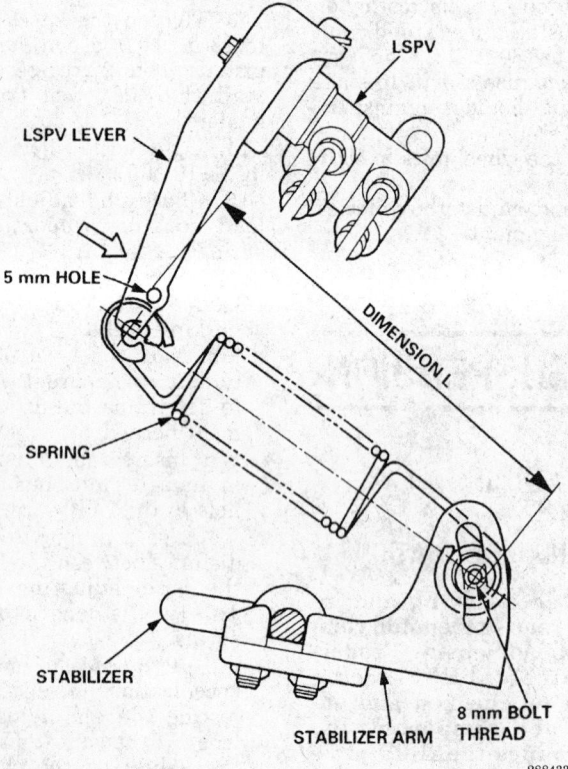

Load sensing proportioning valve spring length and adjusting nut

17. Test drive the vehicle and check for proper brake system operation.

Coil Spring

REMOVAL AND INSTALLATION

NOTE: After removing and installing any rear suspension component, the Load Sensing Proportioning Valve (LSPV) spring length must be checked and adjusted. This step is important: the LSPV determines the fluid pressure for the rear brakes.

1. Raise and safely support the vehicle.
2. Remove the rear wheels.
3. Place a floor jack under the lower control arm spring perch and raise it slightly.
4. Remove the shock absorber flange bolt and the knuckle flange bolt from the lower control arm.
5. Slowly lower the floor jack to release the tension on the coil spring.
6. Remove the coil spring and the upper and lower spring seats.
   To Install:
7. Replace the upper and lower spring seats if they are distorted or have disintegrated.
8. Install the spring seats into position.
9. Install the coil spring. Align the ends of the coil with the notches on the spring seats.
10. Raise the floor jack under the lower control arm to compress the spring.
11. Install the shock absorber and knuckle flange bolts. Hand-tighten them only at this point.
12. Lower the jack and move it under the knuckle. Raise the jack under the knuckle until it is supporting the weight of the vehicle. Tighten each of the two flange bolts to 76 ft. lbs. (103 Nm).
13. Lower the floor jack. Install the rear wheels. Lower the vehicle.
14. Tighten the wheel nuts to 80 ft. lbs. (110 Nm).
15. Check the adjustment of the load sensing proportioning valve spring:
    a. Make sure the vehicle is not loaded with cargo. Release the parking brake.
    b. Note the level of fuel in the tank and compare it with the chart to determine the degree of adjustment needed.
    c. Insert a metal pin 5.0–5.3mm in diameter into the 5mm diameter hole in the LSPV arm.

Example:

Type: LX, 7–passenger
Fuel level: 1/2

The table shows that dimension L of fully-fueled 7–passenger LX 7 is 132 mm (5.197 in). However, because the fuel level is 1/2, dimension L should be compensated by –2 mm.
Therefore, it should be adjusted at 130 mm (5.118 in).

U.S.A. MODEL:

mm (in)

Type	Dimension L	Fuel Level			
		3/4	1/2	1/4	0
LX 6-pass.	131 (5.157)	– 1 (– 0.039)	– 2 (– 0.079)	– 3 (– 0.012)	– 4 (– 0.157)
LX 7-pass.	132 (5.197)	↑	↑	↑	↑
EX	131 (5.157)	↑	↑	↑	↑

CANADA MODEL:

mm (in)

Passenger	Dimension L	Fuel Level			
		3/4	1/2	1/4	0
6-pass.	131 (5.157)	– 1 (– 0.039)	– 2 (– 0.079)	– 3 (– 0.012)	– 4 (– 0.157)
7-pass.	132 (5.197)	↑	↑	↑	↑

NOTE: If the vehicle is equipped with a trailer hitch, add 3 mm to dimension L before compensating for the fuel level.

288435

Load sensing proportioning valve Specifications Chart

d. Use a caliper to measure the distance between the 5mm pin and the 8mm adjusting bolt thread. This is the length of the LSPV spring.

e. If the measurement is out of specification, loosen the 8mm adjusting bolt and adjust the spring length to specification according to the values on the chart. Tighten the adjusting nut to 9 ft. lbs. (12 Nm).

16. Check and adjust the rear wheel alignment.

17. Test drive the vehicle and check for proper brake system operation.

Upper Control Arms

REMOVAL AND INSTALLATION

NOTE: After removing and installing any rear suspension component, the Load Sensing Proportioning Valve (LSPV) spring length must be checked and adjusted. This step is important: the LSPV determines the fluid pressure for the rear brakes.

1. Raise and safely support the vehicle.
2. Remove the rear wheels.

3. Support the lower control arm with a jack.
4. Use a ball joint separator tool to disconnect the upper control arm ball joint from the knuckle.
5. Hold the yoke nut with a box-end wrench while loosening the flange bolt. Remove the flange bolt; then, remove the upper control arm.

To Install:

NOTE: If the ball joint is faulty, the entire upper arm must be replaced. However, the ball joint boot can be replaced if the joint is not worn.

6. Fit the upper control arm into position. Install the flange bolt while holding the yoke nut with a wrench. Only hand-tighten the bolt at this time.
7. Connect the ball joint to the knuckle. Install the castle nut and tighten it to 29–35 ft. lbs. (39–47 ft. lbs.). Then, tighten the castle nut only enough to install a new cotter pin.
8. Place a floor jack under the knuckle and raise it until it is supporting the weight of the vehicle. Tighten the flange bolt to 76 ft. lbs. (103 Nm).
9. Lower the floor jack. Install the rear wheels. Lower the vehicle.
10. Tighten the wheel nuts to 80 ft. lbs. (110 Nm).
11. Check the adjustment of the load sensing proportioning valve spring:
   a. Make sure the vehicle is not loaded with cargo. Release the parking brake.
   b. Note the level of fuel in the tank and compare it with the chart to determine the degree of adjustment needed.
   c. Insert a metal pin 5.0–5.3mm in diameter into the 5mm diameter hole in the LSPV arm.
   d. Use a caliper to measure the distance between the 5mm pin and the 8mm adjusting bolt thread. This is the length of the LSPV spring.
   e. If the measurement is out of specification, loosen the 8mm adjusting bolt and adjust the spring length to specification according to the values on the chart. Tighten the adjusting nut to 9 ft. lbs. (12 Nm).
12. Check and adjust the vehicle's rear wheel alignment.
13. Test drive the vehicle and check for proper brake system operation.

Lower Control Arms

REMOVAL AND INSTALLATION

NOTE: After removing and installing any rear suspension component, the Load Sensing Proportioning Valve (LSPV) spring length must be checked and adjusted. This step is important: the LSPV determines the fluid pressure for the rear brakes.

1. Raise and safely support the vehicle.
2. Remove the rear wheels.
3. Place a floor jack under lower control arm **B's** spring perch and raise it slightly.
4. Remove the shock absorber flange bolt and the knuckle flange bolt from the lower control arm.
5. Slowly lower the floor jack to release the tension on the coil spring.
6. Remove the lower control arm **B** bracket flange bolt. Note the position of the alignment shims. Remove lower control arm B.
7. Remove the nut securing lower arm **A** to the rear knuckle. Then, remove the flange bolt from lower control arm **A's** bracket. Remove lower control arm **A**.

To Install:

NOTE: Use new self-locking nuts when installing both lower control arms.

8. Inspect the lower control arms for signs of corrosion, distortion, or other damage. Replace any damaged parts.
9. Install lower control arm **A** and hand-tighten its nut and flange bolt.
10. Fit lower control arm **B** into position, and install its flange bolt, alignment shims, and a new self-locking nut. Only hand-tighten these fasteners.
11. Install the spring seats into position.
12. Install the coil spring. Align the ends of the coil with the notches on the spring seats.
13. Raise the floor jack under the lower control arm to compress the spring.
14. Install the shock absorber and knuckle flange bolts. Only hand-tighten them at this time.
15. Lower the jack and move it under the knuckle. Raise the jack under the knuckle until it is supporting the weight of the vehicle. Tighten each of the two flange bolts to 76 ft. lbs. (103 Nm). Tighten the control arm **B** self-locking nut to 40 ft. lbs. (54 Nm). Tighten the control arm **A** flange bolt to 69 ft. lbs. (93 Nm); and

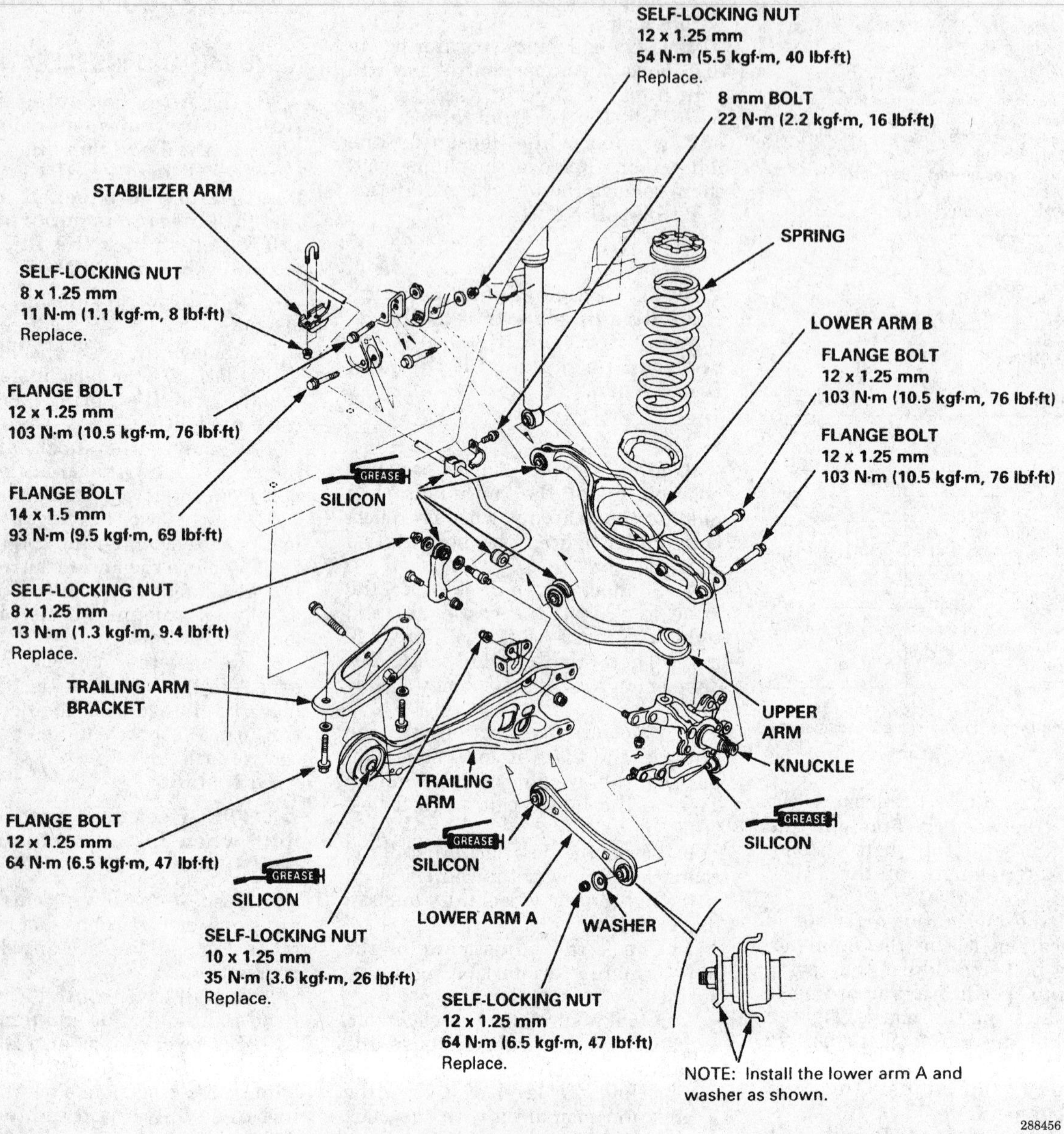

SELF-LOCKING NUT
12 x 1.25 mm
54 N·m (5.5 kgf·m, 40 lbf·ft)
Replace.

8 mm BOLT
22 N·m (2.2 kgf·m, 16 lbf·ft)

STABILIZER ARM

SPRING

SELF-LOCKING NUT
8 x 1.25 mm
11 N·m (1.1 kgf·m, 8 lbf·ft)
Replace.

LOWER ARM B

FLANGE BOLT
12 x 1.25 mm
103 N·m (10.5 kgf·m, 76 lbf·ft)

FLANGE BOLT
12 x 1.25 mm
103 N·m (10.5 kgf·m, 76 lbf·ft)

FLANGE BOLT
12 x 1.25 mm
103 N·m (10.5 kgf·m, 76 lbf·ft)

FLANGE BOLT
14 x 1.5 mm
93 N·m (9.5 kgf·m, 69 lbf·ft)

GREASE
SILICON

SELF-LOCKING NUT
8 x 1.25 mm
13 N·m (1.3 kgf·m, 9.4 lbf·ft)
Replace.

TRAILING ARM
BRACKET

UPPER
ARM

KNUCKLE

GREASE
SILICON

FLANGE BOLT
12 x 1.25 mm
64 N·m (6.5 kgf·m, 47 lbf·ft)

TRAILING
ARM

GREASE
SILICON

GREASE
SILICON

LOWER ARM A

WASHER

SELF-LOCKING NUT
10 x 1.25 mm
35 N·m (3.6 kgf·m, 26 lbf·ft)
Replace.

SELF-LOCKING NUT
12 x 1.25 mm
64 N·m (6.5 kgf·m, 47 lbf·ft)
Replace.

NOTE: Install the lower arm A and
washer as shown.

288456

Rear suspension components

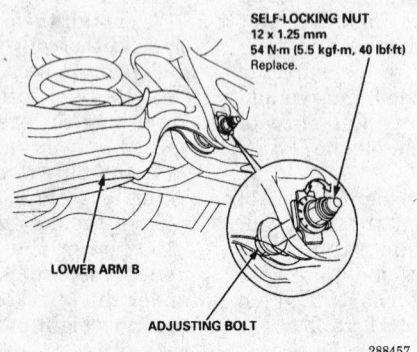

SELF-LOCKING NUT
12 x 1.25 mm
54 N·m (5.5 kgf·m, 40 lbf·ft)
Replace.

LOWER ARM B

ADJUSTING BOLT

288457

Lower control arm B alignment shims

the control arm **A** self-locking nut to
47 ft. lbs. (64 Nm).

16. Lower the floor jack. Install the
rear wheels. Lower the vehicle.

17. Tighten the wheel nuts to 80 ft.
lbs. (110 Nm).

18. Check the adjustment of the
load sensing proportioning valve
spring:

a. Make sure the vehicle is not
loaded with cargo. Release the
parking brake.

b. Note the level of fuel in the
tank and compare it with the chart
to determine the degree of adjustment needed.

c. Insert a metal pin 5.0–5.3mm in diameter into the 5mm diameter hole in the LSPV arm.

d. Use a caliper to measure the distance between the 5mm pin and the 8mm adjusting bolt thread. This is the length of the LSPV spring.

e. If the measurement is out of specification, loosen the 8mm adjusting bolt and adjust the spring length to specification according to the values on the chart. Tighten the adjusting nut to 9 ft. lbs. (12 Nm).

19. Check and adjust the vehicle's rear wheel alignment.

20. Test drive the vehicle and check for proper brake system operation.

Sway Bar

REMOVAL AND INSTALLATION

NOTE: After removing and installing any rear suspension component, the Load Sensing Proportioning Valve (LSPV) spring length must be checked and adjusted. This step is important: the LSPV determines the fluid pressure for the rear brakes.

1. Raise and safely support the vehicle.

2. Remove the rear wheels.

3. Remove the self-locking nuts securing the LSPV arm U-bolt to the stabilizer bar. Remove the U-bolt.

4. Unbolt the stabilizer links from the trailing arms.

5. Unbolt the stabilizer bar brackets. Remove the stabilizer bar

6. Remove the bushings from the stabilizer bar. Inspect them for cracks and distortion. Replace any damaged parts.

To Install:

NOTE: Use new self-locking nuts when installing the stabilizer bar.

7. Assemble the bushings onto the stabilizer bar.

8. Fit the flat part of the stabilizer bar into the recess of the LSPV arm. Install the U-bolt and loosely tighten the nuts.

9. Install the stabilizer bar brackets and tighten the mounting bolts to 16 ft. lbs. (22 Nm).

10. Connect the stabilizer links to the trailing arms. Tighten the self-locking nuts to 26 ft. lbs. (35 Nm).

11. Tighten the LSPV U-bolt nuts to 8 ft. lbs. (11 Nm).

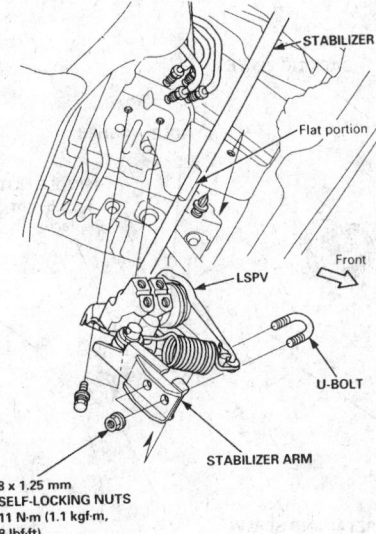

8 x 1.25 mm
SELF-LOCKING NUTS
11 N·m (1.1 kgf·m, 8 lbf·ft)

288418

LSPV assembly, arm, and U-bolt

12. Install the rear wheels and lower the vehicle.

13. Tighten the wheel nuts to 80 ft. lbs. (110 Nm).

14. Check the adjustment of the load sensing proportioning valve spring:

a. Make sure the vehicle is not loaded with cargo. Release the parking brake.

b. Note the level of fuel in the tank and compare it with the chart to determine the degree of adjustment needed.

c. Insert a metal pin 5.0–5.3mm in diameter into the 5mm diameter hole in the LSPV arm.

d. Use a caliper to measure the distance between the 5mm pin and the 8mm adjusting bolt thread. This is the length of the LSPV spring.

e. If the measurement is out of specification, loosen the 8mm adjusting bolt and adjust the spring length to specification according to the values on the chart. Tighten the adjusting nut to 9 ft. lbs. (12 Nm).

15. Test drive the vehicle and check for proper brake system operation.

Rear Wheel Bearings

ADJUSTMENT

1. Raise and support the vehicle safely.

2. Remove the rear wheels.

3. Install the lug nuts and tighten them to 80 ft. lbs. (110 Nm).

4. Use a dial gauge to measure front bearing end play at the hub flange.

5. Use a dial gauge to measure rear bearing end play at the center of the hub's grease cap.

6. Pull the rotor assembly in and out to measure the bearing play. Compare the dial gauge readings.

7. The standard bearing end play is 0–0.002 in. (0–0.05 mm). If the end play measurement exceeds the standard, the wheel bearings must be replaced. The wheel bearings cannot be adjusted.

REMOVAL AND INSTALLATION

1. Loosen the hub spindle nut.

2. Raise and safely support the vehicle and remove the rear wheels.

3. Engage the parking brake.

4. Remove the caliper bracket mounting bolts. Use a piece of wire to hang the caliper out of the way.

5. Remove the two 6mm brake rotor retaining screws. If the rotor has seized onto the hub, screw two 8mm bolts into the threaded holes to pop the rotor loose.

6. Release the parking brake. Remove the brake rotor.

7. Remove the spindle nut and washer. Remove the hub unit.

To Install:

8. Clean the hub unit in solvent. Inspect the hub unit and wheel bearing for damage.

NOTE: If the wheel bearing is faulty, the entire hub unit must be replaced.

9. Clean excess brake dust and grease from the backing plate and brake rotor.

10. Install the hub unit, spindle washer, and spindle nut. Only hand-tighten the spindle nut.

11. Install the brake rotor and tighten the retaining screws to 7 ft. lbs. (10 Nm).

12. Install the caliper and tighten the bracket mounting bolts to 28 ft. lbs. (39 Nm).

13. Install the rear wheels and lower the vehicle.

14. Tighten the spindle nut to 181 ft. lbs. (245 Nm).

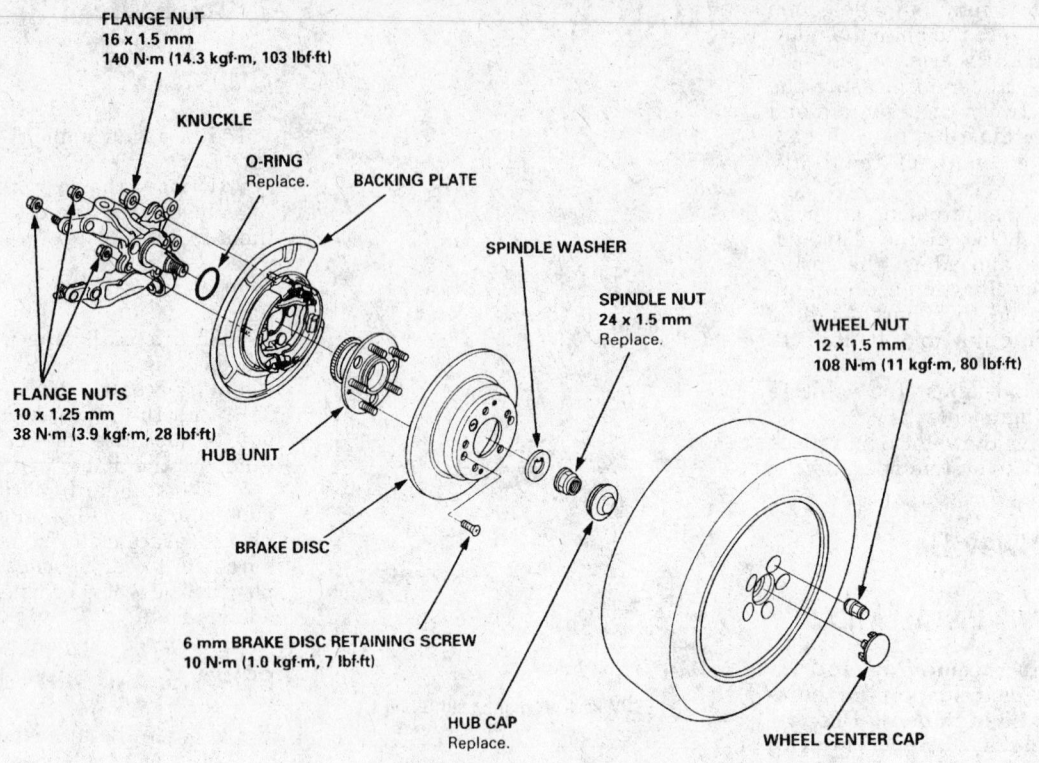

FLANGE NUT
16 x 1.5 mm
140 N·m (14.3 kgf·m, 103 lbf·ft)

KNUCKLE

O-RING
Replace.

BACKING PLATE

SPINDLE WASHER

SPINDLE NUT
24 x 1.5 mm
Replace.

WHEEL NUT
12 x 1.5 mm
108 N·m (11 kgf·m, 80 lbf·ft)

FLANGE NUTS
10 x 1.25 mm
38 N·m (3.9 kgf·m, 28 lbf·ft)

HUB UNIT

BRAKE DISC

6 mm BRAKE DISC RETAINING SCREW
10 N·m (1.0 kgf·m, 7 lbf·ft)

HUB CAP
Replace.

WHEEL CENTER CAP

288496

Rear hub and wheel bearing components

15. Tighten the wheel nuts to 80 ft. lbs. (110 Nm).

16. Road test the vehicle and check the operation of the brakes.

HONDA and ISUZU

HONDA-Passport **Isuzu**-Amigo • Pick-Up • Rodeo

FIRING ORDERS

NOTE: To avoid confusion, always replace spark plug wires one at a time.

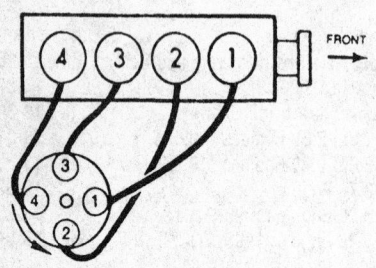

250202

2.3L (VIN L) and 2.6L (VIN E) Engines
Firing Order: 1-3-4-2
Distributor Rotation: Counterclockwise

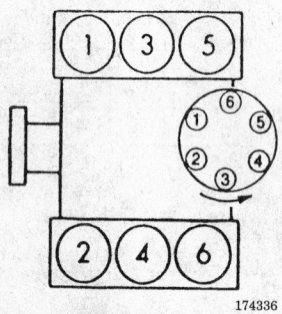

174336

3.1L (VIN Z) Engine
Engine Firing Order: 1-2-3-4-5-6
Distributor Rotation:
Counterclockwise

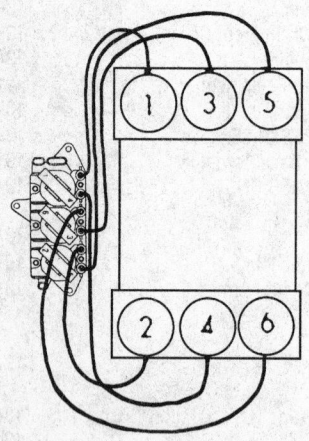

250210

3.2L (VIN V) Engine
Engine Firing Order: 1-2-3-4-5-6
Distributorless Ignition

ENGINE ELECTRICAL

Distributor

REMOVAL AND INSTALLATION

1. Rotate the engine and bring No. 1 cylinder to top dead center of its compression stroke.

NOTE: To bring the engine to TDC of the No. 1 compression stroke, remove the spark plug for the No. 1 cylinder. With the engine cool, turn the crankshaft over until compression is forced out of the spark plug hole. Watch the crankshaft damper while feeling for compression. When compression is felt, align the mark on the crankshaft damper with the O° mark on the timing cover.

2. Remove the air cleaner assembly, if needed.
3. Disconnect the negative battery cable. Disconnect and tag all the electrical connectors along with the spark plug wires from the distributor.

NOTE: Disconnecting the negative battery cable on some vehicles may interfere with the functions of the on-board computer systems and may require the computer to undergo a relearning process once the negative battery cable is reconnected.

4. Disconnect the vacuum hoses from the vacuum advance, if so equipped.
5. Make an alignment mark on the base of the distributor and the cylinder head or engine block. Also mark the position of the rotor to the distributor housing.
6. Remove the distributor holddown bolt and bracket, then lift out the distributor assembly. Remove the distributor housing seal and discard it. Lubricate the new seal with clean engine oil before installation.

To install:

Crankshaft Not Rotated

1. Install the distributor assembly. Be sure to align the scribe marks made during disassembly. Lightly tighten the distributor mounting bolt.

2. Connect the vacuum hoses to the vacuum controller.
3. Connect all the electrical connectors along with the spark plug wires to the distributor.
4. Install the air cleaner, if removed.
5. Connect the negative battery cable. Start the engine, set the timing and check the idle speed. Tighten the distributor mounting bolts to 14 ft. lbs. (19 Nm).

Crankshaft Rotated

1. Remove the No. 1 spark plug.
2. Rotate the crankshaft in the direction of rotation until compression is felt at the spark plug hole.
3. Continue rotating the engine in the same direction until the timing marks line up and when No. 1 cylinder is at TDC.
4. Align the rotor with the No. 1 tower on the distributor cap and install the distributor.
5. Reconnect the distributor wiring, ignition wires, and vacuum hoses.
6. Install the air cleaner, if removed. Connect the negative battery cable.
7. Check and/or adjust the ignition timing when finished.
8. Tighten the distributor mounting bolts to 14 ft. lbs. (19 Nm).

Ignition Timing

ADJUSTMENT

2.3L (VIN L) MFI and 2.6L (VIN E) Engines

NOTE: On carbureted engines, set the air gap in the distributor before timing the engine. The timing marks are located near the front crankshaft pulley and consist of a pointer with graduations attached to the engine block and a mark on the crankshaft pulley.

1. If equipped, check and correct the air gap in the distributor. Proper air gap should be 0.012–0.020 in. (0.30–0.50mm).
2. Connect a timing light to the No. 1 spark plug wire.
3. Set the parking brake and block the wheels.
4. Start the engine and allow it to warm up.
5. Make sure the air conditioner is **OFF**.
6. Disconnect and plug the evaporative emission canister purge line.
7. Disconnect and plug the exhaust gas recirculation vacuum lines.

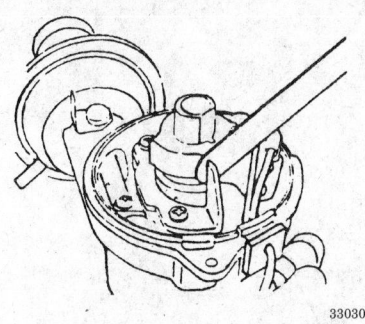

Use a feeler gauge to adjust the air gap — 2.3L (VIN L) carbureted engine

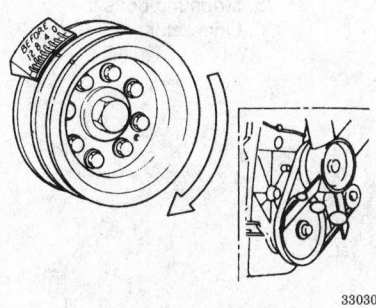

Timing mark location — 2.3L and 2.6L engines

8. While the engine idles, point the timing light at the notched line on the crankshaft pulley.

9. The ignition timing should be;

a. 6° BTDC at 900 rpm for carbureted engines with an automatic transmission.

b. 6° BTDC at 800 rpm for carbureted engines with a manual transmission.

c. 12° BTDC at 900 rpm for fuel injected models.

10. If adjustment is needed, loosen the distributor mounting bolts and turn the distributor counterclockwise to advance the timing or clockwise to retard the timing.

11. Tighten the distributor mounting bolts and recheck the timing and the idle.

NOTE: When tightening the distributor mounting bolt, make sure that the distributor body does not rotate together with the mounting bolt.

12. After everything has been rechecked, reconnect the vacuum lines and remove the timing light.

3.1L (VIN Z) Engine

1. Warm up the engine to normal operating temperature. Connect a

timing light to the No. 1 (left front) spark plug wire. You can also use the No. 6 wire, if it is more convenient.

NOTE: On vehicles equipped with HEI, do not pierce the spark plug wire insulation; doing so will cause the plug to misfire. The best method is an inductive pick-up timing light.

2. Clean off the timing marks and mark the pulley or damper notch and timing scale with white chalk.

3. Disconnect the timing harness which comes out of the harness conduit next to the distributor. This will put the IC system in the bypass mode. Check the underhood emission sticker for any other hoses or wires which may need to be disconnected.

4. Start the engine and adjust the idle speed if needed. With an automatic transmission, set the specified idle speed in **PARK**.

5. Connect a tachometer to the **TACH** terminal on the distributor.

——————— **WARNING** ———————
Never ground the HEI TACH terminal; serious system damage will result.

6. Aim the timing light at the pointer marks. Be careful not to touch the fan. If the pulley or damper notch isn't aligned with the proper timing mark (see the underhood emissions label), the timing will have to be adjusted.

7. Loosen the distributor base clamp locknut. Turn the distributor slowly to adjust the timing. Turn the distributor in the direction of rotor rotation to retard, and against the direction of rotation to advance.

8. Tighten the locknut. Check the timing again, in case the distributor moved slightly as you tightened it.

9. Reconnect the timing connector. Correct the idle speed if necessary.

10. Stop the engine and disconnect the timing light.

3.2L (VIN V) Engine

Vehicles with the (VIN V) engine are equipped with a distributorless ignition system. The ignition timing is controlled by the PCM through the input of engine control system sensors. The ignition timing is set at 5 degrees BTDC for vehicles equipped with manual or automatic transmissions. The ignition timing cannot be adjusted.

Alternator

PRECAUTIONS

Several precautions must be observed with alternator equipped vehicles to avoid damage to the unit.

• If the battery is removed for any reason, make sure it is reconnected with the correct polarity. Reversing the battery connections may result in damage to the 1–way rectifiers.

• When utilizing a booster battery as a starting aid, always connect the positive to positive terminals and the negative terminal from the booster battery to a good engine ground on the vehicle being started.

• Never use a fast charger as a booster to start vehicles.

• Disconnect the battery cables when charging the battery with a fast charger.

• Never attempt to polarize the alternator.

• Do not use test lights of more than 12 volts when checking diode continuity.

• Do not short across or ground any of the alternator terminals.

• The polarity of the battery, alternator and regulator must be matched and considered before making any electrical connections within the system.

• Never separate the alternator on an open circuit. Make sure all connec-

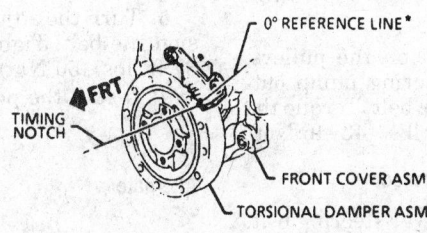

Timing mark on crankshaft damper and timing chain cover — 3.1L engine

tions within the circuit are clean and tight.

• Disconnect the battery ground terminal when performing any service on electrical components.

• Disconnect the battery if arc welding is to be done on the vehicle.

REMOVAL AND INSTALLATION

1. Disconnect the negative battery cable.
2. On the 3.2L (VIN V) engine, raise and support the vehicle. Remove the right front wheel.
3. Remove the alternator belt.
4. Disconnect and label the alternator wiring harnesses.
5. Remove the mounting bolt from the alternator bracket.
6. On 3.1L engines, remove the air pump bracket bolt from the rear of the alternator.
7. Remove the alternator from the engine.

To install:

8. Position the alternator on the bracket and install the pivot bolt.
9. On 3.1L engines, install the air pump bracket bolt to the rear of the alternator.
10. Install the belt and adjust the tension.
11. Reconnect the alternator wiring harnesses.
12. On the 3.2L (VIN V) engine, install the right front wheel. Lower the vehicle.
13. Connect the negative battery cable.

Drive Belt

REMOVAL AND INSTALLATION

All Engines Except 3.1 L (VIN Z)

Power Steering Pump

1. Loosen the lock and pivot bolts.
2. Pivot the power steering pump towards the crankshaft and remove the drive belt.

To install:

3. Install the belt on the pulleys. Pivot the power steering pump outward and tension the belt. Torque the lockbolt to 30–32 ft. lbs. (43–46 Nm).

Air Conditioning Compressor

1. Remove the power steering belt.
2. Loosen the lockbolt in the center of the A/C compressor tensioner pulley.
3. Turn the adjusting bolt to relieve the belt tension.

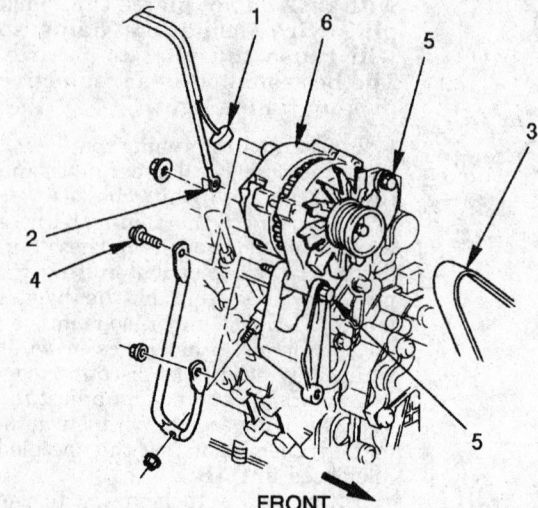

1. Wiring connector
2. Battery lead wire
3. Drive belt
4. Rear bracket fixing bolt
5. Mounting bolts
6. Generator

FRONT

179407

Alternator removal

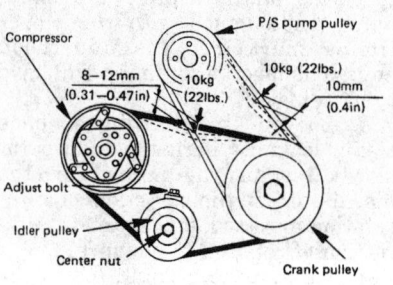

A/C compressor and power steering belts

322251

4. Remove the belt.

To install:

5. Install the A/C compressor belt on pulleys.
6. Turn the adjusting bolt and tension the belt. Tighten the lockbolt to 37 ft. lbs. (50 Nm).
7. Install the power steering belt.

Alternator

1. Remove the power steering and A/C compressor belts.
2. Loosen the adjusting plate and mounting bolts.

3. Pivot the alternator towards the crankshaft to loosen the belt.
4. Remove the belt.

To install:

5. Install the belt on to the pulleys.
6. Pry the alternator outwards and tension the belt. Tighten the mounting bolt to 16 ft. lbs. (22 Nm). Tighten the lockbolt to 17 ft. lbs. (24 Nm).
7. Install the A/C compressor belt and power steering pump belt.

3.1L (VIN Z) Engine

The V6 engines are equipped with a serpentine belt and automatic belt tensioner. The tension is maintained by a spring loaded pulley/tensioner. The indicator mark on the moveable portion of the tensioner must be within the limits of the slotted area on the stationary portion of the tensioner. Any reading outside the limits indicates either a defective belt or tensioner.

To remove the belt, install a ½ in. ratchet handle into the square slot in the tensioner and move far enough to slide the belt off the pulleys. Mark the belt routing for installation.

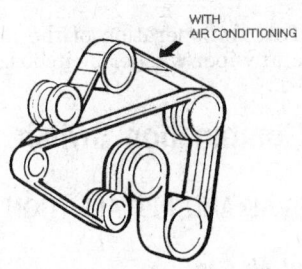

WITH
AIR CONDITIONING

178272

Serpentine belt routing — 3.1L (VIN Z) engine

Starter

REMOVAL AND INSTALLATION

1. Disconnect the negative battery cable.
2. If necessary, raise and safely support the vehicle.
3. If equipped, remove the skid plate.
4. On 3.2L engines, do the following;
   a. Disconnect the heated oxygen sensor connectors from the front exhaust pipe.
   b. Unbolt the front exhaust pipes from the exhaust manifolds. Unbolt the front exhaust crossover pipe from the catalytic converter flange.
   c. Separate the front exhaust crossover pipes from the exhaust manifold and then slide it toward the rear of the vehicle to gain access to the starter motor.
5. Remove the starter heat shield, if equipped.
6. Label and disconnect the starter wiring connectors.

NOTE: It may be necessary to disconnect and remove the EGR pipe.

7. Remove the starter mounting bolt and nut and the shims. Remove the starter from the vehicle.

To install:

8. Install the starter to the engine. Install the shims, the mounting bolt, and nut. Tighten the bolt and nut to 30 ft. lbs. (41 Nm) each.
9. Connect the wiring connectors to the starter.
10. Install the starter heat shield.
11. On 3.2L engines, complete the following;
   a. Install the front exhaust crossover pipe. Tighten the front mounting nuts to 49 ft. lbs. (67 Nm). Tighten the catalytic con-

verter flange bolt and nuts to 20 ft. lbs. (27 Nm).
   b. Connect the heated oxygen sensor connectors.
12. Install the skid plate, if equipped. Tighten the bolts to 27 ft. lbs. (37 Nm).
13. Reconnect the negative battery cable.
14. Verify that the starter operates correctly.

CHASSIS ELECTRICAL

Blower Motor

REMOVAL AND INSTALLATION

NOTE: The heater and A/C blower motor is located under the right side of the dash.

1. Disconnect the negative battery cable.
2. Remove the right-side dashboard lower trim panel.
3. Disconnect the blower motor harness.
4. Remove the blower mounting screws, then the blower motor assembly.
5. Remove the retaining clip or nut; then, remove the cage from the motor.

To install:

6. Install the cage onto the motor, and install the retaining clip or nut.
7. Install the blower motor into the housing and install the mounting screws.
8. Connect the harness to the blower motor.
9. Install the right-side dashboard lower trim panel.
10. Connect the negative battery cable.
11. Test the blower motor operation at all speeds.

Windshield Wiper Motor

REMOVAL AND INSTALLATION

Front

1. Disconnect the negative battery cable.
2. Uncouple the electrical connector from the wiper motor.
3. Unbolt the wiper motor from the bulkhead.

4. On Amigo and Pick-up models, remove the wiper motor bracket bolts and pull the motor out of the cowl far enough to disconnect the motor from the linkage.
5. Disconnect the wiper motor from the wiper linkage arm.

To install:

6. Connect the wiper motor to the wiper linkage arm. Torque the nut to 9–11 ft. lbs. (12–16 Nm).
7. Install the wiper motor mounting bolts.
8. Install the wiper motor and the bracket bolts.
9. Connect the electrical connector to the wiper motor.
10. Connect the negative battery cable.

Rear

1. Disconnect the negative battery cable.
2. Remove the rear door trim panel to expose the wiper motor and disconnect the wiring harness.
3. Loosen the wiper motor mounting bolts and disconnect the wiper linkage from the motor.
4. Remove the motor from the rear door.

To install:

5. Position the wiper motor in the rear door and connect the wiper linkage to the motor.
6. Install the wiper motor mounting bolts.
7. Connect the harness to the wiper motor.
8. Install the rear door trim panel.
9. Connect the negative battery cable.

Wiper Switch

REMOVAL AND INSTALLATION

NOTE: The pushbutton wiper switch is a part of the switch cluster located on the left side of the dash.

1. Disconnect the negative battery cable.
2. Remove the instrument cluster-to-dash screws.
3. Pull the instrument cluster forward and disconnect the harness from the headlight and wiper switches.
4. From the rear of the instrument cluster, loosen the switch cluster-to-instrument cluster screws.
5. Separate the wiper switch from the switch cluster.

To install:

6. Install the wiper switch to the switch cluster.

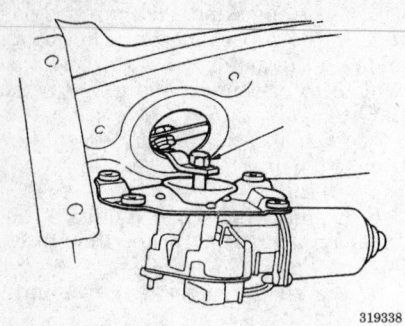

Front wiper motor linkage connection — Passport

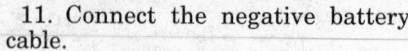

Inner panel and rear wiper motor location

7. Tighten the switch cluster-to-instrument cluster screws.
8. Connect the harness to the switch cluster.
9. Install the instrument cluster.
10. Connect the negative battery cable.

Headlight Switch

REMOVAL AND INSTALLATION

1. Disconnect the negative battery cable.
2. Remove the instrument cluster-to-dash screws from below the cluster shroud.
3. If equipped with tilt steering, move the steering column to its lowest position.
4. Pull the instrument cluster forward and disconnect the harness from the headlight and wiper/washer switches.
5. Tilt the instrument cluster upward to clear the steering wheel and cluster shroud. Tilting the cluster assembly and using a stub or offset screwdriver will allow the switch to

be removed without removing the cluster from the dashboard.
6. From the rear of the instrument cluster, remove the switch cluster-to-instrument cluster shroud screws.
7. Separate the headlight switch from the cluster shroud.
To install:
8. Install the headlight switch to the cluster shroud.
9. Connect the harness to the switch cluster.
10. Install the instrument cluster and its mounting screws.

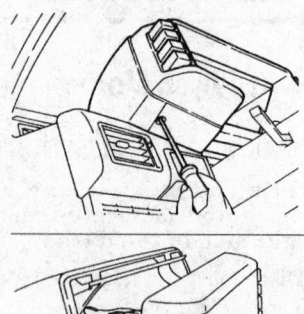

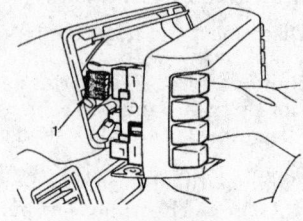

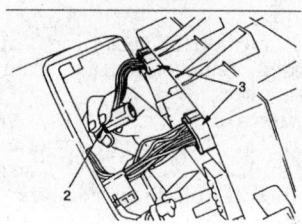

1 Lighting and wiper connections
2 Speedometer connections
3 Meter connections

254214

Removal of instrument cluster assembly

11. Connect the negative battery cable.
12. Test the operation of the headlight and wiper/washer switches.

Combination Switch

REMOVAL AND INSTALLATION

Without Air Bag

1. Disconnect the negative battery cable.
2. Remove the steering wheel.

NOTE: Do not strike the steering column or use air tools to loosen the steering wheel nut. The impact may damage the steering column's energy-absorbing properties.

3. Remove the steering column upper and lower covers and the horn contact ring.
4. Remove the turn signal cancelling sleeve.
5. Disconnect the harnesses from the combination switch.
6. Remove the combination switch-to-steering column screws. Then, remove the switch.

NOTE: On vehicles equipped with cruise control, the cruise control switch is an integral part of the combination switch.

To install:
7. Install the combination switch to the steering column and secure it with screws.
8. Connect the combination switch harness.
9. Install the steering column cover.
10. Install the contact ring.
11. Install the steering wheel and horn pad.
12. Connect the negative battery cable.

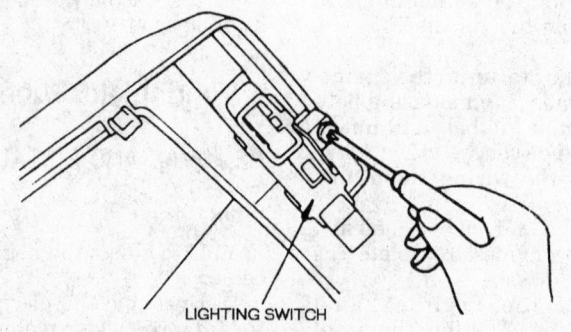

LIGHTING SWITCH

254215

Headlight switch mounting screw

With Air Bag

NOTE: 1995 models are available with and without dual Air Bags. Air Bags were added as standard equipment during the middle of the year's production run. Air Bag equipped vehicles may be referred to as 1995½ models in parts and service listings.

---------- CAUTION ----------

The Air Bag system must be disabled before the steering wheel and combination switch are removed. Failure to disable the system may result in repairs and possible personal injury. Do not damage, cut, or attempt to alter the yellow Air Bag wiring harness.

NOTE: On vehicles with Air Bags, the headlight and wiper switches are integral parts of the combination switch and Air Bag cable reel assembly.

1. Disconnect the negative then positive battery cables.
2. Disarm the Air Bag system.
3. Remove the steering wheel.
4. Remove the lower dashboard cover and disconnect the dimmer switch.
5. Remove the upper and lower steering column covers. Be careful with the wiring harness that runs through the lower cover; it contains the Air Bag wiring.
6. Unscrew and remove the combination switch and Air Bag cable reel assembly.

NOTE: The combination switch and cable reel assembly cannot be disassembled. They are serviced and replaced as one complete assembly.

To install:
7. Install the combination switch and cable reel assembly onto the steering column shaft.
8. Reconnect the switch wiring harnesses.
9. Install the lower steering column cover, making sure that the Air Bag wiring is routed correctly and not pinched. Install the upper steering column cover.
10. Reconnect the dimmer switch and install the dashboard lower cover.
11. Check the alignment of the Air Bag cable reel.
   a. Turn the cable reel clockwise to its fully locked position. Don't turn the cable reel past the point at which you begin to feel resistance to its rotation.
   b. Turn the cable reel about three turns in the opposite direction until the pointer on the cable reel is aligned with the neutral mark.
12. Install the steering wheel and tighten the nut to 25 ft. lbs. (34 Nm).
13. Reconnect the positive and negative battery cables.
14. Turn the ignition to the **ON** position, but don't start the engine. The AIR BAG warning light should turn on and flash on and off for seven seconds, and then turn off. This light sequence indicates that the Air Bag system is functioning normally. If the AIR BAG light doesn't come on, or stays on longer than seven seconds, the system must be diagnosed.
15. Check the operation of the combination switch controls.

Ignition Lock Cylinder

REMOVAL AND INSTALLATION

NOTE: 1995 models are available with and without dual air bags. Air Bags were added as standard equipment during the

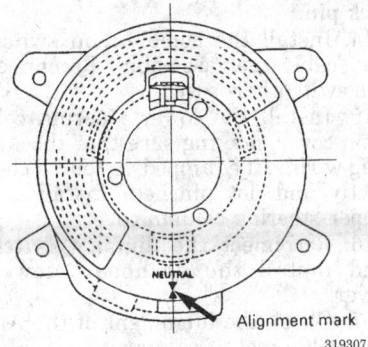

Cable reel alignment marks

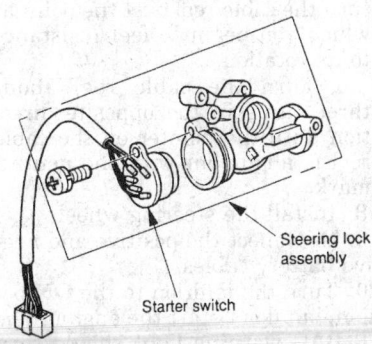

Ignition starter switch and lock cylinder

middle of the year's production run. Air bag equipped vehicles may be referred to as 1995½ models in parts and service listings.

---------- CAUTION ----------

The Air Bag system must be disabled before the steering wheel and ignition switch are removed. Failure to disable the system may result in system repairs and possible personal injury. Do not damage, cut, or attempt to alter the yellow Air Bag wiring harness.

1. Turn the steering wheel so that the vehicle's front wheels are pointing straight ahead.
2. Turn the ignition switch to the **LOCK** position. Remove the key.
3. Disconnect the negative and positive battery cables.
4. If equipped, disarm the Air Bag system. Remove the steering wheel.
5. Remove the lower dashboard cover and disconnect the dimmer switch.
6. Remove the upper and lower steering column covers. Be careful with the wiring harness that runs through the lower cover; it may contain the Air Bag wiring.
7. Unscrew and remove the combination switch and Air Bag cable reel assembly, if equipped. The switch and cable reel assembly cannot be disassembled.
8. If equipped with an automatic transmission, disconnect the shift lock cable from the ignition lock cylinder.
9. Unscrew the starter switch from the ignition lock body and disconnect its wiring harness from the fuse box.
10. Unbolt the ignition lock cylinder from the steering column. Remove the snapring and the bushing to remove the lock cylinder from the steering column.
To install:
11. Install the lock cylinder onto the steering column. Install and evenly tighten the bolts. Install a new bushing and snapring.
12. Reconnect the starter switch to the lock cylinder. Connect the wiring.
13. Connect the A/T shift lock cable to the ignition switch using a new lock pin.
14. Install the combination switch and cable reel assembly. Reconnect the switch harnesses.
15. Install the lower steering column cover, making sure that the Air Bag wiring, if equipped, is routed correctly and not pinched. Install the upper steering column cover.

16. Reconnect the dimmer switch and install the dashboard lower cover.

17. Check the alignment of the Air Bag cable reel, if equipped;

a. Turn the cable reel clockwise to its fully locked position. Don't turn the cable reel past the point at which you begin to feel resistance to its rotation.

b. Turn the cable reel about three turns in the opposite direction until the pointer on the cable reel is alignment with the neutral mark.

18. Install the steering.

19. Reconnect the positive and negative battery cables.

20. Turn the ignition to the **ON** position, but don't start the engine. The AIR BAG warning light should turn on and flash on and off for seven seconds, and then turn off. This indicator light sequence indicates that the Air Bag system is functioning normally. If the AIR BAG light doesn't come on, or stays on longer than seven seconds, the system must be diagnosed.

21. Check the operation of the ignition switch and combination switch controls.

Ignition Switch

REMOVAL AND INSTALLATION

NOTE: 1995 models are available with and without dual air bags. Air Bags were added as standard equipment during the middle of the year's production run. Air bag equipped vehicles may be referred to as 1995½ models in parts and service listings.

------ **CAUTION** ------
The Air Bag system must be disabled before the steering wheel and ignition switch are removed. Failure to disable the system may result in system repairs and possible personal injury. Do not damage, cut, or attempt to alter the yellow air Bag wiring harness.

1. Turn the steering wheel so that the vehicle's front wheels are pointing straight ahead.

2. Turn the ignition switch to the **LOCK** position. Remove the key.

3. Disconnect the negative and positive battery cables. Wait at least five minutes if equipped with an Air Bag.

4. Disarm the Air Bag system, if equipped. Remove the steering wheel.

5. Remove the lower dashboard cover and disconnect the dimmer switch.

6. Remove the upper and lower steering column covers. Be careful with the wiring harness that runs through the lower cover; it may contain the Air Bag wiring.

7. Unscrew and remove the combination switch and Air Bag cable reel assembly, if equipped. The switch and cable reel assembly cannot be disassembled.

8. If equipped with an automatic transmission, disconnect the shift lock cable from the ignition lock cylinder.

9. Unscrew the starter switch from the ignition lock body and disconnect the harness from the fuse box.

10. Unbolt the ignition lock cylinder from the steering column. Remove the snapring and bushing to remove the lock cylinder from the steering column.

To install:

11. Install the lock cylinder onto the steering column. Install and evenly tighten the bolts. Install a new bushing and snapring.

12. Reconnect the starter switch to the lock cylinder. Connect the wiring harness.

13. Connect the A/T shift lock cable to the ignition switch using a new lock pin.

14. Install the combination switch and cable reel assembly. Reconnect the switch harnesses.

15. Install the lower steering column cover, making sure that the Air Bag wiring, if equipped, is routed correctly and not pinched. Install the upper steering column cover.

16. Reconnect the dimmer switch and install the dashboard lower cover.

17. Check the alignment of the Air Bag cable reel, if equipped.

a. Turn the cable reel clockwise to its fully locked position. Don't turn the cable reel past the point at which you begin to feel resistance to its rotation.

b. Turn the cable reel about three turns in the opposite direction until the pointer on the cable reel is alignment with the neutral mark.

18. Install the steering wheel.

19. Reconnect the positive and negative battery cables.

20. Turn the ignition to the **ON** position, but don't start the engine. The AIR BAG warning light should turn on and flash on and off for seven seconds, and then turn off. This indi-

cator light sequence indicates that the Air Bag system is functioning normally. If the AIR BAG light doesn't come on, or stays on longer than seven seconds, the system must be diagnosed.

21. Check the operation of the ignition switch and combination switch controls.

Park/Neutral Safety Switch

REMOVAL AND INSTALLATION

Amigo and Pick-Up

The neutral safety switch is mounted to the transmission at the shift lever. The backup switch is incorporated into this assembly.

1. Disconnect the negative battery cable.

2. Disconnect the harness and remove the shift lever from the transmission.

3. Remove the retaining bolts and switch.

To install:

4. Install the switch to the transmission. Align the groove and the neutral basic line and tighten the mounting bolts to 9 ft. lbs. (12 Nm).

5. Connect the harness to the switch.

6. Install the shifter assembly.

7. Connect the negative battery cable and check that the engine will only start in the park or neutral position.

Rodeo and Passport

1. Disconnect the negative battery cable.

2. Place the gear selector lever in the **N** position.

3. Remove the air cleaner assembly.

4. Raise and support the vehicle safely.

5. Remove the switch cover.

6. Disconnect the selector lever from the switch.

7. Remove the heat protector from the transmission case.

8. Disconnect the black switch harness from the engine harness and remove the mounting clip.

9. Remove the mounting bolts and park/neutral switch from the transmission.

To install:

10. Position the park/neutral switch on the transmission and loosely install the mounting bolts.

11. With the transmission in **N**, rotate the switch until the slot in the switch housing aligns with the selec-

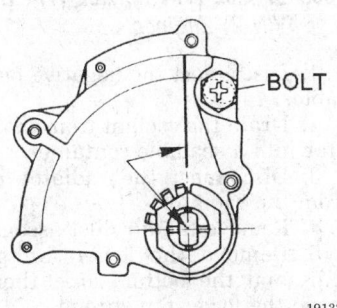

Park/Neutral switch adjustment

tor shaft bushing, and insert a $3/32$ in. drill bit or punch into the slot. Tighten the bolts to 10 ft. lbs. (13 Nm).

12. Remove the drill bit or punch.

13. Connect the harness and install the mounting clip.

14. Install the heat protector.

15. Connect the selector lever to the switch.

16. Install the switch cover.

17. Lower the vehicle.

18. Install the air cleaner assembly.

19. Connect the negative battery cable and check that the engine will only start in the **P** or **N** positions.

Powertrain Control Module

REMOVAL AND INSTALLATION

Without Air Bag

1. Disconnect the negative battery cable.

2. Remove the left kick panel trim.

3. Disconnect the harness at the control module.

4. Unfasten the mounting bolts and remove the Powertrain Control Module (PCM) from the vehicle.

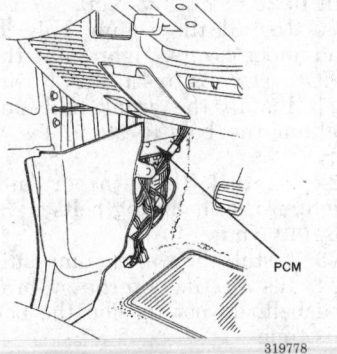

PCM location — vehicles without Air Bags

To install:

5. Position the PCM on the body and install the mounting bolts.

6. Connect the electrical connectors to the control module.

7. Replace the kick panel trim.

8. Connect the negative battery cable.

With Air Bag

NOTE: **1995 models are available with and without dual Air Bags. Air Bags were added as standard equipment during the middle of the year's production run. Air Bag equipped vehicles may be referred to as 1995½ models in parts and service listings.**

─── **CAUTION** ───

The Air Bag system must be disabled before the PCM is removed. Failure to disable the system may result in system repairs and possible personal injury. Do not damage, cut, or attempt to alter the yellow Air Bag wiring harness.

1. Turn the steering wheel so that the vehicle's front wheels are pointing straight ahead.

2. Turn the ignition switch to the **LOCK** position. Remove the key.

3. Disconnect the negative and positive battery cables. Wait at least five minutes before working around the Air Bags.

4. Disarm the Air Bag system.

5. Remove the glove box door and the passenger's Air Bag lower cover.

6. Disconnect the yellow 2–way Air Bag harness from the passenger's side Air Bag.

7. Remove the four console screws. Remove the console.

8. Unscrew and remove the trim panel located below the dashboard cubby hole.

9. Unfasten the three Powertrain Control Module (PCM) mounting bolts.

10. Disconnect and remove the PCM.

NOTE: **Do not touch the connector pins or soldered circuit board.**

To install:

11. Fit the PCM into position and install the three mounting bolts.

12. Reconnect the PCM wiring harnesses. Make sure the connectors are firmly seated into the PCM.

13. Install the trim panel and screws.

14. Install the console and screws.

15. Reconnect the yellow 2–way connector to the passenger's Air Bag.

16. Install the lower cover and glove box door.

17. Reconnect the yellow 3–way connector at the base of the steering column.

18. Reconnect the positive and negative battery cables.

19. Turn the ignition to the **ON** position, but don't start the engine. The AIR BAG indicator light should turn on and flash on and off for seven seconds. The AIR BAG indicator light should then turn off. If the indicator light doesn't turn on, or stays on, the system must be diagnosed.

20. Check the PCM for trouble codes.

ENGINE COOLING

Radiator

REMOVAL AND INSTALLATION

1993–94 Models

1. Disconnect the negative battery cable.

2. Drain the cooling system into a suitable container.

PCM location — vehicles with Air Bags

3. Remove the upper, lower, and reservoir hoses from the radiator.

4. Disconnect the oil cooling lines from the radiator if equipped with an automatic transmission.

5. Remove the fan shroud attaching bolts and the shroud.

6. Remove the radiator attaching bolts and the radiator.

To install:

7. Install the radiator and attaching bolts.

8. Install the fan shroud and the shroud attaching bolts.

9. Reconnect the radiator hoses.

10. Connect the oil cooling lines, if removed.

11. Refill the cooling system.

12. Connect the negative battery cable and check for coolant system leaks.

1995–97 Models

1. Disconnect the negative battery cable.

2. Loosen the drain plug on the bottom of the radiator and drain the coolant into a container.

3. If equipped with an automatic transmission, disconnect and plug the ATF cooler hoses.

4. Remove the upper radiator hose.

5. Remove the two air intake duct mounting bolts; then, move the duct and deflector out of the way.

6. Disconnect the lower radiator hose from the radiator.

7. Disconnect the coolant recovery hose from the radiator.

8. Disconnect the lower fan guide clips and the bottom lock, then remove the lower fan shroud from the vehicle.

9. On 2.6L engines, remove the four upper fan shroud attaching bolts and remove the shroud. Remove the radiator bolts and carefully remove the radiator.

10. On 3.2L engines, remove the radiator bracket. Lift the radiator from the vehicle with the hoses attached. Remove the cushions from the bottom of the radiator.

To install:

11. Carefully lower the radiator into the vehicle, take care not to damage the radiator on the fan blades.

12. On 2.6L engines, position the radiator and install the four bolts. Install the upper fan shroud and its attaching bolts.

13. On 3.2L engines, install the radiator bracket.

14. Install the lower fan guide, make sure that the clips are fully engaged.

15. Connect the coolant recovery hose to the radiator.

16. Connect the lower radiator hose to the radiator.

17. If equipped with a automatic transmission, connect the ATF cooler hoses.

18. Install the air intake duct assembly and the two mounting bolts.

19. Install the upper radiator hose.

20. Refill and bleed the air from the cooling system.

21. Connect the negative battery cable.

Water Pump

REMOVAL AND INSTALLATION

1993–94 2.3L (VIN L), 2.6L (VIN E) and 3.1L (VIN Z) Engines

1. Disconnect the negative battery cable.

2. Remove the undercover, if equipped, and drain the cooling system into a container.

3. Remove the drive belt from the water pump pulley.

4. Remove the fan blade and pulley from the pump hub.

5. Remove the water pump-to-engine bolts, then the water pump and gasket.

6. Clean and inspect the mounting surfaces.

To install:

7. Install a new gasket and water pump. Torque the water pump-to-engine bolts to 14 ft. lbs. (19 Nm).

8. Install the fan blade and pulley.

9. Install and adjust the drive belt.

10. Refill the cooling system and install the undercover, if equipped.

11. Connect the negative battery cable.

12. Operate the engine to normal operating temperatures and check for leaks.

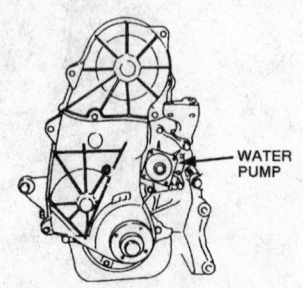

191426

Water pump assembly — 2.3L and 1993–94 2.6L engines

1995–97 2.3L (VIN L), 2.6L (VIN E) and 3.1L (VIN Z) Engines

1. Disconnect the negative battery cable.

2. Drain the coolant from the radiator into a sealable container.

3. Disconnect the radiator hoses from the radiator.

4. Remove the air duct assembly.

5. Remove the lower fan guide clips and the bottom lock, then remove the lower fan shroud.

6. Remove the upper fan shroud bolts and remove the shroud.

7. Remove the nuts attaching the fan to the water pump, then remove the fan.

8. If equipped with power steering, remove the drive belt.

9. If equipped with A/C, loosen the A/C idler pulley nuts, then remove the mounting bolts and idler pulley. Remove the A/C compressor belt.

10. Remove the alternator belt.

11. Remove the pulley from the water pump.

12. On 3.1L engines, remove the alternator upper and lower brackets. Remove the power steering pump lower bracket and swing aside. Remove the bottom radiator hose and heater hose from the pump.

13. Rotate the crankshaft to align the crankshaft pulley timing marks.

14. Remove the starter and install flywheel holder (part No. J-38674) or equivalent.

15. Remove the crankshaft pulley bolt and pulley.

16. Remove the upper and lower timing belt covers.

17. Remove the four bolts and one nut from the water pump and remove the pump from the engine.

To install:

18. Clean the water pump mounting surface.

19. Install the water pump with a new gasket. Tighten the mounting bolts to 14 ft. lbs. (19 Nm), and the nut to 20 ft. lbs. (25 Nm).

20. Install the timing belt lower and upper covers. Tighten the timing belt cover bolts to 4 ft. lbs. (6 Nm).

21. Install the crankshaft pulley, tighten the bolt to 90 ft. lbs. (122 Nm).

22. Install the starter motor. Tighten the mounting bolts to 30 ft. lbs. (40 Nm).

23. Install the water pump pulley.

24. Install the alternator bracket and belt, do not tension the belt at this time.

25. If equipped with A/C, install the and idler pulley, then adjust the belt tension.

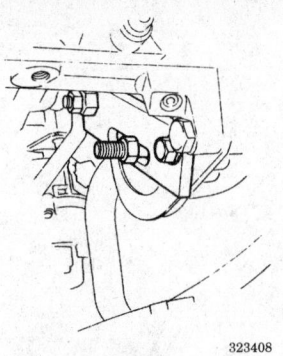

**Locking the flywheel —
Passport (VIN E) engine**

26. If equipped with power steering, install the bracket and belt, then adjust the drive belt.

27. Install the fan pulley to the water pump, and adjust the alternator belt tension. Tighten the fan attaching nuts to 20 ft. lbs. (27 Nm). Install the cooling fan.

28. Install the upper and lower fan shroud.

29. Install the air duct assembly.

30. Connect the radiator hoses.

31. Fill and bleed the cooling system.

32. Connect the negative battery cable.

3.2L (VIN V) Engine

1. Disconnect the negative battery cable.

2. Drain the engine coolant into a sealable container.

3. Remove the upper radiator hose.

4. Remove the timing belt and idler pulley. The timing belt must be replaced if it has been contaminated by oil or coolant.

5. Unbolt and remove the water pump. Clean any gasket material or sealant residue from the water pump mating sealing surfaces.

To install:

6. Install the water pump using a new gasket. Tighten the mounting bolts to 13 ft. lbs. (18 Nm) in a two-step crisscross sequence.

7. Install the idler pulley. Tighten the mounting bolt to 31 ft. lbs. (42 Nm).

8. Install and tension the timing belt.

9. Install the upper radiator hose.

10. Refill and bleed the cooling system.

11. Connect the negative battery cable. Start the engine and check for coolant leaks.

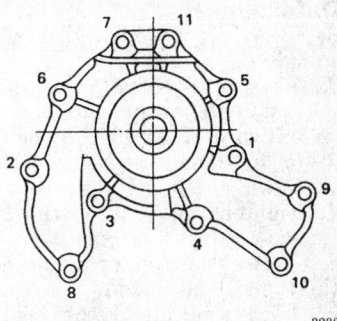

Water pump bolt tightening sequence — Passport (VIN V) engine

Thermostat

REMOVAL AND INSTALLATION

1. Disconnect the negative battery cable.

2. Drain the coolant into a sealable container.

3. Disconnect the upper radiator hose from the thermostat housing.

4. Unbolt and remove the thermostat housing.

5. Remove the thermostat and gasket.

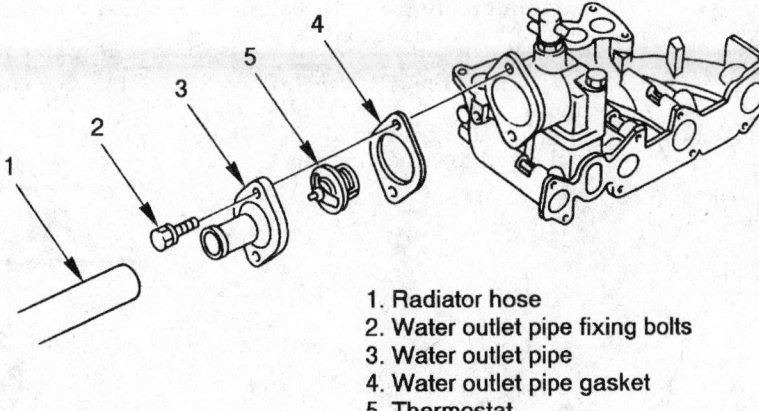

1. Radiator hose
2. Water outlet pipe fixing bolts
3. Water outlet pipe
4. Water outlet pipe gasket
5. Thermostat

Water outlet pipe and thermostat assembly — Passport (VIN E) engine

To install:

6. Clean the sealing surface thoroughly.

7. Install the thermostat and a new gasket. The thermostat's pin faces outward. Tighten the thermostat housing mounting bolts to 18 ft. lbs. (25 Nm) on all engines except the 3.2L. Tighten the bolts on the 3.2L engine to 14 ft. lbs. (19 Nm).

8. Connect the upper radiator hose to the thermostat housing.

9. Refill and bleed the cooling system.

10. Connect the negative battery cable.

11. Warm the engine up to normal operating temperature and test the operation of the thermostat. Check for coolant leaks.

Cooling Fan

REMOVAL AND INSTALLATION

1. Disconnect the negative battery cable.

2. If necessary, remove the intake air duct.

3. Remove the fan shroud.

4. Remove the four nuts attaching the cooling fan to the water pump pulley studs. Be careful not to damage the radiator.

To install:

5. Install the cooling fan on the water pump pulley. Tighten the nuts to 6 ft. lbs. (8 Nm).

6. Install the fan shroud.

7. Install the intake air duct if it was removed.

8. Connect the negative battery cable.

Cooling System Bleeding

1. Fill the radiator with a 50/50 mixture of coolant and water to just

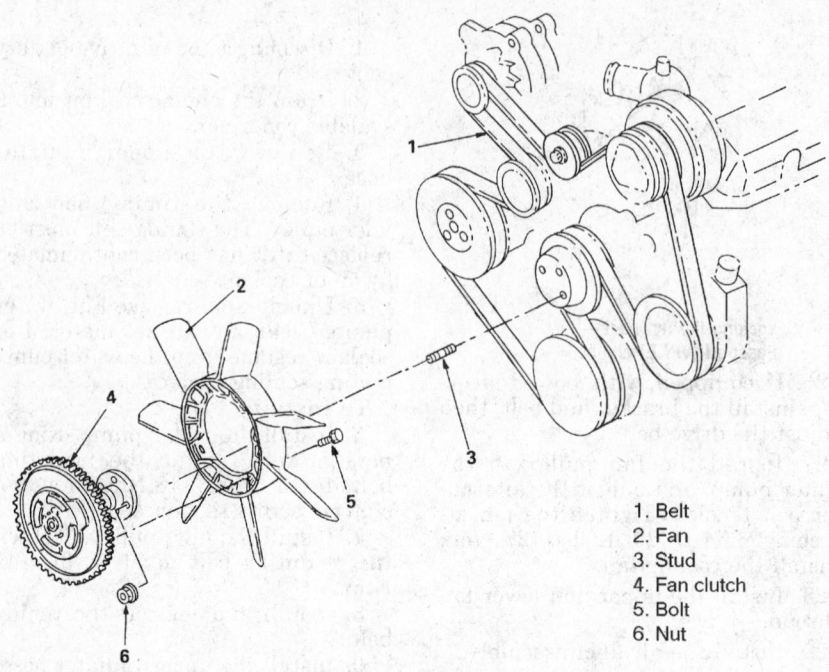

1. Belt
2. Fan
3. Stud
4. Fan clutch
5. Bolt
6. Nut

187615

Cooling fan and clutch assembly

below the radiator filler neck. Do not use more than 60% coolant, or the efficiency of the cooling system will be impaired.

2. Fill the reserve tank to the MAX line with the coolant mixture.

3. With the radiator cap off, warm the engine to operating temperature. The heater controls should be set to maximum heat. Add coolant as needed while the engine is warming up.

4. When the thermostat opens (upper hose hot and visible coolant flow in radiator) top off the coolant and install the radiator cap.

5. Refill the reserve tank to the proper level.

6. Shut off the engine and allow it to cool. As the engine cools it will draw coolant from the reserve tank.

NOTE: Any air in the cooling system will be released through the reservoir bottle during normal engine operation.

7. Refill the reserve tank to the MAX line after the engine has cooled.

FUEL SYSTEM

Fuel System Service Precautions

Safety is the most important factor when performing not only fuel system maintenance but any type of maintenance. Failure to conduct maintenance and repairs in a safe manner may result in serious personal injury or death. Maintenance and testing of the vehicle's fuel system components can be accomplished

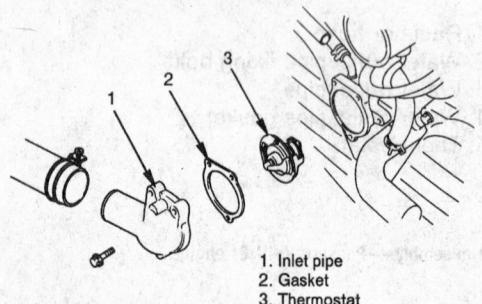

1. Inlet pipe
2. Gasket
3. Thermostat

320577

Thermostat assembly — Passport (VIN V) engine

safely and effectively by adhering to the following rules and guidelines.

• To avoid the possibility of fire and personal injury, always disconnect the negative battery cable unless the repair or test procedure requires that battery voltage be applied.

• Always relieve the fuel system pressure prior to disconnecting any fuel system component (injector, fuel rail, pressure regulator, etc.), fitting or fuel line connection. Exercise extreme caution whenever relieving fuel system pressure to avoid exposing skin, face and eyes to fuel spray. Please be advised that fuel under pressure may penetrate the skin or any part of the body that it contacts.

• Always place a shop towel or cloth around the fitting or connection prior to loosening to absorb any excess fuel due to spillage. Ensure that all fuel spillage (should it occur) is quickly removed from engine surfaces. Ensure that all fuel soaked cloths or towels are deposited into a suitable waste container.

• Always keep a dry chemical (Class B) fire extinguisher near the work area.

• Do not allow fuel spray or fuel vapors to come into contact with a spark or open flame.

• Always use a backup wrench when loosening and tightening fuel line connection fittings. This will prevent unnecessary stress and torsion to fuel line piping. Always follow the proper torque specifications.

• Always replace worn fuel fitting O-rings with new. Do not substitute fuel hose or equivalent, where fuel pipe is installed.

Fuel System Pressure

RELIEVING

Except 3.1L (VIN Z) Engine

1. Remove the fuel filler cap.
2. Remove the fuel pump relay from the underhood relay box.
3. Start the engine and let it run until it stalls, then crank the engine for an additional 30 seconds.
4. Turn the ignition switch to the **OFF** position and remove the key. Disconnect the negative battery cable and then install the fuel pump relay.

3.1 (VIN Z) Engine

NOTE: This fuel system is equipped with an internal constant bleed feature to reduce fuel system pressure when the engine is turned OFF.

1. Disconnect the negative battery cable. Loosen fuel filler cap.
2. Connect a fuel pressure gauge to the fuel pressure connection.
3. Wrap a shop cloth around the fitting while connecting the gauge.
4. Install the bleed hose into an approved container and open the valve. Connect the negative battery cable.
5. When the repair to the fuel system is complete check all of the fittings for leaks.

Idle Speed

ADJUSTMENT

2.3L Carbureted Engine

1. Set the parking brake and block the wheels.
2. Place the transmission in **NEUTRAL**.
3. Let the engine warm up to normal operating temperature.
4. Make sure that the choke is fully open and the air cleaner is installed.
5. Make sure that the A/C is **OFF**.
6. Disconnect and plug the distributor vacuum, canister purge, and EGR vacuum lines.
7. Shut off the vacuum to the idle compensator by bending the hose.
8. Turn the throttle adjusting screw to adjust.
• California: 900 rpm
• Federal: 800 rpm
9. If the vehicle is equipped with A/C, turn it **ON** and set the blower to the highest position.

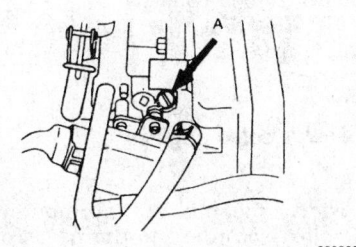

330309

Idle speed adjusting screw — carbureted engine

10. Use the adjusting screw at the tip of the carburetor throttle lever to set the correct fast idle.

1993–95 MFI 2.3L (VIN L) and 2.6L (VIN E) Engines

1. Apply the parking brake.
2. The idle speed must be adjusted with the engine running under the following conditions;
• Allow the engine to reach normal operating temperature.
• Throttle valve must be closed and the throttle valve switch idle contact on.
• Front wheels straight ahead if equipped with power steering.
• Air conditioner must be turned **OFF**
• Manual transmission in **NEUTRAL** or automatic transmission in **PARK** or **NEUTRAL**
• Harness connector for pressure regulator VSV disconnected
• Canister purge vacuum hose disconnected and plugged
• EGR vacuum hose disconnected and plugged
3. Connect a tachometer to the engine.
4. Correct idle speed should be between 850–950 RPM. If the idle speed is incorrect, turn the adjustment screw on the throttle body.
5. Connect the pressure regulator VSV harness and disconnect the tachometer from the engine.
6. Connect the vacuum hoses for the EGR and the charcoal canister.

1996–97 2.6L (VIN E) Engine

The engine control system determines the idle speed for this engine. If the idle speed is out of specification, it may be a symptom of an engine control system problem. The idle speed cannot be adjusted.

3.1L (VIN Z) Engine

Controlled idle speed is programmed into the PCM, which determines the correct IAC valve pintle position to maintain the desired idle speed for all engine operating conditions. No adjustment is necessary. The minimum air rate is set at the factory with a stop screw. The throttle stop screw is covered with a plug at the factory.

3.2L (VIN V) Engine

The idle speed of the engine is controlled by the PCM. It is not adjustable. The correct idle speed is 750 RPM for vehicles equipped with manual or automatic transmissions.

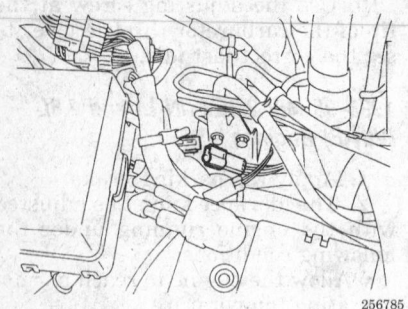

Disconnect the pressure regulator VSV harness connector

256785

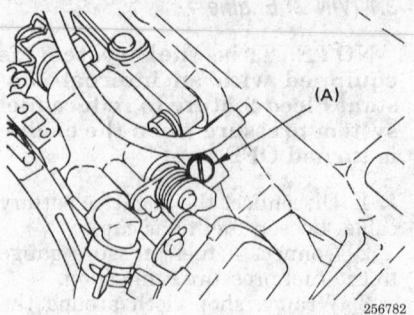

Idle speed adjustment screw — 1993–95 2.3L MFI and 2.6L engines

256782

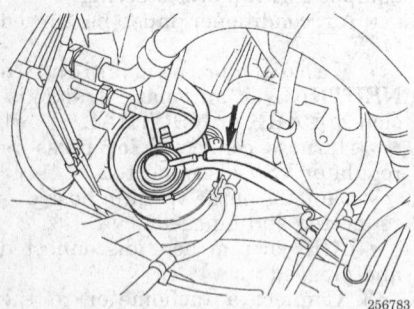

Disconnect the canister purge vacuum hose

256783

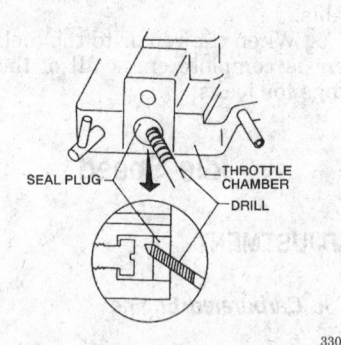

SEAL PLUG   THROTTLE CHAMBER   DRILL

Idle mixture screw — carbureted engine

330024

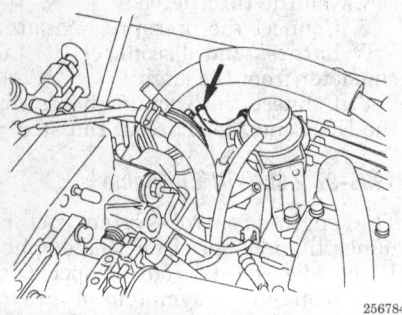

Disconnect the EGR vacuum signal hose

256784

Mixture

ADJUSTMENT

2.3L Carbureted Engine

1. Set the parking brake and block the front wheels.
2. Remove the carburetor.
3. Remove the plug for the idle mixture screw. The plug can be removed by drilling a hole and inserting a screw to pull out the plug.
4. Reinstall the carburetor assembly.

5. Make the idling speed adjustment with the engine at normal operating temperature.
6. Make sure that the air cleaner is installed and the choke valve is open.
7. Disconnect and plug the EGR, the canister purge line, idle compensator line, and the distributor vacuum lines.
8. Adjust the idle speed to 800 rpm.
9. Turn the idle mixture screw all the way in and then back out 3 turns.
10. Adjust the setting of the idle mixture screw to achieve the maximum speed.
11. Reset the throttle screw to 850 rpm.
12. Turn the idle mixture screw clockwise until the engine speed is down to 800 rpm.
13. Insert a new plug for the idle mixture screw.

Fuel Injected Engines

The air/fuel mixture is computer controlled according to the needs of the engine and is not adjustable. If the air/fuel mixture is too lean or too rich, other problems with the engine and/or engine control system exist.

Fuel Filter

REMOVAL AND INSTALLATION

2.3L Carbureted Engine

1. Raise and safely support the vehicle.

——— CAUTION ———
Fuel will spill out when the hoses are removed from the filter. Observe all applicable safety precautions when working around fuel. Do not allow fuel or fuel vapor to come in contact with spark or open flame. Keep a dry chemical (class B) fire extinguisher near the work area. Also, never store or drain fuel into an open container due to the possibility of fire or explosion.

2. Disconnect the fuel hoses from the filter. Plug the ends to prevent fuel spillage.
3. Remove the fuel filter from the mounting bracket.
   To install:
4. Position the filter in the bracket and connect the fuel hoses. Be sure the filter in installed in the proper direction.
5. Lower the vehicle, start the engine and carefully check for leaks.

Fuel Injected Engines

——— CAUTION ———
Fuel injection systems remain under pressure even after the engine has been turned off. The fuel system pressure must be relieved before disconnecting any fuel lines. Failure to do so may result in fire and personal injury.

1. Relieve the fuel system pressure.
2. Disconnect the negative battery cable.
3. Raise and safely support the vehicle.
4. Use block-off clamps to pinch the fuel lines shut. Disconnect the fuel lines from the fuel filter. Clean up any fuel spills.
5. Loosen the filter mounting bolt and remove the fuel filter.
   To install:
6. Install the fuel filter with the label, arrow, or manufacturer's mark facing toward the front of the vehicle. Tighten the bracket bolt.
7. Connect the fuel lines to the fuel filter.
8. Lower the vehicle to the floor and install the filler cap.

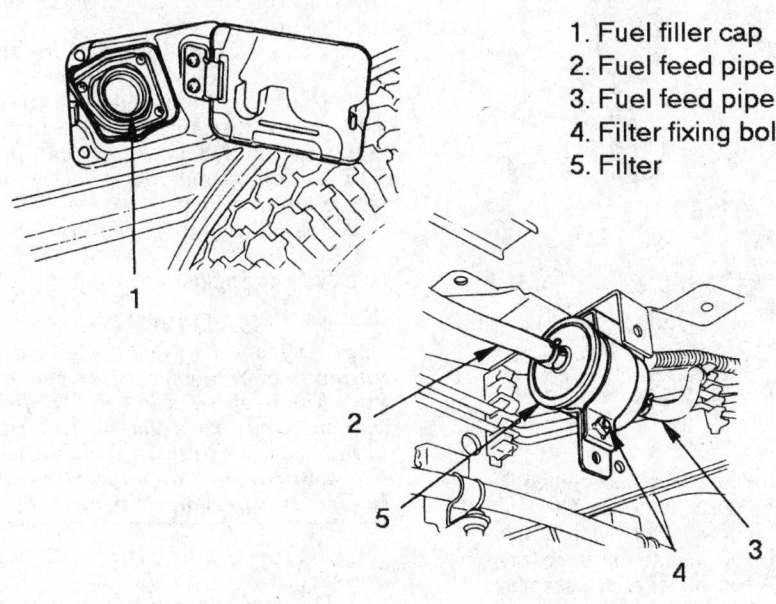

1. Fuel filler cap
2. Fuel feed pipe
3. Fuel feed pipe
4. Filter fixing bolt
5. Filter

320201

Fuel filter mounting — fuel injected engines

9. Reconnect the negative battery cable.

10. Start the engine and inspect the fuel filter connections for leaks.

Fuel Pump

REMOVAL AND INSTALLATION

2.3L Carbureted Engine

The fuel pump is located at the right side of the engine, directly under the intake manifold.

1. Disconnect the negative battery cable.

2. Remove the air cleaner assembly.

3. Remove the intake manifold assembly.

4. Disconnect and plug the fuel lines at the fuel pump.

5. Remove the fuel pump-to-engine bolts, then remove the pump assembly.

To install:

6. Remove the cylinder head cover.

7. Rotate the engine to position the No. 4 cylinder at TDC.

8. Lift the fuel pump pushrod toward the camshaft and hold it in the raised position.

9. Using a new gasket, install the fuel pump on the engine; torque the bolts to 20 ft. lbs. (27 Nm).

10. Connect the fuel hoses to the fuel pump.

11. Using a new gasket, install the intake manifold.

12. Install the air cleaner assembly.

13. Connect the negative battery cable.

14. Start the engine and check for fuel leaks.

Fuel Injected Engines

> **CAUTION**
> *Fuel injection systems remain under pressure even after the engine has been turned off. The fuel system pressure must be relieved before disconnecting any fuel lines. Failure to do so may result in fire and personal injury.*

1. Disconnect the negative battery cable.

2. Relieve the fuel system pressure.

3. Raise and safely support the vehicle.

4. Drain the fuel into an approved sealable container. Install the drain plug with a new washer and tighten it to 14 ft. lbs. (19 Nm) for 8mm bolts, or 22 ft. lbs. (30 Nm) for 14mm bolts.

> **CAUTION**
> *Observe all applicable safety precautions when working around fuel. Do not allow fuel spray or vapor to come in contact with a spark or open flame. Keep a class B dry chemical fire extinguisher near your work area. Never drain or store fuel in an open container due to the risk of fire or explosion.*

5. If equipped, remove the fuel tank undercover.

6. Disconnect the feed, return, and vapor hoses.

7. Disconnect all wiring harness connectors.

8. Disconnect the filler neck.

9. Position a suitable jack under the tank, remove the retainers and lower the tank far enough to disconnect any remaining wiring or hoses.

10. Remove the fuel tank from the vehicle.

To install:

11. Position a floor jack under the tank; raise it far enough to reconnect the wiring and hoses. Install the retainers. Torque the mounting bolts to 27 ft. lbs. (37 Nm).

12. Connect the filler neck.

13. Connect all harness connectors.

14. Connect the feed, return, and vapor hoses.

15. Refill the tank. Install the fuel tank undercover.

16. Connect the negative battery cable and check for leaks.

Fuel Injector

REMOVAL AND INSTALLATION

2.3L (VIN L) and 2.6L (VIN E) MFI Engines

> **CAUTION**
> *Fuel injection systems remain under pressure even after the engine has been turned off. The fuel system pressure must be relieved before disconnecting any fuel lines. Failure to do so may result in fire and/or personal injury.*

1. Relieve the fuel pressure.

2. Disconnect the negative battery cable.

3. Remove the air intake duct.

4. Remove the throttle body for extra working room.

5. Label and disconnect the harnesses from the fuel injectors.

6. Disconnect the fuel lines from the fuel rail.

CAUTION

Do not allow fuel spray or fuel vapors to come in contact with a spark or open flame. Keep a dry chemical fire extinguisher nearby. Never store fuel in an open container due to risk of fire or explosion.

7. Clean up any fuel that may spill on the engine or intake manifold.

8. Remove the fuel rail and injectors from the intake manifold as an assembly.

9. Separate the fuel injectors from the fuel rail.

To install:

10. Lubricate new O-rings with clean engine oil and install them onto the fuel injectors.

11. Install the fuel injectors onto the fuel rail.

12. Lubricate the fuel injector O-rings with clean engine oil and install them, along with the fuel rail, onto the intake manifold.

13. Install the fuel rail mounting bolts and tighten the bolts to 14 ft. lbs. (19 Nm).

14. Connect the fuel lines to the fuel rail using new sealing washers.

15. Connect the harnesses to the injectors.

16. Install the throttle body with a new gasket, and tighten the mounting bolts to 14 ft. lbs. (19 Nm).

17. Install the air intake duct.

18. Connect the negative battery cable.

19. Turn the ignition to the **ON** position to pressurize the fuel system. Then, check the injectors and fuel line fittings for leaks.

20. After warming the engine up to normal operating temperature, check the operation of the throttle cable and adjust it if necessary.

3.1L (VIN Z) Engine

1. Disconnect the negative battery cable.

2. Relieve the fuel system pressure.

3. Remove the air cleaner assembly.

4. Remove the harness to the injectors by squeezing the plastic tabs and pulling straight up.

5. Remove the fuel meter cover attaching screws and the cover assembly.

6. Remove the meter outlet passage gasket and the pressure regulator dust seal. If the fuel meter cover gasket is stuck to the meter body, leave it in place. If it is stuck to the

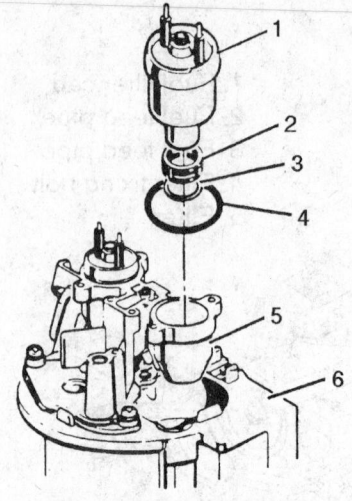

1. Fuel injector assembly
2. Fuel injector inlet filter
3. Fuel injector lower "o" ring
4. Fuel injector upper "o" ring
5. Fuel meter body assembly
6. Throttle body assembly

177752

Fuel injector and O-rings — 3.1L engine

fuel meter cover, remove it, and place it on the fuel meter body.

NOTE: The fuel meter cover gasket will be used to protect the fuel meter body in the next step.

7. Use a screwdriver and fulcrum to carefully pry the injector out of the fuel meter body.

8. Discard the fuel meter cover gasket, upper and lower O-rings.

To install:

NOTE: Be sure to replace the injector with the identical part. Injectors from other models will fit but are calibrated for different flow rates. Also, injector service packages may contain an injector washer (spacer). This washer is not required for this application.

9. Lubricate the new O-rings with engine oil and install the large O-ring in the fuel injector body and the small O-ring on the fuel injector nozzle. Make sure the O-rings are properly seated.

10. Align the raised lug on the injector base with the notch in the fuel metering body and install the fuel injector. The electrical terminals should be parallel with the throttle shaft.

11. Using new gaskets, apply Loctite 262® or equivalent to the screw threads and install the fuel meter cover. Torque the screws to 27 inch lbs. (3 Nm).

12. Connect the harnesses to the fuel injectors.

13. Connect the negative battery cable, turn the ignition switch to the **ON** position for two seconds then **OFF** for ten seconds. Turn the switch **ON** again and inspect for fuel leaks.

14. Install the air cleaner assembly.

3.2L (VIN V) Engine

CAUTION

Fuel injection systems remain under pressure even after the engine has been turned off. The fuel system pressure must be relieved before disconnecting any fuel lines. Failure to do so may result in fire and personal injury.

1. Relieve the fuel system pressure.

2. Disconnect the negative battery cable and reinstall the fuel pump relay.

3. Remove the air cleaner assembly.

4. Disconnect the accelerator pedal cable from the throttle body and bracket.

5. Disconnect the charcoal canister hose from the vacuum pipe.

6. Disconnect the air vacuum hose and booster hose from the common chamber.

7. Disconnect the harnesses from the MAP sensor; charcoal canister vacuum switching valve; exhaust gas recirculation valve; intake air temperature sensor and the engine ground cable.

8. Disconnect the spark plug wires from the valve covers.

9. Remove the ignition module assembly with the spark plug wires attached.

10. Disconnect the vacuum hoses from the throttle body.

11. Remove the throttle body mounting bolts. Then, remove the throttle body.

12. Disconnect the PCV hose, fuel pressure control valve vacuum hose, evaporative emission canister purge hose and EGR valve assembly from the common chamber.

13. Remove the common chamber from the intake manifold (six bolts, two nuts and three brackets).

14. Disconnect the fuel feed and return hoses from the fuel rail. Unbolt the fuel rail brackets from the cylinder head cover.

15. Disconnect the thermo sensor.

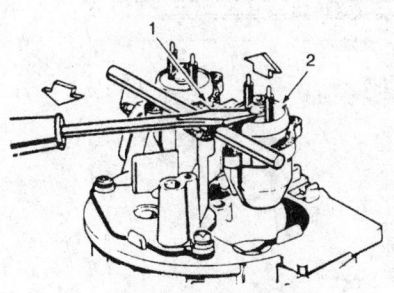

1. Fuel meter cover gasket
2. Fuel injector assembly

177753

Fuel injector removal — 3.1L engine

1. Fuel injector - top view
A. Part identification number
B. Vendor identification

C. Build date code
   D. Year
   E. Day
   F. Month
      1-9 (Jan-Sept)
      O, N, D (Oct, Nov, Dec)

177754

Fuel injector identification

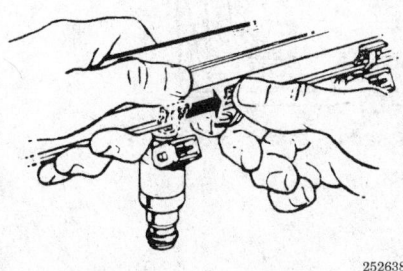

252638

Slide the retainer to remove the fuel injector — 3.2L engine

16. Remove the fuel rail mounting bolts. Lift the fuel rail up with the injectors still attached. Remove the fuel rail and injector assembly from the vehicle.

17. Slide the retainer clip sideways to remove each fuel injector from the fuel rail.

18. Clean up any fuel that might spill.

To install:

NOTE: **Always use new O–rings when assembling the fuel injectors and installing the fuel rail assembly to the intake manifold. Always use new sealing washers when reconnecting fuel lines.**

19. Lubricate new O-rings with engine oil and install them on the fuel injectors.

20. Install the fuel injectors onto the fuel rail and secure with the retaining clips.

21. Install the fuel rail to the intake manifold. Make sure the injectors are securely seated into the manifold ports.

22. Connect the harnesses to the fuel injectors and the thermo sensor.

23. Connect the fuel return and feed hoses to the fuel rail.

24. Install the common chamber. Torque the bolts and nuts in a crisscross pattern to 17 ft. lbs. (23 Nm).

NOTE: **Replace any gaskets between the throttle body, common chamber, and intake manifold to prevent the possibility of leaks.**

25. Install the EGR assembly. Torque the bolts to 78 inch lbs. (9 Nm).

26. Connect the charcoal canister purge, fuel pressure regulator vacuum hoses and PCV hose to the common chamber.

27. Install the throttle body assembly and connect the vacuum hoses to the throttle body. Torque the throttle body mounting bolts to 16 ft. lbs. (22 Nm).

28. Install the ignition module assembly. Torque the mounting bolts to 16 ft. lbs. (22 Nm).

29. Reconnect the spark plug wires.

30. Connect the MAP sensor; charcoal canister vacuum switching valve; exhaust gas recirculation valve; intake air temperature sensor and the engine ground cable.

31. Connect the air vacuum and vacuum booster hoses to the common chamber.

32. Connect the charcoal canister vacuum hose to the vacuum pipe.

33. Connect the accelerator cable to the throttle body and bracket.

34. Install the air cleaner and reconnect the negative battery cable.

35. Turn the ignition key to the **ON** position for two seconds, then **OFF**. Turn the ignition **ON** again to pressurize the fuel system and check for leaks where the fuel system was disconnected.

EMISSION CONTROLS

Service Interval Lamp

RESETTING

Oxygen Sensor Life Indicator light

The oxygen sensor must be replaced after 90,000 miles (144,000 km) of vehicle operation. When the odometer reading reaches 90,000 miles (144,000 km), the oxygen sensor life indicator light (O2) will illuminate to remind the driver to change the oxygen sensor.

After replacing the oxygen sensor, the oxygen sensor life indicator light must be reset to remind the driver to replace the oxygen sensor after the next 90,000 miles (144,000 km). The reset screw is located in the back of the instrument cluster. Perform the reset procedure as follows:

1. Remove the Instrument panel cluster assembly.

2. Remove the masking tape from hole **B**.

3. Remove the screw from hole **A** and install it to hole **B**.

4. Apply new masking tape to hole **A**.

NOTE: **The above procedure assumes that the oxygen sensor is being replaced for the first time (after 90,000 miles/144,000 km). For the subsequent reset procedure (after the next 90,000 miles/144,000 km), the hole positions will be opposite of the above procedure.**

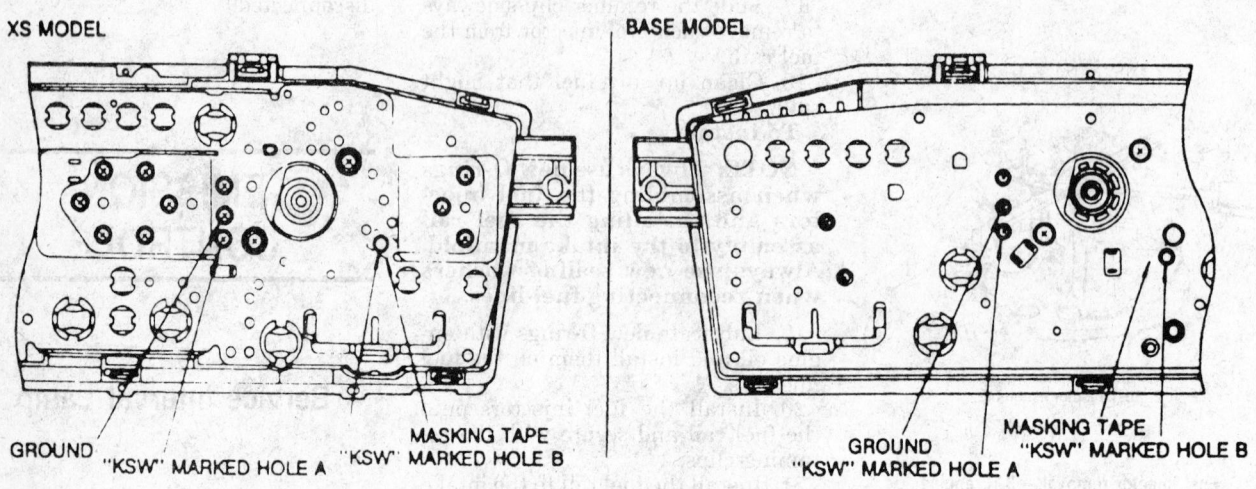

XS MODEL

BASE MODEL

GROUND
"KSW" MARKED HOLE A
"KSW"
MASKING TAPE
MARKED HOLE B

GROUND
"KSW" MARKED HOLE A
MASKING TAPE
"KSW" MARKED HOLE B

277880

Resetting the oxygen sensor life indicator (Pick-up)

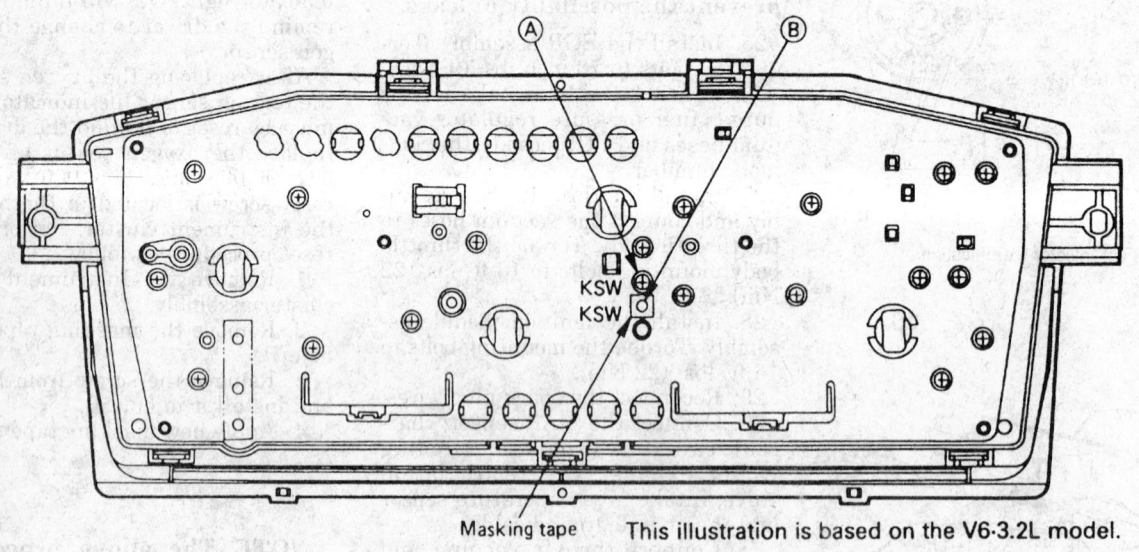

Ⓐ

Ⓑ

KSW
KSW

Masking tape   This illustration is based on the V6-3.2L model.

276485

Resetting the oxygen sensor life indicator — Passport

ENGINE MECHANICAL

Engine Assembly

REMOVAL AND INSTALLATION

1993–94 2.3L (VIN L) and 2.6L (VIN E) Engines

NOTE: Disconnecting the negative battery cable on some vehicles may interfere with the functions of the on-board computer systems and may require the computer to undergo a relearning process once the negative battery cable is reconnected.

1. Relieve the fuel system pressure, if fuel injected.
2. Disconnect both battery cables, negative cable first. Remove the battery.
3. Matchmark the hood-to-hinges and remove the hood.
4. Remove the undercover, if equipped. Open the drain plugs on the radiator and the cylinder block and drain the cooling system.
5. Remove the air cleaner assembly.
6. Label and disconnect the necessary hoses, harnesses, cables and control rods from the engine.
7. Label and disconnect the following items:
    a. Air switching valve hose.
    b. Oxygen sensor wire.
    c. Vacuum switching valve hose.
    d. Thermal vacuum switching valve hose.
    e. Pressure regulator vacuum hose.
    f. Canister hose.
    g. PCM harness.
    h. Fuel hose.
8. Remove the clutch return spring, clutch control cable, if equipped, backup light switch wires and speedometer cable from the transmission.
9. Remove the radiator grille.
10. Disconnect the upper and lower radiator hoses and the reservoir tank hose.
11. Remove the fan shroud, blade and radiator.
12. If equipped with A/C, remove the compressor from the engine and move aside; do not disconnect the pressure hoses.

13. Remove the gear shift lever by performing the following procedures:
    a. Place the gear shift lever in **N**.
    b. Remove the front console from the floor.
    c. Pull the shift lever boot and grommet upward.
    d. Remove the shift lever cover bolts and shift lever.
14. If equipped with 4-wheel drive, remove the transfer shift lever by performing the following procedures:
    a. Place the transfer shift lever in **H**.
    b. Pull the shift lever boot and dust cover upward.
    c. Remove the shift lever retaining bolts.
    d. Pull the shift lever from the transfer case.
15. Raise and safely support the vehicle. Remove the front wheels.
16. Drain the oil from the engine, transmission and transfer case, if equipped.
17. If equipped with an automatic transmission, perform the following procedures:
    a. Remove the oil level gauge (dipstick) and the tube.
    b. Disconnect the shift select control link rod from the select lever.
    c. Disconnect the downshift cable from the transmission.
    d. Disconnect and plug the fluid coolant lines from the transmission.
18. If equipped with a 1-piece driveshaft, remove the driveshaft flange-to-pinion nuts. Lower the driveshaft and remove.
19. If equipped with a 2-piece driveshaft, perform the following procedures:
    a. Remove the rear driveshaft flange-to-pinion nuts.
    b. Remove the rear driveshaft flange-to-front driveshaft flange bolts and the rear driveshaft.
    c. Remove the center bearing-to-chassis bolts, move the front driveshaft rearward away from the transmission.
20. Remove the front driveshaft's splined yoke flange-to-transfer case bolts and separate the front driveshaft from the transfer case; do not allow the splined flange to fall away from the driveshaft.
21. Remove the starter-to-engine bolts. Disconnect the starter wires, and remove.
22. Remove the exhaust pipe-to-exhaust manifold nuts, exhaust pipe bracket-to-transmission bolts, front exhaust pipe-to-2nd exhaust pipe

bolts and the front exhaust pipe from the vehicle.
23. Attach an engine hanger to the rear of the exhaust manifold.
24. Using an engine hoist, connect it to the engine hangers and support the engine.
25. If equipped with a manual transmission, perform the following procedures:
    a. Using a transmission jack, place it under the transmission; do not support.
    b. Remove the rear mount-to-transmission nuts.
    c. Remove the rear mount-to-crossmember nuts/bolts and the mount.

NOTE: Further removal of the transmission may require an assistant.

    d. Remove the clutch cover and the transmission-to-engine bolts.
    e. Move the transmission rearward into the crossmember and floor pan area; the transmission may rest on the crossmember.
    f. Lower the front of the transmission toward the jack.
    g. Firmly, grasp the transmission while someone raises the jack toward the transmission.
    h. Carefully lower the transmission onto the jack and center it.
    i. Lower the jack and move the transmission rearward.
26. If equipped with an automatic transmission, perform the following procedures:

NOTE: Removal of the transmission will require an assistant.

    a. Remove the torque converter-to-flexplate bolts through the starter hole.
    b. Using a transmission jack, place it under the transmission; do not support it.
    c. Remove the rear mount-to-transmission nuts.
    d. Remove the rear mount-to-crossmember nuts/bolts and the mount.
    e. Remove the transmission-to-engine bolts.
    f. Move the transmission rearward into the crossmember and floor pan area; the transmission may rest on the crossmember.
    g. Lower the front of the transmission toward the jack.
    h. Firmly, grasp the transmission while the assistant raises the jack toward the transmission.
    i. Carefully, lower the transmission onto the jack and center it.
    j. Lower the jack and move the transmission rearward.

27. Remove the transmission/transfer case assembly by performing the following procedures:

a. Using a transmission jack, place it under the transmission and support the assembly.

b. Remove the rear mount-to-transmission nuts.

c. Remove the rear mount-to-side mount member nuts/bolts and the mount.

d. Remove the transmission-to-engine bolts.

e. Move the transmission assembly rearward.

f. Carefully lower the transmission.

28. Remove the engine-to-mount nuts/bolts.

29. Using the hoist, slowly, lift the engine; be sure to hold the front of the engine higher than the rear.

30. Place the engine on a workstand.

To install:

31. Using a hoist, slowly lower the engine into the vehicle; be sure to hold the front of the engine higher than the rear.

32. Install the engine-to-mount nuts/bolts.

33. Install the transmission/transfer assembly by performing the following procedures:

a. Raise the transmission into position.

b. Move the transmission forward and engage it with the engine.

c. Install the engine-to-transmission bolts.

d. Install the rear mount and the rear mount-to-side mount member nuts/bolts.

e. Install the rear mount-to-transmission nuts.

f. Remove the transmission jack.

34. If equipped with an automatic transmission, perform the following procedures:

NOTE: Installation of the transmission will require an assistant.

a. Raise the transmission into position.

b. Raise the rear of the transmission and move it into position on the crossmember.

c. Move the transmission forward and engage it with the engine.

d. Install the engine-to-transmission bolts.

e. Install the mount and the rear mount-to-crossmember nuts/bolts.

f. Install the rear mount-to-transmission nuts.

g. Install the torque converter-to-flexplate bolts through the starter hole.

35. If equipped with a manual transmission, perform the following procedures:

NOTE: Installation of the transmission may require an assistant.

a. Raise the transmission into position.

b. Raise the rear of the transmission and move it into position on the crossmember.

c. Move the transmission forward and engage it with the engine.

d. Install the engine-to-transmission bolts.

e. Install the mount and the rear mount-to-crossmember nuts/bolts.

f. Install the rear mount-to-transmission nuts.

36. Remove the engine hoist and the engine hanger from the rear of the exhaust manifold.

37. Install the front exhaust pipe, exhaust pipe-to-exhaust manifold nuts, exhaust pipe bracket-to-transmission bolts and front exhaust pipe-to-2nd exhaust pipe bolts.

38. Install the starter and the starter-to-engine bolts. Reconnect the starter wiring.

39. Install the front driveshaft's splined yoke flange-to-transfer case bolts.

40. If equipped with a 2-piece driveshaft, perform the following procedures:

a. Install the front driveshaft into the transmission and the center bearing-to-chassis bolts.

b. Install the rear driveshaft and the rear driveshaft flange-to-front driveshaft flange bolts.

c. Install the rear driveshaft flange-to-pinion nuts.

41. If equipped with a 1-piece driveshaft, install the driveshaft into the transmission and the driveshaft flange-to-pinion nuts.

42. If equipped with an automatic transmission, perform the following procedures:

a. Connect the fluid coolant lines to the transmission.

b. Connect the downshift cable to the transmission.

c. Connect the shift select control link rod to the select lever.

d. Install the oil level gauge and the tube.

43. Install the front wheels and lower the vehicle.

44. Install the transfer shift lever by performing the following procedures:

a. Position the shift lever into the transfer case.

b. Install the shift lever retaining bolts.

c. Push the dust cover and the shift lever boot downward.

45. Install the gear shift lever by performing the following procedures:

a. Install the shift lever and the shift lever cover bolts.

b. Push the grommet and shift lever boot downward.

c. Install the front console to the floor panel.

46. If equipped with A/C, install the compressor.

47. Install the radiator, the fan blade and the fan shroud.

48. Connect the upper and lower radiator hoses and reservoir tank hose.

49. Install the radiator grille to the deflector panel.

50. Install the clutch return spring, if equipped, the clutch control cable, if equipped, the backup light switch connector and the speedometer cable to the transmission.

51. Connect the following items:

a. Air switch valve hose.

b. Oxygen sensor wire.

c. Vacuum switch valve hose.

d. Thermal vacuum switching valve hose.

e. Pressure regulator vacuum hose.

f. Canister hose.

g. PCM harness.

h. Fuel hose(s).

52. Install the remaining components.

53. Install the battery and connect both battery cables, the positive cable first.

54. Adjust the drive belt tension. Start the engine, check for leaks.

55. Check and/or adjust the idle speed and ignition timing.

1995–97 2.3L (VIN L) Engine

NOTE: The manufacturer recommends removing the transmission from the vehicle before removing the engine. If you choose to leave the transmission in the vehicle, it must be securely supported.

1. Relieve the fuel system pressure.

2. Disconnect the negative and positive battery cables. Remove the battery.

3. Use a felt–tipped marker to matchmark the hood hinge plates to the hood. Remove the hood.

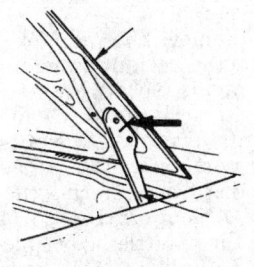

256142

Matchmark the hood and hinges

4. Remove the air cleaner duct and hose assembly. Using a clean shop cloth, cover the air intake to prevent dirt from entering the engine.

5. Drain the coolant from the radiator and engine block.

6. Label and disconnect the following components:
   a. Engine ground cables.
   b. Throttle cable.
   c. Starter motor wires.
   d. Alternator terminals.
   e. Oxygen sensor connector.
   f. Oil pressure switch connector.
   g. Air switching valve hose.
   h. Vacuum switching valve hose.
   i. Thermal vacuum switching valve hose.
   j. Pressure regulator vacuum hose.
   k. Canister hose.
   l. PCM harness.
   m. Fuel inlet and return lines.

7. Disconnect the upper and lower radiator hoses and the reservoir tank hose.

8. Remove the fan shroud and the cooling fan assembly.

9. Remove the radiator.

10. If equipped with air conditioning, remove the compressor from the engine and move it aside; do not disconnect the pressure hoses.

11. Remove the gear shift lever by performing the following procedures:
   a. Place the gear shift lever in **N**.
   b. Remove the front console from the floor panel.
   c. Pull the shift lever boot and grommet upward.
   d. Remove the shift lever cover bolts and the shift lever.

12. Raise and safely support the vehicle. Remove the front wheels.

13. Drain the engine oil and the transmission fluid.

14. If equipped with an automatic transmission, perform the following procedures:
   a. Remove the dipstick and its tube.

   b. Disconnect the shift select control link rod from the select lever.
   c. Disconnect the downshift cable from the transmission.
   d. Disconnect and plug the fluid coolant lines from the transmission.

15. If equipped with a one-piece driveshaft, remove the driveshaft flange-to-pinion attaching bolts. Lower the driveshaft and pull it from the transmission.

16. If equipped with a two-piece driveshaft, perform the following procedures:
   a. Remove the bolts from the axle pinion flange.
   b. Unbolt the rear driveshaft section from the front driveshaft section.
   c. Remove the center bearing-to-chassis bolts, move the front driveshaft section away from the transmission to remove it.

17. Remove the starter.

18. Unbolt the exhaust pipe from the exhaust manifold and the catalytic converter. The exhaust pipe may remain in the vehicle suspended by its bracket.

19. Attach an engine hanger to the rear of the exhaust manifold.

20. Connect a lifting chain to the engine hangers. Raise the chain slightly to support the engine.

21. If equipped with a manual transmission, perform the following procedures:
   a. Place a jack under the transmission. Use the jack to support the transmission in place, not to take up the transmission weight.
   b. Remove the rear mount bracket attaching bolts.
   c. Remove the rear mount attaching nuts and remove the mount.

NOTE: Further removal of the transmission may require an assistant.

   d. Remove the clutch cover and the transmission attaching bolts.
   e. Move the transmission rearward into the crossmember and floor pan area; the transmission may rest on the crossmember.
   f. Lower the front of the transmission toward the jack.
   g. Firmly grasp the transmission while the assistant raises the jack toward the transmission.
   h. Carefully lower the transmission onto the jack and center it.
   i. Lower the jack and move the transmission rearward.

22. If equipped with an automatic transmission, perform the following procedures:

NOTE: Removal of the transmission will require an assistant.

   a. Place a jack under the transmission. Use the jack to support the transmission in place, not to take up the transmission weight.
   b. Remove the rear mount bracket attaching bolts.
   c. Remove the rear mount attaching nuts and remove the mount.
   d. Remove the transmission attaching bolts.
   e. Move the transmission rearward into the crossmember and floor pan area; the transmission may rest on the crossmember.
   f. Lower the front of the transmission toward the jack.
   g. Firmly grasp the transmission while the assistant raises the jack toward the transmission.
   h. Carefully lower the transmission onto the jack and center it.
   i. Lower the jack and move the transmission rearward.

23. Remove the engine mount nuts.

24. Slowly hoist the engine up a few inches. Be sure to hold the front of the engine higher than the rear.

25. Verify that all wires and hoses have been disconnected from the engine.

26. Lift the engine out of the vehicle.

27. Place the engine on a workstand.

To install:

28. Slowly lower the engine into the vehicle; be sure to hold the front of the engine higher than the rear.

29. Install the engine mount nuts and bolts. Only hand–tighten the mounting nuts at this time.

30. If equipped with an automatic transmission, perform the following procedures:

NOTE: Installation of the transmission will require an assistant.

   a. Raise the transmission into position.
   b. Raise the rear of the transmission and move it into position on the crossmember.
   c. Move the transmission forward and engage it with the engine.
   d. Install the torque converter attaching bolts through the starter hole. Tighten the bolts to 13 ft. lbs. (18 Nm).

e. Install the transmission attaching bolts. Tighten the bolts to 28 ft. lbs. (38 Nm).

f. Install the rear mount bracket to the transmission. Tighten the bolts to 28 ft. lbs. (38 Nm).

g. Lower the mount onto the crossmember and tighten the rear mount attaching nuts to 62 ft. lbs. (83 Nm).

31. If equipped with a manual transmission, perform the following procedures:

NOTE: Installation of the transmission may require an assistant.

a. Raise the transmission into position.

b. Raise the rear of the transmission and move it into position on the crossmember.

c. Move the transmission forward and engage it with the engine.

d. Align the input shaft splines of the transmission with the clutch disc splines.

e. Install the transmission attaching bolts. Tighten the bolts to 28 ft. lbs. (38 Nm).

f. Install the rear mount to the transmission. Tighten the bracket attaching bolts to 28 ft. lbs. (38 Nm).

g. Lower the mount onto the crossmember and tighten the mounting nuts to 62 ft. lbs. (83 Nm).

32. Remove the lifting chain and the engine hanger from the rear of the exhaust manifold.

33. Tighten the engine mounting nuts to specification at this time. Tighten the mounting nuts to 62 ft. lbs. (83 Nm). Tighten the mount bolts to 37 ft. lbs. (50 Nm).

34. Connect the front exhaust pipe to the exhaust manifold and the catalytic converter. Tighten the flange nuts to 49 ft. lbs. (67 Nm).

35. Install the starter.

36. Install the driveshaft(s).

37. If equipped with an automatic transmission, perform the following procedures:

a. Connect the fluid coolant lines to the transmission.

b. Connect the downshift cable to the transmission.

c. Connect the shift select control link rod to the select lever.

d. Install the oil level gauge and the tube.

38. Reconnect the clutch cable, if equipped.

39. Install the front wheels and lower the vehicle.

40. Install the gear shift lever:

a. Install the shift lever and the shift lever cover bolts.

b. Push the grommet and shift lever boot downward.

c. Install the front console.

41. If equipped with air conditioning, install the compressor to the engine.

42. Install the cooling fan assembly.

43. Install the radiator and the fan shrouds.

44. Connect the upper and lower radiator hoses and the reservoir tank hose.

45. Install the radiator grille to the deflector panel.

46. Connect the following components:

a. Engine ground cables.

b. Throttle cable.

c. Starter motor wires.

d. Alternator terminals.

e. Oxygen sensor connector.

f. Oil pressure switch connector.

g. Air switching valve hose.

h. Vacuum switching valve hose.

i. Thermal vacuum switching valve hose.

j. Pressure regulator vacuum hose.

k. Canister hose.

l. PCM harness.

m. Fuel inlet and return lines.

47. Install the air cleaner duct and hose assembly.

48. Fill the engine and transmission with fresh fluids.

49. Install or connect the remaining components.

50. Verify that all wires, hoses, and cables have been reconnected properly.

51. Start the engine, check for fluid, air, and fuel leaks.

52. Check and/or adjust the idle speed and ignition timing.

53. Adjust the throttle cable and clutch cable.

1995–97 2.6L (VIN E) Engine

NOTE: The transmission or transmission and transfer case assembly should be completely removed from the vehicle before the engine is removed. If you chose to leave the transmission in the vehicle after separating it from the engine, it must be securely supported. The transfer case shouldn't be separated from the transmission.

1. Relieve the fuel pressure.

2. Disconnect the negative and positive battery cables. Remove the battery.

3. Use a felt–tipped marker to matchmark the hinge plates. Remove the hood.

4. Remove the radiator skid plate. Drain the coolant from the radiator and engine block.

5. Remove the air cleaner box and the intake air duct. Cover the throttle body port with a shop towel to prevent dirt from entering the engine.

6. Disconnect the throttle cable from the throttle body linkage.

7. Label and disconnect the following:

a. Air switch valve hose.

b. Oxygen sensor harness.

c. Power booster vacuum hose.

d. Alternator wiring harness.

e. Fuel pressure regulator vacuum hose.

f. Canister hose.

g. Engine wiring harness connectors located on the right wheel well.

h. Inlet and return fuel lines.

i. Starter motor cables.

j. Engine ground cables.

k. Oil pressure switch connectors.

8. Remove the radiator grille from the deflector panel.

9. Remove the radiator.

10. If equipped with A/C, remove the compressor from the engine and move it aside. Do not disconnect the A/C lines.

11. If equipped with a manual transmission, remove the gear shift lever by performing the following procedures:

a. Place the gear shift lever in **N**.

b. Remove the front console.

c. Pull the shift lever boot and grommet upward.

d. Remove the shift lever cover bolts and the shift lever.

12. If equipped with four–wheel drive, remove the transfer case shift lever by performing the following procedures:

a. Place the transfer shift lever in **2H**.

b. Pull the shift lever boot and dust cover upward.

c. Remove the shift lever retaining bolts.

d. Pull the shift lever from the transfer case.

13. Raise and safely support the vehicle. Remove the front wheels.

14. Disconnect the backup light switch and the vehicle speed sensor cable from the transmission.

15. Remove the transmission and transfer case skid plates.

16. Drain the oil from the engine, transmission and transfer case, if equipped.

17. If equipped with an automatic transmission, perform the following procedures:

a. Remove the dipstick and the tube.

b. Disconnect the shift select control link rod from the select lever.

c. Disconnect the downshift cable from the transmission.

d. Disconnect and plug the fluid coolant lines from the transmission.

18. Remove the driveshaft(s).

19. Remove the starter.

20. If equipped with a clutch slave cylinder, remove it from the transmission and move it aside.

21. Unbolt the front exhaust pipe flanges and separate the front pipe from the exhaust system.

22. Attach engine hangers to the engine and connect a chain hoist to the engine hangers and support the engine.

23. Remove the transmission or transmission and transfer case assembly.

24. Support the weight of the engine with the chain hoist. Unbolt the engine mounts.

25. Using the hoist, slowly lift the engine from the vehicle; be sure to hold the front of the engine higher than the rear.

26. Place the engine on a workstand.

To install:

27. Using the hoist, slowly lower the engine into the vehicle. Be sure to hold the front of the engine higher than the rear.

28. Install the engine mount nuts and bolts. Tighten the engine mount and transmission mount bolts to 30 ft. lbs. (41 Nm). Tighten the engine mount nuts to 62 ft. lbs. (83 Nm). Tighten the transmission mount nuts to 30 ft. lbs. (41 Nm).

29. Install the transmission and/or transfer assembly.

30. Remove the engine hoist and the engine hangers from the engine.

31. Install the front exhaust pipe and reconnect the exhaust system with new self-locking nuts.

32. If equipped with a clutch slave cylinder, install it onto the transmission.

33. Install the starter.

34. Install the driveshafts.

35. If equipped with an automatic transmission, perform the following procedures:

a. Connect the fluid coolant lines to the transmission.

b. Connect the downshift cable to the transmission.

c. Connect the shift select control link rod to the select lever.

d. Install the oil level gauge and the tube.

36. Connect the backup light switch and the vehicle speed sensor cable to the transmission.

37. Install the front wheels and lower the vehicle.

38. Install the transfer case shift lever by performing the following procedures:

a. Position the shift lever into the transfer case.

b. Install the shift lever retaining bolts.

c. Push the dust cover and the shift lever boot downward.

39. Install the gear shift lever by performing the following procedures:

a. Install the shift lever and the shift lever cover bolts.

b. Push the grommet and shift lever boot downward.

c. Install the front console.

40. Install the power steering pump and bracket to the engine.

41. If equipped with A/C, install the compressor to the engine.

42. Install the radiator and grille

43. Reconnect the following:

a. Air switch valve hose.

b. Oxygen sensor harness.

c. Power booster vacuum hose.

d. Alternator wiring harness.

e. Fuel pressure regulator vacuum hose.

f. Canister hose.

g. Engine wiring harness connectors located on the right wheel well.

h. Inlet and return fuel lines.

i. Starter motor cables.

j. Engine ground cables.

k. Oil pressure switch connectors.

44. Reconnect the throttle cable.

45. Install the air cleaner duct and hose.

46. Refill the engine, transmission and transfer case with fresh fluids.

47. Refill and bleed the cooling system.

48. Refill and bleed the power steering system.

49. Install the radiator skid plate and tighten the bolts to 27 ft. lbs. (37 Nm).

50. Align the matchmarks and install the hood.

51. Install the battery. Connect the positive and negative battery cables.

52. Start the engine, check for fuel, coolant, and oil leaks.

53. After the engine has warmed up, check the throttle cable deflection and operation. Check the belt tension and adjust if necessary.

3.1L (VIN Z) Engine

1. Relieve the fuel pressure. Disconnect both battery cables, the negative cable first. Remove the battery.

2. Matchmark the hood-to-hinges and remove the hood.

3. Remove the undercover, if equipped. Open the drain plugs on the radiator and the cylinder case and drain the cooling system.

4. Remove the air cleaner. Using a clean shop cloth, cover the air cleaner port to prevent dirt from entering the engine.

5. Label and disconnect the necessary hoses, electrical connectors, control cables and control rods from the engine.

6. Label and disconnect the following items:

a. Air switch valve hose.

b. Oxygen sensor wire.

c. Vacuum switch valve hose.

d. Thermal vacuum switching valve hose.

e. Pressure regulator vacuum hose.

f. Canister hose.

g. PCM harness.

h. Fuel hose(s).

7. Remove the clutch return spring (if equipped), the clutch control cable (if equipped), the backup light switch connector and the speedometer cable from the transmission.

8. Remove the radiator grille from the deflector panel.

9. Remove the fan shroud, fan blade assembly and the radiator.

10. If equipped with A/C, remove the compressor and place aside. Do not disconnect the pressure hoses.

11. Perform the following procedures:

a. Remove the power steering pump-to-engine brackets and move the pump aside.

b. Remove the spark plug wire from the No. 1 spark plug.

c. Remove the distributor cap with the No. 1 spark plug wire.

d. Remove the ignition coil.

12. Remove the gear shift lever by performing the following procedures:

a. Place the gear shift lever in **N**.

b. Remove the front console from the floor panel.

c. Pull the shift lever boot and grommet upward.

d. Remove the shift lever cover bolts and the shift lever.

13. Remove the transfer shift lever by performing the following procedures:

a. Place the transfer shift lever in **2H**.

b. Pull the shift lever boot and dust cover upward.

c. Remove the shift lever retaining bolts.

d. Pull the shift lever from the transfer case.

14. Raise and safely support the vehicle. Remove the front wheels. Drain the oil from the engine, transmission and transfer case.

15. If equipped with an automatic transmission, perform the following procedures:

a. Remove the oil level gauge and the tube.

b. Disconnect the shift select control link rod from the select lever.

c. Disconnect the downshift cable from the transmission.

d. Disconnect and plug the fluid coolant lines from the transmission.

16. Remove the driveshaft(s).

17. Remove the starter.

18. If equipped with a clutch slave cylinder, remove it from the transmission and move it aside.

19. Remove the exhaust pipe-to-exhaust manifold nuts, the exhaust pipe bracket-to-transmission bolts, the front exhaust pipe-to-2nd exhaust pipe bolts and the front exhaust pipe from the vehicle.

20. Attach an engine hanger to the rear of the exhaust manifold.

21. Using an engine hoist, connect it to the engine hangers and support the engine.

22. Remove the catalytic converter and the parking brake cable bracket.

23. Remove the transmission/transfer case assembly.

24. Remove the engine-to-mount nuts/bolts.

25. Using the hoist, slowly, lift the engine; be sure to hold the front of the engine higher than the rear.

26. Place the engine on a workstand.

To install:

27. Using the hoist, slowly, lower the engine into the vehicle; be sure to hold the front of the engine higher than the rear.

28. Install the engine-to-mount nuts/bolts.

29. Install the transmission/transfer assembly.

30. Install the catalytic converter and the parking brake cable bracket.

31. Remove the engine hoist and the engine hanger from the rear of the exhaust manifold.

32. Install the front exhaust pipe, exhaust pipe-to-exhaust manifold nuts, the exhaust pipe bracket-to-transmission bolts, the front exhaust pipe-to-2nd exhaust pipe bolts.

33. If equipped with a clutch slave cylinder, install it onto the transmission.

34. Install the starter.

35. Install the driveshaft(s).

36. If equipped with an automatic transmission, perform the following procedures:

a. Connect the fluid coolant lines to the transmission.

b. Connect the downshift cable to the transmission.

c. Connect the shift select control link rod to the select lever.

d. Install the oil level dipstick and the tube.

37. Install the front wheels and lower the vehicle.

38. Install the transfer shift lever by performing the following procedures:

a. Position the shift lever into the transfer case.

b. Install the shift lever retaining bolts.

c. Push the dust cover and the shift lever boot downward.

39. Install the gear shift lever by performing the following procedures:

a. Install the shift lever and the shift lever cover bolts.

b. Push the grommet and shift lever boot downward.

c. Install the front console to the floor panel.

40. Perform the following procedures:

a. Install the ignition coil.

b. Install the distributor cap with the No. 1 spark plug wire.

c. Install the spark plug wire from the No. 1 spark plug and reconnect the wires to the distributor cap.

d. Install the power steering pump-to-engine brackets.

41. If equipped with A/C, install the compressor to the engine.

42. Install the radiator, the fan blade assembly and the fan shroud.

43. Install the clutch return spring (if equipped), the clutch control cable (if equipped), the back-up light switch connector and the speedometer cable to the transmission.

44. Connect the following items:

a. Air switch valve hose.

b. Oxygen sensor wire.

c. Vacuum switch valve hose.

d. Thermal vacuum switching valve hose.

e. Pressure regulator vacuum hose.

f. Canister hose.

g. PCM harness.

h. Fuel hoses.

45. Connect the necessary hoses, electrical connectors, control cables and control rods to the engine.

46. Refill the engine, the transmission, the transfer case and the cooling system. Install the undercover, if equipped.

47. Install the hood.

48. Install the battery and connect both battery cables, the positive cable first.

49. Install the drive belt. Start the engine, check for leaks.

50. Check and/or adjust the idle speed and ignition timing.

3.2L (VIN V) Engine

— **WARNING** —

The transmission and transfer case assembly may be completely removed from the vehicle before the engine is removed. If you chose to leave the transmission and transfer case assembly in the vehicle, it must be securely supported.

1. Shift the transmission into the **N** or **NEUTRAL** position. If equipped with four-wheel drive, shift the transfer case into the **2H** position and verify that the front axle and hubs are not engaged. Set the parking brake and securely block the rear wheels while the vehicle is on the ground.

2. Relieve the fuel pressure.

3. Disconnect the negative and positive battery cables. Remove the battery.

4. Use a felt-tipped marker to matchmark the hood hinge plates. Remove the hood.

5. If equipped with a manual transmission, remove the gear shift lever:

a. Verify that the transmission is in **N**.

b. Remove retaining screws from the front console.

c. Remove the shift knob.

d. Pull the shift lever boot and grommet upward.

e. Remove the shift lever cover bolts and the shift lever.

6. If equipped with an automatic transmission:

a. Verify that the transmission is in **N**.

b. Remove the retaining screws from the front console.

c. Disconnect the shift lock cable.

d. Label and uncouple the wiring connectors.

e. After the vehicle has been raised and supported, disconnect the shift control rod from the selector lever linkage.

7. If equipped with four–wheel drive, remove the transfer case shift lever:

a. Verify that the transfer case is in **2H**.

b. Pull the shift lever boot and dust cover upward.

c. Remove the shift lever retaining bolts.

d. Pull the shift lever from the transfer case.

8. If equipped, remove the radiator skid plate.

9. Drain the engine coolant into a container.

10. Remove the air cleaner and the intake air duct. Use a clean shop cloth to plug the throttle body port to prevent dirt from entering the engine.

11. If necessary, remove the vehicle's grille to prevent it from being damaged.

12. Remove the radiator.

13. Remove the drive belts.

14. If equipped with A/C, unbolt the compressor and move it out of the work area. Don't disconnect the A/C lines.

15. Unbolt the power steering pump from its bracket. Move the pump out of the way with the hydraulic line connected.

16. Remove the starter.

17. Disconnect the throttle cable from the throttle body linkage.

18. Label and disconnect the following vacuum hoses from the intake manifold chamber:

a. PCV hose

b. EVAP canister vacuum hose

c. Brake booster hose

19. Label and disconnect the following sensor connectors from the rear of the intake manifold chamber:

a. Ignition control module connectors

b. Linear EGR valve

c. MAP sensor

d. EVAP purge valve

e. Throttle position sensor

f. Idle air control valve

g. Intake air temperature sensor

20. Disconnect the EGR valve supply tube and bracket.

21. Disconnect the MAP sensor tube, and then unbolt the MAP sensor bracket.

22. If necessary, the intake manifold chamber may be removed to avoid damage. If removed, cover the intake ports to keep dirt or foreign objects out.

23. If necessary, the ignition coil assembly may be removed to avoid damaged. Label the spark plug wires to avoid confusion.

24. If necessary, the cruise control actuator and cable brackets may be unbolted to move the actuator and cable out of the work area.

25. Raise and safely support the vehicle. Remove the front wheels.

26. Drain the oil from the engine, transmission and transfer case.

27. Disconnect the backup light switch connector and the speed sensor connector from the transmission.

28. If equipped with an automatic transmission, perform the following procedures:

a. Remove the dipstick and the tube.

b. Disconnect the shift select control link rod from the select lever.

c. Disconnect the downshift cable from the transmission.

d. Disconnect and plug the fluid coolant lines from the transmission.

29. Unbolt and remove the rear driveshaft. Unbolt the center bearing and lower the driveshaft from the vehicle.

30. If equipped, remove the front driveshaft's splined yoke flange-to-transfer case bolts and separate the front driveshaft from the transfer case; do not allow the splined flange to fall away from the driveshaft.

31. If equipped with a clutch slave cylinder, remove it from the transmission and move it aside.

32. Label and disconnect the oxygen sensor connectors.

33. Unbolt the front exhaust pipe flanges from the exhaust manifolds and catalytic converters. Separate the exhaust system from the engine, and move it out of the work area. If necessary, the front part of the exhaust system may be removed from the vehicle.

34. Attach a engine lifting chain to the engine hangers. The engine hangers are located on the right and left sides of the engine below the valve covers.

35. Make sure the engine is safely supported.

NOTE: Make sure that no engine components will be damaged by the lifting chain.

36. Remove the transmission/transfer case assembly.

37. Unbolt the engine mounts.

38. Raise the engine slightly. Verify that all vacuum lines and electrical connectors have been disconnected so that the engine removal is not obstructed.

39. Raise the chain hoist to lift the engine out of the vehicle. If necessary, keep the front of the engine higher than the rear to clear the bulkhead.

40. Secure the engine to a workstand.

To install:

41. Using the chain hoist, slowly lower the engine into the vehicle. Be sure to hold the front of the engine higher than the rear.

42. Install the engine mount nuts and bolts. Tighten the engine mount bolts to 30 ft. lbs. (41 Nm). Tighten the engine mount nuts to 37 ft. lbs. (50 Nm).

—————— **WARNING** ——————
Make sure that the transmission mounting dowels are in the correct locations for the type of transmission (M/T or A/T). Incorrect dowel positioning can crack the transmission case.

43. Install the transmission/transfer case assembly.

44. Install the front exhaust system components. Use new self–locking nuts where necessary.

45. Reconnect the oxygen sensor connectors.

46. If equipped with a clutch slave cylinder, install it onto the transmission.

47. Install the front driveshaft's splined yoke flange-to-transfer case bolts. Tighten the flange bolts to 46 ft. lbs. (63 Nm).

48. Install the driveshaft into the transmission and the driveshaft flange nuts to 46 ft. lbs. (63 Nm).

49. If equipped with an automatic transmission, perform the following procedures:

a. Connect the fluid coolant lines to the transmission.

b. Connect the downshift cable to the transmission.

c. Connect the shift select control link rod to the select lever.

d. Install the dipstick and the tube.

50. Connect the backup light switch connector and the vehicle speed sensor connector to the transmission.

51. Install the front wheels and lower the vehicle.

52. Remove the engine lifting chain.

53. Install the power steering pump and bracket to the engine.

54. Install the A/C compressor to the engine.

55. If removed, install the intake manifold chamber.

56. If removed, install the ignition coil assembly and reconnect the spark plug wires.

57. Reconnect the EGR valve supply tube and bracket.

58. Reconnect the MAP sensor tube, and then install the MAP sensor bracket.

59. Reconnect the following vacuum hoses to the intake manifold chamber:

a. PCV hose

b. EVAP canister vacuum hose

c. Brake booster hose

60. Reconnect the following sensor connectors to the rear of the intake manifold chamber:

a. Ignition control module connectors

b. Linear EGR valve

c. MAP sensor

d. EVAP purge valve

e. Throttle position sensor

f. Idle air control valve

g. Intake air temperature sensor

61. Reconnect the throttle cable from the throttle body linkage.

62. Install the starter.

63. Install and adjust the drive belts.

64. If removed, install the cruise control actuator and cable brackets.

65. Install the radiator.

66. Install the radiator skid plate and tighten the bolts to 27 ft. lbs. (37 Nm).

67. Reconnect the heater hoses.

68. Install the air cleaner duct and hose.

69. If equipped with a manual transmission, install the gear shift lever:

a. Install the shift lever and its mounting cover bolts.

b. Install the shift lever boot and grommet.

c. Install the front console.

d. Install the shift knob.

e. Verify that the transmission is in **N**.

70. If equipped with an automatic transmission:

a. Reconnect the shift control rod from the selector lever linkage.

b. Reconnect the wiring connectors.

c. Reconnect the shift lock cable.

d. Install the front console.

e. Verify that the transmission is in **N**.

71. If equipped with four–wheel drive, install the transfer case shift lever:

a. Install the shift lever onto the transfer case.

b. Install the shift lever retaining bolts.

c. Install the shift lever dust cover and boot.

d. Verify that the transfer case is in **2H**.

72. Refill the engine, transmission and transfer case with fresh fluids.

73. Refill and bleed the cooling system.

74. Refill and bleed the power steering system if the hydraulic lines were opened.

75. Align the matchmarks and install the hood.

76. Verify that all vacuum lines, cooling hoses, control cables, and electrical connections have been routed and connected properly.

77. Install the battery. Reconnect the positive and negative battery cables.

78. Crank the engine until it starts to remove any air from the fuel lines. Carefully check all fuel lines and fittings for signs of leakage.

79. Warm it up to normal operating temperature. Check the adjustment and operation of the throttle cable.

80. Shut the engine off. Recheck the drive belt tensions.

81. Check all fluid levels and refill as necessary.

82. Test drive the vehicle.

Engine Mounts

REMOVAL AND INSTALLATION

2.3L (VIN L) and 2.6L (VIN E) Engines

1. Disconnect the negative battery cable.

2. Raise and safely support the vehicle.

3. Carefully raise the engine slightly to remove the weight from the engine mounts.

4. Remove the engine mount(s) as needed.

To install:

------- WARNING -------

The engine may need to be moved with a prybar to get the bolt holes to line up. Be careful when prying around the engine components. The engine must be securely supported.

To install:

5. Install the engine mount(s). Lower the engine as needed to align the bolt holes. Install the engine mount nuts.

6. Lower the weight of the engine on the mount(s) and remove the engine lifting device.

7. Tighten the engine mount bolts to the following specifications:

a. Right side mount-to-frame nut: 37 ft. lbs (50 Nm).

b. Left side mount-to-frame nut: 37 ft. lbs (50 Nm).

c. Right side mount-to-engine bracket nut: 45 ft. lbs (61 Nm).

d. Left side mount-to-engine bracket nut: 38 ft. lbs (51 Nm).

8. Lower the vehicle and connect the negative battery cable.

3.1L (VIN Z) Engine

1. Disconnect the negative battery cable. Using a hoist, lift up on the engine hoisting tabs to remove weight from the engine mounts.

2. Loosen the engine mount hardware and remove the mounts from the engine and crossmember.

To install:

3. Install the mount and replace the hardware. Torque the bolts and nuts to 38 ft. lbs. (51 Nm).

4. Lower the engine and connect the negative battery cable.

3.2L (VIN V) Engine

1. Disconnect the negative battery cable.

2. Attach a chain hoist to the engine lifting hooks and raise the engine slightly to support the weight.

3. If necessary, raise and safely support the vehicle.

4. Loosen the engine mount fasteners and remove the mounts from the engine block and frame rails.

------- WARNING -------

The engine may need to be moved with a prybar to get the bolt holes to line up. Be careful when prying around the engine components. The engine must be securely supported.

To install:

5. Install the mount and replace the fasteners.

6. Tighten the mount bracket bolts to 30 ft. lbs. (41 Nm). Tighten the bracket nuts to 37 ft. lbs. (50 Nm).

7. Lower the vehicle if it was raised.

8. Remove the engine lifting equipment.

9. Reconnect the negative battery cable.

10. Run the engine and check the mounts for excess motion and any signs of looseness.

Cylinder Head

REMOVAL AND INSTALLATION

2.3L (VIN L) and 2.6L (VIN E) Engines

1. Relieve the fuel system pressure.
2. Disconnect the negative battery cable. Drain the cooling system.
3. Remove the drive belts.
4. Remove the air pump switching valve.
5. Remove the air pump hoses from the manifold and air pump.
6. Remove the MAP sensor hose, charcoal canister hose, vacuum booster and Vacuum Switching Valve (VSV) hoses.
7. Rotate the engine to position the No. 1 cylinder on TDC.
8. Remove the distributor.
9. Remove the exhaust manifold-to-exhaust pipe bolts.
10. Label and disconnect the connectors and vacuum hoses which may be in the way.
11. Remove the linkage to the carburetor or throttle body.
12. Remove the coolant hoses and cooling fan. It is not necessary to remove the radiator.
13. Remove the crankshaft pulley.
14. Remove the timing belt.
15. Remove the camshaft pulley and camshaft boss.
16. Remove the timing belt guide plate and the cylinder head front plate.
17. Remove the rocker arm cover and gasket.

— **WARNING** —

Cylinder head warpage can result if the cylinder head is removed from a hot engine. Allow the engine to cool to ambient temperature before removing the cylinder head.

18. Remove the cylinder head-to-engine bolts slowly and in sequence.

19. Remove the cylinder head and gasket.
20. Clean the gasket mounting surfaces.
   To install:
21. Using a new gasket, install the cylinder head and torque the bolts, in sequence to 58 ft. lbs. (79 Nm) in the 1st step, and to 65–79 ft. lbs. (88–107 Nm) in the final step.
22. Install the cylinder head front plate.
23. Install the camshaft pulley.
24. Align the camshaft pulley mark with the mark on the front plate. Make sure the keyway on the crankshaft if facing upward, aimed at the pointer on the engine block.
25. Install the timing belt. Install the timing belt covers, using a new gasket.
26. Install the crankshaft pulley.
27. Install the cooling fan and coolant hoses. If removed, reinstall the radiator.
28. Install the linkage to the carburetor or throttle body.
29. Reconnect the harnesses and vacuum hoses.
30. Install the exhaust manifold-to-exhaust pipe bolts.
31. Install the distributor.
32. Install the drive belts.
33. Adjust the valve lash to specification.
34. Using a new gasket, install the rocker arm cover. Tighten the bolts evenly in a crisscross pattern.
35. Reconnect the negative battery cable. Refill the cooling system.
36. Start the engine and check for leaks.

3.1L (VIN Z) Engine

Left Side

1. Relieve the fuel pressure. Disconnect the negative battery cable. Drain the cooling system.
2. Remove the intake manifold.
3. Raise and safely support the vehicle.

4. Disconnect the exhaust pipe from the exhaust manifold and remove the exhaust manifold-to-cylinder head bolts.
5. Remove the dipstick tube from the engine.
6. Lower the vehicle.
7. Loosen the rocker arm nuts, turn the rocker arms and remove the pushrods; keep the pushrods in the same order as removed.
8. Remove the cylinder head bolts in stages and in the reverse order of torquing.
9. Remove the cylinder head; do not pry on the head to loosen it.
10. Clean the gasket mounting surfaces thoroughly.
   To install:
11. Position a new cylinder head gasket over the dowel pins with the words **This Side Up** facing upwards. Carefully, guide the cylinder head into place.
12. Torque the cylinder head bolts in sequence to 41 ft. lbs. (55 Nm). Go back and turn each bolt an additional 90 degrees.
13. Install the pushrods; make sure the lower ends are in the lifter heads. Torque the rocker arm nuts to 14–20 ft. lbs. (20–27 Nm).
14. Install the intake manifold.
15. Install the dipstick tube to the engine.
16. Install the exhaust manifold-to-cylinder head bolts and the exhaust pipe-to-exhaust manifold nuts.
17. Refill the cooling system. Start the engine and check for leaks.

Right Side

1. Relieve the fuel pressure. Disconnect the negative battery cable. Drain the cooling system.
2. Remove the intake manifold.
3. If equipped, remove the cruise control servo bracket, the air management valve and hose.
4. Raise and safely support the vehicle.
5. Disconnect the exhaust pipe from the exhaust manifold and remove the exhaust manifold-to-cylinder head bolts.
6. Remove the exhaust pipe, crossover and heat shield, if equipped.
7. Lower the vehicle.
8. Label and disconnect the electrical wiring and vacuum hoses that may interfere with the removal of the right cylinder head.
9. Loosen the rocker arm nuts, turn the rocker arms and remove the pushrods; keep the pushrods in the same order as removed.
10. Remove the cylinder head bolts in stages and in the reverse order of torquing.

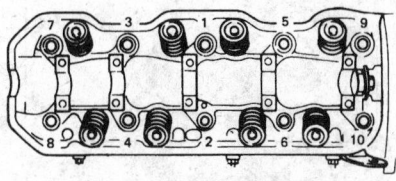

Cylinder head bolt removal sequence — 2.3L and 2.6L engines

191236

Cylinder head bolt torque sequence — 2.3L and 2.6L engines

191237

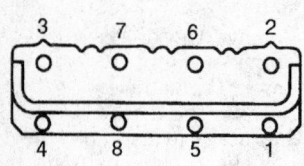

Cylinder head bolt removal sequence — 3.1L engine

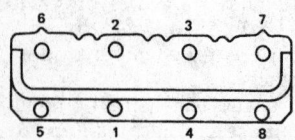

Cylinder head bolt torque sequence — 3.1L engine

11. Remove the cylinder head; do not pry on the head to loosen it.

12. Clean the gasket mounting surfaces thoroughly.

To install:

13. Position a new cylinder head gasket over the dowel pins with the words **This Side Up** facing upwards. Carefully, guide the cylinder head into place.

14. Torque the cylinder head bolts in sequence to 41 ft. lbs. (55 Nm). Go back and turn each bolt an additional 90 degrees.

15. Install the pushrods; make sure the lower ends are in the lifter heads. Torque the rocker arm nuts to 14–20 ft. lbs. (20–27 Nm).

16. Install the intake manifold.

17. Install the exhaust pipe at crossover, the crossover and the heat shield, if equipped.

18. Install the exhaust manifold-to-cylinder head bolts and the exhaust pipe-to-exhaust manifold nuts.

19. Connect the wiring and vacuum hoses to the right cylinder head.

20. If equipped, install the cruise control servo bracket, the air management valve and hose.

21. Refill the cooling system. Start the engine and check for leaks.

1993–95 3.2L (VIN V) Engine

NOTE: Isuzu has issued a recall notice for 1993–94 Rodeos equipped with 3.2L (VIN V) engines. This is campaign number 94V–094, and involves faulty camshaft end plugs. The plugs may dislodge from the cylinder heads and can cause rapid oil loss. Remember this when ordering part; a service kit is available for affected vehicles.

1. Relieve the fuel system pressure and disconnect the negative battery cable.

2. Remove the air cleaner assembly.

3. Remove the upper cooling fan shroud and the cooling fan assembly.

4. Disconnect the accelerator cable from the throttle body and the bracket.

5. Disconnect the canister vacuum hose from the vacuum pipe.

6. Disconnect the air vacuum hose from the common chamber.

7. Disconnect the vacuum booster hose from the common chamber.

8. Disconnect the MAP sensor; Canister Vacuum Switching Valve (VSV); Exhaust Gas Recirculation VSV; Intake Air Temperature sensor and ground connectors.

9. Remove the spark plug wires from the cylinder head cover.

10. Remove the ignition control module assembly.

11. Remove the four bolts and the throttle body from the common chamber.

12. Disconnect the vacuum hoses from the throttle body.

13. Disconnect the positive crankcase ventilation hose from the common chamber.

14. Disconnect the fuel pressure control valve vacuum hose from the common chamber.

15. Disconnect the evaporative emission canister purge hose from the common chamber.

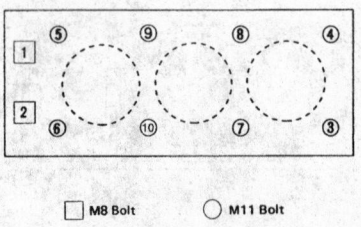

M8 Bolt   M11 Bolt

Cylinder head bolt removal sequence — 1993 3.2L engine

16. Remove the EGR valve assembly.

17. Remove the common chamber from the intake manifold.

18. Disconnect the fuel feed and return hoses from the fuel rail assembly.

19. Disconnect the harnesses to the fuel injectors and the thermo sensor.

20. Remove the intake manifold.

21. Remove the engine coolant manifold by removing the heater hose and mounting bolts.

22. Remove the drive belts.

23. Remove the power steering pump.

24. Remove the fan pulley assembly.

25. Remove the crankshaft pulley and damper.

26. Remove the oil cooler hoses and bracket on the timing belt cover.

27. Remove the timing belt.

28. Remove the cylinder head cover.

29. Remove the power steering pump bracket.

30. Remove the front exhaust pipes from the exhaust manifolds.

31. Remove the dipstick tube bracket from the cylinder head.

——— WARNING ———
The cylinder head and engine block must be at room temperature before removing the cylinder head.

32. Remove the cylinder head bolts in the illustrated sequence, gradually and in two steps.

33. Remove the cylinder head.

To install:

34. Install new camshaft seals and retaining plates onto the cylinder. Tighten the right camshaft seal retaining plate bolts to 65 inch lbs. (7 Nm). Tighten the left camshaft seal retaining plate bolts to 191 inch lbs. (22 Nm).

35. Thoroughly clean the cylinder head and engine block sealing surfaces.

36. Place a new head gasket on the engine block and carefully position the cylinder head on top of the new gasket.

NOTE: Do not reuse or apply oil to the cylinder head bolts.

37. Install new cylinder head bolts and torque them in sequence to 47 ft. lbs. (64 Nm) for the M11 bolts and 15 ft. lbs. (21 Nm) for the M8 bolts.

38. Install the dipstick tube bracket to the cylinder head.

39. Connect the front exhaust pipes to the exhaust manifolds. Tighten the exhaust bolts to 48 ft. lbs. (67 Nm).

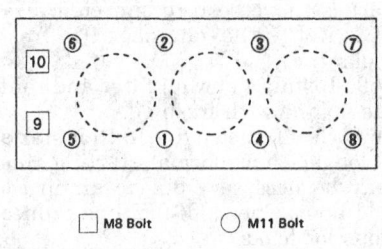

M8 Bolt   M11 Bolt

249520

Cylinder head bolt torque sequence — 1993 3.2L engine

40. Install the power steering pump bracket. Torque the mounting bolts to 34 ft. lbs. (46 Nm).

41. Install the cylinder head covers.

42. Install the timing belt.

43. Install the timing belt cover and oil cooler hoses and bracket.

44. Install the crankshaft pulley. Torque the center bolt to 123 ft. lbs. (167 Nm).

45. Install the fan pulley assembly.

46. Install the power steering pump.

47. Install the drive belts.

48. Install the engine coolant manifold and the heater hose.

49. Install the intake manifold.

50. Install the fuel injector harnesses and the fuel hoses to the fuel rail.

51. Install the common chamber. Torque the nuts and bolts to 17 ft. lbs. (24 Nm).

52. Install the EGR valve assembly. Torque the mounting bolts on the valve side to 69 inch lbs. (8 Nm) and the bolts on the exhaust side to 21 ft. lbs. (28 Nm).

53. Connect the evaporative emission canister purge hose.

54. Connect the fuel pressure control valve vacuum hose.

55. Connect the positive crankcase ventilation hose.

56. Install the throttle body assembly. Torque the mounting bolts to 14 ft. lbs. (19 Nm).

57. Connect the vacuum hoses to the throttle body.

58. Install the ignition control module and the spark plug wires.

59. Connect the MAP sensor; Canister Vacuum Switching Valve (VSV); Exhaust Gas Recirculation (VSV); Intake Air Temperature sensor and ground connectors.

60. Connect the vacuum booster hose.

61. Connect the air vacuum hose.

62. Connect the accelerator cable. Adjust the accelerator cable by pulling the cable housing while closing the throttle valve and tightening the adjusting nut and screw cap by hand temporarily. Now loosen the adjusting nut by three turns and then tightening the screw cap. Make sure the throttle valve reaches the screw stop when the throttle is closed.

63. Install the cooling fan assembly and the upper fan shroud.

64. Install the air cleaner assembly.

65. Connect the negative battery cable.

66. Refill and bleed the cooling system.

67. Refill and bleed the power steering pump if necessary.

68. Refill the engine with fresh oil.

69. Run the engine and check for leaks and proper compression.

1996–97 3.2L (VIN V) Engine

——— WARNING ———

The cylinder head should be cool to the touch before it is removed. If the head bolts are loosened on a hot engine, the cylinder head may warp.

1. Relieve the fuel system pressure.

——— CAUTION ———

Fuel injection systems remain under pressure even after the engine has been turned off. The fuel system pressure must be relieved before disconnecting any fuel lines. Failure to do so may result in fire and personal injury.

2. Disconnect the negative battery cable and reinstall the fuel pump relay.

3. Raise and support the vehicle safely.

4. Disconnect the front exhaust pipes from the exhaust manifolds. If necessary, separate the front exhaust pipes from the crossover pipe. Label and disconnect the oxygen sensors.

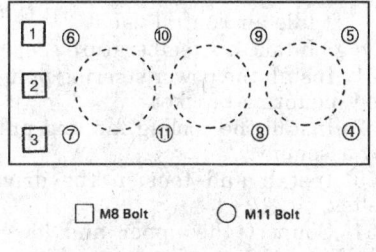

M8 Bolt   M11 Bolt

324691

Cylinder head bolt removal sequence — 1994–97 3.2L engine

5. Lower the vehicle.

6. Drain the coolant into a sealable container. Disconnect and remove the upper and lower radiator hoses.

7. Remove the air intake duct and the air cleaner box.

8. Loosen and remove the drive belts.

9. Remove the cooling fan and pulley assembly.

10. Unbolt the power steering pump mounting bracket. Move the pump and bracket out of the way without disconnecting the hydraulic lines.

11. Disconnect the throttle cable from the throttle body linkage.

12. Label and disconnect the following vacuum hoses from the intake manifold chamber:
   a. PCV hose
   b. EVAP canister vacuum hose
   c. Brake booster hose

13. Label and disconnect the following sensor connectors from the rear of the intake manifold chamber:
   a. Ignition control module connectors
   b. Linear EGR valve
   c. MAP sensor
   d. EVAP purge valve
   e. Throttle position sensor
   f. Idle air control valve
   g. Intake air temperature sensor

14. Disconnect the EGR valve supply tube and bracket.

15. Remove the throttle body, and then the intake manifold chamber.

16. Carefully clean any dirt from the fuel rail and fuel fittings.

17. Disconnect the fuel feed and return lines from the front of the fuel rail. Clean up any spilled fuel.

18. Label and disconnect the fuel injector wiring harness.

19. Remove the fuel injectors and lower intake manifold as an assembly. If desired, the fuel rail and injectors may be removed separately as an assembly. Remove the intake manifold gaskets.

20. Cover the intake openings with a sheet of plastic or clean shop towels to keep out dirt and foreign objects.

21. Label the ignition coil assemblies, then disconnect and remove.

22. Unbolt the oil cooler line brackets from the timing belt covers.

23. Remove the timing belt.

——— WARNING ———

If the timing belt is worn, damaged, or shows signs of oil or coolant contamination, it must be replaced.

24. Disconnect the heater hoses from the engine; then, unbolt and remove the engine coolant manifold.

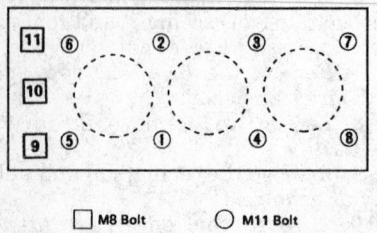

☐ M8 Bolt   ○ M11 Bolt

324741

Cylinder head bolt torque sequence — 1994–97 3.2L engine

25. Unbolt the dipstick tube from the cylinder head.

26. Loosen the valve cover bolts in a crisscross sequence. Remove the valve covers.

27. Loosen the cylinder head bolts in a two-step crisscross pattern, working from the outer bolts to those at the center of the head, as illustrated. First, partially loosen the 11mm bolts, then partially loosen the 8mm bolts. Finally, loosen all the bolts and then remove them.

28. Remove the cylinder head. If it sticks, tap it with a wooden or plastic–faced mallet.

29. Remove the head gasket.

30. Inspect the cylinder head for cracking or warpage. Inspect the engine block for any signs of damage. Carefully clean the head gasket mating surfaces, don't scratch or gouge the machined aluminum surfaces.

To install:

NOTE: Use new head bolts when installing the cylinder head. Do not apply oil to the head bolt threads.

31. Make sure all mating surfaces are clean and free of oil, coolant, or gasket residue.

32. Install new cylinder head gaskets.

33. Install the cylinder head. Install the new head bolts and tighten them by hand only.

34. Follow these steps to tighten the cylinder head bolts to their final torque specification:

  a. Use a two–step crisscross pattern to tighten the 11mm bolts to 47 ft. lbs. (64 Nm). Start tightening with the center bolts, and work toward the outer bolts.

  b. Tighten the 8mm bolts to 15 ft. lbs. (21 Nm). Start with the bolt closest to the exhaust side of the head and work toward the intake side.

35. Apply a 0.7–0.8 in. (2–3mm) bead of sealant to the joint were the camshaft holders meet the cylinder head. Install the valve cover with a new gasket before the sealant cures.

36. Tighten the valve cover bolts to 6 ft. lbs. (8 Nm) in a crisscross pattern.

37. Verify that the camshaft and crankshaft timing marks are properly aligned.

38. Install the timing belt..

39. Fit the oil cooler line brackets onto the timing cover and tighten the bolts to 13 ft. lbs. (18 Nm).

40. Install the engine coolant manifold and reconnect the heater hoses. Tighten the bolts to 16 ft. lbs. (22 Nm).

41. Install the dipstick tube bracket.

42. Raise and safely support the vehicle. Install and reconnect the front exhaust pipes. Reconnect the oxygen sensors. Lower the vehicle.

43. Install the intake manifold with a new gasket. Tighten the bolts and nuts to 17 ft. lbs. (23 Nm).

44. Reconnect the fuel injector wiring harness. Reconnect the fuel feed and return lines.

45. Install and reconnect the ignition coil assemblies.

46. Install the intake manifold chamber and throttle body with new gaskets. Tighten the nuts and bolts to 17 ft. lbs. (23 Nm).

47. Reconnect the throttle cable to the throttle body linkage.

48. Reconnect the EGR valve supply tube and bracket.

49. Reconnect the following vacuum hoses to the intake manifold chamber:

  a. PCV hose

  b. EVAP canister vacuum hose

  c. Brake booster hose

50. Reconnect the following sensor connectors to the rear of the intake manifold chamber:

  a. Ignition control module connectors

  b. Linear EGR valve

  c. MAP sensor

  d. EVAP purge valve

  e. Throttle position sensor

  f. Idle air control valve

  g. Intake air temperature sensor

51. Install the power steering pump and mounting bracket.

52. Install the cooling fan and pulley assembly.

53. Install and tension the drive belts.

54. Connect the upper and lower radiator hoses.

55. Install the air cleaner box and air intake duct.

56. Install the hood.

57. Verify that all fuel lines, vacuum and coolant hoses, and wiring harness have been reconnected.

58. Refill the engine with fresh coolant.

59. Install a new oil filter and refill the engine with fresh oil.

60. Crank the engine until it starts. A longer than normal starting time may be necessary due to air in the fuel lines. Check all fuel line connections for leaks.

61. Bleed any air from the cooling system.

62. Bleed the power steering system if necessary.

63. Warm the engine up to normal operating temperature and check the operation of the thermostat and water pump.

64. Check the throttle cable operation and adjustment.

65. Check the engine oil level and add if necessary.

Lash Adjusters

BLEEDING

3.2L (VIN V) Engine

NOTE: The hydraulic lash adjusters are self–bleeding. The lash adjusters must be primed before installation to purge excess air. If the lash adjusters are worn, they should be replaced.

1. Disassemble the rocker arms and shafts.

2. Remove the hydraulic lash adjuster from the rocker arm. The rocker arms and lash adjusters are a set. Label them so they aren't confused upon reassembly.

3. Inspect the hydraulic lash adjusters for excess movement and replace if necessary. The hydraulic lash adjusters are designed to be self–bleeding, but new lash adjusters must be primed before installation. To prime an adjuster, proceed as follows:

  a. Use a small–diameter rod (0.08 in. or 2mm) to push in the adjuster's check ball.

  b. Submerge the lash adjuster in a tub of clean engine oil.

  c. Hold the lash adjuster and pump the plunger with your finger to fill the adjuster with oil and displace any air.

  d. Keep pumping the plunger until it's hard and no more air bubbles come out. Then, remove the rod to release the check ball.

  e. Leave the lash adjuster in the tub of oil.

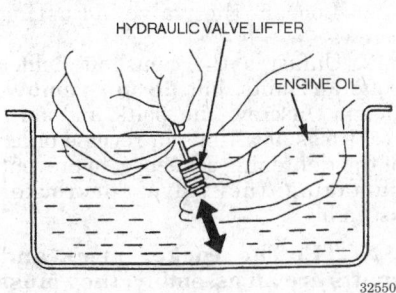

Priming the valve lash adjuster — 3.2L engine

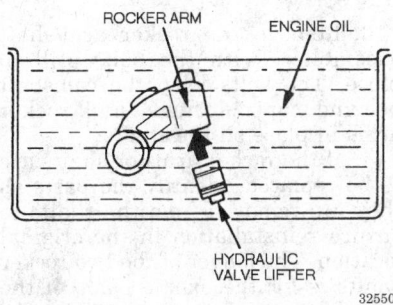

Installing the valve lash adjuster into the rocker arm

4. Submerge the rocker arm into the tub of engine oil. Install the hydraulic lash adjuster into the rocker arm.

5. Reassemble the rocker arm and shaft assemblies and install them.

Valve Lash

ADJUSTMENT

2.3L (VIN L) and 2.6L (VIN E) Engines

NOTE: The valves are adjusted with the engine COLD. It is best to allow an engine to sit overnight before beginning a valve adjustment. While all valve adjustments must be made as accurately as possible, it is better to have the valve adjustment slightly loose rather than slightly tight. A burned valve may result from overly tight valve adjustments.

1. Remove the rocker arm cover and discard the gasket.

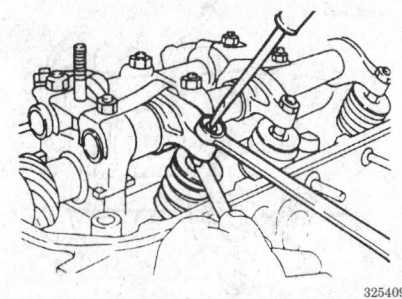

Adjusting the valve lash — 2.3L and 2.6L engines

2. Rotate the crankshaft pulley until the No. 1 piston is at TDC of the compression stroke.

NOTE: To make sure the piston is on the correct stroke, remove the spark plug and place your finger over the hole. Feel for air being forced out of the spark plug hole. Both valves on No. 1 cylinder will be closed. Stop turning the crankshaft when the TDC timing mark on the crankshaft pulley is directly aligned with the timing mark pointer.

3. With the No. 1 piston at TDC of the compression stroke, adjust the clearances of the following valves: Intake 1 and 2; Exhaust 1 and 3.

4. Adjust the clearance by loosening the locknut and turning the adjusting screw. Retightening the locknut when the proper thickness feeler gauge passes between the rocker arm and valve stem and has a slight drag. Clearance is 0.006 in. (0.15mm) for intake; and 0.010 in. (0.25mm) for exhaust.

5. Rotate the crankshaft 1 complete revolution (360 degrees) to position the No. 4 piston at TDC of its compression stroke and adjust the clearances of the following valves: Intake 3 and 4; Exhaust 2 and 4.

6. After each valve is adjusted, tighten its locknut to 10 ft. lbs. (13 Nm).

7. After adjustment, install the rocker arm cover using a new gasket and sealant.

8. Retighten the crankshaft pulley if it was loosened during the valve adjustment.

3.1L (VIN Z) Engine

1. Remove the valve covers.

2. Rotate the crankshaft until the No. 1 cylinder is on the TDC of its compression stroke.

NOTE: When the notch on the damper pulley is aligned with the 0 timing mark and the rocker arms of the No. 1 cylinder do not move, the engine is at the TDC of the compression stroke of the No. 1 cylinder. If the rocker arms move as the timing mark approaches the "0" mark, rotate the crankshaft one full revolution until the timing mark aligns with "0" again.

3. With the engine at TDC of the No. 1 cylinder, adjust the following valves: Exhaust 1, 2 and 3; Intake: 1, 5 and 6.

4. Back out the adjusting nut until lash is felt.

5. Tighten the adjusting nut until the lash is removed, then, turn the nut 1½ additional turns to center the lifter plunger.

6. Rotate the engine 1 complete revolution and reposition the notch on the damper pulley with the 0 mark on the timing tab; this is the No. 4 cylinder firing position.

7. With the engine at TDC of the No. 4 cylinder, adjust the following valves: Exhaust: 4, 5 and 6; Intake 2, 3 and 4.

8. Back out the adjusting nut until lash is felt.

9. Tighten the adjusting nut until the lash is removed, plus, turn the nut 1½ additional turns to center the lifter plunger.

10. Using a new gasket and sealant, install the rocker arm covers.

3.2L (VIN E) Engine

The 3.2L (VIN V) engine is equipped with hydraulic lash adjusters. No valve adjustment is necessary.

Rocker Arms and Shafts

REMOVAL AND INSTALLATION

1993–94 2.3L (VIN L) and 2.6L (VIN E) Engines

1. Disconnect the negative battery cable.

2. Remove the rocker cover.

3. Remove the drive belts, cooling fan, and water pump pulley.

4. Remove the timing belt.

5. Loosen the rocker arm shaft bracket nuts a little at a time, in sequence, starting with the end brackets.

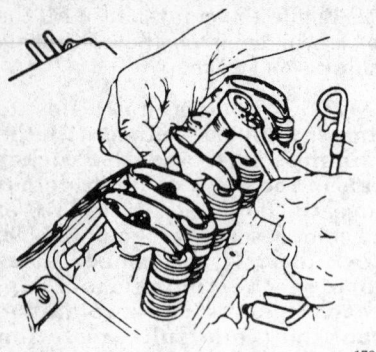

178634

Adjusting the valve lash — 3.1L engine

6. Remove the nuts from the rocker arm shaft brackets. Remove shaft assembly.

7. To disassemble the rockers and shafts; remove the spring from the rocker arm shaft, the rocker brackets and arms. Keep parts in order for reassembly.

To install:

8. Before installing, apply a generous amount of clean engine oil to the rocker arm shaft, rocker arms and valve stems.

9. Install the longer shaft on the exhaust valve side and the shorter shaft on the intake side so the alignment marks on the shafts are turned on the front side of the engine.

10. Assemble the rocker arm shaft brackets and rocker arms to the shafts so the cylinder number, on the upper face of the brackets, points toward the front of the engine.

11. Make certain the amount of projection of the rocker arm shaft beyond the face of the No. 1 rocker arm shaft bracket, is longer on the exhaust side shaft than on the intake shaft when the rocker arm shaft stud holes are aligned with the rocker arm shaft bracket stud holes.

12. Place the rocker arm shaft springs in position between the shaft bracket and rocker arm.

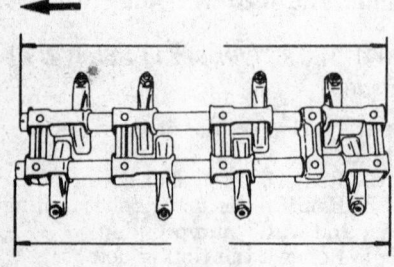

191360

The longer rocker arm shaft is the exhaust side shaft — 2.3L and 2.6L engines

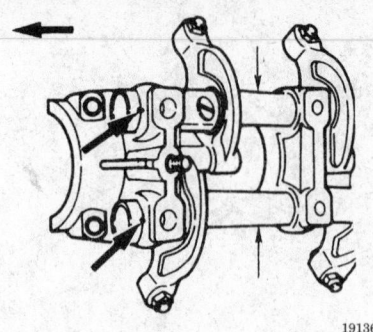

191361

The punch mark must face the front of the engine — 2.3L and 2.6L engines

13. Check that the punch mark on the rocker arm shaft is facing upward, then, install the rocker arm shaft bracket assembly onto the cylinder head studs. Torque the nuts to 16 ft. lbs. (22 Nm) and the bolts to 6 ft. lbs. (8 Nm).

NOTE: Hold the rocker arm springs while torquing the nuts to prevent damage to the spring. Start with the center bracket and work outward.

14. Install the timing belt and cover.

15. Adjust the valves as needed.

16. Install the water pump pulley and cooling fan.

17. Install and adjust the drive belts.

18. Install the rocker cover.

19. Connect the negative battery cable and check the ignition timing.

1995–97 2.3L (VIN L) and 2.6L (VIN E) Engines

1. Disconnect the negative battery cable.

2. Remove the drive belts.

3. Remove the cooling fan and the water pump pulley.

4. Unbolt and remove the power steering pump. Unbolt the hydraulic line brackets from the upper timing cover. Move the pump out without disconnecting the hydraulic lines.

5. Remove the valve cover and the upper timing cover.

6. Rotate the crankshaft to set the engine at TDC/compression for the No. 1 cylinder. The arrow mark on camshaft sprocket aligns with mark on the rear timing cover.

7. Remove the lower timing belt cover.

8. Remove the crankshaft pulley.

9. Verify that the engine is set at TDC/compression for the No. 1 cylinder. The notch on the crankshaft sprocket aligns with the pointer on the oil seal retainer.

10. Remove the timing belt.

11. Loosen the valve adjusting screws.

12. Unfasten the camshaft holder bolts and nuts, but **do not remove** them. Unscrew the bolts and nuts two turns at a time, in reverse order of the tightening sequence, to prevent damaging the valves or rocker assembly.

NOTE: The rocker arms and shafts are an assembly; they must be removed from the engine as a unit. Always follow the torque sequence carefully when removing or installing the rocker shaft assembly.

13. Remove the rocker arm/shaft assemblies, with the bolts still in place. The bolts keep the camshaft bearing caps, springs, and rocker arms in place on the shafts.

14. If the rocker arms or shafts are to be replaced, identify the parts as they are removed from the shafts to ensure reinstallation in the original location. The longer of the two rocker shafts is for the exhaust side of the cylinder head.

To install:

15. Lubricate the camshaft journals and lobes with clean engine oil.

16. Apply sealant to the contact surface of the No. 1 camshaft holder. Set the rocker arm assembly in place and loosely thread the camshaft holder bolts and nuts. The punch mark on each rocker shaft faces the front of the engine.

17. Tighten the rocker arm bolts and nuts to the following specifications in a two–step crisscross pattern:

   a. Tighten the rocker arm bolts to 6 ft. lbs. (8 Nm).

   b. Tighten rocker arm nuts to 16 ft. lbs. (22 Nm).

18. Install the timing belt.

19. Install the lower timing cover and the crankshaft pulley.

20. Adjust the valves and tighten the locknuts to 9 ft. lbs. (12 Nm).

21. Install the timing belt covers.

22. Install the water pump pulley, tightening the nuts to 20 ft. lbs. (27 Nm). Install the cooling fan.

23. Install the engine drive belts.

24. Change the engine oil and filter.

25. Reconnect the negative battery cable.

26. Start the engine. Check and adjust the ignition timing.

27. Check for oil leaks.

3.1L (VIN Z) Engine

1. Disconnect the negative battery cable.

2. Remove the rocker arm covers.

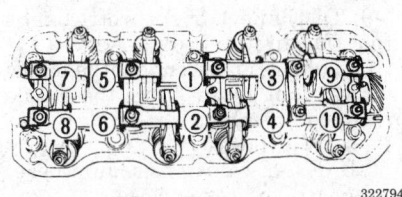

Camshaft holder tightening sequence — 1995–97 2.3L and 2.6L engines

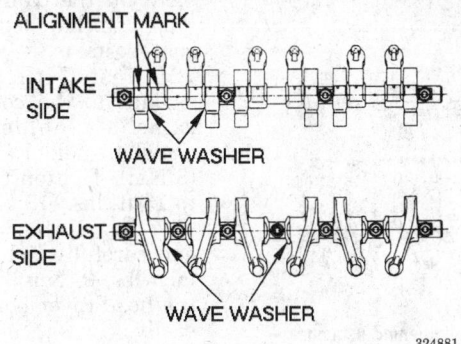

Intake and exhaust rocker arms and shafts — 3.2L engine

3. Remove the rocker arm nut, pivot balls, rocker arm and the pushrods.

---- **WARNING** ----
Keep all components in order so they may be reinstalled in the same location.

To install:

4. Inspect the friction contact surfaces for scoring or abnormal wear and replace as necessary.

5. Install the pushrods in their original location; be sure the lower ends are seated in the lifter.

6. Coat the bearing surfaces of the rocker arms and pivot balls with Molykote® or equivalent.

7. Install the rocker arms, balls and nuts on the studs.

8. Adjust the valve lash.

9. Install the rocker arm covers.

10. Connect the negative battery cable.

1993–95 3.2L (VIN V) Engine

NOTE: Isuzu has issued a recall notice for 1993–94 Rodeos equipped with 3.2L (VIN V) engines. This is campaign number 94V–094, and involves faulty camshaft end plugs. The plugs may dislodge from the cylinder heads and can cause rapid oil loss. Remember this when ordering parts: a service kit is available for affected vehicles.

1. Disconnect negative battery cable.

2. Remove the air cleaner assembly.

3. Disconnect the accelerator pedal cable from the throttle body and cable brackets.

4. Disconnect the canister vacuum hose from the Vacuum Switch Valve (VSV).

5. Disconnect the vacuum booster hose from the common chamber duct.

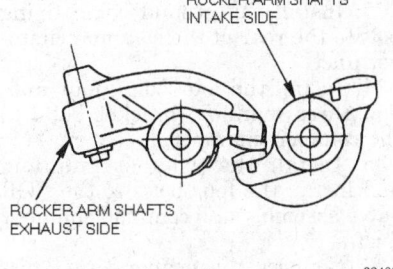

Rocker arm contact areas — 3.2L engine

6. Disconnect the harness from the Idle Air Control Valve, Throttle Position sensor, Manifold Absolute Pressure sensor, canister VSV, EGR VSV, Intake Air Temperature sensor and VSV.

7. Remove the high tension cable from the cylinder heads.

8. Disconnect the harness from the ignition module.

9. Remove the three bolts from the electronic ignition bracket and assembly.

10. Remove the bolts from the throttle body, then remove the assembly.

11. Disconnect the canister VSV and the EGR VSV vacuum hose from the throttle body.

12. Disconnect the fuel pressure control valve vacuum hose from the common chamber duct.

13. Disconnect the PCV hose from the common chamber duct.

14. Disconnect the evaporative emission canister purge hose from the common chamber duct.

15. Remove the four bolts from the EGR valve assembly common chamber duct and remove the exhaust manifold.

16. Remove the four bolts, four nuts and three manifold bracket fixing bolts from the common chamber duct.

17. Remove the ground cable fixing bolt from the rear of the common chamber duct.

18. Remove the six bolts and two nuts from the common chamber duct.

19. Remove the common chamber duct bracket fixing bolts from the rear of the common chamber duct.

20. Remove the timing belt.

21. Remove the cylinder head cover.

22. Remove the camshaft holders and camshaft.

23. Remove the rocker arm shaft bolts. Lift the rocker arm assembly from the cylinder head.

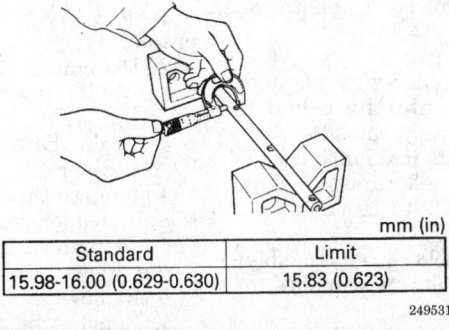

Standard	Limit
15.98–16.00 (0.629–0.630)	15.83 (0.623)

mm (in)

249531

Measure the shaft at the point where the rocker arm moves on the shaft — 3.2L engine

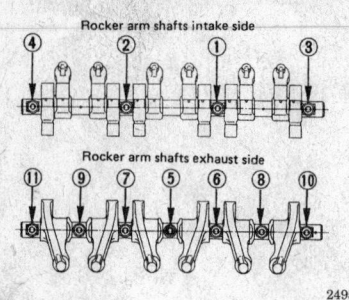

Rocker arm shaft bolt tightening sequence — 1993–95 3.2L engine

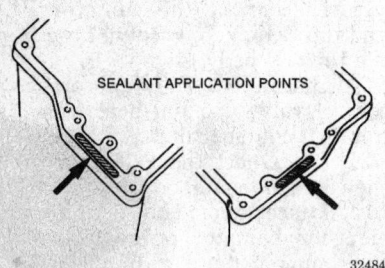

Apply sealant to the cylinder head at the camshaft holder mounts — 3.2L engine

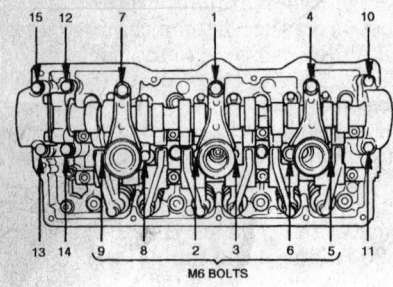

Camshaft holder bolt tightening sequence — 3.2L engine

24. The hydraulic lifters are attached to the rocker arms. Remove them and inspect, bleed, or replace as necessary.

To install:

25. Install new camshaft seals and retaining plates onto the cylinder head. Tighten the right camshaft seal retaining plate bolts to 65 inch lbs. (7 Nm). Tighten the left camshaft seal retaining plate bolts to 191 inch lbs. (21 Nm).

26. Install the rocker arm assembly and tighten the bolts in sequence to 13 ft. lbs. (18 Nm).

27. Oil the camshaft bearing journals, camshaft lobes, and rocker arm contact areas.

28. Install the camshaft. Apply sealant to the contact edges of the camshaft holders. Tighten the camshaft holder bolts to 69 inch lbs. (8 Nm). Tighten the remaining bolts to 13 ft. lbs. (18 Nm).

29. Install the cylinder head covers and carefully tighten the bolts to 69 inch lbs. (8 Nm). Do not overtighten the head cover bolts; they crack very easily.

30. Install the timing belt.

31. Install the common chamber duct bracket bolts to the rear of the common chamber duct.

32. Install the six bolts and two nuts to the common chamber duct.

33. Install the ground cable fixing bolt to the rear of the common chamber duct.

34. Install the four bolts, four nuts and three manifold bracket bolts to the common chamber duct.

35. Install the exhaust manifold and install the four bolts to the EGR valve assembly and common chamber duct.

36. Connect the evaporative emission canister purge hose to the common chamber duct.

37. Connect the PCV hose to the common chamber duct.

38. Connect the fuel pressure control valve vacuum hose to the common chamber duct.

39. Connect the canister VSV and the EGR VSV vacuum hose to the throttle body.

40. Install the remaining components.

41. Verify that all vacuum hoses, lines, and wiring harnesses are reconnected.

42. Install the air cleaner assembly.

43. Connect the negative battery cable.

1996–97 3.2L (VIN V) Engine

1. Relieve the fuel system pressure.

2. Disconnect the negative battery cable and reinstall the fuel pump relay.

3. Drain the coolant into a sealable container.

4. Remove the air intake duct and the air cleaner box.

5. Disconnect and remove the upper and lower radiator hoses.

6. Remove the drive belts.

7. Remove the cooling fan and pulley assembly.

8. Unbolt the power steering pump mounting bracket. Move the pump and bracket out of the way without disconnecting the hydraulic lines.

9. Disconnect the throttle cable from the throttle body linkage.

10. Label and disconnect the following vacuum hoses from the intake manifold chamber:
   a. PCV hose
   b. EVAP canister vacuum hose
   c. Brake booster hose

11. Label and disconnect the following sensor connectors from the rear of the intake manifold chamber:
   a. Ignition control module connectors
   b. Linear EGR valve
   c. MAP sensor
   d. EVAP purge valve
   e. Throttle position sensor
   f. Idle air control valve
   g. Intake air temperature sensor

12. Disconnect the EGR valve supply tube and bracket.

13. Remove the throttle body, then remove the intake manifold chamber.

14. Carefully clean any dirt from the fuel rail and fuel fittings.

CAUTION

Fuel injection systems remain under pressure even after the engine has been turned off. The fuel system pressure must be relieved before disconnecting any fuel lines. Failure to do so may result in fire and personal injury.

---

15. Disconnect the fuel feed and return lines from the front of the fuel rail. Clean up any spilled fuel.

16. Remove the intake manifold gaskets.

17. Label and disconnect the ignition coil from the wiring harness. Remove the coil assemblies.

18. Unbolt the oil cooler line brackets from the timing belt covers.

19. Remove the timing belt.

WARNING

If the timing belt is worn, damaged, or shows signs of oil or coolant contamination, it must be replaced.

---

20. Loosen the valve cover bolts in a crisscross sequence. Remove the valve covers.

21. Remove the camshaft sprockets and back covers.

22. Loosen the camshaft holder bolts in a crisscross sequence to prevent warping.

23. Remove the camshaft and camshaft holders from the cylinder head.

24. Inspect the camshaft lobes and journals for signs of wear or damage.

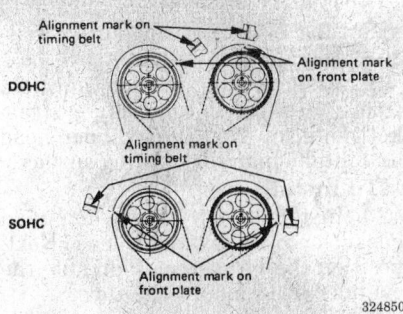

Camshaft sprocket alignment marks — 1996–97 3.2L engine

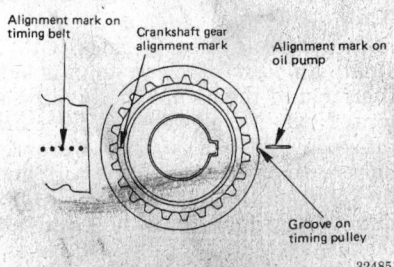

Timing belt dotted mark alignment — 1996–97 3.2L engine

25. Loosen the exhaust and intake rocker shaft bolts in a crisscross sequence to prevent warping.

26. Remove the intake and exhaust rocker shafts from the cylinder head.

27. If the rocker arms and shafts must be disassembled, label the parts and wave washers so that they can be reassembled in the same positions.

28. If necessary, remove the hydraulic valve lash adjusters from the rocker arms.

29. Inspect the hydraulic lash adjusters for excess movement and replace if necessary. The hydraulic lash

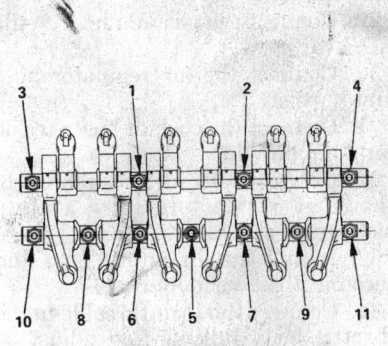

Rocker shaft bolt tightening pattern — 1996–97 3.2L engine

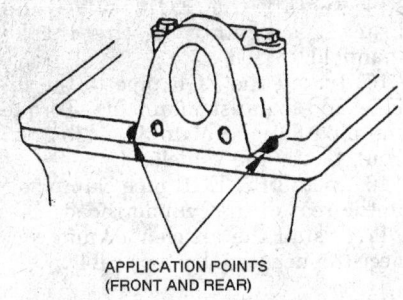

APPLICATION POINTS (FRONT AND REAR)

Apply sealant to the camshaft holders — 3.2L engine

adjusters are designed to be self-bleeding, but new adjusters must be primed before installation.

To install:

30. Reassemble the rocker arm, shaft, and hydraulic lash adjuster components. Assemble the hydraulic lash adjusters to the rockers before removing them from the tub of oil. The intake rocker arms all face the same direction when installed.

31. Lubricate the rocker arms and shafts with clean engine oil.

32. Install the intake and exhaust rocker arms and shaft assemblies. Tighten the shaft holder bolts to 13 ft. lbs. (18 Nm), starting with the intake shaft and then moving to the exhaust shaft. Make sure the intake and exhaust rockers contact each other properly.

33. Make sure all mating surfaces are clean and free of oil, coolant, or gasket residue.

34. Lubricate the camshaft lobes and journals with clean engine oil.

35. Apply a bead of sealant to the front and rear camshaft holder mating surfaces on the cylinder head.

36. Install the camshaft and holder assembly onto the cylinder head before the sealant cures. Install the camshaft holder bolts, but don't tighten them yet.

37. Use a crisscross sequence to tighten the camshaft holder bolts. Tighten the 8mm bolts to 13 ft. lbs. (18 Nm). Tighten the 6mm bolts to 6 ft. lbs. (8 Nm).

38. Use a seal driver to install a new camshaft seal.

39. Install the camshaft sprocket back covers and tighten the bolts to 12 ft. lbs. (17 Nm).

40. Install the camshaft sprockets so that the timing marks are aligned. Tighten the bolts to 46 ft. lbs. (64 Nm).

41. Apply a 0.7–0.8 in. (2–3mm) bead of sealant to the joint were the

camshaft holders meet the cylinder head. Install the valve cover with a new gasket before the sealant cures.

42. Tighten the valve cover bolts to 6 ft. lbs. (8 Nm) in crisscross pattern.

43. Verify that the camshaft and crankshaft timing marks are properly aligned.

44. Install the timing belt.

45. Fit the oil cooler line brackets onto the timing cover and tighten the bolts to 13 ft. lbs. (18 Nm).

46. Reconnect the fuel feed and return lines.

47. Install and reconnect the ignition coil assemblies.

48. Install the intake manifold chamber and throttle body with new gaskets. Tighten the nuts and bolts to 17 ft. lbs. (24 Nm).

49. Reconnect the throttle cable to the throttle body linkage.

50. Reconnect the EGR valve supply tube and bracket.

51. Reconnect the following vacuum to the intake manifold chamber:
   a. PCV hose
   b. EVAP canister vacuum hose
   c. Brake booster hose

52. Reconnect the following sensor connectors to the rear of the intake manifold chamber:
   a. Ignition control module connectors
   b. Linear EGR valve
   c. MAP sensor
   d. EVAP purge valve
   e. Throttle position sensor
   f. Idle air control valve
   g. Intake air temperature sensor

53. Install the power steering pump and mounting bracket.

54. Install the cooling fan and pulley assembly.

55. Install the drive belts.

56. Install and reconnect the upper and lower radiator hoses.

57. Install the air cleaner box and air intake duct.

58. Verify that all fuel lines, vacuum and coolant hoses, and wiring harness have been reconnected.

59. Refill the engine with fresh coolant.

60. Crank the engine until it starts. A longer than normal starting time may be necessary due to air in the fuel lines. Check all fuel line connections for leaks.

61. Bleed any air from the cooling system.

62. Bleed the power steering system if necessary.

63. Check the throttle cable operation and adjustment.

64. Check the engine oil level and add if necessary.

Intake Manifold

REMOVAL AND INSTALLATION

Carbureted Engine

1. Disconnect the negative battery cable and remove the air cleaner assembly.
2. Remove the EGR pipe clamp bolt at the rear of the cylinder head.
3. Raise and support the vehicle safely. Remove the EGR pipe from the intake and exhaust manifolds.
4. Remove the EGR valve and bracket assembly from the intake manifold.
5. Lower the vehicle and drain the cooling system.
6. Remove the upper coolant hoses from the manifold.
7. Disconnect the accelerator linkage, vacuum lines, electrical wiring and fuel line from the intake manifold.
8. Remove the intake manifold mounting nuts and remove the manifold from the cylinder head.
9. Remove the lower heater hose while holding the manifold away from the engine. Remove the manifold from the vehicle.
   To install:
10. Inspect the manifold and the cylinder head sealing surfaces for warpage. If warpage is greater than 0.002 in. (0.05mm), the sealing surface must be refaced.
11. Position the intake manifold near the cylinder head and connect the lower heater hose to the manifold. Using a new gasket, install the intake manifold. Torque the mounting bolts and nuts to 16 ft. lbs. (22 Nm).
12. Connect the accelerator linkage, vacuum lines, electrical wiring and fuel line to the intake manifold.
13. Connect the upper coolant hose to the intake manifold.

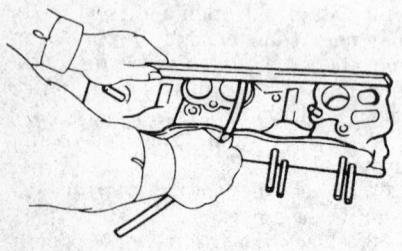

182307

Check the manifold's sealing surface for warpage

14. Install the EGR valve and bracket assembly to the intake manifold.
15. Install the EGR pipe to the intake and exhaust manifolds. Torque the pipe flange nut to 33 ft. lbs. (45 Nm). Lower the vehicle.
16. Install the EGR pipe clamp bolt to the rear of the cylinder head.
17. Install the air cleaner and connect the negative battery cable.

2.3L (VIN L) and 2.6L (VIN E) Engines

1. Relieve the fuel pressure.
2. Disconnect the negative battery cable.
3. Drain the engine coolant.
4. Disconnect the throttle cable from the throttle body linkage.
5. Disconnect and remove the air intake duct.
6. Disconnect the vacuum hose from the EGR valve. If equipped, disconnect the EGR temperature sensor connector.
7. Disconnect the EGR fuel pressure control rubber hose.
8. Use a flare wrench to disconnect the EGR pipe fitting from the intake manifold.

> ——— CAUTION ———
> *Fuel injection systems remain under pressure after the engine has been turned OFF. Properly relieve fuel pressure before disconnecting any fuel lines. Failure to do so may result in fire or personal injury.*

9. Disconnect the fuel feed hose from the fuel rail.
10. Disconnect the coolant hoses from the intake manifold.
11. Disconnect the air regulator hose and harness from the lower rear of the intake manifold chamber.
12. Disconnect the throttle position sensor harness and the coolant hoses from the throttle body.
13. Loosen the throttle body mounting bolts, and remove the throttle body from the intake manifold chamber.
14. Loosen the intake manifold chamber mounting nuts and bolts, then remove the intake manifold chamber from the lower part of the manifold assembly.
15. Remove the fuel injector rail attaching bolts.
16. Label and disconnect the fuel injector harnesses.
17. Carefully lift the fuel rail and injectors from the intake manifold as an assembly.

18. Loosen the intake manifold mounting bolts and nuts in a crisscross sequence.
19. Remove the intake manifold from the engine. Clean any old gasket material from the cylinder head and intake manifold mating surfaces.
   To install:
20. Install a new intake manifold gasket onto the cylinder head. Next, position the intake manifold onto the cylinder head mounting studs.
21. Install the intake manifold attaching bolts and nuts, Tighten the bolts and nuts in a two-step crisscross pattern beginning in the center and working outward. The final torque specification is 16 ft. lbs. (22 Nm).
22. Lubricate new O-rings with a small amount of clean engine oil, then install them onto the fuel injectors. Next, install the fuel injectors into the fuel rail, if they were removed. Install the fuel injectors and fuel rail assembly onto the intake manifold. Tighten the fuel rail attaching bolts to 14 ft. lbs. (19 Nm).
23. Reconnect the fuel injector wiring harness connectors.
24. Install the intake manifold chamber to the lower part of the manifold using a new gasket. Tighten the bolts and nuts to 20 ft. lbs. (27 Nm) in a two-step crisscross pattern.
25. Install a new throttle body gasket, then install the throttle body. Tighten the mounting bolts to 14 ft. lbs. (19 Nm).
26. Reconnect the throttle position sensor connector. Reconnect the coolant hoses to the throttle body.
27. Connect the EGR pipe to the intake manifold, torque the flange nut to 33 ft. lbs. (45 Nm).
28. Connect the EGR fuel pressure control rubber hose. If equipped, reconnect the EGR temperature sensor connector.
29. Connect the vacuum hose to the EGR valve.
30. Connect the air regulator hose and harness.
31. Connect the coolant hoses to the intake manifold.
32. Connect the fuel feed hose to the fuel rail using new sealing washers.
33. Install the air intake duct and reconnect the vacuum hose.
34. Connect the throttle cable to the throttle body linkage and adjust as necessary.
35. Refill and bleed the cooling system.
36. Connect the negative battery cable.

37. Turn the ignition switch **ON** and check for fuel leaks at the fuel rail.

38. Check the manifold coolant hoses for leaks. Check the intake manifold mating surfaces for leaks.

3.1L (VIN Z) Engine

1. Relieve the fuel pressure. Disconnect the negative battery cable.

2. Remove the air cleaner. Drain the cooling system.

3. Label and disconnect the wires and hoses from the TBI unit and the intake manifold.

4. Disconnect and plug the fuel lines from the TBI unit.

5. Disconnect the accelerator cables from the TBI unit.

6. Disconnect the ignition wires from the spark plugs and coil.

7. Remove the distributor hold-down clamp and the distributor.

8. Label and disconnect the EGR vacuum line and the emission hoses.

9. Remove the pipe brackets from the rocker arm covers.

10. Remove the rocker arm covers.

11. Remove the upper radiator hose and the heater hose.

12. Disconnect the electrical connectors from the coolant sensors.

13. Remove the intake manifold nuts/bolts, the manifold and gaskets.

14. Clean the gasket mounting surfaces.

To install:

15. Using RTV sealant, apply a ⅛ in. (3mm) bead to the front and rear of the block; make sure no water or oil is present.

16. Using new gaskets, marked right and left, apply a ¼ in. (6mm) bead of sealant to hold them in place and install onto the cylinder heads. The gaskets may have to be cut to fit around the pushrods.

17. Install the intake manifold and torque the nuts/bolts, in sequence, to

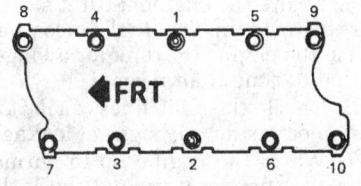

177944

Intake manifold bolt tightening sequence — 3.1L engine

19 ft. lbs. (26 Nm). Retorque using the same sequence.

NOTE: Make sure the areas between the case ridges and the intake manifold are completely sealed.

18. Install the heater and radiator hose to the manifold.

19. Using new gaskets, install the rocker arm covers.

20. Connect the electrical connectors to the coolant sensor.

21. Install the remaining components.

22. Install the air cleaner. Connect the negative battery cable. Refill the cooling system.

1993–95 3.2L (VIN V) Engine

CAUTION

Fuel injection systems remain under pressure even after the engine has been turned off. The fuel system pressure must be relieved before disconnecting any fuel lines. Failure to do so may result in fire and personal injury.

1. Relieve the fuel pressure.

2. Disconnect the negative battery cable.

3. Remove the air cleaner assembly.

4. Disconnect the accelerator pedal cable from the throttle body and bracket.

5. Disconnect the charcoal canister vacuum hose from the vacuum pipe.

6. Disconnect the air vacuum hose and the vacuum booster hose from the common chamber.

7. Disconnect the following electrical connectors:
   a. MAP sensor
   b. charcoal canister vacuum switching valve (VSV)
   c. exhaust gas recirculation VSV
   d. intake air temperature sensor
   e. engine ground cable
   f. fuel injector connectors
   g. thermo sensor connector

8. Disconnect the spark plug wires.

9. Remove the ignition module assembly with the spark plug wires attached.

10. Disconnect the vacuum hoses from the throttle body.

11. Remove the four throttle body mounting bolts. Then, remove the throttle body.

12. Disconnect the PCV hose from the common chamber.

13. Disconnect the fuel pressure control valve vacuum hose from the common chamber.

14. Disconnect the evaporative emission canister purge hose from the common chamber.

15. Remove the EGR valve assembly from the common chamber.

16. Remove the common chamber (six bolts, two nuts, and three brackets).

17. Disconnect the fuel feed and return hoses from the fuel rail. Remove the bracket mounting bolts from the cylinder head cover.

18. Remove the two bolts and four nuts to remove the intake manifold from the engine.

To install:

NOTE: Use new self-locking nuts when installing the intake manifold. Use new manifold gaskets. Use new sealing washers when reconnecting the fuel lines.

19. Install the intake manifold on the engine. Torque the bolts and nuts to 17 ft. lbs. (23 Nm). Tighten the bolts from the center towards the ends.

20. Connect the electrical connectors to the fuel injectors and the thermo sensor.

21. Connect the fuel return and feed hoses to the fuel rail.

22. Install the common chamber. Torque the bolts and nuts to 17 ft. lbs. (23 Nm).

23. Install the EGR assembly. Torque the bolts to 78 inch lbs. (9 Nm).

24. Connect the charcoal canister purge and fuel pressure regulator vacuum hoses to the common chamber.

25. Connect the PCV hose to the common chamber.

26. Install the throttle body assembly and connect the vacuum hoses to the throttle body. Torque the throttle body mounting bolts to 16 ft. lbs. (22 Nm).

27. Install the ignition module assembly. Torque the mounting bolts to 16 ft. lbs. (22 Nm).

28. Reconnect the spark plug wires.

29. Reconnect the following electrical connectors:
   a. MAP sensor
   b. charcoal canister vacuum switching valve (VSV)
   c. exhaust gas recirculation VSV
   d. intake air temperature sensor
   e. engine ground cable

30. Connect the air vacuum hose and the vacuum booster hose to the common chamber.

31. Connect the charcoal canister vacuum hose to the vacuum pipe.

32. Connect the accelerator cable to the throttle body and bracket. Adjust the cable so that the linkage rests

against the stop when moved by hand.

33. Install the air cleaner assembly.

34. Verify that all electrical connectors and vacuum lines have been reconnected.

35. Connect the negative battery cable.

36. Turn the ignition key to the **ON** position for two seconds, then **OFF**. Turn the ignition **ON** again to pressurize the fuel system and check for leaks.

1996–97 3.2L (VIN V) Engine

1. Relieve the fuel system pressure.

2. Disconnect the negative battery cable.

3. Remove the air cleaner and intake duct.

4. Drain the engine coolant to a level below the upper radiator hose.

5. Disconnect the throttle cable from the throttle body linkage.

6. Label and disconnect the following vacuum hoses from the intake manifold chamber:

  a. PCV hose

  b. EVAP canister vacuum hose

  c. Brake booster hose

7. Label and disconnect the following sensor connectors from the rear of the intake manifold chamber:

  a. Ignition control module connectors

  b. Linear EGR valve

  c. MAP sensor

  d. EVAP purge valve

  e. Throttle position sensor

  f. Idle air control valve

  g. Intake air temperature sensor

8. Disconnect the EGR valve supply tube and bracket.

9. Remove the throttle body.

WARNING

Don't use solvent of any type when cleaning the gasket mating surfaces of the throttle body and intake manifold. Solvent may damage the machined surfaces of these components. Be careful not to scratch the mating surfaces.

10. Disconnect the MAP sensor tube, and then unbolt the MAP sensor bracket.

11. Unbolt the intake manifold chamber from the brackets which are located at its front and rear edges.

12. Loosen the manifold mounting bolts and two nuts in a crisscross sequence.

13. Remove the bolts and nuts, and then lift the chamber off of the base of the intake manifold. Note the positions of the long and short bolts.

14. Cover the intake manifold with a sheet of plastic, or clean shop towels to keep out dirt and foreign objects.

15. Carefully clean any dirt from the fuel rail and fuel fittings.

CAUTION

Fuel injection systems remain under pressure even after the engine has been turned off. The fuel system pressure must be relieved before disconnecting any fuel lines. Failure to do so may result in fire and personal injury.

16. Disconnect the fuel feed and return lines from the front of the fuel rail. Clean up any spilled fuel.

17. Label and disconnect the fuel injector wiring harness.

18. Unbolt the fuel return line bracket from the front of the intake manifold.

19. Unbolt the fuel rail from the intake manifold.

20. Carefully lift the fuel rail and injectors off of the intake manifold as an assembly. Move the fuel rail out of the way so it won't be damaged.

21. Remove the fuel rail spacer grommets from the sides of the intake manifold. Replace the spacer grommets if they are cracked or ripped.

22. Loosen the intake manifold nuts and bolts in a crisscross sequence working from the outer edges of the manifold toward the center.

23. Push the engine wiring harnesses aside and lift the intake manifold up and off of the engine block.

24. Remove the intake manifold gaskets. Be careful not to drop any pieces of the gaskets into the engine.

25. Cover the intake openings with a sheet of plastic or clean shop towels to keep out dirt and foreign objects.

To install:

26. Remove the covers from the intake openings. Install new intake manifold gaskets.

27. Fit the intake manifold into position. Move the wiring harness back into position.

28. Install the intake manifold nuts and bolts. Tighten them in a two–step crisscross pattern to 15 ft. lbs. (20 Nm) working from the center of the manifold toward the outer edges.

29. Install the fuel rail spacer grommets.

30. Install the fuel rail assembly onto the intake manifold. Make sure all the fuel injectors are properly seated. Tighten the fuel rail bolts to 5 ft. lbs. (7 Nm).

31. Reconnect the fuel feed and return lines. Install the return line bracket bolt.

32. Reconnect the fuel injector wiring harness.

33. Install a new intake manifold chamber gasket. Install the intake manifold chamber.

34. Install the six intake manifold chamber bolts and two nuts. Tighten the nuts and bolts to 15 ft. lbs. (20 Nm) in a crisscross sequence.

35. Install the intake manifold chamber bracket bolts.

36. Reconnect the MAP sensor tube and bracket.

37. Reconnect the EGR supply tube and bracket.

38. Install the throttle body with a new gasket. Tighten the bolts to 10 ft. lbs. (14 Nm) in a crisscross sequence.

39. Reconnect the following sensor connectors to the rear of the intake manifold chamber:

  a. Ignition control module connectors

  b. Linear EGR valve

  c. MAP sensor

  d. EVAP purge valve

  e. Throttle position sensor

  f. Idle air control valve

  g. Intake air temperature sensor

40. Reconnect the following vacuum hoses to the intake manifold chamber:

  a. PCV hose

  b. EVAP canister vacuum hose

  c. Brake booster hose

41. Reconnect the throttle cable to the throttle body linkage.

42. Install the air cleaner and air intake duct.

43. Reconnect the negative battery cable.

44. Refill and bleed the cooling system.

45. Crank the engine until it starts. Air trapped in the fuel lines may cause the engine to crank for a longer period of time than normal.

46. Check the fuel lines, fuel rail, and injectors for any signs of leakage.

47. Warm the engine up to normal operating temperature and check the operation of the throttle cable and linkage. Adjust if necessary.

48. Check the manifold and throttle body mating surfaces for vacuum leaks.

Exhaust Manifold

REMOVAL AND INSTALLATION

2.3L (VIN L) and 2.6L (VIN E) Engines

——————— CAUTION ———————
Allow the engine to cool to ambient temperature before removing the exhaust manifold.

1. Disconnect the negative battery cable.
2. Remove the intake air duct.
3. Disconnect the hoses from the air pump.
4. Remove the air pump mounting bolts. Slip the drive belt off the pulley and remove the air pump.
5. Remove the manifold heat shield.
6. Remove the EGR pipe clamp bolt from the rear of the cylinder head.
7. Raise and safely support the vehicle.
8. Disconnect the EGR pipe from the exhaust manifold.

NOTE: The dipstick and tube may be removed for extra access to the oxygen sensor and EGR pipe.

9. Disconnect the front exhaust pipe from the exhaust manifold.
10. Uncouple the oxygen sensor harness.
11. Loosen the exhaust manifold nuts in a crisscross pattern.
12. Remove the exhaust manifold from the cylinder head. If it sticks, tap it with a soft-faced mallet.
To install:

NOTE: Install the new exhaust manifold gasket with the stamped mark facing outward.

13. Inspect the exhaust manifold mating surfaces for warpage or other damage. Replace if necessary. The warpage limit is 0.016 in. (0.4mm).

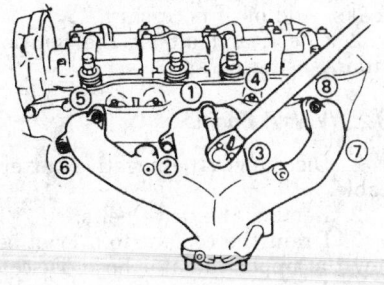

329919

Exhaust manifold tightening sequence — 2.3L and 2.6L engines

14. Using a new gasket, install the exhaust manifold. Tighten the manifold nuts in a crisscross pattern starting in the center and working outward:
  • 2.6 liter (VIN E) engine: 33 ft. lbs. (44 Nm)
  • 2.3 liter (VIN L) engine: 16 ft. lbs. (22 Nm)
15. Install the tube and the dipstick if they were removed.
16. Connect the exhaust pipe to the exhaust manifold. Tighten the nuts to 49 ft. lbs. (67 Nm).
17. If the oxygen sensor was removed, coat its threads with small amount anti-seize compound. Don't get any anti-seize on the sensor's tip. Install the sensor and tighten its fitting to 31 ft. lbs. (42 Nm).
18. Connect the harness to the oxygen sensor.
19. Install the EGR pipe to the intake and exhaust manifolds and tighten the bolts to 17 ft. lbs. (24 Nm).
20. Lower the vehicle.
21. Install the manifold heat shield.
22. Install the EGR pipe clamp bolt to the rear of the cylinder head.
23. Install the air pump and drive belt.
24. Connect the hoses to the air pump.
25. Install the intake air duct.
26. Connect the negative battery cable.

3.1L (VIN Z) Engine

Left or Right Exhaust Manifold

1. Disconnect the negative battery cable.
2. Raise and safely support the vehicle.
3. Remove the front wheels.
4. Remove the dust cover.
5. Disconnect the exhaust pipe from the exhaust manifold.
6. If removing the left manifold, remove the power steering pump bracket.
7. If removing the right manifold, remove the alternator support bracket.
8. Using a 6-point socket, remove the exhaust manifold mounting bolts.
9. Remove the exhaust manifold.
To install:
10. Install the exhaust manifold. Torque the mounting bolts to 25 ft. lbs. (34 Nm).
11. If installing the left manifold, install the power steering pump bracket.
12. If installing the right manifold, install the alternator support bracket.

13. Connect the exhaust pipe to the exhaust manifold.
14. Install the dust cover and the front wheels.
15. Lower the vehicle to the floor.
16. Connect the negative battery cable, start the engine and check for leaks.

3.2L (VIN V) Engine

Left or Right Exhaust Manifold

NOTE: Allow the engine to cool completely before removing the exhaust manifolds.

1. Disconnect the negative battery cable.
2. If removing the left manifold, remove the air duct.
3. If removing the left manifold, remove the EGR pipe mounting bolts from the exhaust manifold.
4. Raise and safely support the vehicle.
5. If necessary to gain extra working room, remove the transfer case skid plate.
6. Label and disconnect the oxygen sensor connectors.
7. Remove the two stud nuts and two bolts and nuts and separate the front exhaust pipes from the exhaust manifold. Be careful not to damage the oxygen sensors when working around the exhaust pipes.
8. Lower the vehicle.
9. Remove the engine hanger and the heat shield.
10. Remove the seven nuts and then remove the exhaust manifold from the cylinder head.
To install:

NOTE: Use new self-locking nuts and new gaskets when installing the exhaust manifolds.

11. Install the exhaust manifold and gasket to the cylinder head using new nuts. Tighten the new nuts to 42 ft. lbs. (57 Nm) in a crisscross sequence.
12. Install the heat shield and the engine hanger.
13. Raise and safely support the vehicle.
14. Install the front exhaust pipes and reconnect the exhaust system. Tighten the exhaust fasteners to the following specifications:
  a. Stud nuts: 49 ft. lbs. (67 Nm)
  b. Flange nuts and bolts: 32–37 ft. lbs. (43–50 Nm).
15. Reconnect the oxygen sensor connectors.
16. Install the skid plate and tighten the bolts to 27 ft. lbs. (37 Nm).
17. Lower the vehicle.

18. If installing the left manifold, install the EGR pipe to the exhaust manifold. Tighten the mounting bolts to 21 ft. lbs. (28 Nm).

19. If installing the left manifold, install the air duct and connect the negative battery cable. Verify that all wires and vacuum lines have been reconnected.

20. Start the engine and check for exhaust leaks.

Front Cover Seal

REMOVAL AND INSTALLATION

3.1L (VIN Z) Engine

1. Disconnect the negative battery cable.
2. Drain the cooling system.
3. Remove the lower radiator hose.
4. Remove the alternator and compressor drive belts.
5. Remove the cooling fan.

— WARNING —
The internal weighted section of the torsional damper (balancer) is assembled to the hub with a rubber sleeve. The removal and installation procedures must be done using the proper tools or damage to the damper can result.

6. Remove the torsional damper retaining bolt. Unbolt and remove the crankshaft pulley. Then, using puller J-24420-B or equivalent, remove the torsional damper.

7. Using a small prybar, remove the oil seal from the front cover. Use care not to damage the crankshaft sealing surface.

To install:

8. Using engine oil, lubricate the new seal and tap it into the front cover using special tool J-35468 or equivalent.

9. Install the torsional damper using tool J-29113 or equivalent. Fasten the crankshaft pulley, then install the torsional damper retaining bolt. Torque this bolt to 70 ft. lbs. (95 Nm).

10. Install the cooling fan and the drive belts.

11. Install the radiator the hose.

12. Refill the cooling system and connect the negative battery cable.

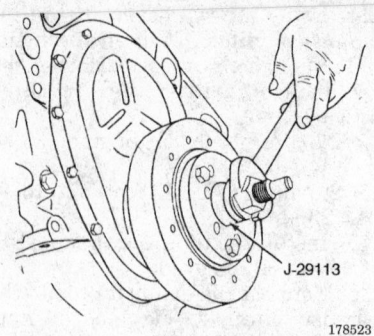

J-29113

178523

Use a special tool to install the torsional damper

Front Crankshaft Seal

REMOVAL AND INSTALLATION

2.3L (VIN L), 2.6L (VIN E) and 3.1L (VIN Z) Engines

1. Disconnect the negative battery cable.
2. Loosen and remove the engine drive belts.
3. Unbolt and remove the cooling fan assembly.
4. If equipped with A/C, remove the belt tensioner.
5. Remove the water pump pulley.
6. Remove the power steering pump from the mount. Don't disconnect the hydraulic lines.
7. On 2.3L and 2.6L engines, remove the upper timing belt cover.
8. Rotate the crankshaft to set the engine at TDC/compression for the No. 1 cylinder. The mark on the camshaft sprocket aligns with the mark on the rear timing cover. The crankshaft sprocket keyway aligns with the mark on the oil seal retainer cover.
9. Remove the crankshaft pulley.
10. On 2.3L and 2.6L engines, remove the lower timing belt cover.
11. On 3.1L engines, remove the timing chain cover.

kg-m(lb.ft.)

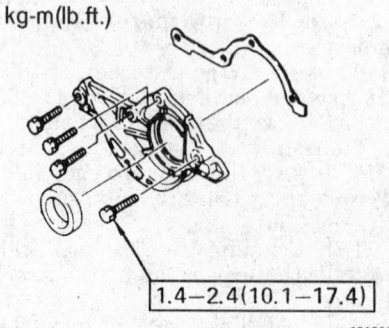

1.4—2.4(10.1—17.4)

191288

Front crankshaft oil seal and retainer

12. Disconnect and remove the starter motor if a flywheel holder tool is to be installed.

13. Remove the timing belt or timing chain.

14. Use tool No. J-22888 or an equivalent puller to remove the oil pump sprocket.

15. If necessary, use a prytool to remove the oil pump seal.

16. Using a small prybar, pry the crankshaft oil seal from the oil seal retainer. Be careful not to damage the crankshaft sealing surface.

To install:

17. Lubricate a new crankshaft oil seal with clean engine oil and tap it into the retainer with an oil seal installation tool. Install a new oil pump oil seal if necessary.

18. Install the crankshaft sprocket so that its keyway and alignment mark point to the alignment mark on the oil seal retainer.

19. Install the oil pump pulley. Apply a small amount of threadlocking compound to the sprocket nut threads. Tighten the nut to 56 ft. lbs. (76 Nm).

20. Verify that all the timing marks are aligned and install the timing belt or chain.

21. On 2.3L and 2.6L engines, install the lower timing cover and tighten the bolts to 4.4 ft. lbs. (6 Nm).

22. On 3.1L engines, install the timing chain cover.

23. Install the crankshaft pulley and tighten the bolt to 87 ft. lbs. (118 Nm).

24. Install the starter motor.

25. On 2.3L and 2.6L engines, install the upper timing cover and tighten the bolts to 4.4 ft. lbs. (6 Nm).

26. Install the power steering pump.

27. Install the cooling fan and tighten the bolts to 20 ft. lbs. (26 Nm).

28. Install the A/C belt tensioner.

29. Install the accessory drive belts.

30. Reconnect the negative battery cable.

31. Start the engine and check for leaks. Add oil if necessary.

32. Check and adjust the ignition timing as necessary.

3.2L (VIN V) Engine

1. Disconnect the negative battery cable.
2. Remove the drive belts.
3. Drain the coolant to a level below the upper radiator hose. Disconnect the upper radiator hose from the coolant inlet.
4. Remove the cooling fan and pulley assembly.

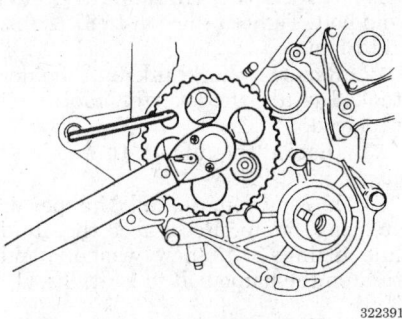

Oil pump sprocket holder

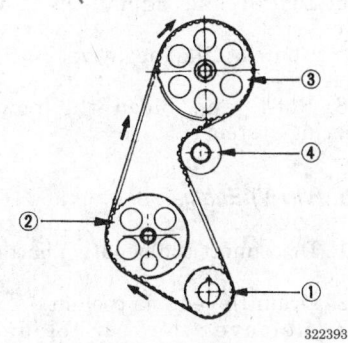

Timing belt installation sequence — 2.3L and 2.6L engines

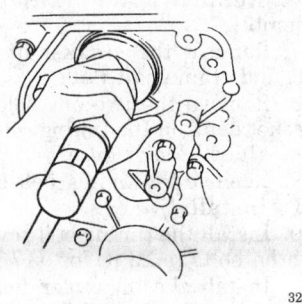

Oil pump seal installation

5. Remove the upper timing cover.
6. Rotate the crankshaft to align the timing marks.
7. Remove the crankshaft pulley and the lower timing belt cover.
8. Remove the timing belt.
9. Remove the crankshaft timing sprocket and key.
10. Use a seal puller and remove the crankshaft oil seal. Be careful not to damage the crankshaft or the oil pump sealing surface.

To install:
11. Apply engine oil to the lip of the seal and install the oil seal using seal installer J-39202 or an equivalent seal driver.
12. Install the crankshaft timing sprocket and key.
13. Install the timing belt.
14. Install the lower timing belt covers and tighten the bolts to 12 ft. lbs. (17 Nm).
15. Install the crankshaft pulley and tighten it to 123 ft. lbs. (167 Nm).
16. Install the upper timing covers.
17. Install the cooling fan and pulley assembly and tighten the bolts to 16 ft. lbs. (22 Nm).
18. Install the accessory drive belts.
19. Connect the upper radiator hose. Refill and bleed the cooling system.
20. Check the engine oil level and top up if necessary.
21. Reconnect the negative battery cable.

Timing Chain and Sprockets

REMOVAL AND INSTALLATION

3.1L (VIN Z) Engine

1. Disconnect the negative battery cable.
2. Remove the timing chain front cover and related components, as described later in this section.
3. Place #1 piston at Top Dead Center (TDC) with the marks on the camshaft and crankshaft sprockets aligned (#4 firing position).
4. Remove the camshaft sprocket mounting bolts, then remove the sprocket with the chain.

NOTE: If the sprocket does not come off easily, a light blow on the lower edge of the sprocket with a plastic mallet should dislodge the sprocket.

5. Remove the crankshaft sprocket using tool J-5825-A, or the equivalent.

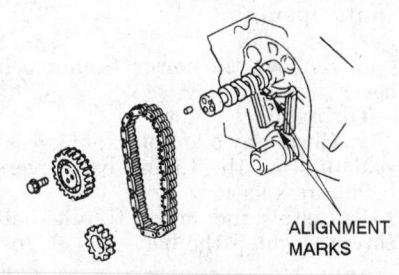

ALIGNMENT MARKS

Timing chain sprocket alignment marks — 3.1L engine

6. Remove the timing chain damper assembly bolts and damper.

To install:
7. Clean all gasket surfaces completely.
8. Install timing chain damper assembly and bolts. Torque bolts to 15 ft. lbs. (21 Nm).
9. Apply Molykote® or equivalent lubricant to the sprocket thrust surface.
10. Align the dowel in the camshaft with the dowel hole in the camshaft sprocket.
11. Hold the camshaft sprocket with the chain hanging down and engage the crankshaft sprocket. Align the marks on the camshaft and crankshaft sprocket.
12. Draw the camshaft sprocket onto the camshaft using the mounting bolts. Torque camshaft mounting bolts to 17 ft. lbs. (23 Nm).
13. Lubricate the timing chain with clean engine oil.
14. Install the front cover and related components, as described in the following procedure.
15. Connect the negative battery cable.

Timing Chain Front Cover

REMOVAL AND INSTALLATION

3.1L (VIN Z) Engine

1. Disconnect the negative battery cable.
2. Drain the cooling system.
3. Raise and safely support the vehicle.
4. Remove the stone guard and the radiator under cover.
5. Remove the serpentine drive belt.
6. Remove the suspension crossmember.
7. Remove the starter and the flywheel inspection cover.
8. Loosen the oil pan bolts.
9. Disconnect the upper radiator hose from the radiator.
10. Remove the A/C pipe bracket from the radiator.
11. Remove the upper and lower fan shrouds.
12. Remove the fan and clutch assembly.
13. Remove the water pump pulley.
14. Remove the power steering pump bracket, if equipped.
15. Insert a prybar into the flywheel to prevent rotation and remove the torsional damper center bolt.
16. Unbolt and remove the crankshaft pulley from the torsional

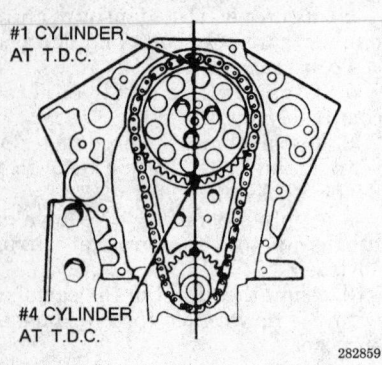

#1 CYLINDER
AT T.D.C.

#4 CYLINDER
AT T.D.C.

282859

Camshaft timing marks — 3.1L engine

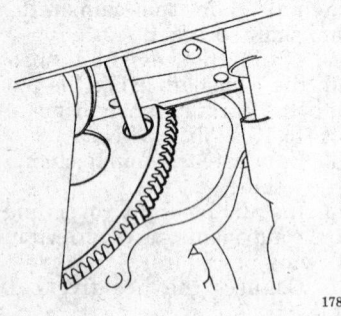

178799

Insert a prybar through the flywheel to hold the crankshaft

damper. Remove the damper using puller J-24420-B or equivalent.

17. Remove the serpentine belt tensioner.

18. Remove the front cover mounting bolts and the front cover.

To install:

19. Thoroughly clean the sealing surface on the front cover and the engine block.

20. Using a new gasket and sealant, install the front cover. Torque the mounting bolts to 20 ft. lbs. (27 Nm).

21. Tighten the oil pan bolts to 18 ft. lbs. (25 Nm) and the nuts to 7 ft. lbs. (10 Nm).

22. Install the starter and the flywheel inspection cover.

23. Install the water pump pulley and the lower radiator hose.

24. Using tool J-29113 or equivalent, install the torsional damper. Attach the crankshaft pulley, then torque the center bolt to 70 ft. lbs. (95 Nm).

25. Install the serpentine belt tensioner.

26. Install the power steering pump bracket, if equipped.

27. Install the fan and clutch assembly.

28. Install the serpentine belt.

29. Install the fan shrouds.

30. Install the A/C pipe bracket on top of the radiator.

31. Connect the upper radiator hose to the radiator.

32. Install the radiator under cover and the stone guard.

33. Install the suspension crossmember.

34. Lower the vehicle to the floor.

35. Connect the negative battery cable and refill the cooling system.

Timing Belt Front Cover

REMOVAL AND INSTALLATION

2.3L (VIN L) and 2.6L (VIN E) Engines

1. Disconnect the negative battery cable.

2. Remove all drive belts.

3. Remove the cooling fan and pulley.

4. If equipped, remove the power steering pump from the mount and unbolt the hydraulic line from the upper timing cover. Move the pump and hydraulics lines out of the work area. If the lines must be disconnect, plug them to prevent fluid loss and contamination.

5. Remove the upper timing cover.

6. Rotate the crankshaft to set the engine at TDC/compression for the No. 1 cylinder. The arrow mark on the camshaft timing sprocket align with the mark on the back timing cover.

7. Use a suitable holder tool to remove the crankshaft pulley bolt and pulley.

NOTE: The starter motor can be removed and a bar-type crankshaft holder tool installed in its opening.

8. Remove the lower timing belt cover.

To install:

9. Clean any oil, coolant, or excess sealant from the timing belt covers before installation.

10. Install the lower timing belt covers. Tighten the bolts to 4 ft. lbs. (6 Nm).

11. Install the crankshaft pulley and bolt. Tighten the bolt to 87 ft. lbs. (118 Nm).

12. Remove the crankshaft holder tool. Install the starter motor, if removed.

13. Install the upper timing belt cover.

14. If removed, install the power steering pump. Reconnect the fluid line fittings with new washers and tighten the banjo bolt to 14 ft. lbs. (19 Nm).

15. Install the cooling fan and pulley and tighten the pulley mounting bolts to 20 ft. lbs. (26 Nm).

16. Install and adjust the drive belts.

17. Connect the negative battery cable.

18. Refill and bleed the power steering system.

3.2L (VIN V) Engine

1. Disconnect the negative battery cable.

2. Drain the engine coolant.

3. Remove the air cleaner assembly.

4. Remove the upper fan shroud and remove the cooling fan.

5. Remove the drive belts.

6. Remove the fan pulley assembly.

7. Remove the crankshaft pulley bolt and crankshaft pulley.

8. Remove the two oil cooler hose bracket bolts on the timing cover. Remove the oil cooler hose.

9. Remove the timing belt cover.

To install:

10. Install the timing belt cover and tighten bolts to 12 ft. lbs. (17 Nm).

11. Install the oil cooler hose and tighten brackets to 16 ft. lbs. (22 Nm).

12. Install the crankshaft pulley. Use tool No. J–8614–01, or an equivalent pulley holder and tighten the bolts to 123 ft. lbs. (167 Nm).

13. Install the fan pulley assembly and tighten the bolts to 16 ft. lbs. (22 Nm).

14. Install the drive belts.

15. Install cooling fan assembly.

16. Install upper fan shroud to the radiator.

17. Install air cleaner assembly.

18. Refill the engine with fresh coolant. Bleed the cooling system.

19. Connect the negative battery cable.

Timing Belt

ADJUSTMENT

WARNING

It is very highly recommended that a timing belt be replaced any time its tension is released. A timing belt should not be viewed as an adjustable component. The timing belt's tension cannot be increased to compensate for wear. If the engine has been disassembled for mechanical work, a new timing belt should be installed. The small cost of a new timing belt is cheap insurance against expensive engine damage which can be caused by the failure of a re-used timing belt.

2.3L (VIN L) and 2.6L (VIN E) Engines

1. Make sure the crankshaft and the camshaft sprockets are aligned with their timing marks, so that the engine is at TDC/compression for the No. 1 cylinder. Install the timing belt onto the sprockets using the following sequence: first, crankshaft sprocket; second, oil pump sprocket; third, camshaft sprocket.

2. Loosen the tensioner mounting bolt. This will allow the tensioner spring to apply pressure to the timing belt.

3. After the spring has pulled the timing belt as far as possible, temporarily tighten the tensioner mounting bolt to 14 ft. lbs. (19 Nm).

4. Rotate the crankshaft counterclockwise two complete revolutions to check the rotation of the belt and the alignment of the timing marks. Listen for any rubbing noises which may mean the belt is binding.

5. Loosen the tensioner pulley bolt to allow the spring to adjust the correct tension. Then, retighten the ten-

sioner pulley bolt to 14 ft. lbs. (19 Nm).

6. Install the timing belt covers and crankshaft pulley.

7. Tighten the crankshaft pulley bolt to 87 ft. lbs. (118 Nm). Tighten the small pulley bolts to 6 ft. lbs. (8 Nm).

3.2L (VIN V) Engine

1. Verify that the camshaft sprocket timing marks are aligned. The groove and the keyway on the crankshaft timing sprocket align with mark on the oil pump. The white pointers on the camshaft timing sprockets align with the dots on the front plate.

2. Install the timing belt. Use clips to secure the belt onto each sprocket until the installation is complete. Align the dotted marks on the timing belt with the timing mark opposite the groove on the crankshaft sprocket.

NOTE: The arrows on the timing belt must follow the belt's direction rotation. The manufacturer's trademark on the belt's spine should be readable left-to-right when the belt is installed.

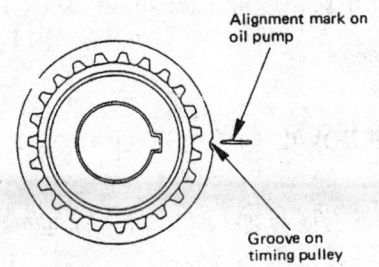

Crankshaft sprocket groove and alignment marks — 3.2L engine

3. Align the white line on the timing belt with the alignment mark on the right bank camshaft timing pulley. Secure the belt with a clip.

4. Rotate the crankshaft counterclockwise to remove the slack between the crankshaft sprocket and the right camshaft timing sprocket.

5. Install the belt around the water pump pulley.

6. Install the belt on the idler pulley.

7. Align the white alignment mark on the timing belt with the alignment mark on the left bank camshaft timing sprocket.

8. Install the crankshaft pulley and tighten the center bolt by hand. Rotate the crankshaft pulley clockwise to give slack between the crankshaft timing pulley and the right bank camshaft timing pulley.

9. Insert a 1.4mm piece of wire through the hole in the pusher to hold the rod in. Install the pusher assembly while pushing the tension pulley toward the belt.

10. Pull the pin out from the pusher to release the rod.

11. Remove the clamps from the sprockets. Rotate the crankshaft pulley clockwise two turns. Measure the rod protrusion to be sure it is between 0.16–0.24 in. (4–6mm).

12. If the tensioner pulley bracket pivot bolt was removed, tighten it to 31 ft. lbs. (42 Nm).

13. Tighten pusher bolts to 14 ft. lbs. (19 Nm).

14. Remove crankshaft pulley. Install the lower and upper timing belt covers and tighten their bolts to 12 ft. lbs. (17 Nm).

15. Install the crankshaft pulley and tighten the pulley bolt to 123 ft. lbs. (167 Nm).

INSPECTION

1. Disconnect the negative and positive battery cables.

2. Label and disconnect the ignition wires and remove the spark plugs. Remove the upper timing belt cover.

3. Rotate the crankshaft to set the engine at TDC/compression for the No. 1 cylinder. Align the pointer on the camshaft sprocket with the mark on the back cover. Verify that the TDC mark on the crankshaft pulley aligns with the pointer on the lower cover.

4. Rotate the crankshaft pulley to cycle the belt through its entire rotation.

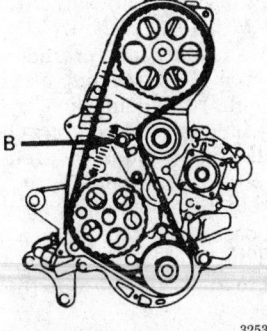

325349

Timing belt tensioner lockbolt

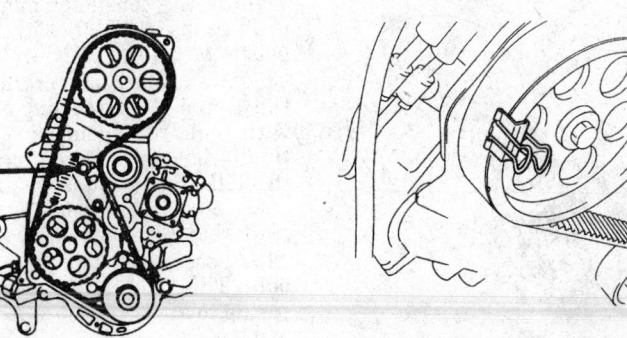

325382

Use clips to hold the timing belt in place during installation

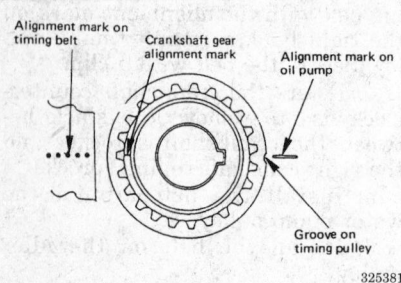

Timing belt dotted mark alignment — 3.2L engine

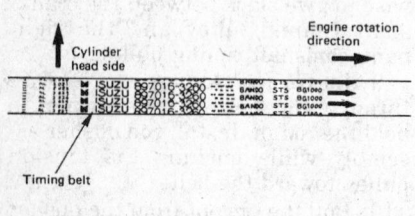

Timing belt — direction of installation and rotation

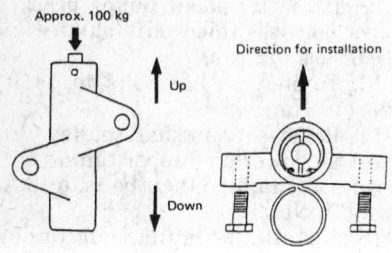

Timing belt pusher (tensioner) with pin installed

5. Inspect the entire length of the timing belt. Look carefully for any signs of the following conditions:

a. Cracked, chipped, or broken teeth.

b. Fraying, separation, or heat damage to the belt's rubber and fiber layers.

c. Oil or coolant leaks which may have contaminated the belt.

d. Make sure the timing marks align. Misaligned marks may indicate that the belt has jumped one

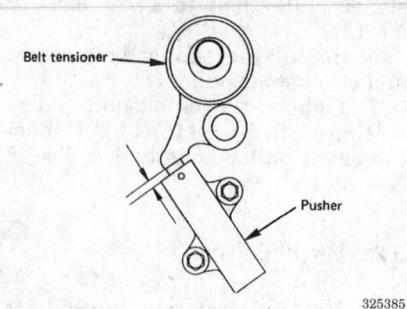

Timing belt tensioner pulley with pusher installed

or more teeth, or has been improperly tensioned.

6. Check the camshaft and crankshaft oil seals for any signs of leakage. Also check the water pump for leakage. The source of any oil or coolant leaks must be found and corrected before a new timing belt is installed.

7. Replace the timing belt if it's damaged in any way, or if its condition or the vehicle's maintenance history is uncertain. Recommended service interval for timing belt replacement is 60,000 miles (96,000 km).

8. After inspection, retighten the crankshaft pulley bolt to 87 ft. lbs. (118 Nm). Install the timing belt covers and valve covers. Install the spark plugs and reconnect the ignition wires. Reconnect the battery cables.

REMOVAL AND INSTALLATION

2.3L (VIN L) and 2.6L (VIN E) Engines

NOTE: With the timing marks aligned, the engine is positioned on the TDC of the No. 1 cylinder's compression stroke.

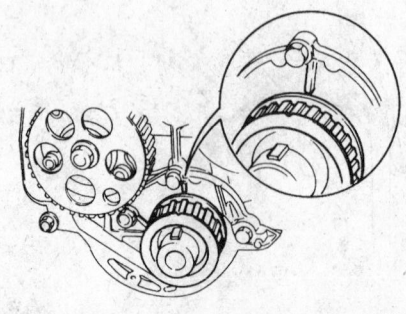

Crankshaft sprocket timing mark — 2.3L and 2.6L engines

1. Disconnect the negative battery cable.

2. Loosen and remove the engine accessory drive belts.

3. Remove the cooling fan assembly and the water pump pulley.

4. Drain the fluid from the power steering reservoir.

5. Unbolt and remove the power steering pump. Unbolt the hydraulic line brackets from the upper timing cover and move the pump out of the work area without disconnecting the hydraulic lines.

6. Disconnect and remove the starter motor if a flywheel holder (part No. J–38674) is to be used.

7. Remove the upper timing belt cover.

8. Rotate the crankshaft to set the engine at TDC/compression for the No. 1 cylinder. The arrow mark on camshaft sprocket aligns with mark on the rear timing cover.

9. Remove the crankshaft pulley.

10. Remove the lower timing belt cover.

11. Verify that the engine is set at TDC/compression for the No. 1 cylinder. The notch on the crankshaft sprocket aligns with the pointer on the oil seal retainer.

12. Release and remove the tensioner spring to release the timing belt's tension.

13. Remove the timing belt.

14. Unbolt the tensioner pulley bracket from the engine's front cover.

15. If necessary, unbolt and remove the camshaft sprockets. Use a puller to remove the crankshaft pulley if necessary. Don't loose the crankshaft sprocket key.

To install:

16. Install the camshaft and crankshaft sprockets if they were removed. Align the timing marks and be sure to install any keys. Tighten the camshaft sprocket bolt to 43 ft. lbs. (59 Nm).

17. Install the tensioner assembly. Tighten the tensioner mounting bolt to 14 ft. lbs. (19 Nm). Tighten the cap bolt to 9 ft. lbs. (13 Nm).

18. Make sure the crankshaft and the camshaft sprockets are aligned with their timing marks. Install the timing belt onto the sprockets using the following sequence: first, crankshaft sprocket; second, oil pump sprocket; third, camshaft sprocket.

19. Loosen the tensioner mounting bolt. This will allow the tensioner spring to apply pressure to the timing belt.

20. After the spring has pulled the timing belt as far as possible, tempo-

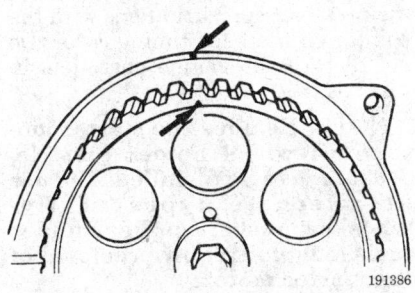

Camshaft sprocket timing mark — 2.3L and 2.6L engines

rarily tighten the tensioner mounting bolt to 14 ft. lbs. (19 Nm).

NOTE: Remove the flywheel holder before rotating the crankshaft. Reinstall the holder to torque the crankshaft pulley bolt.

21. Rotate the crankshaft counterclockwise two complete revolutions to check the rotation of the belt and the alignment of the timing marks. Listen for any rubbing noises which may mean the belt is binding.

22. Loosen the tensioner pulley bolt to allow the spring to adjust the correct tension. Then, retighten the tensioner pulley bolt to 14 ft. lbs (19 Nm).

23. Install the lower timing cover and the crankshaft pulley.

24. Tighten the crankshaft pulley bolt to 87 ft. lbs. (118 Nm). Tighten the small pulley bolts to 6 ft. lbs. (8 Nm).

25. Install the upper timing cover.

26. Install the starter if it was removed. Tighten the bolts to 30 ft. lbs. (40 Nm).

27. Install the power steering pump. If the hydraulic lines were disconnected, refill and bleed the power steering system.

28. Install the water pump pulley and tighten its nut to 20 ft. lbs. (26 Nm).

29. Install the cooling fan assembly.

30. Install and adjust the accessory drive belts.

31. Connect the negative battery cable.

3.2L (VIN V) Engine

1. Disconnect the negative battery cable.

2. Drain the engine coolant into a sealable container.

3. Remove the air cleaner assembly and intake air duct.

4. Disconnect the upper radiator hose from the coolant inlet.

5. Remove the upper fan shroud from the radiator.

6. Remove the four nuts retaining the cooling fan assembly. Remove the cooling fan from the fan pulley.

7. Loosen and remove the power steering drive belt.

8. Loosen and remove the air conditioning compressor drive belt.

9. Loosen and remove the alternator drive belt.

10. Remove the upper timing belt covers.

11. Remove the fan pulley assembly.

12. Rotate the crankshaft to align the camshaft timing marks with the pointer dots on the back covers. Verify that the pointer on the crankshaft aligns with the mark on the lower timing cover.

NOTE: When the timing marks are aligned on 1994–95 vehicles, no cylinders will be at TDC/compression. When the timing marks are aligned on 1995½–97 vehicles, the No. 2 cylinder is at TDC/compression.

------ **WARNING** ------

Align the camshaft and crankshaft sprockets with their alignment marks before removing the timing belt. Failure to align the belt and sprocket marks may result in valve damage.

13. Use tool No. J–8614–01, or a suitable pulley holding tool to remove the crankshaft pulley center bolt. Remove the crankshaft pulley.

14. Disconnect the two oil cooler hose bracket bolts on the timing cover. Move the oil cooler hoses and bracket off of the lower timing cover.

15. Remove the lower timing belt cover.

16. Remove the pusher assembly (tensioner) from below the belt tensioner pulley. The pusher rod must always be facing upward to prevent oil leakage. Push the pusher rod in, and insert a wire pin into the hole to keep the pusher rod retracted.

17. Remove the timing belt.

18. Use tool No. J–41472, or a suitable pulley holding tool to loosen and remove the camshaft sprocket bolt. Remove the camshaft sprockets.

19. Inspect the water pump and replace it if there is any doubt about its condition.

20. Repair any oil or coolant leaks before installing a new timing belt. If the timing belt has been contaminated with oil or coolant, or is damaged, it must be replaced.

To install:

21. Install the camshaft sprockets. Use a holding tool, and tighten their bolts to 41 ft. lbs. (55 Nm).

22. Verify that the sprocket timing marks are still aligned. The groove and the keyway on the crankshaft timing sprocket align with mark on the oil pump. The white pointers on the camshaft timing sprockets align with the dots on the front plate.

23. Install the timing belt. Use clips to secure the belt onto each sprocket until the installation is complete. Align the dotted marks on the timing belt with the timing mark opposite the groove on the crankshaft sprocket.

NOTE: The arrows on the timing belt must follow the belt's direction rotation. The manufacturer's trademark on the belt's spine should be readable left–to–right when the belt is installed.

24. Align the white line on the timing belt with the alignment mark on the right bank camshaft timing pulley. Secure the belt with a clip.

25. Rotate the crankshaft counterclockwise to remove the slack between the crankshaft sprocket and the right camshaft timing sprocket.

26. Install the belt around the water pump pulley.

27. Install the belt on the idler pulley.

28. Align the white alignment mark on the timing belt with the alignment mark on the left bank camshaft timing sprocket.

29. Install the crankshaft pulley and tighten the center bolt by hand. Rotate the crankshaft pulley clockwise to give slack between the crankshaft timing pulley and the right bank camshaft timing pulley.

30. Insert a 1.4mm piece of wire through the hole in the pusher to hold the rod in. Install the pusher assembly while pushing the tension pulley toward the belt.

31. Pull the pin out from the pusher to release the rod.

32. Remove the clamps from the sprockets. Rotate the crankshaft pulley clockwise two turns. Measure the rod protrusion to be sure it is between 0.16–0.24 in. (4–6mm).

33. If the tensioner pulley bracket pivot bolt was removed, tighten it to 31 ft. lbs. (42 Nm).

34. Tighten pusher bolts to 14 ft. lbs. (19 Nm).

35. Remove the crankshaft pulley. Install the lower and upper timing belt covers and tighten their bolts to 12 ft. lbs. (17 Nm).

36. Fit the oil cooler hose onto the timing cover and tighten its mounting bracket bolts to 16 ft. lbs. (22 Nm).

37. Install the crankshaft pulley and tighten the pulley bolt to 123 ft. lbs. (167 Nm).

38. Install fan pulley assembly and tighten bolts to 16 ft. lbs. (22 Nm).

39. Install and adjust the alternator drive belt.

40. Install and adjust the air conditioning drive belt.

41. Install and adjust the power steering pump drive belt.

42. Install cooling fan assembly and tighten bolts to 6 ft. lbs. (8 Nm).

43. Install upper fan shroud.

44. Install air cleaner assembly and intake air duct.

45. Connect the upper radiator hose to the coolant inlet.

46. Refill and bleed the cooling system.

47. Connect the negative battery cable.

Timing Belt Sprockets

REMOVAL AND INSTALLATION

2.3L (VIN L) and 2.6L (VIN E) Engines

Camshaft Sprocket

1. Disconnect the negative battery cable.

2. Remove the timing belt.

3. Remove the camshaft sprocket-to-camshaft bolt and the sprocket.

NOTE: It may be necessary to use a rubber mallet to tap the sprocket from the camshaft.

4. Remove and replace the camshaft oil seal.

To install:

5. Align the camshaft sprocket-to-rear plate timing marks. With the crankshaft sprocket aligned with the timing mark, install the timing belt.

6. Apply the tensioner pulley spring pressure to the timing belt.

7. Rotate the crankshaft 2 complete revolutions in the opposite direction of rotation and realign the timing marks.

8. Loosen the tensioner pulley bolt to allow the spring to adjust the correct tension. Torque the tensioner pulley bolt to 14 ft. lbs. (19 Nm).

9. Install the timing covers and the crankshaft pulley.

10. Install the accessories and the drive belts.

11. Connect the negative battery cable.

Crankshaft Sprocket

1. Disconnect the negative battery cable.

2. Remove the timing belt.

3. Using a puller, press the sprocket from the crankshaft.

4. Remove and replace the crankshaft oil seal.

To install:

5. Align the crankshaft sprocket-to-oil seal retainer plate timing marks.

6. With the camshaft sprocket aligned with its timing mark, install the timing belt.

7. Apply the tensioner pulley spring pressure to the timing belt.

8. Rotate the crankshaft 2 complete revolutions in the opposite direction of rotation and realign the timing marks.

9. Loosen the tensioner pulley bolt to allow the spring to adjust the correct tension. Torque the tensioner pulley bolt to 14 ft. lbs. (19 Nm).

10. Install the timing covers and the crankshaft pulley.

11. Install the accessories and the drive belts.

12. Connect the negative battery cable.

Camshaft

REMOVAL AND INSTALLATION

2.3L (VIN L) and 2.6L (VIN E) Engines

1. Disconnect the negative battery cable.

2. Loosen and remove the engine accessory drive belts.

3. Remove the cooling fan assembly.

4. Remove the water pump pulley.

5. Label and disconnect the ignition wires.

6. Remove the upper timing belt cover.

7. Rotate the crankshaft to position the No. 1 cylinder on the TDC of

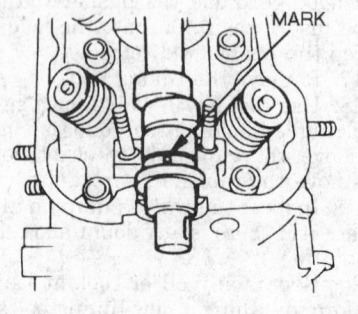

191212

Camshaft installation positioning mark — 2.3L and 2.6L engines

its compression stroke. The mark on the camshaft sprocket aligns with the pointer on the back timing cover. Do not disturb the engine once it is in this position.

NOTE: Remove the starter motor if flywheel holder tool No. J–38674 is to be installed. The use of alternate types of flywheel/crankshaft pulley holder tools may not require removal of the starter motor.

8. Remove the valve cover.

9. Remove the crankshaft pulley and the lower timing belt cover.

10. Remove the timing belt. Mark its direction of rotation if it is to be reinstalled.

11. Unbolt and remove the camshaft pulley.

12. Loosen the camshaft holder bolts in a two–step crisscross sequence to prevent warpage. Remove the bolts and lift the rocker arm assembly off of the cylinder head. Label the positions of the rocker arms and shafts if the rockers and shafts are to be disassembled.

13. Lift the camshaft from the cylinder head. Remove the camshaft oil seals.

14. Inspect the camshaft lobes and journals for signs of wear. Replace parts as necessary.

15. Cover the cylinder head with a sheet of plastic or clean shop towels to keep out dust and foreign objects.

To install:

16. Lubricate the camshaft and its journals with clean engine oil and position it onto the cylinder head. The marks on the camshaft must face upwards when installed to ensure that the engine remains at TDC/compression for the No. 1 cylinder.

17. Assemble the rocker arm and shaft assemblies and camshaft holders.

18. Apply silicon gasket sealer to the front side of the contact surface of the No. 1 camshaft holder prior to installation. Don't allow the sealant to cure before installing the holder.

19. Install the rocker arm and shaft assembly. Install all the retaining bolts, but only hand–tighten them at this point.

20. Starting in the center of the cylinder head and working outward, use a two–step crisscross pattern to tighten the holder bolts to 6 ft. lbs. (8 Nm) and the holder nuts to 16 ft. lbs. (22 Nm).

21. Lubricate a new camshaft oil seal when clean engine oil. Use seal installer J–33183 or equivalent to drive the seal into position.

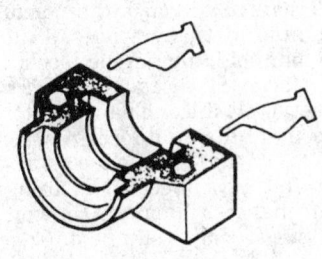

Silicone sealer application points on rocker shaft bracket #1

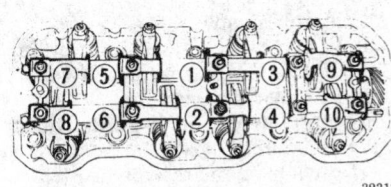

Camshaft holder tightening sequence — 2.3L and 2.6L engines

22. Install the back timing cover.

23. Install the camshaft sprocket and its key. Tighten the bolt to 43 ft. lbs. (59 Nm).

24. Verify that the timing marks are aligned and that the engine is at TDC/compression for the No. 1 cylinder. Install the timing belt and tension the timing belt.

25. Install the lower timing belt cover. Install the crankshaft pulley and tighten the bolt to 87 ft. lbs. (118 Nm).

26. Adjust the valve lash. Tighten the adjusting screw locknuts to 9 ft. lbs. (13 Nm).

27. Rotate the engine manually two times to check that there is no piston-to-valve interference.

28. Apply silicone sealant to the front and rear valve cover mating surfaces of the cylinder head. Immediately install a new gasket and the valve cover. Tighten the nuts to 7 ft. lbs. (10 Nm).

29. Install the upper timing belt cover.

30. Install the starter motor if it was removed. Tighten the mounting bolts to 30 ft. lbs. (41 Nm).

31. Install the water pump pulley and tighten its bolts to 6 ft. lbs. (8 Nm). Install the cooling fan.

32. Install the engine accessory drive belts and adjust their tensions.

33. Change the engine oil and filter.

34. Verify that all electrical connectors and vacuum lines have been reconnected.

35. Reconnect the negative battery cable.

36. Start the engine. Check and adjust the ignition timing.

3.1L (VIN Z) Engine

NOTE: Use long bolts threaded into the camshaft to help remove the shaft without damaging the camshaft bearings. Remove the camshaft slowly while supporting the weight with the long bolt.

1. Relieve the fuel pressure. Disconnect the negative battery cable.

2. Remove the timing cover and the camshaft sprocket.

3. Remove the upper fan shroud and the radiator.

4. Disconnect the fuel line, the accelerator linkage, the vacuum hoses and electrical connectors from the throttle body unit.

5. Remove the rocker arm covers.

6. Loosen the valves, rotate them 90° and remove the pushrods; be sure to keep them aligned so they may be installed in their original positions.

7. Remove the intake manifold.

8. Using a hydraulic lifter removal tool, pull the valve lifters from the engine.

9. Using 3 long bolts, thread them into the camshaft holes. Grasp the bolts and carefully, pull the camshaft from the front of the engine.

NOTE: All the camshaft bearing journals are the same diameter; exercise care in removing the camshaft so the bearings do not become damaged.

To install:

10. Lubricate the camshaft with engine oil and carefully install it into the engine. Take care not to damage the bearings.

11. Using a hydraulic lifter installation tool or equivalent, install the hydraulic lifters into the engine.

12. Using new gaskets and sealant, install the intake manifold.

13. Install the pushrods and the rocker arms.

14. Install the camshaft sprocket, the timing chain and the front cover; be sure the timing marks are aligned.

15. Adjust the valves.

16. Using new gaskets, install the rocker arm covers.

17. Connect the fuel line, accelerator linkage and vacuum hoses to the throttle body.

18. Install the radiator and the fan shroud.

19. Refill the cooling system.

20. Start the engine and allow it to reach normal operating temperatures. Check and/or adjust the timing.

3.2L (VIN V) Engine

NOTE: Isuzu has issued a recall notice for 1993–94 Rodeos equipped with 3.2L VIN V engines. This is campaign number 94V–094, and involves faulty camshaft end plugs. The plugs may dislodge from the cylinder heads and can cause rapid oil loss. Remember this when ordering parts; a service kit is available for affected vehicles.

— CAUTION —
Fuel injection systems remain under pressure even after the engine has been turned off. The fuel system pressure must be relieved before disconnecting any fuel lines. Failure to do so may result in fire and personal injury.

1. Relieve the fuel system pressure.

2. Disconnect the negative battery cable and reinstall the fuel pump relay.

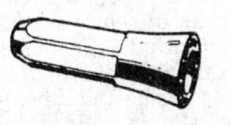

J-33183   Camshaft oil seal installer

Camshaft oil seal installation tool J-33183

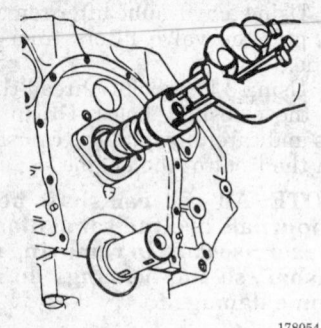

Use long bolts to help remove the camshaft — 3.1L engine

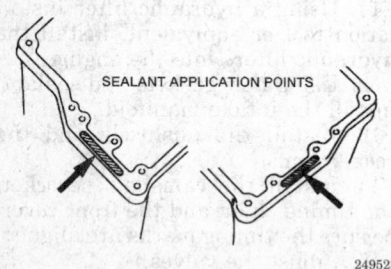

Apply sealant to the cylinder head where the camshaft mounts — 3.2L engine

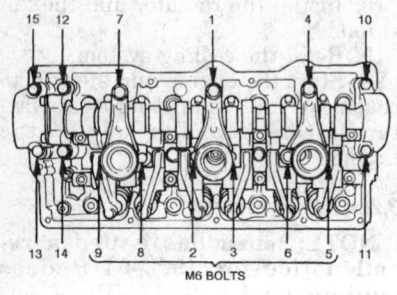

Camshaft holder torque sequence — 3.2L engine

3. Drain the coolant into a sealable container.

4. Support the hood as far open as possible.

5. Remove the air intake duct and the air cleaner box.

6. Disconnect and remove the upper and lower radiator hoses. Catch any coolant that runs out.

7. Loosen and remove the power steering pump, A/C compressor, and alternator drive belts.

8. Remove the cooling fan and its pulley assembly.

9. Unbolt the power steering pump mounting bracket. Move the

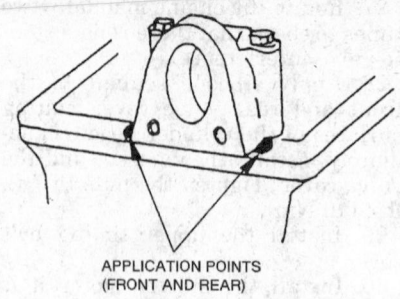

Apply sealant to the camshaft mounting brackets — 3.2L engine

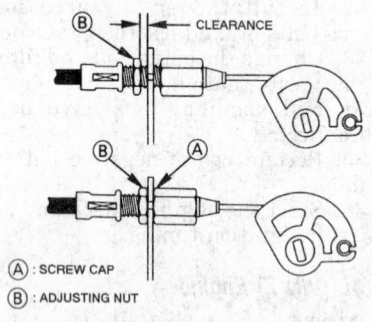

Accelerator cable adjustment — 3.2L engine

pump and bracket out of the way without disconnecting the hydraulic lines.

10. Disconnect the throttle cable from the throttle body linkage.

11. Label and disconnect the following vacuum hoses from the intake manifold chamber:

   a. PCV hose

   b. EVAP canister vacuum hose

   c. Brake booster hose

12. Label and disconnect the following sensor connectors from the rear of the intake manifold chamber:

   a. Ignition control module connectors

   b. Linear EGR valve

   c. MAP sensor

   d. EVAP purge valve

   e. Throttle position sensor

   f. Idle air control valve

   g. Intake air temperature sensor

13. Disconnect the EGR valve supply tube and bracket.

14. First, remove the throttle body, and then remove the intake manifold chamber.

15. Carefully clean any dirt from the fuel rail and fuel fittings.

16. Disconnect the fuel feed and return lines from the front of the fuel rail. Clean up any spilled fuel.

17. Remove the intake manifold gaskets. Be careful not to drop any

pieces of the gaskets into the engine. Don't scratch or gouge the machined aluminum mating surfaces of the intake manifold and engine block.

18. Cover the intake openings with a sheet of plastic or clean shop towels to keep out dirt and foreign objects.

19. Label the ignition coil assemblies and disconnect them from the wiring harness. Remove the coil assemblies so they won't be damaged.

20. Unbolt the oil cooler line brackets from the timing belt covers.

21. Remove the upper timing belt covers.

22. Rotate the crankshaft to align the camshaft timing marks with the pointer dots on the back covers. When the timing marks are aligned, the No. 2 cylinder is at TDC/compression.

23. Remove the crankshaft pulley. Remove the lower timing belt cover.

24. Remove the pusher assembly (tensioner) from below the timing belt tensioner pulley. The pusher rod must always be facing upward to prevent oil leakage. Push the pusher rod in, and insert a wire pin into the hole to keep the pusher rod retracted.

25. Remove the timing belt.

—— WARNING ——
If the timing belt is worn, damaged, or shows signs of oil or coolant contamination, it must be replaced.

26. Loosen the valve cover bolts in a crisscross sequence. Remove the valve covers.

27. Remove the camshaft sprockets and back covers.

28. Loosen the camshaft holder bolts in a crisscross sequence to prevent warping.

29. Remove the camshaft and camshaft holders from the cylinder head.

30. Inspect the camshaft lobes and journals for signs of wear or damage.

To install:

31. Make sure all mating surfaces are clean and free of oil, coolant, or gasket residue.

32. Lubricate the camshaft lobes and journals with clean engine oil.

33. Apply a bead of sealant to the front and rear camshaft holder mating surfaces on the cylinder head.

34. Install the camshaft and holder assembly onto the cylinder head before the sealant cures. Install the camshaft holder bolts, but don't tighten them yet.

35. Use a crisscross sequence to tighten the camshaft holder bolts. Tighten the 8mm bolts to 13 ft. lbs.

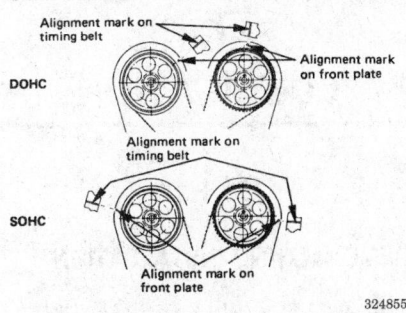

Camshaft sprocket alignment marks — 3.2L engine

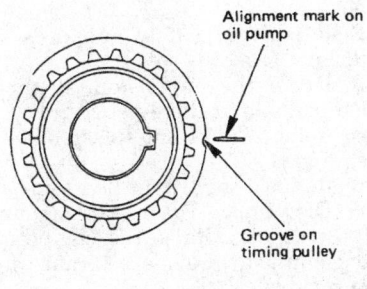

Crankshaft sprocket groove and alignment marks — 3.2L engine

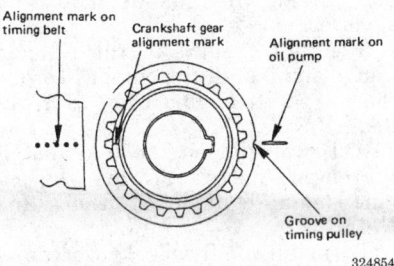

Timing belt dotted mark alignment — 3.2L engine

(18 Nm). Tighten the 6mm bolts to 6 ft. lbs. (8 Nm).

36. Use a seal driver to install a new camshaft seal.

37. Install the camshaft sprocket back covers and tighten their bolts to 12 ft. lbs. (16 Nm).

38. Install the camshaft sprockets so that the timing marks are aligned. Tighten the bolts to 46 ft. lbs. (64 Nm).

39. Apply a 2–3mm bead of sealant to the joint were the camshaft holders meet the cylinder head. Install

the valve cover with a new gasket before the sealant cures.

40. Tighten the valve cover bolts to 6 ft. lbs. (8 Nm) in crisscross pattern.

41. Verify that the camshaft and crankshaft timing marks are properly aligned.

42. Install and tension the timing belt. Tighten the pusher bolts to 14 ft. lbs. (19 Nm).

43. Install the lower timing belt covers and tighten the bolts to 13 ft. lbs. (18 Nm). Install the crankshaft pulley. Tighten the pulley bolt to 123 ft. lbs. (167 Nm).

44. Install the upper timing belt covers and tighten the bolts to 13 ft. lbs. (18 Nm).

45. Fit the oil cooler line brackets onto the timing cover and tighten the bolts to 13 ft. lbs. (18 Nm).

46. Reconnect the fuel feed and return lines.

47. Install and reconnect the ignition coil assemblies.

48. Install the intake manifold chamber and throttle body with new gaskets. Tighten the nuts and bolts to 17 ft. lbs. (24 Nm).

49. Reconnect the throttle cable to the throttle body linkage.

50. Reconnect the EGR valve supply tube and bracket.

51. Reconnect the following vacuum to the intake manifold chamber:
   a. PCV hose
   b. EVAP canister vacuum hose
   c. Brake booster hose

52. Reconnect the following sensor connectors to the rear of the intake manifold chamber:
   a. Ignition control module connectors
   b. Linear EGR valve
   c. MAP sensor
   d. EVAP purge valve
   e. Throttle position sensor
   f. Idle air control valve
   g. Intake air temperature sensor

53. Install the power steering pump and mounting bracket.

54. Install the cooling fan and its pulley assembly.

55. Install and tension the alternator, A/C compressor, and power steering pump drive belts.

56. Install and reconnect the upper and lower radiator hoses.

57. Install the air cleaner box and air intake duct.

58. Verify that all fuel lines, vacuum and coolant hoses, and wiring harness have been reconnected.

59. Refill the engine with fresh coolant.

60. Crank the engine until it starts. A longer than normal starting time

may be necessary due to air in the fuel lines. Check all fuel line connections for leaks.

61. Bleed any air from the cooling system.

62. Bleed the power steering system if necessary.

63. Check the throttle cable operation and adjustment.

64. Check the engine oil level and add if necessary.

Piston and Connecting Rod

POSITIONING

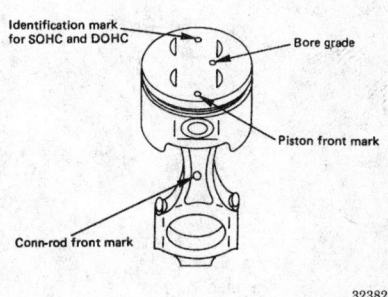

Piston identification marks — 2.3L (VIN L) and 2.6L (VIN E) engines

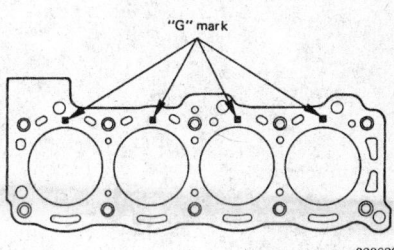

Cylinder bore diameter code location — G mark — 2.3L (VIN L) and 2.6L (VIN E) engines

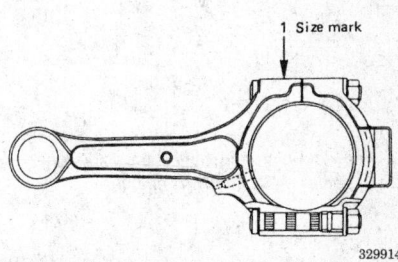

Connecting rod size mark — 2.3L (VIN L) and 2.6L (VIN E) engines

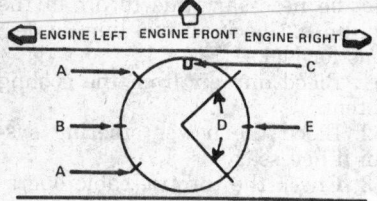

A. Oil Ring Gaps
B. 2nd Compression Ring Gap
C. Notch in Piston
D. Oil Ring Space Gap
   (Tang in Hole Or Slot With Arc)
E. Top Compression Ring Gap

329938

Piston ring gap schematic — 3.1L (VIN Z) engine

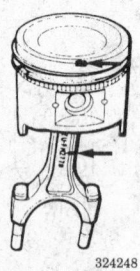

324248

Piston and connecting rod front marks — 3.2L (VIN V) engine

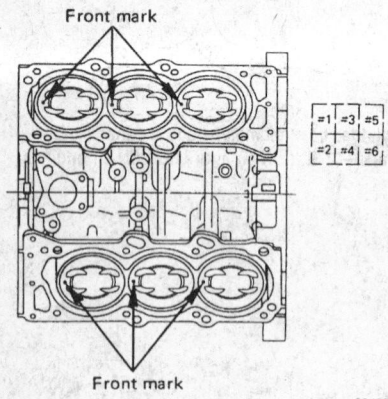

324251

Cylinder and piston identification marks on engine block — 3.2L (VIN V) engine

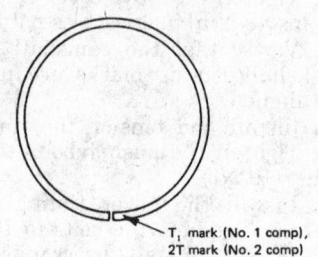

T₁ mark (No. 1 comp), 2T mark (No. 2 comp)

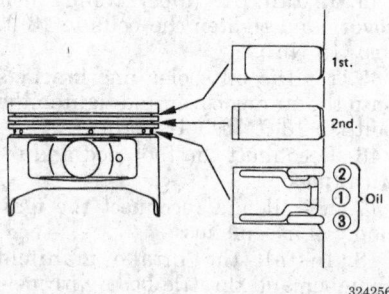

324256

Piston ring positions and markings — 2.3L (VIN L), 2.6L (VIN E) and 3.2L (VIN V) engines

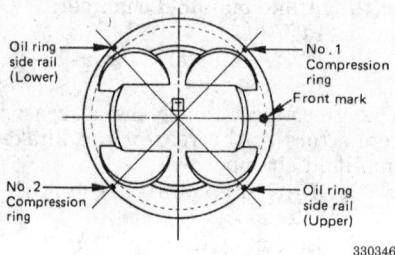

330346

Piston ring end-gap locations — 3.2L (VIN V) engine

ENGINE LUBRICATION

Oil Pan

REMOVAL AND INSTALLATION

1993–94 2.3L (VIN L) and 2.6L (VIN E) Engines

Upper Pan

1. Disconnect the negative battery cable.
2. Raise and safely support the vehicle.
3. Drain the engine oil. Remove the dipstick and the dipstick tube.
4. Remove the front splash shield, if equipped.
5. Remove the flywheel cover.
6. Disconnect the engine mount nuts and bolts. Raise the engine off the mounts to provide clearance for pan removal.
7. Remove the oil pan bolts and remove the pan. Use oil pan seal cutter J–37228 or equivalent if needed to break the oil pan-to-block sealant.
   To install:
8. Clean the gasket mounting surfaces.
9. Apply sealant to the oil pan flange and install the oil pan. Torque the oil pan-to-engine bolts to 13 ft. lbs. (18 Nm).
10. Lower the engine and install the engine mounts. Torque the engine mount bolts to 41 ft. lbs. (55 Nm).
11. Install the flywheel cover.
12. Install the dipstick and tube.
13. Install the engine under cover, if equipped.
14. Refill the crankcase with engine oil.
15. Connect the negative battery cable, start the engine and check for leaks.

Lower Pan

1. Raise and safely support the vehicle.
2. Drain the crankcase.
3. Remove the lower oil pan-to-upper oil pan bolts and the lower pan.
4. Clean the gasket mounting surfaces.
   To install:
5. Using a new gasket and sealant, install the lower oil pan and torque the bolts to 4–8 ft. lbs. (5–11 Nm).
6. Refill the crankcase.
7. Lower the vehicle to the floor.

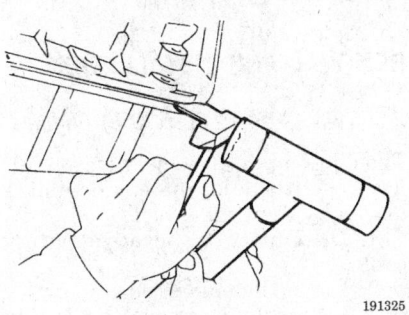

Oil pan seal cutter

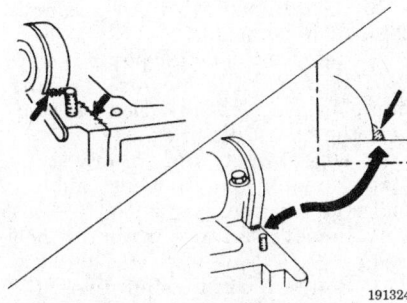

Apply sealant to the corners of the oil seal retainer and the bearing cap — 2.3L and 2.6L engines

8. Start the engine and check for leaks.

1995–97 2.3L (VIN L) and 2.6L (VIN E) Engine

1. Disconnect the negative battery cable.
2. Drain the engine oil.
3. Attach a chain hoist to the engine lifting hooks.
4. Raise and safely support the vehicle.
5. Remove the front wheels.
6. Remove the dipstick from the dipstick tube.
7. Remove the radiator skid plate, if equipped and the radiator lower shroud.
8. Remove the suspension crossmember.
9. Remove the flywheel dust cover.
10. Matchmark the Pitman arm to the steering shaft. Use a puller to remove the Pitman arm.
11. Unbolt the idler arm assembly from the frame.
12. If equipped with 4WD; support the axle assembly with a jack and remove the mounting bolts on both sides of the axle assembly. Lower the axle assembly to gain access to the oil

pan. After lowering the axle, support it with jackstands.

NOTE: The lower section of the oil pan may be separated and removed from the upper section of the oil pan.

13. Raise the chain hoist to take the engine's weight off the mounts.
14. Unbolt the engine mounts from the brackets on either side of the oil pan.
15. Remove the oil pan mounting bolts and bolt retainers. Use a sealer cutter to break the seal and remove the oil pan from the engine block.

To install:
16. Thoroughly clean and dry the sealing surface of the oil pan and engine block. Apply beads of sealant to the front and rear oil seal retainer surfaces. Install the oil pan to the engine block within five minutes of sealer application. Install the bolt retainers and all the mounting bolts. Torque the mounting bolts in sequence to 4 ft. lbs. (5.5 Nm) on 2.3L engine oil pans, and 13 ft. lbs. (18 Nm) for 2.6L engine oil pans.
17. Connect the engine mounts to the brackets.
18. Lower the hoist.
19. On 4WD models; raise the axle housing assembly into position. Tor-

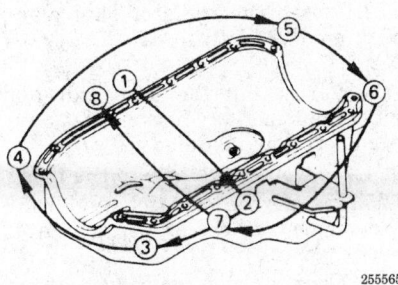

Oil pan bolt tightening sequence — 1995–97 2.3L and 2.6L engines

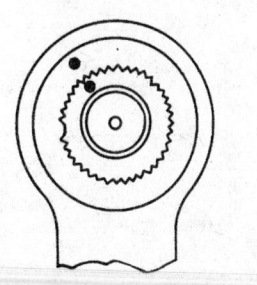

Matchmark the Pitman arm and sector shaft before removal — 3.1L engine

que the axle–to–frame mounting bolts and nuts to 112 ft. lbs. (152 Nm).
20. Install the idler arm bracket. Torque the mounting bolts to 33 ft. lbs. (45 Nm).
21. Align the matchmark and install the Pitman arm on the selector shaft. Torque the nut to 160 ft. lbs. (216 Nm).
22. Install the remaining components.
23. Refill the engine with the proper amount of oil.
24. Connect the negative battery cable.
25. Start the engine and check for leaks.
26. Check and adjust the front wheel alignment and the steering wheel spoke angle.

3.1L (VIN Z) Engine

1. Disconnect the negative battery cable.
2. Remove the dipstick. Raise and safely support the vehicle. Drain the crankcase.
3. Remove the front skid plate and crossmember.
4. Remove the front driveshaft from the front differential.
5. Remove the braces from the flywheel cover.
6. Disconnect the electrical connectors from the starter. Remove the starter-to-engine bolts and the starter.
7. Remove the flywheel inspection cover.
8. Matchmark the Pitman arm-to-Pitman shaft for reassembly. Remove the Pitman arm-to-Pitman arm shaft nut and separate the Pitman arm from the Pitman shaft.
9. Remove the idler arm-to-shaft nut and separate the idler arm from the shaft.
10. Remove the rubber hose from the front axle vent and support the axle housing assembly.
11. Remove both bolts from the left axle housing isolator and the right axle housing isolator, then, lower the front axle housing assembly.
12. Remove the oil pan-to-engine bolts and oil pan. Discard the gasket.
13. Clean the gasket mounting surfaces.
To install:
14. Using a new gasket and sealant, install the oil pan. Torque the rear pan-to-engine bolts to 18 ft. lbs. (24 Nm) and the other bolts/nuts/studs to 7 ft. lbs. (10 Nm).
15. Install the front axle housing assembly. Torque the front drive axle

mounting bolts to 112 ft. lbs. (152 Nm).

16. Install the Pitman arm and the idler arm. Torque the Pitman arm nut to 159 ft. lbs. (215 Nm) and the idler arm nut to 86 ft. lbs. (117 Nm).

17. Install the flywheel dust cover.

18. Install the starter motor. Torque the bolts to 27 ft. lbs. (36 Nm).

19. Install the driveshaft. Torque the driveshaft bolts to 46 ft. lbs. (63 Nm.).

20. Install the suspension crossmember. Torque the bolts to 58 ft. lbs. (78 Nm).

21. Install the under cover and the stone guard.

22. Install the dipstick and refill the crankcase with oil.

23. Connect the negative battery cable, start the engine and check for leaks.

3.2L (VIN V) Engine

1. Disconnect the negative battery cable.

2. Drain the engine oil.

3. Raise and safely support the vehicle.

4. Remove the front wheels.

5. Remove the dipstick from the dipstick tube.

6. Remove the radiator skid plate and the radiator lower shroud.

7. Remove the suspension crossmember from below the oil pan.

8. Remove the flywheel dust cover.

9. Disconnect the tie rod ends from the steering knuckles.

10. Matchmark the Pitman arm to the steering shaft and use a puller, (tool No. J–29107, or equivalent) to remove the Pitman arm.

11. Unbolt the idler arm bracket from the frame. Then, remove the steering linkage from the vehicle as an assembly.

12. If equipped with 4WD, perform the following steps:

    a. Support the axle assembly with a jack and safety stands.

    b. Remove the mounting bolts and nuts from the axle assembly mounting brackets on both sides of the axle assembly.

    c. Lower the axle assembly to gain access to the oil pan. Support the axle assembly so that the half-shafts aren't stressed.

13. Remove the oil pan mounting bolts. Use a sealer cutter to break the seal and separate the oil pan from the engine block.

To install:

14. Thoroughly clean and dry the sealing surface of the oil pan and engine block. Apply a continuous bead of sealant to the oil pan flange and install the oil pan to the engine block. Do not allow the sealant to cure before installation. Tighten the mounting bolts to 74 ft. lbs. (10 Nm) in a two–step crisscross sequence.

15. Install the axle housing assembly. Tighten the axle mounting bracket bolts and nut to 112 ft. lbs. (152 Nm). If the axle housing flange bolts at the mounting brackets were loosened or removed, tighten them to 61 ft. lbs. (82 Nm).

16. Install the idler arm. Tighten the mounting bolts to 33 ft. lbs. (45 Nm).

17. Align the matchmark and install the Pitman arm on the sector shaft. Tighten the nut to 160 ft. lbs. (216 Nm).

18. Install the flywheel dust cover.

19. Install the suspension crossmember. Tighten the mounting bolts to 58 ft. lb. (78 Nm).

20. Install the steering linkage, Pitman arm, and idler arm assembly. Tighten the castle nuts to the following specifications:

    a. Tie rod end castle nuts: 73 ft. lbs. (98 Nm)

    b. Idler arm bracket bolts: 33 ft. lbs. (44 Nm)

    c. Pitman arm nuts: 159 ft. lbs. (216 Nm).

21. Install the lower fan shroud.

22. Install the radiator skid plate. Tighten the bolts to 27 ft. lbs. (37 Nm).

23. Verify that the front axle and any suspension components have been correctly installed.

24. Lower the vehicle to the floor.

25. Install the dipstick and refill the engine with the proper amount of oil.

26. Connect the negative battery cable.

27. Start the engine and check for oil leaks.

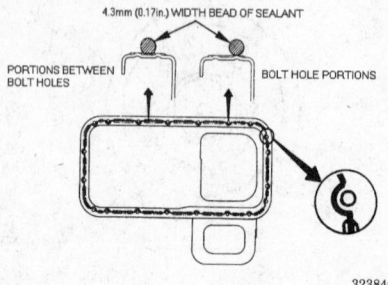

Apply sealant to the oil pan flange — 3.2L engine

Oil Pump

REMOVAL AND INSTALLATION

2.3L (VIN L) and 2.6L (VIN E) Engines

The oil pump is attached to the front, lower right side of the engine and is driven by the timing belt.

1. Disconnect the negative battery cable.

2. Drain the engine oil.

3. Loosen and remove the engine accessory drive belts.

4. Unbolt and remove the cooling fan assembly.

5. If equipped with A/C, remove the belt tensioner.

6. Remove the water pump pulley.

7. Remove the power steering pump from its mount. Don't disconnect the hydraulic lines.

8. Disconnect and remove the starter motor if a flywheel holding tool is to be used.

9. Remove the upper timing belt cover.

10. Rotate the crankshaft to set the engine at TDC/compression for the No. 1 cylinder. The mark on the camshaft sprocket will align with the mark on the rear timing cover.

11. Remove the crankshaft pulley.

12. Remove the lower timing belt cover. Verify that the engine is set at TDC/compression for the No. 1 cylinder; the pointer on the crankshaft sprocket aligns with the pointer on the oil seal retainer.

13. Loosen the timing belt tensioner and relax the tension and remove the timing belt from the crankshaft sprocket.

14. Use tool No. J–22888 or an equivalent puller to remove the oil pump sprocket.

15. Unbolt the oil pump and remove it from the engine.

16. Remove the O–ring seal from the oil pump housing.

To install:

WARNING
The timing belt must be replaced if it is damaged, or has come in contact with oil or coolant.

---

17. Inspect the oil pump and its rotors for signs of scoring and damage. Replace the pump or any damaged parts.

18. Lubricate and install a new O–ring seal.

19. Install the oil pump and tighten the bolts to 14 ft. lbs. (19 Nm).

20. Align the timing marks and install the oil pump sprocket. Apply a small amount of threadlocking com-

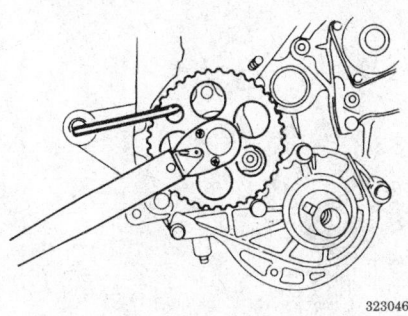

Oil pump sprocket holder — 2.3L and 2.6L engines

pound to the nut threads and tighten it to 56 ft. lbs. (76 Nm).

21. Verify that the engine is at TDC/compression for the No. 1 cylinder.

22. Install the timing belt.

23. If a flywheel holder tool was used, remove it.

24. Apply the tensioner pulley spring pressure to the timing belt.

NOTE: Remove the crankshaft holder before rotating the crankshaft to tension the timing belt.

25. Rotate the crankshaft counterclockwise for two complete revolutions and realign the timing marks.

26. Loosen the tensioner pulley bolt to allow the spring to adjust the correct tension. Torque the tensioner pulley bolt to 14 ft. lbs. (19 Nm).

27. Install the lower timing cover and tighten the bolts to 4.4 ft. lbs. (6 Nm).

28. Install the crankshaft pulley and tighten the bolt to 87 ft. lbs. (118 Nm).

29. Install the upper timing cover.

30. Install the starter motor and tighten the mounting bolts to 30 ft. lbs. (40 Nm).

31. Install the power steering pump. If the hydraulic lines were disconnected, refill and bleed the power steering system.

32. Install the cooling fan and tighten the bolts to 20 ft. lbs. (26 Nm).

33. Install and adjust the accessory drive belts.

34. Refill the engine with fresh oil.

35. Reconnect the negative battery cable.

36. Start the engine and check for leaks.

37. Check the engine's oil pressure.

3.1L (VIN Z) Engine

The oil pump is attached to the cylinder block and is located under the oil pan.

1. Disconnect the negative battery cable. Raise and safely support the vehicle.

2. Drain the crankcase and remove the oil pan.

3. Remove the oil pump-to-engine bolts and the oil pump.

To install:

4. Align the oil pump shaft with the hexagon socket and install the pump. Torque the oil pump-to-engine bolts to 30 ft. lbs. (41 Nm).

5. Install the oil pan.

6. Refill the crankcase with oil.

7. Connect the negative battery cable. Start the engine and check for leaks.

3.2L (VIN V) Engine

1. Disconnect the negative battery cable.

2. Remove the timing belt. If the timing belt is damaged or has been contaminated with oil or coolant, it must be replaced.

3. Remove the crankshaft timing pulley.

4. Raise and safely support the vehicle.

5. Drain the engine oil.

6. Remove the oil pan.

7. Remove the oil pipe and O-ring.

8. Remove the oil strainer and O-ring.

9. Remove the oil cooler assembly.

10. Remove the oil pump mounting bolts, and then remove the oil pump from the engine block.

To install:

11. Install a new oil seal into the oil pump housing.

12. Thoroughly clean the sealing surface of the oil pump and the engine block.

13. Apply sealant to the oil pump. Be careful not to block the oil ports.

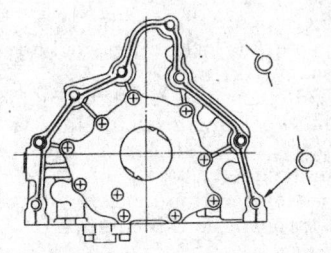

Oil pump sealant application points — 3.2L engine

14. Apply engine oil to the seal lip and install the oil pump on the engine block. Tighten the mounting bolts to 13 ft. lbs. (18 Nm). Take care not to drop the garter spring from the seal lid during installation.

15. Install the oil cooler assembly.

16. Install the oil pipe and O-ring.

17. Install the oil strainer and O-ring.

18. Install the oil pan.

19. Install the crankshaft timing pulley.

20. Install the timing belt.

21. Install the remaining accessories and drive belts.

22. Lower the vehicle.

23. Refill the engine with oil.

24. Connect the negative battery cable, start the engine and check for proper oil pressure.

NOTE: If the oil pressure does not build up almost immediately, stop the engine and investigate the cause.

25. Check for oil leaks.

TRANSMISSION

Manual Transmission Assembly

REMOVAL AND INSTALLATION

MUA5 and MSG5C Transmissions

1. Disconnect the negative and positive battery cables.

2. Remove the battery.

3. Support the hood as far open as possible. If you choose to remove the hood, first matchmark the hood hinge plates with a felt-tipped marker.

4. Remove the console and shift boot. Unbolt the shift lever from the transmission case and remove it. Cover the quadrant box hole to prevent contaminants from entering the transmission.

5. Raise and support the vehicle safely.

6. Drain the transmission oil. Install the drain plug with a new washer.

7. Remove the transfer case skid plates, if equipped.

8. If equipped with a two-piece driveshaft, remove the center bearing retainer bolts.

9. If 4WD equipped, mark the driveshafts to the differential flanges

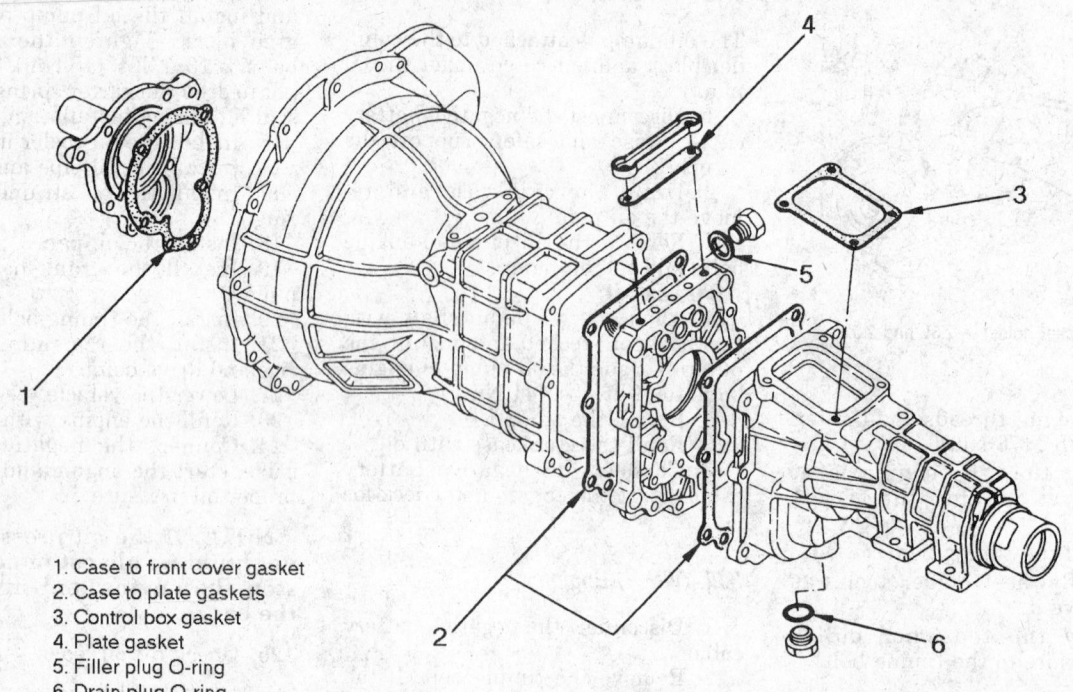

1. Case to front cover gasket
2. Case to plate gaskets
3. Control box gasket
4. Plate gasket
5. Filler plug O-ring
6. Drain plug O-ring

254876

Transmission repair gaskets and O-rings — MUA5 and MSG5C transmissions

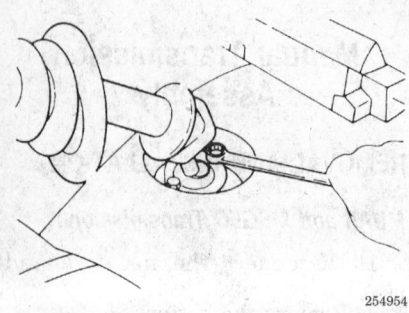

254954

Shift lever mounting bolts — MUA5 and MSG5C transmissions

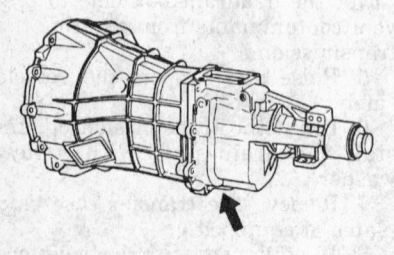

254869

Transmission oil drain plug location (4x2) — MUA5 and MSG5C transmissions

and remove the front and rear driveshafts.

10. Matchmark the driveshaft flanges to the differential and transmission flanges. Remove the driveshaft.

11. Remove the starter.

12. Disconnect the speedometer cable.

13. For MUA5 transmissions, unbolt the slave cylinder from the side of the transmission case. Don't disconnect the hydraulic line.

14. For MSG5C transmissions, disconnect the clutch cable from the release fork.

15. Remove the exhaust pipe bracket from the transmission case. Disconnect the front exhaust pipe from the exhaust manifold and the second exhaust pipe.

16. Support the engine with a lifting chain or jack. Support the transmission with a jack.

17. Remove the rear housing mount from the transmission. Remove the mount bracket from the third crossmember.

18. Remove the quadrant box from the transmission, if equipped.

NOTE: The frame crossmember may interfere with transmission removal. An assis-

tant will be helpful for shifting the transmission back and away from the engine.

19. Position a jack under the transmission and remove the engine-to-transmission bolts. Move the transmission as far to the rear of the vehicle as possible, and then lower the clutch housing end of the transmission toward the jack.

To install:

20. Raise the transmission into position.

21. Using a transmission jack, position the transmission-to-engine and torque the retaining bolts to 28 ft. lbs. (38 Nm). Install the quadrant box.

22. Install the mount bracket and tighten the bolt to 27 ft. lbs. (37 Nm).

23. Install the frame-to-rear housing mount bolts at the number three crossmember, and tighten to 62 ft. lbs. (83 Nm).

24. Tighten the engine mount nuts to 30 ft. lbs. (41 Nm).

25. Install the exhaust pipe and bracket.

26. If equipped with 4WD, install the front and rear driveshafts in the marked locations. Tighten the flange bolts to 43 ft. lbs. (63 Nm).

27. Align the matchmarks and install the driveshaft. Torque the

center bearing bolts to 46 ft. lbs. (65 Nm). Position the retaining bolts with the head facing rearward and torque to 22 ft. lbs. (30 Nm).

28. Install the speedometer cable and starter motor. Tighten the starter motor bolts to 30 ft. lbs. (41 Nm).

29. For MUA5 transmissions, install the slave cylinder. If necessary, refill and bleed the clutch hydraulic system.

30. For MSG5C transmissions, connect the clutch cable to the release fork.

31. Install the transfer case skid plates, if equipped.

32. Refill the transmission with the proper type of oil.

33. Lower the vehicle and install the shift lever, boot, and console.

34. Install the battery and reconnect the positive and negative cables.

35. Install the hood if it was removed.

MUA5C Transmission

NOTE: The transfer case is an integral part of the transmission housing. Although the two cases can be separated, the transfer case should be removed with the transmission.

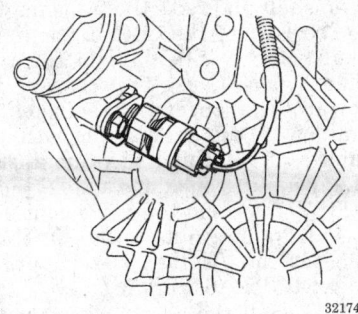

Speed sensor harness connector location — MUA5C transmission

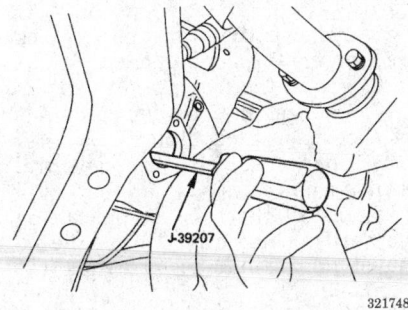

Insert the release bearing remover through the bell housing

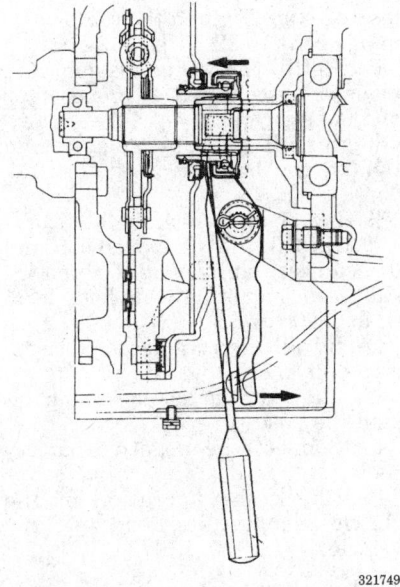

Push the release bearing fork toward the transmission

1. Shift the transmission into **N**. Shift the transfer case into **2H** and drive the vehicle forward and backward a few feet/meters to make sure the front axle is not engaged.

2. Use a felt-tipped marker to matchmark the hood to the hood hinges. Remove the hood.

3. Disconnect the negative battery cable.

4. Remove the shift knobs and the console and shift boots.

5. Remove the shift lever and the transfer case shift lever (if equipped), by unbolting their mounting plates from the transmission case.

6. Raise and safely support the vehicle.

7. Drain the oil from the transmission and transfer case.

8. Remove the exhaust and transfer case skid plates.

9. Disconnect the oxygen sensor connector from the transmission harness.

10. Unbolt the exhaust flanges. Separate and remove the catalytic converter, left front, and center exhaust pipes.

11. Remove the harness heat protector.

12. Remove the slave cylinder heat protector. Unbolt and remove the slave cylinder. Don't disconnect the hydraulic line.

13. Remove the slave cylinder dust covers.

14. Matchmark the driveshafts at the flanges and remove them.

15. Disconnect the reverse light switch, 4WD indicator switch, 1–2 and 3–4 indicator switch harness connectors.

16. Disconnect the speed sensor harness connector.

17. Remove the two harness clamps from the transmission case.

18. Attach a chain hoist to the engine.

19. Using a transmission jack, raise the transmission slightly. Remove the two rear transmission mounting nuts.

20. Remove the center crossmember (eight bolts).

21. On 4WD vehicles, remove the front crossmember and the front driveshaft.

NOTE: Make sure the engine assembly is properly supported when removing the front crossmember.

22. Remove the three flywheel inspection cover bolts.

23. Use clutch release bearing remover J–39207 or an equivalent prytool to release the bearing from the pressure plate. Push the release fork toward the rear of the vehicle. Insert the tool between the release bearing and the pressure plate collar. Move the lever to the rear to pry.

24. Raise the engine slightly with a chain hoist and remove the bolts and nuts securing the transmission to the engine.

25. Carefully pull the transmission rearward. Lower the transmission from the vehicle.

To install:

26. Apply a thin coating of molybdenum grease to the splines of the input shaft, and then slowly raise the transmission into position against the rear of the engine. Align the splines of the input shaft with the grooves of the clutch disc hub and install the transmission to the engine.

NOTE: It may be helpful the put the transmission in gear and rotate the driveshaft flange so the input shaft will turn and engage the grooves in the clutch disc hub.

27. Install the transmission case bolts. Tighten the upper six bolts to 56 ft. lbs. (76 Nm). Tighten the two remaining large bolts to 56 ft. lbs. (76 Nm). Tighten the remaining three bolts to 4.4 ft. lbs. (6 Nm).

28. Push the release bearing fork rearward with a force of 13–18 ft. lbs.

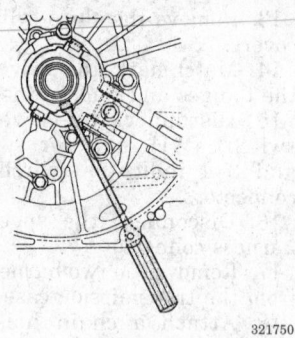

321750

Insert the tool between the wedge collar and the release bearing

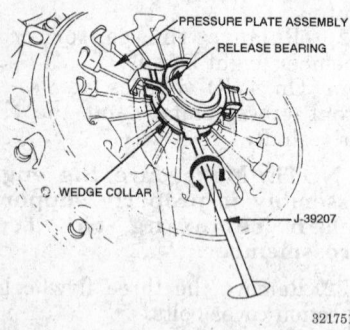

PRESSURE PLATE ASSEMBLY

RELEASE BEARING

WEDGE COLLAR

J-39207

321751

Turn the remover to separate the release bearing

(18–24 Nm) to engage the release bearing with the pressure plate. A click sound will be heard when the bearing engages the pressure plate properly.

29. Install the flywheel inspection cover.

30. On 4WD vehicles, install the front crossmember and the front driveshaft. Tighten the crossmember bolts to 58 ft. lbs. (78 Nm). Tighten the driveshaft flange bolts to 46 ft. lbs. (62 Nm).

31. Install the center crossmember and transmission mount. Tighten the center crossmember mounting bolts to 37 ft. lbs. (50 Nm) and the transmission mounting nuts to 30 ft. lbs. (41 Nm).

32. Remove the transmission jack and the engine hoist.

33. Connect the transmission harness connectors and install the harness clamps.

34. Install the rear driveshaft and tighten the flange bolts to 46 ft. lbs. (62 Nm).

35. Apply grease to the dimple on the end of the release bearing fork and install the slave cylinder, heat protector, and dust covers. Tighten the slave cylinder bolts to 32 ft. lbs. (43 Nm). Tighten the dust cover bolts to 4.4 ft. lbs. (6 Nm).

36. Install the exhaust pipes and heat protectors using new self–locking nuts. Tighten the exhaust manifold flange bolts to 49 ft. lbs. (67 Nm). Tighten the center pipe flange bolts to 32 ft. lbs. (43 Nm).

37. Refill the transmission and transfer case with the proper type and quantity of oil.

38. Install the skid plates and tighten the bolts to 27 ft. lbs. (37 Nm).

39. Lower the vehicle to the floor.

40. Install the transmission and transfer case shift levers. Tighten the shift lever mounting plate bolts to 15 ft. lbs. (20 Nm).

41. Install the console, shift boots, and shift knobs.

42. Align the matchmarks and install the hood.

43. Connect the negative battery cable.

44. Check the operation of the clutch and starter. Road test the vehicle.

T5R Transmission

NOTE: The transfer case is an integral part of the transmission housing. Although the two cases can be separated, the transfer case should be removed with the transmission.

1. Disconnect the negative battery cable.

2. Shift the transmission into neutral. Remove the gearshift and transfer case shift knobs.

3. Remove the four console screws and lift the console and shift boot over the shift levers.

4. Unbolt and remove the shift lever and its cover plate from the transmission case. If equipped, remove the transfer case shift lever.

5. Raise and safely support the vehicle.

6. Drain the transmission oil.

7. Disconnect and remove the starter.

8. Unbolt the slave cylinder from the transmission case. Don't disconnect the hydraulic line.

9. Matchmark the driveshaft U–joints to the transmission and differential flanges.

10. If the vehicle is equipped with 4WD; matchmark the front driveshaft U–joints to the differential and transfer case.

NOTE: On 4WD vehicles, the transfer case skid plates and front exhaust pipe must be removed before removing the front drive shaft.

11. Remove the driveshaft and the center bearing.

12. Disconnect the front exhaust pipe from the manifold and the catalytic converter. It is not necessary to remove the exhaust pipe from the chassis.

13. Disconnect the reverse and neutral switch connectors from the transmission.

14. Disconnect the speedometer cable or speed sensor connector from the transmission.

15. Remove the flywheel inspection cover.

16. Support the transmission with a transmission jack.

17. Raise the transmission slightly so the jack supports its weight.

18. Support the rear of the engine with a jack or chain hoist.

19. Unbolt and remove the center crossmember and the transmission mount.

20. With the engine and transmission supported, remove the bolts securing the transmission case to the engine.

21. Pull the transmission away from the engine so that the mainshaft clears the pressure plate. Remove the transmission from the vehicle.

To install:

22. Apply a thin coating of molybdenum grease to the splines of the mainshaft and raise the transmission to the rear of the engine. Align the shaft splines with the clutch driven plate splines. Push the transmission toward the engine to engage the splines of the mainshaft with the grooves in the clutch disc hub. Install the mounting bolts and nuts.

23. Tighten the 10mm transmission case bolts and nuts to 28–30 ft. lbs. (38–41 Nm). Tighten the 6mm bolts to 4.4 ft. lbs. (6 Nm).

24. Install the center crossmember to the frame. Then, install the transmission mount. Tighten the crossmember bolts to 56 ft. lbs. (76 Nm) and the mount nuts to 30 ft. lbs. (41 Nm).

25. Remove the transmission jack and engine lifting equipment.

26. Install the flywheel dust cover.

27. Connect the speedometer cable or speed sensor connector.

28. Connect the reverse and neutral switch connectors.

29. Install the exhaust pipe using new self–locking nuts. Torque the exhaust flange nuts to 49 ft. lbs. (67 Nm).

30. Install the driveshafts. Tighten the center bearing bolts to 45 ft. lbs.

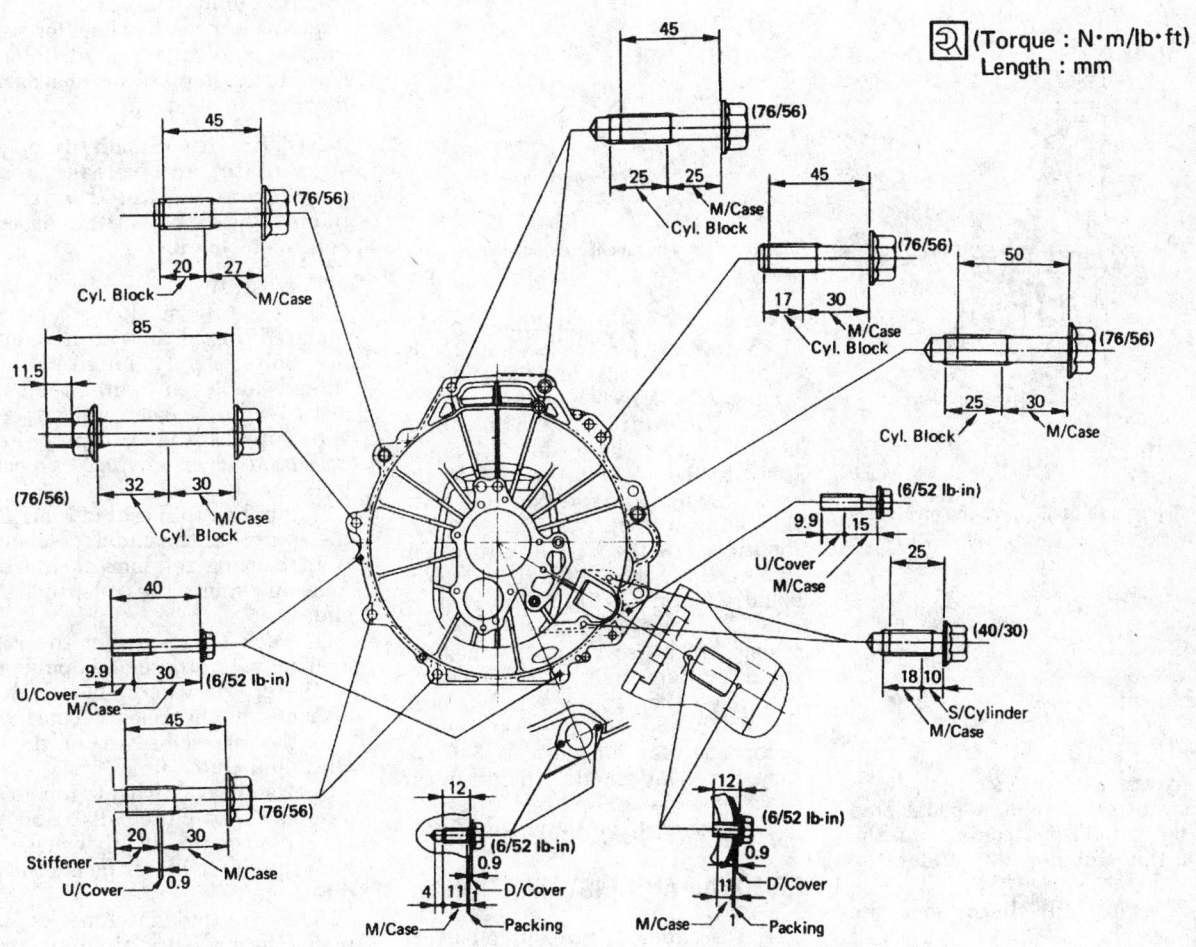

Transmission mounting bolt locations and torque specifications — MUA5C transmission

(60 Nm). Tighten the flange bolts to 46 ft. lbs. (63 Nm).

31. Install the skid plates, if equipped. Tighten the skid plate bolts to 27 ft. lbs. (37 Nm).

32. Coat the tip of the slave cylinder with molybdenum grease. Install the slave cylinder assembly and tighten the bolts to 37 ft. lbs. (50 Nm).

33. Install the starter and tighten the bolts to 30 ft. lbs. (41 Nm).

34. Refill the transmission with the proper type and quantity of oil.

35. Lower the vehicle.

36. Lubricate the shift ball and lower edge of the shift lever. Install the gearshift lever assembly to the transmission case and tighten the bolts to 15 ft. lbs. (20 Nm). On 4WD vehicles, install the transfer case lever.

37. Install the shift boot(s), console, and shift knob(s).

38. Connect the negative battery cable.

Clutch Assembly

PEDAL ADJUSTMENT

Cable Driven

1. Loosen the outer cable adjusting nut. Pull the outer cable, located under the hood, forward as far as possible and secure it.

2. Turn the adjusting nut inward until it touches the damper rubber washer on the fire wall.

3. Depress and release the clutch pedal 3 times.

4. Fully tighten the adjusting nut.

5. Measure the free-play distance in the pedal. Cable—operated clutches should have a play of 0.59–0.98 in. (15–25mm).

6. If the distance is out of specification, loosen the locknut to adjust the distance. Tighten the locknut to secure the adjusting nut.

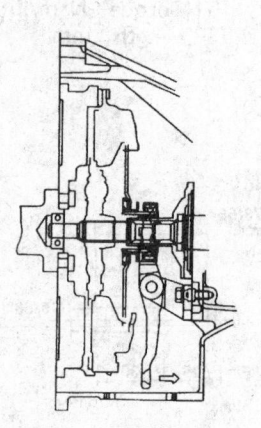

Engage the release bearing with the pressure plate

321753

Hydraulic

1. Locate the clutch switch (with cruise control) or clutch pedal stop bolt (without cruise control) at the top of the clutch pedal under the dash.

2. If equipped, disconnect the clutch switch.

3. Loosen the clutch switch or stop bolt as equipped. Loosen the clutch master cylinder pushrod yoke nut.

4. Adjust the clutch master cylinder pushrod to obtain a clutch pedal height:
- 1993 vehicles: 6.73–7.13 in. (17.1–18.1cm) for the 2.6L (VIN E) engine, or 7.64–8.03 in. (19.4–20.4cm) for vehicles with V6 engines

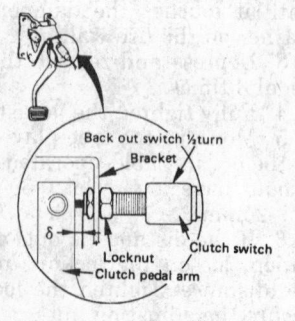

Clutch switch adjustment (with cruise control)

325065

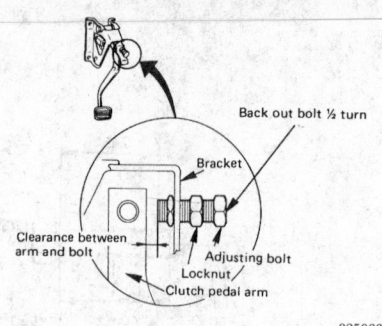

325066

Clutch switch adjustment (without cruise control)

- 1994–97 vehicles: 7.28–7.68 in. (18.5–19.5cm) for the 2.6L (VIN E) engine, or 7.64–8.03 in. (19.4–20.4cm) for the 3.2L (VIN V) engine

5. After adjusting the pedal height, tighten the pushrod yoke nut to 12 ft. lbs. (16 Nm). Screw in the clutch switch (cruise control) until the plunger is fully depressed and then unscrew the switch ½ turn and tighten the locknut. There should be 0.020–0.059 in. (0.5–1.5mm) of clearance. Tighten the pedal stop bolt (no cruise control) so it just touches the pedal and then tighten the locknut.

6. After the pedal height has been completely adjusted, check the clutch pedal free play and adjust if necessary. Free-play should fall within the range of 0.20–0.59 in. (5–15mm).

7. Reconnect the clutch switch.

REMOVAL AND INSTALLATION

1. Disconnect the negative battery cable.

2. Raise and support the vehicle safely.

3. On 2WD vehicles, remove the transmission. On 4WD models, remove the transmission and transfer case as an assembly.

4. If the pressure plate is going to be reused, matchmark the pressure plate to the flywheel so the pressure plate can be reassembled in the same position. Lock the flywheel in place to prevent it from turning.

NOTE: It is recommended that the clutch disc, pressure plate and release bearing be replaced with new parts when the clutch assembly requires service.

5. Loosen the pressure plate mounting bolts, one turn at a time in an crisscross sequence. Loosening in a sequence will relieve the spring tension to avoid distorting or bending the pressure plate.

6. Install a clutch alignment tool through the clutch disc into the pilot

bearing to support the pressure plate and cover assembly. Remove the bolts and clutch assembly.

7. Remove the release bearing from the input shaft and fork.

8. Inspect the flywheel for scoring, grooves and cracks or discoloration from heat. Replace or resurface the flywheel as needed.

NOTE: The clutch disc, pressure plate, and release bearing should be replaced with new parts when the clutch assembly requires service.

To install:

9. If the flywheel was removed or replaced, install it with new mounting bolts. Apply Loctite® to the threads and then tighten the bolts in a two-step crisscross pattern to 40–43 ft. lbs. (54–58 Nm). After installation, clean any excess Loctite® from the flywheel.

10. Apply a thin coat of lubricant to the splines in the clutch disc and the front bearing retainer on the transmission where the release bearing slides.

11. Pack the recess in the release bearing with grease and apply grease to the groove where the clutch fork attaches to the release bearing. Install the release bearing on the input shaft and clutch fork.

12. Using a clutch alignment tool, assemble the clutch disc and pressure plate onto the flywheel. Make sure the clutch disc is facing the right direction.

13. Align the matchmarks, if reusing the pressure plate and install the pressure plate mounting bolts. On four cylinder engines (2.3L (VIN L) and 2.6L (VIN E) torque the bolts to 10–16 ft. lbs. (14–22 Nm), and on the 3.1L (VIN Z) engine torque the bolts to 12–18 ft. lbs. (16–24 Nm) using a star pattern tightening sequence. Remove the aligning tool.

14. On 2WD vehicles, install the transmission. On 4WD models, install the transmission and transfer case as an assembly.

15. Adjust the clutch cable linkage, if equipped. Bleed the clutch hydraulic system, if equipped.

16. Reconnect the negative battery cable.

Clutch Cable

ADJUSTMENT

1. Check the clutch pedal height. If necessary, adjust it so its within specifications.

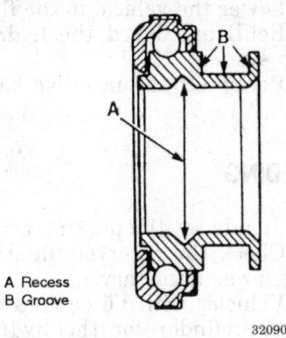

A Recess
B Groove

320907

Apply grease to the release bearing recess and groove

2. Working in the engine compartment, loosen the outer cable adjusting nut. Pull the outer cable forward as far as possible and secure it.

3. Turn the adjusting nut inward it touches the damper rubber washer on the firewall.

4. Depress and release the clutch pedal 3 times.

5. Pull the outer cable forward again and fully tighten the adjusting nut.

6. Loosen the nut to provide a 0.2 in. (5mm) clearance between the adjusting nut and the damper washer.

7. Release the outer cable and tighten the locknut to secure the adjusting nut.

8. Check the clutch cable free-play and adjust if necessary.

REMOVAL AND INSTALLATION

1. Disconnect the negative battery cable.

2. Loosen the clutch cable lock and adjusting nuts. Remove the clutch cable clip in the engine compartment.

3. Raise and safely support the vehicle. Remove the spring from the shift fork end.

4. Remove the assist spring from the pedal.

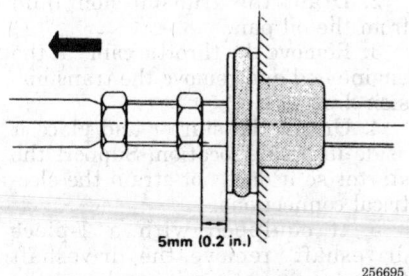

5mm (0.2 in.)

256695

Clutch cable adjustment

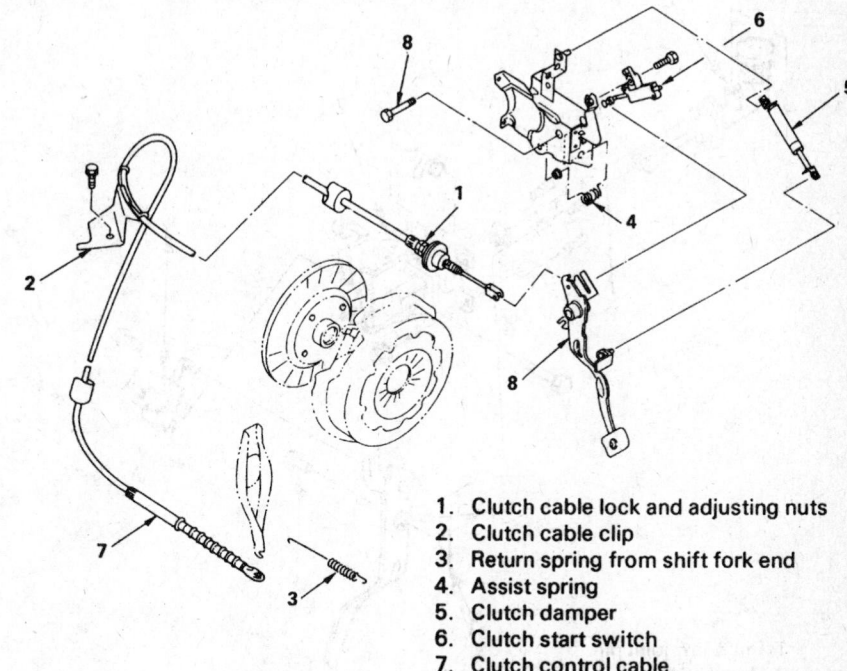

1. Clutch cable lock and adjusting nuts
2. Clutch cable clip
3. Return spring from shift fork end
4. Assist spring
5. Clutch damper
6. Clutch start switch
7. Clutch control cable
8. Clutch pedal and bolt

254771

Clutch cable and related components

5. Remove the clutch damper from the pedal and the pedal bracket.

6. Disconnect the cable end from the clutch release fork and pull the cable assembly through the bracket.

7. Lower the vehicle enough to disengage the hooked part of the clutch pedal from the cable eye. Pull the cable assembly towards the engine compartment and remove the cable from the vehicle.

To install:

8. Pass the cable through the dash panel and hook it on the top of the clutch pedal.

9. Install the clutch damper.

10. Install the assist spring onto the pedal.

11. Connect the cable to the clutch release fork.

12. Install the return spring.

13. Install the clutch cable clip.

14. Adjust the clutch pedal height and free-play.

15. Reconnect the negative battery cable.

16. Test the operation of the clutch pedal and cable.

Clutch Master Cylinder

REMOVAL AND INSTALLATION

─────── **CAUTION** ───────
The Air Bag system wiring is routed behind the dashboard. It is encased in yellow insulation. Do not damage the wiring or its insulation. Do not bump or strike any Air Bag components. If any component must be serviced or disconnected, the system must first be disabled.
────────────────────

1. Disconnect the negative and positive battery cables. Wait at least five minutes before working around the Air Bags.

2. Use a suction pump to remove the hydraulic fluid from the clutch master cylinder reservoir.

3. Disconnect the hydraulic line from the clutch master cylinder.

4. From inside the vehicle, remove the instrument panel lower cover. Then, disconnect the instrument panel dimmer switch from the panel lower cover.

5. Unbolt and remove the knee bolster and the air conditioning duct.

6. Remove the clutch pedal pivot pin.

7. Disconnect the clutch switch.

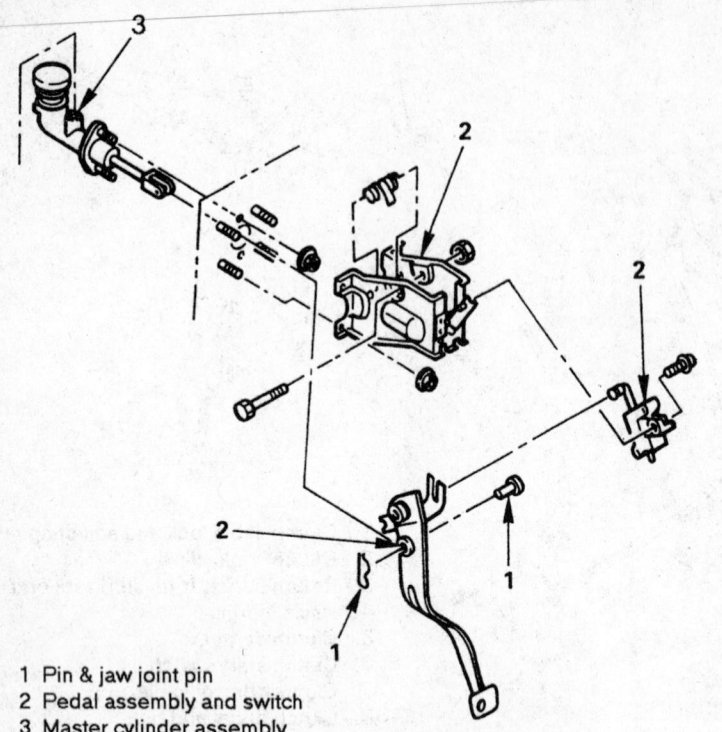

1 Pin & jaw joint pin
2 Pedal assembly and switch
3 Master cylinder assembly

320766

Clutch master cylinder assembly

8. Unbolt and remove the clutch pedal bracket assembly from the bulkhead.

9. Remove the master cylinder mounting nuts.

10. Remove the clutch master cylinder from the engine side of the bulkhead.

To install:

11. Install the clutch master cylinder on the bulkhead. Tighten the mounting nuts to 12 ft. lbs. (16 Nm).

12. Install the clutch pedal bracket and switch. Tighten the mounting nuts to 15 ft. lbs. (20 Nm).

13. Install the clutch pedal and pivot pin using a new lock pin.

14. Adjust the clutch pedal height:
• 1993 Rodeo 4 cylinder: 6.7–7.1 in. (17.1–18.1cm)
• 1994–97 Rodeo and Passport 4-cylinder: 7.28–7.68 in. (18.5–19.5cm)
• 1993–97 Rodeo and Passport V6: 7.64–8.03 in. (19.4–20.4cm)

15. Install the air conditioning duct, the knee bolster, and the instrument panel lower cover.

16. Connect the hydraulic line to the master cylinder. Tighten the line fitting nut to 14 ft. lbs. (19 Nm).

17. Bleed the clutch hydraulic system.

18. Reconnect the positive and negative battery cables.

Clutch Slave Cylinder

REMOVAL AND INSTALLATION

1. Disconnect the negative battery cable.

2. Use a suction pump to remove the hydraulic fluid from the clutch master cylinder reservoir.

3. Raise and safely support the vehicle.

4. Disconnect the hydraulic line from the clutch slave cylinder. Plug the line to prevent fluid loss.

5. Unbolt and remove the slave cylinder from the transmission case.

To install:

6. Install the slave cylinder and connect the hydraulic line. Make sure the tip of the slave cylinder snaps into the release fork. Tighten the slave cylinder mounting bolts to 32 ft. lbs. (43 Nm).

7. On slave cylinders with flare fittings, torque the hydraulic line fitting to 11–17 ft. lbs. (16–25 Nm).

8. On slave cylinders with banjo fittings, use new washers and torque the hydraulic line fitting to 25 ft. lbs. (34 Nm).

9. Lower the vehicle to the floor.

10. Refill and bleed the hydraulic clutch system.

11. Reconnect the negative battery cable.

BLEEDING

1. Firmly set the parking brake.

2. Check the reservoir fluid level and refill as necessary.

3. Vehicles with V6 engines use a damper cylinder for the hydraulic clutch system. Bleed the damper cylinder before bleeding the slave cylinder.

NOTE: The bleeding procedure is the same for both the slave cylinder and the damper cylinder.

4. Connect a vinyl tube to the bleeder screw. Submerge the other end of the tube in a container of brake fluid.

5. Pump the clutch pedal slowly several times and hold it depressed.

6. Loosen the bleeder screw and allow the air bubbles and fluid to flow into the container. Tighten the bleeder screw.

7. Release the clutch pedal.

8. Add fluid to the reservoir if necessary. Don't let the reservoir run dry.

9. Continue to pump the pedal until the fluid draining from the tube is clear of air bubbles.

10. Refill the reservoir to the full mark. Remove the vinyl tube. Tighten the bleeder screw and install its rubber cap.

Automatic Transmission Assembly

REMOVAL AND INSTALLATION

2WD Models

1. Disconnect the negative battery cable. Raise and safely support the vehicle.

2. Drain the transmission fluid from the oil pan.

3. Remove the throttle cable at the engine end and remove the transmission dipstick.

4. Unbolt the starter and place it aside in a safe location. Support the starter so it does not strain the electrical connections.

5. If equipped with a 1-piece driveshaft, remove the driveshaft flange nuts at the pinion, lower the driveshaft and pull it from the transmission.

6. If equipped with a 2-piece driveshaft, perform the following procedures:

    a. Remove the rear driveshaft flange nuts at the pinion.

    b. Remove the rear driveshaft flange bolts from the front driveshaft flange. Remove the rear driveshaft.

    c. Remove the center bearing bolts from the chassis, move the front driveshaft rearward and from the transmission.

7. Disconnect the shift lever at the shifter end.

8. Disconnect the speedometer cable.

9. Disconnect the oil cooler lines and place the cooler bypass line close to the transmission case to prevent damage during transmission removal.

10. Remove the torque converter bolts from the flexplate through the starter hole.

11. Using a transmission jack, place it under the transmission and raise it slightly.

12. Remove the rear mount nuts from the transmission.

13. Remove the rear mount nuts/bolts from the crossmember. Remove the mount.

14. Remove the transmission-to-engine bolts.

15. Move the transmission back and lower the transmission out of the vehicle.

To install:

NOTE: Installation of the transmission will require an assistant.

16. Raise the transmission into position.

17. Raise the rear of the transmission and move it into position on the crossmember.

18. Move the transmission forward and engage it with the engine.

19. Install the engine-to-transmission bolts and torque to 47 ft. lbs. (64 Nm).

NOTE: Make sure the torque converter is seated properly. The transmission should mount flush to the engine block; if the transmission does not mount flush, check the torque convertor. Do not use the mounting bolts to pull the transmission flush to the engine or torque converter damage may occur.

20. Install the mount and the rear mount nuts/bolts to the crossmember.

21. Install the rear mount nuts to the transmission. Torque to 30 ft. lbs. (41 Nm).

22. Install the torque converter bolts to the flexplate through the starter hole, and torque to 22 ft. lbs. (30 Nm).

23. Connect the oil cooler lines, speedometer cable, and shift linkage.

24. If equipped with a 2-piece driveshaft, perform the following procedures:

    a. Install the front driveshaft into the transmission, and the center bearing bolts to the chassis.

    b. Install the rear driveshaft, and the rear driveshaft flange bolts to the front driveshaft flange.

    c. Install the rear driveshaft flange nuts to the pinion.

25. If equipped with a 1-piece driveshaft, install the driveshaft into the transmission and the driveshaft flange nuts to the pinion.

26. Install the starter.

27. Connect the throttle (downshift) cable to the transmission.

28. Install the oil level dipstick and the tube.

29. Refill the transmission.

30. Connect the negative battery cable.

31. Start the engine and check for leaks.

4WD Models

NOTE: The transfer case is an integral part of the transmission housing. Although the two cases can be separated, the transfer case should be removed with the transmission.

1. Use a felt-tipped marker to matchmark the hood to the hood hinges. Remove the hood.

2. Shift the transmission into the **N** position, and the transfer case into the **2H** position.

3. Disconnect the negative battery cable.

4. Remove the air cleaner assembly.

5. Remove the transfer case shift knob. Remove the four console retaining screws.

6. Remove the center console assembly and disconnect the console switch wiring connectors.

7. Disconnect the shift lock cable and the shift control rod from the selector lever assembly.

8. Unbolt and remove the transfer case control lever.

9. Raise and safely support the vehicle.

10. Remove the transmission and transfer case skid plates.

11. Remove the exhaust pipe protectors.

12. Disconnect the oxygen sensor connectors from the transmission harness.

13. Remove the nuts and bolts, and then separate the catalytic converters, center, and front exhaust pipes from exhaust manifold and tailpipe. Remove the exhaust components from the vehicle, or move them out of the way.

14. Matchmark the front and rear driveshafts to the differential and transfer case flanges.

15. Unbolt and remove the front driveshaft.

16. Unbolt the front part of the rear driveshaft from the U-joint behind the driveshaft center bearing. Unbolt the rear driveshaft from the transfer case. Unbolt the center bearing from the frame, and remove the driveshaft.

17. Disconnect the oil cooler lines from the transmission. Plug the lines to prevent fluid loss and contamination.

18. Remove the brackets securing the oil cooler lines to the engine stiffener.

19. Remove the front suspension crossmember.

20. Remove the dipstick and tube. Disconnect the breather hoses from the tube.

21. Place a transmission jack under the transmission and transfer case unit for support.

22. Raise the transmission slightly and remove the eight bolts securing the rear mount and the third crossmember.

NOTE: Make sure the engine and transmission assembly is properly supported before removing the rear mount and third crossmember.

23. Remove the five engine stiffener bolts and the stiffener.

24. Remove the heat protector.

25. Disconnect the transmission harness connectors and the mode switch harness connector from the engine harness.

26. Disconnect the harness clamp from the clamp bracket.

27. Disconnect the ground cable from the engine.

28. Remove the starter.

29. Remove the flexplate (flywheel) inspection cover.

30. Remove the three bolts securing the flexplate to the torque converter.

NOTE: Remove the radiator upper fan shroud and the cooling fan to access the crankshaft center bolt to turn the crankshaft.

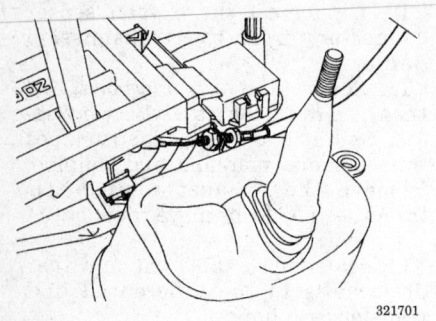

Disconnect the shift lock cable from the selector lever — 4WD models

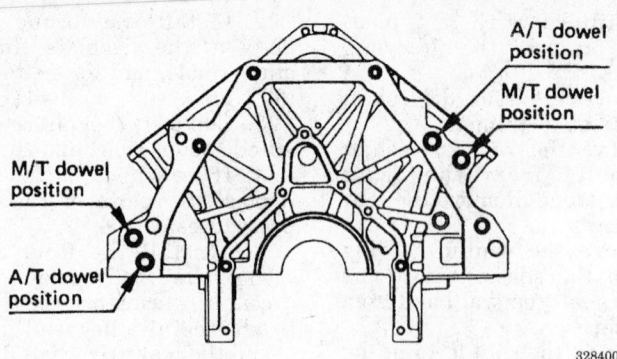

Transmission mounting dowel positions

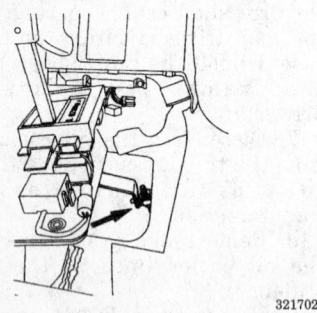

Disconnect the shift control rod from the selector lever — 4WD models

31. Attach a chain hoist to the engine lifting points.
32. Raise the engine slightly with a hoist and remove the transmission-to-engine bolts.
33. Separate the transmission from the engine and lower the transmission from the vehicle.

To install:

NOTE: Use new self-locking nuts when reconnecting the exhaust system. Replace any color-coded self-locking bolts when installing the frame crossmembers.

34. Install the transfer case onto the transmission case if the two were separated.
35. Install a new O-ring on the front pump shaft.

NOTE: Make sure the transmission dowel pins are installed in the correct position. If the dowels are in the wrong hole, the transmission case may crack.

36. Make sure the torque converter is fully seated in the front pump and slowly raise the front of the transmission into position until it is flush with the rear of the engine and install the transmission-to-engine bolts. Tighten the upper six mounting bolts to 56 ft. lbs. (76 Nm). Tighten the lower three mounting bolts to 35 ft. lbs. (47 Nm). Tighten the remaining two bolts to 52 ft. lbs. (70 Nm).

NOTE: If the transmission does not seat flush against the engine block before the bolts are installed, check to see that the torque converter is seated properly in the transmission, do not use the bolts to draw the transmission to the engine block.

37. Install the third crossmember. Tighten the bolts to 37 ft. lbs. (50 Nm).
38. Install the rear mount and lower the engine from the hoist. Tighten the nuts to 30 ft. lbs. (41 Nm).
39. Remove the transmission jack and the engine hoist.
40. Install the flexplate to torque converter bolts. Tighten the bolts in a crisscross pattern to 41 ft. lbs. (55 Nm).
41. Install the flexplate inspection cover.
42. Connect the transmission and mode switch harness connectors to the engine harness.
43. Connect the harness clamp to the clamp bracket and radiator clip.
44. Connect the ground cable to the engine.
45. Install the starter and tighten the mounting bolts to 30 ft. lbs. (41 Nm).
46. Install the heat protector.
47. Install the stiffener. Tighten the bolts to 35 ft. lbs. (47 Nm).
48. Install the transmission dipstick tube and dipstick. Connect the breather hoses to the tube.
49. Install the front suspension crossmember. Tighten the bolts to 58 ft. lbs. (79 Nm).
50. Connect the transmission fluid cooling lines to the transmission. Torque the line fittings to the following specs:
   • 1993–94 Rodeo and Passport: 33 ft. lbs. (45 Nm)
   • 1995–97 Rodeo and Passport: 40 ft. lbs. (54 Nm)
51. Install the front and rear driveshafts. Tighten the flange bolts to 46 ft. lbs. (62 Nm).
52. Install the exhaust pipes and the catalytic converter. Tighten the manifold nuts to 49 ft. lbs. (67 Nm). Tighten the center pipe and converter bolts to 32 ft. lbs. (43 Nm).
53. Reconnect the oxygen sensors.
54. Install the exhaust pipe protectors.
55. Install the skid plates and tighten their mounting bolts to 27 ft. lbs. (37 Nm).
56. Lower the vehicle.
57. Install the transfer case control lever.
58. Connect the shift control rod to the selector lever. Adjust the shift cable so that there is 0.059–0.098 in. (1.5–2.5mm) of slack at the selector lever.
59. Connect the console switch connectors and the shift lock cable to the selector lever.
60. Install the center console assembly and the transfer case lever knob.
61. Install the upper fan shroud and the cooling fan.
62. Install the air cleaner assembly.
63. Align the matchmarks and install the hood.
64. Refill the transmission with Dexron® II ATF.
65. Connect the negative battery cable.
66. Start the engine and shift through the gear range three times to circulate the ATF through the valve body.
67. Recheck the fluid level and top-up if necessary.
68. Road test the vehicle.

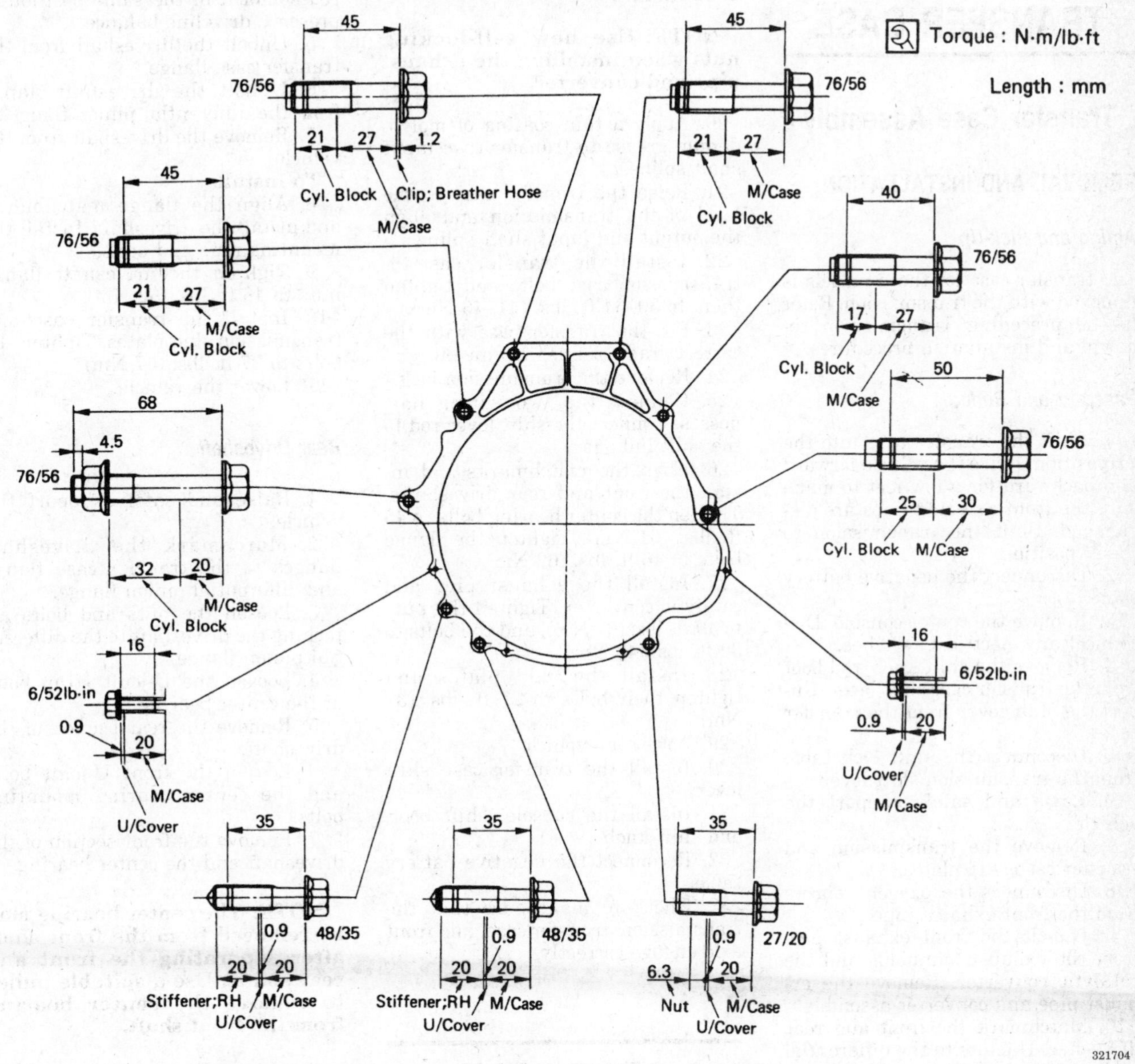

Torque : N·m/lb·ft

Length : mm

45

76/56

21 — 27 — 1.2

Cyl. Block — Clip; Breather Hose

M/Case

45

76/56

21 — 27

M/Case

Cyl. Block

45

76/56

21 — 27

M/Case

Cyl. Block

40

76/56

17 — 27

Cyl. Block

M/Case

68

4.5

76/56

32 — 20

M/Case

Cyl. Block

50

76/56

25 — 30

Cyl. Block   M/Case

16

6/52lb·in

0.9

20

M/Case

U/Cover

16

6/52lb·in

0.9

20

U/Cover

M/Case

35

0.9

48/35

20 — 20

Stiffener;RH   M/Case

U/Cover

35

0.9

48/35

20 — 20

Stiffener;RH   M/Case

U/Cover

35

0.9

27/20

6.3 — 20

Nut   M/Case

U/Cover

321704

Automatic transmission mounting bolt locations and torque specifications

Throttle Cable

ADJUSTMENT

1. Disconnect the negative battery cable.

2. Tighten the adjusting nut with the cable stretched. Allow only enough slack so that the throttle valve is closed. Tighten the nut so that the cable is straight in its stretching direction and is not binding.

3. Confirm that the throttle lever moves freely and returns to rest on the stopper screw.

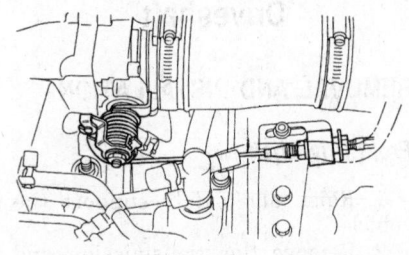

324975

Throttle cable attachment at the throttle body

4. Confirm that the cable and the return spring do not interfere with other.

5. Connect the negative battery cable.

TRANSFER CASE

Transfer Case Assembly

REMOVAL AND INSTALLATION

Amigo and Pick-Up

The transfer case for these models is removed with the transmission. Refer to the preceding Transmission removal and installation procedures.

Passport and Rodeo

1. Shift the transfer case into the **2H** position. Drive the vehicle forward and backward for a few feet to make sure the front axle and hubs are disengaged. Shift the transmission to the **N** position.
2. Disconnect the negative battery cable.
3. Remove the center console. Disconnect any electrical switches.
4. Remove the shift knob and boot from the transfer case shift lever. Unbolt the shift lever from the transfer case.
5. Disconnect the shift lock cable from the transmission shift lever.
6. Raise and safely support the vehicle.
7. Remove the transmission and transfer case skid plates.
8. Disconnect the oxygen sensors from the front exhaust pipe.
9. Unbolt the front exhaust pipe from the exhaust manifolds and the catalytic converter. Remove the exhaust pipe and converter assembly.
10. Matchmark the front and rear driveshaft U-joints to the differential and transfer case flanges.
11. Unbolt and remove the front driveshaft.
12. Unbolt the rear driveshaft from the rear differential and transfer case flanges.
13. Unbolt the center bearing and remove the rear driveshaft.
14. Drain the transfer case oil.
15. Disconnect the transmission shift linkage from the shift lever rod.
16. Label and disconnect the two wiring harnesses from the transfer case.
17. Support the transfer case with a transmission jack.
18. Remove the bolts securing the transfer case to the transmission.
19. Separate the transfer case from the transmission output shaft. Lower the transfer case from the vehicle.

To install:

NOTE: Use new self-locking nuts when installing the exhaust pipe and converter.

20. Apply a thin coating of molybdenum grease to transfer case input shaft splines.
21. Raise the transfer case to the level of the transmission and align the output and input shaft splines.
22. Install the transfer case-to-transmission case bolts and tighten them to 30–34 ft. lbs. (41–46 Nm).
23. Fill the transfer case with the correct grade of fresh engine oil.
24. Remove the transmission jack.
25. Connect the two wiring harnesses. Connect the shift lever rod to the shift linkage.
26. Align the matchmarks and install the front and rear driveshafts. Tighten the center bearing bolts to 45 ft. lbs. (61 Nm). Tighten the flange bolts to 46 ft. lbs. (62 Nm).
27. Install the exhaust pipe and catalytic converter. Tighten the nuts to 49 ft. lbs. (67 Nm), and the bolts to 32 ft. lbs. (43 Nm).
28. Install the skid plates and tighten their bolts to 27 ft. lbs. (37 Nm).
29. Lower the vehicle.
30. Install the transfer case shift lever.
31. Install the console, shift boot, and shift knob.
32. Reconnect the negative battery cable.
33. Check to make sure that the transmission, transfer case, and front axle engage correctly.

DRIVE AXLE

Driveshaft

REMOVAL AND INSTALLATION

Front Driveshaft

1. Raise and safely support the vehicle.
2. Remove the transmission and transfer case skid plates.
3. Matchmark the driveshaft flanges to the transfer case flange and differential pinion flange.
4. Matchmark the front and rear parts of the driveshaft so they can be

reassembled in the same position to preserve driveline balance.
5. Unbolt the driveshaft from the transfer case flange.
6. Unbolt the driveshaft flange from the differential pinion flange.
7. Remove the driveshaft from the vehicle.

To install:

8. Align the flange matchmarks and install the driveshaft. Install the mounting nuts and bolts.
9. Tighten the driveshaft flange nuts to 46 ft. lbs. (62 Nm).
10. Install the transfer case and transmission skid plates. Tighten the bolts to 27 ft. lbs. (37 Nm).
11. Lower the vehicle.

Rear Driveshaft

1. Raise and safely support the vehicle.
2. Matchmark the driveshaft flanges to the transfer case flange and differential pinion flange.
3. Loosen the nuts and bolts attaching the driveshaft to the differential pinion flange.
4. Loosen the U-joint strap bolts at the center bearing.
5. Remove the rear section of the driveshaft.
6. Loosen the front U-joint bolts and the center bearing mounting bolts.
7. Remove the front section of the driveshaft and the center bearing.

NOTE: The center bearing may be removed from the front shaft after separating the front and rear shafts. Use a suitable puller to remove the center bearing from the front shaft.

To install:

8. If the yoke has been removed from the front shaft, install it to the shaft with the washer and locknut. Torque the locknut to 112 ft. lbs. (152 Nm).
9. Align the matchmarks and install the front section of the driveshaft and the center bearing. Tighten the center bearing bolts to 45 ft. lbs. (61 Nm). Tighten the front flange bolts to 46 ft. lbs. (62 Nm).
10. Align the matchmarks and install the rear section of the driveshaft. Install the mounting bolts and nuts, but only hand-tighten them.
11. Tighten the U-joint strap bolts to 15 ft. lbs. (21 Nm). Tighten the flange bolts to 46 ft. lbs. (62 Nm).
12. Lower the vehicle.

U-Joints

REMOVAL AND INSTALLATION

1. Raise and support the vehicle safely.

2. Matchmark and remove the driveshafts.

3. Matchmark the driveshaft yokes so that driveline balance is preserved upon reassembly. Remove the snaprings that retain the bearing caps.

4. Select two press components; one must be small enough to pass through the yoke holes for the bearing caps, the other must be large enough to receive the bearing cap.

5. Use a vise or a press and position the small and large press components on either side of the U-joint. Press in on the smaller press component so it presses the opposite bearing cap out of the yoke and into the larger press component. If the cap does not come all the way out, grasp it with a pair of pliers and work it out.

6. Reverse the position of the press components so that the smaller press component presses on the cross. Press the other bearing cap out of the yoke.

7. Repeat the procedure on the other bearing caps.

To install:

8. Grease the bearing caps and needles thoroughly if they are not pregreased. Start a new bearing cap into a side of the yoke. Position the cross in the yoke.

NOTE: Some U-joints have a grease fitting that must be installed in the joint before assembly. When installing the fitting, make sure that once the driveshaft is installed in the vehicle, the fitting will be accessible.

9. Select two press components small enough to pass through the yoke holes. Put the press components against the cross and the cap and press the bearing cap ¼ in. (6mm) below the surface of the yoke. If there is a sudden increase in the force needed to press the cap into place, or if the cross starts to bind, the bearings are cocked. They must be removed and restarted in the yoke. Failure to do so will cause premature bearing failure.

10. Install a new snapring.

11. Start the new bearing cap into the opposite side. Place a press component on it and press in until the opposite bearing contacts the snapring.

12. Install a new snapring. It may be necessary to grind the facing surface of the snapring slightly to permit easier installation.

13. Install the other bearings in the same manner.

14. Check the joint for free movement. If binding exists, tap the yoke ears with a brass or plastic-faced hammer to seat the bearing needles. If binding still exists, disassemble the joint and check to see if the needles are in place. Do not strike the bearings unless the shaft is supported firmly. Do not install the driveshaft until all joints move freely.

15. Install the driveshafts.

16. Lower the vehicle. Test drive the vehicle and check for driveline vibrations.

Halfshaft

REMOVAL AND INSTALLATION

NOTE: The front axle must be lowered from the vehicle if the inboard CV-joint cases must be removed.

1. Shift the transfer case to the **2H** position and verify that the front axle is not engaged.

2. Raise and safely support the vehicle.

3. Remove the front wheels.

4. Remove the boot bands from the inboard joint boot. Push the boot away from the inboard joint case to expose the circlip. Remove the circlip from the inboard joint case.

5. Separate the halfshaft from the inboard joint case. Remove the halfshaft from the vehicle. The inboard joint case remains attached to the axle.

To install:

6. Fit the inboard joint boot over the edge of the case and into position. Gently squeeze the boot while lifting up the inner edge to release any

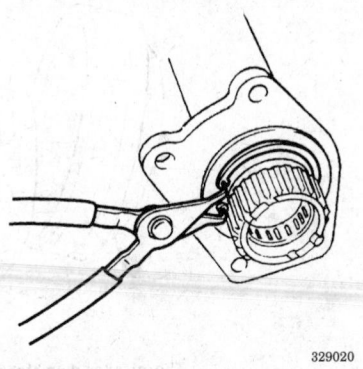

329020

Snapring removal — Passport

trapped air. The joint's extended length from the end of the case to the outer edge of the boot should be 6.5 in. (16.5cm).

7. Install new boot bands and clench them so that the crimp faces the front of the vehicle.

8. Assemble any suspension components that were disconnected to remove the halfshaft.

9. Install the front wheels.

10. Lower the vehicle. Tighten any suspension fasteners to their final torque specifications.

CV-Boot

REMOVAL AND INSTALLATION

1. Remove the halfshaft from the vehicle.

2. Place the halfshaft in a vise with padded jaws.

3. Use a flat prying tool to remove the balls from the inboard joint cage by prying outward from the inside of the cage.

4. Remove the snapring from the inboard spider assembly. Matchmark the spider to the halfshaft and remove it. If a puller is used, be careful not to damage the machined surfaces of the spider.

5. Remove the inboard joint cage from the halfshaft.

6. Wrap tape over the splines of the halfshaft to protect the boot if it is to be reused. Remove the inboard boot from the shaft.

7. Remove the dust seal and the boot bands from the outboard CV-joint.

NOTE: The outboard CV-joint can't be disassembled. Inspect it for rough motion or noise, and replace it if necessary.

8. Remove the outboard CV-boot.

9. Inspect all parts for wear or damage and replace as necessary.

To install:

10. Pack the outboard joint with the grease included with the boot repair kit. Don't mix different types of grease. Install the dust seal and the outboard CV-boot. Install new boot bands and clench them so that the crimp faces the front of the vehicle when the shaft is installed. Make sure the boot isn't twisted.

11. Fit the inboard boot onto the halfshaft. Remove the tape from the halfshaft splines.

12. Install the inboard joint cage onto the halfshaft with its small-diameter side facing the outboard joint.

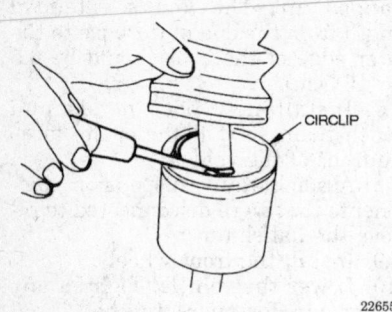

226552

Remove the circlip from the inner joint — Passport

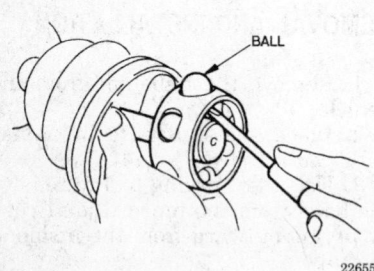

226553

Carefully remove the balls from the cage with a flat bladed tool — Passport

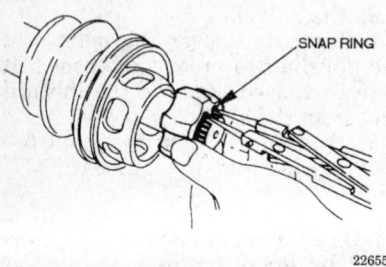

226554

Remove the snapring from the shaft — Passport

13. Install the spider onto the half-shaft, and then install a new snapring.

14. Pack the cage, spider, and inboard joint with CV-joint grease. Grease the balls and push them into the inboard joint cage.

15. Pack the inboard joint case with CV-joint grease. Verify that all the inboard joint components have been assembled properly and greased thoroughly.

16. Fit the inboard joint cage into the inboard joint case. Move the inboard joint back-and-forth several times to make sure it is seated.

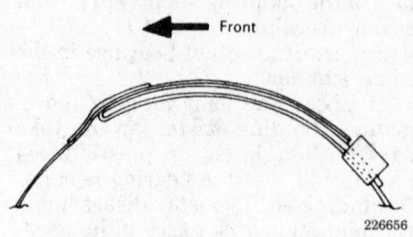

226656

Band installation direction — Passport

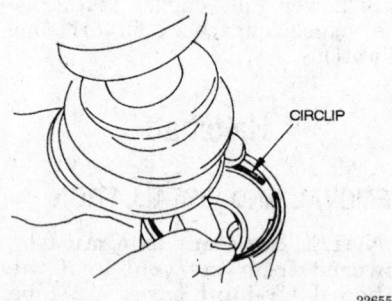

226556

Installing the circlip in the outer race of the inner joint — Passport

17. Install a new circlip into the inboard joint case. The circlip's end gap must be positioned away from the recesses in the edge of the inboard joint case.

18. Fit the inboard joint boot over the edge of the case and into position. Gently squeeze the boot while lifting up the inner edge to release any trapped air. The joint's extended length from the end of the case to the outer edge of the boot should be 6.5 in. (16.5cm).

19. Install new boot bands and clench them so that the crimp faces the front of the vehicle.

20. Install the halfshaft on the vehicle.

Rear Axle Shaft, Bearing and Seal

REMOVAL AND INSTALLATION

Dana Rear Axle

1. Raise and safely support the vehicle.

2. Remove the rear wheels.

3. If equipped with disc brakes, remove the rear brake caliper and support bracket. Use wire hooks to support the calipers.

4. If equipped with drum brakes, remove the brake drums and shoe assemblies.

5. Place a container under the axle flange to catch any dripping gear oil.

6. Remove the four axle retaining nuts and lockwashers.

7. Remove the axle shaft assembly from the axle housing.

8. Remove the snapring and bearing cup.

9. Break the retainer ring with a hammer and chisel.

10. Break the bearing cage with a hammer and chisel, then remove the bearing cage and roller.

11. Remove the oil seal and retainer. Then, remove the parking brake assembly and brake backing plate from the axle.

12. Use a press and a bearing splitter to remove the inner race from the axle shaft.

To install:

13. Install the parking brake assembly and brake backing plate. Then, install the retainer plate.

14. Install the oil seal into the bearing holder. Install the bearing holder onto the axle assembly. The cup side of the bearing faces the inboard side of the axle shaft.

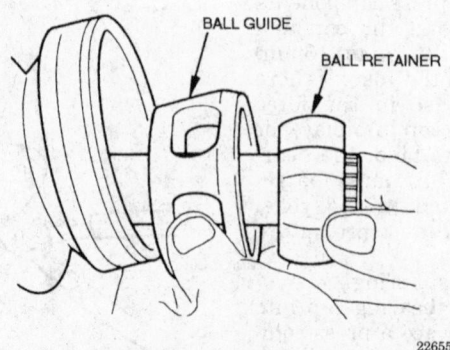

226555

The smaller diameter of the cage faces the outer CV-joint — Passport

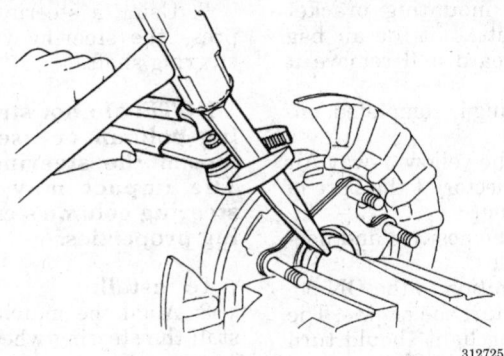

312725

Use a chisel to remove the retainer ring — Dana rear axle

15. Press the bearing assembly on the axle shaft.

16. Install the retainer ring using a press.

17. Install the snapring.

18. Place the axle assembly into the axle housing and install the lockwashers and retaining nuts. Tighten the retaining nuts to 55 ft. lbs. (75 Nm).

19. Install the brake rotors and calipers.

20. Refill the differential with the proper type of gear oil.

21. Install the rear wheels.

Saginaw Rear Axle

1. Raise and safely support the vehicle.

2. Remove the rear wheels. Then, remove the brake drums.

3. Drain the rear axle lubricant and remove the rear axle housing cover.

4. Remove the pinion shaft lockbolt and the pinion shaft.

5. Push the axle shaft slightly inward and remove the C-lock clip from the innermost end of the axle shaft.

6. Remove the axle from the axle housing.

7. Pry the oil seal from the axle housing.

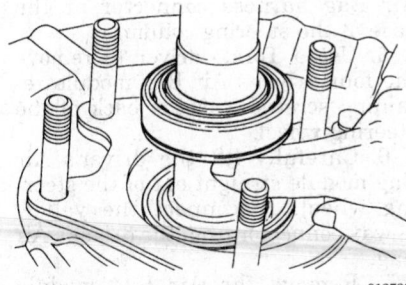

312728

Install the bearing with its cup towards the inboard side — Dana rear axle

8. Remove the axle bearing, use a axle bearing puller (part No. J-22813–01) on a slide hammer or a slide hammer with a hook.

To install:

9. Use a bearing driver (part No. J-23765) and handle (part No. J-8092) to install the bearing into the axle shaft housing.

10. Use an oil seal installer (part No. J-23771 or equivalent) and install the oil seal into the axle housing.

11. Carefully slide the axle shaft into the axle housing taking care not to damage the oil seal. The axle may have to be rotated a little to properly engage the splines in the differential side gear.

12. Install the C-lock clip to the axle shaft then pull the shaft out to seat the clip.

13. Install the pinion shaft and pinion shaft lockbolt. Torque the lockbolt to 25 ft. lbs. (34 Nm).

14. Use a new gasket and install the differential cover. Torque the cover bolts to 20 ft. lbs. (27 Nm).

15. Fill the differential with the proper amount of lubricant.

16. Install the brake drums.

17. Install the rear wheels.

18. Lower the vehicle to the floor.

STEERING

Air Bag

─── **CAUTION** ───

Some vehicles are equipped with an air bag system, also known as the Supplemental Inflatable Restraint (SIR) or Supplemental Restraint System (SRS). The system must be disabled before performing service on or around system components, steering column, instrument panel components, wiring and sensors. Failure to follow safety and disabling procedures could result in accidental air bag deployment, possible personal injury and unnecessary system repairs.

DISARMING

Driver's Side

─── **CAUTION** ───

The Air Bag system must be disarmed before any of its components are disconnected or removed. Failing to disarm the system before servicing components may cause accidental deployment of the Air Bag, resulting in repairs and possible personal injury.

1. Turn the steering wheel so that the vehicle's front wheels are pointing straight ahead.

2. Turn the ignition switch to the **LOCK** position. Remove the key.

3. Disconnect the negative and positive battery cables. Wait at least five minutes before working around the air bags.

NOTE: Removing the SRS-1 and SRS-2 fuses from the under-dash fuse block has the same effect as disconnecting the battery.

4. Disconnect the yellow 3-way Air Bag harness connector at the base of the steering column.

5. After servicing is completed, enable the Air Bag.

6. Reconnect the yellow 3-way Air Bag harness connector at the base of the steering column. Then, install the fuses if they were removed.

7. Reconnect the positive and negative battery cables.

8. Turn the ignition to the **ON** position, but don't start the engine. The AIR BAG warning light should turn on and flash on and off for seven seconds, and then turn off. This light sequence indicates that the Air Bag system is functioning normally. If the AIR BAG light doesn't come on, or stays on longer than seven seconds, the system must be diagnosed.

Passenger's Side

1. Turn the steering wheel so that the vehicle's front wheels are pointing straight ahead.

2. Turn the ignition switch to the **LOCK** position. Remove the key.

3. Disconnect the negative and positive battery cables. Wait at least

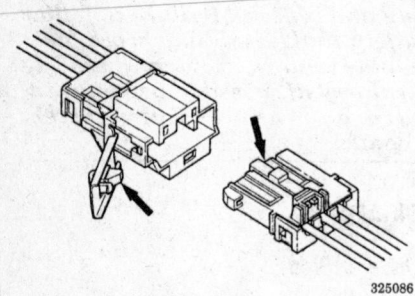

Air Bag wiring harness connectors

325086

five minutes before working around the air bags.

NOTE: Removing the SRS-1 and SRS-2 fuses from the under-dash fuse block has the same effect as disconnecting the battery.

4. Remove the glove box door and the passenger's air bag lower cover.

5. Disconnect the yellow 4-way Air Bag harness connector from the passenger's side air bag.

6. If the passenger's air bag module must be removed; carefully re-

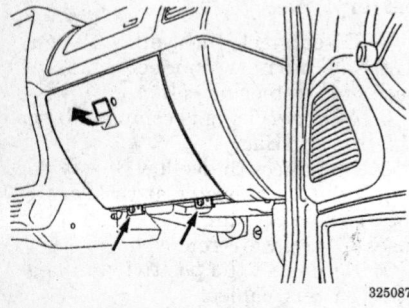

Glove box mounts

325087

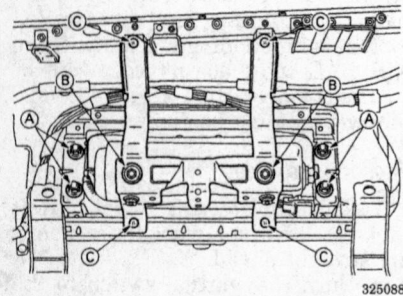

Passenger's air bag mounting bracket

325088

move the lower mounting bracket and mounting nuts. Lift the air bag out of the dashboard and remove it from the vehicle.

7. After servicing is completed, enable the Air Bag.

8. Reconnect the yellow 3-way Air Bag harness connector at the base of the steering column.

9. Reconnect the positive and negative battery cables.

10. Turn the ignition to the **ON** position, but don't start the engine. The AIR BAG warning light should turn on and flash on and off for seven seconds, and then turn off. This light sequence indicates that the Air Bag system is functioning normally. If the AIR BAG light doesn't come on, or stays on longer than seven seconds, the system must be diagnosed.

PRECAUTIONS

Several precautions must be observed when handling the inflator module to avoid accidental deployment and possible personal injury.

- Never carry the inflator module by the wires or connector on the underside of the module.
- When carrying a live inflator module, hold securely with both hands, and ensure that the bag and trim cover are pointed away.
- Place the inflator module on a bench or other surface with the bag and trim cover facing up.
- With the inflator module on the bench, never place anything on or close to the module which may be thrown in the event of an accidental deployment.

Steering Wheel

REMOVAL AND INSTALLATION

Without Air Bag

1. Set the steering wheel so that the front wheels are pointing straight ahead.

2. Disconnect the negative battery cable.

3. Remove the horn pad screws from the rear of the steering wheel. Pry the horn pad upward from the steering wheel to remove it.

4. Disconnect the horn lead.

5. Remove the steering column nut.

6. Matchmark the steering wheel-to-steering shaft position for reassembly.

7. Using a steering wheel puller, press the steering wheel from the steering shaft.

NOTE: Do not strike the steering column or use air tools to loosen the steering wheel nut. The impact may damage the steering column's energy-absorbing properties.

To install:

8. Align the matchmarks and install the steering wheel on the shaft. Torque the steering wheel nut to 22–29 ft. lbs. (30–39 Nm).

9. Reconnect the horn lead and install the horn pad.

10. Connect the negative battery cable.

11. Check the steering wheel spoke angle and test the operation of the horn.

With Air Bag

———— **CAUTION** ————
The Air Bag system must be disabled before the steering wheel and steering column are removed. Failure to disable the system may result in system repairs and possible personal injury. Do not damage, cut, or attempt to alter the yellow Air Bag wiring harness.
————————————

1. Turn the steering wheel so that the vehicle's front wheels are pointing straight ahead.

2. Turn the ignition switch to the **LOCK** position. Remove the key.

3. Disconnect the negative and positive battery cables. Wait at least five minutes before working around the Air Bags.

NOTE: Removing the SRS-1 and SRS-2 fuses from the under-dash fuse block has the same effect as disconnecting the battery.

4. Disconnect the yellow 3-way Air Bag harness connector at the base of the steering column.

5. Use a Torx® driver to remove the four driver's Air Bag module retaining screws from the back of the steering wheel.

6. Carefully lift the driver's Air Bag module straight out of the steering wheel. Disconnect the yellow 2-way connector from the driver's Air Bag.

7. Remove the Air Bag module from the vehicle and place it face-up on a clean bench away from your work area.

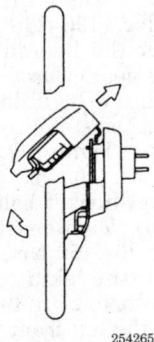

254265

After removing the screw, lift up on the horn pad to remove it

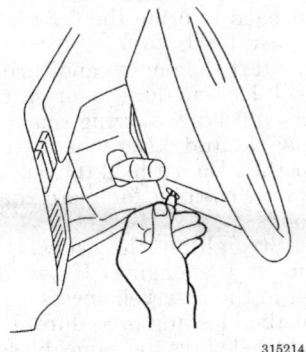

315214

Air Bag retaining bolts

—— **CAUTION** ——

Always carry a live Air Bag with its trim cover and cushion pointed away from your body. When placing a live Air Bag on a bench or other surface, always point the trim cover up, away from the surface. Following these precautions will lessen the chance of personal injury if the Air BAG accidentally deploys.

8. Disconnect the horn leads.
9. Make alignment marks on the steering wheel and steering column shaft.

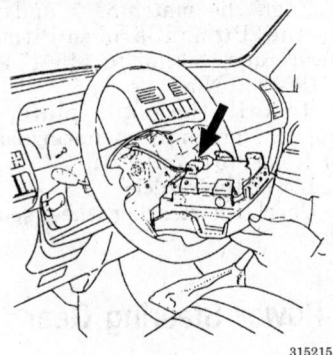

315215

Removing and installing an Air Bag module

10. Loosen and remove the steering wheel nut. Use tool No. J–29752 or an equivalent steering wheel puller to remove the steering wheel.

—— **WARNING** ——

The puller bolts may damage the Air Bag cable reel if they are threaded too deeply into the steering wheel.

To install:
11. Check the alignment of the Air Bag cable reel.
 a. Turn the cable reel clockwise to its fully locked position. Don't turn the cable reel past the point at which you begin to feel resistance to its rotation.
 b. Turn the cable reel about three turns in the opposite direction until the pointer on the cable reel is aligned with the neutral mark.
12. Align the steering wheel matchmarks. Install the steering wheel and tighten the nut to 25 ft. lbs. (34 Nm). Reconnect the horn leads.
13. Carefully connect the yellow 2–way connector to the Air Bag module. Install the Air Bag onto the steering wheel. Install the four

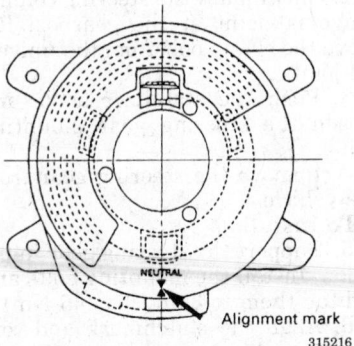

NEUTRAL

Alignment mark

315216

Air Bag cable reel alignment marks

Torx®bolts while supporting the Air Bag module with your hand. Tighten each bolt to 6 ft. lbs. (8 Nm).
14. Reconnect the yellow 3–way Air Bag harness connector at the base of the steering column.
15. Reconnect the positive and negative battery cables.
16. Turn the ignition to the **ON** position, but don't start the engine. The AIR BAG warning light should turn on and flash on and off for seven seconds, and then turn off. This light sequence indicates that the Air Bag system is functioning normally. If the AIR BAG light doesn't come on, or stays on longer than seven seconds, the system must be diagnosed.
17. Check the operation of the horn buttons.

Tie Rod Ends

REMOVAL AND INSTALLATION

1. Set the front wheels and steering wheel in the straight ahead position. Remove the ignition key and lock the steering column in this position.
2. Raise and safely support the vehicle.
3. If necessary, remove the radiator skid plate.
4. Matchmark the tie rod ends to the tie rod shaft for reinstallation.
5. Remove the cotter pin and castle nut from the tie rod end. Loosen the tie rod end locknut.
6. Using a ball joint separator tool, separate the tie rod ends from the steering knuckle and center link.
7. Remove the tie rod.
8. Unscrew the tie rod ends from the tie rod.
9. Check the tie rod ends for damage and excess play. Replace any damaged or worn parts.
To install:
10. Install the tie rod ends onto the tie rod, then align the matchmarks. Torque the locknut (inner or outer) to 87 ft. lbs. (118 Nm).
11. Install the tie rod end onto the steering knuckle and center link. Install the castle nuts and torque them to 72 ft. lbs. (98 Nm). Tighten the castle nuts only enough to install a new cotter pin.
12. Install the radiator skid plate and tighten its bolts to 27 ft. lbs. (37 Nm).
13. Check and adjust the front wheel alignment. Tighten the tie rod locknuts to 87 ft. lbs. (118 Nm).

Manual Steering Gear

REMOVAL AND INSTALLATION

1. Make sure the vehicle's front wheels are pointed straight ahead.

2. Disconnect the negative battery cable.

3. Raise and safely support the vehicle.

4. Remove the skid plate, if equipped.

5. Remove Pitman arm nut and washer. Matchmark the Pitman arm to the Pitman shaft.

6. Use a puller tool to press the Pitman arm from the Pitman shaft.

7. Matchmark the steering shaft joint to the steering gear. Then, remove the steering gear-to-steering shaft joint clamp bolt.

8. Remove the steering gear-to-frame bolts. Remove the steering gear from the vehicle.

To install:

9. Install the steering gear and mounting bolts. Torque the mounting bolts to 29–40 ft. lbs. (39–54 Nm).

10. Align the matchmarks and install the steering gear-to-steering shaft joint clamp bolt. Torque the bolt to 18 ft. lbs. (25 Nm).

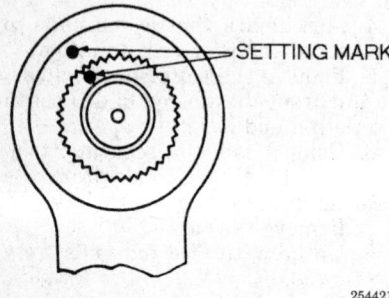

254421

Matchmark the Pitman arm and sector shaft

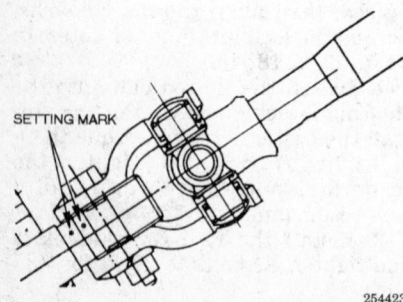

254423

Matchmark the steering shaft and gear

11. Align the matchmark and install the Pitman arm-to-Pitman shaft. Torque the nut to 145–174 ft. lbs. (196–236 Nm).

12. Install the skid plate, if equipped. Tighten the mounting bolts to 27 ft. lbs. (37 Nm).

13. Lower the vehicle.

14. Check the steering wheel spoke angle.

Power Steering Gear

REMOVAL AND INSTALLATION

1. Set the front wheels and the steering wheel in the straight ahead position. Lock the steering column in this position and remove the key.

2. Disconnect the negative and positive battery cables. Wait at least five minutes before working around the Air Bags, if equipped.

――――― **WARNING** ―――――
If the vehicle is equipped with a driver's Air Bag, be careful not to violently jar or strike the steering column.

3. Drain the power steering fluid from the pump reservoir.

4. Raise and safely support the vehicle.

5. Remove the left front wheel.

6. Remove the radiator skid plate, if equipped.

7. Remove the lower fan shroud.

8. Disconnect the stabilizer bar linkages and loosen the stabilizer bar bracket bolts.

9. Matchmark the Pitman arm to the Pitman shaft. Remove the nut and washer from the Pitman arm.

10. Use puller tool to press the Pitman arm from the Pitman shaft.

11. Use a flare wrench to disconnect the power steering lines from the steering gear. Plug the lines to prevent fluid loss and contamination.

12. Matchmark the steering column universal joint at the gearbox. Remove the clamp bolt from the universal joint.

13. Push the stabilizer aside and loosen the steering gear mounting bolts.

14. Remove the steering gear from the vehicle.

To install:

15. Support the steering gear in position. Install the mounting bolts and tighten them to 33 ft. lbs. (45 Nm).

16. Align the matchmark and connect the steering column universal joint to the steering gear.

17. Install the universal joint clamp bolt and tighten it to 18 ft. lbs. (25 Nm).

18. Align the matchmark and install the Pitman arm onto the Pitman shaft. Make sure the notched teeth are aligned. Install a new washer and tighten the Pitman arm nut to 145–174 ft. lbs. (196–236 Nm).

19. Connect the hydraulic lines to the steering gear. Carefully tighten the line fittings to 33 ft. lbs. (45 Nm).

20. Connect the stabilizer bar to the control arms. Tighten the linkage bolts to 8 ft. lbs. (10 Nm), and tighten the mounting bracket bolts to 21 ft. lbs. (28 Nm). Tighten the linkage nuts to 37 ft. lbs. (50 Nm).

21. Install the skid plate, and tighten its bolts to 27 ft. lbs. (37 Nm).

22. Install the left front wheel.

23. Lower the vehicle. Reconnect the negative and positive battery cable.

24. Refill and bleed the power steering system.

   a. Fill the reservoir with Dexron® II ATF.

   b. Raise the front wheels off the ground and block the rear wheels.

   c. Before starting the engine, turn the wheel lock-to-lock several times so the level of fluid in the reservoir drops. Refill the reservoir as needed to bring the fluid to the specified level.

   d. Start the engine and turn the wheel lock-to-lock 3 or 4 more times until any buzzing sound disappears and the wheel turns smoothly. Do not hold the wheel in the lock position for more than 5 seconds.

   e. Straighten the wheels and turn off the engine. If the fluid level in the reservoir increases, repeat the bleeding procedure. If the fluid level stays the same, bleeding is completed.

25. Verify that the steering wheel spokes are centered when the front wheels are in the straight ahead position. Check and adjust the front wheel alignment.

Power Steering Pump

BLEEDING

――――― **WARNING** ―――――
Use only Dexron® IIE ATF in the power steering system. Other types of fluids may cause system damage.

1. Fill the power steering reservoir to the proper level with the engine cold.

2. Start the engine and allow it to run until it reaches normal operating temperature.

3. Turn the engine **OFF** and check the fluid level. If necessary, fill the reservoir to the proper level.

4. Raise and support the vehicle so its front wheels are off the ground. Block the rear wheels.

5. Run the engine and turn the steering wheel from lock-to-lock, in both directions, 3–4 times; do not hold the steering wheel at the lock position for more than 5 seconds or temperature rise will result.

6. Return the steering wheel to center, turn the engine **OFF** and allow the fluid to sit for 5 minutes before adding any more.

7. If necessary, repeat the bleeding procedure until the air bubbles are removed from the system. No buzzing noises should be heard after the bleeding procedure is completed.

8. Fill the system to the proper level when finished.

REMOVAL AND INSTALLATION

1993–94 Models

1. Disconnect the negative battery cable.

2. Disconnect and plug the inlet and outlet fluid lines from the power steering pump.

3. Remove the drive belt from the pump.

4. Remove the pump-to-bracket bolts and the pump from the brackets.

To install:

5. Install the pump to the mounting bracket.

6. Connect the hydraulic lines to the pump.

7. Install and adjust the drive belt.

8. Refill the reservoir with Dexron® II ATF.

9. Connect the negative battery cable, start the engine and bleed the power steering system.

1995–97 Models

1. Disconnect the negative battery cable.

2. Drain the fluid from the power steering pump reservoir.

3. Disconnect and plug the inlet and outlet fluid lines from the power steering pump.

4. Loosen the adjusting bolt to relieve the belt tension.

5. Remove the drive belt from the pump pulley.

6. Loosen and remove the pump pulley nut. Use a puller to remove the pulley if it sticks.

7. Unbolt and remove the pump from its mounting bracket.

To install:

8. Install the pump to the mounting bracket and tighten the mounting bolts to 30 ft. lbs. (41 Nm).

9. Install the pulley and tighten its nut to 58 ft. lbs. (78 Nm).

10. Connect the hydraulic line banjo bolt to the pump using new crush washers. Tighten the banjo bolt to 15 ft. lbs. (20 Nm).

11. Install and adjust the drive belt. Tighten the adjusting bolt to 30 ft. lbs. (41 Nm).

12. Connect the negative battery cable.

13. Refill and bleed the power steering system.

BRAKES

Anti-Lock Brake System Service

PRECAUTIONS

• Certain components within the Anti-Lock Brake System (ABS) are not intended to be serviced or repaired individually. Only those components with removal and installation procedures should be serviced.

• Do not use rubber hoses or other parts not specifically specified for and ABS system. When using repair kits, replace all parts included in the kit. Partial or incorrect repair may lead to functional problems and require the replacement of components.

• Lubricate rubber parts with clean, fresh brake fluid to ease assembly. Do not use lubricated shop air to clean parts; damage to rubber components may result.

• Use only specified brake fluid from an unopened container.

• If any hydraulic component or line is removed or replaced, it may be necessary to bleed the entire system.

• A clean repair area is essential. Always clean the reservoir and cap thoroughly before removing the cap. The slightest amount of dirt in the fluid may plug an orifice and impair the system function. Perform repairs after components have been thoroughly cleaned; use only denatured

alcohol to clean components. Do not allow ABS components to come into contact with any substance containing mineral oil; this includes used shop rags.

• The Anti-Lock control unit is a microprocessor similar to other computer units in the vehicle. Ensure that the ignition switch is **OFF** before removing or installing controller harnesses. Avoid static electricity discharge at or near the controller.

• If any arc welding is to be done on the vehicle, the control unit should be unplugged before welding operations begin.

Master Cylinder

REMOVAL AND INSTALLATION

1. Disconnect the negative battery cable.

2. Use a suction pump to remove the brake fluid from the master cylinder reservoir.

3. Disconnect the reservoir fluid level indicator connector at the master cylinder.

4. Disconnect the brake lines from the master cylinder using a flare nut wrench. Cap the lines while they are disconnected to prevent moisture or dirt from entering the brake system.

NOTE: Brake fluid is highly corrosive to paint. Take care not to spill brake fluid on any painted surface of the vehicle.

5. Remove the master cylinder mounting nuts.

6. Remove the master cylinder by pulling it straight forward off of the power booster studs.

To install:

7. If installing a new power brake booster, apply 19.7 in. Hg (66.5 kPa) of vacuum to the power booster and check the distance from the flange face of the booster to the end of the pushrod. Pushrod length should be 0.709–0.717 in. (18.0–18.2mm).

8. Install a new pushrod seal on the master cylinder.

9. Coat a new dust seal with silicon grease and install it onto the master cylinder. The seal's groove must face downward.

10. Bench–bleed the master cylinder before installing it:

a. Wrap the master cylinder with a shop towel and secure it in a vice.

b. Attach bleeding tubes to the outlet ports, then route the tubes into a container partially filled with clean brake fluid.

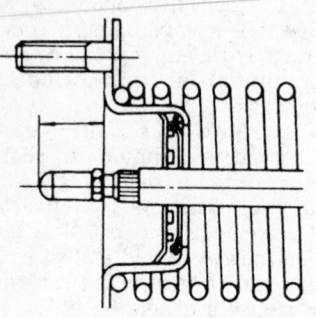

255414

Measure the master cylinder pushrod length

c. Fill the reservoir with clean brake fluid.

d. Slowly depress the master cylinder piston with a wooden dowel. Then, allow the piston to return. Follow this procedure several times until there are no air bubbles in the fluid draining from the bleeding tubes.

11. Position the master cylinder on the power booster and install the mounting nuts. Tighten the master cylinder–to–power booster nuts to 10–16 ft. lbs. (13–21 Nm) for Amigo and Pick-up models, 7–12 ft. lbs. (9–16 Nm) for all other models.

NOTE: The bleeding process can be expedited by bench bleeding a new master cylinder prior to installation.

12. Connect the brake lines to the master cylinder. Torque the fittings to 9–15 ft. lbs. (12–20 Nm) using a flare nut wrench.

13. Connect the fluid level indicator connector to the master cylinder.

14. Fill the master cylinder with clean brake fluid.

15. Bleed the master cylinder and brake system.

16. Reconnect the negative battery cable.

Front Brake Caliper

REMOVAL AND INSTALLATION

1. Raise and safely support the vehicle. Remove the front wheels.

2. Remove some brake fluid from the master cylinder reservoir.

3. Disconnect and plug the brake fluid line from the caliper.

4. Remove the brake caliper mounting bolt and guide bolt and remove the caliper from the mount. The brackets can be remove for additional work space.

5. Remove the brake pads and clips from the caliper. Inspect the brake pads for wear; replace them, if necessary.

To install:

6. Install the brake pads and clips onto the caliper.

7. If the caliper bracket was removed, tighten the bolts to 103–126 ft. lbs. (139–171 Nm).

8. Install the caliper on the mounting bracket. Torque the caliper-to-mounting bracket bolts to 20–27 ft. lbs. (27–37 Nm) for 4-cylinder models, or 54 ft. lbs. (74 Nm) for 6-cylinder models.

9. Connect the fluid line to the caliper using new washers. Torque the brake line banjo fitting to 26 ft. lbs. (35 Nm).

----- **WARNING** -----
Be sure the hook end of the flexible brake line is positioned in the anti-rotation cavity.

10. Refill the master cylinder reservoir and bleed the brake system.

11. Install the front wheels and lower the vehicle.

Rear Brake Caliper

REMOVAL AND INSTALLATION

1. Raise and safely support the vehicle. Remove the rear wheels.

2. Remove some fluid from the master cylinder reservoir.

3. Disconnect and plug the brake fluid line from the caliper.

NOTE: Do not touch the adjustment of the parking brake cable and discard the park cable mounting pin after removal.

4. If equipped with caliper–actuated parking brakes; remove the mounting pin from the parking

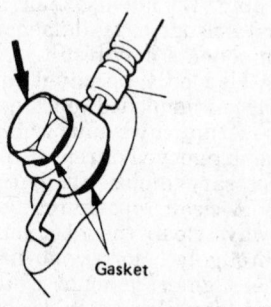

Gasket

255376

Brake line, banjo bolt and anti-rotation hook — front and rear calipers

brake cable and disconnect the parking cable from the disc caliper.

5. Remove the brake caliper mounting bolt and guide bolt and remove the caliper from the mount.

6. Remove the brake pads and clips from the caliper. Inspect the brake pads for wear; replace them, if necessary.

To install:

7. Install the brake pads and clips onto the caliper.

8. If the mounting bracket was removed, tighten the bolts to 69–84 ft. lbs. (93–114 Nm).

9. Install the caliper on the mounting bracket. Torque the caliper-to-mounting bracket bolts to 12–17 ft. lbs. (16–24 Nm), or 32 ft. lbs. (43 Nm) on vehicles with shoe–type parking brakes.

10. Connect the parking brake cable to the caliper and install a new mounting pin.

11. Connect the fluid line to the caliper using new washers. Torque the brake line banjo fitting to 26 ft. lbs. (35 Nm).

12. Refill the master cylinder reservoir and bleed the brake system.

13. Install the rear wheels and lower the vehicle.

Front Disc Brake Pads

REMOVAL AND INSTALLATION

Most disc brake pads are equipped with wear indicators. If a squealing noise occurs from the brakes while driving, check the pad wear indicator plate. If there is evidence of the indicator plate contacting the brake disc, the brake pad should be replaced.

----- **CAUTION** -----
Brake shoes may contain asbestos, which has been determined to be a cancer causing agent. Never clean the brake surfaces with compressed air. Avoid inhaling any dust from any brake surface. Use a commercially available brake cleaning fluid when cleaning brake surfaces.

1. Remove half of the volume of brake fluid from the master cylinder to prevent overflow when the caliper piston is compressed.

2. Raise and safely support the vehicle.

3. Remove the wheel and tire assemblies.

4. Remove the brake caliper without disconnecting the brake line. Support the caliper with a length of

wire. Do not let the caliper hang from the brake hose.

NOTE: On some disc brake systems it is not necessary to remove the caliper when installing new brake pads. Remove the lower slide bolt and rotate the caliper upward to remove the pads.

5. Remove the brake pads and shims. Inspect the brake rotor and machine or replace as necessary. Check the minimum thickness (specification is cast into the rotor) before machining.

To install:

6. Use a suitable tool to push the caliper piston into its bore.

7. Apply a thin coat of grease to the rear face of the brake pad and install the shim. Install the brake pads.

8. Install the calipers. Lubricate the caliper bolts and boots. If equipped with a 4-cylinder engine, tighten the caliper mounting bolts to 24 ft. lbs. (33 Nm). If equipped with a 6-cylinder engine, tighten the caliper mounting bolts to 54 ft. lbs. (74 Nm).

9. Install the wheel and tire assemblies and lower the vehicle.

10. Apply the brakes several times to seat the pads before moving the vehicle. Check the fluid in the master cylinder and add as necessary.

Rear Disc Brake Pads

REMOVAL AND INSTALLATION

1. Raise and safely support the vehicle. Remove the rear wheels.

2. Remove the brake caliper mounting bolts and remove the caliper without disconnecting the brake fluid line. Support the caliper so it does not hang on the brake line.

3. Remove the brake pads and retaining clips from the caliper.

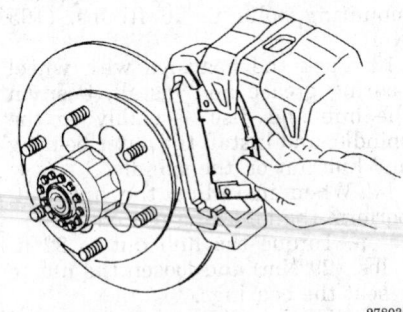

278035

Rotate the caliper upward to remove the pads

4. If equipped with caliper–activated parking brakes; use tool J-37617 or equivalent to rotate the piston clockwise until it retracts into the bore. Align the notches of the piston face so the centerline of the notches is perpendicular to the centerline of the mounting bosses.

To install:

5. Install the new brake pads and clips in the caliper and install the caliper in the mounting bracket.

6. Tighten the caliper mounting bolts to 12–17 ft. lbs. (16–24 Nm), or 32 ft. lbs. (43 Nm) on vehicles with shoe-type parking brakes.

7. Install the rear wheels. Check the brake fluid level.

8. Pump the brake pedal until pressure is felt before moving the vehicle.

Front Brake Rotor

REMOVAL AND INSTALLATION

2WD

1. Raise and safely support the vehicle. Remove the front wheels.

2. Remove the brake caliper without disconnecting the fluid line. Support the caliper aside.

3. Remove the brake caliper mounting bracket.

4. Remove the dust cover, cotter pin and locknut from the rotor.

5. Place a hand over the outer wheel bearing to prevent the bearing from falling on the floor and remove the rotor and hub from the spindle.

6. Matchmark the hub and rotor so it can be assembled in the same position and remove the bolts attaching the hub to the rotor, then pull off the rotor.

7. Thoroughly clean, inspect and repack the bearings with wheel bearing grease. Replace the bearings if needed. Use a brass drift to drive the bearing races out of the hub.

To install:

8. Use a suitably–sized bearing driver to drive new bearing races into the hub.

9. Install the rotor on the hub and torque the bolts to 76 ft. lbs. (103 Nm).

10. Install the inner bearing and the grease seal in the hub if removed and position the hub and rotor assembly on the spindle.

11. Install the outer bearing and spindle nut in the hub. Torque the hub nut to 22 ft. lbs. (29 Nm) to seat the bearings and then loosen the nut. Using a spring gauge, connected to the stud bolt at 90 degrees, retorque

the hub nut until the spring gauge measures a bearing preload of 4.4–5.5 lbs. (2.0–2.5 kg) for a new bearing and grease seal, or 2.6–4.0 lbs. (1.2–1.8 kg) for a used bearing and a new grease seal.

12. Install the caliper mounting bracket and torque the bolts to 115 ft. lbs. (156 Nm).

13. Install the caliper.

14. Install the front wheels and lower the vehicle.

15. Pump the brake pedal several times to test the system. Bleed the brakes if necessary.

4WD Models

With Automatic Hubs

1. Move the transfer case shift lever into **2H** and move the vehicle forward and rearward about 3 ft. (0.91m).

2. Raise and safely support the vehicle. Remove the front wheels.

3. Remove the 4WD hub cap attaching bolts and the cap.

4. Remove the brake caliper and support the caliper on a wire; do not disconnect the brake hose. Remove the brake caliper support bracket from the steering knuckle.

5. Using snapring pliers, remove the snapring and shims from the hub assembly.

6. Remove the drive clutch assembly, the inner cam and lockwasher.

7. Using a hub nut wrench, loosen the hub nut.

8. Pull the hub from the spindle.

9. Scribe matchmarks on the hub, then remove the bolts attaching the hub to the rotor. Separate the disc from the hub.

To install:

10. Install the hub to the rotor and torque the bolts to 76 ft. lbs. (103 Nm).

11. Install the hub nut and adjust the bearing.

12. Install the lockwasher with the larger diameter of the tapered bore facing out. If the bolt holes in the lock plate do not align with the corresponding holes in the nut, reverse the lock plate. If the bolt holes are still out of alignment, turn in the nut just enough to obtain alignment and install the lockscrew tightly so its head is lower than the surface of the washer.

13. Clean the flange surface of the hub, thread holes, the surface of the lockwasher and the splines of the axle shaft.

14. Install the inner cam by aligning the keyway of the inner cam with the groove of the knuckle. If the cam is difficult to install, use the special

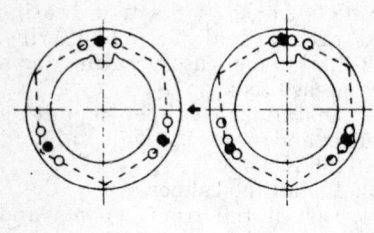

Lockwasher installation — 4WD models with automatic hubs

251135

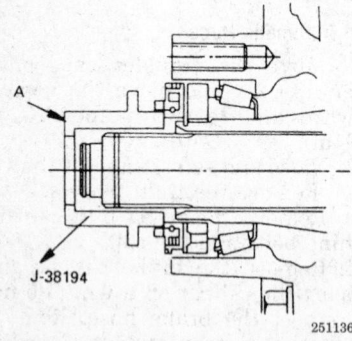

J-38194

251136

Special tool J-38194 or equivalent should be used to install the inner cam

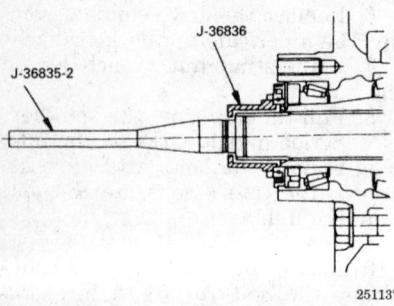

J-36836

J-36835-2

251137

Install special tools to measure shim thickness

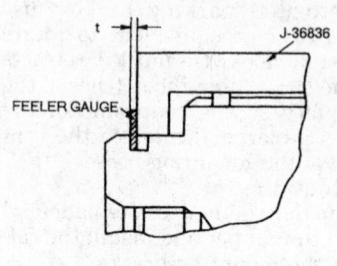

FEELER GAUGE

J-36836

251138

Measure clearance "t" between the special tool and the snapring groove

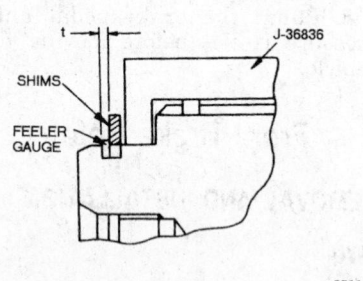

SHIMS

FEELER GAUGE

J-36836

251139

If clearance "t" is larger than snapring groove, selected shims must be used

the hub and the snapring groove on the axle shaft.

    d. If the clearance is larger than the snapring groove, selected shims must be installed so that clearance "t" is 0–0.039 in. (0– 0.1mm). Shims come in thicknesses of 0.2, 0.3, 0.5, and 1.0mm.

    e. Remove the special tool J-36836 or equivalent.

16. Install the drive clutch assembly.

17. Install the shims selected above by hand on the axle and use the fol-

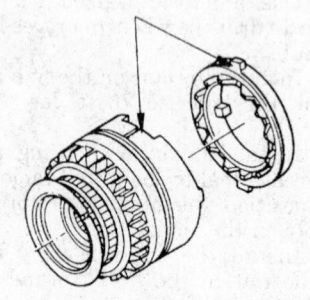

251140

Drive clutch assembly installation — 4WD models with automatic hubs

lowing steps to install a new snapring.

    a. Install special tool J-36835-2 or equivalent onto the axle.

    b. Install the snapring on the tool.

    c. Install tool driver J-36835-1 or equivalent.

    d. Pull out the axle shaft by pulling tool J-36835-2 or equivalent. Install the snapring to the axle by pushing on tool J-36835-1.

    e. Remove tool driver J-36835-2 or equivalent from the axle and check the fit of the snapring.

18. Install the housing assembly and cap. Torque the cap bolts to 43 ft. lbs. (58 Nm).

19. Install the front wheels.

20. Lower the vehicle to the floor.

21. Pump the brake pedal several times to test the system. Bleed the brakes if necessary.

With Manual Hubs

1. Shift the transfer case lever into **2H** and the manual hub into the **FREE** position.

2. Raise and safely support the vehicle.

3. Remove the front wheels.

4. Remove the brake caliper assembly and mounting bracket.

5. Loosen the 6 bolts and remove the housing assembly. Be careful not to lose the detent ball and spring.

6. Use snapring pliers and remove the snapring and shims.

7. Remove 6 bolts and remove housing assembly from the hub.

8. Remove the lockwasher retaining screw and remove the lockwasher.

9. Use a special hub nut wrench and remove the hub nut.

10. Remove the hub and rotor assembly from the spindle.

11. Matchmark the hub and rotor. Place the rotor in a soft jaw vise and remove the bolts attaching the rotor to the hub.

    To install:

12. Align the matchmark and assemble the hub and rotor. Torque the mounting bolts to 76 ft. lbs. (103 Nm).

13. Pack the bearings with wheel bearing grease and install. Position the hub and disc assembly on the spindle, and install the outer bearing and hub nut on the spindle.

14. When installing the hub nut, perform the following procedures:

    a. Torque the hub nut to 22 ft. lbs. (29 Nm) and loosen the nut to seat the bearings.

    b. Use a spring gauge connected to the stud bolt at 90° to measure preload.

tool or equivalent and a plastic hammer to lightly tap the inner cam into place.

15. Do the following steps to select the proper shim:

    a. Lower the vehicle and support the lower control arm with a block of wood and a floor jack to place the axle in the normal horizontal position.

    b. Install the special adjusting tools onto the hub until it comes in contact with the lockwasher.

    c. Pull the axle out as far as possible and using a feeler gauge, measure the clearance "t" between

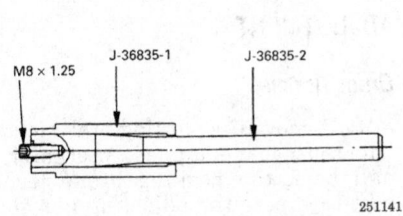

M8 × 1.25   J-36835-1   J-36835-2

251141

Special tools for snapring installation

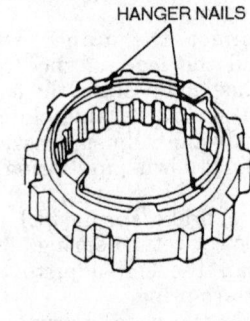

HANGER NAILS

251144

Hook the retainer spring onto the upper portion of the hanger nails

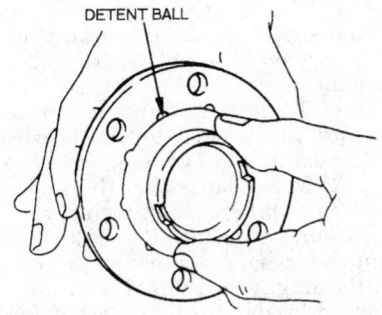

DETENT BALL

251142

Align the detent ball with the groove cut in the cover — 4WD models with manual hubs

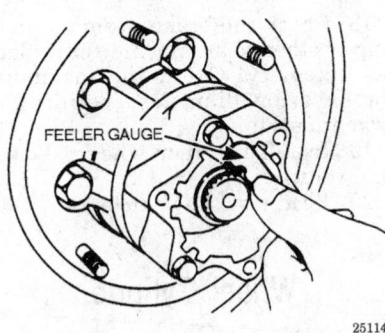

FEELER GAUGE

251145

Use a feeler gauge to measure the snapring clearance

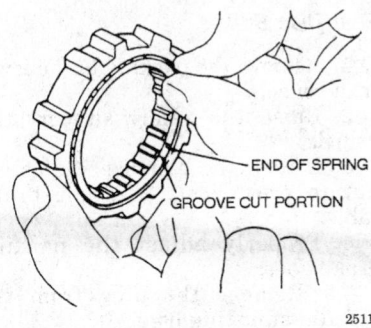

END OF SPRING

GROOVE CUT PORTION

251143

Align the end of the retainer spring with the end of the clutch spring groove's cut portion

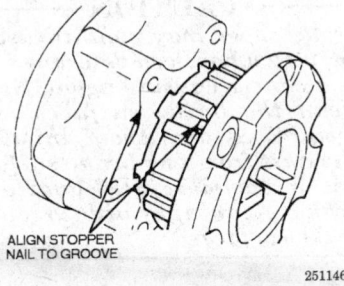

ALIGN STOPPER NAIL TO GROOVE

251146

Install the clutch assembly into the housing, while aligning the stopper nails with the grooves of the body — 4WD models with manual hubs

c. Retorque the hub nut until the spring gauge measures a bearing preload of 4.4–5.5 lbs. (2.0–2.5 kg) for a new bearing and oil seal, or 2.6–4.0 lbs. (1.2–1.8 kg) for a used bearing and new oil seal.

15. Install the lockwasher with the larger diameter of the tapered bore to the outer side of the vehicle. If the bolt holes in the lock plate do not align with the corresponding holes in the nut, reverse the lock plate. If the bolt holes are still out of alignment, turn in the nut just enough to obtain alignment and install the lockscrew

tightly so its head is lower than the surface of the washer.

16. Apply grease to the spacer, ring, snapring and the splined part of the inner assembly.

17. Assemble the inner assembly and the clutch assembly.

18. Install the body assembly on the hub. Torque the bolts to 9 ft. lbs. (12 Nm).

19. Install the selective shim and a new snapring. The clearance between the free wheeling hub body and the snapring should be 0–0.01 in. (0.25mm), shims are available in selective sizes.

20. Install the housing assembly. Torque the bolts to 43 ft. lbs. (58 Nm).

21. Install the front wheels.

22. Lower the vehicle to the floor.

23. Pump the brake pedal several times to test the system. Bleed the brakes if necessary.

Rear Brake Rotor

REMOVAL AND INSTALLATION

1. Raise and safely support the vehicle. Remove the rear wheels.

2. Engage the parking brake.

3. Remove the brake caliper without disconnecting the fluid line. Support the caliper aside with wire. Remove the caliper mounting bracket.

4. Release the parking brake.

5. Remove the rotor from the hub.

To install:

6. Install the brake rotor onto the hub. Install any retaining screws.

7. Install the caliper mounting bracket and attach the caliper.

8. Install the rear wheels and lower the vehicle.

9. Pump the brake pedal to test the system. Bleed the brakes if necessary.

Brake Drums

REMOVAL AND INSTALLATION

———— CAUTION ————
Brake shoes may contain asbestos which has been determined to be a cancer causing agent. Never clean the brake surfaces with compressed air. Avoid inhaling any dust from any brake surface. Use a commercially available brake cleaning fluid to clean brake surfaces.

1. Raise and safely support the vehicle. Release the parking brake.

2. Remove the rear wheels.

3. Use chalk to mark the brake drum to one of the wheel studs as an index mark for reinstallation.

4. Remove the retaining screw that holds the brake drum to the axle flange.

5. Pull the brake drum from the axle flange.

To install:

6. Align the index mark and install the brake drum to the axle flange.

7. Install the retaining screw to secure the brake drum to the axle flange.

8. Install the rear wheels.

Rear Brake Shoes

REMOVAL AND INSTALLATION

—————— CAUTION ——————

Brake shoes may contain asbestos, which has been determined to be a cancer causing agent. Never clean the brake surfaces with compressed air. Avoid inhaling any dust from any brake surface. When cleaning brake surfaces, use a commercially available brake cleaning fluid.

1. Raise and safely support the vehicle.
2. Remove the rear wheels.
3. Remove the brake drums.
4. Remove the brake return springs.
5. Remove the leading shoe holding pin and spring, and then the leading shoe.
6. Remove the self–adjuster and the adjuster lever.
7. Remove the trailing shoe holding pin and spring.
8. Disconnect the parking brake cable from the trailing shoe and remove the trailing shoe. Remove the parking brake lever from the trailing shoe.

To install:
9. Attach the parking brake lever to the trailing shoe.
10. Connect the parking brake cable to the parking brake lever.
11. Apply a thin coat of high temperature grease to the shoe contact points on the brake backing plate (locations A and C in the accompanying illustration), piston contact surface (B), and self-adjuster (D).
12. Position the trailing shoe on the backing plate and install the hold-down pin, spring, and retainer. Be careful not to stretch the return

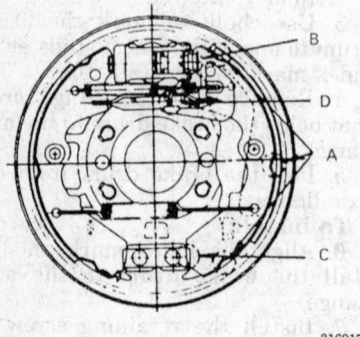

316917

Brake shoe lubrication points

spring when fitting the shoes onto the backing plate.

13. Connect the upper return spring and the leading shoe to the trailing shoe and position the leading brake shoe on the backing plate.
14. Install the adjuster assembly and the hold-down pin, spring, and retainer.
15. Use a brake spring tool to install the lower return spring.
16. Install the self–adjuster lever and adjuster spring.
17. Adjust the shoe-to-drum clearance to 0.0098–0.0157 in. (0.25–0.40mm) and install the brake drum.
18. Check the brake drum for scoring or other wear. Machine or replace as necessary. Check the maximum brake drum diameter specification when machining.
19. Install the rear wheels. Lower the vehicle.
20. Road test the vehicle.

Wheel Cylinder

REMOVAL AND INSTALLATION

—————— CAUTION ——————

Brake shoes may contain asbestos, which has been determined to be a cancer causing agent. Never clean the brake surfaces with compressed air. Avoid inhaling any dust from any brake surface. Use a commercially available brake cleaning fluid to clean brake surfaces.

1. Raise and safely support the vehicle.
2. Remove the rear wheels.
3. Remove the brake drums and shoes.
4. Remove the wheel cylinder from the backing plate by removing the two mounting bolts.

To install:
5. Install the wheel cylinder to the backing plate and secure with bolts. Torque the bolts to 6–9 ft. lbs. (8–12 Nm).
6. Install the brake line and tighten fitting to 9–14 ft. lbs. (13–19 Nm).
7. Install the brake shoes and drums.
8. Install the rear wheels.
9. Bleed the complete brake system and torque the bleeder fittings to 5–9 ft. lbs. (7–12 Nm).
10. Adjust the brake shoes.

Parking Brake Cable

ADJUSTMENT

Drum Brakes

1. Remove the parking brake lever trim console, if equipped. Alternately, pull back the parking brake lever boot to expose the adjusting nut and cable equalizer.
2. Raise and safely support the vehicle.
3. Properly adjust the rear brakes.
   a. Remove the plug from the drum adjusting hole.
   b. Use a small prytool to turn the shoe adjuster downward until the wheel cannot be turned by hand.
   c. Turn the adjuster upward with the prytool until the wheel can be turned by hand.
   d. Make sure the shoes don't drag or bind against the drum.
4. Adjust the parking brake at the equalizer so the parking brake is fully engaged (wheel cannot be turned by hand) when the handle is pulled nine to eleven notches. On 1996–97 vehicles, the handle only needs to be pulled up six notches.

Rear Disc Brakes

1. Remove the parking brake lever trim console.
2. Raise the safely support the vehicle.
3. Loosen the equalizer adjusting nut to provide slack in the brake cable.
4. Properly adjust the parking brake shoes.
   a. Remove the plug from the drum adjusting hole.
   b. Use a small prytool to turn the shoe adjuster downward until the wheel cannot be turned by hand.
   c. Turn the adjuster upward approximately six to eight notches with the prytool until the wheel can be turned by hand.
   d. Make sure the shoes don't drag or bind against inside of the brake rotor.
5. Remove the rubber plug in the backing plate and turn the adjusting screw downward until the rotor will not turn by hand.
6. Slowly turn the adjuster upward just until the rotor begins to turn and install the rubber plug.
7. Adjust the cable adjusting nut so the lever travels six notches when pulled with a force of 66 lbs. (30 kg) and then tighten the cable locknut.

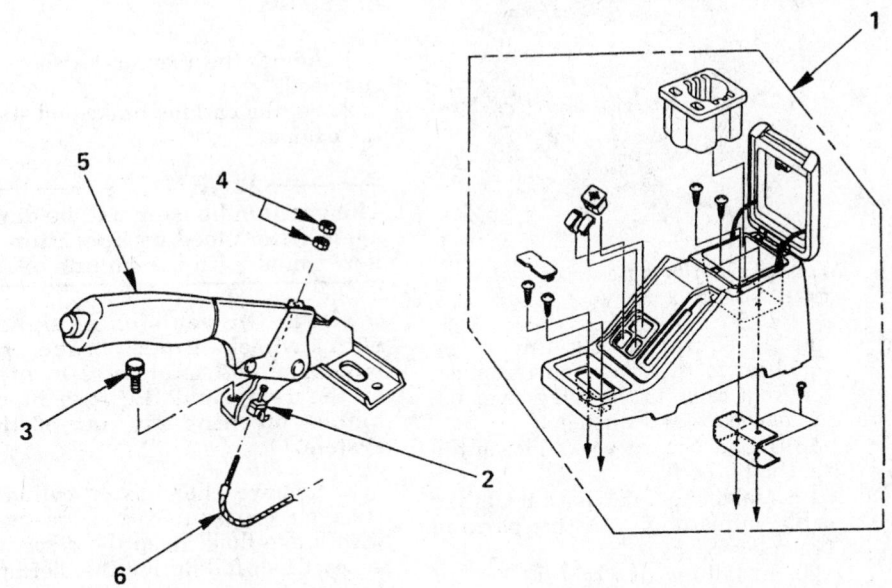

1. Rear console
2. Switch connector
3. Bolt
4. Nut
5. Parking brake lever
6. Parking brake front cable

325002

Brake lever and console components

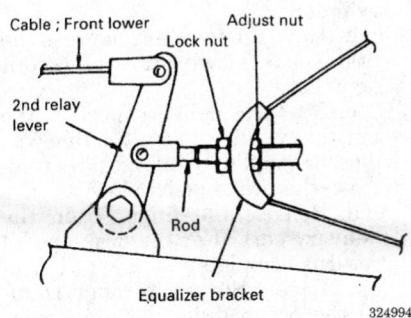

Parking brake adjustment — drum brakes

324994

Tighten the locknut to 8 ft. lbs. (11 Nm).

REMOVAL AND INSTALLATION

Drum Brakes

1. If equipped, remove the rear console.
2. Make sure the parking brake lever or handle is in the fully released position.
3. Raise and safely support the vehicle.
4. If equipped with a dash-board–mounted parking brake lever, disconnect the front cable from the relay lever. Next, remove the nut to

release the equalizer lever and re-move the clip to separate the cable from the lever.
5. Remove the equalizer nut and the equalizer bracket.
6. Remove the rear wheels and brake shoes.
7. Disconnect the parking brake cable from the lever on the brake shoe.
8. Use a 13mm offset wrench to compress the retaining lugs on the cable and work the cable out of the backing plate (toward the inside of the vehicle).
9. Remove any clips or brackets attaching the cable to the frame and remove the cable from the vehicle.

To install:
10. Pass the parking brake cable through the hole in the backing plate until the retaining lugs spread apart and hold the cable securely in place.
11. Install the clips and brackets that attach the cable to the frame.
12. Connect the parking brake cable to the brake shoe lever.
13. Install the brake shoes and the brake drums. Install the rear wheels.
14. Connect the parking brake cable at the equalizer bracket.
15. If equipped, reconnect the equalizer to the lever and reconnect the front cable to the relay lever us-ing new clips and washers.

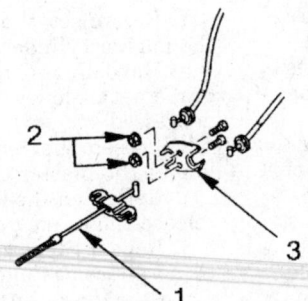

1. Front cable
2. Nuts
3. Retainer assembly

324996

Parking brake adjustment — rear disc brakes

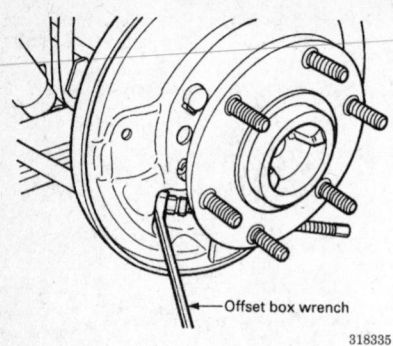

Cable attachment to the backing plate — drum brakes

318335

16. Adjust the parking brake and lower the vehicle to the floor.

17. Install the rear console.

Rear Disc Brakes

1. Remove the rear console.

2. Raise and safely support the vehicle.

3. If equipped with a dashboard-mounted parking brake lever, disconnect the front cable from the relay lever. Next, remove the nut to release the equalizer lever and remove the clip to separate the cable from the lever.

4. Remove the rear wheels.

5. Remove the calipers and support with wire. Don't disconnect the brake lines.

6. Remove the rear brake rotors.

7. Remove the parking brake shoes and remove the cable from the lever on the shoe.

8. Remove the cable bracket bolts and the clips attaching the cable to the frame.

9. Remove the cable from the equalizer assembly.

To install:

10. Apply grease to the front and rear linkages of the cable.

11. Attach the cable to the equalizer assembly. Torque the nuts on the retainer assembly to 10 ft. lbs. (13 Nm).

12. Install the clips holding the cable to the frame.

13. Pass the cable through the backing plate and install the bracket bolts. Torque the bolts to 4 ft. lbs. (6 Nm).

14. Apply grease to the brake shoe contact areas, adjuster assembly, and brake shoe anchors.

15. Install the parking brake shoes and the rotor.

16. Install the caliper assembly and adjust the parking brake shoes.

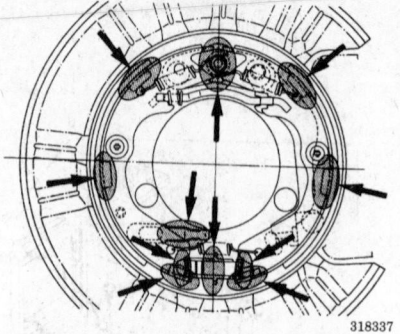

Apply grease lightly to the shaded areas — models with rear disc brakes

318337

17. If equipped, reconnect the equalizer to the lever and reconnect the front cable to the relay lever using new clips and washers.

18. Install the wheels and lower the vehicle to the floor.

19. Make the final parking brake cable adjustment at the parking brake lever.

20. Install the rear console.

Parking Brake Shoes

ADJUSTMENT

NOTE: Adjust the parking brake shoes before adjusting the parking brake lever and cable.

1. Remove the parking brake lever console.

2. Raise and safely support rear of the vehicle. Block the front wheels.

3. Loosen the adjusting nut on the parking brake cable equalizer.

4. Remove the adjusting hole plug from the rear brake backing plate.

5. Use a brake adjusting tool or a small prybar to turn the adjuster wheel downward until the rotor cannot be rotated by hand.

6. Turn the adjuster wheel upward 7 or 8 notches until the rotor can be turned by hand. Make sure the rotor doesn't drag when turned.

7. Inspect the parking brake shoes for excess wear. The minimum lining thickness is 0.039 in. (1.0mm).

8. Make sure the return springs and areas of metal–to–metal contact are lubricated with brake grease. No lubricant or cleaner must contact the brake shoe friction surfaces.

9. Adjust the parking brake cable so the parking brakes are fully engaged when the lever is pulled upward 6 or 7 notches. Tighten the equalizer nut to 4.5 ft. lbs. (6 Nm).

Brake Hydraulic System

BLEEDING

1. Adjust the rear brake shoes if equipped.

2. Set the parking brake and start the engine.

——————— **WARNING** ———————
The vacuum booster will be damaged if the bleeding operation is performed with the engine off.

NOTE: On vehicles equipped with 4 wheel anti-lock brakes, remove the 40A main fuse located at the relay and the fuse block before bleeding air out of the system.

3. Remove the master cylinder reservoir cap and fill the reservoir with brake fluid. Keep the reservoir at least half full during the bleeding operation.

4. If the master cylinder is replaced or overhauled, first bleed the air from the master cylinder and then from each caliper or wheel cylinder. Bleed the master cylinder as follows:

a. Disconnect the rear wheel brake line from the master cylinder.

b. Have an assistant depress the brake pedal slowly once and hold it depressed.

c. Seal the delivery port of the master cylinder where the line was disconnected with a finger, then release the brake pedal slowly.

d. Release the finger from the delivery port after the brake pedal returns completely.

e. Repeat Steps B through D until the brake fluid (not air) comes out of the delivery port during Step B.

NOTE: Do not let the fluid level in the reservoir drop below the halfway mark.

f. Reconnect the brake line to the master cylinder.

g. Have an assistant depress the brake pedal slowly once and hold it depressed.

h. Loosen the rear wheel brake line at the master cylinder.

i. Retighten the brake line, then release the brake pedal slowly.

j. Repeat Steps G through I until no air comes out from the port when the brake line is loosened.

k. Bleed the air from the front wheel brake line connection by repeating Steps A through J.

5. If equipped with a rear wheel ABS proportioning valve, this valve must be bled before the calipers.

6. Bleed the air from each wheel in the following order: Right rear wheel, Left rear wheel, Right front caliper and Left front caliper. Bleed the air as follows:

   a. Place the proper size box wrench over the bleeder screw.

   b. Cover the bleeder screw with a transparent tube and submerge the free end of the tube in a transparent container containing brake fluid.

   c. Have an assistant pump the brake pedal slowly 3 times, then hold it depressed.

   d. Remove the air along with the brake fluid by loosening the bleeder screw.

   e. Retighten the bleeder screw, then release the brake pedal slowly.

   f. Repeat Steps C through E until the air is completely removed. It may be necessary to repeat the bleeding procedure 10 or more times for front wheels and 15 or more times for rear wheels.

   g. Go to the next wheel in sequence after each wheel is bled.

7. Depress the brake pedal several times after the air has been removed from all wheel cylinders and calipers. If the pedal feels spongy, the entire bleeding procedure must be repeated.

8. After the bleeding operation is completed on each individual wheel, check the level of brake fluid in the reservoir and refill up to the **MAX** level, if necessary.

9. Install the master cylinder reservoir cap and shut off the engine.

Brake System Depressurizing

NOTE: A TECH-1, or equivalent scan tool and the appropriate cartridges are needed to depressurize and bleed the four-wheel anti-lock brake system. The TECH-1's MOTOR REHOME function sets the modulator pistons to their uppermost position, allowing the modulator to be properly bled.

1. Diagnose, repair, and clear any ABS diagnostic trouble codes before bleeding the system.

2. Connect the TECH-1 tool to the vehicle. Select the F5 MOTOR REHOME function to set the hydraulic modulator for bleeding.

3. Clean any dirt from the master cylinder reservoir. Be careful not to let any brake fluid or cleaner contact or enter any of the ABS system's electrical connectors.

4. Refill and cap the master cylinder reservoir.

5. Prime the hydraulic modulator unit:

   a. Attach a clear plastic bleeder tube to the modulator's rear bleeder valve. Route the tube into a clear container partially filled with brake fluid.

   b. Loosen the rear bleeder valve ½–¾ turn.

   c. Have an assistant slowly depress the brake pedal and hold it until fluid begins to flow out the bleeder. Close the bleeder and release the brake pedal.

   d. Repeat steps B and C until the fluid is free of air bubbles. Don't allow the reservoir to run dry.

   e. Repeat the priming process for the modulator's front bleeder valve.

WARNING

After the hydraulic modulator unit has been primed (depressurized), it must be bled after the brake calipers are bled. Bleeding the modulator after the calipers

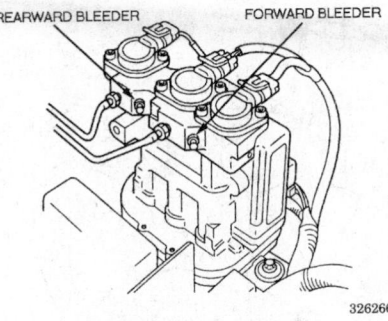

REARWARD BLEEDER   FORWARD BLEEDER

326266

Hydraulic modulator unit

will ensure that any air remaining in the brake system will be purged.

6. Refill and cap the master cylinder reservoir.

7. Bleed the brake system.

Wheel Speed Sensor

REMOVAL AND INSTALLATION

1. Make sure the ignition is turned **OFF**. Remove the key.

2. Raise and safely support the vehicle.

3. Unbolt the speed sensor cable brackets.

4. Uncouple the speed sensor connector from the ABS wiring harness.

5. Unbolt the speed sensor from the steering knuckle and remove it.

To install:

6. Inspect the speed sensor. Replace the sensor if it is damaged or shorted:

   a. Clean any dirt or corrosion from the speed sensor probe.

   b. Check the speed sensor for a short circuit. Bend the cable while checking for continuity.

   c. Check the sensor ring for damaged or chipped teeth.

7. Install the speed sensor to the steering knuckle.

8. Install the speed sensor brackets. Be careful not to bend or twist the cable during installation.

9. Couple the speed sensor and ABS harness connectors. Verify that the white line on the cable insulation is not twisted.

10. Tighten the speed sensor bolts to the following specifications:

   a. Speed sensor bolt: 8 ft. lbs. (11 Nm)

   b. Lower cable bracket: 18 ft. lbs. (24 Nm)

   c. Upper cable bracket: 4 ft. lbs. (6 Nm)

11. Lower the vehicle.

12. Turn the ignition to the **ON** position. Verify that both the ABS or ANTILOCK and ABS ACTIVE (if equipped) indicators come on. Start the engine and verify that all ABS indicator lights turn off.

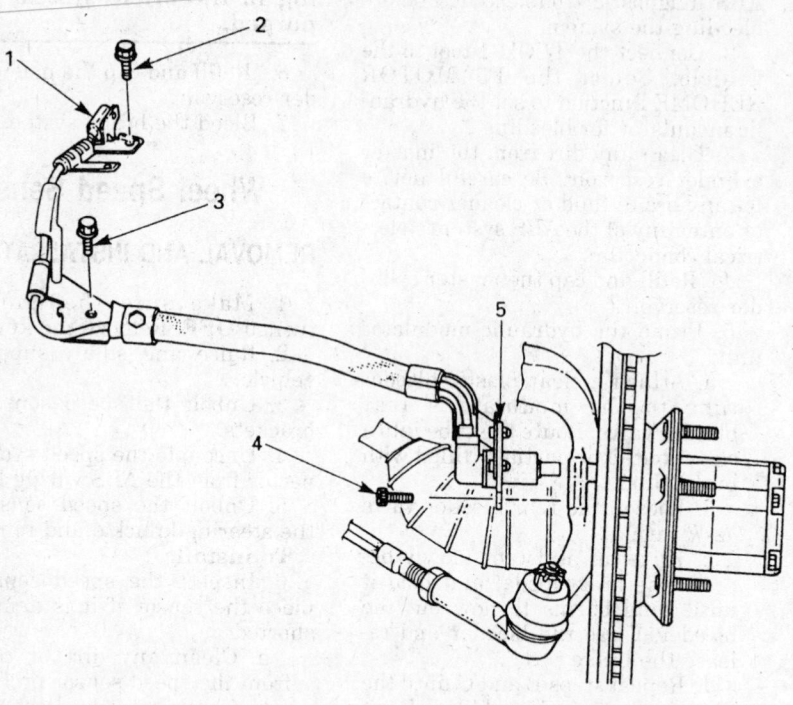

1. Speed sensor connector
2. Sensor cable bolt (Upper side)
3. Sensor cable bolt (Lower side)
4. Sensor bolt
5. Speed sensor

330195

Front wheel speed sensor

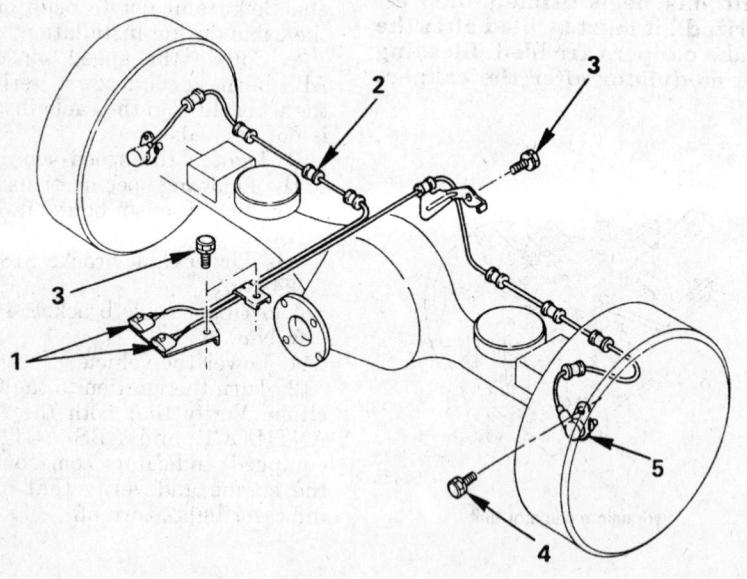

1. Speed sensor connector
2. Clip (11 pieces)
3. Sensor cable bolt
4. Sensor bolt
5. Speed sensor

330196

Rear wheel speed sensors — 4WD models

FRONT SUSPENSION

Shock Absorber

REMOVAL AND INSTALLATION

1. Raise and support the vehicle safely.
2. Remove the front wheels.
3. Support the lower control arm with a floor jack.
4. Hold the shaft of the shock absorber with a wrench to keep it from turning and remove the upper mounting nut, retainer, and rubber grommet.

NOTE: On some vehicles, it may be necessary to remove the bump stops to gain access to the lower mounting bolt.

5. Unbolt the shock absorber from the lower control arm. Remove the shock absorber from the vehicle.

To install:

6. Install the lower retainer and rubber grommet onto the upper shaft of the shock absorber. Then, fully extend the shock and install the upper shaft into the mounting hole in the frame bracket.

7. Install the upper rubber grommet, retainer and attaching nut onto the shock absorber upper shaft. Only hand–tighten the mounting nut at this time.

8. Install the shock absorber to the lower control arm bracket and install the mounting bolt and nut. Torque the mounting bolt to 60–61 ft. lbs. (82–84 Nm).

9. Tighten the upper mounting nut to 14–15 ft. lbs. (19–20 Nm).

10. Install the bump stop if removed, and tighten the bolts to 30 ft. lbs. (41 Nm).

11. Install the front wheels and lower the vehicle.

Torsion Bar

REMOVAL AND INSTALLATION

1. Raise and safely support the vehicle.
2. Remove the front wheels.
3. Mark the location of the height adjustment bolt and remove it from the height control arm.
4. Mark the location of the torsion bar shaft and the height control arm. Remove the height control arm from the torsion bar and the frame crossmember.

5. Mark the location of the torsion bar on the lower control arm bracket.

6. Unbolt the torsion bar bracket, and remove the torsion bar from the lower control arm. The torsion bars are marked with a **L** or **R** for identification.

To install:

7. Inspect the torsion bar and height control arm components for signs of damage. If the rubber torsion bar seat is damaged, replace it.

8. Apply a generous amount of grease to the serrated ends of the torsion bar.

9. Use a floor jack to raise the lower control arm to help hold the rubber seat in contact with the torsion bar bracket while the torsion bar is being installed.

10. Align the matchmark and insert the front end of the torsion bar into the control arm.

11. Align the matchmark and install the height control arm in position so the end is reaching the adjusting bolt. Be sure to grease the part of the height control arm that fits into the chassis.

12. Turn the adjusting bolt so that the matchmark aligns with the mark made on the height control arm.

13. Tighten the torsion bar control arm bracket bolts to 86 ft. lbs. (116 Nm).

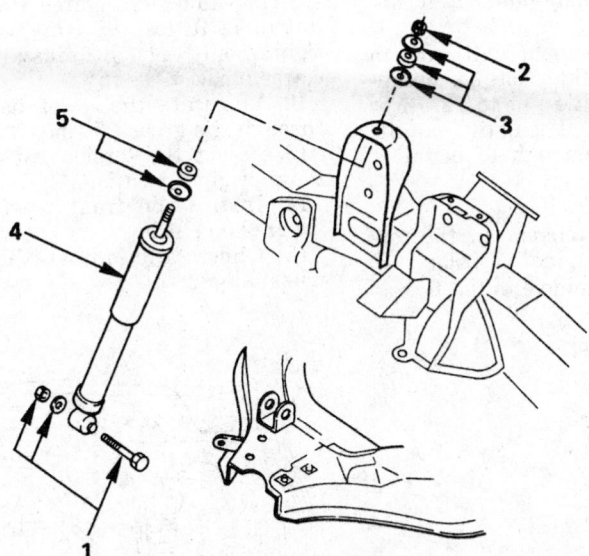

1. Bolt, nut and washer
2. Nut
3. Rubber bushing and washer
4. Shock absober
5. Rubber bushing and washer

312643

Front shock absorber and mounting components

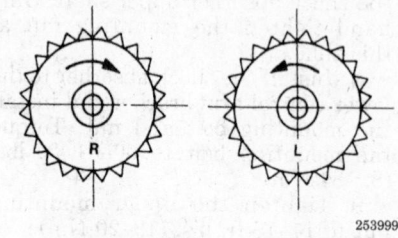

Matchmark the height adjustment bolt and collar

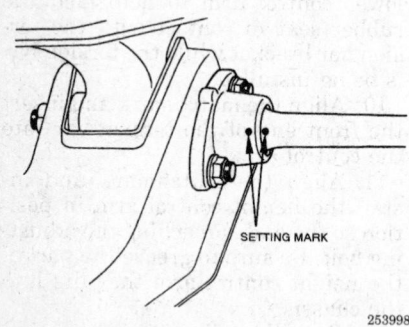

Matchmark the torsion bar to the lower control arm

14. Install the front wheels.
15. Lower the vehicle and check the vehicle height.
16. Check and adjust the front wheel alignment.

Upper Ball Joints

REMOVAL AND INSTALLATION

1. Raise and safely support the vehicle.
2. Remove the front wheels.

Left and right torsion bar identification

3. Mark the position of the torsion bar adjuster. Loosen the adjuster to relieve the torsion bar tension.
4. Support the lower control arm with a floor jack.
5. Remove the upper ball joint castle nut.
6. Using a ball joint separator tool, separate the upper ball joint from the steering knuckle.
7. Remove the upper ball joint from the control arm.
To install:
8. Install the ball joint onto the control arm. Tighten the upper ball joint bolts:
   • Four–bolt–style ball joint: 21–25 ft. lbs. (29–35 Nm)
   • 1993–94 vehicles with three–bolt–style ball joints: 51 ft. lbs. (69 Nm).
   • 1995–97 vehicles with three–bolt–style ball joints: 42 ft. lbs. (57 Nm).
9. Install the upper ball joint to the steering knuckle. Torque the upper ball joint castle nut to 72–73 ft. lbs. (96–98 Nm). Then, tighten the castle nut only enough to install a new cotter pin.
10. Install the front wheels.
11. Adjust the tension on the torsion bar to the original position.
12. Lower the vehicle to the floor.

Lower Ball Joints

REMOVAL AND INSTALLATION

2WD

1. Raise and safely support the vehicle under the frame.
2. Remove the front wheels.
3. Disconnect the outer tie rod from the steering knuckle.
4. Mark the position of the torsion bar adjuster and release the torsion bar tension.
5. Remove the lower ball joint cotter pin and castellated nut.
6. Using J-29107 or equivalent, disconnect the ball joint from the knuckle.
7. Remove the lower ball joint mounting bolts from the lower control arm.
8. Remove the ball joint.
To install:
9. Install the lower ball joint on the lower control arm and torque the bolts to 45–56 ft. lbs. (62–75 Nm) for Amigo and Pick-up models, 68–83 ft. lbs. (93–113 Nm) for Rodeo and Passport models.
10. Install the ball joint stud into the steering knuckle and install the castellated nut. Torque the nut to 87–101 ft. lbs. (117–137 Nm), with just enough additional torque to align the cotter pin hole with a castellation on the nut. Install a new cotter pin.
11. Connect the tie rod end to the steering knuckle. Tighten the castle nut to 72 ft. lbs. (98 Nm), and then tighten only enough to install a new cotter pin.
12. Lubricate the lower ball joint through the grease fitting.
13. Adjust the torsion bar tension to its original position.
14. Install the front wheels and lower the vehicle.
15. Check and adjust the front wheel alignment.

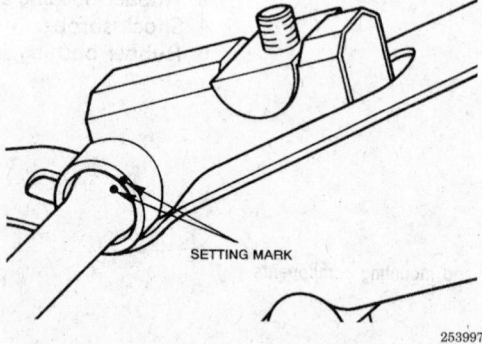

Matchmark the torsion bar and the height control arm

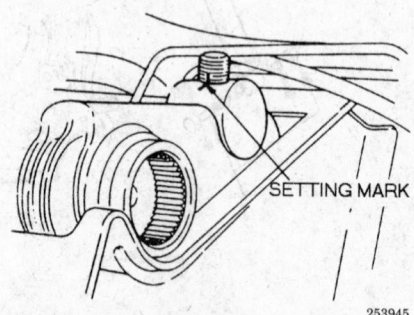

Mark the torsion bar adjusting bolt before loosening

4WD

1. Shift the transfer case into **2H** and make sure the front axle isn't engaged.
2. Raise and safely support the vehicle under the frame.
3. Remove the front wheels.
4. Disconnect the outer tie rod from the steering knuckle.
5. Unbolt and remove the hub cover. Note the positions of the shims.
6. Remove the snapring to release the axle shaft from the hub.
7. Mark the position of the torsion bar adjuster and release the torsion bar tension.
8. Remove the upper and lower ball joint cotter pins and castle nuts.
9. Using tool No. J-29107 or an equivalent ball joint removal tool, disconnect the ball joints from the knuckle.
10. Remove the knuckle/hub assembly from the axle shaft.
11. Remove the lower ball joint mounting bolts from the lower control arm.
12. Remove the ball joint.

To install:

13. Install the lower ball joint on the lower control arm and torque the bolts to 69–83 ft. lbs. (93–113 Nm).
14. Install the ball joint studs into the steering knuckle and install the castellated nut. Torque the lower nut to 87–101 ft. lbs. (117–137 Nm), torque the upper nut to 72 ft. lbs. (98 Nm). Then, tighten the nuts just enough to align the cotter pin hole with a castellation on the nut. Install a new cotter pin.
15. Connect the tie rod end to the steering knuckle. Tighten the castle nut to 72 ft. lbs. (98 Nm), and then tighten only enough to install a new cotter pin.
16. Reinstall the shims and the snapring on the axle shaft.
17. Lubricate the lower ball joint through the grease fitting.
18. Adjust the torsion bar tension to the original position.
19. Install the front wheels and lower the vehicle.
20. Check and adjust the front wheel alignment.

Upper Control Arms

REMOVAL AND INSTALLATION

1. Raise and safely support the vehicle.
2. Remove the front wheels.

3. Mark the position of the torsion bar adjuster. Loosen the adjuster to release the torsion bar tension.
4. Remove the cotter pin and castle nut from the upper control arm ball joint. Use a ball joint separator tool to separate the upper control arm from the steering knuckle.

NOTE: Do not allow the steering knuckle to hang by the flexible brake line. Wire the steering knuckle up to the frame temporarily.

5. Unbolt the upper control arm pivot shaft and remove the upper control arm from the frame bracket. Be sure to note the position and number of shims used for adjusting the camber and caster angles when removing the upper control arm. The shims must be replaced in their original position.

NOTE: It is helpful to wrap each shim pack in electrical tape. This will keep the appropriate shims together and organized. It is optional to remove the tape when the shims are reinstalled.

6. To remove the pivot shaft and bushings from the upper control arm assembly, remove the bushing nuts from the pivot shaft by loosening them alternately, then remove the pivot shaft and bushings using J-29755 or equivalent and a press.

To install:

7. Press new bushings into the control arm and install the pivot shaft.
8. Install the pivot shaft and bushing in the upper control arm. Only hand–tighten the pivot nuts at this time.

NOTE: Tighten the thinner shim pack's nut first for improved shaft-to-frame clamping force and torque retention.

9. Install the control arm to the frame. Replace the shims in the same

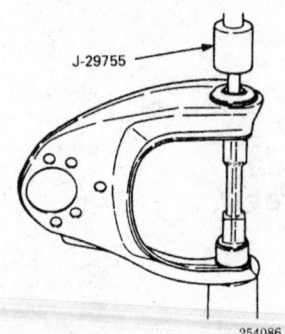

J-29755

254086

Removing the upper control arm bushings — Amigo and Pick-up

position that they were removed from. Torque the bolts to 80 ft. lbs. (108 Nm) on the 2WD Pick-up, and 112 ft. lbs. (152 Nm) on the 4WD and Amigo models.
10. Install the ball joint stud through the steering knuckle. Install the castellated nut and tighten it to 80 ft. lbs. (108 Nm) on the 2WD Pick-up, and 72 ft. lbs. (98 Nm) on the 4WD and Amigo models. Then, tighten the castle nut just enough to install a new cotter pin. Remove the wire used to support the knuckle.
11. Adjust the torsion bar to its original position.
12. Install the wheel assembly and lower the vehicle.
13. Torque the nuts on the upper control arm pivot shaft to 80 ft. lbs. (108 Nm) after the vehicle is on the floor.
14. Check and adjust the front wheel alignment.

Lower Control Arms

REMOVAL AND INSTALLATION

1. Raise and safely support the vehicle.
2. Remove the front wheels.
3. Mark the position of the torsion bar adjuster and release the torsion bar tension.
4. Disconnect the stabilizer bar from the lower control arm.
5. Unbolt the torsion bar bracket from the lower control arm. Then, remove the torsion bar from the lower control arm.
6. Disconnect the shock absorber from the lower control arm.
7. Unbolt the lower ball joint from the lower control arm.

NOTE: On four–wheel drive vehicles, the knuckle/hub should be supported with a jack so that the front axle shafts do not bear the weight of the suspension components.

8. Remove the retaining nuts and use a soft metal drift to drive out the bolts holding the lower control arm to the chassis.
9. Remove the lower control arm from the vehicle.

To install:

10. Mount the lower control arm to the frame. Drive the bolts into position carefully. Do not damage the bolt threads. Do not tighten the bolts until the vehicle is on the ground.
11. Install the lower ball joint to the lower control arm. Tighten the retaining bolts 76 ft. lbs. (103 Nm) on

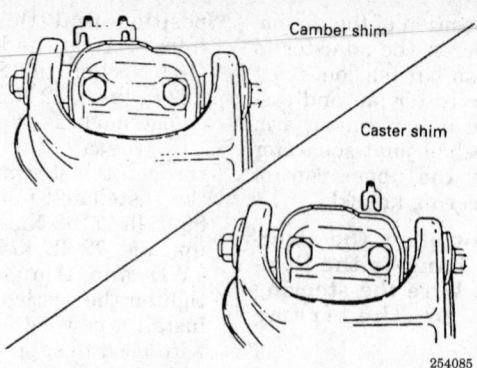

Camber shim

Caster shim

254085

Camber and caster shims

the 4WD and Amigo, and 51 ft. lbs. (69 Nm) on the 2WD vehicles.

12. Install the stabilizer bar to the lower control arm. Tighten the linkage bolt to 18 ft. lbs. (24 Nm).

13. Assemble the lower ball joint to the steering knuckle. Torque the nut to 94 ft. lbs. (127 Nm) on the 4WD and Amigo, and 109 ft. lbs. (147 Nm) on the 2WD vehicles. Then, tighten the nut only enough to install a new cotter pin.

14. Adjust the torsion bar to the original position.

15. Install the front wheels and lower the vehicle.

16. With the weight of the vehicle on the suspension:

a. On 4WD and Amigo models tighten the rear pivot bolt to 145 ft. lbs. (196 Nm), and the front pivot bolt to 116 ft. lbs. (157 Nm)

b. On 2WD Pick-up models tighten the bolt to 93 ft. lbs. (126 Nm).

Sway Bar

REMOVAL AND INSTALLATION

1. Raise and safely support the vehicle.

2. Remove the front wheels.

3. If equipped, remove the radiator skid plate.

4. Unbolt the sway bar linkages from the lower control arms.

5. Unbolt the frame bushing brackets and remove the stabilizer bar.

To install:

6. Position the sway bar and install the bracket bolts and end links. Torque the bracket bolts to 21 ft. lbs. (28 Nm) on the 4WD and Amigo models and 14 ft. lbs. (19 Nm) on the 2WD Pick-up. Torque the end-link nuts to 8 ft. lbs. (10 Nm) on the 4WD and Amigo models and 18 ft. lbs. (25 Nm) on the 2WD Pick-up.

7. Install the radiator skid plate and tighten its mounting bolts to 27 ft. lbs. (37 Nm).

8. Install the front wheels and lower the vehicle.

Front Wheel Bearings

ADJUSTMENT

1. Raise and safely support the vehicle.

2. Remove the front wheels.

3. If equipped, unbolt and remove the hub dust cap.

4. If equipped, remove the locking hub assembly.

5. Use a hub nut wrench to loosen and remove the spindle nut.

6. Pull the hub from the spindle.

7. Hold the outer wheel bearing with your hand to prevent it from falling out of the hub. Remove the hub and rotor assembly from the spindle.

8. Thoroughly clean, inspect and repack the bearings with wheel bearing grease. Replace the bearings if needed. Use a brass drift to drive the bearing races out of the hub.

9. Clean the flange surface of the hub, thread holes, surface of the lockwasher, and the splines of the axle shaft.

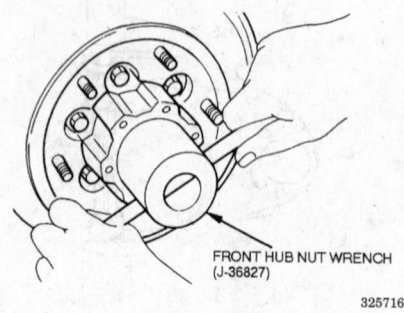

FRONT HUB NUT WRENCH (J-36827)

325716

Hub nut wrench

10. Use bearing drivers to install new inner and outer bearing races and oil seals.

11. Pack the bearings and the hub cavity with wheel bearing grease.

12. Install the outer bearing and spindle nut into the hub. Tighten the hub nut to 22 ft. lbs. (29 Nm) to seat the bearings and then fully loosen the nut. Use a spring scale connected to the wheel stud bolt at 90 degrees to measure bearing preload. Then, re-tighten the hub nut until the spring gauge measures a bearing preload of 4.4–5.5 lbs. (2.0–2.5 kg) for a new bearing and grease seal, or 2.6–4.0 lbs. (1.2–1.8 kg) for a used bearing and a new grease seal.

13. Install the lockwasher on the spindle nut. The larger diameter, tapered side of the lockwasher faces out, and the holes in the lockwasher must align with the holes in the spindle nut. If the holes don't align, reverse the lockwasher, or tighten the spindle just enough to bring them into alignment.

14. If equipped, install the locking hub assembly or the hub dust cap.

15. Install the front wheels.

16. Lower the vehicle.

REMOVAL AND INSTALLATION

2WD Vehicles

1. Raise and safely support the vehicle. Remove the front wheels.

2. Remove the brake caliper without disconnecting the line. Support the caliper aside.

3. Remove the brake caliper mounting bracket.

4. Remove the dust cover, cotter pin and locknut from the rotor.

5. Place a hand over the outer wheel bearing to prevent the bearing from falling on the floor and remove the rotor and hub from the spindle.

6. Matchmark the hub and rotor so it can be assembled in the same position, then remove the bolts attaching the hub to the rotor. Pull off the rotor.

7. Thoroughly clean, inspect and repack the bearings with wheel bearing grease. Replace the bearings if needed. Use a brass drift to drive the bearing races out of the hub.

To install:

8. Use a suitably-sized bearing driver to drive new bearing races into the hub.

9. Install the rotor on the hub and torque the bolts to 76 ft. lbs. (103 Nm).

10. Install the inner bearing and the grease seal in the hub if removed

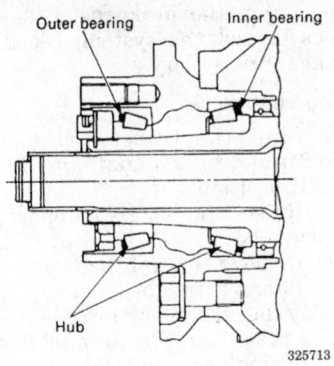

Outer and inner wheel bearings

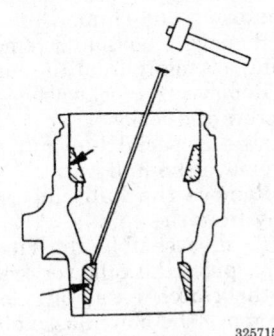

Use a brass drift to remove the inner or outer races

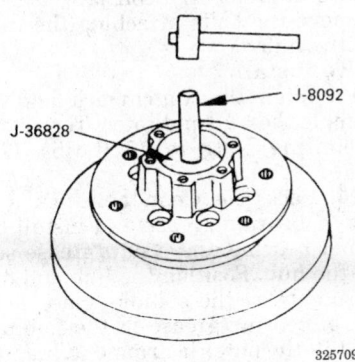

Outer bearing race driver

and position the hub and rotor assembly on the spindle.

11. Install the outer bearing and spindle nut in the hub. Torque the hub nut to 22 ft. lbs. (29 Nm) to seat the bearings and then loosen the nut. Adjust the wheel bearing.

12. Install the caliper mounting bracket and torque the bolts to 115 ft. lbs. (156 Nm).

13. Install the brake caliper.

14. Install the front wheels and lower the vehicle.

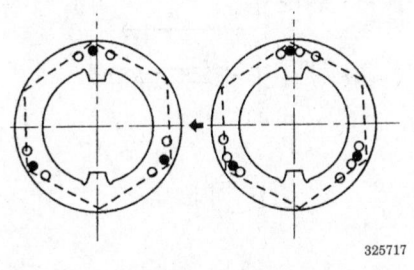

Lockwasher alignment

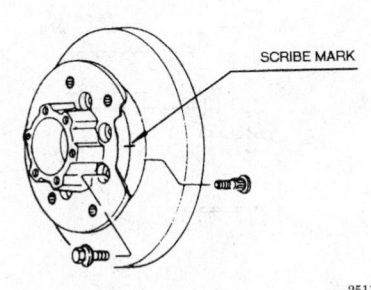

Matchmark the hub and rotor for reassembly

15. Pump the brake pedal several times to test the system. Bleed the brakes if necessary.

4WD

With Automatic Hubs

1. Shift the transfer case shift lever into **2H** and move the vehicle forward and rearward about 3 ft. (0.91m).

2. Raise and safely support the vehicle. Remove the front wheels.

3. Remove the 4WD hub cap attaching bolts and the cap.

4. Remove the brake caliper and support the caliper on a wire; do not disconnect the brake hose. Remove the caliper support bracket from the steering knuckle.

5. Using snapring pliers, remove the snapring and shims from the hub assembly.

6. Remove the drive clutch assembly, inner cam and lockwasher.

7. Using a hub nut wrench, loosen the hub nut and pull the hub from the spindle.

8. Remove the inner and outer bearings from the hub. Clean and inspect the bearings for pitting and scoring and replace if needed.

9. If necessary, use a brass drift to remove the bearing races from the hub.

NOTE: Always replace the bearing and race as a matched set if replacement is needed.

10. If equipped with ABS, unbolt and remove the ABS sensor ring.

11. If removing the disc from the hub, scribe matchmarks, then remove the bolts attaching the hub to the rotor. Separate the disc from the hub.

To install:

12. Use a suitably–sized bearing driver to install new bearing races into the hub.

13. If the hub and rotor were separated, install the hub to the rotor and torque the bolts to 76 ft. lbs. (103 Nm).

14. If equipped, install the ABS sensor ring and tighten the bolts to 13 ft. lbs. (18 Nm).

15. Pack the bearings with wheel bearing grease and install the inner bearing and a new grease seal in the hub. Position the hub and disc assembly on the spindle, place 1 to 1½ oz. of bearing grease in the hub and install the outer bearing and hub nut on the spindle.

16. When installing the hub nut, perform the following procedures:

a. Torque the hub nut to 22 ft. lbs. (30 Nm) and loosen the nut to seat the bearings.

b. Use a spring gauge connected to the stud bolt at 90° to measure preload.

c. Retorque the hub nut until the spring gauge measures a bearing preload of 4.4–5.5 lbs. (2.0–2.5 kg) for a new bearing and new oil seal, or 2.6–4.0 lbs. (1.2–1.8 kg) for a used bearing and new oil seal.

17. Install the lockwasher with the larger diameter of the tapered bore facing out. If the bolt holes in the lock plate do not align with the corresponding holes in the nut, reverse the lock plate. If the bolt holes are still out of alignment, turn in the nut just enough to obtain alignment and install the lockscrew tightly so the head is lower than the surface of the washer.

18. Clean the flange surface of the hub, thread holes, surface of the lockwasher and the splines of the axle shaft.

19. Install the inner cam by aligning the keyway of the inner cam with the groove of the knuckle. If the cam is difficult to install, use the special tool or equivalent and a plastic hammer to lightly tap the inner cam into place.

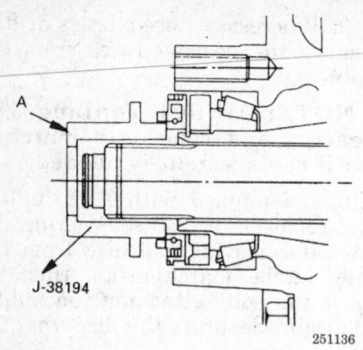

251136

A special tool should be used to install the inner cam

20. Do the following steps to select the proper shim:

a. Lower the vehicle and support the lower control arm with a block of wood and a floor jack to place the axle in the normal ride position.

b. Install the special adjusting tools onto the hub until it comes in contact with the lockwasher.

c. Pull the axle out as far as possible, and using a feeler gauge, measure the clearance "t" between the hub and the snapring groove on the axle shaft.

d. If the clearance is larger than the snapring groove, the selected shims must be installed so that clearance "t" is 0–0.039 in. (0–0.1mm). Shims come in thicknesses of 0.2, 0.3, 0.5, and 1.0mm.

e. Remove the special tool J-36836 or equivalent.

21. Install the drive clutch assembly.

22. Install the shims selected above by hand on the axle and use the following steps to install a new snapring.

a. Install special tool J-36835-2 or equivalent onto the axle.

b. Install the snapring on the tool.

c. Install tool driver J-36835-1 or equivalent.

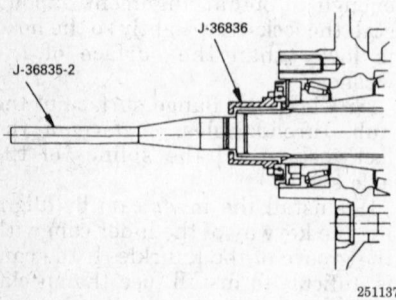

251137

Install special tools to measure shim thickness

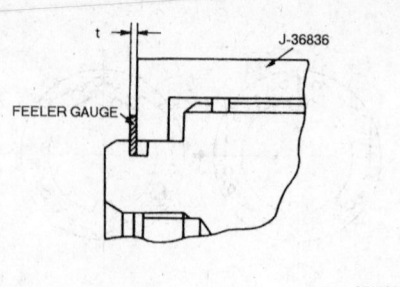

251138

Measure clearance "t" between the special tool and the snapring groove

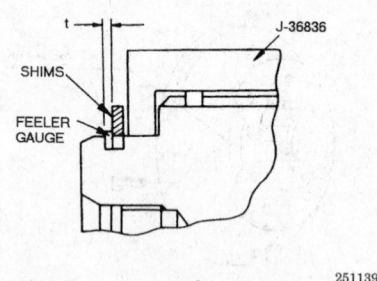

251139

If clearance "t" is larger than snapring groove, selected shims must be used

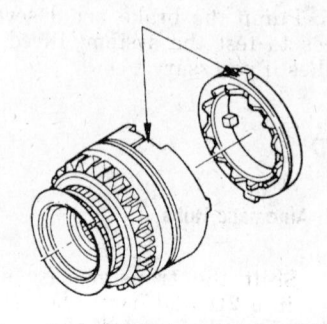

251140

Drive clutch assembly installation — 4WD models with automatic hubs

d. Pull out the axle shaft by pulling tool J-36835-2 or equivalent. Install the snapring to the axle by pushing on tool J-36835-1.

e. Remove tool driver J-36835-2 or equivalent from the axle and check the fit of the snapring.

23. Install the housing assembly and cap. Torque the cap bolts to 43 ft. lbs. (58 Nm).

24. Install the remaining components.

25. Install the front wheels.

26. Lower the vehicle to the floor.

27. Pump the brake pedal several times to test the system. Bleed the brakes if necessary.

With Manual Hubs

1. Shift the transfer case lever into **2H** and the manual hub into the **FREE** position.

2. Raise and safely support the vehicle.

3. Remove the front wheels.

4. Remove the brake caliper assembly and mounting bracket.

5. Loosen the 6 bolts and remove the housing assembly. Be careful not to lose the detent ball and spring.

6. Use snapring pliers and remove the snapring and shims.

7. Remove 6 bolts and remove the housing assembly from the hub.

8. Remove the lockwasher retaining screw and lockwasher.

9. Use a special hub nut wrench and remove the hub nut.

10. Remove the hub and rotor assembly from the spindle.

11. To disassembly the clutch assembly, push the follower knob and turn the clutch assembly clockwise and remove the clutch assembly. Remove the retaining spring by turning it counterclockwise.

12. Matchmark the hub and rotor. Place the rotor in a soft jaw vise and remove the bolts attaching the rotor to the hub.

To install:

13. Align the matchmark and assemble the hub and rotor. Torque the mounting bolts to 76 ft. lbs. (103 Nm).

14. Pack the clean bearings with wheel bearing grease and install the inner bearing and a new grease seal in the hub. Position the hub and disc assembly on the spindle, place 1–1½ oz. of bearing grease in the hub and install the outer bearing and hub nut on the spindle.

15. When installing the hub nut, perform the following procedures:

a. Torque the hub nut to 22 ft. lbs. (29 Nm) and loosen the nut to seat the bearings.

b. Use a spring gauge connected to the stud bolt at 90° to measure preload.

c. Retorque the hub nut until the spring gauge measures a bearing preload of 4.4–5.5 lbs. (2.0–2.5 kg) for a new bearing and oil seal, or 2.6–4.0 lbs. (1.2–1.8 kg) for a used bearing and new oil seal.

16. Install the lockwasher with the larger diameter of the tapered bore to the outer side of the vehicle. If the bolt holes in the lock plate do not align with the corresponding holes in the nut, reverse the lock plate. If the

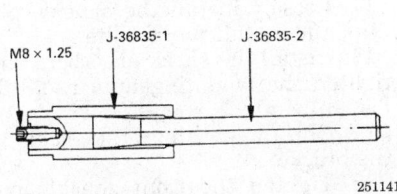

Special tools for snapring installation

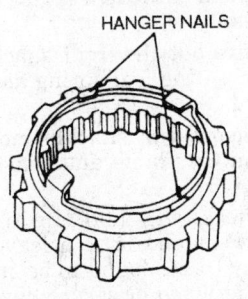

Hook the retainer spring onto the upper portion of the hanger nails

REAR SUSPENSION

Shock Absorber

REMOVAL AND INSTALLATION

1. Raise and safely support the vehicle.
2. Remove the shock absorber-to-lower mount nut, washers, and bushings.
3. Remove the shock absorber-to-chassis nut, washers, and bushings.
4. Remove the shock absorber.
   To install:
5. Install the shock absorber to the upper mount. Do not tighten the nut until the vehicle is on the floor.
6. Install the shock absorber on the lower mount. Do not tighten the nut until the vehicle is on the floor.
7. Lower the vehicle to the floor and torque the mounting nuts to 25–32 ft. lbs. (34–43 Nm.)

bolt holes are still out of alignment, turn in the nut just enough to obtain alignment and install the lockscrew tightly so its head is lower than the surface of the washer.

17. Apply grease to the spacer, ring, snapring and the splined part of the inner assembly.
18. Assemble the inner assembly and the clutch assembly.

19. Install the body assembly on the hub. Torque the bolts to 9 ft. lbs. (12 Nm).

20. Install the selective shim and a new snapring. The clearance between the free wheeling hub body and the snapring should be 0–0.01 in. (0.25mm); shims are available in selective sizes.
21. Install the housing assembly. Torque the bolts to 43 ft. lbs. (58 Nm).
22. Install the front wheels.
23. Lower the vehicle to the floor.
24. Pump the brake pedal several times to test the system. Bleed the brakes if necessary.

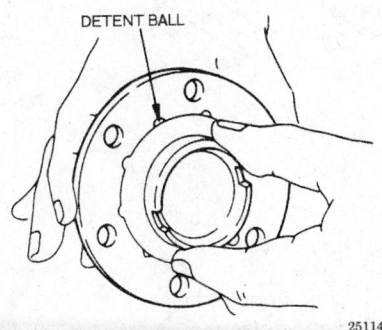

Align the detent ball with the groove cut in the cover — 4WD models with manual hubs

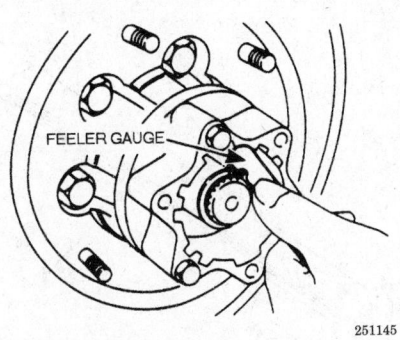

Use a feeler gauge to measure the snapring clearance

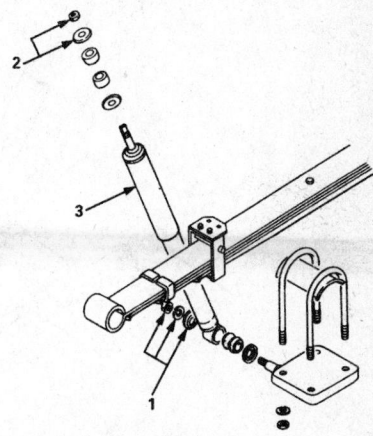

1 Lower nut and washer
2 Upper nut and washer
3 Shock absorber

Rear shock absorber and related components

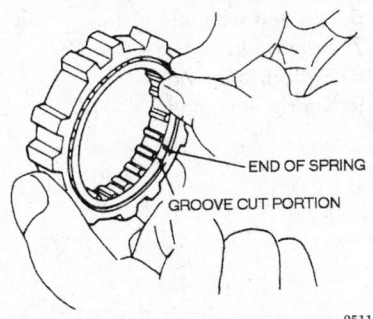

Align the end of the retainer spring with the end of the clutch spring groove's cut portion

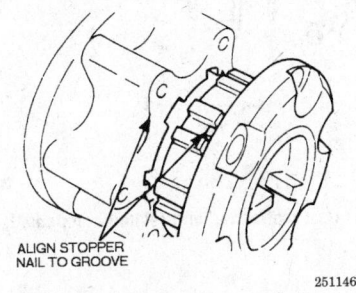

Install the clutch assembly into the housing while aligning the stopper nails with the grooves of the body — 4WD models with manual hubs

Leaf Spring

REMOVAL AND INSTALLATION

1. Raise and safely support the vehicle so the leaf springs are hanging freely.

2. Remove the rear wheels.

3. Support the rear axle housing to remove the weight of the axle housing from the springs.

4. Remove the rear shock absorbers.

5. Remove the parking brake cable clips.

6. Remove the nuts from the U-bolts holding the springs to the axle housing.

7. Remove the front and rear shackle pin nuts.

8. Drive out the rear shackle pin by using a hammer and drift. Lower the rear end of the leaf spring assembly to the floor.

9. Drive out the front shackle pin and remove the leaf spring assembly rearward.

10. Remove the shackle mounting bolt from the frame and remove the shackle.

11. Check the leaf springs for cracks, wear and broken leaves. Replace any leaves found to be cracked, broken, fatigued or seriously worn.

12. Check the shackles for bending and the pins for wear.

13. Check the U-bolts for distortion or other damage.

To install:

14. Mount the shackle to the frame and hand-tighten the mounting nut.

15. Align the front end of the leaf spring assembly with the front bracket and install the shackle pin.

16. Align the rear end of the leaf spring assembly with the shackle and install the shackle pin.

17. Loosely install the shackle pin nuts and install the U-bolts.

18. Install the shock absorbers and tighten the mounting nuts to 28 ft. lbs. (39 Nm).

19. Clip the parking brake cable to the bracket.

20. Tighten the front shackle pin nut to 58–87 ft. lbs. (78–118 Nm).

21. Tighten the rear shackle pin nuts to 101–123 ft. lbs. (137–167 Nm). Tighten the U-bolt nuts to 44–53 ft. lbs. (60–71 Nm).

22. Install the wheels. Remove the axle housing support and lower the vehicle so the weight is on the leaf springs.

23. With the vehicle weight resting on the springs, tighten the rear shackle mounting nut to 101–123 ft. lbs. (137–167 Nm).

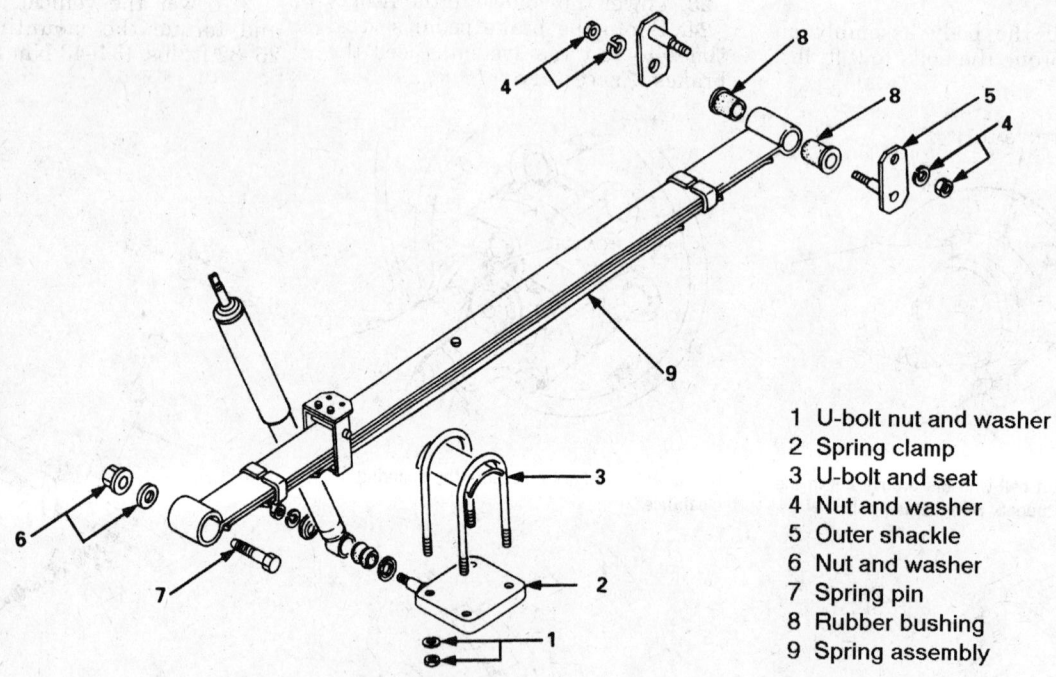

1 U-bolt nut and washer
2 Spring clamp
3 U-bolt and seat
4 Nut and washer
5 Outer shackle
6 Nut and washer
7 Spring pin
8 Rubber bushing
9 Spring assembly

311610

Rear leaf spring and related components

FIRING ORDERS

NOTE: To avoid confusion, always replace spark plug wires one at a time.

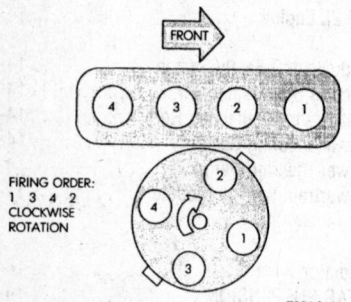

FIRING ORDER:
1 3 4 2
CLOCKWISE
ROTATION

7921dg01

2.5L (VIN P) Engine
Engine Firing Order: 1–3–4–2
Distributor Rotation: Clockwise

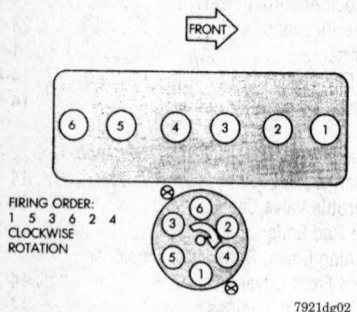

FIRING ORDER:
1 5 3 6 2 4
CLOCKWISE
ROTATION

7921dg02

4.0L (VIN S) Engines
Engine Firing Order: 1–5–3–6–2–4
Distributor Rotation: Clockwise

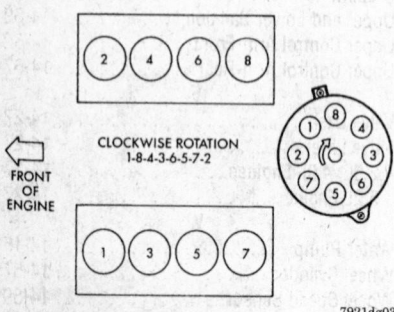

CLOCKWISE ROTATION
1-8-4-3-6-5-7-2

FRONT OF ENGINE

7921dg03

5.2L (VIN Y) Engines
Engine Firing Order: 1–8–4–3–6–5–7–2
Distributor Rotation: Clockwise

ENGINE ELECTRICAL

Distributor

REMOVAL AND INSTALLATION

Except 5.2L (VIN Y) Engines

1. Disconnect the negative battery cable.
2. Unfasten the distributor cap retaining screws. Remove the distributor cap with the coil and spark plug wires attached and position them aside.
3. Disconnect the distributor primary wiring connector.
4. Remove the No. 1 spark plug.
5. Hold a finger over the spark plug hole and rotate the engine until compression pressure is felt. Slowly continue to rotate the engine until the timing index on the vibration damper pulley aligns the Top Dead Center (TDC) mark (0 degree) on the timing degree scale.

NOTE: Always rotate the engine in the direction of normal rotation. Do not turn the engine backward to align the timing marks.

6. Scribe a mark on the distributor housing in line with the tip of the rotor.
7. Note the position of the rotor and distributor housing in relation to the surrounding engine components as reference points for installing the distributor.
8. Remove the distributor hold-down bolt and clamp.
9. Lift the distributor straight up and out of the engine.
To install:
10. If the crankshaft has been rotated after distributor removal, cylinder No. 1 must be returned to TDC of the compression stroke. Refer to the appropriate steps earlier in the procedure if necessary.
11. Check the position of the oil pump slot. On 2.5L (VIN P) engines, it should be slightly before the 10 o'clock position. On 4.0L (VIN S) engines, it should be slightly before the 11 o'clock position. If necessary, rotate the oil pump gear to position using a screwdriver.

NOTE: Factory replacement distributors are equipped with a plastic alignment pin installed. If this pin is in place, the next step may be skipped.

12. Remove the camshaft position sensor from the distributor. Four different alignment holes are provided on the plastic ring. Rotate the distributor shaft, then insert a 3/16 in. pin punch through the proper alignment hole and mating access hole in the distributor. This prevents the shaft from rotating.
13. Clean the distributor mounting area of the cylinder block.
14. Install a new distributor mounting gasket.

NOTE: There is a fork on the distributor housing where the housing seats against the engine block. The slot in the fork aligns with the distributor hold-down bolt hole in the engine block. The distributor is correctly installed when the rotor is correctly positioned. This is the slot in the fork aligned with the hold-down bolt hole in the cylinder block. Because of the fork in the distributor housing initial ignition timing is not adjustable (the distributor cannot be rotated).

15. Position the distributor shaft in the cylinder block, while holding the centerline of the base slot in the 1 o'clock position.
16. When the distributor is fully seated in the block, the centerline of the base slot should be aligned to the clamp bolt mounting hole on the engine. On the 2.5L (VIN P) engines, the rotor should point slightly past the 3 o'clock position. On 4.0L (VIN S) engines it should be pointed at the 5 o'clock position.

NOTE: It may be necessary to move the rotor and shaft (slightly) to engage the distributor shaft with the slot in the oil pump shaft. the same may have to be done to engage the distributor gear with the camshaft gear.

5.2L (VIN Y) Engines

1. Disconnect the negative battery cable.
2. Remove the air cleaner tube at the throttle body.
3. Label and remove the high tension wires from the distributor cap.
4. Remove the distributor cap. Note the position of the rotor and distributor. Scribe a mark on the base of

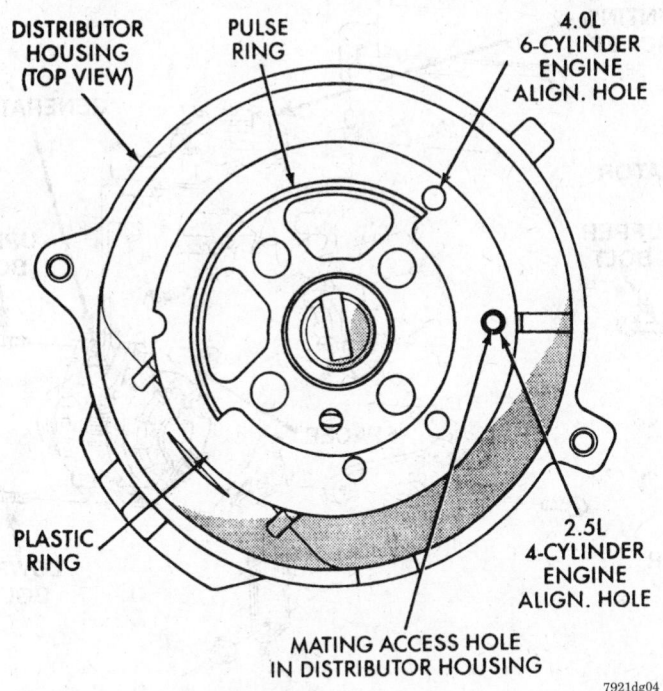

DISTRIBUTOR HOUSING (TOP VIEW)
PULSE RING
4.0L 6-CYLINDER ENGINE ALIGN. HOLE
PLASTIC RING
2.5L 4-CYLINDER ENGINE ALIGN. HOLE
MATING ACCESS HOLE IN DISTRIBUTOR HOUSING
7921dg04

Distributor pin alignment holes

the distributor and the engine as an installation reference.

5. Turn the engine clockwise, using a socket on the end of the crankshaft damper bolt, until the rotor is pointing to the No. 1 spark plug wire post and the timing mark on the damper aligns with the 0 on the timing scale; No. 1 cylinder is at TDC on the compression stroke.

NOTE: The timing mark is on the edge of the vibration damper, closest to the front engine cover.

6. Unplug the camshaft position sensor wiring connector.
7. Remove the rotor.
8. Remove the bolt for the distributor hold-down clamp.
9. Remove the distributor from the engine.
To install:
10. If the crankshaft has been rotated after distributor removal, cylinder No. 1 must be returned to TDC of the compression stroke. Refer to the appropriate steps earlier in the procedure if necessary.
11. Clean the top of the cylinder block.
12. Lubricate the oil seal on the distributor with engine oil.
13. Install the rotor.
14. Position the distributor into the block and engage the tongue of the

distributor shaft with the slot in the distributor oil pump drive gear. Position the rotor to the No. 1 spark plug terminal position.
15. Install the hold-down clamp and loosely install the bolt.
16. Rotate the distributor housing until the rotor is aligned to the **cyl. No. 1** alignment mark on the camshaft position sensor.
17. Tighten the hold-down clamp bolt to 200 inch lbs. (22 Nm).
18. Connect the camshaft position sensor wiring harness to the main engine harness.
19. Install the distributor cap and tighten the screws.
20. Ensure the spark plug wires are firmly connected to their terminals.
21. Connect the negative battery cable. Start the engine and check for proper operation.

Ignition Timing

ADJUSTMENT

Base ignition timing is not adjustable. The distributor does not have a built-in centrifugal or vacuum assisted advance. Base ignition timing and timing advance are controlled by the Powertrain Control Module

(PCM) which monitors inputs from various sensors to determine and adjust correct ignition timing. On the 5.2L (VIN Y) engines, proper distributor position can be checked as follows:

1. Connect the DRB or equivalent scan tool to the data link connector. It is located in the engine compartment.
2. Gain access to the SET SYNC screen on the scan tool.

— WARNING —
The engine will be running. Be careful not to stand in line with the fan blades or belt. Do not wear loose clothing.

3. Follow the directions on the scan tool screen and start the engine. With the engine running, the words IN RANGE should appear on the screen along with 0°.
4. If a plus or minus is displayed and/or the degree displayed is not zero, loosen the distributor hold-down bolt and rotate the distributor until IN RANGE appears. Continue to rotate until achieving as close to 0° as possible. After adjustment, tighten the clamp bolt.

NOTE: The degree scale on the SET SYNC screen is referring to fuel synchronization only. It is not referring to ignition timing. Do not attempt to adjust ignition timing using this method. Rotating the distributor will have no effect on ignition timing.

Alternator

REMOVAL AND INSTALLATION

1. Disconnect the negative battery cable.
2. Loosen and remove the drive belt.
3. On Cherokee models with 4.0L (VIN S) engines, raise and safely support the vehicle.
4. Remove the nuts from the harness hold-down. Disconnect the electrical harness from the alternator.
5. Remove the mounting bolts and the alternator.

NOTE: On 1997 Wrangler models, a spacer is pressed into the rear mounting ear of the alternator. Carefully pry this spacer slightly rearward to loosen the alternator from the upper mounting bracket. This will allow the alternator to be tilted for removal.

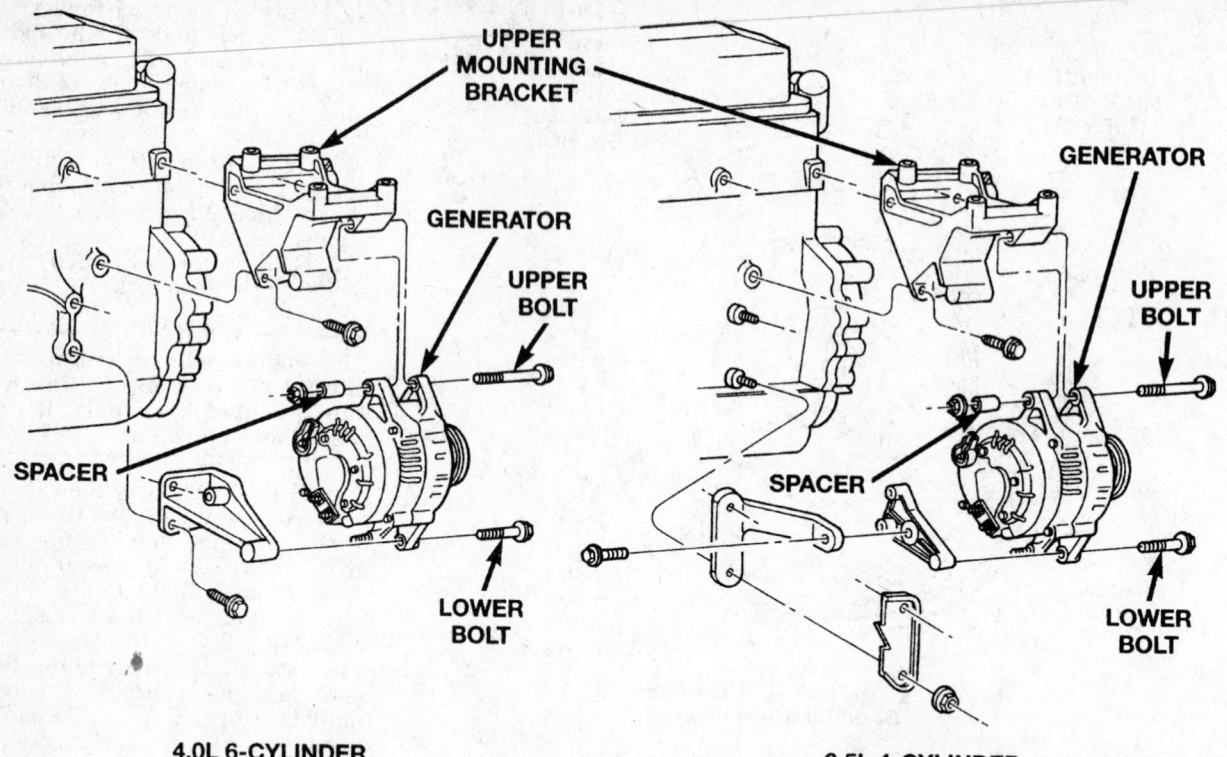

4.0L 6-CYLINDER

2.5L 4-CYLINDER

7921dg05

Alternator mounting — 1997 Wrangler shown

To install:

6. Attach the alternator to the mounting bracket and loosely install the bolts/nuts.

7. Tighten the mounting bolts to 41 ft. lbs. (55 Nm), except on 5.2L (VIN Y) engines. On 5.2L (VIN Y) engines, tighten the bolts to 30 ft. lbs. (41 Nm).

8. Connect the electrical harness to the alternator.

9. Install and tension the drive belt.

──── **WARNING** ────

Never force a belt over a pulley rim using a prybar, as synthetic fiber damage could result. If equipped with a serpentine belt, ensure it is routed correctly.

10. Connect the negative battery cable.

Starter

REMOVAL AND INSTALLATION

Except 5.2L (VIN Y) Engines and 2.5L (VIN P) Cherokee

1. Disconnect the negative battery cable, then the positive cable.

2. Raise and safely support the vehicle.

3. Disconnect the battery cable and solenoid feed wire from the starter solenoid.

4. Remove the starter mounting bolts and remove the starter and any shims from the vehicle.

To install:

5. Install the starter in position with its shims (if applicable).

6. Tighten the mounting bolts on 2.5L (VIN P) engines to 33 ft. lbs. (45 Nm). On 4.0L (VIN S) engines tighten the lower (or front) bolt to 30 ft. lbs. (41 Nm); tighten the upper (or rear) bolt to 40 ft. lbs. (55 Nm).

7. Connect the wire harness to the starter solenoid.

8. Lower the vehicle.

9. Connect the positive, then negative battery cable.

2.5L (VIN P) Cherokee

1. Disconnect the negative battery cable, then the positive cable.

2. Raise and safely support the vehicle.

3. Remove the exhaust clamp from the bracket.

4. On automatic transmissions, remove the nut and bolt from the forward end of the brace rod, then remove the nut from the lower end of the brace rod.

5. Remove the nut, bolt and bracket from the bell housing (manual transmissions).

6. Disconnect the wiring harness from the starter.

7. Remove the starter mounting bolts and remove the starter and any shims from the vehicle.

To install:

8. Install the starter in position with its shims (if applicable).

9. Tighten the mounting bolts to 33 ft. lbs. (45 Nm).

10. Connect the wire harness to the starter solenoid.

11. Tighten the brace bolts to 55 ft. lbs. (74 Nm) and the exhaust clamp bolt to 7 ft. lbs. (10 Nm).

12. Lower the vehicle.

13. Connect the positive, then negative battery cable.

5.2L (VIN Y) Engines

1. Disconnect the negative battery cable, then the positive cable.

2. Raise and safely support the vehicle.

3. Disconnect the wiring harness from the starter.

4. Remove the lower starter mounting bolt and exhaust brace.

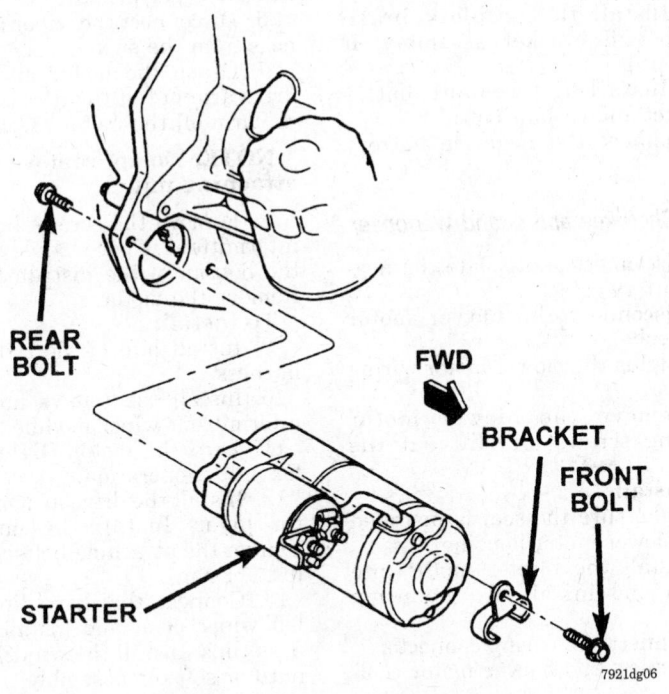

REAR BOLT

FWD

BRACKET

FRONT BOLT

STARTER

7921dg06

Starter mounting — 4.0L (VIN S) engine

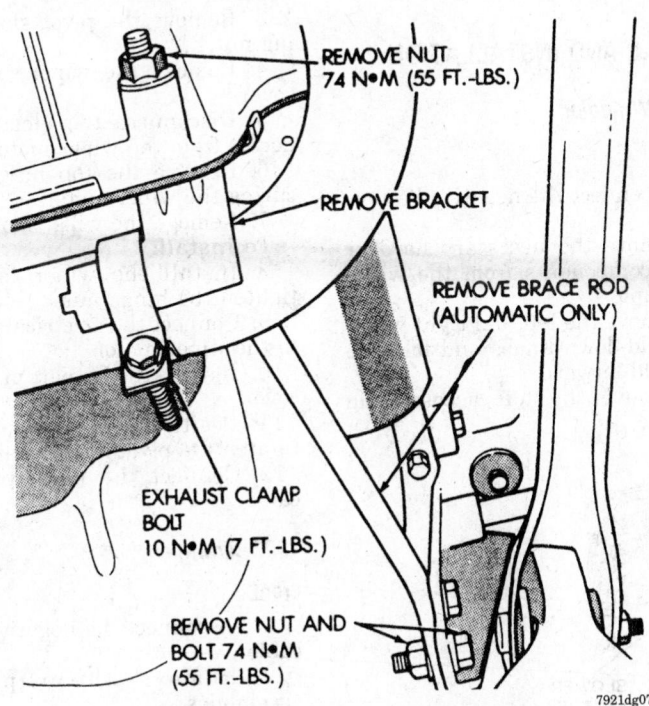

REMOVE NUT
74 N•M (55 FT.-LBS.)

REMOVE BRACKET

REMOVE BRACE ROD
(AUTOMATIC ONLY)

EXHAUST CLAMP
BOLT
10 N•M (7 FT.-LBS.)

REMOVE NUT AND
BOLT 74 N•M
(55 FT.-LBS.)

7921dg07

Exhaust clamp and brace — 2.5L (VIN P) Cherokee

5. Remove the upper starter mounting nut, lock washer and oil cooler line bracket.

6. Move the starter towards the front of the vehicle until the starter gear housing nose clears the bell housing. Now, tilt the nose downwards past the exhaust pipe and remove the starter from the vehicle.

To install:

7. Install the starter in position.

8. Install the mounting nut/bolt with the appropriate bracket, brace and lockwasher. Tighten the mounting hardware to 50 ft. lbs. (60 Nm).

9. Connect the wire harness to the starter solenoid.

10. Lower the vehicle.

11. Connect the positive, then negative battery cable.

CHASSIS ELECTRICAL

Blower Motor

REMOVAL AND INSTALLATION

Wrangler

1993–95 Models

NOTE: It is not necessary to discharge the refrigerant system.

1. Disconnect the negative battery cable.

2. Remove the hose clamps and dash grommet retaining screws.

3. Remove the evaporator housing-to-instrument panel screws and the housing mounting bracket screw.

4. Lower the evaporator housing to gain access to the blower motor attaching screws.

5. Remove the blower motor attaching screws and remove the blower motor. Disconnect the blower motor wiring and remove the blower motor from the vehicle.

To install:

6. Install the blower motor and connect the wiring. Install the blower motor attaching screws.

7. Position the evaporator housing and install the housing-to-instrument panel screws and housing mounting bracket screw.

8. Install the dash grommet retaining screws and hose clamps.

9. Connect the negative battery cable.

1997 Models

1. Disconnect the negative battery cable, then the positive cable.
2. Remove the battery.
3. Remove the Powertrain Control Module (PCM).
4. Unplug the blower motor wiring connector.
5. Remove the three blower motor mounting screws, then remove the assembly from the housing.
6. If necessary, remove the clip retaining the wheel and remove the wheel from the motor.

To install:

7. Install the blower motor and wheel into the housing.
8. Install and tighten the mounting screws.
9. Engage the motor wiring connector, then install the PCM.
10. Install the battery. Connect the positive cable first, then the negative cable.

Cherokee

NOTE: The blower motor is removed from the engine compartment.

1. Disconnect the negative battery cable.
2. If equipped with 4.0L (VIN S) engine, proceed as follows:
   a. Remove the coolant bottle retaining strap and move the bottle aside. Remove the coolant bottle bracket.
   b. If equipped with anti-lock brakes, remove the anti-lock brake pump and bracket and position it aside.
3. Unplug the blower motor wiring connector.
4. Remove the blower motor mounting screws and lift out the motor.

To install:

5. Install the blower motor into position and connect the electrical leads.

6. If equipped with 4.0L (VIN S) engine, proceed as follows:
   a. Install the anti-lock brake pump and bracket assembly, if equipped.
   b. Install the coolant bottle bracket and coolant bottle.
7. Connect the negative battery cable.

Grand Cherokee and Grand Wagoneer

1. Disconnect and isolate the negative battery cable.
2. Disconnect the blower motor cooling tube.
3. Unplug the blower motor wiring connector.
4. Remove the blower motor mounting screws and lift out the motor.

To install:

5. Make sure the seal is installed on the blower motor housing.
6. Install the blower motor into position and install the mounting screws.
7. Connect the wiring connector.
8. Connect the blower motor cooling tube.
9. Connect the negative battery cable.

Wiper Motor

REMOVAL AND INSTALLATION

1993–95 Wrangler

Front

1. Disconnect the negative battery cable.
2. Remove the necessary hard or soft top components from the windshield frame.
3. Remove the left and right windshield hold-down knobs and fold the windshield forward.
4. Remove the left access hole cover.

5. Disconnect the drive link from the left wiper pivot.
6. Disconnect the wiper motor harness from the switch.
7. Grasp the motor and pull the drive arm out of the access hole. Pry the arm off the motor pivot.

NOTE: Do not remove the pivot attaching nut.

8. Remove the screws holding the intermittent wiper module bracket to the bottom of the instrument panel. Remove the motor.

To install:

9. Install and connect the wiring harness.
10. Install the screws holding the intermittent wiper module.
11. Turn the motor **ON** and check for proper operation.
12. Install the drive arm on the motor pivot. Install the motor and tighten the attaching bolts to 96 inch lbs. (11 Nm).
13. Connect the drive link at the left wiper pivot and install the harness clips. Install the windshield and hard or soft top assembly.

Rear

1. Disconnect the negative battery cable.
2. Remove the wiper arm from the wiper motor.
3. Remove the pivot shaft retaining nut.
4. Remove the wiper motor trim cover.
5. Disconnect the electrical connector from the wiper motor.
6. Remove the top hinge nut securing the wiper motor.
7. Remove the wiper motor.

To install:

8. Install the wiper motor and tighten the hinge nut.
9. Connect the electrical connector to the wiper motor.
10. Install the wiper motor trim cover.
11. Position the wiper arm and tighten the pivot shaft retaining nut.
12. Connect the negative battery cable.

1997 Wrangler

Front

1. Disconnect the negative battery cable.
2. Remove the wiper arm assemblies.
3. Open the hood. Pull back each end of the hood seal from the dash panel to the cowl plenum panel pinch weld far enough to remove one screw on each outboard end of the cowl plenum cover/grille panel.

POWERTRAIN CONTROL MODULE

BATTERY TRAY

POWER DISTRIBUTION CENTER

BLOWER MOTOR

FRO

7921dg08

Blower motor — 1997 Wrangler

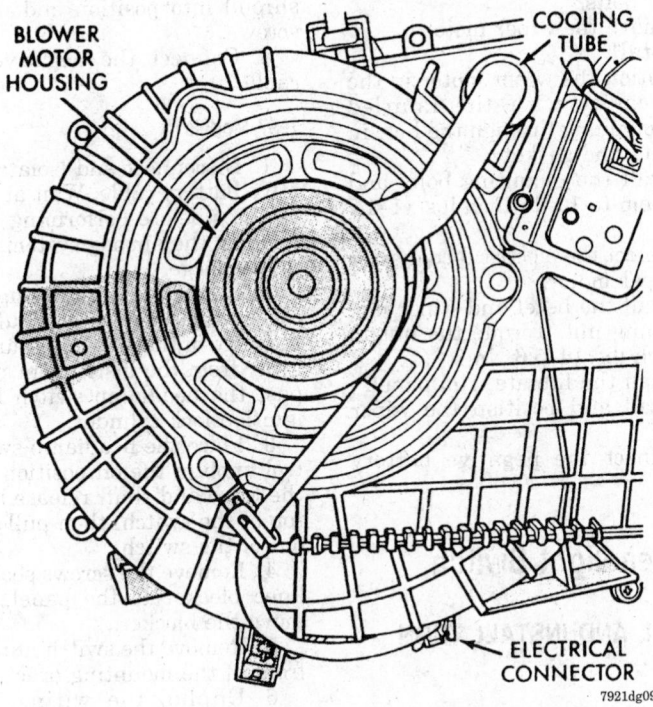

Blower motor on Grand Cherokee

4. Close the hood and remove the one screw in the top center of the cowl plenum cover/grille panel.

5. Remove the four screws securing the cowl plenum cover/grille panel near the base of the windshield.

6. Carefully remove the cowl plenum cover/grille panel from the vehicle, so as not to damage the paint around the pivot openings of the panel.

7. Reach into the cowl plenum and unplug the wiper motor wiring connector.

8. Remove the three wiper linkage cowl mounting bracket screws.

9. Remove the motor and linkage assembly from the cowl plenum as a unit.

10. Release the retainer securing the wiper motor connector to the bracket.

11. Turn the linkage and motor assembly over and remove the nut holding the wiper crank arm to the output shaft.

12. Remove the three screws holding the motor to the linkage assembly bracket and remove the motor.

13. Reverse the removal procedure to install.

Rear

1. Disconnect the negative battery cable.

2. Remove the wiper arm assembly.

3. Remove the motor output shaft nut.

4. Remove the external output shaft bezel and gasket.

5. From inside the liftglass, remove the three screws securing the wiper motor cover.

6. Unplug the wiring connector.

7. Loosen, but do not remove, the right liftglass hinge nut.

8. Pull carefully on the motor until the output shaft clears the hole in the liftglass.

9. Move the motor towards the right side of the vehicle until the slotted hole in the motor mounting bracket clears the grommet under the right liftglass hinge nut.

10. Remove the rear wiper motor from the vehicle.

11. Installation is the reverse of removal.

Cherokee

Front

1. Disconnect the negative battery cable.

2. Remove the wiper arm assemblies.

3. Remove the cowl and trim panel.

4. Disconnect the washer hose.

5. Remove the cowl mounting bracket attaching bolts and the pivot pin attaching screws.

6. Disconnect the wiring harness and remove the assembly.

NOTE: Some motors are protected by a rubber case, care should be used so as not to damage this protective coat.

To install:

7. Install the wiper motor assembly and connect the wiring harness. Take care not to damage the rubber case.

8. Install the pivot pin attaching screws. Install the cowl mounting bracket and washer hose.

9. Install the wiper arm assemblies and test for proper operation. Tighten the wiper motor attaching screws to 35–50 inch lbs. (47–67 Nm).

Rear

1. Remove the wiper arm from the pivot pin by depressing the tab and pulling straight out.

2. Slide the clip along the hose until the clip is off the mounting.

3. Disconnect the washer hose.

4. Remove the pivot pin retaining nut.

5. Remove the external bezel and seal.

6. Remove the liftgate interior trim panel.

7. Disconnect the electrical connector from the wiper motor.

8. Remove the wiper motor mounting screws.

9. Remove the wiper motor.

To install:

10. Position the wiper motor into the liftgate cavity with the pivot pin protruding through the hole in the liftgate.

11. Install the mounting screws.

12. Connect the electrical connector to the wiper motor.

13. Install the pivot pin, seal bezel and attaching nut. Torque the nut to 32 inch lbs. (4 Nm).

14. Lubricate the male end of the bezel with water and connect the washer hose.

15. Install the liftgate panel trim.

16. Install the wiper arm assembly and connect the external washer hose to the bezel.

17. Slide the clip along the hose until it is over the hose mount.

18. Position the arm so the blade is parallel to the window and comes no closer than 5mm to the window seal when operating on a wet surface.

19. Connect the negative battery cable.

Grand Cherokee and Grand Wagoneer

Front

1. Disconnect the negative battery cable.

2. Lift the wiper arms upward, slide the tab up and remove the wiper arms.

3. Remove the cowl grille screws, disconnect the washer hose and remove the grille.

4. Remove the bolts securing the wiper linkage.

5. Turn the linkage over and remove the nut securing the crank arm to the motor.

6. Remove the screws holding the linkage to the wiper motor and remove the motor.

To install:

7. Install the wiper motor and tighten the screws and nut.

8. Install the wiper linkage.

9. Connect the washer hose to the cowl grille and install the grille.

10. Install the wiper arm assemblies.

11. Connect the negative battery cable.

Rear

1. Disconnect the negative battery cable.

2. Lift the wiper arm and insert a ⅛ inch pin into the arm hole.

3. Remove the wiper arm assembly from the pivot pin by depressing the tab and pulling the blade straight out of the arm.

4. Remove the wiper motor retaining nut.

5. Remove the external panel.

6. Remove the 5 screws holding the liftgate interior panel.

7. Remove the panel with a wide flat bladed tool.

8. Disconnect the wiper motor electrical connector.

9. Remove the 2 wiper motor mounting bolts.

10. Remove the wiper motor.

To install:

11. Position the wiper motor in the liftgate cavity with the knurled driver protruding through the hole in the liftgate and gasket.

12. Install the mounting bolts and torque them to 10–15 inch lbs. (1–1.7 Nm).

13. Connect the electrical connector to the wiper motor.

14. Install the bezel and wiper motor retaining nut. Torque the nut to 35–50 inch lbs. (4–5.6 Nm).

15. Install the liftgate trim panel.

16. Install and position the wiper arm.

17. Connect the negative battery cable.

Headlight Switch

REMOVAL AND INSTALLATION

Wrangler

1993–95 Models

1. Disconnect the negative battery cable.

2. Remove the instrument panel shroud retaining screws.

3. Pull the instrument panel shroud outward and upward while applying downward force to the indicator panel and remove the shroud.

4. Remove the screws retaining the switch, pull the switch from the instrument panel cavity and disconnect the electrical connector.

To install:

5. Connect the electrical connectors to the headlight switch and install the retaining screws.

6. Position the instrument panel shroud under the steering column and slide the holding tabs into the shroud notches.

7. Place the instrument panel shroud into position and install the screws.

8. Connect the negative battery cable.

1997 Models

1. Disconnect and isolate the negative battery cable. Wait at least two minutes before performing any work to allow the air bag system capacitor to discharge.

2. Remove the steering column cover attaching screws. Pull the upper edge of the steering column cover away from the instrument panel and past the headlamp switch knob and ignition lock cylinder.

3. Place the headlamp switch control knob in the on position. Depress the knob and shaft release button on top of the switch, then pull the shaft out of the switch.

4. Remove the screws securing the knee blocker to the panel, then remove the blocker.

5. Remove the switch nut from the front of the mounting bracket.

6. Unplug the wiring harness, then remove the switch from the panel.

7. Installation is the reverse of removal.

8. Connect the battery cable.

Cherokee

1. Disconnect the negative battery cable. Disable the air bag system, if equipped.

2. Pull the light switch control knob out as far as it will go.

3. If equipped, remove the instrument panel trim plate.

4. From under the dash depress the headlight switch shaft retainer button and pull the shaft along with the knob from the headlight switch assembly.

5. Remove the headlight switch retaining nut.

6. Disconnect the electrical connector from the switch.

7. Remove the headlight switch from the vehicle.

To install:

8. Connect the electrical connector, install the headlamp switch and tighten the nut.

9. Insert the shaft and knob into the switch.

10. Enable the air bag system and connect the negative battery cable.

Grand Cherokee and Grand Wagoneer

1. Disconnect and isolate the negative battery cable. Wait at least two minutes before performing any work

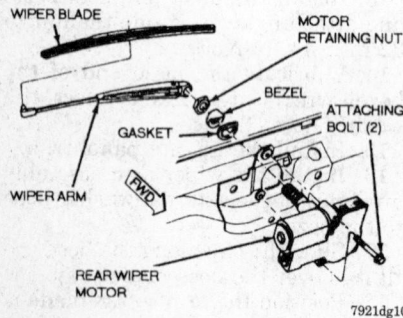

Rear wiper motor — Grand Cherokee and Grand Wagoneer

7921dg10

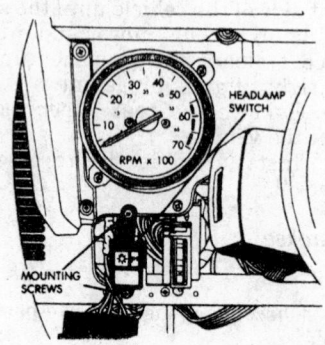

Headlight switch — 1993–95 Wrangler

7921dg11

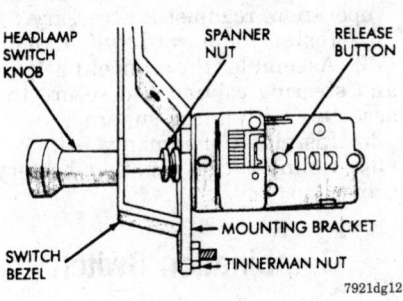

Headlight switch — Cherokee

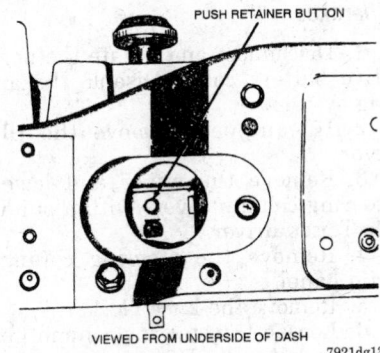

Headlight switch shaft removal

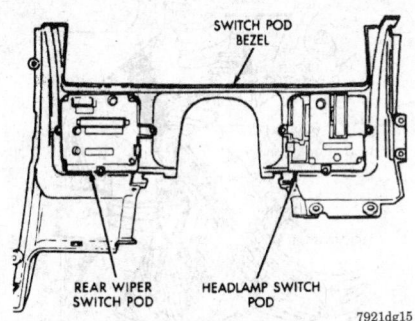

Rear view of the switch pod bezel

Turn Signal Switch

REMOVAL AND INSTALLATION

1993–95 Wrangler and 1993–94 Cherokee

1. Disconnect the negative battery cable.
2. Remove the steering wheel.
3. Remove the lockplate cover by compressing with tool C–4156 or equivalent.
4. Release and discard the steering shaft retaining snapring.
5. Remove the compressor tool.
6. Remove the lockplate, canceling cam, upper bearing preload spring and thrust washer from the steering column.
7. Remove the horn button components from the canceling cam.
8. Remove the hazard warning switch knob.
9. Remove the dimmer switch actuator arm attaching screw.
10. Remove the turn signal switch attaching screws.
11. Remove the lower instrument panel cover trim panel.
12. Remove the lower steering column cover.
13. If equipped with automatic transmission column shift selector, remove the **PRNDL** cable clip.
14. Remove the nuts securing the steering column bracket to the brake sled.
15. Remove the bolts holding the steering column bracket to the column.
16. Loosen the column brace mounting nut at the drivers side kick panel and allow the column to drop.
17. Push the turn signal connector up and out of the steering column connector.
18. Pry up the locking tabs of the steering column connector and remove the connector from the column bracket.
19. Tape the connector flat against the wire harness to prevent it from hanging up during removal.
20. Remove the plastic harness cover by pulling it up and over the weld nuts, then open and slide the cover off the harness.
21. Remove the combination lever by pulling it out straight from the column.
  To install:
22. Position the combination switch into position on the column while guiding the harness. Ensure the wires are laying flat on the bottom of the inside column.

to allow the air bag system capacitor to discharge.

2. Remove the ashtray.
3. Remove the screws holding the center cluster bezel and remove the bezel.
4. Remove the 2 screws holding the dash panel.
5. Gently pry the defroster grille out of the dash panel.
6. Unplug the auto headlamp and sun sensors, if equipped and remove the defroster grille.
7. Remove the screws, through the defroster duct opening, securing the dash panel.

8. Remove the 3 screws above the instrument panel cluster securing the dash panel.
9. Open the glove box and remove the 2 screws holding the dash panel.
10. Pull up on the dash panel, unsnap the end clips and remove the panel.
11. With the left door open, remove the screw from the side of the lower trim panel.
12. Remove the screws securing the steering column covers.
13. Remove the screw from the bottom of the lower trim panel and unsnap it from the instrument panel.
14. Remove the knee blocker.
15. Remove the steering column retaining nuts.
16. Remove the screws holding the bezels.
17. Remove the screws holding the switch pod bezel.
18. Pull the switch pod bezel out far enough to disconnect the electrical connectors and remove the switch pod assemblies.
19. Remove the headlight switch retaining screws and switch.
  To install:
20. Install the switch to the left pod assembly.
21. Connect the electrical connectors and install the switch pod assemblies.
22. Position the steering column and install the retaining nuts.
23. Install the knee blocker.
24. Install the lower trim panel.
25. Install the steering column cover screws.
26. Install the screw securing the left side of the lower trim panel.
27. Position the dash panel and install the screws and defroster grille.
28. Install the center cluster bezel.
29. Install the ashtray.
30. Enable the air bag system and connect the negative battery cable.

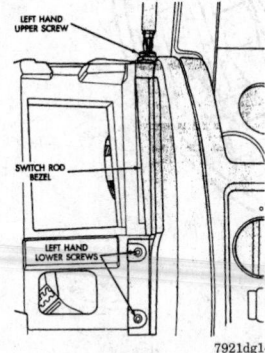

Left switch pod bezel screws

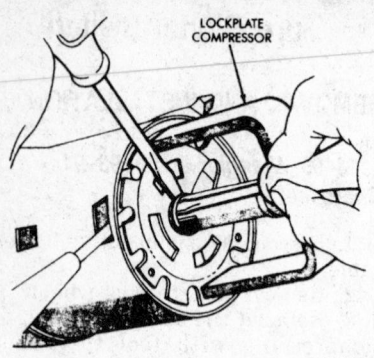

Lockplate removal

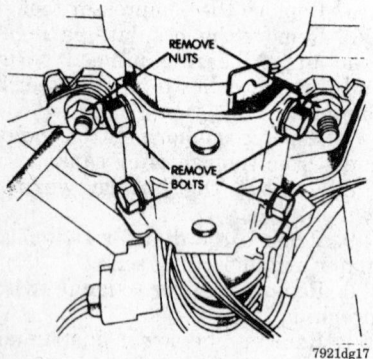

Steering column attaching hardware

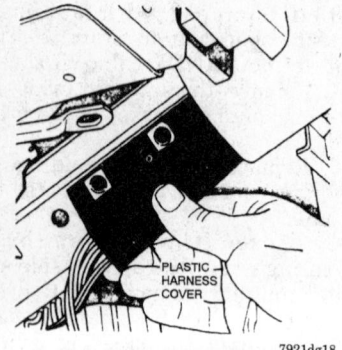

Plastic harness cover

23. Remove the tape and connect the electrical connector.
24. Connect the turn signal connector.
25. Position the steering column and tighten the mounting nuts and bolts.
26. If equipped with a column shift selector, install the **PRNDL** cable clip with the transmission in **N**. Move the selector through all ranges and ensure the indicator is aligned.
27. Install the lower steering column cover.
28. Install the lower instrument panel cover trim panel.

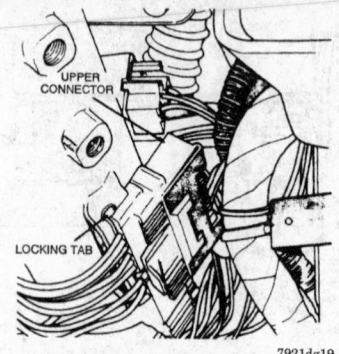

Harness connector

29. Install the turn signal switch attaching screws.
30. Loosely install the dimmer switch actuating screws and adjust the switch as follows:

a. Compress the switch and insert a 3/32 inch diameter drill bit into the adjustment hole.

NOTE: The drill bit will prevent horizontal movement of the switch.

b. Move the switch toward the steering wheel to eliminate rod lash.
c. Tighten the screw.
d. Connect the negative battery cable.

e. Remove the drill bit and test operation, readjust if necessary.
31. Install the hazard switch knob.
32. Assemble the canceling cam and steering column and secure the assembly with a new snapring.
33. Install the steering wheel.
34. Connect the negative battery cable.

Combination Switch

REMOVAL AND INSTALLATION

Except 1993–95 Wrangler and 1993–94 Cherokee

1. Disconnect and isolate the negative battery cable. Disable the air bag system.
2. If equipped, remove the tilt lever.
3. Remove the upper and lower steering column covers with a suitable Torx® driver.
4. Remove the steering column trim panel.
5. Remove the knee blocker.
6. Loosen the steering column upper bracket nuts. Do not remove the nuts.
7. Using Snap-On tamper-proof bit TTXR20B2 or equivalent, remove the multi-function switch screws.

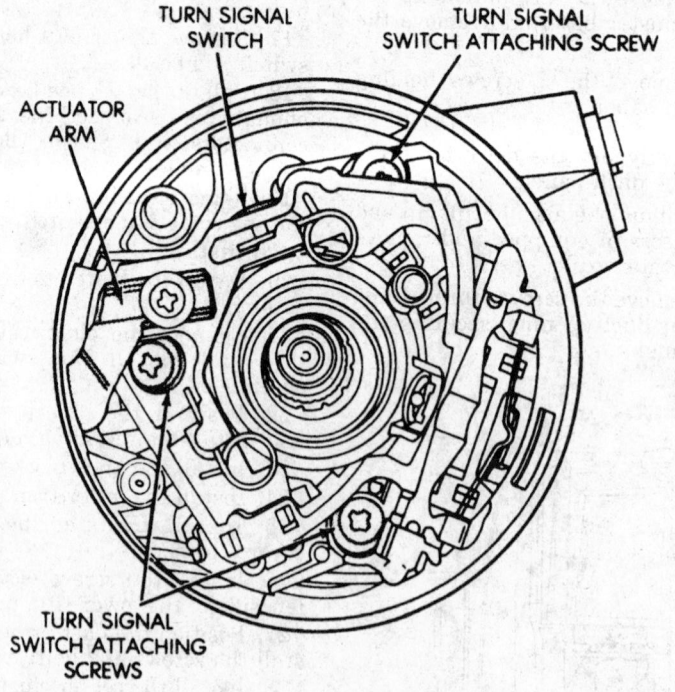

Turn signal switch retaining screws

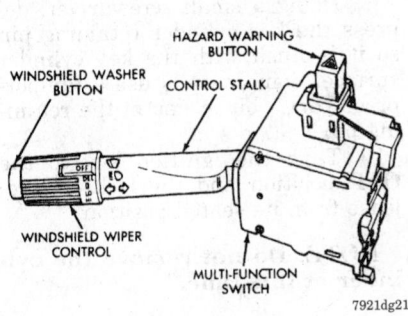

Combination switch

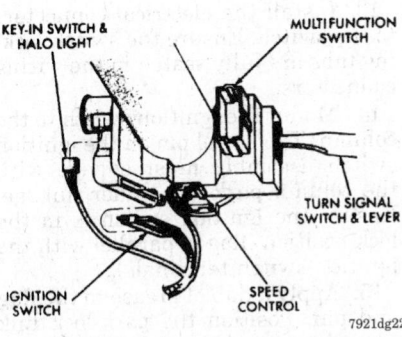

Steering column electrical connections

8. Pull the switch away from the column, loosen the connector screw (which will remain in the connector) and unplug the electrical connector.

To install:

9. Connect the electrical connector to the multi-function switch and tighten the retaining screw.

10. Mount the multi-function switch to the steering column and tighten the screws.

11. Position the steering column and tighten the retaining nuts.

12. Install the knee blocker, lower trim panel and steering column covers.

13. If equipped, install the tilt steering lever.

14. Connect the negative battery cable.

1997 Wrangler

1. Disconnect and isolate the negative battery cable. Wait at least two minutes before performing any work to allow the air bag system capacitor to discharge.

2. Remove the steering column cover.

3. If the vehicle is equipped with tilt steering, move the column to the fully raised position.

4. Turn the ignition switch to the **ON** position.

5. Insert a small screwdriver or pin punch through the access hole in the lower steering column shroud and depress the lock cylinder tumbler.

6. While the tumbler is depressed, pull the lock cylinder out.

7. If equipped with tilt steering, lower the column.

8. Remove the steering column shrouds.

9. Remove the two screws securing the switch water shield and bracket to the top of the steering column.

10. Remove the one screw located below the combination switch lever that secures the switch shield and bracket to the column.

11. Pull the lower mounting tab of the shield bracket far enough to clear the screw boss below the lever.

12. Remove the switch from the column and unplug the switch connectors.

13. Remove the water shield from the switch if necessary.

14. Install the components in the reverse order of removal.

15. Connect the negative battery cable.

Ignition Lock Cylinder

REMOVAL AND INSTALLATION

1993–95 Wrangler and 1993–94 Cherokee

1. Disconnect the negative battery cable.

2. Remove the steering wheel.

3. Remove the lockplate cover by compressing with tool C–4156 or equivalent.

4. Release and discard the steering shaft retaining snapring.

5. Remove the compressor tool.

6. Remove the lockplate, canceling cam, upper bearing preload spring and thrust washer from the steering column.

7. Remove the hazard warning switch knob.

8. Remove the turn signal/wiper/cruise stalk by pulling it straight out. The wiper switch must be in the off position.

9. Unplug the turn signal switch connector.

10. Remove the turn signal switch screws, then guide the switch over the steering column shaft.

11. Turn the ignition switch to the **ON** position.

WARNING

Do not attempt to remove the key warning buzzer switch and contacts separately. If separated, the contacts can detach and drop into the steering column.

---

12. Remove the buzzer switch and contacts with needle-nosed pliers or a paper clip.

13. Insert a thin screwdriver into the slot adjacent to the switch attaching screw boss. Depress the spring latch located at the bottom of the slot, then pull the lock cylinder out.

14. To install, insert the lock cylinder into the column. Push the cylinder inward until it locks in place.

15. Install the remaining components in the reverse order of removal.

Except Wrangler and 1993–94 Cherokee

Please refer to the ignition switch removal and installation procedure to remove the lock cylinder.

1997 Wrangler

1. Disconnect and isolate the negative battery cable. Wait at least two minutes before performing any work to allow the air bag system capacitor to discharge.

2. If the vehicle is equipped with automatic transmission, place the shifter in the PARK position.

3. Turn the ignition switch to the **ON** position.

4. Insert a small screwdriver or pin punch through the access hole in the lower steering column shroud and depress the lock cylinder tumbler.

5. While the tumbler is depressed, pull the lock cylinder out.

6. Installation is the reverse of removal.

Ignition Switch

REMOVAL AND INSTALLATION

1993–95 Wrangler and 1993–94 Cherokee

1. Disconnect the negative battery cable.

2. If equipped, remove the windshield wiper intermittent control module and its bracket.

3. Turn the ignition to the **ACC** position.

4. Remove the headlamp dimmer switch attaching nuts.

5. Lift the switch from the steering column while disengaging the actuator rod.

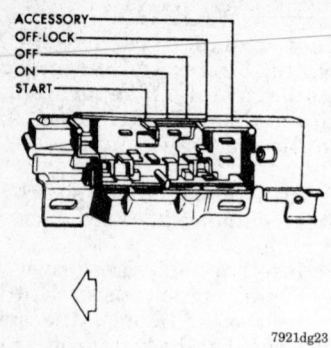

ACCESSORY
OFF-LOCK
OFF
ON
START

7921dg23

Ignition switch — 1993-95 Wrangler and 1993-94 Cherokee

6. Tape the ignition and dimmer switch actuator rods to the steering column to prevent disengagement from the upper position of the steering column.

7. Remove the ignition switch-to-steering column retaining screws.

8. Disengage the ignition switch from the remote actuator rod by lifting straight up and remove the switch from the column.

9. Disconnect the black connector and then the other connector from the switch.

To install:

10. Place the ignition switch in the **ACC** position.

11. Place the slider bar in the ignition switch to the **ACC** detent position.

12. Connect the colored (non-black) connector and then the black connector to the ignition switch.

13. Slip the remote actuator rod into the access hole on the switch.

14. Install the switch to the column, be careful not to move the slider bar out of the detent position.

15. Remove the tape from the rods.

16. Loosely install the ignition switch-to-steering column screws.

17. While holding the key in the **ACC** position, slide the ignition switch up towards the steering wheel (non-tilt steering wheel) or down away from the steering wheel (tilt steering wheel) to remove slack from the switch. Tighten the attaching screws.

NOTE: Do not allow the ignition to move from the ACC position. Because the ignition and dimmer switches share the same 2 mounting screws, 1 screw must be removed from the ignition switch. This must be done after the ignition switch has been adjusted and before the dimmer switch has been installed.

18. Remove 1 screw, but do not remove the stud/nut.

19. Install and adjust the dimmer switch.

20. If equipped, install the intermittent wiper control module and bracket.

21. Connect the negative battery cable.

Except Wrangler and 1993-94 Cherokee

1. Disconnect the negative battery cable.

2. If equipped, remove the tilt lever.

3. Remove the upper and lower steering column covers with a suitable Torx® driver.

4. Using an appropriate tamper-proof Torx® bit, remove the ignition switch screws.

5. Pull the ignition switch away from the column.

6. Release the 2 connector locks on the 7-terminal wiring connector and remove the connector from the ignition switch.

7. Release the connector lock on the key-in-switch and halo light 4-terminal connector and remove the connector from the ignition switch.

8. Insert the key into the ignition lock and ensure it is in the **LOCK** position.

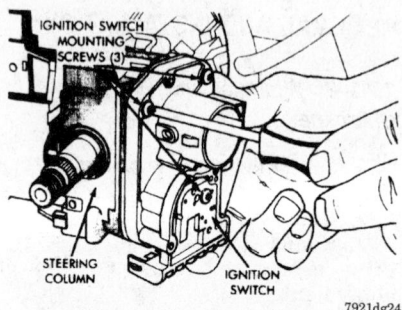

IGNITION SWITCH MOUNTING SCREWS (3)

STEERING COLUMN

IGNITION SWITCH

7921dg24

Ignition switch removal — except Wrangler and 1993-94 Cherokee

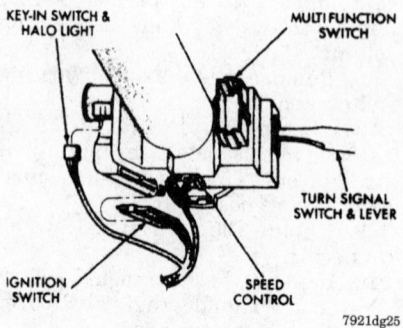

KEY-IN SWITCH & HALO LIGHT

MULTI FUNCTION SWITCH

TURN SIGNAL SWITCH & LEVER

IGNITION SWITCH

SPEED CONTROL

7921dg25

Key-in switch and halo connector — except Wrangler and 1993-94 Cherokee

9. Using a small screwdriver, depress the key cylinder retaining pin so it is flush with the key cylinder surface. Some models use a tamper-proof Torx® bit in lieu of the retaining pin.

10. Turn the ignition key to the **OFF** position and the lock will release from its seated position.

NOTE: Do not remove the cylinder at this time.

11. Turn the key to the **LOCK** position and remove the key.

12. Remove the ignition lock.

To install:

13. Install the electrical connectors to the switch. Ensure the switch locking tabs are fully seated in the wiring connectors.

14. Mount the ignition switch to the column. The dowel pin on the ignition switch assembly must engage with the column park-lock slider linkage. Ensure the ignition switch is in the lock position (flag is parallel with the ignition switch terminals).

15. Apply a dab of grease to the flag and pin. Position the park-lock link and slider to mid-travel. Position the ignition lock against the lock housing face. Ensure the pin is inserted into the park-lock link contour slot and tighten the retaining screw.

16. With the ignition lock and switch in the **LOCK** position, insert the lock into the switch assembly until it bottoms. Install the retaining screw, if equipped.

17. Assemble the column covers.

18. If equipped, install the tilt wheel lever.

19. Connect the negative battery cable.

1997 Wrangler

1. Disconnect and isolate the negative battery cable. Wait at least two minutes before performing any work to allow the air bag system capacitor to discharge.

2. Remove the lock cylinder.

3. Remove the steering column covers.

4. Unplug the electrical connectors at the rear of the ignition switch.

5. Remove the ignition switch mounting screw. A tamper-proof Torx® bit will be necessary.

6. Using a small screwdriver, push on the locking tab and remove the switch from the steering column.

To install:

7. Before installing the switch, rotate the slot in the switch to the **ON** position.

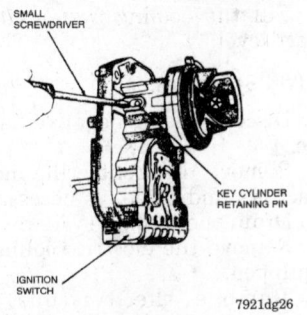

**Key cylinder retaining ring —
except Wrangler and 1993–94
Cherokee**

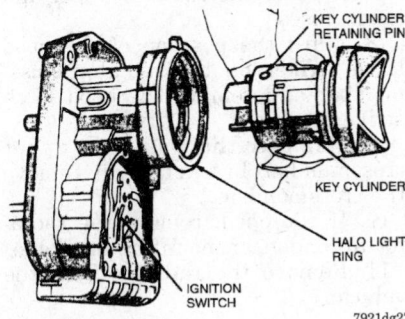

**Key cylinder removal — except Wrangler and
1993–94 Cherokee**

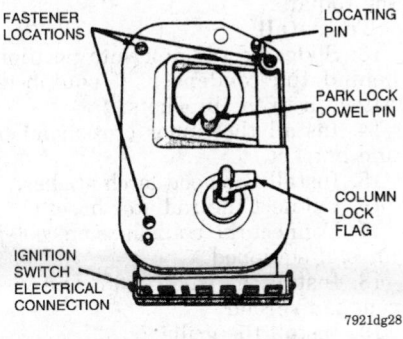

**Column flag position — except Wrangler and
1993–94 Cherokee**

8. Position the switch to the column and install the tamper-proof screw.

9. Engage the two electrical connections. Make sure the locking tabs are fully seated into the connectors.

10. Install the column covers and connect the battery cable.

Park/Neutral Safety Switch

REMOVAL AND INSTALLATION

1. Raise and safely support the vehicle.
2. Unplug the switch wires.
3. Remove the switch from the case.

To install:

4. Move the shift lever to the Park and Neutral positions. Verify that the operating lever fingers are centered in the switch opening.

5. Install a new seal on the switch and install the switch. Tighten to 25 ft. lbs. (34 Nm).

6. Connect the switch and lower the vehicle.

7. Check the transmission fluid level.

Powertrain Control Module

REMOVAL AND INSTALLATION

The Powertrain Control Module (PCM) can be found at the following locations:

• 1993–95 Wrangler: behind the washer fluid reservoir
• 1997 Wrangler: behind the battery
• Cherokee: next to the air cleaner
• Grand Cherokee/Wagoneer: on the right rear side of the cowl panel

1. Disconnect the negative battery cable.

2. Remove any components necessary to access the control module (battery, reservoir tank or air cleaner).

3. On 1997 Wrangler models, remove the plastic caps from over the connectors.

4. Loosen the connector mounting bolt (if equipped).

5. Carefully unplug the connector(s) from the control module.

6. Remove the control module mounting bolts, then remove the module from the vehicle.

7. Installation is the reverse of removal. Be careful when engaging the connectors, so as not to damage the connecting pins.

ENGINE COOLING

Radiator

REMOVAL AND INSTALLATION

Wrangler

1. Disconnect the negative battery cable.
2. Drain the cooling system.
3. Remove the radiator upper and lower hoses.
4. Disconnect the overflow tube from the radiator.
5. If equipped, remove the transmission cooler lines.
6. Remove the fan shroud mounting bolts and pull the fan shroud back to the engine.
7. On some models, the power steering fluid reservoir is mounted on the shroud. Tie the reservoir back to prevent spillage.
8. Remove all attaching bolts and screws that secure the radiator to the radiator support.
9. Remove the condenser to radiator mounting bolts.
10. Pull the radiator out of the vehicle taking care not to damage the radiator fins.

To install:

11. Slide the radiator into position behind the condenser, if equipped and torque the mounting screws 6 ft. lbs. (8 Nm).
12. Close the radiator drain.
13. Install the fan shroud.
14. Connect the transmission cooler lines, if equipped.
15. Connect the hoses to the radiator.
16. Connect the negative battery cable.
17. Fill the cooling system.
18. Connect the reserve bottle hose and install the radiator cap.

Cherokee

2.5L (VIN P) Engines

1. Disconnect the negative battery cable.
2. Remove the front grille mounting screws and grille, as necessary.
3. Drain the cooling system.
4. Remove the radiator upper and lower hoses.
5. If equipped, remove the transmission cooler lines.
6. Remove the fan shroud mounting bolts and pull the fan shroud back to the engine.

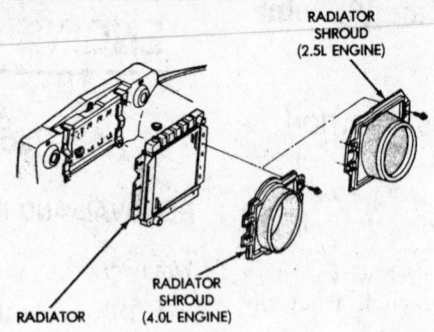

Radiator mounting on Wrangler

7. Remove the alignment dowel E-clip from the lower radiator mounting bracket.

8. Disconnect the overflow tube from the radiator.

9. Remove all attaching bolts and screws that secure the radiator to the radiator support.

10. Remove the condenser to radiator mounting bolts and pull the radiator out of the vehicle.

NOTE: Take care not to damage the radiator fins.

11. Empty the remaining coolant in the radiator.

To install:

12. Slide the radiator into position behind the condenser, if equipped.

13. Align the dowel pin with the bottom mounting bracket and install the E-clip.

14. Tighten the condenser-to-radiator bolts to 55 inch lbs. (6.2 Nm).

15. Install and tighten the radiator mounting bolts.

16. Install the grille.

17. Connect the transmission cooler lines, if equipped.

18. Install the fan shroud.

19. Connect the radiator hoses.

20. Connect the negative battery cable.

21. Fill the cooling system to the correct level.

4.0L (VIN S) Engines

1. Disconnect the negative battery cable.

2. Remove the front grille mounting screws and grille, as necessary.

3. Drain the cooling system.

4. Remove the electric cooling fan if equipped.

5. Remove the (viscous) fan shroud mounting bolts and pull the fan shroud back to the engine.

6. If equipped, remove the transmission cooler lines.

7. Remove the radiator upper and lower hoses.

8. Mark the position of the hood latch striker on the radiator crossmember and remove the hood latch striker.

9. Remove the radiator upper crossmember bracket, then remove the crossmember.

10. If equipped, remove the radiator-to-condenser mounting brackets.

11. Remove the radiator from the vehicle.

NOTE: Take care not to damage the radiator fins.

12. Empty the remaining coolant in the radiator.

To install:

13. Slide the radiator into position behind the condenser, if equipped. Attach it to the brackets.

14. Install the upper crossmember and bracket.

15. Install the hood latch striker.

16. Connect the radiator hoses.

17. Connect the transmission cooler lines, if equipped.

18. Install the electric fan and viscous fan shroud.

19. Install the grille.

20. Connect the negative battery cable.

21. Fill the cooling system to the correct level.

Grand Cherokee and Grand Wagoneer

1. Disconnect the negative battery cable.

2. Open the radiator valve and drain the cooling system.

3. Remove the fan and shroud assembly.

4. If equipped, disconnect the automatic transmission cooling line quick-fit connections.

5. Matchmark the upper radiator crossmember and adjust the crossmember to the left or right.

6. Eight clips are used to retain a rubber seal to the body. Gently pry up the outboard clips (2 per side) un-

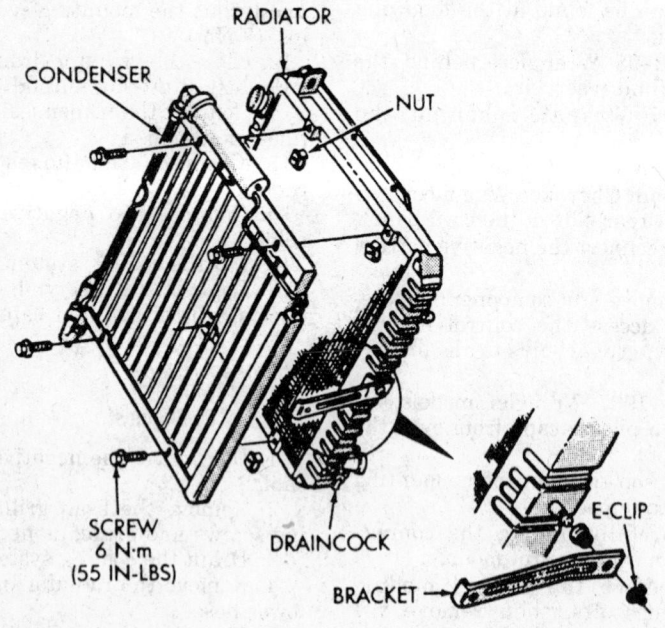

Radiator mounting on 2.5L (VIN P) Cherokee

til the rubber seal can be removed. Do not remove the seals entirely. Fold back the seal on both sides to access the grille opening reinforcement mounting bolts and remove the bolts.

7. Remove the grille.

8. Remove the upper brace bolt from each of the 2 radiator braces.

9. Remove the crossmember-to-radiator mounting nuts.

10. Working through the grille opening, remove the lower bracket bolt securing the lower part of the hood latch or hood latch cable from the crossmember.

11. Lift the crossmember straight up and position it aside.

12. If equipped with A/C, remove the 2 A/C condenser-to-radiator mounting bolts which also retain the side mounted rubber air seals.

13. If not equipped with A/C, remove the bolts retaining the side mounted rubber air seals compressed between the radiator and crossmember.

NOTE: Note the location of the air seals. To prevent overheating, they must be installed in their original position.

14. Disconnect the coolant reservoir/overflow tank hose from the radiator.

15. Disconnect the upper hose from the radiator.

16. Carefully lift the radiator a slight amount and disconnect the lower hose from the radiator.

17. Lift the radiator up and out of the engine compartment, take care not to scrape the fins or disturb the A/C condenser if equipped.

NOTE: If equipped with an auxiliary automatic transmission oil cooler, use caution during radiator removal. The oil cooler lines are routed through a rubber air seal on the left side of the radiator. Do not cut or tear this seal.

To install:

18. Lower the radiator into the vehicle. Guide the alignment dowels into the hoses in the rubber air seals and then through the A/C support brackets, if equipped. Continue to guide the radiator through the rubber grommets located in the lower crossmember.

NOTE: If equipped with A/C, the L-shaped brackets, located on the bottom of the condenser, must be positioned between the bottom of the rubber air seals and top of rubber grommets.

19. Connect the lower radiator hose to the radiator.

20. Connect the upper radiator hose to the radiator.

21. If equipped with A/C, install the bolts condenser-to-radiator mounting bolts.

22. If not equipped with A/C, install the rubber air seal retaining bolts.

23. Connect the reservoir/overflow tank hose to the radiator.

24. If the radiator-to-upper crossmember rubber insulators were removed, install them.

25. Install the hood latch support bracket-to-lower frame crossmember bolt.

26. Install the bolts securing the upper radiator crossmember to the body.

27. Install the radiator-to-upper crossmember nuts.

28. Install a bolt to each upper radiator brace.

29. Install the grille.

30. Position the rubber seal and push down on the clips until seated.

31. If equipped, connect the transmission cooling lines.

32. Install the fan shroud with the fan.

33. Install the fan and shroud.

34. Rotate the fan blades and ensure they do not interfere with the shroud and at least 1.0 inch (25mm) of clearance is allowed. Correct as necessary.

35. Fill the cooling system.

Water Pump

REMOVAL AND INSTALLATION

2.5L (VIN P) and 4.0L (VIN S) Engines

NOTE: Some vehicles use a serpentine drive belt and have a reverse rotating water pump coupled with a viscous fan drive assembly. The components are identified by the words REVERSE stamped on the cover of the viscous drive and on the inner side of the fan. The word REV is also cast into the body of the water pump.

1. Disconnect the negative battery cable.

2. Drain the cooling system.

3. Disconnect the hoses at the pump.

4. Remove the drive belts.

5. Remove the power steering pump bracket.

6. Remove the fan and shroud.

7. If equipped, remove the idler pulley to gain clearance for pump removal.

8. Unbolt and remove the pump.

To install:

9. Clean the mating surfaces thoroughly.

10. Using a new gasket, install the pump and torque the bolts to 22 ft. lbs. (30 Nm).

11. If removed, install the idler pulley.

12. Reconnect the hoses at the pump and install accessory drive belt.

13. Install the power steering pump bracket. Install the fan and shroud.

14. Adjust the belt tension and fill the cooling system to the correct level.

15. Operate the engine with the heater control valve in the **HEAT** position until the thermostat opens to purge air from the system. Check coolant level and fill as required.

5.2L (VIN Y) Engines

1. Disconnect the negative battery cable.

2. Open the radiator valve and drain the cooling system.

3. Remove the cooling fan and shroud as an assembly.

4. Remove the accessory drive belt.

5. Remove the water pump pulley from the hub.

6. Disconnect the hoses from the water pump.

7. Loosen the heater hose coolant return tube mounting bolt and nut and remove the tube. Discard the O-ring.

8. Remove the water pump mounting bolts.

9. Loosen the clamp at the water pump end of the bypass hose. Slip the bypass hose from the water pump while removing the pump from the engine. Discard the gasket.

To install:

10. Clean all gasket mating surfaces.

11. Guide the water pump and new gasket into position while connecting the bypass hose to the pump. Torque the water pump bolts to 30 ft. lbs. (40 Nm).

12. Install the bypass hose clamp.

13. Spin the water pump to ensure the pump impeller does not rub against the timing chain cover.

14. Coat a new O-ring with coolant and install it to the heater hose coolant return tube.

15. Install the coolant return tube to the engine. Ensure the slot in the tube bracket is bottomed to the

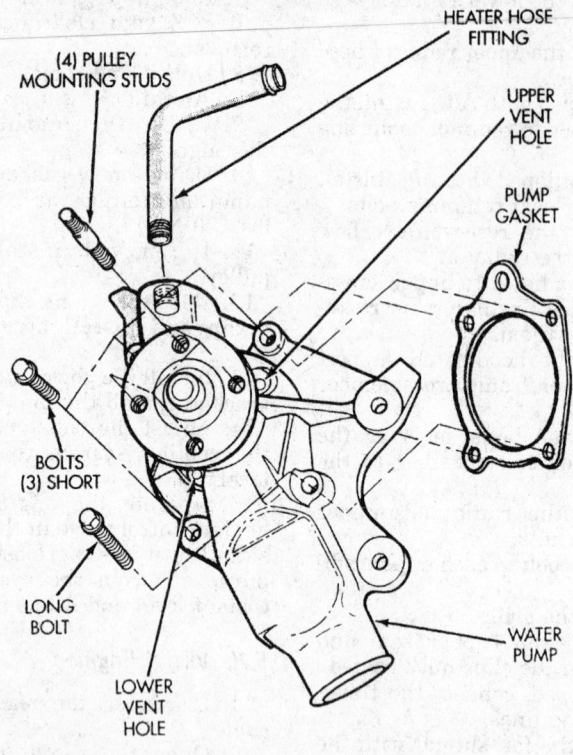

(4) PULLEY
MOUNTING STUDS

HEATER HOSE
FITTING

UPPER
VENT
HOLE

PUMP
GASKET

BOLTS
(3) SHORT

LONG
BOLT

LOWER
VENT
HOLE

WATER
PUMP

7921dgac

Water pump — 2.5L (VIN P) and 4.0L (VIN S) engines

mounting bolt. This will properly position the return tube.

16. Connect the radiator hose to the water pump.

17. Connect the heater hose and clamp to the return tube.

18. Install the water pump pulley and torque the bolts to 20 ft. lbs. (27 Nm).

19. Install the accessory drive belt.

20. Install the cooling fan and shroud.

21. Fill the cooling system.

22. Connect the negative battery cable.

23. Start the engine and check for leaks.

Thermostat

REMOVAL AND INSTALLATION

2.5L (VIN P) and 4.0L (VIN S) Engines

1. Disconnect the negative battery cable.

2. Open the radiator valve and drain the cooling system.

3. Remove the necessary hoses from the thermostat housing.

4. If necessary, disconnect the coolant temperature sensor electrical connector.

5. Remove the 2 attaching screws and lift the housing from the engine.

6. Remove the thermostat and gasket.

To install:

7. Clean all gasket surfaces thoroughly.

8. Place the thermostat in the housing with the spring inside the engine.

9. Install a new gasket with a small amount of sealing compound applied to both sides.

10. Install the water outlet and tighten the attaching bolts to 15 ft. lbs. (20 Nm).

11. Connect the hoses and if disconnected, the coolant temperature sensor connector to the housing.

12. Refill the cooling system.

13. Connect the battery cable.

5.2L (VIN Y) Engines

1. Disconnect the negative battery cable.

2. Open the radiator valve and drain the cooling system.

3. On air conditioned vehicles, remove the alternator support bracket.

Remove the alternator if more room is necessary to access the alternator.

4. Remove the necessary hoses from the thermostat housing.

5. If necessary, disconnect the coolant temperature sensor electrical connector.

6. Remove the 2 attaching screws and lift the housing from the engine.

7. Remove the thermostat and gasket.

To install:

8. Clean all gasket surfaces thoroughly.

9. Place the thermostat in the housing with the spring inside the engine.

10. Install a new gasket with a small amount of sealing compound applied to both sides.

11. Install the water outlet and tighten the attaching bolts to 200 inch lbs. (23 Nm).

12. Connect the hoses and if disconnected, the coolant temperature sensor connector to the housing.

13. If applicable, install the alternator and support bracket.

14. Refill the cooling system.

15. Connect the battery cable.

Electric Cooling Fan

REMOVAL AND INSTALLATION

1. Disconnect the negative battery cable.

2. Unplug the electrical connector.

3. Remove the upper fan shroud bolts.

4. Lift the fan assembly up and out of the engine compartment.

To install:

5. Insert the shroud into its retaining tabs and install the upper bolts.

6. Engage the electrical connector.

7. Connect the negative battery cable.

Cooling System Bleeding

PROCEDURE

As the engine operates, any air trapped in the cooling system gathers under the radiator cap. The next time the engine is operated, thermal expansion of the coolant will push any trapped air past the radiator cap into the overflow tank. Here it escapes into the atmosphere through the tank. When the engine cools, coolant will be drawn into the radiator to replace any removed air.

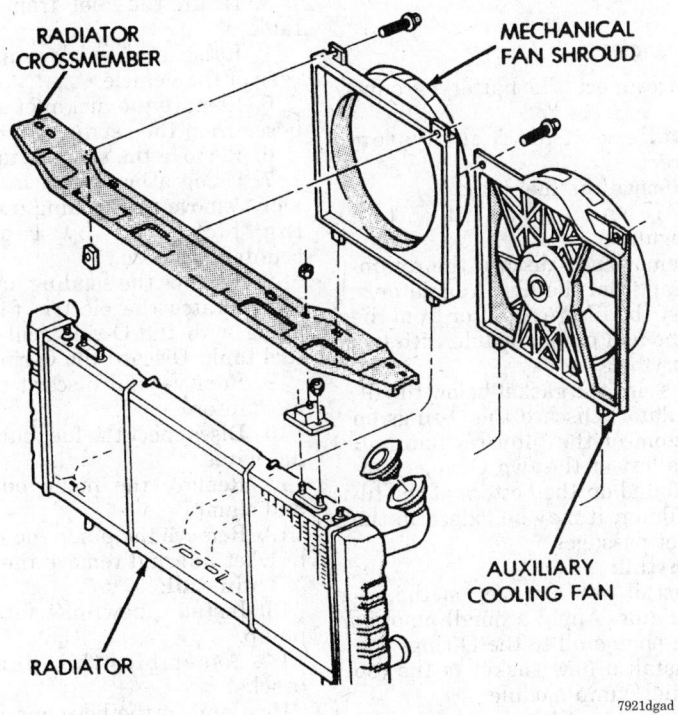

RADIATOR CROSSMEMBER

MECHANICAL FAN SHROUD

AUXILIARY COOLING FAN

RADIATOR

7921dgad

Electric cooling fan mounting

FUEL SYSTEM

Fuel System Service Precautions

Safety is the most important factor when performing not only fuel system maintenance but any type of maintenance. Failure to conduct maintenance and repairs in a safe manner may result in serious personal injury or death. Maintenance and testing of the vehicle's fuel system components can be accomplished safely and effectively by adhering to the following rules and guidelines.

• To avoid the possibility of fire and personal injury, always disconnect the negative battery cable unless the repair or test procedure requires that battery voltage be applied.

• Always relieve the fuel system pressure prior to disconnecting any fuel system component (injector, fuel rail, pressure regulator, etc.), fitting or fuel line connection. Exercise extreme caution whenever relieving fuel system pressure to avoid exposing skin, face and eyes to fuel spray.

Please be advised that fuel under pressure may penetrate the skin or any part of the body that it contacts.

• Always place a shop towel or cloth around the fitting or connection prior to loosening to absorb any excess fuel due to spillage. Ensure that all fuel spillage (should it occur) is quickly removed from engine surfaces. Ensure that all fuel soaked cloths or towels are deposited into a suitable waste container.

• Always keep a dry chemical (Class B) fire extinguisher near the work area.

• Do not allow fuel spray or fuel vapors to come into contact with a spark or open flame.

• Always use a backup wrench when loosening and tightening fuel line connection fittings. This will prevent unnecessary stress and torsion to fuel line piping. Always follow the proper torque specifications.

• Always replace worn fuel fitting O-rings with new. Do not substitute fuel hose or equivalent where fuel pipe is installed.

Fuel System Pressure

RELIEVING

— **CAUTION** —

The fuel system is under constant pressure, even with the engine off. Fuel pressure must be released before servicing any fuel supply or fuel return system component. Do not allow fuel to spill onto the engine intake or exhaust manifolds. Place shop towels under and around the any fittings to be disconnected. This will absorb any fuel spilled when residual pressure is released from the fuel system.

With Test Port

1. Disconnect the negative battery cable.
2. Remove the fuel tank filler cap.
3. Remove the cap from the pressure test port on the fuel rail in the engine compartment.
4. Place one end of the hose from a suitable fuel pressure gauge into an approved gasoline container.
5. Screw the other end of the hose onto the fuel pressure test port to relieve the fuel system pressure.
6. After the fuel pressure has been released, remove the hose from the test port and reinstall the test port cap.

Without Test Port

1. Remove the fuel pump relay.
2. Start the engine and allow it to run until it stalls.
3. Attempt restarting the engine until it no longer runs.
4. Turn the ignition key to the **OFF** position.

Idle Speed

ADJUSTMENT

A (factory adjusted) set screw is used to mechanically limit the position of the throttle body throttle plate. Never attempt to adjust the idle speed using this screw. All idle speed functions are controlled by the Powertrain Control Module (PCM).

Mixture

ADJUSTMENT

The fuel mixture is controlled by the Powertrain Control Module (PCM).

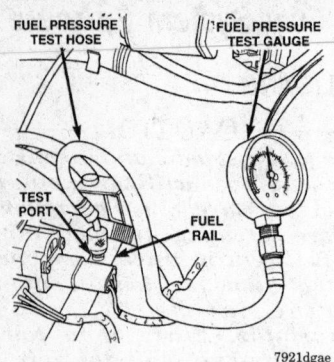

Fuel pressure test with test port

No adjustments are necessary or possible.

Fuel Filter

REMOVAL AND INSTALLATION

Except 1997 Wrangler

1. Disconnect the battery ground cable.
2. Relieve the fuel system pressure.
3. Raise and support the rear of the vehicle safely.
4. Remove the filter shield, if equipped.
5. Disconnect the fuel lines from the filter.
6. Remove the filter strap bolt and remove the filter.

To install:

NOTE: The filter is marked for installation. IN goes towards the fuel tank; OUT towards the engine.

7. Place the new filter on the frame rail and tighten the strap bolt.
8. Connect the fuel lines to the filter.
9. Install the filter shield, if applicable.

10. Connect the negative battery cable.

1997 Wrangler

1. Disconnect the battery ground cable.
2. Relieve the fuel system pressure.
3. Remove the fuel tank.
4. Clean around the filter/regulator.
5. Remove and discard the retaining clamp from the filter/regulator.
6. Pry the filter/regulator from the top of the fuel pump module with two small prytools.
7. Discard the gasket below the filter/regulator. Discard the O-rings on the bottom of the filter/regulator. If the smallest of the two O-rings cannot be found on the bottom of the filter/regulator, it may be lodged in the fuel inlet passage.

To install:

8. Install new O-rings on the filter/regulator. Apply a small amount of clean engine oil to the O-rings.
9. Install a new gasket to the top of the fuel pump module.
10. Press the filter/regulator into the top of the module until it snaps into position (a positive click must be felt or heard).
11. The arrow on top of the fuel pump module must be pointing towards the front of the vehicle. Rotate the filter/regulator until the fuel supply tube is pointed at the 10 o'clock position.
12. Install a new retainer clamp to the top of the filter/regulator.
13. Install the fuel tank.

Fuel Pump

REMOVAL AND INSTALLATION

Cherokee and 1993–95 Wrangler

1. Disconnect the negative battery cable.

2. Remove the fuel tank filler cap.
3. Drain the fuel from the fuel tank.
4. Raise and safely support the rear of the vehicle.
5. Remove the fuel inlet and outlet hoses from the sending unit.
6. Remove the sending unit wires.
7. Using a brass punch and hammer, remove the sending unit retaining lock ring by tapping it counterclockwise.
8. Remove the sending unit, which incorporates the electric fuel pump, along with the O-ring seal from the fuel tank. Discard the O-ring.
9. Remove and discard the pump inlet filter.
10. Disconnect the fuel pump terminal wires.
11. Remove the pump outlet hose and clamp.
12. Remove the pump top mounting bracket nut and remove the pump.

To install:

13. Install a new inlet filter on the pump.
14. Assemble the pump and bracket.
15. Connect the hose and wiring.
16. Install the unit and new O-ring in the tank. The rubber stopper on the end of the fuel return tube must be inserted into the cup in the fuel tank reservoir.
17. Install the lock ring. Carefully tap it into place until it seats against the stop on the tank.
18. Connect the hoses.
19. Connect the wiring.
20. Lower the vehicle.
21. Refill the fuel tank.
22. Run the engine and check for leaks.

1997 Wrangler

NOTE: The fuel pump is not serviceable. If the fuel pump is found to be defective, the entire fuel pump module assembly must be replaced.

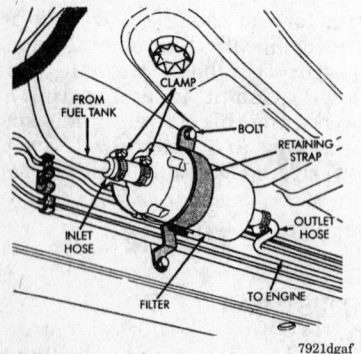

Fuel filter — Cherokee

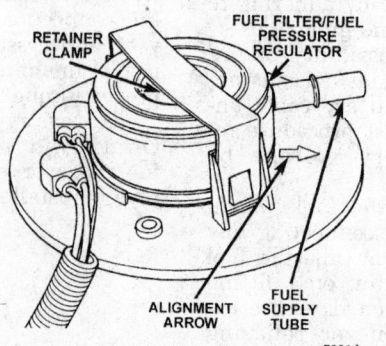

Fuel filter/pressure regulator — 1997 Wrangler

1. Disconnect the battery ground cable.

2. Relieve the fuel system pressure.

3. Remove the fuel tank.

4. Install special tool 6856 to the fuel pump module locknut and remove the locknut. The fuel pump module will spring up when the locknut is removed.

5. Remove the module from the tank.

To install:

6. Using a new gasket, position the fuel pump module into the opening in the fuel tank.

7. Position the locknut over the module. Rotate the module until the arrow is pointed toward the front of the vehicle.

8. Install special tool 6856 to the locknut. Tighten the locknut to 25 ft. lbs. (34 Nm).

9. Rotate the fuel filter/pressure regulator until the fitting is pointed to the 10 o'clock position.

10. Install the fuel tank.

Grand Cherokee and Grand Wagoneer

1. Disconnect the negative battery cable.

2. Remove the fuel tank filler cap.

3. Drain the fuel from the fuel tank.

4. Raise and safely support the rear of the vehicle.

5. Remove the fuel inlet and outlet hoses from the sending unit.

6. Remove the sending unit wires.

7. Using a brass punch and hammer, remove the sending unit retaining lock ring by tapping it counterclockwise.

8. Remove the sending unit, which incorporates the electric fuel pump, along with the O-ring seal from the fuel tank. Discard the O-ring.

NOTE: The fuel pump cannot be replaced separately. The sending unit and pump must be replaced as an assembly.

To install:

9. Install the unit and new O-ring in the tank. The rubber stopper on the end of the fuel return tube must be inserted into the cup in the fuel tank reservoir.

10. Install the lock ring. Carefully tap it into place until it seats against the stop on the tank.

11. Connect the hoses.

12. Connect the wiring.

13. Lower the vehicle.

14. Refill the fuel tank.

15. Run the engine and check for leaks.

Fuel Injector

REMOVAL AND INSTALLATION

2.5L (VIN P) and 4.0L (VIN S) Engines

1. Disconnect the negative battery cable.

2. Relieve fuel system pressure.

3. If equipped, remove the air cleaner crossover tube above the fuel rail.

4. Disconnect the fuel lines at the ends of the fuel rail assembly.

5. Mark and disconnect the injector wire harness connectors.

6. Remove the fuel rail retaining bolts.

7. Disconnect the vacuum line from the fuel pressure regulator.

8. Remove the fuel rail assembly from the engine.

NOTE: On models with automatic transmission, it may be necessary to remove the automatic transmission throttle pressure cable and bracket to remove the fuel rail assembly.

9. Remove the clips that retain the injectors to the fuel rail and remove the injectors.

To install:

10. Install the injectors with new O-rings and clips. Apply a small amount of engine oil to the O-rings to aid in installation.

11. Install the fuel rail and tighten the fuel rail mounting bolts to 20 ft. lbs. (27 Nm), except on 1997 Wrangler. On 1997 Wrangler, tighten the bolts to 100 inch lbs. (11 Nm).

12. Connect the vacuum line to the fuel pressure regulator.

13. Connect the fuel injector electrical connectors.

14. Connect the fuel lines to the injectors.

15. Install the air cleaner crossover tube.

16. Connect the negative battery cable.

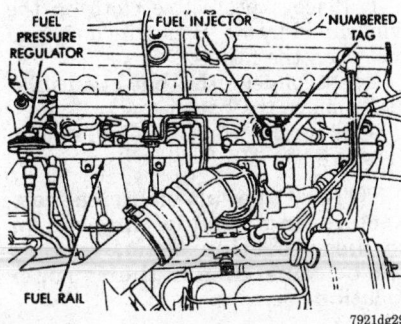

Fuel rail — 2.5L (VIN P) and 4.0L (VIN S) engines

5.2L (VIN Y) Engines

1. Disconnect the negative battery cable.

2. Relieve the fuel system pressure.

3. Remove the air duct from the throttle body.

4. Disconnect the Manifold Absolute Pressure (MAP) sensor, Idle Air Control (IAC) motor and Throttle Position Sensor (TPS) electrical connectors from the throttle body.

5. Disconnect the vacuum line from the throttle body.

6. Disconnect (unsnap) the control cables from the throttle body (lever) arm.

7. Remove the throttle body from the intake manifold. Discard the gasket.

8. If equipped with A/C, disconnect the compressor-to-intake manifold support bracket.

9. Disconnect the electrical connectors from the fuel injectors.

NOTE: The fuel injector wiring harness is numerically tagged (INJ. 1, INJ. 2, etc.) for injector position identification.

10. Remove the EVAP canister purge solenoid/bracket assembly from the intake manifold.

NOTE: Do not attempt to disconnect the fuel line/tubes at the rear of the fuel rail. Fuel rail connections are made under the vehicle at the frame rail.

11. Raise and support the vehicle safely.

12. Disconnect the fuel rail quick-connect fittings at the fuel lines leading to the rear of the vehicle.

13. Lower the vehicle.

14. Remove the remaining fuel rail mounting bolts.

NOTE: Do not attempt to separate the fuel rail halves at the connecting hoses and do not attempt to install clamps to the hoses. When removing the fuel rail do not bend or kink these hoses.

15. Carefully rock the left fuel rail until the fuel injectors start to clear the intake manifold. Repeat the procedure on the right side.

16. Remove the fuel rail, with the fuel injectors attached, from the engine.

17. Remove the clips retaining the injector to the fuel rail and remove the injectors.

To install:

18. Coat each injector O-ring with engine oil.

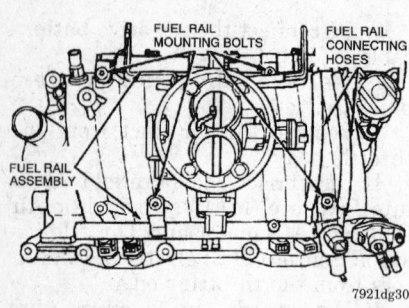

Fuel rail mountng bolts — 5.2L (VIN Y) engines

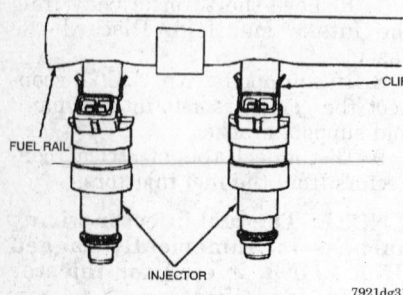

Fuel injector and rail assembly

19. Install the injectors and clips to the fuel rail.

20. Position each injector to the intake manifold.

21. Push the right side of the fuel rail down, taking care not to tear the O-ring, until the injector bottom. Repeat the procedure on the left side.

22. Install the fuel rail mounting bolts.

23. Install the EVAP canister purge solenoid to the intake manifold.

24. Connect the air temperature sensor electrical connector.

25. Connect the electrical connectors to the injectors.

26. If equipped with A/C, install the support bracket.

27. Clean the mating surfaces on the throttle body.

28. Install the throttle body with a new gasket and torque the bolts to 200 inch lbs. (23 Nm).

29. Connect the control cables. If equipped with automatic transmission, the throttle cable must be adjusted.

30. Connect the vacuum line and electrical connectors to the throttle body.

31. Install the air duct to the throttle body.

32. Connect the vacuum line to the fuel pressure regulator.

33. Raise and support the vehicle safely.

34. Connect the fuel rail lines.

35. Lower the vehicle.

36. Connect the negative battery cable.

37. Start the engine and check for leaks.

EMISSION CONTROLS

Emission Warning Lamp

RESETTING

1993 Models

NOTE: 1994 and later models are not equipped with a maintenance reminder lamp.

1. Connect the DRB-II, or equivalent scan tool, to the diagnostic connector located next to the Powertrain Control Module in the engine compartment.

2. Follow the DRB-II Function Flow Diagram to reset the EMR light.

ENGINE MECHANICAL

Engine Assembly

REMOVAL AND INSTALLATION

Wrangler

1. Place a protective cloth on the windshield frame. Raise the hood and rest it on the frame.

2. Disconnect the battery cables and remove the battery.

3. Properly relieve the fuel system pressure.

NOTE: Label all electrical connectors and vacuum lines prior to disconnecting them, so they can be reinstalled in their proper locations.

4. Drain the cooling system.

5. Disconnect the wires from the alternator.

6. Disconnect the ignition coil and distributor wire connections.

7. Disconnect the oil pressure sending unit connector.

8. Disconnect the wires from the starter.

9. Disconnect the fuel injection wires.

10. Disconnect the fuel lines from the fuel rails.

11. Remove the fuel line bracket from the intake manifold.

12. Disconnect the engine ground strap.

13. Remove the air cleaner assembly.

14. Disconnect the canister purge hose from the vapor canister "T" connector.

15. Disconnect the idle speed actuator wire connector.

16. Disconnect the throttle cable and remove it from the bracket.

17. Disconnect the throttle rod from the bellcrank.

18. If equipped, disconnect the cruise control cable.

19. Disconnect the oxygen sensor electrical connector.

20. Disconnect the upper and lower hoses from the radiator.

21. Disconnect the coolant hoses from the rear of the intake manifold and thermostat housing.

22. Disconnect the heater hoses.

23. Remove the fan shroud screws.

24. Remove the radiator and fan shroud.

25. Remove the engine cooling fan.

26. Remove the engine cooling fan and install a 5/16 x 1/2 inch capscrew through the fan pulley into the water pump flange. This will maintain the pulley and water pump in alignment when the crankshaft is rotated.

27. If equipped, disconnect the check valve from the power brake booster.

28. If equipped with power steering, perform the following:

   a. Disconnect the steering hoses from the fittings at the steering gear.

   b. Drain the pump reservoir.

   c. Cap all fittings once removed.

29. Raise and support the vehicle safely.

30. Remove the oil filter.

31. Remove the starter.

32. Remove the flywheel access cover.

33. Remove the engine support cushion-to-bracket through-bolts.

34. Disconnect the exhaust pipe from the manifold.

35. Remove the upper flywheel housing bolts and loosen the bottom bolts.

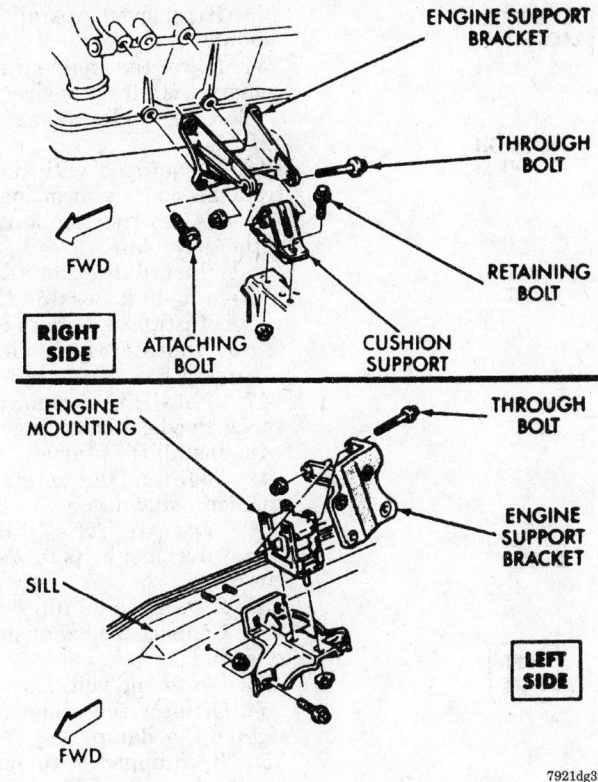

Front engine mounts — Wrangler

36. Remove the engine shock damper bracket from the sill.
37. Lower the vehicle.
38. Attach a lifting device to the engine.
39. Place a support under the bellhousing.
40. Remove the remaining flywheel bolts.
41. Lift the engine from the vehicle.
42. Install the oil filter to keep foreign material out of the engine.
To install:
43. Remove the oil filter.
44. Lower the engine into the vehicle. To ease installation, remove the engine support cushions to aid in engine-to-transmission alignment.
45. Insert the transmission shaft into the clutch spline.
46. Align the flywheel housing with the engine.
47. Install and tighten the flywheel housing bolts finger-tight.
48. If removed, install the engine support cushions.
49. Lower the engine into place and remove the lifting device.
50. Raise and support the vehicle safely.
51. Attach the engine shock damper bracket to the sill.
52. Attach the exhaust pipe to the manifold and torque the nuts to 23 ft. lbs. (31 Nm).

53. Install the flywheel access cover.
54. Install the remaining flywheel bolts and torque them to 28 ft. lbs. (38 Nm).
55. Install the starter.
56. Install the oil filter.
57. Lower the vehicle.
58. Connect the coolant lines and tighten the clamps.
59. If equipped with power steering.
   a. Connect the hoses to the steering gear and torque the nut to 38 ft. lbs. (52 Nm).
   b. Fill the pump reservoir with fluid.
60. Remove the alignment capscrew and install the fan assembly.
61. Install the accessory drive belt.
62. Install the radiator and shroud.
63. Connect the radiator hoses.
64. Connect the oxygen sensor electrical connector.
65. Connect the throttle valve rod and retainer. Connect the throttle cable and install the rod and spring.
66. If equipped, connect the speed control cable.
67. Install the vacuum hose and check valve to the brake booster.
68. Connect the electrical connections disconnected during removal.
69. Connect the fuel lines to the fuel rail.

70. Install the fuel line bracket to the intake manifold.
71. Install the air cleaner.
72. Install the battery and connect the cables.
73. Fill the engine to the proper level with oil.
74. Fill the cooling system.
75. Start the engine and check for leaks.
76. Fill the fluid levels to the proper level.

Cherokee

1. Disconnect the negative battery cable.
2. Properly relieve the fuel system pressure.
3. If equipped with A/C, properly discharge the system using a recovery/recycling machine.
4. Matchmark the hood and hinges and remove the hood.
5. Drain the cooling system.

NOTE: Label all electrical connectors and vacuum lines prior to disconnecting them, so they can be reinstalled in their proper locations.

6. Remove the upper, lower and coolant recovery hoses.
7. Remove the fan shroud.
8. If equipped with an automatic transmission, disconnect the fluid cooler lines.
9. Remove the radiator and if equipped, A/C condenser.
10. Remove the engine cooling fan and install a $5/16$ x $1/2$ inch capscrew through the fan pulley into the water pump flange. This will maintain the pulley and water pump in alignment when the crankshaft is rotated.
11. Disconnect the heater hoses.
12. Disconnect the throttle linkages, speed control cable, if equipped and throttle valve rod.
13. Disconnect the oxygen sensor electrical connector.
14. Disconnect the fuel injection harness connectors.
15. Disconnect the quick-connection fuel lines at the fuel rail and return line.
16. Remove the fuel line bracket from the intake manifold.
17. Remove the air cleaner assembly.
18. If equipped with A/C, remove the service valves and cap the compressor ports.
19. Remove the power brake vacuum check valve from the booster, if equipped.

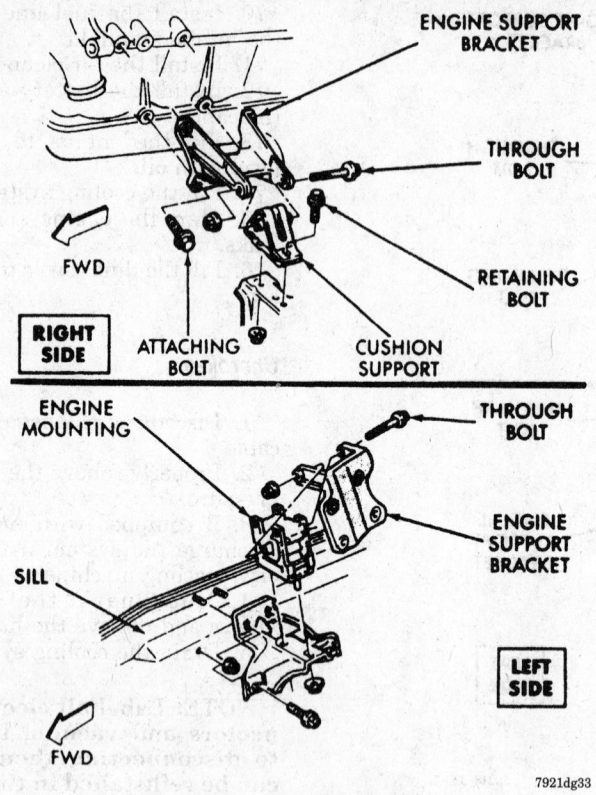

Front engine mounts — Cherokee

20. If equipped with power steering, perform the following:

a. Disconnect the steering hoses from the fittings at the steering gear.

b. Drain the pump reservoir.

c. Cap all fittings once removed.

21. Disconnect the coolant hoses from the rear of the intake manifold.

22. Identify, tag and disconnect all necessary wires and vacuum lines.

23. Raise and support the vehicle safely.

24. Remove the oil filter.

25. Remove the starter.

26. Disconnect the exhaust pipe from the manifold.

27. Remove the flywheel/converter housing access cover.

28. If equipped with an automatic transmission, matchmark the converter to the driveplate and remove the bolts.

29. Remove the upper flywheel/converter housing bolts and loosen the bottoms bolts.

30. Remove the engine mount-to-engine compartment bracket bolts.

31. Remove the engine shock damper bracket from the sill.

32. Lower the vehicle.

33. Attach a lifting device to the engine.

34. Raise the engine slightly off the front supports.

35. Place a support stand under the transmission housing.

36. Remove the remaining flywheel bolts.

37. Lift the engine out of the vehicle.

38. Install the oil filter to keep foreign material out of the engine.

To install:

39. Remove the oil filter.

40. Lower the engine into the vehicle. To ease installation, remove the engine mounts to aid in engine-to-transmission alignment.

41. If equipped with a manual transmission, perform the following.

a. Insert the transmission shaft into the clutch spline.

b. Align the flywheel housing with the engine.

c. Install and tighten the flywheel housing bolts finger-tight.

42. If equipped with an automatic transmission, perform the following.

a. Align the torque converter housing with the engine.

b. Loosely install the converter housing lower bolts and install the next higher nut and bolt on each side.

c. Tighten all 4 bolts finger-tight.

43. If removed, install the engine mounts.

44. Lower the engine into place and remove the lifting device.

45. Raise and support the vehicle safely.

46. If equipped with an automatic transmission, perform the following.

a. Align the torque converter to the driveplate.

b. Install the bolts and torque them to 40 ft. lbs. (54 Nm).

c. Install the access cover.

d. Install the exhaust pipe support.

47. Install the remaining converter/flywheel bolts finger-tight.

48. Install the starter.

49. Tighten the engine support cushion bolts/nuts.

50. Torque the loose converter/flywheel bolts to 28 ft. lbs. (38 Nm).

51. Install the oil filter.

52. Connect the exhaust pipe to the manifold.

53. Lower the vehicle.

54. Connect the coolant hoses and tighten the clamps.

55. If equipped with power steering, perform the following:

a. Unplug the lines and connect them to the steering gear. Torque the fittings to 38 ft. lbs. (52 Nm).

b. Fill the pump reservoir with fluid.

56. Remove the alignment cap screw and install the fan.

57. Install the radiator, condenser, if equipped and fan shroud.

58. Connect the radiator hoses.

59. If equipped with an automatic transmission, connect the cooling lines.

60. Connect the oxygen sensor electrical connector.

61. Connect the throttle valve rod and retainer. Connect the throttle cable and install the rod and spring.

62. If equipped, connect the cruise control cable.

63. Connect the fuel lines to the throttle body.

64. Connect all vacuum lines and electrical connectors disconnected during removal.

65. If equipped with A/C, connect the service valves to the compressor ports.

66. Install the air cleaner.

67. Install the hood.

68. Connect the battery cables.

69. Fill the cooling system.

70. Start the engine and check for leaks.

71. If equipped, recharge the A/C system.

72. Check and top off fluid levels.

Grand Cherokee and Grand Wagoneer

4.0L (VIN S) Engines

1. Matchmark the hood to the hinges and remove the hood.
2. Remove the battery.
3. Properly relieve the fuel system pressure.
4. Drain the cooling system.
5. Remove the air cleaner and tube.
6. Remove the radiator.
7. Remove the heater hoses.
8. Label and disconnect the necessary vacuum lines.
9. Remove the distributor cap and wiring.
10. Disconnect the accelerator linkage.
11. Remove the air duct from the throttle body.
12. Label and disconnect the Manifold Absolute Pressure (MAP) sensor, Idle Air Control (IAC) motor and Throttle Position Sensor (TPS) electrical connectors from the throttle body.
13. Disconnect the vacuum line from the throttle body.
14. Disconnect (unsnap) the control cables from the throttle body (lever) arm.
15. Remove the throttle body from the intake manifold. Discard the gasket.
16. Disconnect the oil pressure electrical connector.
17. If equipped, properly discharge the A/C system.
18. Disconnect the A/C lines from the compressor.
19. If equipped with power steering, disconnect the lines from the pump.
20. Remove the starter.
21. Remove the alternator.
22. Raise and support the vehicle safely.
23. Disconnect the fuel line connections coming from the fuel rail.
24. Disconnect the exhaust pipe from the manifold.

25. Support the transmission with a stand.
26. Remove the bell housing bolts and inspection plate.
27. Attach a C-clamp to the bottom of the torque converter housing to prevent the torque converter from coming out.
28. Matchmark the torque converter to the driveplate and remove the bolts.
29. Disconnect the engine from the torque converter driveplate.
30. Install a suitable lifting device to the engine.

——————— **WARNING** ———————
Do not lift the engine by the intake manifold.

31. Remove the front engine mount thru-bolts.
32. Lower the vehicle.
33. Remove the engine from the vehicle and mount on a suitable workstand.
To install:
34. Remove the engine from the workstand and position it in the engine compartment.
35. Raise and support the vehicle.
36. Position the torque converter and driveplate. Torque the bolts to 271 inch lbs. (31 Nm).
37. Install the front engine mount thru-bolts.
38. Install the bell housing bolts and torque them to 30 ft. lbs. (41 Nm).
39. Remove the C-clamp and install the inspection plate. Remove the stand from the transmission.
40. Connect the exhaust pipe to the manifold.
41. Connect the fuel rail lines.
42. Lower the vehicle.
43. Install the starter.
44. Install the alternator.
45. If equipped, install the power steering hoses.
46. If equipped, connect the A/C hoses.

47. Connect the accelerator linkage.
48. Connect the starter wires.
49. Connect the oil pressure electrical connector.
50. Install the distributor cap and wires.
51. Connect the vacuum lines.
52. Install the radiator, radiator hoses and heater hoses.
53. Install the fan shroud into position.
54. Install the air cleaner.
55. Install the battery.
56. Fill the cooling system.
57. Start the engine and check for leaks.
58. If equipped, recharge the A/C system.
59. Install the hood.
60. Road test the vehicle.

5.2L (VIN Y) Engines

1. Matchmark the hood to the hinges and remove the hood.
2. Remove the battery.
3. Properly relieve the fuel system pressure.
4. Drain the cooling system.
5. Remove the air cleaner and tube.
6. Remove the radiator.
7. Remove the heater hoses.
8. Label and disconnect the necessary vacuum lines.
9. Remove the distributor cap and wiring.
10. Disconnect the accelerator linkage.
11. Remove the air duct from the throttle body.
12. Label and disconnect the Manifold Absolute Pressure (MAP) sensor, Idle Air Control (IAC) motor and Throttle Position Sensor (TPS) electrical connectors from the throttle body.
13. Disconnect the vacuum line from the throttle body.
14. Disconnect (unsnap) the control cables from the throttle body (lever) arm.
15. Remove the throttle body from the intake manifold. Discard the gasket.
16. Disconnect the oil pressure electrical connector.
17. If equipped, properly discharge the A/C system.
18. Disconnect the A/C lines from the compressor.
19. If equipped with power steering, disconnect the lines from the pump.
20. Remove the starter.
21. Remove the alternator.
22. Raise and support the vehicle safely.
23. Disconnect the fuel line connections coming from the fuel rail.

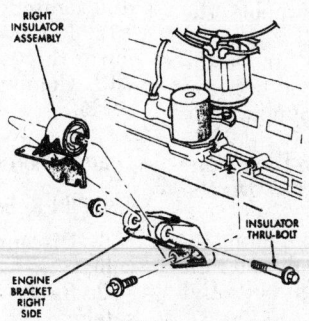

Right side engine mount — 4.0L (VIN S) Grand Cherokee

24. Disconnect the exhaust pipe from the manifold.

25. Support the transmission with a stand.

26. Remove the bell housing bolts and inspection plate.

27. Attach a C-clamp to the bottom of the torque converter housing to prevent the torque converter from coming out.

28. Matchmark the torque converter to the driveplate and remove the bolts.

29. Disconnect the engine from the torque converter driveplate.

30. Install a suitable lifting device to the engine.

-------- WARNING --------

Do not lift the engine by the intake manifold.

31. Remove the front engine mount thru-bolts.

32. Lower the vehicle.

33. Remove the engine from the vehicle and mount on a suitable workstand.

To install:

34. Remove the engine from the workstand and position it in the engine compartment.

35. Raise and support the vehicle.

36. Position the torque converter and driveplate. Torque the bolts to 271 inch lbs. (31 Nm).

37. Install the front engine mount thru-bolts.

38. Install the bell housing bolts and torque them to 30 ft. lbs. (41 Nm).

39. Remove the C-clamp and install the inspection plate. Remove the stand from the transmission.

40. Connect the exhaust pipe to the manifold.

41. Connect the fuel rail lines.

42. Lower the vehicle.

43. Install the starter.

44. Install the alternator.

45. If equipped, install the power steering hoses.

46. If equipped, connect the A/C hoses.

47. Connect the accelerator linkage.

48. Connect the starter wires.

49. Connect the oil pressure electrical connector.

50. Install the distributor cap and wires.

51. Connect the vacuum lines.

52. Install the radiator, radiator hoses and heater hoses.

53. Install the fan shroud into position.

54. Install the air cleaner.

55. Install the battery.

56. Fill the cooling system.

57. Start the engine and check for leaks.

58. If equipped, recharge the A/C system.

59. Install the hood.

60. Road test the vehicle.

Engine Mounts

REMOVAL AND INSTALLATION

1993 Wrangler and Cherokee

1. Disconnect the negative battery cable.

2. Raise and safely support the vehicle.

3. Position a floor jack under the oil pan with a block of wood between the pan and the jack. Support the engine.

4. Remove the nut from the mount through-bolt, but do not remove the through-bolt.

5. Remove the engine mount-to-frame bracket retaining bolt and nut.

6. Remove the through-bolt and the engine mount.

To install:

7. If the engine support bracket was removed, position the support bracket and install the attaching bolts. Tighten to 45 ft. lbs. (61 Nm).

8. Install the engine mount into position and install the through-bolt and nut.

9. Install the mount-to-frame bracket bolt and nut. On Cherokee, tighten to 30 ft. lbs. (41 Nm). On Wrangler, tighten to 48 ft. lbs. (65 Nm).

10. Tighten the through-bolt to 48 ft. lbs. (65 Nm).

11. Remove the engine support and lower the vehicle.

12. Connect the negative battery cable.

1994–97 Wrangler

1. Disconnect the negative battery cable.

2. Raise and safely support the vehicle.

3. Position a floor jack under the oil pan with a block of wood between the pan and the jack. Support the engine.

4. Remove the nut from the mount through-bolt, but do not remove the through-bolt.

5. Remove the engine mount-to-frame bracket retaining bolt and nut.

6. Remove the through-bolt and the engine mount.

To install:

7. If the engine support bracket was removed, position the support

bracket and install the attaching bolts. Tighten to 46 ft. lbs. (62 Nm).

8. Install the engine mount into position.

9. Install the mount-to-frame bracket bolt and nut. Tighten to 38 ft. lbs. (52 Nm).

10. Install the through-bolt and nut and tighten to 51 ft. lbs. (69 Nm).

11. Remove the engine support and lower the vehicle.

12. Connect the negative battery cable.

1994–97 Cherokee

1. Disconnect the negative battery cable.

2. Raise and safely support the vehicle.

3. Position a floor jack under the oil pan with a block of wood between the pan and the jack. Support the engine.

4. Remove the nut from the mount through-bolt, but do not remove the through-bolt.

5. Remove the engine mount-to-frame bracket retaining bolt and nut.

6. Remove the through-bolt and the engine mount.

To install:

7. If the engine support bracket was removed, position the left bracket and the right bracket with generator brace onto the cylinder block. Install the bolts and stud nuts.

8. On the right side, tighten the bolts to 45 ft. lbs. (61 Nm) and the stud nuts to 34 ft. lbs. (46 Nm). On the left side, tighten the bolts to 45 ft. lbs. (61 Nm).

9. If the frame support bracket was removed, position the brackets onto the lower front sill and install the bolts and stud nuts. Tighten the bolts to 40 ft. lbs. (54 Nm) and the stud nuts to 30 ft. lbs. (41 Nm).

10. Install the engine mount into position. Tighten the right mount nuts to 48 ft. lbs. (65 Nm) and the left mount bolt/nut to 30 ft. lbs. (41 Nm).

11. Install the through-bolt and nut and tighten to 48 ft. lbs. (65 Nm).

12. Remove the engine support and lower the vehicle.

13. Connect the negative battery cable.

Grand Cherokee and Grand Wagoneer

4.0L (VIN S) Engines

1. Disconnect the negative battery cable.

2. Raise and support the vehicle safely.

3. Support the engine using a floor jack and a block of wood under the oil pan.

4. Remove the nut from the through-bolt but do not remove the through-bolt.

5. Remove the retaining bolts and nuts from the engine support cushions.

6. Remove the engine support cushions.

To install:

7. If the engine support bracket was removed, position the bracket and install the attaching bolts. Tighten the bolts to 45–48 ft. lbs. (61–65 Nm).

8. Install the engine support cushion and the through-bolt. Install the engine support cushion attaching bolts and tighten to 30–48 ft. lbs. (41–48 Nm). Tighten the through-bolt to 89 ft. lbs. (121 Nm).

9. Remove the engine support and lower the vehicle.

10. Connect the negative battery cable.

5.2L (VIN Y) Engines

1. Disconnect the negative battery cable.

2. Position the fan to assure clearance for the radiator top tank and hose.

3. Raise and support the vehicle safely.

4. Support the engine using a floor jack and a block of wood under the oil pan.

5. Remove the engine support insulator through-bolts and nuts.

6. Raise the engine slightly.

7. Remove the engine support insulator bolts and remove the insulator.

To install:

8. If removed, install the sill bracket assembly.

9. Install the right side bracket onto the sill. Torque the bolts to 48 ft. lbs. (65 Nm).

10. Install the left side bracket onto the sill. Torque the top bolts to 48 ft. lbs. (65 Nm), side bolts to 70 ft. lbs. (96 Nm) and bottom bolts to 89 ft. lbs. (121 Nm).

11. With the engine raised slightly, position the support insulator assembly onto the engine block. Torque the bolts to 65 ft. lbs. (88 Nm).

12. Lower the engine while aligning the engine support insulator into the sill bracket.

13. Install the through-bolt and nut. Torque the right side nut to 48 ft. lbs. (65 Nm) and left side to 89 ft. lbs. (121 Nm).

14. Remove the engine support and lower the vehicle.

15. Connect the negative battery cable.

Cylinder Head

REMOVAL AND INSTALLATION

2.5L (VIN P) Engines

1. Disconnect the negative battery cable.

2. Properly relieve the fuel system pressure.

3. Drain the cooling system.

4. Disconnect the hoses at the thermostat housing.

5. Remove the air cleaner.

6. Remove the rocker arm cover.

7. Remove the rocker arms and the pushrods. Keep them in their original order for installation.

8. Remove the power steering pump bracket. Suspend the pump aside.

9. Remove the intake and exhaust manifolds.

10. If equipped with A/C, perform the following:

    a. Remove the compressor and position it aside with the lines attached.

    b. Remove the compressor bracket bolts from the cylinder head.

    c. Loosen the through-bolt at the bottom of the bracket.

11. Disconnect the fuel lines and vacuum advance hose.

12. Remove the intake and exhaust manifolds.

13. Remove the spark plugs and wires.

14. Remove the ignition coil and bracket assembly.

15. Disconnect the temperature sending unit wire.

16. Remove the cylinder head bolts in the reverse order of the installation torque sequence.

17. Lift the head off the engine and remove the head gasket.

18. Thoroughly clean the gasket mating surfaces. Remove all traces of old gasket material. Remove all car-

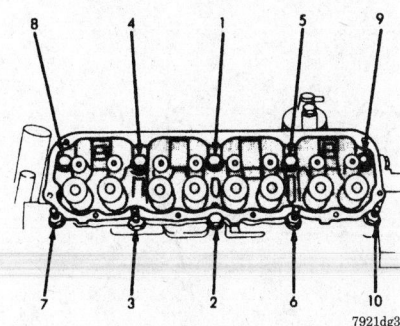

8 4 1 5 9

7 3 2 6 10

7921dg35

2.5L (VIN P) cylinder head torque sequence

bon deposits from the combustion chambers. Lay a straight-edge across the head and check for flatness. Total deviation should not exceed 0.008 in.

To install:

19. Install the head gasket.

NOTE: Do not apply sealer as the cylinder head gaskets are of a composition type.

20. Fabricate 2 cylinder head alignment dowels from used head bolts. Using the longest bolts, trim the hex head off and cut slots into the top.

NOTE: Cylinder head bolts should be reused only once. Replace the head bolts which were previously used or are marked with paint. If head bolts are to be reused, mark each head bolt with paint for future reference.

21. Coat the threads of bolt No. 7 with Loctite® PST sealant or equivalent.

22. Install the cylinder head and tighten the head bolts in the proper sequence to the following torque specifications:

    a. Torque all bolts to 22 ft. lbs. (30 Nm).

    b. Torque all bolts to 45 ft. lbs. (61 Nm).

    c. Torque bolts 1 through 6 to 110 ft. lbs. (150 Nm)

    d. Torque bolt 7 to 100 ft. lbs. (136 Nm).

    e. Torque bolts 8 through 10 to 110 ft. lbs. (150 Nm)

23. Connect the temperature sending unit wire.

24. Install the ignition coil, spark plugs and wires.

25. Install the intake and exhaust manifolds.

26. Install the fuel lines and vacuum advance hose.

27. If equipped, install the power steering pump and bracket.

28. Install the pushrods and rocker arms in their original positions.

29. Install the rocker arm cover.

30. Install the A/C compressor.

31. Install the accessory drive belt.

32. Install the air cleaner.

33. Connect the hoses at the thermostat housing.

34. Fill the cooling system.

35. Connect the negative battery cable.

36. Run the engine to normal operating temperature and check for leaks.

4.0L (VIN S) Engines

1. Disconnect the negative battery cable.

2. Drain the cooling system.

3. Disconnect the hoses at the thermostat housing.

4. Properly relieve the fuel system pressure.

5. Remove the cylinder head cover.

6. Remove the pushrods, bridges, pivots and rocker arms.

NOTE: The valve train components must be replaced in their original positions.

7. Remove the intake and exhaust manifold from the cylinder head.

8. Disconnect the spark plug wires and remove the spark plugs.

9. Disconnect the temperature sending unit wire, ignition coil and bracket assembly from the engine.

10. Remove the accessory drive belt(s).

11. Unbolt and set aside the power steering pump and bracket. Do not disconnect the hoses.

12. Remove the intake and exhaust manifold assembly.

13. If equipped with A/C, perform the following:

   a. Remove the compressor and position it aside with the lines attached.

   b. Remove the compressor bracket bolts from the cylinder head.

   c. Loosen the through-bolt at the bottom of the bracket.

14. Remove the alternator.

15. Remove the cylinder head bolts, the cylinder head and gasket from the block.

NOTE: Bolt No. 14 cannot be removed until the head is moved forward. Pull the bolt out as far as it will go and suspend in place by wrapping with tape.

16. Discard the gasket. Thoroughly clean the head and block mating surfaces. Check them for warpage with a straight-edge. Deviation should not exceed 0.002 in. in a 6 in. span.

To install:

17. Coat a new head gasket with suitable sealing compound and place it on the block. Most replacement gaskets will have the word **TOP** stamped on them.

NOTE: Apply sealing compound only to the cylinder head gasket. Do not allow sealing compound to enter the cylinder bore.

18. Install the cylinder head and bolts. The threads of bolt No. 11 must be coated with Loctite® 592 sealant before installation. Torque the bolts in 3 steps, using the correct sequence:

   a. Torque all bolts to 22 ft. lbs. (30 Nm).

   b. Torque all bolts to 45 ft. lbs. (61 Nm).

   c. Retorque all bolts to 45 ft. lbs. (61 Nm).

   d. Torque bolts 1 through 10 to 110 ft. lbs. (150 Nm).

   e. Torque bolt 11 to 100 ft. lbs. (136 Nm).

   f. Torque bolts 12 through 14 to 110 ft. lbs. (149 Nm)

NOTE: Cylinder head bolts should be reused only once. Replace the head bolts which were previously used or are marked with paint. If head bolts are to be reused, mark each head bolt with paint for future reference. Head bolts should be installed using sealer.

19. Install the ignition coil.

20. Install the air conditioning compressor.

21. Install the alternator.

22. Install the intake and exhaust manifold assembly.

23. Install the power steering pump and bracket.

24. Install the accessory drive belt(s).

25. Connect the temperature sending unit wire, ignition coil and bracket.

26. Install the spark plugs and wires.

27. Install the pushrods, rocker arm assembly, gasket and cylinder head cover.

28. Connect the hoses at the thermostat housing.

29. Fill the cooling system.

30. Connect the negative battery cable.

31. Run the engine to normal operating temperature and check for leaks.

5.2L (VIN Y) Engines

1. Disconnect the negative battery cable.

2. Properly relieve the fuel system pressure.

3. Drain the cooling system.

4. Remove the alternator.

5. Disconnect the PCV valve.

6. Disconnect the EVAP fuel lines.

7. Remove the air cleaner and disconnect the fuel lines.

8. Disconnect the accelerator linkage, speed control cable, if equipped, and transmission kickdown cables.

9. Remove the return spring.

10. Remove the distributor cap and wires.

11. Disconnect the coil wires.

12. Disconnect the heat indicator sending unit wire.

13. Disconnect the heater and by-pass hoses.

14. Remove the cylinder head covers, discard the gaskets.

15. Remove the intake manifold and throttle body.

16. Remove the exhaust manifolds.

17. Remove the rocker arm assemblies and pushrods.

NOTE: Identify the rocker arms and pushrods for installation purposes.

18. Remove the spark plugs.

19. Remove the cylinder head bolts.

20. Remove the heads and discard the gaskets.

21. Thoroughly clean the gasket mating surfaces.

To install:

22. Apply Perfect Sealer No. 5 or equivalent, to the inner corners of the new head gaskets.

23. Position the head gaskets onto the block.

24. Position the cylinder heads onto the cylinder block.

25. Install the cylinder head bolts as follows:

   a. Torque all bolts, in sequence, to 50 ft. lbs. (68 Nm).

   b. Torque all bolts, in sequence, to 105 ft. lbs. (143 Nm).

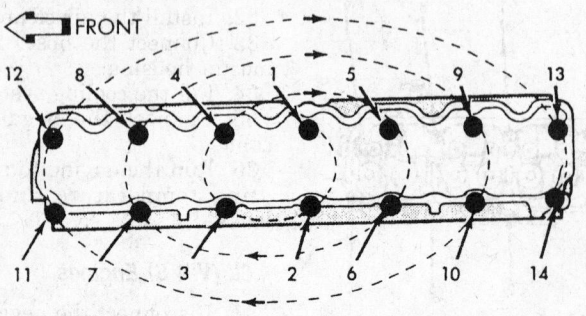

7921dgah

4.0L (VIN S) cylinder head torque sequence

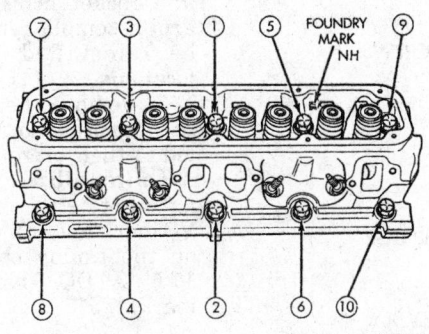

7921dgai

5.2L (VIN Y) cylinder head torque sequence

c. Repeat Step b to ensure the torque is correct.

26. Install the pushrods and rocker arms to their original positions.

27. Install the intake and exhaust manifolds.

28. Install the spark plugs.

29. Install the coil wires.

30. Connect the heat indicating sending unit wire.

31. Connect the heater and bypass hoses.

32. Install the distributor cap and wires.

33. Hook up the return spring.

34. Connect the accelerator linkage, speed control cable, if equipped and transmission kickdown cables.

35. Install the fuel lines.

36. Install the alternator.

37. Install the intake manifold-to-alternator bracket support rod.

38. Place new cylinder head cover gaskets into position and install the cylinder head covers. Torque the bolts to 96 inch lbs. (11 Nm).

39. Install the PCV valve.

40. Connect the EVAP lines.

41. Install the air cleaner.

42. Fill the cooling system.

43. Connect the negative battery cable. Start the engine and check for leaks.

Valve Lifters

BLEEDING

It is not necessary to charge the lifters with engine oil. They will charge themselves within a very short period of engine operation.

REMOVAL AND INSTALLATION

------ **WARNING** ------

To prevent valve mechanism damage, do not run the engine above fast idle until all lifters fill with oil and become quiet.

2.5L (VIN P) and 4.0L (VIN S) Engines

1. Disconnect the negative battery cable.

2. Remove the rocker arm cover.

3. Remove the rocker arm assembly by alternately loosening the bolts 1 turn at a time.

4. Remove the pushrods.

NOTE: Keep all components in the order they were removed so they can be reinstalled in their original positions.

5. Remove the lifter through the pushrod opening in the cylinder head using a hydraulic lifter removal tool.

To install:

6. Dip each lifter in MOPAR engine oil supplement or equivalent, and install into the lifter bore. If reusing the old lifters, make sure they are reinstalled in their original bores.

7. Install the pushrods in their original positions.

8. Install the rocker arm assembly components in their original positions.

9. Pour the remaining engine oil supplement into the engine.

NOTE: The engine oil supplement must remain in the engine for at least 1000 miles but need not be drained until the next scheduled oil change.

10. Install the rocker arm cover.

11. Connect the negative battery cable.

5.2L (VIN Y) Engines

1. Disconnect the negative battery cable.

2. Remove the air cleaner.

3. Remove the cylinder head covers.

4. Remove the rocker assembly and pushrods.

NOTE: Keep all components in the order they were removed.

5. Remove the intake manifold.

6. Remove the yoke retainer and aligning yokes.

7. Install valve lifter tool C-4129-A or equivalent, through the opening in the cylinder head and seat the tool firmly onto the valve lifter.

8. Pull the valve lifter out of the cylinder block using a twisting motion. If all lifters are to be removed, identify each one for installation.

9. If the lifter bore in the cylinder block is scored, scuffed or shows signs of sticking, ream the bore to the next oversize and replace the lifters with oversized.

To install:

10. Lubricate the lifters with clean engine oil or suitable assembly lube.

11. Install the valve lifters and pushrods into their original positions. Ensure the oil feed hole in the side of the valve lifter body faces up away from the crankshaft.

12. Install the aligning yokes with the arrow pointing toward the crankshaft.

13. Install the yoke retainer. Torque the bolts to 200 inch lbs. (23 Nm).

14. Install the rocker arms.

15. Install the intake manifold.

16. Install the cylinder head covers with new gaskets.

17. Connect the negative battery cable.

18. Start the engine and allow it to idle until the valve lifters fill with oil and become quiet. Check for leaks.

Valve Lash

ADJUSTMENT

These engines are equipped with hydraulic valve lifters. No valve clearance adjustments are possible or necessary.

Rocker Arms

REMOVAL AND INSTALLATION

2.5L (VIN P) and 4.0L (VIN S) Engines

1. Disconnect the negative battery cable.

2. Label and disconnect the necessary hoses and vacuum lines from the cylinder head cover.

3. Remove the cylinder head cover retaining bolts and remove the cylinder head cover.

4. Alternately loosen the rocker arm bolts, one turn at a time, to avoid damaging the rocker arm bridge.

5. Remove the rocker arm bolts, bridges, pivots and rocker arms.

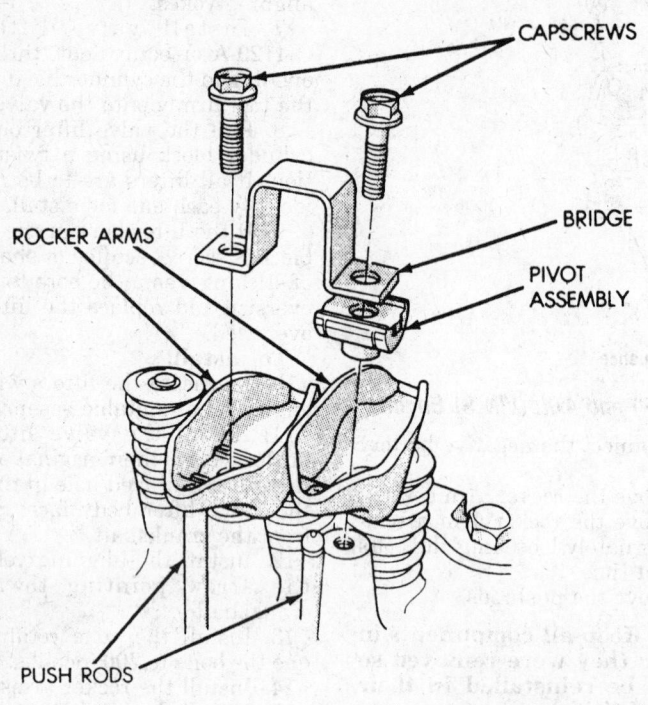

CAPSCREWS

ROCKER ARMS

BRIDGE

PIVOT ASSEMBLY

PUSH RODS

7921dgaj

Rocker arm assembly — 2.5L (VIN P) and 4.0L (VIN S) engines

Keep all parts in order so they can be reinstalled in their original locations.

6. Remove the pushrods. Keep them in order so they can be reinstalled in their original locations.

To install:

7. Clean all parts in solvent and allow to dry. If available, blow out the oil passages in the rocker arms and pushrods with compressed air.

8. Inspect all parts for wear or damage and replace as necessary.

9. Install the pushrods in their original locations.

10. Lubricate the pushrod tips and rocker arm bearing surfaces with clean engine oil.

11. Install the rocker arms, pivots and bridges in their original locations. Loosely install the rocker arm bolts.

12. Tighten the rocker arm bolts alternately and evenly, one turn at a time, to avoid damaging the rocker arm bridges.

13. Tighten the rocker arm bolts to 21 ft. lbs. (28 Nm).

14. Clean the cylinder head cover and cylinder head cover mating surfaces.

NOTE: Some vehicles are equipped with a cylinder head cover containing an integral gasket. This gasket should not be re-moved. If sections of this gasket material are missing or damaged, the cylinder head cover must be replaced. Sections of this type cover with minor damage may be repaired with liquid gasket material.

15. Install the cylinder head cover using a new gasket or sealant, as required.

16. Connect the hoses and vacuum lines.

17. Connect the negative battery cable.

5.2L (VIN Y) Engines

1. Disconnect the negative battery cable.

2. Label and disconnect the necessary hoses from the cylinder head cover.

3. If removing the left cylinder head cover, remove the coolant tube bracket.

4. Remove the spark plug wires from the holders and disconnect them from the spark plugs.

5. Remove the cylinder head cover and gasket. The steel backed silicon gasket can be used again if not damaged.

6. Remove the rocker arm bolts and remove the rocker arm pivots and rocker arms. Keep the rocker arm assemblies in order so they can be reinstalled in their original locations.

7. Remove the pushrods, keeping them in order so they can be reinstalled in their original locations.

To install:

8. Rotate the crankshaft until the **V8** mark lines up with the TDC mark on the timing chain cover (located 17.5° ATDC from the No. 1 firing mark).

— WARNING —
Do not rotate or crank the engine during or immediately after rocker arm installation. Allow about 5 minutes for the hydraulic lifters to bleed down.

---

9. Install the pushrods in their original locations. Make sure they are seated in the lifters.

10. Lubricate the pushrod tips, rocker arm bearing surfaces and rocker arm pivots with clean engine oil.

11. Install the rocker arm and pivot assemblies in their original locations. Tighten the bolts to 21 ft. lbs. (28 Nm).

12. Install the cylinder head cover and gasket. On the left cover, install the coolant tube bracket. Tighten the cylinder head cover retaining bolts to 96 inch lbs. (11 Nm).

13. Connect the spark plug wires to the spark plugs and install the wires in the holders.

14. Connect all hoses that were disconnected during the removal procedure.

15. Connect the negative battery cable.

Intake Manifold

REMOVAL AND INSTALLATION

2.5L (VIN P) Engines

1. Disconnect the negative battery cable.

2. Properly relieve the fuel system pressure.

3. Drain the cooling system.

4. Remove the air cleaner.

5. Remove the power steering pump from its bracket. Position it aside.

6. Disconnect the fuel lines from the fuel rail.

7. Disconnect the accelerator cable from the throttle body and hold-down bracket.

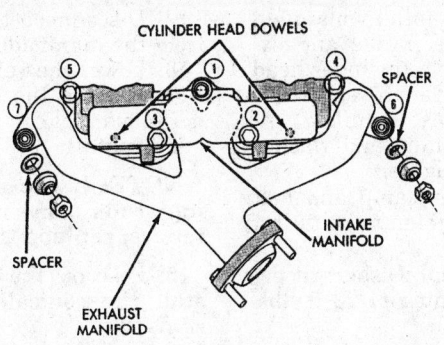

Intake and exhaust manifold — 2.5L (VIN P) engines

7921dgak

CYLINDER HEAD DOWELS
SPACER
INTAKE MANIFOLD
SPACER
EXHAUST MANIFOLD

WARNING

When disconnecting the cruise control connector at the throttle body, do not pry the connector off with pliers or a screwdriver. Use finger pressure only. Prying the connector off could break it.

8. Label and unplug the electrical and vacuum connections from the intake manifold.

9. Remove the molded vacuum harness.

10. Remove bolts 2 through 5 securing the intake manifold to the cylinder head. Slightly loosen bolt No. 1 and nuts 6 and 7.

11. Remove the intake manifold and gaskets.

To install:

12. Clean the gasket mating surfaces.

13. Install a new gasket over the locating dowels.

14. Position the manifold in place and finger-tighten the mounting bolts.

15. Tighten the fasteners in sequence as follows:

a. Tighten fastener No. 1 to 30 ft. lbs. (41 Nm).

b. Tighten fasteners No. 2 through No. 7 to 23 ft. lbs. (31 Nm).

16. Installation of the remaining components is the reverse of removal.

5.2L (VIN Y) Engines

1. Disconnect the negative battery cable.

2. Properly relieve the fuel system pressure.

3. Drain the cooling system.

4. Remove the air cleaner.

5. Remove the alternator.

6. Remove the fuel lines and fuel rail.

7. Disconnect the accelerator linkage and, if equipped, the cruise control and transmission kickdown cables.

8. Remove the return spring.

9. Remove the distributor cap and wires.

10. Disconnect the coil wires.

11. Disconnect the heat indicator sending unit wire.

12. Disconnect the heater and by-pass hoses.

13. Disconnect the PCV and EVAP lines.

14. If equipped with A/C, remove the compressor and position it aside with the lines attached.

15. Remove the support bracket from the mounting bracket and intake manifold.

16. Remove the intake manifold and discard the gaskets.

17. Remove the throttle body and discard the gasket.

18. Turn the intake manifold upside down and support it. Remove the bolts and lift the plenum pan off the manifold. Discard the gasket.

19. Clean all gasket mating surfaces. Clean the intake manifold with solvent and blow dry with compressed air. The plenum pan rail must be clean, dry and free of all foreign material.

To install:

20. Place a new plenum pan gasket onto the seal rail of the intake manifold.

21. Position the pan over the gasket and align the holes. Hand-tighten the bolts.

22. Torque the bolts as follows:

a. Torque all bolts, in sequence, to 24 inch lbs. (2.7 Nm).

b. Torque all bolts, in sequence, to 48 inch lbs. (5.4 Nm).

c. Torque all bolts, in sequence, to 84 inch lbs. (9.5 Nm).

d. Repeat Step c to ensure proper torque.

23. Using a new gasket, install the throttle body onto the intake manifold. Tighten the bolts to 200 inch lbs. (23 Nm).

24. Place the 4 plastic locator dowels into the holes in the block.

25. Apply Mopar rubber adhesive sealant or equivalent, to the 4 corner joints.

NOTE: An excessive amount of sealant is not required to ensure a leak proof seal, however, and excessive amount of sealant may reduce the effectiveness of the flange gasket. The sealant should be slightly higher than the crossover gaskets (approximately 0.2 inch).

26. Install the front and rear crossover gaskets onto the dowels.

27. Install the flange gaskets. Ensure the vertical port alignment tab is resting on the deck face of the block. Also, the horizontal mating alignment tabs must be in position with the mating cylinder head gasket tabs. The words MANIFOLD SIDE should be visible on the center of each flange gasket.

28. Carefully lower the intake manifold into place. Use the alignment dowels in the crossover gaskets to position the manifold. Once in place ensure the gaskets are still in position.

29. Torque the manifold bolts in sequence to the following specifications:

a. Bolts 1–4 — 72 inch lbs. (8 Nm) in 12 inch lbs. (1.4 Nm) intervals

b. Bolts 5–12 — 72 inch lbs. (8 Nm)

c. Repeat Steps a and b to ensure proper torque.

d. All bolts — 12 ft. lbs. (16 Nm)

e. Repeat Step d to ensure proper torque.

30. Connect the PCV and EVAP lines.

31. Install the coil wires.

32. Connect the heat indicator sending unit wire.

33. Connect the heater and bypass hoses.

34. Install the distributor cap and wires.

35. Hook up the return spring.

36. Connect the accelerator linkage and, if equipped, cruise control and transmission kick down cables.

37. Install the fuel lines and fuel rail.

38. Install the support bracket.

39. Install the alternator and drive belt.

40. If equipped with A/C, install the compressor.

41. Install the air cleaner.

42. Fill the cooling system.

43. Connect the negative battery cable.

44. Start the engine and check for leaks.

5.2L – V-8

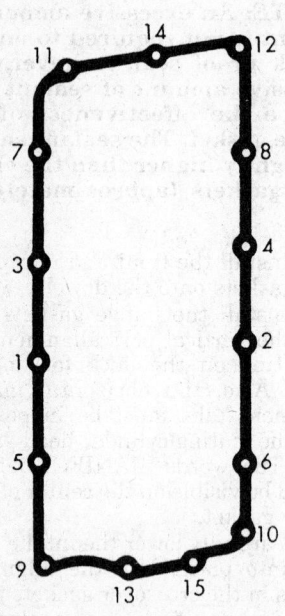

Plenum pan bolt tightening
sequence — 5.2L (VIN Y) engines

7921dgal

Exhaust Manifold

REMOVAL AND INSTALLATION

2.5L (VIN P) Engines

1. Disconnect the negative battery cable.
2. Remove the intake manifold.
3. Disconnect the exhaust pipe at the manifold.
4. Remove the fasteners and exhaust manifold.
   To install:
5. Clean the intake manifold and cylinder head mating surfaces.

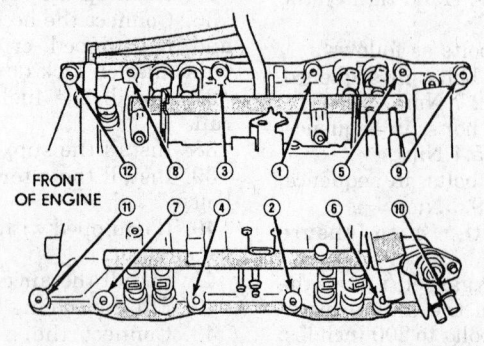

FRONT
OF ENGINE

Intake manifold tightening sequence — 5.2L (VIN Y)
engines

7921dgam

6. Using a new intake manifold gasket, position the intake and exhaust manifolds on the cylinder head and place spacers over the end studs to center the exhaust manifold. Install the end stud nuts and washer clamps but do not tighten.
7. Install washer clamp and bolt at position 1 and tighten to 30 ft. lbs. (41 Nm).
8. Install bolts and washers at positions 2–5 and tighten to 23 ft. lbs. (31 Nm).
9. Tighten end stud nuts (positions 6 and 7) to 23 ft. lbs. (31 Nm).
10. Install all components removed from the intake manifold.
11. Connect the exhaust pipe and tighten the bolts to 23 ft. lbs. (31 Nm).

5.2L (VIN Y) Engines

1. Disconnect the negative battery cable.
2. Remove the exhaust manifold heat shields.
3. Remove the spark plug wire loom and cables from the mounting stud at the rear of the valve cover and position the cables at the top of the valve cover.
4. Label and disconnect the 2 hoses from the EGR valve.
5. Disconnect the electrical connector and hoses from the EGR transducer.
6. Remove the EGR valve and discard the gasket.
7. Disconnect the oil pressure sending unit electrical connector.
8. Using oil pressure sending unit remover C-4597 or equivalent, remove the sending unit.
9. Loosen the EGR mounting nut from the intake manifold.
10. Remove the mounting bolts and EGR tube. Discard the gasket.
11. Raise and safely support the vehicle.

12. Disconnect the exhaust pipes from the manifolds.
13. Lower the vehicle.
14. Remove the fasteners and exhaust manifold.
    To install:

NOTE: If the manifold mounting studs came out with the fasteners, replace the studs.

15. Position the manifold and install the conical washers on the studs.
16. Install new bolt and washer assemblies into the remaining holes. Working from the center outward, torque the fasteners to 20 ft. lbs. (27 Nm).
17. Raise and support the vehicle safely.
18. Connect the exhaust pipes to the manifolds and torque the fasteners to 23 ft. lbs. (31 Nm).
19. Lower the vehicle.
20. Clean the EGR and tube gasket mating surfaces.
21. Install a new gasket onto the exhaust manifold ends of the EGR tube and install the tube. Tighten the tube nut to the intake manifold and torque the tube-to-exhaust manifold bolts to 204 inch lbs. (14 Nm).
22. Coat the threads of the oil pressure sending unit with sealer taking care not to apply sealant to the opening. Install the sending unit and torque it to 130 inch lbs. (14 Nm) and connect the electrical connector.
23. Install the EGR valve and new gasket to the intake manifold and torque the bolts to 200 inch lbs. (23 Nm).
24. Position the EGR transducer and connect the vacuum lines and electrical connector.
25. Position the spark plug cables and loom into place and connect the cables.
26. Install the exhaust heat shields and torque the bolts to 20 ft. lbs. (27 Nm).
27. Connect the negative battery cable.

Combination Manifold

REMOVAL AND INSTALLATION

4.0L (VIN S) Engine

NOTE: The intake and exhaust manifold are mounted externally on the left side of the engine and are attached to the cylinder head. They are removed as a unit.

1. Disconnect the negative battery cable.

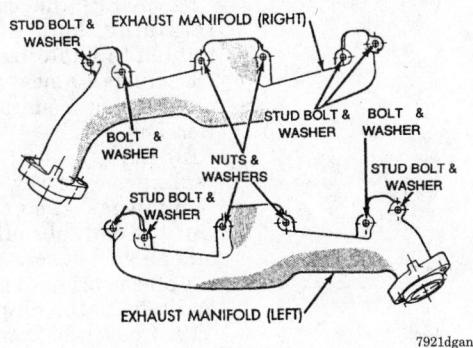

Exhaust manifold — 5.2L (VIN Y) engines

2. Remove the air cleaner assembly.

3. Disconnect the accelerator cable, cruise control cable, if equipped and transmission line pressure cable.

4. Disconnect all electrical connectors on the intake manifold.

5. Disconnect and remove the fuel supply and return lines from the fuel rail assembly.

6. Remove the fuel rail and injectors.

7. Loosen the accessory drive belts.

8. Remove the power steering pump.

9. Disconnect the exhaust pipe from the manifold and discard the seal.

10. Remove the intake and exhaust manifold.

To install:

11. Clean the gasket mating surfaces thoroughly. Install a new gasket over the alignment dowels and position the exhaust manifold to the cylinder head. Install bolt No. 3 finger-tight.

12. Install the intake manifold and the remaining bolts and washers.

13. Tighten bolts, in sequence, to the following torque specifications:

    a. Bolts 1–5 — 24 ft. lbs. (33 Nm)

    b. Bolts 6 and 7 — 23 ft. lbs. (31 Nm)

    c. Bolts 8–11 — 24 ft. lbs. (33 Nm)

14. Install the fuel rail and injectors.

15. Install the power steering pump and tension the accessory belt to specification.

16. Using new O-rings, install the fuel supply and return lines.

17. Connect all electrical connectors, vacuum connectors, throttle cable, cruise control cable and transmission lines pressure cable.

18. Install the air cleaner assembly.

19. Using a new seal, connect the exhaust pipe to the manifold and torque the bolts to 23 ft. lbs. (31 Nm).

20. Connect the negative battery cable.

21. Start the engine and check for leaks.

Front Cover Seal

REMOVAL AND INSTALLATION

2.5L (VIN P) and 4.0L (VIN S) Engines

1. Disconnect the negative battery cable.

2. Remove the drive belts and fan shroud, if equipped.

3. If equipped, remove the crankshaft pulley. Remove the vibration damper retaining bolt.

4. Remove the vibration damper using a suitable puller.

5. Using a suitable tool, pry the oil seal from the front cover. Take care not to damage the front cover or crankshaft.

To install:

6. Position the replacement oil seal on seal installation tool 6139 or equivalent. Install the seal in the cover.

7. Apply a light coat of oil to the seal lip and the vibration damper seal contact surface. Install the vibration damper and tighten the damper bolt to 80 ft. lbs. (108 Nm).

8. Install the crankshaft pulley and torque the bolts to 20 ft. lbs. (27 Nm).

9. Install the drive belts and fan shroud.

10. Connect the negative battery cable.

11. Start the engine and check for leaks.

5.2L (VIN Y) Engines

1. Disconnect the negative battery cable.

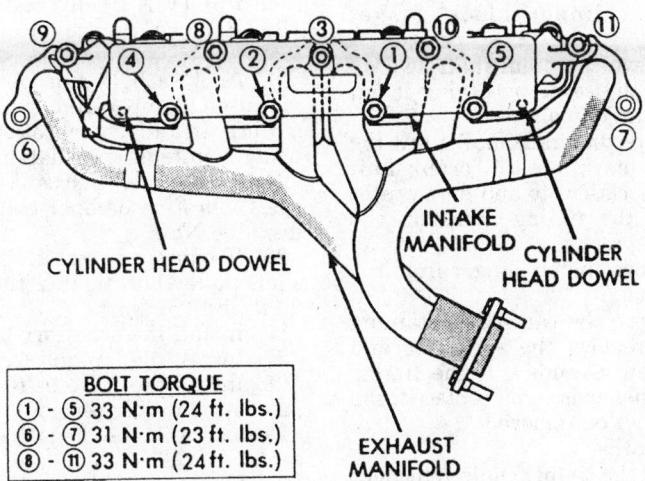

BOLT TORQUE	
①-⑤	33 N·m (24 ft. lbs.)
⑥-⑦	31 N·m (23 ft. lbs.)
⑧-⑪	33 N·m (24 ft. lbs.)

Combination manifold tightening sequence — 4.0L (VIN S) engines

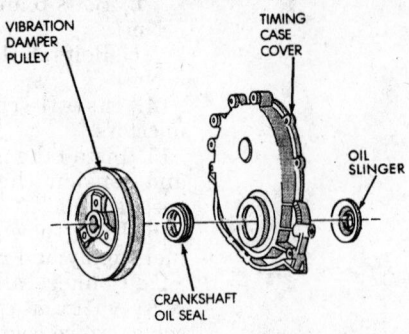

Exploded view of the front cover and seal

2. Remove the fan shroud retaining bolts and set the shroud back over the engine.

3. Remove the cooling fan.

4. Remove the accessory drive belt and the vibration damper pulley.

5. Remove the vibration damper bolt and washer. Remove the vibration damper using puller tool C-3688 or equivalent.

6. Using a suitable tool, pry the oil seal from the front cover. Take care not to damage the front cover.

To install:

7. Position the replacement oil seal on seal installation tool 6635 or equivalent. Install the seal in the cover using the tool.

8. Apply a light coat of oil to the vibration damper hub seal contact surface and install the damper using a suitable installation tool. Tighten the damper bolt to 135 ft. lbs. (183 Nm).

9. Install the crankshaft pulley and torque the bolts to 200 inch lbs. (23 Nm).

10. Install the drive belt, fan and fan shroud.

11. Connect the negative battery cable.

12. Start the engine and check for leaks.

Timing Chain, Sprockets, Tensioner and Front Cover

REMOVAL AND INSTALLATION

2.5L (VIN P) and 4.0L (VIN S) Engines

1. Disconnect the negative battery cable.

2. Remove the drive belt(s), fan and fan shroud. If equipped, remove the accessory drive belt pulley.

3. Remove the vibration damper retaining bolt and washer. Remove the vibration damper using a suitable puller.

4. Remove the accessory drive brackets attached to the timing cover.

5. Remove the A/C compressor, if equipped, and alternator bracket from the cylinder head and move to one side.

6. Remove the oil pan-to-timing case cover bolts and the cover-to-cylinder block bolts.

7. Remove the timing case cover front seal and gasket from the engine.

NOTE: Make sure the tension spring and thrust pin do not fall out of the camshaft sprocket retaining preload bolt.

8. Cut off the oil pan side gasket end tabs and oil pan front seal tabs flush with the front face of the cylinder block. Remove the gasket tabs.

9. Clean the timing case cover, oil pan and cylinder block gasket surfaces.

10. Remove the crankshaft seal oil seal from the cover by prying it out with a suitable tool.

11. Rotate the crankshaft until the **0** timing mark on the crankshaft sprocket is closest to and on a center line with the timing mark on the camshaft sprocket.

12. Remove the oil slinger from the crankshaft.

13. Remove the camshaft retaining bolt and remove the sprockets and chain as an assembly. If the timing chain is to be replaced, the oil pan must be removed.

To install:

14. Turn the timing chain tensioner lever to the unlock (down) position. Pull the tensioner block toward the tensioner lever to compress the spring. Hold the block and turn the tensioner lever to the lock (up) position.

15. Install the sprockets and timing chain. Ensure the timing marks on the sprockets are properly aligned.

16. Install the camshaft sprocket retaining bolt and washer and tighten to 80 ft. lbs. (108 Nm).

17. Install a new seal in the timing cover using a suitable seal installation tool.

18. Apply sealer to both sides of the replacement cover gasket and position the gasket on the cylinder block. Cut the end tabs off the replacement oil pan gasket corresponding to those cut off the original gasket. Attach the end tabs to the oil pan with sealer.

19. Coat the front cover seal end tab recesses generously with sealer and position the seal on the timing cover.

20. Apply engine oil to the seal/oil pan contact surface, then position the cover on the cylinder block.

NOTE: Make sure the tension spring and thrust pin are in place in the preload bolt, before installing the cover.

21. Insert timing case cover alignment tool 6139 or equivalent, in the crankshaft opening. Install the cover bolts and tighten the bolts as follows:

  a. 1/4 in. cover-to-block bolts to 60 inch lbs. (7 Nm).

  b. 5/16 in. cover-to-block bolts to 192 inch lbs. (22 Nm).

  c. 1/4 in. oil pan-to-cover bolts to 84 inch lbs. (10 Nm) on 4.0L (VIN S) engines or 120 inch lbs. (14 Nm) on 2.5L (VIN P) engines.

  d. 5/16 in. oil pan-to-cover bolts to 132 inch lbs. (15 Nm) on 4.0L (VIN S) engines or 84 inch lbs. (10 Nm) on 2.5L (VIN P) engines.

22. Remove the cover alignment tool.

23. Apply a light film of oil to the vibration damper hub seal contact surface. Install the vibration damper using a suitable installation tool.

24. Install and tighten the crankshaft vibration damper bolt to 80 ft. lbs. (108 Nm).

25. If equipped, install the crankshaft pulley and tighten the bolts to 20 ft. lbs. (27 Nm).

26. Install the accessory brackets.

27. Install the fan and fan shroud.

28. Install the drive belt(s) and adjust to the proper tension.

29. Connect the negative battery cable.

30. Start the engine and check for leaks.

5.2L (VIN Y) Engines

1. Disconnect the negative battery cable.

2. Properly relieve the fuel system pressure.

3. Drain the cooling system.

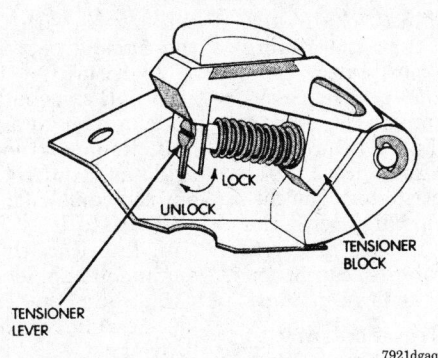

Timing chain tensioner — 2.5L (VIN P) engines

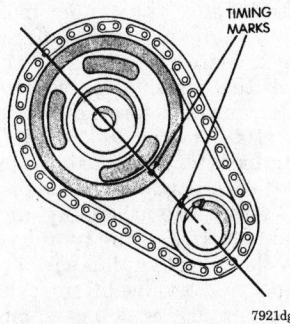

Timing chain alignment — 2.5L (VIN P) and 4.0L (VIN S) engines

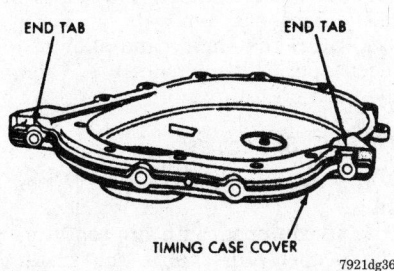

Timing cover end tabs — 2.5L (VIN P) and 4.0L (VIN S) engines

4. Remove the serpentine belt.
5. Remove the cooling fan shroud and position it on the engine.
6. Remove the water pump.
7. Remove the power steering pump.
8. Remove the vibration damper using puller C-3688 or equivalent.
9. Disconnect the fuel lines.
10. Loosen the oil pan bolts and remove the front bolt at each side.
11. Remove the timing chain cover bolts. Remove the chain cover and gasket using extreme caution to avoid damaging the oil pan gasket.

12. Place a scale next to the timing chain so any movement of the chain can be measured.
13. Place a torque wrench and socket over the camshaft sprocket attaching bolt. Apply 30 ft. lbs. (41 Nm) with the cylinder heads installed or 15 ft. lbs. (20 Nm) with the cylinder heads removed.

NOTE: With the torque applied the crankshaft sprocket should not be permitted to move, but it may be necessary to block the crankshaft to prevent rotation.

14. Hold the scale with the dimension reading even with the edge of a chain link. Apply 30 ft. lbs. (41 Nm) with the cylinder heads installed or 15 ft. lbs. (20 Nm) with the cylinder heads removed, in the reverse direction. Note the amount of chain movement.
15. Install a new timing chain if the movement exceeds 1/8 inch (3.175mm).
16. Remove the camshaft sprocket retaining bolt.
17. Remove the timing chain and sprockets.
   To install:
18. Position the camshaft and crankshaft sprockets on a bench with the timing marks facing each other.
19. Position the timing chain onto the sprockets.
20. Turn the crankshaft and camshaft to align with the keyway location in the crankshaft and camshaft sprockets.
21. Keeping tension on the chain, slide the sprocket and chain assembly onto the engine.
22. Ensure the timing marks are still aligned by using a straight edge.
23. Install the camshaft bolt and torque it to 50 ft. lbs. (68 Nm).
24. Install a new timing chain cover gasket to the chain cover. Apply a small amount of Mopar silicone rubber adhesive sealant or equivalent, at the joint where the chain cover and oil pan gasket meet.
25. Install the timing chain cover taking care not to damage to oil pan. Torque the timing chain cover bolts to 30 ft. lbs. (41 Nm) and oil pan bolts to 215 inch lbs. (24 Nm).
26. Install the vibration damper. Torque the crankshaft bolt to 135 ft. lbs. (183 Nm) and pulley bolt to 200 inch lbs. (23 Nm).
27. Connect the fuel lines.
28. Install the water pump.
29. Install the power steering pump.
30. Install the serpentine belt.
31. Install the cooling fan shroud.
32. Fill the cooling system.
33. Connect the negative battery cable.

Camshaft

REMOVAL AND INSTALLATION

2.5L (VIN P) Engines

1. Disconnect the negative battery cable.
2. If equipped with air conditioning, properly discharge the system using a recovery/recycling machine.
3. Drain the cooling system.

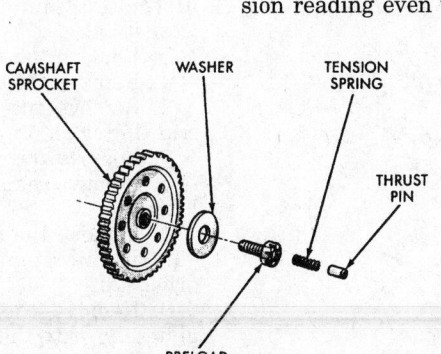

Sprocket preload bolt — 4.0L (VIN S) engines

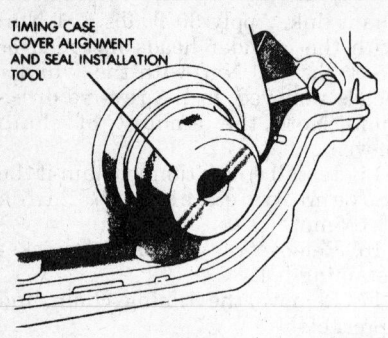

TIMING CASE COVER ALIGNMENT AND SEAL INSTALLATION TOOL

7921dg37

Cover alignment tool — 2.5L (VIN P) and 4.0L (VIN S) engines

4. Remove the radiator and air conditioning condenser, if equipped.

5. Matchmark the distributor and engine for installation.

6. Matchmark the rotor position by marking it on the distributor body.

7. Remove the distributor and wires.

8. Remove the rocker arm cover.

9. Remove the rocker arm assemblies.

10. Remove the pushrods.

NOTE: Keep all valve train components in order for installation.

11. Remove the hydraulic lifters.

12. Remove the timing case cover.

NOTE: If the camshaft sprocket appears to have been rubbing against the cover, check the oil pressure relief holes in the rear cam journal for debris.

13. Remove the timing chain and sprockets.

14. Slide the camshaft from the engine.

To install:

15. Inspect the camshaft for wear and damage.

16. Lubricate the camshaft with an engine oil supplement.

17. Install the camshaft, taking care not to damage the cam bearings.

18. Install the timing chain and sprockets. Ensure that the timing marks are correctly positioned.

19. Install the timing case cover, hydraulic lifters, pushrods, rocker arm assemblies, cylinder head cover, distributor and ignition wires.

20. Install the distributor. The distributor rotor should align with the position of the No. 1 spark plug terminal on the distributor cap when the distributor shaft is down in place.

NOTE: It may be necessary to rotate the oil pump shaft with a long flat-blade screwdriver to engage the oil pump drive tang.

21. Install all other components in reverse order of removal.

22. Start the engine and allow it to reach operating temperature. Check the ignition timing.

4.0L (VIN S) Engines

1. Disconnect the negative battery cable.

2. If equipped with air conditioning, properly discharge the system using a recovery/recycling machine.

3. Drain the cooling system.

4. Remove the radiator and air conditioning condenser, if equipped.

5. Properly relieve the fuel system pressure.

6. Remove the valve cover and gasket, rocker assemblies, pushrods and lifters.

NOTE: The pushrods must be replaced in their original locations.

7. Remove the drive belts, cooling fan, fan hub assembly, vibration damper and timing chain cover.

8. Remove the distributor assembly, including the spark plug wires.

9. Remove the cylinder head.

10. Remove the valve lifters. Keep them in order for installation.

11. Rotate the crankshaft until the timing mark of the crankshaft sprocket is adjacent to and on a center line with the timing mark of the camshaft sprocket.

12. Remove the crankshaft sprocket, camshaft sprocket and timing chain as an assembly.

13. Remove the front bumper or grille as required and carefully slide out the camshaft.

To install:

14. Lubricate the camshaft with an engine oil supplement.

15. Slide the camshaft into the block carefully to avoid damage to the bearings.

16. Install the crankshaft sprocket, camshaft sprocket and the timing chain as an assembly.

17. Make sure the timing mark of the crankshaft sprocket is adjacent to and on a centerline with, the timing mark of the camshaft sprocket. Torque the camshaft sprocket bolt to 80 ft. lbs. (109 Nm).

18. Lubricate the tension spring, thrust pin and pin bore in the preload bolt. Install the assembly on the preload bolt. Install the timing cover. Install the vibration damper.

19. Install the valve lifters, cylinder head, pushrods, rocker assemblies, valve cover and gasket.

20. Install the distributor assembly. The rotor should be aligned with the No. 1 cylinder spark plug terminal on the cap when the distributor is fully seated on the cylinder block.

21. Install the condenser and receiver/drier.

22. Install the radiator and fill the cooling system.

23. Connect the negative battery cable.

24. Start the engine and allow it to reach operating temperature.

25. Check the ignition timing.

5.2L (VIN Y) Engines

1. Disconnect the negative battery cable.

2. Remove the engine from the vehicle and mount on a suitable workstand.

3. Remove the intake manifold.

4. Remove the valve covers.

5. Remove the timing case cover and timing chain.

6. Remove the rocker arms.

7. Remove the pushrods and valve lifters.

8. Remove the distributor and lift out the oil pump and distributor driveshaft.

9. Remove the camshaft thrust plate and note the location of the oil tab.

10. Install a long bolt into the camshaft to facilitate the removal of the camshaft.

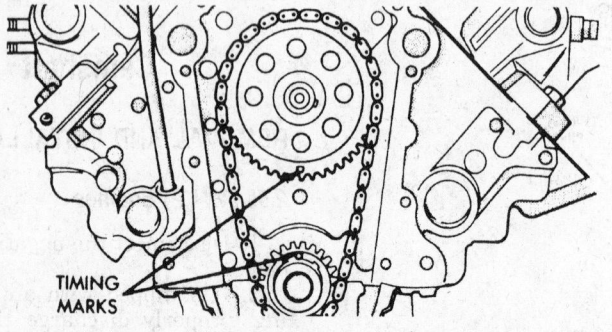

TIMING MARKS

7921dgat

Timing chain alignment marks — 5.2L (VIN Y) engines

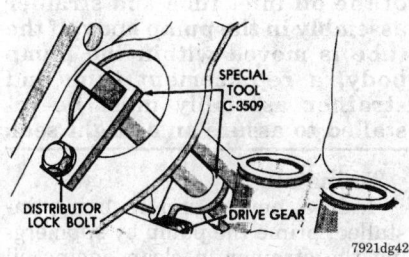

Camshaft — 5.2L (VIN Y) engines

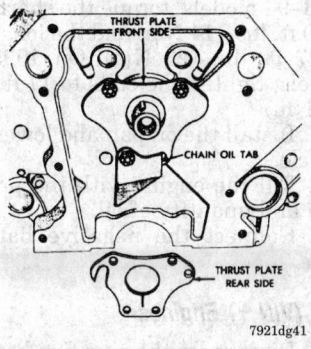

Camshaft holding tool — 5.2L (VIN Y) engines

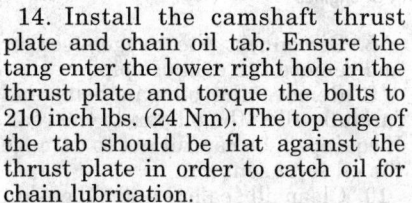

Timing chain oil tab installation —
5.2L (VIN Y) engines

11. Remove the camshaft, being careful not to damage the cam bearings with the cam lobes.

To install:

12. Lubricate the camshaft lobes and bearings with engine oil and insert the camshaft to within 2 inches (51mm) of its final position in the cylinder block.

13. Install camshaft gear installer tool C-3509 or equivalent, with the tool secured in the distributor drive

gear position. Hold the tool in position with a distributor lockplate bolt.

NOTE: The tool will restrict the camshaft from being pushed in too far and prevent knocking out the plug in the rear of the cylinder block. The tool should remain installed until the camshaft and crankshaft sprockets with the chain are installed.

14. Install the camshaft thrust plate and chain oil tab. Ensure the tang enter the lower right hole in the thrust plate and torque the bolts to 210 inch lbs. (24 Nm). The top edge of the tab should be flat against the thrust plate in order to catch oil for chain lubrication.

15. Install the timing chain and sprockets.

16. Remove installer tool C-3509.

17. Ensure the camshaft end-play is 0.002–0.010 inch (0.051–0.76mm); if not, install a new thrust plate.

18. Install the driveshaft and distributor.

19. Install the lifters and pushrods.

20. Install the rocker arms.

21. Install the valve covers with new gaskets.

22. Install the intake manifold.

23. Install the engine.

24. Connect the negative battery cable.

Piston and Connecting Rod

POSITIONING

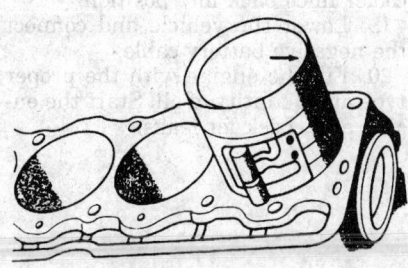

Connecting rod and piston installation

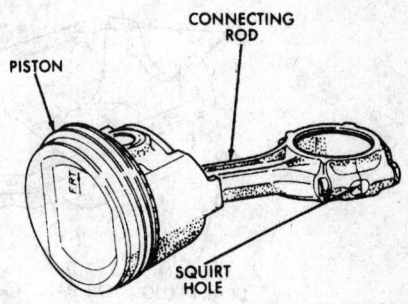

Correct piston and rod alignment

ENGINE LUBRICATION

Oil Pan

REMOVAL AND INSTALLATION

Wrangler and Cherokee

1. Disconnect the negative battery cable.

2. Raise and support the vehicle safely.

3. Drain the engine oil.

4. Disconnect the exhaust pipe at the manifold.

5. Disconnect the exhaust hanger at the catalytic converter and lower the pipe.

6. Remove the starter.

7. Remove the bellhousing access cover.

8. If equipped, disconnect the oil level sensor.

9. Position a jackstand directly under the vibration damper. Place a piece of wood between the jack and the vibration damper.

10. Remove the engine mount through-bolts. Using the jack, raise the engine until there is enough room to remove the oil pan.

11. Remove the oil pan bolts and remove the oil pan.

12. Clean all sealant and old gasket material from the oil pan and cylinder block mating surfaces. Thoroughly clean the oil pan.

To install:

13. Fabricate 4 alignment dowels from 1½ in. x ¼ in. bolts. Cut the heads off the bolts and cut a slot into the top of the dowel to allow installation/removal with a screwdriver.

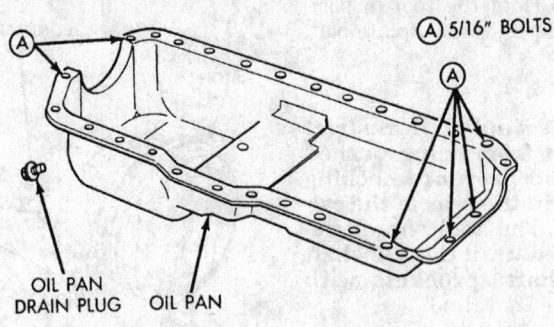

(A) 5/16" BOLTS

OIL PAN
DRAIN PLUG   OIL PAN

7921dgau

Oil pan on 4.0L (VIN S) engines

14. Install 2 dowels in the timing case cover and the other 2 in the cylinder block. Slide the one-piece gasket over the dowels and onto the block and timing case cover.

15. Position the oil pan over the dowels and onto the gasket. Install the ¼ in. pan bolts and tighten to 120 inch lbs. (14 Nm), except on 1997 Wrangler. On 1997 Wrangler, tighten the bolts to 84 inch lbs. (10 Nm). Install the 5/16 in. pan bolts and tighten to 156 inch lbs. (18 Nm), except on 1997 Wrangler. On 1997 Wrangler, tighten the bolts to 132 inch lbs. (15 Nm).

16. Remove the dowels and install the remaining ¼ in. pan bolts.

17. Lower the engine and install the engine mount through-bolts and nuts. Lower the jack and remove the piece of wood.

18. Connect the oil level sensor if equipped.

19. Install the bellhousing access cover.

20. Install the starter.

21. Connect the exhaust pipe to the hanger and the exhaust manifold.

22. Install the oil pan drain plug and tighten to 25 ft. lbs. (34 Nm).

23. Lower the vehicle.

24. Fill the crankcase to the proper level with the recommended oil.

25. Connect the negative battery cable. Start the engine and check for leaks.

Grand Cherokee and Grand Wagoneer

1. Disconnect the negative battery cable.

2. Raise and safely support the vehicle.

3. Remove the oil pan drain plug and drain the engine oil.

4. Remove the oil filter.

5. Remove the starter.

6. If equipped, disconnect the oil level sensor.

7. Position the oil cooler lines out of the way.

8. Disconnect the oxygen sensor and remove the exhaust pipe.

9. Remove the oil pan bolts and carefully slide the oil pan to the rear. If equipped, be careful not to damage the oil level sensor.

10. Clean all sealant and old gasket material from the oil pan and cylinder block mating surfaces. Thoroughly clean the oil pan.

To install:

11. Fabricate 4 alignment dowels from 1½ x 5/16 in. bolts. Cut the heads off the bolts and cut a slot in the dowel to allow installation/removal with a screwdriver.

12. Install the dowels in the cylinder block. Apply a small amount of silicone sealant in the corner of the cap and cylinder block.

13. Slide the one-piece gasket over the dowels and onto the block. Position the oil pan over the dowels and onto the gasket. If equipped, be careful not to damage the oil level sensor.

14. Install the oil pan bolts and tighten to 215 inch lbs. (24 Nm). Remove the dowels and install the remaining bolts. Tighten to 215 inch lbs. (24 Nm).

15. Install the drain plug and tighten to 25 ft. lbs. (34 Nm).

16. Install the exhaust pipe and connect the oxygen sensor.

17. Install the oil filter. If equipped, connect the oil level sensor.

18. Install the starter. Move the oil cooler lines back into position.

19. Lower the vehicle and connect the negative battery cable.

20. Fill the engine with the proper type and quantity of oil. Start the engine and check for leaks.

Oil Pump

REMOVAL AND INSTALLATION

2.5L (VIN P) and 4.0L (VIN S) Engines

1. Disconnect the negative battery cable. Raise and safely support the vehicle.

2. Drain the engine oil and remove the oil pan.

3. Unbolt and remove the pump assembly from the block. Discard the gasket.

———— **WARNING** ————

If the oil pump is not to be serviced, do not disturb the position of the oil inlet tube and strainer assembly in the pump body. If the tube is moved within the pump body, a replacement tube and strainer assembly must be installed to assure an airtight seal.

To install:

4. If a new pump is being installed, prime the pump by submerging the strainer in clean engine oil and turning the pump gears until oil emerges from the pump feed hole.

5. Using a new gasket, install the pump on the cylinder block. On 1993–94 models, torque the short bolt to 10 ft. lbs. (14 Nm) and the long bolt to 17 ft. lbs. (23 Nm). On 1995–97 models, tighten the bolts to 17 ft. lbs. (23 Nm).

6. Install the oil pan and lower the vehicle.

7. Fill the engine with the proper type and quantity of oil.

8. Connect the negative battery cable.

5.2L (VIN Y) Engines

1. Disconnect the negative battery cable.

2. Raise and safely support the vehicle.

3. Drain the engine oil and remove the oil pan.

4. Unbolt and remove the pump assembly from the rear main bearing cap.

To install:

5. If a new pump is being installed, prime the pump by submerging the pickup in clean engine oil and turning the pump gears until oil emerges from the pump feed hole.

6. Install the oil pump. During installation, slowly rotate the pump body to ensure driveshaft-to-pump rotor shaft engagement.

7. Hold the oil pump base flush against the mating surface of the rear main bearing cap and finger

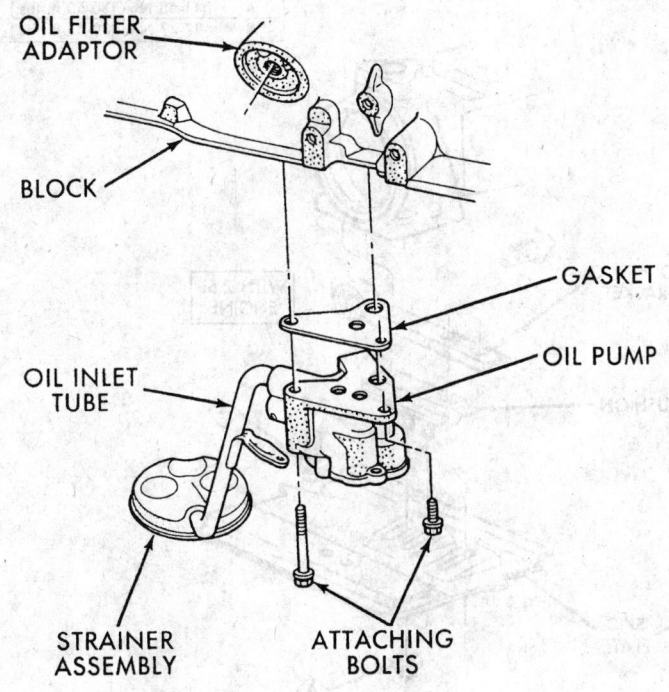

OIL FILTER ADAPTOR

BLOCK

GASKET

OIL PUMP

OIL INLET TUBE

STRAINER ASSEMBLY

ATTACHING BOLTS

7921dgav

Oil pump assembly

tighten the pump mounting bolts. Tighten the mounting bolts to 30 ft. lbs. (41 Nm).

8. Install the oil pan and lower the vehicle. Fill the engine with the proper type and quantity of oil.

9. Connect the negative battery cable.

TRANSMISSION

Manual Transmission Assembly

REMOVAL AND INSTALLATION

Wrangler and Cherokee

1. Disconnect the negative battery cable.

2. Shift the vehicle into first or third gear, then raise and safely support the vehicle.

3. Support the engine with an adjustable jackstand. Position a wood block between the jack and oil pan to avoid damaging the pan.

4. Disconnect the necessary exhaust system components.

5. Remove the skid plate.

6. Disconnect the rear cushion and bracket from the transmission.

7. Remove the rear crossmember.

8. Remove the slave cylinder from the clutch housing.

9. On 4WD models, disconnect the transfer case shift linkage.

10. Unplug the electrical connectors and vent hose.

11. Lower the transmission enough to provide access to the shift lever.

12. Reach up and around the transmission case and unseat the shift lever dust boot from the transmission shift tower. Reposition the boot to access the lever retainer.

13. Press the shift lever retainer downward and turn it counterclockwise to release it.

14. Lift the lever and retainer out of the shift tower.

NOTE: It is not necessary to remove the shift lever from the floor pan boot. Leave the lever in place for installation.

15. Matchmark the front and rear driveshaft and yoke for installation alignment.

16. Unbolt and remove the driveshaft(s).

17. Remove the crankshaft position sensor.

18. Unclip the wiring harnesses from the transmission and transfer case (if equipped).

19. On 1993–95 models, disconnect the clutch master cylinder hydraulic line from the concentric bearing inlet line.

20. Support the transmission/transfer case with a transmission jack. Secure the assembly to the jack with chains.

21. Except on 1997 2.5L (VIN P) engines:

a. Remove the clutch housing brace rod, if equipped.

b. Remove the clutch housing-to-engine bolts and remove the transmission/transfer case assembly.

c. On 4WD models, remove the bolts attaching the transmission to the transfer case and separate the components.

22. On 1997 2.5L (VIN P) engines:

a. On 4WD models, remove the nuts attaching the transfer case to the transmission and remove the transfer case.

b. Remove the clutch housing brace rod.

c. Remove the transmission-to-engine attaching bolts, then remove the transmission from the vehicle.

To install:

23. If necessary, install the clutch housing on the transmission. Tighten the bolts to 27 ft. lbs. (37 Nm). except on 1997 4.0L (VIN S) engines. On 1997 4.0L (VIN S) engines, tighten the 3/8 in. bolts to 27 ft. lbs. (37 Nm) and the 7/16 in. bolts to 43 ft. lbs. (58 Nm).

24. Lubricate the contact surfaces of the release fork pivot ball stud and release stud with high temperature grease. Then install the release bearing, fork and retainer clip.

25. Lubricate the pilot bearing and transmission input splines with high temperature grease.

26. Align the transmission input shaft, release bearing and clutch disc splines, then slide the transmission into place.

27. Install and tighten the mounting bolts to 28 ft. lbs. (38 Nm), except on 1997 2.5L (VIN P) engines. Tighten to 55 ft. lbs. (75 Nm) on 1997 2.5L (VIN P) engines. Be sure the transmission is properly seated before installing the mounting bolts.

28. Reach up around the transmission and insert the shift lever in the shift tower. Press the lever retainer downward and turn it clockwise to lock it in place. Install the dust lever on the dust boot.

29. Install the slave cylinder.

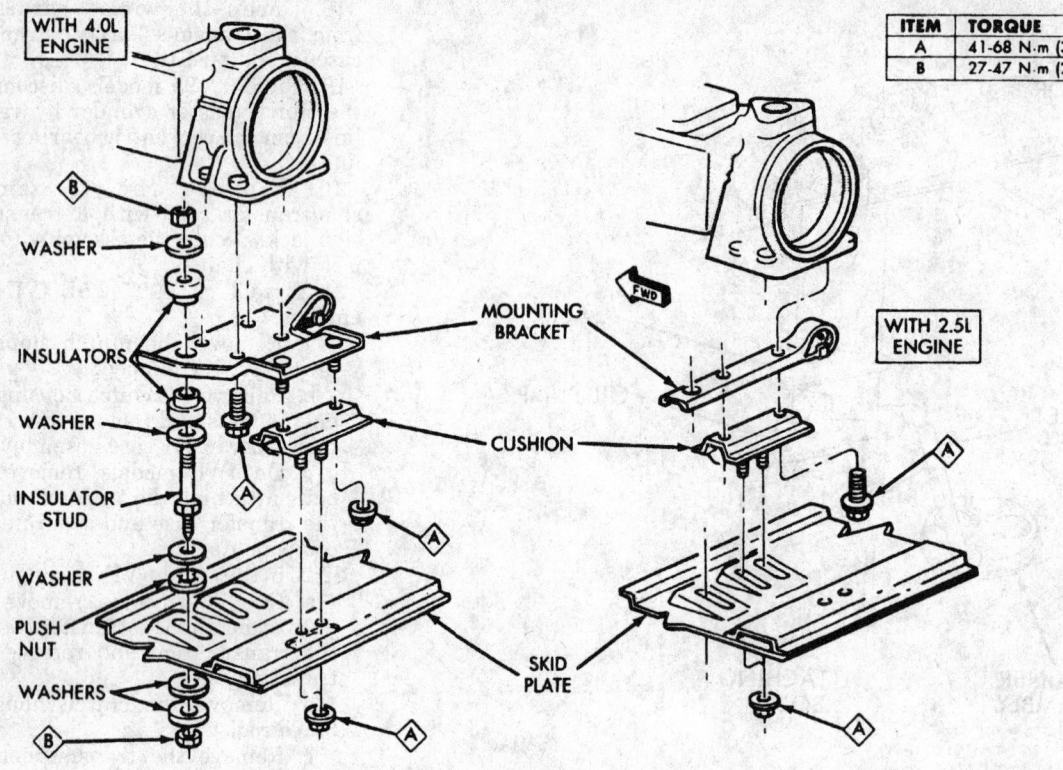

ITEM	TORQUE
A	41-68 N·m (30-50 ft. lbs.)
B	27-47 N·m (20-35 ft. lbs.)

7921dg43

Rear transmission mounting — Wrangler

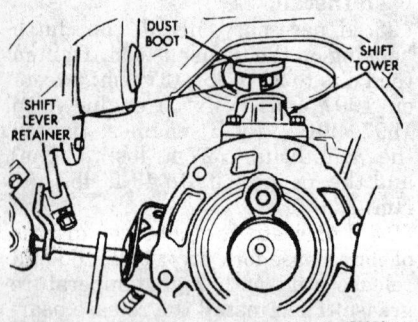

Removing and installting the shift lever

7921dg44

30. On 4WD models, align and install the transfer case on the transmission. Tighten the attaching hardware to 26 ft. lbs. (35 Nm).
31. Connect the vent hoses and shift linkage.
32. Connect the wire harnesses. Secure the wire harnesses in clip/tie straps on the transmission/transfer case.
33. Install the crankshaft position sensor.
34. Install the rear crossmember. Tighten the crossmember-to-frame bolts to 31 ft. lbs. (41 Nm). Tighten the transmission-to-rear support bolts/nuts to 33 ft. lbs. (45 Nm).

35. Remove the jackstand.
36. Align and install the front/rear driveshafts. Tighten the clamp bolts to 170 inch lbs. (19 Nm).
37. Install the skid plate, if equipped. Tighten the bolts to 31 ft. lbs. (42 Nm). Tighten the stud nuts to 150 inch lbs. (17 Nm).
38. Fill the transmission and transfer case (if equipped).
39. Lower the vehicle.
40. Connect the negative battery cable.

Grand Cherokee

1. Disconnect the negative battery cable.
2. Shift the vehicle into neutral, then raise and safely support the vehicle.
3. Remove the skid plate.
4. Matchmark the front and rear driveshaft and yoke for installation alignment.
5. Unbolt and remove the driveshaft(s).
6. Disconnect the transfer case shift linkage.
7. Disconnect the harness from the distance sensor.
8. Unclip the wiring harnesses from the transmission and transfer case.

9. Unplug the electrical connectors and vent hose.
10. Support the transmission/transfer case with a transmission jack. Secure the assembly to the jack with chains.
11. Support the engine with a jack positioned under the clutch housing or oil pan flange.
12. Remove the bolts securing the rear mount to the crossmember.
13. Remove the rear crossmember.
14. Remove the nuts attaching the transfer case to the transmission and remove the transfer case.
15. Reach up and around the transmission case and unseat the shift lever dust boot from the transmission shift tower. Reposition the boot to access the lever retainer.
16. Press the shift lever retainer downward and turn it counterclockwise to release it.
17. Lift the lever and retainer out of the shift tower.

NOTE: It is not necessary to remove the shift lever from the floor pan boot. Leave the lever in place for installation.

18. Remove the crankshaft position sensor.
19. Remove the slave cylinder from the clutch housing.

ITEM	TORQUE
A	54-75 N·m (40-55 FT.LBS.)
B	33-49 N·m (24-36 FT.LBS.)
C	33-60 N·m (24-44 FT.LBS.)

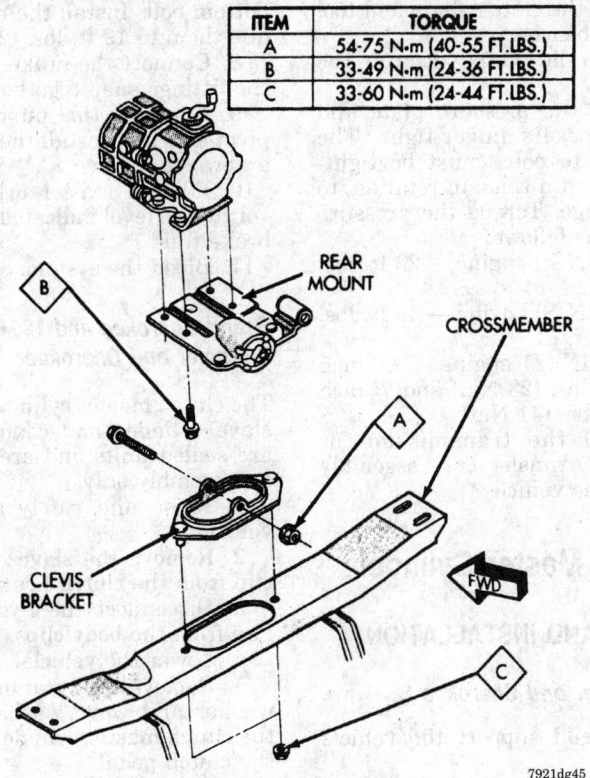

Rear transmission mounting — Grand Cherokee

7921dg45

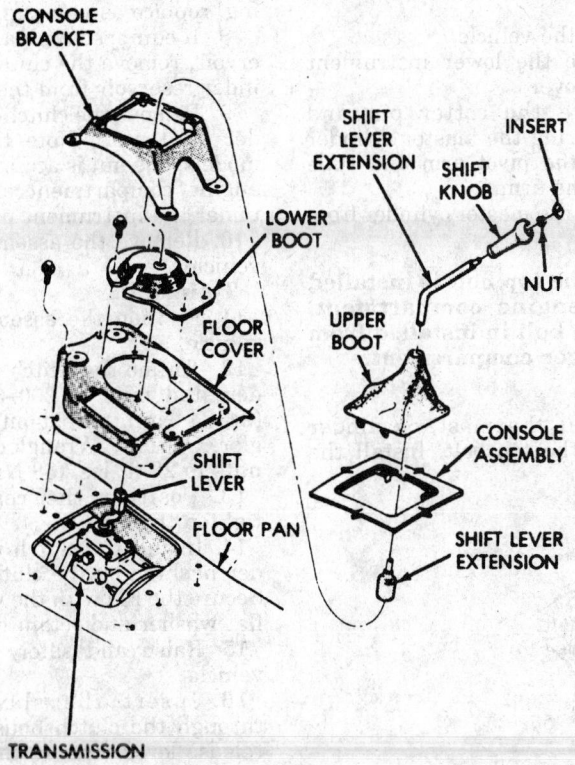

7921dg46

Shift lever attachment — Grand Cherokee

20. Remove the transmission-to-engine attaching bolts, then remove the transmission from the vehicle.

To install:

21. Lubricate the contact surfaces of the release fork pivot ball stud and release stud with high temperature grease. Then install the release bearing, fork and retainer clip.

22. Lubricate the pilot bearing and transmission input splines with high temperature grease.

23. Align the transmission input shaft, release bearing and clutch disc splines, then slide the transmission into place.

24. Install and tighten the mounting bolts to 45 ft. lbs. (61 Nm). Be sure the transmission is properly seated before installing the mounting bolts.

25. Reach up around the transmission and insert the shift lever in the shift tower. Press the lever retainer downward and turn it clockwise to lock it in place. Install the dust lever on the dust boot.

26. Install the slave cylinder.

27. Align and install the transfer case on the transmission. Tighten the attaching hardware to 26 ft. lbs. (35 Nm).

28. Move the support stand from under the engine and reposition it under the transmission. Remove the transmission jack.

29. Install the rear crossmember. Tighten the bolts to 30 ft. lbs. (41 Nm). Tighten the transmission-to-rear support hardware to 33 ft. lbs. (45 Nm).

30. Install the slave cylinder.

31. Install the crankshaft position sensor.

32. Connect the vent hoses and shift linkage.

33. Connect the wire harnesses. Secure the wire harnesses in clip/tie straps on the transmission/transfer case.

34. Align and install the front/rear driveshafts. Tighten the clamp bolts to 170 inch lbs. (19 Nm).

35. Install the skid plate, if equipped. Tighten the bolts to 31 ft. lbs. (42 Nm). Tighten the stud nuts to 150 inch lbs. (17 Nm).

36. Fill the transmission and transfer case (if equipped).

37. Lower the vehicle.

38. Connect the negative battery cable.

Clutch Assembly

REMOVAL AND INSTALLATION

All Models

1. Raise and safely support the vehicle.
2. Remove the transmission or transmission/transfer case assembly.
3. Matchmark the pressure plate and flywheel. Loosen the pressure plate bolts, a little at a time, in rotation, to avoid warpage.
4. Remove the pressure plate and clutch disc.
5. Inspect the flywheel for scoring, cracks, warpage or other wear; resurface or replace as necessary.
6. Inspect the pilot bearing for excessive wear or damage and replace as necessary.

To install:
7. If removed, install the pilot bearing after lubricating with grease. Seat the bearing in the crankshaft with a clutch alignment tool.
8. Check the clutch disc runout by installing the disc on the transmission input shaft. Runout should not exceed 0.020 in. (0.5mm) when measured ¼ inch from the outer edge of the facing.

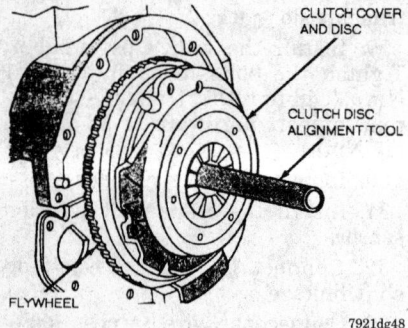

FLYWHEEL

CLUTCH COVER AND DISC

CLUTCH DISC ALIGNMENT TOOL

7921dg48

Clutch disc alignment

9. Install the clutch alignment tool in the pilot bearing.
10. Install the clutch disc on the tool.
11. Install the pressure plate and tighten the bolts finger-tight. The pressure plate bolts must be tightened a little at a time, in rotation, to avoid warpage. Torque the pressure plate bolts as follows:
- 2.5L (VIN P) engine — 23 ft. lbs. (31 Nm)
- 4.0L (VIN S) engine — 40 ft. lbs. (54 Nm)
- 5.2L (VIN Y) engine — ⁵⁄₁₆ inch bolts: 17 ft. lbs. (23 Nm) and ³⁄₈ inch bolts 30 ft. lbs. (41 Nm)
12. Install the transmission or transmission/transfer case assembly and lower the vehicle.

Clutch Master Cylinder

REMOVAL AND INSTALLATION

1993 Wrangler and Cherokee

1. Raise and support the vehicle safely.
2. Disconnect the quick-disconnect fitting by pushing the round disc inward to unsnap and separate the fittings.
3. Lower the vehicle.
4. Remove the lower instrument panel trim cover.
5. Remove the cotter pin and washer securing the master cylinder pushrod to the pivot arm and slide the rod off the arm.
6. Unbolt the master cylinder from the firewall.

NOTE: The top bolt is installed from the engine compartment. The bottom bolt in installed from the passenger compartment.

To install:
7. Position the master cylinder and install the top bolt. Install the

SPACER PLATE

CLUTCH DISC

CLUTCH COVER

HYDRAULIC THROWOUT BEARING

CLUTCH HOUSING

PILOT BEARING

PRESSURE PLATE BOLTS

PIN

RETAINING NUT

7921dg47

Exploded view of the clutch components

bottom bolt. Install the nuts and torque them to 19 ft. lbs. (26 Nm).
8. Connect the brake line. Ensure the fittings snap together securely.
9. Connect the pushrod to the pivot arm and install the washer and a new cotter pin.
10. Fill the master cylinder reservoir to the level indicated with DOT 3 brake fluid.
11. Bleed the system.

Grand Cherokee and 1994–97 Wrangler and Cherokee

The clutch master cylinder, reservoir, slave cylinder and connecting lines are sealed units and are serviced as an assembly only.

1. Raise and safely support the vehicle.
2. Remove the slave cylinder and clip from the clutch housing.
3. Disconnect the hydraulic fluid line from the body clips.
4. Lower the vehicle.
5. Remove the retaining ring, flat washer and wave washer attaching the clutch master cylinder pushrod to the clutch pedal.
6. Slide the master cylinder pushrod off the clutch pedal pin.
7. Inspect the clutch pedal bushing, replace as necessary.
8. If equipped with a separate reservoir, remove the clutch master cylinder reservoir from the dash panel.
9. Remove the clutch master cylinder stud nuts. Note that on some models, one nut is accessible from the engine compartment and one from under the instrument panel.
10. Remove the assembly from the vehicle.

To install:
11. Position the assembly into the vehicle.
12. Torque the clutch master cylinder stud nuts to 200–300 inch lbs. (24–34 Nm), except on 1997 Wrangler. On 1997 Wrangler, tighten the nuts to 28 ft. lbs. (38 Nm).
13. Position the reservoir and tighten the screws.
14. Install the clutch master cylinder pushrod on the clutch pedal pin. Secure the rod with the wave washer, flat washer and retaining ring.
15. Raise and safely support the vehicle.
16. Insert the slave cylinder through the clutch housing into the release lever. Ensure the cap on the end of the rod is securely engaged in the lever. Torque the bolts to 200–300 inch lbs. (24–34 Nm), except on 1997 Wrangler. On 1997 Wrangler, tighten the nuts to 17 ft. lbs. (23 Nm).

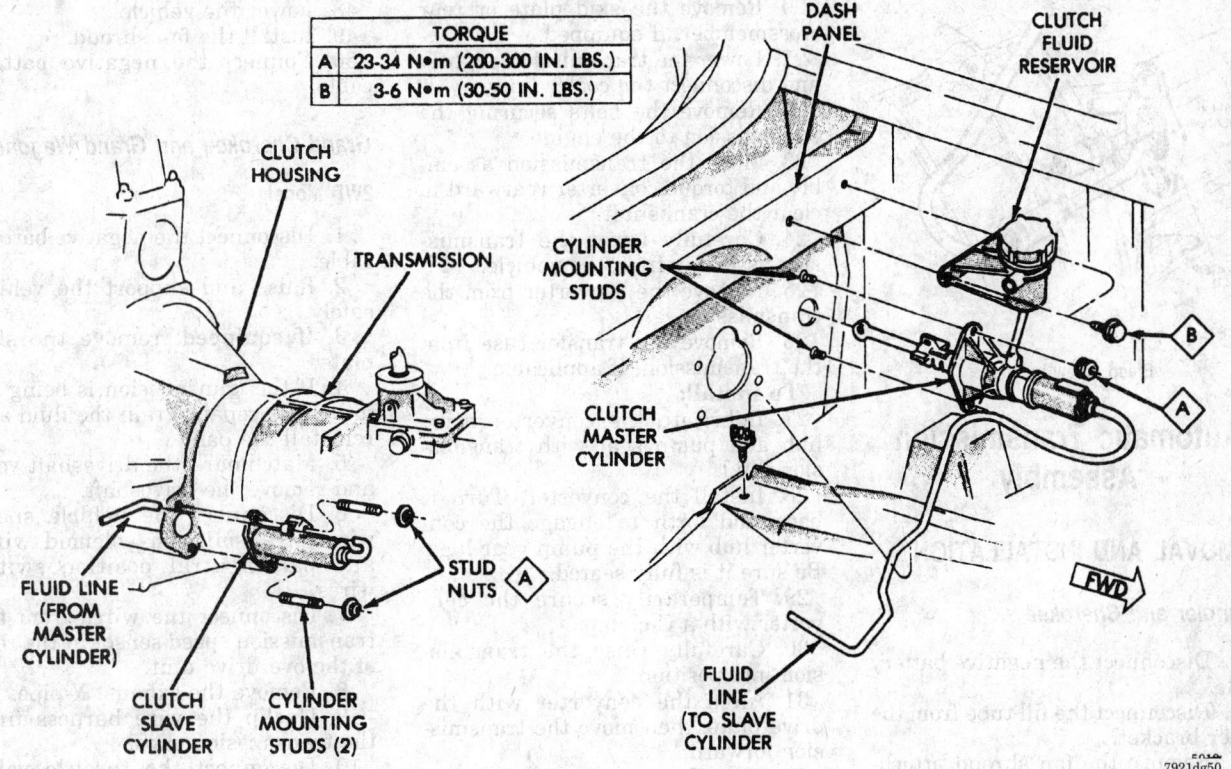

TORQUE	
A	23-34 N•m (200-300 IN. LBS.)
B	3-6 N•m (30-50 IN. LBS.)

Hydraulic system components

17. Insert the fluid line in the body clips.

Clutch Slave Cylinder

REMOVAL AND INSTALLATION

1993 Wrangler and Cherokee

The hydraulic concentric bearing is serviced as an assembly only. The release bearing portion of the assembly is permanently attached to the piston. The hydraulic lines are also permanently attached.

1. Disconnect the negative battery cable.
2. Raise and support the vehicle safely.
3. Remove the transmission.
4. Disconnect the clutch master cylinder fluid line.
5. Remove the insulator plate bolts and slide the plate off the bleed line.
6. Remove the concentric bearing retaining nut.
7. Remove the concentric bearing from the transmission input shaft. If the bearing will be reused, secure the

bearing and piston with rubber bands.

To install:

8. Inspect the bearing mounting pin and replace the front cover if the pin is damaged. Install the concentric bearing on the transmission input shaft.
9. Guide the bearing fluid and bleed lines through the openings in the clutch housing.
10. Position the bearing boss on the mounting pin and seat the bearing against the transmission. Install a new retaining nut and unhook the T-handle straps retaining the bearing.
11. Install the insulator and plate.
12. Install the transmission and transfer case and connect the clutch master cylinder fluid line.
13. Fill and bleed the clutch hydraulic system.

Hydraulic System Bleeding

PROCEDURE

NOTE: This procedure only applies to 1993 Cherokee and Wrangler models. The replacement

reservoir, master cylinder, slave cylinder and fluid line assembly is prefilled with fluid from the factory for Grand Cherokee and 1994-97 Wrangler and Cherokee models. Bleeding is not necessary or possible on these models.

1. Attach a hose to the clutch bleed screw and immerse the end of the hose in a clear container partially filled with clutch fluid.
2. Open the bleed screw.
3. Fill the clutch master cylinder with fluid until the fluid running out is free of air bubbles. Close the bleed screw.
4. Check the pedal freeplay. If it exceeds ½ in. (12.7mm), continue the bleeding procedure.
5. Open the bleeder screw.
6. Have an assistant depress the pedal to the floor.
7. Tighten the bleed valve, then release the pedal. Continue until the fluid running out is free of air bubbles. Do not allow the reservoir to run out of fluid.
8. Check the pedal freeplay again. If it still exceeds ½ in. (12.7mm), some air is still trapped in the system. Repeat the bleeding procedure.

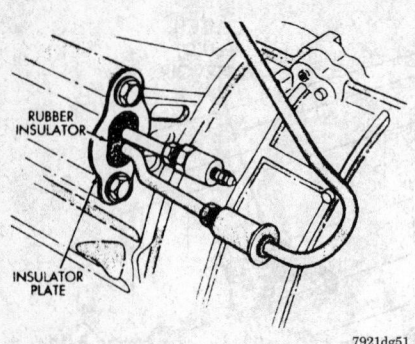

Bleed screw location

Automatic Transmission Assembly

REMOVAL AND INSTALLATION

Wrangler and Cherokee

1. Disconnect the negative battery cable.
2. Disconnect the fill tube from the upper bracket.
3. Remove the fan shroud attaching bolts.
4. Raise and support the vehicle safely.
5. Remove the torque converter inspection cover. Remove the skid plate for access if necessary.
6. Remove the transmission fill tube and O-ring.
7. Remove the starter.
8. Matchmark the driveshaft assemblies for installation.
9. Disconnect the driveshafts and secure the assemblies aside.
10. Disconnect the exhaust pipe at the manifold.
11. Drain the transfer case fluid, if the transfer case is to be serviced as well.
12. Unplug the wiring harnesses, hoses and linkages from the transmission and transfer case.
13. On 1995–97 models, remove the transfer case.
14. Remove the crankshaft position sensor.
15. Matchmark the converter driveplate and converter assembly for reassembly.
16. Remove the bolts attaching the torque converter to the flexplate.
17. Support the engine with a support stand.
18. Support the transmission with a transmission jack. Secure it with a chain.
19. Remove the hardware securing the cushion and torque arm bracket to the skid plate.

20. Remove the skid plate or rear crossmember, if equipped.
21. Lower the transmission slightly and disconnect the cooler lines.
22. Remove the bolts securing the transmission to the engine.
23. Move the transmission assembly and torque converter rearward to clear the crankshaft.
24. Carefully lower the transmission assembly from the vehicle.
25. Remove the converter from the transmission.
26. Remove the transfer case from the transmission, if applicable.

To install:
27. Lubricate the converter drive hub and pump seal with transmission fluid.
28. Install the converter. Turn it back and forth to engage the converter hub with the pump gear lugs. Be sure it is fully seated.
29. Temporarily secure the converter with a C-clamp.
30. Carefully raise the transmission into position.
31. Align the converter with the drive plate, then move the transmission forward.
32. Install two attaching bolts. Tighten the bolts just enough to hold the assembly in place.
33. Install the bolts attaching the converter. On 1997 Wrangler, tighten the bolts as follows:
- 9.5 in. 3 lug converter: 40 ft. lbs. (54 Nm)
- 9.5 in. 4 lug converter: 55 ft. lbs. (74 Nm)
- 10 in. 4 lug converter: 55 ft. lbs. (74 Nm)
- 10.75 in. 4 lug converter: 270 inch lbs. (31 Nm)
34. Install the bolts attaching the transmission assembly to the engine. Torque the bolts as shown in the illustration.
35. Install the crankshaft position sensor.
36. Install the fill tube and O-ring.
37. Connect the cooler lines.
38. Install the transfer case.
39. Install the rear crossmember and attach the transmission rear support to the crossmember.
40. Remove the transmission jack.
41. Connect the wiring harnesses, hoses and linkages.
42. Install the inspection cover on the converter housing.
43. Install the exhaust pipes.
44. Install the starter motor.
45. Install the driveshafts.
46. Install the skid plate and rear cushion, if removed.
47. Fill the transmission and transfer case with the appropriate fluid.

48. Lower the vehicle.
49. Install the fan shroud.
50. Connect the negative battery cable.

Grand Cherokee and Grand Wagoneer

2WD Models

1. Disconnect the negative battery cable.
2. Raise and support the vehicle safely.
3. If equipped, remove the skid plate.
4. If the transmission is being removed for repair, drain the fluid and reinstall the pan.
5. Matchmark the driveshaft yoke and remove the driveshaft.
6. Disconnect the vehicle speed wires, transmission solenoid wires and park-neutral position switch wires.
7. Disconnect the wires from the transmission speed sensor at the rear of the overdrive unit.
8. Remove the exhaust Y-pipe.
9. Unclip the wire harness from the transmission clips.
10. Disconnect the throttle valve and gearshift cables from the levers on the valve body manual shaft. Position the cables aside and secure them to the underbody.
11. Remove the dust cover from the transmission converter housing.
12. Remove the starter.
13. Remove the bolts attaching the converter to the driveplate.
14. Disconnect the cooler fluid lines from the transmission.
15. Support the transmission with a jack.
16. Remove the nuts and bolts securing the rear crossmember to the insulator and remove the crossmember.
17. Lower the jack to gain access to the upper portion of the transmission.
18. Remove the crankshaft position sensor.
19. Remove the transmission fill tube and discard the O-ring.
20. Remove the bolts attaching the transmission to the engine.
21. Slide the transmission back and secure a C-clamp to the converter.
22. Remove the transmission.

To install:
23. Ensure the torque converter hub and hub drive are free from sharp edges, scratches or nicks. Polish with 400 grit sandpaper if necessary.
24. Lubricate the converter hub and pump seal with high temperature grease.

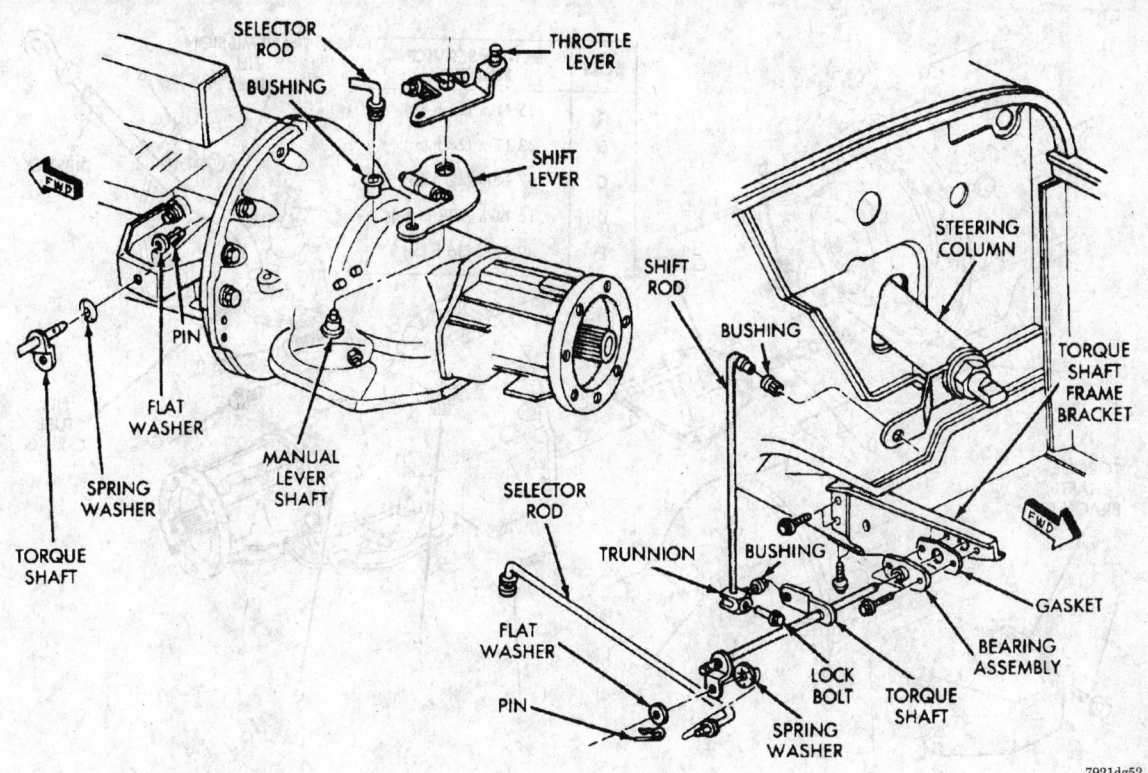

Shift linkage — Wrangler

7921dg52

BOLT	SET-TO SERVICE TORQUE
A	58 N·m (43 ft-lbs)
B	20 N·m (181 in-lbs)

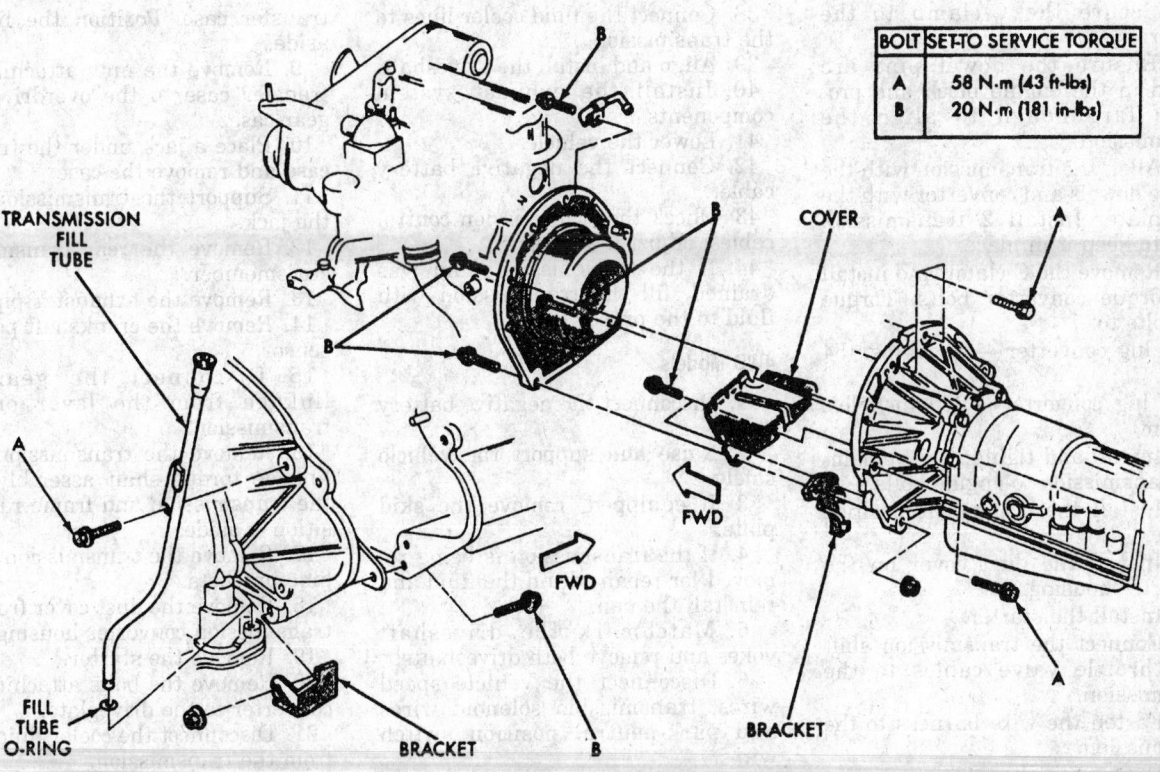

AW-4 transmission installation

7921dgaw

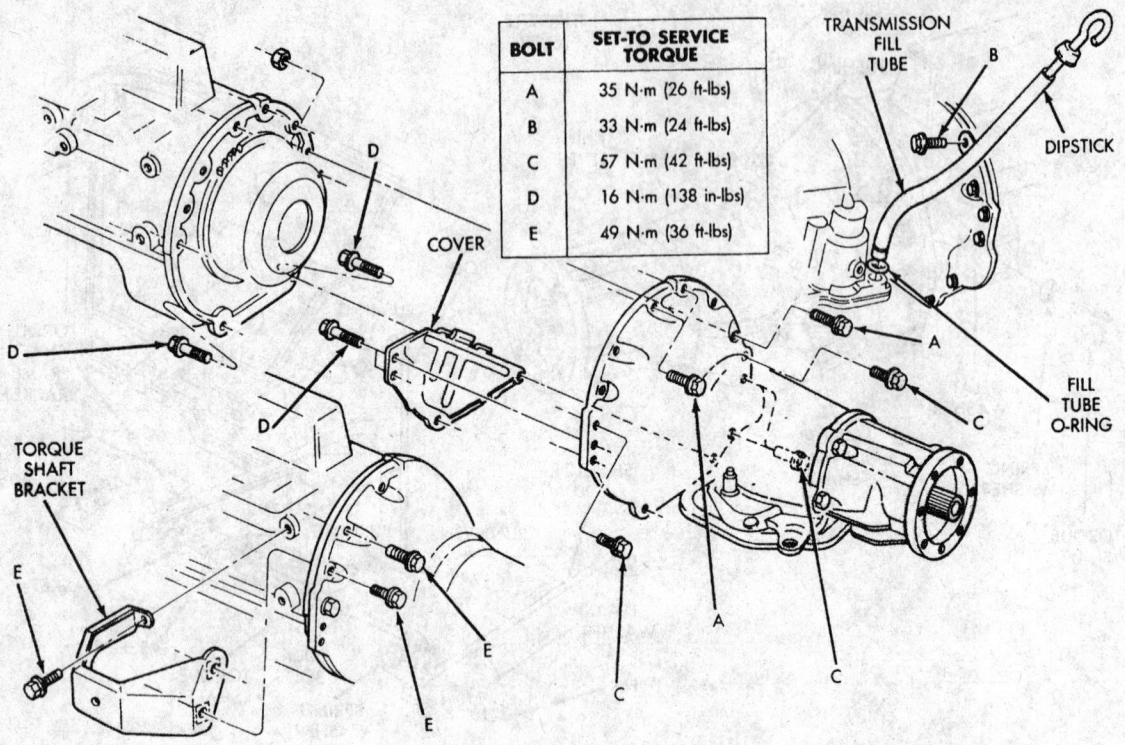

BOLT	SET-TO SERVICE TORQUE
A	35 N·m (26 ft-lbs)
B	33 N·m (24 ft-lbs)
C	57 N·m (42 ft-lbs)
D	16 N·m (138 in-lbs)
E	49 N·m (36 ft-lbs)

30RH/32RH transmission installation

7921dgax

25. Secure the C-clamp to the converter.

26. Ensure the dowel pins are seated in the engine block and protrude far enough to align the transmission.

27. Align the transmission with the engine dowels and converter with the driveplate. Install 2 transmission bolts to keep it in place.

28. Remove the C-clamp and install the torque converter bolts. Torque the bolts to:

- 3 lug converter — 40 ft. lbs. (54 Nm)
- 4 lug converter — 270 inch lbs. (31 Nm)

29. Install and tighten the remaining transmission-to-engine bolts.

30. Install the crankshaft position sensor.

31. Install the dust cover on the converter housing.

32. Install the starter.

33. Connect the transmission shift and throttle valve cables to the transmission.

34. Fasten the wire harness to the transmission.

35. Connect the harness connectors disconnected during removal.

36. Install the transmission filler tube with a new O-ring.

37. Install the rear crossmember.

38. Connect the fluid cooler lines to the transmission.

39. Align and install the driveshaft.

40. Install the exhaust system components.

41. Lower the vehicle.

42. Connect the negative battery cable.

43. Check the transmission control cables; adjust if necessary.

44. If the transmission fluid was drained, fill the transmission with fluid to the proper level.

4WD Models

1. Disconnect the negative battery cable.

2. Raise and support the vehicle safely.

3. If equipped, remove the skid plate.

4. If the transmission is being removed for repair, drain the fluid and reinstall the pan.

5. Matchmark the driveshaft yokes and remove both driveshafts.

6. Disconnect the vehicle speed wires, transmission solenoid wires and park-neutral position switch wires.

7. Unclip the wire harness from the transmission clips.

8. Disconnect the transfer case shift linkage from the lever. Remove the linkage and bracket from the

transfer case. Position the linkage aside.

9. Remove the nuts attaching the transfer case to the overdrive unit gear case.

10. Place a jack under the transfer case and remove the case.

11. Support the transmission with the jack.

12. Remove the rear transmission crossmember.

13. Remove the exhaust Y-pipe.

14. Remove the crankshaft position sensor.

15. Disconnect the gearshift linkage from the lever on the transmission.

16. Remove the transmission shift linkage torque shaft assembly from the transmission and frame rail. Position it aside.

17. Remove the transmission-to-engine brackets.

18. Remove the dust cover from the transmission converter housing.

19. Remove the starter.

20. Remove the bolts attaching the converter to the driveplate.

21. Disconnect the cooler fluid lines from the transmission.

22. Disconnect the solenoid and park/neutral position wire switch wires.

23. Remove the transmission fill tube and discard the O-ring.

24. Lower the jack to gain access to the upper portion of the transmission.

25. Remove the bolts attaching the transmission to the engine.

26. Slide the transmission back and secure a C-clamp to the converter.

27. Move the transmission rearward until it clears the engine block dowels.

NOTE: On some models, part of the flange joining the vehicle cab and dash panel may interfere with transmission removal. If necessary peen this part of the flange over with a mallet.

28. Remove the transmission.

To install:

29. Ensure the torque converter hub and hub drive are free from sharp edges, scratches or nicks. Polish with 400 grit sandpaper if necessary.

30. Lubricate the converter hub and pump seal with high temperature grease.

31. Secure the C-clamp to the converter.

32. Ensure the dowel pins are seated in the engine block and protrude far enough to align the transmission.

33. Align the transmission with the engine dowels and converter with the driveplate. Install 2 transmission bolts to keep it in place.

34. Remove the C-clamp and install the torque converter bolts. Torque the bolts as follows:
- 3 lug converter — 40 ft. lbs. (54 Nm)
- 4 lug, except 10.75 inch converter — 55 ft. lbs. (74 Nm)
- 4 lug, 10.75 inch converter — 270 inch lbs. (31 Nm)

35. Install the starter.

36. Install the strut brackets securing the transmission to the engine and front axle.

37. Install and tighten the remaining transmission–to–engine bolts.

38. Install the crankshaft position sensor.

39. Install the transmission filler tube with a new O-ring.

40. Install the exhaust system components.

41. Install the shift linkage torque bracket.

42. Connect the shift linkage to the transmission.

43. Connect the harness connectors disconnected during removal.

44. Install the rear crossmember.

45. Install the transfer case. Torque the nuts as follows:
- 3/8 stud nuts — 35 ft. lbs. (47 Nm)
- 5/16 stud nuts — 26 ft. lbs. (35 Nm)

46. Install the damper on the transfer case rear retainer if removed. Torque the nuts to 40 ft. lbs. (54 Nm).

47. Connect the transfer case shift linkage.

48. Connect the fluid cooler lines to the transmission.

49. Align and install the driveshafts. Torque the U-joint clamp bolts to 170 inch lbs. (19 Nm).

50. Fill the transfer case to the proper level with fluid.

51. Lower the vehicle.

52. Connect the negative battery cable.

53. Fill the transmission to the proper level with Mopar ATF Plus or equivalent.

54. Check the transmission control cables, adjust if necessary.

55. Check and adjust the transmission and transfer case shift linkage.

Throttle Valve Cable

ADJUSTMENT

AW-4 and 30RH/32RH

1. Turn the ignition switch to **OFF**.

2. Press the throttle cable button all the way down, then, push the cable plunger inward.

3. Rotate the primary throttle lever to the wide open throttle position.

4. Hold the primary throttle lever in this position and let the cable plunger extend.

5. Release the lever when the plunger is fully extended.

6. The cable is now adjusted.

42RE/44RE

1. Turn the ignition switch to **OFF**.

2. Disconnect the cable from the stud.

3. Press the cable lock button inward to release the cable.

4. Center the end of the cable to the stud within 0.039 in. (1mm).

5. Install the cable onto the stud.

46RH

1. Turn the ignition switch to **OFF**.

2. Remove the air cleaner.

3. Position a 0.110–0.120 in. (2.8–3.0mm) feeler gauge between the idle stop and throttle lever.

4. Press the cable lock button to release the cable.

5. Raise the vehicle for access to the throttle valve lever.

6. Rotate the lever towards the front of the vehicle until the ratcheting sound stops.

7. Remove the feeler gauges and check the cable adjustment. The throttle lever should begin to move at the same time the lever on the throttle body moves off idle.

8. Lower the vehicle and install the air cleaner.

DRIVE AXLE

Driveshaft

REMOVAL AND INSTALLATION

All Models

Front

NOTE: These vehicles may come equipped with one of 2 different type front driveshafts. The first type has a conventional universal joint at the axle but a double offset joint at the transfer case. The second type has a conventional universal joint at the axle and a double cardan joint at the transfer case.

1. Place the transmission and transfer case in **N**.

2. Raise and safely support the vehicle.

3. Matchmark the shaft ends, axle and transfer case.

4. Remove the U-joint strap bolts at the front axle yoke.

5. Remove the double offset joint flange nuts at the transfer case.

To install:

6. Position the driveshaft on the transfer case output shaft and axle yoke aligning the marks.

7. Torque the new U-joint strap-to-axle yoke bolts to 14 ft. lbs. (19 Nm).

8. Torque the new U-joint strap-to-transfer case bolts to 20 ft. lbs. (27 Nm).

9. Lower the vehicle.

Rear

1. Place the transmission in **N**.

2. Raise and safely support the vehicle.

3. Matchmark the yokes and flanges.

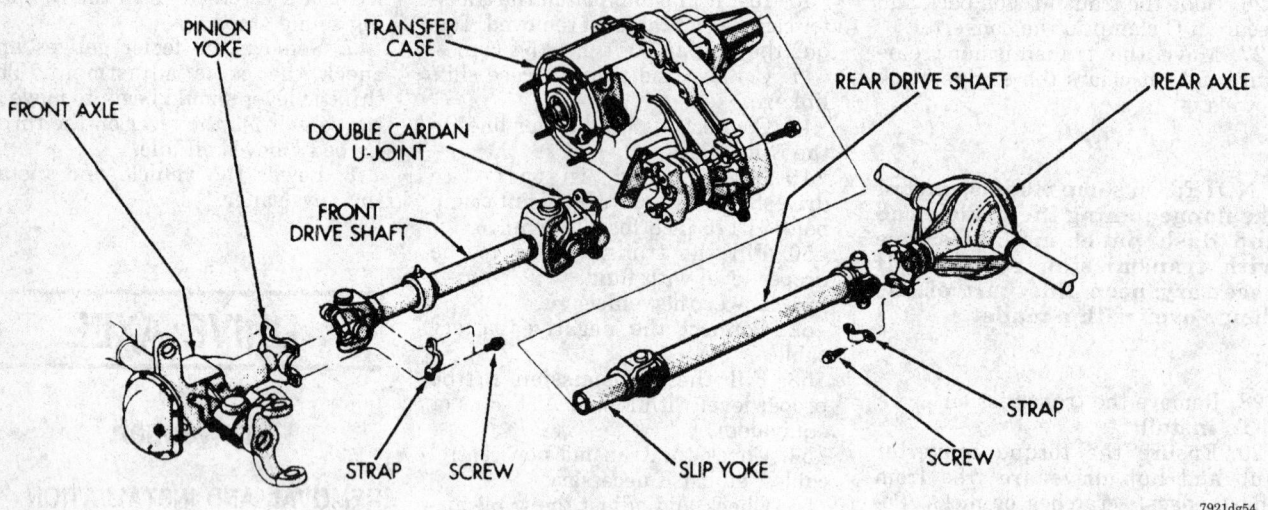

Front and rear driveshafts

7921dg54

4. If equipped with Command-Trac®, the driveshaft may be removed by disconnecting it at the axle and sliding it from the front yoke, leaving the front yoke attached to the transfer case. If done this way, matchmark the driveshaft and front yoke before separation. On some models, it will be necessary to pry the clamp from the dust boot.

5. If equipped with Selec-Trac®, disconnect the yokes from the axle and transfer case. Remove the driveshaft.

6. Installation is the reverse of removal. Torque the U-joint strap nuts to 14 ft. lbs. (19 Nm). If applicable, crimp the clamp of the dust boot.

NOTE: New U-joint straps should be used.

U-Joints

REMOVAL AND INSTALLATION

All Models

Most Jeep vehicles use a conventional universal joint at both ends of both driveshafts. All models have C-type retainer rings on the inside of the bearing caps. The constant velocity joint is also assembled with snaprings on the outside.

The U-joints are not serviceable and if defective, must replaced as a unit. If the socket yokes, balls, springs, bearings, seals, thrust washers, spiders or bearing caps are damaged or worn, replace the complete U-joint.

Single Cross Cardan Joint

1. Clamp the yoke, not the tube, in a vise.
2. Remove the bearing cap C-retainers. Tap on the bearing caps to relieve pressure as necessary.
3. Support the yoke on the vise jaws.
4. Tap one bearing cap in until the opposite one comes out.
5. Turn the yoke around and tap the exposed end of the spider to drive the remaining bearing cap out.

To install:
6. Lubricate all needle bearings, bearing caps, and bearing surfaces with chassis grease.
7. Place the seals on the spider.
8. Install one cap and needle bearing assembly partway into the shaft yoke.
9. Install the spider and the opposite bearings and cap.
10. Support the yoke and seat both caps with a hammer.
11. Install the retainer C-clips. Tap the bearing caps as necessary.

12. Install the other two cap and bearing assemblies. Hold them in place with tape until the shaft is reinstalled.

Double Cardan Joint

1. Remove the bearing cap retainer snaprings.
2. Mark all components for reassembly.
3. Use a 5/8 in. socket as a bearing cap driver and a 1¹/₁₆ in. socket as a bearing cap receiver. Squeeze the assembly in a vise to force out the bearing caps.
4. Repeat the operation of step 3 to remove the bearing caps at the other end of the joint.
5. Clean all parts in solvent and dry.

NOTE: Do not disassemble the socket yoke, centering ball, spring, needle bearings, retainer, and thrust washers. These parts are sold as an assembly only.

To install:
6. Lubricate all bearings and contact surfaces with chassis grease.
7. Install the bearing caps on the transfer case yoke ends of the rear spider. Tape them in place.
8. Assemble the socket yoke and the rear spider.

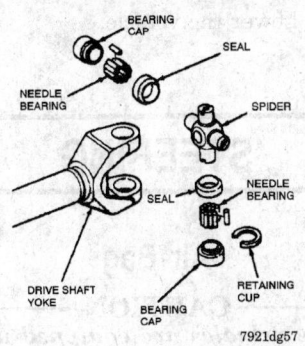

Exploded view fo a single cardan U-joint

9. Place the rear spider in the link yoke and install the bearing caps. Press them into place with the ⅝ in. socket. Install the snaprings.

10. Install the front spider, bearing caps, and snaprings in the driveshaft yoke.

11. Install the thrust washer and socket spring in the ball socket bearing bore. Install the thrust washer on the ball socket bearing boss on the driveshaft yoke. Align the ball socket bearing boss with the ball socket bearing bore and insert the boss into the bore.

12. Align the front spider with the link yoke and install the bearing caps and snaprings.

Front Axle Shaft

REMOVAL AND INSTALLATION

All Models

1. Raise and support the vehicle safely.

2. Remove the wheel.

3. Remove the caliper and wheel.

4. Remove the hub and bearing assembly.

5. If equipped with a shift motor, perform the following:

    a. Disconnect the vacuum and wiring connector from the shift housing.

    b. Remove the indicator switch.

    c. Remove the shift motor housing cover, gasket and shield from the housing.

    d. Remove the E-clips from the shift motor housing and shaft. Remove the shift motor and fork from the housing.

    e. Remove the O-ring seal from the shift motor shaft.

6. Remove the axle from the housing, being careful to avoid damaging the axle shaft oil seals in the differential.

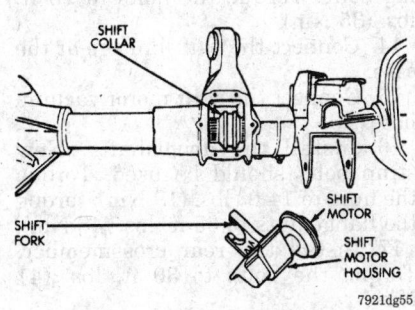

Shift motor housing and shift collar

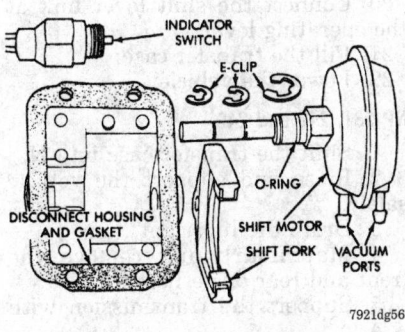

Shift motor components

To install:

7. Thoroughly clean the axle shaft and apply a thin film of wheel bearing grease to the shaft splines, seal contact surface and hub bore.

8. Carefully install the axle shaft, being careful to avoid damaging the differential seals.

9. If equipped with a shift motor, perform the following:

    a. Install a new O-ring seal on the shift motor.

    b. Insert the shift motor shaft through the hole in the housing and shift fork. The shift fork offset should be toward the differential.

    c. Install the E-clips on the shift motor shaft and housing.

    d. Install the shift motor housing gasket and cover. Ensure the shift fork is correctly guided into the shift collar groove.

    e. Install the shift motor shield and torque the bolts to 101 inch lbs. (11 Nm).

    f. Add 5 ounces of API grade 5 hydraulic gear lubricant or equivalent, into the shift motor housing through the indicator switch mounting hole.

    g. Install the indicator switch.

    h. Connect the electrical connector and vacuum hoses.

10. Install the wheel bearing and hub assembly.

11. Install the rotor and caliper.

12. Install the wheel.

Transfer Case Assembly

REMOVAL AND INSTALLATION

All Models

NP 207

1. Shift the case into **4H**.

2. Raise and safely support the vehicle.

3. Drain the transfer case.

4. Matchmark the rear driveshaft and remove it.

5. Disconnect the speedometer cable, vacuum hoses and vent hose from the case.

6. Support the transmission with a transmission jack.

7. Remove the crossmember.

8. Matchmark the front driveshaft and remove it.

9. Disconnect the shift lever linkage rod at the case.

10. Remove the shift lever bracket bolts.

11. Support the transfer case with a jack and remove the attaching bolts.

12. Pull the case out of the vehicle.

To install:

13. Raise the transfer case into position. Torque the attaching bolts to 26 ft. lbs. (35 Nm).

14. Connect the shift lever linkage rod at the case.

15. Install the shift lever bracket bolts.

16. Install the front driveshaft. New strap bolts should be used. Torque the nuts to 14 ft. lbs. (19 Nm). Torque the flange bolts to 35 ft. lbs. (47 Nm).

17. Install the crossmember. Torque the bolts to 30 ft. lbs. (41 Nm).

18. Remove the support jack.

19. Connect the speedometer cable, vacuum hoses and vent hose at the case.

20. Install the rear driveshaft. Use new strap bolts, torqued to 14 ft. lbs. (19 Nm). Torque the flange bolts to 35 ft. lbs. (47 Nm).

21. Fill the transfer case to the correct level.

22. Lower the vehicle.

NP 228

1. Raise and safely support the vehicle.

2. Drain the transfer case.

3. Disconnect the speedometer cable, vacuum hoses and vent hose from the case.

4. Disconnect the shift lever linkage rod at the case.

5. Support the transmission with a jack.

6. Remove the rear crossmember.

7. Matchmark the rear driveshaft and remove it.

8. Matchmark the front driveshaft and remove it.

9. Remove the shift lever bracket bolts.

10. Support the transfer case with a jack and remove the attaching bolts.

11. Pull the case rearward out of the vehicle.

To install:

12. Position the transfer case in the vehicle.

13. Torque the attaching bolts to 40 ft. lbs. (54 Nm).

14. Install the shift lever bracket bolts.

15. Install the front and rear driveshafts. New strap bolts should be used. Torque the strap bolt nuts to 14 ft. lbs. (19 Nm); the flange nuts to 35 ft. lbs. (47 Nm).

16. Install the rear crossmember. Torque the bolts to 30 ft. lbs. (41 Nm).

17. Connect the shift lever linkage rod at the case.

18. Connect the speedometer cable, vacuum hoses and vent hose from the case.

19. Fill the transfer case to the correct level.

20. Lower the vehicle.

NP 229

1. Raise and safely support the vehicle.

2. Drain the transfer case.

3. Disconnect the speedometer cable and vent hose.

4. Disconnect the shift lever link at the operating lever.

5. Support the transmission with a jack.

6. Remove the rear crossmember.

7. Matchmark the driveshafts and remove them.

8. Disconnect the shift motor vacuum hoses.

9. Disconnect the shift linkage at the case.

10. Support the transfer case with a floor jack or transmission jack and remove the attaching bolts.

11. Pull the case rearward and remove it.

12. Clean the gasket mating surfaces and use new gasket material for installation.

To install:

13. Raise the transfer case into position. Make certain the case and transmission are mated without binding, before torquing the attach-

ing bolts. Torque the bolts to 26 ft. lbs. (35 Nm).

14. Connect the shift linkage at the case.

15. Connect the shift motor vacuum hoses.

16. Install the driveshafts. New strap bolts should be used. Torque the nuts to 14 ft. lbs. (19 Nm); torque the flange nuts to 35 ft. lbs. (48 Nm).

17. Install the rear crossmember. Torque the bolts to 30 ft. lbs. (41 Nm).

18. Remove the transmission support jack.

19. Connect the speedometer cable and vent hose.

20. Connect the shift lever link at the operating lever.

21. Fill the transfer case.

22. Lower the vehicle.

NP 231, 242 and 249

1. Shift the transfer case into **N**.

2. Raise and support the vehicle safely.

3. Drain the lubricant.

4. Matchmark and remove the front and rear driveshafts.

5. Support the transmission with a jack.

6. Remove the rear crossmember.

7. Disconnect the speedometer cable or speed sensor connector.

8. Disconnect the linkage.

9. Disconnect the vent and vacuum hoses and the indicator wire.

10. Support the transfer case with a transmission jack.

11. Remove the transfer case-to-transmission bolts.

12. Pull the case rearward to disengage it and lower it from the truck.

To install:

13. Raise the transfer case into position. Make certain the case and transmission are mated without binding, before torquing the attaching bolts. Torque the bolts to 26 ft. lbs. (35 Nm).

14. Connect the shift linkage at the case.

15. Connect the vacuum hoses.

16. Install the driveshafts. New strap bolts should be used. Torque the nuts to 14 ft. lbs. (19 Nm); torque the flange nuts to 35 ft. lbs. (48 Nm).

17. Install the rear crossmember. Torque the bolts to 30 ft. lbs. (41 Nm).

18. Remove the transmission floor jack.

19. Connect the speedometer cable or speed sensor connector and vent hose.

20. Connect the shift lever link at the operating lever.

21. Fill the transfer case to the correct lever.

22. Lower the vehicle.

STEERING

Air Bag

——— **CAUTION** ———

Some vehicles are equipped with an air bag system, also known as the Supplemental Inflatable Restraint (SIR) or Supplemental Restraint System (SRS). The system must be disabled before performing service on or around system components, steering column, instrument panel components, wiring and sensors. Failure to follow safety and disabling procedures could result in accidental air bag deployment, possible personal injury and unnecessary system repairs.

PRECAUTIONS

Several precautions must be observed when handling the inflator module to avoid accidental deployment and possible personal injury.

• Never carry the inflator module by the wires or connector on the underside of the module.

• When carrying a live inflator module, hold securely with both hands, and ensure that the bag and trim cover are pointed away.

• Place the inflator module on a bench or other surface with the bag and trim cover facing up.

• With the inflator module on the bench, never place anything on or close to the module which may be thrown in the event of an accidental deployment.

DISARMING

Cherokee

1. Disconnect the negative battery cable and isolate.

2. Using a small prytool, remove the plastic cover plug from the top outer surface of the steering wheel hub.

3. Exit the vehicle and disarm the air bag by reaching through the driver's side window and turning the arming screw counterclockwise to its travel limit. When the screw has reached its travel limit, it will extend 1 inch above the outer surface of the

wheel hub cover. Use an 8mm socket; do not use power driven tools.

4. To arm the system, exit the vehicle and reach through the driver's side window. Turn the arming screw clockwise to its travel limit. Torque should not exceed 10–15 inch lbs. (1.1–1.7 Nm). Reinstall the plastic cover.

Except Cherokee

1. Disconnect and isolate the negative battery cable. Wait two minutes for the system capacitor to discharge before performing any service.
2. To arm the system, connect the negative battery cable.

Steering Wheel

REMOVAL AND INSTALLATION

——————— **CAUTION** ———————
Before removing the steering wheel, disconnect the negative battery cable and disable the air bag system. Failure to do so could result in accidental air bag deployment and possible injury.

1993–95 Wrangler and 1993–97 Cherokee

Without Air Bag

1. Disconnect the negative battery cable.
2. Set the front tires in a straight-ahead position.
3. Pull the horn button from the steering wheel.
4. If equipped with a standard wheel, remove the trim cover attaching screws and the trim cover. If equipped with a sport wheel, remove the horn contact and flexplate.
5. Remove the steering wheel nut and, if equipped, vibration damper.
6. Scribe a line mark on the steering wheel and steering shaft if there is not one already.

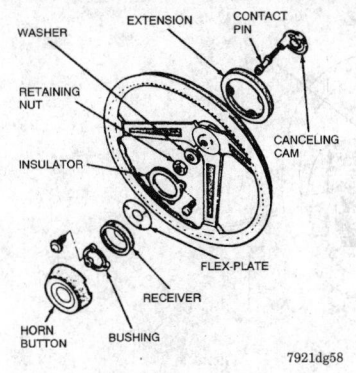

Sport wheel removal

7. Remove the steering wheel using a suitable puller.

To install:

8. To install, align the scribe marks on the steering shaft with the steering wheel.
9. If equipped, install the vibration damper on the hub. Install the steering wheel retaining nut and tighten to 25 ft. lbs. (34 Nm).
10. Install the internal components, horn flexplate and the trim cover/horn button.
11. Connect the negative battery cable.

With Air Bag

1. Set the front tires in a straight-ahead position.
2. Disconnect the negative battery cable. Disable the air bag system.
3. From the backside of the steering wheel, remove the four air bag module attaching nuts.
4. Remove the speed control switch and disconnect the wire feeds.
5. Remove the wheel retaining nut.
6. Scribe a line mark on the steering wheel and steering shaft if there is not one already.
7. Remove the steering wheel using a suitable puller.

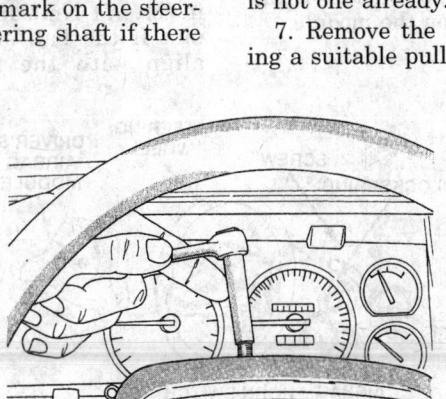

Air bag disarming bolt — Cherokee

To install:

8. To install, align the scribe marks on the steering shaft with the steering wheel.
9. Ensure the wheel compresses the 2 lock tabs on the clockspring.
10. Pull the air bag and if equipped, cruise control wires through the larger hole and horn wire through the smaller hole. Ensure they are not pinched.
11. Install the steering wheel nut and torque it to 45 ft. lbs. (61 Nm) while forcing the steering wheel don the shaft with the nut.
12. Connect the wire feed to the horn buttons.
13. Connect the feed wires and torque the air bag retaining nuts to 90 inch lbs. (10 Nm).
14. Connect the negative battery cable.
15. Check for proper operation of the air bag warning light.

1997 Wrangler

1. Set the front tires in a straight-ahead position.
2. Disconnect the negative battery cable. Wait at least 2 minutes for the reserve capacitor to discharge.
3. From the backside of the steering wheel, remove the two air bag module attaching nuts.
4. Disconnect the wire feeds.
5. Remove the wheel retaining nut.
6. Scribe a line mark on the steering wheel and steering shaft if there is not one already.
7. Remove the steering wheel using a suitable puller.

To install:

8. To install, align the scribe marks on the steering shaft with the steering wheel.
9. Install the steering wheel nut and torque it to 45 ft. lbs. (61 Nm) while forcing the steering wheel down the shaft with the nut.
10. Connect the wire feeds.
11. Tighten the air bag retaining nuts to 90 inch lbs. (10 Nm).
12. Connect the negative battery cable.
13. Check for proper operation of the air bag warning light.

Grand Cherokee and Grand Wagoneer

1. Set the front tires in a straight-ahead position.
2. Disconnect the negative battery cable. Wait at least 2 minutes for the reserve capacitor to discharge.
3. Remove the air bag module retaining nuts from behind the steering wheel.
4. Remove the air bag module.

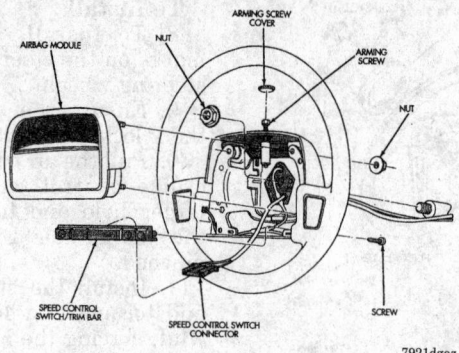

Air bag module removal

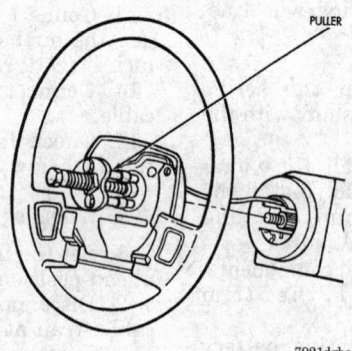

Steering wheel removal

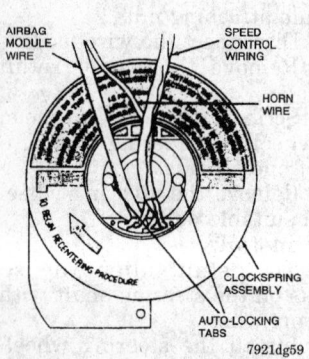

Clockspring assembly

5. If equipped, remove the cruise control switch.

6. Disconnect the horn wiring.

7. Remove the steering wheel retaining nut.

8. Scribe a line mark on the steering wheel and steering shaft if there is not one already.

9. Remove the steering wheel using a suitable puller.

To install:

10. To install, align the scribe marks on the steering shaft with the steering wheel.

11. Ensure the wheel compresses the 2 lock tabs on the clockspring.

12. Pull the air bag and if equipped, cruise control wires through the larger hole and horn wire through the smaller hole. Ensure they are not pinched.

13. Install the steering wheel nut and torque it to 45 ft. lbs. (61 Nm) while forcing the steering wheel don the shaft with the nut.

14. Connect the wire feed to the horn buttons.

15. Connect the feed wires and torque the air bag retaining nuts to 90 inch lbs. (10 Nm). Ensure the air bag is completely seated. the latching clip arms must be visible on top of the connector housing on the module.

16. Connect the negative battery cable.

17. Check for proper operation of the air bag warning light.

Tie Rod Ends

REMOVAL AND INSTALLATION

Wrangler

1. Raise and safely support the vehicle.

2. Remove the cotter pins and nuts from the tie rod end ball studs, drag link end ball stud, and steering damper piston rod ball stud.

3. Use a suitable puller to loosen the ball studs, then remove the tie rod.

4. If necessary, loosen the adjustment sleeve clamp bolts and remove the tie rod end from the tie rod. Count the number of turns required to remove the tie rod end so the replacement can be reinstalled in approximately the same position.

To install:

5. If removed, install the tie rod end to the tie rod. Thread the tie rod end on the same number of turns that was required to remove the old one. Position the adjustment sleeve bolts so the threaded ends of the bolts face rearward and are angled upward.

6. Attach the tie rod ends to the steering knuckles and the drag link ball stud to the tie rod. Install the nuts and tighten to 35 ft. lbs. (47 Nm).

7. Install new cotter pins in the ball studs. If the ball stud hole does not align with the nut castellation, tighten the nut further until the cotter pin can be installed.

8. Attach the steering damper to the tie rod and tighten the nut to 53 ft. lbs. (71 Nm). Install a new cotter pin. If the ball stud hole does not align with the nut castellation,

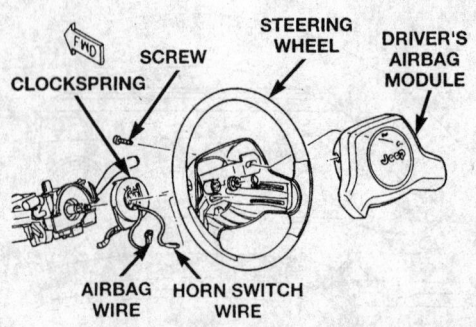

Air bag removal — 1997 Wrangler

tighten the nut further until the cotter pin can be installed.

9. Lower the vehicle and check the toe adjustment.

Cherokee

1. Raise and safely support the vehicle.

2. Remove the cotter pins at the steering knuckle and drag link.

3. Loosen the ball studs using a suitable puller.

4. If necessary, loosen the end clamp bolts and remove the tie rod ends from the tube. Count the number of turns required to remove the

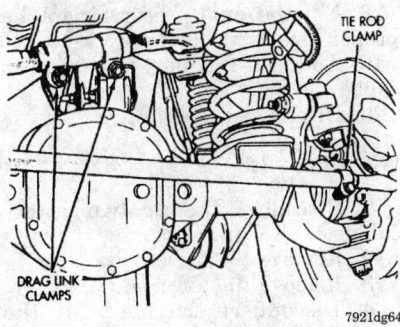

7921dg64

Tie rod and drag link positioning

tie rod ends so the replacement tie rod ends can be reinstalled in the same approximate position.

To install:

5. If removed, install the tie rod ends in the tube. Thread the tie rod ends into the tube the same number of turns required to remove the old ones.

6. Position the tie rod clamp bolts so the bolt heads face the rear of the vehicle and tighten to 20 ft. lbs. (27 Nm).

7. Install the tie rod on the drag link and steering knuckle and install the retaining nuts.

8. Tighten the ball stud-to-steering knuckle nut to 35 ft. lbs. (47 Nm) and the ball stud-to-drag link nut to 55 ft. lbs. (75 Nm). Install new cotter pins. If the ball stud hole does not align with the nut castellation, further tighten the nut in order to install the cotter pin.

9. Lower the vehicle and check the toe adjustment.

Grand Cherokee and Grand Wagoneer

1. Remove the cotter pins at the steering knuckle and drag link.

2. Loosen the ball studs using a suitable puller.

3. If necessary, loosen the end clamp bolts and remove the tie rod

ends from the tube. Count the number of turns required to remove the tie rod ends so the replacement tie rod ends can be reinstalled in the same approximate position.

To install:

4. If removed, install the tie rod ends in the tube. Thread the tie rod ends into the tube the same number of turns required to remove the old ones.

5. Position the tie rod clamps and tighten to 20 ft. lbs. (27 Nm).

6. Install the tie rod on the drag link and steering knuckle and install the retaining nuts.

7. Tighten the ball stud nuts to 55 ft. lbs. (75 Nm) and install new cotter pins. If the ball stud hole does not align with the nut castellation, further tighten the nut in order to install the cotter pin.

8. Check the toe adjustment.

Manual Steering Gear

REMOVAL AND INSTALLATION

Wrangler and Cherokee

1. Place the front wheels in the straight ahead position.

2. Disconnect the steering shaft from the gear.

FASTENER TORQUE			
LETTER	N•m	IN. LBS.	FT. LBS.
◇	251	—	185
◇			
◇	74	—	55
◇			
◇	49	—	36
◇	27	—	20

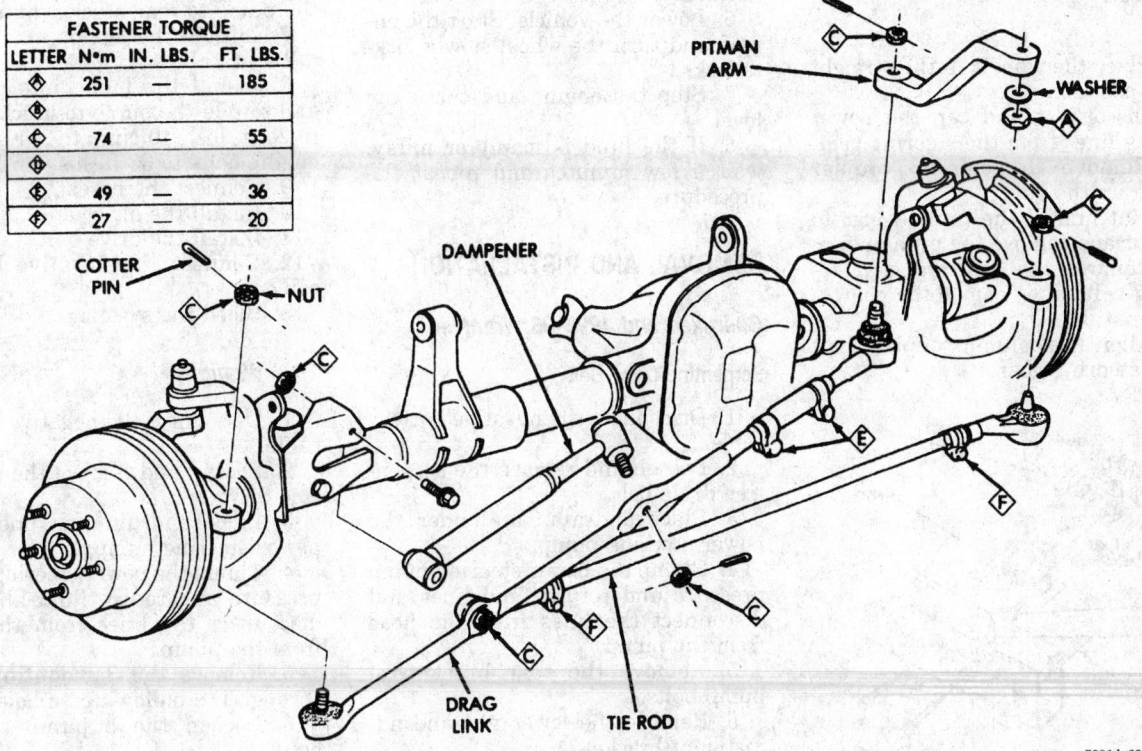

7921dg63

Exploded view fo the steering linkage

3. Raise and safely support the vehicle.

4. Disconnect the center link from the pitman arm.

5. If necessary, remove the front stabilizer bar.

6. Remove the pitman arm nut, matchmark the arm and shaft and remove the arm with a puller.

7. Unbolt and remove the gear.

To install:

8. Install the steering gear. Torque the steering gear to frame bolts to 70 ft. lbs. (95 Nm).

9. Install the pitman arm to its original position. Torque the nut to 185 ft. lbs. (252 Nm) and stake the nut securely.

10. If equipped, install the stabilizer bar and bolts.

11. Connect the center link to the pitman arm and torque the nut to 55 ft. lbs. (75 Nm).

12. Lower the vehicle.

13. Connect the steering shaft to the gear and tighten the nuts.

Power Steering Gear

REMOVAL AND INSTALLATION

All Models

1. Place the wheels in the straight ahead position.

2. Disconnect and cap the power steering lines from the steering gear.

3. Remove the column coupler shaft from the gear.

4. Matchmark the pitman arm to the gear and remove the pitman arm.

5. Remove the steering gear retaining bolts and remove the gear.

To install:

6. Align the column coupler shaft to the steering gear.

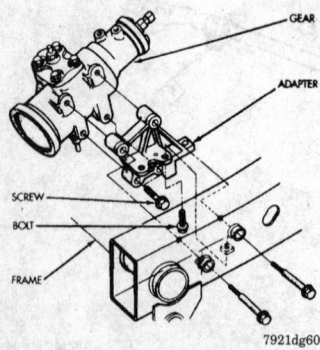

7921dg60

Steering gear installation

7. Position the steering gear and bracket on the frame rail and install the bolts. Torque the bolts to:

1993–95 Wrangler — 78 ft. lbs. (105 Nm)

Cherokee and 1997 Wrangler — 70 ft. lbs. (96 Nm)

Grand Cherokee and 1993 Grand Wagoneer — 65 ft. lbs. (88 Nm)

8. Align and install the pitman arm.

9. Connect the power steering lines.

10. Fill the power steering reservoir to the proper level with fluid and bleed the system.

Power Steering Pump

BLEEDING

1. Fill the pump reservoir to the proper level and let the fluid settle for 2 minutes.

2. Start the engine and let it run for a few seconds, then turn the engine off.

3. Add fluid if necessary. Repeat the procedure until the level remains constant.

4. Raise the wheels off the ground. Slowly turn the wheel right and left, lightly contacting the stops 20 times.

5. Check the fluid level and add if necessary.

6. Lower the vehicle. Start the engine and turn the wheel slowly lock-to-lock.

7. Stop the engine and check the level.

8. If the fluid is foamy or milky, wait a few minutes and repeat the procedure.

REMOVAL AND INSTALLATION

Cherokee and 1993–95 Wrangler

Serpentine Drive Belt

1. Disconnect the negative battery cable.

2. Loosen and remove the serpentine drive belt.

3. Place a drain pan under the power steering pump.

4. Clamp the power steering pump pressure and return fluid lines and disconnect the lines from the hose from the pump.

5. Remove the rear bracket-to-pump bolts.

6. Remove the lower nut and adjustment bracket.

7. Remove the adjuster and pivot bolts.

8. Tilt the pump forward and remove the pump and front bracket assembly from the engine bracket.

9. Remove the bracket from the pump.

To install:

10. Install the bracket to the pump and torque the bolts to 21 ft. lbs. (28 Nm).

11. Position the pump and bracket on the engine bracket.

12. Install the pivot bolt.

13. Install the adjuster bolt.

14. Install the adjuster stud nut.

15. Install the rear bracket-to-pump bolts and torque them to 21 ft. lbs. (28 Nm).

16. Install the serpentine belt.

17. Connect the power steering lines and remove the clamps.

18. Fill the pump reservoir to the proper level with fluid.

19. Connect the negative battery cable.

20. Bleed the system.

V-Type Drive Belt

1. Disconnect the negative battery cable.

2. Remove the drive belt.

3. Remove the air cleaner.

4. Disconnect the hoses at the pump and cap the hose ends.

5. Remove the front bracket-to-engine bolts.

6. Support the pump and remove the pump-to-rear bracket nuts.

7. Lift out the pump.

To install:

8. Install the pump into position and torque the pump-to-bracket nuts to 28 ft. lbs. (40 Nm); the bracket-to-engine bolts to 33 ft. lbs. (45 Nm).

9. Connect the hoses.

10. Install the air cleaner.

11. Install the drive belt.

12. Connect the negative battery cable.

13. Bleed the system.

1997 Wrangler

1. Disconnect the negative battery cable.

2. Loosen and remove the serpentine drive belt.

3. Place a drain pan under the power steering pump.

4. Clamp the power steering pump pressure and return fluid lines and disconnect the lines from the hose from the pump.

5. Remove the 3 mounting bolts through the pulley access holes.

6. Loosen the 3 pump bracket bolts.

7. Tilt pump downward and remove from the engine.

FASTENER TORQUE			
LETTER	N•m	IN. LBS.	FT. LBS.
A	57	—	42
B	28	250	21
C	47	—	35

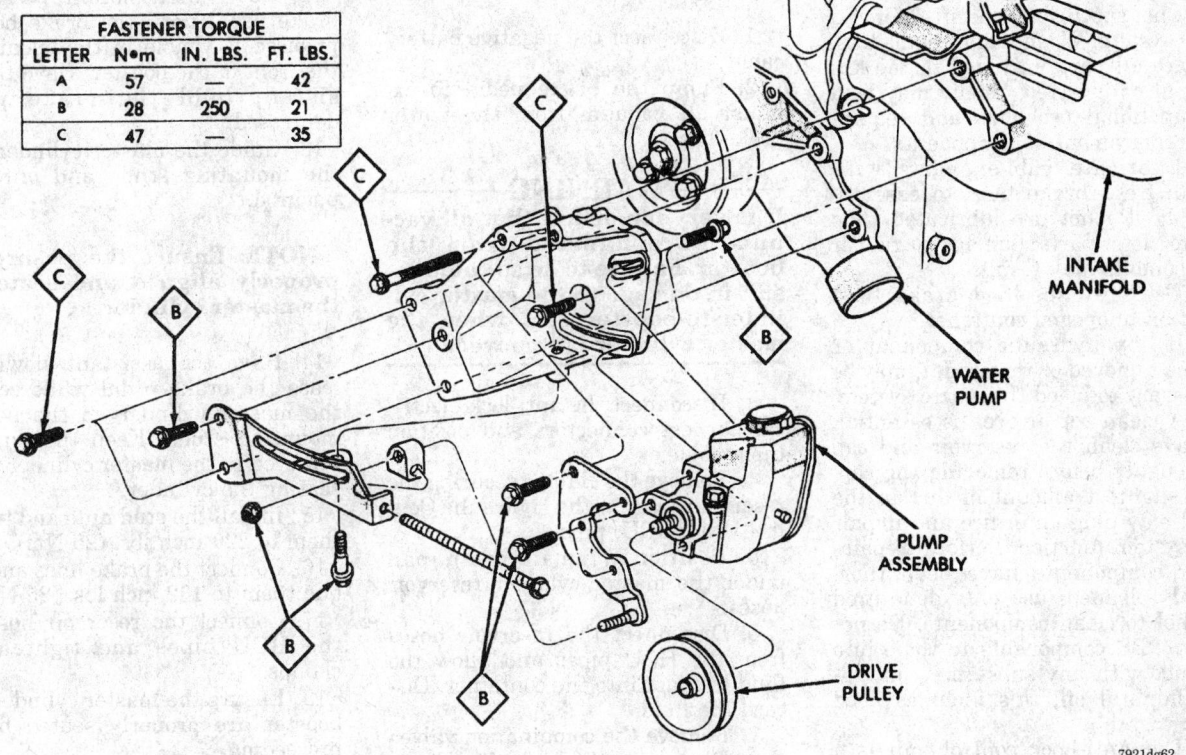

INTAKE MANIFOLD

WATER PUMP

PUMP ASSEMBLY

DRIVE PULLEY

7921dg62

Power steering pump installation

To install:

8. Install the pump on the engine.

9. Tighten the bracket bolts to 35 ft. lbs. (47 Nm).

10. Install the mounting bolts and tighten to 20 ft. lbs. (27 Nm).

11. Install the hoses and drive belt.

Grand Cherokee and Grand Wagoneer

4.0L (VIN S) Engine

1. Disconnect the negative battery cable.

2. Loosen and remove the serpentine drive belt.

3. Place a drain pan under the power steering pump.

4. Clamp the power steering pump pressure and return fluid lines and disconnect the lines from the hose from the pump.

5. Remove the rear bracket-to-pump bolts.

6. Remove the lower nut and adjustment bracket.

7. Remove the adjuster and pivot bolts.

8. Tilt the pump forward and remove the pump and front bracket assembly from the engine bracket.

9. Remove the bracket from the pump.

To install:

10. Install the bracket to the pump and torque the bolts to 21 ft. lbs. (28 Nm).

11. Position the pump and bracket on the engine bracket.

12. Install the pivot bolt.

13. Install the adjuster bolt.

14. Install the adjuster stud nut.

15. Install the rear bracket-to-pump bolts and torque them to 21 ft. lbs. (28 Nm).

16. Install the serpentine belt.

17. Connect the power steering lines and remove the clamps.

18. Fill the pump reservoir to the proper level with fluid.

19. Connect the negative battery cable.

20. Bleed the system.

5.2L (VIN Y) Engine

1. Disconnect the negative battery cable.

2. Loosen and remove the serpentine drive belt.

3. Place a drain pan under the power steering pump.

4. Disconnect the return and pressure lines from the pump.

5. Remove the bolts attaching the pump to the bracket on the engine block.

6. If necessary, remove the bracket-to-engine block bolts.

To install:

7. Install the bracket to the engine block and torque the bolts to 30 ft. lbs. (41 Nm).

8. Mount the pump on the bracket and torque the bolts 20 ft. lbs. (27 Nm).

9. Install the serpentine belt.

10. Connect the fluid lines to the pump and remove the clamps.

11. Fill the power steering reservoir to the proper level with fluid.

12. Connect the negative battery cable.

13. Bleed the system.

BRAKES

Anti-Lock Brake System Service

PRECAUTIONS

• Certain components within the Anti-Lock Brake System (ABS) are not intended to be serviced or repaired individually. Only those components with removal and installation procedures should be serviced.

• Do not use rubber hoses or other parts not specifically specified for and ABS system. When using repair kits, replace all parts included in the kit. Partial or incorrect repair may lead to functional problems and require the replacement of components.

• Lubricate rubber parts with clean, fresh brake fluid to ease assembly. Do not use lubricated shop air to clean parts; damage to rubber components may result.

• Use only specified brake fluid from an unopened container.

• If any hydraulic component or line is removed or replaced, it may be necessary to bleed the entire system.

• A clean repair area is essential. Always clean the reservoir and cap thoroughly before removing the cap. The slightest amount of dirt in the fluid may plug an orifice and impair the system function. Perform repairs after components have been thoroughly cleaned; use only denatured alcohol to clean components. Do not allow ABS components to come into contact with any substance containing mineral oil; this includes used shop rags.

• The Anti-Lock control unit is a microprocessor similar to other computer units in the vehicle. Ensure that the ignition switch is **OFF** before removing or installing controller harnesses. Avoid static electricity discharge at or near the controller.

• If any arc welding is to be done on the vehicle, the control unit should be unplugged before welding operations begin.

Master Cylinder

REMOVAL AND INSTALLATION

Except 1997 Wrangler

Without ABS

1. Disconnect the negative battery cable.
2. Disconnect and plug the brake lines.
3. Disconnect the wires from the stoplight switch.
4. Remove the attaching nuts and lift the assembly from the vehicle.
    To install:
5. Fill the master cylinder with brake fluid and operate the pushrod until fluid squirts from the ports.
6. Install the master cylinder. Torque the mounting nuts to 15–21 ft. lbs. (20–29 Nm).
7. Connect the stoplight switch.
8. Connect the brake lines.
9. Bleed the brake system.

With ABS

1. Disconnect the negative battery cable.
2. Pump the brake pedal to exhaust all vacuum from the power brake booster.

WARNING
It is very important that all vacuum be exhausted from the booster. Failure to do so could result in damage to the master cylinder-to-booster seal when the master cylinder is removed.

---

3. Disconnect the anti-lock electrical harness connectors and position them aside.
4. Remove the clamps securing the reservoir hoses to the Hydraulic Control Unit (HCU) pipes.
5. Position a small drain pan under the master cylinder reservoir hoses.
6. Disconnect the reservoir hoses from the HCU pipes and allow the fluid to drain into the container. Discard the fluid.
7. Remove the combination valve.
8. Disconnect the brake lines from the master cylinder.
9. Remove the nuts attaching the master cylinder to the brake booster mounting studs.
10. Pull the master cylinder forward and out of the studs.
    To install:
11. If installing a new master cylinder, fill the reservoir with fluid and depress the pushrod until brake fluid comes out of the fluid line connections.

WARNING
The seal between the master cylinder and brake booster can be damaged if the master cylinder is improperly installed. To avoid damage, install the master cylinder only as described.

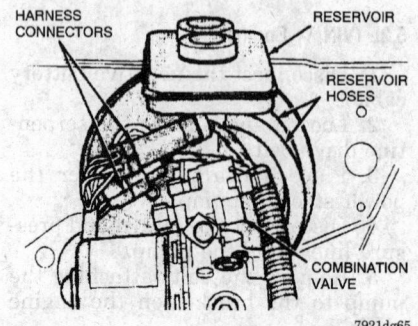

Master cylinder mounting with ABS

12. Have an assistant press the brake pedal until the brake booster pushrod is visible in the opening at the front of the booster. Have the assistant hold the brake pedal depressed.
13. Guide the master cylinder onto the mounting studs and onto the pushrod.

NOTE: Ensure the pushrod is properly aligned and seated in the master cylinder.

14. Have the assistant slowly release the brake pedal while seating the master cylinder on the booster mounting studs. Keep the pushrod centered in the master cylinder while seating the cylinder.
15. Install the stud nuts and torque them to 220 inch lbs. (25 Nm).
16. Connect the brake lines and torque them to 132 inch lbs. (25 Nm).
17. Connect the reservoir hoses to the HCU pipes and tighten the clamps.
18. Ensure the master cylinder and booster are properly seated before proceeding.
19. Install the combination valve.
20. Connect the electrical connectors.
21. Fill the master cylinder reservoir and bleed the brakes.
22. Install the washer reservoir.
23. Install the air cleaner.
24. Connect the negative battery cable.

1997 Wrangler

1. Disconnect the negative battery cable.
2. Remove the evaporative canister.
3. Disconnect and plug the brake lines.
4. Remove the combination valve.
5. Remove the attaching nuts and lift the assembly from the vehicle.
    To install:
6. Fill the master cylinder with brake fluid and operate the pushrod until fluid squirts from the ports.
7. Install the master cylinder. Torque the mounting nuts to 18 ft. lbs. (24 Nm).
8. Install the combination valve. Torque the mounting nuts to 18 ft. lbs. (24 Nm).
9. Connect the brake lines. Tighten the master cylinder lines to 11 ft. lbs. (15 Nm) and the combination valve lines to 15 ft. lbs. (21 Nm).
10. Bleed the brake system.

Brake Caliper

REMOVAL AND INSTALLATION

All Models

——— CAUTION ———

Brake linings may contain asbestos. Asbestos is a known cancer-causing agent. When working on brakes, remember that the dust which accumulates on the brake parts may contain asbestos. Always wear a protective face covering, such as a painter's mask, when working on the brakes. NEVER blow the dust from the brakes or drum! There are solvents made for the purpose of cleaning brake parts. Use them!

1. Drain ⅔ of the brake fluid from the front reservoir. Use the bleeder screw at the front outlet port to drain the fluid. If equipped with anti-lock brakes, relieve the system pressure.
2. Raise and safely support the vehicle.
3. Remove the wheels.
4. Place a C-clamp on the caliper so the solid end contacts the back of the caliper and screw end contacts the metal part of the outboard brake pad.
5. Tighten the clamp until the caliper moves far enough to force the piston to the bottom of the piston bore. This will back the brake pads off of the rotor surface to facilitate the removal and installation of the caliper assembly.
6. Remove the C-clamp.

NOTE: Do not push down on the brake pedal or the piston and brake pads will return to their original positions up against the rotor.

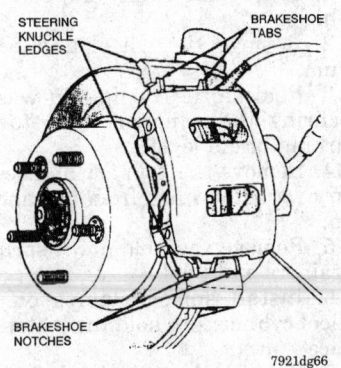

STEERING KNUCKLE LEDGES

BRAKESHOE TABS

BRAKESHOE NOTCHES

7921dg66

Caliper mounting

7. Remove both of the mounting bolts and lift the caliper off the rotor.

NOTE: If just the brake pads are being replaced, it is not necessary to remove the caliper assembly entirely from the vehicle. Do not remove the brake line. Rest the caliper on the front spring or other support. Do not allow the brake hose to support the weight of the caliper.

8. If the caliper is being removed, it is necessary to disconnect the brake fluid hose. Clean the brake fluid hose-to-caliper connection thoroughly. Remove the hose-to-caliper bolt. Cap or tape the open ends to keep dirt out. Discard the copper gaskets.

To install:
9. Connect the brake line to the caliper with new sealing washers and fitting bolt. Hand-tighten the fitting bolt.
10. Position the caliper into place over the rotor.
11. Coat the caliper mounting bolt with silicone grease and torque them to 7–15 ft. lbs. (10–20 Nm).
12. Position the brake line clear of all chassis components, untwisted and free of kinks. Torque the fitting bolt to 23 ft. lbs. (31 Nm).
13. Install the wheels.
14. Fill the master cylinder with fluid and bleed the brake system.
15. Before driving the vehicle, pump the brakes several times to seat the pads.

Disc Brake Pads

REMOVAL AND INSTALLATION

All Models

——— CAUTION ———

Brake linings may contain asbestos. Asbestos is a known cancer-causing agent. When working on brakes, remember that the dust which accumulates on the brake parts may contain asbestos. Always wear a protective face covering, such as a painter's mask, when working on the brakes. NEVER blow the dust from the brakes or drum! There are solvents made for the purpose of cleaning brake parts. Use them!

1. Raise and safely support the vehicle.

2. Drain ⅔ of the brake fluid from the front reservoir. Use the bleeder screw at the front outlet port to drain the fluid.
3. Raise and support the vehicle safely.
4. Remove the wheels.
5. Remove the brake caliper. Use a suitable tool to compress the caliper piston into the bore.
6. Hold the anti-rattle clip against the caliper anchor plate and remove the outboard brake pad.
7. Remove the inboard pad and its anti-rattle clip.

To install:
8. Clean all the mounting holes and bushing grooves in the caliper ears. Clean the mounting bolts. Replace the bolts if they are corroded or if the threads are damaged. Wipe the inside of the caliper clean, including the exterior of the dust boot. Inspect the dust boot for cuts or cracks and for proper seating in the piston bore. If evidence of fluid leakage is noted, the caliper should be rebuilt.

NOTE: Do not use abrasives on the bolts in order not to destroy their protective plating. Do not use compressed air to clean the inside of the caliper, as it may unseat the dust boot seal.

9. Install the inboard anti-rattle clip on the trailing end of the anchor plate. The split end of the clip must face away from the rotor.
10. Install the inboard pad in the caliper. The pad must lay flat against the piston.
11. Install the outboard pad in the caliper while holding the anti-rattle clip.
12. With the pads installed, position the caliper over the rotor. Line up the mounting holes in the caliper and the support bracket and insert the mounting bolts. Make sure the bolts pass under the retaining ears on the inboard shoes. Push the bolts through until they engage the holes of the outboard pad and caliper ears. Thread the bolts into the support bracket and tighten them to 11 ft. lbs. (15 Nm).
13. Fill the master cylinder with brake fluid and pump the brake pedal to seat the pads.
14. Install the wheel assembly and lower the vehicle. Check the level of the brake fluid in the master cylinder and fill as necessary. Test the operation of the brakes before taking the vehicle onto the road.

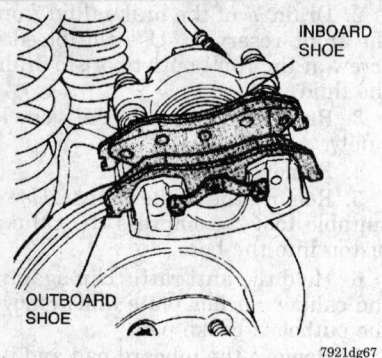

Disc brake pad installation

Brake Rotor

REMOVAL AND INSTALLATION

—— CAUTION ——

Brake linings may contain asbestos. Asbestos is a known cancer-causing agent. When working on brakes, remember that the dust which accumulates on the brake parts may contain asbestos. Always wear a protective face covering, such as a painter's mask, when working on the brakes. NEVER blow the dust from the brakes or drum! There are solvents made for the purpose of cleaning brake parts. Use them!

2WD Models

1. Raise and safely support the vehicle.
2. Remove the wheels.
3. Remove the caliper without disconnecting the brake line. Suspend the caliper aside; do not let it hang by the brake hose.
4. Remove the grease cap, cotter pin, nut cap, nut and washer from the spindle.
5. Pull slowly on the hub and catch the outer bearing as it falls.

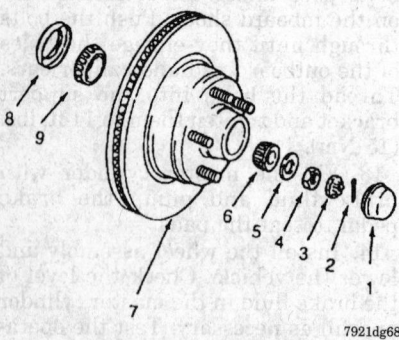

Disc brake rotor — 2WD vehicles

6. Remove the hub and rotor. The inner bearing and seal can be removed by prying out and discarding the inner seal.
To install:
7. Clean and repack the hub and bearings, install the inner bearing and a new seal.
8. Position the hub and rotor on the spindle and install the outer bearing.
9. Install the washer and nut.
10. While turning the rotor, torque the nut to 25 ft. lbs. (34 Nm) to seat the bearings.
11. Back off the nut ½ turn and, while turning the rotor, torque the nut to 19 inch lbs. (2 Nm).
12. Install the nut cap and a new cotter pin. Install the grease cap.
13. Install the caliper.
14. Install the wheels and lower the vehicle. Before moving the vehicle, pump the brake pedal several times to seat the brake pads against the rotor.

4WD Models

1. Raise and support the vehicle.
2. Remove the wheel and tire assembly.
3. Remove the caliper.
4. Remove the retainers securing the rotor to the hub studs.
5. Remove the rotor from the hub.
6. Installation is the reverse of removal.

Brake Drum

REMOVAL AND INSTALLATION

—— CAUTION ——

Brake linings may contain asbestos. Asbestos is a known cancer-causing agent. When working on brakes, remember that the dust which accumulates on the brake parts may contain asbestos. Always wear a protective face cover-

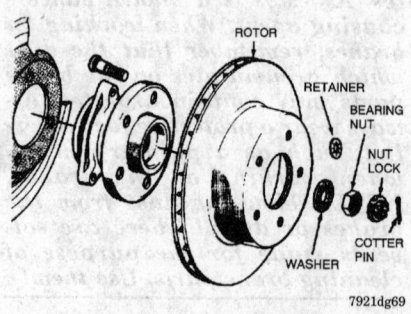

Disc brake rotor — 4WD

ing, such as a painter's mask, when working on the brakes. NEVER blow the dust from the brakes or drum! There are solvents made for the purpose of cleaning brake parts. Use them!

All Models

1. Raise and safely support the vehicle.
2. Remove the wheel.
3. Remove the spring nuts (if installed) from the lug bolts and remove the drum from the vehicle.

NOTE: It may be necessary to back off the brake adjusters to remove the drum.

To install:
4. Ensure the contacting surfaces are clean and flat. Install the drum on the hub.
5. Adjust the brake shoes, if necessary.
6. Install the spring nuts on the lug bolts.
7. Install the wheel.

Brake Shoes

REMOVAL AND INSTALLATION

—— CAUTION ——

Brake linings may contain asbestos. Asbestos is a known cancer-causing agent. When working on brakes, remember that the dust which accumulates on the brake parts may contain asbestos. Always wear a protective face covering, such as a painter's mask, when working on the brakes. NEVER blow the dust from the brakes or drum! There are solvents made for the purpose of cleaning brake parts. Use them!

All Models

1. Raise and safely support the vehicle.
2. Remove the wheel and brake drum.
3. Remove the U-clip and washer securing the adjuster cable to the parking brake lever.
4. Remove the primary and secondary return springs from the anchor pin.
5. Remove the hold-down springs, retainers and pins.
6. Install spring clamps on the wheel cylinders to hold the pistons in place.
7. Remove the adjuster lever, adjuster screw and spring.

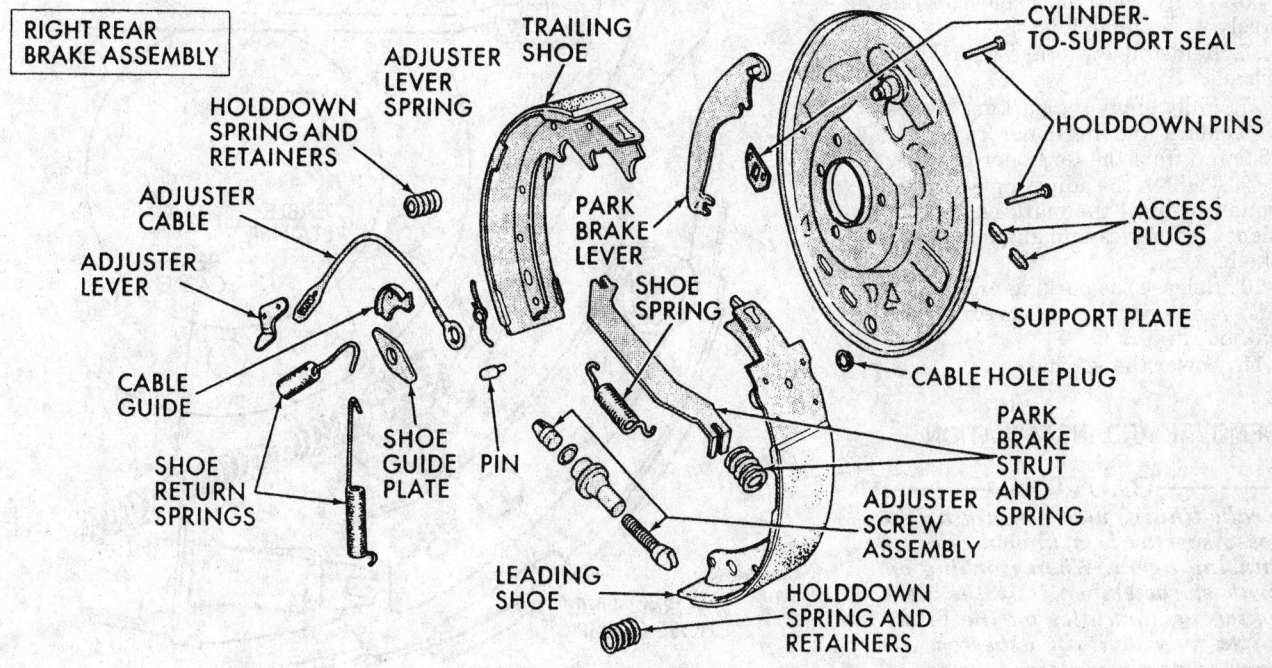

RIGHT REAR BRAKE ASSEMBLY

HOLDDOWN SPRING AND RETAINERS
ADJUSTER LEVER SPRING
TRAILING SHOE
CYLINDER-TO-SUPPORT SEAL
HOLDDOWN PINS
ADJUSTER CABLE
ADJUSTER LEVER
PARK BRAKE LEVER
ACCESS PLUGS
CABLE GUIDE
SHOE SPRING
SUPPORT PLATE
SHOE RETURN SPRINGS
SHOE GUIDE PLATE
PIN
CABLE HOLE PLUG
PARK BRAKE STRUT AND SPRING
ADJUSTER SCREW ASSEMBLY
LEADING SHOE
HOLDDOWN SPRING AND RETAINERS

7921dgbc

Drum brake components

8. Remove the adjuster cable and cable guide.

9. Remove the brake shoes and parking brake strut.

10. Disconnect the cable from the parking brake lever and remove the lever.

To install:

11. Clean the support plate with brake cleaner.

12. Apply multi-purpose grease to the brake shoe contact surfaces on the backing plate.

13. Lubricate the adjuster screw threads.

14. Attach the parking brake lever to the secondary brake shoe. Use a new washer and U-clip.

15. Remove the wheel cylinder clamps.

16. Attach the parking brake cable to the lever.

17. Install the brake shoes on the support plate. Secure the shoes with new hold-down springs, pins and retainers.

18. Install the parking brake strut and spring.

19. Install the guide plate and adjuster cable to the anchor pin.

20. Install the return springs.

21. Install the adjuster cable guide on the secondary shoe.

22. Install the adjuster screw, spring and lever. Connect to the adjuster cable.

23. Adjust the shoes to the drum. Install the drum.

24. Install the wheel/tire assemblies and lower the vehicle.

25. Verify a firm brake pedal before moving the vehicle.

Wheel Cylinder

REMOVAL AND INSTALLATION

——— CAUTION ———
Brake linings may contain asbestos. Asbestos is a known cancer-causing agent. When working on brakes, remember that the dust which accumulates on the brake parts may contain asbestos. Always wear a protective face covering, such as a painter's mask, when working on the brakes. NEVER blow the dust from the brakes or drum! There are solvents made for the purpose of cleaning brake parts. Use them!

All Models

1. Raise and safely support the vehicle.

2. Remove the wheel and brake drum.

3. Disconnect the brake line at the wheel cylinder.

4. Remove the brake shoes.

5. Remove the wheel cylinder attaching bolts and remove the wheel cylinder.

To install:

6. Position the wheel cylinder on the backing plate.

7. Connect the brake line to the cylinder fitting.

8. Install the wheel cylinder mounting bolts. Torque the bolts to 7 ft. lbs. (10 Nm).

9. Tighten the brake line fitting to 140 inch lbs. (16 Nm).

10. Install the brake shoes.

11. Install the brake drum and wheel.

12. Lower the vehicle.

13. Fill the brake system to the correct level and bleed the brakes.

Parking Brake Cable

ADJUSTMENT

1. Raise the vehicle.

2. Back off the tensioner nut to create slack in the cables.

3. Remove the rear wheel assemblies.

4. Check the rear brake shoes adjustment with a brake gauge.

5. Verify that the cables operate freely with no binding.

6. Reinstall the brake drums and wheels.

7. Fully apply the parking brakes.

8. Mark the tensioner rod ¼ in. (6.5mm) from the tensioner bracket.

9. Tighten the adjusting nut at the equalizer until the mark on the tensioner rod moves into alignment with the bracket.

10. Release the parking brake cable and verify that the wheels turn freely without drag.

11. Lower the vehicle.

REMOVAL AND INSTALLATION

------- **CAUTION** -------

Brake linings may contain asbestos. Asbestos is a known cancer-causing agent. When working on brakes, remember that the dust which accumulates on the brake parts may contain asbestos. Always wear a protective face covering, such as a painter's mask, when working on the brakes. NEVER blow the dust from the brakes or drum! There are solvents made for the purpose of cleaning brake parts. Use them!

1. Raise the vehicle and loosen the equalizer nuts until the cables are slack.

2. Disengage the cable from the equalizer and remove the cable.

3. Remove the cable bracket from the suspension arm.

4. Remove the rear wheel and brake drum.

5. Remove the secondary brake shoe and disconnect the cable from the lever.

6. Compress the cable retainer and remove the cable from the backing plate.

To install:

7. Install a new cable in the backing plate. Be sure the cable retainer is seated.

8. Attach the cable to the lever on the brake shoe and install the brake shoe.

9. Adjust the brake shoes to the drum using a brake gauge.

10. Install the drum and wheel.

11. Install the cable/bracket to the suspension arm.

12. Engage the cable in the equalizer and install the equalizer nuts.

13. Adjust the cable.

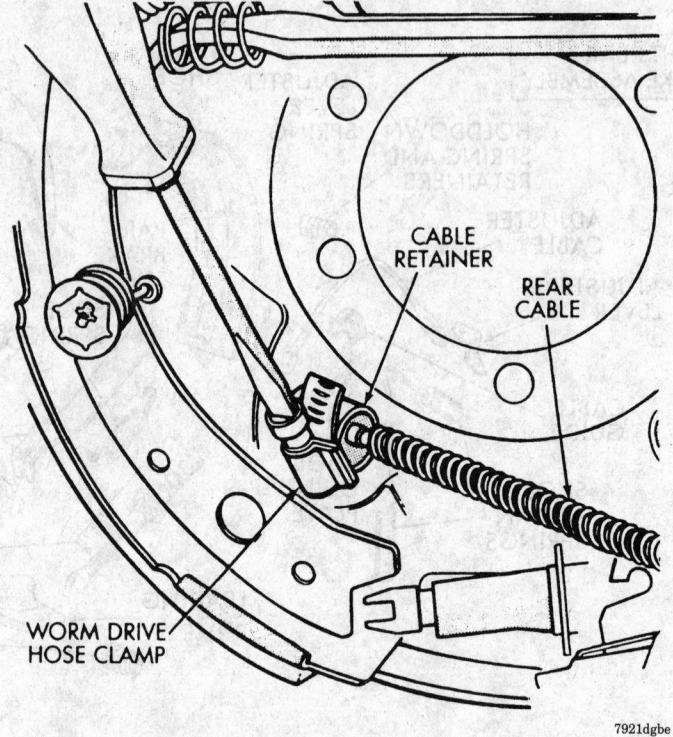

CABLE RETAINER

REAR CABLE

WORM DRIVE HOSE CLAMP

7921dgbe

Compressing the cable retainer

Brake System Bleeding

PROCEDURE

All Models

Bleed the brake system in the following sequence:
- Master cylinder
- Right rear wheel
- Left rear wheel
- Right front wheel
- Left front wheel

Bleeding the anti-lock braking system is basically a 3 step process consisting of a conventional manual brake bleed, a 2nd brake bleed using scan tool DRB-II or equivalent, to run the pump and a repeat of the conventional manual brake bleed.

1. Clean the master cylinder reservoir caps and reservoir exterior.

2. Fill the master cylinder with DOT 3 standard brake fluid.

3. Bleed the master cylinder and HCU at the brake fittings, if equipped with ABS.

4. Attach a bleed hose to the bleed screw on the first wheel brake unit to be bled. Immerse the end of the bleed hose in a container partially filled with brake fluid.

5. Have an assistant apply and hold the brake pedal.

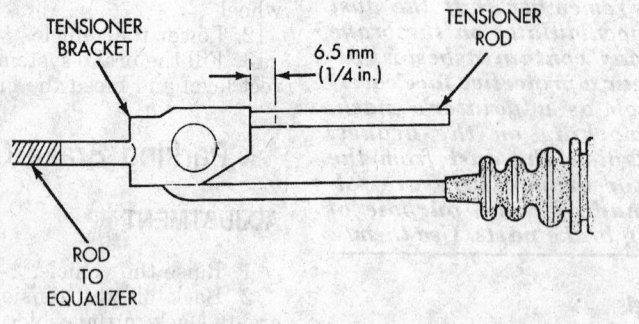

TENSIONER BRACKET

6.5 mm (1/4 in.)

TENSIONER ROD

ROD TO EQUALIZER

7921dgbd

Tensioner rod measurement

6. Open the bleed screw ½ turn and close the screw once the brake pedal reaches the floor.

NOTE: Do not pump the brake pedal at any time while bleeding. This compresses air into small bubbles which are distributed throughout the system.

7. Repeat the bleeding operation 5–7 more times at each wheel or until the fluid entering the container is free of air bubbles. Check the reservoir fluid lever frequently and add fluid if necessary.

NOTE: Do not allow the master cylinder reservoir to run dry while bleeding the system. Running the reservoir dry will allow air to reenter the system necessitating a second bleeding operation.

8. If equipped with ABS, connect the DRB-II scan tool or equivalent, to the diagnostic connector.
9. Perform the "Bleed Brake" operation as described in the scan tool manual.
10. Repeat Steps 4–9.
11. Verify proper brake operation before moving the vehicle.

Wheel Speed Sensor

REMOVAL AND INSTALLATION

────── **CAUTION** ──────
Brake linings may contain asbestos. Asbestos is a known cancer-causing agent. When working on brakes, remember that the dust which accumulates on the brake parts may contain asbestos. Always wear a protective face covering, such as a painter's mask, when working on the brakes. NEVER blow the dust from the brakes or drum! There are solvents made for the purpose of cleaning brake parts. Use them!

Front

1. Raise the vehicle and turn the wheel for access to the sensor.
2. Remove the sensor wire from the mounting brackets.
3. Clean the area surrounding the sensor.
4. Remove the bolt securing the sensor.
5. Disconnect the sensor wire at the harness plug, then remove the sensor and wire.

To install:
6. Apply Loctite® 242 to the attaching bolt.
7. Position the sensor on the steering knuckle. Install the bolt finger-tight.
8. Tighten the sensor to 11 ft. lbs. (14 Nm).
9. Attach the sensor wire to the bracket.
10. Properly route the wire and attach it to the brackets.
11. Engage the electrical connection.

Rear

1. Except on Wrangler, Raise and fold the rear seat. Disconnect the sensors from the rear harness. Push the sensor grommets and wires through the floorpan.
2. Raise the vehicle.
3. Disconnect the sensor wires at the rear axle.
4. Remove the wheel and tire assembly.
5. Remove the clips securing the sensor wire.
6. Remove the bolt securing the sensor and remove the sensor.

To install:
7. Insert the sensor wire through the support plate and seat the grommet in the plate.
8. Apply Loctite® 242 to the attaching bolt.
9. Install the bolt finger-tight.
10. Set the gap as follows:
   a. If the original sensor is being installed, remove any remaining pieces of cardboard spacer from the sensor face. Then adjust the air gap to 0.043 in. (1.1mm) with a brass feeler gauge. Tighten the bolt to 11 ft. lbs. (14 Nm).
   b. If a new sensor is being installed, push the cardboard spacer on the sensor against the tone ring. Tighten the bolt to 6 ft. lbs. (8 Nm). Correct air gap will be established as the tone ring rotates.
   c. Verify the sensor gap adjustment. If adjustment changed after tightening the bolt, readjust.
11. Route the sensor wires and engage the wire harnesses.
12. Install the wire to the clips.
13. Lower the vehicle.

FRONT SUSPENSION

Shock Absorbers

REMOVAL AND INSTALLATION

Cherokee and Wrangler

1. Remove the locknuts and washers from the upper stud.
2. Raise and support the vehicle safely.
3. If necessary for access, remove the wheels.
4. Remove the lower attaching nuts and bolts.
5. Pull the shock absorber eyes and rubber bushings from the mounting pins.

To install:

NOTE: Before installing new shocks, they should be purged of air. To do this, hold the shock upright and fully extend it, then invert and compress it. Do this several times.

6. Position the shock on the vehicle and install the mounting hardware.
7. Tighten the upper end nut to 8 ft. lbs. (11 Nm) and the lower end bolts to 14 ft. lbs. (19 Nm)., except on 1997 Wrangler. On 1997 Wrangler, tighten the lower bolts to 250 inch lbs. (28 Nm) and the upper retainers to 17 ft. lbs. (23 Nm).
8. Install the wheels and lower the vehicle.

Grand Cherokee and Grand Wagoneer

1. Remove the upper nut, retainer and grommet from the engine compartment.
2. Remove the lower bolt and nut from the mounting bracket.
3. Remove the shock absorber.

To install:

NOTE: Before installing new shocks, they should be purged of air. To do this, hold the shock upright and fully extend it, then invert and compress it. Do this several times.

4. Position the shock on the vehicle and install the mounting hardware.
5. Tighten the upper retaining nut to 14 ft. lbs. (19 Nm) and lower bolt and nut to 17 ft. lbs. (23 Nm).
6. Install the wheels and lower the vehicle.

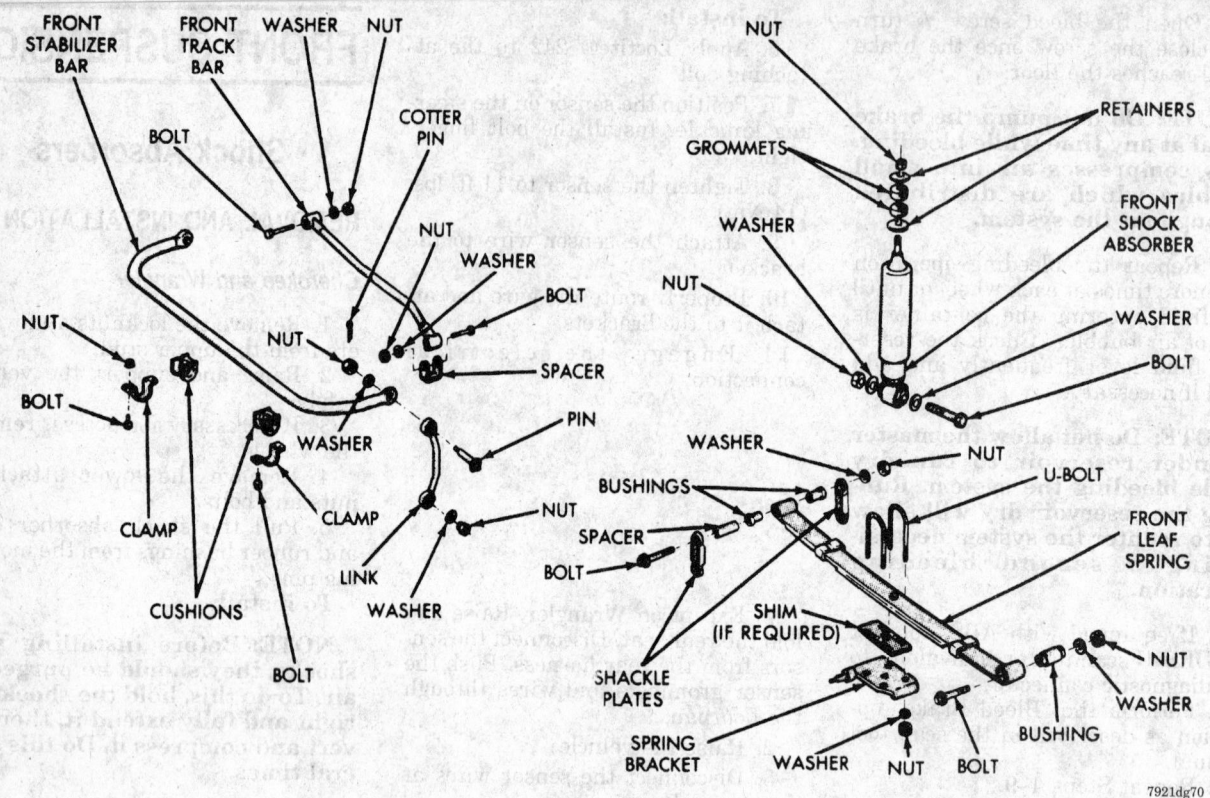

Front suspension — 1993–95 Wrangler

7921dg70

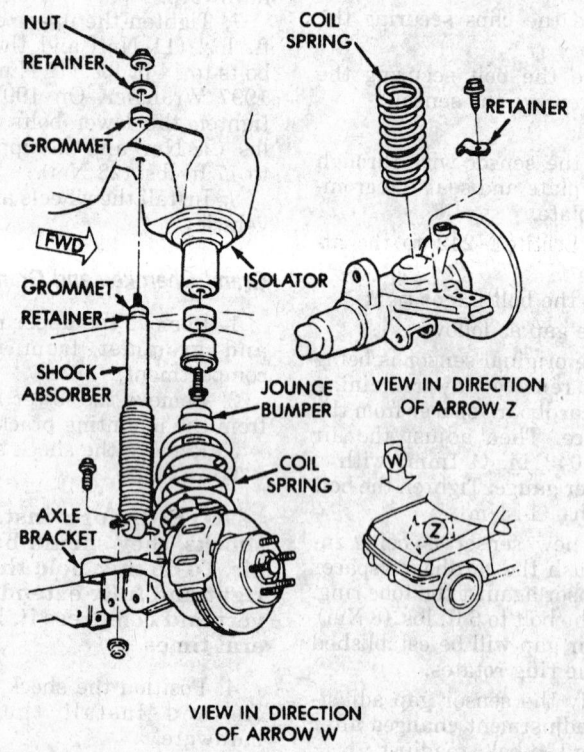

Front shock absorber mounting — Grand Cherokee/Wagoneer

7921dg71

Coil Springs

REMOVAL AND INSTALLATION

Cherokee

— CAUTION —

Coil springs are under a great deal of tension when installed in the vehicle. Serious injury or death may result from being hit by an expanding spring. A piece of chain fastened to the frame and wrapped around the coil spring will keep the spring from flying out if it should slip before it is fully expanded.

1. Raise and support the vehicle safely.
2. Support the axle with a floor jack.
3. Remove the wheels.
4. On 4WD trucks, matchmark and disconnect the front driveshaft from the axle.
5. Disconnect the lower control arm at the axle.
6. Disconnect the stabilizer bar links and the shock absorbers at the axle.
7. Disconnect the track bar at the sill bracket.

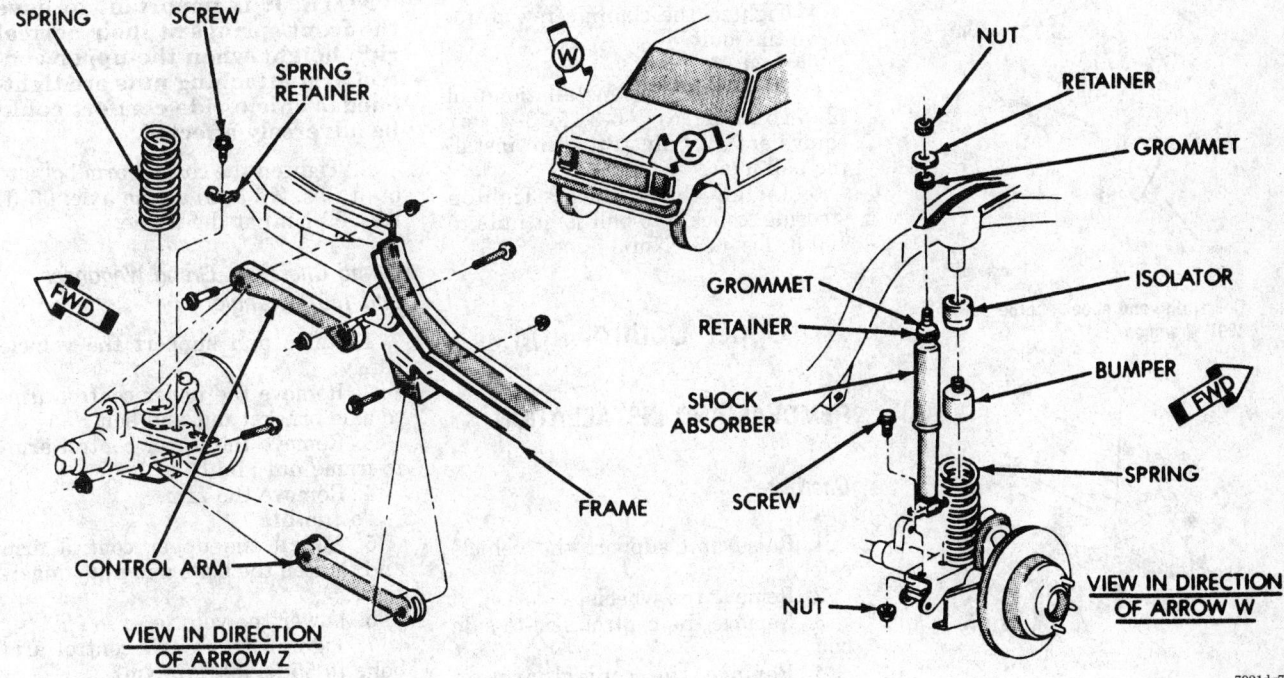

Front coil spring — Cherokee

8. Disconnect the tie rod at the pitman arm.

9. Lower the axle until tension is removed from the spring, then loosen the spring retainer and remove the spring.

To install:

10. Position the replacement spring on the retainer, tighten the spring retainer bracket screw and lift the axle into position.

11. Connect the lower control arm to the axle. Tighten the control arm-to-axle bolt to 133 ft. lbs. (180 Nm).

12. Remove the support jack.

13. Connect the stabilizer bar links and the shock absorbers at the axle. Tighten the shock absorber-to-axle bolt to 14 ft. lbs. (19 Nm) and the stabilizer bar-to-axle bolt to 70 ft. lbs. (95 Nm).

14. Connect the track bar at the sill bracket. Tighten the track bar-to-frame rail bolt to 35 ft. lbs. (47 Nm).

15. Connect the tie rod at the pitman arm. Tighten the center link-to-pitman arm bolt to 35 ft. lbs. (47 Nm).

NOTE: New strap bolts must be used each time the driveshaft is disconnected.

16. On 4WD trucks, connect the front driveshaft. Tighten U-joint-to-axle bolt to 14 ft. lbs. (19 Nm).

17. Install the wheels and lower the vehicle.

Grand Cherokee, Grand Wagoneer and 1997 Wrangler

1. Raise and safely support the vehicle, allowing the front axle to hang.

2. Support the axle with a jack.

3. Paint or scribe alignment marks on the cam adjusters and axle bracket for installation reference.

4. Matchmark and disconnect the front driveshaft from the axle.

5. Disconnect the lower suspension arm nut, cam and cam bolt from the axle.

6. Disconnect the stabilizer bar links and shock absorbers at the axle.

7. Disconnect the track bar at the frame rail bracket.

8. Disconnect the drag link at the pitman arm.

9. Lower the axle until spring is free from the upper mount, then remove the coil spring clip screw and remove the spring.

10. If necessary, remove the jounce bumper from the upper spring mount.

To install:

11. If removed, install the jounce bumper and tighten the bolts to 31 ft. lbs. (42 Nm).

12. Position the replacement spring on axle pad. Install the spring clip and tighten the screw to 16 ft. lbs. (21 Nm).

13. Raise the axle into position until the spring seats in the upper mount.

14. Connect the stabilizer bar links and shock absorbers to the axle bracket. Connect the track bar to the frame rail bracket.

15. Install the lower suspension arm to the axle.

16. Connect the driveshaft to the yoke.

17. Lower the vehicle.

Leaf Spring

REMOVAL AND INSTALLATION

1993–95 Wrangler

1. Raise and support the vehicle safely.

2. Support the front axle so the weight is relieved.

3. If equipped, remove the stabilizer bar link attaching nut.

4. Remove the nuts, U-bolts and bracket from the axle.

5. Remove the nut and bolt attaching the spring front eye to the shackle.

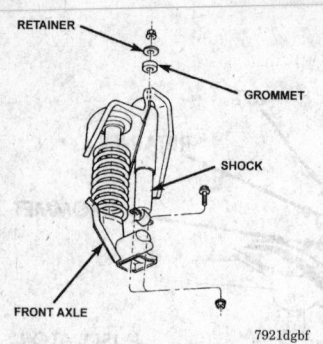

Coil spring and shock aborber —
1997 Wrangler

7921dgbf

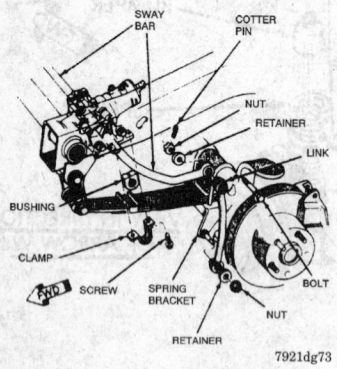

Leaf spring — 1993–95 Wrangler

7921dg73

6. Remove the spring from the vehicle.

To install:

7. Position the spring front eye in the shackle.

8. Position the rear eye in the hanger bracket.

9. Install the spring bracket, U-bolts and nuts. Torque the nuts to 90 ft. lbs. (122 Nm).

10. If equipped, attach the stabilizer bar links.

11. Remove the axle support.

12. Lower the vehicle.

13. Torque the front shackle plate nut to 100 ft. lbs. (135 Nm).

14. Torque the rear eye bracket nut to 105 ft. lbs. (142 Nm).

Upper and Lower Ball Joint

REMOVAL AND INSTALLATION

NOTE: This procedure requires the use of a special ball joint removal tool.

1. Raise and support the vehicle safely.

2. Remove the wheel. Remove the steering knuckle.

3. Position a ball joint removal tool (J-34503-1 and 34503-3 or

equivalent), as illustrated, to remove the ball joint.

4. Tighten the clamp screw to remove the joint.

To install:

5. Use a ball joint installation tool (J-34503-5 or J-34503-4 or equivalent), as illustrated, to install the ball joint.

6. Install the knuckle. Tighten steering knuckle-to-ball joint nuts to 100 ft. lbs. (135 Nm).

Upper Control Arm

REMOVAL AND INSTALLATION

Cherokee

1. Raise and support the vehicle safely.

2. Remove the wheels.

3. Remove the control arm-to-axle bolt.

4. Remove the control arm-to-frame bolt and remove the arm.

To install:

5. Install the replacement control arm and all nuts and bolts finger-tight.

6. Install the wheels.

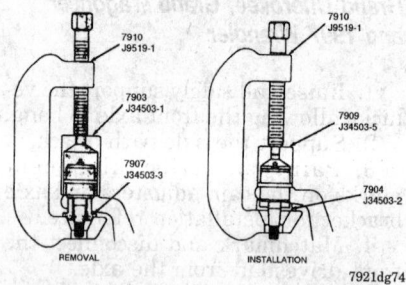

Upper ball joint removal and installation

7921dg74

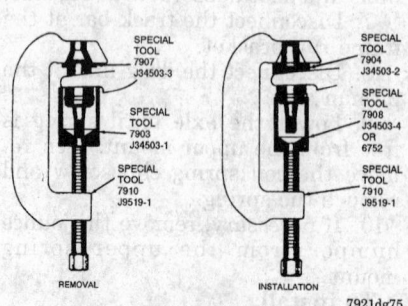

Lower ball joint removal and installation

7921dg75

7. Lower the vehicle.

NOTE: It is important to have the front springs at their normal ride height when the upper control arm attaching nuts are tightened. Vehicle ride comfort could be adversely affected.

8. Tighten the control arm bolts to 55 ft. lbs. (75 Nm) at the axle; 66 ft. lbs. (89 Nm) at the frame.

Grand Cherokee, Grand Wagoneer and 1997 Wrangler

1. Raise and support the vehicle safely.

2. Remove the upper control arm-to-axle bracket nut and bolt.

3. Remove the upper control arm-to-frame nut and bolt.

4. Remove the arm.

To install:

5. Install the upper control arm and tighten the bolts and nuts finger-tight.

6. Lower the vehicle.

7. Tighten the upper control arm bolts to 55 ft. lbs. (75 Nm).

Lower Control Arm

REMOVAL AND INSTALLATION

Cherokee

1. Raise and support the vehicle safely.

2. Remove the wheels.

3. Remove the retainers securing the lower control arm at the axle and rear bracket.

4. Remove the arm.

To install:

5. Install the replacement arm and tighten the attaching bolts finger-tight.

6. Install the wheels, then lower the vehicle.

NOTE: It is important to have the front springs at their normal ride height when the lower control arm attaching nuts are tightened. Vehicle ride comfort could be adversely affected.

7. Tighten the lower control arm attaching bolts to 133 ft. lbs. (180 Nm).

Grand Cherokee, Grand Wagoneer and 1997 Wrangler

1. Raise and safely support the vehicle, allowing the suspension to hang freely.

2. Paint or scribe alignment marks on the cam adjusters and sus-

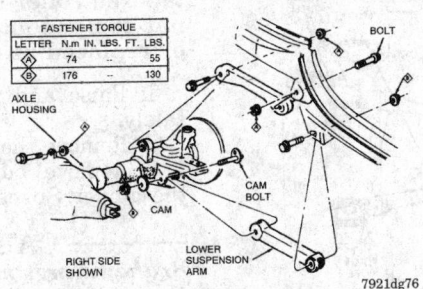

FASTENER TORQUE		
LETTER	N.m IN. LBS.	FT. LBS.
◇	74	55
◈	176	130

AXLE HOUSING

BOLT

CAM

CAM BOLT

RIGHT SIDE SHOWN

LOWER SUSPENSION ARM

7921dg76

Exploded view of the suspension arms — Grand Cherokee/Wagoneer

pension arm for installation reference.

3. Remove the lower control arm nut, cam and cam bolt from the axle.

4. Remove the nut and bolt from the frame rail bracket.

5. Remove the arm.

To install:

6. Install the lower control arm.

7. Tighten the rear bolts finger-tight.

8. Install the cam bolt, cam and nut in the axle while realigning the reference marks.

9. Lower the vehicle.

10. On Grand Cherokee and Grand Wagoneer, tighten the nuts to 130 ft. lbs. (176 Nm). On 1997 Wrangler, tighten the axle bracket nuts to 85 ft. lbs. (115 Nm) and the frame bracket nut to 130 ft. lbs. (176 Nm).

Sway Bar

REMOVAL AND INSTALLATION

1993–95 Wrangler

1. Raise and safely support the vehicle.

2. Remove the retaining nut from the connecting link bolt.

3. Disconnect the bracket retaining nuts and brackets.

4. Remove the stabilizer bar.

To install:

5. Install the sway bar.

6. Install and torque the retaining brackets and fasteners to 30 ft. lbs. (41 Nm).

7. Install and torque the upper link bolts and nuts to 45 ft. lbs. (61 Nm).

8. Tighten the link spring bracket nuts to 45 ft. lbs. (61 Nm).

9. Lower the vehicle.

1997 Wrangler

1. Remove the upper link nuts and separate the links from the bar with tool MB-990635.

2. Remove the front bumper valance.

3. Remove the bar retaining bolts.

4. Remove the bar from the vehicle.

5. Remove the lower link nuts and bolts and remove the links.

To install:

6. Center the stabilizer bar on top of the frame rails and install the retainers and bolts. Tighten to 70 ft. lbs. (95 Nm).

7. Install the upper link nuts and tighten to 45 ft. lbs. (61 Nm).

8. Install the bumper valance.

Cherokee

1. Raise and support the vehicle safely.

2. Disconnect the stabilizer bar at the connecting links. If necessary, disengage the connecting links from the brackets on the frame rail and steering box.

3. Remove the stabilizer bar-to-frame clamps and cushions, and remove the stabilizer bar.

To install:

4. Position the bar on the vehicle and install the clamps and cushions finger-tight.

5. Tighten the clamp-to-frame bolts to 55 ft. lbs. (75 Nm), the stabilizer bar-to-connecting link nuts to 27 ft. lbs. (37 Nm), and the connecting link-to-axle bolts to 70 ft. lbs. (95 Nm).

Grand Cherokee and Grand Wagoneer

1. Raise and safely support the vehicle.

2. Remove the sway bar links from the axle brackets.

3. Disconnect the sway bar from the links.

4. Disconnect the sway bar clamps from the frame rails.

5. Remove the sway bar.

To install:

6. Install the sway bar on the frame rail and install the clamps and

bolts. Ensure the bar is centered with equal spacing on both sides. Tighten the retaining bracket bolts to 55 ft. lbs. (75 Nm).

7. Install the connecting rod link and grommets onto the stabilizer bar and brackets. Tighten the nuts to 70 ft. lbs. (96 Nm).

8. Tighten the sway bar-to-connecting link nut to 27 ft. lbs. (36 Nm).

9. Lower the vehicle.

Front Wheel Bearings

REPLACEMENT

NOTE: Sodium-based grease is not compatible with lithium-based grease. Read the package labels and be careful not to mix the two types. If there is any doubt as to the type of grease used, completely clean the old grease from the bearing and hub before replacing.

Cherokee

2WD Models

1. Raise and support the vehicle safely.

2. Remove the wheels.

3. Remove the caliper without disconnecting the brake line. Suspend it out of the way using a piece of wire to prevent damage.

———— **CAUTION** ————

Brake linings may contain asbestos. Asbestos is a known cancer-causing agent. When working on brakes, remember that the dust which accumulates on the brake parts may contain asbestos. Always wear a protective face covering, such as a painter's mask, when working on the brakes. NEVER blow the dust from the brakes or drum! There are solvents made for the purpose of cleaning brake parts. Use them!

4. Remove the grease cap, cotter pin, nut cap, nut, and washer from the spindle. Discard the cotter pin.

5. Slowly remove the hub and rotor. Catch the outer bearing as it falls.

6. Carefully drive out the inner bearing and seal from the hub, using a wood block.

7. Inspect the bearing races for excessive wear, pitting or grooves. If they are cracked or grooved, or if pitting and excess wear is present, drive them out with a drift or punch.

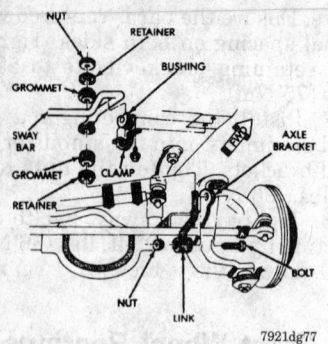

Exploded view of the sway bar mounting — Cherokee

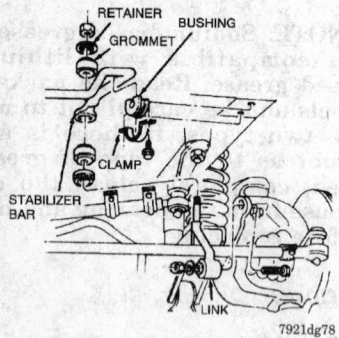

Exploded view of the sway bar — Grand Cherokee/Wagoneer

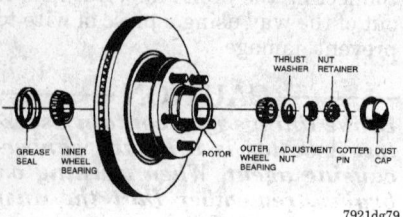

Exploded view of the bearings — 2WD models

8. Check the bearing for excess wear, pitting or cracks, or excess looseness.

NOTE: If it is necessary to replace either the bearing or the race, replace both. Never replace just a bearing or a race. These parts wear in a mating pattern. If just one is replaced, premature failure of the new part will result.

To install:

9. On vehicles with drum brakes, cover the spindle with a cloth and thoroughly brush all dirt from the

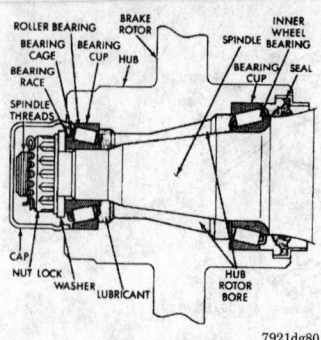

Cut-away view of the hub and bearings — 2WD models

brakes. Never blow the dirt off the brakes, due to the possible presence of asbestos in the dirt, which is harmful to your health when inhaled.

10. Remove the cloth and thoroughly clean the spindle and the inside of the hub.

11. Pack the inside of the hub with high temperature wheel bearing grease. Add grease to the hub until it is flush with the inside diameter of the bearing cup.

12. Pack the bearings with the same grease. A needle-shaped wheel bearing packer is best for this operation. If one is not available, place a large amount of grease in the palm of your hand and slide the edge of the bearing cage through the grease to pick up as much as possible, then work the grease in until it squeezes through the bearing.

13. If a new race is being installed, very carefully drive it into position until it bottoms all around, using a brass drift. Be careful to avoid scratching the surface.

14. Place the inner bearing in the race and install a new grease seal.

15. Clean the rotor contact surface if necessary.

16. Position the hub and rotor on the spindle and install the outer bearing.

17. Install the washer and nut.

18. While turning the rotor, tighten the nut to 25 ft. lbs. (34 Nm) to seat the bearings.

19. Back the nut off ½ a turn. While turning the rotor, tighten the nut to 19 inch lbs. (2 Nm).

20. Install the nut cap and a new cotter pin. Install the grease cap.

21. Install the caliper.

22. Install the wheels.

4WD Models

NOTE: If the hub is equipped with ball bearings, the entire unit must be replaced if defective. If the hub is equipped with

tapered roller bearings, its internal components can be serviced or replaced as necessary.

1. Raise and support the vehicle safely.
2. Remove the wheels.
3. Remove, but do not disconnect, the caliper. Suspend it out of the way.

--- **CAUTION** ---

Brake linings may contain asbestos. Asbestos is a known cancer-causing agent. When working on brakes, remember that the dust which accumulates on the brake parts and/or in the drum may contain asbestos. Always wear a protective face covering, such as a painter's mask, when working on the brakes. NEVER blow the dust from the brakes or drum! There are solvents made for the purpose of cleaning brake parts. Use them!

4. Remove the rotor.
5. Remove the cotter pin, nut retainer, axle nut and washer.
6. Remove the 3 hub-to-steering knuckle attaching bolts.
7. Remove the hub/bearing carrier and the rotor shield.
8. Using an arbor press, press the hub out of the bearing carrier. Special tools 5073 and 5074 are available for this job. Secure the carrier to the press plate with M12x1.75mmx40mm bolts.
9. Cut and remove the plastic cage from the hub inner bearing. Using diagonal pliers or tin snips, cut the bearing cage. Discard the rollers after removing the cage.
10. Remove what remains of the inner bearing as follows:
   a. Install a bearing separator tool on the inner bearing.
   b. Position the separator tool and hub in an arbor press.
   c. Force the hub out of the inner bearing with press pin tool 5074.
11. Remove the bearing carrier outer seal and discard it.
12. Drive the inner bearing seal out and discard it. If you're using tool 5078, make sure that the word JEEP faces downward.
13. Attach press plate tool 5073 to the rear of the carrier. Secure it in the press using M12x1.75mmx40mm bolts.
14. Position bearing race remover 5076 in the carrier bore between the inner and outer bearing races.
15. Position the press pin tool 5074 on tool 5076.
16. Place the bearing carrier in the press and force the inner bearing

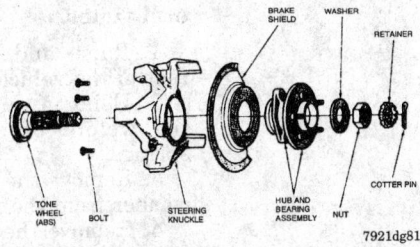

Exploded view of the hub assembly — 4WD models

race from the carrier bore. Reverse the position of the carrier and tools and force the outer bearing race from the bore.

To assemble and install:

17. Thoroughly clean all reusable parts with a safe solvent. Discard any parts that appear worn or damaged.

18. Attach press plate tool 5073 on the bearing carrier. Secure it in the press using M12x1.75mmx40mm bolts.

19. Position the new outer bearing race in the bore.

20. Position bearing race installation tool 5077 on the race. Make sure that the word JEEP faces the downward. Press the race into the bore. The race should be flush with the machined shoulder of the carrier.

21. Position the new inner bearing race in the carrier bore. Reverse the position of the carrier and tools and force the inner race into the bore.

22. Thoroughly pack the new outer bearing with wheel bearing grease. Make sure that the bearing is fully packed.

23. Coat the race with wheel bearing grease and place the bearing in the bore.

24. Place the new outer seal on the bearing and position bearing installation tool 5079 on the seal. Place the carrier in the press and force the seal into the bore. Apply wheel bearing grease to the seal lip.

25. Insert the hub through the seal and outer bearing and into the bearing carrier bore.

26. Install bearing installation tool 5078 into the rear of the bearing carrier bore and place the race installation tool 5077 on the front of the hub. Make sure that the word JEEP on 5077 is facing the hub.

27. Place the assembly in the press and force the hub shaft into the carrier bore.

28. Pack the new inner bearing with wheel bearing grease. Make

sure that the bearing is thoroughly packed.

29. Coat the inner seal lip with wheel bearing grease and place it on the inner bearing.

30. Coat the inner bearing race with wheel bearing grease.

31. Place the carrier in a press along with tool 5077. The word JEEP must face the hub. Position the bearing and seal in the carrier. Place seal installation tool 5080 on the seal.

32. Force the bearing and seal into the bore and onto the hub shaft.

—— **WARNING** ——

Use extreme care when forcing the assembly into position! The carrier must rotate freely after installation of the bearing! Do not attempt to eliminate bearing lash with the press. Final bearing preload is attained by tightening the drive axle nut.

33. Install the new outer seal on the carrier.

34. Thoroughly clean the axle shaft and apply a thin coating of lithium-based grease to the splines and seal contact surfaces.

35. Install the slinger, rotor shield and hub/bearing assembly on the axle shaft.

36. Coat the carrier bolt threads with Loctite®. Install them and tighten to 75 ft. lbs. (102 Nm).

37. Install the rotor and caliper.

38. Install the washer and axle shaft nut. Tighten the nut to 175 ft. lbs. (237 Nm).

39. Install the nut retainer and cotter pin. NEVER back off the nut to install the cotter pin! ALWAYS advance it!

40. Install the wheel.

Grand Cherokee, Grand Wagoneer and Wrangler

1. Raise and support the vehicle safely.

2. Remove the wheel.

3. Remove the caliper and rotor.

4. Remove the cotter pin, nut retainer and axle hub nut.

5. Remove the hub-to-knuckle bolts. Remove the hub from the steering knuckle.

To install:

6. Install the hub to the steering knuckle. Tighten the bolts to 75 ft. lbs. (102 Nm).

7. Install the hub washer and nut. Tighten the nut to 175 ft. lbs. (237 Nm). Install the nut retainer and a new cotter pin.

8. Install the rotor and caliper.

9. Install the wheel and lower the vehicle.

REAR SUSPENSION

Shock Absorber

REMOVAL AND INSTALLATION

Wrangler

NOTE: Before installing new shocks, they should be purged of air. To do this, hold the shock upright and fully extend it, then invert and compress it. Do this several times.

1. Raise and safely support the rear of the vehicle.

2. Using a suitable jack, relieve the axle weight from the springs.

3. Remove the upper locknut and washer from the frame bracket stud.

4. Remove the lower bolt, nut and washers from the axle shaft tube bracket.

5. Remove the shock absorber.

To install:

6. Position the shock absorber upper eye on the frame bracket stud and install the washer and nut.

7. Position the shock absorber lower eye in the axle shaft tube bracket and install the bolt, washers and nut.

8. Tighten the nuts to 44 ft. lbs. (60 Nm), except on 1997 Wrangler. On 1997 Wrangler, tighten the upper bolt to 23 ft. lbs. (31 Nm). Tighten the lower bolts to 74 ft. lbs. (100 Nm).

9. Remove the supports and lower the vehicle.

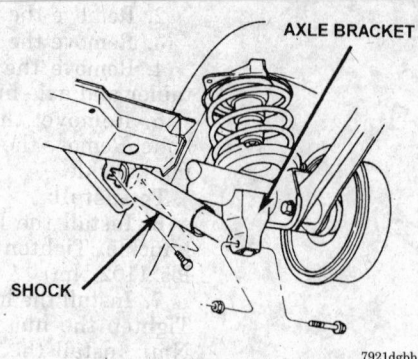

AXLE BRACKET

SHOCK

7921dgbh

Rear shock mounting — 1997 Wrangler

Cherokee

NOTE: Before installing new shocks, they should be purged of air. To do this, hold the shock upright and fully extend it, then invert and compress it. Do this several times.

1. Raise and safely support the vehicle.
2. Using a suitable jack, relieve the weight of the axle from the springs.
3. Remove the upper shock absorber-to-body bolts.
4. Remove the lower stud nut and washer.
5. Remove the shock absorber.

To install:

6. Position the shock absorber on the axle shaft tube bracket stud, then install the upper attaching bolts.
7. Install the washer and nut to the stud and tighten to 44 ft. lbs. (60 Nm).
8. Tighten the upper attaching bolts to 15 ft. lbs. (20 Nm).
9. Remove the supports and lower the vehicle.

Grand Cherokee and Grand Wagoneer

NOTE: Before installing new shocks, they should be purged of air. To do this, hold the shock up-

right and fully extend it, then invert and compress it. Do this several times.

1. Raise and safely support the rear of the vehicle.
2. Using a suitable jack, relieve the weight of the axle from the springs.
3. Remove the upper stud nut and washer from the frame rail stud.
4. Remove the lower nut and bolt from the axle bracket.
5. Remove the shock absorber.

To install:

6. Position the shock absorber on the frame rail stud and axle bracket.
7. Install the nut and retainer on the stud and tighten to 52 ft. lbs. (70 Nm).
8. Install the bolt and nut to the axle bracket and tighten to 68 ft. lbs. (92 Nm).
9. Remove the supports and lower the vehicle.

Coil Spring

REMOVAL AND INSTALLATION

Grand Cherokee, Grand Wagoneer and 1997 Wrangler

1. Raise and safely support the vehicle.

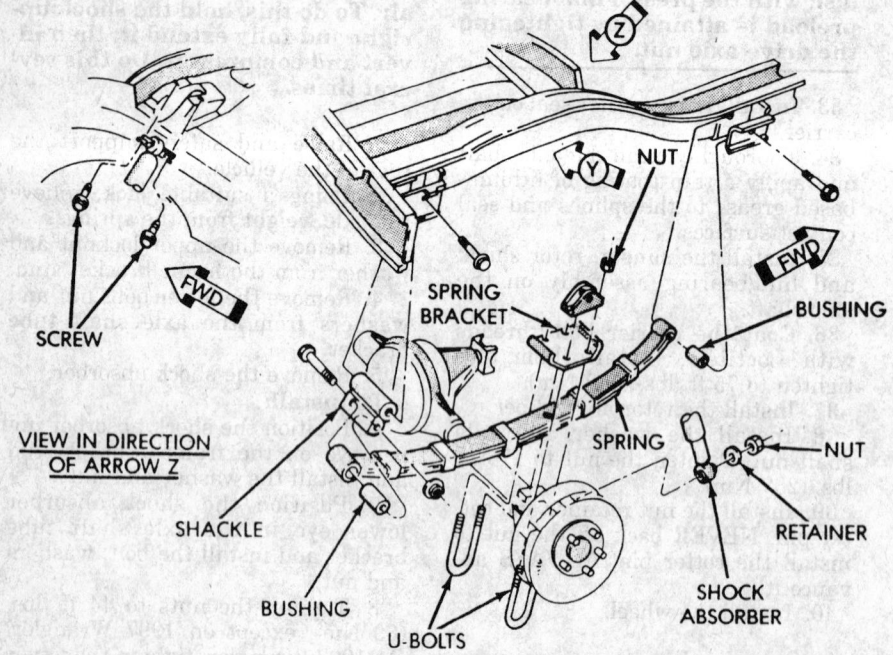

Z

Y NUT

FWD

FWD

SCREW

SPRING BRACKET

BUSHING

VIEW IN DIRECTION OF ARROW Z

SHACKLE

BUSHING

U-BOLTS

SPRING

NUT

RETAINER

SHOCK ABSORBER

Rear suspension components — Cherokee

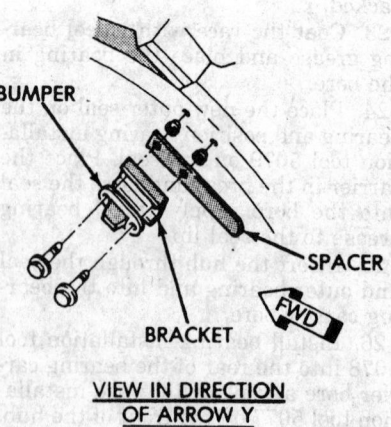

BUMPER

SPACER

BRACKET

FWD

VIEW IN DIRECTION OF ARROW Y

7921dgbg

2. Support the axle with a suitable jack.

3. Disconnect the sway bar links and shock absorbers from the axle bracket.

4. Disconnect the track bar from the frame rail bracket.

5. Lower the axle until the spring is free from the upper mount seat. Remove the coil spring clip screw and remove the spring.

To install:

6. Position the coil spring on the axle pad, then install the spring clip and screw (if equipped). Tighten the screw to 16 ft. lbs. (22 Nm).

7. Raise the axle into position until the spring seats in the upper mount.

8. Connect the sway bar links and shock absorbers to the axle bracket. Connect the track bar to the frame rail bracket.

9. Remove the supports and lower the vehicle.

Leaf Spring

REMOVAL AND INSTALLATION

1993–95 Wrangler

1. Raise the vehicle and support it safely.

2. Position a jack under the axle. Raise the axle to relieve the springs of the axle weight.

3. Remove the wheels.

4. Remove the spring U-bolts and tie plates.

5. Remove the bolt attaching the spring rear eye to the shackle.

6. Remove the bolt attaching the spring front eye to the shackle.

7. Remove the spring from its mounting.

To install:

8. Position the replacement spring front eye in the shackle. Loosely install the attaching bolt. Perform the same procedure for the rear spring eye.

9. Position and align the spring with the axle tube.

10. Lower the rear axle until it is completely supported by the spring.

11. Install the spring bracket, U-bolts and nuts. Torque the nuts to 90 ft. lbs. (122 Nm).

12. Install the wheels.

13. Lower the vehicle and tighten the spring eye bolts to 100 ft. lbs. (136 Nm).

Cherokee

1. Raise the vehicle and support it safely.

2. Position a jack under the axle. Raise the axle to relieve the springs of the axle weight.

3. Remove the wheels.

4. If equipped, disconnect the sway bar links at the spring plates.

5. Remove the spring U-bolts and spring plates.

6. Remove the bolt attaching the spring rear eye to the shackle.

7. Remove the bolt attaching the spring front eye to the shackle.

8. Lower the axle and remove the spring from its mounting.

To install:

9. Position the replacement spring front eye in the shackle. Loosely install the attaching bolt. Perform the same procedure for the rear spring eye.

10. Position and align the spring with the axle tube.

11. Install the spring plates, U-bolts and nuts. Torque the nuts to 52 ft. lbs. (70 Nm).

12. If equipped, connect the sway bar link to the spring bracket and tighten the nut to 70 ft. lbs. (96 Nm).

13. Install the wheels.

14. Lower the vehicle.

15. Torque the spring eye bolts to 65 ft. lbs. (88 Nm).

Upper Control Arm

REMOVAL AND INSTALLATION

Grand Cherokee and Grand Wagoneer

1. Raise and support the vehicle safely.

2. Remove the upper suspension arm nut and bolt at the axle bracket. Remove the ABS wire bracket from the arm.

3. Remove the nut and bolt at the frame rail and remove the suspension arm.

To install:

4. Position the upper arm at the axle and frame rail.

5. Install the fasteners finger-tight. Install the ABS wire onto the arm.

6. Remove the supports and lower the vehicle.

7. Tighten the fasteners to 55 ft. lbs. (75 Nm).

1997 Wrangler

1. Raise and support the vehicle safely.

2. Remove the ABS wire bracket and the parking brake cable bracket from the arm.

3. Remove the upper suspension arm nut and bolt at the axle bracket.

COIL SPRING

FRAME RAIL STUD

RETAINER

NUT

SHOCK ABSORBER

COIL SPRING

CLIP

AXLE BRACKET

VIEW IN DIRECTION OF ARROW Z

7921dgbi

Coil spring — Grand Cherokee/Wagoneer

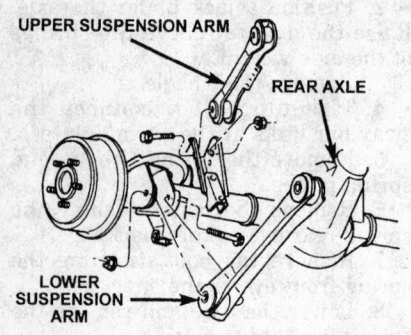

Upper and lower control arms — 1997 Wrangler

4. Remove the nut and bolt at the frame rail and remove the suspension arm.

To install:

5. Position the upper arm at the axle and frame rail.

6. Install the fasteners finger-tight. Install the ABS wire bracket and parking brake cable bracket onto the arm.

7. Remove the supports and lower the vehicle.

8. Tighten the fasteners to 55 ft. lbs. (75 Nm).

Lower Control Arm

REMOVAL AND INSTALLATION

Grand Cherokee, Grand Wagoneer and 1997 Wrangler

1. Raise and support the vehicle safely.

2. Remove the lower control arm nut and bolt at the axle bracket.

3. Remove the lower control arm nut and bolt at the frame rail.

4. Remove the lower control arm.

To install:

5. Install the lower control arm and tighten the bolts finger-tight.

6. Lower the vehicle.

7. Torque the bolts and nuts to 130 ft. lbs. (177 Nm).

Sway Bar

REMOVAL AND INSTALLATION

Cherokee

1. Raise and support the vehicle safely.

2. Disconnect the sway bar links from the springs.

3. Disconnect the sway bar from the frame rails.

4. Remove the sway bar.

To install:

5. Install the sway bar and torque the bar link bolts to 55 ft. lbs. (74 Nm).

6. Connect the sway bar to the frame rail and torque the bolts to 40 ft. lbs. (54 Nm).

7. Lower the vehicle.

Grand Cherokee and Grand Wagoneer

1. Raise and support the vehicle safely.

2. Remove 1 wheel.

3. Disconnect the stabilizer bar links from the axle brackets.

4. Lower the exhaust by disconnecting the muffler and tail pipe hangers.

5. Disconnect the stabilizer bar from the links.

6. Disconnect the stabilizer bar clamps from the frame rails. Remove the stabilizer bar.

To install:

7. Position the stabilizer bar on the frame rail and install the clamps and bolts.

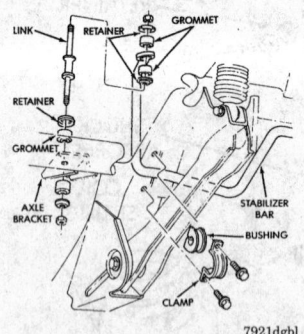

Rear stabilizer bar — Grand Cherokee/Wagoneer

8. Ensure the bar is centered with equal spacing on both sides. Torque the bolts to 40 ft. lbs. (54 Nm).

9. Install the links and grommets onto the stabilizer bar and axle brackets. Install the nuts and torque them to 27 ft. lbs. (36 Nm).

10. Connect the muffler and tail pipe to their hangers.

11. Install the wheel and lower the vehicle.

1997 Wrangler

1. Raise and support the vehicle safely.

2. Remove the stabilizer link bolts from the frame mounts.

3. Remove the link bolts from the bar.

4. Remove the retainer bolts and retainers from the axle, then remove the bar.

To install:

5. Install the stabilizer bar on the axle mounts and install the retainers and bolts.

NOTE: Ensure that the bar is centered with equal spacing on both sides and is positioned above the differential housing.

6. Tighten the retainer bolts to 40 ft. lbs. (54 Nm).

7. Install the links onto the stabilizer bar and frame mounts. Install the bolts and nuts finger-tight.

8. Lower the vehicle.

9. Tighten the links to 40 ft. lbs. (54 Nm).

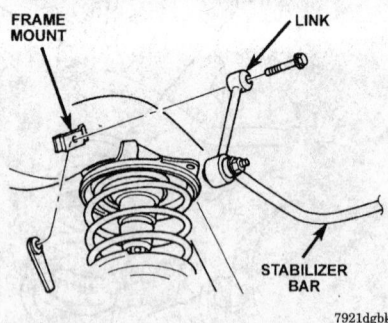

Stabilizer link — 1997 Wrangler

FIRING ORDERS

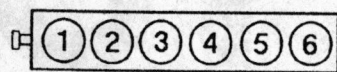

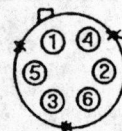

275403

6-4.5L (VIN F) engine
Engine firing order: 1-5-3-6-2-4
Distributor rotation: Counterclockwise

NOTE: To avoid confusion, always replace spark plug wires one at a time.

ENGINE ELECTRICAL

NOTE: Disconnecting the negative battery cable on some vehicles may interfere with the functions of the on board computer systems and may require the computer to undergo a relearning process, once the negative battery cable is reconnected.

Distributor

REMOVAL

1. Disconnect the negative battery cable.
2. Disconnect the distributor connectors.
3. Remove the distributor cap without disconnecting the secondary leads. Position aside.
4. Matchmark the rotor with the distributor housing and housing with the cylinder block.
5. Remove the distributor hold-down bolt and pull the distributor from the cylinder block.

INSTALLATION

Engine Not Disturbed

1. Install a new O-ring to the distributor and lubricate with engine oil.

2. Insert the distributor into the cylinder block, while aligning the matchmarks made during removal. Install the distributor hold-down bolt.
3. Install the distributor cap. Connect the distributor connector.
4. Connect the negative battery cable. Start the engine and allow normal operating temperature to be reached.
5. Check and if necessary, adjust the ignition timing.

Engine Disturbed

1. Install a new O-ring to the distributor and lubricate with engine oil.
2. Remove the No. 1 cylinder spark plug.
   a. Place a finger or compression gauge over the spark plug hole.
   b. Turn the crankshaft until compression starts to build up. Continue turning the crankshaft until the crankshaft pulley groove align with the timing mark "0" of the timing chain.
3. If necessary, remove the valve cover.
   a. Check that the timing marks with 1 and 2 dots are in straight line on the cylinder head surface as shown.
   b. If not, turn the crankshaft one revolution (360 degrees) and align the crankshaft pulley groove with the timing mark "0" of the timing chain.
4. Align the groove of the distributor housing with the protrusion on the driven gear.
5. Insert the distributor into the cylinder block. Install the distributor hold-down bolt.
6. Install the distributor cap. Connect the distributor connector.
7. Connect the negative battery cable. Start the engine and allow normal operating temperature to be reached.
8. Check and if necessary, adjust the ignition timing.

Ignition Timing

ADJUSTMENT

1. Warm up the engine to normal operating temperature.
2. Connect a Toyota or Lexus hand-held tester or the OBD II scan tool. Remove the fuse cover on the instrument panel. Connect the tool to the DLC3.

3. Connect the timing light to the engine.
4. Check the idle speed by racing the engine speed to 2,500 rpm for approximately 90 seconds. Idle speed should be 650±50 rpm.
5. Inspect and adjust ignition timing.
   a. Using tool No. SST 09843-18020 (jumper wire), or equivalent, connect terminals TE1 and E1 of the DLC1.
   b. Using a timing light, check the ignition timing. It should be 3° BTDC at idle.
6. Adjustment can be made by loosening the hold-down bolt and turning the distributor. Tighten the hold-down bolt and recheck the timing.
7. Remove the SST from the DLC1.
8. Disconnect the timing light from the engine.
9. Remove the tester or scan tool.

Alternator

PRECAUTIONS

Several precautions must be observed with alternator equipped vehicles to avoid damage to the unit.

- If the battery is removed for any reason, make sure it is reconnected with the correct polarity. Reversing the battery connections may result in damage to the 1-way rectifiers.
- When utilizing a booster battery as a starting aid, always connect the positive to positive terminals and the negative terminal from the booster battery to a good engine ground on the vehicle being started.
- Never use a fast charger as a booster to start vehicles.
- Disconnect the battery cables when charging the battery with a fast charger.
- Never attempt to polarize the alternator.
- Do not use test lights of more than 12 volts when checking diode continuity.
- Do not short across or ground any of the alternator terminals.
- The polarity of the battery, alternator and regulator must be matched and considered before making any electrical connections within the system.
- Never separate the alternator on an open circuit. Make sure all connections within the circuit are clean and tight.

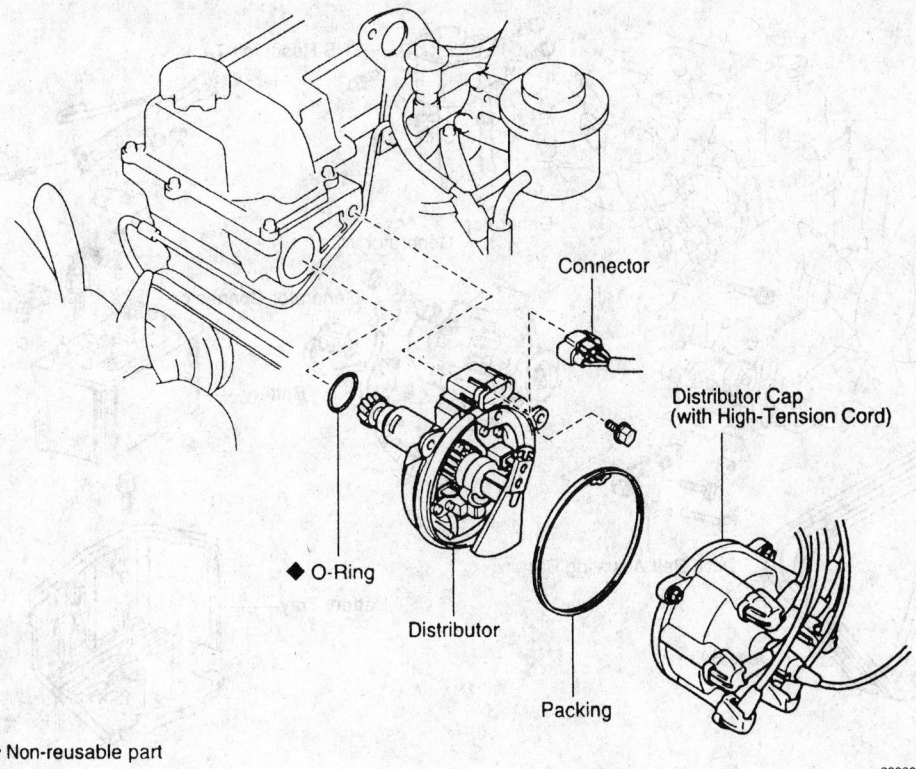

Connector

Distributor Cap
(with High-Tension Cord)

◆ O-Ring

Distributor

Packing

▸ Non-reusable part

299393

Distributor and related components

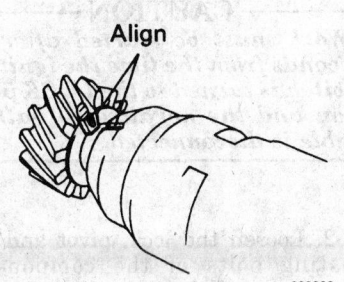

Align

299392

Align the groove of the distributor housing with the protrusion on the driven gear

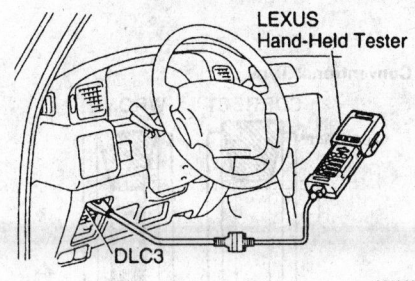

LEXUS
Hand-Held Tester

DLC3

301069

Connection for scan tool on the instrument panel

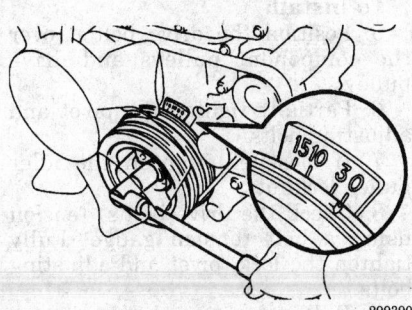

299390

Align the crankshaft pulley groove with the timing mark "0" of the timing chain cover

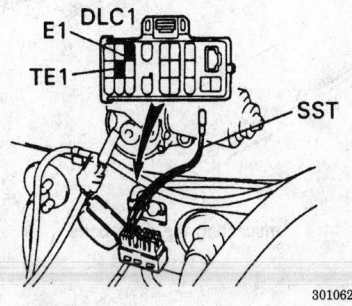

DLC1
E1
TE1
SST

301062

Connect service tool No. SST 9843-18020, or equivalent, to terminals TE1 and E1 of data link connector 1

• Disconnect the battery ground terminal when performing any service on electrical components.
• Disconnect the battery if arc welding is to be done on the vehicle.

REMOVAL AND INSTALLATION

1. Disconnect the battery cables and remove the battery and the battery tray.

— CAUTION —

On air bag equipped vehicles, wait at least 90 seconds from the time that the ignition switch is turned to the LOCK position and the battery is disconnected before performing any further work.

2. Disconnect P/S reservoir tank.
3. Loosen the lockbolt, pivot bolt and the adjusting bolt.
4. Remove the two drive belts.
5. Disconnect the alternator connector. Remove the rubber cap and nut and disconnect the alternator wire.
6. Disconnect the alternator wire clamp from the generator.
7. Remove the lockbolt, pivot bolt, nut and the drive belt adjusting bar.
8. Remove the alternator.

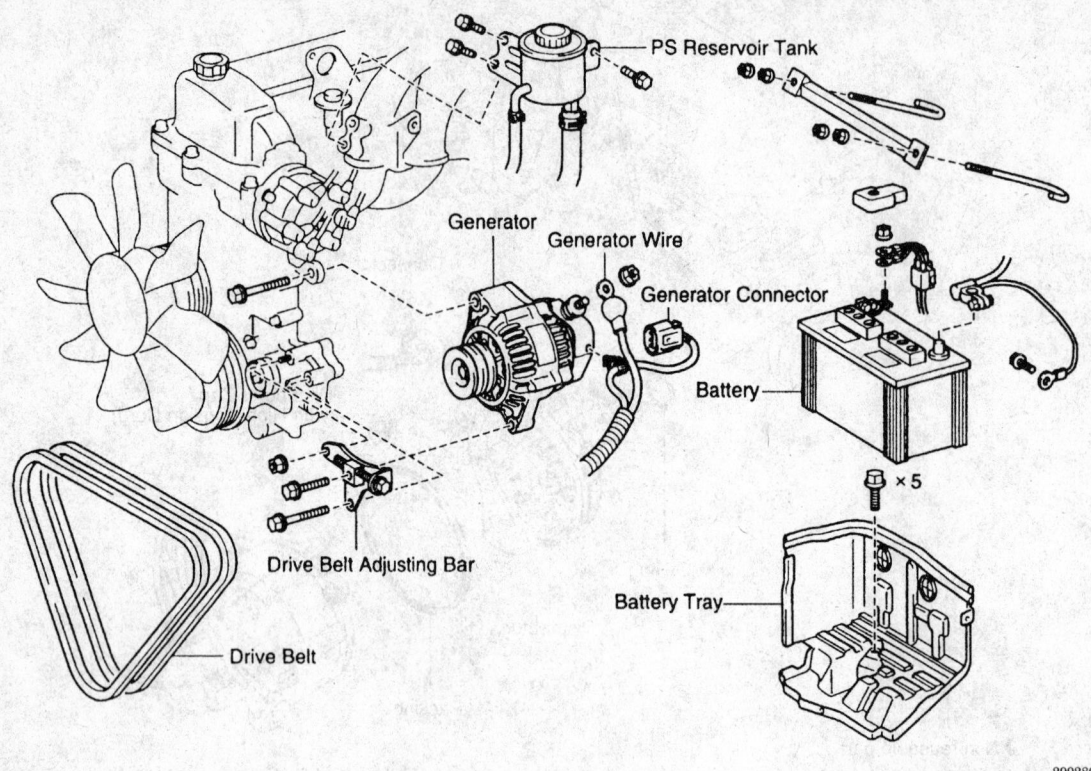

PS Reservoir Tank

Generator

Generator Wire

Generator Connector

Battery

Drive Belt Adjusting Bar

Battery Tray

Drive Belt

299280

Alternator and related components

To install:

9. Mount the alternator on the bracket with the pivot bolt. Do not tighten the bolt yet.

10. Install the drive belt adjusting bar with the bolt and nut. Torque the bolt to 15 ft. lbs. (21 Nm).

11. Temporarily install the lockbolt. Connect the alternator connector and the wire with the nut and rubber cap.

12. Connect the alternator wire clamp to the alternator.

13. Install the drive belts. The drive belt tension should be, 125±25 lbs. for a new belt and 80±20 lbs. for a used belt.

14. Tighten the pivot bolt to 43 ft. lbs. (59 Nm) and the lockbolt to 15 ft. lbs. (21 Nm).

15. Connect the power steering reservoir tank.

16. Install the battery tray, the battery and connect the cables.

17. Start the vehicle and check the alternator operation.

Drive Belt

REMOVAL AND INSTALLATION

1. Disconnect the negative battery cable.

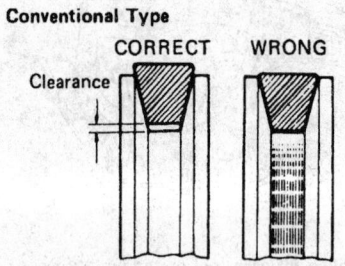

Conventional Type

CORRECT   WRONG

Clearance

V-Ribbed Type

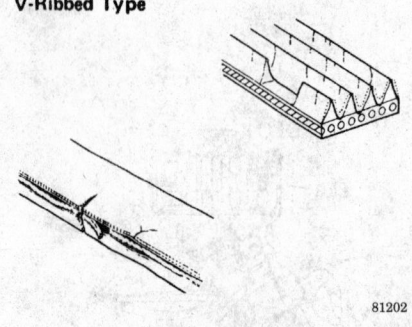

81202

Inspecting the drive belts

CAUTION

Work must be started after 90 seconds from the time the ignition switch is turned to the LOCK position and the negative (-) battery cable is disconnected.

2. Loosen the lock, pivot and adjusting bolts of the component. Loosen the idler pulley, if equipped.

3. Remove the drive belt(s) from its component.

4. Visually check the belt for separation of the ribs, torn or worn ribs, or cracks in the inner ridges of the ribs. If necessary, replace the drive belt.

To install:

5. Position the drive belt(s) over the component pulleys and drive pulley.

6. Partially tighten the pivot and adjusting bolts.

7. If loosened, tighten the idler pulley locknut.

8. Check the drive belt(s) tension using a belt tension gauge. Fully tighten the lock, pivot and adjusting bolts.

9. Belt tensions should be as follows:

   a. Alternator: Used — 80±20 lbs. New — 125±25 lbs.

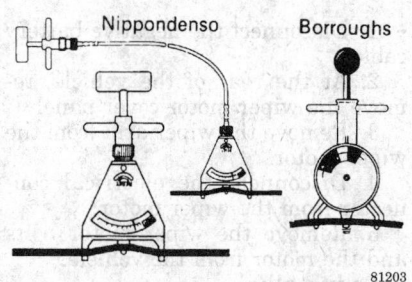

Check the drive belt tension using a belt tension gauge

b. A/C: Used — 80±25 lbs. New — 125±25 lbs.

NOTE: After installing the drive belt(s), check that the belt does not touch the bottom of the pulley groove (conventional type belts) or that it fits properly in the ribbed grooves (V-ribbed type belts).

10. Connect the negative battery cable.

Starter

REMOVAL AND INSTALLATION

1993–94

1. Disconnect the negative battery cable.

WARNING

To avoid possible air bag deployment (if equipped), do not start working on the vehicle until 90 seconds has elapse from the time the ignition switch is turned OFF and the negative battery terminal is disconnected.

---

2. Remove the air cleaner and hose, if necessary.
3. Disconnect and tag all connectors, cables and hoses as required.
4. Remove the battery and battery tray, if necessary.
5. Disconnect the connector and wire from the starter.
6. Remove the fasteners and then remove the starter from the bell-housing.
   To install:
7. Install the starter. Torque the to 29 ft. lbs. (39 Nm).
8. Connect the wiring and starter connector.

9. Install the battery tray and battery, if removed.
10. Connect all hoses, cables and wire connectors.
11. Connect the battery cable(s). Check that the engine starts.

1995–97

1. Disconnect the negative battery cable.

CAUTION

Work must be started after 90 seconds from the time the ignition switch is turned to the LOCK position and the negative (-) battery cable is disconnected.

---

2. Disconnect the starter connector.
3. Remove the nut and disconnect the starter wire.
4. Remove the two bolts and the starter.
   To install:
5. Install the starter and torque the bolts to 29 ft. lbs. (39 Nm).
6. Connect the starter wire and torque the nut to 6.5 ft. lbs. (8.8 Nm).
7. Connect the starter connector.
8. Connect the negative battery cable.

CHASSIS ELECTRICAL

Blower Motor

REMOVAL AND INSTALLATION

1993–94

1. Disconnect the negative battery cable.
2. Disconnect the electrical connectors from the motor.
3. Remove the blower motor fasteners and lower the blower motor out of the air inlet duct.
4. To install, reverse the removal procedure. Check to ensure the seal around the motor flange is in good condition. Be sure to position the motor so the flexible tube can be attached to the motor.

1995–97

Front Blower Motor

1. Disconnect the negative battery cable.

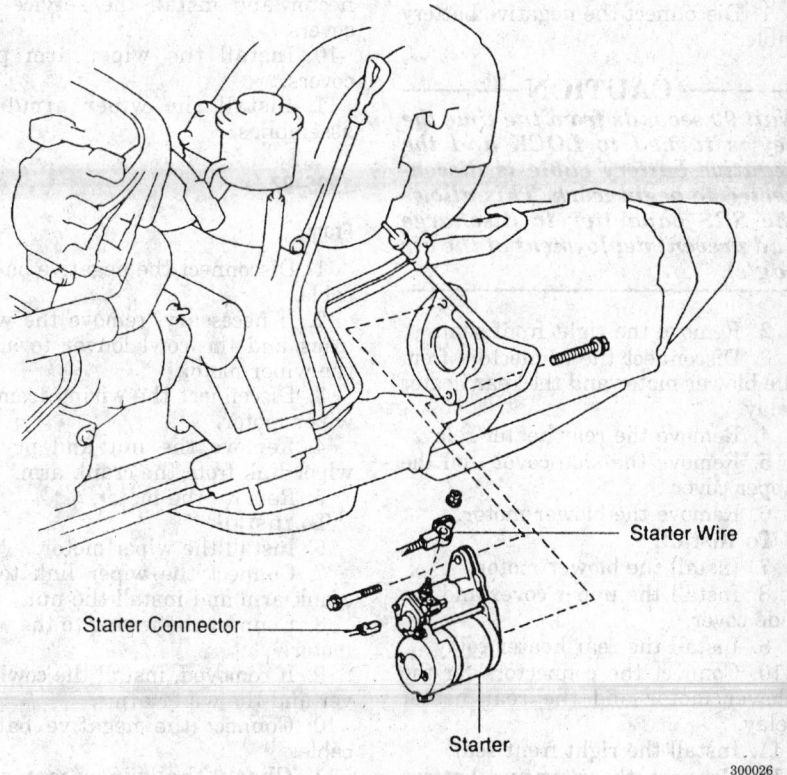

Starter and related components

—————— CAUTION ——————

Wait 90 seconds from the time the key is turned to LOCK and the negative battery cable is disconnected to begin work. This allows the SRS capacitor to discharge and prevent deployment of the air bag(s).

2. Remove the right-hand scuff plate.
3. Disconnect the link from the lower cover of the blower motor.
4. Remove the lower cover while pushing the locking protrusion.
5. Disconnect the electrical connectors from the blower motor and blower resistor.
6. Remove the blower motor to case screws and remove the motor from the case.

To install:

7. Install the blower motor to the case and install the three screws.
8. Connect the electrical connectors to the blower motor and blower resistor.
9. Connect the link to the lower cover of the blower motor.
10. Install the right-hand scuff plate.
11. Connect the negative battery cable to the battery.

Rear Blower Motor

1. Disconnect the negative battery cable.

—————— CAUTION ——————

Wait 90 seconds from the time the key is turned to LOCK and the negative battery cable is disconnected to begin work. This allows the SRS capacitor to discharge and prevent deployment of the air bag(s).

2. Remove the right front seat.
3. Disconnect the connectors from the blower motor and the rear heater relay.
4. Remove the rear heater relay.
5. Remove the side cover and the upper cover .
6. Remove the blower motor.

To install:

7. Install the blower motor.
8. Install the upper cover and the side cover.
9. Install the rear heater relay.
10. Connect the connectors for the blower motor and the rear heater relay.
11. Install the right front seat.
12. Connect the negative battery cable.

Windshield Wiper Motor

REMOVAL AND INSTALLATION

1993–94

NOTE: The wiper motor is removed with the linkage assembly.

1. Disconnect the negative battery cable.
2. Remove the wiper arm retaining nuts, then, the wiper arm/blade assemblies.
3. Remove both wiper arm pivot covers and the pivot-to-cowl attaching screws.
4. Remove the service hole covers from the cowl area of the engine compartment.
5. Disconnect the wiring from the wiper motor.
6. From the engine compartment, remove the wiper motor plate-to-cowl screws. Withdraw the wiper motor and the linkage from the cowl panel as an assembly.
7. Pry the linkage from of the wiper motor.

To install:

8. Attach the linkage to the wiper motor. Place the wiper motor into position and install the mounting screws.
9. Connect the wiper motor connector and install the service hole covers.
10. Install the wiper arm pivot covers.
11. Install the wiper arm/blade assemblies.

1995–97

Front

1. Disconnect the negative battery cable.
2. If necessary, remove the wiper arms and the cowl louver to access the wiper motor.
3. Disconnect the wiring from the wiper motor.
4. Remove the nut and pry the wiper link from the crank arm.
5. Remove the motor.

To install:

6. Install the wiper motor.
7. Connect the wiper link to the crank arm and install the nut.
8. Connect the wiring to the wiper motor.
9. If removed, install the cowl louver and the wiper arms.
10. Connect the negative battery cable.
11. Check the wiper motor for proper operation.

Rear

1. Disconnect the negative battery cable.
2. At the rear of the vehicle, remove the wiper motor cover panel.
3. Remove the wiper arm from the wiper motor.
4. Disconnect the electrical connector from the wiper motor.
5. Remove the wiper motor bolts and the motor from the vehicle.

To install:

6. Install the wiper motor.
7. Connect the electrical connector to the wiper motor.
8. Install the wiper arm to the wiper motor.
9. Install the wiper motor cover panel.
10. Connect the negative battery cable.
11. Check the wiper motor for proper operation.

Combination Switch

REMOVAL AND INSTALLATION

1. Disconnect the negative battery cable and disable the air bag system.

—————— CAUTION ——————

Work must be started after 90 seconds from the time the ignition switch is turned to the LOCK position and the negative (-) battery cable is disconnected.

2. Remove the steering wheel.

—————— CAUTION ——————

The air bag system must be disarmed before removing the steering wheel. Failure to do so may cause accidental deployment, property damage or personal injury.

3. Remove the upper and lower steering column shroud screws and the shrouds.
4. Remove the combination switch screws and the switch from the column.
5. Disconnect the electrical connector from the combination switch. To remove the wires from the electrical connector, perform the following procedures:
   a. Using a small prybar, insert it into the open end between the locking lugs and the terminal.
   b. Pry the locking lugs upward and pull the terminal out from the rear.

Front Wiper

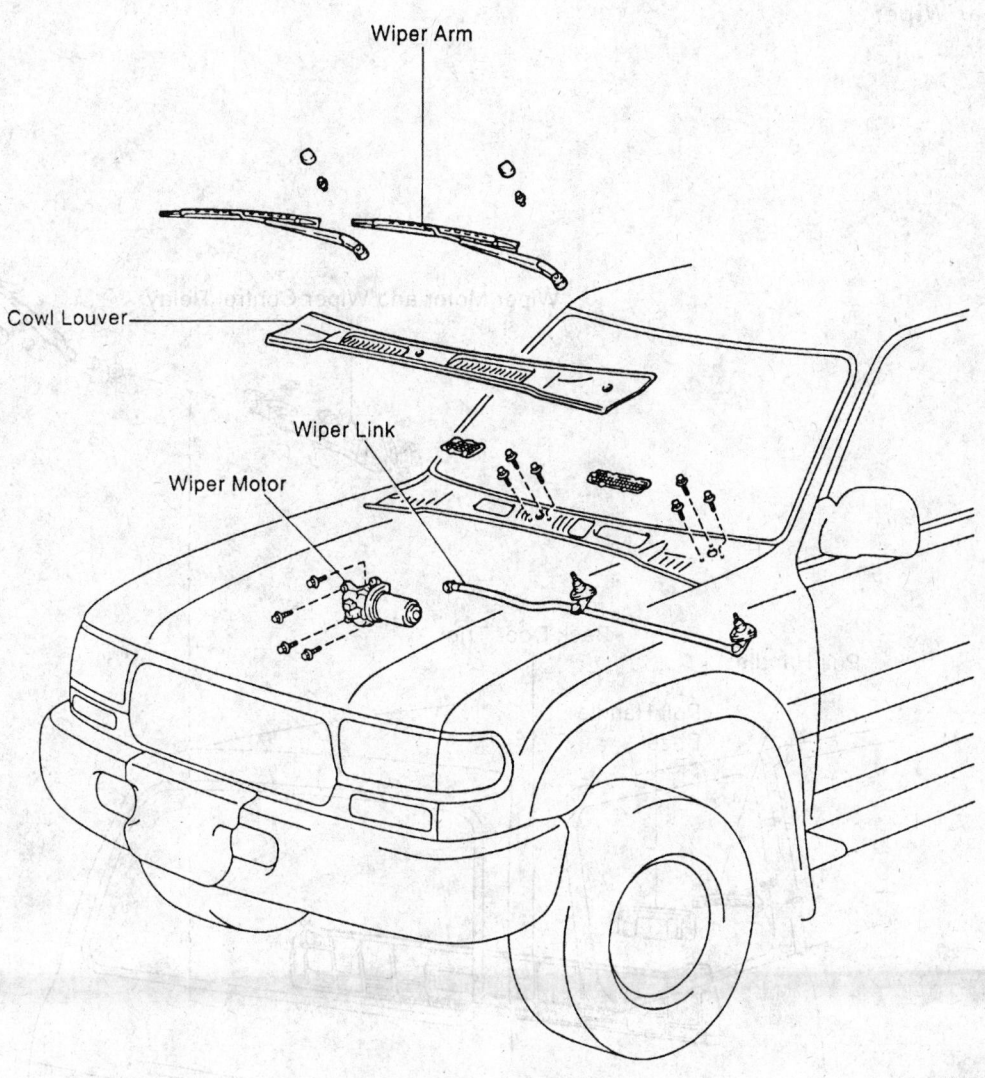

Wiper Arm

Cowl Louver

Wiper Link

Wiper Motor

301020

Front wiper motor and related components

6. Remove the light control switch.

a. Remove the two screws and the ball set plate from the switch body.

b. Remove the ball and slide out the switch from the switch body with the spring.

7. Remove the headlight dimmer, the turn signal and the wiper washer switches by removing the screws on the backside of the combination switch.

To install:

8. Install the headlight dimmer, the turn signal and the wiper washer switches to the combination switch.

9. Install the light control switch. Slide in the switch with the spring to the switch body. Install the ball and the ball set plate with the two screws. Make sure that the switch operates smoothly.

10. Push the wires into the connector until they lock securely in place.

11. Install the switch into the steering column.

12. Install the upper and lower steering column shroud.

13. Install the steering wheel and torque the nut to 25 ft. lbs. (34 Nm).

14. Enable the air bag system.

15. Connect the negative battery cable.

Ignition Lock Cylinder

REMOVAL AND INSTALLATION

1. Disconnect the negative battery terminal.

2. Remove the upper and lower steering column covers.

3. Disconnect the ignition switch from the electrical connector.

4. Using the key in the ignition switch, turn it to the **ACC** position.

5. Using a thin rod, place it into the hole of the cylinder lock housing. Pushing down on the thin rod, pull out the cylinder lock.

Rear Wiper

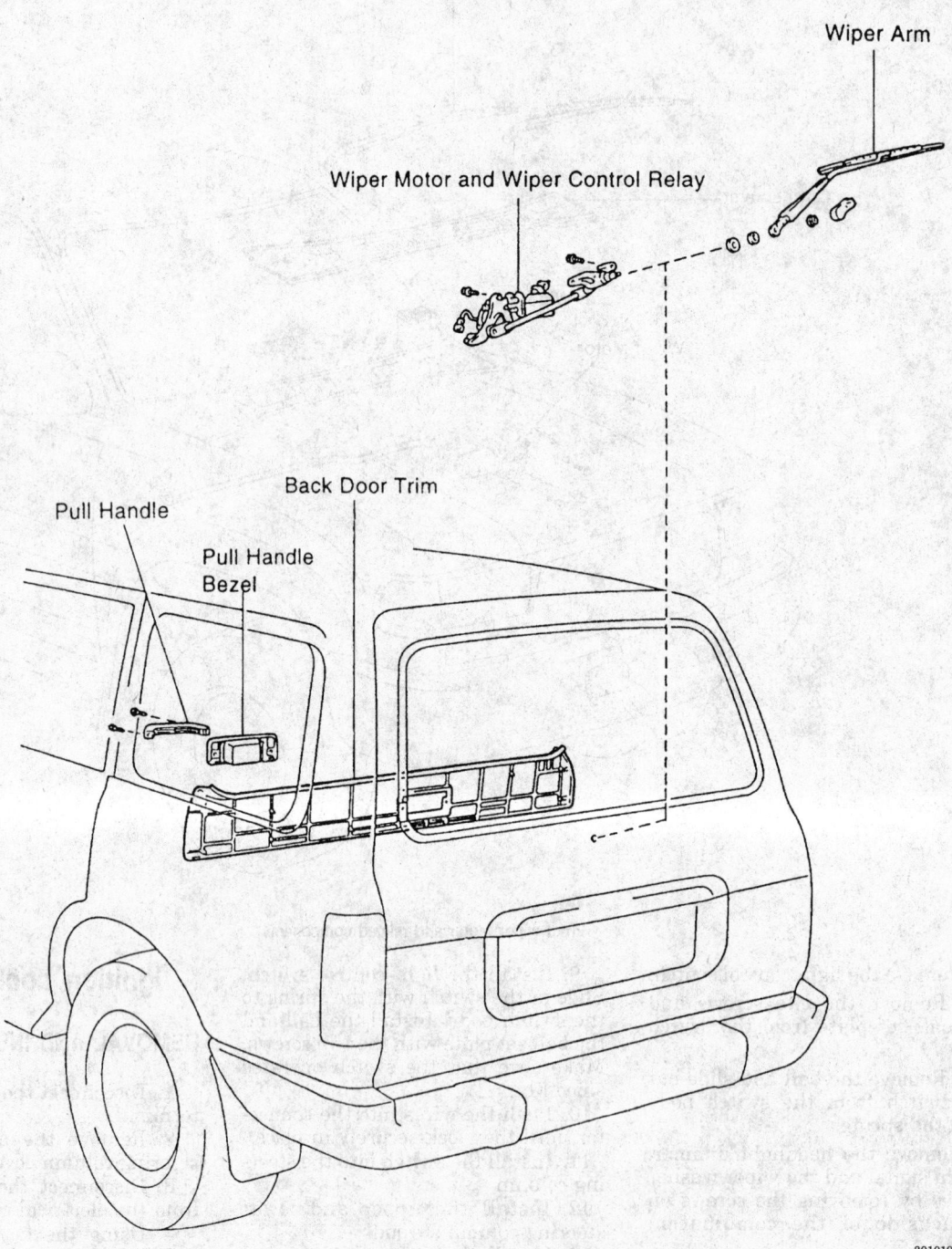

Wiper Arm

Wiper Motor and Wiper Control Relay

Back Door Trim

Pull Handle

Pull Handle Bezel

301019

Rear wiper motor location

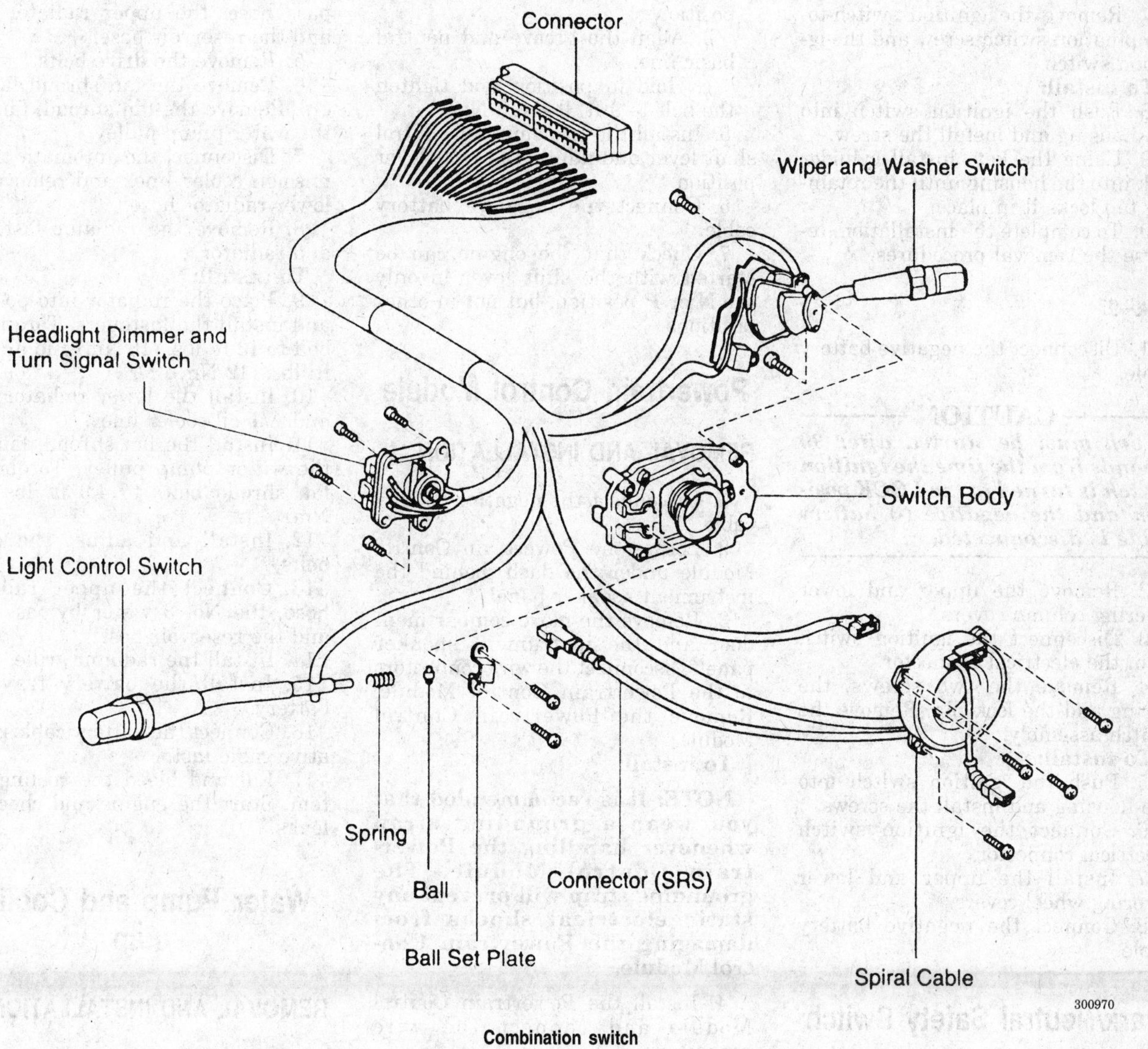

Connector

Wiper and Washer Switch

Headlight Dimmer and
Turn Signal Switch

Switch Body

Light Control Switch

Spring

Ball

Connector (SRS)

Ball Set Plate

Spiral Cable

300970

Combination switch

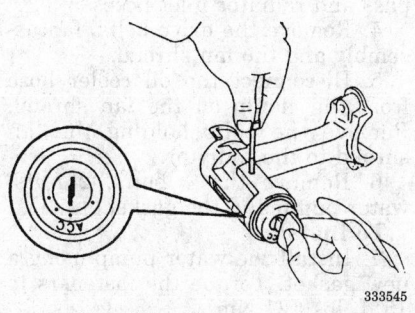

333545

Lock cylinder

To install:

6. Using the key, install the cylinder lock into the housing until the retaining tab locks it in place.

7. Connect the ignition switch electrical connector.

8. Install the upper and lower steering column covers.

9. Connect the negative battery terminal.

Ignition Switch

REMOVAL AND INSTALLATION

1993–94

1. Disconnect the negative battery terminal.

2. Remove the upper and lower steering column covers.

3. Disconnect the ignition switch from the electrical connector.

4. Using the key in the ignition switch, turn it to the **ACC** position.

5. Using a thin rod, place it into the hole of the cylinder lock housing. Pushing down on the thin rod, pull out the cylinder lock.

6. Remove the unlock warning switch-to-combination switch screws and the unlock warning switch.

7. Remove the ignition switch-to-combination switch screw and the ignition switch.

To install:

8. Push the ignition switch into the housing and install the screw.

9. Using the key, install cylinder lock into the housing until the retaining tab locks it in place.

10. To complete the installation, reverse the removal procedures.

1995–97

1. Disconnect the negative battery cable.

------ **CAUTION** ------
Work must be started after 90 seconds from the time the ignition switch is turned to the LOCK position and the negative (-) battery cable is disconnected.
------

2. Remove the upper and lower steering column covers.

3. Disconnect the ignition switch from the electrical connector.

4. Remove the two screws, the spring and the lock pin. Remove the switch assembly.

To install:

5. Push the ignition switch into the housing and install the screws.

6. Connect the ignition switch electrical connector.

7. Install the upper and lower steering wheel covers.

8. Connect the negative battery cable.

Park/Neutral Safety Switch

REMOVAL AND INSTALLATION

1. Disconnect the negative battery cable.

------ **CAUTION** ------
Wait at least 90 seconds from the time the ignition key is turned to the LOCK position and the negative battery cable has been disconnected. This will allow the SRS system capacitor sufficient time to discharge and prevent accidental deployment of the air bag(s).
------

2. Remove the transmission control shaft lever.

3. Disconnect the park/neutral position switch connector. Pry off the lock washer and remove the nut. Remove the bolt and pull out the switch.

To install:

4. Install the park/neutral switch.

   a. Set the shift lever to the **N** position.

   b. Align the groove and neutral basic line.

   c. Hold in position and tighten the bolt to 9 ft. lbs. (13 Nm).

5. Install the transmission control shaft lever and adjust the shift lever position.

6. Connect the negative battery cable.

7. Check that the engine can be started with the shift lever in only the **N** or **P** position, but not in other positions.

Powertrain Control Module

REMOVAL AND INSTALLATION

1. Disconnect the negative battery cable.

2. Locate the Powertrain Control Module under the dash, behind the instrument speaker panel.

3. Remove the glove compartment door and the instrument speaker panel. Disconnect the wire connectors to the Powertrain Control Module. Remove the Powertrain Control Module.

To install:

NOTE: It is recommended that you wear a grounding strap whenever handling the Powertrain Control Module. The grounding strap will prevent any static electrical shocks from damaging the Powertrain Control Module.

4. Install the Powertrain Control Module and connect the wire connectors.

5. Install the instrument speaker panel and the glove compartment door.

6. Connect the negative battery cable.

7. Check for proper operation.

ENGINE COOLING

Radiator

REMOVAL AND INSTALLATION

1. Disconnect the battery cables; negative cable first. Remove the battery and battery tray.

2. Drain the cooling system.

3. Remove the radiator grille.

4. Disconnect the No. 3 water by-pass hose, the upper radiator hose and the reservoir hose.

5. Remove the drive belts.

6. Remove the fan shroud fasteners. Remove the fan shroud, fan and the water pump pulley.

7. Disconnect the automatic transmission cooler lines and remove the lower radiator hose.

8. Remove the radiator fasteners and radiator.

To install:

9. Place the radiator into position and install the fasteners. Torque the bolt to 13 ft. lbs. (18 Nm) and nut to 9 ft. lbs. (12 Nm).

10. Install the lower radiator hose and A/T oil cooler lines.

11. Install the fan shroud, fan and the water pump pulley. Torque the fan shroud bolts to 4.9 ft. lbs. (5.4 Nm).

12. Install and adjust the drive belts.

13. Connect the upper radiator hose, the No. 3 water by-pass hose and the reservoir hose.

14. Install the radiator grille.

15. Install the battery tray and battery.

16. Connect the battery cables; negative cable last.

17. Fill and bleed the cooling system. Start the engine and check for leaks.

Water Pump and Cooling Fan

REMOVAL AND INSTALLATION

1. Disconnect the negative battery cable.

2. Drain the engine coolant.

3. Disconnect the No. 3 water by-pass and radiator inlet hoses.

4. Remove the drive belts, fan assembly and the fan shroud.

5. Disconnect the oil cooler hose from the clamp on the fan shroud. Remove the bolts holding the fan shroud to the radiator.

6. Remove the 4 bolts, 2 nuts, water pump and the gasket.

To install:

7. Install the water pump using a new gasket. Torque the fasteners to 15 ft. lbs. (21 Nm).

8. Install the water pump pulley, fan shroud and the drive belts.

   a. Place the fan with the fluid coupling, water pump pulley and the fan shroud in position.

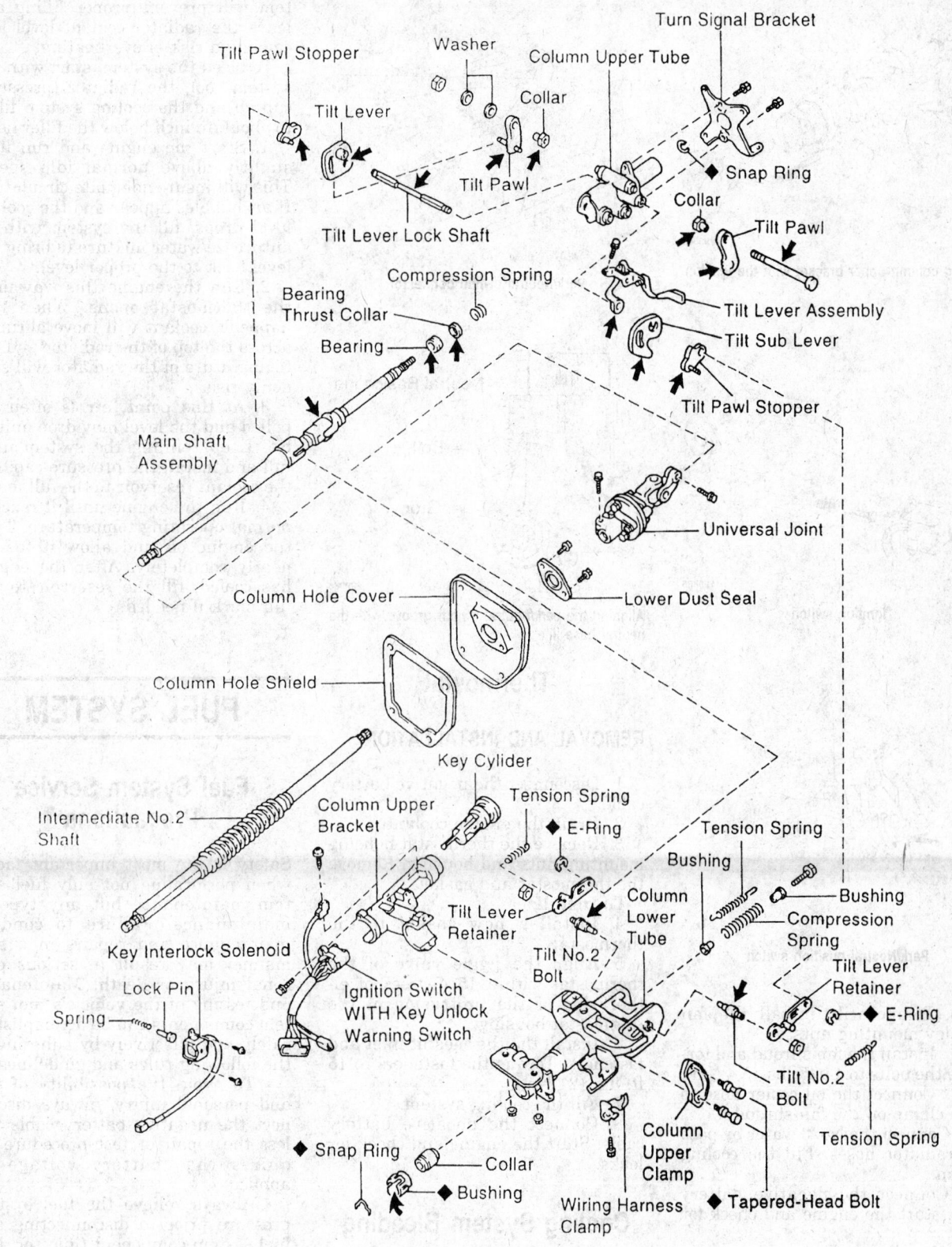

Tilt Pawl Stopper

Washer

Turn Signal Bracket

Tilt Lever

Collar

Column Upper Tube

Tilt Pawl

◆ Snap Ring

Collar

Tilt Pawl

Tilt Lever Lock Shaft

Tilt Lever Assembly

Compression Spring

Tilt Sub Lever

Bearing Thrust Collar

Tilt Pawl Stopper

Bearing

Main Shaft Assembly

Universal Joint

Column Hole Cover

Lower Dust Seal

Column Hole Shield

Key Cylinder

Tension Spring

Intermediate No.2 Shaft

Column Upper Bracket

◆ E-Ring

Tension Spring

Bushing

Bushing

Column Lower Tube

Compression Spring

Tilt Lever Retainer

Tilt No.2 Bolt

Key Interlock Solenoid

Tilt Lever Retainer

Lock Pin

Ignition Switch WITH Key Unlock Warning Switch

◆ E-Ring

Spring

Tilt No.2 Bolt

◆ Snap Ring

Collar

Tension Spring

◆ Bushing

Column Upper Clamp

Wiring Harness Clamp

◆ Tapered-Head Bolt

◆ Non-reusable part
← Molybdenum disulphide lithium base grease

299802

Steering column components

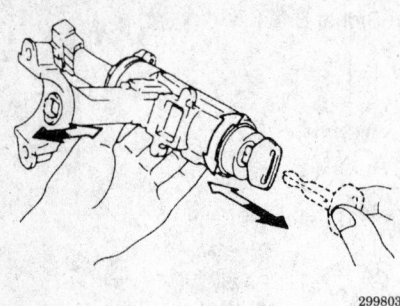

Steering column upper bracket with the ignition switch

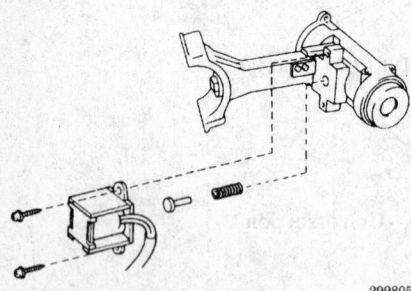

Ignition switch

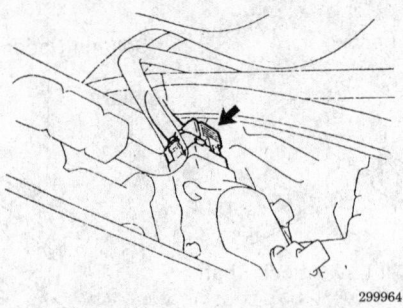

Park/Neutral switch connector

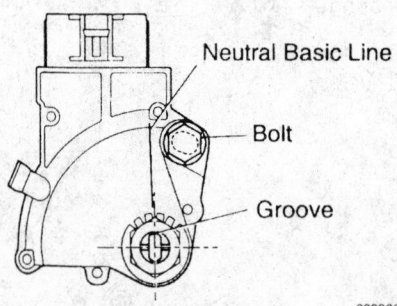

Neutral Basic Line

Bolt

Groove

Aligning the park/neutral switch groove with the neutral base line

Thermostat

REMOVAL AND INSTALLATION

1. Disconnect the negative battery cable.
2. Drain the engine coolant.
3. Remove the thermostat housing mounting nuts and housing. Remove the thermostat and gasket.

To install:

4. Install a new gasket to the thermostat.
5. Align the jiggle valve of the thermostat within 15 degrees of either side of the protrusion of the thermostat housing.
6. Install the thermostat inlet and fasteners. Torque the fasteners to 15 ft. lbs. (21 Nm).
7. Fill the cooling system.
8. Connect the negative battery cable. Start the engine and check for leaks.

Cooling System Bleeding

PROCEDURE

After working on the cooling system, even to replace the thermostat, it must be bled. Air trapped in the sys-

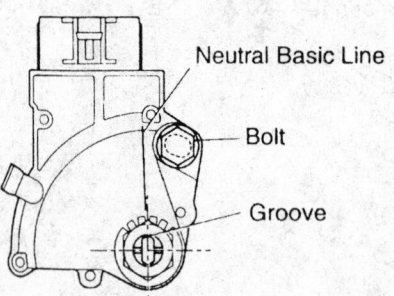

Park/Neutral position switch

b. Temporarily install the fan pulley mounting nuts.

c. Install the fan shroud and torque the bolts to 4.9 ft. lbs. (5.4 Nm).

d. Connect the oil cooler hose to the clamp on the fan shroud.

9. Connect the No. 3 water by-pass and radiator hoses. Fill the cooling system.

10. Connect the negative battery cable, start the engine and check for leaks.

tem will prevent proper filling and leave the radiator coolant level low, causing a risk of overheating.

To bleed the system, start with the system cool, the radiator (pressure) cap off and the cooling system filled to about an inch below the filler neck.

1. Start the engine and run it at slightly above normal idle speed. This will insure adequate circulation. If air bubbles appear and the coolant level drops, fill the system with an antifreeze/water mixture to bring the level back to the proper level.

2. Run the engine this way until the thermostat opens. When this happens, coolant will move abruptly across the top of the radiator and the temperature of the radiator will suddenly rise.

3. At this point, air is often expelled and the level may drop quite a bit. Keep refilling the system until full and install the pressure cap. Fill the coolant reservoir to the full mark.

4. Run the engine until it reaches normal operating temperature. Turn the engine off and allow it to cool nearly completely. After the engine has cooled, fill the reservoir to the full mark if needed.

FUEL SYSTEM

Fuel System Service Precautions

Safety is the most important factor when performing not only fuel system maintenance but any type of maintenance. Failure to conduct maintenance and repairs in a safe manner may result in serious personal injury or death. Maintenance and testing of the vehicle's fuel system components can be accomplished safely and effectively by adhering to the following rules and guidelines.

• To avoid the possibility of fire and personal injury, always disconnect the negative battery cable unless the repair or test procedure requires that battery voltage be applied.

• Always relieve the fuel system pressure prior to disconnecting any fuel system component (injector, fuel rail, pressure regulator, etc.), fitting or fuel line connection. Exercise extreme caution whenever relieving fuel system pressure to avoid exposing skin, face and eyes to fuel spray. Please be advised that fuel under

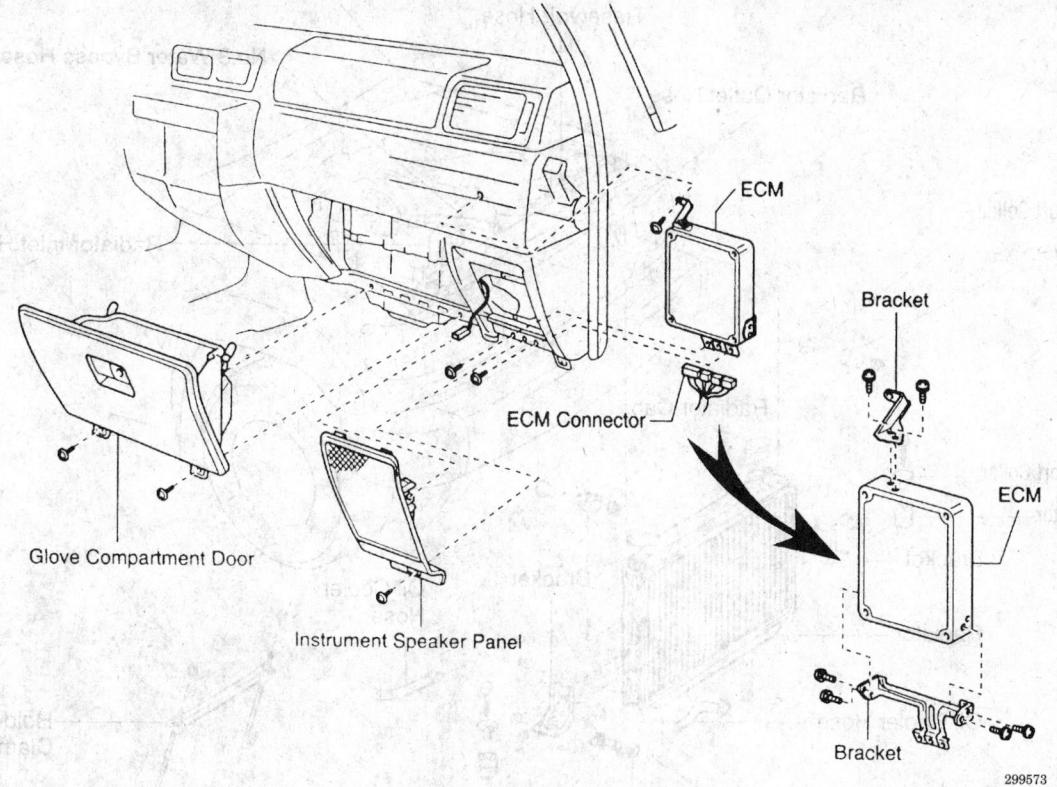

Glove Compartment Door

Instrument Speaker Panel

ECM

Bracket

ECM Connector

Bracket

ECM

299573

Powertrain Control Module location

pressure may penetrate the skin or any part of the body that it contacts.

• Always place a shop towel or cloth around the fitting or connection prior to loosening to absorb any excess fuel due to spillage. Ensure that all fuel spillage (should it occur) is quickly removed from engine surfaces. Ensure that all fuel soaked cloths or towels are deposited into a suitable waste container.

• Always keep a dry chemical (Class B) fire extinguisher near the work area.

• Do not allow fuel spray or fuel vapors to come into contact with a spark or open flame.

• Always use a backup wrench when loosening and tightening fuel line connection fittings. This will prevent unnecessary stress and torsion to fuel line piping. Always follow the proper torque specifications.

• Always replace worn fuel fitting O-rings with new. Do not substitute fuel hose or equivalent, where fuel pipe is installed.

Fuel System Pressure

RELIEVING

—————— **CAUTION** ——————

Fuel injection systems remain under pressure after the engine has been turned OFF. Properly relieve fuel pressure before disconnecting any fuel lines. Failure to do so may result in fire or personal injury.

1. Disconnect the negative battery terminal.
2. Place a catch-pan under the joint to be disconnected. A large quantity of fuel may be released when the joint is opened.
3. Wear eye or full face protection.
4. Place a shop towel over the area and slowly loosen the joint using a wrench of the correct size. Use a back-up wrench if needed.
5. Allow the fuel left in the line to bleed off slowly before fully disconnecting the joint.
6. Plug the opened lines immediately to prevent fuel spillage or the entry of dirt.
7. Dispose of the released fuel properly.

8. After connecting fuel lines, connect the negative battery cable and start the engine.
9. Check for leaks and repair as needed.

Idle Speed

ADJUSTMENT

Idle speed is controlled by the Powertrain Control Module and is not adjustable.

Mixture

ADJUSTMENT

Air/Fuel mixture is controlled by the Powertrain Control Module and is not adjustable.

Fuel Filter

REMOVAL AND INSTALLATION

1. Disconnect the negative battery cable.

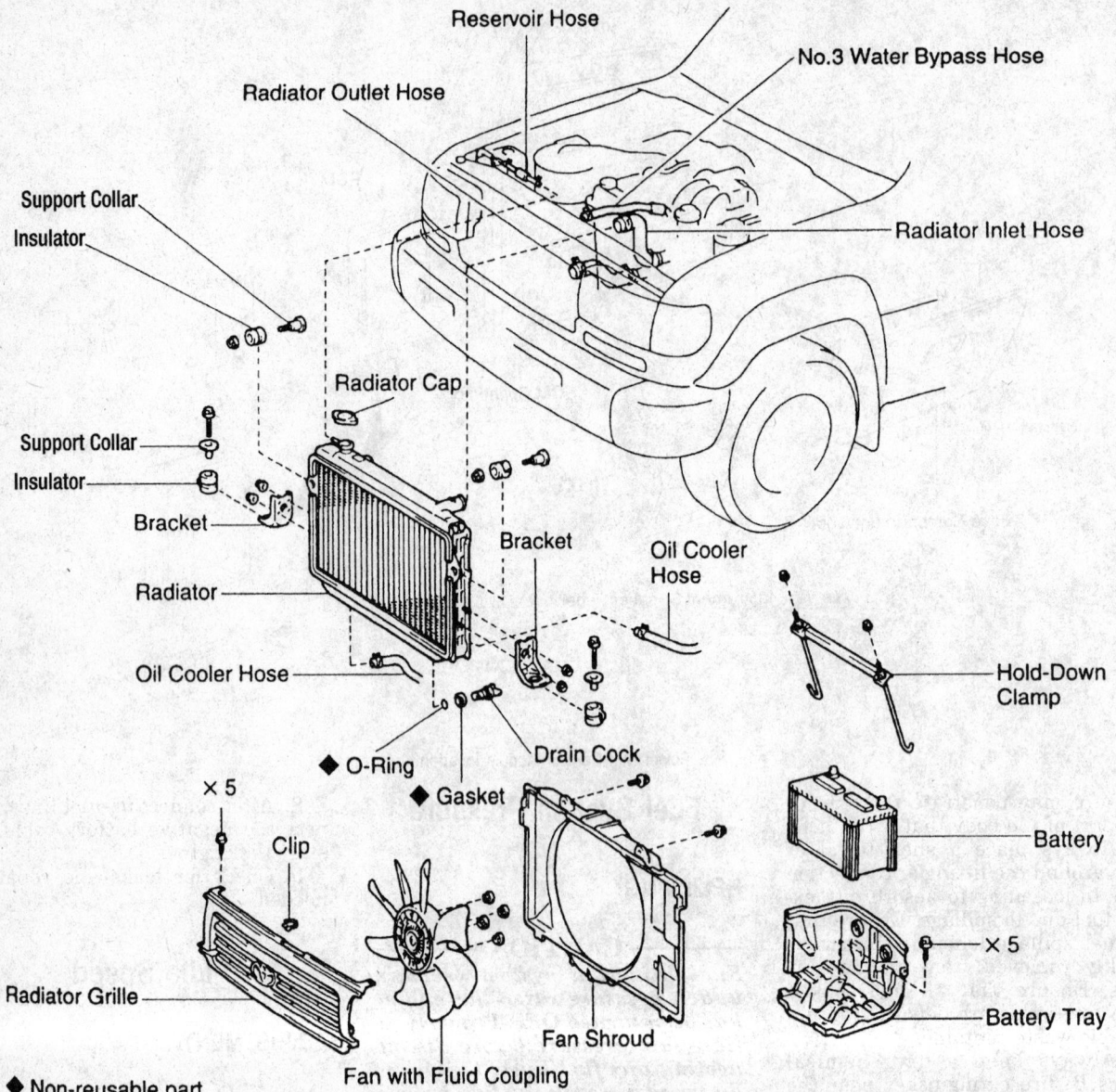

Radiator and related components

◆ Non-reusable part

CAUTION

Work must be started after 90 seconds from the time the ignition switch is turned to the LOCK position and the negative (-) battery cable is disconnected.

2. Relieve the fuel system pressure.

CAUTION

Fuel injection systems remain under pressure after the engine has been turned OFF. Properly relieve fuel pressure before disconnecting any fuel lines. Failure to do so may result in fire or personal injury.

NOTE: The fuel filter is located in the engine compartment, at the inlet line to the fuel rail.

3. Disconnect and plug the inlet and outlet lines from the filter.
4. Remove the fuel filter retaining bolts and remove the filter.

To install:
5. Install the fuel filter.
6. Use new O-rings and tighten the lines to 21 ft. lbs. (29 Nm).
7. Connect the negative battery cable.
8. Start the engine and check for leaks.

Fuel Pump

REMOVAL AND INSTALLATION

1993–94

1. Disconnect the negative battery cable. Relieve the fuel pressure.
2. Remove the fuel tank from the vehicle.
3. Remove the access plate-to-fuel tank bolts, then, pull out the plate/fuel pump assembly.
4. Disconnect the electrical connectors from the fuel pump. Pull the bracket from the lower-side of the fuel pump, then, remove the fuel pump from the fuel hose.

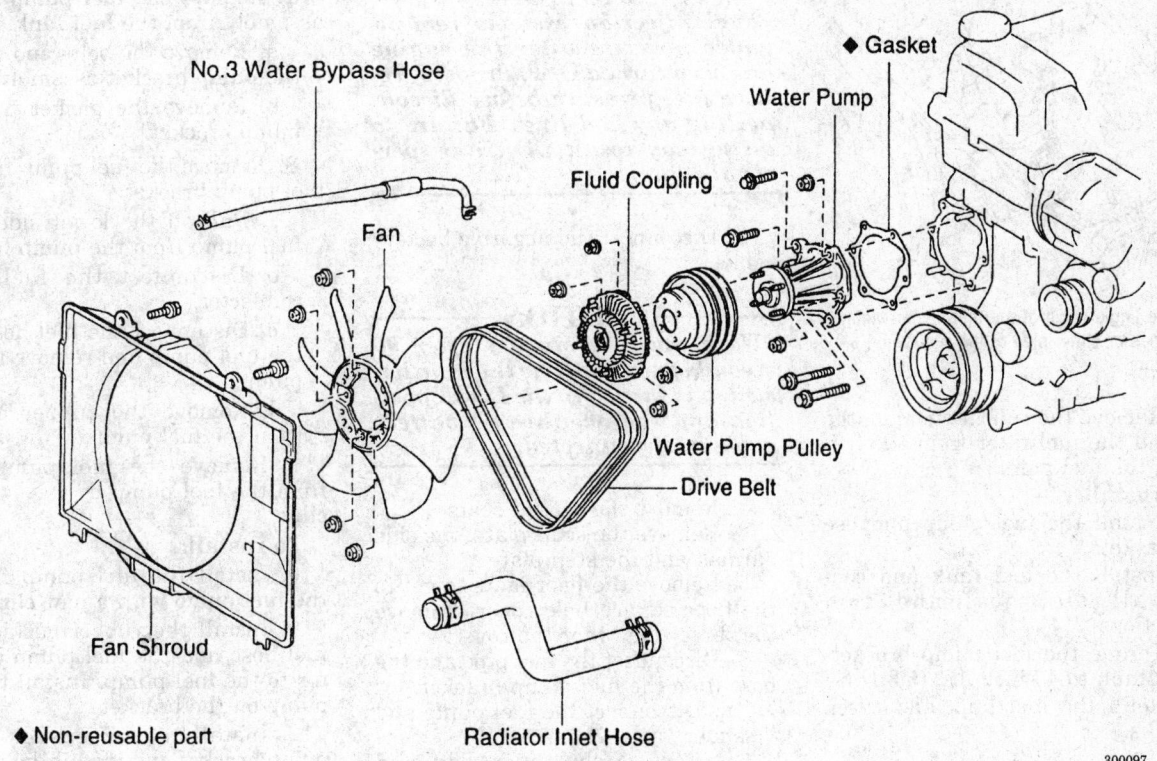

No.3 Water Bypass Hose

Fluid Coupling

◆ Gasket

Water Pump

Fan

Water Pump Pulley

Drive Belt

Fan Shroud

◆ Non-reusable part

Radiator Inlet Hose

300097

Water pump and related components

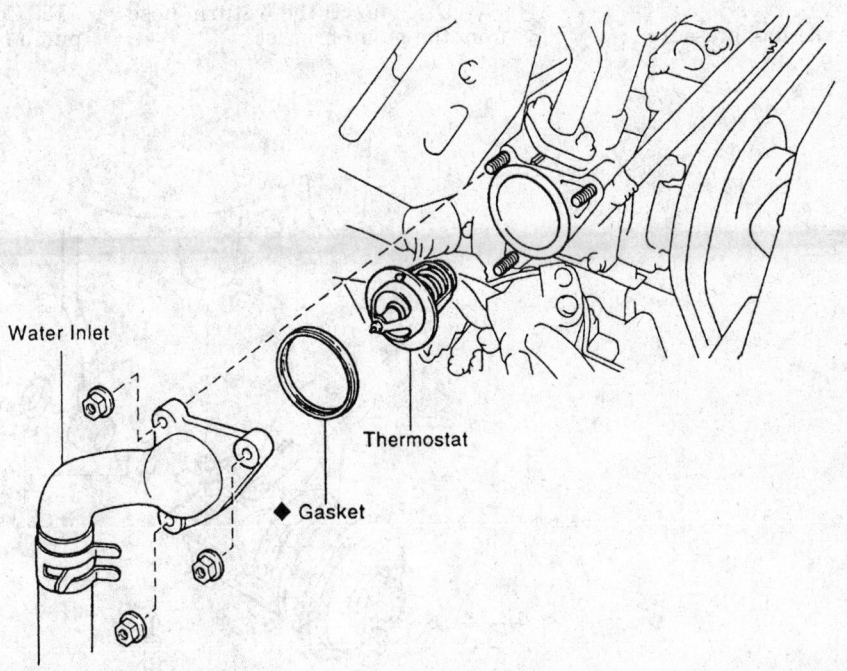

Water Inlet

Thermostat

◆ Gasket

◆ Non-reusable part

300028

Thermostat and related components

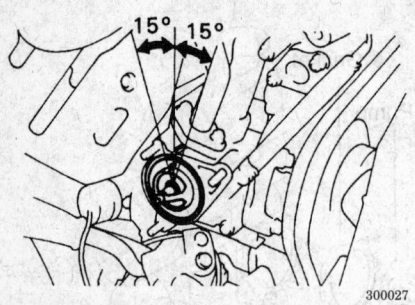

Align the jiggle valve of the thermostat within 15 degrees of either side of the protrusion of the housing

5. Remove the rubber cushion, the clip and the fuel filter from the bottom of the fuel pump.

To install:

6. Install the fuel pump and use new gaskets.

7. Install the fuel tank and connect all electrical and fuel connections.

8. Torque the fuel pump bracket-to-fuel tank to 43 inch lbs. (5.3 Nm).

9. Refill the fuel tank and check for leaks.

1995–97

1. Relieve the fuel pressure.

— CAUTION —

Fuel injection systems remain under pressure after the engine has been turned OFF. Properly relieve fuel pressure before disconnecting any fuel lines. Failure to do so may result in fire or personal injury.

2. Disconnect the negative battery cable.

— CAUTION —

Work must be started after 90 seconds from the time the ignition switch is turned to the LOCK position and the negative (-) battery cable is disconnected.

3. Remove the second seats.

4. Remove the scuff plate, the side garnish and the step plate.

5. Remove the floor mats to get to the floor service hole cover. Remove the three screws for the cover.

6. Disconnect the fuel pipe and the hose from the fuel pump bracket.

a. Disconnect the fuel pump and sender gauge connector.

b. Remove the union bolt and gaskets and disconnect the outlet pipe from the pump bracket.

c. Disconnect the return hose from the pump bracket.

---

7. Remove the fuel pump bracket assembly from the fuel tank.

a. Remove the bolts and pull out the pump bracket assembly.

b. Remove the gasket from the pump bracket.

8. Remove the fuel pump from the fuel pump bracket.

a. Pull off the lower side of the fuel pump from the pump bracket.

b. Disconnect the fuel pump connector.

c. Disconnect the fuel hose from the fuel pump and remove the fuel pump.

d. Remove the rubber cushion from the fuel pump.

9. Remove the fuel pump filter from the fuel pump by removing the clip.

To install:

10. Install the fuel pump filter to the fuel pump with a new clip.

11. Install the rubber cushion, the fuel hose and the fuel pump connector to the fuel pump. Install the fuel pump on the bracket.

12. Install a new gasket to the pump bracket and install the bracket in the fuel tank. Torque the bolts to 2.8 ft. lbs. (3.9 Nm).

13. Connect the return hose to the pump bracket.

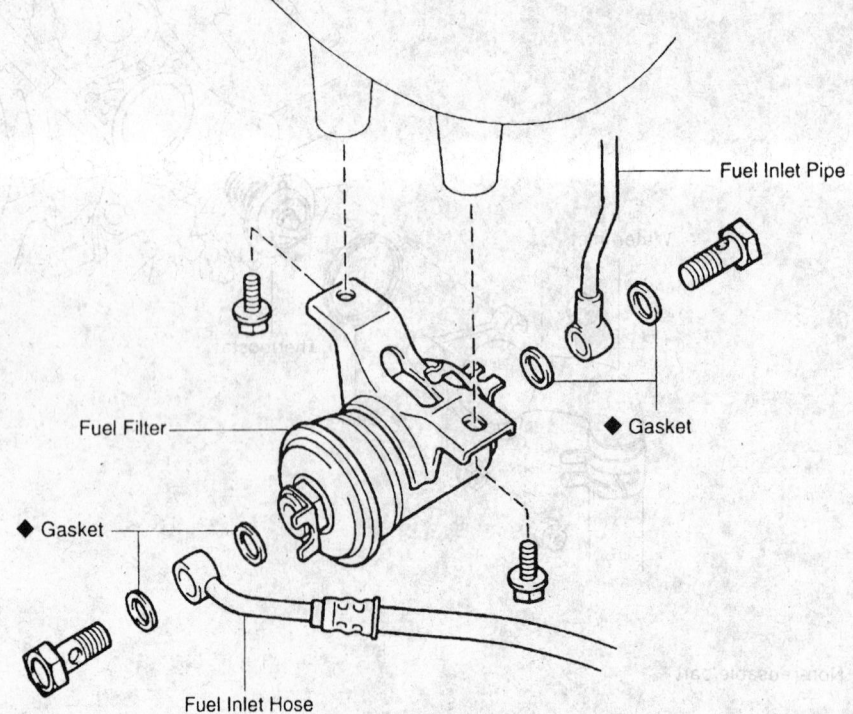

Fuel Inlet Pipe

◆ Gasket

Fuel Filter

◆ Gasket

Fuel Inlet Hose

◆ Non-reusable part

Fuel filter — 1995–97

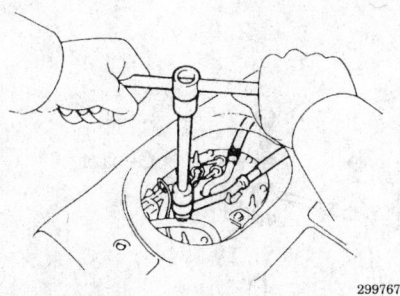

Disconnecting the fuel pipe and hose

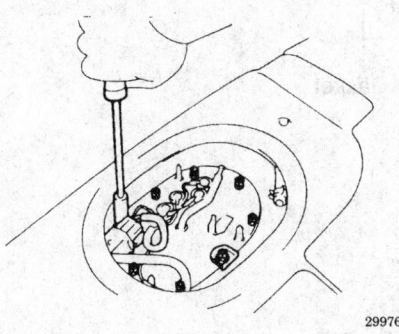

Removing the fuel pump bracket from the fuel pump — 1995–97

14. Connect the outlet pipe to the pump bracket and torque the union bolt to 22 ft. lbs. (29 Nm).

15. Connect the fuel pump and sender gauge connector.

16. Install the service hole cover.

17. Install the floor mats.

18. Install the step plate, the side garnish and the scuff plate.

19. Install the second seats and torque the bolts to 29 ft. lbs. (39 Nm).

20. Connect the negative battery cable.

21. Start the engine and check for leaks.

Fuel Injector

REMOVAL AND INSTALLATION

1. Relieve the fuel pressure.

> **CAUTION**
>
> *Fuel injection systems remain under pressure after the engine has been turned OFF. Properly relieve fuel pressure before disconnecting any fuel lines. Failure to do so may result in fire or personal injury.*

2. Disconnect the negative battery cable.

> **CAUTION**
>
> *Work must be started after 90 seconds from the time the ignition switch is turned to the LOCK position and the negative (-) battery cable is disconnected.*

3. Drain the engine coolant.

4. Disconnect the cruise control actuator cable.

5. Disconnect the accelerator cable.

6. Disconnect the throttle cable.

7. Disconnect the No. 2 PCV hoses.

8. Remove the EGR vacuum modulator and valve.

9. Remove the bolt holding the heater inlet pipe and the air intake chamber.

10. Disconnect the following items:

  a. No. 1 PCV hose.

  b. Vacuum sensing hose.

  c. No. 2 water bypass hose.

  d. Evaporative emission control hose.

  e. Brake booster hose.

  f. Throttle position sensor connector.

  g. Idle air control valve connector.

  h. Connector for the emission control valve set assembly.

  i. EGR gas temperature sensor connector.

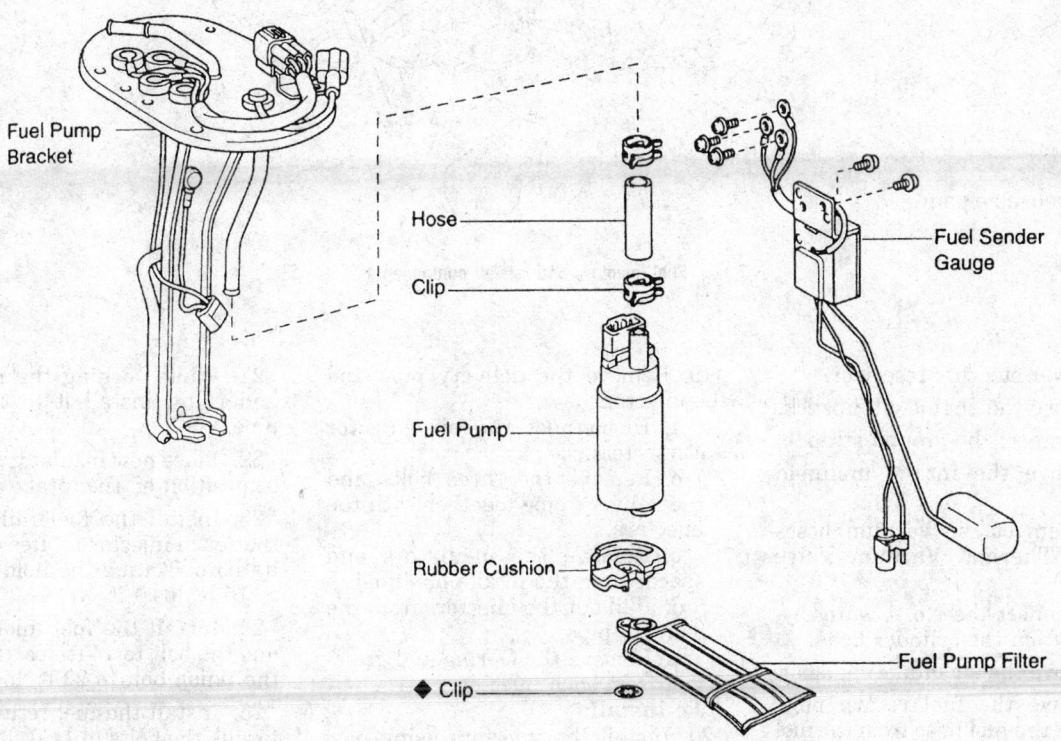

Fuel Pump Bracket

Hose

Clip

Fuel Pump

Rubber Cushion

◆ Clip

Fuel Sender Gauge

Fuel Pump Filter

▶ **Non-reusable part**

Fuel pump and related components — 1995–97

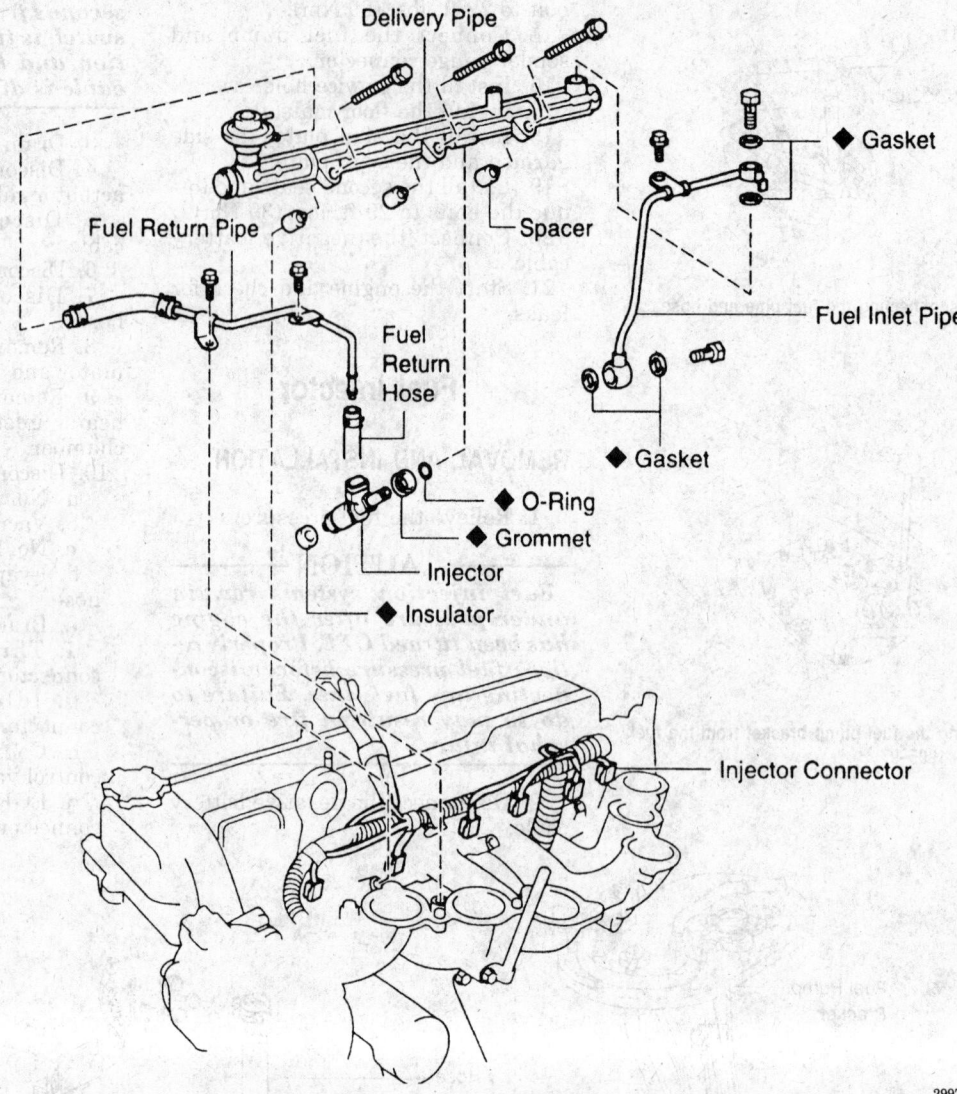

Delivery Pipe

Fuel Return Pipe

Spacer

◆ Gasket

Fuel Inlet Pipe

Fuel Return Hose

◆ Gasket

◆ O-Ring

◆ Grommet

Injector

◆ Insulator

Injector Connector

▶ Non-reusable part

299705

Fuel injectors and related components

j. Power steering reservoir.

11. Remove the engine oil dipstick.

12. Disconnect the ground strap.

13. Remove the intake manifold stay.

14. Disconnect the vacuum hoses from the Thermal Vacuum Valve (TVV).

15. Disconnect the No. 1 water by-pass hose from the cylinder head.

16. Remove the air intake chamber.

17. Remove the fuel return pipe. Disconnect the fuel hose from the fuel pressure regulator.

18. Remove the fuel inlet pipe.

19. Remove the delivery pipe and the injectors.

   a. Disconnect the six injector connectors.

   b. Remove the three bolts and the delivery pipe together with the injectors.

   c. Remove the insulators and spacer from the intake manifold.

   d. Pull out the injectors from the delivery pipe.

   e. Remove the O–ring and grommet from each injector.

   To install:

20. Install the injectors using new O-rings. Lubricate the O-rings with clean gasoline.

21. While turning the injector left and right, install it to the delivery pipe.

22. Place new insulator and spacers in position on the intake manifold.

23. Install the fuel rail and check that each injector rotates smoothly in its bore. Torque the hold-down bolts to 15 ft. lbs. (21 Nm).

24. Install the fuel inlet pipe. Torque the bolt to 14 ft. lbs. (20 Nm) and the union bolt to 22 ft. lbs. (29 Nm).

25. Install the fuel return pipe and torque the bolts to 14 ft. lbs. (20 Nm). Connect the fuel hose to the fuel pressure regulator.

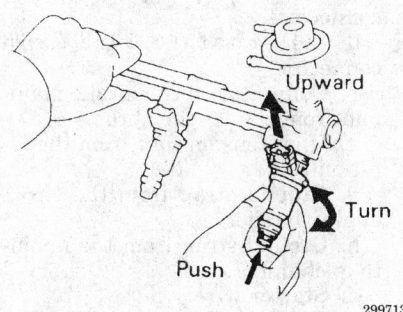

299713

Installing the fuel injectors to the delivery pipe

26. Install the air intake chamber. Torque the bolts to 15 ft. lbs. (21 Nm).

27. Connect the No. 1 water bypass hose to the cylinder head.

28. Connect the vacuum hoses to the TVV.

29. Install the intake manifold stay. Torque the bolts to 26 ft. lbs. (36 Nm).

30. Connect the ground strap.

31. Install the engine oil dipstick.

32. Connect the P/S reservoir tank and torque the bolts to 13 ft. lbs. (18 Nm).

33. Connect the following:

a. EGR gas temperature sensor connector.

b. Connector for the emission control valve set assembly.

c. Idle air control valve connector.

d. Throttle position sensor connector.

e. Brake booster hose.

f. Evaporative emission control hose.

g. No. 2 water bypass hose.

h. Vacuum sensing hose.

i. No. 1 PCV hose.

34. Install the bolt holding the heater inlet pipe and the air intake chamber. Torque the bolts to 14 ft. lbs. (20 Nm).

35. Install the EGR valve and the vacuum modulator.

36. Connect the throttle cable.

37. Connect the accelerator cable.

38. Connect the cruise control actuator cable.

39. Connect the air cleaner hose.

40. Connect the No. 2 PCV hose.

41. Fill the engine with coolant.

42. Connect the negative battery cable. Start the engine and check for leaks.

ENGINE MECHANICAL

Engine Assembly

REMOVAL AND INSTALLATION

1. Properly relieve the fuel system pressure.

2. Disconnect the battery cables and remove the battery and the battery tray.

— CAUTION —

Work must be started after 90 seconds from the time the ignition switch is turned to the LOCK position and the negative (-) battery cable is disconnected.

3. Raise and safely support the vehicle.

4. Drain the engine coolant, transmission oil and the engine oil.

5. Remove the hood.

6. Remove the radiator grille and remove the radiator.

7. Disconnect the oil cooler hose from the oil cooler pipe.

8. Remove the air cleaner hose, cap and the air cleaner case.

9. Disconnect the cruise control actuator cable from the throttle body.

10. Disconnect the accelerator cable from the throttle body.

11. Disconnect the heater hoses.

12. Disconnect the engine wire harness and the heater valve from the cowl panel.

13. Disconnect the brake booster vacuum hose.

— CAUTION —

Fuel injection systems remain under pressure after the engine has been turned OFF. Properly relieve fuel pressure before disconnecting any fuel lines. Failure to do so may result in fire or personal injury.

14. Disconnect the EVAP and fuel hoses.

15. Disconnect the following wires and connectors:

a. Two heated oxygen sensor connectors.

b. DCL1 clamp.

c. Two oil pressure gauge connectors.

d. Alternator wire and connector.

e. Connector on the intake manifold from the fender apron.

f. High tension cord from the ignition coil.

g. Ground strap from the No. 1 engine hanger.

h. Ground strap from the air intake chamber.

i. Starter wire.

j. Ground cable from the cylinder block.

16. Loosen the idler pulley nut and adjusting bolt and remove the A/C drive belt.

17. Disconnect the A/C compressor and remove the bracket.

18. Remove the radiator pipe.

a. Remove the two nuts holding the radiator pipe to the No. 1 oil pan.

b. Disconnect the No. 2 radiator hose from the water inlet and remove the radiator pipe.

19. Remove the union bolt and the two gaskets and disconnect the pressure hose from the P/S pump.

20. Disconnect the return hose from the P/S reservoir tank.

21. Disconnect the engine wire from the cabin.

a. Remove the glove compartment door.

b. Remove the screw and speaker panel.

c. Disconnect the A/C amplifier.

d. Disconnect the connector from the Powertrain Control Module and the cowl wire.

e. Pull out the engine wire from the cabin.

22. Remove the stabilizer bar.

23. Put matchmarks on the flanges and remove the front and rear propeller shafts.

24. Remove the transfer shift lever.

a. Remove the nut and the transmission control rod.

b. Remove the transfer shift lever knob.

c. Lift up the console slightly in order to disconnect the connector.

d. Remove the shifter console.

e. Remove the center console box.

f. Disconnect the connectors and remove the transfer shift lever boot and the transmission shift lever assembly.

g. Pull out the pin and disconnect the shift rod.

h. Remove the hose clamp and the transfer shift lever.

25. Remove the front exhaust pipe.

a. Disconnect the heated oxygen sensor connector.

b. Remove the nuts and bolts holding the exhaust to the rear TWC.

c. Loosen the clamp bolt and disconnect the clamp from the No. 1 support bracket.

d. Remove the No. 1 support bracket.

e. Remove the front exhaust pipe.

26. Disconnect the ground strap from the heat insulator.

27. Place a jack under the transmission. Put a block of wood between the jack and the transmission oil pan to prevent damage to the pan.

28. Remove the frame crossmember.

29. Attach the engine hoist chain to the two engine hangers.

30. Remove the nuts holding the engine front mounting insulators to the frame.

31. Lift the engine with the transmission out of the vehicle slowly and carefully. Make sure that the engine is clear of all wiring and hoses.

To install:

32. Attach the engine hoist chain to the engine hangers and lower the engine and transmission assembly into the engine compartment.

33. Install the nuts holding the engine front mounting insulators to the frame crossmember.

34. Keep the engine level with a jack and remove the chain hoist.

35. Install the frame crossmember and tighten the bolts to 45 ft. lbs. (61 Nm).

36. Tighten the nuts holding the crossmember to the engine rear mounting insulator to 54 ft. lbs. (74 Nm).

37. Tighten the nuts holding the engine front mounting insulators to the frame to 54 ft. lbs. (74 Nm).

38. Connect the ground strap to the heat insulator.

39. Install the front exhaust pipe and torque the nuts to 46 ft. lbs. (63 Nm).

a. Install the No. 1 support bracket and torque the bolts to 17 ft. lbs. (24 Nm).

b. Connect the clamp and tighten the clamp bolt to 14 ft. lbs. (19.5 Nm).

c. Connect the front exhaust pipe to the rear TWC and torque the bolts to 34 ft. lbs. (46 Nm).

40. Install the transfer shift lever and hose clamp. Torque the bolts to 13 ft. lbs. (18 Nm).

a. Connect the shift rod and install the pin.

b. Install the transfer shift lever boot and transmission shift lever assembly and torque the bolts to 4 ft. lbs. (5.4 Nm).

c. Connect the connectors to the transmission shift lever assembly.

d. Install the center console box.

e. Connect the pattern select switch connector.

f. Install the shifter console. Install the transfer shift lever knob.

g. Shift the shift lever to N position.

h. fully turn the control shaft lever back and return two notches. It is now in the neutral position.

i. Connect the transmission control rod and torque the nut to 9 ft. lbs. (13 Nm).

41. Install the front and rear propeller shafts.

a. At the differential side, align the matchmarks on the flanges and torque the front shaft to 54 ft. lbs. (74 Nm) and the rear shaft to 65 ft. lbs. (88 Nm).

b. At the transfer side, align the matchmarks on the flanges and torque the front shaft to 54 ft. lbs. (74 Nm) and the rear shaft to 65 ft. lbs. (88 Nm).

42. Temporarily install the stabilizer bar to the axle housing.

43. Connect the stabilizer bar brackets to 13 ft. lbs. (18 Nm).

44. After the vehicle is lowered and resting on its suspension, torque the bolts holding the stabilizer bar to the axle housing to 19 ft. lbs. (25 Nm).

45. Push the engine wire through the cowl panel and connect the three connectors to the Powertrain Control Module and the two connectors to the cowl wire.

46. Connect the A/C amplifier with its screw.

47. Install the speaker panel and the glove compartment door.

48. Connect the return hose to the P/S reservoir tank.

49. Connect the P/S pressure hose with the union bolt and torque to 42 ft. lbs. (56 Nm).

50. Install the radiator pipe.

a. Connect the No. 2 radiator hose to the water inlet.

b. Install the two nuts holding the radiator pipe to the No. 1 oil pan. Torque the nuts to 15 ft. lbs. (21 Nm).

51. Install the A/C bracket and torque the bolts to 27 ft. lbs. (37 Nm).

52. Install the A/C compressor and torque the bolts to 18 ft. lbs. (25 Nm).

53. Install and adjust the A/C drive belt.

54. Connect the following wires and connectors:

a. Two heated oxygen sensor connectors.

b. DCL1 clamp.

c. Two oil pressure gauge connectors.

d. Alternator wire and connector.

e. Connector on the intake manifold from the fender apron.

f. High tension cord from the ignition coil.

g. Ground strap from the No. 1 engine hanger.

h. Ground strap from the air intake chamber.

i. Starter wire.

j. Ground cable from the cylinder block.

55. Connect the fuel inlet hose to the fuel filter and tighten the union bolt to 22 ft. lbs. (39 Nm). Connect the fuel return hose.

56. Connect the EVAP hose and the brake booster vacuum hose.

57. Connect the heater valve and the engine wire to the cowl panel. Connect the engine wire and the ground strap.

58. Connect the heater hoses.

59. Connect the accelerator cable to the throttle body.

60. Connect the cruise control actuator cable to the throttle body.

61. Install the air cleaner case, the hose and the cap.

62. Connect the oil cooler hose to the oil cooler pipe.

63. Install the radiator.

64. Install the radiator grille.

65. Install the battery tray and the battery. Connect the battery cables.

66. Fill with engine with oil and fill the transmission with the proper amount and type of fluid.

67. Fill the radiator with engine coolant.

68. Start the engine and check for leaks.

69. Check the automatic transmission fluid level.

70. Check the ignition timing.

71. Install the hood and test drive the vehicle.

72. Recheck the engine coolant and oil levels.

Engine Mounts

REMOVAL AND INSTALLATION

1. Disconnect the negative battery cable.

2. Raise and safely support the vehicle.

—————— CAUTION ——————
Make certain vehicle is properly supported when raising the engine.

3. Raise the engine slightly to take the tension off the motor engine mounts.

NOTE: When raising the engine to replace the engine mounts, DO NOT place the jack directly under the oil pan or the crankshaft torsional damper. Use a block of wood between the engine and the jack to prevent damage at the jacking point.

4. Remove the through bolt nuts and through bolts.

5. Remove the engine mount bolts.

6. Raise the engine enough to remove the engine mounts.

NOTE: Only raise the engine enough to remove the mounts. Raising the engine too far could result in damage to some of the engine components.

To install:

7. Install the engine mounts and the engine bolts. Torque the bolts to 54 ft. lbs. (74 Nm).

8. Lower the vehicle and connect the negative battery cable.

Cylinder Head

REMOVAL AND INSTALLATION

1. Properly relieve the fuel system pressure.

2. Disconnect the battery cables and remove the battery and the battery tray.

3. Drain the engine coolant.

4. Remove the air cleaner hose and cap.

5. Disconnect the cruise control actuator cable from the throttle body.

6. Disconnect the accelerator cable from the throttle body.

7. Disconnect the throttle cable from the throttle body.

8. Disconnect the engine ground strap from the No. 1 engine hanger and the ground strap from the air intake chamber.

9. Disconnect the connector on the intake manifold from the left fender apron.

10. Disconnect the brake booster vacuum hose.

11. Disconnect the EVAP hose and disconnect the fuel return hose.

─────── **CAUTION** ───────
Fuel injection systems remain under pressure after the engine has been turned OFF. Properly relieve fuel pressure before discon-necting any fuel lines. Failure to do so may result in fire or personal injury.

12. Disconnect the heater hoses.

13. Disconnect the engine wire and heater valve from the cowl panel.

14. Remove the No. 2 and the No. 3 cylinder head covers

15. Remove the distributor.

16. Disconnect the P/S reservoir tank.

17. Disconnect the radiator inlet hose and the No. 3 water bypass hose.

18. Remove the alternator.

19. Remove the throttle body.

20. Remove the oil dipsticks and guides for the engine and transmission. Pull out the dipstick together with the dipstick guide and remove the O-ring from the dipstick guide.

21. Remove the intake manifold stay.

22. Disconnect the fuel inlet hose from the fuel filter.

23. Disconnect the following connectors:
   a. ECT sender gauge connector, the ECT cut switch connector and the ECT sensor connector.
   b. Knock sensor connector.
   c. Crankshaft position sensor connector.

24. Remove the bolt and disconnect the engine wire harness from the cylinder block.

25. Disconnect the following:
   a. Oil level sensor connector.
   b. Two connectors from the transmission.
   c. Starter connector.
   d. Disconnect the two heated oxygen sensor connectors.
   e. Disconnect the park/neutral position (PNP) switch connector.
   f. Remove the two bolts and disconnect the engine wire from the intake manifold and the cylinder block.
   g. Disconnect the PCV hose from the PCV valve.
   h. Remove the bolt holding the engine wire to the intake manifold.
   i. Disconnect the connector for the emission control valve set assembly and the three injector connectors.
   j. Disconnect the engine wire harness clamp.
   k. Disconnect the EGR gas temperature sensor connector.
   l. Disconnect the clamp of the No. 6 injector wire from the bracket.
   m. Disconnect the engine wire harness from the cylinder head and the intake manifold.

26. Remove the three bolts and disconnect the No. 2 water bypass pipe from the cylinder head.

27. Disconnect the heated oxygen sensor connector.

28. Remove the nuts and bolts holding the front exhaust pipe to the rear TWC.

29. Disconnect the front exhaust pipe and remove the gasket.

30. Remove the clamp from the No. 1 support bracket and remove the bracket.

31. Remove the front exhaust pipe and the gaskets.

32. Remove the No. 1 and No. 2 exhaust manifolds. Remove the No. 1 and No. 2 heat insulators.

33. Remove the ground cable, heater pipe and gasket.

34. Remove the water bypass outlet and the pipe. Remove the three O-rings from the water bypass outlet and the pipe.

35. Remove the cylinder head cover.

36. Remove the semi-circular plug from the cylinder head.

37. Remove the spark plugs.

38. Set the No. 1 cylinder to TDC of the compression stroke.
   a. Turn the crankshaft pulley and align its groove with the **0** mark on the timing chain cover.
   b. Check that the timing marks (one and two dots) of the camshaft drive and driven gears are in straight line on the cylinder head surface. If not, turn the crankshaft one revolution (360°) and align the marks as above.

39. Remove the chain tensioner.

40. Place matchmarks on the camshaft timing gear and the timing chain and remove the camshaft timing gear.
   a. Hold the intake camshaft with a wrench and remove the bolt and the distributor gear.
   b. Remove the camshaft timing gear and chain from the intake camshaft and leave on the slipper and the damper.

41. Remove the camshafts.

NOTE: Since the thrust clearance of the camshaft is small, the camshaft must be kept level while it is being removed. If the camshaft is not kept level, the portion of the cylinder head receiving the shaft thrust may crack or be damaged, causing the camshaft to seize or break. To avoid this, the following steps should be carried out.

42. Remove the exhaust camshaft.
   a. Bring the service bolt hose of the driven sub-gear upward by turning the hexagon wrench head

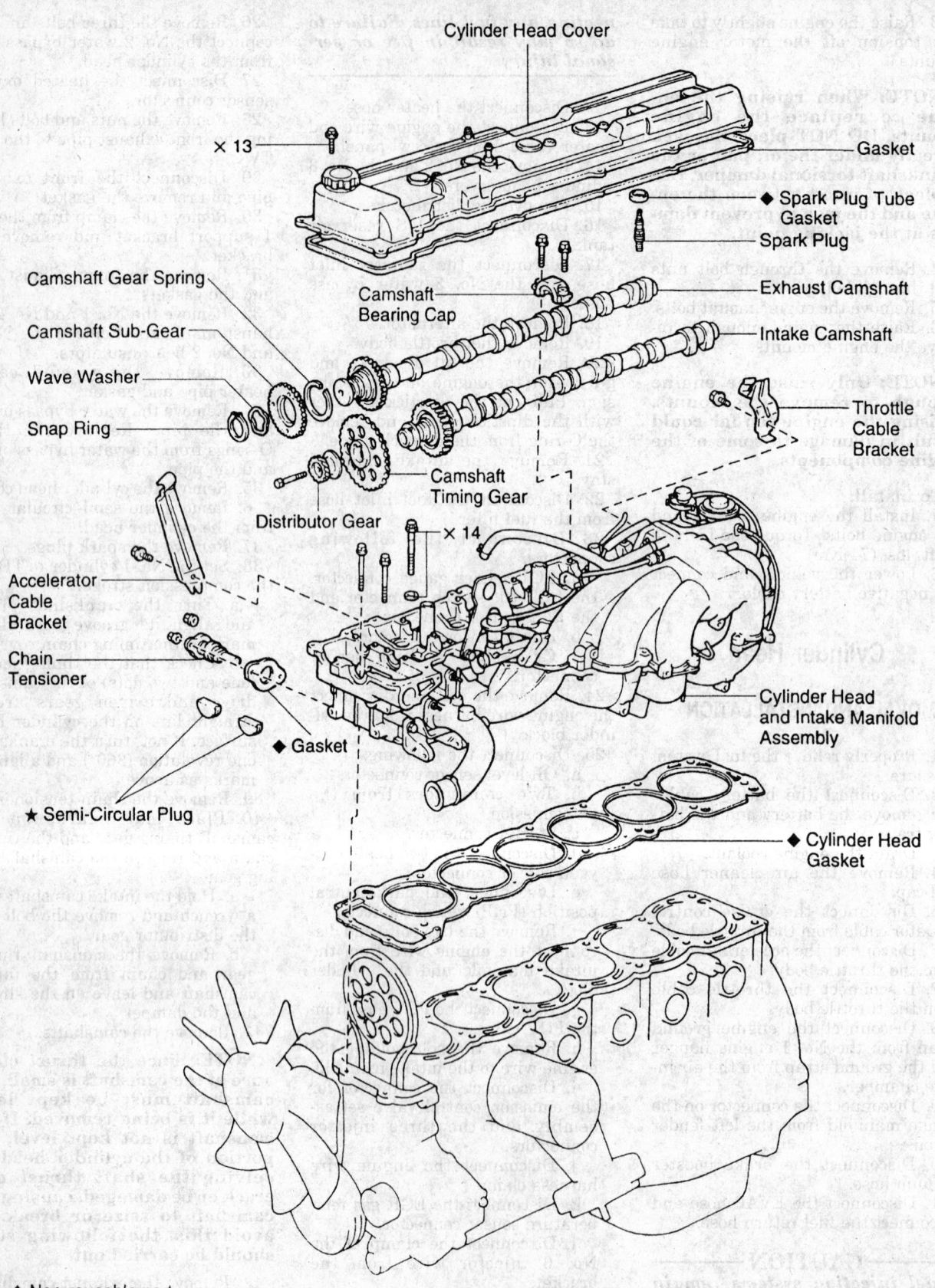

Cylinder Head Cover

× 13

Gasket

◆ Spark Plug Tube Gasket

Spark Plug

Camshaft Gear Spring

Camshaft Sub-Gear

Wave Washer

Snap Ring

Camshaft Bearing Cap

Exhaust Camshaft

Intake Camshaft

Throttle Cable Bracket

Camshaft Timing Gear

Distributor Gear

× 14

Accelerator Cable Bracket

Chain Tensioner

◆ Gasket

★ Semi-Circular Plug

Cylinder Head and Intake Manifold Assembly

◆ Cylinder Head Gasket

◆ Non-reusable part

299333

Cylinder head, camshafts, intake manifold and related components

portion of the exhaust camshaft with a wrench.

b. Secure the exhaust camshaft sub-gear to the main gear with a service bolt. When removing the camshaft, make sure that the torsional spring force of the sub-gear has been eliminated by the above operation.

c. Set the timing mark (two dot marks) of the camshaft driven gear at approximately 35° angle by turning the hexagon wrench head portion of the intake camshaft with a wrench.

d. Lightly push the camshaft towards the rear without applying excessive force.

e. Loosen and remove the No. 1 bearing cap bolts, alternately loosening the left and right bolts uniformly.

f. Loosen and remove the No. 2, No. 3, No. 5 and the No. 7 bearing cap bolts, alternately loosening the left and right bolts uniformly in several passes, in sequence.

NOTE: Do not remove the No. 4 and No. 6 bearing cap bolts at this stage.

g. Remove the four bearing caps.

h. Alternately and uniformly loosen and remove the No. 4 and the No. 6 bearing cap bolts.

i. If the camshaft is not being lifted out straight and level, retighten the four No. 4 and No. 6 bearing cap bolts. Then reverse the order of the above steps from (g) to (e) and repeat steps from (c) to (h) once again.

j. Remove the two bearing caps and exhaust camshaft. Do not pry on or attempt to force the camshaft with a tool or any other object.

43. Remove the intake camshaft.

a. Set the timing mark (two dot marks) of the camshaft drive gear at approximately a 25° angle by turning the hexagon wrench head portion of the intake camshaft with a wrench.

NOTE: The above angle arrows the No. 1 and the No. 4 cylinder cam lobes of the intake camshaft to push their valve lifters evenly.

b. Lightly push the intake camshaft towards the front without applying excessive force.

c. Loosen and remove the No. 1 bearing cap bolts, alternately loosening the left and the right bolts uniformly.

d. Loosen and remove the No. 3, No. 4, No. 6 and the No. 7 bearing cap bolts, alternately loosening the

left and right bolts uniformly in several passes in sequence.

NOTE: Do not remove the No. 2 and No. 5 bearing cap bolts at this stage.

e. Remove the four bearing caps.

f. Alternately and uniformly loosen and remove the No. 2 and the No. 5 bearing cap bolts.

g. If the camshaft is not being lifted out straight and level, retighten the four No. 2 and No. 5 bearing cap bolts. Then reverse the order of the above steps from (e) to (c) and repeat steps from (a) to (f) once again.

h. Remove the two bearing caps and the exhaust camshaft.

44. Remove the cylinder head and the intake manifold assembly.

a. Remove the two bolts in front of the head before the other head bolts are removed.

b. Loosen and remove the 14 cylinder head bolts in sequence using several passes.

CAUTION
Cylinder head warpage or cracking could result from removing bolts in incorrect order.

299347

Front cylinder head bolts

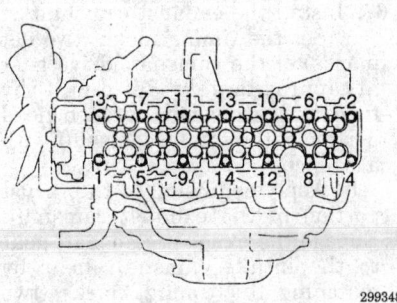

299348

Cylinder head bolt loosening sequence

c. Lift the cylinder head from the dowels on the cylinder block and place the cylinder head on wooden blocks on the bench.

d. If the cylinder head is difficult to lift off, pry between the cylinder head and the cylinder block with a flat prying tool.

45. Remove the alternator bracket.

46. Remove the two nuts, the water outlet and the gasket.

47. Loosen the union nut and remove the EGR pipe and gasket.

48. Remove the heater inlet pipe and hose.

49. Remove the air intake chamber and the intake manifold assembly.

a. Disconnect the vacuum hoses from the TVV.

b. Remove the 10 bolts, the two nuts and the intake manifold and gasket.

50. Remove the No. 1 water bypass hose.

51. Remove the No. 1 and the No. 2 engine hangers.

52. Remove the two engine wire clamp brackets.

53. Remove the accelerator cable bracket and the throttle cable bracket.

54. Remove the valve lifters and shims. Arrange the valve lifters and shims in correct order for reinstallation.

To install:

55. Install the valve lifters and shims. Check to make sure that the valve lifter rotates smoothly by hand.

56. Install the accelerator cable bracket and the throttle cable bracket.

57. Install the engine wire clamp brackets.

58. Install the No. 1 and No. 2 engine hangers and torque the bolts to 30 ft. lbs. (41 Nm).

59. Install the air intake chamber and intake manifold assembly.

a. Place a new gasket so that the rear mark is toward the rear side.

b. Torque the intake manifold bolts to 15 ft. lbs. (21 Nm).

c. Connect the vacuum hoses to the TVV.

60. Install the heater hose to the cylinder head and connect the pipe to the intake manifold. Torque the bolts to 15 ft. lbs. (21 Nm).

61. Temporarily install the union nut to the EGR valve. Install the EGR pipe to the cylinder head. Torque the bolts to 15 ft. lbs. (21 Nm). Torque the union nut to 58 ft. lbs. (78 Nm).

62. Install a new gasket and the water outlet. Torque the nuts to 15 ft. lbs. (21 Nm).

63. Install the alternator bracket and torque the bolts to 32 ft. lbs. (43 Nm).

64. Install the cylinder head and the intake manifold assembly.

a. Apply seal packing on the end of the engine block by the timing belt.

b. Install a new cylinder head gasket on the cylinder block.

c. Install the cylinder head.

65. Install the cylinder head bolts.

a. The cylinder head bolts are tightened in three progressive steps. Apply a light coat of engine oil on the threads and under the heads of the cylinder head bolts.

b. Install the 14 cylinder head bolts and tighten progressively in sequence to 29 ft. lbs. 39 Nm).

c. Mark the front of the cylinder head bolt head with paint.

d. Retighten the cylinder head bolts by 90° in numerical order.

e. Retighten the cylinder head bolts an additional 90° so that the painted mark is now facing to the rear.

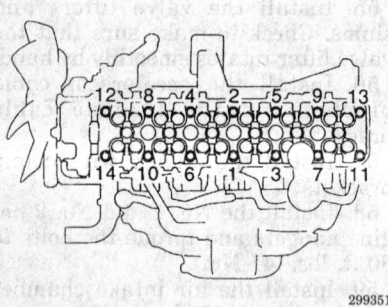

299351

Cylinder head bolt tightening sequence

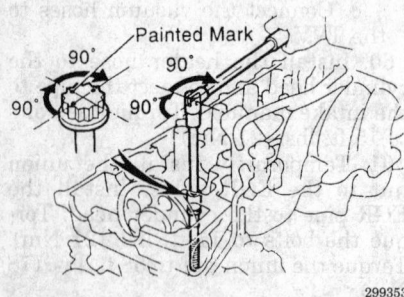

299353

Turning cylinder head bolts 90 degrees

— **WARNING** —

Do not combine steps D and E; the above steps must be followed exactly and in order to prevent cylinder head damage or pre-mature gasket failure.

---

f. Install and torque the two mounting bolts to 15 ft. lbs. (21 Nm).

66. Install the camshafts.

NOTE: Since the thrust clearance of the camshaft is small, the camshaft must be kept level while it is being installed. If the camshaft is not kept level, the portion of the cylinder head receiving the shaft thrust may crack or be damaged, causing the camshaft to seize or break. To avoid this, the following steps should be carried out.

a. Apply engine oil to the thrust portion of the intake camshaft.

b. Lightly place the intake camshaft on top of the cylinder head so that the No. 1 and the No. 4 cylinder cam lobes face downward.

c. Lightly push the camshaft towards the front without applying excessive force. Place the No. 2 and the No. 5 bearing caps in their proper location.

d. Temporarily tighten these bearing cap bolts uniformly and alternately in several passes until the bearing caps are snug with the cylinder head.

e. Place the No. 3, No. 4, No. 6 and the No. 7 bearing caps in their proper location. Temporarily tighten these bearing cap bolts, alternately tightening the left and right bolts uniformly.

f. Place the No. 1 bearing cap in its proper location. Check that there is no gap between the cylinder head and the contact surface of the bearing cap.

g. Uniformly tighten the 14 bearing cap bolts in several passes to 12 ft. lbs. (16 Nm).

67. Install the exhaust camshaft.

a. Set the timing mark (two dot marks) of the camshaft drive gear at approximately 35° angle by turning the hexagon wrench head portion of the intake camshaft with a wrench.

b. Apply engine oil to the thrust portion of the exhaust camshaft. Engage the exhaust camshaft gear to the intake camshaft hear by matching the timing marks (two dot marks) on each gear.

c. Roll down the exhaust camshaft onto the bearing journals

---

while engaging the gears with each other. Lightly push the intake camshaft towards the front without applying excessive force.

d. Install the No. 4 and the No. 6 bearing caps in their proper location. Temporarily tighten these bearing cap bolts, alternately tightening the left and right bolts uniformly.

e. Place the No. 2, No. 3, No. 5 and the No. 7 bearing caps in their proper location. Temporarily tighten these bearing cap bolts, alternately tightening the left and right bolts uniformly.

f. Tighten the 14 bearing cap bolts in several passes to 12 ft. lbs. (16 Nm).

g. Bring the service bolt installed in the driven sub-gear upward by turning the hexagon wrench head portion of the camshaft with a wrench. Remove the service bolt.

h. Check that the intake and the exhaust camshafts turn smoothly.

68. Set the No. 1 cylinder to TDC of the compression stroke. Turn the crankshaft pulley and align its groove with the timing mark 0 of the timing chain cover. Turn the camshaft so that the timing marks with one and two dots will be in straight line on the cylinder head surface.

69. Install the camshaft timing gear.

a. Check that the matchmarks on the camshaft timing gear and the timing chain are aligned. Place the gear over the straight pin of the intake camshaft.

b. Align the straight pin of the distributor gear with the straight pin groove of the intake camshaft gear.

c. Hold the intake camshaft with a wrench, install and torque the bolt to 54 ft. lbs. (74 Nm).

70. Install the chain tensioner. Push the tensioner by hand until it touches the head installation surface, then install and torque the two nuts to 15 ft. lbs. (21 Nm).

71. Check the valve timing.

a. Turn the crankshaft pulley and align its groove with the timing mark **0** of the timing chain cover. Always turn the crankshaft clockwise.

b. Check that the timing marks (one and two dots) of the camshaft drive and driven gears are in straight line on the cylinder head surface. If not, turn the crankshaft one revolution (360°) and align the marks.

72. Check valve clearance and adjust if necessary.

73. Install the spark plugs.

74. Install the semi–circular plug to the cylinder head.

75. Make sure that the No. 1 cylinder is in TDC of the compression stroke.

76. Install the cylinder head cover.

77. Install the water bypass outlet and the pipe.

    a. Install and new O–ring to the water bypass outlet. Install new O–rings to the water bypass pipe.

    b. Assemble the water bypass outlet and the pipe and install with the two bolts torqued to 15 ft. lbs. (21 Nm).

78. Install the heater pipe and the ground cable. Torque the heater pipe bolt to 14 ft. lbs. (20 Nm) and the nut to 15 ft. lbs. (21 Nm).

79. Install the No. 1 and the No. 2 exhaust manifolds. Torque the nuts to 29 ft. lbs. (39 Nm).

80. Install the No. 1 insulator and No. 2 heat insulator and torque the bolts to 14 ft. lbs. (20 Nm).

81. Install the front exhaust pipe. Torque the nuts to 46 ft. lbs. (63 Nm).

82. Install the No. 1 support bracket and torque to 17 ft. lbs. (24 Nm).

83. Connect the clamp and tighten the clamp bolt to 14 ft. lbs. (20 Nm).

84. Connect the front exhaust pipe to the rear TWC and torque the bolts to 34 ft. lbs. (46 Nm).

85. Connect the No. 2 water bypass pipe to the cylinder head and torque the bolts to 14 ft. lbs. (20 Nm).

86. Connect the following:

    a. Connect the engine wire harness from the cylinder head and the intake manifold.

    b. Connect the clamp of the No. 6 injector wire from the bracket.

    c. Connect the EGR gas temperature sensor connector.

    d. Connect the engine wire harness clamp.

    e. Connect the connector for the emission control valve set assembly and the three injector connectors.

    f. Install the bolt holding the engine wire to the intake manifold.

    g. Connect the PCV hose from the PCV valve.

    h. Install the two bolts and connect the engine wire from the intake manifold and the cylinder block.

    i. Connect the park/neutral position (PNP) switch connector.

    j. Connect the two heated oxygen sensor connectors.

    k. Starter connector.

    l. Two connectors from the transmission.

    m. Oil level sensor connector.

87. Install the bolt and connect the engine wire harness from the cylinder block.

88. Connect the following:

    a. ECT sender gauge connector, the ECT cut switch connector and the ECT sensor connector.

    b. Knock sensor connector.

    c. Crankshaft position sensor connector.

89. Connect the fuel inlet hose to the fuel filter with the union bolt. Torque to 22 ft. lbs. (29 Nm).

90. Install the intake manifold stay and torque the bolts to 26 ft. lbs. (36 Nm).

91. Install the oil dipsticks and the guides for the engine and the transmission. Torque the oil dipstick guide bolts to 14 ft. lbs. (20 Nm).

92. Install the throttle body.

93. Install the alternator and the drive belts.

94. Connect the No. 3 water bypass hose.

95. Connect the radiator inlet hose.

96. Connect the P/S reservoir tank and torque the bolts to 14 ft. lbs. (20 Nm).

97. Install the distributor.

98. Install the No. 2 and the No. 3 cylinder head covers.

99. Connect the heater valve and the engine wire harness to the cowl panel.

100. Connect the heater hoses.

101. Connect the fuel return hose, the EVAP hose and the brake booster vacuum hose.

102. Connect the connector on the intake manifold to the left fender apron.

103. Connect the ground strap to the No. 1 engine hanger and the air intake chamber.

104. Connect the throttle cable to the throttle body. Adjust the throttle cable.

105. Connect the accelerator cable to the throttle body.

106. Connect the cruise control actuator cable to the the throttle body.

107. Install the air cleaner hose and cap.

108. Install the battery tray, the battery and connect the cables.

109. Refill the engine coolant.

110. Start the engine and check for leaks.

111. Make necessary engine adjustments.

112. Road test the vehicle and recheck the engine coolant level.

Valve Clearance Adjustment

REMOVAL AND INSTALLATION

1. Disconnect the negative battery cable.

2. Drain the engine coolant.

3. Remove the PCV hoses.

4. Remove the air cleaner cap, MAF meter and the resonator.

5. Disconnect the following connectors:

    a. ECT sensor connector.

    b. Oil pressure sensor connector.

    c. If disconnected, the A/C compressor connector.

6. Remove the throttle body.

7. Disconnect the engine wire and the heater valve from the cowl panel.

8. Disconnect the spark plug wires.

9. Remove the cylinder head cover.

10. Set the No. 1 cylinder to TDC of the compression stroke.

    a. Turn the crankshaft pulley clockwise and align its groove with the 0 mark on the timing chain cover.

    b. Check that the timing marks (one and two dots) of the camshaft drive and driven gears are in a straight line on the cylinder head surface. If not, turn the crankshaft one revolution (360°) and align the marks.

11. Inspect the valve clearance.

    a. Measure the clearance between the valve lifter and the camshaft. Measure the first, second and fourth intake and the first, third and fifth exhaust valves.

    b. Turn the crankshaft pulley one revolution (360°) and align the marks as above. Measure the third, fifth and sixth intake and the second, fourth and sixth exhaust valves.

12. Valve clearance cold should be:

• Intake: 0.006–0.010 in. (0.15–0.25 mm)

• Exhaust: 0.010–0.014 in. (0.25–0.35 mm)

13. Adjust the valve clearance by using adjusting shims.

    a. Turn the equipment driveshaft so that the camshaft lobe for the valve to be adjusted faces up.

    b. Using tool No. SST 09248–55040, or equivalent, press down the valve lifter and place SST 09248–05420 or equivalent, between the camshaft and the valve lifter. Remove SST 09248–55040.

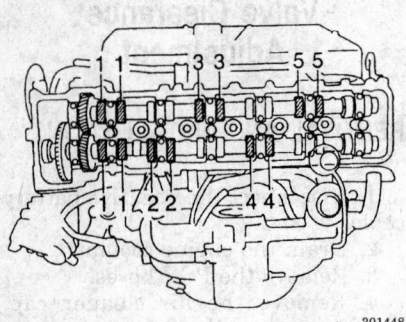

301448

First valve adjustment

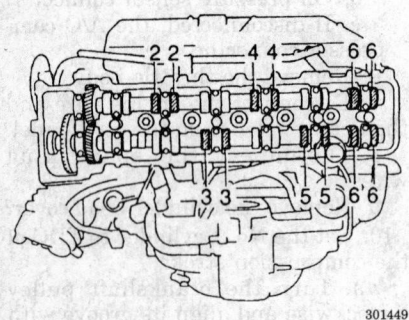

301449

Second valve adjustment

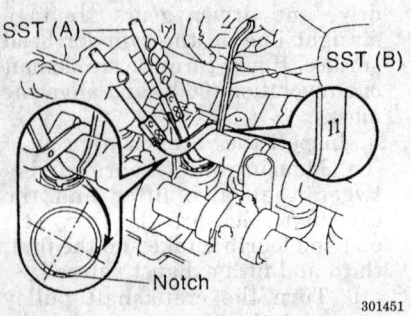

301451

Removing the adjusting shim

c. Remove the adjusting shim with a small flat prying tool and a magnetic finger.

d. Determine the replacement adjusting shim size according to the following formula, or use the adjusting shim charts.

e. Using a micrometer, measure the thickness of the removed shim. Calculate the thickness of a new

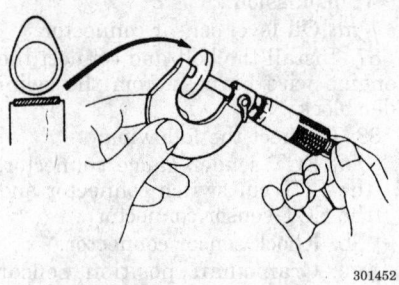

301452

Measuring shim thickness

shim so that the valve clearance comes within the specified value.

• T: Thickness of the removed shim

• A: Measured valve clearance

• N: Thickness of the new shim

f. Intake: N=T+(A-0.008 in. (0.20 mm))

g. Exhaust: N=T+(A-0.012 in. (0.30 mm))

h. Install a new adjusting shim. Place it on the valve lifter. Using the SST 09248–55040, press down the valve lifter and remove SST 09248–05420.

i. Recheck the valve clearance.

14. Reinstall the cylinder head cover.

15. Reconnect the engine wire and clamps.

16. Connect the following:

   a. ECT sensor connector.

   b. Oil pressure sensor connector.

   c. If disconnected, the A/C compressor connector.

17. Install the throttle body.

18. Install the spark plug wires.

19. Install the PCV hoses.

20. Install the air cleaner cap, MAF meter and the resonator.

21. Refill with engine coolant.

22. Check the ignition timing.

Intake Manifold

REMOVAL AND INSTALLATION

1. Properly relieve the fuel system pressure.

—————— **CAUTION** ——————
Fuel injection systems remain under pressure after the engine has been turned OFF. Properly relieve fuel pressure before disconnecting any fuel lines. Failure to do so may result in fire or personal injury.

2. Disconnect the negative battery cable.

—————— **CAUTION** ——————
Work must be started after 90 seconds from the time the ignition switch is turned to the LOCK position and the negative (-) battery cable is disconnected.

3. Drain the engine coolant.

4. Remove the air cleaner hose and cap.

5. Disconnect the cruise control actuator cable from the throttle body.

6. Disconnect the accelerator cable from the throttle body.

7. Disconnect the throttle cable from the throttle body.

8. Disconnect the engine ground strap from the No. 1 engine hanger and the ground strap from the air intake chamber.

9. Disconnect the connector on the intake manifold from the left fender apron.

10. Disconnect the brake booster and EVAP hoses.

11. Disconnect the fuel inlet and return lines from the fuel rail.

12. Remove the heater inlet pipe and hose.

13. Remove the radiator inlet hose, the No. 3 water bypass hose and the alternator.

14. Remove the intake manifold stay.

15. Disconnect the following electrical connectors:

   a. ECT sender gauge connector.

   b. ECT cut switch connector.

   c. ECT sensor connector.

   d. Knock sensor connector.

   e. Crankshaft Position Sensor connector.

16. Remove the bolt that secures the engine harness to the cylinder block.

17. Disconnect the PCV hose from the PCV valve.

18. Remove the bolt that secures the engine harness to the intake manifold.

19. Disconnect the following:

   a. Engine wire clamps.

   b. EGR gas temp. sensor connector.

   c. The fuel injector connectors.

   d. The connector to the emission control valve set.

20. Remove the No. 2 water bypass pipe.

21. Remove the air intake chamber and the intake manifold assembly.

   a. Disconnect the vacuum hoses from the TVV.

   b. Remove the 10 bolts, the two nuts and the intake manifold and gasket.

Adjusting Shim Selection Chart (Intake)

New shim thickness mm (in.)

Shim No.	Thickness	Shim No.	Thickness
1	2.500 (0.0984)	10	2.950 (0.1161)
2	2.550 (0.1004)	11	3.000 (0.1181)
3	2.600 (0.1024)	12	3.050 (0.1201)
4	2.650 (0.1043)	13	3.100 (0.1220)
5	2.700 (0.1063)	14	3.150 (0.1240)
6	2.750 (0.1083)	15	3.200 (0.1260)
7	2.800 (0.1102)	16	3.250 (0.1280)
8	2.850 (0.1122)	17	3.300 (0.1299)
9	2.900 (0.1142)		

HINT: New shims have the thickness in millimeters imprinted on the face.

Intake valve clearance (Cold):
0.15 – 0.25 mm (0.006 – 0.010 in.)

EXAMPLE: The 2.800 mm (0.1102 in.) shim is installed, and the measured clearance is 0.440 mm (0.0173 in.). Replace the 2.800 mm (0.1102 in.) shim with a No. 12 shim.

The chart lists installed shim thickness across the top (from 2.500 (0.0984) to 3.300 (0.1299) mm) and measured clearance down the left side (from 0.000 – 0.030 (0.0000 – 0.0012) to 1.031 – 1.050 (0.0406 – 0.0413)), with shim numbers 1 through 17 at each intersection.

Adjusting shim chart — Intake

301612

Adjusting Shim Selection Chart (Exhaust)

New shim thickness mm (in.)

Shim No.	Thickness	Shim No.	Thickness
1	2.500 (0.0984)	10	2.950 (0.1161)
2	2.550 (0.1004)	11	3.000 (0.1181)
3	2.600 (0.1024)	12	3.050 (0.1201)
4	2.650 (0.1043)	13	3.100 (0.1220)
5	2.700 (0.1063)	14	3.150 (0.1240)
6	2.750 (0.1083)	15	3.200 (0.1260)
7	2.800 (0.1102)	16	3.250 (0.1280)
8	2.850 (0.1122)	17	3.300 (0.1299)
9	2.900 (0.1142)		

HINT: New shims have the thickness in millimeters imprinted on the face.

Exhaust valve clearance (Cold):

0.25 — 0.35 mm (0.010 — 0.014 in.)

EXAMPLE: The 2.800 mm (0.1102 in.) shim is installed, and the measured clearance is 0.440 mm (0.0173 in.). Replace the 2.800 mm (0.1102 in.) shim with a No. 10 shim.

Adjusting shim chart — Exhaust

301613

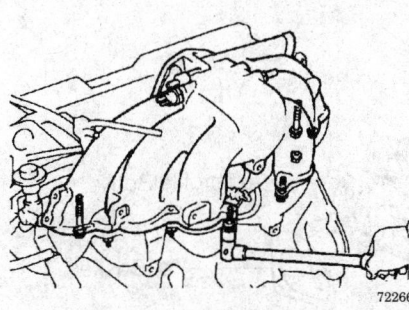

Removing air intake chamber and gaskets

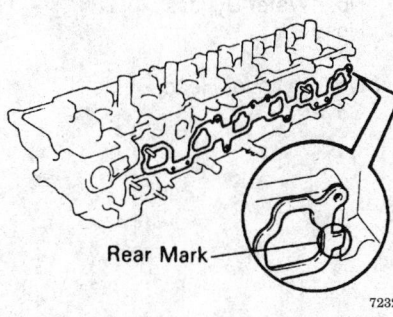

Install the intake manifold gasket so that the rear mark is toward the rear side

To install:

22. Install the air intake chamber and intake manifold assembly.

   a. Place a new gasket so that the rear mark is toward the rear side.

   b. Torque the intake manifold bolts to 15 ft. lbs. (21 Nm).

   c. Connect the vacuum hoses to the TVV.

23. Install the No. 2 water bypass pipe.

24. Connect the following:

   a. Engine wire clamps.

   b. EGR gas temp. sensor connector.

   c. The fuel injector connectors.

   d. The connector to the emission control valve set.

25. Install the bolt that secures the engine wire harness to the intake manifold.

26. Connect the PCV hose to the PCV valve.

27. Install the bolt that secures the engine wire harness to the cylinder block.

28. Connect the following:

   a. ECT sender gauge connector.

   b. ECT cut switch connector.

   c. ECT sensor connector.

   d. Knock sensor connector.

   e. Crankshaft Position Sensor connector.

29. Install the intake manifold stay and torque the bolts to 26 ft. lbs. (36 Nm).

30. Install the radiator inlet hose, the No. 3 water bypass hose and the alternator.

31. Install the heater hose to the cylinder head and connect the pipe to the intake manifold. Torque the bolts to 15 ft. lbs. (21 Nm).

32. Connect the fuel inlet and return hoses to the fuel rail.

33. Connect the brake booster and EVAP hoses.

34. Connect the connector on the intake manifold to the left fender apron.

35. Connect the ground straps to the No.1 engine hanger and the air intake chamber.

36. Connect the throttle cable to the throttle body.

37. Connect the accelerator cable to the throttle body.

38. Connect the cruise control actuator cable to the the throttle body.

39. Install the air cleaner hose and cap.

40. Connect the negative battery cable.

41. Refill the engine coolant.

42. Start the engine and check for leaks.

Exhaust Manifold

REMOVAL AND INSTALLATION

1993–94

1. Disconnect the negative battery cable.

2. Raise and support the vehicle.

3. Remove the heat insulator, if required.

4. Disconnect the No. 1 and No. 2 front exhaust pipe from the exhaust manifold.

5. Lower the vehicle.

6. Remove the dipsticks and guides for the engine and transmission, as required.

7. Remove the heater pipe, air pipe and PAIR reed valve, as required.

8. Remove the engine hanger, if necessary. Remove the No. 1 and No. 2 heat insulators.

9. Remove the fasteners and No. 1 and No. 2 exhaust manifolds and gaskets.

To install:

10. Install the No. 1 and No. 2 exhaust manifolds using new gaskets. Torque the nuts to 29 ft. lbs. (39 Nm).

11. Install the heat insulators. Torque the fasteners to 14 ft. lbs. (19

Nm). Install the engine hanger, if removed.

12. Install the heater pipe, air pipe and PAIR reed valve, as required.

13. Install the dipsticks and guides for the engine and transmission, as required.

14. Raise the vehicle and support it safely.

15. Install the No. 1 and No. 2 front exhaust pipe to the exhaust manifold using new gaskets. Torque the fasteners to 46 ft. lbs. (62 Nm).

16. Install the heat insulators, as required.

17. Lower the vehicle. Connect the negative battery cable. Start the engine and check for leaks.

1995–97

1. Disconnect the negative battery cable.

2. Raise and safely support the vehicle.

3. Working from under the vehicle, disconnect the heated oxygen sensor connector.

4. Remove the nuts and bolts holding the front exhaust pipe to the rear TWC.

5. Loosen the pipe clamp bolt.

6. Remove the two bolts and the pipe bracket.

7. Remove the four nuts and disconnect the front exhaust pipe. Remove the gasket.

8. Lower the vehicle and remove the six bolts and the exhaust manifold heat insulators.

9. Remove the 13 nuts, the No. 1 and the No. 2 exhaust manifolds and the gaskets.

To install:

10. Install the new gaskets and the No. 1 and the No. 2 exhaust manifolds. Uniformly tighten the nuts in several passes. Torque the nuts to 29 ft. lbs. (39 Nm).

11. Install the exhaust manifold heat insulators with the bolts and torque to 14 ft. lbs. (19 Nm).

12. Connect the exhaust pipe to the exhaust manifold with a new gasket and torque the four new nuts to 46 ft. lbs. (63 Nm).

13. Install the No. 1 support bracket and torque the bolts to 17 ft. lbs. (24 Nm).

14. Connect the clamp and tighten the clamp bolt to 14 ft. lbs. (19.5 Nm).

15. Connect the front exhaust pipe to the rear TWC with a new gasket and torque to 34 ft. lbs. (46 Nm).

16. Connect the heated oxygen sensor connector.

17. Connect the negative battery cable.

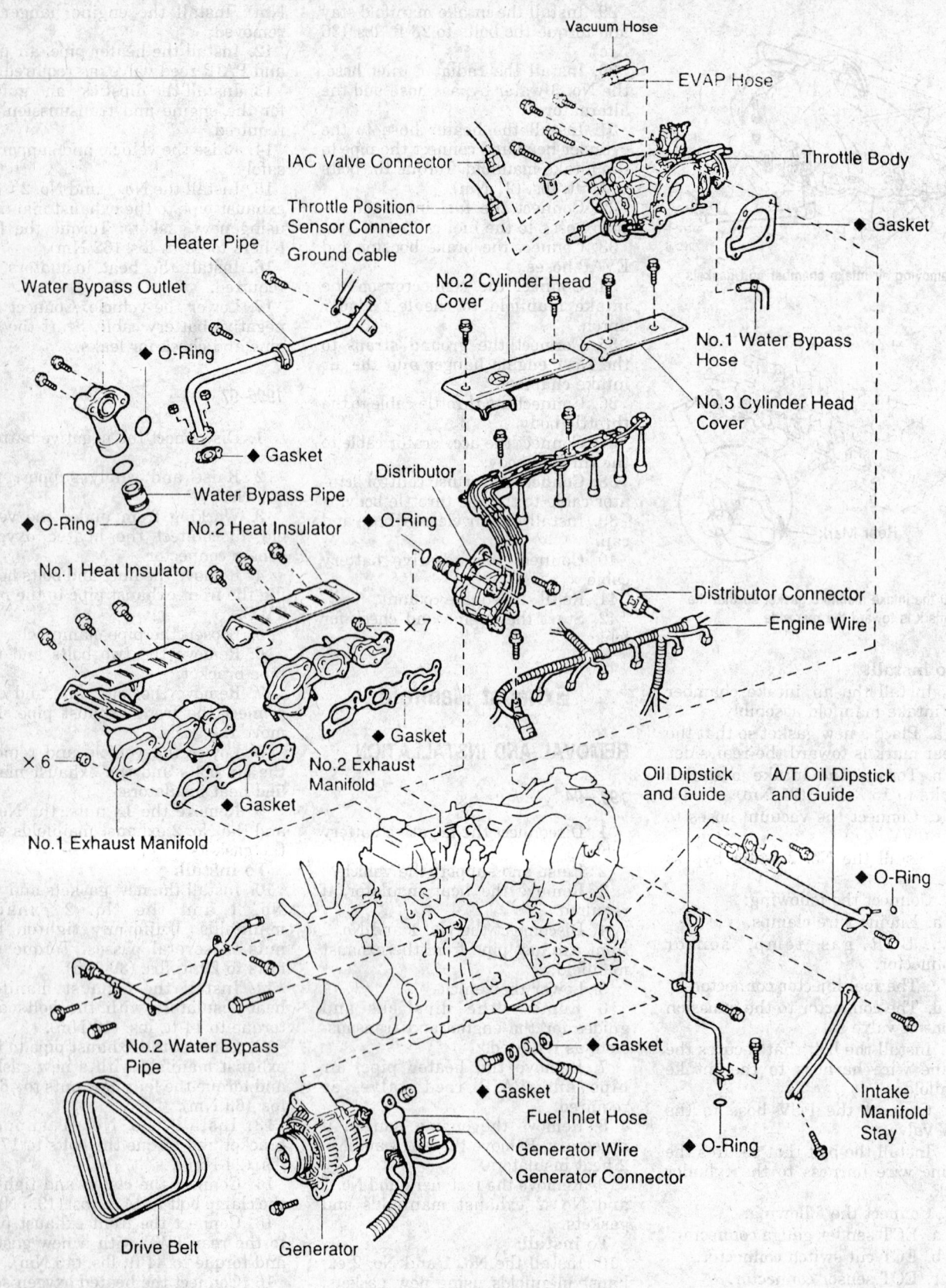

Vacuum Hose

EVAP Hose

IAC Valve Connector

Throttle Body

Throttle Position Sensor Connector

Ground Cable

Heater Pipe

Water Bypass Outlet

◆ O-Ring

◆ Gasket

◆ O-Ring

Water Bypass Pipe

No.2 Heat Insulator

No.1 Heat Insulator

× 7

× 6

No.1 Exhaust Manifold

No.2 Cylinder Head Cover

◆ Gasket

No.1 Water Bypass Hose

No.3 Cylinder Head Cover

Distributor

◆ O-Ring

Distributor Connector

Engine Wire

◆ Gasket

No.2 Exhaust Manifold

◆ Gasket

Oil Dipstick and Guide

A/T Oil Dipstick and Guide

◆ O-Ring

No.2 Water Bypass Pipe

◆ Gasket

◆ Gasket

Fuel Inlet Hose

Generator Wire

Generator Connector

◆ O-Ring

Intake Manifold Stay

Drive Belt

Generator

◆ Non-reusable part

299647

Exhaust system components

18. Start the engine and make sure that there are no exhaust leaks.

Front Cover Seal

REMOVAL AND INSTALLATION

1. Disconnect the negative battery cable.

————— **CAUTION** —————
Work must be started after 90 seconds from the time the ignition switch is turned to the LOCK position and the negative (-) battery cable is disconnected.
————————————————

2. Remove the crankshaft pulley.
3. Using a suitable tool, pry out the oil seal. Be careful not to damage the crankshaft.
To install:
4. Apply MP grease to the new oil seal lip.
5. Using tool No. SST 09316–60010, or equivalent, and a mallet, tap in the new oil seal until its surface is flush with the timing chain cover edge.
6. Install the crankshaft pulley.
   a. Align the pulley set key with the key groove of the pulley and slide on the pulley.

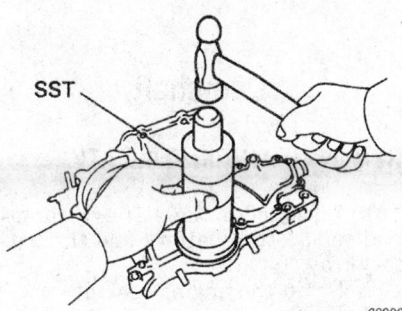

SST

Installing timing chain cover oil seal
68906

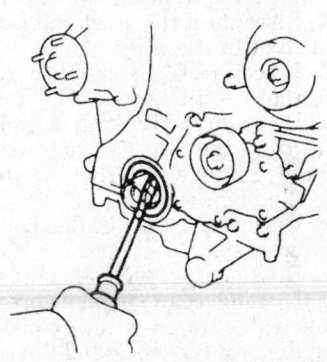

Oil seal removal on the engine
299671

   b. Torque the pulley bolt to 304 ft. lbs. (412 Nm).
7. Install the engine undercover.
8. Connect the negative battery cable.

Timing Chain Front Cover

REMOVAL AND INSTALLATION

1. Disconnect the negative battery cable.

————— **CAUTION** —————
Wait at least 90 seconds from the time the ignition switch is turned to the LOCK position and the negative (-) battery cable is disconnected before starting work. This will allow sufficient time for the SRS capacitor to discharge and prevent accidental deployment of the air bag(s).
————————————————

2. Raise and safely support the vehicle.
3. Drain the engine oil and the engine coolant.
4. Remove the engine undercover.
5. Remove the radiator.

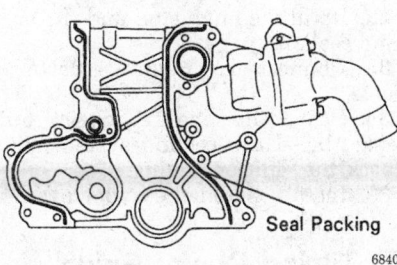

Seal Packing

68406

Apply seal packing to the timing chain cover, as shown

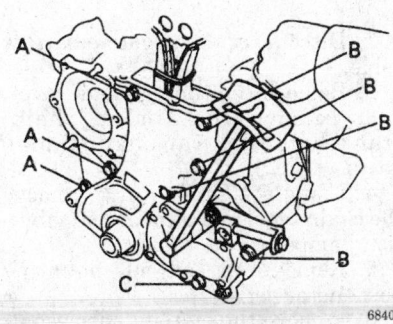

68407

Install the oil pump and drive belt adjusting bar. Each bolt length is indicated: "A" 1.18 inch (30mm), "B" 1.97 inch (50mm), "C" 2.38 inch (60mm)

6. Disconnect and remove the A/C compressor and the bracket.

NOTE: Do not disconnect the A/C refrigerant lines. If the lines are disconnected, be sure to use the proper refrigerant recovery and evacuation procedures.

7. Remove the radiator pipe. Disconnect the No. 2 radiator hose from the water inlet.
8. Remove the water pump and the gasket.
9. Remove the cylinder head.
10. Disconnect the oil cooler pipe bracket from the No. 1 oil pan.
11. Remove the oil level sensor.
12. Remove the bolts holding the No. 1 oil pan to the transmission housing.
13. Remove the No. 2 and No. 1 oil pans.
14. Remove the crankshaft pulley. Using tool No. SST 09213–58012 and 09330–00021, or equivalent, remove the pulley bolt. Remove the crankshaft pulley.
15. Check the thrust clearance of the oil pump driveshaft gear.
   a. Using a dial indicator with a lever type attachment, measure the thrust clearance.
   b. Maximum thrust clearance is 0.0118 in. (0.30 mm).
   c. If the thrust clearance is greater than maximum, replace the oil pump driveshaft gear and/or timing chain cover.
16. Remove the drive belt idler pulley.
17. Remove the timing chain cover.
   a. Remove the nine bolts, two nuts and the drive belt adjusting bar.
   b. Remove the oil pump, the O–rings and the gasket.
To install:
18. Install the timing chain cover. Apply packing seal to the cover before installation.
   a. Engage the gear of the oil pump drive rotor with the gear of the oil pump drive gear and install the oil pump.
   b. Install the oil pump and the drive belt adjusting bar and torque the bolts to 15 ft. lbs. (21 Nm).

NOTE: Install the timing chain cover with the proper length bolts in their correct locations. The bolts measurements are: Bolt A = 1.18 in. (30 mm), Bolt B = 1.97 in. (50 mm) and Bolt C = 2.38 in. (60 mm).

19. Remove the cord securing the timing chain to the sprockets and guides.

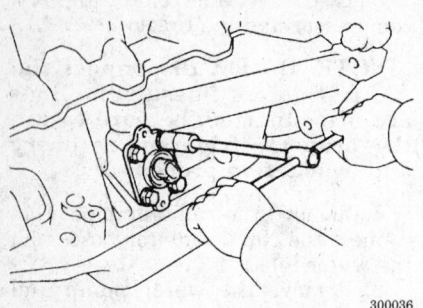

Removing the oil level sensor

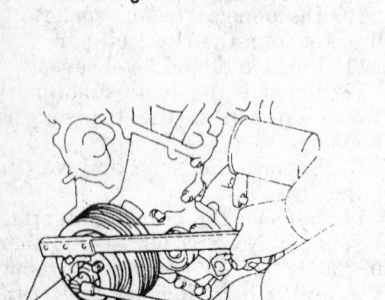

SST

Removing the crankshaft pulley bolt

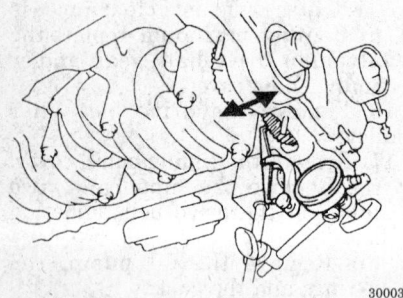

Checking thrust clearance of the oil pump drive

20. Install the drive belt idler pulley and torque the bolt to 32 ft. lbs. (43 Nm).
21. Install the crankshaft pulley.
    a. Align the pulley set key with the key groove of the pulley and slide on the pulley.
    b. Install the pulley bolt and torque to 304 ft. lbs. (412 Nm).
22. Install the No. 1 and the No. 2 oil pans.
23. Install the bolts holding the No. 1 oil pan to the transmission housing. Torque the bolts to 53 ft. lbs. (72 Nm).

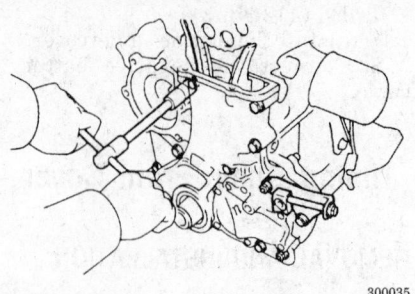

Removing the timing chain cover

24. Install the oil level sensor with a new gasket and torque the bolts to 48 in. lbs. (5.4 Nm).
25. Connect the oil cooler pipe bracket to the No. 1 oil pan.
26. Install the cylinder head.
27. Install the water pump and torque the bolts to 15 ft. lbs. (21 Nm).
28. Connect the No. 2 radiator hose to the water inlet. Install the nuts holding the radiator pipe to the No. 1 oil pan and torque to 15 ft. lbs. (21 Nm).
29. Install the A/C compressor bracket and torque the bolts to 27 ft. lbs. (37 Nm).
30. Install the A/C compressor and torque the bolts to 18 ft. lbs. (25 Nm). Install and adjust the drive belt.
31. Install the radiator.
32. Refill the engine oil and the engine coolant.
33. Connect the negative battery cable.
34. Start the engine, check for leaks, bleed the cooling system and check the ignition timing.
35. Install the engine undercover.

Timing Chain, Guide, Tensioner and Sprocket

REMOVAL AND INSTALLATION

1. Disconnect the negative battery cable.
2. Remove the timing chain cover.
3. Remove the timing chain, crankshaft and camshaft timing gears.
4. Using a 10mm wrench, remove the chain tensioner slipper and vibration damper.
5. Remove the oil jet mounting bolt and oil jet.
6. Remove the oil pump driveshaft gear.
7. Remove the pump drive gear from the crankshaft. If the oil pump driveshaft gear cannot be removed by

hand, use 2 flat-bladed tools and carefully pry it off.
   To install:
8. Turn the crankshaft until the set key is facing downward. Install the pump drive gear.
9. Lubricate the shaft portion of the oil pump driveshaft gear and install the gear.
10. Inspect the oil jet. Replace, if necessary. Install the oil jet and torque the mounting bolt to 14 ft. lbs. (20 Nm).
11. Install the chain tensioner slipper and vibration damper. Torque the damper bolts to 14 ft. lbs. (20 Nm) and the slipper bolt to 51 inch lbs. (69 Nm). Check that the tensioner slipper moves freely.
12. Install the crankshaft timing gear.
    a. Install the timing chain on the camshaft timing gear with the bright link aligned with the timing mark on the camshaft timing gear.
    b. Install the timing chain on the crankshaft timing gear with the other bright link aligned with the timing mark on the crankshaft timing gear.
    c. Wrap the timing chain with a piece of cord and make sure it doesn't come loose.
13. Install the timing chain cover.
14. Remove the drive belt idler pulley.
15. Install the crankshaft timing gear.

Camshaft

REMOVAL AND INSTALLATION

1. Disconnect the battery cables and remove the battery and the battery tray.
2. Drain the engine coolant.
3. Remove the air cleaner hose and cap.
4. Disconnect the cruise control actuator cable from the throttle body.
5. Disconnect the accelerator cable from the throttle body.
6. Disconnect the throttle cable from the throttle body.
7. Disconnect the engine ground strap from the No. 1 engine hanger and the ground strap from the air intake chamber.
8. Disconnect the brake booster vacuum hose.
9. Disconnect the heater hoses.
10. Disconnect the engine wire and heater valve from the cowl panel.
11. Remove the No. 2 and the No. 3 cylinder head covers
12. Remove the distributor.

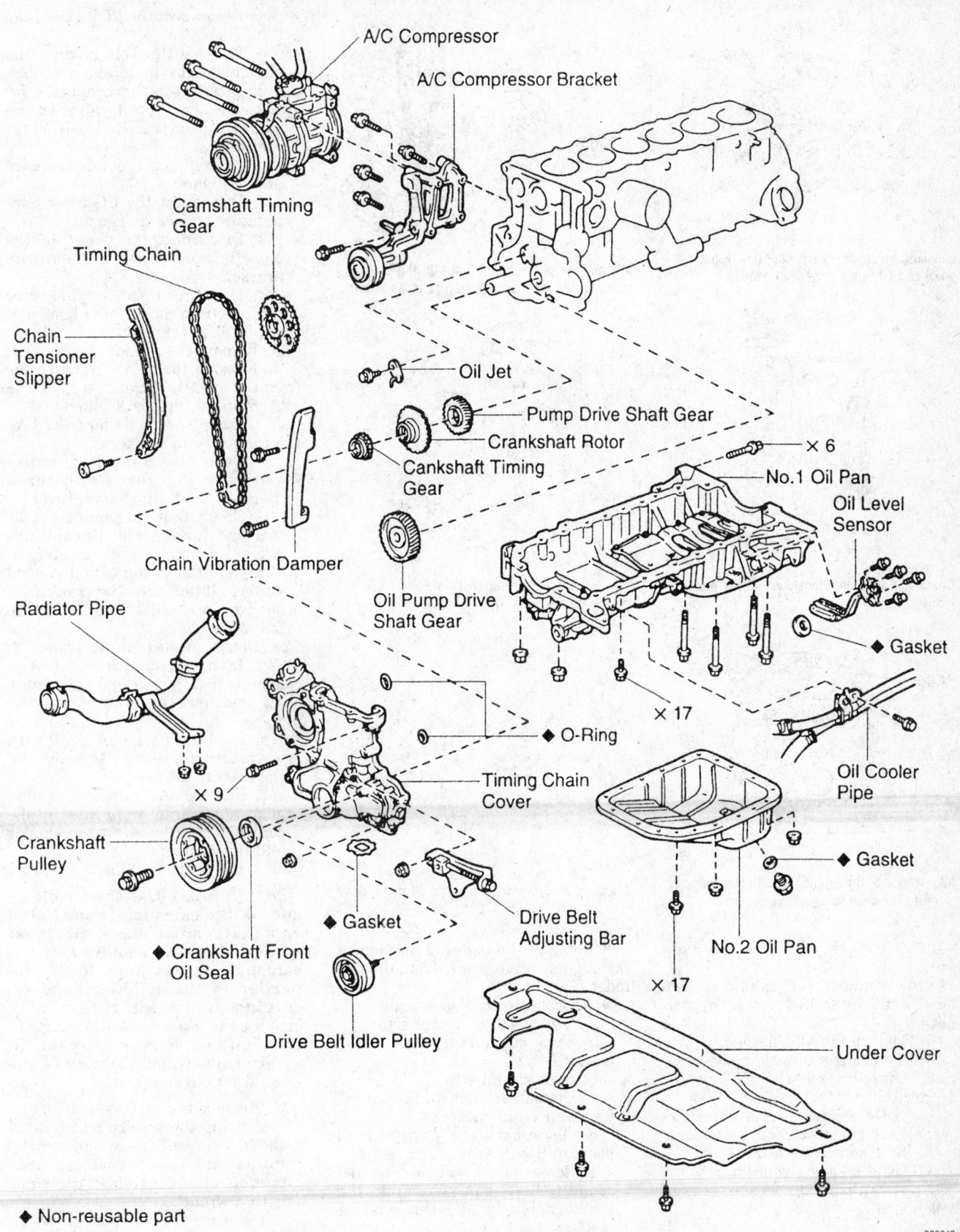

A/C Compressor

A/C Compressor Bracket

Camshaft Timing Gear

Timing Chain

Chain Tensioner Slipper

Oil Jet

Pump Drive Shaft Gear

Crankshaft Rotor

Cankshaft Timing Gear

Oil Pump Drive Shaft Gear

Chain Vibration Damper

× 6

No.1 Oil Pan

Oil Level Sensor

◆ Gasket

× 17

◆ O-Ring

Radiator Pipe

Timing Chain Cover

Oil Cooler Pipe

◆ Gasket

× 9

Crankshaft Pulley

◆ Gasket

Drive Belt Adjusting Bar

No.2 Oil Pan

◆ Crankshaft Front Oil Seal

× 17

Drive Belt Idler Pulley

Under Cover

◆ Non-reusable part

300049

Timing chain, timing cover, oil pump and related components

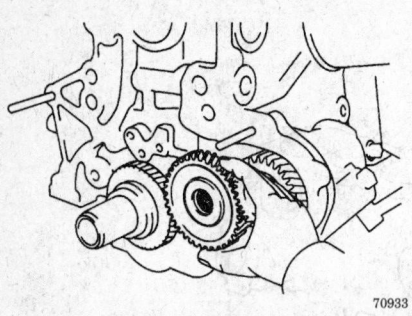

70933

Installing pump driveshaft gear (crankshaft gear) and oil pump driveshaft gear

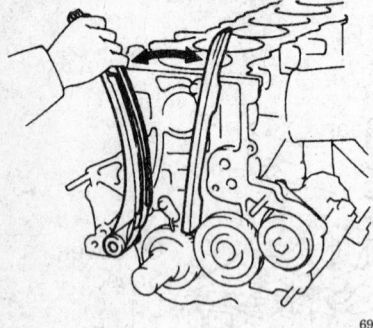

69000

Check that the chain tensioner slipper moves freely

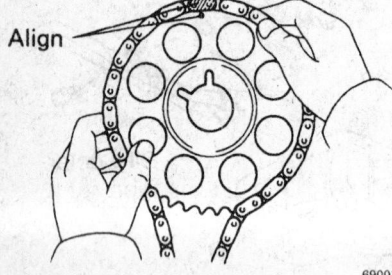

Align

69001

Align either of the bright link with the timing mark on the camshaft gear, as shown

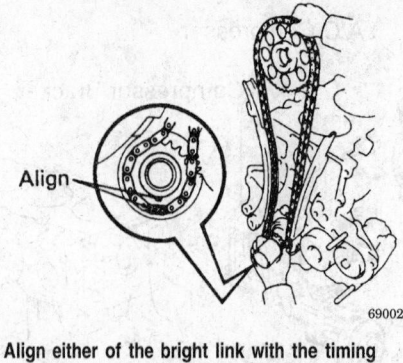

Align

69002

Align either of the bright link with the timing mark on the crankshaft timing gear, as shown

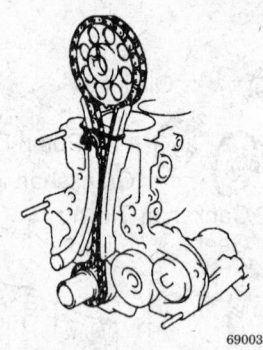

69003

Tie the timing chain with a cord as shown

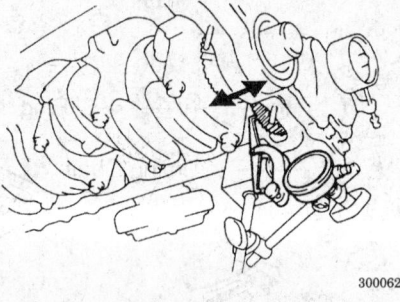

300062

Checking thrust clearance of the oil pump drive

13. Disconnect the P/S reservoir tank.

14. Disconnect the radiator inlet hose and the No. 3 water bypass hose.

15. Remove the alternator.

16. Remove the throttle body.

17. Disconnect the following connectors:

a. ECT sender gauge connector, the ECT cut switch connector and the ECT sensor connector.

b. Knock sensor connector.

c. Crankshaft position sensor connector.

18. Remove the bolt and disconnect the engine wire harness from the cylinder block.

19. Disconnect the following:

a. Oil level sensor connector.

b. Two connectors from the transmission.

c. Starter connector.

d. Disconnect the two heated oxygen sensor connectors.

e. Disconnect the park/neutral position (PNP) switch connector.

f. Remove the two bolts and disconnect the engine wire from the intake manifold and the cylinder block.

g. Disconnect the PCV hose from the PCV valve.

h. Remove the bolt holding the engine wire to the intake manifold.

i. Disconnect the connector for the emission control valve set assembly and the three injector connectors.

j. Disconnect the engine wire harness clamp.

k. Disconnect the EGR gas temperature sensor connector.

l. Disconnect the clamp of the No. 6 injector wire from the bracket.

m. Disconnect the engine wire harness from the cylinder head and the intake manifold.

20. Remove the cylinder head cover.

21. Remove the semi–circular plug from the cylinder head.

22. Remove the spark plugs.

23. Set the No. 1 cylinder to TDC of the compression stroke.

a. Turn the crankshaft pulley and align its groove with the **0** mark on the timing chain cover.

b. Check that the timing marks (one and two dots) of the camshaft drive and driven gears are in straight line on the cylinder head surface. If not, turn the crankshaft one revolution (360°) and align the marks as above.

24. Remove the chain tensioner.

25. Place matchmarks on the camshaft timing gear and the timing chain and remove the camshaft timing gear.

a. Hold the intake camshaft with a wrench and remove the bolt and the distributor gear.

b. Remove the camshaft timing gear and chain from the intake camshaft and leave on the slipper and the damper.

26. Remove the camshafts.

NOTE: Since the thrust clearance of the camshaft is small, the camshaft must be kept level while it is being removed. If the camshaft is not kept level, the portion of the cylinder head receiving the shaft thrust may crack or be damaged, causing the camshaft to seize or break. To avoid this, the following steps should be carried out.

27. Remove the exhaust camshaft.

a. Bring the service bolt hose of the driven sub–gear upward by turning the hexagon wrench head portion of the exhaust camshaft with a wrench.

b. Secure the exhaust camshaft sub–gear to the main gear with a service bolt. When removing the camshaft, make sure that the tor-

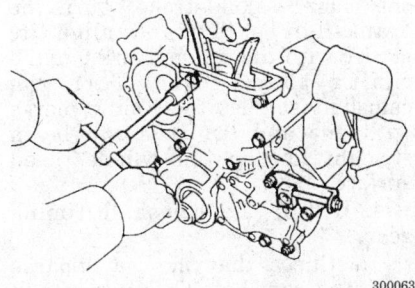

Removing the timing chain cover

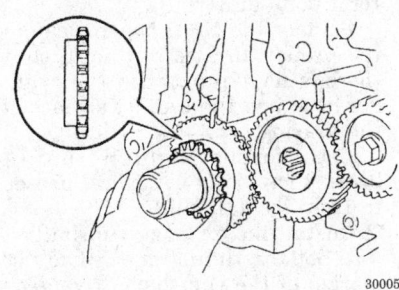

Installing the crankshaft timing gear

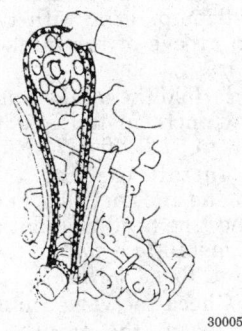

Removing the timing chain and the camshaft timing gear

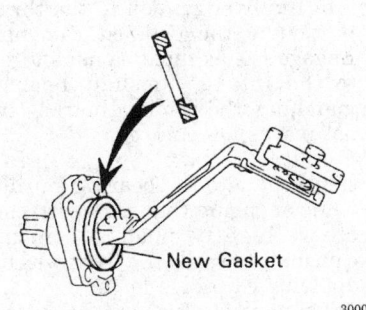

— New Gasket

Installing the oil level sensor

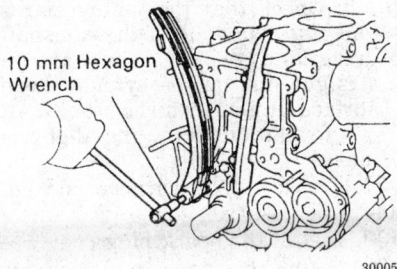

10 mm Hexagon Wrench

Removing the chains tensioner slipper and the vibration damper

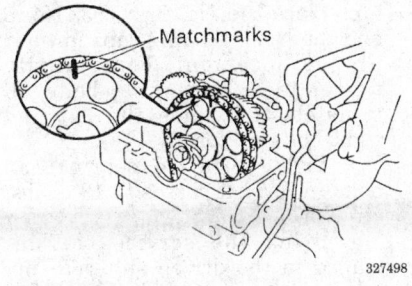

Matchmarks

Matchmarks on the timing chain and the camshaft gear

sional spring force of the sub–gear has been eliminated by the above operation.

c. Set the timing mark (two dot marks) of the camshaft driven gear at approximately 35° angle by turning the hexagon wrench head portion of the intake camshaft with a wrench.

d. Lightly push the camshaft towards the rear without applying excessive force.

e. Loosen and remove the No. 1 bearing cap bolts, alternately loosening the left and right bolts uniformly.

f. Loosen and remove the No. 2, No. 3, No. 5 and the No. 7 bearing cap bolts, alternately loosening the left and right bolts uniformly in several passes, in sequence.

NOTE: Do not remove the No. 4 and No. 6 bearing cap bolts at this stage.

g. Remove the four bearing caps.

h. Alternately and uniformly loosen and remove the No. 4 and the No. 6 bearing cap bolts.

i. If the camshaft is not being lifted out straight and level, retighten the four No. 4 and No. 6 bearing cap bolts. Then reverse the order of the above steps from (g) to (e) and repeat steps from (c) to (h) once again.

j. Remove the two bearing caps and exhaust camshaft. Do not pry on or attempt to force the camshaft with a tool or any other object.

28. Remove the intake camshaft.

a. Set the timing mark (two dot marks) of the camshaft drive gear at approximately a 25° angle by turning the hexagon wrench head portion of the intake camshaft with a wrench.

NOTE: The above angle arrows the No. 1 and the No. 4 cylinder cam lobes of the intake camshaft to push their valve lifters evenly.

b. Lightly push the intake camshaft towards the front without applying excessive force.

c. Loosen and remove the No. 1 bearing cap bolts, alternately loosening the left and the right bolts uniformly.

d. Loosen and remove the No. 3, No. 4, No. 6 and the No. 7 bearing cap bolts, alternately loosening the left and right bolts uniformly in several passes in sequence.

NOTE: Do not remove the No. 2 and No. 5 bearing cap bolts at this stage.

e. Remove the four bearing caps.

f. Alternately and uniformly loosen and remove the No. 2 and the No. 5 bearing cap bolts.

g. If the camshaft is not being lifted out straight and level, retighten the four No. 2 and No. 5 bearing cap bolts. Then reverse the order of the above steps from (e) to (c) and repeat steps from (a) to (f) once again.

h. Remove the two bearing caps and the exhaust camshaft.

29. Remove the valve lifters and shims. Arrange the valve lifters and shims in correct order for reinstallation.

To install:

30. Install the valve lifters and shims. Check to make sure that the valve lifter rotates smoothly by hand.

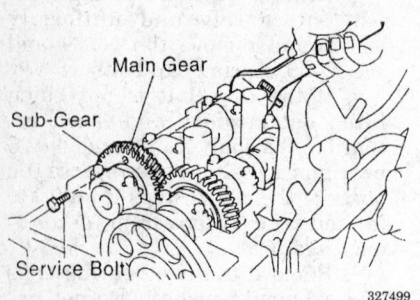

Exhaust camshaft sub-gear

327499

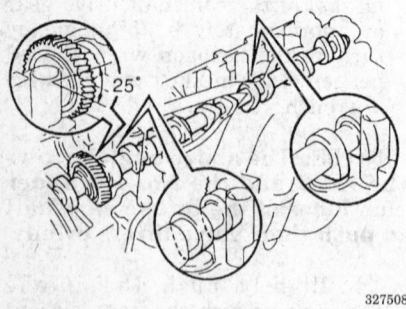

Intake camshaft with No. 1 and No. 4 cam lobes down

327508

31. Install the camshafts.

NOTE: Since the thrust clearance of the camshaft is small, the camshaft must be kept level while it is being installed. If the camshaft is not kept level, the portion of the cylinder head receiving the shaft thrust may crack or be damaged, causing the camshaft to seize or break. To avoid this, the following steps should be carried out.

a. Apply engine oil to the thrust portion of the intake camshaft.

b. Lightly place the intake camshaft on top of the cylinder head so that the No. 1 and the No. 4 cylinder cam lobes face downward.

c. Lightly push the camshaft towards the front without applying excessive force. Place the No. 2 and the No. 5 bearing caps in their proper location.

d. Temporarily tighten these bearing cap bolts uniformly and alternately in several passes until the bearing caps are snug with the cylinder head.

e. Place the No. 3, No. 4, No. 6 and the No. 7 bearing caps in their proper location. Temporarily tighten these bearing cap bolts, al-

ternately tightening the left and right bolts uniformly.

f. Place the No. 1 bearing cap in its proper location. Check that there is no gap between the cylinder head and the contact surface of the bearing cap.

g. Uniformly tighten the 14 bearing cap bolts in several passes to 12 ft. lbs. (16 Nm).

32. Install the exhaust camshaft.

a. Set the timing mark (two dot marks) of the camshaft drive gear at approximately 35° angle by turning the hexagon wrench head portion of the intake camshaft with a wrench.

b. Apply engine oil to the thrust portion of the exhaust camshaft. Engage the exhaust camshaft gear to the intake camshaft hear by matching the timing marks (two dot marks) on each gear.

c. Roll down the exhaust camshaft onto the bearing journals while engaging the gears with each other. Lightly push the intake camshaft towards the front without applying excessive force.

d. Install the No. 4 and the No. 6 bearing caps in their proper location. Temporarily tighten these bearing cap bolts, alternately tightening the left and right bolts uniformly.

e. Place the No. 2, No. 3, No. 5 and the No. 7 bearing caps in their proper location. Temporarily tighten these bearing cap bolts, alternately tightening the left and right bolts uniformly.

f. Tighten the 14 bearing cap bolts in several passes to 12 ft. lbs. (16 Nm).

g. Bring the service bolt installed in the driven sub-gear upward by turning the hexagon wrench head portion of the camshaft with a wrench. Remove the service bolt.

h. Check that the intake and the exhaust camshafts turn smoothly.

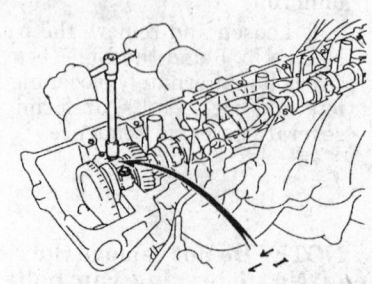

Installing No. 1 bearing cap bolts

327510

33. Set the No. 1 cylinder to TDC of the compression stroke. Turn the crankshaft pulley and align its groove with the timing mark **0** of the timing chain cover. Turn the camshaft so that the timing marks with one and two dots will be in straight line on the cylinder head surface.

34. Install the camshaft timing gear.

a. Check that the matchmarks on the camshaft timing gear and the timing chain are aligned. Place the gear over the straight pin of the intake camshaft.

b. Align the straight pin of the distributor gear with the straight pin groove of the intake camshaft gear.

c. Hold the intake camshaft with a wrench, install and torque the bolt to 54 ft. lbs. (74 Nm).

35. Install the chain tensioner. Push the tensioner by hand until it touches the head installation surface, then install and torque the two nuts to 15 ft. lbs. (21 Nm).

36. Check the valve timing.

a. Turn the crankshaft pulley and align its groove with the timing mark **0** of the timing chain cover. Always turn the crankshaft clockwise.

b. Check that the timing marks (one and two dots) of the camshaft drive and driven gears are in straight line on the cylinder head surface. If not, turn the crankshaft one revolution (360°) and align the marks.

37. Check valve clearance and adjust if necessary.

38. Install the spark plugs.

39. Install the semi-circular plug to the cylinder head.

40. Make sure that the No. 1 cylinder is in TDC of the compression stroke.

41. Install the cylinder head cover.

42. Connect the following:

a. Connect the engine wire harness from the cylinder head and the intake manifold.

b. Connect the clamp of the No. 6 injector wire from the bracket.

c. Connect the EGR gas temperature sensor connector.

d. Connect the engine wire harness clamp.

e. Connect the connector for the emission control valve set assembly and the three injector connectors.

f. Install the bolt holding the engine wire to the intake manifold

g. Connect the PCV hose from the PCV valve.

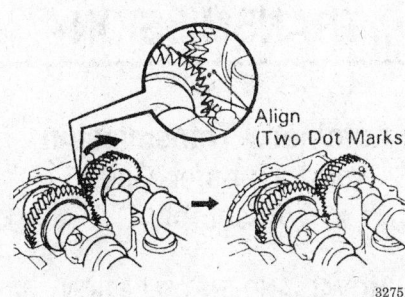

Align
(Two Dot Marks)

327512

Matchmarks for intake and exhaust camshafts

h. Install the two bolts and connect the engine wire from the intake manifold and the cylinder block.

i. Connect the park/neutral position (PNP) switch connector.

j. Connect the two heated oxygen sensor connectors.

k. Starter connector.

l. Two connectors from the transmission.

m. Oil level sensor connector.

43. Install the bolt and connect the engine wire harness from the cylinder block.

44. Connect the following:

a. ECT sender gauge connector, the ECT cut switch connector and the ECT sensor connector.

b. Knock sensor connector.

c. Crankshaft position sensor connector.

45. Install the throttle body.

46. Install the alternator and the drive belts.

47. Connect the No. 3 water bypass hose.

48. Connect the radiator inlet hose.

49. Connect the P/S reservoir tank and torque the bolts to 14 ft. lbs. (20 Nm).

50. Install the distributor.

51. Install the No. 2 and the No. 3 cylinder head covers.

52. Connect the heater valve and the engine wire harness to the cowl panel.

53. Connect the heater hoses.

54. Connect the ground strap to the No. 1 engine hanger and the air intake chamber.

55. Connect the brake booster vacuum hose.

56. Connect the throttle cable to the throttle body. Adjust the throttle cable.

57. Connect the accelerator cable to the throttle body.

58. Connect the cruise control actuator cable to the the throttle body.

59. Install the air cleaner hose and cap.

60. Install the battery tray, the battery and connect the cables.

61. Refill the engine coolant.

62. Start the engine and check for leaks.

63. Make necessary engine adjustments.

64. Road test the vehicle and recheck the engine coolant level.

Piston and Connecting Rod

POSITIONING

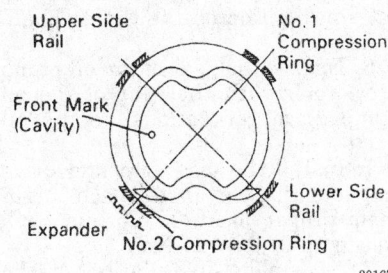

Upper Side Rail

No.1 Compression Ring

Front Mark (Cavity)

Lower Side Rail

Expander

No.2 Compression Ring

301671

Piston ring positioning

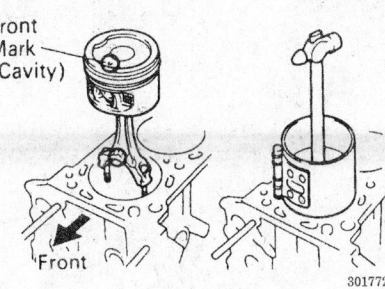

Front Mark (Cavity)

Front

301772

Installing the piston

LUBRICATION

Oil Pan

REMOVAL AND INSTALLATION

1. Disconnect the negative battery cable.

---- **CAUTION** ----

Work must be started after 90 seconds from the time the ignition switch is turned to the LOCK position and the negative (-) battery cable is disconnected.

2. Raise and safely support the vehicle.

3. Drain the engine oil.

4. Remove the engine undercover.

5. Disconnect the oil cooler pipe bracket from the No. 1 oil pan.

6. Remove the oil level sensor.

7. Remove the bolts holding the No. 1 oil pan to the transmission housing.

8. Remove the No. 2 oil pan.

9. Remove the No. 1 oil pan.

To install:

10. Install the No. 1 oil pan.

a. Apply seal packing to the No. 1 oil pan.

b. Install the oil pan and torque the 14 mm bolts to 32 ft. lbs. (44 Nm) and the 12 mm bolts to 14 ft. lbs. (20 Nm).

11. Apply seal packing to the No. 2 oil pan and torque the bolt to 5.7 ft. lbs. (7.8 Nm) and the nuts to 6.5 ft. lbs. (8.8 Nm).

12. Install the bolts holding the No. 1 oil pan to the transmission housing and torque to 53 ft. lbs. (72 Nm).

13. Install the oil level sensor and torque the bolts to 4.9 ft. lbs. (5.4 Nm).

14. Connect the oil cooler pipe bracket to the No. 1 oil pan.

15. Install the engine undercover.

16. Fill with engine oil.

17. Connect the negative battery cable.

18. Start the engine and check for leaks.

Oil Pump

REMOVAL AND INSTALLATION

1. Disconnect the negative battery cable.

2. Remove the timing chain cover.

NOTE: The oil pump is incorporated in the timing chain cover.

3. Remove the oil pump by prying the portions between the cylinder block and oil pump.

NOTE: Be careful not to damage the contact surfaces of the cylinder block and oil pump.

4. Remove the O-rings and gasket from the oil pump (timing chain cover).

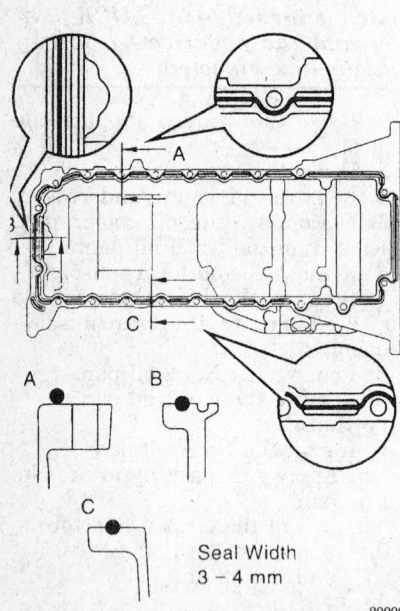

Seal Width
3 – 4 mm

299925

Applying the packing seal to the No. 1 oil pan

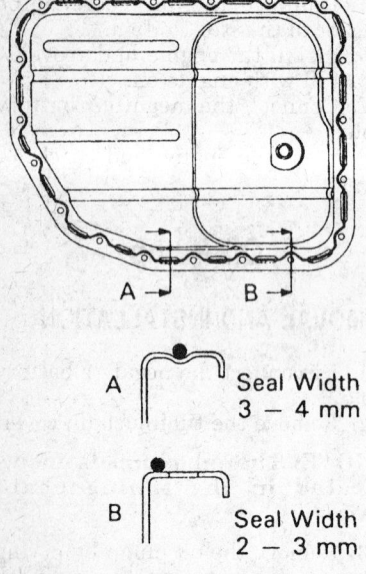

A Seal Width
3 – 4 mm

B Seal Width
2 – 3 mm

299926

Applying the packing seal to the No. 2 oil pan

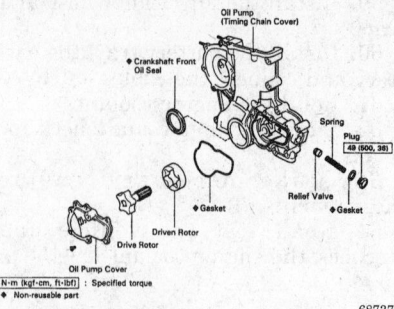

Oil pump (timing chain cover) components, exploded view

68737

To install:

5. Apply sealant PN 08826-00080 or equivalent to the oil pump (timing chain cover), as shown. Place new O-rings in position on the timing chain cover.

6. Engage the gear of the oil pump drive rotor with the gear of the oil pump drive gear and slide the oil pump.

7. Install the oil pump and drive belt adjusting bar. Install each bolt in their proper location. Torque each bolt to 15 ft. lbs. (21 Nm).

8. Complete installation of the timing chain cover.

9. Reconnect the negative battery cable. Start the engine and check for leaks.

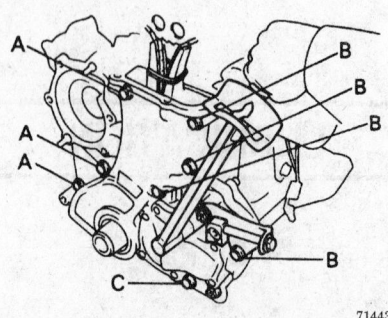

71443

Install the oil pump and drive belt adjusting bar. Each bolt length is indicated: "A" 1.18 inch (30mm), "B" 1.97 inch (50mm), "C" 2.38 inch (60mm)

TRANSMISSION

Automatic Transmission and Transfer Case Assembly

REMOVAL AND INSTALLATION

1993–94

1. Disconnect the negative battery cable.

2. If required, remove the air cleaner assembly.

3. Disconnect the transmission throttle cable from the throttle body.

4. Raise and safely support the vehicle.

5. Drain the transmission fluid.

6. Disconnect the wiring connectors (near the starter) for the neutral start switch and the back-up light switch. If equipped, disconnect the solenoid (overdrive) switch wiring at the same location.

7. If equipped, disconnect the oil level gauge.

8. Disconnect the starter wiring at the starter. Remove the mounting bolts and the starter from the engine.

9. Make matchmarks on the rear driveshaft flange and the differential pinion flange. These marks must be aligned during installation.

10. Unbolt the rear driveshaft flange. If the vehicle has a 2-piece driveshaft, remove the center bearing bracket-to-frame bolts. Remove the driveshaft from the vehicle.

11. Disconnect the speedometer cable (tie it aside). Disconnect the shift linkage from the transmission.

12. Disconnect the transmission oil cooler lines at the transmission.

13. Disconnect the exhaust pipe clamp and remove the oil filler tube, as required.

14. Support the transmission, using a jack with a wooden block placed between the jack and the transmission pan. Raise the transmission, just enough to take the weight off of the rear mount.

15. Remove the rear engine mount with the bracket and the engine undercover, to gain access to the engine crankshaft pulley.

16. Remove the stiffener plates, if equipped.

17. Place a wooden block (or blocks) between the engine oil pan and the front frame crossmember.

18. Slowly, lower the transmission until the engine rests on the wooden block.

19. Remove the rubber plug(s) from the service holes located at the rear of the engine in order to gain access to the torque convertor bolts.

20. Rotate the crankshaft (to remove the torque convertor bolts) to access the bolts through the service holes.

21. Obtain a bolt of the same dimensions as the torque convertor bolts. Cut the head off of the bolt and hacksaw a slot in the bolt opposite the threaded end.

NOTE: This modified bolt is used as a guide pin. Two guides pins are needed to properly install the transmission.

22. Thread the guide pin into one of the torque convertor bolt holes. The guide pin will help keep the convertor with the transmission.

23. Remove the stiffener plates from the transmission.

24. Remove the transmission-to-engine bolts, then carefully move the transmission rearward by prying on the guide pin through the service hole.

25. Pull the transmission rearward and lower it (front end down) out of the vehicle.

To install:

26. Apply a coat of multi-purpose grease to the torque convertor stub shaft and the corresponding pilot hole in the flexplate.

27. Install the torque convertor into the front of the transmission. Push inward on the torque convertor while rotating it to completely couple the torque convertor to the transmission.

28. To make sure the convertor is properly installed, measure the distance between the torque convertor mounting lugs and the front mounting face of the transmission. The proper distance is 0.079 inch (20mm).

29. Install guide pins into 2 opposite mounting lugs of the torque converter.

30. Raise the transmission to the engine, align the transmission with the engine alignment dowels and position the convertor guide pins into the mounting holes of the flexplate.

31. Install and tighten the transmission-to-engine mounting bolts. Torque the bolts to 47 ft. lbs. (63 Nm).

32. Remove the convertor guide pins and install the convertor mounting bolts. Rotate the crankshaft as necessary to gain access to the guide pins and bolts through the service holes. Evenly, tighten the convertor

mounting bolts to 13 ft. lbs. (17 Nm). Install the rubber plugs into the access holes.

33. Install the engine undercover. Raise the transmission slightly and remove the wood block(s) from under the engine oil pan.

34. Install the transmission crossmember. Torque the crossmember-to-frame bolts to 26–36 ft. lbs. (34–48 Nm).

35. Lower the transmission onto the crossmember and install the transmission mounting bolts. Torque the bolts to 19 ft. lbs. (26 Nm).

36. Install the oil filler tube and connect the exhaust pipe clamp.

37. Connect the oil cooler lines to the transmission and torque the fittings to 25 ft. lbs. (34 Nm).

38. Install the remaining components by reversing the removal procedure.

39. Adjust the transmission throttle cable.

40. Refill the transmission.

41. Connect the negative battery cable. Start the engine and check for leaks.

42. Road test the vehicle for proper operation.

43. Recheck all fluid levels.

1995–97

1. Disconnect the battery cables and remove the battery and the battery tray.

CAUTION

Work must be started after 90 seconds from the time the ignition switch is turned to the LOCK position and the negative (-) battery cable is disconnected.
---

2. Loosen the fan shroud of the cooling fan to avoid damage to the fan.

3. Disconnect the throttle cable.

4. Raise and support the vehicle.

5. Drain the transmission and transfer case fluid.

6. Remove the upper side starter mounting bolt.

7. Remove the transmission select lever and the transfer shift lever.

  a. Remove the clip, washer and the wave washer and disconnect the link.

  b. Remove the nut and washer and disconnect the link.

  c. Remove the transfer shift lever knob.

  d. Remove the console and the transfer shift lever boot.

  e. Remove the center console box and disconnect the three connectors.

  f. Remove the transmission shift lever assembly and the transfer shift lever.

8. Disconnect the No. 1 and No. 2 vehicle speed sensors, the park/neutral switch, the solenoid connector and the A/T fluid temperature sensor.

9. Disconnect the connectors and hoses from the transfer.

10. Remove the front and rear propeller (drive) shaft.

11. Remove the oil lever gauge, the upper side mounting bolt and the filler pipe.

12. Loosen the two oil cooler pipe union nuts.

13. Remove the four stabilizer bar bracket mounting bolts.

14. Remove the engine undercover.

15. Remove the torque converter clutch mounting bolts.

  a. Remove the converter hole plug.

  b. Turn the crankshaft to gain access to each bolt.

  c. Hold the crankshaft pulley nut with a wrench and remove the bolts.

16. Remove the front exhaust pipe assembly.

  a. Loosen the clamp bolt and disconnect the clamp from the No. 1 support bracket.

  b. Remove the No. 1 support bracket.

17. Disconnect the starter terminal and the connector and remove the starter.

18. Place a jack under the transmission and remove the crossmember.

19. Lower the rear end of the transmission.

20. Separate the wire harness from the transmission and the transfer case.

21. Remove the oil cooler pipe mounting bolts from the torque converter clutch housing and disconnect the two oil cooler pipes from the elbows.

22. Remove the ten bolts and remove the transmission.

23. Remove the transfer case from the transmission.

To install:

24. Install the transfer case to the transmission.

25. Install the transmission and torque the bolts to 53 ft. lbs. (72 Nm).

26. Connect the two oil cooler pipes and install the oil cooler pipe mounting bolts to the torque converter clutch housing.

27. Install the crossmember and torque the bolts to 45 ft. lbs. (61 Nm) and the nuts to 54 ft. lbs. (74 Nm).

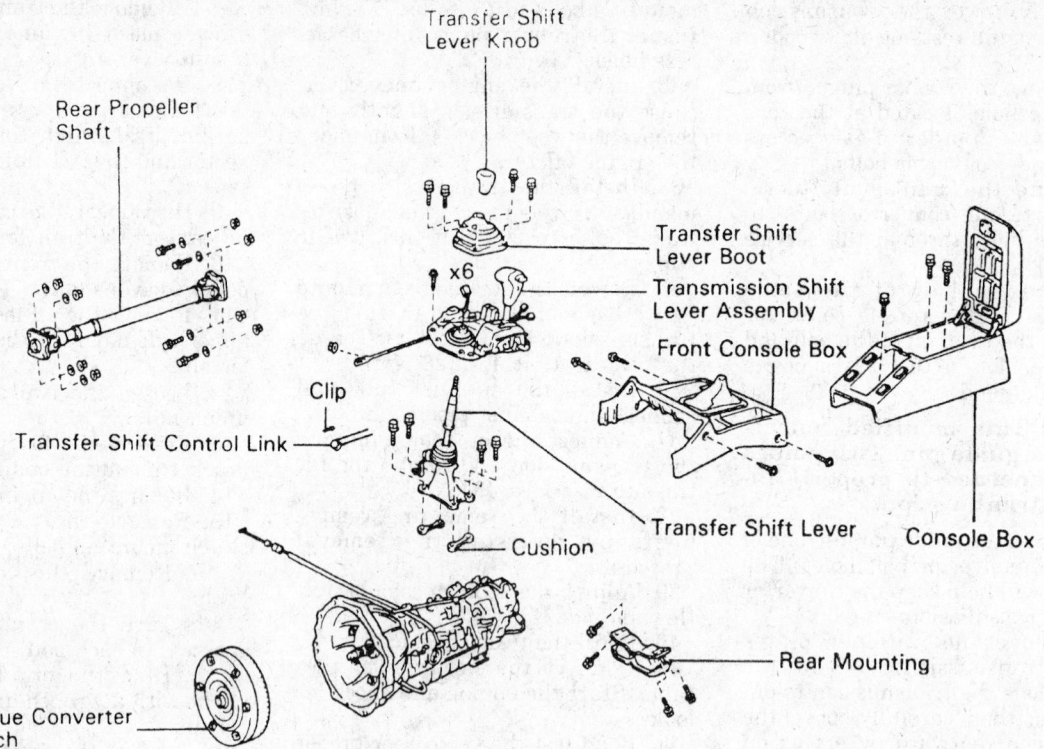

Transfer Shift
Lever Knob

Rear Propeller
Shaft

Transfer Shift
Lever Boot

Transmission Shift
Lever Assembly

Front Console Box

Clip

Transfer Shift Control Link

Transfer Shift Lever

Console Box

Cushion

Torque Converter
Clutch

Rear Mounting

300396

Transmission and related components

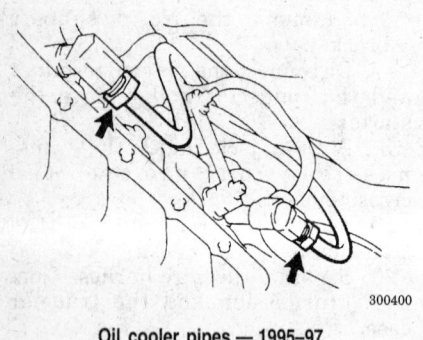

Oil cooler pipes — 1995–97

300400

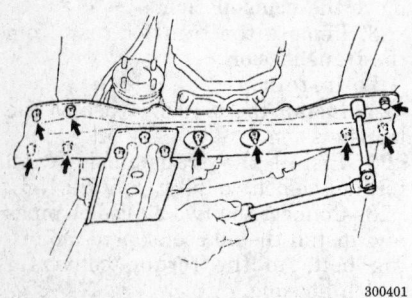

300401

Removing the crossmember — 1995–97

28. Install the starter and torque the bolt to 29 ft. lbs. (39 Nm). Connect the terminal and the connector.

29. Install the No. 1 support bracket and torque the bolts to 29 ft. lbs. (39 Nm).

30. Connect the clamp from the No. 1 support bracket and torque the bolt to 14 ft. lbs. (19 Nm).

31. Install the front exhaust pipe assembly and torque the bolts to 29 ft. lbs. (39 Nm).

32. Install the torque converter clutch mounting bolts and torque to 40 ft. lbs. (55 Nm).

NOTE: First install the gray colored bolt and then the five other bolts.

33. Install the engine undercover. Torque the bolts to 21 ft. lbs. (28 Nm).

34. Install the stabilizer bar bracket mounting and torque the bolts to 13 ft. lbs. (18 Nm).

35. Tighten the oil cooler pipe union nuts.

36. Install a new O–ring and install the filler pipe. Install the upper side mounting bolt and the level gauge.

37. Install the front and rear propeller (drive) shaft.

38. Connect the connectors and hoses to the transfer case.

39. Connect the A/T fluid temperature sensor, the solenoid connector, the park/neutral switch and the No. 1 and No. 2 vehicle speed sensors.

40. Install the transmission select lever and transfer shift lever.

a. Install the transmission shift lever assembly and the transfer shift lever. Torque the bolts for the shift lever to 13 ft. lbs. (18 Nm).

b. Install the center console box and connect the three connectors.

c. Install the console and the transfer shift lever boot.

d. Install the transfer shift lever knob.

e. Install the nut and washer and connect the link.

f. Install the clip, washer and the wave washer and connect the link.

41. Install the upper side starter mounting bolt and torque to 29 ft. lbs. (39 Nm).

42. Lower the vehicle and fill the transfer case and transmission with the proper fluid.

43. Connect the throttle cable.

44. Tighten the fan shroud of the cooling fan.

45. Install the battery tray and the battery. Connect the battery cables.

46. Test drive the vehicle and check the shifting operation.

DRIVE AXLE

Driveshaft

REMOVAL AND INSTALLATION

1. Disconnect the negative battery cable from the battery.

——— CAUTION ———
Work must be started after 90 seconds from the time the ignition switch is turned to the LOCK position and the negative (-) battery cable is disconnected.

2. Raise and safely support the vehicle.
3. Place matchmarks on the flanges of the front driveshaft and the transfer case.
4. Remove the four nuts, bolts and the washers. Disconnect the front driveshaft from the transfer.
5. Place matchmarks on the flanges of the driveshaft and the front differential housing.
6. Remove the four nuts, bolts and the washers. Disconnect the driveshaft from the differential housing.
7. Place matchmarks on the flanges of the rear driveshaft and the transfer case.
8. Remove the four nuts, bolts and the washers. Disconnect the rear driveshaft from the transfer case.
9. Place matchmarks on the flanges of the rear driveshaft and the differential housing.
10. Remove the four nuts, bolts and the washers. Disconnect the rear driveshaft from the differential housing.
   To install:
11. Align the matchmarks on the driveshaft and differential flanges.
12. Install the four washers, bolts and the nuts to the front driveshaft and differential carrier. Torque the bolts to 54 ft. lbs. (74 Nm).
13. Install the four washers, bolts and the nuts to the front driveshaft and transfer case. Torque the bolts to 54 ft. lbs. (74 Nm).
14. Install the four washers, bolts and the nuts to the rear driveshaft and differential carrier. Torque the bolts to 65 ft. lbs. (88 Nm).
15. Install the four washers, bolts and the nuts to the rear driveshaft and transfer case. Torque the bolts to 65 ft. lbs. (88 Nm).

16. Lower the vehicle and connect the negative battery cable to the battery.

U-Joints

REMOVAL AND INSTALLATION

1. Remove the driveshaft from the vehicle.
2. Matchmark the yoke and the driveshaft.
3. Using a brass bar and hammer, slightly tap in the bearing outer race.
4. Remove the four snaprings from the bearings.
5. Using tool No. SST 09332–25010, or equivalent, push out the bearing from the flange.
6. Clamp the bearing outer race in a vise and tap off the flange with a hammer.
7. Repeat Steps 3, 5 and 6 for the other bearings.
8. Check for worn or damaged parts. Inspect the bearing journal surfaces for wear.
   To install:
9. Install the bearing cups, seals and O-rings on the spider.
10. Grease the spider and the bearings.

NOTE: Be sure to hold the bearing caps while greasing the U-joints. The grease will force the bearing caps off the spider when they are not secured in the driveshaft yoke.

11. Remove the bearing cups from the spider.
12. Position the spider in the yoke.
13. Start the bearings in the yoke and then press them into place, using a vise. Install the snaprings to hold the bearing cups in place.
14. Make sure the bearings and snaprings are fully seated by lightly tapping on the yoke with a hammer.
15. If the axial play of the spider is greater than 0.002 inch (0.05 mm), select snaprings which will provide the correct play. Be sure that the snaprings are the same size on both sides or driveshaft noise and vibration will result.
16. Install the driveshaft in the vehicle.

Axle Shaft and Seal

REMOVAL AND INSTALLATION

Front

1. Raise and safely support the vehicle.

2. Remove the front axle hub.
3. Remove the bolt and disconnect the speed sensor from the steering knuckle.
4. Remove the knuckle spindle mounting bolts.
5. Remove the dust seal and the dust cover.
6. Remove the knuckle spindle.
   a. Using a brass bar and hammer, tap the knuckle spindle of the steering knuckle.
   b. Remove the knuckle spindle and the gasket.
7. Position one flat part of the outer shaft upward and remove the axle shaft.
8. Remove the oil seal set.
   a. Remove the bolts from the end retainer.
   b. Remove the oil seal and retainer, the felt dust seal, the rubber seal and the steel ring.
9. Using tool No. SST 09308–00010, or equivalent, remove the axle housing oil seal.
   To install:
10. Using tool No. SST 09618–60010, or equivalent, and a hammer, install the new oil seal into the axle housing.
11. Install the oil seal set.
12. Install the oil seal end retainer to the knuckle and torque the bolts to 4.9 ft. lbs. (5.4 Nm).
13. Position one flat part of the outer shaft upward and install the shaft.
14. Pack molybdenum disulfide lithium base grease NLGI No. 2 into the knuckle to about three fourths of the knuckle.
15. Place the dust cover, dust seal and a new gasket on the spindle. Torque the spindle mounting bolts to 34 ft. lbs. (47 Nm).
16. Connect the speed sensor and the bolts to the steering knuckle. Torque to 13 ft. lbs. (18 Nm).
17. Install the axle hub.
18. Check the front wheel alignment and the ABS speed sensor signal.

Rear

1993–94

1. Raise and safely support the vehicle.
2. Remove the wheels.
3. Place a suitable drain pan under the axle.
4. Remove the plug and drain the oil from the differential.
5. Remove the brake drum and parking brake cable clamp.
6. Remove the cover from the back of the differential housing.

Front Propeller Shaft

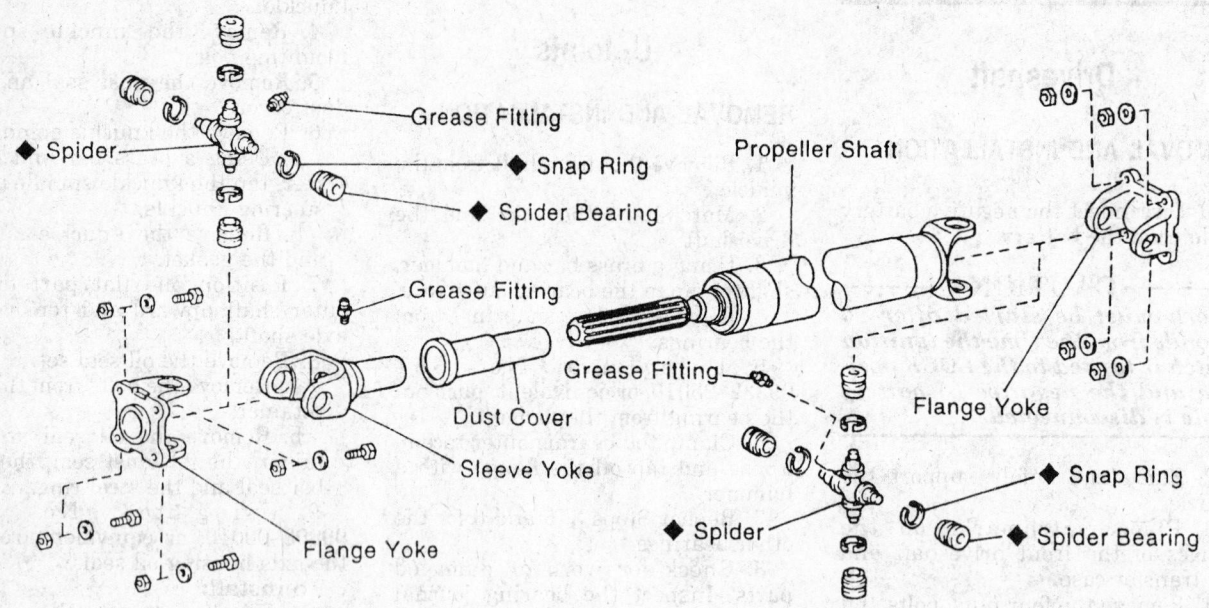

◆ Spider
Grease Fitting
◆ Snap Ring
◆ Spider Bearing

Grease Fitting
Dust Cover
Sleeve Yoke
Flange Yoke

Propeller Shaft

Grease Fitting

Flange Yoke

◆ Spider
Grease Fitting

◆ Snap Ring
◆ Spider Bearing

Rear Propeller Shaft

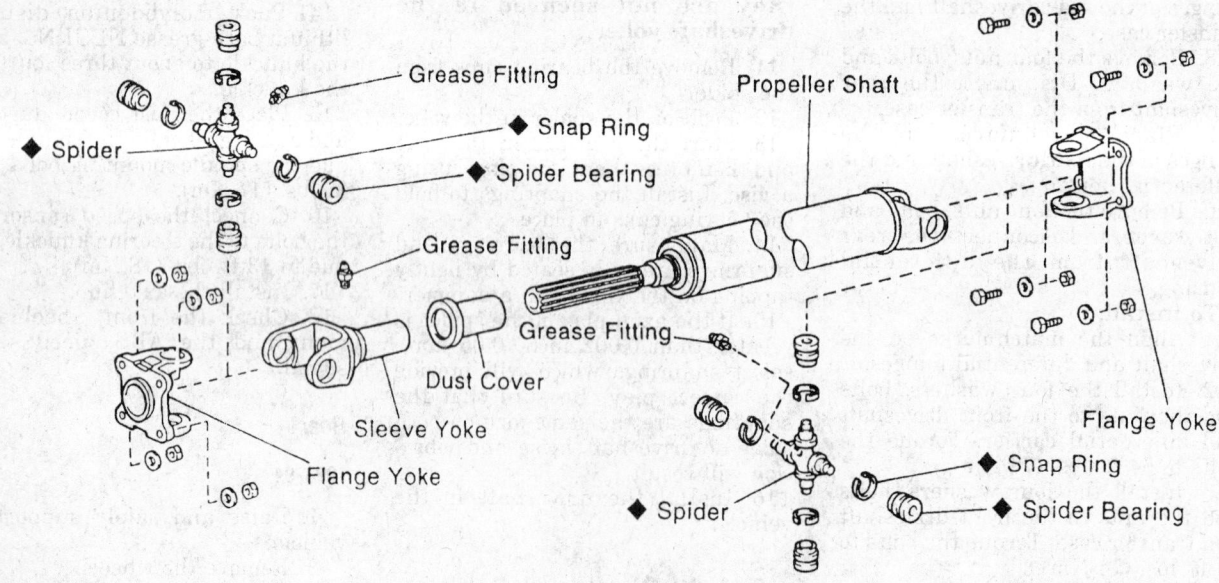

◆ Spider
Grease Fitting
◆ Snap Ring
◆ Spider Bearing

Grease Fitting
Dust Cover
Sleeve Yoke
Flange Yoke

Propeller Shaft

Grease Fitting

Flange Yoke

◆ Spider
Grease Fitting

◆ Snap Ring
◆ Spider Bearing

◆ Non-reusable part

300646

Driveshaft component assembly

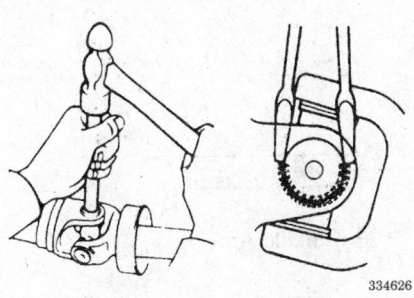

Removing the snaprings

334626

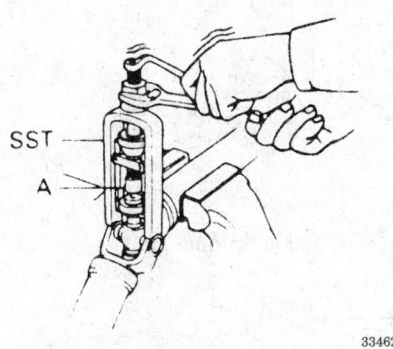

SST

A

334627

Removing the bearing from the flange

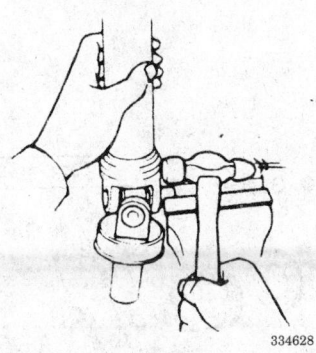

334628

Bearing outer race in a vise, tap off the flange with a hammer

7. Remove the pin from the differential pinion shaft and withdraw the pinion shaft and it's spacer from the case.

8. Push the axle shaft to the differential side and remove the remove the C-lock from the axle shaft.

9. Carefully withdraw the axle shaft from the housing. Be careful not to damage the oil seal when removing the axle shaft.

10. If replacing the bearing and/or seal, use a suitable bearing puller and withdraw the bearing and seal.

To install:

11. Apply MP grease to the bearing and seal lip. Using a suitable bearing installer, drive a new bearing and seal into place.

12. Carefully insert the axle shaft into the axle housing.

13. Install the axle shaft C-lock and pull the axle shaft fully toward the outer side of the vehicle.

14. Install the pinion shaft and spacer. Torque the pinion shaft pin to 20 ft. lbs. (27 Nm).

15. Install the differential cover. Torque the mounting nuts to 9 ft. lbs. (13 Nm).

16. Install the parking brake clamp, brake drum and wheel.

17. Fill the differential with the correct oil.

18. Lower the vehicle.

19. Road test the vehicle for proper operation.

1995–97

1. Raise and support the vehicle safely.

2. Remove the rear axle shaft nuts.

3. Using a brass bar and hammer, tap on the bolt heads and remove the cone washers.

4. Install and gradually tighten two bolts and pull the axle shaft. Remove the bolts from the axle shaft.

5. Remove the axle shaft and the gasket.

6. Using tool No. SST 09308–00010, or equivalent, remove the oil seal.

To install:

7. Using tool No. SST 09517–36010, or equivalent and a hammer, install a new oil seal into the hub. Apply MP grease to the oil seal lip.

8. Install the rear axle shaft.

9. Install the cone washers and torque the nuts to 25 ft. lbs. (34 Nm).

STEERING

Air Bag

———— CAUTION ————

Some vehicles are equipped with an air bag system, also known as the Supplemental Inflatable Restraint (SIR) or Supplemental Restraint System (SRS). The system must be disabled before performing service on or around system components, steering column, instrument panel components, wiring and sensors. Failure to follow safety and disabling procedures could result in accidental air bag deployment, possible personal injury and unnecessary system repairs.

PRECAUTIONS

Several precautions must be observed when handling the inflator module to avoid accidental deployment and possible personal injury.

• Never carry the inflator module by the wires or connector on the underside of the module.

• When carrying a live inflator module, hold securely with both hands and ensure that the bag and trim cover are pointed away.

• Place the inflator module on a bench or other surface with the bag and trim cover facing up.

• With the inflator module on the bench, never place anything on or close to the module which may be thrown in the event of an accidental deployment.

DISARMING

To avoid personal injury when working on vehicles equipped with an air bag, the negative battery cable must be disconnected and at least 90 seconds must elapse before working on the system. Failure to do so may result in deployment of the air bag.

Steering Wheel

REMOVAL AND INSTALLATION

NOTE: To avoid possible unexpected deployment of the air bag (if equipped), work must not be started after approximately 90 seconds or longer from the time the ignition switch is turned OFF and the negative battery cable disconnected.

1. Disconnect the negative battery cable.

2. Position the wheels in a straight ahead position.

3. On 1993–94 models:

a. Loosen and remove the steering wheel center cover (pad) retaining screws, if equipped.

b. Pull the wheel pad out from the steering wheel and disconnect the air bag connector, if equipped.

4. On 1995–97 models:

a. Place the front wheels straight ahead.

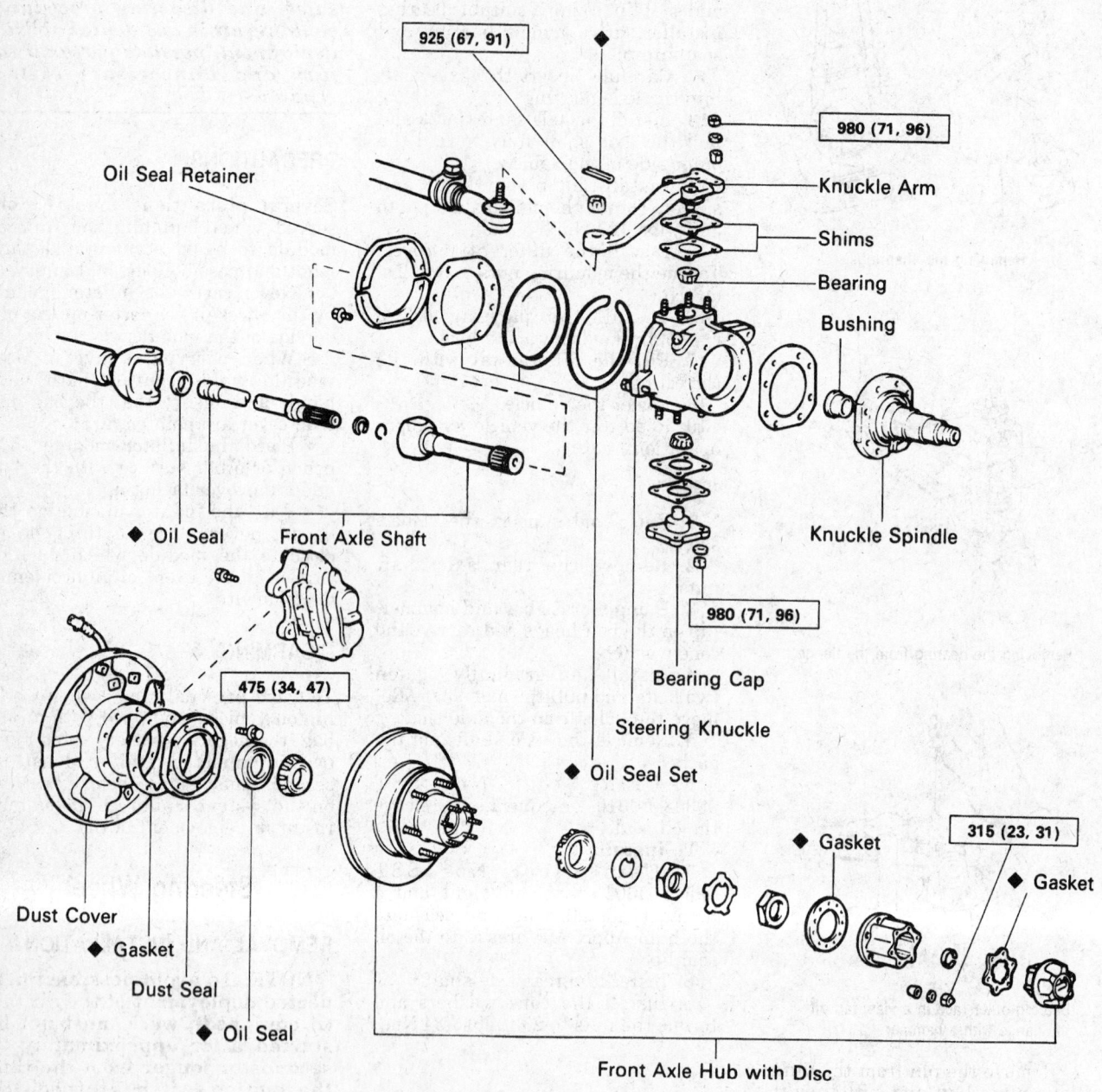

925 (67, 91)

980 (71, 96)

Oil Seal Retainer

Knuckle Arm

Shims

Bearing

Bushing

Oil Seal

Front Axle Shaft

Knuckle Spindle

980 (71, 96)

475 (34, 47)

Bearing Cap

Steering Knuckle

◆ Oil Seal Set

Dust Cover

◆ Gasket

Dust Seal

◆ Gasket

315 (23, 31)

◆ Gasket

◆ Oil Seal

Front Axle Hub with Disc

kg-cm (ft-lb, N·m) : Specified torque

◆ Non-reusable part

Steering knuckle and axle shaft components, exploded view — 1993–94

90531

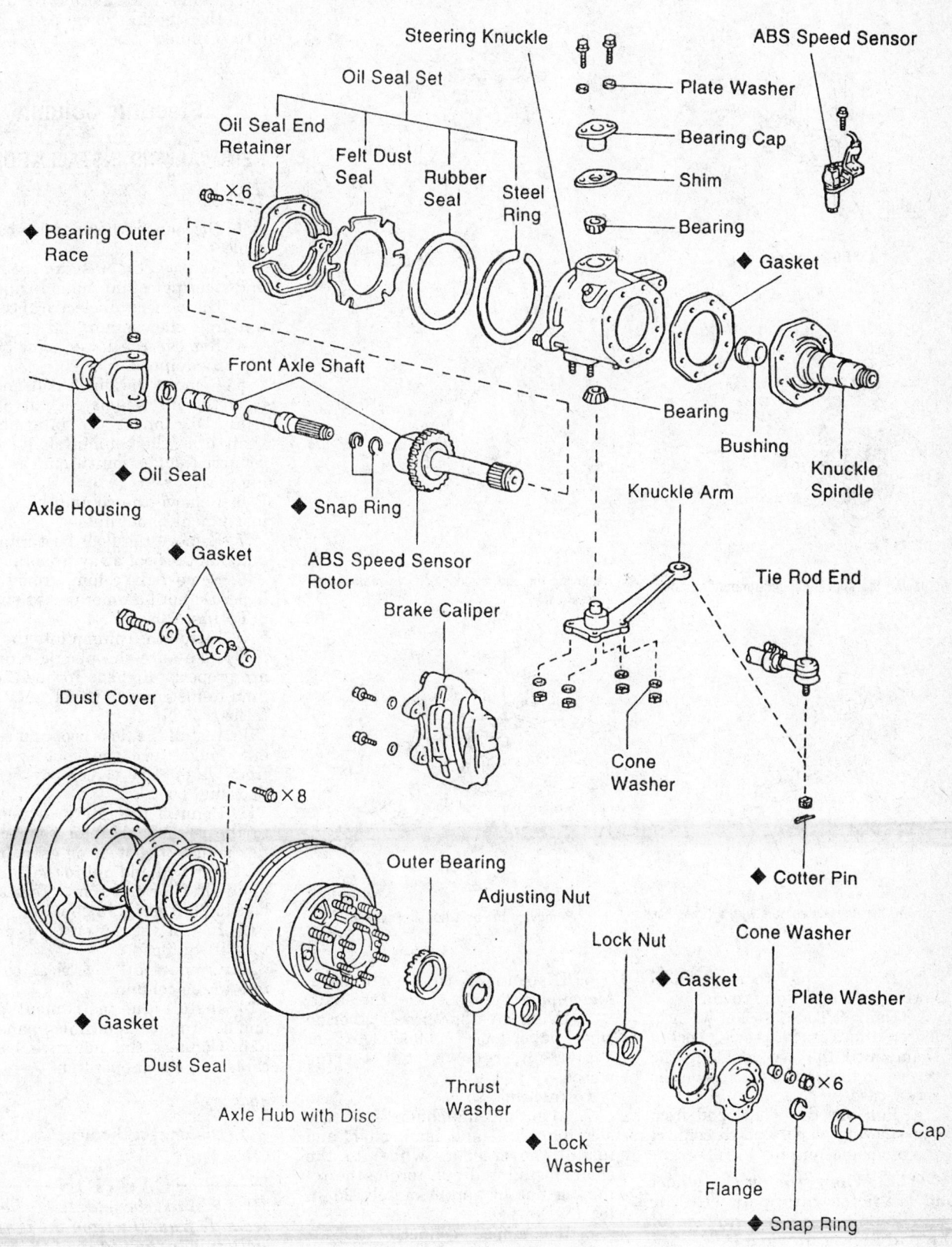

Steering Knuckle

Oil Seal Set

ABS Speed Sensor

Oil Seal End
Retainer

Felt Dust
Seal

Rubber
Seal

Steel
Ring

Plate Washer

Bearing Cap

Shim

Bearing

◆ Gasket

× 6

◆ Bearing Outer
Race

Front Axle Shaft

Bearing

Bushing

Knuckle
Spindle

◆ Oil Seal

Axle Housing

◆ Snap Ring

ABS Speed Sensor
Rotor

Knuckle Arm

Tie Rod End

◆ Gasket

Brake Caliper

Cone
Washer

Dust Cover

Cotter Pin

× 8

Outer Bearing

Adjusting Nut

Lock Nut

Cone Washer

◆ Gasket

Plate Washer

◆ Gasket

Dust Seal

Axle Hub with Disc

Thrust
Washer

◆ Lock
Washer

Flange

Cap

◆ Snap Ring

◆ Non-reusable part

300657

Axle shaft and related components — 1995–97

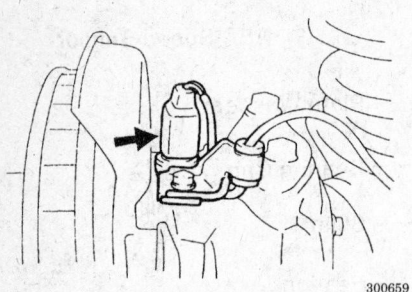

ABS speed sensor

300659

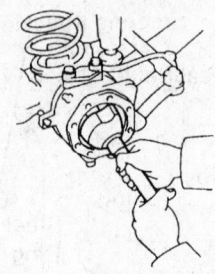

Removing the axle shaft

300663

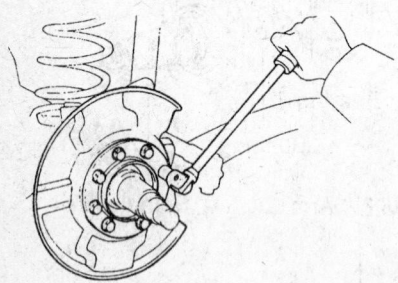

Removing the knuckle spindle mounting bolts

300661

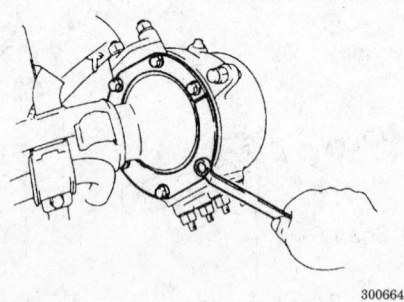

Removing the end retainer bolts to remove the oil seal set

300664

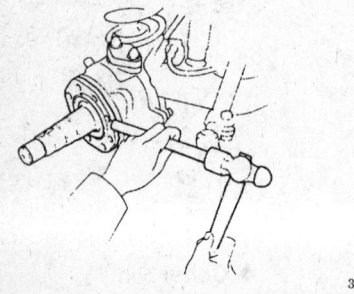

Removing the knuckle spindle with a brass bar

300662

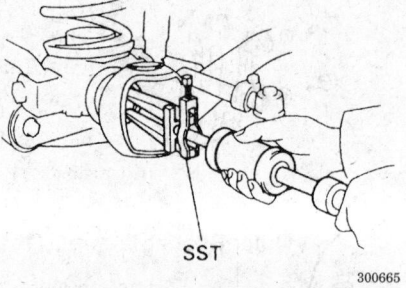

SST

Removing the axle housing oil seal

300665

b. Remove the steering wheel lower No. 2 and No. 3 cover.

c. Using a Torx® socket wrench, loosen the Torx® screws. Loosen them until the groove along the screw circumference catches on the screw case.

d. Pull out the wheel pad from the steering wheel and disconnect the air bag connector.

NOTE: When storing the wheel pad (vehicles equipped with air bag), keep the upper surface of the pad facing upward.

5. Disconnect the horn wire. Matchmark the wheel and the shaft.

6. Using an appropriate steering wheel puller tool No. 0960920011, or equivalent, remove the steering wheel.

To install:

7. Align the matchmarks on the steering wheel and main shaft and install the steering wheel to the shaft. Install and and torque the retaining nut to approximately 25 ft. lbs. (34 Nm).

8. If equipped, connector the air bag connector.

9. Connect the horn wire and install the steering wheel pad.

10. Connect the negative battery cable.

Steering Column

REMOVAL AND INSTALLATION

1993–94

1. Disconnect the negative battery cable.

2. Remove the instrument panel and steering column finish panels.

3. Disconnect all electrical connectors from the column.

4. Remove the lower joint protectors, if equipped.

5. Remove the pinch bolt and nut from the intermediate shaft or flex joint. Disconnect the intermediate shaft from the column shaft or disconnect the flex joint from the steering gear.

6. Remove the column bracket-to-instrument panel nuts.

7. Remove the floor boot retainers and pull the boot away from the floor.

8. Remove the column from the vehicle. Do not hammer on the shaft.

To install:

9. Install the column into the vehicle. Make sure the plastic retainers are properly aligned. Torque the column-to-instrument panel nuts to 18 ft. lbs. (25 Nm).

10. Install the floor boot and retainers. Make sure the boot is sealed from water. Use sealer between the floor and boot.

11. Connect the intermediate shaft to the column shaft or connect the flex joint to the steering gear. Install the pinch bolt and nut to the intermediate shaft or flex joint. Torque the bolt to 26 ft. lbs. (35 Nm).

12. Install the lower joint protectors, if equipped.

13. Connect all electrical connectors to the column.

14. Install the instrument panel and steering column finish panels.

15. Connect the negative battery cable and check operation.

1995–97

1. Disconnect the negative battery cable.

— **CAUTION** —

Work must be started after 90 seconds from the time the ignition switch is turned to the LOCK position and the negative (-) battery cable is disconnected.

2. Remove the steering wheel.

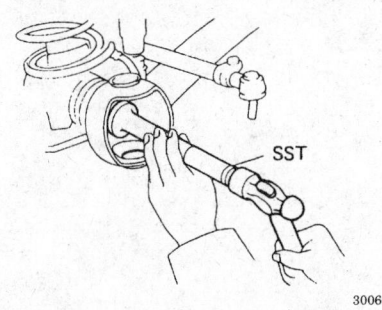

300666

Installing the axle housing oil seal

3. Remove the instrument lower finish panel. Disconnect the hood lock control cable and fuel lid control cable.

4. Remove the No. 2 heater register duct.

5. Remove the upper and lower column covers.

6. Remove the combination switch with the spiral cable. Disconnect the two connectors and the air bag connector.

7. Remove the spiral cable.

8. Remove the four column hole cover set bolts.

9. Remove the link joint protector.

10. Place matchmarks on the sliding yoke sub–assembly and the worm gear valve body shaft. Loosen bolt A and remove bolt B. Remove the yoke sub–assembly.

11. Remove the tilt steering column assembly.

To install:

12. Install the tilt steering column assembly. Torque the bolts to 18 ft. lbs. (25 Nm).

13. Align the matchmarks and connect the sliding yoke sub–assembly. Torque bolts A and B to 25 ft. lbs. (34 Nm).

14. Install the link joint protector. Torque the bolts to 9 ft. lbs. (12 Nm).

15. Torque the four column hole cover set bolts to 9 ft. lbs. (12 Nm).

16. Install the combination switch the with the spiral cable. Connect the air bag connector.

17. Install the upper and lower column covers.

18. Install the instrument lower finish panel. Connect the hood lock control cable and the fuel lid control cable.

19. Center the spiral cable.

a. Check that the front wheels are facing straight ahead.

b. Turn the cable counterclockwise by hand until it becomes harder to turn the cable.

c. Rotate the cable clockwise about three turns to align the red mark. The cable will rotate about three turns to either left or right of the center.

20. Install the steering wheel.

a. Align the matchmarks on the wheel and main shaft assembly.

b. Connect the connector and torque the wheel set nut to 25 ft. lbs. (34 Nm).

21. Connect the air bag connector.

22. Install the wheel pad after confirming that the circumference groove of the Torx screws is caught on the screw case.

23. Using a Torx socket wrench, torque the screws to 6.6 ft. lbs. (9 Nm).

NOTE: If the wheel pad has been dropped or there are other defects, replace the wheel pad with a new one. Make sure that the wires do not interfere or are pinched between other parts when installing.

24. Connect the negative battery cable.

25. Enable the air bag system.

26. Road test the vehicle to check that the steering wheel is straight.

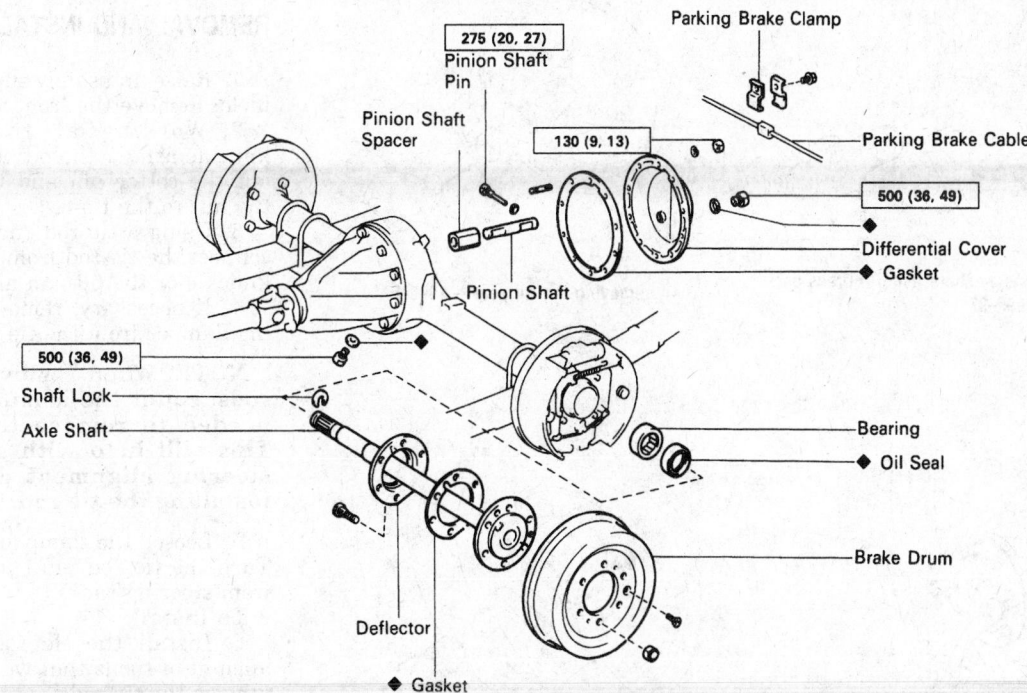

kg-cm (ft-lb, N·m) : Specified torque
◆ Non-reusable part

Removal and installation of the rear axle shaft components — 1993–94

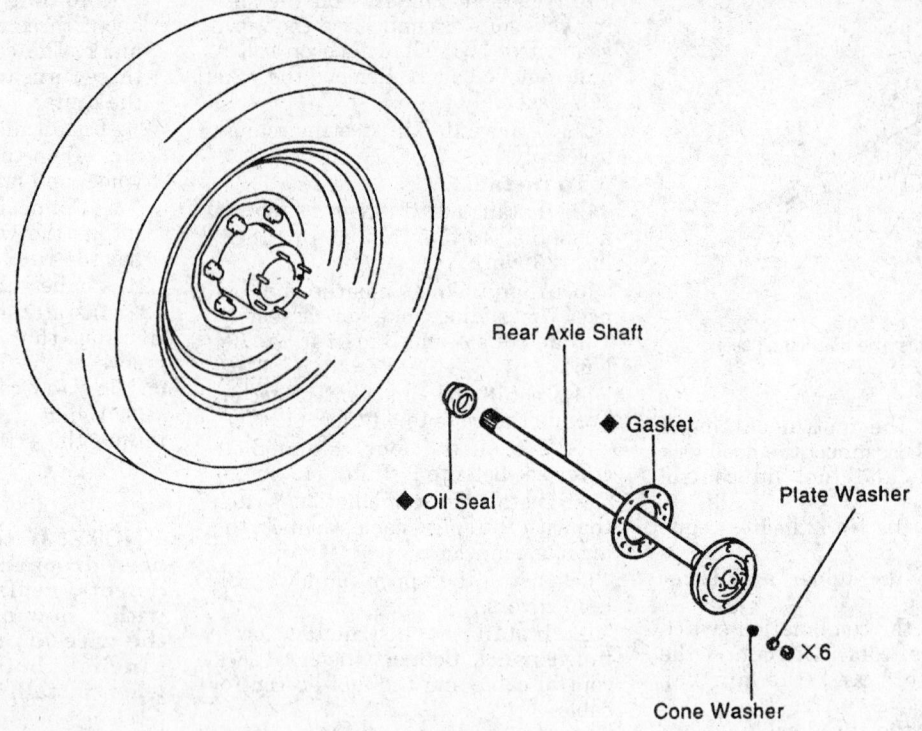

Rear Axle Shaft

◆ Gasket

◆ Oil Seal

Plate Washer

×6

Cone Washer

300684

◆ Non-reusable part

Rear axle and related components — 1995–97

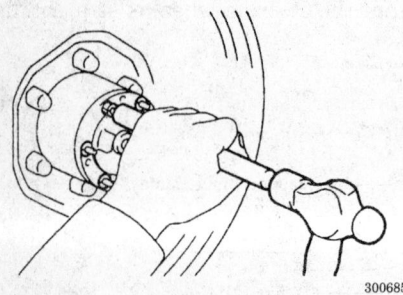

300685

Removing the cone washers using a brass bar and hammer — 1995–97

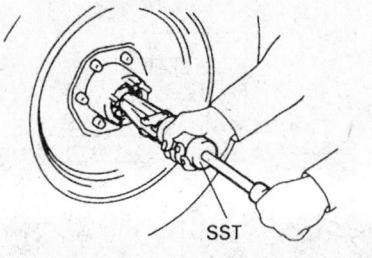

SST

300687

Removing the oil seal — 1995–97

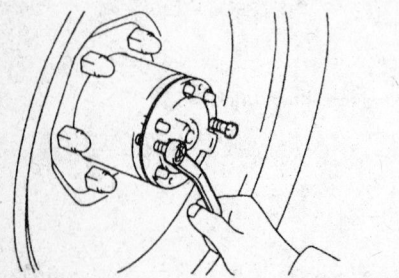

300686

Using bolts to remove the axle shaft — 1995–97

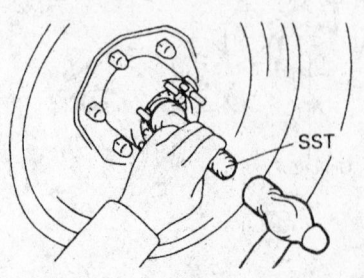

SST

300688

Installing the oil seal — 1995–97

Tie Rod Ends

REMOVAL AND INSTALLATION

1. Raise and safely support the vehicle. Remove the front wheels.

2. Working at the steering knuckle arm or the pitman arm, pull out the cotter pin and then remove the nut to the tie rod.

3. Using a tie rod end puller, disconnect the tie rod from the steering knuckle or the pitman arm.

4. If necessary, remove the steering damper from the tie rod end.

NOTE: When removing the tie rods, count the amount of turns needed to remove the tie rod. This will help with getting the steering alignment close when installing the tie rod.

5. Loosen the clamp for the tie rod. Turn the tie rod until it is removed from steering rack.

To install:

6. Install the tie rod the same amount of turns that were needed to remove the tie rod.

7. Tighten the tie rod end to steering knuckle nut to 67 ft. lbs. (91 Nm) or the tie rod to the pitman arm nut to 67 ft. lbs. (91 Nm). Install a new cotter pin.

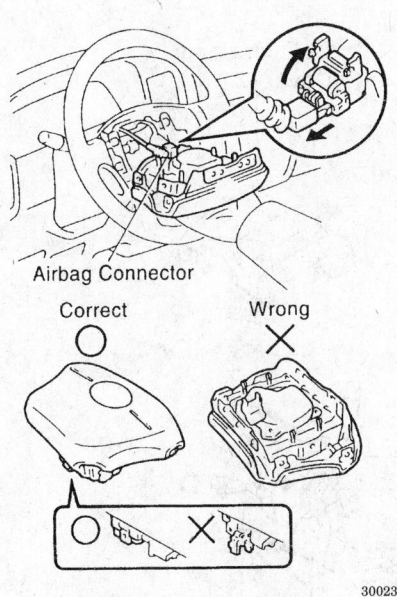

Airbag Connector

Correct   Wrong

Steering wheel with the air bag

300236

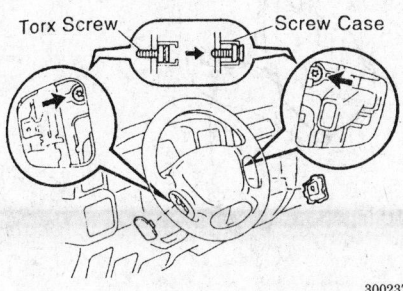

Torx Screw   Screw Case

Torx screws holding the wheel pad — 1995-97

300237

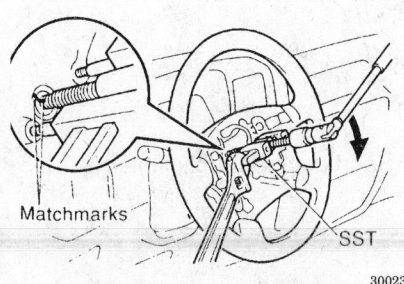

Matchmarks   SST

Steering wheel with a puller

300238

8. If removed, tighten the tie rod end to the steering damper nut to 54 ft. lbs. (74 Nm).

NOTE: If the hole, for the cotter pin, on the tie rod does not line up with the nut, always tighten the nut until the hole lines up.

9. Install the front wheels and lower the vehicle. Check the front end alignment.

10. After the alignment is complete, torque the tie rod clamp to 27 ft. lbs. (37 Nm).

11. Check steering wheel center point.

Power Steering Gear

REMOVAL AND INSTALLATION

1993-94

1. Raise and safely support the vehicle.

2. Disconnect the pressure lines from the steering gear.

3. Remove the intermediate shaft-to-steering gear bolt and the steering column-to-firewall bolts.

4. Loosen the steering column-to-dash bolts. Remove the pitman arm-to-steering gear nut.

5. Using a puller, separate the relay rod from the pitman shaft and the pitman arm from the steering gear.

6. Pull the steering column towards the passenger compartment to uncouple the steering shaft from the steering gear.

7. Remove the steering gear-to-frame bolts and the steering gear from the vehicle.

To install:

8. Place the steering gear into position and install the mounting bolts. Torque the steering gear-to-frame bolts to 40–63 ft. lbs. (54–86 Nm).

9. Connect the relay rod to the pitman shaft and the pitman arm to the steering gear. Torque the pitman arm-to-steering gear nut to 120–141 ft. lbs. (163–190 Nm), the intermediate shaft-to-steering gear bolt to 22–32 ft. lbs. (29–43 Nm).

10. Connect the pressure and return hoses. Torque the pressure hose fitting to 29–36 ft. lbs. (40–48 Nm) and the return hose fitting to 24–30 ft. lbs. (35–41 Nm).

NOTE: During installation of the hydraulic lines, position each line clear of any surrounding components, then tighten the fittings.

11. Fill and bleed the power steering system.

1995-97

1. Drain the power steering fluid.

2. Disconnect the negative battery cable.

———— CAUTION ————

Work must be started after 90 seconds from the time the ignition switch is turned to the LOCK position and the negative (-) battery cable is disconnected.

3. Raise and safely support the vehicle.

4. Remove the frame seal.

5. Remove the link joint protector.

6. Remove the pitman arm set nut and spring washer. Using tool No. SST 09628–62011, or equivalent, disconnect the pitman arm from the gear assembly.

7. Using tool No. SST 09631–22020, or equivalent, disconnect the pressure feed and the return tubes.

8. Disconnect the sliding yoke sub–assembly from the steering gear.

9. Remove the mounting bolts and the power steering gear assembly.

To install:

10. Install the power steering gear assembly and torque the bolts to 105 ft. lbs. (142 Nm).

11. Connect the sliding yoke sub–assembly to the steering gear.

12. Connect the pressure feed and the return tubes. Torque the tube fittings to 26 ft. lbs. (36 Nm).

13. Connect the pitman arm and align the alignment marks on the cross shaft and the pitman arm. Install the spring washer and the pitman arm set nut. Torque the nut to 130 ft. lbs. (177 Nm).

14. Refill and bleed the power steering system.

15. Install the link joint protector.

16. Install the frame seal.

17. Connect the negative battery cable.

18. Test drive the vehicle. Check the steering wheel center point.

Power Steering Pump

BLEEDING

1. Raise and safely support the vehicle.

2. Fill the pump reservoir with power steering fluid.

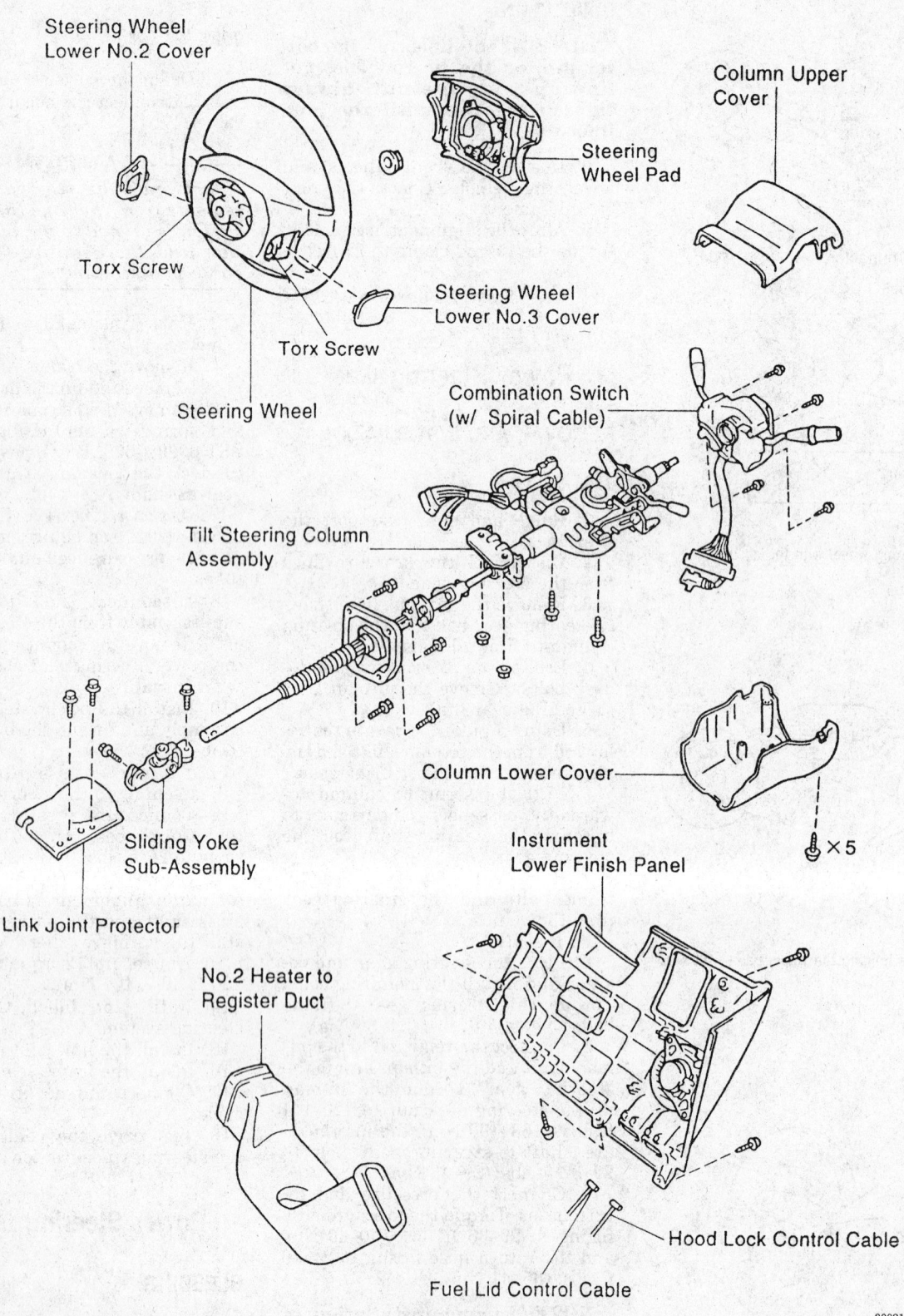

Steering Wheel Lower No.2 Cover

Steering Wheel Pad

Column Upper Cover

Torx Screw

Steering Wheel Lower No.3 Cover

Torx Screw

Steering Wheel

Combination Switch (w/ Spiral Cable)

Tilt Steering Column Assembly

Column Lower Cover

Sliding Yoke Sub-Assembly

Instrument Lower Finish Panel

× 5

Link Joint Protector

No.2 Heater to Register Duct

Hood Lock Control Cable

Fuel Lid Control Cable

300214

Steering column and related components

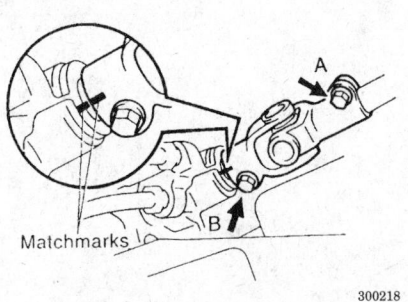

Matchmarks

300218

Sliding yoke sub assembly

3. With the engine running, rotate the steering wheel from lock to lock several times. Add fluid as necessary.

NOTE: Perform the bleeding procedure until all of the air is bled from the system.

4. The fluid should not have a lot of bubbles and the steering wheel should turn smoothly from lock to lock after the air is bleed from the system.

5. The fluid level should not have risen more than 0.20 inch (5mm); if it does, check the pump.

REMOVAL AND INSTALLATION

1993–94

NOTE: Disconnect the air hoses from the air control valve and the high tension wires from the distributor.

1. Disconnect the negative battery cable. Loosen the power steering pump pulley nut.

NOTE: Use the drive belt as a brake to keep the pulley from rotating.

2. Place a container under the pump. Disconnect the return line and the pressure tube, then drain the fluid into the container.

3. Loosen the idler pulley nut and the adjusting bolt, then remove the drive belt.

4. Remove the drive pulley and the Woodruff key from the pump shaft.

5. Remove the mounting bolts and the power steering pump from the vehicle.

To install:

6. Place the power steering pump into position and install the mounting bolts. Torque the pump pulley mounting bolt to 29 ft. lbs. (40 Nm), the pump pulley nut to 32 ft. lbs. (42 Nm) and the pressure hoses to 33 ft. lbs. (45 Nm).

7. Install the drive belt and adjust the belt tension.

8. Bleed the power steering system.

1995–97

1. Disconnect the negative battery cable.

─────── **CAUTION** ───────

Work must be started after 90 seconds from the time the ignition switch is turned to the LOCK position and the negative (-) battery cable is disconnected.

2. Place a container under the pump. Disconnect the return line and the pressure feed tube, then drain the fluid into the container.

3. Remove the power steering pump mounting nuts and remove the pump from the vehicle.

To install:

4. Place the power steering pump into position. Torque the nuts to 27 ft. lbs. (36 Nm).

5. Connect the pressure feed tube, make sure that the stopper of the tube is touching the pump body, then torque the union bolt to 42 ft. lbs. (56 Nm).

6. Connect the return hose.

7. Connect the negative battery cable.

8. Fill and bleed the power steering system.

─────────────────────

BRAKES

Anti-Lock Brake System Service

PRECAUTIONS

• Certain components within the Anti-Lock Brake System (ABS) are not intended to be serviced or repaired individually. Only those components with removal and installation procedures should be serviced.

• Do not use rubber hoses or other parts not specifically specified for and ABS system. When using repair kits, replace all parts included in the kit. Partial or incorrect repair may lead to functional problems and require the replacement of components.

• Lubricate rubber parts with clean, fresh brake fluid to ease assembly. Do not use lubricated shop

air to clean parts; damage to rubber components may result.

• Use only specified brake fluid from an unopened container.

• If any hydraulic component or line is removed or replaced, it may be necessary to bleed the entire system.

• A clean repair area is essential. Always clean the reservoir and cap thoroughly before removing the cap. The slightest amount of dirt in the fluid may plug an orifice and impair the system function. Perform repairs after components have been thoroughly cleaned; use only denatured alcohol to clean components. Do not allow ABS components to come into contact with any substance containing mineral oil; this includes used shop rags.

• The Anti-Lock control unit is a microprocessor similar to other computer units in the vehicle. Ensure that the ignition switch is **OFF** before removing or installing controller harnesses. Avoid static electricity discharge at or near the controller.

• If any arc welding is to be done on the vehicle, the control unit should be unplugged before welding operations begin.

Master Cylinder

REMOVAL AND INSTALLATION

1. Disconnect the negative battery cable from the battery.

─────── **CAUTION** ───────

Wait at least 90 seconds from the time the ignition switch is turned to the LOCK position and the negative (-) battery cable is disconnected before starting work. This will allow the SRS system capacitor sufficient time to discharge.

─────────────────────

2. Disconnect the level warning switch connector.

3. Take out the fluid from the master cylinder with a syringe.

4. Using a line wrench, disconnect the brake lines from the master cylinder.

5. Remove the four nuts holding the master cylinder to the brake booster.

6. Remove the master cylinder, clamp and the gasket from the brake booster.

To install:

7. Adjust the length of the brake booster pushrod as follows:

a. Install a new gasket on the master cylinder.

b. Install tool No. SST 09737–00010, or equivalent, on the

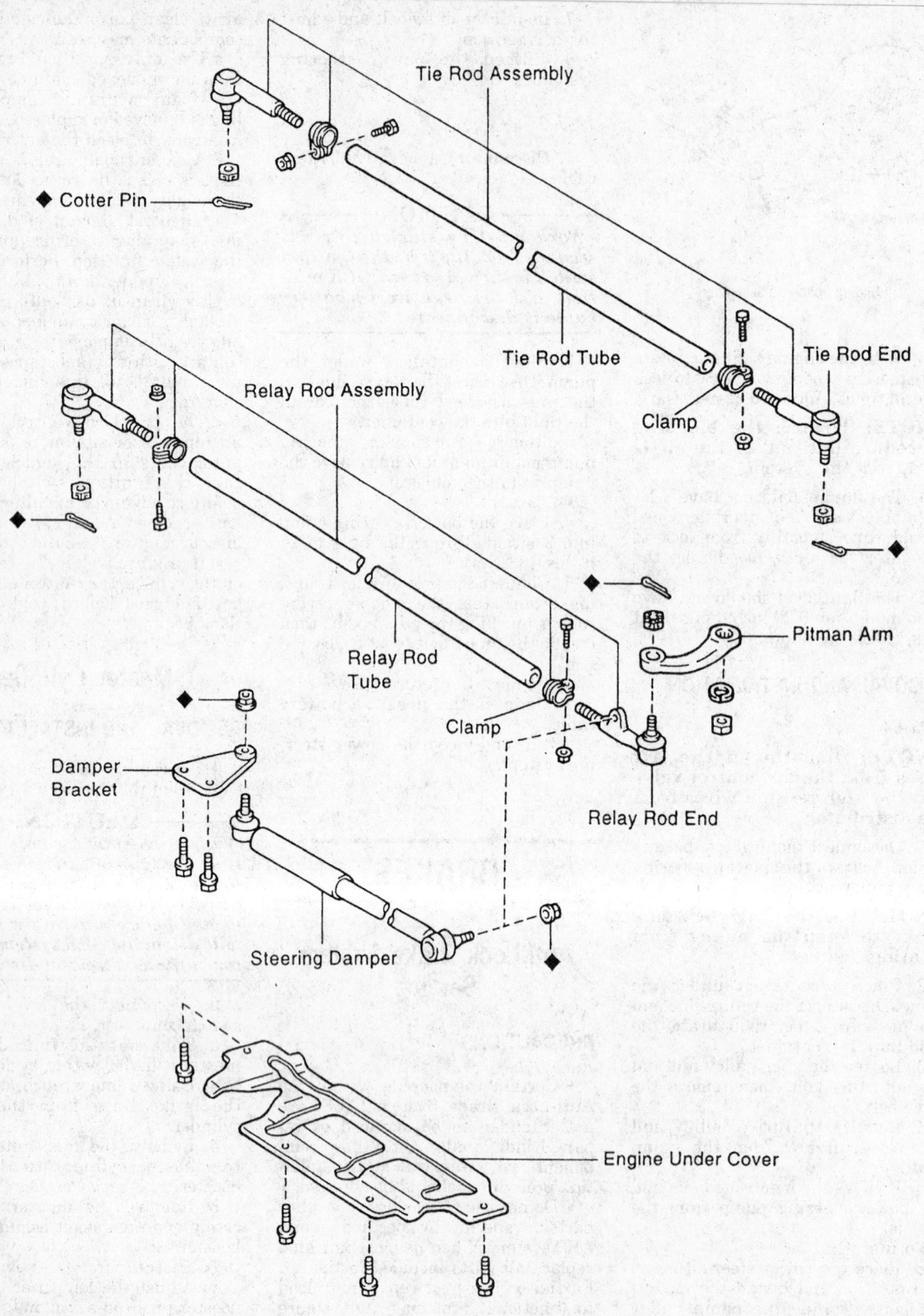

Tie Rod Assembly

◆ Cotter Pin

Tie Rod Tube

Tie Rod End

Clamp

Relay Rod Assembly

◆

Pitman Arm

Relay Rod Tube

Clamp

Damper Bracket

◆

Relay Rod End

Steering Damper

◆

Engine Under Cover

◆ Non-reusable part

333499

Steering linkage components

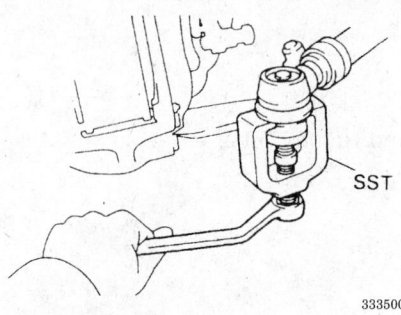

Disconnecting the tie rod with a ball joint remover

master cylinder gasket and lower the pin until its tip slightly touches the piston.

c. Turn the tool upside down and position it on the brake booster.

d. Measure the clearance between the booster pushrod and the pin head on the SST tool.

e. Clearance should be 0 inch (0 mm). If not, adjust the booster pushrod length until the pushrod lightly touches the pin head.

NOTE: When adjusting the pushrod, depress the brake pedal so that the pushrod sticks out.

8. Bench bleed the master cylinder.

9. Remove the SST tool and install the master cylinder to the brake booster using a new gasket.

10. Install the master cylinder four nuts and torque the nuts to 9 ft. lbs. (13 Nm).

11. Using a line wrench, connect the two brake lines to the master cylinder. Torque the union nuts to 11 ft. lbs. (15 Nm).

12. Connect the level warning switch connector.

13. Fill the brake reservoir with brake fluid and bleed the brake system.

14. Check and adjust brake pedal.

15. Connect the negative battery cable to the battery.

Brake Caliper

REMOVAL AND INSTALLATION

1. Disconnect the negative battery cable from the battery.

2. Raise and support the vehicle safely.

3. Remove the wheels.

4. Disconnect the brake hose from the caliper by removing the union bolt and two gaskets. Plug the end of the hose to prevent loss of fluid.

5. Remove the bolts that attach the caliper to the torque plate.

6. Lift the bottom of the caliper up and remove the caliper assembly.

To install:

7. Grease the caliper slides and bolts with lithium grease or equivalent. Install the caliper and secure with the bolts. Torque the bolts to 90 ft. lbs. (123 Nm).

8. Connect the brake hose to the caliper, using two new washers. Make sure the flexible hose lock is securely in the lock hole of the caliper. Torque the union bolt to 22 ft. lbs. (30 Nm).

9. Fill the brake system to the proper level and bleed the brake system.

10. Install the tire and wheel assembly.

11. Top off the brake fluid level in the master cylinder. Check for leaks and proper brake operation.

12. Connect the negative battery cable to the battery.

Disc Brake Pads

REMOVAL AND INSTALLATION

Front

1. Raise the vehicle and support it safely.

2. Remove the wheels.

3. Remove the clip, pins and anti-rattle spring.

4. Withdraw the pads and remove the anti-squeal shims.

To install:

5. Before installing the new pads, check the disc thickness and disc runout.

6. Siphon out a small amount of brake fluid from the reservoir.

7. Press in the pistons with a hammer handle or equivalent.

8. Apply disc brake grease to both sides of the inner anti-squeal shim. Install the anti-squeal shims to the new pads.

9. Install the pads.

10. Install the anti-rattle springs and pins. Install the clip.

11. Install the wheels.

12. Check and adjust the fluid level. Apply the brake pedal several times.

13. Road test the vehicle for proper operation.

Rear

1. Raise the vehicle and support it safely.

2. Remove the wheels.

3. Remove the brake caliper and suspend it so the hose is not stretched.

NOTE: Do not disconnect the brake hose.

4. Remove the brake pads, anti-squeal shim, pad support plates and wear indicators.

To install:

5. Before installing the new pads, check the disc thickness and disc runout.

6. Install the pad support plates.

7. Install the pad wear indicator plate to each pads.

8. Install the anti-squeal shim to the outer pad. Install the pads so the wear indicator plate is facing upward.

9. Install the brake caliper.

10. Install the wheels.

11. Apply the brake pedal several times.

12. Road test the vehicle for proper operation.

Brake Rotor

REMOVAL AND INSTALLATION

1993–94

1. Raise the vehicle and support it safely.

2. Remove the wheels.

3. Remove the brake caliper.

4. On 2WD vehicles, remove the grease cap, cotter pin and nut from the hub. Remove the rotor from the vehicle.

5. On 4WD vehicles, remove the locknut and washer from the free wheeling hub. Remove the adjusting nut and thrust washer. Remove the hub with disc together with the outer bearing.

To install:

6. Before installing the rotor, thoroughly clean and repack the wheel bearings, using MP grease. Coat inside the hub and cap with MP grease.

7. Install a new bearing seal. Coat the seal lip with MP grease.

8. On 2WD vehicles, install the rotor, outer bearing and thrust washer. Adjust the bearing preload.

9. On 4WD vehicles, install the free wheeling hub locknut and washer. Adjust the preload and check that the bearing has no play. Secure the locknut by bending one of the lock washer teeth inward and the other lock washer teeth outward.

10. On 2WD vehicles, install the locknut, cotter pin and hub grease cap.

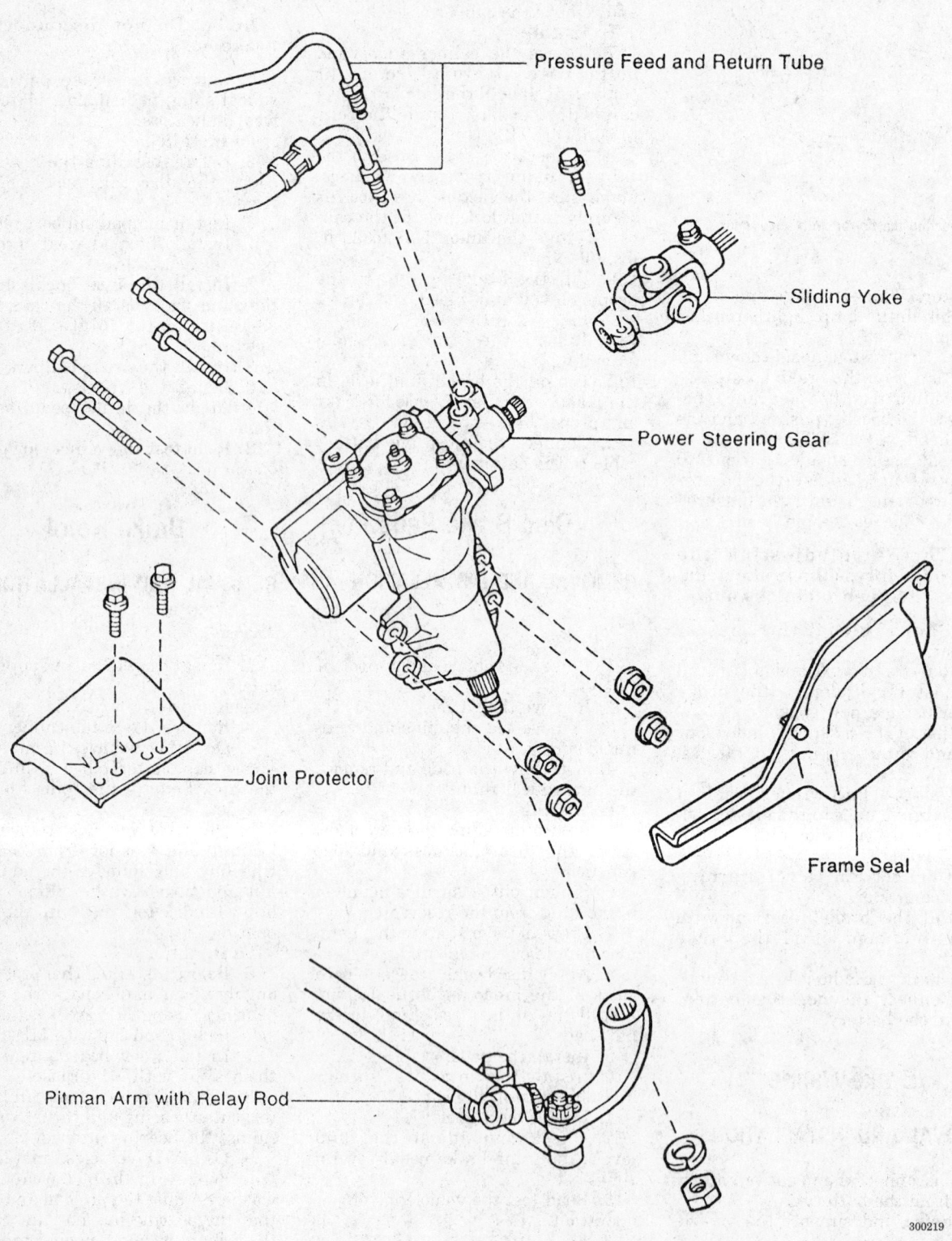

Power steering gear and related components

Pressure Feed and Return Tube

Sliding Yoke

Power Steering Gear

Joint Protector

Frame Seal

Pitman Arm with Relay Rod

300219

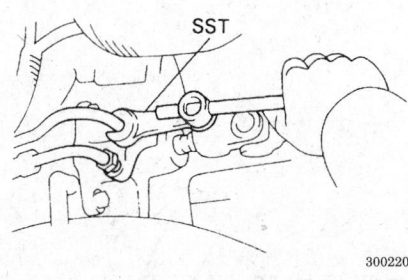

300220

Pressure feed and return tubes

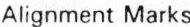

Alignment Marks

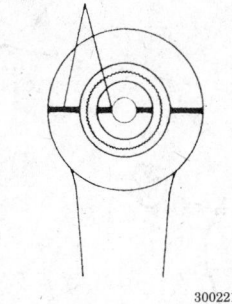

300221

Pitman arm alignment marks

11. On 4WD vehicles, install the free wheeling hub or automatic locking hub.

12. Install the brake caliper.

13. Install the wheels.

14. Check and adjust the brake fluid level.

15. Road test the vehicle for proper operation.

1995–97

1. Disconnect the negative battery cable from the battery.

2. Loosen the wheel lugs slightly, then raise and safely support the vehicle.

3. Remove the wheel(s) and temporarily install two of the wheel lug nuts.

4. Remove the two mounting bolts holding the caliper.

5. Remove the caliper and hang the caliper from a piece of wire. Do not disconnect the brake hose.

――――― **WARNING** ―――――
Do not allow the caliper to hang freely from the vehicle. Always support the caliper with a wire from the vehicle.

6. Remove the flange.

a. Remove the grease cap from the flange.

b. Using a snapring expander, remove the snapring.

c. Loosen the six mounting nuts. Using a hammer and a brass bar, rap on the bolt heads a remove the six cone washers, plate washers and nuts.

d. Remove the flange and the gasket.

7. Remove the axle hub with the disc.

a. Using a flat prying tool, release the lock washer.

b. Using tool No. SST 09607–60020, or equivalent, remove the locknut, lock washer, the adjusting nut and the thrust washer.

c. Remove the hub and disc together with the outer bearing.

To install:

8. Install the hub and the disc together with the outer bearing.

9. Adjust the preload.

a. Using tool No. SST 09607–60020, or equivalent, torque the adjusting nut to 43 ft. lbs. (59 Nm).

b. Turn the hub right and left two or three times and re–torque to 43 ft. lbs. (59 Nm).

c. Loosen the nut until it can be turned by hand. Torque the adjusting nut again to 4 ft.lbs. (5.4 Nm).

d. Using a spring tension gauge, measure the preload. It should read 6.4–12.6 lbs. (2.9–5.7 kg).

10. Install the lock washer and the locknut. Torque the locknut to 47 ft. lbs. (64 Nm). Secure the locknut by bending one of the lock washer teeth inward and the outer lock washer teeth outward.

11. Install the flange.

a. Place a new gasket on the axle hub and apply grease to the inner flange splines. Install the flange to the axle hub.

b. Install the six cone washer, plate washers and nuts.

c. Torque the six nuts to 26 ft. lbs. (35 Nm).

d. Install the bolt in the axle shaft and pull it out. Using a snapring expander, install a new snapring.

e. Remove the bolt and install the cap to the flange.

12. Install the brake caliper.

13. Install the wheels. Secure the wheel lugs.

14. Lower the vehicle and tighten the lug nuts. Before moving the vehicle, make sure to pump the brake pedal to seat the brake pads against the rotors.

15. Connect the negative battery cable to the battery.

Parking Brake Cable

ADJUSTMENT

Pull the parking brake lever all the way up and count the number of clicks. The parking brake lever should travel approximately 7–9 clicks at 44 lbs. (20 kg) of force.

NOTE: Before adjusting the parking brake, make sure that the rear brake shoe clearance has been adjusted.

1. Remove the rear console box.

2. Loosen the locknut and turn the adjusting nut (at the parking brake lever) until the lever travel is correct.

3. Apply the parking brake several times and again check that there is no drag with the brake released.

4. Tighten the locknut to 4.9 ft. lbs. (5.4 Nm).

5. Install the rear console box.

Brake System Bleeding

PROCEDURE

Start the bleeding procedure at the caliper or wheel cylinder the furthest from the master cylinder.

1. Connect a vinyl tube to the bleeder screw on the brake cylinder and submerge the other end of the tube in a transparent container half filled with clean brake fluid.

2. Pump the brake pedal several times and loosen the bleeder screw with the pedal held down.

3. When brake fluid stops coming out of the tube with the brake pedal held to the floor, tighten the bleeder screw and release the brake pedal.

4. Repeat Steps 2 and 3 until no air bubbles can be seen in the container.

5. Repeat the procedure for each wheel.

6. Check the level in the master cylinder. Add fluid as necessary.

Wheel Speed Sensor

REMOVAL AND INSTALLATION

Front

1. Disconnect the negative battery cable.

2. Raise and safely support the vehicle.

3. Disconnect the speed sensor connector.

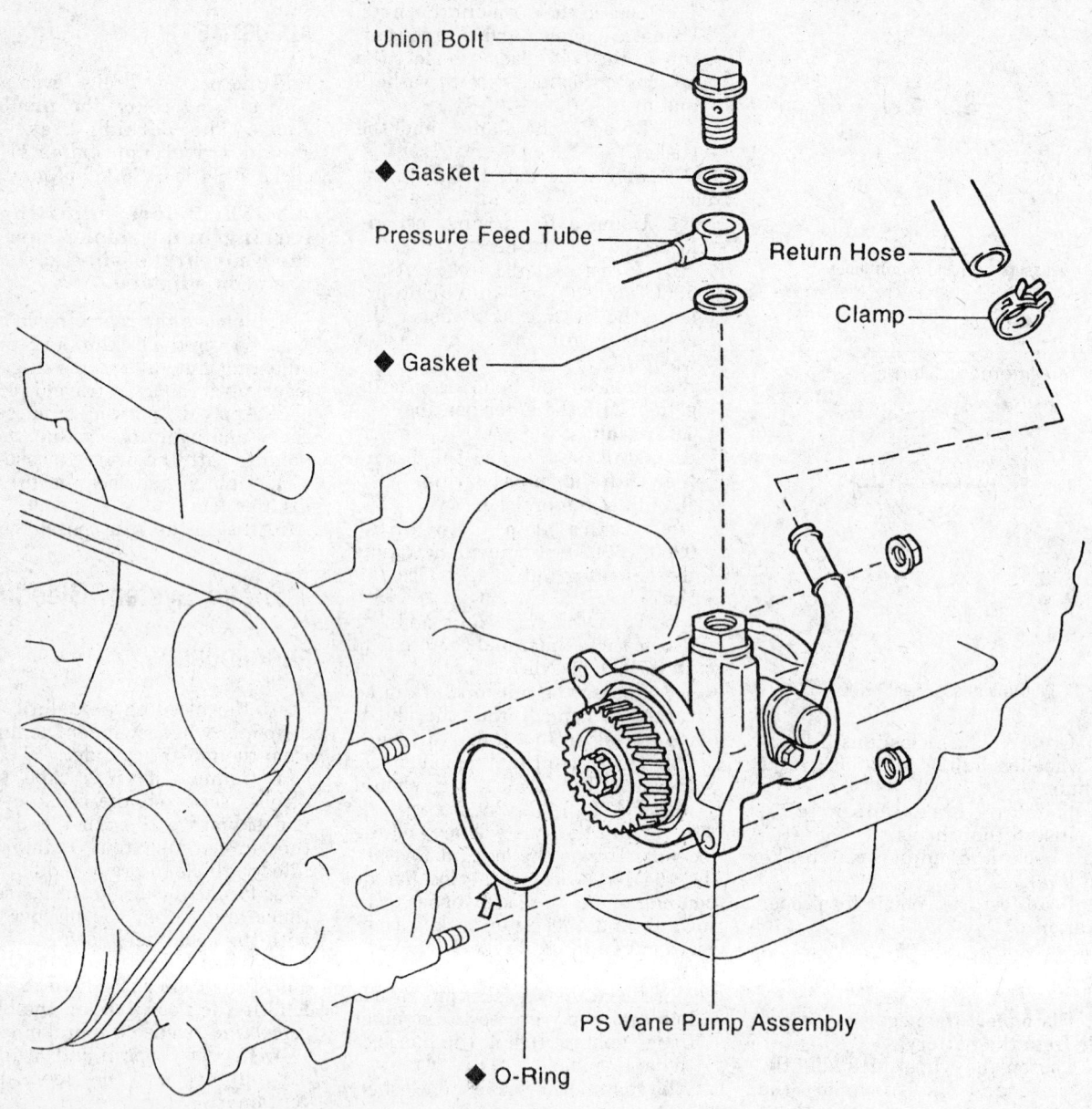

Union Bolt

◆ Gasket

Pressure Feed Tube

Return Hose

Clamp

◆ Gasket

PS Vane Pump Assembly

◆ O-Ring

◆ Non-reusable part
⇦ Power steering fluid

300230

Power steering pump

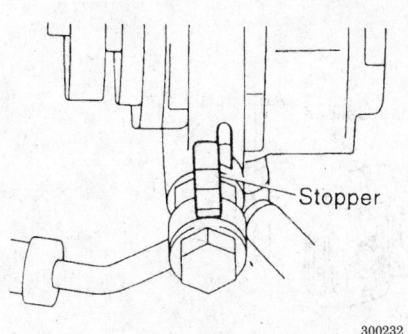

Pressure feed tube connected to the pump

4. Remove the eight clamp bolts and two clips holding the sensor harness to the axle housing.

5. Remove the bolts holding the speed sensor to the steering knuckle.

Remove the speed sensor from the vehicle.

To install:

6. Install the speed sensor to the steering knuckle and install the bolts. Torque the bolt to 13 ft. lbs. (18 Nm).

7. Torque the clamp bolts and the clips holding the sensor harness to the axle housing.

8. Connect the speed sensor connector.

9. Lower the vehicle.

10. Connect the negative battery cable.

11. After installation, check the speed sensor signal.

Rear

1. Disconnect the negative battery cable.

2. Raise and safely support the vehicle.

3. Disconnect the parking brake cable from the lower control arm.

4. Remove the three clamp bolts and two clips holding the sensor wire harness to the suspension arm and the frame.

5. Remove the lockbolt from the axle carrier and remove the speed sensor.

To install:

6. Install the speed sensor and tighten the lockbolt to 13 ft. lbs. (18 Nm).

7. Install the three clamp bolts to hold the sensor wire harness to the suspension arm and the frame.

8. Lower the vehicle.

9. Connect the negative battery cable.

10. After installation, check the speed sensor signal.

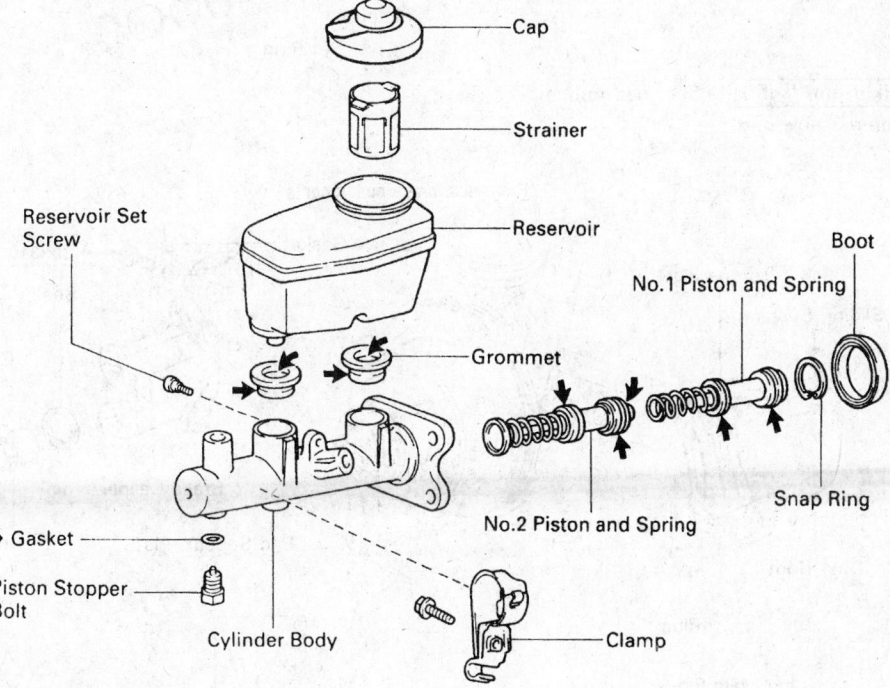

◆ Non-reusable part
← Lithium soap base glycol grease

Master cylinder component assembly

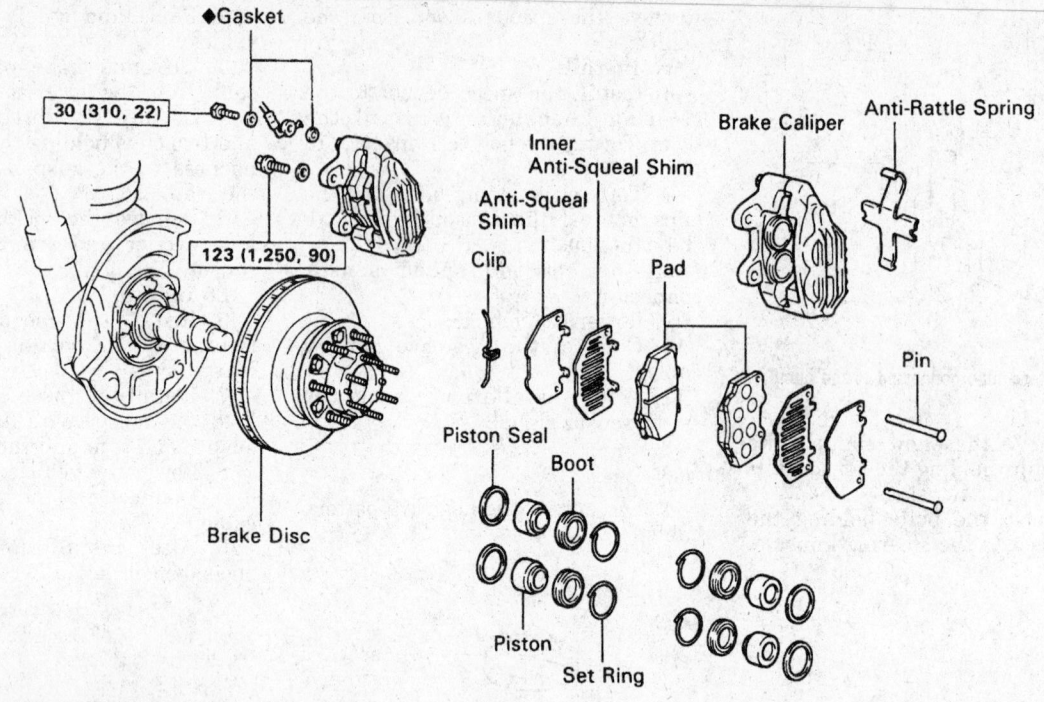

Front disc brake components

◆ Gasket

30 (310, 22)

123 (1,250, 90)

Brake Disc

Inner Anti-Squeal Shim

Anti-Squeal Shim

Clip

Brake Caliper

Anti-Rattle Spring

Pad

Pin

Piston Seal

Boot

Piston

Set Ring

N·m (kgf·cm, ft·lbf) : Specified torque

◆ Non-reusable part

100817

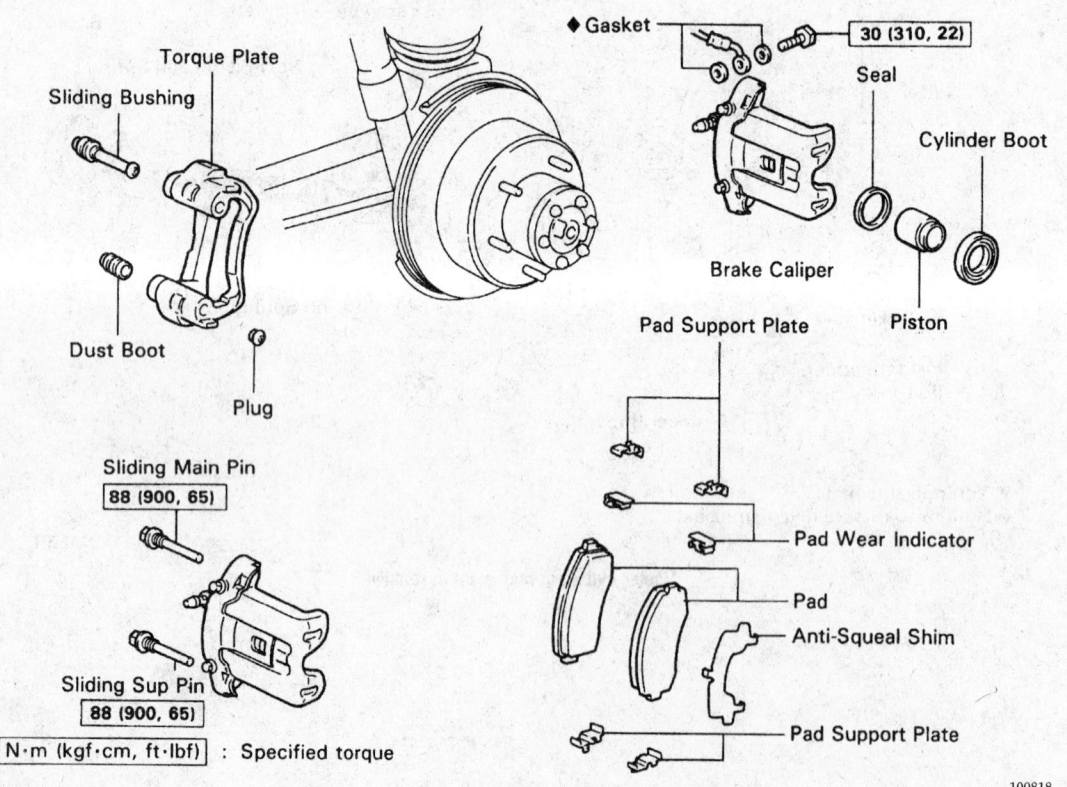

Rear disc brake components

◆ Gasket

30 (310, 22)

Torque Plate

Sliding Bushing

Seal

Cylinder Boot

Dust Boot

Plug

Brake Caliper

Piston

Pad Support Plate

Sliding Main Pin

88 (900, 65)

Pad Wear Indicator

Pad

Anti-Squeal Shim

Sliding Sup Pin

88 (900, 65)

Pad Support Plate

N·m (kgf·cm, ft·lbf) : Specified torque

100818

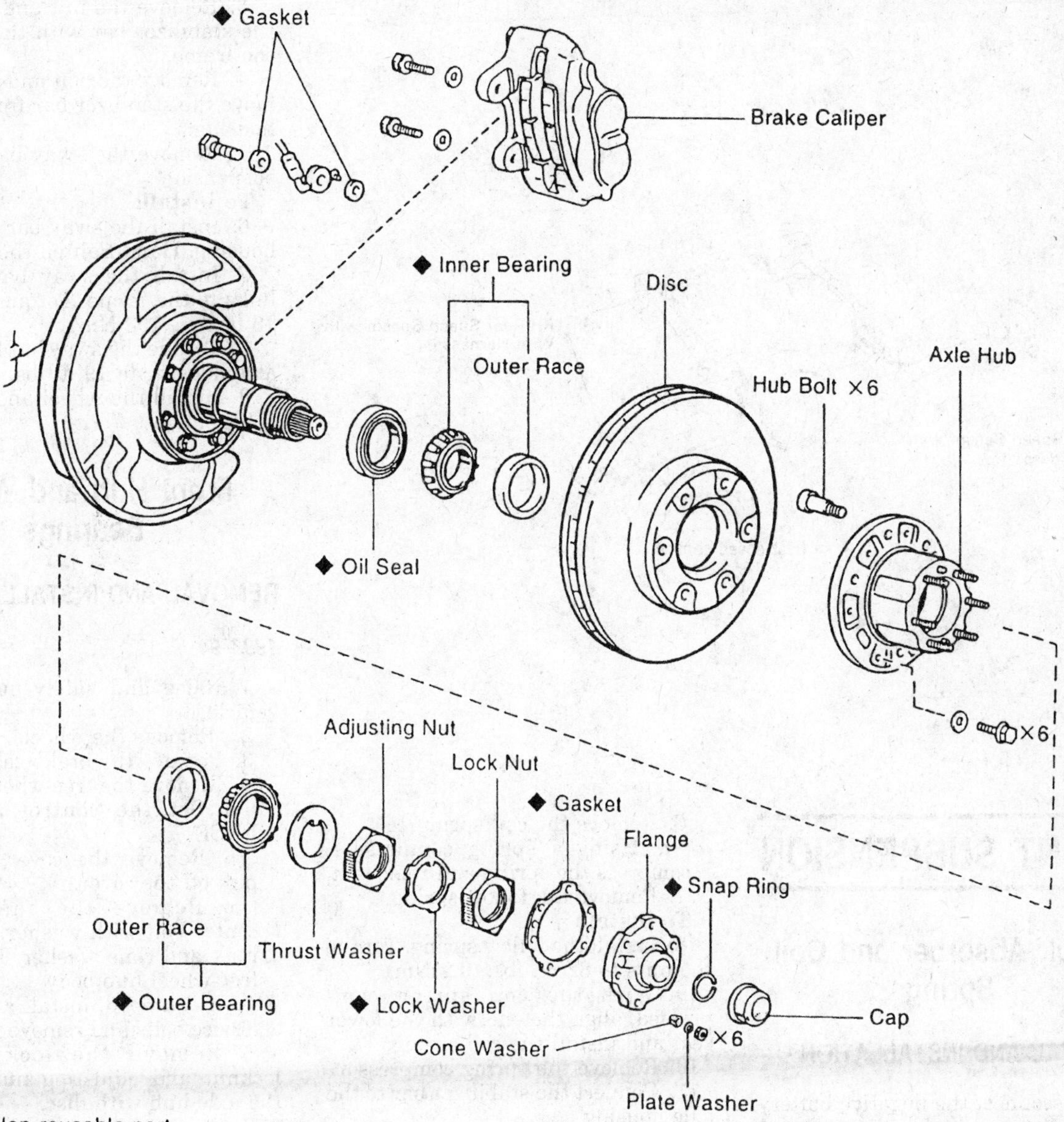

◆ Gasket

Brake Caliper

◆ Inner Bearing

Outer Race

Disc

Hub Bolt ×6

Axle Hub

◆ Oil Seal

◎ ⟿ ×6

Adjusting Nut

Lock Nut

◆ Gasket

Flange

◆ Snap Ring

Outer Race

Thrust Washer

◆ Outer Bearing   ◆ Lock Washer

Cap

Cone Washer

◎⟿ ×6

Plate Washer

◆ Non-reusable part

300455

Front brake rotor removal

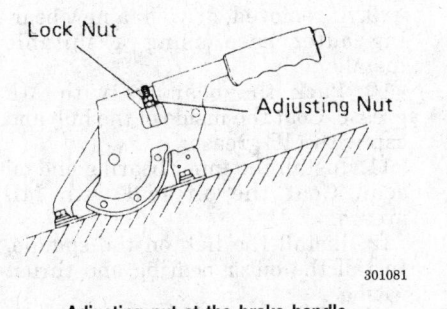

Lock Nut

Adjusting Nut

301081

Adjusting nut at the brake handle

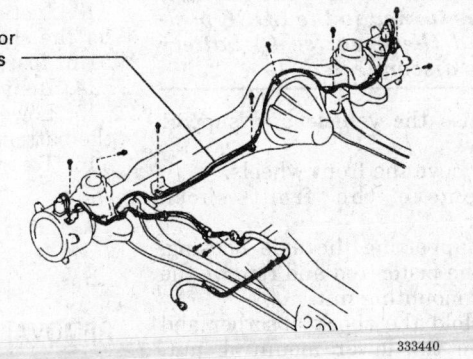

Front Speed Sensor
with Wire Harness

333440

Front speed sensor

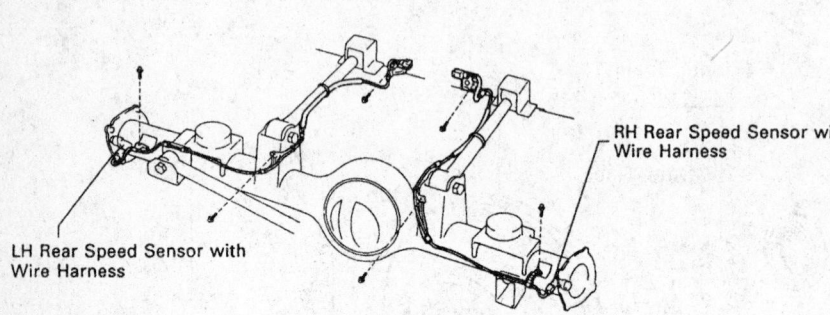

LH Rear Speed Sensor with
Wire Harness

RH Rear Speed Sensor with
Wire Harness

333441

Rear speed sensor

3. Remove the nut and disconnect the stabilizer bar with the link from the frame.

4. Remove the bolt and nut and remove the stabilizer bar from the axle housing.

5. Remove the sway bar from the suspension.

To install:

6. Install the sway bar to the axle housing. Hand tighten the bolts.

7. Install the sway bar with the links to the frame. Torque the nut to 76 ft. lbs. (103 Nm).

8. Torque the sway bar bolts to the axle housing to 19 ft. lbs. (25 Nm).

9. Install the wheel and lower the vehicle.

Front Hub and Wheel Bearings

REMOVAL AND INSTALLATION

1993–94

1. Raise and safely support the vehicle.

2. Remove the wheels.

3. Remove the brake caliper.

4. Remove the free-wheel hub:

a. Set the control handle to FREE.

b. Remove the cover bolts and pull off the cover.

c. Remove the snapring or center bolt with washer, mounting nuts and cone washer. Pull of the free-wheel hub body.

5. If equipped, install and tighten 2 service bolts and remove the flange.

6. Remove the lock washer, locknut and adjusting nut. Remove the axle hub with disc.

7. Remove the oil seal and inner bearing, using a suitable puller.

8. If replacing bearing outer race, drive out the outer bearing race using a brass bar and hammer.

To install:

9. If removed, drive in a new bearing outer race using a suitable installer.

10. Pack the bearings with MP grease. Coat the inside of the hub and cap with MP grease.

11. Install the inner bearing and oil seal. Coat the oil seal with MP grease.

12. Install the hub on the spindle. Install the outer bearing and thrust washer.

13. Adjust the preload:

a. Torque the bearing adjusting nut to 43 ft. lbs. (59 Nm).

b. Turn the hub right and left 2 or 3 times and re-torque.

FRONT SUSPENSION

Shock Absorber and Coil Spring

REMOVAL AND INSTALLATION

1. Disconnect the negative battery cable.

———— **CAUTION** ————

Work must be started after 90 seconds from the time the ignition switch is turned to the LOCK position and the negative (-) battery cable is disconnected.

2. Raise the vehicle and support the body.

3. Remove the front wheels.

4. Remove the front shock absorber.

a. Supporting the axle housing, hold the piston rod and remove the upper mounting nut.

b. Hold the shock absorber and remove the lower mounting nut, shock absorber, cushions and the retainers.

5. Disconnect the stabilizer bar from the axle housing.

6. Remove the coil spring.

a. Using a spring compressor, compress the spring and remove it.

7. Remove the follow spring.

To install:

8. Install the follow spring. Torque the nuts to 82 in. lbs. (9.2 Nm).

9. With the coil spring compressed, align the end with the lower seat and install the coil spring.

10. Remove the spring compressor.

11. Connect the stabilizer bar to the axle housing.

12. Install the retainers, cushions, shock absorber and the lower mounting nut. Torque the nut to 51 ft. lbs. (69 Nm).

13. Install the upper mounting nut for the shock absorber and torque to 51 ft. lbs. (69 Nm).

14. Install the wheels.

15. Lower the vehicle and connect the battery cable.

16. Test drive the vehicle.

Sway Bar

REMOVAL AND INSTALLATION

1. Raise and safely support the vehicle.

2. Remove the wheels.

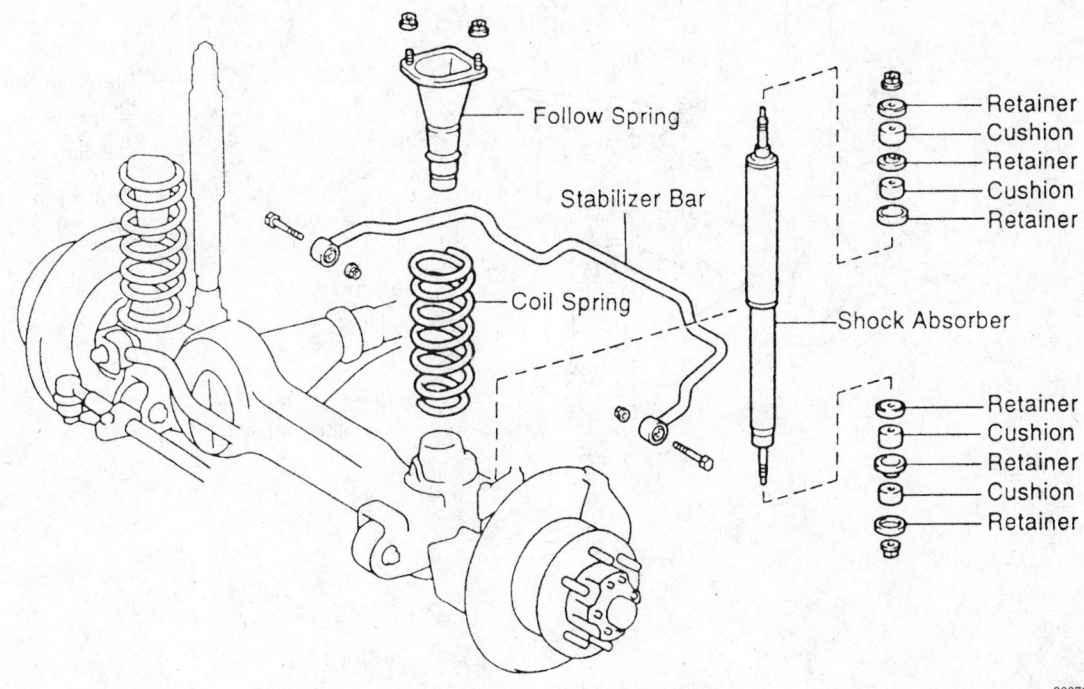

Coil spring and shock absorber components

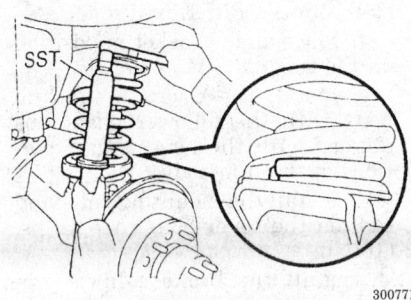

Removing the coil spring

c. Loosen the nut until it can be turned by hand.

d. Re-torque to 48 inch ft. lbs. (5.4 Nm).

14. Install the lock washer and nut. Torque to 65 ft. lbs. (88 Nm) Land Cruiser. Secure the locknut by bending 1 of the lock washer tooth inward and another outward.

15. Install the free-wheel hub body, using a new gasket. If equipped, install the flange with a new gasket. Torque the nuts to 23 ft. lbs. (31 Nm).

16. Install the snapring or center bolt with washer. Apply MP grease to the inner hub splines.

17. Install the free-wheel hub cover with a new gasket.

18. Install the cover to the body with the follower pawl tabs aligned with the no-toothed portions of the body. Torque the cover mounting bolts to 7 ft. lbs. (10 Nm).

19. Bleed the brake system, as required.

1995–97

1. Raise and safely support the vehicle.

2. Remove the wheels.

3. If equipped with ABS brakes, disconnect the ABS speed sensor from the steering knuckle.

4. Remove the two brake caliper support bracket bolts and wire the caliper to the side. Do not allow the caliper to hang from the brake hose.

5. Remove the free-wheel hub:

a. Set the control handle to FREE.

b. Remove the cover bolts and pull off the cover.

c. Remove the center bolt with washer.

d. Remove the mounting nuts and washer to the hub body.

e. Using a brass bar and hammer, tap on the bolts head and remove the cone washer.

f. Pull off the free wheel hub body and gasket.

6. If equipped without a free wheeling hub, remove the flange for the axle hub as follows:

a. Remove the grease cap from the flange.

b. Remove the bolt from the flange.

c. Remove the six mounting nuts to the flange.

d. Using a brass bar and hammer, tap on the bolts head and remove the six cone washers.

e. Install two bolts to the flange. Tighten the bolts to remove the flange.

f. Remove the gasket for the flange.

7. Using a screwdriver, release the taps on the lockwasher.

8. Using tool No. SST 09607–60020 or equivalent, remove the locknut.

9. Remove the lock washer and adjusting nut.

10. Remove the claw washer.

11. Remove the axle hub and rotor as an assembly. Remove the outer bearing with the hub and disc.

12. Remove the oil seal and inner bearing, using a suitable puller.

13. If replacing bearing outer race, drive out the outer bearing race using a brass bar and hammer.

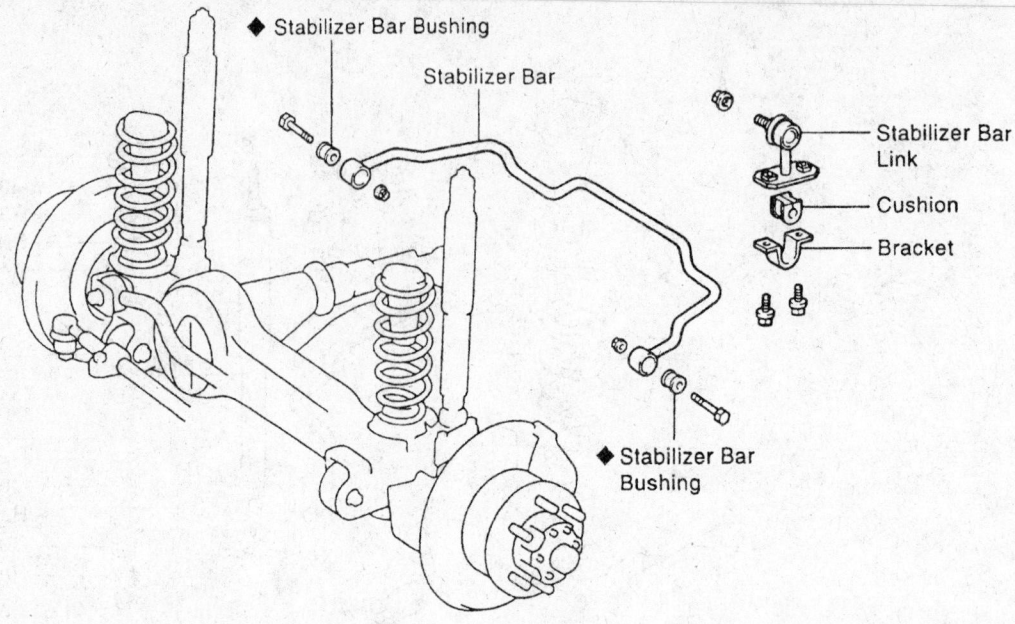

- Stabilizer Bar Bushing
- Stabilizer Bar
- Stabilizer Bar Link
- Cushion
- Bracket
- Stabilizer Bar Bushing

◆ Non-reusable part

300880

Sway bar component assembly

To install:

14. If removed, drive in a new bearing outer race using a suitable installer.

15. Pack the bearings with MP grease. Coat the inside of the hub and cap with MP grease.

16. Install the inner bearing and oil seal. Coat the oil seal with MP grease.

17. Install the hub on the spindle. Install the outer bearing and claw washer.

18. Adjust the preload:

  a. Torque the bearing adjusting nut to 43 ft. lbs. (59 Nm).

  b. Turn the hub right and left 2 or 3 times and re-torque.

  c. Loosen the nut until it can be turned by hand.

  d. Re-torque to 4 ft. lbs. (5.4 Nm)

  e. Check the bearing preload with a spring scale. The preload should be 6.4–12.6 ft. lbs. (28–56 Nm).

19. Install the lock washer and nut. Torque to 47 ft. lbs. (64 Nm)

20. Check that there is no bearing end-play.

21. Secure the locknut by bending one lock washer tooth inward and another outward.

22. If equipped without a free wheeling hub, install the flange as follows:

  a. Place a new gasket in position on the axle hub.

  b. Install the flange to the axle hub.

  c. Install the six cone washers, plate washers and nuts.

  d. Install the six nuts. Torque the nuts to 26 ft. lbs. (35 Nm).

  e. Install the bolt and torque the bolt to 13 ft. lbs. (18 Nm).

  f. Install the grease cap.

23. If equipped with a free wheeling hub, install the hub as follows:

  a. Place a new gasket in position on the front axle hub.

  b. Install the free wheeling hub body with the six cone washers and nuts. Torque the nuts to 23 ft. lbs. (31 Nm).

  c. Install the bolt with the washer. Torque the bolt to 13 ft. lbs. (18 Nm).

  d. Apply multi purpose grease to the inner hub splines.

  e. Set the control handle and clutch to the FREE position.

  f. Place a new gasket in position on the cover.

  g. Install the cover to the hub body with the follower pawl tabs aligned with the non–toothed portions of the hub body.

  h. Install the mounting bolts and tighten the cover bolts to 7 ft. lbs. (10 Nm).

24. Install the brake caliper support bracket to the steering knuckle and torque the two bolts to 90 ft. lbs. (123 Nm).

25. Clean the threads of the bolts and steering knuckle.

26. Apply sealant to the bolt threads and connect the knuckle arm with the brake line bracket to the steering knuckle. Torque the bolts to 135 ft. lbs. (183 Nm).

27. If equipped with ABS brakes, connect the ABS speed sensor to the steering knuckle.

28. Install the front wheel and lower the vehicle.

29. Check the ABS speed sensor signal.

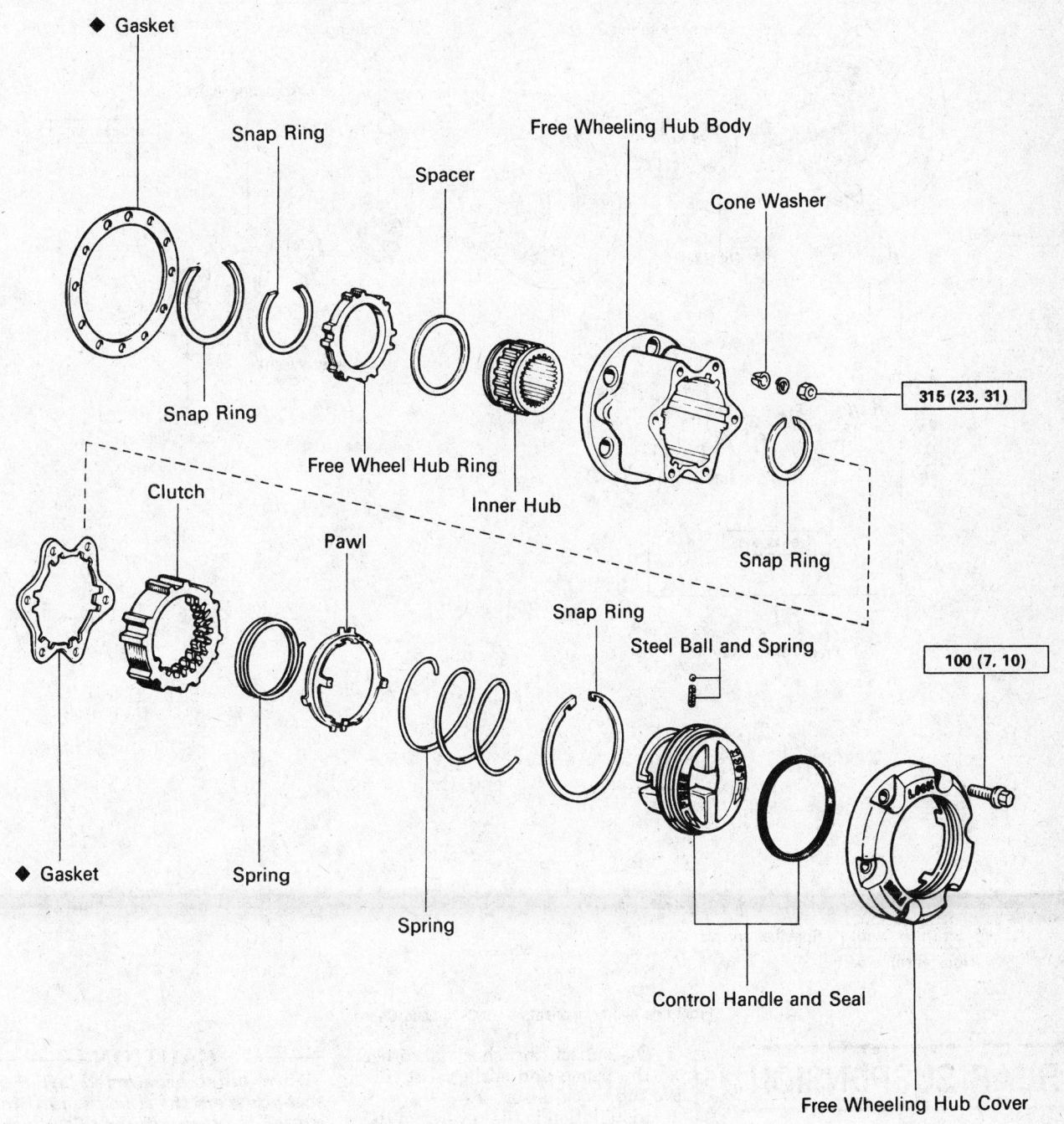

Gasket

Snap Ring

Spacer

Free Wheeling Hub Body

Cone Washer

Snap Ring

Free Wheel Hub Ring

Inner Hub

315 (23, 31)

Snap Ring

Clutch

Pawl

Snap Ring

Steel Ball and Spring

100 (7, 10)

◆ Gasket

Spring

Spring

Control Handle and Seal

Free Wheeling Hub Cover

| kg-cm (ft-lb, N·m) | : Specified torque

◆ Non-reusable part

90529

Free wheel hub components — 1993–94 Land Cruiser

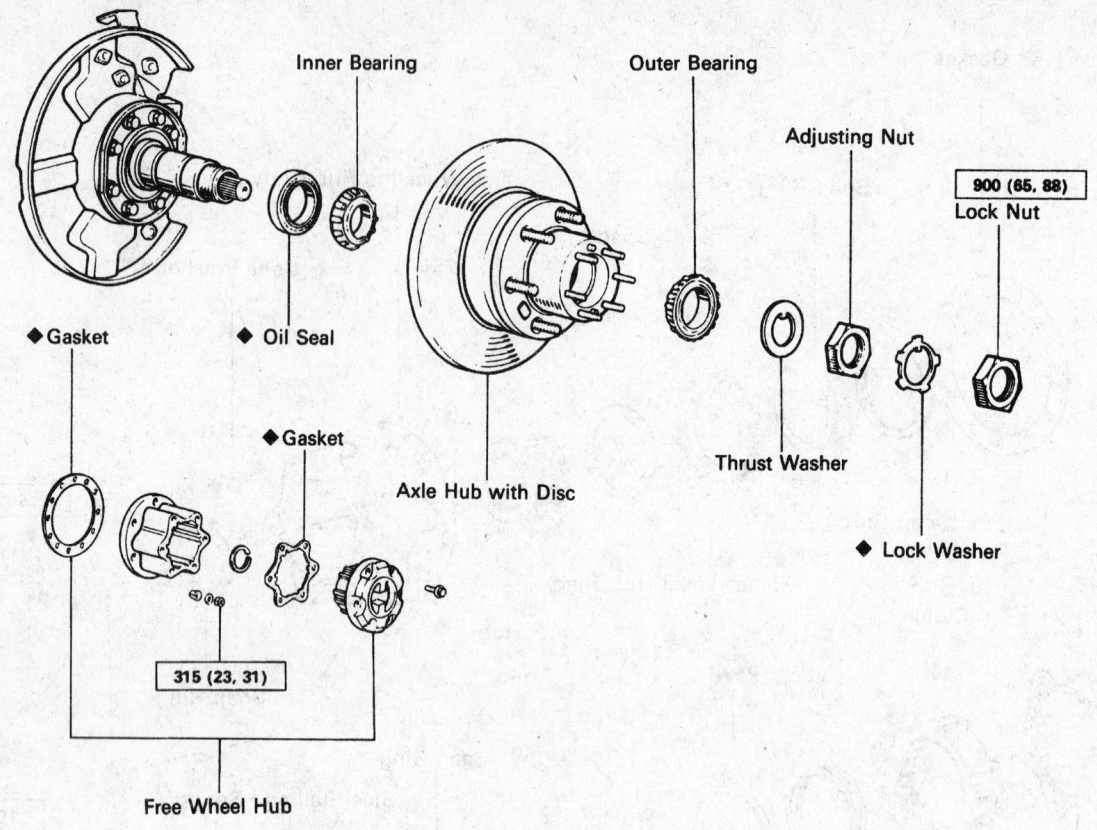

Inner Bearing Outer Bearing

Adjusting Nut

900 (65, 88)
Lock Nut

◆ Gasket ◆ Oil Seal

◆ Gasket

Axle Hub with Disc

Thrust Washer

◆ Lock Washer

315 (23, 31)

Free Wheel Hub

kg-cm (ft-lb, N·m) : Specified torque

◆ Non-reusable part

90530

Front axle hub components — 1993–94 Land Cruiser

REAR SUSPENSION

Shock Absorber

REMOVAL AND INSTALLATION

1. Raise and safely support the frame with stands.
2. Support the axle housing with a floor jack.
3. Remove the wheels.
4. Lower the floor jack to take the tension off of the spring.

5. Disconnect the shock absorber from the frame and spring seat. Remove the shock absorber.
6. Installation is the reverse of the removal procedure. Torque the shock absorber-to-frame mounting bolt to 47 ft. lbs. (64 Nm). Torque the shock absorber-to-spring seat bolt to 27 ft. lbs. (37 Nm).

Coil Spring

REMOVAL AND INSTALLATION

1. Disconnect the negative battery cable.

— CAUTION —
Work must be started after 90 seconds from the time the ignition switch is turned to the LOCK position and the negative (-) battery cable is disconnected.

2. Raise the vehicle and support the body. Hold the rear axle housing with a jack.
3. Remove the rear wheels.
4. Remove the rear shock absorber.
   a. Remove the bolt holding the shock absorber from the rear axle housing and disconnect the shock absorber.

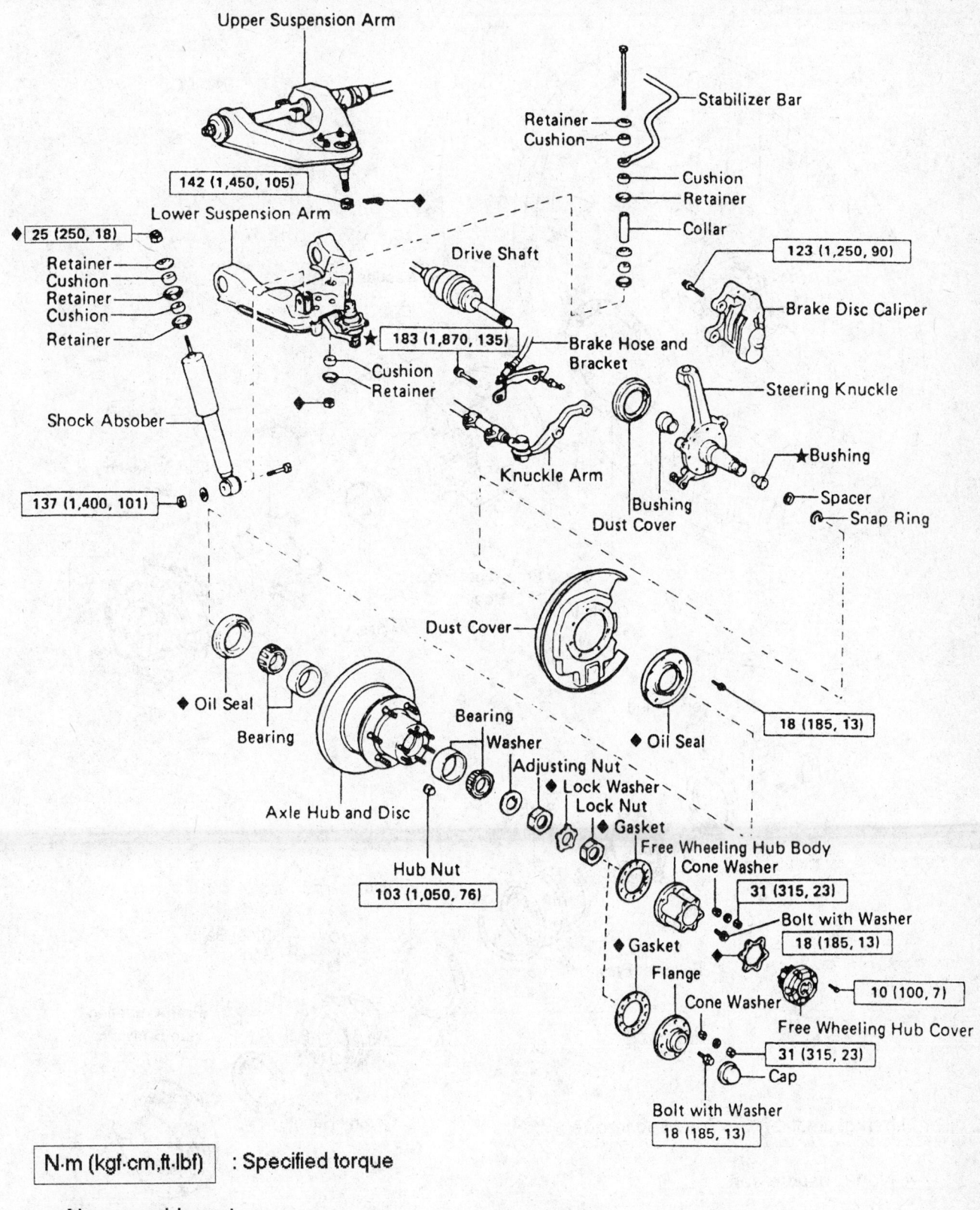

Front axle hub and wheel bearing component assembly — 1995–97 (4WD)

| N·m (kgf·cm,ft·lbf) | : Specified torque |

◆ Non-reusable part

★ Precoated part

335218

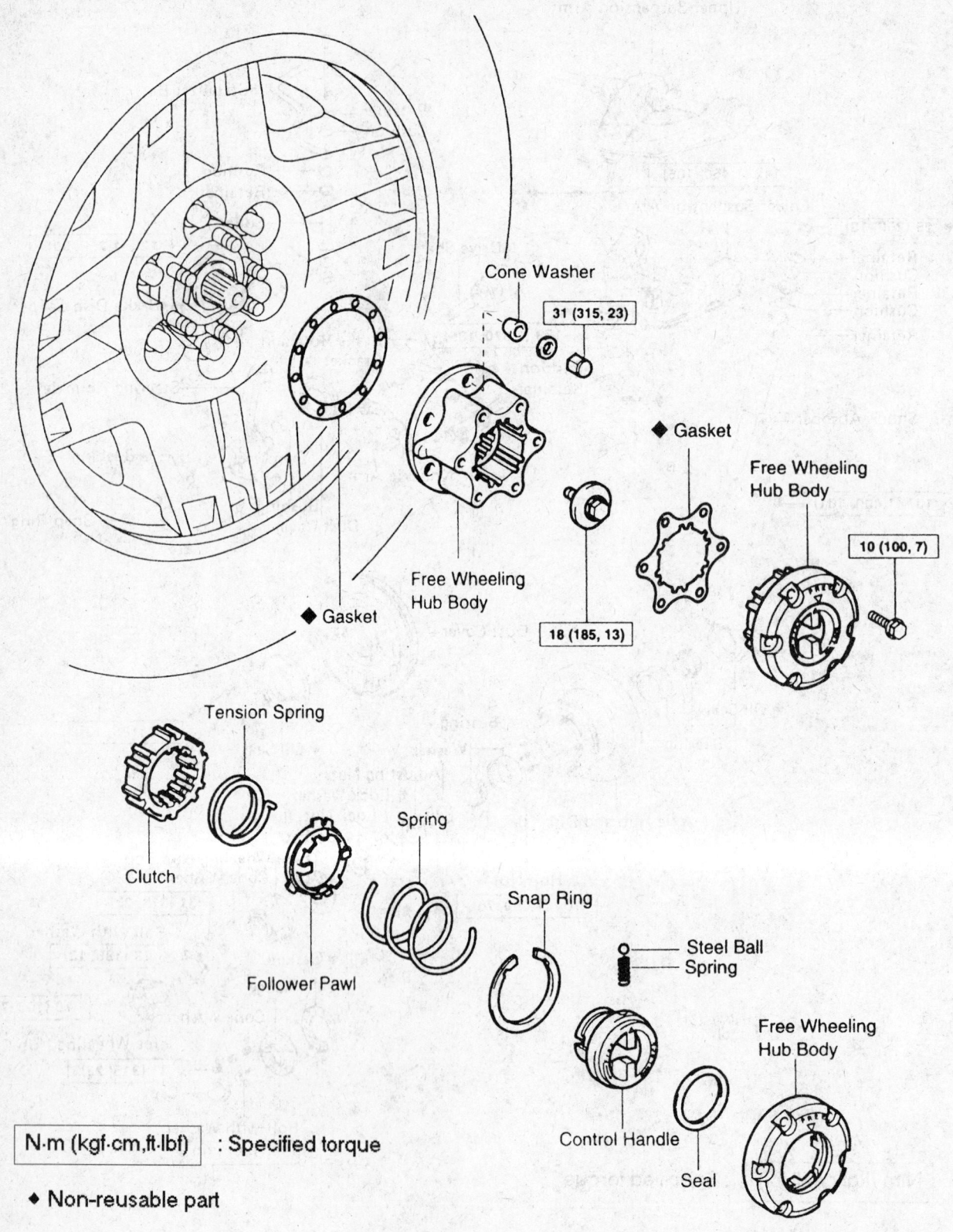

Cone Washer

31 (315, 23)

◆ Gasket

Free Wheeling
Hub Body

Free Wheeling
Hub Body

10 (100, 7)

◆ Gasket

18 (185, 13)

Tension Spring

Clutch

Follower Pawl

Spring

Snap Ring

Steel Ball

Spring

Control Handle

Seal

Free Wheeling
Hub Body

N·m (kgf·cm,ft·lbf) : Specified torque

◆ Non-reusable part

335217

Free wheeling hub component assembly — 1995–97

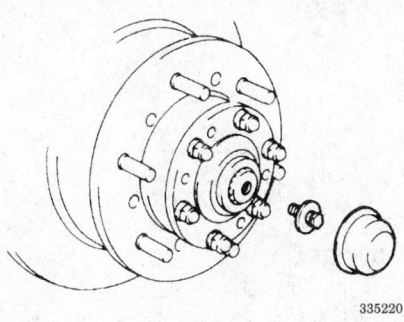

Grease cap and bolt (automatic hub) — 1995–97

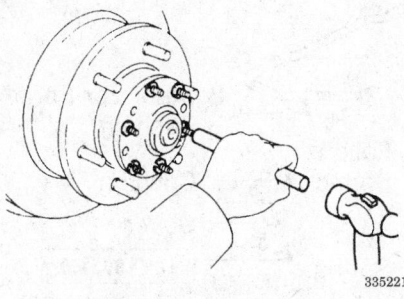

Removing the six cone washer (automatic hub) — 1995–97

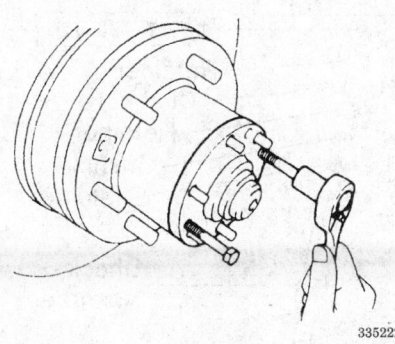

Removing the flange (automatic hub) — 1995–97

b. Remove the two upper bolts and the shock absorber.

5. Disconnect the stabilizer bar brackets from the rear axle housing.

6. Disconnect the lateral control rod from the rear axle housing.

7. Remove the coil spring. Begin to slowly lower the axle housing. While lowering the rear axle housing, remove the coil spring and the insulator.

8. Remove the follow spring from the frame.

To install:

9. Install the follow spring. Torque the nuts to 21 ft. lbs. (28 Nm).

10. Install the coil spring with a jack under the axle housing. Slowly raise the jack.

11. Connect the lateral control rod to the rear axle housing.

12. Connect the stabilizer bar to the axle housing.

13. Install the shock absorber and the upper bolt. Torque the nut to 37 ft. lbs. (50 Nm).

14. Install the shock absorber to the rear axle housing and torque to 47 ft. lbs. (64 Nm).

15. Install the wheels.

16. Lower the vehicle and connect the battery cable.

Upper Control Arms

REMOVAL AND INSTALLATION

1. Safely jack up the rear axle housing and support the frame with stands. Hold the rear axle housing with a jack.

2. Remove the rear wheels.

3. Remove the nuts, plate washers, bolts and the upper control arm from the frame.

4. Remove the nuts, plate washers, bolts and the upper control arm from the axle housing.

5. Remove the control arm.

To install:

6. Install the upper control arm to the axle housing and the frame and torque the bolts to 130 ft. lbs. (177 Nm).

7. Install the rear wheels and lower the vehicle.

Lower Control Arms

REMOVAL AND INSTALLATION

1. Safely jack up the rear axle housing and support the frame with stands. Hold the rear axle housing with a jack.

2. Remove the rear wheels.

3. Remove the nuts, plate washers, bolts and the lower control arm from the frame.

4. Remove the nuts, plate washers, bolts and the lower control arm from the axle housing.

5. Remove the control arm.

To install:

6. Install the lower control arm to the axle housing and the frame and torque the bolts to 130 ft. lbs. (177 Nm).

7. Install the rear wheels and lower the vehicle.

8. Check the vehicle's alignment.

Sway Bar

REMOVAL AND INSTALLATION

1. Raise and safely support the vehicle.

2. Remove the wheels.

3. Remove the nut and disconnect the stabilizer bar with the links from the frame.

4. Remove the two bolts and remove the stabilizer bar from the axle housing with the bracket.

5. Remove the sway bar from the suspension.

To install:

6. Install the sway bar to the axle housing and hand tighten the bolts.

7. Install the sway bar with the links to the frame and torque the nut to 11 ft. lbs. (15 Nm).

8. Torque the sway bar bolts to the axle housing with the brackets to 13 ft. lbs. (18 Nm).

9. Install the wheel and lower the vehicle.

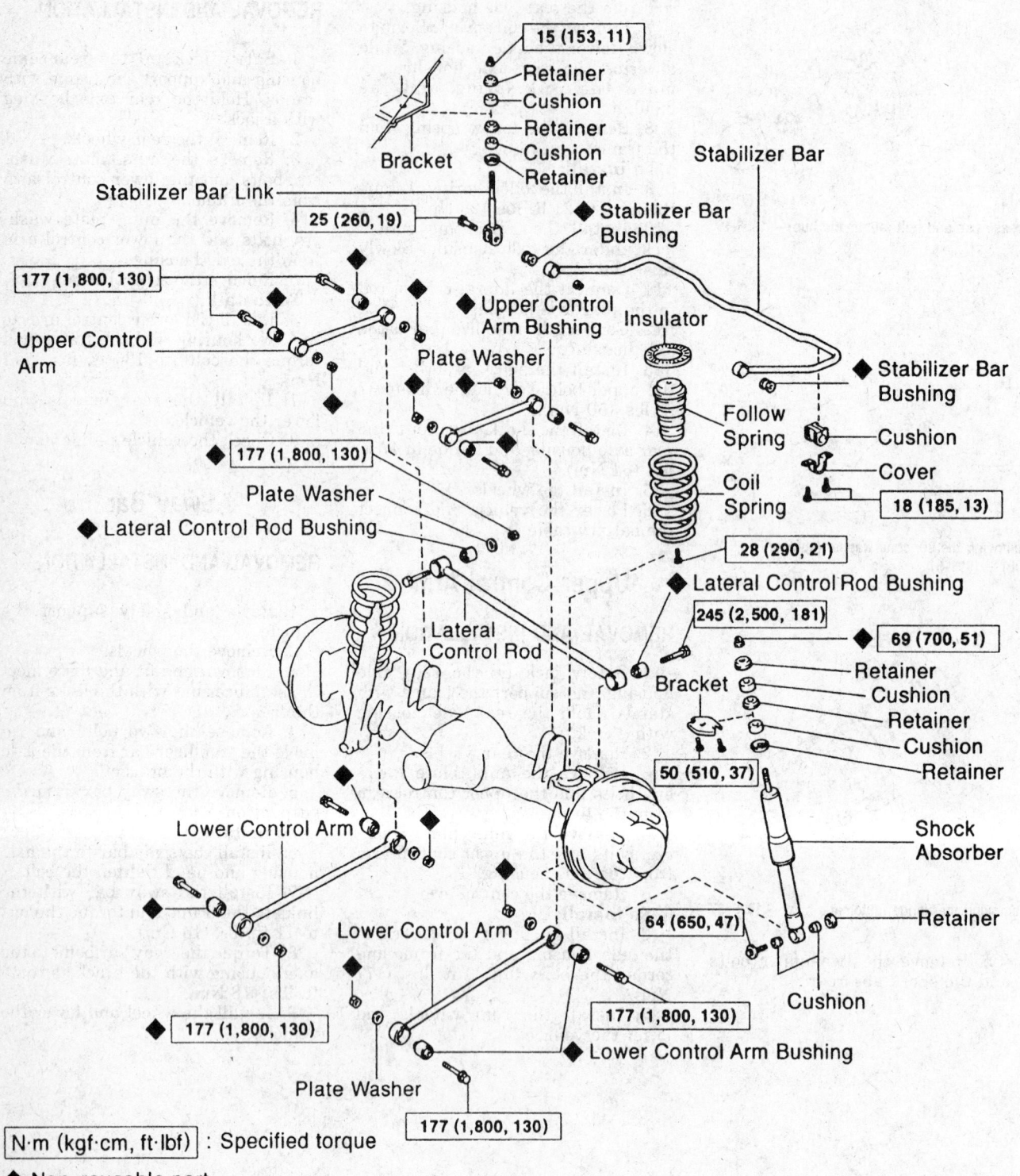

15 (153, 11)
Retainer
Cushion
Retainer
Cushion
Retainer
Bracket
Stabilizer Bar
Stabilizer Bar Link
25 (260, 19)
Stabilizer Bar Bushing
◆ Upper Control Arm Bushing
Insulator
177 (1,800, 130)
◆ Stabilizer Bar Bushing
Upper Control Arm
Plate Washer
Follow Spring
Cushion
◆ Cover
177 (1,800, 130)
Coil Spring
18 (185, 13)
Plate Washer
28 (290, 21)
◆ Lateral Control Rod Bushing
◆ Lateral Control Rod Bushing
Lateral Control Rod
245 (2,500, 181)
◆ 69 (700, 51)
Bracket
Retainer
Cushion
Retainer
Cushion
Retainer
50 (510, 37)
Shock Absorber
Lower Control Arm
Lower Control Arm
64 (650, 47)
Retainer
177 (1,800, 130)
177 (1,800, 130)
Cushion
◆ Lower Control Arm Bushing
Plate Washer
177 (1,800, 130)

N·m (kgf·cm, ft·lbf) : Specified torque

◆ Non-reusable part

300896

Rear suspension components

FIRING ORDERS

NOTE: To avoid confusion, always replace spark plug wires one at a time.

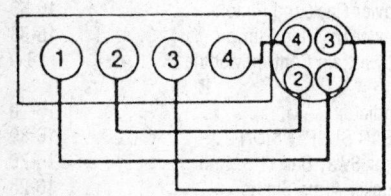

202993

2.2L Engine
Engine Firing Order: 1–3–4–2
Distributor Rotation: Clockwise

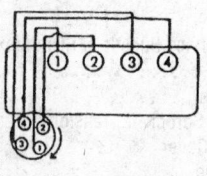

253638

2.6L Engine
Engine Firing Order: 1–3–4–2
Distributor Rotation: Clockwise

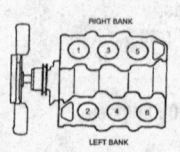

RIGHT BANK

LEFT BANK

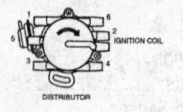

IGNITION COIL

DISTRIBUTOR

287315

3.0L Engine
Engine Firing Order:
1–2–3–4–5–6
Distributor Rotation:
Clockwise

ENGINE ELECTRICAL

NOTE: Disconnecting the negative battery cable on some vehicles may interfere with the functions of the on board computer systems and may require the computer to undergo a relearning process, once the negative battery cable is reconnected.

Distributor

REMOVAL

2.2L and 2.6L Engines

1. Disconnect the negative battery cable.
2. Remove the distributor cap from the distributor, leaving the spark plug wires attached. If spark plug wire removal is necessary to remove the distributor cap, tag the wires prior to removal so they can be reinstalled in the correct position.
3. Disconnect the electrical connectors and vacuum hose(s), if equipped, from the distributor.
4. Mark the position of the rotor in relation to the distributor housing and the position of the distributor housing on the cylinder head.
5. Remove the distributor hold-down bolt(s) and remove the distributor.
6. Check the distributor O-ring for cuts or other damage and replace, if necessary.

3.0L Engine

1. Disconnect the negative battery cable.
2. Label and remove the spark plug wires.

G6

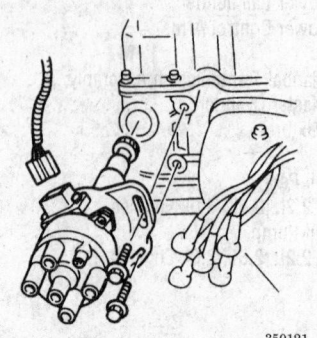

350121

Distributor installation — 2.6L engine

3. Turn the crankshaft so the No. 1 cylinder is at TDC of compression.
4. Disconnect the electrical connector from the distributor.
5. Mark the position of the rotor in relation to the distributor housing and the position of the distributor housing on the cylinder head.
6. Remove the distributor hold-down bolt(s), then remove the distributor.
7. Check the distributor O-ring for cuts or other damage and replace, if necessary.

INSTALLATION

Timing Not Disturbed

2.2L and 2.6L Engines

1. Lubricate the distributor O-ring with clean engine oil.
2. Install the distributor with the hold-down bolt(s), aligning the marks that were made during removal. Tighten the hold-down bolt(s) to 14–19 ft. lbs. (19–25 Nm).
3. Connect the electrical connectors and vacuum hose(s), if equipped.
4. Install the distributor cap on the distributor. Connect the spark plug wires, if removed.
5. Connect the negative battery cable. Start the engine and check the ignition timing.

3.0L Engine

1. Lubricate the distributor O-ring with clean engine oil.
2. On the 1996–97 MPV, align the matchmarks on the distributor shaft and the distributor drive gear.
3. Install the distributor with the hold-down bolt, aligning the marks that were made during removal. Tighten the hold-down bolt to 14–18 ft. lbs. (19–25 Nm).
4. Connect the electrical connector.
5. Connect the spark plug wires.
6. Connect the negative battery cable. Start the engine and check the ignition timing.

Timing Disturbed

2.2L and 2.6L Engines

1. Disconnect the spark plug wire from the No. 1 cylinder spark plug and remove the spark plug. Make sure the engine is cool enough to touch, place a finger over the spark plug hole.
2. Turn the crankshaft in the normal direction of rotation until compression is felt at the spark plug hole.

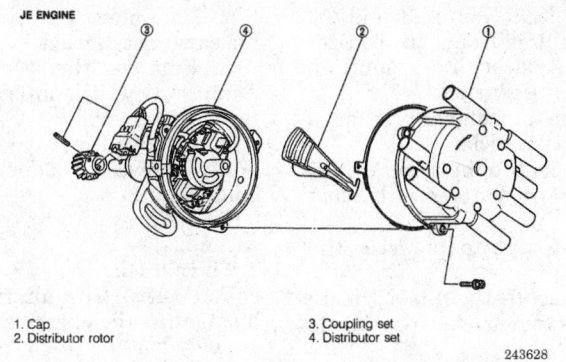

1. Cap
2. Distributor rotor
3. Coupling set
4. Distributor set

243628

Distributor assembly (exploded view) — 1993–95 MPV

3. Align the mark on the crankshaft pulley with the TDC mark on the timing belt cover.

4. Lubricate the distributor O-ring with clean engine oil.

5. Turn the distributor shaft until the rotor points to the No. 1 spark plug tower on the distributor cap and install the distributor. Install the distributor hold-down bolt(s) and align the distributor housing with the mark made on the cylinder head during removal. Snug the bolt(s).

6. Connect the electrical connectors and vacuum hose(s), if equipped.

7. Install the distributor cap on the distributor. Connect the spark plug wires, if removed.

8. Install the spark plug in the No. 1 cylinder and connect the spark plug wire.

9. Connect the negative battery cable. Start the engine and adjust the ignition timing. Tighten the distributor hold-down bolt(s) to 14–19 ft. lbs. (19–25 Nm) after the timing has been set.

3.0L Engine

1. Disconnect the spark plug wire from the No. 1 cylinder spark plug and remove the spark plug. After making sure the engine is cool enough to touch, place a finger over the spark plug hole.

2. Turn the crankshaft in the normal direction of rotation until compression is felt at the spark plug hole.

3. On the 1996–97 MPV, align the mark on the crankshaft pulley with the TDC mark on the timing belt cover.

4. Lubricate the distributor O-ring with clean engine oil.

5. Align the matchmarks on the distributor shaft and the distributor drive gear.

6. Turn the distributor shaft until the rotor points to the No. 1 spark plug tower on the distributor cap and install the distributor. Install the distributor hold-down bolt(s) and align

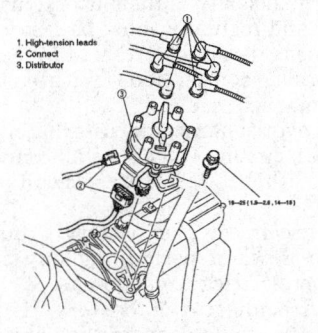

1. High-tension leads
2. Connect
3. Distributor

N·m (kgf·m, ft·lb)

301777

Distributor assembly — 1996–97 MPV

the distributor housing with the mark made on the cylinder head during removal. Snug the bolt(s).

7. Connect the electrical connector.

8. Connect the spark plug wires.

9. Install the spark plug in the No. 1 cylinder and connect the spark plug wire.

10. Connect the negative battery cable. Start the engine and adjust the ignition timing. Tighten the distributor hold-down bolt to 14–18 ft. lbs. (19–25 Nm) after the timing has been set.

Ignition Timing

ADJUSTMENT

2.2L Engine

1. Check the general condition of the engine components, including battery charge, poor spark plug condition, vacuum leaks, clogged filters, etc. These conditions must be cured before setting timing or idle speed.

2. Shift the transmission into the **P** or neutral position and apply the parking brake.

3. Turn all electrical accessories OFF.

4. Start the engine and allow it to warm up to normal operating temperature.

5. Connect an inductive timing light to the engine.

6. Connect a jumper wire between the green, single-pin test connector and a good ground. Connect a test tachometer to the white, single-pin check connector.

7. Check the idle speed and adjust it if needed. Manual transmission vehicles should idle at 730–770 rpm; automatic transmission vehicles should idle at 750–790 rpm.

8. Using the timing light, check the yellow timing mark on the crankshaft pulley against the degree scale. Correct timing is 5–7 degrees BTDC.

9. If the timing is incorrect, loosen the distributor lock bolt. Turn the distributor to make the necessary adjustment.

10. Tighten the distributor lock bolt. Recheck the ignition timing and adjust as needed.

11. Remove the jumper wire and the other test equipment.

2.6L Engine

1. Check the general condition of the engine components, including battery charge, poor spark plug condition, vacuum leaks, clogged filters, etc. These conditions must be cured before setting timing or idle speed.

2. Turn all electrical accessories OFF.

3. Connect a timing light. Place the transmission selector in **P** or neutral.

4. Start the engine and allow it to warm to normal operating temperature.

5. Connect a jumper wire between the green, single-pin test connector and a good ground. Connect a tachometer to the white, single-pin check connector.

6. Check the idle speed and adjust it if needed. Manual transmission vehicles should idle at 730–770 rpm; automatic transmission vehicles should idle at 750–790 rpm.

7. Using the timing light, check the yellow timing mark on the crankshaft pulley against the degree scale. Correct timing is 4–6 degrees BTDC, 5 degrees preferred.

8. If the timing is incorrect, loosen the distributor lock bolt. Turn the distributor to make the necessary adjustment.

9. Tighten the distributor lock bolt. Remove the jumper wire; recheck the idle speed and adjust as needed.

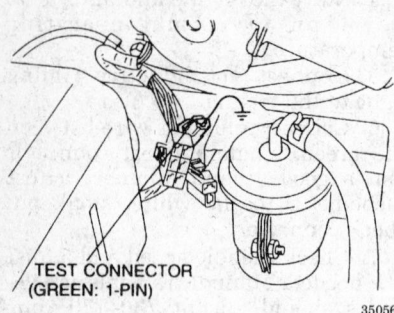

TEST CONNECTOR
(GREEN: 1-PIN)

350560

Test connector — 2.2L and 2.6L engines

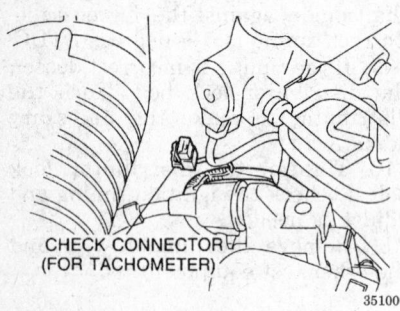

CHECK CONNECTOR
(FOR TACHOMETER)

351000

Tachometer connection — 2.2L and 2.6L engines

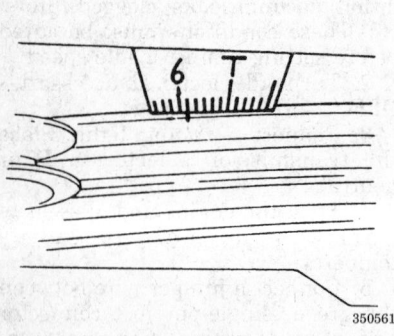

350561

Timing scale — 2.2L and 2.6L engines

Alternator

PRECAUTIONS

Several precautions must be observed with alternator equipped vehicles to avoid damage to the unit.

• If the battery is removed for any reason, make sure it is reconnected with the correct polarity. Reversing the battery connections may result in damage to the 1-way rectifiers.

• When utilizing a booster battery as a starting aid, always connect the positive to positive terminals and the negative terminal from the booster battery to a good engine ground on the vehicle being started.

• Never use a fast charger as a booster to start vehicles.

• Disconnect the battery cables when charging the battery with a fast charger.

• Never attempt to polarize the alternator.

• Do not use test lights of more than 12 volts when checking diode continuity.

• Do not short across or ground any of the alternator terminals.

• The polarity of the battery, alternator and regulator must be matched and considered before making any electrical connections within the system.

• Never separate the alternator on an open circuit. Make sure all connections within the circuit are clean and tight.

• Disconnect the battery ground terminal when performing any service on electrical components.

• Disconnect the battery if arc welding is to be done on the vehicle.

REMOVAL AND INSTALLATION

2.2L Engine

1. Disconnect the negative (ground) cable. On some vehicles, it may be necessary to remove the battery.

2. Remove the nut holding the alternator wire to the terminal at the rear of the alternator.

3. Pull the multiple connector from the rear of the alternator.

4. Remove the alternator adjusting arm bolt. Swing the alternator in and disengage the fan belt.

5. Remove the alternator pivot bolt and remove the alternator from the vehicle.

To install:

6. Place the alternator in position and install the pivot bolt hand tight.

7. Install the adjusting arm bolt. Place the drive belt in position and tension it properly. Secure the adjusting and pivot bolts.

8. Be sure to adjust the drive belt tension and to connect the battery properly.

2.6L Engine

B2600i

1. Disconnect the negative battery cable.

2. Disconnect the wiring and label for easy installation.

3. Remove the power steering pump pulley, if it interferes with alternator removal.

4. Remove the drive belts.

5. Remove the support mounting bolt and nut.

6. Remove the alternator assembly.

To install:

7. Position the alternator assembly against the engine and install the support bolt.

8. Install the adjusting strap mounting bolt.

9. Install the alternator belts and adjust to specification.

10. Install the power steering pump pulley.

11. Tighten all the support bolts and nuts.

12. Connect all alternator terminals.

13. Connect the negative battery cable.

MPV

1. Disconnect the negative battery cable.

2. Tag and disconnect the wire and connector from the alternator.

3. Remove the drive belt.

4. Remove the support mounting bolt and nut.

5. Remove the alternator assembly.

To install:

6. Position the alternator assembly against the engine and install the support bolt.

7. Install the adjusting strap mounting bolt.

8. Install and adjust the alternator belt.

9. Tighten the support mounting bolt to 28–38 ft. lbs. (38–51 Nm), and the adjusting bolt to 14–18 ft. lbs. (19–25 Nm).

10. Connect all alternator terminals.

11. Connect the negative battery cable.

3.0L Engine

1. Disconnect the negative battery cable. Wait at least 90 seconds before performing any work.

2. Tag and disconnect the wire and connector from the alternator.

3. Remove the power steering pulley.

4. Remove the mounting and adjusting bolts.

5. Remove the drive belts.

6. Remove the alternator assembly.

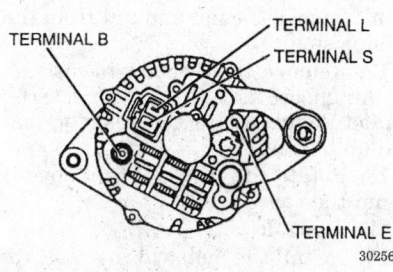

Terminal locations — 3.0L engines

302567

To install:

7. Position the alternator assembly against the engine and install the support bolt.

8. Install the adjusting strap mounting bolt.

9. Install the power steering pulley. Tighten the nut to 29–43 ft. lbs. (40–58 Nm).

10. Install and adjust the drive belts.

11. Tighten the support mounting bolt to 28–38 ft. lbs. (38–51 Nm), and the adjusting bolt to 14–18 ft. lbs. (19–25 Nm).

12. Connect all alternator terminals.

13. Connect the negative battery cable.

Drive Belt

REMOVAL AND INSTALLATION

Power Steering Pump Belt

1. Turn the ignition **OFF** and remove the key. Allow the engine to cool.

2. Loosen the idler pulley locknut to release the drive belt tension.

3. Remove the power steering belt.

To install:

4. Install the power steering belt and make sure it is correctly seated on the pulleys.

5. Adjust the power steering belt tension/deflection by turning the adjusting bolt. A new belt should deflect 0.28–0.31 in. (7–8mm) for 2.2L engine or 0.26–0.28 in. (6.6–7.2mm) except for 2.2L engine, and a used belt should deflect 0.31–0.35 in. (8–9mm) for 2.2L engine or 0.28–0.31 in. (7–8mm) except for 2.2L engine.

6. Tighten the idler pulley locknut to 27–38 ft. lbs. (37–52 Nm).

7. Run the engine for 5 minutes and then recheck the belt deflection.

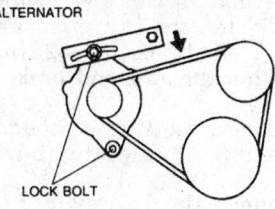

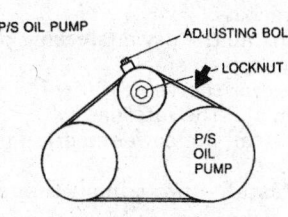

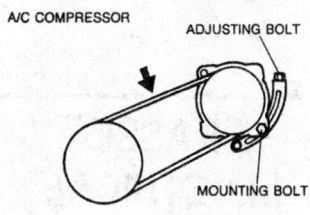

Drive belt routings and component adjusting bolts — 2.2L engine

350795

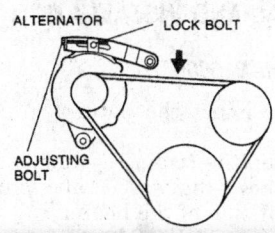

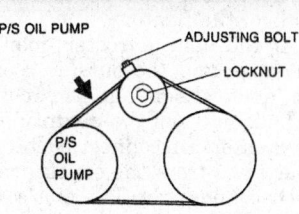

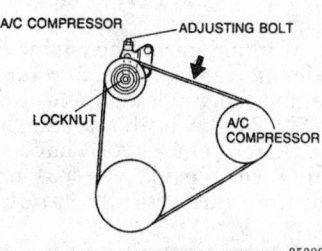

Drive belt routings and component mounting bolts — 2.6L and 3.0L engine

350802

Alternator Belt

1. Turn the ignition **OFF** and remove the key. Allow the engine to cool.

2. Remove the power steering pump belt and A/C compressor belt, if necessary.

3. Loosen the alternator upper bracket adjusting bolt.

4. Loosen the lower through-bolt.

5. Remove the alternator belt.

To install:

6. Install the alternator belt and make sure it is correctly seated on the pulleys.

7. Turn the adjusting bolt to adjust the belt deflection. New belts should deflect 0.28–0.31 in. (7–8mm) for 2.2L engine or 0.39–0.47 in. (10–12mm) except for 2.2L engine. Used belts should deflect 0.31–0.35 in. (9–10mm) for 2.2L engine or 0.43–0.51 in. (11–13mm) except for 2.2L engine.

8. Once the correct deflection has been reached, tighten the upper bracket lock bolt to 14–19 ft. lbs. (19–25 Nm).

9. Tighten the lower through-bolt to 28–38 ft. lbs. (38–51 Nm).

10. Install and tension the power steering pump belt and A/C compressor belt, if necessary.

11. Run the engine for 5 minutes and then recheck the belt deflection.

A/C Compressor Belt

1. Turn the ignition **OFF** and remove the key. Allow the engine to cool.

2. Remove the power steering pump belt and alternator belt, if necessary.

3. For 2.2L engine, loosen the A/C compressor mounting bolts and adjusting bolt to release the drive belt tension.

4. Except for 2.2L engine, loosen the tensioner bracket adjusting bolt and tensioner pulley center locknut to release the drive belt tension.

5. Remove the A/C belt.

To install:

6. Install the A/C belt, and make sure it is correctly seated on the pulleys.

7. Adjust the A/C belt tension/deflection by turning the tensioner bracket adjusting bolt. A new belt should deflect 0.39–0.47 in. (10–12mm) for 2.2L engine or 0.33–0.39 in. (8.5–10mm) except for 2.2L engine, and a used belt should deflect 0.47–0.55 in. (12–14mm) for 2.2L engine or 0.39–0.45 in. (10–11.5mm) except for 2.2L engine.

8. For 2.2L engine, tighten the A/C compressor mounting bolts to 27–38 ft. lbs. (37–52 Nm).

9. Except for 2.2L engine, tighten the tensioner pulley center locknut to 27–38 ft. lbs. (37–52 Nm).

10. Install and tension the alternator belt, if necessary, and the power steering pump belt.

11. Run the engine for 5 minutes and then recheck the belt deflection.

Starter

REMOVAL AND INSTALLATION

2.2L, 2.6L and 3.0L Engines

Except 4WD MPV

1. Disconnect the negative battery cable. Raise and safely support the vehicle.

2. Disconnect the electrical connectors from the starter.

3. Remove the starter mounting bolts and remove the starter.

To install:

4. Place the starter in position, after cleaning the mounting flange surfaces.

5. Install and tighten the mounting bolts. Tighten the starter mounting bolts to 27–38 ft. lbs. (37–52 Nm).

6. Connect all wiring connectors. Connect the negative battery cable.

4WD MPV

1. Disconnect the negative battery cable.

2. Remove the drive belts.

3. Remove the power steering pump pulley.

4. Remove the alternator.

5. Raise and safely support the vehicle. Remove the splash shields.

6. Remove the power steering pump mounting bolts and position the pump aside, without disconnecting the power steering hoses.

7. Remove the automatic transmission cooler line brackets.

8. Mark the position of the driveshaft on the axle flange, and remove the front driveshaft.

9. Remove the wiring harness bracket and the automatic transmission cooler line bracket that is next to the starter.

10. Disconnect the electrical connectors from the starter.

11. Remove the fuel and brake line shield.

12. Remove the starter mounting bolts and remove the starter.

To install:

13. Clean the mounting surface flanges. Place the starter motor into position and install the mounting bolts. Tighten the starter mounting bolts to 27–38 ft. lbs. (37–52 Nm).

14. Install the fuel and brake line shield.

15. Connect the starter motor wiring. Install the wiring harness bracket.

16. Connect the driveshaft.

17. Install the transmission cooler line brackets.

18. Install the power steering pump and the splash shields.

19. Lower the vehicle.

20. Install the alternator.

21. Install the power steering pump pulley.

22. Install and adjust the drive belts.

23. Connect the negative battery cable.

CHASSIS ELECTRICAL

Blower Motor

REMOVAL AND INSTALLATION

B2200 and B2600i

1. Disconnect the battery ground cable.

2. Drain the cooling system.

3. Remove the water valve shield at the left side of the heater.

4. Disconnect the 2 hoses from the left side of the heater.

5. At the heat-defroster door, the water valve and the outside recirculation door, disengage the control cable housing from the mounting clip on the heater. Disconnect each of the 3 cable wires from the crank arms.

6. Disconnect the fan motor electrical lead.

7. Remove the glove compartment for clearance.

8. Working inside the engine compartment, remove the 2 retaining nuts and the single bolt and washer which hold the heater to the bulkhead. Later vehicles also have a retaining bolt inside the passenger compartment which must be removed.

9. Disconnect the 2 defroster ducts from the heater and remove the heater.

10. Remove the 5 screws and separate the halves of the heater assembly.

11. Loosen the fan retaining nut. Lightly tap on the nut to loosen the fan. Remove the fan and nut from the motor shaft.

12. Remove the 3 motor-to-case retaining screws and disconnect the bullet connector to the resistor and ground screw.

13. Rotate the motor and remove it from the case.

To install:

14. Install the motor in the case, rotating it slightly.

15. Install the retaining screws and connect the bullet connector and ground wire.

16. Install the fan on the shaft and install the nut.

17. Assemble the halves together and install the 5 retaining screws.

18. Install the heater on the dash so the heater duct indexes with the air intake duct and the 2 mounting studs enter their respective holes.

19. From the engine side of the bulkhead, install the nuts on the mounting studs. While an assistant holds the heater in position, install the mounting bolt.

20. Connect the defroster ducts.

21. Connect the heat-defrost door control cable to the door crank arm. Set the control lever (upper) in the HEAT position and turn the crank arm toward the mounting clip as far as it will go. Engage the cable housing in the clip and install the screw in the clip.

22. Connect the water valve control cable wire to the crank arm on the water valve lever. Locate the cable housing in the mounting clip. Set the control lever in the HOT position and pull the valve plunger and lever to the full outward position. This will move the lever crank arm toward the cable mounting clip as far as it will go. Tighten the clip and screw.

23. Insert the outside recirculation door control cable into the hole in the door crank arm. Bend the wire over and tighten the screw Set the center control lever in the REC position and turn the door crank arm toward the mounting clip as far as it will go. Engage the cable housing in the clip and install the screw in the clip.

24. Connect the fan motor electrical lead.

25. Connect the 2 hoses to the heater core tubes, at the left side of the heater, and tighten the clamp.

26. Install the water valve shield and tighten the 3 screws (left side of the heater).

27. Refill the cooling system and connect the battery ground cable.

28. Run the engine and check for leaks. Check the operation of the heater.

29. Install the glove compartment.

Windshield Wiper Motor

REMOVAL AND INSTALLATION

B2200 and B2600i

1. Disconnect the negative battery cable. Remove the wiper arm/blade assembly.

2. Remove the rubber seal from the leading edge of the cowl grille.

3. Remove the attaching screws and remove the cowl grille.

4. Remove the access hole covers.

5. Remove the bolts attaching the wiper shaft drives.

6. Matchmark the position of the wiper linkage in relation to the face of the wiper motor. Disconnect the wiper linkage from the wiper motor.

7. Disconnect the electrical connector from the wiper motor.

8. Remove the mounting bolts and the wiper motor.

   To install:

9. Position the wiper motor and install the mounting bolts. Tighten to 61–87 inch lbs. (6.9–9.8 Nm).

10. Connect the electrical connector.

11. Attach the wiper linkage to the motor, aligning the mark that was made during removal.

12. Install the bolts attaching the wiper shaft drives and tighten to 61–87 inch lbs. (6.9–9.8 Nm).

13. Install the access hole covers, cowl grille and seal.

14. Install the wiper arm/blade assemblies. Adjust the arm height to 0.8 in. (20mm) from the lower windshield moulding and tighten the arm retaining nuts to 7.2–10 ft. lbs. (9.8–14 Nm).

15. Connect the negative battery cable and check wiper operation.

1993–95 MPV

1. Disconnect the negative battery cable.

2. Disconnect the electrical connector from the wiper motor.

3. Disconnect the wiper linkage from the motor crank arm. If necessary to remove the crank arm from the motor, matchmark the arm to the bracket prior to removal.

4. Remove the mounting bolts and remove the wiper motor.

   To install:

5. Position the wiper motor and install the mounting bolts. Tighten to 61–87 inch lbs. (6.9–9.8 Nm).

6. If removed, install the motor crank arm, aligning the marks that were made during removal.

7. Connect the wiper linkage to the motor crank arm. Connect the electrical connector.

8. Make sure when in the park position, the wiper blades are 0.98–1.38 in. (25–35mm) from the lower windshield moulding.

1996–97 MPV

1. Make sure the wipers are in the parked position.

2. Disconnect the negative battery cable and wait at least 90 seconds before performing any work.

3. Remove the 2 nuts and remove the 2 wiper arms.

4. Remove the 2 nuts and the 2 seal rings securing the wiper transmission to the cowl.

5. Remove the 2 bolts securing the wiper motor bracket and disconnect the wiper motor connector.

6. Disconnect the wiper transmission from the wiper motor bell crank by carefully prying the transmission link from the bell crank.

7. Remove the nut and the O-ring and remove the bell crank from the wiper motor.

8. Remove the 3 wiper motor mounting bolts and the wiper motor.

   To install:

9. Install the wiper motor with the 3 mounting bolts and torque the bolts to 61–86 inch lbs. (6.9–9.8 Nm).

10. Install the bell crank to the wiper motor with the seal and the nut.

11. Install the wiper transmission to the bell crank.

12. Install the wiper transmission assembly through the openings in the cowl panel. Reconnect the wiper motor connector and install the 2 bolts securing the wiper motor bracket.

13. Install the wiper pivot seal rings and secure with the 2 nuts. Torque the nuts to 61–86 inch lbs. (6.9–9.8 Nm).

14. Reinstall the wiper arms and secure with the 2 nuts. Torque the nuts to 86.8–120 inch lbs. (9.81–13.7 Nm).

15. Reconnect the negative battery cable and check for proper operation.

Combination Switch

REMOVAL AND INSTALLATION

B2200 and B2600i

The combination turn signal, windshield wiper, and headlight switch is mounted on the steering column, and must be replaced as an assembly.

1. Disconnect the negative battery cable.

2. Remove the steering wheel.

3. Remove the "Lights-Hazard" Indicator and the steering column shroud.

4. Unplug the electrical multiple connectors at the base of the steering column.

5. Pull the headlight knob from its shaft.

6. Remove the snapring, which retains the switch, from the steering shaft. Pull the turn indicator canceling cam from the shaft.

7. Remove the single retaining bolt near the bottom of the switch. Remove the complete switch from the column.

   To install:

8. Place the switch in position and secure it.

9. Install the turn indicator cam and snapring. Install the remaining components.

10. Check the operation of the switch before installing the steering wheel.

1993–95 MPV

——— **CAUTION** ———

The air bag system must be disarmed before removing the steering wheel. Failure to do so may cause accidental deployment, property damage or personal injury.

——— **CAUTION** ———

Always carry an air bag assembly with the bag and trim cover away from your body. Store the assembly facing upward; never place the assembly face down on any surface.

1. Disconnect the battery.

2. Remove the steering wheel.

3. Remove the attaching screws, and remove the upper and lower steering column covers.

4. Disconnect the electrical connectors.

5. Remove the combination retaining screw, and remove the switch.

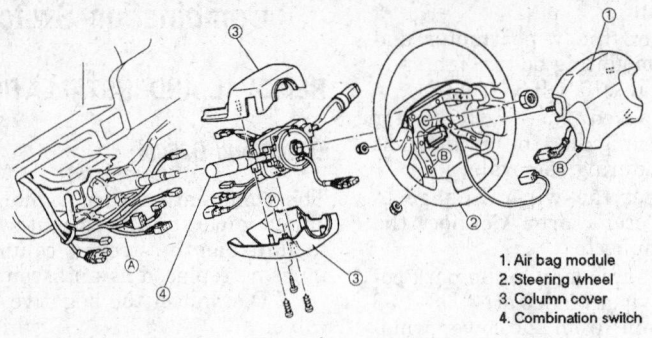

1. Air bag module
2. Steering wheel
3. Column cover
4. Combination switch

241220

Combination switch exploded view — 1993–95 MPV

To install:

6. Install the combination switch and secure with the retaining screw.

7. Connect all combination switch connectors.

8. Install the steering column covers with retaining screws.

9. Make sure the front wheels are aligned straight ahead and set the clock spring connector by turning it clockwise until it stops, then return the connector 2¾ turns. Align the marks on the clock spring connector and the outer housing.

10. Install the steering wheel.

11. Connect the negative battery cable.

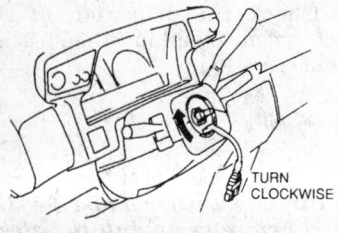

TURN CLOCKWISE

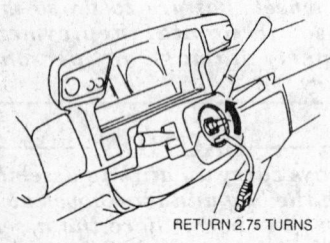

RETURN 2.75 TURNS

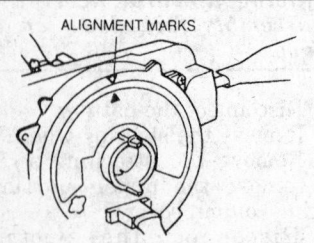

ALIGNMENT MARKS

241238

Steering column clock spring and connector — MPV

12. Check all the functions of the combination switch for proper operation.

1996–97 MPV

———— **CAUTION** ————
The air bag system must be disarmed before removing the steering wheel. Failure to do so may cause accidental deployment, property damage or personal injury.

———— **CAUTION** ————
Always carry an air bag assembly with the bag and trim cover away from your body. Store the assembly facing upward; never place the assembly face down on any surface.

1. Disconnect the negative battery cable and wait at least 90 seconds before performing any work.

2. Remove the service caps and remove the bolts securing the air bag module to the steering wheel. Disconnect the air bag connector and remove the air bag module.

3. Remove the steering wheel nut and remove the steering wheel.

4. Remove the screws and the upper and lower steering column covers.

5. Disconnect the clock spring connector, remove the mounting screws, and remove the clock spring.

6. Disconnect the combination switch connector, remove the mounting screws and remove the combination switch.

To install:

7. Install the combination switch with the mounting screws and reconnect the connector.

8. Install the clock spring with its screws and reconnect the connector. The clock spring must be adjusted in the following manner:

a. Turn the clock spring clockwise until it stops.

b. Turn the clock spring counterclockwise 2¾ turns.

c. Align the marks on the clock spring connector with that on the outer housing.

9. Reinstall the steering column upper and lower covers with the mounting screws.

10. Reinstall the steering wheel and tighten with the nut.

11. Reconnect the air bag module connector and install the air bag with its mounting bolts. Reinstall the service caps.

12. Reconnect the negative battery cable and check for proper operation.

Ignition Lock Cylinder

REMOVAL AND INSTALLATION

B2200 and B2600i

1. Disconnect the negative battery cable.

2. Remove the steering wheel.

3. Remove the upper and lower steering column covers by removing the screws from the lower cover.

4. If equipped with an automatic transmission, disconnect the interlock cable from the ignition lock.

5. Disconnect the wiring harness connectors from the ignition switch

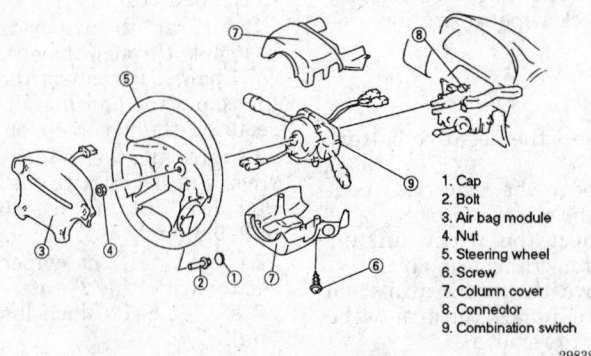

1. Cap
2. Bolt
3. Air bag module
4. Nut
5. Steering wheel
6. Screw
7. Column cover
8. Connector
9. Combination switch

298380

Combination switch and components — 1996–97 MPV

and then remove the ignition electrical switch from the ignition lock cylinder.

6. Use a small chisel to tap a notch into the head of each shear-off bolt. Next, unscrew the shear-off bolts using a standard screwdriver in the chiseled notch. Be careful to avoid damaging the steering column and combination switch.

7. Remove the ignition lock cylinder assembly from the steering column.

To install:

8. Install the ignition lock cylinder assembly onto the steering column.

9. Tighten the new shear-off bolts until their heads break off.

10. Install the ignition electrical switch and reconnect the wiring harness connectors.

11. If equipped, reconnect the A/T interlock cable to the ignition lock.

12. Install the upper and lower steering column covers.

13. Install the steering wheel.

14. Reconnect the negative battery cable.

15. Test the operation of the ignition switch at all positions.

MPV

The ignition lock is an integral part of the ignition switch. The ignition lock is serviced only as the ignition switch is replaced.

Ignition Switch

REMOVAL AND INSTALLATION

MPV

1. Disconnect the negative battery cable. Wait at least 90 seconds before performing any work.

2. Remove the steering column cover and the lower dash panel.

3. Remove the screw and disconnect the connector.

4. Extract the key-reminder switch terminals from the ignition switch and remove the ignition switch.

5. Installation is the reverse of removal.

B2200 and B2600i

1. Disconnect the battery ground cable.

2. Remove the steering column covers.

3. Disconnect the wiring harness connector at the switch.

4. Remove the attaching screw and lift out the switch.

5. Installation is the reverse of removal.

Park/Neutral Safety Switch

REMOVAL AND INSTALLATION

Electronically Controlled Transmission

1. Disconnect the negative battery cable.

2. Raise and safely support the vehicle.

3. Disconnect the control linkage from the transmission manual shaft.

4. Disconnect the park/neutral switch harness connector, remove the switch mounting bolts, and remove the park/neutral switch.

To install and adjust:

5. Install the park/neutral switch over the manual shaft.

6. Install the switch mounting bolts but do not tighten.

7. Make sure the transmission manual shaft lever is positioned at the **L** position (fully forward). Turn the manual shaft lever fully rearward, then return it 2 notches (**N** position).

8. Insert a 0.16 inch (4.0mm) pin through the holes of the switch and the manual shaft lever.

9. Tighten the switch mounting bolts to 22–35 inch lbs. (2.5–3.9 Nm). Remove the pin.

10. Connect the electrical connector.

11. Reinstall the shift control linkage.

12. Safely lower the vehicle, reconnect the negative battery cable, and check for proper operation.

Hydraulically Controlled Transmission

1. Disconnect the negative battery cable.

2. Raise and safely support the vehicle.

3. Disconnect the control linkage from the transmission manual shaft.

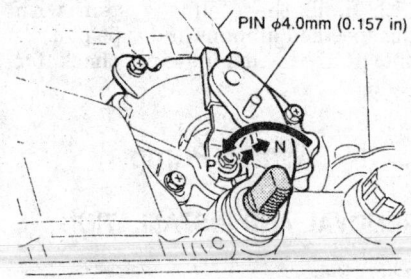
PIN φ4.0mm (0.157 in)

349805

Adjusting park/neutral switch — B2200 and B2600i

4. Disconnect the wiring harness connector and remove the park/neutral switch mounting bolts. Remove the switch from the manual shaft lever.

To install and adjust:

5. Position the park/neutral switch over the manual shaft lever and install the mounting bolts. Do not fully tighten the mounting bolts.

6. Move the manual shaft lever to the **N** position.

7. Remove the screw on the switch body and move the switch so the screw hole is aligned with the small hole inside the switch. Check their alignment by inserting a 0.079 in. (2.0mm) diameter pin through the holes.

8. Tighten the switch mounting bolts to 43–61 inch lbs. (4.9–6.9 Nm).

9. Install the screw in the park/neutral switch body and reconnect the wiring harness connector.

10. Reinstall the control linkage.

11. Safely lower the vehicle, reconnect the negative battery cable, and check for proper operation.

Powertrain Control Module

REMOVAL AND INSTALLATION

B2200 and B2600i

1. Disconnect the negative battery cable.

2. Remove the retaining screws from the right side kick panel.

3. Remove the kick panel.

4. Unplug the wiring connector from the PCM and slide out the PCM.

5. Installation is the reverse of the removal procedure.

NOTE: After the PCM has been disconnected and reconnected the vehicle should be driven a few miles to allow the PCM to relearn the vehicle operating strategy.

MPV

1. Disconnect the negative battery cable.

2. Remove the right side scuff plate and front side trim.

3. Pull back the carpet on the passenger-side floor panel under the dash.

4. Remove the 4 retaining bolts from the floor panel to access the PCM.

5. Unplug the connectors from the PCM. Remove the retaining screws at the bottom of the PCM and remove the module.

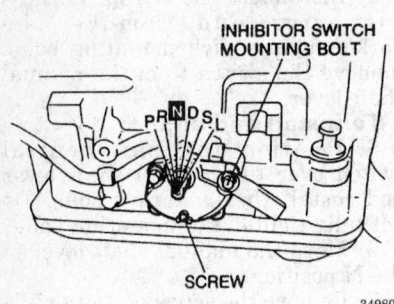

INHIBITOR SWITCH
MOUNTING BOLT

SCREW

349807

Park/neutral switch alignment screw

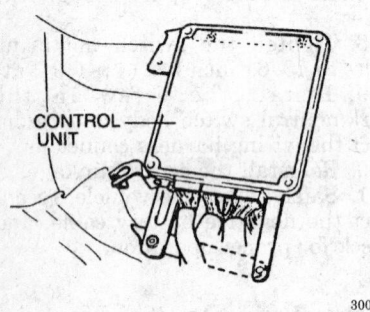

CONTROL
UNIT

300417

Powertrain control module location — MPV

6. Installation is the reverse of the removal procedure.

NOTE: After the PCM has been unplugged and plugged in, the vehicle should be driven a few miles so the PCM can relearn the operating strategy.

ENGINE COOLING

Radiator

REMOVAL AND INSTALLATION

B2200 and B2600i

1. Drain the cooling system.

— CAUTION —
When draining the coolant, always drain the coolant into a sealable container. Coolant should be reused unless it is contaminated or several years old.

2. If equipped, remove the fan shroud.
3. Remove the fan. Check the fan clutch, for looseness or loss of fluid

packing. If so, the fan clutch will have to be replaced.
4. Drain the coolant into a clean container.
5. Disconnect the upper and lower radiator hoses.
6. Disconnect the coolant reservoir hose.
7. On vehicles with automatic transmission, disconnect the cooler lines.
8. Unbolt and remove the radiator.
   To install:
9. Install the radiator against the supports and tighten the mounting bolts.
10. Install the hoses on the radiator. Tighten the clamps.
11. Install the fan.
12. If equipped, install the fan shroud.
13. Refill the cooling system with the specified amount and type of coolant. Run the engine and check for leaks.

MPV

1. Drain the cooling system.
2. Remove the fresh air duct.
3. Disconnect the upper and lower radiator hoses.
4. Disconnect the coolant reservoir hose.
5. If equipped with an automatic transmission, disconnect the cooler lines.
6. Remove the fan shroud.
7. Remove the fan. Check the fan clutch for looseness or leakage. If worn out, the fan clutch will have to be replaced.
8. Unbolt and remove the radiator.
   To install:
9. Install the radiator against the supports, and tighten the mounting bolts to 16–22 ft. lbs. (20–30 Nm).
10. Install the fan shroud.
11. Install the fan.
12. Install the hoses and cooler lines on the radiator. Tighten the clamps.
13. Refill the cooling system with the specified amount and type of coolant. Run the engine and check for leaks.

Water Pump

REMOVAL AND INSTALLATION

2.2L Engine

NOTE: Use special tool No. 49E301060 or equivalent on the engine flywheel gear to stop the

engine from rotating during removal and installation of the crankshaft pulley.

1. Drain the cooling system.

— CAUTION —
When draining the coolant, always drain the coolant into a sealable container. Coolant should be reused unless it is contaminated or several years old.

2. Disconnect the negative battery cable.
3. Remove the alternator belt.
4. Remove the power steering pump belt.
5. Remove the upper and the lower timing belt covers.
6. Turn the crankshaft to position the **A** mark on the camshaft pulley with the mark on the housing.
7. Remove the crankshaft pulley mounting bolts and the pulley.
8. Remove the tensioner pulley lock bolt, the pulley and the spring.
9. Remove the water inlet pipe from the water pump.
10. Remove the water pump retaining bolts. Remove the water pump from its mounting.
11. Thoroughly clean the mounting surfaces of the pump and engine.
    To install:
12. Install the new water pump and torque the pump bolts to 14–19 ft. lbs. (19–25 Nm).
13. Install the idler pulley.
14. Install the timing belt and be sure the timing mark on the timing belt pulley is aligned with the matching mark. Make sure the mark **A** of the camshaft pulley is aligned with the timing mark. If it is not, turn the camshaft to align it.
15. Install the timing belt tensioner and spring. Temporarily secure it as the spring is fully extended.
16. Loosen the tensioner lock bolt. Turn the crankshaft twice in the direction of rotation. Align the timing marks. Tighten the timing belt tensioner lock bolt to 28–38 ft. lbs. Check the timing belt tension. The timing belt deflection should be 11–13mm (0.43–0.51 inch) at 22 lbs.
17. Install the fan and pulley.
18. Install the drive belts.
19. Fill the cooling system and check the timing.

2.6L Engine

1. Disconnect the battery ground.
2. Drain the cooling system.

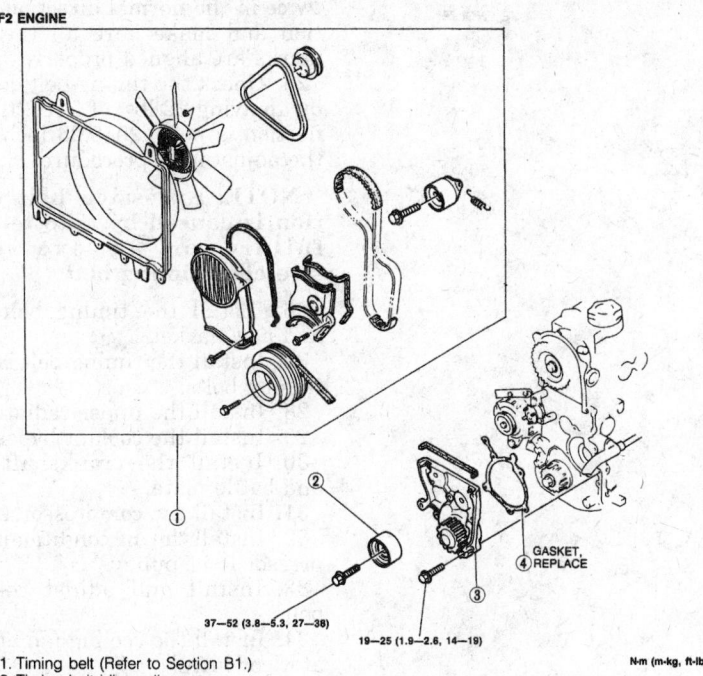

F2 ENGINE

37—52 (3.8—5.3, 27—38)

19—25 (1.9—2.6, 14—19)

GASKET, REPLACE

N·m (m-kg, ft-lb)

1. Timing belt (Refer to Section B1.)
2. Timing belt idler pulley
3. Water pump
   Inspect for body cracks and damaged
   gasket surface
4. Gasket

349901

Water pump assembly — 2.2L engine

— **CAUTION** —

When draining the coolant, always drain the coolant into a sealable container. Coolant should be reused unless it is contaminated or several years old.

3. Remove the accessory drive belts.
4. Remove the fan and shroud.
5. Remove the water pump pulley.
6. Unbolt and remove the water pump.
7. Thoroughly clean the gasket mounting surfaces.

To install:

8. Using a new gasket coated with sealer, position the water pump on the engine. For Pick-Ups, tighten the bolts to 19 ft. lbs. For MPV, tighten the bolts to 14—18 ft. lbs. (19—25 Nm).
9. Install the fan and pulley.
10. Install the drive belts and shroud.
11. Fill the cooling system.
12. Run the engine and check for leaks.

3.0L Engine

1. Position the engine at TDC on the compression stroke.
2. Disconnect the negative battery cable.

3. Remove the air cleaner assembly.
4. Drain the cooling system.
5. Remove the spark plug wires.
6. Remove the fresh air duct assembly.
7. Remove the cooling fan and radiator cowling.
8. Remove the drive belts.
9. Remove the air conditioning compressor idler pulley. Remove the compressor, and position it to the side without disconnecting the hose lines.
10. Remove the crankshaft pulley and baffle plate.
11. Remove the coolant bypass hose.
12. Remove the upper radiator hose.
13. Remove the timing belt cover retaining bolts. Remove the timing belt covers and gasket.
14. Turn the crankshaft to align the mating marks of the pulleys.
15. Remove the upper idler pulley.
16. Remove the timing belt. If reusing the belt be sure to mark the direction of rotation.
17. Remove the timing belt auto tensioner.
18. Unbolt and remove the water pump. Discard the gasket.
19. Thoroughly clean the mating surfaces of the pump and engine.

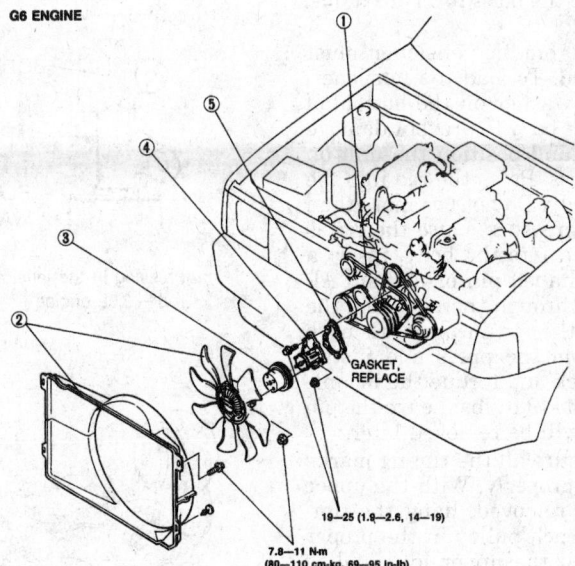

G6 ENGINE

19—25 (1.9—2.6, 14—19)

GASKET, REPLACE

7.8—11 N·m
(80—110 cm-kg, 69—95 in-lb)

1. Drive belt
2. Cooling fan and radiator cowling
3. Water pump pulley
4. Water pump
   Inspect body cracks and damaged gasket surface
5. Gasket

349902

Water pump assembly — 2.6L engine

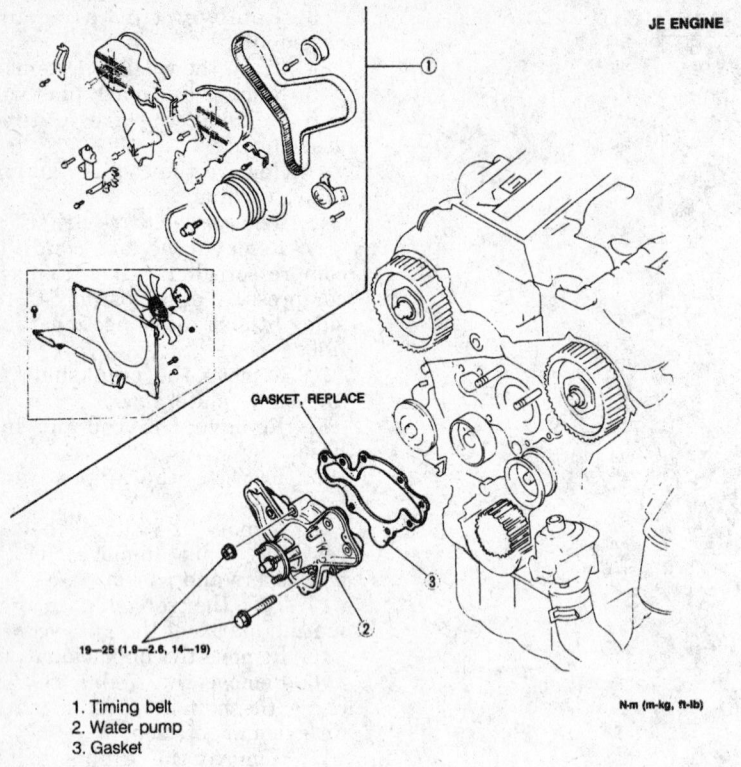

JE ENGINE

GASKET, REPLACE

19—25 (1.9—2.6, 14—19)

N·m (m-kg, ft-lb)

1. Timing belt
2. Water pump
3. Gasket

300513

Water pump assembly — 3.0L engine

To install:

20. Position the pump and a new gasket, coated with sealer, on the engine. Torque the bolts to 14–18 ft. lbs. (19–25 Nm).

21. The automatic tensioner must be pre-loaded. To load the tensioner, place a flat washer on the bottom of the tensioner body to prevent damage to the body and position the unit on an arbor press. Press the rod into the tensioner body. Do not use more than 2000 lbs of pressure. Once the rod is fully inserted into the body, insert a suitable L-shaped pin or a small Allen wrench through the body and the rod to hold the rod in place. Remove the unit from the press and install onto the block and torque the mounting bolt to 14–19 ft. lbs. Leave the pin in place, it will be removed later.

22. Make sure all the timing marks are aligned properly. With the upper idler pulley removed, hang the timing belt on each pulley in the proper order. Install the upper idler pulley and torque the mounting bolt to 27–38 ft. lbs. Rotate the crankshaft twice in the normal direction of rotation to align all the timing marks.

23. Make sure all the marks are aligned correctly. If not, repeat the previous step.

24. Remove the pin from the auto tensioner. Again turn the crankshaft

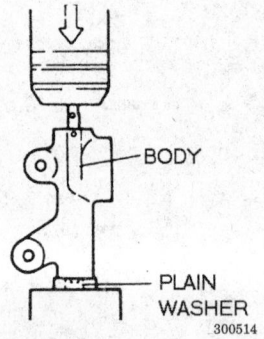

BODY

PLAIN
WASHER

300514

Pressing in the tensioner rod — 3.0L engine

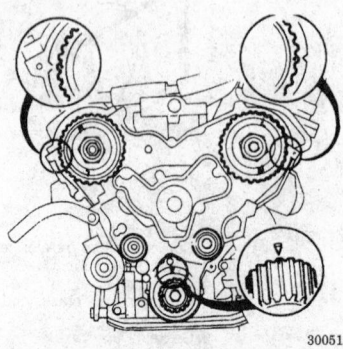

300515

Timing belt sprocket aligning marks — 3.0L engine

twice in the normal direction of rotation and make sure all the timing marks are aligned properly.

25. Check the timing belt deflection by applying 22 lbs. of force. If the deflection is not 0.20–0.28 inch, repeat the adjustment procedure.

NOTE: Excessive belt deflection is caused by auto tensioner failure or an excessively stretched timing belt.

26. Install the timing belt covers and new gasket.

27. Install the timing belt cover retaining bolts.

28. Install the upper radiator hose.

29. Install the coolant bypass hose.

30. Install the crankshaft pulley and baffle plate.

31. Install the compressor.

32. Install the air conditioning compressor idler pulley.

33. Install and adjust the drive belts.

34. Install the cooling fan and radiator cowling.

35. Install the fresh air duct assembly.

36. Install the spark plug wires.

37. Fill the cooling system.

38. Install the air cleaner assembly.

39. Connect the negative battery cable.

Thermostat

REMOVAL AND INSTALLATION

2.2L Engine

1. Drain enough coolant to bring the coolant level down below the thermostat housing. The thermostat housing is located on the left front side of the cylinder block.

2. Disconnect the temperature sending unit wire.

3. Remove the coolant outlet elbow.

4. If so equipped, position the vacuum control valve out of the way. The vacuum control valve is not used on California vehicles.

5. Disconnect the coolant by-pass hose from the thermostat housing.

6. Remove the thermostat and housing from the engine.

7. Note the position of the jiggle pin and remove the thermostat from the housing.

8. Remove all gasket material from the parts.

To install:

9. Position the thermostat in the housing with the jiggle pin up. Coat a new gasket with sealer and install it on the thermostat housing.

10. Install the thermostat housing using a new gasket with water resistant sealer. Torque the bolts to 20 ft. lbs.

11. Install the coolant outlet elbow and vacuum control valve, if equipped.

12. Connect the by-pass and radiator hoses.

13. Connect the temperature sending unit wire.

14. Fill the cooling system with the proper coolant. Operate the engine and check the coolant lever. Check for leaks.

2.6L Engine

The thermostat housing is at the end of the upper hose, on the cylinder head side, above the alternator.

1. Drain the cooling system to a point below the housing.

2. Remove the upper hose.

3. Remove the upper nut and lower bolt and remove the housing.

4. Remove the gasket and thermostat. Discard the gasket.

5. Thoroughly clean the mating surfaces of the head and housing.

To install:

6. Position the new thermostat in the head with the jiggle pin on the upper side.

7. Coat the gasket with an adhesive sealer and stick it in place on the head.

8. For Pick-Ups, install the housing and torque the nut and bolt to 19 ft. lbs. For MPV, torque the nut and bolt to 14–18 ft. lbs. (19–25 Nm).

9. Fill the cooling system.

3.0L Engine

NOTE: The thermostat housing is located at the engine end of the lower radiator hose.

1. Disconnect the negative battery cable.

2. Raise and support the front end on jackstands.

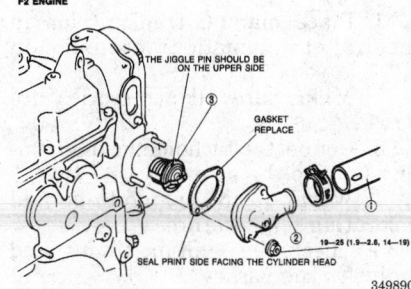

F2 ENGINE

Thermostat — 2.2L engine

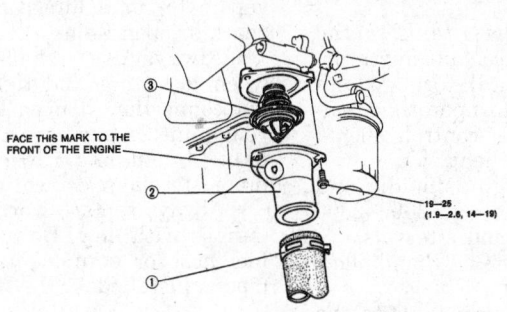

JE ENGINE

FACE THIS MARK TO THE FRONT OF THE ENGINE

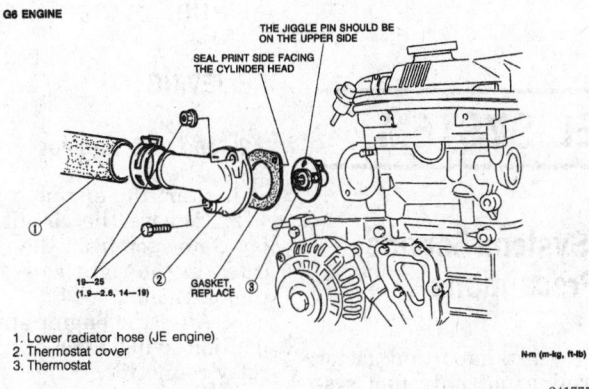

G6 ENGINE

THE JIGGLE PIN SHOULD BE ON THE UPPER SIDE

SEAL PRINT SIDE FACING THE CYLINDER HEAD

GASKET, REPLACE

1. Lower radiator hose (JE engine)
2. Thermostat cover
3. Thermostat

Nm (m-kg, ft-lb)

241775

Thermostat and housing — 2.6L (G6) and 3.0L (JE) engines

3. Drain the cooling system.

4. Remove the lower hose.

5. Unbolt and remove the housing and thermostat.

NOTE: This engine has a housing which incorporates an O-ring, eliminating the need for a gasket. Therefore, don't use sealer when replacing the housing.

6. Install a new thermostat in the housing and position the housing on the engine. Some housings are equipped with a location mark on the side, the mark should face the front of the engine when the housing is installed. Torque the bolts to 14–18 ft. lbs. (19–25 Nm).

7. Install the lower hose.

8. Fill the cooling system.

9. Lower the front end of the vehicle.

10. Connect the negative battery cable.

Electric Cooling Fan

REMOVAL AND INSTALLATION

1. Loosen and remove the drive belts.

2. Remove the bolts that retain the fan shroud to the radiator support. Remove (if there is enough room between the fan blades and radiator) or position the shroud back over the water pump and fan assembly, if necessary to gain working room.

3. Loosen and remove the fan to water pump mounting bolts or nuts and remove the fan assembly. Don't lay the fan, if equipped with a fan clutch, on its side. Fluid could be lost and the fan clutch might require replacement. Inspect the condition of the fan blades, if any are cracked or damaged, replace the fan.

To install:

4. Install the fan assembly in position on the water pump and pulley.

5. Secure the fan assembly with the mounting nuts or bolts. Tighten the nuts to 69–70 inch lbs. (7.8–11 Nm).

6. Install the shroud and drive belts. Adjust the drive belts to the proper tension.

Cooling System Bleeding

PROCEDURE

1. Fill the radiator with a 50/50 mixture of coolant and water to just below the radiator filler neck. Do not use more than 60 percent coolant, or

the efficiency of the cooling system will be impaired.

2. Fill the reserve tank to the MAX line with the coolant mixture.

3. Install the radiator cap and warm the engine to operating temperature. The heater controls should be set to maximum heat.

4. Allow the engine to idle and then run the engine at 2000–2800 rpm for a few 5-second intervals.

5. Shut the engine OFF and allow it to cool.

6. Refill the reserve tank to the proper level.

FUEL SYSTEM

Fuel System Service Precautions

Safety is the most important factor when performing not only fuel system maintenance but any type of maintenance. Failure to conduct maintenance and repairs in a safe manner may result in serious personal injury or death. Maintenance and testing of the vehicle's fuel system components can be accomplished safely and effectively by adhering to the following rules and guidelines.

• To avoid the possibility of fire and personal injury, always disconnect the negative battery cable unless the repair or test procedure requires that battery voltage be applied.

• Always relieve the fuel system pressure prior to disconnecting any fuel system component (injector, fuel rail, pressure regulator, etc.), fitting or fuel line connection. Exercise extreme caution whenever relieving fuel system pressure to avoid exposing skin, face and eyes to fuel spray. Please be advised that fuel under pressure may penetrate the skin or any part of the body that it contacts.

• Always place a shop towel or cloth around the fitting or connection prior to loosening to absorb any excess fuel due to spillage. Ensure that all fuel spillage (should it occur) is quickly removed from engine surfaces. Ensure that all fuel soaked cloths or towels are deposited into a suitable waste container.

• Always keep a dry chemical (Class B) fire extinguisher near the work area.

• Do not allow fuel spray or fuel vapors to come into contact with a spark or open flame.

• Always use a backup wrench when loosening and tightening fuel line connection fittings. This will prevent unnecessary stress and torsion to fuel line piping. Always follow the proper torque specifications.

• Always replace worn fuel fitting O-rings with new. Do not substitute fuel hose or equivalent, where fuel pipe is installed.

Fuel System Pressure

RELIEVING

2.2L and 2.6L Engines

1. Start the engine.

2. Remove the circuit opening relay connector from the relay box, located in the right side of the engine compartment.

3. After the engine stalls, turn the ignition switch OFF and reinstall the relay connector.

3.0L Engine

— CAUTION —
Fuel injection systems remain under pressure after the engine has been turned OFF. Properly relieve fuel pressure before disconnecting any fuel lines. Failure to do so may result in fire or personal injury.

— CAUTION —
Do not allow fuel spray or fuel vapors to come in contact with a spark or open flame. Keep a dry chemical fire extinguisher nearby. Never store fuel in an open container due to risk of fire or explosion.

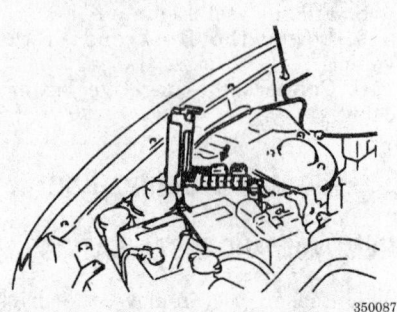

350087

Circuit opening relay connector location

1. Start the engine.

2. Disconnect the fuel pump connector, located at the PCM.

3. After the engine stalls, turn the ignition switch OFF.

4. Connect the fuel pump connector.

Idle Speed

ADJUSTMENT

Carbureted 2.2L Engine

Fast Idle

1. Set the fast idle cam so the fast idle lever rests on the second step of the cam.

2. Adjust the clearance between the air horn wall and the lower edge of the throttle plates, by turning the fast idle adjusting screw. Clearance should be 0.84–1.04mm (0.0331–0.0409 in.). Make sure the choke valve clearance hasn't changed.

Curb Idle Adjustment

Make sure the ignition timing, spark plugs, and carburetor float level are all in normal operating condition. Turn OFF all lights and other unnecessary electrical loads.

1. Connect a tachometer to the engine.

2. Start the engine and allow it to reach normal operating temperature. Make sure the choke valve has fully opened.

3. Check the idle speed. If necessary, adjust it to specification by turning the Throttle Adjusting Screw (TAS). The idle speed should be 800–850 rpm with manual transmission in neutral or automatic transmission in P.

Fuel Injected Engines

Except MPV

1. Place manual transmission in neutral or automatic transmission in P.

2. Make sure all accessories are OFF.

3. Connect a tachometer and timing light to the engine.

4. Warm up the engine to normal operating temperature.

5. Check the ignition timing and adjust, if necessary.

6. Ground the green 1-pin test connector to the body with a jumper wire.

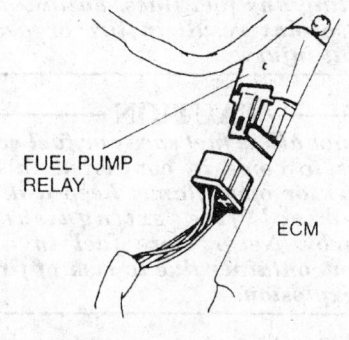

FUEL PUMP RELAY

ECM

304470

Fuel pump connector location

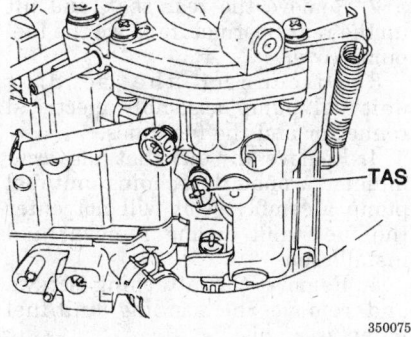

TAS

350075

Throttle adjusting screw (TAS) — carbureted 2.2L engines

7. Check the idle speed. Specifications are as follows:

2.2L and 2.6L engines

Manual transmission: 730–770 rpm

Automatic transmission: 750–790 rpm

3.0L engine: 780–820 rpm.

8. If the idle speed is not within specification, adjust by turning the air adjusting screw.

9. After adjustment, disconnect the jumper wire from the test connector. Recheck the ignition timing.

1995 MPV

1. Apply the parking brake. Place the shift lever in **P**.

2. Start the engine and bring to normal operating temperature. Make sure all accessories are **OFF**.

3. Connect a tachometer to the engine according to the manufacturer's instructions.

4. Check the idle speed and adjust, if necessary; it should be 750–790 rpm for 2.6L engines or 780–820 rpm for 3.0L engines.

5. Connect a jumper wire between the **TEN** terminal and the **GND** terminal of the data link connector.

6. Adjust the idle speed by turning the air adjusting screw.

7. Remove the jumper wire.

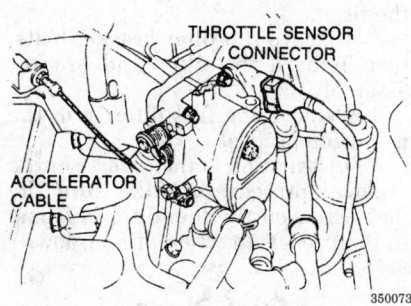

THROTTLE SENSOR CONNECTOR

ACCELERATOR CABLE

350073

Throttle body — fuel injected 2.2L engines

8. Remove all test equipment.

Mixture

ADJUSTMENT

2.2L Engine

Carbureted Engines

1. Start the engine and allow it to reach normal operating temperature. Let the engine run at idle.

2. Connect a dwell meter (90 degrees, 4-cylinder) to the air/fuel check connector (Br/Y).

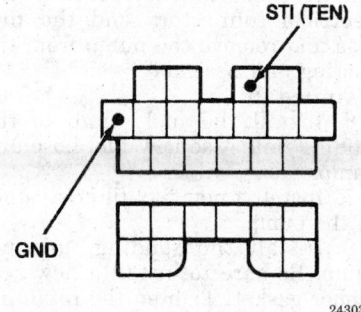

STI (TEN)

GND

243033

Connect a jumper wire to the Data Link Connector

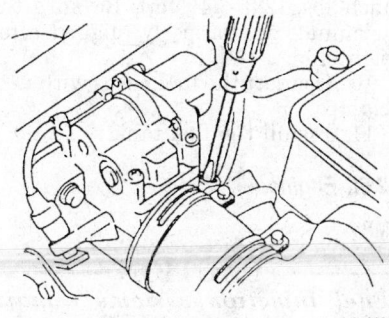

243032

Turn the air adjusting screw

3. Check the air/fuel mixture at the specified idle speed. The mixture should be 20–70 degrees. If the mixture is not as specified, adjust as follows:

a. Remove the carburetor and knock out the spring pin. Reinstall the carburetor.

b. Install the air cleaner and make sure the idle compensator is closed. Make sure all vacuum hoses are properly connected.

c. Connect a tachometer to the engine.

d. Warm up the engine and run it at idle. Make sure the idle speed is correct.

e. Reconnect the dwell meter to the air/fuel check connector.

f. Adjust the mixture to 27–45 degrees by turning the mixture adjusting screw.

g. Tap in the spring pin.

Fuel Injected Engines

The air/fuel mixture is computer controlled according to the needs of the engine and is not adjustable. If the air/fuel mixture is too lean or too rich, other problems with the engine and/or engine control system exist.

Fuel Filter

REMOVAL AND INSTALLATION

2.2L and 2.6L Engines

Except MPV

1. Relieve the fuel system pressure. Disconnect the negative battery cable.

2. Raise and safely support the vehicle, if necessary.

3. Disconnect the fuel lines from the fuel filter.

4. Remove the fuel filter and bracket assembly.

5. Installation is the reverse of the removal procedure. Make sure the flow arrow on the fuel filter is facing in the proper direction of fuel flow.

2.6L and 3.0L Engines

MPV

— **CAUTION** —

Fuel injection systems remain under pressure after the engine has been turned OFF. Properly relieve fuel pressure before disconnecting any fuel lines. Failure to do so may result in fire or personal injury.

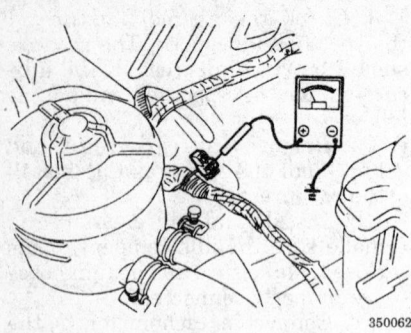

Connect the dwell meter to the check connector — 2.2L engine

350062

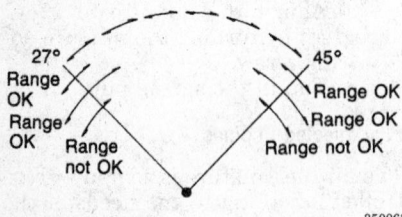

27° Range OK	45° Range OK
Range OK	Range OK
Range not OK	Range not OK

350063

Dwell meter readings for idle mixture — 2.2L engine

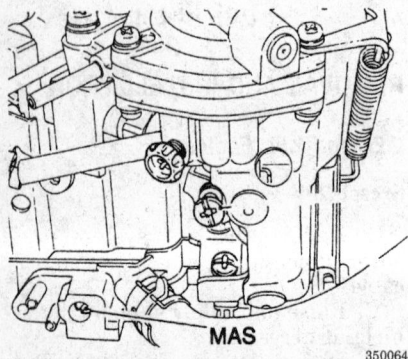

MAS

350064

Mixture adjusting screw — 2.2L engine

CAUTION

Do not allow fuel spray or fuel vapors to come in contact with a spark or open flame. Keep a dry chemical fire extinguisher nearby. Never store fuel in an open container due to risk of fire or explosion.

NOTE: The fuel filter is located in the engine compartment, next to the pulsation damper.

1. Relieve the fuel system pressure. Disconnect the negative battery cable.

2. Disconnect the fuel lines from the filter.

3. Remove the filter bracket bolts, then remove the filter and bracket assembly.

4. Remove the fuel filter from the mounting bracket.

5. Installation is the reverse of the removal procedure. Make sure the flow arrow on the fuel filter is facing in the proper direction of fuel flow.

Fuel Pump

REMOVAL AND INSTALLATION

2.2L and 2.6L Engines

Except MPV

NOTE: The fuel pump is located inside the fuel tank, attached to the tank sending unit assembly.

1. Disconnect the negative battery cable.

2. Remove the fuel tank.

3. Remove any dirt that has accumulated around the sending unit/fuel pump assembly so it will not enter the fuel tank during removal and installation.

4. Remove the attaching screws and remove the sending unit/fuel pump assembly.

5. If necessary, disconnect the electrical connectors and the fuel hose and remove the pump from the sending unit assembly.

To install:

6. Install the fuel pump to the sending unit. Use new fuel hose and clamps.

7. Install a new fuel filter/strainer on the pump.

8. Install the sending unit/fuel pump. Be sure to install a new seal rubber gasket. Tighten the retaining screws.

9. Tighten the positive terminal (blue) to 10–17 inch lbs. (1.2–2.0 Nm). Tighten the negative terminal 20–29 inch lbs. (2.3–3.4 Nm). Be sure the terminals are properly aligned after tightening.

10. Connect the fuel hoses with new clamps.

11. Install the fuel tank.

2.6L Engine

MPV

CAUTION

Fuel injection systems remain under pressure after the engine has been turned OFF. Properly relieve fuel pressure before disconnecting any fuel lines. Failure to do so may result in fire or personal injury.

CAUTION

Do not allow fuel spray or fuel vapors to come in contact with a spark or open flame. Keep a dry chemical fire extinguisher nearby. Never store fuel in an open container due to risk of fire or explosion.

1. Relieve the fuel system pressure, and disconnect the negative battery cable.

2. Remove the rear seat, and lift up the rear floormat. Remove the fuel pump cover.

3. Disconnect the sending unit/fuel pump assembly electrical connector and the fuel lines.

4. Remove any dirt that has accumulated around the sending unit/fuel pump assembly so it will not enter the fuel tank during removal and installation.

5. Remove the attaching screws and remove the sending unit/fuel pump assembly.

6. If necessary, disconnect the electrical connectors and the fuel hose, and remove the pump from the sending unit assembly.

To install:

7. If removed, connect the electrical connectors and the fuel hose, and install the pump to the sending unit assembly.

8. Install a new seal rubber gasket, and position the assembly in the fuel tank. Install the mounting bolts.

9. Connect the sending unit/fuel pump assembly electrical connector and the fuel lines.

10. Install the fuel pump cover, and replace the floormat.

11. Install the rear seat.

12. Connect the negative battery cable.

Fuel Injector

REMOVAL AND INSTALLATION

2.2L and 2.6L Engines

Except MPV

1. Relieve the fuel system pressure.

2. Remove the air chamber.

3. Disconnect the vacuum hose.

4. Disconnect the fuel lines.

5. Remove the pressure regulator and delivery pipe.

6. Disconnect the injector wiring.

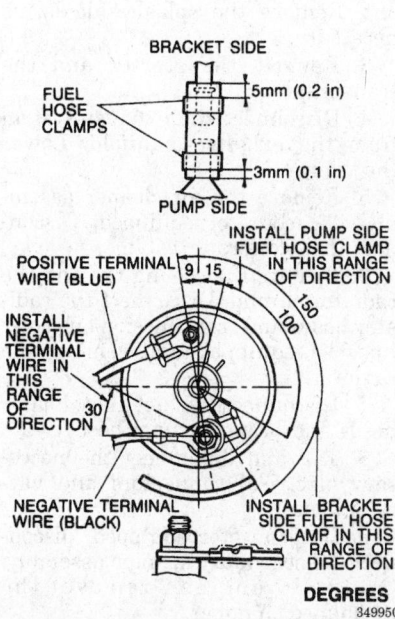

Terminal and clamp alignment on the top of the pump/sender unit — 2.2L and 2.6L engines

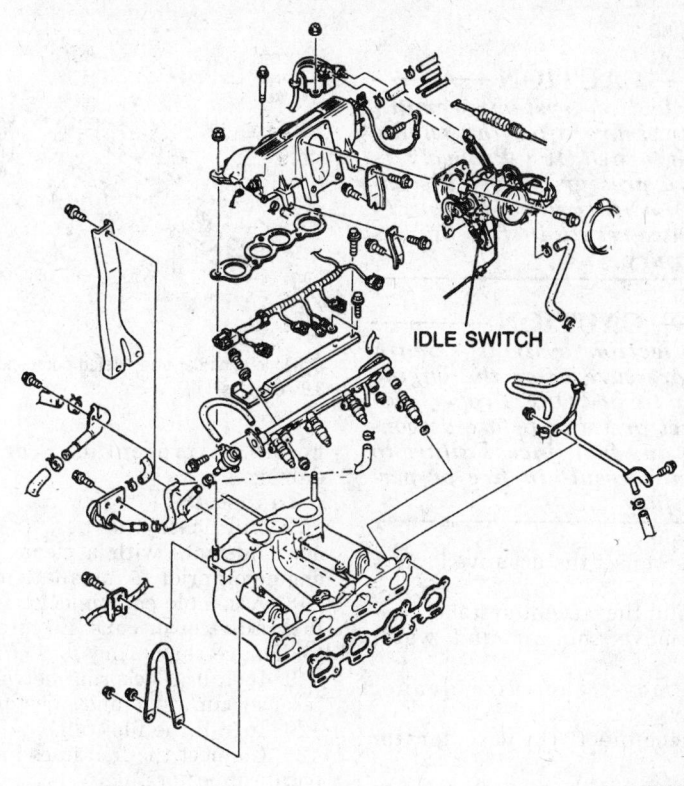

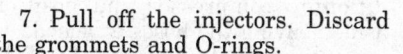

Fuel injection system exploded view — 2.2L and 2.6L engines

7. Pull off the injectors. Discard the grommets and O-rings.

8. Installation is the reverse of removal. Torque the delivery pipe bolts to 15 ft. lbs. Always use new O-rings coated with clean engine oil.

2.6L Engine

MPV

CAUTION

Fuel injection systems remain under pressure after the engine has been turned OFF. Properly relieve fuel pressure before disconnecting any fuel lines. Failure to do so may result in fire or personal injury.

CAUTION

Do not allow fuel spray or fuel vapors to come in contact with a spark or open flame. Keep a dry chemical fire extinguisher nearby. Never store fuel in an open container due to risk of fire or explosion.

1. Relieve the fuel system pressure.

2. Disconnect the negative battery cable.

3. Remove the upper intake manifold (dynamic chamber) as follows:

   a. Remove the manifold brackets.

   b. Disconnect and label the PCV hose, intake temperature sensor and ground wire.

   c. Remove the fuel injector harness bracket, and then remove the upper intake manifold.

4. Label and disconnect the vacuum and fuel hoses from the fuel rail.

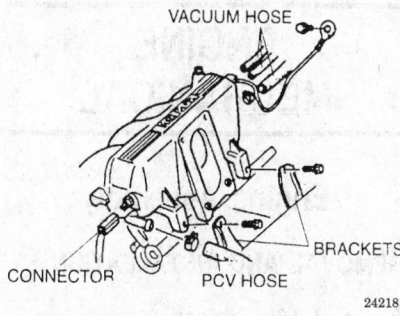

Dynamic chamber and related components — 2.6L (VIN 1) MPV

5. Remove the fuel rail with the pressure regulator attached.

6. Disconnect the injector harness from injectors.

7. Invert fuel injector rail assembly.

8. Remove lock rings securing injectors to fuel rail receiver cups. Pull injectors upward from receiver cups.

9. If injectors are to be reused, place a protective cap on injector nozzle to prevent dirt or other damage.

To install:

10. Lubricate the new O-ring of each injectors with a clean drop of engine oil prior to installation.

11. Assemble each injector into the fuel rail receiver cups. Be careful not to damage the O-rings.

12. Install lock ring between receiver cup ridge and injector slot.

13. Install the fuel rail. Connect the vacuum and fuel hoses.

14. Install the upper intake manifold (dynamic chamber). Tighten the mounting bolts for the manifold and brackets to 14–18 ft. lbs. (19–25 Nm).

15. Connect the negative battery cable.

16. Start the engine and examine for leaks.

3.0L Engine

— CAUTION —

Fuel injection systems remain under pressure after the engine has been turned OFF. Properly relieve fuel pressure before disconnecting any fuel lines. Failure to do so may result in fire or personal injury.

— CAUTION —

Fuel injection systems remain under pressure after the engine has been turned OFF. Properly relieve fuel pressure before disconnecting any fuel lines. Failure to do so may result in fire or personal injury.

---

1. Disconnect the negative battery cable.
2. Drain the coolant system.
3. Remove the air and water hoses.
4. Remove the air cleaner assembly.
5. Disconnect the accelerator cable.
6. Disengage the throttle position sensor connector.
7. Remove the throttle body and gasket.
8. Remove the Bypass Air Control (BAC) valve and intake air pipe.
9. Remove the extension manifolds.
10. Remove the dynamic chamber.
11. Disconnect the fuel lines and electrical connectors.
12. Remove the fuel rail.
13. Disconnect the injector harness from injectors.
14. Invert the fuel injector rail assembly.
15. Remove the lock rings securing the injectors to the fuel rail receiver cups. Pull the injectors upward from the receiver cups.
16. If the injectors are to be reused, place a protective cap on the injector

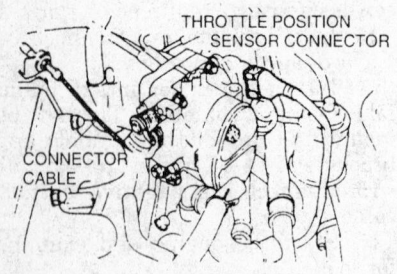

301390

Throttle body and related components — 3.0L engines

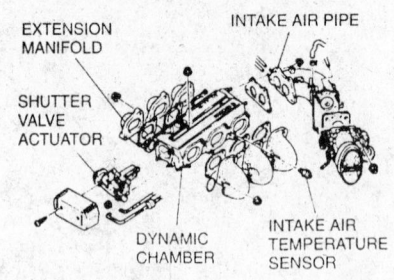

301391

Dynamic chamber and related components — 3.0L engines

nozzle to prevent dirt or other damage.

To install:

17. Lubricate the new O-ring of each injectors with a clean drop of engine oil prior to installation.
18. Assemble each injector into the fuel rail receiver cups. Be careful not to damage the O-rings.
19. Install a lock ring between the receiver cup ridge and injector slot.
20. Install the fuel rail.
21. Connect the fuel lines and electrical connectors.
22. Install the dynamic chamber.
23. Install the extension manifolds.
24. Install the intake air pipe, with a new gasket, and BAC valve.
25. Install the throttle body with a new gasket.
26. Engage the throttle position sensor connector.
27. Connect and adjust the accelerator cable.
28. Install the air cleaner assembly.
29. Install the air and water hoses.
30. Fill and bleed the coolant system.
31. Connect the negative battery cable.

ENGINE MECHANICAL

Engine Assembly

REMOVAL AND INSTALLATION

2.2L and 2.6L Engines

Except MPV

1. Relieve the fuel system pressure on fuel injected models. Disconnect the battery cables and remove the battery.

2. Raise and safely support the vehicle. Drain the engine oil and coolant. Remove the splash shields, as necessary.
3. Remove the starter and the transmission.
4. Disconnect the exhaust system from the exhaust manifold. Lower the vehicle.
5. Remove the air cleaner assembly, if carburetor equipped. Disconnect the accelerator cable.
6. Remove the cooling fan and the radiator shroud. Disconnect the radiator hoses and transmission oil cooler lines, if equipped and remove the radiator.
7. Disconnect the fuel lines, heater hoses and brake vacuum hose.
8. Tag and disconnect the necessary electrical connectors and vacuum hoses.
9. If carburetor equipped, disconnect the secondary air pipe assembly. On 2.6L engine, remove the resonance chamber.
10. Remove the accessory drive belt(s). If equipped, remove the power steering pump pulley and the power steering pump. Position the pump aside, leaving the hoses connected.
11. If equipped, remove the air conditioning compressor and position aside, leaving the hoses attached.
12. Remove the gusset plates, if equipped. Remove the transmission oil cooler line retainers, if equipped.
13. Attach suitable engine lifting equipment to the engine. Remove the engine mount nuts and remove the engine from the vehicle.
14. Install the engine on a workstand.

To install:

15. Remove the engine from the workstand and position in the vehicle. Install the engine mount nuts and tighten to 30–36 ft. lbs. (40–49 Nm).
16. Install the gusset plates, if equipped. Attach the transmission oil cooler line retainers, if equipped.
17. Install the air conditioning compressor, if equipped. Tighten the mounting bolts to 29–40 ft. lbs. (39–54 Nm).
18. Install the power steering pump, if equipped. Tighten the mounting bolts to 23–34 ft. lbs. (31–46 Nm). Install the power steering pump pulley and tighten the nut to 36–43 ft. lbs. (49–59 Nm).
19. Install the secondary air pipe, if equipped. Install the resonance chamber, if equipped.
20. Connect all vacuum lines and electrical connectors. Connect the

brake vacuum hose, heater hoses and fuel lines.

21. Install the accessory drive belt(s) and cooling fan. Install the fan shroud, radiator and radiator hoses.

22. Adjust the accessory drive belt tension.

23. Connect the accelerator cable. Install the air cleaner assembly on carbureted engine.

24. Raise and safely support the vehicle. Connect the exhaust pipe to the exhaust manifold and tighten the attaching nuts to 30–36 ft. lbs. (40–49 Nm).

25. Install the starter and the transmission assembly. Install the splash shields and lower the vehicle.

26. Fill the crankcase with the proper type and quantity of engine oil. Install the battery and connect the battery cables.

27. Fill and bleed the cooling system. Run the engine and check for leaks and proper operation.

2.6L Engine

MPV

─────── **CAUTION** ───────

Fuel injection systems remain under pressure after the engine has been turned OFF. Properly relieve fuel pressure before disconnecting any fuel lines. Failure to do so may result in fire or personal injury.

─────── **CAUTION** ───────

Do not allow fuel spray or fuel vapors to come in contact with a spark or open flame. Keep a dry chemical fire extinguisher nearby. Never store fuel in an open container due to risk of fire or explosion.

1. Relieve the fuel system pressure.

2. Disconnect the battery cables, and remove the battery.

3. Raise and safely support the vehicle. Drain the engine oil and coolant.

4. Remove the splash shield.

5. Remove the starter, and disconnect the transmission.

6. Disconnect the exhaust pipe from the exhaust manifold and lower the vehicle.

7. Remove the fresh air duct and the radiator hoses.

8. Disconnect the transmission oil cooler lines from the radiator, if equipped.

9. Remove the purge control solenoid valve.

10. Remove the radiator, fan shroud and cooling fan.

11. Remove the accessory drive belts.

12. Disconnect the accelerator cable and remove the resonance chamber and air cleaner.

13. Remove the Pressure Regulator Control (PRC) solenoid valve.

14. Tag and disconnect the vacuum hose to vacuum actuator, and the canister hose.

15. Disconnect the brake vacuum hose, heater hoses and fuel lines.

16. Tag and disconnect the engine harness connectors and the ground wire.

17. Remove the shroud upper panel and the air conditioning pipe bracket.

18. If equipped, remove the power steering pump and position aside, leaving the hoses attached. It is necessary to remove the power steering pulley prior to removing the pump.

19. If equipped, remove the air conditioning compressor and position aside, leaving the hoses attached.

20. Remove the lower grille and radiator grille.

21. Remove the shroud upper plate and the additional condenser fan, if equipped.

22. Attach suitable engine lifting equipment to the engine. Remove the engine mount nuts and remove the engine from the vehicle.

23. Install the engine on a workstand.

To install:

24. Remove the engine from the workstand. Lower the engine into the vehicle, being careful not to damage the piping.

NOTE: Lean the air conditioning condenser forward to ease engine installation.

25. Install the engine mount nuts and tighten to 25–36 ft. lbs. (34–49 Nm). Install the additional condenser fan, if equipped.

26. Apply a bead of sealer to each side of the front support, then install the shroud upper plate. Tighten the mounting bolts to 61–87 inch lbs. (6.9–9.8 Nm).

27. Install the radiator grille and lower grille.

28. Install the air conditioning compressor, if equipped. Tighten the mounting bolts to 13–20 ft. lbs. (18–26 Nm). Install the air conditioner pipe bracket and tighten the mounting nuts to 61–87 inch lbs. (6.9–9.8 Nm).

29. Install the power steering pump, if equipped. Tighten the mounting bolts to 23–34 ft. lbs. (31–46 Nm). Install the pump pulley

and tighten the nut to 29–43 ft. lbs. (39–59 Nm).

30. Install the shroud upper panel and tighten the bolts to 69–95 inch lbs. (7.8–11.0 Nm).

31. Connect the engine harness electrical connectors and ground wire.

32. Connect the brake vacuum hose, heater hoses and fuel lines.

33. Connect the vacuum hose to the vacuum actuator, and the canister hose.

34. Install the Pressure Regulator Control (PRC) solenoid valve.

35. Connect the accelerator cable.

36. Install the resonance chamber and air cleaner.

37. Install the cooling fan.

38. Install the fan shroud and the radiator.

39. Install the drive belts. Adjust the drive belt tension.

40. Install the purge control solenoid valve.

41. Install the radiator hoses and fresh air duct. If equipped, connect the transmission oil cooler lines.

42. Raise and safely support the vehicle. Connect the exhaust pipes to the exhaust manifolds. Tighten the nuts to 25–36 ft. lbs. (34–49 Nm).

43. Install the starter and transmission assembly.

44. Install the splash shield, and lower the vehicle.

45. Install the battery, and connect the negative battery cables.

46. Fill the crankcase with the proper type and quantity of engine oil.

47. Fill and bleed the cooling system. Run the engine and check for leaks and proper operation.

3.0L Engine

─────── **CAUTION** ───────

Fuel injection systems remain under pressure after the engine has been turned OFF. Properly relieve fuel pressure before disconnecting any fuel lines. Failure to do so may result in fire or personal injury.

─────── **CAUTION** ───────

Do not allow fuel spray or fuel vapors to come in contact with a spark or open flame. Keep a dry chemical fire extinguisher nearby. Never store fuel in an open container due to risk of fire or explosion.

1. Relieve the fuel system pressure.

2. Disconnect the battery cables and remove the battery.

3. Raise and safely support the vehicle. Drain the engine oil and coolant.

4. Remove the splash shield.

5. Remove the starter, and disconnect the transmission from the engine.

6. Disconnect the exhaust pipes from the exhaust manifolds and lower the vehicle.

7. Remove the fresh air duct and the radiator hoses.

8. Disconnect the transmission oil cooler lines from the radiator, if equipped.

9. Remove the radiator, fan shroud and cooling fan.

10. Remove the accessory drive belts.

11. Tag and disconnect the volume air flow sensor connector. Remove the air cleaner and the volume air flow sensor.

12. Disconnect the brake vacuum hose, heater hoses and fuel lines.

13. Disconnect the vacuum hose to the vacuum actuator and the canister hose.

14. Disconnect the accelerator cable.

15. Remove the alternator.

16. If equipped, remove the power steering pump and position aside, leaving the hoses attached. It is necessary to remove the power steering pulley prior to removing the pump.

17. If equipped, remove the air conditioning compressor and position aside, leaving the hoses attached.

18. Remove the protector cover from the front of the engine.

19. Tag and disconnect the emission harness connectors and the ground wire.

20. Remove the shroud upper panel and the air conditioning pipe bracket.

21. Remove the lower grille and radiator grille. Remove the shroud upper plate.

22. Attach suitable engine lifting equipment to the engine. Remove the engine mount nuts and remove the engine from the vehicle.

23. Install the engine on a workstand.

To install:

24. Remove the engine from the workstand. Lower the engine into the vehicle, being careful not to damage the piping.

NOTE: Lean the air conditioning condenser forward to ease engine installation.

25. Install the engine mount nuts and tighten to 25–36 ft. lbs. (34–49 Nm).

26. Apply a bead of sealer to each side of the front support, then install the shroud upper plate. Tighten the mounting bolts to 61–87 inch lbs. (6.9–9.8 Nm).

27. Install the radiator grille and lower grille.

28. Install the air conditioning compressor, if equipped. Tighten the mounting bolts to 13–20 ft. lbs. (18–26 Nm). Install the air conditioner pipe bracket and tighten the mounting nuts to 61–87 inch lbs. (6.9–9.8 Nm).

29. Install the power steering pump, if equipped. Tighten the mounting bolts to 23–34 ft. lbs. (31–46 Nm). Install the pump pulley and tighten the nut to 29–43 ft. lbs. (39–59 Nm).

30. Install the shroud upper panel and tighten the bolts to 69–95 inch lbs. (7.8–11.0 Nm).

31. Install the alternator.

32. Connect the accelerator cable. Install the air cleaner and volume airflow sensor.

33. Connect all electrical connectors and vacuum hoses.

34. Connect the brake vacuum hose, heater hoses and fuel lines.

35. Connect the vacuum hose to the vacuum actuator and the canister hose.

36. Install the accessory drive belts and the cooling fan. Install the fan shroud and the radiator. Adjust the drive belt tension.

37. Install the radiator hoses and fresh air duct.

38. If equipped, connect the transmission oil cooler lines.

39. Raise and safely support the vehicle. Connect the exhaust pipes to the exhaust manifolds. Tighten the nuts to 25–36 ft. lbs. (34–49 Nm).

40. Install the starter, and connect the transmission assembly.

41. Install the splash shield and lower the vehicle.

42. Install the battery and connect the negative battery cables. Fill the crankcase with the proper type and quantity of engine oil.

43. Fill and bleed the cooling system. Run the engine and check for leaks and proper operation.

Engine Mounts

REMOVAL AND INSTALLATION

2.2L and 2.6L (VIN 4) Engines

Front Mounts

1. Remove the fan shroud attaching bolts.

2. Support the engine using a wood block and a jack under the oil pan.

3. Remove the engine mount bracket nut(s). Raise the engine with the jack just enough to remove the mount.

4. Remove the attaching nuts/bolts and remove the mount.

5. Installation is the reverse of the removal procedure. Tighten the mounting bolts to 27–50 ft. lbs. (37–63 Nm).

Rear Mounts

1. Raise and safely support the vehicle.

2. Support the transmission with a jack.

3. Remove the mount-to-crossmember nuts/bolts and the mount-to-transmission bolts.

4. Raise the transmission just enough to remove the mount.

5. Installation is the reverse of the removal procedure. Tighten the mounting bolts to 60–85 ft. lbs. (86–115 Nm).

2.6L (VIN 1) Engine

MPV

1. Disconnect the negative battery cable.

2. Raise and support the vehicle safely.

3. Attach a hoist to the engine, and lift until the slack in the chain is taken up.

4. Remove the nuts from the engine mounts.

NOTE: Inspect the engine compartment for components that may bind when the engine is raised. Disconnect these components.

5. Lift the engine the exact amount needed to remove the engine mount. Do not lift any higher.

6. Remove the engine mount(s).

To install:

7. Install the engine mount(s) to the engine, tighten the mounting bolts to 29–42 ft. lbs. (39–57 Nm).

8. Lower the engine, and install the engine mount to the frame. Tighten the nuts to 26–36 ft. lbs. (35–49 Nm).

9. Remove the engine hoist and lower the vehicle.

10. Connect the negative battery cable.

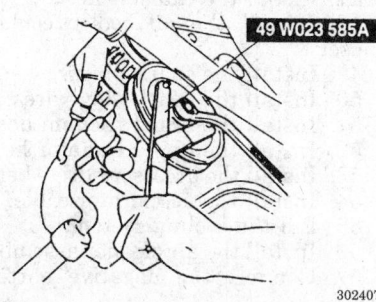

`49 W023 585A`

`302407`

Removing and installing the power steering pump pulley — 1996–97 3.0L engines

Cylinder Head

REMOVAL AND INSTALLATION

2.2L Engine

1. Relieve the fuel system pressure if an injected model. Disconnect the negative battery cable and drain the cooling system.

——————— CAUTION ———————
When draining the coolant, always drain the coolant into a sealable container. Coolant should be reused unless it is contaminated or several years old.

2. Disconnect the spark plug wires and remove the spark plugs.
3. Disconnect the accelerator cable. If equipped with automatic transmission, disconnect the throttle cable.
4. Remove the air intake pipe.
5. Remove the air cleaner and fuel hose. Cover the fuel hose to prevent leakage, if equipped with a carburetor. If injected, remove the air intake hose. If equipped with a carburetor, remove the fuel pump.

——————— CAUTION ———————
Never smoke when working around gasoline! Avoid all sources of sparks or ignition. Gasoline vapors are EXTREMELY volatile!

6. Remove the upper radiator hose, water by-pass hose, heater hose, and brake vacuum hose.
7. Remove the 3-way and EGR solenoid valve assemblies.
8. Disconnect the engine harness connector and ground wire.
9. Remove the vacuum chamber and exhaust manifold insulator.
10. Remove the EGR pipe and exhaust pipe.
11. Remove the exhaust manifold.

12. Remove the intake manifold bracket and the intake manifold.
13. Remove the distributor.
14. Loosen the air conditioning compressor and bracket, position it aside and tie it out of the way. Do not disconnect the refrigerant lines!
15. Remove the upper timing belt cover and the timing belt tensioner spring.
16. To remove the timing belt, perform the following:

   a. Rotate the crankshaft so the **1** on the camshaft pulley is aligned with the timing mark on the front housing.

   b. When timing marks are aligned, loosen the timing belt tensioner lock bolt. Pull the tensioner as far out as will go and then temporarily tighten the lock bolt to hold it there.

   c. Lift the timing belt from the camshaft pulley and position it out of the way.

17. Remove the cylinder head cover and cover gasket.
18. Loosen the cylinder head bolts in the proper sequence and remove the cylinder head and head gasket.

To install:

19. Position a new head gasket on the cylinder block and install the cylinder head. Apply oil to the threads and seat faces of the cylinder head bolts and install.
20. Tighten the cylinder head bolts in 2–3 steps in the proper sequence. The final torque specification is 59–64 ft. lbs. (80–86 Nm).
21. Apply silicone sealer to each side of the front and rear camshaft bearing caps where the cap meets the cylinder head. Install the rocker arm cover and tighten the bolts to 26–35 inch lbs. (2.9–3.9 Nm).
22. Install the timing belt and tensioner. Install the timing belt front cover.
23. Install the intake and exhaust manifolds.
24. Install the upper radiator and water by-pass hoses. If carburetor equipped, install the secondary air pipe.
25. Install the distributor. Install the spark plugs and connect the spark plug wires.
26. Connect all vacuum hoses and electrical connectors. Connect the heater hoses and the brake vacuum hose.
27. If equipped, install the fuel pump. Connect the fuel lines.
28. Install the cooling fan and shroud.

29. Connect the accelerator cable. Install the air cleaner or air intake hose.
30. Install the splash shield. Connect the negative battery cable.
31. Fill and bleed the cooling system. Run the engine and check for leaks.
32. Check and adjust the ignition timing.

2.6L Engine

1. Properly relieve the fuel system pressure.

——————— CAUTION ———————
Never smoke when working around gasoline! Avoid all sources of sparks or ignition. Gasoline vapors are EXTREMELY volatile!

2. Disconnect the negative battery cable.
3. Remove the air cleaner assembly.
4. Drain the coolant.
5. Position the engine at TDC on the compression stroke so all the pulley matchmarks are aligned.
6. Remove the accelerator cable. Remove the air intake pipe and resonance chamber.
7. Remove the accessory drive belts and A/C belt idler.
8. Remove the upper radiator hose.
9. Remove the brake vacuum hose.
10. Remove the spark plug wires.
11. Remove the spark plugs.
12. Remove the oil cooler coolant hose.
13. Remove the canister hose.
14. Remove the fuel lines.
15. Disconnect the oxygen sensor.
16. Remove the solenoid valves.
17. Disconnect the emissions harness.
18. Remove the rocker cover. Check to ensure the engine is set on TDC. The timing mark on the camshaft sprocket should be 90 degrees to the right, parallel to the top of the cylinder head. Make sure the yellow crankshaft pulley timing mark is aligned with the indicator pin.
19. Mark the position of the distributor rotor in relation to the distributor housing, and the distributor housing in relation to the cylinder head. Remove the distributor. Do not rotate the engine after distributor removal.
20. Hold the crankshaft pulley with a suitable tool and remove the distributor drive gear/camshaft pulley retaining bolt and the drive gear. Remove the upper timing cover assembly.

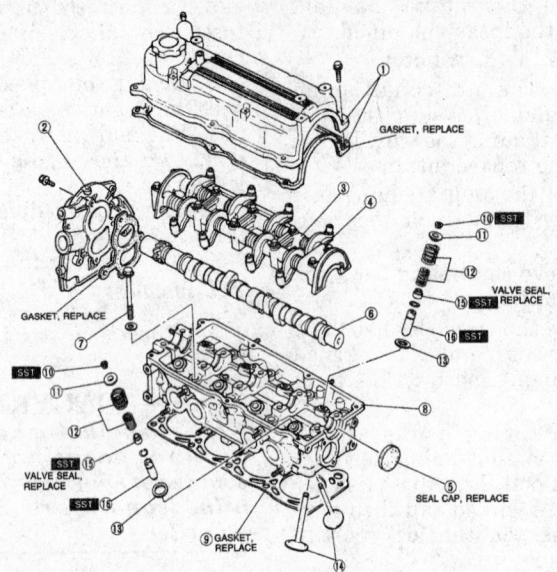

1. Cylinder head cover and gasket
2. Front housing
3. Rocker arm and shaft assembly
4. Hydraulic lash adjuster (HLA)
5. Seal cap
6. Camshaft
7. Cylinder head bolt
8. Cylinder head
9. Cylinder head gasket
10. Valve keepers
11. Upper spring seat
12. Valve spring, outer and inner
13. Lower spring seat
14. Valve
15. Valve seal
16. Valve guide

350195

Upper engine assembly — 2.2L engine

21. Push the timing chain adjuster sleeve in towards the left, and insert a 0.0787 in. (2mm) diameter x 1.77 in. (45mm) long pin into the lever hole to hold it in place.

22. Wire the chain to the pulley and remove the pulley from the camshaft. Do not allow the sprocket and chain to fall down into the engine and cause the chain to become disengaged from the crankshaft sprocket.

23. Remove the intake manifold bracket.

24. Disconnect the exhaust pipe.

25. Remove the 2 front head bolts.

26. Remove the remaining head bolts starting from the middle and working outward toward the ends of the head.

27. Lift off the head.

28. Discard the head gasket.

29. Thoroughly clean the mating surfaces of the head and block.

30. Check the head and block for flatness with a straightedge.

To install:

31. Apply RTV sealer to the top front of the block.

32. Place a new head gasket on the block.

33. Position the head on the block.

34. Clean the head bolts and apply oil to the threads.

35. Tighten the head bolts, in 2 even steps, to 64 ft. lbs.

36. Torque the 2 front bolts, to 17 ft. lbs.

37. Place the camshaft pulley on the camshaft and tighten the bolt to to 95 inch lbs.; the nut to 87 inch lbs.

38. Connect the exhaust pipe.

39. Install the intake manifold bracket.

40. Install the upper timing cover assembly.

41. Install the distributor.

42. Install the rocker cover.

43. Connect the emissions harness.

44. Install the solenoid valves.

45. Connect the oxygen sensor.

46. Install the fuel lines.

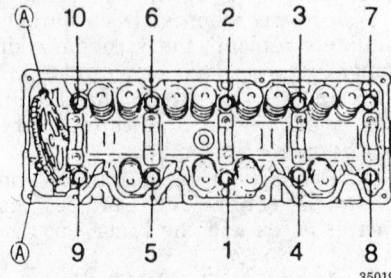

350197

Cylinder head bolt torque sequence — 2.6L engine

47. Install the canister hose.

48. Install the oil cooler coolant hose.

49. Install the spark plugs.

50. Install the spark plug wires.

51. Install the brake vacuum hose.

52. Install the upper radiator hose.

53. Install the accessory drive belts.

54. Install the accelerator cable.

55. Fill the cooling system.

56. Install the air cleaner assembly.

57. Connect the negative battery cable.

3.0L Engine

— **CAUTION** —

Fuel injection systems remain under pressure after the engine has been turned OFF. Properly relieve fuel pressure before disconnecting any fuel lines. Failure to do so may result in fire or personal injury.

— **CAUTION** —

Do not allow fuel spray or fuel vapors to come in contact with a spark or open flame. Keep a dry chemical fire extinguisher nearby. Never store fuel in an open container due to risk of fire or explosion.

1. Position the engine at TDC on the compression stroke.

2. Properly relieve the fuel system pressure.

3. Disconnect the negative battery cable.

4. Remove the air cleaner assembly.

5. Disconnect the accelerator cable.

6. Drain the cooling system.

7. Remove the spark plug wires.

8. Remove the fresh air duct assembly.

9. Remove the cooling fan and radiator cowling.

10. Remove the drive belts.

11. Remove the air conditioning compressor idler pulley. If necessary, remove the compressor and position it to the side.

12. Remove the crankshaft pulley and baffle plate.

13. Remove the coolant bypass hose.

14. Remove the upper radiator hose.

15. Remove the timing belt cover assembly retaining bolts. Remove the timing belt cover assembly and gasket.

16. Turn the crankshaft to align the mating marks of the pulleys.

17. Remove the upper idler pulley.

18. Remove the timing belt. If re-using the belt be sure to mark the direction of rotation.

19. Disconnect and plug canister, brake vacuum and fuel hoses. If equipped with automatic transmission, disconnect the automatic transmission vacuum hose.

20. Remove the 3-way solenoid valve assembly and disconnect all engine harness connector and grounds.

21. If equipped with automatic transmission, remove the dipstick. Disconnect the required vacuum hoses. Disconnect the accelerator linkage.

22. Remove the distributor and the EGR pipe.

23. Remove the 6 extension manifolds. Remove the O-rings from the extension manifolds and replace with new ones. Remove the intake manifold by loosening the retaining bolts in the proper sequence.

24. Remove the cylinder head cover, gasket and seal washers.

25. Remove the center exhaust pipe insulator and pipe. Disconnect the exhaust manifold retaining bolts. Remove the exhaust manifold with insulator.

26. Remove the seal plate.

27. Remove the cylinder head retaining bolts in the proper sequence in 2 or 3 stages. Remove the cylinder head from the vehicle.

28. Thoroughly clean the cylinder head and cylinder block contact surfaces to remove any dirt or oil. Check the cylinder head for warpage and cracks. The maximum allowable warpage is 0.10mm. Inspect the cylinder head bolts for damaged threads and make sure they are free from grease and dirt. After the bolts are cleaned, measure the length of each bolt and replace out of specifications bolts as required.

   a. Length: Intake — 108mm; Exhaust — 138mm

   b. Maximum: Intake — 109mm; Exhaust — 139mm

29. Check the oil control plug projection at the cylinder block. Projection should be 0.53–0.57mm. If correct, apply clean engine oil to a new O-ring and position it on the control plug.

To install:

30. Place the new cylinder head gasket on the left bank with the **L** mark facing up. Place the new cylinder head gasket on the right bank with the **R** mark facing up. Install the cylinder onto the block. Tighten

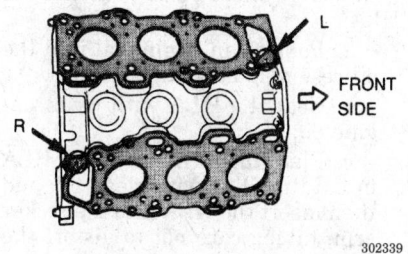

Head gasket positioning — 3.0L engine

the head bolts in the following manner:

   a. Coat the threads and the seating faces of the head bolts with clean engine oil.

   b. Torque the bolts in the proper sequence to 14 ft. lbs.

   c. Paint a mark on the head of each bolt.

   d. Using this mark as a reference, tighten the bolts in the proper sequence an additional 90 degrees.

   e. Repeat the previous step.

31. Install the seal plate.

32. Install the exhaust manifold with insulator.

33. Connect the exhaust manifold retaining bolts.

34. Install the center exhaust pipe insulator and pipe.

35. Install the cylinder head cover, gasket and seal washers.

36. Install the intake manifold by loosening the retaining bolts in the proper sequence.

37. Install the O-rings from the extension manifolds.

38. Install the 6 extension manifolds.

39. Install the distributor and the EGR pipe.

40. If equipped with automatic transmission, install the dipstick.

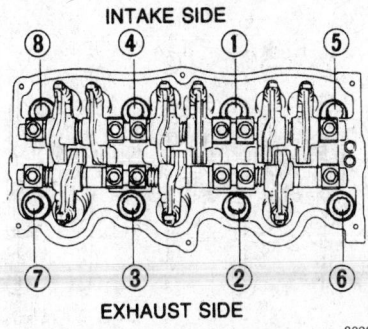

INTAKE SIDE

EXHAUST SIDE

302341

Cylinder head bolt torque sequence — 3.0L engine

Connect the required vacuum hoses. Connect the accelerator linkage.

41. Install the 3-way solenoid valve assembly and connect all engine harness connector and grounds.

42. Connect the canister, brake vacuum and fuel hoses. If equipped with automatic transmission, connect the automatic transmission vacuum hose.

43. To install the timing belt, first the automatic tensioner must be loaded. To load the tensioner:

   a. Place a flat washer on the bottom of the tensioner body to prevent damage to the body and position the unit on an arbor press.

   b. Press the rod into the tensioner body. Do not use more than 2000 lbs. of pressure.

   c. Once the rod is fully inserted into the body, insert a suitable L-shaped pin or a small Allen wrench through the body and the rod to hold the rod in place.

   d. Remove the unit from the press and install onto the block and torque the mounting bolt to 14–19 ft. lbs.

   e. Leave the pin in place, it will be removed later.

44. Make sure all the timing marks are aligned properly. With the upper idler pulley removed, hang the timing belt on each pulley in the order.

45. Install the upper idler pulley and torque the mounting bolt to 27–38 ft. lbs.

46. Rotate the crankshaft twice in the normal direction of rotation to align all the timing marks.

47. Make sure all the marks are aligned correctly.

48. Remove the pin from the auto tensioner. Again turn the crankshaft twice in the normal direction of rotation and make sure all the timing marks are aligned properly.

49. Check the timing belt deflection by applying 22 lbs. of force. If the deflection is not 5–7mm, repeat the adjustment procedure.

NOTE: Excessive belt deflection is caused by auto tensioner failure or an excessively stretched timing belt.

50. Install the upper idler pulley.

51. Install the timing belt cover assembly and new gasket.

52. Install the upper radiator hose.

53. Install the coolant bypass hose.

54. Install the crankshaft pulley and baffle plate.

55. Install the A/C compressor.

56. Install the air conditioning compressor idler pulley.

57. Install the accessory drive belts.

58. Install the cooling fan and radiator cowling.

59. Install the fresh air duct assembly.

60. Install the spark plug wires.

61. Connect and adjust the accelerator cable.

62. Fill the cooling system.

63. Install the air cleaner assembly.

64. Connect the negative battery cable.

Valve Lifters

BLEEDING

2.2L and 2.6L Engines

NOTE: The manufacturer does not recommend that the Hydraulic Lash Adjusters (HLA) be bled. Removing an HLA from its rocker arm will release its oil. If the HLAs are removed from the rocker arms, new ones should be installed using new O-rings.

1. Before installing new HLAs, fill the rocker arm oil reservoir with fresh engine oil.

2. Apply fresh engine oil to the new HLA and its O-ring.

3. Install the new HLA into the rocker arm, taking care not to damage or distort its O-ring.

REMOVAL AND INSTALLATION

2.6L Engine

1. Disconnect the negative battery cable.

2. Remove the air intake hose.

3. Remove the rocker arm cover.

4. Loosen the rocker arm/shaft assembly mounting bolts in 2–3 steps in the proper sequence. Remove the rocker arm/shaft assembly together with the bolts.

5. If necessary, disassemble the rocker arm/shaft assembly, noting the position of each component to ease reassembly.

6. Check for wear or damage to the contact surfaces of the shafts and rocker arms; replace as necessary.

7. Check the surface of the Hydraulic Lash Adjuster (HLA) for wear and damage. If the HLA is worn or damaged, it must be replaced.

8. Remove the HLA from the rocker arm. Don't remove the HLA unless necessary, because oil leakage will occur if the O-ring is damaged.

To install:

9. Follow these steps to install the HLA:

   a. Pour clean engine oil into the oil reservoir in the rocker arm.

   b. Coat the HLA with clean engine oil.

   c. Place the rocker arm and HLA into a tub filled with clean oil, and then insert the HLA into the rocker arm, taking care not to distort the O-ring.

10. Apply clean engine oil to the rocker arm shafts and rocker arms and assemble the rocker arm/shaft assembly in the reverse order of disassembly, noting the following:

   a. The intake side shaft has twice as many oil holes as the exhaust side shaft.

   b. The No. 4 camshaft cap has an oil hole from the cylinder head; make sure it is installed correctly.

11. Apply clean engine oil to the camshaft journals and valve stem tips.

12. Install the rocker arm/shaft assembly and tighten the mounting bolts, in sequence, in 2–3 steps to a maximum torque of 14–19 ft. lbs. (19–25 Nm).

13. Coat a new gasket with silicone sealer and install on the rocker arm

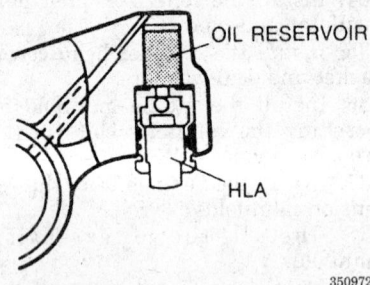

Hydraulic lash adjuster and oil reservoir — 2.6L engine

OIL RESERVOIR

HLA

350972

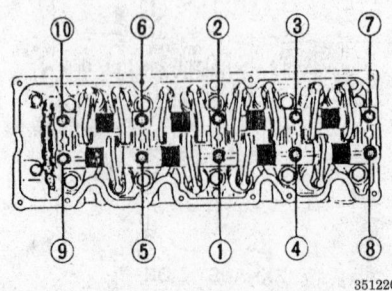

Rocker arm tightening sequence — 2.6L engine

351220

cover. Apply sealer to the cylinder head in the area of the half circle seals and install the rocker arm cover. Install the mounting bolts and tighten to 52–78 inch lbs. (5.9–8.8 Nm).

14. Install the air intake hose. Connect the negative battery cable, start the engine and check for leaks and proper operation.

3.0L Engine

1. Disconnect the negative battery cable.

2. If removing the driver's side rocker arm/shaft assembly, proceed as follows:

   a. Remove the air inlet tube.

   b. Tag and disconnect the necessary electrical connectors and vacuum hoses from the throttle body and intake air pipe.

   c. Disconnect the throttle cable.

   d. Remove the throttle body and intake air pipe.

3. Remove the rocker arm cover.

4. Loosen the rocker arm and shaft assembly mounting bolts in sequence, in 2–3 steps. Remove the assembly with the bolts.

5. If necessary, disassemble the rocker arm/shaft assembly, noting the position of each component to ease reassembly.

6. Remove the valve lifter (Hydraulic Lash Adjuster — HLA) and inspect it. Replace the HLA as necessary.

7. Check for wear or damage to the contact surfaces of the shafts and rocker arms; replace as necessary.

8. Measure the rocker arm inner diameter, it should be 0.7480–0.7493 in. (19.000–19.033mm). Measure the rocker arm shaft diameter, it should be 0.7464–0.7472 in. (18.959–18.980mm).

9. Subtract the shaft diameter from the rocker arm diameter to get the oil clearance. The oil clearance should be 0.0008–0.0029 in. (0.020–0.074mm) and should not exceed 0.004 in. (0.10mm). Replace parts, as necessary, if the oil clearance is not within specification.

To install:

10. To install the HLA, pour engine oil into the rocker arm oil reservoir. Apply engine oil to the HLA, and carefully install the HLA into the rocker arm.

11. Apply clean engine oil to the rocker arm shafts and rocker arms and assemble the rocker arm/shaft assembly. The intake side shaft has twice as many oil holes as the exhaust side shaft.

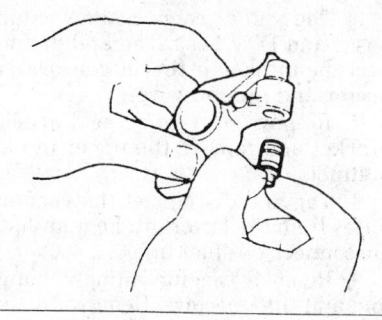

Valve lifter (Hydraulic Lash Adjuster —
HLA) — 3.0L engine

12. Apply clean engine oil to the
camshaft journals and valve stem
tips.

13. Install the rocker arm/shaft assembly and tighten the mounting
bolts, in sequence, in 2–3 steps to a
maximum torque of 14–19 ft. lbs.
(19–25 Nm).

**NOTE: Be careful that the
rocker arm shaft spring does not
get caught between the shaft and
mounting boss during
installation.**

14. Coat a new gasket with silicone
sealant and install on the rocker arm
cover. Install the rocker arm cover
with new seal washers and tighten
the bolts to 30–39 inch lbs. (3.4–4.4
Nm).

15. Install the intake air pipe,
throttle body and air intake tube, if
removed. Connect the throttle cable
and the necessary electrical connectors and vacuum hoses.

16. Connect the negative battery
cable, start the engine and check for
leaks and proper operation.

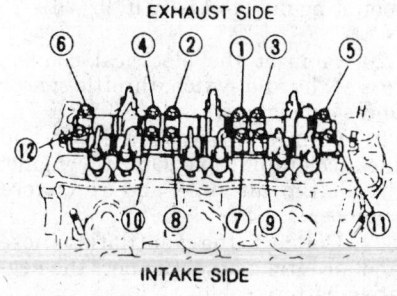

Rocker shaft bolt torque sequence — 3.0L
engine

Valve Lash

ADJUSTMENT

2.2L, 2.6L and 3.0L Engines

The engines covered by this manual
are equipped with Hydraulic Lash
Adjusters (HLAs). Valve clearance
adjustments are not necessary or
possible.

Rocker Arms/Shafts

REMOVAL AND INSTALLATION

2.2L Engine

1. Disconnect the negative battery
cable.
2. Remove the air cleaner assembly or air intake hose, as required.
3. Remove the rocker arm cover.
4. Loosen the rocker arm/shaft assembly mounting bolts in 2–3 steps in
the proper sequence. Remove the
rocker arm/shaft assembly together
with the bolts.
5. If necessary, disassemble the
rocker arm/shaft assembly, noting
the position of each component to
ease reassembly.
6. Check for wear or damage to
the contact surfaces of the shafts and
rocker arms; replace as necessary.
7. Measure the rocker arm inner
diameter, it should be 0.6300–0.6310
in. (16.000–16.027mm). Measure the
rocker arm shaft diameter, it should
be 0.6286–0.6293 in.
(15.966–15.984mm).
8. Subtract the shaft diameter
from the rocker arm diameter to get
the oil clearance. The oil clearance
should be 0.0006–0.0024 in.
(0.016–0.061mm) and should not exceed 0.004 in. (0.10mm). Replace
parts, as necessary, if the oil clearance is not within specification.
To install:
9. Apply clean engine oil to the
rocker arm shafts and rocker arms
and assemble the rocker arm/shaft
assembly in the reverse order of disassembly. Make sure the rocker arm
shaft oil holes in the center camshaft
cap face each other.

**NOTE: Use the mounting bolts
for alignment.**

10. Apply silicone sealant to the
cylinder head on the front and rear
camshaft cap mounting surface. Apply clean engine oil to the camshaft
journals and valve stem tips.
11. Install the rocker arm/shaft assembly and tighten the bolts, in sequence, in 2–3 steps to a maximum
torque of 13–20 ft. lbs. (18–26 Nm).
12. Apply silicone sealant to each
side of the front and rear camshaft
cap and the cylinder head in the area
where the caps meet the cylinder
head.
13. Install the rocker arm cover and
tighten the mounting bolts to 26–35
inch lbs. (2.9–3.9 Nm).
14. Install the air cleaner assembly
or air intake tube. Connect the negative battery cable, start the engine
and check for leaks and proper
operation.

2.6L Engine

1. Disconnect the negative battery
cable.
2. Remove the air intake hose.
3. Remove the rocker arm cover.
4. Loosen the rocker arm/shaft assembly mounting bolts in 2–3 steps in
the proper sequence. Remove the
rocker arm/shaft assembly together
with the bolts.
5. If necessary, disassemble the
rocker arm/shaft assembly, noting
the position of each component to
ease reassembly.
6. Check for wear or damage to
the contact surfaces of the shafts and
rocker arms; replace as necessary.
7. Measure the rocker arm inner
diameter, it should be 0.8268–0.8281
in. (21.000–21.033mm). Measure the
rocker arm shaft diameter, it should
be 0.8252–0.8260 in.
(20.959–20.980mm).
8. Subtract the shaft diameter
from the rocker arm diameter to get
the oil clearance. The oil clearance
should be 0.0008–0.0029 in.
(0.020–0.074mm) and should not exceed 0.004 in. (0.10mm). Replace
parts, as necessary, if the oil clearance is not within specification.
To install:
9. Apply clean engine oil to the
rocker arm shafts and rocker arms
and assemble the rocker arm/shaft
assembly in the reverse order of disassembly, noting the following:
  a. The intake side shaft has
twice as many oil holes as the exhaust side shaft.
  b. The No. 4 camshaft cap has an
oil hole from the cylinder head;
make sure it is installed correctly.
10. Apply clean engine oil to the
camshaft journals and valve stem
tips.
11. Install the rocker arm/shaft assembly and tighten the mounting
bolts, in sequence, in 2–3 steps to a
maximum torque of 14–19 ft. lbs.
(19–25 Nm).

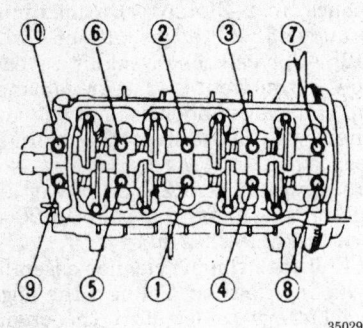

Rocker shaft bolt torque sequence — 2.2L engine

350293

12. Coat a new gasket with silicone sealer and install on the rocker arm cover. Apply sealer to the cylinder head in the area of the half circle seals and install the rocker arm cover. Install the mounting bolts and tighten to 52–78 inch lbs. (5.9–8.8 Nm).

13. Install the air intake hose. Connect the negative battery cable, start the engine and check for leaks and proper operation.

3.0L Engine

1. Disconnect the negative battery cable.

2. If removing the driver's side rocker arm/shaft assembly, proceed as follows:

   a. Remove the air inlet tube.

   b. Tag and disconnect the necessary electrical connectors and vacuum hoses from the throttle body and intake air pipe.

   c. Disconnect the throttle cable.

   d. Remove the throttle body and intake air pipe.

3. Remove the rocker arm cover.

4. Loosen the rocker arm and shaft assembly mounting bolts in sequence, in 2–3 steps. Remove the assembly with the bolts.

5. If necessary, disassemble the rocker arm/shaft assembly, noting the position of each component to ease reassembly.

6. Remove the valve lifter (Hydraulic Lash Adjuster — HLA) and inspect it. Replace the HLA as necessary.

7. Check for wear or damage to the contact surfaces of the shafts and rocker arms; replace as necessary.

8. Measure the rocker arm inner diameter, it should be 0.7480–0.7493 in. (19.000–19.033mm). Measure the rocker arm shaft diameter, it should be 0.7464–0.7472 in. (18.959–18.980mm).

9. Subtract the shaft diameter from the rocker arm diameter to get

the oil clearance. The oil clearance should be 0.0008–0.0029 in. (0.020–0.074mm) and should not exceed 0.004 in. (0.10mm). Replace parts, as necessary, if the oil clearance is not within specification.

To install:

10. To install the HLA, pour engine oil into the rocker arm oil reservoir. Apply engine oil to the HLA, and carefully install the HLA into the rocker arm.

11. Apply clean engine oil to the rocker arm shafts and rocker arms and assemble the rocker arm/shaft assembly. The intake side shaft has twice as many oil holes as the exhaust side shaft.

12. Apply clean engine oil to the camshaft journals and valve stem tips.

13. Install the rocker arm/shaft assembly and tighten the mounting bolts, in sequence, in 2–3 steps to a maximum torque of 14–19 ft. lbs. (19–25 Nm).

NOTE: Be careful that the rocker arm shaft spring does not get caught between the shaft and mounting boss during installation.

14. Coat a new gasket with silicone sealant and install on the rocker arm cover. Install the rocker arm cover with new seal washers and tighten the bolts to 30–39 inch lbs. (3.4–4.4 Nm).

15. Install the intake air pipe, throttle body and air intake tube, if removed. Connect the throttle cable and the necessary electrical connectors and vacuum hoses.

16. Connect the negative battery cable, start the engine and check for leaks and proper operation.

Intake Manifold

REMOVAL AND INSTALLATION

2.2L and 2.6L Engines

1. Relieve the fuel system pressure and disconnect the negative battery cable. Drain the cooling system.

2. Disconnect the air intake tube and ventilation hose. Remove the air pipe and resonance chamber on 2.6L engine.

3. Disconnect the accelerator cable and coolant hoses. Tag and disconnect the electrical connectors to the solenoid valve, throttle sensor and idle switch.

4. Remove the throttle body.

5. Remove the upper intake manifold brackets.

6. Tag and disconnect the vacuum hoses and PCV hose. Tag and disconnect the intake air thermosensor connector and ground wire.

7. Remove the injector harness bracket and remove the upper intake manifold.

8. Tag and disconnect the vacuum hoses from the lower intake manifold. Disconnect the fuel lines.

9. Remove the fuel supply manifold and the injectors. Remove the injector harness and bracket.

10. Remove the pulsation damper and the intake manifold bracket. Remove the attaching nuts and remove the lower intake manifold.

To install:

11. Clean all gasket mating surfaces.

12. Position a new intake manifold-to-cylinder head gasket and install the lower intake manifold. Tighten the nuts to 14–19 ft. lbs. (19–25 Nm).

13. Install the intake manifold bracket and pulsation damper. Install the injector harness and bracket. Tighten the pulsation damper and injector harness bracket bolts to 69–95 inch lbs. (7.8–11.0 Nm).

14. Install the injectors and the fuel supply manifold. Tighten the fuel supply manifold attaching bolts and tighten to 14–19 ft. lbs. (19–25 Nm).

15. Connect the fuel lines. Connect the vacuum hoses to the lower intake manifold.

16. Position a new gasket and install the upper intake manifold. Tighten the attaching bolts/nuts to 14–19 ft. lbs. (19–25 Nm).

17. Install the injector harness bracket. Connect the ground wire and air thermosensor electrical connector. Connect the PCV hose and the vacuum hoses to the upper intake manifold.

18. Install the upper intake manifold brackets.

19. Position a new gasket and install the throttle body. Tighten the mounting nuts to 14–19 ft. lbs. (19–25 Nm).

20. Connect the electrical connectors at the idle switch, throttle sensor and solenoid valve.

21. Connect the coolant hoses and the accelerator cable. On 2.6L engine, install the air pipe and resonance chamber.

22. Connect the ventilation hose and air intake hose. Connect the negative battery cable.

23. Fill and bleed the cooling system. Run the engine and check for leaks and proper operation.

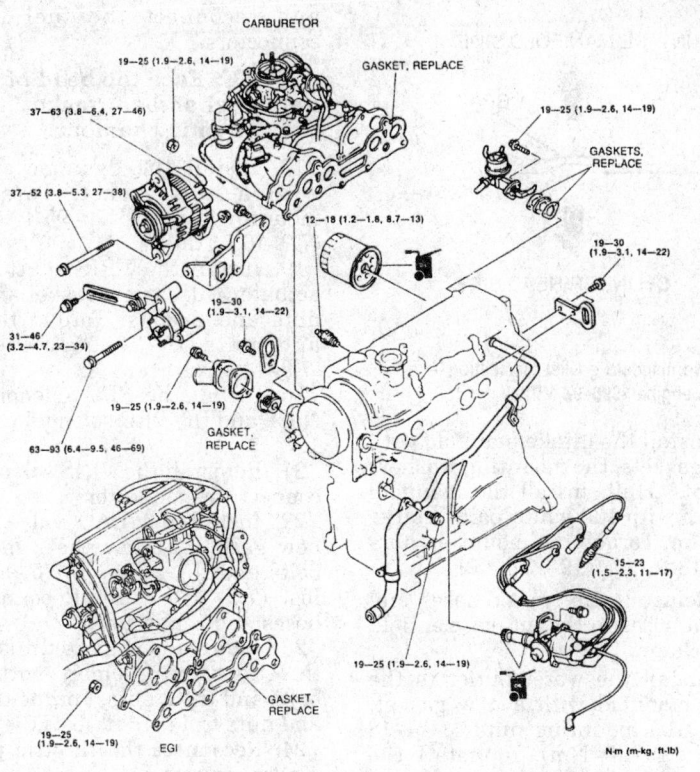

Upper engine assembly — 2.2L and 2.6L engines

3.0L Engine

1993–95 MPV

— **CAUTION** —

Fuel injection systems remain under pressure after the engine has been turned OFF. Properly relieve fuel pressure before disconnecting any fuel lines. Failure to do so may result in fire or personal injury.

— **CAUTION** —

Do not allow fuel spray or fuel vapors to come in contact with a spark or open flame. Keep a dry chemical fire extinguisher nearby. Never store fuel in an open container due to risk of fire or explosion.

1. Relieve the fuel system pressure, and disconnect the negative battery cable. Drain the cooling system.

2. Remove the air intake tube from the throttle body. Disconnect the accelerator cable.

3. Disconnect the throttle sensor connector and the coolant hoses. Remove the throttle body.

4. Tag and disconnect the vacuum hoses. Remove the bypass air control valve and the intake air pipe.

5. Remove the extension manifolds. Remove the dynamic chamber (upper intake plenum) with the shutter valve actuator.

6. Remove the fuel supply manifold and the injectors. Disconnect the coolant hoses.

7. Loosen the lower intake manifold nuts, in sequence, in 2 steps, then remove the lower intake manifold.

To install:

8. Clean all gasket mating surfaces.

9. Position new lower intake manifold gaskets, and install the lower intake manifold.

10. Install the intake manifold washers with the white paint mark upward. Install the nuts and tighten, in sequence, in 2 steps to a maximum torque of 14–19 ft. lbs. (19–25 Nm).

11. Install the injectors and the fuel supply manifold. Tighten the attaching bolts to 14–19 ft. lbs. (19–25 Nm).

12. Connect the coolant hoses.

13. Install a new O-ring on the lower intake manifold and install the upper intake plenum. Apply clean engine oil to new O-rings and install on the extension manifolds. Position new gaskets and install the extension manifolds. Tighten the attaching nuts to 14–19 ft. lbs. (19–25 Nm).

14. Position a new gasket and install the intake air pipe. Install the bypass air control valve. Tighten the attaching bolts/nuts to 14–19 ft. lbs. (19–25 Nm).

15. Position a new gasket and install the throttle body. Tighten the attaching nuts to 14–19 ft. lbs. (19–25 Nm).

16. Connect the coolant and vacuum hoses. Connect the throttle sensor connector and accelerator cable.

17. Adjust the accelerator cable deflection to 0.039–0.118 inch (1–3mm).

18. Connect the air intake tube and the negative battery cable.

19. Fill and bleed the cooling system. Run the engine and check for leaks and proper operation.

1996–97 MPV

— **CAUTION** —

Fuel injection systems remain under pressure after the engine has been turned OFF. Properly relieve fuel pressure before disconnecting any fuel lines. Failure to do so may result in fire or personal injury.

— **CAUTION** —

Do not allow fuel spray or fuel vapors to come in contact with a spark or open flame. Keep a dry chemical fire extinguisher nearby. Never store fuel in an open container due to risk of fire or explosion.

1. Relieve the fuel system pressure, and disconnect the negative battery cable. Drain the cooling system.

2. Remove the clamps and remove the air intake hose from the throttle body and the MAF sensor.

3. Disconnect the accelerator cable and if equipped, the cruise control cable.

4. Disconnect the throttle position sensor connector.

5. Remove the 2 bolts and the 2 nuts and remove the throttle body unit and gasket.

6. Disconnect the hoses and connector and remove the BAC valve.

7. Remove the VRIS solenoid valve.

8. Remove the PRC solenoid valve No. 1 and the PRC solenoid valve No. 2.

9. Remove the 2 bolts and the 2 nuts and remove the VRIS shutter valve actuator and the gasket.

10. Remove the mounting bolts and the dynamic chamber.

11. Disconnect the fuel delivery pipe from the fuel distributor. Dis-

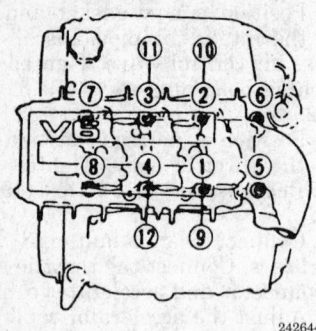

242644

Intake manifold torque sequence tightening — 3.0L engine 1993–95 MPV

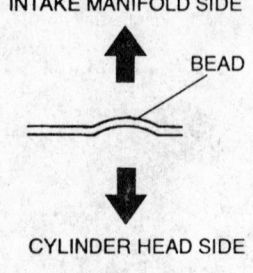

INTAKE MANIFOLD SIDE

BEAD

CYLINDER HEAD SIDE

302042

Intake manifold gasket installation — 3.0L engine 1996–97 MPV

connect the fuel injector connectors, remove the mounting nuts and remove the fuel distributor with the injectors.

12. Remove the water hoses from the water outlet and remove the water outlet.

13. Disconnect any water hoses and vacuum hoses from the intake manifold. Remove the mounting nuts, washers, and remove the intake manifold and gaskets.

To install:

NOTE: Face the bead of the intake manifold gasket toward the intake manifold.

14. Install the intake manifold with 2 new gaskets, the mounting washers and nuts. Hint: install the manifold washers with the white paint marks facing up. Torque the mounting nuts to 14–18 ft. lbs. (19–25 Nm).

15. Reinstall any water hoses and vacuum hoses to the intake manifold and reclamp.

16. Install the water outlet to the intake manifold with a new gasket. Torque the mounting nuts to 14–18 ft. lbs. (19–25 Nm). Reinstall the water hoses with the clamps.

17. Reinstall the fuel delivery pipe with the injectors and torque the mounting nuts to 14–18 ft. lbs. (19–25 Nm). Reconnect the fuel delivery pipe

and reconnect the fuel injector connectors.

NOTE: Face the bead of the dynamic chamber gasket toward the dynamic chamber.

18. Install the dynamic chamber with a new gasket. Torque the mounting bolts to 70–95.4 inch lbs. (7.9–10.7 Nm).

19. Install the VRIS shutter valve actuator with a new gasket and the 2 nuts and 2 bolts. Torque the bolts and nuts to 70–95.4 inch lbs. (7.9–10.7 Nm).

20. Install the PRC solenoid valve No. 2 and the PRC solenoid valve No. 1.

21. Reinstall the VRIS solenoid and connect the connectors.

22. Install the BAC valve with a new gasket. Torque the 2 mounting bolts and the 2 nuts to 70–95.4 inch lbs. (7.9–10.7 Nm). Reconnect the hoses and connector.

23. Using a new gasket, install the throttle body assembly with the 2 bolts and the 2 nuts. Torque the bolts and nuts to 14–16 ft. lbs. (19–25 Nm).

24. Reconnect the throttle position sensor connector.

25. Reconnect the accelerator cable and if equipped, the cruise control cable.

26. Install the air intake hose to the throttle body assembly and the MAF sensor with the 2 clamps.

27. Refill the cooling system, reconnect the negative battery cable, start the engine and operate until normal temperature, check for coolant leaks, and proper operation.

Exhaust Manifold

REMOVAL AND INSTALLATION

2.2L Engine

1. Disconnect the negative battery cable.

2. Remove the air cleaner and fresh air hose assembly.

3. Raise and safely support the vehicle.

4. Remove the 2 attaching nuts from the front exhaust pipe at the manifold and disconnect the exhaust pipe from the manifold. Discard the gasket.

5. Lower the vehicle and remove the mounting screws and remove the exhaust manifold heat shield.

6. On carbureted vehicles, disconnect and remove the secondary air pipe.

7. Remove the nuts and remove the exhaust manifold and gasket.

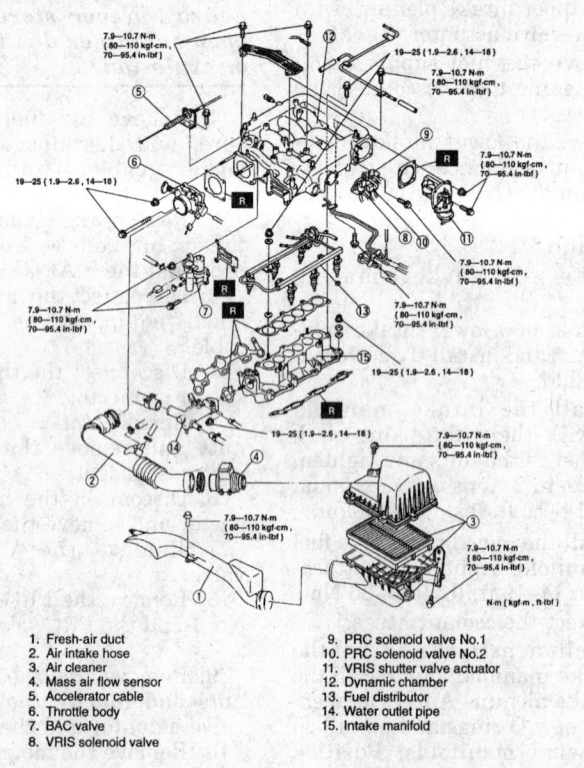

1. Fresh-air duct
2. Air intake hose
3. Air cleaner
4. Mass air flow sensor
5. Accelerator cable
6. Throttle body
7. BAC valve
8. VRIS solenoid valve
9. PRC solenoid valve No.1
10. PRC solenoid valve No.2
11. VRIS shutter valve actuator
12. Dynamic chamber
13. Fuel distributor
14. Water outlet pipe
15. Intake manifold

302039

Intake air system — 3.0L engine 1996–97 MPV

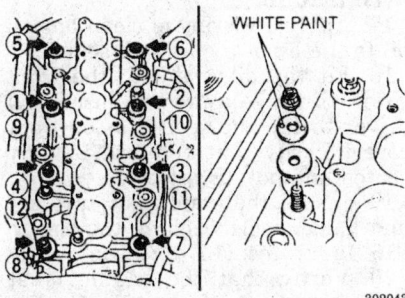

Intake manifold tightening sequence and mounting washers — 3.0L engine 1996–97 MPV

DYNAMIC CHAMBER SIDE

↑

BEAD

↓

INTAKE MANIFOLD SIDE

302040

Dynamic chamber gasket installation — 3.0L engine 1996–97 MPV

To install:

8. Install a new gasket and secure the manifold to the cylinder head mounting studs with the attaching nuts. Torque the exhaust manifold mounting nuts to 16–18 ft. lbs.

9. If equipped with a carburetor, reinstall the secondary air pipe assembly.

10. Replace the exhaust manifold heat shield with its mounting screws.

11. Raise the vehicle and install a new exhaust pipe gasket and reconnect the exhaust pipe to the exhaust manifold. Torque the nuts to 16–21 ft. lbs.

12. Safely lower the vehicle, install the air cleaner and fresh air duct,reconnect the negative battery cable and check for leaks.

2.6L Engine

1. Disconnect the negative battery cable.

2. Drain the cooling system.

─────── **CAUTION** ───────

When draining the coolant, always drain the coolant into a sealable container. Coolant should be reused unless it is contaminated or several years old.

3. Remove the oil dipstick tube.

4. Remove the coolant bypass pipe.

5. Disconnect the exhaust pipe from the manifold.

6. Remove the heat shield.

7. Unbolt and remove the manifold. Discard the gasket.

To install:

8. Clean all mating surfaces. Place a new gasket and exhaust manifold in position and loosely install the mounting bolts. After all of the manifold bolts and attached components are loosely installed, secure all mounting bolts. Torque the bolts to 16–21 ft. lbs. (22–28 Nm).

9. Install the heat shield and tighten the mounting bolts to 14–22 ft. lbs. (19–30 Nm).

10. Connect the exhaust pipe to the manifold. Torque the bolts to 28–38 ft. lbs. (38–51 Nm).

11. Install the coolant bypass pipe.

12. Install the oil dipstick tube.

13. Fill and bleed the cooling system.

14. Connect the negative battery cable.

3.0L Engine

1. Disconnect the negative battery cable. Wait at least 90 seconds before performing any work.

2. Raise and safely support the vehicle.

3. If necessary remove the engine splash shield.

4. Disconnect the oxygen sensor connector.

5. Remove the nuts attaching the 3 way pre-catalytic convertor to the main catalytic convertor and the LH exhaust manifold. Remove the 3 way pre-catalytic convertor and the gaskets.

6. Remove the bolts and the LH and the RH exhaust manifold insulators.

7. Remove the attaching bolts and nuts and the center exhaust pipe insulators.

8. Remove the bolts securing the center exhaust manifold to the LH and RH exhaust manifolds. Remove the center exhaust manifold and the gaskets.

9. Remove the mounting nuts, the LH and the RH exhaust manifolds, and the gaskets.

To install:

10. Before installation make sure all mating surfaces are clean of any gasket material.

11. Install the RH and the LH exhaust manifolds with new gaskets and torque the mounting nuts to 16–20 ft. lbs.

12. Install the center exhaust manifold with 2 new gaskets to the RH and LH exhaust manifolds. Torque the bolts to 14–18 ft. lbs.

13. Reinstall the RH and the LH exhaust manifold insulators with the attaching bolts. Torque the bolts to 14–18 ft. lbs.

14. Install the exhaust insulator to the center exhaust pipe with the bolts and nuts. Torque the bolts and nuts to 14–18 ft. lbs.

15. Install the 3 way pre-catalytic convertor using new gaskets to the LH exhaust manifold and the main catalytic convertor. Torque the nuts to 28–38 ft. lbs.

16. Reconnect the oxygen sensor connector.

17. If necessary, reinstall the engine splash shield.

18. Safely lower the vehicle and reconnect the negative battery cable.

Front Cover Seal

REMOVAL AND INSTALLATION

2.6L Engine

1. Disconnect the negative battery cable.

2. Drain the engine oil.

3. Remove the cooling fan cowling from the radiator.

4. Remove the cooling fan and fan clutch assembly.

5. Remove the engine drive belts.

6. Remove the water pump pulley.

7. Use and pulley holding tool such as 49-S120-710, or equivalent to hold the crankshaft pulley, and then loosen the crankshaft pulley bolt.

8. Remove the crankshaft pulley and spacer. Note the position of the spacer for reassembly.

9. Clean any oil or grime from the front cover.

10. Wrap protective tape or a shop cloth around the crankshaft so it won't be scratched or damaged by seal removal tools.

11. Use a seal removal tool to remove the front cover seal. Work carefully to avoid damaging the crankshaft or the sealing edges of the front cover.

To install:

12. Apply a coating of clean engine oil to the lip of the new oil seal.

13. Fit the oil seal into the front cover; then use a properly-sized seal driver to tap the seal into the front cover. Verify that the seal is fully seated and not distorted.

14. Install the crankshaft pulley and spacer and fit a holding tool onto the pulley. Install the pulley bolt and

tighten it to 130–145 ft. lbs. (177–196 Nm).

15. Install the water pump pulley and then install and tension the engine drive belts.

16. Install the cooling fan assembly.

17. Install the cooling fan cowling.

18. Refill the engine with fresh oil.

19. Reconnect the negative battery cable, and then run the engine and check for oil leaks.

Front Crankshaft Seal

REMOVAL AND INSTALLATION

2.2L Engine

1. Disconnect the negative battery cable.

2. Drain the engine oil.

3. Remove the cooling fan cowling from the radiator.

4. Remove the cooling fan and fan clutch assembly.

5. Remove the engine drive belts.

6. Remove the cooling fan pulley and bracket.

7. If equipped with a carbureted engine, disconnect the secondary air pipe assembly and move it out of the work area.

8. Remove the upper timing cover.

9. Rotate the crankshaft to set the engine at TDC/compression for the No. 1 cylinder. The No. 1 arrow mark on the camshaft sprocket should point straight up and align with the notch on the rear of the timing belt housing.

10. Remove the timing belt. If the timing belt is worn, damaged, or contaminated with oil or coolant, it should be replaced.

11. Remove the crankshaft pulley and spacer. Note the position of the spacer for reassembly.

12. Remove the lower timing belt cover.

13. Use a pulley holder and a crankshaft lock tool 49-S120-710 and 49-H011-101A or their equivalents, to hold and loosen the crankshaft sprocket bolt. After loosening the bolt, remove the crankshaft sprocket.

14. Clean any oil or dirt from the front of the engine.

15. Wrap protective tape or a shop cloth around the crankshaft so it won't be scratched or damaged by seal removal tools.

16. Use a seal removal tool to remove the front cover seal. Work carefully to avoid damaging the crankshaft or the sealing edges of the front cover.

To install:

17. Apply a coating of clean engine oil to the lip of the new oil seal.

18. Fit the oil seal into the front cover; then use a properly-sized seal driver to tap the seal into the front cover. Verify that the seal is fully seated and not distorted.

19. Install the crankshaft sprocket and holder tools. Tighten the bolt to 116–123 ft. lbs. (157–167 Nm).

20. Verify that the engine is at TDC/compression for the No. 1 cylinder, and then install and tension the timing belt.

21. Install the upper and lower timing belt covers, make sure the cover gaskets are fully seated, and not distorted. Make sure no foreign objects or fluids entered the timing belt case during assembly.

22. Install the crankshaft pulley and spacer. Install the pulley bolts and tighten them to 9–13 ft. lbs. (12–17 Nm).

23. If equipped, reconnect the secondary air pipe assembly.

24. Install the cooling fan bracket and pulley and then install and tension the engine drive belts.

25. Install the cooling fan assembly.

26. Install the cooling fan cowling.

27. Refill the engine with fresh oil.

28. Reconnect the negative battery cable, and then run the engine and check for oil leaks.

3.0L Engine

1. Disconnect the negative battery cable.

2. Remove the timing belt covers and the timing belt.

3. If not removed during the timing belt removal procedure, remove the crankshaft sprocket bolt.

4. Remove the crankshaft sprocket, using a suitable puller.

5. Cut the seal lip with a razor knife. Protect the crankshaft with a shop towel and pry the seal from the engine.

To install:

6. Lubricate the seal lip with clean engine oil and push the seal slightly in by hand.

7. Tap the seal in evenly using a seal installer. Install the seal until it is flush with the oil pump body.

8. Install the crankshaft sprocket. Install the sprocket key with the tapered side toward the oil pump body. Tighten the crankshaft bolt to 123 ft. lbs. (167 Nm).

9. Install the timing belt and covers.

10. Connect the negative battery cable.

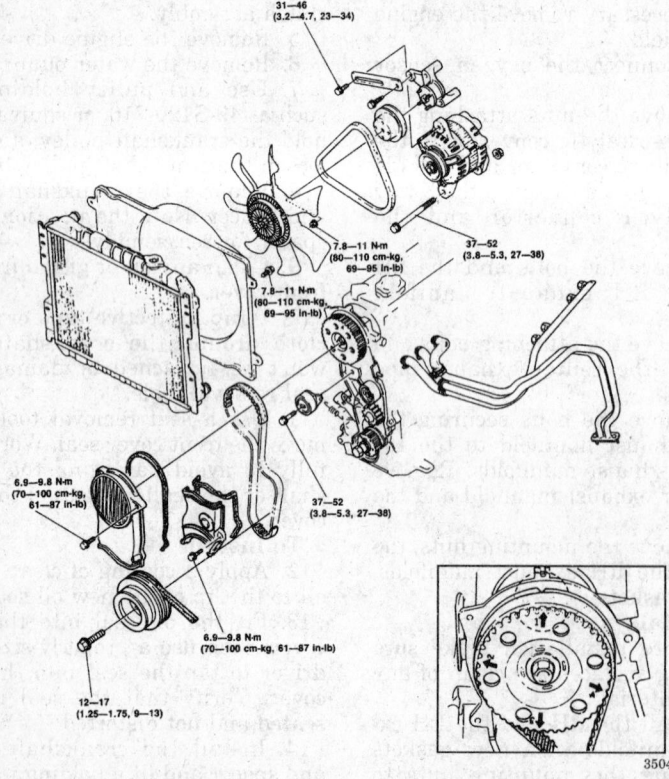

31—46
(3.2—4.7, 23—34)

7.8—11 N·m
(80—110 cm-kg,
69—95 in-lb)

7.8—11 N·m
(80—110 cm-kg,
69—95 in-lb)

37—52
(3.8—5.3, 27—38)

6.9—9.8 N·m
(70—100 cm-kg,
61—87 in-lb)

37—52
(3.8—5.3, 27—38)

6.9—9.8 N·m
(70—100 cm-kg, 61—87 in-lb)

12—17
(1.25—1.75, 9—13)

350482

Timing belt components: inset shows the camshaft TDC mark — 2.2L engine

Timing Belt, Sprockets, Tensioner and Front Cover

ADJUSTMENT

2.2L Engine

> ——— **WARNING** ———
> **It is recommended that a timing belt be replaced any time its tension is released.**

Setting the Tension of a New Timing Belt

1. Make sure the crankshaft and the camshaft sprockets are aligned with their timing marks, so the engine is at TDC/compression for the No. 1 cylinder. Install the timing belt, making sure to keep the tensioner side of the belt as tight as possible.

2. Rotate the crankshaft twice in a clockwise direction to seat the belt. Verify that the timing marks align correctly.

3. Loosen the timing belt tensioner lock bolt. This will allow the tensioner spring to apply pressure to the timing belt.

4. After the spring has pulled the timing belt as far as possible, tempo-

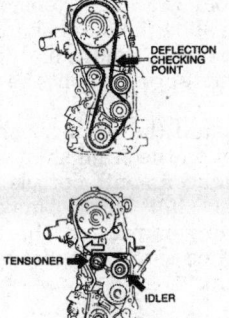

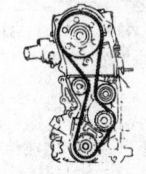

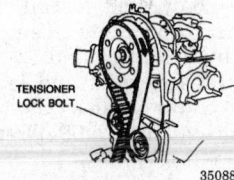

350885

Timing belt routing, tension bolt, and deflection check point — 2.2L engine

rarily tighten the tensioner lock bolt to 27–38 ft. lbs. (37–52 Nm).

5. Rotate the crankshaft clockwise 2 complete revolutions to check the rotation of the belt and the alignment of the timing marks. Listen for any rubbing noises which may mean the belt is binding.

6. Check the deflection of the timing belt at a point above the idler pulley. If a force of 22 lbs. (98 N) is applied; a new belt should deflect 0.31–0.35 in. (8.0–9.0mm), a used belt should deflect 0.35–0.39 in. (9.0–10.0mm). If the deflection isn't correct, repeat steps 2–4 to re-tension the belt. If necessary, replace the tensioner spring.

7. Install the timing belt covers.

Timing Belt Inspection

1. Disconnect the negative and positive battery cables.

2. Label and disconnect the ignition wires and remove the spark plugs. Remove the upper timing belt cover.

3. Rotate the crankshaft to set the engine at TDC/compression for the No. 1 cylinder. Align the No. 1 arrow on the camshaft sprocket with the mark on the back cover. Verify that the TDC mark on the crankshaft pulley aligns with the pointer on the lower cover.

4. Rotate the crankshaft pulley clockwise to cycle the belt through its entire rotation.

5. Inspect the entire length of the timing belt. Look carefully for any signs of the following conditions:

    a. Cracked, chipped, or broken teeth.

    b. Fraying, separation, or heat damage to the belt's rubber and fiber layers.

    c. Oil or coolant leaks which may have contaminated the belt.

    d. Make sure the timing marks align. Misaligned marks may indicate that the belt has jumped one or more teeth, or has been improperly tensioned.

6. Check the camshaft and crankshaft oil seals for any signs of leakage. Also check the water pump for leakage. The source of any oil or coolant leaks must be found and corrected before a new timing belt is installed.

7. Replace the timing belt if it's damaged in any way, or if its condition or the vehicle's maintenance history is uncertain. Recommended service interval for timing belt

replacement is 60,000 miles (96,000 Km).

8. After inspection, install the timing belt covers. Install the spark plugs and reconnect the ignition wires. Reconnect the battery cables.

REMOVAL AND INSTALLATION

2.2L Engine

1. Disconnect the negative battery cable and drain the cooling system.

2. Remove the timing belt cover.

3. Turn the crankshaft to align the mark of the camshaft sprocket with the front housing matching mark.

4. Remove the tensioner and spring. Mark the timing belt direction of rotation if it is to be reused.

5. Remove the timing belt.

To install:

6. Make sure the mark on the crankshaft sprocket is aligned with the matching mark on the oil pump body.

7. Make sure the mark on the camshaft sprocket is aligned with the matching mark on the front housing.

8. Install the timing belt tensioner and spring. Temporarily secure it with the spring fully extended.

9. Install the timing belt so there is no looseness at the water pump pulley and idler pulley side. If the timing belt is being reused, it must be installed in the same direction of rotation.

10. Remove the spark plugs to make engine rotation easier.

11. Turn the crankshaft twice clockwise in the direction of rotation. Make sure the matching marks are correctly aligned; if not repeat the installation procedure.

12. Loosen the tensioner lock bolt and apply tension to the belt. Tighten the tensioner lock bolt to 27–38 ft. lbs. (37–52 Nm).

13. Turn the crankshaft twice clockwise in the direction of rotation and align the matching marks. Check the timing belt deflection. The deflection should be 0.31–0.35 in. (8–9mm) on a new belt or 0.35–0.39 in. (9–10mm) on a used belt. Do not apply tension other than that of the tensioner spring.

14. If the deflection is not correct, repeat Steps 11–13.

15. Install the remaining components in the reverse order of removal. Fill and bleed the cooling system.

16. Run the engine and check for leaks and proper operation. Check the idle speed and the ignition timing.

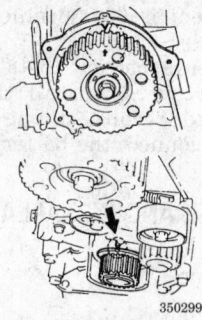

350299

Timing belt sprocket
matching marks — 2.2L
engine

Timing Chain, Sprockets and Front Cover

REMOVAL AND INSTALLATION

2.6L Engine

----- CAUTION -----

Fuel injection systems remain under pressure after the engine has been turned OFF. Properly relieve fuel pressure before disconnecting any fuel lines. Failure to do so may result in fire or personal injury.

----- CAUTION -----

Do not allow fuel spray or fuel vapors to come in contact with a spark or open flame. Keep a dry chemical fire extinguisher nearby. Never store fuel in an open container due to risk of fire or explosion.

1. Relieve the fuel system pressure, and disconnect the negative battery cable.
2. Drain the cooling system and engine oil.
3. Remove the fan shroud and cooling fan.
4. Remove the cylinder head.
5. Remove the oil pan.
6. Remove the drive belts.
7. Remove the water pump pulley.
8. Disconnect the electrical wiring harness and remove the mounting bolts and remove the alternator. Remove the bolts and the alternator bracket.
9. Remove the power steering pump and the pump bracket.
10. Without disconnecting the A/C lines, remove the A/C compressor mounting bolts and remove the A/C compressor. Secure the compressor out of the way. Remove the bolts and the A/C compressor mounting bracket.

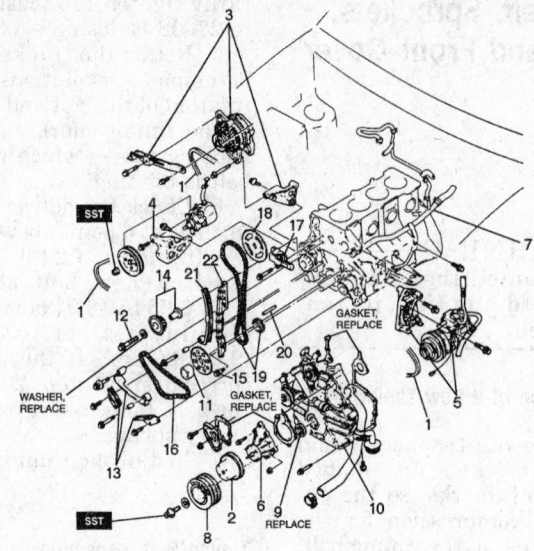

332325

Timing chain and related components — 2.6L engine

1. Drive belt
2. Water pump pulley
3. Generator and bracket
4. P/S oil pump and bracket
5. A/C compressor and bracket
6. Water pump
7. Coolant bypass pipe
8. Crankshaft pulley
9. Front oil seal
10. Chain cover
11. Spacer
12. Idler sprocket assembly lock bolt
13. Chain guides
14. Idler sprocket assembly
15. Crankshaft sprocket
16. Balancer chain
17. Chain adjuster
18. Camshaft pulley
19. Timing chain and timing gear
20. Key
21. Chain lever
22. Chain guide

11. Remove the water pump mounting bolts, the water pump and gasket.
12. Remove the coolant bypass pipe.
13. Remove the bolt, and by using a suitable puller, remove the crankshaft pulley.
14. Remove the timing chain cover bolts and the cover.
15. Remove the spacer, the idler sprocket assembly lock bolt, the chain guides, and the idler sprocket assembly.
16. Remove the balancer chain.
17. Remove the timing chain adjuster and the timing chain.
18. Remove the lock bolts and remove the crankshaft timing sprocket and the camshaft sprocket.
19. Remove the key and the chain lever and guide.
20. Inspect the chain, sprockets and guides for damage and/or wear.

To install:
21. Install the timing chain guides and tighten to 7.2–8.7 ft. lbs. (10–12 Nm).
22. If removed, install the tensioner onto the cylinder block.
23. Align the plated links of the timing chain with the timing marks on the sprockets as the chain and sprockets are assembled. Secure the

pulley and chain with wire to prevent misalignment.
24. Hold the chain tensioner head in, then slide the crankshaft sprocket onto the crankshaft and place the camshaft sprocket on the sprocket holder.
25. Install the balancer shaft drive sprocket on the crankshaft. Assemble the balancer shaft sprockets to the balancer shaft chain, making sure the timing marks on the sprockets are aligned with the polished links on the chain.

NOTE: Be careful not to confuse the right and left sprockets as they are installed in opposite directions.

26. While holding the assembled sprockets and chain, align the timing mark on the crankshaft sprocket with the chain and install the balancer shaft sprockets. Temporarily tighten the bolts by hand.
27. Install the right and left lower balancer chain guides and tighten the mounting bolts to 69–95 inch lbs. (7.8–11.0 Nm).
28. Install the upper chain guide and loosely tighten the mounting and adjusting bolts. Set the chain guide to the fully downward position.

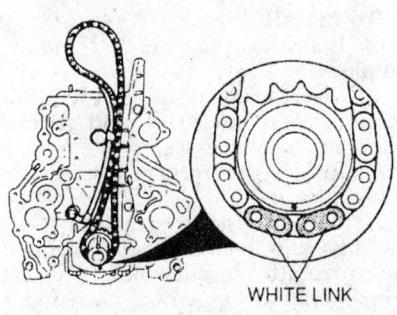

332327

Crankshaft sprocket chain alignment — 2.6L engine

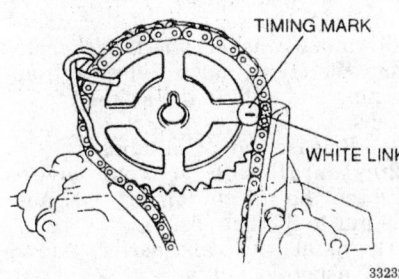

332329

Camshaft sprocket chain alignment — 2.6L engine

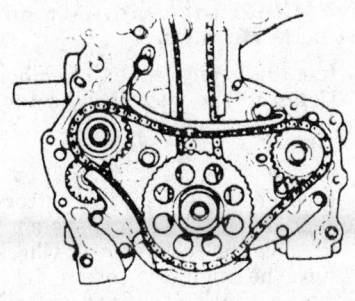

332330

Balancer chain assembly — 2.6L engine

29. Tighten the idler sprocket assembly lock bolt to 27–38 ft. lbs. (37–52 Nm), and install the spacer.

30. Rotate both balancer shaft sprockets slightly to position the chain slack at the center between the left balancer shaft sprocket and the oil pump sprocket.

31. Adjust the balancer chain tension as follows:

　a. Loosen the upper chain guide adjusting bolt.

　b. Push on the chain guide just above the adjusting slot with a force of approximately 11 lbs., then pull back the guide 0.126–0.149

inch (3.2–3.8mm). Tighten the bolt to 69–95 inch lbs. (7.8–11.0 Nm). Tighten the guide pivot bolt to the same specification.

　c. The chain slack at the notch in the guide should be 0.12 inch (3mm) when the guide is properly adjusted.

32. Install the timing cover with a new gasket Torque the timing cover bolts to 14–19 ft. lbs. (19–25 Nm).

33. Install the crankshaft pulley with the lock bolt and washer. Torque the lock bolt to 130–145 ft. lbs. (177–196 Nm).

34. Reinstall the coolant bypass pipe.

35. Install the water pump with a new gasket and torque the mounting nuts and bolts to 14–19 ft. lbs. (19–25 Nm).

36. Install the power steering pump bracket and the power steering pump.

37. Reattach the A/C compressor bracket and install the A/C compressor.

38. Remount the alternator bracket and the alternator. Reconnect the alternator wiring harness connector.

39. Install the water pump pulley with the mounting nuts.

40. Install the drive belts and adjust to specifications.

41. Install the oil pan.

42. Install the cylinder head.

43. Install the fan shroud and cooling fan.

44. Fill the crankcase with the proper type and quantity of engine oil. Fill and bleed the cooling system. Run the engine and check for leaks and proper operation.

45. Check the idle speed and ignition timing and adjust, if necessary.

Camshaft

REMOVAL AND INSTALLATION

2.2L Engine

1. Disconnect the negative battery cable.

2. Remove the air cleaner assembly.

3. Drain the cooling system.

——— **CAUTION** ———

When draining the coolant, always drain the coolant into a sealable container. Coolant should be reused unless it is contaminated or several years old.

4. Remove the front cover assembly.

5. Remove the cam gear.

6. Remove the thermostat housing.

7. Remove the distributor assembly.

8. Remove the rocker cover.

9. Remove the rear housing.

10. Remove the rocker arm assembly.

11. If equipped, remove the thrust plate.

12. Remove the camshaft from the cylinder head.

　To install:

13. Coat the camshaft with clean engine oil and position it on the cylinder head.

14. Install the thrust plate and new gasket.

15. Install the rocker arm assembly.

16. Install the rear housing and new gasket.

17. Install the rocker cover and new gasket.

18. Install the distributor assembly.

19. Install the thermostat housing and new gasket.

20. Install the cam gear.

21. Install the front cover assembly.

22. Fill the cooling system.

23. Install the air cleaner assembly.

24. Connect the negative battery cable.

2.6L Engine

Except MPV

1. Disconnect the negative battery cable.

2. Remove the air cleaner assembly.

3. Remove the rocker cover.

4. Rotate the crankshaft until No. 1 piston is at the top of its compression stroke (both valves closed).

5. Remove the distributor drive gear.

6. Record the position of mating mark on camshaft sprocket and plated link on timing chain.

7. Remove the camshaft sprocket service cover.

8. Push the chain adjuster sleeve towards the left and insert a 0.0787 in. (2mm) diameter by 1.85 in. (47mm) long pin in the lever hole to hold the adjuster.

9. Wire the camshaft sprocket and timing chain together and remove the timing chain and camshaft sprocket assembly from the camshaft, and lay aside.

10. Remove the rocker arms/shafts assembly.

11. Remove the cover end seals.

12. Carefully remove camshaft without cocking to prevent damage to cam.

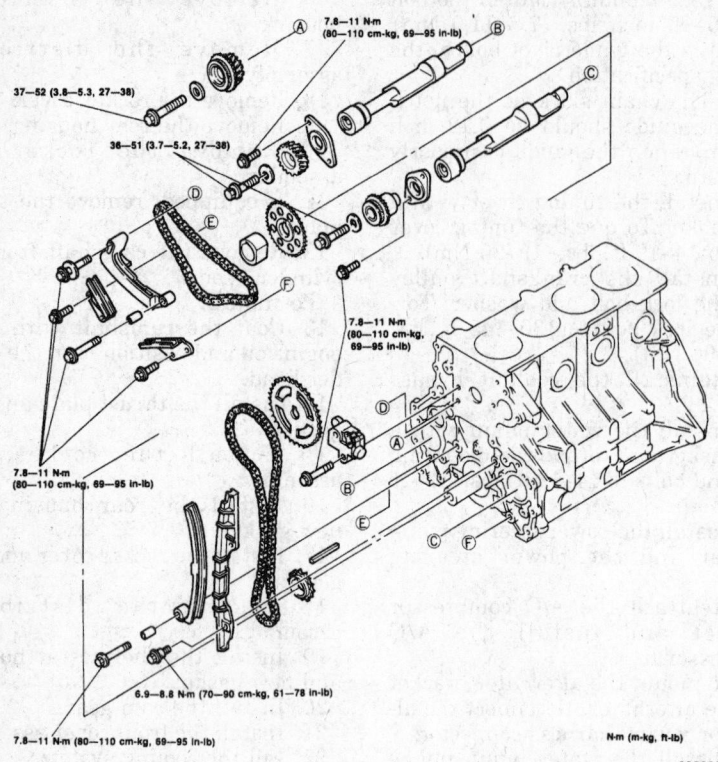

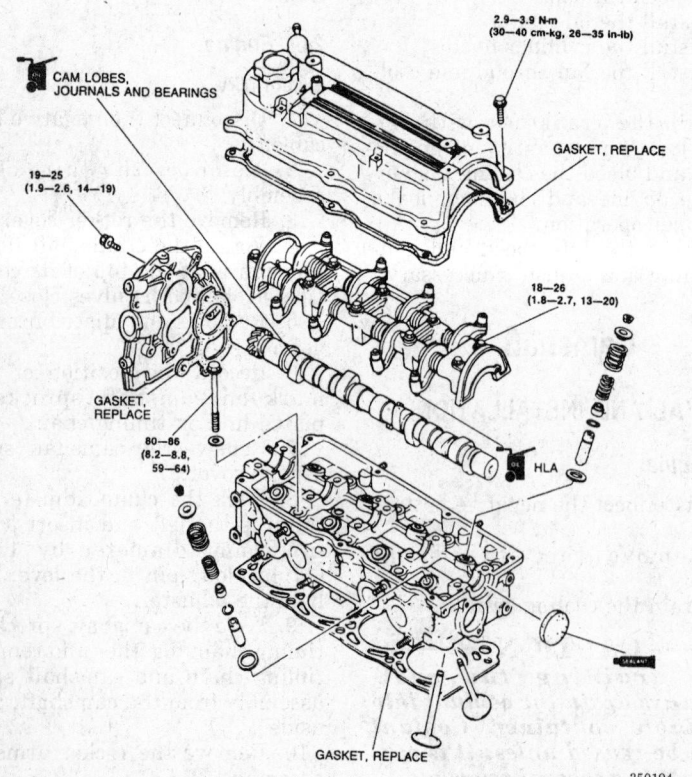

Timing and balancer chain assembly — 2.6L engine

Camshaft and cylinder head assembly — 2.2L engine

To install:

13. Lubricate the camshaft and set in place.

14. Coat the end seals with RTV silicone gasket material and install them in their recesses.

15. Install the rocker arms/shafts assembly.

16. With the rocker arms/shafts assembly and bearing caps torqued down, rotate the camshaft so dowel pin is on the top center of cylinder head.

17. Install the timing chain and camshaft sprocket assembly. Make sure the timing mark and the white link align. Torque the bolt to 45 ft. lbs. Remove the wire.

18. Install the distributor drive gear, a new washer, and the lockbolt. Hold the crankshaft to keep it from turning and tighten the lockbolt to 45 ft. lbs.

19. Release the chain adjuster.

20. Install the service cover. Tighten the bolt to 95 inch lbs. and the nut to 87 inch lbs.

21. Install the valve cover. Use a new gasket coated with sealer. Coat the tops of the end seals with RTV silicone gasket material. Torque the cover bolts to 78 inch lbs.

NOTE: After servicing rocker shaft assembly, Intake/Exhaust Valve Clearance adjustment must be performed.

22. Install the air cleaner assembly.

23. Connect the negative battery cable.

MPV

1. Disconnect the negative battery cable, and drain the cooling system.

2. Remove the rocker cover. Check to ensure the engine is set on TDC. The timing mark on the camshaft sprocket should be 90 degrees to the right, parallel to the top of the cylinder head. Make sure the yellow crankshaft pulley timing mark is aligned with the indicator pin.

3. Remove the seal cover.

4. Mark the position of the distributor rotor in relation to the distributor housing, and the distributor housing in relation to the cylinder head. Remove the distributor. Do not rotate the engine after distributor removal.

5. Hold the crankshaft pulley with a suitable tool and remove the distributor drive gear/camshaft pulley retaining bolt and the drive gear. Remove the upper timing cover assembly.

6. Push the timing chain adjuster sleeve in towards the left, and insert a 2mm diameter x 45mm long pin into the lever hole to hold it in place.

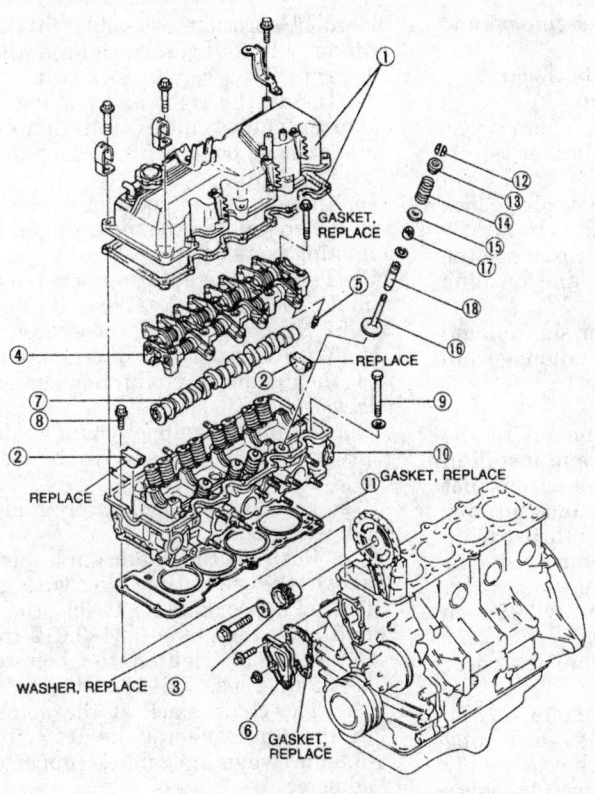

1. Cylinder head cover and gasket
2. Seal cover
3. Distributor drive gear
4. Rocker arm and shaft assembly
5. Hydraulic lash adjuster
6. Service cover
7. Camshaft
8. Timing chain cover attaching bolt
9. Cylinder head bolt
10. Cylinder head
11. Cylinder head gasket
12. Valve keepers
13. Upper spring seat
14. Valve spring
15. Lower spring seat
16. Valve
17. Valve seat
18. Valve guide

GASKET, REPLACE

REPLACE

REPLACE

GASKET, REPLACE

WASHER, REPLACE ③

⑥ GASKET, REPLACE

350193

Camshaft and cylinder head assembly — 2.6L engine

7. Wire the chain to the pulley and remove the pulley from the camshaft. Do not allow the sprocket and chain to fall down into the engine and cause the chain to become disengaged from the crankshaft sprocket.

8. Remove the rocker arm/shaft assembly.

9. Remove the camshaft.

10. Inspect the camshaft for wear and/or damage and replace if necessary.

To install:

11. Apply clean engine oil to the camshaft journals, lobes and bearings. Install the camshaft with the dowel pin facing upwards.

12. Install the rocker arm/shaft assembly and tighten the bolts, in sequence, in 2–3 steps to 14–19 ft. lbs. (19–25 Nm). Make sure the rocker arm shaft spring does not get caught between the shaft and mounting boss during installation.

13. Install the camshaft sprocket and tighten the bolt to 37–44 ft. lbs. (50–60 Nm).

NOTE: Remove all old sealer from the distributor drive gear and apply sealer to the gear face, then seat the gear fully on the camshaft.

14. Install the distributor drive gear and distributor.

15. Install the seal cover.

16. Coat a new gasket with silicone sealant and install on the rocker arm cover. Install the cover and tighten the bolts to 53–78 inch lbs. (6–9 Nm).

17. Remove the timing chain adjuster sleeve pin. Install the upper timing chain cover.

18. Connect the negative battery cable. Fill and bleed the cooling system.

19. Run the engine and check for leaks and proper operation. Check the idle speed and ignition timing.

3.0L Engine

─────── **CAUTION** ───────
Fuel injection systems remain under pressure after the engine has been turned OFF. Properly relieve fuel pressure before disconnecting any fuel lines. Failure to do so may result in fire or personal injury.

─────── **CAUTION** ───────
Do not allow fuel spray or fuel vapors to come in contact with a spark or open flame. Keep a dry chemical fire extinguisher nearby. Never store fuel in an open container due to risk of fire or explosion.

1. Disconnect the negative battery cable.

2. Drain the cooling system.

3. Relieve the fuel system pressure.

4. Remove the PCV valve and blind cover.

5. Remove the cylinder head.

6. If removing the driver's side camshaft, remove the distributor and the distributor spacer.

7. Remove the camshaft sprocket bolt and sprocket.

8. Remove the seal plate. Pry out the camshaft seal, being careful not to damage the seal housing.

9. Remove the rocker arm/shaft assembly.

10. Remove the thrust plate bolts and remove the thrust plate. Slide the camshaft out of the cylinder head. If removing the driver's side camshaft, remove the distributor drive gear.

11. Inspect the camshaft for wear and/or damage and replace if necessary.

To install:

NOTE: If installing the driver's side camshaft, remove all old sealer from the distributor drive gear and apply sealer to the gear face, then seat the gear fully on the camshaft.

12. Apply clean engine oil to the camshaft journals, lobes and bearings. Install the camshaft and the thrust plate. Tighten the thrust plate to 69–95 inch lbs. (7.8–11.0 Nm).

13. Apply clean engine oil to a new camshaft seal lip and press the seal into the cylinder head, using a seal installer.

14. Install the rocker arm/shaft assembly and tighten the bolts, in sequence, in 2–3 steps to 14–19 ft. lbs. (19–25 Nm). Make sure the rocker arm shaft spring does not get caught between the shaft and mounting boss during installation.

15. Install the seal plates and tighten the bolts to 69–95 inch lbs. (7.8–11.0 Nm).

16. Align and install the camshaft sprocket. Tighten the bolt to 52–59 ft. lbs. (71–80 Nm).

17. If installing the driver's side camshaft, apply clean engine oil to a new O-ring and install on the distributor spacer. Install the spacer and tighten the nuts to 69–95 inch lbs. (7.8–11.0 Nm). Install the distributor.

18. Install the cylinder head.

19. Install the blind cover and PCV valve.

20. Connect the negative battery cable.

21. Fill and bleed the cooling system.

22. Run the engine and check for leaks and proper operation. Check the idle speed and ignition timing.

Balance Shaft

REMOVAL AND INSTALLATION

2.6L Engine

——————— CAUTION ———————

Fuel injection systems remain under pressure after the engine has been turned OFF. Properly relieve fuel pressure before disconnecting any fuel lines. Failure to do so may result in fire or personal injury.

——————— CAUTION ———————

Do not allow fuel spray or fuel vapors to come in contact with a spark or open flame. Keep a dry chemical fire extinguisher nearby. Never store fuel in an open container due to risk of fire or explosion.

1. Relieve the fuel system pressure, and disconnect the negative battery cable.

2. Drain the engine engine oil and the cooling system.

3. Remove the cylinder head.

4. Remove the oil pan.

5. Remove the timing chain cover.

6. Remove the balancer shaft chain.

7. Remove the thrust plate lock bolts and remove the balancer shaft(s). Be careful not to damage the balancer shaft journals and bushing during removal.

8. Check the balancer shaft(s) and bushings for wear and/or damage and replace as necessary.

To install:

9. Apply clean engine oil to the balancer shaft journals and install in the cylinder block, being careful not to damage the bushings and journals.

10. Loosely tighten the thrust plate lock bolts and make sure the balancer shaft(s) rotate smoothly. Tighten the lock bolts to 69–95 inch lbs. (7.8–11.0 Nm).

11. Install the crankshaft balancer sprocket.

12. Set the balancer chain on the idler sprocket assembly so the timing mark on the idler sprocket assembly and the brown link of the balancer chain align.

13. Install the balancer chain so the 5 alignment marks on the chain, sprocket and block align, and attach the idler sprocket assembly to the cylinder block. Loosely tighten the idler sprocket assembly lock bolt.

14. Install the right and left lower balancer chain guides and tighten the mounting bolts to 69–95 inch lbs. (7.8–11.0 Nm).

15. Install the upper chain guide and loosely tighten the mounting and adjusting bolts.

16. Tighten the idler sprocket assembly lock bolt to 27–38 ft. lbs. (37–52 Nm), and install the spacer.

17. With manual transmission, adjust the balancer chain tension as follows:

   a. Tighten the upper chain guide pivot bolt to 69–95 inch lbs. (7.8–11.0 Nm).

   b. Loosen the upper chain guide adjusting bolt.

   c. Push on the chain guide just above the adjusting slot with a force of approximately 11 lbs., then pull back the guide 0.24–0.012 in. (6.0–0.3mm). Tighten the bolt to 69–95 inch lbs. (7.8–11.0 Nm).

   d. The chain slack at the notch in the guide should be 0.23 in. (5.8mm) when the guide is properly adjusted.

18. With automatic transmission, adjust the balancer chain tension as follows:

   a. Fabricate a piece of wood, 0.118–0.138 inch (3.0–3.5mm) thick and 0.335–0.374 inch (8.5–9.5mm) wide.

   b. Insert the piece of wood in the notch in the upper chain guide.

   c. Push on the chain guide just above the adjusting slot with a force of 2.9–3.7 lbs. and tighten the adjusting and pivot bolts to 69–95 inch lbs. (7.8–11.0 Nm).

   d. Remove the wood from between the chain and chain guide, making sure no wood shavings are left.

   e. Measure the chain slack. It should be 0.039–0.059 in. (1.0–1.5mm) at the notch in the guide.

NOTE: If the upper chain guide bottoms on the adjusting bolt during the adjustment procedure, the balancer chain must be replaced.

19. Install the timing chain cover.

20. Install the oil pan.

21. Install the cylinder head.

22. Fill the crankcase with the proper type and quantity of oil. Fill and bleed the cooling system.

23. Run the engine and check for leaks and proper operation. Check the idle speed and ignition timing.

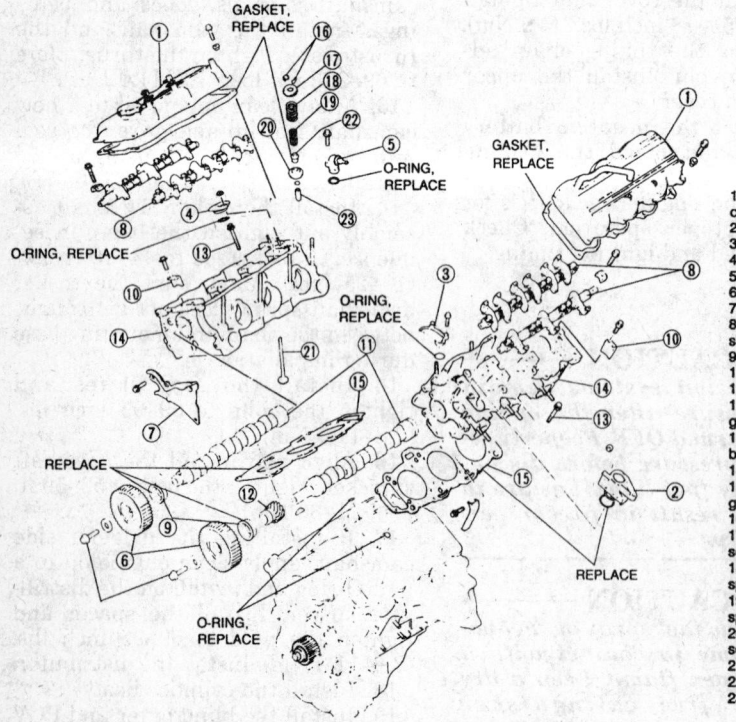

1. Cylinder head cover
2. Distributor spacer
3. Blind cover
4. Blind cover
5. PCV valve
6. Camshaft pulley
7. Seal plate
8. Rocker arm and shaft
9. Camshaft oil seal
10. Thrust plate
11. Camshaft
12. Distributor drive gear
13. Cylinder head bolt
14. Cylinder head
15. Cylinder head gasket
16. Valve keepers
17. Upper spring seat
18. Outer valve spring
19. Inner valve spring
20. Lower spring seat
21. Valve
22. Valve seal
23. Valve guide

302175

Camshaft and cylinder head components — 3.0L engines

Piston and Connecting Rod

POSITIONING

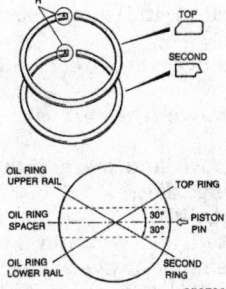

Piston ring UP marks and end-gap positions — 2.2L and 2.6L engines

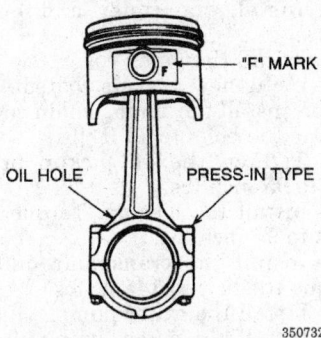

Piston and connecting rod F mark and oil hole: 2.2L engine

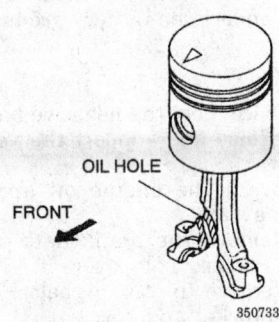

Piston and connecting rod arrow front mark and oil hole: 2.6L engine

ENGINE LUBRICATION

Oil Pan

REMOVAL AND INSTALLATION

2.2L Engine

1. Disconnect the battery ground cable.
2. Raise and support the vehicle on jackstands. Drain the oil.
3. Remove the skid plate.
4. Place a floor jack under the front of the engine at the crankshaft pulley and take up the weight of the engine. Or use a shop crane to support the engine.
5. Remove the crossmember.
6. Remove the cotter pin and nut and, with a puller, disconnect the idler arm from the center link.
7. Remove the engine mount gusset plates from the sides of the engine.
8. Remove the bell housing front cover.
9. Unbolt and remove the oil pan. A flat tipped screwdriver may be used to break the seal between the pan and block.
10. Clean all the gasket surfaces. Straighten and portion of the pan rim that is bent.
11. Clean the oil pan, oil pump pickup tube and oil pump screen.
   To install:
12. If you are using a gasket, install a new oil pan gasket coated with oil resistant sealer. Place RTV silicone sealer at the points. If you are using RTV silicone gasket material in place of a conventional gasket, run a ⅛ in. bead around the rim of the pan, going inboard of each bolt hole. Tighten the pan bolts within 30 minutes of application. Tighten the pan bolts to 5–9 ft. lbs.
13. Install the bell housing front cover. Torque the bell housing cover to 15–20 ft. lbs.
14. Install the engine mount gusset plates from the sides of the engine.
15. Install the idler arm on the center link.
16. Install the cotter pin and nut. Torque the idler arm nut to 25–30 ft. lbs.
17. Install the crossmember.
18. Remove the shop crane.
19. Install the skid plate.
20. Fill the engine with the proper amount of oil.

21. Install the battery ground cable.

2.6L Engine

1. Raise and support the front end on jackstands.
2. Drain the oil.
3. Remove the splash pan.
4. Remove the engine braces.
5. Remove the stabilizer bracket.
6. Unbolt and remove the oil pan.
7. Clean the oil pan and engine block gasket surfaces thoroughly.
8. Install a new pan gasket.
9. Install oil pan and tighten screws to 87 inch lbs.
10. Install all other parts in reverse order of removal.

3.0L Engine

1. Disconnect the negative battery cable.
2. Raise and safely support the vehicle.
3. On 4WD, attach a suitable engine lifting tool to hold the engine.
4. Drain the oil.
5. Remove the splash shield.
6. On 2WD, remove the 2 engine braces (gusset plates).
7. On 4WD, perform the following procedures:
   a. Remove the fresh-air duct and the cooling fan cowling.
   b. Remove the driver's side engine mount.
   c. Remove the transmission lower mount.
   d. Remove the oil cooler hose and pipe.
   e. Remove the stabilizer brackets.
8. Unbolt and remove the oil pan.
   To install:
9. Clean the oil pan and engine block gasket surfaces thoroughly.
10. Install a new pan gasket.
11. Install the oil pan and tighten the bolts to 61–87 inch lbs. (7–10 Nm).
12. On 4WD, perform the following procedures:
   a. Install the stabilizer brackets, then tighten the bolts to 14–18 ft. lbs. (19–25 Nm).
   b. Install the oil cooler hose and pipe.
   c. Install the transmission lower mount. Tighten the mounting bolt to 32–44 ft. lbs. (44–60 Nm), and the nut to 24–33 ft. lbs. (32–46 Nm).
   d. Install the engine mount. Tighten the mounting bolts to 26–36 ft. lbs. (35–49 Nm).
   e. Install the cooling fan cowling and the fresh-air duct.

13. On 2WD, install the engine braces, then tighten the mounting bolts to 28–38 ft. lbs. (38–51 Nm).

14. Install the splash shield and tighten the bolts to 61–87 inch lbs. (7–10 Nm).

15. On 4WD, remove the engine lifting tool.

16. Fill the engine oil, and lower the vehicle.

17. Connect the negative battery cable.

Oil Pump

REMOVAL AND INSTALLATION

2.2L Engine

Due to the inaccessibility of the oil pump, it is recommended that the engine assembly be removed.

1. Disconnect the battery ground.
2. Drain the cooling system and drain the engine oil.

—— CAUTION ——
When draining the coolant, always drain the coolant into a sealable container. Coolant should be reused unless it is contaminated or several years old.

3. Remove the engine assembly from the vehicle and secure it to a suitable engine stand.
4. Remove the timing belt.
5. Remove the crankshaft timing sprocket bolt and using a suitable puller, remove the crankshaft sprocket.
6. Unbolt and remove the oil pan. A thin flat bladed prying tool may be used to break the seal between the oil pan and the cylinder block.
7. Remove the 3 bolts and remove the oil pickup tube.
8. Remove the oil pan baffle and gasket.
9. Remove the oil pump mounting bolts and remove the oil pump from the cylinder block.
   To install:
10. Apply a thin coating of grease to the O-ring and install it in its recess in the oil pump body.
11. Apply a thin bead of RTV silicone sealer to the pump mounting surface.
12. Coat the oil seal lip with clean engine oil and install the pump. Torque the bolts to 14–19 ft. lbs. (19–25 Nm) and 27–38 ft. lbs. (37–52 Nm) respectively.
13. Clean all the gasket surfaces. Straighten any portion of the pan rim that is bent.

Oil pump assembly — 2.2L engine

14. Clean the oil pan and replace the oil pump pickup tube and oil pump screen.
15. If you are using a gasket, install a new oil pan gasket coated with oil resistant sealer. Place RTV silicone sealer at the points. If you are using RTV silicone gasket material in place of a conventional gasket, run a 1/8 in. bead around the rim of the pan, going inboard of each bolt hole. Tighten the pan bolts within 30 minutes of application. Tighten the pan bolts to 5–9 ft. lbs.
16. Reinstall the timing belt crankshaft sprocket with the mounting bolt and torque the bolt to 116–123 ft. lbs. (157–167 Nm).
17. Replace timing the belt if it has been contaminated by oil or grease, or shows any sign of damage, wear, cracks or peeling.
18. Reinstall the engine assembly to the vehicle.
19. Fill the engine with the proper amount of oil.
20. Fill the cooling system.
21. Install the battery ground cable, start the engine, bleed the cooling system, make any necessary adjustments, check for leaks and check for proper operation.

2.6L Engine

1. Disconnect the battery ground.
2. Drain the cooling system.

—— CAUTION ——
When draining the coolant, always drain the coolant into a sealable container. Coolant should be reused unless it is contaminated or several years old.

3. Remove the accessory drive belts.
4. Remove the fan and shroud.
5. Remove the water pump pulley.
6. Unbolt and remove the water pump.
7. Remove the crankshaft pulley.
8. Remove the oil pan.

9. Remove the timing chain cover.

NOTE: The pump is built into the cover

10. Remove the oil pickup tube.
11. Remove the pump cover from the case.
12. Remove the inner and outer rotors.
13. Remove the pressure relief valve.
14. Remove and discard the water inlet pipe gasket.
    To install:
15. Install a new water inlet pipe gasket using adhesive sealer.
16. Install the oil pickup tube using a new gasket. Torque the bolts to 95 inch lbs.
17. Install the pressure relief valve. Torque the plug to 45 ft. lbs.
18. Install the inner and outer rotors.
19. Install the pump cover.
20. Using new gaskets coated with sealer, install the timing chain cover. Torque the bolts to 19 ft. lbs.
21. Tighten the oil pickup brace bolt to 95 inch lbs.
22. Install the oil pan. Torque the bolts to 95 inch lbs.
23. Install the crankshaft pulley. Torque the bolt to 145 ft. lbs.
24. Install the water pump.
25. Install the water pump pulley.
26. Install the fan and shroud.
27. Install the accessory drive belts.
28. Fill the cooling system.
29. Connect the battery ground.

3.0L Engine

1. Disconnect the negative battery cable. Raise and support the vehicle safely.
2. Drain the engine oil and the cooling system.
3. Remove the crankshaft pulley and the timing belt covers.
4. Remove the timing belt, crankshaft sprocket and key.
5. Remove the thermostat and gasket.
6. Remove the oil pan, oil strainer and O-ring.
7. Unbolt and remove the oil pump and gasket.
   To install:
8. Press in a new oil seal and coat the seal lip with clean engine oil. Use a new gasket, O-ring and sealant as required. Torque the oil pump retaining bolts to 14–18 ft. lbs. (19–25 Nm).
9. Install the oil pan and torque the pan bolts 5–8 ft. lbs. (8–11 Nm).
10. Install the crankshaft sprocket and key.
11. Install the timing belt and covers. Install the crankshaft pulley and

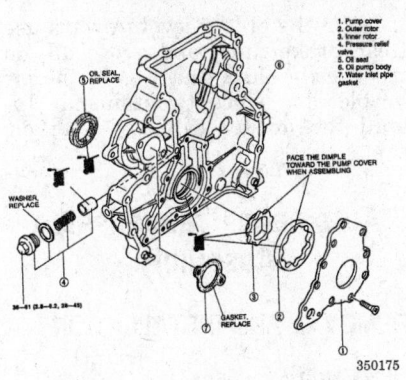

Oil pump assembly — 2.6L engine

tighten the pulley bolt to 116–123 ft. lbs. (157–167 Nm).

12. Install the thermostat and gasket. Torque the thermostat housing bolts 14–18 ft. lbs. (19–25 Nm).

13. Fill the crankcase to the recommended level with fresh oil. Fill the cooling system.

14. Crank the engine to prime the oil pump.

15. Start the engine and check for leaks.

TRANSMISSION

Manual Transmission Assembly

REMOVAL AND INSTALLATION

B2200 and B2600i

NOTE: On 4WD vehicles, the transmission and transfer case are removed as a unit.

1. Disconnect the negative battery cable.

2. Remove the knobs from the transfer case, if equipped and transmission shifters.

3. Remove the console box, if equipped.

4. Remove the insulator plate and shifter boot.

5. Remove the attaching bolts and the shift lever(s).

6. Raise and support the vehicle safely. Drain the transfer case, if equipped and transmission oil.

7. Remove the transmission and transfer case, if equipped, splash shields. Remove the starter.

8. Disconnect and remove the front exhaust pipe.

9. Matchmark and remove the driveshaft(s).

10. Disconnect the speedometer cable, 4WD switch, if equipped and backup light switch wires from the transmission or transmission/transfer case.

11. Remove the slave cylinder without disconnecting the fluid line. Support the slave cylinder aside.

12. Remove the transmission gusset plates and clutch housing cover. Support the transmission and engine with jacks.

13. Raise the transmission or transmission/transfer case and remove the crossmember.

14. Remove the transmission or transmission/transfer case.

To install:

15. Raise the transmission or transmission/transfer case assembly into position. Install the transmission-to-engine bolts and tighten to 51–65 ft. lbs. (69–88 Nm).

16. Install the crossmember and tighten the crossmember-to-chassis bolts to 23–34 ft. lbs. (31–46 Nm).

17. Lower the transmission or transmission/transfer case assembly and remove the jacks. Install the crossmember-to-transmission mount nuts and tighten to 23–34 ft. lbs. (31–46 Nm).

18. Install the gusset plates and clutch housing cover. Install the slave cylinder.

19. Connect the backup light and, if equipped, 4WD switch electrical connectors and the speedometer cable.

20. Install the driveshaft(s), aligning the marks that were made during removal.

21. Install the front exhaust pipe and the starter. Install the splash shields.

22. Fill the transmission and, if equipped, transfer case with the proper type and quantity of fluid. Lower the vehicle.

23. Install the shift lever(s) and tighten the mounting bolts to 25–37 ft. lbs. (34–50 Nm).

24. Install the shifter boot and insulator plate. Tighten the mounting bolts to 25–37 ft. lbs. (34–50 Nm).

25. Install the console box and the shifter knob(s).

26. Connect the negative battery cable. Check the transmission for leaks and proper operation.

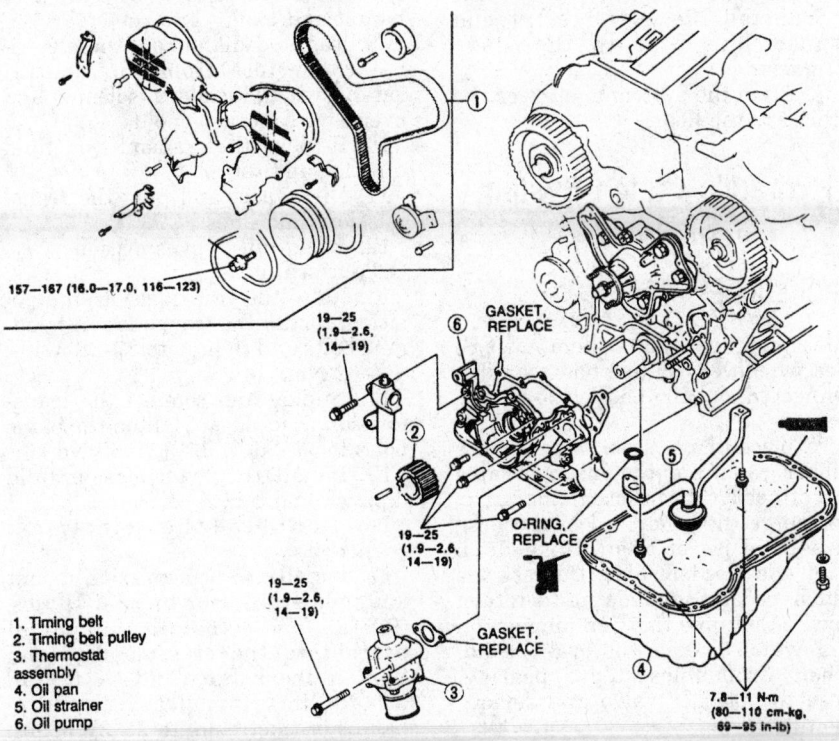

1. Timing belt
2. Timing belt pulley
3. Thermostat assembly
4. Oil pan
5. Oil strainer
6. Oil pump

Oil pump and related components — 3.0L engine

Clutch Assembly

REMOVAL AND INSTALLATION

B2200 and B2600i

———— CAUTION ————

The clutch driven disc contains asbestos, which has been determined to be a cancer causing agent. Never clean clutch surfaces with compressed air! Avoid inhaling any dust from any clutch surface! When cleaning clutch surfaces, use a commercially available brake cleaning fluid.

1. Remove the transmission.
2. Remove the 4 attaching and 2 pilot bolts holding the clutch cover to the flywheel. Loosen the bolts evenly a turn or 2 at a time. If the clutch cover is to be reinstalled, mark the flywheel and clutch cover to show the location of the 2 pilot holes.
3. Remove the clutch disc.

To install:

4. Install the clutch disc on the flywheel. Do not touch the facing or allow the facing to come in contact with grease or oil. The clutch disc can be aligned using a tool made for that purpose, or with an old mainshaft.
5. Install the clutch cover on the flywheel and install the 4 standard bolts and the 2 pilot bolts.
6. To avoid distorting the pressure plate, tighten the bolts evenly a few turns at a time until they are all tight.
7. Torque the bolts to 13–20 ft. lbs. using a star pattern.
8. Remove the aligning tool.
9. Apply a light film of lubricant to the release bearing, release lever contact area on the release bearing hub and to the input shaft bearing retainer.
10. Install the transmission.
11. Check the operation of the clutch and if necessary, adjust the pedal free-play and the release lever.

Clutch Master Cylinder

REMOVAL AND INSTALLATION

B2200 and B2600i

1. Using a flare nut wrench, disconnect and plug the fluid outlet line at the outlet fitting on the master cylinder one-way valve.
2. Remove the nuts and bolts attaching the master cylinder to the firewall.

3. Remove the cylinder straight out away from the firewall.

To install:

4. Start the pedal pushrod into the master cylinder and position the master cylinder on the firewall.
5. Install the attaching nuts and bolts. Torque the nuts to 12–17 ft. lbs.
6. Connect the fluid outlet line to the master cylinder fitting.
7. Bleed the hydraulic system.
8. Check the clutch pedal free-travel and adjust as necessary.

Clutch Slave Cylinder

REMOVAL AND INSTALLATION

B2200 and B2600i

1. Raise and support the front end on jackstands.
2. Back off the flare nut on the fluid pipe to free the slave cylinder hose.
3. Pull off the hose-to-bracket retaining clip and pull the hose from the bracket. Cap the pipe to prevent fluid loss.
4. Unbolt and remove the slave cylinder.

To install:

5. Install the slave cylinder to the clutch housing.
6. Install the hose bracket and connect the hose to the slave cylinder.
7. Bleed the system and lower the vehicle to the floor.

Hydraulic Clutch System Bleeding

PROCEDURE

The clutch hydraulic system must be bled whenever the line has been disconnected or air has entered the system.

To bleed the system, remove the rubber cap from the bleeder valve and attach a rubber hose to the valve. Submerge the other end of the hose in a large jar of clean brake fluid. Open the bleeder valve. Depress the clutch pedal and allow it to return slowly. Continue this pumping action and watch the jar of brake fluid. When air bubbles stop appearing, close the bleeder valve and remove the tube.

During the bleeding process, the master cylinder must be kept at least ¾ full. After the bleeding operation is finished, install the cap on the bleeder valve and fill the master cyl-

inder to the proper level. Always use fresh brake fluid, and above all, do not use the fluid that was in the jar for bleeding, since it contains air. Install the master cylinder reservoir cap.

Automatic Transmission Assembly

REMOVAL AND INSTALLATION

B2200 and B2600i

2WD Vehicles

1. Disconnect the negative battery cable. Raise and safely support the vehicle.
2. Drain the transmission fluid.
3. Mark the position of the driveshaft on the axle flange and remove the driveshaft.
4. Disconnect the speedometer cable, vacuum hose and shift lever. Remove the vacuum line bracket from the transmission.
5. Remove the gusset plates and bellhousing cover.
6. Remove the torque converter attaching bolts from the flywheel.
7. Support the transmission and engine with jacks.
8. Remove the transmission mount and mount bracket.
9. Tag and disconnect the electrical connectors from the neutral safety switch, kickdown solenoid and overdrive cancel solenoid.
10. Remove the transmission fluid dipstick and tube.
11. Disconnect and plug the transmission fluid lines.
12. Remove the transmission.

To install:

13. Raise the transmission into position. Install the transmission-to-engine bolts and tighten to 27–38 ft. lbs. (37–52 Nm).
14. Unplug and connect the transmission fluid lines. Tighten the banjo bolts to 17–26 ft. lbs. (24–35 Nm).
15. Install the transmission fluid dipstick and tube.
16. Connect the electrical connectors.
17. Install the transmission mount and tighten the bolts to 7.2–17 ft. lbs. (9.8–23 Nm). Install the mount bracket to the crossmember and tighten the bolts to 23–34 ft. lbs. (31–46 Nm). Install the mount-to-mount bracket bolts to 23–34 ft. lbs. (31–46 Nm).
18. Remove the support jacks.
19. Install the torque converter attaching bolts and tighten to 25–36 ft. lbs. (34–49 Nm).

20. Install the gusset plates and bellhousing cover. Tighten the gusset plate bolts to 27–38 ft. lbs. (37–52 Nm).

21. Connect the shift lever, vacuum hose and speedometer cable. Attach the vacuum line bracket.

22. Install the driveshaft and lower the vehicle.

23. Connect the negative battery cable. Fill the transmission with the proper type and quantity of fluid.

24. Check the transmission for leaks and proper operation.

4WD Vehicles

1. Disconnect the negative battery cable.

2. Remove the shifter knob and the console box.

3. Remove the insulator plate and boot. Remove the 4WD shift lever.

4. Raise and safely support the vehicle. Remove the splash shields and drain the transmission fluid.

5. Disconnect and remove the front exhaust pipe.

6. Mark the position of the driveshaft on the flanges and remove the driveshafts.

7. Disconnect the speedometer cable and the 4WD indicator switch connector, if equipped.

8. Disconnect the shift cable and vacuum hose, if equipped.

9. Remove the gusset plate, if equipped.

10. Loosen the front differential mounting bolts and remove the No. 2 crossmember.

11. Remove the torque converter attaching bolts.

12. Support the transmission and engine with jacks. Remove the transmission-to-engine bolts.

13. Disconnect and plug the transmission fluid lines at the transmission. Remove the bracket from the transmission.

14. Remove the rear transmission crossmember.

15. Tag and disconnect the electrical connectors at the transmission.

16. Remove the transmission dipstick and tube.

17. Lower the transmission from the vehicle.

To install:

18. Raise the transmission into position and install the transmission-to-engine bolts. Tighten to 27–38 ft. lbs. (37–52 Nm).

19. Install the dipstick and tube.

20. Connect the electrical connectors.

21. Install the rear transmission crossmember and tighten the trans-mission-to-chassis bolts to 23–34 ft. lbs. (31–46 Nm).

22. Lower the transmission to the crossmember and install the mount-to-crossmember nuts. Tighten to 23–34 ft. lbs. (31–46 Nm). Remove the jacks.

23. Connect the transmission fluid lines to the transmission and tighten the banjo bolts to 17–26 ft. lbs. (24–35 Nm). Attach the fluid line bracket.

24. Install the torque converter attaching bolts and tighten to 27–40 ft. lbs. (36–54 Nm). Install the bellhousing cover.

25. Install the No. 2 crossmember.

26. Install the gusset plates, if equipped.

27. Connect the shifter cable and the bracket, if equipped. Connect the speedometer cable.

28. Install the driveshafts, aligning the marks that were made during removal.

29. Install the front exhaust pipe and the splash shields. Lower the vehicle.

30. Install the 4WD shift lever and the insulator plate and boot.

31. Install the console box and the shifter knob.

32. Connect the negative battery cable. Fill the transmission with the proper type and quantity of fluid.

33. Check the transmission for leaks and proper operation.

MPV

2WD

1. Disconnect the negative battery cable.

2. Raise and support the vehicle.

3. Drain the transmission fluid but do not remove the pan.

4. Disconnect the speedometer cable.

5. Label for identification and location and then disconnect all electrical wiring at the transmission.

6. Remove the exhaust pipe and heat shield.

7. Matchmark and disconnect the driveshaft.

8. Remove the selector cable from the transmission shift lever and the cable bracket and remove the shift selector cable.

9. Remove the filler tube.

10. Remove the access cover from the lower front of the converter housing.

11. Matchmark the drive plate (flywheel) and torque converter for reassembly. Remove the 4 bolts holding the torque converter to the drive plate.

12. Remove the starter.

13. Remove the exhaust pipe bracket.

14. Remove the bolts connecting the crossmember to the transmission.

15. Disconnect and plug the cooler lines from the radiator at the transmission.

16. Remove the gusset plates.

17. Support the transmission with a jack. Remove the crossmember-to-frame bolts, and remove the crossmember.

18. Make sure the transmission is securely supported. Secure it to the jack with a safety chain, if necessary.

19. Remove the converter housing-to-engine bolts.

20. With a prybar, exert light pressure between the converter and the drive plate to prevent the converter from disengaging from the transmission as it is removed.

21. Lower the transmission and converter as an assembly. Be careful not to let the converter fall out.

To install:

22. Place the transmission on the jack. Be sure the converter is properly installed.

23. Raise the transmission into place. Install the converter housing-to-engine bolts, and torque in 2 stages to 28–38 ft. lbs. (38–51 Nm).

24. Install the gusset plates. Torque the bolts to 28–38 ft. lbs. (38–51 Nm).

25. Install the exhaust pipe bracket, torque the bolts to 14–18 ft. lbs. (19–25 Nm).

26. Connect the oil cooler pipes.

27. Install the transmission crossmember mounting bolts. Torque to 32–44 ft. lbs. (44–60 Nm).

28. Loosely and evenly tighten the torque converter mounting bolts. Torque to 27–39 ft. lbs. (37–53 Nm).

29. Install the access cover. Remove the jack.

30. Install the starter.

31. Install the fluid filler tube with a new O-ring.

32. Connect the electrical connectors, and replace the wires in the clip. Install the vacuum hose.

33. Reconnect the speedometer cable.

34. Reconnect the shift selector cable to the transmission gear selector lever. Secure the cable end with the washer and the hitchpin. Reinstall the selector cable to the cable bracket with the clip.

35. Install the heat insulator and exhaust pipe.

36. Insert the driveshaft into the transmission. Install the center bearing support. Bolt the driveshaft to the rear of the axle flange.

37. Install a new pan gasket and the fluid pan, if this has not already been done.

38. Lower the vehicle. Connect the negative battery cable. Fill the transmission through the dipstick tube with the specified fluid, being careful not to overfill, and check for leaks.

4WD

1. Disconnect the negative battery cable.

2. Raise and support the vehicle.

3. Drain the transmission fluid but do not remove the pan.

4. Disconnect the speedometer cable.

5. Label for identification and location and then disconnect all electrical wiring at the transmission.

6. Remove the exhaust pipe and heat shield.

7. Matchmark and disconnect the front driveshaft and the rear driveshaft.

8. Remove the selector cable from the transmission shift lever and the cable bracket and remove the shift selector cable.

9. Remove the filler tube.

10. Remove the access cover from the lower front of the converter housing.

11. Matchmark the drive plate (flywheel) and torque converter for reassembly. Remove the 4 bolts holding the torque converter to the drive plate.

12. Remove the starter.

13. Remove the exhaust pipe bracket.

14. Remove the bolts connecting the crossmember to the transmission.

15. Disconnect and plug the cooler lines from the radiator at the transmission.

16. Remove the gusset plates.

17. Support the transmission with a jack. Remove the crossmember-to-frame bolts, and remove the crossmember.

18. Make sure the transmission is securely supported. Secure it to the jack with a safety chain, if necessary.

19. Support the transfer case with a jack and remove the transfer case mounting bolts. Remove the transfer case. It may be necessary to tap the case with a plastic hammer to break it free from the transmission.

20. Remove the converter housing-to-engine bolts.

21. With a prybar, exert light pressure between the converter and the drive plate to prevent the converter from disengaging from the transmission as it is removed.

22. Lower the transmission and converter as an assembly. Be careful not to let the converter fall out.

To install:

23. Place the transmission on the jack. Be sure the converter is properly installed.

24. Raise the transmission into place. Install the converter housing-to-engine bolts, and torque in 2 stages to 28–38 ft. lbs. (38–51 Nm).

25. Apply silicone sealant to the transfer case.

26. Support the transfer case with a jack and install the transfer case. Apply sealant to the bolt threads and tighten to 27–39 ft. lbs. (37–52 Nm).

27. Install the gusset plates. Torque the bolts to 28–38 ft. lbs. (38–51 Nm).

28. Install the exhaust pipe bracket, torque the bolts to 14–18 ft. lbs. (19–25 Nm).

29. Connect the oil cooler pipes.

30. Install the transmission crossmember mounting bolts. Torque to 32–44 ft. lbs. (44–60 Nm).

31. Loosely and evenly tighten the torque converter mounting bolts. Torque to 27–39 ft. lbs. (37–53 Nm).

32. Install the access cover. Remove the jack.

33. Install the starter.

34. Install the fluid filler tube with a new O-ring.

35. Connect the electrical connectors, and replace the wires in the clip. Install the vacuum hose.

36. Reconnect the speedometer cable.

37. Reconnect the shift selector cable to the transmission gear selector lever. Secure the cable end with the washer and the hitchpin. Reinstall the selector cable to the cable bracket with the clip.

38. Install the heat insulator and exhaust pipe.

39. Install the front driveshaft and the rear driveshaft. Install the center bearing support.

40. Install a new pan gasket and the fluid pan, if this has not already been done.

41. Lower the vehicle. Connect the negative battery cable. Fill the transmission through the dipstick tube with the specified fluid, being careful not to overfill, and check for leaks.

Transfer Case Assembly

REMOVAL AND INSTALLATION

B2200 and B2600i

Manual Transmission

1. Disconnect the battery ground cable.

2. Raise and support the vehicle on jackstands.

3. Drain the transmission.

4. Remove the shift knobs and boots.

5. Remove the console.

6. Remove the insulator plate.

7. Remove the shift grommet.

8. Remove the shift levers.

9. Remove the upper right side transmission support plate.

10. Remove the transmission splash shield.

11. Remove the transfer case splash shield.

12. Remove the exhaust pipe.

13. Matchmark and remove the front and rear driveshafts.

14. Disconnect the speedometer cable.

15. Disconnect the 4WD indicator connector.

16. Disconnect the back-up light switch.

17. Remove the front converter spring and nut.

18. Remove the clutch release cylinder.

19. Remove the transmission side support plates.

20. Remove the bellhousing inspection cover.

21. Support the transmission/transfer case with a transmission jack.

22. Remove the crossmember.

23. Lower the transmission assembly until all the transmission-to-engine bolts are available.

24. Support the engine with a jackstand.

25. Remove the transmission-to-engine bolts.

26. Pull the unit backward until the mainshaft clears the clutch.

27. Lower the unit and remove it from under the vehicle.

28. Stand the transmission on end with the transfer case up.

29. Remove the transfer case-to-extension housing bolts.

30. Lift the transfer case straight up and off. It may be necessary to tap it loose with a plastic mallet.

NOTE: Removing the transfer case in this manner will avoid damage to the control rod.

To install:

31. Stand the transmission on end with the rear up.

32. Coat the mating surfaces of the transfer case and extension housing with RTV gasket material.

33. Install the transfer case on the extension housing, indexing the control lever end.

34. Apply RTV gasket material to the bolt threads and torque them to 35 ft. lbs.

35. Secure the control lever end with a spring pin.

36. Coat the mating surfaces of the shift control lever case and transfer case with RTV gasket material and install the shift control case. Apply RTV to the bolts and tighten them to 22 ft. lbs.

37. Raise the unit into position.

38. Push the unit forward until the mainshaft enters the clutch. and the locating lugs engage the bellhousing.

39. Install the transmission-to-engine bolts. Torque the bolts to 45 ft. lbs.

40. Install the crossmember. Torque the crossmember bolts to 50 ft. lbs.; the crossmember-to-transmission bolts to 40 ft. lbs.

41. Remove the engine jackstand.

42. Remove the transmission/transfer case transmission jack.

43. Install the bellhousing inspection cover.

44. Install the transmission side support plates.

45. Install the clutch release cylinder.

46. Install the front converter spring and nut.

47. Connect the back-up light switch.

48. Connect the 4WD indicator connector.

49. Connect the speedometer cable.

50. Install the front and rear driveshafts.

51. Install the exhaust pipe.

52. Install the transmission splash shield.

53. Install the transfer case splash shield.

54. Install the upper right side transmission support plate.

55. Install the shift levers.

56. Install the shift grommet.

57. Install the insulator plate.

58. Install the console.

59. Install the shift knobs and boots.

60. Fill the transmission.

61. Lower the vehicle.

62. Connect the battery ground cable.

Automatic Transmission

1. Disconnect the negative cable from the battery.

2. Raise and support the vehicle.

3. Drain the transmission fluid but do not remove the pan. After the fluid has drained, install a few bolts to hold the pan in place, temporarily.

4. Remove the exhaust pipe bracket bolt from the right side of the converter housing.

5. Remove the splash shield and skid plate.

6. Remove the exhaust pipe flange bolts from the rear of the resonator or catalytic converter, and disconnect the pipe.

7. Matchmark and disconnect the front and or rear the driveshaft(s).

8. Disconnect the speedometer cable.

9. Disconnect the shift rod from the manual lever.

10. Remove the shift knob.

11. Remove the console.

12. Remove the 4WD shift lever plate and transfer case lever.

13. Disconnect the 4WD indicator wiring.

14. Remove the vacuum hose from the diaphragm. Disconnect the electrical connectors from the downshift solenoid and inhibitor switch, and remove their wires from the clip.

15. Disconnect and plug the cooler lines from the radiator at the transmission. Use a flare nut wrench if one is available.

16. Remove the side support plates.

17. Remove the access cover from the lower front of the converter housing.

18. Matchmark the drive plate (flywheel) and torque converter for reassembly. Remove the 4 bolts holding the torque converter to the drive plate.

19. Remove the bolts connecting the crossmember to the transmission.

20. Support the transmission with a jack. Remove the crossmember-to-frame bolts, and remove the crossmember.

21. Make sure the transmission is securely supported. Secure it to the jack with a safety chain, if necessary.

22. Lower the transmission to provide working clearance, and remove the starter.

23. Remove the converter housing-to-engine bolts.

24. Remove the fluid filler tube.

25. With a pry bar, exert light pressure between the converter and the drive plate to prevent the converter from disengaging from the transmission as it is removed.

26. Lower the transmission and converter as an assembly. Be careful not to let the converter fall out.

27. Stand the transmission on end with the transfer case up.

28. Remove the transfer case-to-extension housing bolts.

29. Lift the transfer case straight up and off. It may be necessary to tap it loose with a plastic mallet.

NOTE: Removing the transfer case in this manner will avoid damage to the control rod.

To install:

30. Stand the transmission on end with the rear up.

31. Coat the mating surfaces of the transfer case and extension housing with RTV gasket material.

32. Install the transfer case on the extension housing, indexing the control lever end.

33. Apply RTV gasket material to the bolt threads and torque them to 35 ft. lbs.

34. Secure the control lever end with a spring pin.

35. Coat the mating surfaces of the shift control lever case and transfer case with RTV gasket material and install the shift control case. Apply RTV to the bolts and tighten them to 22 ft. lbs.

36. Place the transmission on the jack. Be sure the converter is properly installed.

37. Raise the transmission into place. Install the converter housing-to-engine bolts, and torque in 2 stages to 23–34 ft. lbs.

38. Lower the transmission on the jack and install the starter.

39. Install the fluid filler tube with a new O-ring.

40. Raise the transmission slightly, and install the crossmember to the frame. Tighten the bolts to 23–34 ft. lbs.

41. Lower the transmission and install the transmission-to-crossmember bolts. Tighten to 23–34 ft. lbs.

42. Align the matchmarks made earlier on the torque converter and drive plate. Install the 4 attaching bolts and torque to 25–36 ft. lbs. in 3 stages.

43. Install the side support plates.

44. Install the access cover. Remove the jack.

45. Connect the cooler lines.

46. Install the 4WD shift lever plate and transfer case lever.

47. Connect the 4WD indicator wiring.

48. Install the electrical connectors to the switch and solenoid, and re-

place the wires in the clip. Install the diaphragm vacuum hose.

49. connect the shift rod to the lever.

50. Reconnect the speedometer cable.

51. Insert the driveshaft into the transmission. Install the center bearing support. Bolt the driveshaft to the rear of the axle flange.

52. Connect the exhaust pipe to the resonator or catalytic converter, using a new gasket. Reinstall the exhaust pipe clamp onto the converter housing, and torque the bolt to 10–15 ft. lbs.

53. Install a new pan gasket and the fluid pan, if this has not already been done.

54. Install the shift knob.

55. Install the console.

56. Lower the vehicle. Connect the battery cable. Fill the transmission through the dipstick tube with the specified fluid, being careful not to overfill, and check for leaks.

MPV

1. Disconnect the negative battery cable. Raise and safely support the vehicle. Drain the transfer case.

2. Mark the position of the driveshafts on the flanges and remove the driveshafts. Push a rag into the double-offset joint to hold the rear driveshaft straight to prevent damaging the boot.

3. Support the transmission with a jack and remove the transmission lower mount. Remove the upper mount.

4. Remove the front exhaust pipe and heat insulator.

5. Disconnect the speedometer and disconnect the electrical connectors.

6. Support the transfer case with a jack and remove the transfer case mounting bolts. Remove the transfer case. It may be necessary to tap the case with a plastic hammer to break it free from the transmission.

To install:

7. Apply silicone sealant to the transfer case.

8. Support the transfer case with a jack and install the transfer case. Apply sealant to the bolt threads and tighten to 27–39 ft. lbs. (37–52 Nm).

9. Connect the electrical connectors and the speedometer cable. Adjust the transfer case shift cable.

10. Install the exhaust pipe and heat insulator.

11. Install the upper transmission mount.

12. Install the lower transmission mount. Loosely install the center

washers and nuts and tighten the outer bolts to 32–44 ft. lbs. (44–60 Nm), then tighten the center nuts to 24–33 ft. lbs. (32–46 Nm).

13. Remove the support jack.

14. Install the driveshafts, aligning the matchmarks. Remove the rag from the double-offset joint and check the boot for damage.

15. Fill the transfer case with the proper type and quantity of fluid.

16. Lower the vehicle and connect the negative battery cable. Check the transfer case for leaks and proper operation.

DRIVE AXLE

Driveshaft

REMOVAL AND INSTALLATION

B2200 and B2600i

1. Matchmark the rear U-joint with the rear companion flange. Remove the bolts attaching the driveshaft to the rear companion flange.

2. On 2-piece units, remove the center support bearing bracket from the underbody.

3. Pull the driveshaft rearward and out of the transmission. Plug the rear seal opening.

4. Installation is the reverse of removal. Make sure you align the matchmarks. Torque the rear companion flange bolts to 39–47 ft. lbs.; the center bearing bracket nuts to 27–38 ft. lbs.

MPV

2WD Vehicles

1. Raise and safely support the vehicle.

2. Matchmark the flanges and remove the mounting nuts and bolts at the differential.

3. Remove the bolts at the center bearing support assembly.

4. Remove the driveshaft from the transmission. Cap the transmission to prevent spillage of fluid.

5. Installation is the reverse of removal. Align the flange marks when installing. Torque the mounting bolts at the differential to 37–43 ft. lbs. (50–58 Nm). Torque the bolts for the center support to 27–39 ft. lbs. (37–53 Nm).

4WD Front Driveshaft

1. Raise and safely support the vehicle.

2. Matchmark the flanges and remove the mounting nuts and bolts at the front differential.

3. Slide the shaft out of the transfer case. Cap the opening to prevent fluid loss.

4. Installation is the reverse of removal. Align the flange marks. Torque the nuts and bolts to 37–43 ft. lbs. (50–58 Nm).

4WD Rear Driveshaft

1. Raise and safely support the vehicle.

2. Matchmark the flanges and remove the mounting nuts and bolts at the differential.

3. Remove the bolts at the center bearing support assembly.

4. Remove the driveshaft from the transfer case. Cap the opening to prevent spillage of fluid.

5. Installation is the reverse of removal. Align the flange marks when installing. Torque the mounting bolts at the differential to 37–43 ft. lbs. (50–58 Nm). Torque the bolts for the center support to 27–39 ft. lbs. (37–53 Nm).

U-Joints

REMOVAL AND INSTALLATION

B2200 and B2600i

Single Cardan U-Joint

1. Remove the driveshaft.

2. If the front yoke is to be disassembled, matchmark the driveshaft and sliding splined yoke (transmission yoke) so driveline balance is preserved upon reassembly. Remove the snaprings which retain the bearing caps.

3. Select 2 sockets, one small enough to pass through the yoke holes for the bearing caps, the other large enough to receive the bearing cap.

4. Using a vise or a press, position the small and large sockets on either side of the U-joint. Press in on the smaller socket so it presses the opposite bearing cap out of the yoke and into the larger socket. If the cap does not come all the way out, grasp it with a pair of pliers and work it out.

5. Reverse the position of the sockets so the smaller socket presses on the cross. Press the other bearing cap out of the yoke.

6. Repeat the procedure on the other bearings.

To install:

7. Grease the bearing caps and needles thoroughly if they are not pregreased. Start a new bearing cap into one side of the yoke. Position the cross in the yoke.

8. Select 2 sockets small enough to pass through the yoke holes. Put the sockets against the cross and the cap, and press the bearing cap 1/4 in. (6mm) below the surface of the yoke. If there is a sudden increase in the force needed to press the cap into place, or if the cross starts to bind, the bearings are cocked, They must be removed and restarted in the yoke. Failure to do so will greatly reduce the life of the bearing.

9. Install a new snapring.

10. Start a new bearing into the opposite side. Place a socket on it and press in until the opposite bearing contacts the snapring.

11. Install a new snapring. It may be necessary to grind the facing surface of the snapring slightly to permit easier installation.

12. Install the other bearings in the same manner.

13. Check the joint for free movement. If binding exists, smack the yoke ears with a brass or plastic faced hammer to seat the bearing needles. Do not strike the bearings, and support the shaft firmly. Do not install the driveshaft until free movement exists at all joints.

14. The nut attaching the yoke and bearing to the front coupling is torqued to 115–130 ft. lbs.

Double Cardan Type U-Joint

1. Place the driveshaft on a suitable workbench.

2. Matchmark the positions of the spiders, the center yoke and the centering socket yoke as related to the stud yoke which is welded to the front of the driveshaft tube.

NOTE: The spiders must be assembled with the bosses in their original position to provide proper clearance.

3. Remove the snaprings that secure the bearings in the front of the center yoke.

4. Position the U-joint tool, T74P-4635-C or equivalent, on the center yoke. Thread the tool clockwise until the bearing protrudes approximately 3/8 in. (10mm) out of the yoke.

5. Position the bearing in a vise and tap on the center yoke to free it from the bearing. Lift the 2 bearing cups from the spider.

6. Re-position the tool on the yoke and move the remaining bearing in the opposite direction so it protrudes approximately 3/8 in. (10mm) out of the yoke.

7. Position the bearing in a vise. Tap on the center yoke to free it from the bearing. Remove the spider from the center yoke.

8. Pull the centering socket yoke off the center stud. Remove the rubber seal from the centering ball stud.

9. Remove the snaprings from the center yoke and from the driveshaft yoke.

10. Position the tool on the driveshaft yoke and press the bearing outward until the inside of the center yoke almost contacts the slinger ring at the front of the driveshaft yoke. Pressing beyond this point can distort the slinger ring interference point.

11. Clamp the exposed end of the bearing in a vise and drive on the center yoke with a soft-faced hammer to free it from the bearing.

12. Reposition the tool and press on the spider to remove the opposite bearing.

13. Remove the center yoke from the spider. Remove the spider form the driveshaft yoke.

14. Clean all serviceable parts in cleaning solvent. If using a repair kit, install all of the parts supplied in the kit.

15. Remove the clamps on the driveshaft boot seal. Discard the clamps.

16. Note the orientation of the slip yoke to the driveshaft tube for installation during assembly. Mark the position of the slip yoke to the driveshaft tube.

17. Carefully pull the slip yoke from the driveshaft. Be careful not to damage the boot seal.

18. Clean and inspect the spline area of the driveshaft.

To install:

19. Lubricate the driveshaft slip splines with Multi-purpose Long-Life lubricant C1AZ-19490-B or equivalent.

20. With the boot loosely installed on the driveshaft tube, install the slip yoke into the driveshaft splines in their original orientation.

21. Using new clamps, install the driveshaft boot in its original position.

22. To assemble the double Cardan joint, position the spider in the driveshaft yoke. Make certain the spider bosses (or lubrication plugs on kits) will be in the same position as originally installed. Press in the bearing using the U-joint tool. Then, install the snaprings.

23. Pack the socket relief and the ball with Multi-purpose Long-Life lubricant C1AZ-19490-B or equivalent, then position the center yoke over the spider ends and press in the bearing. Install the snaprings.

24. Install a new seal on the centering ball stud. Position the centering socket yoke on the stud.

25. Place the front spider in the center yoke. Make certain the spider bosses (or lubrication plugs on kits) are properly positioned.

26. With the spider loosely positioned on the center stop, seat the first pair of bearings into the centering socket yoke. Then, press the second pair into the centering yoke. Install the snaprings.

27. Apply pressure on the centering socket yoke and install the remaining bearing cup.

28. If a kit was used, lubricate the U-joint through the grease fitting, using Multi-purpose Long-Life lubricant C1AZ-19490-B or equivalent.

Constant Velocity (CV) Type U-Joint

1. Place the driveshaft on a suitable workbench.

NOTE: The CV-joint components are matched. Extreme care should be take not to mix or substitute components.

2. Remove the clamp retaining the shroud to the outer bearing race and flange assembly.

3. Carefully tap the shroud lightly with a blunt tool and remove the shroud. Be careful not to damage the shroud, dust boot or outer bearing race and flange assembly.

4. Peel the boot upward and away from the outer bearing race and flange assembly.

5. Remove the wire ring that retains the inner race to the outer race.

6. Remove the inner race and shaft assembly from the outer race and flange assembly. Remove the cap and spring from inside the outer retainer.

7. Remove the circlip retaining the inner race assembly to the shaft, using snapring pliers. Discard the clip and remove the inner race assembly.

8. If required, remove the clamp retaining the boot to the shaft and remove the boot.

9. Carefully pry the ball bearings from the cage. Be careful not to scratch or damage the cage, race or ball bearings.

10. Rotate the inner race to align with the cage windows and remove the inner race through the wider end of the cage.

To install:

11. Install the inner bearing race in the bearing cage. Install the race through the large end of the cage with the counterbore facing the large end of the cage.

12. Push the race to the top of the cage and rotate the race until all the ball slots are aligned with the windows. This will lock the race to the top of the cage.

13. With the bearing cage and inner race properly aligned, install the ball bearings. The bearings can be pressed through the bearing cage with the heel of the hand. Repeat this step until the remaining ball bearings are installed.

14. If removed, install a new dust boot on the shaft, using a new clamp. Make certain the boot is seated in its groove.

NOTE: The clamp is a fixed diameter push-on metal ring.

15. Install the inner bearing assembly on the shaft. Make certain the circlip is exposed.

16. Install a new circlip on the shaft. Do not over-expand or twist the circlip during installation.

17. Install the spring and cap in the outer bearing retainer and flange.

18. Fill the outer bearing retainer with 3 oz. of Constant Velocity Joint Grease, D8RZ-19590-A or equivalent.

19. Insert the inner race and shaft assembly in the outer bearing retainer and flange.

20. Push the inner race down until the wire spring groove is visible and install the wire ring.

21. Fill the top of the outer bearing retainer with Constant Velocity Joint Grease, D8RZ-19590-A or equivalent. Remove all excess grease from the external surfaces.

22. Pull the dust boot over the retainer. Make certain the boot is seated in the groove and that any air pressure which may have built up in the boot is relieved.

NOTE: Insert a dulled screwdriver blade between the boot and outer bearing retainer and allow the trapped air to escape from the boot.

23. Install the shroud over the boot and retainer and install the clamp.

Halfshaft

REMOVAL AND INSTALLATION

B2200 and B2600i

1. Raise and safely support the vehicle. Remove the wheel and tire assembly.

2. Remove the drive flange hub.

3. Remove the caliper, mounting support and knuckle arm. Support the caliper aside with rope or mechanics wire; do not let the caliper hang by the brake hose.

4. Disconnect the stabilizer bar and the tie rod end.

5. Remove the lower mount of the shock absorber.

6. Remove the snapring and spacer.

7. Support the lower control arm with a jack.

8. Disconnect the upper and lower ball joints and the knuckle.

9. Lower the lower control arm and remove the knuckle assembly.

10. Remove the splash shield.

11. Using a suitable prybar, pry out the halfshaft from the differential and remove the halfshaft from the vehicle. Be careful not to damage the dust cover or oil seal.

To install:

12. Install a new clip on the halfshaft. Coat the differential seal with clean transmission fluid.

13. Install the halfshaft in the differential, being careful not to damage the seal. After installation, attempt to pull the halfshaft outward to make sure it does not come out.

14. Install the knuckle and hub to the halfshaft and ball joints. Install the spacer and a new snapring.

15. Install the lower mount of the shock absorber and loosely tighten the bolt.

16. Connect the stabilizer bar and tie rod end.

17. Install the caliper assembly, knuckle arm and wheel and tire assembly. Apply sealant to the drive flange and install it.

18. Install the splash shield and lower the vehicle.

19. Tighten the lower shock absorber mount to 41–59 ft. lbs. (55–80 Nm).

20. Check the front end alignment.

MPV

1. Raise and safely support the vehicle. Remove the wheel and tire assembly.

2. Drain the differential gear oil.

3. Remove and discard the halfshaft locknut.

4. Disconnect the tie rod end from the knuckle.

5. Remove the caliper and brake rotor from the knuckle. Support the caliper aside with rope or mechanics wire; do not let it hang by the brake hose.

6. Remove the nut and bolts and remove the lower ball joint. Remove the bolts and nuts and remove the knuckle/hub assembly from the strut.

NOTE: If the halfshaft is stuck to the hub, install a used locknut so it is flush with the end of the shaft, then tap the nut with a soft mallet.

7. Remove the splash shield.

8. Using a prybar, pry out the halfshaft from the differential and remove the halfshaft from the vehicle. Be careful not to damage the dust cover or oil seal.

To install:

9. Install a new clip on the halfshaft. Coat the differential seal with clean transmission fluid.

10. Install the halfshaft in the differential, being careful not to damage the seal. After installation, attempt to pull the halfshaft outward to make sure it does not come out.

11. Install the knuckle/hub assembly to the strut.

12. Install the lower ball joint.

13. Install the brake rotor and caliper assembly.

14. Connect the tie rod end to the knuckle.

15. Install a new locknut and tighten to 174–231 ft. lbs. (236–313 Nm). After tightening, stake the locknut using a blunt chisel.

16. Install the splash shield, tighten the mounting bolts to 12–16 ft. lbs. (16–22 Nm).

17. Install the wheel and tire assembly and lower the vehicle.

18. Fill the differential with gear oil.

19. Check the front end alignment.

CV-Joint Boot

REPLACEMENT

B2200 and B2600i

NOTE: Do not attempt to disassemble the outer CV-joint. If outer CV-boot replacement is necessary, the inner CV-joint and boot must be first be removed.

Inner Boot

1. Remove the halfshaft from the vehicle and mount it in a vise with protective jaw caps.

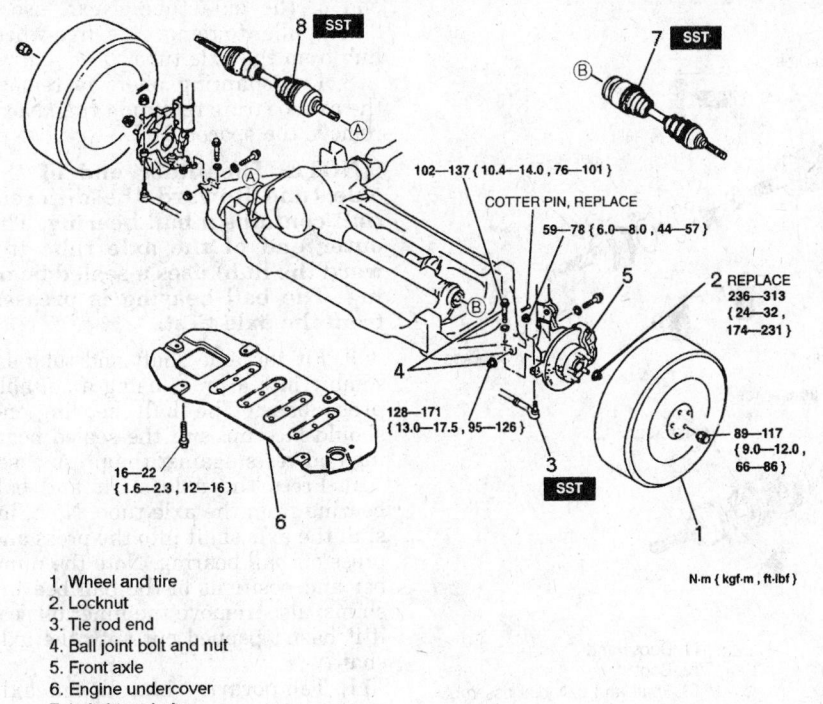

102—137 { 10.4—14.0 , 76—101 }

COTTER PIN, REPLACE

59—78 { 6.0—8.0 , 44—57 }

2 REPLACE
236—313
{ 24—32 ,
174—231 }

128—171
{ 13.0—17.5 , 95—126 }

89—117
{ 9.0—12.0 ,
66—86 }

16—22
{ 1.6—2.3 , 12—16 }

N·m { kgf·m , ft-lbf }

1. Wheel and tire
2. Locknut
3. Tie rod end
4. Ball joint bolt and nut
5. Front axle
6. Engine undercover
7. Left drive shaft
8. Right drive shaft

300245

Halfshaft and related components, exploded view — MPV

2. Pry up the boot band locking clips with a small prybar and remove the bands with pliers.

3. Slide the boot back on the shaft to expose the inner CV-joint.

4. Mark the CV-joint housing and cage for proper reassembly and remove the retaining clip with a small prybar. Remove the housing.

5. Mark the shaft, cage, balls and inner ring for reassembly and remove the snapring. Turn the cage about 30 degrees, remove the balls and remove the cage from the inner ring.

6. Remove the inner ring from the shaft with a press or drive it off with a hammer and brass drift.

7. Wrap the shaft splines with tape and remove the inner boot.

To install:

8. Wrap the shaft splines with tape and slide a new boot onto the shaft. Remove the tape.

NOTE: The inner and outer CV-boots are different and cannot be interchanged.

9. Install the inner ring on the shaft, aligning the marks that were made during removal.

10. Install the cage with the big end facing the snapring groove. Install the cage on the inner ring, aligning the marks made during removal and turn it 30 degrees. Install the balls into their proper positions and install a new snapring into the groove.

11. Fill the CV-joint housing with the proper quantity and type of CV-joint grease and apply the grease thoroughly to the cage, inner ring and ball assembly.

12. Align the marks and install the CV-joint housing on the shaft and install a new retaining clip.

13. Apply about 120 grams (4.2 oz.) of CV-joint grease to the inside of the inner boot and slide the boot over the CV-joint. Carefully lift up the small end of the boot to release any trapped air.

14. Set the halfshaft to the required length before installing the boot bands. On B-Series Pick-Up, the right side halfshaft length should be 24.49 in. (622mm) and the left side should be 21.81 in. (554mm). On MPV, the right side halfshaft length should be 22.30 in. (566.5mm) and the left side should be 19.63 in. (498.5mm).

15. Install the new CV-joint boot bands. Fold the band back by pulling the end with pliers, then lock the end of the band by bending the locking clips.

NOTE: The bands should always be mounted in the direction opposite the forward revolving direction of the halfshaft.

16. Remove the halfshaft from the vise and install it in the vehicle.

Outer Boot

1. Remove the halfshaft from the vehicle and mount it in a vise with protective jaw caps.

2. Remove the inner CV-boot.

3. Remove the dust cover, if equipped, using a hammer and a drift.

4. Pry up the boot band locking clips with a small prybar and remove the bands with pliers.

5. Slide the outer CV-boot off the shaft.

To install:

6. Wrap the shaft splines with tape and slide a new boot onto the shaft. Remove the tape.

NOTE: The inner and outer CV-boots are different and cannot be interchanged.

7. Apply about 120 grams of CV-joint grease to the inside of the outer boot and slide the boot over the CV-joint. Carefully lift up the small end of the boot to release any trapped air.

8. Install new CV-joint boot bands. Fold the band back by pulling the end with pliers, then lock the end of the band by bending the locking clips.

NOTE: The bands should always be mounted in the direction opposite the forward revolving direction of the halfshaft.

9. Press on a new dust cover, if equipped.

10. Install the inner CV-joint boot.

11. Install the halfshaft in the vehicle.

MPV

1. Raise and safely support the vehicle. Remove the wheel and tire assembly.

2. Drain the differential gear oil.

3. Remove and discard the half-shaft locknut.

4. Disconnect the tie rod end from the knuckle.

5. Remove the caliper and brake rotor from the knuckle. Support the caliper aside with rope or mechanics wire; do not let it hang by the brake hose.

6. Remove the nut and bolts and remove the lower ball joint. Remove

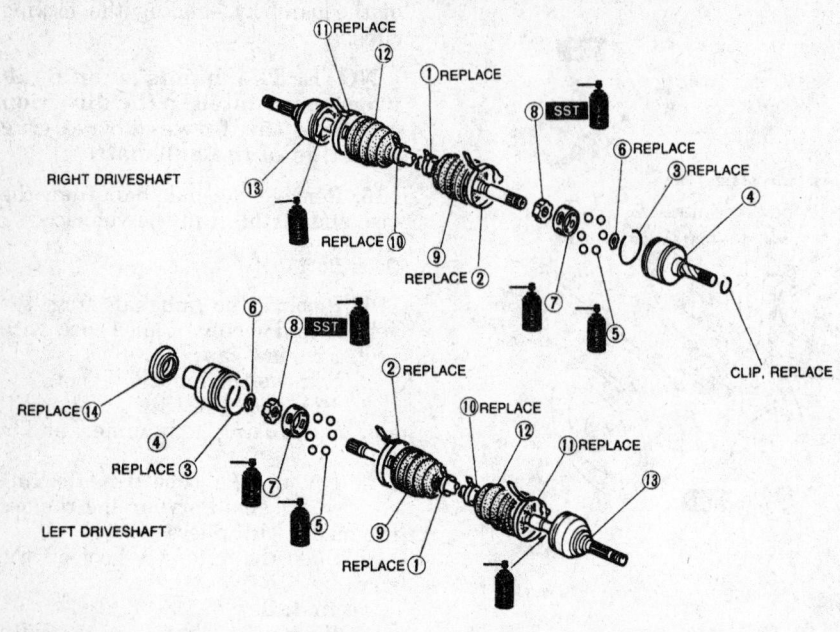

RIGHT DRIVESHAFT

LEFT DRIVESHAFT

1. Boot band
2. Boot band
3. Clip
4. Outer ring
5. Ball
6. Snap ring
7. Cage
8. Inner ring
9. Boot
10. Boot band
11. Boot band
12. Boot
13. Shaft and ball joint assembly
14. Dust cover

350059

Axle shaft and CV-boot assembly — B2200 and B2600i

the bolts and nuts and remove the knuckle/hub assembly from the strut.

NOTE: If the halfshaft is stuck to the hub, install a used locknut so it is flush with the end of the shaft, then tap the nut with a soft mallet.

7. Remove the splash shield.

8. Using a prybar, pry out the halfshaft from the differential and remove the halfshaft from the vehicle. Be careful not to damage the dust cover or oil seal.

To install:

9. Install a new clip on the halfshaft. Coat the differential seal with clean transmission fluid.

10. Install the halfshaft in the differential, being careful not to damage the seal. After installation, attempt to pull the halfshaft outward to make sure it does not come out.

11. Install the knuckle/hub assembly to the strut.

12. Install the lower ball joint.

13. Install the brake rotor and caliper assembly.

14. Connect the tie rod end to the knuckle.

15. Install a new locknut and tighten to 174–231 ft. lbs. (236–313 Nm). After tightening, stake the locknut using a blunt chisel.

16. Install the splash shield, tighten the mounting bolts to 12–16 ft. lbs. (16–22 Nm).

17. Install the wheel and tire assembly and lower the vehicle.

18. Fill the differential with gear oil.

19. Check the front end alignment.

Axles

REMOVAL AND INSTALLATION

B2200 and B2600i

Front

1. Turn the ignition to the **LOCK** position and remove the key.

2. Raise and safely support the vehicle.

3. Remove the halfshafts, and then remove the front axle from the vehicle.

4. If equipped, remove the axle control assembly from the axle case.

5. Remove the axle mounting bracket from the left end of the axle tube.

6. Unbolt the axle shaft and tube assembly from the differential. Note the positions of the gear sleeves and needle bearing for reassembly.

7. Remove the bolts which secure the remote free-wheel hub to the left

end of the axle tube. Next, use a 2-bolt puller to remove the free-wheel hub from the axle tube.

8. Use snapring pliers to remove the axle bearing retaining ring. Next, remove the spacer.

NOTE: The inner end of the axle tube (toward the differential) contains a ball bearing. The outer end of the axle tube (toward the hub) uses a sealed bearing. The ball bearing is pressed from the axle first.

9. Fit the axle shaft and tube assembly into a press using a suitable press base. The ball bearing-end should face up, and the sealed bearing end rests against the press base.

10. Press the axle shaft and ball bearing from the axle tube. Next, install the axle shaft into the press and press off ball bearing. Note the number and positions of the ball bearing shims; also, remove the inner oil seal if it hasn't popped out with the axle shaft.

11. Temporarily install the axle shaft back into the axle tube so it fits into the sealed bearing. Install the axle shaft and tube assembly into the press so the sealed bearing faces down. Install a suitably sized press base to support the axle tube so the splines on the axle shaft won't be damaged.

12. Press the axle shaft out of the axle tube so it bottoms in the press base. The sealed bearing will be pressed out by the axle shaft. Next, install the axle shaft into the press and press off the sealed bearing and dust seal.

13. Use SST 49-UO27-004, or an equivalent seal removal tool to remove the outer oil seal from the axle tube.

14. Inspect the axle shaft for out-of-round and other signs of damage, inspect the splines for wear and distortion. Replace parts as necessary.

To install:

15. Apply clean differential lubricant to a new inner oil seal. Press the seal into position, using SST 49-UO27-006, or an equivalent seal driver.

16. Install the ball bearing along with the shims. No more than 2 shims should be installed. Press these components into position and then install the retaining ring.

17. Measure the clearance between the ball bearing and the retaining ring. If the clearance isn't within the specification of 0.0059 in. (0.15mm), it should be adjusted by adding or removing shims.

18. Remove the retaining ring.

break the collar. Be careful to avoid damaging the shaft.

— **CAUTION** —

Wear eye protection when grinding the collar and breaking the collar from the shaft.

11. Using a press or puller, remove the hub and bearing assembly from the shaft. Remove the spacer from the shaft.

12. Remove the bearing and seal from the hub.

13. Using a drift, tap the race from the hub.

14. Check all parts for wear or damage. If either race is to be replaced, both must be replaced. The race in the axle housing can be removed with a slide hammer and adapter. Replace the bearing and races as a set. Always replace the seals, regardless of what other service is being performed.

To install:

15. The outer race must be installed using an arbor press. The inner race can be driven into place in the axle housing.

16. Pack the hub with lithium wheel bearing grease.

17. Tap a new oil seal into the axle housing until it is flush with the end of the housing. Coat the seal lip with wheel bearing grease.

18. Install a new spacer on the shaft with the larger flat surface up.

19. Install a new seal in the hub.

20. Thoroughly pack the bearing with clean, lithium based, wheel bearing grease. If one is available, use a grease gun adapter meant for packing bearings.

21. Place the bearing in the hub, and, using a press, press the hub and bearing assembly onto the shaft.

22. Press the new collar onto the shaft. The press pressure for the collar is critical. Press pressures should be 9240–13,420 lb. (4200–6100 kg).

23. Install one shaft in the housing, being very careful to avoid damaging the inner seal.

24. If only one shaft was being serviced, the other must now be removed to check bearing play on the serviced axle. If both shafts were removed, leave the other one out for now.

25. Tighten the backing plate bolts on the one installed axle to 80 ft. lbs.

26. Mount a dial indicator on the backing plate, with the pointer resting on the axle shaft flange. Check the axial play. Standard bearing play should be 0.0256–0.0374 in. (0.65–0.95mm).

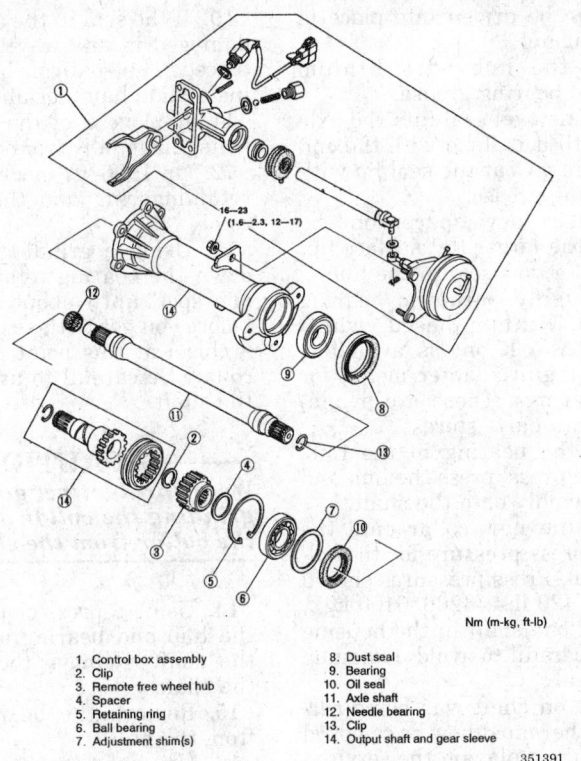

Nm (m-kg, ft-lb)

1. Control box assembly
2. Clip
3. Remote free wheel hub
4. Spacer
5. Retaining ring
6. Ball bearing
7. Adjustment shim(s)
8. Dust seal
9. Bearing
10. Oil seal
11. Axle shaft
12. Needle bearing
13. Clip
14. Output shaft and gear sleeve

351391

Exploded view of the front axle shaft components — B2200 and B2600i

19. Invert the axle tube in the press so the outer (hub) end faces up. Use SST 49-UO27-005 and a press base such as 49-U027-006 to press the axle shaft and sealed bearing into position. The bearing must be installed with its seal side upward.

20. Install the retaining ring.

21. Install the free-wheel hub gear by pressing it into place and then install a new clip.

22. Measure the clearance between the free-wheel hub gear and the clip. If the clearance isn't within specification of 0.0059 in. (0.15mm), it should be adjusted by adding or removing spacers.

23. Use SST 49-M005-795 or an equivalent seal driver to press in a new dust seal.

24. Apply clean differential lubricant to the needle bearing and then install it and install a new clip on the axle shaft.

25. Install the free-wheel hub assembly and the axle mounting bracket. Install the axle control, making sure the the vacuum disc linkage and and the shifting fork fit into place properly and aren't binding.

26. Install the front axle into the vehicle. Install the halfshafts.

27. Reassemble any suspension components removed. Install the front wheels.

28. Refill the axle with the proper differential lubricant.

29. Lower the vehicle and then check and adjust the front wheel alignment and ride height.

30. Verify that the front axle engages correctly and functions properly in 4WD. Road test the vehicle.

Rear

1. Raise and support the rear end on jackstands.

2. Remove the wheel and brake drum.

3. Remove the brake shoes.

4. Remove the parking brake cable retainer.

5. Disconnect and cap the brake lines at the wheel cylinders.

6. Remove the bolts securing the backing plate and bearing housing.

7. Slide the axle shaft from the axle housing. Be careful to avoid damaging the oil seal with the shaft.

8. If the seal in the axle housing is damaged in any way, it must be replaced. The seal can be removed using a slide hammer and adapter.

9. Remove 2 of the backing plate bolts, diagonally from each other.

10. Using a grinding wheel, grind down the bearing retaining collar in one spot, until about 0.197 in. (5mm) remains before you get to the axle shaft. Place a chisel at this point and

27. If play is not within specifications, shims are available for correcting it.

28. Install the other shaft and torque the backing plate bolts. Check the play as on the first shaft. Play should be 0.0019–0.0098 in. (0.05–0.25mm). If not, correct it with shims.

29. Install the brake drums and wheels.

30. Bleed the brake system.

1993 MPV

1. Raise and support the rear end on jackstands.

2. Remove the wheel and brake drum.

3. Remove the brake shoes.

4. Remove the parking brake cable retainer.

5. Disconnect and cap the brake lines at the wheel cylinders.

6. Remove the bolts securing the backing plate and bearing housing.

7. Slide the axle shaft from the axle housing. Be careful to avoid damaging the oil seal with the shaft.

8. If the seal in the axle housing is damaged in any way, it must be replaced. The seal can be removed using a slide hammer and adapter.

9. Remove 2 of the backing plate bolts, diagonally from each other.

10. Using a grinding wheel, grind down the bearing retaining collar in one spot, until about 5mm remains before you get to the axle shaft. Place a chisel at this point and break the collar. Be careful to avoid damaging the shaft.

— **CAUTION** —
Wear some kind of protective goggles when grinding the collar and breaking the collar from the shaft!

11. Using a press or puller, remove the hub and bearing assembly from the shaft. Remove the spacer from the shaft.

12. Remove the bearing and seal from the hub.

13. Using a drift, tap the race from the hub.

14. Check all parts for wear or damage. If either race is to be replaced, both must be replaced. The race in the axle housing can be removed with a slide hammer and adapter. It's a good idea to replace the bearing and races as a set. It's also a good idea to replace the seals, regardless of what other service is being performed.

To install:

15. The outer race must be installed using an arbor press. The inner race can be driven into place in the axle housing.

16. Pack the hub with lithium based wheel bearing grease.

17. Tap a new oil seal into the axle housing until it is flush with the end of the housing. Coat the seal lip with wheel bearing grease.

18. Install a new spacer on the shaft with the larger flat surface up.

19. Install a new seal in the hub.

20. Thoroughly pack the bearing with clean, lithium based, wheel bearing grease. If one is available, use a grease gun adapter meant for packing bearings. These are available at all auto parts stores.

21. Place the bearing in the hub, and, using a press, press the hub and bearing assembly onto the shaft.

22. Press the new collar onto the shaft. The press pressure for the collar is critical. Press pressures should be 9240–13,420 lbs. (4200–6100 kg).

23. Install one shaft in the housing being very careful to avoid damaging the inner seal.

24. If only on shaft was being serviced, the other must now be removed to check bearing play on the serviced axle. If both shafts were removed, leave the other one out for now.

25. Tighten the backing plate bolts on the one installed axle to 80 ft. lbs.

26. Mount a dial indicator on the backing plate, with the pointer resting on the axle shaft flange. Check the axial play. Standard bearing play should be 0.57mm.

27. If play is not within specifications, shims are available for correcting it.

28. Install the other shaft and torque the backing plate bolts. Check the play as on the first shaft.

29. Install the brake drums and wheels. Bleed the brake system.

1994–97 MPV

1. Raise and support the rear end on jackstands.

2. Remove the wheel(s).

3. Remove the brake caliper assembly.

4. Remove the disc/drum plate.

5. Remove the parking brake shoe assembly.

6. Disconnect the parking brake cable.

7. Disconnect and cap the brake line(s).

8. Remove the dust cover and rear axle assembly by using a slide hammer.

9. Slide the axle shaft from the axle housing. Be careful to avoid damaging the oil seal with the shaft.

10. If the seal in the axle housing is damaged in any way, it must be replaced. The seal can be removed using a slide hammer and adapter.

11. Remove 2 of the backing plate bolts, diagonally from each other.

12. On 1996–97 models, remove the retaining ring and the ABS sensor rotor.

13. Using a grinding wheel, grind down the bearing retaining collar in one spot, until about 5mm remains before you get to the axle shaft. Place a chisel at this point and break the collar. Be careful to avoid damaging the shaft.

— **CAUTION** —
Wear protective goggles when grinding the collar and breaking the collar from the shaft.

14. Using a press or puller, remove the hub and bearing assembly from the shaft. Remove the spacer from the shaft.

15. Remove the bearing and seal from the hub.

16. Using a drift, tap the race from the hub.

17. Check all parts for wear or damage. If either race is to be replaced, both must be replaced. The race in the axle housing can be removed with a slide hammer and adapter. It's a good idea to replace the bearing and races as a set. It's also a good idea to replace the seals, regardless of what other service is being performed.

To install:

18. The outer race must be installed using an arbor press. The inner race can be driven into place in the axle housing.

19. Pack the hub with lithium based wheel bearing grease.

20. Tap a new oil seal into the axle housing until it is flush with the end of the housing. Coat the seal lip with wheel bearing grease.

21. Install a new spacer on the shaft with the larger flat surface up.

22. Install a new seal in the hub.

23. Thoroughly pack the bearing with clean, lithium-based, wheel bearing grease.

24. Place the bearing in the hub, and, using a press, press the hub and bearing assembly onto the shaft.

25. Press the new collar onto the shaft. The press pressure for the collar is critical. If 5940 lbs. (26,478 N) or less is required to press the collar, replace the shaft.

26. On 1996–97 models, press a new ABS speed sensor rotor to the axle shaft and install a new retainer.

27. Install one shaft in the housing being very careful to avoid damaging the inner seal.

28. If only on shaft was being serviced, the other must now be removed to check bearing play on the serviced axle. If both shafts were removed, leave the other one out for now.

29. Tighten the backing plate bolts on the one installed axle to 80 ft. lbs.

30. Mount a dial indicator on the backing plate, with the pointer resting on the axle shaft flange. Check the axial play. Standard bearing play should be 0.0224 in. (0.57mm).

31. If play is not within specifications, shims are available for correcting it.

32. Install the other shaft and torque the backing plate bolts. Check the play as on the first shaft.

33. Install the brake line and parking brake cable.

34. Install the parking brake shoe assembly.

35. Install the brake disc/drum.

36. Install the caliper assembly.

37. Install the wheel(s).

38. Lower the vehicle.

STEERING

Air Bag

CAUTION

Some vehicles are equipped with an air bag system, also known as the Supplemental Inflatable Restraint (SIR) or Supplemental Restraint System (SRS). The system must be disabled before performing service on or around system components, steering column, instrument panel components, wiring and sensors. Failure to follow safety and disabling procedures could result in accidental air bag deployment, possible personal injury and unnecessary system repairs.

PRECAUTIONS

Several precautions must be observed when handling the inflator module to avoid accidental deployment and possible personal injury.

• Never carry the inflator module by the wires or connector on the underside of the module.

• When carrying a live inflator module, hold securely with both hands, and ensure that the bag and trim cover are pointed away.

• Place the inflator module on a bench or other surface with the bag and trim cover facing up.

• With the inflator module on the bench, never place anything on or close to the module which may be thrown in the event of an accidental deployment.

DISARMING

1994–95 MPV

1. Turn the ignition switch to **LOCK**.

2. Disconnect the negative battery cable.

3. Remove the left side cover, lower panel, and the column cover.

4. Disconnect the orange and blue clock spring connectors.

5. After servicing, connect the negative battery cable. Turn the ignition switch to **ON**. Verify that the air bag system warning light illuminates for 4–8 seconds and then goes OFF.

1996–97 MPV

Driver's Air Bag

1. Turn the ignition switch to **LOCK**.

2. Disconnect the negative battery cable and wait for more than 1 minute to allow the backup power supply to deplete its stored power.

3. Remove the lower column cover and disconnect the orange and blue clock spring connectors.

4. After servicing, connect the negative battery cable. Turn the ignition switch to **ON**. Verify that the air bag system warning light illuminates for 4–6 seconds and then goes OFF.

Passenger Air Bag

1. Turn the ignition switch to **LOCK**.

2. Disconnect the negative battery cable and wait for more than 1 min-

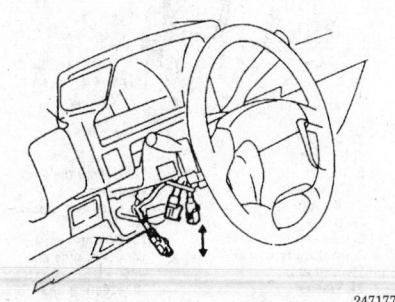

247177

Clock spring or driver's side air bag connector — 1994–97 MPV

ute to allow the backup power supply to deplete its stored power.

3. Remove the glove compartment cover.

4. Disconnect the orange and blue passenger-side air bag module connector.

5. After servicing, connect the negative battery cable. Turn the ignition switch to **ON**. Verify that the air bag system warning light illuminates for 4–6 seconds and then goes OFF.

Steering Wheel

REMOVAL AND INSTALLATION

B2200, B2600i and 1993–94 MPV

1. Disconnect the negative battery cable.

2. Remove the steering wheel pad from the steering wheel.

   a. Pull the horn pad straight up from the steering wheel.

3. Remove the steering wheel attaching bolt or nut. Check to see if the steering wheel and steering shaft have alignment marks or flats. If there are no steering wheel-to-steering column shaft alignment marks or flats, matchmark the steering wheel and column shaft so they can be reassembled in the same position.

4. Using a steering wheel puller, remove the steering wheel from the steering column shaft.

   NOTE: Do not hammer on the steering wheel or steering shaft or use a knock-off type steering wheel puller, as either will damage the steering column.

   To install:

5. Install the steering wheel on the steering column shaft, aligning the marks or flats on the steering wheel with the marks or flats on the steering shaft.

6. Install the steering wheel attaching nut and tighten to 29–36 ft. lbs. (39–49 Nm).

7. Connect the horn switch and, if equipped, cruise control wires and install the steering wheel pad.

8. Connect the negative battery and check the steering column for proper operation.

1995–97 MPV

CAUTION

The air bag system must be disarmed before removing the steering wheel. Failure to do so may cause accidental deployment, property damage or personal injury.

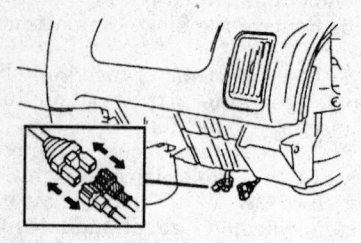

Passenger side air bag connector — 1996–97 MPV

1. Set the steering wheel so the front wheels are straight ahead.
2. Disconnect the negative battery cable.
3. Disarm the air bag.
4. At the back of the steering wheel hub, remove the nuts that hold the air bag assembly and remove the air bag. Place it in a safe place, pad side up.

— CAUTION —

Always carry an air bag assembly with the bag and trim cover away from your body. Store the assembly facing upward; never place the assembly face down on any surface.

5. Matchmark the wheel to the shaft and remove the nut. Use a puller to remove the wheel.

To install:

6. Double check that the front wheels are in the straight-ahead position and mount the steering wheel. Make sure the matchmarks are aligned.
7. Torque the steering wheel nut to 29–36 ft. lbs. (40–49 Nm). Install the air bag unit.
8. Connect the negative battery cable.

Tie Rod Ends

REMOVAL AND INSTALLATION

B2200 and B2600i

1. Raise and support the vehicle.
2. Loosen the tie rod jam nuts.
3. Remove and discard the cotter pin from the ball socket end, and remove the nut.
4. Use a ball joint puller to loosen the ball socket stud from the center link or steering arm, as required.
5. Unscrew the tie rod end from the threaded sleeve, counting the number of threads until it's off. The

threads may be left or right hand threads. Tighten the jam nuts to 58 ft. lbs.

To install:

6. Lightly coat the threads with grease, and turn the new end in as many turns as were required to remove it. This will give the approximate correct toe-in.
7. Install the ball socket studs into center link or steering arm. Tighten the nuts to 43 ft. lbs. Install a new cotter pin. You may tighten the nut to fit the cotter pin, but don't loosen it.
8. Check and adjust the toe-in, and tighten the tie rod clamps or jam nuts.

1994 MPV with Manual Steering

1. Raise and support the vehicle safely. Remove the wheel and tire assembly.
2. Remove the cotter pin and loosen the nut on the tie rod end ball stud. With the nut protecting the ball stud, press the stud from the knuckle using press tool 49-0118-850C or equivalent. Remove the nut and the tie rod end from the knuckle.
3. Paint a reference mark across the tie rod, jam nut and shaft.
4. Loosen the jam nut and unscrew the tie rod end from the shaft.

To install:

5. Thread the tie rod onto the shaft and align the marks made during removal. If installing a new tie rod end, try to assemble it in the same position as the old one.
6. Install the tie rod end into the knuckle. Install the nut and tighten to 47 ft. lbs. (64 Nm).
7. Install a new cotter pin. If the cotter pin cannot be installed because the ball stud hole and the nut castellation do not align, tighten the nut further until the cotter pin can be installed. Never loosen the nut to install the cotter pin.
8. Tighten the jam nut.
9. Install the wheel and tire assembly and lower the vehicle. Check the front wheel alignment.

1993–95 MPV with Power Steering

1. Raise and support the front end on jackstands.
2. Remove the wheels.
3. Loosen the tie rod end ball stud nut and separate the tie rod end from the knuckle arm with a separator tool.
4. Matchmark the tie rod end and tie rod and loosen the locknut.
5. Unscrew the tie rod end, counting the number of turns until it's off, for installation purposes.

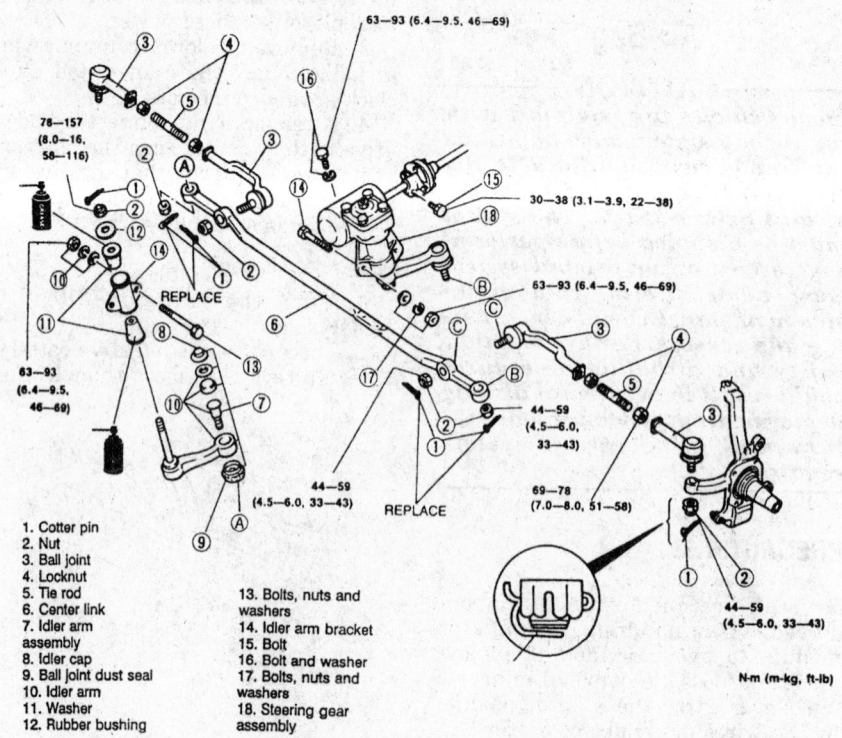

1. Cotter pin
2. Nut
3. Ball joint
4. Locknut
5. Tie rod
6. Center link
7. Idler arm assembly
8. Idler cap
9. Ball joint dust seal
10. Idler arm
11. Washer
12. Rubber bushing
13. Bolts, nuts and washers
14. Idler arm bracket
15. Bolt
16. Bolt and washer
17. Bolts, nuts and washers
18. Steering gear assembly

N·m (m-kg, ft-lb)

Steering linkage components — B2200 and B2600i

To install:

6. Install the tie rod end onto the tie rod the same number of turns that it took to remove it. Tighten the locknut to 51–57 ft. lbs. (69–78 Nm).

7. Install the ball stud into the steering knuckle. Tighten the ball stud nut to 44–57 ft. lbs. (59–78 Nm) and install the cotter pin.

8. Lower the vehicle and install a new cotter pin. Always advance the nut to align the cotter pin hole. Never back it off. Check the front alignment.

Power Rack and Pinion

REMOVAL AND INSTALLATION

MPV

2WD Vehicles

1. Place the front wheels in the straight-ahead position. Raise and safely support the vehicle.

2. Remove the wheel and tire assemblies. Remove the splash shield.

3. Remove the cotter pins and nuts from both tie rod end studs. Separate the tie rod ends from the knuckles.

4. Remove the pinch bolt from the intermediate shaft-to-pinion shaft coupling.

5. Disconnect and plug the pressure line from the rack and pinion assembly. Loosen the clamp and disconnect the return line from the rack and pinion assembly. Plug the line.

6. If equipped with automatic transmission, remove the change counter assembly to remove the protector plate mounting bolt.

7. Remove the steering bracket mounting bolts and remove the rack and pinion assembly and brackets.

8. If necessary, remove the brackets.

To install:

9. If removed, install the brackets and tighten the mounting bolts, in sequence, to 54–69 ft. lbs. (74–93 Nm).

10. Install the rack and pinion assembly and brackets in the vehicle. Tighten the bracket-to-chassis bolts to 46–69 ft. lbs. (63–93 Nm).

11. If equipped with automatic transmission, install the change counter assembly.

12. Connect the return line and tighten the clamp. Connect the pressure line and tighten the nut to 23–35 ft. lbs. (31–47 Nm).

13. Install the pinch bolt in the intermediate shaft-to-pinion shaft coupling and tighten to 13–20 ft. lbs. (18–26 Nm).

14. Position the tie rod end studs in the knuckles and install the nuts. Tighten the nuts to 43–58 ft. lbs. (59–78 Nm) and install new cotter pins.

15. Install the splash shield and the wheel and tire assemblies. Lower the vehicle and bleed the power steering system.

4WD Vehicles

1. Place the front wheels in the straight-ahead position. Raise and safely support the vehicle.

2. Remove the wheel and tire assemblies. Remove the splash shield.

3. Remove the cotter pins and nuts from both tie rod end studs. Separate the tie rod ends from the knuckles.

4. Disconnect and plug the pressure and return hoses at the pressure and return lines.

5. Remove the pressure and return lines from the rack and pinion assembly.

6. Remove the pinch bolt from the intermediate shaft-to-pinion shaft coupling.

7. Working inside the vehicle, remove the lower panel and column cover from under the steering column. Remove the steering column mounting bolts and nuts and pull the column and intermediate shaft rearward to separate the intermediate shaft from the pinion shaft.

8. Mark the position of the front driveshaft on the axle flange and remove the front driveshaft.

9. Remove the rack and pinion assembly mounting bracket bolts and the front differential/joint shaft assembly mounting bolts.

10. Slide the differential/joint shaft assembly rearward. Slide the rack and pinion assembly rearward and turn it 90 degrees, then remove it from the left side of the vehicle.

To install:

11. Install the rack and pinion assembly from the left side of the vehicle, turn it 90 degrees and move it forward into position. Install the mounting bolts and tighten, in sequence, to 54–69 ft. lbs. (74–93 Nm).

12. Move the differential/joint shaft assembly forward, install the mounting bolts and tighten to 49–72 ft. lbs. (67–97 Nm).

13. Install the driveshaft, aligning the marks made during removal.

14. Working inside the vehicle, move the steering column and intermediate shaft forward to engage the intermediate shaft with the pinion shaft. Install and tighten the steering column nuts and bolts to 12–17 ft. lbs. (16–23 Nm). Install the lower panel and column cover.

15. Install the pinch bolt in the intermediate shaft-to-pinion shaft coupling and tighten to 13–20 ft. lbs. (18–26 Nm).

16. Install the pressure and return lines on the rack and pinion assembly. Connect the pressure and return hoses to the lines.

17. Position the tie rod end studs in the knuckles and install the nuts. Tighten the nuts to 43–58 ft. lbs. (59–78 Nm) and install new cotter pins.

18. Install the splash shield and the wheel and tire assemblies. Lower the vehicle and bleed the power steering system.

Power Steering Pump

BLEEDING

B2200 and B2600i

— **WARNING** —
Use only Dexron® II or Type F ATF in the power steering system. Other types of fluids may cause system damage.

1. Fill the reservoir with fresh ATF. Allow the fluid to settle for a few minutes.

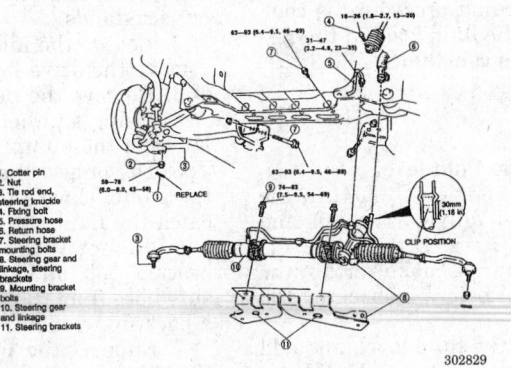

1. Cotter pin
2. Nut
3. Tie rod end, steering knuckle
4. Fixing bolt
5. Pressure hose
6. Return hose
7. Steering bracket mounting bolts
8. Steering gear and linkage, steering brackets
9. Mounting bracket bolts
10. Steering gear and linkage
11. Steering brackets

302829

Rack and pinion mounting, exploded view — 2WD MPV

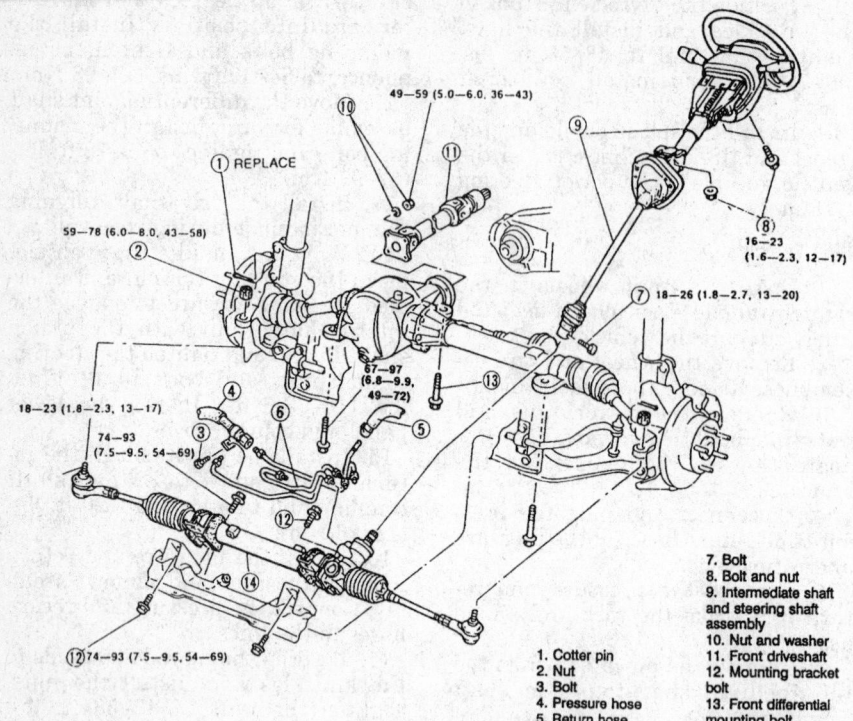

49—59 (5.0—6.0, 36—43)

① REPLACE

59—78 (6.0—8.0, 43—58)

16—23
(1.6—2.3, 12—17)

18—26 (1.8—2.7, 13—20)

87—97
(6.8—9.9,
49—72)

18—23 (1.8—2.3, 13—17)

74—93
(7.5—9.5, 54—69)

74—93 (7.5—9.5, 54—69)

7. Bolt
8. Bolt and nut
9. Intermediate shaft and steering shaft assembly
10. Nut and washer
11. Front driveshaft
12. Mounting bracket bolt
13. Front differential mounting bolt
14. Steering gear and linkage

1. Cotter pin
2. Nut
3. Bolt
4. Pressure hose
5. Return hose
6. Pressure and return pipes

302830

Rack and pinion mounting, exploded view — 4WD MPV

2. Recheck the fluid level and add more if necessary.

3. Raise and support the vehicle so its front wheels are off the ground. Block the rear wheels.

4. Start the engine and slowly turn the steering wheel lock-to-lock 3–4 more times until any buzzing sound disappears and the wheel turns smoothly. Do not hold the wheel in either lock position for more than 5 seconds.

5. Add fluid if necessary.

6. Lower the vehicle.

7. Straighten the wheels and turn OFF the engine. If the fluid level in the reservoir increases, or the fluid foams — repeat the bleeding procedure.

8. The bleeding procedure is completed when the fluid level in the reservoir remains constant.

1995–97 MPV

1. Check the fluid level.

2. With the engine OFF, turn the steering wheel fully to the left and right several times. If the vehicle is elevated to do this, make certain the vehicle is level before rechecking the fluid level.

3. Recheck the fluid level, and add fluid (Dexron® II or M-III) if necessary.

4. Repeat the previous 2 steps until the fluid level stabilizes.

5. Start the engine and let it idle.

6. Turn the steering wheel fully to the left and right several times.

7. Verify that the fluid level has not dropped, and that the fluid is smooth, not foamy.

8. If necessary, repeat the previous 2 steps until the fluid level stabilizes.

REMOVAL AND INSTALLATION

B2200

1. Raise and support the front end on jackstands.

2. Loosen the idler pulley bolt and remove the drive belt.

3. Remove the pump pulley nut.

4. Using a puller, remove the pulley from the pump.

5. Disconnect the return hose from the pump. Have a drain pan ready to catch the fluid.

6. Remove the pressure hose bracket bolt and unscrew the pressure hose from the pump. Always use a back-up wrench.

7. Support the pump and remove the front and rear pump-to-bracket bolts. Remove the pump.

To install:

8. Place the pump in position and install the mounting bolts. Connect the pressure and return lines. Install the drive pulley and belt. Adjust the drive belt to the proper tension. Fill and bleed the system. Observe the following torques:

Pump pulley: 43 ft. lbs.
Pressure line connection: 35 ft. lbs.
Pressure line bracket: 17 ft. lbs.
Pump-to-bracket: 34 ft. lbs.

B2600i

1. Raise and support the front end on jackstands.

2. Remove the pump adjusting bolt and remove the drive belt.

3. Remove the lower pump-to-bracket through-bolt and spacer.

4. Rotate the pump to get to the hoses.

5. Disconnect the return hose from the pump. Have a drain pan ready to catch the fluid.

6. Remove the pressure hose from the pump. Always use a back-up wrench.

7. Support the pump and remove the upper pump-to-bracket bolt. Remove the pump.

To install:

8. Place the pump in position and install the upper mounting bolt loosely. Connect the pressure and return hoses. Install the lower mounting and adjusting bolts. Tighten the mounting bolts enough so the pump can still be move for belt adjustment. Install the drive belt and adjust to proper tension. Tighten the mounting bolts. Fill and bleed the system. Observe the following torques:

Pressure line connection: 35 ft. lbs.
Lower pump-to-bracket: 34 ft. lbs.
Upper pump-to-bracket: 20 ft. lbs.
Adjusting bolt: 34 ft. lbs.

MPV

1. Disconnect the negative battery cable.

2. Loosen the idler pulley locknut.

3. Loosen the pump adjusting bolt.

4. Remove the drive belt.

5. Remove the pump pulley nut.

6. Using a puller, remove the pulley.

7. Disconnect the pressure switch wiring connector.

8. Place a drain pan under the pump.

9. Matchmark the pressure line connection and disconnect it.

10. Remove the pressure line bracket bolt.

11. Disconnect the return line.

12. Remove the pump-to-bracket bolts and lift out the pump.

To install:

13. Place the pump in position and install the mounting bolts. Connect the return and pressure line. Install the drive pulley and belt. Adjust the drive belt to the proper tension. Fill and bleed the system. Connect the negative battery cable. Tighten the mounts and lines to the following torques:

Mounting bolts to 34 ft. lbs.

Pressure line connection to 35 ft. lbs.

Pressure line bracket bolt: 17 ft. lbs.

Pulley nut/bolt: 43 ft. lbs.

Adjusting bolt: 35 ft. lbs.

Idler pulley locknut: 38 ft. lbs.

BRAKES

Anti-Lock Brake System Service

PRECAUTIONS

• Certain components within the Anti-Lock Brake System (ABS) are not intended to be serviced or repaired individually. Only those components with removal and installation procedures should be serviced.

• Do not use rubber hoses or other parts not specifically specified for and ABS system. When using repair kits, replace all parts included in the kit. Partial or incorrect repair may lead to functional problems and require the replacement of components.

• Lubricate rubber parts with clean, fresh brake fluid to ease assembly. Do not use lubricated shop air to clean parts; damage to rubber components may result.

• Use only specified brake fluid from an unopened container.

• If any hydraulic component or line is removed or replaced, it may be necessary to bleed the entire system.

• A clean repair area is essential. Always clean the reservoir and cap thoroughly before removing the cap. The slightest amount of dirt in the fluid may plug an orifice and impair the system function. Perform repairs after components have been thoroughly cleaned; use only denatured alcohol to clean components. Do not allow ABS components to come into contact with any substance containing mineral oil; this includes used shop rags.

• The Anti-Lock control unit is a microprocessor similar to other computer units in the vehicle. Ensure that the ignition switch is **OFF** before removing or installing controller harnesses. Avoid static electricity discharge at or near the controller.

• If any arc welding is to be done on the vehicle, the control unit should be unplugged before welding operations begin.

Master Cylinder

REMOVAL AND INSTALLATION

1. Clean all dirt and grease from the master cylinder and lines. Disconnect and cap the brake lines from the master cylinder.
2. Disconnect the fluid level sensor connector.
3. Unbolt and remove the master cylinder from the power booster.

To install:

4. Place the master cylinder in position and loosely install the mounting nuts.
5. Connect the hydraulic lines to the master cylinder, but do not tighten fully at this time. Tighten the master cylinder mounting nuts to 7.2–12 ft. lbs. (9.8–16 Nm).
6. Tighten the brake lines.
7. Connect the sensor connector.
8. Fill and properly bleed the brake system. Check the system for proper operation.

Brake Caliper

REMOVAL AND INSTALLATION

1. Raise and safely support the vehicle. Remove the wheel and tire assembly.
2. Remove the banjo bolt and disconnect the brake hose from the caliper. Plug the hose to prevent fluid leakage.
3. Remove the caliper mounting bolt and pivot the caliper about the mounting pin and off the brake rotor. Remove the caliper from the pin.
4. Installation is the reverse of the removal procedure. Lubricate the caliper mounting bolts or bolt and pin prior to installation.

5. Tighten the caliper mounting bolt(s) to:

B2200 and B2600i: 23–30 ft. lbs. (31–41 Nm)

MPV

Front caliper: 61–69 ft. lbs. (83–93 Nm)

Rear caliper: 37–50 ft. lbs. (50–68 Nm)

6. Bleed the brake system.

Disc Brake Pads

REMOVAL AND INSTALLATION

1. Raise and support the front end on jackstands.
2. Remove the wheels.
3. Remove the lower lock pin bolt from the caliper.
4. Rotate the caliper upward and remove the brake pads, shims, guide plates and if equipped, the springs.

To install:

5. Remove the master cylinder reservoir cap and remove about half of the fluid from the reservoir.
6. Using a large C-clamp and piece of wood, depress the caliper piston(s) until they bottom in their bores.
7. Install the shims, guide plates (1994–97 MPV), new pads and if removed, the springs (1994–97 MPV).
8. Reposition the caliper and install the lock pin bolt. Torque the lockbolt to:

B2200 and B2600i: 30 ft. lbs.

MPV

1993: 69 ft. lbs.

1994–97

Front caliper: 62–68 ft. lbs. (84–93 Nm)

Rear caliper: 28–36 ft. lbs. (38–49 Nm)

9. Install the wheels, lower the vehicle, refill the master cylinder and depress the brake pedal a few times to restore pressure. Bleed the system if required.

Brake Rotor

REMOVAL AND INSTALLATION

B2200 and B2600i

——— **CAUTION** ———
Brake pads contain asbestos, which has been determined to be a cancer causing agent. Never clean the brake surfaces with compressed air. Avoid inhaling any dust from any brake surface.

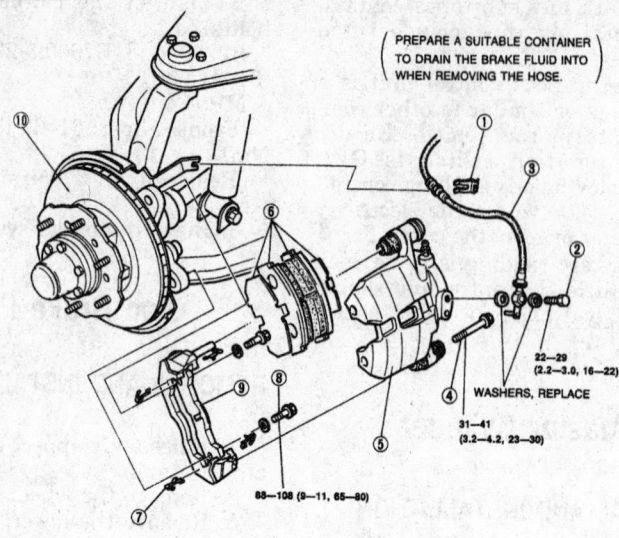

PREPARE A SUITABLE CONTAINER TO DRAIN THE BRAKE FLUID INTO WHEN REMOVING THE HOSE.

22—29 (2.2—3.0, 16—22)

WASHERS, REPLACE

31—41 (3.2—4.2, 23—30)

88—108 (9—11, 65—80)

1. Clip
2. Bolt
3. Brake hose
4. Lock bolts
5. Brake caliper assembly

6. Disc pad
7. Shims
8. Bolts
9. Mounting support
10. Disc

349957

Brake caliper assembly exploded view — B2200 and B2600i

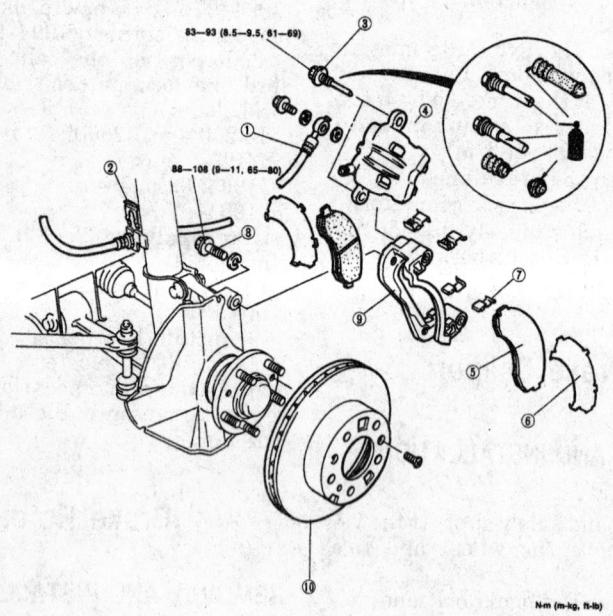

83—93 (8.5—9.5, 61—69)

88—108 (9—11, 65—80)

N·m (m-kg, ft-lb)

1. Brake hose
2. Clip
3. Lock bolt
4. Brake caliper assembly
5. Disc pad

6. Shim
7. Guide plate
8. Bolt
9. Mounting support
10. Disc plate

299211

Front rotor and brake assembly — MPV

1. Raise and safely support the vehicle. Remove the wheel and tire assembly.

2. Remove the caliper and support it aside with mechanic's wire; do not let the caliper hang by the brake hose. Remove the disc brake pads and mounting support.

3. On 2WD, remove the dust cap, cotter pin, nut, washer and outer bearing. Remove the rotor from the spindle.

4. On 4WD, remove the locking hub or drive flange, snapring and spacer, set bolts and bearing set plate.

5. Remove the bearing locknut; remove the hub and rotor assembly, being careful not to let the washer and bearing fall.

6. Inspect the rotor for scoring, wear and runout. Machine or replace as necessary.

To install:

7. If rotor replacement is necessary, remove the attaching bolts and separate the rotor from the hub.

8. Install the rotor. Tighten the hub bolts to 40–51 ft. lbs. (54–69 Nm).

9. Adjust the wheel bearings.

MPV

— **CAUTION** —
Brake pads may contain asbestos, which has been determined to be a cancer causing agent. Never clean the brake surfaces with compressed air. Avoid inhaling any dust from any brake surface.

1. Raise and safely support the vehicle. Remove the wheel and tire assembly.

2. Remove the brake caliper; support the caliper aside with wire. Do not let the caliper hang by the brake hose.

3. Remove the attaching screw.

4. Remove the rotor.

5. Inspect the rotor for scoring, wear and runout. Machine or replace as necessary.

6. Installation is the reverse of removal.

Brake Drums

REMOVAL AND INSTALLATION

1. Raise and support the rear of the vehicle on jackstands.

2. Remove the wheels.

3. Remove the drum attaching screws and insert them in the threaded holes in the drum. Turn the

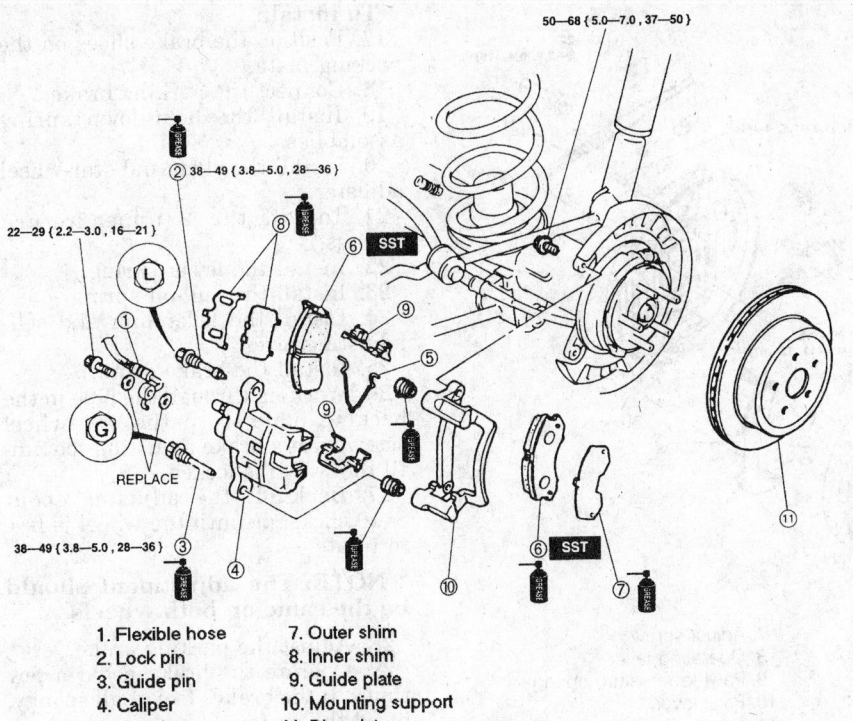

50—68 { 5.0—7.0 , 37—50 }

② 38—49 { 3.8—5.0 , 28—36 }

22—29 { 2.2—3.0 , 16—21 }

⑧

⑥ SST

L

⑨

①

⑤

REPLACE

G

⑨

38—49 { 3.8—5.0 , 28—36 } ③

④

⑩

⑥ SST

⑦

⑪

1. Flexible hose
2. Lock pin
3. Guide pin
4. Caliper
5. V-spring
6. Disc pad
7. Outer shim
8. Inner shim
9. Guide plate
10. Mounting support
11. Disc plate

N·m { kgf·m , ft-lbf }
299213

Rear rotor and brake assembly — MPV

screws inward, evenly, to force the drum off the hub.

4. Thoroughly inspect the drum. Discard a cracked drum. If the drum is suspected of being out of round, or shows signs of wear or has a ridged or rough surface, have it turned on a lathe at a machine shop. The maximum oversize diameter is stamped into the drum.

5. Installation is the reverse of removal. Make sure the holes are aligned for the attaching screws. Tighten the screws evenly to install the drum.

Brake Shoes

REMOVAL AND INSTALLATION

2WD B2200 and B2600i

1. Raise and support the rear end on jackstands.
2. Remove the drums.
3. Remove the retracting springs.
4. Remove the hold-down springs and guide pins by turning the collars 90 degrees with a pliers, or spring tool, releasing the springs.
5. Remove the parking brake link and disconnect the parking brake cable from the lever.

6. Remove the adjusting pawl and spring.
7. Remove the shoes, noting in which place the shoe with the longer lining is installed.
8. Inspect the shoes for cracks, heat checking or contamination by oil or grease. Minimum lining thickness is 0.039 in. (1.00mm). If heat checking or discoloration is noted, the wheel cylinders are probably at fault and will have to be rebuilt or replaced.

NOTE: Never replace the shoes on one side of the vehicle, only! Always replace shoes on both sides!

9. Clean the backing plate with an approved cleaning fluid.
10. Lubricated the threads of the starwheel with lithium based or silicone based grease. Apply a small dab of lithium or silicone based grease to the pads on which the brake shoes ride.

To install:
11. Transfer the parking brake lever to the new shoe.
12. Position the shoes on the backing plate.
13. Connect the parking brake cable to the lever.
14. Install the hold-down springs and guide pins.

15. Install the adjusting pawl and spring.
16. Install the adjusting screw assembly.
17. Install the retracting springs.
18. Install the drums.
19. Working through the 2 holes in the backing plate, reach through the hole in the center with a brake adjusting spoon and turn the star-wheel screw until the wheel is locked, that is, it can't be turned by hand.
20. Reach through the outside hole with a small bar and hold off the adjusting pawl while backing off the star-wheel about 6–7 clicks, or until the wheel is free to rotate.

NOTE: The adjustment should be the same on both wheels.

21. Adjust the parking brake.
22. Operate the brake pedal a few times. If the brakes feel at all spongy, bleed the system.

MPV, 4WD B2200 and B2600i

1. Raise and support the rear end on jackstands.
2. Remove the wheels.
3. Remove the brake drum.
4. Disconnect the parking brake cable.
5. Remove the hold-down spring assemblies.
6. Remove the self-adjusting lever.
7. Remove the lever link.
8. Remove the pull-off spring.
9. Remove the lower shoe-to-shoe spring.
10. Remove the 2 upper return springs.
11. Remove the star-wheel adjuster.
12. Remove the brake shoes and strut.
13. Remove the parking brake cable lever.
14. Inspect the shoes for cracks, heat checking or contamination by oil or grease. Minimum lining thickness is 1.00mm. If heat checking or discoloration is noted, the wheel cylinders are probably at fault and will have to be rebuilt or replaced.

— **CAUTION** —
Never replace the shoes on one side only. Always replace shoes on both sides.

15. Clean the backing plate with an approved cleaning fluid.
16. Lubricate the threads of the starwheel with lithium or silicone grease. Apply a small dab of this grease to the pads on which the brake shoes ride.

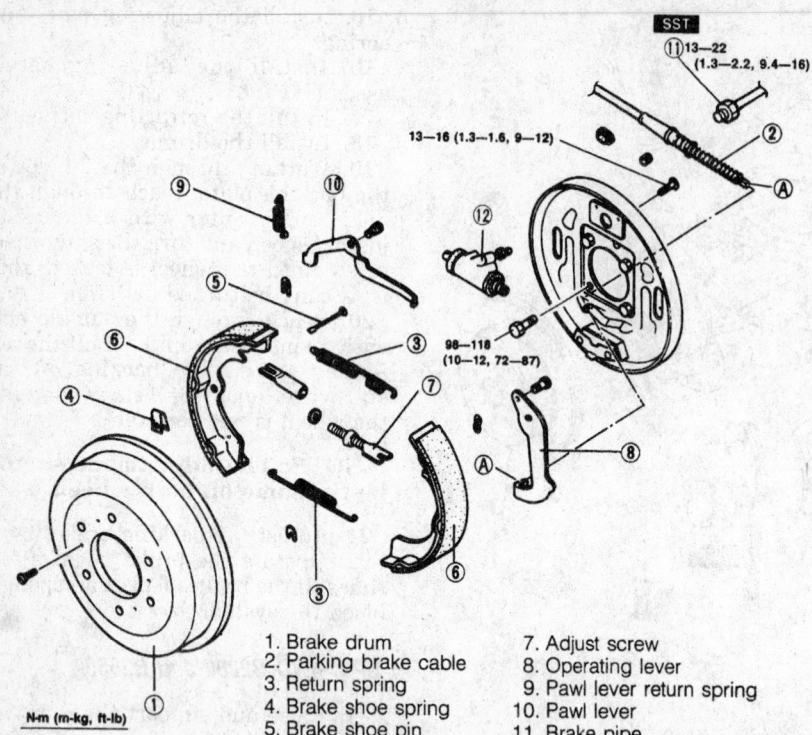

1. Brake drum
2. Parking brake cable
3. Return spring
4. Brake shoe spring
5. Brake shoe pin
6. Brake shoe
7. Adjust screw
8. Operating lever
9. Pawl lever return spring
10. Pawl lever
11. Brake pipe
12. Wheel cylinder assembly

N·m (m-kg, ft-lb)

350044

Drum brake assembly — 2WD B2200 and B2600i

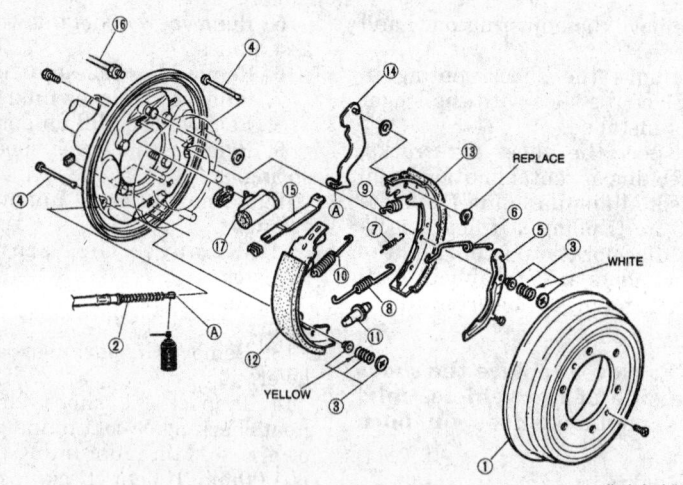

1. Brake drum
2. Parking brake cable
3. Hold spring and sleeve
   Caution
   • **Primary side** .... Yellow
   • **Secondary side** .... White
4. Hold spring pin
5. Adjust lever
6. Link
7. Pull-off spring
8. Shoe spring
9. Return spring
10. Return spring
11. Adjuster
12. Primary brake shoe
13. Secondary brake shoe
14. Operating lever
15. Strut
16. Brake pipe
17. Wheel cylinder assembly

N·m (m-kg, ft-lb)

241569

Rear drum brake assembly — MPV, 4WD B2200 and B2600i

To install:

17. Position the brake shoes on the backing plate.
18. Connect the parking brake.
19. Install the hold-down spring assemblies.
20. Install the strut and star-wheel adjuster.
21. Install the 2 upper return springs.
22. Install the lower spring.
23. Install the pull-off spring.
24. Install the lever link and self-adjusting lever.
25. Install the brake drum.
26. Working through the hole in the backing plate, turn the star-wheel screw with a brake adjusting tool until the wheel is locked.
27. Back off the adjuster about 8–10 clicks, or until the wheel is free to rotate.

NOTE: The adjustment should be the same on both wheels.

28. Adjust the parking brake.
29. Operate the brake pedal a few times. If the brakes feel at all spongy, bleed the system.

Wheel Cylinder

REMOVAL AND INSTALLATION

1. Disconnect the negative battery cable.
2. Raise and safely support the vehicle.
3. Remove the tire and wheel assembly.
4. Remove the brake drum and the brake shoes.
5. Disconnect and plug the brake line(s) at the wheel cylinder.
6. Remove the attaching nuts from behind the backing plate and remove the wheel cylinder.
7. Installation is the reverse of removal.

NOTE: Install the wheel cylinder to the brake backing plate with its mounting bolts. Do not tighten the mounting bolts until the brake line has been installed to the wheel cylinder.

8. Properly bleed the brake system.

Parking Brake Cable

ADJUSTMENT

B2200 and B2600i

1. Make sure the rear brake shoes are properly adjusted.

2. Start the engine and depress the brake pedal several times while the vehicle is moving in reverse.

3. Stop the engine.

4. Loosen the locknut at adjusting portion of the parking brake handle. Next, turn the adjusting nut to adjust the parking brake cable.

5. Turn the adjusting nut until the parking brake is fully applied when the lever is pulled 7–12 notches.

6. Tighten the locknut.

7. Test the operation of the parking brake.

8. Raise and safely support the vehicle, and then verify that the rear wheels will turn freely when the parking brake is off. Make sure the rear brake shoes don't drag or bind. Lower the vehicle.

MPV

1993

1. Make sure the rear brake shoes are properly adjusted.

2. Remove the screw and remove the parking brake lever cover. Remove the adjusting nut clip.

3. Turn the adjusting nut until the parking brake is fully applied when the lever is pulled 5–7 notches.

4. Install the adjusting nut clip and the parking brake lever cover.

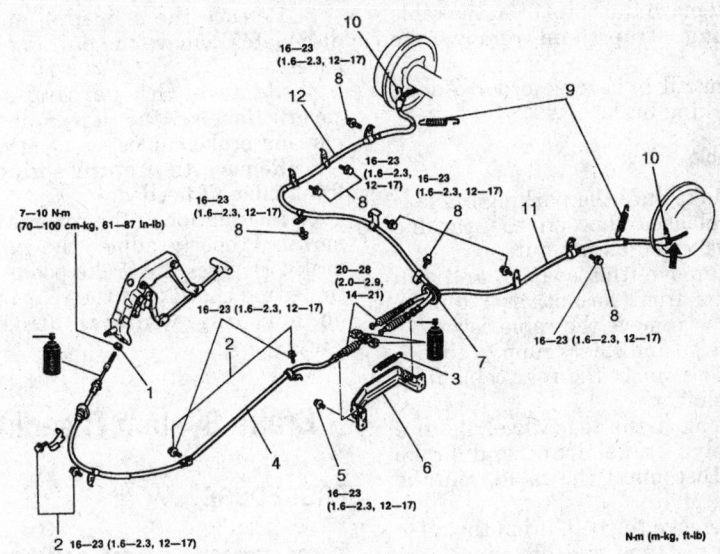

1. Nut	7. Grommet
2. Bolt	8. Bolt
3. Spring	9. Spring
4. Front brake cable	10. Clip
5. Bolt	11. Rear cable, (left)
6. Bracket	12. Rear cable, (right)

351179

Parking brake cable components — B2200 and B2600i (4WD)

1994–97

1. Remove the parking brake lever cover.

2. Remove the adjusting nut clip and turn the adjusting nut at the front of the parking cable to adjust.

3. After adjustment, with the ignition switch in the **ON** position, pull the parking brake lever one notch. Check that the parking brake warning light illuminates.

4. When the cable is properly adjusted, the lever should move 3–6 notches.

5. Replace the adjusting nut clip.

REMOVAL AND INSTALLATION

B2200, B2600i and 1993 MPV

Front Cable

1. Make sure the parking brake is fully released. Remove the parking brake lever adjusting nut from the forward end of the front cable.

2. Remove the seat(s) and roll back the front format, as required. On MPV, remove the cable cover. Raise and safely support the vehicle, as necessary.

3. Disengage the rear cables from the equalizer and remove the spring. Disconnect the front cable from the equalizer.

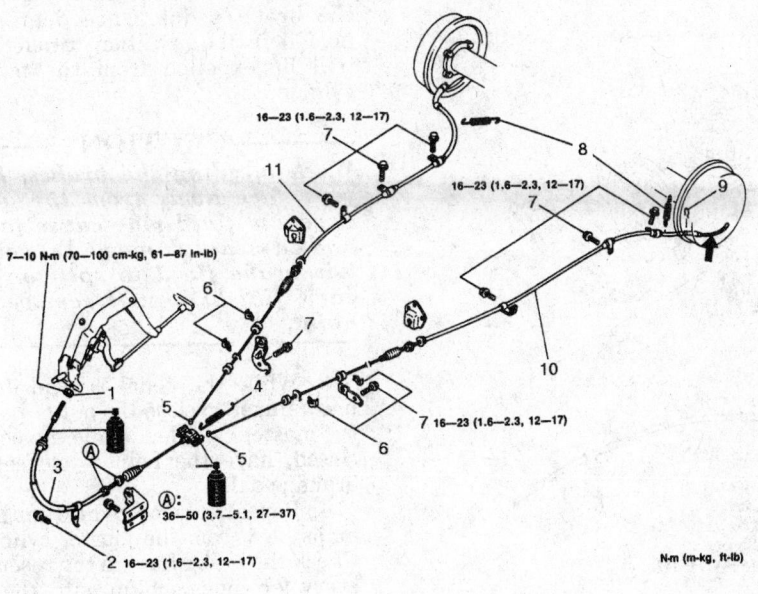

1. Nut	7. Bolts
2. Bolt	8. Spring
3. Front brake cable	9. Clip
4. Spring	10. Rear cable (left)
5. Brake cable connector	11. Rear cable (right)
6. Clip	

351178

Parking brake cable components — B2200 and B2600i (2WD)

4. Remove the bolts from the cable retaining straps and remove the cable.

5. Install in reverse order. Adjust the parking brake.

Rear Cable

1. Make sure the parking brake is fully released. Loosen the parking brake lever adjusting nut.

2. Remove the seat(s) and roll back the front format, as required. On MPV, remove the cable cover.

3. Raise and safely support the vehicle. Disconnect the rear cable from the equalizer.

4. Remove the rear wheel and tire assembly, brake drum and brake shoes. Disconnect the cable from the backing plate.

5. Remove the bolts from the cable retaining straps and disconnect the spring from the cable. Remove the cable.

6. Install in reverse order. Adjust the parking brake.

1994–97 MPV

1. Raise and safely support the vehicle.

2. Release the parking brake and remove the parking brake lever adjusting nut. Remove rear seat No. 1, front floormat and cover.

3. Remove the caliper(s), and the disc plate. Remove the parking brake shoes.

4. Remove the parking brake cover. Remove the left and right parking brake cables, as needed.

5. Remove the return spring and front cable, if needed.

6. Installation is the reverse of removal. Properly adjust the parking brake. Depress the brake pedal a few times and check that the rear brakes do not drag while rotating the wheels.

Brake System Bleeding

PROCEDURE

B2200, B2600i and 1993–94 MPV

1. Raise and safely support the vehicle.

2. Fill the master cylinder reservoir with fresh brake fluid. Don't let the reservoir run dry during the bleeding process.

3. Have an assistant slowly pump the brake pedal several times and then apply a steady pressure to the brake pedal.

4. Attach a clear flexible hose to the bleed screw and place it into a clear container partially filled with brake fluid. Loosen the brake bleeder screw at the brake caliper or wheel cylinder and allow the fluid to flow. Close the bleeder screw.

5. Repeat this procedure for each wheel, until no air bubbles appear in the brake fluid. Begin the bleeding procedure at the wheel furthest (right rear) from the master cylinder and then work toward the wheel closest (left front) to the master cylinder.

6. Lower the vehicle.

7. Check the fluid level in the master cylinder and add fluid, if necessary. Road test the vehicle and check the brake performance.

1995–97 MPV

Master Cylinder

Due to the location of the fluid reservoir, bench bleeding of the master cylinder is not recommended. The master cylinder is to be bled while mounted on the brake booster. If the fluid reservoir runs dry, bleeding of the entire system will be necessary. Two people will be required to bleed the brake system.

1. Fill the brake fluid reservoir with clean brake fluid. Disconnect the brake tube from the master cylinder.

2. Have a helper slowly depress the brake pedal. Once depressed, hold it in that position. Brake fluid will be expelled from the master cylinder.

> **CAUTION**
> *When bleeding the brakes, keep your face away from the area. Spraying fluid may cause facial and/or visual damage. Do not allow brake fluid to spill on the car's finish; it will remove the paint.*

3. While the pedal is held down, use a finger to close the outlet port of the master cylinder. While the port is closed, have the helper release the brake pedal.

4. Repeat this procedure until all air is bled from the master cylinder. Check the brake fluid in the reservoir every 4–5 times, making sure the reservoir does not run dry. Add clean DOT 3 brake fluid to the reservoir as needed. All air is bled from the master cylinder when the fluid expelled from the port is free of bubbles.

5. Connect the brake tube to the port on the master cylinder. Add clean fluid to fill the reservoir to the appropriate level.

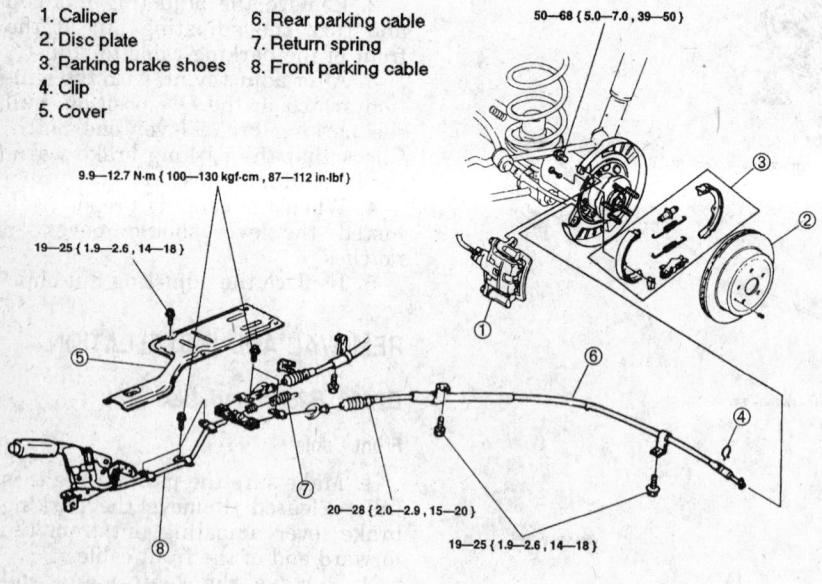

1. Caliper
2. Disc plate
3. Parking brake shoes
4. Clip
5. Cover
6. Rear parking cable
7. Return spring
8. Front parking cable

9.9—12.7 N·m { 100—130 kgf-cm , 87—112 in·lbf }

19—25 { 1.9—2.6 , 14—18 }

50—68 { 5.0—7.0 , 39—50 }

20—28 { 2.0—2.9 , 15—20 }

19—25 { 1.9—2.6 , 14—18 }

N·m (kgf·m , ft·lbf)

242391

Parking brake cable(s), exploded view — 1994–97 MPV

Calipers

1. Fill the master cylinder with fresh brake fluid. Check the level often during this procedure. Raise and safely support the vehicle.

2. Starting with the wheel furthest from the master cylinder, remove the protective cap from the bleeder and place where it will not be lost. Clean the bleeder screw.

3. Start the engine and run at idle.

CAUTION

When bleeding the brakes, keep face away from the brake area. Spewing fluid may cause physical and/or visual damage. Do not allow brake fluid to spill on the car's finish; it will remove the paint.

---

4. If the system is empty, the most efficient way to get fluid down to the wheel is to loosen the bleeder about ½–¾ turn, place a finger firmly over the bleeder and have a helper pump the brakes slowly until fluid comes out the bleeder. Once fluid is at the bleeder, close it before the pedal is released inside the vehicle.

NOTE: If the pedal is pumped rapidly, the fluid will churn and create small air bubbles, which are almost impossible to remove from the system. These air bubbles will accumulate and a spongy pedal will result.

5. Once fluid has been pumped to the caliper, open the bleed screw again, have the helper press the brake pedal to the floor, lock the bleeder and have the helper slowly release the pedal. Wait 15 seconds and repeat the procedure (including the 15 second wait) until no more air comes out of the bleeder upon application of the brake pedal. Remember to close the bleeder before the pedal is released inside the vehicle each time the bleeder is opened. If not, air will be introduced into the system.

6. If a helper is not available, connect a small hose to the bleeder, place the end in a container of brake fluid and proceed to pump the pedal from inside the vehicle until no more air comes out the bleeder. The hose will prevent air from entering the system.

7. Repeat the procedure on the remaining calipers in the following order:
   a. Left front caliper
   b. Left rear caliper
   c. Right front caliper

8. Hydraulic brake systems must be totally flushed if the fluid becomes contaminated with water, dirt or other corrosive chemicals. To flush, bleed the entire system until all fluid has been replaced with the correct type of new fluid.

9. Install the bleeder cap on the bleeder to keep dirt out. Always road test the vehicle after brake work of any kind is done.

Wheel Speed Sensor

REMOVAL AND INSTALLATION

B2200 and B2600i

1. Turn the ignition to the **LOCK** position and remove the key.

2. Raise and safely support the vehicle.

3. The speed sensor is located on top of the differential toward the front of the vehicle. Disconnect the wiring harness connector from the speed sensor.

4. Remove the sensor retaining bolt and then remove the sensor from the differential.

To install:

5. Measure the clearance (air gap) between the sensor's probe tip and the sensor rotor:
   a. Measure the distance between the sensor rotor and the sensor mounting surface on the differential.
   b. Measure the distance between the sensor's probe tip and mounting base.
   c. Subtract the probe tip-to-mounting base measurement (Step B) from the rotor-to-sensor mounting surface measurement (Step A). The difference between the 2 measurements is the sensor clearance. For B2200 trucks, the clearance should fall within the range of 0.020–0.039 in. (0.5–1.0mm); for B2600 trucks, the clearance should be 0.020–0.047 in. (0.5–1.2mm).

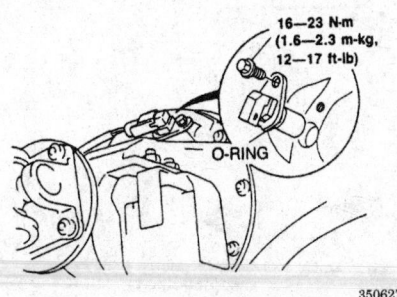

16–23 N·m
(1.6–2.3 m·kg,
12–17 ft·lb)

O-RING

350627

Speed sensor location — B2200, B2600i and 1993–95 MPV

d. If the clearance is less than the specification, use an adjustment shim, No. PO49-27-155 during installation. If the clearance is greater than the specification, the speed sensor should be replaced.

e. If the sensor rotor is damaged, inspect the differential gears for damage.

6. Install a new O-ring onto the sensor and then lubricate it with a small amount of clean engine oil.

7. Install the speed sensor and tighten its retaining bolt to 12 ft. lbs. (16 Nm). Reconnect the wiring harness connector.

8. Lower the vehicle.

9. Turn the ignition **ON** and make sure the ANTI LOCK indicator light flashes and then turns OFF. If the indicator light doesn't turn on, or stays on, the system fault must be traced. Road test the vehicle to verify ABS operation.

1993–95 MPV

1. Disconnect the negative battery cable. Wait at least 90 seconds before performing any work.

2. Raise and safely support the vehicle.

3. Disconnect the speed sensor electrical connector.

4. Remove the speed sensor mounting bolt and remove the sensor from the axle housing with the O-ring.

To install:

5. Before installing the speed sensor, check for proper clearance and adjust if necessary. Adjust as follows:
   a. Measure the clearance between the ring gear speed sensor and the sensor rotor teeth. If the clearance is less than 0.020 in., adjust it during the sensor installation by using the PO49-27-155 adjustment shim. If the clearance is more than specification, replace the speed sensor with a new one.
   b. Measure the distance between the sensor attaching surface and the rotor teeth. Measure the distance between the sensor attaching surface and the sensor pole piece. Subtract this measurement from the first. Specified clearance: 0.020–0.047 in.

6. Clean the ring gear speed sensor attaching surface, lubricate the sensor O-ring with differential oil, and install the speed sensor with the bolt. Torque the bolt to 12–16 ft. lbs. (16–22 Nm).

7. Reconnect the electrical connector. Safely lower the vehicle, and reconnect the negative battery cable.

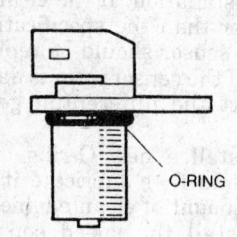

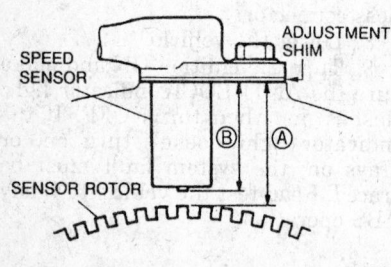

Speed sensor clearance inspection and shim — B2200, B2600i and 1993–95 MPV

1996–97 MPV

Front

1. Disconnect the negative battery cable. Wait at least 90 seconds before performing any work.
2. Raise and safely support the vehicle.
3. Remove the wheel and tire assembly.
4. Disconnect the speed sensor connector.
5. Remove the bolt from the clamp securing the speed sensor harness from the body.
6. Remove the bolt securing the speed sensor bracket to the lower strut mount.

7. Remove the bolt attaching the speed sensor and remove the speed sensor.

To install:

8. Install the speed sensor with the bolt and check that the clearance between the speed sensor and the speed sensor rotor is 0.0197–0.0512 in. (0.5–1.3mm). Torque the bolt to 18 ft. lbs. (25 Nm).
9. Reinstall the speed sensor bracket to the lower strut mount with the bolt and torque the bolt to 18 ft. lbs. (25 Nm).
10. Install the speed sensor wire clamp to the body with the bolt and reconnect the connector.
11. Install the wheel and tire assembly and safely lower the vehicle.
12. Reconnect the negative battery cable.

Rear

1. Disconnect the negative battery cable. Wait at least 90 seconds before performing any work.
2. Raise and safely support the vehicle and remove the tire and wheel assembly.
3. Disconnect the speed sensor electrical connector.
4. Remove the bolt holding the speed sensor clamp to the body.

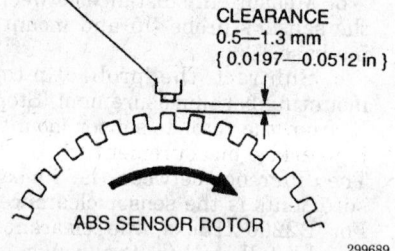

Front ABS speed sensor clearance — 1996–97 MPV

5. Remove the 3 bolts securing the speed sensor harness to the differential and the axle tube.
6. Remove the speed sensor mounting bolt and the speed sensor.

To install:

7. Before installing the speed sensor, measure the distance between the tip of the speed sensor and the underside of the mounting flange. Use calipers or similar measuring device.
8. Measure the distance between the speed sensor mounting surface and the speed sensor rotor. Subtract the first measurement from the second. The clearance should be: 0.0119–0.0433 in. (0.3–1.1mm).
9. Install the speed sensor with the mounting bolt and torque the bolt to 18 ft. lbs. (25 Nm).
10. Reattach the speed sensor to the axle tube and the differential using the 3 bolts. Torque the bolts to 18 ft. lbs. (25 Nm).
11. Reinstall the speed sensor clamp to the body with the bolt and reconnect the electrical connector.
12. Install the tire and wheel assembly and safely lower the vehicle.
13. Reconnect the negative battery cable.

FRONT SUSPENSION

Strut

REMOVAL AND INSTALLATION

MPV

1. Raise and safely support the vehicle. Remove the wheel and tire assembly.
2. Support the lower control arm with a jack.
3. Remove the clip attaching the brake hose to the strut and disconnect the hose from the strut.
4. Remove the strut-to-knuckle attaching bolts and nuts.
5. Working in the engine compartment, remove the 4 attaching nuts from the strut tower and remove the strut assembly from the vehicle.
6. Remove the rubber cap from the upper mounting block. Loosen the upper attaching nut, but do not remove it.
7. Install a suitable spring compressor and compress the coil spring.
8. Remove the upper attaching nut and slowly relieve the tension on

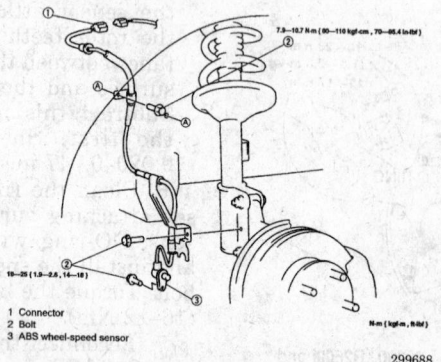

1 Connector
2 Bolt
3 ABS wheel-speed sensor

Front ABS speed sensor — 1996–97 MPV

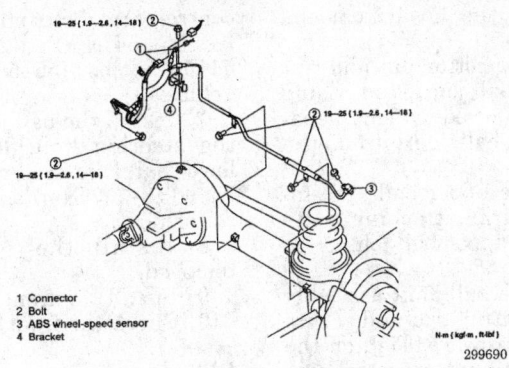

1 Connector
2 Bolt
3 ABS wheel-speed sensor
4 Bracket

Nm (kgf·m, ft·lbf)
299690

Rear ABS speed sensor — 1996–97 MPV

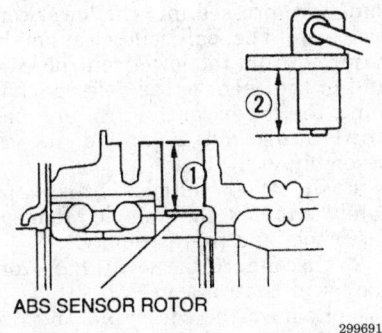

ABS SENSOR ROTOR

299691

Clearance measurement for the rear ABS speed sensor — 1996–97 MPV

the coil spring, using the spring compressor. When the spring is no longer under tension, remove the spring compressor.

9. Remove the upper mounting block, upper spring seat, spring seat, coil spring, bump stopper and ring rubber from the strut.

To install:

10. Secure the strut in a vise equipped with protective jaw covers, so the strut will not be damaged.

11. Apply a suitable rubber grease to the ring rubber and install it on the bump stopper. Install the bump stopper on the strut.

12. Attach the spring compressor to the coil spring and compress the spring.

13. Install the compressed spring on the strut and install the spring seat.

14. Install the upper spring seat. The flat of the strut rod must fit correctly into the upper spring seat.

15. Install the upper mounting block. Install and loosely tighten the upper attaching nut.

16. Remove the spring compressor. Make sure the spring is properly seated in the upper and lower spring seats.

17. Secure the upper spring seat in a vise and tighten the upper attach-

ing nut to 47–59 ft. lbs. (64–80 Nm). Install the rubber cap on the upper mounting block.

18. Install the strut assembly in the strut tower, making sure the white mark on the upper mounting block is in the front-inside direction. Install the attaching nuts and tighten to 34–46 ft. lbs. (47–62 Nm).

19. Install the strut to the knuckle and tighten the attaching bolts and nuts to 69–86 ft. lbs. (93–117 Nm).

20. Position the brake hose on the strut and install the clip. Remove the jack from under the lower control arm.

21. Install the wheel and tire assembly and lower the vehicle. Check the front end alignment.

Shock Absorber

REMOVAL AND INSTALLATION

B2200 and B2600i

1. Raise and safely support the vehicle. Remove the wheel and tire assembly.

2. Remove the upper shock absorber nuts, retainer and bushing.

3. Remove the lower shock absorber-to-lower control arm mounting bolt, nut and washer.

4. Slightly compress the shock absorber and remove it from the vehicle. Remove the remaining retainers and bushing from the upper shock absorber stud.

To install:

5. Install the shock absorber and install the mounting bolts, nuts, washers and bushings. Do not tighten at this time.

6. Install the wheel and tire assembly and lower the vehicle.

7. With the vehicle unladen, tighten the upper shock absorber mounting nuts until the stud protrudes 0.28 in. (7mm) above the upper nut. Tighten the lower mounting

bolt and nut to 41–59 ft. lbs. (55–80 Nm).

Torsion Bar

REMOVAL AND INSTALLATION

B2200 and B2600i

1. Raise and safely support the vehicle. Remove the wheel and tire assembly.

2. Support the lower control arm with a jack.

3. Remove the cotter pin and nut from the lower ball joint stud. Separate the ball joint from the knuckle.

4. Remove the bolt, washer and nut attaching the shock absorber to the lower control arm.

5. Mark the position of the anchor bolt and swivel for reference during reassembly and remove the anchor bolt and swivel.

6. Mark the position of the torsion bar on the anchor arm and remove the anchor arm.

7. Mark the position of the torsion bar on the torque plate and remove the torsion bar. If removing the torsion bar from both sides of the vehicle, mark their positions as the torsion bars are not interchangeable.

8. Remove the attaching bolts and remove the torque plate.

9. Check the torsion bar for bending or for looseness between the serrations of the torsion bar and anchor arm and/or torque plate, replace as necessary.

To install:

10. Install the torque plate and tighten the attaching bolts to 55–69 ft. lbs. (75–93 Nm).

11. Coat the serrations of the torsion bar with grease and install in the torque plate, aligning the marks made during the removal procedure. If both torsion bars were removed, make sure the correct torsion bar is being installed.

12. Coat the serrations on the other end of the torsion bar with grease and install the anchor arm onto the torsion bar, aligning the marks made during the removal procedure.

13. Install the anchor bolt and swivel. Tighten the anchor bolt until the marks made during removal are aligned.

14. Connect the shock absorber to the lower control arm and loosely tighten the nut and bolt. Connect the lower ball joint to the knuckle; install the nut and tighten to 87–116 ft. lbs. (118–157 Nm). Install a new cotter pin.

15. Install the wheel and tire assembly and lower the vehicle. With the vehicle unladen, tighten the shock absorber-to-lower control arm bolt and nut to 41–59 ft. lbs. (55–80 Nm).

16. Check vehicle ride height as follows:

a. Check the front and rear tire pressures and bring to specification.

b. Measure the distance from the center of each front wheel to the fender brim. The difference must not be greater than 0.39 in. (10mm).

c. If the difference is not as specified, turn the necessary torsion spring anchor bolt to adjust.

17. Check the front end alignment.

NOTE: If, for some reason, you didn't matchmark the torsion bar anchor bolt, or the matchmarks were lost, or you're installing a new, unmarked torsion bar, here's a procedure to help you attain the correct ride height:

18. Install the anchor arm on the torsion bar so there is 4.92 in. (125mm) between the lowest point on the arm and the crossmember directly above it.

19. Tighten the anchor bolt until the anchor arm contacts the swivel. Then, tighten the bolt an additional 1.77 in. (45mm) travel.

Upper Ball Joints

REMOVAL AND INSTALLATION

B2200 and B2600i

1. Raise and safely support the vehicle. Remove the wheel and tire assembly.

2. Support the lower control arm with a jack.

3. Remove the clip attaching the brake hose to the upper control arm

and disconnect the hose from the arm.

4. Remove the cotter pin and nut from the upper ball joint stud. Using tool 49-0727-575 or equivalent, separate the upper ball joint from the knuckle.

5. Remove the upper ball joint-to-upper control arm attaching bolts and remove the upper ball joint.

To install:

6. Position the ball joint assembly on the upper control arm and secure the mounting hardware. Tighten the upper ball joint-to-upper control arm attaching bolts to 18–25 ft. lbs. (25–33 Nm) and the upper ball joint stud nut to 22–38 ft. lbs. (29–51 Nm). Install a new cotter pin.

7. Install the clip securing the brake hose to the upper control arm.

8. Install the wheel and tire assembly.

9. Check the front end alignment.

Lower Ball Joints

REMOVAL AND INSTALLATION

B2200 and B2600i

1. Raise and safely support the vehicle. Remove the front wheel.

2. Support the lower control arm with a jack.

3. On 2WD vehicles, remove the bolts attaching the tension rod to the lower control arm.

4. Remove the cotter pin and nut from the lower ball joint stud. Separate the ball joint from the knuckle using tool 49-0727-575 or equivalent.

5. Remove the bolts/nuts attaching the lower ball joint to the lower control arm and remove the lower ball joint.

To install:

6. Position the ball joint assembly to the lower control arm and secure it with the mounting hardware. Tighten the lower ball joint-to-lower

control arm bolts/nuts to 32–40 ft. lbs. (43–54 Nm) on 2WD vehicles or 41–50 ft. lbs. (55–68 Nm) on 4WD vehicles.

7. Install the ball joint stud nut to the steering knuckle. Tighten the lower ball joint stud nut to 87–116 ft. lbs. (118–157 Nm) and install a new cotter pin.

8. Install the tension rod if removed.

9. Install the front wheel.

10. Check the front end alignment.

MPV

On 2WD vehicles, the lower ball joints are pressed into the lower control arm. The ball joints cannot be removed from the lower control arms and in the event of the defective ball joint, the lower control arm and ball joint must be replaced as an assembly.

1. On 4WD vehicles, raise and safely support the vehicle. Remove the wheel and tire assembly.

2. Disconnect the sway bar from the lower control arm.

3. Remove the cotter pin and nut from the lower ball joint stud. Separate the ball joint from the knuckle.

4. Remove the 2 upper bolts, one through-bolt and remove the ball joint.

To install:

5. Install the ball joint to the lower control arm.

6. Install the ball stud into the steering knuckle. Torque the nut to 116–137 ft. lbs. (157–186 Nm) and install the cotter pin.

7. Connect the sway bar to the lower control arm.

8. Install the wheel and tire assembly and lower the vehicle.

Upper Control Arms

REMOVAL AND INSTALLATION

B2200 and B2600i

1. Loosen the wheel lugs nuts slightly. Raise and support the front end on jackstands placed under the frame.

2. Remove the wheels. Support the lower arm with a floor jack.

3. Remove the cotter pin and nut from the upper ball joint and separate the ball joint from the upper arm using a ball joint separator tool.

4. Remove the nuts and bolts that retain the upper arm shaft to the support bracket. Note the number and location of the shims under the nuts. These must be installed in their

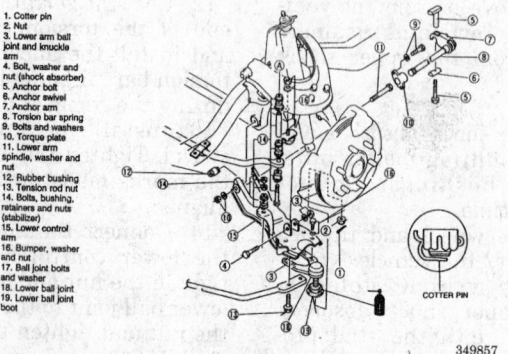

1. Cotter pin
2. Nut
3. Lower arm ball joint and knuckle arm
4. Bolt, washer and nut (shock absorber)
5. Anchor bolt
6. Anchor swivel
7. Anchor arm
8. Torsion bar spring
9. Bolts and washers
10. Torque plate
11. Lower arm spindle, washer and nut
12. Rubber bushing
13. Tension rod nut
14. Bolts, bushing, retainers and nuts (stabilizer)
15. Lower control arm
16. Bumper, washer and nut
17. Ball joint bolts and washer
18. Lower ball joint
19. Lower ball joint boot

COTTER PIN

349857

Front suspension components — B2200 and B2600i

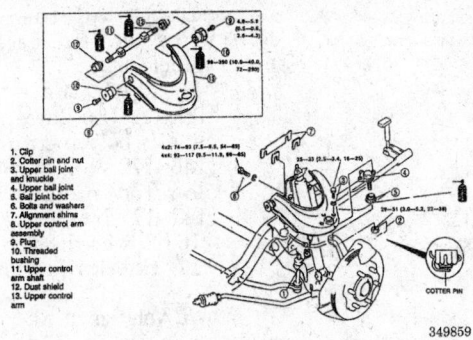

Upper control arm and ball joint mounting — B2200 and B2600i

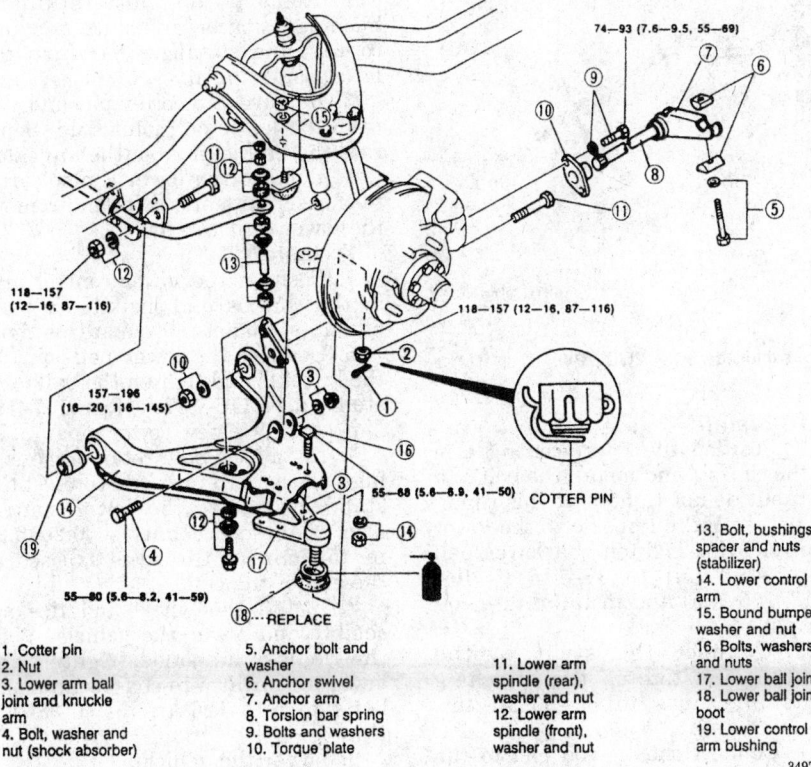

1. Cotter pin
2. Nut
3. Lower arm ball joint and knuckle arm
4. Bolt, washer and nut (shock absorber)
5. Anchor bolt and washer
6. Anchor swivel
7. Anchor arm
8. Torsion bar spring
9. Bolts and washers
10. Torque plate
11. Lower arm spindle (rear), washer and nut
12. Lower arm spindle (front), washer and nut
13. Bolt, bushings, spacer and nuts (stabilizer)
14. Lower control arm
15. Bound bumper, washer and nut
16. Bolts, washers and nuts
17. Lower ball joint
18. Lower ball joint boot
19. Lower control arm bushing

Lower control arm and ball joint mounting — B2200 and B2600i (4WD)

exact locations for proper wheel alignment. Check all parts for wear or damage. Replace any suspect parts.

To install:

5. Place the control arm assembly in position. Install the mounting bolts and nuts with the alignment shims in correct positions. Torque the upper arm shaft mounting bolts to: 60–68 ft. lbs. for 2WD or 69–85 ft. lbs. for 4WD.

6. Install the ball joint stud to the steering knuckle. Tighten the ball joint nut to 30–37 ft. lbs. and install a new cotter pin.

7. Install the wheel and lower the vehicle.

8. Check the wheel alignment and adjust if necessary.

Lower Control Arms

REMOVAL AND INSTALLATION

B2200 and B2600i

1. Raise and safely support the vehicle. Remove the front wheel.

2. Support the lower control arm with a jack.

3. Remove the cotter pin and nut from the lower ball joint stud. Separate the ball joint from the knuckle using tool 49-0727-575 or equivalent.

4. Remove the bolt, washer and nut attaching the shock absorber to the lower control arm.

5. Remove the torsion bar, anchor arm and torque plate assembly.

6. Remove the bolt(s) and nut(s) attaching the lower control arm to the frame.

7. On 2WD vehicles, remove the bolts attaching the tension rod to the lower control arm.

8. Remove the bolts, bushings, retainers, spacer and nuts connecting the sway bar to the lower control arm.

9. Remove the lower control arm from the vehicle. Remove the lower ball joint, if necessary.

To install:

10. Position the lower control arm to the frame and install the attaching bolt(s) and nut(s), but do not tighten at this time.

11. Install the torsion bar, anchor arm and torque plate assembly.

12. On 2WD vehicles, install the tension rod bolt and tighten to 69–86 ft. lbs. (93–117 Nm).

13. Attach the sway bar to the control arm with the bolts, bushings, retainers, spacer and nuts. Tighten the nuts so 0.73 in. (18.5mm) of thread is exposed at the end of the bolt.

14. Install the shock absorber to the lower control arm and loosely tighten the mounting bolt and nut.

15. Install the front wheel and lower the vehicle.

16. With the vehicle unladen, tighten the lower control arm-to-frame bolt and nut on 2WD vehicles and the front side lower control arm-to-frame bolt and nut on 4WD vehicles to 87–116 ft. lbs. (118–157 Nm). Tighten the rear side lower control arm-to-frame bolt and nut on 4WD vehicles to 116–145 ft. lbs. (157–196 Nm).

17. With the vehicle unladen, tighten the shock absorber-to-lower control arm bolt and nut to 41–59 ft. lbs. (55–80 Nm).

18. Check vehicle ride height as follows:

a. Check the front and rear tire pressures and bring to specification.

b. Measure the distance from the center of each front wheel to the fender brim. The difference must not be greater than 0.39 in. (10mm).

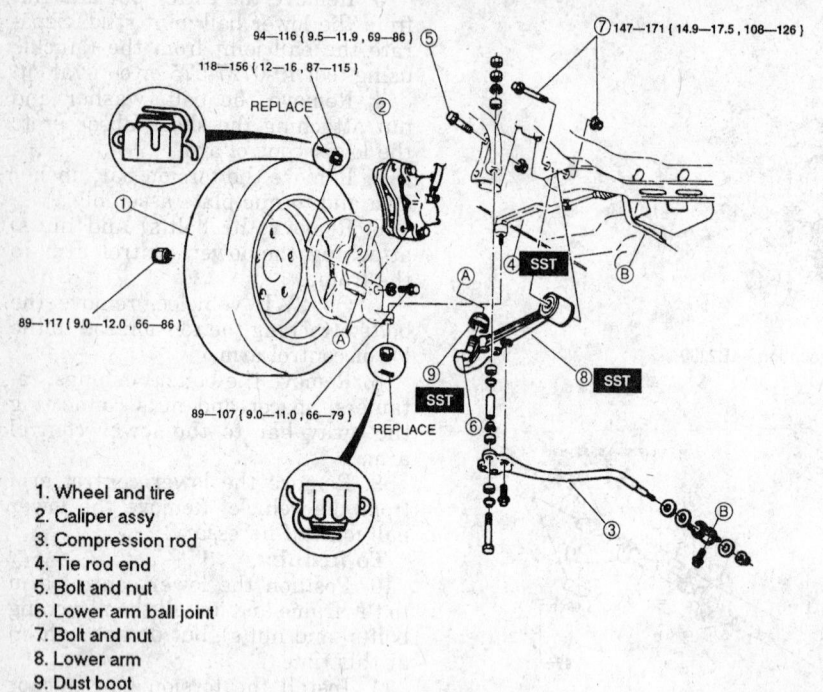

94—116 { 9.5—11.9, 69—86 }

118—156 { 12—16, 87—115 }

REPLACE

147—171 { 14.9—17.5 , 108—126 }

89—117 { 9.0—12.0, 66—86 }

89—107 { 9.0—11.0, 66—79 }

REPLACE

SST

SST

SST

1. Wheel and tire
2. Caliper assy
3. Compression rod
4. Tie rod end
5. Bolt and nut
6. Lower arm ball joint
7. Bolt and nut
8. Lower arm
9. Dust boot

N·m { kgf·m , ft·lbf }

303182

Ball joint and lower control arm assembly, exploded view — 2WD MPV

c. If the difference is not as specified, turn the necessary torsion spring anchor bolt to adjust.

19. Check the front end alignment.

MPV

2WD Vehicles

1. Raise and safely support the vehicle. Remove the wheel and tire assembly.

2. Remove the brake caliper and support it aside with mechanics wire, do not let it hang by the brake hose.

3. Remove the nuts, bolts, spacer, washers and bushings and remove the compression rod from the lower control arm and chassis and disconnect the stabilizer bar from the lower control arm.

4. Remove the cotter pin and nut and separate the tie rod end from the knuckle.

5. Remove the bolts and nuts and disconnect the strut from the knuckle.

6. Remove the cotter pin and nut from the lower ball joint stud and separate the lower ball joint from the knuckle.

7. Remove the mounting bolt and nut and remove the lower control arm from the vehicle.

To install:

8. Position the lower control arm to the chassis and install the bolt and nut, but do not tighten at this time.

9. Install the knuckle to the lower control arm. Tighten the lower ball joint stud nut to 87–115 ft. lbs. (118–156 Nm) and install a new cotter pin.

10. Connect the strut to the knuckle and tighten the attaching bolts and nuts to 69–86 ft. lbs. (94–116 Nm).

11. Connect the tie rod end to the knuckle. Tighten the tie rod end stud nut to 43–58 ft. lbs. (59–78 Nm) and install a new cotter pin.

12. Install the compression rod to the lower control arm and chassis. Tighten the compression rod-to-lower control arm mounting bolts to 76–93 ft. lbs. (103–126 Nm) and the compression rod bushing-to-chassis bolts to 61–76 ft. lbs. (83–103 Nm). Install the compression rod nut but do not tighten at this time.

NOTE: The left-hand compression rod nut has left-hand threads.

13. Connect the stabilizer bar to the control arm with the bolt, washers, bushings, spacer and nuts. Tighten the nuts so 0.24 in. (6mm) of thread is exposed at the end of the bolt.

14. Install the caliper and the wheel and tire assembly. Lower the vehicle.

15. With the vehicle unloaded, tighten the lower control arm-to-chassis bolt and nut to 94–108 ft. lbs. (146–172 Nm). Tighten the compression rod nut to 108–126 ft. lbs. (147–171 Nm).

16. Lower the vehicle.

17. Check the front end alignment.

4WD Vehicles

1. Raise and safely support the vehicle. Remove the wheel and tire assembly.

2. Remove the bolt, retainers, bushings, spacer and nuts and disconnect the stabilizer bar from the lower control arm.

3. Remove the cotter pin and nut from the lower ball joint stud. Separate the ball joint from the knuckle.

4. Remove the lower control arm-to-chassis nuts and bolts and remove the lower control arm.

To install:

5. Position the lower control arm to the chassis and install the bolts and nuts. Do not tighten at this time.

6. Connect the lower ball joint to the knuckle and tighten the ball joint stud nut to 116–137 ft. lbs. (157–186 Nm). Install a new cotter pin.

7. Install the bolt, retainers, bushings, spacer and nuts and connect the stabilizer bar to the lower control arm. Tighten the nuts so 0.20–0.28 in. (5–7mm) of thread is exposed at the end of the bolt.

8. Install the wheel and tire assembly and lower the vehicle. With the vehicle unloaded, tighten the lower control arm-to-chassis nuts and bolts to 102–126 ft. lbs. (138–171 Nm).

9. Lower the vehicle.

10. Check the front end alignment.

Sway Bar

REMOVAL AND INSTALLATION

B2200 and B2600i

1. Raise and support the front end on jackstands.

2. Unbolt the stabilizer bar-to-frame clamps.

3. Unbolt the stabilizer bar from the lower control arms. Keep all the bushings, washers and spacers in order.

4. Check all parts for wear or damage and replace anything which looks suspicious.

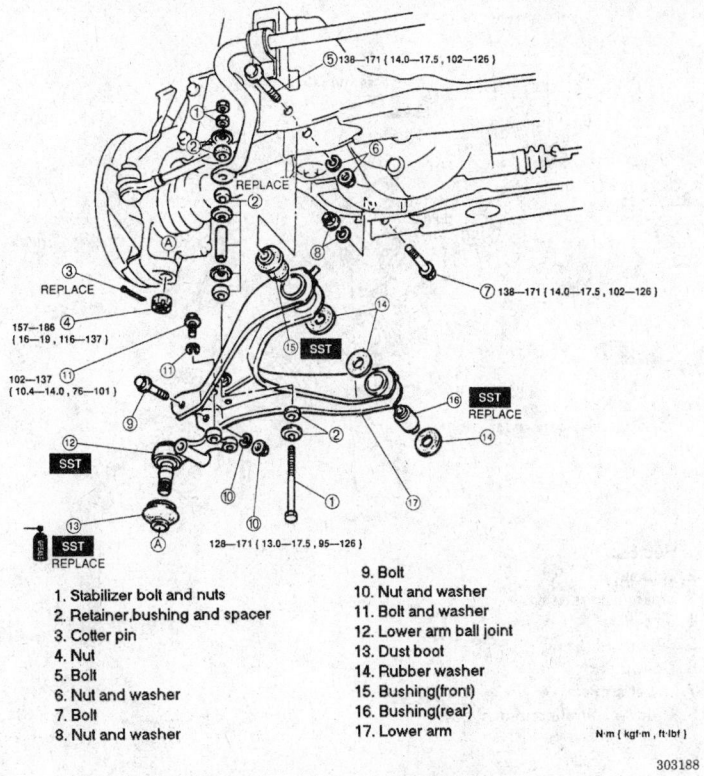

1. Stabilizer bolt and nuts
2. Retainer, bushing and spacer
3. Cotter pin
4. Nut
5. Bolt
6. Nut and washer
7. Bolt
8. Nut and washer
9. Bolt
10. Nut and washer
11. Bolt and washer
12. Lower arm ball joint
13. Dust boot
14. Rubber washer
15. Bushing (front)
16. Bushing (rear)
17. Lower arm

N·m (kgf·m , ft·lbf)

303188

Lower control arm mounting — 4WD MPV

To install:

5. Install the stabilizer bar to the lower control arms and mount the frame brackets. Tighten all fasteners lightly, then torque them to specifications with the wheels on the ground.
Stabilizer-to-control arm nut: 34 ft. lbs.
Stabilizer-to-frame clamp bolts: 19 ft. lbs.

MPV

1. Raise and support the front end on jackstands.
2. Remove the splash shield.
3. Disconnect the end links at the compression rods.
4. Remove the clamp bolts. Lift out the stabilizer bar.
5. Inspect all parts for wear and/or damage. Replace as necessary.
To install:
6. Place the sway bar on the frame and install the U-brackets finger-tight.
7. Install the end links to the control arms and tighten until, on 2WD, 0.47–0.55 in. (12–14mm) of thread is visible above the nut; on 4WD, 0.20–0.28 in. (5–7mm) of thread is visible above the nut
8. Lower the vehicle to the floor and tighten the U-bracket bolts to

37–44 ft. lbs. (51–60 Nm) on 2WD or 14–18 ft. lbs. (19–25 Nm) on 4WD.
9. Install the splash shield.
10. Lower the vehicle.

Front Wheel Bearings

ADJUSTMENT

B2200 and B2600i

1. Raise and safely support the vehicle and then remove the front wheels.
2. Unbolt the brake caliper from the steering knuckle and hang it out of the way with a wire hook. Don't disconnect the brake hose.
3. If equipped with 4WD, remove the locking hub assembly.
4. If equipped with 2WD, remove the hub dust cap and clean the old grease out of it. Remove the cotter pin from the hub castle nut.
5. Tighten the hub castle nut to 14–22 ft. lbs. (20–29 Nm) with a torque wrench. Next, turn the hub and rotor assembly 2–3 times to seat the bearing.
6. Loosen the hub castle nut so the hub and rotor assembly turns freely.
7. Attach a spring scale to one of the wheel lug bolts and measure the force required to move the hub by

pulling on the spring scale; this is the frictional force.
8. Using the frictional force found in step 7, tighten the hub castle nut until the reading on the spring scale reaches the sum of the frictional force plus the bearing preload. The preload added to be added to the frictional force is 1.3–2.4 lbs. (6–11 N) for 2WD vehicles or 1.3–2.6 lbs. (6–12 N) for 4WD vehicles.
9. Make sure the hub and rotor assembly turns smoothly and freely, and the the bearings don't feel loose when the hub is pulled away from the steering knuckle.
10. Install a new cotter pin and then pack the hub dust cap with fresh grease and install it.
11. If equipped with 4WD, install the locking hub assembly and make sure it functions properly.
12. Install the caliper and tighten the mounting bolts to 65 ft. lbs. (88 Nm).
13. Install the front wheels and lower the vehicle.

1995–97 MPV

1. Raise and support the vehicle safely. Remove the tire and wheel assembly.
2. Remove and properly support the caliper assembly.
3. Position a dial indicator gauge against the dust cap. Push and pull the disc brake rotor or brake drum in and out in the axial direction and measure the end-play of the wheel bearing.
4. End-play should not exceed 0.002 in. (0.05mm).
5. If end-play is excessive, check the hub nut torque or replace the bearing.

REMOVAL AND INSTALLATION

B2200 and B2600i

1. Raise and safely support the vehicle. Remove the wheel and tire assembly.
2. Remove the brake caliper and support it with mechanics wire. Do not let the caliper hang by the brake hose.
3. Remove the grease cap, cotter pin, retainer, adjusting nut and washer. Discard the cotter pin.
4. Remove the outer bearing and pull the hub and rotor off the spindle. Remove the grease seal using a seal removal tool. Discard the grease seal.
5. Remove the inner bearing from the hub. Remove all traces of old lubricant from the bearings, hub and

spindle with solvent and dry thoroughly.

6. Inspect the bearings and bearing races for scratches, pits or cracks. If the bearings and/or races are worn or damaged, remove the races with a brass drift.

To install:

7. If the bearing races were removed, install new races in the hub with suitable installation tools. Make sure the races are properly seated.

8. Using a bearing packer, pack the bearings with high-temperature wheel bearing grease. If a packer is not available, work as much grease as possible between the rollers and cages by hand.

9. Place a small amount of grease within the hub and grease the races. Install the inner bearing. Install a new grease seal using a seal installer. Apply grease to the lips of the seal.

10. Install the hub and rotor assembly on the spindle. Install the outer bearing, washer and adjusting nut. Adjust the bearings.

11. Install the retainer, a new cotter pin and the grease cap.

12. Install the caliper and the wheel and tire assembly. Lower the vehicle.

13. Before driving the vehicle, pump the brake pedal several times to restore normal brake travel.

MPV

2WD Vehicles

1. Raise and safely support the vehicle.
2. Remove the wheel assembly.
3. Remove the hub dust cap.
4. Remove the locknut.
5. Remove the brake caliper.
6. Remove the disc plate.
7. Remove the hub assembly.
8. Disconnect the tie-rod end from the knuckle/spindle.
9. Disconnect the lower arm.

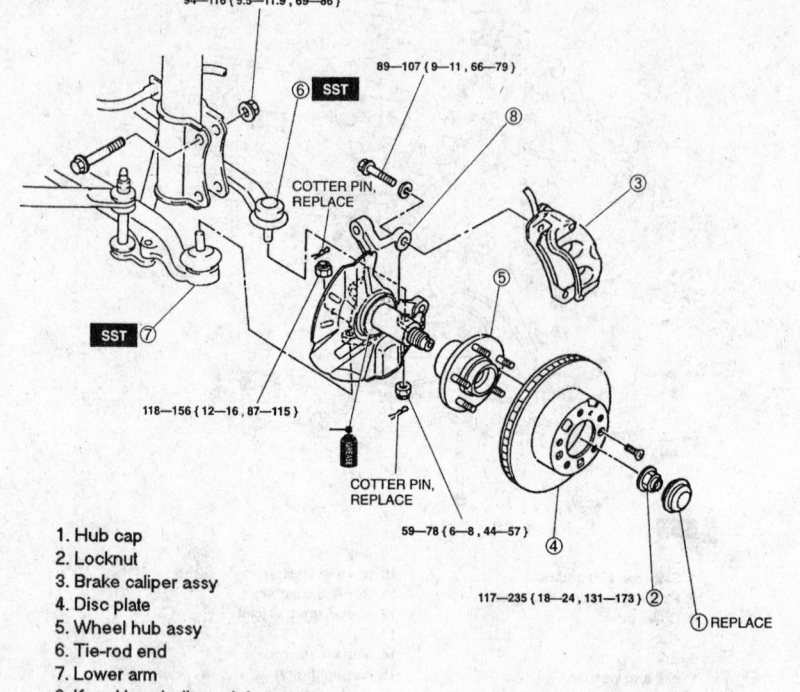

1. Hub cap
2. Locknut
3. Brake caliper assy
4. Disc plate
5. Wheel hub assy
6. Tie-rod end
7. Lower arm
8. Knuckle spindle and dust cover

N·m { kgf·m , ft-lbf }

303176

Front hub and related components, exploded view — 2WD MPV

10. Remove the knuckle/spindle assembly.
11. Remove the wheel bearings from the hub assembly, if needed.

To install:

12. Install wheel bearings to the hub assembly, if removed.
13. Install the knuckle/spindle assembly. Torque the strut mounting nut to 69–86 ft. lbs. (94–116 Nm).
14. Install the lower arm, ball joint to the knuckle/spindle assembly. Torque the ball joint nut to 87–115 ft. lbs. (118–156 Nm).
15. Connect the tie-rod end, torque the nut to 44–57 ft. lbs. (59–78 Nm).
16. Install the hub assembly.
17. Install the disc plate.

18. Install the brake caliper, torque the mounting bolts to 66–79 ft. lbs. (89–107 Nm).
19. Install the locknut, torque to 131–173 ft. lbs. (117–235 Nm).
20. Install the hub dust cap.
21. Install the wheel assembly.
22. Lower the vehicle.

4WD Vehicles

1. Raise and safely support the vehicle.
2. Remove the wheel assembly.
3. Remove the locknut.
4. Remove the brake caliper.
5. Remove the disc plate retaining screw(s).
6. Disconnect the tie-rod end from the knuckle.
7. Disconnect the lower ball joint.
8. Remove the disc plate.
9. Remove the ball joint mounting nuts and bolts.
10. Remove the knuckle, wheel hub and dustplate as an assembly.
11. Remove the wheel bearings from the hub assembly, if needed.

To install:

12. Install wheel bearings to the hub assembly, if removed.
13. Install the knuckle assembly. Torque the strut mounting nut to 69–86 ft. lbs. (94–116 Nm).
14. Install the ball joint mounting nuts and bolts. Torque the upper

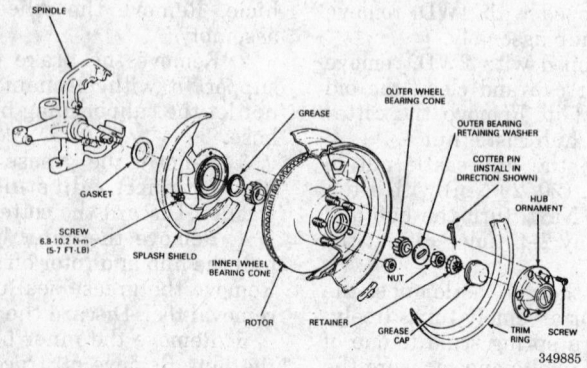

Front wheel bearing and hub assembly — B2200 and B2600i

349885

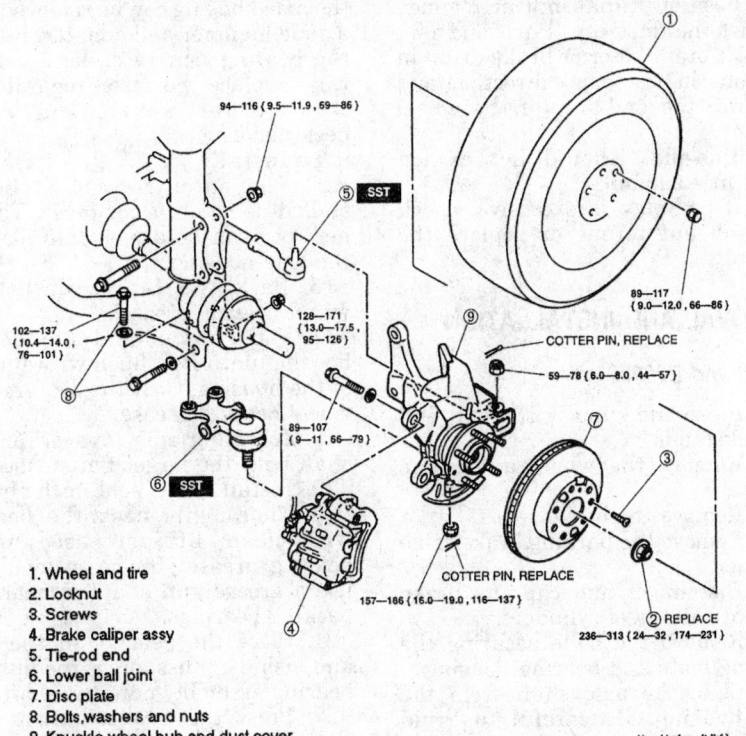

94—116 { 9.5—11.9 , 59—86 }

102—137
{ 10.4—14.0 ,
76—101 }

128—171
{ 13.0—17.5 ,
95—126 }

89—107
{ 9—11 , 66—79 }

157—186 { 16.0—19.0 , 116—137 }

89—117
{ 9.0—12.0 , 66—86 }

COTTER PIN, REPLACE

59—78 { 6.0—8.0 , 44—57 }

COTTER PIN, REPLACE

② REPLACE
236—313 { 24—32 , 174—231 }

1. Wheel and tire
2. Locknut
3. Screws
4. Brake caliper assy
5. Tie-rod end
6. Lower ball joint
7. Disc plate
8. Bolts, washers and nuts
9. Knuckle, wheel hub and dust cover

N·m { kgf·m , ft·lbf }

303177

Front hub and related components, exploded view — 4WD MPV

mounting bolts to 76–101 ft. lbs. (102–137 Nm). Torque the through-bolt nut to 95–106 ft. lbs. (128–171 Nm).

15. Replace the disc plate.

16. Install ball joint to the knuckle assembly. Torque the ball joint nut to 116–137 ft. lbs. (157–186 Nm).

17. Connect the tie-rod end, torque the nut to 44–57 ft. lbs. (59–78 Nm).

18. Install the brake caliper, torque the mounting bolts to 66–79 ft. lbs. (89–107 Nm).

19. Install the disc plate retaining nut.

20. Install the locknut, torque to 174–231 ft. lbs. (236–313 Nm).

21. Install the wheel assembly.

22. Lower the vehicle.

REAR SUSPENSION

Shock Absorber

REMOVAL AND INSTALLATION

B2200 and B2600i

1. Raise and support the rear end on jackstands.

2. Remove the wheels.

3. Unbolt the shock absorber at each end and remove it.

4. Place the shock absorber in position and secure the mounting hardware. Torque each bolt to 58 ft. lbs.

MPV

1. Raise and safely support the vehicle. Remove the splash shield.

2. Remove the stabilizer bar.

3. If equipped, remove the nut and disconnect the height sensor from the rear axle.

4. Remove the bolt attaching the parking brake cable bracket.

5. Support the rear axle housing with a jack. Raise the jack slightly to take the load off the shock absorbers.

6. Remove the attaching bolts and nuts and disconnect the shock absorbers from the lower axle housing.

7. Slowly lower the axle housing until the spring tension is relieved. Remove the coil springs.

8. Remove the spring seats and bump stopper, if equipped.

To install:

9. Install the upper and lower spring seats and the bump stopper, if removed.

10. Install the coil springs, making sure the larger diameter coil is toward the axle housing.

11. Raise the axle housing enough to connect the shock absorbers. Install the attaching bolts and nuts and tighten to 56–75 ft. lbs. (76–102 Nm). Remove the jack.

12. Install the bolt attaching the parking brake cable bracket and the nut attaching the height sensor.

13. Install the stabilizer bar. Tighten the link bolt nut until 0.28 in. (7mm) of thread is exposed at the top of the link bolt. Do not tighten the stabilizer bar bushing bracket bolts at this time.

14. Lower the vehicle. With the vehicle unladen, tighten the stabilizer bar bushing bracket bolts to 26–37 ft. lbs. (35–50 Nm).

15. Connect the height sensor, if disconnected.

16. Install the splash shield.

Leaf Spring

REMOVAL AND INSTALLATION

B2200 and B2600i

1. Raise and support the rear of the vehicle on jackstands under the frame.

— **CAUTION** —
The rear leaf springs are under considerable tension. Be very careful when removing and installing them, they can exert enough force to cause serious injuries.

2. Place a floor jack under the rear axle to take up its weight.

3. Disconnect the lower end of the shock absorbers.

4. Remove the spring U-bolts and plate.

5. Remove the spring front bolt.

6. Remove the rear shackle nuts and the shackle.

7. Lift the spring from the vehicle.

To install:

8. Place the leaf spring in position and install the rear shackle and front mount bolt. Install the spring U-bolts. Secure all fasteners snugly and lower the vehicle. When the vehicle is on its wheels, torque the nuts and bolts. Observe the following torque specifications:

Spring rear shackle nuts: 58 ft. lbs.
2WD U-bolt nuts: 58 ft. lbs.
4WD U-bolt nuts: 101 ft. lbs.
Front spring pin nut: 72 ft. lbs.
Shock absorber: 58 ft. lbs.

Sway Bar

REMOVAL AND INSTALLATION

MPV

1. Raise and safely support the vehicle.
2. Remove the sway bar links from both sides of the sway bar.
3. Remove the sway bar bracket and bushing from the axle and remove the sway bar.

To install:

4. Install the bushing on the bar and install the bar finger-tight on the axle housing.
5. Install the sway bar links to the frame brackets.
6. Lower the vehicle to the floor and tighten the sway bar to axle bolts to 26–37 ft. lbs. (35–50 Nm).

Wheel Bearings

ADJUSTMENT

B2200 and B2600i

1. Raise and safely support the vehicle.
2. Rotate each rear wheel by hand. Make sure there is no abnormal noise and the wheel rotates smoothly.
3. Hold the tire at its upper and lower edges and move the wheel slightly in and out to inspect for wheel bearing play and looseness.
4. The standard bearing play is 0.002–0.010 in. (0.05–0.25mm), and there shouldn't be noticeable noise or play when the wheel is moved in and out by hand. Further inspection of the rear bearings involves removal of the rear axle shafts.
5. If the play is excessive, rear axle bearings should be replaced, or the axle shafts should be removed and a shim of the correct thickness should be installed.

1995–97 MPV

1. Raise and support the vehicle safely. Remove the tire and wheel assembly.
2. Remove and properly support the caliper assembly.

3. Position a dial indicator gauge against the dust cap. Push and pull the disc brake rotor or brake drum in and out in the axial direction and measure the end-play of the wheel bearing.
4. End-play should not exceed 0.002 in. (0.05mm).
5. If end-play is excessive, check the hub nut torque or replace the bearing.

REMOVAL AND INSTALLATION

B2200 and B2600i

1. Raise and support the rear end on jackstands.
2. Remove the wheel and brake drum.
3. Remove the brake shoes.
4. Remove the parking brake cable retainer.
5. Disconnect and cap the brake lines at the wheel cylinders.
6. Remove the bolts securing the backing plate and bearing housing.
7. Slide the axle shaft from the axle housing. Be careful to avoid damaging the oil seal with the shaft.
8. If the seal in the axle housing is damaged in any way, it must be replaced. The seal can be removed using a slide hammer and adapter.
9. Remove 2 of the backing plate bolts, diagonally from each other.
10. Using a grinding wheel, grind down the bearing retaining collar in one spot, until about 0.197 in. (5mm) remains before you get to the axle shaft. Place a chisel at this point and break the collar. Be careful to avoid damaging the shaft.

CAUTION

Wear eye protection when grinding the collar and breaking the collar from the shaft.

---

11. Using a press or puller, remove the hub and bearing assembly from the shaft. Remove the spacer from the shaft.
12. Remove the bearing and seal from the hub.
13. Using a drift, tap the race from the hub.
14. Check all parts for wear or damage. If either race is to be replaced, both must be replaced. The race in

the axle housing can be removed with a slide hammer and adapter. Replace the bearing and races as a set. Always replace the seals, regardless of what other service is being performed.

To install:

15. The outer race must be installed using an arbor press. The inner race can be driven into place in the axle housing.
16. Pack the hub with lithium wheel bearing grease.
17. Tap a new oil seal into the axle housing until it is flush with the end of the housing. Coat the seal lip with wheel bearing grease.
18. Install a new spacer on the shaft with the larger flat surface up.
19. Install a new seal in the hub.
20. Thoroughly pack the bearing with clean, lithium based, wheel bearing grease. If one is available, use a grease gun adapter meant for packing bearings.
21. Place the bearing in the hub, and, using a press, press the hub and bearing assembly onto the shaft.
22. Press the new collar onto the shaft. The press pressure for the collar is critical. Press pressures should be 9240–13,420 lbs. (4200–6100 kg).
23. Install one shaft in the housing, being very careful to avoid damaging the inner seal.
24. If only one shaft was being serviced, the other must now be removed to check bearing play on the serviced axle. If both shafts were removed, leave the other one out for now.
25. Tighten the backing plate bolts on the one installed axle to 80 ft. lbs.
26. Mount a dial indicator on the backing plate, with the pointer resting on the axle shaft flange. Check the axial play. Standard bearing play should be 0.0256–0.0374 in. (0.65–0.95mm).
27. If play is not within specifications, shims are available for correcting it.
28. Install the other shaft and torque the backing plate bolts. Check the play as on the first shaft. Play should be 0.0019–0.0098 in. (0.05–0.25mm). If not, correct it with shims.
29. Install the brake drums and wheels.
30. Bleed the brake system.

FIRING ORDERS

NOTE: To avoid confusion, always replace spark plug wires one at a time.

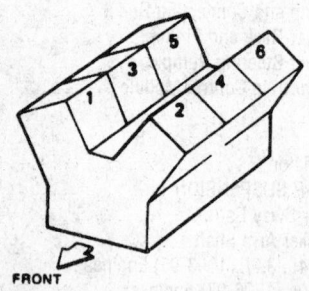

3.5L (VIN M) Engine
Engine Firing Order: 1–2–3–4–5–6
Distributorless Ignition System

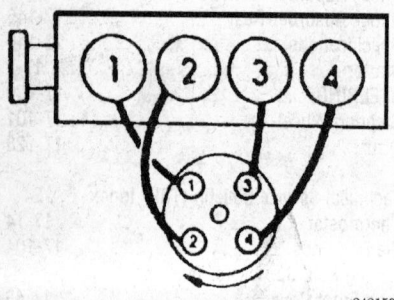

Mighty Max and Montero 2.4L Engines
Engine Firing Order: 1–3–4–2
Distributor Rotation: Clockwise

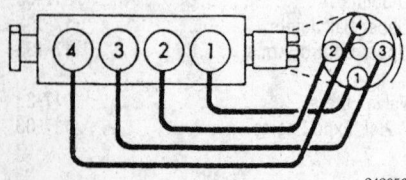

Expo 1.8L and 2.4L Engines
Engine Firing Order: 1–3–4–2
Distributor Rotation: Counterclockwise

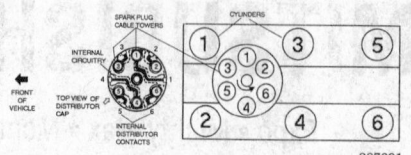

287091

3.0L Engine
Engine Firing Order: 1–2–3–4–5–6
Distributor Rotation: Counterclockwise

ENGINE ELECTRICAL

NOTE: Disconnecting the negative battery cable on some vehicles may interfere with the functions of the on board computer systems and may require the computer to undergo a relearning process, once the negative battery cable is reconnected.

Distributor

REMOVAL AND INSTALLATION

2.4L (VIN G) and 3.0L (VIN H) Engines

Timing Not Disturbed

1. Disconnect the negative battery cable.
2. Disconnect the distributor pickup lead wires and vacuum hose(s), if equipped.
3. Unfasten the distributor cap retaining clips or screws and lift off the distributor cap with all ignition wires connected. Remove the coil wire if necessary. Matchmark the rotor to the distributor housing and the distributor housing to the engine.

NOTE: Do not crank the engine during this procedure. If the engine is cranked, the matchmark must be disregarded.

5. Remove the retaining nut and remove the distributor from the engine.
To install:
6. Install a new distributor housing O-ring.

7. Install the distributor in the engine so the rotor is aligned with the matchmark on the housing and the housing is aligned with the matchmark on the engine. Make sure the distributor is fully seated and the distributor shaft is fully engaged.
8. Install the retaining nut finger-tight only. Connect the vacuum hose(s), if removed.
9. Connect the distributor pickup electrical harness.
10. Install the distributor cap and secure.
11. Connect the negative battery cable.
12. Check the ignition timing and adjust as required. Tighten the retaining nut.

Timing Disturbed

1. Disconnect the negative battery cable.
2. Disconnect the distributor pickup lead wires and vacuum hose(s), if equipped.
3. Unfasten the distributor cap retaining clips or screws and lift off the distributor cap with all ignition wires connected. Remove the coil wire if necessary.
4. Remove the retaining nut and remove the distributor from the engine.
To install:
5. Install a new distributor housing O-ring.
6. Rotate the engine so No. 1 piston is on TDC of compression stroke and the timing mark on the vibration damper is aligned with **T** on the timing indicator.
7. Install the distributor so the rotor is aligned with the No. 1 ignition wire on the distributor cap. Take note that the distributor shaft is fully engaged and the housing is fully seated.

NOTE: Some distributor caps may contain runners inside the cap. If so, make sure the rotor is pointing to where the No. 1 runner originates inside the cap and not where the No. 1 ignition wire plugs into the cap.

8. Install the retaining nut finger-tight only. Connect the vacuum hose(s), if removed.
9. Connect the distributor electrical harness.
10. Install the distributor cap and secure.
11. Connect the negative battery cable.
12. Adjust the ignition timing and tighten the retaining nut.

Ignition Timing

ADJUSTMENT

Montero

3.0L (VIN H) Engine

Before attempting to adjust the ignition timing, make sure of the following:

* The engine should be at normal operating temperature.
* The lights and all accessories should be OFF.
* If equipped with an automatic transmission, the transmission should be in **P** or**N**.

1. Insert a paper clip into the one–pin connector between the primary side of the ignition coil and the noise filter. The connector should not be disconnected.
2. Connect a primary voltage detection tachometer to the paper clip.

NOTE: Do not use the scan tool. When the scan tool is connected to the data link connector, the ignition timing will be unchanged, instead of reverting to the base ignition timing.

3. Connect the timing light and run the engine at idle speed.
4. Verify that the idle speed is 700±100 rpm.
5. Turn the ignition switch **OFF** and disconnect the brown waterproof female connector from the ignition timing adjustment connector.
6. Use a jumper wire to ground the ignition timing adjustment terminal.

NOTE: Grounding this terminal sets the engine to base ignition timing.

7. Start the engine and run it at idle. Check the base timing; it should be 5° BTDC±2°.

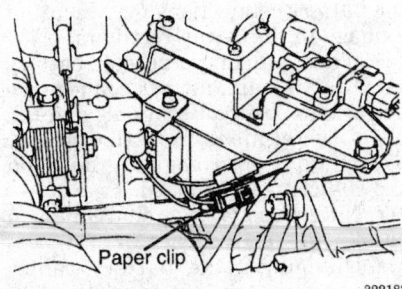

Ignition coil connector — Montero with 3.0L (VIN H) engines

8. If the base timing is out of specification, loosen the hold-down nut and turn the distributor. Turning the distributor clockwise will advance the timing and counter-clockwise will retard the ignition timing. After the base timing is set to specifications, tighten the distributor hold-down nut and recheck the base timing.
9. Turn the ignition switch **OFF** and disconnect the jumper wire from the ignition timing adjustment connector.
10. Start the engine and allow it to idle. The ignition timing at idle should be approximately 15°BTDC.

NOTE: Ignition timing under computer control is variable by ±7°, even under normal operating conditions. The ignition timing is automatically advanced by about 5° from 15°BTDC at higher altitudes.

11. Remove all the test equipment.

24 Valve 3.0L (VIN H) and 3.5L (VIN M) Engines

Before attempting to adjust the ignition timing, make sure of the following:

* The engine should be at normal operating temperature.
* The lights and all accessories should be OFF.
* If equipped with an automatic transmission, the transmission should be in **P** or **N**.

1. Insert a paper clip into the engine speed detection connector (blue), and then connect a tachometer to the paper clip.

NOTE: Do not use the scan tool. When the scan tool is connected to the data link connector, the ignition timing will be unchanged, instead of reverting to the base ignition timing.

2. Connect the timing light and run the engine at idle speed.

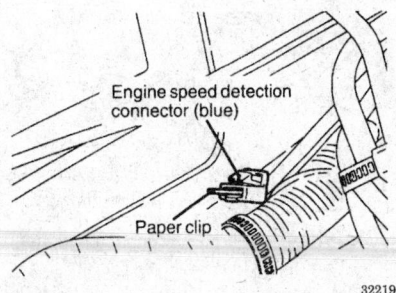

Engine speed detection connector — 24 Valve 3.0L (VIN H) and 3.5L (VIN M) engine

3. Idle speed should be 700±100 rpm.

NOTE: The reading on the tachometer indicates $\frac{1}{3}$ of the actual engine speed. The actual engine speed is really 3 times the tachometer reading.

4. Turn the ignition switch **OFF** and disconnect the waterproof female connector from the ignition timing adjustment connector (brown).
5. Use a jumper wire to ground the ignition timing adjustment terminal.

NOTE: Grounding this terminal sets the engine to the base ignition timing.

6. Start the engine and run it at idle speed. Check the base timing, it should be 5°BTDC ± 3°.

NOTE: The ignition timing is controlled by the ECM and is not adjustable. The ECM determines the timing based on input from the crankshaft position sensor.

7. Turn the ignition switch **OFF** and disconnect the jumper wire from the ignition timing adjustment connector.
8. Start the engine and allow it to idle. The ignition timing at idle should be approximately 15°BTDC.

NOTE: Ignition timing under computer control is variable by ±7°, even under normal operating conditions. The ignition timing is automatically advanced by about 5° from 15°BTDC at higher altitudes.

9. Turn the ignition switch **OFF** and remove all the test equipment.

Mighty Max

1. Firmly apply the parking brake and block the drive wheels.
2. Start the engine and allow to operate until normal operating temperature is reached.
3. Turn all lights and accessories **OFF**.
4. Place the transmission lever in **N**.
5. Without disconnecting the connector, insert a paper clip into the tachometer terminal.
  a. Connect the red lead of a tachometer to the paper clip and connect the black lead to a ground.
  b. Set the idle speed to specifications.
6. Turn the engine **OFF**.
7. Remove the water-proof cover from the ignition timing adjusting connector. Connect a jumper wire

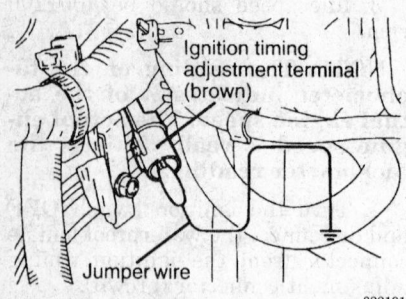

Ignition timing adjustment terminal — 24 Valve 3.0L (VIN H) and 3.5L (VIN M) engine

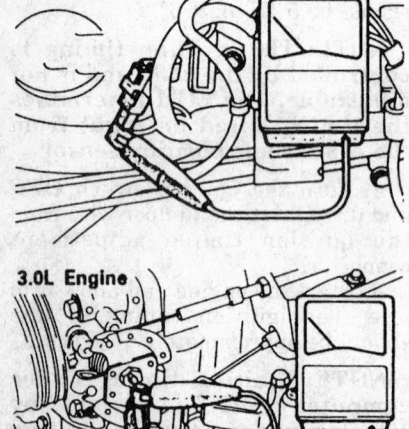

Tachometer connector location — Mighty Max with the 2.4L (VIN G) and 3.0L (VIN H) engines

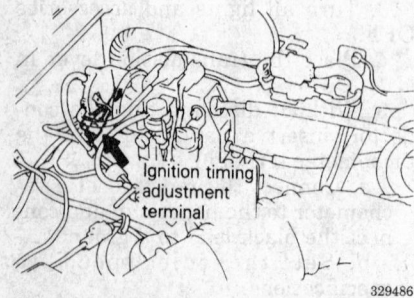

Ignition timing adjustment terminal — Mighty Max with the 2.4L (VIN G) and 3.0L (VIN H) engines

from the ignition timing adjusting terminal to ground.

8. Connect a conventional timing light to No. 1 cylinder spark plug wire. Start the engine and allow to idle.

9. Aim the timing light at the timing scale.

10. Basic ignition timing should be 5°BTDC±2°.

11. Loosen the distributor nut to allow for distributor rotation.

12. Turn the distributor in the proper direction until the specified timing is reached. Tighten the retainer nut and recheck the timing.

13. Turn the engine **OFF**.

14. Remove the jumper wire from the ignition timing adjusting terminal and install the water-proof cover.

15. Start the engine and check the actual ignition timing. This reading should be 8°BTDC for the 2.4L engine or 15° BTDC for the 3.0L engine.

NOTE: The actual timing may fluctuate according to the control mode of the engine control unit; this is a normal condition.

16. Turn the engine **OFF** and remove all test equipment.

Expo 1.8L and 2.4L (VIN-G) Engine

1. Set the parking brake, start and run the engine until normal operating temperature is obtained. Keep all lights and accessories OFF and the front wheels straight-ahead. Place the transaxle in **P** or automatic transaxle or neutral for manual transaxle.

NOTE: On Canadian vehicles the lights will remain on when the vehicle is running, this will not be a problem.

2. Locate the wire connector on the ignition coil connector. Insert a paper clip behind the TACH terminal connector to act as a tachometer adapter. Connect a tachometer to the

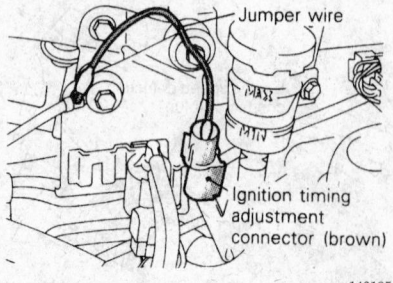

Ignition timing connector location — Expo 1.8L and 2.4L (VIN-G) engine

paper clip. If not at specification, set the idle speed at the correct level.

3. Turn the engine **OFF** and remove the water-proof cover from the ignition timing adjusting connector. This connector is a brown connector located near the center of the firewall. Connect a jumper wire from this terminal to a good ground.

4. Connect a conventional power timing light to the No. 1 cylinder spark plug wire. Start the engine and run at idle.

5. Aim the timing light at the timing scale located near the crankshaft pulley.

6. Loosen the distributor hold-down nut just enough so the housing can be rotated.

7. Turn the housing in the proper direction until the specified timing is reached. Tighten the hold-down nut and recheck the timing. Turn the engine **OFF**.

8. Remove the jumper wire from the ignition timing adjusting terminal and install the water-proof cover.

9. Start the engine and check the actual timing without the terminal grounded. This reading should be approximately 5 degrees more than the basic timing. Actual timing may increase according to altitude. Also, actual timing may fluctuate because of slight variation accomplished by the ECU. As long as the basic timing is correct, the engine is timed correctly.

10. Turn the engine **OFF**. Disconnect the timing equipment and tachometer.

Alternator

PRECAUTIONS

Several precautions must be observed with alternator equipped vehicles to avoid damage to the unit.

• If the battery is removed for any reason, make sure it is reconnected with the correct polarity. Reversing the battery connections may result in damage to the 1-way rectifiers.

• When utilizing a booster battery as a starting aid, always connect the positive to positive terminals and the negative terminal from the booster battery to a good engine ground on the vehicle being started.

• Never use a fast charger as a booster to start vehicles.

• Disconnect the battery cables when charging the battery with a fast charger.

• Never attempt to polarize the alternator.

- Do not use test lights of more than 12 volts when checking diode continuity.
- Do not short across or ground any of the alternator terminals.
- The polarity of the battery, alternator and regulator must be matched and considered before making any electrical connections within the system.
- Never separate the alternator on an open circuit. Make sure all connections within the circuit are clean and tight.
- Disconnect the battery ground terminal when performing any service on electrical components.
- Disconnect the battery if arc welding is to be done on the vehicle.

REMOVAL AND INSTALLATION

Mighty Max and 1993–95 Montero

1. Disconnect the negative battery cable.

--- CAUTION ---
Wait at least 90 seconds after the negative (-) battery cable is disconnected to prevent possible deployment of the air bag.

2. If equipped, remove the alternator cover.
3. Remove the alternator drive belt.
4. Remove the alternator brace bolt(s), nut(s) and applicable spacers.
5. Remove the alternator from the mounting bracket, label and disconnect all wires from the rear of the unit.
To install:
6. Connect all wiring to their proper terminals on the rear of the alternator.
7. Position the alternator in the mounting bracket.
8. Install the alternator brace bolt(s), nut(s) and applicable spacers.
9. Install the alternator belt and adjust the tension as required.
10. If removed, install the alternator cover.
11. Connect the negative battery cable and check the alternator for proper operation.

1996–97 Montero

1. Disconnect the negative battery cable.
2. Remove the alternator cover.
3. Remove the alternator drive belt.
4. Remove the alternator brace bolts, nuts and spacers.
5. Remove the alternator from the mounting bracket, label and discon-

nect all wires from the rear of the unit.
To install:
6. Connect all wiring to their proper terminals on the rear of the alternator.
7. Position the alternator in the mounting bracket.
8. Install the alternator brace bolts, nuts and spacers.
9. Torque the top bolt to:
- 3.0L 12 valve: 14–18 ft. lbs. (20–25 Nm)
- 3.0L 24 valve and 3.5L: 38 ft. lbs. (52 Nm)
10. Torque the bottom bolt to:
- 3.0L 12 valve: 9 ft. lbs. (13 Nm)
- 3.0L 24 valve and 3.5L: 16 ft. lbs. (22 Nm)
11. Install the alternator belt and adjust the tension as required.
12. Install the alternator cover.
13. Connect the negative battery cable and check the alternator for proper operation.

Expo 1.8L (4G93) Engine

1. Disconnect negative battery cable.
2. Remove the accessory drive belts.
3. Disconnect the electrical harness from the alternator.
4. Remove the alternator mounting nut, bolt and upper brace assembly from the vehicle.
To install:
5. Install the alternator and secure using mounting nuts. Make sure the upper brace assembly is in place.
6. Install and adjust drive belts to the proper tension. Secure all mounting hardware.
7. Reconnect the negative battery cable and check system operation.

Expo 2.4L (VIN 4G64) Engine

1. Disconnect the negative battery cable.
2. Remove the left side cover panel under the vehicle.
3. Remove the drive belts.
4. Remove both water pump pulleys.
5. Remove the alternator upper bracket/brace.
6. Disconnect the alternator electrical connectors and remove alternator.
To install:
7. Position the alternator on the lower mounting fixture and install the lower mounting bolt and nut. Tighten nut just enough to allow for movement of the alternator.
8. Install the alternator upper bracket/brace and connect the alternator electrical harness.

9. Install the water pump pulleys.
10. Install the drive belts and adjust to the proper tension.
11. Install the left side cover panel under the vehicle as required.
12. Connect the negative battery cable and check for proper operation.

Drive Belt

REMOVAL AND INSTALLATION

All Vehicles Except Expo

1. Disconnect the negative battery cable.
2. Loosen the lock, pivot and adjusting bolts of the component. Loosen the idler pulley, if equipped.
3. Remove the drive belt(s) from its component.
4. Visually check the belt for separation of the ribs, torn or worn ribs or cracks in the inner ridges of the ribs. If necessary, replace the drive belt.
To install:
5. Position the drive belt(s) over the component pulleys and drive pulley.
6. Partially tighten the pivot and adjusting bolts.
7. Check the drive belt(s) tension using a belt tension gauge. Fully tighten the lock, pivot and adjusting bolts.

NOTE: After installing the drive belt(s), check that the belt does not touch the bottom of the pulley groove (conventional type belts) or that it fits properly in the ribbed grooves (V-ribbed type belts).

8. Connect the negative battery cable.
9. If a new belt was installed, run the engine for approximately 5 minutes and then recheck the tension.

Expo 1.8L (4G93) and 2.4L (4G64) Engines

Excessive belt tension will cause damage to the alternator and water pump pulley bearings, while loose belt tension will produce slip and premature wear on the belt. Be sure to adjust the belt tension to the proper level.

To adjust the tension on a drive belt, loosen the adjusting bolt or fixing bolt locknut on the desired component, bracket or tension pulley. Then move the component or turn the adjusting bolt to adjust belt tension. Once the desired value is reached, secure the bolt or locknut and recheck tension.

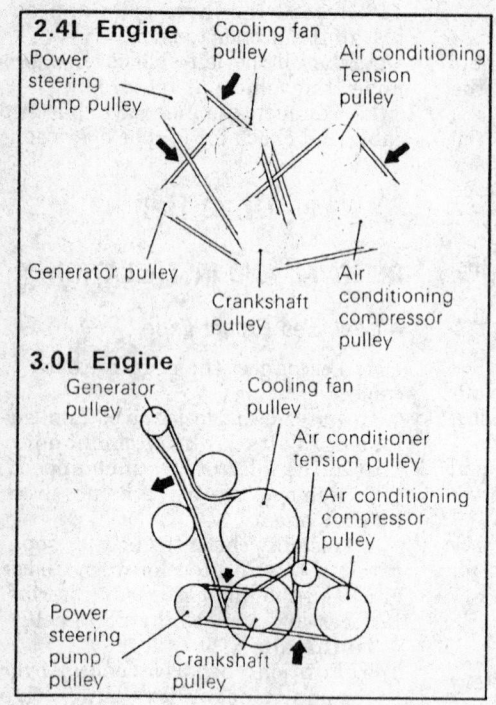

2.4L Engine
Power steering pump pulley
Cooling fan pulley
Air conditioning Tension pulley
Generator pulley
Crankshaft pulley
Air conditioning compressor pulley

3.0L Engine
Generator pulley
Cooling fan pulley
Air conditioner tension pulley
Air conditioning compressor pulley
Power steering pump pulley
Crankshaft pulley

Standard value:
<2.4L Engine>
For water pump and Generator
    7.0 – 10.0 mm (.28 – .39 in.)
For Power steering pump
    6.0 – 9.0 mm (.23 – .35 in.)
For Air conditioning compressor
    8.5 – 10.0 mm (.33 – .39 in.)
<3.0L Engine>
For Generator
    8.0 – 10.0 mm (.32 – .39 in.)
For Power steering pump
    9.0 – 12.0 mm (.35 – .47 in.)
For Air conditioning compressor
    8.5 – 10.0 mm (.33 – .39 in.)

318154

Drive belts — 2.4L (VIN G) and 3.0L (VIN H) engines

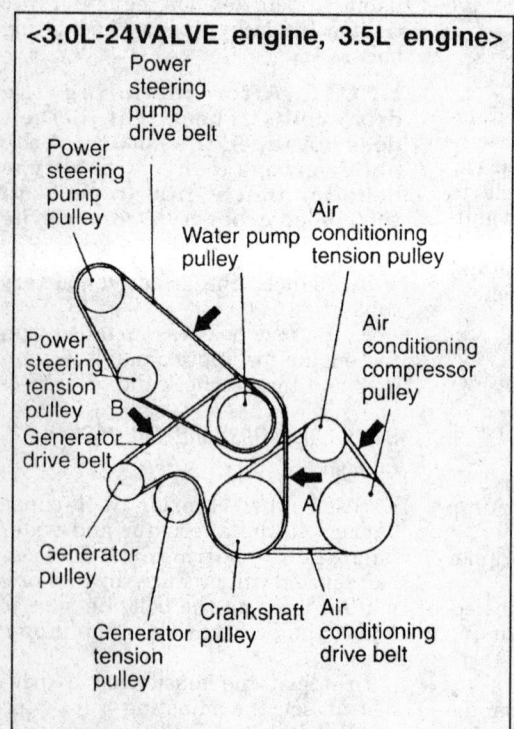

<3.0L-24VALVE engine, 3.5L engine>
Power steering pump drive belt
Power steering pump pulley
Power steering pump pulley
Water pump pulley
Air conditioning tension pulley
Air conditioning compressor pulley
Power steering tension pulley
B
Generator drive belt
Generator pulley
Generator tension pulley
Crankshaft pulley
Air conditioning drive belt
A

<3.0L-24VALVE engine, 3.5L engine>
Standard value:

Item		Check value mm (in.)	
		3.0L-24VALVE engine	3.5L engine
For generator	A	5.0–7.0 (.20–.28)	5.0–7.0 (.20–.28)
	B	8.5–10.5 (.33–.41)	8.5–10.5 (.33–.41)
For power steering		10.5–14.5 (.41–.57)	13.0–17.0 (.51–.67)
For air conditioning		6.5–7.5 (.26–.30)	6.5–7.5 (.26–.30)

A: Measure between the water pump pulley and the crankshaft pulley.
B: Measure between the water pump pulley and the generator.

318155

Drive belts — 24 Valve 3.0L (VIN H) and 3.5L (VIN M) engines

Belt replacement is similar to adjustment, with the exception the belt will have to be properly routed around the pulleys. It is important to note the routing of the belt before removal. For individual belt replacement, start with the outer most belt. If a removed belt is to be reused, be certain to mark the direction of rotation on the belt, to extend belt life.

Starter

REMOVAL AND INSTALLATION

3.0L (VIN H) and 3.5L (VIN M) Engines

1. Disconnect the negative battery cable.
2. Raise the vehicle and support it safely.
3. Remove the starter cover.
4. Label and disconnect the wiring to the starter motor.
5. Remove the starter mounting bolts.
6. Remove the starter motor from the vehicle.

To install:

7. Mount the starter to the engine block. Torque the mounting bolts to 20–25 ft. lbs. (27–34 Nm).
8. Connect the wiring to the starter.
9. Connect the negative battery cable and check the starter for proper operation.

2.4L (VIN G) Engine

1. Disconnect the negative battery cable.
2. Raise the vehicle and support safely.
3. Label and disconnect the wiring to the starter motor.
4. Remove the starter mounting bolts.
5. Remove the starter motor from the vehicle.

To install:

6. Mount the starter to the engine block. Torque the mounting bolts to 20–24 ft. lbs. (27–33 Nm).
7. Connect the wiring to the starter.
8. Connect the negative battery cable and check the starter for proper operation.

Expo 1.8L (4G93) and 2.4L (4G64) Engines

1. Disconnect the negative battery cable.
2. Disconnect and remove the air cleaner assembly as required.

3. Disconnect the starter motor electrical connections.
4. Remove the starter motor mounting bolts and remove the starter.

To install:

5. Install the starter.
6. Tighten starter mounting bolts to 20–25 ft. lbs. (27–34 Nm.)
7. Connect the starter motor electrical connections.
8. Connect the negative battery cable and check the starter for proper operation.

CHASSIS ELECTRICAL

Blower Motor

REMOVAL AND INSTALLATION

Except Expo

1. Disconnect the negative battery cable.
2. Remove the glove box stopper and the glove box from the dash panel.
3. Disconnect the wiring connector to the blower motor.
4. Remove the blower motor mounting screws.
5. Remove the blower motor and fan as an assembly from the heater unit.

To install:

6. Check the condition of the blower motor gasket and replace it if necessary.
7. Install the blower motor assembly to the heater unit.
8. Connect the wiring to the blower motor.
9. Install the glove box to the dash panel.
10. Connect the negative battery cable and check the operation of the blower motor.

Expo

1. Disconnect the negative battery cable.
2. Remove the glove box assembly.
3. Pry off the speaker cover, located on the lower right of the glove box.
4. Remove the passenger side knee protector, which is the panel surrounding the glove box opening.
5. Remove the glove box frame along the top of the glove box.

6. Remove the lap heater duct.
7. Disconnect the electrical connector form the blower motor.
8. Remove the cooling tube form the blower assembly.
9. Remove the blower motor assembly.
10. Remove the fan retaining nut and fan cage, in order to renew the motor.

To install:

11. Check that the blower motor shaft is not bent and that the packing (sealing material) and blower case are in good condition.
12. Assemble the fan and motor.
13. Install the blower assembly. Connect the wiring and cooling tube.
14. Install the lap heater duct.
15. Install the glove box frame, interior trim pieces and glove box assembly.
16. Connect the negative battery cable and check the entire climate control system for proper operation.
17. Connect the negative battery cable and check the blower for proper operation.

Windshield Wiper Motor

REMOVAL AND INSTALLATION

Montero

Front Wiper Motor

1. Disconnect the negative battery cable.

— CAUTION —

Wait at least 90 seconds after the negative (-) battery cable is disconnected to prevent possible deployment of the air bag.

2. Remove the motor retaining bolts.
3. Remove the wiper link from the motor by prying the link from the crank arm pin with a suitable prying tool.
4. Disconnect the electrical connector and remove the motor.

To install:

5. Install the motor and connect the electrical connector.
6. Connect the wiper link.
7. Install the motor retaining bolts.
8. Connect the negative battery cable and check the wiper motor for proper operation.

Rear Wiper Motor

1. Disconnect the negative battery cable.

— CAUTION —
Wait at least 90 seconds after the negative (-) battery cable is disconnected to prevent possible deployment of the air bag.

2. Remove the rear interior hatch panel cover.
3. Remove the wiper arm and the blade assembly.
4. Remove the back door trim and remove the wiper motor and bracket.
To install:
5. Install the wiper motor and bracket assembly.
6. Install the back door trim.
7. Install the wiper arm and the blade assembly.
8. Install the rear panel cover.
9. Connect the negative battery cable.
10. Check the wiper motor for proper operation.

Mighty Max

1. Disconnect the negative battery cable.
2. Remove the motor retaining bolts.
3. Remove the wiper link from the motor by prying the link from the crank arm pin with a suitable prying tool.
4. Disconnect the electrical connector and remove the motor.
To install:
5. Install the motor and connect the electrical connector.
6. Connect the wiper link.
7. Install the motor retaining bolts.
8. Connect the negative battery cable and check the wiper motor for proper operation.

Expo

Front

1. Disconnect the negative battery cable.
2. Remove the windshield wiper arms by unscrewing the cap nuts and lifting the arms from the linkage posts.
3. Remove the front deck garnish panel.
4. Loosen the wiper motor assembly mounting bolts, then using a flat tipped screwdriver to pry the linkage from the motor.
5. Disconnect the motor wiring and remove the motor assembly.

NOTE: The installation angle of the crank arm and motor has been factory set, do not remove them unless it is necessary to do

so. If arm must be removed, remove them only after marking their mounting positions.
To install:
6. Install the windshield wiper motor mounting bolts and connect the linkage.
7. Connect the electrical harness to the motor.
8. Install the front deck garnish panel.
9. Install the wiper arms and tighten nuts to 17 ft. lbs. (24 Nm).
10. Connect the negative battery cable and check the wiper system for proper operation.

REAR

1. Disconnect the negative battery cable.
2. Remove the rear wiper arm by removing the cap nut cover, unscrewing the cap nut and lifting the arm from the linkage post.
3. Remove the large interior trim panel. Use a plastic trim stick to unhook the trim clips of the liftgate trim. There will be a row of metal liftgate clips across the top. There will be two rows of trim clips that retain the rest of the panel.
4. Disconnect the electrical harness at the wiper motor.
5. Remove the rear wiper assembly mounting bolt.

NOTE: Do not loosen the grommet for the wiper post.

To install:
6. Install the motor and torque mounting bolt to 7 ft. lbs. (9 Nm). If the grommet was removed, ensure that it is installed with the arrow pointing downward.
7. Install the wiper arm and tighten mounting nut to 6 ft. lbs. (8 Nm).
8. Connect the negative battery cable and check rear wiper system for proper operation.
9. If operation is satisfactory, fit the tabs on the upper part of the liftgate trim into the liftgate clips and secure the liftgate trim.

Combination Switch

REMOVAL AND INSTALLATION

Montero

1993 Vehicles

— CAUTION —
Wait at least 60 seconds after disconnecting the battery before doing any work on the SRS system.

1. Disconnect negative battery cable.

— CAUTION —
The air bag system (SRS) must be disarmed before removing the steering wheel. Failure to do so may cause accidental deployment, property damage or personal injury.

2. Remove the air bag or the horn pad.
3. Remove the steering wheel nut. Install a steering wheel puller and remove the steering wheel.

— CAUTION —
Always carry an air bag assembly with the bag and trim cover away from your body. Store the assembly with the bag and trim cover facing up; never place the assembly face down on the floor or workbench.

4. If equipped with tilt steering column, put the column in lowest position.
5. Remove the upper and lower column covers.
6. Remove the wiring harness band and disconnect the harness connectors.
7. Remove the combination switch.
To install:
8. Install the switch.
9. Connect the electrical connector and install the harness band.
10. Install the upper and lower covers.
11. Install the steering wheel. Tighten the steering wheel nut to 29 ft. lbs. (39 Nm).
12. Install the air bag or the horn pad.
13. Connect the negative battery cable.

1994–97 Vehicles

1. Disconnect negative battery cable.
2. Remove the upper and lower column covers.
3. Remove the screws holding the combination switch.
4. Remove the combination switch.
To install:
5. Install the switch and tighten the screws.
6. Install the upper and lower covers.
7. Connect the negative battery cable.

Mighty Max

1. Disconnect negative battery cable.

2. Remove the steering wheel pad. Matchmark and remove the steering wheel.

3. If equipped with tilt steering column, put the column in lowest position.

4. Remove the upper and lower column covers.

5. Remove the wiring harness band and disconnect the harness connectors.

6. Remove the combination switch mounting screws and remove the switch.

To install:

7. Install the switch and tighten the retaining screws.

8. Connect the electrical connector and install the harness band.

9. Install the upper and lower covers.

10. Align the marks made during removal and install the steering wheel. Tighten the steering wheel nut to 30 ft. lbs. (41 Nm).

11. Connect the negative battery cable.

1993 Expo

1. Disconnect the negative battery cable.

2. Remove the steering wheel.

3. Remove the four meter (instrument cluster) hood mounting screws and lift the hood off the three retaining clips.

4. Remove the two screws securing the hood release to the lower instrument panel. Remove the lower instrument panel under the steering column, then the upper and lower column covers.

5. Disconnect all connectors and remove the wiring clip. Remove the retaining screw and remove the column switch assembly.

To install:

6. Install the switch assembly and secure all harness connectors with clips if needed. Make sure the wires are not pinched or out of place.

7. Install the steering wheel and torque the retaining nut to 29 ft. lbs. (40 Nm).

8. Install the column covers and lower instrument panel.

9. Connect the hood release to the lower instrument panel.

10. Install the meter hood assembly.

11. Connect the battery cable and check all functions of the combination switch for proper operation.

1994–95 Expo

> **CAUTION**
> *On vehicles equipped with an air bag, be sure to disarm the system before attempting repairs on the vehicle. Failure to do so could result in severe personal injury and damage to vehicle.*

1. To disarm air bag, perform the following steps:

a. Position the front wheels in the straight-ahead position and place the key in the **LOCK** position. Remove the key from the ignition lock cylinder.

b. Disconnect the negative battery cable and insulate the cable end with high-quality electrical tape or similar non-conductive wrapping.

c. Wait at least one minute before working on the vehicle. The air bag system is designed to retain enough voltage to deploy the air bag for a short period of time even after the battery has been disconnected.

2. To gain access to clockspring wiring, perform the following steps:

a. Remove the four meter hood mounting screws and lift the hood off the three retaining clips.

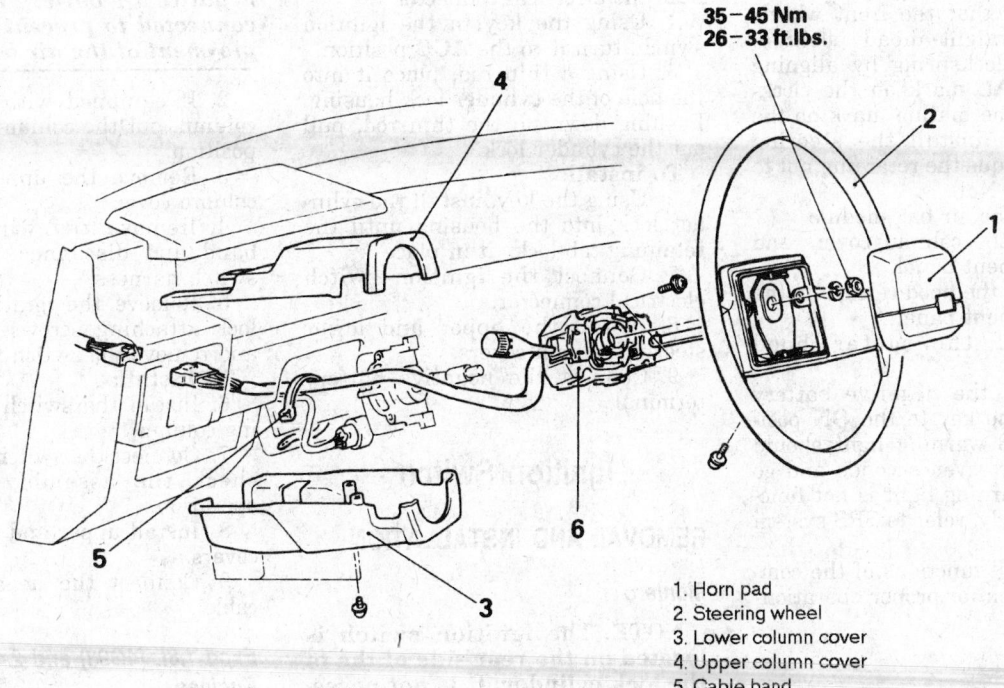

35–45 Nm
26–33 ft.lbs.

1. Horn pad
2. Steering wheel
3. Lower column cover
4. Upper column cover
5. Cable band
6. Column switch

329037

Combination switch mounting — Mighty Max

b. Remove the two screws securing the hood release to the lower instrument panel. Remove the lower instrument panel under the steering column, then the upper and lower column covers.

3. Remove the steering wheel as follows:

a. Remove the air bag module mounting nut from behind the steering wheel.

b. To disconnect the connector of the clockspring from the air bag module, press the air bag's lock toward the module to spread the lock open. While holding lock in this position, use a small tipped prying tool to gently pry the connector from the module.

c. Store the air bag module in a clean, dry place with the pad cover facing up.

d. Remove the steering wheel retaining nut and use a steering wheel puller to remove the wheel. Do not use a hammer or the collapsible mechanism in the column could be damaged.

4. Disconnect all connectors and remove the wiring clip. Remove the three retaining screws and remove the column switch assembly.

To install:

5. Install the switch assembly and secure all harness connectors with clips if needed. Make sure the wires are not pinched or out of place.

6. Confirm that the front wheels are in a straight-ahead position. Center the clockspring by aligning the **NEUTRAL** mark on the clockspring with the mating mark on the casing. Then install the steering wheel and torque the retaining nut to 29 ft. lbs. (40 Nm).

7. Install the air bag module.

8. Install the column covers and lower instrument panel.

9. Connect the hood release to the lower instrument panel.

10. Install the meter hood assembly.

11. Connect the negative battery cable, turn the key to the **ON** position, the SRS warning light should illuminate for seven seconds and go out. If the warning light is not functioning properly, refer to SRS system diagnosis.

12. Check all functions of the combination switch for proper operation.

Ignition Lock Cylinder

REMOVAL AND INSTALLATION

Except Expo

1. Disconnect the negative battery terminal.

2. Remove the upper and lower steering column covers.

3. Disconnect the ignition switch from the electrical connector.

4. Using the key in the ignition switch, turn it to the **ACC** position.

5. Using a thin rod, place it into the hole of the cylinder lock housing. Pushing down on the thin rod, pull out the cylinder lock.

To install:

6. Using the key, install the cylinder lock into the housing until the retaining tab locks it in place.

7. Connect the ignition switch electrical connector.

8. Install the upper and lower steering column covers.

9. Connect the negative battery terminal.

Expo

1. Disconnect the negative battery terminal.

2. Remove the upper and lower steering column covers.

3. Disconnect the ignition switch from the electrical connector.

4. Using the key in the ignition switch, turn it to the **ACC** position.

5. Using a thin rod, place it into the hole of the cylinder lock housing. Pushing down on the thin rod, pull out the cylinder lock.

To install:

6. Using the key, install the cylinder lock into the housing until the retaining tab locks it in place.

7. Connect the ignition switch electrical connector.

8. Install the upper and lower steering column covers.

9. Connect the negative battery terminal.

Ignition Switch

REMOVAL AND INSTALLATION

Montero

NOTE: The ignition switch is located on the rear side of the of the lock cylinder. It is not necessary to remove the lock cylinder to replace the switch.

1. Disconnect the negative battery cable.

2. If equipped with tilt steering column, position the column in alignment with the the column.

3. Remove the upper and lower column covers.

4. Remove the wiring harness band and disconnect the ignition switch harness.

5. Remove the ignition switch to lock cylinder attaching screws and remove the switch from the rear of the lock cylinder.

6. Remove the key reminder switch assembly.

To install:

7. Install the key reminder switch assembly.

8. Install the ignition switch to the rear of the lock cylinder and secure with attaching screws.

9. Connect the wiring harness connections.

10. Install a wiring harness band around the harness.

11. Install upper and lower column covers.

12. Connect the negative battery cable and check the switch for proper operation.

Mighty Max

1. Disconnect the negative battery cable.

--- **CAUTION** ---
Wait at least 90 seconds after the negative (-) battery cable is disconnected to prevent possible deployment of the air bag.
---

2. If equipped with tilt steering column, put the column in its lowest position.

3. Remove the upper and lower column covers.

4. Remove the wiring harness band and disconnect the ignition switch harness.

5. Remove the ignition switch to lock attaching screws, if equipped, and remove the switch from the lock.

To install:

6. Install the switch to the steering column.

7. Connect the switch harness and check the assembly for proper operation.

8. Install upper and lower column covers.

9. Connect the negative battery cable.

Expo 1.8L (4G93) and 2.4L (4G64) Engines

1. Disconnect the negative battery cable.

2. Remove the hood lock release lever from the lower panel.

3. Remove the lower instrument panel knee protector.

NOTE: Use proper steering wheel puller equipment when removing the steering wheel. The use of a hammer for removal could damage the collapsible mechanism within the column.

4. Remove the instrument panel hood.

5. Remove the lower steering column cover.

6. Remove the upper steering column cover.

7. Remove the clip that holds the wiring harness against the steering column.

8. Insert the key into the steering lock cylinder and turn to the **ACC** position. With a small pointed tool, push the lock pin of the steering lock cylinder inward and pull the lock cylinder out.

9. Remove the key reminder switch, if equipped.

10. Unplug the ignition switch harness connector. Remove the ignition switch mounting screws and pull the switch from the steering lock cylinder.

To install:

11. Install the ignition switch into the rear of the lock cylinder housing. Be sure to align the keyway of the ignition switch with interlock cylinder.

12. Connect harness connections and install the wiring clip.

13. Install the steering column upper and lower covers.

14. Install the instrument panel hood.

15. Install the knee protector.

16. Connect the negative battery cable and check the ignition switch and lock for proper operation.

Park/Neutral Safety Switch

REMOVAL AND INSTALLATION

Expo and Montero

1. Disconnect the negative battery cable.

2. Disconnect the control cable from the adjusting lever.

3. Disconnect the harness connector from the park/neutral switch.

4. Remove the nut that secures the adjusting lever and remove the lever.

5. Remove the bolts that secure the switch and remove the switch.

To install:

6. Install the switch and secure with mounting bolts. Do not tighten the bolts at this time.

7. Install adjusting lever and secure with mounting nut. Do not tighten the nut at this time.

8. Set the shift selector to the **N** position.

9. Align the switch and the adjustment lever **N** positions.

10. Hold in position and tighten the bolt to 48 inch lbs. (6 Nm).

11. Gently pull the control cable and tighten the control cable adjusting nut to 17 ft. lbs. (24 Nm).

12. Connect the negative battery cable and check that the engine will only start in the park or neutral position.

13. Check that the reverse lights operate only in the reverse position.

Mighty Max

1993–95 Vehicles

1. Disconnect the negative battery cable.

2. Raise and safely support the vehicle. Position a drain pan under the switch.

3. Disconnect the switch electrical connector.

4. Remove the switch from the case.

To install:

5. Verify that the switch operating lever fingers are centered in the switch opening in the case when in park and neutral.

6. Install a new seal and screw the switch on the case. Tighten to 24 ft. lbs. (33 Nm).

7. Check the continuity of the switch for proper operation. Reconnect the electrical connector.

8. Lower the vehicle and check the transmission fluid level.

9. Reconnect the negative battery cable and check the switch for proper operation.

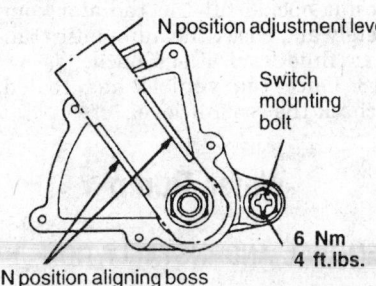

Park/Neutral switch adjusting lever and switch alignment — Montero and Expo

Powertrain Control Module

REMOVAL AND INSTALLATION

Except Expo

NOTE: The engine control module is located under the right hand corner of the dash panel.

1. Disconnect the negative battery cable.

2. Remove the trim cover(s), as necessary, to gain access to the control module.

3. Disconnect the electrical connector.

4. Remove the module mounting bolts.

5. Remove the module from the mounting bracket.

To install:

6. Install the module onto the bracket.

7. Install the module mounting bolts.

8. Connect the electrical connector.

9. Install the trim cover(s).

10. Connect the negative battery cable.

Expo 1.8L (4G93) and 2.0L (4G64) Engines

The engine control module is located behind glove box.

1. Disconnect negative battery cable.

2. Locate and access the module.

3. Unplug wiring connector and remove mounting hardware. Slide out control unit.

4. Installation is the reverse of removal.

ENGINE COOLING

Radiator

REMOVAL AND INSTALLATION

Except Expo

1. Disconnect the negative battery cable.

─── **CAUTION** ───

Wait at least 90 seconds after the negative (-) battery cable is disconnected to prevent possible deployment of the air bag.

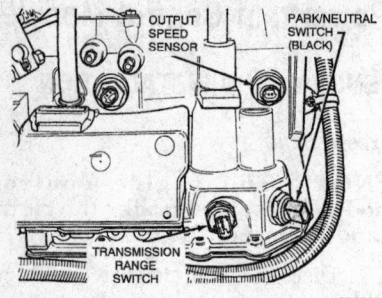

Park/Neutral switch removal and installation — 1993-95 Mighty Max with 4 speed automatic electronic transaxle

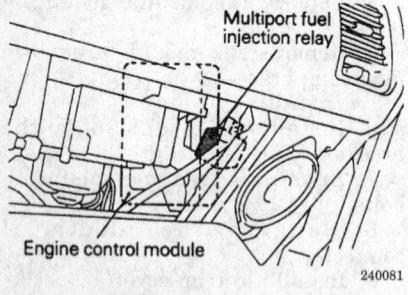

Engine control module location — Expo 1.8L (4G93) and 2.0L (4G64) engines

2. Open the radiator petcock and drain the coolant. After draining, close the petcock.

3. If necessary, remove the air cleaner case.

4. Disconnect the radiator hoses and coolant reserve tank hose from the radiator.

5. Remove the shroud from the radiator.

6. If equipped, remove the automatic transmission oil cooler lines. Cap the lines to minimize contamination.

7. Remove the radiator mounting screws.

8. Carefully lift the radiator out of the engine compartment.

To install:

9. Lower the radiator into position and install the mounting screws.

10. Connect the transmission oil cooler lines, if removed.

11. Install the fan shroud assembly.

12. Connect the radiator hoses and coolant reserve tank hose to the radiator.

13. If removed, install the air cleaner case.

14. Fill the cooling system.

15. Connect the negative battery cable.

16. Start the engine and let run until it reaches normal operating temperature, then check the coolant level and the transmission fluid level, if equipped.

Expo 1.8L (4G93) and 2.4L (4G64) Engines

1. Disconnect the negative battery cable.

— **CAUTION** —

Allow the cooling system to completely cool before attempting any repair or draining the system. Injury from scalding could result if radiator cap or hose connections are removed while system is hot.

2. Drain the cooling system.

3. Disconnect the overflow tube and remove the overflow tank.

4. Disconnect the upper and lower radiator hoses.

5. Disconnect the electrical connectors for the cooling fan and air conditioning condenser fan, if equipped. Remove the fan assembly.

6. Disconnect the thermo sensor wires.

7. Disconnect and plug the automatic transaxle cooler lines, if equipped with automatic transaxle.

8. Remove the upper radiator mounts and lift out the radiator assembly.

9. Service the lower mounts, as required.

To install:

10. Install the radiator and fan assembly, if removed as an assembly.

11. Connect the automatic transaxle cooler lines, if disconnected.

12. Connect the thermo wires.

13. Install the fan if removed separately.

14. Install the radiator hoses.

15. Install the overflow tube and reservoir, if removed.

16. Fill the system with coolant.

17. Connect the negative battery cable, run the vehicle until the thermostat opens, fill the radiator completely and check the automatic transaxle fluid level, if equipped.

18. Once the vehicle has cooled, recheck the coolant level.

Water Pump

REMOVAL AND INSTALLATION

Except Expo

1. Disconnect the negative battery cable.

— **CAUTION** —

Wait at least 90 seconds after the negative (-) battery cable is disconnected to prevent possible deployment of the air bag.

2. Drain the cooling system.

3. If equipped with fuel injection, properly release the fuel pressure.

— **CAUTION** —

Fuel injection systems remain under pressure after the engine has been turned OFF. Properly relieve fuel pressure before disconnecting any fuel lines. Failure to do so may result in fire or personal injury.

4. Remove the upper radiator shroud.

5. Remove all accessory belts. Remove the air conditioning compressor tensioner pulley, if equipped.

6. Remove the cooling fan and clutch assembly and remove the water pump pulley.

7. Disconnect the radiator hose from the water pump.

8. Remove the crankshaft pulley(s).

9. Remove the timing belt covers. If the same timing belt will be reused, mark the direction of the timing belt's rotation, for installation in the same direction. Make sure the engine is positioned so the No. 1 cylinder is at the TDC of its compression stroke and the sprockets timing marks are aligned with the engine's timing mark indicators. Remove the timing belt.

10. The water pump bolts are different lengths, note their positions before removing. Remove the water pump mounting bolts and remove the pump from the block and the water pipe connection. Remove the O-ring from the water pipe connection.

To install:

11. Clean and dry the mating surfaces of the block and water pump. Install a new O-ring to the water pipe connection. Coat the new O-ring with water to aid in installation.

12. Install the water pump with a new gasket to the block and tighten the bolts to:

- 2.4L Mighty Max: 9-11 ft. lbs. (12-15 Nm)
- 3.0L Mighty Max: 14-19 ft. lbs. (20-27 Nm)
- Montero: 17 ft. lbs. (24 Nm)

13. Tighten the alternator bracket bolt to 17 ft. lbs. (23 Nm).

14. Install the timing belt(s) and covers.

15. Install the crankshaft pulley(s).

16. Connect the radiator hose to the water pump.

17. Install the water pump pulley. Install the cooling fan and clutch assembly.

18. Install the air conditioning compressor tensioner pulley, if equipped.

19. Install the accessory belts, adjust if necessary.

20. Install the upper radiator shroud.

21. Fill the radiator with coolant. This cooling system has a self-bleeding thermostat, so system bleeding is not required.

22. Connect the negative battery cable, run the vehicle until the thermostat opens and fill the overflow tank. Check for leaks.

23. Once the vehicle has cooled, recheck the coolant level.

Expo 1.8L (4G93) Engine

1. Disconnect the negative battery cable.

2. Rotate the engine and position the No. 1 piston to TDC of its compression stroke.

3. Drain the cooling system.

4. Remove the engine undercover.

5. Disconnect the clamp bolt from the power steering hose.

6. Support the engine with the appropriate equipment and remove the engine mount bracket.

7. Remove the timing belt from the front of the engine.

8. Remove the timing belt rear cover.

9. Remove the alternator brace if necessary.

NOTE: The water pump mounting bolts are different in length, note their positioning for reassembly.

10. Remove the water pump mounting bolts and remove the pump.
To install:

11. Thoroughly clean and dry both mating surfaces of the water pump and block.

12. Apply a 0.09–0.12 inch (2.5–3.0mm) continuous bead of sealant to water pump and install the pump assembly.

NOTE: Install the water pump within 15 minutes of the application of the sealant. Wait 1 hour after installation of the water pump to refill the cooling system or starting the engine.

13. Properly position the bolts and tighten the bolts to 18 ft. lbs. (24 Nm).

14. Reinstall the timing belt rear cover.

15. Reinstall the timing belt and related parts.

16. Install the engine mount and bracket.

17. Connect the clamp bolt to the power steering hose.

18. Install the engine drive belts and adjust.

19. Install the engine undercover.

20. Fill the system with coolant.

21. Connect the negative battery cable, run the vehicle until the thermostat opens and fill the radiator completely.

22. Once the vehicle has cooled, recheck the coolant level.

Expo 2.4L (4G64) Engine

1. Disconnect the negative battery cable.

2. Drain the cooling system.

3. Remove the engine undercover.

4. Disconnect the clamp bolt from the power steering hose.

5. Support the engine with the appropriate equipment and remove the engine mount bracket.

6. Remove the timing belt from the front of the engine.

7. Disconnect the coolant hoses from the pump, if equipped.

8. Remove the alternator brace.

9. Remove the water pump, gasket and O-ring where the water inlet pipe(s) joins the pump.
To install:

10. Thoroughly clean and dry both gasket surfaces of the water pump and block.

11. Install a new O-ring into the groove on the front end of the water inlet pipe. Do not apply oils or grease to the O-ring. Wet with clean antifreeze only.

12. Install the gasket and pump assembly and tighten the bolts.

13. Connect the hoses to the pump.

14. Reinstall the timing belt and related parts.

15. Install the engine drive belts and adjust.

16. Fill the system with coolant.

17. Connect the negative battery cable, run the vehicle until the thermostat opens and fill the radiator completely.

18. Once the vehicle has cooled, recheck the coolant level.

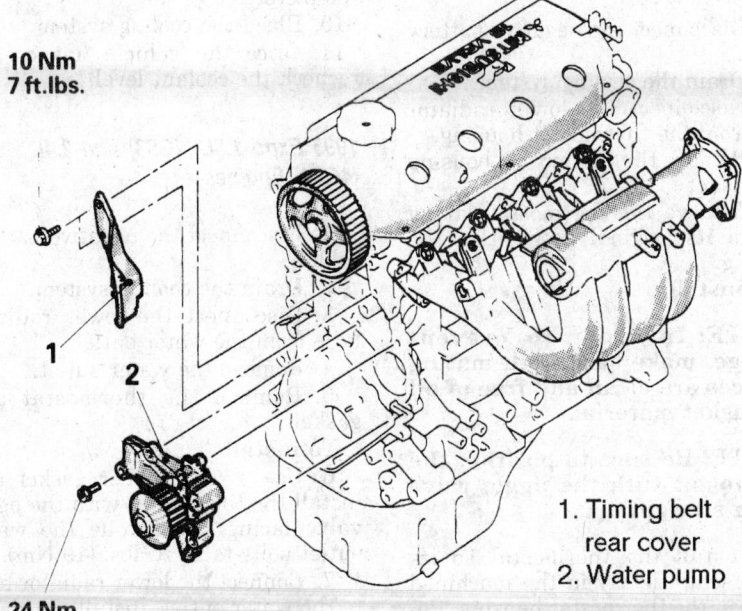

10 Nm
7 ft.lbs.

1

2

24 Nm
18 ft.lbs.

1. Timing belt rear cover
2. Water pump

332344

Water pump and related components — Expo 1.8L (4G93) engine

1. Alternator brace
2. Water pump
3. Gasket
4. O-ring

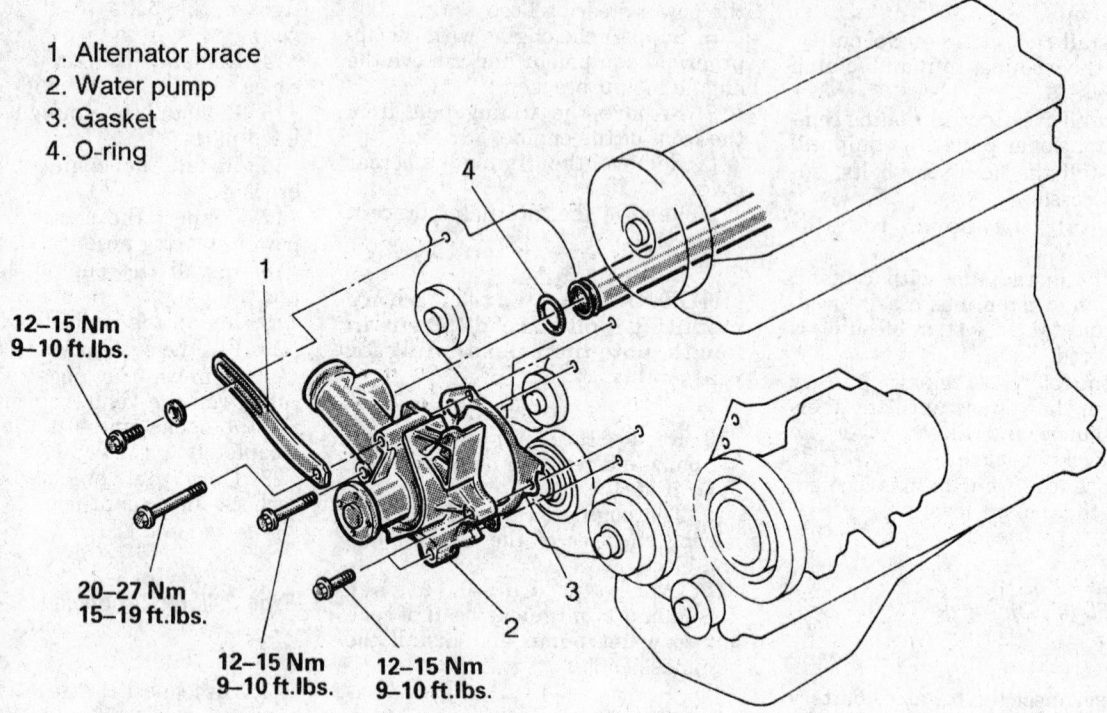

12–15 Nm
9–10 ft.lbs.

20–27 Nm
15–19 ft.lbs.

12–15 Nm
9–10 ft.lbs.

12–15 Nm
9–10 ft.lbs.

241613

Water pump and related components — Expo 2.4L (4G64) Engine

Thermostat

REMOVAL AND INSTALLATION

Except Expo

1. Disconnect the negative battery cable.

——————— CAUTION ———————
Wait at least 90 seconds after the negative (-) battery cable is disconnected to prevent possible deployment of the air bag.

2. Drain the coolant down to thermostat level or below.
3. Remove the upper radiator hose from the thermostat housing, then remove the housing.
4. Remove the thermostat and discard the gasket.
To install:
5. Clean the housing mating surfaces and use a new gasket.
6. Install the thermostat and gasket.
7. Install the housing and tighten the retaining bolts/nuts.
8. Connect the negative battery cable. Fill the radiator completely and start the vehicle.
9. Run the vehicle until the thermostat opens. Check the coolant level

in the overflow tank and fill if necessary.

Expo 1.8L (4G93) and 2.4L (4G64) Engines

1. Disconnect the negative battery cable.
2. Drain the cooling system.
3. Disconnect the lower radiator hose from the thermostat housing.
4. Remove the thermostat housing and gasket.
5. Remove the thermostat taking note of its original position in the housing.
To install:

NOTE: In order to prevent leakage, make sure both mating surfaces are clean and free of all old gasket material.

NOTE: Be sure to position the thermostat with the jiggle valve facing straight up.

6. Install the thermostat so its flange seats tightly in the machined recess in the thermostat housing. Refer to its location prior to removal.
7. Use a new gasket and reinstall the thermostat housing. Torque the housing mounting bolts to 14 ft. lbs. (19 Nm).

8. Connect the lower hose and fill the system with coolant.
9. Connect the negative battery cable, run the vehicle until the thermostat opens and fill the radiator completely.
10. Bleed the cooling system.
11. Once the vehicle has cooled, recheck the coolant level.

1995 Expo 1.8L (4G93) and 2.4L (4G64) Engines

1. Disconnect the negative battery cable.
2. Drain the cooling system.
3. Disconnect the lower radiator hose from the water outlet.
4. Remove the water outlet.
5. Remove the thermostat and gasket.
To install:
6. Use a new rubber gasket and install the thermostat with the jiggle valve facing up. Torque the water outlet bolts to 14 ft. lbs. (19 Nm).
7. Connect the lower radiator hose to the water outlet. Install the clamp in the original position.
8. Refill the cooling system and connect the negative battery cable.
9. Start the engine and check for leaks.

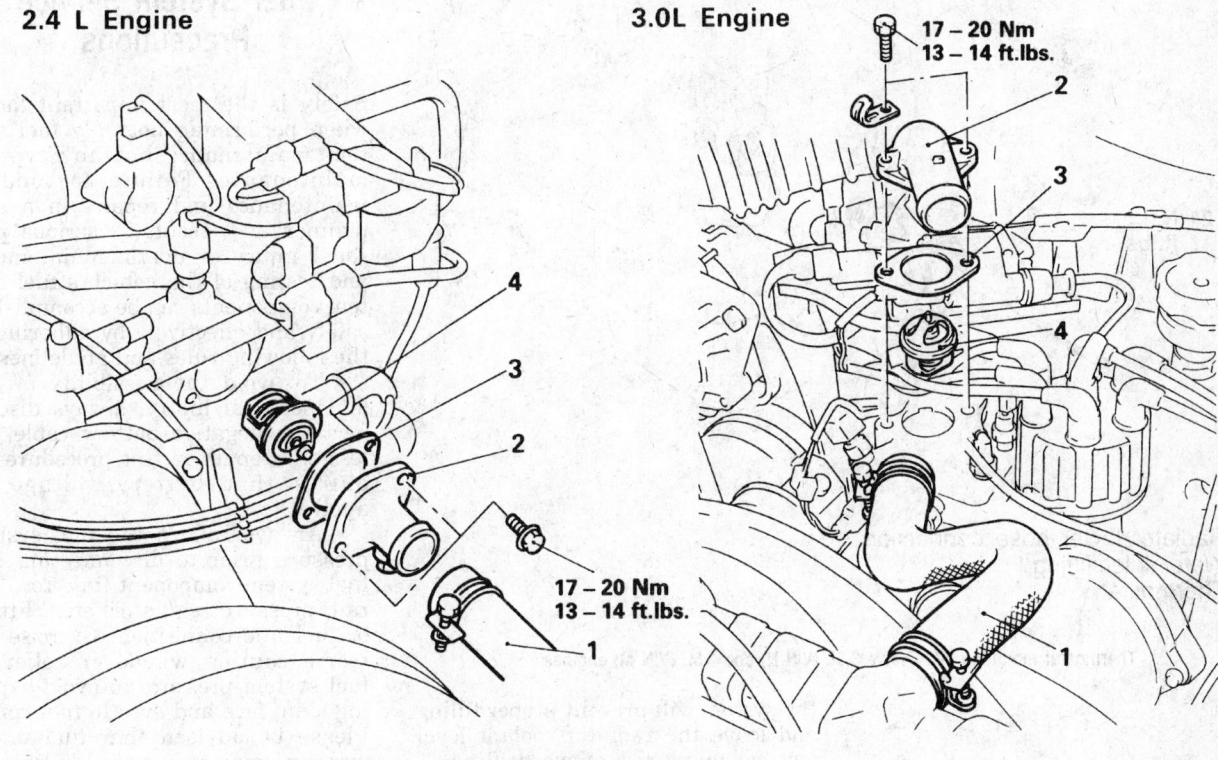

2.4 L Engine

3.0L Engine

17 – 20 Nm
13 – 14 ft.lbs.

17 – 20 Nm
13 – 14 ft.lbs.

329753

Removal steps

1. Radiator upper hose connection
2. Water outlet fitting
3. Water outlet fitting gasket
4. Thermostat

Thermostat mounting — 2.4L (VIN G) and 3.0L (VIN H) engines

Electric Cooling Fan

REMOVAL AND INSTALLATION

Except Expo

1. Disconnect the negative battery cable.
2. Remove the fan shroud.
3. Remove the retaining nuts and remove the fan and clutch assembly.
4. If necessary, remove the retaining bolts and remove the fan from the fan clutch.

To install:

5. If separated, attach the fan to the fan clutch and tighten the bolts to 7–9 ft. lbs. (10–12 Nm).
6. Install the fan and clutch assembly and tighten the retaining nuts to 7–9 ft. lbs. (10–12 Nm).
7. Install the fan shroud.
8. Connect the negative battery cable.

Expo 1.8L (4G93) and 2.4L (4G64) Engines

1. Disconnect the negative battery cable.
2. Disconnect the electrical connectors from the cooling fan.

3. Remove the fan and shroud bolts. Remove the fan shroud assembly.

NOTE: If equipped with air conditioning, it may be necessary to remove the condenser fan during this procedure.

4. With fan removed from the vehicle, remove the center fan nut and the three attaching bolts.

To install:

5. Install the fan blade to the motor, then connect the motor to the shroud.
6. Align the fan assembly to the radiator and secure with mounting bolts.

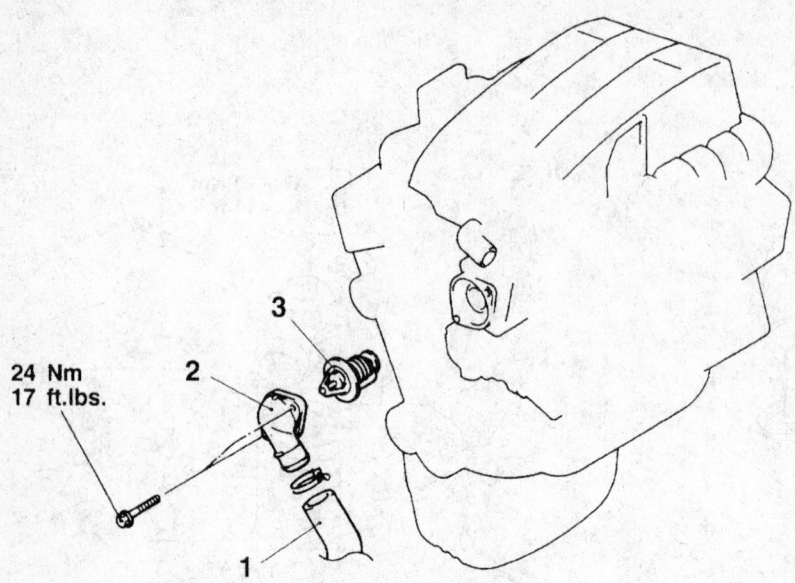

1. Radiator lower hose connection
2. Water inlet fitting
3. Thermostat

329755

Thermostat mounting — 24 valve 3.0L (VIN H) and 3.5L (VIN M) engines

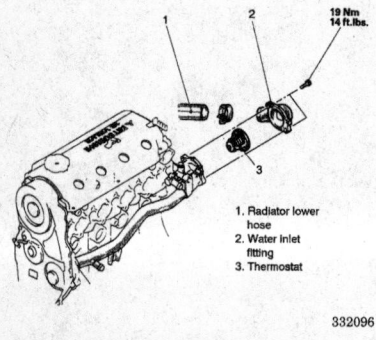

1. Radiator lower hose
2. Water inlet fitting
3. Thermostat

332096

Thermostat and related components — Expo 1.8L (4G93) and 2.4L (4G64) engines

7. Connect all fan wiring and secure away from abrasion.

8. Connect the negative battery cable. Start the engine and check fan operation.

Cooling System Bleeding

After working on the cooling system, even to replace the thermostat, the system must be bled. Air trapped in the system will prevent proper filling and leave the radiator coolant level low, causing a risk of overheating.

To bleed the system, start with the system cool, the radiator cap off and the radiator filled to about an inch below the filler neck.

1. Start the engine and run it at slightly above normal idle speed. This will insure adequate circulation. If air bubbles appear and the coolant level drops, fill the system with an antifreeze/water mixture to bring the level back to the proper level.

2. Run the engine this way until the thermostat opens. When this happens, coolant will move abruptly across the top of the radiator and the temperature of the radiator will suddenly rise.

3. At this point, air is often expelled and the level may drop quite a bit. Keep refilling the system until the level is near the top of the radiator and remains constant.

4. If the vehicle has an overflow tank, fill the radiator right up to the filler neck. Replace the radiator filler cap.

FUEL SYSTEM

Fuel System Service Precautions

Safety is the most important factor when performing not only fuel system maintenance but any type of maintenance. Failure to conduct maintenance and repairs in a safe manner may result in serious personal injury or death. Maintenance and testing of the vehicle's fuel system components can be accomplished safely and effectively by adhering to the following rules and guidelines.

• To avoid the possibility of fire and personal injury, always disconnect the negative battery cable unless the repair or test procedure requires that battery voltage be applied.

• Always relieve the fuel system pressure prior to disconnecting any fuel system component (injector, fuel rail, pressure regulator, etc.), fitting or fuel line connection. Exercise extreme caution whenever relieving fuel system pressure to avoid exposing skin, face and eyes to fuel spray. Please be advised that fuel under pressure may penetrate the skin or any part of the body that it contacts.

• Always place a shop towel or cloth around the fitting or connection prior to loosening to absorb any excess fuel due to spillage. Ensure that all fuel spillage (should it occur) is quickly removed from engine surfaces. Ensure that all fuel soaked cloths or towels are deposited into a suitable waste container.

• Always keep a dry chemical (Class B) fire extinguisher near the work area.

• Do not allow fuel spray or fuel vapors to come into contact with a spark or open flame.

• Always use a backup wrench when loosening and tightening fuel line connection fittings. This will prevent unnecessary stress and torsion to fuel line piping. Always follow the proper torque specifications.

• Always replace worn fuel fitting O-rings with new. Do not substitute fuel hose or equivalent, where fuel pipe is installed.

Fuel System Pressure

RELIEVING

———— **CAUTION** ————

The fuel system is under a constant pressure, even with the engine off. This pressure must be relieved before disconnecting any fuel system component, fitting or fuel line connection. Failure to do so may result in personal injury.

1. Disconnect the fuel pump electrical connector, located at the rear side of the fuel tank.
2. Start the engine.
3. After the engine stalls, turn the ignition switch **OFF** and reconnect the fuel pump connector.
4. Disconnect the negative battery cable, then continue with the service procedure.

Idle Speed

ADJUSTMENT

Montero

———— **WARNING** ————

The electrical system and computer(s) are easily damaged through careless work habits. Since electricity travels at close to the speed of light, damage is instantaneous and expensive. Always observe the following rules when connecting the tachometer.

- Do not leave the ignition switch ON for more than 5 minutes when the engine is not running.
- The tachometer lead must only be connected to the correct terminal; accidental contact with another terminal can cause great damage.
- Once connected to the terminal, the tachometer lead and connectors must be protected against grounding to any metal surface of the car. It is highly recommended that any adapter have fully insulated connectors or be wrapped in dry cloth or tape for protection; otherwise, the igniter and/or ignition coil can be damaged.
- Never disconnect the battery cables while the engine is running.
- Make certain the engine and ignition wiring is connected properly before connecting the tachometer. It is particularly important that ground circuits be clean and tight.
1. The air cleaner should be in place and all wires and vacuum hoses connected. All accessories should be

OFF, the transmission in neutral and the drive wheels blocked.

2. SOHC 12 valve engine: Insert a paper clip into the female side of the 1–pin connector. Do not disconnect the connector.
3. DOHC, SOHC 24 valve engines: Insert a paper clip into the 1–pin blue connector.
4. Connect a tachometer to the paper clip.
5. Use a jumper wire to connect terminal 1 to a ground.
6. Disconnect the waterproof female connector (brown) from the ignition timing adjusting connector.

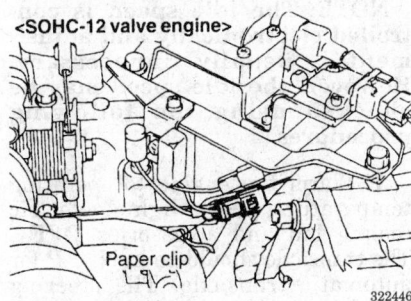

<SOHC-12 valve engine>

Paper clip

322446

Pin connector for 3.0L (VIN H) 12 valve engines — Montero

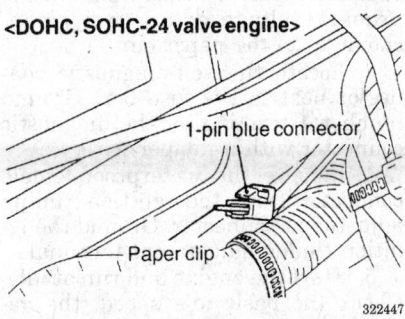

<DOHC, SOHC-24 valve engine>

1-pin blue connector

Paper clip

322447

Pin connector for 24 valve 3.0L (VIN H) and 3.5L (VIN M) engines — Montero

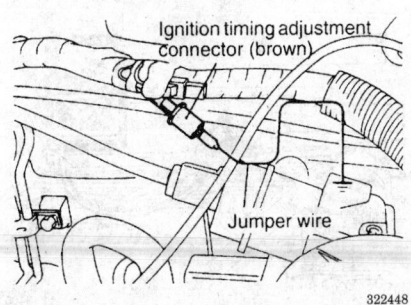

Ignition timing adjustment connector (brown)

Jumper wire

322448

Waterproof ignition timing adjusting connector — Montero

7. Use a jumper wire to ground the ignition timing adjusting terminal.
8. Start the engine and run at idle.
9. Set the idle speed by turning the idle adjusting screw to obtain the proper idle speed.
10. Idle speed should be 700 ± 50 r/min.
11. Disconnect the jumper wires and recheck idle.
12. Disconnect the tachometer and replace any covers or plugs which were removed from the service connector.

Mighty Max

———— **WARNING** ————

The electrical system and computer(s) are easily damaged through careless work habits. Since electricity travels at close to the speed of light, damage is instantaneous and expensive. Always observe the following rules when connecting the tachometer.

- Do not leave the ignition switch ON for more than 5 minutes when the engine is not running.
- The tachometer lead must only be connected to the correct terminal; accidental contact with another terminal can cause great damage.
- Once connected to the terminal, the tachometer lead and connectors must be protected against grounding to any metal surface of the car. It is highly recommended that any adapter have fully insulated connectors or be wrapped in dry cloth or tape for protection; otherwise, the igniter and/or ignition coil can be damaged.
- Never disconnect the battery cables while the engine is running.
- Make certain the engine and ignition wiring is connected properly before connecting the tachometer. It is particularly important that ground circuits be clean and tight.
1. The air cleaner should be in place and all wires and vacuum hoses connected. All accessories should be **OFF**, the transmission in neutral and the drive wheels blocked.
2. Insert a paper clip into the female side of the 1–pin connector. Do not disconnect the connector.
3. Connect a tachometer to the paper clip.
4. Use a jumper wire to ground the diagnostic test mode control terminal of the data link connector.
5. Use a jumper wire to ground the ignition timing adjusting terminal.
6. Start the engine and run at idle.

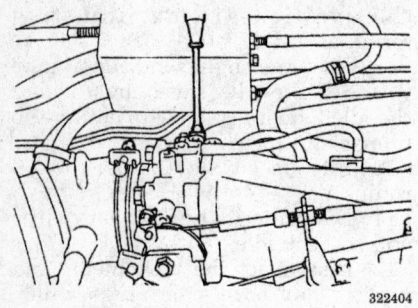

Idle speed adjustment screw — Montero

Pin connector — Mighty Max

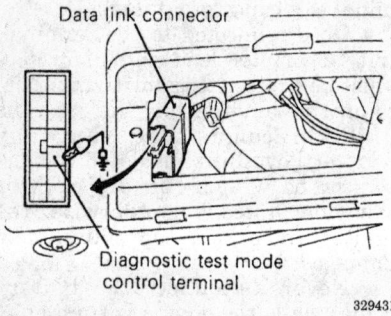

Diagnostic test mode control terminal — Mighty Max

7. Set the idle speed by turning the idle adjusting screw to obtain the proper idle speed.

8. Idle speed should be 700±50 r/min.

9. Disconnect the jumper wires and recheck idle.

10. Disconnect the tachometer and replace any covers or plugs which were removed from the service connector.

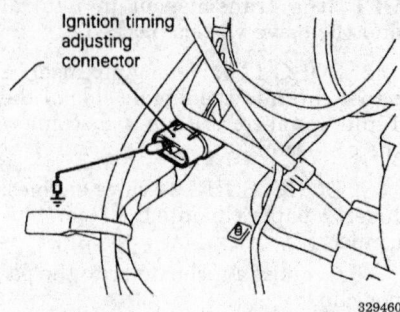

Terminal for adjustment of the ignition timing — Mighty Max

Expo 1.8L (4G93) and 2.4L (4G64) Engines

NOTE: The idle speed is controlled electronically and adjustment is usually unnecessary. However, the idle speed may be checked using the following procedures.

1. Warm the engine to operating temperature, leave lights, electric cooling fan and accessories OFF. The transaxle should be in N or P for automatic transaxle. The steering wheel in a neutral position for vehicles with power steering.

2. Insert the paper clip into the single terminal rpm connector in the engine compartment, and connect the primary voltage detection type tachometer to the paper clip.

3. Locate the self-diagnostic connector next to the fuse box. Ground number 1 terminal of the diagnostic connector with a jumper wire.

4. Remove the waterproof female connector from the ignition timing adjustment connector. Ground the ignition timing adjustment terminal.

5. Start the engine and run at idle. Check the basic idle speed, the desired value is 700–800 rpm.

6. If the value is not within specifications, turn the Speed Adjusting

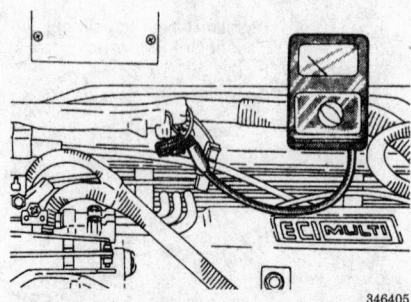

Tachometer connection location — Expo 1.8L (4G93) and 2.4L (4G64) engines

Screw (SAS) to make the necessary adjustment.

NOTE: If the idle speed is higher than the standard value, inspect the SAS screw for evidence of movement. If there is evidence that the SAS screw has been adjusted, readjust to the proper setting. If the screw does not look as though it has been adjusted, it is possible that there is leakage as a result of deterioration of the Fast Idle Air Valve (FIAV) and, if so the throttle body should be replaced.

7. Turn the ignition OFF. Disconnect and remove the jumper wires from the diagnosis control terminal and the ignition timing adjustment terminal.

8. Start the engine and let run at idle speed for about 10 minutes, check to be sure the idling condition is normal.

Mixture

ADJUSTMENT

Air/Fuel mixture is controlled by the ECU and is not adjustable.

Fuel Filter

REMOVAL AND INSTALLATION

Mighty Max and 1993–95 Montero

——— CAUTION ———
Fuel injection systems remain under pressure after the engine has been turned OFF. Properly relieve fuel pressure before disconnecting any fuel lines. Failure to do so may result in fire or personal injury.

——— CAUTION ———
Do not allow fuel spray or fuel vapors to come in contact with a spark or open flame. Keep a dry chemical fire extinguisher nearby. Never store fuel in an open container due to risk of fire or explosion.

1. Relieve the fuel system pressure.

2. Disconnect the negative battery cable.

3. Remove the fuel filter protector if equipped.

4. Using a back-up wrench disconnect the fuel lines from the filter.

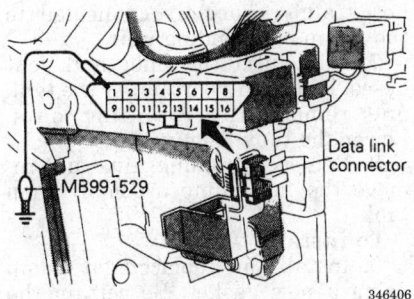

Diagnostic connector and terminal identification — Expo 1.8L (4G93) and 2.4L (4G64) engines

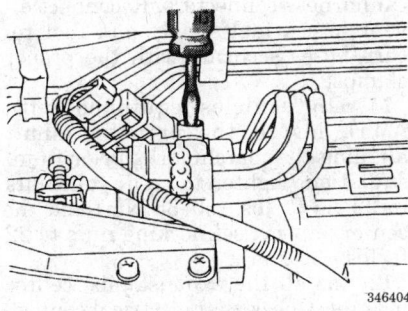

Engine speed adjusting screw location — Expo 1.8L (4G93) and 2.4L (4G64) engines

5. Remove the filter from the mounting bracket.

To install:

6. Install the fuel filter to the mounting bracket in the proper direction.

7. Using new gaskets, connect the fuel lines to the fuel filter. Use a back-up wrench to hold the fuel filter. Torque the eye bolts to 18–25 ft. lbs. (25–35 Nm).

8. Install the fuel filter protector if equipped.

9. Connect the negative battery cable.

10. Start the engine and check for leaks.

1996–97 Montero

————— CAUTION —————
Fuel injection systems remain under pressure after the engine has been turned OFF. Properly relieve fuel pressure before disconnecting any fuel lines. Failure to do so may result in fire or personal injury.

————— CAUTION —————
Do not allow fuel spray or fuel vapors to come in contact with a spark or open flame. Keep a dry

chemical fire extinguisher nearby. Never store fuel in an open container due to risk of fire or explosion.

1. Relieve the fuel system pressure.

2. Disconnect the negative battery cable.

3. Remove the fuel filter protector if equipped.

4. Using a back-up wrench disconnect the fuel lines from the filter.

5. Remove the filter from the mounting bracket.

To install:

6. Install the fuel filter to the mounting bracket in the proper direction.

7. Using new gaskets, connect the fuel lines to the fuel filter. Use a back-up wrench to hold the fuel filter. Torque the eye bolts to 22 ft. lbs. (30 Nm)

8. Install the fuel filter protector if equipped.

9. Connect the negative battery cable.

10. Start the engine and check for leaks.

Expo 1.8L (4G93) and 2.4L (4G64) Engines

A replaceable fuel filter is located in the engine compartment.

————— CAUTION —————
Do not use conventional fuel filters, hoses or clamps when servicing fuel injection systems. They are not compatible with the injection system and could fail, causing personal injury or damage to the vehicle. Use only hoses and clamps specifically designed for fuel injection.

————— CAUTION —————
Fuel injection systems remain under pressure, after the engine has been turned OFF. Properly relieve fuel pressure before disconnecting any fuel lines. Failure to do so may result in fire or personal injury.

1. Relieve the fuel system pressure.

NOTE: Wrap shop towels around the fitting that is being disconnected to absorb residual fuel in the lines.

2. Hold the fuel filter nut securely with a backup or spanner wrench. Cover the hoses with shop towels and remove the eye bolt. Discard the gaskets.

3. Separate the flare nut connection at the filter. Discard the gaskets.

4. Remove the mounting bolts and the fuel filter from the vehicle.

To install:

5. If equipped with flare fitting, tighten the fitting by hand before installing the filter to the vehicle.

6. Install the filter to its bracket only finger-tight. Movement of the filter will ease attachment of the fuel lines.

7. Install new gaskets and connect the high pressure hose and eye bolt, then the main pipe. While holding the fuel filter nut, tighten the eye bolts to 22 ft. lbs. (30 Nm). Tighten the flare nut to 27 ft. lbs. (37 Nm).

8. Tighten the filter mounting bolts fully.

9. Install the air cleaner assembly, if removed.

10. Connect the negative battery cable, install the fuel filler cap, turn the key to the **ON** position to pressurize the fuel system and check for leaks.

11. Release the fuel pressure and repair leaks as required.

Fuel Pump

REMOVAL AND INSTALLATION

Montero

1. Disconnect the negative battery cable.

————— CAUTION —————
Fuel injection systems remain under pressure after the engine has been turned OFF. Properly relieve fuel pressure before disconnecting any fuel lines. Failure to do so may result in fire or personal injury.

2. Relieve the fuel system pressure.

NOTE: The manufacturer recommends draining of the fuel tank.

————— CAUTION —————
Do not allow fuel spray or fuel vapors to come in contact with a spark or open flame. Keep a dry chemical fire extinguisher nearby. Never store fuel in an open container due to risk of fire or explosion.

3. Remove the rear floor carpeting.

4. Remove the fuel pump cover.

5. Disconnect the fuel pump connector and the fuel hoses.

6. Remove the fuel pump assembly.

To install:

7. Install the fuel pump assembly into the fuel tank.

8. Tighten the nuts to 24 in. lbs. (2.5 Nm).

9. Connect the fuel lines and the fuel pump connector.

10. Install the fuel pump cover and torque the bolts to 9 ft. lbs. (12 Nm).

11. Install the rear floor carpeting.

12. If drained, refill the fuel tank.

13. Connect the negative battery cable.

14. Start the vehicle; check for leaks and proper operation.

Mighty Max

1. Properly relieve the fuel system pressure.

─────── **CAUTION** ───────

Fuel injection systems remain under pressure after the engine has been turned OFF. Properly relieve fuel pressure before disconnecting any fuel lines. Failure to do so may result in fire or personal injury.

2. Disconnect the negative battery cable.

─────── **CAUTION** ───────

Wait at least 90 seconds after the negative (-) battery cable is disconnected to prevent possible deployment of the air bag.

3. Raise and safely support the vehicle.

4. If equipped, remove the fuel tank protector.

5. Position a suitable container under the fuel tank. Remove the fuel tank drain plug and drain the fuel from the tank.

─────── **CAUTION** ───────

Observe all applicable safety precautions when working around gasoline. Do not allow fuel spray or fuel vapors to come in contact with a spark or open flame. Keep a dry chemical (Class B) fire extinguisher near the work area.

6. Remove the fuel tank from the vehicle.

7. Remove the fuel pump retaining screws and remove the pump from the tank.

To install:

8. Clean the seal area of the tank. Install a new gasket.

9. Install the fuel pump in the same position as originally installed.

10. Install the fuel pump retaining screws and tighten to 1.8 ft. lbs. (2.5 Nm).

11. Install the fuel tank and torque to 18–22 ft. lbs. (25–30 Nm).

12. Install the fuel tank drain plug and the fuel tank protector, if equipped.

13. Lower the vehicle. Refill the fuel tank and install the cap.

14. Connect the negative battery cable.

15. Connect a jumper wire from the fuel pump check connector on the back of the fuse block to the positive battery terminal and operate the fuel pump. Check for leaks.

Expo 1.8L (4G93) and 2.4L (4G64) Engines

1. Relieve fuel system pressure. Remove the fuel filler cap.

─────── **CAUTION** ───────

Fuel injection systems remain under pressure after the engine has been turned OFF. Properly relieve fuel pressure before disconnecting any fuel lines. Failure to do so may result in fire or personal injury.

2. Disconnect the negative battery cable.

3. Raise and safely support vehicle.

4. The fuel pump is located in the fuel tank. Drain the fuel from the fuel tank.

─────── **CAUTION** ───────

Do not allow fuel spray or fuel vapors to come in contact with a spark or open flame. Keep a dry chemical fire extinguisher nearby. Never store fuel in an open container due to risk of fire or explosion.

5. On vehicles equipped with AWD, remove the rear propeller shaft from the vehicle as follows:

a. Remove the center exhaust pipe bracket.

b. Matchmark the differential companion flange to the propeller flange yoke.

c. Remove the bolts, washers and nuts from the center support. Remove the propeller shaft assembly in a straight and level manner to avoid damage to the boot caused by pinching.

d. Install cover into the rear end of the transfer case to prevent the entry of foreign materials.

6. Disconnect the return hose, high pressure hose and all other hoses and connectors connected to the pump and sending unit.

7. Disconnect the filler and vent hoses. Place a support under the tank and remove the retaining nuts. Lower the tank from vehicle.

8. Remove retaining nuts and remove the fuel pump assembly from tank.

To install:

9. Install the replacement pump using a new gasket. Be certain the pump is installed in the same location, facing the same direction as before.

10. Install the fuel tank and secure the retainer nuts. Connect all electrical harness connectors. Reconnect all vent hoses, fuel supply and fuel return hoses securing with the proper clamps.

11. On vehicles equipped with AWD, install the propeller shaft aligning the matchmarks prior to installation. Tighten the rear yoke nuts to 22–25 ft. lbs. (30–35 Nm) and the center support self-locking nuts to 22 ft. lbs. (30 Nm).

12. Install the exhaust pipe center bracket. Check that electrical connectors are properly installed and all fuel hose connections are tight.

13. Connect the negative battery cable and check the entire fuel system for proper operation and leaks. If repairing of a fuel leak is required, release the fuel system pressure prior to repairing system.

Fuel Injector

REMOVAL AND INSTALLATION

1993–95 3.0L (VIN H) and 3.5L (VIN M) Engines

─────── **CAUTION** ───────

Fuel injection systems remain under pressure after the engine has been turned OFF. Properly relieve fuel pressure before disconnecting any fuel lines. Failure to do so may result in fire or personal injury.

1. Properly relieve the fuel system pressure.

2. Disconnect the negative battery cable.

3. Partially drain the cooling system.

4. Remove the air intake hose from the throttle body.

5. Label and disconnect the electrical connectors and vacuum hoses from the throttle body and air intake plenum.

6. Disconnect the accelerator cable and, if equipped, throttle control cable.

7. Disconnect the coolant hoses.

8. Disconnect the EGR temperature sensor connector.

9. Disconnect the coil wire and remove the ignition coil.

10. Disconnect the engine oil filler neck bracket from the air intake plenum.

11. Unbolt the EGR tube from the air intake plenum.

12. Remove the plenum-to-engine brackets.

13. Unbolt the air intake plenum assembly from the intake manifold and remove. Note the position of the mounting bolts as they are removed.

14. Cover the high pressure fuel hose with a clean shop towel to prevent fuel spray due to residual pressure in the line. Disconnect the high pressure fuel hose from the fuel rail.

15. Remove the fuel return line and vacuum hose from the fuel pressure regulator.

16. Label and disconnect the electrical connectors from the injectors.

17. Remove the fuel rail retaining bolts.

18. Lift the rail, with injectors attached, up and away from the engine.

19. Remove the injectors from the fuel rail pulling straight out away from the rail. Remove the lower insulators.

To install:

20. Install new insulators to the intake manifold.

21. Install a new grommet and O-ring onto the injector. Coat the O-ring lightly with clean engine oil to aid in assembly.

22. Install the injectors into the rail, making sure the injector turns freely. If they do not, check for a misaligned O-ring and reinstall.

23. Install the assembled fuel rail with injectors onto the manifold.

24. Tighten the fuel rail bolts to 96 inch lbs. (10 Nm).

25. Connect the electrical connectors to the injectors.

26. Connect the fuel return hose and the vacuum hose to the pressure regulator.

27. Using a new O-ring coated lightly with clean engine oil, install the high pressure fuel line. Tighten the attaching bolts to 12–24 inch lbs. (2–3 Nm).

28. Install the air intake plenum to the intake manifold using a new gasket.

29. Install the plenum-to-intake manifold brackets.

30. Attach the EGR tube to the plenum using a new gasket.

31. Attach the engine oil filler neck bracket to the plenum.

32. Install the ignition coil and connect the coil wire.

33. Connect the EGR temperature sensor connector. Connect all remaining hoses, vacuum lines and electrical connectors to the plenum and throttle body.

34. Connect the accelerator cable and, if equipped, throttle control cable.

35. Connect the air intake hose.

36. Fill the cooling system and connect the negative battery cable.

37. Connect a jumper wire from the fuel pump activation terminal to the positive battery post and inspect the system for leaks.

1996–97 3.0L (VIN H) Engine

— **CAUTION** —

Fuel injection systems remain under pressure after the engine has been turned OFF. Properly relieve fuel pressure before disconnecting any fuel lines. Failure to do so may result in fire or personal injury.

1. Properly relieve the fuel system pressure.

2. Disconnect the negative battery cable.

— **CAUTION** —

Work must be started after 90 seconds from the time the ignition switch is turned to the LOCK position and the negative (-) battery cable is disconnected.

3. Partially drain the cooling system.

4. Remove the air intake hose from the throttle body.

5. Label and disconnect the electrical connectors and vacuum hoses from the throttle body and air intake plenum.

6. Disconnect the accelerator cable and the throttle control cable.

7. Disconnect the coolant hoses.

8. Disconnect the engine oil filler neck bracket from the air intake plenum.

9. Unbolt the EGR tube from the air intake plenum.

10. Remove the plenum brackets.

11. Unbolt the air intake plenum assembly from the intake manifold and remove. Note the position of the mounting bolts as they are removed.

12. Cover the high pressure fuel hose with a clean shop towel to prevent fuel spray due to residual pressure in the line. Disconnect the high pressure fuel hose from the fuel rail.

13. Remove the fuel return line and vacuum hose from the fuel pressure regulator.

14. Label and disconnect the electrical connectors from the injectors.

15. Remove the fuel rail retaining bolts.

16. Lift the rail, with injectors attached, up and away from the engine.

17. Remove the injectors from the fuel rail pulling straight out away from the rail. Remove the lower insulators.

To install:

18. Install new insulators to the intake manifold.

19. Install a new grommet and O-ring onto the injector. Coat the O-ring lightly with clean engine oil to aid in assembly.

20. Install the injectors into the rail, making sure the injector turns freely. If they do not, check for a misaligned O-ring and reinstall.

21. Install the assembled fuel rail with injectors onto the manifold.

22. Tighten the fuel rail bolts to 9 ft. lbs. (12 Nm).

23. Connect the electrical connectors to the injectors.

24. Connect the fuel return hose and the vacuum hose to the pressure regulator.

25. Using a new O-ring coated lightly with clean engine oil, install the high pressure fuel line. Tighten the attaching bolts to 12–24 inch lbs. (2–3 Nm).

26. Install the air intake plenum to the intake manifold using a new gasket.

27. Install the plenum to the intake manifold brackets.

28. Attach the EGR tube to the plenum using a new gasket.

29. Connect the EGR temperature sensor connector. Connect all remaining hoses, vacuum lines and electrical connectors to the plenum and throttle body.

30. Connect the accelerator cable and the throttle control cable.

31. Connect the air intake hose.

32. Fill the cooling system and connect the negative battery cable.

33. Connect a jumper wire from the fuel pump activation terminal to the positive battery post and inspect the system for leaks.

1996–97 3.5L (VIN M) Engines

---- **CAUTION** ----

Fuel injection systems remain under pressure after the engine has been turned OFF. Properly relieve fuel pressure before disconnecting any fuel lines. Failure to do so may result in fire or personal injury.

1. Properly relieve the fuel system pressure.
2. Disconnect the negative battery cable.

---- **CAUTION** ----

Work must be started after 90 seconds from the time the ignition switch is turned to the LOCK position and the negative (-) battery cable is disconnected.

3. Partially drain the cooling system.
4. Remove the air intake hose from the throttle body.
5. Label and disconnect the electrical connectors and vacuum hoses from the throttle body and air intake plenum.
6. Disconnect the accelerator cable and the throttle control cable.
7. Disconnect the coolant hoses.
8. Disconnect the EGR temperature sensor connector.
9. Unbolt the EGR tube from the air intake plenum.
10. Remove the intake manifold plenum cover.
11. Remove the intake manifold plenum stay brackets.
12. Unbolt the air intake plenum assembly from the intake manifold and remove. Note the position of the mounting bolts as they are removed.
13. Remove the induction control valve assembly.
14. Cover the high pressure fuel hose with a clean shop towel to prevent fuel spray due to residual pressure in the line. Disconnect the high pressure fuel hose from the fuel rail.
15. Remove the fuel return line and vacuum hose from the fuel pressure regulator.
16. Label and disconnect the electrical connectors from the injectors.
17. Remove the fuel rail retaining bolts.
18. Lift the rail, with injectors attached, up and away from the engine.
19. Remove the injectors from the fuel rail pulling straight out away from the rail. Remove the lower insulators.
To install:
20. Install new insulators to the intake manifold.

21. Install a new grommet and O-ring onto the injector. Coat the O-ring lightly with clean engine oil to aid in assembly.
22. Install the injectors into the rail, making sure the injector turns freely. If they do not, check for a misaligned O-ring and reinstall.
23. Install the assembled fuel rail with injectors onto the manifold.
24. Tighten the fuel rail bolts to 9 ft. lbs. (12 Nm).
25. Connect the electrical connectors to the injectors.
26. Connect the fuel return hose and the vacuum hose to the pressure regulator.
27. Using a new O-ring coated lightly with clean engine oil, install the high pressure fuel line. Tighten the attaching bolts to 7 ft. lbs. (9 Nm).
28. Install the induction control valve assembly.
29. Install the air intake plenum to the intake manifold using a new gasket.
30. Install the intake manifold plenum stay brackets.
31. Attach the EGR tube to the plenum using a new gasket.
32. Connect the EGR temperature sensor connector. Connect all remaining hoses, vacuum lines and electrical connectors to the plenum and the throttle body.
33. Connect the accelerator cable and the throttle control cable.
34. Connect the air intake hose.
35. Fill the cooling system and connect the negative battery cable.
36. Connect a jumper wire from the fuel pump activation terminal to the positive battery post and inspect the system for leaks.

2.4L (VIN G) Engine

1. Properly relieve the fuel system pressure.

---- **CAUTION** ----

Fuel injection systems remain under pressure after the engine has been turned OFF. Properly relieve fuel pressure before disconnecting any fuel lines. Failure to do so may result in fire or personal injury.

2. Disconnect the negative battery cable.

---- **CAUTION** ----

Wait at least 90 seconds after the negative (-) battery cable is disconnected to prevent possible deployment of the air bag.

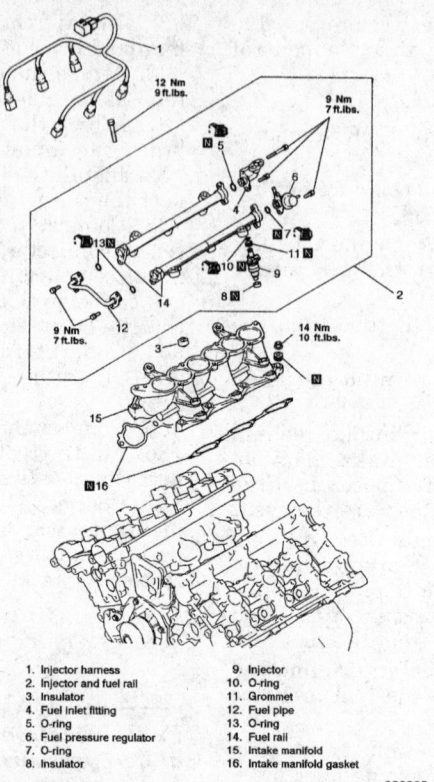

1. Injector harness
2. Injector and fuel rail
3. Insulator
4. Fuel inlet fitting
5. O-ring
6. Fuel pressure regulator
7. O-ring
8. Insulator
9. Injector
10. O-ring
11. Grommet
12. Fuel pipe
13. O-ring
14. Fuel rail
15. Intake manifold
16. Intake manifold gasket

320985

Fuel rail and injectors — 1996–97 3.5L (VIN M) engines

3. Partially drain the cooling system.

4. Remove the air intake hose, breather hose, air intake pipe and air hose.

5. Label and disconnect the electrical connectors, water hoses and vacuum hoses from the throttle body.

6. Disconnect the accelerator and, if equipped, kickdown cables.

7. Remove the throttle body from the intake manifold.

8. Cover the high pressure fuel hose with a clean shop towel to prevent fuel spray from residual pressure in the line. Disconnect the high pressure fuel hose from the fuel rail.

9. Remove the fuel return line and vacuum hose from the fuel pressure regulator.

10. Label and disconnect the electrical connectors from the injectors.

11. Remove the fuel rail retaining bolts.

12. Lift the rail, with the injectors attached, up and away from the engine.

13. Remove the injector from the rail by pulling straight out away from rail.

14. Remove the fuel rail insulators.

To install:

15. Install a new grommet and O-rings onto the injector. Coat the O-rings lightly with clean engine oil.

16. Install the injector to the fuel rail, making sure the injector turns freely. If it does not turn, check for a damaged or misaligned O-ring and reinstall.

17. Install new fuel rail insulators to the intake manifold. Install the fuel rail with injectors onto the manifold.

18. Tighten the fuel rail retaining bolts to 72–108 inch lbs. (6–9 Nm).

19. Install the fuel return line and vacuum hose to the fuel pressure regulator.

20. Connect the electrical harness to the injectors.

21. Replace the O-ring and connect the high pressure fuel line to the fuel rail.

22. Install the throttle body using a new gasket.

23. Connect all hoses, vacuum lines and electrical connectors.

24. Connect the accelerator and, if equipped, kickdown cables.

25. Install the air hose, air intake pipe, breather hose and air intake hose.

26. Refill the cooling system. Connect the negative battery cable.

27. Connect a jumper wire from the fuel pump activation terminal to the positive battery post and inspect the system for leaks.

Expo 1.8L (4G93) and 2.4L (4G64) Engines

1. Relieve the fuel system pressure following proper procedures.

── CAUTION ──
Fuel injection systems remain under pressure after the engine has been turned OFF. Properly relieve fuel pressure before disconnecting any fuel lines. Failure to do so may result in fire or personal injury.

2. Disconnect the PCV hose from the valve cover. Also disconnect the breather hose at the opposite end of the valve cover.

3. Remove the bolts holding the high pressure fuel line to the fuel rail and disconnect the line. Be prepared to contain fuel spillage; plug the line to keep out dirt and debris.

4. Remove the vacuum hose from the fuel pressure regulator.

5. Disconnect the fuel return hose from the pressure regulator. Remove the fuel pressure regulator mounting bolts and remove from the fuel rail.

6. Label and disconnect the electrical connector from each injector.

7. Remove the bolt(s) holding the fuel rail to the manifold. Carefully lift the rail up and remove it with the injectors attached. Take great care not to drop an injector. Place the rail and injectors in a safe location on the workbench; protect the tips of the injectors from dirt and/or impact.

8. Remove and discard the injector insulators from the intake manifold. The insulators are not reusable.

9. Remove the injectors from the fuel rail by pulling gently in a straight outward motion. Make certain the grommet and O-ring come off with the injector.

To install:

10. Install a new insulator in each injector port in the manifold.

11. Remove the old grommet and O-ring from each injector. Install a new grommet and O-ring; coat the O-ring lightly with clean, thin oil.

12. If the fuel pressure regulator was removed, replace the O-ring with a new one and coat it lightly with clean, thin oil. Insert the regulator straight into the rail, then check that it can be rotated freely. If it does not rotate smoothly, remove it and inspect the O-ring for deformation or jamming. When properly installed, align the mounting holes and tighten the retaining bolts to 7 ft. lbs. (9 Nm).

This procedure must be followed even if the fuel rail was not removed.

13. Install the injector into the fuel rail, constantly turning the injector left and right during installation. When fully installed, the injector should still turn freely in the rail. If it does not, remove the injector and inspect the O-ring for deformation or damage.

14. Install the delivery pipe and injectors to the engine. Make certain that each injector fits correctly into its port and that the rubber insulators for the fuel rail mounts are in position.

15. Install the fuel rail retaining bolts and tighten them to 9 ft. lbs. (12 Nm).

16. Connect the wiring harnesses to the appropriate injector.

17. Connect the fuel return hose to the pressure regulator, then connect the vacuum hose.

18. Replace the O-ring on the high pressure fuel line, coat the O-ring lightly with clean, thin oil and install the line to the fuel rail. Tighten the mounting bolts to specifications.

19. Connect the PCV hose and the breather hose if they were disconnected.

20. Connect the negative battery cable. Pressurize the fuel system and inspect all connections for leaks.

ENGINE MECHANICAL

Engine Assembly

REMOVAL AND INSTALLATION

3.0L (VIN H) and 3.5L (VIN M) Engines

1. Relieve the fuel system pressure.

── CAUTION ──
Fuel injection systems remain under pressure after the engine has been turned OFF. Properly relieve fuel pressure before disconnecting any fuel lines. Failure to do so may result in fire or personal injury.

2. Disconnect the negative battery cable.

3. Matchmark and remove the hood.

4. Remove the oil dipstick.

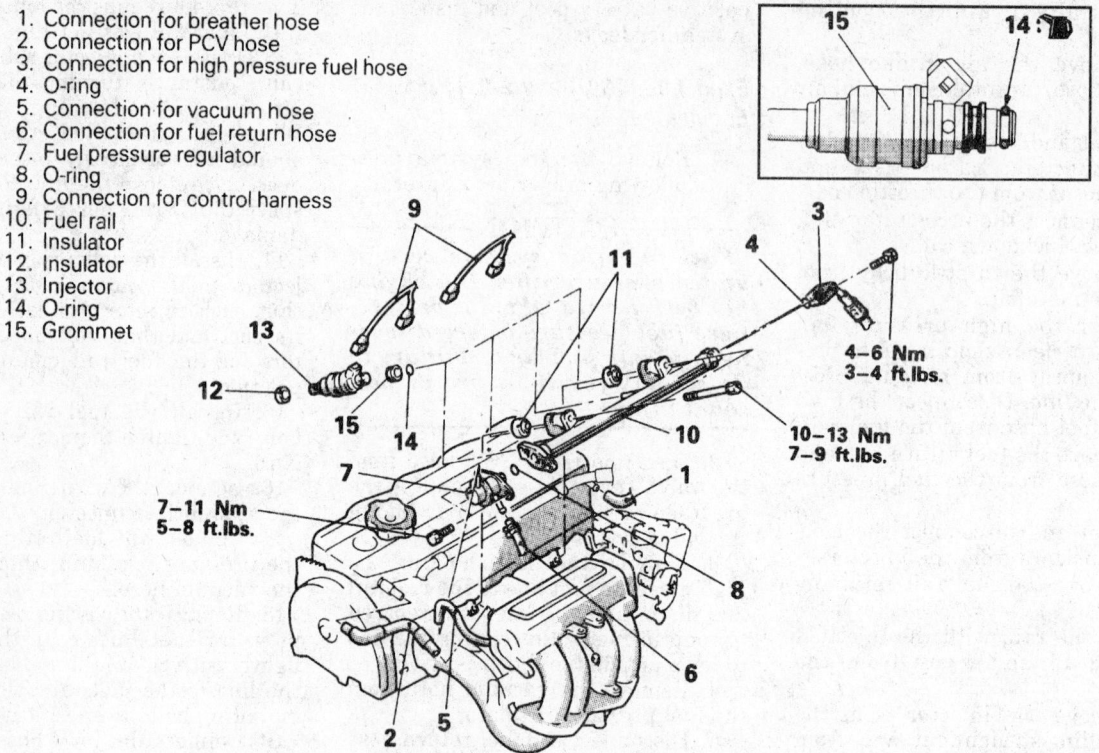

1. Connection for breather hose
2. Connection for PCV hose
3. Connection for high pressure fuel hose
4. O-ring
5. Connection for vacuum hose
6. Connection for fuel return hose
7. Fuel pressure regulator
8. O-ring
9. Connection for control harness
10. Fuel rail
11. Insulator
12. Insulator
13. Injector
14. O-ring
15. Grommet

4–6 Nm
3–4 ft.lbs.

10–13 Nm
7–9 ft.lbs.

7–11 Nm
5–8 ft.lbs.

330405

Fuel injector and related parts — Expo 1.8L (4G93) and 2.4L (4G64) engines

5. Raise the vehicle and support safely.

6. Remove the engine under cover.

7. Drain the engine oil and coolant. Remove the lower radiator hose.

8. Remove the starter.

9. Remove the exhaust pipe from the exhaust manifolds.

10. If equipped with 4WD, remove transfer case from vehicle.

11. If equipped with a manual transmission, remove the transmission and all related parts.

12. If equipped with an automatic transmission and 2WD:

  a. Remove the inspection plate.

  b. Matchmark the flexplate to the converter; remove the torque converter bolts and move the torque converter back as far as it will go.

  c. Remove the lower bell housing bolts.

  d. Lower the vehicle.

13. Remove all intake ducts and air intake hoses.

14. Disconnect all linkages and cables from the throttle body.

15. Cover the fuel line connections with a clean shop rag and disconnect and plug the fuel lines.

16. If equipped with air conditioning, unbolt the air conditioning compressor from the engine and position

it aside. It is not necessary to remove the lines from the compressor.

17. Remove the radiator and shroud. Remove the fan and all related parts.

18. Disconnect the heater hoses.

19. Unbolt the power steering pump from its brackets and position it to the side. Do not remove the hoses from the pump.

20. Remove the alternator.

21. If equipped, remove the ignition coil and power transistor assembly.

22. Label and disconnect all remaining electrical connectors, vacuum hoses and check for any other items preventing engine removal.

23. Attach an engine removal device to the engine support eyes on the engine.

24. If equipped with an automatic transmission, support the transmission with a floor jack or equivalent. Remove the remaining bell housing bolts.

25. Remove the engine mount nuts and remove the engine from the vehicle.

To install:

26. Lower the engine into position and install the engine mount nuts. Torque the nuts to 20 ft. lbs. (27 Nm), on the Montero torque to 33 ft. lbs. (44 Nm).

27. Install the upper bell housing bolts.

28. Remove the engine removal device and the transmission support. Install the oil dipstick.

29. Raise the vehicle and support safely. Install the remaining bell housing bolts.

30. Install transfer case, if equipped.

31. If equipped with a manual transmission, install the transmission and all related parts.

32. If equipped with an automatic transmission, align the torque converter and flexplate and install the bolts.

33. Install the inspection plate and starter motor.

34. Install the exhaust pipe to the exhaust manifolds using new gaskets. Install the lower radiator hose.

35. Lower the vehicle.

36. Connect the heater hoses.

37. Connect the negative battery cable.

38. Install the alternator, power steering pump and all brackets.

39. Install the air conditioning compressor.

40. Connect all linkages and cables to the carburetor or throttle body.

41. Install the ignition coil and power transistor assembly, if equipped.

42. Connect all electrical connectors and vacuum hoses.

43. Install the fan and all related parts. Adjust all belt tensions, as required.

44. Install the radiator, shroud and upper hose.

45. Install the air cleaner assembly, ducts and air intake hose.

46. Fill the engine with the specified amount of oil and fill the radiator with coolant.

47. Connect the negative battery cable.

48. Connect the jumper wire from the fuel pump activation terminal to the positive battery post. Inspect the system for leaks.

49. Check the automatic transmission fluid level, if equipped.

50. Recheck all engine adjustments.

51. Install and align the hood.

2.4L (VIN G) Engine

1. Relieve the fuel system pressure.

CAUTION

Fuel injection systems remain under pressure after the engine has been turned OFF. Properly relieve fuel pressure before disconnecting any fuel lines. Failure to do so may result in fire or personal injury.

---

2. Disconnect the negative battery cable.

CAUTION

Wait at least 90 seconds after the negative (-) battery cable is disconnected to prevent possible deployment of the air bag.

---

3. Matchmark and remove the hood.

4. Remove the oil dipstick.

5. Raise the vehicle and support safely.

6. Remove the engine under cover.

7. Drain the engine oil and coolant. Remove the lower radiator hose.

8. Remove the starter.

9. Remove the exhaust pipe from the exhaust manifold.

10. If equipped with 4WD, remove transfer case from vehicle.

11. If equipped with a manual transmission, remove the transmission and all related parts.

12. If equipped with an automatic transmission and 2WD:

   a. Remove the inspection plate.

   b. Matchmark the flexplate to the converter; remove the torque converter bolts and move the tor-

que converter back as far as it will go.

   c. Remove the lower bell housing bolts.

   d. Lower the vehicle.

13. Remove all intake ducts and air intake hoses.

14. Disconnect all linkages and cables from the throttle body.

15. Cover the fuel line connections with a clean shop rag and disconnect and plug the fuel lines.

16. If equipped with air conditioning, unbolt the air conditioning compressor from the engine and position it aside. It is not necessary to remove the lines from the compressor.

17. Remove the radiator and shroud. Remove the fan and all related parts.

18. Disconnect the heater hoses.

19. Unbolt the power steering pump from its brackets and position it to the side. Do not remove the hoses from the pump.

20. Remove the alternator. Remove the ignition coil and power transistor assembly, if equipped.

21. Label and disconnect all remaining electrical connectors, vacuum hoses and check for any other items preventing engine removal.

22. Attach an engine removal device to the engine support eyes on the engine.

23. If equipped with an automatic transmission, support the transmission with a floor jack or equivalent. Remove the remaining bell housing bolts.

24. Remove the engine mount nuts and remove the engine from the vehicle.

To install:

25. Lower the engine into position and install the engine mount nuts. Torque the nuts to 14–22 ft. lbs. (30–40 Nm).

26. Install the upper bell housing bolts.

27. Remove the engine removal device and the transmission support. Install the oil dipstick.

28. Raise the vehicle and support safely. Install the remaining bell housing bolts.

29. Install transfer case, if equipped.

30. If equipped with a manual transmission, install the transmission and all related parts.

31. If equipped with an automatic transmission, align the torque converter and flexplate and install the bolts.

32. Install the inspection plate and starter motor.

33. Install the exhaust pipe to the exhaust manifold using new gaskets. Install the lower radiator hose.

34. Lower the vehicle.

35. Connect the heater hoses.

36. Connect the negative battery cable.

37. Install the alternator, power steering pump and all brackets.

38. Install the air conditioning compressor.

39. Connect all linkages and cables to the throttle body.

40. Install the ignition coil and power transistor assembly.

41. Connect all electrical connectors and vacuum hoses.

42. Install the fan and all related parts. Adjust all belt tensions, as required.

43. Install the radiator, shroud and upper hose.

44. Install the air cleaner assembly, ducts and air intake hose.

45. Fill the engine with the specified amount of oil and fill the radiator with coolant.

46. Connect the negative battery cable.

47. Connect the jumper wire from the fuel pump activation terminal to the positive battery post. Inspect the system for leaks.

48. Check the automatic transmission fluid level, if equipped.

49. Recheck all engine adjustments.

50. Install and align the hood.

Expo 1.8L (4G93) Engine

1. Relieve fuel system pressure.

2. Disconnect the negative battery cable. Remove the under cover if equipped.

3. Matchmark the hood and hinges and remove the hood assembly. Remove the air cleaner assembly and all adjoining air intake duct work.

4. Drain the engine coolant and remove the radiator assembly, coolant reservoir and intercooler.

5. Remove the transaxle assembly.

6. Disconnect the accelerator cable, breather hose and heater hose connections from the engine.

7. Note the locations and remove vacuum hoses from engine. Be sure to disconnect brake booster vacuum supply.

8. Disconnect the fuel feed and return hoses.

9. Disconnect the oxygen sensor connection, coolant temperature gauge and coolant temperature sensor connections.

10. Disconnect the oil pressure switch connection.

11. On models with automatic transmissions, disconnect the thermo switch.

12. Disconnect the harness connections for the idle speed control motor and throttle position sensor.

13. Disconnect the EGR temperature sensor (California).

14. Note locations for reassembly and disconnect injector connections.

15. Disconnect the power transistor and ignition coil connections.

16. Disconnect the alternator and power steering switch wiring.

17. Remove the air conditioner drive belt and the air conditioning compressor. Leave the hoses attached. Do not discharge the system. Wire the compressor aside.

18. Remove the power steering pump and wire aside.

19. Remove the starter and alternator harness clamp.

20. Remove the exhaust manifold to head pipe nuts. Discard the gasket.

21. Attach a hoist to the engine and support the engine weight. Remove the engine mount bracket. Remove any torque control brackets (roll stoppers).

22. Remove the engine assembly from the vehicle.

To install:

23. Install the engine and secure in position. The front lower mount through bolt nut should not be tightened until the full weight of the engine is on the mount. Tighten through bolt to 72 ft. lbs. (100 Nm) and bracket mounting bolts to 42 ft. lbs. (58 Nm). The bracket mounting nut is tightened to 38 ft. lbs. (53 Nm).

24. Using a new gasket, position exhaust pipe onto the manifold and tighten the flange nuts to 33 ft. lbs. (45 Nm).

25. Install the power steering pump, alternator and air conditioner compressor. Install and adjust drive belts, tighten all mounting bolts.

26. Connect the alternator and power steering wiring.

27. Secure the alternator and starter harness clamp.

28. Connect the ignition coil and power transistor connections.

29. Connect the fuel injector harness connections.

30. On California models, connect EGR temperature sensor plug

31. Connect the wiring for idle speed control motor and throttle position sensor.

32. On automatic transmission models, connect the thermo switch.

33. Connect the oil pressure switch wiring.

34. Connect the oxygen sensor, coolant temperature gauge and coolant temperature sensor.

35. Using a new O-ring connect fuel feed hose and tighten the bolts to 44 inch lbs. (5 Nm).

36. Using a new hose clamp, connect the fuel return hose.

37. Connect the noted vacuum hoses and connect brake booster vacuum supply.

38. Connect the breather hose, heater hoses and accelerator cable. Inspect accelerator cable for proper adjustment.

39. Install the transaxle assembly.

40. Install the radiator assembly and refill the cooling system, engine oil and transmission oil.

41. Install the air cleaner and hood assembly.

42. Connect the negative battery cable and run engine.

43. Inspect all the connections and check all fluid levels.

Expo 2.4L (4G64) Engines

1. Relieve the fuel system pressure.

——————— CAUTION ———————

Fuel injection systems remain under pressure after the engine has been turned OFF. Properly relieve fuel pressure before disconnecting any fuel lines. Failure to do so may result in fire or personal injury.

2. Disconnect the negative battery cable. Remove the under cover if equipped.

3. Matchmark the hood and hinges and remove the hood assembly.

4. Remove the air cleaner assembly and all adjoining air intake duct work.

5. Drain the engine coolant and remove the radiator assembly and coolant reservoir.

6. Remove the transaxle assembly.

7. Note locations and remove the vacuum hoses from engine. Be sure to disconnect brake booster vacuum supply.

8. Disconnect the heater hoses at the cylinder head and coolant inlet pipe.

9. Disconnect the fuel feed and return hoses.

10. Disconnect the accelerator cable and breather hose.

11. Disconnect the oxygen sensor connection, coolant temperature gauge and coolant temperature sensor connections.

12. Disconnect the wiring at the distributor.

13. Disconnect the harness connections for the idle speed control and throttle position sensor.

14. Note locations for reassembly and disconnect injector connections.

NOTE: At this point all wiring to the intake should be disconnected. Position the main wiring harness aside.

15. Disconnect the alternator and oil pressure switch wiring.

16. Remove the air conditioner drive belt and the air conditioning compressor. Leave the hoses attached. Do not discharge the system. Wire the compressor aside.

17. Remove the power steering pump and wire aside.

18. Remove the exhaust manifold to head pipe nuts. Discard the gasket.

19. Attach a hoist to the engine and support the engine weight. Remove the engine mount bracket. Remove any torque control brackets (roll stoppers).

20. Remove the engine assembly from the vehicle.

To install:

21. Install the engine and secure in position. The front lower mount through bolt nut should not be tightened until the full weight of the engine is on the mount. Tighten through bolt to 51 ft. lbs. (69 Nm) and bracket mounting bolts to 42 ft. lbs. (58 Nm).

22. Using a new gasket, position exhaust pipe onto the manifold and tighten 3 hole flange nuts to 33 ft. lbs. (44 Nm).

23. Install the power steering pump, alternator and air conditioner compressor. Install and adjust drive belts, tighten all mounting bolts.

24. Connect the alternator and oil pressure switch wiring.

25. Connect the noise filter, ignition coil and power transistor connections, at the distributor.

26. Connect the fuel injector harness connections.

27. Connect the wiring for the idle speed control and throttle position sensor.

28. Connect the oxygen sensor, coolant temperature gauge and coolant temperature sensor.

29. Using new O-rings, connect fuel feed and return hoses and tighten bolts to 3–4 ft. lbs. (4–6 Nm).

30. Connect the vacuum hoses and connect the brake booster vacuum supply.

31. Connect the breather hose, heater hoses and accelerator cable.

Inspect the accelerator cable for proper adjustment.

32. Install the transaxle assembly.

33. Install the radiator assembly and refill the cooling system, engine oil and transmission oil.

34. Install the air cleaner and hood assembly.

35. Connect the negative battery cable and run engine.

36. Inspect all the connections and check all fluid levels.

Engine Mounts

REMOVAL AND INSTALLATION

1993–95 Vehicles

Front

1. Disconnect negative battery cable.

2. Install engine support fixture in place. Raise and safely support the vehicle.

3. Remove the engine front support insulator.

4. Remove the front insulator stopper.

5. Remove the heat protector.

6. Unbolt the rear crossmember or mount as required.

7. Raise engine with support fixture far enough to remove mounts, remove remaining bolts and mounts. Transfer insulator and stopper to new mount.

To install:

8. Install mounts to engine block. Install stopper and heat protector to the mount if equipped.

9. Lower the engine to original position and insert bolts through mounting brackets.

10. Install heat protector to insulator if equipped.

11. Lower the vehicle and remove the engine support fixture.

12. Reconnect the negative battery cable.

Rear

1. Disconnect negative battery cable.

2. Raise and safely support the vehicle.

3. Install transmission jack into position and raise transmission slightly.

4. Remove rear mount attaching bolts and crossmember attaching bolts. Remove the engine mount from the vehicle.

To install:

5. Position the mount on the crossmember and install the mounting bolts.

6. Lower the transmission onto the mount and install the nuts and bolts.

7. Lower the vehicle to the floor and connect the negative battery cable.

1996–97 3.0L (VIN H) and 3.5L (VIN M) Engines

Front

1. Install an engine support fixture in place. Raise and safely support the vehicle.

2. Remove the engine front support insulator.

3. Remove the front insulator stopper.

4. Remove the heat protector.

5. Unbolt the rear crossmember or mount as required.

6. Raise engine with support fixture far enough to remove mounts, remove remaining bolts and mounts. Transfer insulator and stopper to new mount.

To install:

7. Install mounts to engine block. Install stopper and heat protector to the mount if equipped.

8. Lower the engine to original position and insert bolts through mounting brackets.

9. Torque the right and left through bolts to:
- 3.0L 12 valve: 12 ft. lbs. (17 Nm)
- 3.0L 24 valve and 3.5L: 20 ft. lbs. (26 Nm)

10. Torque the right and left mounts to the frame to:
- 3.0L 12 valve: 25 ft. lbs. (34 Nm)
- 3.0L 24 valve and 3.5L: 33 ft. lbs. (44 Nm)

11. Install heat protector to insulator if equipped.

12. Lower the vehicle and remove the engine support fixture.

Rear

1. Raise and safely support the vehicle.

2. Install a transmission jack into position and raise transmission slightly.

3. Remove rear mount attaching bolts and crossmember attaching bolts. Remove the engine mount from the vehicle.

To install:

4. Position the mount on the crossmember and torque the mounting bolts:
- 3.0L: 16 ft. lbs. (22 Nm)
- 3.5L: 18 ft. lbs. (24 Nm)

5. Lower the transmission onto the mount and install the nuts and bolts.

1996–97 2.4L (VIN G) Engine

Front

1. Disconnect negative battery cable.

———— **CAUTION** ————

Wait at least 90 seconds after the negative (-) battery cable is disconnected to prevent possible deployment of the air bag.

2. Install engine support fixture in place. Raise and safely support the vehicle.

3. Remove the engine front support insulator.

4. Remove the front insulator stopper.

5. Remove the heat protector.

6. Unbolt the rear crossmember or mount as required.

7. Raise engine with support fixture far enough to remove mounts, remove remaining bolts and mounts. Transfer insulator and stopper to new mount.

To install:

8. Install mounts to engine block. Install stopper and heat protector to the mount if equipped.

9. Lower the engine to original position and insert bolts through mounting brackets and torque to 14–22 ft. lbs. (20–30 Nm).

10. Install heat protector to insulator.

11. Lower the vehicle and remove the engine support fixture.

12. Reconnect the negative battery cable.

Rear

1. Disconnect negative battery cable.

———— **CAUTION** ————

Wait at least 90 seconds after the negative (-) battery cable is disconnected to prevent possible deployment of the air bag.

2. Raise and safely support the vehicle.

3. Install transmission jack into position and raise transmission slightly.

4. Remove rear mount attaching bolts and crossmember attaching bolts. Remove the engine mount from the vehicle.

To install:

5. Position the mount on the crossmember and install the mounting bolts.

6. Lower the transmission onto the mount and install the nuts and bolts and torque to:
- 4WD: 13–18 ft. lbs. (18–25 Nm)
- RWD: 14–18 ft. lbs. (20–25 Nm)

7. Lower the vehicle to the floor and connect the negative battery cable.

Expo 1.8L (4G93) Engine

Upper mount

1. Disconnect the negative battery cable.

2. Raise and safely support the engine so it is not resting on the engine mount. One suggested way is a block of wood between a floor jack and the oil pan. Use care not to bend or damage any components.

3. Remove the retainer bolt from the clamp securing the power steering pressure hose and the air conditioning low pressure hose, if equipped.

4. Remove the engine mount bracket and body connection through bolt.

5. Disconnect the engine mounting bracket, by removing the two nuts and bolts securing it to the engine and remove.

To install:

6. Install the engine mounting bracket and stopper plate. Align the through bolt hole to the notch in the body bracket, make sure they are installed properly. Torque upper mount to engine nuts to 35 ft. lbs., bolts to 42 ft. lbs. (58 Nm) and the upper mount through bolt nut to 72 ft. lbs. (100 Nm). There is a second nut that secures the through bolt, tighten the nut to 38 ft. lbs. (53 Nm).

Lower mount

1. Disconnect the negative battery cable.

2. Raise and safely support the engine so it is not resting on the engine mount. One suggested way is a block of wood between a floor jack and the oil pan. Use care not to bend or damage any components.

3. Remove the stopper through bolt.

4. Remove the stopper mount frame bolts and pry mount out.

To install

5. Position the lower front roll stopper so the part of the bracket with the hole in it is facing away from the engine. Install the frame mounting bolts and tighten.

6. Install the front lower mount through bolt nut, but do not fully tighten until the full weight of the engine is on the mount. Torque lower

mount through bolt nut 38 ft. lbs. (53 Nm).

7. Connect the negative battery cable.

Expo 2.4L (4G64) Engines

Upper mount

1. Disconnect the negative battery cable.

2. Raise and safely support the engine so it is not resting on the engine mount. Use care not to bend or damage any components.

3. Remove the retainer bolt from the clamp securing the power steering pressure hose and the air conditioning low pressure hose, if equipped.

4. Remove the engine mount bracket and body connection through bolt.

5. Disconnect the engine mounting bracket, by removing the two nuts and bolts securing it to the engine and remove.

To install:

6. Install the engine mounting bracket and stopper plate. Align the through bolt hole to the notch in the body bracket, make sure they are installed properly. Torque upper mount to engine nut and bolt to 42 ft. lbs. (58 Nm) and the upper mount through bolt nut to 51 ft. lbs. (70 Nm).

Lower

1. Disconnect the negative battery cable.

2. Raise and safely support the engine so it is not resting on the engine mount. Use care not to bend or damage any components.

3. Remove the stopper through bolt.

4. Remove the stopper mount frame bolts and pry mount out.

To install

5. Position the lower front roll stopper so the part of the bracket with the hole in it is facing away from the engine. Install the frame mounting bolts and tighten.

6. Install the front lower mount through bolt nut, but do not fully tighten until the full weight of the engine is on the mount. Torque lower mount through bolt nut 42 ft. lbs. (58 Nm).

7. Connect the negative battery cable.

Cylinder Head

REMOVAL AND INSTALLATION

2.4L (VIN G) Engine

1. Disconnect negative battery cable.

——————— **CAUTION** ———————
Wait at least 90 seconds after the negative (-) battery cable is disconnected to prevent possible deployment of the air bag.

2. Properly relieve the fuel system pressure.

——————— **CAUTION** ———————
Fuel injection systems remain under pressure after the engine has been turned OFF. Properly relieve fuel pressure before disconnecting any fuel lines. Failure to do so may result in fire or personal injury.

3. Drain the cooling system.

4. Remove the upper radiator hose. Disconnect the heater hoses.

5. Remove the air intake hose and pipe.

6. Disconnect the accelerator, and if equipped, kickdown cable.

7. Disconnect and plug the fuel lines.

8. If equipped with power steering, unbolt the power steering pump from its brackets and position it to the side.

NOTE: Do not disconnect the power steering lines.

9. Remove the timing belt upper cover and valve cover.

10. Rotate the crankshaft clockwise and align the timing mark on the camshaft sprocket with the timing mark on the cylinder head.

11. Remove the camshaft bolt.

12. Remove the sprocket from the camshaft (with the timing belt attached) and allow it to rest on the lower cover. Secure the belt to the sprocket so that they do not become disengaged.

——————— **WARNING** ———————
Do not rotate the crankshaft after the camshaft sprocket is removed from the camshaft. Secure the sprocket and timing belt so there is no slack in the belt. Make sure the sprocket does not become disengaged from the timing belt. If the engine is disturbed or the timing belt moved, the camshaft timing will have to be reset.

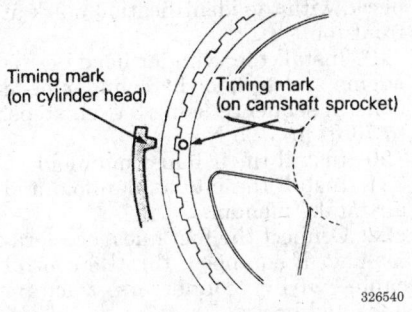

Timing mark
(on cylinder head)

Timing mark
(on camshaft sprocket)

326540

Align the camshaft sprocket timing mark with the timing mark on the cylinder head — 2.4L (VIN G) engines

13. Label and disconnect the spark plug wires from the spark plugs. Remove the distributor cap and wires.

14. Mark the position of the rotor and distributor housing in relation to the cylinder head and remove the distributor.

15. Label and disconnect all vacuum lines, hoses and wiring connectors from the manifolds and cylinder head.

16. Raise the vehicle and support safely.

17. Remove the exhaust pipe from the exhaust manifold.

18. Lower the vehicle.

19. Remove cylinder head bolts, starting from the outside and working inward. Remove the cylinder head from the engine.

20. If necessary, remove the intake and exhaust manifolds from the cylinder head.

21. Clean the cylinder head gasket mating surfaces.

To install:

22. If removed, install the intake and exhaust manifolds to the cylinder head.

23. Install a new head gasket to the block and position the cylinder head assembly with all head bolts and washers. Torque the bolts in sequence, in 2 or 3 steps, to 72 ft. lbs. (100 Nm).

24. Install the camshaft sprocket to the camshaft and tighten the bolt to 72 ft. lbs. (100 Nm).

25. Install the distributor, aligning the marks made during removal. Install the distributor cap and spark plug wires.

26. Install the power steering pump and adjust the belt tension.

27. Connect the heater hoses and upper radiator hose.

28. Install the valve cover and upper timing belt cover.

29. Connect the fuel lines. Connect the accelerator, and if equipped, kickdown cable.

30. Connect all remaining wiring connectors, vacuum lines and hoses in their proper locations.

31. Install the air intake pipe and hose.

32. Refill the cooling system.

33. Connect the negative battery cable.

34. Start the engine and check for leaks. Check the ignition timing.

3.0L (VIN H) Engines

12 Valve Heads

— **CAUTION** —
Fuel injection systems remain under pressure after the engine has been turned OFF. Properly relieve fuel pressure before disconnecting any fuel lines. Failure to do so may result in fire or personal injury.

1. Relieve the fuel system pressure.

2. Disconnect the negative battery cable.

— **CAUTION** —
Work must be started after 90 seconds from the time the ignition switch is turned to the LOCK position and the negative (-) battery cable is disconnected.

3. Drain the cooling system. Remove the upper radiator hose.

4. Remove the air intake hose.

5. Remove the accessory drive belts, fan and pulleys.

6. Remove the air conditioning compressor, power steering pump and mounting brackets and position them to the side, without disconnecting the lines.

7. If removing the right cylinder head, disconnect the wiring and remove the alternator cover, alternator and alternator stay.

8. Remove the timing belt covers.

9. Remove the timing belt as follows:

   a. Rotate the crankshaft and bring the piston in No. 1 cylinder to TDC on the compression stroke. Align the camshaft and crankshaft sprocket timing marks.

   b. Mark the timing belt in the direction of rotation for reinstallation purposes.

   c. Loosen the timing belt tensioner bolt and turn the tensioner counterclockwise. Remove the timing belt.

— **WARNING** —
Do not rotate the crankshaft or camshaft sprockets after the timing belt has been removed.

10. Label and disconnect the spark plug wires from the spark plugs.

11. If removing the left cylinder head, remove the distributor cap. Mark the position of the rotor and the distributor housing in relation to the cylinder head, then remove the distributor.

12. Remove the valve cover.

13. Remove the EGR pipe and the air intake plenum stays.

14. Disconnect the accelerator and, if equipped, throttle control cable. Disconnect and plug the fuel lines.

15. Label and disconnect the wiring connectors, vacuum lines and hoses from the air intake plenum, intake manifold and cylinder head.

16. Remove the air intake plenum and intake manifold.

17. Remove the exhaust manifold.

18. If removing the right cylinder head, remove the dipstick tube.

19. If necessary, remove the camshaft sprocket bolt and camshaft sprocket. Remove the alternator bracket and/or timing belt rear cover.

20. Remove the cylinder head bolts starting from the outside and working inward.

21. Remove the cylinder head from the engine.

22. Clean the gasket mounting surfaces.

To install:

23. Install the new cylinder head gasket over the dowels on the engine block, with the identification mark at front top.

24. Install the cylinder head on the engine and torque the cylinder head bolts in sequence using 3 even steps, to 70 ft. lbs. (95 Nm).

25. If removed, install the timing belt rear cover and/or alternator bracket. Torque the alternator bracket bolts in the proper sequence.

26. If removed, install the dipstick tube using a new O-ring coated with clean engine oil.

27. Install the exhaust manifold.

28. Install the intake manifold and air intake plenum.

29. Install the EGR pipe and air intake plenum stays.

30. Connect the fuel lines, accelerator and, if equipped, throttle control cable, wiring connectors, vacuum lines and hoses.

31. Install the valve cover.

32. Install the distributor, if removed, aligning the marks made dur-

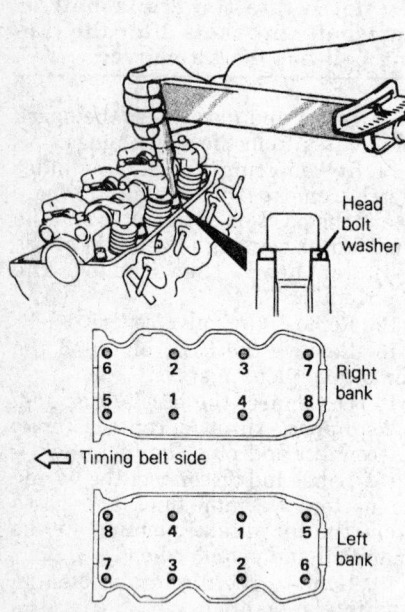

Tighten the cylinder head bolts in the order shown, in 2 or 3 passes — 12 valve 3.0L (VIN H) engines

Head bolt washer

Right bank

⇐ Timing belt side

Left bank

319331

ing removal. Install the distributor cap and spark plug wires.

33. Make sure the camshaft and crankshaft sprocket timing marks are aligned.

34. Turn the timing belt tensioner to the extreme counterclockwise position and temporarily tighten the bolt.

35. Install the timing belt in the original rotation direction. Loosen the timing belt tensioner bolt and allow the spring force of the tensioner to tension the belt.

36. Turn the crankshaft 2 turns in the normal direction of rotation and check the timing mark alignment.

37. If the timing is correct, tighten the tensioner bolt to 21 ft. lbs. (30 Nm). If the timing is incorrect, repeat the belt installation procedure.

38. Install the timing belt covers.

39. If removed, install the alternator, alternator cover and alternator stay.

40. Install the air conditioning compressor and power steering pump with the brackets.

41. Install the pulleys, cooling fan and accessory drive belts. Tighten the crankshaft pulley bolt to 137 ft. lbs. (190 Nm).

42. Install the upper radiator hose and the air intake hose.

43. Fill the cooling system.

44. Connect the negative battery cable.

45. Start the engine and check for leaks. Check the ignition timing.

24 Valve Heads

1. Relieve the fuel system pressure.

2. Disconnect the negative battery cable.

3. Drain the cooling system. Remove the upper radiator hose.

4. Remove the accessory drive belts, fan and pulleys.

5. Remove the air conditioning compressor, power steering pump and mounting brackets and position them to the side, without disconnecting the lines.

6. Remove the timing belt covers.

7. Remove the timing belt as follows:

   a. Rotate the crankshaft and bring the piston in No. 1 cylinder to TDC on the compression stroke. Align the camshaft and crankshaft sprocket timing marks.

   b. Mark the timing belt in the direction of rotation for reinstallation purposes.

   c. Loosen the timing belt tensioner bolt and turn the tensioner counterclockwise. Remove the timing belt.

—— **WARNING** ——

Do not rotate the crankshaft or camshaft sprockets after the timing belt has been removed.

8. Label and disconnect the spark plug wires from the spark plugs.

9. Remove the valve cover.

10. Disconnect and plug the fuel lines.

11. Label and disconnect the wiring connectors, vacuum lines and hoses from the air intake plenum, intake manifold and cylinder head.

12. Remove the air intake plenum and intake manifold.

13. Remove the exhaust manifold.

14. If necessary, remove the camshaft sprocket bolt and camshaft sprocket. Remove the alternator bracket and/or timing belt rear cover.

15. Remove the cylinder head bolts starting from the outside and working inward.

16. Remove the cylinder head from the engine.

17. Clean the gasket mounting surfaces.

To install:

18. Install the new cylinder head gasket over the dowels on the engine block, with the identification mark at front top.

19. Install the cylinder head on the engine and torque the cylinder head bolts in sequence using 3 even steps, to 70 ft. lbs. (95 Nm).

20. Install the exhaust manifold.

21. Install the intake manifold and air intake plenum.

22. Connect the fuel lines, accelerator and, if equipped, throttle control cable, wiring connectors, vacuum lines and hoses.

23. Install the valve cover.

24. Make sure the camshaft and crankshaft sprocket timing marks are aligned.

25. Turn the timing belt tensioner to the extreme counterclockwise position and temporarily tighten the bolt.

26. Install the timing belt in the original rotation direction. Loosen the timing belt tensioner bolt and allow the spring force of the tensioner to tension the belt.

27. Turn the crankshaft 2 turns in the normal direction of rotation and check the timing mark alignment.

28. If the timing is correct, tighten the tensioner bolt to 21 ft. lbs. (30 Nm). If the timing is incorrect, repeat the belt installation procedure.

29. Install the timing belt covers.

30. If removed, install the alternator, alternator cover and alternator stay.

31. Install the air conditioning compressor and power steering pump with the brackets.

32. Install the pulleys, cooling fan and accessory drive belts. Tighten the crankshaft pulley bolt to 137 ft. lbs. (190 Nm).

33. Install the upper radiator hose and the air intake hose.

34. Fill the cooling system.

35. Connect the negative battery cable.

36. Start the engine and check for leaks. Check the ignition timing.

3.5L (VIN M) Engines

—— **CAUTION** ——

Fuel injection systems remain under pressure after the engine has been turned OFF. Properly relieve fuel pressure before disconnecting any fuel lines. Failure to do so may result in fire or personal injury.

1. Relieve fuel system pressure. Disconnect the negative battery cable.

CAUTION

Work must be started after 90 seconds from the time the ignition switch is turned to the LOCK position and the negative (-) battery cable is disconnected.

2. Drain the cooling system.

3. Remove the air intake hoses.

4. Remove air intake plenum and intake manifold.

5. Remove the exhaust manifold.

6. Remove the timing belt.

7. Remove the breather hose.

8. Remove the spark plug cable center cover and remove the spark plug cables.

9. Remove the valve cover.

10. To remove the intake camshaft sprocket, hold the camshaft with a wrench on the hexagon near the end of the camshaft and remove the sprocket bolt.

11. Remove the rear timing belt cover.

12. Remove the ignition coil.

13. Disconnect all water hoses from the thermostat housing and remove the housing.

14. Disconnect the water inlet from the front head and discard O-ring.

15. Loosen the cylinder head mounting bolts in three steps, starting from the outside and working inward. Lift off the cylinder head assembly and remove the head gasket.

To install:

16. Thoroughly clean and dry the mating surfaces of the head and block. Check the cylinder head for cracks, damage or engine coolant leakage. Remove scale, sealing compound and carbon. Clean oil passages thoroughly. Check the head for flatness. End to end, the head should be within 0.0012 in. , normally with 0.008 in. the maximum allowed out of true. The total thickness allowed to be removed from the head and block is 0.008 in. maximum.

17. Place a new head gasket on the cylinder block with the identification marks in the front top (upward) position. Do not use sealer on the gasket.

18. Carefully install the cylinder head on the block. Make sure the head bolt washers are installed with the chamfered edge upward. Using three even steps, torque the head bolts in sequence, to 76–83 ft. lbs. (105–115 Nm) for non-turbocharged cold engine or 87–94 ft. lbs. (120–130 Nm) for turbocharged cold engine.

19. Install new O-ring and connect the water inlet to the front head.

20. Replace the gaskets and install the thermostat housing and connect the hoses.

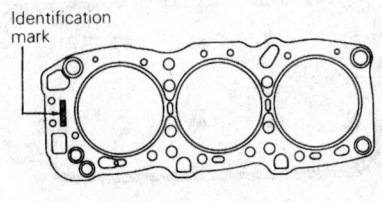

Identification mark

Cylinder head gasket identification marks — 3.5L (VIN M) engines

319309

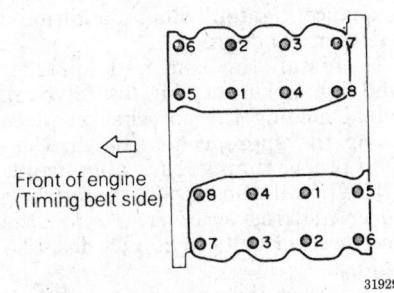

Front of engine (Timing belt side)

Cylinder head bolt tightening sequence — 3.5L (VIN M) engines

319291

21. Install the ignition coil and center rear timing belt cover.

22. Install the intake camshaft sprocket. Use hex flange on camshaft to secure and torque the retaining bolt to 65 ft. lbs. (90 Nm).

23. Apply sealer to the lower edges of the half-round portions of the belt-side of the new gasket and install the valve cover. Torque the bolts to 7 ft. lbs. (10 Nm).

24. Connect the spark plug cables and install the center cover.

25. Install the breather hose.

26. Install the timing belt and all related items.

27. Using all new gaskets, install the intake manifold, air intake plenum, turbocharger and exhaust manifold, following the proper torque sequences.

28. Install the air intake hoses.

29. Change the engine oil and oil filter.

30. Fill the system with coolant.

31. Connect the negative battery cable, run the vehicle until the thermostat opens, fill the radiator completely.

32. Adjust the accelerator cable. Check and adjust the idle speed and ignition timing.

33. Once the vehicle has cooled, recheck the coolant level.

Expo 1.8L (4G93) Engine

1. Relieve fuel system pressure. Disconnect the negative battery cable.

2. Position the No. 1 piston to TDC of its compression stroke.

3. Remove the air cleaner assembly.

4. Drain the cooling system.

5. Disconnect the brake booster vacuum hose and PVC valve connection.

6. Note the locations and disconnect the vacuum hoses from the intake and throttle body.

7. Remove the upper radiator hose, overflow tube and the water hose from the thermostat to the throttle body.

8. Wrap the connection with a shop towel and disconnect the high pressure fuel line at the fuel rail. Discard the O-ring.

9. Disconnect the fuel return hose from the fuel pressure regulator.

10. Disconnect the accelerator cable connection from the throttle body.

11. Disconnect the electrical harnesses at the oil pressure switch, oxygen sensor, water temperature sensor connectors and distributor.

12. Disconnect the electrical harnesses at the idle air control motor and the EGR temperature sensor.

13. Disconnect the wiring from condenser, idle speed control, throttle position sensor and knock sensor.

14. Note harness plug connections for reassembly and disconnect fuel injectors.

15. Disconnect the spark plug cables from each spark plug.

16. Unbolt the control harness assembly and position aside.

17. Remove the thermostat housing, thermostat and the thermostat case with O-ring from the engine.

18. Remove the rocker cover.

19. Remove the timing belt upper cover.

20. If not already done, rotate the crankshaft in the clockwise direction to align the camshaft timing marks. Matchmark the camshaft sprocket and the timing belt. Tie the camshaft sprocket and the timing belt together so the sprocket will not move with respect to the timing belt.

21. While holding the camshaft sprocket in position using the appropriate wrench, remove the camshaft sprocket and with the belt attached. Wire the sprocket and belt aside making sure constant tension is maintained on the belt. Do not allow

the belt to slacken or engine timing may be altered.

NOTE: When removing the camshaft sprocket, do not allow the crankshaft to rotate. Confirm proper engine timing during installation.

22. Loosen the cylinder head bolts in two or three steps in the the proper sequence.

23. Remove the cylinder head from the engine.

——— **CAUTION** ———
When removing the cylinder head, take care not to bend or damage the plug guide. The plug guide can not be replaced.

24. Remove the cylinder head gasket from the block.

To install:
25. Thoroughly clean and dry the mating surfaces of the head and block. Check the cylinder head for cracks, damage or engine coolant leakage. Remove scale, sealing compound and carbon. Clean oil passages thoroughly. Check the head for flatness. end to end, the head should be within 0.002 inch with 0.008 inch the maximum allowed out of true. The total thickness allowed to be removed from the head and block is 0.008 in. maximum.

26. Place a new head gasket on the cylinder block with the identification marks facing upward. Make sure the gasket has the proper identification mark for the engine. Do not use sealer on the gasket.

27. Carefully install the cylinder head on the block.

28. Inspect the cylinder head bolt prior to installation. The length below the head of the bolts should not exceed 3.795 inches (96.4mm). If bolt shank length exceeds limit, bolt must be replaced. New bolts are always recommended.

29. Apply a small amount of engine oil to the thread section of the bolt and install so the chamfer of the washer faces upward.

30. Tighten the cylinder head bolts in the proper order as follows:
   a. In the proper tightening sequence, torque bolts to 54 ft. lbs. (75 Nm).
   b. In the reverse order of the tightening sequence, fully loosen bolts.
   c. In the proper tightening sequence, torque bolts to 14 ft. lbs. (20 Nm).
   d. In the proper tightening sequence, tighten bolts an additional ¼turn (90 degrees).

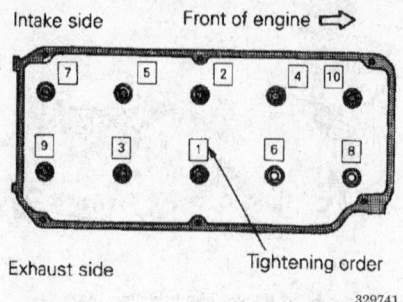

Cylinder head bolt torque sequence — 1.8L and 2.4L engines

   e. In the proper tightening sequence, tighten bolts an additional ¼turn (90 degrees).

31. Install the camshaft sprocket and tighten bolt to 65 ft. lbs. (90 Nm), while holding the sprocket in place using the appropriate wrench. Confirm proper timing mark alignment.

32. Install the upper timing belt cover and rocker cover. Torque the rocker cover bolts to 29 inch lbs. (3.3 Nm).

33. Loosen the water pipe mounting bolt for ease of thermostat housing installation.

34. Apply a thin bead of sealant MD970389 or equivalent, to the water tube connection on the thermostat case.

35. Apply a small amount of water to the O-ring of the water inlet pipe and press the thermostat case assembly onto the water inlet pipe. Install the thermostat case assembly mounting bolt tightening to 16 ft. lbs. (22 Nm).

36. Tighten the water pipe mounting bolt.

37. Install the thermostat into the housing so the jiggle valve is located at the top. Tighten the housing bolts to 10 ft. lbs. (14 Nm).

38. Connect the upper radiator hose to the thermostat housing.

39. Connect or install all previously disconnected hoses, cables and electrical connections. Adjust the throttle cable(s).

40. Replace the O-ring for the high pressure hose and install a new clamp on the return hose and reconnect the fuel lines.

41. Install the air intake hose. Connect the breather hose and air cleaner case cover.

42. Reconnect the brake booster and the PCV vacuum hoses.

43. Change the engine oil and oil filter.

44. Fill the system with coolant.

45. Connect the negative battery cable, run the vehicle until the thermostat opens, fill the radiator completely.

46. Check and adjust the idle speed and ignition timing.

47. Check all systems for leaks. Allow the engine to cool and recheck the coolant level.

Expo 2.4L (4G64) Engines

1. Relieve fuel system pressure. Disconnect the negative battery cable.

2. Drain the cooling system.

3. Disconnect the accelerator cable.

4. Remove the radiator.

5. Disconnect the air flow sensor connector and the air intake hose. Remove the air cleaner cover.

6. Disconnect the PCV hose.

7. Disconnect the water hose connection at the throttle body to water inlet pipe.

8. Disconnect the water hose connection at the throttle body to thermostat hose.

9. Wrap the connection with a shop towel and disconnect the high pressure fuel line at the fuel rail.

10. Disconnect the fuel return hose and remove the O-ring.

11. Disconnect the accelerator cables connection at the throttle body.

12. Disconnect the spark plug cables from the spark plugs.

13. Disconnect the electrical connectors from the oxygen sensor, water temperature gauge unit, engine coolant temperature sensor, TPS, power transistor connector, fuel injectors, ignition coil, distributor, and air conditioner compressor. Label prior to disconnecting to assure correct relocation on assembly.

14. Remove the bolt retaining the power steering hose and air conditioner hose clamp.

15. Remove the coolant reservoir. Remove the bolt holding the ground wire to the manifold.

16. Place a jack and wood block under the oil pan and carefully lift just enough to take the weight off the engine mounting bracket. Then remove the engine mounting bracket taking note of the position of the mount stopper.

17. Remove the valve cover, gasket and half-round seal.

18. Remove the timing belt front upper cover.

19. If possible, rotate the crankshaft clockwise until the timing marks on the cam sprocket and belt align. Matchmark the timing sprocket to the belt. Remove the sprocket bolt and remove the

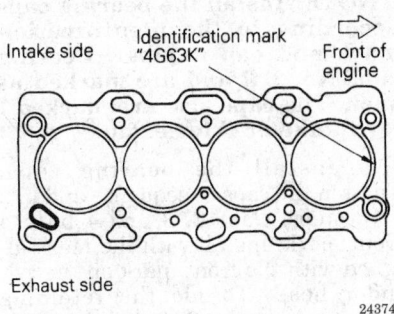

Intake side — Identification mark "4G63K" — Front of engine

Exhaust side

243749

Cylinder head identification marks — Expo 2.4L (4G64) engines

sprocket with the timing belt attached. Attach a flexible cord to the hood and suspend the sprocket so it cannot turn and there is no slack in the belt. Remove the timing belt rear upper cover.

20. Loosen the head bolts in the correct sequence in 2 or 3 steps. Remove the cylinder head bolts and head assembly from the block.

To install:

21. Thoroughly clean and dry the mating surfaces of the head and block. Remove scale, sealing compound and carbon. Clean oil passages thoroughly.

22. Perform the following checks before reassembly:

• Check the cylinder head for cracks, damage or engine coolant leakage.

• Check the head for flatness. End to end, the head should be within 0.002 in. (0.051mm) normally with 0.008 in. (0.200mm) the maximum allowed out of true. The total thickness allowed to be removed from the head and block is 0.008 in. (0.200mm) maximum.

• Check the cylinder head bolts for stretching or necking. The shank length should not exceed 3.91 in. (99.4 mm).

23. Place a new head gasket on the cylinder block with the identification marks at the top (upward) position. Make sure the gasket has the proper identification mark for the engine. Do not use sealer on the gasket. Replace the turbo gasket and ring, if equipped.

24. Carefully install the cylinder head on the block. Torque the cylinder head using the following procedure:

a. Torque all bolts in sequence to 58 ft. lbs. (78 Nm).

b. Using loosening sequence, fully loosen all bolts.

c. Torque all bolts in sequence to 14 ft. lbs. (20 Nm).

d. Tighten all bolts in sequence ¼ turn (90 °).

e. Once again, tighten all bolts in sequence ¼ turn (90 °).

NOTE: Install the head bolt washer so the sagging side made by tapping out the washer is facing upward.

25. Install the camshaft sprocket and tighten bolt to 65 ft. lbs. (90 Nm), while holding the sprocket in place using the appropriate wrench. Confirm proper timing mark alignment.

26. Apply sealer to the perimeter of the half-round seal and to the lower edges of the half-round portions of the belt-side of the new gasket. Install the valve cover.

27. Install the engine mount positioning the stopper in the same direction as it was prior to removal.

28. Install the power steering and air conditioning compressor hose clamp in position and secure with the retainer bolt. Tighten the bolt to 9 ft. lbs. (12 Nm).

29. Install the coolant reservoir tank.

30. Reconnect all electrical harness connectors disconnect during disassembly. Connect the ground wire to the manifold.

31. Connect the accelerator cables and the spark plug cables.

32. Replace the O-rings and reconnect the fuel lines.

33. Reconnect the water hoses to throttle body, thermostat and the heater assembly.

34. Install the air intake case cover, air flow sensor connector and the radiator.

35. Fill the system with coolant. Adjust the accelerator cable.

36. Firmly set the parking brake. Start the engine and allow to idle until the thermostat opens, add coolant as required to fill system to the appropriate level.

37. Check all systems for leaks. Allow the engine to cool and recheck the coolant level.

Valve Lifters

REMOVAL AND INSTALLATION

Expo 2.4L (4G64) Engines

1. Disconnect the negative battery cable.

2. Disconnect the PCV and breather hoses. Remove the valve cover.

3. Install lash adjuster retainer tools MD998443 or equivalent, to pre-

vent the auto-lash adjuster from falling out of the rocker arm.

4. Loosen rocker arm and shaft assembly evenly in several steps. Remove the rocker arm and shaft assembly as a complete unit.

NOTE: It is essential that all parts be kept in the same order and orientation for reinstallation. Be sure to mark and separate parts, so not to be confused during reassembly.

5. Remove the lash retaining tools and remove the lash adjusters from the rocker bores.

6. Visually inspect the rocker arm roller and replace if damaged or seizure is evident. Check the roller for smooth rotation. Replace if excess play or binding is present. Also, inspect valve contact surface for possible damage or seizure. It is recommended that all rocker arms and lash adjusters be replaced together.

To install:

7. Immerse the lash adjusters in clean diesel fuel. Using a small wire, move the plunger of the lash adjuster up and down 4 or 5 times while pushing down lightly on the check ball in order to bleed out the air.

8. Install the lash adjusters in the rocker arms and secure with lash retainers.

9. Assemble the rocker assembly components in their original locations. Lubricate the rocker shaft with heavy engine oil and position on the cylinder head.

10. Install the rocker assembly and remove the lash adjuster retainers.

11. Install the valve cover, with a new gasket and semi-circular packing in place.

12. Connect the negative battery cable.

Lash Adjusters

BLEEDING

NOTE: The hydraulic lash adjuster is a precision component that relies on a clean operating environment. When handling the lash adjuster, make certain that no dirt or foreign particles are allowed to get inside the unit. DO NOT try to take the unit apart as it is not rebuildable. The adjuster is filled with diesel fuel. Hold the adjuster in the upright position so that the diesel fuel does not spill out. When cleaning the unit, only use clean diesel fuel.

To bleed the hydraulic lash adjusters when disassembling the rocker arms or as part of an engine overhaul, perform the following:

1. Mount the adjuster into special bleeding tool 09246–32100 and immerse the tool and adjuster in a container of clean diesel fuel.
2. With air bleed wire 09246–32200 inserted in the adjuster, lightly press down on the steel ball and compress the plunger four or five times.
3. Remove the air bleed wire from the adjuster and push down firmly on the plunger. If the plunger moves even slightly, repeat Steps 1 and 2 until the plunger stops moving. If the plunger continues to move, replace it.

Rocker Arms

REMOVAL AND INSTALLATION

3.5L (VIN M) Engine

1. Relieve the fuel system pressure.
2. Disconnect negative battery cable.
3. Remove the intake manifold plenum.
4. Remove the timing belt cover and the timing belt.

—— **WARNING** ——
DO NOT rotate the crankshaft or camshafts after the timing belt has been removed. If rotated, severe internal engine damage will result from the pistons hitting the valves.

5. Remove the center cover, breather, PCV hoses, and the spark plug cables.
6. Remove the rocker cover and the semi-circular packing.
7. Matchmark the positioning of the crankshaft position sensor at the rear of the camshaft and remove the sensor.
8. If equipped with a camshaft sensor, remove the sensor from the front of the engine.
9. Being sure to hold the flats of the camshaft, loosen the camshaft sprocket bolts.
10. Noting the positioning and location of the sprockets, remove the sprockets from the camshafts.

NOTE: Be sure to note the positioning of the knock pin at the end of the camshafts for reinstallation purposes.

NOTE: Be sure to keep the valve train components labeled and in proper order for reassembly.

11. Loosen the bearing cap bolts in 2–3 steps. Label and remove all camshaft bearing caps.

NOTE: If the bearing caps are difficult to remove, use a plastic hammer to gently tap the components.

12. Mark the components and remove the intake and the exhaust camshafts.
13. Remove the rocker arms and the lash adjusters. Be sure to note the location of the valve train components for reinstallation purposes.
14. Check the camshaft journals for wear or damage. Check the cam lobes for damage. Also, check the cylinder head oil holes for clogging.

To install:

NOTE: Lubricate the valve train components with clean engine oil.

15. Bleed and install the the lash adjusters to the to the original bores in the cylinder head.
16. Install the rocker arms to the cylinder head.
17. Lubricate the camshafts with clean engine oil and position the camshafts on the cylinder head.

—— **WARNING** ——
Be sure to properly position the knock pins of the camshaft to prevent valve to piston interference.

NOTE: Do not confuse the intake camshaft with the exhaust camshaft. The intake camshaft on the Montero has a P stamped on the hexagon. The exhaust camshaft on the Montero has a K stamped on the hexagon.

NOTE: Install the bearing caps according to the identification mark and cap number. Bearing caps No. 2, 3 and are marked as such. The caps also are marked I for intake or E for exhaust.

18. Install the bearing caps. Tighten the caps in sequence and in 2 or 3 steps. Caps 2, 3 and 4 have a front mark. Install with the mark aligned with the front mark on the cylinder head. Torque the retaining bolts for caps No. 2, 3 and 4 to 8 ft. lbs. (11 Nm) and torque the retaining bolts for the front and rear caps to 14 ft. lbs. (20 Nm).
19. Apply a coating of engine oil to the oil seals and install the oil seals to the front and rear of the camshafts.
20. Holding the flats of the camshaft, install and tighten the sprocket bolts to 65 ft. lbs. (90 Nm).
21. If removed, install the camshaft position sensor and tighten the mounting bolts to 78 inch lbs. (9 Nm).
22. Aligning the matchmark, install the crankshaft position sensor at the rear of the camshaft and tighten the mounting nut to 7 ft. lbs. (12 Nm).
23. Align the marks on the camshaft and crankshaft sprockets. Install the timing belt assembly.
24. Install the rocker cover and the semi-circular packing.
25. Install the intake manifold plenum.
26. Install the spark plug cables, center cover, breather and PCV hoses.
27. Connect the negative battery cable and check for leaks.

Expo 1.8L (4G93) Engine

1. Disconnect the negative battery cable.
2. Rotate the engine and position the No. 1 piston to TDC of its compression stroke.
3. Label and disconnect the spark plug cables.

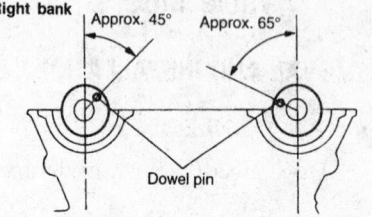
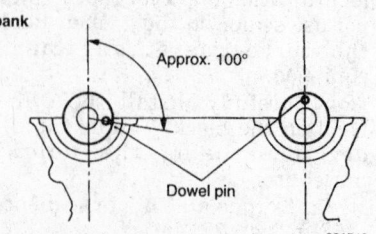

Proper positioning of the camshaft knock pins — 3.5L (VIN M) engines

321543

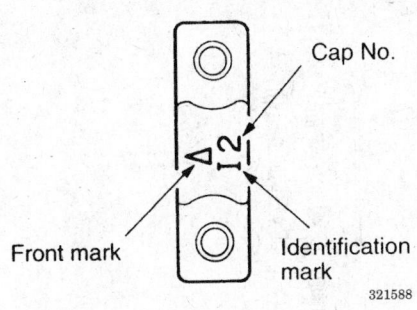

Identification of the camshaft bearing caps — 3.5L (VIN M) engines

4. Disconnect the air flow sensor connector and remove the air cleaner case cover.

5. Disconnect the accelerator cable, breather hose and PCV hose connections.

6. Remove the rocker cover and discard the gasket.

7. Loosen both rocker arm shaft assemblies gradually and evenly and remove the rocket shafts from the vehicle. Do not disassembly rocker arms and rocker arm shaft assemblies.

8. If disassembly is required, keep all parts in the exact order of removal. Inspect the roller surfaces of the rockers. Replace if there are any signs of damage or if the roller does not turn smoothly. Check the inside bore of the rockers and the adjuster tip for wear.

To install:

9. Lubricate the rocker shaft with clean engine oil and install the rockers and springs in their proper places.

10. Install the rocker arm and shaft assemblies. Tighten the rocker arm shaft retainer bolts to 23 ft. lbs. (32 Nm).

11. Check valve adjustment and install valve cover with a new gasket. Tighten the valve cover bolts to 29 inch lbs. (3.3 Nm).

12. Connect the spark plug cables.

13. Connect the accelerator cable, breather hose and PCV hose.

14. Connect the air flow sensor connector and install the air cleaner case cover.

15. Connect the negative battery cable. Run the engine at idle until normal operating temperature is reached. Check idle speed and ignition timing and adjust as required.

Expo 2.4L (4G64) engines

1. Disconnect the negative battery cable.

2. Disconnect the PCV and breather hoses. Remove the valve cover.

3. Install lash adjuster retainer tools MD998443 or equivalent, to prevent the auto-lash adjuster from falling out of the rocker arm.

4. Loosen rocker arm and shaft assembly evenly in several steps. Remove the rocker arm and shaft assembly as a complete unit.

NOTE: If any parts are to be reused, it is essential that all parts be kept in the same order and orientation for reinstallation. Be sure to mark and separate parts, so parts won't be mixed during reassembly.

5. Carefully disassemble the shaft assembly. Visually inspect the rocker arm roller and replace if damage or seizure is evident. Check the roller for smooth rotation. Replace if excess play or binding is present. Also, inspect valve contact surface for possible damage or seizure. It is recommended that all rocker arms and lash adjusters be replaced together.

To install:

6. Immerse the lash adjusters in clean diesel fuel. Using a small wire, move the plunger of the lash adjuster up and down 4 or 5 times while pushing down lightly on the check ball in

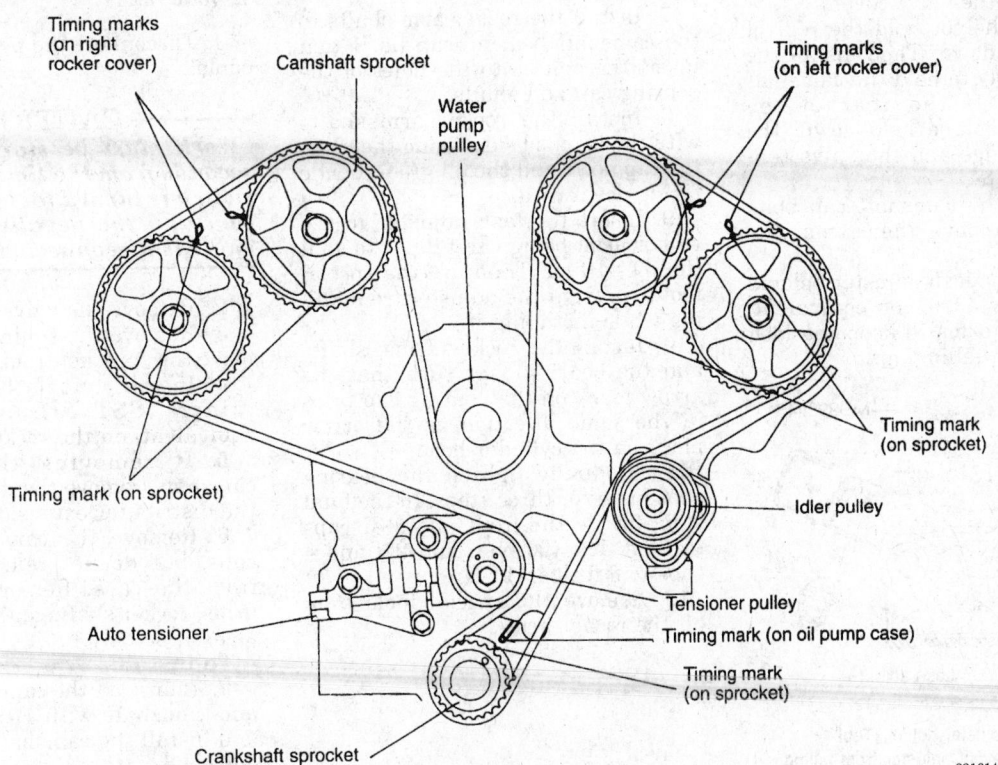

Sprocket alignment for timing belt installation — 3.5L (VIN M) engines

order to bleed out the air. Install the lash adjusters in the rocker arms.

7. Assemble the rocker assembly components in their original locations, lubricate the rocker shaft with heavy engine oil and position on the cylinder head.

8. Install the shaft mounting bolts and tighten all bolts evenly and gradually to 21-25 ft. lbs. (28-34 Nm).

9. Remove the lash adjuster retainers.

10. Install the valve cover, with a new gasket and semi-circular packing in place.

11. Connect the negative battery cable.

Rocker Arm Shaft

REMOVAL AND INSTALLATION

2.4L (VIN G) and 1993–95 3.0L (VIN H) Engines

1. Disconnect the negative battery cable.

—————— CAUTION ——————
Wait at least 90 seconds after the negative (-) battery cable is disconnected to prevent possible deployment of the air bag.

2. Remove the valve cover.

3. Have a helper hold the rear of the camshaft down. Then install the rear cap loosely to hold the camshaft in position. If the rear of the camshaft cannot be held down, the belt will dislodge and the valve timing will be lost.

4. Loosen the camshaft cap bolts but do not remove them from the caps.

5. Install the lash adjuster holders (Tools MD998443–01 or equivalent) on the rocker arms to keep the lash adjusters from falling out.

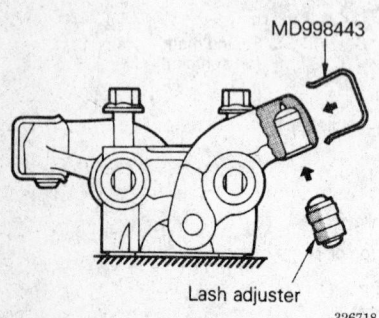

Insert the lash adjuster holder (Tool MD998443) to prevent adjuster from falling out — 2.4L (VIN G) and 1993–95 3.0L (VIN H) engines

6. Remove the caps, arms, shafts and bolts all as an assembly.

To install:

7. Install the rocker arm shafts to the camshaft bearing cap no. 1 and insert the bolts into the holes of the bearing cap and shafts.

8. Install the rocker arm shafts with the notched side facing the bearing cap no. 1 and the oil grooved side facing downward.

9. Insert the lash adjuster to the rocker arm, being careful not to spill the diesel fuel. Then use the special tool to prevent the adjuster from falling while installing it.

10. Install the rocker arms, shafts and the bearing caps such that the arrow mark on the bearing cap faces in the same direction as the arrow mark on the cylinder head.

11. Gradually tighten the bearing caps in two or three steps. In the final step, torque the front and rear caps to 14 ft. lbs. (20 Nm) and 2,3 and 4 caps to 8 ft. lbs. (11 Nm).

12. Remove the special tools from all the rocker arms.

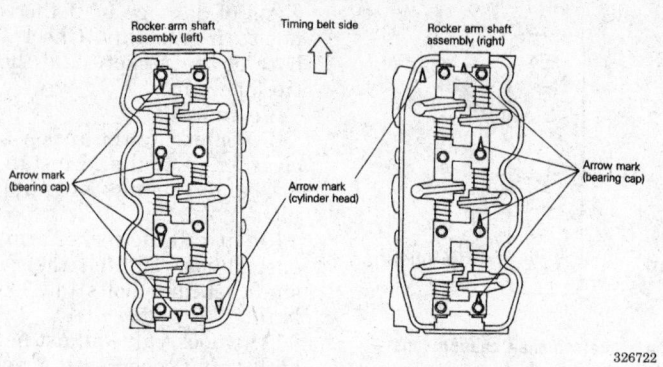

Bearing caps installed correctly — 3.0L (VIN H) engine

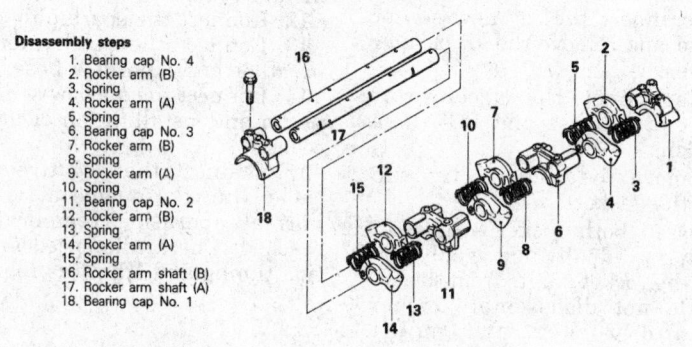

Disassembly steps
1. Bearing cap No. 4
2. Rocker arm (B)
3. Spring
4. Rocker arm (A)
5. Spring
6. Bearing cap No. 3
7. Rocker arm (B)
8. Spring
9. Rocker arm (A)
10. Spring
11. Bearing cap No. 2
12. Rocker arm (B)
13. Spring
14. Rocker arm (A)
15. Spring
16. Rocker arm shaft (B)
17. Rocker arm shaft (A)
18. Bearing cap No. 1

Rocker arms and shafts — 3.0L (VIN H) engine

1996–97 3.0L (VIN H) Engines

12 Valve Heads

1. Disconnect the negative battery cable.

—————— CAUTION ——————
Work must be started after 90 seconds from the time the ignition switch is turned to the LOCK position and the negative (-) battery cable is disconnected.

2. Remove the valve cover.

3. Remove the timing belt and remove the sprocket from the camshaft.

4. Install auto lash adjuster retainers SST MD998443–01 or equivalent, on the rocker arms.

5. If removing the left side camshaft, remove the distributor and the distributor extension.

6. Remove the camshaft bearing caps, but do not remove the bolts from the caps. Remove the rocker arms, rocker shafts and bearing caps, as an assembly.

To install:

7. Lubricate the camshaft journals and camshaft with clean engine oil and install the camshaft in the cylinder head.

8. Align the camshaft bearing caps with the arrow mark (depending on

cylinder numbers) and in numerical order.

NOTE: The arrow marks on the left rocker shaft bearing caps face in same direction as the arrow marked on the left cylinder head which is away from the timing belt. The arrows on the right cylinder head and the right rocker bearing caps face toward the timing belt.

9. Apply sealer at the ends of the bearing caps and install the assembly.

10. Torque the front and rear bearing cap bolts to 14 ft. lbs. (20 Nm) and bearing caps 2,3 and 4 to 8 ft. lbs. (11 Nm).

11. Remove the SST from the rocker arms.

12. If removed, install the distributor.

13. Install the sprockets, timing belt and the timing belt cover.

14. Install the valve cover and connect the negative battery cable.

15. Start the engine and check for leaks and proper operation.

24 Valve Heads

1. Disconnect the negative battery cable.

––––––––––– **CAUTION** –––––––––––
Work must be started after 90 seconds from the time the ignition switch is turned to the LOCK position and the negative (-) battery cable is disconnected.
––––––––––––––––––––––––––––––––––

2. Remove the valve cover.

3. Remove the timing belt and remove the sprocket from the camshaft.

4. Install auto lash adjuster retainers SST MD998443 or equivalent, on the rocker arms.

5. If removing the left side camshaft, remove the distributor and the distributor extension.

6. Remove the rocker arms, rocker shafts and bearing caps, as an assembly.

7. Remove the camshaft from the cylinder head.

8. Inspect the bearing journals on the camshaft and the cylinder head.

To install:

9. Lubricate the camshaft journals and camshaft with clean engine oil and install the camshaft in the cylinder head.

10. Install the rocker arms, rocker arm shaft and the rocker shaft spring.

   a. Temporarily tighten the rocker shaft with the bolts at the condition that all the intake valve rocker arms do not push the valves.

   b. Insert the rocker shaft spring from above and mount it at right angles to the plug guide.

   c. Before installing the exhaust rocker arms and the rocker arm shaft, mount the rocker shaft spring.

   d. Remove the SST used to hold the lash adjuster in position.

   e. Check to ensure that the flat side of the rocker shaft is perpendicular to the cylinder head, and facing the valves.

   f. Gradually tighten the bearing caps in two or three steps. In the final step torque to 23 ft. lbs. (31 Nm).

11. If removed, install the distributor.

12. Install the sprockets, timing belt and timing belt cover.

13. Install the valve cover and connect the negative battery cable.

14. Start the engine and check for leaks and proper operation.

Intake Manifold

REMOVAL AND INSTALLATION

2.4L (VIN G) Engine

1. Relieve the fuel pressure.

––––––––––– **CAUTION** –––––––––––
Fuel injection systems remain under pressure after the engine has been turned OFF. Properly relieve fuel pressure before disconnecting any fuel lines. Failure to do so may result in fire or personal injury.
––––––––––––––––––––––––––––––––––

2. Disconnect the negative battery cable.

––––––––––– **CAUTION** –––––––––––
Wait at least 90 seconds after the negative (-) battery cable is disconnected to prevent possible deployment of the air bag.
––––––––––––––––––––––––––––––––––

3. Drain the engine coolant. Disconnect the upper radiator hose from the thermostat housing.

4. Remove the air intake hoses, breather hose and the air intake pipe.

5. Disconnect all wires, hoses and linkages to the throttle body.

6. Remove the ignition coil.

7. Disconnect the brake booster hose and vacuum hose cluster from the air intake plenum.

8. Unbolt the air intake plenum from the intake manifold and remove the plenum from the engine.

9. Cover the fuel line with a clean shop rag and disconnect the fuel lines from the fuel rail. Keep the line covered or plugged.

10. Remove the fuel rail assembly with injectors intact.

11. Disconnect the heater hose from the manifold.

12. Disconnect the wires to the engine coolant switches.

13. Matchmark the rotor to the housing and the housing to the cylinder head and remove the distributor.

14. Unbolt the intake manifold from the cylinder head and remove from the engine.

15. Clean and dry the mating surfaces of the manifold and cylinder head.

To install:

16. Using a new gasket, install the intake manifold to the head. Starting from the middle and working outward, torque the retaining nuts to 10 ft. lbs. (14 Nm).

17. Connect the wires to the engine coolant switches. Install the distributor with matchmarks aligned.

18. Install the fuel rail assembly to the manifold and connect the fuel line using a new O-ring.

19. Connect the heater hose to the manifold.

20. Install the air intake plenum with a new gasket. Torque the retaining bolts to 12 ft. lbs. (16 Nm).

21. Connect the vacuum hoses cluster, brake booster hose and all wires, hoses and linkages to the throttle body.

22. Install the ignition coil.

23. Install the air intake pipe and hoses.

24. Connect the upper radiator hose.

25. Fill the radiator with coolant.

26. Connect the negative battery cable and connect the jumper wire from the fuel pump activation terminal to the positive battery post to inspect the system for leaks.

27. Set all adjustments to specifications.

1993–95 3.0L (VIN H) and 3.5L (VIN M) Engines

1. Relieve the fuel pressure.

––––––––––– **CAUTION** –––––––––––
Fuel injection systems remain under pressure after the engine has been turned OFF. Properly relieve fuel pressure before disconnecting any fuel lines. Failure to do so may result in fire or personal injury.
––––––––––––––––––––––––––––––––––

2. Disconnect the negative battery cable.

3. Drain the engine coolant.

4. Disconnect the upper radiator hose from the thermostat housing.

5. Remove the air intake hose from the throttle body.

6. Disconnect all wires, hoses and linkages to the throttle body.

7. Disconnect the EGR temperature sensor wire.

8. Remove the ignition coil.

9. Remove the engine oil filler neck bracket.

10. Unbolt the EGR tube from the air intake plenum.

11. Disconnect the PCV hose and vacuum hose cluster from the plenum.

12. Remove the plenum to engine brackets.

13. Unbolt the air intake plenum assembly from the intake manifold and remove.

14. Cover the fuel lines with a clean shop rag and disconnect the fuel lines from the fuel rail. Keep the lines covered.

15. Remove the fuel rail with injectors in place.

16. Disconnect the bypass hose and the upper radiator hose from the thermostat housing. Disconnect the wires to the coolant temperature switches.

17. Remove the intake manifold retaining nuts and remove the manifold from the cylinder heads.

18. Remove the gaskets and thoroughly clean and dry the mating surfaces of the manifold and heads.

To install:

19. Position the manifold over the studs and install the retaining nuts. Torque to 10 ft. lbs. (14 Nm) for the 3.0L 12 valve and the 3.5L engines and 16 ft. lbs. (21 Nm) for the 3.0L 24 valve engine. Start from the center and working outward.

20. Connect the hoses and connect the wires to the coolant switches.

21. Install the fuel rail assembly and connect the fuel hoses.

22. Using a new gasket, install the air intake plenum to the intake manifold. Torque the nuts and bolts to 12 ft. lbs. (16 Nm). Install the plenum to engine brackets.

23. Connect the PCV hose and vacuum hose cluster to the plenum.

24. Connect the EGR tube.

25. Install the engine oil filler neck bracket.

26. Install the ignition coil assembly.

27. Connect the EGR temperature sensor wire.

28. Connect all wires, hoses and linkages to the throttle body.

29. Install the air intake hose to the throttle body.

30. Connect the upper radiator hose to the thermostat housing.

31. Refill the radiator with coolant.

32. Connect the negative battery cable.

33. Connect the jumper wire from the fuel pump activation terminal to the positive battery post; check the system for fuel leaks.

34. Set all adjustments to specifications.

1996–97 3.0L (VIN H) Engine

1. Relieve the fuel pressure.

— **CAUTION** —

Fuel injection systems remain under pressure after the engine has been turned OFF. Properly relieve fuel pressure before disconnecting any fuel lines. Failure to do so may result in fire or personal injury.

2. Disconnect the negative battery cable.

— **CAUTION** —

Work must be started after 90 seconds from the time the ignition switch is turned to the LOCK position and the negative (-) battery cable is disconnected.

3. Drain the engine coolant.

4. Remove the air intake hose from the throttle body.

5. Label and disconnect the electrical connectors and vacuum hoses from the throttle body and air intake plenum.

6. Disconnect the accelerator cable and the throttle control cable.

7. Disconnect the coolant hoses.

8. Disconnect the engine oil filler neck bracket from the air intake plenum.

9. Unbolt the EGR tube from the air intake plenum.

10. Remove the plenum brackets.

11. Unbolt the air intake plenum assembly from the intake manifold and remove. Note the position of the mounting bolts as they are removed.

12. Cover the high pressure fuel hose with a clean shop towel to prevent fuel spray due to residual pressure in the line. Disconnect the high pressure fuel hose from the fuel rail.

13. Remove the fuel return line and vacuum hose from the fuel pressure regulator.

14. Label and disconnect the electrical connectors from the injectors.

15. Remove the fuel rail retaining bolts.

16. Lift the rail, with injectors attached, up and away from the engine.

17. Remove the intake manifold retaining nuts and remove the manifold from the cylinder heads.

18. Remove the gaskets and thoroughly clean and dry the mating surfaces of the manifold and heads.

To install:

19. Position the manifold over the studs and install the retaining nuts. Torque to 10 ft. lbs. (14 Nm) for the 3.0L 12 valve and 16 ft. lbs. (21 Nm) for the 3.0L 24 valve engine. Start from the center and working outward.

20. Connect the hoses and connect the wires to the coolant switches.

21. Install the fuel rail assembly and connect the fuel hoses.

22. Using a new gasket, install the air intake plenum to the intake manifold. Torque the nuts and bolts to 10 ft. lbs. (13 Nm). Install the plenum to engine brackets.

23. Connect the PCV hose and vacuum hose cluster to the plenum.

24. Connect the EGR tube.

25. Connect the EGR temperature sensor wire.

26. Connect all wires, hoses and linkages to the throttle body.

27. Install the air intake hose to the throttle body.

28. Connect the upper radiator hose to the thermostat housing.

29. Refill the radiator with coolant.

30. Connect the negative battery cable.

31. Connect the jumper wire from the fuel pump activation terminal to the positive battery post; check the system for fuel leaks.

32. Set all adjustments to specifications.

1996–97 3.5L (VIN M) Engines

1. Relieve the fuel pressure.

— **CAUTION** —

Fuel injection systems remain under pressure after the engine has been turned OFF. Properly relieve fuel pressure before disconnecting any fuel lines. Failure to do so may result in fire or personal injury.

2. Disconnect the negative battery cable.

— **CAUTION** —

Wait at least 90 seconds after the negative (-) battery cable is disconnected to prevent possible deployment of the air bag.

3. Partially drain the cooling system.

4. Remove the air intake hose from the throttle body.

5. Label and disconnect the electrical connectors and vacuum hoses from the throttle body and air intake plenum.

6. Disconnect the accelerator cable and the throttle control cable.

7. Disconnect the coolant hoses.

8. Disconnect the EGR temperature sensor connector.

9. Unbolt the EGR tube from the air intake plenum.

10. Remove the intake manifold plenum cover.

11. Remove the intake manifold plenum stay brackets.

12. Unbolt the air intake plenum assembly from the intake manifold and remove. Note the position of the mounting bolts as they are removed.

13. Remove the induction control valve assembly.

14. Cover the high pressure fuel hose with a clean shop towel to prevent fuel spray due to residual pressure in the line. Disconnect the high pressure fuel hose from the fuel rail.

15. Remove the fuel return line and vacuum hose from the fuel pressure regulator.

16. Label and disconnect the electrical connectors from the injectors.

17. Remove the fuel rail retaining bolts.

18. Lift the rail, with injectors attached, up and away from the engine.

19. Remove the intake manifold retaining nuts and remove the manifold from the cylinder heads.

20. Remove the gaskets and thoroughly clean and dry the mating surfaces of the manifold and heads.

To install:

21. Position the manifold over the studs and install the retaining nuts. Torque to 10 ft. lbs. (14 Nm). Start from the center and work outward.

22. Connect the hoses and connect the wires to the coolant switches.

23. Install the fuel rail assembly and connect the fuel hoses.

24. Install the induction control valve assembly and torque to 6 ft. lbs. (9 Nm).

25. Using a new gasket, install the air intake plenum to the intake manifold. Torque the nuts and bolts to 13 ft. lbs. (18 Nm). Install the plenum to engine brackets.

26. Connect the PCV hose and vacuum hose cluster to the plenum.

27. Connect the EGR tube.

28. Connect the EGR temperature sensor wire.

29. Connect all wires, hoses and linkages to the throttle body.

30. Install the air intake hose to the throttle body.

31. Connect the upper radiator hose to the thermostat housing.

32. Refill the radiator with coolant.

33. Connect the negative battery cable.

34. Connect the jumper wire from the fuel pump activation terminal to the positive battery post; check the system for fuel leaks.

35. Set all adjustments to specifications.

Expo 1.8L (4G93) and 2.4L (4G64) Engines

1. Relieve the fuel system pressure.

————— **CAUTION** —————
Fuel injection systems remain under pressure after the engine has been turned OFF. Properly relieve fuel pressure before disconnecting any fuel lines. Failure to do so may result in fire or personal injury.

2. Disconnect battery negative cable and drain the cooling system.

3. Disconnect the accelerator cable and the air intake hose.

4. Tag and disconnect the electrical connectors from the oxygen sensor, coolant temperature sensor, idle speed control connection, EGR temperature sensor, oil pressure switch, spark plug wires and distributor connectors.

5. Disconnect the wiring from the throttle position sensor, fuel injectors and disconnect the ground cables.

6. Remove all vacuum hoses and pipes as necessary, including the brake booster and PCV vacuum lines.

7. Disconnect the upper radiator hose, heater hose and water bypass hose.

8. Disconnect the high pressure fuel line and the fuel return hose.

9. Remove the fuel rail, fuel injectors, pressure regulator and insulators.

10. Remove the intake manifold support bracket.

11. If the thermostat housing is preventing removal of the intake manifold, remove it.

12. Remove the intake manifold mounting bolts/nuts and remove the intake manifold assembly.

To install:

13. Clean all gasket material from the cylinder head intake mounting surface and intake manifold assembly. Check both surfaces for cracks or other damage. Check the intake manifold water passages and jet air passages for clogging. Clean if necessary.

14. Using a straight edge, measure the distortion of the intake manifold-to-cylinder head. Total distortion or warpage should be 0.006 inches (0.15mm or less).

15. Install a new intake manifold gasket to the head and install the manifold. Torque the manifold in a criss-cross pattern, starting from the inside and working outwards to 14 ft. lbs. (20 Nm).

16. If removed, install the thermostat housing.

17. Install the intake manifold brace bracket.

18. Install the fuel delivery pipe, injectors and pressure regulator to the engine. Torque the retaining bolts to 108 inch lbs. (12 Nm).

19. Using a new O-ring for the feed pipe and a new clamp for the return pipe, install the fuel hoses.

20. Connect the upper radiator hose, heater hose and water bypass hoses.

21. Install the vacuum hoses and pipes as necessary. Be sure to connect the brake booster and PCV vacuum lines.

22. Connect the wiring to the throttle position sensor, fuel injectors and connect the ground cables.

23. Connect the electrical wiring to the oxygen sensor, coolant temperature sensor, idle speed control connection, EGR temperature sensor, oil pressure switch, spark plug wires and distributor connectors.

24. Connect and adjust the accelerator cable and install the air intake hose.

25. Fill the system with coolant.

26. Connect the negative battery cable, run the vehicle until the thermostat opens, fill the radiator completely.

27. Check and adjust the idle speed and ignition timing.

28. Once the vehicle has cooled, recheck the coolant level.

Exhaust Manifold

REMOVAL AND INSTALLATION

2.4L (VIN G) Engine

1. Disconnect the negative battery cable.

————— **CAUTION** —————
Wait at least 90 seconds after the negative (-) battery cable is disconnected to prevent possible deployment of the air bag.

2. Remove the heat cowl from the exhaust manifold.

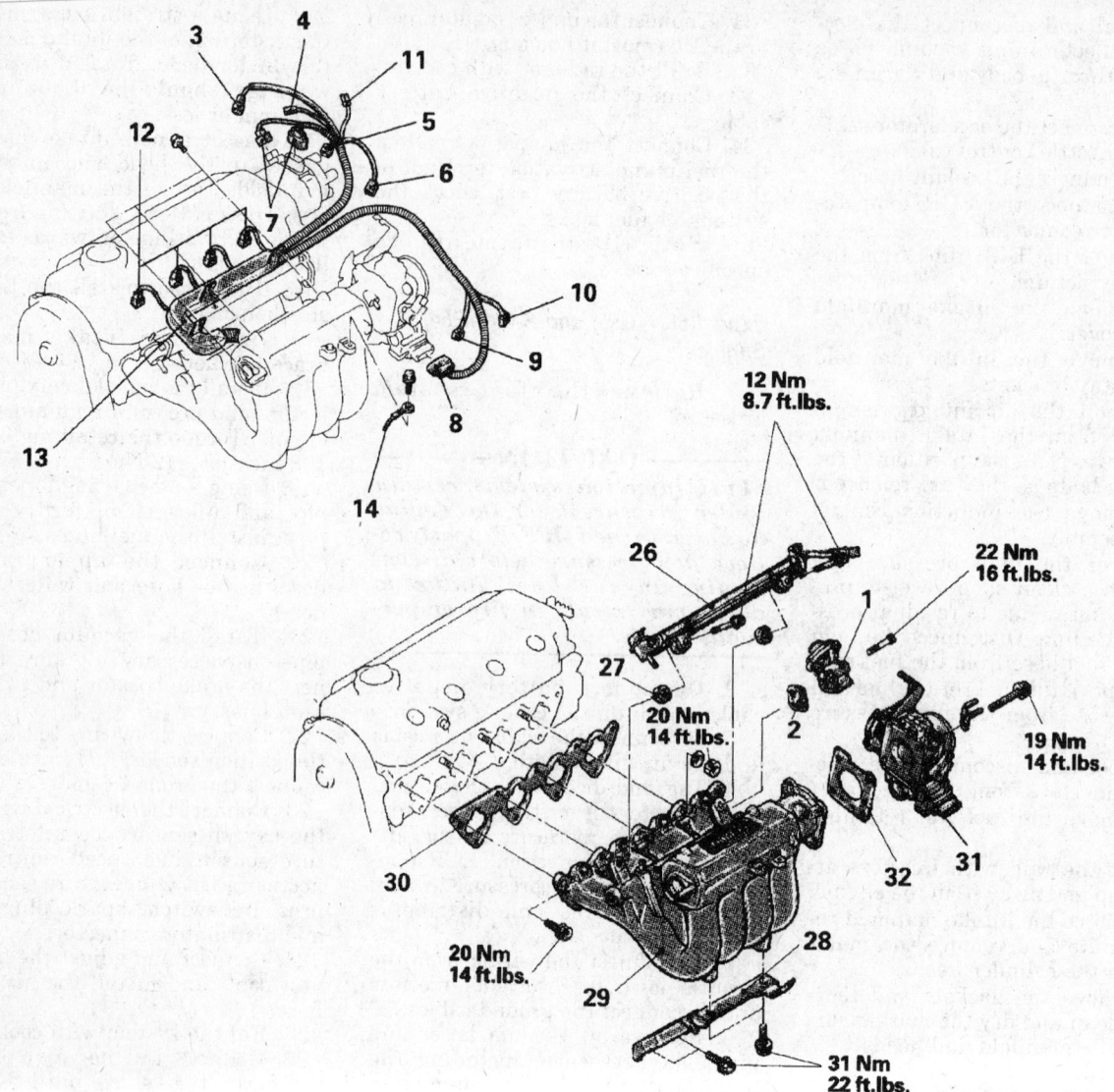

1. EGR valve
2. EGR valve gasket
3. Heated oxygen sensor connector
4. Oil pressure switch connector
5. Engine temperature gauge unit connector
6. Engine coolant temperature sensor connector
7. Distributor connector
8. Idle air control motor connector
9. Heated oxygen sensor connector (front) <Vehicles for California>
10. EGR temperature sensor connector <Vehicles for California>
11. Throttle position sensor connector
12. Injector connector
13. Control harness assembly
14. Ground wire
15. Breather hose connection
16. PCV hose connection
17. Vacuum hose connection
18. Vacuum pipe
19. Water hose connection (Thermostat case → Throttle body)

20. Water hose connection (Throttle body → Water inlet fitting)
21. High-pressure fuel hose connection
22. O-ring
23. Fuel return hose connection
24. Heater hose connection
25. Brake booster vacuum hose connection
26. Delivery pipe, injector and pressure regulator assembly
27. Insulator
28. Intake manifold stay
29. Intake manifold
30. Intake manifold gasket
31. Throttle body
32. Throttle body gasket

330730

Intake manifold and related components — Expo 1.8L (4G93) and 2.4L (4G64) engines

3. If equipped, remove the aspirator valve assembly.

4. Raise and safely support the vehicle.

5. Disconnect the exhaust pipe from the manifold. Lower the vehicle.

6. If equipped, disconnect the oxygen sensor connector and ground cable.

7. Remove the manifold mounting nuts and remove the manifold and gasket from the engine.

To install:

8. Install the exhaust manifold and tighten the mounting nuts to 13 ft. lbs. (18 Nm), starting from the middle and working outward.

9. If removed, connect the oxygen sensor connector and ground cable.

10. Raise and safely support the vehicle.

11. Connect the exhaust pipe to the manifold. Lower the vehicle.

12. If removed, install the aspirator valve assembly.

13. Install the heat cowl to the exhaust manifold.

14. Connect the negative battery cable.

15. Start the engine and check for exhaust leaks.

12 Valve 3.0L (VIN H) Engines

1993–95 Vehicles

1. Disconnect the negative battery cable. Raise and safely support the vehicle.

2. Disconnect the oxygen sensor.

3. Disconnect the exhaust pipe from the exhaust manifolds.

4. If removing the left manifold:

   a. Disconnect the EGR tube.

   b. Remove the air intake plenum bracket.

   c. Remove the bracket from the exhaust manifold.

5. If removing the right manifold:

   a. Remove the alternator bracket.

   b. Remove the engine hanger.

   c. Remove the air duct.

6. Remove the heat shield.

7. Remove the manifold to cylinder head nuts and remove the manifolds.

8. Clean the gasket mounting surfaces. Inspect the manifolds for cracks, flatness and/or damage.

To install:

9. Install the new gasket and manifold. Tighten the manifold-to-cylinder head nuts to 14 ft. lbs. (19 Nm).

**NOTE: The numbers 2–4–6 on the gasket indicate that the gasket should be installed on the left cylinder head. The numbers

1–3–5 indicate that the gasket should be installed on the right cylinder head.**

10. Connect the heat shield.

11. Connect the EGR tube to the manifold, exhaust manifold bracket and air intake plenum bracket, if removed.

12. Install the alternator bracket, engine hanger and air duct, if removed.

13. Connect the exhaust pipe to the exhaust manifolds.

14. Connect the oxygen sensor.

15. Lower the vehicle and connect the negative battery cable.

16. Start the engine and check for exhaust leaks.

1996–97 Vehicles

1. Disconnect the negative battery cable. Raise and safely support the vehicle.

2. Remove the oil dipstick and the guide.

3. Remove the right heat protector, the engine hanger and the right exhaust manifold and gasket.

4. Remove the left heat protector, the bracket and the exhaust manifold and gasket.

5. Remove the water hoses, the heater pipe and gasket, the water pipe and O–ring and the water inlet pipe.

To install:

6. Install the water inlet pipe, the water pipe and gasket, the heater pipe and gasket and the water hoses.

7. Install the new gasket and manifold. Tighten the exhaust manifold head nuts to 14 ft. lbs. (19 Nm).

NOTE: The numbers 2–4–6 on the gasket indicate that the gasket should be installed on the left cylinder head. The numbers 1–3–5 indicate that the gasket should be installed on the right cylinder head.

8. Connect the heat shield.

9. Install the bracket and the engine hanger.

10. Install the oil dipstick and guide with a new O–ring.

11. Lower the vehicle and connect the negative battery cable.

12. Start the engine and check for exhaust leaks.

24 Valve 3.0L (VIN H) and 3.5L (VIN M) Engines

1995 Vehicles

1. Disconnect the battery cables and remove the battery and tray.

2. Raise and safely support the vehicle.

3. Remove the air duct and the air cleaner cover.

4. Disconnect the exhaust pipe from the exhaust manifolds.

5. Remove the heat shield.

6. Remove the EGR tube.

7. Remove the manifold to cylinder head nuts and remove the manifolds.

8. Clean the gasket mounting surfaces. Inspect the manifolds for cracks, flatness and/or damage.

To install:

9. Install the new gasket and manifold. Tighten the manifold head nuts to 22 ft. lbs. (29 Nm).

10. Connect the heat shield.

11. Connect the EGR tube to the manifold and the exhaust manifold bracket.

12. Install the air duct.

13. Connect the exhaust pipe to the exhaust manifolds.

14. Lower the vehicle, install the battery and connect the cables.

15. Start the engine and check for exhaust leaks.

1996–97 Vehicles

1. Disconnect the negative battery cable.

2. Raise and safely support the vehicle.

3. Disconnect the exhaust pipe from the exhaust manifolds.

4. Remove the oil dipstick guide and O–ring.

5. Remove the heat shields.

6. Remove the exhaust manifolds.

7. Clean the gasket mounting surfaces. Inspect the manifolds for cracks, flatness and/or damage.

To install:

8. Install the new gasket and manifold. Tighten the manifold head nuts to:
   - 3.0L: 22 ft. lbs. (29 Nm)
   - 3.5L: 33 ft. lbs. (45 Nm)

9. Connect the heat shield.

10. Connect the exhaust pipe to the exhaust manifolds.

11. Install the oil dipstick guide with a new O–ring.

12. Lower the vehicle, and connect the negative battery cable.

13. Start the engine and check for exhaust leaks.

Expo 1.8L (4G93) Engines

1. Disconnect battery negative cable.

2. Raise the vehicle and support safely.

3. Remove the exhaust pipe to exhaust manifold nuts and separate exhaust pipe. Discard the gasket.

4. Lower the vehicle.

5. Remove the outer exhaust manifold heat shield and engine hanger.

6. Disconnect the electrical connector and remove the oxygen sensor.

7. Remove the exhaust manifold mounting bolts, the inner heat shield and the exhaust manifold.

To install:

8. Clean all gasket material from the mating surfaces and check the manifold for damage.

9. Install a new gasket and install the manifold. Tighten the nuts to in a crisscross pattern to 13-22 ft. lbs. (18-30 Nm).

10. Install the heat shield.

11. Install a new flange gasket and connect the exhaust pipe.

12. Connect the negative battery cable and check for exhaust leaks.

Expo 2.4L (4G64) Engines

1. Disconnect battery negative cable.

2. Raise the vehicle and support safely.

3. Leaving hoses connected, disconnect the power steering pump and position aside.

4. Remove the exhaust pipe to exhaust manifold nuts and separate exhaust pipe. Discard gasket.

5. Lower the vehicle.

6. Remove the electric cooling fan assembly.

7. Remove the outer exhaust manifold heat shield and engine hanger.

8. Disconnect the electrical connector and remove the oxygen sensor.

9. Remove the exhaust manifold mounting bolts, the inner heat shield and the exhaust manifold.

To install:

10. Clean all gasket material from the mating surfaces and check the manifold for damage.

11. Install a new gasket and install the manifold. Tighten the nuts to in a crisscross pattern to 22 ft. lbs. (30 Nm).

12. Install the heat shields.

13. Install the electric cooling fan assembly as required.

14. Install a new flange gasket and connect the exhaust pipe.

15. Install the power steering pump and adjust the drive belt for proper tension.

16. Connect the negative battery cable and check for exhaust leaks.

Front Crankshaft Seal

REMOVAL AND INSTALLATION

2.4L (VIN G) Engines

1. Disconnect the negative battery cable.

——— **CAUTION** ———
Wait at least 90 seconds after the negative (-) battery cable is disconnected to prevent possible deployment of the air bag.

2. Drain the engine coolant into a suitable container so it can be reused.

3. Remove the radiator and cooling fan.

4. Remove the accessory drive belts and the air conditioner tension pulley if equipped.

5. Remove the crankshaft pulley and the timing belt front covers.

6. Reinstall the crankshaft pulley bolt and use it to rotate the engine clockwise until the timing marks are aligned.

7. Remove the timing belt and crankshaft sprocket.

8. Remove the inner timing belt and the inner crankshaft sprocket.

9. Carefully pry out the crankshaft seal without scratching the crankshaft.

To install:

10. Place the seal guide over the crankshaft and coat it with engine oil.

11. Slide the seal over the guide until it touches the front case assembly then use the appropriate seal driver to install the seal. Remove the seal guide.

12. Install the inner crankshaft sprocket and the inner timing belt. Adjust the belt tension.

13. Install the flange and the timing belt sprocket on the crankshaft.

14. Install the main timing belt. Adjust the belt tension.

15. Install the timing belt front covers and the crankshaft pulleys.

16. Install the air conditioner belt tension pulley if equipped.

17. Install the accessory drive belts. Adjust the belts to specifications.

18. Install the cooling fan, radiator and the fan shroud.

19. Fill the cooling system to the proper level.

20. Connect the negative battery cable, start the engine and check for leaks.

3.0L (VIN H) and 3.5L (VIN M) Engines

1993-95 Vehicles

1. Disconnect the negative battery cable.

2. Remove the accessory drive belts.

3. Remove the crankshaft pulley.

4. Remove the timing belt covers and the timing belt.

5. Remove the crankshaft sprocket.

6. Pry out the oil seal with a flat prying tool, being careful not to damage the crankshaft.

To install:

7. Coat the lip of the new seal with oil and install the seal using the proper seal driver.

8. Install the crankshaft sprocket and the timing belt.

9. Install the timing belt covers.

10. Install the crankshaft pulley.

11. Install the accessory drive belts.

12. Connect the negative battery cable.

13. Start the engine and check for proper operation.

1996-97 Vehicles

1. Disconnect the negative battery cable.

——— **CAUTION** ———
Wait at least 90 seconds after the negative (-) battery cable is disconnected to prevent possible deployment of the air bag.

2. Remove the accessory drive belts.

3. Remove the crankshaft pulley.

4. Remove the timing belt covers and the timing belt.

5. Remove the crankshaft sprocket.

6. Cut out a portion in the crankshaft oil seal lip and pry out the oil seal with a flat prying tool, being careful not to damage the crankshaft.

To install:

7. Coat the lip of the new seal with oil and install the seal using the proper seal driver.

8. Install the crankshaft sprocket and the timing belt.

9. Install the timing belt covers.

10. Install the crankshaft pulley.

11. Install the accessory drive belts.

12. Connect the negative battery cable.

13. Start the engine and check for proper operation.

Expo 1.8L (4G93) Engine

1. Disconnect the negative battery cable.

2. Remove the timing belt.

3. Drain the oil.

4. Remove the crankshaft pulley retainer bolts and remove the pulley.

5. Remove the vibration damper retainer bolt and washer and remove damper. If difficult to remove, the appropriate puller may be used.

6. Remove the crankshaft sprocket. If sprocket is difficult to remove, the appropriate puller may be used.

7. Pry out the oil seal from front of engine.

To install:

8. Using proper size driver, install new front seal.

---WARNING---

Small nicks and burrs on crankshaft surface will damage the oil seal. Use care when installing the oil seal not to damage crankshaft surface.

9. Lubricate the lips of the new seal with clean engine oil.

10. Install the crankshaft sprocket and vibration damper. Torque the retaining bolt to specifications.

11. Install the timing belt, timing covers, valve cover and remaining components.

12. Install the engine under cover and connect the negative battery cable.

13. Fill engine oil, start the engine and check for leaks.

Expo 2.4L (4G64) Engines

1. Disconnect the negative battery cable.

2. Remove the timing belt.

3. Drain the engine oil.

4. Remove the crankshaft pulley retainer bolts and remove the pulley.

5. Remove the crankshaft sprocket retainer bolt and washer from the sprocket, if used, and remove the sprocket. If no bolts are used on the sprocket, use the appropriate puller to remove.

6. Pry out the oil seal from front of engine.

To install:

7. Using proper size driver, install new front seal.

---WARNING---

Small nicks and burrs on crankshaft surface will damage oil seal. Use care when installing oil seal not to damage crankshaft surface.

8. Install the crankshaft sprocket and torque the retaining bolt to 80–94 ft. lbs. (110–130 Nm).

9. Install the timing belt, timing covers, valve cover and remaining components.

10. Install the engine under cover and connect the negative battery cable.

11. Fill engine oil, start the engine and check for leaks.

Timing Belt, Sprockets, Tensioner and Front Cover

ADJUSTMENT

2.4L (VIN G) Engine

1. Disconnect the negative battery cable.

2. Remove the timing belt cover.

3. Adjust the silent shaft (inner) belt first:

    a. Loosen the pulley center bolt so the pulley may be moved.

    b. Move the pulley up by hand so the center span of the long side of the belt deflects 1/4 inch.

    c. Hold the pulley tight so the pulley itself does not rotate when the bolt is tightened. Tighten the bolt to 15 ft. lbs. (20 Nm). If the pulley has moved, the belt will be too tight.

4. Check the timing (outer) belt tension.

5. To adjust the timing (outer) belt:

    a. First loosen the tensioner pulley bolts.

    b. Turn the crankshaft and camshaft sprocket forward 2 teeth.

    c. Allow the spring to take up any slack.

    d. Check that the the deflection of the longest span (between the camshaft and oil pump sprockets) is 1/2 inch. Do not overtighten the belt or it will howl.

6. First tighten the lower pulley bolts to 35 ft. lbs. (47 Nm) and then the upper bolt to the same value.

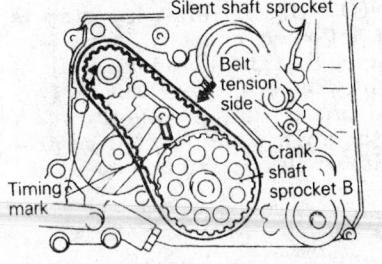

329675

Inner belt timing marks — 2.4L (VIN G) engines

9. Install the timing belt, timing covers, valve cover and remaining components.

10. Install the engine under cover and connect the negative battery cable.

11. Fill engine oil, start the engine and check for leaks.

7. Install the covers and all related parts.

3.0L (VIN H) Engine

1. Disconnect the negative battery cable.

2. Remove the air conditioner compressor belt to make work easier.

3. Loosen the bolt that holds the timing belt tensioner in place through the access hole in the timing belt cover.

4. Turn the crankshaft 2 turns in the normal direction. Allow the spring to pull the tensioner in automatically.

5. Tighten the tensioner locking bolt.

6. Install the access hole cover and the air compressor belt.

7. Connect the negative battery cable.

REMOVAL AND INSTALLATION

2.4L (VIN G) Engine

1. Make sure that the engines No. 1 piston is at TDC in the compression stroke.

---CAUTION---

Wait at least 90 seconds after the negative (-) battery cable is disconnected to prevent possible deployment of the air bag.

2. Disconnect the negative battery cable.

3. Remove the spark plug wires from the tree on the upper cover.

4. Drain the cooling system.

5. Remove the shroud, fan and accessory drive belts.

6. Remove the radiator as required.

7. Remove the power steering pump, alternator, air conditioning compressor, tension pulley and accompanying brackets, as required.

8. Remove the upper front timing belt cover.

9. Remove the water pump pulley and the crankshaft pulley(s).

10. Remove the lower timing belt cover mounting screws and remove the cover.

11. If the belt(s) are to be reused, mark the direction of rotation on the belt.

12. Remove the timing (outer) belt tensioner and remove the belt. Unbolt the tensioner from the block and remove.

13. Remove the outer crankshaft sprocket and flange.

14. Remove the silent shaft (inner) belt tensioner and remove the inner

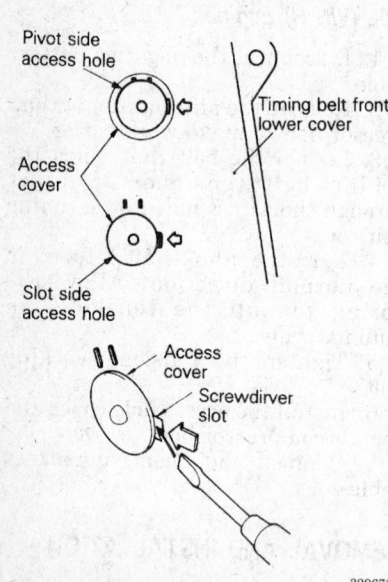

Access holes in the lower timing belt cover for tension adjustment — 2.4L (VIN G) engines

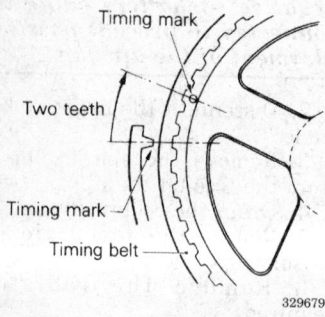

Turn the sprocket and belt forward 2 teeth — 2.4L (VIN G) engines

belt. Unbolt the tensioner from the block and remove.

15. To remove the camshaft sprockets, use SST MB990767–01 and MIT308239 or equivalent.

To install:

16. Install the camshaft sprockets and torque to 65 ft. lbs. (90 Nm).

17. Align the timing mark of the silent shaft belt sprockets on the crankshaft and silent shaft with the marks on the front case. Wrap the silent shaft belt around the sprockets so there is no slack in the upper span of the belt and the timing marks are still in line.

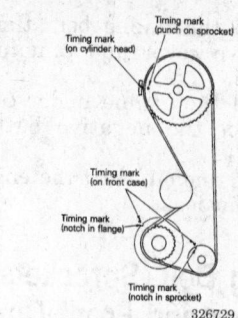

Alignment of timing marks — 2.4L (VIN G) engines

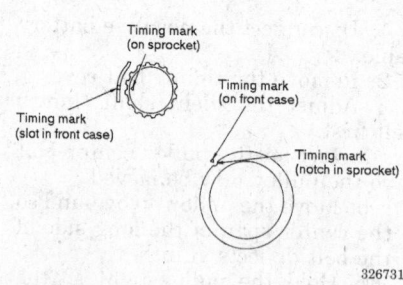

Alignment of inner belt timing marks — 2.4L (VIN G) engines

18. Install the tensioner initially so the actual center of the pulley is above and to the left of the installation bolt.

19. Move the pulley up by hand so the center span of the long side of the belt deflects about ¼ inch.

20. Hold the pulley tightly so it does not rotate when the bolt is tightened. Tighten the bolt to 15 ft. lbs. (20 Nm). If the pulley has moved, the belt will be too tight.

21. Install the timing belt tensioner fully toward the water pump and temporarily tighten the bolts. Place the upper end of the spring against the water pump body. Align the timing marks of the cam, crankshaft and oil pump sprockets with the corresponding marks on the front case or head.

NOTE: If the following step is not followed exactly, there is a chance that the silent shaft alignment will be 180 degrees off. This will cause a noticeable vibration in the engine and the entire procedure will have to be repeated.

22. Before installing the timing belt, ensure that the left side silent shaft is in the correct position.

NOTE: It is possible to align the timing marks on the camshaft sprocket, crankshaft sprocket and the oil pump sprocket with the left balance shaft out of alignment.

23. With the timing mark on the oil pump pulley aligned with the mark on the front case, check the alignment of the left balance shaft to assure correct shaft timing.

    a. Remove the plug located on the left side of the block in the area of the starter.

    b. Insert a tool having a shaft diameter of 0.3 in. (8mm) into the hole.

    c. With the timing marks still aligned, the tool must be able to go in at least 2⅓ in. If it can only go in about 1 in. turn the oil pump sprocket 1 complete revolution.

    d. Recheck the position of the balance shaft with the timing marks realigned. Leave the tool in place to hold the silent shaft while continuing.

24. Install the belt to the crankshaft sprocket, oil pump sprocket and the camshaft sprocket, in that order. While doing so, make sure there is no slack between the sprockets except where the tensioner will take it up when release.

25. Recheck the timing marks' alignment.

26. If all are aligned, loosen the tensioner mounting bolt and allow the tensioner to apply tension to the belt.

27. Remove the tool that is holding the silent shaft in place and turn the crankshaft clockwise a distance equal to 2 teeth of the camshaft sprocket. This will allow the tensioner to automatically tension the belt the proper amount.

NOTE: Do not manually apply pressure to the tensioner. This will overtighten the belt and will cause a howling noise.

28. First tighten the lower mounting bolt and then tighten the upper spacer bolt.

29. To verify that belt tension is correct, check that the deflection of the longest span (between the camshaft and oil pump sprockets) is ½ inch.

30. Install the lower timing belt cover. Make sure the packing is positioned in the inner grooves of the covers properly when installing.

31. Install the water pump pulley and the crankshaft pulley(s).

32. Install the upper front timing belt cover.

33. Install the power steering pump, alternator, air conditioning compressor, tension pulley and accompanying brackets, as required.

34. Install the radiator, shroud, fan and accessory drive belts.

35. Install the spark plug wires to the tree on the upper cover.

36. Refill the cooling system.

37. Connect the negative battery cable. Start the engine and check for leaks.

3.0L (VIN H) Engines

1993–95 Vehicles

1. If possible, position engine with No. 1 cylinder at TDC.

2. Disconnect the negative battery cable.

3. Drain the cooling system. Remove the drive belts.

4. Remove the upper radiator shroud.

5. Remove the fan and fan pulley.

6. Without disconnecting the lines, remove the power steering pump from its bracket and position it to the side. Remove the pump brackets.

7. Remove the belt tensioner pulley bracket.

8. Without releasing the refrigerant remove the air conditioning compressor from its bracket and position it to the side. Remove the bracket.

9. Remove the cooling fan bracket.

10. On some vehicles it may be necessary to remove the pulley from the crankshaft to access the lower cover bolts.

11. Remove the timing belt cover bolts and the upper and lower covers from the engine.

12. If the same timing belt will be reused, mark the direction of timing belt's rotation, for installation in the same direction. Make sure engine is positioned so No. 1 cylinder is at the TDC of it's compression stroke and the sprockets timing marks are aligned with the engine's timing mark indicators.

13. Loosen the timing belt tensioner bolt and remove the belt. If not removing the tensioner, position it as far away from the center of the engine as possible and tighten the bolt.

14. If tensioner is being removed, mark outside of the spring to ensure that it is not installed backwards. Unbolt the tensioner and remove it along with the spring.

15. Slide the timing belt off of the sprockets.

To install:

16. Install the tensioner, if removed, and hook the upper end of the spring to the water pump pin. Install the lower end of the spring to the tensioner in exactly the same position as originally installed.

17. If not already done, position both camshafts so the timing marks line up with those on the alternator bracket (rear bank) and inner timing cover (front bank). Rotate the crankshaft so the timing mark aligns with the mark on the oil pump.

18. Install the timing belt on the crankshaft sprocket and while keeping the belt tight on the tension side (right side), install the belt on the front camshaft sprocket.

19. Install the belt on the water pump pulley, then the rear camshaft sprocket and the tensioner.

20. Rotate the front camshaft counterclockwise to tension the belt between the front camshaft and the crankshaft. If the timing marks came out of line, repeat the procedure.

21. Install the crankshaft sprocket flange.

22. Loosen the tensioner bolt and allow the spring to tension the belt.

23. Slowly turn the crankshaft 2 full turns in the clockwise direction until the timing marks align. Now that the belt is properly tensioned, torque the tensioner lock bolt to 35 ft. lbs. (48 Nm).

24. Install the upper and lower covers to the engine and secure with the retaining screws. Make sure the packing is positioned in the inner grooves of the covers properly when installing.

25. Install the crankshaft pulley if it was removed. Torque the bolt to 110 ft. lbs. (150 Nm).

26. Install the air conditioning bracket and compressor to the engine. Install the belt tensioner.

27. Install the power steering pump into position. Install the fan pulley and fan.

28. Install the fan shroud to the radiator.

29. Refill the cooling system.

30. Connect the negative battery cable. Start the engine and check for fluid leaks.

1996–97 Vehicles

1. If possible, position engine with No. 1 cylinder at TDC.

2. Disconnect the negative battery cable.

3. Drain the cooling system. Remove the drive belts.

4. Remove the upper radiator shroud.

5. Remove the fan and fan pulley.

6. Without disconnecting the lines, remove the power steering pump from its bracket and position it to the side. Remove the pump brackets.

7. Remove the belt tensioner pulley bracket.

8. Without releasing the refrigerant remove the air conditioning compressor from its bracket and position it to the side. Remove the bracket.

9. Remove the cooling fan bracket.

10. On some vehicles it may be necessary to remove the pulley from the crankshaft to access the lower cover bolts.

11. Remove the timing belt cover bolts and the upper and lower covers from the engine.

12. If the same timing belt will be reused, mark the direction of timing belt's rotation, for installation in the same direction. Make sure engine is positioned so No. 1 cylinder is at the TDC of it's compression stroke and the sprockets timing marks are aligned with the engine's timing mark indicators.

13. Loosen the timing belt tensioner bolt and remove the belt. If not removing the tensioner, position it as far away from the center of the engine as possible and tighten the bolt.

14. If tensioner is being removed, mark outside of the spring to ensure

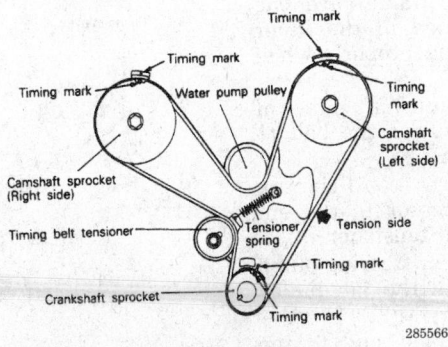

Installation of the timing belt. Align the timing marks as shown — 1993–95 3.0L (VIN H) engines

that it is not installed backwards. Unbolt the tensioner and remove it along with the spring.

15. Using SST MB990767–01 and MIT308239 or equivalent, remove the camshaft sprockets.

To install:

16. Hold the hexagonal portion of the camshaft with a wrench when tightening the camshaft sprocket bolt and torque to:

- 3.0L 12 valve: 66 ft. lbs. (90 Nm)
- 3.0L 24 valve: 64 ft. lbs. (88 Nm)

17. If removed, install the tensioner and hook the upper end of the spring to the water pump pin. Install the lower end of the spring to the tensioner in exactly the same position as originally installed.

18. Position both camshafts so the timing marks line up with those on the alternator bracket (rear bank) and inner timing cover (front bank). Rotate the crankshaft so the timing mark aligns with the mark on the oil pump.

19. Install the timing belt on the crankshaft sprocket and while keeping the belt tight on the tension side (right side), install the belt on the front camshaft sprocket.

20. Install the belt on the water pump pulley, then the rear camshaft sprocket and the tensioner.

21. Rotate the front camshaft counterclockwise to tension the belt between the front camshaft and the crankshaft. If the timing marks came out of line, repeat the procedure.

22. Install the crankshaft sprocket flange.

23. Loosen the tensioner bolt and allow the spring to tension the belt.

24. Slowly turn the crankshaft 2 full turns in the clockwise direction until the timing marks align. Now that the belt is properly tensioner, torque the tensioner lock bolt to 35 ft. lbs. (48 Nm).

25. Install the upper and lower covers to the engine and secure with the retaining screws. Make sure the packing is positioned in the inner grooves of the covers properly when installing.

26. Install the crankshaft pulley if it was removed. Torque the bolt to 110 ft. lbs. (150 Nm).

27. Install the air conditioning bracket and compressor to the engine. Install the belt tensioner.

28. Install the power steering pump into position. Install the fan pulley and fan.

29. Install the fan shroud to the radiator.

30. Refill the cooling system.

31. Connect the negative battery cable. Start the engine and check for fluid leaks.

3.5L (VIN M) Engines

1994–95 Vehicles

1. If possible, position engine with No. 1 cylinder at TDC.
2. Disconnect the negative battery cable.
3. Drain the cooling system. Remove the drive belts.
4. Remove the upper radiator shroud.
5. Remove the fan and fan pulley.
6. Without disconnecting the lines, remove the power steering pump from its bracket and position it to the side. Remove the pump brackets.
7. Remove the belt tensioner pulley bracket.
8. Without releasing the refrigerant remove the air conditioning compressor from its bracket and position it to the side. Remove the bracket.
9. Remove the cooling fan bracket.
10. On some vehicles it may be necessary to remove the pulley from the crankshaft to access the lower cover bolts.
11. Remove the timing belt cover bolts and the upper and lower covers from the engine.
12. If the same timing belt will be reused, mark the direction of timing belt's rotation, for installation in the same direction. Make sure engine is positioned so No. 1 cylinder is at the TDC of it's compression stroke and the sprockets timing marks are aligned with the engine's timing mark indicators.
13. Loosen the timing belt tensioner bolt and remove the belt. If not removing the tensioner, position it as far away from the center of the engine as possible and tighten the bolt.
14. Remove the sprockets by holding the hexagonal portion of the

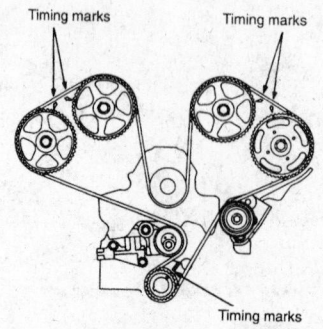

Timing belt and components — 3.5L (VIN M) engines

camshaft with a wrench while removing the sprocket bolt.

To install:

15. Install the crankshaft pulley and turn the crankshaft sprocket timing mark forward 3 teeth to move the piston slightly past No. 1 cylinder top dead center.

16. If removed, install the camshaft sprockets and torque the bolts to 64 ft. lbs. (88 Nm).

17. Align the timing mark of the left bank side camshaft sprocket.

18. Align the timing mark of the right bank side camshaft sprocket and hold the sprocket with a wrench so that it doesn't turn.

19. Set the timing belt onto the water pump pulley.

20. Check that the camshaft sprocket timing mark of the left bank side is aligned and clamp the timing belt with double clips.

21. Set the timing belt onto the idler pulley.

22. Turn the crankshaft one turn counterclockwise and set the timing belt onto the crankshaft sprocket.

23. Set the timing belt on the tension pulley.

24. Place the tension pulley pin hole so that it is towards the top. Press the tension pulley onto the timing belt, and then provisionally tighten the fixing bolt. Torque the bolt to 35 ft. lbs. (48 Nm).

25. Slowly turn the crankshaft 2 full turns in the clockwise direction until the timing marks align. Remove the four double clips.

26. Install the upper and lower covers to the engine and secure with the retaining screws. Make sure the packing is positioned in the inner grooves of the covers properly when installing.

27. Install the crankshaft pulley if it was removed. Torque the bolt to 110 ft. lbs. (150 Nm).

28. Install the air conditioning bracket and compressor to the engine. Install the belt tensioner.

29. Install the power steering pump into position. Install the fan pulley and fan.

30. Install the fan shroud to the radiator.

31. Refill the cooling system.

32. Connect the negative battery cable. Start the engine and check for fluid leaks.

1996–97 Vehicles

1. Disconnect the negative battery cable.
2. Drain the cooling system. Remove the drive belts.
3. Remove the upper radiator shroud.

4. Remove the fan and fan pulley.

5. Without disconnecting the lines, remove the power steering pump from its bracket and position it to the side. Remove the pump brackets.

6. Remove the belt tensioner pulley bracket.

7. Without releasing the refrigerant remove the air conditioning compressor from its bracket and position it to the side. Remove the bracket.

8. Remove the cooling fan bracket.

9. On some vehicles it may be necessary to remove the pulley from the crankshaft to access the lower cover bolts.

10. Remove the timing belt cover bolts and the upper and lower covers from the engine.

11. Remove the crankshaft position sensor connector.

12. Using SST MB990767-01 and MD998754 or equivalent, remove the crankshaft pulley from the crankshaft.

13. Remove the timing belt.

a. Align the timing marks.

b. Loosen the center bolt on the tension pulley and remove the belt.

14. If the same timing belt will be reused, mark the direction of timing belt's rotation, for installation in the same direction. Make sure engine is positioned so No. 1 cylinder is at the TDC of it's compression stroke and the sprockets timing marks are aligned with the engine's timing mark indicators.

15. Remove the auto–tensioner, the tension pulley and the tension arm assembly.

16. Remove the sprockets by holding the hexagonal portion of the camshaft with a wrench while removing the sprocket bolt.

To install:

17. Install the crankshaft pulley and turn the crankshaft sprocket timing mark forward 3 teeth to move the piston slightly past No. 1 cylinder top dead center.

18. If removed, install the camshaft sprockets and torque the bolts to 64 ft. lbs. (88 Nm).

19. Align the timing mark of the left bank side camshaft sprocket.

20. Align the timing mark of the right bank side camshaft sprocket and hold the sprocket with a wrench so that it doesn't turn.

21. Set the timing belt onto the water pump pulley.

22. Check that the camshaft sprocket timing mark of the left bank side is aligned and clamp the timing belt with double clips.

23. Set the timing belt onto the idler pulley.

24. Turn the crankshaft one turn counterclockwise and set the timing belt onto the crankshaft sprocket.

25. Set the timing belt on the tension pulley.

26. Place the tension pulley pin hole so that it is towards the top. Press the tension pulley onto the timing belt, and then provisionally tighten the fixing bolt. Torque the bolt to 35 ft. lbs. (48 Nm).

27. Slowly turn the crankshaft 2 full turns in the clockwise direction until the timing marks align. Remove the four double clips.

28. Install the crankshaft position sensor connector.

29. Install the upper and lower covers to the engine and secure with the retaining screws. Make sure the packing is positioned in the inner grooves of the covers properly when installing.

30. Install the crankshaft pulley if it was removed. Torque the bolt to 110 ft. lbs. (150 Nm).

31. Install the air conditioning bracket and compressor to the engine. Install the belt tensioner.

32. Install the power steering pump into position. Install the fan pulley and fan.

33. Install the fan shroud to the radiator.

34. Refill the cooling system.

35. Connect the negative battery cable. Start the engine and check for fluid leaks.

Expo 1.8L (4G93) Engines

1. Disconnect the negative battery cable. Remove the engine under cover.

2. Rotate crankshaft clockwise and position the No. 1 piston at TDC compression stroke.

3. Raise and safely support the weight of the engine using the appropriate equipment. Remove the front engine mount bracket, A/C hose clamp and accessory drive belts.

4. Remove the crankshaft pulley.

5. Remove timing belt upper and lower covers.

6. Make a mark on the timing belt indicating the direction of rotation so it may be reassembled in the same direction if it is to be reused. Loosen the timing belt tensioner, insert a screwdriver into the tensioner and release tension by prying screwdriver against spring tension. Temporarily

tighten tensioner bolt to provide slack and remove the timing belt.

NOTE: Coolant or engine oil drastically shortens the life of the timing belt. Do not allow engine oil or coolant to contact the timing belt, sprockets or tensioner assembly.

7. Remove the tensioner spacer, tensioner spring and tensioner assembly.

NOTE: It is recommended that the timing belt is replaced every 60,000 miles (96,000 Km).

8. Inspect the timing belt for cracks or wear. Check the tensioner pulley for smooth rotation.

To install:

9. Position the tensioner, tensioner spring and tensioner spacer on engine block.

10. Align the timing marks on the camshaft sprocket and crankshaft sprocket. This will position No. 1 piston on TDC on the compression stroke.

11. Position the timing belt on the crankshaft sprocket, water pump sprocket, camshaft sprocket and the tensioner keeping the tension side of the belt tight.

12. Apply slight counterclockwise force to the camshaft sprocket to give tension to the belt and make sure all timing marks are aligned.

13. Loosen the tensioner bolt and allow the spring to remove the slack.

14. Turn the crankshaft clockwise two rotations and tighten the adjuster bolt to 18 ft. lbs. (24 Nm) and tighten the pivot (spring) bolt to 35 ft. lbs. (45 Nm).

15. Install the timing belt covers and torque the cover bolts to 84 inch lbs. (10 Nm).

16. Install the crankshaft pulley and torque the retaining bolt to 134 ft. lbs. (185 Nm).

17. Install the front engine mount bracket and mount.

18. Connect the A/C hose clamp and install the accessory drive belts.

19. Install the engine undercover.

20. Connect the negative battery cable.

Expo 2.4L (4G64) engines

1. If possible, position the engine so the No. 1 piston is at TDC.

2. Disconnect the negative battery cable.

3. Remove the timing belt covers.

4. To loosen the timing (outer) belt tensioner, install special tool MD998738 or equivalent, to the slot and screw inward to move tensioner

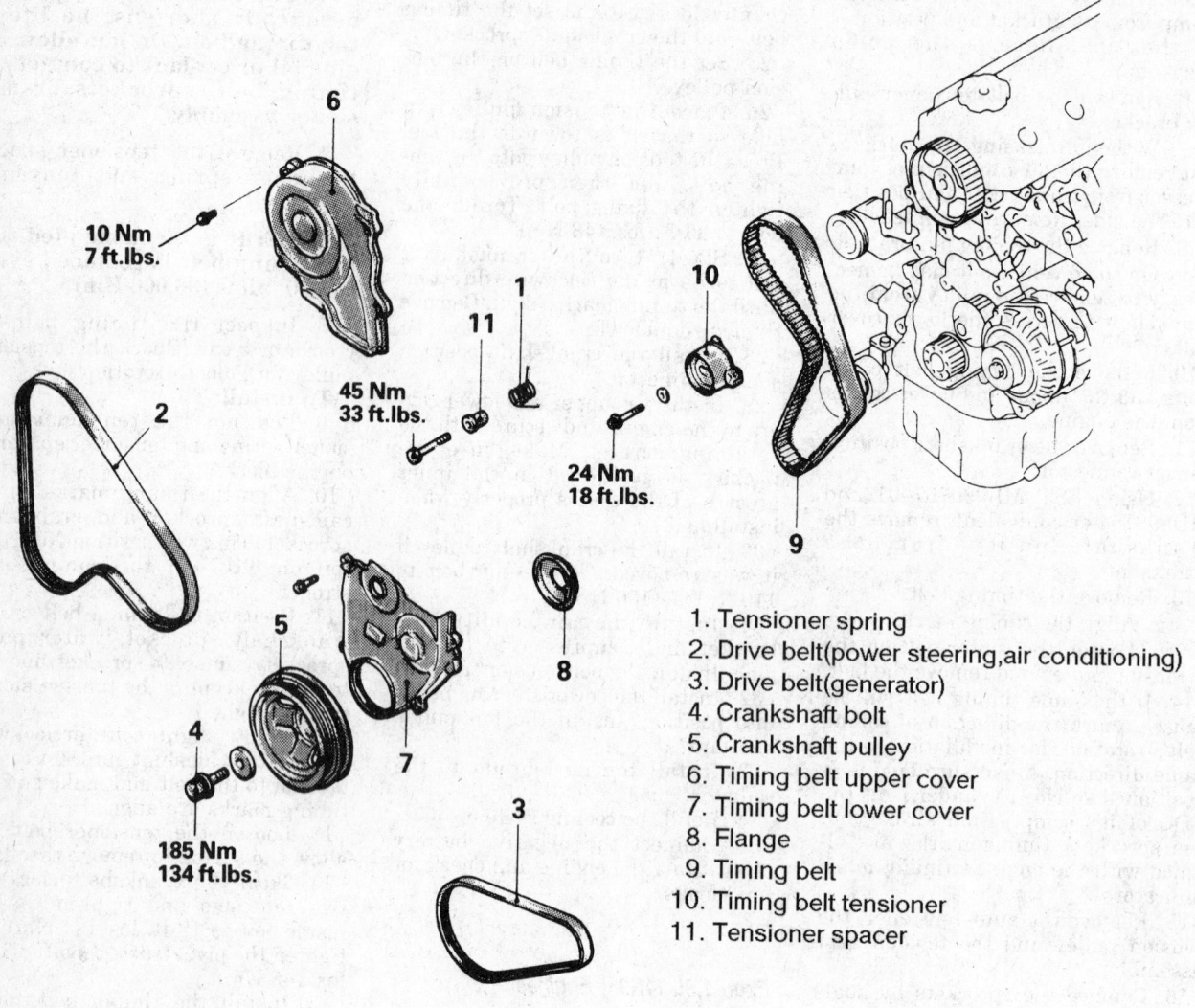

1. Tensioner spring
2. Drive belt(power steering,air conditioning)
3. Drive belt(generator)
4. Crankshaft bolt
5. Crankshaft pulley
6. Timing belt upper cover
7. Timing belt lower cover
8. Flange
9. Timing belt
10. Timing belt tensioner
11. Tensioner spacer

332161

Timing belt and related components — Expo 1.8L (4G93) engines

toward the water pump. Once the tension has been relieved, remove the outer timing belt.

NOTE: If timing belts are going to be reused, mark the direction of rotation on the belt. This will ensure the belt is reinstalled in same direction, extending belt life.

5. Remove the outer crankshaft sprocket and flange.
6. Loosen the silent shaft (inner) belt tensioner and remove the belt.
To install:
7. Turn both tensioner pulleys and check for any signs of bearing wear.

8. Align the timing marks of the silent shaft sprockets and the crankshaft sprocket with the timing marks on the front case. Wrap the timing belt around the sprockets so there is no slack in the upper span of the belt and the timing marks are still aligned.
9. Install the tensioner pulley and move the pulley by hand so the long side of the belt deflects about ¼ inch.
10. Hold the pulley tightly so the pulley cannot rotate when the bolt is tightened. Tighten the bolt to 14 ft. lbs. (19 Nm) and recheck the deflection amount.
11. Align the timing marks of the camshaft, crankshaft and oil pump

sprockets with their corresponding marks on the front case or rear cover.

NOTE: There is a possibility to align all timing marks and have the oil pump sprocket and silent shaft out of time, causing an engine vibration during operation. If the following step is not followed exactly, there is a 50 percent chance that the silent shaft alignment will be 180 degrees off.

12. Before installing the timing belt, ensure that the left side (rear) silent shaft (oil pump sprocket) is in the correct position as follows:
    a. Remove the plug from the rear side of the block and insert a tool

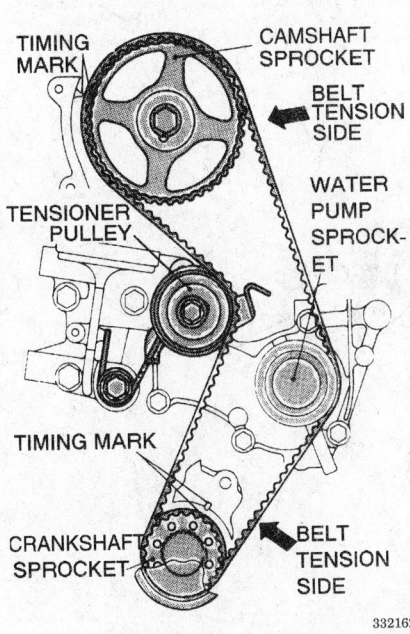

Timing belt routing and mark identification — Expo 1.8L (4G93) engines

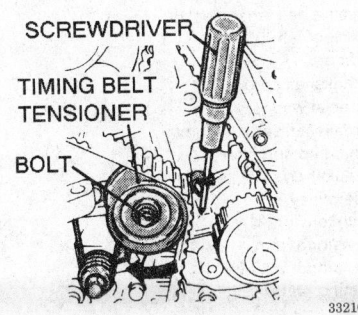

Proper method of releasing belt tension — Expo 1.8L (4G93) engines

with shaft diameter of 0.31 inches. (8mm) into the hole.

b. With the timing marks still aligned, the shaft of the tool must be able to go in at least 2 1/2 inches. If the tool can only go in about 1 in., the shaft is not in the correct orientation and will cause a vibration during engine operation. Remove the tool from the hole and turn the oil pump sprocket 1 complete revolution. Realign the timing marks and insert the tool. The shaft of the tool must go in at least 2 1/2 inches.

c. Recheck and realign the timing mark.

d. Leave the tool in place to hold the silent shaft while continuing.

13. Install the belt to the crankshaft sprocket, oil pump sprocket, then camshaft sprocket, in that order. While doing so, make sure there is no slack between the sprocket except where the tensioner is installed.

14. To adjust the timing (outer) belt perform the following steps:

a. Turn the crankshaft 1/4-turn counterclockwise, then turn it clockwise to move No. 1 cylinder to TDC.

b. Loosen the center bolt. Using tool MD998752 or equivalent and a torque wrench, apply a torque of 2.6 ft. lbs. (3.6 Nm). Tighten the center bolt.

c. Screw the special tool into the engine left support bracket until its end makes contact with the tensioner arm. At this point, screw the special tool in some more and remove the set wire attached to the auto tensioner, if the wire was not previously removed. Then remove the special tool.

d. Rotate the crankshaft two complete turns clockwise and let it sit for approximately 15 minutes. Then, measure the auto tensioner protrusion (the distance between the tensioner arm and auto tensioner body) to ensure that it is within 0.15–0.18 inch (3.8–4.5mm). If out of specification, repeat Step 1–4 until the specified value is obtained.

NOTE: Do not manually overtighten the belt or it will howl.

15. Install the timing belt covers and all related items.

16. Connect the negative battery cable.

17. Run the engine until the thermostat opens. Check and adjust ignition timing.

Expo 1.8L (4G93) Engine

This vehicle uses a 1.8L engine in various models. A common trait shared among these engines is their belt driven Overhead Camshaft (OHC) design. The following basic procedure applies to all models. Note that different options may mean slightly different procedures may be required. As is true with all OHC design engines, valve timing is critical or engine damage may occur. Pay close attention to all timing marks. Many technicians will set an engine up to TDC of the No. 1 cylinder's firing stroke before beginning disassembly. This should bring all timing marks into alignment. If the timing indica-

tors on the sprockets are hard to see, take the time to mark them with a small amount of white paint. A few minutes spent prior to tear-down will save much time at reassembly and should help assure that the engine is properly timed. In addition, many vehicle manufacturers advise that a new replacement timing belt be installed anytime the belt has been removed or new replacement sprockets are installed.

1. Disconnect the negative battery cable.

2. Remove the valve cover.

3. Remove timing belt.

4. Remove the crankshaft pulley retainer bolts and remove the pulley.

5. Remove the crankshaft sprocket retainer bolt and washer from the sprocket, if used, and remove sprocket. If sprocket is difficult to remove, the appropriate puller may be used. If no bolts are used on the sprocket, use the appropriate puller to remove.

6. Remove the camshaft sprocket using the appropriate spanner wrench if available. If not, note that most camshafts are made with a hexagon-shaped piece near the forward cam lobes so that a wrench can be used to hold the camshaft to keep it from turning when the sprocket bolt is being loosened or tightened. In no case should a tool be inserted through the sprocket to jam the sprocket to keep it from turning. This will damage the sprocket so that it should not be reused.

To install:

7. Install the sprockets to their appropriate shafts. Install the retainer bolts and torque the camshaft sprocket bolt to specification.

8. Torque the crankshaft sprocket retaining bolt to specification.

9. Install the timing belt, timing covers, valve cover and remaining components.

10. Install the engine under cover and connect the negative battery cable.

1995 Expo 1.8L (4G93) Engine

1. Disconnect the negative battery cable.

2. Rotate the engine and position the No. 1 piston to TDC of its compression stroke.

3. Label and disconnect the spark plug cables.

4. Matchmark the positioning of the distributor housing and the positioning of the distributor rotor to the engine block and remove the distributor.

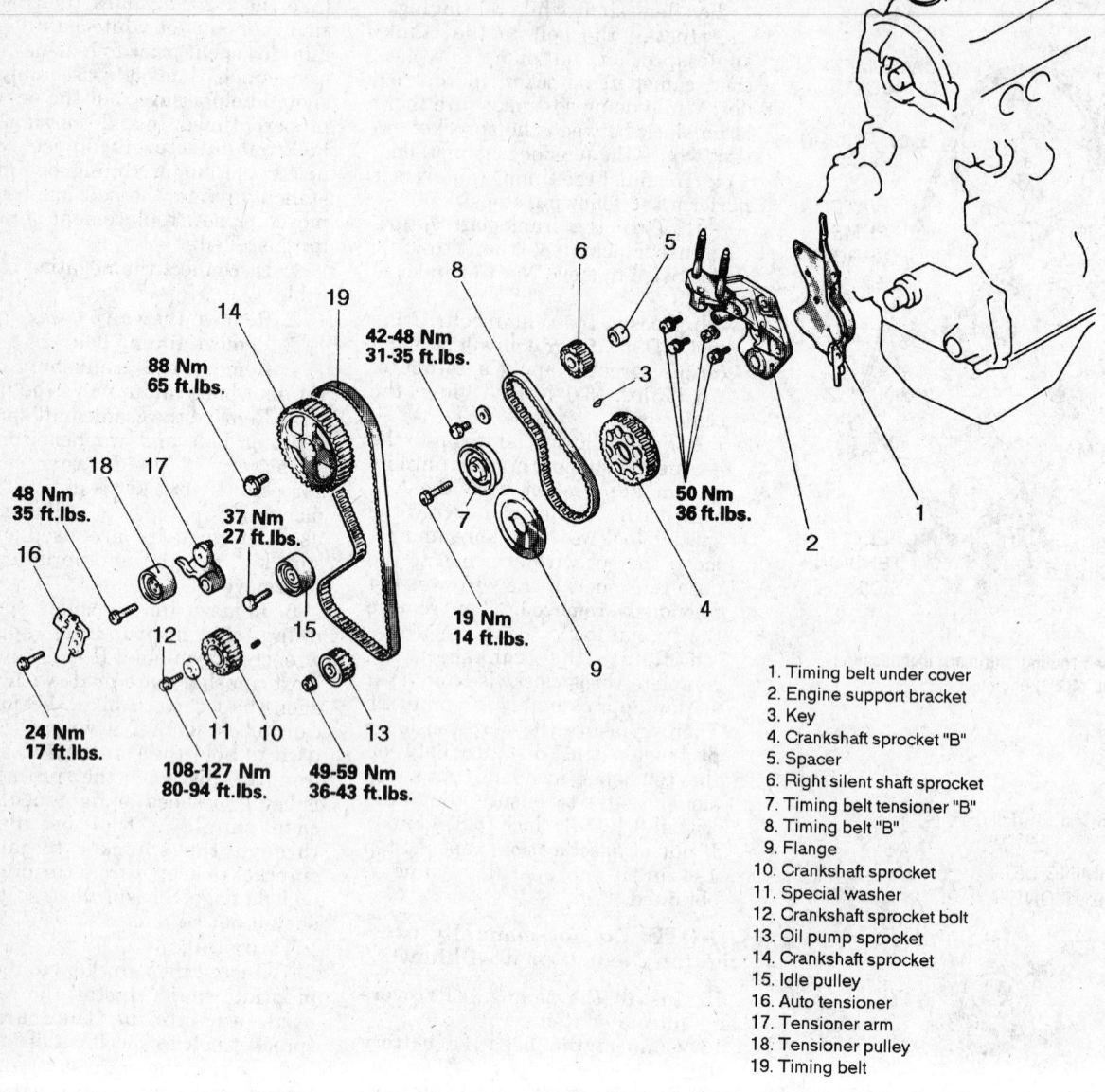

88 Nm
65 ft.lbs.

42-48 Nm
31-35 ft.lbs.

48 Nm
35 ft.lbs.

37 Nm
27 ft.lbs.

50 Nm
36 ft.lbs.

19 Nm
14 ft.lbs.

24 Nm
17 ft.lbs.

108-127 Nm
80-94 ft.lbs.

49-59 Nm
36-43 ft.lbs.

1. Timing belt under cover
2. Engine support bracket
3. Key
4. Crankshaft sprocket "B"
5. Spacer
6. Right silent shaft sprocket
7. Timing belt tensioner "B"
8. Timing belt "B"
9. Flange
10. Crankshaft sprocket
11. Special washer
12. Crankshaft sprocket bolt
13. Oil pump sprocket
14. Crankshaft sprocket
15. Idle pulley
16. Auto tensioner
17. Tensioner arm
18. Tensioner pulley
19. Timing belt

241427

Timing belt component identification — Expo 2.4L (4G64) engines

5. Disconnect the air flow sensor connector and remove the air cleaner case cover.

6. Disconnect the accelerator cable, breather hose and PCV hose connections.

7. Remove the rocker cover and discard the gasket.

8. Loosen both rocker arm shaft assemblies gradually and evenly and remove the rocker shafts from the vehicle. Do not disassembly rocker arms and rocker arm shaft assemblies.

9. Remove the timing belt covers.

NOTE: DO NOT allow the camshaft or the crankshaft to rotate after the timing belt is removed.

10. Remove the timing belt assembly.

11. Holding the camshaft sprocket from turning, loosen and remove the bolt that secures the sprocket.

12. Remove the camshaft sprocket from the camshaft. Note the positioning of the dowel pin at the end of the camshaft.

13. Remove the camshaft oil seal from the front of the cylinder head.

14. Remove the camshaft from the head.

15. Carefully check all parts for damage and wear.

To install:

16. Lubricate the camshaft journals and camshaft with clean engine oil and install the camshaft in the cylinder head. Be sure to position the dowel pin at the end of the camshaft as noted during the removal procedure.

17. Check the camshaft end-play between the thrust case and camshaft. The camshaft end-play should be 0.002–0.008 in. (0.05–0.20mm). If the end-play is not

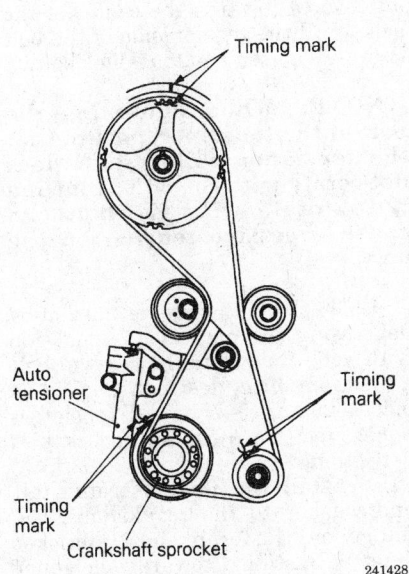

Camshaft timing belt timing mark identification — Expo 2.4L (4G64) engines

241428

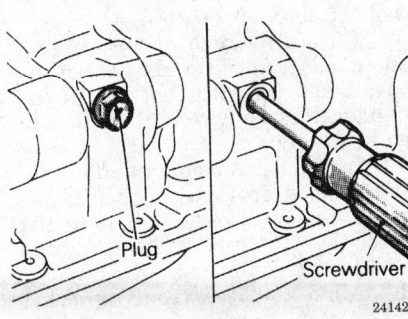

Proper method of aligning oil pump sprocket — Expo 2.4L (4G64) engines

241429

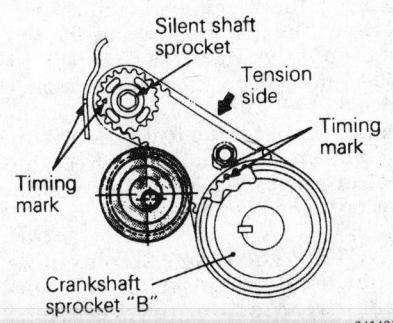

Oil pump timing belt timing mark identification — Expo 2.4L (4G64) engines

241431

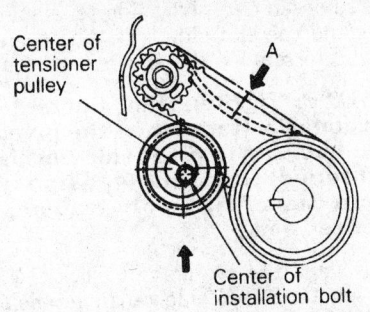

Proper method of adjusting oil pump timing belt — Expo 2.4L (4G64) engines

241432

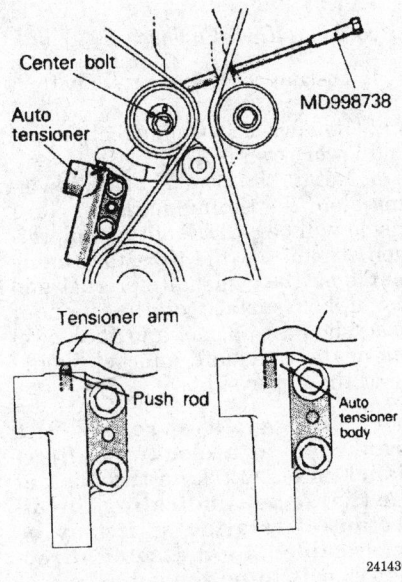

Auto tensioner adjustment — Expo 2.4L (4G64) engines

241430

within specification, replace the camshaft thrust bearing.

18. Install a new camshaft oil seal. Be sure to lubricate the lips of the seal with clean engine oil.

19. Install camshaft sprocket and torque the retainer bolt to 65 ft. lbs. (90 Nm). Be sure to secure the sprocket while tightening the bolt.

20. Install the timing belt assembly.

21. Install the timing belt covers.

22. Install the rocker arm and shaft assemblies. Tighten the rocker arm shaft retainer bolts to 23 ft. lbs. (32 Nm).

23. Check valve adjustment and install valve cover with a new gasket. Tighten the valve cover bolts to 29 inch lbs. (3.3 Nm).

24. Align the distributor marks and install the distributor.

25. Connect the spark plug cables.

26. Connect the accelerator cable, breather hose and PCV hose.

27. Connect the air flow sensor connector and install the air cleaner case cover.

28. Connect the negative battery cable. Run the engine at idle until normal operating temperature is reached. Check idle speed and ignition timing and adjust as required.

Expo 2.4L (4G64) Engines

1. Disconnect the negative battery cable.

2. Remove timing belts.

3. Remove the crankshaft pulley retainer bolts and remove the pulley.

4. Remove the crankshaft sprocket retainer bolt and washer from the sprocket, if used, and remove sprocket. If sprocket is difficult to remove, the appropriate puller may be used. If no bolts are used on the sprocket, use the appropriate puller to remove.

5. Remove the camshaft sprocket using the appropriate spanner wrench; hold the shaft in position while removing the bolt.

To install:

6. Install the sprockets to their appropriate shafts. Install the retainer bolts and torque the camshaft sprocket bolt to 65 ft. lbs. (88 Nm).

7. Torque the crankshaft sprocket retaining bolt to 80–94 ft. lbs. (110–130 Nm).

8. Install the timing belt, timing covers, and remaining components.

9. Install the engine under cover and connect the negative battery cable.

Expo 1.8L (4G93) Engine

1. Disconnect the negative battery cable. Remove the engine under cover.

2. Rotate crankshaft clockwise and position engine at TDC compression stroke.

3. Raise and safely support the weight of the engine using the appropriate equipment. Remove the front engine mount bracket and accessory drive belts.

4. Remove timing belt upper and lower covers.

5. Make a mark on the back of the timing belt indicating the direction of rotation so it may be reassembled in the same direction if it is to be re-

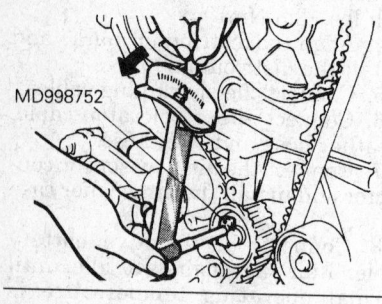

MD998752

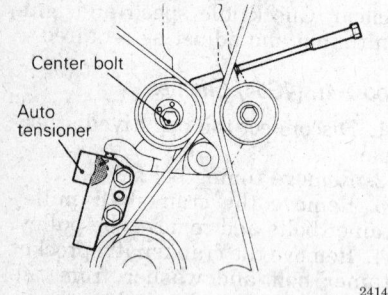

Center bolt

Auto tensioner

241433

Using special tools to adjust belt tension — Expo 2.4L (4G64) engines

used. Loosen the timing belt tensioner and remove the timing belt.

NOTE: Coolant and engine oil will drastically shorten the timing belts's life. Do not allow engine oil or coolant to contact the timing belt, the sprockets or tensioner assembly.

6. Remove the tensioner spacer, tensioner spring and tensioner assembly.

7. Inspect the timing belt for cracks on back surface, sides, bottom and check for separated canvas. Check the tensioner pulley for smooth rotation.

To install:

8. Position the tensioner, tensioner spring and tensioner spacer on engine block.

9. Align the timing marks on the camshaft sprocket and crankshaft sprocket. This will position No. 1 piston on TDC on the compression stroke.

10. Position the timing belt on the crankshaft sprocket and keeping the tension side of the belt tight, set it on the camshaft sprocket.

11. Apply counterclockwise force to the camshaft sprocket to give tension to the belt and make sure all timing marks are aligned.

12. Loosen the pivot side tensioner bolt and the slot side bolt. Allow the spring to remove the slack.

NOTE: Tighten the slot side tensioner bolt and then the pivot side bolt. If the pivot side bolt is tightened first, the tensioner could turn with bolt, causing over tension.

13. Turn the crankshaft clockwise. Loosen the pivot side tensioner bolt and then the slot side bolt to allow the spring to take up any remaining slack. Tighten the adjuster bolt to 18 ft. lbs. (24 Nm).

14. Install the timing belt covers and all related items.

15. Connect the negative battery cable.

Expo 2.4L (4G64) Engines

1. Disconnect the negative battery cable.

2. Remove the timing belt upper and lower covers.

3. Rotate the crankshaft clockwise and align the timing marks so No. 1 piston will be at TDC of the compression stroke. At this time the timing marks on the camshaft sprocket and the upper surface of the cylinder head should coincide, and the dowel pin of the camshaft sprocket should be at the upper side.

NOTE: Always rotate the crankshaft in a clockwise direction. Make a mark on the back of the timing belt indicating the direction of rotation so it may be reassembled in the same direction if it is to be reused.

4. Remove the auto tensioner cylinder assembly two mounting bolts and remove the outermost timing belt.

CAUTION

Release tension gradually when removing cylinder assembly to prevent damage and possible injury.

5. Remove the silent shaft (inner) belt tensioner center bolt and the pulley.

To install:

6. Check both tensioner and idler pulley for bearing wear, and replace if needed.

7. Align the timing marks on the crankshaft sprocket and the silent shaft sprocket. Fit the inner timing belt over the crankshaft and silent shaft sprocket. Ensure that there is no slack in the belt.

8. While holding the inner timing belt tensioner with your fingers, adjust the timing belt tension by applying a force towards the center of the belt, until the tension side of the belt is taut. Tighten the tensioner bolt.

NOTE: When tightening the bolt of the tensioner, ensure that the tensioner pulley shaft does not rotate with the bolt. Allowing it to rotate with the bolt can cause excessive tension on the belt.

9. Check belt for proper tension by depressing the belt on its' long side with your finger and noting the belt deflection. The desired reading is 0.20–0.28 in. (5–7mm). If tension is not correct, readjust and check belt deflection.

10. Install the flange, crankshaft and washer to the crankshaft. The flange on the crankshaft sprocket must be installed towards the inner timing belt sprocket. Tighten bolt to 80–94 ft. lbs. (110–130 Nm).

11. To install the oil pump sprocket, insert a Phillips screwdriver with a shaft 0.31 in. (8mm) in diameter into the plug hole in the left side of the cylinder block to hold the left silent shaft. Tighten the nut to 36–43 ft. lbs. (50–60 Nm).

12. Carefully push the auto tensioner rod in until the set hole in the rod aligned up with the hole in the cylinder. Place a wire into the hole to retain the rod.

13. Install the tensioner pulley onto the tensioner arm. Locate the pinhole in the tensioner pulley shaft to the left of the center bolt. Then, tighten the center bolt finger-tight.

14. Align the crankshaft sprocket and oil pump sprocket timing marks.

15. After alignment of the oil pump sprocket timing marks, remove the plug on the cylinder block and insert a Phillips screwdriver with a shaft diameter of 0.31 in. (8mm) through the hole. If the shaft can be inserted 2.4 in. deep, the silent shaft is in the correct position. If the shaft of the tool can only be inserted 0.8–1.0 in. (20–25mm) deep, turn the oil pump sprocket 1 turn and realign the marks. Reinsert the tool making sure it is inserted 2.4 in. deep. Keep the tool inserted in hole for the remainder of this procedure.

NOTE: The above step assures that the oil pump socket is in correct orientation to the silent shafts. This step must not be skipped or a vibration may develop during engine operation.

16. Install the timing belt as follows:

a. Install the timing belt around the camshaft idler pulley, oil pump sprocket, crankshaft sprocket and the tensioner pulley. Remove the 2 spring clips.

b. Lift upward on the tensioner pulley in a clockwise direction and tighten the center bolt. Make sure all timing marks are aligned.

c. Rotate the crankshaft 1/4 turn counterclockwise. Then, turn in clockwise until the timing marks are aligned again.

17. To adjust the timing (outer) belt, turn the crankshaft 1/4 turn counterclockwise, then turn it clockwise to move No. 1 cylinder to TDC.

18. Loosen the center bolt. Using tool MD998738 or equivalent and a torque wrench, apply a torque of 1.88–2.03 ft. lbs. (2.6–2.8 Nm). Tighten the center bolt.

19. Screw the special tool into the engine left support bracket until its end makes contact with the tensioner arm. At this point, screw the special tool in some more and remove the set wire attached to the auto tensioner, if the wire was not previously removed. Then remove the special tool.

20. Rotate the crankshaft 2 complete turns clockwise and let it sit for approximately 15 minutes. Then, measure the auto tensioner protrusion (the distance between the tensioner arm and auto tensioner body) to ensure that it is within 0.15–0.18 in. (3.8–4.5mm). If out of specification, repeat Step 1–4 until the specified value is obtained.

21. If the timing belt tension adjustment is being performed with the engine mounted in the vehicle, and clearance between the tensioner arm and the auto tensioner body cannot be measured, the following alternative method can be used:

a. Screw in special tool MD998738 or equivalent, until its end makes contact with the tensioner arm.

b. After the special tool makes contact with the arm, screw it in some more to retract the auto tensioner pushrod while counting the number of turns the tool makes until the tensioner arm is brought into contact with the auto tensioner body. Make sure the number of turns the special tool makes conforms with the standard value of 2½–3 turns.

c. Install the rubber plug to the timing belt rear cover.

22. Install the timing belt covers and all related items.

23. Connect the negative battery cable.

24. Run engine until thermostat opens. Check ignition timing and adjust as necessary.

Expo 2.4L (4G64) Engines

1. Disconnect the negative battery cable.

2. Remove the timing belt upper and lower covers.

3. Rotate the crankshaft clockwise and align the timing marks so No. 1 piston will be at TDC of the compression stroke. At this time the timing marks on the camshaft sprocket and the upper surface of the cylinder head should coincide, and the dowel pin of the camshaft sprocket should be at the upper side.

NOTE: Always rotate the crankshaft in a clockwise direction. Make a mark on the back of the timing belt indicating the direction of rotation so it may be reassembled in the same direction if it is to be reused.

4. Remove the auto tensioner cylinder assembly two mounting bolts and remove the outermost timing belt.

─────── CAUTION ───────
Release tension gradually when removing cylinder assembly to prevent damage and possible injury.

5. Remove the silent shaft (inner) belt tensioner center bolt and the pulley.

To install:

6. Check both tensioner and idler pulley for bearing wear, and replace if needed.

7. Align the timing marks on the crankshaft sprocket and the silent shaft sprocket. Fit the inner timing belt over the crankshaft and silent shaft sprocket. Ensure that there is no slack in the belt.

8. While holding the inner timing belt tensioner with your fingers, adjust the timing belt tension by applying a force towards the center of the belt, until the tension side of the belt is taut. Tighten the tensioner bolt.

NOTE: When tightening the bolt of the tensioner, ensure that the tensioner pulley shaft does not rotate with the bolt. Allowing it to rotate with the bolt can cause excessive tension on the belt.

9. Check belt for proper tension by depressing the belt on its' long side with your finger and noting the belt

deflection. The desired reading is 0.20–0.28 in. (5–7mm). If tension is not correct, readjust and check belt deflection.

10. Install the flange, crankshaft and washer to the crankshaft. The flange on the crankshaft sprocket must be installed towards the inner timing belt sprocket. Tighten bolt to 80–94 ft. lbs. (110–130 Nm).

11. To install the oil pump sprocket, insert a Phillips screwdriver with a shaft 0.31 in. (8mm) in diameter into the plug hole in the left side of the cylinder block to hold the left silent shaft. Tighten the nut to 36–43 ft. lbs. (50–60 Nm).

12. Carefully push the auto tensioner rod in until the set hole in the rod aligned up with the hole in the cylinder. Place a wire into the hole to retain the rod.

13. Install the tensioner pulley onto the tensioner arm. Locate the pinhole in the tensioner pulley shaft to the left of the center bolt. Then, tighten the center bolt finger-tight.

14. Align the crankshaft sprocket and oil pump sprocket timing marks.

15. After alignment of the oil pump sprocket timing marks, remove the plug on the cylinder block and insert a Phillips screwdriver with a shaft diameter of 0.31 in. (8mm) through the hole. If the shaft can be inserted 2.4 in. deep, the silent shaft is in the correct position. If the shaft of the tool can only be inserted 0.8–1.0 in. (20–25mm) deep, turn the oil pump sprocket 1 turn and realign the marks. Reinsert the tool making sure it is inserted 2.4 in. deep. Keep the tool inserted in hole for the remainder of this procedure.

NOTE: The above step assures that the oil pump socket is in correct orientation to the silent shafts. This step must not be skipped or a vibration may develop during engine operation.

16. Install the timing belt as follows:

a. Install the timing belt around the camshaft idler pulley, oil pump sprocket, crankshaft sprocket and the tensioner pulley. Remove the 2 spring clips.

b. Lift upward on the tensioner pulley in a clockwise direction and tighten the center bolt. Make sure all timing marks are aligned.

c. Rotate the crankshaft 1/4 turn counterclockwise. Then, turn in clockwise until the timing marks are aligned again.

17. To adjust the timing (outer) belt, turn the crankshaft 1/4 turn coun-

terclockwise, then turn it clockwise to move No. 1 cylinder to TDC.

18. Loosen the center bolt. Using tool MD998738 or equivalent and a torque wrench, apply a torque of 1.88–2.03 ft. lbs. (2.6–2.8 Nm). Tighten the center bolt.

19. Screw the special tool into the engine left support bracket until its end makes contact with the tensioner arm. At this point, screw the special tool in some more and remove the set wire attached to the auto tensioner, if the wire was not previously removed. Then remove the special tool.

20. Rotate the crankshaft 2 complete turns clockwise and let it sit for approximately 15 minutes. Then, measure the auto tensioner protrusion (the distance between the tensioner arm and auto tensioner body) to ensure that it is within 0.15–0.18 in. (3.8–4.5mm). If out of specification, repeat Step 1–4 until the specified value is obtained.

21. If the timing belt tension adjustment is being performed with the engine mounted in the vehicle, and clearance between the tensioner arm and the auto tensioner body cannot be measured, the following alternative method can be used:

 a. Screw in special tool MD998738 or equivalent, until its end makes contact with the tensioner arm.

 b. After the special tool makes contact with the arm, screw it in some more to retract the auto tensioner pushrod while counting the number of turns the tool makes until the tensioner arm is brought into contact with the auto tensioner body. Make sure the number of turns the special tool makes conforms with the standard value of 2½–3 turns.

 c. Install the rubber plug to the timing belt rear cover.

22. Install the timing belt covers and all related items.

23. Connect the negative battery cable.

24. Run engine until thermostat opens. Check ignition timing and adjust as necessary.

Expo 1.8L (4G93) Engines

1. Disconnect the negative battery cable.

2. Remove the engine undercover.

3. Remove the coolant recovery tank.

4. Using the proper equipment, slightly raise the engine to take the weight off of the side engine mount. Support the engine in this position.

5. Disconnect the power steering/air conditioning hose retaining clamp and remove the engine mount bracket.

6. Remove the drive belts, tension pulley brackets, water pump pulley and crankshaft pulley.

NOTE: If drive belts are going to be reused, mark the direction of rotation on the belt. This will ensure the belt will be reinstalled with the same rotation, extending belt life.

7. Remove all attaching screws and remove the upper and lower timing belt covers.

NOTE: Take notice of the locations of each timing belt cover fastener during the removal procedure. Due to the difference in lengths, it is important that they are installed in their original locations.

To install

8. Make sure all pieces of packing are positioned in the inner grooves of the covers and install front covers.

9. Install both pulleys and tensioner. Properly route drive belts and adjust.

10. Connect engine mount bracket and lower engine.

11. Connect the hose retaining clamp to the engine mount.

12. Install engine undercover.

13. Install the coolant recovery tank and top off coolant.

14. Reconnect the battery cable.

Expo 2.4L (4G64) Engines

1. Disconnect the negative battery cable.

2. Remove the engine undercover.

3. Using the proper equipment, slightly raise the engine to take the weight off of the side engine mount. Support the engine in this position.

4. Remove the engine mount bracket.

5. Remove the drive belts, tension pulley brackets, water pump pulley and crankshaft pulley.

NOTE: If drive belts are going to be reused, mark the direction of rotation on the belt. This will ensure the belt will be reinstalled with the same rotation, extending belt life.

6. Remove all attaching screws and remove the upper and lower timing belt covers.

NOTE: Take notice of the locations of each timing belt cover fastener during the removal procedure. Due to the difference in

lengths, it is important that they are installed in their original locations.

To install

7. Make sure all pieces of packing are positioned in the inner grooves of the covers and install front covers.

8. Install both pulleys and tensioner. Properly route drive belts and adjust.

9. Connect engine mount bracket and lower engine.

10. Install engine undercover.

11. Reconnect the battery cable.

Camshaft

REMOVAL AND INSTALLATION

2.4L (VIN G) Engine

1. Disconnect the negative battery cable.

------- CAUTION -------
Wait at least 90 seconds after the negative (-) battery cable is disconnected to prevent possible deployment of the air bag.

2. Remove the valve cover and the upper timing belt cover.

3. Rotate the crankshaft clockwise and align the camshaft sprocket timing mark with the timing mark on the cylinder head.

4. Matchmark the rotor to the distributor housing and remove the distributor.

5. Remove the camshaft sprocket retaining bolt. Pull the sprocket from the camshaft, with the timing belt attached, and place it on top of the timing belt front lower cover. Secure the sprocket in place so the belt remains taut.

------- WARNING -------
Do not rotate the crankshaft after the sprocket is removed from the camshaft. Make sure there is no slack in the timing belt. Make sure the timing belt does not disengage from the sprocket. If the crankshaft is rotated or the timing belt position is disturbed, timing belt and sprocket alignment will have to be set.

6. Remove the camshaft cap bolts evenly and gradually.

7. Install auto lash adjuster retainer MD998443 or equivalent, to each rocker arm to hold the auto lash adjusters in place.

8. Remove the caps, shafts, rocker arms and bolts together as an assembly.

9. Remove the camshaft with the front seal from the engine.

To install:

10. Install a new roll pin to the camshaft.

11. Lubricate the camshaft and install with the front seal in place.

12. Install the camshaft so the hole in the sprocket will line up with the roll pin.

13. Install the caps, shafts and arms assembly. Tighten the camshaft bearing cap bolts to 85 inch lbs. (10 Nm), in the following order:

    a. No. 3, No. 2, No. 4, front cap, rear cap.

    b. Repeat the sequence increasing the torque to 14 lbs. (20 Nm).

14. Install the sprocket to the camshaft, engaging the roll pin. Torque the bolt to 65 ft. lbs. (90 Nm).

15. Install the distributor, aligning the mark made during removal.

16. Install the valve cover and upper timing belt cover. Connect the negative battery cable.

17. Start the engine and check for leaks. Check the ignition timing.

3.0L (VIN H) Engines

1993–95 12 Valve Vehicles

1. Disconnect the negative battery cable.

2. Remove the valve cover.

3. Remove the timing belt and remove the sprocket from the camshaft.

4. Install auto lash adjuster retainers MD998443 or equivalent, on the rocker arms.

5. If removing the left side camshaft, remove the distributor and the distributor extension.

6. Remove the camshaft bearing caps, but do not remove the bolts from the caps. Remove the rocker arms, rocker shafts and bearing caps, as an assembly.

7. Remove the camshaft from the cylinder head.

8. Inspect the bearing journals on the camshaft, cylinder head and bearing caps.

To install:

9. Lubricate the camshaft journals and camshaft with clean engine oil and install the camshaft in the cylinder head.

10. Align the camshaft bearing caps with the arrow mark (depending on cylinder numbers) and in numerical order.

NOTE: The arrow marks on the left rocker shaft bearing caps face in same direction as the arrow marked on the left cylinder head which is away from the tim-

ing belt. The arrows on the right cylinder head and the right rocker bearing caps face toward the timing belt.

11. Apply sealer at the ends of the bearing caps and install the assembly.

12. Torque the bearing cap bolts to 85 inch lbs. (10 Nm), in the following sequence: No. 3, No. 2, No. 1 and No. 4.

13. Repeat the sequence increasing the torque to 175 inch lbs. (18 Nm).

14. Install the distributor, if removed.

15. Install the sprockets, timing belt and timing belt cover.

16. Install the valve cover and connect the negative battery cable.

17. Start the engine and check for leaks and proper operation.

1996–97 12 Valve Vehicles

1. Disconnect the negative battery cable.

CAUTION

Work must be started after 90 seconds from the time the ignition switch is turned to the LOCK position and the negative (-) battery cable is disconnected.

2. Remove the valve cover.

3. Remove the timing belt and remove the sprocket from the camshaft.

4. Install auto lash adjuster retainers SST MD998443-01 or equivalent, on the rocker arms.

5. If removing the left side camshaft, remove the distributor and the distributor extension.

6. Remove the camshaft bearing caps, but do not remove the bolts from the caps. Remove the rocker arms, rocker shafts and bearing caps, as an assembly.

7. Remove the camshaft from the cylinder head.

8. Inspect the bearing journals on the camshaft, cylinder head and bearing caps.

To install:

9. Lubricate the camshaft journals and camshaft with clean engine oil and install the camshaft in the cylinder head.

10. Align the camshaft bearing caps with the arrow mark (depending on cylinder numbers) and in numerical order.

NOTE: The arrow marks on the left rocker shaft bearing caps face in same direction as the arrow marked on the left cylinder head which is away from the timing belt. The arrows on the right

cylinder head and the right rocker bearing caps face toward the timing belt.

11. Apply sealer at the ends of the bearing caps and install the assembly.

12. Torque the front and rear bearing cap bolts to 14 ft. lbs. (20 Nm) and bearing caps 2,3 and 4 to 8 ft. lbs. (11 Nm).

13. Remove the SST from the rocker arms.

14. If removed, install the distributor.

15. Install the sprockets, timing belt and the timing belt cover.

16. Install the valve cover and connect the negative battery cable.

17. Start the engine and check for leaks and proper operation.

24 Valve Vehicles

1. Disconnect the negative battery cable.

CAUTION

Work must be started after 90 seconds from the time the ignition switch is turned to the LOCK position and the negative (-) battery cable is disconnected.

2. Remove the valve cover.

3. Remove the timing belt and remove the sprocket from the camshaft.

4. Install auto lash adjuster retainers SST MD998443 or equivalent, on the rocker arms.

5. If removing the left side camshaft, remove the distributor and the distributor extension.

6. Remove the rocker arms, rocker shafts and bearing caps, as an assembly.

7. Remove the camshaft from the cylinder head.

8. Inspect the bearing journals on the camshaft and the cylinder head.

To install:

9. Lubricate the camshaft journals and camshaft with clean engine oil and install the camshaft in the cylinder head.

10. Install the rocker arms, rocker arm shaft and the rocker shaft spring.

    a. Temporarily tighten the rocker shaft with the bolts at the condition that all the intake valve rocker arms do not push the valves.

    b. Insert the rocker shaft spring from above and mount it at right angles to the plug guide.

    c. Before installing the exhaust rocker arms and the rocker arm shaft, mount the rocker shaft spring.

d. Remove the SST used to hold the lash adjuster in position.

e. Check to ensure that the flat side of the rocker shaft is perpendicular to the cylinder head, and facing the valves.

f. Gradually tighten the bearing caps in two or three steps. In the final step torque to 23 ft. lbs. (31 Nm).

11. If removed, install the distributor.

12. Install the sprockets, timing belt and timing belt cover.

13. Install the valve cover and connect the negative battery cable.

14. Start the engine and check for leaks and proper operation.

3.5L (VIN M) Engine

1. Relieve the fuel system pressure.

2. Disconnect negative battery cable.

3. Remove the intake manifold plenum.

4. Remove the timing belt cover and the timing belt.

——— **WARNING** ———
DO NOT rotate the crankshaft or camshafts after the timing belt has been removed. If rotated, severe internal engine damage will result from the pistons hitting the valves.

5. Remove the center cover, breather, PCV hoses, and the spark plug cables.

6. Remove the rocker cover and the semi-circular packing.

7. Matchmark the positioning of the crankshaft position sensor at the rear of the camshaft and remove the sensor.

8. If equipped with a camshaft sensor, remove the sensor from the front of the engine.

9. Being sure to hold the flats of the camshaft, loosen the camshaft sprocket bolts.

10. Noting the positioning and location of the sprockets, remove the sprockets from the camshafts.

NOTE: Be sure to note the positioning of the knock pin at the end of the camshafts for reinstallation purposes.

NOTE: Be sure to keep the valve train components labeled and in proper order for reassembly.

11. Loosen the bearing cap bolts in 2–3 steps. Label and remove all camshaft bearing caps.

NOTE: If the bearing caps are difficult to remove, use a plastic hammer to gently tap the components.

12. Mark the components and remove the intake and the exhaust camshafts.

13. Remove the rocker arms and the lash adjusters. Be sure to note the location of the valve train components for reinstallation purposes.

14. Check the camshaft journals for wear or damage. Check the cam lobes for damage. Also, check the cylinder head oil holes for clogging.

To install:

NOTE: Lubricate the valve train components with clean engine oil.

15. Bleed and install the the lash adjusters to the to the original bores in the cylinder head.

16. Install the rocker arms to the cylinder head.

17. Lubricate the camshafts with clean engine oil and position the camshafts on the cylinder head.

——— **WARNING** ———
Be sure to properly position the knock pins of the camshaft to prevent valve to piston interference.

NOTE: Do not confuse the intake camshaft with the exhaust camshaft. The intake camshaft on the Montero has a P stamped on the hexagon. The exhaust camshaft on the Montero has a K stamped on the hexagon.

NOTE: Install the bearing caps according to the identification mark and cap number. Bearing caps No. 2, 3 and are marked as such. The caps also are marked I for intake or E for exhaust.

18. Install the bearing caps. Tighten the caps in sequence and in 2 or 3 steps. Caps 2, 3 and 4 have a front mark. Install with the mark aligned with the front mark on the cylinder head. Torque the retaining bolts for caps No. 2, 3 and 4 to 8 ft. lbs. (11 Nm) and the retaining bolts for the front and rear caps to 14 ft. lbs. (20 Nm).

19. Apply a coating of engine oil to the oil seals and install the oil seals to the front and rear of the camshafts.

20. Holding the flats of the camshaft, install and tighten the sprocket bolts to 65 ft. lbs. (90 Nm).

21. If removed, install the camshaft position sensor and tighten the mounting bolts to 78 inch lbs. (9 Nm).

22. Aligning the matchmark, install the crankshaft position sensor at the rear of the camshaft and tighten the mounting nut to 7 ft. lbs. (12 Nm).

23. Align the marks on the camshaft and crankshaft sprockets. Install the timing belt assembly.

24. Install the rocker cover and the semi-circular packing.

25. Install the intake manifold plenum.

26. Install the spark plug cables, center cover, breather and PCV hoses.

27. Connect the negative battery cable and check for leaks.

Expo 2.4L (4G64) Engine

1. Disconnect the negative battery cable.

2. Remove the breather hose. Disconnect the PCV hose.

3. Matchmark and remove the distributor.

4. Remove the timing belt.

5. Disconnect and tag the spark plug wires.

6. Remove the rocker cover.

7. Remove the camshaft sprocket retainer bolt while holding shaft stationary with appropriate spanner wrench. Remove the sprocket from the shaft.

8. Remove the camshaft oil seal.

9. Remove both rocker arm shaft assemblies from the head. Do not disassemble rocker arms from the rocker arm shaft unless worn or damaged.

10. Remove the camshaft from the cylinder.

11. Inspect the bearing journals on the camshaft, cylinder head, and bearing caps.

To install:

12. Lubricate the camshaft journals and camshaft with clean engine oil and install the camshaft in the cylinder head.

13. Install the rocker arm and shaft assemblies. Tighten the rocker arm shaft retainer bolts to 21–25 ft. lbs. (29–35 Nm).

14. Install new camshaft oil seal.

15. Install camshaft sprocket and retainer bolt torquing to 65 ft. lbs. (90 Nm).

16. Install the timing belt.

17. Install the distributor.

18. Install the rocker cover using new gasket material on mating surfaces.

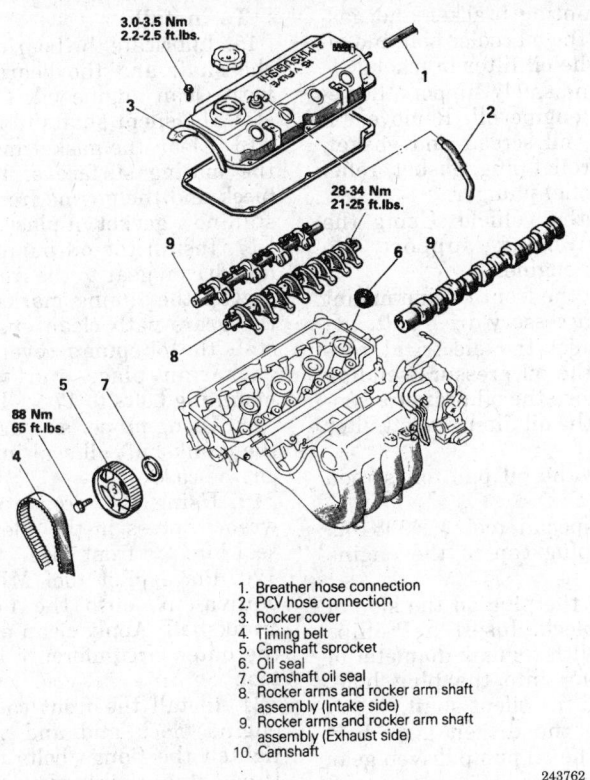

1. Breather hose connection
2. PCV hose connection
3. Rocker cover
4. Timing belt
5. Camshaft sprocket
6. Oil seal
7. Camshaft oil seal
8. Rocker arms and rocker arm shaft assembly (Intake side)
9. Rocker arms and rocker arm shaft assembly (Exhaust side)
10. Camshaft

243762

Camshaft and related components — Expo 2.4L (4G64) engine

19. Connect the spark plug cables.

20. Install the breather hose and connect the PCV hose.

21. Connect the negative battery cable. Run the engine at idle until normal operating temperature is reached. Check idle speed and ignition timing and adjust as required.

Expo 1.8L (4G93) Engine

1. Disconnect the negative battery cable.

2. Rotate the engine and position the No. 1 piston to TDC of its compression stroke.

3. Label and disconnect the spark plug cables.

4. Matchmark the positioning of the distributor housing and the positioning of the distributor rotor to the engine block and remove the distributor.

5. Disconnect the air flow sensor connector and remove the air cleaner case cover.

6. Disconnect the accelerator cable, breather hose and PCV hose connections.

7. Remove the rocker cover and discard the gasket.

8. Loosen both rocker arm shaft assemblies gradually and evenly and remove the rocket shafts from the vehicle. Do not disassembly rocker

arms and rocker arm shaft assemblies.

9. Remove the timing belt covers.

NOTE: DO NOT allow the camshaft or the crankshaft to rotate after the timing belt is removed.

10. Remove the timing belt assembly.

11. Holding the camshaft sprocket from turning, loosen and remove the bolt that secures the sprocket.

12. Remove the camshaft sprocket from the camshaft. Note the positioning of the dowel pin at the end of the camshaft.

13. Remove the camshaft oil seal from the front of the cylinder head.

14. Remove the camshaft from the head.

15. Carefully check all parts for damage and wear.

To install:

16. Lubricate the camshaft journals and camshaft with clean engine oil and install the camshaft in the cylinder head. Be sure to position the dowel pin at the end of the camshaft as noted during the removal procedure.

17. Check the camshaft end-play between the thrust case and camshaft. The camshaft end-play should be 0.002–0.008 in.

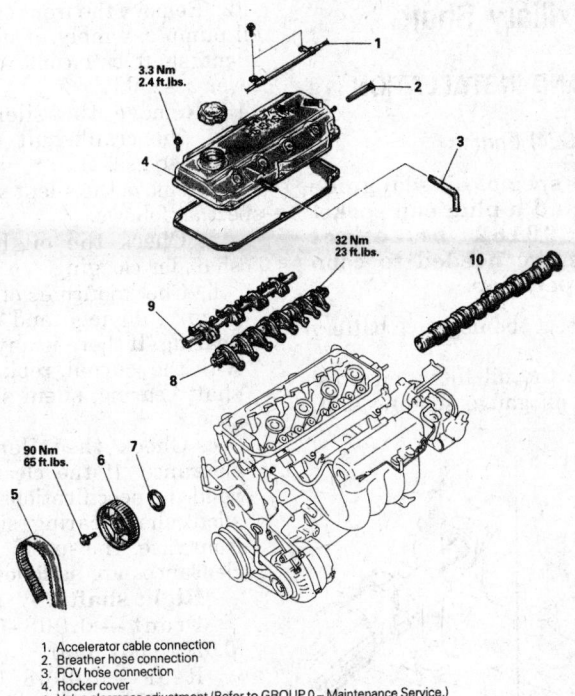

1. Accelerator cable connection
2. Breather hose connection
3. PCV hose connection
4. Rocker cover
● Valve clearance adjustment (Refer to GROUP 0 – Maintenance Service.)
5. Timing belt (Refer to P. 9B-41.)
6. Camshaft sprocket
7. Camshaft oil seal
8. Rocker arms and rocker arm shaft assembly (Intake side)
9. Rocker arms and rocker arm shaft assembly (Exhaust side)
10. Camshaft

329170

Camshaft and related components — Expo 1.8L (4G93) engines

(0.05–0.20mm). If the end-play is not within specification, replace the camshaft thrust bearing.

18. Install a new camshaft oil seal. Be sure to lubricate the lips of the seal with clean engine oil.

19. Install camshaft sprocket and torque the retainer bolt to 65 ft. lbs. (90 Nm). Be sure to secure the sprocket while tightening the bolt.

20. Install the timing belt assembly.

21. Install the timing belt covers.

22. Install the rocker arm and shaft assemblies. Tighten the rocker arm shaft retainer bolts to 23 ft. lbs. (32 Nm).

23. Check valve adjustment and install valve cover with a new gasket. Tighten the valve cover bolts to 29 inch lbs. (3.3 Nm).

24. Align the distributor marks and install the distributor.

25. Connect the spark plug cables.

26. Connect the accelerator cable, breather hose and PCV hose.

27. Connect the air flow sensor connector and install the air cleaner case cover.

28. Connect the negative battery cable. Run the engine at idle until normal operating temperature is reached. Check idle speed and ignition timing and adjust as required.

Auxiliary Shaft

REMOVAL AND INSTALLATION

Expo 2.4L (4G64) Engine

NOTE: A special oil seal guide MD998285 and a plug cap socket tool MD998162 or exact equivalents are needed to complete this operation.

1. Disconnect the negative battery cable.

2. Remove the oil filter, oil pressure switch, oil gauge sending unit, oil filter mounting bracket and gasket. Remove the oil cooler bolt and oil cooler from the oil filter bracket.

3. Raise and safely support the vehicle. Drain engine oil. Remove engine oil pan, oil screen and gasket. Remove the relief plug, gasket, relief spring and relief plunger.

4. Lower the vehicle. Using the proper equipment, support the weight of the engine.

5. Remove the front engine mount bracket and accessory drive belt.

6. Disconnect the electrical connector from the oil pressure sending unit and remove the oil pressure sensor. Remove the oil filter and oil filter bracket.

7. Remove the oil pan, oil screen and gasket.

8. Using special tool MD998162, remove the plug cap in the engine front cover.

9. Remove the plug on the side of the engine block. Insert a Phillips screwdriver with a shank diameter of 0.32 in. (8mm) into the plug hole. This will hold the silent shaft.

10. Remove the driven gear bolt that secures the oil pump driven gear to the silent shaft.

11. Remove and tag the front cover mounting bolts. Note the lengths of the mounting bolts as they are removed for proper installation.

12. Remove the front case cover and oil pump assembly. If necessary, the silent shaft can come out with the cover assembly.

13. Remove the silent shaft oil seals, the crankshaft oil seal and front case gasket.

14. Remove the silent shafts and inspect as follows:

  a. Check the oil holes in the shaft for clogging.

  b. Check journals of the shaft for seizure, damage and contact with bearing. If there is anything wrong with the journal, replace the silent shaft bearing, silent shaft or front case.

  c. Check the silent shaft oil clearance. If the clearance is beyond the specifications, replace the silent shaft bearing, silent shaft or front case. The specifications for oil clearances are as follows:

    Right shaft
    Front — 0.008–0.0024 in. (0.020–0.061 mm)
    Rear — 0.0008–0.0021 in. (0.02–0.05 mm)
    Left shaft
    Front — 0.008–0.0021 in. (0.020–0.054 mm)
    Rear — 0.0020–0.0036 in. (0.050–0.091mm)

To install:

15. Lubricate the bearing surface of the shaft and the bearing journals with clean engine oil. Carefully install the silent shafts to the block.

16. Clean the gasket material from the mating surface of the cylinder block and the engine front cover. Install new gasket in place.

17. Install the oil pump drive gear and driven gear to the front case, lining up the timing marks. Lubricate the gears with clean engine oil. Install the oil pump cover, with new gasket in place and tighten the mounting bolts to 17 ft. lbs. (24 Nm).

18. Using proper size driver, install the crankshaft oil seal into the front engine case.

19. Using the proper size socket wrench, press in the silent shaft oil seal into the front case.

20. Place pilot tool MD998285 or equivalent, onto the nose of the crankshaft. Apply clean engine oil to the outer circumference of the pilot tool.

21. Install the front case onto the engine block and and temporarily tighten the flange bolts (other than those for tightening the filter bracket). Mount the oil filter bracket with new gasket in place. Install the three bolts with washers and tighten to 16 ft. lbs. (22 Nm).

22. Insert the Phillips screwdriver into the hole on the side of the engine block. Secure the oil pump driven gear onto the left silent shaft by tightening the driven gear flange bolt to 29 ft. lbs. (40 Nm).

23. Install a new O-ring onto the groove in the front case. Using special socket tool, install and tighten the plug cap to 20 ft. lbs. (27 Nm).

24. Install the oil pump relief plunger and spring into the bore in the oil filter bracket and tighten to 36 ft. lbs. (50 Nm). Make sure a new gasket is in place.

25. Clean both mating surfaces of the oil pan and the cylinder block. Apply sealant in the groove in the oil pan flange, keeping towards the inside of the bolt holes. The width of the sealant bead applied is to be about 0.016 in. (4mm) wide.

NOTE: After applying sealant to the oil pan, do not exceed 15 minutes before installing the oil pan.

26. Install the oil pan to the engine and secure with the retainers. Tighten bolts to 6 ft. lbs. (8 Nm).

27. Install the oil pressure gauge unit and the oil pressure switch. Connect the electrical harness connector.

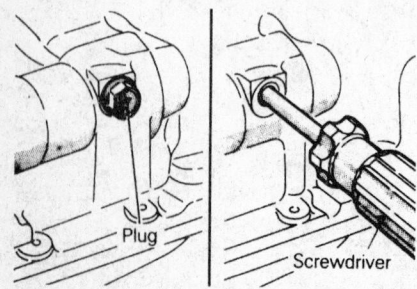

139954

Holding silent shaft for oil pump gear removal — Expo 2.4L (4G64) Engine

28. Install the oil cooler secure with oil cooler bolt tightened to 33 ft. lbs. (45 Nm).

29. Install new oil filled with clean engine oil.

NOTE: The timing of the oil pump sprocket and connected silent shaft can be incorrect, even with the timing mark aligned. Incorrect orientation of the silent shaft will result in engine vibration during operation. Follow the alignment procedure for timing belt installation.

30. Install the timing belts and all related items. Make sure the timing and orientation of the silent shafts is correct by inserting the alignment tool in the hole in the left side of the engine block, as specified in the timing belt section of this chapter.

31. Install any remaining components removed during disassembly.

32. Connect the negative battery cable and start the engine. Check for proper timing and inspect for leaks.

Balance Shaft

REMOVAL AND INSTALLATION

2.4L (VIN G) Engine

1. Disconnect the negative battery cable.

——— **CAUTION** ———
Wait at least 90 seconds after the negative (-) battery cable is disconnected to prevent possible deployment of the air bag.

2. Release fuel pressure, if applicable.

——— **CAUTION** ———
Fuel injection systems remain under pressure after the engine has been turned OFF. Properly relieve fuel pressure before disconnecting any fuel lines. Failure to do so may result in fire or personal injury.

3. Raise and safely support the vehicle.

4. Drain the engine oil.

5. Remove the oil filter and the oil pan from the engine.

6. Remove the oil pressure electrical connector and the oil pressure switch.

7. Lower the vehicle.

8. Remove the timing belt cover(s).

9. Remove the timing belts.

10. Remove the timing belt sprockets.

11. Remove the front case from the block.

12. Remove the oil pump components.

13. Using SST MIT304204 and MD998371–01 or equivalent, remove the front bearing for the balance shaft.

14. Using SST MIT304204, MB991603 and MD998372–01 or equivalent, and remove the rear balance shaft bearing.

15. Remove the balance shafts from the block.

To install:

16. Coat the balance shafts with clean engine oil and install into the block.

17. Install the front and rear balance shaft bearings.

18. Install the drive and driven gears into place so the timing marks are mated with each other. Coat gears with clean engine oil.

19. Attach special tool MD998285 or equivalent, (used to locate the front case assembly and protect the oil seal) to the end of the crankshaft and coat the outer surface with engine oil.

20. Install the front case assembly with a new front case gasket to the block. Install the retaining bolts.

21. Install the oil filter bracket together with the oil filter bracket gasket and install retaining bolts with washers. Torque bolts to 15 ft. lbs. (22 Nm).

22. Install the timing belts, covers and related parts.

23. Connect the negative battery cable, add clean oil to correct level, start engine and check for leaks.

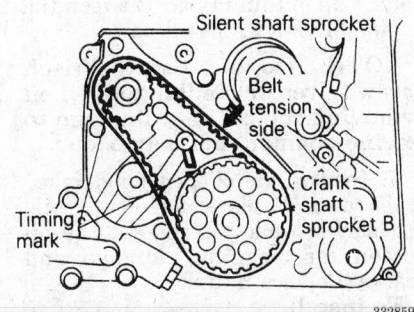

Timing belt "B" timing marks — 2.4L (VIN G) engines

Piston and Connecting Rod

POSITIONING

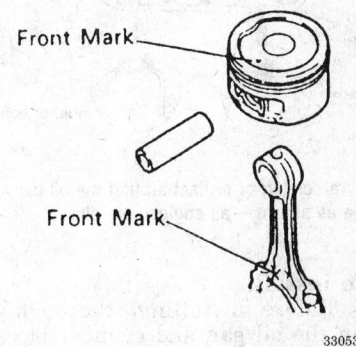

330537

Aligning the piston and the connecting rod

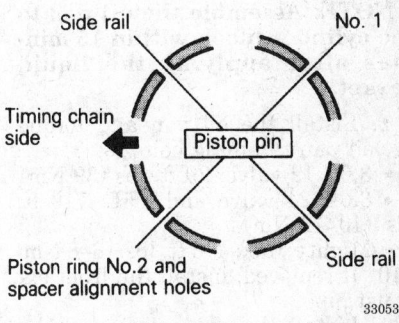

330538

Arrange the piston ring and oil ring gaps (side rail and spacer) as shown

---

ENGINE LUBRICATION

Oil Pan

REMOVAL AND INSTALLATION

Except Expo

1. Disconnect the negative battery cable.

2. Raise the vehicle and support it safely.

3. Remove the skid plate and the engine undercover.

4. Place a suitable container into position. Remove the oil drain plug and drain the engine oil.

5. If necessary, remove the front exhaust pipe.

6. Remove the oil pan.

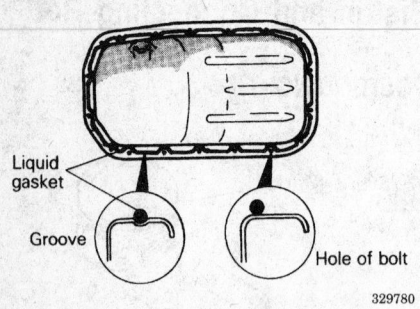

Liquid gasket

Groove

Hole of bolt

329780

Apply a coating of sealant around the oil pan flange as shown — all engines are similar

To install:

7. Before installing, thoroughly clean the oil pan and cylinder block mating surfaces.

8. Apply liquid gasket around the surface of the oil pan.

NOTE: Assemble the oil pan to the cylinder block within 15 minutes after applying the liquid gasket.

9. Install the oil pan and torque the oil pan retaining bolts to:
- 3.0L 12 valve: 29 ft. lbs (39 Nm)
- 3.0L 24 valve and 3.5L: 7–9 ft. lbs. (10–12 Nm)
- Mighty Max: 4–6 ft. lbs. (6–8 Nm)

10. If removed, install the front exhaust pipe.

11. Install the skid plate and the engine undercover.

12. Fill with engine oil.

13. Connect the negative battery cable.

14. Start the vehicle and check for leaks.

Expo 1.8L (C-4G93) Engine

1. Disconnect the negative battery cable.

2. Raise the vehicle and support safely.

3. Remove the oil pan drain plug and drain the engine oil.

4. Disconnect and lower the exhaust pipe from the engine manifold.

5. Remove the bell housing lower cover.

6. Remove the oil pan retainer bolts. Tap the oil pan with a rubber mallet to break the seal.

NOTE: Do not use a chisel, screwdriver or similar tool when removing the oil pan. Damage to engine components may occur. If available, oil pan remover tool MD998727 or equivalent may be used break the seal.

7. Inspect the oil pan for damage and cracks. Replace if faulty. While the pan is removed, inspect the oil screen for clogging, damage and cracks. Replace if faulty.

To install:

8. Using a wire brush or other tool, scrape clean all gasket surfaces of the cylinder block and the oil pan so that all loose material is removed. Clean sealing surfaces of all dirt and oil.

9. Apply sealant around the gasket surfaces of the oil pan in such a manner that all bolt holes are circled and there is a continuous bead of sealer around the entire outside edge of the oil pan.

NOTE: The continuous bead of sealer should be applied in a bead approximately 0.16 in. (4mm) in diameter.

10. Install the oil pan onto the cylinder block within 15 minutes after applying sealant. Install the fasteners and tighten to 60 inch lbs. (5 Nm).

11. Install the bellhousing cover.

12. Connect the exhaust pipe from the engine manifold with new gasket in place. Tighten the exhaust pipe to manifold flange nuts to 33 ft. lbs. (45 Nm). Install and tighten the support bolt to 18 ft. lbs. (25 Nm).

13. Install the oil drain plug and tighten to 29 ft. lbs. (40 Nm).

14. Lower the vehicle and fill the crankcase to the proper level with clean engine oil.

15. Connect the negative battery cable. Start the engine and check for leaks.

Expo 2.4L (4G64) Engine

FWD

1. Disconnect the negative battery cable.

2. Raise the vehicle and support safely.

3. Remove the oil pan drain plug and drain the engine oil.

4. Disconnect and lower the exhaust pipe from the engine manifold.

5. Remove the left side axle shaft.

6. Remove the oil pan retainer bolts. Tap in thin prybar between the engine block and the oil pan.

NOTE: Do not use a chisel, screwdriver or similar tool when removing the oil pan. Damage to engine components may occur.

7. Inspect the oil pan for damage and cracks. Replace if faulty. While the pan is removed, inspect the oil screen for clogging, damage and cracks. Replace if faulty.

To install:

8. Using a wire brush or other tool, scrape clean all gasket surfaces of the cylinder block and the oil pan so that all loose material is removed. Clean sealing surfaces of all dirt and oil.

9. Apply sealant around the gasket surfaces of the oil pan in such a manner that all bolt holes are circled and there s a continuous bead of sealer around the entire perimeter of the oil pan.

NOTE: The continuous bead of sealer should be applied in a bead approximately 0.16 in. (4mm) in diameter.

10. Install the oil pan onto the cylinder block within 15 minutes after applying sealant. Install the fasteners and tighten to 4–6 ft. lbs. (6–8 Nm).

11. Connect the exhaust pipe from the engine manifold with new gasket in place. Tighten the exhaust pipe to manifold flange nuts to 29 ft. lbs. (40 Nm). Install and tighten the support bolt to 29 ft. lbs. (40 Nm).

12. Install the axle shaft and check the transxle fluid level.

13. Install the oil drain plug and tighten to 29 ft. lbs. (40 Nm), If not already done.

14. Lower the vehicle and fill the crankcase to the proper level with clean engine oil.

15. Connect the negative battery cable. Start the engine and check for leaks.

AWD

1. Disconnect the negative battery cable.

2. Raise the vehicle and support safely.

3. Remove the oil pan drain plug and drain the engine oil.

4. Disconnect and lower the exhaust pipe from the engine manifold.

5. Remove the transfer assembly.

6. Remove the oil pan retainer bolts. Tap in thin prybar between the engine block and the oil pan.

NOTE: Do not use a chisel, screwdriver or similar tool when removing the oil pan. Damage to engine components may occur.

7. Inspect the oil pan for damage and cracks. Replace if faulty. While the pan is removed, inspect the oil screen for clogging, damage and cracks. Replace if faulty.

To install:

8. Using a wire brush or other tool, scrape clean all gasket surfaces of the cylinder block and the oil pan so that all loose material is removed. Clean sealing surfaces of all dirt and oil.

9. Apply sealant around the gasket surfaces of the oil pan in such a

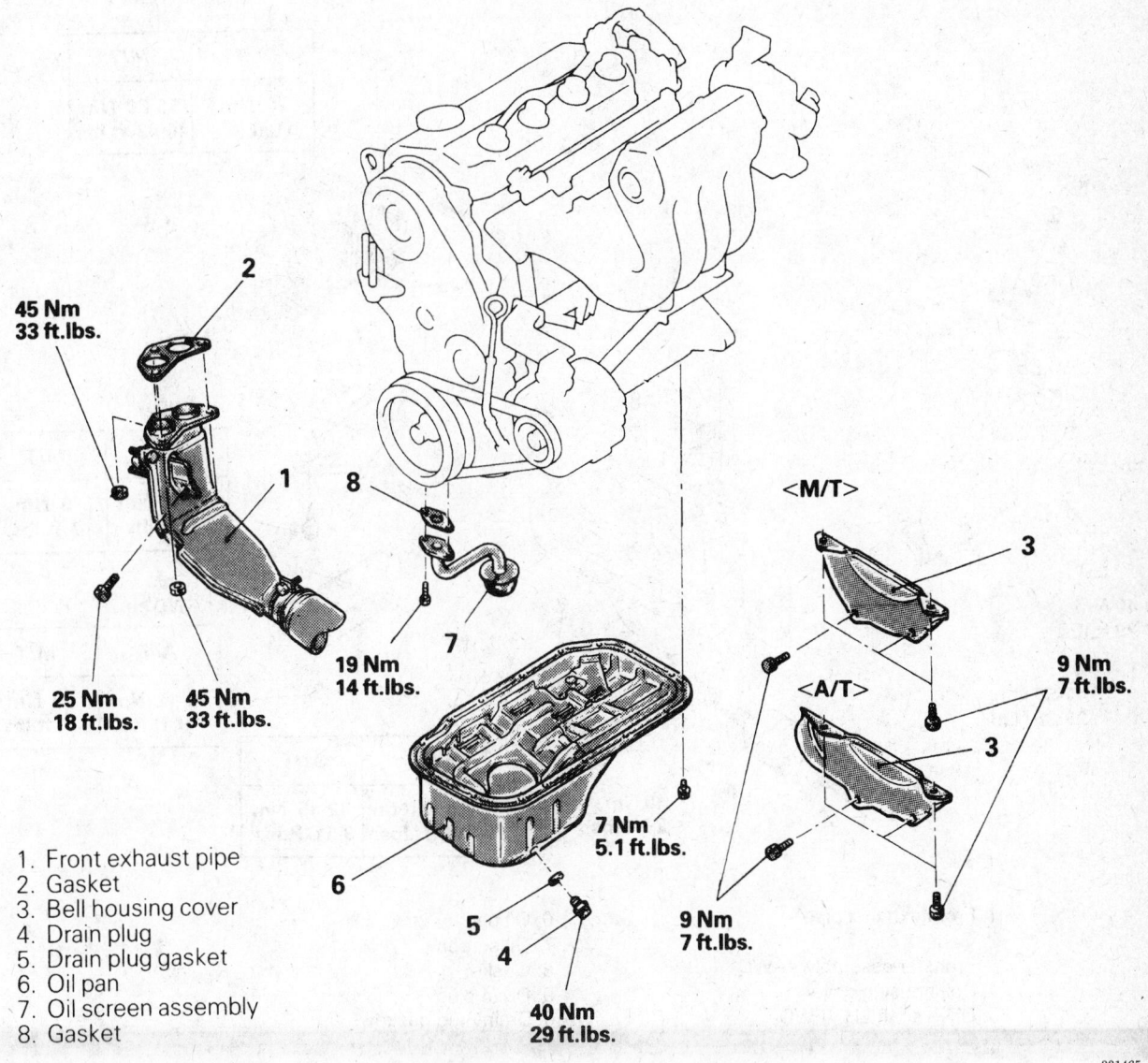

1. Front exhaust pipe
2. Gasket
3. Bell housing cover
4. Drain plug
5. Drain plug gasket
6. Oil pan
7. Oil screen assembly
8. Gasket

45 Nm
33 ft.lbs.

25 Nm
18 ft.lbs.

45 Nm
33 ft.lbs.

19 Nm
14 ft.lbs.

7 Nm
5.1 ft.lbs.

40 Nm
29 ft.lbs.

9 Nm
7 ft.lbs.

9 Nm
7 ft.lbs.

9 Nm
7 ft.lbs.

\<M/T\>

\<A/T\>

331465

Oil pan and related components — Expo 1.8L (C-4G93) engine

manner that all bolt holes are circled and there s a continuous bead of sealer around the entire perimeter of the oil pan.

NOTE: The continuous bead of sealer should be applied in a bead approximately 0.16 in. (4mm) in diameter.

10. Install the oil pan onto the cylinder block within 15 minutes after applying sealant. Install the fasteners and tighten to 4–6 ft. lbs. (6–8 Nm).
11. Install the transfer assembly and check fluid level.

12. Connect the exhaust pipe from the engine manifold with new gasket in place. Tighten the exhaust pipe to manifold flange nuts to 29 ft. lbs. (40 Nm). Install and tighten the support bolt to 29 ft. lbs. (40 Nm).
13. Install the oil drain plug and tighten to 29 ft. lbs. (40 Nm), If not already done.
14. Lower the vehicle and fill the crankcase to the proper level with clean engine oil.
15. Connect the negative battery cable. Start the engine and check for leaks.

Oil Pump

REMOVAL AND INSTALLATION

2.4L (VIN G) Engine

1. Disconnect the negative battery cable.

— **CAUTION** —
Wait at least 90 seconds after the negative (-) battery cable is disconnected to prevent possible deployment of the air bag.

2. Remove the timing belt covers, timing belts and sprockets.

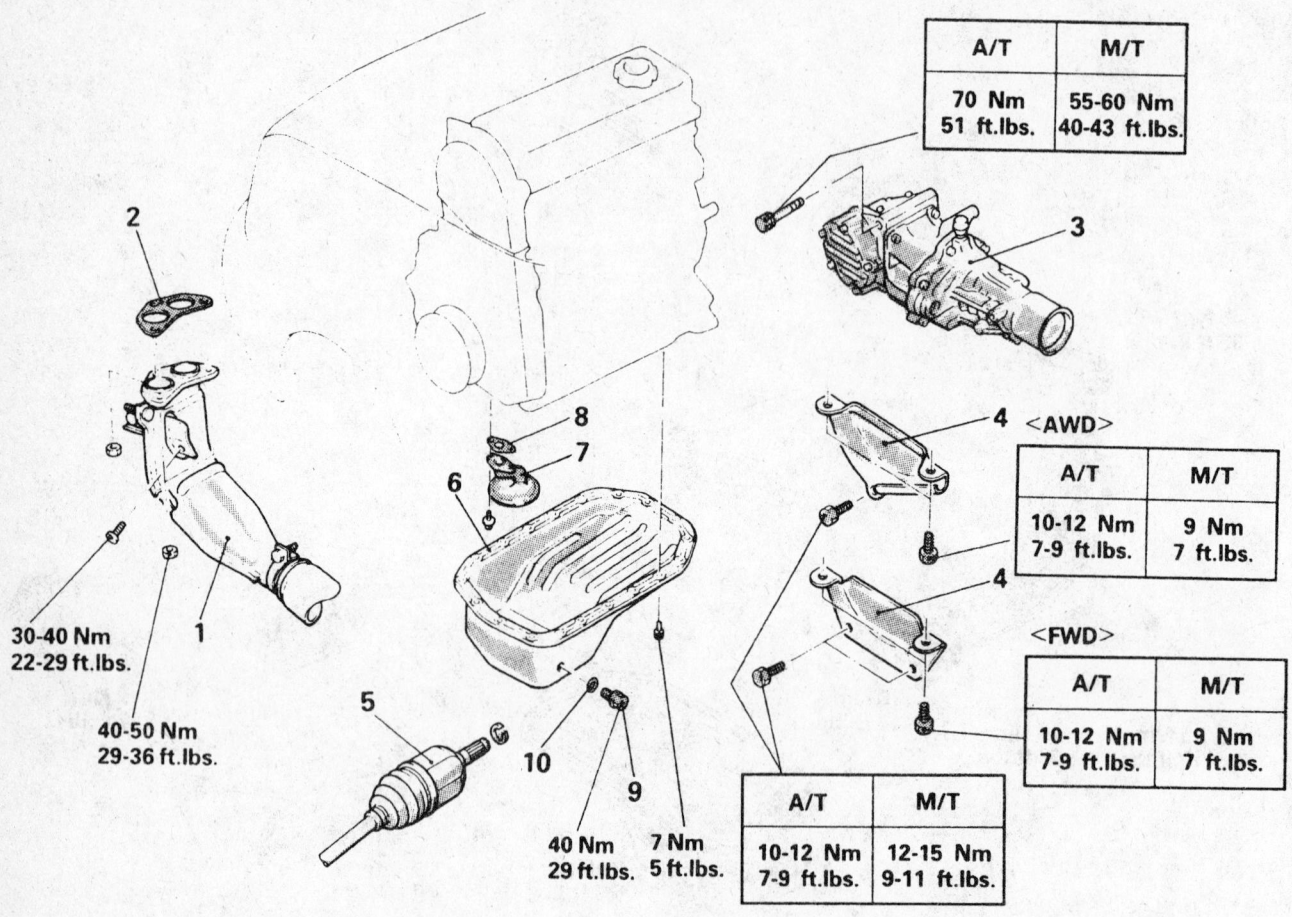

A/T	M/T
70 Nm 51 ft.lbs.	55-60 Nm 40-43 ft.lbs.

30-40 Nm
22-29 ft.lbs.

40-50 Nm
29-36 ft.lbs.

4 <AWD>

A/T	M/T
10-12 Nm 7-9 ft.lbs.	9 Nm 7 ft.lbs.

<FWD>

A/T	M/T
10-12 Nm 7-9 ft.lbs.	9 Nm 7 ft.lbs.

40 Nm **7 Nm**
29 ft.lbs. **5 ft.lbs.**

A/T	M/T
10-12 Nm 7-9 ft.lbs.	12-15 Nm 9-11 ft.lbs.

1. Front exhaust pipe
2. Gasket
3. Transfer assembly <AWD>
4. Bell housing cover
5. Drive shaft LH <FWD>
6. Oil pan
7. Oil screen
8. Gasket
9. Drain plug
10. Drain plug gasket

240807

Oil pan and related components — Expo 2.4L (4G64) engine

3. Raise the vehicle and support safely.

4. Drain the oil and remove the oil filter.

5. Remove the oil pan and gasket. Remove the oil pump pickup and gasket.

6. Remove the oil pressure relief plunger plug and gasket. Remove the spring and plunger from the oil filter bracket.

7. Remove the 4 bracket mounting bolts and remove the oil filter mount and gasket.

8. Using special tool MD998162, remove the plug cap and gasket that covers the oil pump driven gear shaft. This is located on the right side of the front case at the front of the engine, just above the protruding drive gear shaft.

9. Using a long socket, remove the retaining bolt from the oil pump driven gear located behind the plug removed earlier.

10. Remove the front case mounting bolts and remove the case from the block.

11. Remove the case gasket from the block.

To install:

12. Prime the pump by pouring fresh oil into the pump intake and turning the driveshaft until oil comes out the pressure port. Repeat a few times until no air bubbles are present. Replace all seals on the case assembly.

13. Install a special seal guide to the crankshaft, MD998285 or equivalent so the smaller diameter faces outward. Coat the outer diameter of the seal with clean engine oil.

14. Install a new front case gasket and install the front case by carefully positioning the crankshaft seal over the seal guide and lining up all bolt holes. Install and tighten the bolts to 17 ft. lbs. (23 Nm).

15. Remove the plug from the left side of the block. Hold the left side silent shaft by inserting a tool in the plug hole and torque the driven gear

bolt to 26 ft. lbs. (35 Nm). Using a new O-ring, install the plug cover.

16. Install the oil filter mounting bracket gasket. Install the mounting bracket and bolts tightening the oil filter mounting bracket bolts to 12 ft. lbs. (16 Nm).

17. Clean or replace the oil pickup screen and install with a new gasket.

18. Install the oil pan using a new gasket.

19. Install the timing sprockets, belts and covers.

20. Fill the engine with the proper amount of engine oil.

21. Connect the negative battery cable and check for proper oil pressure and leaks.

3.0L (VIN H) and 3.5L (VIN M) Engines

1993–95 Vehicles

1. Disconnect the negative battery cable.
2. Remove the dipstick.
3. Raise the vehicle and support safely.
4. Remove the timing belt.
5. Drain the engine oil and remove the oil pan from the engine. Remove the oil pickup.

6. Remove the oil pump mounting bolts and remove the pump from the front of the engine.

NOTE: Note the position of each oil pump case retaining bolts to facilitate installation. The bolts are of different length.

To install:

7. Clean the gasket mounting surfaces of the pump and engine block.

8. Prime the pump by pouring fresh oil into the inlet and turning the rotors or by packing pump with petroleum jelly. Using a new gasket, install the oil pump on the engine and torque all bolts to 10 ft. lbs. (14 Nm).

9. Install the balancer and crankshaft sprockets.

10. Clean out the oil pickup or replace as required. Replace the oil pickup gasket ring and install the pickup to the pump.

11. Install the timing belt, oil pan and all related parts.

12. Install the dipstick. Fill the engine with the proper amount of oil.

13. Connect the negative battery cable.

14. Start the engine and check for proper oil pressure. Check for leaks.

1996–97 Vehicles

1. Disconnect the negative battery cable.
2. Remove the timing belt.
3. Remove the oil dipstick.
4. Drain the engine oil.
5. Remove the oil pan from the engine.
6. Remove the oil pump mounting bolts and remove the pump from the front of the engine.
7. Remove the oil filter and the bracket.

NOTE: Note the position of each oil pump case retaining bolts to facilitate installation. The bolts are of different length.

To install:

8. Clean the gasket mounting surfaces of the pump and engine block.

9. Prime the pump by pouring fresh oil into the inlet and turning the rotors or by packing pump with petroleum jelly. Using a new gasket, install the oil pump on the engine and torque all bolts to 10 ft. lbs. (14 Nm).

10. Clean out the oil pickup or replace as required. Replace the oil pickup gasket ring and install the pickup to the pump.

11. Install the oil filter and the bracket.

12. Install the oil pan.

13. Install the timing belt.

14. Install the dipstick. Fill the engine with the proper amount of oil.

15. Connect the negative battery cable.

16. Start the engine and check for proper oil pressure. Check for leaks.

Expo 1.8L (C-4G93) Engines

NOTE: Whenever the oil pump is disassembled or the cover removed, it is suggested the gear cavity is filled with petroleum jelly for priming purposes. Do not use grease.

1. Disconnect the negative battery cable.

2. Remove the front engine mount bracket and accessory drive belts.

3. Remove timing belt upper and lower covers.

4. Remove the timing belt and crankshaft sprocket.

5. Remove the oil pan.

6. Remove the oil screen and gasket.

7. Remove and tag the front cover mounting bolts. Note the lengths of the mounting bolts as they are removed for proper installation.

8. Remove the front case cover and oil pump assembly.

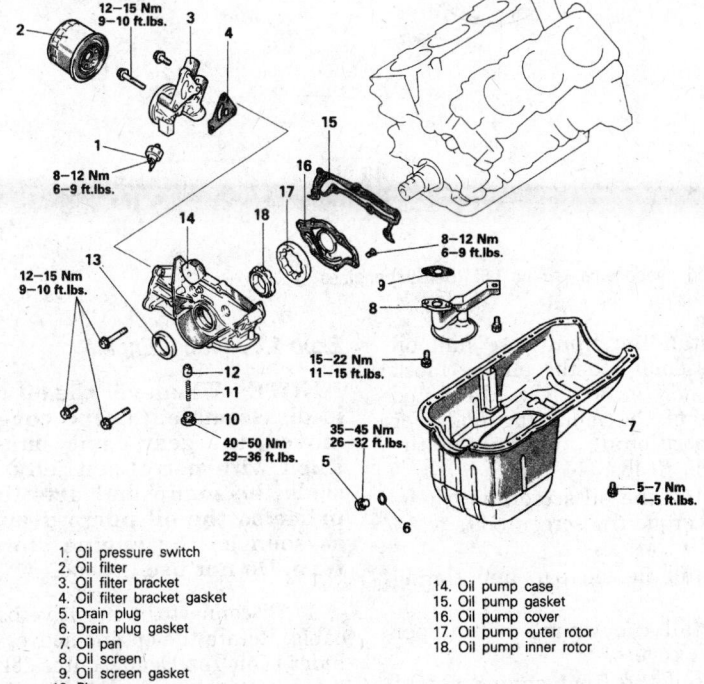

1. Oil pressure switch
2. Oil filter
3. Oil filter bracket
4. Oil filter bracket gasket
5. Drain plug
6. Drain plug gasket
7. Oil pan
8. Oil screen
9. Oil screen gasket
10. Plug
11. Relief spring
12. Relief plunger
13. Crankshaft front oil seal
14. Oil pump case
15. Oil pump gasket
16. Oil pump cover
17. Oil pump outer rotor
18. Oil pump inner rotor

285431

Oil pan and oil pump components — 3.0L (VIN H) and 3.5L (VIN M) engines

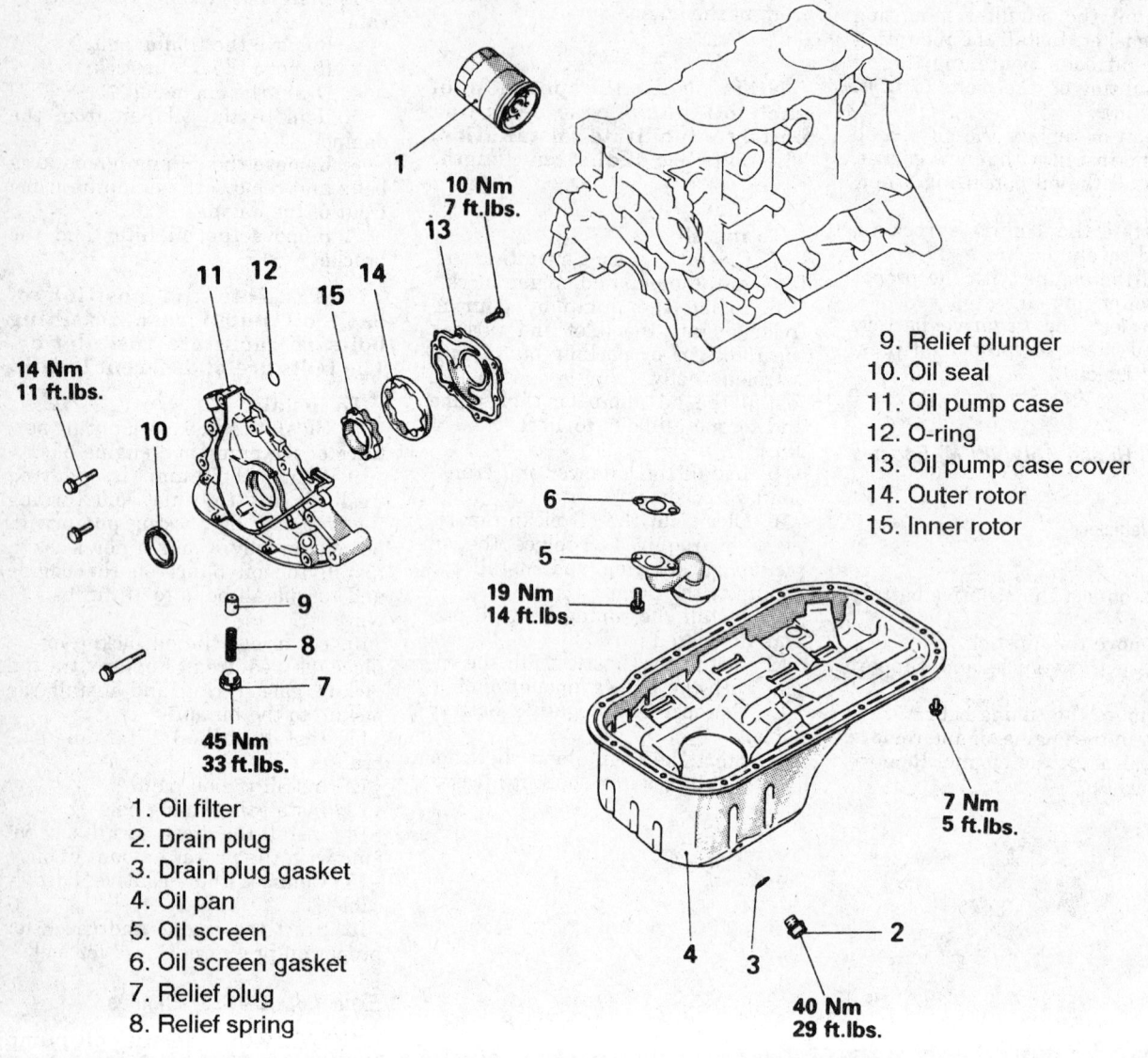

10 Nm
7 ft.lbs.

14 Nm
11 ft.lbs.

45 Nm
33 ft.lbs.

19 Nm
14 ft.lbs.

7 Nm
5 ft.lbs.

40 Nm
29 ft.lbs.

9. Relief plunger
10. Oil seal
11. Oil pump case
12. O-ring
13. Oil pump case cover
14. Outer rotor
15. Inner rotor

1. Oil filter
2. Drain plug
3. Drain plug gasket
4. Oil pan
5. Oil screen
6. Oil screen gasket
7. Relief plug
8. Relief spring

331447

Oil pump and related components — Expo 1.8L (C-4G93) engines

9. Check the front case for damage or cracks. Replace the front seal. Replace the oil screen O-ring. Clean all parts thoroughly with a safe solvent. Check the pump gears for wear or damage, and ensure that the relief valve can slide freely in the case.

To install:

10. Thoroughly clean all gasket material from all mounting surfaces.

11. Apply engine oil to the entire surface of the gears or rotors.

12. Install the gears and rotor to the front case and install the oil pump case cover. Torque the oil pump case cover bolts to 84 inch lbs. (10 Nm).

13. Install the front case and oil pump assembly to the engine block using a new gasket. Use the noted locations of the mounting bolts for proper positioning and tighten the bolts to 11 ft. lbs. (14 Nm).

14. Install the oil screen with new gasket. Torque the screen bolts to 14 ft. lbs. (19 Nm).

15. Install the oil pan and timing belt.

16. Install the timing belt upper and lower covers.

17. Install the front engine mount bracket and accessory drive belts.

18. Connect the negative battery cable and refill the engine oil. Check for adequate oil pressure.

Expo 2.4L (4G64) Engine

NOTE: Whenever the oil pump is disassembled or the cover removed, the gear cavity must be filled with petroleum jelly. This seals the pump and acts lime a prime so the oil pump draws oil as soon as the engine starts to turn. Do not use grease.

1. Disconnect the negative battery cable. Rotate the engine so No. 1 cylinder is on Top Dead Center (TDC) of its compression stroke. The timing marks should be aligned at this point.

2. Raise and safely support the vehicle.

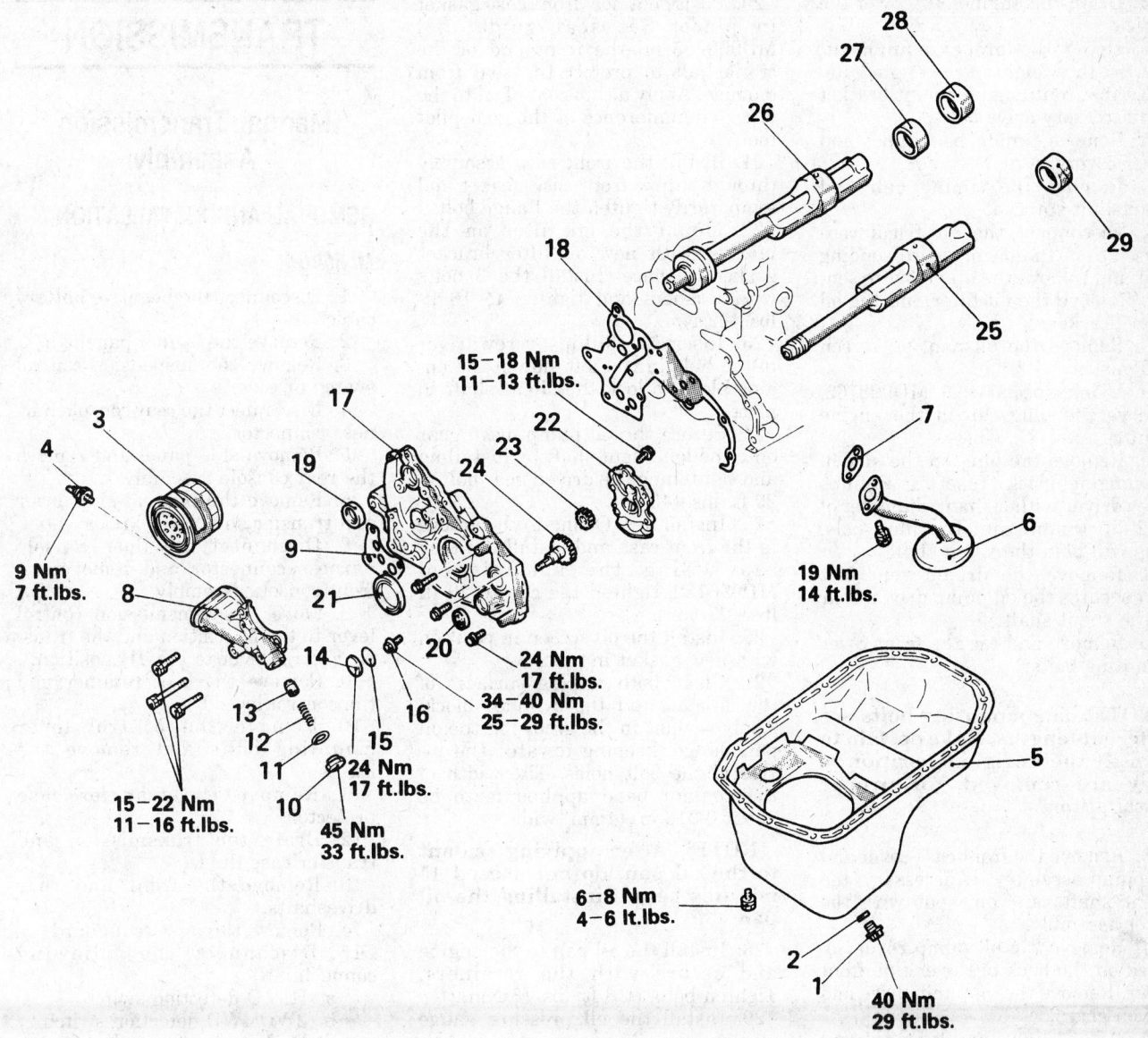

15—18 Nm
11—13 ft.lbs.

9 Nm
7 ft.lbs.

19 Nm
14 ft.lbs.

24 Nm
17 ft.lbs.

34—40 Nm
25—29 ft.lbs.

15—22 Nm
11—16 ft.lbs.

24 Nm
17 ft.lbs.

45 Nm
33 ft.lbs.

6—8 Nm
4—6 lt.lbs.

40 Nm
29 ft.lbs.

1. Drain plug
2. Gasket
3. Oil filter
4. Oil pressure switch
5. Oil pan
6. Oil screen
7. Gasket
8. Oil filter bracket
9. Gasket
10. Relief plug
11. Gasket
12. Relief spring
13. Relief plunger
14. Plug cap
15. O-ring
16. Driven gear bolt
17. Front case
18. Gasket
19. Oil seal
20. Oil seal
21. Crankshaft front oil seal
22. Oil pump cover
23. Oil pump driven gear
24. Oil pump drive gear
25. Left silent shaft
26. Right silent shaft
27. Silent shaft front bearing
28. Right silent shaft rear bearing
29. Left silent shaft rear bearing

325177

Oil pump and related components — Expo 2.4L (4G64) engine

3. Drain the engine oil. Lower the vehicle.

4. Using the proper equipment, support the weight of the engine. Remove the front engine mount bracket and accessory drive belts.

5. Remove timing belt upper and lower covers.

6. Remove the timing belt and crankshaft sprocket.

7. Disconnect the electrical connector from the oil pressure sending unit and remove the oil pressure sensor. Remove the oil filter and the oil filter bracket.

8. Remove the oil pan, oil screen and gasket.

9. Using special tool MD998162, remove the plug cap in the engine front cover.

10. Remove the plug on the side of the engine block. Insert a Phillips screwdriver with a shank diameter of 0.32 in. (8mm) into the plug hole. This will hold the silent shaft.

11. Remove the driven gear bolt that secures the oil pump driven gear to the silent shaft.

12. Remove and tag the front cover mounting bolts.

NOTE: The mounting bolts are different lengths, make certain to identify their original location as they are removed, for proper installation.

13. Remove the front case cover and oil pump assembly. If necessary, the silent shaft can come out with the cover assembly.

14. Remove the oil pump cover, located on the back of the engine front cover. Remove the oil pump drive and driven gears.

15. After disassembling the oil pump, clean all components and remove gasket material from mating surfaces.

16. Assemble the oil pump gears into the front case and rotate it to ensure smooth rotation and no looseness. Make sure there is no ridge wear on the contact surface between the front case and the gear surface of the oil pump front cover.

  To install

17. Align the timing mark on the oil pump drive gear with that on the driven gear and install them into the engine front case. Apply engine oil to the gears.

18. Install the oil pump cover and tighten the retainer bolts to 17 ft. lbs. (24 Nm).

19. Using the appropriate driver, install a new crankshaft seal into the front case.

20. Position a new front case gasket in place. Set seal guide tool MD998285 on the front end of the crankshaft to protect the seal from damage. Apply a thin coat of oil to the outer circumference of the seal pilot tool.

21. Install the front case assembly through a new front case gasket and temporarily tighten the flange bolts.

22. Mount the oil filter on the bracket with new oil filter bracket gasket in place. Install the 3 bolts with washers and tighten to 16 ft. lbs. (22 Nm).

23. Insert a Phillips screwdriver into a hole in the left side of the engine block to lock the silent shaft in place.

24. Secure the oil pump drive gear onto the left silent shaft by installing and tightening the driven gear bolt to 29 ft. lbs. (40 Nm).

25. Install new O-ring to the groove in the front case and install the plug cap. Using the special tool MD998162, tighten the cap to 20 ft. lbs. (27 Nm).

26. Install the oil screen in position with new gasket in place.

27. Clean both mating surfaces of the oil pan and the cylinder block. Apply sealant in the groove in the oil pan flange, keeping towards the inside of the bolt holes. The width of the sealant bead applied is to be about 0.016 in. (4mm) wide.

NOTE: After applying sealant to the oil pan, do not exceed 15 minutes before installing the oil pan.

28. Install the oil pan to the engine and secure with the retainers. Tighten bolts to 6 ft. lbs. (8 Nm).

29. Install the oil pressure gauge unit and the oil pressure switch. Connect the electrical harness connector.

30. Refill crankcase with oil. Install new oil filter.

31. Install the timing belts and timing belt covers. Assemble the remaining components to the front of the engine.

32. Connect the negative battery cable and start the engine. Verify oil pressure. Inspect for leaks.

TRANSMISSION

Manual Transmission Assembly

REMOVAL AND INSTALLATION

Montero

1. Disconnect the negative battery cable.

2. Remove the switch panel.

3. Remove the suspension control switch or cover.

4. Disconnect the rear console harness connector.

5. Remove side panel and remove the rear console assembly.

6. Remove the manual shift lever and transfer shift lever knobs.

7. Disconnect the floor console harness connector and remove the front console assembly.

8. Move the transmission control lever to the **N** position and the transfer control lever to the **4H** position.

9. Remove the boot retainer and the control lever boot.

10. Remove the control lever mounting bolts and remove the levers.

11. Remove the transfer case protector.

12. Drain the transmission and transfer case fluid.

13. Remove the front and rear driveshafts.

14. Remove the dust seal guard.

15. Disconnect the following connectors:
   a. HI/LO detection switch
   b. 2WD/4WD detection switch
   c. Back-up light switch
   d. Center differential lock detection switch
   e. Center differential lock operation switch
   f. 4WD operation detection switch

16. Disconnect the speedometer cable.

17. Remove the heat protector.

18. Remove the clutch release cylinder (with the clutch hose attached) from the transmission and suspend it from the body using wire.

19. Remove the starter motor and cover.

20. Remove the heat protector.

21. Remove the transmission stays.

22. Remove the bell housing cover.

23. Support the transmission with a jack and remove the transfer roll stopper and the transfer mounting bracket.

24. Remove the No. 2 crossmember.

25. Remove the engine mounting rear insulator.

26. Remove the transfer case protector bracket.

27. Remove the mass damper.

28. Remove the transmission mounting bolts and pull the transmission slowly towards the rear.

29. When the transmission and transfer assembly are lowered, tilt the front end of the transmission downward and slowly lower forward, while using care to make sure that the rear of the transmission does not interfere with the No. 4 crossmember.

To install:

30. Replace the transmission to the engine and install the mounting bolts.

31. Install the mass damper. Torque the bolts to 51 ft. lbs. (69 Nm).

32. Install the transfer case protector bracket.

33. Install the engine mounting rear insulator.

34. Install the No. 2 crossmember. Torque the mounting bolts to 47 ft. lbs. (64 Nm).

35. Install the transfer mounting bracket and then the transfer roll stopper.

36. Install the bell housing cover.

37. Install the transmission stays.

38. Install the heat protector.

39. Install the starter cover and starter motor.

40. Install the clutch release cylinder. Torque the mounting bolts to 25 ft. lbs. (34 Nm).

41. Install the heat protector.

42. Connect the speedometer cable.

43. Connect all switch connectors.

44. Install the dust seal guard.

45. Install the driveshafts.

46. Fill the transmission and the transfer case with the proper fluids.

47. Install the transfer case protector.

48. Install the shift levers. Torque the mounting bolts to 14 ft. lbs. (19 Nm).

49. Install the control lever boot and install the retainer.

50. Connect the harness connector and install the front console assembly.

51. Install the shift lever knobs.

52. Install the rear console assembly and install the side panel.

53. Connect the rear console harness connector.

54. Install the suspension control switch or cover.

55. Install the switch panel.

56. Connect the negative battery cable.

Mighty Max

1. Disconnect the negative battery cable.

2. Place the shifter(s) in the neutral position.

3. Unscrew the shift knob from the control lever. Remove the retainer screws from the dust cover retaining plate and slide plate and boot off of the lever.

4. Remove the retainer screws from the stopper plate and remove the lever assembly from the transmission. Cover the opening with a clean towel to prevent dirt from entering the transmission.

5. Raise the vehicle and support safely. Remove the skid plate, if equipped.

6. Drain the transmission fluid. If equipped with 4WD, drain the oil from the transfer case.

7. Matchmark and remove the driveshaft(s) from the vehicle.

8. Disconnect the speedometer cable from the transmission or transfer case.

9. Disconnect the clutch cable connection.

10. Disconnect the reverse light switch harness connector.

11. Remove the starter and the bell housing cover.

12. Support the weight of the engine using a jack stand with a block of wood to protect the oil pan.

13. Support the transmission with a transmission jack.

14. Remove the transfer case bracket, if equipped.

15. Remove the rear crossmember.

16. Remove the transmission to bell housing bolts.

17. Slide the transmission backwards until the input shaft clears the clutch disc. Remove the transmission from the vehicle.

To install:

18. Lubricate the pilot bushing and input shaft splines very lightly with high temperature lubricant.

19. Mount the transmission securely on a transmission jack and lift it in place until the input shaft is centered in the bell housing opening. Roll the transmission forward until the input shaft splines fully engage with the clutch disc.

20. Install the transmission to bell housing bolts.

21. Lift the transmission using the transmission jack and install the rear crossmember into position. Install the transfer case bracket, if equipped. Torque the frame bolts. Remove the transmission and engine support fixtures.

22. Install the starter. Install the bell housing cover.

23. Connect the reverse light switch and clip all wiring to the transmission case.

24. Connect the speedometer cable.

25. Install the driveshaft(s) making sure to align matchmarks.

26. Connect the clutch cable, using a new cotter pin.

27. Fill the transmission and transfer case with the proper amount of gear oil.

28. Install the skip plate, if equipped.

29. Lower the vehicle.

30. Install the shift lever assembly, boot and console.

31. Connect the negative battery cable. Road test the vehicle for proper operation. Check operation of the reverse lights.

2-Wheel Drive Expo

1. Disconnect the battery cables, negative cable first. Remove the battery and tray.

2. Remove the coolant reservoir.

3. Remove the air cleaner.

4. Disconnect the clutch cable, speedometer cable and backup light wiring from the transaxle.

5. Remove the upper engine-to-transaxle bolts.

6. Disconnect the select control lever and switch harness.

7. Remove the starter.

8. Disconnect and tag all wiring from the transaxle.

9. Raise and support the vehicle safely.

10. Remove the front wheels.

11. Drain the transaxle fluid.

12. Remove the extension and shift rod from the engine compartment.

13. Remove the stabilizer and strut bar from the lower control arm.

14. Remove the left and right halfshafts.

15. Support the transaxle with a suitable floor jack, taking care to avoid damaging the pan.

16. Remove the bellhousing cover.

17. Remove the remaining transaxle-to-engine bolts.

18. Remove the transaxle mounting bolt.

19. Lower the jack and slide the transaxle from under the vehicle.

To install:

20. Secure the transaxle on a transaxle jack and position it to the engine.

21. Carefully guide the transaxle input shaft into the clutch assembly. Make sure the transaxle is seated properly and is flush to the engine

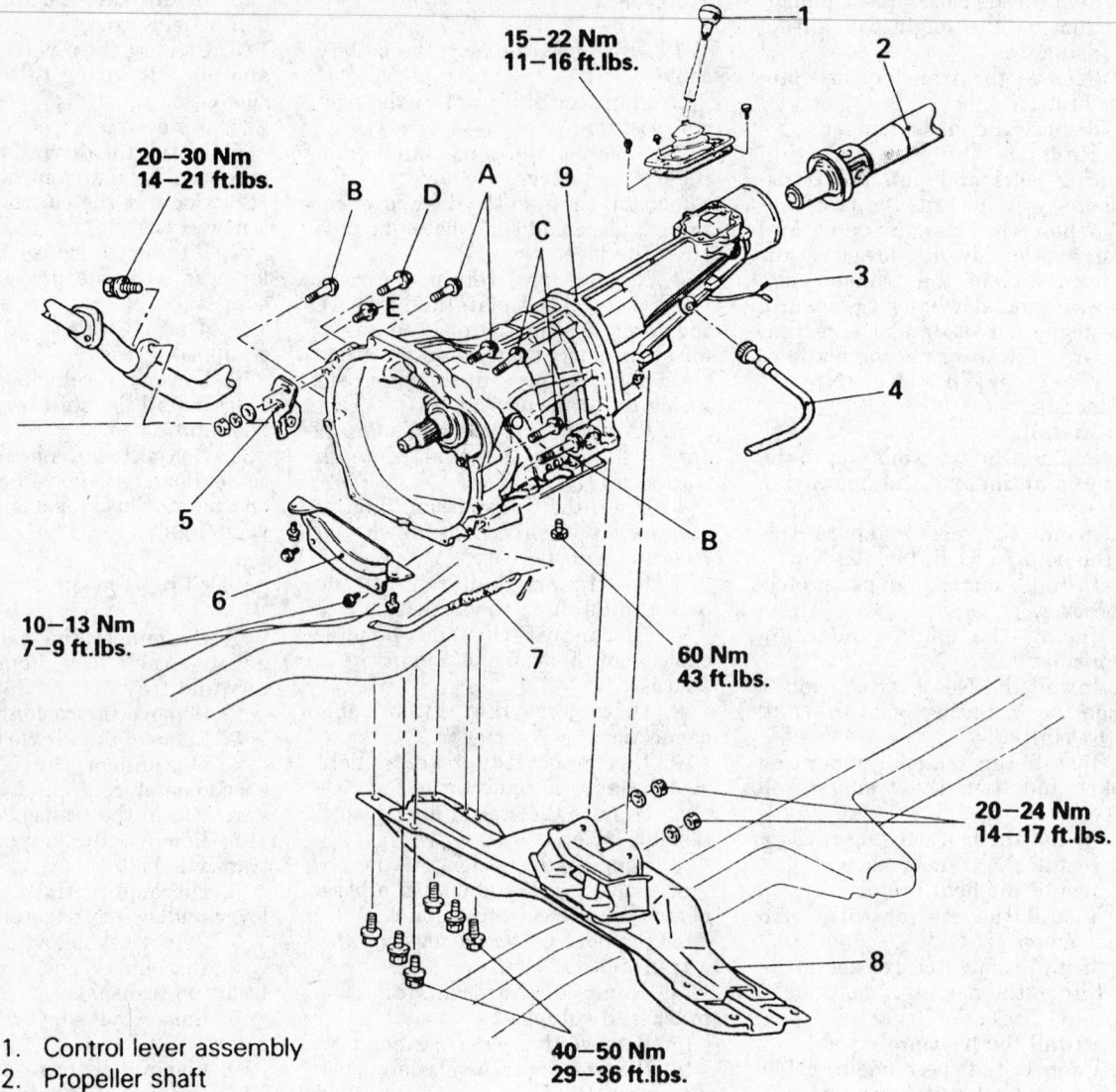

15–22 Nm
11–16 ft.lbs.

20–30 Nm
14–21 ft.lbs.

10–13 Nm
7–9 ft.lbs.

60 Nm
43 ft.lbs.

20–24 Nm
14–17 ft.lbs.

40–50 Nm
29–36 ft.lbs.

1. Control lever assembly
2. Propeller shaft
3. Backup light switch connector connection
4. Speedometer cable connection
5. Exhaust pipe mounting bracket
6. Bell housing cover
7. Clutch cable connection
8. No. 2 crossmember
9. Transmission assembly

Items	Nm	ft.lbs.	O.D. × Length mm (in.)		Bolt identification
A	43–55	31–40	⑦	10 × 40 (.4 × 1.6)	
B	43–55	31–40	⑦	10 × 65 (.4 × 2.6)	⑦ D × L
C	22–32	16–23	⑦	10 × 60 (.4 × 2.4)	
D	20–27	15–20	⑦	8 × 55 (.3 × 2.2)	
E	20–27	15–20	⑦	8 × 25 (.3 × 1.0)	

328969

Transmission assembly — 2WD Mighty Max

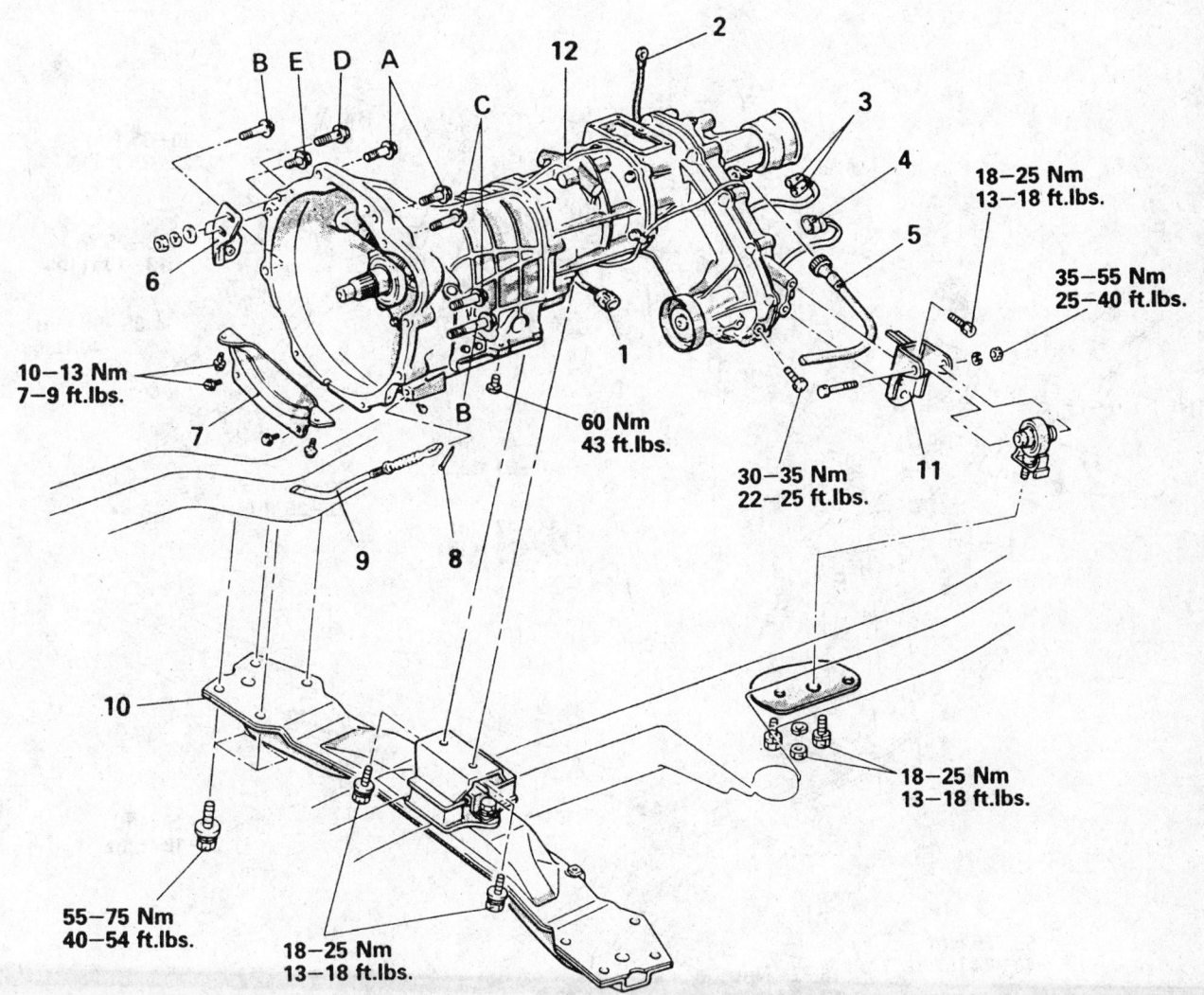

**10–13 Nm
7–9 ft.lbs.**

**18–25 Nm
13–18 ft.lbs.**

**35–55 Nm
25–40 ft.lbs.**

**60 Nm
43 ft.lbs.**

**30–35 Nm
22–25 ft.lbs.**

**18–25 Nm
13–18 ft.lbs.**

**55–75 Nm
40–54 ft.lbs.**

**18–25 Nm
13–18 ft.lbs.**

1. Backup light switch harness connection
2. Gound cable
3. 4WD indicator harness connection
4. Pulse generator connector
5. Speedometer cable connection
6. Exhaust pipe mounting bracket
7. Bell housing cover
8. Cotter pin
9. Clutch cable connection
10. No. 2 crossmember
11. Transfer mounting bracket
12. Transmission and transfer assy

Items	Nm	ft.lbs.	O.D. × Length mm (in.)	Bolt identification
A	43 – 55	31 – 40	⑦ 10 × 40 (.4 × 1.6)	⑦ D × L
B	43 – 55	31 – 40	⑦ 10 × 65 (.4 × 2.6)	
C	22 – 32	16 – 23	⑦ 10 × 60 (.4 × 2.4)	
D	20 – 27	15 – 20	⑦ 8 × 55 (.3 × 2.2)	
E	20 – 27	15 – 20	⑦ 8 × 25 (.3 × 1.0)	

328974

Transmission assembly — 1993–94 4WD Mighty Max with 2.4L (VIN G) engines

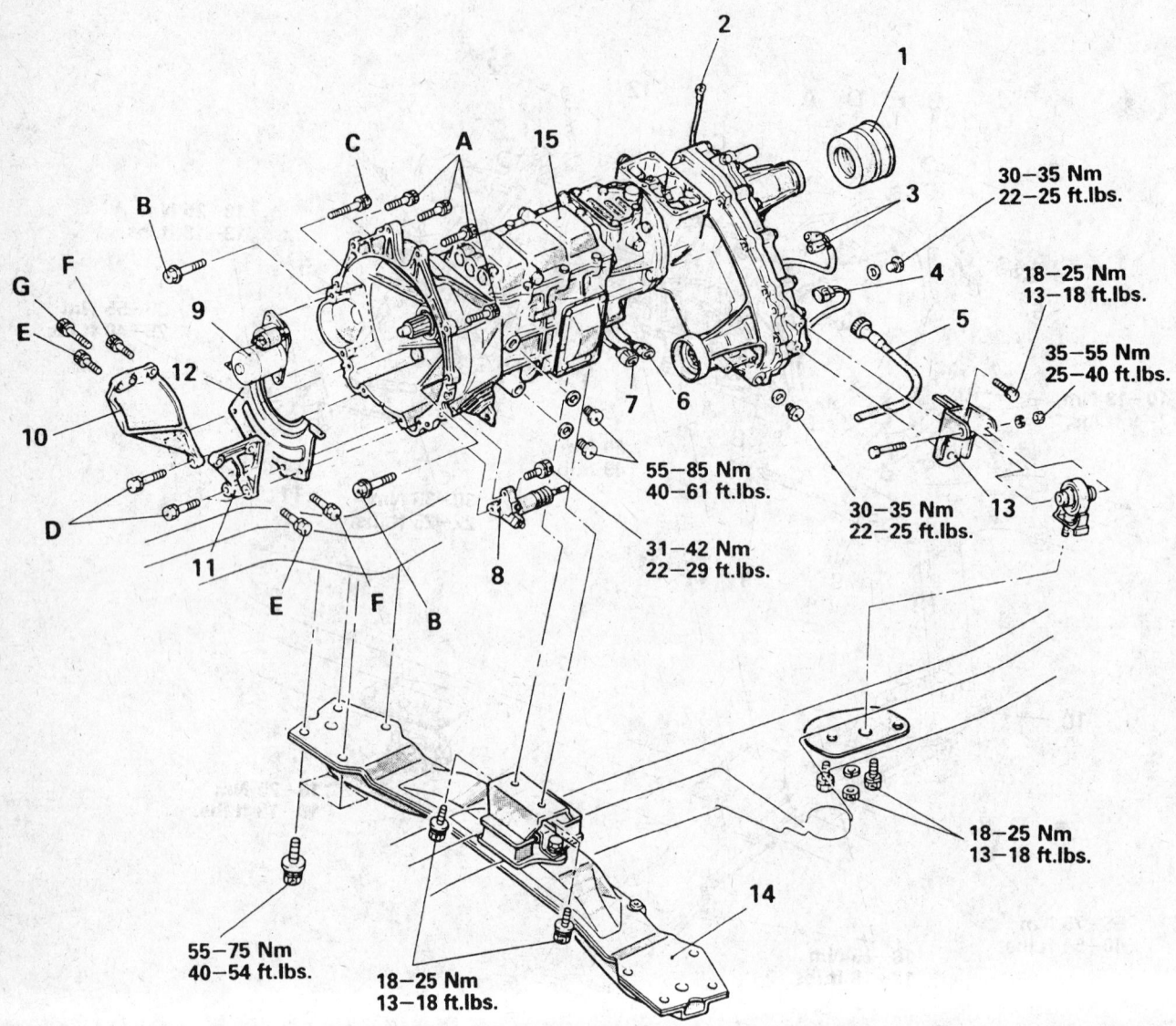

1. Dust seal guard
2. Gound cable
3. 4WD indicator light switch connection
4. Pulse generator connector
5. Speedometer cable
6. Oxygen sensor connector
7. Back-up light switch connector
8. Clutch release cylinder
9. Starter motor
10. Transmission stay (R.H.)
11. Transmission stay (L.H.)
12. Bell housing cover
13. Transfer mounting bracket
14. No. 2 crossmember
15. Transmission and transfer assembly

	Nm	ft.lbs.	O.D. × Length mm (in.)	Bolt identification
A	65 – 85	47 – 61	⑦ 12 × 40 (.5 × 1.6)	⑦ D × L
B	80 – 100	58 – 72	⑦ 12 × 55 (.5 × 2.2)	
C	27 – 34	20 – 25	⑦ 10 × 55 (.4 × 2.2)	
D	30 – 42	22 – 30	⑦ 10 × 40 (.4 × 1.6)	
E	65 – 85	47 – 61	⑦ 12 × 35 (.5 × 1.4)	
F	33 – 50	24 – 36	⑦ 10 × 30 (.4 × 1.2)	
G	65 – 85	47 – 61	⑦ 12 × 50 (.5 × 2.0)	

328975

Transmission assembly — 4WD Mighty Max with 3.0L (VIN H) engines

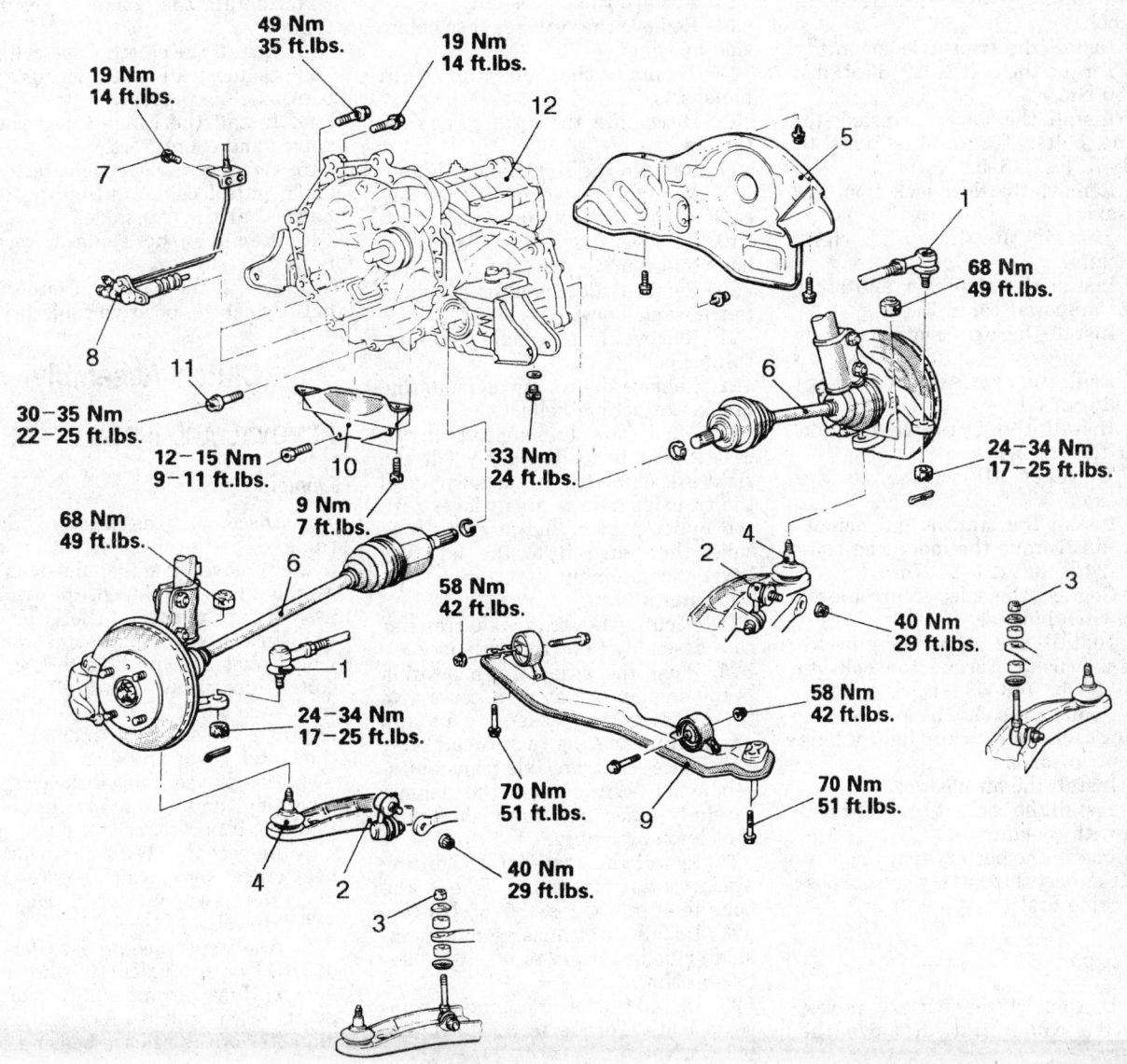

1. Connection for tie rod end
2. Connection for stabilizer bar
3. Self locking nut
4. Connection for lower arm ball joint
5. Under cover (R.H.)
   Draining of the transaxle oil
6. Drive shaft connection
7. Clutch oil line bracket bolt
8. Connection for release cylinder
9. Center member
10. Bell housing cover
11. Transaxle assembly lower part coupling bolts
12. Transaxle assembly

241506

Underside connection points — 2-Wheel Drive Expo

flange. Install two transaxle-to-engine bolts.

22. Install the transaxle mounting bolt. Torque the bolt to 29–36 ft. lbs. (40–50 Nm).

23. Install the lower transaxle-to-engine bolts. Torque the bolts to 32–39 ft. lbs. (43–53 Nm).

24. Remove the floor jack from the transaxle.

25. Install the left and right halfshafts.

26. Install the stabilizer and strut bar to the lower control arm.

27. Install the extension and shift rod.

28. Refill the transaxle with an approved gear oil.

29. Install the front wheels and lower the vehicle.

30. Connect all wiring to the transaxle.

31. Install the starter and mounting bolts. Torque the mounting bolts to 20–24 ft. lbs. (27–34 Nm).

32. Connect the select control lever and switch harness.

33. Install the upper engine-to-transaxle bolts. Torque the bolts to 32–39 ft. lbs. (43–53 Nm).

34. Connect the clutch cable, speedometer cable and backup light wiring to the transaxle.

35. Install the air cleaner.

36. Install the coolant reservoir and refill with coolant.

37. Install the battery tray and battery. Connect the battery cables, positive cable first.

4WD Expo

1. Disconnect the battery cables, negative cable first. Remove the battery.

2. Remove the coolant reserve tank.

3. Disconnect the speedometer cable, shift control cable and backup light harness at the transaxle.

4. Remove the range select control valves and connectors.

5. Tag and disconnect all other wiring attached to the transaxle.

6. Remove the clutch slave cylinder.

7. Remove the vacuum reservoir tank.

8. Disconnect the starter wiring and remove.

9. Remove the upper engine-to-transaxle bolts.

10. Raise and support the vehicle safely.

11. Remove the front wheels, lower engine cover and skid plate.

12. Drain the transaxle and transfer case.

13. Remove the driveshaft.

14. Remove the transfer case extension housing.

15. Remove the left and right halfshafts.

16. Disconnect the right strut from the lower arm.

17. Remove the right fender liner.

18. Take up the weight of the transaxle with a suitable floor jack.

19. Remove the bellhousing cover bolts and remove the cover.

20. Remove the remaining engine-to-transaxle bolts.

21. Remove the transaxle mount insulator bolt.

22. Remove the transaxle mounting bracket attaching bolts.

23. Move the transaxle/transfer case assembly to the right. Tilt the right side of the transaxle down, until the transfer case is about level with the upper part of the steering rack tube, then turn it to the left and lower the assembly.

To install:

24. Secure the transaxle/transfer case assembly to a transaxle jack.

25. Raise the assembly in position to the engine. It may be necessary to tilt or angle the assembly in and around the steering rack tube.

26. Once the transaxle/transfer assembly is positioned to the engine, carefully guide the input shaft into the clutch assembly.

27. Install the transaxle mounting bracket attaching bolts. Torque the bolts to 40–43 ft. lbs. (55–60 Nm).

28. Install the transaxle mount insulator bolt. Torque to 40–43 ft. lbs. (55–60 Nm).

29. Install the lower engine-to-transaxle bolts. Torque to 31–40 ft. lbs. (43–55 Nm).

30. Install the bellhousing cover bolts and install the cover.

31. Remove the the floor jack from the transaxle.

32. Install the right fender liner.

33. Connect the right strut to the lower arm.

34. Install the left and right halfshafts.

35. Install the transfer case extension housing and the driveshaft.

36. Refill the transaxle and transfer case with an approved gear oil.

37. Install the front wheels, lower engine cover and skid plate.

38. Lower the vehicle.

39. Install the upper engine-to-transaxle bolts. Torque the bolts to 40–43 ft. lbs. (55–60 Nm).

40. Install the starter motor and mounting bolts. Torque the bolts to 22–25 ft. lbs. (30–35 Nm). Connect the starter wiring.

41. Install the vacuum reservoir tank.

42. Install the clutch slave cylinder.

43. Connect all other wiring to the transaxle, as tagged.

44. Install the range select control valves and connectors.

45. Connect the speedometer cable, shift control cable and backup light harness at the transaxle.

46. Remove the coolant reserve tank.

47. Install the battery. Connect the battery cables, positive cable first.

Clutch Assembly

REMOVAL AND INSTALLATION

Montero

1. Disconnect the negative battery cable.

2. Remove the transmission assembly and if equipped, the transfer assembly from the vehicle.

3. Insert a suitable tool in the flywheel pilot bearing hole to keep the clutch disc from falling off. Loosen the clutch cover retainer bolts gradually in a crisscross fashion. Remove the clutch cover and disc.

4. Check the release bearing for scorching, damage or strange noise. Replace, if necessary.

5. Inspect the flywheel surface for heat cracks or scoring. Reface or replace the flywheel as required.

To install:

6. Apply specified grease (Part No. 0101011 or equivalent) to the clutch disc splines, input shaft, contact points of the release fork and inside diameter of the release bearing.

NOTE: Do not allow oil or grease to contact the clutch facing and pressure plate.

7. Align the clutch disc to the flywheel, using a suitable tool.

NOTE: When installing the clutch disc, be sure that the surface having the manufacturer's stamped mark is on the pressure plate side.

8. Install the clutch cover with the dowel pin holes in alignment with the dowel pins in the flywheel and tighten the bolts gradually in a crisscross fashion. Tighten the bolts to 14 ft. lbs (19 Nm).

9. Install the transmission assembly and if equipped, the transfer assembly to the vehicle.

10. Connect the negative battery cable. Road test the vehicle for proper operation.

1. Connection for tie rod end
2. Connection for stabilizer bar
3. Connection for lower arm ball joint
4. Under cover (R.H.)
5. Drive shaft nut (R.H.)
6. Drive shaft (R.H.)
7. Connection for drive shaft and inner shaft
8. Clutch oil line bracket bolts
9. Connection for clutch release cylinder
10. Front exhaust pipe
11. Transfer assembly
12. Center member
13. Bell housing cover
14. Transaxle assembly lower part coupling bolts
15. Transaxle assembly

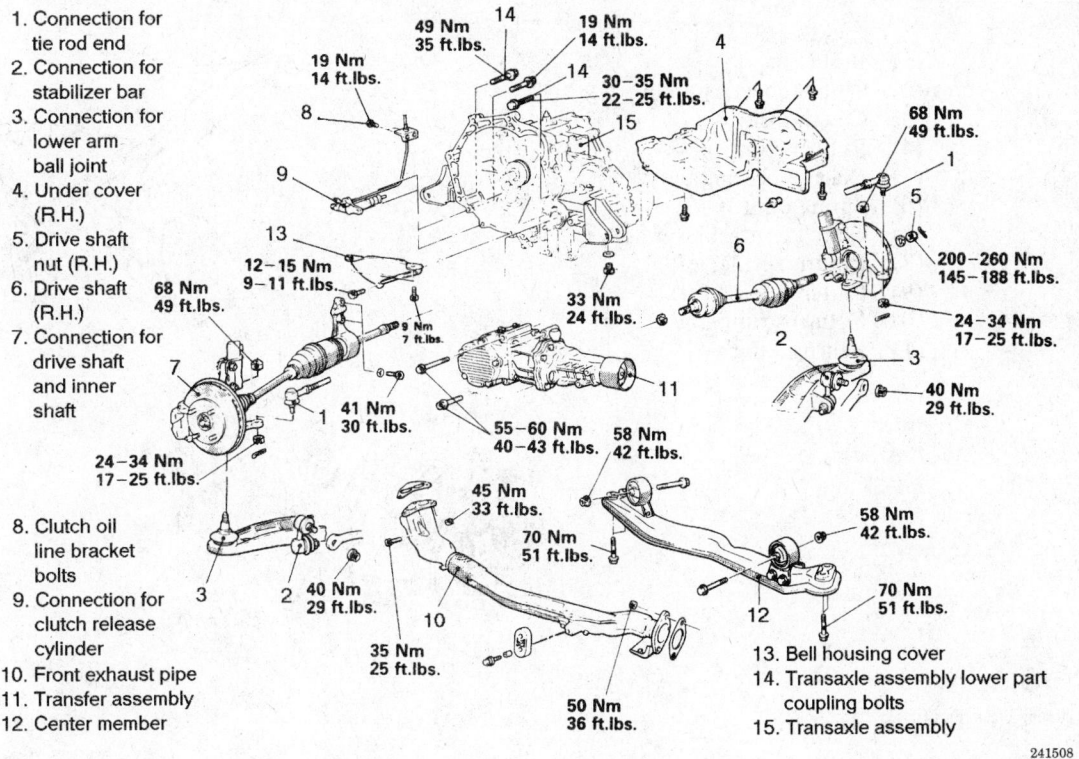

241508

Underside connection points — 4-Wheel Drive Expo

Mighty Max

1. Disconnect the negative battery cable.

2. Remove the transmission assembly and if equipped, the transfer assembly from the vehicle.

3. Insert a suitable tool in the flywheel pilot bearing hole to keep the clutch disc from falling off. Loosen the clutch cover retainer bolts gradually in a crisscross fashion. Remove the clutch cover and disc.

4. Check the release bearing for scorching, damage, or strange noise. Replace, if necessary.

5. Inspect the flywheel surface for heat cracks or scoring. Resurface or replace the flywheel as required.

To install:

6. Apply specified grease (Part No. 0101011 or equivalent) to the clutch disc splines, input shaft, contact points of the release fork and the inside diameter of the release bearing.

NOTE: Do not allow oil or grease to contact the clutch facing and pressure plate.

7. Align the clutch disc to the flywheel, using a suitable tool.

NOTE: When installing the clutch disc, be sure that the surface having the manufacturer's stamped mark is on the pressure plate side.

8. Install the clutch cover with the dowel pin holes in alignment with the dowel pins in the flywheel and tighten the bolts gradually in a crisscross fashion. Tighten the bolts to 11–15 ft. lbs (15–22 Nm).

9. Install the transmission assembly and if equipped, the transfer assembly to the vehicle.

10. Check the clutch pedal height/free-play adjustment.

11. Connect the negative battery cable. Road test the vehicle for proper operation.

Expo with F5M22, F5M31, F5M33, W5M31 and W5M33 Manual Transmissions

1. Disconnect the negative battery cable. Raise and safely support the vehicle.

2. Remove the transaxle assembly from the vehicle.

3. Remove the pressure plate attaching bolts, pressure plate and clutch disc. If the pressure plate is to be reused, loosen the bolts in a diagonal pattern, 1 or 2 turns at a time. This will prevent warping the the clutch cover assembly.

4. Remove the return clip and the pressure plate release bearing. Do not use solvent to clean the bearing.

5. Inspect the clutch release fork and fulcrum for damage or wear. If necessary, remove the release fork and unthread the fulcrum from the transaxle.

6. Carefully inspect the condition of the clutch components and replace any worn or damaged parts.

To install:

7. Inspect the flywheel for heat damage or cracks. Resurface or replace the flywheel as required. Install the flywheel using new bolts.

8. Install the fulcrum and tighten to 25 ft. lbs. (35 Nm). Install the release fork. Apply a coating of multi-purpose grease to the point of contact with the fulcrum and the point of contact with the release bearing. Apply a coating of multi-purpose grease to the end of the release cylinder's

(1) Bolt (6)
(2) Clutch cover assembly
(3) Clutch disc
(4) Return clip (2)
(5) Release bearing
(6) Spring pin (2)
(7) Clutch control lever ass'y
(8) Return spring, left
(9) Clutch release fork
(10) Return spring, right
(11) Felt packing

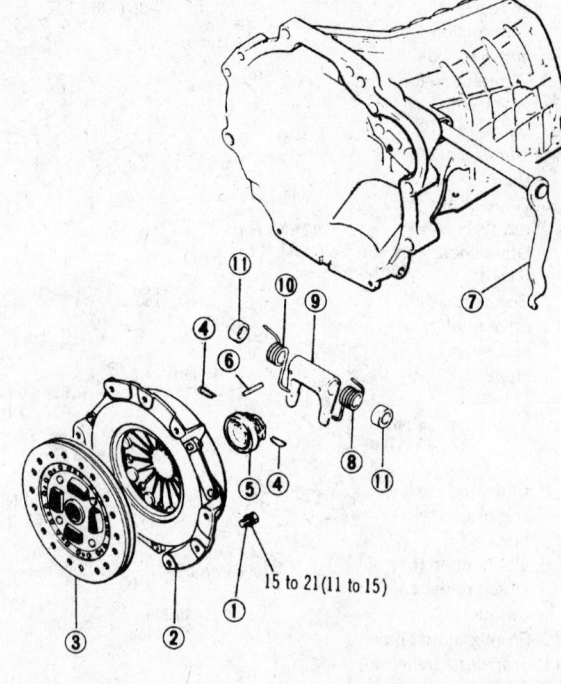

Tightening torque : Nm (ft-lbs.)

328871

Clutch components — Mighty Max

push rod and the push rod hole in the release fork.

NOTE: When installing the clutch, apply grease to each part, but be careful not to apply excessive grease. Excessive grease will cause clutch slippage and shudder.

9. Apply multi-purpose grease to the clutch release bearing. Pack the bearing inner surface and the groove with grease. Do not apply grease to the resin portion of the bearing. Place the bearing in position and install return clip.

10. Apply a coating of grease to the clutch disc splines and then use a brush to rub it in the grooves. Using a universal clutch disc alignment tool, position the clutch disc on the flywheel. Install the retainer bolts and tighten a little at a time, in a diagonal sequence. Tighten them to a final torque of 16 ft. lbs. (22 Nm). Remove the aligning tool.

11. Install the transaxle assembly and check fluid level.

12. Check for proper clutch operation.

Clutch Cable

ADJUSTMENT

Mighty Max

1. Measure the clutch pedal height from the face of the pedal pad to the floorboard. This distance should be 6.5–6.7 in. (166–171 mm).

2. If the pedal height is not correct, turn the pedal stopper bolt or clutch switch until the pedal height is adjusted to its standard value.

3. Measure the clutch pedal free-play. The standard value should be 0.8–1.4 in. (20–35 mm).

4. If the clutch pedal free-play is not within the standard value, adjust it as follows:

a. Pull the clutch cable lightly at the toeboard and turn adjusting nut until the adjusting jut clearance is adjusted to 0.12–0.16 in. (3–4 mm).

b. After making the adjustment, depress the clutch pedal several times and check that the clutch pedal play is with the standard value.

5. Measure the distance between the clutch pedal and floorboard when the clutch is disengaged. The standard value should be 2.4 in. (60 mm) or more.

Clutch Master Cylinder

REMOVAL AND INSTALLATION

Except Expo

1. Disconnect the negative battery cable.

2. Remove as much fluid as possible from the clutch master cylinder reservoir.

3. Remove the cotter pin and remove the clevis pin from the clutch pedal.

4. Disconnect and plug the fluid line from the clutch master cylinder.

5. Remove the mounting nuts and remove the cylinder from the firewall. Remove the reservoir from the cylinder.

To install:

6. Install the brake cylinder to the firewall and torque the nuts to 7 ft. lbs. (9 Nm) for 1985–91 and 9 ft. lbs. (13 Nm) for 1993–95.

7. Connect the fluid line to the master cylinder.

8. Connect to the clutch pedal with a new cotter pin and clevis pin.

9. Fill with DOT 3 or 4 brake fluid.

10. Bleed the clutch system.

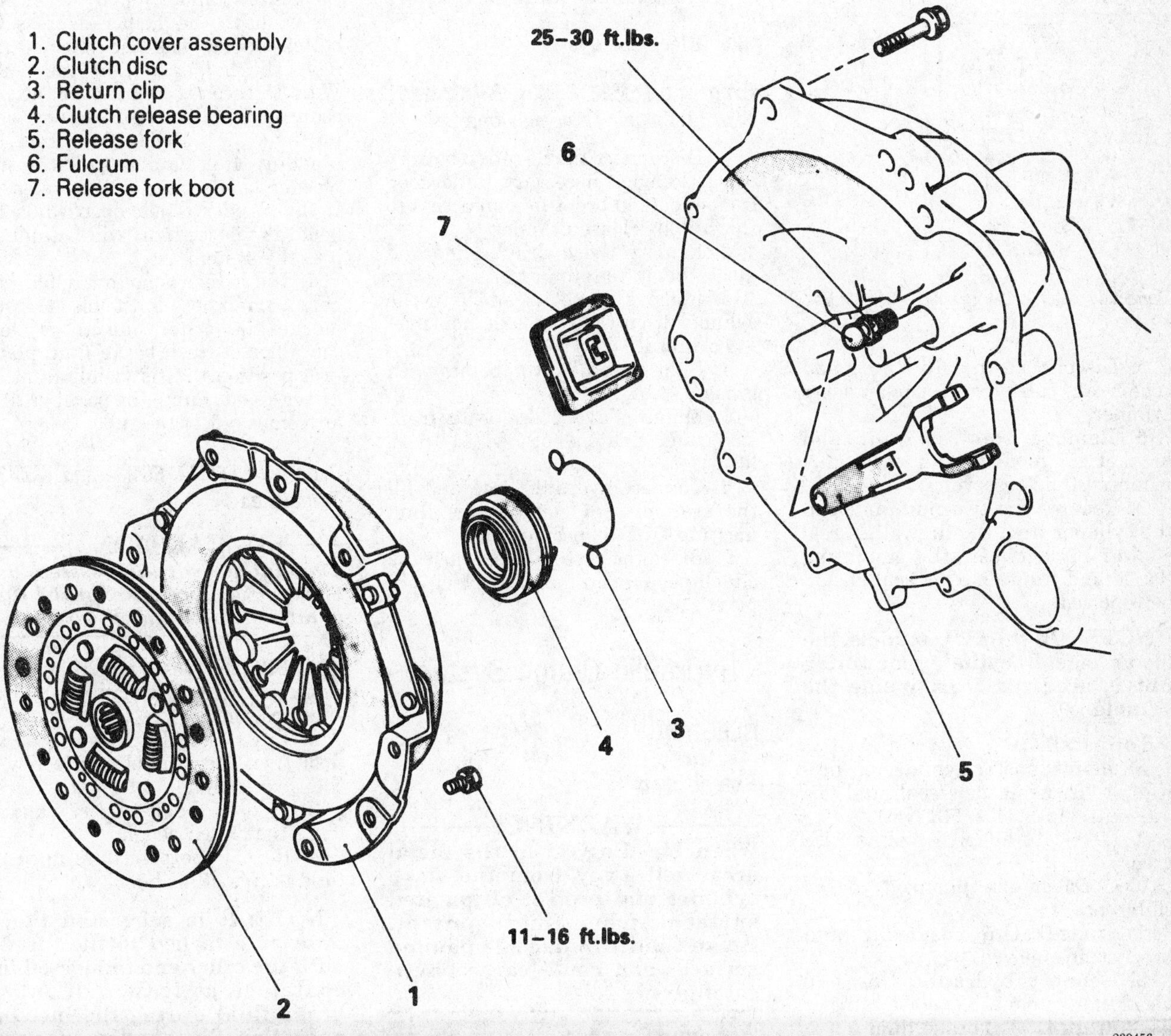

1. Clutch cover assembly
2. Clutch disc
3. Return clip
4. Clutch release bearing
5. Release fork
6. Fulcrum
7. Release fork boot

25–30 ft.lbs.

7

6

3

4

5

11–16 ft.lbs.

2

1

239456

Clutch assembly exploded view — 4-Wheel Drive Expo

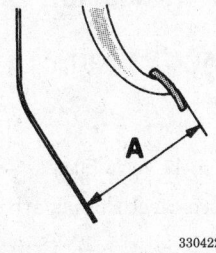

330422

**Checking clutch pedal
height — Mighty Max**

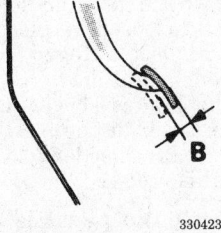

330423

**Checking clutch pedal free-
play — Mighty Max**

***Expo with F5M22, F5M31 and W5M33
Transaxles***

1. Remove necessary underhood
components in order to gain access to
the clutch master cylinder.

——— **WARNING** ———
**Clutch hydraulic system uses
brake fluid which is harmful to
painted surfaces.**

2. Loosen the bleeder screw on the
slave cylinder and drain the system.
3. Disconnect the pushrod from
the clutch pedal.
4. Disconnect the clutch pedal
from the pedal bracket.

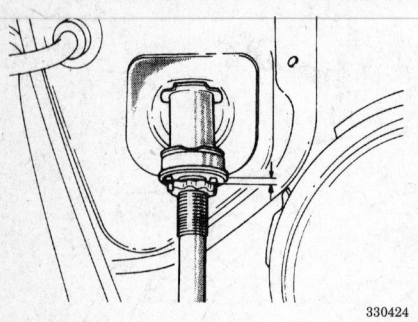

330424

Correcting clutch pedal free-play — Mighty Max

5. Disconnect the fluid line and reservoir tube from the master cylinder.

6. Remove the reservoir and bracket on models with externally mounted fluid reservoirs.

7. Remove the two nuts and pull the cylinder from the firewall. A seal should be between the mounting flange and firewall. This seal should be replaced.

NOTE: On the 4WD models, the lower master cylinder mounting nut is accessed from inside the vehicle.

To install

8. Mount master cylinder on firewall studs, using new seal, and torque nuts to 9 ft. lbs. (13 Nm).

9. Lubricate all pivot points with grease.

10. Connect the pushrod to the clutch pedal.

11. Install the reservoir and bracket, if removed.

12. Connect hydraulic line and bleed the system at the slave cylinder using fresh DOT 3 brake fluid.

13. Check the adjustment of the clutch pedal for proper freeplay.

Clutch Slave Cylinder

REMOVAL AND INSTALLATION

Except Expo

1. Disconnect the negative battery cable.

2. Raise the vehicle and support safely.

3. Remove the eye-bolt and washer from the clutch hose at the release cylinder.

4. Remove the mounting bolts and remove the slave cylinder from the transmission case.

To install:

5. Install the slave cylinder with the mounting bolts.

6. Replace the eye-bolt washer.

7. Connect the negative battery cable.

8. Bleed the system.

Expo with F5M22, F5M31, F5M33 and W5M33 Manual Transmissions

1. Disconnect the negative battery cable. Remove necessary underhood components in order to gain access to the clutch release cylinder.

2. Remove the hydraulic line and allow the system to drain.

3. Remove the bolts and pull the cylinder from the transaxle housing.

To install

4. Lubricate all pivot points with grease.

5. Mount slave cylinder to transaxle and tighten bolts to 11-16 ft. lbs.

6. Connect hydraulic line and fill the system with clean brake fluid meeting DOT 3 specifications.

7. Bleed the system and adjust the clutch pedal height and the clevis pin play.

Hydraulic Clutch System

BLEEDING

Except Expo

—————— WARNING ——————
When bleeding, keep the facial area well away from the slave cylinder and protect all painted surfaces from fluid contact. Brake fluid will damage painted surfaces and could cause physical injury.

1. Fill the clutch master cylinder with fresh DOT 3 brake fluid.

2. Have a helper sit in the vehicle.

3. Raise the vehicle and support it safely.

4. Remove the bleeder screw cap.

5. If the system is empty, the most efficient way to get fluid down to the cylinder is to loosen the bleeder about ½–¾ turn; place a finger firmly over the bleeder and have the helper pump the brakes slowly until fluid pressure is felt at the bleeder. Once fluid is at the bleeder, close before the pedal is released.

NOTE: If the pedal is pumped rapidly, the fluid will churn and create small air bubbles, which are difficult and time consuming to remove from the system. These air bubbles will eventually congregate and will result in a spongy pedal.

6. Once fluid has been pumped to the slave cylinder, open the bleeder screw, have the helper depress the clutch pedal, lock the bleeder and have the helper release the pedal. Wait 15 seconds and repeat the procedure (including the 15 second wait) until no air bubbles flow from the bleeder. Remember to close the bleeder before the pedal is released. If the bleeder is left open when the pedal is released, air will be induced into the system.

7. If a helper is not available, connect a small hose to the bleeder, submerge the other end in a clean container of fresh brake fluid placed in a position that is visible from the driver's seat. Pump the pedal until no air comes out of the tube.

Expo with F5M22, F5M31 and W5M33 Transaxles

—————— CAUTION ——————
The clutch hydraulic system uses brake fluid. Use care; brake fluid is harmful to painted surfaces.

1. Fill the reservoir with clean brake fluid meeting DOT 3 specifications.

2. Press the clutch pedal to the floor then open the bleeder screw on the slave cylinder.

3. Tighten the bleed screw and release the clutch pedal.

4. Repeat the procedure until the fluid is free of air bubbles.

NOTE: It is suggested that a hose be attached to the bleeder with the other end immersed in a container at least half full of brake fluid during the bleeding operation. Do not allow the reservoir to run out of fluid during bleeding.

Automatic Transmission Assembly

REMOVAL AND INSTALLATION

Montero

3.0L (VIN H) 12 Valve Engine

1. Disconnect the negative battery cable.

2. Remove the switch panel.

3. Remove the suspension control switch or cover.

4. Disconnect the rear console harness connector.

5. Remove side panel and the rear console assembly.

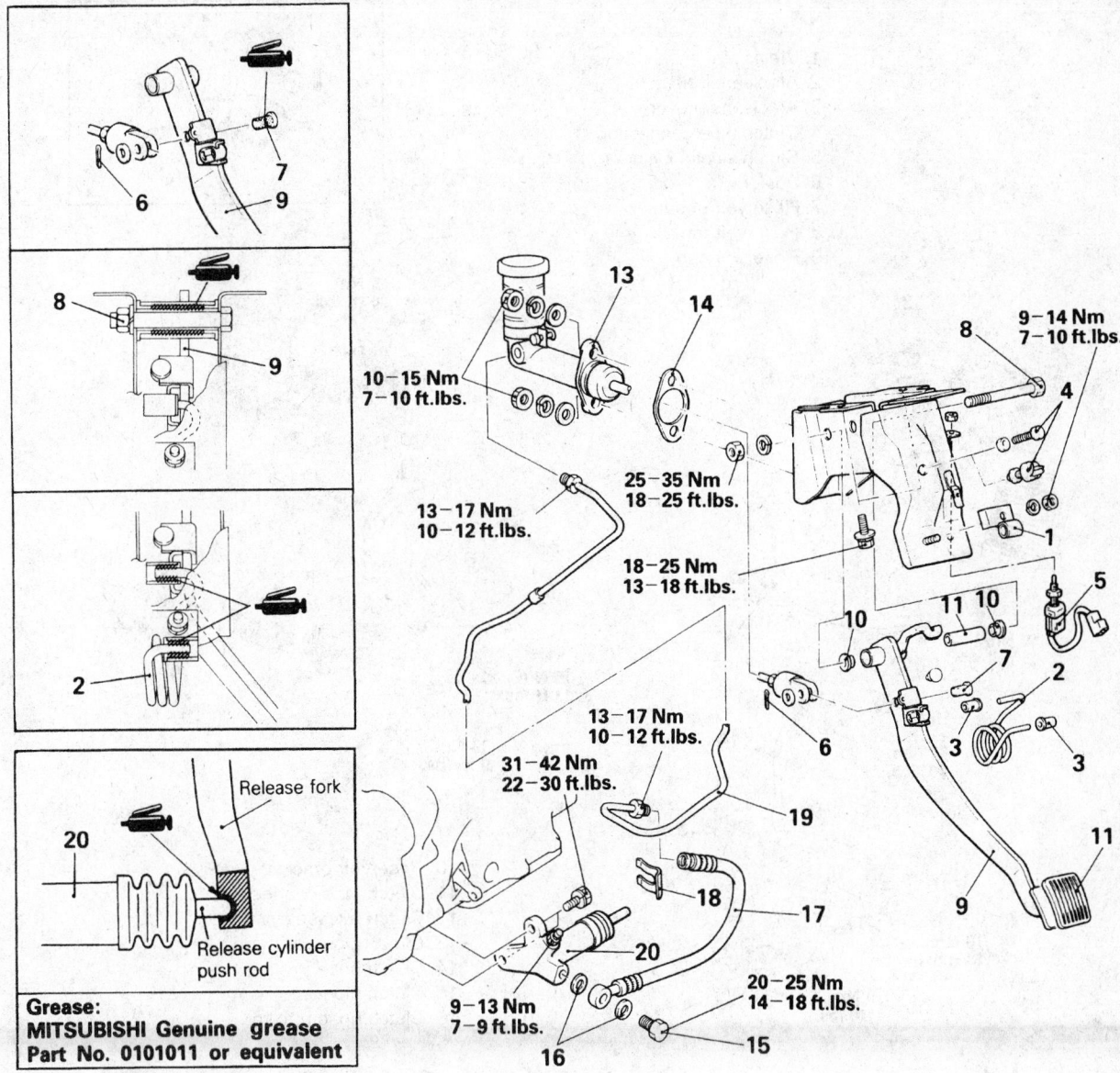

10−15 Nm
7−10 ft.lbs.

13−17 Nm
10−12 ft.lbs.

25−35 Nm
18−25 ft.lbs.

18−25 Nm
13−18 ft.lbs.

13−17 Nm
10−12 ft.lbs.

31−42 Nm
22−30 ft.lbs.

9−14 Nm
7−10 ft.lbs.

9−13 Nm
7−9 ft.lbs.

20−25 Nm
14−18 ft.lbs.

Release fork

Release cylinder
push rod

Grease:
MITSUBISHI Genuine grease
Part No. 0101011 or equivalent

Clutch pedal

1. Bracket
2. Turnover spring
3. Bushing
4. Stopper bolt <Vehicles without cruise control system> or Clutch pedal position switch <Vehicles with cruise control system>
5. Clutch pedal position switch (used for starter interlock system)
6. Cotter pin
7. Clevis pin
8. Clutch pedal mounting bolt
9. Clutch pedal
10. Bushing

11. Spacer
12. Pedal pad

Clutch master cylinder

6. Cotter pin
7. Clevis pin
13. Clutch master cylinder
14. Sealer

Clutch line

15. Eye bolt
16. Gasket
17. Clutch hose
18. Hose clip
19. Clutch tube

Clutch release cylinder

15. Eye bolt
16. Gasket
17. Clutch hose
20. Clutch release cylinder

328894

Hydraulic clutch system components — except Expo

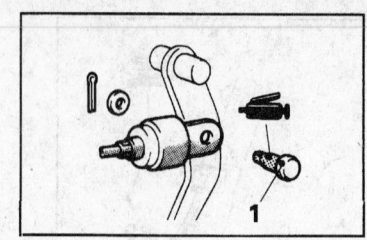

1. Clevis pin
2. Air cleaner element
3. Air cleaner cover
4. Clutch pipe connection
5. Clutch master cylinder
6. Sealer
7. Reservoir hose
8. Reservoir tank
9. Reservoir cap

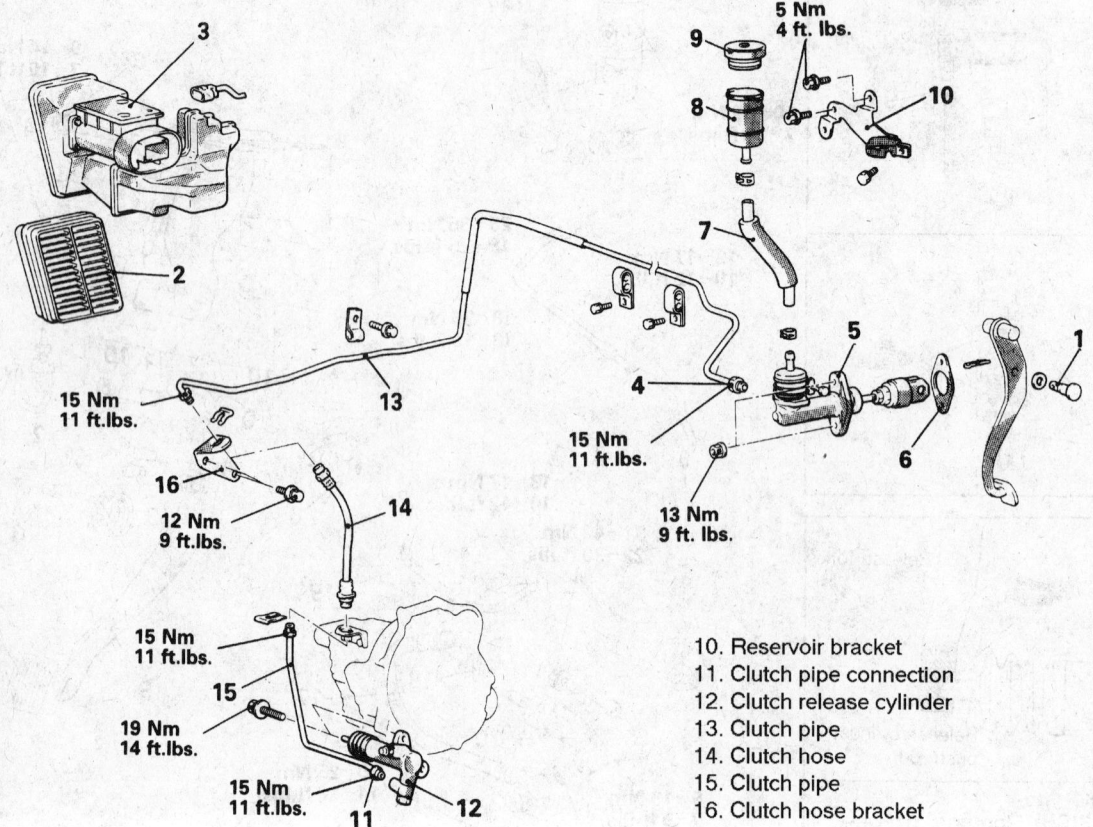

10. Reservoir bracket
11. Clutch pipe connection
12. Clutch release cylinder
13. Clutch pipe
14. Clutch hose
15. Clutch pipe
16. Clutch hose bracket

Clutch hydraulic system component identification — Expo

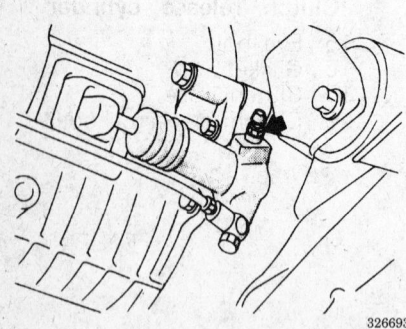

Clutch hydraulic system bleeder screw — Expo with F5M22, F5M31 and W5M33 transaxles

6. Remove the transfer shift lever knob.

7. Disconnect the floor console harness connector.

8. Set the A/T selector lever to L and remove the front console assembly.

9. Disconnect the key-interlock cable connection (selector lever assembly side).

10. Disconnect the shift-lock cable connection (selector lever assembly side).

11. Disconnect the transmission control cable connection (selector lever assembly side).

12. Remove the transmission selector lever assembly.

13. Remove the transfer control lever boot retainer and the boot.

14. Set the transfer control lever to 2H. Remove the transfer control lever mounting bolts and remove the assembly.

15. Remove the transfer case protector.

16. Remove the front exhaust pipe.

17. Drain the transmission fluid and the transfer case fluid.

18. Remove the front and rear driveshafts.

19. Remove the fluid filler pipe.

20. Disconnect the throttle cable connection.

21. Remove the dust seal guard.

22. Disconnect the transmission control cable connection.
23. Disconnect:
   a. Speed sensor connector
   b. High range / Low range detection switch connector
   c. 4WD operation detection switch connector
   d. Center differential lock operation detection switch connector
   e. Center differential lock detection switch connector
   f. 2WD/4WD detection switch connector
   g. Park/Neutral position switch connector
   h. Fluid cooler pipe connection
24. Remove the starter motor and cover.
25. Remove the heat protector.
26. Remove the left hand and right hand transmission stays.
27. Remove the bell housing cover.
28. Remove the transfer roll stopper and the transfer mounting bracket.
29. Remove the No. 2 crossmember.
30. Remove the engine mount rear insulator.
31. Remove the transfer case protector bracket.
32. Remove the mass damper.
33. Remove the torque converter connecting bolt.
34. Using a jack stand, support the transmission and remove the mounting bolts. Remove the transmission assembly.

To install:
35. Using a jack stand for support, mount the transmission to the engine and install the mounting bolts.
36. Install the torque converter connecting bolt.
37. Install the transfer case protector bracket.
38. Install the engine mount rear insulator.
39. Install the No. 2 crossmember.
40. Install the transfer mounting bracket and the transfer roll stopper.
41. Install the bell housing cover.
42. Install the left and right transmission stays.
43. Install the heat protector.
44. Install the starter motor cover and the starter.
45. Connect:
   a. Fluid cooler pipe connection
   b. Park/Neutral position switch connector
   c. 2WD/4WD detection switch connector
   d. Center differential lock detection switch connector
   e. Center differential lock operation detection switch connector

   f. 4WD operation detection switch connector
   g. High range / Low range detection switch connector
   h. Speed sensor connector
46. Connect the transmission control cable.
47. Install the dust seal guard.
48. Connect the throttle guard connection.
49. Install the fluid filler pipes.
50. Install the front and rear driveshafts.
51. Fill the transmission and the transfer case with the proper type and volume of fluid.
52. Install the front exhaust pipe.
53. Install the transfer case protector.
54. Install the transfer control lever, lever boot and boot retainer.
55. Install the transmission selector lever assembly and connect the transmission control cable.
56. Connect the shift-lock cable connection.
57. Connect the key-interlock cable connection.
58. Install the front console assembly.
59. Connect the floor console harness connector.
60. Install the transfer shift lever knob.
61. Install the rear console assembly and the side panel.
62. Connect the rear console harness connector.
63. Install the suspension control switch or cover.
64. Install the switch panel.
65. Connect the negative battery cable.

3.0L (VIN H) 24 Valve and 3.5L (VIN M) Engines

1. Disconnect the negative battery cable.
2. Remove the switch panel.
3. Remove the suspension control switch or cover.
4. Disconnect the rear console harness connector.
5. Remove side panel and the rear console assembly.
6. Remove the transfer shift lever knob.
7. Disconnect the floor console harness connector.
8. Set the A/T selector lever to L and remove the front console assembly.
9. Disconnect the key-interlock cable connection (selector lever assembly side).
10. Disconnect the shift-lock cable connection (selector lever assembly side).

11. Disconnect the transmission control cable connection (selector lever assembly side).
12. Remove the transmission selector lever assembly.
13. Remove the transfer control lever boot retainer and the boot.
14. Set the transfer control lever to 2H. Remove the transfer control lever mounting bolts and remove the assembly.
15. Remove the transfer case protector.
16. Remove the front exhaust pipe.
17. Drain the transmission fluid and the transfer case fluid.
18. Remove the front and rear driveshafts.
19. Remove the fluid filler pipe.
20. Disconnect the throttle cable connection.
21. Remove the dust seal guard.
22. Disconnect the transmission control cable connection.
23. Disconnect:
   a. Speed sensor connector
   b. High range / Low range detection switch connector
   c. Low range operation detection switch
   d. 4WD operation detection switch connector
   e. Center differential lock operation detection switch connector
   f. Center differential lock detection switch connector
   g. 2WD/4WD detection switch connector
   h. Park/Neutral position switch connector
   i. Solenoid valve connector
   j. Fluid cooler pipe connection
24. Remove the starter motor and cover.
25. Remove the heat protector.
26. Remove the No. 2 crossmember.
27. Remove the engine rear mount bracket.
28. Remove the mass damper.
29. Remove the torque converter connecting bolt.
30. Using a jack stand, support the transmission and remove the mounting bolts. Remove the transmission assembly.

To install:
31. Using a jack stand for support, mount the transmission to the engine and install the mounting bolts.
32. Install the torque converter connecting bolt.
33. Install the mass damper.
34. Install the engine rear mount bracket.
35. Install the No. 2 crossmember.
36. Install the heat protector.
37. Install the starter motor cover and the starter.

38. Connect:
   a. Fluid cooler pipe connection
   b. Solenoid valve connector
   c. Park/Neutral position switch connector
   d. 2WD/4WD detection switch connector
   e. Center differential lock detection switch connector
   f. Center differential lock operation detection switch connector
   g. 4WD operation detection switch connector
   h. Low range operation detection switch
   i. High range / Low range detection switch connector
   j. Speed sensor connector

39. Connect the transmission control cable.
40. Install the dust seal guard.
41. Connect the throttle cable connection.
42. Install the fluid filler pipes.
43. Install the front and rear driveshafts.
44. Fill the transmission and the transfer case with the proper type and volume of fluid.
45. Install the front exhaust pipe.
46. Install the transfer case protector.
47. Install the transfer control lever, lever boot and boot retainer.
48. Install the transmission selector lever assembly and connect the transmission control cable.
49. Connect the shift-lock cable connection.
50. Connect the key-interlock cable connection.
51. Install the front console assembly.
52. Connect the floor console harness connector.
53. Install the transfer shift lever knob.
54. Install the rear console assembly and the side panel.
55. Connect the rear console harness connector.
56. Install the suspension control switch or cover.
57. Install the switch panel.
58. Connect the negative battery cable.

Mighty Max

1. Disconnect the negative battery cable.
2. If equipped with 4WD, disconnect the 4WD indicator light switch connector and the ground cable at the transfer case.

3. Remove the following components from the 4WD floor control console:
   a. Remove the shifter knob from the 4WD control.
   b. Remove the top trim panel around the outside of the boot, retaining screws and slide the boot(s) up and off of the shaft.
   c. Remove the center storage compartment and the rear cover plate. Remove the retaining screws the console from the vehicle.
4. Disconnect the pulse generator connector.
5. Raise and safely support the vehicle.
6. Remove the skid plate, if equipped. Drain the transmission and transfer case, as equipped.
7. Matchmark and remove the driveshaft(s).
8. Disconnect the speedometer cable from the transmission or transfer case.
9. Disconnect the shifter linkage or cable.
10. Unplug all transmission electrical connectors.
11. Remove the exhaust pipe from the vehicle. Remove the exhaust bracket from the transmission case.
12. Remove the filler neck and dipstick.
13. Remove the torque converter inspection plate. Matchmark the flexplate to the torque converter and remove the torque converter bolts.
14. Remove the starter assembly.
15. Disconnect and plug the oil cooler lines.
16. Using a transmission jack, support the transmission. Remove the retaining bolts at the crossmember.
17. Remove the rear crossmember.
18. Lower the transmission down slightly and unbolt the 4WD shifter from the transfer case, if equipped.
19. Remove the bell housing bolts and mounting brackets.
20. Pull the transmission assembly rearward to clear the aligning dowels and remove from the vehicle.

To install:
21. Install the transmission assembly to the engine using dowels as guides. Install the bell housing bolts and torque to 35 ft. lbs. (47 Nm).
22. Install the 4WD shifter, if equipped. Raise the assembly up into position and install the rear crossmember and mounting hardware. Torque the crossmember to frame bolts to 36 ft. lbs. (50 Nm).
23. Align the flexplate to torque converter. Apply Loctite to the threads and install the torque converter bolts. Torque the bolts to 25 ft.

lbs. (34 Nm). Install the inspection plate.
24. Install the filler tube with a new O-ring and the dipstick.
25. Install the exhaust and the support brackets.
26. Connect all switch connectors that were removed.
27. Connect the shifter linkage or cable and the throttle cable.
28. Align and install the driveshaft(s).
29. Install the ground wire and the 4WD indicator light switch connector to the transfer case, if equipped.
30. Fill the transfer case with hypoid gear oil with an API classification of GL-4 or higher.
31. Connect the throttle cable.
32. Install the skid plate, if equipped. Lower the vehicle.
33. Install the transfer case shifter boot and console, if equipped.
34. Fill the transmission with the proper amount of Dexron II.
35. Connect the negative battery cable, start the engine and run through all gears. Add fluid until the transmission is properly filled.
36. Check the operation of the neutral safety switch and the reverse lights.
37. Road test the vehicle and check for leaks.

Expo with F4A23 and W4A32 Automatic Transaxles

NOTE: On both Front Wheel Drive (FWD) and All Wheel Drive (AWD) vehicles, the transaxle and converter must be removed and installed as an assembly.

FWD Vehicles

1. Disconnect negative battery cable.
2. Remove the air cleaner assembly.
3. Disconnect the transaxle control lever. Disconnect and plug the oil cooler lines.
4. Disconnect the pulse generator connector, oil temperature connector, kickdown servo switch connector, inhibitor switch connector and solenoid valve connection.
5. Disconnect the speedometer cable connection. Remove the oil level dipstick and tube.
6. Install holding fixture to the top of the engine to support engine weight.
7. Remove the top transaxle upper coupling bolts.
8. Raise and safely support the vehicle.
9. Remove the starter motor leaving wire harness attached.

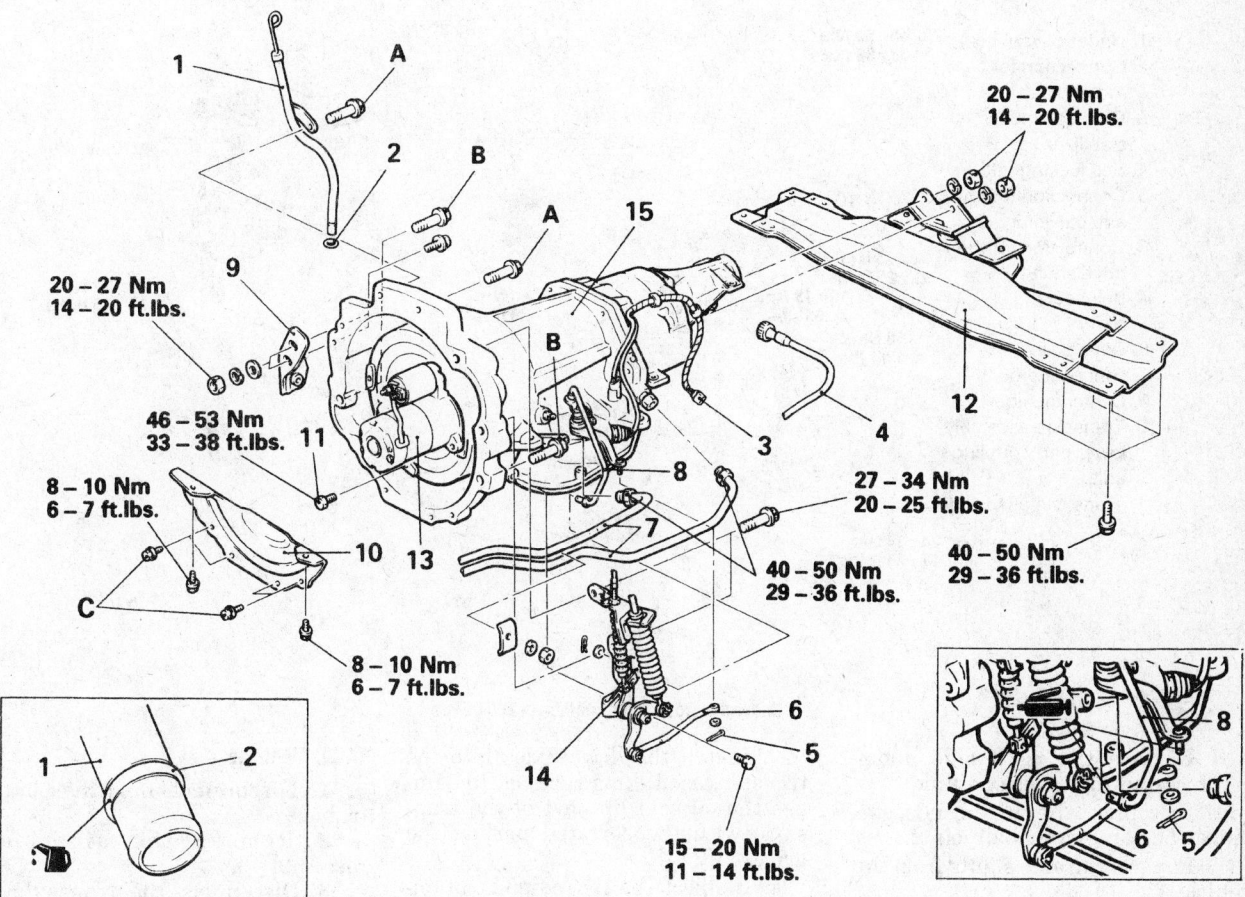

Removal steps

1. Oil filler tube
2. O-ring
3. Transmission harness connector
4. Speedometer cable
5. Cotter pin
6. Transmission control rod (Transmission side)
7. Automatic transmission cooler tube
8. Transmission throttle lever (Bell crank bracket side)
9. Exhaust pipe mounting bracket
10. Bell housing cover
11. Special bolt
12. No.2 crossmember
13. Starter motor
14. Bell crank bracket assembly
15. Transmission assembly

	Nm	ft.lbs.	O.D.×Length mm (in.)	Bolt identification
A	43 – 55	31 – 40	⑦ 10×50 (.4×2.0)	⑦ D×L
B	43 – 55	31 – 40	⑦ 10×70 (.4×2.8)	
C	30 – 42	21 – 30	⑦ 10×16 (.4×.6)	

328948

Automatic transmission assembly and related components — Mighty Max

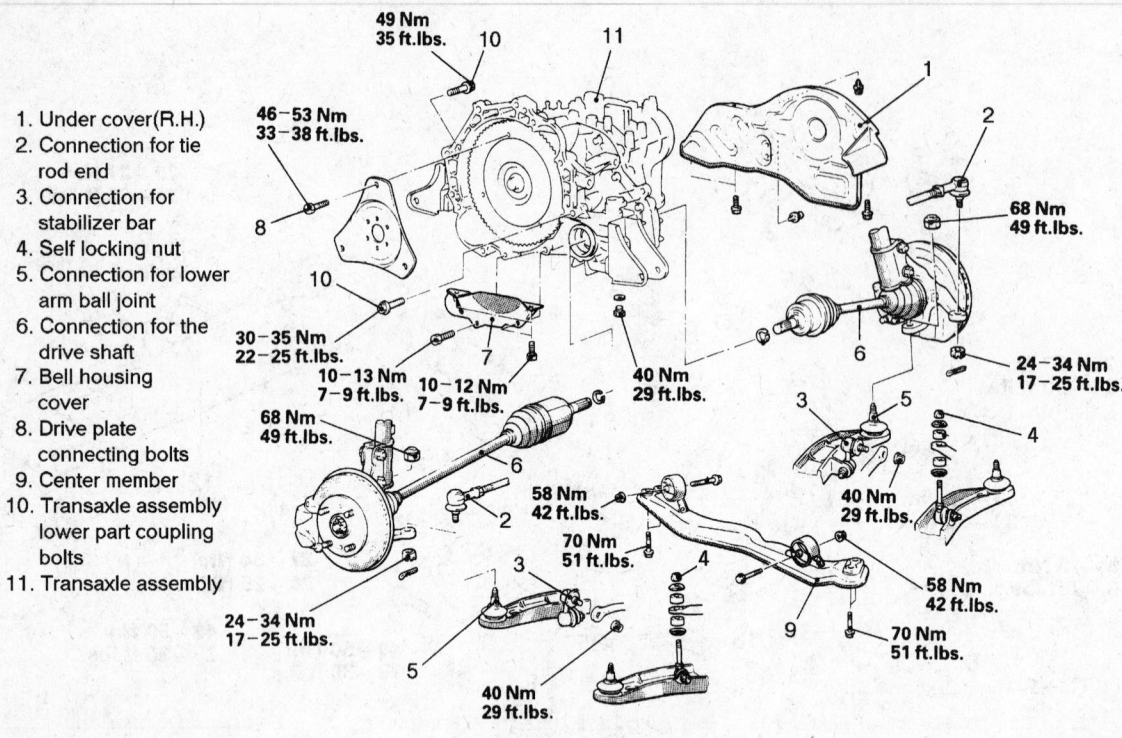

1. Under cover(R.H.)
2. Connection for tie rod end
3. Connection for stabilizer bar
4. Self locking nut
5. Connection for lower arm ball joint
6. Connection for the drive shaft
7. Bell housing cover
8. Drive plate connecting bolts
9. Center member
10. Transaxle assembly lower part coupling bolts
11. Transaxle assembly

Underside connection points — 2WD Expo

10. Remove the right side under cover. Drain the transaxle fluid.

11. Disconnect the tie rod ends, stabilizer bar and lower ball joints.

12. Remove the axle shafts from the vehicle.

13. Remove the lower bellhousing cover. Scribe a mark on the driveplate and transaxle converter face using chalk. Remove the driveplate connecting bolts while turning the crankshaft.

14. Support the transaxle using a transmission jack. Remove the center support.

15. Remove the transaxle mount bolt and bracket.

16. Remove the lower transaxle case coupling bolts, press the torque converter towards the transfer case to prevent separation during removal and lower the transfer case from the vehicle.

To install:

17. Install the transaxle into the vehicle and secure using the lower case coupling bolts.

18. Install the transaxle mount bolt and bracket, torque through bolt nut to 51 ft. lbs. (70 Nm).

19. Align the scribe marks on the converter and the driveplate. Install the driveplate connecting bolts torquing to 33–38 ft. lbs. (46–53 Nm).

20. Install the drive axles into the transfer case taking care not to damage the oil seal lip part of the transaxle with the serrated part of the driveshaft.

21. Connect the tie rod ends, stabilizer bar and lower ball joints.

22. Install the right side under cover.

23. Lower the vehicle. Install the upper transaxle coupling bolts.

24. Connect the speedometer cable, and the electrical harness connectors disconnected during the removal procedure.

25. Install the starter motor torquing the retainer bolts to 35 ft. lbs. (49 Nm).

26. Connect the transaxle cooler hoses and the connections for the manual controls.

27. Install the air cleaner assembly and the oil level dipstick and tube.

28. Refill with Dexron II, Mopar ATF Plus type 7176, or equivalent automatic transaxle fluid.

29. Start the engine and allow to idle for 2 minutes. Apply parking brake and move selector through each gear position, ending in **N**. Recheck fluid level and add if necessary. Fluid level should be between the marks in the **HOT** range. Check operation of all gauges and meters.

AWD Vehicles

1. Disconnect negative battery cable.

2. Remove the air cleaner assembly.

3. Disconnect the transaxle control lever. Disconnect and plug the oil cooler lines.

4. Disconnect the pulse generator connector, oil temperature connector, kickdown servo switch connector, inhibitor switch connector and solenoid valve connection.

5. Disconnect the speedometer cable connection. Remove the oil level dipstick and tube.

6. Install holding fixture to the top of the engine to support engine weight.

7. Remove the top transaxle upper coupling bolts.

8. Raise and safely support the vehicle.

9. Remove the starter motor leaving wire harness attached.

10. Remove the right side under cover. Drain the transaxle fluid.

11. Disconnect the tie rod ends, stabilizer bar and lower ball joints.

12. Remove the axle shafts from the vehicle.

13. Remove the driveshaft from the transfer case, insert a prybar between the driveshaft and the transaxle case and pry the shaft from the

1. Under cover (R.H.)
2. Connection for tie rod end
3. Connection for stabilizer bar
4. Connection for lower arm ball joint
5. Drive shaft nut (R.H.)
6. Drive shaft (R.H.)
7. Connection for drive shaft and inner shaft
8. Front exhaust pipe
9. Transfer assembly
10. Bell housing cover
11. Drive plate connecting bolts
12. Center member
13. Transaxle assy lower part coupling bolts
14. Transaxle assembly

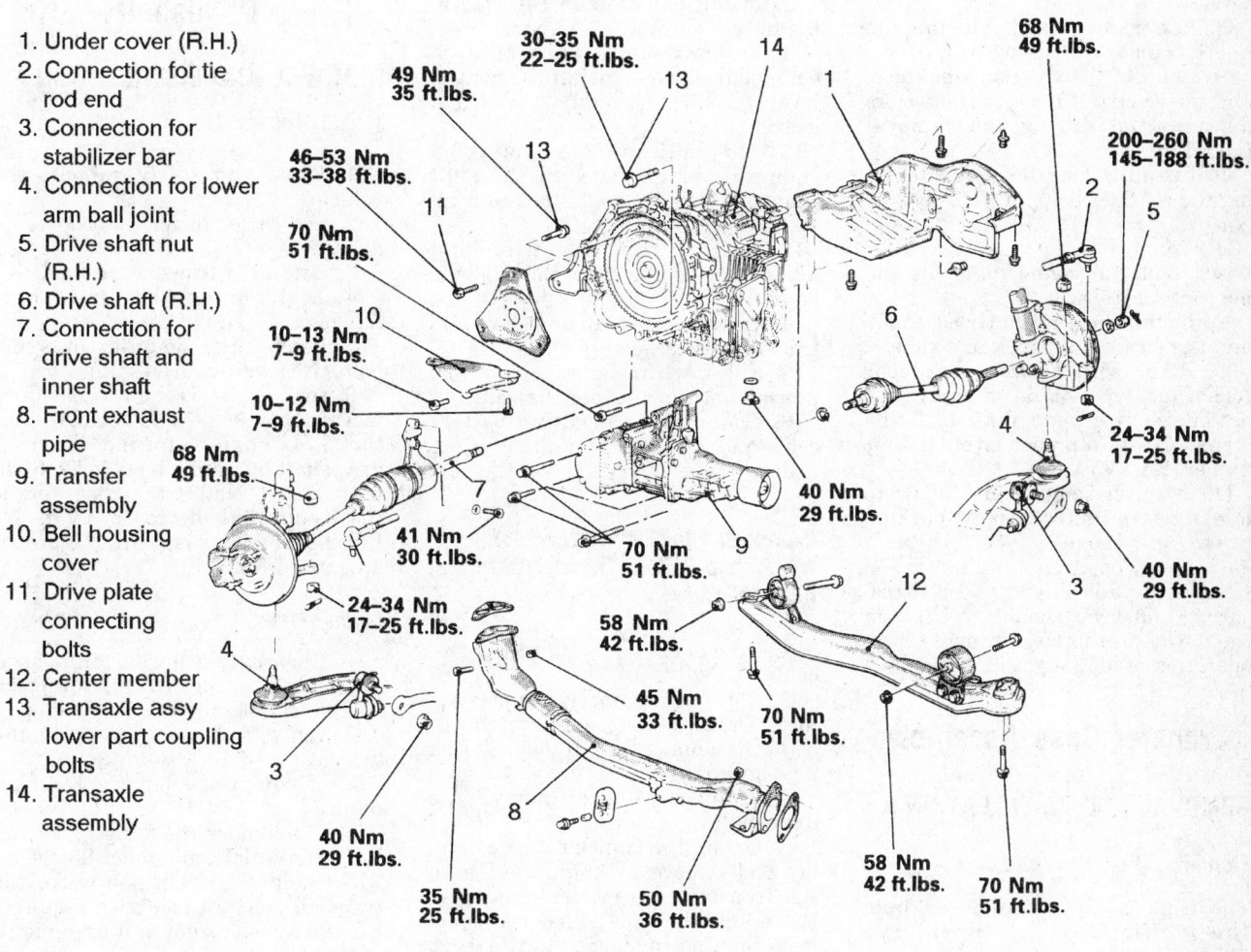

30–35 Nm
22–25 ft.lbs.

49 Nm
35 ft.lbs.

46–53 Nm
33–38 ft.lbs.

70 Nm
51 ft.lbs.

10–13 Nm
7–9 ft.lbs.

10–12 Nm
7–9 ft.lbs.

68 Nm
49 ft.lbs.

41 Nm
30 ft.lbs.

24–34 Nm
17–25 ft.lbs.

40 Nm
29 ft.lbs.

35 Nm
25 ft.lbs.

50 Nm
36 ft.lbs.

58 Nm
42 ft.lbs.

45 Nm
33 ft.lbs.

70 Nm
51 ft.lbs.

70 Nm
51 ft.lbs.

58 Nm
42 ft.lbs.

70 Nm
51 ft.lbs.

40 Nm
29 ft.lbs.

68 Nm
49 ft.lbs.

200–260 Nm
145–188 ft.lbs.

24–34 Nm
17–25 ft.lbs.

40 Nm
29 ft.lbs.

239251

Underside connection points — 4WD Expo

transaxle housing. Swing the shafts out of the way keeping the joints straight, and suspend using wire. Turn the right shaft 90 degrees toward the front of the vehicle so it will be out of the way.

NOTE: Do not pull on the shaft during removal from the transaxle. This will damage the inboard joint. Do not insert the prybar so deep as to damage the oil seal.

14. Remove the lower bellhousing cover. Scribe a mark on the driveplate and transaxle converter face using chalk. Remove the

driveplate connecting bolts while turning the crankshaft.

15. Support the transaxle using a transmission jack. Remove the center support.

16. Remove the transaxle mount bolt and bracket.

17. Disconnect the front exhaust pipe.

18. Drain and remove the transfer assembly.

19. Remove the lower transaxle case coupling bolts, press the torque converter towards the transfer case to prevent separation during removal and lower the transfer case from the vehicle.

To install:

20. Install the transaxle into the vehicle and secure using the lower case coupling bolts.

21. Install the transaxle mount bolt and bracket, torque through bolt nut to 51 ft. lbs. (70 Nm).

22. Align the scribe marks on the converter and the driveplate. Install the driveplate connecting bolts torquing to 33–38 ft. lbs. (46–53 Nm).

23. Install the transfer assembly and the center crossmember. Remove the transmission jack.

24. Install the center exhaust pipe.

25. Install the drive axles into the transfer case taking care not to damage the oil seal lip part of the tran-

saxle with the serrated part of the driveshaft.

26. Connect the tie rod ends, stabilizer bar and lower ball joints.

27. Install the right side under cover.

28. Lower the vehicle. Install the upper transaxle coupling bolts.

29. Connect the speedometer cable, and the electrical harness connectors disconnected during the removal procedure.

30. Install the starter motor torquing the retainer bolts to 35 ft. lbs. (49 Nm).

31. Connect the transaxle cooler hoses and the connections for the manual controls.

32. Install the air cleaner assembly and the oil level dipstick and tube.

33. Refill with Dexron II, Mopar ATF Plus type 7176, or equivalent automatic transaxle fluid. Fill the transfer case to proper level GL-4 or higher, SAE 75W-90W.

34. Start the engine and allow to idle for 2 minutes. Apply parking brake and move selector through each gear position, ending in **N**. Recheck fluid level and add if necessary. Fluid level should be between the marks in the **HOT** range. Check operation of all gauges and meters.

Transfer Case Assembly

REMOVAL AND INSTALLATION

Montero

The transfer case is removed from the vehicle along with the transmission.

Mighty Max

1. Disconnect the negative battery cable.

2. Raise and safely support the vehicle.

3. Remove the transmission and transfer case as an assembly.

4. Remove the plug from right side of the transfer case, under the control housing.

5. Remove the select spring and plunger from from the housing bore.

6. On automatic transmission equipped vehicles, remove the control lever housing assembly, cover and gasket.

7. On manual transmission equipped vehicles, remove the spring pin that retains the shift changer to the control shaft using a pin punch.

8. Remove the transfer case to transmission retaining nuts and sep-

arate the transfer case from the transmission.

To install:

9. Align and seat transfer case on transmission.

10. Install and tighten the attaching nuts to 30 ft. lbs. (42 Nm).

11. On automatic transmission equipped vehicles, install the control lever housing assembly, cover and gasket.

12. On manual transmission equipped vehicles, install the shift changer to the control shaft and install a new roll pin.

13. Install the select spring, plunger and plug into the housing bore.

14. Install the transmission and transfer case assembly.

15. Fill the transfer case with the proper lubricant. Lower the vehicle.

16. Connect the negative battery cable and check the operation of the transfer case.

Expo with F4A23, F5M22, F5M31, W4A32 and W5M33 Transfer Case Assemblies

1. Disconnect the battery negative cable.

2. Raise and properly support vehicle. Drain the transfer assembly.

3. Disconnect the front exhaust pipe and hanger.

4. Matchmark and remove the driveshaft.

5. Unbolt the transfer case assembly and remove by sliding out from the transaxle. Cover the opening in the transaxle and transfer case to keep oil from dripping and to keep dirt out.

To install:

6. Install the transfer case assembly to the transaxle. Tighten the transfer case to transaxle bolts to proper specification.

7. Lubricate the driveshaft sleeve yoke and oil seal lip on the transfer extension housing. Install the drive shaft.

NOTE: Use care when installing the rear propeller shaft to the transfer case, not to damage the output shaft seal.

8. Connect the exhaust pipe, using a new gasket.

9. Refill the transfer case with gear oil of correct classification. Check fluid level in transaxle and add as required.

10. Lower the vehicle and connect the negative battery cable.

DRIVE AXLE

Driveshaft

REMOVAL AND INSTALLATION

Montero

1. Raise and safely support the vehicle.

2. Set the transfer shift lever to 2H.

3. Drain the transfer case oil.

4. Matchmark the driveshaft flanges.

5. Remove the mounting nuts and bolts, remove the driveshaft.

6. Installation is the reverse of removal. On 1994–1996 vehicles with the 3.5L engine, torque the rear driveshaft mounting bolts to 72–80 ft. lbs. (98–108 Nm). Otherwise, torque the mounting bolts to 36–43 ft. lbs. Fill the transfer case with the proper amount of fluid.

Mighty Max

1. Ensure the vehicle is in neutral and front hubs are free, if equipped. Disconnect negative battery cable.

2. Raise and safely support the vehicle.

3. If equipped, Remove the skid plate.

4. Matchmark the flange yoke and the differential companion flange.

5. Position a drain pan under the transmission tail shaft, as required. Fluid may leak from unit on removal of the driveshaft.

6. If equipped with a 2 piece driveshaft, remove the rear shaft, then remove the center bearing retainer.

7. Remove the retaining bolts or nuts from the flange(s) and remove the shaft from the vehicle.

To install:

8. If equipped with a 2 piece driveshaft, install the front shaft, then install the center bearing retainer and torque to 22–29 ft. lbs. (30–40 Nm).

9. Align the mating marks and install the driveshaft into the transfer case/transmission and the companion flange. Torque the bolts to 36–43 ft. lbs. (50–60 Nm).

10. Install the mounting nuts or bolts.

11. Install the skid plate, if removed.

12. Lower the vehicle and connect the negative battery cable.

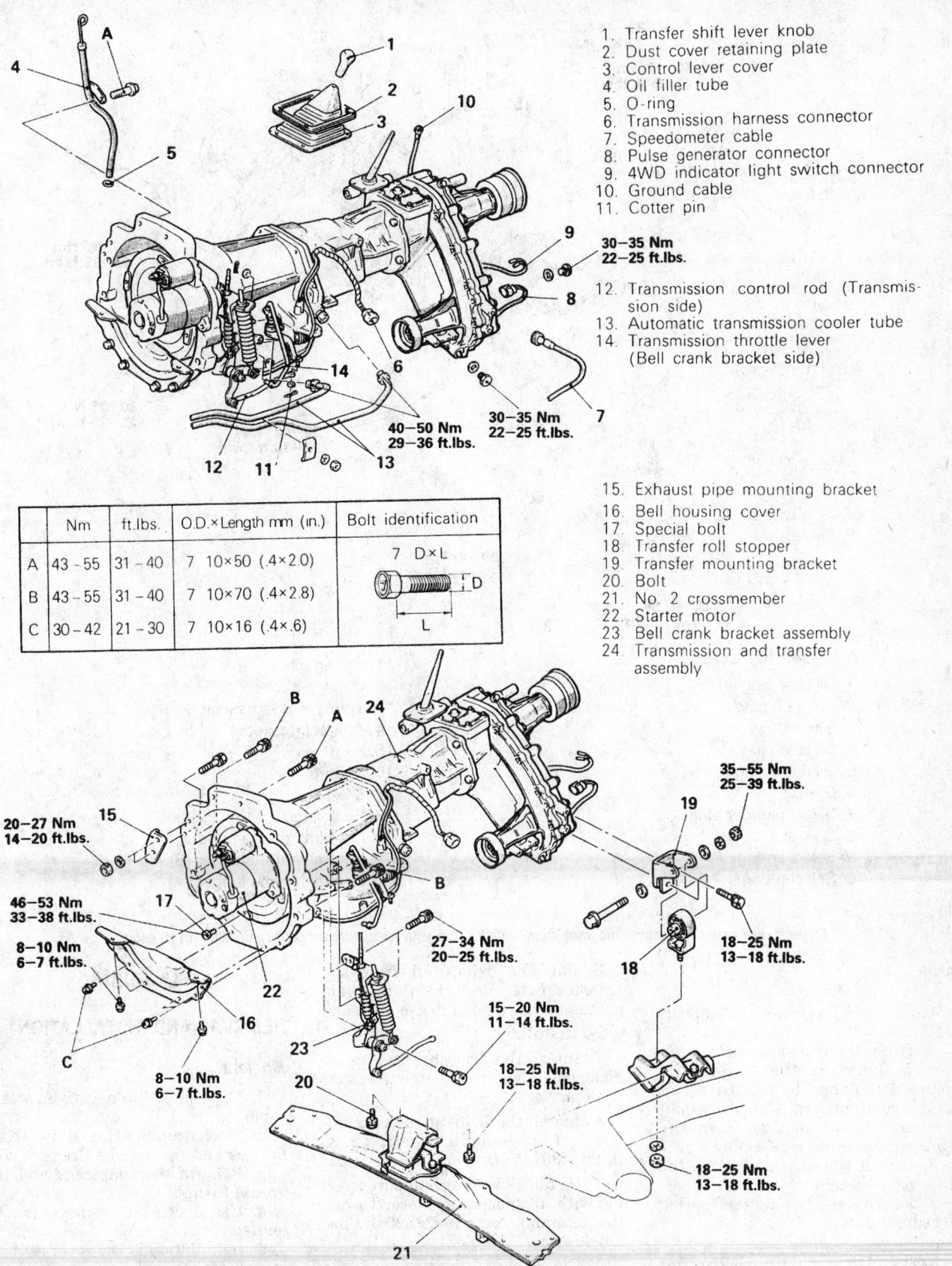

1. Transfer shift lever knob
2. Dust cover retaining plate
3. Control lever cover
4. Oil filler tube
5. O-ring
6. Transmission harness connector
7. Speedometer cable
8. Pulse generator connector
9. 4WD indicator light switch connector
10. Ground cable
11. Cotter pin

30–35 Nm
22–25 ft.lbs.

12. Transmission control rod (Transmission side)
13. Automatic transmission cooler tube
14. Transmission throttle lever (Bell crank bracket side)

30–35 Nm
22–25 ft.lbs.

40–50 Nm
29–36 ft.lbs.

15. Exhaust pipe mounting bracket
16. Bell housing cover
17. Special bolt
18. Transfer roll stopper
19. Transfer mounting bracket
20. Bolt
21. No. 2 crossmember
22. Starter motor
23. Bell crank bracket assembly
24. Transmission and transfer assembly

	Nm	ft.lbs.	O.D.×Length mm (in.)	Bolt identification
A	43 – 55	31 – 40	7 10×50 (.4×2.0)	7 D×L
B	43 – 55	31 – 40	7 10×70 (.4×2.8)	
C	30 – 42	21 – 30	7 10×16 (.4×.6)	

35–55 Nm
25–39 ft.lbs.

20–27 Nm
14–20 ft.lbs.

46–53 Nm
33–38 ft.lbs.

8–10 Nm
6–7 ft.lbs.

27–34 Nm
20–25 ft.lbs.

18–25 Nm
13–18 ft.lbs.

15–20 Nm
11–14 ft.lbs.

8–10 Nm
6–7 ft.lbs.

18–25 Nm
13–18 ft.lbs.

18–25 Nm
13–18 ft.lbs.

Transfer case assembly removal and installation — Mighty Max with automatic transmissions

327545

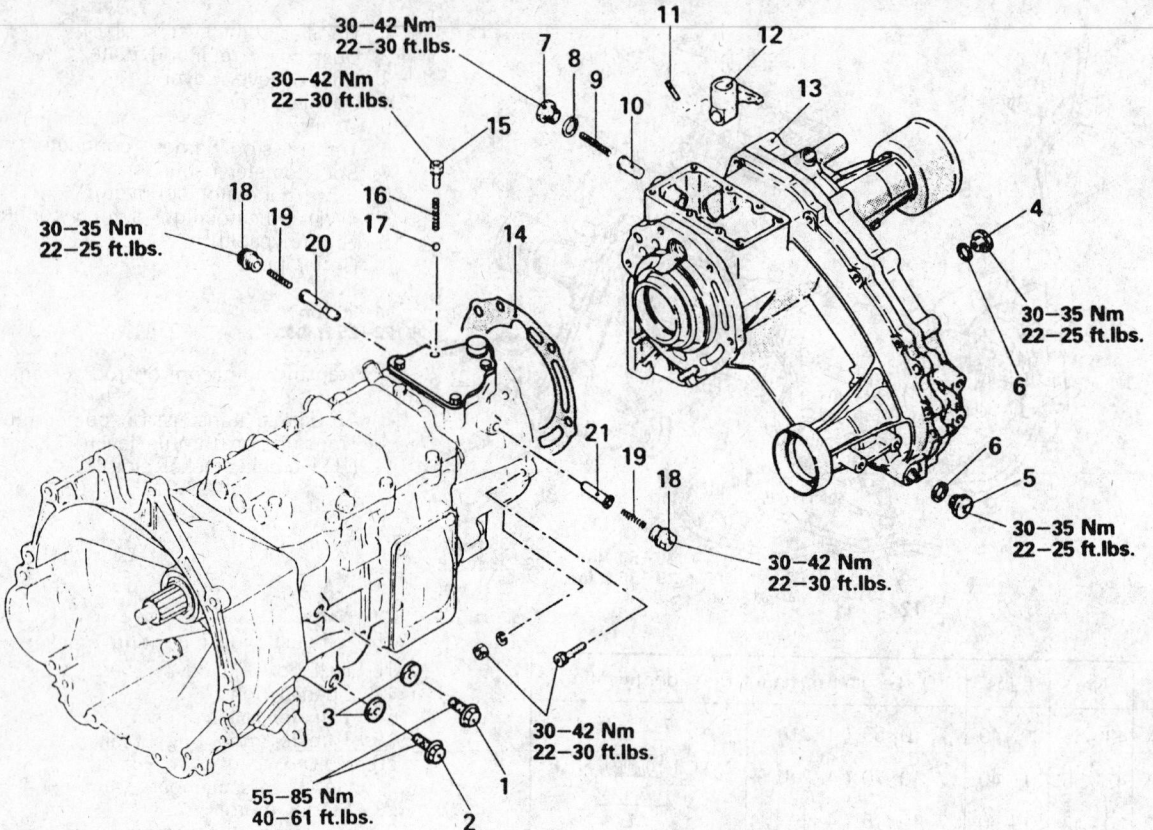

30—42 Nm
22—30 ft.lbs.

30—42 Nm
22—30 ft.lbs.

30—35 Nm
22—25 ft.lbs.

30—35 Nm
22—25 ft.lbs.

30—35 Nm
22—25 ft.lbs.

30—42 Nm
22—30 ft.lbs.

30—42 Nm
22—30 ft.lbs.

55—85 Nm
40—61 ft.lbs.

327547

Disassembly steps

1. Oil filler plug
2. Oil drain plug
3. Gasket
4. Oil filler plug
5. Oil drain plug
6. Gasket
7. Select plunger plug
8. Gasket
9. Select spring
10. Select plunger
11. Spring pin
12. Change shifter
13. Transfer case assembly
14. Adapter gasket
15. Plug
16. Spring
17. Steel ball
18. Seal plug
19. Neautral return spring
20. Neutral return plunger (B)
21. Neutral return plunger (A)

Transfer case assembly removal and installation — Mighty Max with manual transmission and 3.0L (VIN H) engine

Expo

1. Raise the vehicle and support safely.

2. Drain the transfer case.

3. Matchmark the differential companion flange to the driveshaft flange yoke, and the two driveshaft sections to one another to ensure proper phasing on reassembly.

4. Unbolt the driveshaft from the differential flange.

5. Remove the two center bearing attaching nuts.

NOTE: Do not confuse the flat washer and the adjusting spacer. Keep them separate for assembly.

6. Pull the driveshaft from the transfer case. Be careful to avoid damaging the transfer case oil seal.

To install:

7. Install the driveshaft to the vehicle and align the matchmarks at the rear yoke.

8. Install the bolts at the rear differential flange and torque to 22–25 ft. lbs. (30–35 Nm).

9. Install the center support bearing with all spacers in place. Torque the retaining nuts to 22–29 ft. lbs. (30-40 Nm).

10. Check the fluid levels in transfer case and rear differential case.

U-Joints

REMOVAL AND INSTALLATION

Montero

1. Disconnect the negative battery cable.

2. Matchmark the driveshaft flanges and remove the driveshaft.

3. Remove the snaprings and the grease fittings.

4. Use a press to remove the U-joints.

5. Installation is the reverse of removal. Measure the clearance of the snapring grooves with a thickness gauge. If the clearance exceeds the

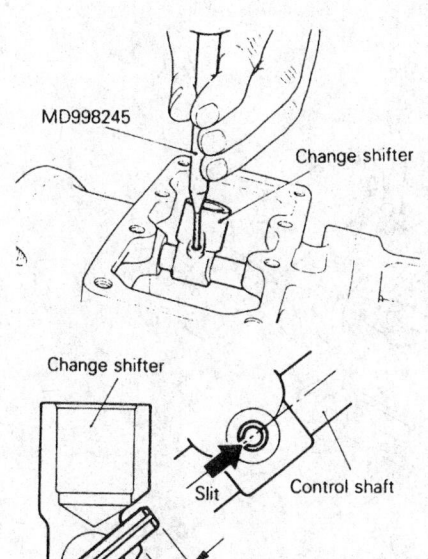

MD998245

Change shifter

Change shifter

Slit

Control shaft

3 – 3.5 mm
(.12 – .13 in.)

327548

Spring pin removal and installation — Mighty Max

standard value, adjust by changing the thickness of the snapring. Standard value: 0.0024 in. (0.06 mm) or less.

Mighty Max

Single Cardan Universal Joint

1. Matchmark the yokes before disassembling so they will be installed in their original locations to retain driveshaft balance.
2. Remove the driveshaft and slip yoke, if equipped, from the vehicle.

NOTE: Do not clamp the driveshaft tube in a vise. Clamp only the forged portion of the welded yoke or the slip yoke in a vise. Do not overtighten the vise jaws.

3. Clamp the yoke in a vise and remove the bearing cap retainers.
4. If equipped with a grease fitting, remove the fitting.
5. Place a socket which has an inside diameter larger than the outside diameter of the bearing cap, against the yoke around the perimeter of the first cap to be removed. Place a socket which is slightly smaller than the cap, on the cap opposite the cap to

be removed. Then position the yoke in a vise.
6. Compress the jaws until the smaller socket has driven the other cap into the larger socket.
7. Release the jaws and remove the cap that is partially out of the yoke.
8. Repeat the procedure for the remaining cap(s).
To install:
9. Clean and remove any rust from the yoke bores and lubricate lightly with suitable lithium based grease.
10. Position the spider cylinders in the yoke bores. Insert the seals into the yoke bores and against the spider cylinders. Tap the bearing caps into the yoke bores far enough to keep the spider in place.
11. Place the socket that is slightly smaller than the cap against the first cap and position the assembly in a vise.
12. Compress the jaws to force the bearing caps into the yoke bores far enough so the retainer grooves are visible.
13. Repeat the procedure for the remaining caps, if necessary.
14. Install the grease fitting, if removed.
15. Install the retaining clips.
16. Install the driveshaft assembly to the vehicle.

Double Cardan Universal Joint

NOTE: The Double Cardan U-joint is not serviceable and must be replaced as a unit.

1. Remove the front driveshaft from the vehicle.
2. Matchmark the yokes before disassembling so they will be installed in their original locations to retain driveshaft balance.
3. Remove the external snaprings.
4. Press the bearing assembly out of the flange yoke far enough to grasp the bearing with vise jaws.
5. Grasp the bearing in the vise jaw and tap the flange yoke with a plastic hammer to remove the bearing.

——— **WARNING** ———
Do not strike the bearing holes as this may damage the snapring groves.

6. Repeat steps 4 and 5 to remove the opposite side bearing.
7. Press the bearing assembly out of the link yoke and remove the bearings.

8. Remove the flange yoke from the link yoke. Press the remaining bearings out of the link yoke and driveshaft yoke.
To install:
9. Fit the cross into the driveshaft yoke.
10. Press the bearings into the driveshaft yoke and install the snaprings.

NOTE: Keep the needle bearings upright in the bearing assembly.

11. Press the bearings into the link yoke at the driveshaft end and install the snaprings.
12. Install the spring in the centering ball. Install the cross in the link yoke and flange yoke.
13. Press the bearings into the link yoke and flange yoke and install the snaprings.
14. Make sure all the snaprings are firmly installed in the grooves.
15. Install the driveshaft. Lower the vehicle and check for proper operation.

Expo

1. Raise and properly support vehicle.

NOTE: Matchmark the rear yoke to shaft and/or the center yoke to yoke for proper installation reference.

2. Remove driveshaft and secure in vise.
3. Remove the snaprings which retain the bearing caps in the slip yoke and the driveshaft.
4. Use a large punch or an arbor press and drive one of the bearing caps in toward the center of the universal joint. The joint will be forced through the opposite side of the yoke.
5. As the opposite side bearing cap is forced from the yoke, grip it with a pair of pliers and pull it, in a twisting motion, out of the yoke.
6. Press the spider cross toward the side you just pushed to force the cap back into the yoke. When the bearing cap starts to clear the yoke, pull it free with a pair of pliers. Repeat the procedure with the other side bearing caps.
7. After removing the bearing caps, lift the bearing (spider) cross from the yoke. Thoroughly clean all dirt and foreign matter from the yoke area on both ends of the driveshaft.

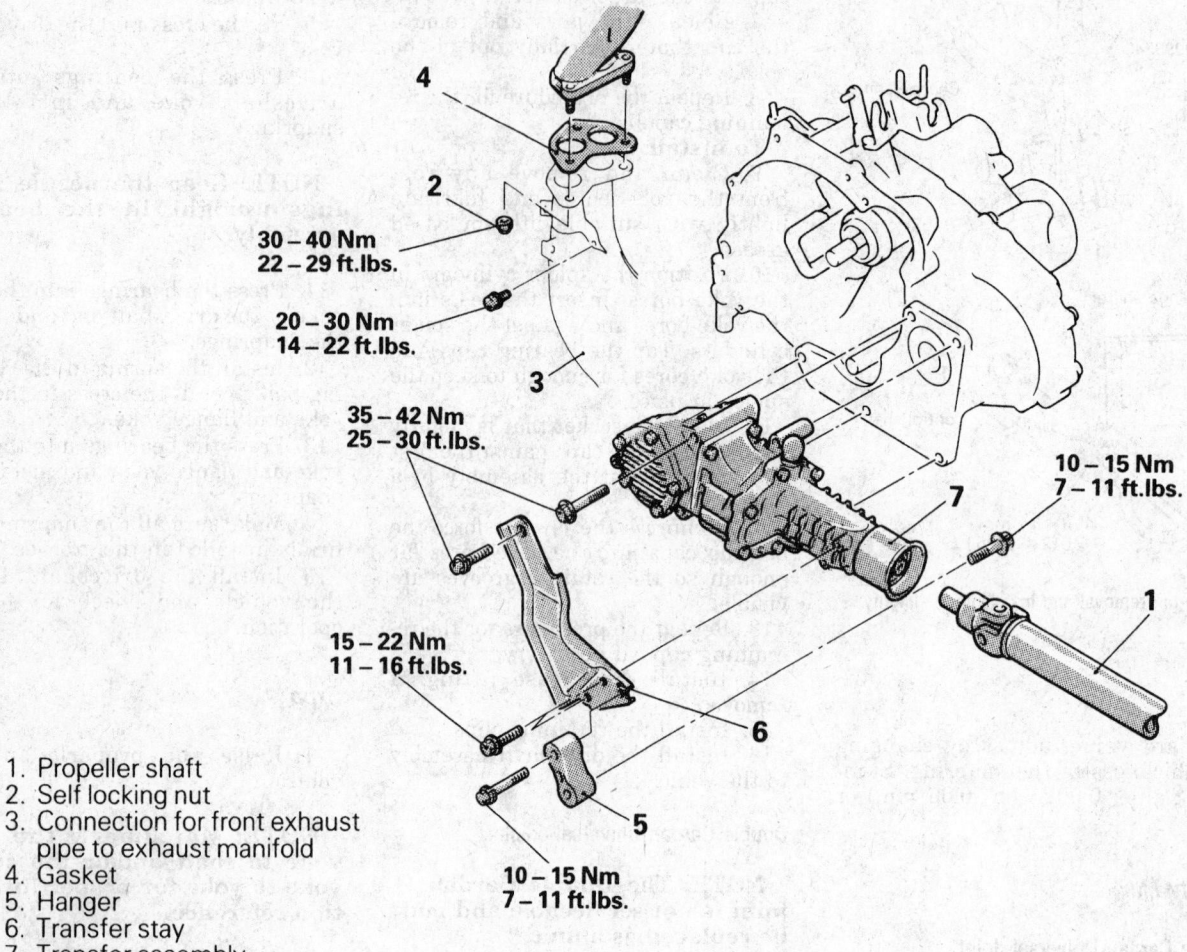

30 – 40 Nm
22 – 29 ft.lbs.

20 – 30 Nm
14 – 22 ft.lbs.

35 – 42 Nm
25 – 30 ft.lbs.

10 – 15 Nm
7 – 11 ft.lbs.

15 – 22 Nm
11 – 16 ft.lbs.

10 – 15 Nm
7 – 11 ft.lbs.

1. Propeller shaft
2. Self locking nut
3. Connection for front exhaust
   pipe to exhaust manifold
4. Gasket
5. Hanger
6. Transfer stay
7. Transfer assembly

241578

Transfer case and related components — Expo with F4A23, F5M22, F5M31, W4A32 and W5M33 transfer case assemblies

To install:

NOTE: When installing new bearing caps within the yokes, it is advisable to use an arbor press. However, if a press is not available, the bearings should be driven into position with extreme care. A heavy jolt on the needle bearing, in the cap, can easily damage or misalign them. A large vise and correct size drivers and spacers can sometimes be used, in place of a punch, to push the bearing caps in or out.

8. Start a bearing cap into the yoke bore.
9. Position the spider into the yoke and into the bearing cap. Push the cap the rest of the way into the yoke bore until it is about 6mm below the outside surface of the yoke. Install a new snapring.
10. Start a bearing cap into the yoke on the opposite side of the one just installed. Carefully press it into the yoke while aligning the spider cross with the bearing center.

11. Continue to pry the cap in until the opposite side bearing cap contacts the snapring. Install a new snapring on the side just installed. Using a feeler gauge, check the clearance between the bearing cap face and the snapring. Clearance should be between 0.0008 — 0.0024 in. (0.02 — 0.06 mm), if not within specification, install appropriate size snapring. Once proper clearance is obtained, check the joint for free movement. Complete the installation of the yoke and bearing caps.

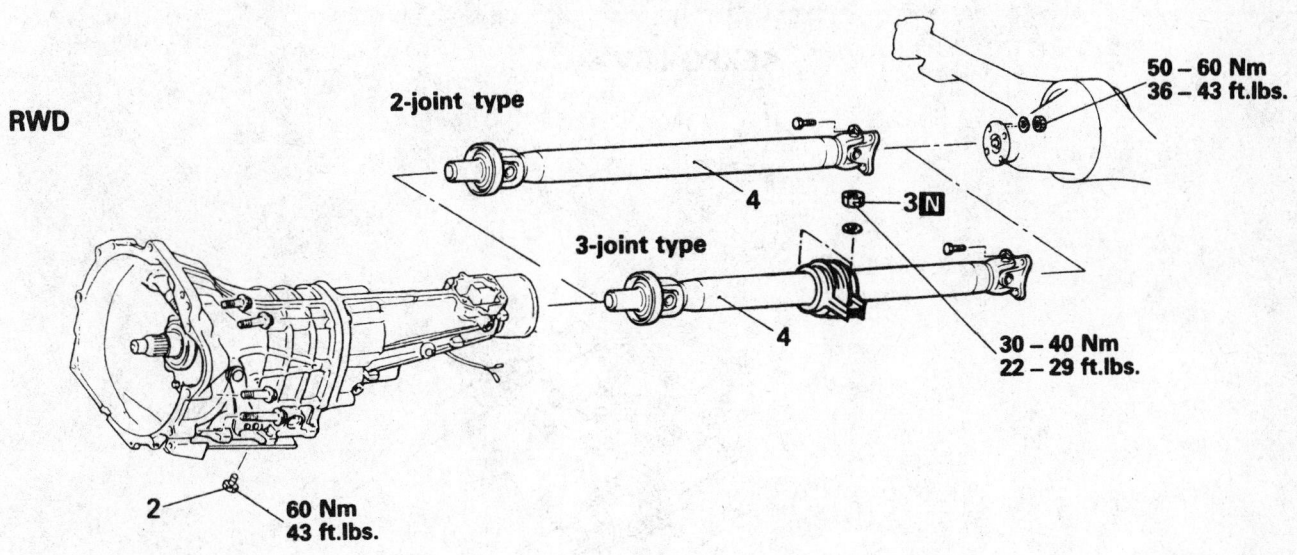

RWD

2-joint type

50 – 60 Nm
36 – 43 ft.lbs.

4

3

3-joint type

4

30 – 40 Nm
22 – 29 ft.lbs.

2   60 Nm
43 ft.lbs.

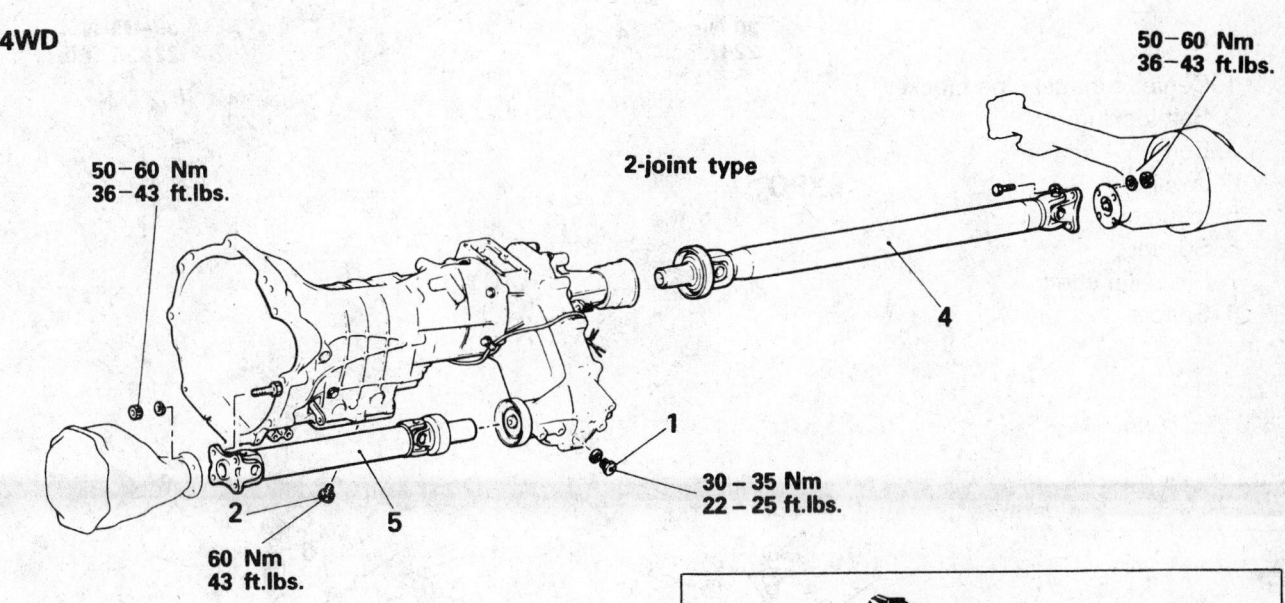

4WD

50 – 60 Nm
36 – 43 ft.lbs.

50 – 60 Nm
36 – 43 ft.lbs.

2-joint type

4

1

2   5

30 – 35 Nm
22 – 25 ft.lbs.

60 Nm
43 ft.lbs.

Rear propeller shaft removal steps

1. Drain plug
2. Drain plug
3. Nut
4. Rear propeller shaft

Front propeller shaft removal steps

1. Drain plug
5. Front propeller shaft

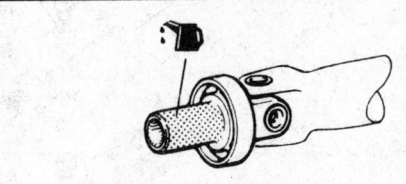

Gear oil:	
2WD MT, 4WD	Hypoid gear oil API classi-fication GL-4 or higher/ SAE viscosity 75W-90 or 75W-85W
2WD A/T	DIAMOND ATF SP or eauivalent

327424

Driveshaft components — Mighty Max

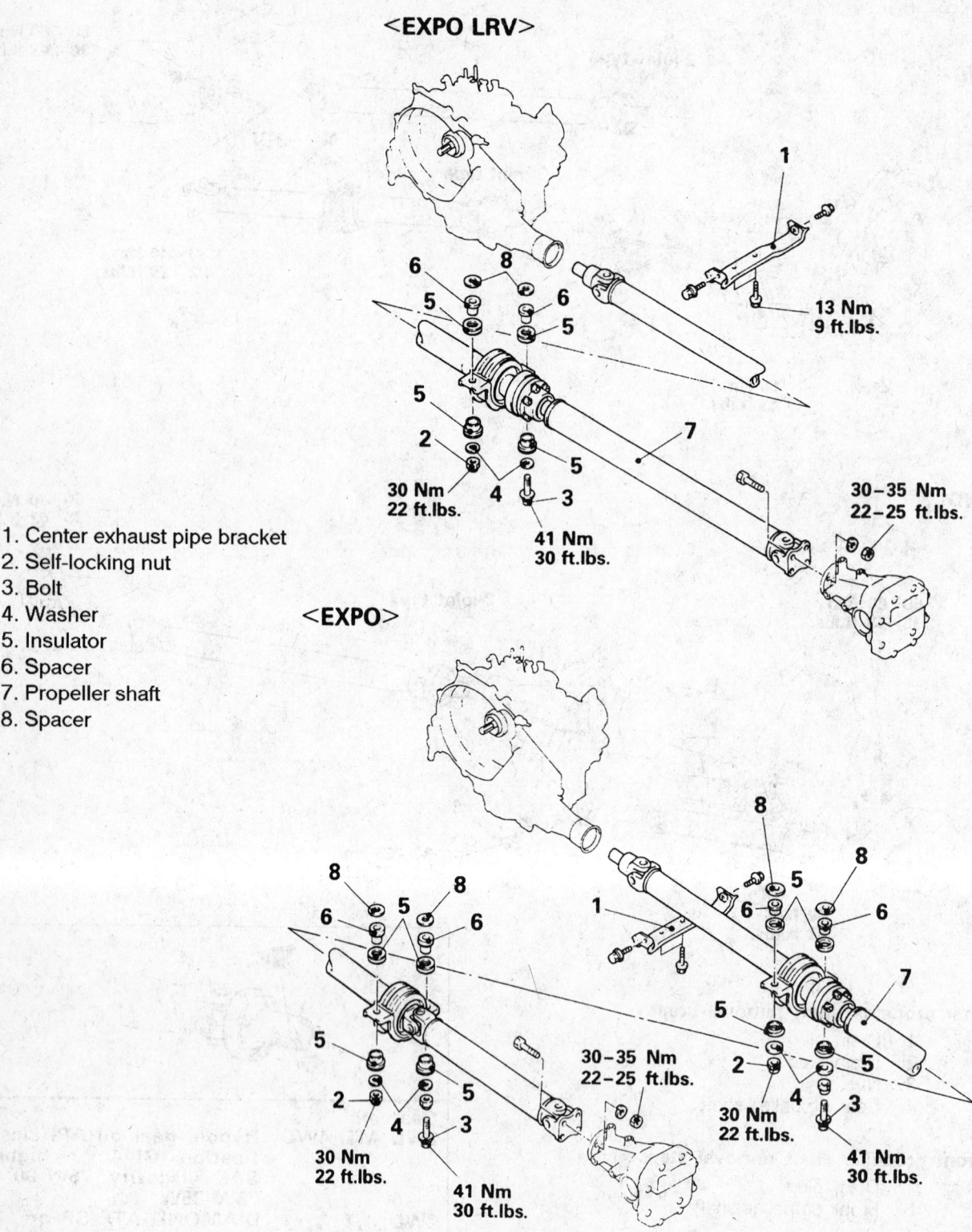

\<EXPO LRV\>

1 — 13 Nm / 9 ft.lbs.

6   8
5   6
    5
5
2
    5
4   3
30 Nm / 22 ft.lbs.
41 Nm / 30 ft.lbs.
7
30 – 35 Nm / 22 – 25 ft.lbs.

1. Center exhaust pipe bracket
2. Self-locking nut
3. Bolt
4. Washer
5. Insulator
6. Spacer
7. Propeller shaft
8. Spacer

\<EXPO\>

8
8   8
6   5
6   6
5
5
2
4   3
30 Nm / 22 ft.lbs.
41 Nm / 30 ft.lbs.

8
1   5
6   8
    6
7
5
2
5
4   3
30 – 35 Nm / 22 – 25 ft.lbs.
30 Nm / 22 ft.lbs.
41 Nm / 30 ft.lbs.

239963

Driveshaft and related components — Expo

Front propeller shaft | **Rear propeller shaft**

Snap ring

Snap ring

321554

Snapring clearance location — Montero

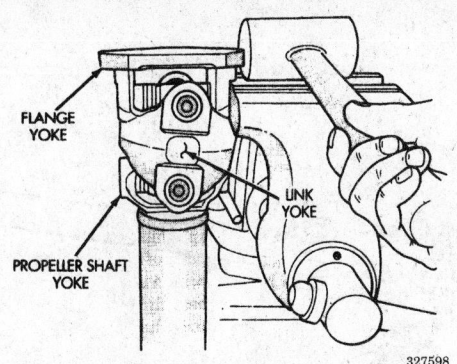

327598

Universal joint bearing removal — Mighty Max

Snap ring thickness mm (in.)		Identification color
Front propeller shaft	1.28 (.050)	–
	1.31 (.052)	Yellow
	1.34 (.053)	Blue
	1.37 (.054)	Purple
Rear propeller shaft	1.50 (.059)	–
	1.55 (.061)	Yellow
	1.60 (.063)	Blue
	1.65 (.065)	Purple

321555

Snapring chart — Montero

12. Position the driveshaft and work on the other end if service is required.

NOTE: After service is completed, check the assembled joints and yokes for freedom of movement. If misalignment of any part causes it to bind, a sharp rap on the side of the yoke with a brass hammer should seat the needle bearings, and provide the desired freedom of movement. Care should be exercised to firmly support the shaft end during this operation, as well as to prevent blows to the bearing caps themselves. Under no circumstances should a driveshaft be installed in a vehicle if there is any bind in the U-joints. If the binding remains, disassemble the yoke and joint a check the needle bearings for correctly vertical alignment.

13. Align the matchmarks and install the driveshaft.

Halfshaft

REMOVAL AND INSTALLATION

Montero

Outer Axle Shaft, Left and Right

1. Disconnect the negative battery cable.
2. Raise and safely support the vehicle. Remove the undercover.
3. Remove the wheels.
4. Remove the hub cover dust cap.
5. Remove the snapring from the inside of the hub. Remove the shim.
6. Remove the front brake caliper assembly. Do not allow the caliper to hang from the brake hose, support with mechanics wire.
7. If equipped with ABS, remove the speed sensor.
8. Separate the tie rod from the steering knuckle assembly.

PRESS

PROPELLER SHAFT YOKE

FLANGE YOKE

LINK YOKE

327597

Universal joint bearing press — Mighty Max

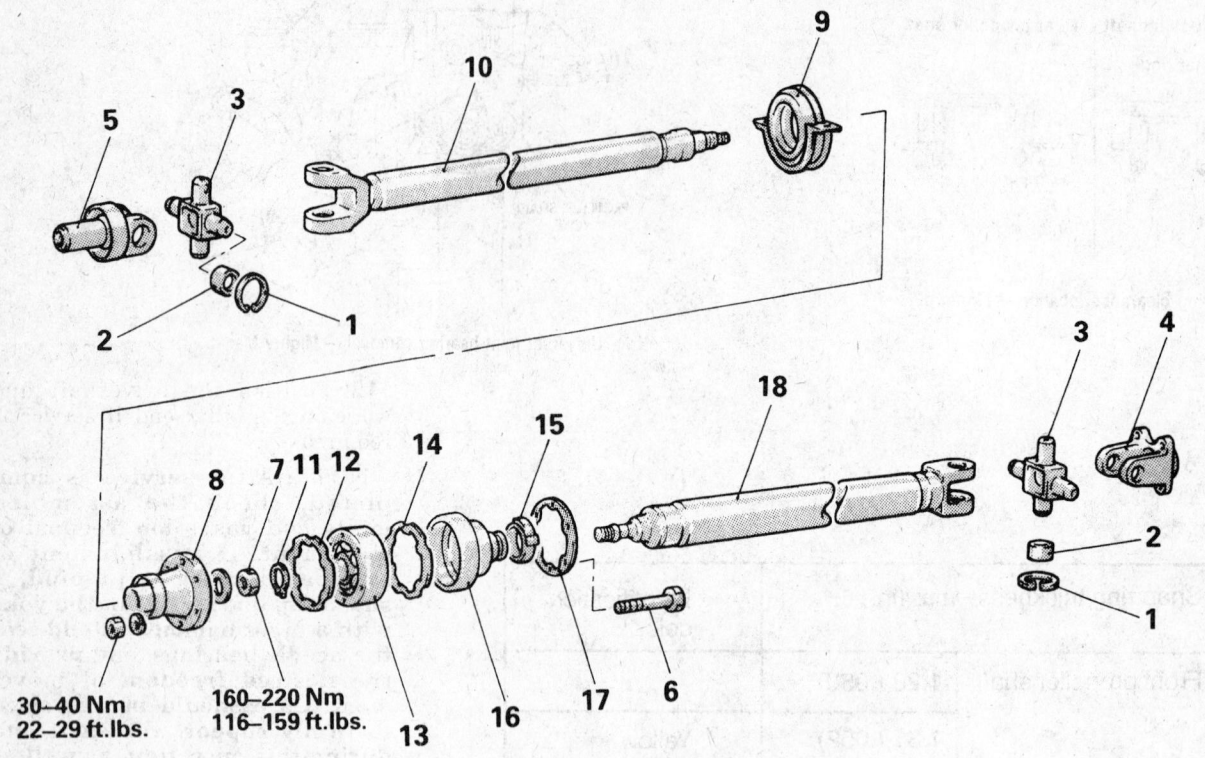

**30–40 Nm
22–29 ft.lbs.**

**160–220 Nm
116–159 ft.lbs.**

1. Snap ring
2. Journal bearing
3. Journal
4. Flange yoke
5. Sleeve yoke
6. Löbro joint mounting bolt
7. Self-locking nut
8. Companion flange
9. Center bearing assembly
10. Front propeller shaft assembly

11. Snap ring
12. Rubber packing
13. Löbro joint assembly
14. Rubber packing
15. Boot band
16. Löbro joint boot
17. Washer
18. Rear propeller shaft assembly

241581

Driveshaft assembly exploded view — Expo

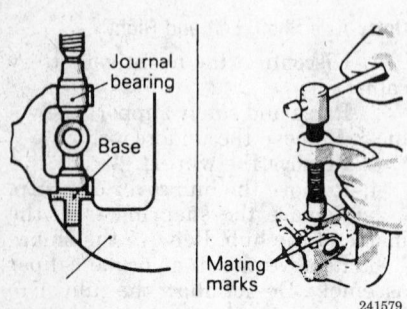

Pressing out universal joint — Expo

241579

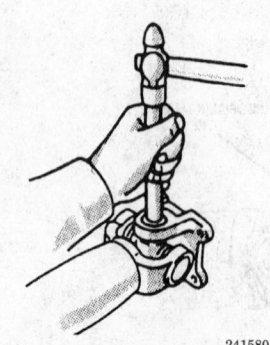

241580

Installing new universal joint — Expo

9. Separate the upper and lower ball joints from the steering knuckle assembly.

10. Remove the front hub/knuckle assembly with the inner and outer bearings intact.

11. Left side, pull the halfshaft from the differential carrier. For the right side and remove the retaining nuts and remove the halfshaft from the vehicle.

To install:

12. Using a NEW circlip, install the left side halfshaft. For the right side halfshaft, install to the inner shaft and tighten the retaining nuts to 36–43 ft. lbs. (49–59 Nm).

Snap ring thickness mm (in.)	Identification colour
1.28 (.0503)	–
1.31 (.0516)	Yellow
1.34 (.0528)	Blue
1.37 (.0539)	Purple
1.40 (.0551)	Brown

241583

Snapring thickness chart — Expo

13. Install the front hub/knuckle and bearing assembly.

14. Install the upper ball joint to the knuckle and torque retaining nut to 54 ft. lbs. (74 Nm). Install the lower ball joint to knuckle and torque retaining nut to 108 ft. lbs. (147 Nm). Install new cotter pins.

15. Install the tie rod end to the steering knuckle and torque to 33 ft. lbs. (44 Nm). Install new cotter pin.

16. Install the speed sensor, if removed.

17. Install the front brake assembly.

18. Install the shim and snapring to the axle shaft. Install the front hub dust cover.

19. Install the wheels and the undercover.

20. Lower the vehicle and connect the negative battery cable.

Inner Axle Shaft

1. Disconnect the negative battery cable.

2. Raise and safely support the vehicle.

3. Remove the undercover.

4. Remove the right side wheel.

5. Remove the right outer halfshaft.

6. Remove the lower shock absorber mounting bolts.

7. Install slide hammer to inner shaft flange and pull the shaft from housing.

To install:

8. Install a NEW circlip to the inner halfshaft and install into the housing. Drive the axle into position.

9. Install the lower shock absorber mounting bolts. Torque the bolts to 65–76 ft. lbs. (88–103 Nm).

10. Install the right halfshaft assembly.

11. Install the undercover.

12. Install the wheel.

13. Lower the vehicle and connect the negative battery cable.

Mighty Max

Right Halfshaft — Outer Axle Shaft

1. Place the free-wheeling hub in the free condition by placing the transfer lever in the **2H** position and moving in reverse for about 6 or 7 feet.

2. Disconnect negative battery cable.

3. Raise and safely support the vehicle. Remove the skid plate, if equipped.

4. Remove the tire and wheel assembly.

5. Remove the hub cover with the use of an oil filter wrench. Install a protective cloth between the wrench and the cover to avoid damage to the cover.

6. Remove the snapring from the inside of the hub. Remove the shim.

7. Remove the front brake caliper and brake pads from the vehicle. Do not allow the caliper to hang from the brake hose, support with mechanics wire.

8. Separate the tie rod from the steering knuckle.

9. Separate the upper and lower ball joints from the steering knuckle.

10. Remove the front hub/knuckle assembly with the inner and outer bearings intact.

11. Remove the halfshaft to axle housing retaining nuts and remove the halfshaft from the vehicle.

To install:

12. Install the halfshaft and the retaining nuts and tighten to 43 ft. lbs. (60 Nm).

13. Install the front hub/knuckle and bearing assembly.

14. Install the upper ball joint to the knuckle and torque retaining nut to 130 ft. lbs. (180 Nm). Install the lower ball joint to knuckle and torque retaining nut to 65 ft. lbs. (90 Nm). Install new cotter pins.

15. Install the tie rod end to the steering knuckle and torque to 33 ft. lbs. (45 Nm). Install new cotter pin.

16. Install the shim and snapring to the axle shaft. Install the front hub cover.

17. Install front brake caliper assembly.

18. Install the tire and wheel assembly. Install skid plate, if removed.

19. Lower the vehicle and connect the negative battery cable.

Right Halfshaft — Inner Axle Shaft

1. Disconnect negative battery cable.

2. Raise and safely support the vehicle.

3. Remove the skid plate, if equipped.

4. Remove the tire and wheel assembly.

5. Remove the right outer halfshaft.

6. Remove the lower shock absorber mounting bolts.

7. Install slide hammer to inner shaft flange and pull from housing. Press the bearing from the axle, as required.

To install:

8. Press new bearing and seal on axle, as required. Install new circlip to inner halfshaft and install into housing. Drive the axle into position.

9. Install the lower shock absorber mounting bolts.

10. Install the right outer halfshaft and related parts.

11. Install the skid plate, if equipped.

12. Install the tire and wheel assembly.

13. Lower the vehicle and connect the negative battery cable.

Left Halfshaft

1. Place the free-wheeling hub in the free condition by placing the transfer lever in the **2H** position and moving in reverse for about 6 or 7 feet.

2. Disconnect negative battery cable.

3. Raise and safely support the vehicle. Remove the skid plate, if equipped.

4. Remove the tire and wheel assembly.

5. Remove the hub cover with the use of an oil filter wrench. Install a protective cloth between the wrench and the cover to avoid damage to the cover.

6. Remove the snapring from the inside of the hub. Remove the shim.

7. Remove the front brake caliper and brake pads from the vehicle. Do not allow the caliper to hang from the brake hose, support with mechanics wire.

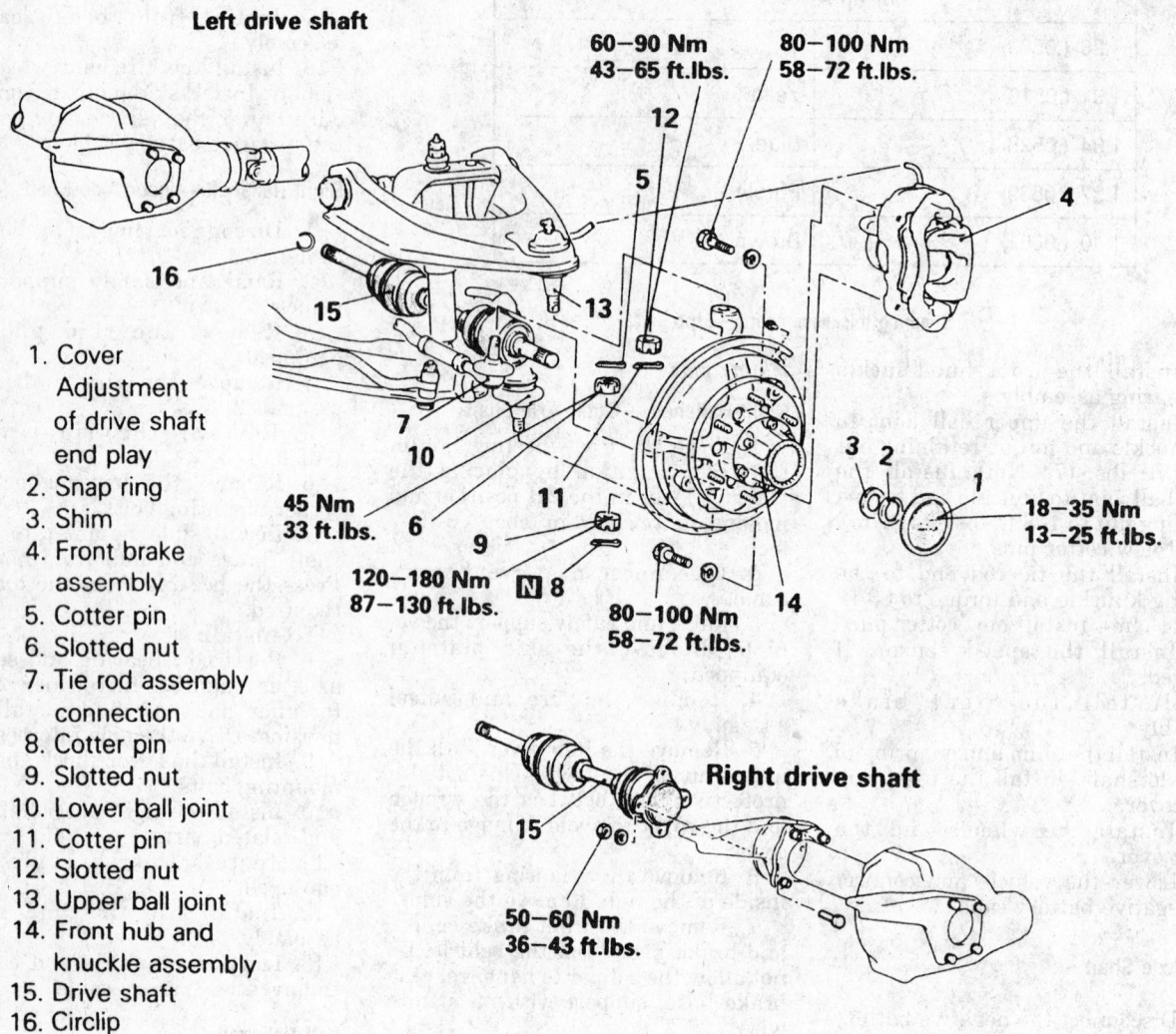

Left drive shaft

60–90 Nm
43–65 ft.lbs.

80–100 Nm
58–72 ft.lbs.

45 Nm
33 ft.lbs.

120–180 Nm
87–130 ft.lbs.

80–100 Nm
58–72 ft.lbs.

18–35 Nm
13–25 ft.lbs.

Right drive shaft

50–60 Nm
36–43 ft.lbs.

1. Cover
   Adjustment
   of drive shaft
   end play
2. Snap ring
3. Shim
4. Front brake
   assembly
5. Cotter pin
6. Slotted nut
7. Tie rod assembly
   connection
8. Cotter pin
9. Slotted nut
10. Lower ball joint
11. Cotter pin
12. Slotted nut
13. Upper ball joint
14. Front hub and
    knuckle assembly
15. Drive shaft
16. Circlip

327454

Halfshaft removal and installation — Mighty Max

8. Separate the tie rod from the steering knuckle.

9. Separate the upper and lower ball joints from the steering knuckle.

10. Remove the front hub/knuckle assembly with the inner and outer bearings intact.

11. Pull the left halfshaft out from the differential carrier. Use care not to damage the oil seal with the splines of the shaft.

To install:

12. Replace circlip on the end of the shaft. Install the halfshaft into the front differential case and drive into position using a plastic hammer.

13. Install the front hub/knuckle and bearing assembly.

14. Install the upper ball joint to the knuckle and tighten retaining nut to 130 ft. lbs. (180 Nm). Install the lower ball joint to knuckle and tighten retaining nut to 65 ft. lbs. (90 Nm). Install new cotter pins.

15. Install the tie rod end to the steering knuckle and tighten to 33 ft. lbs. (45 Nm). Install new cotter pin.

16. Install the shim and snapring to the axle shaft. Install the front hub cover.

17. Install front brake caliper assembly.

18. Install the tire and wheel assembly. Install skid plate, if removed.

19. Lower the vehicle and connect the negative battery cable.

Expo

1. Raise the vehicle and support safely.

2. Remove the bolts that attach the rear halfshaft to the companion flange.

3. Remove the cotter pin and axle nut from the outer shaft.

4. Remove the rear driveshaft from the vehicle.

5. If the differential companion shaft is to be removed, connect a slide hammer to the flange and pull the shaft from the differential.

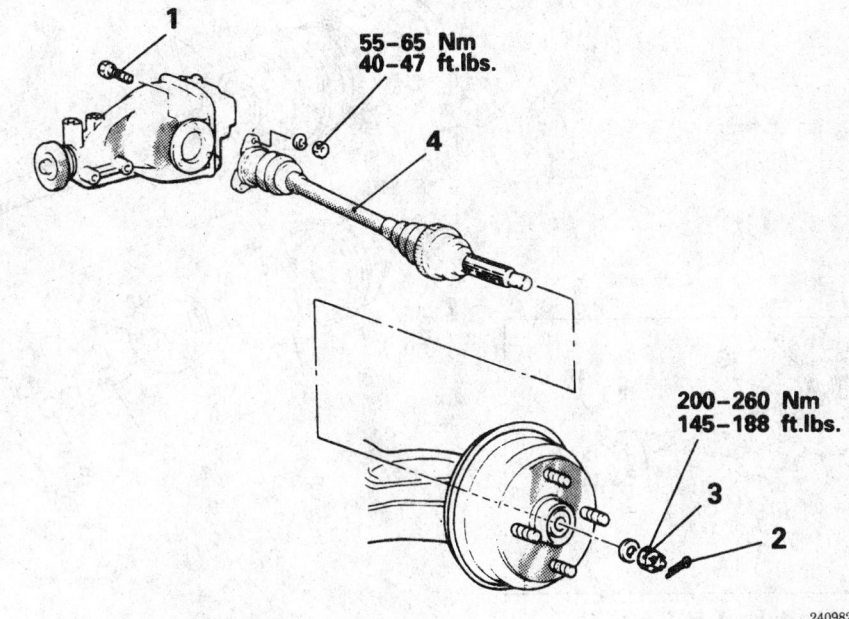

1. Bolt
2. Cotter pin
3. Drive shaft nut
4. Drive shaft

55–65 Nm
40–47 ft.lbs.

200–260 Nm
145–188 ft.lbs.

Axle shaft and related components — Expo

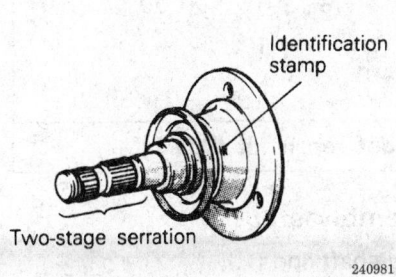

Identification stamp

Two-stage serration

Companion shaft identification — Expo

To install:

NOTE: If the companion shaft is being replaced or both shafts were removed together, it is important to properly identify the companion shaft. The right and left side companion shafts are different, as are limited slip differentials, which use a two stage serration on the companion shaft.

6. Replace the circlip and install the companion flange to the differential case. Make sure it snaps in place. On limited slip differentials, ensure

that both serrations are fully engaged to the differential.

7. Install the axle shaft through the hub and install the axle nut. Torque the nut to 145 to 188 ft. lbs. (200 to 260 Nm) and secure with cotter pin.

8. Install the companion flange bolts and tighten to 40 to 47 ft. lbs. (55 to 65 Nm).

9. Check the fluid level in the rear differential.

CV-Joint Boot

REPLACEMENT

1. Remove the halfshaft from the vehicle and secure in a soft-jawed vise.

2. Remove the boot bands on the inner Double Offset Joint (DOJ). Remove the circlip.

3. Remove the DOJ outer race from the shaft.

4. Remove the balls from the DOJ cage, prying from the inside of the cage outward.

5. Rotate the cage while pushing toward the Birfield Joint (BJ). The cage will drop down to expose a snapring on the halfshaft. Remove the snapring.

6. Remove the DOJ cage and inner race from the halfshaft.

7. Remove the circlip from the shaft.

8. Wrap tape over the threads of the halfshaft. Remove the boot from the shaft.

9. Remove the dust cover and boot protector from the BJ.

NOTE: Do not disassemble the Birfield Joint (BJ).

10. Remove the boot bands. Remove the BJ boot from the DOJ end of the halfshaft.

11. Inspect all parts for wear or damage and replace, as necessary.

To install:

12. Install the BJ boot and bands over the DOJ end of the shaft. Install the DOJ boot and bands onto the halfshaft. Apply ½ of supplied grease into the BJ boot and install the bands onto the boot.

13. Install the DOJ cage onto the halfshaft with the smaller diameter side of the cage facing the installed BJ boot. Install the circlip.

14. Install the DOJ inner race onto the halfshaft and secure with the snapring.

15. Apply the proper grease to the DOJ inner race, the DOJ cage and the balls. Insert the balls into the

LEFT DRIVE SHAFT

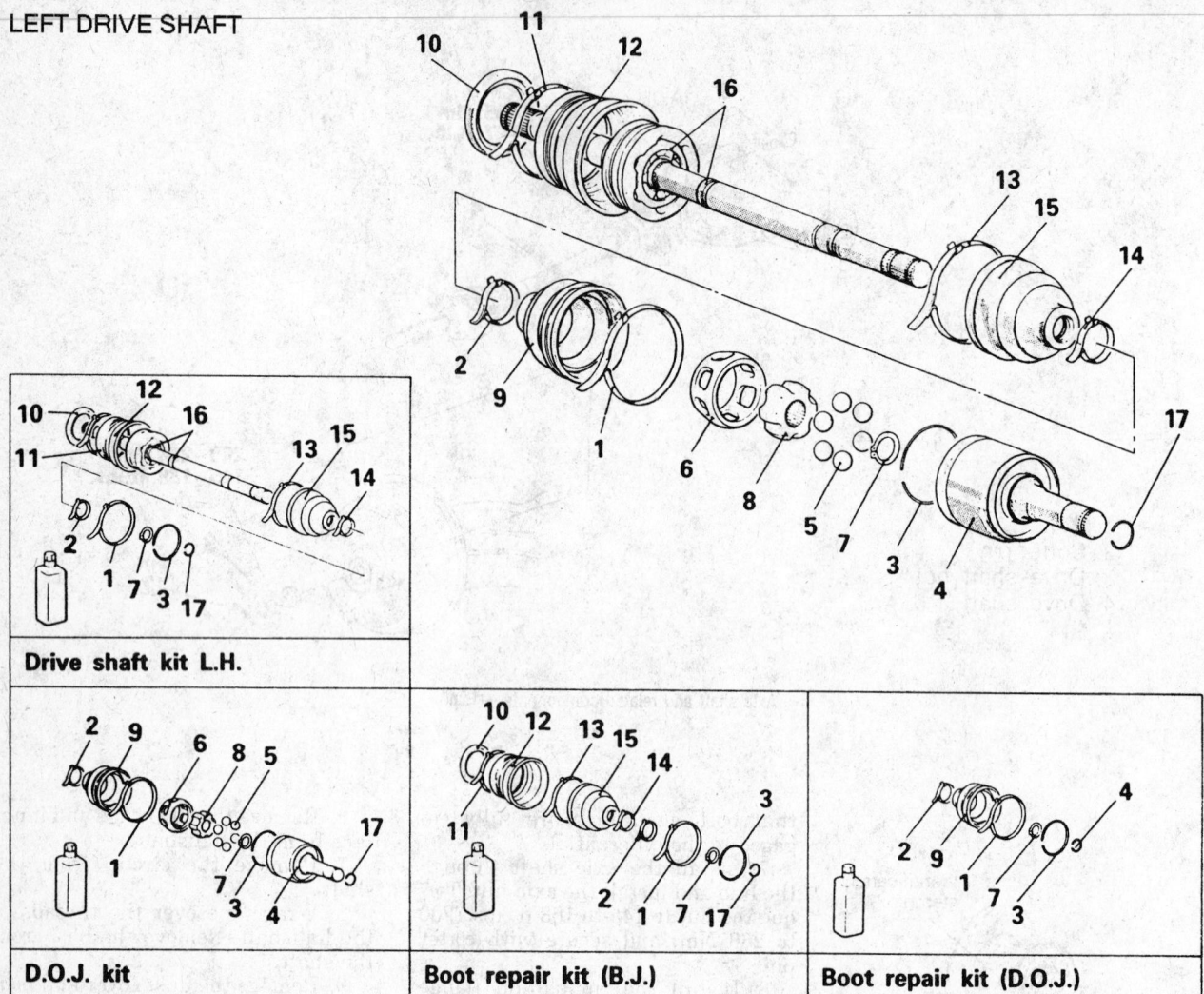

Drive shaft kit L.H.

D.O.J. kit

Boot repair kit (B.J.)

Boot repair kit (D.O.J.)

Disassembly steps

1. Boot band A
2. Boot band B
3. Circlip
4. D.O.J. outer race
5. Ball
6. D.O.J. cage
7. Snap ring
8. D.O.J. inner race
9. D.O.J. boot
10. Dust cover
11. Boot protector band
12. Boot protector
13. Boot band A
14. Boot band B
15. B.J. boot
16. Drive shaft and B.J.
17. Circlip

Reassembly steps

16. Drive shaft and B.J.
15. B.J. boot
13. Boot band A
14. Boot band B
2. Boot band B
9. D.O.J. boot
1. Boot band A
6. D.O.J. cage
8. D.O.J. inner race
7. Snap ring
5. Ball
4. D.O.J. outer race
3. Circlip
17. Circlip
12. Boot protector
11. Boot protector band
10. Dust cover

327369

Disassembled view of the left side halfshaft

RIGHT DRIVE SHAFT

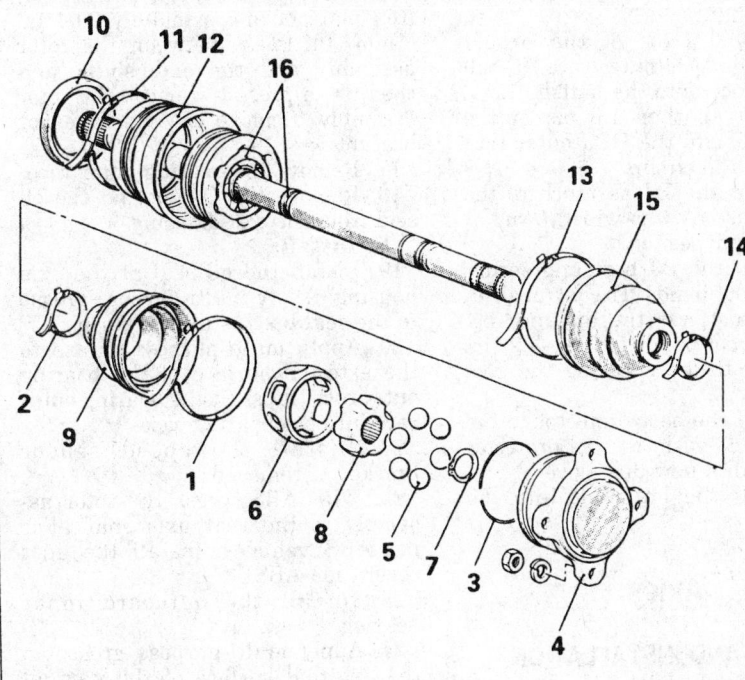

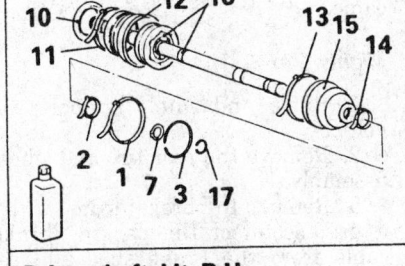

Drive shaft kit R.H.

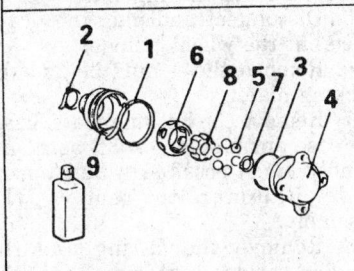

D.O.J. kit

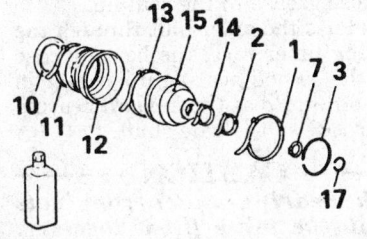

Boot repair kit (B.J.)

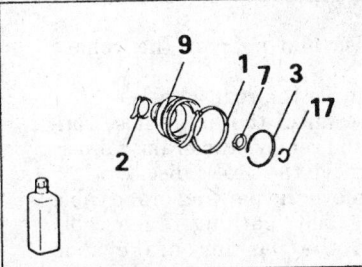

Boot repair kit (D.O.J.)

Disassembly steps

1. Boot band A
2. Boot band B
3. Circlip
4. D.O.J. outer race
5. Ball
6. D.O.J. cage
7. Snap ring
8. D.O.J. inner race
9. D.O.J. boot
10. Dust cover
11. Boot protector band
12. Boot protector
13. Boot band A
14. Boot band B
15. B.J. boot
16. Drive shaft and B.J.

Reassembly steps

16. Drive shaft and B.J.
15. B.J. boot
13. Boot band A
14. Boot band B
2. Boot band B
9. D.O.J. boot
1. Boot band A
6. D.O.J. cage
8. D.O.J. inner race
7. Snap ring
5. Ball
4. D.O.J. outer race
3. Circlip
12. Boot protector
11. Boot protector band
10. Dust cover

327370

Disassembled view of the right side halfshaft

cage from the outside pushing in toward the shaft.

16. Apply 1.9 oz. of the proper grease to the DOJ outer race. Install the outer race onto the halfshaft.

17. Apply another 1.9 oz. of proper grease to the DOJ outer race and install the circlip.

18. Add to the BJ as much of the proper grease as was wiped away at the time of inspection.

19. Install the BJ boot and secure with new boot bands. To control air in the DOJ boot, set the distance between the boot bands to 3.03–3.27 in. (77–83mm) before securing the boot bands.

20. Install the boot protector to the BJ and secure with the boot protector band. Install a new dust cover.

21. Install the halfshaft into the vehicle.

Axles

REMOVAL AND INSTALLATION

Montero

1. Disconnect the negative battery cable.
2. Raise and support the vehicle safely.
3. Remove the rear wheels.
4. Disconnect the brake hose connection and remove the brake caliper.
5. Remove the brake disc.
6. Remove the parking brake cable attaching bolt, parking brake cable end and the parking brake shoe assembly.
7. If equipped with ABS, disconnect the speed sensor.
8. Using an axle puller, remove the rear axle shaft assembly.

NOTE: Be careful not to damage the oil seal when removing the rear axle shaft.

9. Remove the snapring.
10. Remove one retainer bolt from the backing plate.
11. Partially grind the retainer ring; then, using a chisel, cut and remove the retainer ring.

NOTE: Be careful not to damage the bearing case and axle shaft.

12. Remove the inboard inner bearing.
13. Using a suitable tool, remove the axle shaft from the bearing case assembly. Remove the outboard inner bearing from the axle shaft.
14. Remove the oil seal.
15. If not equipped with ABS, remove the dust cover.

16. If equipped with ABS, insert an iron plate of approximately 0.04 in. (1mm) thickness between the rotor assembly and the axle shaft, and then use a press to remove the rotor assembly. Remove the speed sensor bracket.
17. Remove the bearing outer race.
18. Remove the O-ring and the oil seal from the axle housing.

To install:

19. Install the oil seal at the axle housing. Apply multi-purpose grease to the seal lip. Install the O-ring.
20. Apply multi-purpose grease to the external surface of the bearing outer race. Press fit the bearing outer race into the bearing case.
21. Install the speed sensor bracket, if removed.
22. With ABS, press the rotor assembly to the rear axle shaft. For non-ABS vehicles, install the dust cover (non-ABS).
23. Install the outboard inner bearing.
24. Apply multi-purpose grease to the external surface of the new oil seal. Press fit the new seal into the bearing case until it is flush with the face of the bearing case. Apply multi-purpose grease to the seal lip.
25. Pass the axle shaft through the bearing inner race, the bearing case and the second bearing inner race in that order. Press the inboard bearing inner race to the axle shaft.

——— **CAUTION** ———
Both bearing inner race sets should be press fitted together. The left and right lengths of the axle shaft are different in vehicles with rear differential lock. The right side is longer; be careful when installing.

26. Press the retainer onto the axle shaft, while checking that the press-fitting force is at the standard value. If the initial press-fitting force is less than the standard value, replace the axle shaft.
- 1993–95 initial press-fitting force: 11,023 lbs. (50,000 N) or more.
- 1996–97 initial press-fitting force: 11,016 lbs. (49,000 N) or more.
- 1993–95 final press-fitting force: 22,046–24,251 lbs. (100,000–110,000 N).
- 1996–97 final press-fitting force: 22,031–24,279 lbs. (98,000–108,000 N).

27. Install the snapring. Measure the clearance between the snapring and the retainer with a thickness gauge, and check that it is 0.0065 in. (0.166 mm) or less. If the clearance exceeds this, change the snapring so

that the clearance is at the standard value.
28. Install the axle shaft assembly to the housing.
29. Install the speed sensor.
30. Install the parking brake assembly.
31. Install the parking brake cable end.
32. Install the brake disc and the caliper assembly.
33. Connect the brake tube.
34. Properly bleed the brake system.
35. Check for proper gear lubricant level.
36. Install the wheels.
37. Lower the vehicle.
38. Connect the negative battery cable.

Mighty Max

1. Raise and safely support the vehicle.
2. Remove the rear tire and wheel assembly.
3. Remove the brake drum.
4. Disconnect the parking brake cable from the brake shoe and remove from the backing plate.
5. Disconnect and plug the brake line(s) at the wheel cylinder.
6. Remove the 4 nuts behind the backing plate.
7. Remove the backing plate, bearing case and the axle shaft as an assembly. If not possible by hand, use a slide hammer to remove the assembly.
8. Remove the O-ring and the bearing preload shims. Save the preload shims for reassembly. Remove the snapring.
9. Remove the oil seal from the axle tube with a hooked slide hammer.
10. To remove the axle shaft bearing, remove the retaining ring:
    a. Remove 1 retaining bolt from the backing plate.
    b. Push the bearing case all the way to the side of the dust cover.
    c. Protect the bearing case with adhesive tape.
    d. Grind through the retaining ring in 1 spot until the axle shaft is exposed.
    e. Cut into the retaining ring with a chisel and remove the ring.

——— **CAUTION** ———
Always wear eye protection and appropriate clothing when grinding or chiseling.

11. Screw the locknut onto the axle shaft about 3 turns. If not equipped with a locknut, it may be necessary to

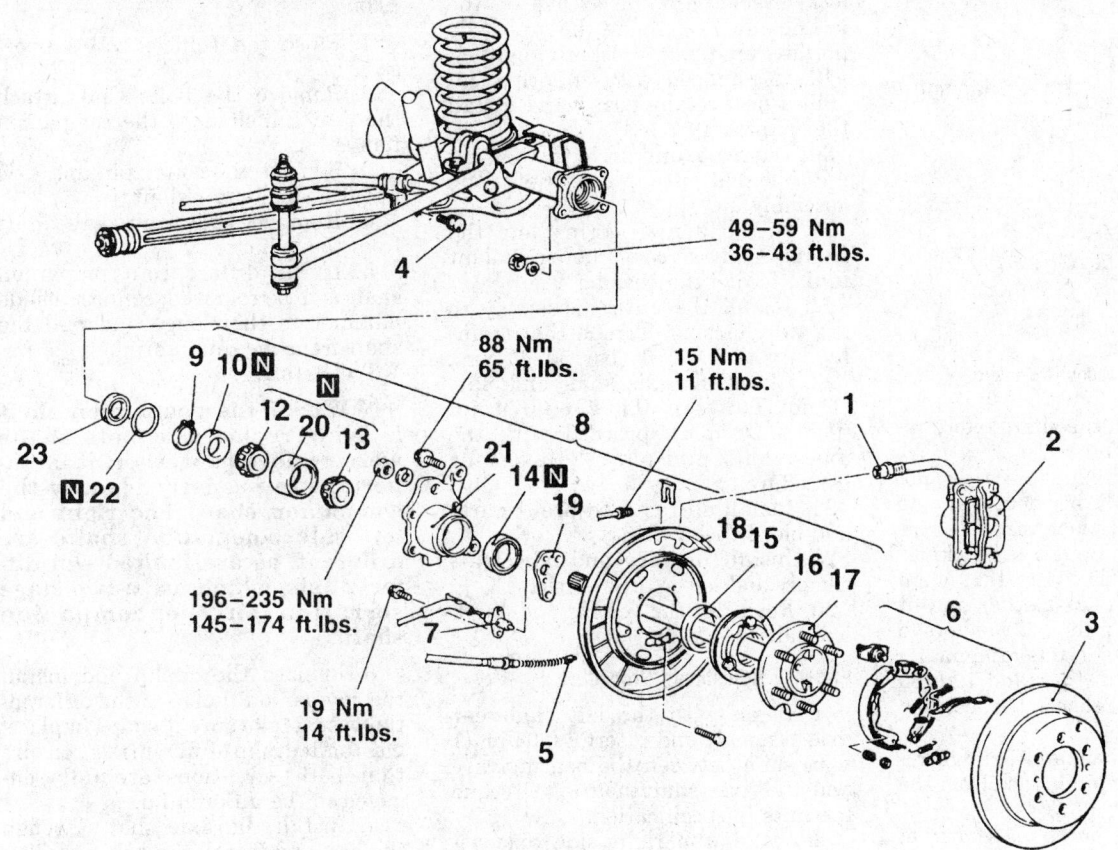

49–59 Nm
36–43 ft.lbs.

88 Nm
65 ft.lbs.

15 Nm
11 ft.lbs.

196–235 Nm
145–174 ft.lbs.

19 Nm
14 ft.lbs.

1. Connection for brake pipe
2. Rear brake assembly
3. Brake disc
4. Parking brake cable attaching bolt
5. Parking cable end
6. Parking brake assembly
7. Speed sensor <Vehicles with ABS>
8. Axle shaft assembly
9. Snap ring
10. Retainer
11. Axle shaft sub assembly
    (Parts from step 13 to step 17)
12. Bearing inner race (inner)
13. Bearing inner race (outer)
14. Oil seal
15. Dust cover <Vehicles without ABS>
16. Rotor assembly <Vehicles with ABS>
17. Axle shaft
18. Backing plate
19. Speed sensor bracket
    <Vehicles with ABS>
20. Bearing outer race
21. Bearing case
22. O-ring
23. Oil seal

Rear axle shaft assembly — Montero

321498

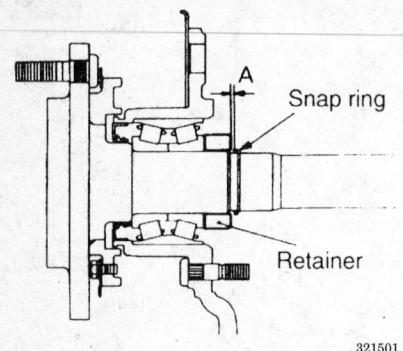

Snap ring location — Montero

install spacers on the shaft before removing the bearing.

12. If tool MB990787-01 is not available, it will be necessary to fabricate a metal plate that fits over the axle shaft and butts the locknut. Drill 4 holes in the plate that align with the 4 bearing case studs and fit the plate. Refit 2 nuts and washers to the bearing case studs diagonally across from each other and tighten them evenly to release the bearing case and bearing.

13. Use a hammer and drift to remove the bearing outer race from the bearing case.

14. Remove the outer oil seal from the bearing case.

To install:

15. Apply grease to the outer surface on the bearing outer race and the lip of the outer oil seal. Drive into the bearing case.

16. Slide the bearing case and bearing over the rear axle shaft. Apply grease on the bearing rollers and install the inner race by pressing into place. Be careful not to damage the dust cover.

17. Pack the bearing with grease.

18. If equipped with locknut, install the washer, the crowned lock washer and the locknut in that order and torque the locknut to 130–159 ft. lbs. (176–220 Nm). Bend the tab on the

lock washer into the groove on the locknut. If the tab and the groove do not line up, tighten locknut slightly.

19. If equipped with snapring, install a new retainer ring and install the snapring.

20. Lubricate and drive the new inner oil seal into place. Refit the assembly.

21. Install a new O-ring and the shims. Apply silicone rubber sealant to the face of the bearing case.

22. Install the entire assembly to the axle housing. Torque the retaining nuts to 36–43 ft. lbs. (50–60 Nm).

23. Check the axle shaft end-play. If not between 0.002–0.0079 in. (0.05–0.20 mm), proceed with the axle shaft end-play adjustment procedure.

24. Install all removed brake parts and bleed the system.

25. Install the tire and wheel assembly and lower the vehicle.

26. Road test the vehicle and check for leaks.

End-play Adjustment Procedure

1. Begin with the left side rear axle assembly and insert a 0.04 in. (1 mm) shim between the bearing case and the axle shaft housing. Torque the nuts to specification.

2. Install the right side axle assembly into the right side housing without its shim and O-ring. Torque the 4 nuts to about 50 inch. lbs.

3. Using a feeler gauge, measure the gap between the bearing case and the axle housing face.

4. Remove the axle shaft and select a shim or shims that is the equal to the sum of the clearance measured in Step 3 plus 0.002–0.0079 in. (0.05–0.20 mm) and install them on the housing. Install the O-ring and apply sealant.

5. Install the axle assembly and torque the nuts to 36–43 ft. lbs. (50–60 Nm).

6. Measure the end-play and complete the installation procedure.

Expo

1. Raise the vehicle and support safely.

2. Remove the bolts that attach the rear halfshaft to the companion flange.

3. Remove the cotter pin and axle nut from the outer shaft.

4. Remove the rear driveshaft from the vehicle.

5. If the differential companion shaft is to be removed, connect a slide hammer to the flange and pull the shaft from the differential.

To install:

NOTE: If the companion shaft is being replaced or both shafts were removed together, it is important to properly identify the companion shaft. The right and left side companion shafts are different, as are limited slip differentials, which use a two stage serration on the companion shaft.

6. Replace the circlip and install the companion flange to the differential case. Make sure it snaps in place. On limited slip differentials, ensure that both serrations are fully engaged to the differential.

7. Install the axle shaft through the hub and install the axle nut. Torque the nut to 145 to 188 ft. lbs. (200 to 260 Nm) and secure with cotter pin.

8. Install the companion flange bolts and tighten to 40 to 47 ft. lbs. (55 to 65 Nm).

9. Check the fluid level in the rear differential.

STEERING

Air Bag

—CAUTION—

Some vehicles are equipped with an air bag system, also known as the Supplemental Inflatable Restraint (SIR) or Supplemental Restraint System (SRS). The system must be disabled before performing service on or around system components, steering column, instrument panel components, wiring and sensors. Failure to follow safety and disabling procedures could result in accidental air bag deployment, possible personal injury and unnecessary system repairs.

Thickness of snap ring mm (in.)	Identification color
2.17 (.0854)	–
2.01 (.0791)	Yellow
1.85 (.0728)	Blue
1.69 (.0665)	Purple
1.53 (.0602)	Red

Snap ring size chart — Montero

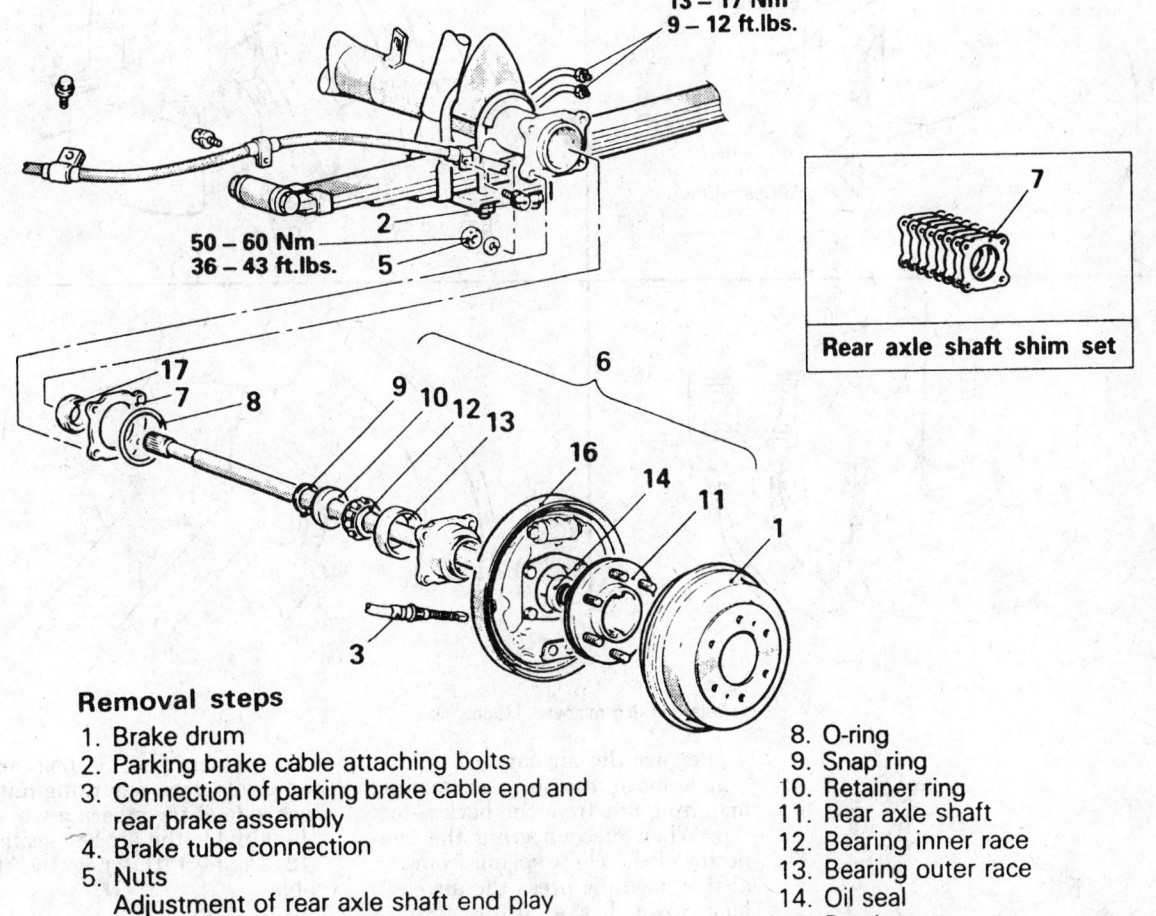

13 – 17 Nm
9 – 12 ft.lbs.

50 – 60 Nm
36 – 43 ft.lbs.

Rear axle shaft shim set

Removal steps

1. Brake drum
2. Parking brake cable attaching bolts
3. Connection of parking brake cable end and rear brake assembly
4. Brake tube connection
5. Nuts
   Adjustment of rear axle shaft end play
6. Rear axle shaft assembly (with parking brake cable)
7. Shim

8. O-ring
9. Snap ring
10. Retainer ring
11. Rear axle shaft
12. Bearing inner race
13. Bearing outer race
14. Oil seal
15. Bearing case
16. Backing plate
17. Oil seal

327523

Rear axle shaft exploded view — Mighty Max

PRECAUTIONS

Several precautions must be observed when handling the inflator module to avoid accidental deployment and possible personal injury.

• Never carry the inflator module by the wires or connector on the underside of the module.

• When carrying a live inflator module, hold securely with both hands, and ensure that the bag and trim cover are pointed away.

• Place the inflator module on a bench or other surface with the bag and trim cover facing up.

• With the inflator module on the bench, never place anything on or close to the module which may be thrown in the event of an accidental deployment.

DISARMING

Montero

To avoid personal injury when working on vehicles equipped with an air bag, the negative battery cable must be disconnected and at least 60 seconds must elapse before working on the system. Failure to do so may result in deployment of the air bag.

Steering Wheel

REMOVAL AND INSTALLATION

Mighty Max

1. Remove the center pad.
2. Remove the steering wheel retaining nut. Matchmark the steering wheel to the shaft.
3. Using a suitable steering wheel puller, pull the steering wheel off of the shaft.
   To install:
4. Install the steering wheel and torque the nut to 33 ft. lbs. (45 Nm).
5. Install the center pad.

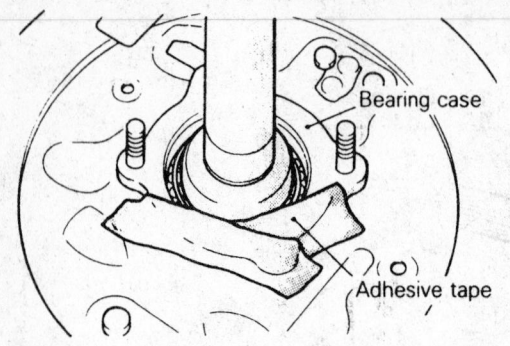

Bearing case

Adhesive tape

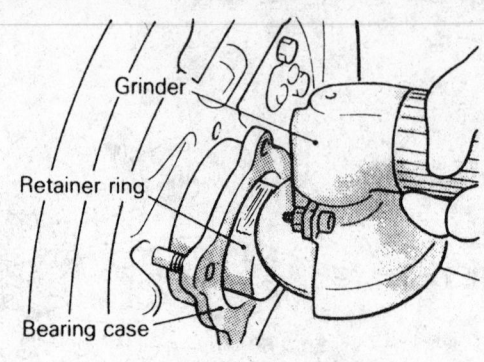

Grinder

Retainer ring

Bearing case

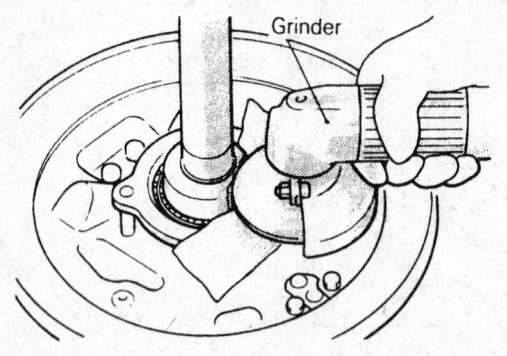

Grinder

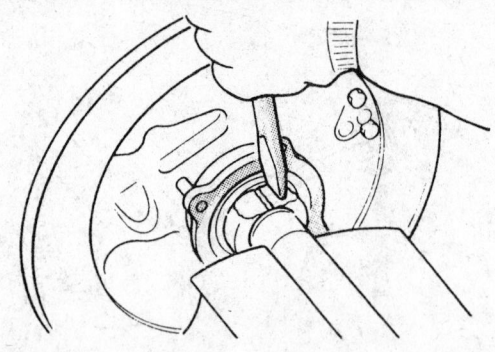

327520

Retaining ring removal — Mighty Max

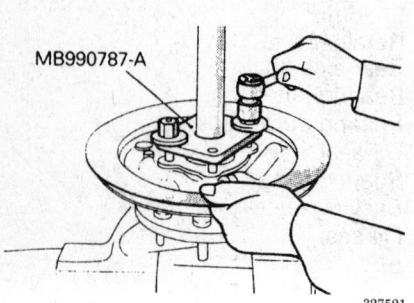

MB990787-A

327521

Axle shaft removal with tool MB990787-A — Mighty Max

Montero

CAUTION

The air bag system (SRS) must be disarmed before removing the steering wheel. Failure to do so may cause accidental deployment, property damage or personal injury.

1. Disable the air bag system
2. Disconnect the negative battery cable.
3. Wait at least 60 seconds after disconnecting the battery before working on the vehicle.

4. Remove the air bag.
  a. Remove the air bag module mounting nut from the back side.
  b. When disconnecting the connector of the clock spring from the air bag module, press the air bag's lock toward the outer side to spread it open. Carefully remove the connector.

CAUTION

Always carry an air bag assembly with the bag and trim cover away from your body. Store the assembly with the bag and trim cover facing up; never place the assembly face down on the floor or workbench.

5. Remove the steering wheel retaining nut. Matchmark the steering wheel to the shaft.
6. Using a suitable steering wheel puller, pull the steering wheel off of the shaft.
  To install:
7. Install the clock spring of the air bag by aligning the mating mark and the neutral position indicator of the clock spring.
8. Install the steering wheel and torque the nut for the Montero to 29 ft. lbs. (39 Nm) and all others to 33 ft. lbs. (45 Nm).
9. Install the air bag assembly.

10. Connect the air bag connector and tighten the mounting nut on the back side of the steering wheel.
11. Enable the air bag system.
12. Connect the negative battery cable.

1993 Expo

1. Disconnect the negative battery cable.
2. Remove the horn pad from the steering wheel by, pulling the lower end of the pad upward. Disconnect horn button connector.
3. Remove steering wheel retaining nut.
4. Matchmark the steering wheel to the shaft.
5. Use a steering wheel puller to remove the steering wheel.

WARNING

Do not hammer on steering wheel to remove it. The collapsible column mechanism may be damaged.

  To install:
6. Line up the matchmarks and install the steering wheel to the shaft.
7. Torque the steering wheel attaching nut to 29 ft. lbs. (40 Nm).
8. Reconnect the horn connector and install the horn pad.

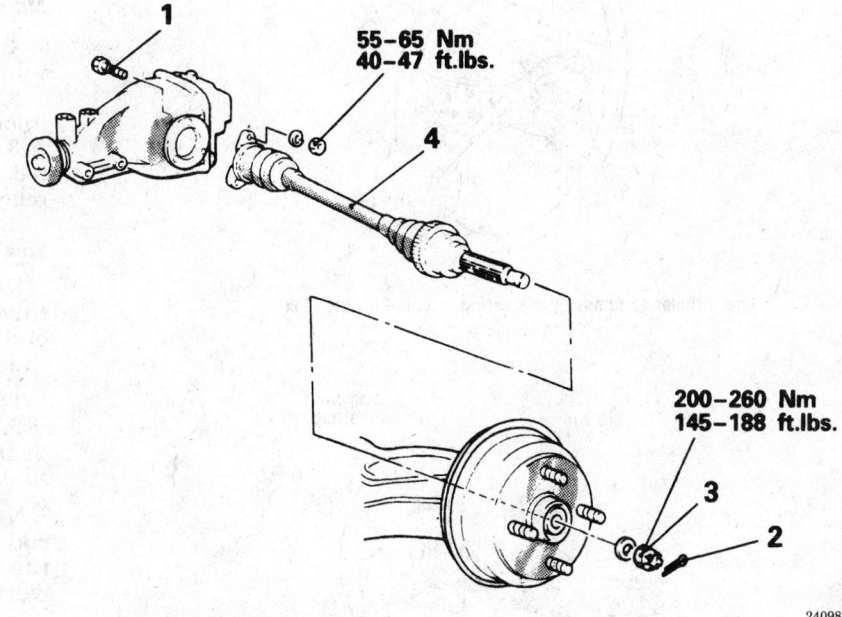

1. Bolt
2. Cotter pin
3. Drive shaft nut
4. Drive shaft

55–65 Nm
40–47 ft.lbs.

200–260 Nm
145–188 ft.lbs.

240982

Axle shaft and related components — Expo

327813

Remove the horn pad from the steering wheel — Mighty
Max

9. Connect the negative battery cable.

1994–95 Expo

------ **CAUTION** ------

If equipped with an air bag, be sure to disarm it before starting repairs on the vehicle. Failure to do so could result in severe personal injury and damage to vehicle.

1. Disarm the air bag as follows:
a. Position the front wheels in the straight-ahead position and place the key in the **LOCK** position. Remove the key from the ignition lock cylinder.
b. Disconnect the negative battery cable and insulate the cable end with high-quality electrical tape or similar non-conductive wrapping.
c. Wait at least 90 seconds before working on the vehicle. The air bag system is designed to retain enough voltage to deploy the air bag for a short period of time even after the battery has been disconnected.

2. Remove the air bag module mounting nut from behind the steering wheel.
3. To disconnect the connector of the clockspring from the air bag module, press the air bag's lock toward the module to spread the lock open. While holding lock in this position, use a small tipped prying tool to gently pry the connector from the module.
4. Remove the air bag module and store in a clean, dry place with the pad cover facing up.
5. Matchmark the steering wheel to the shaft.
6. Remove the steering wheel retaining nut and use a steering wheel puller to remove the wheel. Do not use a hammer or the collapsible mechanism in the column could be damaged.
To install:
7. Confirm that the front wheels are in a straight-ahead position. Center the clockspring by aligning the **NEUTRAL** mark on the clockspring with the mating mark on the casing. Then install the steering wheel and torque the new retaining nut to 29 ft. lbs. (40 Nm).
8. Install the air bag module.
9. Connect the negative battery cable, turn the key to the **ON** position, the SRS warning light should

327814

Use a puller to remove the steering wheel — Mighty Max and Montero

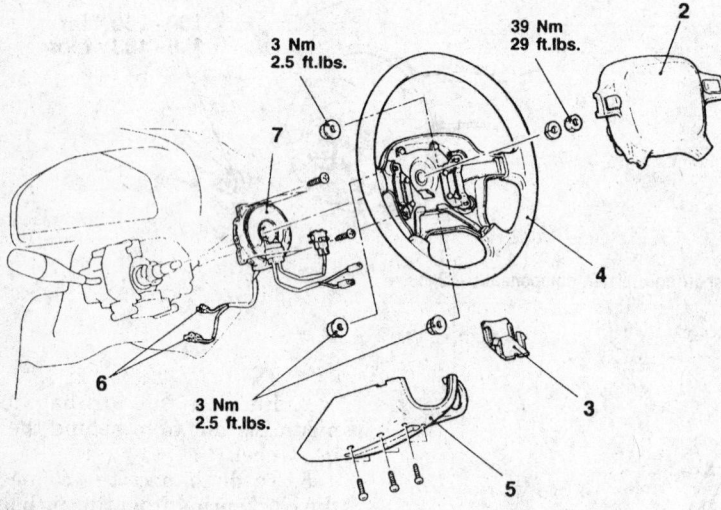

- Post-installation inspection
1. Connection of the negative (–) battery cable to the battery
2. Air bag module
3. Cap
4. Steering wheel
5. Column cover lower
6. Clock spring and body wiring harness connection
7. Clock spring
- Pre-installation inspection

327799

Steering wheel and air bag — Montero

illuminate for seven seconds and go out. If the warning light is not functioning properly, refer to SRS system diagnosis.

Tie Rod Ends

REMOVAL AND INSTALLATION

Mighty Max

1. Raise and safely support the vehicle.
2. Remove the cotter pin and nut from the tie rod end stud.

3. Using a puller, separate the tie rod from the steering knuckle or center link.
4. Loosen the sleeve clamp nut and remove the tie rod end counting the number of turns for re–installation.
   To install:
5. Install the tie rod ends, turning the same amount of times as during the removal.
6. Install the tie rod to the center link or the steering knuckle and torque the stud nuts to 33 ft. lbs. (45 Nm) and install a new cotter pin.

7. Lubricate the front end.
8. Check wheel alignment.

Montero

1. Raise and safely support the vehicle.
2. Remove the cotter pin and nut from the tie rod end stud.
3. Using a puller, separate the tie rod from the steering knuckle or the relay rod.
4. Loosen the sleeve clamp nut and remove the tie rod end.

NOTE: Count the number of turns while removing the tie rod end so it can be turned back on the same amount of turns.

To install:
5. Install the tie rod end and tighten the nut on the tie rod end stud.

NOTE: Be sure to install the tie rod end into the adjusting sleeve the same number of turns that were required to remove it.

6. Torque the stud nuts to 33 ft. lbs. (45 Nm) and install a new cotter pin.
7. Lubricate the front end.
8. Check wheel alignment.

Expo

Outer

1. Raise the front of the vehicle and support it on jackstands. Remove the wheel.
2. Remove the cotter pin and the tie rod ball joint stud nut. Note the position of the steering linkage.
3. Wire brush the threads on the tie rod shaft and lubricate with penetrating oil.
4. Using a suitable ball joint separator tool, remove the tie rod ball joint from the steering knuckle.
5. Loosen the locknut and remove the outer tie rod end from the tie rod. Count the number of complete turns it takes to completely remove it.
   To install:
6. Install the new tie rod end, turning it in exactly as many turns as it was to remove the old one. Make sure it is correctly positioned in relationship to the steering linkage.
7. Connect the outer tie rod end to the steering knuckle and install the castle nut. Torque the nut to 25 ft. lbs. (34 Nm).
8. Install a new cotter pin to the castle nut.
9. Tighten the tie rod end locking nut to 30 ft. lbs. (42 Nm).
10. Install the wheel and tire assembly.

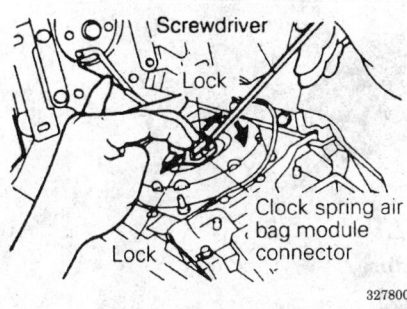

Disconnecting the clock spring air bag module connector — Montero

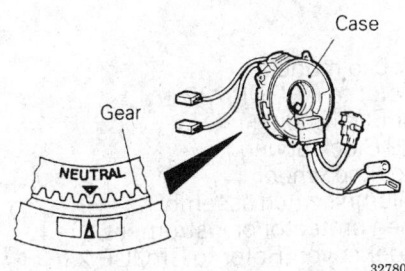

Aligning the mating mark and the neutral mark for the clock spring — Montero

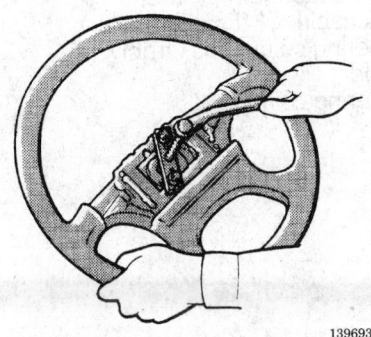

Using a puller to remove the steering wheel — 1993 Expo

11. Lower the vehicle and perform a front end alignment.

Inner

1. Raise the front of the vehicle and support it on jackstands. Remove the wheel.
2. Remove the cotter pin and the outer tie rod ball joint stud nut. Note the position of the steering linkage.
3. Wire brush the threads on the tie rod shaft and lubricate with penetrating oil.
4. Using a suitable ball joint separator tool, remove the tie rod ball joint from the steering knuckle.

5. Loosen the locknut and remove the tie rod end from the tie rod. Count the number of complete turns it takes to completely remove it.
6. Remove the tie rod-to-steering gear locknut.
7. Remove the clamps that secure the flexible boot to the steering gear.
8. Slide the boot from the inner tie rod and remove the boot.
9. Bend the lock plate tabs from the inner tie rod end nut.
10. Loosen the inner tie rod end nut from the steering gear and remove the inner tie rod end.

To install:
11. Using a new lock plate, install the tie rod end and torque the tie rod to 65 ft. lbs. (90 Nm).
12. Bend the tabs of the new lock plate to secure the inner tie rod end.
13. Slide the boot onto the steering gear and secure it with new clamps.
14. Install the outer tie rod end to the steering gear locknut.
15. Install the outer tie rod end, turning it in exactly as many turns as it was to remove the old one. Make sure it is correctly positioned in relationship to the steering linkage.
16. Connect the outer tie rod end to the steering knuckle and install the castle nut. Torque the nut to 25 ft. lbs. (34 Nm.).
17. Install a new cotter pin to the castle nut.
18. Tighten the tie rod end locking nut to 30 ft. lbs. (42 Nm).
19. Install the wheel and tire assembly.
20. Lower the vehicle and perform a front end alignment.

Power Rack and Pinion

REMOVAL AND INSTALLATION

1. Disconnect the negative battery cable.
2. Remove the pressure switch connector from the side of the pump.
3. Disconnect the return fluid line. Remove the reservoir cap and allow the return line to drain the fluid from the reservoir. If the fluid is contaminated, disconnect the ignition high tension cable and crank the engine several times to drain the fluid from the gearbox.
4. Disconnect the pressure line.
5. Remove the pump drive belt and unbolt the pump from its bracket.
To install:
6. Install the pump, wrap the belt around the pulley and tighten the bolts.

7. Replace the O-rings and connect the pressure line. Connect the pressure line so the notch in the fitting aligns and contacts the pump's guide bracket.
8. Connect the return line.
9. Connect the pressure switch connector.
10. Adjust the belt tension and tighten the adjusting bolts.
11. Connect the negative battery cable.
12. Refill the reservoir and bleed the system.

Expo

———— **WARNING** ————
If equipped with air bag, prior to removal of the steering gear box, center the front wheels and remove the ignition key. Failure to do so may damage the SRS clockspring and render SRS system inoperative, risking serious driver injury.

1. Drain power steering system:
   a. Disconnect the return hose at the reservoir and place into a suitable container.
   b. Disable the ignition system. While cranking the engine, turn the wheels several times, until system has been drained.
2. Disconnect the battery negative cable. Raise the vehicle and support safely.
3. On AWD vehicles, properly support the engine and remove the rear engine mount bracket.
4. Remove the pinch bolt holding the lower steering column joint to the rack and pinion input shaft.
5. Remove the cotter pins and disconnect the tie rod ends from the steering knuckle.
6. Disconnect the power steering fluid pressure pipe and return hose from the rack fittings.
7. Remove the rack and pinion steering assembly and its rubber mounts from the right side of the vehicle.
To install:
8. Align steering shaft and install the steering gear into the vehicle. Secure rack assembly using the retainer clamps and bolts.
9. Install the pinch bolt and torque to 13 ft. lbs. (18 Nm).
10. Install the engine mount bracket, if removed.
11. Connect the power steering fluid lines to the rack fittings.
12. Connect the tie rod ends to the steering knuckles.

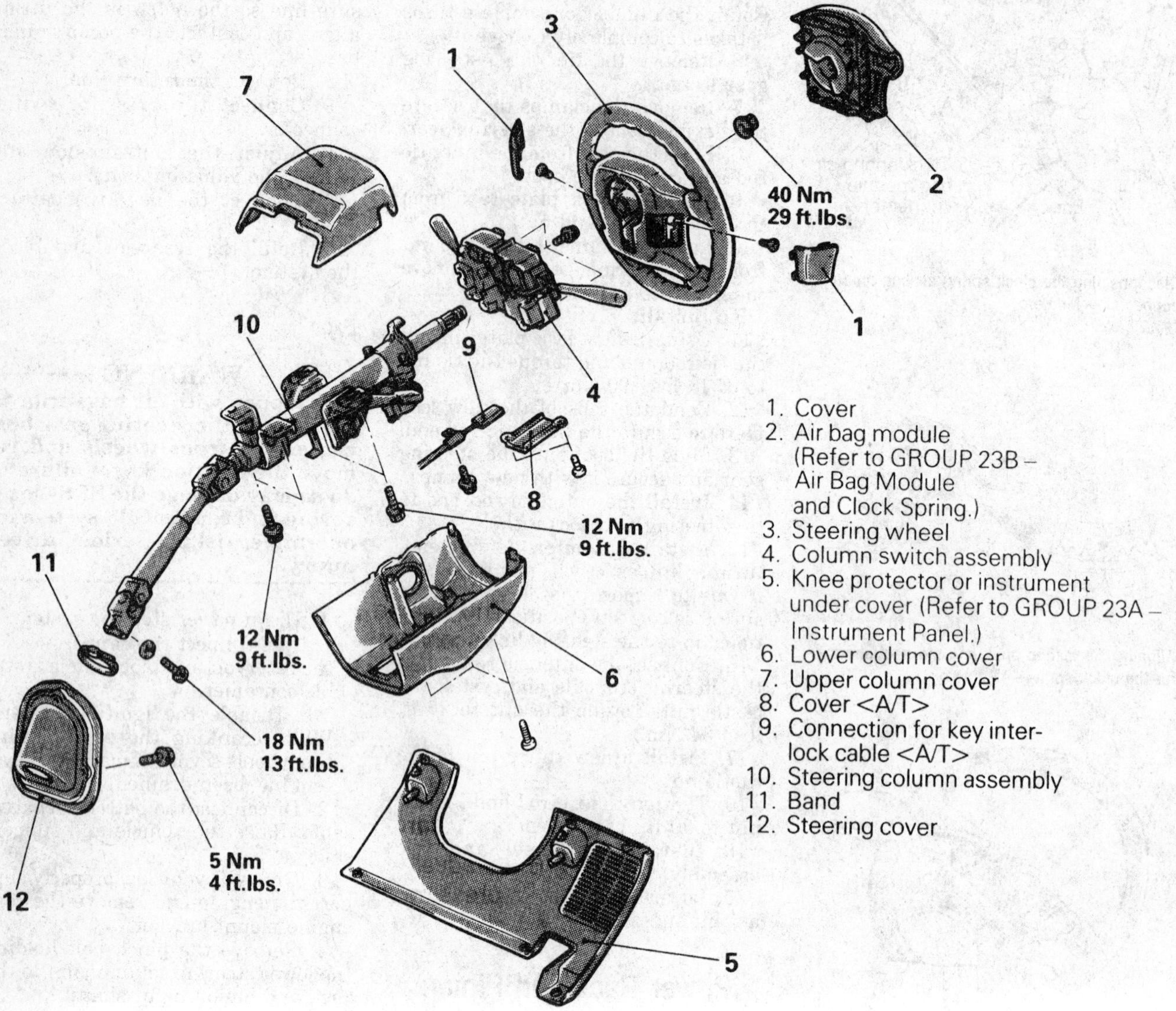

40 Nm
29 ft.lbs.

12 Nm
9 ft.lbs.

12 Nm
9 ft.lbs.

18 Nm
13 ft.lbs.

5 Nm
4 ft.lbs.

1. Cover
2. Air bag module
   (Refer to GROUP 23B –
   Air Bag Module
   and Clock Spring.)
3. Steering wheel
4. Column switch assembly
5. Knee protector or instrument
   under cover (Refer to GROUP 23A –
   Instrument Panel.)
6. Lower column cover
7. Upper column cover
8. Cover <A/T>
9. Connection for key interlock cable <A/T>
10. Steering column assembly
11. Band
12. Steering cover

332027

Steering column exploded view — 1994–95 Expo

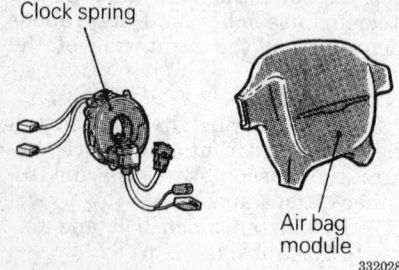

Clock spring

Air bag
module

332028

Air bag parts identification — 1994–95 Expo

13. Connect the negative battery cable. Refill the reservoir and bleed the system.
14. Perform a front end alignment.

Power Steering Pump

BLEEDING

1. Fill the reservoir with Dexron®II.
2. Raise the front end of the vehicle.
3. Disconnect the ignition coil wire.
4. Simultaneously crank the engine and turn the steering wheel from lock to lock. Repeat this several times.

NOTE: If the bleeding procedure is done with the engine running, high speed rotation of the pump will churn the fluid and create air bubbles filling the system with air. Bleed the system while cranking the engine only.

5. Lower the front end.
6. Connect one end of a tube to the breather plug on the steering box and place the other a container.
7. Start the engine and allow it to idle.
8. Loosen the breather plug and turn the steering wheel from lock to

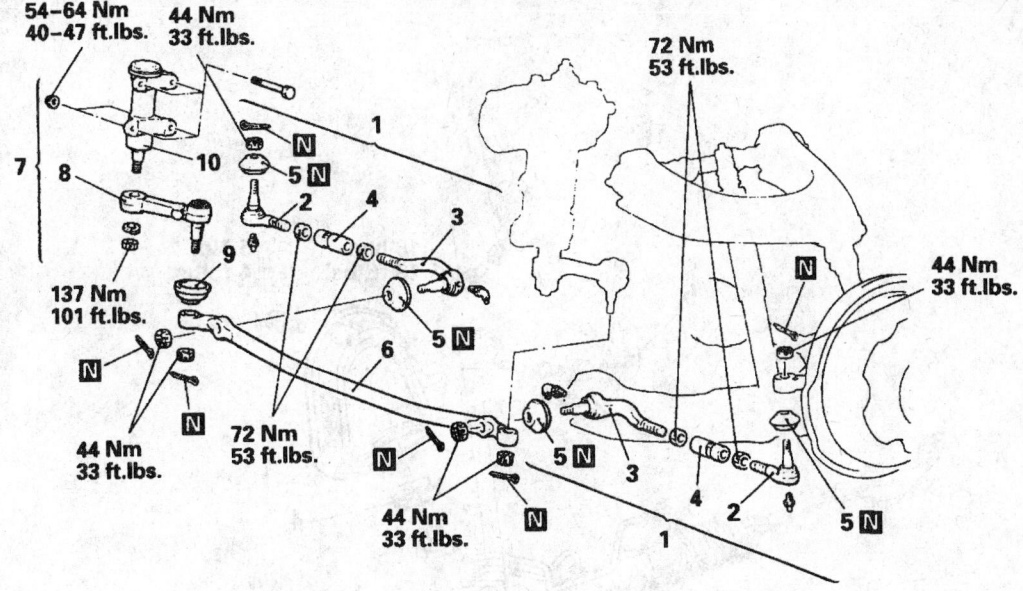

1. Tie rod assembly
2. Tie rod end, outer
3. Tie rod end, inner
4. Pipe
5. Dust cover
6. Relay rod
7. Idler arm (complete)
8. Idler arm
9. Dust cover
10. Idler arm support

321609

Steering linkage and related components — Montero

lock continuously until no more air bubbles appear in the fluid coming out the tube.

NOTE: Do not hold the steering wheel all the way against the stop for more than 5 seconds.

9. After the bleeding is done, tighten the breather plug and refill the reservoir.

REMOVAL AND INSTALLATION

Except Expo

1. Disconnect the negative battery cable.
2. Remove the pressure switch connector from the side of the pump.
3. Disconnect the return fluid line. Remove the reservoir cap and allow the return line to drain the fluid from the reservoir. If the fluid is contaminated, disconnect the ignition high tension cable and crank the engine several times to drain the fluid from the gearbox.
4. Disconnect the pressure line.
5. Remove the pump drive belt and unbolt the pump from its bracket.

To install:
6. Install the pump, wrap the belt around the pulley and tighten the bolts.
7. Replace the O-rings and connect the pressure line. Connect the pressure line so the notch in the fitting aligns and contacts the pump's guide bracket.
8. Connect the return line.
9. Connect the pressure switch connector.
10. Adjust the belt tension and tighten the adjusting bolts.
11. Connect the negative battery cable.
12. Refill the reservoir and bleed the system.

Expo

1. Disconnect the battery negative cable.
2. Loose and remove the alternator drive belt.
3. Remove the pressure switch connector from the side of the pump.

NOTE: If the alternator is located under the oil pump, cover it with a shop towel to protect it from oil.

4. Disconnect the return fluid line. Remove the reservoir cap and allow the return line to drain the fluid from the reservoir. If the fluid is contaminated, disconnect the ignition high tension cable and crank the engine several times to drain the fluid from the gearbox.
5. Disconnect the pressure line.
6. Remove the pump drive belt and unbolt the pump from its bracket.

To install:
7. Install the pump, wrap the belt around the pulley and tighten the mounting bolts.
8. Replace the O-rings and connect the pressure line. Connect the pressure line so the notch in the fitting aligns and contacts the pump's guide bracket.
9. Connect the return line. Connect the pressure switch connector.
10. Adjust both the power steering and alternator belts for proper tension and tighten the adjusting bolts.
11. Refill the reservoir and bleed the system.

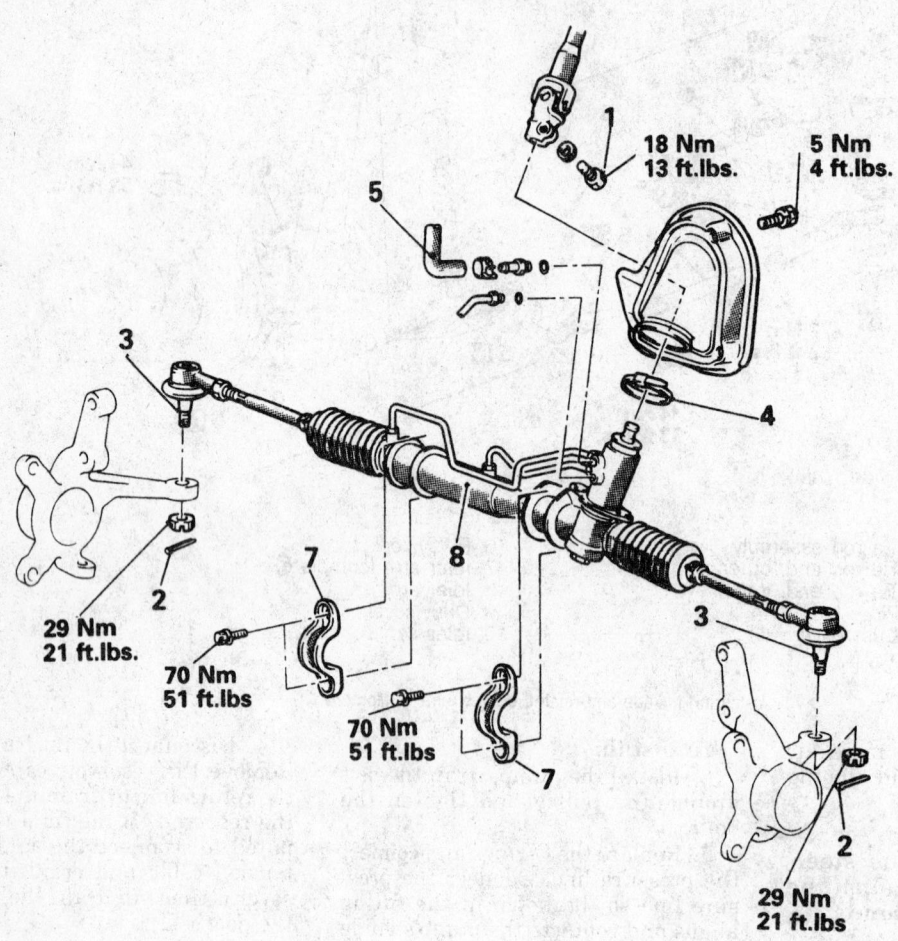

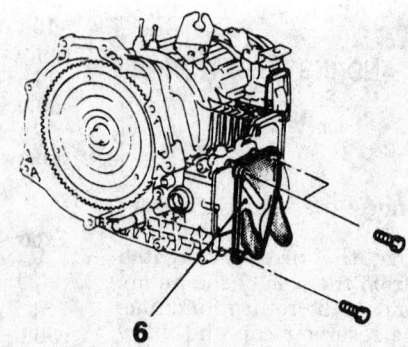

1. Joint assembly and gear box connecting bolt
2. Cotter pin
3. Connection for tie-rod end and knuckle
4. Band
5. Connection for return hose
6. Bracket <AWD>
7. Clamp
8. Gear box assembly

240942

Steering rack assembly and related components — Expo

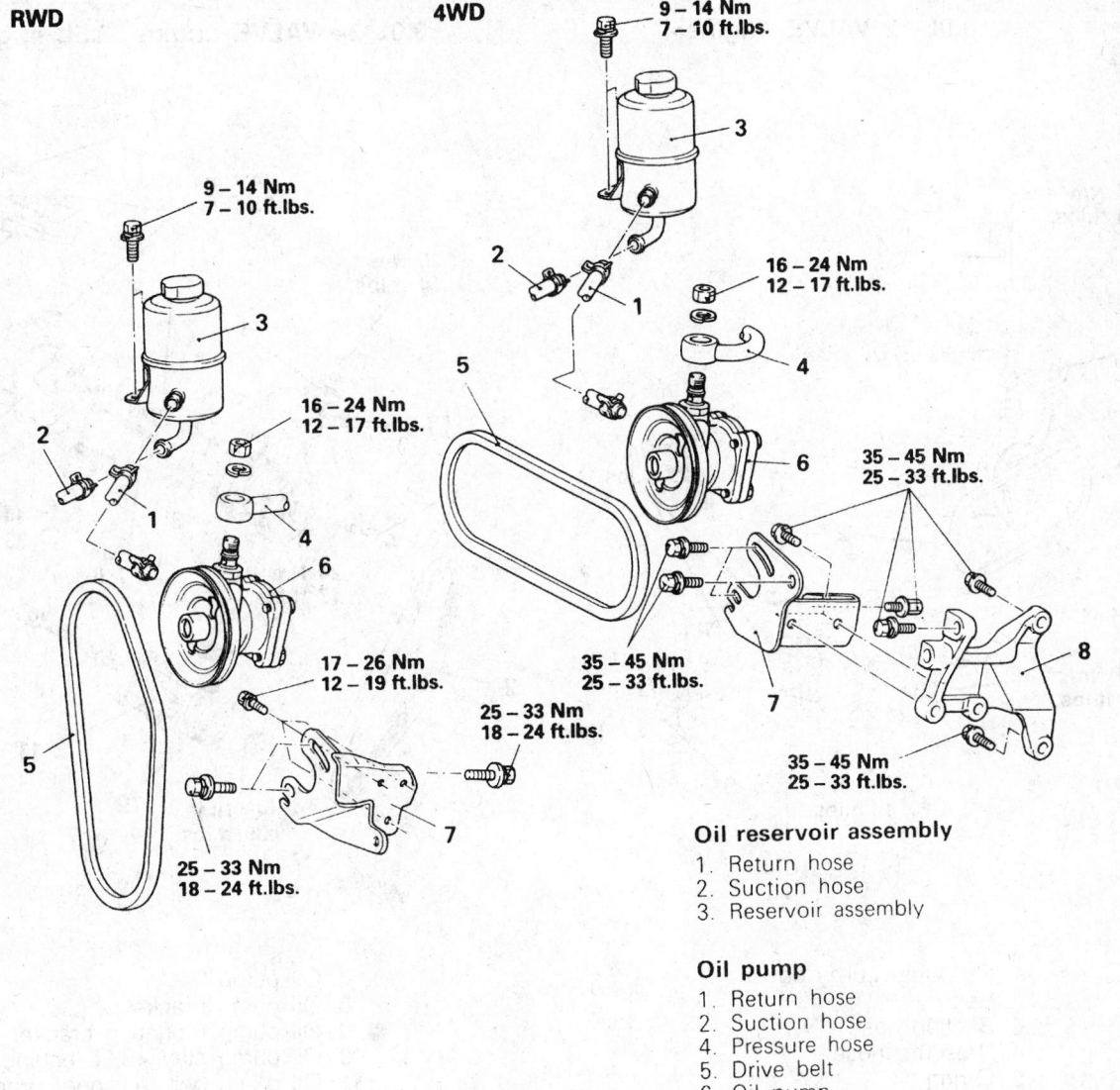

RWD

4WD

9 – 14 Nm
7 – 10 ft.lbs.

9 – 14 Nm
7 – 10 ft.lbs.

16 – 24 Nm
12 – 17 ft.lbs.

16 – 24 Nm
12 – 17 ft.lbs.

35 – 45 Nm
25 – 33 ft.lbs.

17 – 26 Nm
12 – 19 ft.lbs.

35 – 45 Nm
25 – 33 ft.lbs.

35 – 45 Nm
25 – 33 ft.lbs.

25 – 33 Nm
18 – 24 ft.lbs.

35 – 45 Nm
25 – 33 ft.lbs.

25 – 33 Nm
18 – 24 ft.lbs.

Oil reservoir assembly

1. Return hose
2. Suction hose
3. Reservoir assembly

Oil pump

1. Return hose
2. Suction hose
4. Pressure hose
5. Drive belt
6. Oil pump
7. Oil pump bracket
8. Oil pump mount bracket

327658

Power steering pump — Mighty Max

BRAKES

Anti-Lock Brake System Service

PRECAUTIONS

• Certain components within the Anti-Lock Brake System (ABS) are not intended to be serviced or repaired individually. Only those components with removal and installation procedures should be serviced.

• Do not use rubber hoses or other parts not specifically specified for and ABS system. When using repair kits, replace all parts included in the kit. Partial or incorrect repair may lead to functional problems and require the replacement of components.

• Lubricate rubber parts with clean, fresh brake fluid to ease assembly. Do not use lubricated shop air to clean parts; damage to rubber components may result.

• Use only specified brake fluid from an unopened container.

• If any hydraulic component or line is removed or replaced, it may be necessary to bleed the entire system.

• A clean repair area is essential. Always clean the reservoir and cap thoroughly before removing the cap. The slightest amount of dirt in the fluid may plug an orifice and impair the system function. Perform repairs after components have been thoroughly cleaned; use only denatured alcohol to clean components. Do not allow ABS components to come into contact with any substance containing mineral oil; this includes used shop rags.

• The Anti-Lock control unit is a microprocessor similar to other computer units in the vehicle. Ensure that the ignition switch is **OFF** before removing or installing control-

<3.0L-12 VALVE engine>

<3.0L-24 VALVE engine, 3.5L engine>

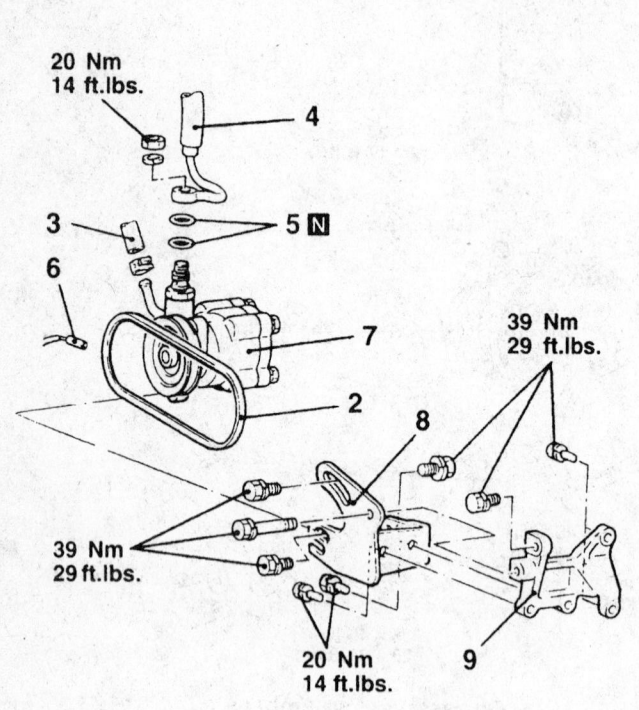

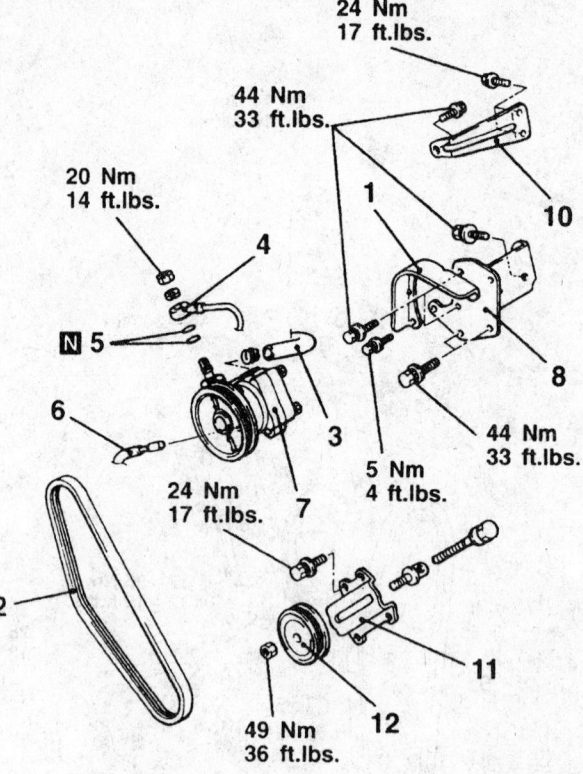

1. Oil pump pulley cover
2. Belt
3. Suction hose
4. Pressure hose
5. O-ring
6. Pressure switch connector <Except 3.0L-12 VALVE engine M/T>

7. Oil pump
8. Oil pump bracket
9. Oil pump mounting bracket
10. Oil pump stay <3.5L engine>
11. Oil pump belt tensioner bracket
12. Oil pump belt tension pulley

327660

Power steering pump — Montero

ler harnesses. Avoid static electricity discharge at or near the controller.

• If any arc welding is to be done on the vehicle, the control unit should be unplugged before welding operations begin.

Master Cylinder

REMOVAL AND INSTALLATION

Except Expo

1. Disconnect the negative battery cable.

2. Disconnect the brake fluid level sensor connector.

3. Disconnect and plug the brake lines at the master cylinder.

4. Remove the nuts attaching the master cylinder to the power booster.

5. Remove the master cylinder from the mounting studs.

To install:

6. Check the clearance between the brake booster pushrod and the master cylinder piston as follows:

a. Measure the distance (1) between the rear edge of the master cylinder body and the bottom of the recess in the master cylinder piston.

b. Measure the distance (2) between the rear edge of the master

cylinder body and the master cylinder flange.

c. Subtract distance (2) from distance (1).

d. Measure the projected length (3) of the pushrod from the master cylinder attaching flange on the brake booster.

e. The value obtained when the value in Step d is subtracted from the value in Step c, is the clearance between the master cylinder piston and the pushrod. Clearance should be 0.026–0.035 in. (0.65–0.90 mm) for the Montero 1993–95, all other vehicles should be 0.1–0.5mm (0.004–0.020 in.).

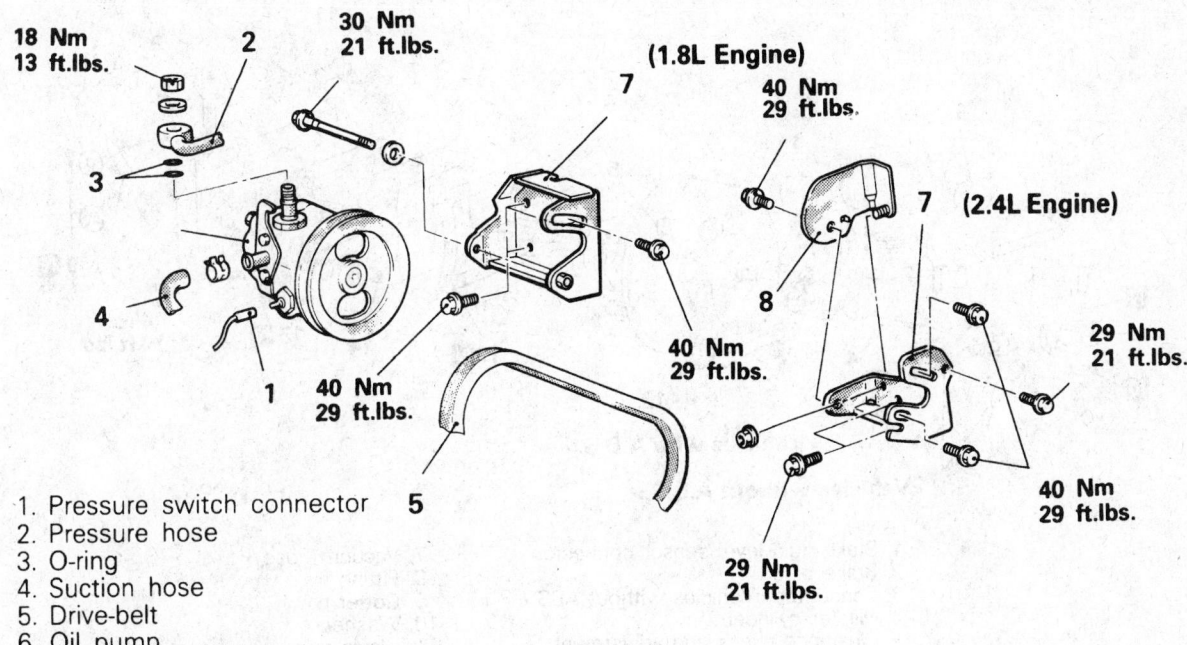

1. Pressure switch connector
2. Pressure hose
3. O-ring
4. Suction hose
5. Drive-belt
6. Oil pump
7. Oil pump bracket
8. Heat protector (2.4L Engine)

240938

Power steering pump mounting — Expo

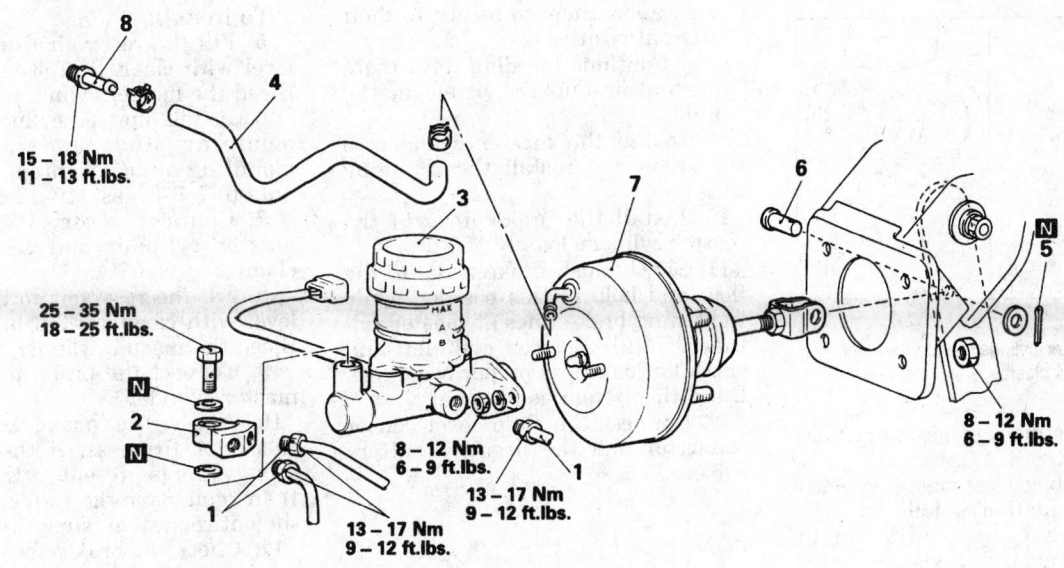

1. Brake tube connector
2. Connector
3. Master cylinder assembly
   Adjusting brake booster push rod and
   primary piston clearance
4. Vacuum hose
5. Cotter pin
6. Clevis pin
7. Brake booster
8. Fitting

327090

Master cylinder and booster assembly — Mighty Max

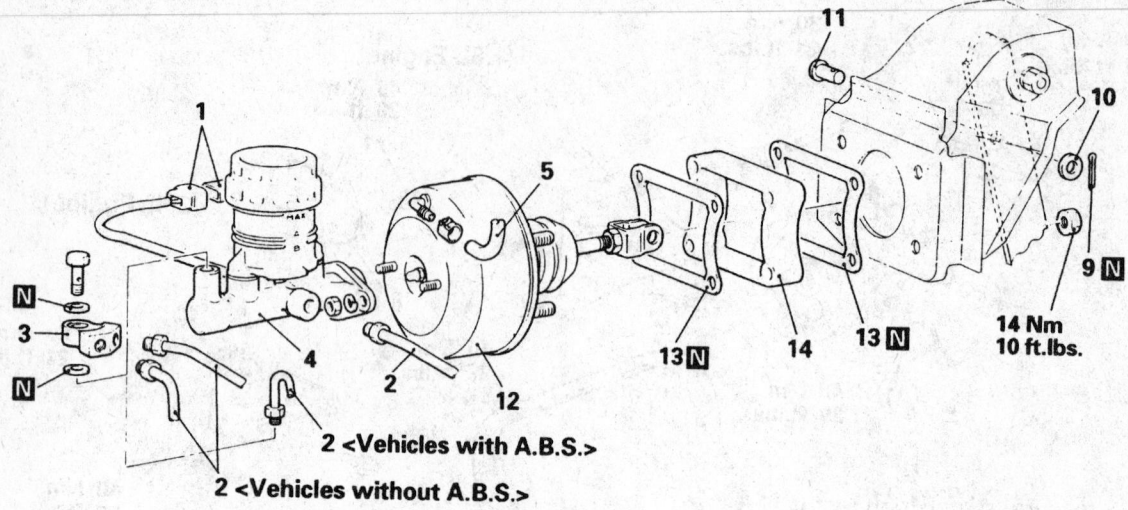

1
3

2 <Vehicles with A.B.S.>

2 <Vehicles without A.B.S.>

11
10
9
14 Nm
10 ft.lbs.
13   14   13

1. Brake fluid level sensor connector
2. Brake pipe
3. Connector <Vehicles without ABS>
4. Master cylinder
• Clearance check and adjustment between primary piston and push rod
5. Vacuum hose (with built-in check valve)
6. Vacuum pipe

7. Vacuum hose
8. Fitting
9. Cotter pin
10. Washer
11. Clevis pin
12. Brake booster
13. Spacer
14. Sealant

327091

Master cylinder and booster assembly — Montero

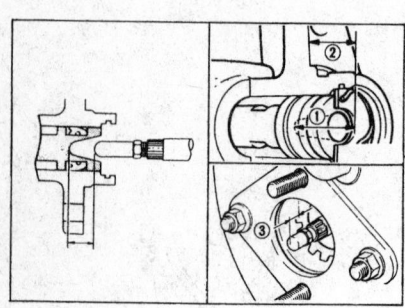

327089

Measuring master cylinder piston-to-brake booster pushrod clearance

7. If adjustment is necessary, turn the pushrod.

8. Bench bleed the master cylinder prior to installation as follows:

a. Mount the replacement master cylinder in a vise.

b. Attach bleed tubes to the outlet ports and insert the tubes in the reservoir. Bleed tubes can be fabricated from brake line and the appropriate fittings.

c. Fill the reservoir with clean brake fluid.

d. Press the master cylinder pistons inward using a suitable tool,

then allow them to return to their original positions.

e. Continue bleeding until there are no air bubbles visible in the fluid.

9. Install the master cylinder to the studs and install the retaining nuts.

10. Install the brake lines to the master cylinder loosely.

11. Slowly push brake pedal to the floor and hold in this position while tightening brake lines at master cylinder. Refill master cylinder and check for leaks and proper pedal feel. Bleed the system as necessary.

12. Connect the fluid level sensor connector and the negative battery cable.

Expo

1. Disconnect the negative battery cable.

2. Disconnect the fluid level sensor connector, if equipped.

3. Disconnect the brake lines from the master cylinder. If a separate fluid reservoir is used, plug the lines to prevent drainage.

4. Remove the two nuts securing the master cylinder to the brake booster and remove the master cylinder.

To install:

5. Fill the reservoir to the proper level with clean DOT 3 brake fluid. Bleed the master cylinder.

6. Install master cylinder to the mounting studs and install the mounting nuts. Tighten mounting nuts to 7–9 ft. lbs. (10–12 Nm).

7. Connect reservoir hoses to master cylinder and secure with clamps.

8. Fill the reservoir to the proper level with clean DOT 3 brake fluid. Bleed the master cylinder.

9. Connect the brake lines to the master cylinder.

10. Apply the brake pedal and check for firmness. If the pedal is spongy, air is present in the system. If air remains in the system, bleeding the entire system is required.

11. Check the brakes for proper operation and leaks.

Brake Caliper

REMOVAL AND INSTALLATION

Front Brake Caliper

1. Raise and safely support the vehicle.

2. Remove the wheel and tire assembly.

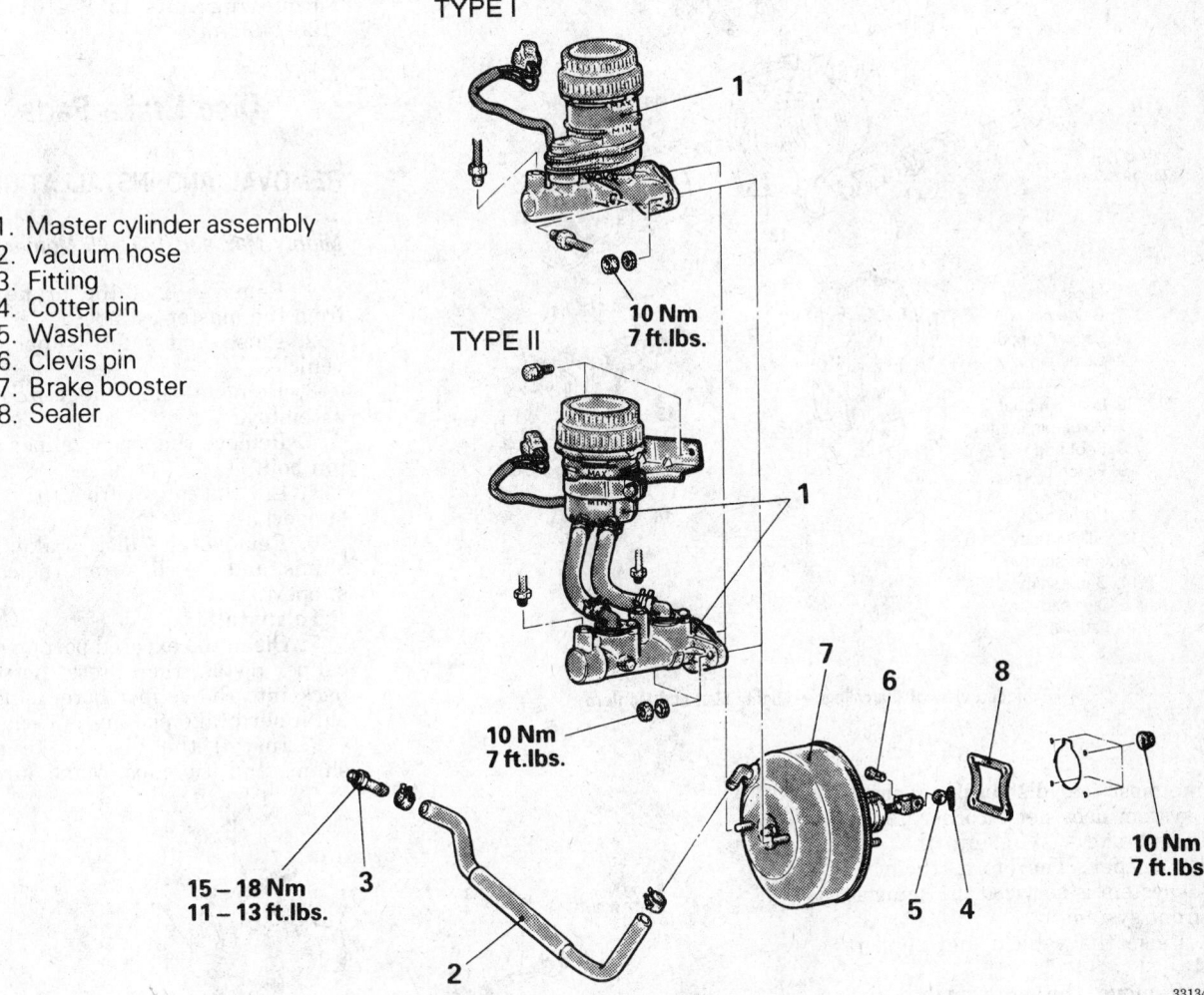

TYPE I

TYPE II

1. Master cylinder assembly
2. Vacuum hose
3. Fitting
4. Cotter pin
5. Washer
6. Clevis pin
7. Brake booster
8. Sealer

10 Nm
7 ft.lbs.

10 Nm
7 ft.lbs.

10 Nm
7 ft.lbs.

15 – 18 Nm
11 – 13 ft.lbs.

331346

Master cylinder and booster assembly — Expo

3. Disconnect the brake hose from the caliper brake line and remove the retaining clip.

4. Remove the caliper guide pin bolts.

5. Lift the caliper from the caliper support.

To install:

6. Make sure the disc brake pad shims and clips are properly positioned.

7. Position the caliper over the rotor so the caliper engages the adaptor correctly. Install the mounting pins and tighten them as follows:

- Montero: 54 ft. lbs. (74 Nm)
- Mighty Max: 72 ft. lbs. (100 Nm)

8. Connect the brake hose to the caliper brake line and install the retaining clip.

9. Bleed the brake system.

10. Install the wheel and tire assembly.

Rear Brake Caliper

1. Raise and safely support the vehicle.

2. Remove the wheel and tire assembly.

3. Disconnect the brake hose from the caliper brake line and remove the retaining clip.

4. Remove the caliper guide pin bolts.

5. Lift the caliper from the caliper support.

To install:

6. Make sure the disc brake pad shims and clips are properly positioned.

7. Position the caliper over the rotor so the caliper engages the adaptor correctly. Install the mounting pins and tighten them to 32 ft. lbs. (44 Nm).

8. Connect the brake hose to the caliper brake line and install the retaining clip.

9. Bleed the brake system.

10. Install the wheel and tire assembly.

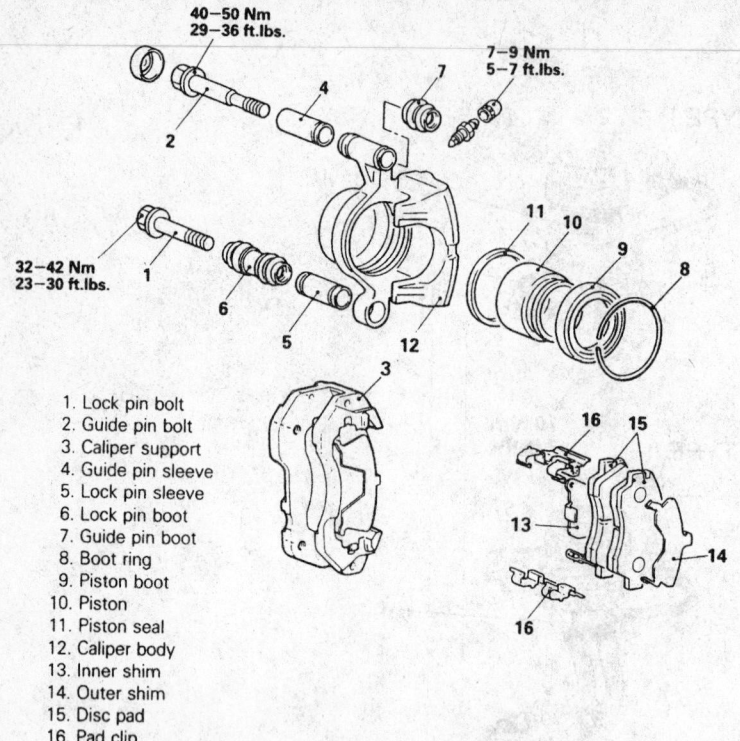

1. Lock pin bolt
2. Guide pin bolt
3. Caliper support
4. Guide pin sleeve
5. Lock pin sleeve
6. Lock pin boot
7. Guide pin boot
8. Boot ring
9. Piston boot
10. Piston
11. Piston seal
12. Caliper body
13. Inner shim
14. Outer shim
15. Disc pad
16. Pad clip

326994

Exploded view of the caliper — Mighty Max and Montero

Expo

Unlike most rear disc brake designs, this system does not incorporate the parking brake system into the rear brake caliper. Therefore, the rear brake system is serviced the same as the front system.

1. Raise the vehicle and support safely.
2. Remove the appropriate tire and wheel assembly.

NOTE: Do not allow the reservoir to empty or air will enter and complete system bleeding will be required.

3. To disconnect the front brake hose, hold the nut on the brake hose side and loosen the flared brake line nut. Remove the brake hose from the caliper.
4. Remove the caliper guide and lock pins and lift the caliper assembly from the caliper support.
To install
5. Position caliper onto the caliper support. Install the guide pin and lock pin. Tighten to specification.
6. Reconnect brake hose.

NOTE: Use caution not to twist brake hose during installation.

7. Bleed the brake system.

8. Apply brake pedal and inspect system. Ensure proper operation and no leakage.
9. Install tire and wheel assembly. Torque lug nuts to 87-101 ft.lbs. (120-140 Nm).

Disc Brake Pads

REMOVAL AND INSTALLATION

Mighty Max and 1993–95 Montero

1. Remove ½ of the brake fluid from the master cylinder.
2. Raise and safely support the vehicle.
3. Remove the wheel and tire assembly.
4. Remove the lower caliper guide pin bolt.
5. Lift the caliper from the caliper support.
6. Remove the disc brake pads, shims, and the clips from the caliper support.
To install:
7. Clean the exposed portion of the caliper piston, then press the piston back into the caliper bore using the old inner brake pad and a C-clamp.
8. Install the disc brake pads, shims, and the clips. Make sure the

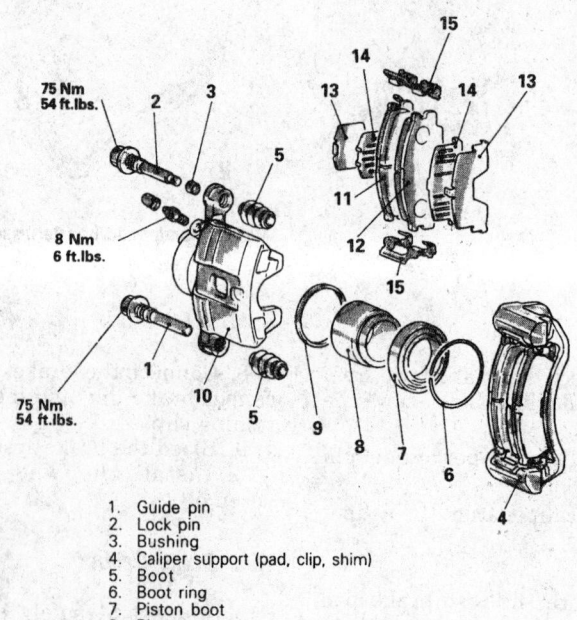

1. Guide pin
2. Lock pin
3. Bushing
4. Caliper support (pad, clip, shim)
5. Boot
6. Boot ring
7. Piston boot
8. Piston
9. Piston seal
10. Caliper body
11. Pad assembly (with wear indicator)
12. Pad assembly
13. Outer shim
14. Inner shim
15. Clip

239359

Front brake caliper and related components — Expo

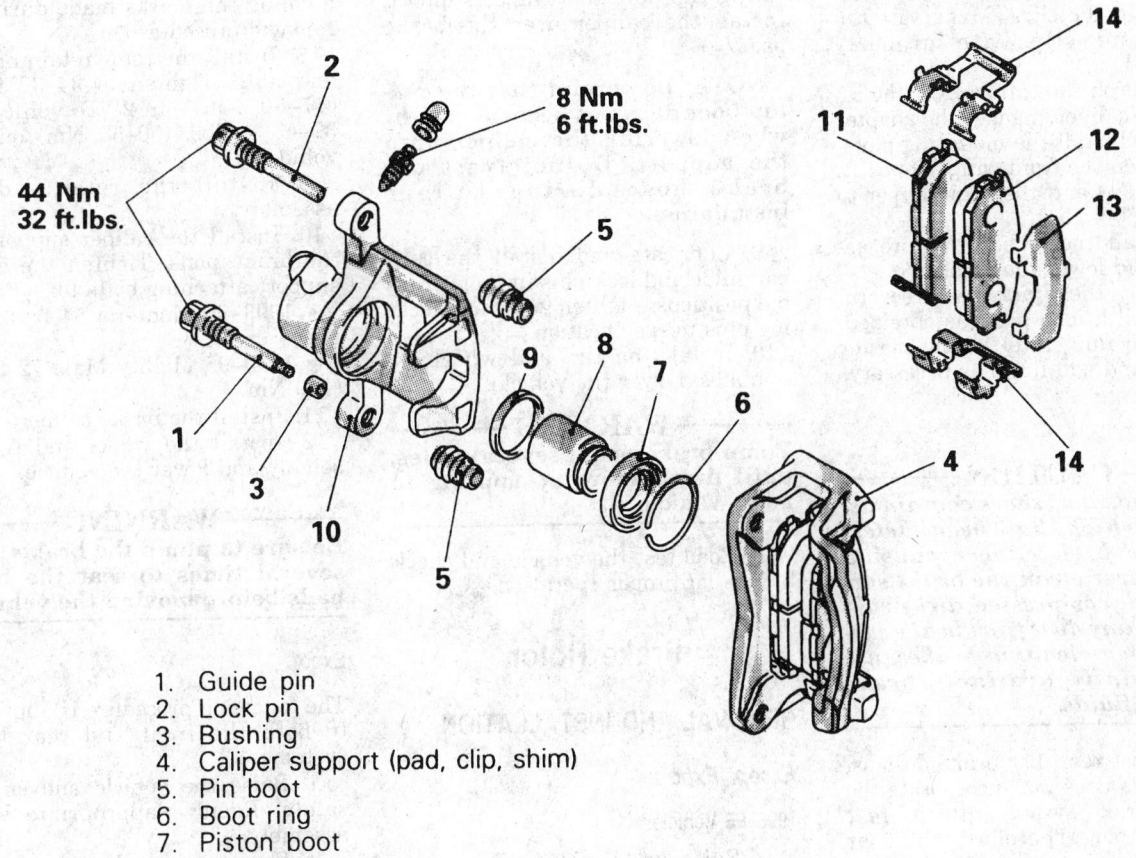

44 Nm 32 ft.lbs.

8 Nm 6 ft.lbs.

1. Guide pin
2. Lock pin
3. Bushing
4. Caliper support (pad, clip, shim)
5. Pin boot
6. Boot ring
7. Piston boot
8. Piston
9. Piston seal
10. Caliper body
11. Pad and wear indicator assembly
12. Pad assembly
13. Outer shim
14. Clip

239360

Rear brake caliper and related components — Expo

Pad clip

Inner shim

Outer shim

Pad clip

327065

Correct position of the clips and shims — Mighty Max and Montero

shims and clips are properly positioned.

9. Position the caliper over the rotor so the caliper engages the adaptor correctly. Install the mounting pin(s) and tighten as follows:

• 1993–95 Montero: 54 ft. lbs. (74 Nm)

• Mighty Max: 72 ft. lbs. (100 Nm)

10. Install the wheel and tire assembly and lower the vehicle.

11. Apply the brake pedal several times until a firm pedal is obtained. Check the fluid level in the master cylinder and add fluid, as necessary.

1996–97 Montero

1. Remove ½ of the brake fluid from the master cylinder.

2. Raise and safely support the vehicle.

3. Remove the wheel and tire assembly.

4. Remove the lower caliper guide pin bolt.

5. Lift the caliper from the caliper support.

6. Remove the disc brake pads, shims, and the clips from the caliper support.

To install:

7. Clean the exposed portion of the caliper piston, then press the piston

back into the caliper bore using the old inner brake pad and a C-clamp.

8. Install the disc brake pads, shims, and the clips. Make sure the shims and clips are properly positioned.

9. Position the caliper over the rotor so the caliper engages the adaptor correctly. Install the mounting pin(s) and tighten the front caliper to 54 ft. lbs. (74 Nm) and the rear caliper to 32 ft. lbs. (44 Nm).

10. Install the wheel and tire assembly and lower the vehicle.

11. Apply the brake pedal several times until a firm pedal is obtained. Check the fluid level in the master cylinder and add fluid, as necessary.

Expo

---— CAUTION ———

Brake pads and shoes contain asbestos, which has been determined to be a cancer causing agent. Never clean the brake surfaces with compressed air! Avoid inhaling any dust from brake surfaces! When cleaning brakes, use commercially available brake cleaning fluids.

Unlike most rear disc brake designs, this system does not incorporate the parking brake system into the rear brake caliper. Therefore, the rear brake system is serviced the same as the front system.

1. Remove some of the brake fluid from the master cylinder reservoir. The reservoir should be no more than ½ full. When the pistons are depressed into the calipers, excess fluid will flow up into the reservoir.

2. Raise the vehicle and support safely.

3. Remove the appropriate tire and wheel assemblies.

4. Remove the caliper guide and lock pins and lift the caliper assembly from the caliper support. Tie the caliper out of the way using wire. Do not allow the caliper to hang by the brake line.

NOTE: On some vehicles, the caliper can be flipped up by leaving the upper pin in place and using it as a pivot point.

5. Remove the brake pads, spring clip and shims. Take note of positioning to aid installation.

6. Install the wheel lug nuts onto the studs and lightly tighten. This is done to hold the disc on the hub.

To install:

7. Use a large C-clamp to compress piston back into caliper bore.

8. Lubricate slide points and install the brake pads, shims and spring clip onto the caliper support. Install the caliper over the brake pads.

NOTE: Be careful that the piston boot does not become caught when lowering the caliper onto the support. Do not twist the brake hose during caliper installation.

9. Lubricate and install the caliper guide and lock pins in their original positions. Tighten guide and locking pins to specification.

10. Install the tire and wheel assemblies. Lower the vehicle.

---— WARNING ———

Pump brake pedal several times, until firm, before attempting to move vehicle.

---

11. Road test the vehicle and check brakes for proper operation.

Brake Rotor

REMOVAL AND INSTALLATION

Except Expo

1993–95 Vehicles

1. Raise and safely support the vehicle.

2. Remove the wheel and tire assembly.

3. Remove the brake caliper. Support the caliper out of the way with wire; do not let the caliper hang by the brake hose.

4. Remove the brake pads and the caliper support.

5. Remove the hub and rotor assembly.

6. Matchmark the rotor to the hub. Remove the rotor-to-hub retaining bolts and nuts and separate the rotor and hub.

To install:

7. Attach the rotor to the hub, aligning the marks made during the removal procedure.

8. Install the rotor retaining bolts and nuts. Tighten to 34–37 ft. lbs. (47–50 Nm) on 2WD vehicles or 37–43 ft. lbs. (50–58 Nm) on 4WD vehicles.

9. Install the rotor and hub assembly.

10. Install the caliper support and the brake pads. Tighten the caliper support attaching bolts to:

- 1993–95 Montero: 54 ft. lbs. (74 Nm)
- 1993–95 Mighty Max: 72 ft. lbs. (100 Nm)

11. Install the brake caliper.

12. Install the wheel and tire assembly and lower the vehicle.

---— WARNING ———

Be sure to pump the brake pedal several times to seat the brake pads before moving the vehicle.

Expo

The following procedure is applicable to both the front and rear brake systems.

1. Raise the vehicle and support safely. Remove appropriate wheel assembly.

2. Remove the caliper and brake pads. Support the caliper out of the way using wire.

3. The rotor on most models is held to the hub by two small threaded screws. Remove screws, if equipped, and pull off the rotor.

To install

4. Position the rotor on the hub and install mounting screws.

5. Install caliper holder and brake pads. Slide caliper over brake pads and tighten guide pins.

6. Install wheel and torque lug nuts. Check brake pedal before attempting to move vehicle.

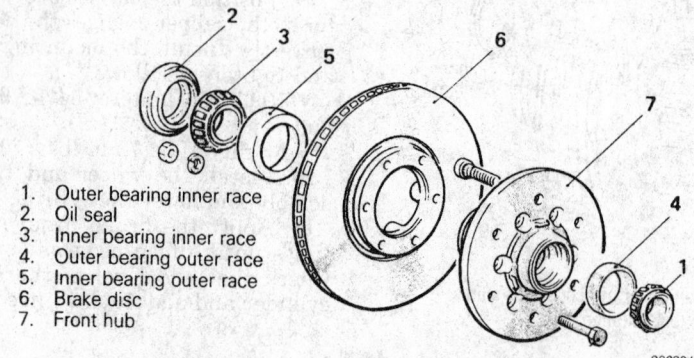

1. Outer bearing inner race
2. Oil seal
3. Inner bearing inner race
4. Outer bearing outer race
5. Inner bearing outer race
6. Brake disc
7. Front hub

286234

2WD hub and rotor assembly — 1993–95 vehicles

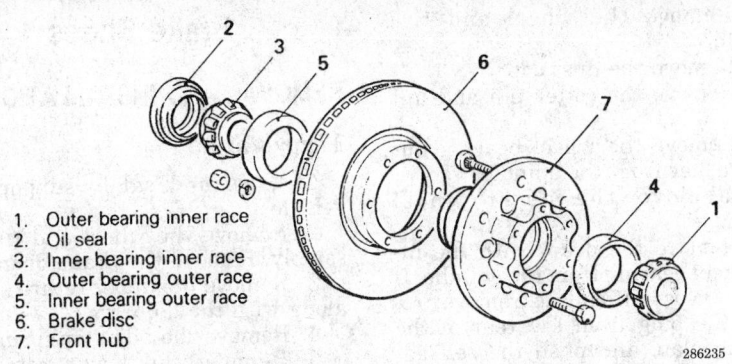

1. Outer bearing inner race
2. Oil seal
3. Inner bearing inner race
4. Outer bearing outer race
5. Inner bearing outer race
6. Brake disc
7. Front hub

286235

4WD hub and rotor assembly — 1993–95 vehicles

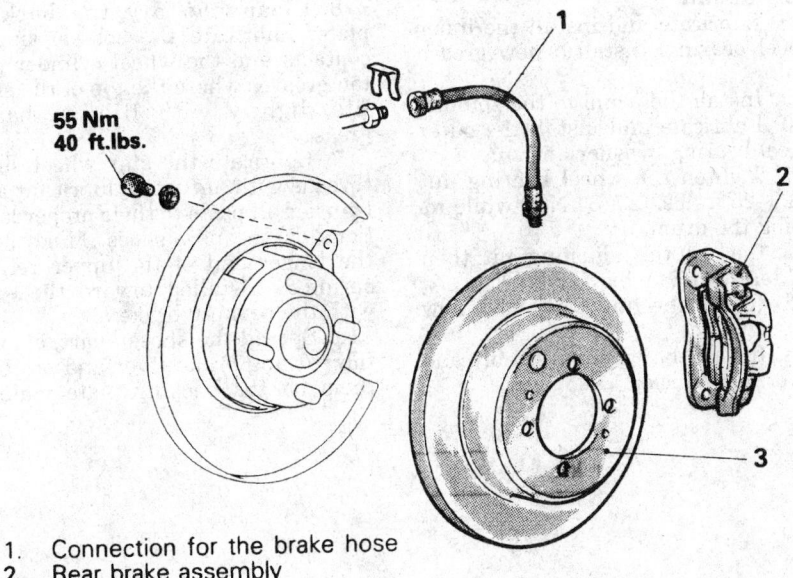

55 Nm
40 ft.lbs.

1. Connection for the brake hose
2. Rear brake assembly
3. Brake disc

239395

Rear brake rotor and related components — Expo

1996–97 Vehicles

1. Raise and safely support the vehicle.

2. Remove the wheel and tire assembly.

3. Remove the brake caliper. Support the caliper out of the way with wire; do not let the caliper hang by the brake hose.

4. Remove the brake pads and the caliper support.

5. Remove the hub cap and the snapring.

6. If equipped with ABS, remove the speed sensor.

7. After removing the lock washer, remove the lock nut with SST MB990954–01 or equivalent.

8. Remove the hub assembly.

9. Matchmark the rotor to the hub. Remove the retaining bolts and nuts and separate the rotor and hub.

To install:

10. Attach the rotor to the hub, aligning the marks made during the removal procedure.

11. Install the rotor retaining bolts and nuts and tighten to 35–43 ft. lbs. (40–59 Nm).

12. Install the rotor and hub assembly.

13. Montero: Install the lock nut.
   a. Tighten to 119 ft. lbs. (162 Nm).
   b. Loosen to 0 ft. lbs. (0 Nm).
   c. Re–tighten to 18 ft. lbs. (25 Nm) and then loosen 30–40°.

14. Mighty Max RWD: Install the lock nut.
   a. Tighten to 22 ft. lbs. (30 Nm).
   b. Loosen to 0 ft. lbs. (0 Nm).
   c. Re–tighten to 6 ft. lbs. (8 Nm) and then loosen 30–40°.

15. Mighty Max 4WD: Install the lock nut.
   a. Tighten to 94–145 ft. lbs. (130–300 Nm).
   b. Loosen to 0 ft. lbs. (0 Nm).
   c. Re–tighten to 18 ft. lbs. (25 Nm) and then loosen 30–40°.

16. If removed, install the ABS speed sensor.

17. Install the snapring and the hub cap.

18. Install the caliper support and the brake pads. Tighten the caliper support attaching bolts to 54 ft. lbs. (74 Nm).

19. Install the brake caliper and torque the bolts to 65 ft. lbs. (88 Nm) for the Montero and 72 ft. lbs. (100 Nm).

20. Install the wheel and tire assembly and lower the vehicle.

— WARNING —

Be sure to pump the brake pedal several times to seat the brake pads before moving the vehicle.

Brake Drums

REMOVAL AND INSTALLATION

Mighty Max

— CAUTION —

Some brake shoes contain asbestos, which has been determined to be a cancer causing agent. Never clean the brake surfaces with compressed air! Avoid inhaling any dust from any brake surface. When cleaning brake surfaces, use a commercially available brake cleaning fluid and the appropriate respiratory protection.

1. Raise and safely support the vehicle.

2. Remove the wheel and tire assembly.

3. Remove the brake drum.

4. If the brake drum is difficult to remove, proceed as follows:
   a. Remove the plug from the brake adjustment access hole on the backing plate.

b. Insert a thin bladed screwdriver through the adjusting hole and hold the adjusting lever away from the star wheel.

c. Retract the brake shoes by turning the star wheel with a brake adjusting tool.

5. If the brake drum will not release from the axle flange, thread M8x1.25 bolts into the holes provided in the brake drum flange surface.

To install:

6. Install the brake drum.

7. Adjust the brake shoes.

8. Install the wheel and tire assembly.

9. Lower the vehicle and check for proper brake operation.

Expo

With Rear Hub Assembly

1. Raise the vehicle and support safely.

2. Remove the wheel and tire assembly.

3. Remove drum retaining screw and remove the brake drum from the vehicle.

4. The installation is the reverse of the removal procedure.

Without Rear Hub Assembly

1. Raise the vehicle and support safely.

2. Remove the wheel and tire assembly.

3. Remove the dust cap.

4. Remove the cotter pin and nut lock.

5. Remove the wheel bearing nut and washer from the spindle.

6. Remove the outer wheel bearing.

7. Remove the drum with the inner wheel bearing from the spindle. If the drum is difficult to remove, remove the plug from the rear of the backing plate and push the self adjuster lever away from the star wheel. Rotate the star wheel with an upward motion to retract the shoes and remove the drum. Remove the grease seal.

To install:

8. Lubricate and install the inner wheel bearing. Install a new grease seal.

9. Install the drum to the spindle.

10. Lubricate and install the outer wheel bearing, washer and nut.

11. Tighten the wheel bearing nut to 20–25 ft. lbs. (27–34 Nm) while rotating the drum.

12. Back off the adjusting nut, then tighten it to 7 ft. lbs. (10 Nm).

13. Install the nut lock and a new cotter pin.

14. Install the wheel assembly and lower the vehicle.

Brake Shoes

REMOVAL AND INSTALLATION

Mighty Max

1. Raise and safely support the vehicle.

2. Remove the wheel and tire assembly. Remove the brake drum.

3. Remove the upper return spring along with the adjuster.

4. Remove the adjuster spring.

5. Remove the lower retaining spring.

6. Remove the hold-down springs.

7. Remove the shoes, disengaging the parking brake lever.

To install:

8. Clean and dry the backing plate. Lubricate the bosses, anchor contacts and the wheel cylinder piston grooves where the top of the shoe fits, lightly with lithium based grease.

9. Lubricate the star wheel shaft threads with anti-seize lubricant and transfer all parts to their proper locations on the new shoes. Make sure the longer end of the upper return spring is installed toward the shoe with the parking brake lever.

10. Spread the shoes apart, engage the parking brake lever and position them on the backing plate making

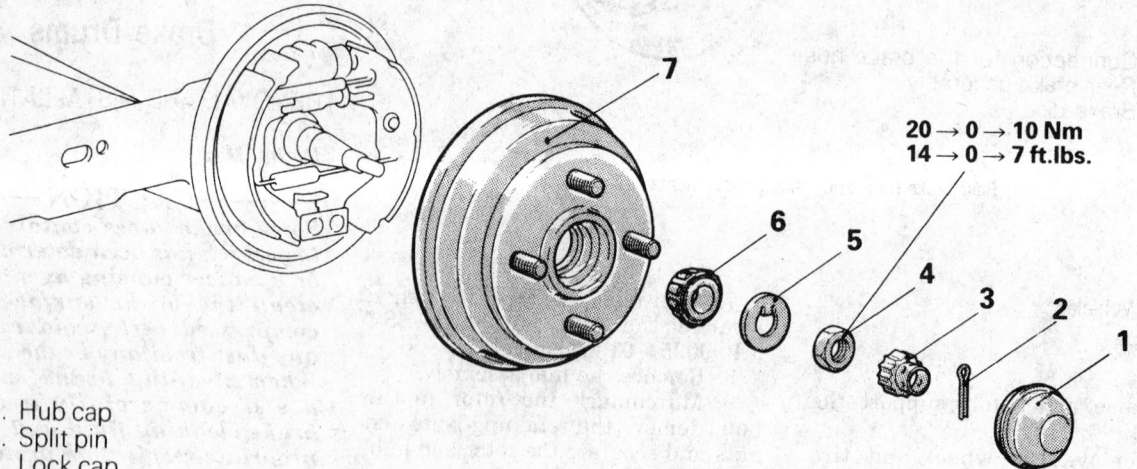

20 → 0 → 10 Nm
14 → 0 → 7 ft.lbs.

1. Hub cap
2. Split pin
3. Lock cap
4. Lock nut
   Adjustment of wheel bearing end play
5. Tongued washer
6. Outer wheel bearing inner race
7. Rear hub assembly

239388

Hubless rear brake drum and related components — Expo

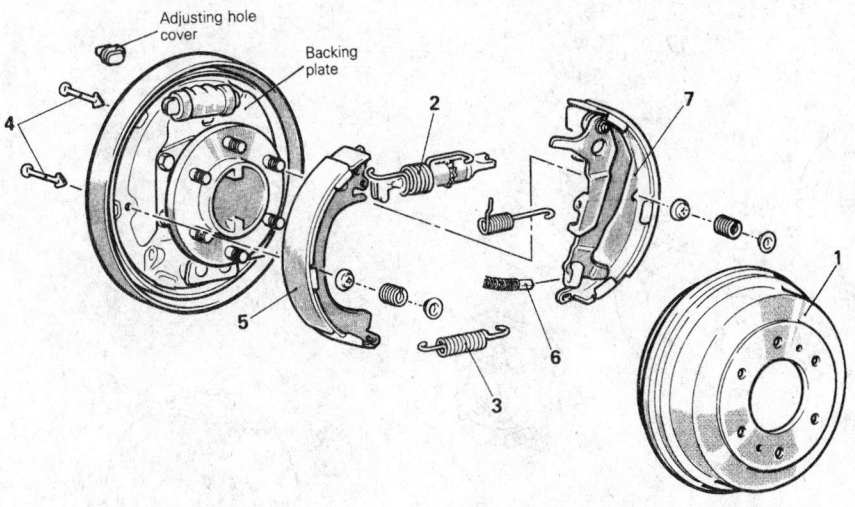

1. Brake drum
2. Shoe return spring with brake shoe adjuster
3. Shoe retainer spring
4. Shoe hold-down pin
5. Shoe and lining assembly
6. Parking brake rear cable end connection
7. Shoe and lever assembly

327056

Brake shoe assembly — Mighty Max

sure the wheel cylinder pins engage the webs of the brake shoes.

11. Install the lower retaining spring and the hold-down springs.

12. Adjust the star wheel until the brake shoes are in contact with the brake drum when installed.

13. Make sure there is no grease on the brake shoe linings and install the brake drums.

14. Install the wheel and tire assembly. Complete the brake adjustment then lower the vehicle and check the brakes for proper operation.

Expo

------ **WARNING** ------
Brake shoes contain asbestos, which has been determined to be a cancer causing agent. Never clean the brake surfaces with compressed air! Avoid inhaling any dust from brake surfaces! When cleaning brakes, use commercially available brake cleaning fluids.
---

1. Raise vehicle and support safely. Remove the appropriate tire and wheel assembly.

2. Remove the brake drum.

3. Remove the shoe to shoe spring.

4. Remove the shoe to lever spring and remove the adjuster assembly.

NOTE: Note the location of all springs and clips for proper reassembly.

5. Remove the shoe hold-down clips and the brake shoes.

6. Disconnect the parking brake cable from the rear shoes by spreading the horseshoe clip apart.

To install

7. Thoroughly clean and dry the backing plate. Ensure the backing plate bosses are smooth, so not to cause binding. If shoes are being replaced due to contamination from brake fluid. repair the leaking wheel cylinder as required.

8. Lubricate backing plate bosses, anchor pin, and parking brake actuating mechanism with a lithium-based grease.

9. Remove, clean and dry all remaining parts. Apply anti-seize to the star wheel threads and transfer all parts to the new shoes.

10. Connect the parking brake arm to the appropriate brake shoe. Attach shoes to backing plate and install all remaining hardware in the reverse order it was removed.

11. Pre-adjust the shoes so the drum slides on with a light drag and

install brake drum. Properly adjust the wheel bearings, if required.

12. Adjust rear brake shoes and install the wheel assemblies.

Wheel Cylinder

REMOVAL AND INSTALLATION

Mighty Max

1. Raise and safely support the vehicle.

2. Remove the wheel assembly, brake drum, and brake shoes.

3. Remove the brake line from the wheel cylinder.

4. Remove the wheel cylinder bolts and remove the cylinder from the backing plate.

To install:

5. Apply a small bead of silicone sealant around the wheel cylinder mounting surface of the backing plate.

6. Start the brake line into the wheel cylinder.

7. Install the wheel cylinder on the backing plate and tighten the retaining bolts.

8. Tighten the brake line.

9. Install the brake shoes and brake drum. Install the wheel assembly and adjust the brakes.

10. Bleed the brake system and lower the vehicle.

Expo

NOTE: It is important not let the master cylinder reservoir run dry, at any time during this procedure, or the entire system will have to be bleed.

1. Raise the vehicle and support it safely.

2. Remove the wheel and the brake drum.

3. Remove the shoe–to–lever spring and the upper shoe-to-shoe spring. Spread the upper portion of the brake shoes slightly.

4. Remove and plug the brake line from the wheel cylinder.

5. Remove the wheel cylinder retaining bolts and remove the cylinder from the backing plate.

To install:

6. Apply a very thin coating of silicone sealer to the cylinder mounting surface, install the cylinder to the backing plate and install the retaining bolts.

7. Connect the brake line to the wheel cylinder.

8. Install brake springs and the brake drum.

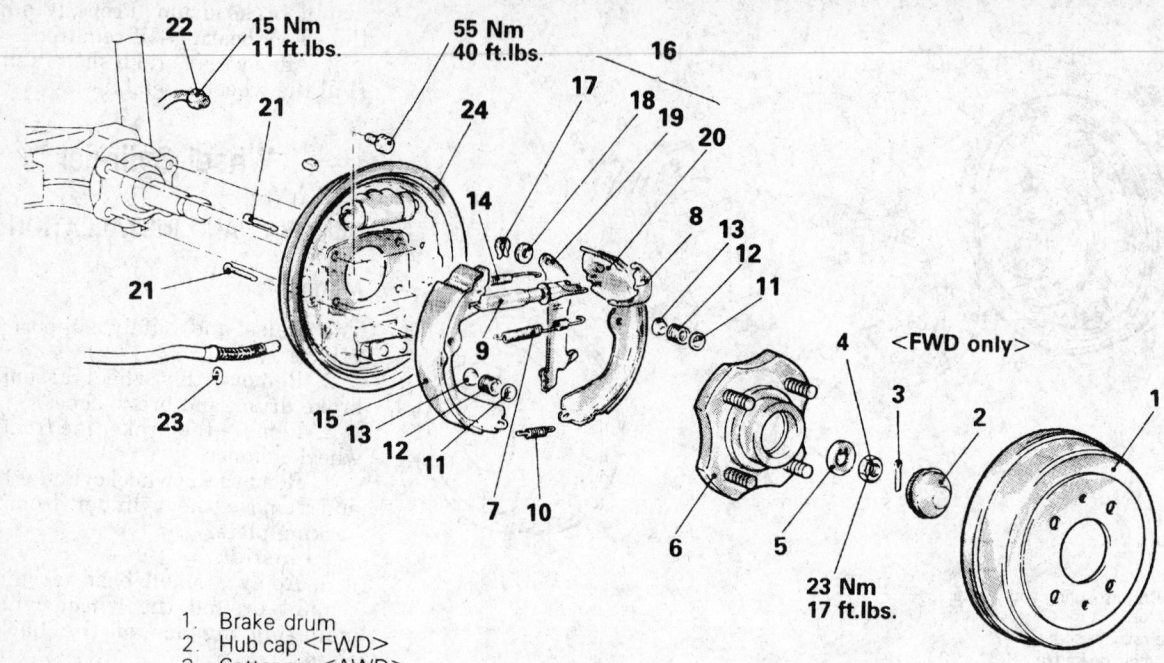

1. Brake drum
2. Hub cap <FWD>
3. Cotter pin <AWD>
4. Lock nut
5. Washer
6. Rear hub assembly
7. Shoe-to-lever spring
8. Adjuster lever
9. Auto adjuster assembly
10. Retainer spring
11. Shoe hold-down cup
12. Shoe hold-down spring
13. Shoe hold-down cup
14. Shoe-to-shoe spring
15. Shoe and lining assembly
16. Shoe and lever assembly
17. Retainer
18. Wave washer
19. Parking lever
20. Shoe and lining assembly
21. Shoe hold-down pin
22. Connection for the brake tube
23. Snap ring
24. Backing plate

239396

Rear drum brake system component identification — Expo

9. To install and adjust the wheel bearing:

a. Lubricate and install the outer wheel bearing, washer and nut.

b. Tighten the wheel bearing nut to 20–25 ft. lbs. (27–34 Nm) while rotating the drum.

c. Back off the adjusting nut, then tighten it to 7 ft. lbs. (10 Nm).

d. Install the nut lock and a new cotter pin.

e. Install the grease cap.

10. Install the tire and wheel assembly.

11. Fill the system with clean brake fluid and bleed the rear brakes.

Parking Brake Cable

ADJUSTMENT

Montero

The parking brake pull rod stroke should be 4–6 notches when pulled. If adjustment is necessary, proceed as follows:

1. Raise and safely support the vehicle.

2. Loosen the adjusting nut to slacken the parking brake cable.

3. Remove the wheel and brake drum.

4. Remove the adjustment hole plug and use a flat–tipped tool to turn the adjuster in the direction of the arrow (the direction which expands the shoe) so that the disc will not rotate.

5. Return the adjuster 3–4 notches in the direction opposite to the direction of the arrow.

6. Tighten the adjusting nut until the pull rod stroke is correct.

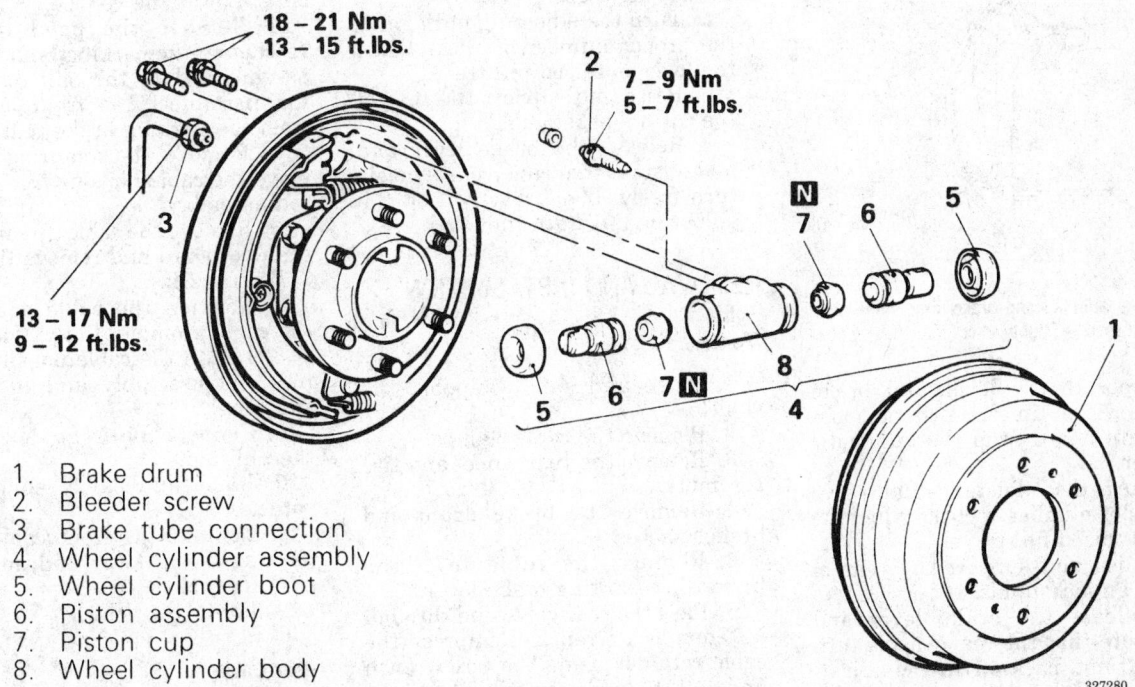

18 – 21 Nm
13 – 15 ft.lbs.

2 7 – 9 Nm
5 – 7 ft.lbs.

13 – 17 Nm
9 – 12 ft.lbs.

327280

1. Brake drum
2. Bleeder screw
3. Brake tube connection
4. Wheel cylinder assembly
5. Wheel cylinder boot
6. Piston assembly
7. Piston cup
8. Wheel cylinder body

Wheel cylinder mounting — all vehicles are similar

WARNING

If the number of pull rod notches engaged is less than specified, the cable has been pulled excessively and failure of the automatic adjuster mechanism will result. Make sure the joint and equalizer are at right angles to each other.

7. Install the brake drum and the wheel and tire assembly.
8. With the pull rod in the released position, turn the rear wheel to confirm that the rear brakes are not dragging.
9. Lower the vehicle.

Mighty Max

The parking brake pull rod stroke should be 16–17 notches when pulled with a force of 66 lbs. If adjustment is necessary proceed as follows:
1. Raise and safely support the vehicle.
2. Loosen the adjusting nut to slacken the parking brake cable.
3. Tighten the adjusting nut slightly, repeating pulling and releasing the parking brake pull rod, to adjust the brake shoe clearance.
4. Tighten the adjusting nut until the pull rod stroke is correct.

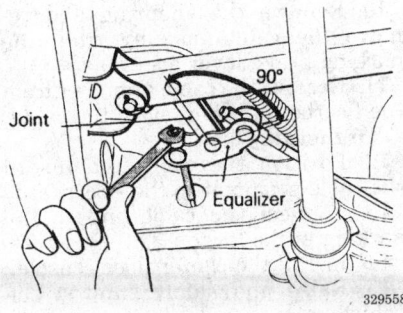

Joint   90°

Equalizer

329558

Parking brake cable adjustment — Mighty Max

WARNING

If the number of pull rod notches engaged is less than specified, the cable has been pulled excessively and failure of the automatic adjuster mechanism will result. Make sure the joint and equalizer are at right angles to each other.

5. Release the parking brake pull rod, remove the brake drum and make sure the brake lever adjuster is touching the shoe.

WARNING

If the parking brake cable is pulled too far, the adjuster lever does not fit the adjuster, resulting in faulty operation of the brake shoe adjuster.

6. Install the brake drum and the wheel and tire assembly.
7. With the pull rod in the released position, turn the rear wheel to confirm that the rear brakes are not dragging.
8. Lower the vehicle.

Expo

With Drum Brakes

NOTE: Make certain that the brake shoes are properly adjusted before attempting to adjust the parking brake.

1. Pull the parking brake lever up with a force of about 45 lbs. If that value cannot be determined, just pull it up as far as possible. The total number of clicks heard should be 5–7.
2. If the number of clicks was not within that range, release the lever and back off the cable adjuster locknut at the base of the lever and tighten the adjusting nut until there is no more slack in the cable.

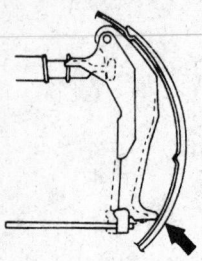

329559

Brake lever adjuster and brake shoe when pull rod is released — Mighty Max

3. Operate the lever and brake pedal several times, until no more clicks are heard from the automatic adjuster.

4. Turn the adjusting nut to give the proper number of clicks when the lever is raised full travel.

5. Raise and support the rear of the car on jackstands.

6. Release the brake lever and make sure that the rear wheels turn freely. If not, back off on the adjusting nut until they do.

With Disc Brakes

1. Pull the parking brake lever up with a force of about 45 lbs. (61 N). The total number of clicks heard should be 3–5. If the number of clicks was not within that range, system requires adjustment.

NOTE: The parking brake shoes must be adjusted before attempting to adjust the cable mechanism

2. To adjust the parking brake shoes perform the following steps.

a. remove the floor console, release the lever and back off the cable adjuster locknut at the base of the lever.

b. Raise the vehicle, support safely and remove the wheel. Remove the hole plug in the brake rotor.

c. Remove the brake caliper and hang out of the way with wire.

d. Use a suitable prybar to pry up on the self-adjuster wheel until the rotor will not turn.

e. Return the adjuster 5 notches in the opposite direction. Make sure the rotor turns freely with a slight drag.

f. Install the caliper and check operation.

3. Once the parking brake shoes have been properly adjusted, adjust

the cable mechanism, by performing the following steps:

a. Turn the adjusting nut to give the proper number of clicks when the lever is raised full travel.

b. Raise and support the rear of the car on jackstands.

c. Release the brake lever and make sure that the rear wheels turn freely. If not, back off on the adjusting nut until they do.

REMOVAL AND INSTALLATION

Montero

1. Raise and safely support the vehicle.

2. Remove the cable clamps.

3. Remove the rear wheel and tire assembly.

4. Remove the brake drum and brakes shoes.

5. Remove the cable from the brake shoe parking brake lever.

6. Pass the rear cable end through a 12mm box wrench. Compress the cable retainer using the box wrench and pull the cable out from the brake backing plate.

7. Lower the vehicle.

8. Remove the center floor console.

9. Remove the parking brake shaft cover.

10. Remove the snapring and remove the cable housing from the parking brake lever assembly.

11. Remove the cable grommet from the floorboard and remove the cable.

To install:

12. Position the cable and install the cable grommet to the floorboard.

13. Connect the cable end to the parking brake lever.

14. Install the center floor console.

15. Raise and safely support the vehicle.

16. Install the cable to the brake backing plate. Make sure the retainer is fully seated.

17. Connect the cable to the brake shoe lever.

18. Install the brake shoes. Adjust the brake shoes and install the brake drum.

19. Install the wheel and tire assembly.

20. Adjust the parking brakes and lower the vehicle.

Mighty Max

Front Cable

1. Raise and safely support the vehicle.

2. Loosen the cable adjustment to allow the front cable to be removed from the parking brake lever.

3. Remove the cable attaching clips. Lower the vehicle.

4. Remove the pawl from the ratchet on the parking brake pull rod assembly. With the pull rod pushed in, disconnect the front cable end from the parking brake pull rod.

5. Remove the snapring and remove the cable housing from the pull rod assembly.

6. Remove the cable grommet from the floorboard and remove the cable.

To install:

7. Position the cable and install the cable grommet to the floorboard.

8. Install the cable housing to the pull rod assembly and install the snapring.

9. Connect the front cable end to the pull rod.

10. Raise and safely support the vehicle.

11. Connect the front cable to the parking brake lever and adjust the parking brakes.

12. Lower the vehicle.

Rear Cable

1. Raise and safely support the vehicle.

2. Disconnect the rear cable from the parking brake lever.

3. Remove the cable clamps.

4. Remove the rear wheel and tire assembly.

5. Remove the brake drum and brakes shoes.

6. Remove the cable from the brake shoe parking brake lever.

7. Pass the rear cable end through a 12mm box wrench. Compress the cable retainer using the box wrench and pull the cable out from the brake backing plate.

To install:

8. Install the cable to the brake backing plate. Make sure the retainer is fully seated.

9. Connect the cable to the brake shoe lever.

10. Install the brake shoes. Adjust the brake shoes and install the brake drum.

11. Install the wheel and tire assembly.

12. Connect the front of the rear cable to the parking brake lever.

13. Adjust the parking brakes and lower the vehicle.

Expo

With Rear Drum Brakes

1. Disconnect the negative battery cable.

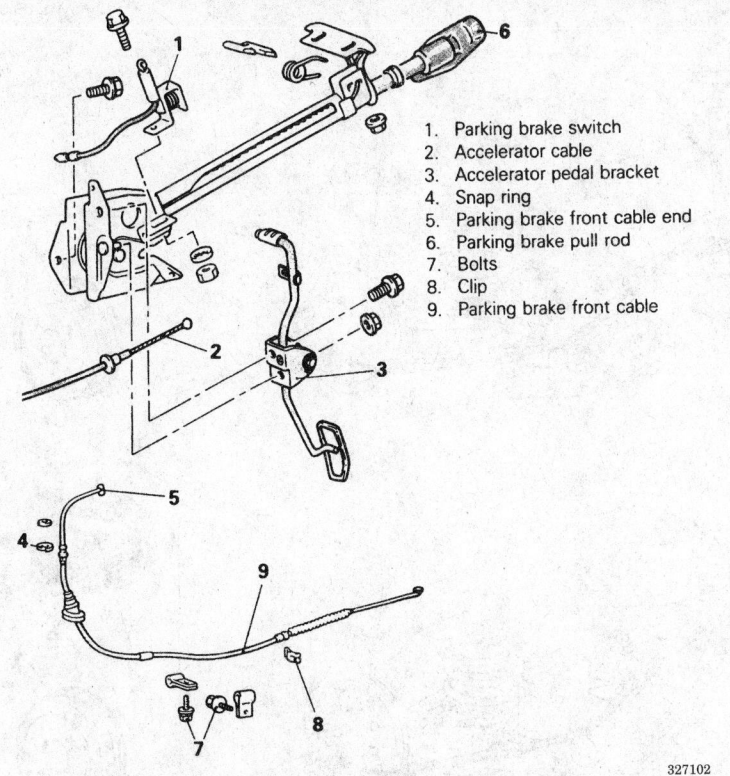

1. Parking brake switch
2. Accelerator cable
3. Accelerator pedal bracket
4. Snap ring
5. Parking brake front cable end
6. Parking brake pull rod
7. Bolts
8. Clip
9. Parking brake front cable

327102

Parking brake pull rod and parking brake front cable — Mighty Max

2. Remove the shifter knob, boot and cover assembly.

NOTE: If equipped with SRS, when removing the floor console, don't allow any impact or shock to the SRS diagnostic unit.

3. Remove the interior rear seat and carpet, in order to gain access to the cables.

4. Loosen the cable adjuster nut and then remove the parking brake cable, by pulling it from the passenger compartment.

5. Raise the vehicle and support safely.

6. At the rear wheel, remove the brake drum and shoes.

7. Disconnect the cable end from the parking brake strut lever. Compress the retaining strips to remove the cable from the backing plate.

8. Unfasten any other frame retainers and remove the cables.

To install

9. Install the cable to the rear actuator. Secure in place with the parking brake cable clip and retainer spring.

10. Install the brake shoes and drum.

11. Position the cable under the vehicle and install retainers loose.

12. Reattach the parking brake cables to the actuator inside the vehi-

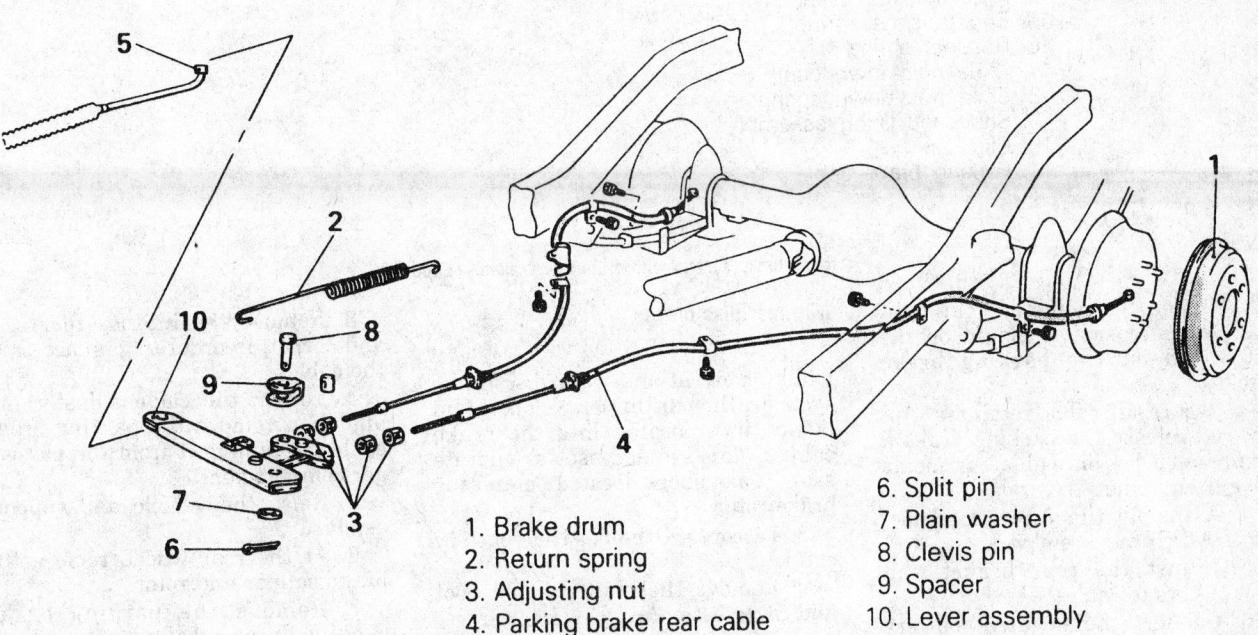

1. Brake drum
2. Return spring
3. Adjusting nut
4. Parking brake rear cable
5.
6. Split pin
7. Plain washer
8. Clevis pin
9. Spacer
10. Lever assembly

327105

Rear parking brake cable and lever assembly — Mighty Max

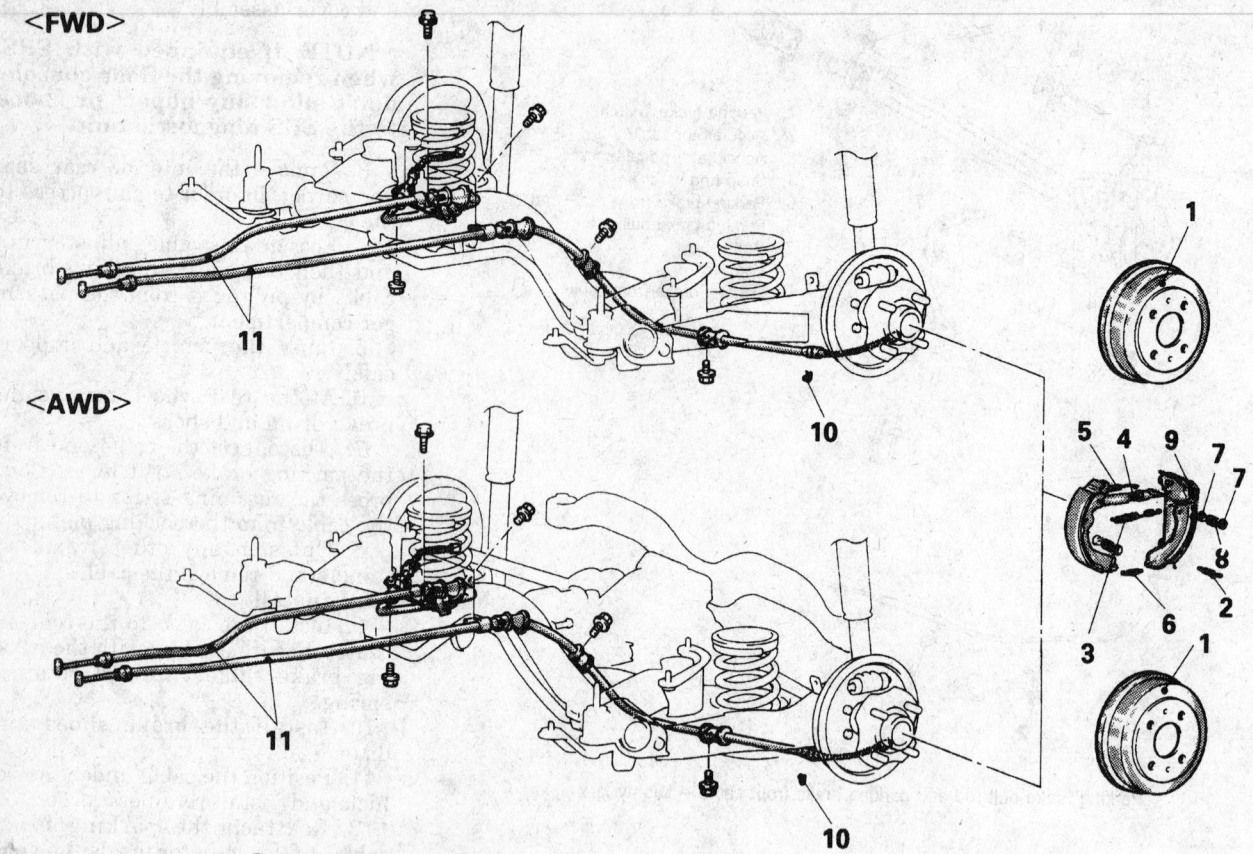

<FWD>

<AWD>

1. Rear brake drum
2. Lever return spring
3. Shoe-to-lever spring
4. Auto adjuster assembly
5. Shoe-to-shoe spring
6. Retainer spring
7. Shoe hold-down cup
8. Shoe hold-down spring
9. Shoe and lining assembly
10. Clip
11. Parking brake cable

240924

Parking brake system with drum rear brakes — Expo

cle. Tighten the adjusting nut until the proper tension is placed on the cable. Adjust the parking brake stroke.

13. Secure all cable retainers. Apply and release the parking brake a number of times once all adjustments have been made.

14. Assemble the interior components which were removed.

15. Adjust the rear brakes and parking brake cables.

16. Connect the negative battery cable and check the rear wheels to confirm that the rear brakes are not dragging.

17. Check that the parking brake holds the vehicle on an incline.

With rear disc brakes

Unlike conventional rear disc brake systems, the parking brake operation is **not** incorporated into the brake caliper. This system, uses a separate set of brake shoes, located behind the brake rotor.

1. Disconnect the negative battery cable.

2. Remove the shifter knob, boot and cover assembly.

NOTE: If equipped with SRS, when removing the floor console, don't allow any impact or shock to the SRS diagnostic unit.

3. Remove the interior rear seat and carpet, in order to gain access to the cables.

4. Loosen the cable adjuster nut and then remove the parking brake cable, by pulling it from the passenger compartment.

5. Raise the vehicle and support safely.

6. At the rear wheel, remove the brake caliper and rotor.

7. Remove the parking brake shoes, following the same procedures as conventional drum brake shoes.

8. Disconnect the cable end from the parking brake strut lever. Compress the retaining strips to remove the cable from the backing plate.

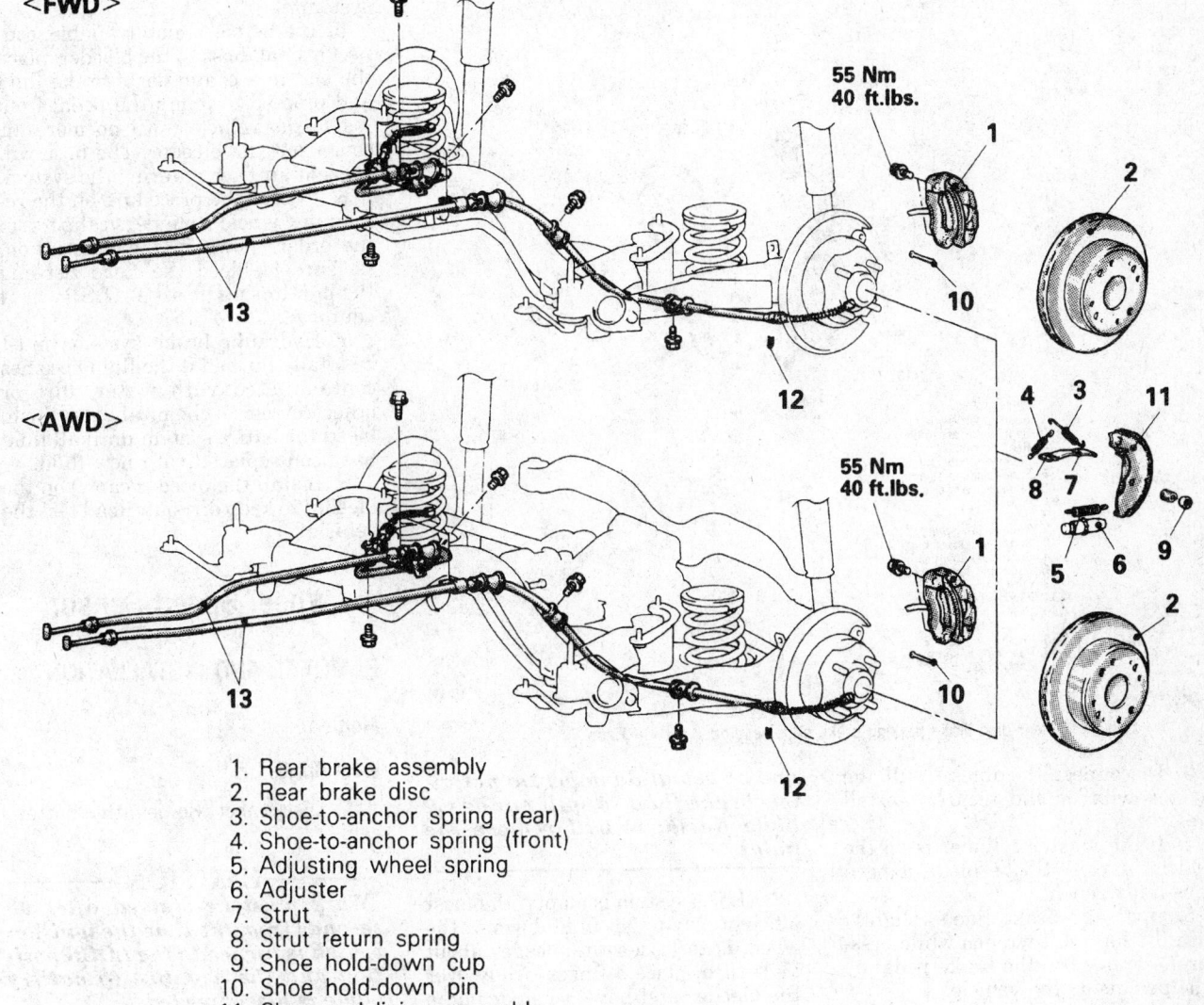

\<FWD\>

55 Nm
40 ft.lbs.

\<AWD\>

55 Nm
40 ft.lbs.

1. Rear brake assembly
2. Rear brake disc
3. Shoe-to-anchor spring (rear)
4. Shoe-to-anchor spring (front)
5. Adjusting wheel spring
6. Adjuster
7. Strut
8. Strut return spring
9. Shoe hold-down cup
10. Shoe hold-down pin
11. Shoe and lining assembly
12. Clip
13. Parking brake cable

240925

Parking brake system with rear disc brakes — Expo

9. Unfasten any other frame retainers and remove the cables.

To install

10. Install the cable to the rear actuator. Secure in place with the parking brake cable clip and retainer spring.

11. Install the parking brake shoes.

12. Install the brake rotor and caliper assembly.

13. Position the cable in the under the vehicle and install retainers loose.

14. Reattach the parking brake cables to the actuator inside the vehicle. Tighten the adjusting nut until the proper tension is placed on the cable. Adjust the parking brake stroke.

15. Secure all cable retainers. Apply and release the parking brake a number of times once all adjustments have been made.

16. Assemble the interior components which were removed.

17. Adjust the parking brake shoes and parking brake cables.

18. Connect the negative battery cable and check the rear wheels to confirm that the rear brakes are not dragging.

19. Check that the parking brake holds the vehicle on an incline.

Brake System Bleeding

MASTER CYLINDER

If the master cylinder is off the vehicle it can be bench bled.

1. Connect short piece(s) of brake line to the outlet fitting(s), bend them until the free end is below the fluid level in the master cylinder reservoirs.

2. Fill the reservoir with fresh brake fluid. Pump the piston slowly until no more air bubbles appear in the reservoir(s).

<FWD>

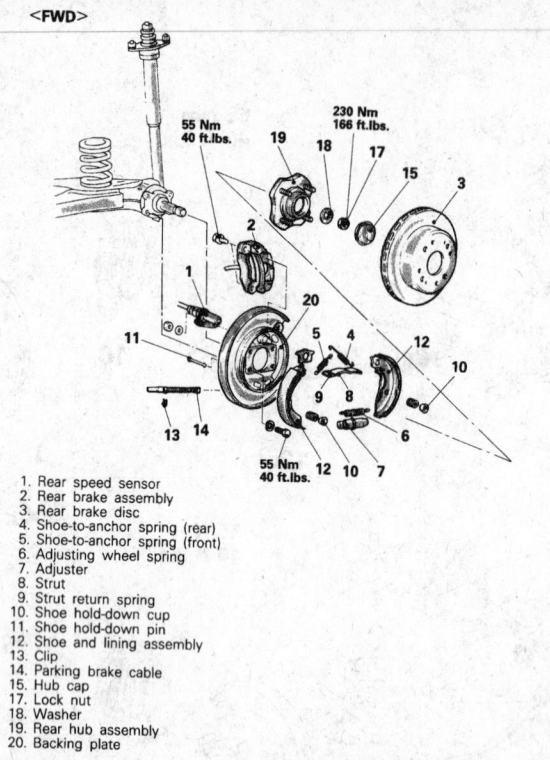

230 Nm
166 ft.lbs.

55 Nm
40 ft.lbs.

55 Nm
40 ft.lbs.

1. Rear speed sensor
2. Rear brake assembly
3. Rear brake disc
4. Shoe-to-anchor spring (rear)
5. Shoe-to-anchor spring (front)
6. Adjusting wheel spring
7. Adjuster
8. Strut
9. Strut return spring
10. Shoe hold-down cup
11. Shoe hold-down pin
12. Shoe and lining assembly
13. Clip
14. Parking brake cable
15. Hub cap
16. Lock nut
17. Lock nut
18. Washer
19. Rear hub assembly
20. Backing plate

240926

Rear disc brake parking brake shoes-exploded view — Expo

3. Disconnect the lines, refill the master cylinder and securely install the cylinder cap.

4. If the master cylinder is on the vehicle, it can still be bled, using a flare nut wrench.

5. Open the brake line(s) slightly with the flare nut wrench while pressure is applied to the brake pedal by a helper inside the vehicle.

6. Be sure to tighten the line before the brake pedal is released.

7. Repeat the process with both lines until no air bubbles come out.

CALIPERS AND WHEEL CYLINDERS

1. Fill the master cylinder with fresh brake fluid. Check the level often during the procedure.

NOTE: Vehicles with ABS, start the engine before bleeding the air.

2. Starting with the right rear wheel, remove the protective cap from the bleeder and place where it will not be lost. Clean the bleed screw.

—— **CAUTION** ——
When bleeding the brakes, keep face away from the brake area. Spewing fluid may cause facial

and/or visual damage. Do not allow brake fluid to spill on the vehicle finish; it will remove the paint.

3. If the system is empty, the most efficient way to get fluid down to the wheel is to loosen the bleeder about $1/2$–$3/4$ turn, place a finger firmly over the bleeder and have a helper pump the brakes slowly until fluid comes out the bleeder. Once fluid is at the bleeder, close it before the pedal is released inside the vehicle.

NOTE: If the pedal is pumped rapidly, the fluid will churn and create small air bubbles, which are almost impossible to remove from the system. These air bubbles will eventually congregate and a spongy pedal will result.

4. Once fluid has been pumped to the caliper or wheel cylinder, open the bleed screw again, have the helper press the brake pedal to the floor, lock the bleeder and have the helper slowly release the pedal. Wait 15 seconds and repeat the procedure (including the 15 second wait) until no more air comes out of the bleeder upon application of the brake pedal. Remember to close the bleeder before the pedal is released inside the vehicle each time the bleeder is opened. If

not, air will be induced into the system.

5. If a helper is not available, connect a small hose to the bleeder, place the end in a container of brake fluid and proceed to pump the pedal from inside the vehicle until no more air comes out the bleeder. The hose will prevent air from entering the system.

6. Repeat the procedure on the remaining wheel cylinders/calipers in the order shown in the illustration. Be sure to bleed the Load Sensing Proportioning Valve (LSPV), if equipped.

7. Hydraulic brake systems must be totally flushed if the fluid becomes contaminated with water, dirt or other corrosive chemicals. To flush, bleed the entire system until all fluid has been replaced with new fluid.

8. Install the bleeder cap(s) on the bleeder to keep dirt out. Road test the vehicle.

Wheel Speed Sensor

REMOVAL AND INSTALLATION

Montero

Front Brakes

1. Disconnect the negative battery cable.

—— **CAUTION** ——
Work must be started after 90 seconds from the time the ignition switch is turned to the LOCK position and the negative (-) battery cable is disconnected.

2. Raise and safely support the vehicle.

3. Disconnect the speed sensor connector.

4. Remove the four clamp bolts holding the sensor harness to the frame, strut and steering knuckle.

5. Remove the bolt holding the speed sensor to the steering knuckle. Remove the speed sensor from the vehicle.

To install:

6. Install the speed sensor to the steering knuckle and install the bolt.

7. Install the clamp bolts holding the sensor harness to the frame, steering knuckle and the strut.

8. Connect the speed sensor connector.

9. Lower the vehicle.

10. Connect the negative battery cable.

11. After installation, check speed sensor signal.

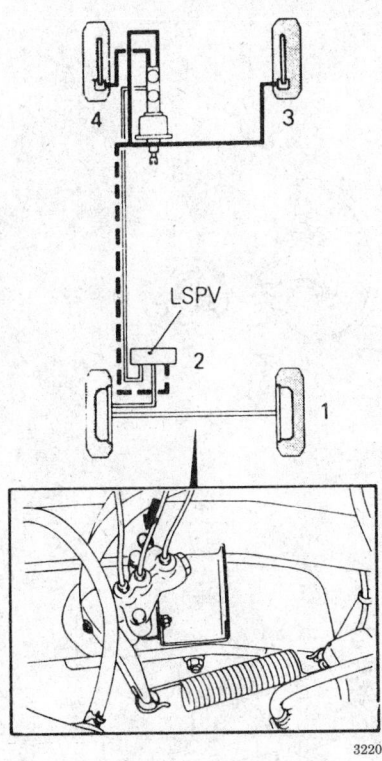

Brake system bleeding sequence — Mighty Max

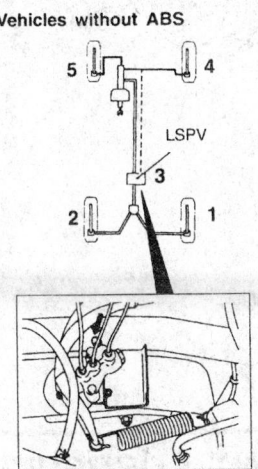

Vehicles without ABS

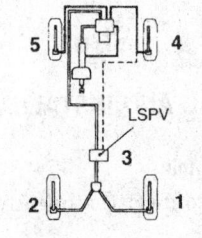

Vehicles with ABS

Brake system bleeding sequence — Montero

Rear Brakes

1. Disconnect the negative battery cable.

─── **CAUTION** ───

Work must be started after 90 seconds from the time the ignition switch is turned to the LOCK position and the negative (-) battery cable is disconnected.

2. Raise and safely support the vehicle.
3. Disconnect the speed sensor connector.
4. Remove the five clamp bolts holding the sensor wire to the frame and remove the speed sensor.
   To install:
5. Install the speed sensor and the five clamp bolts holding the sensor wire to the frame.
6. Connect the speed sensor connector.
7. Lower the vehicle and connect the negative battery cable.
8. Check the speed sensor signal.

Mighty Max

1. Disconnect the negative battery cable.
2. Raise and safely support the vehicle.
3. Disconnect the speed sensor connector at the rear differential.
   To install:
4. Connect the speed sensor connector at the differential.
5. Lower the vehicle and connect the negative battery cable.
6. Check the speed sensor signal.

Expo

Front-FWD And AWD Models

1. Disconnect the negative battery cable.

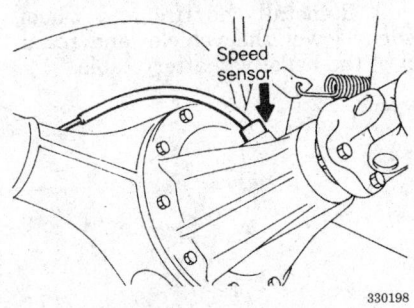

Rear speed sensor — Mighty Max

─── **CAUTION** ───

Wait at least 90 seconds after the negative (-) battery cable is disconnected to prevent possible deployment of the air bag.

2. Raise and safely support the vehicle. Remove the necessary tire and wheel assembly.
3. Remove the fender splash shield.
4. Disconnect the ABS speed sensor connector.
5. Remove the sensor harness clamp bolts and clamps.
6. Remove the ABS speed sensor mounting bolt and the sensor.
   To Install:
7. Install the ABS speed sensor with its mounting bolt.

NOTE: The clearance between the wheel speed sensor and the rotor's toothed surface is not adjustable, but measure the distance between the sensor installation surface and the rotor's toothed surface. Standard value is: 0.012–0.035 in. If not within specifications, replace the speed sensor or the toothed rotor.

8. Reinstall the sensor harness with its clamps and bolts.
9. Reconnect the speed sensor connector.
10. Install the fender splash shield.
11. Reinstall the tire and wheel, safely lower the vehicle, and reconnect the negative battery cable.

Rear-FWD And AWD

1. Disconnect the negative battery cable.

─── **CAUTION** ───

Wait at least 90 seconds after the negative (-) battery cable is disconnected to prevent possible deployment of the air bag.

2. Raise and safely support the vehicle. Remove the necessary tire and wheel assembly.
3. Disconnect the ABS speed sensor connector.
4. Remove the sensor harness clamp bolts and clamps.
5. Remove the ABS speed sensor mounting bolt and the sensor.

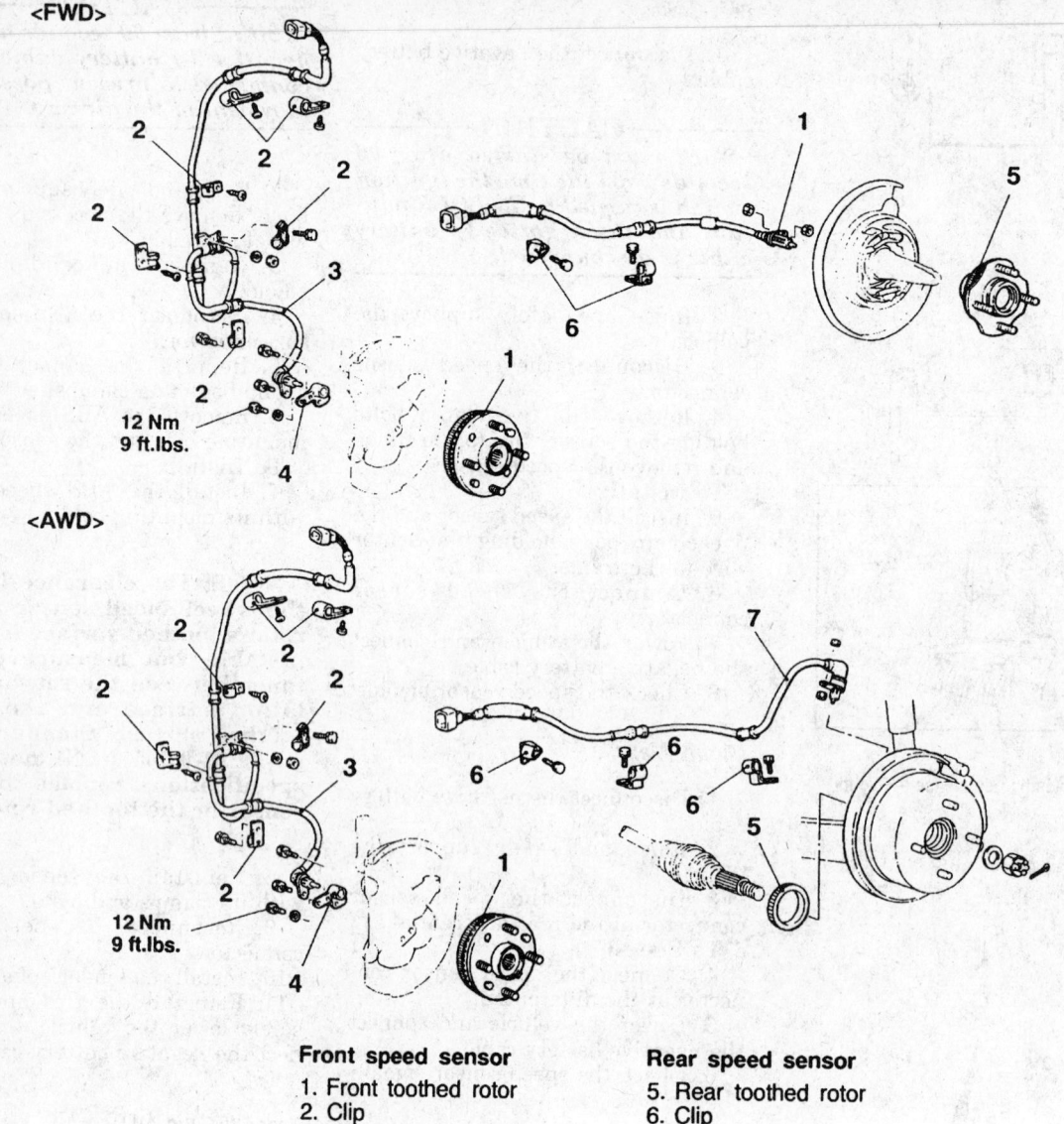

<FWD>

12 Nm
9 ft.lbs.

<AWD>

12 Nm
9 ft.lbs.

Front speed sensor

1. Front toothed rotor
2. Clip
3. Front speed sensor
4. Front speed sensor bracket

Rear speed sensor

5. Rear toothed rotor
6. Clip
7. Rear speed sensor

346331

Speed sensor and related components — Expo

To Install:

6. Install the ABS speed sensor with its mounting bolt.

NOTE: The clearance between the wheel speed sensor and the rotor's toothed surface is not adjustable, but measure the distance between the sensor installation surface and the rotor's toothed surface. Standard value is: 0.008–0.028 in. for the FWD and 0.012–0.035 in. for the AWD. If not within specifications, replace the speed sensor or the toothed rotor.

7. Reinstall the sensor harness with its clamps and bolts.

8. Reconnect the speed sensor connector.

9. Reinstall the tire and wheel, safely lower the vehicle, and reconnect the negative battery cable.

FRONT SUSPENSION

Strut

REMOVAL AND INSTALLATION

Strut assembly

1. Disconnect the negative battery cable.

2. Remove the daytime running lamp relay mounting bracket and position relay assembly aside.

3. Raise and safely support the vehicle.

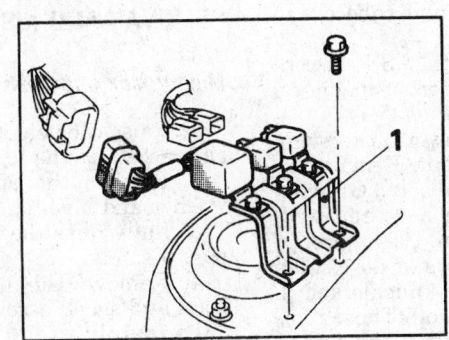

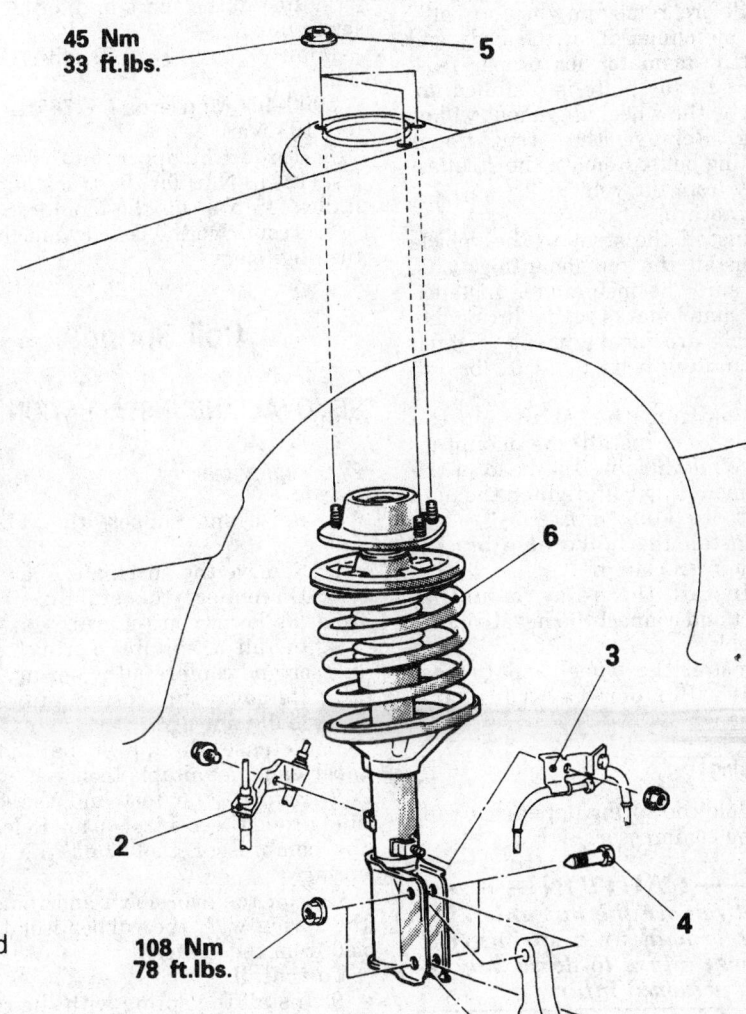

45 Nm
33 ft.lbs. — 5

108 Nm
78 ft.lbs.

1. Daytime running lamp relay and
   control unit
2. Brake tube clamp
3. Front speed sensor clamp
   <Vehicles with ABS>
4. Bolts
5. Flange nut
6. Strut assembly

241280

Front strut assembly and related components — Expo

4. Remove the brake hose and tube bracket retainer bolt and bracket from the front strut. Do not pry the brake hose and tube clamp away when removing.

5. If equipped with ABS, disconnect the front speed sensor mounting clamp from the strut.

6. Support the lower arm using floor jack or equivalent. Remove the lower strut to knuckle bolts. Once the mounting bolts have been removed, jack up the lower arm. Use a piece of wire to attach the brake hose, tube and driveshaft to the knuckle and to help keep the weight off. These components are not to be pulled.

7. Before removing the top bolts, make matchmarks on the body and the strut insulator for proper reassembly. If this plate is installed improperly, the wheel alignment will be wrong. Remove the strut upper mounting bolts. Remove the strut assembly from the vehicle.

To install:

8. Install the strut to the vehicle and install the top mounting bolts. Make sure the insulator is installed so the matchmarks made during disassembly are in alignment. Tighten the mounting bolts to 33 ft. lbs. (45 Nm).

9. Position the strut on the knuckle and install the mounting bolts. While holding the head of the lower mounting bolt, tighten the nuts to 78 ft. lbs. (108 Nm).

10. Install the brake hose bracket and the ABS clamp.

11. Install the relay mounting bracket and connect the negative battery cable.

12. Install the wheel and tire assembly. Perform a front end alignment.

Coil Spring

1. Hold the spring upper seat with a spring compressor.

────── **CAUTION** ──────
Do not remove the nut unless the spring is held by a spring compressor. Failure to do so may result in personal injury.

2. Compress the spring and then remove the self-locking nut holding the strut insulator.

3. Remove the spring.

To install:

4. With the spring being held in the spring compressor, align the spring in the grooves in the upper and lower seats.

5. Install the self-locking nut and tighten to 43–51 ft. lbs. (60–70 Nm).

Shock Absorber

REMOVAL AND INSTALLATION

Mighty Max and 1993–95 Montero

1. Raise the vehicle just enough to gain access to the upper shock nut.

2. Remove the upper shock nut, washer and bushing.

3. Fully raise the vehicle and support it safely.

4. Remove the lower mounting bolts and shock assembly.

To install:

5. Install the shock and torque the lower nut to:

Mighty Max: 7–10 ft. lbs. (9–14 Nm)

1993–95 Montero: 65–76 ft. lbs. (88–103 Nm)

6. Torque the upper nut to 9–13 ft. lbs. (12–18 Nm) for the truck and 11 ft. lbs. (15 Nm) for the Montero.

7. Test drive the vehicle and check the alignment.

Coil Spring

REMOVAL AND INSTALLATION

2WD Mighty Max

1. Raise and support the vehicle safely.

2. Remove the shock absorber.

3. Disconnect the stabilizer bar from the lower control arm.

4. Install a suitable spring compressor and compress the spring.

5. Remove the cotter pin and lower ball joint nut.

6. Release the lower ball joint taper using a suitable tool.

7. Remove the tool and the ball stud from the control arm. Release the compressor tool from the coil spring.

8. Pull the arm down and remove the spring with the rubber isolation pad from the vehicle.

To install:

9. Install the spring with the rubber isolator. Install the compressor tool and compress spring so the lower ball joint can be inserted through the knuckle.

10. Torque the lower ball joint nut to 87–130 ft. lbs. (120–180 Nm). Install a new cotter pin. Remove the spring compressor.

11. Connect the sway bar to the lower control arm, if equipped.

12. Install the shock absorber. Install the wheel and tire assembly.

Torsion Bar

REMOVAL AND INSTALLATION

Montero and 4WD Mighty Max

1. Raise the vehicle and support it safely.

2. Remove skid plate, if equipped.

3. Support the lower control arm at a point away from the torsion bar mounting point to the arm.

4. Fold the dust covers back and slide them away from the ends of the bar.

5. If the bars are to be reused, matchmark the torsion bar at both ends to the anchor and identify left from right.

6. Paint or measure the distance of the exposed threads of the rear mounting bolt down to the nut to aid in adjustment when installing. Remove the rear anchor arm mounting nut and bolt.

7. Remove the torsion bar from the front anchor arm.

To install:

8. If the bar is being reused, lubricate the ends and install the torsion bar aligning the matchmarks. If a new bar is being used, align the white stripe on the front splines with the mark on the anchor.

NOTE: There is a mark on the front of the torsion bar to differentiate between left and right. Do not install the bar with the mark facing the rear.

9. Install the torsion bar to the rear anchor so the length of the mounting bolt from the nut to the head of the bolt is the specified length with the rebound bumper in contact with the crossmember. The specifications are as follows:

a. Raider and Montero left side — 5.3–5.6 in.

b. Raider and Montero right side — 4.9–5.2 in.

c. Ram 50 and Mighty Max left side — 5.5–5.9 in.

d. Ram 50 and Mighty Max right side — 5.3–5.8 in.

10. To initially set the riding height, tighten the rear anchor mounting nut to the same point at which it was removed if the old bar is being reused. If a new bar has been installed, tighten the nut so the exposed length of the bolt threads is within specifications:

a. Raider left side — 2.4 in.

b. Raider right side — 2.8 in.

c. Ram 50 and Mighty Max left side — 3.9 in.

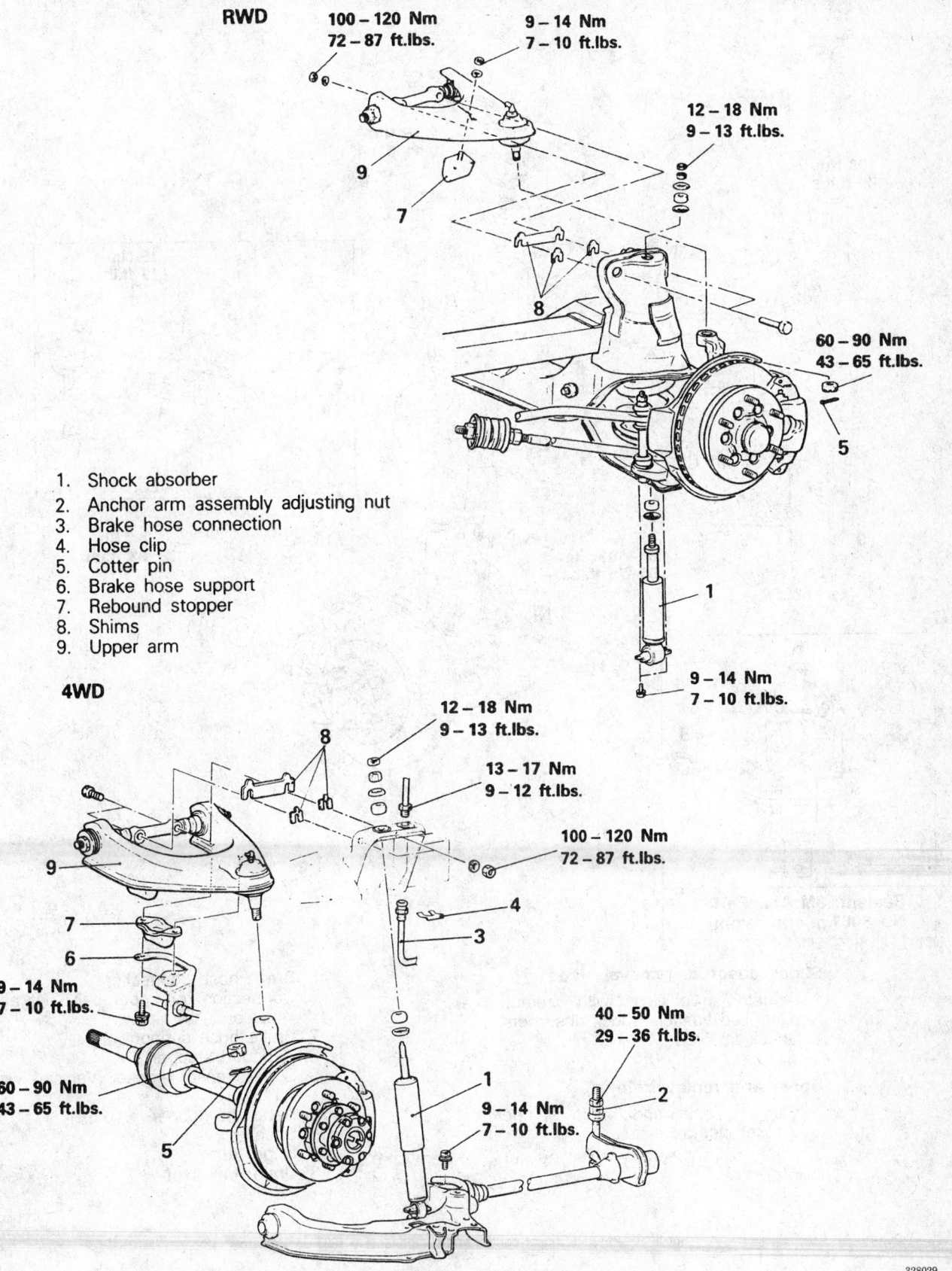

RWD

100 – 120 Nm
72 – 87 ft.lbs.

9 – 14 Nm
7 – 10 ft.lbs.

12 – 18 Nm
9 – 13 ft.lbs.

60 – 90 Nm
43 – 65 ft.lbs.

9 – 14 Nm
7 – 10 ft.lbs.

1. Shock absorber
2. Anchor arm assembly adjusting nut
3. Brake hose connection
4. Hose clip
5. Cotter pin
6. Brake hose support
7. Rebound stopper
8. Shims
9. Upper arm

4WD

12 – 18 Nm
9 – 13 ft.lbs.

13 – 17 Nm
9 – 12 ft.lbs.

100 – 120 Nm
72 – 87 ft.lbs.

9 – 14 Nm
7 – 10 ft.lbs.

60 – 90 Nm
43 – 65 ft.lbs.

40 – 50 Nm
29 – 36 ft.lbs.

9 – 14 Nm
7 – 10 ft.lbs.

328029

Shock absorber and upper control arm components — Mighty Max

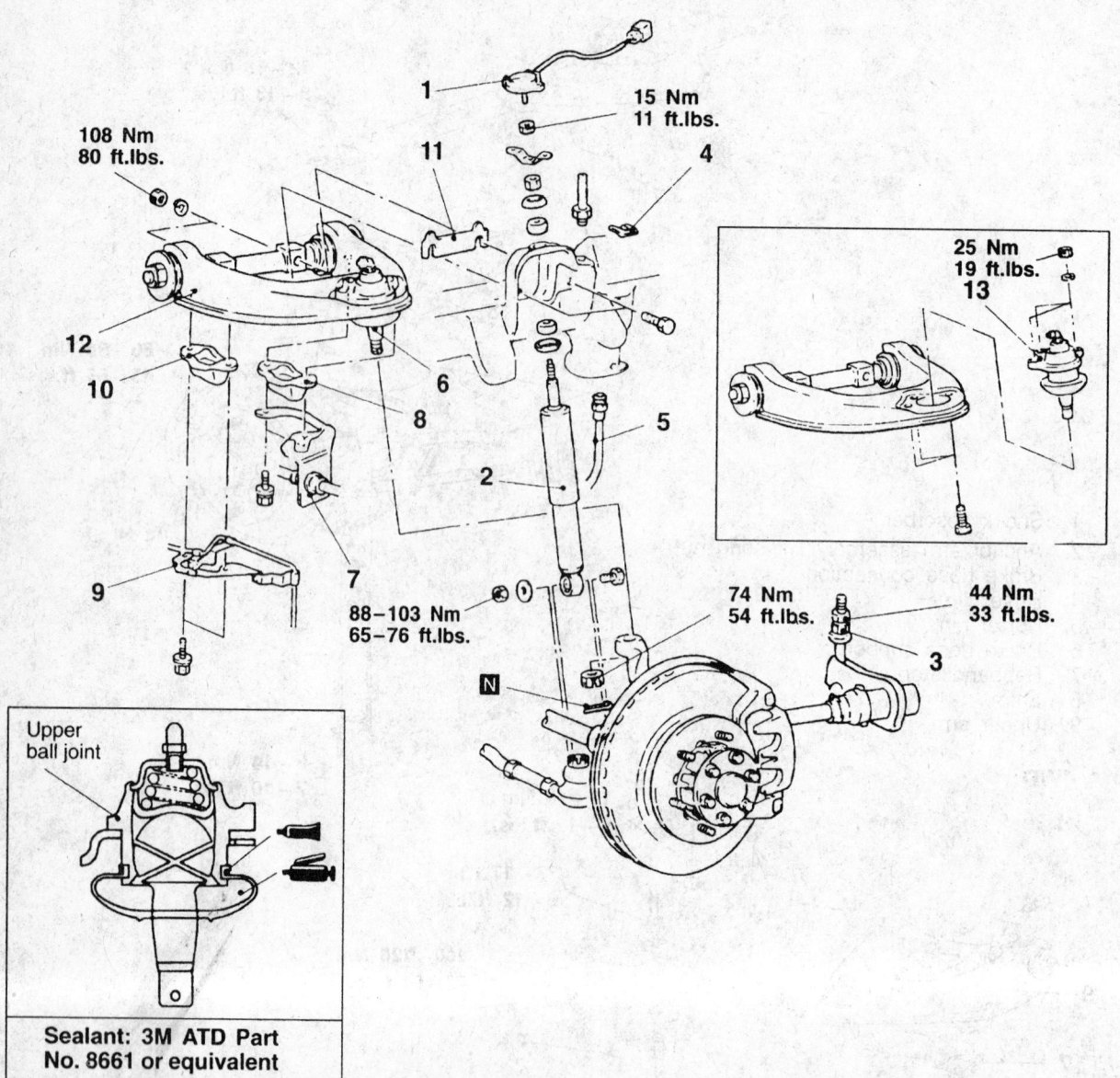

108 Nm
80 ft.lbs.

15 Nm
11 ft.lbs.

25 Nm
19 ft.lbs.

74 Nm
54 ft.lbs.

44 Nm
33 ft.lbs.

88–103 Nm
65–76 ft.lbs.

Upper
ball joint

**Sealant: 3M ATD Part
No. 8661 or equivalent**

Shock absorber removal steps

1. Actuator (Vehicles with remote controlled variable shock absorbers)
2. Shock absorber

Upper arm removal steps

- Bump stopper and bump stopper bracket clearance adjustment
3. Anchor arm assembly adjusting nut
4. Hose clip

5. Brake hose connection
6. Connection for upper ball joint and knuckle
7. Brake hose support
8. Rebound stopper
9. Speed sensor bracket (Vehicles with A.B.S.)
10. Rebound stopper
11. Shim
12. Upper arm
13. Upper ball joint

328030

Shock absorber and upper control arm components — 1993–95 Montero

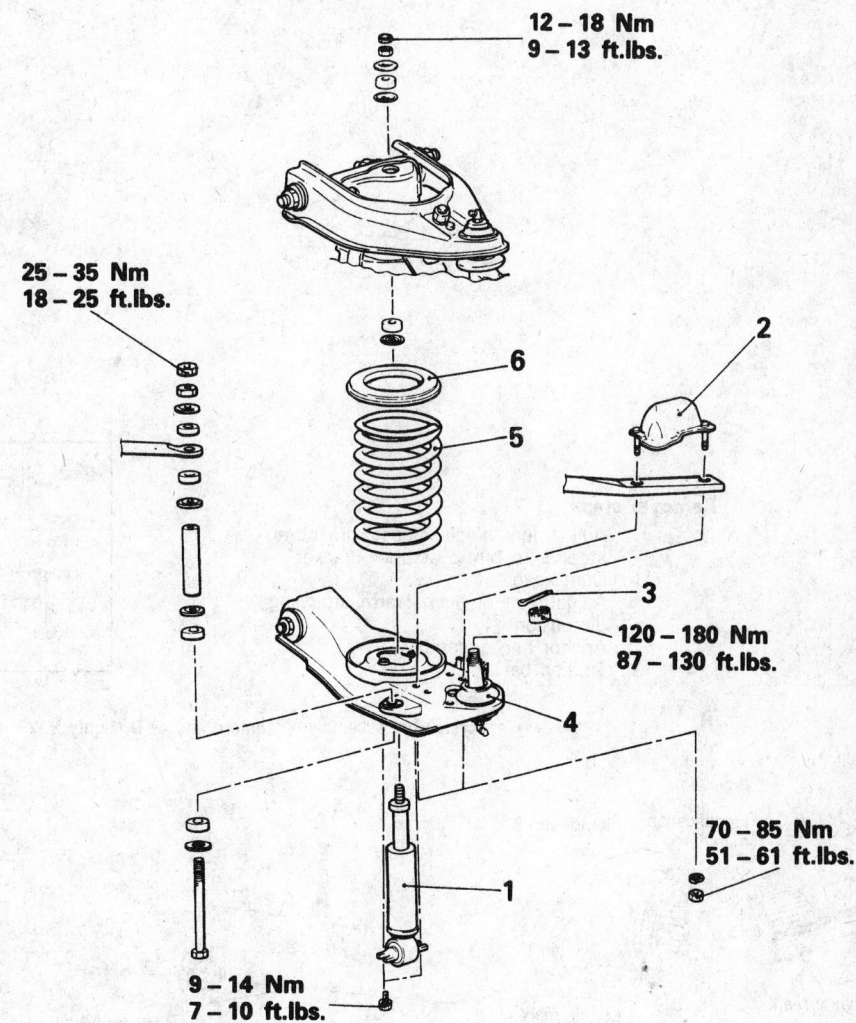

1. Shock absorber
2. Bump stopper
3. Cotter pin
4. Lower ball joint
5. Coil spring
6. Spring seat

Coil spring and related components — 2WD Mighty Max

327958

d. Ram 50 and Mighty Max right side — 3.4 in.

e. Montero — 3.15 in.

11. Fill the dust covers with grease and fold them back into position.

12. Adjust the torsion to the correct riding height.

Upper Ball Joints

REMOVAL AND INSTALLATION

Montero

1. Remove the upper control arm from the vehicle.

2. Remove the dust boot and the snapring from the ball joint.

3. Use the special tools to press the ball joint out of the upper control arm.

To install:

4. Line up the mating marks on the ball joint with the mark on the upper control arm.

5. Use the special tool to press the ball joint into the upper control arm.

6. Install the snapring and the dust boot on the ball joint.

7. Install the upper control arm on the vehicle.

8. Apply grease to the upper ball joint and all other suspension components with a grease fitting.

9. Tighten the ball joint stud nut to 54 ft. lbs. (74 Nm)

10. Install the new cotter pin.

Mighty Max

If the upper ball joint is bad, the upper control arm must be replaced. The ball joint is not removable from the control arm.

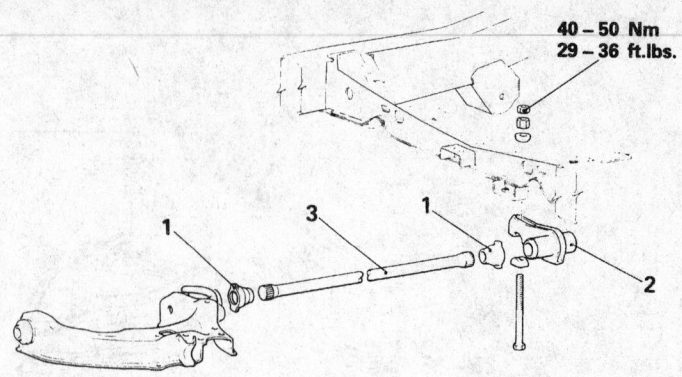

40 – 50 Nm
29 – 36 ft.lbs.

Removal steps

Adjustment of clearance from bump stopper to bump stopper bracket
1. Dust cover
   Adjustment of anchor arm attaching dimension
2. Anchor arm assembly
3. Torsion bar

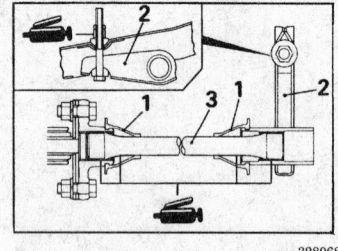

Torsion bar assembly — Montero and 4WD Mighty Max

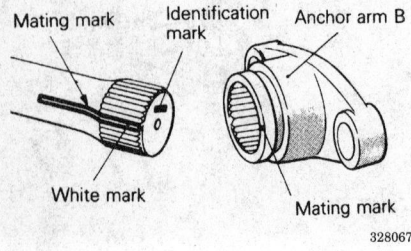

Torsion bar identification marks — Montero and 4WD Mighty Max

Mating mark
Identification mark
Anchor arm B
White mark
Mating mark

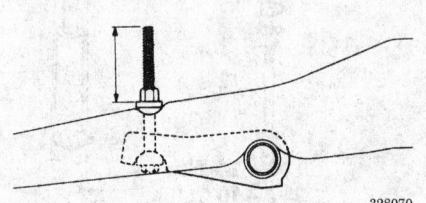

Exposed length of bolt dimension — Montero and 4WD Mighty Max

Lower Ball Joints

REMOVAL AND INSTALLATION

Montero

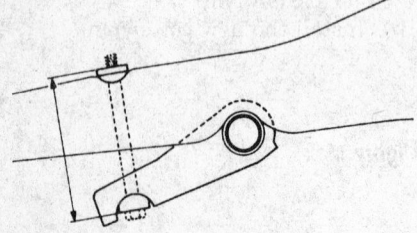

Anchor arm to crossmember dimension — Montero and 4WD Mighty Max

1. Raise the vehicle and support it safely.
2. Apply upward pressure to the lower control arm with a jack or an adjustable stand.

— **CAUTION** —

Do not disconnect the lower ball joint stud from the steering knuckle unless the lower control arm has a stand or a jack under it.

3. Remove the ball joint stud nut and disconnect the stud from the steering knuckle.
4. Remove the ball joint retaining nuts and bolts and remove the ball joint from the arm.

To install:
5. Install the lower ball joint to the control arm. Torque the ball joint retaining nuts and bolts to 60 ft. lbs. (81 Nm).
6. Torque the ball stud nut to 108 ft. lbs. (147 Nm). Install a new cotter pin.
7. Lubricate the ball joint with a grease gun.
8. If equipped with a torsion bar, adjust the riding height.
9. Check and adjust the alignment if necessary.

Mighty Max

1. Raise the vehicle and support it safely.
2. Apply upward pressure to the lower control arm with a jack or an adjustable stand.

CAUTION

Do not disconnect the lower ball joint stud from the steering knuckle unless the lower control arm has a stand or a jack under it.

---

3. Remove the ball joint stud nut and disconnect the stud from the steering knuckle.

4. Remove the ball joint retaining nuts and bolts and remove the ball joint from the arm.

To install:

5. Install the lower ball joint to the control arm. Torque the ball joint retaining nuts and bolts to:
- RWD: 30 ft. lbs. (42 Nm)
- 4WD: 54 ft. lbs. (74 Nm)

6. Torque the ball stud nut to 87–130 ft. lbs. (120–180 Nm) and install a new cotter pin.

7. Lubricate the ball joint with a grease gun.

8. Adjust the riding height, if equipped with a torsion bar.

9. Check and adjust the alignment if necessary.

Expo

The lower ball joint is an integral part of the lower control arm assembly, and can not be serviced separately. A worn or damaged ball joint, requires replacement of lower control arm assembly.

Upper Control Arms

REMOVAL AND INSTALLATION

1993–95 Montero

2WD Vehicles

1. Raise the vehicle and support it safely.

2. Remove the tire and wheel assembly.

3. Remove the shock absorber.

4. Compress the coil spring using a suitable coil spring compressor tool.

5. Remove the cotter pin and upper ball joint nut.

6. Suspend the rotor assembly so there is not excessive pull on the brake hose.

7. Release the upper ball joint from the knuckle using a suitable ball joint separator tool.

8. Loosen the pivot bar retaining nuts and bolts, identify and remove the alignment shims, remove the nuts and bolts and remove the arm from the vehicle.

To install:

9. Install the arm to the frame rail bracket, install the shims in their original locations and install the retaining nuts and bolts. Torque the nuts initially to 40 ft. lbs. (54 Nm).

10. Torque the ball joint nut to 60 ft. lbs. (81 Nm). Install a new cotter pin. Remove the spring compressor.

11. Install the shock absorber.

12. Align the front end. When all settings are at specifications, torque the pivot bar retaining bolts to 72–87 ft. lbs. (100–120 Nm) for 1985–91 and 80 ft. lbs. (109 Nm) for 1993–95.

4WD Vehicles

1. Raise the vehicle and support safely. Remove the skid plate.

2. Remove the shock absorber.

3. Turn the torsion bar adjustment nut counterclockwise to relieve all tension from the torsion bar.

4. Disconnect the brake hose from the brake line and remove the hose from the bracket.

5. Remove the cotter pin from the upper ball stud.

6. Release the upper ball joint from the knuckle using a suitable ball joint separator tool.

7. Loosen the pivot bar retaining nuts and bolts, identify and remove the alignment shims, remove the nuts and bolts.

8. Remove the control arm from the vehicle.

To install:

9. Position the arm at the frame rail bracket.

10. Install the shims in their original locations and install the retaining nuts and bolts. Torque the nuts initially to 40 ft. lbs. (54 Nm).

11. Insert the upper ball stud in the steering knuckle arm bore and install the nut. Torque the nut to 60 ft. lbs. (81 Nm) and install a new cotter pin. Install the shock absorber and attach the brake hose.

12. Turn the torsion bar adjustment nut clockwise to apply a load on the bar.

13. Lower the vehicle.

14. Set the riding height. Align the front end. When all settings are at specifications, torque the pivot bar retaining bolts to 80 ft. lbs. (109 Nm).

1996–97 Montero

1. Raise the vehicle and support safely.

2. Remove the shock absorber.

3. Turn the torsion bar adjustment nut counterclockwise to relieve all tension from the torsion bar.

4. Disconnect the brake hose from the brake line and remove the hose from the bracket.

5. Remove the cotter pin from the upper ball stud.

6. Release the upper ball joint from the knuckle using a suitable ball joint separator tool.

7. Loosen the pivot bar retaining nuts and bolts, identify and remove the alignment shims, remove the nuts and bolts.

8. Remove the control arm from the vehicle.

To install:

9. Position the arm at the frame rail bracket.

10. Install the shims in their original locations and install the retaining nuts and bolts. Do not tighten at this time.

11. Insert the upper ball stud in the steering knuckle arm bore and install the nut. Torque the nut to 54 ft. lbs. (74 Nm) and install a new cotter pin. Install the shock absorber and attach the brake hose.

12. Turn the torsion bar adjustment nut clockwise to apply a load on the bar.

13. Lower the vehicle.

14. Set the riding height. Align the front end. When all settings are at specifications, torque the upper control arm retaining bolts to 80 ft. lbs. (109 Nm).

Mighty Max

2WD Vehicles

1. Raise the vehicle and support it safely.

2. Remove the tire and wheel assembly.

3. Remove the shock absorber.

4. Compress the coil spring using a suitable coil spring compressor tool.

5. Remove the cotter pin and upper ball joint nut.

6. Suspend the rotor assembly so there is not excessive pull on the brake hose.

7. Release the upper ball joint from the knuckle using a suitable ball joint separator tool.

8. Loosen the pivot bar retaining nuts and bolts, identify and remove the alignment shims, remove the nuts and bolts and remove the arm from the vehicle.

To install:

9. Install the arm to the frame rail bracket, install the shims in their original locations and install the retaining nuts and bolts. Torque the nuts initially to 40 ft. lbs. (54 Nm).

10. Torque the ball joint nut to 65 ft. lbs. (90 Nm). Install a new cotter pin. Remove the spring compressor.

11. Install the shock absorber.

12. Align the front end. When all settings are at specifications, torque the pivot bar retaining bolts to 72–87 ft. lbs. (100–120 Nm).

4WD Vehicles

1. Raise the vehicle and support safely. Remove the skid plate.
2. Remove the shock absorber.
3. Turn the torsion bar adjustment nut counterclockwise to relieve all tension from the torsion bar.
4. Disconnect the brake hose from the brake line and remove the hose from the bracket.
5. Remove the cotter pin from the upper ball stud.
6. Release the upper ball joint from the knuckle using a suitable ball joint separator tool.
7. Loosen the pivot bar retaining nuts and bolts, identify and remove the alignment shims, remove the nuts and bolts.
8. Remove the control arm from the vehicle.

To install:

9. Position the arm at the frame rail bracket.
10. Install the shims in their original locations and install the retaining nuts and bolts. Torque the nuts initially to 40 ft. lbs. (54 Nm).
11. Insert the upper ball stud in the steering knuckle arm bore and install the nut. Torque the nut to 65 ft. lbs. (90 Nm) and install a new cotter pin. Install the shock absorber and attach the brake hose.
12. Turn the torsion bar adjustment nut clockwise to apply a load on the bar.
13. Lower the vehicle.
14. Set the riding height. Align the front end. When all settings are at specifications, torque the pivot bar retaining bolts to 72–87 ft. lbs. (100–120 Nm).

Lower Control Arms

REMOVAL AND INSTALLATION

Montero

1. Raise the vehicle and support it safely.
2. Remove the skid plate.
3. Remove the torsion bar, anchors and lower arm shaft.
4. Remove the shock absorber lower attaching bolt.
5. Disconnect the stabilizer bar from the lower control arm.
6. Remove the cotter pin and the nut from the lower ball stud. Separate the lower ball stud from the steering knuckle using a suitable puller.
7. Remove the pivot bolts and remove the arm from the vehicle.

To install:

8. Install the new control arm to the vehicle.
9. Install the pivot bolts, but do not torque.
10. Insert the ball stud into the steering knuckle bore. Install the slotted nut, torque to 100 ft. lbs. (136 Nm). Install a new cotter pin.
11. Attach the stabilizer bar to the control arm. Install the shock mount bolts.
12. Install the torsion bar and turn the adjustment bolt clockwise to apply a load to the bar.
13. Lower the vehicle so weight of the vehicle is on suspension.
14. Torque the pivot nuts to 108 ft. lbs. (147 Nm).
15. Set the riding height.
16. Align the front suspension.

Mighty Max

2WD Vehicles

1. Raise the vehicle and support safely.
2. Remove the shock absorber.
3. Disconnect the sway bar and strut bar from the lower control arm.
4. Compress the spring using a suitable spring compressor tool.
5. Remove the cotter pin and lower ball joint nut.
6. Release the lower ball joint from the knuckle using a suitable ball joint separator.
7. Pull the arm down and remove the spring with the rubber isolation pad from the vehicle. Remove the lower arm shaft mounting nuts and remove the arm from the vehicle.

To install:

8. Install the arm to the crossmember finger-tight. Install the spring with the rubber isolators. Install the compressor tool and compress spring so lower ball joint can be inserted through the knuckle.
9. Torque the lower ball joint slotted nut to 87–130 ft. lbs. (120–180 Nm). Install a new cotter pin. Remove the spring compressor.
10. Connect the sway bar and strut bar to the lower control arm.
11. Install the shock absorber.
12. Lower the vehicle completely. With the weight of the vehicle on the suspension and torque the lower control arm to crossmember mounting nut to 40–54 ft. lbs. (55–75 Nm).
13. Align the front suspension.

4WD Vehicles

1. Raise the vehicle and support it safely.
2. Remove the skid plate.
3. Remove the torsion bar, anchors and lower arm shaft.

4. Remove the shock absorber lower attaching bolt.
5. Disconnect the stabilizer bar from the lower control arm.
6. Remove the cotter pin and the nut from the lower ball stud. Separate the lower ball stud from the steering knuckle using a suitable puller.
7. Remove the pivot bolts and remove the arm from the vehicle.

To install:

8. Install the new control arm to the vehicle.
9. Install the pivot bolts, but do not torque.
10. Insert the ball stud into the steering knuckle bore. Install the slotted nut, torque to 87–130 ft. lbs. (120–180 Nm) and install a new cotter pin.
11. Attach the stabilizer bar to the control arm. Install the shock mount bolts.
12. Install the torsion bar and turn the adjustment bolt clockwise to apply a load to the bar.
13. Lower the vehicle so weight of the vehicle is on suspension.
14. Torque the pivot nuts to 101–116 ft. lbs. (140–160 Nm).
15. Set the riding height.
16. Align the front suspension.

Expo

1. Raise the vehicle and support safely.
2. To disconnect the sway link, hold ball stud with a hex wrench and remove the self-locking nut with a box wrench.
3. Disconnect the ball joint stud from the steering knuckle.
4. Remove the lower arm inner mounting bolt.
5. Remove the rear mount clamp bolts and remove the rear retainer clamp.
6. Remove the arm from the vehicle.

To install:

7. Install the control arm to the vehicle. Temporarily secure with bolt and a new nut.
8. Install the rear mount clamp and temporarily tighten the clamp mounting bolts. Once the weight of the vehicle is on the suspension, the bolts will be tightened to 51 ft. lbs. (70 Nm).
9. Connect the ball joint stud to the knuckle. Install a new nut and torque to 49 ft. lbs. (68 Nm).
10. Connect the stabilizer links and tighten the locknut, using the same procedure as removal.
11. Lower the vehicle to the floor.

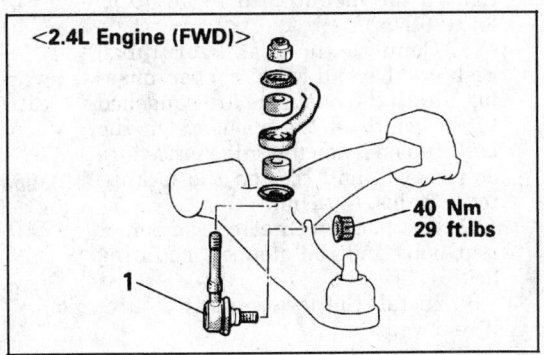

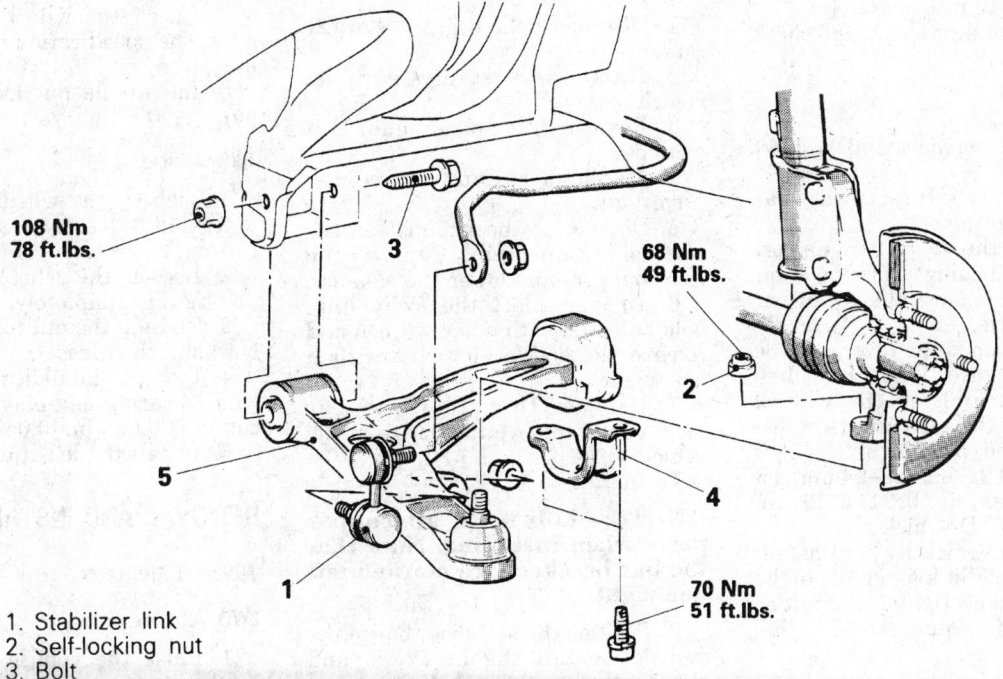

1. Stabilizer link
2. Self-locking nut
3. Bolt
4. Clamp
5. Lower arm

240791

Lower control arm and related components — Expo

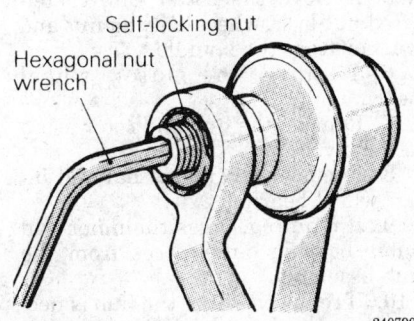

240790

Proper method of disconnecting the ball and socket style link — Expo

12. Once the full weight of the vehicle is on the suspension, torque the inner lower arm mounting bolt nut to 78 ft. lbs. (108 Nm).

13. Inspect all suspension bolts, making sure they all have been fully tightened.

Sway Bar

REMOVAL AND INSTALLATION

Montero

1993 Vehicles

1. Raise the vehicle and support it safely.

2. Remove the front sway bar brackets and retainers.

3. Remove the sway bar support brackets and bushings from the lower control arm. Remove the sway bar from the vehicle.

To install:

4. Install the sway bar.

5. Install the stabilizer bar to the link assembly and torque to 11 ft. lbs. (15 Nm).

6. Install the stabilizer link to the lower arm and torque the nuts to 25 ft. lbs. (33 Nm).

7. Install the brackets and torque the screws to 17 ft. lbs. (24 Nm).

8. Test drive the vehicle and check the alignment.

1994–97 Vehicles

1. Raise the vehicle and support it safely.

2. Remove the front sway bar brackets and retainers.

3. Remove the sway bar support brackets and bushings from the lower control arm. Remove the sway bar from the vehicle.

To install:

4. Install the sway bar.

5. Torque the nuts to the stabilizer link to 69 ft. lbs. (93 Nm).

6. Install the stabilizer link to the lower arm and torque the nuts to 25 ft. lbs. (33 Nm).

7. Install the brackets and torque the screws to 17 ft. lbs. (24 Nm).

8. Test drive the vehicle and check the alignment.

Mighty Max

1. Raise the vehicle and support safely.

2. Remove the front sway bar brackets and retainers.

3. Remove the sway bar support brackets and bushings from the lower control arm. Remove the sway bar from the vehicle.

4. Installation is the reverse of the removal procedure. When installing the stabilizer link to the control arms, tighten the nuts until the exposed threaded portion of the link bolt is 0.87–0.94 in. (22–24mm) for 2WD vehicles or 0.31–0.35 in. (8–9mm) for 4WD vehicles.

5. On 4WD vehicles when installing the stabilizer link assembly to the stabilizer bracket, tighten the nut so as to obtain 16–18mm (0.63-0.71 in.) of exposed threads.

Expo

2WD Vehicles

1. Disconnect the negative battery cable.

2. Raise and safely support vehicle.

3. Remove the splash guard, if equipped.

4. Remove the center cross-member rear installation bolts and roll stopper mounting bolt.

5. Remove the stabilizer link bolts.

6. Remove the stabilizer bar mounts and remove the bar from the vehicle.

To install

NOTE: Lubricate all rubber parts when installing. Note that the bar brackets are marked left and right.

7. Position the stabilizer bar in the vehicle, install the brackets, and

tighten the mounting bolts to 19 ft. lbs. (26 Nm).

8. Connect the stabilizer links. Tighten the end with rubber bushings, until the bushings are squashed to the width of the washer. On the ball stud end, use the same procedure as removal, and tighten the locknut to 29 ft. lbs. (40 Nm).

9. Install the center crossmember rear bolts and roll stopper mounting bolt.

10. Install the lower splash guard, if removed.

11. Lower the vehicle and connect the negative battery cable.

4WD Vehicles

1. Disconnect the negative battery cable.

2. Raise and safely support vehicle.

3. Remove the splash guard, if equipped.

4. Matchmark and remove the driveshaft.

5. Disconnect the exhaust at the manifold connection. Remove the front hangers and lower the exhaust.

6. To disconnect the sway link, hold ball stud with a hex wrench and remove the self-locking nut with a box wrench.

7. Remove the stabilizer bar mounts and remove the bar from the vehicle.

To install

NOTE: Lubricate all rubber parts when installing. Note that the bar brackets are marked left and right.

8. Position the stabilizer bar in the vehicle, install the brackets, and tighten the mounting bolts to 16 ft. lbs. (22 Nm).

9. Connect the stabilizer links and tighten the locknut, using the same procedure as removal.

10. Align the driveshaft and install.

11. Using a new gasket, connect the exhaust pipe.

12. Install the lower splash guard, if removed.

13. Lower the vehicle and connect the negative battery cable.

Front Wheel Bearings

ADJUSTMENT

Montero

1. Tighten the wheel bearing nut to 119 ft. lbs. (162 Nm) while turning the rotor.

2. Loosen the wheel bearing adjusting nut completely.

3. Tighten the nut to 18 ft. lbs. (25 Nm) and then loosen 30°–40°.

4. Using a dial indicator, check the wheel bearing end-play. The specification is 0.002 in. (0.05mm).

5. Install the lock nut.

Mighty Max

2WD Vehicles

1. Tighten the wheel bearing nut to 22 ft. lbs. (30 Nm) while turning the rotor.

2. Loosen the wheel bearing adjusting nut completely.

3. Tighten the nut to 6 ft. lbs. (8 Nm).

4. Check the wheel bearing end-play. The specification is 0.001–0.003 in.

5. Install the nut lock and cotter pin.

4WD Vehicles

1. Tighten the wheel bearing nut to 94–145 ft. lbs. (130–200 Nm) while turning the rotor.

2. Loosen the wheel bearing adjusting nut completely.

3. Tighten the nut to 18 ft. lbs. (25 Nm) and then loosen 30°–40°.

4. Using a dial indicator, check the wheel bearing end-play. The specification is 0.002 in. (0.05mm).

5. Install the lock nut.

REMOVAL AND INSTALLATION

1993–95 Montero

2WD Vehicles

1. Raise the vehicle and support safely.

2. Remove the tire and wheel assembly.

3. Remove the caliper assembly and suspend it from the upper arm.

4. Remove the dust cap.

5. Remove the cotter pin, castellated nut lock, wheel bearing nut and washer from the spindle.

6. Remove the outer wheel bearing.

7. Remove the hub and rotor as an assembly.

8. Remove the grease seal and inner wheel bearing.

9. If required, press the inner and outer bearing outer races from the hub assembly.

10. If replacement of the hub is necessary, matchmark the brake disc with the hub and then separate the hub from the disc.

To install:

11. If removed, assemble the brake disc to the hub, while aligning the

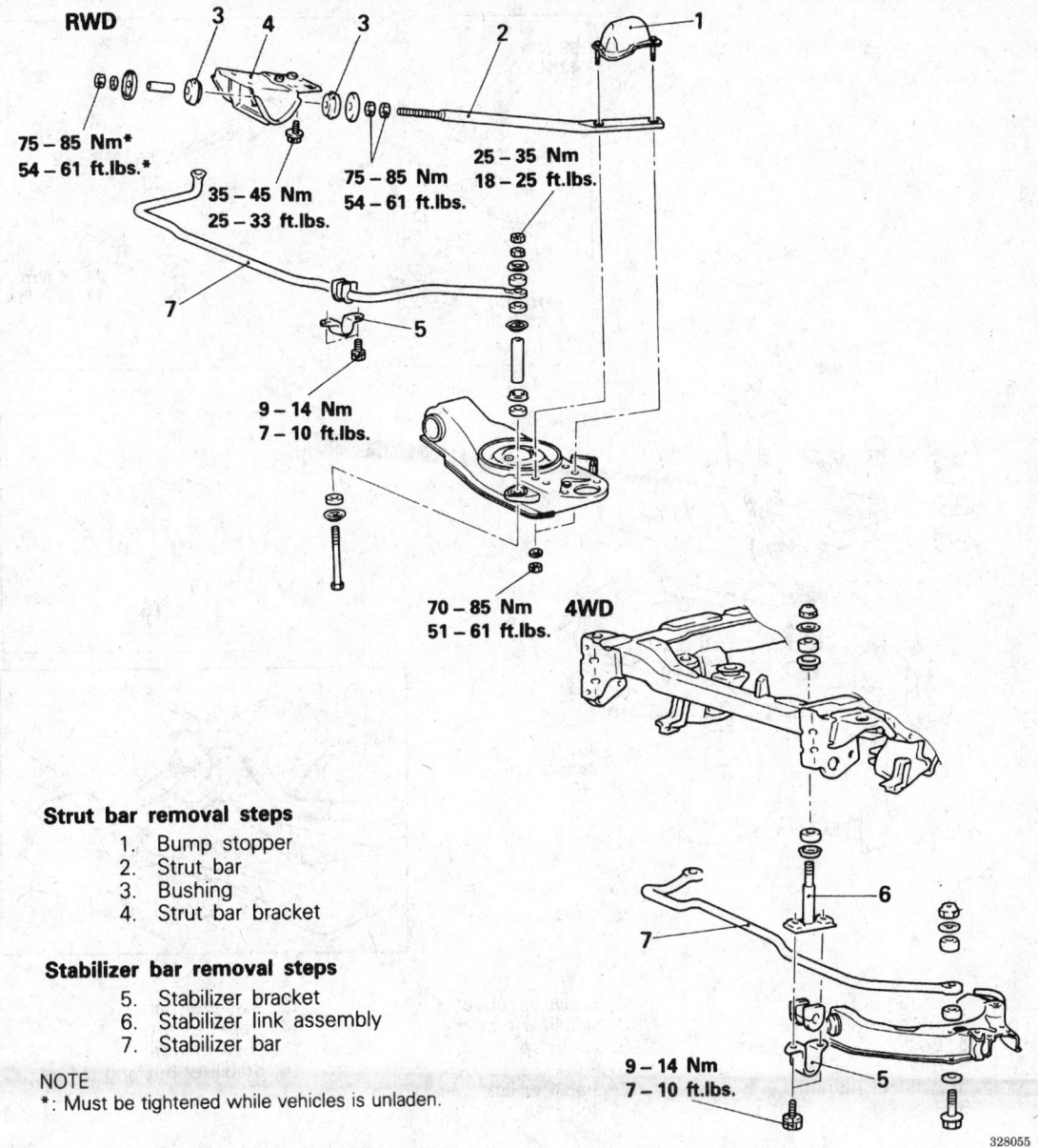

RWD

75 – 85 Nm*
54 – 61 ft.lbs.*

35 – 45 Nm
25 – 33 ft.lbs.

75 – 85 Nm
54 – 61 ft.lbs.

25 – 35 Nm
18 – 25 ft.lbs.

9 – 14 Nm
7 – 10 ft.lbs.

70 – 85 Nm
51 – 61 ft.lbs.

4WD

9 – 14 Nm
7 – 10 ft.lbs.

328055

Strut bar removal steps

1. Bump stopper
2. Strut bar
3. Bushing
4. Strut bar bracket

Stabilizer bar removal steps

5. Stabilizer bracket
6. Stabilizer link assembly
7. Stabilizer bar

NOTE
*: Must be tightened while vehicles is unladen.

Stabilizer bar and strut bar components — Mighty Max

matchmarks. Torque the mounting bolts to 34–38 ft. lbs (47–52 Nm).

12. If removed, press-fit the inner and outer bearing outer races into the hub assembly.

13. Lubricate the seal lip and inside surface of the front hub with MP grease. Re-pack and install the inner wheel bearing. Install a new grease seal.

14. Install the hub assembly to the spindle.

15. Lubricate and install the outer wheel bearing, washer and nut. When the bearing preload is properly set, install the nut lock and a new cotter pin.

16. Install the grease cap.

17. Install the brake pads and caliper.

18. Install the tire and wheel assembly.

4WD Vehicles

1. Raise the vehicle and support safely.

2. Remove the tire and wheel assembly.

3. If equipped with free-wheeling hub, place the free-wheeling hub in the free condition.

NOTE: The free condition can be obtained by shifting the transfer shift lever to the 2H position and then moving in reverse for approximately 3.3–6.5 ft.

a. Remove the hub cover.

b. Remove the snapring from the driveshaft.

c. Remove the bolts and remove the automatic free-wheeling hub.

4. Remove the caliper assembly and suspend it from the upper arm.

5. Remove the lock washer and lock nut.

6. Remove the hub and rotor as an assembly from the knuckle together with the inner and outer bearings.

7. Remove the outer bearing, grease seal and inner wheel bearing.

8. If required, press the inner and outer bearing outer races from the hub assembly.

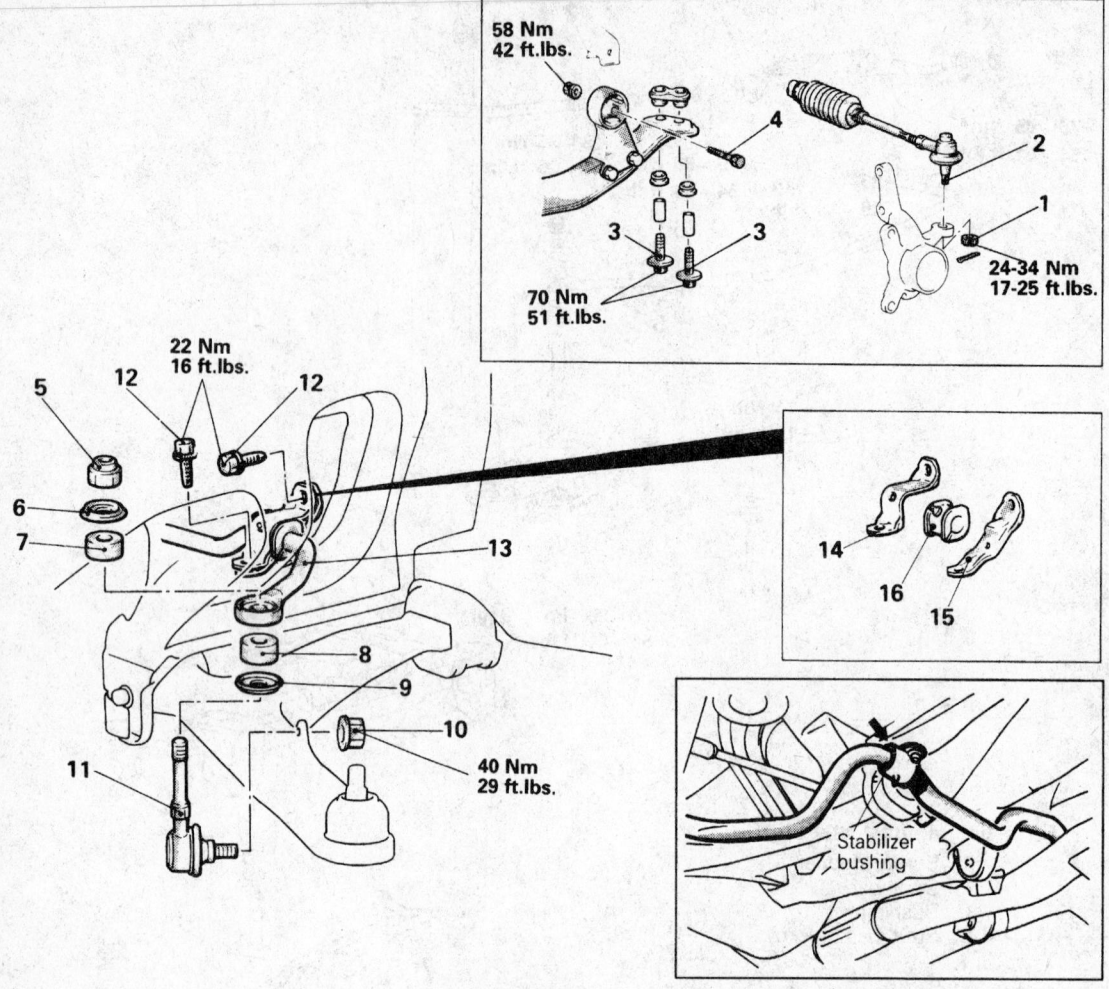

1. Castle nut
2. Tie rod end
3. Bolt
4. Bolt
5. Self-locking nut
6. Joint cup
7. Stabilizer rubber
8. Stabilizer rubber
9. Joint cup
10. Nut
11. Stabilizer link
12. Bolt
13. Stabilizer bar
14. Upper fixture
15. Lower fixture
16. Stabilizer bushing

241334

2WD Front sway bar and related components — Expo

9. If replacement of the hub is necessary, matchmark the brake disc with the hub and then separate the hub from the disc.

To install:

10. If removed, assemble the brake disc to the hub, while aligning the matchmarks.

11. If removed, press-fit the inner and outer bearing outer races into the hub assembly.

12. Lubricate the seal lip and inside surface of the front hub with MP grease. Re-pack and install the inner wheel bearing. Install a new grease seal.

13. Install the hub assembly to the spindle.

14. Lubricate and install the outer wheel bearing and lock nut. When the bearing preload is properly set, install the lock washer.

15. If equipped with free-wheeling hub:

   a. Apply a coating of semi-drying sealant to the free-wheeling hub body and front hub contact surfaces.

   b. Align the key of the brake "B" and the keyway of the knuckle spindle and loosely install the automatic free-wheeling hub assembly. Torque the mounting bolts to 36–43 ft. lbs (50–60 Nm).

16. Install the wheel and tire assembly. Lower the vehicle.

Mighty Max

2WD Vehicles

1. Raise the vehicle and support safely.

2. Remove the tire and wheel assembly.

3. Remove the caliper assembly and suspend it from the upper arm.

4. Remove the dust cap.

5. Remove the cotter pin, castellated nut lock, wheel bearing nut and washer from the spindle.

6. Remove the outer wheel bearing.

7. Remove the hub and rotor as an assembly.

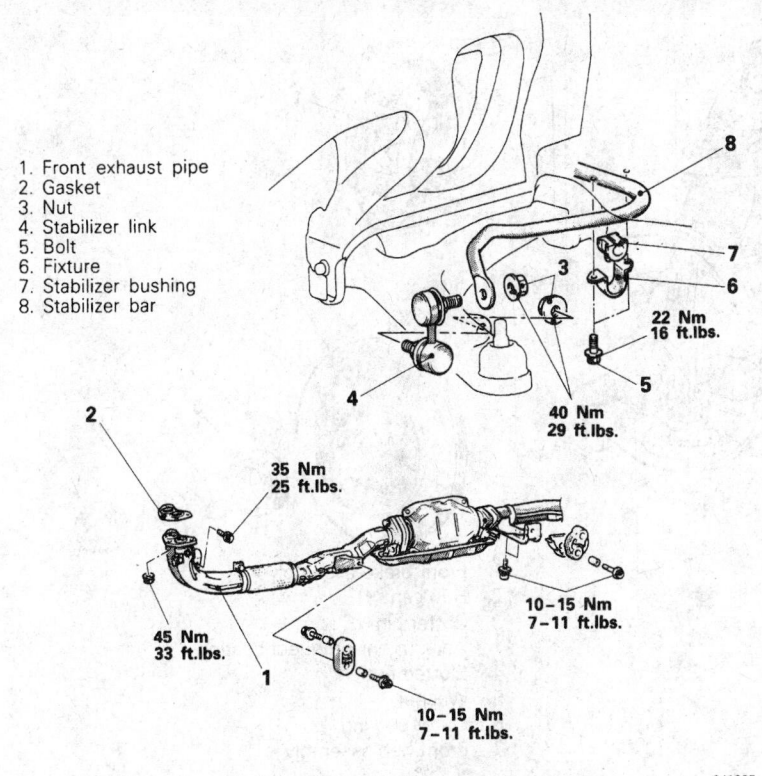

1. Front exhaust pipe
2. Gasket
3. Nut
4. Stabilizer link
5. Bolt
6. Fixture
7. Stabilizer bushing
8. Stabilizer bar

22 Nm
16 ft.lbs.

40 Nm
29 ft.lbs.

35 Nm
25 ft.lbs.

45 Nm
33 ft.lbs.

10-15 Nm
7-11 ft.lbs.

10-15 Nm
7-11 ft.lbs.

241335

4WD Front sway bar and related components — Expo

8. Remove the grease seal and inner wheel bearing.

9. If required, press the inner and outer bearing outer races from the hub assembly.

10. If replacement of the hub is necessary, matchmark the brake disc with the hub and then separate the hub from the disc.

To install:

11. If removed, assemble the brake disc to the hub, while aligning the matchmarks. Torque the mounting bolts to 58–72 ft. lbs (80–100 Nm).

12. If removed, press-fit the inner and outer bearing outer races into the hub assembly.

13. Lubricate the seal lip and inside surface of the front hub with MP grease. Re-pack and install the inner wheel bearing. Install a new grease seal.

14. Install the hub assembly to the spindle.

15. Lubricate and install the outer wheel bearing, washer and nut. When the bearing preload is properly set, install the nut lock and a new cotter pin.

16. Install the grease cap.

17. Install the brake pads and caliper.

18. Install the tire and wheel assembly.

4WD Vehicles

1. Raise the vehicle and support safely.

2. Remove the tire and wheel assembly.

3. If equipped with free-wheeling hub, place the free-wheeling hub in the free condition.

NOTE: The free condition can be obtained by shifting the transfer shift lever to the 2H position and then moving in reverse for approximately 3.3–6.5 ft.

a. Remove the hub cover.

b. Remove the snapring from the driveshaft.

c. Remove the bolts and remove the automatic free-wheeling hub.

4. Remove the caliper assembly and suspend it from the upper arm.

5. Remove the lock washer and lock nut.

6. Remove the hub and rotor as an assembly from the knuckle together with the inner and outer bearings.

7. Remove the outer bearing, grease seal and inner wheel bearing.

8. If required, press the inner and outer bearing outer races from the hub assembly.

9. If replacement of the hub is necessary, matchmark the brake disc with the hub and then separate the hub from the disc.

To install:

10. If removed, assemble the brake disc to the hub, while aligning the matchmarks.

11. If removed, press-fit the inner and outer bearing outer races into the hub assembly.

12. Lubricate the seal lip and inside surface of the front hub with MP grease. Re-pack and install the inner wheel bearing. Install a new grease seal.

13. Install the hub assembly to the spindle.

14. Lubricate and install the outer wheel bearing and lock nut. When the bearing preload is properly set, install the lock washer.

15. If equipped with free-wheeling hub:

a. Apply a coating of semi-drying sealant to the free-wheeling hub body and front hub contact surfaces.

b. Align the key of the brake "B" and the keyway of the knuckle spindle and loosely install the automatic free-wheeling hub assembly. Torque the mounting bolts to 36–43 ft. lbs (50–60 Nm).

16. Install the wheel and tire assembly. Lower the vehicle.

Expo

Removal

1. Disconnect the negative battery cable.

2. Remove the cotter pin from the driveshaft nut. With the brakes applied, loosen the halfshaft nut.

3. Raise the vehicle and support safely. Remove the halfshaft nut.

4. If equipped with ABS, remove the front wheel speed sensor.

5. If equipped with Active-ECS, disconnect the height sensor from the lower control arm.

6. Remove the caliper assembly and brake pads. Suspend the caliper with a wire.

7. Using tool MB991113 or equivalent, disconnect the ball joint and tie rod end from the steering knuckle.

NOTE: It is important to use proper methods of joint separation. Use of unapproved techniques can result in damage to joint and possible failure.

8. Remove the halfshaft by setting up a puller on the outside wheel hub and pushing the halfshaft from the front hub. After pressing the outer shaft, insert a prybar between the transaxle case and the halfshaft and pry the shaft from the transaxle.

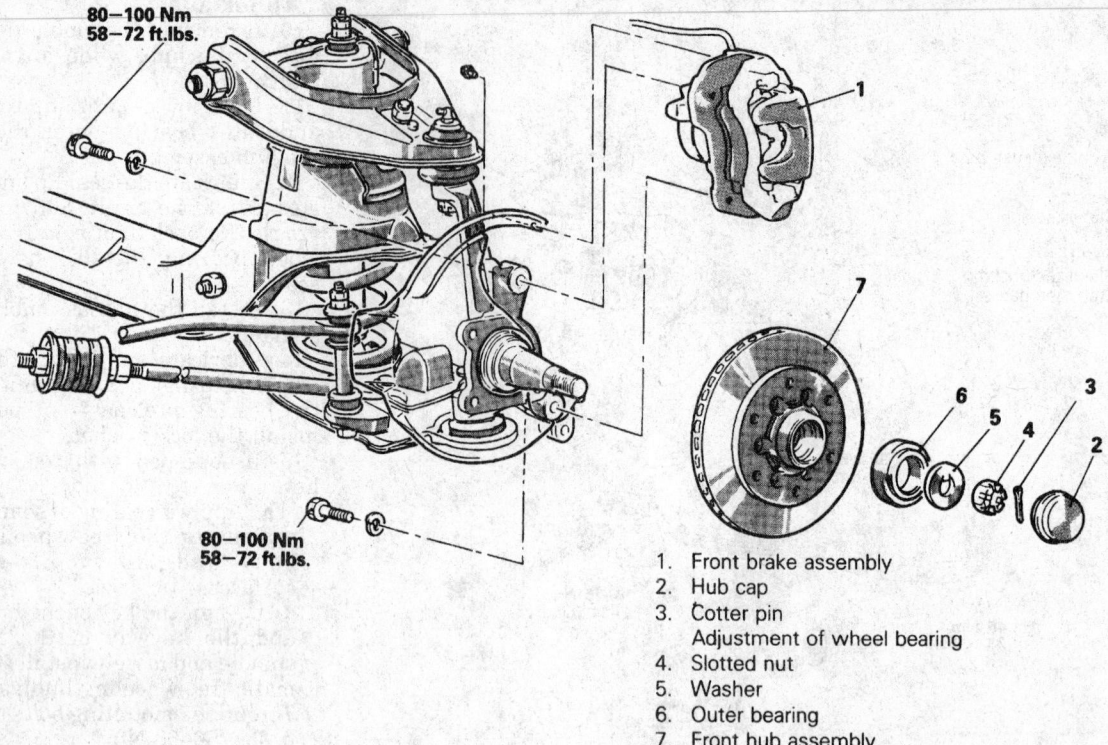

80—100 Nm
58—72 ft.lbs.

80—100 Nm
58—72 ft.lbs.

1. Front brake assembly
2. Hub cap
3. Cotter pin
   Adjustment of wheel bearing
4. Slotted nut
5. Washer
6. Outer bearing
7. Front hub assembly

328802

Front axle hub components — Mighty Max 2WD vehicles

9. Unbolt the lower end of the strut and remove the hub and steering knuckle assembly from the vehicle.

10. Install the hub/knuckle assembly in a vise. Using puller MB991056 or equivalent, remove the hub from the knuckle.

Disassembly

NOTE: Do not use a hammer to accomplish this or the bearing will be damaged.

1. Remove the oil seal from the axle side of the knuckle using a small prying tool.

2. Remove the wheel bearing inner race from the front hub using a puller.

NOTE: Be careful that the front hub does not fall when the inner race is removed.

3. Remove the snapring from the axle side of the knuckle. Remove the bearing from the knuckle using a puller.

4. Once the bearing is removed, the bearing outer race can be removed by tapping out with a brass drift pin and a hammer.

Assembly

1. Fill the wheel bearing with multipurpose grease. Apply a thin coat-

ing of multipurpose grease to the knuckle and bearing contact surfaces.

2. Press the wheel bearing into the knuckle using appropriate pressing tool. Once the bearing is installed, install the inner race using the proper driving tool.

3. Drive the oil seal into the knuckle by using the proper size driver. Drive seal into knuckle until it is flush with the knuckle end surface.

4. Using pressing tool MB990998 or equivalent, mount the front hub assembly into the knuckle. Tighten the nut of the pressing tool to 144—188 ft. lbs. (200—260 Nm). Rotate the hub to seat the bearing.

5. Mount the knuckle assembly in a vise. Check the hub assembly turning torque and end-play as follows:

a. Using a torque wrench and socket MB990998 or equivalent, turn the hub in the knuckle assembly. Note the reading on the torque wrench and compare to the desired reading of 16 inch lbs. (1.8 Nm) or less. This is known as the breakaway torque.

b. Check for roughness when turning the bearing.

c. Mount a dial indicator on the hub so the pointer contacts the machined surface on the hub.

d. Check the end-play.

e. Compare the reading to the limit of 0.002 in. (0.05mm).

6. If the starting torque or the hub end-play are not within specifications while the nut is tightened to 144—188 ft. lbs. (200—260 Nm), the bearing, hub or knuckle have probably not been installed correctly. Repeat the disassembly and assembly procedure and recheck starting torque and end-play.

Installation

1. Install the hub and knuckle assembly onto the vehicle. Install the lower ball joint stud into the steering knuckle and install new nut. Tighten to 52 ft. lbs. (72 Nm).

2. Install the halfshaft into the transaxle extension housing and guide the outer end through the hub/knuckle assembly.

3. Install the two front strut lower mounting bolts and tighten to 80—94 ft. lbs. (110—130 Nm).

4. Install the connection for the tie rod end and tighten nut to 25 ft. lbs. (34 Nm). Install new cotter pin and bend to lock nut in position.

5. Install the brake disc and caliper assembly.

6. If equipped with Active-ECS, connect the height sensor and tighten

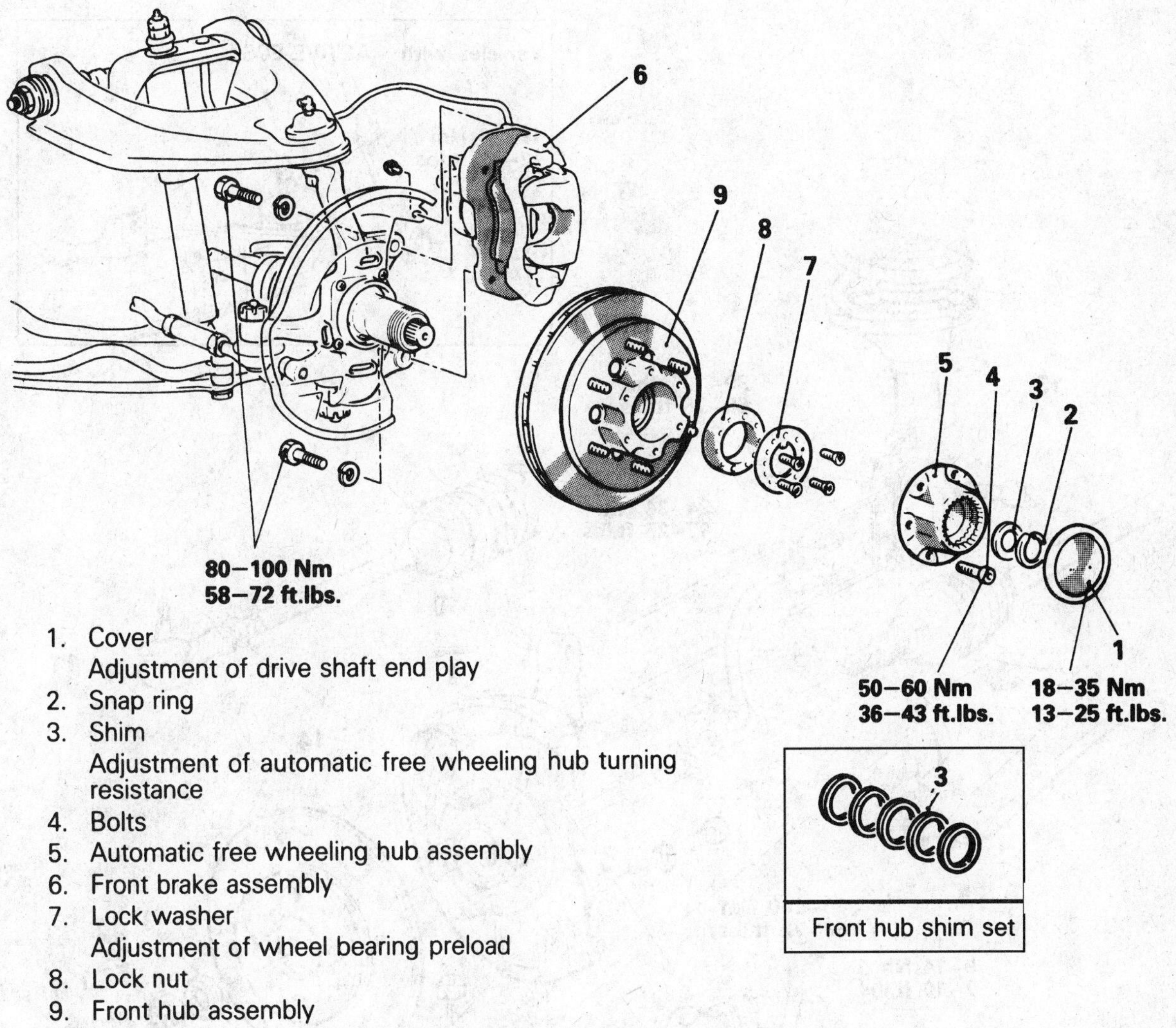

**80—100 Nm
58—72 ft.lbs.**

**50—60 Nm   18—35 Nm
36—43 ft.lbs.   13—25 ft.lbs.**

Front hub shim set

1. Cover
   Adjustment of drive shaft end play
2. Snap ring
3. Shim
   Adjustment of automatic free wheeling hub turning resistance
4. Bolts
5. Automatic free wheeling hub assembly
6. Front brake assembly
7. Lock washer
   Adjustment of wheel bearing preload
8. Lock nut
9. Front hub assembly

328801

Front axle hub and free-wheeling hub components — Mighty Max 4WD vehicles

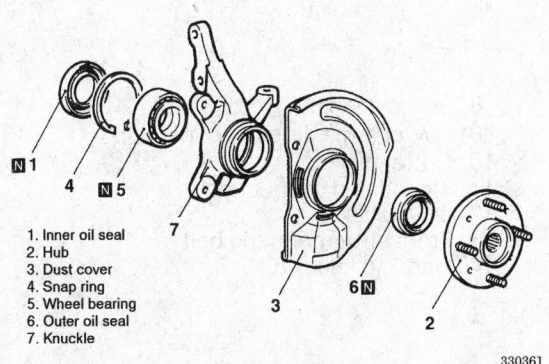

1. Inner oil seal
2. Hub
3. Dust cover
4. Snap ring
5. Wheel bearing
6. Outer oil seal
7. Knuckle

330361

Wheel bearing assembly exploded view — Expo

the mounting bolt to 15 ft. lbs. (20 Nm).

7. Install the front speed sensor, if removed.

NOTE: When installing front speed sensor, make sure harness is routed in the original position and that it is not twisted.

8. Install the washer and new locknut to the end of the halfshaft. Tighten the locknut snugly.

9. Install the tire and wheel assembly onto the vehicle. Lower the vehicle to the ground.

10. With the weight of the vehicle on the ground and the brakes ap-

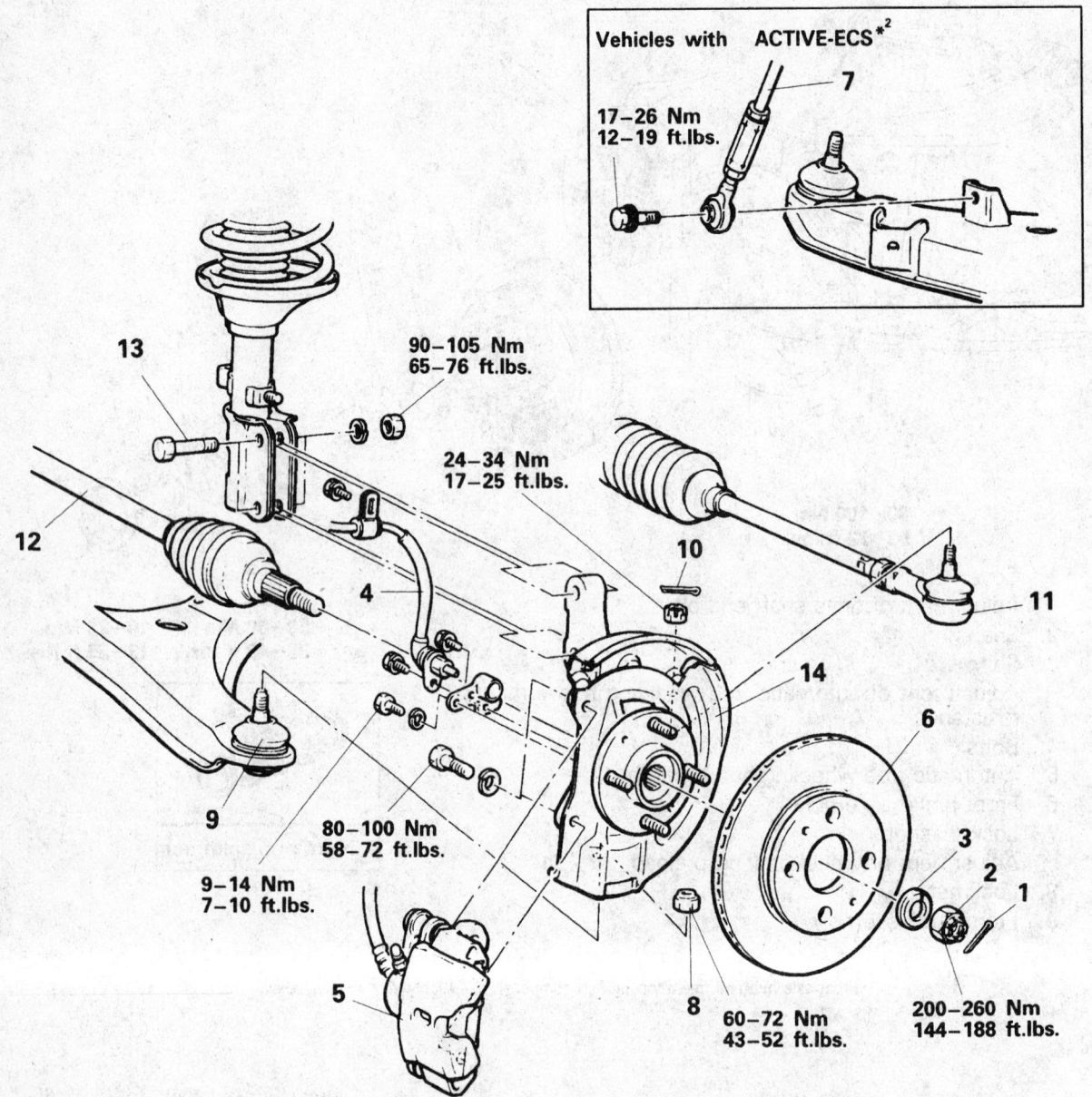

Vehicles with ACTIVE-ECS*²

17–26 Nm
12–19 ft.lbs.

90–105 Nm
65–76 ft.lbs.

24–34 Nm
17–25 ft.lbs.

80–100 Nm
58–72 ft.lbs.

9–14 Nm
7–10 ft.lbs.

60–72 Nm
43–52 ft.lbs.

200–260 Nm
144–188 ft.lbs.

1. Cotter pin
2. Drive shaft nut
3. Washer
4. Front speed sensor connection
5. Caliper assembly
6. Brake disc
7. Front height sensor connection
8. Self locking nut
9. Lower arm ball joint connection
10. Cotter pin
11. Tie rod end connection
12. Drive shaft
13. Front strut mounting bolt
14. Hub and knuckle

330366

Exploded view of the hub and knuckle assembly — Expo

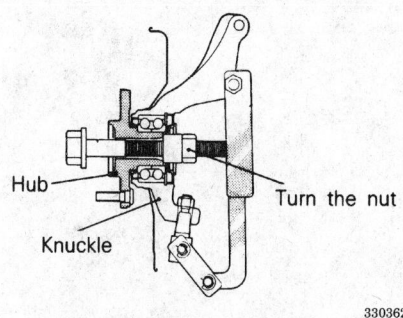

Hub

Turn the nut

Knuckle

330362

Use of press tool for hub removal — Expo

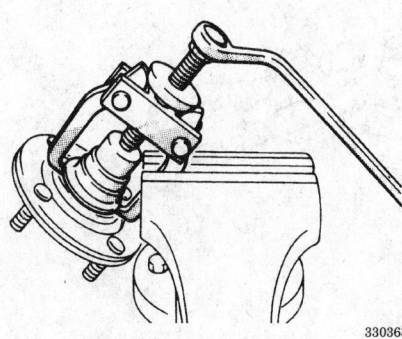

330363

Removing inner race from hub — Expo

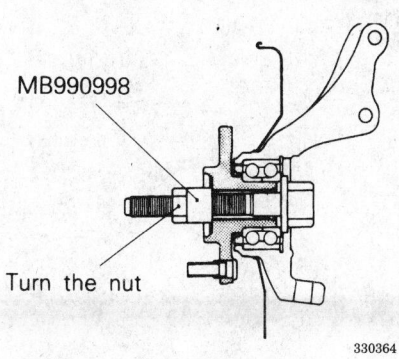

MB990998

Turn the nut

330364

Pressing new bearing assembly into knuckle — Expo

plied, tighten the locknut to 144–188 ft. lbs. (200–260 Nm).

11. Install the cotter pin in the first matching holes and bend it securely.

REAR SUSPENSION

Shock Absorber

REMOVAL AND INSTALLATION

Montero

1. Raise the vehicle and support it safely.
2. Support the rear axle assembly with a hydraulic floor jack, so that the shock absorber may be removed.
3. Remove the upper and lower mounting nuts and bolts that attach the shock to the frame and bracket.
4. Remove the shock from the vehicle.
   To install:
5. Install the shock.
6. Torque the lower bolt to 181 ft. lbs. (245 Nm)
7. Torque the upper nut to 11 ft. lbs. (15 Nm).
8. Remove the floor jack from under the axle assembly.
9. Lower the vehicle.

Mighty Max

1. Raise the vehicle and support it safely.
2. Remove the upper and lower mounting nuts and bolts that attach the shock to the frame and bracket.
3. Remove the shock from the vehicle.
   To install:
4. Install the shock.
5. Torque the upper and lower nuts to 18 ft. lbs. (25 Nm).
6. Lower the vehicle.

Expo

NOTE: The strut assembly is a load bearing component, therefore the vehicle chassis and axle weight must be supported separately, requiring the use of two separate lifting devices.

1. Raise and support vehicle chassis.
2. Raise and support arm assembly slightly.
3. Remove the trunk interior trim to gain access to the top mounting nuts.
4. Remove the top cap and upper shock mounting nuts.
5. Remove the shock lower retaining nut and remove the assembly from the vehicle.

To install

6. Position strut assembly so that lower mounting nut can be installed and lightly tightened.
7. Use jack to raise or lower arm, so that top strut plate studs aligns through body. Raise jack to hold strut assembly in position.
8. Install top plate nuts on studs. Tighten the upper shock mounting nuts to 33 ft. lbs. (45 Nm).
9. With the full weight of the vehicle on the suspension, torque the lower mounting bolt to 72 ft. lbs. (100 Nm).
10. Install top cap and interior trim.

Coil Spring

REMOVAL AND INSTALLATION

Montero

1. Raise the vehicle and support it safely.
2. Remove the parking brake cable attaching bolt.
3. Using the proper equipment, support the weight of the axle.
4. Remove the bolt that attaches the lateral rod to the body.
5. Remove the lower shock mounting bolts.
6. Lower the axle and remove the coil springs with their seats.
   To install:
7. Install the coil to the lower axle.
8. Install the lower shock mounting bolts and torque to 181 ft. lbs. (245 Nm).
9. Support the axle and install the lateral rod bolt. Torque to 170 ft. lbs. (230 Nm).
10. Attach the parking brake cable.
11. Test drive the vehicle and check the alignment.

Expo

FWD Vehicle

1. Raise and properly support the vehicle.
2. Remove the rear stabilizer bar.

NOTE: Perform the following steps, working on one side at a time.

3. Using a jack to support the lower arm, remove the rear shock absorber lower mounting bolt.
4. If equipped with ABS, remove the speed sensor clamp bolt and relocate out of the way. Do not apply tension to the wire harness of the connector.
5. Scribe mating marks on the lower control arm shaft (inner

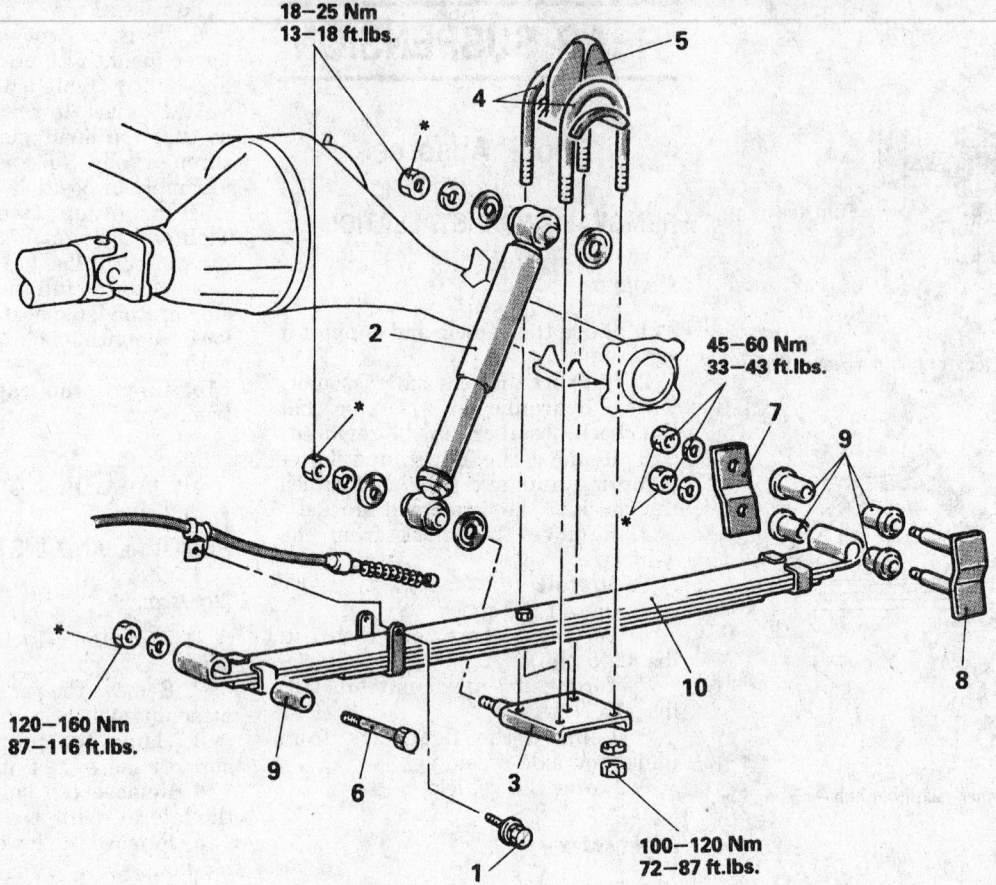

18–25 Nm
13–18 ft.lbs.

45–60 Nm
33–43 ft.lbs.

120–160 Nm
87–116 ft.lbs.

100–120 Nm
72–87 ft.lbs.

1. Parking brake cable attaching bolt
2. Shock absorber
3. U-bolt seat
4. U-bolts
5. Bump stopper
6. Bolt
7. Shackle plate
8. Shackle assembly
9. Rubber bushings
10. Rear spring

328827

Rear leaf spring suspension components — Mighty Max

mounting bolt) and the crossmember. To remove the coil spring, loosen the shaft assembly nut and flange bolt nut (outer mounting bolt), then slowly lower the rear end of the lower arm. It is not necessary to remove the nuts, only to loosen them.

To install:

6. Install the coil spring into the seats making sure both ends of the spring are correctly aligned with the spring seat groove.

7. Slowly raise the rear the rear end of the lower arm and align the scribe marks made during disassembly. Once the full weight of the vehi-

cle is on the ground, tighten shaft and flange mounting nuts to 69 ft. lbs. (95 Nm).

8. Install the speed sensor clamp to it's original location and secure the wire harness making.

9. Reconnect the lower portion of the shock and tighten the retaining bolt to 72 ft. lbs. (100 Nm).

10. Install the stabilizer bar.

11. Lower the arm and remove the jack.

12. Check rear alignment.

AWD Vehicle

1. Raise and properly support the vehicle.

2. Remove the rear stabilizer bar.

NOTE: Perform the following steps, working on one side at a time.

3. Remove the rear driveshaft mounting bolts at the carrier flange and hang the driveshaft from the vehicle body using wire.

4. Using a jack to support the lower arm, remove the rear shock absorber lower mounting bolt.

5. If equipped with ABS, remove the speed sensor clamp bolt and relocate out of the way. Do not apply tension to the wire harness of the connector.

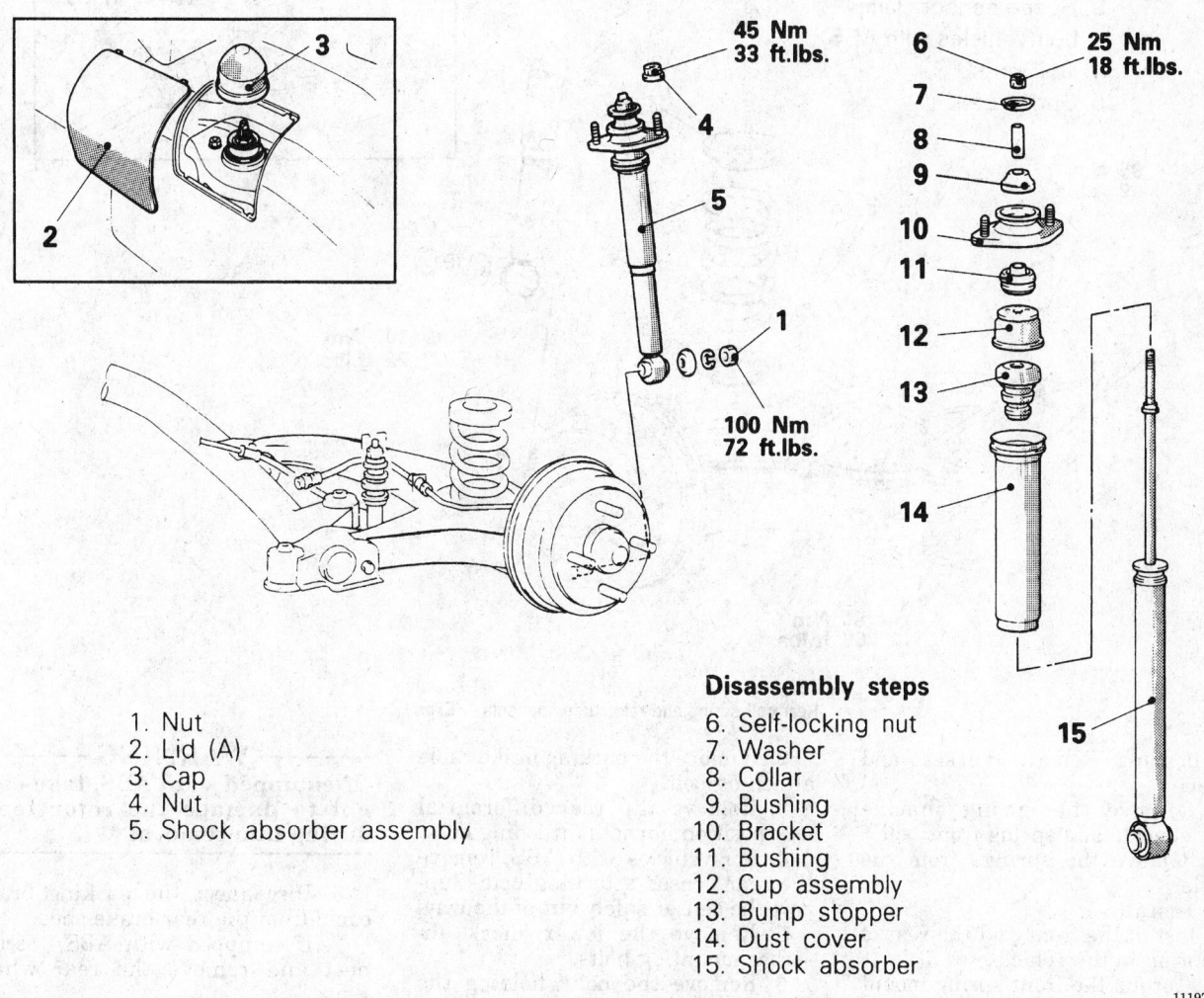

1. Nut
2. Lid (A)
3. Cap
4. Nut
5. Shock absorber assembly

Disassembly steps

6. Self-locking nut
7. Washer
8. Collar
9. Bushing
10. Bracket
11. Bushing
12. Cup assembly
13. Bump stopper
14. Dust cover
15. Shock absorber

45 Nm
33 ft.lbs.

100 Nm
72 ft.lbs.

25 Nm
18 ft.lbs.

111978

Rear shock absorber mounting points — Expo

6. Scribe mating marks on the lower control arm shaft (inner mounting bolt) and the crossmember. To remove the coil spring, loosen the shaft assembly nut and flange bolt nut (outer mounting bolt), then slowly lower the rear end of the lower arm. It is not necessary to remove the nuts, only to loosen them.

To install:

7. Install the coil spring into the seats making sure both ends of the spring are correctly aligned with the spring seat groove.

8. Slowly raise the rear the rear end of the lower arm and align the scribe marks made during disassembly. Once the full weight of the vehicle is on the ground, tighten shaft and flange mounting nuts to 69 ft. lbs. (95 Nm).

9. Install the speed sensor clamp to it's original location and secure the wire harness making.

10. Reconnect the lower portion of the shock and tighten the retaining bolt to 72 ft. lbs. (100 Nm).

11. Install the rear driveshaft to the flange and secure tightening mounting bolts to 40–47 ft. lbs. (55–65 Nm).

12. Install the stabilizer bar.

13. Lower the arm and remove the jack.

14. Check rear alignment.

Leaf Spring

REMOVAL AND INSTALLATION

Mighty Max

1. Raise the vehicle and support it safely.

2. Remove the parking bake cable attaching bolt.

3. Disconnect the lower shock mounting from the U-bolt seats on both sides.

4. Using the proper equipment, support the weight of the axle.

5. Remove the nuts, washers and U-bolts attaching the springs to the

1. Shock absorber
2. Bolt(AWD)
3. Speed sensor clamp
   bolt(vehicles with ABS)
4. Coil spring
5. Spring seat

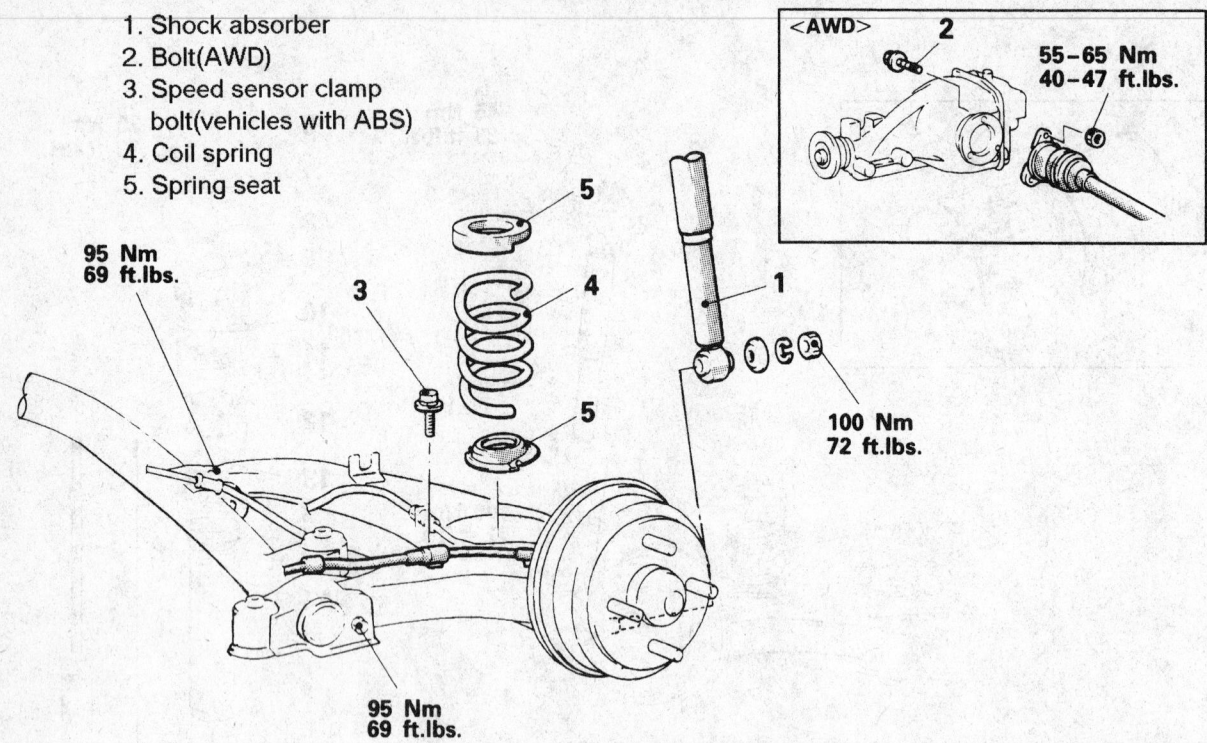

95 Nm
69 ft.lbs.

100 Nm
72 ft.lbs.

95 Nm
69 ft.lbs.

\<AWD\>

55–65 Nm
40–47 ft.lbs.

239574

Rear coil spring and related components — Expo

axle housing. Remove the seat and spacer.

6. Remove the spring shackle bolts, shackle and spring front bolt.

7. Remove the springs from the vehicle.

To install:

8. Install the front and the rear of the spring to the vehicle.

9. Torque the front spring mount nut to 43 ft. lbs. (60 Nm).

10. Torque the rear spring mount nut to 43 ft. lbs. (60 Nm).

11. Place the axle housing on the spring and install the U-bolts and related parts. Torque U-bolt nuts to 87 ft. lbs. (120 Nm).

12. Connect the lower shock mounts to the U-bolt seats.

13. Install the parking brake cable attaching bolt.

14. Make a visual inspection to be sure all components are installed and lower the vehicle to the floor.

Lower Control Arms

REMOVAL AND INSTALLATION

Montero

1. Raise and safely support the vehicle.

2. Remove the parking brake cable attaching bolt.

3. Remove the rear differential lock position harness attaching bolt.

4. For vehicles with ABS, remove the rear sensor attaching bolt. Support the sensor safely out of the way.

5. Remove the lower shock absorber mounting bolts.

6. Remove the bolts holding the lower arm.

7. Remove the lower control arm.

To install:

8. Install the lower arm and torque the nuts to 181 ft. lbs. (230 Nm).

9. Install the lower shock absorber bolt. Torque to 181 ft. lbs. (245 Nm).

10. If removed, install the ABS sensor bracket.

11. Attach the rear differential lock position harness.

12. Attach the parking brake cable.

13. Test drive the vehicle and check the rear alignment.

Expo

1. Remove the rear stabilizer bar.

2. If equipped with AWD, remove the rear axle shaft.

3. Remove the rear brake drum.

4. If equipped with ABS, remove the rear caliper assembly and brake disc.

5. Remove the rear hub assembly.

WARNING

If equipped with ABS, take care not to damage the rotor teeth during hub removal.

6. Disconnect the parking brake cable from the rear brake shoe.

7. If equipped with ABS, disconnect and remove the rear wheel sensor.

NOTE: The speed sensor has a pole piece projecting from it. This exposed tip must be protected from impact or damage. Do not allow the pole piece to contact the toothed wheel during removal or installation.

8. If equipped with a link bracket, remove the mounting bolts from the control arm.

9. Remove the rear shock and coil spring.

10. Remove the brake line and parking brake mounting bolts from the lower control arm.

11. Matchmark and remove the inboard lower arm pivot bolt. Remove the flange bolt and the arm from the vehicle.

To install:

12. Install the arm on the vehicle and secure with the flange bolt, temporarily tighten the nut. Install the

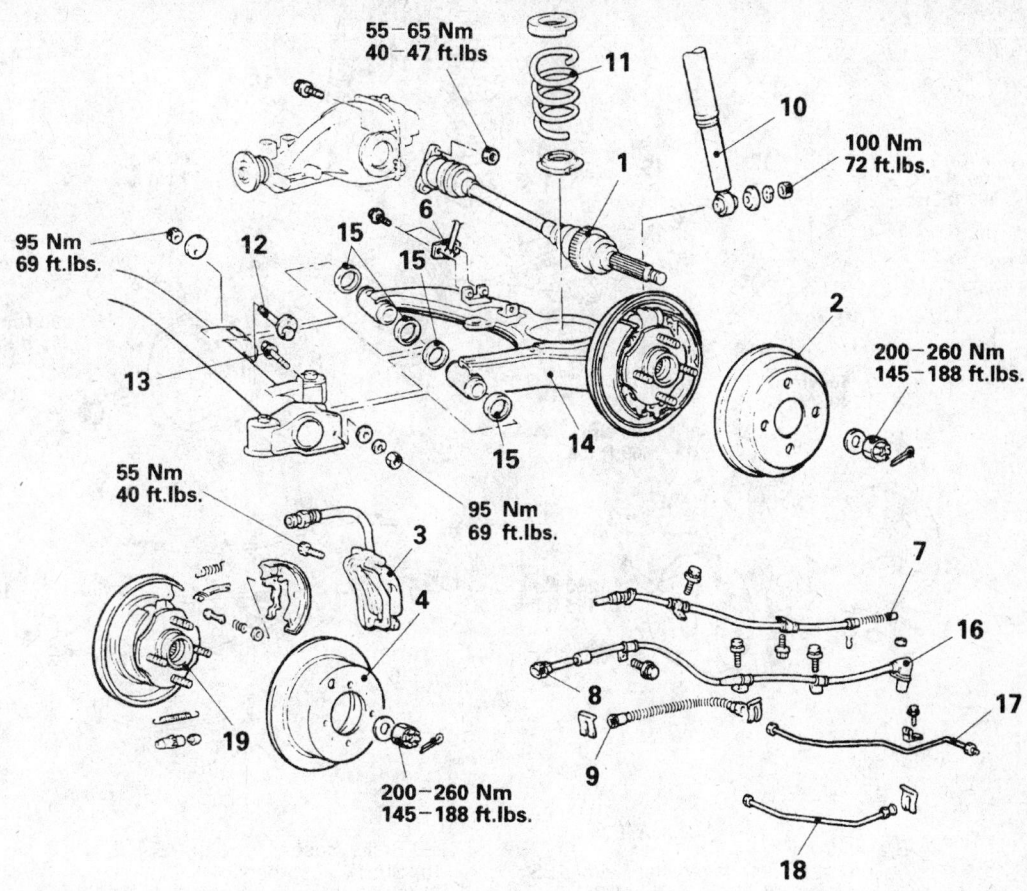

1. Drive shaft
2. Brake drum
3. Caliper assembly <Vehicles with ABS>
4. Brake disk <Vehicles with ABS>
6. Link bracket <EXPO>
7. Connection for parking brake cable and brake shoe
8. Rear sensor connector <Vehicles with ABS>
9. Brake hose
10. Shock absorber
11. Coil spring
12. Shaft assembly
13. Flange bolt
14. Lower arm assembly
15. Stopper
16. Rear speed sensor <Vehicles with ABS>
17. Brake pipe
18. Brake pipe <Vehicles with ABS>
19. Hub assembly

139787

AWD rear suspension component identification — Expo

arm pivot bolt and temporarily tighten the nut.

13. Install the rear shock and coil spring.

14. Connect the link bracket, if removed.

15. Install the brake line and parking brake mounting bolts to the lower control arm.

16. Connect the parking brake cable to the rear brake shoe.

17. Install the rear hub assembly.

18. Install the rear brake drum or, if equipped with ABS, install the rear caliper assembly and brake disc.

19. Install the rear axle shaft.

20. Install the rear stabilizer bar.

21. Install and connect the rear wheel speed sensor. Use a brass or other non-magnetic feeler gauge to check the air gap between the tip of the pole piece and the toothed wheel. Correct gap is 0.012–0.035 in. (0.3–0.9mm). Tighten the 2 sensor bracket bolts to 10 ft. lbs. (14 Nm) with the sensor located so the gap is the same at several points on the toothed wheel. If the gap is incorrect, it is likely that the toothed wheel is worn or improperly installed.

22. Lower the vehicle and tighten the lower arm flange bolt nut and the arm pivot bolt to 69 ft. lbs. (95 Nm).

23. Bleed the brake system if any lines where opened. Adjust the park-

ing brake and perform a rear wheel alignment.

Sway Bar

REMOVAL AND INSTALLATION

Montero

1. Raise and support the vehicle safely.

2. Remove the lower shock absorber bolts.

3. Remove the sway bar bushings, nuts and mounting bolts.

4. Remove the brackets and sway bar.

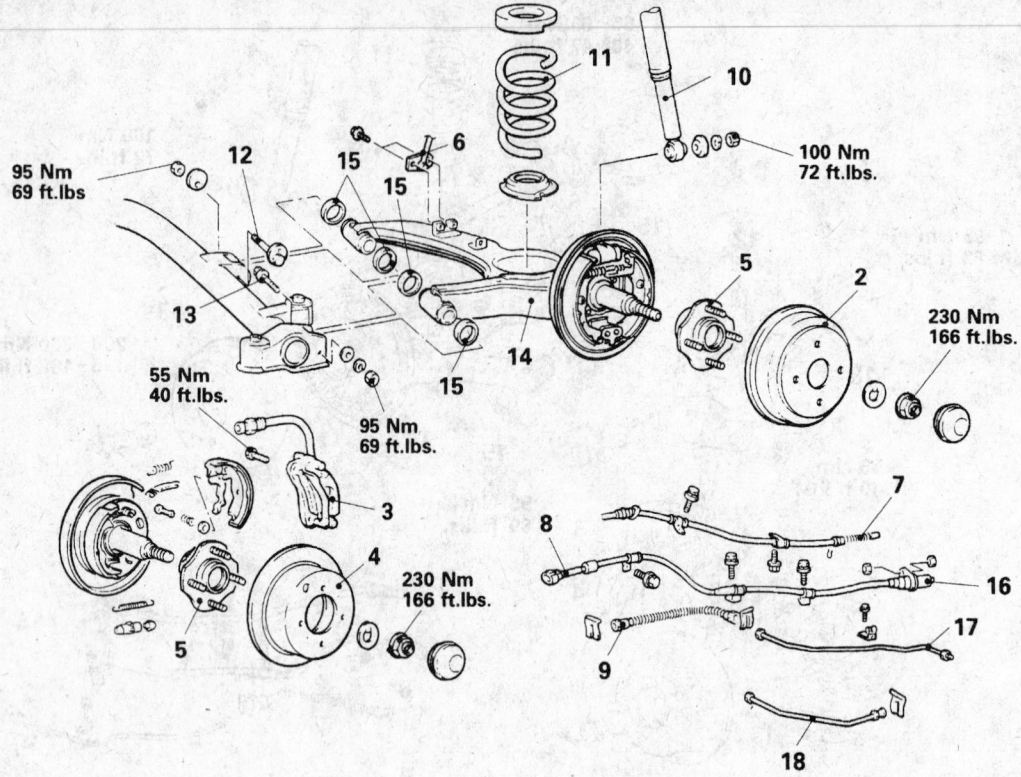

2. Brake drum
3. Caliper assembly <Vehicles with ABS>
4. Brake disk <Vehicles with ABS>
5. Hub assembly
6. Link bracket <EXPO>
7. Connection for parking brake cable and brake shoe
8. Rear sensor connector <Vehicles with ABS>
9. Brake hose
10. Shock absorber
11. Coil spring
12. Shaft assembly
13. Flange bolt
14. Lower arm assembly
15. Stopper
16. Rear speed sensor <Vehicles with ABS>
17. Brake pipe
18. Brake pipe <Vehicles with ABS>

139786

FWD rear suspension component identification — Expo

To install:

5. Install the sway bar.

6. Torque the bolts for the brackets to 25 ft. lbs. (34 Nm).

7. Install the bushings, nuts and mounting bolts and torque the nuts to 11 ft. lbs. (15 Nm).

8. Install the lower shock absorber bolts and torque to 181 ft. lbs. (245 Nm).

9. Lower the vehicle and test drive.

Expo

1. Raise and support the vehicle safely.

2. Remove the self-locking nuts at the sway link. Once the stabilizer bar nut is removed, remove the joint cups and stabilizer rubber bushings.

3. Remove the retainer bolts and the stabilizer bar brackets. Remove the bushing.

4. Remove the stabilizer bar.

5. Inspect the bar for damage, wear and deterioration and replace as required.

To install:

6. Install the stabilizer bar into the vehicle.

7. Install the center stabilizer bar bushings, brackets and bolts. Tighten the bolts to 17 ft. lbs. (23 Nm).

8. Assemble the joint cups and stabilizer rubber to the link. Install a new self-locking nut onto the link.

Tighten the self-locking nut so the protrusion of the stabilizer link from the top of the joint cup is within 0.98 to 1.06 in. (25 to 27 mm).

9. Lower the vehicle.

Wheel Bearings

REMOVAL AND INSTALLATION

Montero

1. Disconnect the negative battery cable.

2. Raise and support the vehicle safely.

3. Remove the rear wheels.

4. Disconnect the brake hose connection and remove the brake caliper.

5. Remove the brake disc.

6. Remove the parking brake cable attaching bolt, parking brake cable end and the parking brake shoe assembly.

7. If equipped with ABS, disconnect the speed sensor.

8. Using an axle puller, remove the rear axle shaft assembly.

NOTE: Be careful not to damage the oil seal when removing the rear axle shaft.

9. Remove the snapring.

10. Remove one retainer bolt from the backing plate.

11. Partially grind the retainer ring; then, using a chisel, cut and remove the retainer ring.

NOTE: Be careful not to damage the bearing case and axle shaft.

12. Remove the inboard inner bearing.

13. Using a suitable tool, remove the axle shaft from the bearing case assembly. Remove the outboard inner bearing from the axle shaft.

14. Remove the oil seal.

15. If not equipped with ABS, remove the dust cover.

16. If equipped with ABS, insert an iron plate of approximately 0.04 in. (1mm) thickness between the rotor assembly and the axle shaft, and then use a press to remove the rotor assembly. Remove the speed sensor bracket.

17. Remove the bearing outer race.

18. Remove the O-ring and the oil seal from the axle housing.

To install:

19. Install the oil seal at the axle housing. Apply multi-purpose grease to the seal lip. Install the O-ring.

20. Apply multi-purpose grease to the external surface of the bearing outer race. Press fit the bearing outer race into the bearing case.

21. Install the speed sensor bracket, if removed.

22. With ABS, press the rotor assembly to the rear axle shaft. For non-ABS vehicles, install the dust cover (non-ABS).

23. Install the outboard inner bearing.

24. Apply multi-purpose grease to the external surface of the new oil seal. Press fit the new seal into the bearing case until it is flush with the face of the bearing case. Apply multipurpose grease to the seal lip.

25. Pass the axle shaft through the bearing inner race, the bearing case and the second bearing inner race in that order. Press the inboard bearing inner race to the axle shaft.

CAUTION

Both bearing inner race sets should be press fitted together. The left and right lengths of the axle shaft are different approximately 0.28 in. (7mm)] in vehicles with rear differential lock. The right side is longer; be careful when installing.

---

26. Press the retainer onto the axle shaft, while checking that the press-fitting force is at the standard value. If the initial press-fitting force is less than the standard value, replace the axle shaft.

• 1993–95 vehicles initial press-fitting force: 11,023 lbs. (50,000 N) or more.

• 1996–97 vehicles initial press-fitting force: 11,016 lbs. (49,000 N) or more.

• 1993–95 vehicles final press-fitting force: 22,046–24,251 lbs. (100,000–110,000 N).

• 1996 — 97 vehicles final press-fitting force: 22,031–24,279 lbs. (98,000–108,000 N).

27. Install the snapring. Measure the clearance between the snapring and the retainer with a thickness gauge, and check that it is 0.0065 in. (0.166 mm) or less. If the clearance exceeds the standard value, change the snapring so that the clearance is at the standard value.

28. Install the axle shaft assembly to the housing.

29. Install the speed sensor.

30. Install the parking brake assembly.

31. Install the parking brake cable end.

32. Install the brake disc and the caliper assembly.

33. Connect the brake tube.

34. Properly bleed the brake system.

35. Check for proper gear lubricant level.

36. Install the wheels.

37. Lower the vehicle.

38. Connect the negative battery cable.

Mighty Max

1. Raise and safely support the vehicle.

2. Remove the rear tire and wheel assembly.

3. Remove the brake drum.

4. Disconnect the parking brake cable from the brake shoe and remove from the backing plate.

5. Disconnect and plug the brake line(s) at the wheel cylinder.

6. Remove the 4 nuts behind the backing plate.

7. Remove the backing plate, bearing case and the axle shaft as an assembly. If not possible by hand, use a slide hammer to remove the assembly.

8. Remove the O-ring and the bearing preload shims. Save the preload shims for reassembly. Remove the snapring.

9. Remove the oil seal from the axle tube with a hooked slide hammer.

10. To remove the axle shaft bearing, remove the retaining ring:

   a. Remove 1 retaining bolt from the backing plate.

   b. Push the bearing case all the way to the side of the dust cover.

   c. Protect the bearing case with adhesive tape.

   d. Grind through the retaining ring in 1 spot until the axle shaft is exposed.

   e. Cut into the retaining ring with a chisel and remove the ring.

CAUTION

Always wear eye protection and appropriate clothing when grinding or chiseling.

---

11. Screw the locknut onto the axle shaft about 3 turns. If not equipped with a locknut, it may be necessary to install spacers on the shaft before removing the bearing.

12. If tool MB990787–01 is not available, it will be necessary to fabricate a metal plate that fits over the axle shaft and butts the locknut. Drill 4 holes in the plate that align with the 4 bearing case studs and fit the plate. Refit 2 nuts and washers to the bearing case studs diagonally across from each other and tighten them evenly to release the bearing case and bearing.

13. Use a hammer and drift to remove the bearing outer race from the bearing case.

14. Remove the outer oil seal from the bearing case.

To install:

15. Apply grease to the outer surface on the bearing outer race and the lip of the outer oil seal. Drive into the bearing case.

16. Slide the bearing case and bearing over the rear axle shaft. Apply grease on the bearing rollers and install the inner race by pressing into place. Be careful not to damage the dust cover.

17. Pack the bearing with grease.

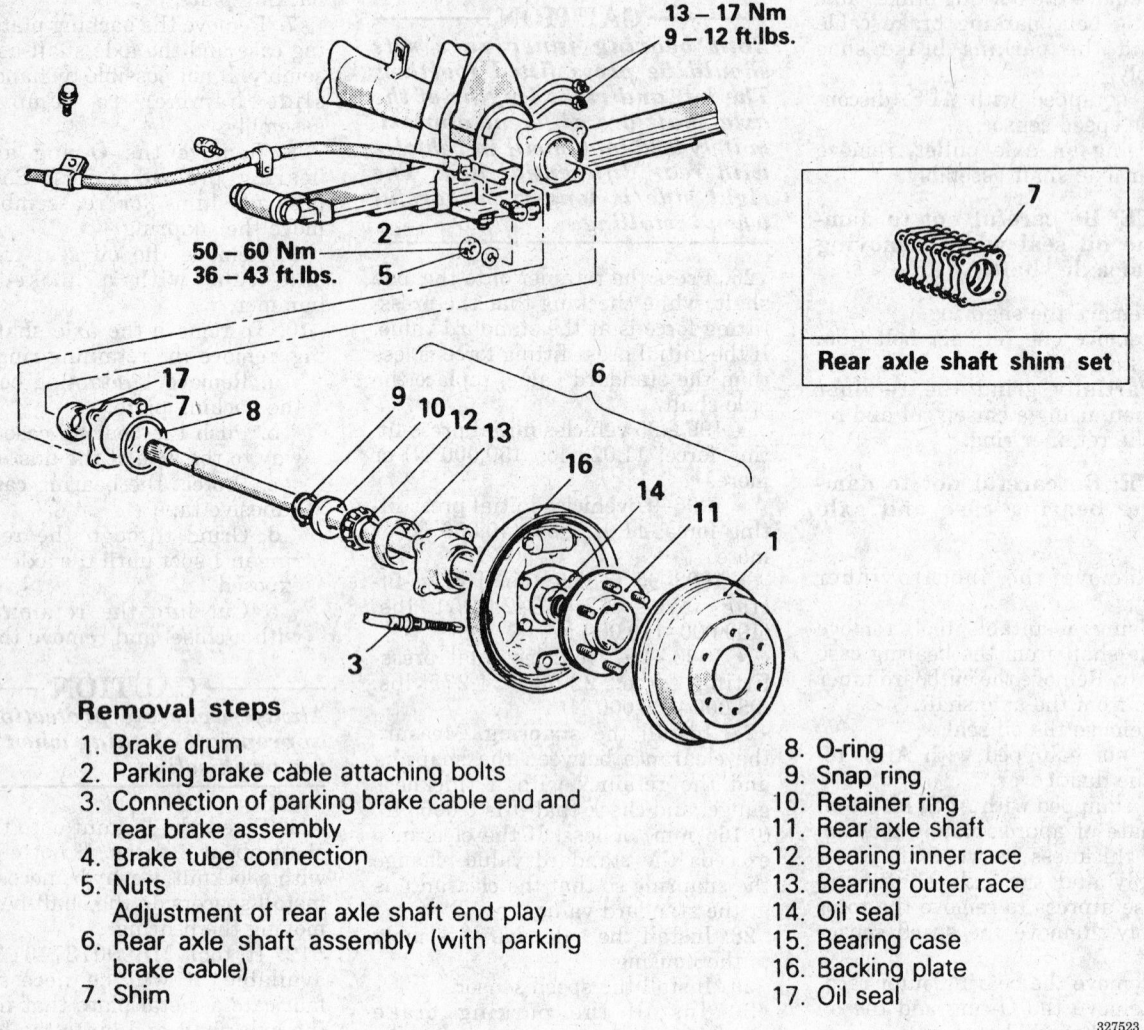

13 – 17 Nm
9 – 12 ft.lbs.

50 – 60 Nm
36 – 43 ft.lbs.

Rear axle shaft shim set

Removal steps

1. Brake drum
2. Parking brake cable attaching bolts
3. Connection of parking brake cable end and rear brake assembly
4. Brake tube connection
5. Nuts
   Adjustment of rear axle shaft end play
6. Rear axle shaft assembly (with parking brake cable)
7. Shim

8. O-ring
9. Snap ring
10. Retainer ring
11. Rear axle shaft
12. Bearing inner race
13. Bearing outer race
14. Oil seal
15. Bearing case
16. Backing plate
17. Oil seal

327523

Exploded view of the rear axle shaft — Mighty Max

18. If equipped with locknut, install the washer, the crowned lock washer and the locknut in that order and torque the locknut to 130–159 ft. lbs. (176–220 Nm). Bend the tab on the lock washer into the groove on the locknut. If the tab and the groove do not line up, tighten locknut slightly.

19. If equipped with snapring, install a new retainer ring and install the snapring.

20. Lubricate and drive the new inner oil seal into place. Refit the assembly.

21. Install a new O-ring and the shims. Apply silicone rubber sealant to the face of the bearing case.

22. Install the entire assembly to the axle housing. Torque the retaining nuts to 36–43 ft. lbs. (50–60 Nm).

23. Check the axle shaft end-play. If not between 0.002–0.0079 in. (0.05–0.20 mm), proceed with the axle shaft end-play adjustment procedure.

24. Install all removed brake parts and bleed the system.

25. Install the tire and wheel assembly and lower the vehicle.

26. Road test the vehicle and check for leaks.

End-play Adjustment Procedure

1. Begin with the left side rear axle assembly and insert a 0.04 in. (1 mm) shim between the bearing case and the axle shaft housing. Torque the nuts to specification.

2. Install the right side axle assembly into the right side housing without its shim and O-ring. Torque the 4 nuts to about 50 inch. lbs.

3. Using a feeler gauge, measure the gap between the bearing case and the axle housing face.

4. Remove the axle shaft and select a shim or shims that is the equal to the sum of the clearance measured in Step 3 plus 0.002–0.0079 in. (0.05–0.20 mm) and install them on the housing. Install the O-ring and apply sealant.

5. Install the axle assembly and torque the nuts to 36–43 ft. lbs. (50–60 Nm).

6. Measure the end-play and complete the installation procedure.

Expo

FWD vehicles

1. Raise the vehicle and support safely.

2. Remove the tire and wheel assembly.

3. Remove the bolt(s) holding the speed sensor bracket to the knuckle and remove the assembly from the vehicle.

NOTE: The speed sensor has a pole piece projecting from it. This exposed tip must be protected from impact or scratches. Do not allow the pole piece to contact the toothed wheel during removal or installation.

4. Remove the brake drums. If equipped with rear disc brakes, remove the caliper from the brake disc and suspend with a wire. Remove the brake rotor.

5. Remove the grease cap, locking nut and tongued washer.

6. Remove the rear hub and bearing assembly.

NOTE: The rear hub assembly can not be disassembled. If bearing replacement is required, replace the assembly as a unit.

To install:

7. Install the hub and bearing assembly.

8. Install the tongued washer and a new locking nut. Torque the lock nut to 166 ft. lbs. (230 Nm). Once the lock nut has been properly torqued, crimp the nut flange over the slot in the spindle.

9. Install the grease cap and brake parts.

10. Temporarily install the speed sensor to the knuckle; tighten the bolts only finger-tight.

11. Route the speed sensor cable correctly and loosely install the clips and retainers. All clips must be in their original position and the sensor cable must not be twisted. Improper installation may cause cable damage or system failure.

NOTE: The wiring in the harness is easily damaged by twisting and flexing. Use the white stripe on the outer insulation to keep the sensor harness properly placed.

12. Use a brass or other non-magnetic feeler gauge to check the air gap between the tip of the pole piece and the toothed wheel. Correct gap is 0.012–0.035 in. (0.3–0.9mm). Tighten the 2 sensor bracket bolts to 10 ft. lbs. (14 Nm) with the sensor located so the gap is the same at several points on the toothed wheel. If the gap is incorrect, it is likely that the toothed wheel is worn or improperly installed.

13. Install the tire and wheel assembly. Be sure to pump brake pedal until firm before moving vehicle.

AWD vehicles

1. Raise and safely support the rear of the vehicle, with the suspension hanging free.

2. Remove the rear wheels.

3. Remove the brake drums. If equipped with rear disc brakes, remove the caliper and rotor assemblies.

4. Remove the bolts that attach the rear halfshaft to the rear carrier.

5. Remove the cotter pin, driveshaft nut cover and nut from the rear driveshaft.

—————— **WARNING** ——————
Do not apply the vehicle weight to the wheel bearing while loosening the driveshaft nut or bearing damage may occur.

6. Use a slide hammer puller and proper adapter to remove the hub assembly from the axle shaft.

7. Remove the lower control arm.

8. Using a hydraulic press and the appropriate adapters, press the inner race from the hub assembly. Remove the outer snap ring and press the outer race from the lower control arm.

To install:

9. Press the new bearing into the lower control arm.

10. Using special adapters MB991400, MB991401 to properly support the bearing races, press the hub into the bearing.

11. Install a wheel bearing preload tool MB990998 to the hub and bearing. Torque the tool nut to 145 to 188 ft. lbs. (200 to 260 Nm). With preload tool in place, use a torque wrench and socket to measure the rotating torque of the bearings. The torque should be 9 inch lbs. (1.1 Nm) or less.

12. Install the lower control arm.

13. Install the rear brake components.

14. Install the axle shaft and torque the retainers on the rear carrier to 40 to 47 ft. lbs. (55 to 65 Nm) and the shaft end nut to 145 to 188 ft. lbs. (200 to 260 Nm).

15. Install the rear wheel assemblies and lower the vehicle.

NOTE: Be sure to pump brake pedal until firm before moving vehicle.

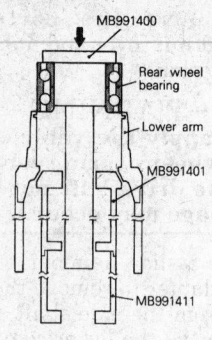

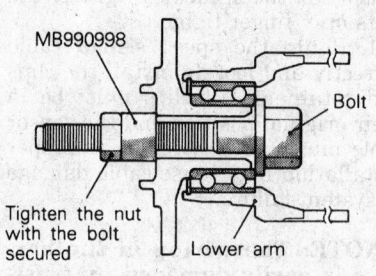

Tighten the nut with the bolt secured

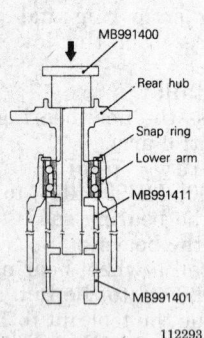

Proper method of assembling bearing and hub — Expo

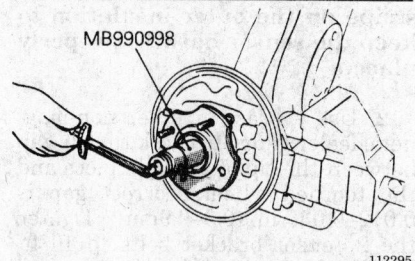

Wheel bearing preload adjustment method — Expo

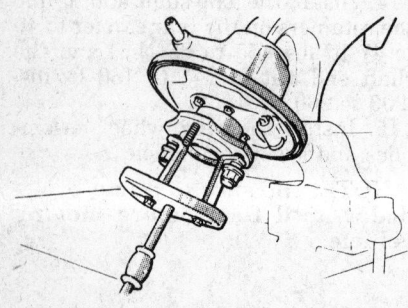

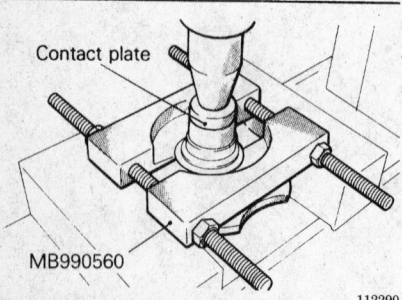

Proper method of removing bearing and hub — Expo

NISSAN 18

Pathfinder • Pick-Up

FIRING ORDERS

NOTE: To avoid confusion, always replace spark plug wires one at a time.

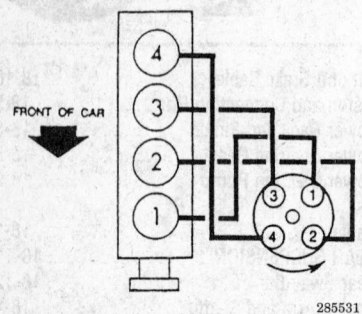

FRONT OF CAR

285531

2.4L Engine
Engine Firing Order: 1–3–4–2
Distributor Rotation: Counterclockwise

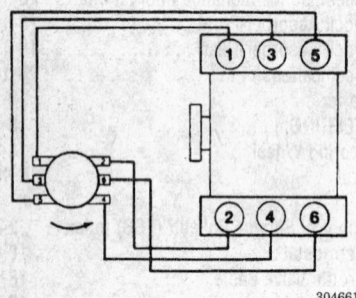

304661

3.0L and 3.3L Engines
Engine Firing Order: 1–2–3–4–5–6
Distributor Rotation: Counterclockwise

ENGINE ELECTRICAL

NOTE: Disconnecting the negative battery cable on some vehicles may interfere with the functions of the on-board computer systems and may require the computer to undergo a relearning process, once the negative battery cable is reconnected.

Distributor

REMOVAL AND INSTALLATION

1. Disconnect the negative battery cable.
2. Disconnect and remove the distributor cap with the plug wires attached.

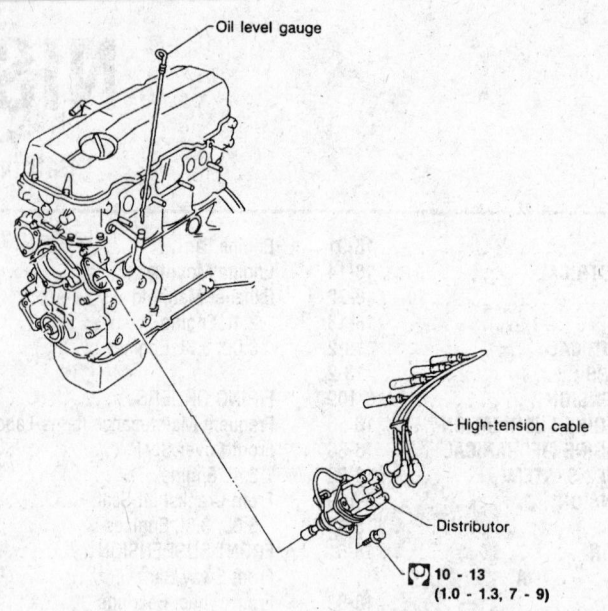

Oil level gauge

High-tension cable

Distributor

10 - 13
(1.0 - 1.3, 7 - 9)

: N·m (kg-m, ft-lb)

: Apply liquid gasket.

309621

Distributor assembly and related components — 2.4L engine

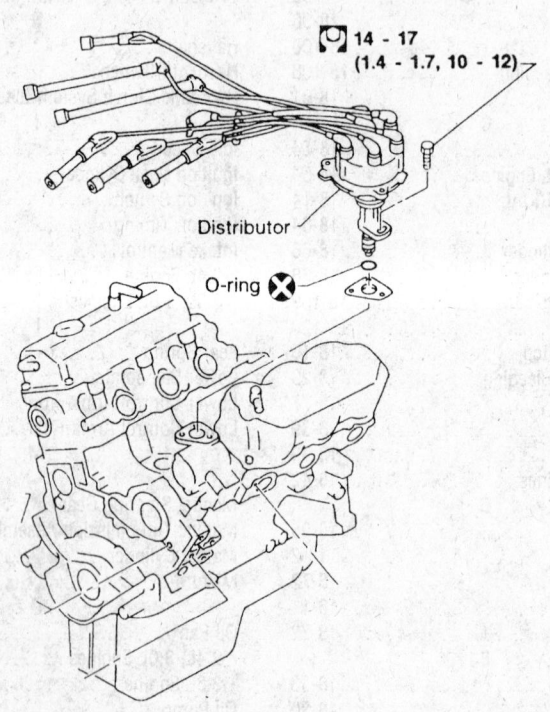

14 - 17
(1.4 - 1.7, 10 - 12)

Distributor

O-ring

309623

Distributor assembly and related components — 3.0L and 3.3L engines

18-2

3. Using a piece of chalk, make alignment marks on the distributor-to-engine and rotor-to-distributor locations; the alignment marks are used for reinstallation.

4. Disconnect the distributor electrical harness connector.

5. Remove the distributor hold-down bolt(s) and lift the distributor assembly from the engine.

To install:

6. If the engine was undisturbed, install the distributor, align the matchmarks and reverse the removal procedures. Start the engine; check and/or adjust the timing.

7. If the crankshaft was turned, the engine disturbed in any manner (while the distributor was removed), or the alignment marks were not drawn, perform the following procedures:

   a. Remove the No. 1 cylinder spark plug.

   b. Turn the crankshaft until the No. 1 piston is positioned on the Top Dead Center (TDC) of the compression stroke.

NOTE: To determine the TDC of the compression stroke, place your thumb over the spark plug hole and feel the air being forced from the cylinder. Stop turning the crankshaft when the timing marks, used to time the engine, are aligned.

   c. Oil the distributor housing-to-cylinder block surface.

   d. Install the distributor so the rotor points toward the No. 1 spark plug terminal tower of the distributor cap (when installed).

   e. When the distributor shaft has reached the bottom of the hole, move the rotor back and forth slightly until the driving lug on the end of the distributor shaft enters the slots cut in the end of the oil pump shaft and the distributor assembly slides down into place.

8. Install the distributor cap.

9. Tighten the distributor hold-down bolt(s).

10. Connect the negative battery cable.

11. Start the engine, check and/or adjust the ignition timing.

Ignition Timing

ADJUSTMENT

1. Visually inspect:
   a. The air cleaner for clogging
   b. Check the hoses/ducts for leaks

   c. Check the EGR valve operation
   d. Check all the electrical connectors
   e. Check the gaskets
   f. Check the throttle valve and throttle sensor operation.

2. Locate the timing marks on the crankshaft pulley and the front of the engine.

3. Clean the timing marks.

NOTE: The ignition timing specification is 13–17° BTDC.

4. Using chalk or white paint, color the mark on the crankshaft pulley and the mark on the scale which will indicate the correct timing when aligned with the notch on the crankshaft pulley.

5. Attach a tachometer to the engine.

6. Attach a timing light to the engine, to number one cylinder ignition wire.

7. Check to make sure all of the wires clear the fan, then, start the engine and allow it to reach normal operating temperatures.

8. Block the front wheels and set the parking brake. Shift the transmission into **NEUTRAL**; do not stand in front of the vehicle when making adjustments.

9. Perform the following procedures:

   a. Race the engine at 2000 rpm for about two minutes under a no-load condition; make sure all of the accessories are turned off.

   b. Perform on-board engine diagnostics and repair any fault code.

   c. Race the engine 2–3 times under no-load, then run the engine for one minute at idle.

   d. Stop the engine and disconnect the throttle position sensor.

   e. Race the engine at 2000 rpm for about two minutes under a no-load condition; make sure all of the accessories are turned off.

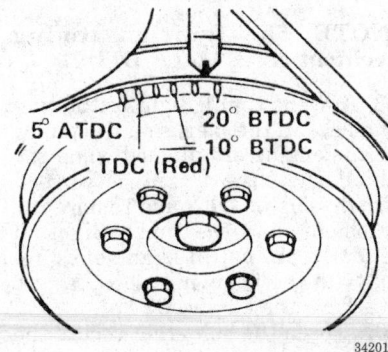

342014

Timing marks — 2.4L and 3.0L engines

   f. Run the engine at idle speed.

NOTE: The ignition timing specification is 13–17° BTDC.

10. Aim the timing light at the timing marks. If the marks on the pulley and the engine are aligned when the light flashes, the timing is correct. Turn the engine **OFF** and remove the tachometer and the timing light. If the marks are not in alignment, proceed with the following steps.

11. Turn the engine **OFF**.

12. Loosen the bolts that secure the distributor just enough so it can be turned.

13. Start the engine. Keep the wires of the timing light clear of the cooling fan.

14. With the timing light aimed at the pulley and the marks on the engine, turn the distributor for the proper adjustment.

15. Race the engine 2–3 times under no-load, then run the engine it at idle.

16. Aim the timing light at the timing marks. If the marks on the pulley and the engine are aligned when the light flashes, the timing is correct.

17. Tighten the bolt that secures the distributor and recheck the timing.

18. Turn the engine **OFF**, then remove the tachometer and the timing light.

19. Connect the throttle position sensor.

Visually check the air cleaner, intake hoses, ducts, EGR valve operation and electrical connections prior to the adjustment of the ignition timing. Correct or repair any problem as required. Be sure to inspect the throttle valve and the throttle position sensor for proper operation.

20. Locate the timing marks on the crankshaft pulley and the front of the engine.

21. Clean the timing marks.

NOTE: The ignition timing specification is 13–17° BTDC.

22. Using chalk or white paint, color the mark on the crankshaft pulley and the mark on the scale which will indicate the correct timing when aligned with the notch on the crankshaft pulley.

23. Attach a tachometer to the engine.

24. Attach a timing light to the engine, to number one cylinder ignition wire.

25. Check to make sure all of the wires clear the fan, then, start the engine and allow it to reach normal operating temperatures.

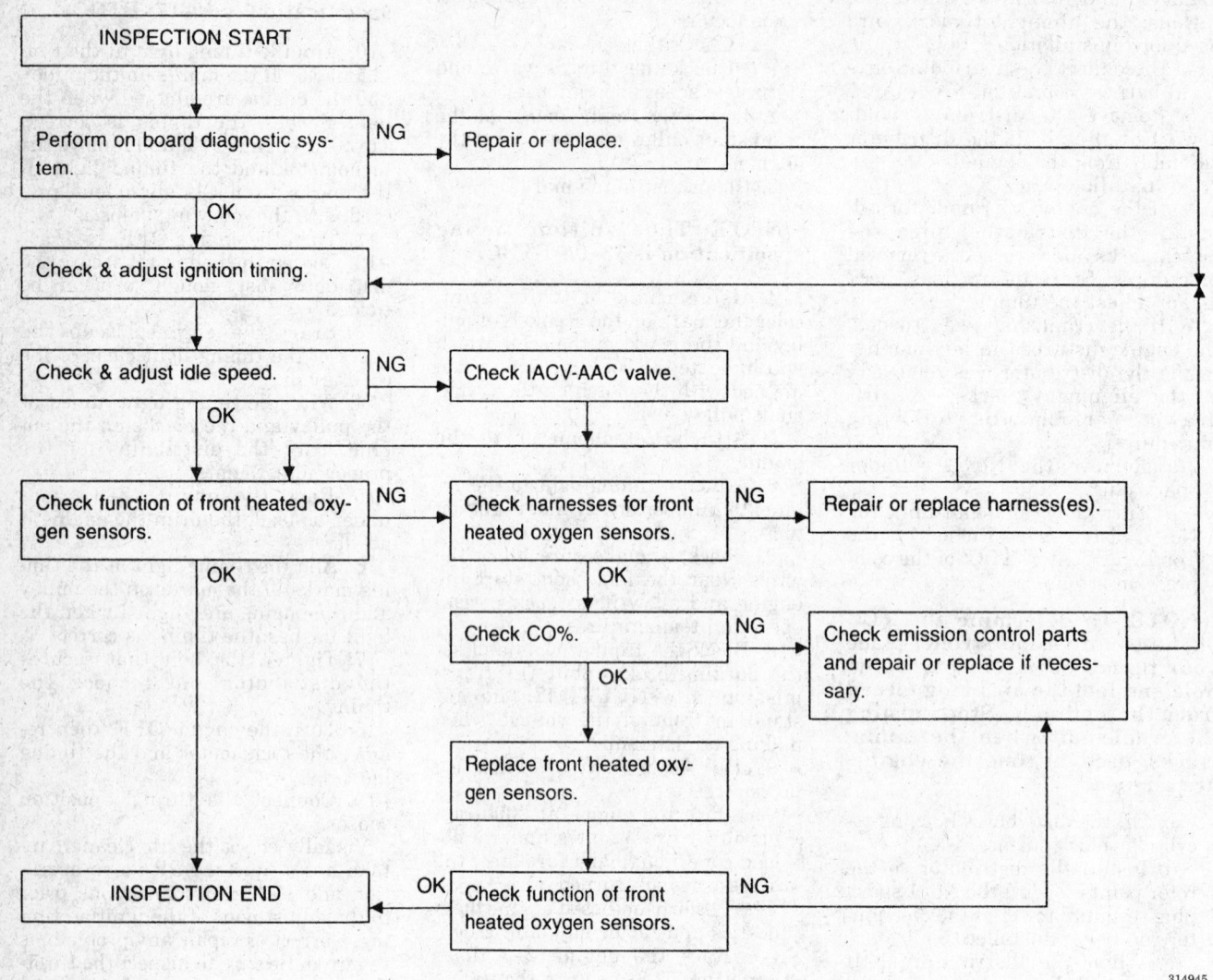

Idle speed/ignition timing/idle mixture inspection

314945

26. Block the front wheels and set the parking brake. Shift the transmission into **NEUTRAL** for automatic and manual transmissions; do not stand in front of the vehicle when making adjustments.

27. Perform the following procedures:

a. Race the engine at 2000 rpm for about two minutes under a no-load condition; make sure all of the accessories are turned off.

b. Perform on board engine diagnostics and repair any fault code.

c. Race the engine 2–3 times under no-load.

d. Stop the engine and disconnect the throttle position sensor.

e. Race the engine 2–3 times under no-load.

f. Run the engine at idle speed.

NOTE: The ignition timing specification is 13–17° BTDC.

28. Aim the timing light at the timing marks. If the marks on the pulley and the engine are aligned when the light flashes, the timing is correct. Turn the engine **OFF** and remove the tachometer and the timing light. If the marks are not in alignment, proceed with the following steps.

29. Turn the engine **OFF**.

30. Loosen the bolts that secure the distributor just enough so it can be turned.

31. Start the engine. Keep the wires of the timing light clear of the cooling fan.

32. With the timing light aimed at the pulley and the marks on the engine, turn the distributor for the proper adjustment.

33. Race the engine 2–3 times under no-load then run the engine at idle.

34. Aim the timing light at the timing marks. If the marks on the pulley and the engine are aligned when the light flashes, the timing is correct.

35. Tighten the bolt that secures the distributor and recheck the timing.

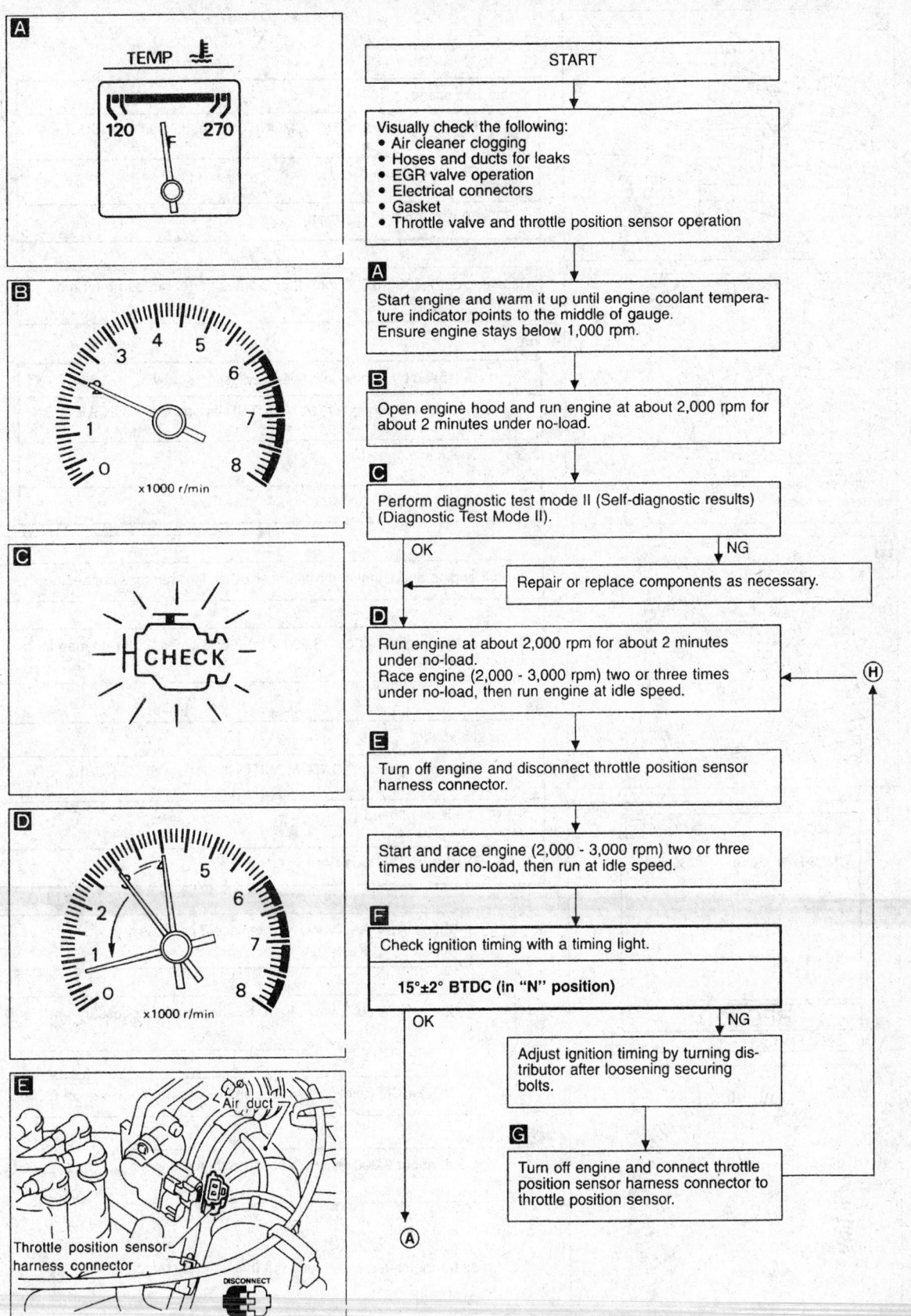

A

TEMP

120 F 270

B

0 1 2 3 4 5 6 7 8

x1000 r/min

C

CHECK

D

0 1 2 3 4 5 6 7 8

x1000 r/min

E

Air duct

Throttle position sensor harness connector

DISCONNECT

START

Visually check the following:
- Air cleaner clogging
- Hoses and ducts for leaks
- EGR valve operation
- Electrical connectors
- Gasket
- Throttle valve and throttle position sensor operation

A
Start engine and warm it up until engine coolant temperature indicator points to the middle of gauge.
Ensure engine stays below 1,000 rpm.

B
Open engine hood and run engine at about 2,000 rpm for about 2 minutes under no-load.

C
Perform diagnostic test mode II (Self-diagnostic results) (Diagnostic Test Mode II).

OK ← → NG

Repair or replace components as necessary.

D
Run engine at about 2,000 rpm for about 2 minutes under no-load.
Race engine (2,000 - 3,000 rpm) two or three times under no-load, then run engine at idle speed.

(H)

E
Turn off engine and disconnect throttle position sensor harness connector.

Start and race engine (2,000 - 3,000 rpm) two or three times under no-load, then run at idle speed.

F
Check ignition timing with a timing light.

15°±2° BTDC (in "N" position)

OK → NG

Adjust ignition timing by turning distributor after loosening securing bolts.

G
Turn off engine and connect throttle position sensor harness connector to throttle position sensor.

(A)

314946

Idle speed/ignition timing/idle mixture adjustment

18-5

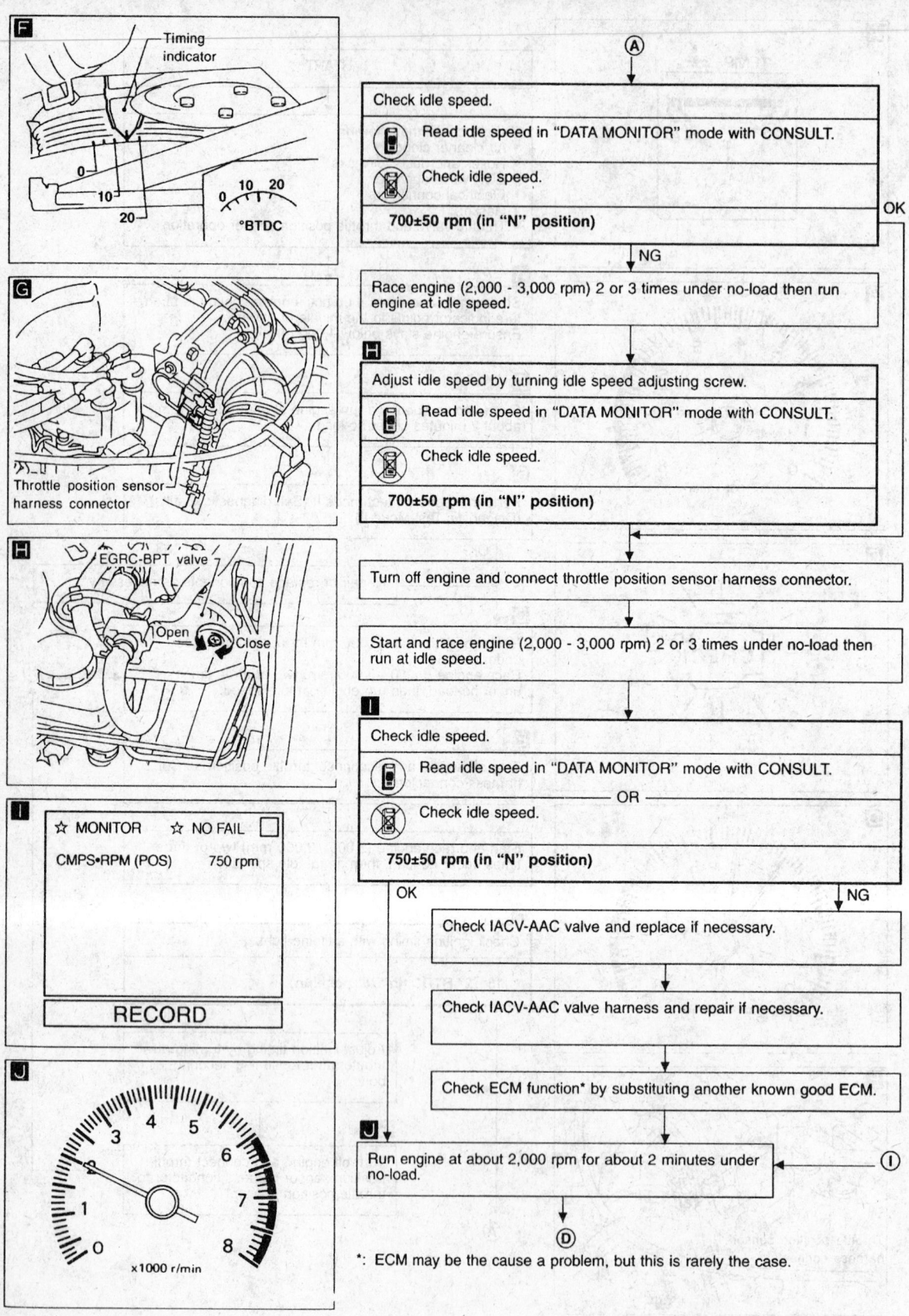

F Timing indicator

0
10
20
0 10 20
0 °BTDC

G Throttle position sensor harness connector

H EGRC-BPT valve
Open
Close

I
☆ MONITOR   ☆ NO FAIL   □

CMPS•RPM (POS)   750 rpm

RECORD

J
x1000 r/min

(A)

Check idle speed.

- Read idle speed in "DATA MONITOR" mode with CONSULT.
- Check idle speed.

700±50 rpm (in "N" position)

→ OK

↓ NG

Race engine (2,000 - 3,000 rpm) 2 or 3 times under no-load then run engine at idle speed.

H

Adjust idle speed by turning idle speed adjusting screw.

- Read idle speed in "DATA MONITOR" mode with CONSULT.
- Check idle speed.

700±50 rpm (in "N" position)

Turn off engine and connect throttle position sensor harness connector.

Start and race engine (2,000 - 3,000 rpm) 2 or 3 times under no-load then run at idle speed.

I

Check idle speed.

- Read idle speed in "DATA MONITOR" mode with CONSULT.

— OR —

- Check idle speed.

750±50 rpm (in "N" position)

OK ↓   NG ↓

Check IACV-AAC valve and replace if necessary.

Check IACV-AAC valve harness and repair if necessary.

Check ECM function* by substituting another known good ECM.

J

Run engine at about 2,000 rpm for about 2 minutes under no-load. ◄— **(I)**

↓ **(D)**

*: ECM may be the cause a problem, but this is rarely the case.

314948

Idle speed/ignition timing/idle mixture adjustment — continued

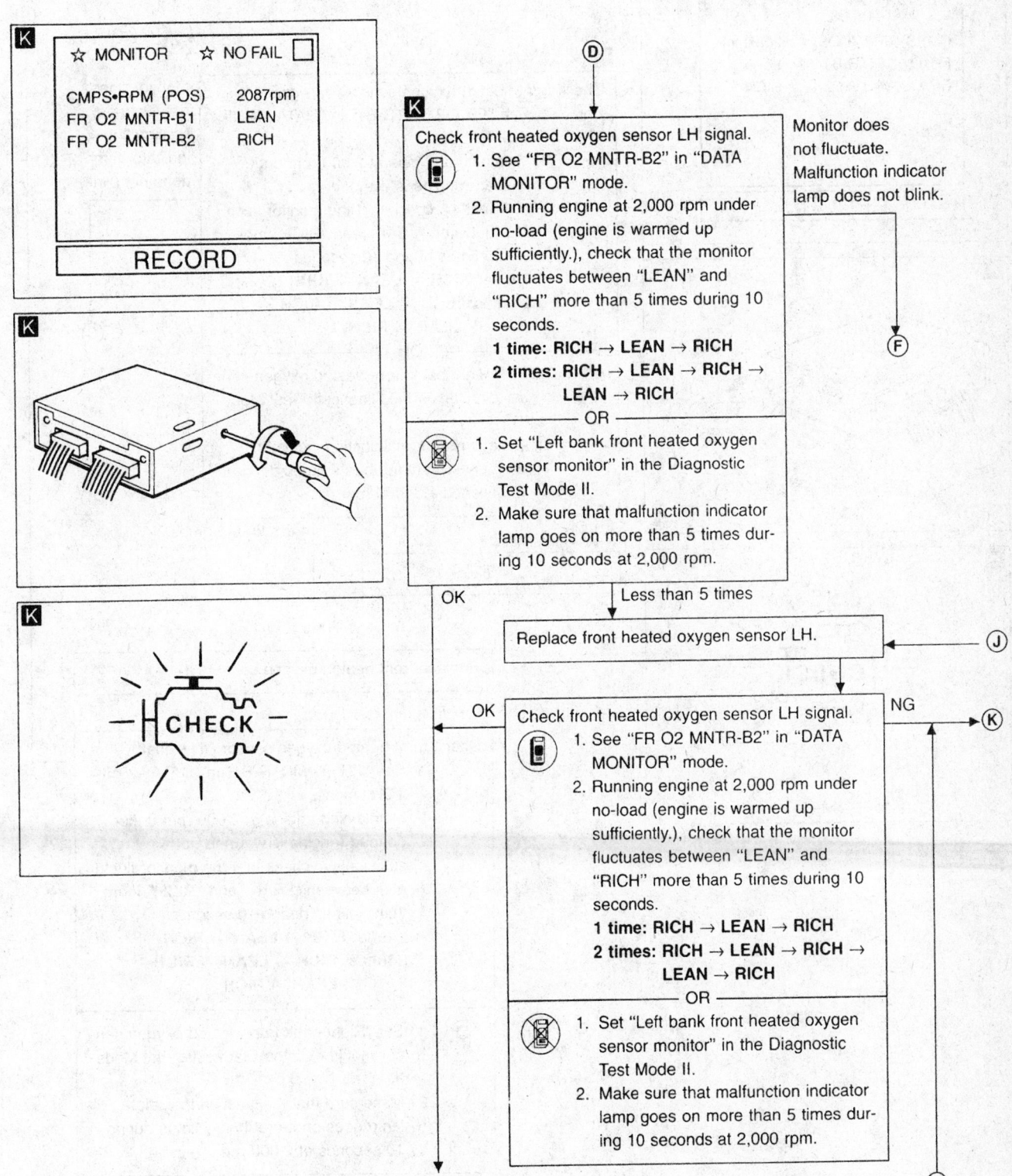

☆ MONITOR   ☆ NO FAIL ☐

CMPS•RPM (POS)   2087rpm
FR O2 MNTR-B1   LEAN
FR O2 MNTR-B2   RICH

RECORD

CHECK

D

Check front heated oxygen sensor LH signal.

1. See "FR O2 MNTR-B2" in "DATA MONITOR" mode.
2. Running engine at 2,000 rpm under no-load (engine is warmed up sufficiently.), check that the monitor fluctuates between "LEAN" and "RICH" more than 5 times during 10 seconds.
   1 time: RICH → LEAN → RICH
   2 times: RICH → LEAN → RICH → LEAN → RICH

— OR —

1. Set "Left bank front heated oxygen sensor monitor" in the Diagnostic Test Mode II.
2. Make sure that malfunction indicator lamp goes on more than 5 times during 10 seconds at 2,000 rpm.

Monitor does not fluctuate. Malfunction indicator lamp does not blink.

F

OK   Less than 5 times

Replace front heated oxygen sensor LH.   J

OK   Check front heated oxygen sensor LH signal.   NG   K

1. See "FR O2 MNTR-B2" in "DATA MONITOR" mode.
2. Running engine at 2,000 rpm under no-load (engine is warmed up sufficiently.), check that the monitor fluctuates between "LEAN" and "RICH" more than 5 times during 10 seconds.
   1 time: RICH → LEAN → RICH
   2 times: RICH → LEAN → RICH → LEAN → RICH

— OR —

1. Set "Left bank front heated oxygen sensor monitor" in the Diagnostic Test Mode II.
2. Make sure that malfunction indicator lamp goes on more than 5 times during 10 seconds at 2,000 rpm.

E

L

314949

Idle speed/ignition timing/idle mixture adjustment — continued

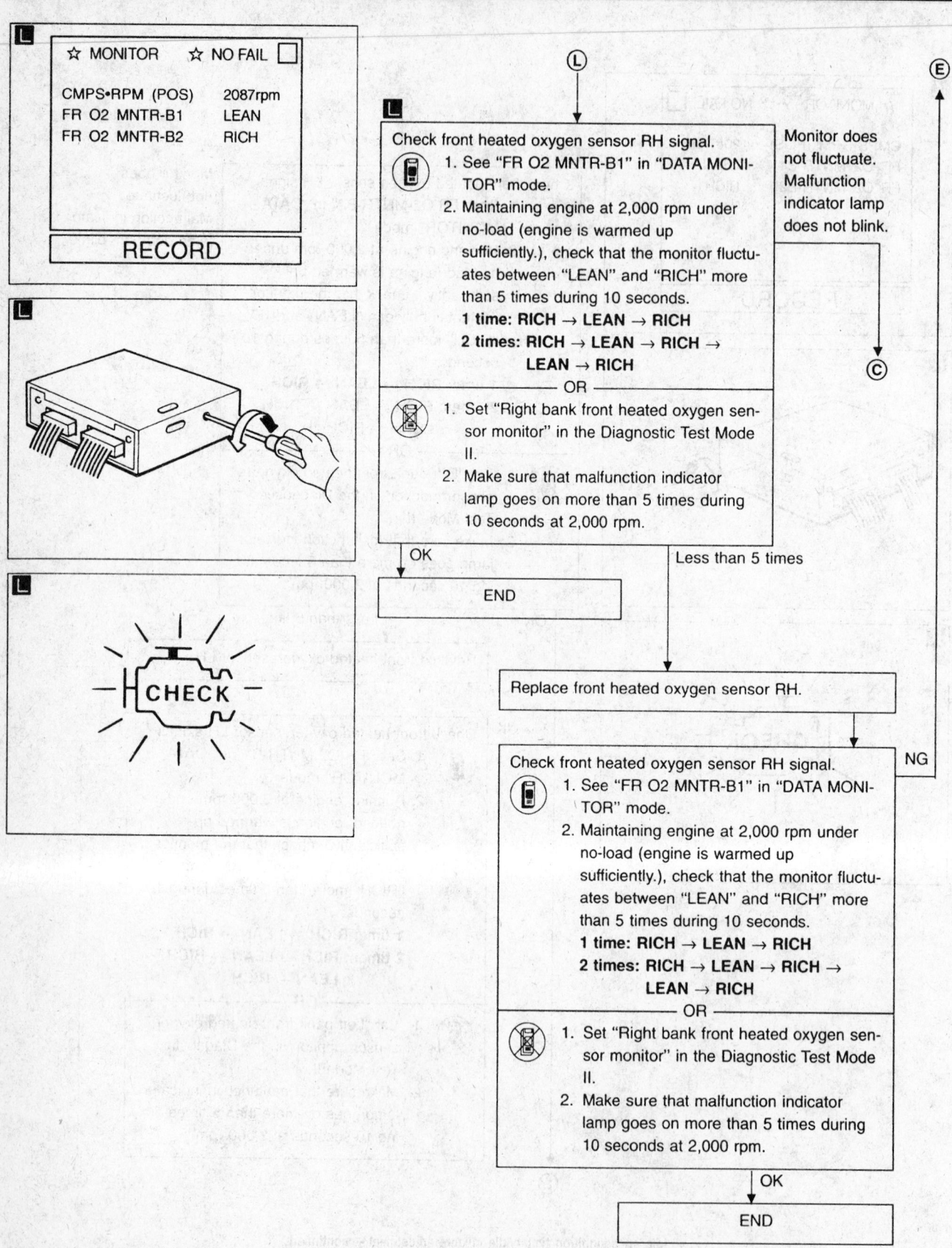

⊕ MONITOR ⊕ NO FAIL ☐

CMPS•RPM (POS)	2087rpm
FR O2 MNTR-B1	LEAN
FR O2 MNTR-B2	RICH

RECORD

CHECK

Ⓛ

Check front heated oxygen sensor RH signal.

1. See "FR O2 MNTR-B1" in "DATA MONI-TOR" mode.
2. Maintaining engine at 2,000 rpm under no-load (engine is warmed up sufficiently.), check that the monitor fluctuates between "LEAN" and "RICH" more than 5 times during 10 seconds.
   1 time: RICH → LEAN → RICH
   2 times: RICH → LEAN → RICH → LEAN → RICH

— OR —

1. Set "Right bank front heated oxygen sensor monitor" in the Diagnostic Test Mode II.
2. Make sure that malfunction indicator lamp goes on more than 5 times during 10 seconds at 2,000 rpm.

OK → END

Less than 5 times

Monitor does not fluctuate. Malfunction indicator lamp does not blink.

Ⓔ

Ⓒ

Replace front heated oxygen sensor RH.

NG

Check front heated oxygen sensor RH signal.

1. See "FR O2 MNTR-B1" in "DATA MONI-TOR" mode.
2. Maintaining engine at 2,000 rpm under no-load (engine is warmed up sufficiently.), check that the monitor fluctuates between "LEAN" and "RICH" more than 5 times during 10 seconds.
   1 time: RICH → LEAN → RICH
   2 times: RICH → LEAN → RICH → LEAN → RICH

— OR —

1. Set "Right bank front heated oxygen sensor monitor" in the Diagnostic Test Mode II.
2. Make sure that malfunction indicator lamp goes on more than 5 times during 10 seconds at 2,000 rpm.

OK → END

314951

Idle speed/ignition timing/idle mixture adjustment — continued

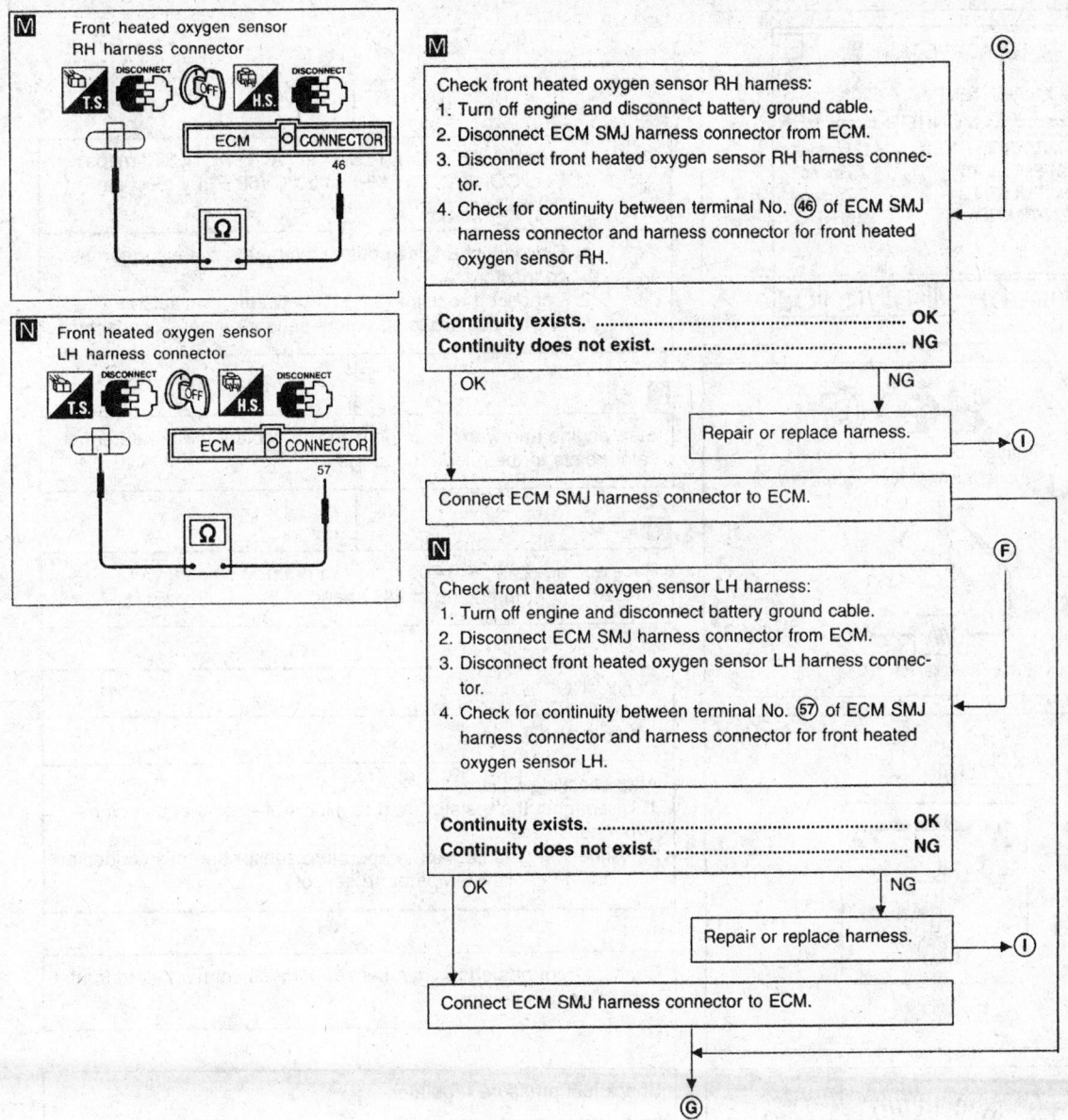

Front heated oxygen sensor RH harness connector

ECM CONNECTOR 46

Front heated oxygen sensor LH harness connector

ECM CONNECTOR 57

M

Check front heated oxygen sensor RH harness:
1. Turn off engine and disconnect battery ground cable.
2. Disconnect ECM SMJ harness connector from ECM.
3. Disconnect front heated oxygen sensor RH harness connector.
4. Check for continuity between terminal No. ㊻ of ECM SMJ harness connector and harness connector for front heated oxygen sensor RH.

Continuity exists. .... OK
Continuity does not exist. .... NG

OK · NG

Repair or replace harness. → Ⓘ

Connect ECM SMJ harness connector to ECM.

N

Check front heated oxygen sensor LH harness:
1. Turn off engine and disconnect battery ground cable.
2. Disconnect ECM SMJ harness connector from ECM.
3. Disconnect front heated oxygen sensor LH harness connector.
4. Check for continuity between terminal No. ㊺ of ECM SMJ harness connector and harness connector for front heated oxygen sensor LH.

Continuity exists. .... OK
Continuity does not exist. .... NG

OK · NG

Repair or replace harness. → Ⓘ

Connect ECM SMJ harness connector to ECM.

Ⓖ

Ⓒ

Ⓕ

314955

Idle speed/ignition timing/idle mixture adjustment — continued

36. Turn the engine **OFF**, then remove the tachometer and the timing light.
37. Connect the throttle position sensor.

Alternator

PRECAUTIONS

Several precautions must be observed with alternator equipped vehicles to avoid damage to the unit.
 • If the battery is removed for any reason, make sure it is reconnected with the correct polarity. Reversing the battery connections may result in damage to the 1-way rectifiers.
 • When utilizing a booster battery as a starting aid, always connect the positive to positive terminals and the negative terminal from the booster battery to a good engine ground on the vehicle being started.
 • Never use a fast charger as a booster to start vehicles.
 • Disconnect the battery cables when charging the battery with a fast charger.
 • Never attempt to polarize the alternator.
 • Do not use test lights of more than 12 volts when checking diode continuity.
 • Do not short across or ground any of the alternator terminals.
 • The polarity of the battery, alternator and regulator must be matched and considered before making any

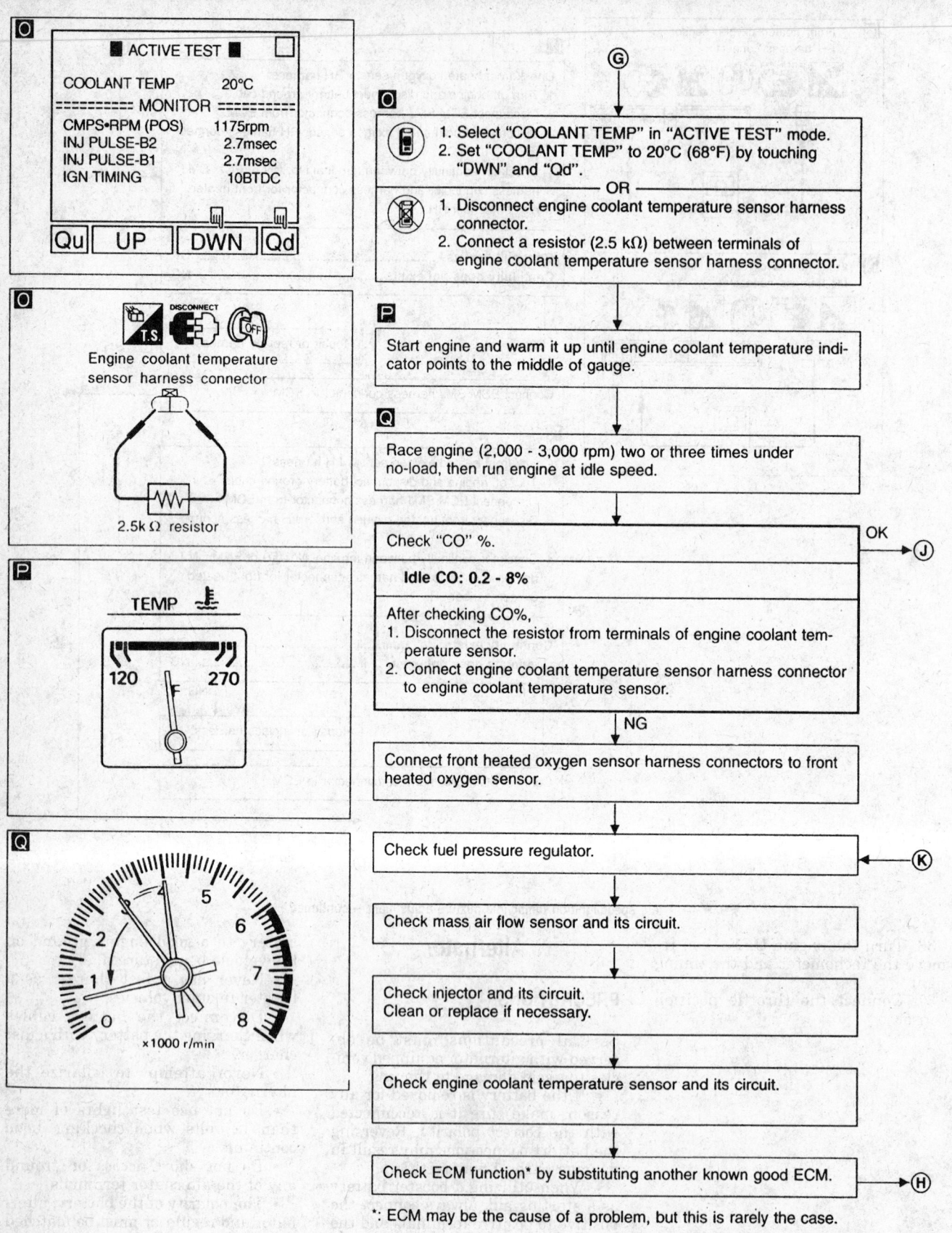

ACTIVE TEST

COOLANT TEMP	20°C
========= MONITOR =========	
CMPS•RPM (POS)	1175rpm
INJ PULSE-B2	2.7msec
INJ PULSE-B1	2.7msec
IGN TIMING	10BTDC

| Qu | UP | DWN | Qd |

Engine coolant temperature
sensor harness connector

2.5k Ω resistor

TEMP

120   270
F

×1000 r/min

G

O
1. Select "COOLANT TEMP" in "ACTIVE TEST" mode.
2. Set "COOLANT TEMP" to 20°C (68°F) by touching "DWN" and "Qd".

— OR —

1. Disconnect engine coolant temperature sensor harness connector.
2. Connect a resistor (2.5 kΩ) between terminals of engine coolant temperature sensor harness connector.

P
Start engine and warm it up until engine coolant temperature indicator points to the middle of gauge.

Q
Race engine (2,000 - 3,000 rpm) two or three times under no-load, then run engine at idle speed.

Check "CO" %.   OK → **J**

Idle CO: 0.2 - 8%

After checking CO%,
1. Disconnect the resistor from terminals of engine coolant temperature sensor.
2. Connect engine coolant temperature sensor harness connector to engine coolant temperature sensor.

NG

Connect front heated oxygen sensor harness connectors to front heated oxygen sensor.

Check fuel pressure regulator.   ← **K**

Check mass air flow sensor and its circuit.

Check injector and its circuit.
Clean or replace if necessary.

Check engine coolant temperature sensor and its circuit.

Check ECM function* by substituting another known good ECM. → **H**

*: ECM may be the cause of a problem, but this is rarely the case.

315002

Idle speed/ignition timing/idle mixture adjustment — continued

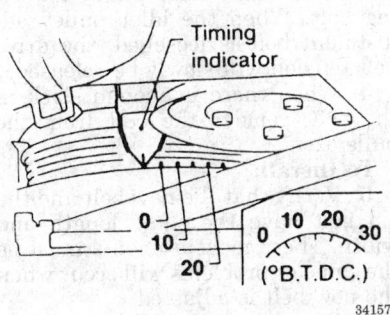

Timing indicator — 3.3L engine

341576

electrical connections within the system.

- Never separate the alternator on an open circuit. Make sure all connections within the circuit are clean and tight.
- Disconnect the battery ground terminal when performing any service on electrical components.
- Disconnect the battery if arc welding is to be done on the vehicle.

REMOVAL AND INSTALLATION

2.4L and 3.0L Engines

NOTE: On some models, the alternator is mounted very low on the engine. On these models, it may be necessary to remove the gravel shield and work from beneath the truck in order to gain access to the alternator.

1. Disconnect the negative battery cable.
2. Remove the alternator pivot bolt. Push the alternator in and remove the drive belt.
3. Pull back the rubber boots and disconnect the wiring from the back of the alternator.

NOTE: To ease installation, tag wiring before disconnecting.

4. Remove the alternator mounting bolt and then withdraw the alternator from its bracket.
To install:
5. Position the alternator in its mounting bracket and lightly tighten the mounting and adjusting bolts.
6. Connect the electrical leads at the rear of the alternator.
7. Properly adjust the belt tension.
8. Connect the negative battery cable. Run engine and check for charging system operation.

3.3L Engine

1. Disconnect the negative battery cable.

2. Loosen the alternator belt idler pulley and remove the belt from the alternator pulley.
3. Raise and safely support the vehicle.
4. Remove the engine undercover.
5. Disconnect the wiring from the alternator.
6. Remove the alternator mounting bolts and remove the alternator.
To install:
7. Position the alternator on the brackets and install the mounting bolts. Torque the upper mounting bolt to 32–43 ft. lbs. (43–58 Nm) and the lower mounting bolt to 33–44 ft. lbs. (45–60 Nm).
8. Connect the wiring to the alternator.
9. Install the engine undercover.
10. Safely lower the vehicle to the floor.
11. Install the alternator drive belt.
12. Connect the negative battery cable.
13. Check the charging system for proper operation.

Drive Belt

REMOVAL AND INSTALLATION

2.4L and 3.0L Engines

NOTE: When replacing more than one belt, it is a good idea to make note or mark what belt goes around what pulley. This will make the installation fast and easy. Also, when a new belt is installed, the manufacturer recommends rechecking the drive belt tension after the vehicle has been driven 1,000 miles.

---**WARNING**---
If a removed belt is to be reused, be certain to mark the direction of rotation on the belt. This will extend the belt's life.

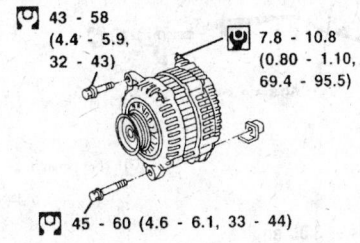

43 - 58
(4.4 - 5.9,
32 - 43)

7.8 - 10.8
(0.80 - 1.10,
69.4 - 95.5)

45 - 60 (4.6 - 6.1, 33 - 44)

N·m (kg-m, ft-lb)   N·m (kg-m, in-lb)

312492

Alternator mounting — 3.3L engine

Alternator — With Drive Belt Adjusting Bolt

1. Disconnect the negative battery cable.
2. Loosen the pivot and mounting bolts of the alternator.
3. Loosen the locking bolt on the alternator adjusting bolt. The alternator adjusting bolt will loosen and tighten the alternator drive belt tension.
4. When there is enough slack in the belt, remove the belt from the alternator pulley.
To install:
5. Verify that the new belt and the old belt have the same length and width. These measurements must be the same or problems will occur when the new belt is adjusted.
6. Correctly route the belt around the pulleys.
7. After new belt is installed correctly, adjust the tension of the new belt.
8. Tighten the mounting, pivot, and lockbolts.
9. Reconnect the negative battery cable.

Alternator — Without Drive Belt Adjusting Bolt

1. Disconnect the negative battery cable.
2. Loosen the pivot and mounting bolts of the alternator.
3. Using a proper tool, pry the component inward to relieve the tension on the drive belt.

NOTE: Always be careful where using the pry bar not to damage the alternator or surrounding components.

4. When there is enough slack in the belt, remove the belt from the alternator pulley.
To install:
5. Verify that the new belt and the old belt have the same length and width. These measurements must be the same or problems will occur when the new belt is adjusted.
6. Correctly route the belt around the pulleys.
7. After new belt is installed correctly, adjust the tension of the new belt.
8. Tighten the mounting and pivot bolts.
9. Reconnect the negative battery cable.

Air Conditioning Compressor

1. Disconnect the negative battery cable.
2. Loosen the lockbolt for the idler pulley.

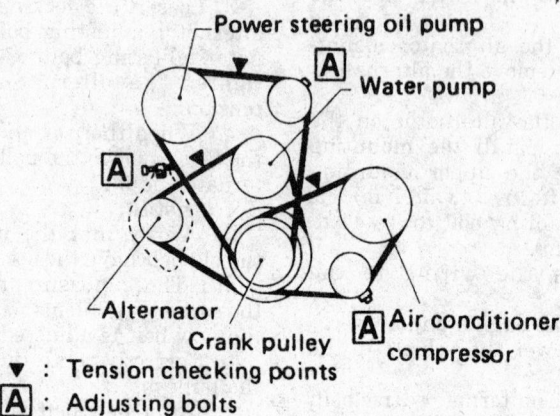

Power steering oil pump

Water pump

Alternator

Crank pulley

Air conditioner compressor

▼ : Tension checking points

A : Adjusting bolts

125935

Drive belt adjustment points — 2.4L engine

▼ : Check point

31 - 42 (3.2 - 4.3, 23 - 31)

Idler pulley

Water pump pulley

Compressor pulley

Power steering pump pulley

14 - 17 (1.4 - 1.7, 10 - 12)

16 - 21 (1.6 - 2.1, 12 - 15)

Alternator pulley

Crank pulley

N·m (kg-m, ft-lb)

125937

Drive belt adjustment points — 3.0L engine

3. Loosen the idler pulley adjusting bolt. When the idler pulley adjustment bolt is loosened, the drive belt tension will slowly be released.

4. When there is enough slack in the belt, remove the belt from the pulleys.

To install:

5. Verify that the new belt and the old belt have the same length and width. These measurements must be the same or problems will occur when the new belt is adjusted.

6. Correctly route the belt around the pulleys.

7. After new belt is installed correctly, adjust the tension of the new belt.

8. Tighten the mounting and pivot bolts.

9. Reconnect the negative battery cable.

Power Steering Pump With Adjustable Idler Pulley

1. Disconnect the negative battery cable.

2. Loosen the lockbolt for the idler pulley.

3. Loosen the idler pulley adjusting bolt. When the idler pulley adjustment bolt is loosened, the drive belt tension will slowly be released.

4. When there is enough slack in the belt, remove the belt from the pulleys.

To install:

5. Verify that the new belt and the old belt have the same length and width. These measurements must be the same or problems will occur when the new belt is adjusted.

6. Correctly route the belt around the pulleys.

7. After new belt is installed correctly, adjust the tension of the new belt.

8. Tighten the mounting, pivot, and lockbolts.

9. Reconnect the negative battery cable.

Power Steering Pump With Non-Adjustable Idler Pulley

1. Disconnect the negative battery cable.

2. Loosen the power steering oil pump mounting and pivot bolts.

3. Loosen the drive belt adjustment locking bolt. The drive belt adjustment bolt is located on the power steering oil pump. The drive belt adjustment bolt will move the power steering oil pump and increase or decrease belt tension.

4. Turn the power steering oil pump adjusting bolt until there is

enough slack in the drive belt to remove it.

5. Remove the drive belt from around the pulleys.

To install:

6. Verify that the new belt and the old belt have the same length and width. These measurements must be the same or problems will occur when the new belt is adjusted.

7. Correctly route the belt around the pulleys.

8. After the drive belt is installed correctly, adjust the drive belt tension.

9. Tighten the mounting, pivot, and lockbolts.

10. Reconnect the negative battery cable.

3.3L Engine

Alternator

1. Remove the A/C compressor belt.

2. Loosen the nut in the center of the idler pulley.

3. Turn the adjusting bolt to loosen the belt.

4. Remove the belt.

To install:

5. Install the belt on the proper pulleys.

6. Turn the adjusting bolt to properly tension the belt.

7. Tighten the nut in the center of the idler pulley to 22–29 ft. lbs. (30–39 Nm) to lock the pulley in place.

8. Install the A/C compressor belt.

Air Conditioning Compressor

1. Loosen the nut in the center of the idler pulley.

2. Turn the adjusting bolt to loosen the belt.

3. Remove the belt.

To install:

4. Install the belt on the proper pulleys.

5. Turn the adjusting bolt to properly tension the belt.

6. Tighten the nut in the center of the idler pulley to 22–29 ft. lbs. (30–39 Nm) to lock the pulley in place.

Power Steering Pump

1. Remove the A/C compressor and alternator belts.

2. Loosen the locknut on the pump bracket.

3. Turn the adjusting bolt to loosen the belt.

4. Remove the belt.

To install:

5. Install the belt on the proper pulleys.

6. Turn the adjusting bolt to properly tension the belt.

7. Tighten the locknut on the bracket to 12–15 ft. lbs. (16–21 Nm).

8. Install the A/C compressor and alternator belts.

Starter

REMOVAL AND INSTALLATION

2.4L Engine

NOTE: On some models with automatic transmission, it may be necessary to disconnect the throttle rod.

1. Disconnect the negative (-) battery cable at the battery, then disconnect the positive (+) battery cable at the starter.

NOTE: The starter is located on the right side of the engine.

2. Remove the front wheel assembly.

3. Remove the front splash shield.

4. Disconnect the remaining electrical connections at the starter solenoid.

5. Remove the two bolts holding the starter to the bell housing and remove the starter by pulling it toward the front of the vehicle.

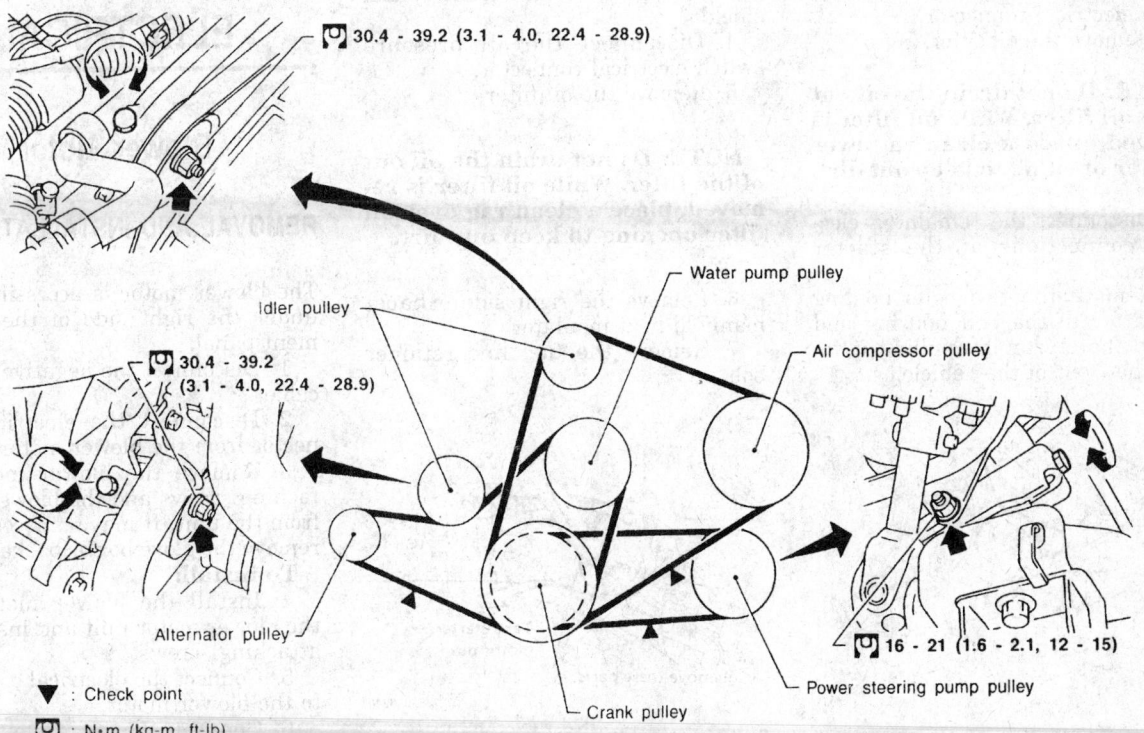

30.4 - 39.2 (3.1 - 4.0, 22.4 - 28.9)

30.4 - 39.2 (3.1 - 4.0, 22.4 - 28.9)

Idler pulley

Water pump pulley

Air compressor pulley

Alternator pulley

Crank pulley

Power steering pump pulley

16 - 21 (1.6 - 2.1, 12 - 15)

▼ : Check point

⬚ : N·m (kg-m, ft-lb)

311477

Drive belt removal/installation — 3.3L engine

To install:

6. Insert the starter into the bell housing and tighten the attaching bolts to 22–29 ft. lbs. (29–39 Nm).

7. Connect all the electrical connections that were removed at the starter.

8. Install the front splash shield and wheel assembly.

9. If removed, connect the throttle rod.

10. Connect the negative battery cable and check the starter for proper operation.

3.0L and 3.3L Engines

2WD Models

NOTE: On some models with automatic transmission, it may be necessary to disconnect the throttle rod.

1. Disconnect the negative (-) battery cable at the battery, then disconnect the positive (+) battery cable at the starter.

2. Remove the right front wheel.

3. Remove the right front splash shield.

4. Remove the right side exhaust manifold heat insulator.

5. Remove the right side exhaust manifold.

6. Disconnect the oil pressure switch electrical connector.

7. Remove the oil filter.

NOTE: Do not drain the oil out of the oil filter. While oil filter is removed, place a clean rag over oil filter opening to keep out dirt.

8. Disconnect the remaining electrical connections at the starter solenoid.

9. Remove the two bolts holding the starter to the bell housing and remove the starter by pulling it toward the front of the vehicle.

To install:

10. Insert the starter into the bell housing and tighten the attaching bolts to 22–29 ft. lbs. (29–39 Nm).

11. Connect all the electrical connections that were disconnected at the starter.

12. Install the oil filter.

13. Connect the oil pressure switch electrical connector.

14. Install the exhaust manifold and the heat insulator.

15. Install the right front splash shield and wheel assembly.

16. If disconnected, connect the throttle rod.

17. Connect the negative battery cable and check the starter for proper operation.

18. Check the oil level and top off as necessary.

4WD Models

NOTE: On some models with automatic transmission, it may be necessary to disconnect the throttle rod.

1. Disconnect the negative (-) battery cable at the battery, then disconnect the positive (+) battery cable at the starter.

2. Remove the right front wheel.

3. Remove the right front splash shield.

4. Disconnect the oil pressure switch electrical connector.

5. Remove the oil filter.

NOTE: Do not drain the oil out of the filter. While oil filter is removed, place a clean rag over oil filter opening to keep out dirt.

6. Remove the right side exhaust manifold heat insulator.

7. Remove the fuel line retainer bolt.

8. Disconnect the remaining electrical connections at the starter solenoid.

9. Remove the two bolts holding the starter to the bell housing and remove the starter by pulling it toward the front of the vehicle.

To install:

10. Insert the starter into the bell housing and tighten the attaching bolts to 22–29 ft. lbs. (29–39 Nm).

11. Connect all the electrical connections that were removed at the starter.

12. Connect the oil pressure switch electrical connector.

13. Install the oil filter.

14. Install the fuel line retaining bolt.

15. Install the exhaust manifold heat insulator.

16. Install the right front splash shield and wheel assembly.

17. Connect the throttle rod, if removed.

18. Connect the negative battery cable and check starter for proper operation.

19. Check the oil level and top off as necessary.

CHASSIS ELECTRICAL

Blower Motor

REMOVAL AND INSTALLATION

The blower motor is accessible from under the right side of the instrument panel.

1. Disconnect the negative battery cable.

2. Disconnect the electrical connector from the blower motor.

3. Remove the blower motor attaching screws and the blower motor from the unit. It may be necessary to remove the glove box or package tray.

To install:

4. Install the blower motor into the blower motor unit and install the attaching screws.

5. Connect the electrical connector to the blower motor.

6. Connect the negative battery cable.

7. Test the operation of the blower motor.

256241

Starter removal through the cylinder head side area on 2WD 3.0L Pick-ups

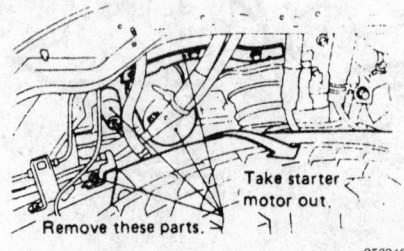

256240

Starter removal between the body and frame on 4WD 3.0L Pick-ups

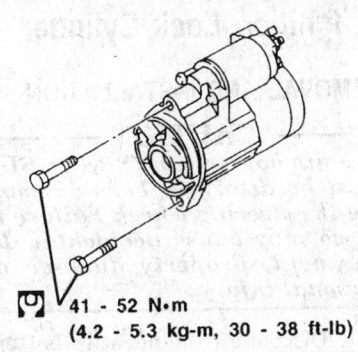

41 - 52 N•m
(4.2 - 5.3 kg-m, 30 - 38 ft-lb)

312544

Starter mounting on 3.3L engine

Windshield Wiper Motor

REMOVAL AND INSTALLATION

Front Wipers

1. Disconnect the negative battery cable.

2. Remove the wiper blades and arms as an assembly from the pivots. The arms are retained to the pivots by nuts; remove the nuts and pull the arms straight off.

3. Remove the cowl top grille screws (from the front edge) and pull the grille forward to disengage the rear tabs.

4. Remove the wiper motor arm-to-connecting rod stop ring.

5. From under the hood, disconnect the electrical connector from the wiper motor harness.

6. Remove the wiper motor mounting bolts and the wiper motor from the vehicle.

To install:

7. Install the motor and secure the bolts.

8. Install the wiper motor arm-to-connecting rod stop ring.

9. Connect the wiper motor electrical connector.

10. Install the cowl grille and the attaching screws.

11. Before installing the wiper arms, be sure the motor is in the PARK position. To do this, connect the battery and wiper motor wiring and turn the ignition switch **ON**. Turn the wiper switch on and cycle the motor 3–4 times, then turn the wiper switch off. The motor should stop in the correct, parked position.

12. Install the wiper arms, the blades should be 0.93 in. (25mm) above the lower windshield molding on the driver's side, or 0.20 in. (5mm) on the passenger's side.

Rear Wiper

1. Disconnect the negative battery cable.

2. From the rear door, remove the wiper blade/arm as an assembly from the pivot. The arm is retained to the pivots by a nut; remove the nut and pull the arm straight off.

3. From inside the rear door, remove wiper motor cover plate.

4. Remove the wiper motor arm-to-connecting rod stop ring.

5. Disconnect the wiring and remove the screws to remove the motor from the door.

To install:

6. Installation is the reverse of removal. Before installing the wiper arm, be sure the motor is in the parked position. To do this, connect the battery and wiper motor wiring and turn the ignition switch **ON**. Turn the wiper switch on and cycle the motor 3–4 times, then turn the wiper switch off. The motor should stop in the correct, parked position.

Headlight Switch

REMOVAL AND INSTALLATION

Refer to the Combination Switch procedure.

Turn Signal Switch

REMOVAL AND INSTALLATION

Refer to the Combination Switch procedure.

Combination Switch

REMOVAL AND INSTALLATION

1993–95 Models

NOTE: The lighting switch, wiper and washer switch and the ASCD switch can be replaced without removing the combination switch base.

─── **CAUTION** ───
The air bag system (SRS or SIR) must be disarmed before removing the steering wheel. Failure to do so may cause accidental deployment, property damage or personal injury.

1. Disconnect the negative battery cable.

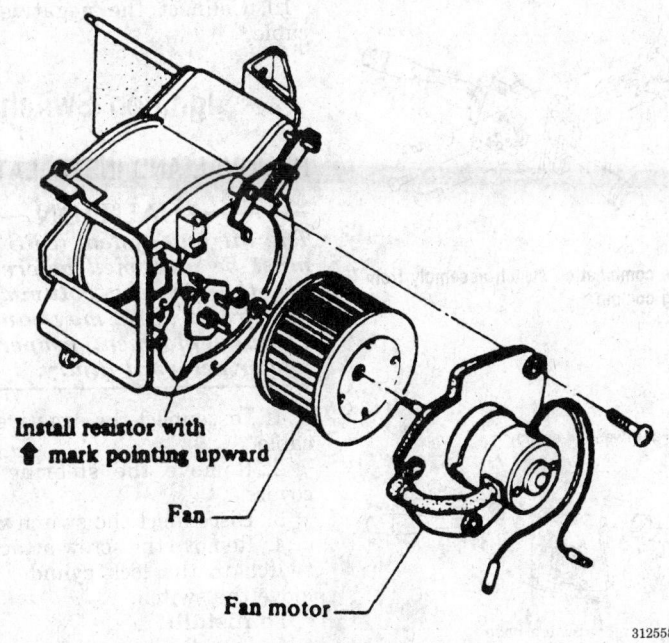

Install resistor with
↑ mark pointing upward

Fan

Fan motor

312558

Blower motor assembly

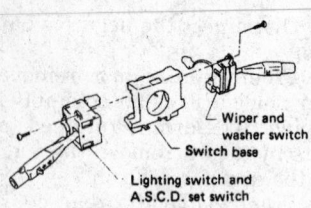

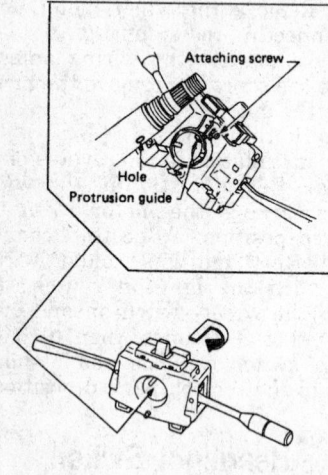

Combination switch component identification

2. Unscrew the two retaining bolts and remove the steering column garnish.

3. Remove the upper and lower steering column covers.

4. Following proper procedures, remove the steering wheel.

5. Trace the switch wiring harness to the multi-connector. Push in the lock levers and pull apart the connector.

6. Unscrew the mounting screws and remove the switch.

To install:

7. Install the switch and tighten the mounting screws. Be sure to align the protrusion on the switch body with the hole in the steering column.

8. Connect the switch multi-connector.

9. Install the steering wheel and column covers.

10. Connect the negative battery cable and check all switch functions.

1996–97 Models

───── **CAUTION** ─────
The air bag system (SRS or SIR) must be disarmed before removing the steering wheel. Failure to

do so may cause accidental deployment, property damage or personal injury.

1. Disconnect and insulate the negative battery cable. Wait at least 10 minutes before doing any work to allow the air bag to disarm.

2. Remove the steering column covers.

3. Remove the mounting screw and remove the switch from the base.

NOTE: The steering wheel must be removed to remove the switch base.

4. Disconnect the electrical connections.

5. To remove the base, first remove the three screws then push and turn the base.

To install:

6. Install the base on the steering column.

7. Install the combination switch to the base.

8. Connect the electrical connections.

9. Install the steering column covers and the steering wheel if removed.

10. Connect the negative battery cable.

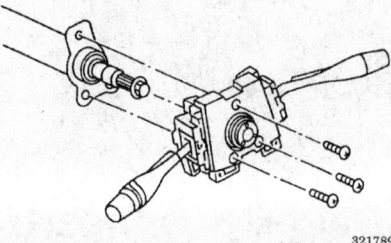

Remove the combination switch assembly from the steering column

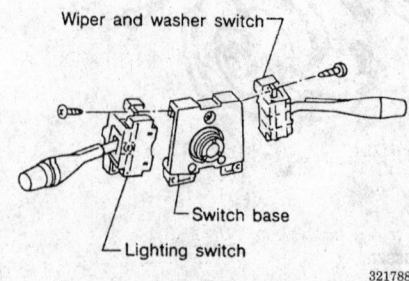

Remove the switches from the base

Ignition Lock Cylinder

REMOVAL AND INSTALLATION

───── **CAUTION** ─────
The air bag system (SRS or SIR) must be disarmed before removing the steering wheel. Failure to do so may cause accidental deployment, property damage or personal injury.

1. Disconnect the negative battery cable.

2. Remove the steering wheel and steering column cover.

3. Disconnect the switch wiring.

4. Drill out the shear bolts holding the lock assembly in place and remove the steering lock from the column.

5. Unscrew the retaining bolts in the lock cylinder and separate the switch from the lock.

To install:

6. Connect the electrical and mechanical portions of the switch.

7. Install the switch onto the column with new self-shear bolts or screws. Make sure the lock mechanism works properly before breaking the heads from the bolts. Tighten the bolts until the heads shear off.

8. Connect the electrical lead to the switch.

9. Install steering column cover and steering wheel.

10. Connect the negative battery cable.

Ignition Switch

REMOVAL AND INSTALLATION

───── **CAUTION** ─────
The air bag system (SRS or SIR) must be disarmed before removing the steering column covers. Failure to do so may cause accidental deployment, property damage or personal injury.

1. Disconnect the negative battery cable.

2. Remove the steering column covers.

3. Disconnect the switch wiring.

4. Remove the screw attaching the switch to the lock cylinder and remove the switch.

To install:

5. Install the switch to the lock cylinder.

6. Connect the harness connector to the switch.

7. Connect the electrical lead to the switch.

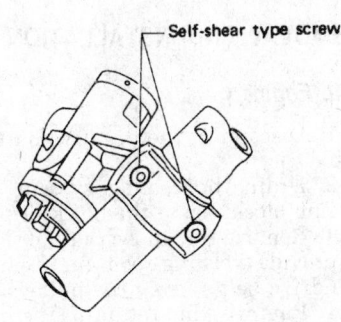

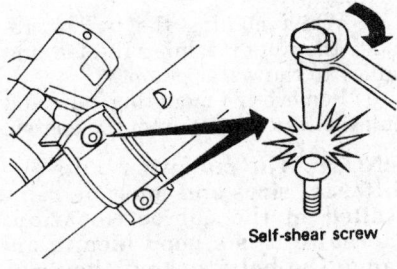

Ignition lock mounting bolt identification and installation

8. Install steering column covers.
9. Connect the negative battery cable.

Park/Neutral Safety Switch

REMOVAL AND INSTALLATION

The switch unit is bolted to the left side of the transmission shift lever. The switch prevents the engine from

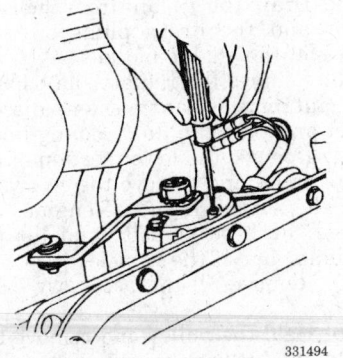

Park/neutral switch removal and installation

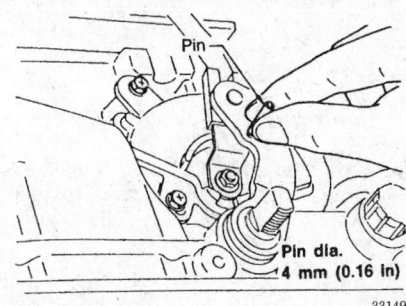

Installation of park/neutral switch adjustment pin

being started in any position except **P** or **N**. It also controls the back-up lights.

1. Raise and support the vehicle safely.
2. Remove the transaxle control linkage connection.
3. Disconnect the electrical harness, then unfasten the switch retaining screws. Remove the switch from the vehicle.
To install:
4. Position switch on transaxle.
5. Install the retaining screws and connect the wiring.
6. Disconnect the manual control linkage from the manual shaft.
7. Set the manual shaft to the **N** position.
8. Loosen the inhibitor switch mounting screws enough to allow for movement of the switch.
9. Insert a 0.16 in. (4mm) diameter pin and move the switch until the pin falls through the locating holes in the inhibitor switch and manual shaft. Tighten the switch screws equally.
10. Remove the pin and connect the manual control linkage to the shaft.
11. Make sure while holding the brakes on, that the engine will start only in **P** or **N** and that the back-up lights only illuminate in reverse.
12. Connect the transaxle control linkage.

Engine Control Module

REMOVAL AND INSTALLATION

Except 1996–97 Pathfinder

NOTE: The Engine Control Module (ECM) is located under the passenger seat.

1. Disconnect negative battery cable.
2. Remove the passenger seat.
3. Unplug wiring connector and remove mounting hardware. Slide out the ECM.

— CAUTION —
When disconnecting and connecting the pin connectors into or from the ECM, take care not to damage, bend or break the pin terminals.

To install:

NOTE: It is recommended that a technician wear a grounding strap whenever servicing the ECM. A grounding strap prevents any static electrical shocks. A static electrical shock can severely damage an ECM.

4. Install the ECM and connect the wiring connector. If there is a bolt holding the connector to the ECM, align the red projection with the connector face.

— CAUTION —
Make sure that there are no breaks or bends on the ECM pin terminals when connecting the wiring.

5. Install the passenger seat.
6. Connect the negative battery cable.

1996–97 Pathfinder

NOTE: The Engine Control Module (ECM) is located behind the glove box.

1. Disconnect the negative battery cable.
2. Remove the glove box bucket.
3. Remove the lower finish panel by reaching through the glove box and releasing the spring clips.
4. Remove the ECM harness protector.
5. Disconnect the ECM connectors.
6. Remove the ECM.
To install:
7. Install the ECM.
8. Connect the harness connectors to the ECM and install the harness protector.
9. Install the lower finish panel.
10. Install the glove box bucket.
11. Connect the negative battery cable.

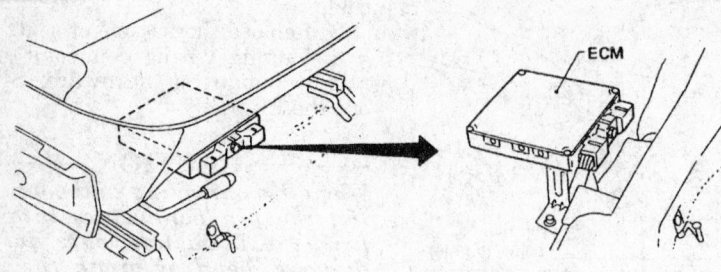

The ECM is located under the passenger seat — except 1996–97 Pathfinder

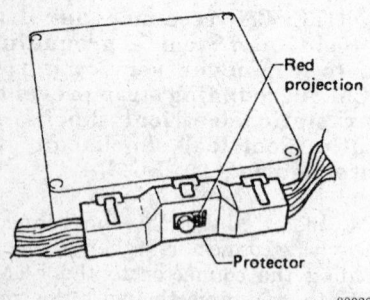

300387

On bolt-retained connectors, align the the red projection with the connector face during installation

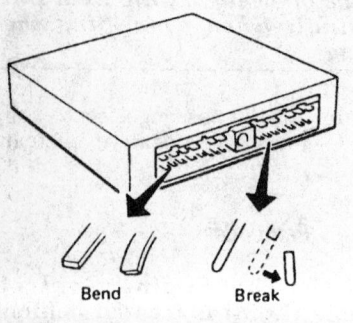

300388

Check the connector for bent or broken pins

ENGINE COOLING

Radiator

REMOVAL AND INSTALLATION

NOTE: The cooling system can be drained by opening the draincock at the bottom of the radiator or by removing the lower radiator hose. Be careful not to damage the fins or core tubes when removing and installing the radiator. Never open the radiator cap when hot! On some models, it may be necessary to remove the grille before removing the radiator.

1. Drain the engine coolant into a clean container. Remove the air cleaner inlet pipe.
2. Disconnect the upper and lower radiator hoses and the coolant reserve tank hose.
3. If equipped, disconnect the automatic transmission oil cooler lines. Plug the lines to keep dirt from entering them.
4. If the radiator has a fan shroud, unbolt the shroud and move it back. Hang it over the fan.

NOTE: Remove the fan coupling with the fan, then remove the fan shroud on 1996–97 Pick-up trucks.

5. Remove the radiator mounting bolts and the radiator.
To install:
6. Install the radiator in the vehicle and tighten the mounting bolts evenly.
7. Check that the rubber mounting legs are in good shape.
8. If equipped with an automatic transmission, connect the cooling lines at the radiator.
9. If applicable, install the radiator fan shroud. If removed, install the fan and coupling.
10. Connect the upper and lower hoses and the coolant reserve tank hose.
11. Refill the cooling system and if necessary, the automatic transmission fluid. Operate the engine until warm, bleed the cooling system, and check the coolant level. Also, check the cooling system for any leaks.

Water Pump

REMOVAL AND INSTALLATION

2.4L Engine

1. Disconnect the negative battery cable.
2. Drain the cooling system and engine block, using the block drain.
3. Remove the upper radiator hose to provide working room and remove the drive belt(s) from the pulleys.
4. Remove the retaining screws, and lift the fan shroud from the engine.
5. While holding the pulley, remove the nuts retaining the fan and pulley to the water pump.
6. Remove the mounting bolts and pull the water pump from the engine.

NOTE: The mounting bolts are different sizes and must be reinstalled in the correct location, therefore it is a good idea to arrange the bolts so that they can be easily identified during installation.

To install:
7. Make sure all gasket surfaces are clean and properly apply silicone sealer to the pump. Install the pump to the engine and torque the bolts to 12–15 ft. lbs. (16–21 Nm).
8. Install the fan clutch, fan, and pulley and torque the nuts or bolts to 5–6 ft. lbs. (7–8 Nm).
9. Install the fan shroud and drive belt(s).
10. Connect the upper hose then fill and bleed the cooling system.
11. Connect the negative battery cable.
12. Start the engine to check for leaks.

3.0L Engine

1. Disconnect the negative battery cable.
2. Drain the coolant from the radiator and the drain plugs on both sides of the engine block.
3. Remove the radiator hoses and, on automatic transmission, disconnect and plug the fluid cooling lines.
4. Remove the lower section of the fan shroud and remove the screws to lift the shroud from the engine. Remove the bracket bolts and lift the radiator out of the vehicle.
5. Remove all the accessory drive belts.
6. Hold the pulley and remove the nuts to remove the fan and pulley from the water pump.
7. Remove the timing belt covers.

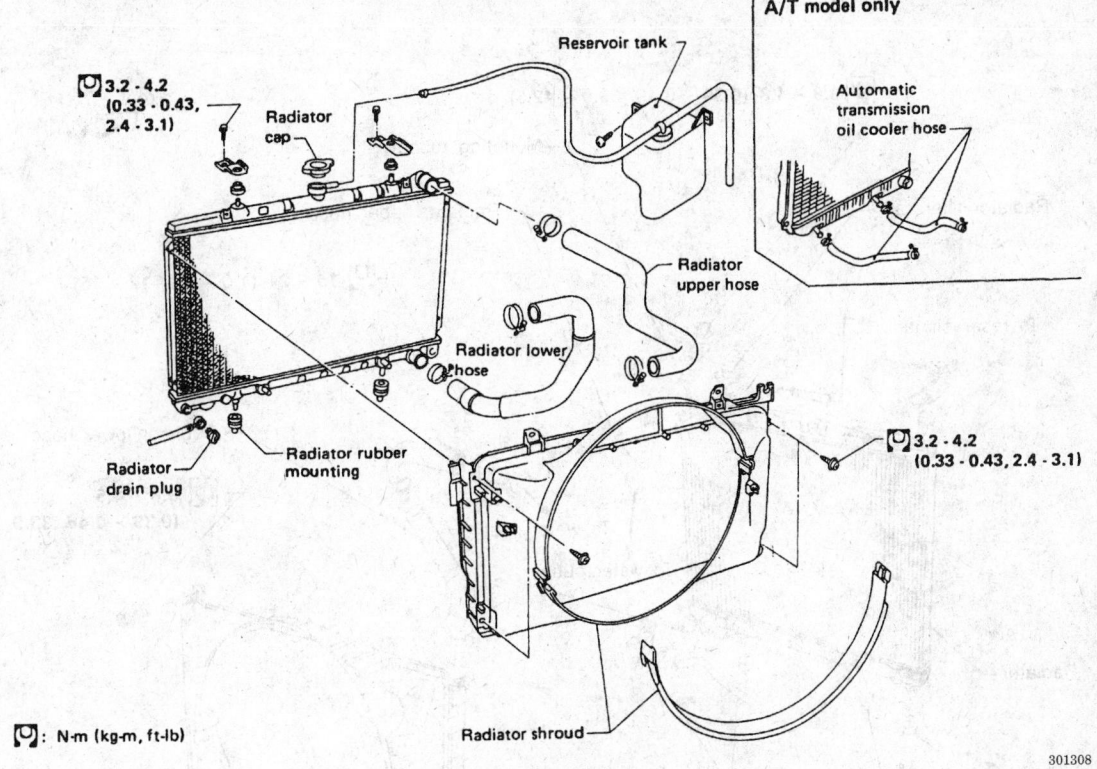

Radiator assembly and related components — KA24E engine

301308

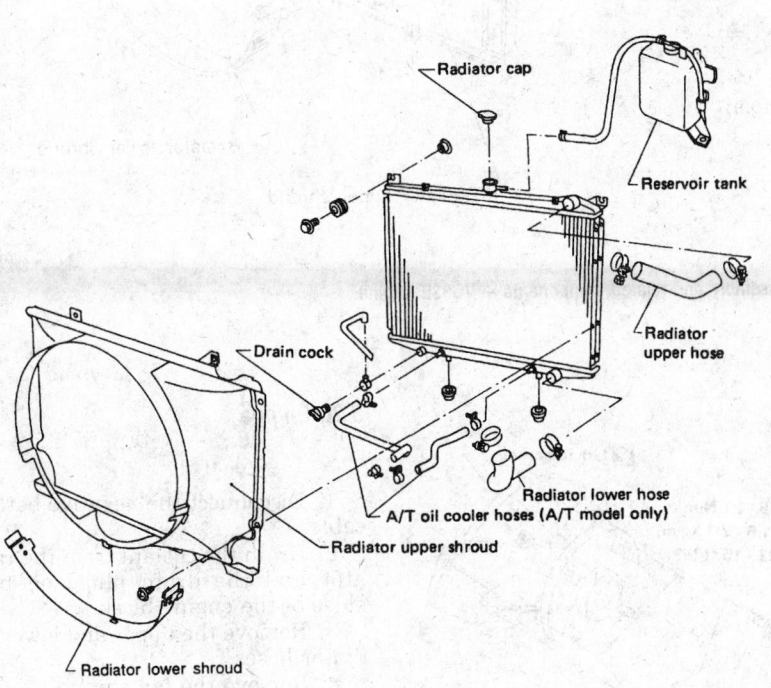

Radiator assembly and related components — VG30E engine

301310

8. Remove the bolts to remove the water pump from the engine.

NOTE: Water pump mounting bolts are different sizes and must be reinstalled in their original locations.

To install:

9. Make sure all gasket surfaces are clean and use a new gasket or silicone sealer when installing the pump to the engine. Torque the bolts to 15 ft. lbs. (21 Nm).

10. Install the timing belt covers. On 4WD models, make sure the sealing surfaces are clean and carefully install the rubber seal when installing the cover. The timing belt must be properly protected from dirt and oil.

11. Install the pulley, fan clutch, and the fan.

12. Install the accessory drive belts and adjust the tension.

13. Install the radiator and fan shroud; connect the cooling system hoses.

14. If equipped with an automatic transmission, connect the A/T oil cooler lines.

15. Connect the negative battery cable.

16. Fill and bleed the cooling system and check for leaks.

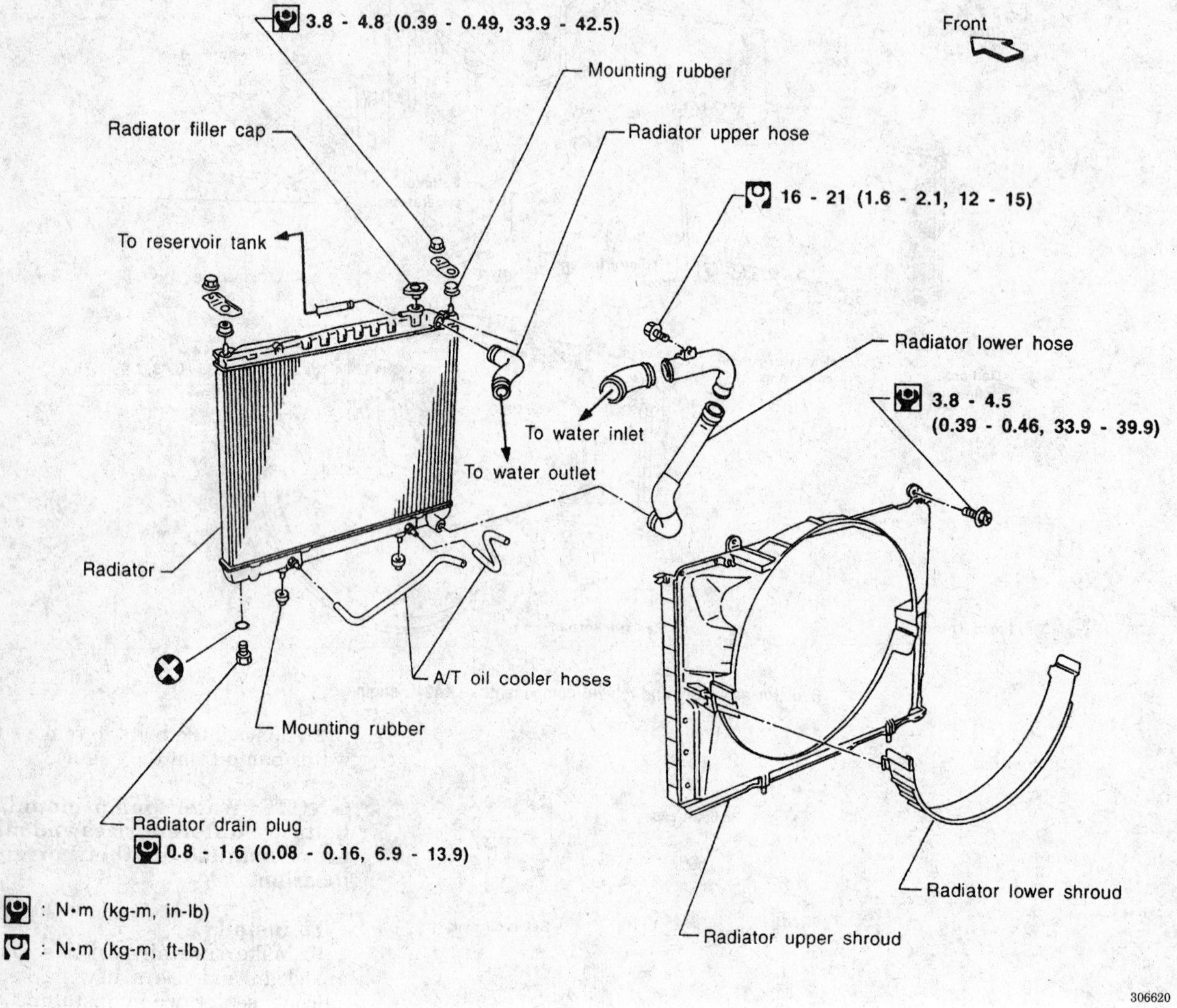

3.8 - 4.8 (0.39 - 0.49, 33.9 - 42.5)

Front

Mounting rubber

Radiator filler cap

Radiator upper hose

16 - 21 (1.6 - 2.1, 12 - 15)

To reservoir tank

Radiator lower hose

3.8 - 4.5
(0.39 - 0.46, 33.9 - 39.9)

To water inlet

To water outlet

Radiator

A/T oil cooler hoses

Mounting rubber

Radiator drain plug
0.8 - 1.6 (0.08 - 0.16, 6.9 - 13.9)

Radiator lower shroud

Radiator upper shroud

: N•m (kg-m, in-lb)

: N•m (kg-m, ft-lb)

306620

Radiator assembly and related components — VG33E engine

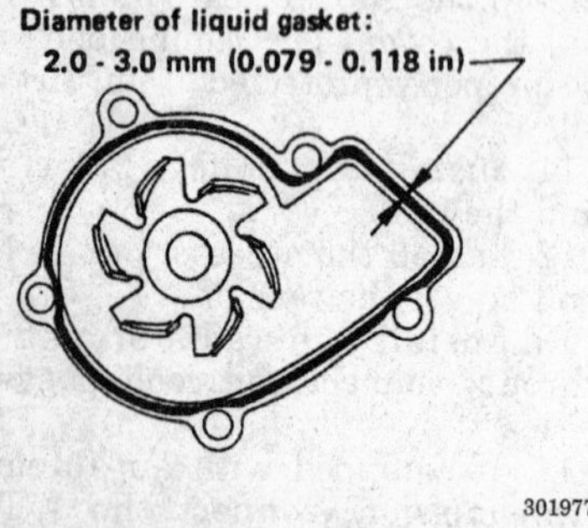

Diameter of liquid gasket:
2.0 - 3.0 mm (0.079 - 0.118 in)

Liquid gasket

16 - 21 N•m
(1.6 - 2.1 kg-m,
12 - 15 ft-lb)

301977

Proper method of applying liquid gasket to the pump assembly

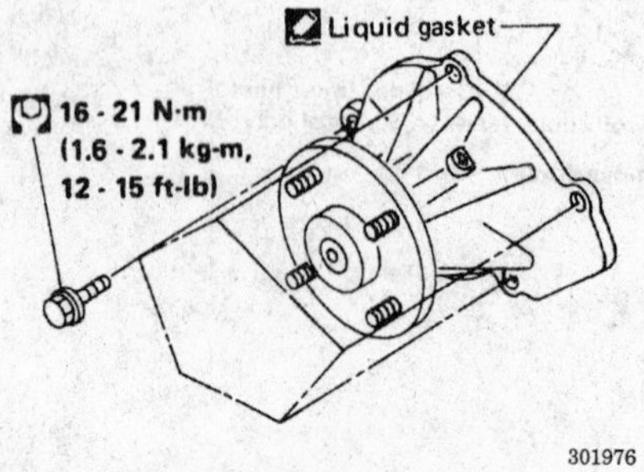

301976

Water pump assembly — 2.4L engine

3.3L Engine

1. Disconnect the negative battery cable.

2. Drain the coolant from the radiator and the drain plugs on both sides of the engine block.

3. Remove the upper and lower radiator hoses.

4. Remove the fan shroud.

5. Remove the drive belts.

6. Remove the cooling fan and the water pump pulley.

7. Remove the crankshaft pulley.

For 2WD   **For 4WD**

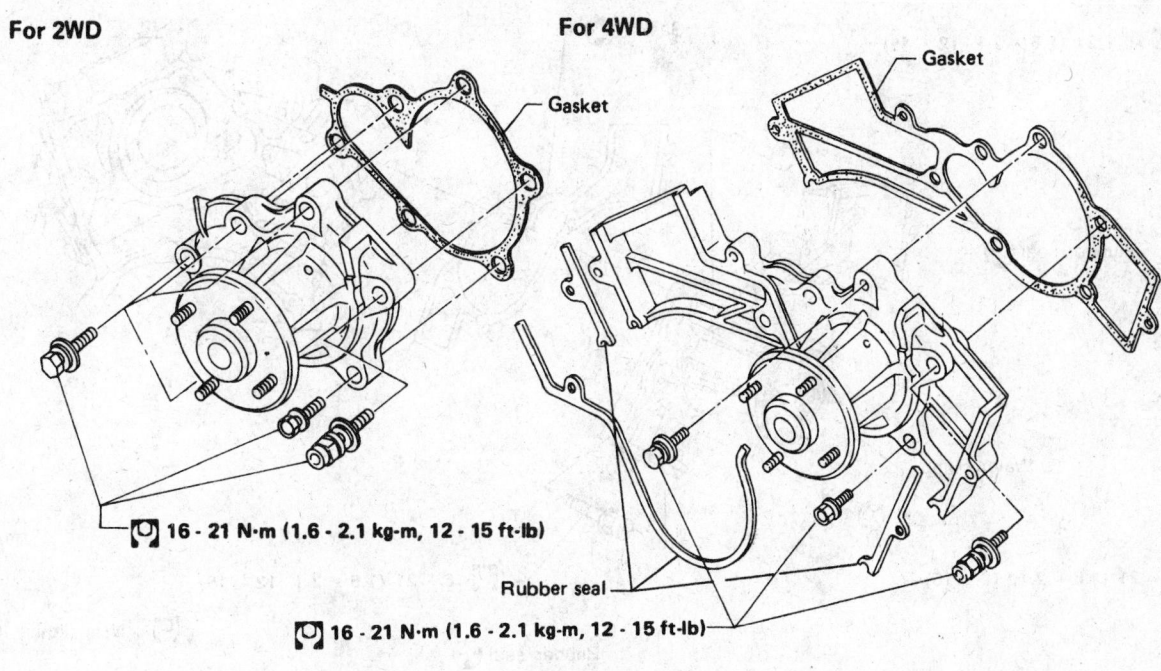

Water pump assembly — 3.0L engine

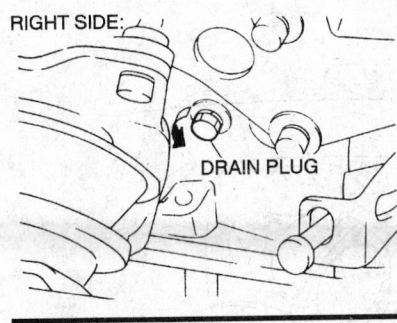

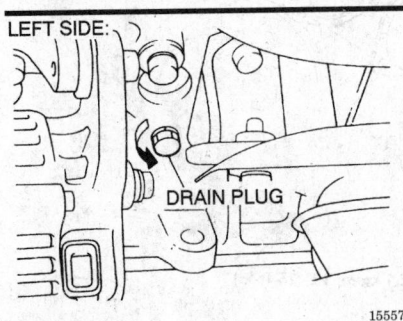

Cooling system drain plug locations —
6-cylinder engines

8. Remove the upper and lower timing belt covers.

NOTE: Water pump mounting bolts are different sizes and must be reinstalled in their original locations.

9. Remove the water pump. Don't let the engine coolant get on the timing belt.
To install:
10. Clean the gasket mating surfaces on the water pump and engine block.
11. Using a new gasket, install the water pump. Torque the mounting bolts to 12–15 ft. lbs. (16–21 Nm).
12. Install the timing belt covers.
13. Install the crankshaft pulley.
14. Install the water pump pulley and the cooling fan.
15. Install the drive belts.
16. Install the fan shroud and the radiator hoses.
17. Connect the negative battery cable.
18. Refill the engine with coolant and bleed the system. Check for leaks.

Thermostat

REMOVAL AND INSTALLATION

1. Disconnect the negative battery cable.
2. Drain the engine coolant to a level below the thermostat housing.
3. Disconnect the coolant hose from the thermostat water outlet.
4. Remove the water outlet-to-thermostat housing bolts, gasket and thermostat.
5. Clean the gasket mounting surfaces.
To install:

NOTE: If the thermostat is equipped with an air bleed or jiggle valve, be sure to position it in the upward direction.

6. Using a new gasket or sealant, position the thermostat/housing assembly on engine.

NOTE: The thermostat spring must face the inside of the engine.

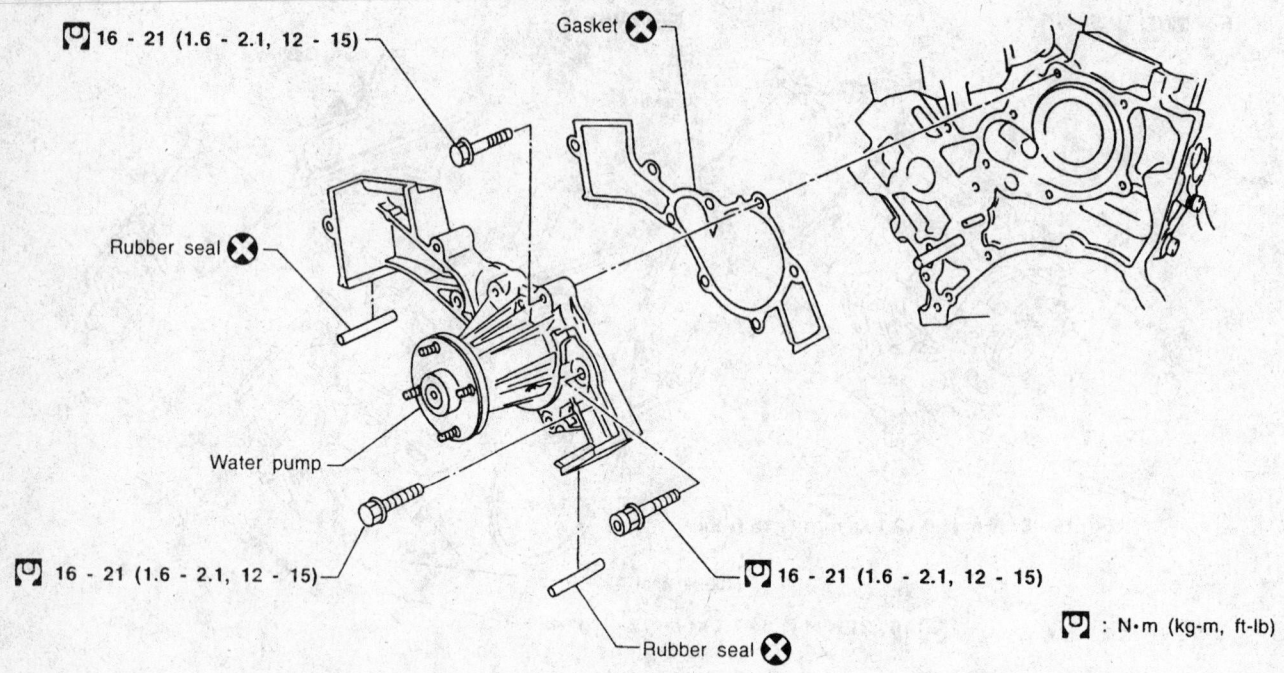

Water pump assembly — 3.3L engine

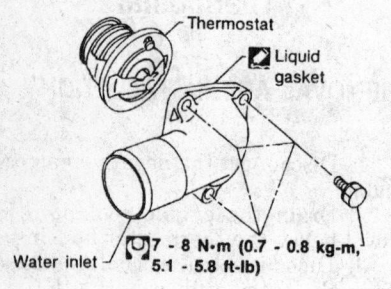

Thermostat and related components — KA24E engine

7. Torque the thermostat housing bolts as follows:
- KA24E engine — 5 ft. lbs. (8 Nm)
- VG30E — 12–15 ft. lbs. (16–21 Nm)
- VG33E — 12–15 ft. lbs. (16–21 Nm)

8. Connect the coolant hose to the thermostat housing.

9. Connect the negative battery cable.

10. Refill the coolant and bleed system as required.

11. Start the engine and warm it to normal operating temperature. Verify proper cooling system operation.

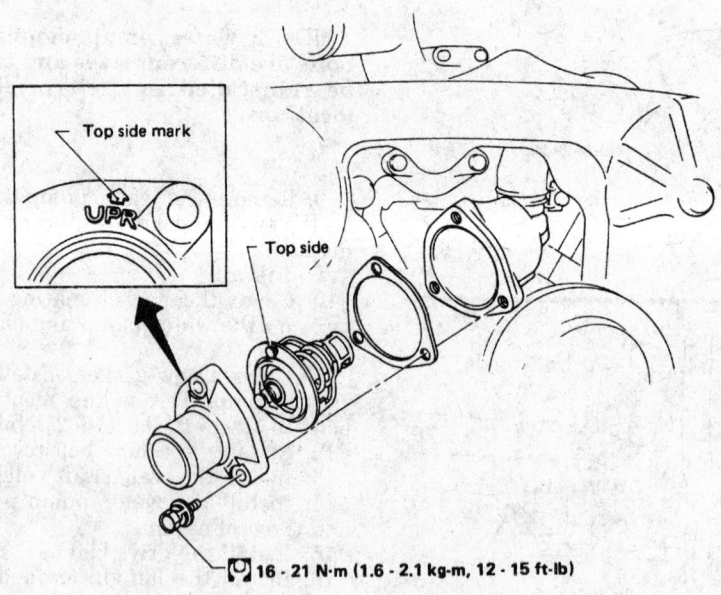

Thermostat and related components — VG30E engine

Engine Fan

REMOVAL AND INSTALLATION

1. Disconnect the negative battery cable.

2. Unfasten the fan shroud securing bolts and remove the fan shroud, if so equipped.

3. Loosen and remove the engine drive belts.

4. Loosen the fan-to-fluid coupling bolts and remove the fan.

To install:

5. Install the fan on the fluid coupling and tighten the bolts to 4–7 ft. lbs. (6–10 Nm).

6. Install all drive belts and adjust as required.

7. Install the fan shroud.

8. Start the engine and check for proper fan operation.

9. Connect the negative battery cable.

Cooling System Bleeding

EXCEPT 3.3L ENGINE

1. Set the heater temperature control to HOT and open the air relief plug.

2. Pour the coolant into the radiator until it comes out the air relief plug, then close the plug.

3. Fill the reservoir and run the engine until it warms to normal operating temperature.

4. Run the engine at 2000 RPM for one minute.

5. Stop the engine and allow it to cool.

6. When the engine is cool again, check the coolant level in the reservoir.

7. Top off coolant level in the radiator and fill reservoir to the MAX line.

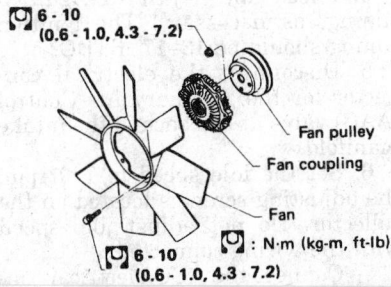

6 - 10
(0.6 - 1.0, 4.3 - 7.2)

Fan pulley
Fan coupling
Fan

6 - 10
(0.6 - 1.0, 4.3 - 7.2)

: N·m (kg-m, ft-lb)

306614

Cooling fan and related components

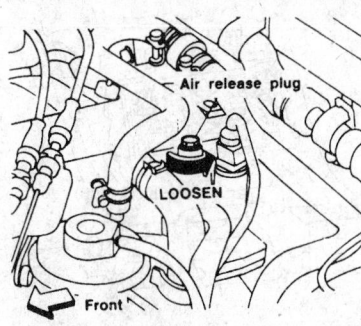

Air release plug

LOOSEN

Front

VG30E engines

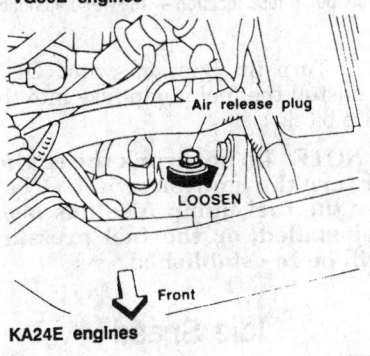

Air release plug

LOOSEN

Front

KA24E engines

300391

Bleeder screw locations — 2.4L and 3.0L engines

8. Check for leaks.

3.3L ENGINE

NOTE: Watch the coolant temperature gauge so as not to overheat the engine.

1. Set the heater temperature control to HOT and open the air relief plug.

2. Pour the coolant into the radiator until it comes out the relief plug, then close the plug. Tighten the relief plug to 61–69 inch lbs. (7–8 Nm).

3. Top off coolant level in the radiator and fill reservoir to the MAX line.

4. Install the radiator cap, start the engine, and run it until it warms to normal operating temperature.

5. Run the engine at 3000 RPM for ten seconds.

6. Stop the engine and allow it to cool.

7. When the engine is cool, check the coolant level in the radiator and reservoir.

8. Repeat Steps 1–7 two more times.

9. Check for leaks.

NOTE: If sound is present at the heater watercock, repeat the bleeding procedure.

GASOLINE FUEL SYSTEM

Fuel System Service Precautions

Safety is the most important factor when performing not only fuel system maintenance but any type of maintenance. Failure to conduct maintenance and repairs in a safe manner may result in serious personal injury or death. Maintenance and testing of the vehicle's fuel system components can be accomplished safely and effectively by adhering to the following rules and guidelines.

• To avoid the possibility of fire and personal injury, always disconnect the negative battery cable unless the repair or test procedure requires that battery voltage be applied.

• Always relieve the fuel system pressure prior to disconnecting any fuel system component (injector, fuel rail, pressure regulator, etc.), fitting or fuel line connection. Exercise extreme caution whenever relieving fuel system pressure to avoid exposing skin, face and eyes to fuel spray. Please be advised that fuel under pressure may penetrate the skin or any part of the body that it contacts.

• Always place a shop towel or cloth around the fitting or connection prior to loosening to absorb any excess fuel due to spillage. Ensure that all fuel spillage (should it occur) is quickly removed from engine surfaces. Ensure that all fuel soaked cloths or towels are deposited into a suitable waste container.

• Always keep a dry chemical (Class B) fire extinguisher near the work area.

• Do not allow fuel spray or fuel vapors to come into contact with a spark or open flame.

• Always use a backup wrench when loosening and tightening fuel line connection fittings. This will prevent unnecessary stress and torsion to fuel line piping. Always follow the proper torque specifications.

• Always replace worn fuel fitting O-rings with new. Do not substitute fuel hose or equivalent, where fuel pipe is installed.

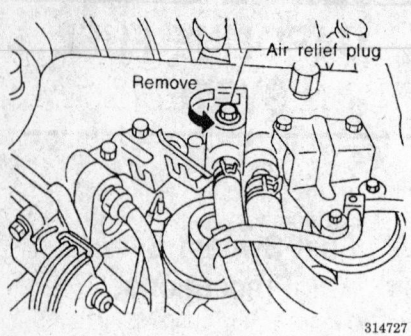

314727

Air relief plug location — 3.3L engine

Fuel System Pressure

RELIEVING

1. Remove the fuel pump fuse.
2. Start the engine.
3. Start the engine and run until the engine stalls.
4. After the engine stalls, try to restart the engine; if the engine will not start, the fuel pressure has been released.

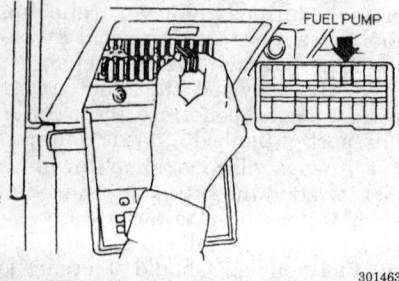

301463

Fuel pump fuse location — 1993–95 Pathfinder and Pick-up

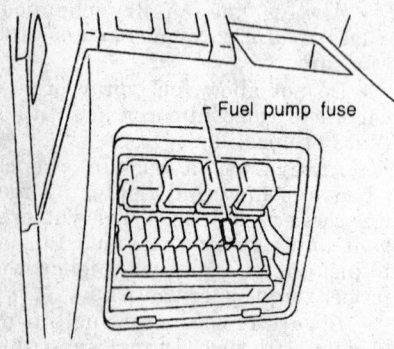

301468

Fuel pump fuse location — 1996–97 Pick-up

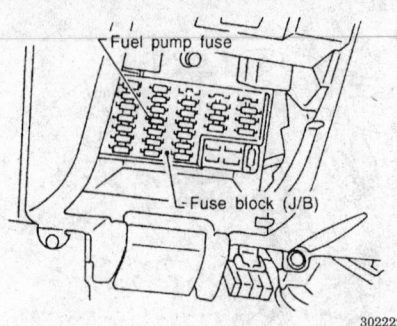

302222

Fuel pump fuse location — 1996–97 Pathfinder

5. Turn the ignition switch **OFF**. Reinstall the fuel pump fuse into the fuse block.

NOTE: Do not crank the engine or turn the ignition switch ON after the fuel pump fuse has been reinstalled, or the fuel pressure will be re-established.

Idle Speed

ADJUSTMENT

2.4L Engine

NOTE: The engine should be in good mechanical condition and all electrical connectors and vacuum hoses connected before making this adjustment. Block the engine drive wheels and apply the parking brake.

1. Start the engine and let it warm up to normal operating temperature.
2. Check the ignition timing.

NOTE: Idle speed and ignition timing must be checked together.

3. Stop the engine and disconnect the throttle position sensor connector.
4. Run the engine under no load at 2,000 rpm for about two minutes. Rev

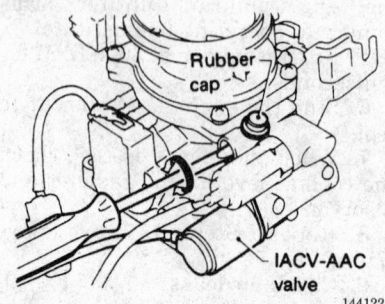

144122

Idle speed adjusting screw location — 1993–94 2.4L Pick-up

the engine two or three times and let it idle.

5. Adjust the idle speed using the idle speed adjusting screw to 700–800 rpm for M/T and A/T vehicles with the transmission in the N position.
6. Stop the engine and connect the throttle position sensor connector.
7. Run the engine under no load at 2,000 rpm for about two minutes. Rev the engine two or three times and let it idle.
8. Check that the the idle speed is 750–850 rpm.
9. If the idle speed is not correct, check the idle air control circuit and repair as necessary.
10. If the idle speed is still not correct, substitute a known good ECM.

NOTE: The ECM may be the cause of the problem, but this is rarely the case.

3.0L Engine

NOTE: On vehicles equipped with an automatic transmission the parking brake should be set and the wheels chocked, then place the shift lever in the N position.

1. Visually inspect:
   a. The air cleaner for clogging
   b. Check the hoses/ducts for leaks
   c. Check the EGR valve operation
   d. Check all the electrical connectors
   e. Check the gaskets
   f. Check the throttle valve and throttle sensor operation and the AIV hose.
2. Start the engine and allow it to reach normal operating temperature. Ensure that the engine speed is below 1000 rpm.
3. Operate the engine under no–load for 2 minutes at about 2000 rpm. Make sure all accessories and lights are **OFF**. Race the engine 2–3 times, then let engine return to idle speed for 1 minute.
4. Check and adjust the ignition timing, as necessary. The ignition timing should be 13–17° BTDC.
5. Disconnect the electrical connector for the Auxiliary Air Control (AAC) valve at the back of the intake manifold.
6. Set the idle speed to 700 rpm, the adjusting screw is located on the collector. Do not adjust idle speed with the wiring connected.
7. Connect the ACC electrical connector, idle speed should be about 700–800 rpm with the ignition timing at 13–17° BTDC. If idle speed is not

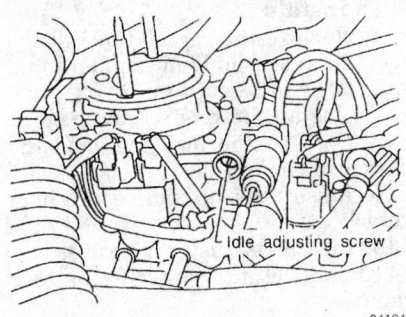

Idle speed adjusting screw — 1995–97 2.4L Pick-up

341843

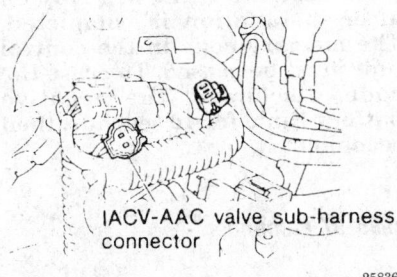

IACV-AAC valve sub-harness connector

258364

IACV-AAC sub-harness connector location — 3.0L engine

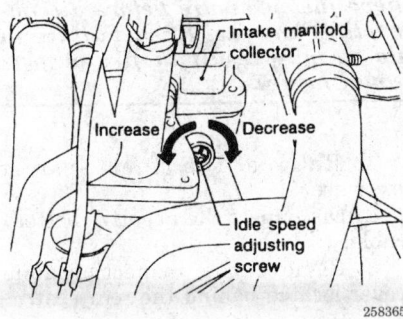

258365

Idle speed adjusting screw location — 3.0L engine

correct, look for a problem in the engine control system.

3.3L Engine

NOTE: The engine should be in good mechanical condition and all electrical connectors and vacuum hoses connected before making this adjustment. Block the engine drive wheels and apply the parking brake.

1. Start the engine and let it warm up to normal operating temperature.

2. Check the ignition timing.
3. Stop the engine and disconnect the throttle position sensor connector.
4. Run the engine under no load at 2,000 rpm for about two minutes. Rev the engine two or three times and let it idle for one minute.
5. Adjust the idle speed using the idle speed adjusting screw to 650–750 rpm for M/T and A/T vehicles with the transmission in the N position.
6. Stop the engine and connect the throttle position sensor connector.
7. Run the engine under no load at 2,000 rpm for about two minutes. Rev the engine two or three times and let it idle for one minute.
8. Check that the the idle speed is 700–800 rpm.
9. If the idle speed is not correct, check the idle air control circuit and repair as necessary.
10. If the idle speed is still not correct, substitute a known good ECM.

NOTE: The ECM may be the cause of the problem, but this is rarely the case.

Mixture

ADJUSTMENT

The air/fuel mixture is computer controlled according to the needs of the engine, and is not adjustable. If the air/fuel mixture is too lean or too rich, other problems with the engine/engine control system exist.

Fuel Filter

REMOVAL AND INSTALLATION

—— CAUTION ——
Fuel injection systems remain under pressure after the engine has been turned OFF. Properly re-

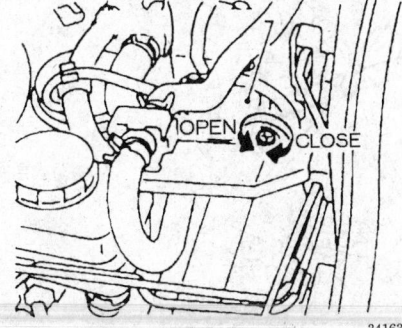

341638

Idle speed adjusting screw — 3.3L engine

lieve fuel pressure before disconnecting any fuel lines. Failure to do so may result in fire or personal injury.

1. Release the fuel pressure from the fuel line as follows:
    a. Remove the fuel pump fuse at the fuse box.
    b. Start the engine.
    c. After the engine stalls, crank the engine two or three times to make sure that the fuel pressure is released.
    d. Turn the ignition switch off and reinstall the fuel pump fuse.

NOTE: After the fuse is reinstalled, do not turn the ignition switch on or fuel pressure will be reestablished.

2. Locate the fuel filter.
3. Loosen the hose clamps at the fuel inlet and outlet lines and slide each line off the filter nipples.
4. Remove the fuel filter from the mounting bracket.
    To install:
5. Mount the filter in the mounting bracket and secure.

NOTE: Always use a high pressure-type fuel filter.

6. Connect the fuel line hoses and secure them with new hose clamps.
7. Start the engine and check for leaks.

Fuel Pump

REMOVAL AND INSTALLATION

Pick-Up

—— CAUTION ——
Fuel injection systems remain under pressure after the engine has been turned OFF. Properly relieve fuel pressure before disconnecting any fuel lines. Failure to do so may result in fire or personal injury.

The fuel pump, equipped with a damper, is located in the fuel tank.

1. Properly relieve the fuel system pressure and disconnect the negative battery cable.
2. Disconnect the fuel gauge electrical connector.
3. Disconnect the fuel outlet and the return hoses.
4. Remove the fuel tank assembly.
5. Remove the ring retaining bolts and the O-ring, then lift the fuel pump assembly from the fuel tank. Plug the opening with a clean rag to

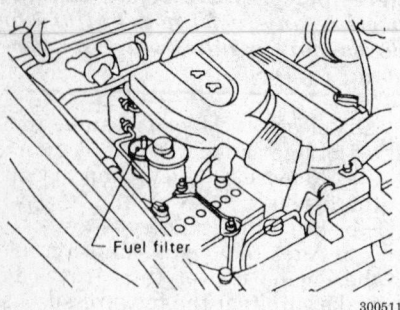

Fuel filter location — KA24E engine

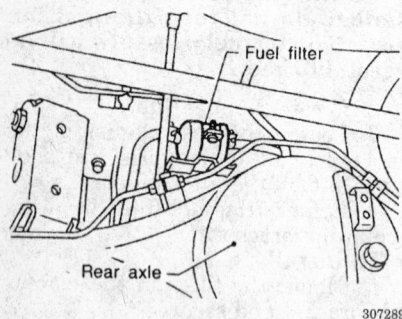

Fuel filter location — VG33E engine

prevent dirt from entering the system.

NOTE: When removing or installing the fuel pump assembly, be careful not to damage or deform it and always install a new O-ring.

To install:

6. Remove the rag and install the fuel pump assembly in the fuel tank; remember to use a new O-ring. Install the ring retaining bolts.

7. Install the fuel tank assembly.

8. Reconnect the fuel lines and the electrical connections

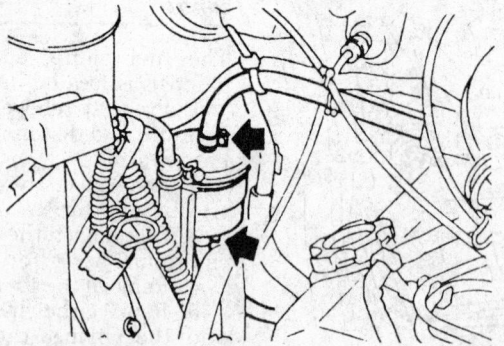

Fuel filter location — VG30E engine

9. Reconnect the negative battery cable.

10. Start the engine and check for fuel leaks.

NOTE: On some models, the Check Engine Light will stay on after installation is completed. The memory code in the control unit must be erased. To erase the code, disconnect the battery cable for 10 seconds, then reconnect it.

1993–95 Pathfinder

The fuel pump, equipped with a damper, is located in the fuel tank.

1. Properly relieve the fuel system pressure and disconnect the negative battery cable.

2. Disconnect the fuel gauge electrical connector and remove the fuel tank inspection cover.

— CAUTION —
Fuel injection systems remain under pressure after the engine has been turned OFF. Properly relieve fuel pressure before disconnecting any fuel lines. Failure to do so may result in fire or personal injury.

3. Disconnect the fuel outlet and the return hoses.

4. Support and lower the fuel tank. Only lower the tank enough to remove the fuel pump.

5. Remove the ring retaining bolts and the O-ring, then lift the fuel pump assembly from the fuel tank. Plug the opening with a clean rag to prevent dirt from entering the system.

NOTE: When removing or installing the fuel pump assembly, be careful not to damage or deform it and always install a new O-ring.

To install:

6. Remove the rag and install the fuel pump assembly in the fuel tank; remember to use a new O-ring.

7. Install the ring retaining bolts.

8. Connect the fuel lines and the electrical connections.

9. Raise tank up and secure it in vehicle.

10. Install the inspection cover.

11. Connect the negative battery cable.

12. Start engine and check for fuel leaks.

NOTE: On some models, the Check Engine Light will stay on after installation is completed. The memory code in the control unit must be erased. To erase the code, disconnect the negative battery cable for 10 seconds, then reconnect it.

1996–97 Pathfinder

— CAUTION —
Fuel injection systems remain under pressure after the engine has been turned OFF. Properly relieve fuel pressure before disconnecting any fuel lines. Failure to do so may result in fire or personal injury.

1. Relieve the fuel system pressure.

2. Disconnect the negative battery cable.

3. Remove the inspection hole cover located behind the rear seat.

4. Disconnect the harness connectors and the fuel tubes from the upper plate of the fuel gauge.

5. Remove the fuel gauge retainer and fuel gauge.

6. Remove the fuel pump with bracket while lifting the pawl of the pump bracket upward.

7. Separate the pump from the sender if necessary.

To install:

8. Connect the pump to the sender if removed.

9. Install the fuel pump assembly in the tank.

10. Install the fuel gauge retainer.

11. Connect the harness connector and the fuel tubes.

12. Install the inspection hole cover.

13. Connect the negative battery cable and check for leaks.

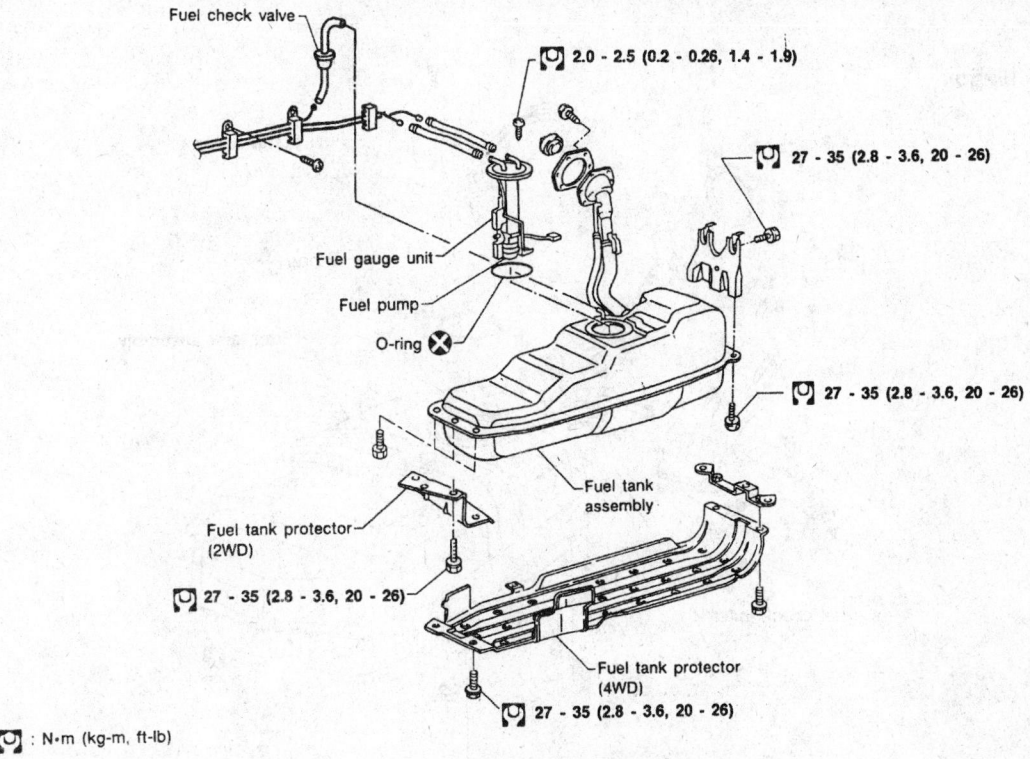

Fuel check valve

2.0 - 2.5 (0.2 - 0.26, 1.4 - 1.9)

27 - 35 (2.8 - 3.6, 20 - 26)

Fuel gauge unit

Fuel pump

O-ring

27 - 35 (2.8 - 3.6, 20 - 26)

Fuel tank assembly

Fuel tank protector (2WD)

27 - 35 (2.8 - 3.6, 20 - 26)

Fuel tank protector (4WD)

27 - 35 (2.8 - 3.6, 20 - 26)

: N•m (kg-m, ft-lb)

300543

Fuel pump assembly and related components — Pick-up

Fuel Injector

REMOVAL AND INSTALLATION

6-Cylinder Engines

1993–94 Models

The engine is equipped 6 fuel injectors, with one located at each cylinder.

1. Relieve the fuel pressure from the system.

2. Remove the 2 drain plugs (from both sides of cylinder block) and drain the engine coolant into a suitable container.

3. Disconnect the ASCD and accelerator control wire from the intake manifold collector.

4. Disconnect the following from the intake collector:
   a. AAC valve
   b. Throttle sensor and throttle valve switch
   c. Ignition coil
   d. EGR control solenoid
   e. Air regulator
   f. Exhaust gas temperature sensor (California models)
   g. PCV hose from RH rocker cover
   h. Air duct hose
   i. Ground wire
   j. EGR Tube

k. Purge hose from canister

5. Disconnect the brake master cylinder, pressure regulator and carbon canister vacuum hoses.

6. Remove the intake collector assembly.

7. Remove the fuel hoses from the injector fuel tube (fuel rail) assembly.

8. Disconnect the injector electrical connectors, and then remove the injectors with fuel tube as an assembly.

9. Remove the clamps securing the fuel injector-to-fuel tube. Pull the fuel injector from the fuel tube assembly. Remove and discard the O-rings.

To install:

NOTE: Replace all fuel injector and fuel tube O-rings. To ease installation, as well as to prevent damage to the seals during assembly, lubricate the seals with a light oil, such as automatic transmission fluid.

10. Lubricate the fuel injector O-rings and press them into the fuel rail assembly.

11. Install the injector clamps and tighten the retaining bolts to 36 inch lbs. (4 Nm).

12. Lubricate the fuel rail O-rings and press them into the intake manifold.

13. Assemble the fuel rail to the intake manifold and tighten the retaining bolts to 12–15 ft. lbs. (16–21 Nm).

14. Connect the electrical connectors to the fuel injectors.

15. Connect the fuel hoses to the fuel rail and secure with new clamps.

16. Using a new gasket, install the intake manifold collector-to-intake manifold bolts. Torque the bolts to 13–16 ft. lbs. (18–22 Nm).

17. Connect the electrical connectors, the vacuum lines and remaining components to the collector assembly.

18. Fill the coolant and properly bleed the cooling system.

19. Connect the negative battery cable.

20. Turn the ignition switch **ON** and check for fuel leaks at the fuel rail.

1995 and 1997 Pick-Up; 1995–97 Pathfinder

The engine is equipped with six fuel injectors, one at each cylinder.

1. Disconnect the negative battery cable.

2. Relieve the pressure from the fuel system.

3. Remove the two drain plugs (from both sides of cylinder block) and drain the engine coolant into a suitable container.

Van and Wagon

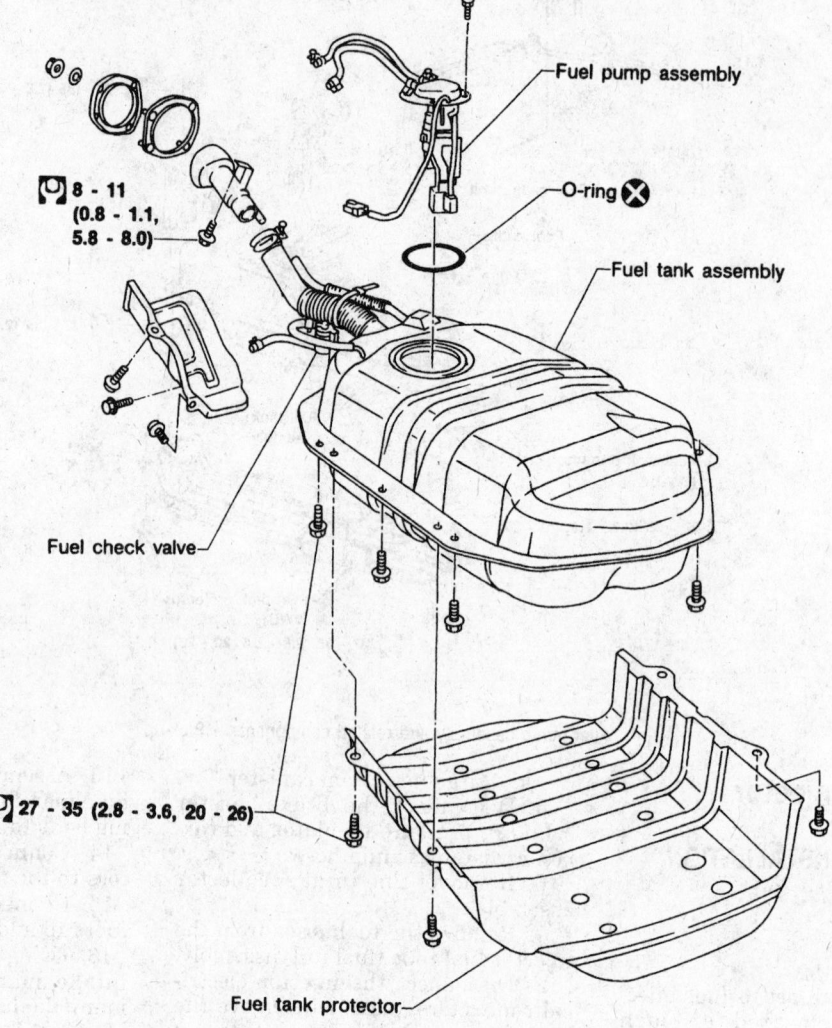

⟐ 8 - 11
(0.8 - 1.1,
5.8 - 8.0)

Fuel pump assembly

O-ring ⊗

Fuel tank assembly

Fuel check valve

⟐ 27 - 35 (2.8 - 3.6, 20 - 26)

⟐ : N·m (kg-m, ft-lb)

Fuel tank protector

253612

Fuel pump assembly and related components — 1993–95 Pathfinder

4. Disconnect the ASCD and accelerator control wire from the intake manifold collector.

5. Disconnect the following electrical connectors from the intake collector (upper intake manifold):
  a. IACV–AAC valve
  b. Throttle position sensor and the closed throttle position switch
  c. Ignition coil
  d. Power transistor
  e. EGRC solenoid valve
  f. IACV air regulator
  g. EGR temperature sensor

6. Remove the coolant hoses from the collector.

7. Disconnect the PCV hose from RH rocker cover

8. Tag, then disconnect the vacuum hoses from the canister, power brake booster, and the fuel pressure regulator.

9. Disconnect the purge hose from the canister.

10. Disconnect the EGR tube from the collector.

11. Remove the ground harness from the collector.

12. Remove the air duct hose.

13. Remove the intake collector assembly.

—CAUTION—
Fuel injection systems remain under pressure after the engine has been turned OFF. Properly relieve fuel pressure before disconnecting any fuel lines. Failure to do so may result in fire or personal injury.

14. Disconnect the fuel feed and return hose from the fuel injector fuel tube assembly.

15. Disconnect the injector electrical connectors, and then remove the injectors with fuel tube as an assembly.

16. Remove the clamps securing the fuel injectors to the fuel tube. Pull the fuel injector from the fuel tube assembly. Remove and discard the O-rings and insulators.

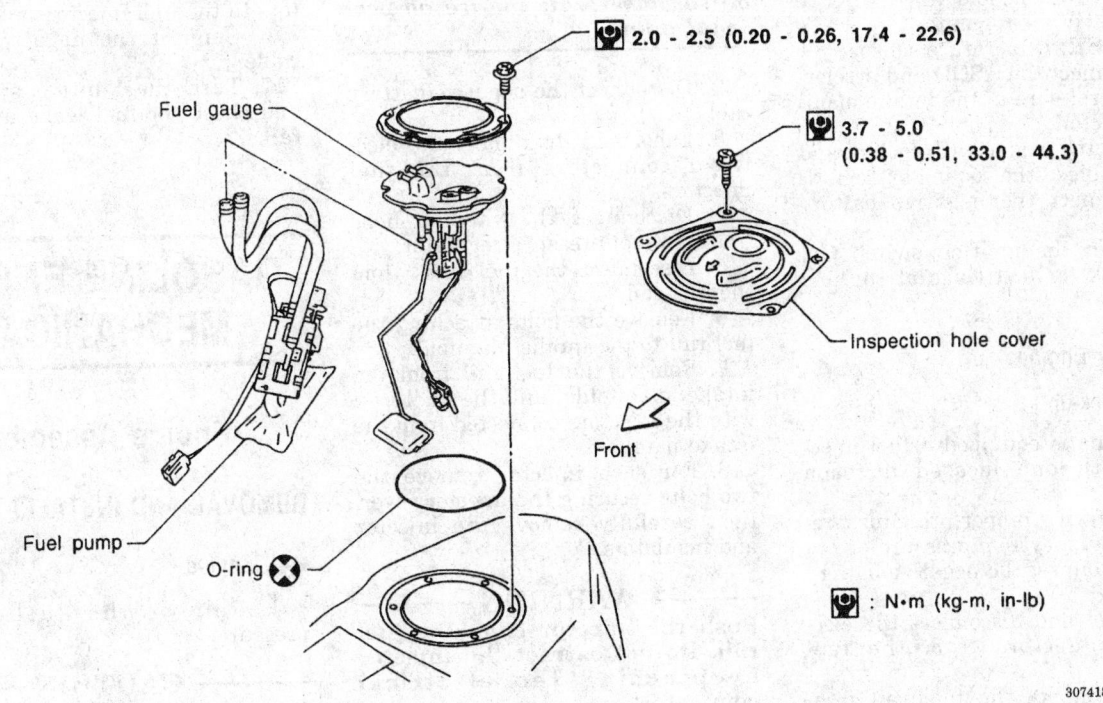

🔧 2.0 - 2.5 (0.20 - 0.26, 17.4 - 22.6)

🔧 3.7 - 5.0
(0.38 - 0.51, 33.0 - 44.3)

Fuel gauge

Inspection hole cover

Front

Fuel pump

O-ring ⊗

🔧 : N•m (kg-m, in-lb)

307418

Fuel pump assembly and related components — 1996–97 Pathfinder

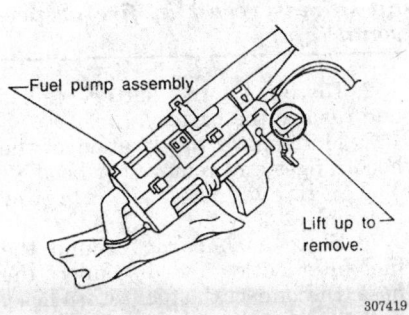

Fuel pump assembly

Lift up to remove.

307419

Lift the pawl of the fuel pump bracket to remove the pump — 1996–97 Pathfinder

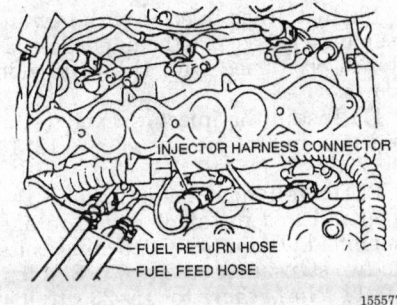

INJECTOR HARNESS CONNECTOR

FUEL RETURN HOSE
FUEL FEED HOSE

155577

Fuel rail connection points — 6-cylinder engines

To install:

NOTE: Replace all fuel injector and fuel tube O-rings. To ease installation, as well as to prevent damage to the seals during assembly, lubricate the seals with a light oil, such as automatic transmission fluid.

17. Lubricate the fuel injector O-rings and press the O-rings and the insulators into the fuel rail assembly.

18. Install the injectors then the injector clamps and tighten the retaining bolts to 36 inch lbs. (4 Nm).

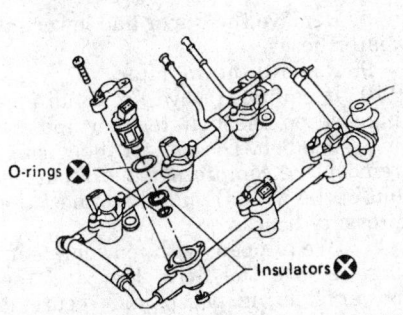

O-rings ⊗

Insulators ⊗

128343

Exploded view of the fuel rail — 6-cylinder engines

19. Lubricate the fuel rail insulators and press them into the intake manifold.

20. Assemble the fuel rail to the intake manifold and tighten the retaining bolts to 12–15 ft. lbs. (16–21 Nm).

21. Connect the electrical connectors to the fuel injectors.

22. Connect the fuel hoses to the fuel rail and secure them with new clamps.

23. Using a new gasket, install the intake manifold collector and mounting bolts. Torque the bolts to 13–16 ft. lbs. (18–22 Nm).

24. Install the air duct hose.

25. Connect the ground harness to the collector.

26. Connect the EGR tube to the collector.

27. Connect the purge hose to the canister.

28. Connect the vacuum hoses to the canister, power brake booster, and the fuel pressure regulator.

29. Connect the PCV hose to the right-side rocker cover

30. Connect the coolant hoses to the collector.

31. Connect the following electrical connectors to the intake collector:

   a. IACV–AAC valve

   b. Throttle position sensor and the closed throttle position switch

   c. Ignition coil

d. Power transistor

e. EGRC solenoid valve

f. IACV air regulator

g. EGR temperature sensor

32. Connect the ASCD and accelerator control wire to the intake manifold collector.

33. Fill the engine with coolant and properly bleed the cooling system.

34. Connect the negative battery cable.

35. Turn the ignition switch **ON** and check for fuel leaks at the fuel rail.

4-Cylinder Engines

1993–94 Pick-Up

The engine is equipped 4 fuel injectors, with one located at each cylinder.

1. Following proper procedure, relieve the fuel system pressure.

2. Disconnect the negative battery cable.

3. Label and disconnect the electrical connectors from the fuel injectors.

4. Disconnect the fuel rail from the fuel system.

5. Remove the fuel rail from the intake manifold; pull the fuel rail with the injectors connected from the intake manifold.

6. Separate the fuel injectors from the fuel rail.

To install:

7. Replace the fuel injector O-rings.

8. Install the fuel injectors to the fuel rail.

9. Lubricate the fuel injector O-rings with automatic transmission fluid and press them, with the fuel rail, into the intake manifold.

10. Install the fuel rail-to-intake manifold bolts.

11. Connect the fuel rail to the fuel system.

12. Connect the electrical connectors to the fuel injectors.

13. Connect the negative battery cable.

14. Turn the ignition switch **ON** and check for fuel leaks at the fuel rail.

1995–97 Pick-Up

The engine is equipped with 4 fuel injectors, one at each cylinder.

1. Following proper procedure, relieve the fuel system pressure.

— CAUTION —

Fuel injection systems remain under pressure after the engine has been turned OFF. Properly relieve fuel pressure before discon-

necting any fuel lines. Failure to do so may result in fire or personal injury.

2. Disconnect the negative battery cable.

3. Label and disconnect the electrical connectors from the fuel injectors.

4. Disconnect the vacuum hose from the fuel pressure regulator.

5. Disconnect the fuel lines from the fuel rail.

6. Remove the bolts attaching the fuel rail to the intake manifold.

7. Remove the fuel rail from the intake manifold; pull the fuel rail with the injectors connected from the intake manifold.

8. For each injector remove the two bolts securing the injector cover, then carefully remove the injector and insulators.

— WARNING —

Push the injector from the fuel rail. Do not extract the injector by pinching the electrical connector.

To install:

9. Lubricate a new O-ring for the injector with silicone oil and install it to the injector.

10. Install the injector with new insulators, make sure the upper insulator is positioned with the plate side down.

11. Install the injector cover to the injector and torque the bolts to 12–15 ft. lbs. (16–21 Nm).

12. Install new insulators to the fuel rail and install it to the intake manifold. Torque the attaching bolts in two steps: first step to 6.5–8 ft. lbs. (9–11 Nm), then to 15–20 ft. lbs. (21–26 Nm).

13. Connect the fuel lines to the fuel rail.

14. Connect the vacuum hose to the fuel pressure regulator.

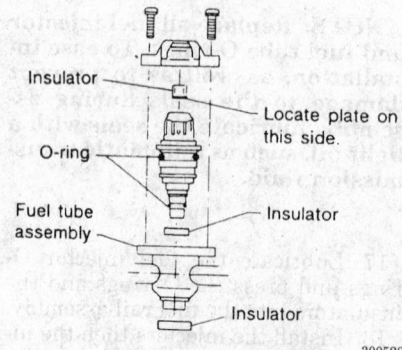

Fuel injector components — 1995–97 Pick-Up

15. Connect the electrical connectors to the fuel injectors.

16. Connect the negative battery cable.

17. Turn the ignition switch **ON** and check for fuel leaks at the fuel rail.

GASOLINE ENGINE MECHANICAL

Engine Assembly

REMOVAL AND INSTALLATION

2.4L Engine

1. Relieve the fuel system pressure.

— CAUTION —

Fuel injection systems remain under pressure after the engine has been turned OFF. Properly relieve fuel pressure before disconnecting any fuel lines. Failure to do so may result in fire or personal injury.

2. Disconnect the battery cables and remove the battery.

3. Matchmark the location of the hood hinges and remove the hood.

4. Remove the air cleaner assembly.

5. Wrap a shop rag around the fuel filter outlet and disconnect the hose. Disconnect the fuel return hose.

6. Raise and safely support the vehicle. If equipped, remove the splash pan from under the engine.

7. Drain the engine oil and the cooling system, including the block drains. Dispose of the old fluids properly.

8. Remove the upper and lower radiator hoses.

9. Remove the radiator.

10. If equipped with air conditioning, loosen the belt tension and remove the belt. Disconnect the wiring, remove the compressor and secure it out of the way. Do not disconnect the pressure hoses.

11. If equipped with power steering, remove the drive belt and the power steering pump and secure it out of the way. Do not disconnect the pressure hoses.

12. Label and disconnect all wiring and vacuum hoses.

13. Disconnect the heater hoses from the engine and disconnect the throttle cable.

14. Remove the starter.

15. Matchmark the driveshaft flange at the rear pinion flange and remove the driveshaft. Plug the extension housing opening to prevent the oil from draining out.

16. If equipped with 4WD, matchmark both front driveshaft flanges so the shaft can be installed in the same position. Remove the front driveshaft.

17. Disconnect the exhaust pipe from the manifold and from the catalytic converter and remove the pipe.

18. Disconnect the speedometer cable and the wiring from the transmission.

19. If equipped an automatic transmission perform the following:

a. Disconnect the selector lever and throttle cables from the transmission.

b. Remove the dipstick tube and disconnect the cooler lines.

c. Remove the torque converter housing dust cover. Matchmark the converter with the driveplate for reassembly; these are balanced together at the factory. Remove the torque converter–to–driveplate (flywheel) bolts. Use a wrench on the crankshaft pulley bolt to rotate the crankshaft to expose the hidden torque converter bolts.

20. If equipped with a manual transmission perform the following:

a. Remove the shifter knob and boot and remove the snapring to lift the shift lever out of the transmission. Stuff a rag in the opening to keep dirt out of the transmission.

b. Without disconnecting the hydraulic hose, remove the clutch slave cylinder from the transmission and secure it aside.

21. If equipped with 4WD perform the following:

a. Remove the torsion bars and the second crossmember

b. Remove the transfer case shift lever assembly from the transfer case.

22. Using a chain hoist, attach it to the engine and lift the engine slightly to take the weight off the mounts. Using an appropriate transmission jack, properly support the transmission and transfer case, if equipped. Remove the transmission mount and crossmember.

23. Remove the bolts securing the transmission to the engine and move the transmission back from the engine. If equipped with an automatic

transmission, secure the torque converter to the transmission. Lower the transmission from the vehicle.

NOTE: When removing the engine mounts, do not loosen the 4 mount cover nuts. The mount is fluid filled and will not function properly if the fluid leaks out.

24. Check to make sure all wires and hoses have been disconnected. Remove the front engine mount bolts and carefully lift the engine out.

To install:

25. Carefully guide the engine into place and start the mount bolts. Tighten the bolts temporarily.

26. If equipped with a manual transmission, perform the following steps:

a. Lightly grease the input shaft splines. On 4WD, apply a silicone sealant to the engine block and rear plate to seal the engine to the transmission.

b. Fit the transmission into place and start the bolts attaching the engine to the transmission. Make sure the input shaft fits properly into the clutch disc and pilot bearing.

c. Torque the bolts to specification.

27. If equipped with an automatic transmission perform the following:

a. Use a dial indicator to check the driveplate run-out while turning the crankshaft. Maximum allowable run-out is 0.020 in. (0.5mm); if beyond specifications, replace the driveplate.

b. Measure and adjust how far the torque converter is recessed into the transmission housing. The distance between the front mounting surface of the transmission and the torque converter–to–driveplate bolt boss should be at least 1.024 in. (26mm).

c. Install the transmission and start the bolts that attach the transmission to the engine. Torque the transmission attaching bolts to specification.

28. If equipped with an automatic transmission, align the matchmarks on the driveplate and torque converter, install the bolts and torque to 33–43 ft. lbs. (44–59 Nm). Turn the crankshaft after tightening the bolts to make sure there is no binding at the driveplate.

29. Install the transmission crossmember and the transmission mount.

30. Loosen the engine mount bolts, tighten the transmission mount bolts and then the engine mounts.

31. If equipped with a manual transmission, perform the following steps:

a. Install the clutch slave cylinder and torque the bolts to 22–30 ft. lbs. (30–40 Nm).

b. Remove the rag from the transmission and install the shifter, then install the snapring to hold the shifter in position.

c. Install the shifter boot and knob.

32. If equipped with an automatic transmission perform the following:

a. Install the torque converter housing dust cover.

b. Install the dipstick tube and connect the cooler lines.

c. Connect the selector lever and throttle cables to the transmission.

33. If the torsion bars were removed, install them in their original location. Make sure the splines are in their original position and set the adjustment to its original position.

34. When installing the driveshafts, be sure to align the matchmarks. Torque the bolts on the front driveshaft (4WD) to 29–33 ft. lbs. (39–44 Nm). If equipped with a one-piece driveshaft, torque the flange bolts to 58–65 ft. lbs. (78–88 Nm). If equipped with a two-piece driveshaft, torque the flange bolts to 29–33 ft. lbs. (39–44 Nm) and torque the center bearing bracket bolts to 12–16 ft. lbs. (16–22 Nm).

35. When installing the exhaust pipe, use new gaskets and torque the flange bolts to 20–27 ft. lbs. (26–36 Nm). Torque the bolts attaching the pipe to the catalytic converter to 23–31 ft. lbs. (31–42 Nm).

36. Connect the speedometer cable. Connect any electrical leads to the transmission.

37. Install the starter and connect the starter leads.

38. Connect the throttle cable and the heater hoses to the engine.

39. Connect the wiring and hoses.

40. Install the power steering pump and its bracket, if equipped; be sure to connect the ground strap to the bracket.

41. Install the air conditioning compressor and drive belt.

42. Adjust the drive belts.

43. Install the radiator and shroud. If equipped with an automatic transmission, unplug oil cooler lines and connect them to the radiator.

44. Install the upper and lower radiator hoses.

45. Install the splash pan, if equipped.

46. Connect the fuel return hose and the hose to the filter outlet.

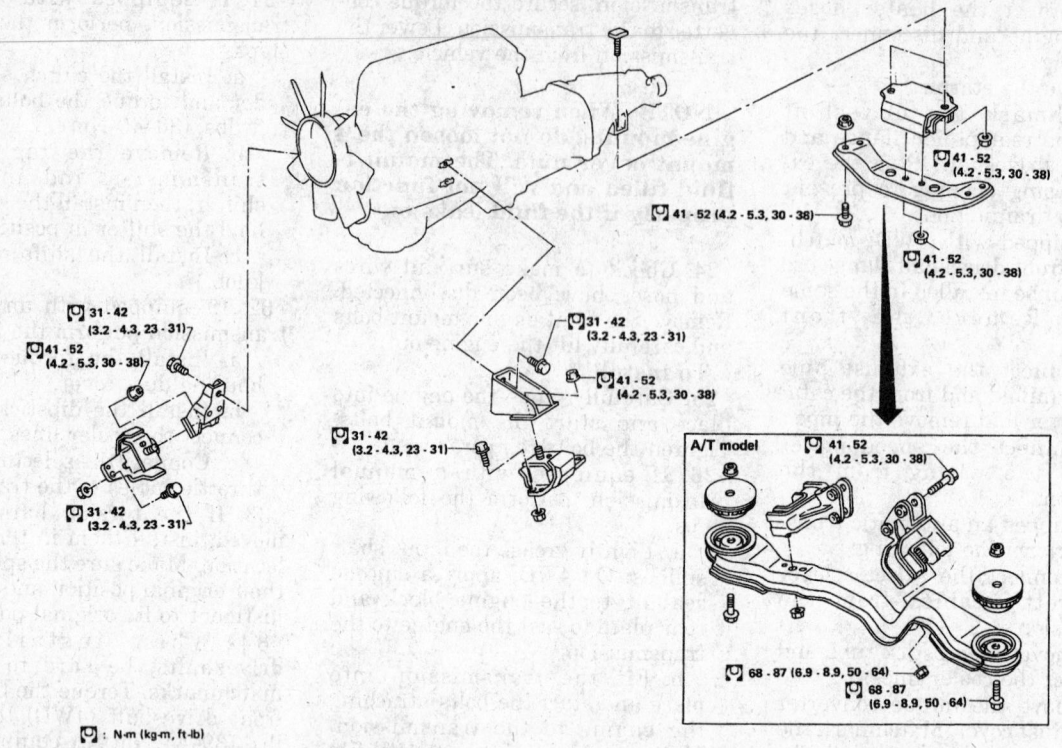

⌷ : N·m (kg-m, ft-lb)

300377

Engine mounts and related components — 2.4L engine

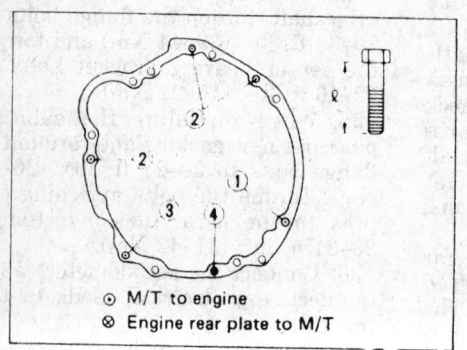

- ⊙ M/T to engine
- ⊗ Engine rear plate to M/T

Bolt No.	Tightening torque N·m (kg-m, ft-lb)	ℓ mm (in)
1	39 - 49 (4.0 - 5.0, 29 - 36)	65 (2.56)
2	39 - 49 (4.0 - 5.0, 29 - 36)	60 (2.36)
3*	19 - 25 (1.9 - 2.5, 14 - 18)	25 (0.98)
4	19 - 25 (1.9 - 2.5, 14 - 18)	16 (0.63)

*: With nut

300378

Manual transmission bolt specifications — 2.4L engine

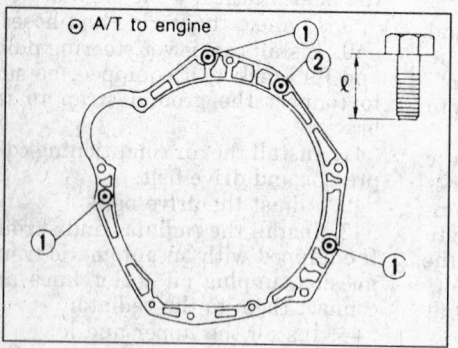

- ⊙ A/T to engine

Tightening torque N·m (kg-m, ft-lb)	Bolt length "ℓ" mm (in)
① 39 - 49 (4.0 - 5.0, 29 - 36)	45 (1.77)
② 39 - 49 (4.0 - 5.0, 29 - 36)	40 (1.57)

300379

Automatic transmission bolt specifications — 2.4L engine

47. Install the air cleaner assembly.
48. Fill the transmission and, if equipped, the transfer case.
49. Fill the engine with fresh oil.
50. Install and adjust the hood. Install the undercover.
51. Connect the negative battery cable, start the engine, allow it to reach normal operating temperatures and check for leaks.

3.0L Engine

1. Relieve the fuel system pressure.

---- **CAUTION** ----

Fuel injection systems remain under pressure after the engine has been turned OFF. Properly relieve fuel pressure before disconnecting any fuel lines. Failure to do so may result in fire or personal injury.

2. Disconnect the battery cables and remove the battery.
3. Matchmark the location of the hood hinges and remove the hood.
4. Remove the air cleaner assembly.
5. Wrap a shop rag around the fuel filter outlet and disconnect the hose. Disconnect the fuel return hose.
6. Raise and safely support the vehicle. If equipped, remove the splash pan from under the engine.
7. Drain the engine oil and the cooling system, including the block drains. Dispose of old fluids properly.
8. Remove the upper and lower radiator hoses.
9. Remove the radiator.
10. If equipped with air conditioning, loosen the belt tension and remove the belt. Disconnect the wiring, remove the compressor and secure it out of the way. Do not disconnect the pressure hoses.
11. If equipped with power steering, remove the drive belt and the power steering pump and secure it out of the way. Do not disconnect the pressure hoses.
12. Label and disconnect all wiring and vacuum hoses.
13. Disconnect the heater hoses from the engine and disconnect the throttle cable.
14. Remove the starter.
15. Matchmark the driveshaft flange at the rear pinion flange and remove the driveshaft. Plug the extension housing opening to prevent the oil from draining out.
16. If equipped with 4WD, matchmark both front driveshaft flanges so the shaft can be installed in the same

position. Remove the front driveshaft.
17. Disconnect the exhaust pipe from the manifolds and from the catalytic converter and remove the pipe.
18. Disconnect the speedometer cable and the wiring from the transmission.
19. Remove the bracket securing the transmission to the engine.
20. If equipped an automatic transmission perform the following:
    a. Disconnect the selector lever and throttle cables from the transmission.
    b. Remove the dipstick tube and disconnect the cooler lines.
    c. Remove the torque converter housing dust cover. Matchmark the converter with the driveplate for reassembly; these are balanced together at the factory. Remove the torque converter-to-driveplate (flywheel) bolts. Use a wrench on the crankshaft pulley bolt to rotate the crankshaft to expose the hidden torque converter bolts.
21. If equipped with a manual transmission perform the following:
    a. Remove the shifter knob and boot and remove the snapring to lift the shift lever out of the transmission. Stuff a rag in the opening to keep dirt out of the transmission.
    b. Without disconnecting the hydraulic hose, remove the clutch slave cylinder from the transmission and secure it aside.
22. If equipped with 4WD perform the following:
    a. Remove the torsion bars and the second crossmember
    b. Remove the transfer case shift lever assembly from the transfer case.
23. Using a chain hoist, attach it to the engine and lift the engine slightly to take the weight off the mounts. Using an appropriate transmission jack, properly support the transmission and transfer case, if equipped. Remove the transmission mount and crossmember.
24. Remove the bolts securing the transmission to the engine and move the transmission back from the engine. If equipped with an automatic transmission, secure the torque converter to the transmission. Lower transmission from the vehicle.

NOTE: When removing the engine mounts, do not loosen the 4 mount cover nuts. The mount is fluid filled and will not function properly if the fluid leaks out.

25. Check to make sure all wires and hoses have been disconnected. Remove the front engine mount bolts and carefully lift the engine out.

To install:

26. Carefully guide the engine into place and start the mount bolts. Tighten the bolts temporarily.
27. If equipped with a manual transmission, perform the following steps:
    a. Lightly grease the input shaft splines. On 4WD, apply a silicone sealant to the engine block and rear plate to seal the engine to the transmission.
    b. Fit the transmission into place and start the bolts attaching the engine to the transmission. Make sure the input shaft fits properly into the clutch disc and pilot bearing.
    c. Torque the 2.36 in. (60mm) and 2.56 in. (65mm) transmission bolts to 29–36 ft. lbs. (39–49 Nm).
    d. Torque the remaining bolts to 22–29 ft. lbs. (29–39 Nm).
28. If equipped with an automatic transmission perform the following:
    a. Use a dial indicator to check the driveplate run-out while turning the crankshaft. Maximum allowable run-out is 0.020 in. (0.5mm); if beyond specifications, replace the driveplate.
    b. Measure and adjust how far the torque converter is recessed into the transmission housing. The distance between the front mounting surface of the transmission and the torque converter-to-driveplate bolt boss should be at least 1.024 in. (26mm).
    c. Install the transmission. Install the transmission attaching bolts and torque the 1.77 in. (45 mm) and the 1.97 in. (50mm) bolts to 29–36 ft. lbs. (39–49 Nm), and torque the 0.98 in. (25mm) bolts to 22–29 ft. lbs. (29–39 Nm).
29. If equipped with an automatic transmission, align the matchmarks on the driveplate and torque converter, install the bolts and torque to 33–43 ft. lbs. (44–59 Nm). Turn the crankshaft after tightening the bolts to make sure there is no binding at the driveplate.
30. Install the bracket securing the transmission to the engine and torque the attaching bolts to 22–29 ft. lbs. (29–39 Nm).
31. Install the transmission crossmember and the transmission mount.
32. Loosen the engine mount bolts, tighten the transmission mount bolts and then the engine mounts.

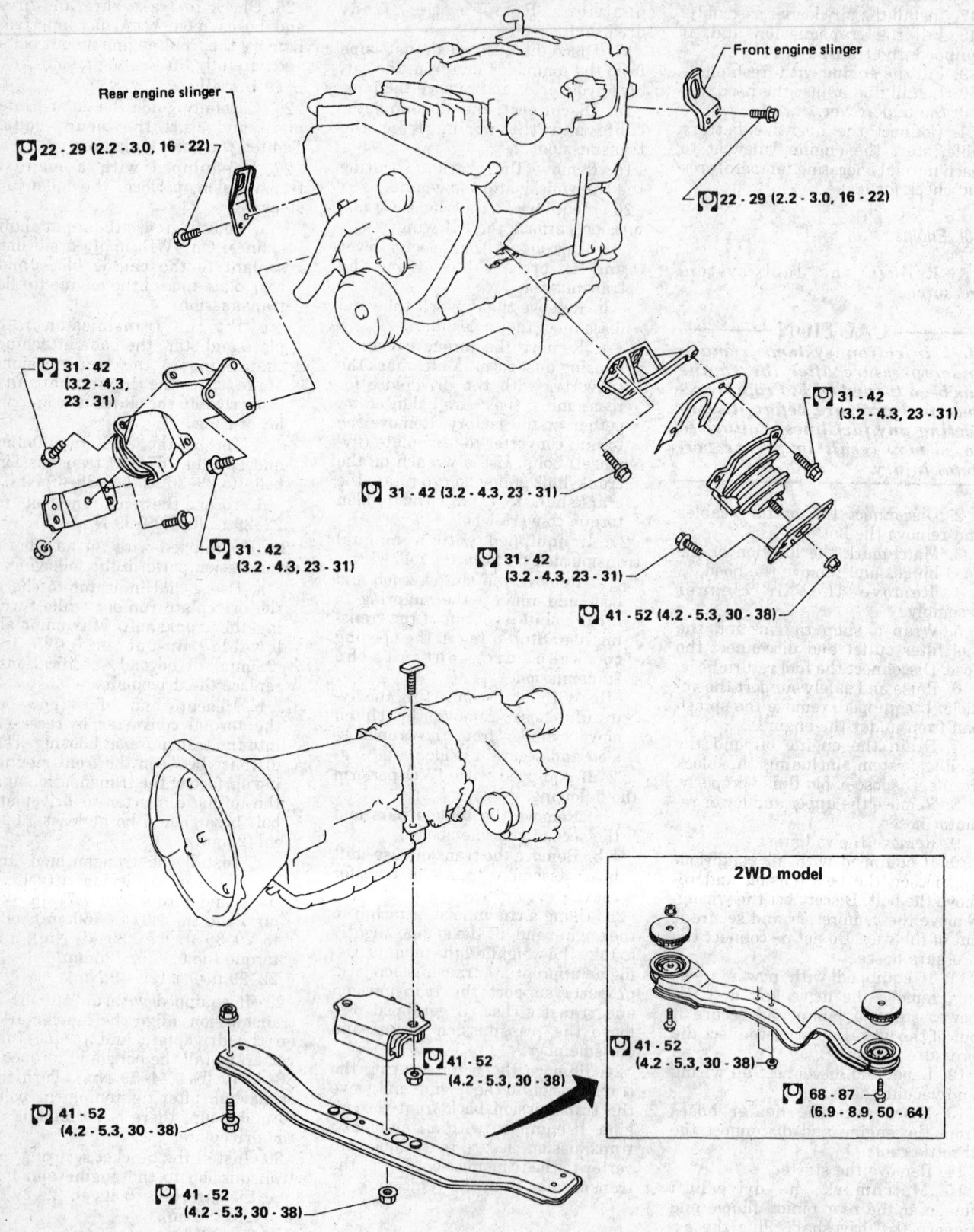

Rear engine slinger

22 - 29 (2.2 - 3.0, 16 - 22)

Front engine slinger

22 - 29 (2.2 - 3.0, 16 - 22)

31 - 42
(3.2 - 4.3,
23 - 31)

31 - 42
(3.2 - 4.3, 23 - 31)

31 - 42 (3.2 - 4.3, 23 - 31)

31 - 42
(3.2 - 4.3, 23 - 31)

31 - 42
(3.2 - 4.3, 23 - 31)

41 - 52 (4.2 - 5.3, 30 - 38)

2WD model

41 - 52
(4.2 - 5.3, 30 - 38)

68 - 87
(6.9 - 8.9, 50 - 64)

41 - 52
(4.2 - 5.3, 30 - 38)

41 - 52
(4.2 - 5.3, 30 - 38)

41 - 52
(4.2 - 5.3, 30 - 38)

: N·m (kg-m, ft-lb)

264983

Engine mounts and related components — 3.0L engine

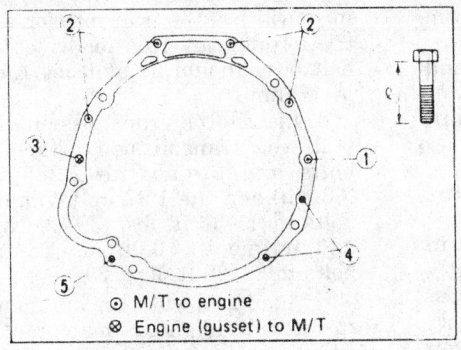

Bolt No.	Tightening torque N·m (kg·m, ft-lb)	ℓ mm (in)
1	39 - 49 (4.0 - 5.0, 29 - 36)	65 (2.56)
2	39 - 49 (4.0 - 5.0, 29 - 36)	60 (2.36)
3	29 - 39 (3.0 - 4.0, 22 - 29)	55 (2.17)
4	29 - 39 (3.0 - 4.0, 22 - 29)	30 (1.18)
5	29 - 39 (3.0 - 4.0, 22 - 29)	25 (0.98)

265678

Manual transmission bolt specifications — 3.0L engine

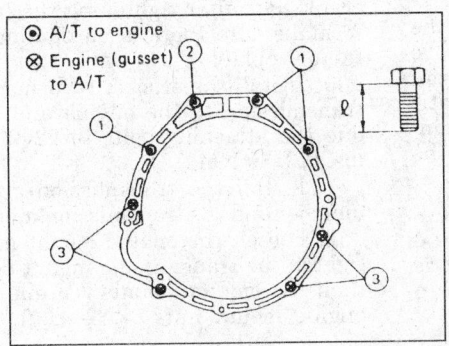

Bolt No.	Tightening torque N·m (kg-m, ft-lb)	Bolt length "ℓ" mm (in)
1	39 - 49 (4.0 - 5.0, 29 - 36)	45 (1.77)
2	39 - 49 (4.0 - 5.0, 29 - 36)	50 (1.97)
3	29 - 39 (3.0 - 4.0, 22 - 29)	25 (0.98)
Gusset to engine	29 - 39 (3.0 - 4.0, 22 - 29)	20 (0.79)

265679

Automatic transmission bolt specifications — 3.0L engine

33. If equipped with a manual transmission, perform the following steps:

   a. Install the clutch slave cylinder and torque the bolts to 22–30 ft. lbs. (30–40 Nm).

   b. Remove the rag from the transmission and install the shifter, then install the snapring to hold the shifter in position.

   c. Install the shifter boot and knob.

34. If equipped with an automatic transmission perform the following:

   a. Install the torque converter housing dust cover.

   b. Install the dipstick tube and connect the cooler lines.

   c. Connect the selector lever and throttle cables to the transmission.

35. If the torsion bars were removed, install them in their original location. Make sure the splines are in their original position and set the adjustment to its original position.

36. When installing the driveshafts, be sure to align the matchmarks. Torque the bolts on the front driveshaft (4WD) to 29–33 ft. lbs. (39–44 Nm). On the rear driveshaft, torque the bolts to 58–65 ft. lbs. (78–88 Nm). If equipped with a center bearing bracket torque the bolts to 12–16 ft. lbs. (16–22 Nm).

37. When installing the exhaust pipe, use new gaskets and torque the flange bolts to 20–27 ft. lbs. (26–36 Nm). Torque the bolts attaching the pipe to the catalytic converter to 32–41 ft. lbs. (43–55 Nm).

38. Connect the speedometer cable. Connect any electrical leads to the transmission.

39. Install the starter and connect the starter leads.

40. Connect the throttle cable and the heater hoses to the engine.

41. Connect the wiring and hoses.

42. Install the power steering pump and its bracket, if equipped; be sure to connect the ground strap to the bracket.

43. Install the air conditioning compressor and drive belt.

44. Adjust the drive belts.

45. Install the radiator and shroud. If equipped with an automatic transmission, unplug oil cooler lines and connect them to the radiator.

46. Install the upper and lower radiator hoses.

47. Install the splash pan, if equipped.

48. Connect the fuel return hose and the hose to the filter outlet.

49. Install the air cleaner assembly.

50. Fill the transmission and, if equipped, the transfer case.

51. Fill the engine with fresh oil.

52. Install and adjust the hood. Install the undercover.

53. Connect the negative battery cable, start the engine, allow it to reach normal operating temperatures and check for leaks.

3.3L Engine

> **CAUTION**
> *Fuel injection systems remain under pressure after the engine has been turned OFF. Properly relieve fuel pressure before disconnecting any fuel lines. Failure to do so may result in fire or personal injury.*

1. Relieve the fuel system pressure.

2. Disconnect the battery cables and remove the battery.

3. Matchmark the location of the hood hinges and remove the hood.

4. Remove the air cleaner assembly.

5. Raise and safely support the vehicle. If equipped, remove the splash pan from under the engine.

6. Drain the engine oil and the cooling system, including the block drains. Dispose of old fluids properly.

7. Remove the upper and lower radiator hoses.

8. Remove the radiator, shroud and cooling fan.

9. If equipped with air conditioning, loosen the belt tension and remove the belt. Disconnect the wiring, remove the compressor and secure it out of the way. Do not disconnect the pressure hoses.

10. If equipped with power steering, remove the drive belt and the power steering pump and secure it out of the way. Do not disconnect the pressure hoses.

11. Label and disconnect all wiring and vacuum hoses.

12. Disconnect the heater hoses from the engine and disconnect the throttle cable.

13. Remove the starter.

14. Matchmark the driveshaft flange at the rear pinion flange and remove the driveshaft. Plug the extension housing opening to prevent the oil from draining out.

15. If equipped with 4WD, matchmark both front driveshaft flanges so the shaft can be installed in the same position. Remove the front driveshaft.

16. Disconnect the exhaust pipe from the manifolds and from the catalytic converter and remove the pipe.

17. Disconnect the speedometer cable and the wiring from the transmission.

18. Remove the bracket securing the transmission to the engine.

19. If equipped with an automatic transmission perform the following:

a. Disconnect the selector lever and throttle cables from the transmission.

b. Remove the dipstick tube and disconnect the cooler lines.

c. Remove the torque converter housing dust cover. Matchmark the converter with the driveplate for reassembly; these are balanced together at the factory. Remove the torque converter-to-driveplate (flywheel) bolts. Use a wrench on the crankshaft pulley bolt to rotate the crankshaft to expose the hidden torque converter bolts.

20. If equipped with a manual transmission perform the following:

a. Remove the shifter knob and boot and remove the snapring to lift the shift lever out of the transmission. Stuff a rag in the opening to keep dirt out of the transmission.

b. Without disconnecting the hydraulic hose, remove the clutch slave cylinder from the transmission and secure it aside.

21. If equipped with 4WD perform the following:

a. Remove the torsion bars and the second crossmember

b. Remove the transfer case shift lever assembly from the transfer case.

22. Attach engine slingers to the engine.

23. Using a chain hoist, attach it to the engine and lift the engine slightly to take the weight off the mounts. Using an appropriate transmission jack, properly support the transmission and transfer case, if equipped. Remove the transmission mount and crossmember.

24. Remove the bolts securing the transmission to the engine and move the transmission back from the engine. If equipped with an automatic transmission, secure the torque converter to the transmission. Lower the transmission from the vehicle.

NOTE: When removing the engine mounts, do not loosen the 4 mount cover nuts. The mount is fluid filled and will not function if the fluid leaks out.

25. Check to make sure all wires and hoses have been disconnected. Remove the front engine mount nuts and carefully lift the engine out.

To install:

26. Carefully guide the engine into place and start the mount bolts. Tighten the bolts temporarily.

27. If equipped with a manual transmission, perform the following steps:

a. Lightly grease the input shaft splines. On 4WD, apply a silicone sealant to the engine block and rear plate to seal the engine to the transmission.

b. Fit the transmission into place and start the bolts attaching the engine to the transmission. Make sure the input shaft fits properly into the clutch disc and pilot bearing.

c. Torque the 2.56 in. (65mm) and 2.28 in. (58mm) transmission bolts to 29–36 ft. lbs. (39–49 Nm).

d. Torque the remaining bolts to 22–29 ft. lbs. (29–39 Nm).

28. If equipped with an automatic transmission perform the following:

a. Use a dial indicator to check the driveplate run-out while turning the crankshaft. Maximum allowable run-out is 0.020 in. (0.5mm); if beyond specifications, replace the driveplate.

b. Measure and adjust how far the torque converter is recessed into the transmission housing. The distance between the front mount-ing surface of the transmission and the torque converter–to–driveplate bolt boss should be at least 1.024 in. (26mm).

c. Install the transmission. Install the transmission attaching bolts and torque the 2.283 in. (58mm) and the 1.87 in. (47.5mm) bolts to 29–36 ft. lbs. (39–49 Nm), and torque the 0.98 in. (25mm) bolts to 22–29 ft. lbs. (29–39 Nm).

29. If equipped with an automatic transmission, align the matchmarks on the driveplate and torque converter, install the bolts and torque to 33–43 ft. lbs. (44–59 Nm). Turn the crankshaft after tightening the bolts to make sure there is no binding at the driveplate.

30. Install the bracket securing the transmission to the engine and torque the attaching bolts to 22–29 ft. lbs. (29–39 Nm).

31. Install the transmission crossmember and the transmission mount.

32. Loosen the engine mount nuts, tighten the transmission mount bolts then the engine mounts. Torque the engine mount nuts to 32–41 ft. lbs. (43–55 Nm).

33. If equipped with a manual transmission, perform the following steps:

a. Install the clutch slave cylinder and torque the bolts to 22–30 ft. lbs. (30–40 Nm).

b. Remove the rag from the transmission and install the shifter, then install the snapring to hold the shifter in position.

c. Install the shifter boot and knob.

34. If equipped with an automatic transmission perform the following:

a. Install the torque converter housing dust cover.

b. Install the dipstick tube and connect the cooler lines.

c. Connect the selector lever and throttle cables to the transmission.

35. If the torsion bars were removed, install them in their original location. Make sure the splines are in their original position and set the adjustment to its original position.

36. When installing the driveshafts, be sure to align the matchmarks. Torque the bolts on the front driveshaft (4WD) to 29–33 ft. lbs. (39–44 Nm). On the rear driveshaft, torque the bolts to 58–65 ft. lbs. (78–88 Nm). If equipped with a center bearing bracket torque the bolts to 12–16 ft. lbs. (16–22 Nm).

37. When installing the exhaust pipe, use new gaskets and torque the flange bolts to 20–27 ft. lbs. (26–36 Nm). Torque the bolts attaching the

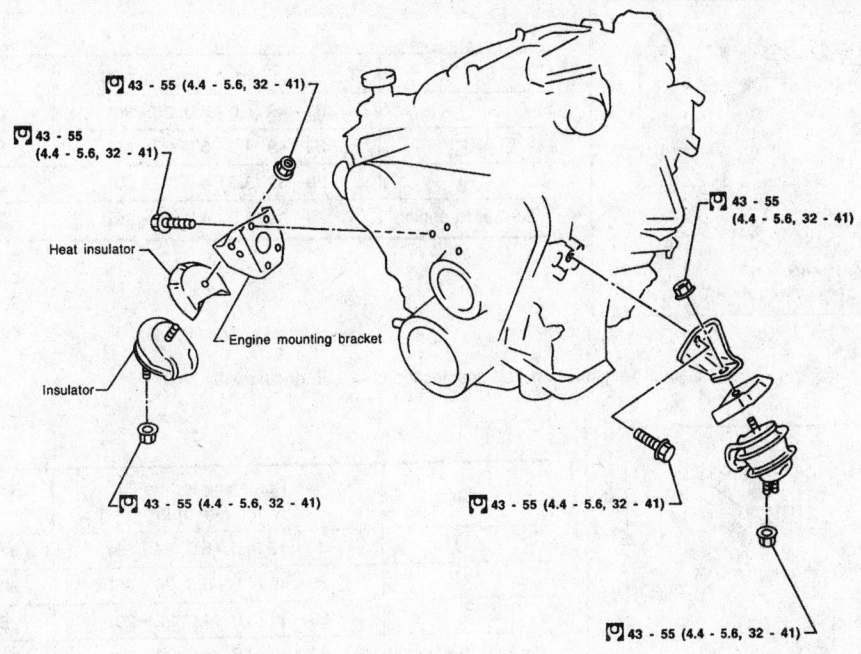

43 - 55 (4.4 - 5.6, 32 - 41)

43 - 55
(4.4 - 5.6, 32 - 41)

Heat insulator

Engine mounting bracket

Insulator

43 - 55
(4.4 - 5.6, 32 - 41)

43 - 55 (4.4 - 5.6, 32 - 41)

43 - 55 (4.4 - 5.6, 32 - 41)

43 - 55 (4.4 - 5.6, 32 - 41)

: N·m (kg-m, ft-lb)

311559

Engine mounts and related components — 3.3L engine

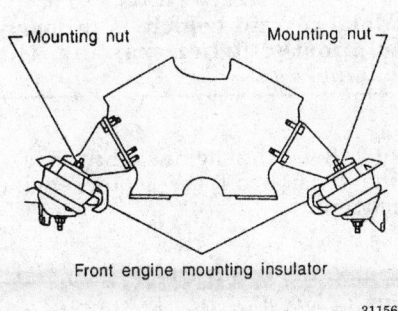

Mounting nut   Mounting nut

Front engine mounting insulator

311560

Engine mounting nuts — 3.3L engine

311562

Engine removal or installation — 3.3L engine

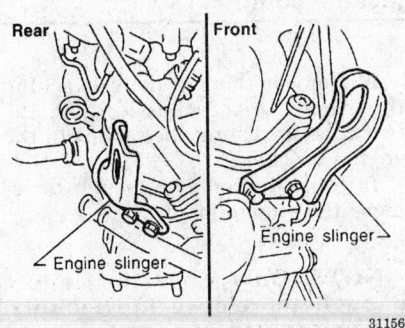

Rear   Front

Engine slinger

Engine slinger

311561

Engine slingers installed — 3.3L engine

pipe to the catalytic converter to 32–41 ft. lbs. (43–55 Nm).

38. Connect the speedometer cable. Connect any electrical leads to the transmission.

39. Install the starter and connect the starter leads.

40. Connect the throttle cable and the heater hoses to the engine.

41. Connect the wiring and hoses.

42. Install the power steering pump and its bracket, if equipped; be sure to connect the ground strap to the bracket.

43. Install the air conditioning compressor and drive belt.

44. Adjust the drive belts.

45. Install the radiator and shroud. If equipped with an automatic transmission, unplug oil cooler lines and connect them to the radiator.

46. Install the upper and lower radiator hoses.

47. Install the splash pan, if equipped.

48. Connect the fuel return hose and the hose to the filter outlet.

49. Install the air cleaner assembly.

50. Fill the transmission and, if equipped, the transfer case.

51. Fill the engine with fresh oil and fill the cooling system with the proper coolant/water mixture.

52. Install and adjust the hood. Install the undercover.

53. Connect the negative battery cable, start the engine, allow it to reach normal operating temperatures and check for leaks. Make all the necessary adjustments.

Engine Mounts

REMOVAL AND INSTALLATION

1993–94 Models

1. Disconnect negative battery cable.

2. Remove the upper fan shroud.

3. Raise and support the vehicle.

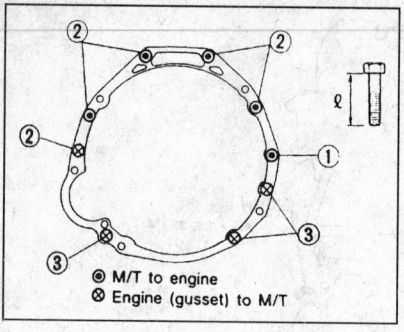

Bolt No.	Tightening torque N·m (kg-m, ft-lb)	ℓ mm (in)
①	39 - 49 (4.0 - 5.0, 29 - 36)	65 (2.56)
②	39 - 49 (4.0 - 5.0, 29 - 36)	58 (2.28)
③	29 - 39 (3.0 - 4.0, 22 - 29)	25 (0.98)
Gusset to engine	29 - 39 (3.0 - 4.0, 22 - 29)	20 (0.79)

◎ M/T to engine
⊗ Engine (gusset) to M/T

311626

Manual transmission bolt specifications — 3.3L engine with 2WD

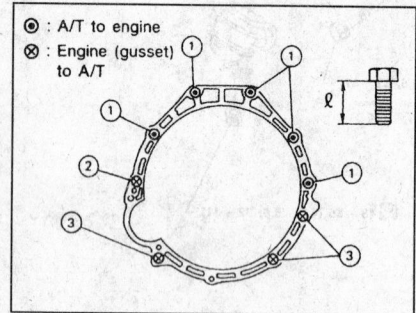

◎ : A/T to engine
⊗ : Engine (gusset) to A/T

Bolt No.	Tightening torque N·m (kg-m, ft-lb)	Bolt length "ℓ" mm (in)
①	39 - 49 (4.0 - 5.0, 29 - 36)	47.5 (1.870)
②	39 - 49 (4.0 - 5.0, 29 - 36)	58.0 (2.283)
③	29 - 39 (3.0 - 4.0, 22 - 29)	25.0 (0.984)
Gusset to engine	29 - 39 (3.0 - 4.0, 22 - 29)	20.0 (0.787)

310289

Automatic transmission bolt specifications — 3.3L engine with 2WD

— WARNING —

Make certain the vehicle is properly supported when raising the engine.

4. Raise the engine slightly to take the tension off the engine mounts.

NOTE: When raising the engine to replace the engine mounts, DO NOT place the jack directly under the oil pan or the crankshaft torsional damper. Use a block of wood between engine and jack to prevent damage at jacking point.

5. Remove the through-bolt nut(s) and through-bolt(s).

6. Remove the engine mount-to-engine block bolts.

NOTE: Some models are equipped with hydraulic mounts. When removing these engine mounts, do not loosen the 4 mount cover nuts. The mount is fluid filled and will not function properly if the fluid leaks out.

7. Raise the engine enough to remove the engine mount(s), then remove the mount(s).

NOTE: Only raise the engine enough to remove the mounts. Raising the engine too far could result in damage to some engine components.

To install:

8. Install the engine mount(s) and engine mount-to-engine block bolts. Tighten the bolts to correct specification.

9. Lower the engine enough to install the engine mount through-bolt(s) and nut(s). Tighten the through-bolts to correct specification.

10. Lower the vehicle and install the upper fan shroud.

11. Connect the negative battery cable.

1995 Models With 3.0L Engine

1. Disconnect negative battery cable.

2. Remove the upper fan shroud.

3. Raise and support the vehicle.

— WARNING —

Make certain vehicle is properly supported when raising the engine.

4. Raise the engine slightly to take the tension off the motor engine mounts.

— WARNING —

When raising the engine to replace the engine mounts, DO NOT place the jack directly under the oil pan or the crankshaft torsional damper. Use a block of wood between engine and jack to prevent damage at jacking point.

5. Remove the four bolts attaching the mount to the chassis, then the two bolts attaching the mount to the engine bracket.

6. Raise the engine enough to remove the engine mount.

NOTE: Only raise the engine enough to remove the mounts. Raising the engine too far could result in damage to some engine components.

7. Remove the nut attaching the mount to the bracket that attaches the mount to the chassis.

NOTE: The vehicle is equipped with hydraulic mounts, when removing these engine mounts, do not loosen the 4 mount cover nuts. The mount is fluid filled and will not function properly if the fluid leaks out.

To install:

8. Install the chassis bracket to the engine mount and torque the nut to 30–38 ft. lbs. (41–52 Nm).
9. With the engine raised enough to install the mount, position the mount in the vehicle with the mount shield installed to the engine bracket.
10. Lower the engine into position, then install the bolts attaching the mount to the chassis and the engine. Torque all the bolts to 23–31 ft. lbs. (31–42 Nm).
11. Remove the support from the engine.
12. Lower the vehicle and install the upper fan shroud.
13. Connect the negative battery cable.

1995–97 Pick-Up With 2.4L Engine

1. Disconnect negative battery cable.
2. Remove the upper fan shroud.
3. Raise and support the vehicle.

——— **WARNING** ———
Make certain the vehicle is properly supported when raising the engine.

4. Raise the engine slightly to take the tension off the motor engine mounts.

——— **WARNING** ———
When raising the engine to replace the engine mounts, DO NOT place the jack directly under the oil pan or the crankshaft torsional damper. Use a block of wood between engine and jack to prevent damage at jacking point.

5. Remove the nut attaching the mount to the engine bracket, then remove the nuts and bolts attaching the mount to the chassis.
6. Raise the engine enough to remove the engine mount.

NOTE: Only raise the engine enough to remove the mounts. Raising the engine too far could result in damage to some engine components.

To install:

7. With the engine raised enough to install the mount, position the mount in the vehicle.
8. Lower the engine into position, then install the nut attaching the mount to the engine. Do not tighten the nut at this time.
9. Install the nuts and bolts attaching the mount to the chassis and torque the nuts and bolts to 23–31 ft. lbs. (31–42 Nm).
10. Torque the nut attaching the mount to the engine bracket to 23–31 ft. lbs. (31–42 Nm).
11. Remove the support from the engine.
12. Lower the vehicle and install the upper fan shroud.
13. Connect the negative battery cable.

1996–97 Pathfinder

1. Disconnect negative battery cable.
2. Remove the upper fan shroud.
3. Raise and support the vehicle.

——— **WARNING** ———
Make certain the vehicle is properly supported when raising the engine.

4. Raise the engine slightly to take the tension off the engine mounts.

——— **WARNING** ———
When raising the engine to replace the engine mounts, DO NOT place the jack directly under the oil pan or the crankshaft torsional damper. Use a block of wood between engine and jack to prevent damage at jacking point.

5. Remove the nut attaching the mount to the engine bracket, then the nuts attaching the mount to the crossmember.
6. Raise the engine enough to remove the engine mount.

NOTE: Only raise the engine enough to remove the mounts. Raising the engine too far could result in damage to some engine components.

NOTE: The vehicle is equipped with hydraulic mounts. When removing these engine mounts, do not loosen the 4 mount cover nuts. The mount is fluid filled and will not function properly if the fluid leaks out.

To install:

7. Install the engine mount on the crossmember and torque the nut to 32–41 ft. lbs. (43–55 Nm).
8. Lower the engine into position, then install the nuts attaching the mount to the engine bracket. Torque the nuts to 32–41 ft. lbs. (43–55 Nm).
9. Remove the support from the engine.
10. If removed, install the engine undercover.
11. Lower the vehicle and install the upper fan shroud.
12. Connect the negative battery cable.

Cylinder Head

REMOVAL AND INSTALLATION

2.4L Engine

——— **CAUTION** ———
Fuel injection systems remain under pressure after the engine has been turned OFF. Properly relieve fuel pressure before disconnecting any fuel lines. Failure to do so may result in fire or personal injury.

NOTE: After completing this procedure, allow the rocker cover-to-cylinder head rubber plugs to dry for 30 minutes before starting the engine. This will allow the liquid gasket sealer to cure properly.

1. Release the fuel system pressure.
2. Disconnect the negative battery cable.
3. Drain the cooling system into a sealable container.
4. Remove the air cleaner assembly.
5. Remove the power steering drive belt, power steering pump, idler pulley, and the power steering pump brackets.
6. Tag and disconnect all vacuum hoses, water hoses, fuel tubes, and wiring harnesses necessary to gain access to cylinder head.
7. Detach the accelerator bracket. If necessary mark the position and remove the accelerator cable wire end from the throttle drum.
8. Tag and disconnect the high tension wires from the spark plugs on the exhaust side of the engine.
9. Disconnect the oxygen sensor electrical connector then, remove the exhaust manifold cover.
10. Disconnect the EGR tube from the exhaust manifold.

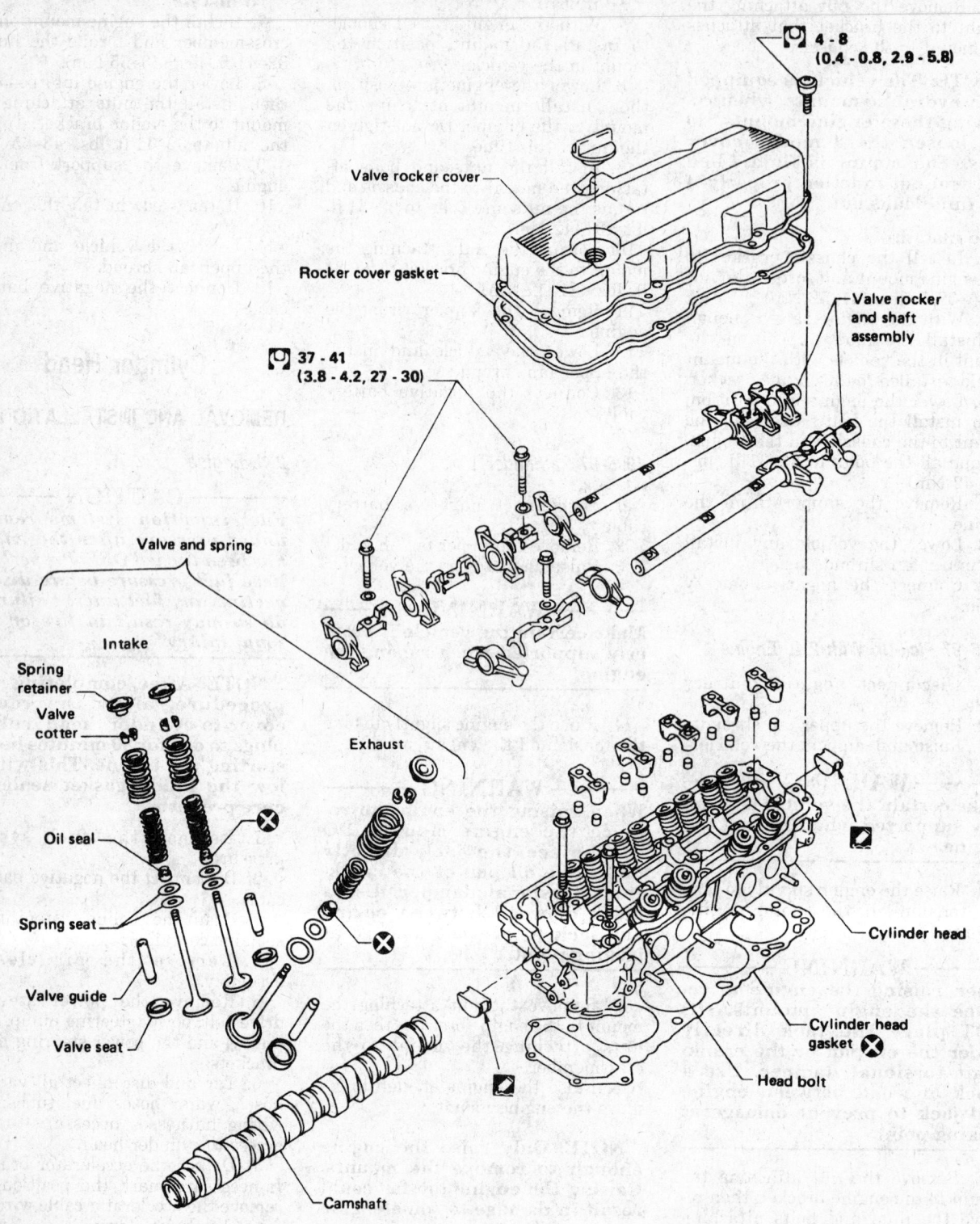

4 - 8
(0.4 - 0.8, 2.9 - 5.8)

Valve rocker cover

Rocker cover gasket

Valve rocker
and shaft
assembly

37 - 41
(3.8 - 4.2, 27 - 30)

Valve and spring

Intake

Spring
retainer

Valve
cotter

Exhaust

Oil seal

Spring seat

Valve guide

Valve seat

Camshaft

Cylinder head

Cylinder head
gasket

Head bolt

: N·m (kg-m, ft-lb)

300167

Exploded view of the cylinder head assembly — 2.4L engine

11. Remove the exhaust manifold mounting nuts and bolts, then remove the manifold from the cylinder head. Discard the gaskets.

12. Remove the intake manifold.

13. Remove the rocker cover. If cover sticks to the cylinder head, tap it with a rubber hammer. Be careful not to strike the rocker arms when removing the rocker arm cover.

14. Remove the spark plugs to protect them from damage.

15. Set No. 1 cylinder piston at TDC on its compression stroke. The No. 1 will be at TDC when the timing pointer is aligned with the red timing mark on the crankshaft pulley.

16. Mark the relationship of the camshaft sprocket to the timing chain with paint or chalk. If this is done, it will not be necessary to locate the factory timing marks. Before removing the camshaft sprocket, it will be necessary to wedge the chain in place so that it will not fall down into the front cover, using special tool KV10105800 or equivalent.

17. Remove the camshaft sprocket bolt and carefully remove the camshaft sprocket.

18. Remove the bolts securing the cylinder head to the front cover assembly.

19. Remove the cylinder head bolts in the correct sequence. Lift the cylinder head off the engine block. It may be necessary to tap the head lightly with a rubber mallet to loosen it.

NOTE: The cylinder head bolts should be loosened in two or three steps, in the correct order to prevent head warpage or cracking.

To install:

20. Thoroughly clean the cylinder block and head surfaces and check both for warpage.

21. Fit the new head gasket. Don't use sealant. Make sure that no open valves are in the way of raised pis-

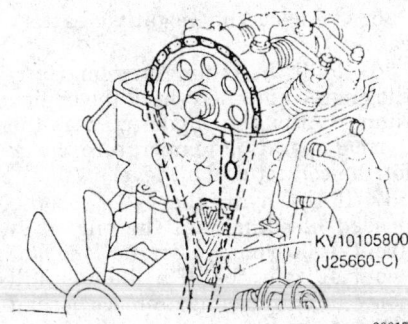

KV10105800 (J25660-C)

300171

Timing chain support installed

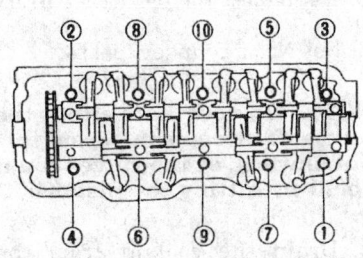

300169

Sequence for cylinder head bolt removal — 2.4L engine

tons, and **never** rotate the crankshaft or camshaft separately because of possible damage which might occur to the valves and/or pistons.

22. Confirm that the No. 1 piston is at TDC on its compression stroke as follows: Align timing mark with the red (0 degree) mark on the crankshaft pulley. Make sure the distributor rotor head is set at No. 1 on the distributor cap. Confirm that the knock pin on the camshaft is set at the top position.

23. Install the cylinder head and torque the head bolts in sequence using the following 5 step procedure:

   a. Torque all the bolts to 22 ft. lbs. (29 Nm).

   b. Torque all the bolts to 58 ft. lbs. (78 Nm).

   c. Loosen all the bolts completely.

   d. Torque all the bolts to 22 ft. lbs. (29 Nm).

   e. Torque all the bolts to 54–61 ft. lbs. (74–83 Nm), or if an angle wrench is used, turn all bolts 80–85 degrees clockwise.

────── **WARNING** ──────
Do not rotate crankshaft and camshaft separately, or valves will hit the tops of the pistons.

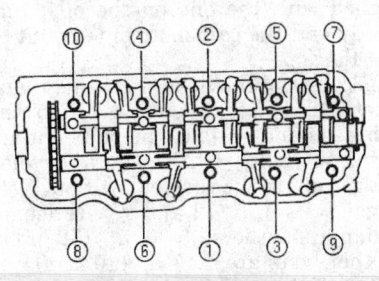

300168

Cylinder head torque sequence — 2.4L engine

24. Install the bolts securing the cylinder head to the front cover assembly.

25. Position the timing chain on the camshaft sprocket by aligning each matchmark. Install the camshaft sprocket to the camshaft, then remove the wedge from the timing chain.

26. Hold the camshaft sprocket stationary, and tighten the sprocket bolt to 87–116 ft. lbs. (118–157 Nm).

27. Install the spark plugs and torque them to 14–22 ft. lbs. (20–29 Nm).

28. Install the intake manifold with new gaskets.

29. Install the exhaust manifold onto the engine with new gaskets.Tighten the nuts/bolts working from the center out, torque the nuts/bolts to 12–15 ft. lbs. (16–21 Nm).

30. Install the EGR tube to the exhaust manifold.

31. Install the exhaust manifold cover and torque the bolts to 3–4 ft. lbs. (4–5 Nm), then connect the oxygen sensor electrical connector.

32. Apply liquid gasket to the rubber plugs and install the rubber plugs in the correct location in the cylinder head. The seating surface of the rubber plugs must be clean and dry. The rubber plugs should be installed within 5 minutes of the sealant application. After the sealant is applied and the rubber plugs are in place, rock the plugs back and forth a few times to distribute the sealant evenly. Wipe the excess sealant from the cylinder head with a clean rag.

33. Check each valve lifter in its free position by forcefully pushing it with your finger. If the lifter moves more than 0.04 in. (1mm) air may be in it.

34. Position the rocker cover and install two of the attaching bolts. Torque the two bolts to 2 ft. lbs. (3 Nm).

35. Torque the rocker cover bolts in the proper sequence to 5–8 ft. lbs. (7–11 Nm).

36. Attach the accelerator bracket and cable.

37. Connect all the vacuum hoses, water hoses, fuel tubes, and electrical connections that were removed to gain access to cylinder head.

38. Install the spark plugs wires in the correct location.

39. Install the power steering bracket and torque the bracket attaching bolts to 16–22 ft. lbs. (22–29 Nm).

40. Install the idler pulley and the power steering pump.

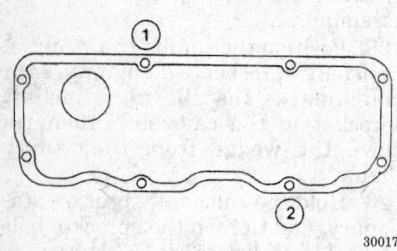

Initially install the two indicated rocker cover bolts — 2.4L engine

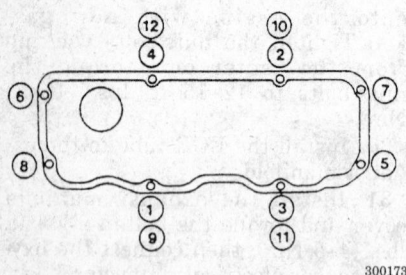

Rocker cover torque sequence — 2.4L engine

41. Install and adjust the drive belts.

42. Drain the engine oil into a sealable container then refill the engine with fresh oil.

43. Fill and bleed the cooling system.

44. Install the air cleaner assembly.

45. Connect the negative battery cable.

46. Run the engine at 1000 rpm with no load for approximately 20 minutes to bleed the air from the valve lifters. If the lifters continue to make noise they need to be replaced.

47. Check the engine for any leaks.

48. Check and/or adjust the ignition timing.

3.0L and 3.3L Engines

1. Release the fuel pressure.

──────── **CAUTION** ────────

Fuel injection systems remain under pressure after the engine has been turned OFF. Properly relieve fuel pressure before disconnecting any fuel lines. Failure to do so may result in fire or personal injury.

2. Disconnect the negative battery cable.

3. Set No. 1 cylinder to TDC.

NOTE: Do not rotate either the crankshaft or camshaft from this point onward, or the valves could be bent by hitting the pistons.

4. Drain the coolant from the engine.

5. Remove the air duct hose.

6. Tag and separate the ASCD and accelerator control wire from the intake manifold collector.

7. Remove the intake collector and intake manifold assembly.

8. Remove the timing belt covers and the timing belt.

9. Remove the camshaft pulleys and the rear timing belt cover.

10. Tag then disconnect the ignition wires from the spark plugs. Disconnect the ignition wire from the ignition coil.

11. Match mark the distributor position, then remove the distributor attaching bolt and the distributor.

12. Remove the harness clamp from the right side cylinder head cover.

13. Unbolt the forward exhaust pipe at the manifold and move it out of the way.

14. Remove the drive belts and the A/C compressor and alternator. Remove the five mounting bolts and then remove the compressor bracket.

15. Remove the cylinder head covers.

16. Loosen the cylinder head bolts, following proper sequence, in three steps.

17. Remove the cylinder head with the exhaust manifold attached.

18. Remove the exhaust manifold from the cylinder head as necessary.

To install:

19. Install the exhaust manifold to the cylinder head, if removed.

20. Check the positions of the timing marks and camshaft sprockets to make sure they have not shifted. The mark on the crankshaft should be aligned with the one on the oil pump body and the camshaft pin should be at the top.

21. Install the head with a new gasket. Apply clean engine oil to the threads and seats of the bolts and install the bolts with washers (beveled edges up) in the correct position. Note that bolts 4, 5, 12 and 13 are longer than the others, 5.00 in. (127mm). Other bolts are 4.17 in. (10.6cm).

22. Tighten the bolts in the proper sequence, in the following stages:

  a. Tighten all bolts, in order, to 22 ft. lbs. (29 Nm)

  b. Tighten all bolts, in order, to 43 ft. lbs. (59 Nm)

  c. Loosen all bolts completely.

  d. Tighten all bolts, in order, to 22 ft. lbs. (29 Nm)

  e. Turn all bolts, in order, 60–65 degrees clockwise. If an angle torque wrench is not available, torque the bolts in order to 40–47 ft. lbs. (54–64 Nm).

23. Check the hydraulic valve lifter(s) by pushing the plunger forcefully with your finger (be sure that the rocker arm is in the free position, NOT on the lobe). If the lifter moves more than 0.04 in. (1mm), it must be bled, as described later.

24. Install the compressor bracket and then install the A/C compressor and alternator.

25. Connect the forward exhaust pipe to the manifold.

26. Install the rear timing belt cover and then install the camshaft pulleys. Make sure the pulley marked **R3** goes on the right and that marked **L3** goes on the left. Torque the camshaft sprockets attaching bolts to 58–65 ft. lbs. (78–88 Nm).

27. Align the timing marks if necessary and then install the timing belt and adjust the belt tension.

28. Install the distributor by aligning the mark on the distributor shaft with the protruding mark on the housing. Confirm that the ignition rotor is pointing toward No. 1 ignition wire on the distributor cap.

29. Install the intake manifold and intake collector assembly.

30. Install the ignition wires to the correct spark plugs and the ignition coil.

31. Reconnect the ASCD and accelerator control wire.

32. Connect all the vacuum hoses and water hoses to the intake collector.

33. Install the air duct hose.

34. Drain the engine oil into a sealable container then refill the engine with new oil.

35. Connect the negative battery cable.

36. Refill the cooling system. Start the engine and then check the engine timing. After the engine reaches the normal operating temperature, check for the correct coolant level.

37. If the hydraulic valve lifter(s) needed bleeding, run the engine at about 1000 rpm, under no load, for about 10 minutes. If a lifter is still noisy after bleeding, replace it and bleed it again.

38. Road test the vehicle for proper operation.

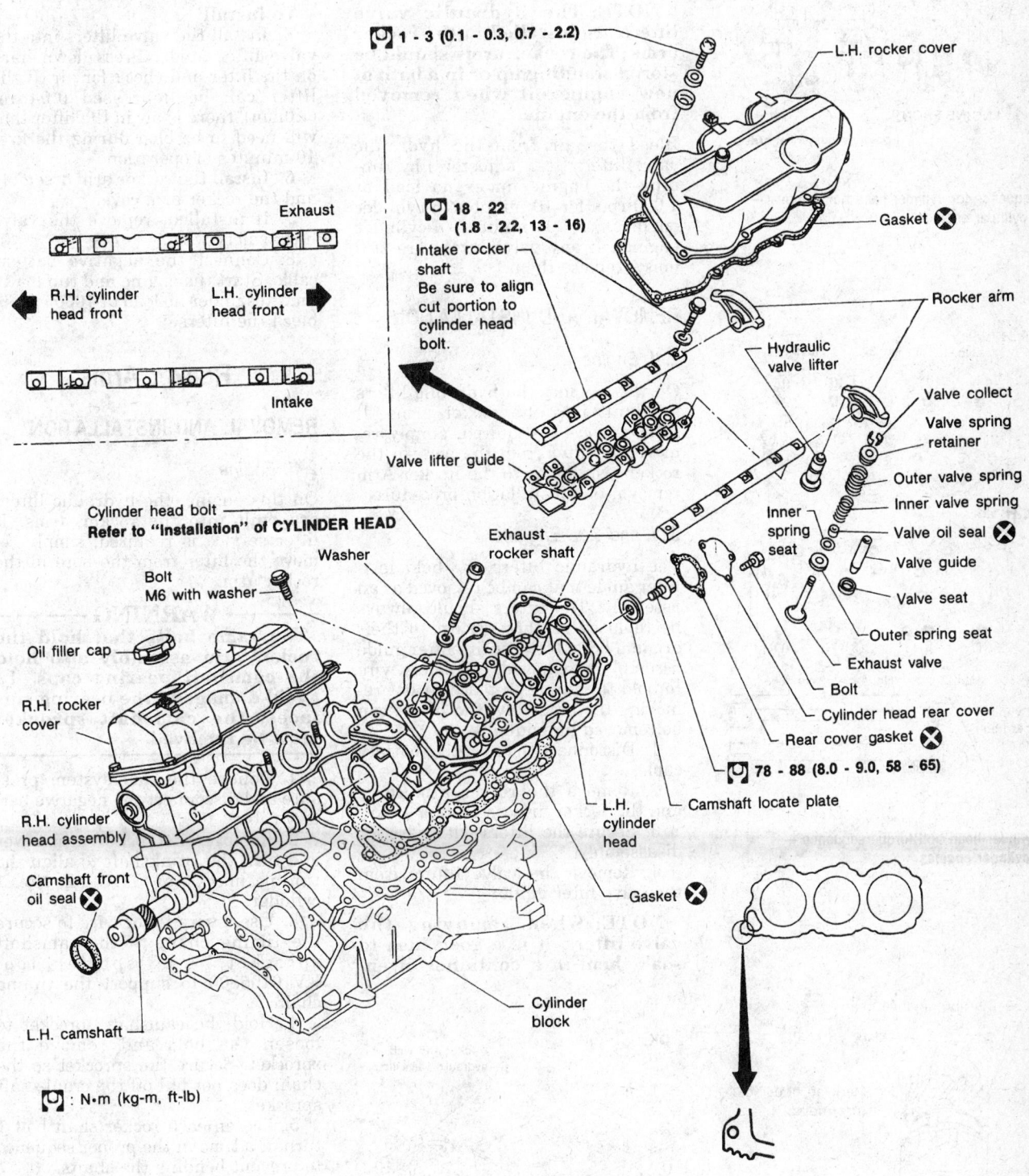

1 - 3 (0.1 - 0.3, 0.7 - 2.2)

L.H. rocker cover

Gasket

18 - 22
(1.8 - 2.2, 13 - 16)

Intake rocker
shaft
Be sure to align
cut portion to
cylinder head
bolt.

Rocker arm

Hydraulic
valve lifter

Valve collect

Valve spring
retainer

Outer valve spring

Inner valve spring

Valve oil seal

Valve lifter guide

Valve guide

Inner
spring
seat

Valve seat

Exhaust
rocker shaft

Outer spring seat

Exhaust valve

Exhaust

R.H. cylinder
head front

L.H. cylinder
head front

Intake

Cylinder head bolt
Refer to "Installation" of CYLINDER HEAD

Washer

Bolt
M6 with washer

Oil filler cap

Bolt

Cylinder head rear cover

Rear cover gasket

78 - 88 (8.0 - 9.0, 58 - 65)

R.H. rocker
cover

Camshaft locate plate

R.H. cylinder
head assembly

L.H.
cylinder
head

Camshaft front
oil seal

Gasket

L.H. camshaft

Cylinder
block

[] : N·m (kg-m, ft-lb)

Exploded view of the cylinder head assembly — 6-cylinder engines

311384

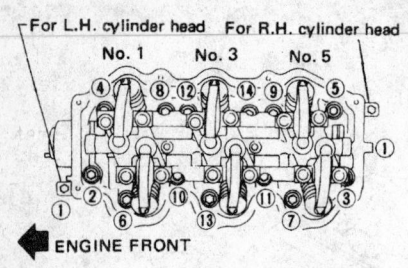

For L.H. cylinder head   For R.H. cylinder head

◀ ENGINE FRONT

Loosen in numerical order.

311386

Sequence for cylinder head bolt removal — 6-cylinder engines

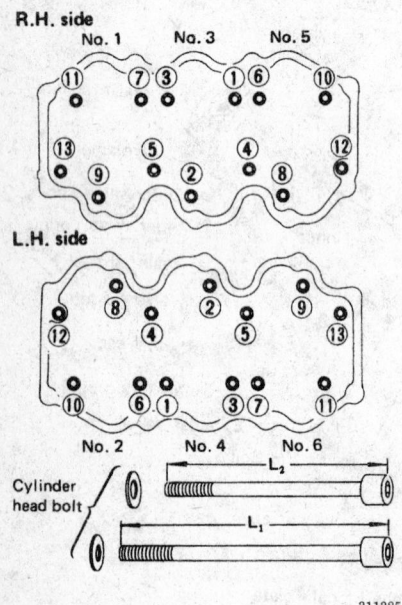

R.H. side

L.H. side

Cylinder head bolt

311385

Cylinder head bolt torque sequence — 6-cylinder engines

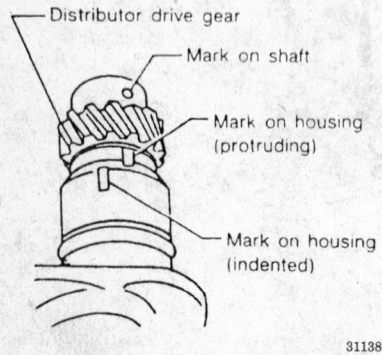

- Distributor drive gear
- Mark on shaft
- Mark on housing (protruding)
- Mark on housing (indented)

311387

Distributor marks — 6-cylinder engines

Valve Lifters

BLEEDING

NOTE: The hydraulic valve lifters are installed in the rocker arms. The rocker arms should be stored standing up or in a bath of new engine oil when removed from the engine.

Bleed the air from the hydraulic valve lifters (lash adjusters) by running the engine under no load at 1,000 rpm for 10 minutes (6-cylinder engines) or 20 minutes (4-cylinder engine). If any valve lifters are still noisy, replace them.

REMOVAL AND INSTALLATION

2.4L Engine

On this engine, the hydraulic lifters are built into the rocker arms. If lifter service is required, simply remove the lifter from the bore in the rocker arm. Refer to the Rocker Arm removal and installation procedure.

3.0L and 3.3L Engines

The hydraulic lifters are held in a lifter guide that can be removed as an assembly. The lifters should always be replaced so they return to their original bore. If the entire lifter guide assembly is removed, use safety wire to hold the lifters in place. If not removing the assembly, the lifters can be removed individually.

1. Disconnect the negative battery cable.
2. Remove the rocker arm cover and the rocker arm assembly.
3. Secure the lifters with wire for disassembly.
4. Remove the valve lifters from the valve lifter guide.

NOTE: When removing the valve lifters, it is a good idea to soak them in a container of en-

OK

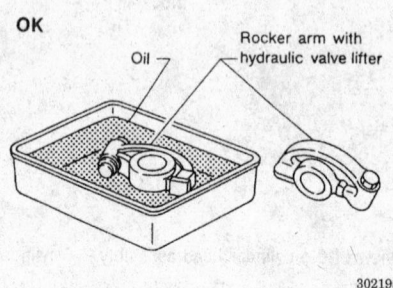

Oil

Rocker arm with hydraulic valve lifter

302195

Proper storage of hydraulic valve lifters

gine oil; this will keep air from entering the lifters. Do not invert the lifters as air can enter them in this manner.

To install:

5. Install the valve lifters into the valve lifter guide. Press down hard on the lifter and check for air. If the lifter can be depressed 0.04 in. (1.0mm), there is air in the lifter that will need to be bled during the first 10 minutes of operation.
6. Install the rocker arm assembly and the rocker arm cover.
7. If installed, remove the valve lifter safety wire.
8. Connect the negative battery cable. Start the engine and run for 10 minutes under no load at 1000 rpm to bleed the lifters.

Rocker Arms

REMOVAL AND INSTALLATION

2.4L Engine

On this engine, the hydraulic lifters are built into the rocker arms. If lifter service is required, simply remove the lifter from the bore in the rocker arm.

——— WARNING ———

The same bolts that hold the rocker arm assembly also hold the camshaft bearing caps. To avoid damage to the bearing surfaces, the camshaft sprocket must be removed.

1. Relieve the fuel system pressure and disconnect the negative battery cable.
2. Remove the rocker arm cover and turn the crankshaft to align the timing marks at TDC on No. 1 cylinder.
3. Use a wire tie or wire to secure the timing chain to the camshaft sprocket. Use special tool KV10105800 to support the timing chain.
4. Hold the camshaft sprocket to loosen the bolt and remove the sprocket. Secure the sprocket so the chain does not fall off the crankshaft sprocket.
5. Loosen each rocker shaft bolt 1 turn at a time in the proper sequence to prevent bending the shafts.
6. When all the bolts are loose, remove the rocker arm shafts with the bolts still in the shafts. This will hold the assembly together.
7. If the rocker arms are to be removed from the shafts, mark them so they can be returned into their origi-

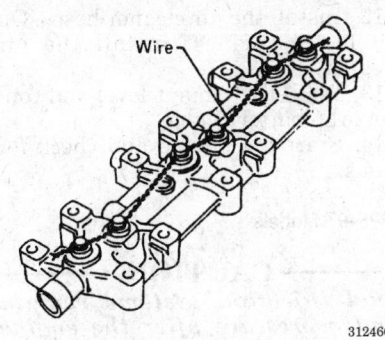

Securing hydraulic valve lifters with wire — 6-cylinder engines

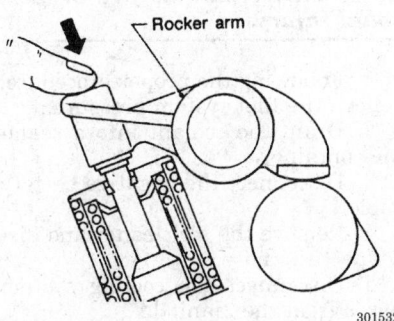

Checking hydraulic lifter operation — 2.4L engine

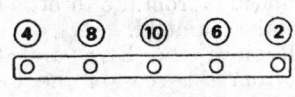

Rocker arm shaft bolt loosening sequence — 2.4L engine

nal position. Remove the bolts from the shaft assembly and remove the parts.

NOTE: Do not allow the arms to lay on their side or they will become air bound. Keep the rocker arms upright or lay them in a pan of new engine oil.

To install:

8. Lubricate the shafts with engine oil and assemble them with the punch marks facing up. Use the bolts to hold the assembly together. Make sure the camshaft and the bearing surfaces are in good condition and lu-

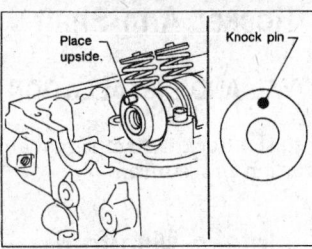

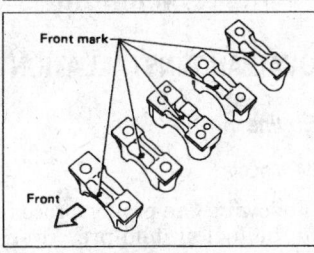

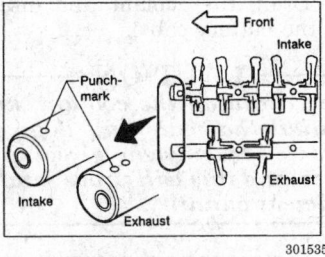

Rocker arm shaft identification marks — 2.4L engine

bricate with engine oil. Make sure the pin on the camshaft sprocket end is up.

NOTE: Punch marks on the front of each shaft that tell which shaft is for the intake side and which is for the exhaust side. This is important for correct rocker arm oiling.

9. Install the rocker arm shafts and tighten the bolts in the proper sequence 1 turn at a time to draw the shafts down evenly against the valve springs without bending the shafts. Tighten the bolts in reverse order of the loosening sequence, tighten bolts from the inside out. Torque the bolts to 27–30 ft. lbs. (37–41 Nm).

10. Install the camshaft sprocket and remove the tie securing the chain. Install the sprocket bolt but don't torque it yet. Rotate the crankshaft 2 full turns to make sure the timing marks line up. When the valve timing is correct, torque the sprocket bolt to 87–116 ft. lbs. (118–157 Nm).

11. Use a silicone sealer on the rubber end plugs and install the rocker arm cover with a new gasket. Install the two indicated attaching bolts and torque them to 2 ft. lbs. (3 Nm).

12. Torque the rocker cover bolts in the proper sequence to 5–8 ft. lbs. (7–11 Nm).

13. Connect the negative battery cable.

14. When the engine is first started, the hydraulic valve lifters may be noisy. Run the engine for 20 minutes at about 1000 rpm. If the noise has not subsided, the lifter will probably never pump up and must be replaced.

3.0L and 3.3L Engines

1. Disconnect the negative battery cable.

2. Align the timing marks to bring No. 1 cylinder to TDC.

3. Remove the valve covers.

4. Loosen the rocker shaft bolts in two or three stages, then remove the rocker shafts with rocker arms, as an assembly.

5. Separate the rocker arms from the shaft.

NOTE: When separating the rocker arms from the rocker arm shafts, be sure to keep the parts in order for reinstallation purposes.

6. Check the rocker arms and the shafts for damage. If necessary, replace the damaged components.

7. Attach a wire to the top of the lifters so that they will not drop from the lifter guide. Carefully remove the lifter guide and lifters from the cylinder head. Put an identification mark on the lifters to avoid mixing them up if they are removed from the guide and to be reused. If the lifters are damaged replace them as necessary.

To install:

8. Install new lifters if replacing them or install the old lifters to their original locations.

——— CAUTION ———
When installing the rocker arm shafts, be certain that they are installed in their original positions.

9. Slide the rocker arms onto the shafts in their proper positions.

10. Make sure that cylinder No. 1 is at TDC.

11. Install the left cylinder head valve lifter guide assembly and remove the wire from the lifters. Install the rocker arm shaft assemblies and attaching bolts, coat the bolt threads and seat surfaces before installing them. Tighten the bolts gradually in three steps to 13–16 ft. lbs. (18–22 Nm).

12. Rotate the crankshaft clockwise 180°, to bring cylinder No. 4 to TDC. Install the right cylinder head valve

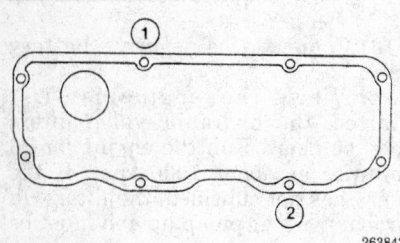

Initially install the two indicated rocker cover bolts

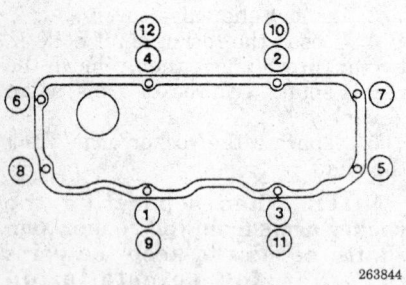

Rocker cover bolt torque sequence

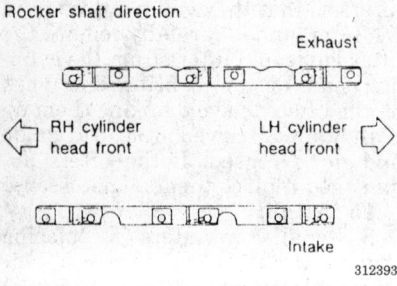

Rocker arm shafts — V6 engines

lifter guide assembly and remove the wire from the lifters. Install the rocker arm shaft assemblies and attaching bolts, coat the bolt threads and seat surfaces before installing them. Tighten the bolts gradually in three steps to 13–16 ft. lbs. (18–22 Nm).

13. Install the rocker covers with new gaskets, torque the rocker cover bolts to 9–26 inch lbs. (1–3 Nm).

14. Install the valve cover.

15. Connect the negative battery cable.

Rocker Arm Shaft

REMOVAL AND INSTALLATION

Refer to the Rocker Arm removal and installation procedure.

Intake Manifold

REMOVAL AND INSTALLATION

2.4L Engine

1993–94 Models

1. Following the proper procedure, relieve the fuel system pressure.

2. Drain the coolant and disconnect the battery cable.

—————— CAUTION ——————
When draining the coolant, keep in mind that cats and dogs are attracted by ethylene glycol antifreeze, and this will prove fatal in sufficient quantity.

3. Remove the air cleaner hoses.

4. Remove the radiator hoses from the manifold.

5. Remove the throttle cable and disconnect the fuel pipe and the return fuel line. Plug the fuel pipe to prevent spilling fuel.

NOTE: When unplugging wires and hoses, mark each hose and its connection with a piece of masking tape, then mark the two pieces of tape with the numbers 1, 2, 3, etc. When assembling, simply match the pieces of tape.

6. Remove all remaining wires, tubes and the EGR and PCV tubes from the rear of the intake manifold. Remove the manifold supports.

7. Unbolt and remove the intake manifold. Remove the manifold with injectors, EGR valve, fuel tubes, etc. still attached.

To install:

8. Clean the gasket mounting surfaces then install the intake manifold on the engine. Always use a new intake manifold gasket. Tighten the mounting bolts from the center out, to 12–15 ft. lbs. (16–21 Nm).

9. Connect all electrical connections, tubes and the EGR and PCV tubes to the rear of the intake manifold. Install the manifold supports.

10. Install the throttle cable and reconnect the fuel pipe and the return fuel line.

11. Install the radiator hoses to the intake manifold.

12. Install the air cleaner hoses. On all other engines, install the air cleaner.

13. Refill the coolant level and connect the battery cable.

14. Start the engine and check for leaks.

1995–97 Models

—————— CAUTION ——————
Fuel injection systems remain under pressure after the engine has been turned OFF. Properly relieve fuel pressure before disconnecting any fuel lines. Failure to do so may result in fire or personal injury.

1. Following the proper procedure, relieve the fuel system pressure.

2. Drain the coolant into a sealable container.

3. Disconnect the negative battery cable.

4. Remove the air cleaner and disconnect the hoses.

5. Disconnect the cooling system hoses from the manifold.

6. Remove the throttle cable and disconnect the fuel feed and return fuel lines. Plug the fuel lines to prevent spilling fuel.

7. Tag, then disconnect the electrical connectors from the throttle body and intake manifold.

8. Remove the EGR and PCV tubes from the rear of the intake manifold.

9. Remove the intake manifold stay (support bracket).

10. Unbolt and remove the intake manifold. Remove the manifold with the fuel injectors, EGR valve, and the throttle body still attached.

To install:

11. Clean the gasket mounting surfaces then install the intake manifold on the engine with a new intake manifold gasket. Tighten the mounting bolts following the proper sequence to 12–15 ft. lbs. (16–21 Nm). The bolts should be torqued from the center towards the ends.

12. Install the intake manifold stay.

13. Connect the EGR and PCV tubes to the rear of the intake manifold.

14. Connect the electrical connections to the throttle body and intake manifold.

15. Connect the fuel feed and return lines.

16. Connect the throttle cable to the throttle body.

17. Connect the cooling system hoses to the intake manifold.

18. Install the air cleaner and the air cleaner hoses.

19. Fill and bleed the air from the cooling system.

20. Connect the negative battery cable.

21. Start the engine and check for leaks.

3.0L and 3.3L Engines

1. Release the fuel system pressure and disconnect the battery cables.

— **CAUTION** —

Fuel injection systems remain under pressure after the engine has been turned OFF. Properly relieve fuel pressure before disconnecting any fuel lines. Failure to do so may result in fire or personal injury.

2. Drain the engine coolant into a sealable container.

3. Remove the air duct hose.

4. Tag then disconnect the spark plug wires from the spark plugs.

5. Disconnect the ASCD and the accelerator control wire from the throttle body.

6. Disconnect all the electrical connectors and the ground wire from the intake manifold and the collector (intake manifold plenum).

7. Disconnect the coolant hoses from the intake manifold and collector.

8. Remove the PCV hose from the right rocker cover.

9. Tag then disconnect the vacuum hoses for the canister, power brake booster, and the fuel pressure regulator.

10. Disconnect the purge hose from the canister.

11. Disconnect the EGR tube from the collector.

12. Remove the collector attaching bolts.

13. Disconnect the fuel feed and fuel return lines from the injector fuel tube assembly.

— **CAUTION** —

Do not allow fuel spray or fuel vapors to come in contact with a spark or open flame. Keep a dry chemical fire extinguisher nearby. Never store fuel in an open container due to risk of fire or explosion.

14. Disconnect all the fuel injector harness connectors.

15. Remove the injector fuel tube assembly.

16. Tag and disconnect the engine temperature switch harness connector and the thermal transmitter harness connector. Remove the coolant hose at the thermostat housing.

17. Remove the intake manifold. Loosen the intake manifold bolts in the proper sequence.

To install:

18. Install the intake manifold and a new gasket to the engine. Tighten the manifold bolts and nuts in two stages until reaching a total torque of 12–14 ft. lbs. (16–20 Nm) on all bolts and 17–20 ft. lbs. (24–27 Nm) on all nuts. On 1996 models, torque the nuts and bolts in sequence to 2.2–3.6 ft. lbs. (3–5 Nm), then torque them to 13–16 ft. lbs. (18–22 Nm).

19. Connect the engine temperature switch harness connector and the thermal transmitter harness connector. Install the coolant line at the thermostat housing.

20. Install the injector fuel tube assembly.

21. Connect all fuel injector harness connectors.

22. Connect the fuel feed and fuel return lines from the injector fuel tube assembly.

23. Install the collector with new gaskets. Tighten collector attaching bolts in two stages, to 13–16 ft. lbs. (18–22 Nm).

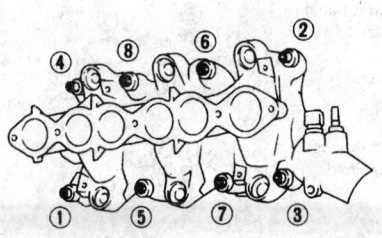

Loosen bolts in numerical order.

309710

Intake manifold bolt removal sequence — V6 engines

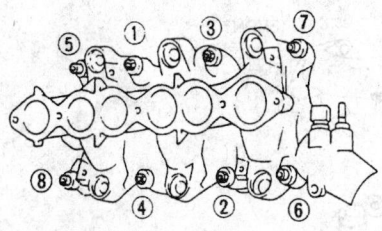

Tighten bolts in numerical order.

309722

Intake manifold bolt installation sequence — V6 engines

24. Connect the EGR tube to the collector.

25. Connect the purge hose to the canister.

26. Connect the vacuum hoses for the canister, power brake booster, and the fuel pressure regulator.

27. Install the PCV hose to the right rocker cover.

28. Connect the coolant hoses to the intake manifold and collector.

29. Connect the electrical connectors and the ground wire to the intake manifold and the collector.

30. Connect and adjust the ASCD and the accelerator control wire to the throttle body.

31. Connect the spark plug wires to the correct spark plugs.

32. Install the air duct hose.

33. Fill and bleed the air from the cooling system.

34. Connect the battery cables.

35. Check the fluid levels, start the engine, and check for leaks.

Exhaust Manifold

REMOVAL AND INSTALLATION

2.4L Engine

1. Disconnect the negative battery cable.

2. Remove the air cleaner assembly. Remove the exhaust manifold heat shield.

3. Tag and disconnect the high tension wires from the spark plugs on the exhaust side of the engine.

4. Disconnect the air induction and/or the EGR tubes from the exhaust manifold.

5. Disconnect the oxygen sensor electrical connector.

6. Disconnect the front exhaust pipe from the exhaust manifold.

NOTE: Soak the exhaust pipe retaining bolts with penetrating oil if necessary to loosen them.

7. Remove the exhaust manifold mounting nuts and then remove the manifold from the cylinder head.

To install:

8. Using a gasket scraper, clean the gasket mounting surfaces.

9. Install the manifold onto the engine with new gaskets. Tighten the nuts/bolts working from the center out, torque the exhaust manifold nuts/bolts to 12–15 ft. lbs. (16–21 Nm).

10. Install the air induction and/or the EGR tubes to the exhaust manifold.

11. Connect the oxygen sensor electrical connector.

12. Connect the exhaust pipe to the manifold.

13. Connect spark plug wires, the air cleaner, and any related hoses.

14. Connect the negative battery cable.

15. Start engine and check for exhaust leaks.

3.0L Engine

1. Disconnect the negative battery cable.

2. Remove the exhaust manifold sub-cover and manifold cover.

3. Remove the EGR tube from the right side exhaust manifold.

4. Remove the exhaust manifold stay.

5. Disconnect the left side exhaust manifold at the exhaust pipe by removing retaining nuts and then disconnect the right side manifold from the connecting pipe.

NOTE: Soak the exhaust pipe retaining bolts with penetrating oil if necessary to loosen them.

6. Remove bolts for each manifold in the order shown.

To install:

7. Clean all gasket surfaces. Install new gaskets.

8. Install the manifold to the engine, tightening the mounting bolts alternately, in two stages, in sequence. Tighten the left side bolts to 13–16 ft. lbs. (18–22 Nm); tighten the right side bolts to 16–20 ft. lbs. (22–27 Nm).

9. Connect the exhaust pipe and the connecting pipe. Be careful not to break these bolts.

10. Install the exhaust manifold stay and the EGR tube to the right side manifold.

11. Install the exhaust manifold covers.

12. Connect the negative battery cable.

13. Start the engine and check for exhaust leaks.

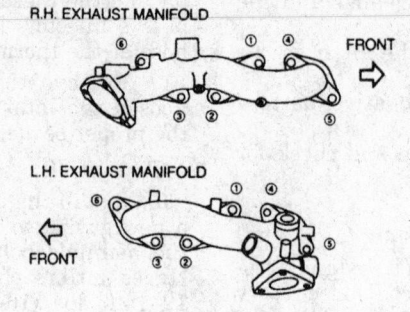

253450

Exhaust manifold bolt tightening sequence — 3.0L engine

3.3L Engine

1. Disconnect the negative battery cable.

2. Remove the exhaust manifold cover.

3. Remove the EGR tube from the left side exhaust manifold.

4. Remove the exhaust manifold stay.

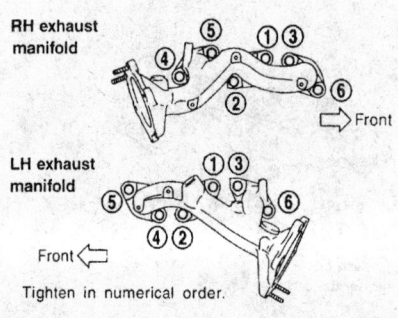

Tighten in numerical order.

307106

Exhaust manifold bolt tightening sequence — 3.3L engine

5. Disconnect the left and right exhaust pipes from the catalytic converters.

NOTE: Soak the exhaust pipe retaining bolts with penetrating oil if necessary to loosen them.

6. Remove the nuts for each manifold. Remove the outer nuts first and the center nuts last.

7. Remove the manifold with the catalytic converter attached.

To install:

8. Clean all gasket surfaces. Install new gaskets.

9. Install the manifold and converter to the engine, tightening the mounting nuts in two stages, in sequence. Tighten the nuts to 21–25 ft. lbs. (28–33 Nm).

10. Connect the exhaust pipe to the converter. Be careful not break these studs.

11. Install the EGR tube to the left side manifold. Torque the tube nut to 29–36 ft. lbs. (39–49 Nm).

12. Install the exhaust manifold covers.

13. Connect the negative battery cable.

14. Start the engine and check for exhaust leaks.

Front Cover Seal

REMOVAL AND INSTALLATION

2.4L Engine

1. Remove the timing chain front cover.

2. Pry the old seal from the cover. Avoid scratching the seal mounting surface.

3. Oil the lip of the new seal. Do not use grease. Press it into place, making sure the flat side faces forward and the lip faces the engine.

4. Install the timing chain front cover.

253449

Exhaust manifold bolt removal sequence — 3.0L engine

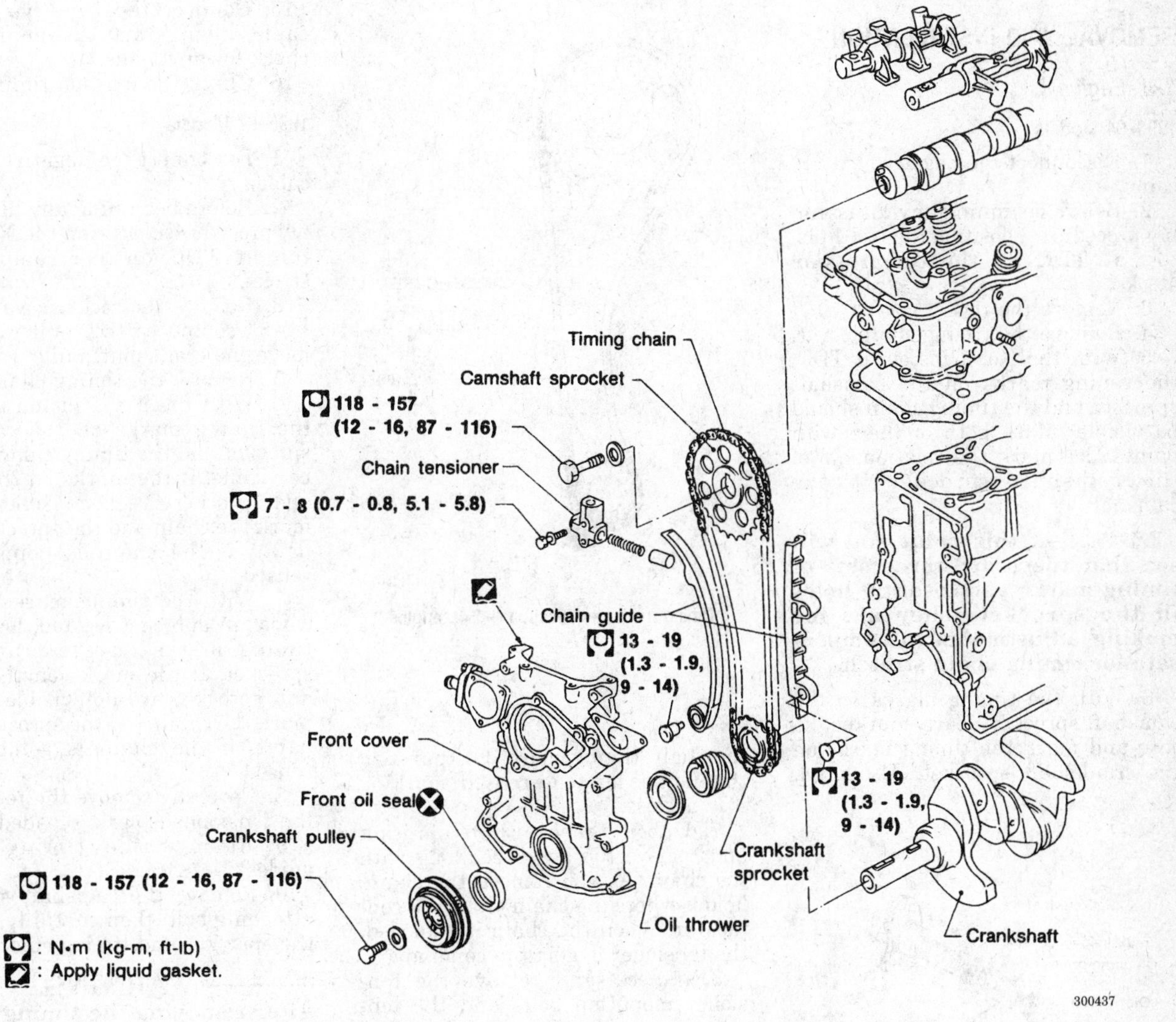

Timing chain

Camshaft sprocket

⌷ 118 - 157
(12 - 16, 87 - 116)

Chain tensioner

⌷ 7 - 8 (0.7 - 0.8, 5.1 - 5.8)

Chain guide

⌷ 13 - 19
(1.3 - 1.9,
9 - 14)

Front cover

Front oil seal ⊗

Crankshaft pulley

⌷ 118 - 157 (12 - 16, 87 - 116)

⌷ 13 - 19
(1.3 - 1.9,
9 - 14)

Crankshaft
sprocket

Oil thrower

Crankshaft

⌷ : N•m (kg-m, ft-lb)
⊘ : Apply liquid gasket.

300437

Timing chain front cover and related components — 2.4L engine

Front Crankshaft Seal

REMOVAL AND INSTALLATION

3.0L and 3.3L Engines

NOTE: The front oil seal is a part of the oil pump body.

1. Disconnect the negative battery cable.
2. Remove the timing belt and the crankshaft sprocket.
3. Remove the oil pump assembly.
4. Remove the oil seal from the oil pump body using a pry tool. Be careful not to damage the oil pump body during seal removal.

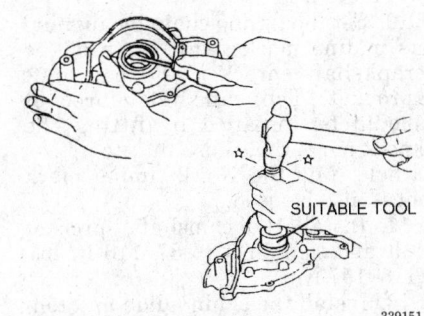

SUITABLE TOOL

339151

Removing and installing the front crankshaft oil seal — 6-cylinder engines

To install:

5. Apply clean engine oil to the new oil seal. Install the seal using proper size driver.
6. Install the oil pump assembly to the engine.
7. Install the remaining components in reverse order of removal.
8. Connect the negative battery cable.

Timing Chain and Sprockets

REMOVAL AND INSTALLATION

2.4L Engine

1993-94 Models

1. Disconnect the negative battery cable.

2. Before beginning any disassembly procedures, position the No. 1 piston at TDC on the compression stroke.

3. Remove the rocker cover.

4. Remove the timing chain cover.

5. With the No. 1 piston at TDC, the timing marks on the camshaft sprocket and the timing chain should be visible. Mark both of them with paint. Also mark the relationship of the camshaft sprocket to the camshaft.

NOTE: At this point you will see that there are three sets of timing marks and locating holes in the sprocket. They are for making adjustments to compensate for timing chain stretch.

6. With the timing marks on the camshaft sprocket clearly marked, locate and mark the timing marks on the crankshaft sprocket. Also mark

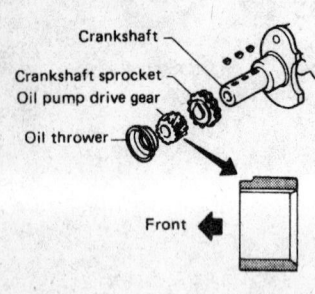

Timing chain sprocket removal and installation — 1993-94 2.4L engine

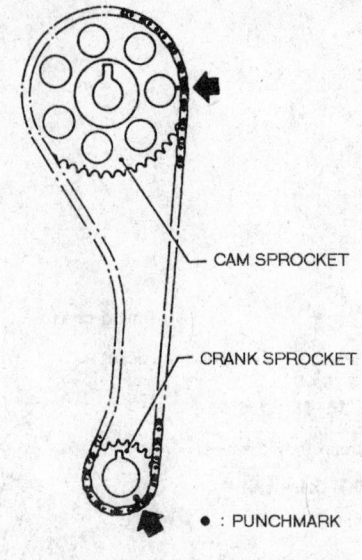

Timing mark identification — 2.4L engine

the chain timing mark. Of course, if the chain is not to be reused, marking it is useless.

7. Unbolt the camshaft sprocket and remove the sprocket along with the chain. As you remove the chain, hold it where the chain tensioner contacts it. When the chain is removed, the tensioner is going to come apart.

8. If necessary, remove the tensioner mounting bolts and the tensioner assembly. Also the crankshaft spocket can be removed, using an appropriate puller.

To install:

9. Install the crankshaft sprocket, if removed.

10. Set the timing chain by aligning its mating marks with those of the crankshaft sprocket and camshaft sprocket. The camshaft sprocket should be installed by fitting the knock pin of the camshaft into its No. 2 hole. And the No. 2 timing mark must also be used.

11. Install the camshaft sprocket bolt and tighten it to 87–116 ft. lbs. (118–157 Nm).

12. Install the chain guide and tensioner. Adjust the protrusion of the chain tensioner spindle to zero clearance. Tighten the bolts to 4–7 ft. lbs. (6–10 Nm).

13. Lightly oil the new seal and install the timing chain cover.

14. Install the rocker cover (as described previously in the Rocker Arm procedure).

15. Connect the negative battery cable, then start the engine and check for any leaks.

16. Check the ignition timing.

1995-97 Models

1. Disconnect the negative battery cable.

2. Before beginning any disassembly procedures, position the No. 1 piston at TDC on the compression stroke.

3. Remove the rocker cover.

4. Use the correct puller and remove the crankshaft pulley.

5. Remove the timing chain cover.

6. With the No. 1 piston at TDC, the timing marks on the camshaft sprocket and the timing chain should be visible. If the marks on the chain and sprocket are not visible match-mark the chain and the sprocket with paint, if the chain is going to be reused.

7. With the timing marks on the camshaft sprocket located, locate the timing marks on the crankshaft sprocket. If the marks on the chain and sprocket are not visible match-mark the chain and the sprocket with paint, if the chain is going to be reused.

8. Carefully remove the tensioner, the tensioner is spring loaded.

9. Remove the timing chain guides.

10. Remove the camshaft sprocket attaching bolt, then carefully remove the sprocket and timing chain.

——— WARNING ———
After removing the timing chain, do not turn the crankshaft or camshaft separately or the pistons will hit the valves.

11. The timing chain sprocket can be removed from the crankshaft after removing the oil thrower and oil pump drive gear. It may be necessary to use a puller to remove the oil pump drive gear and the timing chain sprocket from the crankshaft. Do not lose the keys from the crankshaft when removing the oil thrower, oil pump drive gear, and the timing chain sprocket.

To install:

12. If removed, install the crankshaft sprocket, oil pump drive gear, and the oil thrower.

13. Set the timing chain by aligning its mating marks with those of the crankshaft sprocket first then the camshaft sprocket. If a new chain is being used the chain has links that

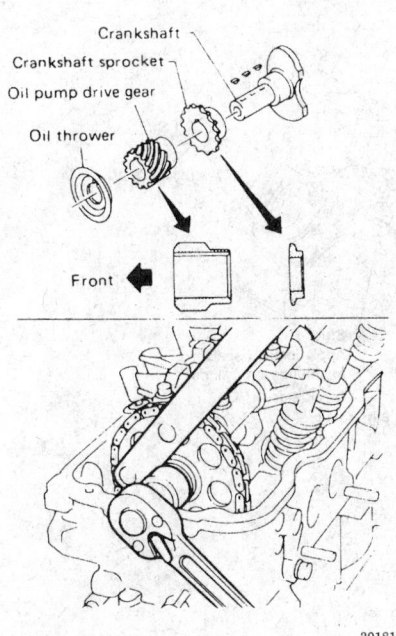

Timing chain sprocket removal and installation — 1995–97 2.4L engine

301815

are marked and should be used for the timing chain alignment. The camshaft sprocket should be installed by fitting the knock pin into the camshaft sprockets hole.

14. Install the camshaft sprocket bolt and tighten it to 101–116 ft. lbs. (137–157 Nm).

15. Install the chain guide and torque the bolts to 9–14 ft. lbs. (13–19 Nm).

16. Install the tensioner guide and torque the bolt to 9–14 ft. lbs. (13–19 Nm). Carefully install the tensioner and torque the mounting bolts to 5–6 ft. lbs. (7–8 Nm).

17. Lightly oil the new seal and install the timing chain cover.

18. Install the crankshaft pulley. Torque the bolt to 105–112 ft. lbs. (142–152 Nm).

19. Install the rocker cover (as described previously in the Rocker Arm procedure).

20. Connect the negative battery cable, then start the engine and check for any leaks.

21. Check the ignition timing.

Timing Chain Front Cover

REMOVAL AND INSTALLATION

2.4L Engine

1. Disconnect the negative battery cable from the battery.

2. Drain the cooling system into a sealable container, then remove the radiator together with the upper and lower radiator hoses.

3. Remove the cooling fan.

4. Remove the drive belts.

5. Remove all of the spark plugs and then set the No. 1 cylinder to TDC of its compression stroke.

6. Remove the power steering pump and mounting brackets from the engine.

7. Remove the A/C compressor idler pulley.

8. Remove the crankshaft pulley bolt and then remove the crankshaft pulley.

9. Remove the distributor.

10. Remove the oil pump attaching screws, and take out the pump and its drive spindle.

11. Remove the rocker cover.

12. Remove the oil pan.

13. Remove the bolts holding the front cover to the front of the cylinder block, then carefully pry the front cover off the front of the engine.

To install:

14. Apply sealer to all of the gaskets and position them on the engine in their proper places.

15. Apply a light coating of oil to the crankshaft oil seal and carefully mount the front cover to the front of the engine and install all of the mounting bolts. Torque the cover attaching bolts to 5–6 ft. lbs. (7–8 Nm).

—— **WARNING** ——
When installing the cover be careful to not damage the head gasket.

16. Install the oil pan.

17. Install the rocker cover (as described previously in the Rocker Arm procedure).

18. Before installing the oil pump, place the gasket over the shaft and make sure that the mark on the drive spindle faces (aligned with) the oil pump hole. Install the oil pump so that the projection on the top of the shaft is located in the exact position as when it was removed (or pointing just beyond the 11 o'clock and 5 o'clock positions when the piston in the No. 1 cylinder is placed at TDC on the compression stroke, if the engine was disturbed since disassem-

bly). Tighten the oil pump attaching screws to 8–10 ft. lbs. (11–15 Nm).

19. Install the distributor in the correct position.

20. Install the crankshaft pulley and bolt. Tighten the bolt to 87–116 ft. lbs. (118–157 Nm).

21. Install the A/C compressor idler pulley.

22. Install the power steering mounting brackets and the pump.

23. Install the fan pulley and cooling fan.

24. Install and adjust the drive belts.

25. Install the spark plugs and torque them to 14–22 ft. lbs. (20–29 Nm).

26. Install the radiator; reconnect the upper and lower radiator hoses.

27. Fill and bleed the air from the cooling system.

28. Connect the negative battery cable.

29. Start the engine, check ignition timing, and check for leaks.

Timing Belt and Tensioner

REMOVAL AND INSTALLATION

3.0L and 3.3L Engines

1. Disconnect the negative battery cable.

2. Remove the engine under cover.

3. Remove the radiator shroud, fan and water pump pulley.

4. Drain the coolant from the radiator into a container and remove the hose from the water pump.

5. Remove the radiator.

6. Remove the power steering, A/C compressor, and the alternator drive belts.

7. Remove the spark plugs.

8. Remove the distributor protector (dust shield).

9. Remove the A/C compressor drive belt idler pulley bracket.

10. Remove the fresh air intake tube at the cylinder head cover.

11. Disconnect the radiator hose from the thermostat housing.

12. Remove the crankshaft pulley bolt and then pull off the pulley with a suitable puller.

13. Remove the bolts and then remove the front upper and lower timing belt covers.

14. Set the No. 1 piston at TDC of its compression stroke. Align the punchmark on the left camshaft sprocket with the punchmark on the timing belt upper rear cover. Align the punchmark on the crankshaft sprocket with the notch on the oil pump housing. Temporarily install

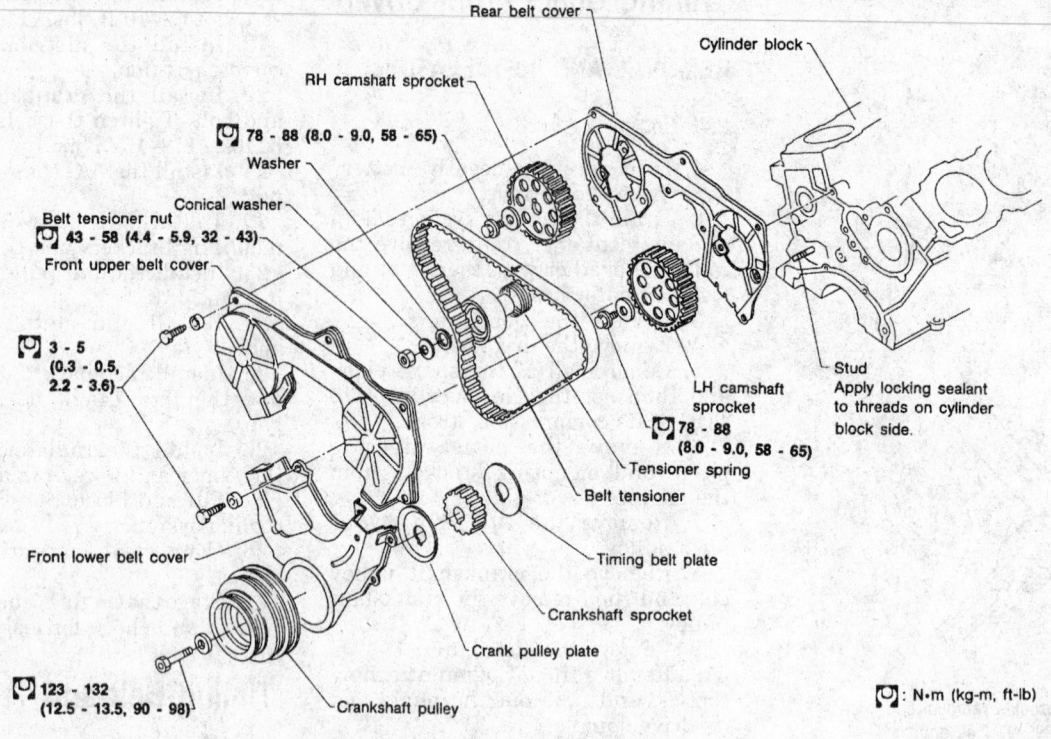

Rear belt cover

Cylinder block

RH camshaft sprocket

⊡ 78 - 88 (8.0 - 9.0, 58 - 65)

Washer

Conical washer

Belt tensioner nut
⊡ 43 - 58 (4.4 - 5.9, 32 - 43)

Front upper belt cover

⊡ 3 - 5
(0.3 - 0.5,
2.2 - 3.6)

LH camshaft
sprocket

⊡ 78 - 88
(8.0 - 9.0, 58 - 65)

Stud
Apply locking sealant
to threads on cylinder
block side.

Tensioner spring

Belt tensioner

Front lower belt cover

Timing belt plate

Crankshaft sprocket

Crank pulley plate

⊡ 123 - 132
(12.5 - 13.5, 90 - 98)

Crankshaft pulley

⊡ : N•m (kg-m, ft-lb)

312410

Exploded view of front engine components — V6 engines

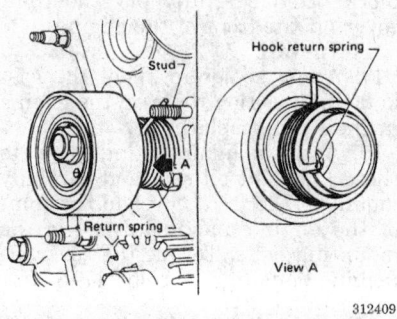

Stud

Hook return spring

Return spring

A

View A

312409

Timing belt tensioner assembly — V6 engines

the crank pulley bolt so the crank-shaft can be rotated if necessary.

15. Loosen the timing belt idler pulley bolt. Using a hexagon wrench, rotate the idler pulley to release its tension on the timing belt and tighten the idler pulley bolt.

─── **WARNING** ───
Once the timing belt is removed, do not rotate crankshaft and camshaft separately, because valves will hit piston heads.

16. Remove the timing belt.

NOTE: Check the tensioner pulley bearing, if any sign of wear is detected, replace the pulley.

17. If tensioner removal is required, release spring tension and remove tensioner center bolt.

To install:
18. Install the tensioner and the return spring. Using a hexagon wrench, turn the tensioner clockwise and temporarily tighten the locknut.
19. Make sure that the timing belt is clean and free from oil or water.
20. When installing the timing belt align the white lines on the belt with the punchmarks on the camshaft and crankshaft sprockets. Have the arrow

on the timing belt pointing toward the front belt covers.

NOTE: To verify that the timing belt is properly installed, count the number of belt teeth between the timing marks. There are 133 teeth on the belt; there should be 40 teeth between the timing marks on the left and right side camshaft sprockets, and 43 teeth between the timing marks on the left side camshaft sprocket and the crankshaft sprocket.

21. To adjust the timing belt tension, perform the following steps:

a. Loosen the tensioner pulley locknut and allow the spring to hold the pulley against the belt.

b. Turn the tensioner 70–80° clockwise and tighten the tensioner locknut.

c. Turn the crankshaft clockwise two to three times and set No. 1 cylinder to TDC.

d. Loosen the tensioner locknut then push on the belt halfway between the tensioner pulley and the camshaft sprocket with a force of 22 lbs. (98 N) and hold the tensioner steady with a hexagon wrench.

e. Set a 0.014 in. (0.35mm) feeler gauge between the belt and pulley on the crankshaft side of the pulley. The feeler gauge should be at least ½ in. (12.7mm) wide.

f. Have an assistant rotate the crankshaft clockwise to make the feeler gauge roll up between the tensioner pulley and the belt. Make sure the gauge is centered on the tensioner pulley.

g. Hold the feeler gauge in place and tighten the tensioner locknut to 32–43 ft. lbs. (43–58 Nm).

h. Rotate the crankshaft to remove the feeler gauge.

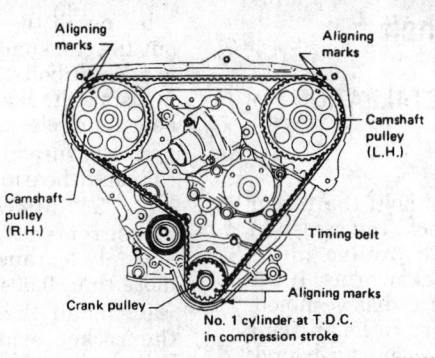

Timing mark identification — V6 engines

i. Turn the crankshaft clockwise two to three times and set No. 1 cylinder to TDC.

j. Check the timing belt deflection by pushing on the belt between the camshaft sprockets with a force of 22 lbs. (98 N). The belt should deflect 0.51–0.59 in. (13–15mm), if out of specification readjust the belt.

——————— WARNING ———————
Before completing the assembly, make certain the belt is properly aligned, by turning the crankshaft two full revolutions and verifying the timing marks are still aligned. If the marks do not align, repeat the belt installation steps, as well as the tensioner adjustment procedure.

22. Install the lower and upper timing belt covers and torque the mounting bolts to 2–4 ft. lbs. (3–5 Nm).
23. Press the crankshaft pulley onto the shaft and then tighten the bolt to 90–98 ft. lbs. (123–132 Nm).
24. Connect the radiator hose to the thermostat housing.
25. Connect the fresh air intake tube at the cylinder head cover.
26. Install the A/C compressor drive belt idler pulley bracket.

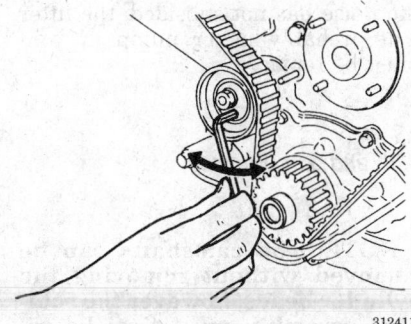

Adjusting timing belt tensioner — V6 engines

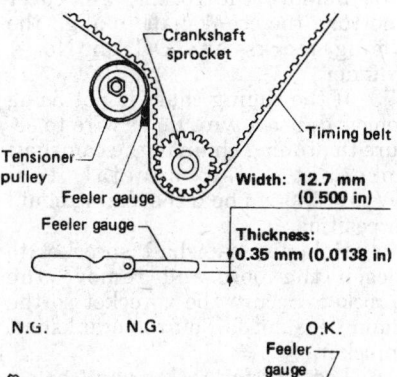

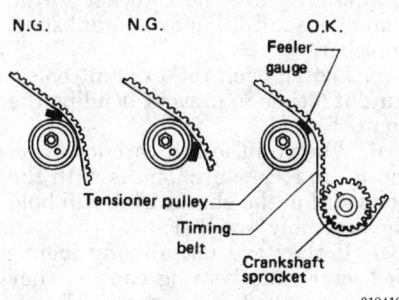

Proper method of adjusting timing belt tension — V6 engines

27. Install the distributor protector (dust shield).
28. Install the spark plugs.
29. Install the power steering, A/C compressor, and the alternator drive belts.
30. Install the radiator.
31. Connect the hose to the water pump.
32. Install the fan shroud and pulleys.
33. Install the engine under cover.
34. Fill the engine with coolant and bleed the air from the cooling system.
35. Connect the negative battery cable.

36. Start the engine and check ignition timing.

Timing Belt Sprockets

REMOVAL AND INSTALLATION

3.0L and 3.3L Engines

1. Disconnect the negative battery cable.
2. Remove timing belt covers.
3. Remove the timing belt.
4. Using an adjustable spanner wrench (to hold the camshaft sprockets) and a socket wrench, remove the camshaft sprocket bolt and washer.
5. Pull the camshaft sprockets from the camshafts. Be careful not to lose the key.

NOTE: The right and left camshaft pulleys are different parts. Install them in their correct positions. The right pulley has an R3 identification mark and the left pulley has an L3.

6. Crankshaft sprocket can pulled off crankshaft. If sprocket is difficult to remove, use an appropriate puller to remove.
To install:
7. Install the camshaft sprockets. Hold the sprocket and torque mounting bolt to 65 ft. lbs. (88 Nm).
8. Align woodruff key and tap crankshaft sprocket onto crankshaft, using a plastic hammer or mallet.
9. Install and adjust the timing belt.
10. Install the timing belt covers.
11. Torque crankshaft pulley bolt to 90 ft. lbs. (123 Nm).
12. Connect the negative battery cable.

Timing Belt Front Cover

REMOVAL AND INSTALLATION

3.0L and 3.3L Engines

1. Disconnect the negative battery cable.
2. Remove the engine under cover.
3. Remove the radiator shroud, the fan, and the pulleys.
4. Drain the coolant from the radiator into a sealable container and remove the hose from the water pump.
5. Remove the radiator.
6. Remove the power steering, A/C compressor, and the alternator drive belts.
7. Remove the distributor protector (dust shield).

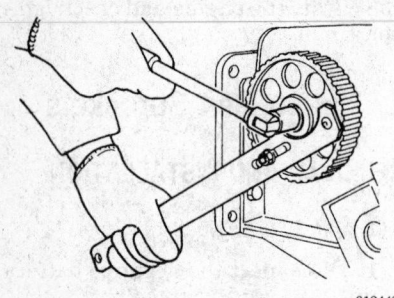

Use a spanner wrench to hold the camshaft sprocket during removal and installation — V6 engines

8. Remove the A/C compressor drive belt idler pulley bracket.

9. Disconnect the radiator hose from the thermostat housing.

10. Remove the crankshaft pulley bolt and then pull off the pulley with a suitable puller.

11. Remove the bolts and then remove the front upper and lower timing belt covers.

To install:

12. Install the lower and upper timing belt covers and torque the mounting bolts to 2–4 ft. lbs. (3–5 Nm).

13. Press the crankshaft pulley onto the shaft and then tighten the bolt to 90–98 ft. lbs. (123–132 Nm).

14. Connect the radiator hose to the thermostat housing.

15. Install the A/C compressor drive belt idler pulley bracket.

16. Install the distributor protector (dust shield).

17. Install the power steering, A/C compressor, and the alternator drive belts.

18. Install the radiator.

19. Connect the water pump hose and fill the engine with coolant. Install the fan shroud and pulleys.

20. Connect the negative battery cable.

21. Start the engine and check for any leaks.

22. Install the engine under cover.

Camshaft

REMOVAL AND INSTALLATION

2.4L Engine

The same bolts that hold the rocker arm assembly also hold the camshaft bearing caps. The hydraulic lifters are built into the rocker arms. If the rocker arm shafts are disassembled, do not allow the arms to lie on their side or they will become air-bound. Keep the rocker arms upright or lay them in a pan of new engine oil.

1. Relieve the fuel system pressure and disconnect the negative battery cable.

2. Remove the rocker arm cover and turn the crankshaft to align the timing marks at TDC on No. 1 cylinder.

3. If the timing chain is not being removed, use a wire tie or wire to secure the timing chain to the camshaft sprocket. Use special tool KV10105800 to hold the timing chain in position.

4. Hold the camshaft sprocket to loosen the bolt and remove the sprocket. Secure the sprocket so the chain does not fall off the crankshaft sprocket.

5. Loosen each rocker shaft bolt 1 turn at a time to prevent bending the shafts.

6. When all the bolts are loose, remove the rocker arm shafts with the bolts still in the shafts. This will hold the assembly together.

7. If they are not already identified, mark the bearing caps so they can be installed in their original position facing the same direction. Lift the caps off and lift the camshaft out.

To install:

8. Inspect the camshaft and the bearings:

   a. Make sure the camshaft and the bearing surfaces are in good condition.

b. Install the bearing caps without the camshaft, torque the rocker arm shaft bolts to specification and measure the inside diameter of the bearing circle.

c. Measure the diameter of the camshaft bearings.

d. The difference between the measurements is the camshaft journal clearance; it should be no more than 0.0047 in. (0.12mm)

e. Install the camshaft without the rocker arms and torque the bolts to specification. The camshaft end-play should be no more than 0.008 in. (0.2mm).

9. Lubricate the camshaft with engine oil and set it in place. Make sure the pin on the sprocket end is up.

10. Install the rocker arm shafts and tighten the bolts in the proper sequence 1 turn at a time to draw the shafts down evenly against the valve springs without bending the shafts. Torque the bolts to 27–30 ft. lbs. (37–41 Nm).

11. Install the camshaft sprocket and remove the tie securing the chain. Install the sprocket bolt but don't torque it yet. Rotate the crankshaft 2 full turns to make sure the timing marks line up. When the valve timing is correct, torque the sprocket bolt to 87–116 ft. lbs. (118–157 Nm).

12. Use a silicone sealer on the rubber end plugs and install the rocker arm cover with a new gasket. Install two of the attaching bolts and torque the two bolts to 2 ft. lbs. (3 Nm).

13. Torque the rocker cover bolts in the proper sequence to 5–8 ft. lbs. (7–11 Nm).

14. Drain the engine oil and refill the engine with fresh oil.

15. Connect the negative battery cable. Run the engine and check for proper operation.

16. When the engine is first started, the hydraulic valve lifters may be noisy. Run the engine for 10–20 minutes at about 1000 rpm. If the noise has not subsided, the lifter will probably never pump up and must be replaced.

3.0L and 3.3L Engines

NOTE: The camshafts can be removed without removing the cylinder heads, however the radiator assembly must first be removed, to provide enough clearance.

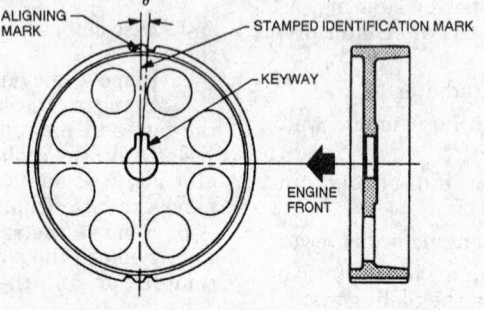

Camshaft sprocket identification — V6 engines

— CAUTION —

Fuel injection systems remain under pressure after the engine has been turned OFF. Properly relieve fuel pressure before disconnecting any fuel lines. Failure to do so may result in fire or personal injury.

1. Following the proper procedures, relieve the fuel system pressure.

2. Disconnect the negative battery cable.

3. Drain cooling system and remove the radiator assembly.

4. Align the timing marks to bring No. 1 cylinder to TDC.

5. Remove the timing covers and belt.

6. Tag then disconnect the ignition wires from the spark plugs. Disconnect the ignition wire from the ignition coil.

7. Match mark the distributor position, then remove the distributor attaching bolt and the distributor.

8. Remove the rocker covers.

9. Remove the camshaft sprockets then the rear timing belt cover.

10. Loosen the rocker shaft bolts in two or three stages, then remove the rocker shafts with rocker arms, as an assembly. If the rocker arm and shaft assembly needs to be disassembled for service, note the location of the components as they are removed. The rocker arms must be installed in the same position if reused.

11. Attach a wire to the top of the lifters so that they will not drop from the lifter guide. Carefully remove the lifter guide and lifters from the cylinder head. Put an identification mark on the lifters to avoid mixing them up if they are removed from the guide and are being reused.

12. Measure the camshaft end-play, it should be 0.0012–0.0024 in. (0.03–0.06mm). If the end-play is out of specification, the locate plate will have to be replaced with a plate of the correct thickness.

13. At the rear of the cylinder head, remove the cylinder head rear cover, the camshaft bolt and the locating plate.

14. Remove the camshaft front oil seal and then slide the camshaft out the front of the cylinder head assembly.

To install:

15. Coat the camshafts with engine oil and carefully install them. Install the locating plates and bolts to the camshafts, torque the bolt to 58–65 ft. lbs. (78–88 Nm). Turn the camshafts so the pin on the sprocket end is up.

16. Install the rear camshaft end covers with new gaskets.

17. Lubricate a new camshaft seal with grease and use a seal driver to install the seal. Make sure that the seals properly seat into the cylinder heads.

18. Install the rear timing belt cover and the camshaft sprockets. Torque the sprocket bolts to 58–65 ft. lbs. (78–88 Nm).

NOTE: The right and left camshaft sprockets are different parts. Install them in their correct positions. The right sprocket has an R3 identification mark and the left has an L3.

19. Install and adjust the timing belt, then install the timing belt covers.

20. Set cylinder No. 1 to TDC.

21. Install the left cylinder head valve lifter guide assembly and remove the wire from the lifters. Install the rocker arm shaft assemblies and attaching bolts, coat the bolt threads and seat surfaces before installing them. Tighten the bolts gradually in three steps to 13–16 ft. lbs. (18–22 Nm).

22. Rotate the crankshaft clockwise 360°, to bring cylinder No. 4 to TDC. Install the right cylinder head valve lifter guide assembly and remove the wire from the lifters. Install the rocker arm shaft assemblies and attaching bolts, coat the bolt threads and seat surfaces before installing them. Tighten the bolts gradually in three steps to 13–16 ft. lbs. (18–22 Nm).

23. Install the rocker covers with new gaskets, torque the rocker cover bolts to 1–2 ft. lbs. (1–3 Nm).

24. Set cylinder No. 1 to TDC by rotating the engine 360° and aligning the timing marks. Install the distributor assembly and connect the wires to their proper locations. Do not tighten the distributor attaching bolt until the timing has been checked and adjusted as necessary.

25. Install the radiator assembly, then refill and bleed the cooling system.

26. Connect the negative battery cable and run the engine for 10–20 minutes, at about 1000 rpm, to pump up the lifter assemblies.

NOTE: If the hydraulic valve lifters are still noisy, replace them and bleed the air from them.

Piston and Connecting Rod

POSITIONING

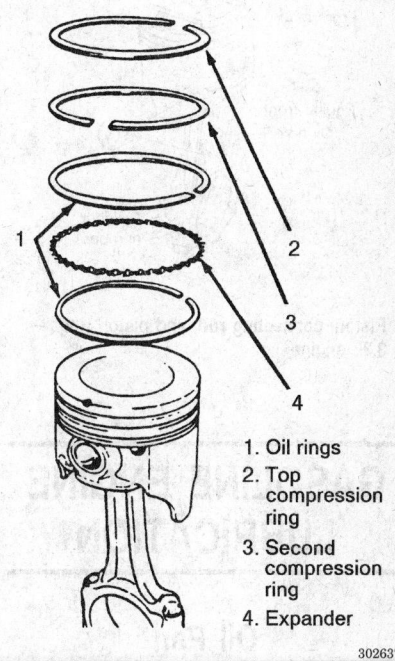

1. Oil rings
2. Top compression ring
3. Second compression ring
4. Expander

302637

Piston, connecting rod, and piston rings — 2.4L engine

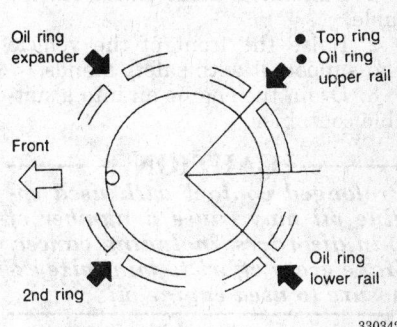

330342

Piston ring positioning — 3.0L engine

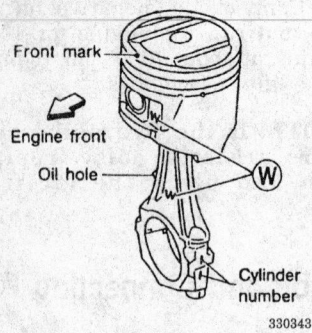

Piston and connecting rod front marks — 3.0L engine

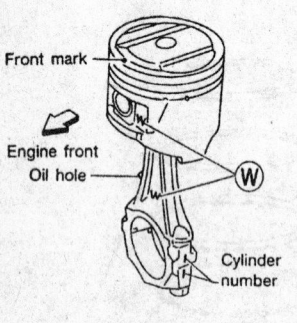

Piston, connecting rod, and piston rings — 3.3L engine

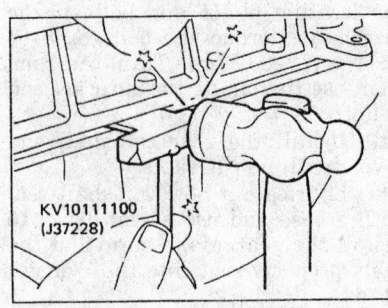

KV10111100 (J37228)

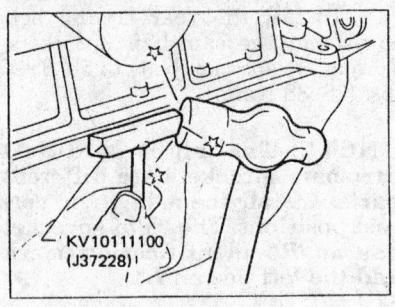

KV10111100 (J37228)

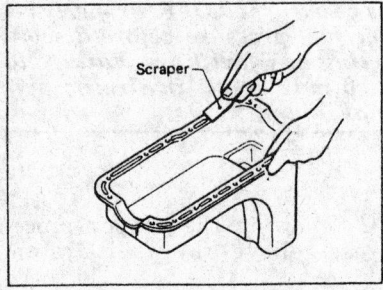

Scraper

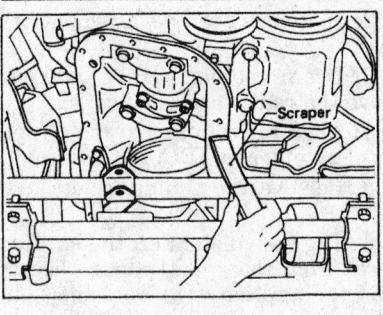

Scraper

Oil pan service points of removal — 2.4L, 3.0L and 3.3L engines

GASOLINE ENGINE LUBRICATION

Oil Pan

REMOVAL AND INSTALLATION

2.4L Engine

1. Disconnect the negative battery cable.
2. Raise the front of the vehicle and support it with safety stands.
3. Drain the engine oil into a suitable container.

> **CAUTION**
> *Prolonged contact with used engine oil may cause a number of skin disorders, including cancer. Make every effort to minimize exposure to used engine oil.*

4. Remove the front stabilizer bar mounting nuts and bolts from the side member.
5. Remove the nuts from the engine mounts and raise the engine.

6. Loosen the oil pan bolts in the proper sequence. Insert a seal cutter tool between the cylinder block and the oil pan and tap it around the circumference of the pan with a hammer. Remove the oil pan. Pull it out from the front side.

NOTE: Be careful not to drive the seal cutter into the oil pump or rear oil seal retainer as you will damage the aluminum mating surface.

To install:

7. Remove all traces of gasket material from the pan and block mating surfaces.
8. Apply a continuous bead of sealant 0.138–0.177 in. (3.5–4.5mm) to the oil pan mating surface. Be sure to trace the sealant bead to the inside of the bolt holes where there is no groove.
9. Install the pan within 5 minutes and tighten all bolts in the proper sequence. Tighten the bolts to 5–6 ft. lbs. (7–8 Nm).
10. Lower the engine and install the engine mount nuts. Torque the nuts to 30–38 ft. lbs. (41–52 Nm).
11. Connect the stabilizer bar.
12. Lower the vehicle.
13. Connect the negative battery cable.

14. Wait at least 30 minutes and then refill the engine with oil. Run the engine until it reaches normal operating temperature and then check for leaks.

3.0L Engine

> **CAUTION**
> *The EPA warns that prolonged contact with used engine oil may cause a number of skin disorders, including cancer! You should make every effort to minimize your exposure to used engine oil. Protective gloves should be worn when changing the oil. Wash your hands and any other exposed skin areas as soon as possible after exposure to used engine oil. Soap and water, or waterless hand cleaner, should be used.*

2WD Vehicles

1. Disconnect the negative battery cable.
2. Raise the vehicle and support safely.
3. Remove the engine under cover and drain the engine oil into a sealable container. Install the drain plug and torque to 22–29 ft. lbs. (29–39 Nm).
4. Remove the bolts attaching the stabilizer bar to the crossmember.

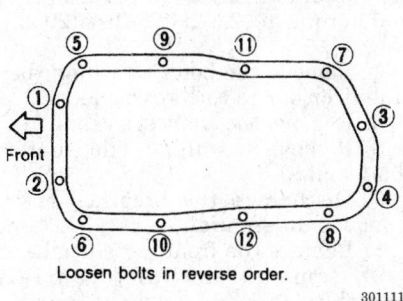

Loosen bolts in reverse order.

301111

Oil pan bolt removal sequence — 2.4L engine

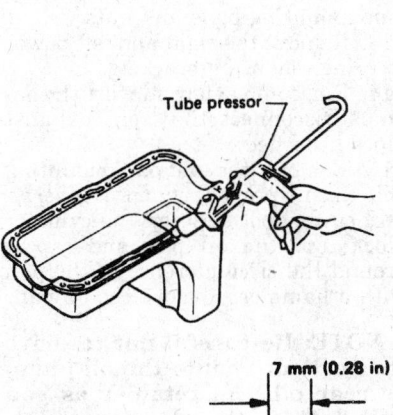

Tube pressor

7 mm (0.28 in)

Groove   Bolt hole

301112

Proper method of installing sealant to oil pan

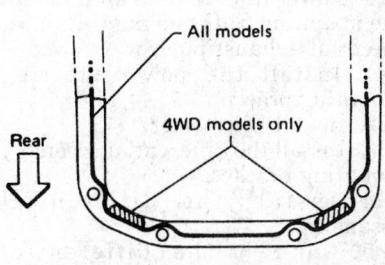

All models

4WD models only

Rear

301113

Liquid gasket placement on the oil pan rail

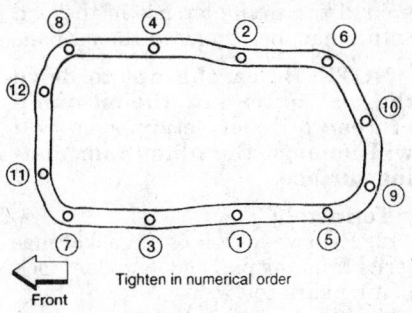

Tighten in numerical order

Front

301115

Oil pan bolt installation sequence — 2.4L engine

5. Remove the front crossmember.
6. Remove the idler arm.
7. Remove the starter motor.
8. Remove the bracket securing the engine to the transmission.
9. Remove the oil pan mounting bolts in the proper sequence. Insert a seal cutter tool between the cylinder block and the oil pan and tap it around the circumference of the pan with a hammer. Remove the oil pan.

NOTE: Be careful not to drive the seal cutter into the oil pump or rear oil seal retainer as you will damage the aluminum mating surface.

To install:
10. Remove all traces of gasket material from the pan and block mating surfaces.
11. Apply sealant to the oil pump and oil seal retainer gasket.
12. Apply a continuous bead of sealant 0.138–0.177 (3.5–4.5 mm) to the oil pan mating surface. Be sure to trace the sealant bead to the inside of the bolt holes where there is no groove.
13. Install the pan within five minutes of sealant application and tighten all bolts in the reverse order of removal. Tighten the bolts to 5 ft. lbs. (7 Nm).

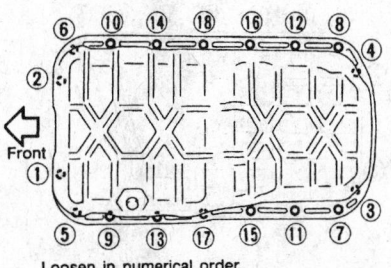

Front

Loosen in numerical order.

253840

Oil pan bolt removal sequence — 3.0L engine with 2WD

14. Install the bracket securing the engine to the transmission.
15. Install the starter.
16. Install the idler arm.
17. Install the front crossmember.
18. Attach the stabilizer bar to the crossmember.
19. Install the engine under cover, then lower the vehicle.
20. Connect the negative battery cable.
21. Wait at least 30 minutes and then refill the engine with oil. Start the engine and run it until it reaches normal operating temperature, then check for leaks.

4WD Vehicles

1. Disconnect the negative battery cable.
2. Raise the vehicle and support safely.
3. Remove the engine under cover and drain the engine oil into a sealable container. Install the drain plug and torque to 22–29 ft. lbs. (29–39 Nm).
4. Remove the front driveshaft.
5. Remove the front drive axle from the vehicle.
6. Remove the idler arm.
7. Remove the starter motor.
8. Remove the transmission mount bracket nuts.
9. Remove the bolts attaching the engine mount.
10. Remove the bracket securing the engine to the transmission.
11. Attach an engine hoist and raise the engine slightly.

—————— WARNING ——————
It may be necessary to disconnect the exhaust to avoid damaging it when raising the engine. When lifting the engine be careful not to contact any adjacent parts, especially the accelerator wire casing end, brake tubes and the brake master cylinder.

12. Remove the oil pan mounting bolts in the proper sequence. Insert a seal cutter tool between the cylinder block and the oil pan and tap it around the circumference of the pan with a hammer. Remove the oil pan.

NOTE: Be careful not to drive the seal cutter into the oil pump or rear oil seal retainer as you will damage the aluminum mating surface.

To install:
13. Remove all traces of gasket material from the pan and block mating surfaces.
14. Apply sealant to the oil pump and oil seal retainer gasket.

15. Apply a continuous bead of sealant 0.138–0.177 in. (3.5–4.5mm) to the oil pan mating surface. Be sure to trace the sealant bead to the inside of the bolt holes where there is no groove.

16. Install the pan within five minutes of sealant application and tighten all bolts in the reverse order of removal. Tighten the bolts to 5 ft. lbs. (7 Nm).

17. Install the bracket securing the engine to the transmission.

18. Install the engine mount bolts and torque the bolts to 23–31 ft. lbs. (31–42 Nm).

19. Install the transmission mount bracket nuts and torque the nuts to 30–38 ft. lbs. (41–52 Nm).

20. Install the starter.

21. Install the idler arm.

22. Install the front drive axle.

23. Install the front drive shaft.

24. Install the engine under cover, then lower the vehicle.

25. Connect the negative battery cable.

26. Wait at least 30 minutes and then refill the engine with oil. Start the engine and run it until it reaches normal operating temperature, then check for leaks.

3.3L Engine

2WD Models

1. Disconnect the negative battery cable.

2. Raise the vehicle and support safely.

3. Remove the engine under cover and drain the engine oil into a sealable container. Install the drain plug and torque to 22–29 ft. lbs. (29–39 Nm).

4. Remove the bolts attaching the stabilizer bar to the crossmember.

5. Remove the front crossmember.

6. Remove the starter motor.

7. Remove the nuts attaching the transmission mount to the crossmember.

8. Remove the right and left engine mounting bolts and nuts.

9. Remove the right and left power steering mounting brackets.

10. Raise and safely support the engine. Disconnect the front exhaust pipes if needed.

11. Remove the oil pan mounting bolts in the proper sequence. Insert a seal cutter tool between the cylinder block and the oil pan and tap it

around the circumference of the pan with a hammer. Remove the oil pan.

NOTE: Be careful not to drive the seal cutter into the oil pump or rear oil seal retainer as you will damage the aluminum mating surface.

To install:

12. Remove all traces of gasket material from the pan and cylinder block mating surfaces.

13. Apply sealant to the oil pump and oil seal retainer gasket.

14. Apply a continuous bead of sealant 0.138–0.177 in. (3.5–4.5mm) to the oil pan mating surface. Be sure to trace the sealant bead to the inside of the bolt holes where there is no groove.

15. Install the pan within five minutes of sealant application and tighten all bolts in the reverse order of removal. Tighten the bolts to 5 ft. lbs. (7 Nm).

16. Lower the engine and install the mounting bolts and nuts. Connect the front exhaust pipes if removed.

17. Install the power steering mounting brackets.

18. Install the starter.

19. Install the front crossmember.

20. Attach the stabilizer bar to the crossmember.

21. Install the engine under cover, then lower the vehicle.

22. Connect the negative battery cable.

23. Wait at least 30 minutes for the sealant to cure and then refill the engine with oil. Start the engine and run it until it reaches normal operating temperature, then check for leaks.

4WD Models

1. Disconnect the negative battery cable.

2. Raise the vehicle and support safely.

3. Remove the engine under cover and drain the engine oil into a sealable container. Install the drain plug and torque to 22–29 ft. lbs. (29–39 Nm).

4. Remove the bolts attaching the stabilizer bar to the crossmember.

5. Remove the front driveshaft.

6. Remove the front axle shafts (half shafts).

7. Disconnect the breather hose from the differential.

8. Remove the front crossmember.

9. Remove the differential assembly.

10. Remove the starter motor.

11. Remove the nuts attaching the transmission mount to the crossmember.

12. Remove the right and left engine mounting bolts and nuts.

13. Remove the right and left power steering mounting brackets.

14. Raise and safely support the engine. Disconnect the front exhaust pipes if needed.

15. Remove the oil pan mounting bolts in the proper sequence. Insert a seal cutter tool between the cylinder block and the oil pan and tap it around the circumference of the pan with a hammer. Remove the oil pan.

NOTE: Be careful not to drive the seal cutter into the oil pump or rear oil seal retainer as you will damage the aluminum mating surface.

To install:

16. Remove all traces of gasket material from the pan and cylinder block mating surfaces.

17. Apply sealant to the oil pump and oil seal retainer gasket.

18. Apply a continuous bead of sealant 0.138–0.177 (3.5–4.5 mm) to the oil pan mating surface. Be sure to trace the sealant bead to the inside of the bolt holes where there is no groove.

19. Install the pan within five minutes of sealant application and tighten all bolts in the reverse order of removal. Tighten the bolts to 5 ft. lbs. (7 Nm).

20. Lower the engine and install the mounting bolts and nuts. Connect the front exhaust pipes if removed.

21. Install the power steering mounting brackets.

22. Install the starter.

23. Install the differential assembly mounting bracket.

24. Install the differential assembly.

25. Connect the differential breather hose.

26. Install the front half shafts.

27. Install the front crossmember.

28. Install the driveshaft.

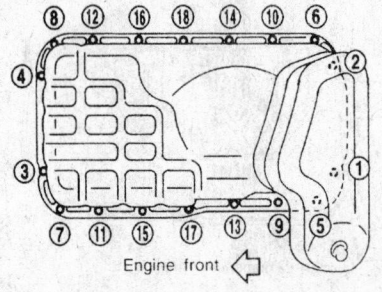

Engine front

310418

Oil pan bolt removal sequence — 3.3L engine with 2WD

29. Attach the stabilizer bar to the crossmember.

30. Install the engine under cover, then lower the vehicle.

31. Connect the negative battery cable.

32. Wait at least 30 minutes for the sealant to cure and then refill the engine with oil. Start the engine and run it until normal operating temperature, then check for leaks.

Oil Pump

REMOVAL AND INSTALLATION

2.4L Engine

The oil pump is mounted externally on the engine, eliminating the need to remove the oil pan in order to remove the oil pump.

1. Turn the engine to TDC. Matchmark and remove the distributor.

2. Disconnect the negative battery cable.

3. Drain the engine oil.

4. Remove the front stabilizer bar.

5. Remove the splash shield.

6. Loosen the mounting bolts and remove the oil pump body with the drive spindle assembly.

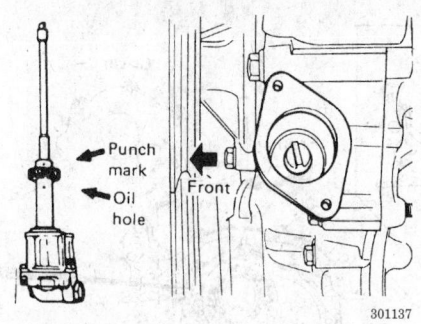

Align the punch mark with the oil hole before oil pump installation — 2.4L engine

To install:

7. If the crankshaft has been moved, turn the crankshaft so that the No. 1 piston is at TDC of the compression stroke before installing the oil pump in the engine.

8. Fill the pump housing with engine oil, then align the punch mark on the spindle with the hole in the oil pump.

9. With a new gasket placed over the drive spindle, install the oil pump and drive spindle assembly so that the projection on the top of the drive

spindle is located in the 11 o'clock position.

10. Align matchmarks and install the distributor with the metal tip of the rotor pointing toward the No. 1 spark plug tower, of the distributor cap.

11. Install the splash shield and front stabilizer bar.

12. Refill the engine with oil.

13. Connect the negative battery cable.

14. Start the engine, ensure proper oil pressure and check for oil leaks.

15. Check ignition timing.

3.0L Engine

1. Disconnect the negative battery cable.

2. Remove the oil pan.

3. Remove the timing belt cover and the timing belt.

4. Remove the crankshaft timing sprocket (it may be necessary to use a puller) and the timing belt plate.

5. Remove the oil pump strainer and pick-up tube from the oil pump.

6. Loosen the oil pump retaining bolts and then remove the oil pump.

To install:

7. Use new gaskets and install a new oil seal. Tighten the 6mm bolts

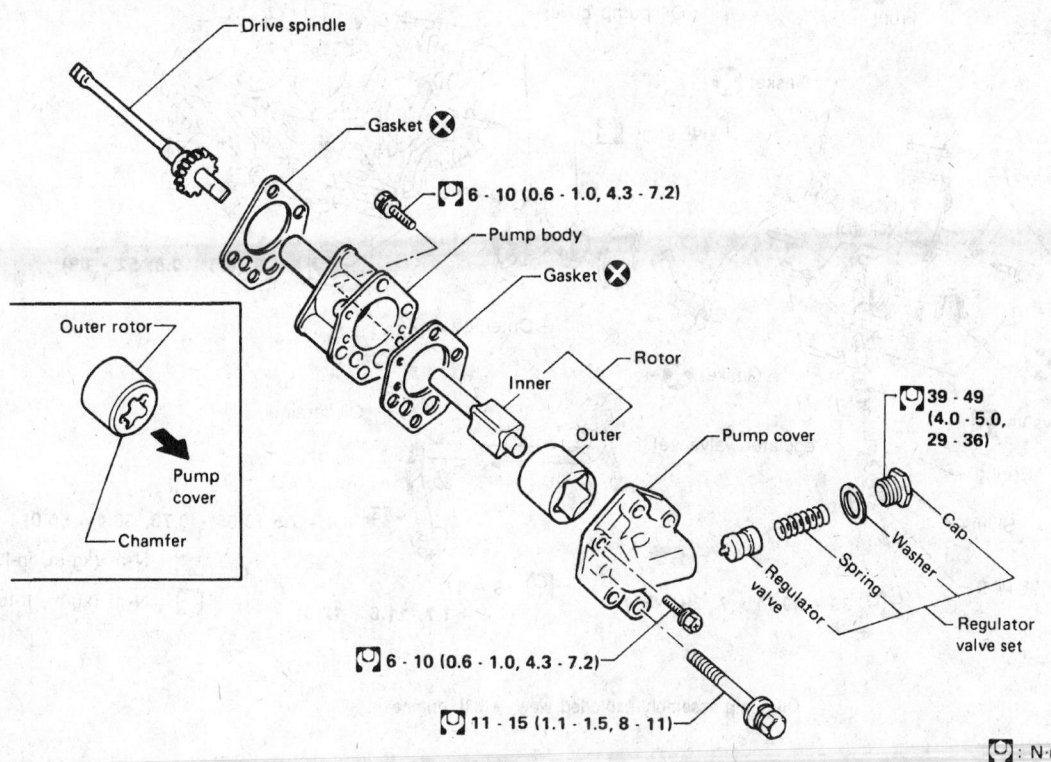

Exploded view of oil pump assembly — 2.4L engine

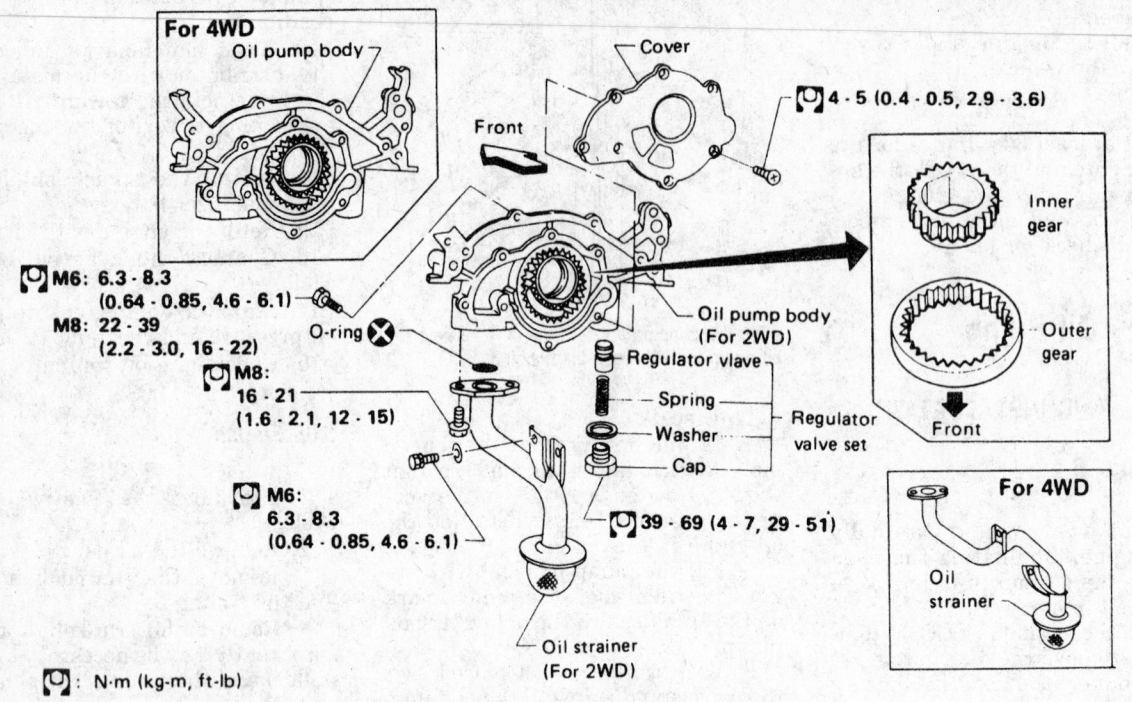

For 4WD
Oil pump body

Cover

Front

4 - 5 (0.4 - 0.5, 2.9 - 3.6)

Inner gear

Outer gear

Front

M6: 6.3 - 8.3 (0.64 - 0.85, 4.6 - 6.1)
M8: 22 - 39 (2.2 - 3.0, 16 - 22)

O-ring

Oil pump body (For 2WD)

Regulator vlave
Spring
Washer
Cap
} Regulator valve set

M8: 16 - 21 (1.6 - 2.1, 12 - 15)

M6: 6.3 - 8.3 (0.64 - 0.85, 4.6 - 6.1)

39 - 69 (4 - 7, 29 - 51)

For 4WD
Oil strainer

Oil strainer (For 2WD)

: N·m (kg-m, ft-lb)

254090

Exploded view of oil pump assembly — 3.0L engine

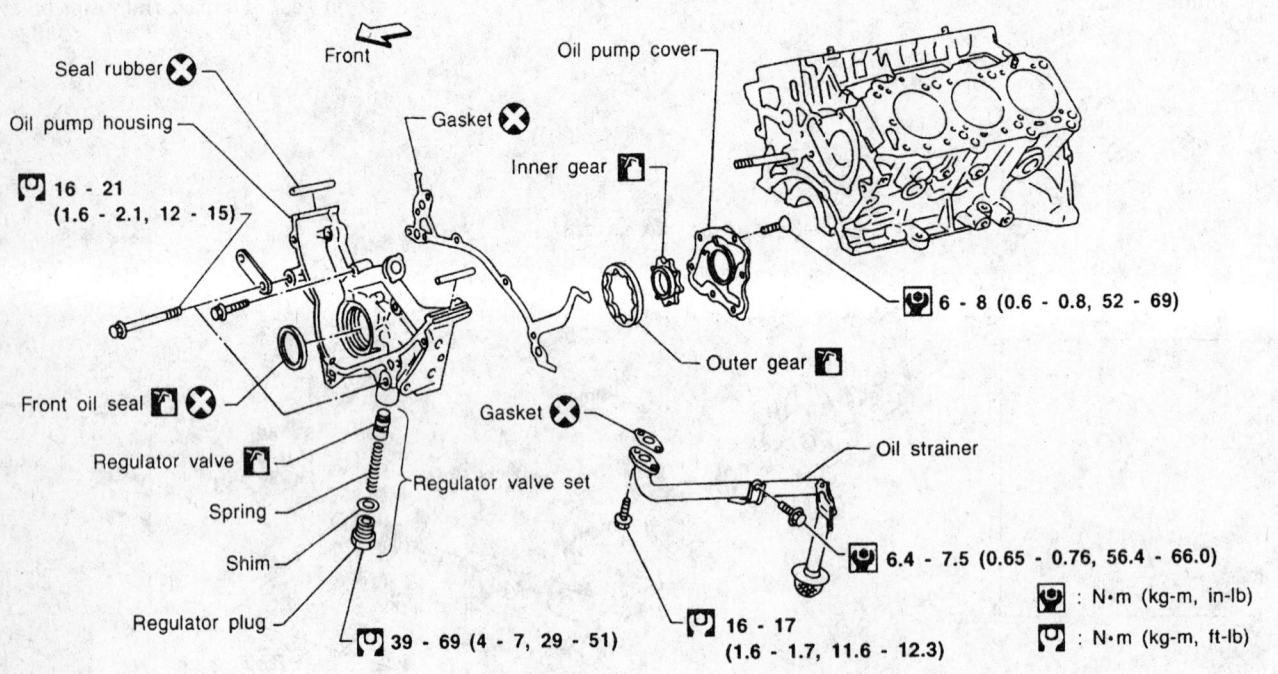

Seal rubber

Front

Oil pump cover

Oil pump housing

Gasket

Inner gear

16 - 21 (1.6 - 2.1, 12 - 15)

6 - 8 (0.6 - 0.8, 52 - 69)

Front oil seal

Gasket

Outer gear

Regulator valve

Oil strainer

Spring

Regulator valve set

Shim

6.4 - 7.5 (0.65 - 0.76, 56.4 - 66.0)

Regulator plug

39 - 69 (4 - 7, 29 - 51)

16 - 17 (1.6 - 1.7, 11.6 - 12.3)

: N·m (kg-m, in-lb)

: N·m (kg-m, ft-lb)

310513

Oil pump assembly exploded view — 3.3L engine

to 4–6 ft. lbs. (6–8 Nm) and the 8mm
bolts to 16–22 ft. lbs. (22–39 Nm).

**NOTE: Before installing the oil
pump, be sure to pack the pump's
cavity with petroleum jelly, then
make sure the O-ring is properly
fitted.**

8. Connect the oil strainer and
pick-up tube to the pump body.
9. Clean gasket surfaces and install the oil pan.
10. Install the timing belt plate and
the crankshaft sprocket.
11. Install the timing belt and front
cover.
12. Connect the negative battery
cable.
13. Refill the engine with oil, start
the engine, and check for any leaks.

3.3L Engine

1. Disconnect the negative battery
cable.
2. Drain the engine oil and the
coolant from the radiator. Save the
coolant so it can be reused.
3. Remove the air duct between
the mass air flow sensor and the
throttle body.
4. Remove the cooling fan.
5. Remove the upper and lower radiator hoses and the fan shroud.
6. Remove the drive belts.
7. Remove the crankshaft pulley.
8. Remove the upper and lower
timing belt covers.
9. Remove the oil pan.
10. Remove the oil strainer (pick-
up).
11. Remove the oil pump mounting
bolts and the oil pump.
To install:
12. Replace the oil seal in the
pump.
13. Using a new gasket, install the
oil pump. Torque the mounting bolts
to 12–15 ft. lbs. (16–21 Nm).
14. Install the oil strainer and the
oil pan.
15. Install the timing belt covers
and the crankshaft pulley.
16. Install the drive belts.
17. Install the fan shroud and the
radiator hoses.
18. Install the cooling fan.
19. Install the air duct between the
throttle body and mass air flow
sensor.
20. Refill the radiator with the coolant that was removed unless the coolant is being changed.
21. Fill the engine with the proper
amount of new engine oil.
22. Connect the negative battery
cable, start the engine and check for
leaks.

TRANSMISSION

Manual Transmission Assembly

REMOVAL AND INSTALLATION

Except 1996–97 Pathfinder

2WD Models

1. Disconnect the negative battery
cable.
2. Remove the shifter knob and
boot, then remove the snapring to lift
the shift lever out of the transmission. Stuff a rag in the opening to
keep dirt out of the transmission.
3. Raise and safely support the vehicle. If equipped, remove the splash
pan or skid plate.
4. If equipped with a V6 engine,
disconnect the exhaust pipe from the
manifolds and from the catalytic converter. Remove the pipe.
5. Remove the starter and drain
the oil from the transmission.
6. On 1996 models, remove the
crankshaft position sensor from the
upper side of the transmission.
7. Matchmark the driveshaft
flange at the rear pinion flange and
remove the driveshaft. Plug the extension housing opening to prevent
dirt from getting in.
8. Disconnect the speedometer
cable and the wiring from the
transmission.
9. Without disconnecting the hydraulic hose, remove the clutch slave
cylinder from the transmission and
secure it aside.
10. Support the engine by placing a
jack under the oil pan, do not place
the jack under the drain plug.
11. Using an appropriate transmission jack, properly support the trans-

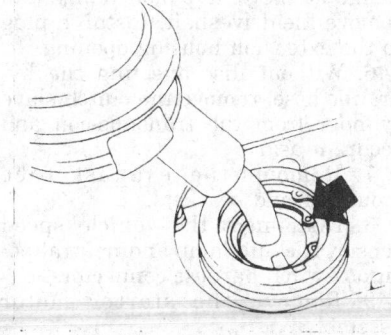

300846

Shifter removal — 2WD models

mission and remove the transmission
mount and crossmember.
12. Remove the transmission
mounting bolts and move the transmission back away from the engine.
Lower the transmission carefully
from the vehicle.

**NOTE: Keep the transmission
mounting bolts in order because
they vary in size. This will expedite the installation process.**

To install:
13. Lightly grease the input shaft
splines. Fit the transmission into
place and start all the transmission
mounting bolts. Make sure the input
shaft fits properly into the clutch disc
and pilot bearing. Torque the transaxle bolts to specification.
14. Install the crossmember and
torque the crossmember mounting
bolts to 30–38 ft. lbs. (41–52 Nm).
15. Install the transmission mount
and torque the nuts to 30–38 ft. lbs.
(41–52 Nm).
16. Remove the transmission jack
from the transmission and the jack
from under the oil pan.
17. Install the clutch slave cylinder
and torque the mounting bolts to
22–30 ft. lbs. (30–40 Nm).
18. Connect the wiring to the transmission and the speedometer cable.
19. Install the drive shaft and align
the matchmarks on the pinion flange.
Install the attaching bolts.
20. If the exhaust pipe was removed, install it with new gaskets.
Torque the flange nuts to 20–27 ft.
lbs. (26–36 Nm) and the bolts attaching the pipe to the catalytic converter
to 32–41 ft. lbs. (43–55 Nm).
21. Install the crankshaft position
sensor if it was removed.
22. Install the starter.
23. If removed, install the splash
pan or skid plate.
24. Refill the transmission with oil
and lower the vehicle.
25. Install the shift lever into the
transmission and install the snapring. Install the shifter boot and the
shifter knob.
26. Connect the negative battery
cable.

4WD Models

1. Disconnect the negative battery
cable.
2. Remove the shifter knob and
boot and remove the snapring to lift
the shift lever out of the transmission. Stuff a rag in the opening to
keep dirt out of the transmission.
3. Raise and safely support the vehicle. If equipped, remove the splash
pan or skid plate.
4. Remove the starter.

5. Remove the crankshaft position sensor from the upper side of the transmission.

6. Drain the oil from the transmission and transfer case.

7. If equipped with a V6 engine, disconnect the exhaust pipe from the manifolds and from the catalytic converter then remove the pipe.

8. Matchmark the rear driveshaft flange at the rear pinion flange and remove the driveshaft. Plug the extension housing opening to prevent dirt from getting in.

9. Matchmark both front driveshaft flanges so the shaft can be installed in the same position. Remove the driveshaft.

10. Disconnect the speedometer cable from the transfer case and the wiring from the transmission.

11. Without disconnecting the hydraulic hose, remove the clutch slave cylinder from the transmission and secure it aside.

12. Remove the torsion bars from the vehicle.

13. Remove the transfer case shift lever.

14. Support the engine by placing a jack under the oil pan, do not place the jack under the drain plug.

15. Using an appropriate transmission jack, properly support the transmission and transfer case, then remove the transmission mount and crossmember.

16. Remove the transmission mounting bolts and move the transmission back away from the engine. Lower the transmission carefully from the vehicle.

NOTE: Keep the transmission mounting bolts in order because they vary in size. This will expedite the installation process.

To install:

NOTE: Apply a silicone sealant to the engine block or rear plate to seal the engine to the transmission.

17. Lightly grease the input shaft splines then fit the transmission into place and start all the transmission mounting bolts. Make sure the input shaft fits properly into the clutch disc and pilot bearing. Torque the transaxle bolts to specification.

18. Install the crossmember and torque the crossmember mounting bolts to 30–38 ft. lbs. (41–52 Nm).

19. Install the rear transmission mount bolts, then torque the mount bolts to 30–38 ft. lbs. (41–52 Nm).

20. Remove the transmission jack from the transmission and the jack from under the oil pan.

21. Install the transfer case shift lever.

22. Install the torsion bars in their original locations. Make sure the splines are in their original position and set the adjustment to its original position.

23. Install the clutch slave cylinder and torque the mounting bolts to 22–30 ft. lbs. (30–40 Nm).

24. Connect the wiring to the transmission and the speedometer cable to the transfer case.

25. Install the driveshafts, be sure to align the matchmarks.

26. If the exhaust pipe was removed, install it with new gaskets. Torque the flange nuts to 20–27 ft. lbs. (26–36 Nm) and the bolts attaching the pipe to the catalytic converter to 32–41 ft. lbs. (43–55 Nm).

27. Install the crankshaft position sensor if it was removed.

28. Install the starter.

29. Refill the transmission and transfer case.

30. If removed, install the splash pan or skid plate then lower the vehicle.

31. Install the shift lever into the transmission and install the snapring. Install the shifter boot and the shifter knob.

32. Connect the negative battery cable.

1996–97 Pathfinder

2WD Models

1. Disconnect the negative battery cable.

2. Remove the shifter knob and boot, then remove the snapring to lift the shift lever out of the transmission. Place a rag in the opening to keep dirt out of the transmission.

3. Raise and safely support the vehicle. If equipped, remove the splash pan or skid plate.

4. Remove the crankshaft position sensor from the upper side of the transmission.

5. Matchmark the driveshaft flange at the rear pinion flange and remove the driveshaft. Install a plug to the extension housing opening.

6. Without disconnecting the hydraulic hose, remove the clutch slave cylinder from the transmission and secure it aside.

7. Remove the exhaust tube mounting and brackets.

8. Disconnect the vehicle speed sensor, back-up lamp and neutral position switch harness connectors.

9. Remove the starter motor assembly.

10. Support the engine by placing a jack with a block of wood under the oil pan. Do not place the jack under the drain plug.

11. Using an appropriate transmission jack, properly support the transmission and remove the transmission mount and crossmember.

12. Remove the transmission mounting bolts and move the transmission back away from the engine. Lower the transmission carefully from the vehicle.

NOTE: Keep the transmission mounting bolts in order because they vary in size. This will expedite the installation process.

To install:

13. Lightly grease the input shaft splines. Fit the transmission into place and start all the transmission mounting bolts. Make sure the input shaft fits properly into the clutch disc and pilot bearing. Torque the transaxle bolts as follows:

   a. Bolts No. 1 and No. 2 torque to 29–39 ft. lbs. (39–49 Nm).

   b. Bolts No. 3 torque to 22–29 ft. lbs. (29–39 Nm).

   c. Gusset-to-engine bolts torque to 22–29 ft. lbs. (29–39 Nm).

14. Install the crossmember and torque the crossmember-to-frame mounting bolts to 57–77 ft. lbs. (77–105 Nm).

15. Install the transmission mount and torque the nuts to 32–41 ft. lbs. (43–55 Nm).

16. Remove the transmission jack from the transmission and the jack from under the oil pan.

17. Install the starter motor assembly.

18. Connect the vehicle speed sensor, back-up lamp and neutral position switch harness connectors.

19. Install the exhaust tube mounting and brackets.

20. Install the clutch slave cylinder and torque the mounting bolts to 22–30 ft. lbs. (30–40 Nm).

21. Install the drive shaft and align the matchmarks on the pinion flange. Install the attaching bolts.

22. Install the crankshaft position sensor.

23. If removed, install the splash pan or skid plate.

24. Refill the transmission with oil and lower the vehicle.

25. Install the shift lever into the transmission and install the snapring. Install the shifter boot and the shifter knob.

26. Connect the negative battery cable and road test the vehicle.

4WD Models

1. Disconnect the negative battery cable.

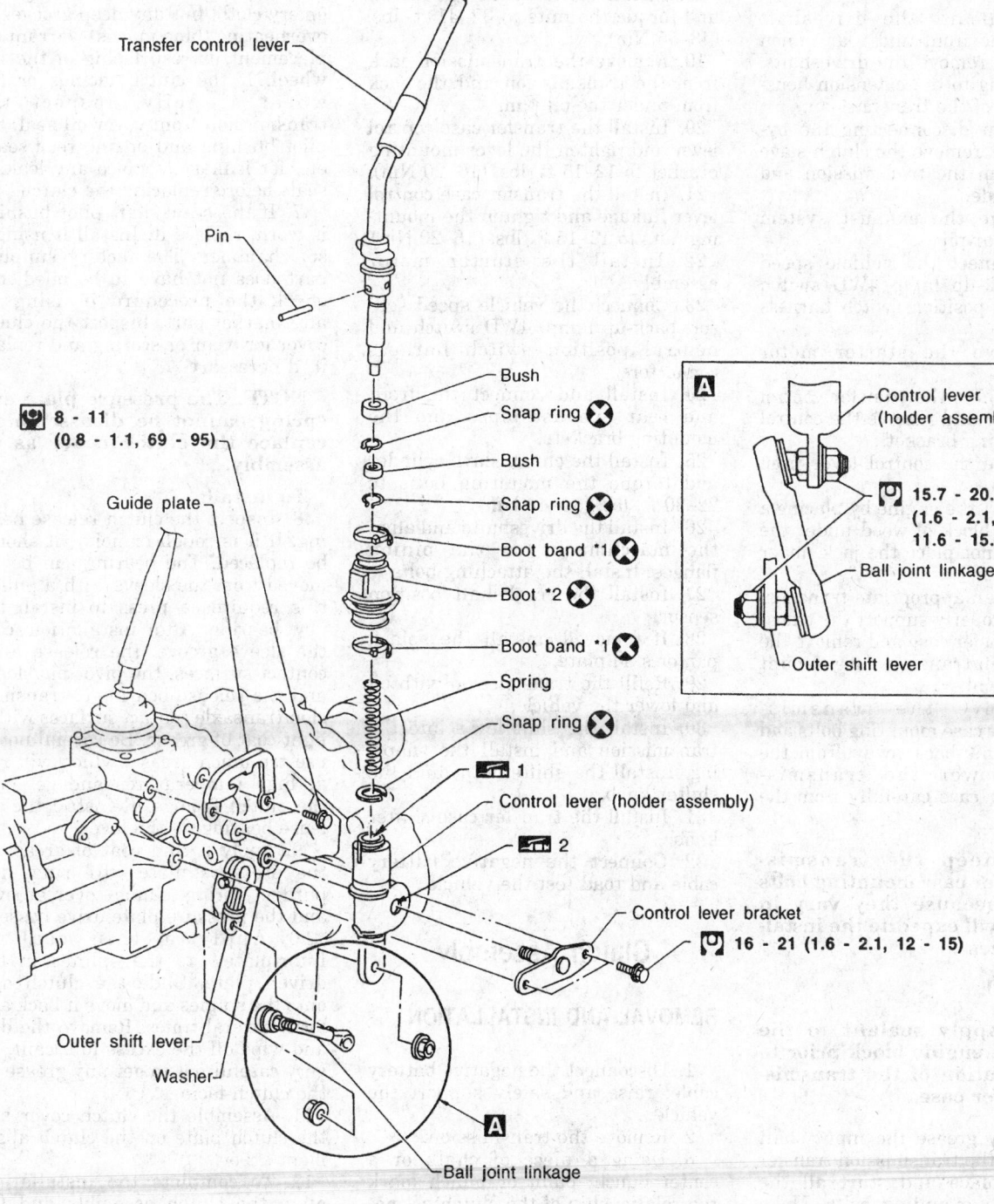

: N·m (kg-m, in-lb)

: N·m (kg-m, ft-lb)

1 : Fill multi-purpose grease up.

2 : Apply multi-purpose grease.

*1 : Securely bend pawls during assembly.
Be careful not to damage boot.

*2 : Do not touch boot with a sharp-pointed
or a hard tool as it breaks easily.

Control knob

Transfer control lever

Pin

8 - 11
(0.8 - 1.1, 69 - 95)

Guide plate

Bush
Snap ring
Bush
Snap ring
Boot band *1
Boot *2
Boot band *1
Spring
Snap ring

1

Control lever (holder assembly)

2

Control lever bracket

16 - 21 (1.6 - 2.1, 12 - 15)

Outer shift lever
Washer

Ball joint linkage

A
Control lever
(holder assembly)

15.7 - 20.6
(1.6 - 2.1,
11.6 - 15.2)

Ball joint linkage

Outer shift lever

311627

Transfer case shifter lever — 1996–97 Pathfinder with 4WD

2. Remove the shifter knob and boot, then remove the snapring to lift the shift lever out of the transmission. Place a rag in the opening to keep dirt out of the transmission.

3. Remove the control knob from the transfer case control lever.

4. Raise and safely support the vehicle. If equipped, remove the splash pan or skid plate.

5. Remove the crankshaft position sensor from the upper side of the transmission.

6. Matchmark the driveshaft flange at the front and rear pinion flanges and remove the driveshafts. Install a plug to the extension housing opening of the transmission.

7. Without disconnecting the hydraulic hose, remove the clutch slave cylinder from the transmission and secure it aside.

8. Remove the exhaust system front and rear pipes.

9. Disconnect the vehicle speed sensor, back-up lamp 4WD switch and neutral position switch harness connectors.

10. Remove the starter motor assembly.

11. Disconnect the transfer control lever linkage and remove the control lever mounting bracket.

12. Remove the control lever from the vehicle.

13. Support the engine by placing a jack with a block of wood under the oil pan. Do not place the jack under the drain plug.

14. Using an appropriate transmission jack, properly support the transmission/transfer case and remove the transmission/transfer case mount and crossmember.

15. Remove the transmission/transfer case mounting bolts and move the unit back away from the engine. Lower the transmission/transfer case carefully from the vehicle.

NOTE: Keep the transmission/transfer case mounting bolts in order because they vary in size. This will expedite the installation process.

To install:

NOTE: Apply sealant to the rear of the engine block prior to the installation of the transmission/transfer case.

16. Lightly grease the input shaft splines. Fit the transmission/transfer case into place and start all the transmission mounting bolts. Make sure the input shaft fits properly into the clutch disc and pilot bearing. Torque the transaxle bolts as follows:

    a. Bolts No. 1 and No. 2 torque to 29–39 ft. lbs. (39–49 Nm).

    b. Bolts No. 3 torque to 22–29 ft. lbs. (29–39 Nm).

    c. Gusset-to-engine bolts torque to 22–29 ft. lbs. (29–39 Nm).

17. Install the crossmember and torque the crossmember-to-frame mounting bolts to 57–77 ft. lbs. (77–105 Nm).

18. Install the transmission mount and torque the nuts to 32–41 ft. lbs. (43–55 Nm).

19. Remove the transmission jack from the transmission and the jack from under the oil pan.

20. Install the transfer case control lever and tighten the lever mounting bracket to 12–15 ft. lbs. (16–20 Nm).

21. Install the transfer case control lever linkage and tighten the mounting nuts to 12–15 ft. lbs. (16–20 Nm).

22. Install the starter motor assembly.

23. Connect the vehicle speed sensor, back-up lamp, 4WD switch and neutral position switch harness connectors.

24. Install and connect the front and rear exhaust pipes and the mounting brackets.

25. Install the clutch slave cylinder and torque the mounting bolts to 22–30 ft. lbs. (30–40 Nm).

26. Install the driveshafts and align the matchmarks on the pinion flanges. Install the attaching bolts.

27. Install the crankshaft position sensor.

28. If removed, install the splash pan or skid plate.

29. Refill the transmission with oil and lower the vehicle.

30. Install the shift lever into the transmission and install the snapring. Install the shifter boot and the shifter knob.

31. Install the transfer case shifter knob.

32. Connect the negative battery cable and road test the vehicle.

Clutch Assembly

REMOVAL AND INSTALLATION

1. Disconnect the negative battery cable; raise and safely support the vehicle.

2. Remove the transmission.

3. Using a piece of chalk or a center punch, paint or punch mark the relationship of the clutch assembly to the flywheel so it can be reassembled in the same position from which it is removed.

4. Using a clutch aligning tool, insert it into the clutch disc hub.

5. Loosen the bolts attaching the clutch cover to the flywheel, a turn at a time in an alternating sequence, until the spring tension is relieved to avoid distorting or bending the clutch cover. Remove the clutch assembly.

6. Inspect the flywheel for scoring, roughness or signs of overheating. Light scoring may be cleaned up with emery cloth, but any deep grooves or overheating (blue marks) warrant replacement or resurfacing of the flywheel. If the clutch facings or flywheel are oily, inspect the transmission front cover oil seal, the pilot bushing and engine rear seals, etc. for leakage; replace any leaking seals before replacing the clutch.

7. If the crankshaft pilot bushing is worn, replace it. Install it using a soft hammer. The factory supplied part does not have to be oiled, but check the procedure if using an aftermarket part. Inspect the clutch cover for wear or scoring and replace it, if necessary.

NOTE: The pressure plate and spring cannot be disassembled; replace the clutch cover as an assembly.

To install:

8. Inspect the clutch release bearing. If it is rough or noisy, it should be replaced. The bearing can be removed from the sleeve with a puller; this requires a press to install the new bearing. After installation, coat the sleeve groove, the release lever contact surfaces, the pivot pin/sleeve and the release bearing-to-transmission/transaxle contact surfaces with a light coat of grease. Be careful not to use too much grease, which will run at high temperatures and get onto the clutch facings. Reinstall the release bearing on the lever.

9. Apply a thin coat of grease to the pressure plate wire ring, diaphragm spring, clutch cover grooves and the pressure plate drive bosses.

10. Apply a thin coat of Lubriplate® to the splines in the driven plate. Slide the clutch disc onto the splines and move it back and forth several times. Remove the disc and wipe off the excess lubricant. Be very careful not to get any grease on the clutch facings.

11. Assemble the clutch cover and the clutch plate on the clutch alignment arbor.

12. To complete the installation, align the clutch assembly and flywheel alignment marks and install

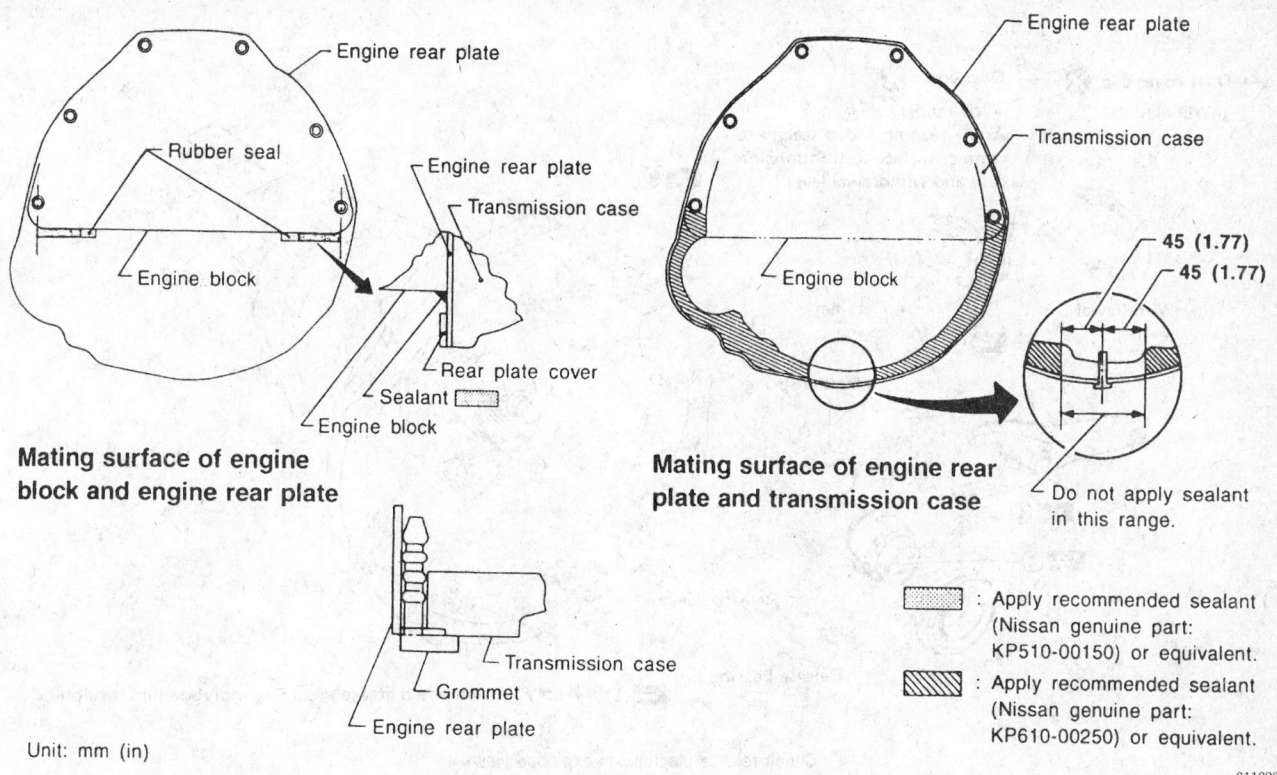

Mating surface of engine block and engine rear plate

Mating surface of engine rear plate and transmission case

45 (1.77)
45 (1.77)

Do not apply sealant in this range.

Unit: mm (in)

Application of sealant to the engine block

311628

: Apply recommended sealant (Nissan genuine part: KP510-00150) or equivalent.

: Apply recommended sealant (Nissan genuine part: KP610-00250) or equivalent.

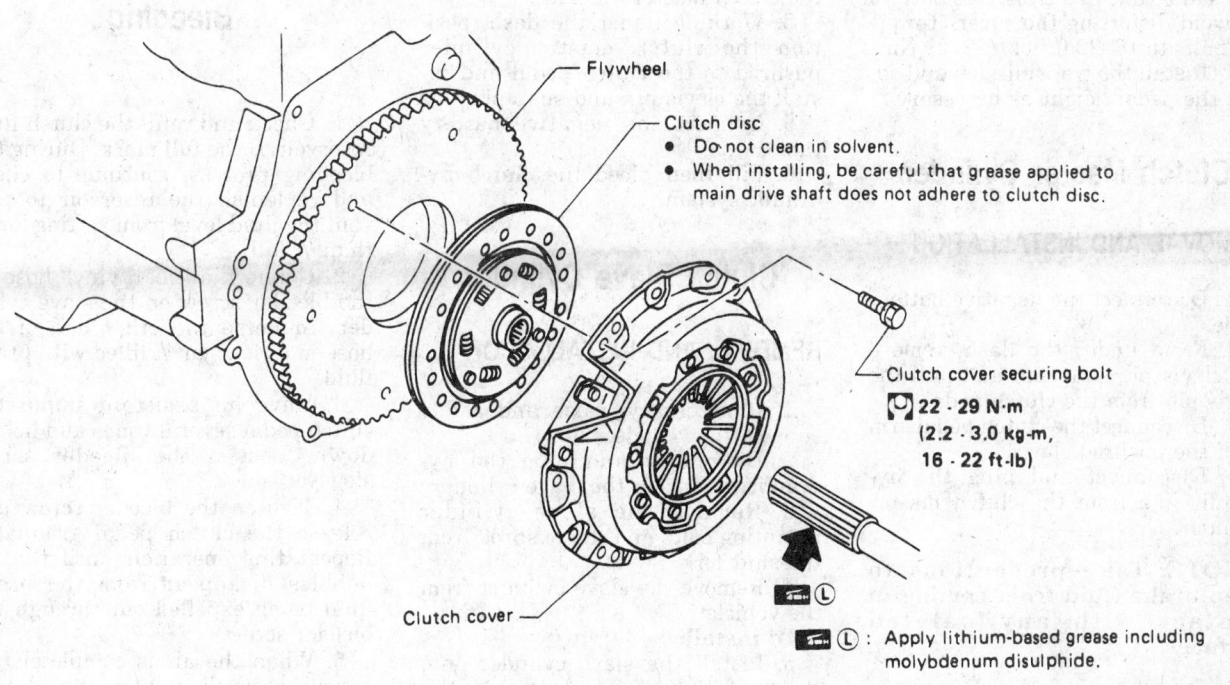

Flywheel

Clutch disc
- Do not clean in solvent.
- When installing, be careful that grease applied to main drive shaft does not adhere to clutch disc.

Clutch cover securing bolt
22 - 29 N·m
(2.2 - 3.0 kg-m,
16 - 22 ft-lb)

Clutch cover

: Apply lithium-based grease including molybdenum disulphide.

311459

Clutch assembly exploded view

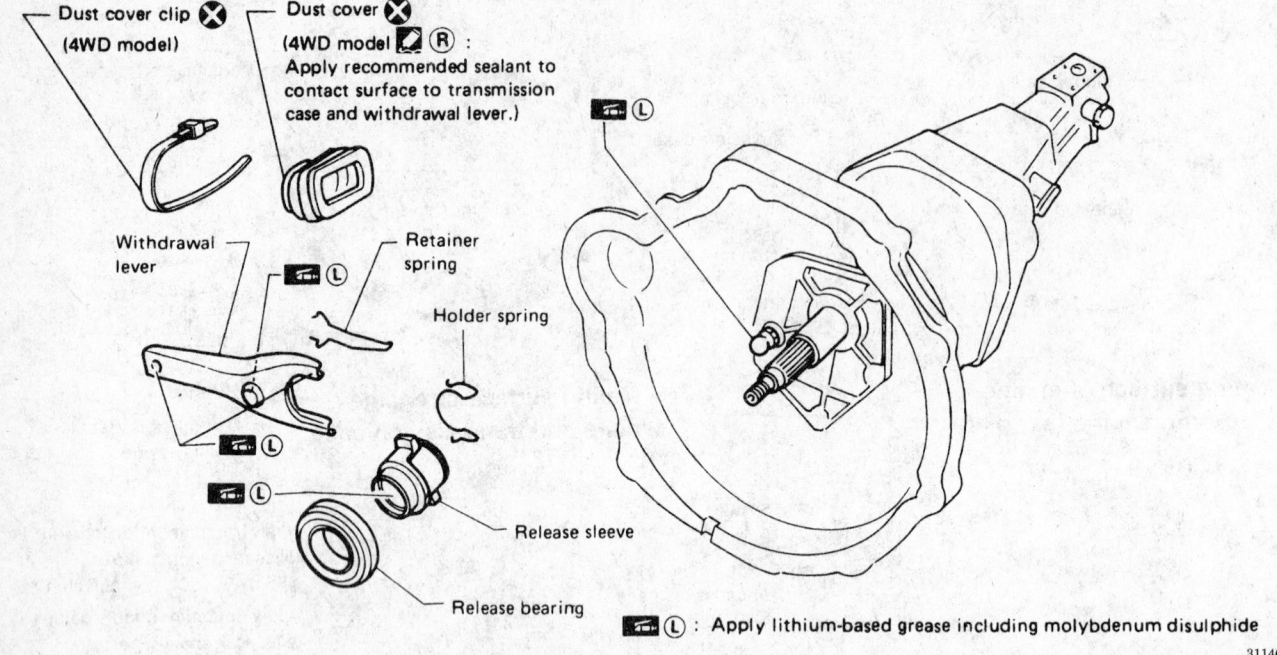

Clutch release mechanism exploded view

the bolts. Tighten the bolts 1 or 2 turns at a time in a crisscross pattern to avoid distorting the cover. Torque the bolts to 16–22 ft. lbs. (22–29 Nm).

13. Install the transmission and adjust the pedal height as necessary.

Clutch Master Cylinder

REMOVAL AND INSTALLATION

1. Disconnect the negative battery cable.

2. From under the dash, remove the clevis pin snap pin and pull the clevis pin from the clutch pedal.

3. Disconnect the clutch pedal arm from the pushrod clevis.

4. Disconnect and plug the hydraulic line from the clutch master cylinder.

NOTE: Take precautions to keep brake fluid from coming in contact with any painted surfaces.

5. Remove the clutch master cylinder.

To install:

6. Position the clutch master cylinder to the bulkhead and install the attaching nuts. Torque the nuts to 6–8 ft. lbs. (8–11 Nm).

7. Connect the hydraulic line to the clutch master cylinder.

8. Working under the dash, position the clutch master cylinder pushrod to the clutch pedal and install the clevis pin and snap pin.

9. Connect the negative battery cable.

10. Fill then bleed the clutch hydraulic system.

Clutch Slave Cylinder

REMOVAL AND INSTALLATION

1. If necessary, raise and safely support the vehicle.

2. Disconnect and plug the hydraulic hose from the slave cylinder.

3. Remove the slave cylinder mounting bolts and the pushrod from the shift fork.

4. Remove the slave cylinder from the vehicle.

To install:

5. Install the slave cylinder onto the bell housing and torque the mounting bolts to 22–30 ft. lbs. (30–40 Nm).

6. Connect the hydraulic hose to the slave cylinder.

7. Fill then bleed the clutch hydraulic system.

Hydraulic Clutch System Bleeding

1. Check and refill the clutch fluid reservoir to the full mark. During the bleeding process, continue to check and replenish the reservoir to prevent the fluid level from getting lower than ½ full.

2. Connect a clear vinyl hose to the bleeder screw on the slave cylinder. Immerse the other end of the hose in a clear jar ½ filled with brake fluid.

3. Have an assistant pump the clutch pedal several times and hold it down. Loosen the bleeder screw slowly.

4. Tighten the bleeder screw and release the clutch pedal gradually. Repeat this operation until the air bubbles disappear from the brake fluid being expelled out through the bleeder screw.

5. When the air is completely removed, securely tighten the bleeder screw and replace the dust cap.

6. Check and refill the master cylinder reservoir as necessary.

7. Depress the clutch pedal several times to check the operation of the clutch and check for leaks.

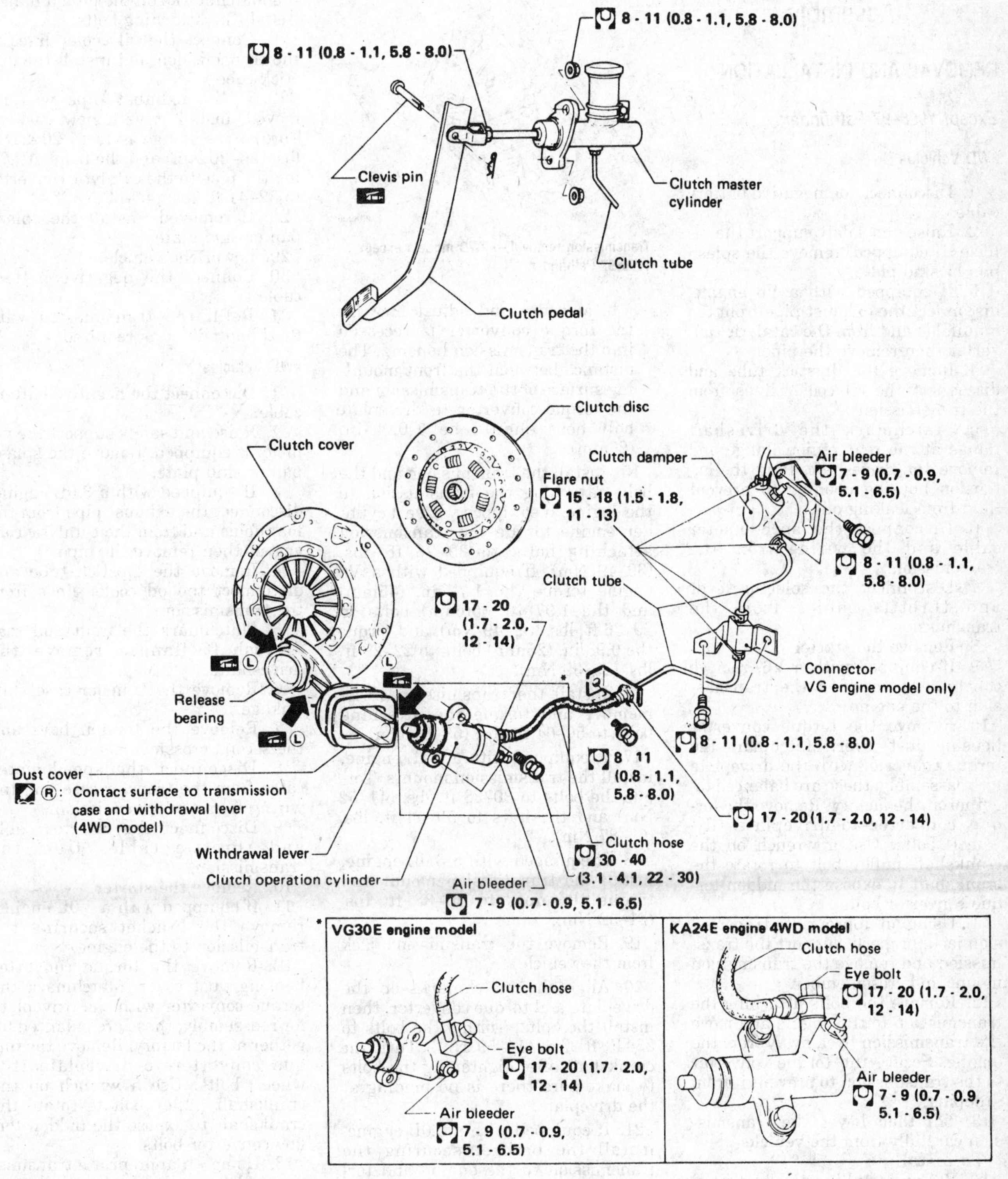

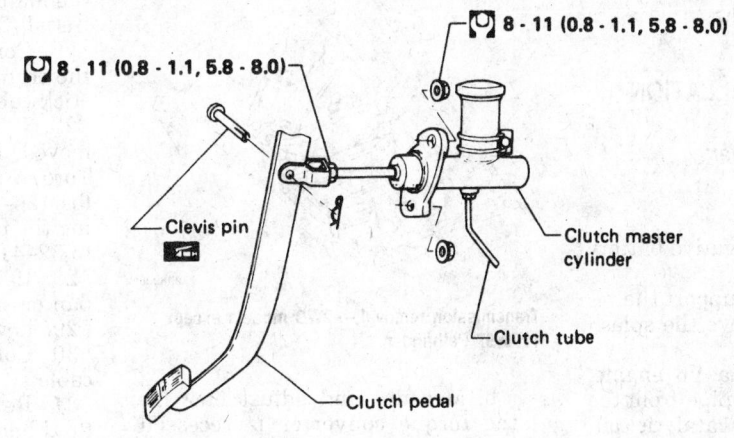

8 - 11 (0.8 - 1.1, 5.8 - 8.0)

8 - 11 (0.8 - 1.1, 5.8 - 8.0)

Clevis pin

Clutch master cylinder

Clutch tube

Clutch pedal

Clutch cover

Clutch disc

Clutch damper

Air bleeder
7 - 9 (0.7 - 0.9, 5.1 - 6.5)

Flare nut
15 - 18 (1.5 - 1.8, 11 - 13)

8 - 11 (0.8 - 1.1, 5.8 - 8.0)

Clutch tube

17 - 20 (1.7 - 2.0, 12 - 14)

Connector
VG engine model only

Release bearing

8 - 11 (0.8 - 1.1, 5.8 - 8.0)

Dust cover
Ⓡ : Contact surface to transmission case and withdrawal lever (4WD model)

17 - 20 (1.7 - 2.0, 12 - 14)

8 - 11
(0.8 - 1.1, 5.8 - 8.0)

Withdrawal lever

Clutch operation cylinder

Clutch hose
30 - 40
(3.1 - 4.1, 22 - 30)

Clutch hose

Air bleeder
7 - 9 (0.7 - 0.9, 5.1 - 6.5)

* VG30E engine model

Clutch hose

Eye bolt
17 - 20 (1.7 - 2.0, 12 - 14)

Air bleeder
7 - 9 (0.7 - 0.9, 5.1 - 6.5)

KA24E engine 4WD model

Clutch hose

Eye bolt
17 - 20 (1.7 - 2.0, 12 - 14)

Air bleeder
7 - 9 (0.7 - 0.9, 5.1 - 6.5)

Ⓡ : Apply recommended sealant (Nissan genuine part: KP115-00100) or equivalent.

Ⓛ : Apply lithium-based grease including molybdenum disulphide.

: N·m (kg-m, ft-lb)

311490

Clutch hydraulic system component identification

Automatic Transmission Assembly

REMOVAL AND INSTALLATION

Except 1996–97 Pathfinder

2WD Vehicles

1. Disconnect the negative battery cable.

2. Raise and safely support the vehicle. If equipped, remove the splash pan or skid plate.

3. If equipped with a V6 engine, disconnect the exhaust pipe from the manifolds and from the catalytic converter then remove the pipe.

4. Remove the dipstick tube and disconnect the oil cooler lines from the transmission.

5. Matchmark the driveshaft flange at the rear pinion flange and remove the driveshaft. Plug the extension housing opening to prevent fluid from leaking out.

6. Disconnect the speedometer cable and the wiring from the transmission.

7. Disconnect the selector lever and throttle cables from the transmission.

8. Remove the starter.

9. If equipped with a V6, remove the bracket securing the transmission to the engine.

10. Remove the torque converter housing dust cover. Matchmark the torque converter with the driveplate for reassembly; these are balanced together at the factory. Remove the torque converter–to–driveplate (flywheel) bolts. Use a wrench on the crankshaft pulley bolt to rotate the crankshaft to expose the hidden torque converter bolts.

11. Using an appropriate transmission jack, properly support the transmission and remove the transmission mount and crossmember.

12. Remove the bolts attaching the transmission to the engine and move the transmission back away from the engine. Secure the torque converter to the transmission to prevent it from dropping.

13. Tilt then lower the transmission carefully from the vehicle.

To install:

14. Before installing the transmission, perform the following checks:

 a. Use a dial indicator to check the driveplate run-out while turning the crankshaft. Maximum allowable run-out is 0.020 in. (0.5mm); if beyond specification, replace the driveplate.

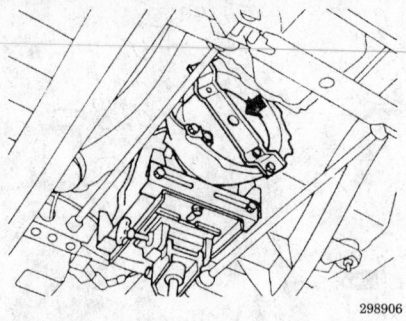

298906

Transmission removal — 2WD models except 1996–97 Pathfinder

 b. Measure and adjust how far the torque converter is recessed into the transmission housing. The distance between the front mounting surface of the transmission and the torque converter–to–driveplate bolt boss should be 1.024 in. (26mm).

15. Install the transmission and the bolts attaching the transmission to the engine. If equipped with a 4 cylinder engine torque the transmission attaching bolts to 29–36 ft. lbs. (39–49 Nm). If equipped with a V6 engine torque the 1.77 in. (45mm) and the 1.97 in. (50mm) bolts to 29–36 ft. lbs. (39–49 Nm), and torque the 0.98 in. (25mm) bolts to 22–29 ft. lbs. (29–39 Nm).

16. Install the transmission crossmember and torque the attaching bolts to 50–64 ft. lbs. (68–87 Nm).

17. If equipped with a 2.4L engine, install the transmission mounts. Torque the bolts to 30–38 ft. lbs. (41–52 Nm) and the nuts to 50–64 ft. lbs. (68–87 Nm).

18. If equipped with a 3.0L engine, install the transmission mount and torque the nuts to 30–38 ft. lbs. (41–52 Nm).

19. Remove the transmission jack from the vehicle.

20. Align the matchmarks on the driveplate and torque converter, then install the bolts. Torque the bolts to 33–43 ft. lbs. (44–59 Nm). Turn the crankshaft after tightening the bolts to make sure there is no binding at the driveplate.

21. If equipped with a 3.0L engine, install the bracket securing the transmission to the engine and torque the attaching bolts to 22–29 ft. lbs. (29–39 Nm).

22. Install the starter.

23. Connect the selector lever and throttle cables to the transmission.

24. Connect the wiring to the transmission and the speedometer cable.

25. Install the driveshaft and align the matchmarks on the pinion flange. Install the attaching bolts.

26. Connect the oil cooler lines to the transmission and install the dipstick tube.

27. If the exhaust pipe was removed, install it with new gaskets. Torque the flange nuts to 20–27 ft. lbs. (26–36 Nm) and the bolts attaching the pipe to the catalytic converter to 32–41 ft. lbs. (43–55 Nm).

28. If removed, install the splash pan or skid plate.

29. Lower the vehicle.

30. Connect the negative battery cable.

31. Refill the transmission with fluid and adjust as required.

4WD Vehicles

1. Disconnect the negative battery cable.

2. Raise and safely support the vehicle. If equipped, remove the splash pan or skid plate.

3. If equipped with a 3.0L engine, disconnect the exhaust pipe from the manifolds and from the catalytic converter, then remove the pipe.

4. Remove the dipstick tube and disconnect the oil cooler lines from the transmission.

5. Matchmark the front and rear driveshaft flanges remove the driveshafts.

6. Remove the transfer case shift linkage.

7. Remove the torsion bars and the second crossmember.

8. Disconnect the speedometer cable from the transfer case and the wiring from the transmission.

9. Disconnect the selector cable and throttle cables from the transmission.

10. Remove the starter.

11. If equipped with a 3.0L engine, remove the bracket securing the transmission to the engine.

12. Remove the torque converter housing dust cover. Matchmark the torque converter with the driveplate for reassembly; these are balanced together at the factory. Remove the torque converter–to–driveplate (flywheel) bolts. Use a wrench on the crankshaft pulley bolt to rotate the crankshaft to expose the hidden torque converter bolts.

13. Using an appropriate transmission jack, properly support the transmission and transfer case and remove the transmission mount and crossmember.

14. Remove the bolts attaching the transmission to the engine and remove the transmission and transfer case from the vehicle.

To install:

15. Before installing the transmission, perform the following checks:

a. Use a dial indicator to check the driveplate run-out while turning the crankshaft. Maximum allowable run-out is 0.020 in. (0.5mm); if beyond specification, replace the driveplate.

b. Measure and adjust how far the torque converter is recessed into the transmission housing. The distance between the front mounting surface of the transmission and the torque converter–to–driveplate bolt boss should be at least 1.024 in. (26mm).

16. Install the transmission and the bolts attaching the transmission to the engine. If equipped with a 2.4L engine, torque the transmission attaching bolts to 29–36 ft. lbs. (39–49 Nm). If equipped with a 3.0L engine, torque the 1.77 in. (45mm) and the 1.97 in. (50mm) bolts to 29–36 ft. lbs. (39–49 Nm), and torque the 0.98 in. (25mm) bolts to 22–29 ft. lbs. (29–39 Nm).

17. Install the transmission crossmember. If equipped with a 2.4L engine, torque the bolts to 50–64 ft. lbs. (68–87 Nm). If equipped with a 3.0L engine, torque the attaching bolts to 30–38 ft. lbs. (41–52 Nm).

18. If equipped with a 2.4L engine, install the transmission mounts. Torque the bolts to 30–38 ft. lbs. (41–52 Nm) and the nuts to 50–64 ft. lbs. (68–87 Nm).

19. If equipped with a 3.0L engine, install the transmission mount and torque the nuts to 30–38 ft. lbs. (41–52 Nm).

20. Remove the transmission jack from the vehicle.

21. Align the matchmarks on the driveplate and torque converter, then install the bolts. Torque the bolts to 33–43 ft. lbs. (44–59 Nm). Turn the crankshaft after tightening the bolts to make sure there is no binding at the driveplate.

22. If equipped with a 3.0L engine, install the bracket securing the transmission to the engine and torque the attaching bolts to 22–29 ft. lbs. (29–39 Nm).

23. Connect the selector cable and throttle cables to the transmission.

24. Connect the speedometer cable to the transfer case and the wiring to the transmission.

25. Install the second crossmember into the vehicle.

26. Install the torsion bars to their original location. Make sure the splines are in their original position

and set the adjustment to its original position.

27. Install the transfer case shift lever.

28. Install the front and rear driveshafts, making sure to align the matchmarks.

29. Connect the oil cooler lines to the transmission and install the dipstick tube.

30. If the exhaust pipe was removed, install it with new gaskets. Torque the flange nuts to 20–27 ft. lbs. (26–36 Nm) and the bolts attaching the pipe to the catalytic converter to 32–41 ft. lbs. (43–55 Nm).

31. If removed, install the splash pan or skid plate.

32. Lower the vehicle.

33. Connect the negative battery cable.

34. Refill the transmission with fluid and adjust as required.

1996–97 Pathfinder

2WD Vehicles

1. Disconnect the negative battery cable.

2. Raise and safely support the vehicle. If equipped, remove the splash pan or skid plate.

3. Remove the dipstick tube and disconnect the oil cooler lines from the transmission. Be sure to plug the fluid line openings.

4. Matchmark the driveshaft flange at the rear pinion flange and remove the driveshaft. Plug the extension housing opening to prevent fluid from leaking out.

5. Disconnect the speedometer cable and the wiring from the transmission.

6. Disconnect the selector lever and throttle cables from the transmission.

7. Remove the starter.

8. Remove the torque converter housing access cover.

9. Matchmark the torque converter with the driveplate for reas-

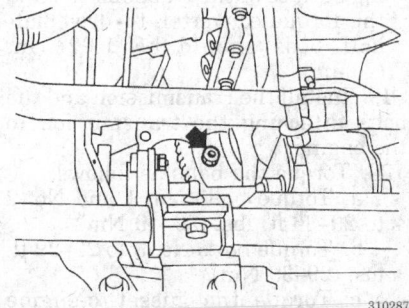

Location of the torque converter bolts — 1996–97 Pathfinder with 2WD

sembly; these are balanced together at the factory.

10. Remove the torque converter–to–driveplate (flywheel) bolts. Use a wrench on the crankshaft pulley bolt to rotate the crankshaft to expose the hidden torque converter bolts.

11. Using an appropriate transmission jack, properly support the transmission and remove the transmission mount and crossmember.

NOTE: The bolts that secure the transmission are different in length. Be sure to note the proper positioning for reinstallation.

12. Remove the bolts attaching the transmission to the engine and move the transmission back away from the engine.

———— **CAUTION** ————

Secure the torque converter to the transmission to prevent it from falling out.

13. Tilt, then lower the transmission carefully from the vehicle.

To install:

14. Before installing the transmission, perform the following checks:

a. Use a dial indicator to check the driveplate run-out while turning the crankshaft. Maximum allowable run-out is 0.020 in. (0.5mm); if beyond specification, replace the driveplate.

b. Measure and adjust how far the torque converter is recessed into the transmission housing. The distance between the front mounting surface of the transmission and the torque converter–to–driveplate bolt boss should be 1.024 in. (26mm).

15. Install the transmission and the bolts attaching the transmission to the engine.

16. Torque the bolts as follows:

a. Torque bolts No. 1 and No. 2 to 29–36 ft. lbs. (39–49 Nm).

b. Torque bolts No. 3 to 22–29 ft. lbs. (29–39 Nm).

c. Torque the gusset-to-engine bolts to 22–29 ft. lbs. (29–39 Nm).

17. Install the transmission crossmember and torque the crossmember-to-frame attaching bolts to 57–77 ft. lbs. (77–105 Nm). Torque the nuts that secure the transmission-to-crossmember to 32–41 ft. lbs. (43–55 Nm).

18. Remove the transmission jack from the vehicle.

19. Align the matchmarks on the driveplate and torque converter, then install the bolts. Torque the bolts to 33–43 ft. lbs. (44–59 Nm). Turn the

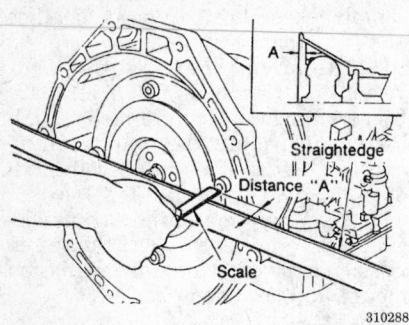

Measuring the torque converter mounting depth — 1996–97 Pathfinder with 2WD

310288

crankshaft after tightening the bolts to make sure there is no binding at the driveplate.

20. Install the torque converter access cover.

21. Install the starter assembly.

22. Connect the selector lever and throttle cables to the transmission.

23. Connect the wiring to the transmission and the speedometer cable.

24. Install the driveshaft and align the matchmarks on the pinion flange. Install the attaching bolts.

25. Connect the oil cooler lines to the transmission and install the dipstick tube.

26. If removed, install the splash pan or skid plate.

27. Lower the vehicle.

28. Connect the negative battery cable.

29. Refill the transmission with fluid and road test.

4WD Vehicles

1. Disconnect the negative battery cable.

2. Raise and safely support the vehicle. If equipped, remove the splash pan or skid plate.

3. Disconnect and remove the front and rear exhaust pipe from the vehicle.

4. Remove the dipstick tube and disconnect the oil cooler lines from the transmission. Be sure to plug the fluid line openings.

5. Matchmark the driveshaft flange at the rear pinion flange and remove the driveshaft. Plug the extension housing opening to prevent fluid from leaking out.

6. Matchmark the driveshaft flange at the front pinion flange and remove the driveshaft.

7. Disconnect the control linkage from the transfer case.

8. Disconnect the speedometer cable and the wiring from the transmission.

9. Disconnect the selector lever and throttle cables from the transmission.

10. Remove the starter.

11. Remove the torque converter housing access cover.

12. Matchmark the torque converter with the driveplate for reassembly; these are balanced together at the factory.

13. Remove the torque converter–to–driveplate (flywheel) bolts. Use a wrench on the crankshaft pulley bolt to rotate the crankshaft to expose the hidden torque converter bolts.

14. Using an appropriate transmission jack, properly support the transmission and transfer case.

15. Remove the transmission mount and crossmember.

NOTE: The bolts that secure the transmission are different in length. Be sure to note the proper positioning for reinstallation.

16. Remove the bolts attaching the transmission and transfer case to the engine and move the transmission back away from the engine.

--- CAUTION ---
Secure the torque converter to the transmission to prevent it from falling out.

17. Tilt then lower the transmission and transfer case carefully from the vehicle.

To install:

18. Before installing the transmission, perform the following checks:

   a. Use a dial indicator to check the driveplate run-out while turning the crankshaft. Maximum allowable run-out is 0.020 in. (0.5mm); if beyond specification, replace the driveplate.

   b. Measure and adjust how far the torque converter is recessed into the transmission housing. The distance between the front mounting surface of the transmission and the torque converter–to–driveplate bolt boss should be 1.024 in. (26mm).

19. Install the transmission and the bolts attaching the transmission to the engine.

20. Torque the bolts as follows:

   a. Torque bolts No. 1 and No. 2 to 29–36 ft. lbs. (39–49 Nm).

   b. Torque bolts No. 3 to 22–29 ft. lbs. (29–39 Nm).

   c. Torque the gusset-to-engine bolts to 22–29 ft. lbs. (29–39 Nm).

21. Install the transmission crossmember and torque the cross-member-to-frame attaching bolts to 57–77 ft. lbs. (77–105 Nm). Torque the nuts that secure the transmission-to-crossmember to 32–41 ft. lbs. (43–55 Nm).

22. Remove the transmission jack from the vehicle.

23. Align the matchmarks on the driveplate and torque converter, then install the bolts. Torque the bolts to 33–43 ft. lbs. (44–59 Nm). Turn the crankshaft after tightening the bolts to make sure there is no binding at the driveplate.

24. Install the torque converter access cover.

25. Install the starter assembly.

26. Connect the selector lever and throttle cables to the transmission.

27. Connect the wiring to the transmission and the speedometer cable.

28. Connect the control linkage to the transfer case.

29. Install the driveshafts and align the matchmarks on the pinion flanges. Install the attaching bolts.

30. Connect the oil cooler lines to the transmission and install the dipstick tube.

31. Using new gaskets, install the exhaust pipes.

32. If removed, install the splash pan or skid plate.

33. Lower the vehicle.

34. Connect the negative battery cable.

35. Refill the transmission with fluid and road test.

Throttle Valve Cable

ADJUSTMENT

NOTE: If so equipped, make sure the Automatic Speed Control Device (ASCD) wire is not pulling the throttle drum.

1. Loosen the locknut, then tighten the adjusting nut until the throttle drum starts to move.

2. From that position, turn back the adjusting nut 1½–2 turns, then secure the locknut.

Transfer Case Assembly

REMOVAL AND INSTALLATION

1. Disconnect the negative battery cable.

2. Raise and safely support the vehicle. If equipped, remove the splash pan or skid plate.

3. Remove the starter. Drain the oil from both the transmission and the transfer case.

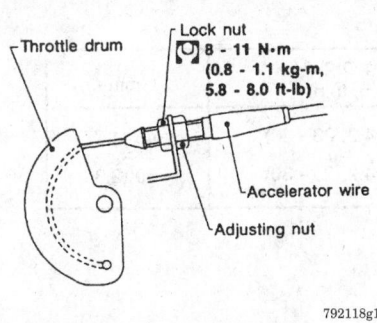

Loosen the locknut before turning the cable's adjusting nut

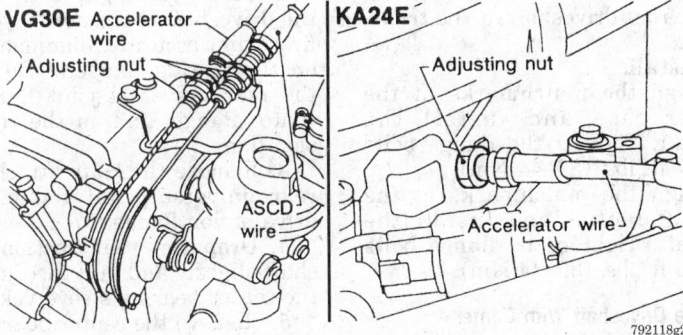

Be careful not to confuse the ASCD and accelerator cables

4. Matchmark the driveshaft flange at the rear differential pinion flange and at both front driveshaft flanges. Remove both driveshafts.

5. Disconnect the selector lever assembly from the transfer case.

6. The torsion bars must be removed:

a. Working under the vehicle, measure and record the length of the threads on the torsion bar adjustment.

b. At the front of the bar, pull the boot back and matchmark the bar to the mounting plate. The spline on the bar must be re-installed in the same position on the plate.

c. Remove the locknut and adjustment nut and remove the 3 nuts at the mounting plate to remove each bar. Mark the bars left and right side for proper installation.

7. Using an appropriate transmission jack, properly support the transmission and remove the transmission mount and crossmember.

8. Remove the transfer case-to-transmission bolts and move the unit back away from the transmission.

NOTE: Mounting bolts are different lengths on manual transmission models.

To install:

9. Clean the mating surfaces and apply a bead of silicone sealant to the transfer case mounting flange.

10. Carefully fit the case into place and start all the mounting bolts. Torque the bolts to 30 ft. lbs. (41 Nm).

11. Install the crossmember and torque the bolts to 58 ft. lbs. (78 Nm). Install the mount bolts and torque to 38 ft. lbs. (52 Nm).

12. Install the driveshafts and make sure to align the matchmarks:

a. On the front driveshaft, torque the bolts to 33 ft. lbs. (44 Nm).

b. On two-piece rear driveshafts, torque the flange bolts to 33 ft. lbs. (44 Nm) and the center bearing bracket bolts to 16 ft. lbs. (22 Nm).

c. On single-piece rear driveshafts, torque the flange bolts to 65 ft. lbs. (88 Nm).

13. Install the selector lever assembly.

14. Install the torsion bars in their original location. Make sure the splines are in their original position and set the adjustment to its original position.

15. Install the remaining components and fill the transfer case and transmission with oil. Check and adjust front suspension height.

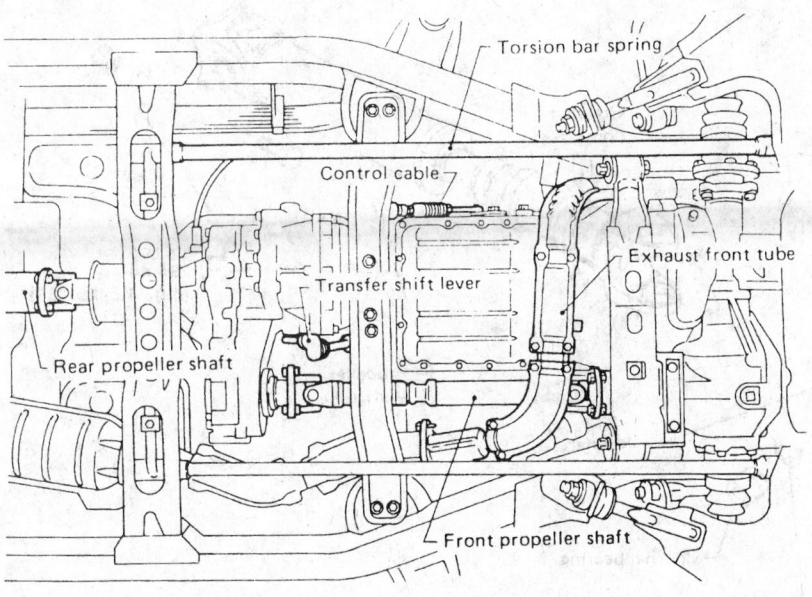

Transfer case and related component locations — typical 4WD model

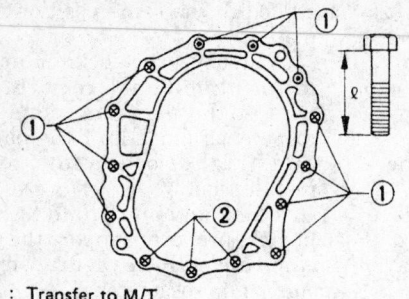

Bolt No.	Tightening torque N·m (kg-m, ft-lb)	l mm (in)
①	31 - 41 (3.2 - 4.2, 23 - 30)	45 (1.77)
②	31 - 41 (3.2 - 4.2, 23 - 30)	60 (2.36)

⊙ : Transfer to M/T
⊗ : M/T to transfer

301903

Transfer case mounting bolt specifications

DRIVE AXLE

Driveshaft

REMOVAL AND INSTALLATION

Except 1996–97 Pathfinder

Front Driveshaft

1. Matchmark the flanges and unbolt the front driveshaft at the front differential.

2. Matchmark the flanges and unbolt the front driveshaft at the transfer case.

To install:

3. Align the matchmarks at the transfer case and install the driveshaft. Tighten the flange bolts to 29–33 ft. lbs. (39–44 Nm).

4. Align the matchmarks at the front differential and install the driveshaft. Tighten the flange bolts to 29–33 ft. lbs. (39–44 Nm).

Two-Piece Driveshaft With Center Bearing — Rear 2WD Models

1. Raise the rear of the truck and support the rear axle housing on jackstands.

2. Before you begin to disassemble the driveshaft components, you must first paint accurate alignment marks on the mating flangers. Do this on the rear universal joint flange, the center flange, and on the transmission flange.

3. Remove the bolts attaching the rear universal joint flange to the drive pinion flange.

4. Drop the rear section of the shaft slightly and pull the unit out of the center bearing sleeve yoke.

5. Remove the center bearing support from the crossmember.

6. Separate the transmission output flange and remove the front half

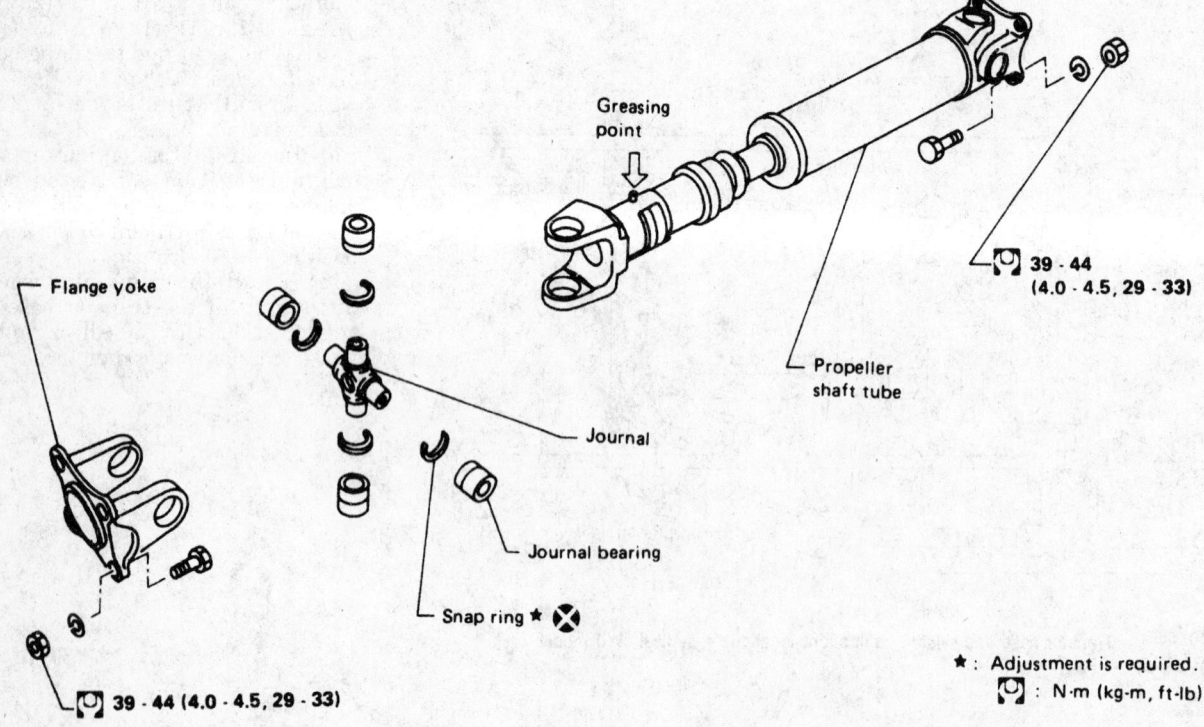

Exploded view of model 2F71H front driveshaft — 4WD models

300372

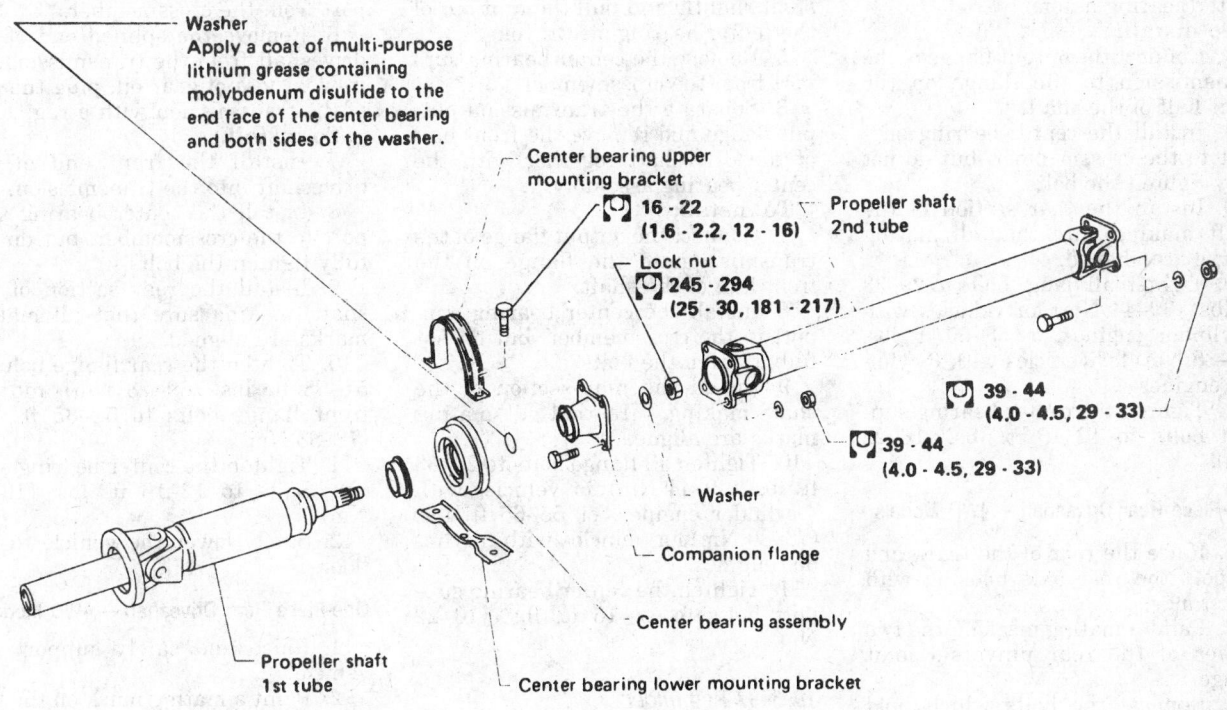

— Washer
Apply a coat of multi-purpose
lithium grease containing
molybdenum disulfide to the
end face of the center bearing
and both sides of the washer.

— Center bearing upper
mounting bracket

🔧 16 - 22
(1.6 - 2.2, 12 - 16)

— Lock nut
🔧 245 - 294
(25 - 30, 181 - 217)

Propeller shaft
2nd tube

🔧 39 - 44
(4.0 - 4.5, 29 - 33)

🔧 39 - 44
(4.0 - 4.5, 29 - 33)

— Washer

— Companion flange

— Center bearing assembly

— Propeller shaft
1st tube

— Center bearing lower mounting bracket

300373

Exploded view of model 3S71A rear driveshaft — 4-cylinder 2WD Pick-up

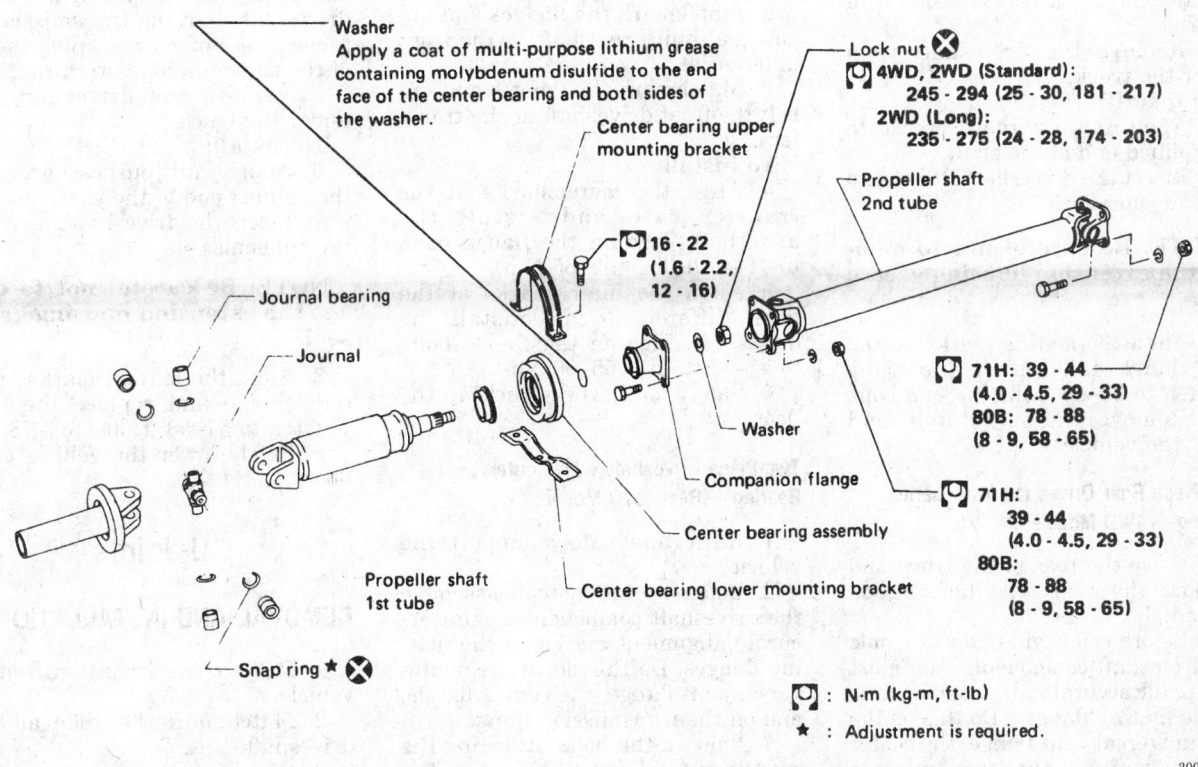

— Washer
Apply a coat of multi-purpose lithium grease
containing molybdenum disulfide to the end
face of the center bearing and both sides of
the washer.

— Lock nut ⊗
🔧 **4WD, 2WD (Standard):**
245 - 294 (25 - 30, 181 - 217)
2WD (Long):
235 - 275 (24 - 28, 174 - 203)

— Center bearing upper
mounting bracket

— Propeller shaft
2nd tube

🔧 16 - 22
(1.6 - 2.2,
12 - 16)

— Journal bearing

— Journal

— Washer

— Companion flange

🔧 **71H:** 39 - 44
(4.0 - 4.5, 29 - 33)
80B: 78 - 88
(8 - 9, 58 - 65)

🔧 **71H:**
39 - 44
(4.0 - 4.5, 29 - 33)
80B:
78 - 88
(8 - 9, 58 - 65)

— Center bearing assembly

— Propeller shaft
1st tube

— Center bearing lower mounting bracket

— Snap ring ★ ⊗

🔧 : N·m (kg-m, ft-lb)

★ : Adjustment is required.

300374

Exploded view of model 3S80B rear driveshaft — 2WD long bed Pick-ups and 6-cylinder 2WD/4WD vehicles

of the driveshaft together with the center bearing assembly.

To install:

7. Connect the output flange of the transmission to the flange on the front half of the shaft.

8. Install the center bearing support to the crossmember, but do not fully tighten the bolts.

9. Install the rear section of the shaft making sure that all mating marks are aligned.

10. Tighten all flange bolts to 29–33 ft. lbs. (39–44 Nm) for vehicles with 4-cylinder engines, or 58–65 ft. lbs. (78–88 Nm) for vehicles with 6-cylinder engines.

11. Tighten the center bearing support bolts to 12–16 ft. lbs. (16–22 Nm).

One-Piece Rear Driveshaft — 4WD Models

1. Raise the rear of the truck and support the rear axle housing with jackstands.

2. Paint a mating mark on the two halves of the rear universal joint flange.

3. Remove the bolts which hold the rear flange together.

4. Remove the splined end of the driveshaft from the transmission. If you don't want to lose a lot of gear oil, plug the end of the transmission with a rag.

5. Remove the driveshaft from under the truck.

To install:

6. Apply multi-purpose grease to the splined end of the shaft.

7. Insert the driveshaft sleeve into the transmission.

NOTE: Be careful not to damage the extension housing grease seal.

8. Align the mating marks on the rear flange and replace the bolts. Tighten to 58–65 ft. lbs. (78–88 Nm).

9. Remove the jackstands and lower the vehicle.

Two-Piece Rear Driveshaft With Center Bearing — 4WD Models

1. Raise the rear of the truck and support the rear axle housing on jackstands.

2. Before you begin to disassemble the driveshaft components, you must first paint accurate alignment marks on the mating flanges. Do this on the rear universal joint flange, the center flange, and on the transmission flange.

3. Remove the bolts attaching the rear universal joint flange to the drive pinion flange.

4. Drop the rear section of the shaft slightly and pull the unit out of the center bearing sleeve yoke.

5. Remove the center bearing support from the crossmember.

6. Separate the transmission output flange and remove the front half of the driveshaft together with the center bearing assembly.

To install:

7. Connect the output flange of the transmission to the flange on the front half of the shaft.

8. Install the center bearing support to the crossmember, but do not fully tighten the bolts.

9. Install the rear section of the shaft making sure that all mating marks are aligned.

10. Tighten all flange bolts to 29–33 ft. lbs. (39–44 Nm) for vehicles with 4-cylinder engines, or 55–65 ft. lbs. (78–88 Nm) for vehicles with 6-cylinder engines.

11. Tighten the center bearing support bolts to 12–16 ft. lbs. (16–22 Nm).

1996–97 Pathfinder

Front Driveshaft

1. Raise and safely support the vehicle.

2. Matchmark the flanges and unbolt the front driveshaft at the front differential.

3. Matchmark the flanges and unbolt the front driveshaft at the transfer case.

To install:

4. Align the matchmarks at the transfer case and install the driveshaft. Tighten the flange bolts to 41–48 ft. lbs. (55–65 Nm).

5. Align the matchmarks at the front differential and install the driveshaft. Tighten the flange bolts to 41–48 ft. lbs. (55–65 Nm).

6. Safely lower the vehicle to the floor.

Two-Piece Driveshaft With Center Bearing — Rear 2WD Models

1. Raise and safely support the vehicle.

2. Before beginning to disassemble the driveshaft components, paint accurate alignment marks on the mating flanges. Do this on the rear universal joint flange, the center flange, and on the transmission flange.

3. Remove the bolts attaching the rear universal joint flange to the drive pinion flange.

4. Drop the rear section of the shaft slightly and pull the unit out of the center bearing sleeve yoke.

5. Remove the center bearing support from the crossmember.

6. Remove the splined end of the driveshaft from the transmission. To prevent loss of gear oil, plug the end of the transmission with a rag.

To install:

7. Install the front end of the driveshaft into the transmission.

8. Install the center bearing support to the crossmember, but do not fully tighten the bolts.

9. Install the rear section of the shaft, making sure that all mating marks are aligned.

10. Tighten the rear flange bolts to 51–58 ft. lbs. (69–78 Nm) and the front flange bolts to 58–65 ft. lbs. (78–88 Nm).

11. Tighten the center bearing support bolts to 12–16 ft. lbs. (16–22 Nm).

12. Safely lower the vehicle to the floor.

One-Piece Rear Driveshaft — 4WD Models

1. Raise and safely support the vehicle.

2. Paint a mating mark on the two halves of the rear universal joint flange.

3. Remove the bolts which hold the rear flange together.

4. Remove the splined end of the driveshaft from the transmission. To prevent loss of gear oil, plug the end of the transmission with a rag.

5. Remove the driveshaft from under the truck.

To install:

6. Apply multipurpose grease to the splined end of the shaft.

7. Insert the driveshaft sleeve into the transmission.

NOTE: Be careful not to damage the extension housing grease seal.

8. Align the mating marks on the rear flange and replace the bolts. Tighten to 51–58 ft. lbs. (69–78 Nm).

9. Safely lower the vehicle to the floor.

U-Joints

REMOVAL AND INSTALLATION

1. Remove the driveshaft from the vehicle.

2. Matchmark the yoke and the driveshaft.

3. Remove the snaprings from the bearings.

4. Position the yoke on vise jaws. Using a bearing remover and a hammer, gently tap the remover until the

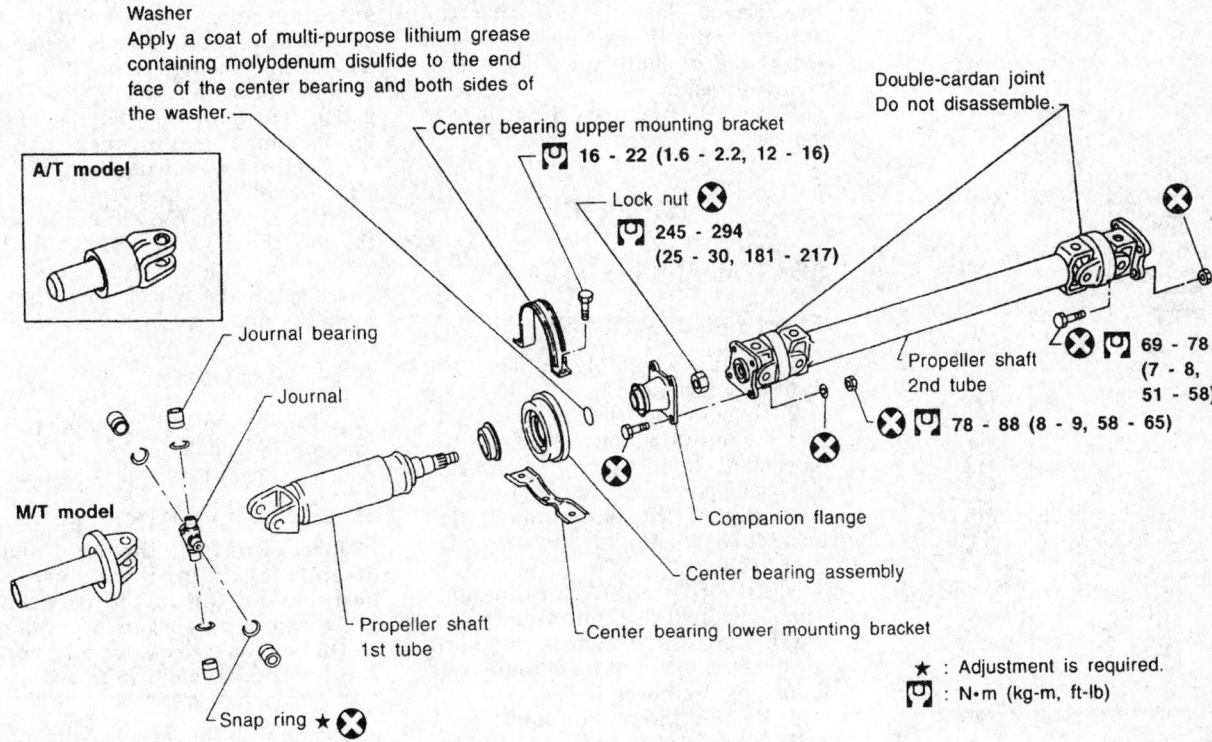

Washer
Apply a coat of multi-purpose lithium grease containing molybdenum disulfide to the end face of the center bearing and both sides of the washer.

A/T model

Journal bearing

Journal

M/T model

Propeller shaft 1st tube

Snap ring ★

Center bearing upper mounting bracket
16 - 22 (1.6 - 2.2, 12 - 16)

Lock nut
245 - 294
(25 - 30, 181 - 217)

Double-cardan joint
Do not disassemble.

Propeller shaft 2nd tube

69 - 78
(7 - 8, 51 - 58)

78 - 88 (8 - 9, 58 - 65)

Companion flange

Center bearing assembly

Center bearing lower mounting bracket

★ : Adjustment is required.
[□] : N·m (kg-m, ft-lb)

306120

Rear driveshaft model 3S80B-D — 1996–97 2WD Pathfinder

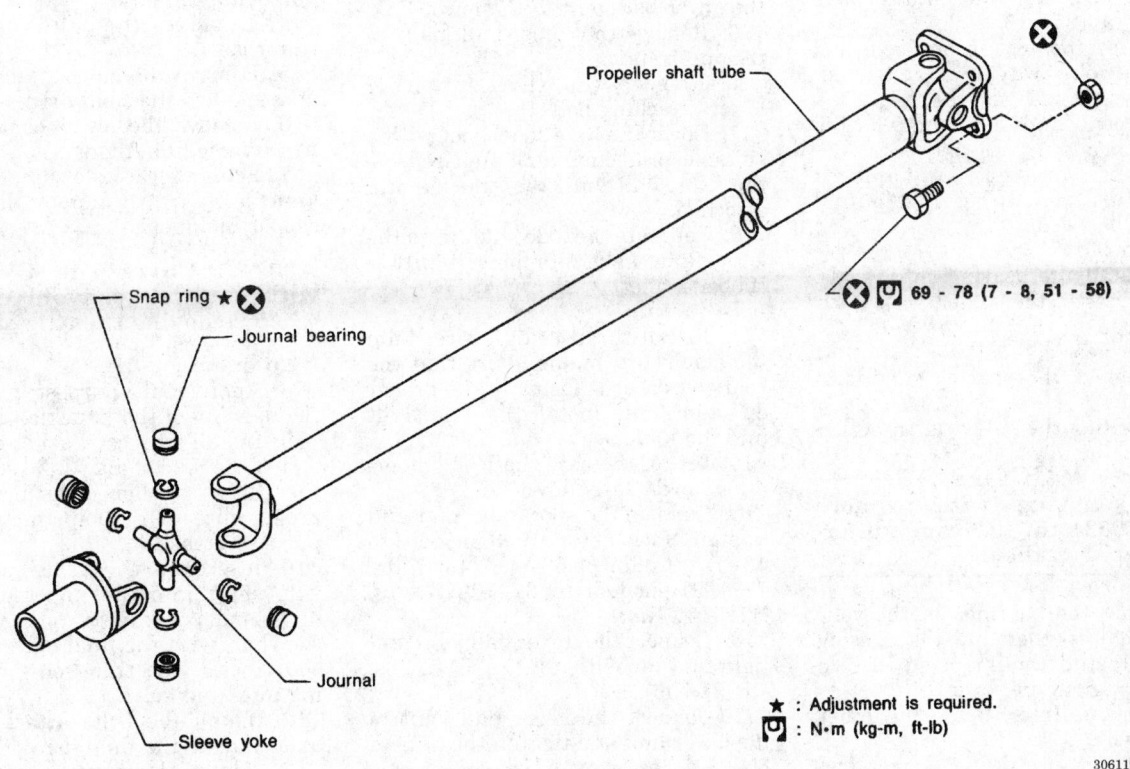

Propeller shaft tube

69 - 78 (7 - 8, 51 - 58)

Snap ring ★

Journal bearing

Journal

Sleeve yoke

★ : Adjustment is required.
[□] : N·m (kg-m, ft-lb)

306119

Rear driveshaft model 2S80B — 4WD vehicles

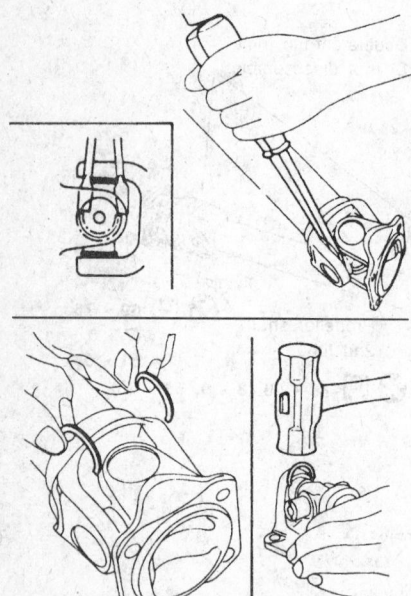

301915

U-Joint service points of removal

bearing is driven out of the yoke about 1 in. (25mm).

5. Place the tool in the vise and drive the yoke away from the tool until the bearing is removed.

6. Repeat Steps 3, 4, and 5 for the other bearings.

7. Check for worn or damaged parts. Inspect the bearing journal surfaces for wear.

To install:

8. Install the bearing cups, seals, and O-rings in the spider.

9. Grease the spider and the bearings.

10. Remove the bearing cups from the spider.

11. Position the spider in the yoke.

— WARNING —
Select snaprings with no more than 0.0024 in. (0.06mm) difference between them.

12. Start the bearings in the yoke and then press them into place, using a vise. Install the snapring to hold the bearing cup in place.

13. Repeat Step 10 for the other bearings.

14. Make sure the bearings and snaprings are fully seated by lightly tapping on the yoke with a hammer.

15. Axial play of the spider is should be 0.008 in. (0.02mm) or less.

Select snaprings which will provide the correct play. Be sure that the snaprings are the same size on both sides or driveshaft noise and vibration will result.

16. Install the driveshaft in the vehicle.

Halfshaft

REMOVAL AND INSTALLATION

Except 1996–97 Pathfinder

1. Raise and safely support the front of the vehicle.

2. Remove the wheel.

3. Remove the bolts attaching the axle shaft to the differential while the brake pedal is being depressed.

4. Remove the free running hub assembly with the brake pedal depressed.

5. Remove the brake caliper assembly without disconnecting the hydraulic brake line. Support or hang the brake caliper with a wire to avoid damaging the hose.

6. Remove the brake rotor.

7. Disconnect the tie rod ball joint from the steering knuckle.

8. Support the lower link with a jack and remove the nuts attaching the lower ball joint to the lower link.

9. Remove the upper ball joint attaching bolts.

10. Remove the shock absorber lower attaching bolt.

11. Cover the axle shaft boot with a suitable protector, and then remove the axle shaft with the knuckle still attached.

12. Separate the axle shaft from the knuckle by lightly tapping it with a rubber mallet.

To install:

13. Install the bearing spacer onto the axle shaft, making sure that the bearing spacer is facing in the proper direction, then install the axle shaft into the knuckle.

14. Install the axle shaft and steering knuckle assembly.

15. Connect the shock absorber and tighten the bolt on 1993 models to 43–58 ft. lbs. (59–78 Nm) and 1994–97 models to 87–108 ft. lbs. (118–147 Nm).

16. Connect the upper ball joint and tighten the bolts to 12–15 ft. lbs. (16–21 Nm).

17. Connect the lower ball joint to the lower link and tighten the nuts to 35–45 ft. lbs. (47–61 Nm).

18. Connect the tie rod ball joint to the steering knuckle.

19. Install the brake rotor and caliper.

20. Temporarily install a new snapring on the axle shaft at the same thickness as it was before removal and then measure the axial end-play of the axle shaft with a dial gauge. The axial end-play should be 0.1–0.3mm. Select another snapring if not within specifications.

21. Install the hub, then connect the axle shaft to the differential and tighten the bolts to 25–33 ft. lbs. (34–44 Nm).

22. Install the wheel, then remove the jackstands and lower the vehicle.

1996–97 Pathfinder

1. Raise and safely support the front of the vehicle.

2. Remove the wheel.

— WARNING —
Before removing the axle shaft, disconnect the ABS wheel sensor and move it out of the way. Failure to do so may result in damage to the sensor wires, which would render the sensor inoperative.

3. Remove the brake caliper assembly without disconnecting the hydraulic brake line. Support or hang the brake caliper with a wire to avoid damaging the hose.

4. Remove the hub cap and snapring.

5. Remove the bolts attaching the axle shaft to the final drive.

6. Remove the lower control arm (transverse link) fixing nut and bolts.

7. Separate the axle shaft from the knuckle by lightly tapping it with a copper hammer.

NOTE: Cover the CV-boots with a towel to avoid damage when removing the axle shaft.

To install:

8. Apply multi-purpose grease to the opening of the knuckle.

9. Install a thrust washer onto the end of the axle shaft. Make sure that the thrust washer is facing in the proper direction, then apply multi-purpose grease.

10. Insert the wheel side end of the axle shaft into the knuckle. Then, align and position the other end of the shaft with the final drive.

11. Install the transverse link fixing nuts and bolts.

12. Install the bolts attaching the axle shaft to the final drive.

13. Adjust the shaft's axial end-play by selecting a suitable snapring as follows:

   a. Temporarily install a new snapring (of the same thickness

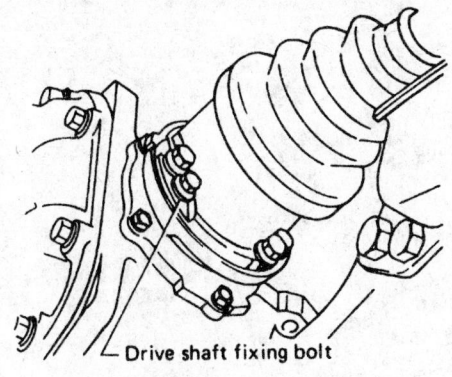

Drive shaft fixing bolt

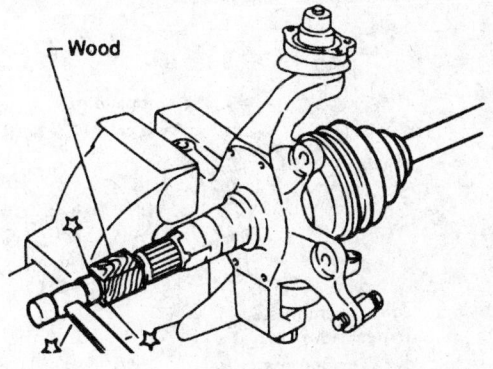

Wood

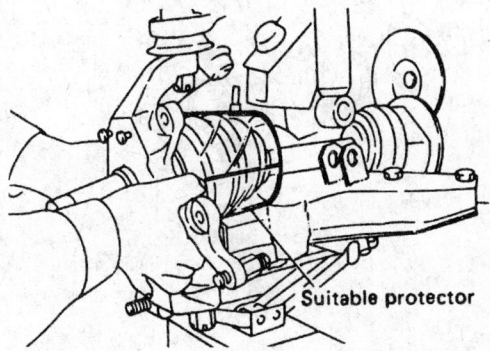

Suitable protector

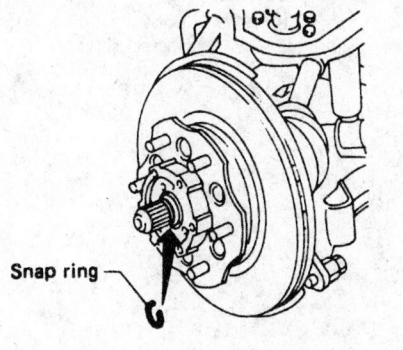

Snap ring

300430

Front axle shaft service points of removal

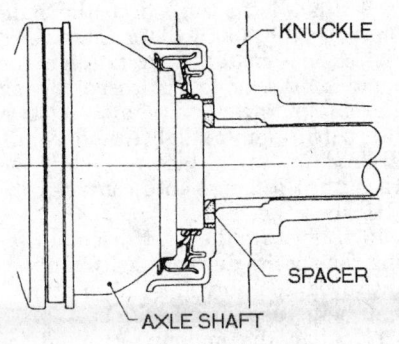

KNUCKLE

SPACER

AXLE SHAFT

300433

Axle shaft installed into the steering knuckle

which was previously removed) on the end of the axle shaft.

NOTE: Do not reuse the old snapring.

    b. Attach a dial gauge to the end of the axle shaft.
    c. Measure the axial end-play. If it is not 0.0177 in. (0.45mm) or less, select a thicker snapring.
14. Install the hub cap.
15. Install the brake caliper and connect the ABS wheel sensor.
16. Install the wheel, then remove the jackstands and lower the vehicle.

CV-Joint Boot

REPLACEMENT

    NOTE: Use only clamps provided with the replacement package when servicing. Plastic wire ties and other straps will not clamp tightly enough and grease will sling out, causing damage to the joint.

TS82F Type

1. Raise and properly support the vehicle. Remove the halfshaft from the vehicle.
2. Remove the plug seal from the slide joint housing by lightly tapping the outer housing.
3. Mark the position of the boot on axle shaft, to ensure the proper boot length on reassembly. Remove the boot bands from the boot, using side cutter pliers.
4. Slide the boot and the joint outer housing toward the wheel side of the shaft. Mark the joint to the shaft so that it can be reinstalled in the same position.
5. Using snapring pliers, remove the snapring securing the tripod joint spider assembly to the halfshaft.
6. Using a press, separate the spider assembly from the shaft. Do not

disassemble the spider and use care in handling.
7. Slide the boot and slide joint housing off the shaft.

    NOTE: If the boot is be reused, wrap vinyl tape around the spline part of the shaft so the boot will not be damaged when removed.

    To install:
8. Clean old grease from joint with solvent and blow dry.
9. Double check that the correct replacement parts are being installed. Wrap vinyl tape around the splines to protect the boot. Install the boot and slide joint housing in the correct order. Lubricate the joint assembly using the specified grease, with half being used to lubricate the joint and half being used inside the boot.
10. Align the matchmarks and press the spider assembly onto the shaft. Secure the spider assembly with the snapring.
11. Secure the boot bands with the halfshaft in a horizontal position. Make sure the boot span on the halfshaft is the proper length.
12. Install the plug seal.
13. Install the halfshaft into the vehicle.

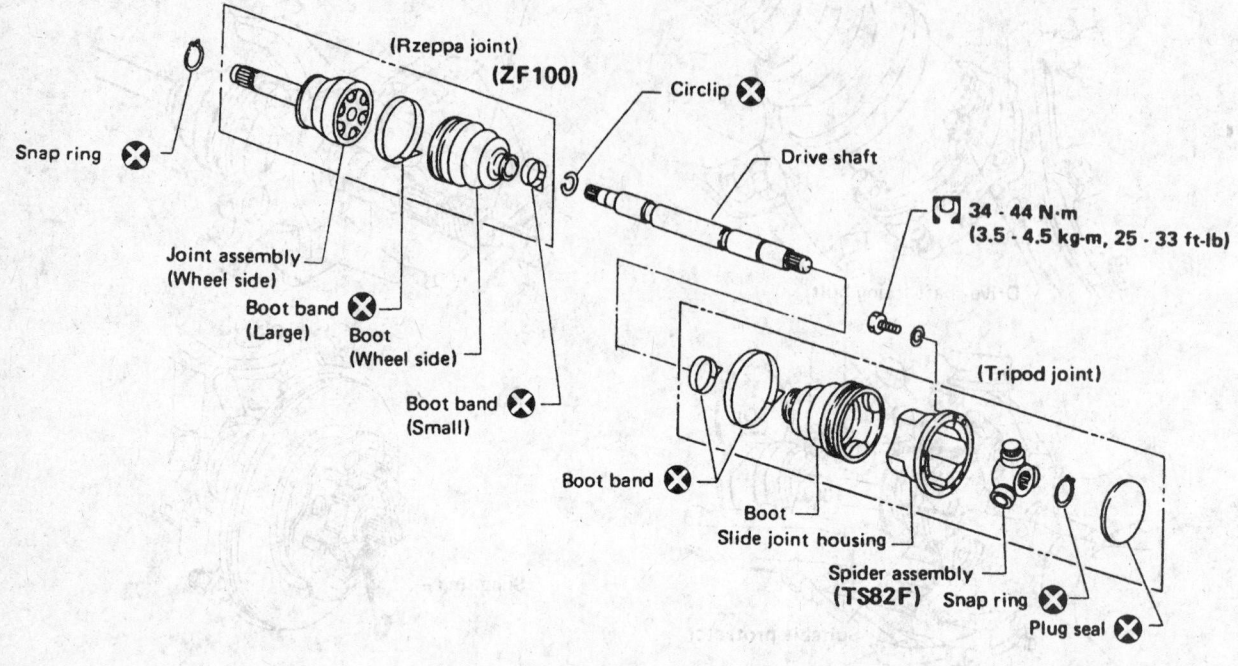

Axle shaft exploded view — 4-cylinder engines

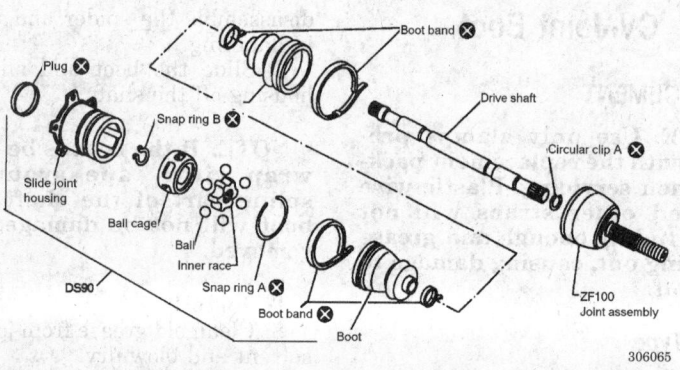

Axle shaft exploded view — 6-cylinder engines

DS90 Type

1. Raise and properly support the vehicle. Remove the halfshaft from the vehicle.

2. Mark the position of the boot on axle shaft, to ensure the proper boot length on reassembly. Remove the boot bands from the boot, using side cutter pliers.

3. Mark the slide joint housing to the shaft so that it can be reinstalled in the same position.

4. Locate and remove the large circlip at the base of the joint. Remove the slide joint housing.

5. Matchmark the inner race and cage to the shaft. Remove the small snapring from the shaft. With a brass drift pin, tap lightly and evenly around the inner race to remove the race and the inner cage from the shaft.

6. If the boot is to be reused, wipe the grease from the splines and wrap the splines in vinyl tape before sliding the boot from the shaft.

To install:

7. Clean old grease from joint with solvent and blow dry.

8. Tape the shaft splines and slide the boot onto the shaft.

9. Install the cage onto the halfshaft so the small diameter side of the cage is installed first. Align the matchmarks made at disassembly on the inner race and shaft. With a brass drift pin, tap lightly and evenly around the inner race to install the race until it comes into contact with the rib of the shaft.

10. Lubricate the joint assembly using the specified grease, with half being used to lubricate the joint and half being used inside the boot.

11. Align the matchmarks and install the outer race (slide joint housing). Install the outer snapring.

12. Secure the boot bands with the halfshaft in a horizontal position. Make sure the boot span on the halfshaft is the proper length.

13. Install the halfshaft into the vehicle.

ZF100 Type

NOTE: This type joint cannot be disassembled.

1. Raise and properly support the vehicle. Remove the halfshaft from the vehicle.

2. Mark the position of the boot on axle shaft, to ensure the proper boot length on reassembly. Remove the boot bands from the boot, using side cutter pliers.

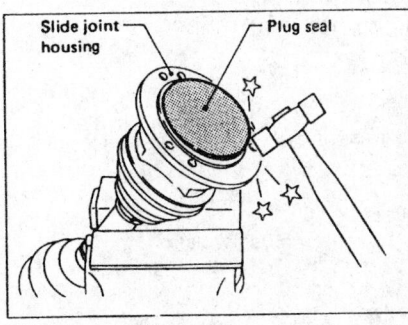

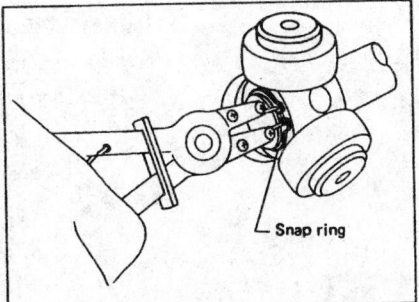

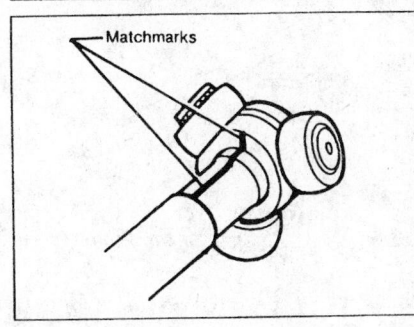

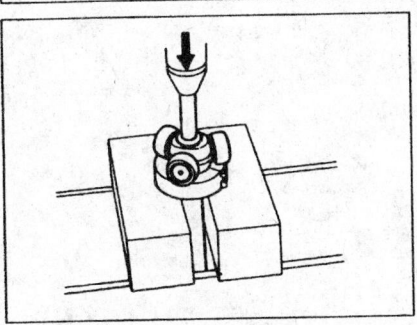

306072

TS82F joint service procedures

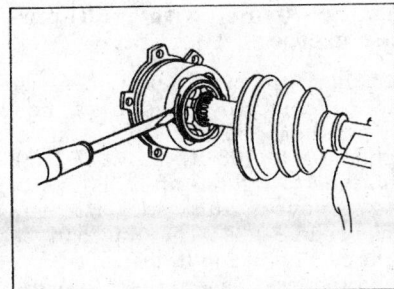

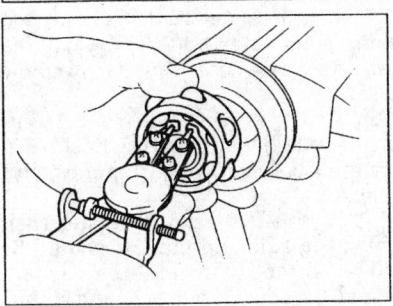

306068

DS90 joint disassembly

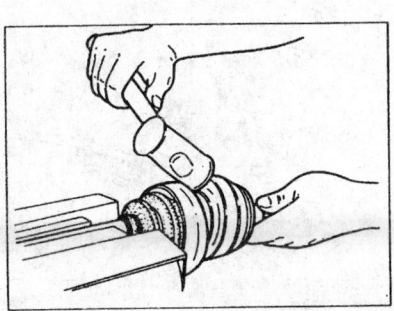

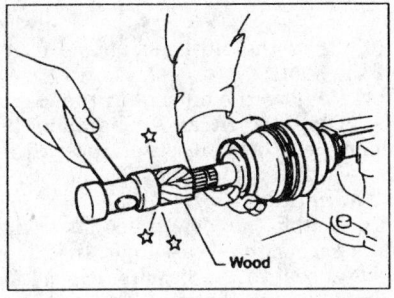

306070

ZF100 joint removal and installation

3. Mark the slide joint housing to the shaft so that it can be reinstalled in the same position.

4. Using a plastic hammer or block of wood, tap the joint off the shaft.

5. If the boot is to be reused, wipe the grease from the splines and wrap the splines in vinyl tape before sliding the boot from the shaft.

To install:

6. Clean the old grease from the joint with solvent and blow dry.

7. Tape the shaft splines and slide the boot onto the shaft.

8. Lubricate the joint assembly using the specified grease, with half being used to lubricate the joint and half being used inside the boot.

9. Install a new outer circular clip to the axle shaft.

10. Align the matchmarks and install the joint assembly to the shaft, using the same method as removal.

11. Secure the boot bands with the halfshaft in a horizontal position. Make sure the boot span on the halfshaft is the proper length.

12. Install the halfshaft into the vehicle.

Axles

REMOVAL AND INSTALLATION

Front

Refer to the Halfshaft procedure.

Rear

Single Rear Wheels With Drum Brakes

1. Raise the rear of the vehicle and support it. Remove the rear wheel.

2. Disconnect the rear parking brake cable.

3. If equipped, remove the ABS wheel speed sensor from the axle assembly.

4. Disconnect the brake tube at the rear brake backing plate. Plug the end of the brake tube to prevent loss of brake fluid.

5. Remove the brake drum.

6. Remove the nuts securing the wheel bearing retainer to the brake backing plate.

7. Pull out the axle shaft assembly together with the brake backing plate using a slide hammer.

8. Remove the oil seal in the axle housing. It can be pried out with a small prybar.

9. To replace the bearing, unbend and discard the lockwasher. Remove the locknut using SST38020000 (spanner wrench) or equivalent.

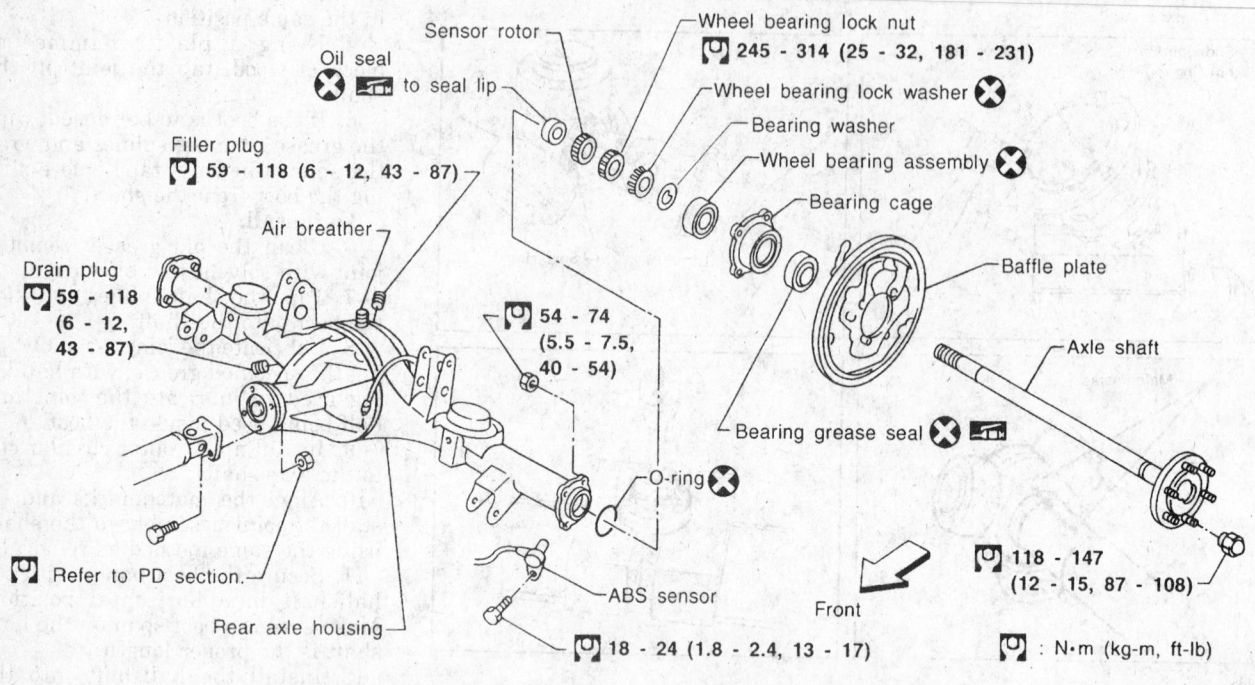

Oil seal ⊗ ▣ to seal lip

Sensor rotor

Filler plug ▣ 59 - 118 (6 - 12, 43 - 87)

Air breather

Drain plug ▣ 59 - 118 (6 - 12, 43 - 87)

Wheel bearing lock nut ▣ 245 - 314 (25 - 32, 181 - 231)

Wheel bearing lock washer ⊗

Bearing washer

Wheel bearing assembly ⊗

Bearing cage

▣ 54 - 74 (5.5 - 7.5, 40 - 54)

Baffle plate

Axle shaft

Bearing grease seal ⊗ ▣

O-ring ⊗

▣ Refer to PD section.

Rear axle housing

ABS sensor

Front

▣ 118 - 147 (12 - 15, 87 - 108)

▣ 18 - 24 (1.8 - 2.4, 13 - 17)

▣ : N•m (kg-m, ft-lb)

306524

Rear axle assembly — single wheels with drum brakes (1996–97 Pathfinder shown)

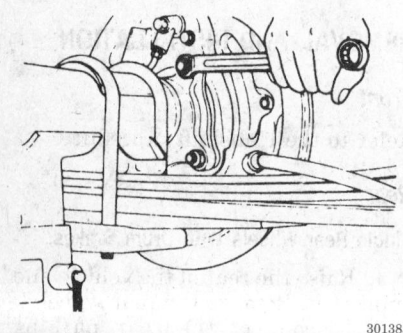

301383

Disconnecting the backing plate from the axle housing

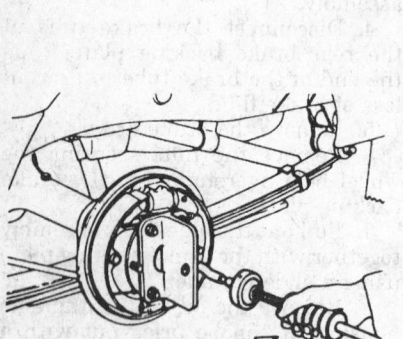

301384

Using a slide hammer to remove the axle shaft from the housing

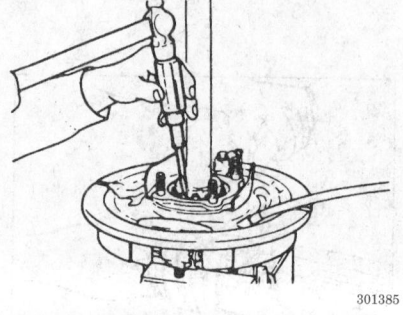

301385

Unbending the lockwasher to remove the bearing locknut

10. Press the old bearing and cage off the shaft.

11. Remove the oil seal in the cage. Use a brass drift and a hammer to remove the bearing outer race after the seal has been removed.

To install:

12. Install the new wheel bearing outer race with a brass drift. Install a new oil seal in the bearing cage. Lubricate the area between the seal lips with grease after installation.

13. Place the bearing spacer on the axle shaft with the chamfered side facing the wheel. Install the wheel bearing inner race using a brass drift

and lightly tapping it onto the axle shaft.

NOTE: Coat each wheel bearing race (cone) with multi-purpose grease.

14. Place the plain flat washer over the bearing, then install a new lockwasher. Install the locknut, tightening to 108 ft. lbs. (147 Nm). Continue to tighten after that until the grooves line up with the lockwasher tabs. The nut can be tightened up to 145 ft. lbs. (196 Nm). Bend the lockwasher tabs into place.

15. Install a new axle grease seal using a hammer and drift. Coat the sealing lip with multi-purpose grease.

16. Lubricate the axle case recess in the axle housing with wheel bearing grease. Coat the axle splines with gear oil.

17. Install the axle shaft and then check the axle end-play. It should be 0.02–0.15mm. The end-play is adjusted by adding or removing shims behind the brake backing plate. Tighten the backing plate attaching nuts to 39–46 ft. lbs. (53–63 Nm) for 1993–94 vehicles and to 40–54 ft. lbs. (54–74 Nm) for 1995–97 vehicles.

18. Reconnect the parking brake cable and the rear brake line. Install the brake drum.

19. Install the ABS wheel speed sensor if removed.

20. Bleed the brake system and install the wheel. Lower the vehicle.

Single Rear Wheels With Disc Brakes

1. Raise the rear of the vehicle and support it. Remove the rear wheel.

2. Disconnect the rear parking brake cable.

3. If equipped, remove the ABS wheel speed sensor from the axle assembly.

4. Disconnect the brake tube on the rear axle housing at the wheel bearing housing attachment point. Plug the end of the brake tube to prevent loss of brake fluid.

5. Remove the brake caliper assembly and the rotor.

6. Remove the nuts securing the wheel bearing retainer to the brake backing plate.

7. Pull out the axle shaft assembly together with the brake backing plate using a slide hammer.

8. Remove the oil seal in the axle housing. It can be pried out with a small prybar.

9. To replace the bearing, unbend and discard the lockwasher. Remove the locknut with a spanner wrench (SST38020000) or equivalent.

10. Remove the wheel bearing together with the bearing housing and the backing plate from the axle shaft.

11. Remove the wheel bearing outer side inner race from the axle shaft.

12. Pry out the old grease seal from the bearing housing.

13. Press the wheel bearing outer race from the bearing housing.

To install:

14. Press the new bearing into the housing until it bottoms.

NOTE: Always press on the outer race of the wheel bearing during installation.

15. Press a new seal into the bearing housing until it bottoms.

NOTE: After installing the new seal, coat the seal lip with multipurpose grease.

16. Install the backing plate over the bearing housing and press the axle shaft into the inner bearing race. Be careful not to damage the seal.

17. Install a flat and a lockwasher.

18. Grease the seat of the locknut and then tighten it to 181–217 ft. lbs. (245–294 Nm).

19. Turn the bearing housing two or three revolutions; it must rotate smoothly. Bend part of the lockwasher in order to lock the nut.

20. Oil the lips of the new axle grease seal with multi-purpose grease and press it into the axle housing with a drift and mallet.

21. Coat the axle splines with gear oil. Coat the seal surface of the shaft with grease.

22. Install the axle shaft and then check the axle end-play. It should be 0.0mm. The end-play is non adjustable; if the end-play is out of specification, replace the wheel bearing.

23. Install the backing plate nuts and torque the nuts to 39–46 ft. lbs. (53–63 Nm) for 1993–94 vehicles and to 40–54 ft. lbs. (54–74 Nm) for 1995–97 vehicles.

24. Install the rotor and caliper assembly.

25. Connect the parking brake cable and the rear brake line.

26. Install the ABS wheel speed sensor if removed.

27. Bleed the brake system and install the wheel. Lower the vehicle.

Dual Rear Wheels

1. Raise the rear of the vehicle and support it. Remove the rear wheel(s).

2. Remove the six bolts and remove the axle from the vehicle.

3. Remove the attaching screws and detach the lockwasher from the rear wheel bearing locknut.

4. Remove the rear wheel bearing locknut using tool KV40105400 or equivalent.

5. Remove the wheel bearing and the wheel hub with the brake drum. Be careful not to drop the outer bearing.

6. Remove the oil seal from the axle housing and discard it. Remove the wheel bearing grease seal from the wheel hub and discard it.

7. Remove the wheel bearings from the wheel hub.

8. Using a brass drift and a hammer, remove the bearing outer races from the wheel hub.

To install:

9. Using a bearing race installing tool, install the new bearing outer races in the wheel hub.

10. Pack the wheel hub with wheel bearing grease.

11. Pack each of the cone bearings with wheel bearing grease and install them in the wheel hub.

12. Install a new grease seal to the wheel hub and lubricate the seal lip with multi-purpose grease.

13. Install a new oil seal in the axle housing and lubricate the seal lip with multi-purpose grease.

14. Install the wheel hub on the vehicle. Install the wheel bearing locknut and torque it to 123–145 ft. lbs. (167–196 Nm).

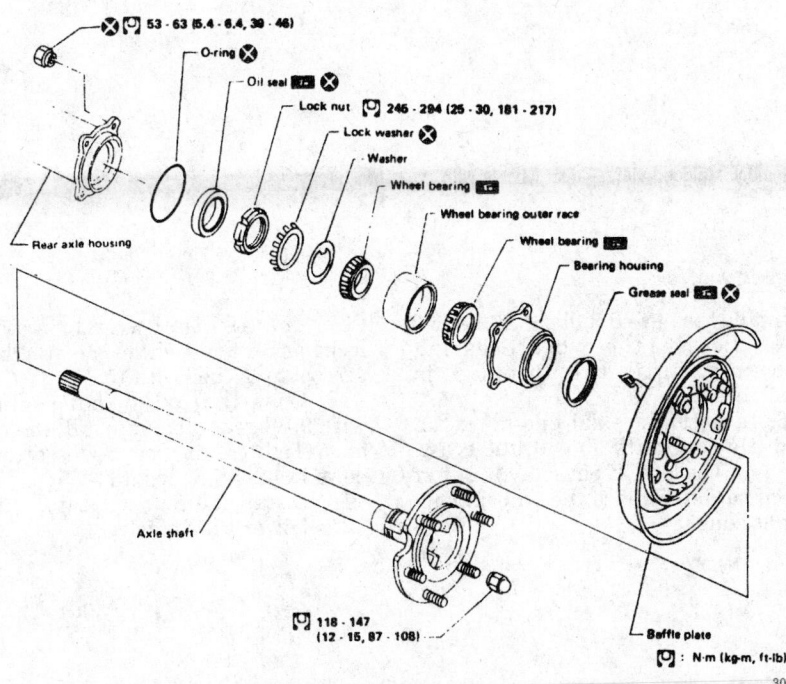

301382

Rear axle assembly exploded view — disc brake models

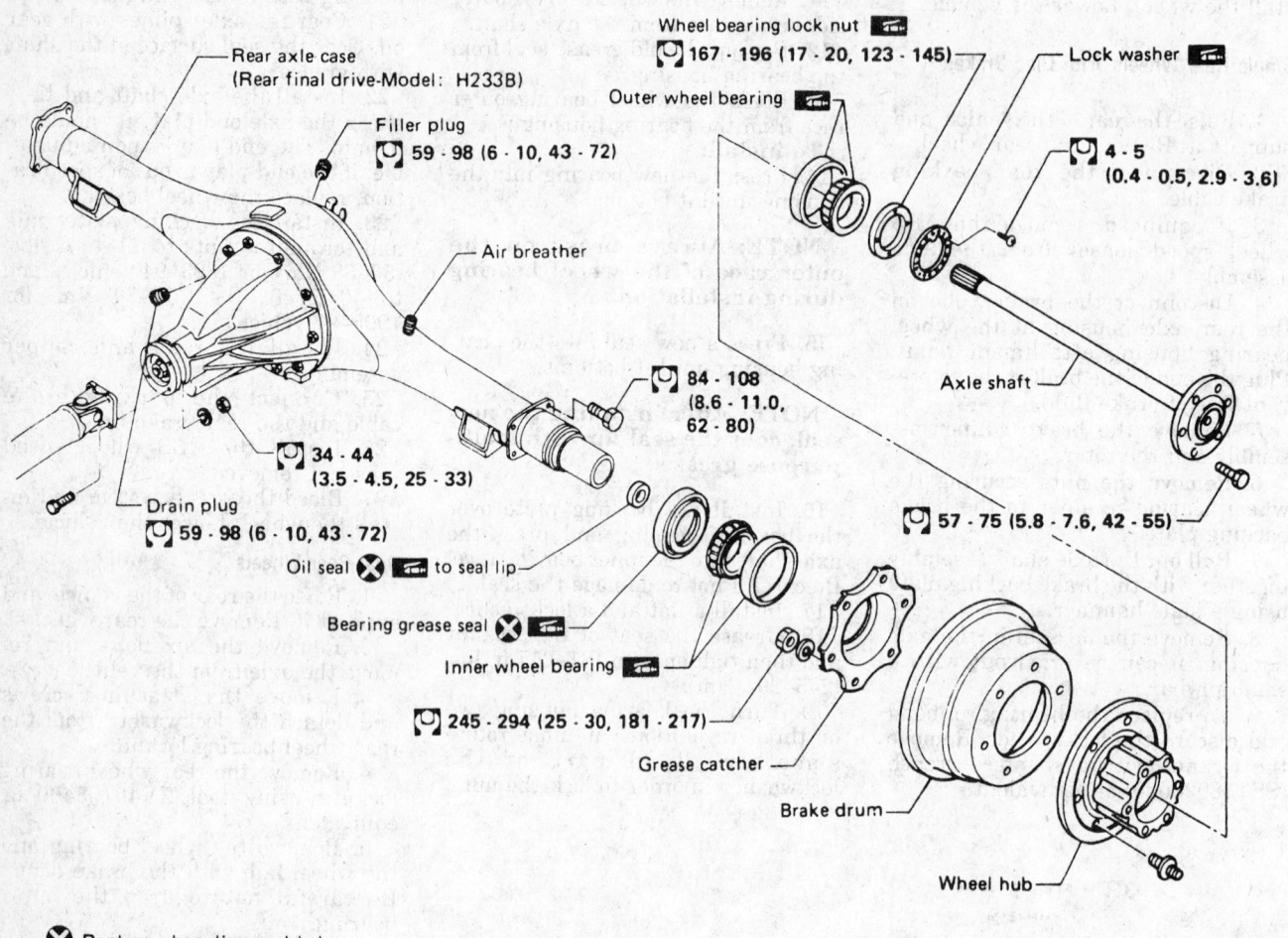

— Rear axle case
(Rear final drive-Model: H233B)

— Filler plug
🔧 59 - 98 (6 - 10, 43 - 72)

Wheel bearing lock nut 🔩
🔧 167 - 196 (17 - 20, 123 - 145)

Outer wheel bearing 🔩

Lock washer 🔩

🔧 4 - 5
(0.4 - 0.5, 2.9 - 3.6)

— Air breather

Axle shaft

🔧 84 - 108
(8.6 - 11.0,
62 - 80)

🔧 34 - 44
(3.5 - 4.5, 25 - 33)

🔧 57 - 75 (5.8 - 7.6, 42 - 55)

— Drain plug
🔧 59 - 98 (6 - 10, 43 - 72)

Oil seal ✖ 🔩 to seal lip

Bearing grease seal ✖ 🔩

Inner wheel bearing 🔩

🔧 245 - 294 (25 - 30, 181 - 217)

Grease catcher —

Brake drum —

Wheel hub —

✖ Replace when disassembled.
🔧 : N·m (kg-m, ft-lb)

301381

Rear axle assembly exploded view — dual wheels with drum brakes

15. Check the wheel bearing preload as follows:
   a. Turn the wheel hub several times in both directions.
   b. Retighten the wheel bearing locknut to 123–145 ft. lbs. (167–196 Nm).
   c. The wheel bearing preload, with new grease and oil seals, as measured at the wheel lug stud

should be 4.6–8.2 lbs. (20.6–36.3 N). Loosen the wheel bearing locknut until this preload is achieved.

16. Inspect the axial end-play; the end-play should be 0.0031 in. (0.08mm) or less. If end-play is out of specification, replace the wheel bearing assembly.

17. Install the wheel bearing lockwasher and tighten the attaching screws to 36 inch lbs. (5 Nm).

18. Coat the axles shaft splines with 90W gear oil and install the rear axle. Tighten the rear axle securing bolts to 42–55 ft. lbs. (57–75 Nm).

19. Install the rear wheel(s).

20. Lower the vehicle.

STEERING

Air Bag

── CAUTION ──

Some vehicles are equipped with an air bag system, also known as the Supplemental Inflatable Restraint (SIR) or Supplemental Restraint System (SRS). The system must be disabled before performing service on or around system components, steering column, instrument panel components, wiring and sensors. Failure to follow safety and disabling procedures could result in accidental air bag deployment, possible personal injury and unnecessary system repairs.

PRECAUTIONS

Several precautions must be observed when handling the inflator module to avoid accidental deployment and possible personal injury.

• Never carry the inflator module by the wires or connector on the underside of the module.

• When carrying a live inflator module, hold securely with both hands, and ensure that the bag and trim cover are pointed away.

• Place the inflator module on a bench or other surface with the bag and trim cover facing up.

• With the inflator module on the bench, never place anything on or close to the module which may be thrown in the event of an accidental deployment.

DISARMING

── CAUTION ──

To avoid rendering the SRS inoperative, which could lead to personal injury or death in the event of a severe frontal collision, extreme caution must be taken when servicing the electrical related systems.

NOTE: All SRS electrical wiring harnesses and connectors are covered with YELLOW outer insulation. Do not use electrical test equipment on any circuit related to the SRS (air bag) sensors. When installing SRS components, always install with the arrow marks facing the front of the vehicle.

To disarm the **SRS** system turn the ignition switch to the **OFF** position. Then disconnect both battery cables starting with the negative cable first and wait at least 10 minutes after the cables are disconnected. Be sure to insulate the battery terminal ends.

To re-arm the **SRS** system, turn the ignition switch to the **OFF** position. Connect both battery cables starting with the positive cable first.

NOTE: The SRS or air bag system is equipped with a self-diagnostic operation. After turning the ignition key to the ON or START position, the AIR BAG warning lamp will illuminate for 7 seconds. After 7 seconds, the AIR BAG lamp will extinguish if no malfunction is detected. If the AIR BAG lamp does not extinguish after 7 seconds, check the SRS self diagnostic system for a malfunction.

Steering Wheel

REMOVAL AND INSTALLATION

1993–94 Models

Position the steering wheel in the straight-ahead position.

1. Disconnect the negative battery cable.

2. Remove the horn pad by removing the screws from the rear of the steering wheel crossbar.

3. Matchmark the top of the steering column shaft and the steering wheel flange.

4. Remove the attaching nut and remove the steering wheel with a puller.

── WARNING ──

Do not strike the shaft with a hammer, as the steering column collapse mechanism may be damaged.

To install:

5. Install the steering wheel so the punchmarks are aligned. Torque the steering wheel nut to 22–29 ft. lbs. (29–39 Nm).

6. Install the horn pad and connect the negative battery cable.

1995–97 Models

── CAUTION ──

The air bag system (SRS or SIR) must be disarmed before removing the steering wheel. Failure to

do so may cause accidental deployment, property damage or personal injury.

Without Air Bag

1. Disconnect the negative battery cable.

2. Ensure that the steering wheel and front tires are positioned in the straight-ahead position.

3. Using an appropriate tool, pry the horn pad off the steering wheel.

4. Remove the steering wheel locknut.

5. Using an appropriate puller, remove the steering wheel.

To install:

6. Apply multi-purpose grease to the entire surface of the turn signal cancel pin and the horn contact slip-ring.

7. Install the steering wheel and tighten the locknut to 22–29 ft. lbs. (29–39 Nm).

8. Install the horn pad. Reconnect the negative battery cable.

With Air Bag

1. Make sure the wheels are pointing straight-ahead and the steering wheel is centered. Disconnect and insulate the negative battery cable and wait at least 10 minutes for the power supply to discharge.

2. Remove the access panel from the rear of the steering wheel and disconnect the air bag module connector.

3. Remove the side access lids, remove the left and right T50H Torx® bolts and discard them. These bolts are specially coated and should not be reused.

4. Carefully remove the air bag module and place in a safe location with the pad side facing upward.

NOTE: The air bag module is a fragile component. Always place it with the pad side facing upward. Do not allow oil, grease or water to come in contact with the module. Do not drop the module; if it is damaged in any way, do not reinstall it to the steering wheel.

5. Disengage the spiral cable and disconnect the horn connector. Remove the steering wheel hold-down nut.

6. Using an appropriate puller, remove the steering wheel.

7. Install the spiral cable to the stopper tool to keep the spiral cable in its correct alignment.

8. To remove the spiral cable, remove the steering column covers. Unplug the connector, remove the four

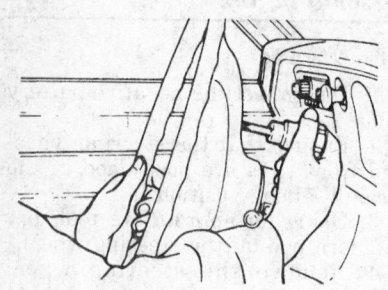

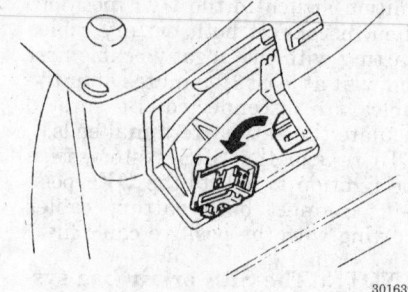

Air bag module connector location

301639

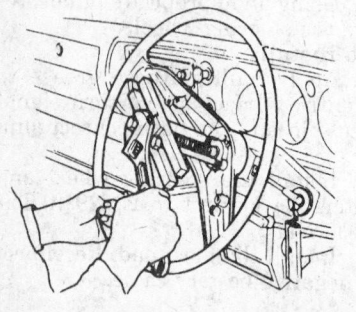

122349

Steering wheel removal — 1993 vehicles

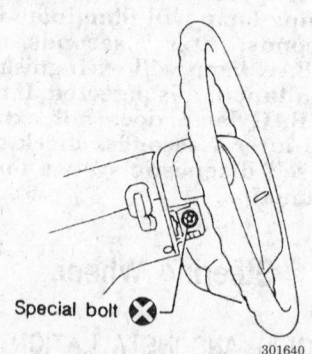

Special bolt ⊗

301640

Air bag's Torx® mounting bolts

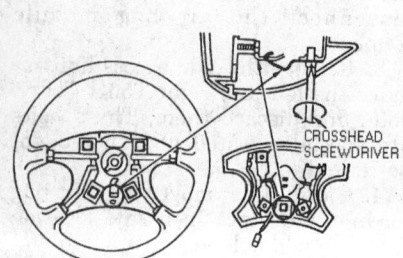

CROSSHEAD SCREWDRIVER

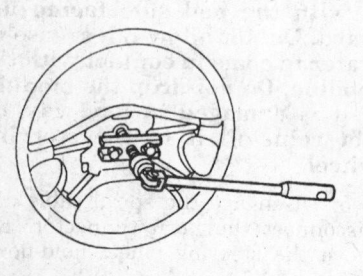

122359

Steering wheel removal — 1994 models

mounting screws and remove the spiral cable.

To install:

9. Connect the spiral cable connectors and install it onto the column. If a stopper tool was not used, align the spiral cable now.

    a. Turn the spiral cable clockwise until it stops. Do not force it.

    b. Turn it back about 2½ turns for power steering and about 4 turns for non-power steering.

    c. Align the white pin with the alignment mark.

10. Pull the spiral cable through the steering wheel opening and install the steering wheel, engaging the spiral cable pin guides.

11. Connect the horn connector and engage the spiral cable with the pawls in the steering wheel.

12. Install the hold-down nut and torque to 22–29 ft. lbs. (29–39 Nm).

13. Carefully position the air bag module. Install new Torx® bolts and torque to 15 ft. lbs. (20 Nm). Connect the air bag module connector.

14. Install the three access lids and the column covers.

15. Make sure no one is in the vehicle and connect the negative battery cable.

16. If the Nissan Consult tool is not available, perform the following:

    a. From the passenger seat, turn the ignition switch to the **ON** position.

    b. Observe the **AIR BAG** warning light on the instrument cluster. The warning light should illuminate for about seven seconds, then go out.

    c. If the warning light illuminates in any sequence except the above, the control unit has detected a fault in the air bag system and it will not operate.

17. The Nissan Consult or equivalent scan tool is required to check the fault diagnosis memory and turn the **AIR BAG** warning light off.

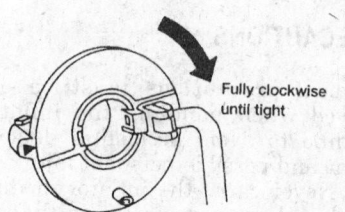

Fully clockwise until tight

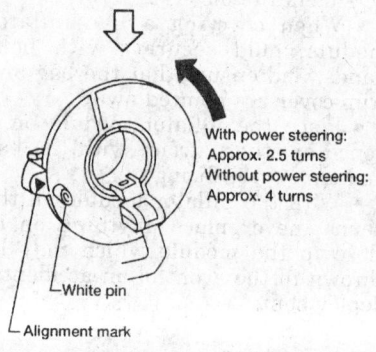

With power steering: Approx. 2.5 turns
Without power steering: Approx. 4 turns

White pin

Alignment mark

301700

Spiral cable alignment marks

Tie Rod Ends

REMOVAL AND INSTALLATION

Except 1996–97 Pathfinder

1. Raise and safely support the vehicle. Remove the wheel/tire assembly.

2. If removing the tie rod as an assembly:

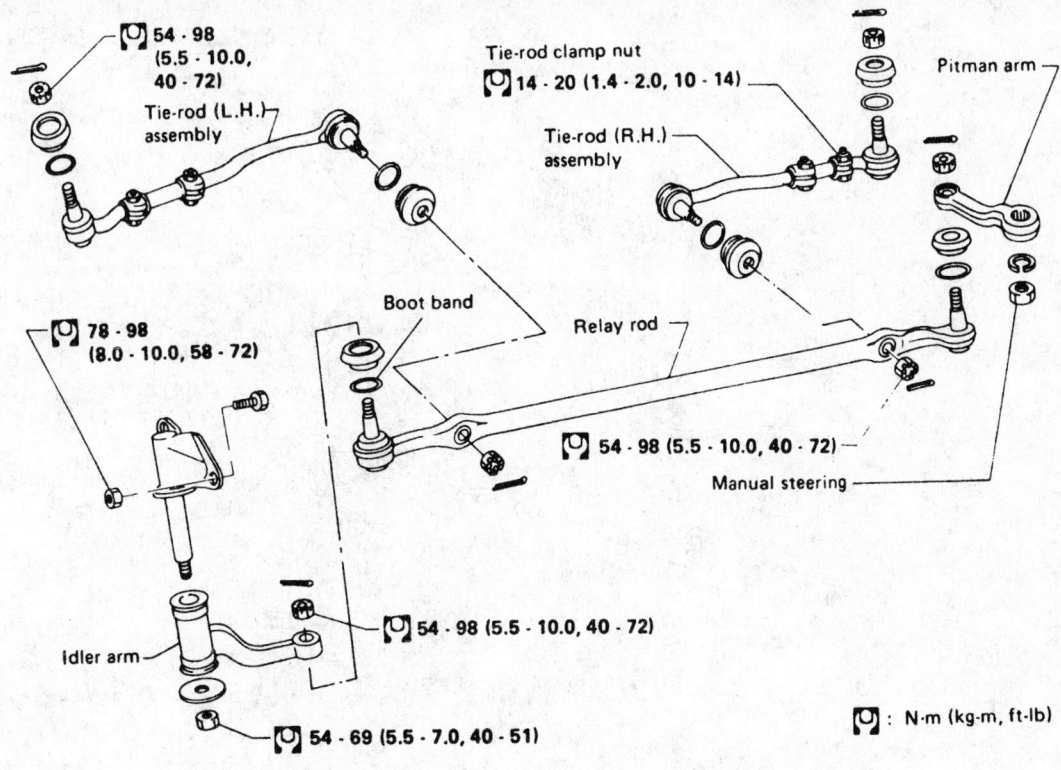

54 - 98
(5.5 - 10.0,
40 - 72)

Tie-rod clamp nut
14 - 20 (1.4 - 2.0, 10 - 14)

Pitman arm

Tie-rod (L.H.)
assembly

Tie-rod (R.H.)
assembly

78 - 98
(8.0 - 10.0, 58 - 72)

Boot band

Relay rod

54 - 98 (5.5 - 10.0, 40 - 72)

Manual steering

54 - 98 (5.5 - 10.0, 40 - 72)

: N·m (kg-m, ft-lb)

Idler arm

54 - 69 (5.5 - 7.0, 40 - 51)

301812

Steering linkage exploded view — 2WD models, except 1996–97 Pathfinder

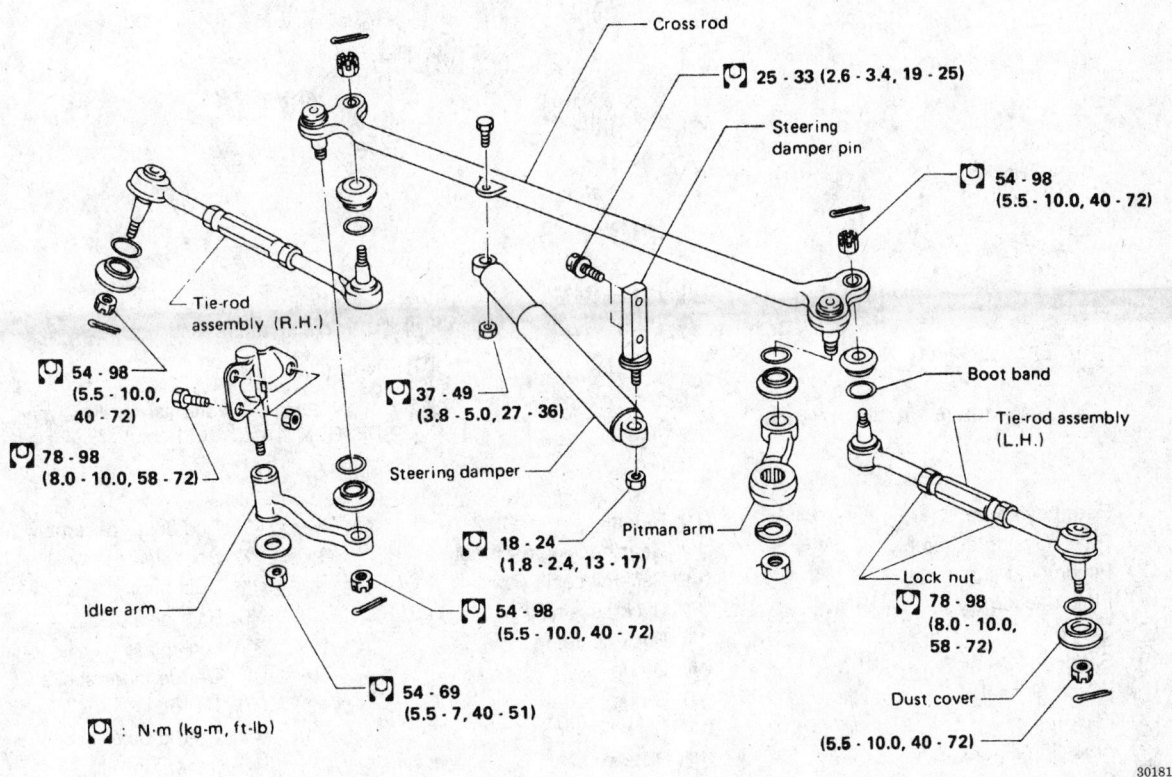

Cross rod

25 - 33 (2.6 - 3.4, 19 - 25)

Steering
damper pin

54 - 98
(5.5 - 10.0, 40 - 72)

Tie-rod
assembly (R.H.)

54 - 98
(5.5 - 10.0,
40 - 72)

78 - 98
(8.0 - 10.0, 58 - 72)

37 - 49
(3.8 - 5.0, 27 - 36)

Steering damper

Boot band

Tie-rod assembly
(L.H.)

Pitman arm

18 - 24
(1.8 - 2.4, 13 - 17)

54 - 98
(5.5 - 10.0, 40 - 72)

Idler arm

Lock nut

78 - 98
(8.0 - 10.0,
58 - 72)

54 - 69
(5.5 - 7, 40 - 51)

Dust cover

: N·m (kg-m, ft-lb)

(5.5 - 10.0, 40 - 72)

301811

Steering linkage exploded view — 4WD models, except 1996–97 Pathfinder

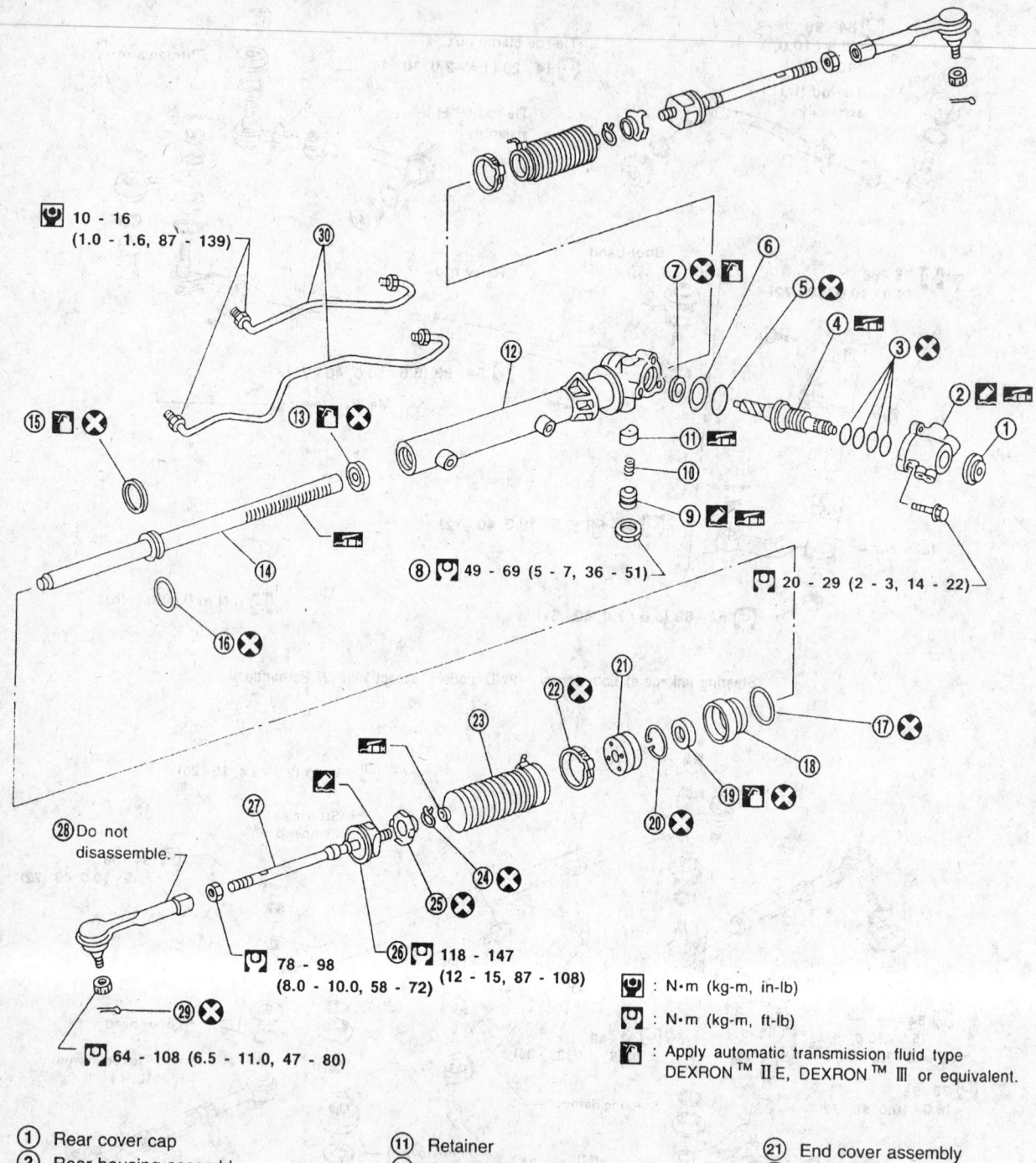

10 - 16
(1.0 - 1.6, 87 - 139)

8 49 - 69 (5 - 7, 36 - 51)

20 - 29 (2 - 3, 14 - 22)

28 Do not disassemble.

78 - 98
(8.0 - 10.0, 58 - 72)

118 - 147
(12 - 15, 87 - 108)

64 - 108 (6.5 - 11.0, 47 - 80)

: N·m (kg-m, in-lb)

: N·m (kg-m, ft-lb)

: Apply automatic transmission fluid type DEXRON™ II E, DEXRON™ III or equivalent.

① Rear cover cap
② Rear housing assembly
③ Pinion seal ring
④ Pinion assembly
⑤ O-ring
⑥ Shim
⑦ Pinion oil seal
⑧ Lock nut
⑨ Adjusting screw
⑩ Spring

⑪ Retainer
⑫ Gear housing assembly
⑬ Rack oil seal
⑭ Rack assembly
⑮ Rack seal ring
⑯ O-ring
⑰ O-ring
⑱ Rack bushing
⑲ Rack oil seal
⑳ Snap ring

㉑ End cover assembly
㉒ Boot clamp
㉓ Dust boot
㉔ Boot clamp
㉕ Lock plate
㉖ Tie-rod inner socket
㉗ Tie-rod
㉘ Tie-rod outer socket
㉙ Cotter pin
㉚ Cylinder tube

314384

Exploded view of the power rack and pinion, tie rods and tie rod ends — 1996–97 Pathfinder

a. Remove the cotter pin and nut attaching the tie rod to the cross rod.

b. Remove the cotter pin and nut attaching the tie rod to steering knuckle.

c. Using a joint separator tool, press the tie rod from the steering knuckle and the tie rod from the cross rod.

3. If removing a defective tie rod end:

a. Remove the cotter pin and nut from the end being removed.

b. Loosen the tie rod clamp or locknut.

c. Using the joint separator tool, press the tie rod from the cross rod or steering knuckle.

d. Measure the tie rod end–to–tie rod clamp distance.

e. Unscrew the tie rod end from the tie rod, while counting the number of turns to remove.

f. Using a new tie rod end, screw the new tie rod end into the tie rod clamp, the proper number of turns. The measured distance should be the same, then, torque the tie rod clamp bolt on 2WD models to 10–14 ft. lbs. (14–20 Nm) or locknut on 4WD models to 58–72 ft. lbs. (78–98 Nm).

To install:

4. Torque the nut attaching the tie rod to the steering knuckle to 40–72 ft. lbs. (54–98 Nm) and install a new cotter pin. If removed, torque the nut attaching the tie rod to the cross rod to 40–72 ft. lbs. (54–98 Nm).

5. Lower the vehicle.

6. Check and/or adjust the front end alignment.

1996–97 Pathfinder

Tie Rod Ends

1. Raise the front of the vehicle and support it on jackstands. Remove the wheel.

2. Remove the cotter pin and the tie rod ball joint stud nut. Note the position of the steering linkage.

3. Using a suitable ball joint separator tool, remove the tie rod ball joint from the steering knuckle.

4. Loosen the locknut and remove the tie rod end from the tie rod. Count the number of complete turns it takes to completely remove it.

To install:

5. Install the new tie rod end, turning it exactly as many turns as were required to remove the old one. Make sure it is correctly positioned in relationship to the steering knuckle.

6. Connect the tie rod end to the steering knuckle and install the cas-

tle nut. Torque the nut to 47–80 ft. lbs. (64–108 Nm.).

7. Install a new cotter pin to the castle nut.

8. Tighten the tie rod end locking nut to 58–72 ft. lbs. (78–98 Nm).

9. Install the wheel and tire assembly.

10. Lower the vehicle and perform a front end alignment.

Tie Rods

1. Raise the front of the vehicle and support it on jackstands. Remove the wheel.

2. Remove the cotter pin and the tie rod ball joint stud nut. Note the position of the steering linkage.

3. Using a suitable ball joint separator tool, remove the tie rod ball joint from the steering knuckle.

4. Loosen the locknut and remove the tie rod end from the tie rod. Count the number of complete turns it takes to completely remove it.

5. Remove the clamps that secure the flexible boot to the steering gear.

6. Slide the boot from the tie rod and remove the boot.

7. Bend the tabs from the lock plate.

8. Loosen the tie rod inner socket (nut) from the steering gear and remove the tie rod.

To install:

9. Using a new lock plate, install the tie rod and torque the inner socket (nut) to 87–108 ft. lbs. (118–147 Nm).

10. Bend the tabs of the new lock plate to secure the inner nut.

11. Slide the boot onto the steering gear and secure it with new clamps.

12. Install the tie rod end, turning it exactly as many turns as were required to remove the old one. Make sure it is correctly positioned in relationship to the steering knuckle.

13. Connect the tie rod end-to-steering knuckle and install the castle nut. Torque the nut to 47–80 ft. lbs. (64–108 Nm).

14. Install a new cotter pin to the castle nut.

15. Tighten the tie rod end locknut to 58–72 ft. lbs. (78–98 Nm).

16. Install the wheel and tire assembly.

17. Lower the vehicle and perform a front end alignment.

Manual Steering Gear

REMOVAL AND INSTALLATION

1. Raise and safely support the vehicle.

2. Remove the steering gear-to-rubber coupling bolt.

3. Matchmark the Pitman arm and sector shaft and, with the wheels in a straight-ahead position, remove the idler arm-to-sector shaft nut.

4. Using the steering gear arm puller tool, press the arm from the steering gear.

5. Remove the steering gear-to-chassis bolts and the steering gear from the vehicle.

6. To install, reverse the removal procedures. Torque as follows:

Steering gear-to-coupling bolt — 17–22 ft. lbs. (24–29 Nm)

Steering gear-to-Pitman arm nut — 94–108 ft. lbs. (127–147 Nm)

Steering gear-to-frame bolts — 62–71 ft. lbs. (84–96 Nm)

ADJUSTMENT

Worm Gear Preload

For this procedure, the steering gear must be removed from the vehicle and placed in a vise.

1. Using the locknut wrench tool, loosen the locknut.

2. Rotate the worm shaft a few times, in both directions, to settle the worm bearing and check the preload.

3. Using the adjusting plug wrench, the torque wrench and an adapter socket, check the worm bearing preload; it should be 1.7–5.2 inch lbs. (0.20–0.59 Nm).

4. To adjust the worm gear preload, turn the adjusting plug with the special pin wrench and recheck the preload.

5. With the worm gear preload set, hold the adjusting plug and tighten the locknut. Check the preload again.

Steering Gear Preload

1. Loosen the adjusting screw locknut.

2. Rotate the worm shaft a few times in both directions to settle the worm bearing and check the preload.

3. Set the worm gear in the straight-ahead position.

4. Using the torque wrench tool and an adapter socket, check the worm gear preload; it should be 7.4–10.9 inch lbs. (0.83–1.23 Nm) for new parts, or 5.2–8.7 inch lbs. (0.59–0.98 Nm) for used parts.

5. If necessary, loosen the locknut and turn the adjusting screw to obtain the correct preload.

6. With the preload set, tighten the locknut.

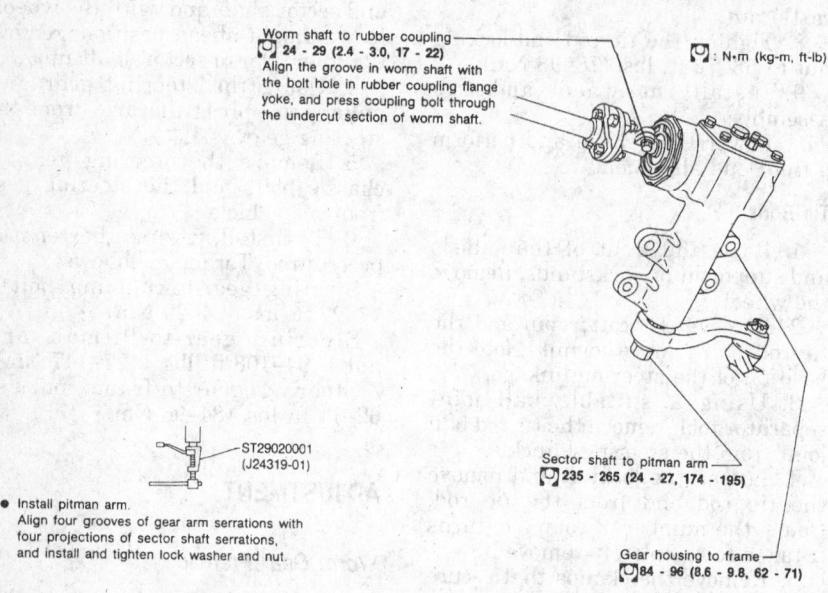

Worm shaft to rubber coupling —
24 - 29 (2.4 - 3.0, 17 - 22)
Align the groove in worm shaft with
the bolt hole in rubber coupling flange
yoke, and press coupling bolt through
the undercut section of worm shaft.

: N·m (kg-m, ft-lb)

ST29020001
(J24319-01)

● Install pitman arm.
Align four grooves of gear arm serrations with
four projections of sector shaft serrations,
and install and tighten lock washer and nut.

Sector shaft to pitman arm —
235 - 265 (24 - 27, 174 - 195)

Gear housing to frame —
84 - 96 (8.6 - 9.8, 62 - 71)

792118g3

Model VB66K manual steering gear installation

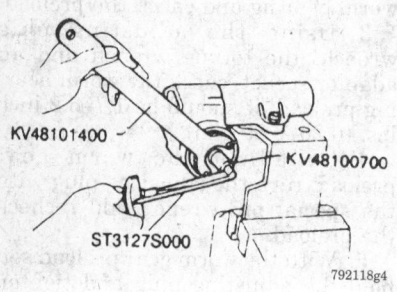

KV48101400
KV48100700
ST3127S000

792118g4

Use the special wrench to adjust worm gear preload

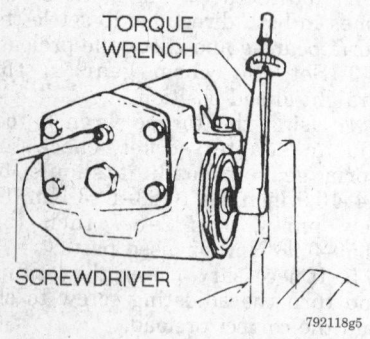

TORQUE WRENCH

SCREWDRIVER

792118g5

Adjusting steering gear preload

Power Steering Gear

REMOVAL AND INSTALLATION

1. Raise and safely support the vehicle.
2. Remove the wormshaft-to-rubber coupling bolt.
3. Matchmark the Pitman arm and sector shaft and, with the wheels in a straight-ahead position, remove the Pitman arm-to-sector shaft nut.
4. Disconnect the fluid lines from the gear, then cap the lines and openings in the gear.
5. Using the steering gear arm puller, press the gear arm from the steering knuckle.
6. Remove the steering gear-to-chassis bolts and the steering gear from the vehicle.
7. To install, reverse the removal procedures. Torque as follows:
Steering gear coupling bolt — 17-22 ft. lbs. (49-51 Nm)
Steering gear-to-Pitman arm nut — 101-130 ft. lbs. (137-177 Nm)
Steering gear-to-frame bolts — 62-71 ft. lbs. (84-96 Nm)
8. Refill the power steering pump reservoir and bleed the system.

ADJUSTMENT

1. Remove the power steering gear and position it in a vise.
2. Loosen the adjusting screw locknut.
3. Set the worm gear in the straight-ahead position.
4. Using the torque wrench and an adapter socket, check the turning torque; it should be 0.9–3.5 inch lbs. (0.1–0.4 Nm).
5. If necessary, use a screwdriver and turn the adjusting screw to obtain the correct preload.
6. With the preload set, tighten the adjusting screw nut.

Power Rack and Pinion

REMOVAL AND INSTALLATION

1996–97 Pathfinder

CAUTION
The air bag system (SRS) must be disarmed before removing the steering wheel. Failure to do so may cause accidental deployment, property damage or personal injury.

1. Disconnect and insulate the negative battery cable. Wait at least 10 minutes before doing any work to allow the air bag system time to disarm.
2. Turn the front wheels to the straight-ahead position and remove the steering wheel to prevent damage to the SRS spiral cable.
3. Raise and safely support the vehicle.
4. Remove the front wheels.
5. Disconnect the steering column lower joint from the transfer gear and steering rack.
6. Using the proper tool, disconnect the tie rod ends from the steering knuckles.
7. Disconnect the pressure and return lines from the rack assembly.
8. Remove the steering rack mounting brackets and remove the rack assembly.
To install:
9. Position the rack in the vehicle and install the insulators and brackets. Torque the bracket bolts to 87–101 ft. lbs. (118–137 Nm).

NOTE: The O-rings for the fluid lines are two different sizes. Be careful not to mix them up.

10. Using new O-rings, connect the pressure and return lines to the rack assembly. Torque the bolts to 22–26 ft. lbs. (30–35 Nm).

11. Center the rack assembly and install the lower steering column joint. Torque the pinch bolts to 17–22 ft. lbs. (24–29 Nm).

12. Connect the lower steering column joint to the transfer gear. Align the lower steering column joint's alignment groove with the matchmark on the transfer steering gear. Torque the pinch bolt to 17–22 ft. lbs. (24–29 Nm).

13. Connect the tie rod ends to the steering knuckle. Torque the castle nuts to 47 ft. lbs. (64 Nm).

14. Lower the vehicle to the ground.

15. Install the steering wheel and air bag assembly.

16. Connect the negative battery cable and enable the SRS system.

17. Refill and bleed the power steering hydraulic system.

18. Check the vehicle's alignment.

ADJUSTMENT

On Vehicle

1. Drive the vehicle on a flat road and turn the steering wheel about 20 degrees. If the wheel returns to center when released, no adjustment is required.

2. If the wheel does not self-center from a slight turn, loosen the locknut and loosen the adjusting screw.

3. If there is excessive play in the steering that is definitely in the rack, loosen the locknut and tighten the adjusting screw.

4. Road test the vehicle again. All adjustments should be made in very small increments. Under normal use, the steering rack should not require any adjustment. If adjustment does not cure the symptom, look for other problems in the steering system such as contaminated fluid, pump or suspension failure or incorrect wheel alignment.

Off Vehicle

NOTE: For a complete adjustment, the power steering rack must be removed from the vehicle and positioned in a vise.

1. Without fluid in the rack, set the gears in the neutral position (wheels straight-ahead).

2. Lubricate the adjusting screw with locking sealant and screw it in.

3. Lightly tighten the locknut.

4. Torque the adjusting screw to 43–52 inch lbs. (4.9–5.9 Nm).

5. Loosen the adjusting screw and retorque it to 0.43–1.74 inch lbs. (0.05–0.20 Nm).

6. Move the rack over its entire stroke several times.

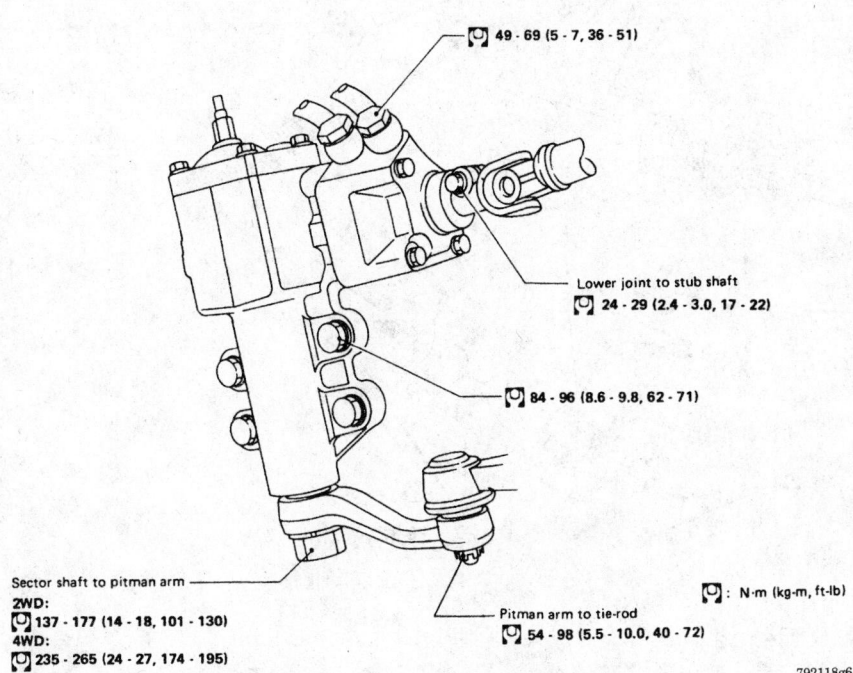

Model PB56S/PB59K power steering gear installation

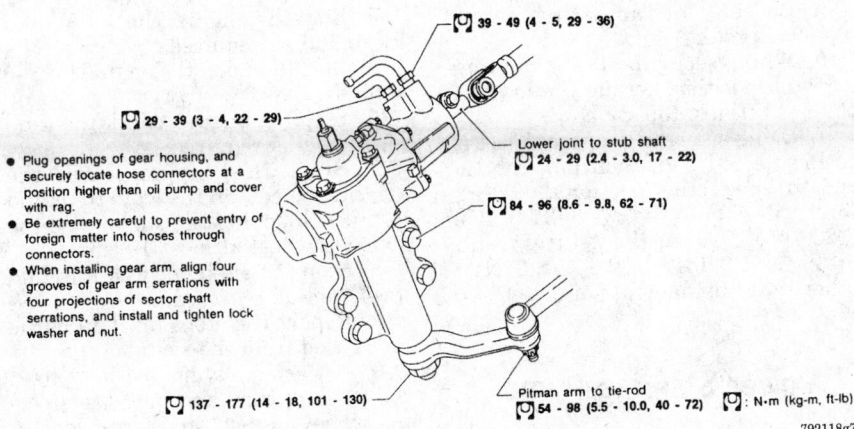

Model PB48S power steering gear installation

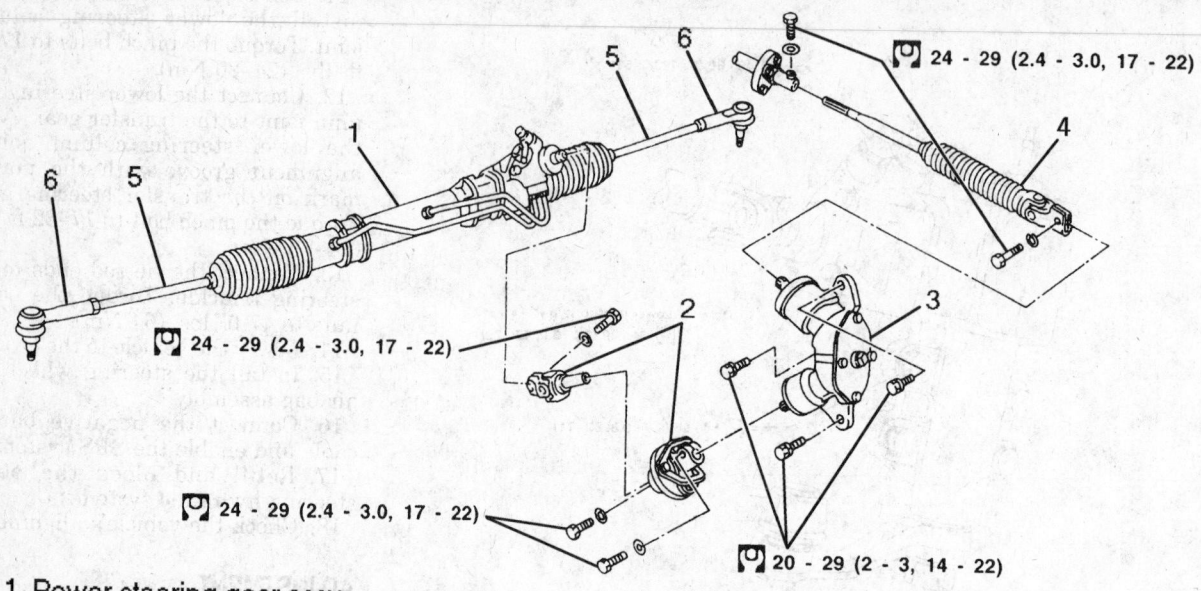

24 - 29 (2.4 - 3.0, 17 - 22)

24 - 29 (2.4 - 3.0, 17 - 22)

24 - 29 (2.4 - 3.0, 17 - 22)

20 - 29 (2 - 3, 14 - 22)

1. Power steering gear assy
2. Steering column lower joint
3. Transfer gear assy
4. Steering column upper joint

5. Tie rod
6. Tie rod end

: N·m (kg-m, ft-lb)

312644

Steering components — 1996–97 Pathfinder

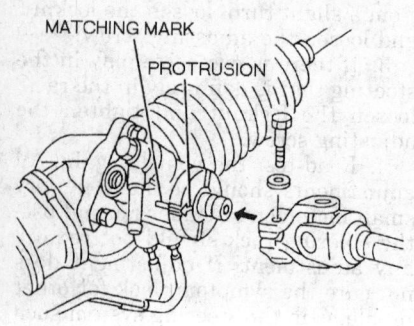

MATCHING MARK

PROTRUSION

312646

Steering rack in centered position — matchmarks aligned

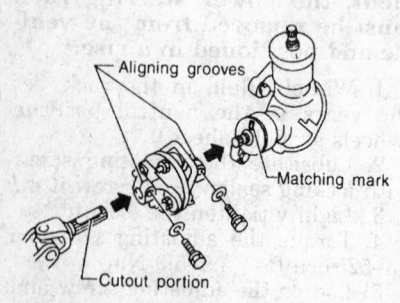

Aligning grooves

Matching mark

Cutout portion

319334

Transfer gear alignment grooves and matchmarks

7. Using an inch lb. torque wrench, measure the pinion rotating torque within the range of 180 degrees from the neutral position.
8. Loosen the adjusting screw and retorque it to 43–52 inch lbs. (4.9–5.9 Nm).
9. Loosen the adjusting screw 40–60 degrees.
10. While securing the adjusting screw in position, torque the locknut to 29–43 ft. lbs. (39–59 Nm).
11. Using a spring gauge, connect it to the tie rod end, then pull the tie rod to check the frictional sliding force; it should be 27.6–37.5 lbs. (122.6–166.7 N) at the neutral point, or 27.6–41.9 lbs. (122.6–186.3 N) at other than the neutral point.

Power Steering Pump

BLEEDING

1. Raise and support the vehicle safely.
2. Check and add fluid to the reservoir, if necessary.
3. Start the engine. Turn the steering wheel quickly (all the way), right and left, just touching the stops; turn the steering wheel at least 10 times.

NOTE: When bleeding the system, make sure the temperature of the fluid reaches 140–176°F (60–80°C).

4. Stop the engine, check the fluid level, add as required.
5. Start and run the engine for 3–5 seconds.
6. Stop the engine, check the fluid level, add as required.
7. Start the engine. Turn the steering wheel (all the way) right and left, just touching the stops; turn the steering wheel at least 10 times.
8. Stop the engine, check the fluid level, add as required.
9. Repeat the steps until all of the air is bled from the system.
10. If the air cannot be bled from the system, turn and hold the steering wheel at each stop for at least 5 seconds, but never more than 15 seconds.

REMOVAL AND INSTALLATION

1. Disconnect the negative battery cable.
2. Remove the drive belt from the power steering pump.
3. Place a container under the power steering pump, disconnect and

plug the pressure lines and drain the fluid into the container.

4. Unfasten the bolts and remove the pump from the vehicle.

5. To install, reverse the removal procedures. Adjust the drive belt tension. Bleed the power steering system.

BRAKES

Anti-Lock Brake System Service

PRECAUTIONS

• Certain components within the Anti-Lock Brake System (ABS) are not intended to be serviced or repaired individually. Only those components with removal and installation procedures should be serviced.

• Do not use rubber hoses or other parts not specifically specified for ABS system. When using repair kits, replace all parts included in the kit. Partial or incorrect repair may lead to functional problems and require the replacement of components.

• Lubricate rubber parts with clean, fresh brake fluid to ease assembly. Do not use lubricated shop air to clean parts; damage to rubber components may result.

• Use only specified brake fluid from an unopened container.

• If any hydraulic component or line is removed or replaced, it may be necessary to bleed the entire system.

• A clean repair area is essential. Always clean the reservoir and cap thoroughly before removing the cap. The slightest amount of dirt in the fluid may plug an orifice and impair the system function. Perform repairs after components have been thoroughly cleaned; use only denatured alcohol to clean components. Do not allow ABS components to come into contact with any substance containing mineral oil; this includes used shop rags.

• The Anti-Lock control unit is a microprocessor similar to other computer units in the vehicle. Ensure that the ignition switch is **OFF** before removing or installing controller harnesses. Avoid static electricity discharge at or near the controller.

• If any arc welding is to be done on the vehicle, the control unit should be unplugged before welding operations begin.

DEPRESSURIZING

1. Turn the ignition key to the **OFF** position.

2. Connect a vinyl tube to each air bleeder valve.

3. Drain the brake fluid from each bleeder valve by depressing the brake pedal.

4. After repairs, refill the reservoir with new **DOT 3** brake fluid until brake fluid comes out of each bleeder valve and close the valve. Bleed the brake system.

Master Cylinder

REMOVAL AND INSTALLATION

Manual Brakes

1993–94 Models

—————— CAUTION ——————
Be careful not to spill brake fluid on the painted surfaces of the vehicle; it will damage the paint.

1. Unfasten the hydraulic lines from the master cylinder.

2. Disconnect the hydraulic fluid pressure differential switch wiring connectors. On models with fluid level sensors, disconnect the fluid level sensor wiring connectors, as well.

3. Remove the master cylinder securing bolts and the clevis pin from the brake pedal. Remove the master cylinder.

To install:

NOTE: Certain models may have an UP mark on the cylinder boot, make sure this is in the correct position.

4. Before tightening the master cylinder mounting bolts, screw the hydraulic line into the cylinder body a few turns, then tighten the mounting hardware.

5. Reinstall the clevis and cotter pin on the brake pedal arm.

6. Bleed the master cylinder and the brake system. Make certain that the brake lines are fully tightened and the fluid is full.

7. Check and adjust the brake pedal if necessary. Pressurize the brake system and check for leaks.

Power Brakes

1993–95 Models

—————— CAUTION ——————
Be careful not to spill brake fluid on the painted surfaces of the vehicle; it will damage the paint.

1. Disconnect the negative battery cable.

2. Disconnect the hydraulic lines from the master cylinder.

3. Disconnect the hydraulic fluid pressure differential switch wiring connectors. On models with fluid level sensors, disconnect the fluid level sensor wiring connectors, as well.

4. Remove the master cylinder mounting nuts and the master cylinder assembly from the power brake unit.

To install:

NOTE: Certain models may have an UP mark on the cylinder boot, make sure this is in the correct position.

5. Position the master cylinder on the brake booster, then screw the hydraulic line into the cylinder body a few turns. Install the master cylinder mounting nuts and torque them to 5.8–8 ft. lbs. (8–11 Nm). Torque the brake line fittings to 11–13 ft. lbs. (15–18 Nm).

6. Bleed the master cylinder and the brake system. Make certain that the brake lines are fully tightened and the fluid is full.

7. Connect the negative battery cable.

8. Check and adjust the brake pedal if necessary. Pressurize the brake system and check for leaks.

1996–97 Pathfinder

—————— CAUTION ——————
Be careful not to spill brake fluid on the painted surfaces of the vehicle; it will damage the paint.

1. Disconnect the negative battery cable.

2. Disconnect the hydraulic lines from the master cylinder.

3. Disconnect the hydraulic fluid pressure differential switch wiring connectors. On models with fluid level sensors, disconnect the fluid level sensor wiring connectors, as well.

4. Remove the master cylinder mounting nuts and the master cylinder assembly from the power brake unit.

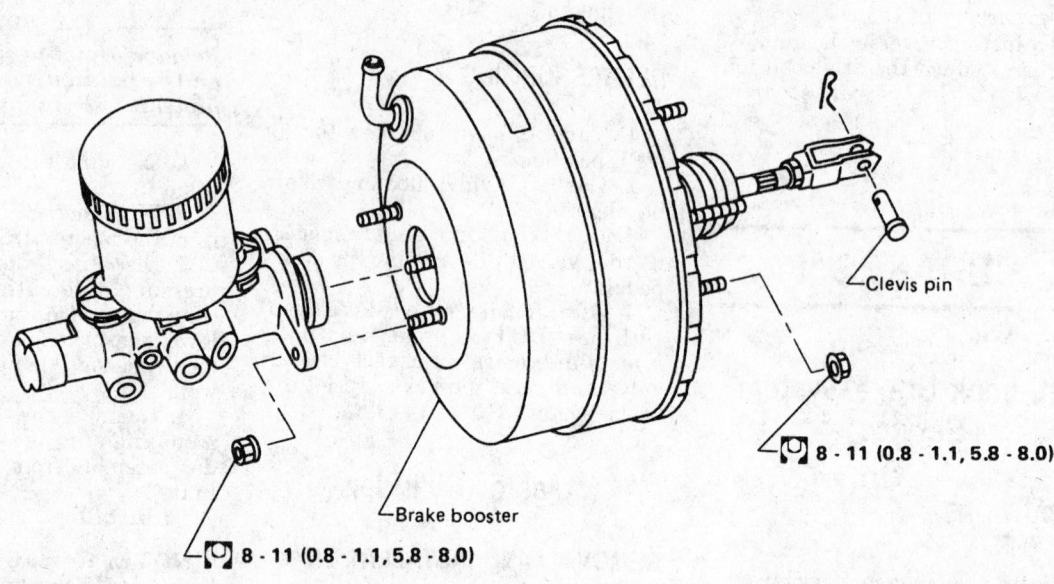

$\boxed{\textcircled{Q}}$ 8 - 11 (0.8 - 1.1, 5.8 - 8.0)

—Clevis pin

—Brake booster

$\boxed{\textcircled{Q}}$ 8 - 11 (0.8 - 1.1, 5.8 - 8.0)

$\boxed{\textcircled{Q}}$: N·m (kg-m, ft-lb)

Typical master cylinder and power brake booster

252534

To install:

NOTE: Certain models may have an UP mark on the cylinder boot, make sure this is in the correct position.

5. Position the master cylinder on the brake booster, then screw the hydraulic line into the cylinder body a few turns. Install the master cylinder mounting nuts and torque them to 9–11 ft. lbs. (9–15 Nm). Torque the brake line fittings to 11–13 ft. lbs. (15–18 Nm).

6. Bleed the master cylinder and the brake system. Make certain that the brake lines are fully tightened and the fluid is full.

7. Connect the negative battery cable.

8. Check and adjust the brake pedal if necessary. Pressurize the brake system and check for leaks.

1996–97 Pick-Up

— CAUTION —

Be careful not to spill brake fluid on the painted surfaces of the vehicle; it will damage the paint.

1. Disconnect the negative battery cable.

2. Disconnect the hydraulic lines from the master cylinder.

3. Disconnect the hydraulic fluid pressure differential switch wiring connectors. On models with fluid level sensors, disconnect the fluid level sensor wiring connectors, as well.

4. Remove the master cylinder mounting nuts and the master cylinder assembly from the power brake unit.

To install:

NOTE: Certain models may have an UP mark on the cylinder boot, make sure this is in the correct position.

5. Position the master cylinder on the brake booster, then screw the hydraulic line into the cylinder body a few turns. Install the master cylinder mounting nuts and torque them to 9–11 ft. lbs. (12–15 Nm). Torque the brake line fittings to 11–13 ft. lbs. (15–18 Nm).

6. Bleed the master cylinder and the brake system. Make certain that the brake lines are fully tightened and the fluid is full.

7. Connect the negative battery cable.

8. Check and adjust the brake pedal if necessary. Pressurize the brake system and check for leaks.

Brake Caliper

REMOVAL AND INSTALLATION

Front

1. Raise the vehicle and support safely.

2. Remove the appropriate tire and wheel assembly.

NOTE: Do not let air into the the master cylinder by allowing the reservoir to empty, or complete system bleeding will be required.

3. Remove the bolt attaching the brake hose to the caliper. Plug the brake hose to prevent brake fluid loss.

4. Remove the caliper support mounting bolts and lift the caliper assembly from the knuckle.

To install

5. Position caliper assembly onto the knuckle and install the bolts. Make sure the rotor fits between the brake pads. Torque the bolts to 53–72 ft. lbs (72–97 Nm).

6. Use new copper washers and connect the brake hose to the caliper.

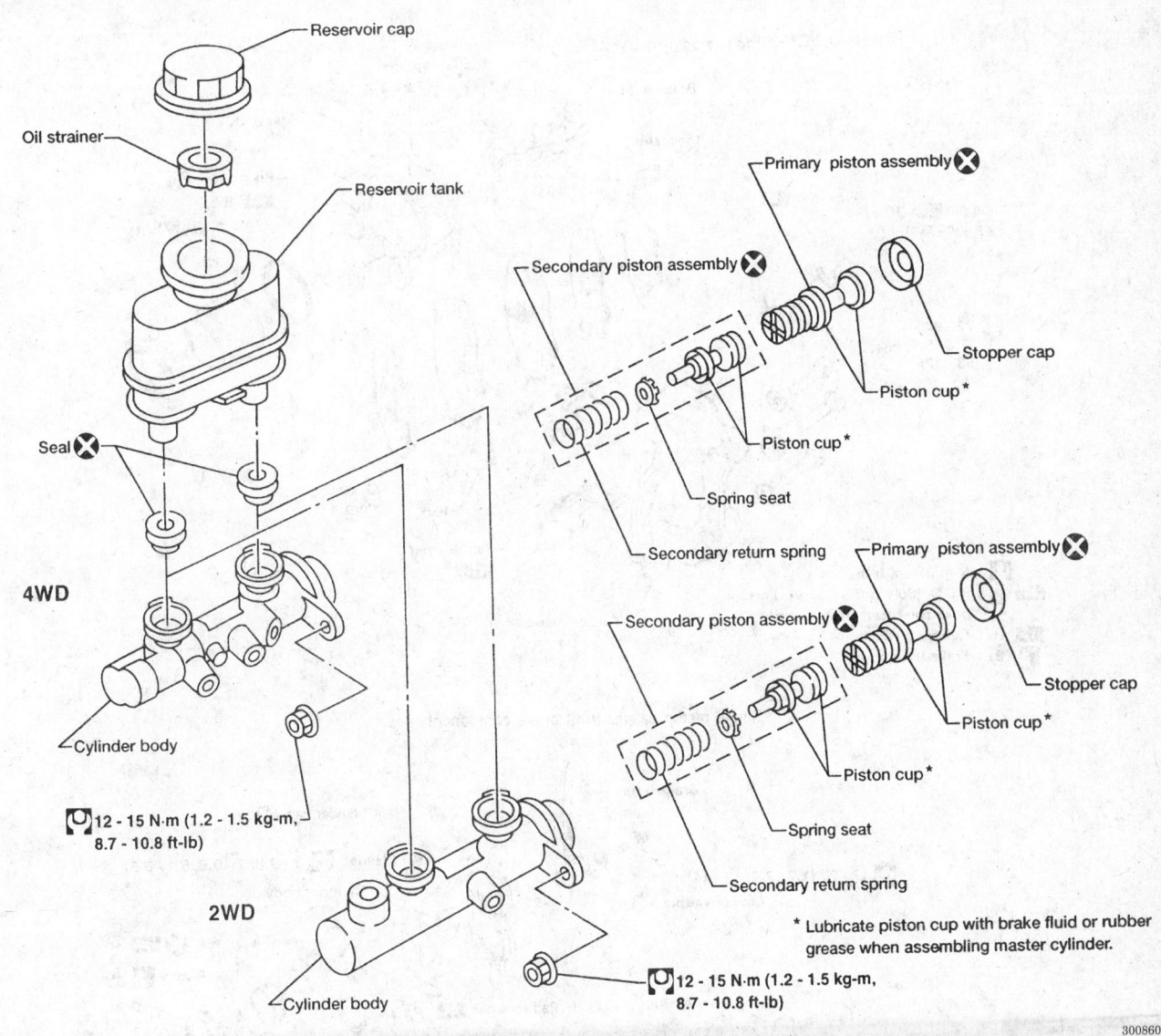

Master cylinder and related components — 1996 Pick-up shown; Pathfinder similar

Torque the brake hose attaching bolt to 12–14 ft. lbs. (17–20 Nm).

NOTE: Use caution not to twist the brake hose during installation.

7. Bleed the brake system.
8. Apply the brake pedal and inspect the system. Ensure proper operation and no leakage.
9. Install tire and wheel assembly. Lower the vehicle and road test.

Rear

NOTE: Unlike most rear disc brake designs, this system does not incorporate the parking brake system into the rear brake caliper; therefore, the rear brake system is serviced in the same manner as the front system.

1. Raise the vehicle and support safely.
2. Remove the appropriate tire and wheel assembly.

NOTE: Do not let air into the the master cylinder by allowing the reservoir to empty, or complete system bleeding will be required.

3. Remove the caliper support mounting bolts and lift the caliper assembly from the baffle plate.
4. Loosen the brake fluid hose with a wrench and turn the caliper to disconnect it from the brake hose. Plug the brake hose to prevent brake fluid loss.
To install:
5. Use a new copper washer and connect the brake hose to the caliper. Torque the hose fitting to 11 ft. lbs. (15 Nm).

NOTE: Use caution not to twist the brake hose during installation.

6. Position the caliper assembly over the baffle plate and install the bolts. Make sure the rotor fits between the brake pads. Torque the bolts to 28–38 ft. lbs. (38–52 Nm).
7. Bleed the brake system.

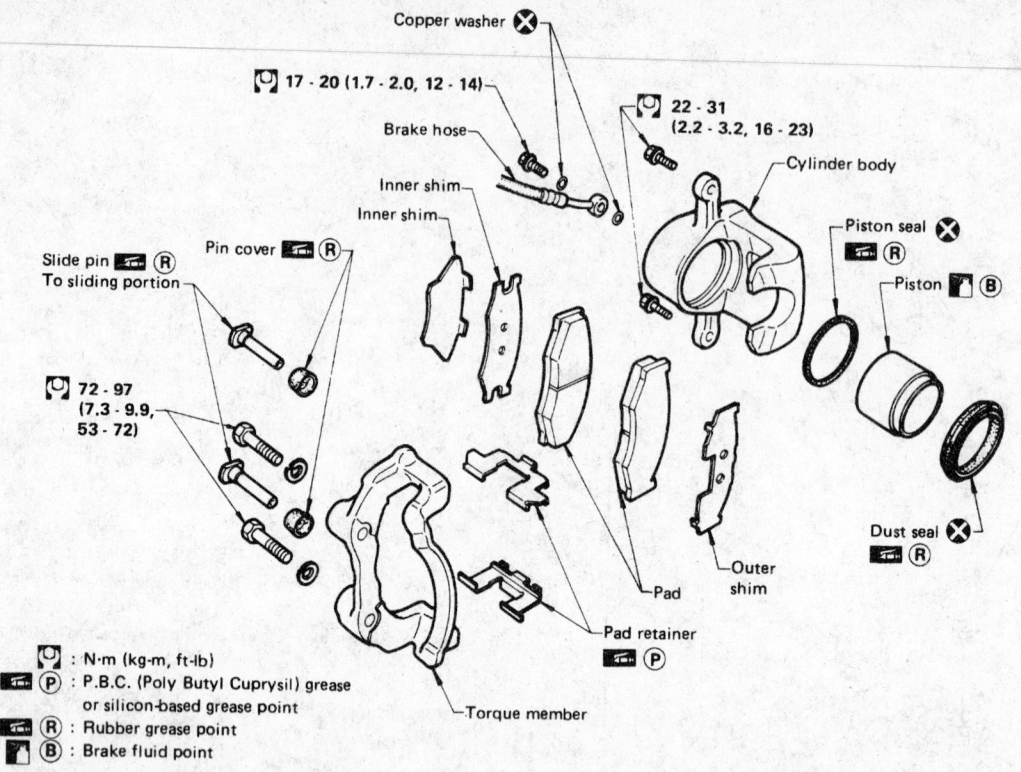

Single piston caliper front brake components

120780

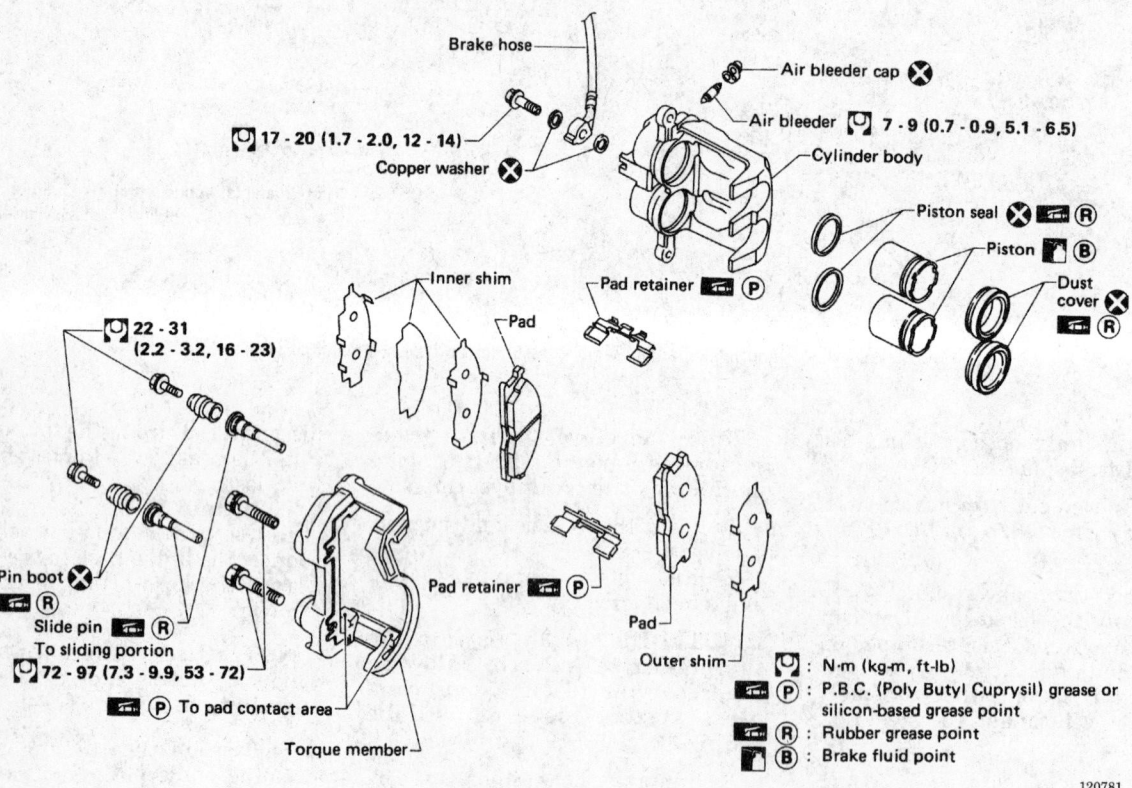

Dual piston caliper front brake components

120781

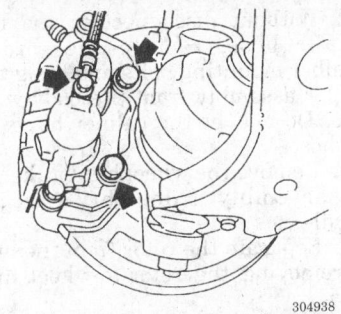

Front brake caliper mounting bolts and brake hose connection — dual piston caliper shown

304938

8. Apply the brake pedal and inspect the system. Ensure proper operation and no leakage.

9. Install tire and wheel assembly. Lower the vehicle and road test.

Disc Brake Pads

REMOVAL AND INSTALLATION

NOTE: **Both the front and rear disc brake pads can be serviced using the same procedure.**

1. Using a syringe, siphon brake fluid from the reservoir, leaving reservoir approximately ½ full.

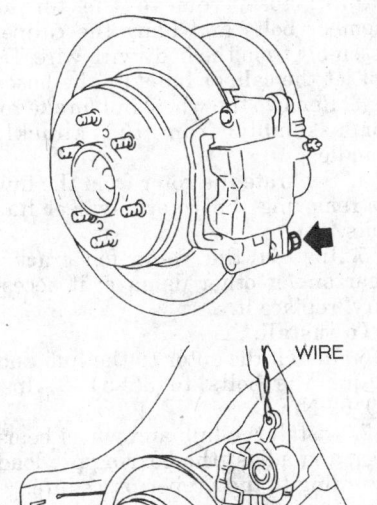

Proper method of accessing rear disc brake pads — 1993–95 Pathfinder

120737

2. Raise and properly support the vehicle.

3. Remove the wheel assemblies.

4. Remove the lower pin bolt from the brake caliper.

5. Swivel the caliper up and away from the torque member. Tie the caliper to a suspension member so that it is out of the way.

6. Lift the 2 brake pads out of the torque member.

7. Remove the inner and outer shims. Remove the 2 pad retainers if they are not still attached to the pads.

8. Check the pad thickness and replace the pads if they are less than 0.079 in. (2mm) thick.

To install:

9. Install the inner and outer shims into the torque member.

10. Install a pad retainer to the bottom of each pad.

11. Install the pads into the torque member.

— CAUTION —

When installing new brake pads, make sure your hands are clean. Do not allow any grease or oil to touch the contact face of the pads.

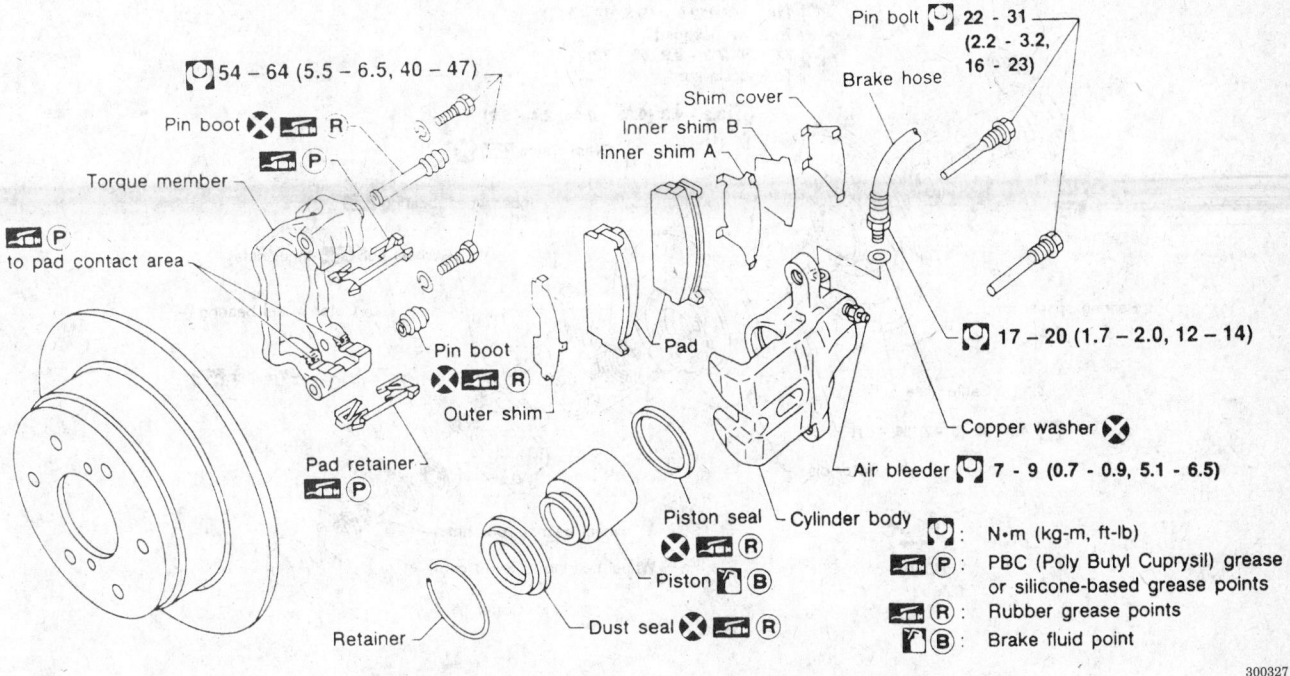

Rear disc brake caliper, rotor and related components — AD14VB system (optional on 1993–95 Pathfinder SE)

300327

12. Use a C–clamp or hammer handle and press the caliper piston(s) back into the housing.

NOTE: Never press the piston into the caliper when the pads are out on both sides of the vehicle.

13. Untie the caliper and swivel it back into position over the torque plate so that the dust boot is not pinched. Install the pin bolt and torque it to 16–23 ft. lbs. (22–31 Nm).

14. Check the condition of the pin boot. Gently pull on it to expel any trapped air.

15. Install the wheel and lower the vehicle.

16. Pump the brakes until the pedal is firm and check the level of brake fluid. Road test the vehicle.

Brake Rotor

REMOVAL AND INSTALLATION

Pick-Up

1. Raise and safely support the vehicle. Remove the wheel assembly.

2. Without disconnecting the hydraulic hose, remove the torque member bolts and hang the caliper assembly from the body with wire. Do not let the caliper hang by the hose.

3. Remove the wheel hub/brake rotor assembly from the knuckle spindle.

4. Separate the rotor from the hub by removing the rotor-to-wheel hub bolts.

5. Inspect the rotor for cracks, wear and/or other damage; if necessary, replace it.

To install:

6. Install the rotor to the hub and torque the bolts to 36–51 ft. lbs. (49–69 Nm).

7. Install the hub and wheel bearings, and adjust the bearing pre-load according to the proper procedure.

8. Install the caliper and pump the brakes until the pedal is firm.

1993–95 Pathfinder

Front

1. Raise and safely support the vehicle. Remove the wheel assembly.

2. Without disconnecting the hydraulic hose, remove the torque member mounting bolts and hang the caliper assembly from the body with wire. Do not let the caliper hang by the hose.

3. Remove the wheel hub/brake rotor assembly from the knuckle spindle.

4. Separate the rotor from the hub by removing the rotor-to-wheel hub bolts.

5. Inspect the rotor for cracks, wear and/or other damage; if necessary, replace it.

To install:

6. Install the rotor to the hub and torque the bolts to 36–51 ft. lbs. (49–69 Nm).

7. Install the hub and wheel bearings and adjust the bearing pre-load according to the proper procedure.

8. Install the caliper and tighten the torque member mounting bolts to 53–72 ft. lbs. (72–97 Nm). Pump the brakes until the pedal is firm.

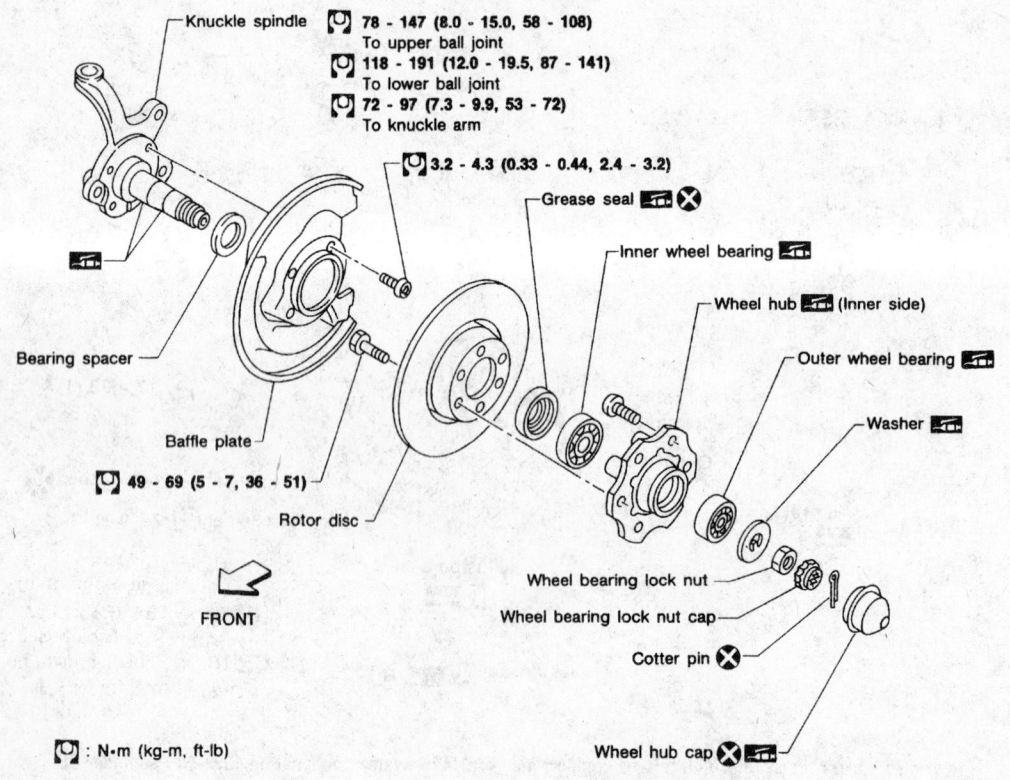

Knuckle spindle — 78 - 147 (8.0 - 15.0, 58 - 108) To upper ball joint
118 - 191 (12.0 - 19.5, 87 - 141) To lower ball joint
72 - 97 (7.3 - 9.9, 53 - 72) To knuckle arm
3.2 - 4.3 (0.33 - 0.44, 2.4 - 3.2)
Grease seal
Inner wheel bearing
Wheel hub (Inner side)
Bearing spacer
Outer wheel bearing
Baffle plate
Washer
49 - 69 (5 - 7, 36 - 51)
Rotor disc
Wheel bearing lock nut
FRONT
Wheel bearing lock nut cap
Cotter pin
: N·m (kg-m, ft-lb)
Wheel hub cap

120637

Wheel hub, bearing and brake rotor assembly — 1993–95 2WD Pick-up

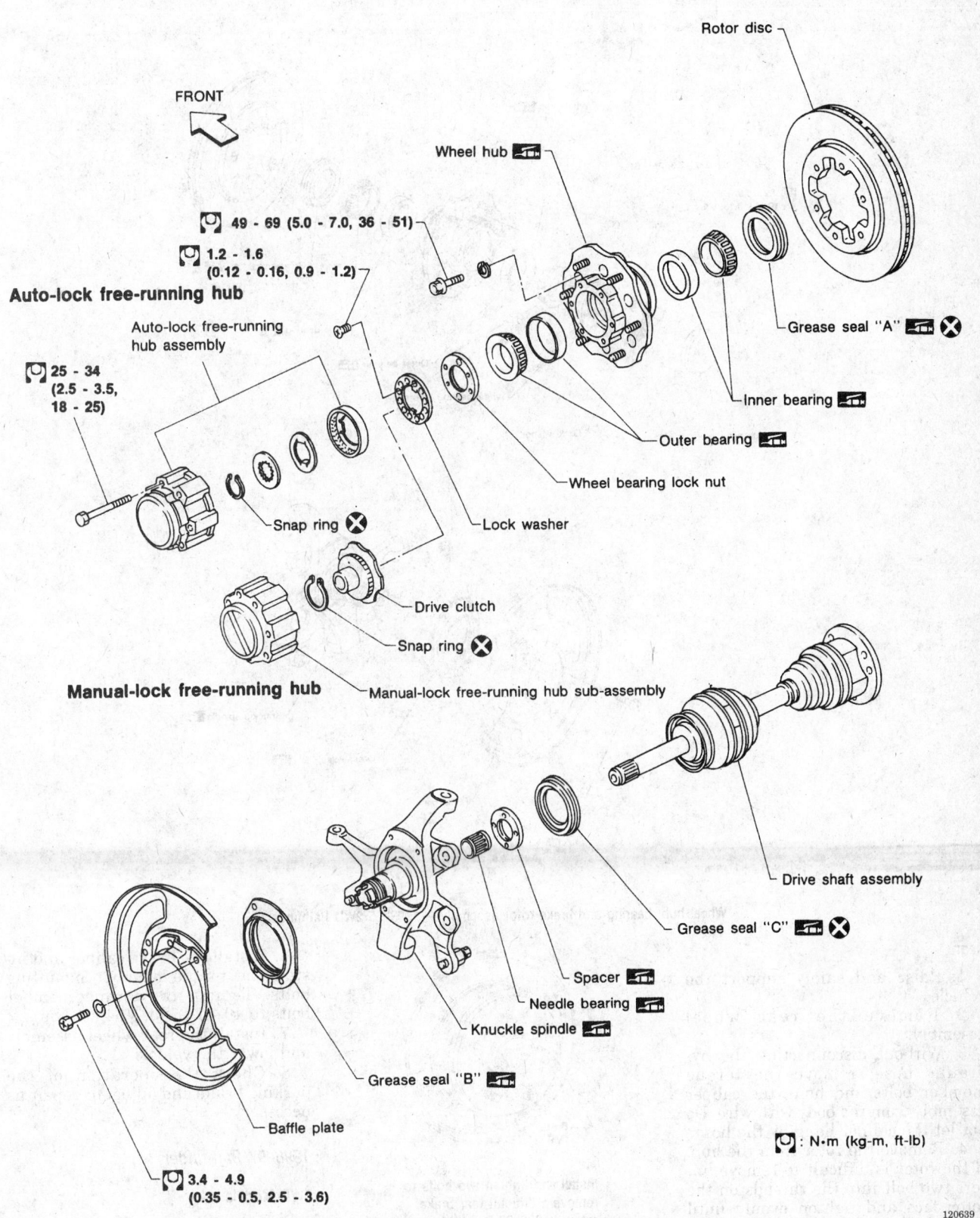

FRONT

Rotor disc

Wheel hub

49 - 69 (5.0 - 7.0, 36 - 51)

1.2 - 1.6
(0.12 - 0.16, 0.9 - 1.2)

Grease seal "A"

Auto-lock free-running hub

Auto-lock free-running
hub assembly

Inner bearing

25 - 34
(2.5 - 3.5,
18 - 25)

Outer bearing

Wheel bearing lock nut

Snap ring

Lock washer

Drive clutch

Snap ring

Manual-lock free-running hub

Manual-lock free-running hub sub-assembly

Drive shaft assembly

Grease seal "C"

Spacer

Needle bearing

Knuckle spindle

Grease seal "B"

Baffle plate

N•m (kg-m, ft-lb)

3.4 - 4.9
(0.35 - 0.5, 2.5 - 3.6)

120639

Wheel hub, bearing and brake rotor assembly — 1993–95 4WD Pick-up and Pathfinder

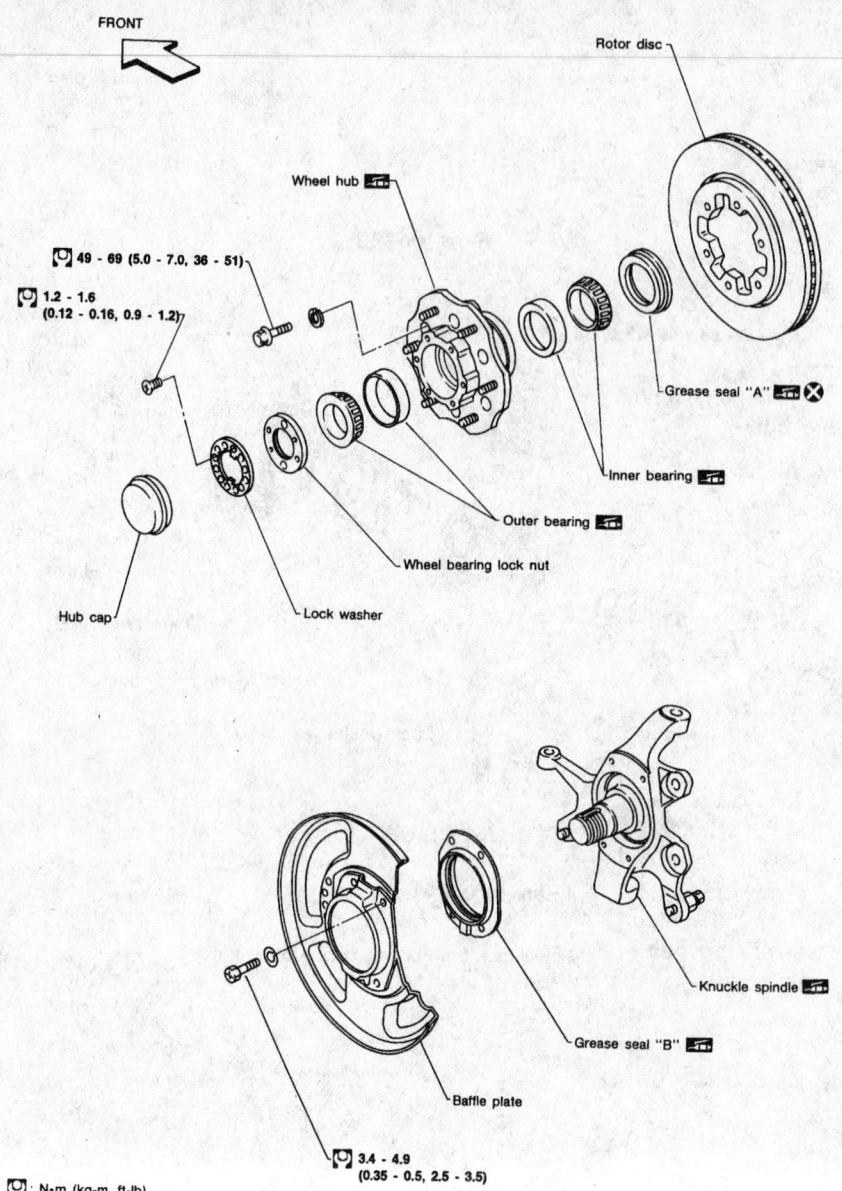

FRONT

Rotor disc

Wheel hub

49 - 69 (5.0 - 7.0, 36 - 51)

1.2 - 1.6
(0.12 - 0.16, 0.9 - 1.2)

Grease seal "A"

Inner bearing

Outer bearing

Wheel bearing lock nut

Hub cap

Lock washer

Knuckle spindle

Grease seal "B"

Baffle plate

3.4 - 4.9
(0.35 - 0.5, 2.5 - 3.5)

: N·m (kg-m, ft-lb)

120638

Wheel hub, bearing and brake rotor assembly — 1993–95 2WD Pathfinder

Rear

1. Raise and safely support the vehicle.
2. Remove the rear wheel assembly.
3. Without disconnecting the hydraulic hose, remove the torque member bolts and hang the caliper assembly from the body with wire. Do not let the caliper hang by the hose.
4. Remove the rotor from the hub. If the rotor is difficult to remove, insert two bolt into the threads on the rotor face and tighten evenly until the rotor is separated from the hub.

To install:

5. Install the rear brake rotor to the hub.

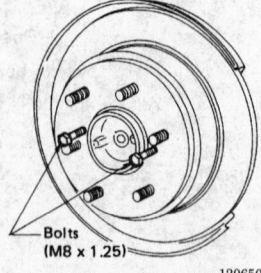

Bolts
(M8 x 1.25)

120650

Install and tighten two bolts to remove a difficult rear brake rotor — 1993–95 Pathfinder with optional rear disc brakes

6. Install the brake caliper and install the torque member mounting bolts. Tighten the torque member bolts to 40–47 ft. lbs. (54–64 Nm).
7. Install the rear wheel assembly and lower the vehicle.
8. Check the operation of the parking brake and adjust or repair as necessary.

1996–97 Pathfinder

2WD Models

1. Raise and safely support the vehicle.
2. Remove the front wheel assembly.

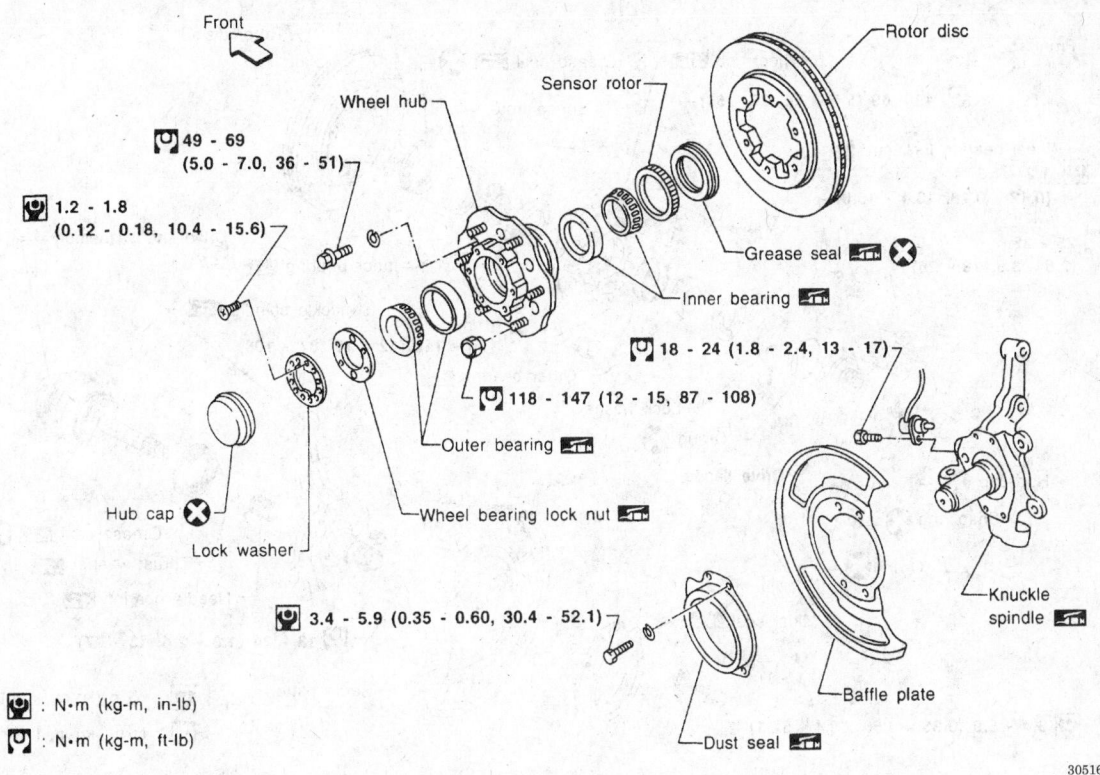

Front

49 - 69
(5.0 - 7.0, 36 - 51)

1.2 - 1.8
(0.12 - 0.18, 10.4 - 15.6)

Wheel hub

Sensor rotor

Rotor disc

Grease seal

Inner bearing

18 - 24 (1.8 - 2.4, 13 - 17)

118 - 147 (12 - 15, 87 - 108)

Outer bearing

Hub cap

Lock washer

Wheel bearing lock nut

Knuckle spindle

3.4 - 5.9 (0.35 - 0.60, 30.4 - 52.1)

Baffle plate

: N·m (kg-m, in-lb)

: N·m (kg-m, ft-lb)

Dust seal

305160

Wheel hub, bearing and brake rotor assembly — 2WD 1996–97 Pathfinder

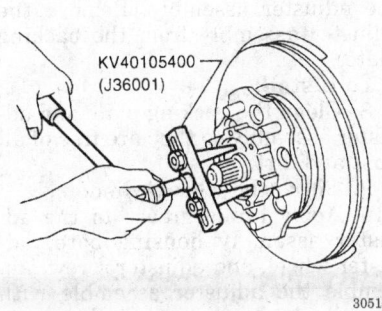

KV40105400
(J36001)

305162

Wheel bearing locknut removal — 2WD 1996–97 Pathfinder

3. Remove the brake caliper with the support. Hang the caliper out of the way with a piece of wire.

4. Remove the bearing cap.

5. Remove the lockwasher.

6. Using special tool J36001 or equivalent, remove the wheel bearing locknut.

7. Remove the hub and rotor assembly and separate the rotor from the hub.

To install:

8. Install the rotor on the hub assembly. Torque the mounting bolts to 36–51 ft. lbs. (49–69 Nm).

9. Install the hub and adjust the wheel bearing.

10. Install the lockwasher and the bearing cap.

11. Install the brake caliper assembly.

12. Install the front wheel and safely lower the vehicle to the floor.

4WD

1. Raise and safely support the vehicle.

2. Remove the front wheel assembly.

3. Remove the brake caliper with the support. Hang the caliper out of the way with a piece of wire.

4. Remove the snapring using the proper tool.

5. Remove the drive flange.

6. Remove the lockwasher.

7. Using special tool J36001 or equivalent, remove the wheel bearing locknut.

8. Remove the hub and rotor assembly and separate the rotor from the hub.

To install:

9. Install the rotor on the hub assembly. Torque the mounting bolts to 36–51 ft. lbs. (49–69 Nm).

10. Install the hub and adjust the wheel bearing.

11. Install the lockwasher.

12. Pack the drive flange with grease and install it. Torque the nuts to 18–26 ft. lbs. (25–35 Nm).

13. Install a new snapring and cap.

14. Install the brake caliper assembly.

15. Install the front wheel and safely lower the vehicle to the floor.

Brake Drums

REMOVAL AND INSTALLATION

1. Remove the hub cap and loosen the lug nuts.

2. Raise the rear of the vehicle and support it on jackstands.

3. Remove the lug nuts, tire and wheel.

4. Release the parking brake.

5. Pull the brake drum from the hub. If difficult to remove try the following:

   a. Strike the face of the drum with a plastic or rubber mallet. This will break free any rust which may develop between the drum and the hub.

   b. Install two 8 x 1.25mm bolts into the holes in the drum and gradually tighten them to pull the drum off the hub.

— **CAUTION** —
Do not depress the brake pedal with the brake drum removed.

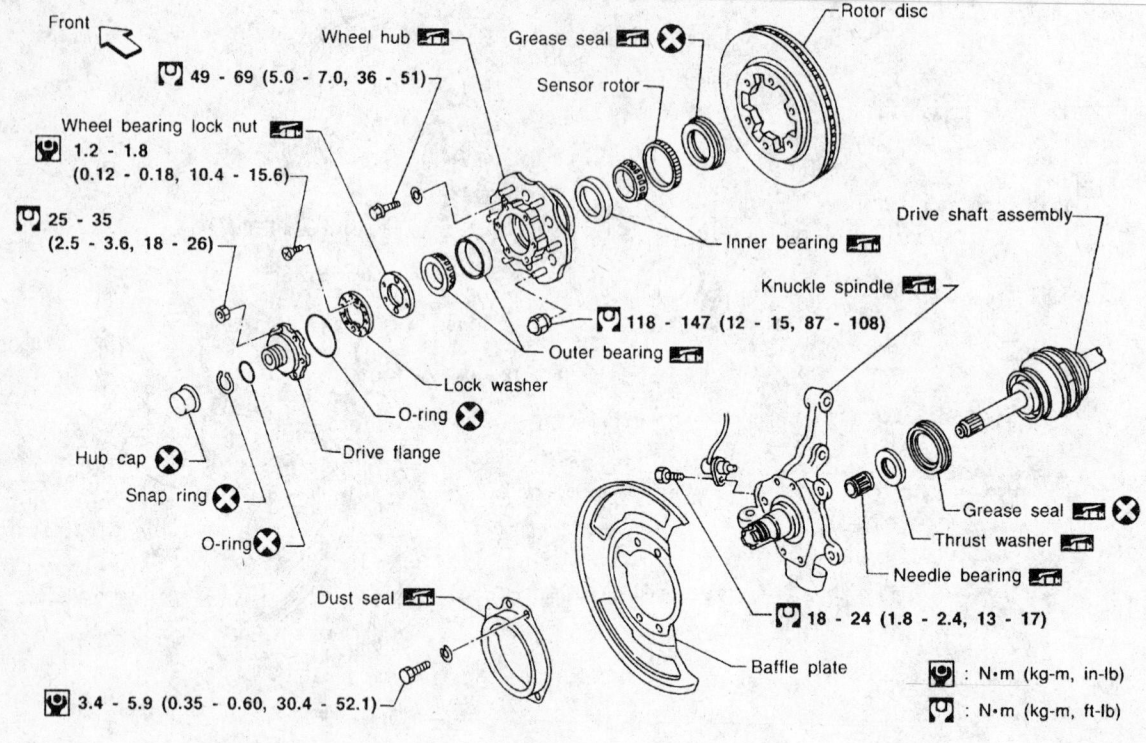

Front

Wheel hub

Grease seal

Rotor disc

Sensor rotor

$\boxed{49 - 69 \ (5.0 - 7.0, \ 36 - 51)}$

Wheel bearing lock nut
$1.2 - 1.8$
$(0.12 - 0.18, \ 10.4 - 15.6)$

$25 - 35$
$(2.5 - 3.6, \ 18 - 26)$

Inner bearing

Drive shaft assembly

Knuckle spindle

$118 - 147 \ (12 - 15, \ 87 - 108)$

Outer bearing

Lock washer

O-ring

Hub cap

Drive flange

Snap ring

O-ring

Grease seal

Thrust washer

Needle bearing

Dust seal

$18 - 24 \ (1.8 - 2.4, \ 13 - 17)$

Baffle plate

: N·m (kg-m, in-lb)

: N·m (kg-m, ft-lb)

$3.4 - 5.9 \ (0.35 - 0.60, \ 30.4 - 52.1)$

305161

Wheel hub, bearing and brake rotor assembly — 4WD 1996–97 Pathfinder

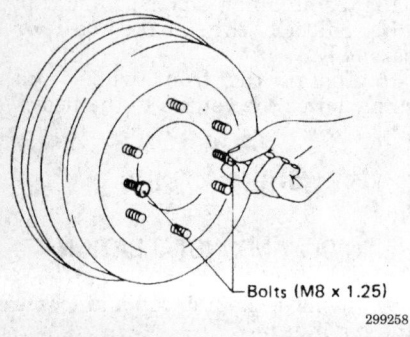

Bolts (M8 × 1.25)

299258

Install and tighten two bolts to remove a stubborn brake drum

To install:

6. Install the brake drum to the hub.

7. Install the tire, wheel and lug nuts.

8. Remove the jackstands and lower the vehicle.

9. Road test the vehicle to ensure that the brakes are working properly.

Brake Shoes

REMOVAL AND INSTALLATION

Except 1996–97 Pathfinder

1. Raise and support the vehicle until the axle to be serviced is off the ground.

2. Remove the wheels and brake drums.

3. With a pair of pliers, remove the brake shoe hold-down anti-rattle spring retainers. Depress the retainer while rotating it 90° to align the slot in the retainer with the flanged end of the pin. Remove the retainers, springs, spring seats, and pins.

4. Open the brake shoes outward against the return springs and remove the parking brake extension link.

5. Disconnect the brake shoe return springs.

6. Remove the brake shoes from the backing plate. The secondary (after) brake shoe must be disconnected from the parking brake toggle lever after withdrawing the toggle lever clevis pin.

7. Remove the rubber boot from behind the brake backing plate and slide the adjuster shim, lockplate, and adjuster springs off the back of the adjuster assembly. Remove the adjuster assembly from the backing plate.

To install:

8. Clean the backing plate and adjuster assembly so they are free of all dust and dirt.

9. Check the wheel cylinders.

10. Apply brake grease to the adjuster assembly housing bore, adjuster wheel, and adjuster screw. Assemble the adjuster assemble with the adjuster screw turned all the way in. Apply brake grease to the sliding surfaces of the adjuster assembly, brake backing plate, and the retaining spring. Install the adjuster assembly to the backing plate.

11. On models with 4WD, after installing the crank lever on the back plate, make sure there is no play between the crank lever and the back plate when pulling the crank lever. If play exists, adjust bolt **A** and locknut **B**.

12. Assemble the secondary (after) brake shoe to the parking brake toggle lever and adjust the clearance between the toggle lever and the brake shoe.

13. Before assembling the brake shoes to the backing plate apply brake grease to the following areas: the brake shoe grooves in the parking brake extension link, the inside sur-

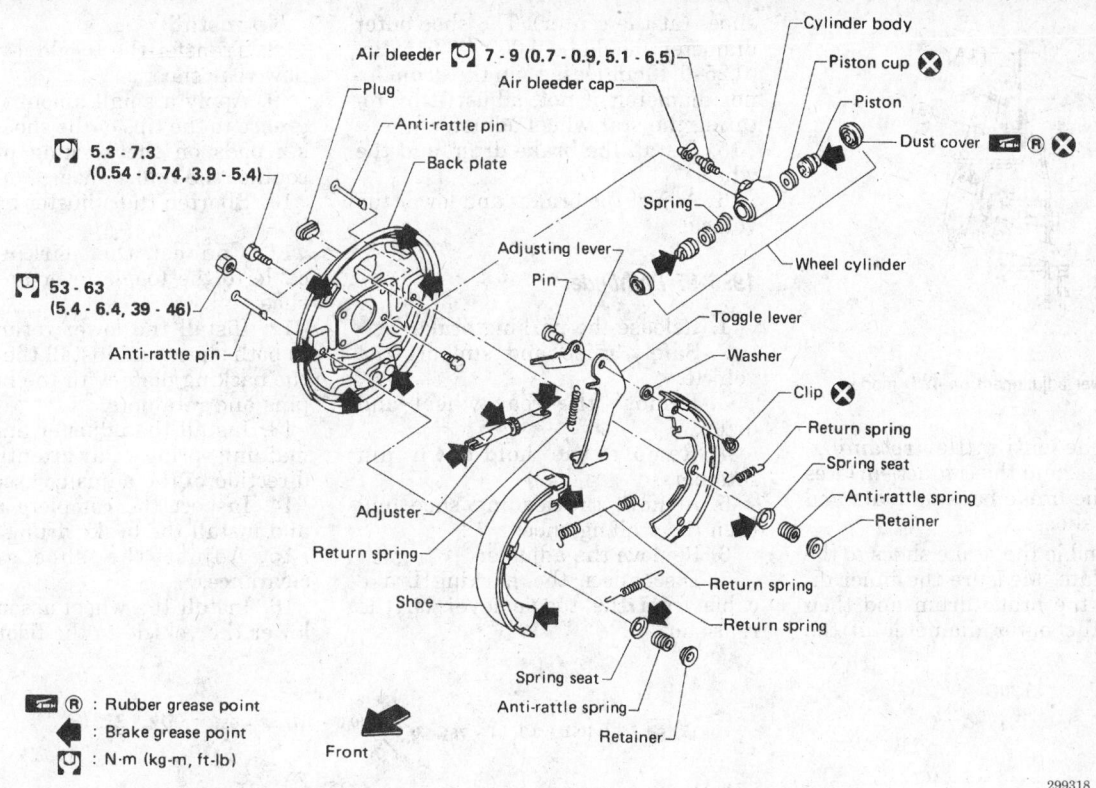

Rear brake shoes and related components — LT26B (2WD) system

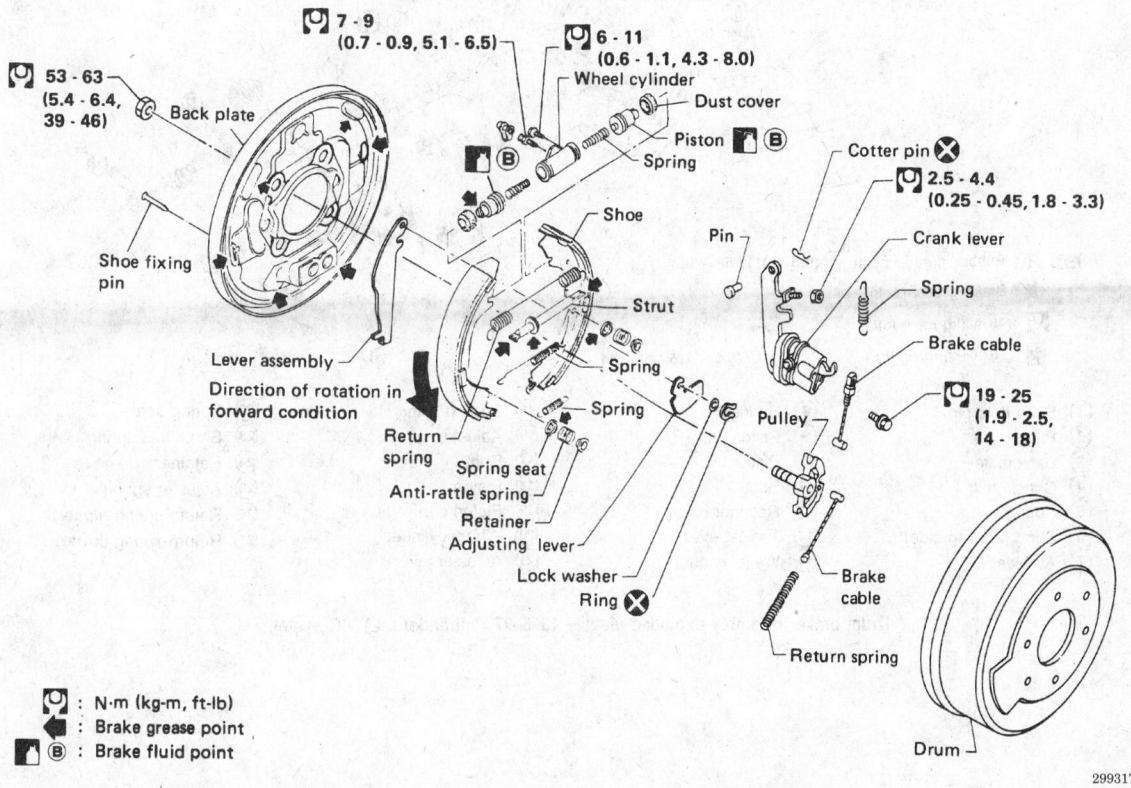

Rear brake shoes and related components — LT30A (4WD) system

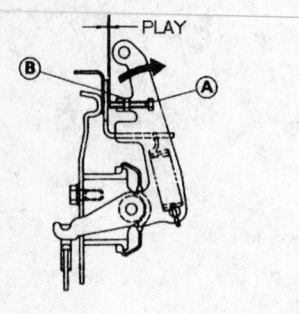

299319

Crank lever adjustment on 4WD models

faces of the anti-rattle (retaining) spring seats, and the contact surfaces between the brake backing plate and the brake shoes.

14. Assemble the brake shoes to the backing plate. Measure the inner diameter of the brake drum and then measure the outer diameter of the shoes (at the center). The shoe outer diameter should be 0.0098–0.0157 in. (0.25–0.40mm) less than the drum inner diameter; if not, adjust it by rotating the star wheel adjuster.

15. Install the brake drum and the wheel.

16. Adjust the brakes and lower the vehicle.

1996–97 Pathfinder

1. Release the parking brake.
2. Safely raise and support the vehicle.
3. Remove the rear wheel and drum.
4. Remove the hold-down pin retainers.
5. Remove the leading shoe and then the trailing shoe.
6. Remove the adjuster.
7. Disconnect the parking brake cable from the toggle lever on the rear shoe.

To install:

8. Transfer the toggle lever to the new rear shoe.
9. Apply a small amount of brake grease to the tips of the shoes and the six pads on the backing plate that contact the brake shoe.
10. Shorten the adjuster by turning it.
11. Connect the parking brake cable to the toggle lever on the rear shoe.
12. Install the lower return spring to both shoes and install the shoes on the backing plate with the hold down pins and retainers.
13. Install the adjuster and the remaining springs. Pay attention to the direction of the adjuster assembly.
14. Inspect the complete assembly and install the brake drum.
15. Adjust the shoe to drum clearance.
16. Install the wheel assembly and lower the vehicle to the floor.

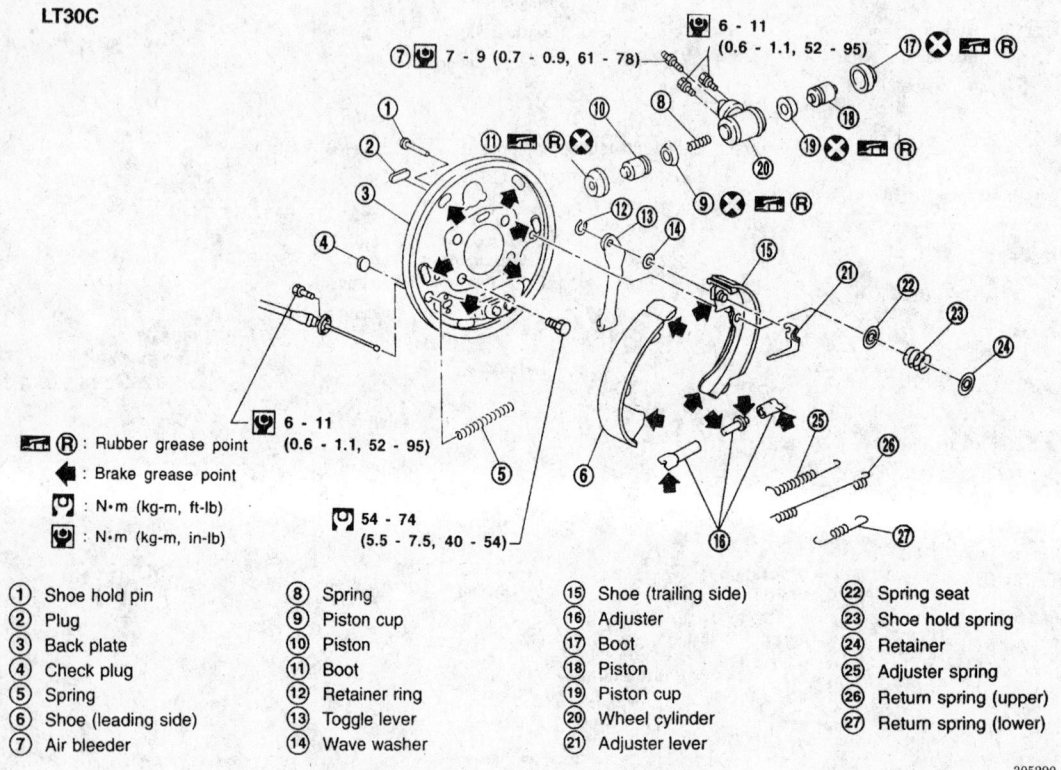

305290

Drum brake assembly exploded view — 1996–97 Pathfinder's LT30C system

① Shoe hold pin
② Plug
③ Back plate
④ Check plug
⑤ Spring
⑥ Shoe (leading side)
⑦ Air bleeder
⑧ Spring
⑨ Piston cup
⑩ Piston
⑪ Boot
⑫ Retainer ring
⑬ Toggle lever
⑭ Wave washer
⑮ Shoe (trailing side)
⑯ Adjuster
⑰ Boot
⑱ Piston
⑲ Piston cup
⑳ Wheel cylinder
㉑ Adjuster lever
㉒ Spring seat
㉓ Shoe hold spring
㉔ Retainer
㉕ Adjuster spring
㉖ Return spring (upper)
㉗ Return spring (lower)

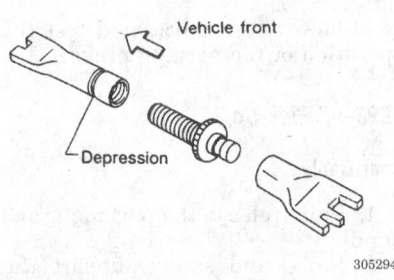

Correct direction of brake shoe adjuster —
1996–97 Pathfinder

Wheel Cylinder

REMOVAL AND INSTALLATION

NOTE: It is important not to let the master cylinder reservoir run dry, at any time during this procedure, or the entire system will have to be bled.

1. Raise the vehicle and support it safely.
2. Remove the wheel and the brake drum.
3. Remove the shoe-to-lever springs or the upper shoe-to-shoe spring. Spread the upper portion of the brake shoes slightly.
4. Remove and plug the brake line from the wheel cylinder.
5. Remove the wheel cylinder retaining bolts and remove the cylinder from the backing plate.

To install:

6. Apply a very thin coating of silicone sealer to the cylinder mounting surface, install the cylinder to the backing plate and install the retaining bolts. Torque the bolts to 5–8 ft. lbs. (6–11 Nm).
7. Connect the brake line to the wheel cylinder.
8. Install the brake springs and the brake drum.
9. Install the tire and wheel assembly.
10. Fill the system with clean brake fluid and bleed the rear brakes.

Parking Brake Cable

ADJUSTMENT

Except 1996–97 Pathfinder

With Drum Brakes

NOTE: Make certain that the brake shoes are properly adjusted before attempting to adjust the parking brake.

1. Pull the parking brake lever up with a force of about 45 lbs. If that value cannot be determined, just pull it up as far as you can. The total number of clicks heard should be 10–12.
2. If the number of clicks was not within that range, release the lever and back off the cable adjuster locknut at the base of the lever and tighten the adjusting nut until there is no more slack in the cable.
3. Operate the lever and brake pedal several times, until no more clicks are heard from the automatic adjuster.
4. Turn the adjusting nut to give the proper number of clicks when the lever is raised full travel.
5. Raise and support the rear of the car on jackstands.
6. Raise the parking brake lever 10–12 notches, the rear wheels should not turn. Release the brake lever and make sure that the rear wheels turn freely. If not, back off on the adjusting nut until they do.

With Disc Brakes

1. Pull the parking brake lever up with a force of about 45 lbs. (61 N). The total number of clicks heard should be 10–12. If the number of clicks was not within that range, system requires adjustment.

NOTE: The parking brake shoes must be adjusted before attempting to adjust the cable mechanism.

2. To adjust the parking brake shoes perform the following steps.
  a. Release the lever and back off the cable adjuster locknut at the base of the lever.
  b. Raise the vehicle, support safely and remove the wheel. Remove the hole plug in the backing plate.
  c. Use a suitable adjusting spoon to pry up on the self-adjuster wheel until the rotor will not turn.
  d. Return the adjuster 7–8 notches in the opposite direction. Make sure the rotor turns freely.
3. Once the parking brake shoes have been properly adjusted, adjust the cable mechanism, by performing the following steps:
  a. Turn the adjusting nut to give the proper number of clicks when the lever is raised full travel.
  b. Raise and support the rear of the car on jackstands.
  c. Raise the parking brake lever 10–12 notches, the rear wheels should not turn. Release the brake lever and make sure that the rear

wheels turn freely. If not, back off on the adjusting nut until they do.

1996–97 Pathfinder

1. Release the parking brake lever and loosen the adjusting nut under the lever.
2. Depress the brake pedal at least 10 times with the engine running.
3. Pull the brake lever up 4–5 notches and tighten the adjusting nut to take the slack out of the cable.
4. Pull the brake lever up with a force of 44 lbs. (196 N). The lever should move up 7–9 notches and the rear brakes should be applied. If not, repeat the adjustment procedure.
5. If needed, bend the warning lamp switchplate so the warning lamp comes on when the lever is pulled one notch.

REMOVAL AND INSTALLATION

1993–95 Models

Rear Cable

1. Fully release the parking brake handle.
2. Raise and safely support the vehicle.
3. Loosen the adjusting nut at the adjuster cable lever.
4. Disconnect the cable from the balance lever or adjuster.
5. Disconnect the rear parking brake cable(s) from the parking brake toggle levers of the rear service brake assemblies.
6. Remove the rear parking brake cable brackets-to-chassis bracket screws.
7. Remove parking brake cable(s) from the vehicle.

To install:

8. Apply a light coat of grease to the cable to make sure it slides properly.
9. Position the parking brake cable in the vehicle and install the brackets-to-chassis bracket screws.
10. Connect the rear parking brake cable(s) to the parking brake toggle levers of the rear service brake assemblies.
11. Connect the cable to the adjuster and install the adjusting nut.
12. Adjust the parking brake cables.
13. Lower the vehicle and test the operation of the parking brake.

Front Cable

1. Fully release the parking brake control lever.
2. Raise and safely support the vehicle.

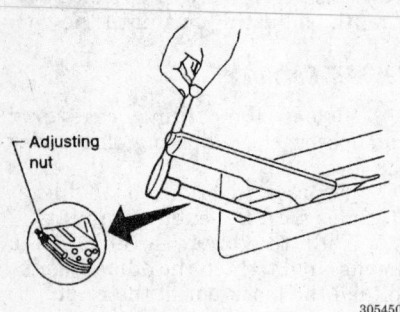

Adjusting nut

305450

Brake lever adjusting nut — 1996–97 Pathfinder

3. Loosen the adjusting nut at the adjuster cable lever.

4. Disconnect the cable from the balance lever or adjuster.

5. Remove the front cable bracket-to-chassis bolt(s) and the cable from the vehicle.

To install:

6. Apply a light coat of grease to the cable to make sure it slides properly.

7. Position the parking brake cable in the vehicle and install the brackets-to-chassis bracket screws.

8. Connect the cable to the balance lever or adjuster.

9. Adjust the parking brake cables.

10. Lower the vehicle and test the operation of the parking brake.

1996–97 Pathfinder

1. Fully release the parking brake handle.

2. Raise and safely support the vehicle.

3. Loosen the adjusting nut at the adjuster cable lever.

4. Disconnect the cable from the balance lever or adjuster.

5. Disconnect the rear parking brake cable(s) from the parking brake toggle levers of the rear service brake assemblies.

6. Remove the rear parking brake cable brackets-to-chassis bracket screws.

7. Remove parking brake cable(s) from the vehicle.

To install:

8. Apply a light coat of grease to the cable to make sure it slides properly.

9. Position the parking brake cable in the vehicle and install the brackets-to-chassis bracket screws.

10. Connect the rear parking brake cable(s) to the parking brake toggle levers of the rear service brake assemblies.

11. Connect the cable to the adjuster and install the adjusting nut.

12. Adjust the parking brake cables.

13. Lower the vehicle and test the operation of the parking brake.

1996–97 Pick-Up

Rear Cable

1. Fully release the parking brake handle.

2. Raise and safely support the vehicle.

3. Loosen the adjusting nut at the adjuster cable lever.

4. Disconnect the cable from the balance lever or adjuster.

5. Disconnect the rear parking brake cable(s) from the parking brake toggle levers of the rear service brake assemblies.

6. Remove the screws attaching the rear parking brake cable brackets to the chassis.

7. Remove parking brake cable(s) from the vehicle.

To install:

8. Apply a light coat of grease to the cable to make sure it slides properly.

9. Position the parking brake cable in the vehicle and install the cable brackets to the chassis.

10. Connect the rear parking brake cable(s) to the parking brake toggle

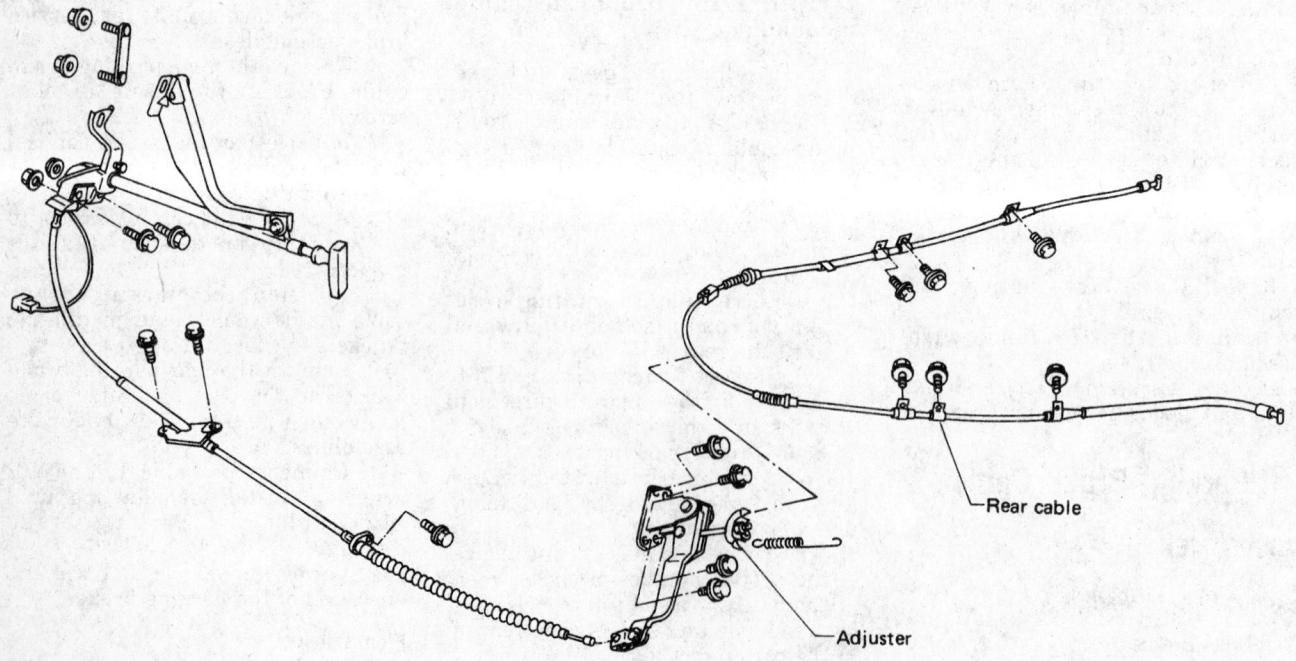

Rear cable

Adjuster

300987

1993–97 2WD Pick-up parking brake cable routing

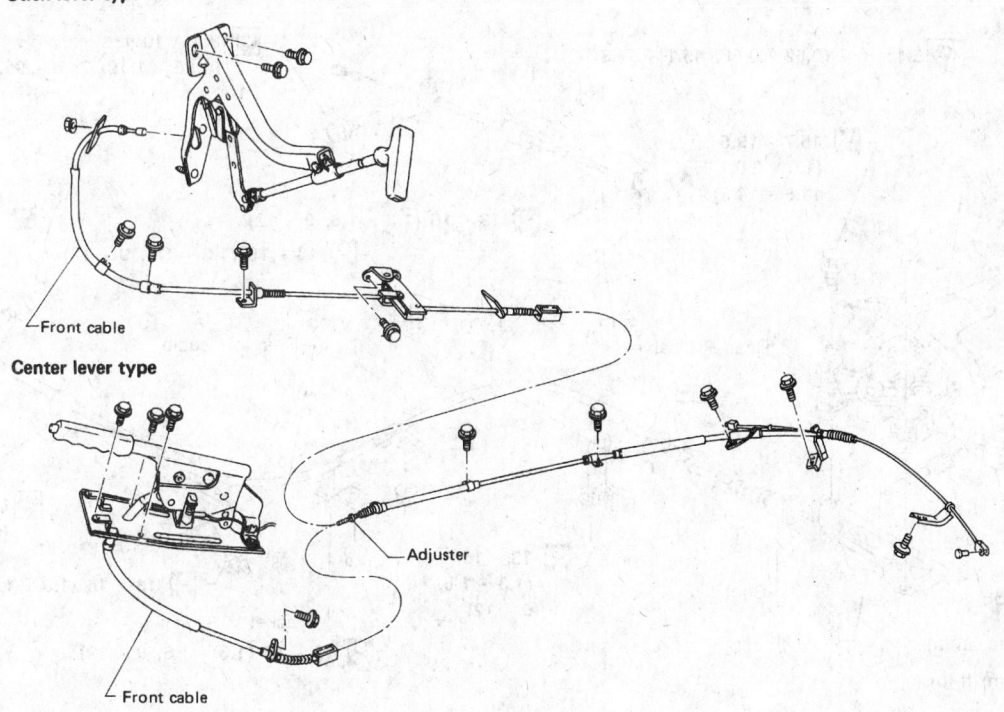

Stick lever type

Center lever type

300988

1993–97 4WD Pick-up parking brake cable routing

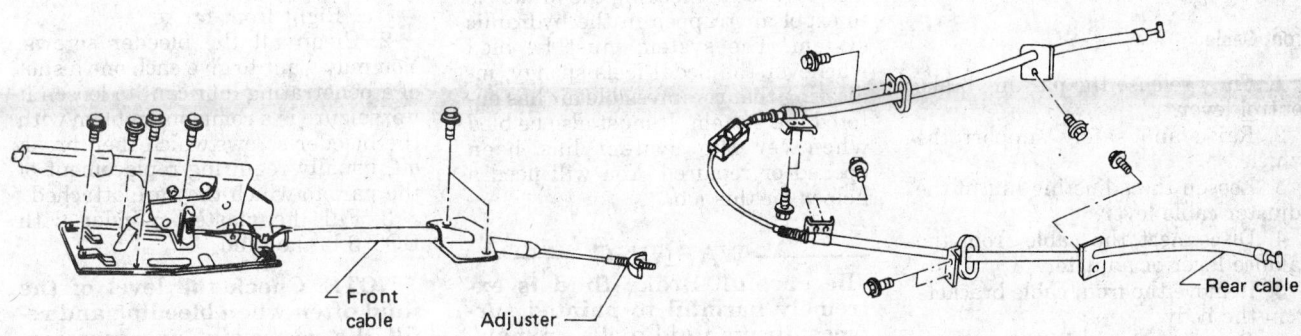

300989

1993–95 Pathfinder parking brake cable routing

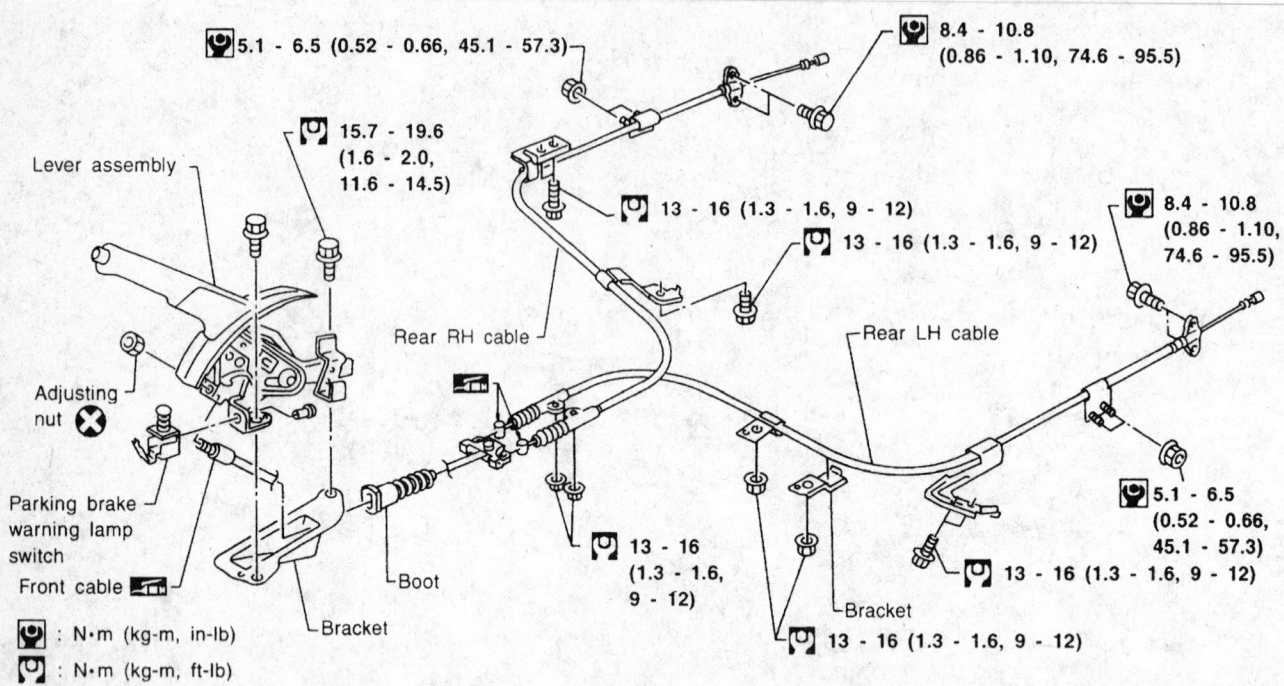

5.1 - 6.5 (0.52 - 0.66, 45.1 - 57.3)

15.7 - 19.6 (1.6 - 2.0, 11.6 - 14.5)

8.4 - 10.8 (0.86 - 1.10, 74.6 - 95.5)

13 - 16 (1.3 - 1.6, 9 - 12)

13 - 16 (1.3 - 1.6, 9 - 12)

8.4 - 10.8 (0.86 - 1.10, 74.6 - 95.5)

Lever assembly

Adjusting nut

Parking brake warning lamp switch

Front cable

Rear RH cable

Rear LH cable

Boot

Bracket

Bracket

5.1 - 6.5 (0.52 - 0.66, 45.1 - 57.3)

13 - 16 (1.3 - 1.6, 9 - 12)

13 - 16 (1.3 - 1.6, 9 - 12)

13 - 16 (1.3 - 1.6, 9 - 12)

: N·m (kg-m, in-lb)

: N·m (kg-m, ft-lb)

Parking brake cable mounting — 1996–97 Pathfinder

levers of the rear service brake assemblies.

11. Connect the cable to the adjuster and install the adjusting nut.

12. Adjust the parking brake cables.

13. Lower the vehicle and test the operation of the parking brake.

Front Cable

1. Fully release the parking brake control lever.

2. Raise and safely support the vehicle.

3. Loosen the adjusting nut at the adjuster cable lever.

4. Disconnect the cable from the balance lever or adjuster.

5. Remove the front cable brackets from the body.

6. Remove the cable from the vehicle.

   To install:

7. Apply a light coat of grease to the cable to make sure it slides properly.

8. Position the parking brake cable in the vehicle and install the brackets to the body with the attaching screws.

9. Connect the cable to the balance lever or adjuster.

10. Adjust the parking brake cables.

11. Lower the vehicle and test the operation of the parking brake.

Brake System Bleeding

The purpose of bleeding the brakes is to expel air trapped in the hydraulic system. The system must be bled whenever the pedal feels spongy, indicating that compressible air has entered the system. It must also be bled whenever the system has been opened or repaired. You will need a helper for this job.

— **WARNING** —

Be careful! Brake fluid is extremely harmful to painted surfaces. Brake fluid picks up moisture from the air. Do not leave the master cylinder or the fluid container uncovered any longer than necessary. Never reuse brake fluid which has been bled from the brake system. Always use proper safety equipment bleeding the brake system.

NOTE: On models with ABS, be sure to turn the ignition OFF and disconnect the ABS actuator connector.

1. The sequence for bleeding is as follows:

   a. Load Sensing Valve (LSV) air bleeder (only on models equipped with LSV)

   b. Left rear brake

   c. Right rear brake

   d. Left front brake

   e. Right front brake

2. Clean all the bleeder screws. You may want to give each one a shot of a penetrating lubricant to loosen it up; seizure is a common problem with the bleeder screws, which then break off, usually requiring replacement of the part to which they are attached.

3. Fill the master cylinder with DOT 3 brake fluid.

NOTE: Check the level of the fluid often when bleeding and refill the reservoirs as necessary. Do not let the reservoir run dry.

4. Attach a length of clear vinyl tubing to the bleeder screw on the wheel cylinder or disc caliper. Insert the other end of the tube into a clear, clean jar half filled with brake fluid.

5. Fully depress the brake pedal several times with all bleeders closed.

6. With the brake pedal depressed, open the bleeder screw ⅓–½ of a turn and allow the fluid to run through the tube. Then close the bleeder screw before the pedal

reaches the end of its travel. Have your assistant slowly release the pedal. Repeat this process until no air bubbles appear in the expelled fluid.

NOTE: If the brake pedal is depressed too fast, small air bubbles will form in the brake fluid.

7. Repeat the procedure on the other remaining bleeder screws, checking the level of fluid in the cylinder reservoirs often.

8. When all the air has been bleed from the system, perform the following steps:

 a. If disconnected, reconnect the actuator.

 b. Pressurize the system and check for leaks.

 c. Check and fill the fluid reservoir.

 d. Road test the vehicle.

Wheel Speed Sensor

REMOVAL AND INSTALLATION

Except 1996–97 Pathfinder

1. Raise and safely support the vehicle.

2. Matchmark and remove the driveshaft from the companion flange at the differential.

3. Matchmark the companion flange to the pinion shaft and remove the flange.

4. Remove the speed sensor from the differential housing.

To install:

5. Install the speed sensor on the differential housing. Torque the mounting bolts to 6–8 ft. lbs. (8–11 Nm).

6. Align the matchmark and install the companion flange. Torque

the nut to 145–210 ft. lbs. (196–284 Nm).

7. Align the matchmark and install the driveshaft.

8. Safely lower the vehicle to the floor.

1996–97 Pathfinder

1. Raise and safely support the vehicle.

2. Remove the wheel speed sensor mounting bolt.

3. Remove the brackets holding the wiring and disconnect the harness connector.

4. Remove the wheel speed sensor.

5. Installation is the reverse of removal. Torque the front sensor mounting bolts to 8–11 ft. lbs. (11–15 Nm) and the rear sensor mounting bolts to 13–17 ft. lbs. (18–24 Nm).

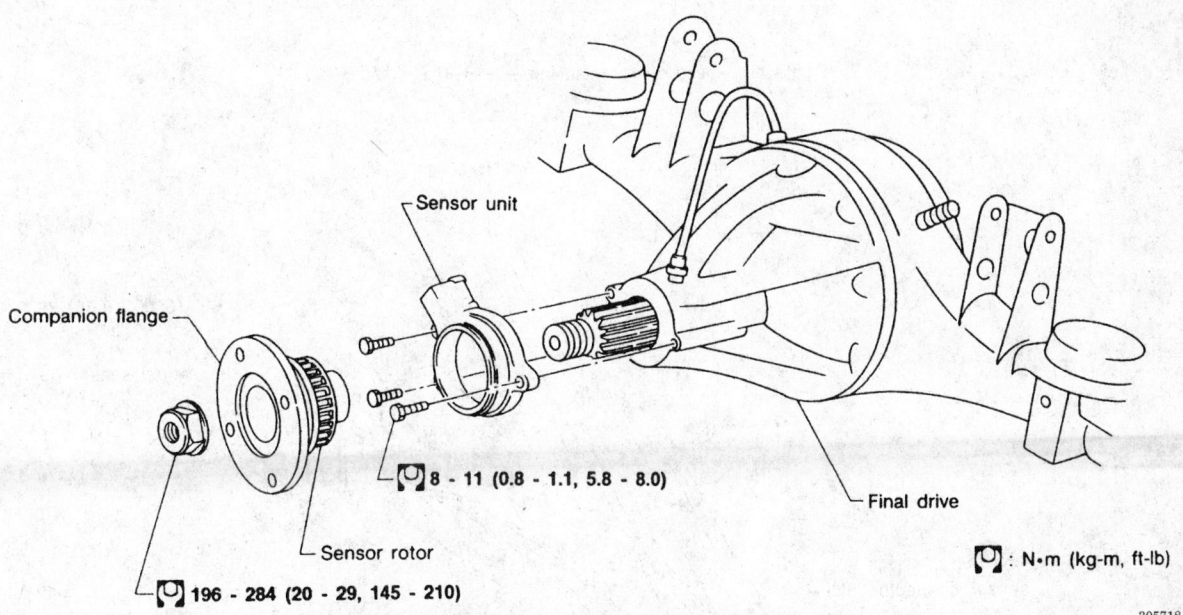

Sensor unit

Companion flange

Sensor rotor

8 - 11 (0.8 - 1.1, 5.8 - 8.0)

196 - 284 (20 - 29, 145 - 210)

Final drive

: N•m (kg-m, ft-lb)

305718

Speed sensor mounting (RWAL brakes) — except 1996–97 Pathfinder

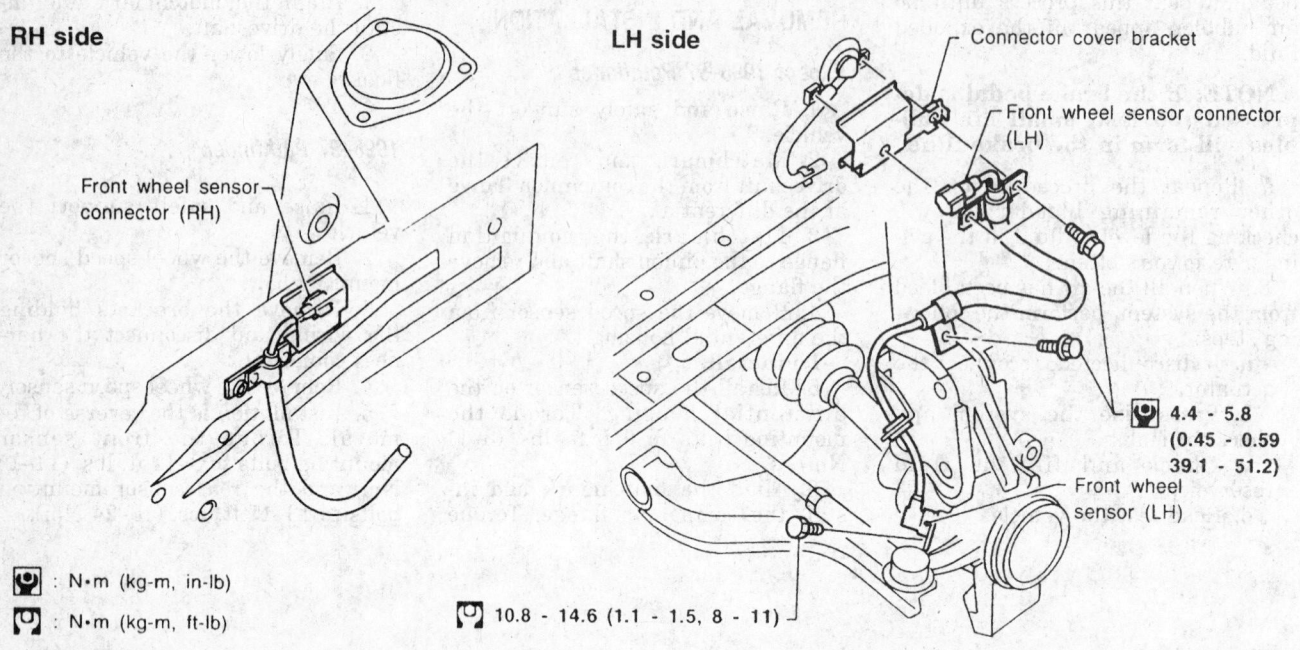

RH side

Front wheel sensor connector (RH)

: N•m (kg-m, in-lb)

: N•m (kg-m, ft-lb)

LH side

Connector cover bracket

Front wheel sensor connector (LH)

4.4 - 5.8
(0.45 - 0.59,
39.1 - 51.2)

Front wheel sensor (LH)

10.8 - 14.6 (1.1 - 1.5, 8 - 11)

305635

Front wheel speed sensor mounting — 1996–97 Pathfinder

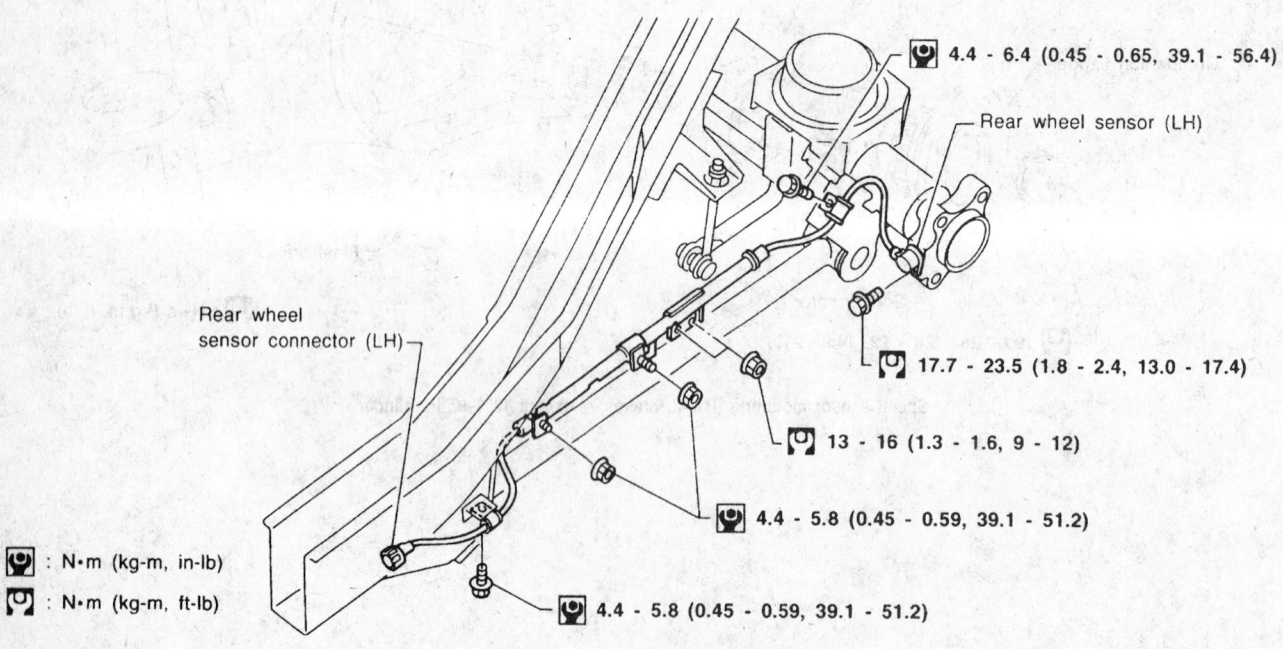

4.4 - 6.4 (0.45 - 0.65, 39.1 - 56.4)

Rear wheel sensor (LH)

Rear wheel sensor connector (LH)

17.7 - 23.5 (1.8 - 2.4, 13.0 - 17.4)

13 - 16 (1.3 - 1.6, 9 - 12)

4.4 - 5.8 (0.45 - 0.59, 39.1 - 51.2)

4.4 - 5.8 (0.45 - 0.59, 39.1 - 51.2)

: N•m (kg-m, in-lb)

: N•m (kg-m, ft-lb)

305636

Rear wheel speed sensor mounting — 1996–97 Pathfinder

FRONT SUSPENSION

Strut

REMOVAL AND INSTALLATION

1996–97 Pathfinder

1. Raise and support the vehicle on jackstands.
2. Remove the front wheel.
3. Detach the brake tube from the strut.
4. Disconnect the ABS wiring from the strut.
5. Disconnect the stabilizer link from the strut.
6. Support the transverse link with a jackstand.
7. Remove the two through bolts and detach the steering knuckle from the strut.

NOTE: Note the positioning of the strut alignment (cutout) mark for reassembly purposes.

8. Support the strut and remove the three upper attaching nuts. Remove the strut from the vehicle.

———— CAUTION ————
Never loosen the center spring retaining nut until the coil spring is compressed, or serious injury or vehicle damage may occur.

9. Place the strut assembly in a vise with the special holding tool (part # ST35652000) or in a spring compressor.
10. Loosen but do not remove the piston rod locknut.
11. Compress the spring with the spring compressor then remove the piston rod locknut.

NOTE: Before removing the strut from the coil spring, note the positioning of the strut in relationship to the coil spring for reassembly.

12. Remove the strut mounting insulator bracket, strut mounting bearing and upper spring seat.
13. Remove the strut, leaving the coil spring compressed.
14. Remove the piston boot and rebound bumper from the strut.
To install:
15. Install the rebound bumper and the boot to the strut piston.
16. Install the strut into the coil spring, make sure the strut and spring are properly positioned.
17. Install the upper spring seat, strut mounting bearing, and the strut mounting insulator bracket. Make sure that the cutout on the upper spring seat is facing the inside of the vehicle.
18. Install the piston rod locknut then remove the spring compressor.
19. Torque the piston rod locknut to 30–39 ft. lbs. (41–53 Nm).

NOTE: When installing the strut, be sure to position the alignment mark toward the inside of the vehicle.

20. Position the strut to the vehicle and install the 3 upper attaching nuts. Tighten the upper mounting nuts to 29–40 ft. lbs. (39–54 Nm).
21. Connect the steering knuckle to the strut and tighten mounting nuts of the mounting bolts to 111–122 ft. lbs. (151–165 Nm).
22. Connect the stabilizer link to the strut and tighten the new mounting nut to 61–76 ft. lbs. (83–103 Nm).
23. Connect the brake tube to the strut and connect the ABS wiring to the strut.
24. Bleed the brake system and install the wheel.
25. Perform a front end alignment.

Shock Absorber

REMOVAL AND INSTALLATION

Except 1996–97 Pathfinder

1. Raise and safely support the vehicle. Remove the wheel assembly.
2. While holding the upper stem of the shock absorber, remove the shock absorber attaching nut, washer and rubber bushing.
3. Remove the bolt attaching the shock absorber to the lower control arm and remove the shock absorber from the vehicle.
To install:
4. Install the shock with a new bushing on the upper shock mounting stud.
5. With the shocks upper stud positioned in the chassis, install the lower attaching bolt. Torque the lower bolt on 2WD vehicles to 47 ft. lbs. (64 Nm) on 1993 models, and to 65 ft. lbs. (88 Nm) on 1994–95 models. Torque the lower bolt on 4WD vehicles to 58 ft. lbs. (78 Nm) on 1993 vehicles, and to 108 ft. lbs. (147 Nm) on 1994–95 vehicles.
6. Install the bushing, washer and attaching nut to the shock stud. Torque the nut to 16 ft. lbs. (22 Nm).
7. Install the front wheels and lower the vehicle.

Coil Spring

REMOVAL AND INSTALLATION

1996–97 Pathfinder

Refer to the Strut removal and installation procedure.

Torsion Bar

REMOVAL AND INSTALLATION

Except 1996–97 Pathfinder

1. Raise and safely support the vehicle with supports placed under the frame. Remove the wheel assemblies.
2. Remove the torsion bar spring adjusting nuts.
3. Remove the dust cover and the snapring from the anchor arm.
4. On 2WD Pick-ups, pull the anchor arm off rearward and remove the torsion bar spring. Remove the torque arm from the control arm.
5. On 4WD vehicles and 2WD Pathfinders remove the nuts attaching the torque arm to the lower control arm. Remove the torsion bar from the front of the vehicle.
6. Keep the torsion bars separated, left and right; they are not interchangeable.
To install:
7. Check the torsion bars for wear, cracks or other damage; replace them if necessary.
8. On 2WD Pick-ups, install the torque arm on the control arm and torque the bolts to 37–50 ft. lbs. (50–68 Nm).
9. If removed, install the snapring and dust cover to the torsion bar.
10. Coat the splines on the inner end of the torsion bar with chassis lube.
11. Raise the lower control arm until it contacts the suspension bumper.
12. Install the torsion bar to its proper location.

NOTE: The torsion bars are marked L and R and are not interchangeable.

13. If equipped, install the nuts attaching the torque arm and torque the nuts to 33–44 ft. lbs. (45–60 Nm).
14. Set the anchor arm position on 2WD trucks so there is 0.24–0.71 in. (6–18mm) of threads exposed for the adjusting nuts.
15. Set the anchor position on 4WD Pick-ups and Pathfinders so there is 1.97–2.36 in. (50–60mm) from the crossmember to the tip of the anchor arm.

When installing rubber parts, final tightening must be
carried out under unladen condition* with tires on ground.
* Fuel, radiator coolant and engine oil full.
  Spare tire, jack, hand tools and mats in designated positions.

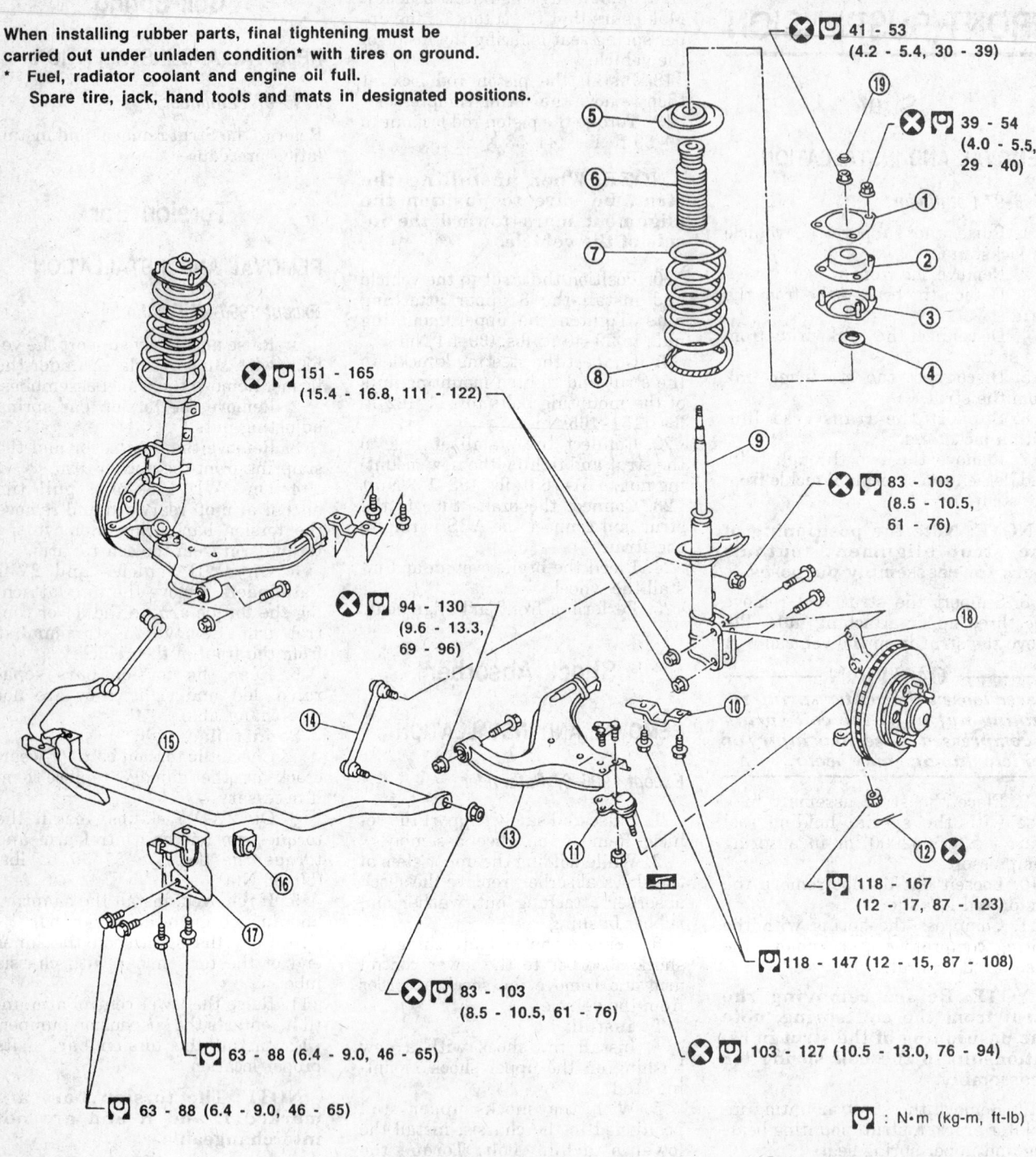

⊗ 🔧 41 - 53
(4.2 - 5.4, 30 - 39)

⊗ 🔧 39 - 54
(4.0 - 5.5,
29 - 40)

⊗ 🔧 151 - 165
(15.4 - 16.8, 111 - 122)

⊗ 🔧 83 - 103
(8.5 - 10.5,
61 - 76)

⊗ 🔧 94 - 130
(9.6 - 13.3,
69 - 96)

🔧 118 - 167
(12 - 17, 87 - 123)

🔧 118 - 147 (12 - 15, 87 - 108)

⊗ 🔧 83 - 103
(8.5 - 10.5, 61 - 76)

🔧 63 - 88 (6.4 - 9.0, 46 - 65)

🔧 63 - 88 (6.4 - 9.0, 46 - 65)

⊗ 🔧 103 - 127 (10.5 - 13.0, 76 - 94)

🔧 : N•m (kg-m, ft-lb)

① Spacer
② Strut mounting insulator
③ Bracket
④ Strut mounting bearing
⑤ Spring upper seat
⑥ Bound bumper
⑦ Coil spring

⑧ (Polyurethane tube)
⑨ Strut assembly
⑩ Bracket
⑪ Lower ball joint assembly
⑫ Cotter pin
⑬ Transverse link

⑭ Stabilizer connecting rod
⑮ Stabilizer bar
⑯ Bushing
⑰ Bracket
⑱ Knuckle spindle
⑲ Cap

312678

Exploded view of the front suspension — 1996–97 2WD Pathfinder; 4WD similar

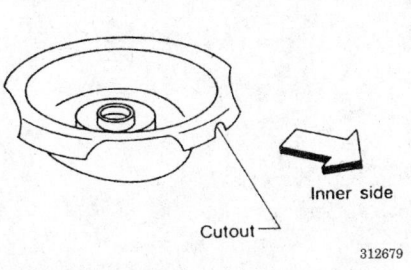

Positioning of strut alignment mark

Inner side

Cutout

312679

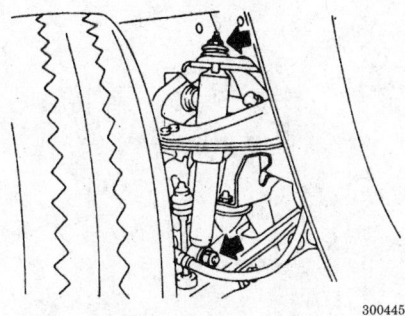

Shock absorber mounting points

300445

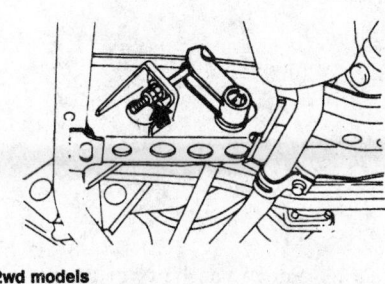

2wd models

4wd models

301866

Anchor arm and adjusting bolt identification

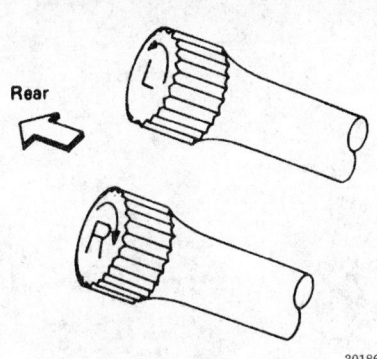

Rear

L

R

301867

Torsion bar identification mark locations

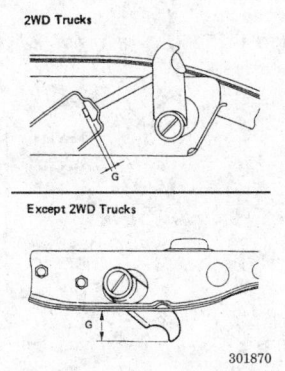

2WD Trucks

G

Except 2WD Trucks

G

301870

Anchor arm positioning

16. Install the snapring and dust cover to the anchor arm.

17. Install the wheel assemblies and lower the vehicle.

18. Following proper procedures, adjust the torsion bars.

Upper Ball Joints

REMOVAL AND INSTALLATION

Except 1996–97 Pathfinder

1. Raise and safely support the vehicle.

2. Remove the wheel/tire assembly.

3. Place a floor jack under the steering knuckle and support it.

4. Remove and discard the cotter pin from the ball joint stud, then loosen the nut. Using the ball joint removal tool, press the upper ball joint from the steering knuckle. Remove the upper ball joint nut.

5. Remove the bolts attaching the upper ball joint to the upper control arm and remove the ball joint from the vehicle.

To install:

6. Install a new ball joint and torque the bolts attaching the ball joint

to the upper control arm to 12–17 ft. lbs. (16–23 Nm).

7. Install the ball joint into the steering knuckle and torque the nut to 58–108 ft. lbs. (78–147 Nm). Install a new cotter pin to secure the nut.

8. Remove the jack from under the steering knuckle.

9. Install the tire and wheel assembly then lower the vehicle.

10. Check and/or adjust the ride height and the front end alignment.

Lower Ball Joints

REMOVAL AND INSTALLATION

Except 1996–97 Pathfinder

NOTE: The lower ball joint on 2WD models is integral with the lower control arm. They are removed and replaced as an assembly.

1. Raise and support the front of the vehicle on jackstands under the frame rails.

2. Remove the front wheels.

3. Make matching marks on the anchor arm crossmember when loosening the adjusting nut until there is no tension on the torsion bar. Remove the torsion bar assembly.

4. Unbolt the shock absorber from the lower arm.

5. Remove the ball joint nut.

6. Using a ball joint separator, disconnect the ball joint from the knuckle.

7. Unbolt the ball joint from the lower arm.

To install:

8. Install the ball joint to the lower arm and tighten the nuts to 45 ft. lbs. (61 Nm).

9. Press the ball stud into the knuckle and tighten the nut to 87–141 ft. lbs. (118–191 Nm). Make sure you use a new cotter pin!

10. Connect the lower end of the shock absorber.

11. Install the torsion bar to the lower arm and align the matchmarks made during disassembly.

12. Install the wheels and lower the vehicle.

13. Check the front end alignment.

1996–97 Pathfinder

1. Raise and support the front of the vehicle on jackstands under the frame rails.

2. Remove the front wheels.

3. Remove the cotter pin from the lower ball joint castle nut.

4. Remove the ball joint nut.

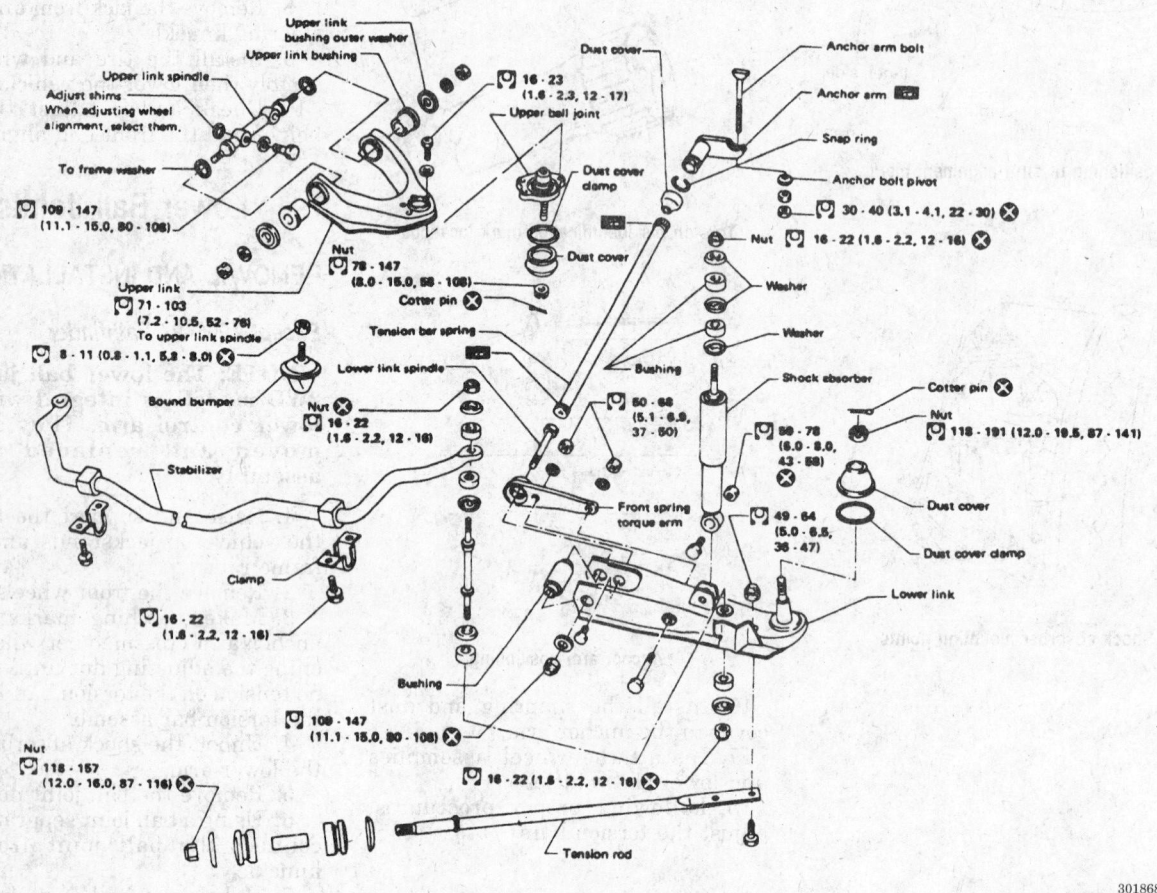

Front suspension exploded view — 2WD Pick-ups

301868

Upper Control Arms

REMOVAL AND INSTALLATION

Except 1996–97 Pathfinder

1. Raise and safely support the vehicle.
2. Remove the wheels.
3. Using a floor jack, raise the lower control arm.
4. Remove the nut attaching the top of the shock absorber to the chassis and compress the shock absorber.
5. Remove the bolts attaching the upper ball joint to the upper control arm.

6. Remove the bolts attaching the upper control arm to the chassis and the upper control arm from the vehicle.

NOTE: If alignment shims are used, be sure to keep them in order for reinstallation purposes.

7. Inspect the ball joint, replace as required.

To install:

8. Install the shims, if used, in their original locations and connect the control arm to the chassis. Torque the bolts attaching the upper control arm to the chassis to 80–108 ft. lbs. (109–146 Nm).

5. Using a ball joint separator, disconnect the ball joint from the knuckle.

6. Unbolt the ball joint from the lower arm and remove the ball joint.

To install:

7. Install the ball joint to the lower control arm and tighten the nuts and bolts to 76–94 ft. lbs. (103–127 Nm).

8. Position the ball joint stud into the knuckle and tighten the nut to 87–123 ft. lbs. (118–167 Nm). Be sure to install a new cotter pin.

9. Install the wheels and lower the vehicle.

10. Check the front end alignment.

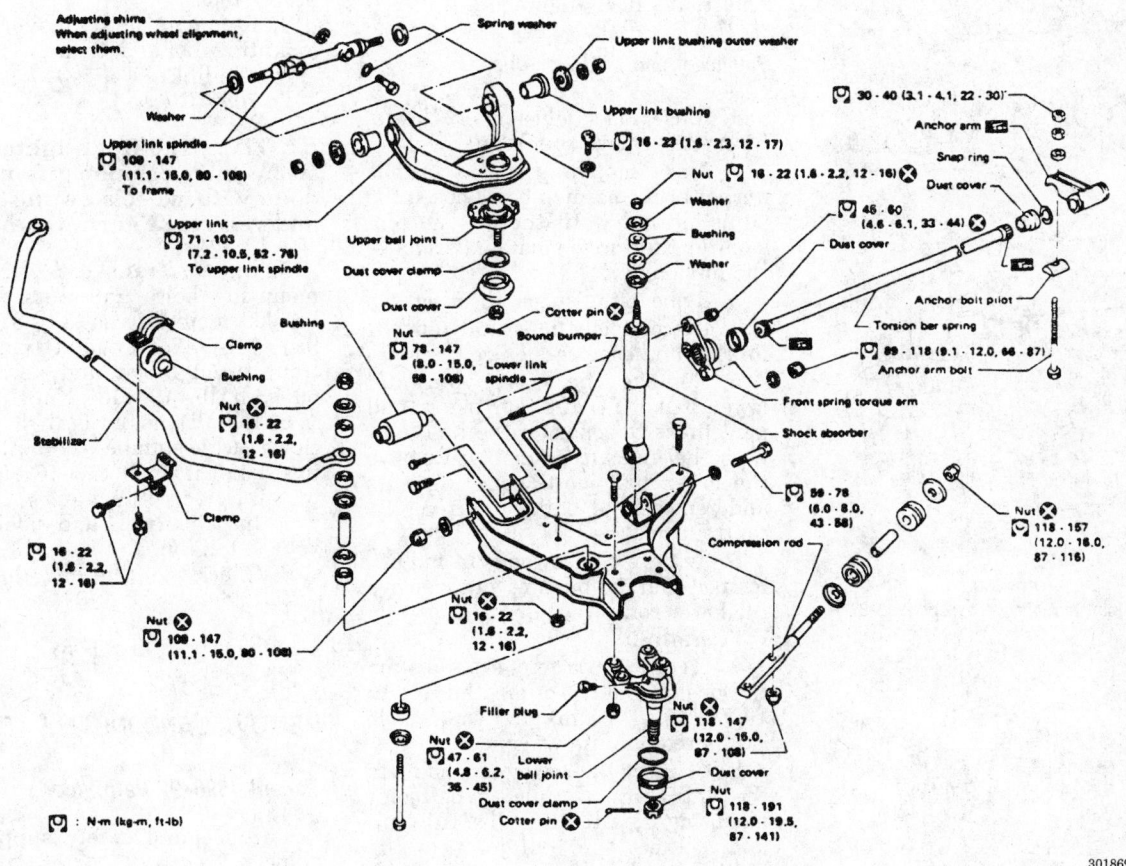

Adjusting shims
When adjusting wheel alignment, select them.

Spring washer

Upper link bushing outer washer

Washer

Upper link spindle

[U] 108 - 147
(11.1 - 15.0, 80 - 108)
To frame

Upper link bushing

[U] 16 - 23 (1.6 - 2.3, 12 - 17)

[U] 30 - 40 (3.1 - 4.1, 22 - 30)

Anchor arm

Snap ring

Dust cover

Upper link

[U] 71 - 103
(7.2 - 10.5, 52 - 76)
To upper link spindle

Nut [U] 16 - 22 (1.6 - 2.2, 12 - 16)

Upper ball joint

Washer

Dust cover clamp

Bushing

Dust cover

Washer

Dust cover

[U] 46 - 60
(4.6 - 6.1, 33 - 44)

Dust cover

Anchor bolt pilot

Torsion bar spring

[U] 89 - 118 (9.1 - 12.0, 66 - 87)

Anchor arm bolt

Bushing

Clamp

Bushing

Cotter pin

Nut [U] 78 - 147
(8.0 - 15.0,
58 - 108)

Sound bumper

Lower link spindle

Front spring torque arm

Stabilizer

Nut [U] 16 - 22
(1.6 - 2.2,
12 - 16)

Shock absorber

Clamp

[U] 59 - 78
(6.0 - 8.0,
43 - 58)

[U] 16 - 22
(1.6 - 2.2,
12 - 16)

Nut [U] 118 - 157
(12.0 - 16.0,
87 - 116)

Compression rod

Nut [U] 108 - 147
(11.1 - 15.0, 80 - 108)

Nut [U] 16 - 22
(1.6 - 2.2,
12 - 16)

Nut [U] 118 - 147
(12.0 - 16.0,
87 - 108)

Filler plug

Nut [U] 47 - 61
(4.8 - 6.2,
36 - 45)

Lower
ball joint

Dust cover

Dust cover clamp

Nut

Cotter pin

Nut [U] 118 - 191
(12.0 - 19.5,
87 - 141)

[U] : N·m (kg-m, ft-lb)

301869

Front suspension exploded view — 4WD Pick-ups and 1993–95 Pathfinder

9. Install the bolts attaching ball joint to the upper control arm and torque to 12–17 ft. lbs. (16–23 Nm).

10. Connect the shock absorber to the chassis and install the nut, torque the nut to 12–16 ft. lbs. (16–22 Nm).

11. Install the wheel assemblies and lower the vehicle.

12. Check and adjust the ride height, if required.

13. Check and/or adjust the front end alignment.

Lower Control Arms

REMOVAL AND INSTALLATION

Except 1996–97 Pathfinder

2WD Pick-Ups

NOTE: The lower ball joint on 2WD models is integral with the lower control arm. They are removed and replaced as a unit.

1. Raise and safely support the vehicle. Remove the front wheels.
2. Remove the shock absorber.
3. Make matching marks on the anchor arm crossmember. Loosen the adjusting nut until there is no tension on the torsion bar and remove the torsion bar.

4. Disconnect the stabilizer bar linkage from the lower control arm.

5. Disconnect the tension rod arm bolts.

6. Remove the cotter pin from the ball joint and use a ball joint press to separate the ball joints from the knuckle assembly.

7. Remove the lower control arm-to-chassis nut/bolt, tap the pivot shaft from the bushing. Push down on the tension rod and remove the lower control arm.

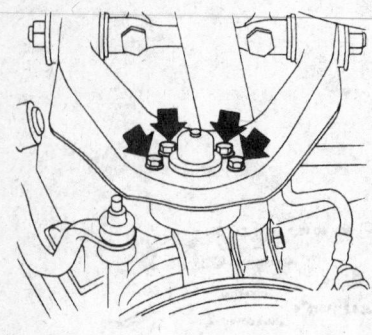

Upper ball joint mounting — except 1996–97
Pathfinder

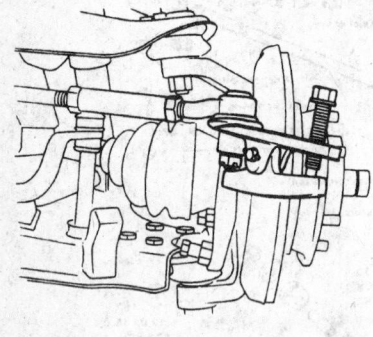

301933

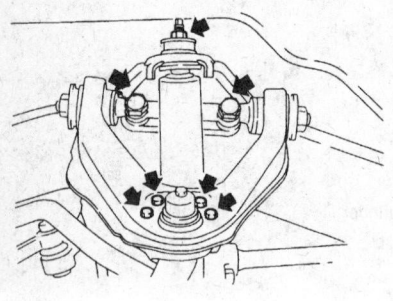

301946

Upper control arm mounting points — except
1996–97 Pathfinder

To install:

8. Install the control arm and the pivot bolt and torque the bolt to 108 ft. lbs. (147 Nm).

9. Connect the lower ball joint and sway bar. Torque the ball joint nut to 141 ft. lbs. (191 Nm) and tighten as required to install a new cotter pin.

10. Connect the tension rod and torque the bolts to 47 ft. lbs. (64 Nm). Install the stabilizer bar linkage and torque to 16 ft. lbs. (22 Nm).

11. Install the torsion bar and shock absorber. Torque the shock ab-

sorber lower mounting bolt on 1986–93 vehicles to 58 ft. lbs. (78 Nm), and on 1994–95 vehicles to 65 ft. lbs. (88 Nm).

12. Install the wheel assembly.

13. Adjust the torsion bar height and front wheel alignment.

Pathfinder and 4WD Pick-Ups

1. Raise and safely support the vehicle and remove the front wheels.

2. Make matching marks on the anchor arm crossmember. Loosen the adjusting nut until there is no tension on the torsion bar and remove the torsion bar.

3. Remove the shock absorber.

4. Disconnect the stabilizer bar linkage and compression rod.

5. Remove the cotter pin and lower ball joint nut and use a ball joint press to separate the ball joint from the knuckle assembly. Remove the lower ball joint mounting nuts and remove the ball joint from the control arm.

6. Remove the lower arm mounting bolt and the bushing nut; remove the lower control arm.

To install:

7. Fit the arm into place and start the bushing nut. Torque the nut to 108 ft. lbs. (147 Nm) after the vehicle is lowered onto its wheels.

8. Connect the lower ball joint to the control arm. Torque the ball joint mounting nuts to 45 ft. lbs. (61 Nm) and connect the ball joint to the knuckle assembly. Torque the ball joint to knuckle mounting nut to 141 ft. lbs. (191 Nm) and tighten as required to install a new cotter pin.

9. Connect the stabilizer bar linkage, compression rod, and install the shock absorber. Torque the stabilizer bar linkage to 16 ft. lbs. (22 Nm) and the compression rod mounting bolts to 108 ft. lbs. (147 Nm) after the vehicle is resting on its wheels. Torque the shock absorber lower mounting bolt on 1993 vehicles to 58 ft. lbs. (78 Nm), and on 1994–97 vehicles to 108 ft. lbs. (147 Nm).

10. Install the torsion bar and align the matchmarks.

11. Install the wheel assembly and lower the vehicle.

12. Check the ride height and the front end alignment.

1996–97 Pathfinder

NOTE: The lower control arm is also known as a transverse link.

1. Raise and safely support the vehicle.

2. Remove the front wheels.

3. Remove lower ball joint to knuckle cotter pin and nut. Separate ball joint stud from knuckle using the proper tool.

4. Remove the transverse link mounting bolts and nuts, then remove the link.

To install:

NOTE: The final tightening of suspension components must be done with wheels on the ground and vehicle at curb weight.

5. Install transverse link with mounting bolts and nuts. Tighten bracket mounting bolts to 87–108 ft. lbs. (118–147 Nm) and tighten the front through-bolt and new nut to 69–96 ft. lbs. (94–130 Nm).

6. Install the lower ball joint to the knuckle/spindle, tighten the nut to 87–123 ft. lbs. (118–167 Nm) and install a new cotter pin.

7. Install wheels and safely lower vehicle to ground.

8. Check the front end alignment.

Sway Bar

REMOVAL AND INSTALLATION

Except 1996–97 Pathfinder

1. Raise and safely support the vehicle.

2. From both sides of the vehicle, remove the stabilizer bar-to-lower control arm connecting rod nut, bushings, and tube.

3. Remove the bracket bolts and brackets and remove the stabilizer bar.

To install:

4. Position the sway bar in its mounting brackets, make sure that the white marks are visible and set toward the center of the vehicle.

5. Connect the sway bar to the lower control arms, make sure that the bushings are in good condition, if not replace them.

6. Torque the bolts mounting the sway bar brackets to 12–16 ft. lbs. (16–22 Nm).

7. Torque the nuts attaching the sway bar links to 12–16 ft. lbs. (16–22 Nm).

8. Lower the vehicle.

NOTE: With the full weight of the vehicle on the ground the white marks should be visible on both sides of the sway bar.

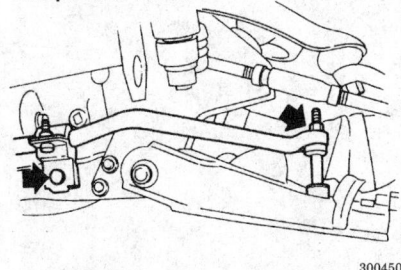

2WD Trucks

Except 2WD Trucks

300450

Front sway bar mounting points — except 1996–97 Pathfinder

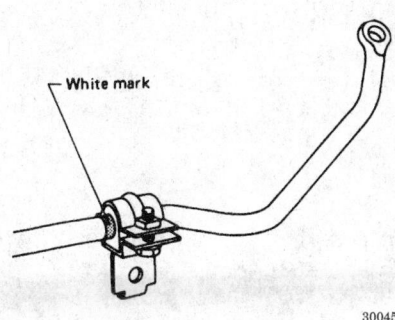

← White mark

300451

Proper positioning of bracket on sway bar — except 1996–97 Pathfinder

1996–97 Pathfinder

1. Raise and safely support the vehicle.
2. From both sides of the vehicle, remove the stabilizer bar-to-strut assembly connecting rod.

NOTE: Be aware of the paint marks on the stabilizer bar for reassembly.

3. Remove the bracket bolts and brackets and remove the stabilizer bar from the vehicle.

To install:

NOTE: Do not fully tighten the mounting bolts until the weight of the vehicle is resting on the ground.

4. Position the stabilizer bar in its mounting brackets, make sure that the paint marks are properly aligned.
5. Connect the stabilizer bar control rods to the strut assembly.
6. Torque the bolts mounting the sway bar brackets to 46–65 ft. lbs. (63–88 Nm).
7. Torque the nuts attaching the stabilizer bar links to 61–76 ft. lbs. (83–103 Nm).
8. Safely lower the vehicle to the ground.

Front Wheel Bearings

ADJUSTMENT

2WD Models

NOTE: Adjust the wheel bearing after the bearing has been replaced or the front axle has been reassembled.

1. Raise the front wheel(s) off the ground and support the vehicle safely.
2. Remove the front wheel(s).
3. Using a suitable tool, spread the brake pads to reduce the drag on the brake rotor.
4. Remove the bearing cap and the cotter pin.
5. Tighten the locknut to 25–29 ft. lbs. (34–39 Nm).
6. Turn the hub assembly several times in both directions to seat the bearings.
7. Again tighten the locknut to 25–29 ft. lbs. (34–39 Nm). Turn the locknut counterclockwise 45°. Install the locking cap and a new cotter pin. If the cotter pin cannot be installed, loosen the nut no more than 15° and install the cotter pin.
8. Measure the turning torque using a spring scale hooked to a hub bolt. Turning torque should be 2.2–6.4 lbs. (9.8–28.4 N) with a new grease seal or 2.2–5.3 lbs. (9.8–23.5 N) with a used grease seal.
9. Repeat the procedure until the correct specification is achieved.
10. Install the wheel(s) and lower the vehicle to the floor.

──────── **CAUTION** ────────
Pump the brakes to reposition the pads against the rotor before moving the vehicle.

4WD Models

NOTE: Adjust the wheel bearing after the bearing has been replaced or the front axle has been reassembled.

1. Raise the front wheel(s) off the ground and support the vehicle safely.
2. Remove the front wheel(s).
3. Using a suitable tool, spread the brake pads to reduce the drag on the brake rotor.
4. Remove the bearing cap and the cotter pin.
5. Tighten the locknut with a torque wrench and special tool KV40105400, or equivalent, to 58–72 ft. lbs. (78–98 Nm).
6. Turn the hub assembly several times in both directions to seat the bearings.
7. Loosen the locknut completely.
8. Tighten the locknut with special tool to 0.4–1.1 ft. lbs. (0.5–1.5 Nm).
9. Turn the hub assembly several times in both directions to seat the bearings.
10. Again tighten the locknut with special tool to 0.4–1.1 ft. lbs. (0.5–1.5 Nm).
11. Measure the starting force A using a spring scale hooked to a hub bolt and record the reading.
12. Install the lockwasher by tightening the locknut 15–30°. Turn the hub assembly several times in both directions to seat the bearings.
13. Measure the starting force with a spring scale after the lockwasher has been installed. Record the measurement and call his measurement B.
14. Calculate wheel bearing preload C using the equation $C = B - A$. Preload C should be 1.59–4.72 lbs. (7.06–20.99 N).
15. Repeat the procedure until the correct specification is achieved.
16. Install the wheel(s) and lower the vehicle to the floor.

──────── **CAUTION** ────────
Pump the brakes to reposition the pads against the rotor before moving the vehicle.

REMOVAL AND INSTALLATION

1993–95 Pathfinder

1. Raise and safely support the vehicle and remove the front wheels.
2. On 4WD models, have an assistant hold the brake pedal and loosen the locking front hub housing bolts.

When installing rubber parts, final tightening must be carried out under unladen condition* with tires on ground.
* Fuel, radiator coolant and engine oil full.
  Spare tire, jack, hand tools and mats in designated positions.

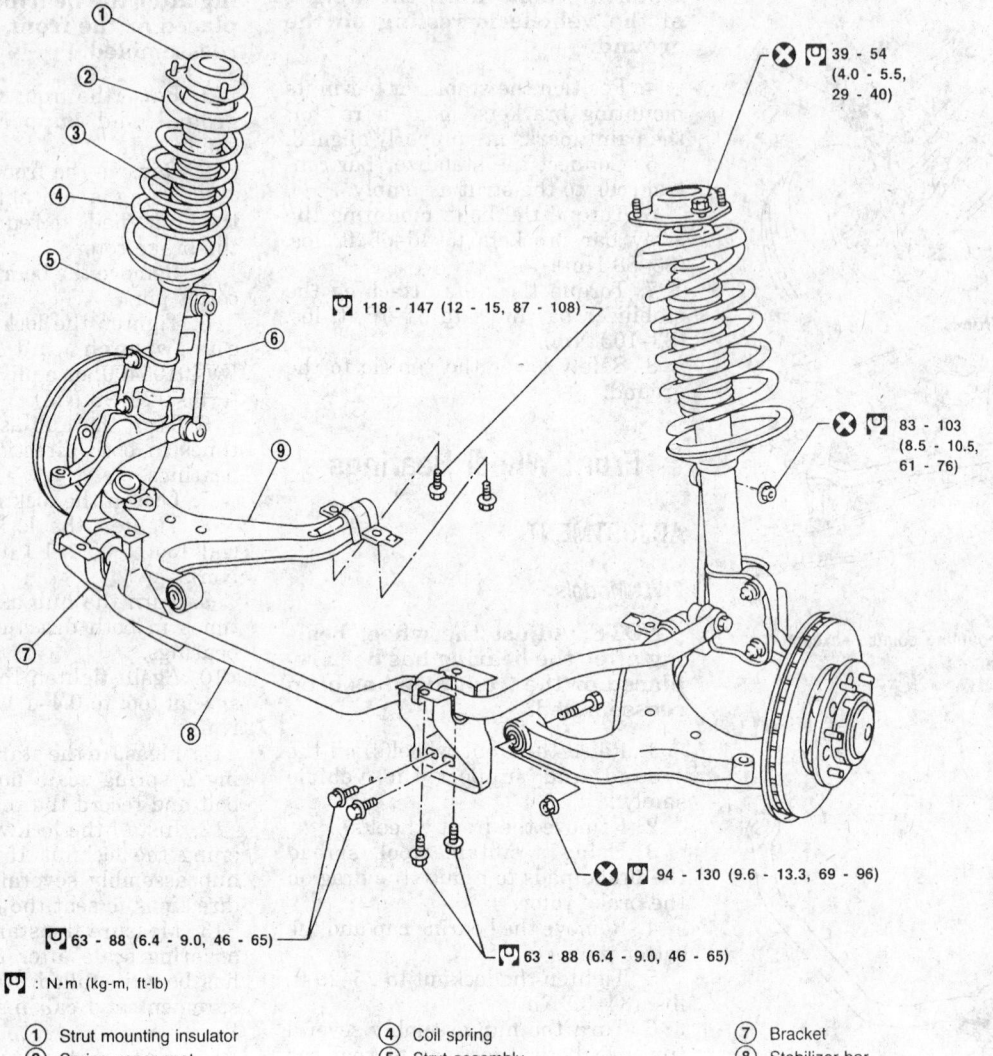

118 - 147 (12 - 15, 87 - 108)

39 - 54
(4.0 - 5.5,
29 - 40)

83 - 103
(8.5 - 10.5,
61 - 76)

94 - 130 (9.6 - 13.3, 69 - 96)

63 - 88 (6.4 - 9.0, 46 - 65)

63 - 88 (6.4 - 9.0, 46 - 65)

: N·m (kg-m, ft-lb)

① Strut mounting insulator
② Spring upper seat
③ Bound bumper
④ Coil spring
⑤ Strut assembly
⑥ Stabilizer connecting rod
⑦ Bracket
⑧ Stabilizer bar
⑨ Transverse link

312737

Front sway (stabilizer) bar assembly — 1996–97 2WD Pathfinder; 4WD similar

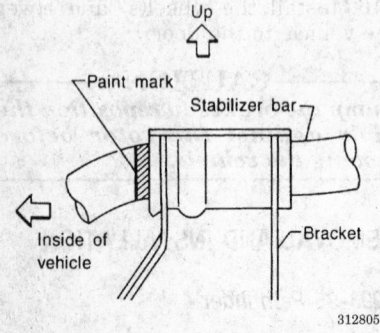

Up

Paint mark

Stabilizer bar

Inside of vehicle

Bracket

312805

Sway bar paint mark alignment — 1996–97 Pathfinder

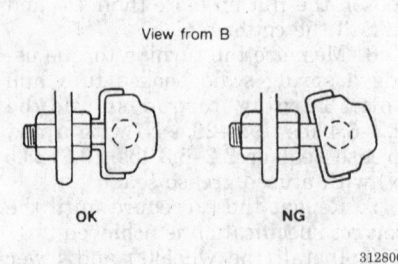

View from B

OK   NG

312800

Alignment of the sway bar control rod links — 1996–97 Pathfinder

Remove the hub assembly housing, the snapring and the hub assembly.

3. Without disconnecting the hydraulic line, remove the brake caliper and hang it from the body with wire. Do not allow the caliper to hang by the hose.

4. Remove the locking screw and remove the lockwasher. Use a pin wrench to loosen the wheel bearing locknut. The torque may be fairly high, **do not** use a hammer and drift pin.

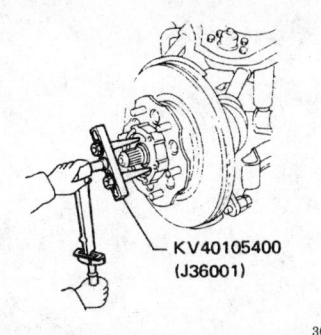

KV40105400
(J36001)

307262

Tighten the locknut with the special tool

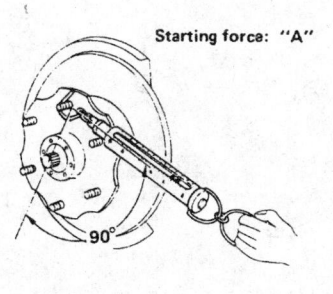

Starting force: "A"

90°

307264

Measuring the starting force

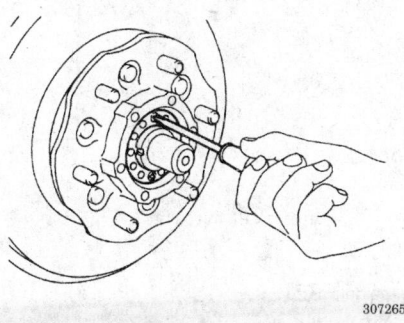

307265

Install the lockwasher on the bearing locknut

5. Remove the locknut and pull the hub off the spindle with the bearings.

NOTE: On 4WD models, use a block of wood and hammer to tap on the end of the halfshaft to break it loose from the hub spline.

6. Pry the inner grease seal out to remove the inner bearing. Discard the seal.

7. Clean and inspect the wheel bearings and replace if worn or heat damaged.

NOTE: Always replace wheel bearings and races together as sets.

8. Using a hammer and punch, drive the bearing races from the hub.
To install:
9. Carefully install the new inner races using the proper size driver, making sure they are fully seated in the hub.
10. Pack the bearings with new grease and pack grease into the hub. Install the inner bearing and press a new inner seal into the hub.
11. Slip the hub assembly onto the spindle and install the outer bearing. Install the halfshaft and grease the locknut. Thread the locknut into place.
12. To set the bearing pre-load, perform the following steps:
   a. Use a pin wrench to torque the locknut to 58–72 ft. lbs. (78–98 Nm). Turn the hub in both directions several times while torquing the nut.
   b. Loosen the locknut, then torque again to 13 inch lbs. (1.5 Nm).
   c. Turn the hub several times and check the nut torque again.
   d. Install the lockwasher. When installing the screw, make sure the locknut turns no more than 30 degrees in either direction.
   e. When bearing pre-load is properly set, there will be no end-play in the hub and it will require no more than 4.7 lbs. of pull at the wheel stud to turn the hub.
13. On 4WD models, install the locking hub and torque the hub bolts to 25 ft. lbs. (34 Nm).
14. Install the brake caliper, disc brake pads, and wheel assembly.
15. Lower the vehicle and pump the brakes until the pedal is firm.

1996–97 Pathfinder

1. Raise and safely support the vehicle and remove the front wheels.
2. Remove the ABS sensor from the steering knuckle/spindle.
3. Without disconnecting the hydraulic line, remove the brake caliper and hang it from the body with wire. Do not allow the caliper to hang by the hose.
4. Remove the hub cap using a suitable tool.
5. On 4WD models, remove the snapring and O-ring.
6. On 4WD models, remove the nuts that secure the drive flange and remove the flange assembly.

7. Remove the screw that secures the lockwasher and remove the lockwasher.
8. Use a pin wrench to loosen the wheel bearing locknut. The torque may be fairly high, **do not** use a hammer and drift pin.
9. Remove the locknut and pull the hub off the spindle with the bearings.

NOTE: On 4WD models, use a block of wood and hammer to tap on the end of the halfshaft to break it loose from the hub spline.

10. Pry the inner grease seal out to remove the inner bearing. Discard the seal.
11. Clean and inspect the wheel bearings and replace if worn or heat damaged.

NOTE: Always replace wheel bearings and races together as sets.

12. Using a hammer and punch, drive the bearing races from the hub.
To install:
13. Carefully install the new inner races using the proper size driver, making sure they are fully seated in the hub.
14. Pack the bearings with new grease and pack grease into the hub. Install the inner bearing and press a new inner seal into the hub.
15. Slip the hub assembly onto the spindle and install the outer bearing.
16. Install the wheel bearing locknut. Use a pin wrench to tighten the locknut. The torque may be fairly high, **do not** use a hammer and drift pin.
17. To set the bearing pre-load, perform the following steps:
   a. Use a pin wrench to torque the locknut to 58–72 ft. lbs. (78–98 Nm). Turn the hub in both directions several times while torquing the nut.
   b. Loosen the locknut, then torque again to 4.3–13 inch lbs. (0.5–1.5 Nm).
   c. Turn the hub several times and check the nut torque again.
   d. Install the lockwasher. When installing the screw, make sure the locknut turns no more than 30 degrees in either direction.
   e. Tighten the lockwasher mounting screw to 11–15 inch lbs. (1.2–1.8 Nm).
   f. When bearing pre-load is properly set, there will be no end-play in the hub and it will require no more than 4.7 lbs. of pull at the wheel stud to turn the hub.

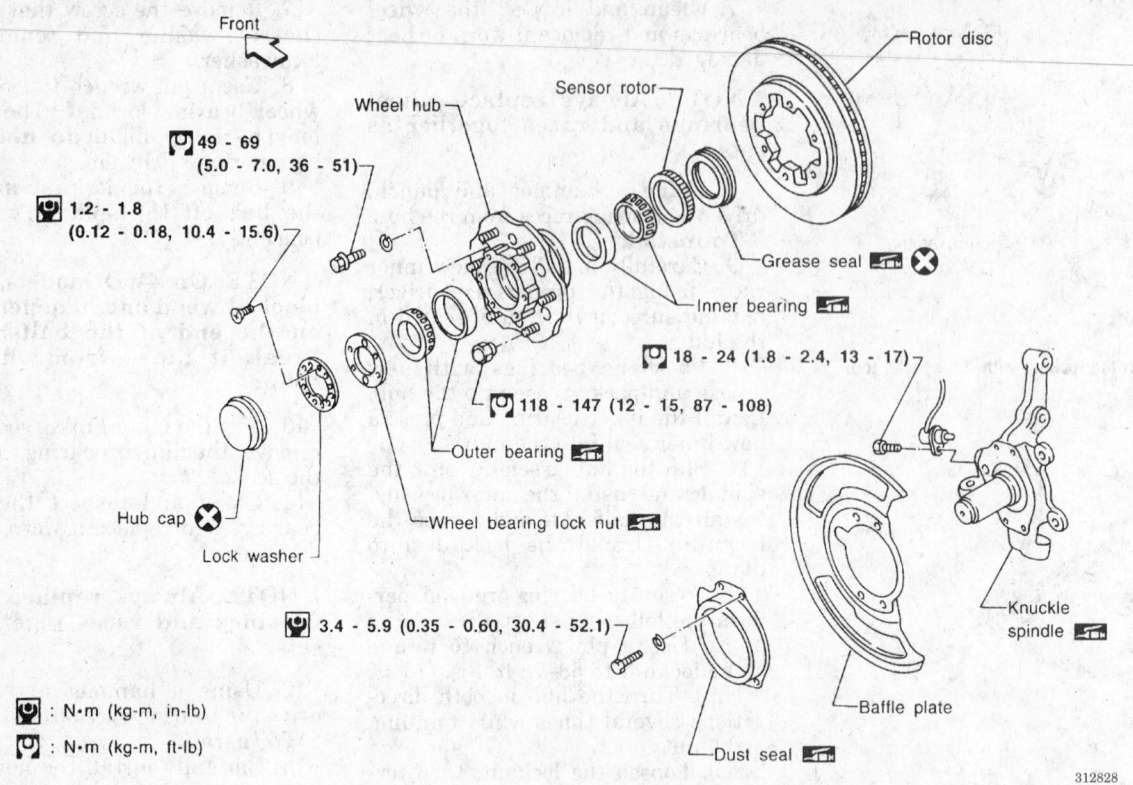

Exploded view of the wheel bearing assembly on 2WD vehicles — 1996–97 Pathfinder

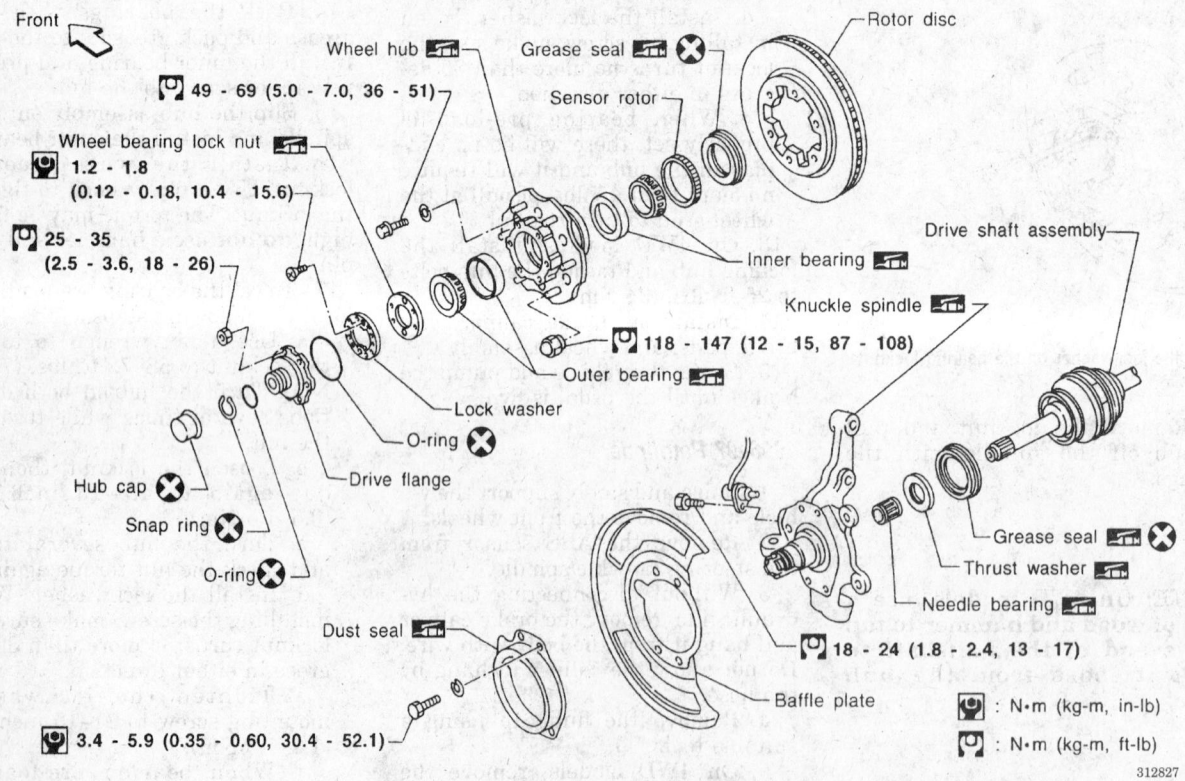

Exploded view of the wheel bearing assembly on 4WD vehicles — 1996–97 Pathfinder

18. On 4WD models, install the drive flange and torque the mounting nuts to 18–26 ft. lbs. (25–35 Nm). Be sure to install new O-rings and pack the flange groves with grease. Also apply grease to the O-rings.

19. On 4WD models, install the snapring.

20. Using a new hub cap, install the cap using a suitable tool.

21. Install the brake caliper assembly.

22. Install the ABS sensor to the steering knuckle/spindle and tighten the mounting bolt to 13–17 ft. lbs. (18–24 Nm).

23. Install the wheel assembly.

24. Lower the vehicle and pump the brakes until the pedal is firm.

Pick-Up

2WD Vehicles

1. Raise and safely support the vehicle and remove the front wheels.

2. Without disconnecting the hydraulic line, remove the brake caliper and hang it from the body on a wire. Do not let it hang by the hose.

3. Remove the wheel hub cup, the cotter pin, the adjusting cap and hub nut.

4. Remove the wheel hub and brake disc assembly. Be careful not to drop the outer wheel bearing.

5. To remove the inner bearing, pry out the grease seal. Discard the seal.

6. Clean and inspect the wheel bearings and replace if worn or heat damaged.

NOTE: Always replace wheel bearings and races together as sets.

7. Using a hammer and punch, drive the bearing races from the hub.
To install:

8. Carefully install the new inner races using the proper size driver, making sure they are fully seated in the hub.

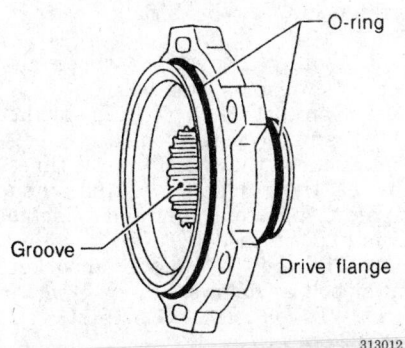

Drive flange assembly — 1996 4WD Pathfinder

313012

9. Pack the bearings with new grease and pack grease into the hub. Install the inner bearing and press a new inner seal into the hub.

10. Slip the hub assembly onto the spindle and install the outer bearing. Grease the locknut, thread it into place.

11. To set the bearing pre-load, perform the following steps:

   a. Install the nut and torque it to 25–29 ft. lbs. (34–39 Nm).

   b. Spin the hub several times in both directions, then torque the nut to 25–29 ft. lbs. (34–39 Nm).

   c. Loosen the nut 45 degrees. Install the locknut cap and a new cotter pin.

12. Install the brake caliper and wheel assembly.

13. Lower the vehicle and pump the brakes until the pedal is firm.

4WD Vehicles

1. Raise and safely support the vehicle and remove the front wheels.

2. Have an assistant hold the brake pedal and loosen the locking front hub housing bolts. Remove the hub assembly housing, the snapring and the hub assembly.

3. Without disconnecting the hydraulic line, remove the brake caliper and hang it from the body with wire. Do not allow the caliper to hang by the hose.

4. Remove the locking screw and remove the lockwasher. Use a pin wrench to loosen the wheel bearing locknut. The torque may be fairly high, **do not** use a hammer and drift pin.

5. Remove the locknut and pull the hub off the spindle with the bearings. Pry the inner grease seal out to remove the inner bearing. Discard the seal.

NOTE: Use a block of wood and hammer to tap on the end of the halfshaft to break it loose from the hub spline.

6. Clean and inspect the wheel bearings and replace if worn or heat damaged.

NOTE: Always replace wheel bearings and races together as sets.

7. Using a hammer and punch, drive the bearing races from the hub.
To install:

8. Carefully install the new inner races using the proper size driver, making sure they are fully seated in the hub.

9. Pack the bearings with new grease and pack grease into the hub.

Install the inner bearing and press a new inner seal into the hub.

10. Slip the hub assembly onto the spindle and install the halfshaft. Install the outer bearing and grease the locknut. Thread the locknut into place.

11. To set the bearing pre-load, perform the following steps:

   a. Use a pin wrench to torque the locknut to 58–72 ft. lbs. (78–98 Nm). Turn the hub in both directions several times while torquing the nut.

   b. Loosen the locknut, then torque again to 13 inch lbs. (1.5 Nm).

   c. Turn the hub several times and check the nut torque again.

   d. Install the lockwasher. When installing the screw, make sure the locknut turns no more than 30 degrees in either direction.

   e. When bearing pre-load is properly set, there will be no end–play in the hub and it will require no more than 4.7 lbs. of pull at the wheel stud to turn the hub.

12. Install the locking hub and torque the hub bolts to 18–25 ft. lbs. (25–34 Nm).

13. Install the brake caliper, brake pads, and wheel assembly.

14. Lower the vehicle and pump the brakes until the pedal is firm.

REAR SUSPENSION

Shock Absorber

REMOVAL AND INSTALLATION

Pathfinder

NOTE: For rear shock absorber replacement the vehicle chassis and axle weight must be supported separately, requiring the use of two separate lifting devices.

1. Raise and properly support vehicle. Remove both rear wheels.

2. If equipped, disconnect the electrical connector from the shock absorber.

3. Working on one side at a time, jack one side of the rear axle and remove upper and lower attaching nuts. While supporting the rear, remove the shock absorber.
To install:

4. Align shock and install both attaching nuts. Do not tighten nuts

completely until the full weight of the vehicle is on the ground.

5. If equipped, connect the electrical connector to the shock absorber.

6. Install wheels, remove the jack from under the rear axle and lower the vehicle.

7. For 1993–95 models, tighten attaching nuts to 22–30 ft. lbs. (30–40 Nm), or for 1996 models tighten the nuts to 36–49 ft. lbs. (49–67 Nm).

Pick-Up

NOTE: For rear shock absorber replacement the vehicle chassis and axle weight must be supported separately, requiring the use of two separate lifting devices.

1. Raise and properly support vehicle. Remove both rear wheels.

2. Working on one side at a time, jack one side of the rear axle and remove upper and lower attaching nuts. While supporting the rear, remove the shock absorber.

To install:

3. Align shock and install both attaching nuts. Do not tighten bolts completely until the full weight of the vehicle is on the ground.

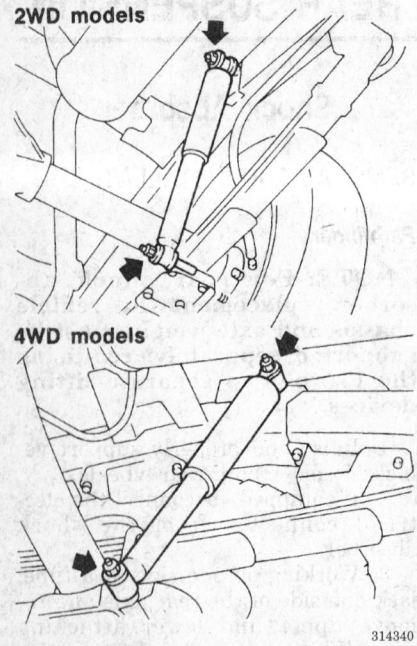

2WD models

4WD models

314340

Rear shock absorber mounting points — Pick-up

4. Install wheels, remove the jack from under the rear axle, and lower vehicle.

5. Tighten attaching nuts to 22–30 ft. lbs. (30–40 Nm) on all models.

Coil Spring

REMOVAL AND INSTALLATION

Pathfinder

The spring removal procedure for the 2WD and 4WD models is the same for both vehicles.

NOTE: The coil spring is a load bearing component, therefore the vehicle chassis and axle weight must be supported separately, requiring the use of two separate lifting devices.

1. Raise the vehicle and support it safely.

2. Using the proper equipment, support the weight of the rear axle.

3. Disconnect the panhard rod at the axle assembly and secure it to the body.

NOTE: Working on one side at a time, complete the following steps for removal and installation of the coil spring.

4. Remove the nut that attaches the shock to the lower mounting bracket.

NOTE: Mark the spring installation direction for reinstallation.

5. Lower the axle assembly until the spring and upper insulator can be removed. Remove the coil spring and lower insulator.

--- **WARNING** ---

Do not stretch the brake hose or parking brake cable.

To install:

NOTE: Check the spring for identification marks and properly align.

6. Position the insulator and install the spring.

7. Raise the axle and loosely attach the lower shock absorber mounting nut.

8. Loosely connect the panhard rod to the mount.

9. Remove the rear axle support and lower the vehicle.

10. With the suspension supporting the weight of the vehicle, torque the lower shock retaining nut. On 1993–95 vehicles, torque the nut to

22–30 ft. lbs. (30–40 Nm), or on 1996 vehicles, torque the nut to 49–65 ft. lbs. (67–88 Nm).

11. On 1993 models, torque the panhard rod mounting nut to 94–123 ft. lbs. (127–167 Nm). On 1994–95 models, torque the mounting nut to 67–90 ft. lbs. (91–122 Nm). On 1996 models, torque the mounting nut to 103–116 ft. lbs. (140–157 Nm).

Leaf Spring

REMOVAL AND INSTALLATION

Pick-Up

--- **CAUTION** ---

The leaf springs are under considerable tension. Be very careful when removing or installing them; they can exert enough force to cause serious injuries.

1. Raise the rear of the truck and support it with jackstands placed under the frame.

2. Disconnect the parking brake cables from the springs.

3. Place a jack under the rear axle housing and raise the housing just enough to remove the weight from the springs.

4. Disconnect the shock absorbers at their lower end.

5. Remove the nuts securing the U–bolts around the axle housing.

6. Remove the nuts from the spring shackles, drive out the shackle pins and remove the spring from the vehicle.

To install:

NOTE: The full weight of the truck must be on the rear wheels before tightening the front pin, shackle, and shock absorber attaching nuts.

7. Install the springs and loosely connect the shackles.

8. Connect the U–bolt nuts and tighten nuts to 65–72 ft. lbs. (88–98 Nm).

9. Loosely connect the shock absorber lower mounting bolt.

10. Connect the parking brake cables and lower the vehicle.

11. Bounce the truck several times to set the suspension and then tighten the front pin and shackle nuts to 94 ft. lbs. (127 Nm).

12. Tighten the shock absorber lower bolt as follows:
- 2WD Models — 12–16 ft. lbs. (16–22 Nm)
- 4WD Models — 22–30 ft. lbs. (30–40 Nm)

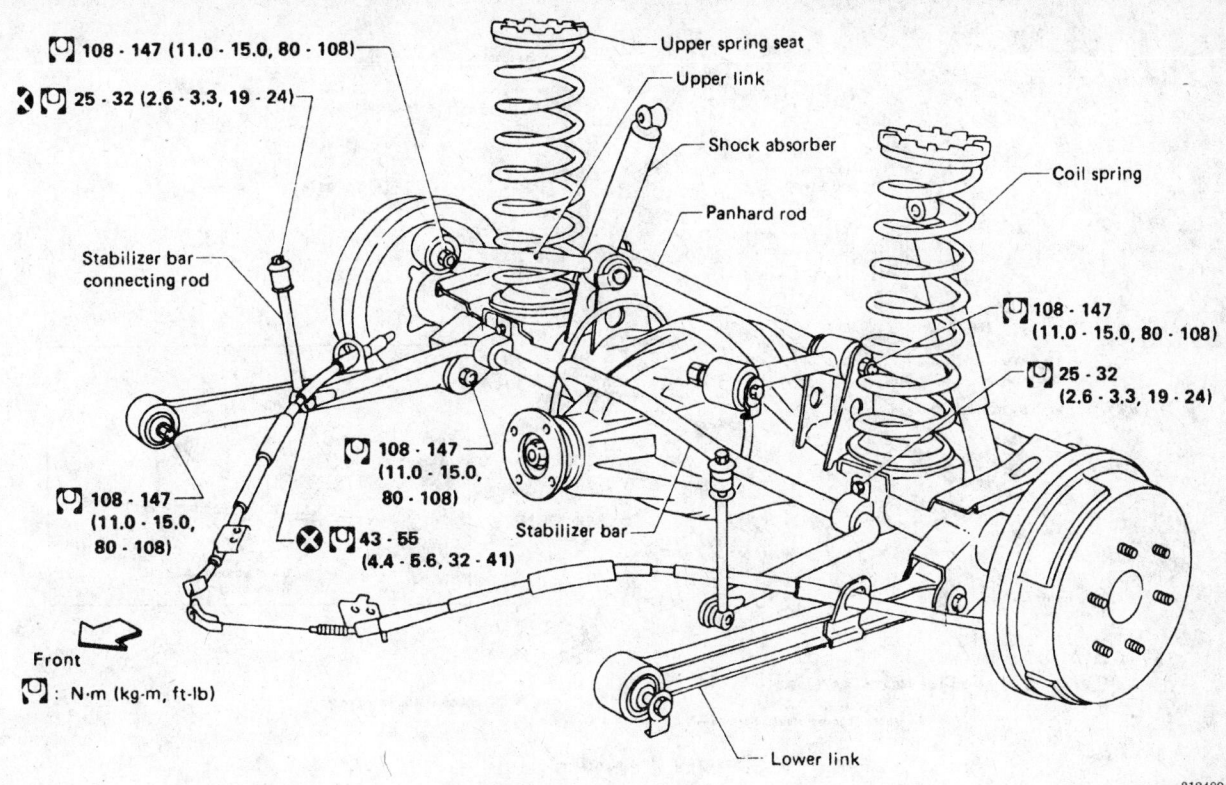

108 · 147 (11.0 · 15.0, 80 · 108)

25 · 32 (2.6 · 3.3, 19 · 24)

Upper spring seat

Upper link

Shock absorber

Panhard rod

Coil spring

108 · 147 (11.0 · 15.0, 80 · 108)

25 · 32 (2.6 · 3.3, 19 · 24)

Stabilizer bar connecting rod

108 · 147 (11.0 · 15.0, 80 · 108)

108 · 147 (11.0 · 15.0, 80 · 108)

43 · 55 (4.4 · 5.6, 32 · 41)

Stabilizer bar

Front

: N·m (kg·m, ft·lb)

Lower link

312489

Rear suspension component identification — Pathfinder

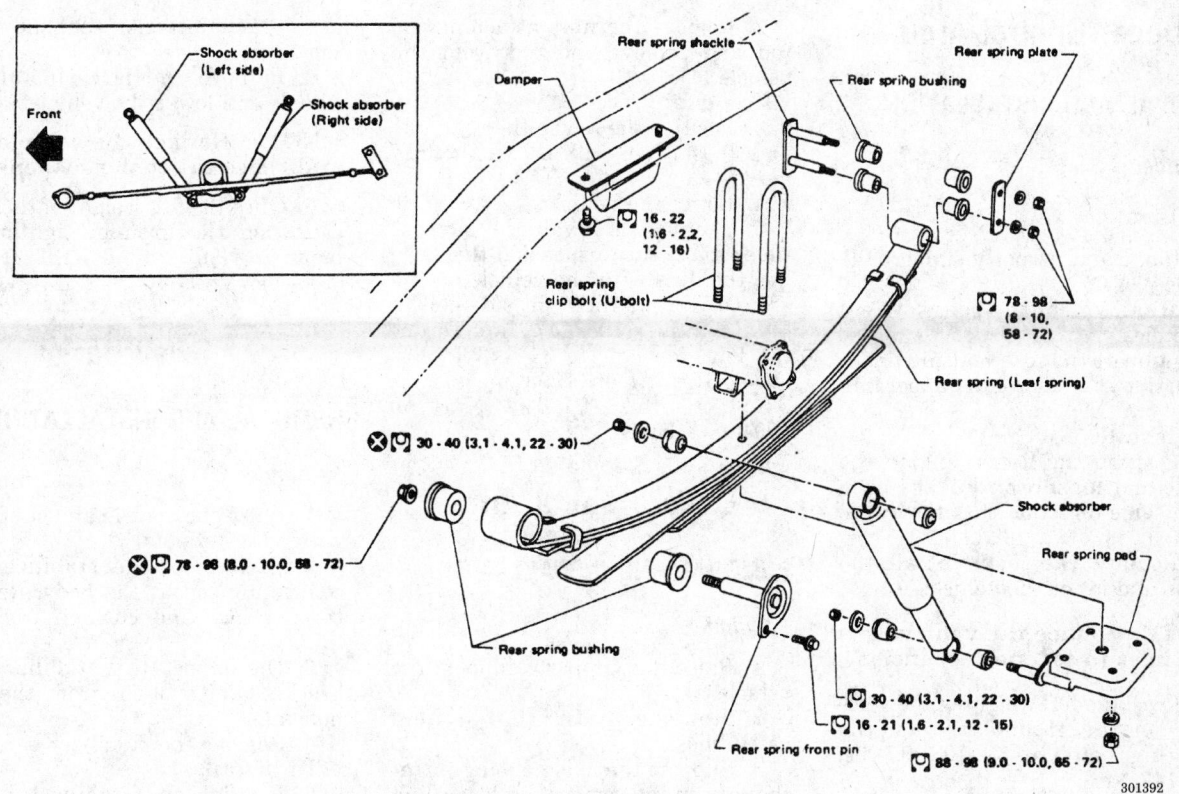

Shock absorber (Left side)

Shock absorber (Right side)

Front

Damper

Rear spring shackle

Rear spring bushing

Rear spring plate

16 · 22 (1.6 · 2.2, 12 · 16)

Rear spring clip bolt (U-bolt)

78 · 98 (8 · 10, 58 · 72)

Rear spring (Leaf spring)

30 · 40 (3.1 · 4.1, 22 · 30)

Shock absorber

78 · 98 (8.0 · 10.0, 58 · 72)

Rear spring pad

Rear spring bushing

Rear spring front pin

30 · 40 (3.1 · 4.1, 22 · 30)

16 · 21 (1.6 · 2.1, 12 · 15)

88 · 98 (9.0 · 10.0, 65 · 72)

301392

Exploded view of rear suspension — 2WD Pick-up

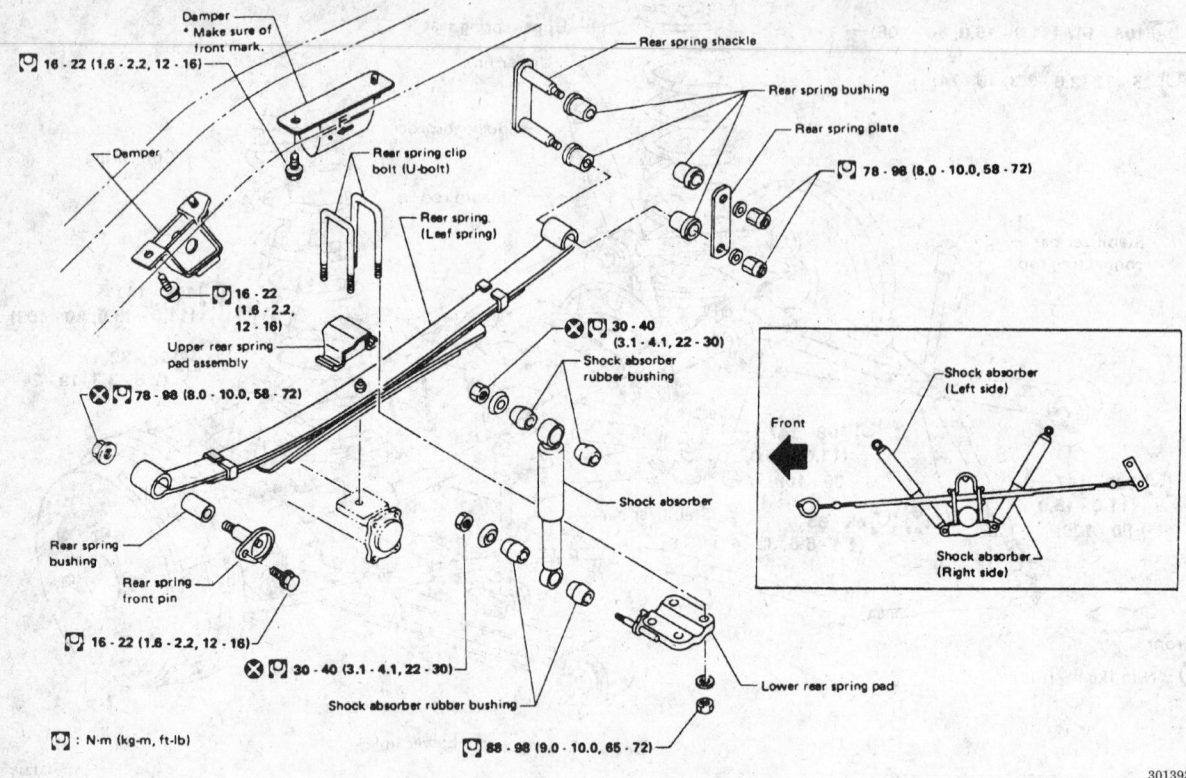

Damper
* Make sure of front mark.

16 - 22 (1.6 - 2.2, 12 - 16)

Rear spring shackle

Rear spring bushing

Rear spring plate

78 - 98 (8.0 - 10.0, 58 - 72)

Damper

Rear spring clip bolt (U-bolt)

16 - 22 (1.6 - 2.2, 12 - 16)

Rear spring (Leaf spring)

Upper rear spring pad assembly

78 - 98 (8.0 - 10.0, 58 - 72)

30 - 40 (3.1 - 4.1, 22 - 30)

Shock absorber rubber bushing

Shock absorber

Rear spring bushing

Rear spring front pin

16 - 22 (1.6 - 2.2, 12 - 16)

30 - 40 (3.1 - 4.1, 22 - 30)

Shock absorber rubber bushing

Lower rear spring pad

: N·m (kg-m, ft-lb)

88 - 98 (9.0 - 10.0, 65 - 72)

Front

Shock absorber (Left side)

Shock absorber (Right side)

301393

Exploded view of rear suspension — 4WD Pick-up

Upper Control Arms

REMOVAL AND INSTALLATION

Pathfinder

Upper Links

1. Raise and properly support the vehicle.
2. Remove both rear wheel assemblies.
3. Support the axle housing under the carrier and remove the upper link bolts.
   To install:
4. Position the upper link in the vehicle and **loosely** install the bolts in the same direction that they were removed.
5. Remove the jack, install the wheels and lower the vehicle.

NOTE: Bounce the vehicle several times to set the suspension.

6. Once the full weight of the vehicle is on the ground, tighten the mounting bolts to 80–108 ft. lbs. (..147 Nm).

vehicle and support ....ing.

2. Remove the nuts attaching the rod and remove the rod from the vehicle.
   To install:
3. Position the rod on the rear axle and loosely install the attaching nuts.
4. Lower the vehicle.
5. Bounce the vehicle several times to set the suspension then torque the nuts with the vehicle on the ground. Torque the right side bolt to 80–108 ft. lbs. (108–147 Nm). Torque the left side bolt to 94–123 ft. lbs. (127–167 Nm) on 1993 models, or 67–90 ft. lbs. (91–122 Nm) on 1994–95 models.

Lower Control Arms

REMOVAL AND INSTALLATION

Pathfinder

1. Raise and properly support the vehicle.
2. Remove both rear wheel assemblies.
3. Support the axle housing under the carrier and remove the lower link bolts.
   To install:
4. Position the lower link in the vehicle and **loosely** install the bolts

in the same direction that they were removed.
5. Remove the jack, install the wheels and lower the vehicle.

NOTE: Bounce the vehicle several times to set the suspension.

6. Once the full weight of the vehicle is on the ground, tighten the mounting bolts to 80–108 ft. lbs. (108–147 Nm).

Sway Bar

REMOVAL AND INSTALLATION

Pathfinder

1. Raise the vehicle and support it with safety stands.
2. Disconnect the stabilizer bar connecting rod at the body. Remove the retainer and cushion from the link.
3. Disconnect the stabilizer bar bracket and cushion from the axle housing.
4. Remove the stabilizer bar.
   To install:
5. Position the stabilizer bar at the axle housing and install the bracket and cushion. Tighten the mounting bolts to 19–24 ft. lbs. (25–32 Nm).

6. Install the connecting rod to the body with the retainers and cushion. Tighten the mounting bolts to 19–24 ft. lbs. (25–32 Nm).

7. Lower the vehicle.

Wheel Bearings

REMOVAL AND INSTALLATION

Single Rear Wheels With Drum Brakes

1. Raise the rear of the vehicle and support it. Remove the rear wheel.

2. Disconnect the rear parking brake cable.

3. If equipped, remove the ABS wheel speed sensor from the axle assembly.

4. Disconnect the brake tube at the rear brake backing plate. Plug the end of the brake tube to prevent loss of brake fluid.

5. Remove the brake drum.

6. Remove the nuts securing the wheel bearing retainer to the brake backing plate.

7. Pull out the axle shaft assembly together with the brake backing plate using a slide hammer.

8. Remove the oil seal in the axle housing. It can be pried out with a small prybar.

9. To replace the bearing, unbend and discard the lockwasher. Remove the locknut using SST38020000 (spanner wrench) or equivalent.

10. Press the old bearing and cage off the shaft.

11. Remove the oil seal in the cage. Use a brass drift and a hammer to remove the bearing outer race after the seal has been removed.

To install:

12. Install the new wheel bearing outer race with a brass drift. Install a new oil seal in the bearing cage. Lubricate the area between the seal lips with grease after installation.

13. Place the bearing spacer on the axle shaft with the chamfered side facing the wheel. Install the wheel bearing inner race using a brass drift and lightly tapping it onto the axle shaft.

NOTE: Coat each wheel bearing race (cone) with multi-purpose grease.

14. Place the plain flat washer over the bearing, then install a new lockwasher. Install the locknut, tightening to 108 ft. lbs. (147 Nm). Continue to tighten after that until the grooves line up with the lockwasher tabs. The nut can be tightened up to 145 ft. lbs. (196 Nm). Bend the lockwasher tabs into place.

15. Install a new axle grease seal using a hammer and drift. Coat the sealing lip with multi-purpose grease.

16. Lubricate the axle case recess in the axle housing with wheel bearing grease. Coat the axle splines with gear oil.

17. Install the axle shaft and then check the axle end-play. It should be 0.02–0.15mm. The end-play is adjusted by adding or removing shims behind the brake backing plate. Tighten the backing plate attaching nuts to 39–46 ft. lbs. (53–63 Nm) for 1993–94 vehicles and to 40–54 ft. lbs. (54–74 Nm) for 1995–97 vehicles.

18. Reconnect the parking brake cable and the rear brake line. Install the brake drum.

19. Install the ABS wheel speed sensor, if removed.

20. Bleed the brake system and install the wheel. Lower the vehicle.

Single Rear Wheels With Disc Brakes

1. Raise the rear of the vehicle and support it. Remove the rear wheel.

2. Disconnect the rear parking brake cable.

3. If equipped, remove the ABS wheel speed sensor from the axle assembly.

4. Disconnect the brake tube on the rear axle housing at the wheel bearing housing attachment point. Plug the end of the brake tube to prevent loss of brake fluid.

5. Remove the brake caliper assembly and the rotor.

6. Remove the nuts securing the wheel bearing retainer to the brake backing plate.

7. Pull out the axle shaft assembly together with the brake backing plate using a slide hammer.

8. Remove the oil seal in the axle housing. It can be pried out with a small prybar.

9. To replace the bearing, unbend and discard the lock washer. Remove the locknut with a spanner wrench (SST38020000) or equivalent.

10. Remove the wheel bearing together with the bearing housing and the backing plate from the axle shaft.

11. Remove the wheel bearing outer side inner race from the axle shaft.

12. Pry out the old grease seal from the bearing housing.

13. Press the wheel bearing outer race from the bearing housing.

To install:

14. Press the new bearing into the housing until it bottoms.

NOTE: Always press on the outer race of the wheel bearing during installation.

15. Press a new seal into the bearing housing until it bottoms.

NOTE: After installing the new seal, coat the seal lip with multi-purpose grease.

16. Install the backing plate over the bearing housing and press the axle shaft into the inner bearing race. Be careful not to damage the seal.

17. Install a flat washer and a lockwasher.

18. Grease the seat of the locknut and then tighten it to 181–217 ft. lbs. (245–294 Nm).

19. Turn the bearing housing two or three revolutions; it must rotate smoothly. Bend part of the lockwasher in order to lock the nut.

20. Oil the lips of the new axle grease seal with multi-purpose grease and press it into the axle housing with a drift and mallet.

21. Coat the axle splines with gear oil. Coat the seal surface of the shaft with grease.

22. Install the axle shaft and then check the axle end play. It should be 0.0mm. The end-play is non adjustable; if the end-play is out of specification, replace the wheel bearing.

23. Install the backing plate nuts and torque the nuts to 39–46 ft. lbs. (53–63 Nm) for 1993–94 vehicles, and to 40–54 ft. lbs. (54–74 Nm) for 1995–97 vehicles.

24. Install the rotor and caliper assembly.

25. Connect the parking brake cable and the rear brake line.

26. Install the ABS wheel speed sensor if removed.

27. Bleed the brake system and install the wheel. Lower the vehicle.

Dual Rear Wheels

1. Raise the rear of the vehicle and support it. Remove the rear wheel(s).

2. Remove the six bolts and remove the axle from the vehicle.

3. Remove the attaching screws and detach the lockwasher from the rear wheel bearing locknut.

4. Remove the rear wheel bearing locknut using tool # KV40105400 or equivalent.

5. Remove the wheel bearing and the wheel hub with the brake drum. Be careful not to drop the outer bearing.

6. Remove the oil seal from the axle housing and discard it. Remove the wheel bearing grease seal from the wheel hub and discard it.

7. Remove the wheel bearings from the wheel hub.

8. Using a brass drift and a hammer, remove the bearing outer races from the wheel hub.

To install:

9. Using a bearing race installing tool, install the new bearing outer races in the wheel hub.

10. Pack the wheel hub with wheel bearing grease.

11. Pack each of the cone bearings with wheel bearing grease and install them in the wheel hub.

12. Install a new grease seal to the wheel hub and lubricate the seal lip with multi-purpose grease.

13. Install a new oil seal in the axle housing and lubricate the seal lip with multi-purpose grease.

14. Install the wheel hub on the vehicle. Install the wheel bearing locknut and torque it to 123–145 ft. lbs. (167–196 Nm).

15. Check the wheel bearing preload as follows:

a. Turn the wheel hub several times in both directions.

b. Retighten the wheel bearing locknut to 123–145 ft. lbs. (167–196 Nm).

c. The wheel bearing preload, with new grease and oil seals, as measured at the wheel lug stud should be 4.6–8.2 lbs. (20.6–36.3 N). Loosen the wheel bearing locknut until this preload is achieved.

16. Inspect the axial end-play; the end-play should be 0.0031 in. (0.08mm) or less. If end-play is out of specification, replace the wheel bearing assembly.

17. Install the wheel bearing lockwasher and tighten the attaching screws to 36 inch lbs. (5 Nm).

18. Coat the axles shaft splines with 90W gear oil and install the rear axle. Tighten the rear axle securing bolts to 42–55 ft. lbs. (57–75 Nm).

19. Install the rear wheel(s).

20. Lower the vehicle.

FIRING ORDERS

NOTE: To avoid confusion, always replace spark plug wires one at a time.

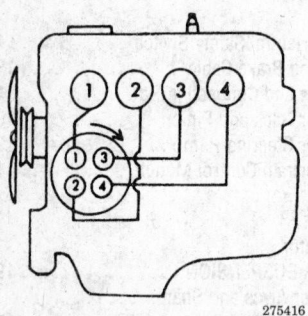

2.4L (22R-E) Engine
Engine Firing Order: 1–3–4–2
Distributor Rotation: Clockwise

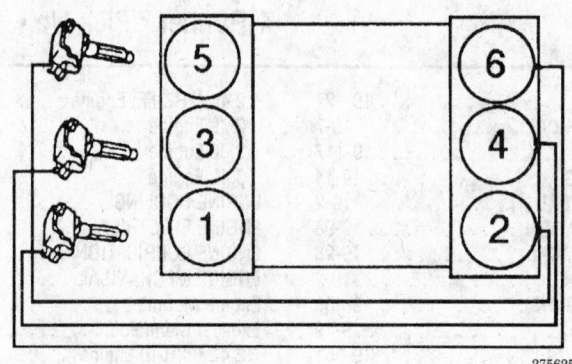

3.4L (5VZ-FE) Engine
Engine firing order: 1–2–3–4–5–6
Distributorless Ignition System

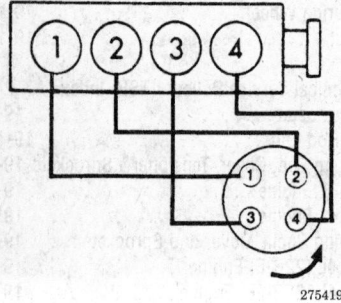

2.4L (2RZ-FE) and 2.7L (3RZ-FE) Engines
Engine firing order: 1–3–4–2,
Distributor rotation: Counterclockwise

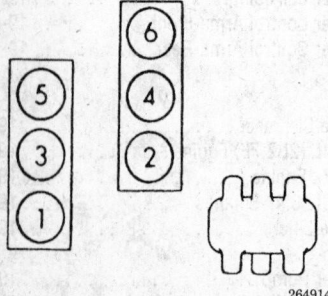

3.0L Engine (3VZ-E) Engine
Engine firing order: 1–2–3–4–5–6
Distributor rotation: counterclockwise

ENGINE ELECTRICAL

NOTE: Disconnecting the negative battery cable on some vehicles may interfere with the functions of the on board computer systems and may require the computer to undergo a relearning process, once the negative battery cable is reconnected.

Distributor

REMOVAL AND INSTALLATION

2.4L (22R-E) Engine

Engine Not Disturbed

1. Disconnect the negative battery cable.

— CAUTION —
Wait 90 seconds from the time the key is turned to LOCK and the negative battery cable is disconnected to begin work. This allows the SRS capacitor to discharge and prevent deployment of the air bag(s).

2. Remove the high tension cable from the coil.

3. Remove the primary wire from the distributor. Remove the distributor cap spring clips or screws, then the cap.

4. Matchmark the rotor to the distributor housing and the distributor housing to the engine block. This will aid in correct positioning of the distributor during installation.

5. Remove the distributor hold-down clamp bolt and the distributor from the engine.

To install:

6. Install a new O-ring to the distributor and lubricate it with engine oil.

7. Insert the distributor into the engine block by aligning the matchmarks made during removal.

8. Engage the distributor drive with the oil pump driveshaft.

9. Install the distributor hold-down clamp, the cap, the high tension wire, the primary wire or the electrical connector.

10. Install the spark plugs cables.

11. Connect the negative battery cable.

12. Start the engine. Check and adjust the ignition timing.

Engine Disturbed

1. Disconnect the negative battery cable.

2. Disconnect the distributor connectors.

3. Remove the distributor cap without disconnecting the secondary leads and position it aside.

4. Remove the distributor hold-down bolt and pull the distributor from the cylinder block.

To install:

5. Install a new O-ring to the distributor and lubricate it with engine oil.

6. Install the distributor and set the timing.

   a. Turn the crankshaft pulley until the timing mark is aligned with 5° BTDC mark.

   b. Check that the rocker arms on the No. 1 cylinder are loose. If not, turn the crankshaft one full turn.

   c. Temporarily install the rotor.

   d. Begin the insertion of the distributor with the rotor pointing upward and the distributor mounting hole approximately at the center position of the bolt hole.

   e. When fully installed, the rotor will rotate to the 10 o'clock position.

   f. Align the rotor tooth with the signal generator (Pick-up coil) projection and torque the bolt to 14 ft. lbs. (19 Nm).

   g. Install the rotor and the distributor cap with the wires.

7. Connect the high tension cords and the wiring connector.

8. Connect the negative battery cable.

9. Start the engine and adjust the ignition timing.

3.0L (3VZ-E), 2.4L (2RZ-FE) and 2.7L (3RZ-FE) Engines

Engine Not Disturbed

1. Disconnect the negative battery cable.

——— **WARNING** ———

The air bag system is equipped with a backup power source. To avoid possible air bag deployment, do not start working on the vehicle until 90 seconds has elapse from the time the ignition switch is turned OFF and the negative battery terminal is disconnected.

2. Disconnect the distributor connectors.

3. Remove the distributor cap without disconnecting the secondary leads and position it aside.

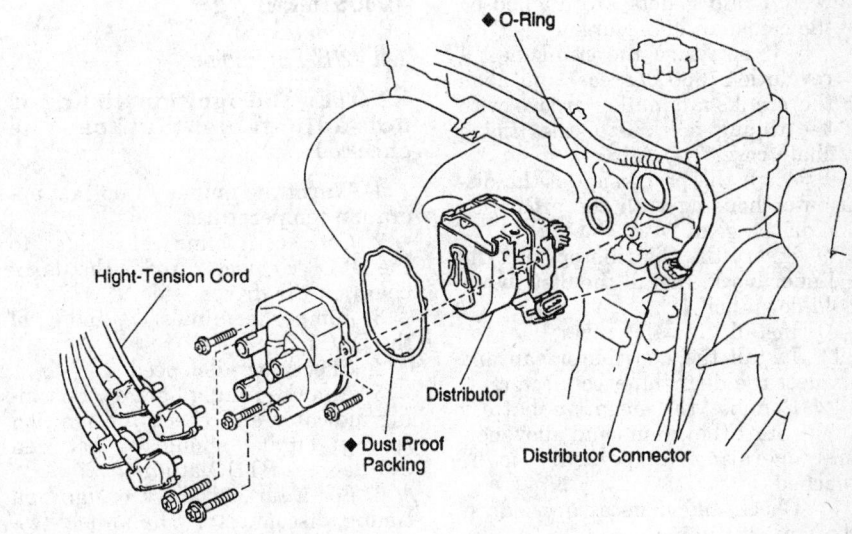

◆ Non-reusable part

291051

Distributor and related components — Tacoma and T-100 With 2.4L (2RZ-FE) and 2.7L (3RZ-FE) engines

4. Matchmark the rotor with the distributor housing and housing with the cylinder block.

5. Remove the distributor hold-down bolt and pull the distributor from the cylinder block.

To install:

6. Install a new O-ring to the distributor and lubricate it with engine oil.

7. Install the distributor into the cylinder block, while aligning the matchmarks made during removal. Install the distributor hold-down bolt.

341831

Align the protrusion on the driven gear with the groove of the distributor housing — 3.0L (3VZ-E) engine

8. Install the distributor cap and connect the distributor connector.

9. Connect the negative battery cable. Start the engine and allow normal operating temperature to be reached.

10. Check and if necessary, adjust the ignition timing.

Engine Disturbed

1. Disconnect the negative battery cable.

2. Disconnect the distributor connectors.

3. Remove the distributor cap without disconnecting the secondary leads and position it aside.

4. Remove the distributor hold-down bolt and pull the distributor from the cylinder block.

To install:

5. Install a new O-ring to the distributor and lubricate it with engine oil.

6. Remove the No. 1 cylinder spark plug.

   a. Place a finger or compression gauge over the spark plug hole.

   b. Turn the crankshaft until compression starts to build up. Continue turning the crankshaft until the crankshaft pulley groove align with the timing mark **0** of the timing chain cover.

7. If necessary, remove the valve cover.

 a. Check that the timing marks with 1 and 2 dots are aligned on the camshaft sub-gears.

 b. If not, turn the crankshaft 1 revolution (360 degrees) and align the crankshaft pulley groove with the timing mark **0** of the timing chain cover.

8. Align the protrusion of the distributor housing with the groove on the driven gear.

9. Insert the distributor into the cylinder block. Install the distributor hold-down bolt.

10. Install the valve cover.

11. Install the distributor cap and connect the distributor connector.

12. Connect the negative battery cable. Start the engine and allow normal operating temperature to be reached.

13. Check and if necessary, adjust the ignition timing.

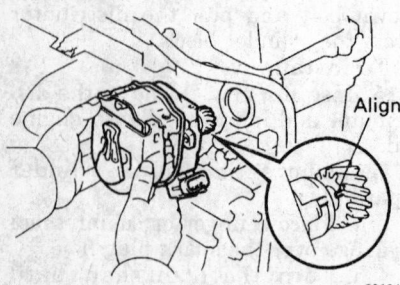

Check that the timing marks with 1 and 2 dots are aligned on camshaft sub-gears — 4Runner, Tacoma and T-100 with 2.4L (2RZ-FE) and 2.7L (3RZ-FE) engines

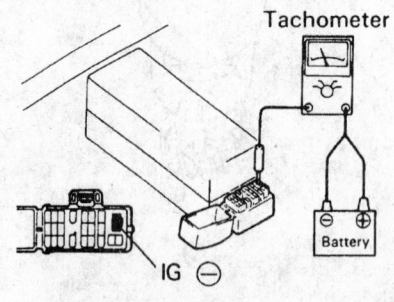

Align the protrusion of the distributor housing with the groove on the driven gear — 4Runner Tacoma and T-100 With 2.4L (2RZ-FE) and 2.7L (3RZ-FE) engines

Ignition Timing

ADJUSTMENT

2.4L (2RZ-FE) Engine

NOTE: The ignition timing is not adjustable, but can be checked.

1. Warm the engine to normal operating temperature.

2. Connect a hand-held tester to the DLC3 connector under the dashboard on the drivers side.

3. Jumper terminals T_{E1} and E_1 of the DLC1.

4. Check the idle speed.

5. Aim the timing light at the timing indicator and check the ignition timing. Timing should be between 3–7 degrees BTDC at idle.

6. For a further check on ignition timing, disconnect the hand-held tester from the DLC3 and disconnect the jumper wire from the DLC1.

7. Point the timing light at the crankshaft pulley and read the timing. Timing should be between 7–18 degrees BTDC at idle.

8. Remove timing light from the engine.

2.4L (22R-E) and 3.0L (3VZ-E) Engines

1. Warm the engine to normal operating temperature.

2. Connect a tachometer and timing light to the engine.

3. Jumper terminals T_{E1} and E_1 of the DLC1.

4. Check the idle speed.

5. Aim the timing light at the timing indicator and check the ignition timing. Timing should be 5° BTDC for the 22R–E engine and 10° BTDC for the 3VZ–E engine at idle.

6. If adjustment is necessary, loosen the distributor hold-down bolt and adjust by turning. Tighten the hold-down bolt and recheck the timing.

Connect a tachometer to the engine as shown — 2.4L (22R-E) and 3.0L (3VZ–E) engines

7. Remove the jumper connector. Check that the ignition timing advances.

8. Disconnect the tachometer and timing light from the engine.

2.7L (3RZ-FE) and 3.4L (5VZ-FE) Engines

Ignition timing is controlled by the ECM and is not adjustable.

Alternator

PRECAUTIONS

Several precautions must be observed with alternator equipped vehicles to avoid damage to the unit.

• If the battery is removed for any reason, make sure it is reconnected with the correct polarity. Reversing the battery connections may result in damage to the 1-way rectifiers.

• When utilizing a booster battery as a starting aid, always connect the positive to positive terminals and the negative terminal from the booster battery to a good engine ground on the vehicle being started.

• Never use a fast charger as a booster to start vehicles.

• Disconnect the battery cables when charging the battery with a fast charger.

• Never attempt to polarize the alternator.

• Do not use test lights of more than 12 volts when checking diode continuity.

• Do not short across or ground any of the alternator terminals.

• The polarity of the battery, alternator and regulator must be matched and considered before making any electrical connections within the system.

• Never separate the alternator on an open circuit. Make sure all connections within the circuit are clean and tight.

• Disconnect the battery ground terminal when performing any service on electrical components.

• Disconnect the battery if arc welding is to be done on the vehicle.

REMOVAL AND INSTALLATION

2.4L (2RZ-FE) Engine

1. Disconnect the negative battery cable.

2. If equipped with power steering, drain the coolant. Remove the engine undercover and remove the water inlet hose.

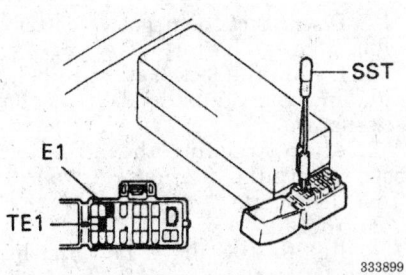

Jumper terminals of the DLC1 as shown — 2.4L (22R-E) and 3.0L (3VZ-E) engines

3. Disconnect the wiring from the alternator.

4. If equipped with air conditioning, remove the No. 2 fan shroud.

5. Loosen the alternator pivot and remove the adjusting bolt. Remove the drive belt.

6. Hold the alternator and remove the pivot. Remove the alternator.

To install:

7. Install the alternator and drive belt.

8. Adjust the belt to:
• New: 125± 25 lbs.
• Used: 80± 20 lbs.

9. After installing and adjusting the alternator belt, tighten the pivot and adjusting bolts.

10. If removed, install the water inlet hose.

11. Install the engine undercover and the No. 2 fan shroud.

12. Connect the wire to the generator and install the nut. Connect the connector to the generator.

13. Close the radiator drain plug and fill with coolant.

14. Connect the negative battery cable.

15. Start the vehicle and check for proper operation.

2.7L (3RZ-FE), 3.0L (3VZ-E) and 3.4L (5VZ-FE) Engines

1. Disconnect the negative battery cable from the battery.

2. Loosen the lock bolt, pivot bolt, nut, and the adjusting bolt at the alternator.

3. After loosening the adjusting bolt, remove the alternator drive belt from the engine.

4. Disconnect the alternator electrical connector.

5. Remove the nut and disconnect the alternator wire from the alternator.

6. Disconnect the wire harness with the clip.

7. Remove the lock bolt, pivot bolt, nut, and the alternator from the engine.

To install:

8. Install the alternator to the engine and install the nut, pivot bolt, and the lock bolt. Do not torque the bolts or nut at this time.

9. Connect the wire harness with the clip to the alternator.

10. Install the alternator wire with the nut.

11. Connect the alternator connector.

12. Install the drive belt to the engine.

13. Tighten the drive belt with the adjusting bolt. Belt tension should be as follows:
• New belt: 165± 10 lbs. — 3RZ-FE and 5VZ-FE T-100 and Tacoma
• Old belt: 100± 25 lbs. — 3RZ-FE and 5VZ-FE T-100 and Tacoma
• New belt: 160± 20 lbs. — 3VZ-E and 5VZ-FE 4Runner
• Old belt: 100± 20 lbs. — 3VZ-E and 5VZ-FE 4Runner

14. Once the belt is tight, torque the pivot bolt to 43 ft. lbs. (59 Nm) (T-100 and Tacoma), 38 ft. lbs. (51 Nm) (4Runner) and the lock bolt to 21 ft. lbs. (29 Nm) (T-100 and Tacoma), 18.5 ft. lbs. (25 Nm) (4Runner).

15. Connect the negative battery to the battery.

Drive Belt

REMOVAL AND INSTALLATION

2.4L (22R-E) and 3.0L (3VZ-E) Engines

1. Disconnect the negative battery cable.

———— **CAUTION** ————

Wait 90 seconds from the time the key is turned to LOCK and the negative battery cable is disconnected to begin work. This allows the SRS capacitor to discharge and prevent deployment of the air bag(s).

2. Loosen the lock, pivot and adjusting bolts of the component. Loosen the idler pulley, if equipped.

3. Remove the drive belt(s) from its component.

4. Visually check the belt for separation of the ribs, torn or worn ribs or cracks in the inner ridges of the ribs. If necessary, replace the drive belt.

To install:

5. Position the drive belt(s) over the component pulleys and drive pulley.

6. Partially tighten the pivot and adjusting bolts.

7. Check the drive belt(s) tension using a belt tension gauge. Fully tighten the lock, pivot and adjusting bolts.

8. Belt tensions should be as follows:
  a. Alternator 2.4L: Used — 80± 20 lbs. New — 125± 25 lbs.
  b. Alternator 3.0L: Used — 100± 20 lbs. New — 160± 20 lbs.
  c. Power steering and air conditioning: Used — 80± 25 lbs. New — 125± 25 lbs.

NOTE: After installing the drive belt(s), check that the belt does not touch the bottom of the pulley groove (conventional type belts) or that it fits properly in the ribbed grooves (V-ribbed type belts).

9. Connect the negative battery cable.

2.4L (2RZ-FE) and 2.7L (3RZ-FE) Engines

Power Steering Drive Belt

1. Loosen the power steering idler pulley set nut.

2. Loosen the belt adjusting bolt.

3. Remove the drive belt from the engine.

To install:

4. Install the drive belt. Tighten the adjusting bolt to adjust the drive belt tension. Drive belt tension is as follows:
• New belt: 135–185 lbs.
• Old belt: 80–120 lbs.

5. Torque the power steering idler pulley set nut to 29 ft. lbs. (39 Nm).

Air conditioning Compressor Drive Belt

1. Remove the power steering drive belt.

2. Raise and safely support the vehicle.

3. Remove the engine undercover.

4. Loosen the idler pulley locknut.

5. Loosen the adjusting bolt to the idler pulley and remove the compressor drive belt.

To install:

6. Install the compressor drive belt to the engine and tighten the adjusting bolt to the idler pulley. Drive belt tension is as follows:
• New belt: 160± 25 lbs.
• Old belt: 100± 20 lbs.

7. Torque the idler pulley locknut to 29 ft. lbs. (39 Nm).

8. Recheck the belt tension.

9. Install the engine undercover.

10. Lower the vehicle.

11. Install the power steering drive belt.

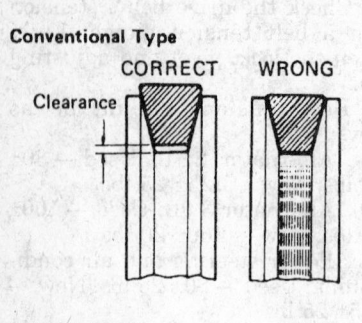

Conventional Type

CORRECT   WRONG

Clearance

V-Ribbed Type

341861

Inspecting the drive belts

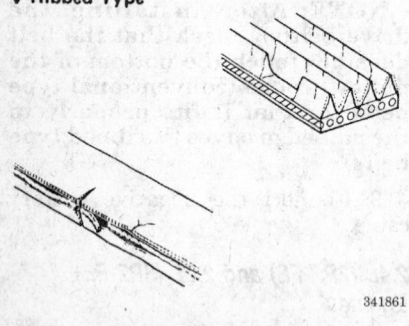

Nippondenso   Borroughs

341862

Check the drive belt tension using a belt tension gauge

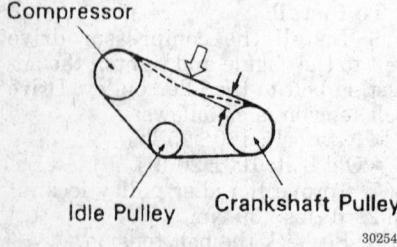

Compressor

Idle Pulley   Crankshaft Pulley

302545

Air conditioning compressor belt — 2.4L (2RZ-FE) and 2.7L (3RZ-FE) 3.4L (5VZ-FE) engines

Alternator Drive Belt

1. Disconnect the negative battery cable from the battery.
2. Loosen the lock bolt, pivot bolt, nut, and the adjusting bolt at the alternator.
3. After loosening the adjusting bolt, remove the alternator drive belt from the engine.

To install:

4. Install the drive belt to the engine.
5. Tighten the drive belt with the adjusting bolt. Belt tension should be as follows:
- New belt: 165± 10 lbs.
- Old belt: 115± 20 lbs.
6. Once the belt is tight, torque the pivot bolt to 43 ft. lbs. (59 Nm) and the lock bolt to 21 ft. lbs. (29 Nm).
7. Connect the negative battery to the battery.

3.4L (5VZ-FE) Engine

Power Steering Drive Belt

1. Loosen the pivot bolt, bracket nut, and the adjusting bolt to the power steering pump.
2. Remove the drive belt.

To install:

3. Install the drive belt to the engine.
4. Tighten the adjusting bolt to adjust the drive belt. The drive belt tension should be as follows:
- New belt tension — 135–180 lbs.
- Used belt tension — 85–120 lbs.
5. Torque the pivot bolt and the bracket nut to 29 ft. lbs. (39 Nm).

Air conditioning Compressor Drive Belt

1. Remove the power steering drive belt.
2. Raise and safely support the vehicle.
3. Remove the engine undercover.
4. Loosen the idler pulley locknut.
5. Loosen the adjusting bolt to the idler pulley and remove the compressor drive belt.

To install:

6. Install the compressor drive belt to the engine and tighten the adjusting bolt to the idler pulley. Drive belt tension is as follows:
- New belt: 160± 25 lbs.
- Old belt: 100± 20 lbs.
7. Torque the idler pulley locknut to 29 ft. lbs. (39 Nm).
8. Recheck the belt tension.
9. Install the engine undercover.
10. Lower the vehicle.
11. Install the power steering belt.

Alternator Drive Belt

1. Disconnect the negative battery cable from the battery.
2. Loosen the lock bolt, pivot bolt, nut, and the adjusting bolt at the alternator.
3. After loosening the adjusting bolt, remove the alternator drive belt from the engine.

To install:

4. Install the drive belt to the engine.
5. Tighten the drive belt with the adjusting bolt. Belt tension should be as follows:
- New belt: 160± 20 lbs.
- Old belt: 100± 20 lbs.
6. Once the belt is tight, torque the pivot bolt to 38 ft. lbs. (51 Nm) and the lock bolt to 25 ft. lbs. (34 Nm).
7. Connect the negative battery to the battery.

Starter

REMOVAL AND INSTALLATION

2.4L (22R-E) and 3.0L (3VZ-E) Engines

1. Disconnect the negative battery cable.
2. Disconnect the wires from the starter.
   a. Remove the nut and disconnect the battery cable from the magnetic switch on the starter motor.
   b. Disconnect the other wire from terminal 50.
3. For the 22R–E engine, remove the nut and bolts and remove the starter motor from the flywheel bell housing.
4. For the 3VZ–E engine, remove the two mounting bolts and remove the starter motor from the flywheel bell housing.

To install:

5. Place the starter motor in the flywheel bell housing and torque the mounting bolts to 29 ft. lbs. (39 Nm).
6. Connect the connector to the terminal on the magnetic switch. Connect the cable from the battery to the terminal on the switch and install the nut.
7. Connect the negative battery cable.

2.4L (2RZ-FE), 2.7L (3RZ-FE) and 3.4L (5VZ-FE) Engines

1. Disconnect the negative battery cable from the battery.
2. Remove the nut and disconnect the starter wire from the starter.

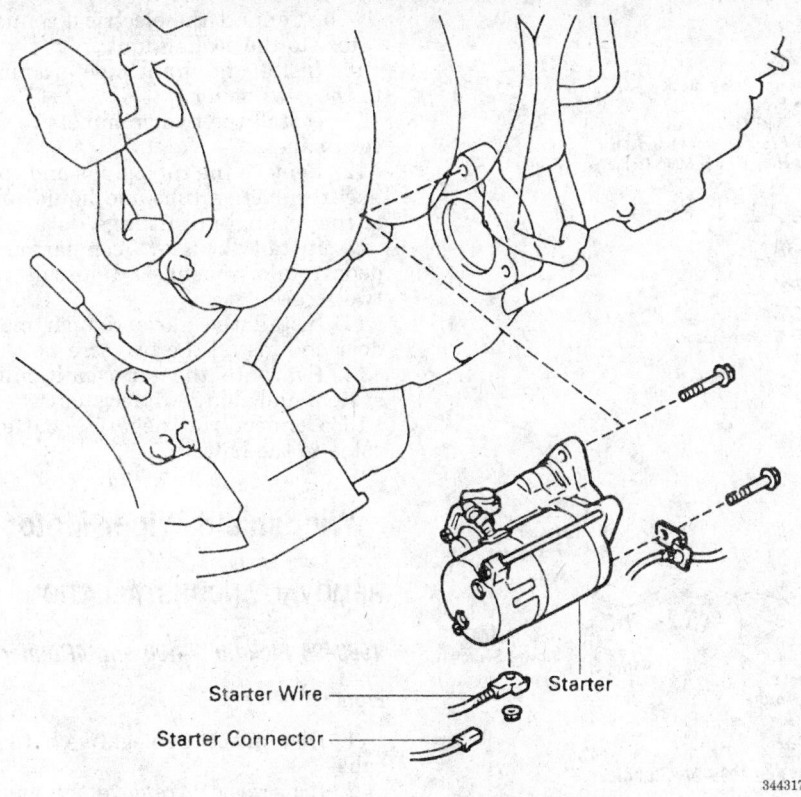

Starter Wire

Starter Connector

Starter

344317

Starter motor connection points — 3RZ-FE and 2RZ-FE engines shown, others similar

3. Disconnect the starter connector from the starter.

4. Remove the starter from the engine by removing the two bolts.

To install:

5. Position the starter on the engine and install the two mounting bolts. Torque the bolts to 29 ft. lbs. (39 Nm).

6. Connect the electrical connector to the starter.

7. Connect the starter wire to the starter and install the nut. Torque the nut to 70 inch lbs. (8.8 Nm).

8. Connect the negative battery cable to the battery.

CHASSIS ELECTRICAL

Blower Motor

REMOVAL AND INSTALLATION

1993–94 Pick-up and 4Runner

1. Disconnect the negative battery cable.

2. Remove the air duct, if equipped.

3. Disconnect the electrical connectors from the motor.

4. Remove the blower motor-to-case screws and lift the motor from the case.

To install:

5. Install the blower motor and tighten the mounting screws.

6. Connect the electrical connector to the blower motor.

7. Install the air duct, if removed.

8. Connect the negative battery cable.

9. Start the engine and turn on the heater to check for proper blower motor operation.

1995 Pick-up and 1995–97 4Runner

Front Blower Motor

1. Disconnect the negative battery cable.

2. Disconnect the connectors from the blower motor and the blower resistor located under the dash on the passenger side of the vehicle.

3. Remove the three screws holding the blower motor and remove the blower motor.

To install:

4. Install the blower motor and tighten the three holding screws.

5. Connect the wire connectors to the blower motor and the blower resistor.

6. Connect the negative battery cable.

7. Start the engine and turn on the heater to check for proper blower motor operation.

Rear Blower Motor

1. Disconnect the negative battery cable.

2. Remove the center console box and remove the rear blower motor.

To install:

3. Install the rear blower box and install the center console box.

4. Connect the negative battery cable.

5. Start the engine and turn on the heater to check for proper blower motor operation.

T-100

1. Disconnect the negative battery cable from the battery.

2. Disconnect the electrical connectors from the blower motor and blower resistor.

3. Remove the blower motor to case screws and remove the motor from the case.

To install:

4. Install the blower motor to the case and install the three screws.

5. Connect the electrical connectors to the blower motor and blower resistor.

6. Connect the negative battery cable to the battery.

Tacoma

1. Disconnect the negative battery cable from the battery.

2. Remove the glove compartment door by removing the two screws.

3. Remove the glove compartment door reinforcement by removing the two screws.

4. Remove the cooling unit as follows:

   a. Properly discharge the refrigerant from the air conditioning system. Make sure to recycle the freon. Do not allow the freon to enter the atmosphere.

   b. Disconnect the suction tube and liquid tube from the cooling unit fittings. Disconnect the fittings at the cowl wall. Plug the cooling unit tubes to prevent any contaminates from entering the air conditioning system.

   c. Remove the two grommets at the cowl wall.

   d. Remove the drain pipe grommet.

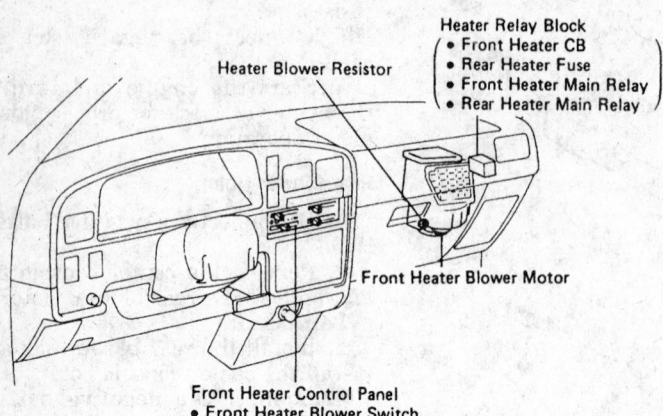

Heater Blower Resistor

Heater Relay Block
• Front Heater CB
• Rear Heater Fuse
• Front Heater Main Relay
• Rear Heater Main Relay

Front Heater Blower Motor

Front Heater Control Panel
• Front Heater Blower Switch

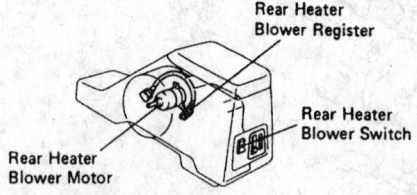

Rear Heater Blower Register

Rear Heater Blower Switch

Rear Heater Blower Motor

Blower motor location — 1995 Pick-up and 1995–97 4Runner

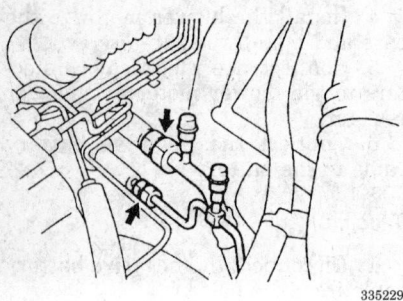

Suction and liquid tubes connection points — Tacoma

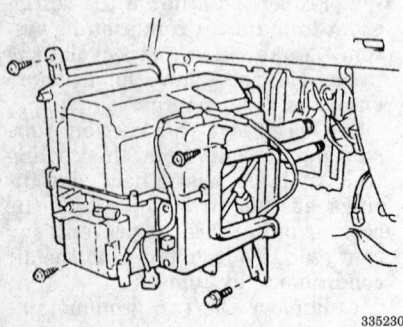

Cooling unit removal — Tacoma

e. Disconnect the electrical connectors from the cooling unit.

f. Remove the cooling unit from the vehicle by removing the bolt and three screws.

5. Remove the blower unit as follows:

a. Disconnect the connector from the blower motor.

b. Disconnect the air inlet damper control cable from the blower motor.

c. Remove the nut and bolt and remove the blower unit from the vehicle.

6. Remove the blower motor from the blower unit by removing the three screws.

To install:

7. Connect the blower motor to the blower unit by installing the three screws.

8. Install the blower unit as follows:

a. Install the blower unit to the vehicle and install the bolt and nut.

b. Connect the air inlet damper control cable to the blower motor.

c. Connect the electrical connector to the blower motor.

9. Install the cooling unit to the vehicle as follows:

a. Install the cooling unit and install the bolt and three screws.

b. Connect the electrical connectors to the cooling unit.

10. Install the drain pipe grommet to the cowl panel.

11. Install the two grommets to the cowl panel.

12. Remove the tube plugs and connect the suction tube and liquid tube to the cooling unit fittings.

13. Install the glove compartment door reinforcement and install the two screws.

14. Install the glove compartment door and install the two screws.

15. Evacuate the air conditioning system and charge the system.

16. Connect the negative battery cable to the battery.

Windshield Wiper Motor

REMOVAL AND INSTALLATION

1993–95 Pick-up, T-100 and 4Runner

Front

1. Disconnect the negative battery cable.

2. If necessary, remove the wiper arms and the cowl louver to access the wiper motor.

3. Disconnect the wiring from the wiper motor.

4. Remove the nut, then, pry the wiper link from the crank arm.

5. Remove the motor.

To install:

6. Install the wiper motor.

7. Connect the wiper link to the crank arm and install the nut.

8. Connect the wiring to the wiper motor.

9. If removed, install the cowl louver and the wiper arms.

10. Connect the negative battery cable.

Rear

1. Disconnect the negative battery cable.

2. At the rear of the vehicle, remove the wiper motor cover panel.

3. Remove the wiper arm from the wiper motor.

4. Disconnect the electrical connector from the wiper motor.

5. Remove the wiper motor bolts and the motor from the vehicle.

To install:

6. Install the wiper motor.

7. Connect the electrical connector to the wiper motor.

8. Install the wiper arm to the wiper motor.

9. Install the wiper motor cover panel.

10. Connect the negative battery cable.

1996–97 4Runner

Front

1. Disconnect the negative battery cable.
2. If necessary, remove the wiper arms and the cowl louver to access the wiper motor.
3. Disconnect the wiring from the wiper motor.
4. Remove the nut and pry the wiper link from the crank arm.
5. Remove the motor.

To install:

6. Install the wiper motor.
7. Connect the wiper link to the crank arm and install the nut.
8. Connect the wiring to the wiper motor.
9. If removed, install the cowl louver and the wiper arms.
10. Connect the negative battery cable.
11. Check the wiper motor for proper operation.

Rear

1. Disconnect the negative battery cable.
2. At the rear of the vehicle, remove the following:
   - Back door trim
   - Plate
   - Back door glass run
   - Outer weatherstrip
   - Back door glass
3. Remove the wiper arm from the wiper motor.
4. Disconnect the electrical connector from the wiper motor.
5. Remove the washer nozzle.
6. Remove the wiper motor bolts and the motor from the vehicle.

To install:

7. Install the washer nozzle.
8. Install the wiper motor. Torque the three set bolts to 48 inch lbs. (5.4 Nm).
9. Install the packing holder and nut and torque to 48 inch lbs. (5.4 Nm).
10. Connect the electrical connector to the wiper motor.
11. Install the wiper arm to the wiper motor. When tightening the rear wiper arm, slide the slider for 0.39 in. (10 mm) to the turning slide and tighten the wiper arm set nut with the step liner on the block. Torque to 48 inch lbs. (5.4 Nm).
12. Install the back door glass, outer weatherstrip, back door glass run, plate and the back door trim.
13. Connect the negative battery cable.

14. Check the wiper motor for proper operation.

1995–97 T-100 and Tacoma

1. Disconnect the negative battery cable from the battery.
2. Remove the wiper arm caps and the nuts.
3. Remove the wiper arms from the vehicle.
4. Remove the cowl louver by removing the screws and clips.
5. Disconnect the electrical connector to the wiper motor.
6. Remove the four bolts to the wiper motor.
7. Connect the claw of the wiper link to the panel.
8. Disconnect the motor from the wiper link and remove the motor from the vehicle.

To install:

9. Install the wiper motor to the vehicle and connect the wiper link to the motor.
10. Disconnect the claw of the wiper link from the panel.
11. Install the four wiper motor bolts to the wiper motor.
12. Connect the electrical connector.
13. Install the cowl louver with the screws and clips.
14. Install the wiper arms with the two bolts. Torque the bolts to 15 ft. lbs. (20 Nm).
15. Install the covers to the wiper arms.
16. Connect the negative battery cable to the battery.

Combination Switch

REMOVAL AND INSTALLATION

T-100 and Tacoma

The combination switch incorporates the headlight switch, turn signal switch, dimmer switch and the windshield wiper switch.

1. Disconnect the negative battery cable.
2. Position the front wheels in a straight-ahead position.

—————— **CAUTION** ——————
Wait 90 seconds from the time the key is turned to LOCK and the negative battery cable is disconnected to begin work. This allows the SRS capacitor to discharge and prevent deployment of the air bag(s). If the wiring connector of the air bag system is disconnected

with the ignition switch at ON or ACC, diagnostic codes will be recorded.

3. If equipped with an air bag, properly disable the air bag system.
4. Remove the steering wheel center pad.

NOTE: When removing the wheel pad, take care not to pull the air bag harness connector. When storing the wheel pad, keep the upper surface of the pad facing upward.

5. Remove the steering wheel assembly.
6. Remove the steering column covers by removing the screws.
7. Remove the combination switch with the spiral cable as follows:
   a. Remove the two screws.
   b. Disconnect the electrical connectors.
   c. If equipped with a air bag, disconnect the air bag connector.
   d. Remove the four screws.
   e. Remove the combination switch from the vehicle.
8. Remove the spiral cable from the combination switch by removing the four screws.
9. Loosen the mounting screws and remove the dimmer switch.
10. Remove the mounting screws and disconnect the headlight/turn signal switch from the combination switch body.
11. Separate the bracket from the wiper switch body by removing the screws. Remove the wiper switch from the switch body.

To install:

12. Install the wiper switch to the switch body and connect the mounting bracket.
13. Install the dimmer and headlight/turn signal switch to the switch body and tighten the mounting screws.
14. Install the spiral cable to the combination switch by installing the four screws.
15. Install the combination switch assembly to the steering column and tighten the mounting screws.
16. Connect the electrical connectors. Push in the terminals until they are securely locked in the connector lug.
17. Install both steering column covers and install the screws.
18. If equipped with a air bag, center the spiral cable as follows:
   a. Check that the front wheels are facing straight ahead.
   b. Turn the cable counterclockwise by hand until it becomes harder to turn the cable.

w/ Intermittent Wiper or Mist Wiper

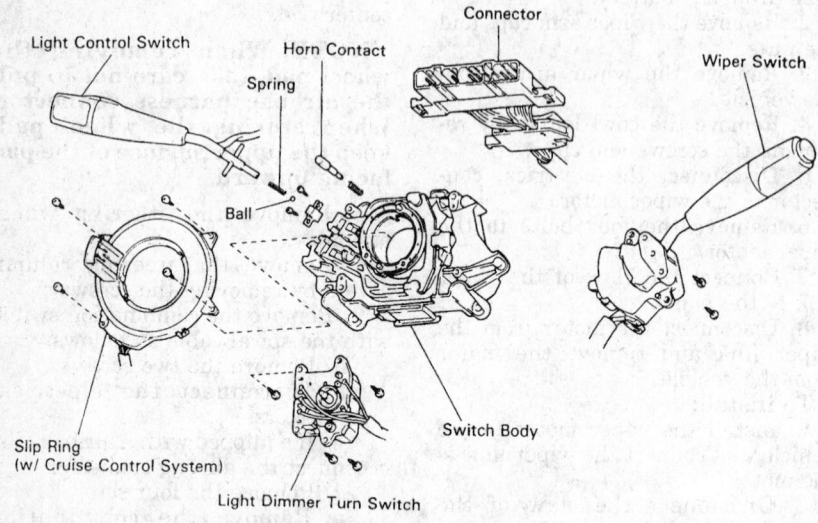

Exploded view of the combination switch assembly — T-100

b. Remove the ball and slide out the switch from the switch body with the spring.

6. Remove the headlight dimmer, the turn signal and the wiper washer switches by removing the screws on the backside of the combination switch.

To install:

7. Install the headlight dimmer, the turn signal and the wiper washer switches to the combination switch.

8. Install the light control switch.

a. Slide in the switch with the spring to the switch body. Install the ball and the ball set plate with the two screws. Make sure that the switch operates smoothly.

9. Push the wires into the connector until they lock securely in place.

10. Install the switch into the steering column.

11. Install the upper and lower steering column shroud.

12. Install the steering wheel and torque the nut to 25 ft. lbs. (34 Nm).

13. Enable the air bag system.

14. Connect the negative battery cable.

Ignition Lock Cylinder

REMOVAL AND INSTALLATION

——— **CAUTION** ———

The air bag system (SRS or SIR) must be disarmed before removing the steering column covers. Failure to do so may cause accidental deployment, property damage or personal injury.

1. Disconnect the negative battery terminal.

2. Remove the upper and lower steering column covers.

3. Disconnect the ignition switch from the electrical connector.

4. Using the key in the ignition switch, turn it to the **ACC** position.

5. Using a thin rod, place it into the hole of the cylinder lock housing. Pushing down on the thin rod, pull out the cylinder lock.

To install:

6. Using the key, install the cylinder lock into the housing until the retaining tab locks it in place.

7. Connect the ignition switch electrical connector.

8. Install the upper and lower steering column covers.

9. Connect the negative battery terminal.

c. Rotate the cable clockwise about three turns to align the red mark.

19. Install the steering wheel to the steering column.

20. Install the steering wheel center pad.

21. Connect the negative battery cable.

22. Enable the air bag system.

Pick-up and 4Runner

The combination switch is composed of the turn signal, the headlight con-

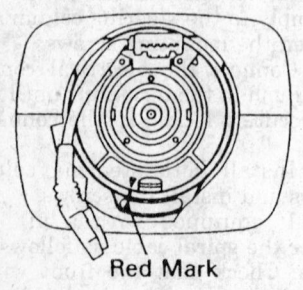

Red Mark

Spiral cable alignment mark — T-100 and Tacoma

trol, the dimmer switch, the wiper, and the washer switch

1. Disconnect the negative battery cable. Remove the steering wheel.

——— **CAUTION** ———

The air bag system must be disarmed before removing the steering wheel. Failure to do so may cause accidental deployment, property damage or personal injury.

2. Remove the upper and lower steering column shroud screws and the shrouds.

3. Remove the combination switch screws and the switch from the column.

4. Disconnect the electrical connector from the combination switch. To remove the wires from the electrical connector, perform the following procedures.

a. Using a small prybar, insert it into the open end between the locking lugs and the terminal.

b. Pry the locking lugs upward and pull the terminal out from the rear.

5. Remove the light control switch.

a. Remove the two screws and the ball set plate from the switch body.

Ignition Switch

REMOVAL AND INSTALLATION

All 1993–94 Models

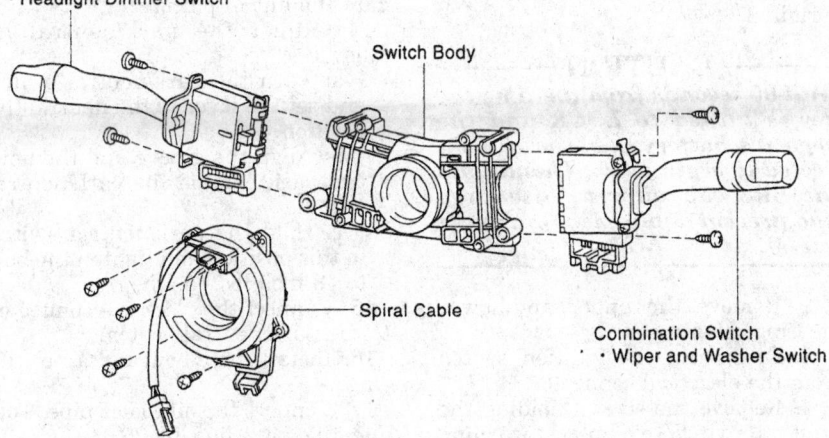

Combination switch component identification — 1995 Pick-up and 4Runner

333498

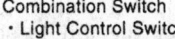

Combination switch component identification — 1996–97 4Runner

333332

—— **CAUTION** ——

The air bag system (SRS or SIR) must be disarmed before removing the steering column covers. Failure to do so may cause accidental deployment, property damage or personal injury.

1. Disconnect the negative battery terminal.
2. Remove the upper and lower steering column covers.
3. Disconnect the ignition switch from the electrical connector.
4. With the key in the ignition switch, turn the lock cylinder to the **ACC** position.
5. Using a thin rod, press in the stop pin and pull out the lock cylinder.
6. Remove the key warning switch from the side of the cylinder housing.
7. Remove the ignition switch from the end of the cylinder housing.
   To install:
8. Position the ignition switch on the end of the housing and install the screw.
9. Using the key, install the lock cylinder into the housing until the stop pin locks it in place.
10. Install the key warning switch on the side of the cylinder housing.
11. Connect the harness connector to the ignition switch and install the steering column covers.
12. Connect the negative battery cable.

1995 Pick-up and 4Runner

1. Disconnect the negative battery terminal.

—— **CAUTION** ——

Wait 90 seconds from the time the key is turned to LOCK and the negative battery cable is disconnected to begin work. This allows the SRS capacitor to discharge and prevent deployment of the air bag(s).

2. Remove the upper and lower steering column covers.
3. Disconnect the ignition switch electrical connector.
4. For manual transmission, remove the three screws holding the ignition switch assembly and remove the switch.
5. For automatic transmission, remove the five screws, the spring, and

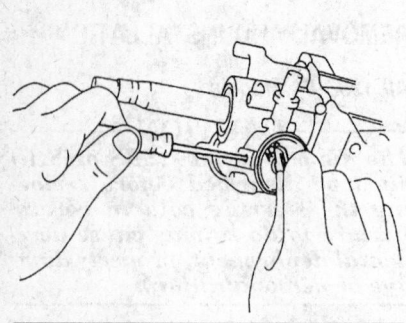

Push the stop pin to remove the lock cylinder — 1993–94 models

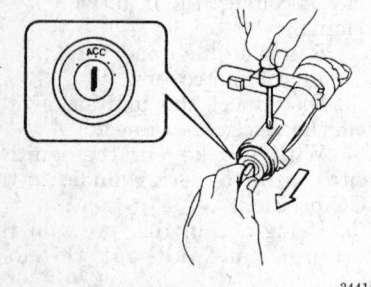

344144

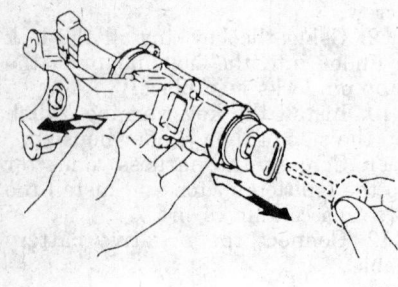

311392

Manual transmission ignition switch with steering column upper bracket — 1995 Pick-up and 4Runner

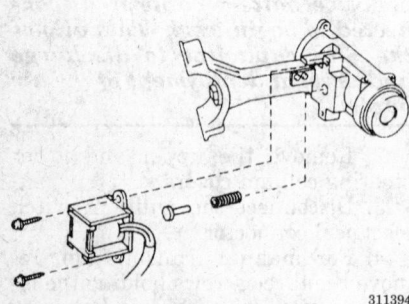

311394

Automatic transmission ignition switch — 1995 Pick-up and 4Runner

the lock pin. Remove the switch assembly.

6. If it's necessary to replace the key cylinder, insert the key in the ignition switch, turn it to the **ACC** position.

7. Using a thin rod, place it into the hole of the cylinder lock housing. Pushing down on the thin rod, pull out the cylinder lock.

8. Remove the unlock warning switch screws and the unlock warning switch.

To install:

9. Install the unlock warning switch and tighten the screws.

10. Using the key, install cylinder lock into the housing until the retaining tab locks it in place. Make sure that the ignition key is in the **ACC** position.

11. For vehicles with manual transmission, install the ignition switch with the three screws. For vehicles with automatic transmission, install the five screws, the spring, and the lock pin.

12. Connect the key unlock warning switch connector.

13. Connect the ignition switch to the electrical connector.

14. Install the upper and lower steering columns.

15. Connect the negative battery cable and check the ignition switch operation.

1995–97 T-100, Tacoma and 1996–97 4Runner

1. Disconnect the negative battery terminal.

────── **CAUTION** ──────
Wait 90 seconds from the time the key is turned to LOCK and the negative battery cable is disconnected to begin work. This allows the SRS capacitor to discharge and prevent deployment of the air bag(s).

2. Remove the upper and lower steering column covers.

3. Disconnect the ignition switch from the electrical connector.

4. Remove the screws holding the ignition switch/key unlock warning switch assembly and remove the assembly from the steering column.

To install:

5. Install the ignition switch/key unlock warning switch assembly with the screws.

6. Connect the ignition switch to the electrical connector.

7. Install the upper and lower steering columns.

8. Connect the negative battery cable and check the ignition switch operation.

Park/Neutral Safety Switch

REMOVAL AND INSTALLATION

All except 1996–97 4Runner

1. Disconnect the negative battery cable to the battery.

2. Apply the parking brake fully and block the rear wheels.

3. Raise and safely support the vehicle.

4. On vehicles equipped with the A43D transmission, remove the front exhaust pipe and heat insulator.

5. Disconnect the oil cooler pipe.

6. Remove the elbow and O-ring.

7. Disconnect the electrical connector from the park/neutral position switch.

8. Using a flat bladed tool, bend the lock washer back away from the nut.

9. Remove the nut and lock washer from the park/neutral switch.

10. Remove the bolt and pull out the park/neutral position switch.

To install:

11. Install the park/neutral position switch to the manual valve shaft. Leave the bolt loose at this time.

12. Install the lock washer and nut. Torque the nut to 61 inch lbs. (6.9 Nm).

13. Using a flat bladed tool, bend the lock washer over the nut to secure the nut in place.

14. Adjust the park/neutral as follows:

   a. Turn the park/neutral switch and set the lever to the neutral (N) position.

   b. Align the groove and the neutral basic line on the park/neutral switch.

   c. Hold the park/neutral switch in this position and tighten the bolt to 48 inch lbs. (5.4 Nm).

15. Connect the electrical connector to the park/neutral switch.

16. Install the elbow and a new O-ring.

17. Connect the oil cooler pipe. Torque the pipe line to 25 ft. lbs. (34 Nm).

18. On vehicles equipped with the A43D transmission, install the front exhaust pipe and heat insulator.

19. Lower the vehicle and check the vehicles shifting positions.

20. Connect the negative battery cable to the battery.

21. Verify that the vehicle only starts in PARK or NEUTRAL.

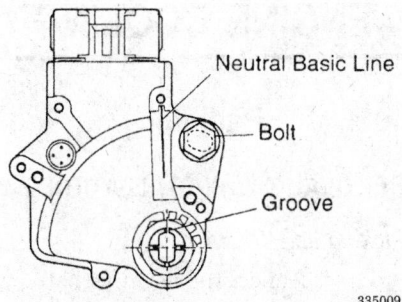

Park/neutral switch alignment

Neutral Basic Line

Bolt

Groove

335009

1996–97 4Runner

1. Disconnect the negative battery cable.

CAUTION

Wait 90 seconds from the time the key is turned to LOCK and the negative battery cable is disconnected to begin work. This allows the SRS capacitor to discharge and prevent deployment of the air bag(s).

2. Raise and safely support the vehicle.
3. Disconnect the park/neutral switch connector.
4. Pry off the lock washer and remove the nut.
5. Remove the bolt and the park/neutral position switch.
To install:
6. Install the park/neutral switch with the bolt and torque to 9 ft. lbs. (13 Nm).
7. Install a new lock plate and nut. Torque the nut to 2.9 ft. lbs. (3.9 Nm).
8. Stake the nut with the lock plate.
9. Connect the negative battery cable.
10. Adjust the park/neutral switch.

Powertrain Control Module

REMOVAL AND INSTALLATION

NOTE: It is recommended that a grounding strap be worn when handling the ECM. The grounding strap will prevent a static electricity discharge from damaging the ECM.

T-100

1. Disconnect the negative battery cable from the battery.

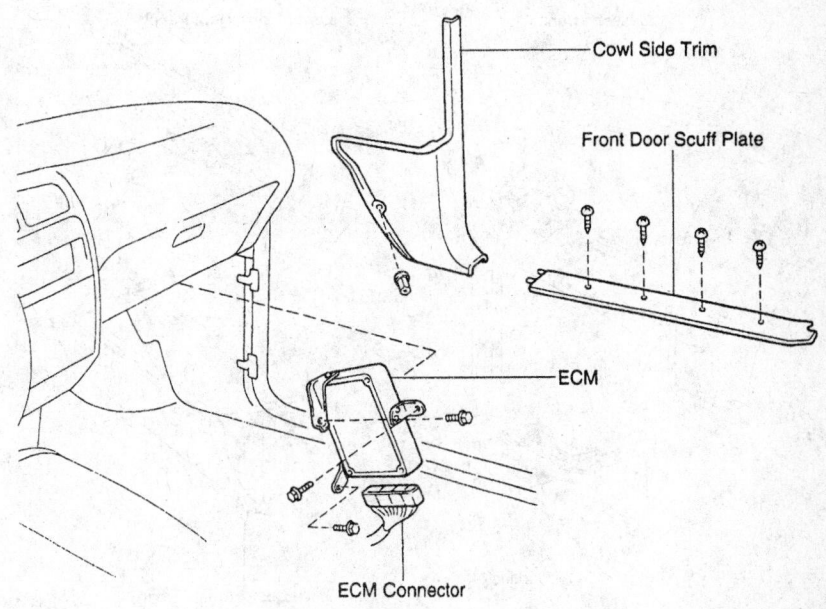

Cowl Side Trim

Front Door Scuff Plate

ECM

ECM Connector

334634

ECM mounting location — T-100

2. Remove the four screws to the right front door scuff plate. Remove the scuff plate from the vehicle.
3. Remove the cowl panel side trim by removing the clip.
4. Disconnect the four ECM electrical connectors.
5. Remove the ECM bolts and remove the ECM from the vehicle.
To install:
6. Install the ECM to the vehicle and install the three bolts.
7. Connect the ECM electrical connectors.
8. Install the cowl side trim and clip.
9. Install the front door scuff plate and four screws.
10. Connect the negative battery cable to the battery.

Tacoma

1. Disconnect the negative battery cable from the battery.
2. Remove the glove compartment door.
3. Remove the lower finish No. 2 panel.
4. Disconnect the ECM electrical connectors.

5. Remove the ECM bolts and remove the ECM from the vehicle.
To install:
6. Install the ECM to the vehicle and install the bolts.
7. Connect the ECM electrical connectors.
8. Install the lower finish No. 2 panel.
9. Install the glove compartment door.
10. Connect the negative battery cable to the battery.

Pick-up and 4Runner

1. Disconnect the negative battery cable.
2. Locate the ECM under the dash, behind the right front kick panel.
3. Remove the kick panel and disconnect the wire connectors to the ECM. Remove the ECM.
To install:
4. Install the ECM and connect the wire connectors.
5. Install the kick panel and connect the negative battery cable.
6. Check for proper operation.

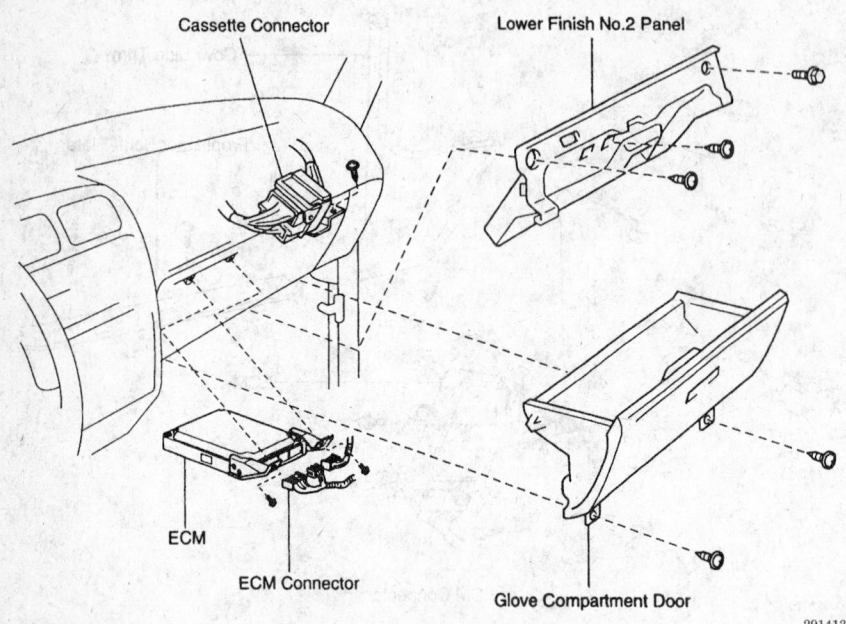

ECM mounting location — Tacoma

291413

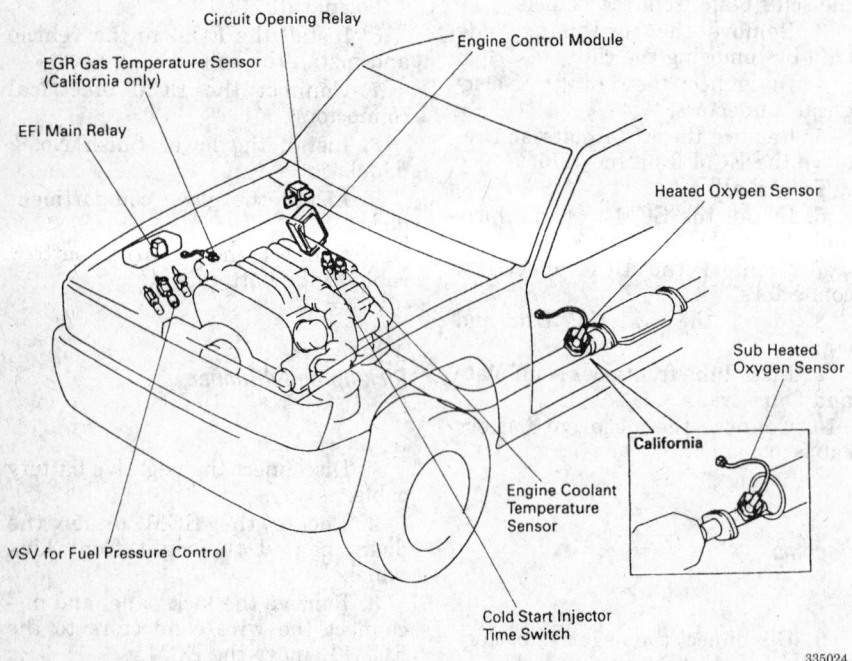

ECM mounting location — Pick-up and 4Runner

335024

ENGINE COOLING

Radiator

REMOVAL AND INSTALLATION

Pick-up and 4Runner

1. Disconnect the negative battery cable.

— **CAUTION** —

Work must be started after 90 seconds from the time the ignition switch is turned to the LOCK position and the negative (-) battery cable is disconnected.

2. Drain the engine coolant.
3. Remove the engine undercover.
4. On the 2.4L (22R-E) engine, remove the air intake connector.
5. Remove the radiator.
    a. Disconnect the reservoir hose.
    b. Remove the radiator hoses.
    c. On the 2.4L (22R-E) engine, with air conditioning, remove the No. 2 fan shroud.
    d. On the 2.4L (22R-E) engine, remove the No. 1 fan shroud.
    e. On the 3.0L (3VZ-E) engine, with air conditioning, remove the No. 2 fan shroud.
    f. With automatic transmission, disconnect the oil cooler hoses. Plug the hose to prevent oil from escaping.
    g. Remove the radiator.
6. Remove the cooling fan(s) from the radiator.

To install:

7. Install the cooling fan(s) to the radiator.
8. Install the No. 1 and No. 2 fan shrouds to the radiator as necessary.
9. Install the radiator.
10. If disconnected, connect the oil cooler hoses.
11. Install the radiator hoses and the reservoir hose.
12. On the 2.4L (22R-E) engine, install the air intake connector.
13. Install the engine under covers.
14. Fill the engine with coolant.
15. Connect the negative battery cable and start the engine.
16. Check for any leaks and bleed the cooling system. Also, verify that the cooling fan is operating correctly.

1996–97 4Runner

1. Disconnect the negative battery cable.
2. Drain the engine coolant.
3. Remove the radiator grille.

4. On the 2.7L (3RZ-FE) engine, disconnect the air pipe.

5. Disconnect the upper radiator hose.

6. Disconnect the radiator reservoir hose.

7. Disconnect the lower radiator hose.

8. Remove the No. 2 fan shroud.

9. Automatic transmission: Disconnect the oil cooler hoses.

10. Remove the radiator.

To install:

11. Install the radiator and torque the bolts to 9 ft. lbs. (12.5 Nm).

12. If disconnected, connect the oil cooler hoses.

13. Install the fan shroud.

14. Connect the lower radiator hose.

15. Connect the radiator reservoir hose.

16. Connect the upper radiator hose.

17. On the 2.7L (3RZ-FE) engine, connect the air pipe.

18. Install the radiator grille.

19. Fill and bleed the cooling system.

20. Connect the negative battery cable.

21. Start the engine and check for leaks.

T-100

3.0L (3VZ-E) and 2.7L (3RZ-FE) Engines

1. Drain the engine coolant from the radiator and engine.

2. Disconnect the negative and positive battery cable from the battery.

3. Remove the battery hold down clamp and remove the battery.

4. Remove the clearance lights from the grille by removing the four screws from each light.

5. Remove the radiator grille from the vehicle by removing the 4 screws and 11 clips.

6. Disconnect the upper radiator hose from the radiator.

7. Disconnect the radiator reservoir hose from the radiator.

8. Loosen the lock bolt and adjusting bolt to the power steering pump. Remove the power steering drive belt from the engine.

9. If equipped with air conditioning, loosen the idler pulley nut and adjusting bolt and remove the air conditioning drive belt.

10. Remove the alternator drive belt, fan (with fan clutch), water pump pulley, and fan shroud as follows:

a. Stretch the belt and loosen the water pump pulley mounting nuts.

b. Loosen the lock, pivot and adjusting bolts for the alternator and remove the alternator drive belt from the engine.

c. Remove the four water pump pulley mounting nuts.

d. Remove the fan (with fan clutch) and the water pump pulley.

e. Remove the No. 2 fan shroud.

f. Remove the No. 1 fan shroud by removing the four bolts.

11. If the vehicle is equipped with an automatic transmission, disconnect the oil cooler hoses from the radiator.

12. Disconnect the lower radiator hose from the radiator.

13. Remove the four bolts to the radiator.

14. Remove the radiator from the vehicle.

15. If the vehicle is equipped with a manual transmission, remove the four bolts and two radiator supports.

16. If the vehicle is equipped with a automatic transmission, remove the eight bolts and two radiator supports.

To install:

17. If the vehicle is equipped with a manual transmission, install the two radiator supports to the radiator and install the four bolts. Torque the bolts to 9 ft. lbs. (13 Nm).

18. If the vehicle is equipped with a automatic transmission, install the two radiator supports with the eight bolts. Torque the bolts to 9 ft. lbs. (13 Nm).

19. Install the radiator to the vehicle with the tabs on the supports through the radiator service holes. Install the four bolts and torque the bolts to 13 ft. lbs. (18 Nm).

20. Connect the lower radiator hose to the radiator.

21. If equipped with automatic transmission, connect the oil cooler hoses to the radiator.

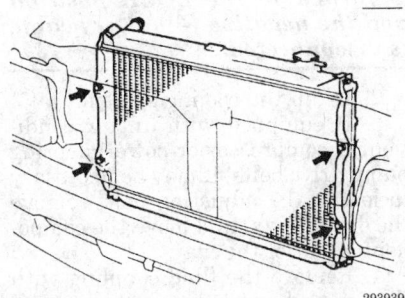

Common location of the radiator bolts on Toyota trucks

22. Install the water pump pulley, fan shroud, fan (with fan clutch), and alternator drive belt as follows:

a. Place the fan (with the fan clutch), water pump pulley and fan shroud in position.

b. Install the water pump pulley mounting nuts but do not torque the nuts at this time.

c. Install the No. 1 fan shroud with the four bolts.

d. Install the No. 2 fan shroud with the two clips.

e. Install the alternator drive belt to the engine.

f. Stretch the alternator belt tight and torque the fan nuts to 16 ft. lbs. (21 Nm).

g. Adjust the drive belt for the alternator.

23. If equipped with air conditioning, install and adjust the air conditioning compressor drive belt.

24. Install and adjust the power steering drive belt.

25. Connect the radiator reservoir hose to the radiator.

26. Connect the upper radiator hose to the radiator.

27. Install the radiator grille to the vehicle with the. 11 clips and four screws.

28. Install the clearance lights to the grille with the four bolts to each light.

29. Install the battery and battery clamp.

30. Connect the negative and positive battery cables.

31. Fill the engine and radiator with coolant.

32. Start the engine and check for leaks.

3.4L (5VZ-FE) Engine

1. Drain the engine coolant from the radiator and engine.

2. Remove the clearance lights from the grille by removing the four screws at each light.

3. Remove the radiator grille from the vehicle by removing the 4 screws and 11 clips.

4. Disconnect the upper radiator hose from the radiator.

5. Disconnect the radiator reservoir hose from the radiator.

6. Remove the No. 2 fan shroud.

7. If the vehicle is equipped with a automatic transmission, disconnect the oil cooler hoses from the radiator.

8. Disconnect the lower radiator hose from the radiator.

9. Remove the four bolts to the radiator.

10. Remove the radiator from the vehicle.

11. Remove the No. 1 fan shroud from the radiator by removing the four bolts.

12. Remove the two radiator supports by removing the eight screws.

To install:

13. Connect the supports to the radiator and install the eight screws.

14. Connect the No. 1 fan shroud to the radiator and install the four bolts.

15. Install the radiator to the vehicle with the tabs on the supports through the radiator service holes. Install the four bolts and torque the bolts to 8.7 ft. lbs. (12 Nm).

16. Connect the No. 2 fan shroud to the radiator.

17. Connect the lower radiator hose to the radiator.

18. If equipped with automatic transmission, connect the oil cooler hoses to the radiator.

19. Connect the radiator reservoir hose to the radiator.

20. Connect the upper radiator hose to the radiator.

21. Install the radiator grille to the vehicle with the 11 clips and four screws.

22. Install the clearance lights to the grille with the four bolts to each light.

23. Fill the engine and radiator with coolant.

24. Start the engine and check for leaks.

Tacoma

1. Drain the engine coolant from the radiator and engine.

2. Remove the clearance lights from the grille by removing the four screws and two clips.

3. Remove the two fillers.

4. Remove the radiator grille from the vehicle by removing the 11 clips.

5. For the California vehicles with 3RZ-FE engine, remove the two bolts and disconnect the air pipe.

6. Disconnect the upper radiator hose from the radiator.

7. Disconnect the radiator reservoir hose from the radiator.

8. Remove the No. 2 fan shroud.

9. If the vehicle is equipped with an automatic transmission, disconnect the oil cooler hoses from the radiator.

10. Disconnect the lower radiator hose from the radiator.

11. Remove the four bolts to the radiator.

12. Remove the radiator from the vehicle.

13. Remove the four bolts and the fan shroud.

14. Remove the four screws and remove the two radiator supports.

To install:

15. Install the two radiator supports to the radiator with the four screws.

16. Install the fan shroud with the four bolts.

17. Install the radiator to the vehicle with the tabs on the supports through the radiator service holes. Install the four bolts and torque the bolts to 9 ft. lbs. (12.5 Nm).

18. Connect the lower radiator hose to the radiator.

19. If equipped with automatic transmission, connect the oil cooler hoses to the radiator.

20. Install the No. 2 fan shroud.

21. Connect the radiator reservoir hose to the radiator.

22. Connect the upper radiator hose to the radiator.

23. If removed, connect the air pipe with the two bolts.

24. Install the radiator grille to the vehicle with the 11 clips.

25. Install the two fillers.

26. Install the clearance lights to the grille with the four bolts and two clips.

27. Connect the negative and positive battery cables.

28. Fill the engine and radiator with coolant.

29. Start the engine and check for leaks.

Water Pump

REMOVAL AND INSTALLATION

2.4L (22R-E) Engine

Pick-up and 4Runner

1. Disconnect the negative battery cable.

―――――― **CAUTION** ――――――

On 1995 and later models, work must be started after 90 seconds from the time the ignition switch is turned to the LOCK position and the negative (-) battery cable is disconnected.

2. Drain the cooling system.

3. If equipped with an air conditioning compressor or power steering pump drive belts, it may be necessary to loosen the adjusting bolt, remove the drive belt(s) and move the component(s) out of the way.

4. Remove the fluid coupling with the fan and water pump pulley.

5. Remove the water pump.

6. Clean the gasket mounting surfaces.

To install:

7. Install the replacement water pump using a new gasket.

8. Install the water pump pulley and fluid coupling with the fan.

9. Install the removed engine drive belts.

10. Fill the cooling system.

11. Connect the negative battery cable. Start the engine and check for leaks.

12. Bleed the cooling system.

3.0L (3VZ-E) and 3.4L (5VZ-FE) Engines

Pick-up and 4Runner

1. Disconnect the negative battery cable.

―――――― **CAUTION** ――――――

On 1995 and later models, work must be started after 90 seconds from the time the ignition switch is turned to the LOCK position and the negative (-) battery cable is disconnected.

2. Drain the cooling system.

3. Remove the timing belt.

4. Remove the thermostat.

5. Disconnect the No. 2 oil cooler hose from the water pump.

6. Remove the water pump by removing the bolts.

7. Thoroughly clean the mating surfaces.

To install:

8. Apply sealant (PN 08826-00100 or equivalent) to the water pump. Parts must be assembled within five minutes of application. Otherwise the material must be removed and reapplied.

9. Install the water pump and torque the bolts to 14 ft. lbs. (20 Nm).

10. Connect the No. 2 oil cooler hose.

11. Install the thermostat.

12. Install the timing belt.

13. Connect the negative battery cable.

14. Fill the cooling system.

15. Start the engine and check for leaks.

2.4L (2RZ-FE) and 2.7L (3RZ-FE) Engines

T-100, Tacoma and 4Runner

1. Disconnect the negative battery cable from the battery.

2. Remove the engine undercover.

3. Drain the cooling system.

4. For the California vehicles with 3RZ-FE engine, remove the two bolts and disconnect the air pipe.

5. Disconnect the upper radiator hose from the radiator.

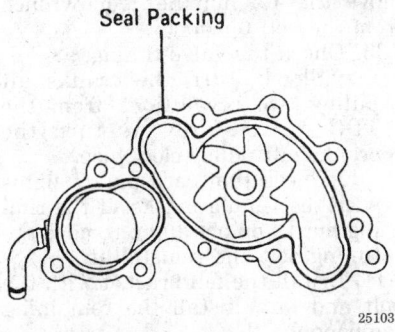

Seal Packing

251033

Water pump with seal packing — 1995 Pick-up and 4Runner with 3.0L (3VZ-E) engine

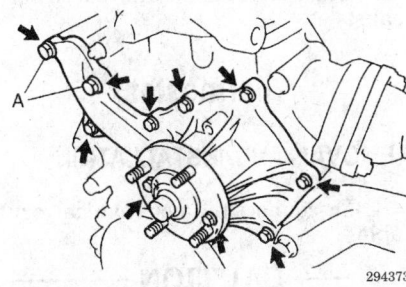

A

294373

Water pump mounting bolt identification — 2.7L (3RZ-FE) engine

6. Remove the oil dipstick guide by removing the bolt.

7. If equipped with power steering, remove the power steering drive belt by loosening the lock bolt and adjusting bolt to the idler pulley.

8. Remove the No. 2 fan shroud by removing the two clips.

9. Remove the No. 1 fan shroud by removing the four bolts.

10. If equipped with air conditioning, loosen the idler pulley nut and adjusting bolt. Remove the air conditioning drive belt from the engine.

11. Remove the alternator drive belt, fan (with fan clutch), water pump pulley, and the fan shroud as follows:

a. Stretch the belt and loosen the water pump pulley mounting nuts.

b. Loosen the lock, pivot, and the adjusting bolts for the alternator and remove the alternator drive belt from the engine.

c. Remove the four water pump pulley mounting nuts.

d. Remove the fan (with fan clutch) and the water pump pulley.

12. Remove the 10 bolts and remove the water pump and gasket from the engine.

To install:

13. Clean all surfaces and apply a thin layer of liquid sealant to a new gasket.

14. Place the gasket and water pump into position. Torque the 14 mm head bolts **A** to 18 ft. lbs. (25 Nm) and the 12 mm head bolts to 78 inch lbs. (9 Nm).

15. Install the water pump pulley, fan shroud, fan (with fan clutch), and the alternator drive belt as follows:

a. Place the fan (with the fan clutch), water pump pulley, and the fan shroud in position.

b. Install the water pump pulley mounting nuts but do not torque the nuts at this time.

c. Install the alternator drive belt to the engine.

d. Stretch the alternator belt tight and torque the fan nuts to 16 ft. lbs. (21 Nm).

e. Adjust the drive belt for the alternator.

16. If equipped with air conditioning, install and adjust the drive belt.

17. Install the No. 1 fan shroud by installing the four bolts.

18. Install the No. 2 fan shroud with the two clips.

19. Install and adjust the power steering drive belt.

20. Install the oil dipstick guide with the bolt.

21. Connect the upper radiator hose to the radiator.

22. If removed, connect the air pipe with the two bolts.

23. Fill and bleed the cooling system.

24. Connect the negative battery cable to the battery.

25. Start the engine and check for leaks.

26. Install the engine undercover.

3.4L (5VZ-FE) Engine

T-100 and Tacoma

1. Disconnect the negative battery cable.

CAUTION

Work must be started after 90 seconds from the time the ignition switch is turned to the LOCK position and the negative (-) battery cable is disconnected.

---

2. Raise and safely support the vehicle.

3. Remove the engine undercover.

4. Drain the engine coolant.

5. Disconnect the upper radiator hose from the engine.

6. Remove the power steering drive belt as follows:

a. Stretch the belt and loosen the fan pulley mounting nuts.

b. Loosen the lock bolt, pivot bolt, and the adjusting bolt and remove the drive belt from the engine.

7. Remove the air conditioning drive belt by loosening the idler pulley nut and adjusting bolt.

8. Loosen the lock bolt, pivot bolt, and the adjusting bolt. Remove the alternator drive belt.

9. Remove the No. 2 fan shroud by removing the two clips.

10. Remove the fan with the fluid coupling and fan pulleys.

11. Disconnect the power steering pump from the engine and set aside. Do not disconnect the lines from the pump.

12. If equipped with air conditioning, disconnect the compressor from the engine and set aside. Do not disconnect the lines from the compressor.

13. If equipped with air conditioning, disconnect the air conditioning bracket.

14. Remove the No. 2 timing belt cover as follows:

a. Disconnect the camshaft position sensor connector from the No. 2 timing belt cover.

b. Disconnect the three spark plug wire clamps from the No. 2 timing belt cover.

c. Remove the six bolts and remove the timing belt cover.

15. Remove the fan bracket as follows:

a. Remove the power steering adjusting strut by removing the nut.

b. Remove the fan bracket by removing the bolt and nut.

16. Set the No. 1 cylinder at TDC of the compression stroke.

a. Turn the crankshaft pulley and align its groove with the timing mark **O** of the No. 1 timing belt cover.

b. Check that the timing marks of the camshaft timing pulleys and the No. 3 timing belt cover are aligned. If not, turn the crankshaft pulley one revolution (360°).

NOTE: If re-using the timing belt, make sure that you can still read the installation marks. If not, place new installation marks on the timing belt to match the timing marks of the camshaft timing pulleys.

17. Remove the timing belt tensioner by alternately loosening the two bolts.

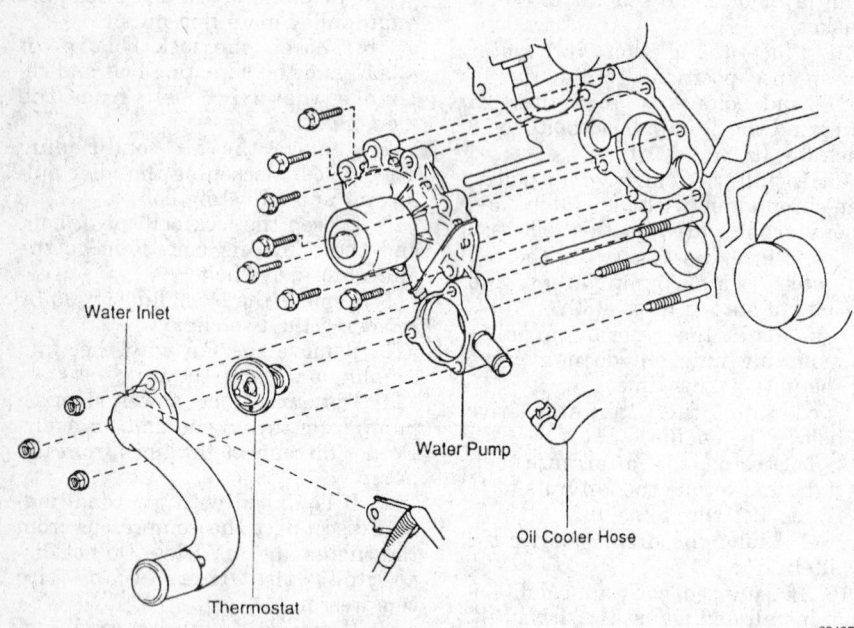

Water pump assembly and related components — 3.4L (5VZ-FE) engine

294375

18. Remove the camshaft timing pulleys.

a. Using SST 09960–10010 or equivalent, remove the pulley bolt, the timing pulley, and the knock pin. Remove the two timing pulleys with the timing belt.

19. Remove the thermostat.

20. Disconnect the No. 2 oil cooler hose from the water pump.

21. Remove the water pump by removing the seven bolts.

22. Thoroughly clean the mating surfaces.

To install:

23. Apply sealant (PN 08826–00100 or equivalent) to the water pump. Parts must be assembled within five minutes of application. Otherwise the material must be removed and reapplied.

24. Install the water pump. Torque the bolts to 14 ft. lbs. (20 Nm).

25. Connect the No. 2 oil cooler hose.

26. Install the thermostat.

27. Install the left camshaft timing pulley. Torque the pulley bolt to 81 ft. lbs. (110 Nm).

28. Set the No. 1 cylinder to TDC of the compression stroke.

29. Connect the timing belt to the left camshaft timing pulley. Check that the installation mark on the tim-ing belt is aligned with the end of the No. 1 timing belt cover.

a. Using SST 09960–01000 or equivalent, slightly turn the left camshaft timing pulley clockwise. Align the installation mark on the timing belt with the timing mark of the camshaft timing pulley, and hang the timing belt on the left camshaft timing pulley.

b. Align the timing marks of the left camshaft pulley and the No. 3 timing belt cover.

c. Check that the timing belt has tension between the crankshaft timing pulley and the left camshaft timing pulley.

30. Install the right camshaft tim-ing pulley and the timing belt.

31. Set the timing belt tensioner as follows:

a. Using a press, slowly press in the pushrod using 220–2,205 lbs. (981–9,807 N) of force.

b. Align the holes of the pushrod and housing, pass a 1.5 mm hexa-gon wrench through the holes to keep the setting position of the pushrod.

c. Release the press and install the dust boot to the tensioner.

32. Install the timing belt tensioner and alternately tighten the bolts to 20 ft. lbs. (28 Nm). Using pliers, re-move the 1.5 mm hexagon wrench from the belt tensioner.

33. Check the valve timing.

a. Slowly turn the crankshaft pulley two revolutions from the TDC to TDC. Always turn the crankshaft pulley clockwise.

b. Check that each pulley aligns with the timing marks. If the tim-ing marks do not align, remove the timing belt and reinstall it.

34. Install the fan bracket with the bolt and nut. Install the remaining components.

35. Fill with engine coolant.

36. Connect the negative battery cable.

37. Start the engine and check for leaks.

Thermostat

REMOVAL AND INSTALLATION

1. Disconnect the negative battery cable.

---- **CAUTION** ----

On 1995 vehicles, work must be started after 90 seconds from the time the ignition switch is turned to the LOCK position and the neg-ative (-) battery cable is disconnected.

2. Drain the cooling system.

3. Disconnect the radiator outlet hose.

4. Remove the water inlet fasten-ers, thermostat, and the gasket.

To install:

5. Place a new gasket to the ther-mostat. Install the thermostat with the jiggle valve upward.

6. Install the water inlet and tor-que the fasteners to 14 ft. lbs. (20 Nm).

7. Connect the radiator hose.

8. Fill and the cooling system.

9. Connect the negative battery cable.

10. Start the engine, bleed the cool-ing system, and check for leaks.

Cooling Fan

REMOVAL AND INSTALLATION

1. Disconnect the negative battery cable from the battery.

2. Remove the fan shroud.

3. Stretch the drive belt and re-move the water pump pulley mount-ing nuts.

4. Remove the cooling fan assem-bly from the engine.

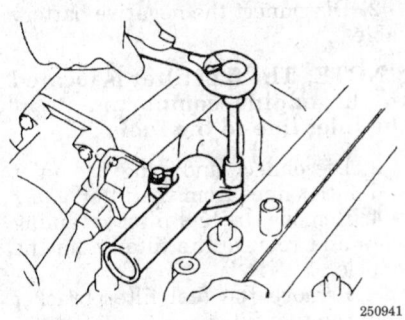

Removing the thermostat housing — 2.4L (22R-E) engine

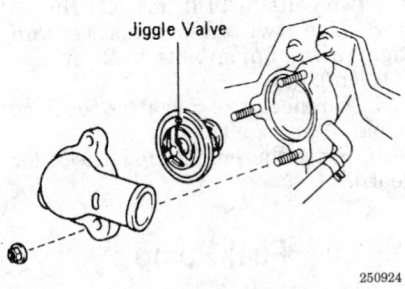

Exploded view of the thermostat and housing — 3.0L (3VZ-E) engine

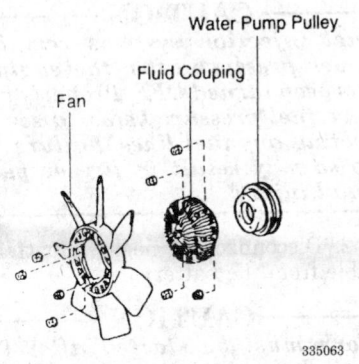

Common cooling fan component assembly

5. Disconnect the cooling fan from the fan clutch by removing the four nuts.

To install:

6. Connect the fan clutch to the cooling fan by installing the four nuts.

7. Install the cooling fan assembly to the engine.

8. Install the nuts to the water pump pulley and torque the nuts to:
• 5VZ-FE engine: 6 ft. lbs. (8 Nm)
• 2RZ-FE and 3RZ-FE engine: 16 ft. lbs. (21 Nm)
• T-100 with 5VZ-FE engine: 4 ft. lbs. (5 Nm)

9. Install the drive belt to the water pump pulley.

10. Install the fan shroud.

11. Connect the negative battery cable to the battery.

Cooling System Bleeding

After working on the cooling system, the system must be bled. Air trapped in the system will prevent proper filling and leave the radiator coolant level low, causing a risk of overheating.

To bleed the system, start with the system cool, the radiator (pressure) cap off and the cooling system filled to about an inch below the filler neck.

1. Start the engine and run it at slightly above normal idle speed. This will insure adequate circulation. If air bubbles appear and the coolant level drops, fill the system with an antifreeze/water mixture to bring the level back to the proper level.

2. Run the engine this way until the thermostat opens. When this happens, coolant will move abruptly across the top of the radiator and the temperature of the radiator will suddenly rise.

3. At this point, air is often expelled and the level may drop quite a bit. Keep refilling the system until full and install the pressure cap. Fill the coolant reservoir to the full mark.

4. Run the engine until it reaches normal operating temperature. Turn the engine off and allow it to cool nearly completely. After the engine has cooled, fill the reservoir to the full mark if needed.

FUEL SYSTEM

Fuel System Service Precautions

Safety is the most important factor when performing not only fuel system maintenance but any type of maintenance. Failure to conduct maintenance and repairs in a safe manner may result in serious personal injury or death. Maintenance

and testing of the vehicle's fuel system components can be accomplished safely and effectively by adhering to the following rules and guidelines.

• To avoid the possibility of fire and personal injury, always disconnect the negative battery cable unless the repair or test procedure requires that battery voltage be applied.

• Always relieve the fuel system pressure prior to disconnecting any fuel system component (injector, fuel rail, pressure regulator, etc.), fitting or fuel line connection. Exercise extreme caution whenever relieving fuel system pressure to avoid exposing skin, face and eyes to fuel spray. Please be advised that fuel under pressure may penetrate the skin or any part of the body that it contacts.

• Always place a shop towel or cloth around the fitting or connection prior to loosening to absorb any excess fuel due to spillage. Ensure that all fuel spillage (should it occur) is quickly removed from engine surfaces. Ensure that all fuel soaked cloths or towels are deposited into a suitable waste container.

• Always keep a dry chemical (Class B) fire extinguisher near the work area.

• Do not allow fuel spray or fuel vapors to come into contact with a spark or open flame.

• Always use a backup wrench when loosening and tightening fuel line connection fittings. This will prevent unnecessary stress and torsion to fuel line piping. Always follow the proper torque specifications.

• Always replace worn fuel fitting O-rings with new. Do not substitute fuel hose or equivalent, where fuel pipe is installed.

Fuel System Pressure

RELIEVING

— **CAUTION** —
Fuel injection systems remain under pressure after the engine has been turned OFF. Properly relieve fuel pressure before disconnecting any fuel lines. Failure to do so may result in fire or personal injury.

1. Disconnect the negative battery terminal.

2. Place a catch-pan under the joint to be disconnected. A large quantity of fuel may be released when the joint is opened.

3. Wear eye or full face protection.

4. Place a shop towel over the area and slowly loosen the joint using a wrench of the correct size. Use a backup wrench if needed.

5. Allow the fuel left in the line to bleed off slowly before fully disconnecting the joint.

6. Plug the opened lines immediately to prevent fuel spillage or the entry of dirt.

7. Dispose of the released fuel properly.

8. After connecting fuel lines, connect the negative battery cable and start the engine.

9. Check for leaks and repair as needed.

Idle Speed

ADJUSTMENT

2.4L (22R-E) and 3.0L (3VZ-E) Engines

1. Warm up the engine until it reaches normal operating temperature.

2. The air cleaner should be in place and all wires and vacuum hoses connected. All accessories should be **OFF**, the transmission in neutral and the drive wheels blocked.

3. Connect a tachometer to the battery and the IG — terminal of the DLC1.

4. Run the engine at 2,500 rpm for two minutes. This insures that the oxygen sensor and other sensors are fully up to temperature and stabilized.

5. Let the engine return to idle.

6. Idle speed should be:
- 22R-E engine: 2WD: 750 rpm
- 22R-E engine 4WD: 850 rpm
- 3VZ-E engine: 800 rpm

7. If the idle is not at specification, set the idle speed by turning the idle speed adjusting screws.

8. Disconnect the tachometer.

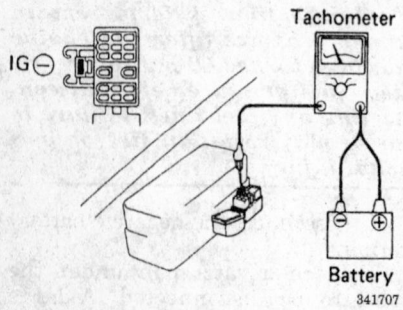

IG ⊖

Tachometer

Battery
341707

Connection for the tachometer — 2.4L (22R-E) and 3.0L (3VZ-E) engines

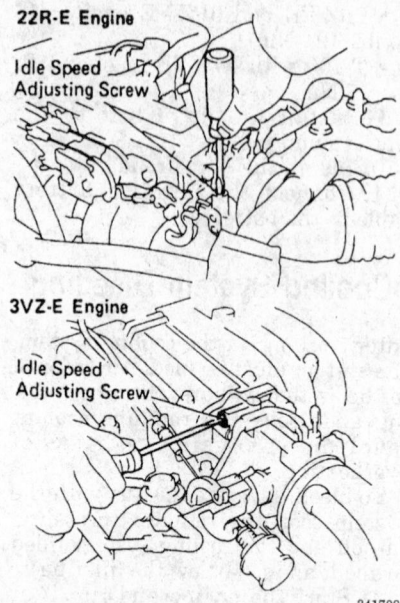

22R-E Engine

Idle Speed Adjusting Screw

3VZ-E Engine

Idle Speed Adjusting Screw

341708

Idle speed adjusting screw — 2.4L (22R-E) and 3.0L (3VZ-E) engines

2.4L (2RZ-FE), 2.7L (3RZ-FE) and 3.4L (5VZ-FE) Engines

Idle speed is controlled by the ECM and is not adjustable.

Mixture

ADJUSTMENT

Fuel mixture on all engines is controlled by the ECM and is not adjustable.

Fuel Filter

REMOVAL AND INSTALLATION

1. Relieve the fuel system pressure.

——— **CAUTION** ———
Fuel injection systems remain under pressure after the engine has been turned OFF. Properly relieve fuel pressure before disconnecting any fuel lines. Failure to do so may result in fire or personal injury.

2. Disconnect the negative battery cable.

NOTE: The fuel filter is located in the engine compartment, at the inlet line to the fuel rail.

3. Disconnect and plug the inlet and outlet lines from the filter.

4. Remove the fuel filter retaining bolts and remove the filter from the vehicle.

5. Remove the fuel filter bracket from the fuel filter.

To install:

6. Install the fuel filter bracket to the fuel filter.

7. Install the fuel filter and torque the two bolts to 14 ft. lbs. (20 Nm).

8. Use two new gaskets and tighten the union bolts to 22 ft. lbs. (30 Nm).

9. Connect the negative battery cable.

10. Start the engine and check for leaks.

Fuel Pump

REMOVAL AND INSTALLATION

1. Relieve the fuel pressure.

——— **CAUTION** ———
Fuel injection systems remain under pressure after the engine has been turned OFF. Properly relieve fuel pressure before disconnecting any fuel lines. Failure to do so may result in fire or personal injury.

2. Disconnect the negative battery cable from the battery.

——— **CAUTION** ———
Work must be started after 90 seconds from the time the ignition switch is turned to the LOCK position and the negative (-) battery cable is disconnected.

3. Drain the fuel from the fuel tank.

——— **CAUTION** ———
Do not allow fuel spray or fuel vapors to come in contact with a spark or open flame. Keep a dry chemical fire extinguisher nearby. Never store fuel in an open container due to risk of fire or explosion.

4. Remove the fuel tank from the vehicle.

5. Disconnect the fuel pump connector from the clamp.

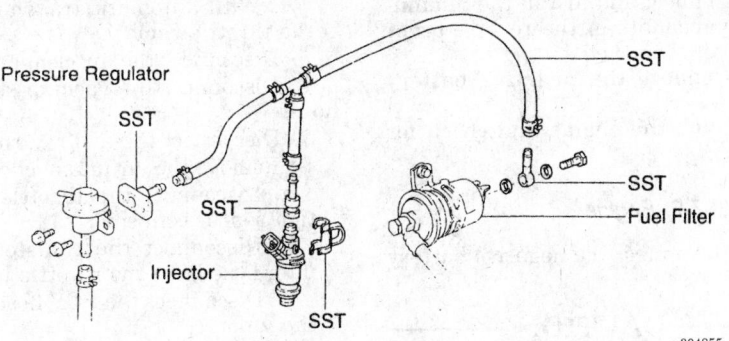

Fuel delivery components — 2.4L (2RZ-FE) and 2.7L (3RZ-FE) engines

304855

Reference (2WD)

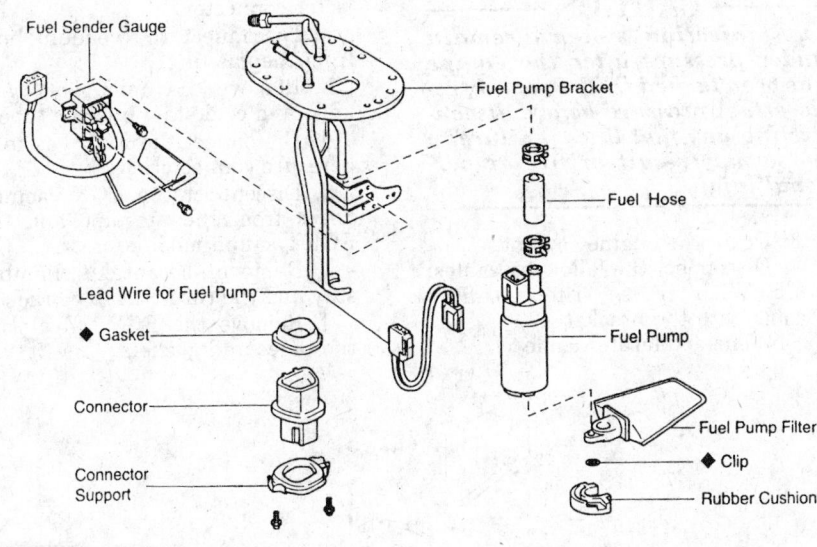

◆ Non-reusable part

291800

Fuel pump assembly and related components — Tacoma shown

6. Remove the access plate bolts, then pull out the fuel pump assembly from the fuel tank.

7. Remove the gasket(s) from the pump bracket.

8. Disconnect the fuel pump connector.

9. Pull the bracket from the lower side of the fuel pump and remove the fuel pump from the fuel hose.

10. Remove the rubber cushion, the clip, and the fuel filter from the bottom of the fuel pump.

To install:

11. Install the fuel pump filter to the fuel pump with a new clip.

12. Install the fuel pump to the fuel pump bracket.

13. Connect the fuel hose to the outlet port of the fuel pump.

14. Connect the fuel pump connector.

15. Install the fuel pump assembly with a new gasket(s). Torque the bolts to 31 inch lbs. (4 Nm).

16. Connect the fuel pump connector to the clamp.

17. Install the fuel tank and connect all electrical and fuel connections.

18. Connect the negative battery cable.

19. Refill the fuel tank and check for leaks.

Fuel Injector

REMOVAL AND INSTALLATION

2.4L (22R-E) Engine

1. Disconnect the negative battery cable.

2. Relieve the fuel pressure.

———— **CAUTION** ————

Fuel injection systems remain under pressure after the engine has been turned OFF. Properly relieve fuel pressure before disconnecting any fuel lines. Failure to do so may result in fire or personal injury.

3. Disconnect the ground strap from the rear side of the engine.

4. Disconnect the accelerator cable.

5. If equipped with an automatic transmission, disconnect the throttle cable from the bracket and the clamp.

6. Disconnect the No. 1 and No. 2 PCV hoses.

7. Disconnect the following items:

a. Brake power booster hose

b. Air control valve hoses

c. Vacuum Switching Valve (VSV)

d. Evaporative emission control hose

e. EGR vacuum hose and modulator

f. Pressure regulator hose for 2-Wheel Drive

g. Fuel pressure-up (VSV) and hose

h. No. 1 and No. 2 air valve hose from the throttle body

i. No. 2 and No. 3 water bypass hoses from the throttle body

j. Cold start injector wire

k. Throttle position wire

8. Remove the following items:

a. Bolt holding the cold start injector to the plenum chamber

b. Bolts holding the No. 1 EGR pipe to the plenum chamber

c. Bolts holding the manifold stay to the plenum chamber

d. Fuel hose clamp, four bolts, two nuts, and the bond strap

e. The plenum chamber with the throttle body and gaskets

9. Disconnect the fuel return hose.

10. Disconnect the following wires:

a. Knock sensor wire

b. Oil pressure sender gauge/switch

c. Starter wire (terminal 50)

d. Transmission wires

e. Air conditioning compressor wires

f. Injector wires

g. Water temperature sender gauge wire

h. Overdrive temperature switch wire (air conditioning)

i. Igniter wire

j. Vacuum Switching Valve (VSV) wire (air conditioning)

k. Cold start injector time switch wire

l. Water temperature sensor wire

11. Disconnect the fuel inlet hose from the delivery pipe with the pulsation damper and gaskets.

12. Remove the injectors from the engine. Take care in handling the injectors.

To install:

13. Lubricate the O-rings with clean gasoline. Install the injectors using new O-rings. Check that each injector rotates smoothly in its bore. Install the fuel rail with the injectors. Torque the fuel rail hold-down bolts to 13 ft. lbs. (17 Nm).

14. Connect the fuel hose with the pulsation damper to the delivery pipe. Torque the bolt to 33 ft. lbs. (44 Nm).

15. Connect and install the remaining components in the reverse order they were removed.

16. Connect the negative battery cable.

17. Start the engine and check for leaks.

3.0L (3VZ-E) Engine

1. Disconnect the negative battery cable.

――――― **CAUTION** ―――――
Wait at least 90 seconds after the negative (-) battery cable is disconnected to prevent possible deployment of the air bag.

2. Relieve the fuel pressure.

――――― **CAUTION** ―――――
Fuel injection systems remain under pressure after the engine has been turned OFF. Properly relieve fuel pressure before disconnecting any fuel lines. Failure to do so may result in fire or personal injury.

3. Drain the engine coolant.

4. Disconnect the following cables:

a. With cruise, the actuator cable with the bracket

b. The accelerator cable

c. With automatic transmission, the throttle cable

5. Disconnect the air cleaner hose.

6. Disconnect the vacuum sensing hose.

7. Disconnect the fuel return hose.

8. Remove the air intake chamber.

a. Disconnect the throttle position sensor connector

b. Disconnect the canister vacuum hose from the throttle body

c. Disconnect the PCV hose from the union.

d. Disconnect the No. 4 water bypass hose from the union of the intake manifold.

e. Disconnect the No. 5 water bypass hose from the water bypass pipe.

f. Disconnect the cold start injector connector.

g. Disconnect the vacuum hose from the gas filter.

h. Remove the union bolt, gaskets, and cold start injector tube.

i. Disconnect the EGR gas temperature connector.

j. Disconnect the EGR vacuum hoses from the air pipe and the EGR vacuum modulator.

k. Remove the intake chamber stay and the throttle cable bracket.

l. Remove the EGR valve with the pipes and gaskets.

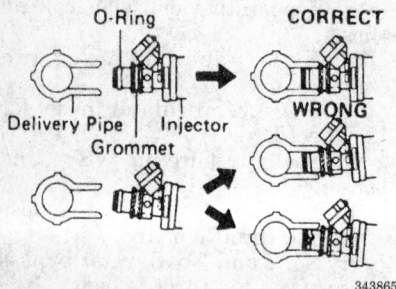

Installing the fuel injectors to the delivery pipe — 2.4L (22R-E) and 3.0L (3VZ-E) engines

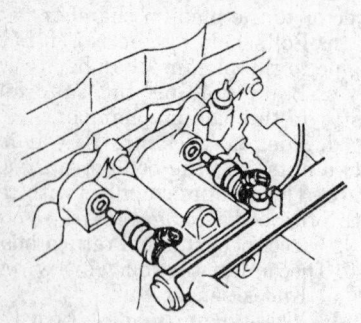

Installing the fuel delivery pipe and injectors — 2.4L (22R-E) engine

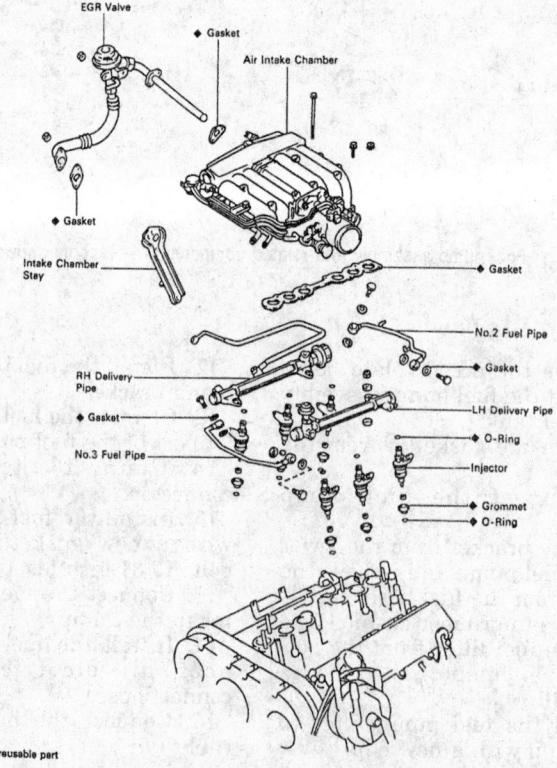

Fuel injectors and related components — 3.0L (3VZ-E) engine

m. Disconnect the No. 1 air hose from the PAIR reed valve.

n. Disconnect the vacuum hoses from the air pipes.

o. Remove the accelerator cable bracket.

p. Remove the intake chamber and gasket.

9. Disconnect the following connectors:

a. Knock sensor

b. Cold start injector time switch connector

c. Engine coolant temperature sensor connector

d. Engine coolant temperature sender gauge connector

e. Right ground strap from the No. 3 camshaft bearing cap

f. Injector connectors

g. Engine wire

10. Remove the No. 2 and the No. 3 fuel pipes.

11. Remove the delivery pipes and the injectors.

12. Remove the insulators, spacers and the O-rings from the cylinder head.

13. Remove the injectors from the delivery pipe.

To install:

14. Install a new grommet and O-ring to the injector.

15. While turning the injector clockwise and counterclockwise, push the injector into the delivery pipe. Make sure that the injectors rotate smoothly.

16. Install the injectors and the fuel rail. Tighten the nuts holding the delivery pipes to the intake manifold to 9 ft. lbs. (13 Nm).

17. Install the No. 2 and No. 3 fuel pipes and torque the bolts to 25 ft. lbs. (34 Nm).

18. Install the connectors removed prior.

19. Install the air intake chamber and torque the bolts to 13 ft. lbs. (18 Nm).

20. Connect the fuel return hose.

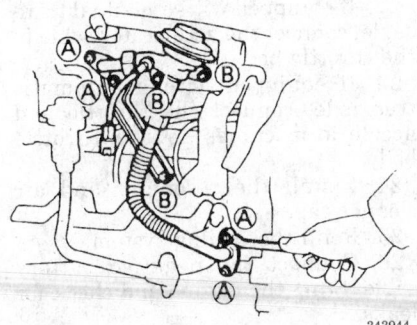

343944

Air intake chamber stay bolt tightening — 3.0L (3VZ-E) engine

21. Connect the vacuum sensing hose.

22. Connect the air cleaner hose.

23. Connect the following cables:

a. With automatic transmission: throttle cable

b. Accelerator cable

c. With cruise: actuator cable with the bracket

24. Fill the engine coolant.

25. Connect the negative battery cable.

26. Start the engine and check for fuel leaks.

2.7L (3RZ-FE) Engine

4Runner

1. Disconnect the negative battery cable.

---- **CAUTION** ----

Wait 90 seconds from the time the key is turned to LOCK and the negative battery cable is disconnected to begin work. This allows the SRS capacitor to discharge and prevent deployment of the air bag(s).

2. Relieve the fuel system pressure.

---- **CAUTION** ----

Fuel injection systems remain under pressure after the engine has been turned OFF. Properly relieve fuel pressure before disconnecting any fuel lines. Failure to do so may result in fire or personal injury.

3. Remove the throttle body.

4. Disconnect the following connectors:

• Four injector connectors

• Crankshaft position sensor connector

• Knock sensor connector

5. Disconnect the DLC1 and the wire clamp from the brackets.

6. Disconnect the engine wire.

7. Remove the delivery pipe and the injectors.

a. Disconnect the vacuum sensing hose from the fuel pressure regulator.

b. Disconnect the fuel return hose from the fuel pressure regulator.

c. Remove the union bolt and the two gaskets and disconnect the fuel inlet pipe from the delivery pipe.

d. Remove the two bolts and the delivery pipe together with the four injectors.

e. Remove the insulators from the spacers.

f. Pull out the four injectors from the delivery pipe.

g. Remove the O-ring and grommet from each injector.

To install:

8. Install the injectors to the delivery pipe.

a. Install a new grommet to the injector.

b. Apply a light coat of gasoline to a new O-ring and install it to the injector.

c. While turning the injector left and right, install it to the delivery pipe. Install the four injectors.

d. Position the injector connector upward.

9. Install the injectors and delivery pipe.

a. Place the four new insulators in position on the spacers.

b. Place the injectors together with the delivery pipe in the position on the cylinder head.

c. Temporarily install the two bolts holding the delivery pipe to the cylinder head.

d. Check that the injectors rotate smoothly.

e. Tighten the bolts holding the delivery pipe to the cylinder head. Torque to 15 ft. lbs. (21 Nm).

f. Connect the fuel inlet pipe to the delivery pipe with two new gaskets and the union bolt. Torque to 22 ft. lbs. (29 Nm).

g. Connect the fuel return hose to the fuel pressure regulator.

h. Connect the vacuum sensing hose to the fuel pressure regulator.

10. Connect the engine wire.

11. Connect the following connectors:

• Four injector connectors

• Crankshaft position sensor connector

• Knock sensor connector

12. Connect the DLC1 and the wire clamp from the brackets.

13. Install the throttle body.

14. Connect the negative battery cable.

15. Start the vehicle and check for leaks.

Tacoma

1. Disconnect the negative battery cable.

2. Drain the engine coolant.

3. Remove the air cleaner cap and resonator.

4. If equipped with a manual transaxle, disconnect the accelerator cable from the throttle body.

5. If equipped with a automatic transaxle, disconnect the accelerator and throttle cables from the throttle body.

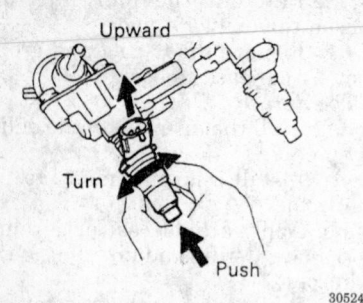

Installing the injector to the delivery pipe — 1996-97 4Runner with 2.7L (3RZ-FE) engine

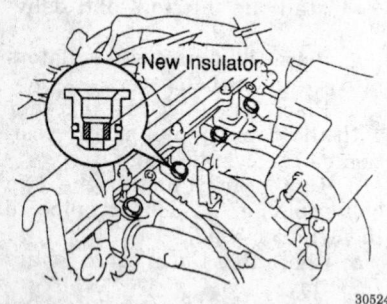

Insulators on the cylinder head — 1996-97 4Runner with 2.7L (3RZ-FE) engine

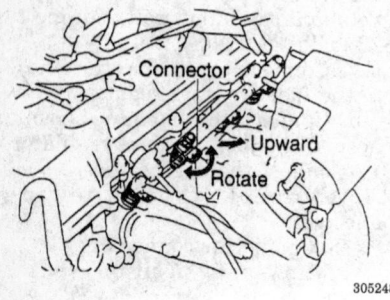

Install the fuel injector with the connector facing upward — 1996-97 4Runner with 2.7L (3RZ-FE) engine

6. Remove the intake air connector.

7. If equipped with air conditioning, remove the air conditioning idle-up valve.

8. Remove the PCV hose.

9. Remove the throttle body as follows:

a. Disconnect the throttle position sensor connector.

b. Disconnect the idle air control valve connector.

c. Disconnect the EVAP hose.

d. Disconnect the three vacuum hoses.

e. If equipped with power steering, remove the air hose for the power steering idle up.

f. Remove the two bolts, two nuts and disconnect the throttle body from the air intake chamber.

g. Remove the throttle body gasket.

h. Disconnect the No. 1 and No. 2 water bypass hoses from the throttle body and remove the throttle body.

i. Remove the air hose from the IAC valve.

10. Remove the VSV for the fuel pressure control as follows:

a. Disconnect the vacuum sensing hoses from the fuel filter and fuel pressure regulator.

b. Disconnect the VSV connector.

c. Remove the bolt, two ground straps and the VSV for the fuel pressure control.

11. Disconnect the engine wire as follows:

a. Disconnect the fuel injectors.

b. Crankshaft position sensor connector.

c. Knock sensor connector.

d. Disconnect the DLC1 from the bracket.

12. Remove the delivery pipe and injectors as follows:

a. Disconnect the vacuum hose from the fuel pressure regulator.

b. Remove the union bolt and two gaskets and disconnect the fuel inlet pipe from the delivery pipe.

c. Remove the two bolts and delivery pipe together with the four injectors.

d. Remove the four insulators from the four spacers.

e. Pull out the four injectors from the delivery pipe.

f. Remove the O-ring and grommet from each injector.

NOTE: Be careful not to drop the injectors when removing the delivery pipe.

To install:

13. Install the injectors to the delivery pipe as follows:

a. Install a new grommet to the injector.

b. Apply a light coat of gasoline to a new O-ring and install it to the injector.

c. While turning the injector left and right, install the injector to the delivery pipe. Install all four injectors.

d. Position the injector connector upward.

14. Install the injectors and delivery pipe. Tighten the bolts holding the delivery pipe to 15 ft. lbs. (21 Nm). Check that the injectors rotate smoothly.

15. Install the fuel inlet pipe to the delivery pipe with two new gaskets. Torque the union bolts to 22 ft. lbs. (29 Nm).

16. Connect the fuel return hose to the fuel pressure regulator.

17. Install the engine wire as follows:

a. Connect the DLC1 to the bracket.

b. Connect the knock sensor connector

c. Connect the crankshaft position sensor connector.

d. Connect the injector connectors.

18. Install the VSV for the fuel pressure control as follows:

a. Install the VSV, two ground straps and bolt.

b. Connect the VSV connector.

c. Connect the vacuum sensing hoses to the fuel filter and fuel pressure regulator.

19. Install the throttle body as follows:

a. Connect the air hose to the IAC valve.

b. Install the throttle body to the engine and connect the No. 1 and No. 2 water bypass hoses to the throttle body.

c. Connect the throttle body to the air intake chamber with a new gasket. Install the two bolts and two nuts. Torque to 14 ft. lbs. (20 Nm).

d. Connect the three vacuum hoses.

e. Connect the EVAP hose.

f. Connect the IAC valve connector.

g. Connect the throttle position sensor connector.

h. If equipped with power steering, connect the air hose for the idle up.

20. Install the PCV hose.

21. Install the air intake connector.

22. If equipped with a manual transaxle, connect the accelerator cable to the throttle body.

23. If equipped with a automatic transaxle, connect the throttle and accelerator cables to the throttle body.

24. Install the resonator and air cleaner cap.

25. Refill the cooling system.

26. Connect the negative battery cable. Start the engine and check for leaks.

27. Check the ignition timing. Road test the vehicle for proper operation.

28. Recheck all fluid levels.

T-100

1. Disconnect the negative battery cable.

2. Drain the engine coolant.

3. Remove the air cleaner cap, MAF meter and resonator.

4. If equipped with a manual transaxle, disconnect the accelerator cable from the throttle body.

5. If equipped with a automatic transaxle, disconnect the accelerator and throttle cables from the throttle body.

6. Remove the intake air connector.

7. If equipped with air conditioning, remove the air conditioning idle-up valve.

8. Remove the No. 1 and No. 2 PCV hoses.

9. Remove the spark plug wires from the spark plugs.

10. Remove the throttle body as follows:

   a. Disconnect the throttle position sensor connector.

   b. Disconnect the idle air control valve connector.

   c. Disconnect the EVAP hose.

   d. Disconnect the three vacuum hoses.

   e. Remove the two bolts, two nuts and disconnect the throttle body from the air intake chamber.

   f. Remove the throttle body gasket.

   g. Disconnect the No. 1 and No. 2 water bypass hoses from the throttle body and remove the throttle body.

11. Disconnect the following connectors:

   • If equipped with air conditioning, disconnect the air conditioning compressor connector.

   • Disconnect the oil pressure sensor connector.

   • Disconnect the ECT sensor connector.

   • Disconnect the EGR gas temperature sensor connector.

   • Disconnect the EGR VSV connector.

12. Disconnect the engine wire as follows:

   a. Remove the two bolts and disconnect the engine wire from the intake chamber.

   b. Disconnect the five engine wire clamps and engine wire.

   c. Disconnect the following connectors:

   • Knock sensor connector
   • Crankshaft position sensor connector
   • Fuel pressure control VSV connector

   d. Disconnect the DLC1 from the bracket.

   e. Disconnect the two engine wire clamps.

   f. Remove the bolt and disconnect the engine wire from the engine.

13. Disconnect the fuel injectors.

14. Remove the EGR valve and vacuum modulator as follows:

   a. Remove the bolt, four nuts, EGR pipe and two gaskets.

   b. Disconnect the two vacuum hoses from the EGR VSV.

   c. Disconnect the water hose from the water bypass pipe.

   d. Remove the bolt, two nuts, EGR valve, vacuum modulator and gasket.

15. Remove the intake chamber stay by removing the two bolts.

16. Remove the fuel return pipe by removing the hoses and two bolts.

17. Remove the intake chamber as follows:

   a. Disconnect the vacuum hose from the gas filter.

   b. Disconnect the brake booster vacuum hose from the intake chamber.

   c. Remove the three bolts, two nuts, air intake chamber and gasket.

18. Remove the fuel inlet tube by removing the union bolts.

19. Remove the delivery pipe and injectors as follows:

   a. Disconnect the vacuum hose from the fuel pressure regulator.

   b. Remove the two bolts and delivery pipe together with the four injectors.

   c. Remove the four insulators from the four spacers.

   d. Pull out the four injectors from the delivery pipe.

   e. Remove the O-ring and grommet from each injector.

NOTE: Be careful not to drop the injectors when removing the delivery pipe.

To install:

20. Install the injectors to the delivery pipe as follows:

   a. Install a new grommet to the injector.

   b. Apply a light coat of gasoline to a new O-ring and install it to the injector.

   c. While turning the injector left and right, install the injector to the delivery pipe. Install all four injectors.

   d. Position the injector connector upward.

21. Install the injectors and delivery pipe. Tighten the bolts holding the delivery pipe to ft. lbs. (21 Nm). Check that the injectors rotate smoothly.

22. Install the fuel tube with four new gaskets. Torque the union bolts to 22 ft. lbs. (29 Nm).

23. Install the air intake chamber as follows:

   a. Install a new gasket and the air intake chamber with the three bolts and two nuts. Torque the bolts and nuts to 15 ft. lbs. (21 Nm).

   b. Connect the vacuum hose to the gas filter.

   c. Connect the brake booster vacuum hose to the intake chamber.

24. Install the fuel return pipe by installing the two bolts and hoses.

25. Install the intake chamber stay by installing the two bolts. Torque the bolts to 14 ft. lbs. (20 Nm).

26. Install the EGR valve and vacuum modulator as follows:

   a. Install a new gasket, EGR valve and vacuum modulator with the bolt and two nuts. Torque the bolt to 74 inch lbs. (8.5 Nm) and the nuts to 14 ft. lbs. (19 Nm).

   b. Connect the following hoses:

   • Two vacuum hoses to the EGR VSV

   • Water bypass hose to the water bypass pipe

   c. Install two new gaskets and EGR pipe with the bolt and four nuts. Torque the bolts and nuts as follows:

   • Bolt to 14 ft. lbs. (18 Nm)
   • Nuts to intake manifold to 14 ft. lbs. (19 Nm)
   • Nuts to cylinder head to 15 ft. lbs. (20 Nm).

27. Connect the injector connectors.

28. Connect the engine wire to the engine as follows:

   a. Connect the engine wire to the intake manifold with the bolt.

   b. Connect the two engine wire clamps.

   c. Connect the DLC1 to the bracket.

   d. Connect the following connectors:

   • Fuel pressure control VSV connector
   • Knock sensor connector
   • Crankshaft position sensor connector

   e. Connect the five engine wire clamps.

   f. Connect the engine wire to the intake chamber with the two bolts.

29. Connect the following connectors to the engine:
- EGR VSV connector
- EGR gas temperature sensor connector
- ECT sensor connector
- Oil pressure sensor connector
- If equipped with air conditioning, connect the compressor connector

30. Install the throttle body as follows:

a. Install the throttle body to the engine and connect the No. 1 and No. 2 water bypass hoses to the throttle body.

b. Connect the throttle body to the air intake chamber with a new gasket. Install the two bolts and two nuts. Torque to 14 ft. lbs. (20 Nm).

c. Connect the three vacuum hoses.

d. Connect the EVAP hose.

e. Connect the IAC valve connector.

f. Connect the throttle position sensor connector.

31. Connect the spark plug wires to the spark plugs.

32. Install the No. 1 and No. 2 PCV hoses.

33. If equipped with air conditioning, install the air conditioning idle up valve.

34. Install the air intake connector by installing the two bolts, hose clamp, and two air hoses.

35. If equipped with a manual transaxle, connect the accelerator cable to the throttle body.

36. If equipped with a automatic transaxle, connect the throttle and accelerator cables to the throttle body.

37. Install the MAF meter, resonator and air cleaner cap.

38. Refill the cooling system.

39. Connect the negative battery cable. Start the engine and check for leaks.

40. Check the ignition timing. Road test the vehicle for proper operation.

41. Recheck all fluid levels.

3.4L (5VZ-FE) Engine

4Runner

1. Disconnect the negative battery cable.

---- CAUTION ----

Wait 90 seconds from the time the key is turned to LOCK and the negative battery cable is disconnected to begin work. This allows

the SRS capacitor to discharge and prevent deployment of the air bag(s).

2. Relieve the fuel system pressure.

---- CAUTION ----

Fuel injection systems remain under pressure after the engine has been turned OFF. Properly relieve fuel pressure before disconnecting any fuel lines. Failure to do so may result in fire or personal injury.

3. Remove the air cleaner hose.
4. Remove the intake air connector.
5. Remove the fuel pressure regulator.
6. Disconnect the fuel inlet pipe.
7. Remove the fuel pipe.
8. Disconnect the injector connectors.
9. Remove the delivery pipes and injectors.

a. Remove the four bolts and the delivery pipe together with the six injectors.

b. Remove the spacers from the intake manifold.

c. Pull out the six injectors from the delivery pipe.

d. Remove the O-ring and grommet from each injector.

To install:

10. Install the injectors to the delivery pipe.

a. Install two new grommets to the injector.

b. Apply a light coat of gasoline to a new O-rings and install it to the injector.

c. While turning the injector left and right, install it to the delivery pipe. Install the six injectors.

d. Position the injector connector upward.

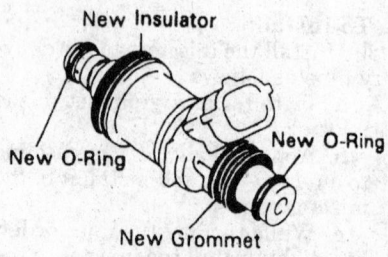

New Insulator

New O-Ring

New O-Ring

New Grommet

305462

Fuel injector seal identification — 1996–97 4Runner with 3.4L (5VZ-FE) engine

11. Install the injectors and delivery pipe.

a. Place the four new spacers in position on the intake manifold.

b. Place the injectors together with the delivery pipe in the position on the intake manifold.

c. Temporarily install the four bolts holding the delivery pipe to the intake manifold.

d. Check that the injectors rotate smoothly.

e. Connect the fuel pipe to the delivery pipe with two new gaskets and the union bolt. Torque to 25 ft. lbs. (34 Nm).

f. Tighten the bolts holding the delivery pipe to the intake manifold. Torque to 10 ft. lbs. (13 Nm).

12. Connect the fuel inlet pipe.

a. Temporarily install the union and the two new gaskets and connect the fuel pipe.

b. Install the clamp and torque the bolt to 6 ft. lbs. (8 Nm).

c. Tighten the union bolt to 25 ft. lbs. (34 Nm).

13. Install the fuel pressure regulator.

14. Visually inspect the air assist lines and the connections.

15. Install the intake air connector.

16. Install the air cleaner hose.

17. Connect the negative battery cable.

18. Check for fuel leaks.

T-100 and Tacoma

1. Disconnect the negative battery cable.

---- CAUTION ----

Work must be started after 90 seconds from the time the ignition switch is turned to the LOCK position and the negative (-) battery cable is disconnected.

2. Relieve the fuel pressure.

---- CAUTION ----

Fuel injection systems remain under pressure after the engine has been turned OFF. Properly relieve fuel pressure before disconnecting any fuel lines. Failure to do so may result in fire or personal injury.

3. Drain the engine coolant.
4. Remove the air cleaner hose.
5. Disconnect the following cables:
- If equipped with cruise control, disconnect the actuator cable with the bracket.
- Accelerator cable
- With automatic transmission, Throttle cable

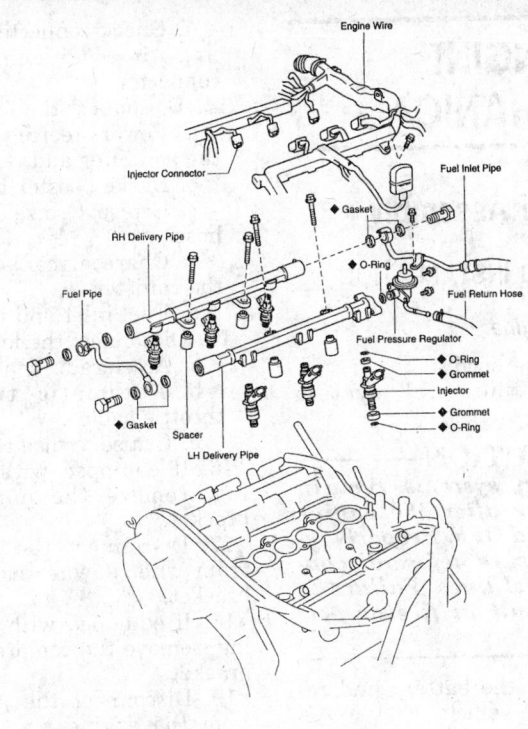

◆ Non-reusable part

291737

Fuel injector component assembly — 1995–97 T-100 and Tacoma with 3.4L (5VZ-FE) engine

6. If equipped with an EGR valve, remove the nuts and remove the EGR pipe and two gaskets.

7. Remove the intake chamber stay as follows:

a. Remove the oil filler tube and No. 1 throttle cable clamp by removing the bolt and two nuts.

b. Remove the intake chamber stay by removing the two bolts.

8. Remove the following connectors:

• VSV connector for the fuel pressure control.

• Disconnect the throttle position sensor connector.

• IAC valve connector

• If equipped with an EGR valve, disconnect the EGR gas temperature connector

• If equipped with an EGR valve, disconnect the VSV connector for the EGR valve

9. Disconnect the following hoses:

• Disconnect the PCV hoses.

• Disconnect the water bypass hoses.

• Disconnect the air assist hose from the intake manifold.

• Two vacuum sensing hoses from the VSV

• EVAP hose

• Air hose from the power steering

• If equipped with air conditioning, disconnect the air hose from the air conditioning idle up valve.

10. Remove the four bolts, two nuts and remove the air intake chamber assembly from the engine.

11. Remove the intake manifold.

12. Disconnect the fuel return hose from the fuel pressure regulator.

13. Remove the fuel pressure regulator from the engine.

14. Disconnect the fuel inlet pipe from the fuel rail by removing the union bolt. Remove the fuel inlet pipe from the intake manifold by removing the clamp bolt.

15. Remove the fuel pipe from the injector rails by removing the two union bolts.

16. Disconnect the injector connectors from the injectors.

17. Remove the delivery pipes and injectors as follows:

a. Remove the four bolts and the delivery pipes together with the six injectors.

b. Remove the four spacers from the intake manifold.

c. Pull out the six injectors from the delivery pipes.

d. Remove the two O-rings and two grommets from each injector.

To install:

18. Install a new grommet and O-ring to the injector.

19. While turning the injector clockwise and counterclockwise, push the injector into the delivery pipe. Make sure that the injectors rotate smoothly.

20. Place the four spacers in position on the intake manifold.

21. Place the delivery pipes with the six injectors in position on the intake manifold.

22. Temporarily install the four bolts to hold the delivery pipes to the intake manifold.

23. Position the injector connector outward.

24. Connect the injector connectors.

25. Torque the bolts to hold the delivery pipes to the intake manifold to 10 ft. lbs. (13 Nm).

26. Install the fuel pipe with four new gaskets. Torque the union bolts to 25 ft. lbs. (34 Nm).

27. Connect the fuel inlet pipe as follows:

a. Connect the fuel pipe with the union bolt and two new gaskets. Do not torque the bolt at this time.

b. Install the fuel pipe inlet to the intake manifold by installing the bolt. Torque the bolt to 71 inch lbs. (8 Nm).

c. Torque the union bolt to 25 ft. lbs. (34 Nm).

28. Install the fuel pressure regulator.

29. Install the intake air connector as follows:

a. Install the intake manifold to the engine by installing the three bolts and two nuts. Torque the bolts and nuts to 14 ft. lbs. (18.5 Nm).

b. Connect the DLC1 to the bracket on the intake manifold.

c. Connect the ground strap to the intake manifold by installing the bolt.

d. Connect the brake booster vacuum hose to the intake air connector.

e. Connect the two fuel return hoses.

f. Connect the engine wire to the intake manifold by installing the bolt.

g. If equipped with air conditioning, connect the idle up valve connector.

30. Install the air intake chamber assembly to the engine by installing the four bolts and two nuts. Torque the bolts and nuts to 14 ft. lbs. (18.5 Nm).

31. Connect the following hoses:
   • Connect the PCV hoses.
   • Connect the water bypass hoses.
   • Connect the air assist hose to the intake manifold.
   • Two vacuum sensing hoses to the VSV
   • EVAP hose
   • Air hose to the power steering
   • If equipped with air conditioning, connect the air hose to the air conditioning idle up valve.
32. Connect the following connectors:
   • VSV connector for the fuel pressure control.
   • Connect the throttle position sensor connector.
   • IAC valve connector
   • If equipped with an EGR valve, connect the EGR gas temperature connector
   • If equipped with an EGR valve, connect the VSV connector for the EGR valve.
33. Install the intake chamber stay as follows:
   a. Install the intake chamber stay and install the two bolts. Torque the bolts to 30 ft. lbs. (40 Nm)
   b. Install a new O-ring to the oil filler tube.
   c. Push in the oil filler tube end into the tube hole in the oil pan.
   d. Install the oil filler tube and No. 1 throttle cable clamp and install the bolt and two nuts.
34. Install two new gaskets and the EGR pipe with the nuts. Torque the clamp nuts to 71 inch lbs. (8 Nm) and the EGR pipe nuts to 14 ft. lbs. (18 Nm).
35. Connect the following cables:
   • If equipped with cruise control, connect the actuator cable with the bracket.
   • Accelerator cable
   • With automatic transmission, connect the throttle cable
36. Install the air cleaner hose.
37. Fill the radiator with engine coolant.
38. Connect the negative battery cable to the battery.
39. Start the engine and check for leaks.

ENGINE MECHANICAL

Engine Assembly

REMOVAL AND INSTALLATION

2.4L (22R-E) Engine

1. Remove the hood.
2. Release the fuel system pressure.

CAUTION

Fuel injection systems remain under pressure after the engine has been turned OFF. Properly relieve fuel pressure before disconnecting any fuel lines. Failure to do so may result in fire or personal injury.

3. Disconnect the battery and remove it from the vehicle.

CAUTION

Wait 90 seconds from the time the key is turned to LOCK and the negative battery cable is disconnected to begin work. This allows the SRS capacitor to discharge and prevent deployment of the air bag(s).

4. Raise and support the vehicle safely.
5. Remove the engine under covers.
6. Drain the engine coolant from the radiator and the cylinder block.
7. Drain the engine oil.
8. Remove the air cleaner assembly.
9. Lower the vehicle and remove the radiator.
10. Remove the power steering belt.
11. If equipped, remove the air conditioning belt.
12. Remove the alternator belt, the fluid coupling, and the fan pulley.
13. Disconnect the following wires and connectors:
   a. Ground strap form the left fender apron
   b. Alternator connector and wire
   c. Ignitor connector
   d. High tension cord for the ignition coil
   e. Distributor wire from the igniter
   f. Ground strap from the engine rear side
   g. ECM connectors
   h. Manual transmission: starter relay connector
   i. Check connector
   j. Air conditioning: compressor connector
14. Disconnect the following hoses:
   a. Power steering air hoses from the gas filter and the air pipe
   b. Brake booster hose
   c. Cruise: Cruise control vacuum hose
   d. Charcoal canister hose from the canister
   e. Fuel inlet and return lines.
15. Disconnect the following cables:
   a. Accelerator cable
   b. Automatic transmission: throttle cable
   c. Cruise: cruise control cable
16. If equipped with power steering, remove the pump from the bracket.
17. Disconnect the ground strap from the power steering pump bracket.
18. If equipped with air conditioning, remove the compressor from the bracket.
19. Disconnect the ground straps from the engine rear side and the right side.
20. Manual transmission: Remove the shift lever(s) from the inside of the vehicle.
21. Remove the rear propeller shaft.
22. 2-Wheel Drive automatic transmission: Disconnect the manual shift linkage from the Park Neutral Position (PNP) switch.
23. 4-Wheel Drive automatic transmission: Disconnect the transfer shift linkage.
24. Disconnect the speedometer cable.
25. 4-Wheel Drive: Remove the transfer undercover.
26. Remove the stabilizer bar.
27. 4-Wheel Drive: Remove the front propeller shaft.
28. Disconnect the oxygen sensor connector.
29. Manual transmission: Remove the clutch release cylinder with the bracket from the transmission.
30. 4-Wheel Drive: Remove the No. 1 front floor heat insulator and brake tube heat insulator.
31. 2-Wheel Drive: Remove the engine rear mounting and the bracket.
32. 4-Wheel Drive: Remove the No. 2 frame crossmember from the side frame.
33. Attach the engine hoist chain to the lift brackets to the engine.
34. Remove the mounting nuts and bolts. Lift the engine/transmission out of the vehicle slowly. Make sure that the engine/transmission is clear of all wiring and hoses.

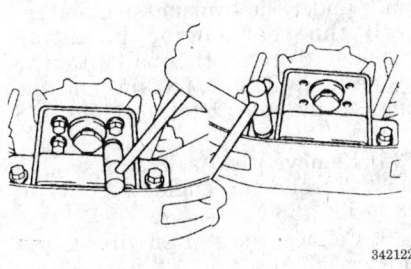

2-Wheel Drive engine rear mounting — 2.4L (22R-E) engine

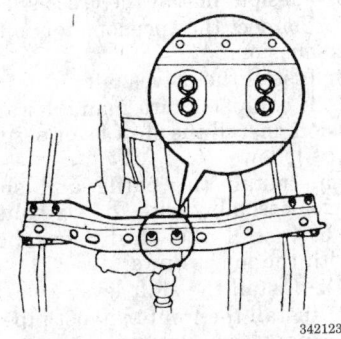

4-Wheel Drive frame crossmember — 2.4L (22R-E) engine

35. Remove the engine with the transmission from the vehicle.

36. Remove the transmission from the engine.

a. Automatic transmission: remove the automatic transmission oil cooler pipes.

b. Remove the starter.

c. Remove the two stiffener plates and the exhaust pipe bracket from the engine.

d. Remove the transmission from the engine.

37. For manual transmission: remove the clutch cover and the disc.

To install:

38. If removed, install the clutch disc and the cover to the flywheel.

39. Connect the transmission to the engine.

40. Lower the engine with the transmission into the engine compartment.

41. 4-Wheel Drive: Place a jack under the transmission. Use a wooden block between the jack and the transmission pan.

42. Jack the transmission up and put it onto the member.

43. Install the engine mount to the frame bracket.

44. Install the engine mount bolts on each side of the engine and remove the chain hoist.

45. 2-Wheel Drive: Install the engine rear mount and bracket. Torque the bolts holding the rear engine mount bracket to the support member to 9 ft. lbs. (13 Nm).

46. Lower the transmission and rest it on the extension housing. Install the rear engine mounting bracket to the mount and torque the bolts to 19 ft. lbs. (25 Nm).

47. 4-Wheel Drive: Install the No. 2 frame crossmember to the side frame and torque the bolts to 70 ft. lbs. (95 Nm).

48. Lower the transmission and the transfer and secure the transmission mount to the crossmember. Torque the bolts to 9 ft. lbs. (13 Nm).

49. 4-Wheel Drive: Install the brake tube heat insulator and the No. 1 front floor heat insulator.

50. Manual transmission: Install the clutch release cylinder with the bracket to the transmission. Torque the bracket to 29 ft. lbs. (39 Nm) and the release cylinder to 9 ft. lbs. (12 Nm).

51. Install the front exhaust pipe. Connect the oxygen sensor.

52. 4-Wheel Drive: Install the front propeller shaft, stabilizer bar, transfer case undercover, and connect the transfer shift linkage.

53. Connect the speedometer cable.

54. 2-Wheel Drive: Connect the manual shift linkage to the PNP switch.

55. Install the rear propeller shaft.

56. Manual transmission: Install the shift lever(s).

57. Connect the ground straps to the engine rear side and the right side.

58. If removed, install the compressor to the bracket.

59. Connect the ground strap for the power steering pump bracket.

60. If removed, install the power steering pump with the bracket.

61. Reconnect all cables, hoses and wires:

62. Install the fan pulley, belt guide, fluid coupling, and the alternator drive belt.

63. If removed, install the air conditioning belt.

64. Install the power steering belt. Torque the power steering pump pulley locknut to 32 ft. lbs. (43 Nm).

65. Install the radiator.

66. Install the air cleaner case and the intake air connector.

67. Fill with engine oil and coolant. If drained, refill the transmission with the appropriate fluid.

68. Install the battery and connect the cables.

69. Install the hood.

70. Start the engine, check for leaks, and perform all engine adjustments.

71. Install the engine undercover.

72. Road test the vehicle.

73. Recheck the coolant, engine oil, and transmission fluid levels.

2.4L (2RZ-FE) Engine

Tacoma

1. Turn the ignition switch **OFF**. Disconnect the battery cables; negative cable first.

—————— **WARNING** ——————

The air bag system is equipped with a backup power source. To avoid possible air bag deployment, do not start working on the vehicle until 90 seconds has elapsed from the time the ignition switch is turned OFF and the negative battery terminal is disconnected.

2. Matchmark the hood hinges and remove the hood.

3. Remove the engine undercover.

4. Drain the engine oil, transmission oil and cooling system.

5. Remove the radiator as outlined in this section.

6. If equipped with power steering, loosen the lock bolt and adjusting bolt to the idler pulley and remove the drive belt.

7. If equipped with air conditioning, loosen the idler pulley nut and adjusting bolt. Remove the air conditioning drive belt from the engine.

8. Remove the alternator drive belt, fan (with fan clutch), water pump pulley, and fan shroud as outlined in this section.

9. If equipped with a manual transaxle, disconnect the accelerator cable from the throttle body.

10. If equipped with a automatic transaxle, disconnect the accelerator and throttle cables from the throttle body.

11. If equipped with cruise control, remove the actuator cover and disconnect the cruise control cable from the actuator.

12. Remove the air cleaner cap, MAF meter and resonator. Remove the air cleaner case.

13. Remove the intake air connector.

14. If equipped with air conditioning, disconnect the air conditioning compressor and bracket as outlined in this section.

15. Disconnect the alternator wires from the alternator.

16. Disconnect the heater hoses at the cowl panel.

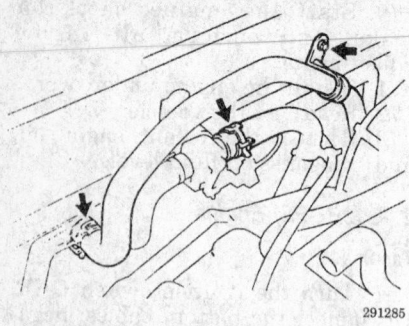

Disconnecting heater hoses — Tacoma with 2.4L (2RZ-FE) engine

17. Disconnect the following hoses:
- Brake booster vacuum hose
- EVAP hose
- If equipped with 4-Wheel Drive with A.D.D., disconnect the vacuum hose.
- If equipped with power steering, disconnect the two power steering hoses
- Fuel return hose
- Fuel inlet hose

18. Remove the power steering pump as follows:

a. Using SST 09960–10010 or equivalent, remove the nut and power steering pulley.

b. Remove the two bolt and disconnect the power steering pump.

19. Disconnect the ECM wiring from the ECM as follows:

a. Remove the four screws to the right front door scuff plate. Remove the scuff plate from the vehicle.

b. Remove the cowl panel side trim by removing the clip.

c. Disconnect the four ECM electrical connectors.

20. Disconnect the engine wire and connectors from the vehicle as follows:

a. Disconnect the igniter connector.

b. Disconnect the ground strap from the cowl top panel.

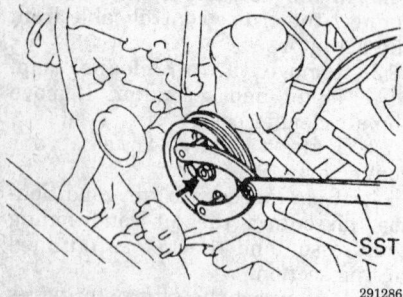

Removing the power steering pulley — Tacoma with 2.4L (2RZ-FE) engine

c. Disconnect the two engine wire clamps.

d. Remove the nuts holding the engine wire retainer to the cowl panel and pull out the engine wire from the vehicle cabin.

21. Disconnect the front exhaust pipe from the exhaust manifold and catalytic converter.

22. If equipped with manual transmission, remove the shift lever assembly as follows:

a. Remove the shift lever knob.

b. Remove the four screws and shift lever boot.

c. Remove the six bolts, shift lever assembly and baffle.

23. Remove the driveshaft from the vehicle.

24. Disconnect the speedometer cable from the transmission.

25. If equipped with manual transmission, remove the clutch release cylinder.

26. If equipped with automatic transmission, remove the cross-shaft.

27. Disconnect the wires at the starter.

28. Position a jack and wooden block under the transmission and remove the rear engine mounting bracket.

29. Attach a suitable engine hoist to the engine hangers.

30. Remove the nuts and bolts from the engine mounts.

NOTE: Make sure the engine/transmission assembly is clear of all wiring and hoses.

31. Carefully lift the engine/transmission assembly out of the vehicle.

To install:

32. Attach the engine hoist to the engine hangers. Carefully lower the engine/transmission assembly into the vehicle. Keep the engine level, while aligning the engine mounts.

33. Install the engine mount fasteners, but do not fully tighten them.

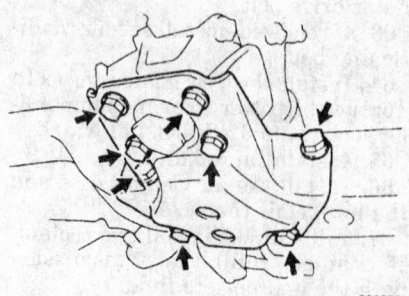

Engine rear mount

34. Position a jack and wooden block under the transmission and install the rear engine mounting bracket. Torque the bolts to the frame to 19 ft. lbs. (58 Nm) and the bolts to the mount to 13 ft. lbs. (18 Nm).

35. Remove the jack and engine hoist. Torque the engine mounts to 28 ft. lbs. (38 Nm).

36. Connect the starter wires to the starter.

37. If equipped with manual transmission, install the clutch release cylinder.

38. If equipped with automatic transmission, install the cross-shaft.

39. Connect the speedometer to the transmission.

40. Install the driveshaft.

41. If equipped with manual transmission, install the shift lever assembly as follows:

a. Install the baffle and shift lever assembly with the six bolts.

b. Install the shift lever boot with the four screws.

c. Install the shift lever knob.

42. Install the front exhaust pipe to the exhaust manifold.

43. Connect all wires and connectors.

a. Install the cowl side trim and clip.

b. Install the front door scuff plate and four screws.

44. Connect the alternator wires to the alternator.

45. Install the power steering pump as follows:

a. Connect the power steering pump to the bracket with the two bolts. Torque the bolts to 43 ft. lbs. (58 Nm).

b. Using SST 09960–10010 or equivalent, install the power steering pulley with the nut. Torque the nut to 32 ft. lbs. (43 Nm).

46. Connect all hoses previously removed.

47. If equipped with air conditioning, install the compressor.

48. Install the intake air connector. Torque the two bolts to 13 ft. lbs. (18 Nm).

49. Install the water pump pulley, fan shroud, fan (with fan clutch), and alternator drive belt.

50. If equipped with air conditioning, install and adjust the drive belt.

51. Install and adjust the power steering drive belt.

52. If equipped with manual transmission, connect the accelerator cable to the throttle body.

53. If equipped with automatic transmission, connect the accelerator

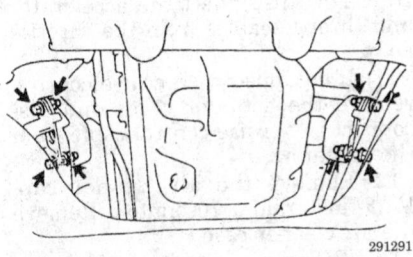

Engine left and right mounts — Tacoma with 2.4L (2RZ-FE) engine

and throttle cables to the throttle body.

54. Install the air cleaner case.

55. Install the MAF meter, resonator and air cleaner cap.

56. Install the radiator to the vehicle with the tabs on the supports through the radiator service holes. Install the four bolts and torque the bolts to 9 ft. lbs. (13 Nm).

57. Connect the lower radiator hose to the radiator.

58. If equipped with automatic transmission, connect the oil cooler hoses to the radiator.

59. Install the No. 2 fan shroud.

60. Connect the radiator reservoir hose to the radiator.

61. Connect the upper radiator hose to the radiator.

62. If removed, connect the air pipe with the two bolts.

63. Install the radiator grille to the vehicle with the 11 clips.

64. Install the two fillers.

65. Install the clearance lights to the grille with the four bolts and two clips.

66. Fill the engine oil, engine coolant, and transmission oil.

67. Connect the negative and positive cables to the battery.

68. Start the engine and check for leaks.

69. Check ignition timing.

70. Install the engine undercover.

71. Install the hood.

72. Road test the vehicle and check all fluids.

2.7L (3RZ-FE) Engine

T-100

1. Turn the ignition switch **OFF**. Disconnect the battery cables; negative cable first.

WARNING

The air bag system is equipped with a backup power source. To avoid possible air bag deployment, do not start working on the vehicle until 90 seconds has elapse from the time the ignition switch is turned OFF and the negative battery terminal is disconnected.

2. Matchmark the hood hinges and remove the hood.

3. Remove the battery and battery tray.

4. Drain the engine oil, transmission oil and cooling system.

5. Remove the expansion tank.

6. Remove the radiator as outlined in this section.

7. Remove the air cleaner cap, MAF meter and resonator. Remove the air cleaner case.

8. If equipped with a manual transaxle, disconnect the accelerator cable from the throttle body.

9. If equipped with a automatic transaxle, disconnect the accelerator and throttle cables from the throttle body.

10. Remove the intake air connector.

11. If equipped with air conditioning, disconnect the air conditioning compressor and bracket.

12. Disconnect the following hoses:
- Brake booster vacuum hose
- EVAP hose
- Two power steering hoses
- Fuel return hose
- Fuel inlet hose

13. Remove the power steering pump as follows:

a. Using SST 09960–10010 or equivalent, remove the nut and power steering pulley.

b. Remove the two bolt and disconnect the power steering pump.

14. Disconnect the alternator wires from the alternator.

15. Disconnect the ECM wiring from the ECM as follows:

a. Remove the four screws to the right front door scuff plate. Remove the scuff plate from the vehicle.

b. Remove the cowl panel side trim by removing the clip.

c. Disconnect the four ECM electrical connectors.

16. Disconnect the engine wire and connectors from the vehicle as follows:

a. Disconnect the igniter connector.

b. Disconnect the ground strap from the cowl top panel.

c. Disconnect the four engine wire clamps.

d. Pull out the engine wire from the vehicle cabin.

17. If equipped with manual transmission, remove the shift lever assembly as follows:

a. Remove the shift lever knob.

b. Remove the four screws and shift lever boot.

c. Remove the six bolts, shift lever assembly and baffle.

18. Remove the sway bar as follows:

a. Remove the nuts and cushions holding the sway bar to the lower control arms.

b. Remove the sway bar bolts and brackets and remove the sway bar from the suspension.

19. Remove the driveshaft from the vehicle.

20. Disconnect the speedometer cable from the transmission.

21. Disconnect the front exhaust pipe from the exhaust manifold and catalytic converter.

22. If equipped with manual transmission, remove the clutch release cylinder.

23. If equipped with automatic transmission, remove the cross-shaft.

24. Disconnect the wires at the starter.

25. Position a jack and wooden block under the transmission and remove the rear engine mounting bracket.

26. Attach a suitable engine hoist to the engine hangers.

27. Remove the nuts and bolts from the engine mounts.

NOTE: Make sure the engine/transmission assembly is clear of all wiring and hoses.

28. Carefully lift the engine/transmission assembly out of the vehicle.

To install:

29. Attach the engine hoist to the engine hangers. Carefully lower the engine/transmission assembly into the vehicle. Keep the engine level, while aligning the engine mounts.

30. Install the engine mount fasteners, but do not fully tighten.

31. Position a jack and wooden block under the transmission and install the rear engine mounting bracket. Torque the bolts to the frame to 42 ft. lbs. (58 Nm) and the bolts to the mount to 13 ft. lbs. (18 Nm).

32. Remove the jack and engine hoist. Torque the engine mounts to 28 ft. lbs. (38 Nm).

33. Connect the starter wires to the starter.

34. If equipped with manual transmission, install the clutch release cylinder.

35. If equipped with automatic transmission, install the cross-shaft.

36. Install the front exhaust pipe to the exhaust manifold.

37. Connect the speedometer to the transmission.

38. Install the driveshaft.

39. Install the sway bar as follows:

a. Place the sway bar in position and install both sway bar bushings and brackets to the frame.

b. Torque the sway bar mounting bolts to 22 ft. lbs. (930 Nm).

c. Connect the sway bar to the lower control arms with the brackets and cushions. Torque the nuts to 9 ft. lbs. (13 Nm).

40. If equipped with manual transmission, install the shift lever assembly as follows:

a. Install the baffle and shift lever assembly with the six bolts.

b. Install the shift lever boot with the four screws.

c. Install the shift lever knob.

41. Reconnect all engine wires.

42. Install the cowl side trim and clip.

a. Install the front door scuff plate and four screws.

43. Install the power steering pump as follows:

a. Connect the power steering pump to the bracket with the two bolts. Torque the bolts to 43 ft. lbs. (58 Nm).

b. Using SST 09960–10010 or equivalent, install the power steering pulley with the nut. Torque the nut to 32 ft. lbs. (43 Nm).

44. Connect all hoses previously removed.

45. Connect the engine heater hoses at the cowl panel.

46. If equipped with air conditioning, install the compressor as follows:

a. Install the air conditioning compressor bracket with the four bolts. Torque the bolts to 32 ft. lbs. (44 Nm).

b. Connect the air conditioning compressor to the bracket with the four bolts. Torque the bolts to 18 ft. lbs. (25 Nm).

47. Install the intake air connector. Torque the two bolts to 13 ft. lbs. (18 Nm).

48. If equipped with manual transmission, connect the accelerator cable to the throttle body.

49. If equipped with automatic transmission, connect the accelerator and throttle cables to the throttle body.

50. Install the air cleaner case.

51. Install the MAF meter, resonator and air cleaner cap.

52. Install the radiator as outlined.

53. If equipped with air conditioning, install and adjust the air conditioning compressor drive belt.

54. Install and adjust the power steering drive belt.

55. Connect the radiator reservoir hose to the radiator.

56. Connect the upper radiator hose to the radiator.

57. Install the radiator grille to the vehicle with the 11 clips and four screws.

58. Install the clearance lights to the grille with the four bolts to each light.

59. Install the battery and battery clamp.

60. Install the radiator expansion tank.

61. Fill the engine oil, engine coolant, and transmission oil.

62. Connect the negative and positive cables to the battery.

63. Start the engine and check for leaks.

64. Check ignition timing.

65. Install the engine undercover.

66. Install the hood.

67. Road test the vehicle and check all fluids.

Tacoma

1. Turn the ignition switch **OFF**. Disconnect the battery cables; negative cable first.

WARNING
The air bag system is equipped with a backup power source. To avoid possible air bag deployment, do not start working on the vehicle until 90 seconds has elapsed from the time the ignition switch is turned OFF and the negative battery terminal is disconnected.

2. Matchmark the hood hinges and remove the hood.

3. Remove the engine undercover.

4. Drain the engine oil, transmission oil and cooling system.

5. Remove the radiator as outlined in this section.

6. If equipped with power steering, loosen the lock bolt and adjusting bolt to the idler pulley and remove the drive belt.

7. If equipped with air conditioning, loosen the idler pulley nut and adjusting bolt. Remove the air conditioning drive belt from the engine.

8. Remove the alternator drive belt, fan (with fan clutch), water pump pulley, and fan shroud.

9. If equipped with a manual transaxle, disconnect the accelerator cable from the throttle body.

10. If equipped with a automatic transaxle, disconnect the accelerator and throttle cables from the throttle body.

11. If equipped with cruise control, remove the actuator cover and disconnect the cruise control cable from the actuator.

12. Remove the air cleaner cap, MAF meter and resonator. Remove the air cleaner case.

13. Remove the intake air connector.

14. If equipped with air conditioning, disconnect the air conditioning compressor and bracket.

15. Disconnect the alternator wires from the alternator.

16. Disconnect the heater hoses at the cowl panel.

17. Disconnect the following hoses:
• Brake booster vacuum hose
• EVAP hose
• If equipped with 4-Wheel Drive with A.D.D., disconnect the vacuum hose.
• If equipped with power steering, disconnect the two power steering hoses
• Fuel return hose
• Fuel inlet hose

18. Remove the power steering pump as follows:

a. Using SST 09960–10010 or equivalent, remove the nut and power steering pulley.

b. Remove the two bolt and disconnect the power steering pump.

19. Disconnect the ECM wiring from the ECM as follows:

a. Remove the four screws to the right front door scuff plate. Remove the scuff plate from the vehicle.

b. Remove the cowl panel side trim by removing the clip.

c. Disconnect the four ECM electrical connectors.

20. Disconnect the engine wire and connectors from the vehicle as follows:

a. Disconnect the igniter connector.

b. Disconnect the ground strap from the cowl top panel.

c. Disconnect the two engine wire clamps.

d. Remove the nuts holding the engine wire retainer to the cowl panel and pull out the engine wire from the vehicle cabin.

21. Disconnect the front exhaust pipe from the exhaust manifold and catalytic converter.

22. If equipped with manual transmission, remove the shift lever assembly as follows:

a. Remove the shift lever knob.

b. Remove the four screws and shift lever boot.

c. Remove the six bolts, shift lever assembly and baffle.

23. Remove the driveshaft from the vehicle.

24. Disconnect the speedometer cable from the transmission.

25. If equipped with manual transmission, remove the clutch release cylinder.

26. If equipped with automatic transmission, remove the cross-shaft.

27. Disconnect the wires at the starter.

28. Position a jack and wooden block under the transmission and remove the rear engine mounting bracket.

29. Attach a suitable engine hoist to the engine hangers.

30. Remove the nuts and bolts from the engine mounts.

NOTE: Make sure the engine/transmission assembly is clear of all wiring and hoses.

31. Carefully lift the engine/transmission assembly out of the vehicle.

To install:

32. Attach the engine hoist to the engine hangers. Carefully lower the engine/transmission assembly into the vehicle. Keep the engine level, while aligning the engine mounts.

33. Install the engine mount fasteners, but do not fully tighten them.

34. Position a jack and wooden block under the transmission and install the rear engine mounting bracket. Torque the bolts to the frame to 19 ft. lbs. (58 Nm) and the bolts to the mount to 13 ft. lbs. (18 Nm).

35. Remove the jack and engine hoist. Torque the engine mounts to 28 ft. lbs. (38 Nm).

36. Connect the starter wires to the starter.

37. If equipped with manual transmission, install the clutch release cylinder.

38. If equipped with automatic transmission, install the cross-shaft.

39. Connect the speedometer to the transmission.

40. Install the driveshaft.

41. If equipped with manual transmission, install the shift lever assembly as follows:

a. Install the baffle and shift lever assembly with the six bolts.

b. Install the shift lever boot with the four screws.

c. Install the shift lever knob.

42. Install the front exhaust pipe to the exhaust manifold.

43. Connect all engine wires.

a. Install the cowl side trim and clip.

b. Install the front door scuff plate and four screws.

44. Connect the alternator wires to the alternator.

45. Install the power steering pump as follows:

a. Connect the power steering pump to the bracket with the two bolts. Torque the bolts to 43 ft. lbs. (58 Nm).

b. Using SST 09960–10010 or equivalent, install the power steering pulley with the nut. Torque the nut to 32 ft. lbs. (43 Nm).

46. Connect all hoses previously removed.

47. Connect the engine heater hoses at the cowl panel.

48. If equipped with air conditioning, install the compressor.

49. Install the intake air connector. Torque the two bolts to 13 ft. lbs. (18 Nm).

50. Install the water pump pulley, fan shroud, fan (with fan clutch), and alternator drive belt as follows:

a. Place the fan (with the fan clutch), water pump pulley and fan shroud in position.

b. Install the water pump pulley mounting nuts but do not torque the nuts at this time.

c. Install the alternator drive belt to the engine.

d. Stretch the alternator belt tight and torque the fan nuts to 16 ft. lbs. (21 Nm).

e. Adjust the drive belt for the alternator.

51. If equipped with air conditioning, install and adjust the drive belt.

52. Install and adjust the power steering drive belt.

53. If equipped with manual transmission, connect the accelerator cable to the throttle body.

54. If equipped with automatic transmission, connect the accelerator and throttle cables to the throttle body.

55. Install the air cleaner case.

56. Install the MAF meter, resonator and air cleaner cap.

57. Install the radiator to the vehicle with the tabs on the supports through the radiator service holes. Install the four bolts and torque the bolts to 9 ft. lbs. (12.5 Nm).

58. Connect the lower radiator hose to the radiator.

59. If equipped with automatic transmission, connect the oil cooler hoses to the radiator.

60. Install the No. 2 fan shroud.

61. Connect the radiator reservoir hose to the radiator.

62. Connect the upper radiator hose to the radiator.

63. If removed, connect the air pipe with the two bolts.

64. Install the radiator grille to the vehicle with the 11 clips.

65. Install the two fillers.

66. Install the clearance lights to the grille with the four bolts and two clips.

67. Fill the engine oil, engine coolant, and transmission oil.

68. Connect the negative and positive cables to the battery.

69. Start the engine and check for leaks.

70. Check ignition timing.

71. Install the engine undercover.

72. Install the hood.

73. Road test the vehicle and check all fluids.

1996–97 4Runner (2-Wheel Drive)

1. Disconnect the negative battery cable.

—————— **CAUTION** ——————

Wait 90 seconds from the time the key is turned to LOCK and the negative battery cable is disconnected to begin work. This allows the SRS capacitor to discharge and prevent deployment of the air bag(s).

2. Relieve the fuel pressure.

—————— **CAUTION** ——————

Fuel injection systems remain under pressure after the engine has been turned OFF. Properly relieve fuel pressure before disconnecting any fuel lines. Failure to do so may result in fire or personal injury.

3. Raise and safely support the vehicle.

4. Remove the engine undercover.

5. Drain the engine coolant.

6. Drain the engine oil and the transmission oil.

7. Remove the hood.

8. Remove the radiator.

9. Remove the drive belt for the alternator and the water pump pulley.

10. Disconnect the accelerator cable from the throttle body.

11. If equipped with cruise control, remove the actuator cover and disconnect the cruise control cable from the actuator.

12. Remove the air cleaner assembly.

a. Disconnect the IAT sensor and the MAF meter connectors.

b. Disconnect the three wire clamps and the engine wire.

c. Loosen the air cleaner hose clamp.

d. Remove the three bolts and the MAF meter, resonator and the air cleaner assembly.

13. If equipped with air conditioning, disconnect the air conditioning compressor.

14. Disconnect the alternator connector.

15. Disconnect the heater hoses.

16. Disconnect the following hoses:
• Brake booster vacuum hose
• EVAP hose
• Two air hoses for the power steering idle-up
• Fuel return hose
• Fuel inlet hose

17. Remove the power steering pump from the engine.

18. Disconnect the engine wire from the cabin.

a. Remove the glove box door.

b. Lower the finish No. 2 panel.

c. Disconnect the four ECM connectors.

d. Disconnect the two cassette connectors and the two wire clamps from the lower finish panel.

e. Disconnect the igniter connector.

f. Disconnect the ground strap from the cowl top panel.

g. Disconnect the two engine wire clamps.

h. Remove the two nuts holding the engine wire retainer to the cowl panel and pull out the engine wire from the cabin.

19. Disconnect the heated oxygen sensor connector and remove the front exhaust pipe.

20. Manual transmission: Remove the shift lever assembly.

a. Remove the shift lever knob.

b. Remove the four screws and the shift lever boot.

c. Remove the six bolts, the shift lever assembly and baffle.

21. Remove the driveshaft.

22. Disconnect the speedometer cable.

23. Manual transmission: Remove the clutch release cylinder.

24. Automatic transmission: Remove the cross-shaft.

25. Disconnect the starter wire.

26. Place a jack under the transmission and remove the engine rear mounting bracket.

27. Install a rear engine hanger in the correct direction.

28. Attach the engine hoist chain to the two engine hangers.

Right and left engine mounts — 1995–97 4Runner (2-Wheel Drive) with 2.7L (3RZ-FE) engine

29. Remove the four bolts and nuts holding the engine front mounting insulators to the frame.

NOTE: Make sure that the engine is clear of all wiring and hoses.

30. Lift the engine and the transmission assembly onto the stand.

31. Separate the engine from the transmission.

To install:

32. Install the transmission to the engine.

33. Attach a chain hoist to the engine hangers.

34. Lower the engine and transmission assembly into the engine compartment.

35. Keep the engine level and align the right and left mounting and body mountings.

36. Attach the right and left mounting insulators to the body mountings and temporarily install the bolts and nuts.

37. Jack up and put the transmission onto the frame.

38. Remove the chain hoist.

39. Remove the bolt and the rear engine hanger.

40. Install the engine rear mounting bracket and torque to:
• Bolt A: 13 ft. lbs. (19 Nm)
• Bolt B: 19 ft. lbs. (26 Nm)

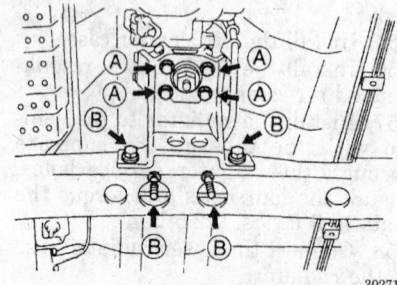

Bolt tightening pattern for the engine rear mounting bracket — 1995–97 4Runner (2-Wheel Drive) with 2.7L (3RZ-FE) engine

41. Tighten the left and right engine mounting insulator bolts and nuts to 28 ft. lbs. (38 Nm).

42. Connect the starter wire.

43. Manual transmission: Install the clutch release cylinder. Torque the clutch line bolt to 29 ft. lbs. (39 Nm) and the clutch release cylinder bolts to 9 ft. lbs. (13 Nm).

44. Automatic transmission: Install the cross-shaft and torque the bolt to 29 ft. lbs. (39 Nm) and the nut to 13 ft. lbs. (18 Nm).

45. Connect the speedometer cable.

46. Install the driveshaft.

47. Manual transmission: Install the shift lever assembly.

48. Install the front exhaust pipe.

a. Install the new gaskets and the front exhaust pipe assembly and torque the three new nuts to 46 ft. lbs. (62 Nm).

b. Install the support bracket and torque the bolts to 29 ft. lbs. (39 Nm).

c. Connect the TWC with a new gasket to the tail pipe and torque to 29 ft. lbs. (39 Nm).

d. Connect the heated oxygen sensor connector.

49. Connect the engine wire to the cabin.

50. Install the power steering pump.

51. Connect all hoses previously removed.

52. Connect the alternator wire.

53. If removed, install the air conditioning compressor and torque the bolts to 18 ft. lbs. (25 Nm).

54. Install the intake air connector and torque the bolts to 13 ft. lbs. (18 Nm).

55. Install the air cleaner assembly.

56. Connect the throttle cable to the throttle body.

57. If disconnected, connect the cruise control cable to the actuator and install the actuator cover.

58. Install the drive belt for the alternator and the water pump pulley.

59. Install the radiator.

60. Refill the engine oil, coolant, and transmission oil.

61. Connect the negative battery cable, start the engine, and check for leaks.

62. Check the ignition timing.

63. Install the engine undercover.

64. Install the hood.

65. Road test the vehicle and recheck the fluid levels.

1996–97 4Runner (4-Wheel Drive)

1. Disconnect the negative battery cable.

— CAUTION —

Wait 90 seconds from the time the key is turned to LOCK and the negative battery cable is disconnected to begin work. This allows the SRS capacitor to discharge and prevent deployment of the air bag(s).

---

2. Relieve the fuel pressure.

— CAUTION —

Fuel injection systems remain under pressure after the engine has been turned OFF. Properly relieve fuel pressure before disconnecting any fuel lines. Failure to do so may result in fire or personal injury.

---

3. Raise and safely support the vehicle.
4. Remove the transmission.
5. Remove the engine undercover.
6. Drain the engine coolant.
7. Drain the engine oil.
8. Remove the hood.
9. Remove the radiator.
10. Remove the drive belt for the alternator and the water pump pulley.
11. Disconnect the accelerator cable from the throttle body.
12. If equipped with cruise control, remove the actuator cover and disconnect the cruise control cable from the actuator.
13. Remove the air cleaner assembly.
14. Remove the intake air connector.
15. If equipped with air conditioning, remove the air conditioning compressor.
16. Disconnect the alternator connector.
17. Disconnect the heater hoses.
18. Disconnect the following hoses:
• Brake booster vacuum hose
• EVAP hose
• Two air hoses for the power steering idle-up
• With A.D.D. — Vacuum hose
• Fuel return hose
• Fuel inlet hose
19. Remove the power steering pump from the engine.
20. Disconnect the engine wire from the cabin.
   a. Remove the glove box door.
   b. Lower the finish No. 2 panel.
   c. Disconnect the ECM connectors.
   d. Disconnect the two cassette connectors and the two wire clamps from the lower finish panel.
   e. VSV connector for the EVAP and clamp.
   f. Disconnect the igniter connector.

   g. Disconnect the ground strap from the cowl top panel.
   h. Disconnect the two engine wire clamps.
   i. Remove the two nuts holding the engine wire retainer to the cowl panel and pull out the engine wire from the cabin.
21. Install a rear engine hanger in the correct direction.
22. Attach the engine hoist chain to the two engine hangers.
23. Remove the bolts and nuts holding the engine front mounting insulators to the frame.
24. Lift the engine out of the vehicle slowly and carefully. Make sure that the engine is clear of all wiring and hoses.
25. Remove the engine from the vehicle.

To install:

26. Attach a chain hoist to the engine hangers.
27. Lower the engine assembly into the engine compartment.
28. Keep the engine level and align the right and left mounting and body mountings.
29. Attach the right and left mounting insulators to the body mountings and temporarily install the bolts and nuts.
30. Remove the chain hoist.
31. Remove the bolt and the rear engine hanger.
32. Tighten the right and left engine mounting insulator bolts and nuts to 28 ft. lbs. (38 Nm).
33. Connect the engine wire to the cabin.
34. Connect the alternator wire.
35. Install the power steering pump.
36. Connect all hoses:
37. If removed, install the air conditioning compressor and torque the bolts to 18 ft. lbs. (25 Nm).
38. Install the intake air connector and torque the bolts to 13 ft. lbs. (18 Nm).

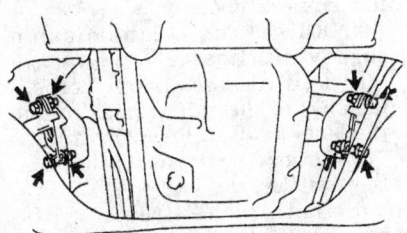

302714

Right and left engine mounts — 1995–97 4Runner (4-Wheel Drive) with 2.7L (3RZ-FE) engine

39. Install the air cleaner assembly.
40. Connect the throttle cable to the throttle body.
41. If removed, install the cruise control cable to the actuator and install the actuator cover.
42. Install the radiator.
43. Install the hood.
44. Fill with engine with oil and the coolant system with coolant.
45. Install the transmission.
46. Install the engine undercover.
47. Connect the negative battery cable.
48. Fill the transmission fluid.
49. Check the ignition timing.
50. Test drive the vehicle and check for leaks.
51. Recheck fluid levels.

3.0L (3VZ-E) Engine

1993–95 4Runner and Pick-up

1. Remove the hood.
2. Relieve the fuel system pressure.

— CAUTION —

Fuel injection systems remain under pressure after the engine has been turned OFF. Properly relieve fuel pressure before disconnecting any fuel lines. Failure to do so may result in fire or personal injury.

---

3. Disconnect the battery and remove it from the vehicle.

— CAUTION —

Wait 90 seconds from the time the key is turned to LOCK and the negative battery cable is disconnected to begin work. This allows the SRS capacitor to discharge and prevent deployment of the air bag(s).

---

4. Raise and safely support the vehicle.
5. Remove the engine under covers.
6. Drain the engine coolant.
7. Drain the engine oil.
8. Lower the vehicle and remove the air cleaner and hose.
9. Disconnect the hoses, fan shrouds, and remove the radiator. On the automatic transmission, disconnect the oil cooler hoses.
10. Manual transmission: Disconnect the clutch release cylinder hose.
11. Remove the power steering drive belt and pump pulley.
12. Disconnect the power steering pump from the engine.
13. Remove the cooling fan.
14. Remove the alternator belt.

15. Disconnect the following straps, wires, and connectors:

    a. Ground strap from the left fender apron

    b. Generator connector and wire

    c. Igniter connector

    d. Oil pressure sender gauge connector

    e. Ground strap from the engine rear side

    f. ECM connectors

    g. VSV connectors

    h. Air conditioning compressor connector

    i. Manual transmission: Starter relay connector

    j. Solenoid resister connector

    k. DLC1

    l. With Automatic Disconnecting Differential (ADD): ADD switch connector

16. Disconnect the following hoses:

    a. Power steering air hoses from the gas filter and air pipe

    b. Brake booster hose

    c. Cruise control: Cruise control vacuum hose

    d. Charcoal canister hose from the canister

    e. VSV vacuum hoses

17. Disconnect the following cables:

    a. Accelerator cable

    b. Automatic transmission: Throttle cable

    c. Cruise: Cruise control cable

18. Disconnect the heater hoses.

19. Disconnect the fuel inlet and the outlet hoses.

20. Disconnect the air conditioning compressor from the engine.

21. Disconnect the heated oxygen sensor and remove the front exhaust pipe.

22. Manual transmission: Remove the shift levers.

23. Remove the rear propeller shaft.

24. 4-Wheel Drive: Remove the front propeller shaft.

25. 2-Wheel Drive automatic transmission: Disconnect the manual shift linkage.

26. 4-Wheel Drive automatic transmission: Disconnect the transfer case shift linkage.

27. Disconnect the speedometer connector.

28. 4-Wheel Drive: Remove the transfer undercover and the stabilizer bar.

29. On the Pick-up, remove the No. 1 frame crossmember.

30. On 4-Wheel Drive vehicles, remove the No. 1 front floor heat insulator and the brake tube heat insulator.

31. 2-Wheel Drive: Remove the engine rear mounting bracket.

32. 4-Wheel Drive: Remove the No. 2 frame crossmember:

    a. Remove the four bolts holding the engine rear mounting insulator to the frame crossmember.

    b. Raise the transmission slightly with a jack.

    c. Remove the eight bolts holding the frame crossmember to the side frame. Remove the frame crossmember.

33. Attach the engine chain hoist to the engine hangers.

34. Remove the four bolts holding the right and left engine mounting insulators to the body mountings.

35. Lift the engine and transmission assembly out of the vehicle slowly and carefully. Make sure that the engine is clear of all wiring, hoses, and cables.

36. Remove the transmission from the engine.

37. Manual transmission: Remove the clutch cover and the disc.

To install:

38. If removed, install the clutch cover and the disc.

39. Install the transmission to the engine.

40. Attach the engine hoist to the engine hangers and slowly lower the engine and transmission assembly into the engine compartment.

41. Keep the engine level and align the right and left mountings to the body mounts. Jack up and put the transmission onto the member.

42. 2-Wheel Drive: Install the engine rear mounting bracket.

    a. Raise the transmission slightly by raising the engine with a jack and a wooden block under the transmission.

    b. Install the engine rear mounting bracket to the support member and torque to 19 ft. lbs. (25 Nm).

    c. Lower the transmission and rest it on the extension housing.

    d. Install the mounting bracket to the mounting insulator. Torque the bolts to 9 ft. lbs. (13 Nm).

43. 4-Wheel Drive: Install the No. 2 frame crossmember.

    a. Raise the transmission slightly with a jack.

    b. Install the frame crossmember to the side frame. Torque the bolts to 70 ft. lbs. (95 Nm).

    c. Lower the transmission and the transfer.

    d. Install the frame crossmember to the engine rear mounting insulator and torque the bolts to 9 ft. lbs. (13 Nm).

44. Tighten the left and right engine mounting insulator bolts to 27 ft. lbs. (37 Nm).

45. 4-Wheel Drive: Install the No. 1 front floor and brake tube heat insulator.

46. On Pickups, install the No. 1 frame crossmember.

47. 4-Wheel Drive: Install the stabilizer bar and the transfer undercover.

48. Connect the speedometer connector.

49. 2-Wheel Drive automatic transmission: Connect the manual shift linkage.

50. 4-Wheel Drive automatic transmission: Connect the transfer shift linkage.

51. 4-Wheel Drive: Install the front propeller shaft.

52. Install the rear propeller shaft.

53. Manual transmission: Install the shift levers.

54. Install the front exhaust pipe to the exhaust manifold and torque to 46 ft. lbs. (62 Nm). Torque the bolts connecting the exhaust to the catalytic converter to 29 ft. lbs. (39 Nm).

55. Connect the heated oxygen sensor connector.

56. Manual transmission: Connect the clutch release cylinder hose.

57. Install the air conditioning compressor. Install the remaining components.

58. Fill the engine with coolant and with oil. If drained, refill the transmission with the appropriate fluid.

59. Start the engine and check for leaks.

60. Perform all engine adjustments.

61. Install the engine undercover.

62. Install the hood.

63. Road test the vehicle.

64. Recheck the engine coolant, oil, and transmission fluid levels.

1993–94 T-100

1. Turn the ignition **OFF**. Disconnect the battery cables; negative first.

2. Remove the battery.

3. Matchmark the hood hinges to facilitate installation; remove the hood assembly.

4. Drain the engine coolant and engine oil.

5. Remove the air cleaner assembly with hose. Remove the radiator. Remove the engine drive belts.

6. Disconnect and tag all wires, connectors and vacuum hoses from the engine.

7. Disconnect the accelerator cable, transmission cable (automatic transmission) and cruise control cable.

8. Remove the power steering pump with hoses attached. Remove the air conditioning compressor with hoses attached. Position each component aside.

9. If equipped with manual transmission, disconnect the clutch release cylinder hose. Remove the shift levers.

10. Disconnect the fuel hoses and heater hoses.

11. Remove the rear driveshaft. Disconnect the speedometer connector.

12. If equipped with automatic transmission, disconnect the manual shift linkage.

13. If equipped with 4-Wheel Drive, remove the transfer undercover. Remove the stabilizer bar and front driveshaft.

14. Remove the frame crossmember. Remove the front exhaust pipe.

15. Remove the front floor heat insulator and brake tube heat insulator (4-Wheel Drive).

16. Remove the transmission mounting and crossmember from side frame.

NOTE: Make sure the engine is clear of all wiring and hoses.

17. Attach a suitable chain to the engine hangers. Remove the mounting nuts and bolts and carefully lift the engine/transmission assembly from the vehicle. Place the engine/transmission assembly onto a suitable stand.

To install:

18. Attach the engine chain and hoist. Carefully lower the engine/transmission assembly into the vehicle.

19. Place a wooden block on the service jack and position the jack under the transmission. Keep the engine level with the jack.

20. Install the engine mounting bolts and nuts.

21. Install the rear mounting and bracket. Torque the bolts to 19 ft. lbs. (25 Nm).

22. Install the frame crossmember(s).

23. If equipped with 4-Wheel Drive, front floor heat insulator and brake tube heat insulator.

24. Install the exhaust pipe.

25. If equipped with 4-Wheel Drive, install the front driveshaft, stabilizer bar and transfer undercover.

26. Connect the manual shift linkage (automatic transmission) or shift levers (manual transmission).

27. Connect the speedometer, fuel hoses and heater hoses.

28. If equipped with manual transmission, connect the clutch release cylinder hose.

29. Install the air conditioning compressor and power steering pump.

30. Reconnect all wiring, connectors and vacuum hoses to the engine.

31. Install the drive belts, radiator and air cleaner assembly.

32. Fill the engine with oil. Fill the cooling system.

33. Install the battery and hood.

34. Start the engine and check for leaks. Perform engine adjustment.

35. Road test the vehicle for proper operation. Recheck all fluid levels.

3.4L (5VZ-FE) Engine

2 Wheel Drive — 1995 Tacoma and 1995–97 T-100

1. Remove the hood.
2. Disconnect the battery and remove it from the vehicle.

─── **CAUTION** ───

Work must be started after 90 seconds from the time the ignition switch is turned to the LOCK position and the negative battery cable is disconnected.

─────────────────

3. Raise and safely support the vehicle.

4. Remove the engine under covers.

5. Drain the engine coolant.

6. Drain the engine oil.

7. Remove the radiator from the vehicle.

8. Remove the power steering (PS) drive belt as follows:

  a. Stretch the belt and loosen the fan pulley mounting nuts.

  b. Loosen the lock bolt, pivot bolt and adjusting bolt and remove the drive belt from the engine.

9. If equipped with air conditioning, remove the air conditioning drive belt by loosening the idle pulley nut and adjusting bolt.

10. Loosen the lock bolt, pivot bolt and adjusting bolt and the alternator drive belt.

11. Remove the fan with the fluid coupling and fan pulleys.

12. Disconnect the PS pump from the engine and set aside. Do not disconnect the lines from the pump.

13. If equipped with air conditioning, disconnect the compressor from the engine and set aside. Do not disconnect the lines from the compressor.

14. Remove the air cleaner cap, MAF meter and resonator.

15. Remove the air cleaner case and filter.

16. Disconnect the following cables:

  • If equipped with cruise control, disconnect the actuator cable with the bracket.

  • Accelerator cable

  • With automatic transmission, Throttle cable

17. Disconnect the heater hoses.

18. Disconnect the following hoses:

  • Brake booster vacuum hose

  • EVAP hose

  • Fuel return hose

  • Fuel inlet hose

19. Disconnect the starter wire and connectors as follows:

  a. Remove the ground strap by removing the bolt.

  b. Remove the nuts and disconnect the positive cable from the battery.

  c. Disconnect the three starter wire clamps and connector.

20. Disconnect the alternator connector and wire.

21. Disconnect the engine wire and connectors as follows:

  a. Remove the four screws to the right front door scuff plate. Remove the scuff plate from the vehicle.

  b. Remove the cowl panel side trim by removing the clip.

  c. Disconnect the ECM electrical connectors.

  d. Disconnect the two connectors from the cowl wire.

  e. Disconnect the igniter connector.

  f. Disconnect the ground strap.

  g. Disconnect the six engine wire clamps.

  h. Pull out the engine wire from the cabin.

22. If equipped with manual transmission, remove the shift lever assembly as follows:

  a. Remove the shift lever knob.

  b. Remove the four screws and the shift lever boot.

  c. Remove the shift lever assembly and gasket by removing the six bolts.

23. Remove the stabilizer bar.

24. Remove the driveshaft from the transmission.

25. Disconnect the speedometer cable.

26. Remove the front exhaust pipe.

27. If equipped with a manual transmission, remove the clutch release cylinder.

28. If equipped with an automatic transmission, remove the cross-shaft.

29. Place a jack under the transmission.

30. Remove the transmission rear mounting bracket by removing the eight bolts.

31. If equipped with air conditioning, remove the bolt and disconnect the air conditioning compressor wire clamp.

32. If necessary, install a No. 2 engine hanger with two bolts. Torque the two bolts to 30 ft. lbs. (40 Nm).

33. Attach the engine hoist chain to the two engine hangers.

34. Remove the four bolts and nuts holding the engine front mounting insulators to the frame.

35. Lift the engine and transmission out of the vehicle.

To install:

36. Install the engine assembly to the vehicle. Attach the engine mounts to the body mountings. Install the bolts and nuts but do not torque at this time.

37. Remove the engine chain hoist the No. 2 engine hanger.

38. If equipped with air conditioning, connect the air conditioning wire with the bolt.

39. Raise the transmission slightly and install the transmission mounting bracket. Torque the bolts to the frame to 43 ft. lbs. (58 Nm) and the bolts to the mounting insulator to 13 ft. lbs. (18 Nm).

40. Torque the engine mounting nuts and bolts to 28 ft. lbs. (38 Nm).

41. If equipped with an automatic transmission, install the cross-shaft.

42. If equipped with a manual transmission, install the clutch release cylinder. Torque the bolts to 9 ft. lbs. (12 Nm).

43. Install the front exhaust pipe.

44. Connect the speedometer cable.

45. Install the driveshaft.

46. Install the stabilizer bar.

47. Install the shift lever assembly as follows:

   a. Install a new gasket and shift lever assembly with the six bolts.

   b. Install the shift lever boot with the four screws.

   c. Install the shift lever knob.

48. Connect all engine wires, hoses and cables.

49. Install the air cleaner case and air filter.

50. Install the MAF meter, resonator and air cleaner cap.

51. If equipped, install the air conditioning compressor. Install the remaining components.

52. Fill the engine with oil.

53. Fill the engine and radiator with coolant.

54. Install the engine undercover.

55. Start the engine and check for leaks.

4-Wheel Drive — 1995 Tacoma and 1995–97 T-100

1. Remove the transmission from the vehicle.
2. Remove the hood.
3. Disconnect the battery and remove it from the vehicle.

— **CAUTION** —

Work must be started after 90 seconds from the time the ignition switch is turned to the LOCK position and the negative battery cable is disconnected.

---

4. Raise and safely support the vehicle.
5. Remove the engine under covers.
6. Drain the engine coolant.
7. Drain the engine oil.
8. Remove the radiator from the vehicle.
9. Remove the power steering (PS) drive belt as follows:

   a. Stretch the belt and loosen the fan pulley mounting nuts.

   b. Loosen the lock bolt, pivot bolt and adjusting bolt and remove the drive belt from the engine.

10. If equipped with air conditioning, remove the air conditioning drive belt by loosening the idle pulley nut and adjusting bolt.

11. Loosen the lock bolt, pivot bolt and adjusting bolt and the alternator drive belt.

12. Remove the fan with the fluid coupling and fan pulleys.

13. Disconnect the PS pump from the engine and set aside. Do not disconnect the lines from the pump.

14. If equipped with air conditioning, disconnect the compressor from the engine and set aside. Do not disconnect the lines from the compressor.

15. Remove the air cleaner cap, MAF meter and resonator.

16. Remove the air cleaner case and filter.

17. Disconnect the following cables:

   • If equipped with cruise control, disconnect the actuator cable with the bracket.

   • Accelerator cable

   • With automatic transmission, Throttle cable

18. Disconnect the heater hoses.

19. Disconnect the following hoses:
   • Brake booster vacuum hose
   • EVAP hose
   • A.D.D. vacuum hose
   • Fuel return hose
   • Fuel inlet hose

20. Disconnect the starter wire and connectors as follows:

   a. Remove the ground strap by removing the bolt.

   b. Remove the nuts and disconnect the positive cable from the battery.

   c. Disconnect the three starter wire clamps and connector.

   d. Disconnect the A.D.D. indicator switch connector.

21. Disconnect the alternator connector and wire.

22. Disconnect the engine wire and connectors as follows:

   a. Remove the four screws to the right front door scuff plate. Remove the scuff plate from the vehicle.

   b. Remove the cowl panel side trim by removing the clip.

   c. Disconnect the ECM electrical connectors.

   d. Disconnect the two connectors from the cowl wire.

   e. Disconnect the igniter connector.

   f. Disconnect the ground strap.

   g. Disconnect the six engine wire clamps.

   h. Pull out the engine wire from the cabin.

23. If equipped with air conditioning, remove the bolt and disconnect the air conditioning compressor wire clamp.

24. If necessary, install a No. 2 engine hanger with two bolts. Torque the two bolts to 30 ft. lbs. (40 Nm).

25. Attach the engine hoist chain to the two engine hangers.

26. Remove the four bolts and nuts holding the engine front mounting insulators to the frame.

27. Lift the engine out of the vehicle.

To install:

28. Install the engine assembly to the vehicle. Attach the engine mounts to the body mountings. Install the bolts and nuts but do not torque at this time.

29. Remove the engine chain hoist the No. 2 engine hanger.

30. If equipped with air conditioning, connect the air conditioning wire with the bolt.

31. Torque the engine mounting nuts and bolts to 28 ft. lbs. (38 Nm).

32. Connect all engine wires, hoses and cables.

33. Install the air cleaner case and air filter.

34. Install the MAF meter, resonator and air cleaner cap.

35. If equipped, install the air conditioning compressor.

36. Install the fan with the fluid coupling and fan pulleys. Torque the nuts to 48 inch lbs. (5.4 Nm).
37. Install the alternator drive belt.
38. If equipped, install and adjust the air conditioning drive belt.
39. Install the PS pump, pump pulley and the drive belt.
40. Install the radiator to the vehicle.
41. Fill the engine with oil.
42. Fill the engine and radiator with coolant.
43. Install the hood.
44. Install the engine undercover.
45. Install the transmission to the vehicle.
46. Start the engine and check for leaks.

1996–97 4Runner (2-Wheel Drive)

1. Remove the hood.
2. Disconnect the battery and remove it from the vehicle.

--- CAUTION ---

Wait 90 seconds from the time the key is turned to LOCK and the negative battery cable is disconnected to begin work. This allows the SRS capacitor to discharge and prevent deployment of the air bag(s).

3. Raise and safely support the vehicle.
4. Remove the engine under covers.
5. Drain the engine coolant.
6. Drain the engine oil.
7. Remove the radiator from the vehicle.
8. Remove the fan with the fluid coupling and fan pulleys.
9. Remove the air cleaner cap, MAF meter, and the resonator.
10. Remove the air cleaner case and filter.
11. Disconnect the heater hoses.
12. Disconnect the following hoses:
   • Brake booster vacuum hose
   • EVAP hose
   • Fuel return hose
   • Fuel inlet hose
13. Disconnect the starter wire and connectors as follows:
   a. Remove the ground strap by removing the bolt.
   b. Disconnect the three starter wire clamps and connector.
14. Disconnect the alternator connector and wire.
15. Disconnect the engine wire from the cabin.
   a. Remove the glove box door.
   b. Lower the finish No. 2 panel.

c. Disconnect the four ECM connectors.
   d. Disconnect the two cassette connectors and the two wire clamps from the lower finish panel.
   e. Disconnect the engine wire clamp.
   f. Disconnect the following:
      • Igniter connector
      • Ground strap
      • VSV connector for the EVAP
      • Vapor pressure sensor connector and clamp
      • Vapor connector for the vapor pressure sensor and clamp
   g. Remove the bolt and wire bracket.
   h. Remove the two nuts holding the engine wire retainer to the cowl panel and pull out the engine wire from the cabin.
16. Remove the driveshaft from the transmission.
17. Disconnect the speedometer cable.
18. Remove the front exhaust pipe.
19. Remove the nut and the control cable.
20. Place a jack under the transmission.
21. Remove the transmission rear mounting bracket by removing the eight bolts.
22. If equipped with air conditioning, remove the bolt and disconnect the air conditioning compressor wire clamp.
23. If necessary, install a No. 2 engine hanger with two bolts. Torque the two bolts to 30 ft. lbs. (40 Nm).
24. Attach the engine hoist chain to the two engine hangers.
25. Remove the four bolts and nuts holding the engine front mounting insulators to the frame.
26. Lift the engine and transmission out of the vehicle.

   To install:
27. Install the engine assembly to the vehicle. Attach the engine mounts to the body mountings. Install the bolts and nuts but do not torque at this time.
28. Remove the engine chain hoist the No. 2 engine hanger.
29. If equipped with air conditioning, connect the air conditioning wire with the bolt.
30. Raise the transmission slightly and install the transmission mounting bracket. Torque the bolts to the frame to 43 ft. lbs. (58 Nm) and the bolts to the mounting insulator to 13 ft. lbs. (18 Nm).
31. Torque the engine mounting nuts and bolts to 28 ft. lbs. (38 Nm).
32. Install the control cable.

33. Install the front exhaust pipe.
34. Connect the speedometer cable.
35. Install the driveshaft.
36. Connect all engine wire, hoses and cables.
37. Install the fan with the fluid coupling and fan pulleys. Torque the nuts to 48 inch lbs. (5.4 Nm).
38. Install the air cleaner case and air filter.
39. Install the MAF meter, resonator, and the air cleaner cap.
40. Install the radiator to the vehicle.
41. Fill the engine with oil.
42. Fill the engine and radiator with coolant.
43. Install the engine undercover.
44. Install the battery and connect the cables.
45. Start the engine and check for leaks.
46. Make any necessary adjustments, install the hood, and road test the vehicle.

1996–97 4Runner (4-Wheel Drive)

1. Remove the transmission from the vehicle.
2. Remove the hood.
3. Release the fuel system pressure.

--- CAUTION ---

Fuel injection systems remain under pressure after the engine has been turned OFF. Properly relieve fuel pressure before disconnecting any fuel lines. Failure to do so may result in fire or personal injury.

4. Disconnect the battery and remove it from the vehicle.

--- CAUTION ---

Wait 90 seconds from the time the key is turned to LOCK and the negative battery cable is disconnected to begin work. This allows the SRS capacitor to discharge and prevent deployment of the air bag(s).

5. Raise and safely support the vehicle.
6. Remove the engine under covers.
7. Drain the engine coolant.
8. Drain the engine oil.
9. Remove the radiator from the vehicle.
10. Remove the fan with the fluid coupling and fan pulleys.
11. Remove the air cleaner cap, MAF meter, and the resonator.
12. Disconnect the heater hoses.

13. Disconnect the following hoses:
- Brake booster vacuum hose
- EVAP hose
- A.D.D. vacuum hose
- Fuel return hose
- Fuel inlet hose

14. Disconnect the starter wire and connectors as follows:
    a. Remove the ground strap by removing the bolt.
    b. Disconnect the three starter wire clamps and connector.

15. Disconnect the alternator connector and wire.

16. Disconnect the engine wire from the cabin.
    a. Remove the glove box door.
    b. Lower the finish No. 2 panel.
    c. Disconnect the four ECM connectors.
    d. Disconnect the two cassette connectors and the two wire clamps from the lower finish panel.
    e. Disconnect the engine wire clamp.
    f. Disconnect the following:
      - Igniter connector
      - Ground strap
      - VSV connector for the EVAP
      - Vapor pressure sensor connector and clamp
      - Vapor connector for the vapor pressure sensor and clamp
    g. Remove the bolt and wire bracket.
    h. Remove the two nuts holding the engine wire retainer to the cowl panel and pull out the engine wire from the cabin.

17. If equipped with air conditioning, remove the bolt and disconnect the air conditioning compressor wire clamp.

18. If necessary, install a No. 2 engine hanger with two bolts. Torque the two bolts to 30 ft. lbs. (40 Nm).

19. Attach the engine hoist chain to the two engine hangers.

20. Remove the four bolts and nuts holding the engine front mounting insulators to the frame.

21. Lift the engine out of the vehicle.

To install:

22. Install the engine assembly to the vehicle. Attach the engine mounts to the body mountings. Install the bolts and nuts but do not torque at this time.

23. Remove the engine chain hoist the No. 2 engine hanger.

24. If equipped with air conditioning, connect the air conditioning wire with the bolt.

25. Torque the engine mounting nuts and bolts to 28 ft. lbs. (38 Nm).

26. Connect the engine wire to the cabin.
    a. Push the engine wire through the cowl panel.
    b. Install the bolt and wire bracket.
    c. Connect the engine wire clamp.
    d. Connect all wires, hoses and cables.

27. Install the fan with the fluid coupling and fan pulleys. Torque the nuts to 48 inch lbs. (5.4 Nm).

28. Install the air cleaner case and air filter.

29. Install the MAF meter, resonator, and the air cleaner cap.

30. Install the radiator to the vehicle.

31. Fill the engine with oil.

32. Fill the engine and radiator with coolant.

33. Install the transmission and refill it with transmission oil.

34. Install the engine undercover.

35. Install the battery and connect the battery cables.

36. Start the engine, make any necessary adjustments, install the hood, and check for leaks.

Engine Mounts

REMOVAL AND INSTALLATION

4Runner, 1993–94 T-100 and 1993–95 Pick-up

1. Disconnect the negative battery cable.
2. Raise and safely support the vehicle.

——— **WARNING** ———
Make certain the vehicle is properly supported when raising the engine.

3. Raise the engine slightly to take the tension off the motor engine mounts.

NOTE: When raising the engine to replace the engine mounts, DO NOT place the jack directly under the oil pan or the crankshaft torsional damper. Use a block of wood between the engine and the jack to prevent damage at the jacking point.

4. Remove the through-bolt nuts and the through-bolts.
5. Remove the engine mount bolts.
6. Raise the engine enough to remove the engine mounts.

NOTE: Only raise the engine enough to remove the mounts. Raising the engine too far could result in damage to some of the engine components.

To install:

7. Install the engine mounts and the engine bolts. Torque the bolts to:
- RH and LH mounts: 29 ft. lbs. (39 Nm)
- Rear mount 2-Wheel Drive: 19 ft. lbs. (25 Nm)
- Rear mount 4-Wheel Drive: 9 ft. lbs. (13 Nm)

8. Lower the vehicle and connect the negative battery cable.

1996–97 4Runner (2-Wheel Drive)

1. Disconnect the negative battery cable.
2. Raise and safely support the vehicle.

——— **WARNING** ———
Make certain the vehicle is properly supported when raising the engine.

3. Raise the engine slightly to take the tension off the motor engine mounts.

NOTE: When raising the engine to replace the engine mounts, DO NOT place the jack directly under the oil pan or the crankshaft torsional damper. Use a block of wood between the engine and the jack to prevent damage at the jacking point.

4. Remove the through-bolt nuts and the through-bolts.
5. Remove the engine mount bolts.
6. Raise the engine enough to remove the engine mounts.

NOTE: Only raise the engine enough to remove the mounts. Raising the engine too far could result in damage to some of the engine components.

To install:

7. Install the engine mounts and the engine bolts. Torque the bolts for the RH and LH mounts to 28 ft. lbs. (38 Nm).

8. Torque the rear engine mount bracket to:
- Bolt A: 13 ft. lbs. (18.5 Nm)
- Bolt B: 19 ft. lbs. (26 Nm)

9. Remove the engine support.
10. Lower the vehicle and connect the negative battery cable.

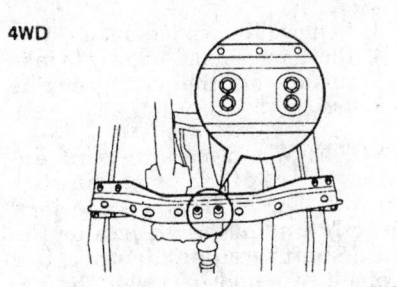

Rear engine mount — 1993–95 with 2WD

210690

Rear engine mount — 1993–95 4WD

210692

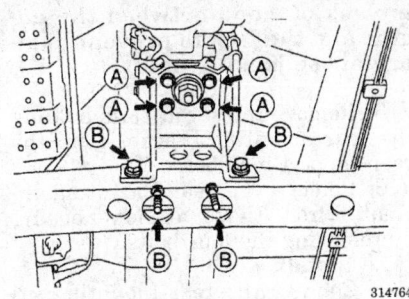

Bolt tightening pattern for the engine rear mounting bracket–1996–97 4Runner with 2.7L engine

314764

314722

Right and left engine mounts–1996–97 4Runner

1996–97 4Runner (4-Wheel Drive)

1. Disconnect the negative battery cable.
2. Raise and safely support the vehicle.

——— **WARNING** ———
Make certain the vehicle is properly supported when raising the engine.

3. Raise the engine slightly to take the tension off the motor engine mounts.

NOTE: When raising the engine to replace the engine mounts, DO NOT place the jack directly under the oil pan or the crankshaft torsional damper. Use a block of wood between the engine and the jack to prevent damage at the jacking point.

4. Remove the through-bolt nuts and the through-bolts.
5. Remove the engine mount bolts.
6. Raise the engine enough to remove the engine mounts.

NOTE: Only raise the engine enough to remove the mounts. Raising the engine too far could result in damage to some of the engine components.

To install:
7. Install the engine mounts and the engine bolts. Torque the bolts for the RH and LH mounts to 28 ft. lbs. (38 Nm).
8. Install the rear engine mount and torque the bolts to 48 ft. lbs. (65 Nm).
9. Remove the engine support.
10. Lower the vehicle and connect the negative battery cable.

1995–97 T-100

2.7L (3RZ-FE) — Left Hand Mount

1. Disconnect the negative battery cable.
2. Raise and safely support the vehicle.

——— **WARNING** ———
Make certain the vehicle is properly supported when raising the engine.

3. Remove the engine undercover.

4. Raise the engine slightly to take the tension off the motor engine mounts.

NOTE: When raising the engine to replace the engine mounts, DO NOT place the jack directly under the oil pan or the crankshaft torsional damper. Use a block of wood between the engine and the jack to prevent damage at the jacking point.

5. Remove the two bolts and two nuts holding the engine mount to the frame.
6. Remove the five bolts holding the engine mount to the engine.
7. Remove the engine mount from the engine.

To install:
8. Install the engine mount to the engine and install the five bolts. Torque the four bolts (A) to the engine to 34 ft. lbs. (45 Nm). Torque the bolt (B) facing sideways to 14 ft. lbs. (20 Nm).
9. Install the two bolts and two nuts to hold the engine mount to the vehicle frame. Torque the nuts and bolts to 28 ft. lbs. (38 Nm).
10. Install the engine undercover.
11. Lower the engine, lower the vehicle, and connect the negative battery cable.

2.7L (3RZ-FE) — Right Hand Mount

1. Disconnect the negative battery cable.
2. Raise and safely support the vehicle.

——— **WARNING** ———
Make certain the vehicle is properly supported when raising the engine.

3. Remove the engine undercover.
4. Raise the engine slightly to take the tension off the motor engine mounts.

NOTE: When raising the engine to replace the engine mounts, DO NOT place the jack directly under the oil pan or the crankshaft torsional damper. Use a block of wood between the engine and the jack to prevent damage at the jacking point.

5. Remove the two bolts and two nuts holding the engine mount to the frame.
6. Remove the four bolts holding the engine mount to the engine.
7. Remove the engine mount from the engine.
To install:
8. Install the engine mount to the engine and install the four bolts. Tor-

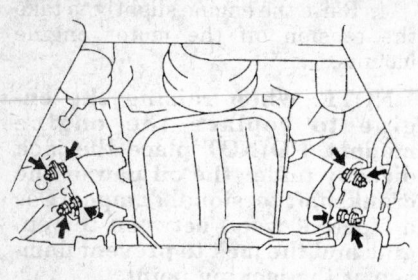

Engine mount to frame bolts and nuts — 1995–97 T-100 with 2.7L (3RZ-FE) engine (left side)

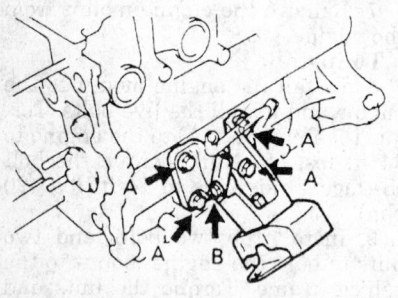

Left engine mount — 1995–97 T-100 with 2.7L (3RZ-FE) engine

que the four bolts to the engine to 34 ft. lbs. (45 Nm).

9. Install the two bolts and two nuts to hold the engine mount to the vehicle frame. Torque the nuts and bolts to 28 ft. lbs. (38 Nm).

10. Install the engine undercover.

11. Lower the engine, lower the vehicle, and connect the negative battery cable.

2.7L (3RZ-FE) — Transmission Mount

1. Disconnect the negative battery cable.

2. Raise and safely support the vehicle.

—— WARNING ——
Make certain the vehicle is properly supported when raising the engine.

3. Remove the engine undercover.

4. Raise the engine and transmission slightly to take the tension off the motor engine mounts.

NOTE: When raising the engine to replace the engine mounts, DO NOT place the jack directly under the oil pan or the crankshaft torsional damper. Use

a block of wood between the engine and the jack to prevent damage at the jacking point.

5. Remove the seven bolts holding the rear mounting bracket from the rear support member.

6. Remove the rear mounting insulator from the extension housing by removing the four bolts.

To install:

7. Connect the rear mounting insulator to the extension housing by installing the four bolts. Torque the four bolts to 19 ft. lbs. (25 Nm).

8. Install the rear mounting bracket to the rear support member by installing the seven bolts. Torque the frame side bolts to 19 ft. lbs. (25 Nm) and the rear mounting side bolts to 13 ft. lbs. (18 Nm).

9. Install the engine undercover.

10. Lower the engine/transmission, lower the vehicle, and connect the negative battery cable.

1995–97 T-100 and Tacoma

3.4L (5VZ-FE) — Left Hand Mount

1. Disconnect the negative battery cable.

2. Raise and safely support the vehicle.

—— WARNING ——
Make certain the vehicle is properly supported when raising the engine.

3. Remove the engine undercover.

4. Raise the engine slightly to take the tension off the motor engine mounts.

NOTE: When raising the engine to replace the engine mounts, DO NOT place the jack directly under the oil pan or the crankshaft torsional damper. Use a block of wood between the engine and the jack to prevent damage at the jacking point.

5. Remove the two bolts and two nuts holding the engine mount to the frame.

6. Remove the three bolts holding the engine mount to the engine.

7. Remove the engine mount from the engine.

To install:

8. Install the engine mount to the engine and install the three bolts. Torque the three bolts to the engine to 32 ft. lbs. (44 Nm).

9. Install the two bolts and two nuts to hold the engine mount to the

vehicle frame. Torque the nuts and bolts to 28 ft. lbs. (38 Nm).

10. Install the engine undercover.

11. Lower the vehicle and connect the negative battery cable.

3.4L (5VZ-FE) — Right Hand Mount

1. Disconnect the negative battery cable.

2. Raise and safely support the vehicle.

—— WARNING ——
Make certain the vehicle is properly supported when raising the engine.

3. Remove the engine undercover.

4. Raise the engine slightly to take the tension off the motor engine mounts.

NOTE: When raising the engine to replace the engine mounts, DO NOT place the jack directly under the oil pan or the crankshaft torsional damper. Use a block of wood between the engine and the jack to prevent damage at the jacking point.

5. Remove the two bolts and two nuts holding the engine mount to the frame.

6. Remove the four bolts holding the engine mount to the engine.

7. Remove the engine mount from the engine.

To install:

8. Install the engine mount to the engine and install the four bolts. Torque the four bolts to the engine to 32 ft. lbs. (44 Nm).

9. Install the two bolts and two nuts to hold the engine mount to the vehicle frame. Torque the nuts and bolts to 28 ft. lbs. (38 Nm).

10. Install the engine undercover.

11. Lower the vehicle and connect the negative battery cable.

Tacoma

2.4L (2RZ-FE) and 2.7L (3RZ-FE) — Left Hand Mount

1. Disconnect the negative battery cable.

2. Raise and safely support the vehicle.

—— WARNING ——
Make certain the vehicle is properly supported when raising the engine.

3. Remove the engine undercover.

4. Raise the engine slightly to take the tension off the motor engine mounts.

NOTE: When raising the engine to replace the engine mounts, DO NOT place the jack directly under the oil pan or the crankshaft torsional damper. Use a block of wood between the engine and the jack to prevent damage at the jacking point.

5. Remove the two bolts and two nuts holding the engine mount to the frame.
6. Remove the five bolts holding the engine mount to the engine.
7. Remove the engine mount from the engine.
To install:
8. Install the engine mount to the engine and install the five bolts. Torque the four bolts (A) to the engine to 34 ft. lbs. (45 Nm). Torque the bolt (B) facing sideways to 14 ft. lbs. (20 Nm).
9. Install the two bolts and two nuts to hold the engine mount to the vehicle frame. Torque the nuts and bolts to 28 ft. lbs. (38 Nm).
10. Install the engine undercover.
11. Lower the engine, lower the vehicle, and connect the negative battery cable.

2.4L (2RZ-FE) and 2.7L (3RZ-FE) — Right Hand Mount

1. Disconnect the negative battery cable.
2. Raise and safely support the vehicle.

——— **WARNING** ———
Make certain the vehicle is properly supported when raising the engine.

3. Remove the engine undercover.
4. Raise the engine slightly to take the tension off the motor engine mounts.

NOTE: When raising the engine to replace the engine mounts, DO NOT place the jack directly under the oil pan or the crankshaft torsional damper. Use a block of wood between the engine and the jack to prevent damage at the jacking point.

5. Remove the two bolts and two nuts holding the engine mount to the frame.
6. Remove the four bolts holding the engine mount to the engine.
7. Remove the engine mount from the engine.

To install:
8. Install the engine mount to the engine and install the four bolts. Torque the four bolts to the engine to 34 ft. lbs. (45 Nm).
9. Install the two bolts and two nuts to hold the engine mount to the vehicle frame. Torque the nuts and bolts to 28 ft. lbs. (38 Nm).
10. Install the engine undercover.
11. Lower the engine, lower the vehicle, and connect the negative battery cable.

2.4L (2RZ-FE) and 2.7L (3RZ-FE) — Transmission Mount

1. Disconnect the negative battery cable.
2. Raise and safely support the vehicle.

——— **WARNING** ———
Make certain the vehicle is properly supported when raising the engine.

3. Remove the engine undercover.
4. Raise the engine and transmission slightly to take the tension off the motor engine mounts.

NOTE: When raising the engine to replace the engine mounts, DO NOT place the jack directly under the oil pan or the crankshaft torsional damper. Use a block of wood between the engine and the jack to prevent damage at the jacking point.

5. Remove the eight bolts holding the rear mounting bracket from the rear support member.
6. Remove the rear mounting insulator from the extension housing by removing the four bolts.
To install:
7. Connect the rear mounting insulator to the extension housing by installing the four bolts. Torque the four bolts to 48 ft. lbs. (65 Nm).
8. Install the rear mounting bracket to the rear support member by installing the eight bolts. Torque the frame side bolts to 19 ft. lbs. (25 Nm) and the rear mounting side bolts to 13 ft. lbs. (18 Nm).
9. Install the engine undercover.
10. Lower the engine/transmission, lower the vehicle, and connect the negative battery cable.

Cylinder Head

REMOVAL AND INSTALLATION

2.4L (22R-E) Engine

NOTE: The rocker arms and rocker arm shaft are secured by the cylinder head bolts. The cylinder head, camshaft, and valve train should be removed as an assembly and then disassembled off the vehicle.

1. Release the fuel system pressure.
2. Disconnect the negative battery cable.

——— **CAUTION** ———
Wait at least 90 seconds after the negative (-) battery cable is disconnected to prevent possible deployment of the air bag.

3. Drain the engine coolant.
4. Remove the intake air connector.
5. Disconnect the exhaust pipe from the exhaust manifold.
6. Remove the oil dipstick.
7. Disconnect the spark plug wires from the spark plugs. Make note of the proper firing order for installation.
8. Remove the distributor and the spark plugs.
9. Remove the radiator inlet hose.
10. Disconnect the heater water inlet hose from the heater water inlet pipe.
11. Disconnect the accelerator cable.
12. If equipped with automatic transmission, disconnect the throttle cable.
13. Disconnect the ground strap from the engine rear side.
14. Disconnect the following:
  a. No. 1 and No. 2 PCV hose
  b. Brake booster hose
  c. With power steering, the air control valve hoses
  d. With air conditioning, the VSV hoses
  e. EVAP hose
  f. EGR vacuum modulator and EGR valve hoses
  g. Fuel pressure up hose
  h. PAIR valve hose
  i. Pressure regulator hose
  j. Vacuum hoses from the throttle body
  k. No. 2 and No. 3 water bypass hoses from the throttle body
  l. With oil cooler, disconnect the No. 1 oil cooler hose from the intake manifold

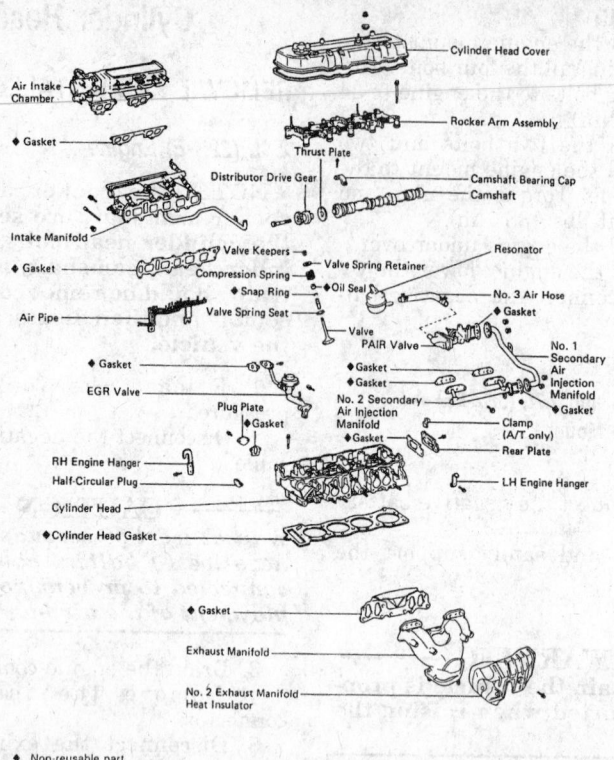

◆ Non-reusable part

339498

Cylinder head and related components — 2.4L (22R-E) engine

m. Without oil cooler, disconnect the No. 1 water bypass hose from the intake manifold

15. Remove the EGR vacuum modulator.

16. Disconnect the following wires:
   a. Cold start injector wire
   b. Throttle position wire
   c. California only, EGR gas temperature sensor wire

17. Remove the air intake chamber with the throttle body.

18. Disconnect the fuel return hose.

19. Disconnect the following wires:
   a. Knock sensor wire
   b. Oil pressure sender gauge wire
   c. Starter wire from terminal 50.
   d. Transmission wires
   e. With air conditioning, the compressor wires
   f. Fuel injector wires
   g. Engine coolant temperature sender gauge wire
   h. With automatic transmission, the OD temperature switch wire
   i. Igniter wire
   j. VSV wires
   k. Cold start injector time switch wire
   l. ECT sensor wire

20. Disconnect the fuel hose from the delivery pipe.

CAUTION

Fuel injection systems remain under pressure after the engine has been turned OFF. Properly relieve fuel pressure before disconnecting any fuel lines. Failure to do so may result in fire or personal injury.

---

21. Disconnect the bypass hose from the intake manifold.

22. If equipped with power steering, remove the drive belt.

23. Remove the power steering bracket.

24. Remove the nuts, grommets, cylinder head cover, and the gasket.

25. Set the No. 1 cylinder to TDC of the compression stroke. Place matchmarks on the sprocket and the chain. Remove the camshaft sprocket bolt.

26. Remove the distributor drive gear and the camshaft thrust plate.

27. Remove the camshaft sprocket.

28. Remove the timing chain cover bolt.

WARNING

The timing chain cover bolt must be removed prior to loosening any of the cylinder head / rocker arm shaft bolts.

---

29. Remove the rocker arm assembly by removing the cylinder head bolts in the reverse order of tightening sequence. Loosen the bolts in two to three stages.

30. Remove the cylinder head rear cover.

31. Remove the cylinder head.

32. Remove the rocker shaft assembly, camshaft caps, and the camshaft.

33. To remove the rocker arms, remove the screws from the rocker shafts and slide the rocker shaft stands, rocker arms, and springs from the rocker shafts.

NOTE: Keep all parts in order as they are removed. All parts must be reinstalled in their original position.

34. Remove the intake manifold, exhaust manifold, the right engine hanger, the left engine hanger, the EGR valve, and the ground strap connector.

To install:

35. Place the camshaft in the cylinder head and install the bearing caps in the numbered order from the front with the arrows pointed toward the front. Torque the bolts to 14 ft. lbs. (20 Nm). Turn the camshaft to position the dowel at the top.

36. Install the cylinder head rear cover.

37. Install the left engine hanger and ground strap.

38. Install the right engine hanger.

39. Install the exhaust manifold. Torque the nuts to 33 ft. lbs. (44 Nm). Install the heat insulator and torque the bolts to 14 ft. lbs. (19 Nm).

40. Install the EGR valve.

41. Install the intake manifold. Torque the nuts and bolts to 14 ft. lbs. (19 Nm).

42. Apply seal packing and to the cylinder block. Only apply sealant at the two areas where the timing chain cover meets the cylinder block.

43. Install the cylinder head on the cylinder block. Be sure to align the dowel pins on the block with the cylinder head.

44. Assemble the rocker arms, springs, and the rocker shaft stands in the same order that they were removed.

45. Place the rocker arm assembly over the dowels on the cylinder head.

46. Install and tighten the head bolts. Following the proper sequence, torque the bolts in three stages to 58 ft. lbs. (78 Nm).

47. Align matchmarks made during removal and install the camshaft sprocket and chain.

48. Install the chain cover bolt and torque it to 9 ft. lbs. (13 Nm).

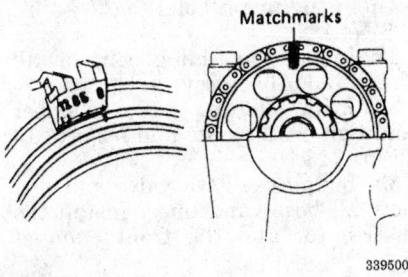

Matchmark the camshaft sprocket to the timing chain — 2.4L (22R-E) engine

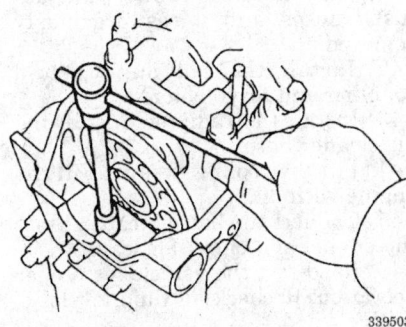

Removing the timing chain cover bolt — 2.4L (22R-E) engine

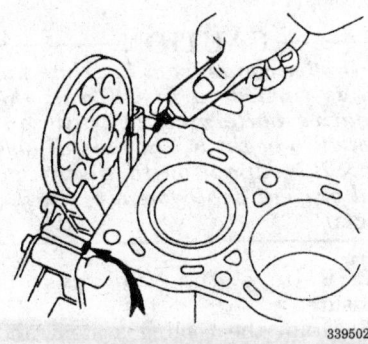

Applying sealant to the cylinder block — 2.4L (22R-E) engine

49. Install the distributor drive gear and camshaft thrust plate. Torque the bolt to 58 ft. lbs. (78 Nm).

50. Check and adjust valve clearance.

51. Install the cylinder head cover with the grommets and the four nuts. Torque the nuts to 4 ft. lbs. (5 Nm).

52. If removed, install the power steering bracket and tighten the bolts to 33 ft. lbs. (44 Nm). Install the drive belt and adjust the belt tension. Install the remaining components.

53. Check the engine oil level and add as necessary.

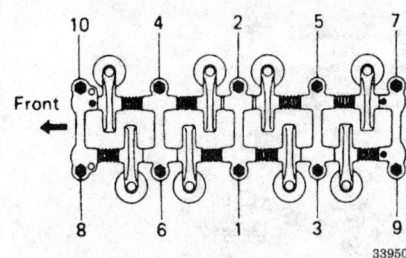

Cylinder head bolt tightening sequence — 2.4L (22R-E) engine

54. Fill the engine coolant.

55. Connect the negative battery cable.

56. Start the engine, check for leaks, and bleed the cooling system.

57. Perform engine adjustments (ignition timing, etc.).

58. Road test the vehicle for proper operation.

3.0L (3VZ-E) Engine

1993–94 Pick-up, 4Runner and T-100

1. Disconnect the negative battery cable.

2. Remove the air cleaner hose and case.

3. Drain the engine coolant.

4. Remove the radiator.

5. Unbolt the power steering pump and position it out of the way with the hoses still attached.

6. Remove all drive belts and then remove the fluid coupling and fan pulley.

7. Tag and disconnect all wires and connectors that will interfere with cylinder head removal.

8. Disconnect the following hoses:
  a. Power steering air hoses
  b. Brake booster hose
  c. Cruise control vacuum hose
  d. Charcoal canister has at the canister
  e. VSV vacuum hose.

9. Disconnect the accelerator, throttle and cruise control cables.

10. Disconnect the clutch release cylinder hose (manual transmission only).

11. Disconnect the heater hoses and the fuel lines.

12. Remove the left side scuff plate and disconnect the O_2 sensor and then remove the front exhaust pipe.

13. Remove the timing belt.

14. Remove the distributor with the spark plug leads attached; position it out of the way.

15. Remove the air intake chamber.

16. Disconnect the connectors and then remove the engine wire.

17. Remove the Nos. 2 and 3 fuel pipes.

18. Remove the No. 4 timing belt cover.

19. Remove the No. 2 idler pulley and the No. 3 timing belt cover.

20. Disconnect the hose and remove the water bypass outlet.

21. Remove the intake manifold. Remove the exhaust crossover pipe.

22. Remove the following from right-hand side:
  a. Remove the reed valve with the No. 1 air injection manifold.
  b. Remove the water bypass pipe mounting bolt.
  c. Remove the cylinder head cover.
  d. Remove the camshaft.
  e. Loosen the cylinder head bolts in several stages, in the opposite order of the tightening sequence.
  f. Remove the air pump bracket and engine hanger.
  g. Lift the cylinder head off of its mounting dowels, do not pry it off.

23. Remove the following from the left-hand side:
  a. Remove the alternator.
  b. Remove the oil dipstick guide tube.
  c. Remove the cylinder head cover.
  d. Remove the camshaft.
  e. Loosen the cylinder head bolts in several stages, in the opposite order of the tightening sequence.
  f. Remove the air pump bracket and engine hanger.
  g. Lift the cylinder head off of its mounting dowels, do not pry it off.

To install:

24. Install the cylinder head on the cylinder block using a new gasket.

25. Lightly coat the threads of the cylinder head bolts with engine oil and then install them into the head.
  a. Alternately tighten them in several passes, in the correct sequence to 33 ft. lbs. (44 Nm).
  b. After the initial tightening, mark the front side the top of the bolt with paint. Tighten the bolts an additional 90 degrees (¼ turn) and check that the mark is now facing the side of the head.
  c. Tighten the bolts an additional 90 degrees and check that the mark is now facing the rear of the head. Install the bolt (A) and tighten it to 27 ft. lbs. (37 Nm).

26. Install the camshaft.

27. Install the alternator and the water bypass pipe mounting bolt.

28. Install the reed valve with the No. 1 injection manifold.

29. Install the oil dipstick tube.

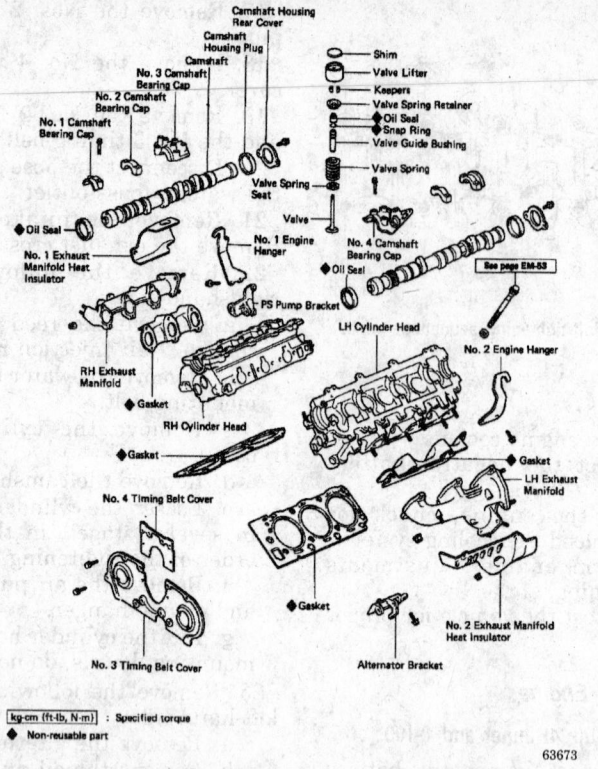

Cylinder head components — 1993–94 Pick-up, 4Runner and T-100 with 3.0L (3VZ-E) engine

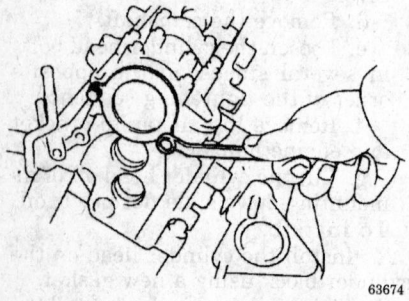

Removing camshaft housing rear cover — 1993–95 Pick-up, 4Runner and T-100 with 3.0L (3VZ-E) engine

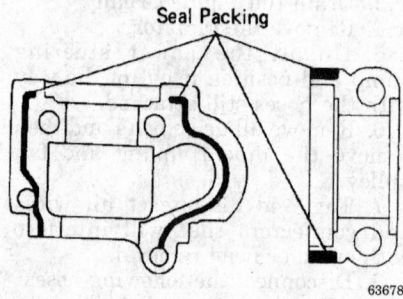

Apply sealant to No. 1 and No. 3 bearing caps as shown — 1993–94 3.0L (3VZ-E) engine

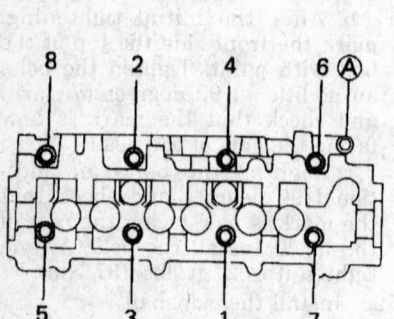

Tighten cylinder head bolts in numerical sequence shown — 1993–94 Pick-up, 4Runner and T-100 with 3.0L (3VZ-E) engine

30. Install the crossover pipe and tighten it to 29 ft. lbs. (39 Nm).

31. Connect the oxygen sensor wire.

32. Install the intake manifold with new gaskets and tighten the mounting bolts to 29 ft. lbs. (39 Nm).

33. Install the water bypass outlet and tighten the bolts to 13 ft. lbs. (18 Nm).

34. Install the fuel delivery pipes and injectors.

35. Install the No. 2 idler pulley. Install the Nos. 3 and 4 timing belt covers and tighten the bolts to 74 inch lbs. (8 Nm).

36. Install the fuel pipes and tighten the union bolts to 22 ft. lbs. (29 Nm).

37. Install the timing belt. Install the cylinder head covers.

38. Install the air intake chamber and tighten the nuts and bolts to 13 ft. lbs. (18 Nm).

39. Install the EGR valve and connect all hoses and lines. Install the distributor and the front exhaust pipe.

40. Connect the fuel lines and heater hoses. Connect the clutch release cylinder hose.

41. Install the power steering pump. Connect all cables (and adjust), hoses and wires previously removed.

42. Install the fan pulley, fluid coupling and drive belts.

43. Install the radiator. Install the air cleaner hose.

44. Fill the cooling system. Fill the engine with oil.

45. Connect the battery cable. Start the engine and check for leaks.

46. Road test the vehicle for proper operation. Recheck all fluid levels.

1995 Pick-up and 4Runner

1. Disconnect the negative battery cable.

— **CAUTION** —

Wait 90 seconds from the time the key is turned to LOCK and the negative battery cable is disconnected to begin work. This allows the SRS capacitor to discharge and prevent deployment of the air bag(s).

2. Relieve the fuel system pressure.

3. Drain the cooling system and the engine oil.

4. Disconnect the air cleaner and the hose.

5. Remove the radiator.

6. On manual transmissions, disconnect the clutch release cylinder hose.

7. Remove the power steering belt and remove the pump.

8. Remove the air conditioning belt.

9. Remove the cooling fan.

10. Remove the alternator belt.

11. Tag and disconnect all of the wires and connectors that interfere in the cylinder head removal.

12. Disconnect the following hoses:

   a. Power steering air hoses from the gas filter and the air pipe

   b. Brake booster hose

   c. Cruise control vacuum hose, if equipped

d. Charcoal canister hose from the canister

e. VSV vacuum hoses

13. Disconnect the following cables:
 a. Accelerator cable
 b. Automatic transmission: Throttle cable
 c. W/Cruise: Cruise control cable

14. Disconnect the heater hoses.

15. Disconnect the fuel inlet and the outlet hoses.

CAUTION

Fuel injection systems remain under pressure after the engine has been turned OFF. Properly relieve fuel pressure before disconnecting any fuel lines. Failure to do so may result in fire or personal injury.

---

16. Remove the front exhaust pipe.

17. Remove the spark plug wires and the distributor.

18. Remove the timing belt.

19. Remove the air intake chamber.

20. Remove the following connectors and wires:
 a. Knock sensor
 b. Cold start injector time switch connector
 c. ECT sensor and sender gauge connector
 d. No. 1 ECT switch connector
 e. Right ground strap from the No. 3 camshaft bearing cap
 f. Injector connectors

21. Remove the engine harness.

22. Remove the union bolts and then remove the No. 2 and 3 fuel pipes.

23. Remove the No. 4 timing belt cover. Remove the No. 2 idler pulley and the No. 3 timing belt cover.

24. Remove the VSV bracket and the VSV from the PAIR reed valve.

25. Remove the PAIR reed valve and the No. 1 air injection manifold.

26. Remove the fuel delivery pipes with their injectors.

27. Remove the water bypass outlet.

28. Remove the intake manifold.

29. Remove the knock sensor wire.

30. Remove the exhaust crossover pipe.

31. Disconnect the water bypass pipe from the right cylinder head.

32. Remove the alternator.

33. Remove the oil dipstick guide and the dipstick.

34. Remove the No. 2 engine hanger from the left side cylinder head.

35. Remove the cylinder head covers.

36. Remove the camshaft bearing cap bolts in reverse order of the tight-ening sequence. Remove the camshafts.

NOTE: Arrange the bearing caps in correct order for installation.

37. Remove the cylinder head bolts in the reverse order of tightening sequence, in several passes. Remove the cylinder heads.

38. Remove the alternator bracket.

39. Remove the exhaust manifold from the right side cylinder head. Remove the exhaust manifold from the left cylinder head.

40. Remove the camshaft housing plugs.

41. Remove the valve lifters and shims by hand. Arrange the valve lifters and shims in correct order for re-installation.

To install:

42. Install the valve lifters and shims. Check that the valve lifter rotates smoothly by hand.

43. Install the camshaft housing plugs with the cup side facing inward. Torque the rear plate bolts to 4 ft. lbs. (5 Nm).

44. Install the exhaust manifolds to the right and left cylinder heads. Torque the nuts to 29 ft. lbs. (39 Nm).

45. Install the alternator bracket and torque the bolts to 27 ft. lbs. (37 Nm).

46. Install the cylinder heads with new gaskets. Tighten the bolts in several passes and in sequence. Torque the bolts to 33 ft. lbs. (44 Nm).

47. Mark the front of the cylinder head bolt head with paint. Retighten the cylinder head bolts in numerical order an additional 90°.

48. Retighten the cylinder head bolts by an additional 90°. Check that the painted mark is now facing rearward.

WARNING

Do not attempt to combine steps 49 and 50. The correct sequence is 33 ft. lbs. + 90° + 90°. It is very important that these instructions are followed precisely to avoid damage to the cylinder head.

---

49. Install the cylinder head six pointed head bolt to each head and torque to 30 ft. lbs. (41 Nm).

50. Install the camshafts. Install the bearing caps in their proper locations and torque the bearing cap bolts to 12 ft. lbs. (16 Nm).

51. Check and adjust valve clearance, if necessary.

52. Install the cylinder head covers and torque to 4 ft. lbs. (5.4 Nm).

53. Install the water bypass pipe to the right side cylinder head.

54. Install the No. 2 engine hanger and torque the bolts to 30 ft. lbs. (40 Nm).

55. Install the oil dipstick guide and the dipstick. Torque the holding bolt to 27 ft. lbs. (37 Nm)

56. Install the alternator.

57. Install the exhaust crossover pipe and torque the bolts to 29 ft. lbs. (39 Nm).

58. Install the knock sensor wire.

59. Install the intake manifold with new gaskets and tighten the mounting bolts to 13 ft. lbs. (18 Nm).

60. Install the water bypass outlet and tighten the 2 bolts to 13 ft. lbs. (18 Nm). Connect the No. 3 water bypass hose to the No. 1 water bypass pipe.

61. Install the fuel delivery pipes and injectors. Torque the delivery pipe holding nuts to 9 ft. lbs. (13 Nm).

62. Install the PAIR reed valve and the No. 1 injection manifold. Torque the bolts to 27 ft. lbs. (37 Nm) and the nuts to 22 ft. lbs. (29 Nm).

63. Install the VSV bracket and the VSV to the PAIR reed valve.

64. Install the No. 2 idler pulley. Torque the bolts to 13 ft. lbs. (18 Nm).

65. Install the No. 3 and No. 4 timing belt covers and tighten the bolts to 74 inch lbs. (8.3 Nm).

66. Install the fuel pipes and tighten the union bolts to 25 ft. lbs. (34 Nm). Connect the vacuum hose to the TVV.

67. Install the engine harness. Connect the following connectors and straps:
 a. Injector connectors
 b. Right ground strap from the No. 3 camshaft bearing cap
 c. No. 1 ECT switch connector
 d. ECT sensor and sender gauge connector
 e. Cold start injector time switch connector
 f. Knock sensor

68. Install the air intake chamber and tighten the nuts and bolts to 13 ft. lbs. (18 Nm).

69. Install the timing belt.

70. Install the distributor and the spark plug wires.

71. Install the front exhaust pipe. Torque the nuts holding the pipe to the left exhaust manifold to 46 ft. lbs. (62 Nm) and the bolts holding the pipe to the catalytic converter to 29 ft. lbs. (39 Nm).

72. Connect the fuel inlet and outlet hoses.

73. Connect the heater hoses.

74. Connect the cables, straps, wiring, connectors and hose removed prior.

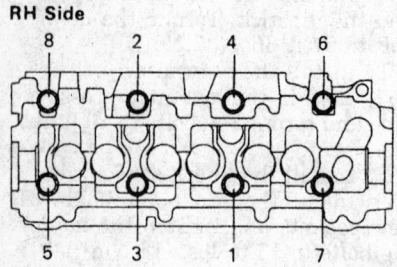

RH Side

LH Side

Cylinder head bolt tightening sequence — 3.0L (3VZ-E) engine

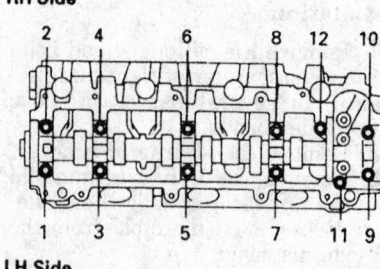

RH Side

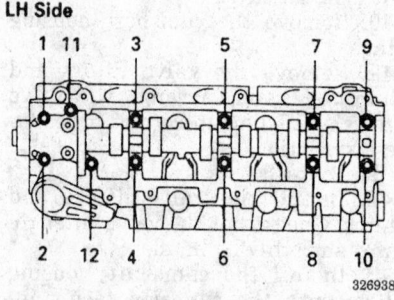

LH Side

Tighten bearing cap bolts in numerical sequence shown — 3.0L (3VZ-E) engine

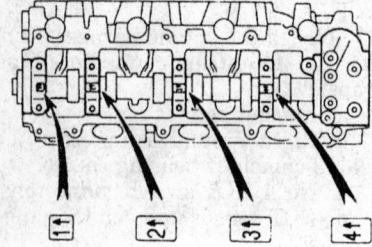

RH Side

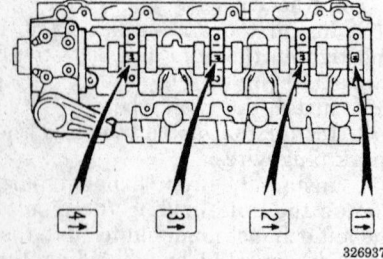

LH Side

Position bearing caps with arrows pointing toward the front (RH side) or rear (LH side) — 3.0L (3VZ-E) engine

6. Remove the intake air connector.

7. If equipped with air conditioning, remove the air conditioning idle-up valve.

8. Remove the power steering drive belt, idler pulley, pump and bracket.

9. Remove the No. 1 and No. 2 PCV hoses.

10. Remove the distributor connector, hold-down bolts, and the distributor.

11. Remove the water housing.

12. Remove the throttle body.

13. Disconnect the following connectors:
- If equipped with air conditioning, disconnect the air conditioning compressor connector.
- Disconnect the oil pressure sensor connector.
- Disconnect the ECT sensor connector.
- Disconnect the EGR gas temperature sensor connector.
- Disconnect the EGR VSV connector.

14. Disconnect the engine wire as follows:

   a. Remove the two bolts and disconnect the engine wire from the intake chamber.

   b. Disconnect the five engine wire clamps and engine wire.

   c. Disconnect the following connectors:
- Knock sensor connector
- Crankshaft position sensor connector
- Fuel pressure control VSV connector

   d. Disconnect the DLC1 from the bracket.

   e. Disconnect the two engine wire clamps.

   f. Remove the bolt and disconnect the engine wire from the engine.

15. Disconnect the fuel injectors.

16. Remove the cylinder head rear cover by disconnecting the heater by-pass hose and removing the three bolts.

17. Remove the EGR valve and vacuum modulator.

18. Remove the intake chamber stay by removing the two bolts.

75. Install the alternator drive belt.

76. Install the cooling fan and torque the nuts to 4 ft. lbs. (5.9 Nm).

77. Install the air conditioning belt.

78. Install the power steering pump and drive belt.

79. If removed, connect the clutch release cylinder hose.

80. Install the radiator.

81. Install the air cleaner and the hose.

82. Fill the cooling system and fill the engine with oil.

83. Connect the battery cable, start the engine, and check for leaks.

84. Road test the vehicle for proper operation and recheck all the fluid levels.

2.4L (2RZ-FE) and 2.7L (3RZ-FE) Engines

1995 Tacoma and 1994–95 T-100

1. Disconnect the negative battery cable.

2. Drain the engine coolant.

3. Remove the air cleaner cap, MAF meter, and the resonator.

4. If equipped with a manual transaxle, disconnect the accelerator cable from the throttle body.

5. If equipped with a automatic transaxle, disconnect the accelerator and throttle cables from the throttle body.

— CAUTION —

Fuel injection systems remain under pressure after the engine has been turned OFF. Properly relieve fuel pressure before disconnecting any fuel lines. Failure to do so may result in fire or personal injury.

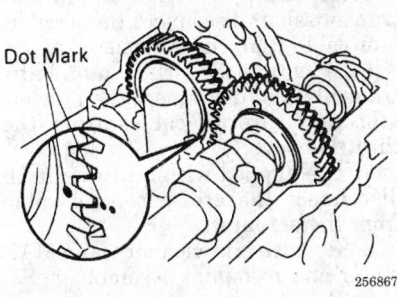

Camshafts TDC/compression timing marks —
2.4L (2RZ-FE) and 2.7L (3RZ-FE) engines

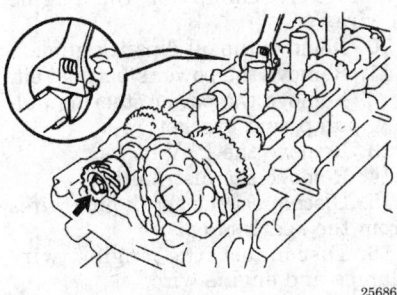

Hold the camshaft at the hexagon wrench head
portion — 2.4L (2RZ-FE) and 2.7L (3RZ-FE)
engines

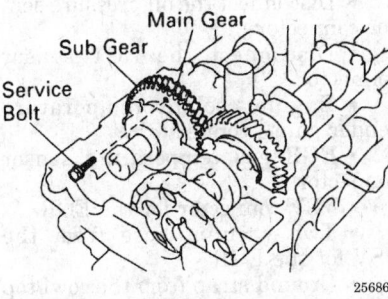

Secure the exhaust camshaft sub-gear to the
main gear with a service bolt — 2.4L (2RZ-FE)
and 2.7L (3RZ-FE) engines

19. Remove the fuel return pipe by removing the hoses and two bolts.

20. Remove the intake chamber assembly.

21. Remove the fuel inlet tube by removing the union bolts.

22. Remove the delivery pipe and injectors.

NOTE: Be careful not to drop the injectors when removing the delivery pipe.

23. Remove the intake manifold by removing the three bolts and two nuts.

24. Disconnect the front exhaust pipe from the exhaust manifold.

25. Remove the heat insulator by removing the two bolts and two nuts.

26. Remove the exhaust manifold and gasket by removing the six nuts.

27. Remove the No. 1 and No. 2 engine hangers.

28. Remove the cylinder head cover by removing the ten bolts.

29. Remove the spark plug wires and plugs from the engine.

30. Set No. 1 cylinder to TDC compression stroke. The groove on the crankshaft pulley should align with the **0** mark on the timing chain cover and the timing marks (one and two dots) of the camshaft gears should form a straight line in respect to the cylinder head surface. If not, turn the crankshaft 1 revolution (360 degrees).

31. Remove the chain tensioner and gasket by removing the two nuts.

32. Remove the camshaft timing gear.

NOTE: Since the thrust clearance of the camshaft is small, the camshaft must be kept level while it is being removed. If the camshaft is not kept lever, the portion of the cylinder head receiving the shaft thrust may crack or be damaged, causing the camshaft to seize or break.

33. Remove exhaust and intake camshafts.

34. Remove the 2 bolts in the front of the head before the other head bolts are removed. Uniformly loosen and remove the remaining head bolts, in the reverse order of the tightening several passes, in the sequence shown.

35. Lift the cylinder head from the block and place the head on wooden blocks on a bench.

To install:

36. Before installing, thoroughly clean the gasket mating surfaces and check for warpage.

37. Apply sealant (PN 08826–00080 or equivalent) to the 2 locations, as shown. Place a new head gasket on the block and install the cylinder head.

38. Lightly coat the cylinder head bolts with engine oil. Install the bolts and tighten in several passes in the sequence shown:

   a. Torque all bolts to 29 ft. lbs. (39 Nm)

   b. Mark the front of the bolt with paint and retighten bolts 90 degrees in the proper sequence.

   c. Retighten an additional 90 degrees. Check that the painted mark is now facing rearward.

39. Install and tighten the 2 front mounting bolts to 15 ft. lbs. (21 Nm).

NOTE: If any of the bolts break, deform or do not meet the torque specification, replace them.

40. Install the intake and exhaust camshafts.

41. Set No. 1 cylinder to TDC compression stroke: Crankshaft pulley groove align with **0** mark on timing cover and camshafts timing marks with one dot and two dots will be straight line on the cylinder head surface.

42. Install the timing gear. Place the gear over the straight pin of the intake camshaft.

43. Install the chain tensioner, using a new gasket (mark toward the front).

44. Check and adjust the valve clearance. Intake valve clearance is 0.006–0.010 inch (0.15–0.25 mm) and exhaust valve clearance is 0.010–0.014 inch (0.25–0.35 mm).

45. Recheck the engine for proper valve timing. Check and adjust the valve clearance.

46. Install the spark plugs and the semi-circular plug.

47. Recheck the engine for proper valve timing. Install the valve cover and engine hangers. Torque the engine hanger bolts to 30 ft. lbs. (42 Nm).

48. Install the exhaust manifold and gasket to the engine and install the six nuts. Torque the nuts to 36 ft. lbs. (49 Nm).

49. Install the heat insulator with the two bolts and two nuts. Torque the bolts and nuts to 48 inch lbs. (5.5 Nm).

50. Install the front exhaust pipe to the exhaust manifold.

51. Install the intake manifold using a new gasket. Torque the bolts and nuts to 22 ft. lbs. (29 Nm).

52. Install the injectors and delivery pipe. Tighten the bolts holding the delivery pipe to 15 ft. lbs. (21 Nm).

53. Install the fuel tube with four new gaskets. Torque the union bolts to 22 ft. lbs. (29 Nm).

54. Install the air intake chamber.

55. Install the fuel return pipe by installing the two bolts and hoses.

56. Install the intake chamber stay by installing the two bolts. Torque the bolts to 14 ft. lbs. (20 Nm).

57. Install the EGR valve, vacuum modulator and any other hoses associated with the units.

58. Install the cylinder head rear cover with a new gasket. Torque the three bolts to 10 ft. lbs. (14 Nm).

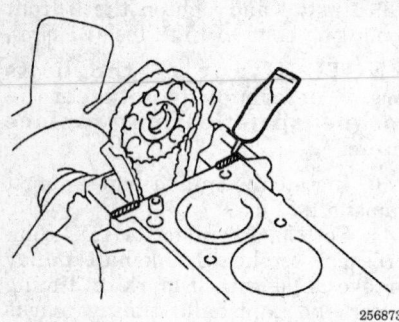

Apply sealant to the 2 locations as shown — 2.4L (2RZ-FE) and 2.7L (3RZ-FE) engines

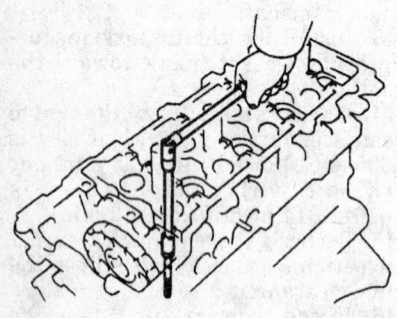

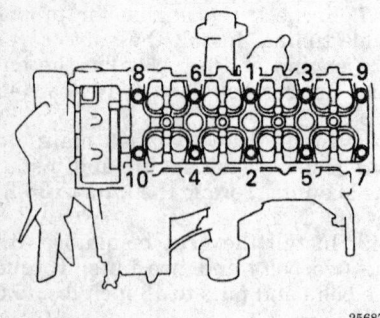

Uniformly tighten the cylinder head bolts in the sequence shown — 2.4L (2RZ-FE) and 2.7L (3RZ-FE) engines

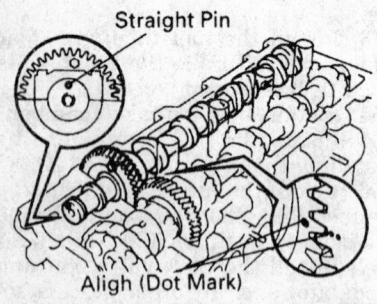

Straight Pin

Aligh (Dot Mark)

Engage both camshaft gears while matching the timing marks as shown — 2.4L (2RZ-FE) and 2.7L (3RZ-FE) engines

59. Connect the heater water by-pass pipe.

60. Connect the injector connectors.

61. Connect the engine wire to the engine as follows:

  a. Connect the engine wire to the intake manifold with the bolt.

  b. Connect the two engine wire clamps.

  c. Connect the DLC1 to the bracket.

  d. Connect the following connectors:

  • Fuel pressure control VSV connector

  • Knock sensor connector

  • Crankshaft position sensor connector

  e. Connect the five engine wire clamps.

  f. Connect the engine wire to the intake chamber with the two bolts.

62. Attach the connectors removed prior.

63. Install the throttle body.

64. Install the water outlet with a new gasket. Install the two bolts and torque the bolts to 14 ft. lbs. (20 Nm). Connect the ECT sender gauge connector and radiator inlet hose.

65. Install the distributor.

66. Install the No. 1 and No. 2 PCV hoses.

67. Install the power steering pump bracket and pump.

68. Install the power steering drive belt and idler pulley.

69. If equipped with air conditioning, install the air conditioning idle up valve.

70. Install the air intake connector by installing the two bolts, hose clamp, and two air hoses.

71. If equipped with a manual transaxle, connect the accelerator cable to the throttle body.

72. If equipped with a automatic transaxle, connect the throttle and accelerator cables to the throttle body.

73. Install the MAF meter, resonator, and the air cleaner cap.

74. Refill the cooling system. Drain and refill the engine oil, if required.

75. Connect the negative battery cable. Start the engine and check for leaks.

76. Check the ignition timing. Road test the vehicle for proper operation.

77. Recheck all fluid levels.

1996–97 4Runner

1. Release the fuel system pressure.

2. Disconnect the negative battery cable.

3. Drain the engine coolant.

4. Remove the air cleaner cap, MAF meter and resonator.

5. If equipped with a manual transmission, disconnect the accelerator cable from the throttle body.

6. If equipped with a automatic transmission, disconnect the accelerator and throttle cables from the throttle body.

7. If equipped with cruise control, disconnect the cruise control cable from the actuator.

8. Remove air cleaner cap, MAF meter and resonator assembly.

9. Remove the intake air connector. Disconnect the following:

  a. Air hose for IAC

  b. Vacuum sensing hose

  c. Wire clamp for the engine wire

10. Remove the oil dipstick guide.

11. Remove the power steering belt.

12. Remove the power steering pulley, pump, and bracket.

13. Remove the PCV hoses.

14. Remove the distributor.

15. Disconnect the spark plug wires from the spark plugs.

16. Disconnect the engine wire clamps and engine wire.

17. Disconnect the following connectors:

  • If equipped with air conditioning, disconnect the air conditioning compressor connector

  • Disconnect the oil pressure sensor connector

  • Disconnect the ECT sensor connector

  • Engine coolant temperature sender gauge connector

  • EGR gas temperature sensor connector

  • VSV connector for the EGR

  • Two vacuum hose from the VSV for the EGR

  • Ground strap from the cowl top panel

  • Engine wire from the air intake chamber

  • Throttle position sensor connector

  • IAC valve connector

  • Crankshaft position sensor connector

  • Knock sensor connector

  • DLC1 from the bracket

  • Engine wire clamp

18. Remove the EGR pipe.

19. Remove the intake chamber stay.

20. Remove the air intake chamber assembly.

21. Disconnect the following hoses:

  a. EVAP hose from the throttle body

  b. Brake booster vacuum hose from the union

  c. Water bypass hose from the water bypass pipe

d. Water bypass hose from the cylinder head rear cover

22. Disconnect the injector connectors.

CAUTION

Fuel injection systems remain under pressure, even after the engine has been turned OFF. The fuel system pressure must be released before disconnecting any fuel lines. Failure to do so may result in fire and/or personal injury.

---

23. Remove the fuel inlet pipe.
24. Disconnect the hoses and remove the fuel return pipe.
25. Remove the delivery pipe and injectors.
26. Remove the intake manifold.
27. Remove the front exhaust pipe.
28. Remove the exhaust manifold and gasket.
29. Remove the water outlet.
30. Remove the cylinder head rear cover.
31. Remove the spark plugs.
32. Remove the front engine hanger.
33. Remove the engine wire brackets.
34. Remove the cylinder head cover.
35. Set No. 1 cylinder to TDC compression stroke. The groove on the crankshaft pulley should align with the **0** mark on the timing chain cover and the timing marks (one and two dots) of the camshaft gears should form a straight line in respect to the cylinder head surface. If not, turn the crankshaft 1 revolution (360 degrees).
36. Remove the chain tensioner and gasket.
37. Remove the camshaft timing gear.
38. Remove exhaust camshafts.
39. To remove the intake camshaft, uniformly loosen and remove the bearing cap bolts in the reverse order of the tightening in several passes, in the sequence shown. Remove the bearing caps and camshaft. Make a note of the bearing cap positions for proper installation.

NOTE: If the camshaft is not being lifted out straight and level, reinstall the No. 3 bearing cap with the two bolts. Then alternately loosen and remove the two bearing cap bolts with the camshaft gear pulled up.

40. Remove the valve lifters and shims from the cylinder head. Arrange the valve lifters and shims in correct order.

41. Remove the cylinder head, uniformly loosen and remove the cylinder head bolts in the reverse order of the tightening in the sequence shown, in several passes.

To install:

42. Before installing, thoroughly clean the gasket mating surfaces and check for warpage.
43. Apply sealant (PN 08826–00080 or equivalent) to the 2 locations, as shown. Place a new head gasket on the block and install the cylinder head.
44. Lightly coat the cylinder head bolts with engine oil. Install the bolts and tighten in several passes in the sequence shown:
    a. Torque all bolts to 29 ft. lbs. (39 Nm)
    b. Mark the front of the bolt with paint and retighten bolts 90 degrees in the proper sequence.
    c. Retighten an additional 90 degrees. Check that the painted mark is now facing rearward.
45. Install and tighten the 2 front mounting bolts to 15 ft. lbs. (21 Nm).

NOTE: If any of the bolts break, deform or do not meet the torque specification, replace them.

46. Install the valve lifters and shims in their proper locations. Check that the valve lifter rotates smoothly by hand.
47. Install the intake and exhaust camshafts.
48. Set No. 1 cylinder to TDC compression stroke: Crankshaft pulley groove align with **0** mark on timing cover and camshafts timing marks with one dot and two dots will be straight line on the cylinder head surface.
49. Install the timing gear. Place the gear over the straight pin of the intake camshaft.
    a. Hold the intake camshaft with a wrench. Install and tighten the bolt to 54 ft. lbs. (74 Nm).
    b. Hold the exhaust camshaft and install the bolt and distributor gear. Torque the bolt to 34 ft. lbs. (46 Nm).
50. Install the chain tensioner, using a new gasket (mark toward the front).
51. Recheck the engine for proper valve timing. Check and adjust the valve clearance.
52. Install the spark plugs.
53. Install the semi-circular plug.
54. Recheck the engine for proper valve timing.
55. Install the cylinder head cover with a new gasket.

56. Install the engine wire brackets.
57. Install the front engine hanger and torque the mounting bolts to 30 ft. lbs. (42 Nm).
58. Install the cylinder head rear cover. Torque the bolts to 10 ft. lbs. (13.5 Nm).
59. Install the water outlet with a new gasket. Torque the bolts to 14 ft. lbs. (20 Nm). Connect the upper radiator hose.
60. Install the exhaust manifold. Torque the bolts to 36 ft. lbs. (49 Nm). Install the remaining components.
61. Fill the engine and radiator with engine coolant.
62. Connect the negative battery cable. Start the engine and check for leaks.
63. Check the ignition timing. Road test the vehicle for proper operation.
64. Recheck all fluid levels.

3.4L (5VZ-FE) Engine

1995–97 T-100 and Tacoma

1. Disconnect the negative battery cable.

CAUTION

Wait 90 seconds from the time the key is turned to LOCK and the negative battery cable is disconnected to begin work. This allows the SRS capacitor to discharge and prevent deployment of the air bag(s).

---

2. Relieve the fuel system pressure.
3. Remove the engine undercover.
4. Drain the cooling system.
5. Remove the front exhaust pipe.
6. Disconnect the air cleaner cap, MAF meter, and the resonator.
7. Disconnect the following cables:
   • If equipped with cruise control, disconnect the actuator cable with the bracket.
   • Accelerator cable
   • With automatic transmission, Throttle cable
8. Disconnect the heater hose.
9. Disconnect the upper radiator hose from the engine.
10. Remove the power steering drive belt.
11. Remove the air conditioning drive belt by loosening the idle pulley nut and adjusting bolt.
12. Loosen the lock bolt, pivot bolt and adjusting bolt and the alternator drive belt.
13. Remove the No. 2 fan shroud by removing the two clips.
14. Remove the fan with the fluid coupling and fan pulleys.

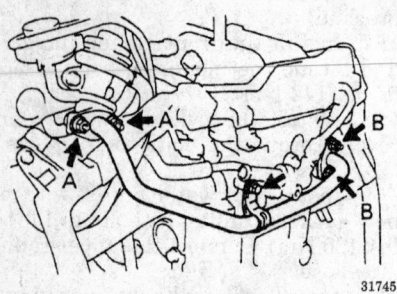

EGR pipe and related components — 1996–97
4Runner with 2.7L (3RZ-FE) engine

15. Disconnect the power steering pump from the engine and set aside. Do not disconnect the lines from the pump.

16. If equipped with air conditioning, disconnect the compressor from the engine and set aside. Do not disconnect the lines from the compressor.

17. If equipped with air conditioning, disconnect the air conditioning bracket.

18. Remove the spark plug wires with the ignition coils.

19. Remove the spark plugs.

20. Remove the No. 2 timing belt cover.

21. Remove the fan bracket as follows:

    a. Remove the power steering adjusting strut by removing the nut.

    b. Remove the fan bracket by removing the bolt and nut.

22. Set the No. 1 cylinder at TDC of the compression stroke.

    a. Turn the crankshaft pulley and align its groove with the timing mark 0 of the No. 1 timing belt cover.

    b. Check that the timing marks of the camshaft timing pulleys and the No. 3 timing belt cover are aligned. If not, turn the crankshaft pulley one revolution (360°).

NOTE: If re-using the timing belt, make sure that you can still read the installation marks. If not, place new installation marks on the timing belt to match the timing marks of the camshaft timing pulleys.

23. Remove the timing belt tensioner by alternately loosening the two bolts.

24. Remove the camshaft timing pulleys.

    a. Using SST 09960–10010 or equivalent, remove the pulley bolt, the timing pulley and the knock

pin. Remove the two timing pulleys with the timing belt.

25. Remove the bolt and the No. 2 idler pulley.

26. Remove the alternator from the engine.

27. If equipped with an EGR valve, remove the nuts and remove the EGR pipe and two gaskets.

28. Remove the intake chamber stay as follows:

    a. Remove the oil filler tube and No. 1 throttle cable clamp by removing the bolt and two nuts.

    b. Remove the intake chamber stay by removing the two bolts.

29. Remove the following connectors:

    • VSV connector for the fuel pressure control.

    • Disconnect the throttle position sensor connector.

    • IAC valve connector

    • If equipped with an EGR valve, disconnect the EGR gas temperature connector

    • If equipped with an EGR valve, disconnect the VSV connector for the EGR valve

30. Disconnect the following hoses:

    • Disconnect the PCV hoses.

    • Disconnect the water bypass hoses.

    • Disconnect the air assist hose from the intake air connector.

    • Two vacuum sensing hoses from the VSV

    • EVAP hose

    • Air hose from the power steering

    • If equipped with air conditioning, disconnect the air hose from the air conditioning idle up valve.

31. Remove the four bolts, two nuts and remove the air intake chamber assembly from the engine.

32. Remove the intake air connector.

33. Disconnect the engine wire from the intake manifold as follows:

    a. Disconnect the following connectors:

    • Oil pressure sensor connector

    • Crankshaft position sensor connector

    • Six injector connectors

    • ECT sender gauge connector

    • ECT sensor connector

    • Knock sensor connector

    • Camshaft position sensor connector

    b. Disconnect the three engine wire clamps.

    c. Remove the three bolts and disconnect the engine wire from the cylinder head.

34. Remove the camshaft position sensor.

35. Remove the No. 3 (rear) timing belt cover by removing the six bolts.

36. Remove the fuel pressure regulator.

— **CAUTION** —

Fuel injection systems remain under pressure after the engine has been turned OFF. Properly relieve fuel pressure before disconnecting any fuel lines. Failure to do so may result in fire or personal injury.

37. Remove the intake manifold assembly.

38. Remove the power steering pump bracket.

39. Remove the oil dipstick and guide.

40. Remove the exhaust crossover pipe and gaskets by removing the six nuts.

41. Remove the left hand exhaust manifold by removing the heat insulator and six nuts for the exhaust manifold.

42. Remove the right hand exhaust manifold by removing the heat insulator and six nuts for the exhaust manifold.

43. Remove the eight bolts, seal washers, cylinder head cover and gasket. Remove both cylinder head covers.

44. Remove the semi circular plugs.

45. Remove the right exhaust camshafts.

46. Remove the right hand intake camshaft.

47. Remove the left exhaust camshafts.

48. Remove the left hand intake camshaft.

49. Remove the valve lifters and shims from the cylinder head. Arrange the valve lifters and shims in correct order.

50. Remove the cylinder heads as follows:

    a. Remove the bolt and disconnect the ground strap.

    b. Using a 8mm hexagon wrench, remove the cylinder head (recessed head) bolt on each cylinder head, then repeat for the other side.

    c. Uniformly loosen and remove the eight cylinder head (12 pointed head) bolts on each cylinder head. Loosen the bolts in several passes and in the reverse order of the tightening sequence shown.

    d. Remove the 16 cylinder head bolts and plate washers.

    e. Lift the cylinder head from the dowels on the cylinder block.

To install:

51. Clean all surfaces.

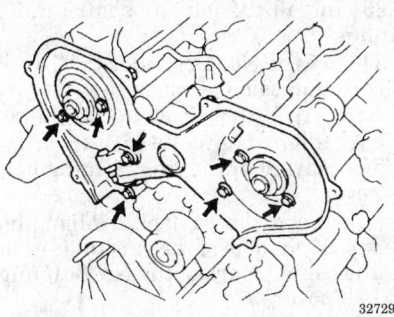

No. 3 (rear) timing belt cover — 1995–97 3.4L (5VZ-FE) engine

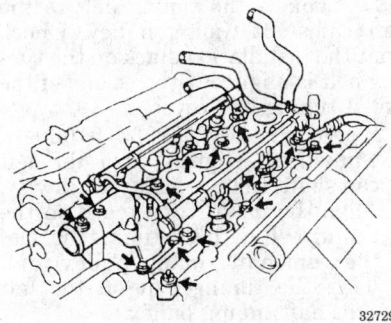

Intake manifold bolts and nuts — 1995–97 T-100 and Tacoma with 3.4L (5VZ-FE) engine

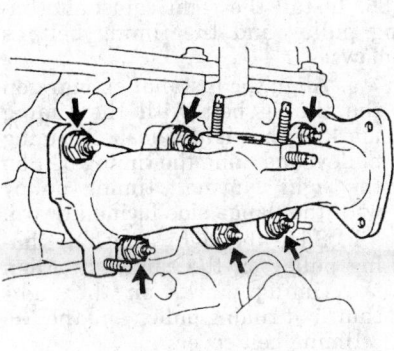

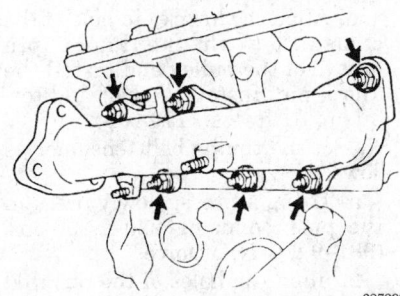

Exhaust manifold nuts — 1995–97 3.4L (5VZ-FE) engine

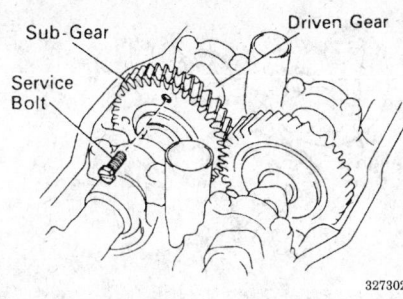

Drive gear service bolt (right side) — 1995–97 3.4L (5VZ-FE) engine

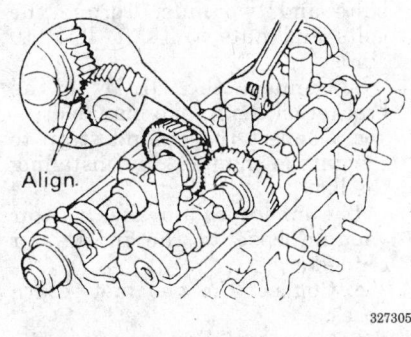

Aligning the timing mark (1 dot mark) of the left camshafts — 1995–97 3.4L (5VZ-FE) engine

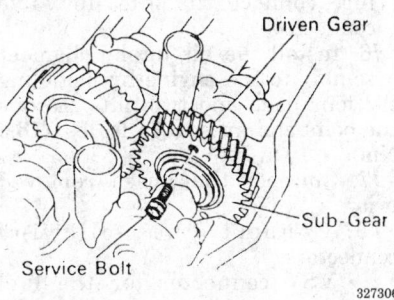

Drive gear service bolt (left side) — 1995–97 3.4L (5VZ-FE) engine

52. Place two new cylinder head gaskets in position on the cylinder block.

53. Place the two cylinder heads on the dowels of the cylinder block.

54. Apply a light coat of engine oil on the threads and under the heads of the cylinder head bolts.

55. Install and uniformly tighten the cylinder head bolts on each cylinder as follows:

   a. In several passes and in the sequence shown, torque the cylinder bolts to 25 ft. lbs. (34 Nm).

   b. Mark the front of the cylinder head bolt with paint.

c. Retighten the cylinder head bolts by 90° in order.

   d. Check that the painted mark is now at a 90° angle to the front.

56. Install the recessed head cylinder head bolts as follows:

   a. Apply a light coat of engine oil on the threads and under the heads of the cylinder head bolts.

   b. Using a 8 mm hexagon wrench, install the cylinder head bolt on each cylinder head, then repeat for the other side, as shown. Torque the bolts to 13 ft. lbs. (18 Nm).

   c. Install the bolt and connect the ground strap.

57. Install the valve lifters and shims. Check that the valve lifter rotates smoothly by hand.

58. Following proper procedures, install the camshafts.

59. Check and adjust the valve clearance.

60. Install the semi circular plugs.

61. Install the cylinder head covers. Uniformly tighten the bolts in several passes to 53 inch lbs. (6 Nm).

62. Install the exhaust manifolds with new gaskets. Torque the nuts to 30 ft. lbs. (40 Nm).

63. Install the exhaust manifold heat insulators with the nuts. Torque the nuts to 71 inch lbs. (8 Nm).

64. Install the exhaust crossover pipe and torque the nuts to 33 ft. lbs. (45 Nm).

65. Install the alternator bracket and torque to 14 ft. lbs. (18 Nm).

66. Install the oil dipstick and guide using a new O-ring.

67. Install the power steering bracket and torque to 14 ft. lbs. (18 Nm).

68. Install two new gaskets and the intake manifold assembly. Install the four plate washers, eight bolts and four nuts. Torque the bolts and nuts to 13 ft. lbs. (18 Nm).

69. Install the intake manifold stay with the two bolts. Torque the bolts to 14 ft. lbs. (18 Nm).

70. Connect the fuel inlet hose.

71. Install the fuel pressure regulator.

72. Install the No. 3 timing belt cover with the six bolts. Torque the bolts to 80 inch lbs. (9 Nm).

73. Install the camshaft position sensor and torque to 71 inch lbs. (8 Nm).

74. Connect the engine wire to the intake manifold in the reverse order of removal.

75. Install the intake air connector as follows:

   a. Install the intake manifold to the engine by installing the three

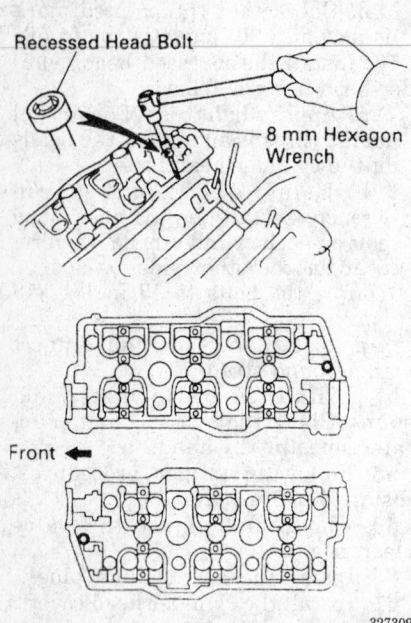

Cylinder head recessed bolts — 1995–97 3.4L (5VZ-FE) engine

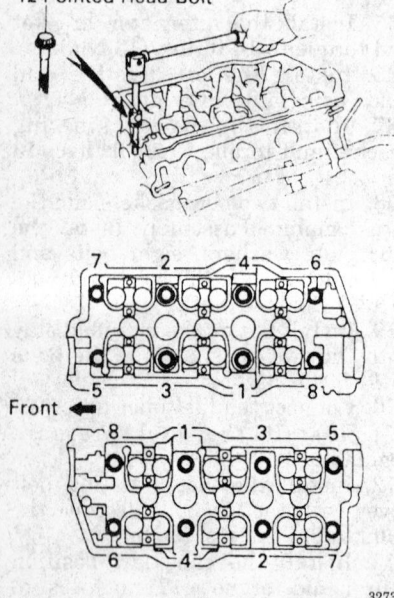

Cylinder head bolt installation sequence — 1995–97 3.4L (5VZ-FE) engine

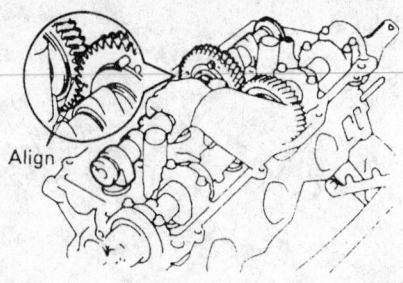

Aligning the right camshafts for installation — 1995–97 3.4L (5VZ-FE) engine

bolts and two nuts. Torque the bolts and nuts to 14 ft. lbs. (19 Nm).

b. Connect the DLC1 to the bracket on the intake manifold.

c. Connect the ground strap to the intake manifold by installing the bolt.

d. Connect the brake booster vacuum hose to the intake air connector.

e. Connect the two fuel return hoses.

f. Connect the engine wire to the intake manifold by installing the bolt.

g. If equipped with air conditioning, connect the idle up valve connector.

76. Install the air intake chamber assembly to the engine by installing the four bolts and two nuts. Torque the bolts and nuts to 14 ft. lbs. (18.5 Nm).

77. Connect the hoses removed prior.

78. Connect the following connectors:
• VSV connector for the fuel pressure control.
• Connect the throttle position sensor connector.
• IAC valve connector
• If equipped with an EGR valve, connect the EGR gas temperature connector
• If equipped with an EGR valve, connect the VSV connector for the EGR valve

79. Install the intake chamber stay.

80. Install two new gaskets and the EGR pipe with the nuts. Torque the clamp nuts to 71 inch lbs. (8 Nm) and the EGR pipe nuts to 14 ft. lbs. (18 Nm).

81. Install the alternator put do not torque the bolts and nuts at this time.

82. Install the No. 2 timing belt idler with the bolt. Torque the bolt to 30 ft. lbs. (40 Nm). Check that the pulley bracket moves smoothly.

83. Install the left camshaft timing pulley.

84. Set the No. 1 cylinder to TDC of the compression stroke.

a. Turn the crankshaft pulley, and align its groove with the timing mark **0** of the No. 1 timing belt cover.

b. Turn the camshaft, align the knock pin hole of the camshaft with the timing mark of the No. 3 timing belt cover.

c. Turn the camshaft timing pulley, align the timing marks of the camshaft timing pulley and the No. 3 timing belt cover.

85. Connect the timing belt to the left camshaft timing pulley. Check that the installation mark on the timing belt is aligned with the end of the No. 1 timing belt cover.

a. Using SST 09960–01000 or equivalent, slightly turn the left camshaft timing pulley clockwise. Align the installation mark on the timing belt with the timing mark of the camshaft timing pulley, and hang the timing belt on the left camshaft timing pulley.

b. Align the timing marks of the left camshaft pulley and the No. 3 timing belt cover.

c. Check that the timing belt has tension between the crankshaft timing pulley and the left camshaft timing pulley.

86. Install the right camshaft timing pulley and the timing belt as follows:

a. Align the installation mark on the timing belt with the timing mark of the right camshaft timing pulley, and hang the timing belt on the right camshaft timing pulley with the flange side facing inward.

b. Slide the right camshaft timing pulley on the camshaft. Align the timing marks on the right camshaft timing pulley and the No. 3 timing belt cover.

c. Align the knock pin hole of the camshaft with the knock pin groove of the pulley and install the knock pin. Install the bolt and torque to 81 ft. lbs. (110 Nm).

87. Set the timing belt tensioner as follows:

a. Using a press, slowly press in the pushrod using 220–2,205 lbs. (981–9,807 N) of force.

b. align the holes of the pushrod and housing, pass a 1.5 mm hexagon wrench through the holes to keep the setting position of the pushrod.

c. Release the press and install the dust boot to the tensioner.

88. Install the timing belt tensioner and alternately tighten the bolts to 20 ft. lbs. (28 Nm). Using pliers, remove the 1.5 mm hexagon wrench from the belt tensioner.

89. Check the valve timing. Install the remaining components.

90. Fill the radiator with engine coolant.

91. Connect the negative battery cable to the battery.

92. Start the engine and check for leaks.

93. Check the ignition timing.

94. Install the engine undercover.

95. Road test the vehicle.

96. Recheck all fluid levels.

1996–97 4Runner

1. Disconnect the negative battery cable.

CAUTION

Wait at least 90 seconds from the time the ignition switch is turned to the LOCK position and the negative (-) battery cable is disconnected before starting the repair procedure.

---

2. Relieve the fuel system pressure.

3. Remove the engine undercover.

4. Drain the cooling system.

5. Remove the front exhaust pipe.

6. Disconnect the air cleaner cap, MAF meter, and the resonator.

7. Disconnect the following cables:
• If equipped with cruise control, disconnect the actuator cable with the bracket.
• Accelerator cable
• With automatic transmission, Throttle cable

8. Disconnect the heater hose.

CAUTION

Fuel injection systems remain under pressure after the engine has been turned OFF. Properly relieve fuel pressure before disconnecting any fuel lines. Failure to do so may result in fire or personal injury.

---

9. Disconnect the following hoses:
• Brake booster vacuum hose
• EVAP hose
• 4-Wheel Drive: A.D.D. vacuum hose
• Fuel inlet and fuel return hose

10. Remove the spark plug wires with the ignition coils.

11. Remove the spark plugs.

12. Remove the intake chamber stay.

13. Remove the No. 2 timing belt cover.

14. Remove the air intake chamber assembly.

15. Remove the following connectors and hoses:
• Throttle position sensor connector
• IAC valve connector
• PCV hoses
• Water bypass hoses
• Air assist hose from the throttle body

16. Remove the intake air connector.

17. Disconnect the engine wire protector.
a. Disconnect the six injector connectors.
b. Disconnect the ECT sensor and sender gauge connectors.
c. Disconnect the engine wire protector from the cylinder head.

18. Remove the fuel pressure regulator.

19. Remove the intake manifold assembly.

20. Set the No. 1 cylinder at TDC of the compression stroke.
a. Turn the crankshaft pulley and align its groove with the timing mark **O** of the No. 1 timing belt cover.
b. Check that the timing marks of the camshaft timing pulleys and the No. 3 timing belt cover are aligned. If not, turn the crankshaft pulley one revolution (360°).

NOTE: If re-using the timing belt, make sure that you can still read the installation marks. If not, place new installation marks on the timing belt to match the timing marks of the camshaft timing pulleys.

21. Remove the timing belt tensioner by alternately loosening the two bolts.

22. Remove the timing belt.

23. Remove the camshaft timing pulleys.
a. Using SST 09960–10010 or equivalent, remove the pulley bolt,

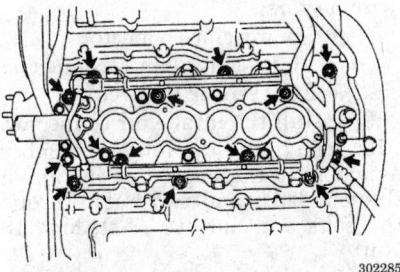

302285

Intake manifold bolts and nuts — 1996–97 4Runner with 3.4L (5VZ-FE) engine

the timing pulley and the knock pin. Remove the two timing pulleys with the timing belt.

24. Remove the bolt and the No. 2 idler pulley.

25. Remove the camshaft position sensor.

26. Remove the No. 3 timing belt cover.

27. Remove the alternator from the engine.

28. Remove the alternator bracket.

29. Disconnect the power steering pump from the engine and set aside. Do not disconnect the lines from the pump.

30. Remove the exhaust crossover pipe and gaskets by removing the six nuts.

31. Remove the left hand exhaust manifold by removing the heat insulator and six nuts for the exhaust manifold.

32. Remove the right hand exhaust manifold by removing the heat insulator and six nuts for the exhaust manifold.

33. Remove the eight bolts, seal washers, cylinder head cover and gasket. Remove both cylinder head covers.

34. Remove the semi circular plugs.

35. Remove the right exhaust and intake camshafts.

36. Remove the left exhaust and intake camshafts.

37. Remove the valve lifters and shims from the cylinder head. Arrange the valve lifters and shims in correct order.

38. Remove the cylinder heads as follows:
a. Remove the bolt and disconnect the ground strap.
b. Using a 8mm hexagon wrench, remove the cylinder head (recessed head) bolt on the cylinder head, then repeat the procedure for the other side.
c. Uniformly loosen and remove the eight cylinder head (12 pointed head) bolts on each cylinder head. Loosen the bolts in several passes and in the reverse order of the tightening sequence shown.
d. Remove the 16 cylinder head bolts and plate washers.
e. Lift the cylinder head from the dowels on the cylinder block.

To install:

39. Clean all surfaces.

40. Place two new cylinder head gaskets in position on the cylinder block.

41. Place the two cylinder heads on the dowels of the cylinder block.

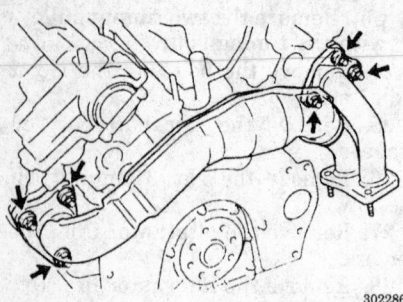

302286

Exhaust crossover pipe connection points

42. Apply a light coat of engine oil on the threads and under the heads of the cylinder head bolts.

43. Install and uniformly tighten the cylinder head bolts on each cylinder as follows:

   a. In several passes and in the sequence shown, torque the cylinder bolts to 25 ft. lbs. (34 Nm).

   b. Mark the front of the cylinder head bolt with paint.

   c. Retighten the cylinder head bolts by 90° in order.

   d. Check that the painted mark is now at a 90° angle to the front.

44. Install the recessed head cylinder head bolts as follows:

   a. Apply a light coat of engine oil on the threads and under the heads of the cylinder head bolts.

   b. Using a 8 mm hexagon wrench, install the cylinder head bolt on each cylinder head, then repeat for the other side, as shown. Torque the bolts to 13 ft. lbs. (18 Nm).

   c. Install the bolt and connect the ground strap.

45. Install the valve lifters and shims. Check that the valve lifter rotates smoothly by hand.

46. Install the right intake and exhaust camshafts.

47. Install the left intake and exhaust camshaft.

48. Check and adjust the valve clearance.

49. Install the semi circular plugs.

50. Install the cylinder head covers. Uniformly tighten the bolts in several passes to 53 inch lbs. (6 Nm).

51. Install the exhaust manifolds with new gaskets. Torque the nuts to 30 ft. lbs. (40 Nm).

52. Install the exhaust manifold heat insulators with the nuts. Torque the nuts to 71 inch lbs. (8 Nm).

53. Install the exhaust crossover pipe and torque the nuts to 33 ft. lbs. (45 Nm).

54. Install the power steering pump.

55. Install the alternator bracket and torque to 14 ft. lbs. (18 Nm).

56. Install the alternator.

57. Install the No. 3 timing belt cover with the six bolts. Torque the bolts to 80 inch lbs. (9 Nm).

58. Install the camshaft position sensor and torque to 71 inch lbs. (8 Nm).

59. Install the timing belt.

60. Install the No. 2 timing belt idler with the bolt. Torque the bolt to 30 ft. lbs. (40 Nm). Check that the pulley bracket moves smoothly.

61. Install the left camshaft timing pulley.

62. Set the No. 1 cylinder to TDC of the compression stroke. Connect the timing belt to the left camshaft timing pulley. Check that the installation mark on the timing belt is aligned with the end of the No. 1 timing belt cover. Install the right camshaft timing pulley and the timing belt. Set the timing belt tensioner.

63. Install the timing belt tensioner and alternately tighten the bolts to 20 ft. lbs. (28 Nm). Using pliers, remove the 1.5 mm hexagon wrench from the belt tensioner.

64. Check the valve timing.

65. Install two new gaskets and the intake manifold assembly. Install the four plate washers, eight bolts and four nuts. Torque the bolts and nuts to 13 ft. lbs. (18 Nm).

66. Install the intake manifold stay with the two bolts. Torque the bolts to 14 ft. lbs. (18 Nm).

67. Install the fuel pressure regulator.

68. Connect the engine wire to the intake manifold as follows:

   a. Install the engine wire to the cylinder head by installing the three bolts.

   b. Connect the three engine wire clamps.

   c. Connect the following connectors:

   • Six injector connectors
   • ECT sender gauge connector
   • ECT sensor connector

69. Install the intake air connector.

70. Install the air intake chamber assembly to the engine by installing the four bolts and two nuts. Torque the bolts and nuts to 13 ft. lbs. (18 Nm).

71. Install the intake chamber stay.

72. Install the No. 2 timing belt cover. Torque the bolts to 80 inch lbs. (9 Nm).

73. Connect the following:
   • Connect the PCV hoses.
   • Connect the water bypass hoses.
   • Connect the air assist hose to the throttle body.
   • IAC valve connector.
   • Throttle position sensor connector.
   • The camshaft position sensor connector to the No. 2 timing belt cover.
   • The three spark plug wire clamps.

74. Connect the following hoses:
   • Brake booster vacuum hose
   • EVAP hose
   • 4-Wheel Drive: A.D.D. vacuum hose
   • Fuel inlet and fuel return hose

75. Install the oil dipstick and guide using a new O-ring.

76. Install the spark plugs.

77. Install the spark plug wires with the ignition coils.

78. Install the alternator drive belt.

79. Install the drive belt.

80. Connect the heater hose.

81. Connect the following cables:
   • If equipped with cruise control, connect the actuator cable with the bracket.
   • Accelerator cable
   • With automatic transmission, connect the throttle cable

82. Install the MAF meter, resonator, and the air cleaner cap.

83. Install the front exhaust pipe.

84. Fill the radiator with engine coolant.

85. Connect the negative battery cable to the battery.

86. Start the engine and check for leaks.

87. Check the ignition timing.

88. Install the engine undercover.

89. Road test the vehicle.

90. Recheck all fluid levels.

Valve Lifters

REMOVAL AND INSTALLATION

For valve lifter replacement procedures, please refer to the Camshaft removal and installation procedure outlined later in this section.

Valve Clearance

ADJUSTMENT

2.4L (2RZ-FE) Engine

1. Disconnect the negative battery cable.

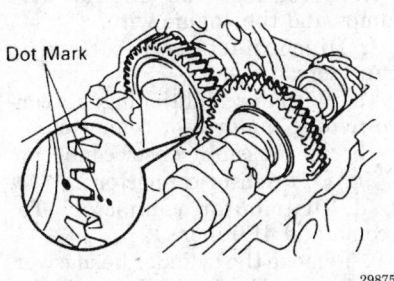

Aligning the timing marks — Tacoma with 2.4L (2RZ-FE) and 2.7L (3RZ-FE) engine

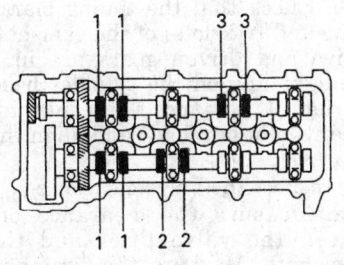

First valve adjustment — 2.4L (2RZ-FE) and 2.7L (3RZ-FE) engines

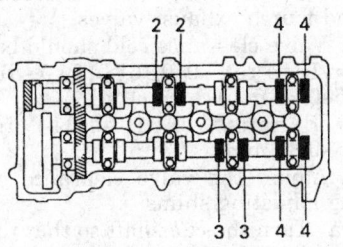

Second valve adjustment — 2.4L (2RZ-FE) and 2.7L (3RZ-FE) engines

2. Remove the air intake connector.

3. Remove the PCV hoses.

4. Disconnect the spark plug wires.

5. Disconnect the four engine wire clamps and the engine wire.

6. Disconnect the following connectors:

   a. With air conditioning: air conditioning compressor connector

   b. Oil pressure sensor connector

   c. ECT sensor connector

   d. Distributor connector

7. Remove the cylinder head cover.

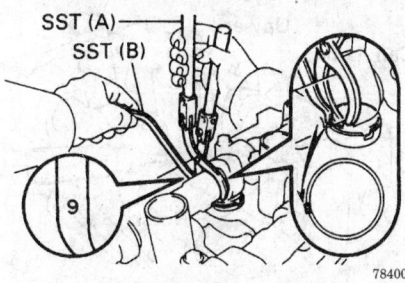

Removing adjusting shim using the special tools shown above — 1995–97 2.4L (2RZ-FE) and 2.7L (3RZ-FE) engines

8. Set the No. 1 cylinder to TDC of the compression stroke.

   a. Turn the crankshaft pulley clockwise and align its groove with the O mark on the timing chain cover.

   b. Check that the timing marks (one and two dots) of the camshaft drive and driven gears are in a straight line on the cylinder head surface. If not, turn the crankshaft one revolution (360°) and align the marks.

9. Inspect the valve clearance.

   a. Measure the clearance between the valve lifter and the camshaft. Measure the first and second intake and the first and third exhaust valves.

   b. Turn the crankshaft pulley one revolution (360°) and align the marks as above. Measure the third and fourth intake and the second and fourth exhaust valves.

10. Valve clearance cold should be:

   • Intake: 0.006–0.010 in. (0.15–0.25 mm)

   • Exhaust: 0.010–0.014 in. (0.25–0.35 mm)

11. Adjust the valve clearance by using adjusting shims.

   a. Turn the equipment driveshaft so that the cam lobe for the valve to be adjusted faces up.

   b. Using SST 09248–55040 or equivalent, press down the valve lifter and place SST 09248–05420 or equivalent, between the camshaft and the valve lifter. Remove SST 09248–55040.

   c. Remove the adjusting shim with a small flat prying tool and a magnetic finger.

   d. Determine the replacement adjusting shim size according to the following formula, or use the adjusting shim charts.

   e. Using a micrometer, measure the thickness of the removed shim. Calculate the thickness of a new

shim so that the valve clearance comes within the specified value.

   • T: Thickness of the removed shim

   • A: Measured valve clearance

   • N: Thickness of the new shim

   f. Intake: $N = T + (A - 0.008$ in. $(0.20$ mm$))$

   g. Exhaust: $N = T + (A - 0.012$ in. $(0.30$ mm$))$

   h. Install a new adjusting shim. Place it on the valve lifter. Using the SST 09248–55040, press down the valve lifter and remove SST 09248–05420.

   i. Recheck the valve clearance.

12. Reinstall the cylinder head cover.

13. Reconnect the engine wire and clamps.

14. Connect the following connectors:

   a. Distributor connector

   b. ECT sensor connector

   c. Oil pressure sensor connector

   d. If disconnected, the air conditioning compressor connector

15. Install the spark plug wires.

16. Install the PCV hoses.

17. Install the air intake connector.

18. Check the ignition timing.

19. Connect the negative battery cable.

2.7L Engine (3RZ-FE) Engine

1994 T-100

NOTE: Inspect and adjust the valve clearance when the engine is cold.

1. Disconnect the negative battery cable. Drain the engine coolant.

2. Remove the air cleaner, MAF meter and resonator.

3. Disconnect the throttle cable and intake air connector.

4. Remove the air conditioning idle-up valve, if equipped. Remove the PCV hoses and spark plug leads.

5. Disconnect all connectors, vacuum hoses and cooling system hoses from the throttle body. Remove the throttle body.

6. Remove the valve cover (cylinder head cover).

7. Set No. 1 cylinder to TDC compression stroke. The groove on the crankshaft pulley should align with the "0" mark on the timing chain cover and the timing marks (one and two dots) of the camshaft gears should form a straight line in respect to the cylinder head surface. If not, turn the crankshaft 1 revolution (360 degrees).

8. Measure the clearance between the valve lifter and camshaft, using a thickness gauge. Record the clear-

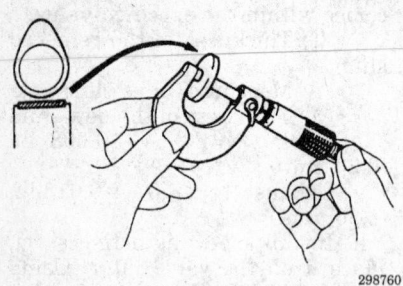

Measuring shim thickness — 2.4L (2RZ-FE) and 2.7L (3RZ-FE) engines

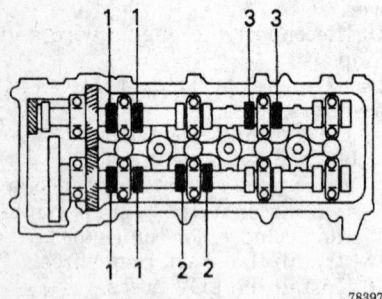

Check only the valves indicated, with the No. 1 cylinder to TDC/compression stroke — 1995–97 2.7L (3RZ-FE) engine

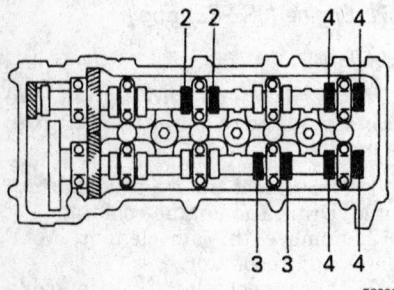

Check only the valves indicated, with the No. 1 cylinder on its exhaust stroke — 1995–97 2.7L (3RZ-FE) engine

ance; it will be used later to determine the required shim. Check only the following valves, as shown.

9. Turn the crankshaft 1 revolution (360 degrees) and align the timing marks of the crankshaft pulley and timing chain cover. Record the clearance; it will be used later to determine the required shim. Check only the following valves, as shown.

10. To adjust:

a. Remove the adjusting shim. Turn the crankshaft to position the lobe of the camshaft on the adjusting valve upward. Position the

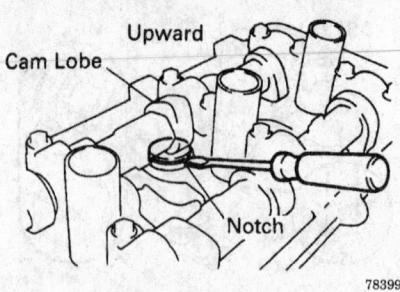

Position the notch of the valve lifter toward the spark plug side — 1995–97 2.7L (3RZ-FE) engine

notch of the valve lifter toward the spark plug.

b. Using SST (A), press down the valve lifter and place SST (B) between the camshaft and valve lifter flange. Remove SST (A).

c. Remove the adjusting shim using a small screw driver and magnet. (SST 09248-55040, SST 09248-05410, SST 09248-05420 or equivalent)

11. Determine the replacement shim size using the following formula:

a. Measure the thickness of the shim removed, using a micrometer.

b. Calculate the thickness of the replacement shim by adding the measured clearance (recorded earlier), plus the thickness of the shim removed.

c. Select a new shim with a thickness as close as possible to the calculated value.

12. Install the new adjusting shim.

13. Install the valve cover.

14. Reconnect all connectors, vacuum hoses and cooling system hoses.

15. Install the throttle body, throttle cable, MAF meter, resonator and air cleaner.

16. Refill the cooling system.

17. Connect the negative battery cable.

18. Check and adjust the ignition timing and idle speed, as required.

Tacoma, 1995–97 T100 and 1996–97 4Runner

1. Disconnect the negative battery cable.

2. Drain the engine coolant.

3. On the Tacoma and 4Runner, remove the intake air connector.

4. On the 1995–97 T100, remove the air cleaner cap, MAF meter, and the resonator.

5. Remove the PCV hoses.

6. Disconnect the spark plug wires.

7. Disconnect the engine wire clamps and the engine wire.

8. Disconnect the following connectors:

a. With air conditioning: air conditioning compressor connector

b. Oil pressure sensor connector

c. ECT sensor connector

d. Distributor connector (Tacoma and 4Runner)

9. Remove the cylinder head cover.

10. Set the No. 1 cylinder to TDC of the compression stroke.

a. Turn the crankshaft pulley clockwise and align its groove with the 0 mark on the timing chain cover.

b. Check that the timing marks (one and two dots) of the camshaft drive and driven gears are in a straight line on the cylinder head surface. If not, turn the crankshaft one revolution (360°) and align the marks.

11. Inspect the valve clearance.

a. Measure the clearance between the valve lifter and the camshaft. Measure the first and second intake and the first and third exhaust valves.

b. Turn the crankshaft pulley one revolution (360°) and align the marks as above. Measure the third and fourth intake and the second and fourth exhaust valves.

12. Valve clearance cold should be:

• Intake: 0.006–0.010 in. (0.15–0.25 mm)

• Exhaust: 0.010–0.014 in. (0.25–0.35 mm)

13. Adjust the valve clearance by using adjusting shims.

a. Turn the camshaft so that the cam lobe for the valve to be adjusted faces up.

b. Using SST 09248–55040 or equivalent, press down the valve lifter and place SST 09248–05420 or equivalent, between the camshaft and the valve lifter. Remove SST 09248–55040.

c. Remove the adjusting shim with a small flat prying tool and a magnetic finger.

d. Determine the replacement adjusting shim size according to the following formula, or use the adjusting shim charts.

e. Using a micrometer, measure the thickness of the removed shim. Calculate the thickness of a new shim so that the valve clearance comes within the specified value.

• T: Thickness of the removed shim

• A: Measured valve clearance

• N: Thickness of the new shim

f. Intake: N=T+ (A — 0.008 in. (0.20 mm))

g. Exhaust: N=T+ (A — 0.012 in. (0.30 mm))

h. Install a new adjusting shim. Place it on the valve lifter. Using the SST 09248–55040, press down the valve lifter and remove SST 09248–05420.

i. Recheck the valve clearance.

14. Reinstall the cylinder head cover.

15. Reconnect the engine wire and clamps.

16. Connect the following connectors:

a. Distributor connector

b. ECT sensor connector

c. Oil pressure sensor connector

d. If disconnected, the air conditioning compressor connector

17. Install the spark plug wires.

18. Install the PCV hoses.

19. On the 1995–97 T100, install the air cleaner cap, MAF meter, and the resonator.

20. On the 4Runner and Tacoma, install the intake air connector.

21. Refill with engine coolant.

22. Check the ignition timing.

23. Connect the negative battery cable.

3.0L (3VZ-E) Engine

NOTE: Inspect and adjust the valve clearance when the engine is cold.

1. Disconnect the negative battery cable.

2. Remove the air intake chamber, valve cover and spark plugs.

3. Set the No. 1 cylinder to TDC/compression. Rotate the crankshaft until the groove on the crankshaft pulley align with "0" on the No. 1 timing belt cover. Check that the lifters on the No. 1 cylinder are loose and valve lifters on the No. 4 are tight.

4. Measure the clearance between the valve lifter and camshaft, using a thickness gauge. Record the clearance; it will be used later to determine the required shim. Valve clearance cold should be:

• Intake: 0.007–0.011 in. (0.18–0.28 mm)

• Exhaust: 0.009–0.013 in. (0.22–0.32 mm)

a. Measure the clearance of No. 6 (intake) and No. 2 (exhaust) valves.

b. Turn the crankshaft pulley ⅓ revolution (120 degrees). Measure the clearance of No. 1 (intake) and No. 3 (exhaust) valves.

c. Turn the crankshaft pulley ⅓ revolution (120 degrees). Measure the clearance of No. 2 (intake) and No. 4 (exhaust) valves.

d. Turn the crankshaft pulley ⅓ revolution (120 degrees). Measure the clearance of No. 3 (intake) and No. 5 (exhaust) valves.

e. Turn the crankshaft pulley ⅓ revolution (120 degrees). Measure the clearance of No. 4 (intake) and No. 6 (exhaust) valves.

f. Turn the crankshaft pulley ⅓ revolution (120 degrees). Measure the clearance of No. 5 (intake) and No. 1 (exhaust) valves.

5. To adjust:

a. Turn the crankshaft to position the lobe of the camshaft on the adjusting valve upward.

b. Using SST (A), press down the valve lifter and place SST (B) between the camshaft and valve lifter flange. Remove SST (A).

c. Remove the adjusting shim using a small screw driver and magnet. **(SST 09248-55020, SST 09248-05011, SST 09248-05021 or equivalent)**

6. Determine the replacement shim size using the following formula:

a. Measure the thickness of the shim removed, using a micrometer.

b. Calculate the thickness of the replacement shim by adding the measured clearance (recorded earlier), plus the thickness of the shim removed.

c. Select a new shim with a thickness as close as possible to the calculated value.

7. Install the new adjusting shim.

8. Install the spark plugs, valve cover and air intake chamber.

9. Connect the negative battery cable.

10. Check and adjust the ignition timing and idle speed, as required.

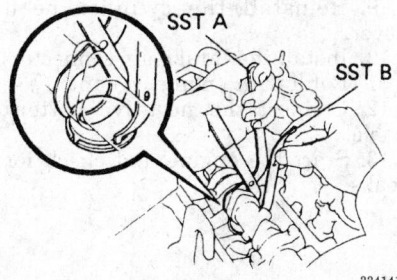

Remove adjusting shim using SST 09248-55020 (09248-05011, 09248-05021) — 3.0L (3VZ-E) engine

3.4L (5VZ-FE) Engine

1. Disconnect the negative battery cable.

2. Drain the engine coolant.

3. Remove the air intake connector.

4. Remove the cylinder head cover.

5. Set the No. 1 cylinder to TDC of the compression stroke.

a. Turn the crankshaft pulley clockwise and align its groove with the **0** mark on the timing chain cover.

b. Check that the timing marks (one and two dots) of the camshaft drive and driven gears are in a straight line on the cylinder head surface. If not, turn the crankshaft one revolution (360°) and align the marks.

6. Inspect the valve clearance.

a. Measure the clearance between the valve lifter and the camshaft. Measure the first intake and the third exhaust valves on the right head and the sixth intake and the second exhaust valves on the left head.

b. Turn the crankshaft ⅔ of a revolution (240°) and adjust the third intake and the fifth exhaust valves on the right head and the second intake and the fourth exhaust valves on the left head.

c. Turn the crankshaft ⅔ of a revolution (240°) and adjust the fifth intake and the first exhaust valves on the right head and the fourth intake and the sixth exhaust valves on the left head.

7. Valve clearance cold should be:

• Intake: 0.006–0.009 in. (0.13–0.23 mm)

• Exhaust: 0.011–0.014 in. (0.27–0.37 mm)

8. Adjust the valve clearance by using adjusting shims.

a. Turn the equipment camshaft so that the cam lobe for the valve to be adjusted faces up.

b. Turn the valve lifter so that the notches are perpendicular to the camshaft.

c. Using SST 09248–55040 or equivalent, press down the valve lifter and place SST 09248–05420 or equivalent, between the camshaft and the valve lifter. Remove SST 09248–55040.

d. Remove the adjusting shim with a small flat prying tool and a magnetic finger.

e. Determine the replacement adjusting shim size according to the following formula, or use the adjusting shim charts.

f. Using a micrometer, measure the thickness of the removed shim.

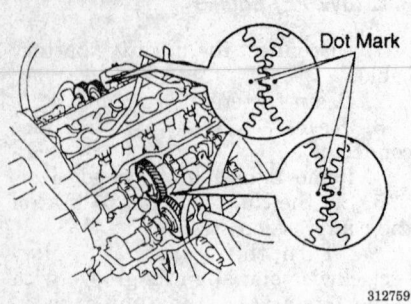

Aligning the timing marks — 3.4L (5VZ-FE) engine

312759

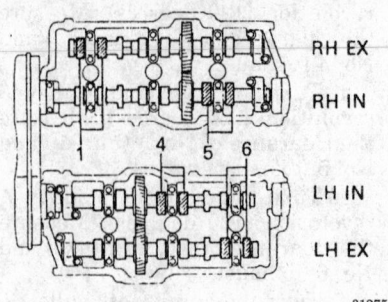

Third valve adjustment — 3.4L (5VZ-FE) engine

312758

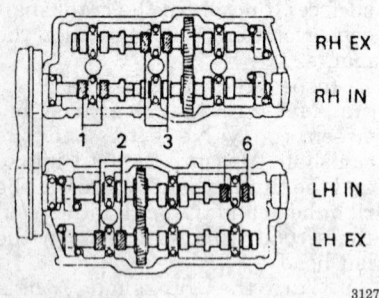

First valve adjustment — 3.4L (5VZ-FE) engine

312760

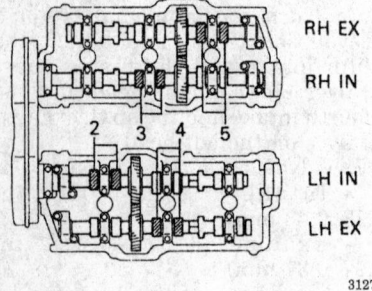

Second valve adjustment — 3.4L (5VZ-FE) engine

312761

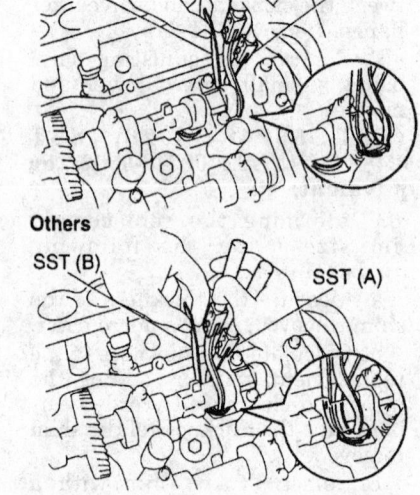

Front of No.1 and Rear of No.6 Cylinders

Others

Removing the adjusting shim — 3.4L (5VZ-FE) engine

312762

Rocker Arms and Shaft

REMOVAL AND INSTALLATION

2.4L (22R-E) Engine

NOTE: Only the 2.4L (22R-E) engine uses rocker arms. All other engines operate the valves through direct action of the camshaft.

NOTE: The rocker arms and rocker arm shaft are secured by the cylinder head bolts. The cylinder head, camshaft, and valve train should be removed as an assembly and then disassembled off the vehicle.

1. Release the fuel system pressure.
2. Disconnect the negative battery cable.

— CAUTION —
Wait at least 90 seconds after the negative (-) battery cable is disconnected to prevent possible deployment of the air bag.

3. Drain the engine coolant.
4. Following proper procedures, remove all cylinder head components necessary for removal.
5. Remove the rocker arm assembly by removing the cylinder head bolts in sequence. Loosen the bolts in two to three stages in the reverse order of the tightening sequence.
6. Remove the cylinder head rear cover.
7. Remove the cylinder head.
8. Remove the rocker shaft assembly, camshaft caps, and the camshaft.
9. To remove the rocker arms, remove the screws from the rocker shafts and slide the rocker shaft stands, rocker arms, and springs from the rocker shafts.

NOTE: Keep all parts in order as they are removed. All parts must be reinstalled in their original position.

10. Remove the intake manifold, exhaust manifold, the right engine hanger, the left engine hanger, the EGR valve, and the ground strap connector.

To install:
11. Place the camshaft in the cylinder head and install the bearing caps in the numbered order from the front with the arrows pointed toward the front. Torque the bolts to 14 ft. lbs. (20 Nm). Turn the camshaft to position the dowel at the top.
12. Install the cylinder head rear cover.

Calculate the thickness of a new shim so that the valve clearance comes within the specified value.
• T: Thickness of the removed shim
• A: Measured valve clearance
• N: Thickness of the new shim
g. Intake: N=T+ (A — 0.007 in. (0.18 mm))
h. Exhaust: N=T+ (A — 0.013 in. (0.32 mm))
i. Install a new adjusting shim. Place it on the valve lifter. Using the SST 09248–55040, press down

the valve lifter and remove SST 09248–05420.
j. Recheck the valve clearance.
9. Reinstall the cylinder head cover.
10. Install the intake air connector.
11. Refill with engine coolant.
12. Connect the negative battery cable.
13. Start the engine and check for leaks.

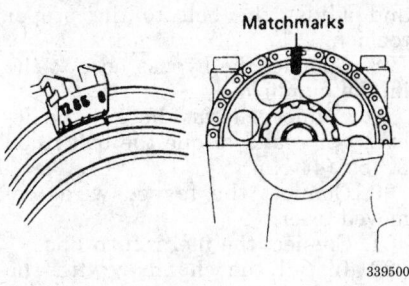

Matchmark the camshaft sprocket to the timing chain — 2.4L (22R-E) engine

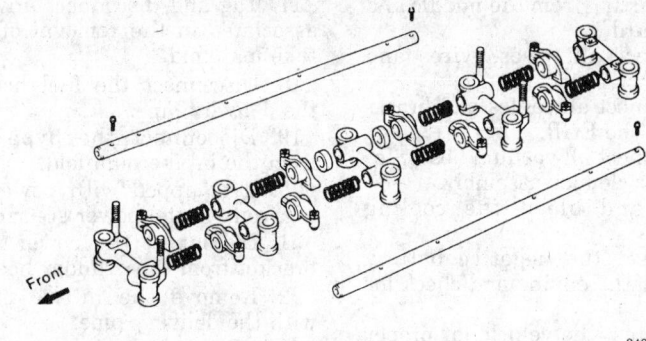

Exploded view of the rocker shaft assembly — 2.4L (22R-E) engine

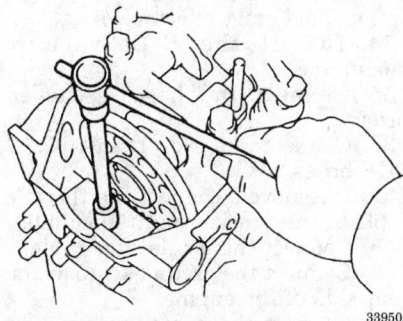

Removing the timing chain cover bolt — 2.4L (22R-E) engine

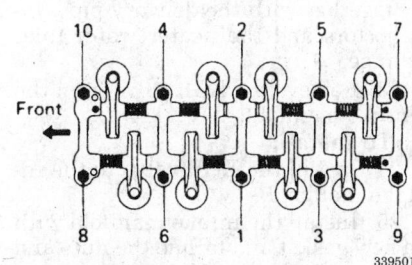

Cylinder head bolt tightening sequence — 2.4L (22R-E) engine

13. Apply seal packing and to the cylinder block. Only apply sealant at the two areas where the timing chain cover meets the cylinder block.

14. Install the cylinder head on the cylinder block. Be sure to align the dowel pins on the block with the cylinder head.

15. Assemble the rocker arms, springs, and the rocker shaft stands in the same order that they were removed.

16. Place the rocker arm assembly over the dowels on the cylinder head.

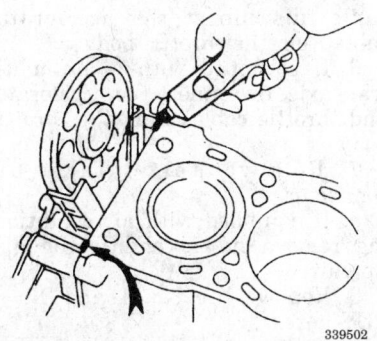

Applying sealant to the cylinder block — 2.4L (22R-E) engine

17. Install and tighten the head bolts. Following the proper sequence, torque the bolts in three stages to 58 ft. lbs. (78 Nm).

18. Align matchmarks made during removal and install the camshaft sprocket and chain.

19. Install the chain cover bolt and torque it to 9 ft. lbs. (13 Nm).

20. Install the distributor drive gear and camshaft thrust plate. Torque the bolt to 58 ft. lbs. (78 Nm).

21. Check and adjust valve clearance.

22. Install the cylinder head cover with the grommets and the four nuts. Torque the nuts to 4 ft. lbs. (5 Nm). Install the remaining components.

23. Check the engine oil level and add as necessary.

24. Fill the engine coolant.

25. Connect the negative battery cable.

26. Start the engine, check for leaks, and bleed the cooling system.

27. Perform engine adjustments (ignition timing, etc.).

28. Road test the vehicle for proper operation.

Intake Manifold

REMOVAL AND INSTALLATION

2.4L (22R-E) Engine

1993–94 Models

1. Relieve the fuel system pressure.

2. Disconnect the negative battery cable.

3. Drain the cooling system.

4. Disconnect the air intake hose from both the air cleaner assembly on one end and the air intake chamber on the other.

5. Tag and disconnect all vacuum lines attached to the intake chamber and manifold.

6. Tag and disconnect the wires to the cold start injector, throttle position sensor and the water hoses from the throttle body.

7. Remove the EGR valve from the intake chamber.

8. Tag and disconnect the actuator cable, accelerator cable and throttle valve cable, if equipped, from the cable bracket on the intake chamber.

9. Unbolt the air intake chamber from the intake manifold and remove the chamber with the throttle body attached.

10. Disconnect the fuel hose from the fuel delivery pipe. Plug hose to prevent fuel leakage.

11. Tag and disconnect the air valve hose from the intake manifold.

12. Make sure all hoses, lines and wires are tagged for later installation and disconnected from the intake manifold. Unbolt the manifold from the cylinder head, removing the delivery pipe and injection nozzle with the manifold.

To install:

13. Clean the gasket mating surfaces and check for warpage.

14. Install the manifold with new gasket. Torque the bolts to 13 ft. lbs.

(18 Nm) starting from the middle and move outward.

15. Connect all hoses, wires and connectors.

16. Reconnect all cables and brackets. Install the EGR.

17. Reconnect all vacuum hoses. Install the air cleaner assembly.

18. Fill and bleed the cooling system.

19. Connect the negative battery cable. Start the engine and check for leaks.

20. Road test the vehicle for proper operation.

1995 Models

1. Drain the cooling system.
2. Disconnect the negative battery cable.

---- **CAUTION** ----

Work must be started after 90 seconds from the time the ignition switch is turned to the LOCK position and the negative (-) battery cable is disconnected.

3. Relieve the fuel pressure.

---- **CAUTION** ----

Fuel injection systems remain under pressure after the engine has been turned OFF. Properly relieve fuel pressure before disconnecting any fuel lines. Failure to do so may result in fire or personal injury.

4. Remove the intake air connector.

5. Remove the distributor cap and spark plug wires.

6. Remove the radiator inlet hose.

7. Remove the heater water inlet hose from the heater water inlet pipe.

8. Disconnect the accelerator cable.

9. If equipped with an automatic transmission, disconnect the throttle cable from the bracket and the clamp.

10. Disconnect the ground strap from the rear side of the engine.

11. Disconnect the No. 1 and No. 2 PCV hoses.

12. Tag and disconnect the all hoses in the way of removing the intake manifold.

13. Remove the EGR vacuum modulator.

14. Disconnect the following wires:
   a. Cold start injector wire
   b. Throttle position wire
   c. California only: EGR gas temperature sensor wire

15. Remove the chamber with the throttle body.

16. Disconnect the fuel return hose.

17. Tag and disconnect any wiring associated in the removal of the intake manifold.

18. Disconnect the fuel hose from the delivery pipe.

19. Disconnect the bypass hose from the intake manifold.

20. If equipped with power steering, remove the power steering belt.

21. Disconnect the power steering bracket from the cylinder head.

22. Remove the intake manifold with the delivery pipe.
   a. Remove the heater inlet pipe from the cylinder head.
   b. Remove the No. 1 air pipe.
   c. Remove the intake manifold together with the delivery pipe, injectors and the heater water inlet pipe.

23. Remove the EGR valve from the intake manifold.

To install:

24. Install the EGR valve to the intake manifold.

25. Install the intake manifold with a new gasket and torque the nuts and bolts to 14 ft. lbs. (19 Nm).
   a. Install the No. 1 air pipe.
   b. Install the heater inlet pipe to the cylinder head.

26. Connect the power steering bracket to the cylinder head. Torque the bolts to 33 ft. lbs. (44 Nm).

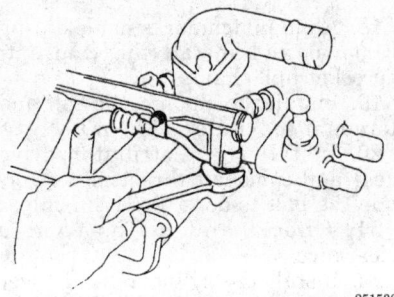

Disconnecting fuel hose from the delivery pipe — 1995 with 2.4L (22R-E) engine

251538

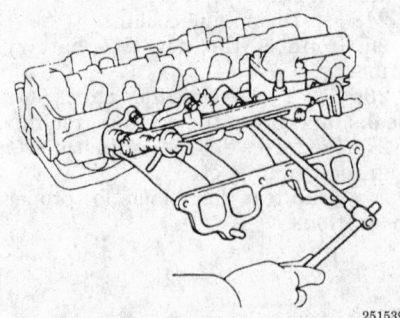

Removing the intake manifold — 1995 with 2.4L (22R-E) engine

251539

27. Install the power steering belt and adjust the belt to the proper tension.

28. Connect the bypass hose to the intake manifold.

29. Connect the fuel hose to the delivery pipe and torque the bolt to 33 ft. lbs. (44 Nm).

30. Connect the tagged wires removed prior.

31. Connect the fuel return line.

32. Install the chamber with the throttle body.

33. Connect the following wires:
   a. California models only, EGR gas temperature sensor wire.
   b. Throttle position wire.
   c. Cold start injector wire.

34. Install the EGR vacuum modulator.

35. Connect the hoses removed prior.

36. Connect the No. 1 and No. 2 PCV hoses.

37. If removed, connect the throttle cable to the bracket and the clamp.

38. Connect the accelerator cable.

39. Connect the ground strap at the rear side of the engine.

40. Connect the heater water inlet hose to the heater water inlet pipe.

41. Connect the radiator inlet pipe.

42. Install the distributor cap and wires.

43. Install the intake air connector.

44. Connect the negative battery cable.

45. Refill and bleed the cooling system.

46. Start the engine and check for leaks.

2.4L (2RZ-FE) Engine

1. Relieve the fuel system pressure.

2. Disconnect the negative battery cable.

3. Drain the engine coolant.

4. Remove the air cleaner cap, MAF meter, and the resonator.

5. If equipped with a manual transaxle, disconnect the accelerator cable from the throttle body.

6. If equipped with a automatic transaxle, disconnect the accelerator and throttle cables from the throttle body.

7. Remove the intake air connector.

8. If equipped with air conditioning, remove the air conditioning idle-up valve.

9. Remove the No. 1 and No. 2 PCV hoses.

10. Remove the spark plug wires from the spark plugs.

11. Remove the throttle body.

Removing the EGR valve — 1995 with 2.4L (22R-E) engine

12. Disconnect the following connectors:
- If equipped with air conditioning, disconnect the air conditioning compressor connector.
- Disconnect the oil pressure sensor connector.
- Disconnect the ECT sensor connector.
- Disconnect the EGR gas temperature sensor connector.
- Disconnect the EGR VSV connector.

13. Disconnect the engine wire as follows:

a. Remove the two bolts and disconnect the engine wire from the intake chamber.

b. Disconnect the five engine wire clamps and engine wire.

c. Disconnect the following connectors:
- Knock sensor connector
- Crankshaft position sensor connector
- Fuel pressure control VSV connector

d. Disconnect the DLC1 from the bracket.

e. Disconnect the two engine wire clamps.

f. Remove the bolt and disconnect the engine wire from the engine.

14. Disconnect the fuel injectors.

15. Remove the EGR valve and vacuum modulator.

16. Remove the intake chamber stay by removing the two bolts.

17. Remove the fuel return pipe by removing the hoses and two bolts.

18. Remove the intake chamber as follows:

a. Disconnect the vacuum hose from the gas filter.

b. Disconnect the brake booster vacuum hose from the intake chamber.

c. Remove the three bolts, two nuts, air intake chamber and gasket.

CAUTION

Fuel injection systems remain under pressure after the engine has been turned OFF. Properly relieve fuel pressure before disconnecting any fuel lines. Failure to do so may result in fire or personal injury.

---

19. Remove the fuel inlet tube by removing the union bolts.

20. Remove the delivery pipe and injectors as follows:

a. Disconnect the vacuum hose from the fuel pressure regulator.

b. Remove the two bolts and delivery pipe together with the four injectors.

c. Remove the four insulators from the four spacers.

d. Pull out the four injectors from the delivery pipe.

e. Remove the O-ring and grommet from each injector.

NOTE: Be careful not to drop the injectors when removing the delivery pipe.

21. Remove the intake manifold by removing the three bolts and two nuts. Remove the gasket from the intake manifold.

To install:

22. Clean the intake manifold surfaces.

23. Install a new gasket and intake manifold with the three bolts and two nuts. Torque the bolts and nuts to 22 ft. lbs. (29 Nm).

24. Install the injectors to the delivery pipe as follows:

a. Install a new grommet to the injector.

b. Apply a light coat of gasoline to a new O-ring and install it to the injector.

c. While turning the injector left and right, install the injector to the delivery pipe. Install all four injectors.

d. Position the injector connector upward.

25. Install the injectors and delivery pipe. Tighten the bolts holding the delivery pipe to 15 ft. lbs. (21 Nm). Check that the injectors rotate smoothly.

26. Install the fuel tube with four new gaskets. Torque the union bolts to 22 ft. lbs. (29 Nm).

27. Install the air intake chamber as follows:

a. Install a new gasket and the air intake chamber with the three bolts and two nuts. Torque the bolts and nuts to 15 ft. lbs. (21 Nm).

b. Connect the vacuum hose to the gas filter.

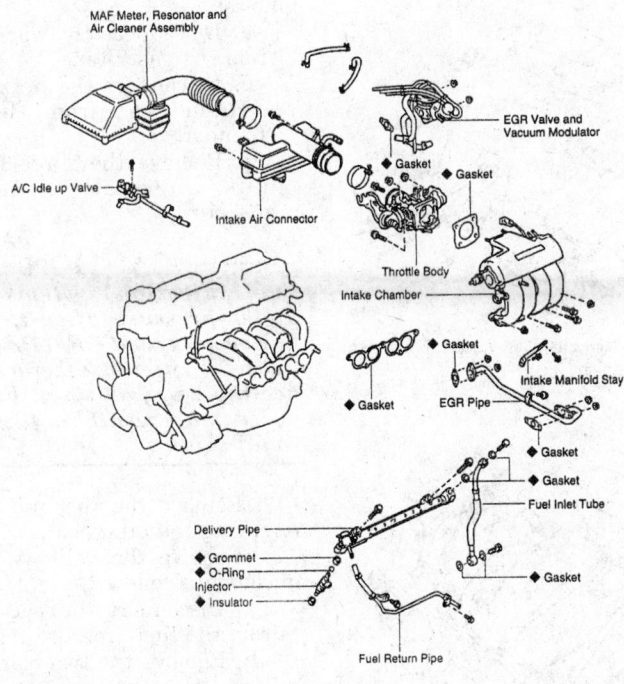

Intake manifold component assembly — 2.4L (2RZ-FE) and 2.7L (3RZ-FE) engines

c. Connect the brake booster vacuum hose to the intake chamber.

28. Install the fuel return pipe by installing the two bolts and hoses.

29. Install the intake chamber stay by installing the two bolts. Torque the bolts to 14 ft. lbs. (20 Nm).

30. Install the EGR valve and vacuum modulator.

31. Connect the injector connectors.

32. Connect the engine wire to the engine as follows:

a. Connect the engine wire to the intake manifold with the bolt.

b. Connect the two engine wire clamps.

c. Connect the DLC1 to the bracket.

d. Connect the following connectors:

• Fuel pressure control VSV connector

• Knock sensor connector

• Crankshaft position sensor connector

e. Connect the five engine wire clamps.

f. Connect the engine wire to the intake chamber with the two bolts.

33. Connect the following connectors to the engine:

• EGR VSV connector

• EGR gas temperature sensor connector

• ECT sensor connector

• Oil pressure sensor connector

• If equipped with air conditioning, connect the compressor connector

34. Install the throttle body.

35. Connect the spark plug wires to the spark plugs.

36. Install the No. 1 and No. 2 PCV hoses.

37. If equipped with air conditioning, install the air conditioning idle up valve.

38. Install the air intake connector by installing the two bolts, hose clamp, and two air hoses.

39. If equipped with a manual transaxle, connect the accelerator cable to the throttle body.

40. If equipped with a automatic transaxle, connect the throttle and accelerator cables to the throttle body.

41. Install the MAF meter, resonator, and the air cleaner cap.

42. Refill the cooling system.

43. Connect the negative battery cable. Start the engine and check for leaks.

44. Check the ignition timing. Road test the vehicle for proper operation.

45. Recheck all fluid levels.

2.7L (3RZ-FE) Engine

1994–97 T-100 and Tacoma

1. Relieve the fuel system pressure.

2. Disconnect the negative battery cable.

3. Drain the engine coolant.

4. Remove the air cleaner cap, MAF meter, and the resonator.

5. If equipped with a manual transaxle, disconnect the accelerator cable from the throttle body.

6. If equipped with a automatic transaxle, disconnect the accelerator and throttle cables from the throttle body.

7. Remove the intake air connector.

8. If equipped with air conditioning, remove the air conditioning idle-up valve.

9. Remove the No. 1 and No. 2 PCV hoses.

10. Remove the spark plug wires from the spark plugs.

11. Remove the throttle body.

12. Disconnect the following connectors:

• If equipped with air conditioning, disconnect the air conditioning compressor connector.

• Disconnect the oil pressure sensor connector.

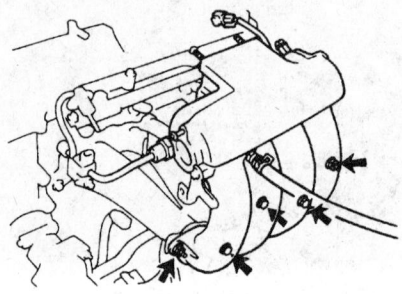

Removing air intake chamber — 1994 T-100 with 2.7L (3RZ-FE) engine

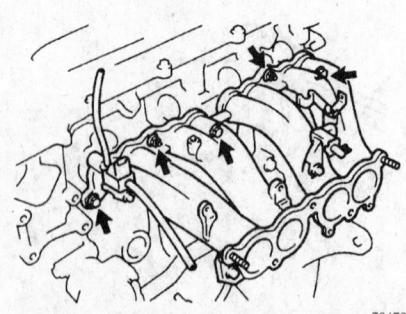

Removing intake manifold — 2.7L (3RZ-FE) engine

• Disconnect the ECT sensor connector.

• Disconnect the EGR gas temperature sensor connector.

• Disconnect the EGR VSV connector.

13. Disconnect the engine wire as follows:

a. Remove the two bolts and disconnect the engine wire from the intake chamber.

b. Disconnect the five engine wire clamps and engine wire.

c. Disconnect the following connectors:

• Knock sensor connector

• Crankshaft position sensor connector

• Fuel pressure control VSV connector

d. Disconnect the DLC1 from the bracket.

e. Disconnect the two engine wire clamps.

f. Remove the bolt and disconnect the engine wire from the engine.

14. Disconnect the fuel injectors.

15. Remove the EGR valve and vacuum modulator.

16. Remove the intake chamber stay by removing the two bolts.

17. Remove the fuel return pipe by removing the hoses and two bolts.

18. Remove the intake chamber as follows:

a. Disconnect the vacuum hose from the gas filter.

b. Disconnect the brake booster vacuum hose from the intake chamber.

c. Remove the three bolts, two nuts, air intake chamber and gasket.

— CAUTION —

Fuel injection systems remain under pressure after the engine has been turned OFF. Properly relieve fuel pressure before disconnecting any fuel lines. Failure to do so may result in fire or personal injury.

19. Remove the fuel inlet tube by removing the union bolts.

20. Remove the delivery pipe and injectors as follows:

a. Disconnect the vacuum hose from the fuel pressure regulator.

b. Remove the two bolts and delivery pipe together with the four injectors.

c. Remove the four insulators from the four spacers.

d. Pull out the four injectors from the delivery pipe.

e. Remove the O-ring and grommet from each injector.

NOTE: Be careful not to drop the injectors when removing the delivery pipe.

21. Remove the intake manifold by removing the three bolts and two nuts. Remove the gasket from the intake manifold.

To install:

22. Clean the intake manifold surfaces.

23. Install a new gasket and intake manifold with the three bolts and two nuts. Torque the bolts and nuts to 22 ft. lbs. (29 Nm).

24. Install the injectors to the delivery pipe as follows:

a. Install a new grommet to the injector.

b. Apply a light coat of gasoline to a new O-ring and install it to the injector.

c. While turning the injector left and right, install the injector to the delivery pipe. Install all four injectors.

d. Position the injector connector upward.

25. Install the injectors and delivery pipe. Tighten the bolts holding the delivery pipe to 15 ft. lbs. (21 Nm). Check that the injectors rotate smoothly.

26. Install the fuel tube with four new gaskets. Torque the union bolts to 22 ft. lbs. (29 Nm).

27. Install the air intake chamber as follows:

a. Install a new gasket and the air intake chamber with the three bolts and two nuts. Torque the bolts and nuts to 15 ft. lbs. (21 Nm).

b. Connect the vacuum hose to the gas filter.

c. Connect the brake booster vacuum hose to the intake chamber.

28. Install the fuel return pipe by installing the two bolts and hoses.

29. Install the intake chamber stay by installing the two bolts. Torque the bolts to 14 ft. lbs. (20 Nm).

30. Install the EGR valve and vacuum modulator as follows:

a. Install a new gasket, EGR valve and vacuum modulator with the bolt and two nuts. Torque the bolt to 74 inch lbs. (9 Nm) and the nuts to 14 ft. lbs. (19 Nm).

b. Connect the following hoses:

• Two vacuum hoses to the EGR VSV

• Water bypass hose to the water bypass pipe

c. Install two new gaskets and EGR pipe with the bolt and four

nuts. Torque the bolts and nuts as follows:

• Bolt to 14 ft. lbs. (18 Nm)

• Nuts to intake manifold to 14 ft. lbs. (19 Nm)

• Nuts to cylinder head to 15 ft. lbs. (20 Nm).

31. Connect the injector connectors.

32. Connect the engine wire to the engine as follows:

a. Connect the engine wire to the intake manifold with the bolt.

b. Connect the two engine wire clamps.

c. Connect the DLC1 to the bracket.

d. Connect the following connectors:

• Fuel pressure control VSV connector

• Knock sensor connector

• Crankshaft position sensor connector

e. Connect the five engine wire clamps.

f. Connect the engine wire to the intake chamber with the two bolts.

33. Connect the following connectors to the engine:

• EGR VSV connector

• EGR gas temperature sensor connector

• ECT sensor connector

• Oil pressure sensor connector

• If equipped with air conditioning, connect the compressor connector

34. Install the throttle body.

35. Connect the spark plug wires to the spark plugs.

36. Install the No. 1 and No. 2 PCV hoses.

37. If equipped with air conditioning, install the air conditioning idle up valve.

38. Install the air intake connector by installing the two bolts, hose clamp, and two air hoses.

39. If equipped with a manual transaxle, connect the accelerator cable to the throttle body.

40. If equipped with a automatic transaxle, connect the throttle and accelerator cables to the throttle body.

41. Install the MAF meter, resonator, and the air cleaner cap.

42. Refill the cooling system.

43. Connect the negative battery cable. Start the engine and check for leaks.

44. Check the ignition timing. Road test the vehicle for proper operation.

45. Recheck all fluid levels.

1996–97 4Runner

1. Disconnect the negative battery cable.

CAUTION

Wait 90 seconds from the time the key is turned to LOCK and the negative battery cable is disconnected to begin work. This allows the SRS capacitor to discharge and prevent deployment of the air bag(s).

2. Release the fuel system pressure.

3. Drain the engine coolant.

4. Remove the air cleaner cap, MAF meter, and the resonator.

5. If equipped with a manual transmission, disconnect the accelerator cable from the throttle body.

6. If equipped with a automatic transmission, disconnect the accelerator and throttle cables from the throttle body.

7. If equipped with cruise control, disconnect the cruise control cable from the actuator.

8. Remove the intake air connector. Disconnect the following:

a. Air hose for IAC

b. Vacuum sensing hose

c. Wire clamp for the engine wire

9. Remove the PCV hoses.

10. Disconnect the engine wire clamps and engine wire.

11. Disconnect the following connectors:

• If equipped with air conditioning, disconnect the air conditioning compressor connector

• Disconnect the oil pressure sensor connector

• Disconnect the ECT sensor connector

• Engine coolant temperature sender gauge connector

• EGR gas temperature sensor connector

• VSV connector for the EGR

• Two vacuum hoses from the VSV for the EGR

• Ground strap from the cowl top panel

• Engine wire from the air intake chamber

• Throttle position sensor connector

• IAC valve connector

• Crankshaft position sensor connector

• Knock sensor connector

• DLC1 from the bracket

• Engine wire clamp

12. Remove the EGR pipe.

13. Remove the intake chamber stay.

14. Remove the air intake chamber assembly.

15. Disconnect the following hoses:
  a. EVAP hose from the throttle body
  b. Brake booster vacuum hose from the union
  c. Water bypass hose from the water bypass pipe
  d. Water bypass hose from the cylinder head rear cover
16. Disconnect the injector connectors.

CAUTION

Fuel injection systems remain under pressure after the engine has been turned OFF. Properly relieve fuel pressure before disconnecting any fuel lines. Failure to do so may result in fire or personal injury.

---

17. Remove the fuel inlet pipe.
18. Disconnect the hoses and remove the fuel return pipe.
19. Remove the delivery pipe and injectors.
  a. Remove the two bolts and the delivery pipe together with the four injectors.
  b. Remove the four insulators form the four spacers.
  c. Pull out the four injectors from the delivery pipe.
  d. Remove the O-ring and grommets from each injector.
  e. Carefully pry out the four spacers.
20. Remove the intake manifold.
  To install:
21. Install the intake manifold and torque the bolts to 22 ft. lbs. (29 Nm).
22. Install the injectors and the delivery pipe.
23. Install the fuel return pipe.
24. Install the fuel inlet pipe with a new gasket and torque the bolts to 22 ft. lbs. (29 Nm).
25. Connect the injector connectors.
26. Install the air intake chamber assembly. Torque the bolts to 15 ft. lbs. (21 Nm). Connect the following hoses:
  a. EVAP hose to the throttle body
  b. Brake booster vacuum hose to union
  c. Water bypass hose to water bypass pipe
  d. Water bypass hose to cylinder head rear cover
27. Install the air intake chamber stay and torque the bolts to 15 ft. lbs. (20 Nm).
28. Install the EGR pipe. Torque the nuts and bolts to:
  • Bolt: 13 ft. lbs. (18 Nm)
  • Nut A: 14 ft. lbs. (19 Nm)
  • Nut B: 15 ft. lbs. (20 Nm)

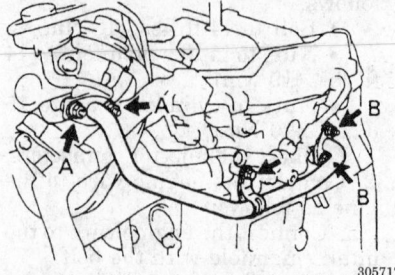

305717

EGR pipe and related components — 1996–97 4Runner with 2.7L (3RZ-FE) engine

29. Connect the engine wire.
  • If disconnected, connect the air conditioning compressor connector
  • Connect the oil pressure sensor connector
  • Connect the ECT sensor connector
  • Engine coolant temperature sender gauge connector
  • EGR gas temperature sensor connector
  • VSV connector for the EGR
  • Two vacuum hose to the VSV for the EGR
  • Ground strap to the cowl top panel
  • Engine wire to the air intake chamber
  • Throttle position sensor connector
  • IAC valve connector
  • Crankshaft position sensor connector
  • Knock sensor connector
  • DLC1 to the bracket
  • Engine wire clamp
30. Install the PCV hoses.
31. Install the intake air connector and torque the bolts to 13 ft. lbs. (18 Nm).
32. If equipped with cruise control, connect the cruise control cable to the actuator.
33. If equipped with a manual transmission, connect the accelerator cable to the throttle body.
34. If equipped with a automatic transmission, connect the accelerator and throttle cables to the throttle body.
35. Fill the engine and radiator with engine coolant.
36. Install the air cleaner cap, MAF meter, and the resonator assembly.
37. Connect the negative battery cable, start the engine, and check for leaks.
38. Road test the vehicle for proper operation.
39. Recheck all fluid levels.

3.0 (3VZ-E) Engine

1993–94 Models

1. Relieve the fuel system pressure. Disconnect the negative battery cable.

WARNING

The air bag system is equipped with a backup power source. To avoid possible air bag deployment, do not start working on the vehicle until 90 seconds has elapse from the time the ignition switch is turned OFF and the negative battery terminal is disconnected.

---

2. Drain the cooling system.
3. Disconnect the air intake hose from both the air cleaner assembly on one end and the air intake chamber on the other.
4. Tag and disconnect all vacuum lines attached to the intake chamber and manifold.
5. Disconnect the throttle position sensor connector at the air chamber. Disconnect the PCV hose at the union.
6. Disconnect the No. 4 water bypass hose at the manifold. Remove the No. 5 bypass hose at the water bypass pipe.
7. Disconnect the cold start injector and the vacuum hose at the fuel filter.
8. Remove the union bolt and gaskets, then remove the cold start injector tube.
9. Disconnect the EGR gas temperature sensor and the EGR vacuum hoses from the air pipe and the vacuum modulator.
10. Remove the EGR valve.
11. Disconnect the No. 1 air hose at the reed valve.
12. Remove the air intake chamber and then remove the engine wire.
13. Remove the union bolts and then remove the No. 2 and 3 fuel pipes.
14. Remove the No. 4 timing belt cover. Remove the No. 2 idler pulley and the No. 3 timing belt cover.
15. Remove the fuel delivery pipes with their injectors.
16. Remove the water bypass outlet and then remove the intake manifold.
  To install:
17. Install the intake manifold with new gaskets and tighten the mounting bolts to 29 ft. lbs. (39 Nm).
18. Install the water bypass outlet and tighten the 2 bolts to 13 ft. lbs. (18 Nm).
19. Install the fuel delivery pipes and injectors.

20. Install the No. 2 idler pulley. Install the No. 3 and 4 timing belt covers and tighten the bolts to 74 inch lbs. (8.3 Nm).

21. Install the fuel pipes and tighten the union bolts to 22 ft. lbs. (29 Nm).

22. Install the engine wire.

23. Install the air intake chamber and tighten the nuts and bolts to 13 ft. lbs. (18 Nm).

24. Install the EGR valve and connect all hoses and lines.

25. Install the air cleaner hose, refill the engine with coolant and connect the battery cable.

1995 Models

1. Relieve the fuel system pressure.

——— **CAUTION** ———
Fuel injection systems remain under pressure after the engine has been turned OFF. Properly relieve fuel pressure before disconnecting any fuel lines. Failure to do so may result in fire or personal injury.

2. Disconnect the negative battery cable. See the air bag warning!

3. Drain the cooling system.

4. Disconnect the air intake hose from both the air cleaner assembly on one end and the air intake chamber on the other.

5. If equipped with manual transmission, disconnect the clutch release cylinder hose.

6. Disconnect the following straps, wires, and connectors:

   a. The ground strap from the LH fender apron.

   b. Alternator connector and wire.

   c. Igniter connector.

   d. Oil pressure sender gauge connector.

   e. The ground strap from the rear of the engine.

   f. The ECM and VSV connectors.

   g. The air conditioning compressor connector.

   h. The Data Link Connector (DLC) No. 1.

   i. With manual transmission: the starter relay and solenoid resistor connectors.

   j. If equipped with Automatic Disconnecting Differential (ADD), unplug the ADD switch connector.

7. Disconnect the following hoses:

   a. With power steering: the air hoses.

   b. Power brake booster hose.

   c. With Cruise: cruise control vacuum hose.

   d. The charcoal canister vacuum hose from the canister and the throttle body.

   e. The VSV vacuum hoses.

   f. The fuel pressure regulator vacuum hose.

8. Disconnect the following cables:

   a. Accelerator cable.

   b. With automatic transmission: Throttle cable.

   c. With Cruise: Cruise control cable.

9. Remove the spark plug wires and the distributor.

10. Disconnect the heater hoses, the fuel inlet hose, and the fuel return hose.

11. Disconnect the throttle position sensor connector at the air chamber. Disconnect the PCV hose at the union.

12. Disconnect the No. 4 water bypass hose at the manifold. Remove the No. 5 bypass hose at the water bypass pipe.

13. Disconnect the cold start injector connector and remove the vacuum hose from the gas filter.

14. Remove the union bolt and gaskets, then remove the cold start injector tube.

15. Disconnect the EGR gas temperature sensor connector and the EGR vacuum hoses from the air pipe and the vacuum modulator.

16. Remove the intake chamber stay and the throttle cable bracket.

17. Remove the No. 1 engine hanger.

18. Remove the power steering pump and bracket.

19. Remove the EGR valve.

20. Disconnect the No. 1 air hose at the PAIR reed valve.

21. Disconnect the vacuum hoses from the air pipes.

22. Remove the accelerator cable bracket.

23. Remove the air intake chamber and then remove the engine wire. Disconnect the following connectors:

   a. Knock sensor

   b. Cold start injector time switch connector

   c. ECT sensor and sender gauge connector

   d. No. 1 ECT switch connector

   e. Right ground strap from the No. 3 camshaft bearing cap

   f. Fuel injector connectors

24. Remove the union bolts and then remove the No. 2 and No. 3 fuel pipes.

25. Remove the fuel delivery pipes with their injectors.

26. Remove the water bypass outlet and then remove the intake manifold.

Exploded view of the intake manifold and related components — 1995 with 3.0L (3VZ-E) engine

To install:

27. Install the intake manifold with new gaskets and tighten the mounting bolts to 13 ft. lbs. (18 Nm).

28. Install the water bypass outlet and tighten the 2 bolts to 13 ft. lbs. (18 Nm).

29. Install the fuel delivery pipes and injectors. Torque the delivery pipe holding nuts to 9 ft. lbs. (13 Nm).

30. Install the No. 2 and No. 3 fuel pipes; tighten the union bolts to 25 ft. lbs. (34 Nm).

31. Install the engine wire and connect the following:
   a. Fuel injector connectors
   b. Right ground strap from the No. 3 camshaft bearing cap
   c. No. 1 ECT switch connector
   d. ECT sensor and sender gauge connector
   e. Cold start injector time switch connector
   f. Knock sensor

32. Install the air intake chamber and tighten the nuts and bolts to 13 ft. lbs. (18 Nm).

33. Install the accelerator cable bracket.

34. Connect the vacuum hoses to the air pipes.

35. Connect the No. 1 air hose to the reed valve.

36. Install the EGR valve and pipes assembly, the air intake chamber stay, and the throttle cable bracket. Torque bolt A to 22 ft. lbs. (29 Nm) and bolt B to 13 ft. lbs. (18 Nm).

37. Install the power steering pump and bracket.

38. Install the No. 1 engine hanger. Torque the bolt to 30 ft. lbs. (40 Nm).

39. Connect the EGR temperature sensor connector and the EGR vacuum hoses to the air pipe and the vacuum modulator.

40. Connect the cold start injector tube with new gaskets and the union bolt. Torque to 11 ft. lbs. (15 Nm).

41. Connect the vacuum hose to the gas filter.

42. Connect the cold start injector connector.

43. Install the No. 5 water bypass hose to the water bypass pipe and connect the No. 4 water bypass hose to the union of the intake manifold.

44. Connect the PCV hose.

45. Connect the throttle position sensor connector to the air chamber.

46. Connect the heater hoses, fuel inlet hose, and the fuel return hose.

47. Install the distributor and the spark plug wires.

48. Connect the following cables:
   a. Accelerator cable.
   b. With automatic transmission: Throttle cable.

   c. With Cruise: Cruise control cable.

49. Connect the following hoses:
   a. With power steering: the air hoses.
   b. Power brake booster hose.
   c. With Cruise: cruise control vacuum hose.
   d. The charcoal canister vacuum hose from the canister and the throttle body.
   e. The VSV vacuum hoses.
   f. The fuel pressure regulator vacuum hose.

50. Connect the following straps, wires, and connectors:
   a. The ground strap from the LH fender apron.
   b. Alternator connector and wire.
   c. Igniter connector.
   d. Oil pressure sender gauge connector.
   e. The ground strap from the rear of the engine.
   f. The ECM and VSV connectors.
   g. The air conditioning compressor connector.
   h. The Data Link Connector (DLC) No. 1.
   i. With manual transmission: the starter relay and solenoid resistor connectors.
   j. With Automatic Disconnecting Differential (ADD): the ADD switch connector.

51. If removed, connect the clutch release cylinder hose.

52. Connect the air intake hose to both the air cleaner assembly on one end and the air intake chamber on the other end.

53. Refill the engine with coolant.

54. Connect the negative battery cable.

55. Start the engine, check for leaks, check the ignition timing, and bleed the cooling system.

56. Perform any necessary adjustments and road test the vehicle.

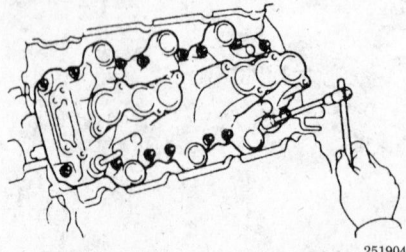

Intake manifold mounting bolts — 1995 with 3.0L (3VZ-E) engine

251904

3.4 (5VZ-FE) Engine

1996–97 4Runner

1. Disconnect the negative battery cable.

— CAUTION —

Wait 90 seconds from the time the key is turned to LOCK and the negative battery cable is disconnected to begin work. This allows the SRS capacitor to discharge and prevent deployment of the air bag(s).

2. Relieve the fuel system pressure.

3. Remove the engine undercover.

4. Drain the cooling system.

5. Disconnect the air cleaner cap, MAF meter, and the resonator.

6. Disconnect the following cables:
• If equipped with cruise control, disconnect the actuator cable with the bracket.
• Accelerator cable
• If equipped with automatic transmission, remove the throttle cable

7. Disconnect the heater hose.

8. Disconnect the following hoses:
• Brake booster vacuum hose
• EVAP hose
• 4-Wheel Drive: A.D.D. vacuum hose
• Fuel inlet and fuel return hose

— CAUTION —

Fuel injection systems remain under pressure after the engine has been turned OFF. Properly relieve fuel pressure before disconnecting any fuel lines. Failure to do so may result in fire or personal injury.

9. Remove the spark plug wires with the ignition coils.

10. Remove the intake chamber stay.

11. Remove the No. 2 timing belt cover.

12. Remove the air intake chamber assembly.

13. Remove the following connectors and hoses:
• Throttle position sensor connector
• IAC valve connector
• PCV hoses
• Water bypass hoses
• Air assist hose from the throttle body

14. Remove the intake air connector.
   a. Remove the bolt and disconnect the engine wire.
   b. Disconnect the fuel return hose.

Fuel delivery pipe mounting bolts — 1995 with 3.0L (3VZ-E) engine

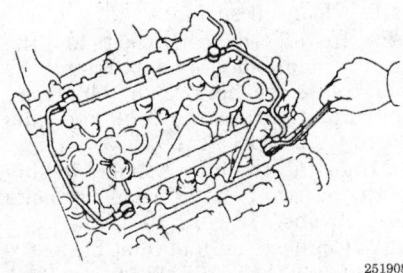

Location of the No. 2 and No. 3 fuel pipes — 1995 with 3.0L (3VZ-E) engine

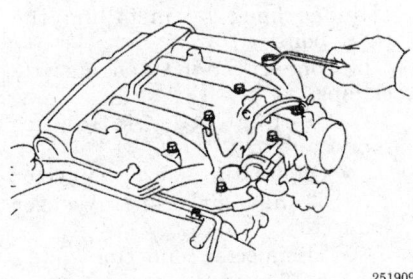

Air intake chamber mounting bolts — 1995 with 3.0L (3VZ-E) engine

c. Disconnect the vacuum hose from the fuel pressure regulator.

d. Disconnect the ground strap from the intake air connector.

e. Disconnect the DLC1 from the bracket.

15. Disconnect the engine wire protector.

a. Disconnect the six injector connectors.

b. Disconnect the ECT sensor and sender gauge connectors.

c. Disconnect the engine wire protector from the cylinder head.

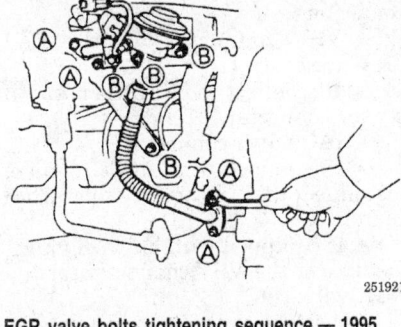

EGR valve bolts tightening sequence — 1995 with 3.0L (3VZ-E) engine

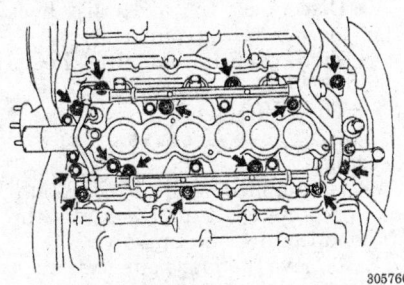

Intake manifold bolts and nuts — 1996–97 4Runner with 3.4 (5VZ-FE) engine

16. Remove the fuel pressure regulator.

17. Remove the intake manifold assembly.

a. Remove the intake manifold stay.

b. Remove the eight bolts, four nuts, four plate washers, the intake manifold, delivery pipes and the injectors assembly with the gaskets.

To install:

18. Install two new gaskets and the intake manifold assembly. Install the four plate washers, eight bolts and four nuts. Torque the bolts and nuts to 13 ft. lbs. (18 Nm).

19. Install the intake manifold stay with the two bolts. Torque the bolts to 14 ft. lbs. (18 Nm).

20. Install the fuel pressure regulator.

21. Connect the engine wire to the intake manifold as follows:

a. Install the engine wire to the cylinder head by installing the three bolts.

b. Connect the three engine wire clamps.

c. Connect the following connectors:

• Six injector connectors
• ECT sender gauge connector
• ECT sensor connector

22. Install the intake air connector as follows:

a. Install the intake manifold to the engine by installing the three bolts and two nuts. Torque the bolts and nuts to 14 ft. lbs. (18.5 Nm).

b. Connect the DLC1 to the bracket on the intake manifold.

c. Connect the ground strap to the intake manifold by installing the bolt.

d. Connect the brake booster vacuum hose to the intake air connector.

e. Connect the two fuel return hoses.

f. Connect the engine wire to the intake manifold by installing the bolt.

23. Install the air intake chamber assembly to the engine by installing the four bolts and two nuts. Torque the bolts and nuts to 14 ft. lbs. (18.5 Nm).

24. Install the intake chamber stay as follows:

a. Install the intake chamber stay and install the two bolts. Torque the bolts to 30 ft. lbs. (40 Nm)

b. Install a new O-ring to the oil filler tube.

c. Push in the oil filler tube end into the tube hole in the oil pan.

d. Install the oil filler tube and No. 1 throttle cable clamp and install the bolt and two nuts.

25. Install the No. 2 timing belt cover and torque the bolts to 80 inch lbs. (9 Nm).

26. Connect the following:

• Connect the PCV hoses.
• Connect the water bypass hoses.
• Connect the air assist hose to the throttle body.
• IAC valve connector.
• Throttle position sensor connector.

27. Connect the following hoses:

• Brake booster vacuum hose
• EVAP hose
• 4-Wheel Drive: A.D.D. vacuum hose
• Fuel inlet and fuel return hose

28. Connect the three clamps for the spark plug wires to the No. 2 timing belt cover.

29. Connect the camshaft position sensor connector to the No. 2 timing belt cover.

30. Install the spark plug wires with the ignition coils.

31. Connect the heater hose.

32. Connect the following cables:
• If equipped with cruise control, connect the actuator cable with the bracket.
• Accelerator cable
• If equipped with automatic transmission, connect the throttle cable
33. Install the MAF meter, resonator, and the air cleaner cap.
34. Fill the radiator with engine coolant.
35. Connect the negative battery cable to the battery.
36. Start the engine and check for leaks.
37. Install the engine undercover.
38. Road test the vehicle.
39. Recheck all fluid levels.

1995–97 T-100 and Tacoma

1. Disconnect the negative battery cable.

— CAUTION —

Wait 90 seconds from the time the key is turned to LOCK and the negative battery cable is disconnected to begin work. This allows the SRS capacitor to discharge and prevent deployment of the air bag(s).

2. Relieve the fuel system pressure.

— CAUTION —

Fuel injection systems remain under pressure after the engine has been turned OFF. Properly relieve fuel pressure before disconnecting any fuel lines. Failure to do so may result in fire or personal injury.

3. Drain the engine coolant.
4. Disconnect the spark plug wires from the spark plugs.
5. Remove the air cleaner cap, MAF meter, and the resonator.
6. Disconnect the following cables:
• If equipped with cruise control, disconnect the actuator cable with the bracket.
• Accelerator cable
• If equipped with automatic transmission, disconnect the throttle cable
7. If equipped with an EGR valve, remove the nuts and remove the EGR pipe and two gaskets.
8. Remove the intake chamber stay as follows:
a. Remove the oil filler tube and No. 1 throttle cable clamp by removing the bolt and two nuts.
b. Remove the intake chamber stay by removing the two bolts.

9. Remove the following connectors:
• VSV connector for the fuel pressure control.
• Disconnect the throttle position sensor connector.
• IAC valve connector
• If equipped with an EGR valve, disconnect the EGR gas temperature connector
• If equipped with an EGR valve, disconnect the VSV connector for the EGR valve
10. Disconnect the following hoses:
• Disconnect the PCV hoses.
• Disconnect the water bypass hoses.
• Disconnect the air assist hose from the intake air connector.
• Two vacuum sensing hoses from the VSV
• The EVAP hose
• Air hose from the power steering
• If equipped with air conditioning, disconnect the air hose from the air conditioning idle up valve.
11. Remove the four bolts, two nuts, and remove the air intake chamber assembly from the engine.
12. Remove the intake air connector as follows:
a. Disconnect the engine wire from the intake air connector by removing the bolt.
b. Disconnect the two fuel return hoses.
c. Disconnect the brake booster vacuum hose from the intake air connector.
d. Remove the bolt and disconnect the ground strap from the intake air connector.
e. Disconnect the DLC1 from the bracket of the intake air connector.
f. If equipped with air conditioning, disconnect idle up valve connector.
g. Remove the intake air connector from the engine by removing the three bolts and two nuts.
13. Disconnect the upper radiator hose from the engine.
14. Disconnect the engine wire from the intake manifold as follows:
a. Disconnect the following connectors:
• Oil pressure sensor connector
• Crankshaft position sensor connector
• Six injector connectors
• ECT sender gauge connector
• ECT sensor connector
• Knock sensor connector
• Camshaft position sensor connector
b. Disconnect the three engine wire clamps.

c. Remove the three bolts and disconnect the engine wire from the cylinder head.
15. Remove the fuel pressure regulator.
16. Disconnect the heater hose.
17. Remove the camshaft position sensor.
18. Remove the intake manifold assembly as follows:
a. Disconnect the fuel inlet hose.
b. Remove the two bolts and the intake manifold stay.
c. Remove the eight bolts, four nuts, four plate washers, and the intake manifold assembly.
To install:
19. Clean all surfaces.
20. Install two new gaskets and the intake manifold assembly. Install the four plate washers, eight bolts and four nuts. Torque the bolts and nuts to 13 ft. lbs. (18 Nm).
21. Install the intake manifold stay with the two bolts. Torque the bolts to 13 ft. lbs. (18 Nm).
22. Connect the fuel inlet hose.
23. Install the camshaft position sensor and torque to 71 inch lbs. (8 Nm).
24. Connect the engine wire to the intake manifold as follows:
a. Install the engine wire to the cylinder head by installing the three bolts.
b. Connect the three engine wire clamps.
c. Connect the following connectors:
• Oil pressure sensor connector
• Crankshaft position sensor connector
• Six injector connectors
• ECT sender gauge connector
• ECT sensor connector
• Knock sensor connector
• Camshaft position sensor connector
25. Install the heater hose.
26. Install the intake air connector as follows:
a. Install the intake manifold to the engine by installing the three bolts and two nuts. Torque the bolts and nuts to 14 ft. lbs. (18.5 Nm).
b. Connect the DLC1 to the bracket on the intake manifold.
c. Connect the ground strap to the intake manifold by installing the bolt.
d. Connect the brake booster vacuum hose to the intake air connector.
e. Connect the two fuel return hoses.

f. Connect the engine wire to the intake manifold by installing the bolt.

g. If equipped with air conditioning, connect the idle up valve connector.

27. Install the air intake chamber assembly to the engine by installing the four bolts and two nuts. Torque the bolts and nuts to 14 ft. lbs. (18.5 Nm).

28. Connect the following hoses:
• Connect the PCV hoses.
• Connect the water bypass hoses.
• Connect the air assist hose to the intake manifold.
• Two vacuum sensing hoses to the VSV
• The EVAP hose
• Air hose to the power steering
• If equipped with air conditioning, connect the air hose to the air conditioning idle up valve.

29. Connect the following connectors:
• VSV connector for the fuel pressure control.
• Connect the throttle position sensor connector.
• IAC valve connector
• If equipped with an EGR valve, connect the EGR gas temperature connector
• If equipped with an EGR valve, connect the VSV connector for the EGR valve

30. Install the intake chamber stay as follows:
a. Install the intake chamber stay and install the two bolts. Torque the bolts to 30 ft. lbs. (40 Nm)
b. Install a new O-ring to the oil filler tube.
c. Push in the oil filler tube end into the tube hole in the oil pan.
d. Install the oil filler tube and No. 1 throttle cable clamp and install the bolt and two nuts.

31. Install two new gaskets and the EGR pipe with the nuts. Torque the clamp nuts to 71 inch lbs. (8 Nm) and the EGR pipe nuts to 14 ft. lbs. (18 Nm).

32. Install the fuel pressure regulator.

33. Connect the three clamps for the spark plug wires to the No. 2 timing belt cover.

34. Connect the camshaft position sensor connector to the No. 2 timing belt cover.

35. Connect the upper radiator hose.

36. Fill with engine coolant.

37. Connect the spark plug wires to the spark plugs.

38. Connect the following cables:
• If equipped with cruise control, connect the actuator cable with the bracket.
• Accelerator cable
• If equipped with automatic transmission, connect the throttle cable

39. Install the air cleaner hose.

40. Fill the radiator with engine coolant.

41. Connect the negative battery cable to the battery.

42. Start the engine and check for leaks.

Exhaust Manifold

REMOVAL AND INSTALLATION

2.4L (22R-E) Engine

1. Disconnect the negative battery cable.

2. Raise and safely support the vehicle.

3. Working from under the vehicle, remove the two bolts holding the front exhaust pipe to the mounting bracket.

4. Remove the three nuts and disconnect the exhaust pipe.

5. Disconnect the main oxygen sensor and the sub oxygen sensor connectors.

6. Remove the three bolts and the exhaust manifold heat insulator.

7. Remove the eight nuts, the exhaust manifold, if equipped, the secondary air injection manifold.

To install:

8. Install a new gasket and the exhaust manifold. Uniformly tighten the nuts in several passes. Torque the nuts to 33 ft. lbs. (44 Nm).

9. Install the exhaust manifold heat insulator with the three bolts and torque to 14 ft. lbs. (19 Nm).

10. Connect the main oxygen and the sub oxygen sensor connectors.

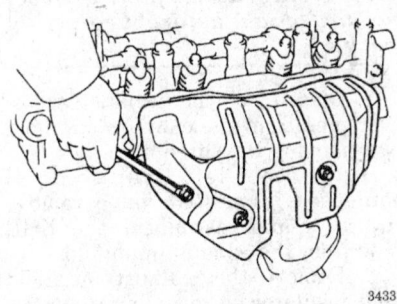

343315

Exhaust manifold heat shield — 2.4L (22R-E) engine

11. Install the front exhaust pipe with a new gasket to the exhaust manifold with the three nuts.

12. Secure the front exhaust pipe to the exhaust pipe clamp.

13. Lower the vehicle safely and connect the negative battery cable.

14. Start the engine and make sure that there are no exhaust leaks.

2.4L (2RZ-FE) and 2.7L (3RZ-FE) Engine

1. Raise and safely support the vehicle.

2. Disconnect the front exhaust pipe from the exhaust manifold as follows:
a. Loosen the clamp bolt and disconnect the clamp from the support bracket.
b. Remove the support bracket by removing the two bolts.
c. Disconnect the three nuts, front exhaust pipe, and the gaskets from the exhaust manifold.

3. Lower the vehicle.

4. Remove the heat insulator by removing the two bolts and two nuts.

5. Remove the exhaust manifold and gasket by removing the six nuts.

To install:

6. Install the exhaust manifold and gasket to the engine and install the six nuts. Torque the nuts to 36 ft. lbs. (49 Nm).

7. Install the heat insulator with the two bolts and two nuts. Torque the bolts and nuts to 48 inch lbs. (5.5 Nm).

8. Raise and safely support the vehicle.

9. Install the front exhaust pipe to the exhaust manifold as follows:
a. Install the two gaskets and the front exhaust pipe assembly to the exhaust manifold. Install the three nuts and torque the nuts to 46 ft. lbs. (62 Nm).
b. Install the support bracket with the two bolts. Torque the brackets to 29 ft. lbs. (39 Nm).
c. Connect the clamp and tighten the clamp bolt. Torque the bolt to 14 ft. lbs. (19 Nm).

10. Lower the vehicle and start the engine.

11. Check for exhaust leaks.

3.0L (3VZ-E) Engine

1. Disconnect the negative battery cable.

2. Raise and safely support the vehicle.

3. Working from under the vehicle, disconnect the heated oxygen sensor connector.

4. Loosen the pipe clamp bolt.

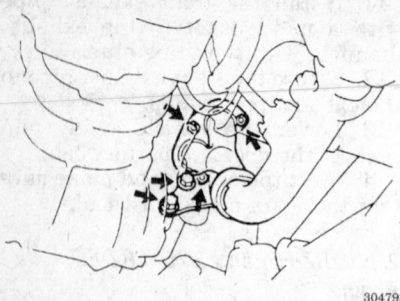

Front exhaust pipe to exhaust manifold — 2.4L (2RZ-FE) and 2.7L (3RZ-FE) engine

Exhaust manifold nuts — 2.4L (2RZ-FE) and 2.7L (3RZ-FE) engine

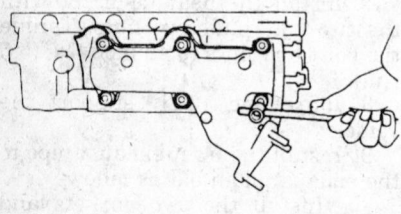

Left side exhaust manifold — 3.0L (3VZ-E) engine

5. Remove the two bolts and the pipe bracket.

6. Remove the three nuts and disconnect the exhaust pipe from the exhaust manifold. Remove the gasket.

7. Remove the two bolts, the joint retainer, the exhaust pipe and the gasket from the catalytic converter.

8. Remove the six nuts, two gaskets, and the exhaust crossover pipe.

9. Remove the bolts, nuts, and the exhaust manifold heat insulators. Remove the six nuts, the left and the right exhaust manifolds, and the gaskets.

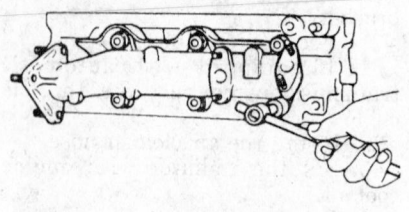

Right side exhaust manifold — 3.0L (3VZ-E) engine

To install:

10. Install the new gaskets and the left and the right exhaust manifolds. Uniformly tighten the nuts in several passes. Torque the nuts to 29 ft. lbs. (39 Nm).

11. Install the exhaust crossover pipe with two new gaskets and six nuts. Torque the nuts to 29 ft. lbs. (39 Nm).

12. Install the left and right exhaust manifold heat insulators with the nuts and bolts.

13. Connect the exhaust pipe to the left exhaust manifold with a new gasket and torque the three new nuts to 46 ft. lbs. (62 Nm).

14. Connect the exhaust pipe to the catalytic converter with a new gasket and torque the two bolts to 29 ft. lbs. (39 Nm).

15. Connect the heated oxygen sensor connector.

16. Secure the front exhaust pipe to the exhaust pipe clamp.

17. Lower the vehicle safely and connect the negative battery cable.

18. Start the engine and make sure that there are no exhaust leaks.

3.4L (5VZ-FE) Engine

1996–97 4Runner, 1995–97 T-100 and Tacoma

— **CAUTION** —
Make certain all surfaces are cool to the touch before beginning work.

1. Disconnect the exhaust crossover pipe from the exhaust manifold by removing the three nuts.

2. On the left manifold, if equipped with an EGR valve, remove the nuts and disconnect the EGR pipe from the exhaust manifold.

3. Remove the exhaust manifold heat insulator by removing the three nuts.

4. Remove the exhaust manifold by removing the six nuts.

To install:

5. Install the exhaust manifold with a new gasket. Torque the six nuts to 30 ft. lbs. (40 Nm).

6. Install the exhaust heat insulator by installing the three nuts. Torque the nuts to 71 inch lbs. (8 Nm).

7. If removed, with an EGR valve, connect the EGR pipe to the exhaust manifold. Torque the nuts to the manifold to 14 ft. lbs. (18 Nm). Torque the clamp nuts to 71 inch lbs. (8 Nm).

8. Connect the crossover pipe to the exhaust manifold with the three bolts and a new gasket. Torque the nuts to 33 ft. lbs. (45 Nm).

Front Cover Seal

REMOVAL AND INSTALLATION

2.4L (22R-E) Engine

Timing Chain Cover Removed

1. Disconnect the negative battery cable.

— **CAUTION** —
Wait 90 seconds from the time the key is turned to LOCK and the negative battery cable is disconnected to begin work. This allows the SRS capacitor to discharge and prevent deployment of the air bag(s).

2. Remove the timing chain front cover.

3. Using a hammer and a flat tipped tool, tap out the oil seal from the inside out.

To install:

4. Using tool SST 09223–50010, or equivalent, and a mallet, tap in the new oil seal until its surface is flush with the oil pump case edge.

5. Apply MP grease to the new oil seal lip.

6. Install the timing chain front cover.

7. Connect the negative battery cable.

Timing Chain Cover Installed

1. Remove the crankshaft pulley.

2. Using a knife, cut off the oil seal lip.

3. Using a flat tipped tool, pry out the oil seal.

NOTE: Tape the end of the prying tool to prevent damage to the crankshaft.

To install:

4. Apply MP grease to a new oil seal lip.

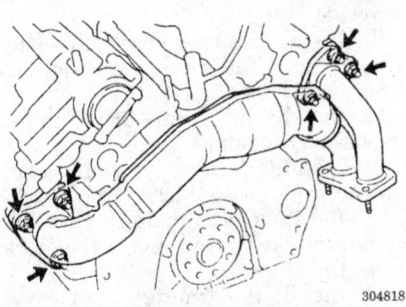

Exhaust crossover pipe — 3.4L (5VZ-FE) engine

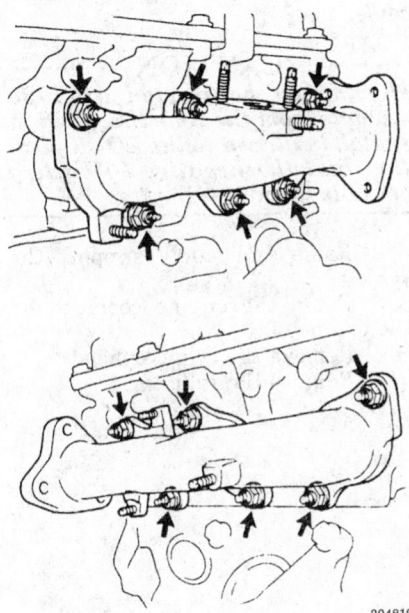

Exhaust manifold nuts — 3.4L (5VZ-FE) engine

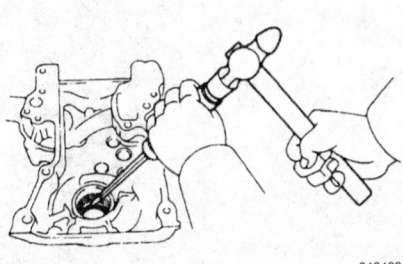

Removing the front cover seal with the timing chain cover removed — 2.4L (22R-E) engine

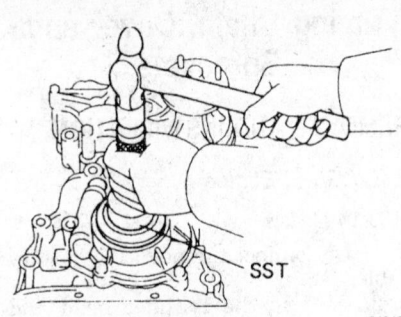

Installation of the timing chain front cover seal with the cover removed — 2.4L (22R-E) engine

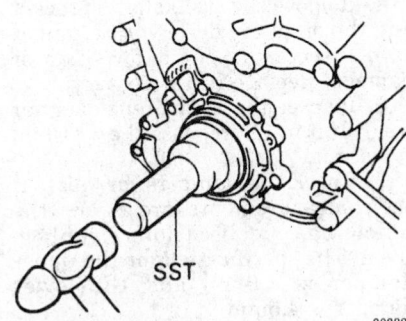

Oil seal installation — 2.4L (2RZ-FE) and 2.7L (3RZ-FE) engines

5. Using SST 09223–50010, or equivalent, and a hammer, tap in the oil seal until its surface is flush with the timing chain cover edge.

6. Install the crankshaft pulley and torque the bolt to 116 ft. lbs. (157 Nm).

2.4L (2RZ-FE) and 2.7L (3RZ-FE) Engines

Timing Chain Cover Installed

1. Disconnect the negative battery cable.
2. Remove the engine drive belts and crankshaft pulley.
3. Using a suitable tool, pry out the oil seal. Be careful not to damage the crankshaft.
   To install:
4. Apply Multi-Purpose (MP) grease to the new oil seal lip.
5. Using tool SST 09223–50510, or equivalent, and a mallet, tap in the new oil seal until its surface is flush with the oil pump cover edge.
6. Install the crankshaft pulley and engine drive belts.

Timing Chain Cover Removed

1. Carefully tap the oil seal out the timing chain cover.
2. Apply Multi-Purpose (MP) grease to the new oil seal lip.

3. Using tool SST 09223–50510, or equivalent, and a mallet, tap in the new oil seal until its surface is flush with the oil pump cover edge.

Front Crankshaft Seal

REMOVAL AND INSTALLATION

2.4L (2RZ-FE) Engine

Timing Chain Cover Installed

1. Disconnect the negative battery cable.
2. Remove the engine drive belts and crankshaft pulley.
3. Using a suitable tool, pry out the oil seal. Be careful not to damage the crankshaft.
   To install:
4. Apply Multi-Purpose (MP) grease to the new oil seal lip.
5. Using tool SST 09223–50510, or equivalent, and a mallet, tap in the new oil seal until its surface is flush with the oil pump cover edge.
6. Install the crankshaft pulley and engine drive belts.

Timing Chain Cover Removed

1. Carefully tap the oil seal out the timing chain cover.
2. Apply Multi-Purpose (MP) grease to the new oil seal lip.
3. Using tool SST 09223–50510, or equivalent, and a mallet, tap in the new oil seal until its surface is flush with the oil pump cover edge.

2.7L (3RZ-FE), 3.0L (3VZ-E) and 3.4L (5VZ-FE) Engines

NOTE: There are 2 methods to replace the oil seal which are as follows:

Oil Pump Body Installed

1. Disconnect the negative battery cable.

--- **CAUTION** ---
Wait 90 seconds from the time the key is turned to LOCK and the negative battery cable is disconnected to begin work. This allows the SRS capacitor to discharge and prevent deployment of the air bag(s).

2. On the 3RZ-FE and 3VZ-E engines, remove the engine undercover (4-Wheel Drive vehicles).
3. Remove the timing belt and crankshaft pulley.
4. Using a knife, cut off the oil seal lip.

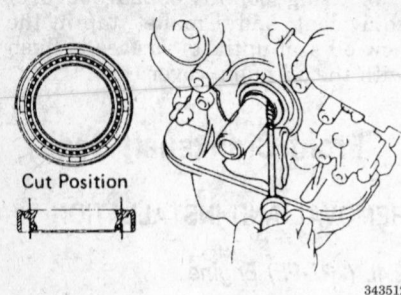

Cut Position

343512

Cutting the oil seal lip — 2.7L (3RZ-FE) and 3.0L (3VZ-E) engines

5. Using a suitable tool, pry out the oil seal. Be careful not to damage the crankshaft.

To install:

6. Apply Multi-Purpose (MP) grease to the new oil seal lip.

7. Using tool SST 09309–37010, or equivalent, and a mallet, tap in the new oil seal until its surface is flush with the oil pump case edge.

8. Install the crankshaft pulley and the timing belt.

9. If removed, install the engine undercover.

10. Connect the negative battery cable.

Oil Pump Body Removed

1. Carefully pry out the seal using a suitable tool.

2. Apply Multi-Purpose (MP) grease to the new oil seal lip.

3. Using tool SST 09309–37010, or equivalent, drive the new seal into place.

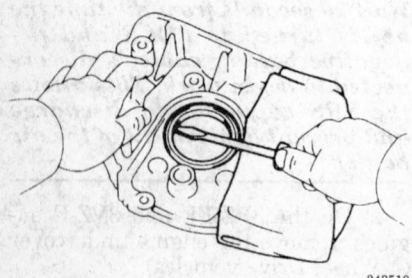

343510

Oil seal removal — 2.7L (3RZ-FE) and 3.0L (3VZ-E) engines

Timing Chain, Cover and Sprockets

REMOVAL AND INSTALLATION

2.4L (22R-E) Engine

1993–94 Models

1. Disconnect the negative battery cable.

2. Remove the cylinder head and timing chain cover.

3. Separate the chain from the damper and remove the chain, complete with the camshaft sprocket.

4. Remove the crankshaft sprocket and the oil pump drive with a puller.

5. Inspect the chain for wear or damage. Replace it if necessary.

6. Inspect the chain tensioner for wear. If it measures less than 11mm, replace it.

7. Check the dampers for wear. If their measurements are below the following specifications, replace them. The specification for the upper damper is 5.0mm and the lower damper is 4.5mm.

To install:

8. Rotate the crankshaft until its key is at TDC. Slide the sprocket in place over the key.

9. Place the chain over the sprocket so its single bright link aligns with the mark on the camshaft sprocket.

10. Install the cam sprocket so the timing mark falls between the two bright links on the chain.

11. Fit the oil pump drive spline over the crankshaft key.

12. Install the timing cover gasket on the front of the block.

13. Rotate the camshaft sprocket counterclockwise to remove the slack from the chain.

14. Install the timing chain cover and cylinder head.

15. Connect the negative battery cable.

1995 Models

1. Disconnect the negative battery cable.

CAUTION
Work must be started after 90 seconds from the time the ignition switch is turned to the LOCK position and the negative (-) battery cable is disconnected.
---

2. Raise and safely support the vehicle.

3. Drain the engine coolant and the engine oil.

4. Remove the cylinder head.

5. Remove the radiator.

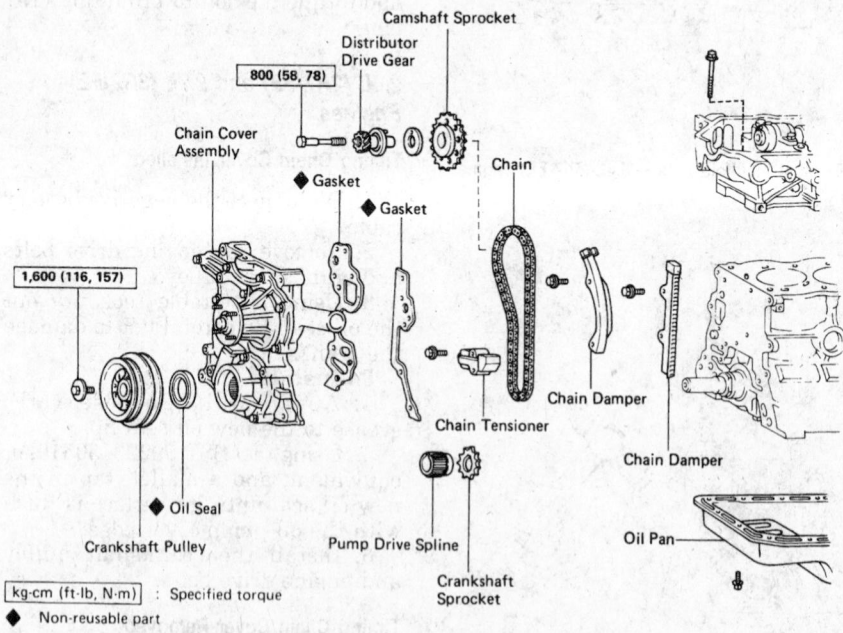

97669

Exploded view of the timing chain components — 2.4L (22R-E) engine

6. On 4-Wheel Drive vehicles, remove the front differential.

7. Remove the oil pan.

8. If equipped with power steering, remove the power steering belt.

9. If equipped with air conditioning, remove the air conditioning belt, compressor, and the bracket.

10. Remove the fluid coupling with the fan and the water pump pulley.

11. Remove the crankshaft pulley.

12. Remove the No. 1 water bypass pipe.

13. Remove the fan belt adjusting bar. With power steering, remove the lower power steering bracket.

14. Disconnect the heater water outlet pipe.

15. Remove the chain cover assembly. Remove the bolts shown by the arrows.

16. Remove the chain and the camshaft sprocket.

17. Remove the pump drive spline and the crankshaft sprocket.

18. Remove the gasket material on the cylinder block.

To install:

19. Install the crankshaft sprocket and the chain.

    a. Turn the crankshaft until the shaft key is on top.

    b. Slide the sprocket over the key on the crankshaft.

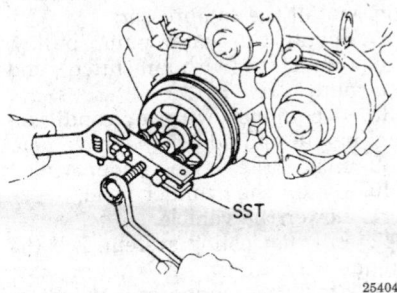

Crankshaft pulley removal — 2.4L (22R-E) engine

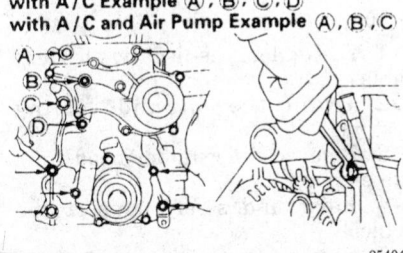

with A/C Example Ⓐ, Ⓑ, Ⓒ, Ⓓ
with A/C and Air Pump Example Ⓐ, Ⓑ, Ⓒ

Timing chain cover — 2.4L (22R-E) engine

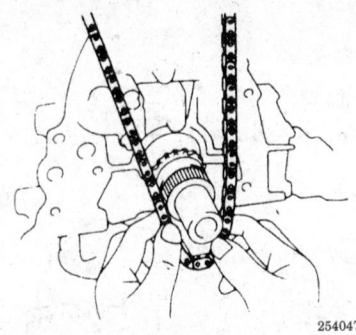

Crankshaft timing sprocket — 2.4L (22R-E) engine

    c. Place the timing chain on the sprocket with the single bright chain link aligned with the timing mark on the sprocket.

20. Place the chain on the camshaft sprocket.

    a. Place the timing chain on the sprocket so that the bright chain link is aligned with the timing mark on the sprocket.

    b. Make sure the chain is positioned between the dampers.

    c. Turn the camshaft sprocket counterclockwise to take the slack out of the chain.

21. Install the oil pipe drive spline by sliding it over the crankshaft key.

22. Install the timing chain cover assembly.

    a. Remove the old cover gaskets and install new gaskets.

    b. Slide the cover assembly over the dowels and the pump spline. Torque the 8 mm bolts to 9 ft. lbs. (13 Nm) and the 10 mm bolts to 29 ft. lbs. (13 Nm).

23. Install the fan belt adjusting bar to the chain cover and the cylinder head. Torque to 9 ft. lbs. (13 Nm).

24. Install the heater water outlet pipe.

25. Install the No. 1 water bypass pipe.

26. Install the crankshaft pulley and torque the bolt to 116 ft. lbs. (157 Nm).

27. Install the water pump pulley and the fluid coupling with the fan. Place the belt onto each pulley. While pulling the belt tight, tighten the four nuts.

28. Adjust the drive belt tension.

29. If removed, install the air conditioning compressor bracket, the compressor and the belt.

30. If removed, install the power steering belt.

31. Apply seal packing to the joint part of the cylinder block, the chain cover, the cylinder block, the rear oil seal retainer and the oil pan. Install the oil pan over the studs on the

block and torque the nuts and bolts to 9 ft. lbs. (13 Nm).

32. Install the radiator.

33. Install the cylinder head.

34. If removed, install the front differential.

35. Fill with engine coolant and engine oil.

36. Connect the negative battery cable.

37. Start the engine and check the ignition timing.

2.4L (2RZ-FE) Engine

1. Disconnect the negative battery cable.

2. Drain the oil and cooling system.

3. Remove the cylinder head from the engine.

4. Raise and safely support the vehicle.

5. Remove the engine undercover by removing the four bolts.

6. If equipped with air conditioning, loosen the idler pulley nut and the adjusting bolt. Remove the drive belt from the engine.

7. Remove the alternator drive belt, fan (with fan clutch), water pump pulley, and the fan shroud.

8. If equipped with air conditioning, disconnect the air conditioning compressor and bracket.

9. Remove the alternator, adjusting bar and bracket.

10. Remove the crankshaft position sensor by removing the two bolts.

11. Remove the oil pan by removing the 16 mounting bolts and 2 nuts.

NOTE: Be careful not to damage the flanges of the oil pan and cylinder block.

12. Remove the two bolts, two nuts, oil strainer, and the gasket.

13. Remove the crankshaft pulley as follows:

    a. If equipped with air conditioning, remove the No. 2 and No. 3 crankshaft pulleys by removing the four bolts.

    b. Using SST 09213–54015 and 09330–00021, or equivalents, remove the crankshaft pulley bolts.

    c. Remove the crankshaft pulley.

14. Remove the timing chain cover as follows:

    a. Remove the two water bypass pipe mounting nuts.

    b. Remove the two timing chain cover mounting bolts.

    c. Remove the nine mounting bolts and two mounting nuts from the timing chain cover.

    d. Using a rubber hammer, loosen the chain cover and remove

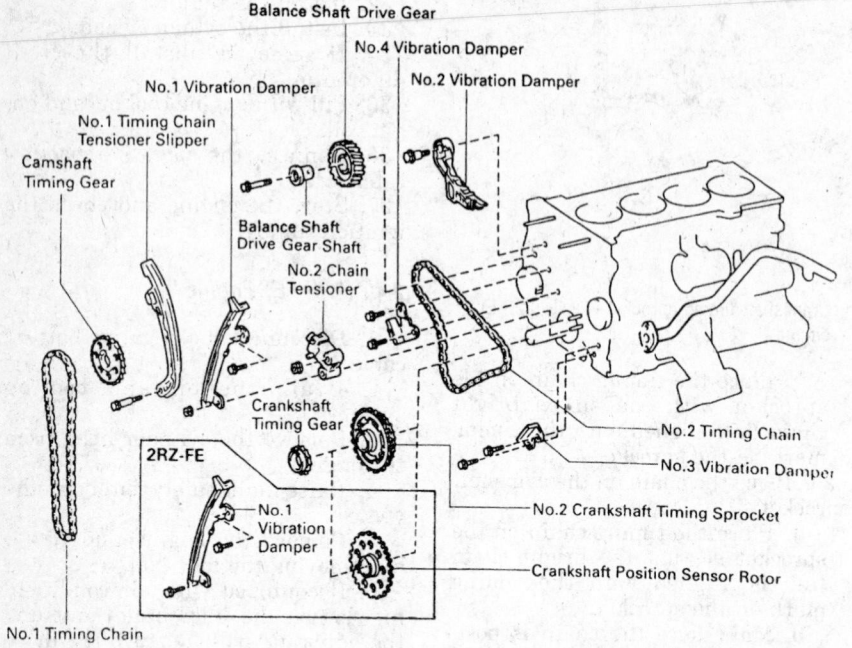

Timing chain component assembly — 2.4L (2RZ-FE) and 2.7L (3RZ-FE) engines

the timing chain cover and three gaskets.

15. Remove the No. 1 timing chain and camshaft timing gear.

16. Remove the crankshaft timing gear.

17. Remove the No. 1 timing chain tensioner slipper and the No. 1 vibration damper by removing the two bolts.

18. Remove the crankshaft position sensor rotor.

19. Remove the oil jet and gasket by removing the bolt.

To install:

20. Before installing, check the timing chain components for wear and other damages. Replace if necessary.

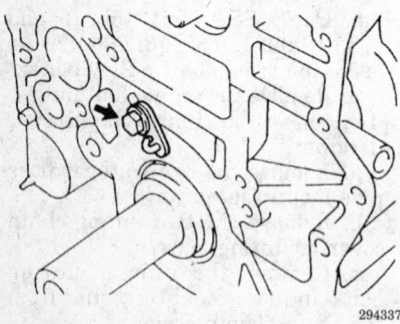

Timing chain oil jet — 2.4L (2RZ-FE) engine

21. Install the crankshaft position sensor rotor with the front mark of the rotor facing forward.

22. Install the oil jet with a new gasket. Torque the bolt to 13 ft. lbs. (18 Nm).

23. Install the No. 1 timing chain tensioner slipper and No. 1 vibration damper. Torque the damper bolts 22 ft. lbs. (29 Nm). Torque the slipper bolt to 20 ft. lbs. (27 Nm).

24. Check that the slipper moves smoothly.

25. Install the crankshaft timing gear.

26. Install the No. 1 timing chain and camshaft timing gear:

　a. Align the timing mark between the mark link of the No. 1 timing chain and install the No. 1 timing chain to the timing gear.

　b. Align the timing mark of the crankshaft timing gear with the mark link of the No. 1 timing chain and install the No. 1 timing chain.

　c. Tie (squeeze) the No. 1 timing chain with a cord and make sure it doesn't come loose.

27. Install the timing chain cover as follows:

　a. Install three new gaskets to the cylinder block and No. 1 bypass pipe.

　b. Install the timing chain cover with the nine bolts and two nuts as follows:
　　• Torque bolts labeled **A** and **D** to 14 ft. lbs. (20 Nm).
　　• Torque bolts labeled **B** to 18 ft. lbs. (25 Nm).
　　• Torque bolts labeled **C** to 32 ft. lbs. (44 Nm).

28. Remove the cord from the chain.

29. Install the 2 rear timing chain cover mounting bolts and water by-pass pipe mounting nuts. Torque to 13 ft. lbs. (18 Nm).

30. Align the pulley set key with the key groove of the pulley and slide on the pulley. Install and torque the pulley bolt to 193 ft. lbs. (260 Nm).

31. If equipped with air conditioning, install the No. 3 and No. 2 crankshaft pulleys with the four bolts. Torque the bolts to 18 ft. lbs. (25 Nm).

32. Install the oil strainer. Torque the fasteners to 13 ft. lbs. (18 Nm).

33. Clean the oil pan and cylinder block mating surfaces. Apply seal packing to the oil pan. Install the oil pan and torque the fasteners to 9 ft. lbs. (13 Nm).

34. Using a new O-ring, install the crankshaft position sensor with the two bolts. Torque the bolts to 74 inch lbs. (8.5 Nm).

35. Install the alternator to the engine.

36. If equipped with air conditioning, install the compressor.

37. Install the water pump pulley, fan shroud, fan (with fan clutch), and alternator drive belt.

38. If equipped with air conditioning, install and adjust the drive belt.

39. Install the engine undercover.

40. Install the cylinder head.

41. Lower the vehicle.

42. Fill the cooling system. Fill the engine with oil.

43. Start the engine and check for leaks.

44. Adjust ignition timing and road test the vehicle for proper operation.

45. Recheck all fluid levels.

2.7L (3RZ-FE) Engine

T-100

1. Disconnect the negative battery cable.

2. Drain the oil and cooling system.

3. Remove the cylinder head from the engine.

4. Raise and safely support the vehicle.

5. Remove the engine undercover by removing the four bolts.

6. If equipped with air conditioning, loosen the idler pulley nut and

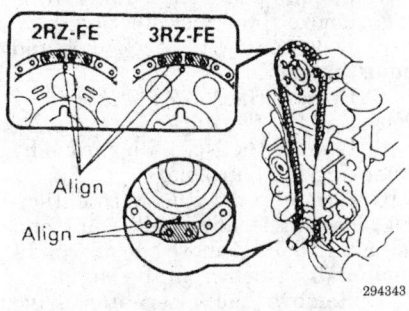

No. 1 timing chain timing marks — 2.4L (2RZ-FE) and 2.7L (3RZ-FE) engines

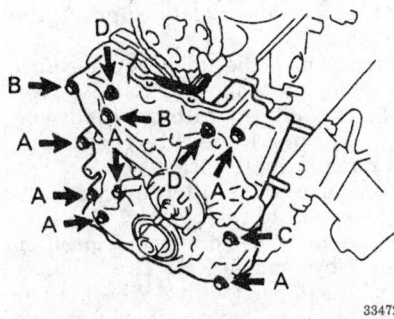

Timing chain cover bolts — 2.4L (2RZ-FE) and 2.7L (3RZ-FE) engines

adjusting bolt and remove the drive belt from the engine.

7. Remove the alternator drive belt, fan (with fan clutch), water pump pulley, and fan shroud.

8. If equipped with air conditioning, disconnect the air conditioning compressor and bracket.

9. Remove the alternator, adjusting bar, and the bracket.

10. Remove the crankshaft position sensor by removing the two bolts.

11. Remove the oil pan by removing the 16 mounting bolts and 2 nuts.

NOTE: Be careful not to damage the flanges of the oil pan and cylinder block.

12. Remove the two bolts, two nuts, oil strainer, and the gasket.

13. Remove the crankshaft pulley as follows:

    a. If equipped with air conditioning, remove the No. 2 and No. 3 crankshaft pulleys by removing the four bolts.

    b. Using SST 09213–54015 and 09330–00021, or equivalents, remove the crankshaft pulley bolts.

    c. Remove the crankshaft pulley.

14. Remove the timing chain cover as follows:

    a. Remove the two water bypass pipe mounting nuts.

    b. Remove the two timing chain cover mounting bolts.

    c. Remove the nine mounting bolts and two mounting nuts from the timing chain cover.

    d. Using a rubber hammer, loosen the chain cover and remove the timing chain cover and three gaskets.

15. Remove the No. 1 timing chain and camshaft timing gear.

16. Remove the crankshaft timing gear.

17. Remove the No. 1 timing chain tensioner slipper and No. 1 vibration damper by removing the nut and two bolts.

18. Install a pin to the No. 2 chain tensioner and lock the plunger.

19. Remove bolt and the No. 2 damper.

20. Remove the two bolts and the No. 3 damper.

21. Remove the nut and No. 2 chain tensioner.

22. Remove the No. 2 timing chain:

    a. Remove the bolt from the balance shaft drive gear.

    b. Remove the balance shaft drive gear with shaft.

    c. Remove the No. 2 timing chain with the No. 2 crankshaft timing sprocket.

23. Remove the No. 4 damper by removing the two bolts.

To install:

24. Before installing, check the timing chain components for wear and other damages. Replace if necessary.

25. Check that No. 1 cylinder is at TDC and that the weights of No. 1 and No. 2 balance shafts are at the bottom side. Install No. 4 vibration damper and install the two bolts.

26. Install No. 2 timing chain as follows:

    a. Match the mark links with the timing marks on No. 2 crankshaft timing sprocket and balance shaft timing sprocket.

    b. Install the other mark link of No. 2 timing chain onto the sprocket behind the large timing mark of the balance shaft drive gear.

    c. Insert the balance shaft drive gear shaft through the balance shaft drive gear so that it fits into the thrust plate hole. Then align the small timing mark of the balance shaft drive gear with the timing mark of the balance shaft timing gear.

    d. Install the bolt to the balance shaft drive gear and torque to 18 ft. lbs. (25 Nm).

    e. Check that the timing mark is matched with the corresponding mark link.

27. Install the No. 2, No. 3 vibration dampers and No. 2 chain tensioner as follows:

    a. Assemble the chain tensioner with the pin installed, then remove the pin after assembly. Avoid pushing the No. 2 vibration damper against the chain.

    b. Install the No. 2 chain tensioner with the nut and torque the nut to 13 ft. lbs. (18 Nm).

    c. Install the No. 3 damper and torque the bolts to 13 ft. lbs. (18 Nm).

    d. Install the No. 2 damper and torque the bolt to 20 ft. lbs. (27 Nm).

    e. Remove the pin from the No. 2 chain tensioner and free the plunger.

28. Install the No. 1 timing chain tensioner slipper and No. 1 vibration damper. Torque the damper bolt and nut to 13 ft. lbs. (18 Nm). Torque the slipper bolt to 20 ft. lbs. (27 Nm).

29. Check that the slipper moves smoothly.

30. Install the crankshaft timing gear.

31. Install the No. 1 timing chain and camshaft timing gear:

    a. Align the timing mark between the mark link of the No. 1 timing chain and install the No. 1 timing chain to the timing gear.

    b. Align the timing mark of the crankshaft timing gear with the mark link of the No. 1 timing chain and install the No. 1 timing chain.

    c. Tie (squeeze) the No. 1 timing chain with a cord and make sure it doesn't come loose.

32. Install the timing chain cover as follows:

    a. Install three new gaskets to the cylinder block and No. 1 bypass pipe.

    b. Install the timing chain cover with the nine bolts and two nuts as follows:

    • Torque bolts labeled **A** and **D** to 14 ft. lbs. (20Nm).

    • Torque bolts labeled **B** to 18 ft. lbs. (25 Nm).

    • Torque bolts labeled **C** to 32 ft. lbs. (44 Nm).

33. Remove the cord from the chain.

34. Install the 2 rear timing chain cover mounting bolts and water bypass pipe mounting nuts. Torque to 13 ft. lbs. (18 Nm).

35. Align the pulley set key with the key groove of the pulley and slide

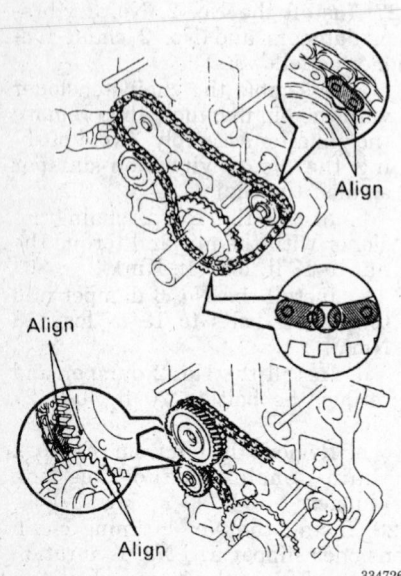

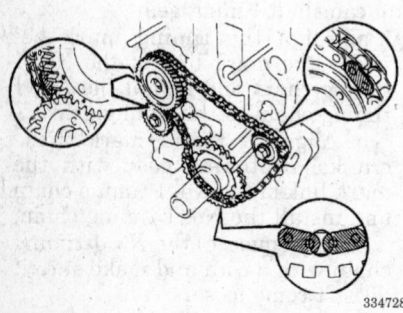

Installing No. 2 timing chain: Match the timing marks as shown — 2.7L (3RZ-FE) engine

Align the timing chain marks as shown — 2.7L (3RZ-FE) engine

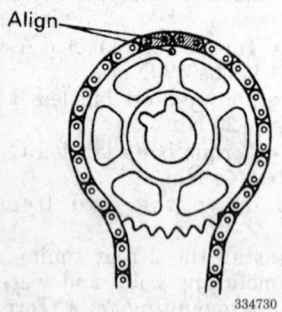

Align the No. 1 timing chain to the spocket, as shown — 2.7L (3RZ-FE) engine

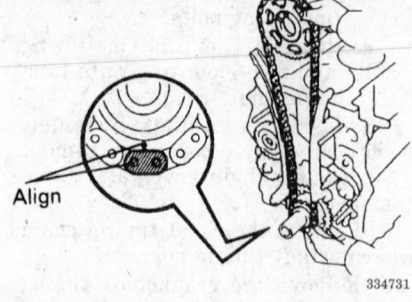

Align the timing mark of the crankshaft timing gear with the mark link of No. 1 timing chain — 2.7L (3RZ-FE) engine

on the pulley. Install and torque the pulley bolt to 193 ft. lbs. (260 Nm).

36. If equipped with air conditioning, install the No. 3 and No. 2 crankshaft pulleys with the four bolts. Torque the bolts to 18 ft. lbs. (25 Nm).

37. Install the oil strainer. Torque the fasteners to 13 ft. lbs. (18 Nm).

38. Clean the oil pan and cylinder block mating surfaces. Apply seal packing to the oil pan. Install the oil pan and torque the fasteners to 9 ft. lbs. (13 Nm).

39. Using a new O-ring, install the crankshaft position sensor with the two bolts. Torque the bolts to 74 inch lbs. (8.5 Nm).

40. Install the alternator to the engine.

41. If equipped with air conditioning, install the compressor.

42. Install the water pump pulley, fan shroud, fan (with fan clutch), and alternator drive belt.

43. If equipped with air conditioning, install and adjust the drive belt.

44. Install the engine undercover.

45. Install the cylinder head.

46. Lower the vehicle.

47. Fill the cooling system. Fill the engine with oil.

48. Start the engine and check for leaks.

49. Adjust ignition timing. Road test the vehicle for proper operation.

50. Recheck all fluid levels.

Tacoma

1. Disconnect the negative battery cable.

2. Drain the oil and the cooling system.

3. If equipped with 4-Wheel Drive, remove the front differential and the halfshafts assembly.

4. Remove the two bolts and disconnect the air pipe.

5. Disconnect the upper radiator hose from the radiator.

6. Remove the oil dipstick guide by removing the bolt.

7. If equipped with power steering, remove the power steering drive belt by loosening the lock bolt and adjusting bolt.

8. Remove the No. 2 fan shroud by removing the two clips.

9. Remove the No. 1 fan shroud by removing the four bolts.

10. If equipped with air conditioning, loosen the idler pulley nut and adjusting bolt. Remove the air conditioning drive belt from the engine.

11. Remove the alternator drive belt, fan (with fan clutch), water pump pulley, and fan shroud.

12. Remove the cylinder head from the engine.

13. Raise and safely support the vehicle.

14. Remove the engine undercover by removing the four bolts.

15. If equipped with air conditioning, disconnect the air conditioning compressor and bracket.

16. Remove the alternator, adjusting bar and bracket.

17. Remove the crankshaft position sensor by removing the two bolts.

18. If equipped with 2-Wheel Drive, remove the stiffener plates by removing the eight bolts.

19. Remove the oil pan by removing the 16 mounting bolts and 2 nuts.

NOTE: Be careful not to damage the flanges of the oil pan and cylinder block.

20. Remove the two bolts, two nuts, oil strainer, and the gasket.

21. Remove the crankshaft pulley as follows:

a. If equipped with air conditioning, remove the No. 2 and No. 3 crankshaft pulleys by removing the four bolts.

b. Using SST 09213–54015 and 09330–00021, or equivalents, remove the crankshaft pulley bolts.

c. Remove the crankshaft pulley.

22. Remove the timing chain cover as follows:

a. Remove the two water bypass pipe mounting nuts.

b. Remove the two timing chain cover mounting bolts.

c. Remove the nine mounting bolts and two mounting nuts from the timing chain cover.

d. Using a rubber hammer, loosen the chain cover and remove the timing chain cover and three gaskets.

23. Remove the No. 1 timing chain and camshaft timing gear.

24. Remove the crankshaft timing gear.

25. Remove the No. 1 timing chain tensioner slipper and No. 1 vibration

damper by removing the nut and two bolts.

26. Install a pin to the No. 2 chain tensioner and lock the plunger.

27. Remove bolt and the No. 2 damper.

28. Remove the two bolts and the No. 3 damper.

29. Remove the nut and No. 2 chain tensioner.

30. Remove the No. 2 timing chain:

a. Remove the bolt from the balance shaft drive gear.

b. Remove the balance shaft drive gear with shaft.

c. Remove the No. 2 timing chain with the No. 2 crankshaft timing sprocket.

31. Remove the No. 4 damper by removing the two bolts.

To install:

32. Before installing, check the timing chain components for wear and other damages. Replace if necessary.

33. Check that No. 1 cylinder is at TDC and that the weights of No. 1 and No. 2 balance shafts are at the bottom side. Install No. 4 vibration damper and install the two bolts.

34. Install No. 2 timing chain as follows:

a. Match the mark links with the timing marks on No. 2 crank-

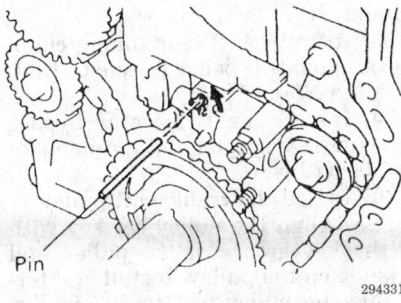

Pin

294331

No. 2 timing chain (installing a pin to No. 2 chain tensioner) — 2.7L (3RZ-FE) engine

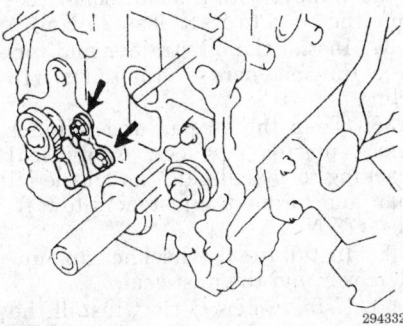

294332

No. 4 vibration damper bolts — 2.7L (3RZ-FE) engine

shaft timing sprocket and balance shaft timing sprocket.

b. Install the other mark link of No. 2 timing chain onto the sprocket behind the large timing mark of the balance shaft drive gear.

c. Insert the balance shaft drive gear shaft through the balance shaft drive gear so that it fits into the thrust plate hole. Then align the small timing mark of the balance shaft drive gear with the timing mark of the balance shaft timing gear.

d. Install the bolt to the balance shaft drive gear and torque to 18 ft. lbs. (25 Nm).

e. Check that the timing mark is matched with the corresponding mark link.

35. Install the No. 2, No. 3 vibration dampers and No. 2 chain tensioner as follows:

a. Assemble the chain tensioner with the pin installed, then remove the pin after assembly. Avoid pushing the No. 2 vibration damper against the chain.

b. Install the No. 2 chain tensioner with the nut and torque the nut to 13 ft. lbs. (18 Nm).

c. Install the No. 3 damper and torque the bolts to 13 ft. lbs. (18 Nm).

d. Install the No. 2 damper and torque the bolt to 20 ft. lbs. (27 Nm).

e. Remove the pin from the No. 2 chain tensioner and free the plunger.

36. Install the No. 1 timing chain tensioner slipper and No. 1 vibration damper. Torque the damper bolt and nut to 13 ft. lbs. (18 Nm). Torque the slipper bolt to 20 ft. lbs. (27 Nm).

37. Check that the slipper moves smoothly.

38. Install the crankshaft timing gear.

39. Install the No. 1 timing chain and camshaft timing gear:

a. Align the timing mark between the mark link of the No. 1 timing chain and install the No. 1 timing chain to the timing gear.

b. Align the timing mark of the crankshaft timing gear with the mark link of the No. 1 timing chain and install the No. 1 timing chain.

c. Tie (squeeze) the No. 1 timing chain with a cord and make sure it doesn't come loose.

40. Install the timing chain cover as follows:

a. Install three new gaskets to the cylinder block and No. 1 bypass pipe.

b. Install the timing chain cover with the nine bolts and two nuts as follows:

• Torque bolts labeled **A** and **D** to 14 ft. lbs. (20 Nm).

• Torque bolts labeled **B** to 18 ft. lbs. (25 Nm).

• Torque bolts labeled **C** to 32 ft. lbs. (44 Nm).

41. Remove the cord from the chain.

42. Install the 2 rear timing chain cover mounting bolts and water bypass pipe mounting nuts. Torque to 13 ft. lbs. (18 Nm).

43. Align the pulley set key with the key groove of the pulley and slide on the pulley. Install and torque the pulley bolt to 193 ft. lbs. (260 Nm).

44. If equipped with air conditioning, install the No. 3 and No. 2 crankshaft pulleys with the four bolts. Torque the bolts to 18 ft. lbs. (25 Nm).

45. Install the oil strainer. Torque the fasteners to 13 ft. lbs. (18 Nm).

46. Clean the oil pan and cylinder block mating surfaces. Apply seal packing to the oil pan. Install the oil pan and torque the fasteners to 9 ft. lbs. (13 Nm).

47. Using a new O-ring, install the crankshaft position sensor with the two bolts. Torque the bolts to 74 inch lbs. (8.5 Nm).

48. Install the alternator to the engine.

49. If equipped with air conditioning, install the compressor.

50. Install the water pump pulley, fan shroud, fan (with fan clutch), and alternator drive belt.

51. If equipped with air conditioning, install and adjust the drive belt.

52. Install the No. 1 fan shroud by installing the four bolts.

53. Install the No. 2 fan shroud with the two clips.

54. Install and adjust the power steering drive belt.

55. Install the oil dipstick guide with the bolt.

56. Connect the upper radiator hose to the radiator.

57. Connect the air pipe with the two bolts.

58. If removed, install the front differential and the halfshafts.

59. Install the engine undercover.

60. Install the cylinder head.

61. Lower the vehicle.

62. Fill the cooling system. Fill the engine with oil.

63. Start the engine and check for leaks.

64. Adjust ignition timing and road test the vehicle for proper operation.

65. Recheck all fluid levels.

1996–97 4Runner

1. Disconnect the negative battery cable.
2. Drain the oil and cooling system.
3. With 4-Wheel Drive: Remove the front differential and halfshafts assembly.
4. Remove the drive belt for the generator, fan with the fluid coupling, and the water pump pulley.
5. Remove the cylinder head assembly.
6. If equipped with air conditioning, disconnect the air conditioning compressor and bracket.
7. Remove the alternator, adjusting bar, and the bracket.
8. Remove the crankshaft position sensor by removing the two bolts.
9. Remove the flywheel housing undercover and the dust seal.
10. Remove the oil pan by removing the 16 mounting bolts and 2 nuts.

NOTE: Be careful not to damage the flanges of the oil pan and cylinder block.

11. Remove the two bolts, two nuts, oil strainer, and the gasket.
12. Remove the crankshaft pulley as follows:
    a. If equipped with air conditioning, remove the No. 2 and No. 3 crankshaft pulleys by removing the four bolts.
    b. Using SST 09213–54015 and 09330–00021, or equivalents, remove the crankshaft pulley bolts.
    c. Remove the crankshaft pulley.
13. Remove the timing chain cover as follows:
    a. Remove the two water bypass pipe mounting nuts.
    b. Remove the two timing chain cover mounting bolts.
    c. Remove the nine mounting bolts and two mounting nuts from the timing chain cover.
    d. Using a rubber hammer, loosen the chain cover and remove the timing chain cover and three gaskets.
14. Remove the No. 1 timing chain and camshaft timing gear.
15. Remove the crankshaft timing gear.
16. Remove the No. 1 timing chain tensioner slipper and No. 1 vibration damper by removing the nut and two bolts.
17. Install a pin to the No. 2 chain tensioner and lock the plunger.
18. Remove bolt and the No. 2 damper.
19. Remove the two bolts and the No. 3 damper.

20. Remove the nut and No. 2 chain tensioner.
21. Remove the No. 2 timing chain:
    a. Remove the bolt from the balance shaft drive gear.
    b. Remove the balance shaft drive gear with shaft.
    c. Remove the No. 2 timing chain with the No. 2 crankshaft timing sprocket.
22. Remove the No. 4 damper by removing the two bolts.

To install:

23. Before installing, check the timing chain components for wear and other damages. Replace if necessary.
24. Check that No. 1 cylinder is at TDC and that the weights of No. 1 and No. 2 balance shafts are at the bottom side. Install No. 4 vibration damper and install the two bolts.
25. Install No. 2 timing chain as follows:
    a. Match the mark links with the timing marks on No. 2 crankshaft timing sprocket and balance shaft timing sprocket.
    b. Install the other mark link of No. 2 timing chain onto the sprocket behind the large timing mark of the balance shaft drive gear.
    c. Insert the balance shaft drive gear shaft through the balance shaft drive gear so that it fits into the thrust plate hole. Then align the small timing mark of the balance shaft drive gear with the timing mark of the balance shaft timing gear.
    d. Install the bolt to the balance shaft drive gear and torque to 18 ft. lbs. (25 Nm).
    e. Check that the timing mark is matched with the corresponding mark link.
26. Install the No. 2, No. 3 vibration dampers and No. 2 chain tensioner as follows:
    a. Assemble the chain tensioner with the pin installed, then remove the pin after assembly. Avoid pushing the No. 2 vibration damper against the chain.
    b. Install the No. 2 chain tensioner with the nut and torque the nut to 13 ft. lbs. (18 Nm).
    c. Install the No. 3 damper and torque the bolts to 13 ft. lbs. (18 Nm).
    d. Install the No. 2 damper and torque the bolt to 20 ft. lbs. (27 Nm).
    e. Remove the pin from the No. 2 chain tensioner and free the plunger.
27. Install the No. 1 timing chain tensioner slipper and No. 1 vibration

damper. Torque the damper bolt and nut to 13 ft. lbs. (18 Nm). Torque the slipper bolt to 20 ft. lbs. (27 Nm).
28. Check that the slipper moves smoothly.
29. Install the crankshaft timing gear.
30. Install the No. 1 timing chain and camshaft timing gear:
    a. Align the timing mark between the mark link of the No. 1 timing chain and install the No. 1 timing chain to the timing gear.
    b. Align the timing mark of the crankshaft timing gear with the mark link of the No. 1 timing chain and install the No. 1 timing chain.
    c. Tie (squeeze) the No. 1 timing chain with a cord and make sure it doesn't come loose.
31. Install the timing chain cover as follows:
    a. Install three new gaskets to the cylinder block and No. 1 bypass pipe.
    b. Install the timing chain cover and torque the nine bolts and two nuts as follows:
    • 12 mm head **A**: 14 ft. lbs. (20 Nm).
    • 12 mm head **B**: 18 ft. lbs. (24.5 Nm).
    • 14 mm head: 32 ft. lbs. (44 Nm).
    • Nut: 14 ft. lbs. (20 Nm).
32. Remove the cord from the chain.
33. Install the 2 rear timing chain cover mounting bolts. Torque to 13 ft. lbs. (18 Nm).
34. Install the two water bypass pipe mounting nuts and torque to 14 ft. lbs. (20 Nm).
35. Install the crankshaft pulley.
    a. Align the pulley set key with the key groove of the pulley and slide on the pulley. Install and torque the pulley bolt to 193 ft. lbs. (260 Nm).
36. If equipped with air conditioning, install the No. 3 and No. 2 crankshaft pulleys with the four bolts. Torque the bolts to 18 ft. lbs. (25 Nm).
37. Install the oil strainer and torque the fasteners to 13 ft. lbs. (18 Nm).
38. Clean the oil pan and cylinder block mating surfaces. Apply seal packing to the oil pan. Install the oil pan and torque the fasteners to 9 ft. lbs. (13 Nm).
39. Install the flywheel housing undercover and the dust seal.
40. Using a new O-ring, install the crankshaft position sensor with the two bolts. Torque the bolts to 74 inch lbs. (9 Nm).

41. Install the alternator, the adjusting bar and the bracket.

42. If equipped with air conditioning, install the compressor.

43. Install the cylinder head assembly.

44. Install the water pump pulley, fan with the fluid coupling, and the drive belt for the alternator.

45. If removed, install the front differential and the halfshafts.

46. Install the engine undercover.

47. Refill the cooling system and the engine with oil.

48. Start the engine and check for leaks.

49. Check and adjust the ignition timing.

50. Road test the vehicle for proper operation.

51. Recheck all fluid levels.

Timing Belt, Cover, Tensioner and Sprockets

REMOVAL AND INSTALLATION

3.0L (3VZ-E) Engine

1993–94 Models

1. Disconnect the negative battery cable.

2. Remove the timing belt covers.

3. Draw a directional arrow on the timing belt and matchmark the belt to each of the pulleys. Remove the timing belt guide and then remove the tension spring.

4. Loosen the idler pulley bolt and shift it left as far as it will go. Tighten the set bolt and relieve the tension on the timing belt. Remove the belt.

5. Remove the crankshaft and camshaft sprocket timing pulleys. Remove the No. 1 idler pulley.

To install:

6. Align the groove in the crankshaft pulley with the key on the

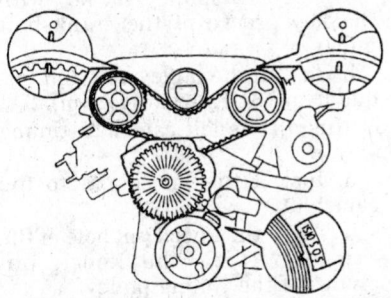

84771

Set No. 1 cylinder at TDC/compression. Check that the matchmarks are aligned as shown — 1993–94 with 3.0L (3VZ-E) engine

crankshaft and press the pulley onto the shaft.

7. Install the idler pulley. Align the groove on the pulley with the cavity of the oil pump and then force it to the left as far as it will go. Temporarily tighten it to 27 ft. lbs. (37 Nm).

8. Position the camshaft pulleys on the camshafts so the match holes in each pulley are in alignment with those on the No. 3 (upper rear) timing cover. Align the pulley matchmark with the one on the cover.

NOTE: Do not install the match pin. Check that the bolt head is not touching the pulley.

9. Install the timing belt around the timing pulleys. If reusing the old belt, make sure the arrow and matchmarks all line up with those made earlier on the pulleys.

10. Move the idler pulley to the right as far as it will go. Install the tension spring and then loosen the pulley bolt until the pulley moves lightly with the tension spring force.

11. Check the valve timing and belt tension by turning the crankshaft 2 complete revolutions clockwise. Check that each pulley aligns with its timing marks. Retighten the idler pulley bolt to 27 ft. lbs. (37 Nm).

12. Remove the camshaft timing pulley bolts and align the match pin hole with the match pin hole in the camshaft. Install the pin and bolt and tighten to 80 ft. lbs. (108 Nm).

13. Remove the crankshaft timing pulley bolt and position the belt guide over the crankshaft pulley so the cupped side is out.

14. Install the timing covers.

15. Connect the negative battery cable.

16. Fill and bleed the cooling system. Start the engine and check for leaks. Adjust engine specifications.

17. Road test the vehicle. Recheck all fluid levels.

1995 Models

1. Disconnect the negative battery cable.

——— **CAUTION** ———
Work must be started after 90 seconds from the time the ignition switch is turned to the LOCK position and the negative (-) battery cable is disconnected.

2. Raise and safely support the vehicle.

3. Remove the engine undercover.

4. Drain the engine coolant.

5. Disconnect the hoses and remove the radiator.

6. Disconnect the No. 2 and No. 3 air hoses from the air pipe.

7. Disconnect the plug wires from the spark plugs.

8. Remove the spark plugs.

9. Remove the power steering drive belt and the pump pulley.

10. Disconnect the power steering pump from the engine.

11. Remove the air conditioning drive belt.

12. Remove the cooling fan.

13. Remove the alternator drive belt.

14. Remove the water outlet.

15. Disconnect the spark plug wire clamps from the mounting bolts and remove the No. 2 timing belt cover.

16. If re-using the timing belt, check that there are installation marks on the timing belt.

17. Set the No. 1 cylinder at TDC of the compression stroke.

   a. Turn the crankshaft pulley and align its groove with the timing mark **0** of the No. 1 timing belt cover.

   b. Check that the timing marks of the camshaft timing pulleys and the No. 3 timing belt cover are aligned. If not, turn the crankshaft pulley one revolution (360°).

18. Remove the timing belt tensioner.

19. Remove the fan bracket.

20. Disconnect the timing belt from the camshaft timing pulleys.

NOTE: If re-using the timing belt, make sure that you can still read the installation marks. If not, place new installation marks on the timing belt to match the timing marks of the camshaft timing pulleys.

21. Remove the camshaft timing pulleys.

   a. Using SST 09960–10010, or equivalent, remove the pulley bolt, the timing pulley, and the knock pin. Remove the two timing pulleys.

22. Remove the crankshaft pulley.

   a. Using SST 09213–58012 and 09330–00021, or equivalent, loosen the pulley bolt.

   b. Remove the SST and the pulley bolt. Remove the crankshaft pulley.

23. Remove the No. 1 timing belt cover.

24. Remove the timing belt guide and remove the timing belt.

25. Remove the pivot bolt, the No. 1 idler pulley, and the plate washer.

26. Remove the crankshaft timing pulley.

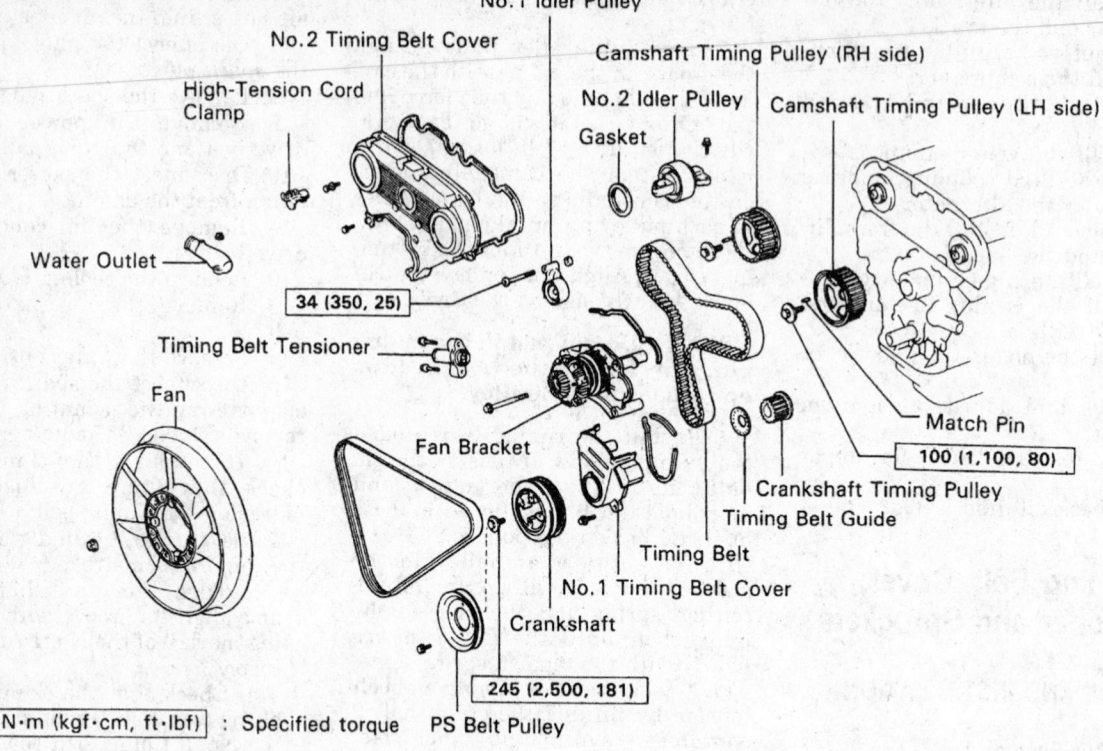

Timing belt and components, exploded view — 1993-95 3.0L (3VZ-E) engine

No.1 Idler Pulley
No.2 Timing Belt Cover
Camshaft Timing Pulley (RH side)
High-Tension Cord Clamp
No.2 Idler Pulley
Camshaft Timing Pulley (LH side)
Gasket
Water Outlet
34 (350, 25)
Timing Belt Tensioner
Match Pin
100 (1,100, 80)
Fan
Fan Bracket
Crankshaft Timing Pulley
Timing Belt Guide
Timing Belt
No.1 Timing Belt Cover
Crankshaft
245 (2,500, 181)
N·m (kgf·cm, ft·lbf) : Specified torque
PS Belt Pulley

90012

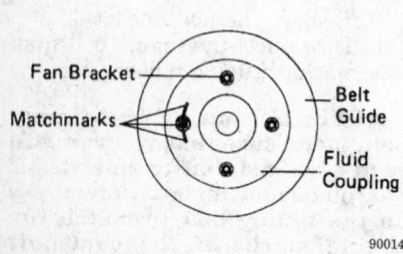

Fan Bracket
Matchmarks
Belt Guide
Fluid Coupling

90014

Place matchmarks on the fluid coupling and fan bracket as shown — 1993-94 3.0L (3VZ-E) engine

253935

Crankshaft pulley at timing mark 0 — 1995 3.0L (3VZ-E) engine

To install:

27. Install the crankshaft timing pulley.

a. Align the timing pulley set key with the key groove of the pulley.

b. Using SST 09214-60010, or equivalent, and a hammer, tap in the timing pulley with the flange side facing inward.

28. Install the plate washer and the No. 1 idler pulley with the pivot bolt and torque to 25 ft. lbs. (34 Nm). Check that the pulley bracket moves smoothly.

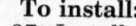

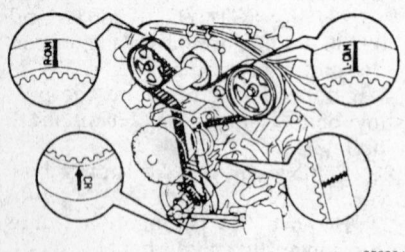

253934

Timing belt installation marks if re-using the belt

29. Temporarily install the timing belt.

a. Using the crankshaft pulley bolt, turn the crankshaft and align the timing marks of the crankshaft timing pulley and the oil pump body.

b. Align the installation mark on the timing belt with the dot mark of the crankshaft timing pulley.

c. Install the timing belt on the crankshaft timing pulley and the water pump pulley.

30. Install the timing belt guide with the cup side facing outward.

31. Install the No. 1 timing belt cover.

32. Install the crankshaft pulley.

a. Align the pulley set key with the key groove of the crankshaft pulley.

b. Install the pulley bolt and torque it to 181 ft. lbs. (245 Nm).

33. Install the left camshaft timing pulley.

a. Install the knock pin to the camshaft.

b. Align the knock pin hole of the camshaft with the knock pin groove of the timing pulley.

c. Slide the timing pulley on the camshaft with the flange side facing outward. Torque the pulley bolt to 80 ft. lbs. (108 Nm).

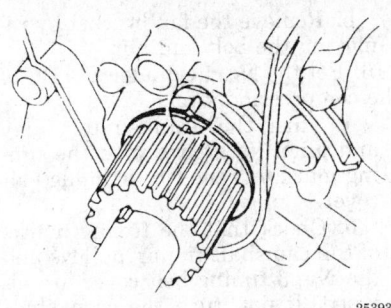

Alignment marks for the crankshaft timing pulley and the oil pump — 1995 3.0L (3VZ-E) engine

253936

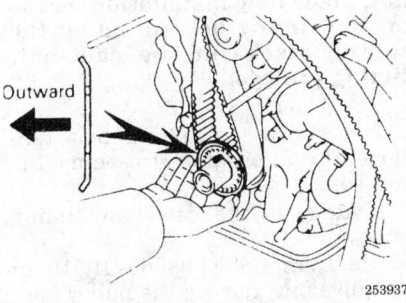

Outward

Timing belt guide — 1995 3.0L (3VZ-E) engine

253937

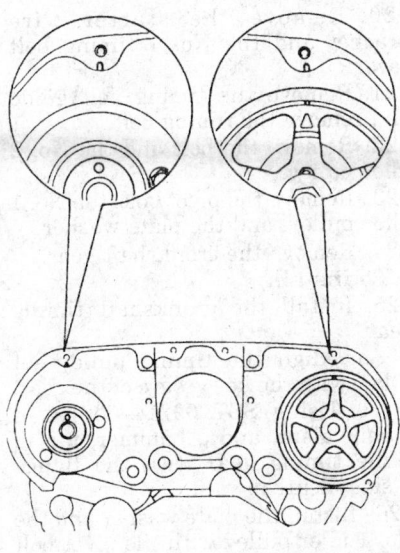

Right and left camshaft pulley alignment marks — 1995 3.0L (3VZ-E) engine

253938

34. Set the No. 1 cylinder to TDC of the compression stroke.

a. Turn the crankshaft pulley, and align its groove with the timing mark **0** of the No. 1 timing belt cover.

b. Turn the camshaft, align the knock pin hole of the camshaft with the timing mark of the No. 3 timing belt cover.

c. Turn the camshaft timing pulley, align the timing marks of the camshaft timing pulley and the No. 3 timing belt cover.

35. Connect the timing belt to the left camshaft timing pulley. Check that the installation mark on the timing belt is aligned with the end of the No. 1 timing belt cover.

a. Using SST 09960–01000, or equivalent, slightly turn the left camshaft timing pulley clockwise. Align the installation mark on the timing belt with the timing mark of the camshaft timing pulley, and hang the timing belt on the left camshaft timing pulley.

b. Align the timing marks of the left camshaft pulley and the No. 3 timing belt cover.

c. Check that the timing belt has tension between the crankshaft timing pulley and the left camshaft timing pulley.

36. Install the right camshaft timing pulley and the timing belt.

a. Align the installation mark on the timing belt with the timing mark of the right camshaft timing pulley, and hang the timing belt on the right camshaft timing pulley with the flange side facing inward.

b. Slide the right camshaft timing pulley on the camshaft. Align the timing marks on the right camshaft timing pulley and the No. 3 timing belt cover.

c. Align the knock pin hole of the camshaft with the knock pin groove of the pulley and install the knock pin. Install the bolt and torque to 80 ft. lbs. (108 Nm).

Alignment of the right camshaft timing pulley and the No. 3 cover — 1995 3.0L (3VZ-E) engine

253939

37. Install the fan bracket and torque the bolts to 30 ft. lbs. (41 Nm).

38. Set the timing belt tensioner.

a. Using a press, slowly press in the pushrod using 200–2,205 lbs. (981–9,807 N) of force.

b. Align the holes of the pushrod and housing, pass a 1.5 mm hexagon wrench through the holes to keep the setting position of the pushrod.

c. Release the press and install the dust boot to the tensioner.

39. Install the timing belt tensioner and alternately tighten the bolts to 20 ft. lbs. (28 Nm). Using pliers, remove the 1.5 mm hexagon wrench from the belt tensioner.

40. Check the valve timing.

a. Slowly turn the crankshaft pulley two revolutions from the TDC to TDC. Always turn the crankshaft pulley clockwise.

b. Check that each pulley aligns with the timing marks. If the timing marks do not align, remove the timing belt and reinstall it.

41. Install the No. 2 timing belt cover. Connect the four clamps on the spark plug wires to the mounting bolts of the No. 2 timing belt cover.

42. Remove the old packing material from the water outlet and apply new seal packing before installation. Torque the nuts to 11 ft. lbs. (15 Nm).

43. Install the alternator drive belt.

44. Install the cooling fan and torque the nuts to 4 ft. lbs. (5.4 Nm).

45. Install the air conditioning drive belt.

46. Install the power steering pump, pump pulley, and the drive belt.

47. Install the spark plugs and torque to 13 ft. lbs. (18 Nm).

48. Connect the spark plug wires to the spark plugs.

49. Connect the No. 2 and the No. 3 air hoses to the air pipe.

50. Install the radiator and connect the hoses.

51. Fill with engine coolant.

52. Connect the negative battery cable.

53. Start the engine and check for leaks.

54. Check the ignition timing and install the engine undercover.

3.4L (5VZ-FE) Engine

1995–97 T-100 and Tacoma

1. Disconnect the negative battery cable.

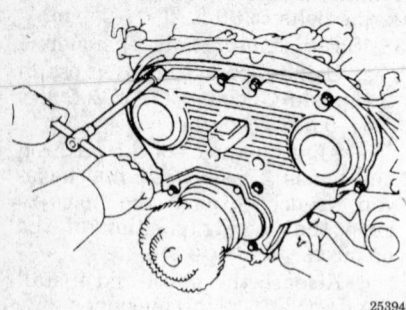

Timing belt cover — 1995 3.0L (3VZ-E) engine

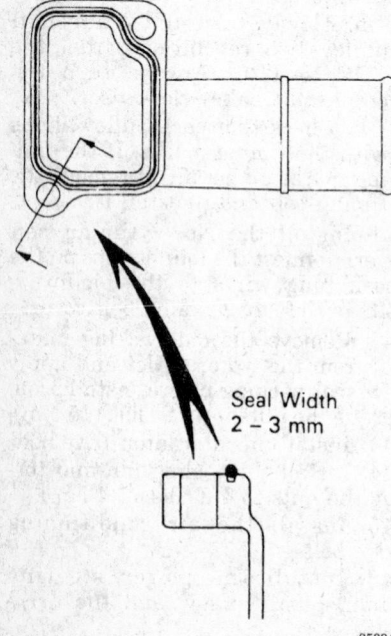

Seal Width
2 – 3 mm

Applying the gasket to the water outlet — 1995
3.0L (3VZ-E) engine

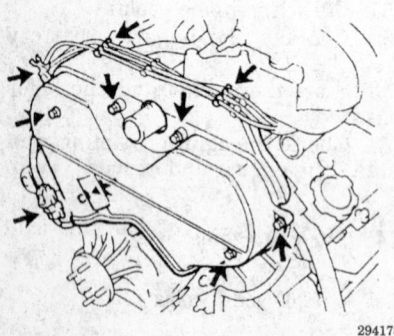

No. 2 timing belt cover — 3.4L (5VZ-FE) engine

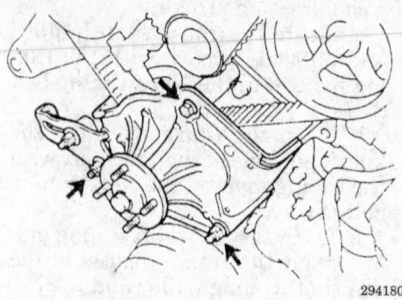

Fan bracket — 3.4L (5VZ-FE) engine

—— CAUTION ——
Work must be started after 90 seconds from the time the ignition switch is turned to the LOCK position and the negative (-) battery cable is disconnected.

2. Raise and safely support the vehicle.

3. Remove the engine undercover.

4. Drain the engine coolant.

5. Disconnect the upper radiator hose from the engine.

6. Remove the power steering drive belt.

7. Remove the air conditioning drive belt by loosening the idler pulley nut and the adjusting bolt.

8. Loosen the lock bolt, pivot bolt, and the adjusting bolt and the alternator drive belt.

9. Remove the No. 2 fan shroud by removing the two clips.

10. Remove the fan with the fluid coupling and fan pulleys.

11. Disconnect the power steering pump from the engine and set aside. Do not disconnect the lines from the pump.

12. If equipped with air conditioning, disconnect the compressor from the engine and set aside. Do not disconnect the lines from the compressor.

13. If equipped with air conditioning, disconnect the air conditioning bracket.

14. Remove the No. 2 timing belt cover as follows:

   a. Disconnect the camshaft position sensor connector from the No. 2 timing belt cover.

   b. Disconnect the three spark plug wire clamps from the No. 2 timing belt cover.

   c. Remove the six bolts and remove the timing belt cover.

15. Remove the fan bracket as follows:

   a. Remove the power steering adjusting strut by removing the nut.

   b. Remove the fan bracket by removing the bolt and nut.

16. Set the No. 1 cylinder at TDC of the compression stroke.

   a. Turn the crankshaft pulley and align its groove with the timing mark **0** of the No. 1 timing belt cover.

   b. Check that the timing marks of the camshaft timing pulleys and the No. 3 timing belt cover are aligned. If not, turn the crankshaft pulley one revolution (360°).

NOTE: If re-using the timing belt, make sure that you can still read the installation marks. If not, place new installation marks on the timing belt to match the timing marks of the camshaft timing pulleys.

17. Remove the timing belt tensioner by alternately loosening the two bolts.

18. Remove the camshaft timing pulleys.

   a. Using SST 09960–10010 or equivalent, remove the pulley bolt, the timing pulley and the knock pin. Remove the two timing pulleys with the timing belt.

19. Remove the crankshaft pulley as follows:

   a. Using SST 09213–54015 and 09330–00021, or equivalent, loosen the pulley bolt.

   b. Remove the SST tool, the pulley bolt, and the pulley.

20. Remove the starter wire bracket and the No. 1 timing belt cover.

21. Remove the timing belt guide and remove the timing belt.

22. Remove the bolt and the No. 2 idler pulley.

23. Remove the pivot bolt, the No. 1 idler pulley, and the plate washer.

24. Remove the crankshaft gear.

To install:

25. Install the crankshaft timing gear.

   a. Align the timing pulley set key with the key groove of the gear.

   b. Using SST 09214–60010, or equivalent, and a hammer, tap in the timing gear with the flange side facing inward.

26. Install the plate washer and the No. 1 idler pulley with the pivot bolt and torque to 26 ft. lbs. (35 Nm). Check that the pulley bracket moves smoothly.

27. Install the No. 2 timing belt idler with the bolt. Torque the bolt to 30 ft. lbs. (40 Nm). Check that the pulley bracket moves smoothly.

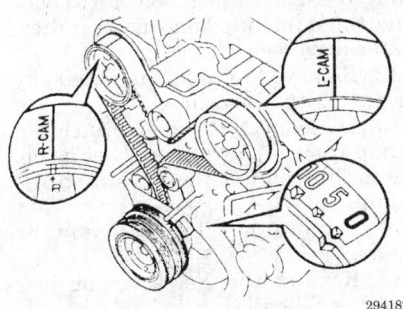

Aligning the crankshaft pulley groove with the timing mark — 3.4L (5VZ-FE) engine

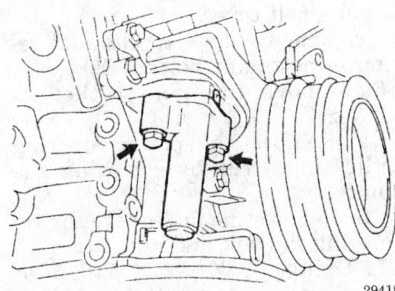

Timing belt tensioner — 3.4L (5VZ-FE) engine

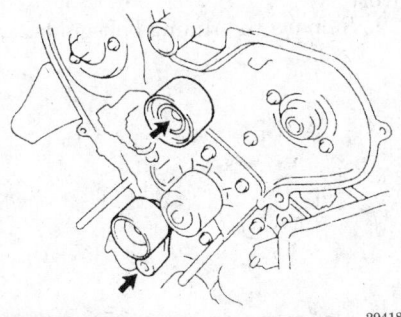

No. 1 and 2 timing belt idlers — 3.4L (5VZ-FE) engine

28. Temporarily install the timing belt.

   a. Using the crankshaft pulley bolt, turn the crankshaft and align the timing marks of the crankshaft timing pulley and the oil pump body.

   b. Align the installation mark on the timing belt with the dot mark of the crankshaft timing pulley.

   c. Install the timing belt on the crankshaft timing pulley, No. 1 idler, and the water pump pulleys.

29. Install the timing belt guide with the cup side facing outward.

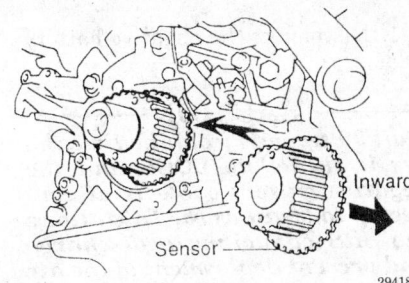

Installation of crankshaft gear — 3.4L (5VZ-FE) engine

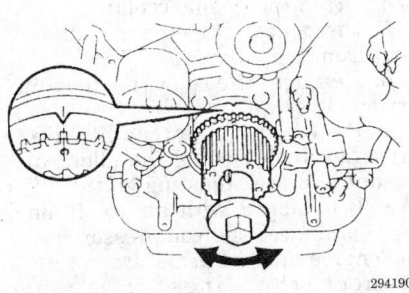

Aligning the timing mark for the crankshaft gear — 3.4L (5VZ-FE) engine

30. Install the No. 1 timing belt cover and starter wire bracket. Torque the timing belt cover bolts to 80 inch lbs. (9 Nm).

31. Install the crankshaft pulley as follows:

   a. Align the pulley set key with the key groove of the crankshaft pulley.

   b. Install the pulley bolt and torque it to 184 ft. lbs. (250 Nm).

32. Install the left camshaft timing pulley.

   a. Install the knock pin to the camshaft.

   b. Align the knock pin hose of the camshaft with the knock pin groove of the timing pulley.

   c. Slide the timing pulley on the camshaft with the flange side facing outward. Torque the pulley bolt to 81 ft. lbs. (110 Nm).

33. Set the No. 1 cylinder to TDC of the compression stroke.

   a. Turn the crankshaft pulley, and align its groove with the timing mark **0** of the No. 1 timing belt cover.

   b. Turn the camshaft, align the knock pin hole of the camshaft with the timing mark of the No. 3 timing belt cover.

   c. Turn the camshaft timing pulley, align the timing marks of the

camshaft timing pulley and the No. 3 timing belt cover.

34. Connect the timing belt to the left camshaft timing pulley. Check that the installation mark on the timing belt is aligned with the end of the No. 1 timing belt cover.

   a. Using SST 09960–01000 or equivalent, slightly turn the left camshaft timing pulley clockwise. Align the installation mark on the timing belt with the timing mark of the camshaft timing pulley, and hang the timing belt on the left camshaft timing pulley.

   b. Align the timing marks of the left camshaft pulley and the No. 3 timing belt cover.

   c. Check that the timing belt has tension between the crankshaft timing pulley and the left camshaft timing pulley.

35. Install the right camshaft timing pulley and the timing belt as follows:

   a. Align the installation mark on the timing belt with the timing mark of the right camshaft timing pulley, and hang the timing belt on the right camshaft timing pulley with the flange side facing inward.

   b. Slide the right camshaft timing pulley on the camshaft. Align the timing marks on the right camshaft timing pulley and the No. 3 timing belt cover.

   c. Align the knock pin hole of the camshaft with the knock pin groove of the pulley and install the knock pin. Install the bolt and torque to 81 ft. lbs. (110 Nm).

36. Set the timing belt tensioner as follows:

   a. Using a press, slowly press in the pushrod using 220–2,205 lbs. (981–9,807 N) of force.

   b. align the holes of the pushrod and housing, pass a 1.5 mm hexagon wrench through the holes to keep the setting position of the pushrod.

   c. Release the press and install the dust boot to the tensioner.

37. Install the timing belt tensioner and alternately tighten the bolts to 20 ft. lbs. (28 Nm). Using pliers, remove the 1.5 mm hexagon wrench from the belt tensioner.

38. Check the valve timing.

   a. Slowly turn the crankshaft pulley two revolutions from the TDC to TDC. Always turn the crankshaft pulley clockwise.

   b. Check that each pulley aligns with the timing marks. If the timing marks do not align, remove the timing belt and reinstall it.

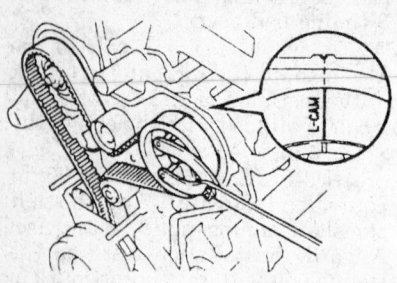

Left hand camshaft timing mark — 3.4L (5VZ-FE) engine

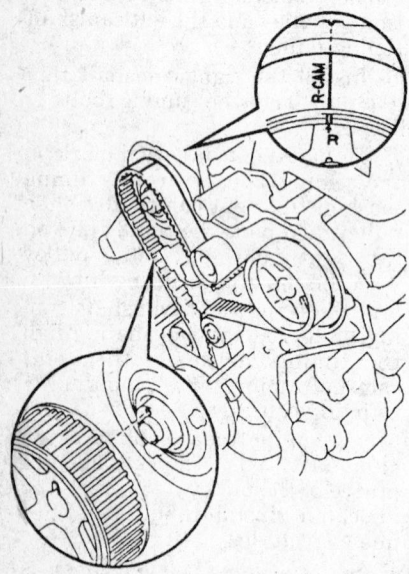

Right hand camshaft timing mark — 3.4L (5VZ-FE) engine

1996–97 4Runner

1. Disconnect the negative battery cable.

—— **CAUTION** ——
Wait 90 seconds from the time the key is turned to LOCK and the negative battery cable is disconnected to begin work. This allows the SRS capacitor to discharge and prevent deployment of the air bag(s).

2. Raise and safely support the vehicle.
3. Remove the engine undercover.
4. Drain the engine coolant.
5. Disconnect the upper radiator hose from the engine.
6. Remove the power steering drive belt.
7. Remove the air conditioning drive belt by loosening the idler pulley nut and the adjusting bolt.
8. If equipped with air conditioning, disconnect the compressor from the engine and set aside. Do not disconnect the lines from the compressor.
9. If equipped with air conditioning, disconnect the air conditioning bracket.
10. Remove the fan with the fluid coupling and fan pulleys.

11. Loosen the lock bolt, pivot bolt, and the adjusting bolt and the alternator drive belt.
12. Remove the No. 2 fan shroud by removing the two clips.
13. Disconnect the power steering pump from the engine and set aside. Do not disconnect the lines from the pump.
14. Remove the oil dipstick and the guide.
15. Remove the No. 2 timing belt cover as follows:
   a. Disconnect the camshaft position sensor connector from the No. 2 timing belt cover.
   b. Disconnect the four spark plug wire clamps from the No. 2 timing belt cover.
   c. Remove the six bolts and remove the timing belt cover.
16. Remove the fan bracket as follows:
   a. Remove the power steering adjusting strut by removing the nut.
   b. Remove the fan bracket by removing the bolt and nut.
17. Using SST 09213–54015 or equivalent, remove the crankshaft pulley.
18. Remove the starter wire bracket and the No. 1 timing belt cover.
19. Remove the timing belt guide.

39. Install the fan bracket with the bolt and nut.
40. Install the power steering adjusting strut with the nut.
41. Install the No. 2 timing belt cover. Torque the bolts to 80 inch lbs. (9 Nm). Install the remaining components.
42. Fill the cooling system with coolant.
43. Connect the negative battery cable.
44. Start the engine and check for leaks.

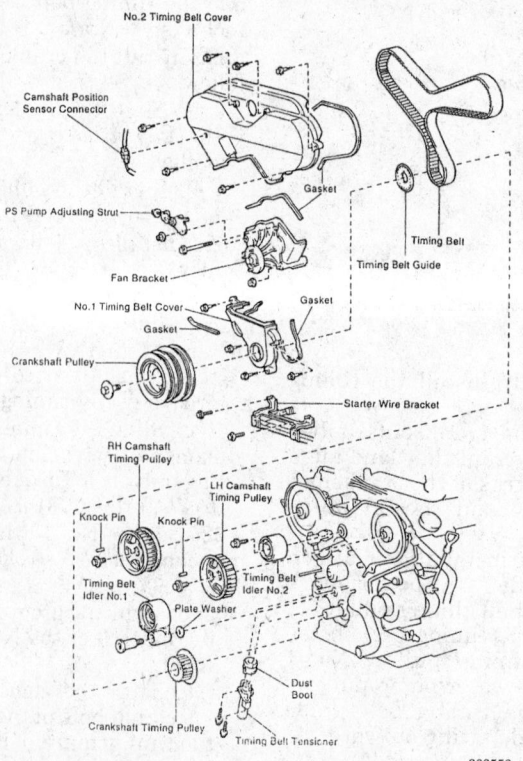

Timing belt component assembly — 1996–97 4Runner with 3.4L (5VZ-FE) engine

20. Set the No. 1 cylinder at TDC of the compression stroke.

   a. Temporarily install the crankshaft pulley bolt to the crankshaft.

   b. Turn the crankshaft and align the timing marks of the crankshaft timing pulley and the oil pump body.

   c. Check that the timing marks of the camshaft timing pulleys and the No. 3 timing belt cover are aligned. If not, turn the crankshaft pulley one revolution (360°).

NOTE: If re-using the timing belt, make sure that you can still read the installation marks. If not, place new installation marks on the timing belt to match the timing marks of the camshaft timing pulleys.

21. Remove the timing belt tensioner by alternately loosening the two bolts.

22. Remove the right and left camshaft pulleys.

23. Remove the No. 2 idler pulley.

24. Using a 10 mm hexagon wrench, remove the pivot bolt, No.1 idler pulley and the plate washer.

25. Remove the timing belt guide and remove the timing belt.

26. Remove the crankshaft timing pulley.

   To install:

27. Install the crankshaft timing pulley.

   a. Align the timing pulley set key with the key groove of the timing pulley and slide on the timing pulley.

   b. Slide on the timing pulley with the flange side facing inward.

28. Install the plate washer and the No. 1 idler pulley with the pivot bolt and torque to 26 ft. lbs. (35 Nm). Check that the pulley bracket moves smoothly.

29. Install the No. 2 timing belt idler with the bolt. Torque the bolt to 30 ft. lbs. (40 Nm). Check that the pulley bracket moves smoothly.

30. Install the left and right camshaft timing pulleys.

31. Set the No. 1 cylinder to TDC of the compression stroke.

   a. Using the crankshaft pulley bolt, turn the crankshaft and align the timing marks of the crankshaft timing pulley and the oil pump body.

   b. Using SST 09960-10010 or equivalent, turn the camshaft pulley, align the timing marks of the timing pulley and the No. 3 timing belt cover.

32. Install the timing belt. The engine should be cold.

   a. Face the front mark on the timing belt forward.

   b. Align the installation mark on the timing belt with the timing mark of the crankshaft timing pulley.

   c. Align the installation marks on the timing belt with the timing marks of the camshaft timing pulleys.

33. Install the timing belt in the following order:

   • Left camshaft timing pulley
   • No. 2 idler pulley
   • Right camshaft timing pulley
   • Water pump pulley
   • Crankshaft timing pulley
   • No. 1 idler pulley

34. Set the timing belt tensioner as follows:

   a. Using a press, slowly press in the pushrod using 220–2,205 lbs. (981–9,807 N) of force.

   b. Align the holes of the pushrod and housing, pass a 1.27 mm hexagon wrench through the holes to keep the setting position of the pushrod.

   c. Release the press and install the dust boot to the tensioner.

35. Install the timing belt tensioner and alternately tighten the bolts to 20 ft. lbs. (27 Nm). Using pliers, remove the 1.27 mm hexagon wrench from the belt tensioner.

36. Check the valve timing.

   a. Slowly turn the crankshaft and align the timing marks of the crankshaft timing pulley and the oil pump body. Always turn the crankshaft pulley clockwise.

   b. Check that the timing marks of the right and left timing pulleys align with the timing marks of the No. 3 timing belt cover. If the marks do not align, remove the timing belt and reinstall it.

37. Install the timing belt guide with the cup side facing outward.

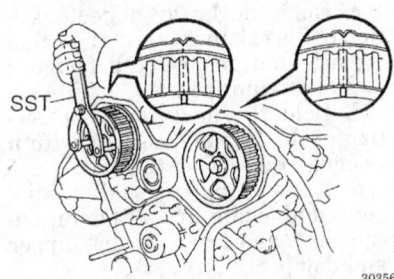
SST

303561

Left hand and right hand camshaft timing marks — 1996–97 4Runner with 3.4L (5VZ-FE) engine

38. Install the No. 1 timing belt cover and starter wire bracket. Torque the timing belt cover to 80 inch lbs. (9 Nm).

39. Install the crankshaft pulley.

   a. Align the pulley set key with the key groove of the pulley and slide the pulley.

   b. Using SST 09213-54014 or equivalent, torque the bolt to 184 ft. lbs. (250 Nm).

40. Install the fan bracket with the bolt and nut.

41. Install the No. 2 timing belt cover and torque the bolts to 80 inch lbs. (9 Nm). Install the remaining components.

42. Fill the cooling system with coolant.

43. Connect the negative battery cable.

44. Start the engine and check for leaks.

45. Check the ignition timing.

Camshaft

REMOVAL AND INSTALLATION

2.7L (3RZ-FE) Engine

1. Disconnect the negative battery cable.

2. Drain the engine coolant.

3. Remove the air cleaner cap, MAF meter, and the resonator.

4. If equipped with a manual transmission, disconnect the accelerator cable from the throttle body.

5. If equipped with a automatic transmission, disconnect the accelerator and throttle cables from the throttle body.

6. If equipped with cruise control, disconnect the cruise control cable from the actuator.

7. Remove the intake air connector. Disconnect the following:

   a. Air hose for IAC
   b. Vacuum sensing hose
   c. Wire clamp for the engine wire

8. Remove the PCV hoses.

9. Disconnect the spark plug wires from the spark plugs.

10. Disconnect the engine wire clamps and engine wire.

11. Disconnect the following connectors:

   • If equipped with air conditioning, disconnect the air conditioning compressor connector
   • Disconnect the oil pressure sensor connector
   • Disconnect the ECT sensor connector
   • Engine coolant temperature sender gauge connector

- EGR gas temperature sensor connector
- VSV connector for the EGR
- Two vacuum hoses from the VSV for the EGR
- Ground strap from the cowl top panel
- Engine wire from the air intake chamber
- Throttle position sensor connector
- IAC valve connector
- Crankshaft position sensor connector
- Knock sensor connector
- DLC1 from the bracket
- Engine wire clamp

12. Remove the EGR pipe.
13. Remove the intake chamber stay.
14. Remove the air intake chamber assembly.
15. Disconnect the following hoses:
 a. EVAP hose from the throttle body
 b. Brake booster vacuum hose from the union
 c. Water bypass hose from the water bypass pipe
 d. Water bypass hose from the cylinder head rear cover
16. Remove the front engine hanger.
17. Remove the engine wire brackets.
18. Remove the cylinder head cover.
19. Set No. 1 cylinder to TDC compression stroke. The groove on the crankshaft pulley should align with the **0** mark on the timing chain cover and the timing marks (one and two dots) of the camshaft gears should form a straight line in respect to the cylinder head surface. If not, turn the crankshaft 1 revolution (360 degrees).
20. Remove the chain tensioner and gasket.
21. Remove the camshaft timing gear as follows:
 a. Remove the 2 semi-circular plugs.

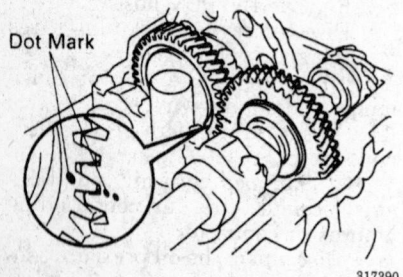

Dot Mark

317390

Camshafts TDC/compression timing marks. Marks with 1 and 2 dots will be in straight line on cylinder head surface — 2.7L (3RZ-FE) engine

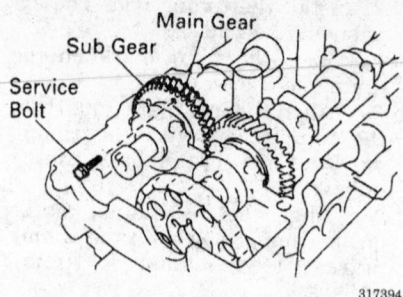

Main Gear
Sub Gear
Service Bolt

317394

Secure the exhaust camshaft sub-gear to the main gear with a service bolt — 2.7L (3RZ-FE) engine

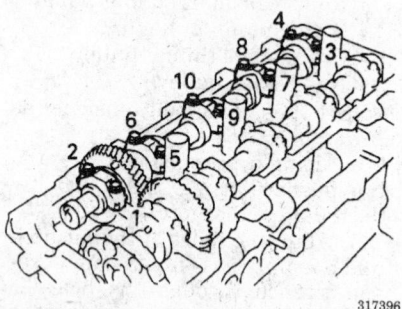

317396

Loosen and remove the exhaust camshaft bearing cap bolts in the sequence shown — 2.7L (3RZ-FE) engine

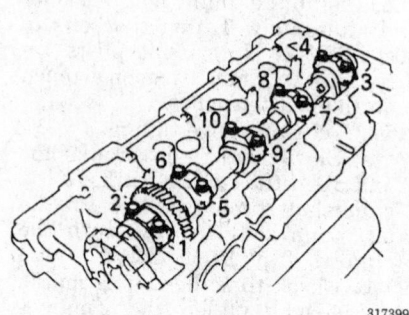

317399

Loosen and remove the intake camshaft bearing cap bolts in the sequence shown — 2.7L (3RZ-FE) engine

 b. Place matchmarks on the camshaft timing gear and No. 1 timing chain.
 c. Hold the hexagon head portion of the exhaust camshaft with a wrench and remove the fastener and distributor gear.
 d. Hold the hexagon head portion of the intake camshaft with a wrench and remove the bolt.
 e. Remove the camshaft timing gear and chain from the intake camshaft and leave on the slipper and damper.

22. Remove exhaust camshafts:
 a. Bring the service bolt hole of the driven sub-gear upward by turning the hexagon head portion of the exhaust camshaft with a wrench.
 b. Secure the exhaust camshaft sub-gear to the driven gear with a service bolt (6mm diameter, 0.63–0.79 inches in length and 1.0mm in thread pitch).

NOTE: When removing the camshaft, make sure that the torsional spring force of the sub-gear has been eliminated by the above operation.

 c. Uniformly loosen and remove the bearing cap bolts in several passes, in the sequence shown.
 d. Remove the bearing caps and camshaft. Make a note of the bearing cap positions for proper installation.

NOTE: Do not pry on or attempt to force the camshaft with a tool or other object.

23. To remove the intake camshaft, uniformly loosen and remove the bearing cap bolts in several passes, in the sequence shown. Remove the bearing caps and camshaft. Make a note of the bearing cap positions for proper installation.

NOTE: If the camshaft is not being lifted out straight and level, reinstall the No. 3 bearing cap with the two bolts. Then alternately loosen and remove the two bearing cap bolts with the camshaft gear pulled up.

24. Remove the valve lifters and shims from the cylinder head. Arrange the valve lifters and shims in correct order.
 To install:
25. Install the valve lifters and shims in their proper locations. Check that the valve lifter rotates smoothly by hand.
26. Install the intake camshaft:
 a. Apply engine oil to the thrust portion of the intake camshaft.
 b. Position the intake camshaft with the knock pin facing upward.
 c. Install the bearing caps in their proper locations. Apply a light coat of engine oil to the threads and install the cap bolts. Uniformly tighten the cap bolts in the sequence shown to 12 ft. lbs. (16 Nm).
27. Install the exhaust camshaft:
 a. Apply engine oil to the thrust portion of the intake camshaft.
 b. Engage the exhaust camshaft gear to the intake camshaft gear by

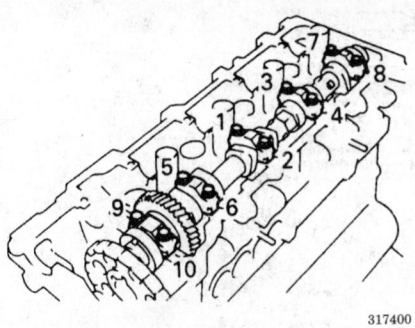

317400

Tighten the intake camshaft bearing cap bolts in the sequence shown — 2.7L (3RZ-FE) engine

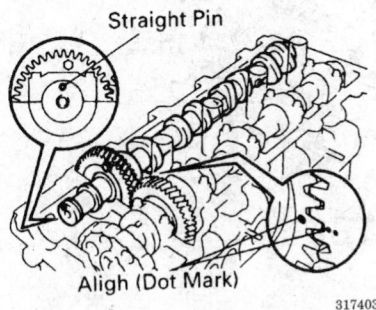

317403

Engage both camshaft gears while matching the timing marks as shown — 2.7L (3RZ-FE) engine

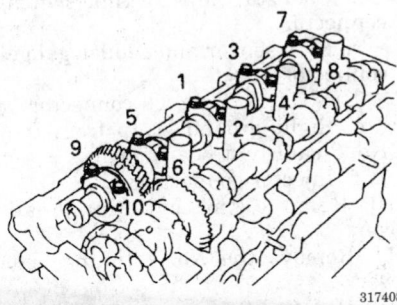

317408

Tighten the exhaust camshaft bearing cap bolts in the sequence shown — 2.7L (3RZ-FE) engine

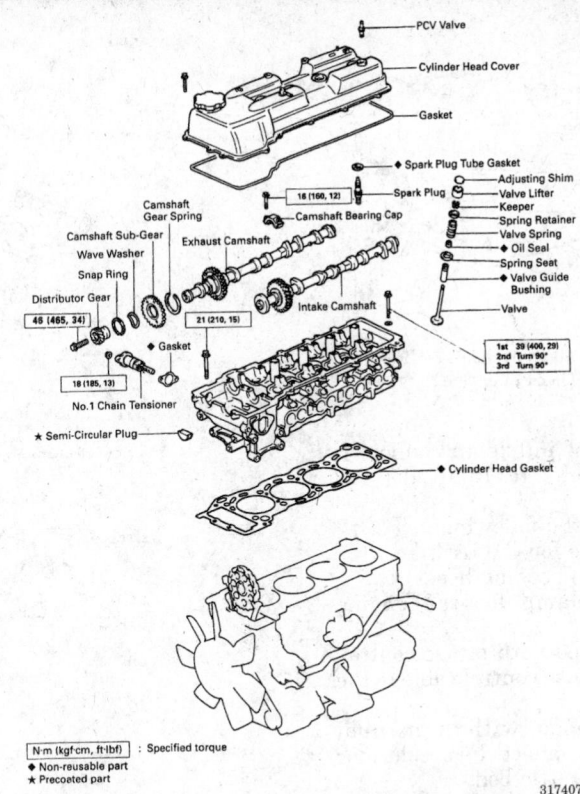

317407

N·m (kgf·cm, ft-lbf) : Specified torque
◆ Non-reusable part
★ Precoated part

Cylinder head components — 2.7L (3RZ-FE) engine

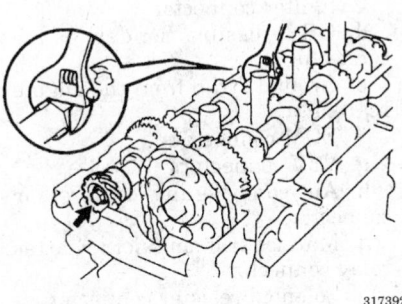

317392

Hold the camshaft at the hexagon wrench head portion — 2.7L (3RZ-FE) engine

matching the timing marks (one and two dots) on each other.

c. Roll down the exhaust camshaft onto the bearing journals while engaging the gears with each other. Install the bearing caps in their proper locations.

d. Apply a light coat of engine oil to the threads and install the cap bolts. Uniformly tighten the cap bolts in the sequence shown to 12 ft. lbs. (16 Nm).

e. Remove the service bolt from the driven sub-gear. Check that the intake and exhaust camshafts turns smoothly.

28. Set No. 1 cylinder to TDC compression stroke: Crankshaft pulley groove align with **0** mark on timing cover and camshafts timing marks with one dot and two dots will be straight line on the cylinder head surface.

29. Install the timing gear. Place the gear over the straight pin of the intake camshaft.

a. Hold the intake camshaft with a wrench. Install and tighten the bolt to 54 ft. lbs. (74 Nm).

b. Hold the exhaust camshaft and install the bolt and distributor gear. Torque the bolt to 34 ft. lbs. (46 Nm).

30. Install the chain tensioner, using a new gasket (mark toward the front).

31. Recheck the engine for proper valve timing. Check and adjust the valve clearance.

32. Install the semi-circular plug.

33. Recheck the engine for proper valve timing.

34. Install the cylinder head cover with a new gasket.

35. Install the engine wire brackets.

36. Install the front engine hanger and torque the mounting bolts to 30 ft. lbs. (42 Nm).

37. Install the air intake chamber assembly. Torque the bolts to 15 ft. lbs. (21 Nm). Connect the hoses removed prior.

38. Install the intake chamber stay.

39. Install the air intake chamber stay and torque the bolts to 15 ft. lbs. (20 Nm).

40. Install the EGR pipe. Torque the nuts and bolts to:
- Bolt: 13 ft. lbs. (18 Nm)
- Nut A: 14 ft. lbs. (19 Nm)
- Nut B: 15 ft. lbs. (20 Nm)

41. Connect the engine wires detached prior.

42. Connect the spark plug wires to the spark plugs.

43. Install the PCV hoses.

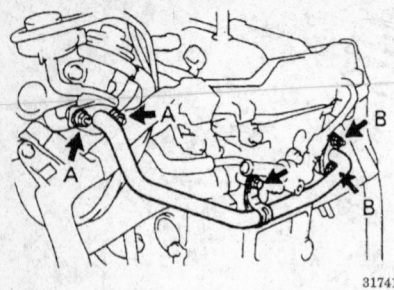

EGR pipe and related components — 1996–97 4Runner with 2.7L (3RZ-FE) engine

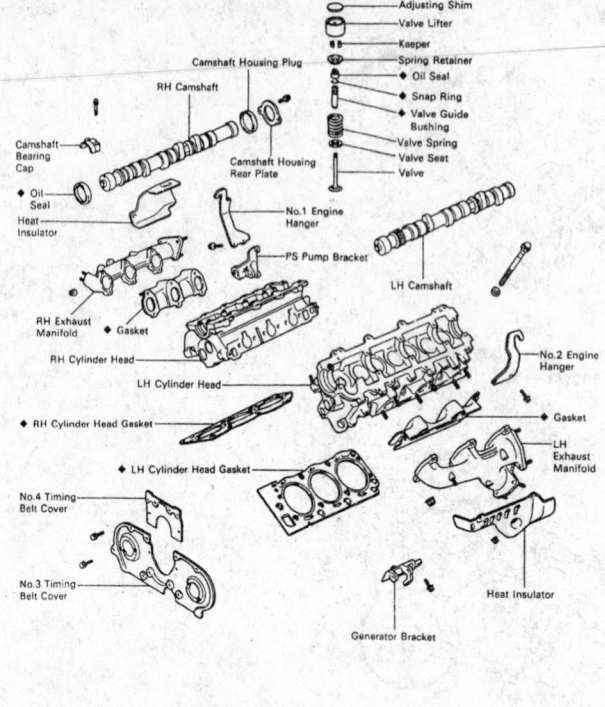

Exploded view of the camshaft and related components — 3.0L (3VZ-E) engine

◆ Non-reusable part

44. Install the intake air connector. Torque the bolts to 13 ft. lbs. (18 Nm).

45. Connect the following:
   a. Air hose for the IAC.
   b. Vacuum sensing hose.
   c. Wire clamp for the engine wire.

46. If equipped with cruise control, connect the cruise control cable to the actuator.

47. If equipped with a manual transmission, connect the accelerator cable to the throttle body.

48. If equipped with a automatic transmission, connect the accelerator and throttle cables to the throttle body.

49. Install the air cleaner cap, MAF meter, and the resonator assembly.

50. Fill the engine and radiator with engine coolant.

51. Connect the negative battery cable. Start the engine and check for leaks.

52. Check the ignition timing. Road test the vehicle for proper operation.

53. Recheck all fluid levels.

3.0L (3VZ-E) Engine

1. Disconnect the negative battery cable.

CAUTION

Wait 90 seconds from the time the key is turned to LOCK and the negative battery cable is disconnected to begin work. This allows the SRS capacitor to discharge and prevent deployment of the air bag(s).

2. Drain the cooling system and the engine oil.

3. Disconnect the air cleaner and the hose.

4. Disconnect the following strap, wires, and connectors:
   a. Ground strap for the left fender apron
   b. Alternator connector and wire

   c. Igniter connector
   d. Oil pressure sender gauge connector
   e. Ground strap from the engine rear side
   f. ECM connectors
   g. VSV connectors
   h. Air conditioning compressor connector.
   i. Manual transmission: Starter relay connector
   j. Solenoid resister connector
   k. W/ Automatic Disconnecting Differential (ADD): ADD switch connector

5. Disconnect the following hoses:
   a. Power steering air hoses from the gas filter and the air pipe
   b. Brake booster hose
   c. W/Cruise: Cruise control vacuum hose
   d. Charcoal canister hose from the canister
   e. VSV vacuum hoses

6. Disconnect the following cables:
   a. Accelerator cable
   b. Automatic transmission: Throttle cable
   c. W/Cruise: Cruise control cable

7. Remove the spark plug wires.

8. Remove the timing belt.

9. Remove the following connectors and wires:
   a. Knock sensor

   b. Cold start injector time switch connector
   c. ECT sensor and sender gauge connector
   d. No. 1 ECT switch connector
   e. Right ground strap from the No. 3 camshaft bearing cap
   f. Injector connectors

10. Remove the cylinder head covers.

11. Remove the No. 4 timing belt cover.

12. Remove the camshaft bearing cap bolts in sequence. Remove the camshafts.

NOTE: Arrange the bearing caps in correct order for installation.

13. Remove the valve lifters and shims by hand. Arrange the valve lifters and shims in correct order for re-installation.

To install:

14. Install the valve lifters and shims. Check that the valve lifter rotates smoothly by hand.

15. Install the camshafts. Install the bearing caps in their proper locations and torque the bearing cap bolts to 12 ft. lbs. (16 Nm).

16. Install the No. 4 timing belt cover.

17. Install the timing belt.

RH Side

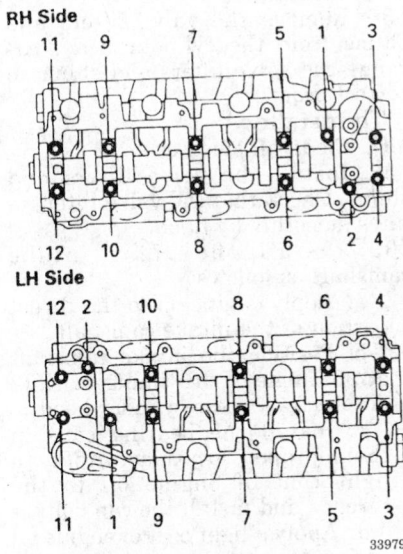

LH Side

Loosen camshaft bearing cap bolts in numerical sequence shown — 3.0L (3VZ-E) engine

RH Side

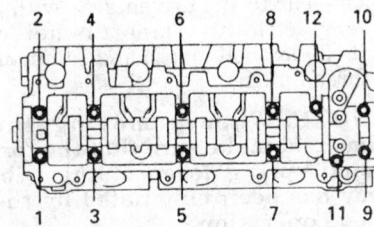

LH Side

Tighten bearing cap bolts in numerical sequence shown — 3.0L (3VZ-E) engine

RH Side

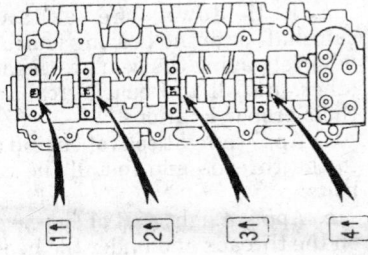

LH Side

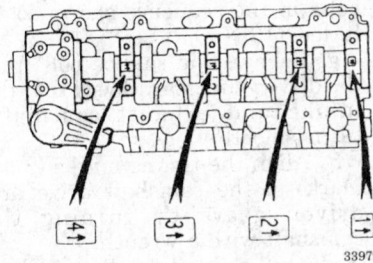

Position bearing caps with arrows pointing toward the front (RH side) or rear (LH side) — 3.0L (3VZ-E) engine

18. Check and adjust valve clearance to:

   a. Intake (Cold): 0.007–0.011 in. (0.18–0.28 mm)

   b. Exhaust (Cold): 0.009–0.013 in. (0.22–0.32 mm)

19. Install the cylinder head covers and torque to 4 ft. lbs. (5 Nm).

20. Install the engine harness. Attach the connectors and straps removed prior.

21. Install the spark plug wires.

22. Connect the cables and hoses removed prior.

23. Install the air cleaner and the hose.

24. Fill the cooling system and fill the engine with oil.

25. Connect the battery cable, start the engine, and check for leaks.

26. Road test the vehicle for proper operation and recheck all the fluid levels.

3.4L (5VZ-FE) Engine

1995–97 T-100 and Tacoma

1. Disconnect the negative battery cable.

—— **CAUTION** ——

Wait 90 seconds from the time the key is turned to LOCK and the negative battery cable is disconnected to begin work. This allows the SRS capacitor to discharge and prevent deployment of the air bag(s).

2. Remove the engine undercover.

3. Drain the cooling system.

4. Disconnect the air cleaner cap, MAF meter, and the resonator.

5. Disconnect the following cables:

• If equipped with cruise control, disconnect the actuator cable with the bracket.

• Accelerator cable

• With automatic transmission, Throttle cable

6. Disconnect the heater hose.

7. Disconnect the upper radiator hose from the engine.

8. Remove the power steering drive belt as follows:

   a. Stretch the belt and loosen the fan pulley mounting nuts.

   b. Loosen the lock bolt, pivot bolt and adjusting bolt and remove the drive belt from the engine.

9. Remove the air conditioning drive belt by loosening the idle pulley nut and adjusting bolt.

10. Loosen the lock bolt, pivot bolt and adjusting bolt and the alternator drive belt.

11. Remove the No. 2 fan shroud by removing the two clips.

12. Remove the fan with the fluid coupling and fan pulleys.

13. Disconnect the power steering pump from the engine and set aside. Do not disconnect the lines from the pump.

14. If equipped with air conditioning, disconnect the compressor from the engine and set aside. Do not disconnect the lines from the compressor.

15. If equipped with air conditioning, disconnect the air conditioning bracket.

16. Remove the spark plug wires with the ignition coils.

17. Remove the spark plugs.

18. Remove the No. 2 timing belt cover as follows:

   a. Disconnect the camshaft position sensor connector from the No. 2 timing belt cover.

   b. Disconnect the three spark plug wire clamps from the No. 2 timing belt cover.

   c. Remove the six bolt and remove the timing belt cover.

19. Remove the fan bracket as follows:

   a. Remove the power steering adjusting strut by removing the nut.

   b. Remove the fan bracket by removing the bolt and nut.

20. Set the No. 1 cylinder at TDC of the compression stroke.

a. Turn the crankshaft pulley and align its groove with the timing mark **0** of the No. 1 timing belt cover.

b. Check that the timing marks of the camshaft timing pulleys and the No. 3 timing belt cover are aligned. If not, turn the crankshaft pulley one revolution (360°).

NOTE: If re-using the timing belt, make sure that you can still read the installation marks. If not, place new installation marks on the timing belt to match the timing marks of the camshaft timing pulleys.

21. Remove the timing belt tensioner by alternately loosening the two bolts.

22. Remove the camshaft timing pulleys.

a. Using SST 09960–10010 or equivalent, remove the pulley bolt, the timing pulley and the knock pin. Remove the two timing pulleys with the timing belt.

23. Remove the bolt and the No. 2 idler pulley.

24. Remove the alternator from the engine.

25. Disconnect the following hoses:
• Disconnect the PCV hoses.
• Disconnect the water bypass hoses.
• Disconnect the air assist hose from the intake air connector.
• Two vacuum sensing hoses from the VSV
• EVAP hose
• Air hose from the power steering
• If equipped with air conditioning, disconnect the air hose from the air conditioning idle up valve.

26. Remove the four bolts, two nuts and remove the air intake chamber assembly from the engine.

27. Remove the intake air connector.

28. Remove the camshaft position sensor.

29. Remove the No. 3 (rear) timing belt cover by removing the six bolts.

30. Remove the eight bolts, seal washers, cylinder head cover and gasket. Remove both cylinder head covers.

31. Remove the semi circular plugs.

32. Remove the right exhaust camshafts as follows:

a. Bring the service bolt hole of the driven sub-gear upward by turning the hexagon head portion of the exhaust camshaft with a wrench.

b. Align the timing mark (2 dot marks) of the camshaft drive and driven gears by turning the camshaft with a wrench.

c. Secure the exhaust camshaft sub-gear to the driven gear with a service bolt (6mm diameter, 16–20mm bolt length and 1.0mm in thread pitch).

NOTE: When removing the camshaft, make sure that the torsional spring force of the sub-gear has been eliminated by the above operation.

d. Uniformly loosen and remove the bearing cap bolts in several passes, in the sequence shown.

e. Remove the bearing caps and camshaft. Make a note of the bearing cap positions for proper installation.

NOTE: Do not pry on or attempt to force the camshaft with a tool or other object.

33. Remove the right hand intake camshaft as follows:

a. Uniformly loosen and remove the bearing cap bolts in several passes, in the sequence shown.

b. Remove the bearing caps, oil seal and camshaft. Make a note of the bearing cap positions for proper installation.

34. Remove the left exhaust camshafts as follows:

a. Align the timing mark (1 dot mark) of the camshaft drive and driven gears by turning the camshaft with a wrench.

b. Secure the exhaust camshaft sub-gear to the driven gear with a service bolt (6mm diameter, 16–20mm bolt length and 1.0mm in thread pitch).

NOTE: When removing the camshaft, make sure that the torsional spring force of the sub-gear has been eliminated by the above operation.

c. Uniformly loosen and remove the bearing cap bolts in several passes, in the sequence shown.

d. Remove the bearing caps and camshaft. Make a note of the bearing cap positions for proper installation.

NOTE: Do not pry on or attempt to force the camshaft with a tool or other object.

35. Remove the left hand intake camshaft as follows:

a. Uniformly loosen and remove the bearing cap bolts in several passes, in the sequence shown.

b. Remove the bearing caps, oil seal and camshaft. Make a note of the bearing cap positions for proper installation.

36. Remove the valve lifters and shims from the cylinder head. Arrange the valve lifters and shims in correct order.

To install:

37. Clean all surfaces.

38. Install the valve lifters and shims. Check that the valve lifter rotates smoothly by hand.

39. Install the right intake camshaft as follows:

a. Apply engine oil to the thrust portion of the intake camshaft.

b. Position the intake camshaft at 90° angle of the timing mark (2 dot marks) on the cylinder head.

c. Install the bearing caps in their proper locations. Apply a light coat of engine oil to the threads and install the cap bolts.

d. Apply a light coat of engine oil on the threads and under the heads of the bearing cap bolts.

e. Uniformly tighten the cap bolts in the sequence shown to 12 ft. lbs. (16 Nm).

40. Install the right exhaust camshaft:

a. Apply engine oil to the thrust portion of the intake camshaft.

b. Align the timing marks (2 dot marks) of the camshaft drive and driven gears.

c. Roll down the exhaust camshaft onto the bearing journals while engaging the gears with each other. Install the bearing caps in their proper locations.

d. Apply a light coat of engine oil to the threads and install the cap bolts.

e. Apply a light coat of engine oil on the threads and under the heads of the bearing cap bolts.

f. Uniformly tighten the cap bolts in the sequence shown to 12 ft. lbs. (16 Nm).

g. Remove the service bolt from the driven sub-gear. Check that the intake and exhaust camshafts turns smoothly.

h. Align the timing marks (2 dot mark) of the camshaft drive and driven gears by turning the camshaft with a wrench.

41. Install the left intake camshaft as follows:

a. Apply engine oil to the thrust portion of the intake camshaft.

b. Position the intake camshaft at 90° angle of the timing mark (1 dot marks) on the cylinder head.

c. Install the bearing caps in their proper locations. Apply a light coat of engine oil to the threads and install the cap bolts.

d. Apply a light coat of engine oil on the threads and under the heads of the bearing cap bolts.

e. Uniformly tighten the cap bolts in the sequence shown to 12 ft. lbs. (16 Nm).

42. Install the left exhaust camshaft:

a. Apply engine oil to the thrust portion of the intake camshaft.

b. Align the timing marks (1 dot marks) of the camshaft drive and driven gears.

c. Roll down the exhaust camshaft onto the bearing journals while engaging the gears with each other. Install the bearing caps in their proper locations.

d. Apply a light coat of engine oil to the threads and install the cap bolts.

e. Apply a light coat of engine oil on the threads and under the heads of the bearing cap bolts.

f. Uniformly tighten the cap bolts in the sequence shown to 12 ft. lbs. (16 Nm).

g. Remove the service bolt.

43. Check and adjust the valve clearance.

44. Install the semi circular plugs.

45. Install the cylinder head covers and uniformly tighten the bolts in several passes to 53 inch lbs. (6 Nm).

46. If removed, install the alternator bracket and torque the bolts to 14 ft. lbs. (18 Nm).

47. Install the No. 3 timing belt cover with the six bolts. Torque the bolts to 80 inch lbs. (9 Nm).

48. Install the camshaft position sensor and torque to 71 inch lbs. (8 Nm).

49. Install the intake air connector.

50. Connect the hoses removed prior.

51. Install the alternator put do not torque the bolts and nuts at this time.

52. Install the No. 2 timing belt idler with the bolt. Torque the bolt to 30 ft. lbs. (40 Nm). Check that the pulley bracket moves smoothly.

53. Install the left camshaft timing pulley.

a. Install the knock pin to the camshaft.

b. Align the knock pin hose of the camshaft with the knock pin groove of the timing pulley.

c. Slide the timing pulley on the camshaft with the flange side facing outward. Torque the pulley bolt to 81 ft. lbs. (110 Nm).

54. Set the No. 1 cylinder to TDC of the compression stroke.

a. Turn the crankshaft pulley, and align its groove with the timing mark 0 of the No. 1 timing belt cover.

b. Turn the camshaft, align the knock pin hole of the camshaft with the timing mark of the No. 3 timing belt cover.

c. Turn the camshaft timing pulley, align the timing marks of the camshaft timing pulley and the No. 3 timing belt cover.

55. Connect the timing belt to the left camshaft timing pulley. Check that the installation mark on the timing belt is aligned with the end of the No. 1 timing belt cover.

a. Using SST 09960–01000 or equivalent, slightly turn the left camshaft timing pulley clockwise. Align the installation mark on the timing belt with the timing mark of the camshaft timing pulley, and hang the timing belt on the left camshaft timing pulley.

b. Align the timing marks of the left camshaft pulley and the No. 3 timing belt cover.

c. Check that the timing belt has tension between the crankshaft timing pulley and the left camshaft timing pulley.

56. Install the right camshaft timing pulley and the timing belt as follows:

a. Align the installation mark on the timing belt with the timing mark of the right camshaft timing pulley, and hang the timing belt on the right camshaft timing pulley with the flange side facing inward.

b. Slide the right camshaft timing pulley on the camshaft. Align the timing marks on the right camshaft timing pulley and the No. 3 timing belt cover.

c. Align the knock pin hole of the camshaft with the knock pin groove of the pulley and install the knock pin. Install the bolt and torque to 81 ft. lbs. (110 Nm).

57. Set the timing belt tensioner as follows:

a. Using a press, slowly press in the pushrod using 220–2,205 lbs. (981–9,807 N) of force.

b. Align the holes of the pushrod and housing, pass a 1.5 mm hexagon wrench through the holes to keep the setting position of the pushrod.

c. Release the press and install the dust boot to the tensioner.

58. Install the timing belt tensioner and alternately tighten the bolts to 20 ft. lbs. (28 Nm). Using pliers, remove the 1.5 mm hexagon wrench from the belt tensioner.

59. Check the valve timing.

a. Slowly turn the crankshaft pulley two revolutions from the TDC to TDC. Always turn the crankshaft pulley clockwise.

b. Check that each pulley aligns with the timing marks. If the timing marks do not align, remove the timing belt and reinstall it.

60. Install the fan bracket with the bolt and nut.

61. Install the power steering adjusting strut with the nut.

62. Install the No. 2 timing belt cover and torque the bolts to 80 inch lbs. (9 Nm). Install the remaining components.

63. Fill the radiator with engine coolant.

64. Connect the negative battery cable to the battery.

65. Start the engine and check for leaks.

66. Check the ignition timing.

67. Install the engine undercover.

68. Road test the vehicle.

69. Recheck all fluid levels.

1996–97 4Runner

1. Release the fuel pressure.
2. Disconnect the negative battery cable.

─── **CAUTION** ───

Wait at least 90 seconds from the time the ignition switch is turned to the LOCK position and the negative (-) battery cable is disconnected before starting the repair procedure.

3. Remove the engine undercover.
4. Drain the cooling system.
5. Disconnect the air cleaner cap, MAF meter, and the resonator.
6. Disconnect the following cables:
• If equipped with cruise control, disconnect the actuator cable with the bracket.
• Accelerator cable
• With automatic transmission, Throttle cable
7. Disconnect the heater hose.

─── **CAUTION** ───

Fuel injection systems remain under pressure, even after the engine is turned OFF. The fuel pressure must be relieved before disconnecting any fuel lines. Failure to do so may result in fire and/or personal injury.

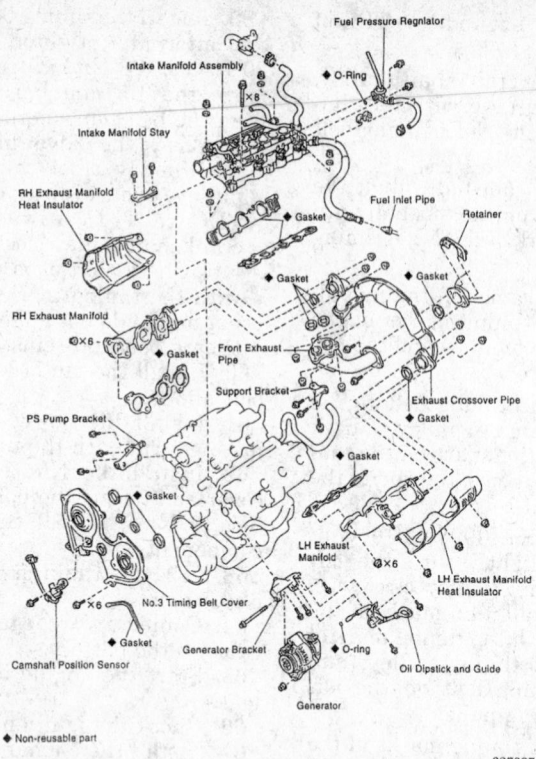

Cylinder head component assembly — 1995–97 T-100 and Tacoma with 3.4L (5VZ-FE) engine

327397

◆ Non-reusable part

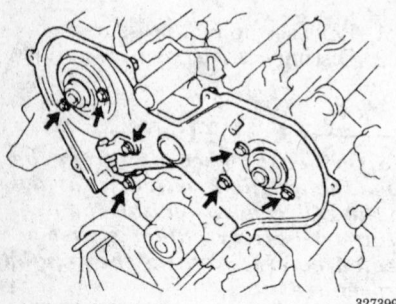

No. 3 (rear) timing belt cover — All with 3.4L (5VZ-FE) engine

327399

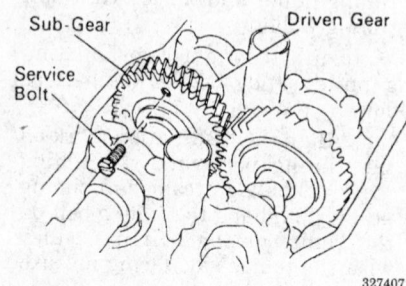

Drive gear service bolt (right side) — All with 3.4L (5VZ-FE) engine

327407

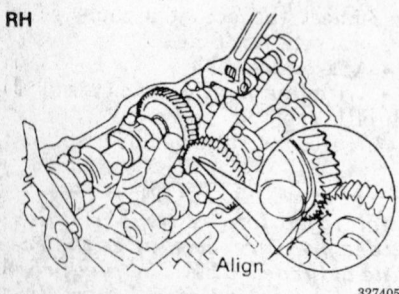

RH

Aligning the timing marks (2 dot marks) of the right camshafts — All with 3.4L (5VZ-FE) engine

327405

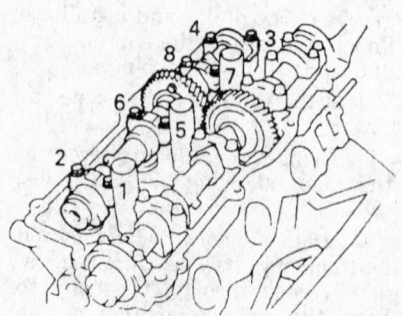

Right exhaust camshaft bolts removal sequence — All with 3.4L (5VZ-FE) engine

327410

8. Disconnect the following hoses:
- Brake booster vacuum hose
- EVAP hose
- 4-Wheel Drive: A.D.D. vacuum hose
- Fuel inlet and fuel return hose

9. Remove the spark plug wires with the ignition coils.

10. Remove the intake chamber stay.

11. Remove the No. 2 timing belt cover as follows:

   a. Disconnect the camshaft position sensor connector from the No. 2 timing belt cover.

   b. Disconnect the three spark plug wire clamps from the No. 2 timing belt cover.

   c. Remove the six bolt and remove the timing belt cover.

12. Remove the air intake chamber assembly.

13. Remove the following connectors and hoses:
- Throttle position sensor connector
- IAC valve connector
- PCV hoses
- Water bypass hoses
- Air assist hose from the throttle body

14. Remove the intake air connector.

15. Disconnect the engine wire protector.

   a. Disconnect the six injector connectors.

   b. Disconnect the ECT sensor and sender gauge connectors.

   c. Disconnect the engine wire protector from the cylinder head.

16. Set the No. 1 cylinder at TDC of the compression stroke.

   a. Turn the crankshaft pulley and align its groove with the timing mark **0** of the No. 1 timing belt cover.

   b. Check that the timing marks of the camshaft timing pulleys and the No. 3 timing belt cover are aligned. If not, turn the crankshaft pulley one revolution (360°).

NOTE: If re-using the timing belt, make sure that you can still read the installation marks. If not, place new installation marks on the timing belt to match the timing marks of the camshaft timing pulleys.

17. Remove the timing belt tensioner by alternately loosening the two bolts.

18. Remove the timing belt.

19. Remove the camshaft timing pulleys.

   a. Using SST 09960–10010 or equivalent, remove the pulley bolt, the timing pulley and the knock

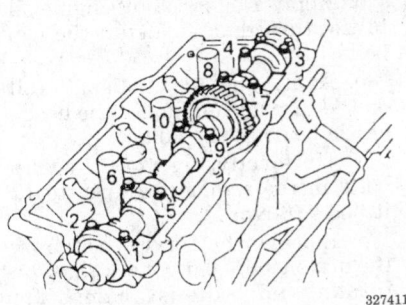

Right intake camshaft bolts removal sequence — All with 3.4L (5VZ-FE) engine

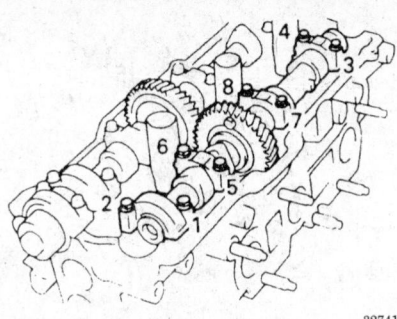

Left exhaust camshaft bolts removal sequence — All with 3.4L (5VZ-FE) engine

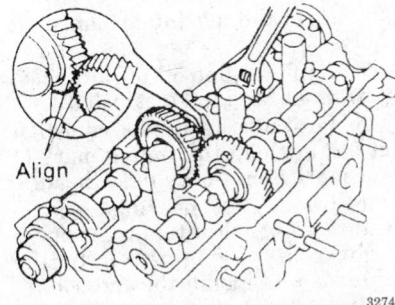

Aligning the timing mark (1 dot mark) of the left camshafts — All with 3.4L (5VZ-FE) engine

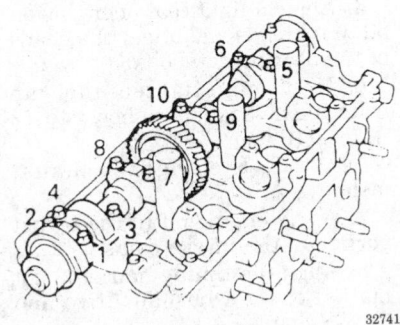

Left intake camshaft bolts removal sequence — All with 3.4L (5VZ-FE) engine

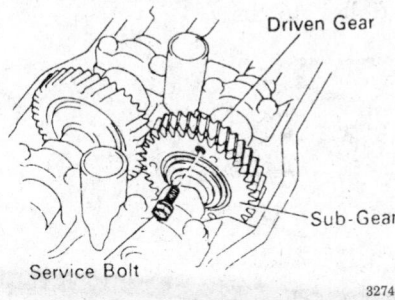

Drive gear service bolt (left side) — All with 3.4L (5VZ-FE) engine

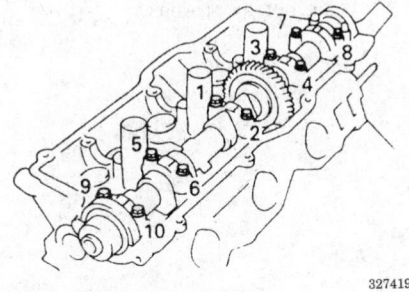

Right intake camshaft tightening sequence — All with 3.4L (5VZ-FE) engine

pin. Remove the two timing pulleys.

20. Remove the bolt and the No. 2 idler pulley.

21. Remove the camshaft position sensor.

22. Remove the No. 3 timing belt cover.

23. Remove the eight bolts, seal washers, cylinder head cover and gasket. Remove both cylinder head covers.

24. Remove the semi circular plugs.

25. Remove the right exhaust camshafts as follows:

a. Bring the service bolt hole of the driven sub-gear upward by turning the hexagon head portion of the exhaust camshaft with a wrench.

b. Align the timing mark (2 dot marks) of the camshaft drive and driven gears by turning the camshaft with a wrench.

c. Secure the exhaust camshaft sub-gear to the driven gear with a service bolt (6mm diameter,

16–20mm bolt length and 1.0mm in thread pitch).

NOTE: When removing the camshaft, make sure that the torsional spring force of the sub-gear has been eliminated by the above operation.

d. Uniformly loosen and remove the bearing cap bolts in several passes, in the sequence shown.

e. Remove the bearing caps and camshaft. Make a note of the bearing cap positions for proper installation.

NOTE: Do not pry on or attempt to force the camshaft with a tool or other object.

26. Remove the right hand intake camshaft as follows:

a. Uniformly loosen and remove the bearing cap bolts in several passes, in the sequence shown.

b. Remove the bearing caps, oil seal and camshaft. Make a note of the bearing cap positions for proper installation.

27. Remove the left exhaust camshafts as follows:

a. Align the timing mark (1 dot mark) of the camshaft drive and driven gears by turning the camshaft with a wrench.

b. Secure the exhaust camshaft sub-gear to the driven gear with a service bolt (6mm diameter, 16–20mm bolt length and 1.0mm in thread pitch).

NOTE: When removing the camshaft, make sure that the torsional spring force of the sub-gear has been eliminated by the above operation.

c. Uniformly loosen and remove the bearing cap bolts in several passes, in the sequence shown.

d. Remove the bearing caps and camshaft. Make a note of the bearing cap positions for proper installation.

NOTE: Do not pry on or attempt to force the camshaft with a tool or other object.

28. Remove the left hand intake camshaft as follows:

a. Uniformly loosen and remove the bearing cap bolts in several passes, in the sequence shown.

b. Remove the bearing caps, oil seal and camshaft. Make a note of the bearing cap positions for proper installation.

29. Remove the valve lifters and shims from the cylinder head. Arrange the valve lifters and shims in correct order.

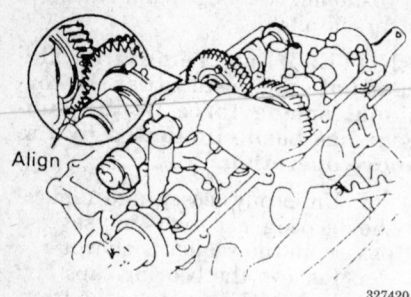

Aligning the right camshafts for installation — All with 3.4L (5VZ-FE) engine

327420

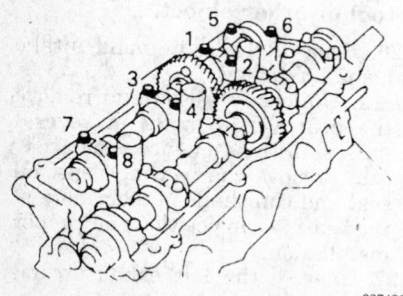

327421

Right exhaust camshaft bolts tightening sequence — All with 3.4L (5VZ-FE) engine

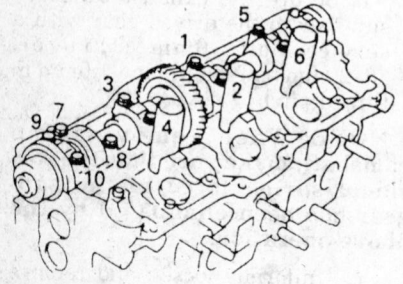

327422

Left intake camshaft bolts tightening sequence — All with 3.4L (5VZ-FE) engine

To install:

30. Install the valve lifters and shims. Check that the valve lifter rotates smoothly by hand.

31. Install the right intake camshaft as follows:

a. Apply engine oil to the thrust portion of the intake camshaft.

b. Position the intake camshaft at 90° angle of the timing mark (2 dot marks) on the cylinder head.

c. Install the bearing caps in their proper locations. Apply a light coat of engine oil to the threads and install the cap bolts.

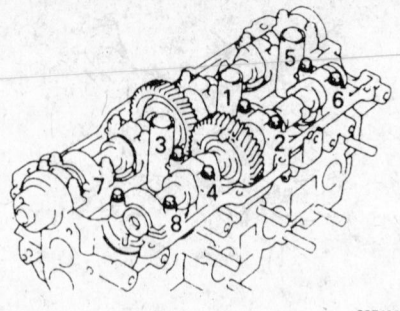

327423

Left exhaust camshaft bolts tightening sequence — All with 3.4L (5VZ-FE) engine

d. Apply a light coat of engine oil on the threads and under the heads of the bearing cap bolts.

e. Uniformly tighten the cap bolts in the sequence shown to 12 ft. lbs. (16 Nm).

32. Install the right exhaust camshaft:

a. Apply engine oil to the thrust portion of the intake camshaft.

b. Align the timing marks (2 dot marks) of the camshaft drive and driven gears.

c. Roll down the exhaust camshaft onto the bearing journals while engaging the gears with each other. Install the bearing caps in their proper locations.

d. Apply a light coat of engine oil to the threads and install the cap bolts.

e. Apply a light coat of engine oil on the threads and under the heads of the bearing cap bolts.

f. Uniformly tighten the cap bolts in the sequence shown to 12 ft. lbs. (16 Nm).

g. Remove the service bolt from the driven sub-gear. Check that the intake and exhaust camshafts turns smoothly.

h. Align the timing marks (2 dot mark) of the camshaft drive and driven gears by turning the camshaft with a wrench.

33. Install the left intake camshaft as follows:

a. Apply engine oil to the thrust portion of the intake camshaft.

b. Position the intake camshaft at 90° angle of the timing mark (1 dot marks) on the cylinder head.

c. Install the bearing caps in their proper locations. Apply a light coat of engine oil to the threads and install the cap bolts.

d. Apply a light coat of engine oil on the threads and under the heads of the bearing cap bolts.

e. Uniformly tighten the cap bolts in the sequence shown to 12 ft. lbs. (16 Nm).

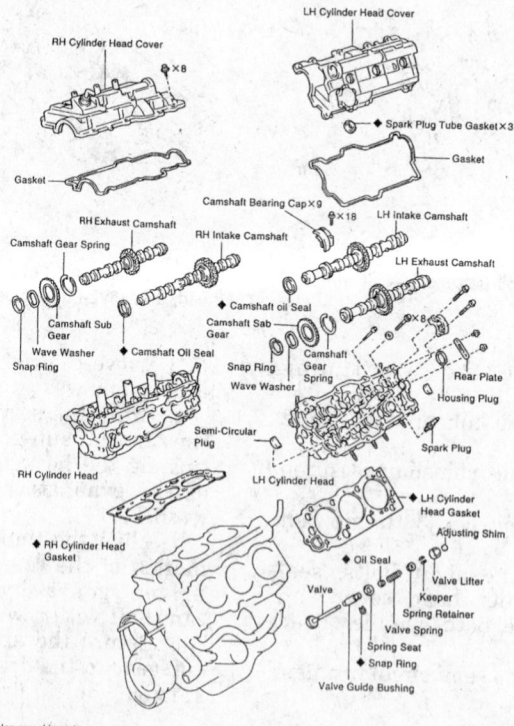

Cylinder head component assembly — All with 3.4L (5VZ-FE) engine

317413

34. Install the left exhaust camshaft:

a. Apply engine oil to the thrust portion of the intake camshaft.

b. Align the timing marks (1 dot marks) of the camshaft drive and driven gears.

c. Roll down the exhaust camshaft onto the bearing journals while engaging the gears with each other. Install the bearing caps in their proper locations.

d. Apply a light coat of engine oil to the threads and install the cap bolts.

e. Apply a light coat of engine oil on the threads and under the heads of the bearing cap bolts.

f. Uniformly tighten the cap bolts in the sequence shown to 12 ft. lbs. (16 Nm).

g. Remove the service bolt.

35. Check and adjust the valve clearance.

36. Install the semi circular plugs.

37. Install the cylinder head covers. Uniformly tighten the bolts in several passes to 53 inch lbs. (6 Nm).

38. Install the No. 3 timing belt cover with the six bolts. Torque the bolts to 80 inch lbs. (9 Nm).

39. Install the camshaft position sensor and torque to 71 inch lbs. (8 Nm).

40. Install the No. 2 timing belt idler with the bolt. Torque the bolt to 30 ft. lbs. (40 Nm). Check that the pulley bracket moves smoothly.

41. Install the left camshaft timing pulley.

a. Install the knock pin to the camshaft.

b. Align the knock pin hose of the camshaft with the knock pin groove of the timing pulley.

c. Slide the timing pulley on the camshaft with the flange side facing outward. Torque the pulley bolt to 81 ft. lbs. (110 Nm).

42. Set the No. 1 cylinder to TDC of the compression stroke.

a. Turn the crankshaft pulley, and align its groove with the timing mark **0** of the No. 1 timing belt cover.

b. Turn the camshaft, align the knock pin hole of the camshaft with the timing mark of the No. 3 timing belt cover.

c. Turn the camshaft timing pulley, align the timing marks of the camshaft timing pulley and the No. 3 timing belt cover.

43. Connect the timing belt to the left camshaft timing pulley. Check

that the installation mark on the timing belt is aligned with the end of the No. 1 timing belt cover.

44. Install the right camshaft timing pulley and the timing belt.

45. Set the timing belt tensioner as follows:

a. Using a press, slowly press in the pushrod using 220–2,205 lbs. (981–9,807 N) of force.

b. align the holes of the pushrod and housing, pass a 1.5 mm hexagon wrench through the holes to keep the setting position of the pushrod.

c. Release the press and install the dust boot to the tensioner.

46. Install the timing belt tensioner and alternately tighten the bolts to 20 ft. lbs. (28 Nm). Using pliers, remove the 1.5 mm hexagon wrench from the belt tensioner.

47. Check the valve timing.

a. Slowly turn the crankshaft pulley two revolutions from the TDC to TDC. Always turn the crankshaft pulley clockwise.

b. Check that each pulley aligns with the timing marks. If the timing marks do not align, remove the timing belt and reinstall it.

48. Connect the engine wire to the intake manifold as follows:

a. Install the engine wire to the cylinder head by installing the three bolts.

b. Connect the three engine wire clamps.

c. Connect the following connectors:
- Six injector connectors
- ECT sender gauge connector
- ECT sensor connector

49. Install the intake air connector.

50. Install the air intake chamber assembly to the engine by installing the four bolts and two nuts. Torque the bolts and nuts to 13 ft. lbs. (18 Nm).

51. Install the intake chamber stay as follows:

a. Install the intake chamber stay and install the two bolts. Torque the bolts to 30 ft. lbs. (40 Nm)

b. Install a new O-ring to the oil filler tube.

c. Push in the oil filler tube end into the tube hole in the oil pan.

d. Install the oil filler tube and No. 1 throttle cable clamp and install the bolt and two nuts.

52. Install the No. 2 timing belt cover. Torque the bolts to 80 inch lbs. (9 Nm). Install the remaining components.

53. Fill the radiator with engine coolant.

54. Connect the negative battery cable to the battery.

55. Start the engine and check for leaks.

56. Check the ignition timing.

57. Install the engine undercover.

58. Road test the vehicle.

59. Recheck all fluid levels.

Piston and Connecting Rod

POSITIONING

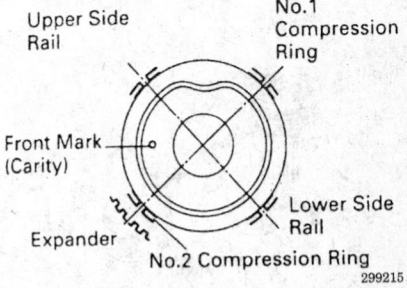

Positioning of the piston rings — 2.4L (2RZ-FE) and 2.7L (3RZ-FE) engines

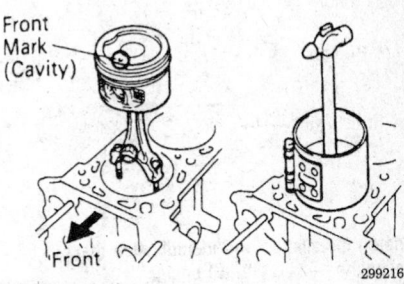

Installing the piston — 2.4L (2RZ-FE), 2.7L (3RZ-FE) and 3.4L (5VZ-FE) engines

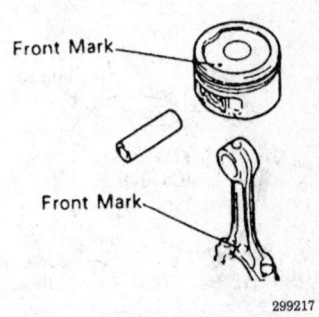

Aligning the piston to the connecting rod — 2.4L (2RZ-FE), 2.7L (3RZ-FE) and 3.4L (5VZ-FE) engines

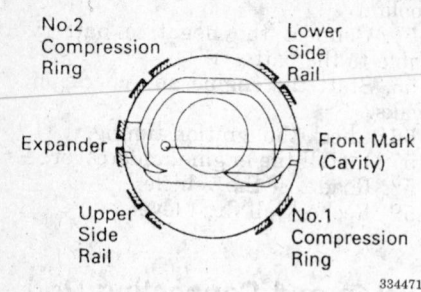

Piston ring positioning — 3.0L (3VZ-E) engine

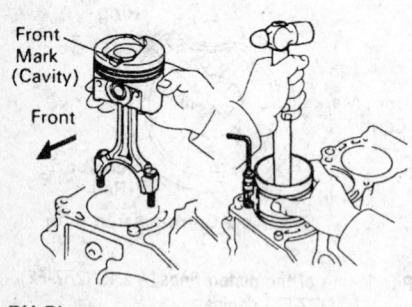

Piston installation and identification mark locations — 3.0L (3VZ-E) engine

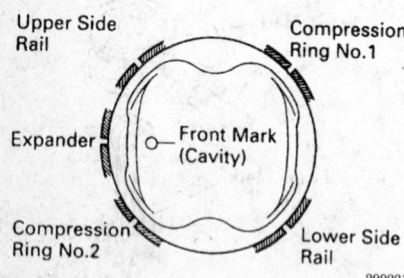

Positioning of the piston rings — 3.4L (5VZ-FE) engine

ENGINE LUBRICATION

Oil Pan

REMOVAL AND INSTALLATION

1993–94 Pick-up and 4Runner

1. Disconnect the negative battery cable.
2. Raise and safely support the vehicle.
3. Drain the engine oil.
4. Remove the steering relay rod and the tie rods from the idler arm, pitman arm, and steering knuckles.
5. Remove the engine stiffening plates.
6. Remove the splash pans from under the engine.
7. Position a floor jack under the transmission and raise the engine/transmission assembly slightly.
8. Remove the front motor mount attaching bolts.
9. Remove the oil pan bolts and remove the oil pan.
To install:
10. Before installing, thoroughly clean the cylinder block and oil pan mating surfaces of any old sealing material. Apply a 5mm bead of gasket sealer to the oil pan.

NOTE: Assemble parts within 5 minutes of applying the sealer.

11. Install the oil pan. Alternately tighten the oil pan bolts to 9 ft. lbs. (12 Nm), starting in the middle of the pan and working out towards the ends.
12. Lower the engine and tighten the motor mount bolts. Install the splash shields and stiffening plate.
13. Install the remaining components.
14. Lower the vehicle.
15. Fill the engine with oil. Start the engine and check for leaks.

1993–94 T-100

3.0L (3VZ-E) Engine

1. Disconnect the negative battery cable.
2. Remove the engine undercover (4-Wheel Drive vehicles).
3. Remove the front differential.
4. Remove the oil pan fasteners and separate the oil pan from the baffle plate.

To install:
5. Apply seal packing (PN 08826–00080 or equivalent) to the oil pan. Install the fasteners and torque to 54 inch lbs. (5.9 Nm).
6. Install the front differential. Install the engine undercover, if required.

1994 T-100

2.7L (3RZ-FE) Engine

1. Disconnect the negative battery cable from the battery.
2. Raise and support the vehicle safely.
3. Remove the engine undercover.
4. Drain the engine oil.
5. Remove the 16 mounting bolts and two nut to the oil pan.
6. Remove the oil pan from the engine.

NOTE: Be careful not to damage the oil pan flanges of the oil pan and cylinder block.

To install:
7. Before installing the oil pans, thoroughly clean the contact surfaces.
8. Apply sealant (PN 08826–00080 or equivalent) to the oil pan as shown.

NOTE: Parts must be assembled within 5 minutes of application. Otherwise the material must be removed and reapplied.

9. Install the oil pan and mounting bolts. Torque the bolts and nuts to 9 ft. lbs. (13 Nm).
10. Install the engine undercover. Lower the vehicle.
11. Fill the engine with oil. Connect the negative battery cable, start the engine and check for leaks.

1995 Pick-up, 1995–97 T-100, Tacoma and 4Runner

1. Disconnect the negative battery cable.

— **CAUTION** —
Wait 90 seconds from the time the key is turned to LOCK and the negative battery cable is disconnected to begin work. This allows the SRS capacitor to discharge and prevent deployment of the air bag(s).

2. Raise and safely support the vehicle.
3. Drain the engine oil.
4. Remove the engine undercover.
5. If equipped with 4-Wheel Drive, remove the front differential.
6. Remove the oil pan by removing the bolts and nuts.

Oil Pump

REMOVAL AND INSTALLATION

2.4L (22R-E) Engine

1993–94 Models

1. Raise and safely support the vehicle.
2. Drain the engine oil. Remove the oil pan, oil strainer and pick-up tube.
3. Remove the drive belts from the crankshaft pulley.
4. Remove the crankshaft bolt, and remove the pulley with a gear puller.
5. Remove the 5 bolts from the oil pump and remove the oil pump assembly.
6. Inspect the drive spline, driven gear, pump body, and timing chain cover for excessive wear or damage. If necessary, replace the gears or pump body or cover. Unbolt the relief valve (the vertical bolt on the pump body) when attached to the engine) and check the pistons, oil passages, and sliding surfaces for burrs or scoring. Inspect the crankshaft front oil seal and replace if worn or damaged.

To install:

7. Install the pump assembly, using a new O-ring.
8. Apply a sealer to the upper bolt and install the 5 bolts. Install the oil strainer and oil pan. Be sure to apply sealer to the corners of the oil pan gasket before installing the pan.
9. Install the crankshaft pulley and drive belts.
10. Connect the negative battery cable. Start the engine and check for leaks.

1995 Models

1. Disconnect the negative battery cable.

———— **CAUTION** ————
Work must be started after 90 seconds from the time the ignition switch is turned to the LOCK position and the negative (-) battery cable is disconnected.

2. Raise and safely support the vehicle.
3. Drain the engine oil.
4. Remove the oil pan, oil strainer, and the pick-up tube.
5. Remove the drive belts.
6. Remove the crankshaft bolt, and remove the pulley with a gear puller.
7. If equipped with air conditioning, remove the air conditioning compressor and the bracket.

8. Loosen the oil pump relief valve plug and remove the oil pump assembly.
9. Remove the oil pump drive spline. If the drive spline cannot be removed by hand, use SST 09213–36020, or equivalent, and remove the spline and the crankshaft timing sprocket together.

NOTE: If the SST is required to remove the oil pump drive spline, the timing cover and timing chain will have to be removed first.

To install:

10. Slide the pump drive spline onto the crankshaft and place a new O-ring into the groove.

NOTE: Prior to installing the oil pump, lubricate the pump drive and driven gears with clean engine oil.

11. Install the pump assembly. Torque the bolts to:
   a. A: 18 ft. lbs. (25 Nm)
   b. B: 14 ft. lbs. (19 Nm)
   c. C: 9 ft. lbs. (13 Nm)
12. Tighten the relief valve plug to 27 ft. lbs. (37 Nm).
13. Install the crankshaft pulley and torque the bolt to 116 ft. lbs. (157 Nm).
14. If removed, install the air conditioning compressor and the bracket.
15. Install and adjust the drive belts.
16. Install the oil strainer.
17. Install the oil pan and torque the nuts and bolts to 3.8 ft. lbs. (5.9 Nm).
18. Refill the engine with oil.
19. Connect the negative battery cable.
20. Start the engine, check oil pressure, and check for leaks.

2.4L (2RZ-FE) Engine

Tacoma

NOTE: The oil pump assembly is mounted in the timing chain cover. To properly service the oil pump, the timing chain cover should be removed from the cylinder block.

1. Disconnect the negative battery cable.
2. Drain the oil and the cooling system.
3. Raise and safely support the vehicle.
4. Remove the engine undercover.
5. If equipped with 4-Wheel Drive, remove the front differential and halfshaft assembly.

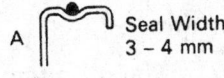

A ⎍ Seal Width 3 – 4 mm

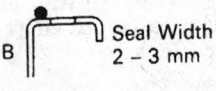

B ⎍ Seal Width 2 – 3 mm

293821

Apply sealant to the oil pan as shown — 1994 T-100 with 2.7L (3RZ-FE) engine

7. Using SST 09032–00100 or equivalent and a brass bar, separate the oil pan from the cylinder block.

To install:

8. Apply seal packing to the oil pan and install the pan to the cylinder block. Torque the nuts and bolts to:
   • Tacoma and T-100: 5.6 ft. lbs. (7.6 Nm)
   • 1995 4Runner and Pick-up: 4.3 ft. lbs. (5.9 Nm)
   • 1996–97 4Runner: 9 ft. lbs. (12.5 Nm)
9. If parts are not assembled within five minutes of applying time, the effectiveness of the seal packing is lost and must be removed and reapplied.
10. If removed, install the front differential.
11. Install the engine undercover.
12. Lower the vehicle.
13. Fill with engine oil.
14. Connect the negative battery cable.
15. Start the engine and check for leaks.

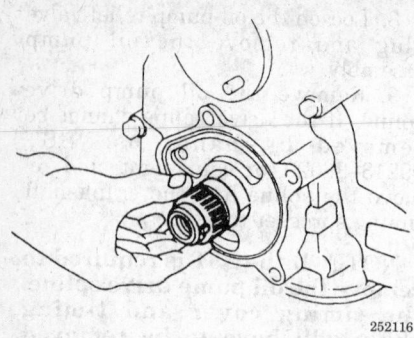

Oil pump drive spline — 1995 with 2.4L (22R-E) engine

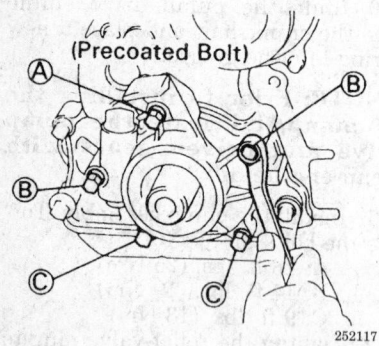

Oil pump bolt tightening — 1995 with 2.4L (22R-E) engine

6. Disconnect the upper radiator hose from the radiator.

7. Remove the oil dipstick guide by removing the bolt.

8. If equipped with power steering, remove the power steering drive belt by loosening the lock bolt and adjusting bolt.

9. Remove the No. 2 fan shroud by removing the two clips.

10. Remove the No. 1 fan shroud by removing the four bolts.

11. If equipped with air conditioning, loosen the idler pulley nut and adjusting bolt and remove the drive belt from the engine.

12. Remove the alternator drive belt, fan (with fan clutch), water pump pulley, and the fan shroud.

13. Remove the cylinder head from the engine.

14. If equipped with air conditioning, disconnect the air conditioning compressor and bracket.

15. Remove the alternator, adjusting bar and bracket.

16. Remove the crankshaft position sensor by removing the two bolts.

17. If equipped with 2-Wheel Drive, remove the stiffener plates by removing the eight bolts.

18. Remove the flywheel housing undercover and dust seal.

19. Remove the oil pan by removing the 16 mounting bolts and 2 nuts.

NOTE: Be careful not to damage the flanges of the oil pan and cylinder block.

20. Remove the two bolts, two nuts, oil strainer, and gasket.

21. Remove the crankshaft pulley.

22. Remove the timing chain cover.

23. Disassemble the oil pump from the front cover by removing the nine screws, pump cover, drive rotor, driven rotor and O-ring.

24. Remove the relief valve as follows:

    a. Using snapring pliers, remove the snapring for the relief valve.

    b. Remove the retainer, spring(s) and relief valve from the front cover.

To install:

25. Install the relief valve as follows:

    a. Install the relief valve, spring(s) and retainer to the valve cover.

    b. Using snapring pliers, install the snapring to hold the relief valve.

26. Install the drive and driven rotors as follows:

    a. Place the drive and driven rotors into the pump body.

    b. Place a new O-ring to the pump body.

    c. Install the pump cover with the nine screws.

27. Install the remaining components in the reverse order of removal.

28. Lower the vehicle.

29. Fill the cooling system and fill the engine with oil.

30. Connect the negative battery cable.

31. Start the engine and check for leaks.

32. Adjust ignition timing. Road test the vehicle for proper operation.

33. Recheck all fluid levels.

2.7L (3RZ-FE) Engine

1996–97 4Runner

1. Disconnect the negative battery cable.

--- **CAUTION** ---
Wait 90 seconds from the time the key is turned to LOCK and the negative battery cable is disconnected to begin work. This allows the SRS capacitor to discharge and prevent deployment of the air bag(s).

2. Remove the cylinder head assembly.

3. Remove the water inlet and housing.

4. Remove the timing chain cover.

5. Remove the nine screws and separate the oil pump from the timing chain cover.

To install:

6. Install the oil pump assembly to the timing chain cover and tighten the nine screws.

7. Install the timing chain cover.

8. Install the two water bypass pipe mounting nuts and torque to 14 ft. lbs. (20 Nm).

9. Install the cylinder head assembly.

1995–97 T-100

NOTE: The oil pump assembly is mounted in the timing chain cover. To properly service the oil pump, the timing chain cover should be removed from the cylinder block.

1. Disconnect the negative battery cable.

2. Drain the oil and cooling system.

3. Remove the cylinder head from the engine.

4. Raise and safely support the vehicle.

5. Remove the engine undercover by removing the four bolts.

6. If equipped with air conditioning, loosen the idler pulley nut and adjusting bolt and remove the drive belt from the engine.

7. Remove the alternator drive belt, fan (with fan clutch), water pump pulley, and fan shroud.

8. If equipped with air conditioning, remove the air conditioning compressor and bracket. Set the compressor aside with the lines attached.

9. Remove the alternator, adjusting bar, and bracket.

10. Remove the crankshaft position sensor by removing the two bolts.

11. Remove the oil pan by removing the 16 mounting bolts and 2 nuts.

NOTE: Be careful not to damage the flanges of the oil pan and cylinder block.

12. Remove the two bolts, two nuts, oil strainer and gasket.

13. Remove the crankshaft pulley.

14. Remove the timing chain cover.

15. Disassemble the oil pump from the front cover by removing the nine screws, pump cover, drive rotor, driven rotor and O-ring.

16. Remove the relief valve as follows:

    a. Using snapring pliers, remove the snapring for the relief valve.

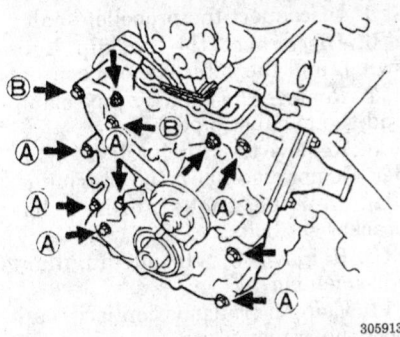

Timing cover bolt pattern — 1996–97 4Runner and T-100 with 2.7L (3RZ-FE) engine

b. Remove the retainer, spring and relief valve from the front cover.

To install:

17. Install the relief valve as follows:

a. Install the relief valve, spring and retainer to the valve cover.

b. Using a snapring pliers, install the snapring to hold the relief valve.

18. Install the drive and driven rotors as follows:

a. Place the drive and driven rotors into the pump body.

b. Place a new O-ring to the pump body.

c. Install the pump cover with the nine screws.

19. Install the timing chain cover.

20. Install the 2 rear timing chain cover mounting bolts and water by-pass pipe mounting nuts. Torque to 13 ft. lbs. (18 Nm).

21. Install the remaining components in the reverse order of the removal sequence.

22. Install the cylinder head.

23. Lower the vehicle.

24. Fill the cooling system. Fill the engine with oil.

25. Connect the negative battery cable.

26. Start the engine and check for leaks.

27. Adjust ignition timing. Road test the vehicle for proper operation.

28. Recheck all fluid levels.

Tacoma

NOTE: The oil pump assembly is mounted in the timing chain cover. To properly service the oil pump, the timing chain cover should be removed from the cylinder block.

1. Disconnect the negative battery cable.

2. Drain the oil and the cooling system.

3. Raise and safely support the vehicle.

4. Remove the engine undercover.

5. If equipped with 4-Wheel Drive, remove the front differential and halfshaft assembly.

6. For the California vehicles with 3RZ-FE engine, remove the two bolts and disconnect the air pipe.

7. Disconnect the upper radiator hose from the radiator.

8. Remove the oil dipstick guide by removing the bolt.

9. If equipped with power steering, remove the power steering drive belt by loosening the lock bolt and adjusting bolt.

10. Remove the No. 2 fan shroud by removing the two clips.

11. Remove the No. 1 fan shroud by removing the four bolts.

12. If equipped with air conditioning, loosen the idler pulley nut and adjusting bolt and remove the drive belt from the engine.

13. Remove the alternator drive belt, fan (with fan clutch), water pump pulley, and the fan shroud.

14. Remove the cylinder head from the engine.

15. If equipped with air conditioning, remove the air conditioning compressor and bracket. Set the compressor aside with the lines attached.

16. Remove the alternator, adjusting bar and bracket.

17. Remove the crankshaft position sensor by removing the two bolts.

18. If equipped with 2-Wheel Drive, remove the stiffener plates by removing the eight bolts.

19. Remove the flywheel housing undercover and dust seal.

20. Remove the oil pan by removing the 16 mounting bolts and 2 nuts.

NOTE: Be careful not to damage the flanges of the oil pan and cylinder block.

21. Remove the two bolts, two nuts, oil strainer, and gasket.

22. Remove the crankshaft pulley.

23. Remove the timing chain cover.

24. Disassemble the oil pump from the front cover by removing the nine screws, pump cover, drive rotor, driven rotor and O-ring.

25. Remove the relief valve as follows:

a. Using snapring pliers, remove the snapring for the relief valve.

b. Remove the retainer, spring(s) and relief valve from the front cover.

To install:

26. Install the relief valve as follows:

a. Install the relief valve, spring(s) and retainer to the valve cover.

b. Using snapring pliers, install the snapring to hold the relief valve.

27. Install the drive and driven rotors as follows:

a. Place the drive and driven rotors into the pump body.

b. Place a new O-ring to the pump body.

c. Install the pump cover with the nine screws.

28. Install the remaining components in the reverse order of removal.

29. Fill the cooling system and fill the engine with oil.

30. Connect the negative battery cable.

31. Start the engine and check for leaks.

32. Adjust ignition timing. Road test the vehicle for proper operation.

33. Recheck all fluid levels.

3.0L (3VZ-E) Engine

1. Disconnect the negative battery cable.

— **CAUTION** —

Work must be started after 90 seconds from the time the ignition switch is turned to the LOCK position and the negative (-) battery cable is disconnected.

2. Remove the engine undercover.

3. If equipped with 4-Wheel Drive, remove the front differential.

4. Remove the timing belt and crankshaft pulley.

5. Remove the oil pan, strainer, and the baffle plate. Be careful not to damage the baffle plate flange.

6. Remove the oil pump body.

7. Remove the O-ring from the cylinder block.

To install:

8. Apply seal packing (PN 08826–00080 or equivalent) to the oil pump. Place a new O-ring into the groove of the cylinder block.

9. Install the oil pump to the crankshaft with the spline teeth of the drive rotor engaged with the large teeth of the crankshaft. Torque the oil pump bolts to 14 ft. lbs. (20 Nm).

10. Apply seal packing to the baffle plate before installation.

11. Install the oil strainer and torque the bolts to 5 ft. lbs. (6.9 Nm).

12. Apply seal packing to the oil pan before installation, torque the nuts and bolts to 4.3 ft. lbs. (5.9 Nm).
13. Install the timing belt.
14. Install the crankshaft pulley and torque the bolt to 181 ft. lbs. (245 Nm).
15. If removed, install the differential.
16. Install the engine undercover.
17. Fill with engine oil.
18. Connect the negative battery cable.
19. Start the engine and check for leaks.

3.4L (5VZ-FE) Engine

1. Disconnect the negative battery cable.

---- **CAUTION** ----

Wait 90 seconds from the time the key is turned to LOCK and the negative battery cable is disconnected to begin work. This allows the SRS capacitor to discharge and prevent deployment of the air bag(s).

---

2. Remove the engine undercover.
3. Remove the crankshaft timing pulley.
4. If equipped with 4-Wheel Drive, remove the front differential.
5. Drain the engine oil from the engine.
6. Remove the timing belt and crankshaft gear.
7. If equipped with automatic transmission, remove the oil cooler tube and clamp.
8. Remove the stiffener plate.
9. Remove the flywheel housing undercover and dust cover.
10. Remove the rear end cover and dust cover.
11. Disconnect the starter wire clamp.
12. Remove the crankshaft position sensor.

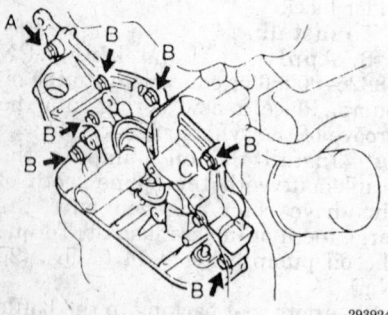

Oil pump bolts — 4Runner, Tacoma and T-100 with 3.4L (5VZ-FE) engine

13. Remove the oil pan by removing the 15 bolts and 4 nuts. Be careful not to damage the baffle plate flange.
14. Remove the oil strainer by removing the bolt and three nuts.
15. Remove the oil baffle plate by removing the nut and two bolts.
16. Remove the oil pump body by removing the eight bolts.
17. Remove the O-ring from the cylinder block.

To install:
18. Apply seal packing (PN 08826–00080 or equivalent) to the oil pump. Place a new O-ring into the groove of the cylinder block.
19. Install the oil pump to the crankshaft with the spline teeth of the drive rotor engaged with the large teeth of the crankshaft. Torque the oil pump bolts to:
• Bolt A to 15 ft. lbs. (20 Nm)
• Bolt B to 31 ft. lbs. (42 Nm)
20. Install the crankshaft position sensor.
21. Install the oil pan baffle plate.
22. Install the oil strainer with a new gasket and torque the bolts to 13 ft. lbs. (18 Nm).
23. Install the remaining components in the reverse order of removal.
24. Fill with engine oil.
25. Connect the negative battery cable.
26. Start the engine and check for leaks.

TRANSMISSION

Manual Transmission Assembly

REMOVAL AND INSTALLATION

1993–95 4Runner and Pick-up

2-Wheel Drive Models

1. Disconnect the negative battery cable.
2. Remove the fan shroud bolts.
3. Remove the transmission shift lever from the inside of the vehicle.
   a. Remove the screws holding the shift lever boot retainer and remove the shift lever boot.
   b. Cover the shift lever cap with a cloth. Pressing down on the shift lever cap, rotate it counterclockwise to remove it.
4. Raise and safely support the vehicle. Drain the transmission fluid.

5. Disconnect the propeller shaft.
6. Disconnect the backup light switch and the vehicle speed sensor.
7. Remove the exhaust pipe clamp and the exhaust pipe.
8. Remove the clutch release cylinder. Do not disconnect the clutch line.
9. Remove the stabilizer bar bracket set bolts.
10. Remove the frame auxiliary crossmember.
11. Using a transmission jack, support the transmission.
12. Remove the four bolts from the engine rear mounting.
13. Remove the mounting bracket.
14. Remove the engine rear mounting from the transmission.
15. Place a piece of wood between the engine oil pan and the front crossmember.
16. Lower the transmission.
17. Disconnect the wiring at the starter. Remove the starter mounting bolts and lower the starter out of the vehicle.
18. Remove the transmission bolts, draw the transmission rearward and down, away from the engine.

To install:
19. Raise the transmission into position under the vehicle.
   a. Install the extension housing between the member and the floor and then slide the transmission forward. Align the input shaft spline with the clutch disc, and install the transmission to the engine.
20. Install and torque the transmission, stiffener, and the starter bolts to:
   a. Transmission bolt: 53 ft. lbs. (72 Nm)
   b. Stiffener plate bolt: 27 ft. lbs. (37 Nm)
   c. Starter bolt: 29 ft. lbs. (39 Nm)
21. Install the engine rear mounting and the bracket. Torque the rear mounting bolts to 19 ft. lbs. (25 Nm).
   a. Raise the transmission slightly by raising the engine with a jack and a wooden block under the transmission.
   b. Install the engine rear mounting bracket to the support member and torque the four bolts to 43 ft. lbs. (59 Nm).
   c. Lower the transmission and rest it on the extension housing.
   d. Install the bracket to the mounting and torque the four bolts to 22 ft. lbs. (29 Nm).
22. Remove the piece of wood from the front crossmember.

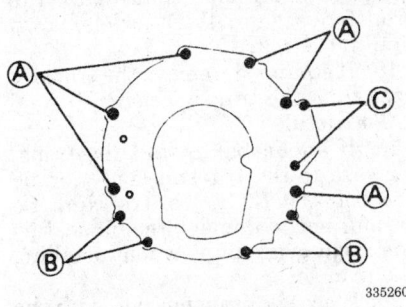

Transmission, stiffener and starter bolt locations — 1993–95 4Runner and Pick-up

23. Install the exhaust pipe, bracket, and the clamp.

a. Install the exhaust pipe and a new gasket to the exhaust manifold. Torque the three new nuts to 46 ft. lbs. (62 Nm).

b. Connect the exhaust pipe to the catalytic converter front side with a new gasket. Torque the two bolts to 29 ft. lbs. (39 Nm).

c. Install the pipe bracket to the clutch housing and torque the two bolts to 29 ft. lbs. (39 Nm).

d. Install and torque the exhaust pipe clamp set bolt to 14 ft. lbs. (19 Nm).

24. Install the clutch release cylinder and torque the two bolts to 9 ft. lbs. (12 Nm).

25. Install the stabilizer bar bracket set bolts and torque to 22 ft. lbs. (29 Nm).

26. Install the frame auxiliary crossmember and torque the bolts to 70 ft. lbs. (95 Nm).

27. Connect the vehicle speed sensor and the backup light switch connector.

28. Connect the propeller shaft.

29. Lower the vehicle.

30. Refill the transmission to the correct level.

31. Install the shift lever.

a. Apply MP grease to the shift lever.

b. Align the groove of the shift lever cap and the pin part of the case cover.

c. Cover the shift lever cap with a cloth. Pressing down on the shift lever cap, rotate it clockwise to install.

d. Install the shift lever boot and retainer with the screws.

32. Install the fan shroud set bolts.

33. Connect the negative battery cable. Start the engine and check for leaks.

34. Road test the vehicle for proper operation. Recheck all fluid levels.

4-Wheel Drive Models

1. Disconnect the negative battery cable.

2. Remove fan shroud set bolts.

3. Remove the heater hose clamp.

4. Remove the transmission shift lever from the inside of the vehicle.

a. Remove the screws holding the shift lever boot retainer and remove the shift lever boot.

b. Cover the shift lever cap with a cloth. Pressing down on the shift lever cap, rotate it counterclockwise to remove it.

5. Remove the transfer shift lever from the inside of the vehicle.

a. Using pliers, remove the snapring and pull out the shift lever from the transfer.

6. Raise the vehicle and drain the transmission and the transfer oil.

7. Remove the propeller shaft dust cover sub-assembly.

8. Disconnect the propeller shaft.

9. Disconnect the vehicle speed sensor, backup light switch connector, and the transfer indicator switch connector.

10. Remove the exhaust pipe bracket.

11. Remove the two bolts holding the clutch release cylinder and lay it along side the engine. Do not disconnect the clutch line.

12. Remove the front differential assembly.

13. Remove the stabilizer bar bracket set bolts.

14. Remove the No. 2 frame crossmember from the side frame.

a. Remove the four bolts from the engine rear mounting.

b. Using a transmission jack, support the transmission.

c. Remove the eight bolts and the No. 2 frame crossmember from the side frame.

15. Lower the transmission with the transfer case.

16. Remove the starter.

17. Remove the exhaust pipe bracket and the stiffener plate bolts.

18. Remove the remaining transmission bolts.

19. Remove the transmission with the transfer toward the rear of the vehicle.

20. Remove the four bolts and the engine rear mounting from the transfer.

21. For the regular cab with the planetary gear type transfer, remove the dynamic damper.

22. Remove the propeller shaft upper dust cover and the transfer from the transmission.

a. Remove the dust cover bolt from the bracket.

b. Remove the transfer adapter rear mounting bolts.

c. Pull the transfer straight up and remove it from the transmission.

NOTE: Take care not to damage the adapter rear oil seal with the transfer input gear spline.

To install:

23. Install the transfer and propeller shaft upper dust cover to the transmission with a new gasket.

a. Shift the two shift fork shafts to the high–four position.

b. Apply MP grease to the adapter oil seal.

c. Place a new gasket to the transfer adapter. Install the transfer to the transmission. Be careful not to damage the oil seal by the input gear spline when installing the transfer.

d. Install and torque the bolts with the propeller shaft upper dust cover to 27 ft. lbs. (37 Nm).

e. Install the dust cover bolt to the bracket. Torque to 17 ft. lbs. (23 Nm).

24. Install the engine rear mounting and torque the four bolts to 19 ft. lbs. (25 Nm).

25. If removed, install the dynamic damper and torque to 27 ft. lbs. (37 Nm).

26. Place the transmission with the transfer at the installation position.

a. Support the transmission with a jack. Align the input shaft spline with the clutch disc, and push the transmission with the transfer fully into position.

27. Install and torque the transmission, stiffener and the starter bolts. Torque to:

a. Transmission bolt: 53 ft. lbs. (72 Nm)

b. Stiffener plate bolt: 27 ft. lbs. (37 Nm)

c. Starter bolt: 29 ft. lbs. (39 Nm)

28. Install the No. 2 frame crossmember.

a. Raise the transmission slightly with a jack.

b. Install the No. 2 frame crossmember to the side frame and torque the eight bolts to 70 ft. lbs. (95 Nm).

c. Lower the transmission and transfer.

d. Install the four mounting bolts to the engine rear mounting and torque the bolts to 9 ft. lbs. (13 Nm).

29. Install the stabilizer bar bracket bolts and torque to 22 ft. lbs. (29 Nm).

30. Install the front differential assembly.

a. Install and torque the three bolts holding the differential carrier cover to the frame to 108 ft. lbs. (147 Nm) and the others to 123 ft. lbs. (167 Nm).

31. Install a new gasket and the exhaust pipe to the exhaust manifold. Torque the three new nuts to 46 ft. lbs. (62 Nm).

32. Install the exhaust pipe bracket to the clutch housing and torque the two bolts to 29 ft. lbs. (39 Nm).

33. Install the exhaust pipe clamp set bolt to 14 ft. lbs. (19 Nm).

34. Install the exhaust pipe and a new gasket to the catalytic converter front side. Torque the two bolts to 29 ft. lbs. (39 Nm).

35. Install the clutch release cylinder and torque the two bolts to 9 ft. lbs. (12 Nm).

36. Install the propeller shaft dust cover sub-assembly.

a. Install the cover and torque the bolt **A** to 27 ft. lbs. (36 Nm) and bolt **B** to 17 ft. lbs. (23 Nm).

37. Connect the vehicle speed sensor, the backup light switch connector and the transfer indicator switch connector.

38. Connect the propeller shaft.

39. Refill the transmission and transfer case to the correct level with the appropriate fluid.

40. Lower the vehicle.

41. Apply MP grease to the transfer shift lever and install the lever. Using pliers, install the snapring.

42. Install the transmission shift lever.

a. Apply MP grease to the transmission shift lever.

b. Align the groove of the shift lever cap and the pin par of the case cover. Cover the shift lever cap with a cloth. Pressing down on the shift lever cap, rotate it clockwise to install.

43. Install the heater hose clamp.

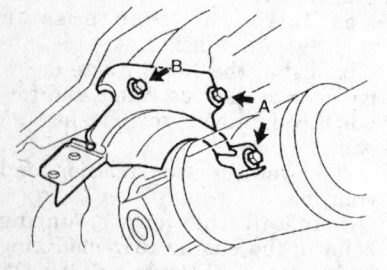

335269

Propeller shaft dust cover sub-assembly attaching bolts — 1993–95 4Runner and Pick-up

44. Install the fan shroud set bolts.

45. Connect the negative battery cable. Start the engine and check for leaks.

46. Road test the vehicle for proper operation. Recheck all fluid levels.

1996–97 4Runner

W59 and R150 Transmissions

1. Disconnect the negative battery cable from the battery.

2. Remove the transmission shift lever from the inside of the vehicle.

a. Remove the four screws and front console box.

b. Remove the screws holding the shift lever boot retainer and remove the shift lever boot.

c. Cover the shift lever cap with a cloth. Pressing down on the shift lever cap, rotate it counterclockwise to remove it.

3. On 4-wheel drive models, remove the transfer shift lever from the inside of the vehicle.

a. Using snapring pliers, remove the snapring and pull out the shift lever from the transfer case.

4. Raise the vehicle and drain the transmission and the transfer oil.

5. Remove the No. 1 and No. 2 engine undercover.

6. Disconnect the front and rear driveshafts.

7. Disconnect the vehicle speed sensor, backup light switch, and the 4-Wheel Drive position switch connectors.

8. If equipped with ABS and/or Differential lock: Disconnect the L4 position switch connector.

9. Remove the two bolts holding the clutch release cylinder and lay it along side the engine. Do not disconnect the clutch line.

10. Remove the exhaust pipe bracket.

11. Remove the rear end plate by removing the nuts and two bolts.

12. Remove the crossmember.

a. Support the transmission rear side.

b. Remove the four bolts from the engine rear mounting.

c. Remove the four bolts, nuts and the crossmember.

13. Remove the four bolts and the engine rear mounting from the transfer case.

14. Using a transmission jack, support the transmission.

15. Disconnect the wiring and the connector and remove the starter.

16. Disconnect the wire harness from the transmission.

17. Remove the transmission mounting bolts from the engine and

lower the transmission with the transfer case (on 4WD models) down and to the rear.

18. If equipped, remove the transfer case from the transmission.

To install:

19. If applicable, install the transfer case to the transmission. Torque the bolts to 17 ft. lbs. (24 Nm). Be careful not to damage the oil seal by the input gear spline when installing the transfer.

20. Place the transmission with the transfer case (4WD models) at the installation position.

21. Support the transmission with a jack. Align the input shaft spline with the clutch disc, and push the transmission with the transfer case fully into position.

22. Install the engine to transmission bolts and torque the bolts to 53 ft. lbs. (72 Nm)

23. Install the starter by installing the two bolts and connecting the electrical wiring. Torque the two bolts to 29 ft. lbs. (39 Nm).

24. Install the engine rear mounting and torque the four bolts to 48 ft. lbs. (65 Nm).

25. Install the crossmember.

a. Raise the transmission slightly with a jack.

b. Install the crossmember and torque the four bolts to 48 ft. lbs. (65 Nm).

c. Lower the transmission and transfer.

d. Install the four mounting bolts to the engine rear mounting and torque the bolts to 14 ft. lbs. (19 Nm).

26. Install the rear end plate by installing the four bolts and nuts. Torque the bolts to 27 ft. lbs. (37 Nm).

27. Install the front exhaust pipe. Tighten the bracket bolts to 52 ft. lbs. (71 Nm), the exhaust pipe-to-catalytic converter bolts to 35 ft. lbs. (48 Nm) and the support bracket bolts to 14 ft. lbs. (19 Nm).

28. Install the clutch release cylinder and torque the two bolts to 9 ft. lbs. (12 Nm).

29. If disconnected, connect the L4 position switch connector.

30. Connect the vehicle speed sensor, backup light switch, and the 4-Wheel Drive position switch connectors.

31. Connect the front and rear driveshafts.

32. Install the No. 1 and No. 2 engine under covers.

33. Refill the transmission to the correct level.

34. Apply MP grease to the transfer shift lever and install the lever. Us-

ing snapring pliers, install the snapring.

35. Install the transmission shift lever.

   a. Apply MP grease to the transmission shift lever.

   b. Align the groove of the shift lever cap and the pin par of the case cover. Cover the shift lever cap with a cloth. Pressing down on the shift lever cap, rotate it clockwise to install.

   c. Install shift lever boot retainer with the four screws.

   d. Install the front console box with the four screws.

36. Connect the negative battery cable, start the engine, and check for leaks.

37. Road test the vehicle for proper operation and recheck all the fluid levels.

1993–94 T-100

2-Wheel Drive Models

1. Disconnect the negative battery cable from the battery.
2. Remove the engine and transmission as an assembly.
3. Remove the right side stiffener plate by removing the four bolts.
4. Remove the left side stiffener plate by removing the four bolts.
5. Remove the transmission mounting bolts from the engine.
6. Pull out the transmission toward the rear.

To install:

7. Connect the transmission to the engine and install the bolts. Torque the bolts to 53 ft. lbs. (72 Nm).
8. Install the left stiffener plate and install the four bolts. Torque the bolts to 27 ft. lbs. (37 Nm).
9. Install the right stiffener plate and install the four bolts. Torque the bolts to 27 ft. lbs. (37 Nm).
10. Install the engine and transmission to the vehicle.
11. Connect the negative battery cable to the battery.
12. Check all fluids.

4-Wheel Drive Models

1. Disconnect the negative battery cable from the battery.
2. Remove the transmission shift lever from the inside of the vehicle as follows:

   a. Remove the four screws and remove the shift lever boot retainer.

   b. Pull up the shift lever boot.

   c. Cover the shift lever cap with a cloth.

   d. Press down on the shift lever cap and rotate it counterclockwise to remove it.

   e. Remove the shift lever from the vehicle.

3. Remove the transfer shift lever from inside the vehicle by pulling up the boot and removing the snapring.
4. Raise and safely support the vehicle.
5. Disconnect the front and rear driveshaft from the vehicle.
6. Disconnect the vehicle speed sensor connector and remove the wire harness from the transmission.
7. Disconnect the backup light switch connector.
8. Remove the two bolts and push the clutch release cylinder away from the transmission.
9. Remove the wire bracket from the transmission by removing the bolt.
10. Remove the starter by removing the wiring and two bolts.
11. Remove the front exhaust pipe from the vehicle.
12. Remove the right stiffener plate by removing the four bolts.
13. Remove the left stiffener plate by removing the four bolts.
14. Remove the sway bar bracket set bolts.
15. Remove the crossmember as follows:

   a. Remove the four bolts from the engine rear mounting.

   b. Raise the transmission slightly with a jack.

   c. Remove the eight bolts and the crossmember.

16. Remove the transmission mounting bolts from the engine.
17. Pull out the transmission toward the rear and remove it from the vehicle.

To install:

18. Install the transmission to the engine and install the six bolts. Torque the bolts to 53 ft. lbs. (72 Nm).
19. Lift the transmission slightly with a jack and install the crossmember with the eight bolts. Torque the bolts to 9 ft. lbs. (13 Nm).
20. Install the sway bar bracket set bolts and torque the bolts to 22 ft. lbs. (30 Nm).
21. Install the right side stiffener plate with the four bolts. Torque the bolts to 27 ft. lbs. (37 Nm).
22. Install the left side stiffener plate with the four bolts. Torque the bolts to 27 ft. lbs. (37 Nm).
23. Install the front exhaust pipe to the vehicle.
24. Install the starter motor.
25. Install the remaining components in the reverse order of removal.

26. Connect the negative battery cable to the battery.
27. Check all fluids and start the engine.
28. Perform a road test.

1995–97 T-100 with Model R150 and R150F Transmission

2-Wheel Drive Models

NOTE: The transmission is removed with the engine.

1. Turn the ignition switch **OFF**. Disconnect the battery cables; negative cable first.

--------- **WARNING** ---------

The air bag system is equipped with a backup power source. To avoid possible air bag deployment, do not start working on the vehicle until 90 seconds has elapsed from the time the ignition switch is turned OFF and the negative battery terminal is disconnected.

2. Matchmark the hood hinges and remove the hood.
3. Remove the battery and battery tray.
4. Drain the engine oil, transmission oil and cooling system.
5. Remove the expansion tank.
6. Remove the radiator.
7. Remove the air cleaner cap, MAF meter and resonator. Remove the air cleaner case.
8. If equipped with a manual transaxle, disconnect the accelerator cable from the throttle body.
9. If equipped with a automatic transaxle, disconnect the accelerator and throttle cables from the throttle body.
10. Remove the intake air connector.
11. If equipped with air conditioning, remove the air conditioning compressor and bracket. Position the compressor aside with the A/C lines attached.
12. Disconnect the heater hoses at the cowl panel.
13. Disconnect the following hoses:
- Brake booster vacuum hose
- EVAP hose
- Two power steering hoses
- Fuel return hose
- Fuel inlet hose
14. Remove the power steering pump.
15. Disconnect the alternator wires from the alternator.
16. Disconnect the ECM wiring from the ECM.

17. Disconnect the engine wire and connectors from the vehicle as follows:

a. Disconnect the igniter connector.

b. Disconnect the ground strap from the cowl top panel.

c. Disconnect the four engine wire clamps.

d. Pull out the engine wire from the vehicle cabin.

18. If equipped with manual transmission, remove the shift lever assembly as follows:

a. Remove the shift lever knob.

b. Remove the four screws and shift lever boot.

c. Remove the six bolts, shift lever assembly and baffle.

19. Remove the sway bar.

20. Remove the driveshaft from the vehicle.

21. Disconnect the speedometer cable from the transmission.

22. Disconnect the front exhaust pipe from the exhaust manifold and catalytic converter.

23. If equipped with manual transmission, remove the clutch release cylinder.

24. If equipped with automatic transmission, remove the cross-shaft.

25. Disconnect the wires at the starter.

26. Position a jack and wooden block under the transmission and remove the rear engine mounting bracket.

27. Attach a suitable engine hoist to the engine hangers.

28. Remove the nuts and bolts from the engine mounts.

NOTE: Make sure the engine/transmission assembly is clear of all wiring and hoses.

29. Carefully lift the engine/transmission assembly out of the vehicle.

30. Safely support the engine/transmission assembly.

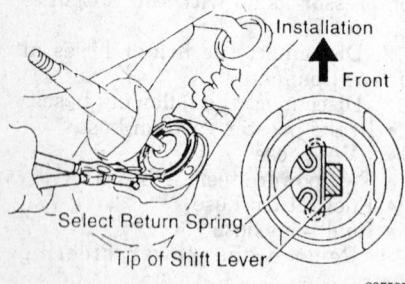

Removing the transfer shift lever — Tacoma, 1996–97 4Runner and 1995–97 T-100 with R150 and R150F transmissions

31. Remove the rear end plate by removing the nuts and four bolts.

32. Remove the starter by removing the two bolts.

33. Remove the transmission from the engine by removing the six mounting bolts from the engine.

34. Pull out the transmission toward the rear.

35. Remove the transmission mount by removing the four bolts.

To install:

36. Install the transmission mount by installing the four bolts. Torque the bolts to 18 ft. lbs. (25 Nm).

37. Connect the transmission to the engine by installing the six bolts. Torque the bolts to 53 ft. lbs. (72 Nm).

38. Install the starter and torque the two bolts to 29 ft. lbs. (39 Nm).

39. Install the rear end plate and torque the four nuts and bolts to 27 ft. lbs. (37 Nm).

40. Attach the engine hoist to the engine hangers. Carefully lower the engine/transmission assembly into the vehicle. Keep the engine level, while aligning the engine mounts.

41. Install the engine mount fasteners, but do not fully tighten.

42. Position a jack and wooden block under the transmission and install the rear engine mounting bracket. Torque the bolts to the frame to 42 ft. lbs. (58 Nm) and the bolts to the mount to 13 ft. lbs. (18 Nm).

43. Remove the jack and engine hoist. Torque the engine mounts to 28 ft. lbs. (38 Nm).

44. Connect the starter wires to the starter.

45. If equipped with manual transmission, install the clutch release cylinder.

46. If equipped with automatic transmission, install the cross-shaft.

47. Install the front exhaust pipe to the exhaust manifold. Tighten the exhaust pipe-to-manifold bolts to 46 ft. lbs. (62 Nm), the support bracket bolts and the exhaust pipe-to-catalytic converter bolts to 29 ft. lbs. (39 Nm), and the exhaust pipe clamp nuts to 14 ft. lbs. (19 Nm).

48. Install the remaining components in the reverse order of removal.

49. Fill the engine oil, engine coolant, and transmission oil.

50. Connect the negative and positive cables to the battery.

51. Start the engine and check for leaks.

52. Install the engine undercover.

53. Install the hood.

54. Road test the vehicle and check all fluids.

4-Wheel Drive Models

1. Disconnect the negative battery cable from the battery.

2. Remove the transmission shift lever from the inside of the vehicle.

a. Remove the four screws and front console box.

b. Remove the screws holding the shift lever boot retainer and remove the shift lever boot.

c. Cover the shift lever cap with a cloth. Pressing down on the shift lever cap, rotate it counterclockwise to remove it.

3. Remove the transfer shift lever from the inside of the vehicle.

a. Using snapring pliers, remove the snapring and pull out the shift lever from the transfer case.

4. Raise the vehicle and drain the transmission and the transfer oil.

5. Disconnect the driveshafts from the vehicle.

6. Disconnect the speedometer cable, backup light switch connector and the transfer indicator switch connector.

7. Remove the two bolts holding the clutch release cylinder and lay it along side the engine. Do not disconnect the clutch line.

8. Remove the exhaust pipe bracket.

9. Remove the starter by disconnecting the starter wires and removing the two bolts.

10. Remove the stiffener plate by removing the four bolts.

11. Remove the rear end plate by removing the nuts and two bolts.

12. Remove the stabilizer bar from the suspension.

13. Using a transmission jack, support the transmission.

14. Remove the frame crossmember from the side frame as follows:

a. Remove the four bolts from the engine rear mounting.

b. Remove the eight bolts and the frame crossmember from the side frame.

15. Remove the six transmission mounting bolts from the engine.

16. Disconnect the three wire clamps from the transmission.

17. Remove the transmission with the transfer toward the rear of the vehicle.

18. Remove the four bolts and the engine rear mounting from the transfer.

19. Remove the transfer adapter rear mounting bolts.

20. Pull the transfer straight up and remove it from the transmission.

To install:

21. Apply MP grease to the adapter oil seal and shift the two shift fork shafts to the high 4 position.

22. Install the transfer to the transmission. Install the bolts and torque the bolts to 27 ft. lbs. (37 Nm). Be careful not to damage the oil seal by the input gear spline when installing the transfer.

23. Install the engine rear mounting and torque the four bolts to 18 ft. lbs. (25 Nm).

24. Place the transmission with the transfer at the installation position.

25. Support the transmission with a jack. Align the input shaft spline with the clutch disc, and push the transmission with the transfer fully into position.

26. Install the engine to transmission bolts. Torque the bolts to 53 ft. lbs. (72 Nm)

27. Install the No. 2 frame crossmember.

   a. Raise the transmission slightly with a jack.

   b. Install the No. 2 frame crossmember to the side frame and torque the eight bolts to 70 ft. lbs. (95 Nm).

   c. Lower the transmission and transfer.

   d. Install the four mounting bolts to the engine rear mounting and torque the bolts to 9 ft. lbs. (13 Nm).

28. Install the stabilizer bar.

29. Install the rear end plate by installing the two bolts and nuts. Torque to 27 ft. lbs. (37 Nm).

30. Install the stiffener plate with the four bolts. Torque the bolts to 27 ft. lbs. (37 Nm).

31. Install the remaining components in the reverse order of removal.

32. Install the transmission shift lever.

   a. Apply MP grease to the transmission shift lever.

   b. Align the groove of the shift lever cap and the pin par of the case cover. Cover the shift lever cap with a cloth. Pressing down on the shift lever cap, rotate it clockwise to install.

   c. Install shift lever boot retainer with the four screws.

   d. Install the front console box with the four screws.

33. Connect the negative battery cable. Start the engine and check for leaks.

34. Road test the vehicle for proper operation. Recheck all fluid levels.

1995–97 T-100 and 2-Wheel Drive Tacoma

1. Remove the transmission with the engine.

2. Remove the left and right side stiffener plates.

3. Remove the rear end plate.

4. Disconnect the connector and remove the starter.

5. Place a stand under the transmission.

6. Remove the transmission mounting bolts and pull the transmission toward the rear.

7. Remove the rear engine mount by removing the four bolts.

To install:

8. Install the rear engine mount by installing the four bolts. Tighten the bolts to 48 ft. lbs. (65 Nm).

9. Install the transmission to the engine.

   a. Align the input shaft spline with the clutch disc and install the transmission to the engine. Torque the three bolts to 53 ft. lbs. (72 Nm).

10. Install the starter and torque the bolts to 29 ft. lbs. (39 Nm). Connect the connectors.

11. Install the rear end plate and torque the bolts to 27 ft. lbs. (37 Nm).

12. Install the left and right side stiffener plates.

13. Install the transmission with the engine assembly.

4-Wheel Drive Tacoma

1. Disconnect the negative battery cable from the battery.

2. Remove the transmission shift lever from the inside of the vehicle.

   a. Remove the four screws and front console box.

   b. Remove the screws holding the shift lever boot retainer and remove the shift lever boot.

   c. Cover the shift lever cap with a cloth. Pressing down on the shift lever cap, rotate it counterclockwise to remove it.

3. Remove the transfer shift lever from the inside of the vehicle.

   a. Using snapring pliers, remove the snapring and pull out the shift lever from the transfer case.

4. Raise the vehicle and drain the transmission and the transfer oil.

5. Disconnect the front and rear driveshafts.

6. Disconnect the speedometer cable and the backup light switch connector.

7. On the Standard cab: disconnect the 4-Wheel Drive position switch connector.

8. On the Extra cab: Disconnect the L4 position switch connector.

9. Remove the two bolts holding the clutch release cylinder and lay it along side the engine. Do not disconnect the clutch line.

10. Remove the exhaust pipe bracket.

11. Remove the starter by disconnecting the starter wires and removing the two bolts.

12. Remove the rear end plate by removing the nuts and two bolts.

13. Remove the crossmember.

   a. Support the transmission rear side.

   b. Remove the four bolts from the engine rear mounting.

   c. Disconnect the O-ring and remove the four bolts, nuts and the crossmember.

14. Using a transmission jack, support the transmission.

15. Remove the six transmission mounting bolts from the engine.

16. Disconnect the three wire clamps from the transmission.

17. Remove the transmission with the transfer toward the rear of the vehicle.

18. Remove the four bolts and the engine rear mounting from the transfer.

19. Remove the transfer adapter rear mounting bolts.

20. Pull the transfer straight up and remove it from the transmission.

To install:

21. Apply MP grease to the adapter oil seal and shift the two shift fork shafts to the high 4 position.

22. Install the transfer to the transmission. Install the bolts and torque the bolts to 17 ft. lbs. (24 Nm). Be careful not to damage the oil seal by the input gear spline when installing the transfer.

23. Place the transmission with the transfer at the installation position.

24. Support the transmission with a jack. Align the input shaft spline with the clutch disc, and push the transmission with the transfer fully into position.

25. Install the engine to transmission bolts. Torque the bolts to 53 ft. lbs. (72 Nm)

26. Install the engine rear mounting and torque the four bolts to 48 ft. lbs. (65 Nm).

27. Install the crossmember.

   a. Raise the transmission slightly with a jack.

   b. Install the crossmember and torque the four bolts to 48 ft. lbs. (65 Nm).

   c. Lower the transmission and transfer.

d. Install the four mounting bolts to the engine rear mounting and torque the bolts to 14 ft. lbs. (19 Nm).

28. Install the rear end plate by installing the four bolts and nuts. Torque to 13 ft. lbs. (18 Nm) on R150 and R150F transmissions, or to 27 ft. lbs. (37 Nm) on W59 transmissions.

29. Install the starter by installing the two bolts and connecting the electrical wiring. Torque the two bolts to 29 ft. lbs. (39 Nm).

30. Install the front exhaust pipe. Tighten the exhaust pipe-to-manifold bolts to 46 ft. lbs. (62 Nm), the exhaust bracket bolts to 33 ft. lbs. (44 Nm) and the exhaust pipe-to-catalytic converter bolts to 35 ft. lbs. (48 Nm).

31. Install the clutch release cylinder and torque the two bolts to 9 ft. lbs. (12 Nm).

32. Connect the L4 position switch connector on the extra cab and the 4-Wheel Drive position switch connector on the standard cab.

33. Connect the vehicle speed sensor and the backup light switch connector.

34. Connect the front and rear driveshafts.

35. Lower the vehicle.

36. Refill the transmission to the correct level.

37. Apply MP grease to the transfer shift lever and install the lever. Using pliers, install the snapring.

38. Install the transmission shift lever.

　a. Apply MP grease to the transmission shift lever.

　b. Align the groove of the shift lever cap and the pin par of the case cover. Cover the shift lever cap with a cloth. Pressing down on the shift lever cap, rotate it clockwise to install.

　c. Install shift lever boot retainer with the four screws.

　d. Install the front console box with the four screws.

39. Connect the negative battery cable. Start the engine and check for leaks.

40. Road test the vehicle for proper operation. Recheck all fluid levels.

Clutch Assembly

REMOVAL AND INSTALLATION

1993–94 Models

1. Raise and safely support the vehicle.
2. Remove the transmission from the vehicle.

3. Make matchmarks on the clutch cover and flywheel, indicating their relationship.

4. Loosen the clutch cover-to-flywheel retaining bolts one turn at a time. The pressure on the clutch disc must be released gradually.

5. Remove the clutch cover-to-flywheel bolts. Remove the clutch cover and the clutch disc.

6. If the clutch release bearing is to be replaced, perform the following:

　a. Remove the bearing retaining clip(s), the bearing and hub.

　b. Remove the release fork and the boot.

　c. The bearing is press fitted to the hub.

　d. Clean all parts and lightly grease the input shaft splines and all of the contact points.

　e. Install the bearing/hub assembly, the fork, the boot and the retaining clip(s) in their original locations.

To install:

7. Inspect the flywheel surface for cracks, heat scoring (blue marks) and warpage. Replace or resurface the flywheel, if any damage is present.

NOTE: Before installing any new parts, make sure they are clean. During installation, do not get grease or oil on any of the components, as this will shorten clutch life considerably.

8. Using an clutch alignment tool, position the clutch disc against the flywheel. The raised center section of the disc faces the transmission.

9. Install the clutch cover over the disc and install the bolts loosely. Align the pressure plate-to-flywheel matchmarks. If a new or rebuilt clutch cover assembly is installed, use the matchmark on the old cover assembly as a reference. Torque the pressure plate-to-flywheel bolts to 14 ft. lbs. (19 Nm), using a crisscross pattern.

10. Install the transmission into the vehicle.

1995–97 Models

1. Disconnect the negative battery cable from the battery.
2. Remove the transmission assembly from the vehicle.
3. Matchmark the clutch cover to the flywheel.
4. At the clutch cover, loosen each bolt one turn until spring tension is released.
5. Remove the set bolts to the clutch cover and pull off the clutch cover with the clutch disc.

6. Remove the retaining clip and withdraw the release bearing.

7. Remove the release fork and boot assembly.

To install:

8. Using a suitable clutch disc alignment tool, install the clutch disc onto the flywheel.

9. Position the clutch cover onto the flywheel and if reusing the old pressure plate, align the matchmarks.

10. Install the clutch cover retaining bolts. Torque the bolts in a crisscross pattern to 14 ft. lbs. (19 Nm).

11. Lubricate the release fork pivot and contact points, the release bearing, bearing hub, and input shaft spline surfaces with a suitable molybdenum disulfide lithium based or multi-purpose grease.

NOTE: Be careful not to apply too much grease or else it will get on the clutch disc and cause it to slip or grab.

12. Install the boot, release fork, hub, and the bearing assemblies.
13. Install the transmission to the vehicle.
14. Connect the negative battery cable to the battery.

Clutch Master Cylinder

REMOVAL AND INSTALLATION

All Models

1. Disconnect the negative battery cable.
2. Draw out fluid with a syringe from the reservoir tank.
3. Disconnect the master cylinder pushrod pin from the top of the clutch pedal.
4. Remove the hydraulic line from the master cylinder, being careful not to damage the compression fitting.
5. Remove the master cylinder mounting nuts.

To install:

6. Set the master cylinder into position and torque the nuts to 9 ft. lbs. (12 Nm).
7. Connect the hydraulic line to the master cylinder and torque to 11 ft. lbs. (15 Nm).
8. Install the pushrod and install the pin.
9. Fill reservoir tank with brake fluid.
10. Bleed the clutch system.
11. Connect the negative battery cable.
12. Adjust the pushrod play clearance.

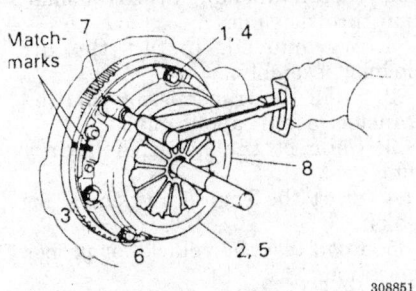

Tightening sequence for the clutch cover — 1995–97 models

Clutch Slave Cylinder

REMOVAL AND INSTALLATION

1993–94 Models

1. Raise and safely support the vehicle.
2. If equipped, remove the tension spring on the clutch fork.
3. Remove the hydraulic line from the release cylinder. Be careful not to damage the fitting.
4. Turn the release cylinder pushrod in sufficiently to gain clearance from the fork.
5. Remove the mounting bolts and withdraw the slave cylinder.
6. To install, reverse the removal procedures. Bleed the clutch system. Adjust the fork tip clearance.

1995–97 Models

1. Disconnect the negative battery cable from the battery.
2. Disconnect the fluid line with a line wrench. Plug the line to prevent oil from dripping. Dispose of fluid properly.
3. Remove the two slave cylinder retaining bolts.
4. Remove the clutch slave cylinder from the vehicle.
To install:
5. Place the clutch slave cylinder in place and install the two bolts. Torque the bolts to 9 ft. lbs. (12 Nm).
6. Connect the fluid line with a line wrench to the slave cylinder. Torque to 11 ft. lbs. (15 Nm).
7. Bleed the slave cylinder to remove any air present in the line.
8. Connect the negative battery cable to the battery.

Hydraulic Clutch System Bleeding

NOTE: If any maintenance on the clutch system was performed or the system is suspected of containing air, bleed the system. Use care; brake fluid will remove the paint from any surface. If the brake fluid spills onto any painted surface, wash it off immediately with soap and water.

1. Fill the clutch reservoir with brake fluid. Check the reservoir level frequently and add fluid as needed.
2. Connect one end of a vinyl tube to the bleeder plug on the slave cylinder and submerge the other end into a clear container half-filled with brake fluid.
3. Slowly pump the clutch pedal several times.
4. Have an assistant hold the clutch pedal down and loosen the bleeder plug until fluid and/or air starts to run out of the bleeder plug. Close the bleeder plug while the pedal is held to the floor.

NOTE: Do not allow the pedal to rise back up while the bleeder is still open. If this happens, it will allow air to re-enter the slave cylinder and cause the clutch system not to work properly.

5. Repeat Steps 2 and 3 until all the air bubbles are removed from the system.
6. Tighten the bleeder plug when all the air is gone.
7. Refill the master cylinder to the proper level as required.
8. Check the system for leaks.

Automatic Transmission Assembly

REMOVAL AND INSTALLATION

Pick-up and 4Runner

Model A340D, A340E, A340H and 1993–94 A340F Transmissions

NOTE: The transfer case and the transmission should be removed as an assembly.

1. Disconnect the negative battery cable.
2. If required, remove the air cleaner assembly.
3. Disconnect the transmission throttle cable from the throttle body.
4. Raise and safely support the vehicle.

5. Remove the engine undercover.
6. Drain the transmission and transfer case (if applicable) fluid.
7. Disconnect the wiring connectors from the transmission and transfer case (if applicable).
8. Disconnect the starter wiring at the starter. Remove the mounting bolts and the starter from the engine.
9. Make matchmarks on the front and rear driveshaft flanges and the differential pinion flanges. These marks must be aligned during installation.
10. Unbolt the front and rear driveshaft flanges. If the vehicle has a two piece driveshaft, remove the center bearing bracket bolts. Remove the driveshaft from the vehicle.
11. Disconnect the speedometer cable.
12. Remove the front exhaust pipe and the bracket.
13. Disconnect the transmission oil cooler lines at the transmission.
14. Disconnect the oil cooler lines bracket and remove the transmission oil filler tube, as required.
15. Support the transmission, using a jack with a wooden block placed between the jack and the transmission pan. Raise the transmission, just enough to take the weight off of the rear mount.
16. Remove the rear engine mount with the bracket, the rear crossmember, and the transfer case (if applicable) undercover.
17. Remove the dynamic damper (Regular Cab only) and the No. 2 cross-shaft bracket.
18. Place a wooden block(s) between the engine oil pan and the front frame crossmember.
19. Slowly, lower the transmission until the engine rests on the wooden block(s).
20. Remove the torque converter cover at the rear of the engine in order to gain access to the converter bolts.
21. Rotate the crankshaft to access the bolts through the service holes and remove the torque converter bolts.
22. Remove the stiffener plates from the transmission.
23. Disconnect the shift control rod and the transfer case shift lever.
24. For the A340H transmissions, perform the following:
  a. Remove the cross-shaft and the No. 2 shifting rod.
  b. Remove the front stabilizer bar.
  c. Support the front differential with a jack and remove the differential mounting bolts.

d. Slowly lower the front differential so there is enough clearance to remove the transmission and transfer case (if applicable). If enough clearance can't be obtained, remove the differential from the vehicle.

25. Remove the stabilizer bar and, if equipped, the auxiliary frame crossmember.

26. For A340D transmissions, obtain a bolt of the same dimensions as the torque converter bolts. Cut the head off of the bolt and hacksaw a slot in the bolt opposite the threaded end. Thread the guide pin into one of the torque converter bolt holes. The guide pin will help keep the converter with the transmission.

NOTE: This modified bolt is used as a guide pin. Two guides pins are needed to properly install the transmission.

27. Remove the transmission bolts, then carefully move the transmission rearward by prying on the dowel pins through the service hole.

28. Pull the transmission rearward and lower it out of the vehicle.

NOTE: Be careful not to drop the torque converter.

29. Separate the transfer case from the transmission

To install:

30. Connect the transfer case to the transmission.

31. Apply a coat of multi-purpose grease to the torque converter stub shaft and the corresponding pilot hole in the flexplate.

32. Install the torque converter into the front of the transmission. Push inward on the torque converter while rotating it to completely couple the torque converter to the transmission.

33. To make sure the converter is properly installed, measure the distance between the torque converter mounting lugs and the front mounting face of the transmission. The proper distance is 0.71 in. (18mm) for the A340H transmission, or 0.79 in. (20mm) for the A340D, A340E and A340F transmissions.

34. For A340D transmissions, install guide pins into 2 opposite mounting lugs of the torque converter.

35. Raise the transmission to the engine and align the transmission with the dowel pins.

36. Install and tighten the transmission mounting bolts to 47 ft. lbs. (63 Nm).

37. Rotate the crankshaft and install the torque converter mounting bolts. Evenly, tighten the converter

mounting bolts to 30 ft. lbs. (41 Nm) for the A340H, A3430D and A340E transmissions, or to 20 ft. lbs. (27 Nm) for the A340F transmission.

38. Install the torque converter access cover.

39. Raise the transmission slightly and remove the wood block(s) from under the engine oil pan.

40. Install the transmission crossmember. Torque the crossmember bolts to 70 ft. lbs. (95 Nm).

41. Install the rear mount and the mounting bracket. Tighten the bracket mounting bolts to 43 ft. lbs. (58 Nm) and tighten the bracket to the rear mount bolts to 9 ft. lbs. (13 Nm).

42. Lower the transmission onto the crossmember and install the transmission to mount bolts. Torque the bolts to 18 ft. lbs. (25 Nm).

43. Remove the wooden blocks from between the frame and the engine. Remove the support from under the transmission.

44. For the A340H transmission, install the front differential. Tighten the two rear mounting bolts to 123 ft. lbs. (167 Nm). Tighten the front mounting through-bolt to 108 ft. lbs. (147 Nm).

NOTE: If the differential oil was drained, refill it at this time.

45. Connect the shift control rod and the transfer case shift lever.

46. Install the front stabilizer bar, if applicable.

47. Install the cross-shaft and the No. 2 shifting rod, if applicable.

48. Install the stiffener plates and tighten the bolts to 27 ft. lbs. (37 Nm).

49. If equipped, install the transfer case undercover and the dynamic damper. Torque the dynamic damper mounting bolts to 27 ft. lbs. (37 Nm).

50. Install the No. 2 cross-shaft bracket.

51. Install the oil filler tube and the oil cooler pipe bracket.

52. Connect the oil cooler lines to the transmission and torque the fittings to 25 ft. lbs. (34 Nm).

53. Install the front exhaust pipe and the support bracket.

54. Connect the speedometer cable.

55. Align the matchmarks of the front and rear driveshaft flanges and the differential pinion flanges and torque the bolts to 54 ft. lbs. (74 Nm).

56. Install the starter and connect the wiring.

57. Connect the wiring connectors to the transmission and the transfer case (if applicable).

58. Install the engine undercover.

59. Lower the vehicle.

60. Install and adjust the transmission throttle cable.

61. If removed, install the air cleaner assembly.

62. Refill the transmission and the transfer case (if applicable).

63. Connect the negative battery cable.

64. Start the engine and check for leaks.

65. Road test the vehicle for proper operation.

66. Recheck all fluid levels.

1995–97 4Runner

Model A340F Transmission

1. Disconnect the negative battery cable.

2. Disconnect the throttle cable from the engine compartment.

3. Remove the ATF level gauge.

4. 3RZ-FE engine: Remove the oil filler pipe upper side bolt.

5. 5VZ-FE engine: Remove the oil filler pipe.

6. Remove the transmission shift lever assembly and transfer shift lever.

   a. Remove the rear console upper panel and disconnect the connectors.

   b. Pull off the heater control knobs.

   c. Remove the center cluster finish panel and disconnect the connectors.

   d. Without 2–4 selector: Remove the transfer shift lever knob.

   e. With 2–4 selector: Remove the bolt and disconnect the transfer shift lever knob.

   f. Remove the front console upper panel.

   g. If equipped, disconnect the 2–4 selector connector and remove the transfer shift lever knob.

   h. Remove the shift control rod.

   i. Disconnect the connector and remove the eight screws and the transmission shift lever assembly.

   j. Using pliers, remove the snapring and pull out the shift lever from the transfer.

7. Remove the engine undercover.

8. Remove the front and rear driveshafts.

9. Remove the exhaust pipe.

10. 3RZ-FE engine: Remove the oil filler pipe.

11. Disconnect the following connectors from the transmission:

• No. 2 vehicle speed sensor connector

• Solenoid connector

• ATF temperature sensor connector

• Park/neutral position switch connector

12. Disconnect the following connectors from the transfer:
- No. 1 vehicle speed sensor connector (3RZ-FE)
- Transfer neutral position switch connector
- Transfer L4 position switch connector
- Transfer 4-Wheel Drive position switch connector
- Actuator connector (w/ 2–4 selector only)

13. Separate the wire harness from the transmission and the transfer.

14. Disconnect the two oil cooler pipes.

15. Remove the rear end plate and torque converter clutch mounting bolt.

16. Support the transmission with a jack stand and remove the engine rear mounting bolts.

17. Remove the four bolts and the crossmember.

18. Disconnect the starter wire and remove the starter.

19. Remove the transmission.

To install:

20. Install the transmission in the vehicle. Torque the bolts to 53 ft. lbs. (71 Nm).

21. Install the starter and torque the bolts to 29 ft. lbs. (39 Nm). Connect the wiring for the starter.

22. Install the crossmember and torque the four bolts to 48 ft. lbs. (65 Nm). Install the four engine rear mounting bolts and torque to 14 ft. lbs. (19 Nm).

23. Install the clutch converter bolts by installing the green colored bolt before the other five. Torque them to 30 ft. lbs. (41 Nm).

24. Install the rear end plate and torque the bolts to 13 ft. lbs. (18 Nm).

25. Install the two oil cooler pipes and torque to 25 ft. lbs. (34 Nm).

26. Install the three bolts for the oil cooler pipe clamps and torque to:
- 10mm head bolt: 4 ft. lbs. (5 Nm)
- 12mm head bolt: 9 ft. lbs. (12 Nm)

27. Connect the harness from the transmission and the transfer.

28. Install the remaining components in the reverse order of removal.

29. Fill the transmission and transfer case with transmission fluid.

30. Connect the throttle cable to the cable clamps in the engine compartment.

31. Connect the negative battery cable.

1993–94 T-100

Model A340E Transmission

1. Disconnect the negative battery cable from the battery.

2. Remove the transmission level gauge.

3. Remove the engine undercover.

4. Disconnect the throttle cable from the engine.

5. Remove the transmission upper filler tube mounting bolt.

6. Raise and safely support the vehicle.

7. Remove the front exhaust pipe from the exhaust manifold and center pipe.

8. Remove the transmission lower filler tube by removing the bolt and pulling the tube from the transmission.

9. Remove the driveshaft from the vehicle.

10. Disconnect the No. 1 vehicle speed sensor electrical connector. After disconnecting the connector. remove the wire harness from the transmission.

11. Disconnect the No. 2 vehicle speed sensor connector and solenoid connector.

12. Remove the cross-shaft as follows:
   a. Remove the clip and disconnect the shifting rod from the transmission.
   b. Remove the nut holding the shift linkage to the cross-shaft.
   c. Remove the four bolts and the cross-shaft from the vehicle.

13. Remove the oil cooler tube as follows:
   a. Disconnect the transmission fluid temperature sensor.
   b. Disconnect the two oil cooler tubes from the transmission.
   c. Remove the bolt and front side tube clamp.
   d. Remove the bolt and the rear side clamp from the transmission.

14. Remove the control shaft lever.

15. Disconnect the park/neutral position switch connector.

16. Remove the starter by disconnecting the electrical connectors and removing the two bolts.

17. Remove the sway bar from the vehicle suspension.

18. Remove the left and right stiffener plates by removing the eight bolts.

19. Remove the converter cover.

20. Remove the throttle cable clamp.

21. Disconnect the rear mounting by removing the four bolts.

22. Using a transmission jack, raise the transmission slightly.

23. Remove the six torque converter bolts from the torque converter.

24. Disconnect the wire harness and connectors from the transmission.

25. Remove the engine components and remove the engine and transmission as an assembly.

26. Remove the six bolts and remove the transmission from the engine.

To install:

27. Connect the transmission to the engine with the six bolts. Torque the six bolts to 47 ft. lbs. (64 Nm).

28. Install the engine and transmission assembly into the vehicle. Install and connect all components for the engine.

29. Slightly jack up the transmission.

30. Connect the rear transmission mounting with the four bolts. Torque the bolts to 9 ft. lbs. (13 Nm).

31. Remove the transmission jack.

32. Install the torque converter bolts to the torque converter. Install the gray bolt first, then the five black bolts. Tighten the bolts evenly and then torque all the bolts to 30 ft. lbs. (41 Nm).

33. Install the converter cover.

34. Install the left and right stiffener plates with the eight bolts. Torque the bolts to 27 ft. lbs. (37 Nm).

35. Install the remaining components in the reverse order of removal. Make certain to tighten the oil cooler tube fittings to 25 ft. lbs. (34 Nm).

36. Fill the transmission with transmission fluid.

37. Connect the negative battery cable to the battery.

38. Start the vehicle and check all fluids.

Model A340F Transmission

1. Disconnect the negative battery cable.

2. Disconnect the transmission throttle cable and clamp from the throttle body.

3. Remove the engine undercover.

4. Drain the transmission fluid.

5. Remove the transfer shift lever from the inside of the vehicle as follows:
   a. Remove the shift lever knob.
   b. Remove the four screws and the boot.
   c. Using pliers, remove the snapring, and pull out the shift lever from the transfer case.

6. Raise and safely support the vehicle.

7. Remove the transmission oil filler tube.

8. Remove the front exhaust pipe.

9. Remove the front and rear driveshafts.

10. Disconnect the No. 1 vehicle speed sensor connector.

11. Disconnect the No. 2 vehicle speed sensor connector.

12. Disconnect the solenoid connector by removing the electrical connector and bolt.

13. Remove the clip and disconnect the No. 2 gear shifting rod from the transmission.

14. Remove the cross-shaft by removing the nut and four bolts.

15. Remove the transmission oil cooler pipes by removing the bolts and clamps.

16. Remove the control shaft lever.

17. Disconnect the park/neutral position switch connector.

18. Remove the left and right stiffener plates by removing the eight bolts.

19. Disconnect the starter wires.

20. Remove the starter from the engine by removing the two bolts.

21. Remove the stiffener plate and rear end plate by removing the eight bolts.

22. Remove the torque converter cover.

23. Remove the throttle clamp.

24. Remove the sway bar.

25. Support the transmission, using a jack with a wooden block placed between the jack and the transmission pan. Raise the transmission just enough to take the weight off of the rear mount.

26. Remove the rear engine mounting bracket.

27. Remove the rear support member by removing the eight bolts.

28. Remove the six bolts from the torque converter.

29. Disconnect or remove any component that will get in the way of removing the transmission.

30. Remove the transmission bolts, then carefully move the transmission rearward.

31. Pull the transmission rearward and lower it out of the vehicle.

To install:

32. Raise the transmission into place and install the bolts. Torque the transmission bolts to 47 ft. lbs. (64 Nm).

33. Install the torque converter bolts and torque the bolts to 30 ft. lbs. (41 Nm). Install the gray bolt first and then install the five black bolts.

34. Install the rear support member and torque the eight bolts to 70 ft. lbs. (97 Nm).

35. Install the rear mounting bracket and install the four bolts. Torque the bolts to 13 ft. lbs. (18 Nm).

36. Remove the jack supporting the transmission.

37. Install the sway bar.

38. Install the throttle cable clamp.

39. Install the torque converter plate.

40. Install the stiffener plate and rear end plate and torque the bolts to 27 ft. lbs. (37 Nm).

41. Install the starter and torque the bolts to 29 ft. lbs. (39 Nm).

42. Connect the starter wires.

43. Connect the park/neutral position switch connector.

44. Install the control shaft lever.

45. Install the oil cooler pipes and clamps. Torque the two oil cooler pipes to 25 ft. lbs. (34 Nm).

46. Install the cross-shaft with the four bolts, washer, and nut. Torque the nut to 9 ft. lbs. (13 Nm) and the bolts as follows:

• Transmission side: 9 ft. lbs. (13 Nm)

• Frame side: 21 ft. lbs. (28 Nm)

47. Install the remaining components in the reverse order of removal. Make sure to tighten the exhaust pipe-to-catalytic converter bolts to 29 ft. lbs. (39 Nm) and the 3 exhaust pipe nuts to 46 ft. lbs. (62 Nm).

48. Fill the transmission with the proper fluid.

49. Connect the negative battery cable to the battery.

50. Road test the vehicle and check for leaks.

51. Check all fluids.

1995–97 T-100

Model A340E Transmission with 2.7L (3RZ-FE) Engine

1. Turn the ignition switch **OFF**. Disconnect the battery cables; negative cable first.

WARNING

The air bag system is equipped with a backup power source. To avoid possible air bag deployment, do not start working on the vehicle until 90 seconds has elapse from the time the ignition switch is turned OFF and the negative battery terminal is disconnected.

2. Matchmark the hood hinges and remove the hood.

3. Remove the battery and battery tray.

4. Drain the engine oil, transmission oil, and the cooling system.

5. Remove the expansion tank.

6. Remove the radiator.

7. Remove the air cleaner cap, MAF meter, and the resonator. Remove the air cleaner case.

8. Disconnect the accelerator and throttle cables from the throttle body.

9. Remove the intake air connector.

10. If equipped with air conditioning, remove the air conditioning compressor and bracket. Position the compressor aside with the A/C lines attached.

11. Disconnect the heater hoses at the cowl panel.

12. Disconnect the following hoses:
• Brake booster vacuum hose
• EVAP hose
• Two power steering hoses
• Fuel return hose
• Fuel inlet hose

13. Remove the power steering pump.

14. Disconnect the alternator wires from the alternator.

15. Disconnect the ECM wiring from the ECM.

16. Disconnect the engine wire and connectors from the vehicle as follows:

a. Disconnect the igniter connector.

b. Disconnect the ground strap from the cowl top panel.

c. Disconnect the four engine wire clamps.

d. Pull out the engine wire from the vehicle cabin.

17. Remove the sway bar.

18. Remove the driveshaft from the vehicle.

19. Disconnect the speedometer cable from the transmission.

20. Disconnect the front exhaust pipe from the exhaust manifold and catalytic converter.

21. Remove the cross-shaft as follows:

a. Remove the clip and disconnect the No. 2 gear shifting rod.

b. Remove the nut, washer, four bolts, and the cross-shaft.

22. Disconnect the wires at the starter.

23. Position a jack and wooden block under the transmission and remove the rear engine mounting bracket.

24. Attach a suitable engine hoist to the engine hangers.

25. Remove the nuts and bolts from the engine mounts.

NOTE: Make sure the engine/transmission assembly is clear of all wiring and hoses.

26. Make sure all connectors, wires, and hoses are disconnected and away from the engine and transmission.

27. Carefully lift the engine/transmission assembly out of the vehicle.

28. Remove the rear end plate by removing the two nuts and four bolts.

29. Turn the crankshaft to gain access to the torque converter bolts.

30. Remove the torque converter bolts.

31. Remove the starter by removing the two bolts.

32. Remove the three mounting bolts from the engine.

33. Pull out the transmission toward the rear.

To install:

34. Connect the engine to the transmission. Install the three bolts and torque the bolts to 53 ft. lbs. (71 Nm).

35. Install the torque converter bolts to the torque converter. Torque the bolts to 30 ft. lbs. (41 Nm).

36. Install the rear end plate and torque the nuts and bolts to 27 ft. lbs. (37 Nm).

37. Connect the starter and bolts. Torque the bolts to 29 ft. lbs. (39 Nm).

38. Attach the engine hoist to the engine hangers. Carefully lower the engine/transmission assembly into the vehicle. Keep the engine level, while aligning the engine mounts.

39. Install the engine mount fasteners, but do not fully tighten.

40. Position a jack and wooden block under the transmission and install the rear engine mounting bracket. Torque the bolts to the frame to 42 ft. lbs. (58 Nm) and the bolts to the mount to 13 ft. lbs. (18 Nm).

41. Remove the jack and engine hoist. Torque the engine mounts to 28 ft. lbs. (38 Nm).

42. Connect the starter wires to the starter.

43. Install the cross-shaft.

44. Install the remaining components in the reverse order of removal. Make sure to tighten the exhaust pipe-to-exhaust manifold nuts to 46 ft. lbs. (62 Nm), the exhaust pipe support bracket bolts to 29 ft. lbs. (39 Nm), the exhaust pipe clamp nuts to 14 ft. lbs. (19 Nm), the exhaust pipe-to-catalytic converter bolts to 29 ft. lbs. (39 Nm), the sway bar mounting bolts to 22 ft. lbs. (30 Nm), the power steering pump-to-bracket bolts to 43 ft. lbs. (58 Nm), the A/C compressor bracket-to-engine bolts to 32 ft. lbs. (44 Nm), and the A/C compressor-to-bracket bolts to 18 ft. lbs. (25 Nm).

45. Fill the engine oil, engine coolant, and the transmission oil.

46. Connect the negative and positive cables to the battery.

47. Start the engine and check for leaks.

48. Check the ignition timing.

49. Install the engine undercover.

50. Install the hood.

51. Road test the vehicle and check all fluids.

Model A340E Transmission with 3.4L (5VZ-FE) Engine

1. Remove the hood.

2. Disconnect the battery and remove it from the vehicle.

——— CAUTION ———
Work must be started after 90 seconds from the time the ignition switch is turned to the LOCK position and the negative (-) battery cable is disconnected.

3. Raise and safely support the vehicle.

4. Remove the engine under covers.

5. Drain the engine coolant.

6. Drain the engine oil.

7. Remove the radiator from the vehicle.

8. Remove the power steering (PS) drive belt.

9. If equipped with air conditioning, remove the air conditioning drive belt by loosening the idler pulley nut and adjusting bolt.

10. Loosen the lock bolt, pivot bolt, and the adjusting bolt. Remove the alternator drive belt.

11. Remove the fan with the fluid coupling and fan pulleys.

12. Disconnect the PS pump from the engine and set aside. Do not disconnect the lines from the pump.

13. If equipped with air conditioning, disconnect the compressor from the engine and set aside. Do not disconnect the lines from the compressor.

14. Remove the air cleaner cap, MAF meter, and the resonator.

15. Remove the air cleaner case and filter.

16. Disconnect the following cables:
- If equipped with cruise control, disconnect the actuator cable with the bracket.
- Accelerator cable
- Throttle cable

17. Disconnect the heater hoses.

18. Disconnect the following hoses:
- Brake booster vacuum hose
- EVAP hose
- Fuel return hose
- Fuel inlet hose

19. Disconnect the starter wire and connectors.

20. Disconnect the alternator connector and wire.

21. Disconnect the engine wire and connectors as follows:
  a. Remove the four screws to the right front door scuff plate. Remove the scuff plate from the vehicle.
  b. Remove the cowl panel side trim by removing the clip.
  c. Disconnect the ECM electrical connectors.
  d. Disconnect the two connectors from the cowl wire.
  e. Disconnect the igniter connector.
  f. Disconnect the ground strap.
  g. Disconnect the six engine wire clamps.
  h. Pull out the engine wire from the cabin.

22. Remove the stabilizer bar.

23. Remove the driveshaft from the transmission.

24. Disconnect the speedometer cable.

25. Remove the front exhaust pipe.

26. Remove the cross-shaft.

27. Place a jack under the transmission.

28. Remove the transmission rear mounting bracket by removing the eight bolts.

29. If equipped with air conditioning, remove the bolt and disconnect the air conditioning compressor wire clamp.

30. If necessary, install a No. 2 engine hanger with two bolts. Torque the two bolts to 30 ft. lbs. (40 Nm).

31. Attach the engine hoist chain to the two engine hangers.

32. Remove the four bolts and nuts holding the engine front mounting insulators to the frame.

33. Lift the engine and transmission out of the vehicle.

34. Remove the starter by removing the two bolts.

35. Remove the transmission six mounting bolts from the engine.

36. If equipped, remove the rear end plate by removing the four bolts and four nuts.

37. Turn the crankshaft to gain access to the torque converter bolts.

38. Remove the torque converter bolts.

39. Pull out the transmission toward the rear.

40. Remove the three mounting bolt from the engine.

41. Pull out the transmission toward the rear.

To install:

42. Connect the engine to the transmission. Install the three bolts and torque the bolts to 53 ft. lbs. (71 Nm).

43. Install the torque converter bolts to the torque converter. Torque the bolts to 30 ft. lbs. (41 Nm).

44. Install the rear end plate and torque the nuts and bolts to 27 ft. lbs. (37 Nm).

45. Connect the starter and bolts. Torque the bolts to 29 ft. lbs. (39 Nm).

46. Install the transmission to the engine and install the six mounting bolts. Torque the bolts to 53 ft. lbs. (71 Nm).

47. Install the starter and install the two bolts. Torque the bolts to 29 ft. lbs. (39 Nm).

48. Install the engine assembly to the vehicle. Attach the engine mounts to the body mounts. Install the bolts and nuts but do not torque at this time.

49. Remove the engine chain hoist the No. 2 engine hanger.

50. If equipped with air conditioning, connect the air conditioning wire with the bolt.

51. Raise the transmission slightly and install the transmission mounting bracket. Torque the bolts to the frame to 43 ft. lbs. (58 Nm) and the bolts to the mounting insulator to 13 ft. lbs. (18 Nm).

52. Torque the engine mounting nuts and bolts to 28 ft. lbs. (38 Nm).

53. Install the cross-shaft.

54. Install the remaining components in the reverse order of removal. Make certain to tighten the exhaust pipe-to-exhaust manifold bolts to 46 ft. lbs. (62 Nm), the exhaust pipe support bracket bolts to 33 ft. lbs. (44 Nm), the exhaust pipe-to-catalytic converter bolts to 35 ft. lbs. (48 Nm) and the cooling fan-to-fluid clutch nuts to 48 inch lbs. (5.4 Nm).

55. Fill the engine with oil and fill the transmission with fluid.

56. Fill the engine and radiator with coolant.

57. Install the engine undercover.

58. Install the battery and connect the cables.

59. Start the engine and check for leaks.

60. Install the hood.

Model A340F Transmission

1. Disconnect the negative battery cable.

2. Disconnect the transmission throttle cable and clamp from the throttle body.

3. Raise and safely support the vehicle.

4. Remove the engine undercover.

5. Drain the transmission fluid.

6. Remove the transfer shift lever front the inside of the vehicle as follows:

  a. Remove the shift lever knob.

  b. Remove the four screws and the boot.

  c. Using pliers, remove the snapring, and pull out the shift lever from the transfer case.

7. Remove the transmission oil filler tube.

8. Remove the front and rear driveshafts.

9. Remove the front exhaust pipe.

10. Disconnect the speedometer cable.

11. Disconnect the No. 2 vehicle speed sensor connector.

12. Disconnect the solenoid connector by removing the electrical connector and bolt.

13. Disconnect the transfer case neutral position switch.

14. Disconnect the transfer case L4 position switch.

15. Remove the clip and disconnect the No. 2 gear shifting rod.

16. Remove the nut, four bolts, and the cross-shaft.

17. Disconnect the starter wires.

18. Remove the oil cooler pipe by removing the bolts and clamps.

19. Disconnect the ATF temperature sensor connector.

20. Disconnect the park/neutral position switch connector.

21. Remove the starter from the engine by removing the two bolts.

22. Remove the stiffener plate and rear end plate by removing the eight bolts.

23. Remove the sway bar.

24. Support the transmission, using a jack with a wooden block placed between the jack and the transmission pan. Raise the transmission just enough to take the weight off of the rear mount.

25. Remove the rear engine mounting bracket.

26. Remove the rear support member by removing the eight bolts.

27. Rotate the crankshaft to access the torque converter bolts. Remove the six bolts from the torque converter.

28. Disconnect or remove any component that will get in the way of removing the transmission.

29. Remove the transmission bolts, then carefully move the transmission rearward.

30. Pull the transmission rearward and lower it out of the vehicle.

To install:

31. Raise the transmission into place and install the bolts. Torque the transmission bolts to 53 ft. lbs. (71 Nm).

32. Install the torque converter bolts and torque the bolts to 30 ft. lbs. (41 Nm).

33. Install the rear support member and torque the eight bolts to 70 ft. lbs. (97 Nm).

34. Install the rear mounting bracket and install the four bolts. Torque the bolts to 13 ft. lbs. (18 Nm).

35. Install the dynamic damper with the two bolts. Torque the bolts to 44 ft. lbs. (61 Nm).

36. Remove the jack supporting the transmission.

37. Install the sway bar.

38. Install the stiffener plate and rear end plate and torque the bolts to 27 ft. lbs. (37 Nm).

39. Install the starter and torque the bolts to 29 ft. lbs. (39 Nm).

40. Connect the park/neutral position switch connector.

41. Connect the ATF temperature sensor connector.

42. Install the oil cooler pipes and clamps. Torque the two oil cooler pipes to 25 ft. lbs. (34 Nm).

43. Connect the starter wires.

44. Install the cross-shaft with the four bolts, washer, and nut. Torque the nut to 9 ft. lbs. (13 Nm) and the bolts as follows:

• Transmission side: 9 ft. lbs. (13 Nm)

• Frame side: 21 ft. lbs. (28 Nm)

45. The balance of installation is the reverse of the removal procedure.

46. Fill the transmission with the proper fluid.

47. Connect the negative battery cable to the battery.

48. Road test the vehicle and check for leaks.

49. Check all fluids.

Tacoma

Model A340F Transmission

1. Remove the ATF level gauge.

2. Remove the engine undercover.

3. Drain the transmission fluid.

4. Disconnect the throttle cable.

5. Remove the No. 1 fan shroud.

6. Remove the transmission shift lever assembly and the transfer shift lever.

  a. Remove the two bolts and the four screws and remove the rear console box.

  b. Remove the front console box with the transfer shift lever knob.

  c. Disconnect the connectors.

  d. Remove the nut and washer and disconnect the shift control rod.

  e. Disconnect the connector and remove the eight screws and the transmission shift lever assembly.

  f. Using snapring pliers, remove the snapring and pull out the shift lever from the transfer.

7. Remove the oil filler pipe with the O-ring.

8. Remove the front and rear driveshaft.

9. Remove the exhaust pipe.

10. Disconnect the speedometer cable.

11. Disconnect the No. 2 vehicle speed sensor connector.

12. Disconnect the solenoid connector.

13. Disconnect the transfer neutral position switch connector.

14. Disconnect the transfer L4 position switch connector.

15. Disconnect the transfer indicator switch.

16. Disconnect the oil cooler pipe.

a. Remove the three bolts and clamps.

b. Loosen the two union nuts and disconnect the two oil cooler pipes.

17. Disconnect the ATF temperature sensor connector.

18. Disconnect the park/neutral position switch connector.

19. Disconnect the connector and remove the starter.

20. Remove the four stabilizer bar bracket mounting bolts.

21. Remove the torque converter clutch mounting bolt.

a. Remove the nuts and bolts and the flywheel housing undercover.

b. While turning the crankshaft to gain access, remove the six bolts.

22. Remove the front differential rear mounting cushion.

a. Using a hexagon wrench, remove the nut.

b. Lift up the front differential. Be careful not to touch the torque converter clutch housing and the front differential companion flange.

c. Remove the two rear mounting cushion mounting bolts.

23. Remove the crossmember.

a. Support the transmission rear side.

b. Remove the four engine rear mounting bolts.

c. Supporting the transmission with a jack, remove the four nuts, bolts and the crossmember.

24. Lower the transmission rear side, separate the wire harness and remove the bolts and the transmission.

To install:

25. Install the transmission to the engine and install the transmission to engine bolts. Torque the bolts to 53 ft. lbs. (71 Nm).

26. Install the crossmember and torque the bolts to 48 ft. lbs. (65 Nm).

27. Install the engine rear mounting bolts and torque to 14 ft. lbs. (19 Nm).

28. Install the front differential rear mounting cushion and torque the nut to 64 ft. lbs. (41 Nm).

29. Install the torque converter clutch mounting bolt. First install the green colored bolt and then the five others. Torque to 30 ft. lbs. (41 Nm).

30. Install the flywheel housing undercover and torque to:

• 3.4L (5VZ-FE): 13 ft. lbs. (18 Nm)
• 2.7L (3RZ-FE): 27 ft. lbs. (37 Nm)

31. Install the four stabilizer bar bracket mounting bolts and torque to 19 ft. lbs. (25 Nm).

32. Install the starter and torque the bolts to 29 ft. lbs. (39 Nm). Connect the connector and terminal.

33. Install the remaining components in the reverse order of removal.

34. Install the ATF level gauge.

35. Fill and check the fluid level.

36. Test drive and check for proper shifting.

1995 Model A340E Transmission

1. Remove the hood.

2. Disconnect the battery and remove it from the vehicle.

—— **CAUTION** ——

Work must be started after 90 seconds from the time the ignition switch is turned to the LOCK position and the negative battery cable is disconnected.

3. Raise and safely support the vehicle.

4. Remove the engine under covers.

5. Drain the engine coolant.

6. Drain the engine oil.

7. Remove the radiator from the vehicle.

8. Remove the power steering drive belt as follows:

a. Stretch the belt and loosen the fan pulley mounting nuts.

b. Loosen the lock bolt, pivot bolt and adjusting bolt and remove the drive belt from the engine.

9. If equipped with air conditioning, remove the air conditioning drive belt by loosening the idle pulley nut and adjusting bolt.

10. Loosen the lock bolt, pivot bolt, and the adjusting bolt and the alternator drive belt.

11. Remove the fan with the fluid coupling and fan pulleys.

12. Disconnect the power steering pump from the engine and set aside. Do not disconnect the lines from the pump.

13. If equipped with air conditioning, disconnect the compressor from the engine and set aside. Do not disconnect the lines from the compressor.

14. Remove the air cleaner cap, MAF meter, and the resonator.

15. Remove the air cleaner case and filter.

16. Disconnect the following cables:

• If equipped with cruise control, disconnect the actuator cable with the bracket.

• Accelerator cable

• With automatic transmission, Throttle cable

17. Disconnect the heater hoses.

18. Disconnect the following hoses:

• Brake booster vacuum hose

• A.D.D. vacuum hose

• EVAP hose

• Fuel return hose

• Fuel inlet hose

19. Disconnect the starter wire and connectors as follows:

a. Remove the ground strap by removing the bolt.

b. Remove the nuts and disconnect the positive cable from the battery.

c. Disconnect the three starter wire clamps and connector.

20. Disconnect the alternator connector and wire.

21. Disconnect the engine wire and connectors as follows:

a. Remove the four screws to the right front door scuff plate. Remove the scuff plate from the vehicle.

b. Remove the cowl panel side trim by removing the clip.

c. Disconnect the ECM electrical connectors.

d. Disconnect the two connectors from the cowl wire.

e. Disconnect the igniter connector.

f. Disconnect the ground strap.

g. Disconnect the six engine wire clamps.

h. Pull out the engine wire harness from the cabin.

22. Remove the stabilizer bar.

23. Remove the driveshaft from the transmission.

24. Disconnect the speedometer cable.

25. Remove the front exhaust pipe.

26. Remove the cross-shaft.

27. Place a jack under the transmission.

28. Remove the transmission rear mounting bracket by removing the eight bolts.

29. If equipped with air conditioning, remove the bolt and disconnect the air conditioning compressor wire clamp.

30. If necessary, install a No. 2 engine hanger with two bolts. Torque the two bolts to 30 ft. lbs. (40 Nm).

31. Attach the engine hoist chain to the two engine hangers.

32. Remove the four bolts and nuts holding the engine front mounting insulators to the frame.

33. Lift the engine and transmission out of the vehicle.

34. Place the engine and transmission on a stand.

35. Remove the ATF level gauge.

36. Remove the transmission oil filler pipe.

37. Loosen the two oil cooler pipe union nuts.

38. Remove the three bolts, three clamps, and the two oil cooler pipes.

39. Disconnect the park/neutral position switch, solenoid connector and No. 2 vehicle speed sensor.

40. Separate the wire harness from the transmission.

41. Remove the flywheel housing under-cover by removing the four bolts.

42. Turn the crankshaft to gain access to the torque converter bolts. Remove the torque converter bolts.

43. Remove the starter mounting bolts and the starter.

44. Remove the six transmission mounting bolts from the engine.

45. Pull out the transmission towards the rear.

To install:

46. Install the transmission to the engine and install the transmission to engine bolts. Torque the bolts to 53 ft. lbs. (71 Nm).

47. Install the starter with the two bolts. Torque the bolts to 29 ft. lbs. (39 Nm).

48. Install the green colored torque converter bolt and then install the other five bolts. Torque the torque converter bolts to 30 ft. lbs. (41 Nm).

49. Install the flywheel housing cover by installing the four bolts. Torque the bolts to 13 ft. lbs. (18 Nm).

50. Install the wire harness to the transmission.

51. Connect the No. 2 vehicle speed sensor, solenoid connector, and the park/neutral position switch.

52. Install the oil cooler pipes.

53. Install the oil filler pipe.

54. Install the ATF gauge.

55. Install the engine assembly to the vehicle. Attach the engine mounts to the body mounts. Install the bolts and nuts but do not torque at this time.

56. Remove the engine chain hoist the No. 2 engine hanger.

57. If equipped with air conditioning, connect the air conditioning wire with the bolt.

58. Raise the transmission slightly and install the transmission mounting bracket. Torque the bolts to the frame to 43 ft. lbs. (58 Nm) and the bolts to the mounting insulator to 13 ft. lbs. (18 Nm).

59. Torque the engine mounting nuts and bolts to 28 ft. lbs. (38 Nm).

60. The balance of installation is the reverse of the removal procedure. Make certain to tighten the exhaust pipe-to-exhaust manifold bolts to 46 ft. lbs. (62 Nm), the exhaust pipe support bracket bolts to 33 ft. lbs. (44 Nm), the exhaust pipe-to-catalytic converter bolts to 35 ft. lbs. (48 Nm) and the cooling fan-to-fluid clutch nuts to 48 inch lbs. (5.4 Nm).

61. Fill the engine with oil.

62. Fill the engine and radiator with coolant.

63. Install the engine undercover.

64. Start the engine and check for leaks.

1996–97 Model A340D and A340E Transmissions

1. Remove the transmission with the engine.

2. Place the engine/transmission assembly on a stand.

3. Remove the bolts, two stiffener plates and rear end plate.

4. Turn the crankshaft to gain access to the torque converter bolts. Remove the torque converter bolts.

5. Remove the starter by removing the two bolts.

6. Remove the transmission mounting bolts from the engine.

7. Pull out the transmission toward the rear and separate the engine from the transmission.

To install:

8. Connect the transmission to the engine and install the bolts. Torque the bolts to 53 ft. lbs. (71 Nm).

9. Install the starter with the two bolts. Torque the bolts to 29 ft. lbs. (39 Nm).

10. Install the torque converter bolts and tighten to 30 ft. lbs. (41 Nm).

11. Install the stiffener plate and rear end plate with the bolts. Torque the bolts to 27 ft. lbs. (37 Nm).

12. Connect the starter wires to the starter.

13. Install the transmission with the engine.

Transfer Case Assembly

REMOVAL AND INSTALLATION

All Models

1. Disconnect the negative battery cable.

2. Raise and safely support the vehicle.

3. Drain the transmission and the transfer case.

4. Remove the transfer case with the transmission.

5. If equipped with an automatic transmission, disconnect the breather hose from the transfer upper cover and the transmission control retainer.

6. Remove the rear engine mounting.

7. Remove the dynamic damper.

8. Remove the propeller shaft upper dust cover and the transfer from the transmissions follows:

   a. Remove the dust cover bolt from the bracket.

   b. Remove the transfer adapter rear mounting bolts.

   c. Pull the transfer straight up and remove it from the transmission. Be careful not to damage the adapter rear oil seal with the transfer input gear spline.

To install:

9. Install the transfer and the propeller shaft upper dust cover to the transmission with a new gasket as follows:

   a. Shift the two shift fork shafts to the high four position.

   b. Apply MP grease to the adapter oil seal.

   c. Place a new gasket to the transfer adapter.

   d. Install the transfer to the transmission. Take care not to damage the oil seal by the input gear spline.

   e. Install the transfer adapter rear mounting bolts and torque the bolts to 27 ft. lbs. (37 Nm).

   f. Install the dust cover bolt to the bracket. Torque the bolt to 17 ft. lbs. (23 Nm).

10. Install the engine rear mounting and torque the bolts to 19 ft. lbs. (25 Nm).

11. Install the dynamic damper and torque the bolts to 27 ft. lbs. (37 Nm).

12. If equipped with an automatic transmission, install the breather hose.

13. Install the transfer case with the transmission to the engine.

14. Fill the transmission and the transfer case with oil.

15. Test drive the vehicle and check the abnormal noise and smooth operation.

16. Recheck the fluid levels.

DRIVE AXLE

Driveshaft

REMOVAL AND INSTALLATION

All Models

2-Wheel Drive Models

1. Disconnect the negative battery cable from the battery.
2. Raise and safely support the vehicle.
3. Place matchmarks on the flanges of the driveshaft and the differential carrier.
4. Remove the four nuts, bolts, and washers. Disconnect the driveshaft from the differential housing.
5. If equipped with a 3 joint driveshaft, remove the two bolts and the center support bearing from the frame crossmember.
6. Pull out the driveshaft yoke from the transmission.
7. Install a transmission output shaft plug into the transmission to prevent fluid loss.

To install:
8. Remove the transmission output shaft plug.
9. Slide the driveshaft into the transmission.
10. If equipped with a 3 joint driveshaft, connect the center support bearing to the vehicle by installing the two bolts. Do not torque the two bolts at this time.

NOTE: Make sure the bearing is installed with the drain hole downwards.

11. Align the matchmarks on the driveshaft and differential flanges.
12. Install the four washers, bolts and nuts to the driveshaft and differential carrier. Torque the bolts to 54 ft. lbs. (74 Nm).
13. With no weight on the vehicle, adjust the center support bearing to keep the bearing perpendicular to the driveshaft.
14. Torque the bolts for the center support bearing to 27 ft. lbs. (36 Nm).
15. Lower the vehicle and connect the negative battery cable to the battery.

4-Wheel Drive Models

1. Disconnect the negative battery cable from the battery.
2. Raise and safely support the vehicle.

3. Place matchmarks on the flanges of the driveshaft and the differential carrier.
4. Remove the four nuts, bolts, and washers. Disconnect the driveshaft from the differential housing.
5. Remove the front propeller shaft No. 2 dust cover.
6. Remove the front propeller shaft dust cover sub-assembly.
7. Suspend the front side of the propeller shaft. Place matchmarks on the flanges. Remove the front driveshaft.
8. If equipped, remove the two mounting bolts for the center support bearing and disconnect the center support bearing from the vehicle.
9. Place matchmarks on the flanges of the driveshaft and the transfer case.
10. Disconnect the driveshaft from the transfer case by removing the four nuts and washers.
11. Remove the rear driveshaft from the vehicle.

To install:
12. Align the matchmarks on the transfer case and driveshaft.
13. Install the four washers and nuts to hold the driveshaft to the transfer case. Torque the nuts to 54 ft. lbs. (74 Nm).
14. Install the front propeller shaft dust cover sub-assembly. Torque the A bolts to 27 ft. lbs. (36 Nm) and the B bolt to 17 ft. lbs. (23 Nm).
15. Connect the center support bearing to the vehicle by installing the two bolts. Do not torque the two bolts at this time.

NOTE: Make sure the bearing is installed with the drain hole downwards.

16. Align the matchmarks on the driveshaft and differential flanges.
17. Install the four washers, bolts and nuts to the driveshaft and differential carrier. Torque the bolts to 54 ft. lbs. (74 Nm).
18. With no weight on the vehicle, adjust the center support bearing to keep the bearing perpendicular to the driveshaft.
19. Torque the bolts for the center support bearing to 27 ft. lbs. (36 Nm).
20. Install the driveshaft protector to the vehicle and install the four bolts and torque to 22 ft. lbs. (29 Nm).
21. Connect the rear driveshaft flange to the companion flange on the transfer and torque to:
• 3VZ-E manual transmission: 56 ft. lbs. (76 Nm)
• Except 3VZ-E: 54 ft. lbs. (74 Nm)

22. Connect the rear driveshaft to the rear differential and torque to:
• 3VZ-E manual transmission: 56 ft. lbs. (76 Nm)
• Except 3VZ-E: 54 ft. lbs. (74 Nm)
23. Lower the vehicle and connect the negative battery cable to the battery.

U-Joints

REMOVAL AND INSTALLATION

All Models

1. Remove the driveshaft from the vehicle.
2. Matchmark the yoke and the driveshaft.
3. Using a brass bar and hammer, slightly tap in the bearing outer race.
4. Remove the four snaprings from the bearings.
5. Using SST 09332–25010 or equivalent, push out the bearing from the flange.
6. Clamp the bearing outer race in a vise and tap off the flange with a hammer.
7. Repeat Steps 3, 5, and 6 for the other bearings.
8. Check for worn or damaged parts. Inspect the bearing journal surfaces for wear.

To install:
9. Install the bearing cups, seals, and O-rings on the spider.
10. Grease the spider and the bearings.

NOTE: Be sure to hold the bearing caps while greasing the U-joints. The grease will force the bearing caps off the spider when they are not secured in the driveshaft yoke.

11. Remove the bearing cups from the spider.
12. Position the spider in the yoke.
13. Start the bearings in the yoke and then press them into place, using a vise. Install the snaprings to hold the bearing cups in place.
14. Make sure the bearings and snaprings are fully seated by lightly tapping on the yoke with a hammer.
15. If the axial play of the spider is greater than 0.002 in. (0.05mm), select snaprings which will provide the correct play. Be sure that the snaprings are the same size on both sides or driveshaft noise and vibration will result.
16. Install the driveshaft in the vehicle.

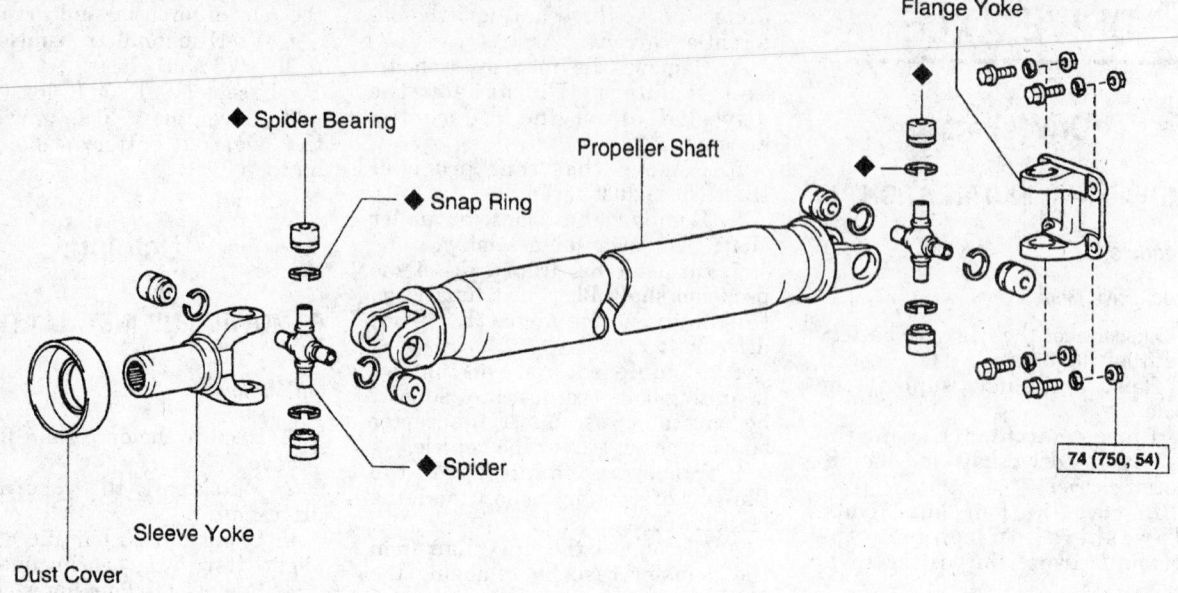

N·m (kgf·cm, ft·lbf) : Specified torque

◆ Non-reusable part

Exploded view of a common driveshaft and U-joints

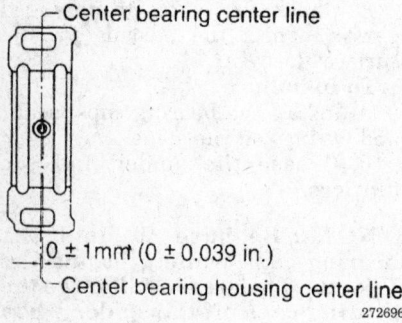

Centering the center bearing — 1995 Pick-up and Tacoma

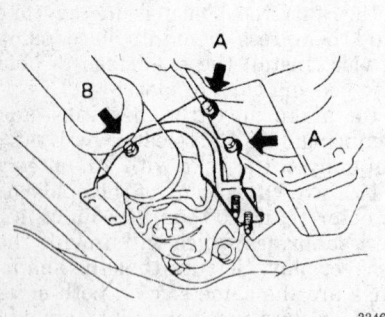

Torquing the shaft dust cover bolts — T-100 (4-Wheel Drive)

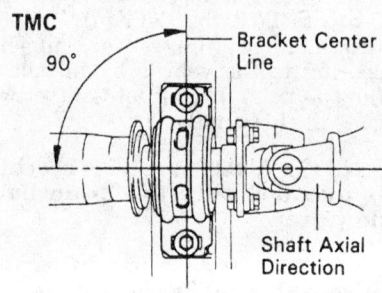

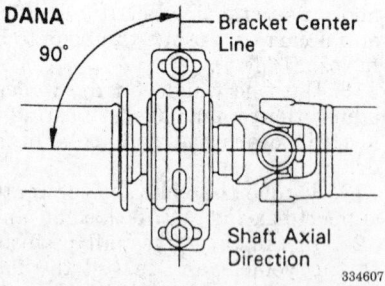

Centering the center bearing — T-100 models

Halfshaft

REMOVAL AND INSTALLATION

1993–94 Pick-up, T-100 and 4Runner

1. Remove the 4-Wheel Drive hub (with the flange) from the axle hub.

2. Raise and safely support the vehicle. Remove the wheel and tire assembly.

3. Disconnect and plug the brake line from the caliper. Remove the caliper from the axle hub.

4. Using a drift punch and a hammer, drive the lock washer tabs away from the locknut.

5. Remove the locknut from the halfshaft. Remove the lock washer, the adjusting nut, the thrust washer, the outer bearing and the axle hub/disc assembly from the vehicle.

6. Remove the knuckle spindle bolts, the dust seal and the dust cover. Using a brass bar and a hammer, tap the steering spindle from the steering knuckle.

7. Turn the halfshaft until a flat spot on the outer shaft is in the upper position, then pull the halfshaft from the steering knuckle.

8. Using a slide hammer, pull the oil seal from the axle housing.

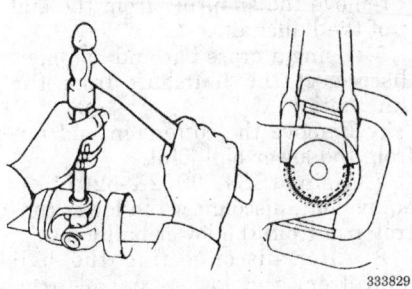

Removing the snaprings (Toyota driveshaft)

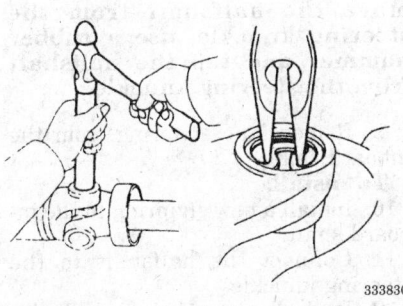

Removing the snaprings (Dana driveshaft)

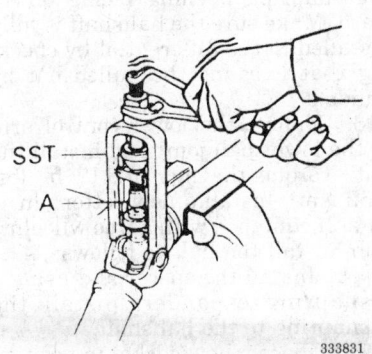

Removing the bearing from the flange

9. Using a clean shop towel, wipe the grease from inside the steering knuckle housing and the halfshaft.

To install:

10. Using an oil seal installation tool, drive a new oil seal into the axle housing until it seat.

11. Install the halfshaft into the axle housing.

12. Using multi-purpose grease, fill the steering knuckle cavity to about ³/₄ full.

13. Install the remaining components by reversing the removal procedure. Use new gaskets and seals.

14. Torque the steering spindle-to-steering knuckle bolts to 38 ft. lbs. (52 Nm), the axle hub adjusting nut to 18 ft. lbs. (25 Nm), the axle hub locknut to 33 ft. lbs. (44 Nm), the free wheel/locking hub nuts to 23 ft. lbs. (33 Nm) and the brake caliper to 65 ft. lbs. (88 Nm).

NOTE: To install the wheel bearings with the axle hub, torque the adjusting nut to 43 ft. lbs. (58 Nm), turn the axle hub (back and forth, several times), loosen the nut and re-torque the adjusting nut to 18 ft. lbs. (24 Nm).

15. Install the wheel and tire assembly. Lower the vehicle.

1995 Pick-up, 4Runner and 1995–97 T-100

1. Raise and safely support the vehicle.

2. Remove the wheel(s) and tire assembly.

3. While having an assistant hold the brake pedal, remove the six nuts holding the halfshaft to the differential.

4. If equipped with a free wheeling hub, remove the free-wheel hub as follows:

    a. Set the control handle to FREE.

    b. Remove the cover bolts and pull off the cover.

    c. Remove the center bolt with washer.

    d. Remove the mounting nuts and washer to the hub body.

    e. Using a brass bar and hammer, tap on the bolts head and remove the cone washer.

    f. Pull off the free wheel hub body and gasket.

5. If equipped without a free wheeling hub, remove the flange for the axle hub as follows:

    a. Remove the grease cap from the flange.

    b. Remove the bolt from the flange.

    c. Remove the six mounting nuts to the flange.

    d. Using a brass bar and hammer, tap on the bolts head and remove the six cone washers.

    e. Install the two bolts to the flange. Tighten the bolts to remove the flange.

    f. Remove the gasket for the flange.

6. Using a snapring expander, remove the snapring from the end of the halfshaft and then remove the spacer.

7. Remove the halfshaft from the differential and then pull the halfshaft from the steering knuckle.

NOTE: It may be necessary to tap the end of the halfshaft with a rubber hammer.

To install:

8. Install the halfshaft to the steering knuckle and differential.

9. Install the six nuts to the differential but do not torque the nuts at this time.

10. Install the spacer and using a snapring expander, install a new snapring to the halfshaft.

11. If equipped without a free wheeling hub, install the flange as follows:

    a. Place a new gasket in position on the axle hub.

    b. Install the flange to the axle hub.

    c. Install the six cone washers, plate washers and nuts.

    d. Install the six nuts. Torque the nuts to 23 ft. lbs. (31 Nm).

    e. Install the bolt and torque the bolt to 13 ft. lbs. (18 Nm).

    f. Install the grease cap.

12. If equipped with a free wheeling hub, install the hub as follows:

    a. Place a new gasket in position on the front axle hub.

    b. Install the free wheeling hub body with the six cone washers and nuts. Torque the nuts to 23 ft. lbs. (31 Nm).

    c. Install the bolt with the washer. Torque the bolt to 13 ft. lbs. (18 Nm).

    d. Apply multi purpose grease to the inner hub splines.

    e. Set the control handle and clutch to the FREE position.

    f. Place a new gasket in position on the cover.

    g. Install the cover to the hub body with the follower pawl tabs aligned with the non-toothed portions of the hub body.

    h. Install the mounting bolts and tighten the cover bolts to 7 ft. lbs. (10 Nm).

13. With an assistant holding down the brake pedal, torque the six nuts holding the halfshaft to the differential. Torque the nuts to 61 ft. lbs. (83 Nm).

14. Install the front wheel(s) and lower the vehicle.

1996–97 4Runner

1. Raise and safely support the vehicle.

2. Remove the front wheel.

3. Drain the differential oil from the differential.

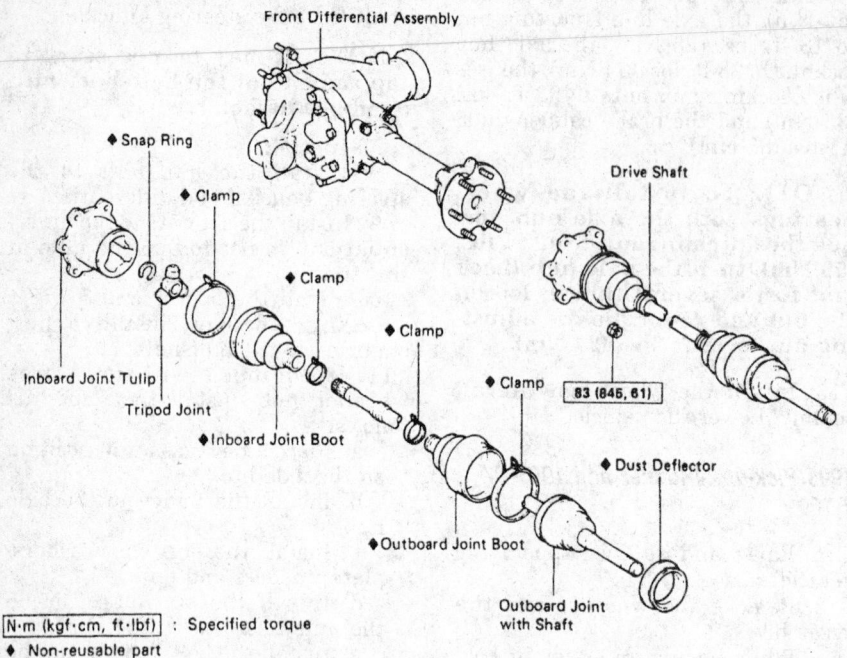

N·m (kgf·cm, ft·lbf) : Specified torque
◆ Non-reusable part

334612

Halfshaft component assembly — 1995 Pick-up and 4Runner and 1995–97 T-100

4. Remove the half shaft locknut.

a. Remove the grease cap.

b. Remove the cotter pin and the lock cap. While applying the brakes, remove the locknut.

5. Using a brass bar and a hammer, disconnect the halfshaft.

6. Disconnect the lower control arm.

7. Push the steering knuckle outward and remove the halfshaft.

8. Remove the snapring from the inboard shaft.

To install:

9. Install the snapring to the inboard shaft.

10. Install the halfshaft and install the steering knuckle.

11. Connect the lower control arm and torque the nut to 105 ft. lbs. (142 Nm).

12. Connect the halfshaft.

a. Set the snapring opening side facing downward.

b. Using SST 09631–10030 and a hammer, strike the inboard joint into the differential.

c. Check that the halfshaft cannot be pulled out by hand.

13. While applying the brakes install the locknut. Torque to 174 ft. lbs. (235 Nm).

14. Install the grease cap.

15. Fill the differential with oil.

16. Install the front wheel.

Tacoma

1. Raise and safely support the vehicle.

2. Drain the differential oil from the differential.

3. If not equipped with a freewheeling hub, disconnect the halfshaft from the steering knuckle as follows:

a. Using a screwdriver, remove the grease cap.

b. Remove the cotter pin and lock cap from the halfshaft.

c. While having an assistant apply the brakes, remove the locknut from the halfshaft.

4. If equipped with free-wheeling hub, remove the free wheel hub as follows:

a. Set the control handle to FREE.

b. Remove the cover bolts and pull off the cover.

c. Remove the center bolt with washer.

d. Remove the mounting nuts and washer to the hub body.

e. Using a brass bar and hammer, tap on the bolts head and remove the cone washer.

f. Pull off the free wheel hub body and gasket.

g. Using a snapring expander, remove the snapring from the end of the halfshaft.

5. Using a brass bar and hammer, disconnect the halfshaft from the differential.

6. Remove the cotter pin and nut from the lower ball joint.

7. Using SST 09628–62011 or equivalent, disconnect the lower control arm from the lower ball joint.

8. After disconnecting the ball joint from the lower control arm, push the steering knuckle outwards and remove the halfshaft from the steering knuckle and vehicle.

NOTE: If it is difficult to remove the halfshaft from the steering knuckle, use a rubber hammer and tap the halfshaft from the steering knuckle.

9. Remove the snapring from the inboard shaft.

To install:

10. Install a new snapring to the inboard shaft.

11. Connect the halfshaft to the steering knuckle.

12. Push the steering knuckle inwards and at the same time, push the halfshaft into the differential with the snapring opening facing downward. Make sure the halfshaft is fully installed to the differential by checking that it cannot be pulled out by hand.

13. Connect the lower control arm to the lower ball joint and install the nut. Torque the nut to 112 ft. lbs. (152 Nm). Install a new cotter pin.

14. If equipped with a free wheeling hub, install the hub as follows:

a. Install the spacer and using a snapring expander, install the snapring to the halfshaft.

b. Place a new gasket in position on the front axle hub.

c. Install the free wheeling hub body with the six cone washers and nuts. Torque the nuts to 23 ft. lbs. (31 Nm).

d. Install the bolt with the washer. Torque the bolt to 13 ft. lbs. (18 Nm).

e. Apply multi purpose grease to the inner hub splines.

f. Set the control handle and clutch to the FREE position.

g. Place a new gasket in position on the cover.

h. Install the cover to the hub body with the follower pawl tabs aligned with the non-toothed portions of the hub body.

i. Install the mounting bolts and tighten the cover bolts to 7 ft. lbs. (10 Nm).

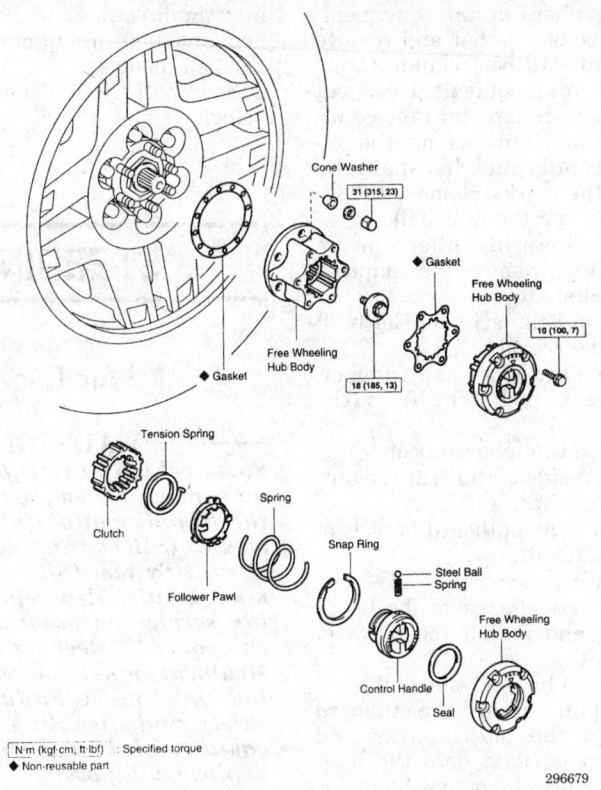

N·m (kgf·cm, ft lbf) : Specified torque
◆ Non-reusable part

296679

Exploded view of the free wheeling hub assembly — Tacoma

15. If equipped without a free wheeling hub, install the halfshaft to the steering knuckle as follows:

   a. Install the locknut to the halfshaft. Torque the nut to 174 ft. lbs. (235 Nm).

   b. Install the lock cap and cotter pin to the halfshaft.

   c. Install the grease cab to the hub.

16. Fill the differential with gear oil.

17. Install the wheels and lower the vehicle.

CV-Joint Boot

REPLACEMENT

1993–95 Pick-up, 4Runner and 1993–94 T-100

1. Disconnect the negative battery cable.

2. Raise and safely support the vehicle.

3. Remove the wheel(s) and tire assembly.

4. If equipped with a free wheeling hub, remove the free-wheel hub as follows:

   a. Set the control handle to FREE.

   b. Remove the cover bolts and pull off the cover.

   c. Remove the center bolt with washer.

   d. Remove the mounting nuts and washer to the hub body.

   e. Using a brass bar and hammer, tap on the bolts head and remove the cone washer.

   f. Pull off the free wheel hub body and gasket.

5. If equipped without a free wheeling hub, remove the flange for the axle hub as follows:

   a. Remove the grease cap from the flange.

   b. Remove the bolt from the flange.

   c. Remove the six mounting nuts to the flange.

   d. Using a brass bar and hammer, tap on the bolts head and remove the six cone washers.

   e. Install two bolts to the flange. Tighten the bolts to remove the flange.

   f. Remove the gasket for the flange.

6. Using a snapring expander, remove the snapring from the end of the halfshaft and then remove the spacer.

7. Remove the halfshaft from the differential and then pull the halfshaft from the steering knuckle.

NOTE: It may be necessary to tap the end of the halfshaft with a rubber hammer.

8. Remove the inboard joint boot clamps.

9. Slide the inboard joint boot toward the outboard joint.

10. Place matchmarks on the inboard joint tulip and the shaft. Remove the inboard joint tulip from the driveshaft.

11. Using a snapring expander remove the snapring and disassemble the tripod joint.

12. Using a brass bar and hammer, remove the tripod joint from the driveshaft. Do not punch the roller.

13. Remove the inboard joint boot.

14. Remove the outboard joint boot clamps and boot. Do not disassemble the outboard joint.

15. Remove the dust deflector.

To install:

16. Install the dust deflector.

17. Temporarily install the new boot and new boot clamps to the outboard joint.

NOTE: Before installing the boot, wrap vinyl tape around the spline of the shaft to prevent damaging the boot.

18. Temporarily install the new boot and the new boot clamps for the inboard joint to the driveshaft.

19. Assemble the tripod joint.

   a. Place the beveled side of the tripod axial spline toward the outboard joint.

   b. Align the matchmarks placed before disassembly.

   c. Using a brass bar and hammer, tap in the tripod joint onto the driveshaft. Do not punch the roller.

20. Using a snapring expander, install a new snapring.

21. Before assembling the boot to the outboard joint, pack the boot with grease.

22. Assemble the inboard joint to the inboard joint tulip. Pack in grease to the inboard tulip and the boot.

23. Assemble the boot clamps to both boots. Be sure that the boot is on the shaft groove and that it is not stretched or contracted.

24. Install the halfshaft to the steering knuckle and differential.

25. Install the six nuts to the differential but do not torque the nuts at this time.

26. Install the spacer and using a snapring expander, install a new snapring to the halfshaft.

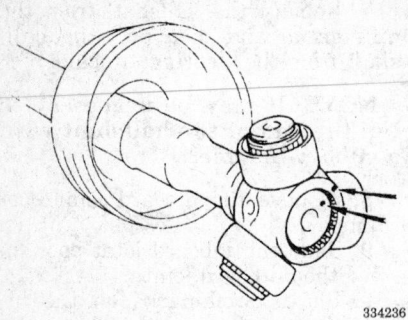

Matchmarks on the tripod joint — Pick-up, 4Runner and T-100

27. If equipped without a free wheeling hub, install the flange as follows:

    a. Place a new gasket in position on the axle hub.

    b. Install the flange to the axle hub.

    c. Install the six cone washers, plate washers and nuts.

    d. Install the six nuts and torque the nuts to 23 ft. lbs. (31 Nm).

    e. Install the bolt and torque the bolt to 13 ft. lbs. (18 Nm).

    f. Install the grease cap.

28. If equipped with a free wheeling hub, install the hub as follows:

    a. Place a new gasket in position on the front axle hub.

    b. Install the free wheeling hub body with the six cone washers and nuts. Torque the nuts to 23 ft. lbs. (31 Nm).

    c. Install the bolt with the washer. Torque the bolt to 13 ft. lbs. (18 Nm).

    d. Apply multi purpose grease to the inner hub splines.

    e. Set the control handle and clutch to the FREE position.

    f. Place a new gasket in position on the cover.

    g. Install the cover to the hub body with the follower pawl tabs aligned with the non-toothed portions of the hub body.

    h. Install the mounting bolts and tighten the cover bolts to 7 ft. lbs. (10 Nm).

29. With an assistant holding down the brake pedal, torque the six nuts holding the halfshaft to the differential. Torque the nuts to 61 ft. lbs. (83 Nm).

30. Install the front wheel(s) and lower the vehicle.

31. Connect the negative battery cable.

Tacoma and 1996–97 4Runner

1. Remove the halfshaft(s) and secure to a suitable holding device.

2. Using pliers or an equivalent, draw the hooks together and remove the large inboard boot clamp. Using side cutters or an equivalent tool, cut the small boot clamp and remove it.

3. Place matchmarks on the inboard joint tulip and halfshaft. Do not punch the marks. Remove the inboard tulip from the halfshaft.

4. Using snapring pliers or an equivalent tool, remove the snapring from the halfshaft.

5. Place matchmarks on the halfshaft and the tripod.

6. Using a brass bar and hammer, remove the tripod joint from the halfshaft.

7. Remove the inboard boot.

8. Using a side cutter, cut the outboard boot clamps.

9. Remove the outboard joint boot from the halfshaft.

To install:

10. Place new clamps to the boot's small ends and install the boots to the halfshaft.

11. Place the beveled side of the tripod axial spline toward the outboard joint. Align the matchmarks and push the tripod joint onto the halfshaft with a brass bar and hammer. Install a new snapring.

12. Pack the outboard tulip and boot with grease. Use the grease supplied with the boot kit. The grease capacity is: 6.20–6.56 oz. (175.76–185.97 g).

13. Pack the inboard joint tulip with grease. The grease capacity is: 8.99–9.34 oz. (254.86–264.78 g) for 1996–97 4Runner models, or 7.58–8.29 oz. (214.89–235.02 g) for Tacoma models. Align the matchmarks and install the inboard tulip to the halfshaft. Install the boot to the inboard tulip.

14. Check that the boot is on the halfshaft groove. Make sure that the boot is not stretched, contracted, or distorted when the shaft is at standard length. The halfshaft standard length is 20.701–21.095 in. (525.8–535.8mm) for 1996–97 4Runner, or 17.094–17.252 in. (434.2–438.2mm) for Tacoma models.

15. Holding the boot large clamp near the closing hooks, use pliers to position the holes in the clamp's free end over the closing hook. Secure the clamp by drawing the closing hook together.

16. Secure the small clamp onto the boot by placing SST 09521–24010 or equivalent, on the clamp and pinching the clamp. Do not over tighten.

17. Use SST 09240–00020 or equivalent, to adjust the clamp clearance. The clearance should be 0.031 in. (0.8mm) or less for 1996–97 4Runner, or 0.059 in. (1.5mm) or less for Tacoma models.

18. Install the halfshaft(s) into the vehicle.

STEERING

Air Bag

—— **CAUTION** ——

Some vehicles are equipped with an air bag system, also known as the Supplemental Inflatable Restraint (SIR) or Supplemental Restraint System (SRS). The system must be disabled before performing service on or around system components, steering column, instrument panel components, wiring and sensors. Failure to follow safety and disabling procedures could result in accidental air bag deployment, possible personal injury and unnecessary system repairs.

PRECAUTIONS

Several precautions must be observed when handling the inflator module to avoid accidental deployment and possible personal injury.

• Never carry the inflator module by the wires or connector on the underside of the module.

• When carrying a live inflator module, hold securely with both hands, and ensure that the bag and trim cover are pointed away.

• Place the inflator module on a bench or other surface with the bag and trim cover facing up.

• With the inflator module on the bench, never place anything on or close to the module which may be thrown in the event of an accidental deployment.

DISARMING

To avoid personal injury when working on vehicles equipped with an air bag, the negative battery cable must be disconnected and at least 90 seconds must elapse before working on the system. Failure to do so may result in deployment of the air bag.

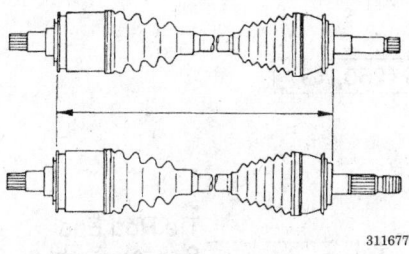

Measuring halfshaft length — 1996–97 4Runner

Steering Wheel

REMOVAL AND INSTALLATION

1993–94 Models

— CAUTION —

The air bag system (SRS) must be disarmed before removing the steering wheel. Failure to do so may cause accidental deployment, property damage or personal injury. Always store the air bag with the pad facing upward.

1. Disconnect the negative battery cable.

— CAUTION —

Wait 90 seconds from the time the key is turned to LOCK and the negative battery cable is disconnected to begin work. This allows the SRS capacitor to discharge and prevent deployment of the air bag(s).

2. Position the wheels in a straight ahead position.
3. Loosen and remove the steering wheel center cover (pad or air bag module) retaining screws.
4. Pull the wheel pad out from the steering wheel and disconnect the air bag connector, if equipped.

NOTE: When storing the wheel pad (vehicles equipped with air bag), keep the upper surface of the pad facing upward.

5. Disconnect the horn wire. Matchmark the wheel and the shaft.
6. Using an appropriate steering wheel puller tool, remove the steering wheel.

To install:

7. Align the matchmarks on the steering wheel and main shaft and install the steering wheel to the shaft. Install and torque the retaining nut to approximately 25 ft. lbs. (34 Nm).

8. If equipped, connector the air bag connector.

NOTE: If the wheel pad has been dropped or there are other defects, replace the wheel pad with a new one. Make sure that the wires do not interfere or are pinched between other parts when installing.

9. Connect the horn wire and install the steering wheel pad.
10. Connect the negative battery cable.

Tie Rod Ends

REMOVAL AND INSTALLATION

1993–94 Models

1. Raise and safely support the vehicle.
2. Remove the wheel and tire assembly. Remove the cotter pin and nut.
3. Using a tie rod end puller, disconnect the tie rod from the relay rod.
4. Using a tie rod end puller remove the tie rod from the steering knuckle.
5. Count the number of turns and remove the tie rod end from the vehicle.

To install:

6. Install the tie rod the same amount of turns as was needed to remove it. Temporarily tighten the adjusting sleeve locknut(s).
7. Connect the tie rod end to the steering knuckle arm and torque the nut to 67 ft. lbs. (90 Nm). Always install a new cotter pin.
8. Install the wheel and lower the vehicle.
9. Perform an alignment on the vehicle. After adjusting the vehicle's toe adjustment, tighten the tie rod adjusting sleeve clamp nuts to 19 ft. lbs. (25 Nm).

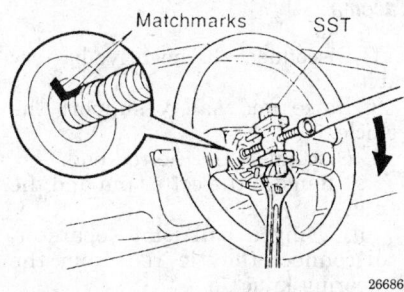

Typical method of removing the steering wheel — 1995 4Runner and Pick-up

1995 Pick-up and 1995–97 4Runner

1. Disconnect the negative battery cable.
2. Raise and safely support the vehicle.
3. Remove the wheel.
4. Remove the nut securing the tie rod to the knuckle or relay rod. Using a ball joint separator, disconnect the tie rod assembly from the relay rod and the steering knuckle arm.
5. Remove the tie rod end from the adjusting sleeve by loosening the locknut and counting the number of turns to remove it.

To install:

6. Install the tie rod the same amount of turns as was needed to remove it. Temporarily tighten the adjusting sleeve locknut(s).
7. Connect the tie rod end to the steering knuckle arm and torque the nut to 67 ft. lbs. (90 Nm).
8. Connect the tie rod end to the relay rod and torque the nut to 67 ft. lbs. (90 Nm).
9. Install the wheel.
10. Lower the vehicle and connect the negative battery cable.
11. Perform an alignment on the vehicle. After adjusting the vehicle's toe adjustment, tighten the tie rod adjusting sleeve locknuts.

1995–97 T-100

1. Disconnect the negative battery cable.

— CAUTION —

Work must be started after 90 seconds from the time the ignition switch is turned to the LOCK position and the negative (-) battery cable is disconnected.

2. Raise and support the vehicle safely.
3. Remove the engine undercover.
4. Remove the nuts securing the tie rod to the relay rod and the steering knuckle. Using a ball joint separator, remove the right and left tie rod assemblies.
5. Remove the bolts holding the right and left tie rod sub-assemblies and the two clamps.
6. Counting the number of turns, remove the tie rod ends from the tie rod (adjusting sleeve).

To install:

7. Install the two tie rod sub-assemblies, tie rod end sub-assemblies, tie rod (adjusting sleeve), and the two clamps.

  a. Screw the tie rod end into the tie rod (adjusting sleeve). Make sure the tie rod is installed the

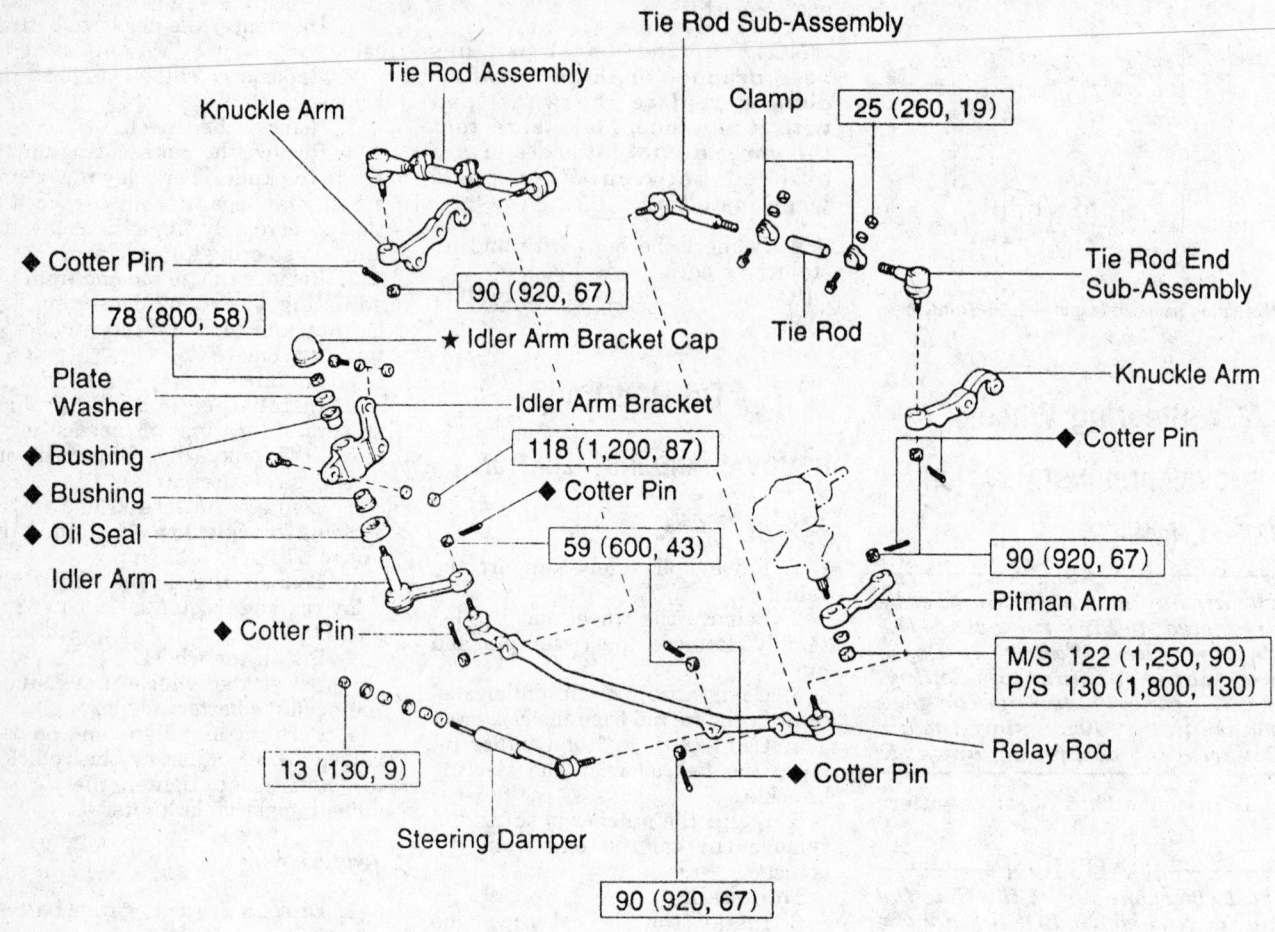

N·m (kgf·cm, ft·lbf) : Specified torque

◆ Non-reusable part

★ Precoated part

254571

Exploded view of steering linkage components — 1995 2WD 4Runner and Pick-up, 4WD models similar

same number of turns as was required to remove it.

b. Temporarily tighten the tie rod locking clamps that secure the tie rod end and sub assembly to the tie rod (adjusting sleeve).

8. Install the right and left tie rod assemblies.

a. Connect the tie rod assembly to the knuckle arm. Install the nut and tighten it to 67 ft. lbs. (90 Nm).

b. Connect the tie rod to the relay rod and torque the nut to 67 ft. lbs. (90 Nm).

9. Install the engine undercover.

10. Connect the negative battery cable.

11. Check the front wheel alignment.

Tacoma

1. Disconnect the negative battery cable.

2. Raise and safely support the vehicle.

3. Disconnect the tie rod end.

a. Remove the cotter pin and the nut.

b. Using a ball joint separator, disconnect the tie rod from the steering knuckle.

4. Loosen the locknut and unscrew the tie rod end from the steering

rack. Count the number of turns for installation.

To install:

5. Install the tie rod end to the steering rack using the amount of turns counted upon removal.

6. Install the ball joint to the steering knuckle and torque the nut to 53 ft. lbs. (72 Nm).

7. Lower the vehicle and connect the negative battery cable.

8. Perform an alignment.

9. Tighten the locknut on the steering rack after all adjustments have been made.

10. Test drive the vehicle.

Manual Rack and Pinion

REMOVAL AND INSTALLATION

Tacoma

1. Disconnect the negative battery cable.

——————— **CAUTION** ———————

Work must be started after 90 seconds from the time the ignition switch is turned to the LOCK position and the negative (-) battery cable is disconnected.

2. Raise and safely support the vehicle.
3. Disconnect the right and left tie rod ends from the knuckle.
4. Disconnect the intermediate No. 2 shaft from the steering rack.
5. Remove the manual steering rack.

To install:

6. Install the manual steering rack and torque the bolts to 148 ft. lbs. (201 Nm).
7. Connect the intermediate No. 2 shaft to the steering rack.
8. Connect the right and left tie rod ends to the steering knuckle. Tighten the castle nuts to 53 ft. lbs. (72 Nm) and install new cotter pins.
9. Connect the negative battery cable.
10. Check the steering wheel center point.
11. Bleed the power steering system.
12. Check the front wheel alignment. Torque the tie rod end locknuts to 67 ft. lbs. (90 Nm)

Power Rack and Pinion

REMOVAL AND INSTALLATION

1993–94 T-100

1. Position the front wheels facing straight ahead.
2. Secure the steering wheel so that it does not turn. The driver's seat belt can be used to secure the steering wheel.
3. Place matchmarks on the intermediate shaft and control valve shaft. Remove the lower and upper joint bolts from the intermediate shaft. Disconnect the shaft.
4. Remove the cotter pin and nut from the tie rod ends. Disconnect the tie rod ends using a separator tool.
5. Disconnect the pressure and return pipes using flare nut wrenches.

6. Remove the bracket bolts and grommets.
7. Remove the steering rack assembly from the vehicle.

To install:

8. Install the gear housing and torque the mounting bolts to 65 ft. lbs. (88 Nm).
9. Install the pressure and return tubes using new O-rings. Torque the fittings to 15 ft. lbs. (20 Nm).
10. Connect the intermediate shaft. Install the lower and upper joint bolts. Torque the bolts to 26 ft. lbs. (35 Nm).
11. Connect the tie rod ends, torque to 67 ft. lbs. (90 Nm) and install a new cotter pin.
12. Connect the negative battery cable.
13. Fill and bleed the steering system.
14. Check and adjust toe-in.

1995–97 T-100

1. Disconnect the negative battery cable.

——————— **CAUTION** ———————

Work must be started after 90 seconds from the time the ignition switch is turned to the LOCK position and the negative (-) battery cable is disconnected.

2. Raise and safely support the vehicle.
3. Disconnect the right and left tie rod ends from the knuckle.
4. Matchmark and disconnect the intermediate shaft from the steering rack.
5. Using SST 09631–22020, or equivalent, remove the pressure feed and the return tubes.
6. Remove the mount bracket and the grommet from the power steering rack assembly.
7. Remove the power steering rack and pinion.

To install:

8. Install the power steering rack and pinion. Torque the mounting bolts to 65 ft. lbs. (88 Nm).
9. Install the grommet and the mount bracket to the gear assembly. Torque the bolts to 65 ft. lbs. (88 Nm).
10. Install a new O-ring and install the pressure feed and return tubes. Torque the line fittings to 14 ft. lbs. (19 Nm).
11. Align the matchmarks and connect the intermediate shaft to the steering rack.

NOTE: If installing a new rack assembly, be sure the steering wheel and the rack are centered.

12. Connect the right and left tie rod ends and tighten nuts to 67 ft. lbs. (90 Nm). Install new cotter pins.
13. Connect the negative battery cable.
14. Check the steering wheel center point.
15. Check the fluid level and bleed the power steering system.
16. Check the front wheel alignment.

Tacoma

1. Disconnect the negative battery cable.

——————— **CAUTION** ———————

Work must be started after 90 seconds from the time the ignition switch is turned to the LOCK position and the negative (-) battery cable is disconnected.

2. Raise and safely support the vehicle.
3. Disconnect the right and left tie rod ends from the knuckle.
4. Disconnect the intermediate No. 2 shaft from the steering rack.
5. Using SST 09631–22020 or equivalent, remove the pressure feed and the return tubes.
6. Remove the power steering rack.

To install:

7. Install the power steering rack and torque the bolts to 148 ft. lbs. (201 Nm) for 2-wheel drive models, or to the following values for 4-wheel drive models:
   • Rack assembly bolt — 123 ft. lbs. (167 Nm)
   • Rack assembly nut — 141 ft. lbs. (191 Nm)
   • Bracket nut and bolt — 123 ft. lbs. (167 Nm)
8. Install a new O-ring and install the pressure feed tube and torque to 33 ft. lbs. (45 Nm). Install the return tube and torque to 36 ft. lbs. (49 Nm) for 2-wheel drive models, or to 29 ft. lbs. (40 Nm) for 4-wheel drive models.
9. Connect the intermediate No. 2 shaft to the steering rack.
10. Connect the right and left tie rod ends to the steering knuckle. Tighten the castle nuts to specification and install new cotter pins.
11. Connect the negative battery cable.
12. Check the steering wheel center point.
13. Bleed the power steering system.
14. Check the front wheel alignment. Torque the tie rod end locknuts to 67 ft. lbs. (90 Nm)

Power Steering Pump

BLEEDING

All Models

1. Raise and safely support the front of the vehicle on jackstands.
2. Check that the fluid level in the reservoir tank is at the maximum level. Fill with DEXRON II® or equivalent.
3. Start the engine.
4. With the engine idle below 1000 rpm, turn the steering wheel from lock to lock. Hold the steering wheel at each lock for 2–3 seconds. Repeat this procedure until all the air bubbles are removed from the fluid.
5. Stop the engine and measure the fluid level.
6. Make sure the rise of the fluid is not over 0.020 in. (5mm).

REMOVAL AND INSTALLATION

1993–94 Models

NOTE: Disconnect the air hoses from the air control valve and the high tension wires from the distributor.

1. Disconnect the negative battery cable. Loosen the power steering pump pulley nut.

NOTE: Use the drive belt as a brake to keep the pulley from rotating.

2. Place a container under the pump. Disconnect the return line and the pressure tube, then drain the fluid into the container.
3. Loosen the idler pulley nut and the adjusting bolt, then remove the drive belt.
4. Remove the drive pulley and the Woodruff key from the pump shaft.

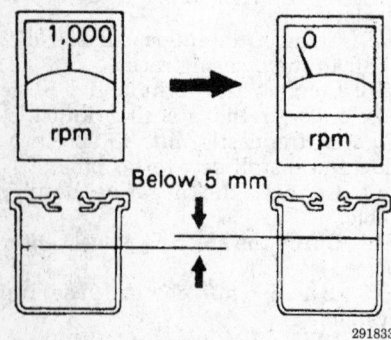

Checking the fluid level once the engine is turned off — 1995–97 models

5. Remove the mounting bolts and the power steering pump from the vehicle.
To install:
6. Place the power steering pump into position and install the mounting bolts. Torque the pump pulley mounting bolt to 29 ft. lbs. (40 Nm), the pump pulley nut to 32 ft. lbs. (42 Nm) and the pressure hoses to 33 ft. lbs. (45 Nm).
7. Install the drive belt and adjust the belt tension.
8. Bleed the power steering system.

1995 4Runner and Pick-up

1. Disconnect the negative battery cable.
2. Disconnect the pressure feed tube from the power steering vane pump on the 22R-E engine, and from the oil reservoir on the 3VZ-E engine.
3. Loosen the power steering pump pulley nut.
4. Place a container under the pump. Disconnect the return line and the pressure tube, then drain the fluid into the container.
5. Loosen the idler pulley nut and the adjusting bolt, then remove the drive belt.
6. Remove the drive pulley and the Woodruff key from the pump shaft.
7. Remove the mounting bolts and the power steering pump from the vehicle.
To install:
8. Place the power steering pump into position and install the mounting bolts. Torque the pump pulley mounting bolt to 29 ft. lbs. (40 Nm), the pump pulley nut to 32 ft. lbs. (43 Nm).
9. Install the power steering pressure and return hoses. Tighten the line fittings to 34 ft. lbs. (47 Nm).
10. Install the drive belt and adjust the belt tension.
11. Connect the negative battery cable.
12. Fill and bleed the power steering system.

1995–97 T-100, Tacoma and 1996–97 4Runner

2.7L (3RZ-FE) and 2.4L (2RZ-FE) Engines

1. Disconnect the negative battery cable.

— **CAUTION** —
Work must be started after 90 seconds from the time the ignition switch is turned to the LOCK position and the negative (-) battery cable is disconnected.

2. Loosen the power steering idle pulley set bolt and the belt adjusting bolt and remove the drive belt.
3. Remove the pressure feed tube.
   a. Using a spanner (24mm) to hold the pressure port union, remove the union bolt and the two gaskets. Drain the fluid into a clean container.
4. Disconnect the return hose.
5. Remove the power steering vane pump mounting bolts and remove the pump assembly.
To install:
6. Install the pump assembly and torque the mounting bolts to 29 ft. lbs. (39 Nm).
7. Connect the return hose.
8. Connect the pressure feed tube and torque the union bolt on each side of the tube to 34 ft. lbs. (47 Nm).
9. Install the drive belt. Adjust the belt tension and torque the power steering idle pulley set nut to 29 ft. lbs. (39 Nm).
10. Fill and bleed the power steering system.
11. Connect the negative battery cable.
12. Test drive the vehicle to check the steering operation.

3.4L (5VZ-FE) Engine

1. Disconnect the negative battery cable.

— **CAUTION** —
Work must be started after 90 seconds from the time the ignition switch is turned to the LOCK position and the negative (-) battery cable is disconnected.

2. Remove the air cleaner assembly.
3. Disconnect the return hose from the power steering vane pump.
4. Remove the union bolt, two gaskets and disconnect the pressure feed tube from the pump assembly. Drain the fluid into a container.
5. Loosen bolts A, B, and C and remove the drive belt.
6. Remove the power steering vane pump mounting bolts and remove the pump assembly.
To install:
7. Install the power steering vane pump assembly.
8. Install the drive belt. Tightening bolt C, adjust the drive belt tension. Torque bolt A and nut B to 29 ft. lbs. (39 Nm).
9. Connect the pressure feed tube, torque the union bolt with a new gasket on each side of the tube to 34 ft. lbs. (47 Nm).
10. Connect the return hose.

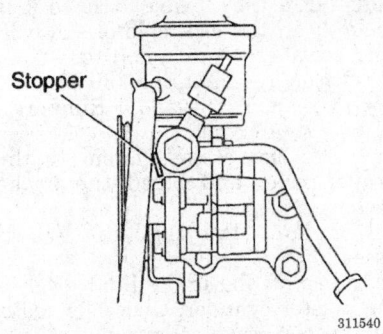

Stopper for the pressure feed tube —
1996–97 4Runner with 2.7L (3RZ-FE) engine

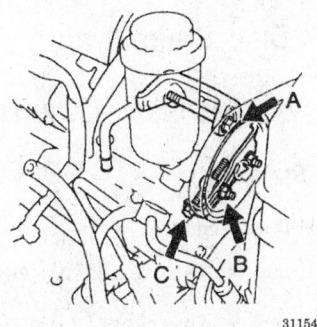

Steering pump torque bolt pattern —
All with 3.4L (5VZ-FE) engine

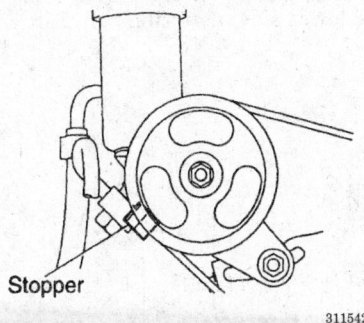

Stopper for the pressure feed tube —
1996–97 4Runner with 3.4L (5VZ-FE) engine

11. Fill and bleed the power steering system.
12. Install the air cleaner assembly.
13. Connect the negative battery cable.
14. Road test the vehicle to check for proper steering operation.

BRAKES

Anti-Lock Brake System Service

PRECAUTIONS

• Certain components within the Anti-Lock Brake System (ABS) are not intended to be serviced or repaired individually. Only those components with removal and installation procedures should be serviced.

• Do not use rubber hoses or other parts not specifically specified for and ABS system. When using repair kits, replace all parts included in the kit. Partial or incorrect repair may lead to functional problems and require the replacement of components.

• Lubricate rubber parts with clean, fresh brake fluid to ease assembly. Do not use lubricated shop air to clean parts; damage to rubber components may result.

• Use only specified brake fluid from an unopened container.

• If any hydraulic component or line is removed or replaced, it may be necessary to bleed the entire system.

• A clean repair area is essential. Always clean the reservoir and cap thoroughly before removing the cap. The slightest amount of dirt in the fluid may plug an orifice and impair the system function. Perform repairs after components have been thoroughly cleaned; use only denatured alcohol to clean components. Do not allow ABS components to come into contact with any substance containing mineral oil; this includes used shop rags.

• The Anti-Lock control unit is a microprocessor similar to other computer units in the vehicle. Ensure that the ignition switch is **OFF** before removing or installing controller harnesses. Avoid static electricity discharge at or near the controller.

• If any arc welding is to be done on the vehicle, the control unit should be unplugged before welding operations begin.

Master Cylinder

REMOVAL AND INSTALLATION

1993–94 Models

1. Disconnect the negative battery cable.

2. Using a syringe, remove the brake fluid from the master cylinder.
3. Disconnect and plug the hydraulic lines at the master cylinder.
4. If equipped, disconnect the level warning switch connector from the master cylinder.
5. Remove the master cylinder-to-power booster mounting nuts. Remove the master cylinder assembly and gasket from the power brake unit.
6. To install, reverse the removal procedures. Torque the master cylinder mounting bolts to 9 ft. lbs. (12 Nm) and the brake lines-to-master cylinder to 11 ft. lbs. (15 Nm).
7. Refill the master cylinder with new brake fluid and bleed the brake system.

1995–97 Models

1. Disconnect the negative battery cable from the battery.

───────── CAUTION ─────────
Wait at least 90 seconds from the time the ignition switch is turned to the LOCK position and the negative (-) battery cable is disconnected before starting work. This will allow the SRS system capacitor sufficient time to discharge.

2. Disconnect the level warning switch connector.
3. Take out the fluid from the master cylinder with a syringe.
4. Using a line wrench, disconnect the brake lines from the master cylinder.
5. Remove the four nuts holding the master cylinder to the brake booster.
6. Remove the master cylinder, clamp, and the gasket from the brake booster.
To install:
7. Adjust the length of the brake booster pushrod as follows:
   a. Install a new gasket on the master cylinder.
   b. Install SST 09737–00010 or equivalent on the master cylinder gasket and lower the pin until its tip slightly touches the piston.
   c. Turn the tool upside down and position it on the brake booster.
   d. Measure the clearance between the booster pushrod and the pin head on the SST tool.
   e. Clearance should be zero. If not, adjust the booster pushrod length until the pushrod lightly touches the pin head.

NOTE: When adjusting the pushrod, depress the brake pedal so that the pushrod sticks out.

8. Bench bleed the master cylinder.

9. Remove the SST tool and install the master cylinder to the brake booster using a new gasket.

10. Install the master cylinder four nuts and torque the nuts to 9 ft. lbs. (13 Nm).

11. Using a line wrench, connect the two brake lines to the master cylinder. Torque the union nuts to 11 ft. lbs. (15 Nm).

12. Connect the level warning switch connector.

13. Fill the brake reservoir with brake fluid and bleed the brake system.

14. Check and adjust brake pedal.

15. Connect the negative battery cable to the battery.

Brake Caliper

REMOVAL AND INSTALLATION

1993–94 Models

1. Raise the vehicle and support it safely.

2. Remove the wheel and tire assembly.

3. Place a suitable container into position to catch the brake fluid.

4. Disconnect the brake hose.

5. Remove the caliper mounting bolts and remove the caliper.

6. Remove the anti-rattle springs, pads, anti-squeal springs, pad guide plates and support plate.

To install:

7. Install the anti-rattle springs, pads, anti-squeal springs, pad guide plates and support plate.

8. Install the caliper. Torque the caliper mounting bolts, as follows:

  a. Previa — Torque the front caliper mounting bolts to 27 ft. lbs. (36 Nm) and rear caliper mounting bolts to 18 ft. lbs. (25 Nm).

  b. T-100 (2-Wheel Drive w/1 ton), Pickup and 4Runner (2-Wheel Drive w/PD60 type disc) — 29 ft. lbs. (39 Nm).

  c. T-100 (2-Wheel Drive w/0.5 ton), Pickup and 4Runner (2-Wheel Drive w/FS17 type disc) — 65 ft. lbs. (88 Nm).

  d. Van — 14 ft. lbs. (20 Nm).

  e. 4-Wheel Drive vehicles to 90 ft. lbs. (123 Nm).

9. Connect the brake hose.

10. Fill the brake reservoir and bleed the brake system.

11. Check for fluid leakage.

12. Install the wheel and tire assembly.

13. Road test the vehicle for proper operation.

1995–97 Models

1. Disconnect the negative battery cable from the battery.

2. Raise and support the vehicle safely.

3. Remove the wheels.

4. Disconnect the brake hose from the caliper by removing the union bolt and two gaskets. Plug the end of the hose to prevent loss of fluid.

5. Remove the bolts that attach the caliper to the torque plate.

6. Lift the bottom of the caliper up and remove the caliper assembly.

To install:

7. Grease the caliper slides and bolts with lithium grease or equivalent. Install the caliper and secure with the bolts. Torque the bolts to:

- T-100 (2-Wheel Drive w/1 ton) — 29 ft. lbs. (39 Nm)
- T-100 (2-Wheel Drive w/0.5 ton), Tacoma, Pickup — 65 ft. lbs. (88 Nm)
- 4Runner — 90 ft. lbs. (123 Nm)
- 4-Wheel Drive vehicles — 65 ft. lbs. (88 Nm)

8. Connect the brake hose to the caliper, using two new washers.

Make sure the flexible hose lock is securely in the lock hole of the caliper. Torque the union bolt to:

- Tacoma — 22 ft. lbs. (30 Nm).
- Pick-up, T-100, and 4 Runner — 11 ft. lbs. (15 Nm)

9. Fill the brake system to the proper level and bleed the brake system.

10. Install the tire and wheel assembly.

11. Top off the brake fluid level in the master cylinder. Check for leaks and proper brake operation.

12. Connect the negative battery cable to the battery.

Disc Brake Pads

REMOVAL AND INSTALLATION

Front Brake Pads

2-Wheel Drive Models

1. Raise the vehicle and support it safely.

2. Remove the wheel and tire assembly.

3. When servicing the front pads, loosen the brake caliper upper side mounting bolt. Loosen and remove the lower side mounting bolt. Lift the

Single-Piston Type:

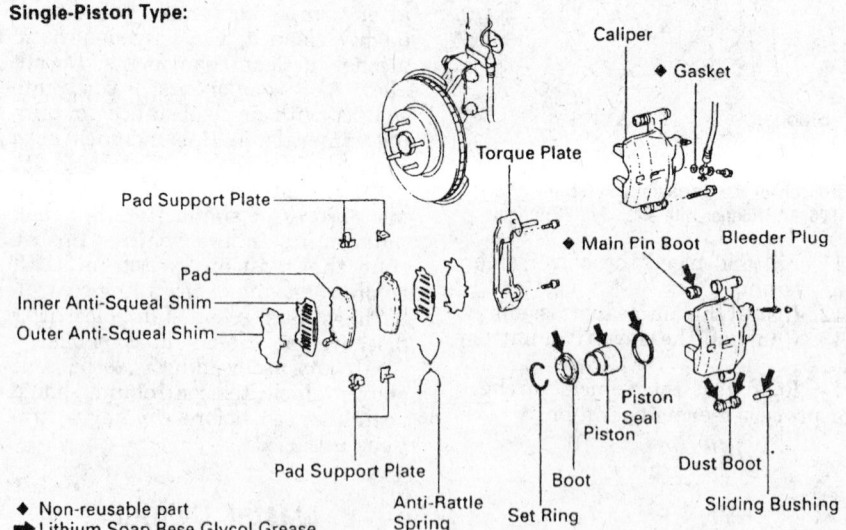

Brake caliper assembly (single piston type) — 1995–97 Models

◆ Non-reusable part
➡ Lithium Soap Bese Glycol Grease

300440

2-Piston Type:

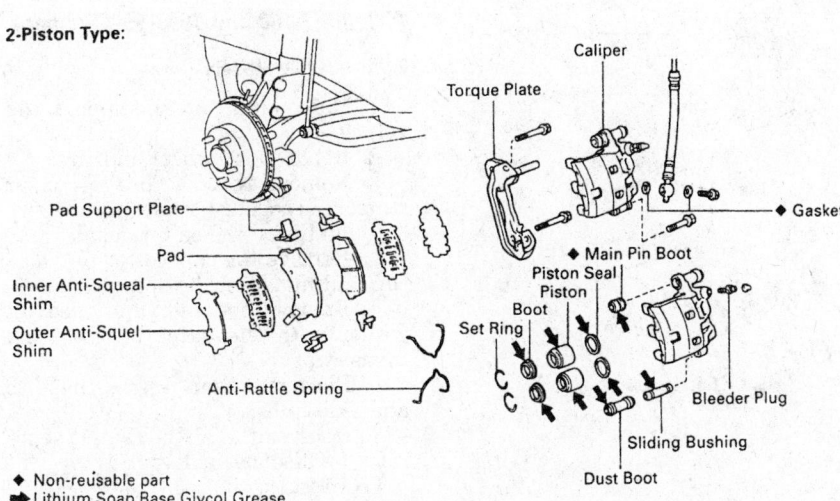

◆ Non-reusable part
➡ Lithium Soap Base Glycol Grease

Brake caliper assembly (dual piston type) — 1995–97 Models

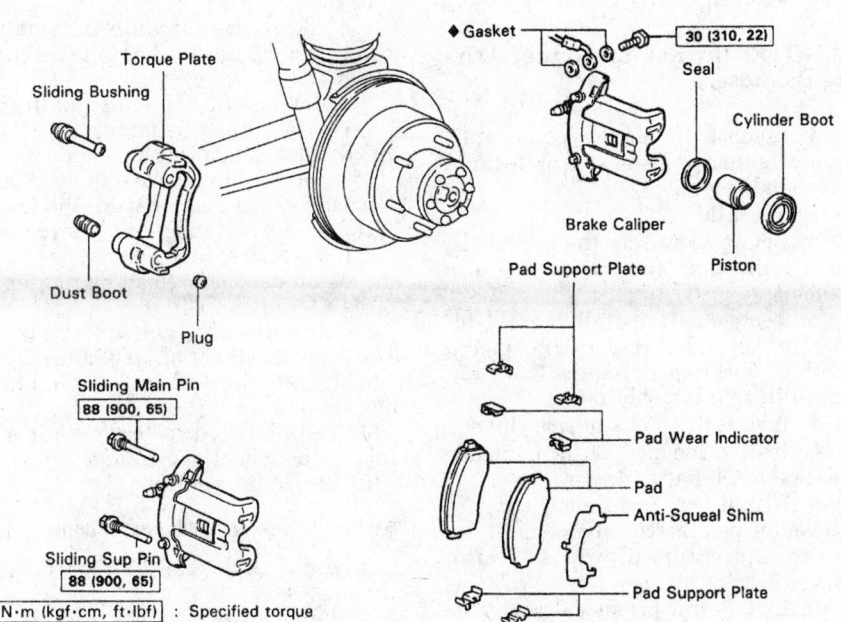

N·m (kgf·cm, ft·lbf) : Specified torque

Rear disc brake components — 1993–94 4-Wheel Drive models

cylinder and suspend it so the hose is not stretched.

NOTE: Do not disconnect the brake hose.

4. If equipped, remove the anti-squeal spring.

5. Remove the brake pads.

To install:

6. Siphon a small amount of brake fluid from the reservoir. Press in the brake caliper piston with a hammer handle or equivalent.

NOTE: Always change the pad on one wheel at a time as there is a possibility of the opposite piston coming out.

7. Before installing the new pads, check the disc thickness and disc runout.

8. Install the pad support plates.

9. Install the anti-squeal shims to each pad.

NOTE: Apply disc brake grease to both side of the inner anti-squeal shims.

10. Install the disc pads so the wear indicator plate is facing downward.

11. If removed, install the anti-squeal springs.

12. Carefully install the brake caliper so the boot is not wedged. Torque the caliper mounting bolts, as follows:

• T-100 (2-Wheel Drive w/1 ton), Pick-up and 4Runner (2-Wheel drive w/PD60 type disc) — 29 ft. lbs. (39 Nm)

• T-100 (2-Wheel Drive w/0.5 ton), Tacoma, Pickup and 4Runner (2-Wheel drive w/FS17 type disc) — 65 ft. lbs. (88 Nm)

13. Install the wheel and tire assembly.

14. Check and adjust the fluid level. Apply the brake pedal several times.

15. Road test the vehicle for proper operation.

4-Wheel Drive Models

1. Raise the vehicle and support it safely.

2. Remove the wheel and tire assembly.

3. Remove the clip, pins, and the anti-rattle spring.

4. Remove the pads and the anti-squeal shims.

5. Remove the caliper, but do not disconnect the brake hose.

— **WARNING** —
Support the caliper with a wire. Do not allow the caliper to hang freely from the brake hose.

Brake Rotor

REMOVAL AND INSTALLATION

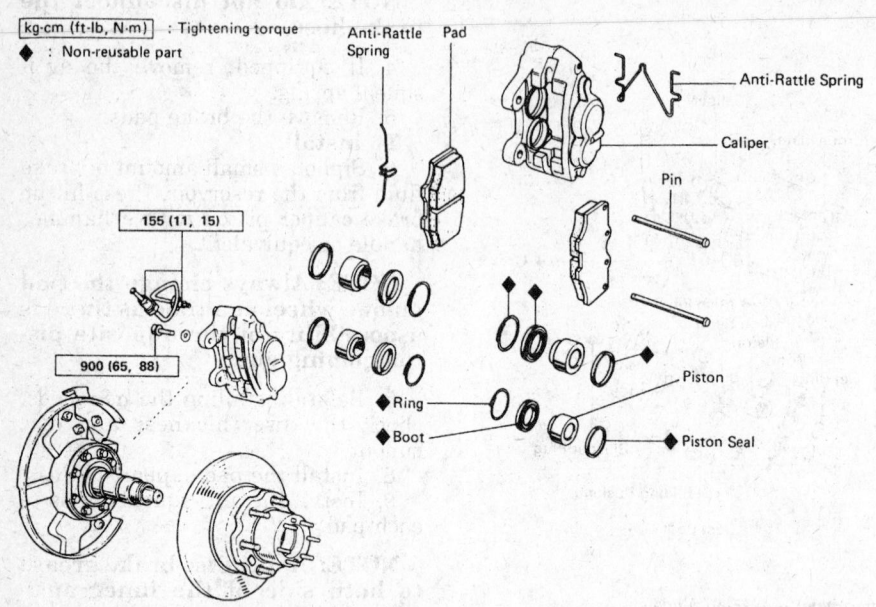

kg·cm (ft-lb, N·m) : Tightening torque
◆ : Non-reusable part

Anti-Rattle Spring — Pad — Anti-Rattle Spring — Caliper — Pin — Piston — Piston Seal — ◆ Ring — ◆ Boot

155 (11, 15)

900 (65, 88)

140897

Front disc brake components — 1993-94 Pick-up and 4Runner 4-Wheel Drive

Pick-up, T-100 and 1993-95 4Runner

2-Wheel Drive Models

1. Raise and safely support the vehicle.
2. Remove the wheel and tire.
3. Remove the disc brake caliper support bracket by removing the two bolts and wire the caliper aside.
4. Remove the cap, cotter pin, lock cap and nut from the spindle.
5. Remove the hub and disc together with the outer bearing and thrust washer.
6. Place matchmarks on the disc and axle hub.
7. Remove the six bolts and separate the disc and axle hub.
   To install:
8. Repack the wheel bearings.
9. Align the marks on the disc and axle hub. Install the six bolts and torque the bolts to 47 ft. lbs. (64 Nm).
10. Install the disc/hub assembly onto the axle shaft, the outer bearing, the thrust washer and the adjusting nut.
11. To adjust the bearing preload, perform the following:
    a. Torque the adjusting nut to 26 ft. lbs. (35 Nm).
    b. Turn the disc/hub assembly 2-3 times, from the left to the right.
    c. Loosen the adjusting nut until it can be turned by hand.
    d. Attach a spring tension gauge to 1 lug on the hub assembly. Pull on the gauge and measure the frictional force. Frictional force = 1.1-31. ft. lbs. (5.0-14.0 N).
    e. Adjust the preload by tightening the nut.
12. Measure the hub axial play. The limit is 0.0020 in. (0.05mm).
13. Install the locknut, cotter pin, and the grease cap.
14. Install the disc brake caliper. Install the wheel and tire assembly.
15. Lower the vehicle.

4-Wheel Drive — Pick-up and 4Runner

1. Raise and safely support the vehicle.
2. Remove the wheel and tire assembly.
3. If equipped with ABS brakes, disconnect the ABS speed sensor from the steering knuckle.
4. Remove the two brake caliper support bracket bolts and wire the caliper to the side. Do not allow the caliper to hang from the brake hose.

To install:

6. Before installing the new pads, check the disc thickness and disc runout.
7. Siphon out a small amount of brake fluid from the reservoir.
8. Temporarily install the old inner brake pad. Press in the pistons with a C-clamp or equivalent. Remove the old inner brake pad.
9. Apply disc brake grease to both sides of the inner anti-squeal shim. Install the anti-squeal shims to the new pads.
10. Install the pads.
11. Install the anti-rattle springs and pins. Install the clip.
12. Install the caliper and the mounting bolts. Torque the mounting bolts to 90 ft. lbs. (123 Nm).
13. Install the wheel and tire assembly.
14. Check and adjust the fluid level. Apply the brake pedal several times.
15. Road test the vehicle for proper operation.

Rear Brake Pads

1. Raise the vehicle and support it safely.
2. Remove the wheel and tire assembly.

3. Remove the brake caliper and suspend it with a wire so the hose is not stretched or stressed.

NOTE: Do not disconnect the brake hose.

4. Remove the brake pads, anti-squeal shim, pad support plates and wear indicators.
   To install:
5. Before installing the new pads, check the disc thickness and disc runout.
6. Temporarily install the old inner brake pad. Press in the piston with a C-clamp or equivalent. Remove the old inner brake pad.
7. Install the pad support plates.
8. Install the pad wear indicator plate to each pads.
9. Install the anti-squeal shim to the outer pad. Install the pads so the wear indicator plate is facing upward.
10. Install the brake caliper. Torque the main sliding pin and the sub pin to 65 ft. lbs. (88 Nm).
11. Install the wheel and tire assembly.
12. Apply the brake pedal several times.
13. Road test the vehicle for proper operation.

5. Remove the free-wheel hub:

a. Set the control handle to FREE.

b. Remove the cover bolts and pull off the cover.

c. Remove the center bolt with washer.

d. Remove the mounting nuts and washer to the hub body.

e. Using a brass bar and hammer, tap on the bolts head and remove the cone washer.

f. Pull off the free wheel hub body and gasket.

6. If equipped without a free wheeling hub, remove the flange for the axle hub as follows:

a. Remove the grease cap from the flange.

b. Remove the bolt from the flange.

c. Remove the six mounting nuts to the flange.

d. Using a brass bar and hammer, tap on the bolts head and remove the six cone washers.

e. Install two bolts to the flange. Tighten the bolts to remove the flange.

f. Remove the gasket for the flange.

7. Using a flat bladed tool, release the taps on the lock washer.

8. Using SST 09607–60020 or equivalent, remove the locknut.

9. Remove the lock washer and adjusting nut.

10. Remove the claw washer.

11. Remove the axle hub and disc as an assembly with the outer wheel bearing.

12. Place matchmarks on the hub and rotor assembly. Remove the six bolts and separate the hub and rotor.

13. Remove the inner grease seal and the inner wheel bearing from the hub.

To install:

14. Repack the wheel bearings and install the inner wheel bearing with a new grease seal.

15. Install the axle hub to the disc and torque the bolts to 47 ft. lbs. (64 Nm).

16. Install the hub on the spindle. Install the outer bearing and claw washer.

17. Adjust the preload:

a. Torque the bearing adjusting nut to 43 ft. lbs. (59 Nm).

b. Turn the hub right and left 2 or 3 times and re-torque.

c. Loosen the nut until it can be turned by hand.

d. Re-torque to 18 ft. lbs. (25 Nm).

18. Check the bearing preload with a spring tension gauge. The preload should be 6.4–12.8 lbs. (28–57 N).

19. Install the lock washer and nut and torque to 35 ft. lbs. (47 Nm). Make sure the bearing has no endplay. Recheck the bearing preload.

20. Secure the locknut by bending one lock washer tooth inward and another outward.

21. If equipped without a free wheeling hub, install the flange as follows:

a. Place a new gasket in position on the axle hub.

b. Install the flange to the axle hub.

c. Install the six cone washers, plate washers and nuts.

d. Install the six nuts and torque the nuts to 23 ft. lbs. (31 Nm).

e. Install the bolt and torque the bolt to 13 ft. lbs. (18 Nm).

f. Install the grease cap.

22. If equipped with a free wheeling hub, install the hub as follows:

a. Place a new gasket in position on the front axle hub.

b. Install the free wheeling hub body with the six cone washers and nuts. Torque the nuts to 23 ft. lbs. (31 Nm).

c. Install the bolt with the washer. Torque the bolt to 13 ft. lbs. (18 Nm).

d. Apply multi purpose grease to the inner hub splines.

e. Set the control handle and clutch to the FREE position.

f. Place a new gasket in position on the cover.

g. Install the cover to the hub body with the follower pawl tabs aligned with the non-toothed portions of the hub body.

h. Install the mounting bolts and tighten the cover bolts to 7 ft. lbs. (10 Nm).

23. Install the brake caliper support bracket to the steering knuckle and torque the two bolts to 90 ft. lbs. (123 Nm).

24. If equipped with ABS brakes, connect the ABS speed sensor to the steering knuckle.

25. Install the front wheel and lower the vehicle.

26. Check the ABS speed sensor signal.

4-Wheel Drive — T-100

1. Raise and safely support the vehicle.

2. Remove the wheel and tire assembly.

3. If equipped with ABS brakes, disconnect the ABS speed sensor from the steering knuckle.

4. Disconnect the knuckle arm with the brake line bracket from the steering knuckle by removing the two bolts.

5. Remove the two brake caliper support bracket bolts and support the caliper to the side with a wire. Do not allow the caliper to hang from the brake hose.

6. Remove the flange for the axle hub as follows:

a. Remove the grease cap from the flange.

b. Remove the bolt from the flange.

c. Remove the six mounting nuts to the flange.

d. Using a brass bar and hammer, tap on the bolts head and remove the six cone washers.

e. Install two bolts to the flange. Tighten the bolts to remove the flange.

f. Remove the gasket for the flange.

7. Using a flat prying tool, release the tabs on the lock washer.

8. Using SST 09607–60020 or equivalent, remove the locknut.

9. Remove the lock washer and adjusting nut.

10. Remove the claw washer.

11. Remove the axle hub and disc as an assembly. Remove the outer bearing with the hub and disc.

To install:

12. Repack the wheel bearings.

13. Install the hub and rotor on the spindle. Install the outer bearing and claw washer.

14. Adjust the preload:

a. Torque the bearing adjusting nut to 43 ft. lbs. (59 Nm).

b. Turn the hub right and left 2 or 3 times and re-torque.

c. Loosen the nut until it can be turned by hand.

d. Re-torque to 18 ft. lbs. (25 Nm).

e. Check that there is no bearing end-play.

15. Install the lock washer and nut. Torque to 35 ft. lbs. (47 Nm). Secure the locknut by bending one of the lock washer tooth inward and another outward.

16. Install the flange as follows:

a. Place a new gasket in position on the axle hub.

b. Install the flange to the axle hub.

c. Install the six cone washers, plate washers, and nuts.

d. Install the six nuts and torque the nuts to 23 ft. lbs. (31 Nm).

e. Install the bolt and torque the bolt to 13 ft. lbs. (18 Nm).

f. Install the grease cap.

17. Install the brake caliper support bracket to the steering knuckle and torque the two bolts to 90 ft. lbs. (123 Nm).

18. Clean the threads of the bolts and steering knuckle.

19. Apply sealant to the bolt threads and connect the knuckle arm with the brake line bracket to the steering knuckle. Torque the bolts to 135 ft. lbs. (183 Nm).

20. If equipped with ABS brakes, connect the ABS speed sensor to the steering knuckle.

21. Install the front wheel and lower the vehicle.

22. Check the ABS speed sensor signal.

1996–97 4Runner

1. Disconnect the negative battery cable from the battery.

2. Remove the wheel(s).

3. Remove the caliper.

4. Hang the caliper from a piece of wire. Do not disconnect the brake hose.

─── **WARNING** ───
Do not allow the caliper to hang freely from the vehicle. Always support the caliper with a wire from the vehicle.

5. Remove the disc from the axle hub.

To install:

6. Position the new rotor disc onto the hub.

7. Install the caliper. Torque the installation bolts to 90 ft. lbs. (123 Nm).

8. Install the wheels.

9. Lower the vehicle and tighten the lug nuts to 83 ft. lbs. (110 Nm).

10. Before moving the vehicle, make sure to pump the brake pedal to seat the brake pads against the rotors.

11. Connect the negative battery cable to the battery.

Tacoma

2-Wheel Drive Models

1. Raise and safely support the front of the vehicle. Place the jackstands under the frame of the vehicle.

2. Remove the brake caliper support bracket by removing the two bolts. Support the brake caliper with a piece of wire. Do not allow the caliper to hang from the brake hose.

3. Remove the cotter pin, lock cap, nut, and the claw washer from the axle hub and disc.

4. Remove the axle hub with the disc from the steering knuckle. Do not drop the outer bearing when removing the hub.

5. If it is necessary to separate the hub and rotor, place matchmarks on the hub and rotor. Remove the five bolts and remove the hub from the rotor.

To install:

6. Connect the hub to the rotor and install the five bolts. Torque the five bolts to 47 ft. lbs. (64 Nm).

7. Clean all parts.

8. Repack the bearings with multi purpose grease and apply the same grease to the outer bearings.

9. Install the hub to the steering knuckle.

10. Install the claw washer and nut to holding the axle hub to the steering knuckle.

11. To adjust the bearing preload, perform the following:

   a. Torque the adjusting nut to 25 ft. lbs. (34 Nm).

   b. Turn the disc/hub assembly 2–3 times, from the left to the right.

   c. Loosen the adjusting nut until it can be turned by hand.

   d. Attach a spring tension gauge to 1 lug on the hub assembly. Pull on the gauge and measure the frictional force. The frictional force should be 1.3–4.0 ft. lbs. (6.0–18.0 N).

   e. Adjust the preload by tightening the nut.

12. Measure the hub axial play. Limit 0.0020 in. (0.05mm).

13. Install the locknut, cotter pin, and the grease cap.

14. Install the disc brake caliper by installing the two bolts. Torque the two bolts to 80 ft. lbs. (108 Nm).

15. Install the wheel and tire assembly.

16. Lower the vehicle.

4-Wheel Drive Models

1. Raise and safely support the vehicle.

2. Remove the wheel and tire assembly.

3. Without a FREE wheeling hub, remove the grease cap, the cotter pin, the lock cap and while applying the brakes remove the locknut.

4. If equipped with FREE wheeling hub, remove the free wheel hub as follows:

   a. Set the control handle to FREE.

   b. Remove the cover bolts and pull off the cover.

   c. Remove the center bolt with washer.

   d. Remove the mounting nuts and washer to the hub body.

   e. Using a brass bar and hammer, tap on the bolts head and remove the cone washer.

   f. Pull off the free wheel hub body and gasket.

   g. Using a snapring expander, remove the snapring from the end of the halfshaft.

5. Remove the two brake caliper support bracket bolts and wire the caliper to the side. Do not allow the caliper to hang from the brake hose.

6. Remove the rotor from the steering knuckle.

To install:

7. Install the brake rotor to the wheel hub.

8. Install a new snapring on the end of the halfshaft.

9. If equipped with a free wheeling hub, install the hub as follows:

   a. Place a new gasket in position on the front axle hub.

   b. Install the free wheeling hub body with the six cone washers and nuts. Torque the nuts to 23 ft. lbs. (31 Nm).

   c. Install the bolt with the washer. Torque the bolt to 13 ft. lbs. (18 Nm).

   d. Apply multi purpose grease to the inner hub splines.

   e. Set the control handle and clutch to the FREE position.

   f. Place a new gasket in position on the cover.

   g. Install the cover to the hub body with the follower pawl tabs aligned with the non-toothed portions of the hub body.

   h. Install the mounting bolts and tighten the cover bolts to 7 ft. lbs. (10 Nm).

10. If equipped without a free wheeling hub, Install the locknut to the halfshaft and torque the nut to 174 ft. lbs. (235 Nm).

11. Install the lock cap and cotter pin and the grease cab to the hub.

12. Install the front wheels and lower the vehicle.

Brake Drums

REMOVAL AND INSTALLATION

1993–94 Models

1. Raise the vehicle and support it safely.

2. Remove the rear wheel and tire assembly.

3. Remove the brake drum retaining screws, if equipped. Remove the brake drum.

4. If equipped with double tire, remove the rear axle shaft and remove the drum with axle hub.

NOTE: If the brake drum cannot be removed easily, insert a screwdriver through the hole in the backing plate. Hold the adjuster lever away from the adjuster and reduce the brake shoe clearance by turning the adjuster wheel.

5. Installation is the reverse of the removal procedure.

1995–97 Models

1. Raise and safely support the vehicle.
2. Remove the rear wheel(s).
3. Remove the brake drum from the axle hub. If there is difficulty in removing the drum, insert a suitable tool through the hole in the rear of the backing plate, and hold the automatic adjusting lever away from the adjuster. Using another suitable tool at the same time, reduce the brake shoe adjuster by turning the adjusting wheel.

To install:

4. Install the brake drum and pull the parking brake lever all the way up until a clicking sound can no longer be heard.

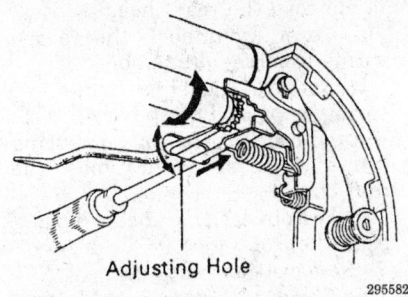

Adjusting Hole

295582

Use a brake adjusting tool (brake spoon) and a prytool to adjust the brake shoes through the adjusting hole — 1995–97 Models

5. Verify that the rear wheels will not turn. If the rear wheels turn, adjust the parking brake cable as necessary.
6. Release the parking brake and remove the brake drum. Measure the brake drum inside diameter and diameter of the brake shoes. Check that the difference between the diameters is the correct shoe clearance. Clearance is: 0.024 in. (6mm).
7. If the brake shoe clearance is not correct, adjust the brake shoes until the clearance is correct.
8. Install the brake drum, replace the wheel(s), and safely lower the vehicle.

9. Road test the vehicle for proper brake operation.

Brake Shoes

REMOVAL AND INSTALLATION

1993–94 Models

1. Raise the vehicle and support it safely.
2. Remove the rear wheel and tire assembly.
3. Remove the brake drum retaining screws, if equipped. Remove the brake drum.
4. If equipped with double tire, remove the rear axle shaft and remove the drum with axle hub.

NOTE: If the brake drum cannot be removed easily, insert a screwdriver through the hole in the backing plate. Hold the adjuster lever away from the adjuster and reduce the brake shoe clearance by turning the adjuster wheel.

5. Remove the front shoe return spring.
   a. Remove the hold-down spring, cups and pin.
   b. Remove the front shoe and anchor spring.
6. Remove the rear shoe hold-down spring, cups and pin.
   a. Remove the rear shoe with strut.
   b. Disconnect the parking brake cable from the lever.
   c. Remove the adjusting lever, spring and strut from the rear shoe.

To install:

7. Before installing the brake shoes, apply high temperature grease to the backing plate shoe contact surfaces. Apply high temperature grease to the adjuster bolt threads and ends.
8. Install the strut, adjusting lever and spring to the rear shoe.
   a. Install the parking brake cable to the lever.
   b. Set the rear shoe in place with the end of the shoe inserted in the wheel cylinder and the other end in the anchor plate.
   c. Install the pin and shoe hold-down springs.
9. Install the anchor spring between the front and rear shoes.
   a. Set the front shoe in place with the end of the shoe inserted in the wheel cylinder and the strut in place.
   b. Install the shoe hold-down spring and pin.
10. Install the return spring.

Bleeder Plug
11 (110, 8)
Piston Cup
Boot
C-Washer
Piston
Adjusting Lever
Spring
Wheel Cylinder
Pin
Hole Plug
E-Ring
Parking Brake Lever
Adjuster
Rear
Cup
Front Shoe
Return Spring
Shoe Hold Down Spring
Anchor Spring
Adjusting Lever Spring
Cup
Gasket
Shoe Hold-Down Spring
Brake Drum

295583

N·m (kgf·cm, ft·lbf) : Specified torque
◆ Non-reusable part

Rear brake drums and related components — 1995–97 2-Wheel Drive models shown, others similar

11. Check the operation of the automatic adjuster mechanism:

a. Apply the parking brake lever and verifying the adjusting bolt turns.

b. Adjust the strut to where it is the shortest possible length.

c. Install the brake drum.

d. Apply the parking brake lever until the clicking sound can no longer be heard.

12. Check the clearance between the brake shoes and drum:

a. Remove the brake drum.

b. Measure the brake drum inside diameter and diameter of the brake shoes. The difference is "Shoe-to-drum clearance" and should be approximately 0.024 in. (0.6mm). If incorrect, check the parking brake system.

13. Install the brake drum.

14. Install the rear wheel and tire assembly.

15. Road test the vehicle for proper operation.

1995–97 Models

1. Loosen the rear wheel lug nuts slightly.

2. Block the front wheels, raise the rear of the vehicle, and safely support it with jackstands.

3. Remove the wheel lug nuts and the wheel.

4. Remove the brake drum.

5. If the drum is difficult to remove, perform the following:

a. Insert a flat prying tool through the hole in the brake drum and hold the automatic adjusting lever away from the adjuster.

b. Reduce the brake shoe adjustment by turning the adjuster bolt with a brake tool.

c. The drum should now be loose enough to remove without much effort.

NOTE: Do not depress the brake pedal once the brake drum has been removed.

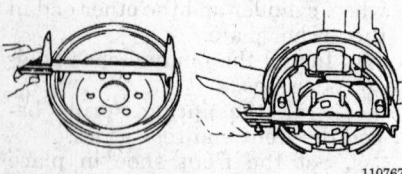

Checking shoe-to-drum clearance — 1993–94 Models

110767

6. Remove the rear shoe.

a. Carefully unhook the return spring from the brake shoe.

b. Remove the shoe hold-down spring, cups and the pin.

c. Disconnect the anchor spring from the rear shoe and remove the rear shoe.

d. Disconnect the anchor spring from the front shoe.

7. Remove the front shoe.

a. Remove the shoe hold-down spring, cups and pin.

b. Remove the return spring from the front shoe.

c. Remove the front shoe with the adjuster.

d. Disconnect the parking brake cable from the front shoe.

To install:

8. Inspect the shoes for signs of unusual wear or scoring.

9. Check the wheel cylinder for any sign of fluid seepage or frozen pistons.

10. Clean and inspect the brake backing plate and all other components. Check that the brake drum inner diameter is within specified limits. Lubricate the backing plate at the positions the brakes come in contact with the backing plate. Also lubricate the anchor plate.

11. Mount the automatic adjuster assembly onto a new rear brake shoe.

12. Install the front shoe.

a. Install the parking brake cable to the front shoe.

b. Install the front shoe with the adjuster.

c. Install the return spring to the front shoe.

d. Install the shoe hold-down spring, cups and pin.

13. Install the rear shoe.

a. Install the anchor spring to the front shoe.

b. Install the anchor spring to the rear shoe and install the rear shoe.

c. Install the shoe hold-down spring, cups and the pin.

d. Hook the return spring to the brake shoe.

14. Install the brake drum.

15. Adjust the brake shoes until a slight drag is felt when the drum is spun by hand.

16. Remove the brake drum and check the clearance between brake shoes and brake drum. Adjust the clearance to specification.

17. Pull the parking lever all the way up until a clicking sound can no longer be heard. Verify that the drum doesn't turn. If the drum turns, adjust the parking brake cable.

18. Install the rear wheels, tighten the wheel lug nuts and lower the vehicle.

19. Retighten the wheel lug nuts and pump the brake pedal a few times before moving the vehicle. Adjust the rear brakes again if necessary.

20. Check the level of brake fluid in the master cylinder, then perform a test drive.

21. Connect the negative battery cable to the battery.

Wheel Cylinder

REMOVAL AND INSTALLATION

All Models

1. Raise and safely support the vehicle.

2. Remove the wheel and tire assemblies.

3. Remove the brake drum and the brake shoes.

NOTE: Do not depress the brake pedal once the brake drum has been removed.

4. Place a suitable container into position to catch the brake fluid. Disconnect and plug the brake line from the wheel cylinder.

5. Remove the wheel cylinder mounting bolts and remove the wheel cylinder from the vehicle.

6. Check the wheel cylinder for any sign of fluid seepage or frozen pistons. Replace the wheel cylinder, if necessary.

To install:

7. Position the wheel cylinder onto the brake backing plate, then install the mounting bolts to 7 ft. lbs. (10 Nm).

8. Attach the hydraulic brake line to the wheel cylinder and tighten the brake line fitting to 11 ft. lbs. (15 Nm).

9. Install the rear brake shoes and brake drum.

10. Install the rear wheel and tire assemblies.

11. Lower the vehicle.

12. Bleed the brake system and road test the vehicle for proper operation.

Parking Brake Cable

ADJUSTMENT

1993–94 Models and 1995–97 T-100

Pull the parking brake lever all the way up and count the number of clicks. The parking brake lever should travel approximately 11–17 clicks at 44 lbs. (196 N).

NOTE: Before adjusting the parking brake, make sure that the rear brake shoe clearance has been adjusted.

2-Wheel Drive Models

1. Working under the vehicle, tighten the adjusting nut at the equalizer until the travel is within limits and there is no drag at the rear shoes.
2. Apply the parking brake several times and again check that there is no drag with the brake released.

4-Wheel Drive Models

1. Working under the vehicle, tighten the bell crank stopper screw until the play at the rear brake links is gone, then loosen the nut one full turn. Tighten the locknut.
2. Tighten one of the adjusting nuts on the intermediate lever while loosening the other, until the travel is correct. Tighten the locknuts.
3. Confirm that the bell crank is in contact with the backing plate.

1996–97 4Runner

Pull the parking brake lever all the way up and count the number of clicks. The parking brake lever should travel approximately 7–9 clicks at 44 lbs. (196 N) of force.

NOTE: Before adjusting the parking brake, make sure that the rear brake shoe clearance has been adjusted.

1. Remove the console box.
2. Loosen or tighten the adjusting nut (at the parking brake lever) until the lever travel is correct.
3. Apply the parking brake several times and again check that there is no drag with the brake released.
4. Install the console box.

1995 4Runner, Pick-up and 1995–97 Tacoma

2-Wheel Drive Models

Pull the parking brake lever all the way up and count the number of clicks. The parking brake lever should travel approximately 12–18 clicks at 44 lbs. (196 N) of force.

NOTE: Before adjusting the parking brake, make sure that the rear brake shoe clearance has been adjusted.

1. Loosen the locknut and turn the adjusting nut until the travel is correct.

2. Then tighten the locknut to 9 ft. lbs. (13 Nm).
3. After adjusting the parking brake, confirm that the rear brakes are not dragging.

4-Wheel Drive Models

Pull the parking brake lever all the way up and count the number of clicks. The parking brake lever should travel approximately 12–18 clicks at 44 lbs. (196 N) of force.

NOTE: Before adjusting the parking brake, make sure that the rear brake shoe clearance has been adjusted.

1. Tighten one of the adjusting nuts of the intermediate lever while loosening the other until the travel of the brake lever is correct.
2. Then tighten the two adjusting nuts to 9 ft. lbs. (13 Nm).
3. After adjusting the parking brake, confirm that the rear brakes are not dragging.

Brake System Bleeding

Start the bleeding procedure at the caliper or wheel cylinder the furthest from the master cylinder.

1. Connect a vinyl tube to the bleeder screw on the brake cylinder and submerge the other end of the tube in a transparent container half filled with clean brake fluid.
2. Pump the brake pedal several times and loosen the bleeder screw with the pedal held down.
3. When brake fluid stops coming out of the tube with the brake pedal held to the floor, tighten the bleeder screw and release the brake pedal.
4. Repeat Steps 2 and 3 until no air bubbles can be seen in the container.
5. Repeat the procedure for each wheel.
6. Check the level in the master cylinder. Add fluid as necessary.

FRONT SUSPENSION

Strut

REMOVAL AND INSTALLATION

1996–97 4Runner

1. Raise and safely support the front of the vehicle.
2. Remove the front wheel.

3. Disconnect the strut from the lower control arm by removing the bolt.
4. Remove the three nuts and the strut assembly.

To install:

5. Install the strut assembly and torque the three nuts to 47 ft. lbs. (64 Nm).
6. Install the lower bolt to hold the strut assembly to the lower control arm. Torque the bolt to 101 ft. lbs. (135 Nm).
7. Install the wheels and lower the vehicle.
8. Check the vehicle alignment.

4-Wheel Drive Tacoma

1. Raise and safely support the vehicle. Place the jack stands under the frame of the vehicle.
2. Remove the nut and bolt holding the strut to the lower control arm.
3. Remove the three nuts holding the strut to the strut tower.
4. Remove the strut from the vehicle.

———— **WARNING** ————
Never loosen the center nut on the strut unless a spring compressor is installed. Serious injury or vehicle damage may result.

To install:

5. Install the strut to the vehicle.
6. Install the three nuts to hold the strut to the strut tower. Torque the nuts to 47 ft. lbs. (64 Nm).
7. Install the lower bolt and nut to hold the strut to the lower control arm. Torque the nut to 101 ft. lbs. (135 Nm).
8. Install the front wheels and lower the vehicle.
9. Check the vehicle alignment.

Shock Absorber

REMOVAL AND INSTALLATION

T-100, Tacoma, Pick-up and 1993–95 4Runner

2-Wheel Drive Models

1. Raise and safely support the front of the vehicle.
2. Remove the front wheel.
3. Disconnect the shock absorber from the lower control arm by removing the two bolts.
4. Remove the nut, retainers, and the cushion from the top of the shock absorber.
5. Remove the shock absorber from the vehicle.

6. Remove the retainers and cushion from the shock absorber.

To install:

7. Install the retainers and cushion to the shock absorber.

8. Install the shock absorber to the vehicle

9. Install the retainers, cushion and nut to the top of the shock absorber. Torque the nut to 18 ft. lbs. (25 Nm).

10. Install the lower bolts to hold the shock absorber to the lower control arm. Torque the bolts to 13 ft. lbs. (19 Nm).

11. Install the wheels and lower the vehicle.

4-Wheel Drive Models

1. Raise and safely support the front of the vehicle.

2. Remove the front wheel.

3. Disconnect the shock absorber from the lower control arm by removing the nut, washer, and the through-bolt.

4. Remove the nut, retainers, and cushion from the top of the shock absorber.

5. Remove the shock absorber from the vehicle.

6. Remove the retainers and cushion from the shock absorber.

To install:

7. Install the retainers and cushion to the shock absorber.

8. Install the shock absorber to the vehicle

9. Install the retainers, cushion, and the nut to the top of the shock absorber. Torque the nut to 18 ft. lbs. (25 Nm).

10. Install the lower through-bolt, washer, and the nut to hold the shock absorber to the lower control arm. Torque the bolt to 101 ft. lbs. (137 Nm).

11. Install the wheels and lower the vehicle.

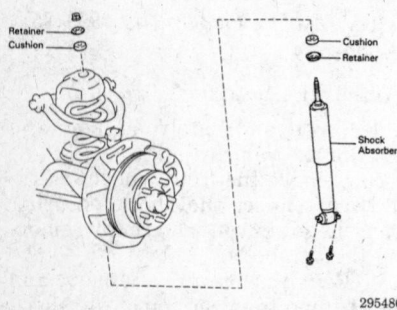

Shock absorber component assembly (2-Wheel Drive) — Tacoma

Coil Spring

REMOVAL AND INSTALLATION

1996–97 4Runner

1. Raise and safely support the vehicle.

2. Remove the strut assembly.

3. Using SST 09727–30030 or equivalent, compress the coil spring until there is clearance on both ends.

4. Remove the support center nut.

— **CAUTION** —
Do not remove the center nut without compressing the spring. Failure to do so may cause property damage or personal injury.

5. Remove the two retainers, cushion, suspension support and the coil spring.

To install:

6. Compress the coil spring and install to the strut.

7. Fit the lower end of the coil spring into the gap of the spring seat of the strut.

8. Install the suspension support.

   a. Install the two retainers, suspension support and the cushion to the rod.

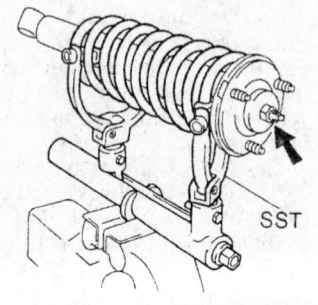

309685

Using a spring compressor to remove the coil spring — 4Runner

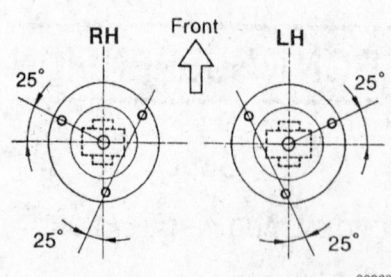

309686

Aligning the strut support to the strut's lower bushing — 4Runner

b. Temporarily tighten the support center nut.

c. Align the suspension support with the strut lower bushing.

d. Face the lower end of the coil spring to the outside of the vehicle.

e. Remove the spring compressor and torque the center nut to 18 ft. lbs. (25 Nm).

9. Install the shock absorber.

Tacoma

2-Wheel Drive Models

1. Raise and safely support the front of the vehicle. Place the jack stands under the frame of the vehicle.

2. Remove the shock absorber from the suspension by removing the two bottom bolts and top nut.

3. Using SST 09727–22011, or equivalent (spring compressor), compress the coil spring.

— **CAUTION** —
The proper tools for this procedure must be used. The spring is under high pressure and can cause serious injury if not properly removed and installed.

4. Remove the nut and disconnect the sway bar link from the lower control arm.

5. Remove the two sway bar bracket bolts on the side of the suspension that the lower control arm is being removed. This will allow access to the lower control arm through-bolt.

6. Support the steering knuckle and upper control arm.

— **WARNING** —
Before removing the lower control arm, make sure the coil spring is safely compressed.

7. Remove the cotter pin and nut from the lower ball joint.

8. Using SST 09628–62011, or equivalent, disconnect the lower ball joint from the lower control arm.

9. Remove the nut from the lower control arm set bolt.

10. Remove the nut from the strut bar front set bolt.

11. Pull out the two bolts and remove the lower control arm and strut bar as an assembly. When the lower control arm is removed, set the coil spring aside.

To install:

12. Place the end of the coil spring in contact with the lower control arm seat.

13. Install the lower control arm, spring, and strut arm to the suspen-

sion. Install the strut arm bolt and lower control arm bolt.

14. Install the nuts for the strut arm bolt and lower control arm bolt. Do not torque the bolts at this time.

15. Connect the lower control arm to the lower ball joint. Install the nut and torque the nut to 80 ft. lbs. (110 Nm). Install a new cotter pin.

16. Remove the support from the upper control arm and steering knuckle.

17. Connect the sway bar bracket to the suspension and install the two bolts. Torque the bolts to 22 ft. lbs. (29 Nm).

18. Connect the sway bar link to the lower control arm and install the nut. Torque the nut to 29 ft. lbs. (39 Nm).

19. Making sure the coil spring is in its correct position, slowly remove the spring compressor from the coil.

20. Install the shock absorber. Torque the top nut to 18 ft. lbs. (25 Nm) and the bottom two bolts to 29 ft. lbs. (39 Nm).

21. Install the wheel and lower the vehicle.

22. Stabilize the suspension by pushing up and down on the vehicle.

23. Torque the strut bar nut and bolt to 221 ft. lbs. (300 Nm) and the lower control arm bolt and nut to 148 ft. lbs. (200 Nm).

24. Check the front wheel alignment.

4-Wheel Drive Models

1. Raise and safely support the vehicle. Place the jack stands under the frame of the vehicle.

2. Remove the nut and bolt holding the strut to the lower control arm.

3. Remove the three nuts holding the strut to the strut tower.

4. Remove the strut from the vehicle.

------ CAUTION ------

The proper tools for this procedure must be used. The spring is under high pressure and can cause serious injury if not properly removed and installed.

5. Using SST 09727–30030, or equivalent, compress the coil spring until there is a clearance on both ends.

6. Remove the strut center nut.

7. Remove the suspension support and coil spring.

8. Remove the insulator from the suspension support.

To install:

9. Install the insulator to the suspension support.

NOTE: Match the bolt of the suspension support with the cut out part of the insulator.

10. Using the coil spring compressor, compress the coil spring and install the coil spring to the strut.

NOTE: Fit the lower end of the coil spring into the gap of the spring seat of the strut.

11. Install the suspension support to the strut rod.

12. Temporarily tighten a new suspension support center nut.

13. Position the suspension support so that a line drawn between the two bolts would be parallel to the direction of the lower bushing.

14. Remove the compressor from the spring.

15. Torque the strut center nut to 22 ft. lbs. (29 Nm).

16. Install the strut to the vehicle.

17. Install the three nuts to hold the strut to the strut tower. Torque the nuts to 47 ft. lbs. (64 Nm).

18. Install the lower bolt and nut to hold the strut to the lower control arm. Torque the nut to 101 ft. lbs. (135 Nm).

19. Install the front wheels and lower the vehicle.

Torsion Bar

REMOVAL AND INSTALLATION

1993–95 4Runner, Pick-up and 1993–97 T-100

2-Wheel Drive Models

NOTE: Great care must be taken to make sure springs are not mixed after removal. It is strongly suggested that before removal, each spring be marked with paint, showing front and

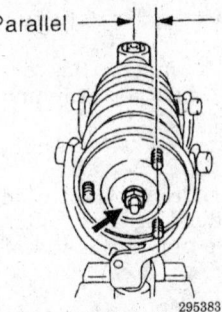

Parallel

295383

Aligning the suspension support with lower bushing (4-Wheel Drive) — Tacoma

rear of spring and from which side of the vehicle it was taken. If the springs are installed backwards or on the wrong sides of the vehicle, they could fracture. If replacing the springs, it is not necessary to mark them.

1. Raise the front of the vehicle and support it safely.

2. Remove the front wheel and tire assembly.

3. Remove the dust cover, if equipped.

4. Place matchmarks on the torsion bar spring, anchor arm, and torque arm.

5. Before removing, measure the distance of protruding bolt end (anchor arm adjusting bolt). This distance is from the top of the anchor arm adjusting locknut to the top of the adjusting bolt end.

NOTE: Use this measurement for reference when adjusting the chassis ground clearance.

6. Remove the anchor arm locknut.

7. Loosen the adjusting nut until all tension on the torsion bar is removed.

8. Remove the torque arm mounting nuts.

9. Remove the anchor arm from the adjusting bolt.

10. Remove the tension bar, torque arm, and the anchor arm.

To install:

NOTE: There are left and right identification marks on the rear end of the torsion bar springs. Be careful not to interchange the torsion bar springs.

11. Apply a light coat of MP grease to the splines of the torsion bar spring.

12. Align the toothless portion of the spring and install the torque arm to the torsion bar spring.

13. Align the toothless portion of the spring and install the anchor arm to the torsion bar spring.

14. Install the torsion bar spring torque arm side and install the anchor arm to the adjusting bolt. Torque the torque arm nuts to 36 ft. lbs. (49 Nm).

15. Tighten the adjusting nut so that the bolt protrusion is equal to that during removal.

16. Install the wheel and tire assembly.

17. Lower the vehicle and bounce the vehicle several times to stabilize the suspension. Check and adjust the chassis ground clearance by turning the adjusting nut.

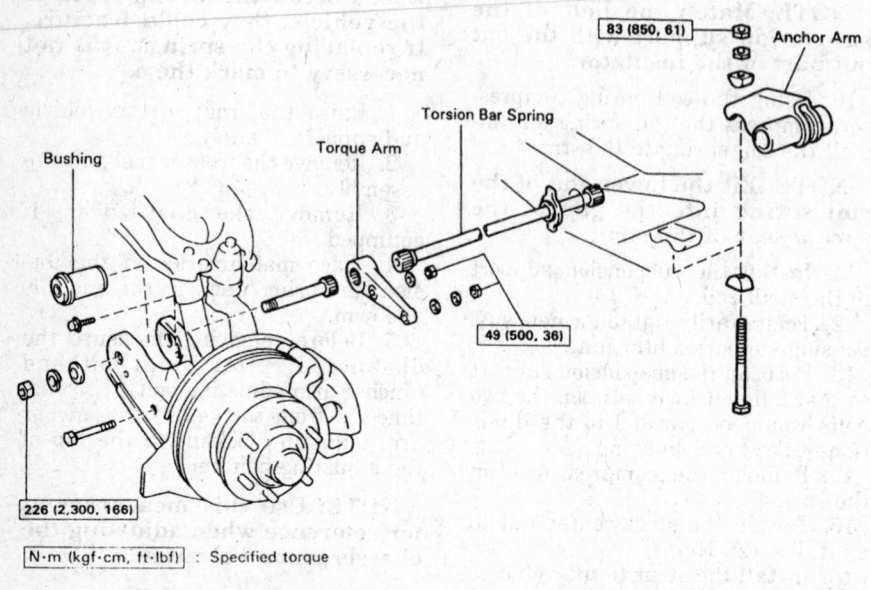

Torsion bar spring component assembly (2-Wheel Drive) — 4Runner, Pick-up and T-100

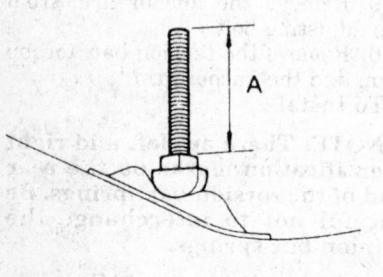

Anchor arm locknut measurement — 1995 4Runner, Pick-up and 1995–97 T-100

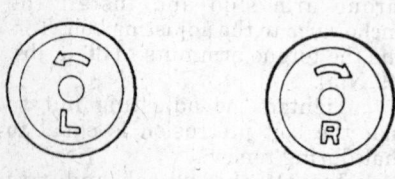

Left and right matchmarks on the end of the torsion bar springs — 1995 4Runner, Pick-up and 1995–97 T-100

18. Torque the locknut to 61 ft. lbs. (83 Nm).

19. Install the dust cover.

4-Wheel Drive Models

NOTE: Great care must be taken to make sure springs are not mixed after removal. It is strongly suggested that before removal, each spring be marked with paint, showing front and rear of spring and from which side of the vehicle it was taken. If the springs are installed backwards or on the wrong sides of the vehicle, they could fracture. If replacing the springs, it is not necessary to mark them.

1. Raise and safely support the vehicle.

2. Matchmark the torsion bar spring, the anchor arm, and the torque arm.

3. Remove the anchor arm locknut.

4. Measure the protruding length of the adjusting arm bolt (from the top of the nut to the end of the bolt threads).

NOTE: The adjusting arm bolt measurement is used as a reference to establish the chassis ground clearance.

5. Remove the adjusting nut, the anchor arm, and the torsion bar spring.

NOTE: When installing the torsion bar springs, be sure to check the left/right indicating marks on the rear end of the springs; be careful not to interchange the springs.

6. Remove the two torque arm nuts and remove the torque arm from the upper control arm.

To install:

7. Install the torque arm to the upper control arm and install the two nuts. Torque the nuts to 64 ft. lbs. (87 Nm).

8. Using molybdenum disulfide lithium base grease, apply a coat to the torsion bar spring splines.

9. If installing the old torsion bar spring, perform the following procedures:

   a. Align the matchmarks, install the torsion bar spring to the torque arm.

   b. Align the matchmarks and install the anchor arm to the torsion bar spring.

   c. Tighten the adjusting nut until the bolt protrusion is the same as it was before.

NOTE: There is one spline on the torsion bar spring that is larger than the others. Install the torsion bar spring into the anchor arm by slowly turning the anchor arm until you feel the larger spline enter the matching point in the anchor arm.

10. If installing a new torsion bar spring, perform the following procedures:

   a. Install one end of the torsion bar spring to the torque arm.

   b. Install the torsion bar spring onto the opposite end of the anchor arm.

   c. Finger tighten the adjusting nut until the adjusting bolt protrudes about 1.570 in.

   d. Tighten the adjusting nut until the adjusting bolt protrudes about 3.430 in.

   e. Install the wheel(s) and remove the jackstands. Bounce the front of the vehicle to stabilize the suspension.

11. To adjust the ground clearance, turn the adjusting nut until the center of the cam plate nut (located of the front end of the lower suspension arm) about 11.220 in. (28.5 cm) above the ground.

12. After adjusting the ground clearance, torque the locknut to 61 ft. lbs., then, install the boots.

Upper Ball Joints

REMOVAL AND INSTALLATION

1993–95 4Runner, Pick-up and
1993–97 T-100

2-Wheel Drive Models

1. Raise and safely support the vehicle. Place the jack stands under the frame of the vehicle.
2. Remove the wheel and tire assembly.
3. Support the lower control arm with a floor jack.
4. Remove the brake caliper and support it out of the way with a wire.
5. Remove the cotter pin and nut from the upper ball joint.
6. Using a ball joint removal tool, separate the ball joint from the knuckle arm.
7. Remove the four nuts, washers and bolts holding the ball joint to the upper control arm.
8. Remove the ball joint from the upper control arm.

To install:

9. Install the ball joint to the upper control arm. Install the bolts, washers, and nuts to hold the upper ball joint to the upper control arm. Tighten the ball joint-to-upper control arm bolts and ball joint-to-steering knuckle castle nuts to the following specifications:

   a. On 1993–94 Pickup and 4Runner, torque the ball joint-to-upper control arm bolts 20 ft. lbs. (26 Nm) for 2-Wheel Drive vehicles or 25 ft. lbs. (33 Nm) for 4-Wheel Drive vehicles. Torque the ball joint-to-steering knuckle castle nut to 80 ft. lbs. (108 Nm) and install a new cotter pin.

   b. On 1995 Pick-up and 4Runner, tighten the ball joint-to-upper control arm bolts to 23 ft. lbs. (31 Nm). Tighten the ball joint-to-steering knuckle castle nut to 80 ft. lbs. (108 Nm) and install a new cotter pin.

   c. On 1993–97 T-100, torque the ball joint-to-upper control arm bolts 23 ft. lbs. (31 Nm) for 2-Wheel Drive vehicles or 25 ft. lbs. (33 Nm) for 4-Wheel Drive vehicles. Torque the ball joint-to-control arm bolts to 80 ft. lbs. (108 Nm) and install a new cotter pin.

NOTE: Be sure to grease the ball joints before moving vehicle.

10. Install the wheel and lower the vehicle.

4-Wheel Drive Models

1. Raise and safely support the vehicle. Place the jack stands under the frame of the vehicle.
2. Remove the wheel and tire assembly.
3. Support the lower control arm with a floor jack.
4. Remove the steering knuckle.
5. Remove the four nuts and disconnect the upper ball joint from the upper control arm.

To install:

6. Install the upper ball joint to the upper control arm and install the four nuts. Torque the nuts to 25 ft. lbs. (33 Nm).
7. Install the steering knuckle.
8. Install the wheel and tire assembly.
9. Lower the vehicle.

1996–97 4Runner

1. Raise and safely support the vehicle.
2. Remove the front wheels.
3. Remove the strut assembly.
4. Remove the grease cap.
5. 4-Wheel Drive: Disconnect the halfshaft.

   a. Remove the cotter pin and lock cap.

   b. While applying the brakes, remove the locknut.
6. With ABS: Remove the ABS speed sensor and wire harness clamp from the steering knuckle.
7. Remove the brake line bracket from the steering knuckle.
8. Remove the front brake caliper and the rotor.
9. Remove the four bolts and disconnect the lower ball joint.
10. Remove the steering knuckle with the axle hub.

   a. Remove the cotter pin and loosen the nut.

   b. Using SST 09950–40010 or equivalent, disconnect the steering knuckle from the upper control arm.

   c. Remove the nut and the steering knuckle.
11. Remove the upper ball joint.

   a. Remove the wire and the boot.

   b. Remove the snapring.

   c. Using SST 09950–40010 or equivalent, and a deep socket wrench, remove the upper ball joint.

To install:

12. Install the upper ball joint.

   a. Install a new ball joint with a new snapring.

   b. Install a new boot and fix it with a new wire.
13. Install the steering knuckle with the axle hub to the upper control

arm. Torque the nut to 80 ft. lbs. (108 Nm). Install a new cotter pin.
14. Install the lower ball joint and torque the four bolts to 59 ft. lbs. (80 Nm).
15. Install the rotor and the caliper. Torque the caliper bolts to 90 ft. lbs. (123 Nm).
16. Install the brake line bracket to the steering knuckle and torque to 21 ft. lbs. (28 Nm).
17. If removed, install the ABS speed sensor and wire harness clamp to the steering knuckle. Torque the bolts to 6 ft. lbs. (8 Nm).
18. If disconnected, install the halfshaft and torque the nut to 174 ft. lbs. (235 Nm).
19. Install the grease cap.
20. Install the strut assembly.
21. Install the front wheel.
22. Lower the vehicle and check the alignment.

Tacoma

2-Wheel Drive Models

1. Raise and safely support the front of the vehicle. Place jackstands under the frame of the vehicle.
2. Remove the wheel.
3. Support the lower control arm with a floor jack.
4. Disconnect the ABS speed sensor wire from the upper control arm.
5. Remove the two bolts and camber adjusting shims from the upper control arm.

NOTE: Before removing the shims from the upper control arm, make a note of each shim size and position. It is important to replace the shims into their original position.

6. Remove the cotter pin and nut from the upper control arm. Using SST 09628–62011 or equivalent, disconnect the upper ball joint from the steering knuckle.
7. Remove the upper control arm from the vehicle.
8. Remove the four nuts and bolts connecting the upper control arm to the upper ball joint. Disconnect the upper control arm from the upper ball joint.

To install:

9. Connect the upper control arm to a new ball joint and install the four bolts and nuts. Torque the nuts to 29 ft. lbs. (39 Nm).

NOTE: Make sure to grease the ball joint.

10. Install the upper control arm to the vehicle.
11. Install the camber adjusting shims and two bolts to the upper con-

trol arm. Torque the two bolts to 94 ft. lbs. (130 Nm).

12. Connect the upper ball joint to the steering knuckle. Install the ball joint nut and torque the nut to 80 ft. lbs. (110 Nm).

13. Connect the ABS speed sensor wire to the upper control arm. Install and torque the ABS bolt to 71 inch lbs. (8 Nm).

14. Install the wheels and lower the vehicle.

15. Check the wheel alignment.

4-Wheel Drive Models

1. Raise and safely support the vehicle.

2. Remove the wheel and tire assembly.

3. Remove the strut from the suspension.

4. If not equipped with a FREE wheeling hub, disconnect the halfshaft from the steering knuckle as follows:

  a. Using a screwdriver, remove the grease cap.

  b. Remove the cotter pin and lock cap from the halfshaft.

  c. While having an assistant apply the brakes, remove the locknut from the halfshaft.

5. If equipped with FREE wheeling hub, remove the free wheel hub as follows:

  a. Set the control handle to FREE.

  b. Remove the cover bolts and pull off the cover.

  c. Remove the center bolt with washer.

  d. Remove the mounting nuts and washer to the hub body.

  e. Using a brass bar and hammer, tap on the bolts head and remove the cone washer.

  f. Pull off the free wheel hub body and gasket.

  g. Using a snapring expander, remove the snapring and spacer from the end of the halfshaft.

6. If equipped with ABS brakes, disconnect the ABS speed sensor from the steering knuckle.

7. Disconnect the brake hose from the steering knuckle by removing the bolt.

8. Remove the two brake caliper support bracket bolts and wire the caliper to the side. Do not allow the caliper to hang from the brake hose.

9. Remove the rotor from the steering knuckle.

10. Remove the four bolts and disconnect the lower ball joint from the steering knuckle.

11. Remove the cotter pin and nut to the upper control arm.

12. Using SST 09950–40010 or equivalent, disconnect the steering knuckle from the upper control arm.

13. Remove the steering knuckle from the vehicle.

NOTE: If it is difficult to remove the halfshaft from the steering knuckle, use a rubber hammer and tap the halfshaft from the steering knuckle.

14. Remove the wire and boot from the upper ball joint.

15. Using a snapring expander, remove the snapring from the ball joint.

16. Using SST 09950–40010 (puller set) or equivalent and a deep socket wrench, remove the upper ball joint from the steering knuckle.

To install:

17. Using SST 09309–37010 or equivalent and a socket wrench, press in a new upper ball joint.

18. Using a snapring expander, install a new snapring.

19. Install a new boot and hold it down with a new piece of wire.

NOTE: Make sure to grease the ball joint.

20. Install the steering knuckle to the halfshaft.

21. Connect the steering knuckle to the lower ball joint by installing the four bolts. Do not torque the bolts at this time.

22. Push down on the upper control arm and connect the upper ball joint to the steering knuckle.

23. Install the upper ball joint nut. Torque the nut to 80 ft. lbs. (105 Nm). Install a new cotter pin.

24. Torque the lower ball joint to steering knuckle bolts to 59 ft. lbs. (80 Nm).

25. Install the brake rotor.

26. Connect the caliper support bracket to the steering knuckle and install the two bolts. Torque the bolts to 90 ft. lbs. (123 Nm).

27. Connect the brake hose clamp to the steering knuckle by installing the bolt. Torque the bolt to 13 ft. lbs. (18 Nm).

28. If equipped with ABS brakes, connect the ABS speed sensor and wire harness to the steering knuckle.

29. Install the spacer and using a snapring expander, install the snapring to the halfshaft.

30. If equipped with a free wheeling hub, install the hub as follows:

  a. Place a new gasket in position on the front axle hub.

  b. Install the free wheeling hub body with the six cone washers and nuts. Torque the nuts to 23 ft. lbs. (31 Nm).

  c. Install the bolt with the washer. Torque the bolt to 13 ft. lbs. (18 Nm).

  d. Apply multi purpose grease to the inner hub splines.

  e. Set the control handle and clutch to the FREE position.

  f. Place a new gasket in position on the cover.

  g. Install the cover to the hub body with the follower pawl tabs aligned with the non-toothed portions of the hub body.

  h. Install the mounting bolts and tighten the cover bolts to 7 ft. lbs. (10 Nm).

31. If equipped without a free wheeling hub, install the halfshaft to the steering knuckle as follows:

  a. Install the locknut to the halfshaft. Torque the nut to 174 ft. lbs. (235 Nm).

  b. Install the lockcap and cotter pin to the halfshaft.

  c. Install the grease cab to the hub.

32. Install the strut to the vehicle. Torque the nut holding the strut to the lower control arm to 101 ft. lbs. (135 Nm). Torque the upper three nuts to 47 ft. lbs. (64 Nm)

33. Install the front wheels and lower the vehicle.

34. Check the wheel alignment.

Lower Ball Joints

REMOVAL AND INSTALLATION

1993–95 4Runner, Pick-up and 1993–97 T-100

2-Wheel Drive Models

1. Raise and safely support the vehicle.

2. Remove the wheel and tire assembly.

3. Support the lower control arm with a floor jack.

4. Remove the cotter pin and nut from the lower ball joint.

5. Using a ball joint removal tool, separate the lower ball joint from the steering knuckle.

6. Remove the ball joint from the lower control arm by removing the three mounting nuts, washers, and bolts.

7. Remove the ball joint from the vehicle.

To install:

8. Install the lower ball joint to the vehicle.

9. Install the lower ball joint bolts, washers, and nuts, then install the ball joint to the steering knuckle.

Tighten the fasteners to the following specifications:

a. On 1993–94 Pickup and 4Runner, torque the ball joint-to-control arm bolts 51 ft. lbs. (69 Nm) for 2-Wheel Drive vehicles or 43 ft. lbs. (58 Nm) for 4-Wheel Drive vehicles. Torque the ball joint-to-steering knuckle bolts to 105 ft. lbs. (142 Nm). Install a new cotter pin.

b. On 1993–94 T-100, torque the ball joint-to-control arm bolts 55 ft. lbs. (75 Nm) for 2-Wheel Drive vehicles or 43 ft. lbs. (58 Nm) for 4-Wheel Drive vehicles. Torque the ball joint-to-steering knuckle bolts to 105 ft. lbs. (142 Nm). Install a new cotter pin.

c. 1995 Pick-up and 4Runner and 1995–97 T-100, tighten the ball joint-to-control arm nuts to 55 ft. lbs. (75 Nm). Tighten the ball joint-to-steering knuckle bolts to 105 ft. lbs. (142 Nm). Install a new cotter pin.

NOTE: Be sure to grease the ball joints before moving the vehicle.

10. Install the wheel and lower the vehicle.

4-Wheel Drive Models

1. Raise and safely support the vehicle. Place the jack stands under the frame of the vehicle.
2. Remove the wheel and tire assembly.
3. Support the lower control arm with a floor jack.
4. Remove the steering knuckle.
5. Remove the four nuts, washers and bolts and disconnect the lower ball joint from the lower control arm.
To install:
6. Install the lower ball joint to the lower control arm and install the four bolts, washers and nuts. Torque the nuts to 25 ft. lbs. (33 Nm).
7. Install the steering knuckle.
8. Install the wheel and tire assembly.
9. Lower the vehicle.

1996–97 4Runner

1. Raise and safely support the vehicle. Place the jack stands under the frame of the vehicle.
2. Remove the wheel and tire assembly.
3. Disconnect the tie rod end.
   a. Loosen the four bolts.
   b. Remove the cotter pin and nut from the tie rod end.

c. Using SST 09610–20012 or equivalent, disconnect the tie rod end from the steering knuckle.
4. Remove the lower ball joint.
   a. Remove the cotter pin and the nut from the lower ball joint.
   b. Using a ball joint separator, disconnect the lower ball joint from the lower suspension arm.
   c. Remove the four bolts. While lifting the upper suspension arm and the steering knuckle, remove the lower ball joint.
To install:
5. Install the lower ball joint to the lower control arm. Torque the four bolts to 59 ft. lbs. (80 Nm).
6. Install the nut and torque to 105 ft. lbs. (142 Nm).
7. Connect the tie rod end to the steering knuckle and torque the nut to 66 ft. lbs. (90 Nm).
8. Install the wheel and tire assembly.

Tacoma

2-Wheel Drive Models

1. Raise and safely support the vehicle. Place the jacks under the frame.
2. Remove the wheel from the vehicle.
3. Support the lower control with a floor jack.

— **WARNING** —
Do not remove the floor jack from the lower control arm until all components are replaced. The coil spring is under high pressure and if not removed properly, the spring could cause damage and/or injury.

4. Loosen the two lower ball joint set bolts.
5. Remove the cotter pin and nut from the tie rod.
6. Using SST 09628–62011 or equivalent, disconnect the tie rod from the ball joint bracket.
7. Remove the cotter pin and nut from the lower ball joint.
8. Using SST 09628–62011 or equivalent, disconnect the lower ball joint from the lower control arm.
9. Remove the two lower ball joint set bolts and remove the ball joint from the suspension.
To install:
10. Install the lower ball joint to the steering knuckle and lower control arm.
11. Install the two lower ball joint set bolts. Do not torque the bolts at this time.

12. Install the lower ball joint nut to hold the lower ball joint to the lower control arm. Torque the nut to 80 ft. lbs. (110 Nm). Install a new cotter pin to the lower ball joint.
13. Connect the tie rod end to the ball joint bracket. Install the nut to hold the tie rod end to the lower ball joint bracket. Torque the nut to 53 ft. lbs. (72 Nm). Install a new cotter pin to the tie rod end.
14. Torque the two lower ball joint set bolts to 116 ft. lbs. (160 Nm).
15. Remove the floor jack from the lower control arm.
16. Install the wheel and lower the vehicle.
17. Check the wheel alignment.

4-Wheel Drive Models

1. Raise and safely support the vehicle. Place the jacks under the frame.
2. Remove the wheel from the vehicle.
3. Loosen the four lower ball joint set bolts.
4. Remove the cotter pin and nut from the tie rod.
5. Using SST 09610–20012 or equivalent, disconnect the tie rod from the ball joint bracket.
6. Remove the cotter pin and nut from the lower ball joint.
7. Using SST 09628–62011 or equivalent, disconnect the lower ball joint from the lower control arm.
8. Remove the four lower ball joint set bolts.
9. While lifting the upper control arm and steering knuckle, remove the ball joint from the suspension.
To install:
10. Install the lower ball joint to the steering knuckle and lower control arm.
11. Install the four lower ball joint set bolts. Do not torque the bolts at this time.
12. Install the lower ball joint nut to hold the lower ball joint to the lower control arm. Torque the nut to 112 ft. lbs. (152 Nm). Install a new cotter pin to the lower ball joint.
13. Connect the tie rod end to the ball joint bracket. Install the nut to hold the tie rod end to the lower ball joint bracket. Torque the nut to 67 ft. lbs. (90 Nm). Install a new cotter pin to the tie rod end.
14. Torque the two lower ball joint set bolts to 83 ft. lbs. (113 Nm).
15. Remove the floor jack from the lower control arm.
16. Install the wheel and lower the vehicle.
17. Check the wheel alignment.

Upper Control Arms

REMOVAL AND INSTALLATION

1993–95 4Runner, Pick-up and 1993–97 T-100

2-Wheel Drive Models

1. Raise and safely support the front of the vehicle. Place jackstands under the frame of the vehicle.
2. Remove the wheel.
3. If equipped with four wheel ABS brakes, remove the front ABS speed sensor bracket from the upper control arm.
4. Support the lower control arm with a floor jack.
5. Remove the caliper support bracket.
6. Remove the four nuts and bolts connecting the upper control arm to the upper ball joint. Disconnect the upper control arm from the upper ball joint.
7. Remove the two bolts and camber adjusting shims from the upper control arm.

NOTE: Before removing the shims from the upper control arm, make a note of each shim size and position. It is important to replace the shims into their original position.

8. Remove the upper control arm from the vehicle.
To install:
9. Install the upper control arm to the vehicle.
10. Install the camber adjusting shims and two bolts to the upper control arm. Torque the two bolts to 71 ft. lbs. (96 Nm).
11. Connect the upper control arm to the ball joint and install the four nuts. Torque the nuts to 23 ft. lbs. (31 Nm).
12. Connect the ABS speed sensor wire to the upper control arm. Install and torque the ABS bolt to 35 inch lbs. (4 Nm).
13. Install the caliper support bracket to the vehicle and torque the two bolts to 80 ft. lbs. (108 Nm).
14. Install the wheels and lower the vehicle.
15. Check the wheel alignment.

4-Wheel Drive Models

1. Raise and safely support the front of the vehicle. Place jackstands under the frame of the vehicle.
2. Remove the wheel.
3. Remove the torsion bar spring and torque arm from the upper control arm.
4. Support the lower control arm with a jack.
5. Remove the four nuts and disconnect the upper ball joint from the upper control arm.
6. If equipped with ABS, remove the front ABS speed sensor bracket from the upper control arm.
7. Disconnect the intermediate shaft from the steering gear housing.
8. Remove the three bolts and remove the upper control arm from the frame.
To install:
9. Install the upper control arm to the frame and install the three bolts. Torque the bolts to 131 ft. lbs. (178 Nm).
10. Connect the intermediate shaft to the steering gear housing.
11. If removed, install the front ABS speed sensor bracket to the upper control arm. Torque the bolt to 48 inch lbs. (5.4 Nm).
12. Connect the upper ball joint to the upper control arm and install the nuts. Torque the nuts to 26 ft. lbs. (33 Nm).
13. Install the torque arm to the upper control arm and torque the nuts to 64 ft. lbs. (87 Nm).
14. Install the torsion bar spring.
15. Install the wheel and lower the vehicle.
16. Check the front wheel alignment.

1996–97 4Runner

1. Raise and safely support the vehicle.
2. Remove the front wheels.
3. Remove the strut.
4. With ABS: Remove the ABS speed sensor and wire harness clamp from the steering knuckle.
5. Disconnect the upper ball joint from the upper control arm.
6. Support the steering knuckle securely.
7. Remove the nut, bolt, washers and the upper control arm.
To install:
8. Install the upper control arm, washers, bolt and the nut. Torque the fasteners to 87 ft. lbs. (115 Nm).
9. Connect the upper ball joint to the upper control arm. Torque the nut to 80 ft. lbs. (108 Nm).
10. If removed, install the ABS speed sensor and wire harness clamp to the steering knuckle. Torque the bolts to 6 ft. lbs. (8 Nm).
11. Install the strut; torque the bottom bolt to 101 ft. lbs. (135 Nm) and the top nuts to 47 ft. lbs. (64 Nm).
12. Install the wheels.
13. Lower the vehicle and check the alignment.

Tacoma

2-Wheel Drive Models

1. Raise and safely support the front of the vehicle. Place jackstands under the frame of the vehicle.
2. Remove the wheel.
3. Support the lower control arm with a floor jack.
4. Remove the two bolts and camber adjusting shims from the upper control arm.

NOTE: Before removing the shims from the upper control arm, make a note of each shim size and position. It is important to replace the shims into their original position.

5. Remove the cotter pin and nut from the upper control arm. Using SST 09628–62011 or equivalent, disconnect the upper ball joint from the steering knuckle.
6. Remove the upper control arm from the vehicle.
7. Remove the four nuts and bolts connecting the upper control arm to the upper ball joint. Disconnect the upper control arm from the upper ball joint.
To install:
8. Connect the upper control arm to the ball joint and install the four bolts and nuts. Torque the nuts to 29 ft. lbs. (39 Nm).
9. Install the upper control arm to the vehicle.
10. Install the camber adjusting shims and two bolts to the upper control arm. Torque the two bolts to 94 ft. lbs. (130 Nm).
11. Connect the upper ball joint to the steering knuckle. Install the ball joint nut and torque the nut to 80 ft. lbs. (110 Nm).
12. Connect the ABS speed sensor wire to the upper control arm. Install and torque the ABS bolt to 71 inch lbs. (8 Nm).
13. Install the wheels and lower the vehicle.
14. Check the wheel alignment.

4-Wheel Drive Models

1. Raise and safely support the front of the vehicle. Place jackstands under the frame of the vehicle.
2. Remove the wheel.
3. Remove the strut from the suspension as follows:
   a. Remove the nut and bolt holding the strut to the lower control arm.
   b. Remove the three nuts holding the strut to the strut tower.
   c. Remove the strut from the vehicle.

4. If equipped with ABS brakes, remove the front ABS speed sensor bracket from the upper control arm.

5. Support the lower control arm with a jack.

6. Remove the cotter pin and nut from the upper control arm. Using SST 09950–40010 or equivalent, disconnect the upper ball joint from the steering knuckle. Make sure not to damage the threads to the upper ball joint.

7. Remove the nut and bolt holding the upper control arm to the strut tower.

8. Remove the upper control arm from the vehicle.

To install:

9. Install the upper control arm to the vehicle.

10. Install the bolt and nut to hold the upper control arm to the strut tower. Do not torque the nut at this time.

11. Connect the upper ball joint to the upper control arm and install the nut. Torque the nut to 80 ft. lbs. (105 Nm). Install a new cotter pin.

12. Connect the ABS speed sensor wire to the upper control arm. Install and torque the ABS bolt to 71 inch lbs. (8 Nm).

13. Install the strut to the vehicle as follows:

   a. Install the strut to the vehicle.

   b. Install the three nuts to hold the strut to the strut tower. Torque the nuts to 47 ft. lbs. (64 Nm).

   c. Install the lower bolt and nut to hold the strut to the lower control arm. Torque the nut to 101 ft. lbs. (135 Nm).

14. Install the wheel and lower the vehicle.

15. Stabilize the suspension by pushing up and down on the vehicle.

16. Torque the upper control arm bolt and nut to 87 ft. lbs. (115 Nm).

17. Check the wheel alignment.

Lower Control Arms

REMOVAL AND INSTALLATION

1993–94 4Runner and Pick-up

2-Wheel Drive Models

1. Raise and safely support the vehicle. Remove the torsion bar spring.

2. Remove the shock absorber, the stabilizer bar and the strut bar from the lower arm.

3. Remove the shock absorber from the lower arm.

4. From the lower ball joint, remove the cotter pin and the nut. Us-

ing a ball joint removal tool, press the ball joint from the lower control arm.

NOTE: If the lower ball joint is not to be replaced, simply unbolt it from the lower control arm. It is not necessary to separate the ball joint from the steering knuckle.

5. Remove the lower control arm shaft nut. Remove the spring torque arm from the other side of the lower control arm, then, remove the lower arm shaft bolt and the lower arm.

To install:

6. Install the components in the reverse order of the removal procedures.

7. Tighten the bolt(s) holding the lower control arm to the frame but do not torque them until the vehicle is on the ground.

8. Torque the ball joint-to-lower control arm nuts/bolts to 51 ft. lbs., the strut bar-to-lower control arm bolts to 70 ft. lbs., the stabilizer bar-to-lower control arm bolts to 9 ft. lbs., the lower shock absorber bolt to 13 ft. lbs., upper shock absorber bolt to 18 ft. lbs. and the lower arm mounting nuts to 166 ft. lbs.

9. Check and/or adjust the front end alignment.

NOTE: Do not torque the control arm bolts fully until the vehicle is lowered and bounced several times; if the bolts are tightened with the control arm(s) hanging, excessive bushing wear will result.

4-Wheel Drive Models

1. Raise and safely support the vehicle. Remove the shock absorber.

2. Disconnect the stabilizer bar from the lower suspension arm.

3. Remove the lower ball joint-to-lower control arm bolts, then, separate the control arm from the ball joint.

4. Using a piece of chalk, place matchmarks on the front/rear adjusting cams.

5. Remove the nuts and adjusting cams and the lower control arms.

To install:

6. Install the components in the reverse order of the removal procedures.

7. Torque the lower ball joint-to-lower control arm bolts to 20 ft. lbs., the stabilizer bar-to-lower control arm bolts to 19 ft. lbs., the shock absorber-to-lower control arm nut/bolt to 101 ft. lbs.

8. Lower the vehicle to the ground, bounce it a few times, align the matchmarks and torque the adjust-

ing cam nuts to 203 ft. lbs. Check and/or adjust the front wheel alignment.

1995 Pick-up

2-Wheel Drive Models

1. Raise and safely support the vehicle. Place the jack stands under the frame of the vehicle.

2. Remove the front wheel.

3. Remove the engine undercover.

4. Remove the torsion bar spring from the vehicle. Leave the torsion torque bar connected to the lower control arm.

5. Remove the cotter pin and nut to the tie rod end.

6. Using SST 09610–20012 or equivalent (tie rod puller), disconnect the tie rod from the steering knuckle.

7. Disconnect the shock absorber from the lower control arm by removing the two bolts.

8. Disconnect the sway bar from the lower control arm by removing the nut, retainers, and the cushion.

9. Disconnect the strut bar from the lower control arm by removing the two bolts and bracket.

10. Disconnect the lower ball joint from the lower control arm by removing the nuts and bolts.

11. Remove the nut to the lower control arm.

12. Remove the two nuts, bolts, and the washers holding the torque arm to the lower control arm.

13. Remove the lower control arm shaft from the lower control arm.

14. Remove the lower control arm from the vehicle.

To install:

15. Install the lower control arm to the vehicle.

16. Install the lower control arm shaft to hold the lower control arm to the frame.

17. Install the torque arm to the lower control arm and install the two bolts, washers, and the nuts. Torque the nuts to 36 ft. lbs. (49 Nm).

18. Install the lower control arm nut to the lower control arm shaft. Do not torque the bolt at this time.

19. Connect the lower ball joint to the lower control arm by installing the bolts and nuts. Torque the bolts to 94 ft. lbs. (127 Nm).

20. Connect the strut bar to the lower control arm and install the bracket two nuts. Torque the two bolts 70 ft. lbs. (95 Nm).

21. Connect the sway bar to the lower control arm and install the retainers, cushion, and the nut. Torque the nut to 9 ft. lbs. (13 Nm).

22. Connect the shock absorber to the lower control arm and install the

two bolts. Torque the bolts to 18 ft. lbs. (25 Nm).

23. Connect the tie rod end to the steering knuckle and install the nut. Torque the nut to 67 ft. lbs. (90 Nm). Install a new cotter pin to the tie rod.

24. Install the torsion bar spring.

25. Install the engine undercover.

26. Install the front wheel and lower the vehicle.

27. Stabilize the vehicle and torque the nut holding the lower control arm to the body to 152 ft. lbs. (206 Nm).

28. Check the front wheel alignment.

4-Wheel Drive Models

1. Raise and safely support the vehicle. Place the jack stands under the frame of the vehicle.

2. Remove the front wheel.

3. Remove the engine undercover.

4. Disconnect the shock absorber from the lower control arm by removing the nut and the through-bolt.

5. Disconnect the sway bar from the lower control arm by removing the nut, retainers, and the cushion.

6. Remove the cotter pin and nut from the ball joint.

7. Using SST 09628-62011 or equivalent (ball joint separator), disconnect the lower control arm from the lower ball joint.

8. Place matchmarks on the front and rear adjusting cams and frame.

9. Remove the nuts and the adjusting cams from the lower control arm.

10. Remove the lower control arm from the vehicle.

To install:

11. Install the lower control arm to the vehicle.

12. Install the adjusting cams and nuts to hold the lower control arm to the vehicle. Align the matchmarks on the frame and adjusting cams but do not torque the nuts at this time.

13. Connect the lower control arm to the lower ball joint.

14. Install the nut to the ball joint and torque the nut to 105 ft. lbs. (142 Nm). Install a new cotter pin to the ball joint.

15. Connect the sway bar to the lower control arm and install the retainers, cushion, and the nut. Torque the nut to 19 ft. lbs. (25 Nm).

16. Connect the shock absorber to the lower control arm and install the through-bolt and nut. Torque to 101 ft. lbs. (137 Nm).

17. Install the engine undercover.

18. Install the wheel and lower the vehicle.

19. Stabilize the vehicle and torque the cam nuts and bolts to 145 ft. lbs. (196 Nm).

20. Check the wheel alignment.

1995 4Runner

1. Raise and safely support the vehicle. Place the jack stands under the frame of the vehicle.

2. Remove the front wheel.

3. Disconnect the shock absorber from the lower control arm by removing the nut and the through-bolt.

4. Disconnect the sway bar from the lower control arm by removing the nut, retainers and cushion.

5. Remove the cotter pin and nut from the ball joint.

6. Using SST 09628-62011 or equivalent (ball joint separator), disconnect the lower control arm from the lower ball joint.

7. Place matchmarks on the front and rear adjusting cams and frame.

8. Remove the nuts and the adjusting cams from the lower control arm.

9. Remove the lower control arm from the vehicle.

To install:

10. Install the lower control arm to the vehicle.

11. Install the adjusting cams and nuts to hold the lower control arm to the vehicle. Align the matchmarks on the frame and adjusting cams but do not torque the nuts at this time.

12. Connect the lower control arm to the lower ball joint.

13. Install the nut to the ball joint and torque the nut to 105 ft. lbs. (142 Nm). Install a new cotter pin to the ball joint.

14. Connect the sway bar to the lower control arm and install the retainers, cushion and nut. Torque the nut to 19 ft. lbs. (25 Nm).

15. Connect the shock absorber to the lower control arm and install the through-bolt and nut. Torque to 101 ft. lbs. (137 Nm).

16. Install the wheel and lower the vehicle.

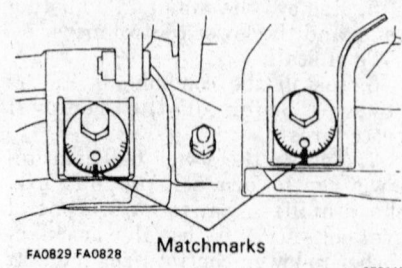

FA0829 FA0828   Matchmarks

252115

Matchmark the adjusting cams as shown — 1995 4Runner and Pick-up, and 1995-97 T-100

17. Stabilize the vehicle and torque the cam nuts and bolts to 145 ft. lbs. (196 Nm).

18. Check the wheel alignment.

1996-97 4Runner

1. Raise and safely support the vehicle.

2. Remove the front wheel.

3. Disconnect the tie rod end.

a. Loosen the four bolts.

b. Remove the cotter pin and nut from the tie rod end.

c. Using SST 09610-20012 or equivalent, disconnect the tie rod end from the steering knuckle.

4. Disconnect the stabilizer bar link.

5. Disconnect the shock absorber from the lower control arm.

6. Remove the lower ball joint.

a. Remove the cotter pin and the nut from the lower ball joint.

b. Using a ball joint separator, disconnect the lower ball joint from the lower suspension arm.

c. Remove the four bolts. While lifting the upper suspension arm and the steering knuckle, remove the lower ball joint.

7. Remove the lower control arm.

a. Place matchmarks on the front and rear adjusting cams.

b. Remove the two bolts, nuts, adjusting cams and the lower control arm.

To install:

8. Install the lower control arm.

9. Align the matchmarks. Install the adjusting cams, nuts and the two bolts. Torque to 96 ft. lbs. (130 Nm).

10. Install the lower ball joint to the lower control arm. Torque the four bolts to 59 ft. lbs. (80 Nm).

11. Install the nut and torque to 105 ft. lbs. (142 Nm).

12. Connect the tie rod end to the steering knuckle and torque the nut to 66 ft. lbs. (90 Nm).

13. Connect the shock absorber to the lower control arm. Torque to 101 ft. lbs. (141 Nm).

14. Connect the stabilizer bar link. Torque to 51 ft. lbs. (69 Nm).

15. Connect the tie rod end to the steering knuckle and torque the nut to 66 ft. lbs. (90 Nm).

16. Install the wheel and tire assembly.

1995-97 T-100

2-Wheel Drive Models

1. Raise and safely support the vehicle. Place the jack stands under the frame of the vehicle.

2. Remove the front wheel.

3. Remove the engine undercover.

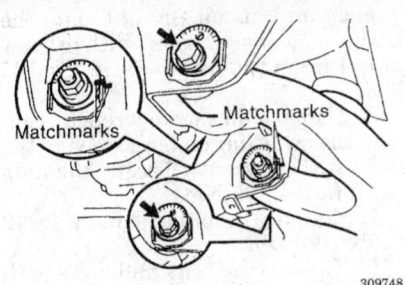

Matchmarks on adjusting cams — 1996–97 4Runner

309748

4. Remove the torsion bar spring from the vehicle. Leave the torsion bar torque bar connected to the lower control arm.

5. Disconnect the shock absorber from the lower control arm by removing the two bolts.

6. Disconnect the sway bar from the lower control arm by removing the nut, retainers and cushion.

7. Disconnect the strut bar from the lower control arm by removing the two nuts. These two nuts also help hold the lower ball joint to the lower control arm.

8. Disconnect the lower ball joint from the lower control arm by removing the two nuts.

9. Remove the nut from the lower control arm.

10. Remove the two nuts, bolts and washers holding the torque arm to the lower control arm.

11. Remove the lower control arm shaft from the lower control arm.

12. Remove the lower control arm from the vehicle.

To install:

13. Install the lower control arm to the vehicle.

14. Install the lower control arm shaft to hold the lower control arm to the frame.

15. Install the torque arm to the lower control arm and install the two bolts, washers, and nuts. Torque to 36 ft. lbs. (49 Nm).

16. Install the lower control arm nut to the lower control arm shaft. Do not torque the bolt at this time.

17. Connect the lower ball joint to the lower control arm by installing the two nuts. Do not torque the bolts at this time.

18. Connect the strut bar to the lower control arm and install the two nuts. The strut bar bolts help hold the lower ball joint to the lower control arm. Torque all the ball joint (four bolts) nuts to 55 ft. lbs. (75 Nm).

19. Connect the sway bar to the lower control arm and install the re-

tainers, cushion, and the nut. Torque the nut to 9 ft. lbs. (13 Nm).

20. Connect the shock absorber to the lower control arm and install the two bolts. Torque the bolts to 18 ft. lbs. (25 Nm).

21. Install the torsion bar spring.

22. Install the engine undercover.

23. Install the front wheel and lower the vehicle.

24. Stabilize the vehicle and torque the nut holding the lower control arm to the body to 152 ft. lbs. (206 Nm).

25. Check the front wheel alignment.

4-Wheel Drive Models

1. Raise and safely support the vehicle. Place the jack stands under the frame of the vehicle.

2. Remove the front wheel.

3. Remove the engine undercover.

4. Disconnect the shock absorber from the lower control arm by removing the nut and the through-bolt.

5. Disconnect the sway bar from the lower control arm by removing the nut, retainers, and the cushion.

6. Remove the cotter pin and nut from the ball joint.

7. Using SST 09628–62011 or equivalent (ball joint separator), disconnect the lower control arm from the lower ball joint.

8. Place matchmarks on the front and rear adjusting cams and frame.

9. Remove the nuts and the adjusting cams from the lower control arm.

10. Remove the lower control arm from the vehicle.

To install:

11. Install the lower control arm to the vehicle.

12. Install the adjusting cams and nuts to hold the lower control arm to the vehicle. Align the matchmarks on the frame and adjusting cams but do not torque the nuts at this time.

13. Connect the lower control arm to the lower ball joint.

14. Install the nut to the ball joint and torque the nut to 105 ft. lbs. (142 Nm). Install a new cotter pin to the ball joint.

15. Connect the sway bar to the lower control arm and install the retainers, cushion and nut. Torque the nut to 19 ft. lbs. (25 Nm).

16. Connect the shock absorber to the lower control arm and install the through-bolt and nut. Torque to 101 ft. lbs. (137 Nm).

17. Install the engine undercover.

18. Install the wheel and lower the vehicle.

19. Stabilize the vehicle and torque the cam nuts and bolts to 145 ft. lbs. (196 Nm).

20. Check the wheel alignment.

Tacoma

2-Wheel Drive Models

1. Raise and safely support the front of the vehicle. Place the jack stands under the frame of the vehicle.

2. Remove the shock absorber from the suspension by removing the two bottom bolts and top nut.

3. Using SST 09727–22011 or equivalent (spring compressor), compress the coil spring.

—————— **CAUTION** ——————

The proper tools for this procedure must be used. The spring is under high pressure and can cause serious injury if not properly removed and installed.

4. Remove the nut and disconnect the sway bar link from the lower control arm.

5. Remove the two sway bar bracket bolts on the side of the suspension that the lower control arm is being removed. This will allow access to the lower control arm through-bolt.

6. Support the steering knuckle and upper control arm.

—————— **WARNING** ——————

Before removing the lower control arm, make sure the coil spring is safely compressed.

7. Remove the cotter pin and nut from the lower ball joint.

8. Using SST 09628–62011 or equivalent, disconnect the lower ball joint from the lower control arm.

9. Remove the nut from the lower control arm set bolt.

10. Remove the nut from the strut bar front set bolt.

11. Pull out the two bolts and remove the lower control arm and strut bar as an assembly. When the lower control arm is removed, set the coil spring aside.

12. Remove the strut bar from the lower control arm by removing the two nuts.

13. Remove the spring bumper from the lower control arm by removing the nut.

14. Remove the No. 3 lower control arm from the lower control arm by removing the two nuts.

To install:

15. Connect the No. 3 lower control arm to the lower control arm and install the two nuts. Torque the nuts to 111 ft. lbs. (150 Nm).

16. Connect the spring bumper to the lower control arm by installing the nut. Torque the nut to 32 ft. lbs. (43 Nm).

17. Connect the strut bar to the lower control arm and install the two nuts. Torque the nuts to 111 ft. lbs. (150 Nm).

18. Place the end of the coil spring and lower control arm seat in contact.

19. Install the lower control arm, spring, and strut arm to the suspension. Install the strut arm bolt and lower control arm bolt.

20. Install the nuts for the strut arm bolt and lower control arm bolt. Do not torque the bolts at this time.

21. Connect the lower control arm to the lower ball joint. Install the nut and torque the nut to 80 ft. lbs. (110 Nm). Install a new cotter pin.

22. Remove the support from the upper control arm and steering knuckle.

23. Connect the sway bar bracket to the suspension and install the two bolts. Torque the bolts to 22 ft. lbs. (29 Nm).

24. Connect the sway bar link to the lower control arm and install the nut. Torque the nut to 29 ft. lbs. (39 Nm).

25. Making sure the coil spring is in its correct position, slowly remove the spring compressor from the coil.

26. Install the shock absorber. Torque the top nut to 18 ft. lbs. (25 Nm) and the bottom two bolts to 29 ft. lbs. (39 Nm).

27. Install the wheel and lower the vehicle.

28. Stabilize the suspension by pushing up and down on the vehicle.

29. Torque the strut bar nut and bolt to 221 ft. lbs. (300 Nm) and the lower control arm bolt and nut to 148 ft. lbs. (200 Nm).

30. Check the front wheel alignment.

4-Wheel Drive Models

1. Raise and safely support the vehicle. Place the jack stands under the frame of the vehicle.

2. Remove the front wheel.

3. Remove the rack and pinion steering from the vehicle.

4. Disconnect the sway bar links from the lower control arm.

5. Remove the nut and bolt holding the strut to the lower control arm.

6. Support the steering knuckle and upper control arm.

7. Remove the cotter pin and nut from the lower ball joint.

8. Using SST 09610–20012 or equivalent, disconnect the lower ball joint from the lower control arm.

9. Place matchmarks on the front and rear adjusting cams and frame.

10. Remove the nuts and adjusting cams from the lower control arm.

11. Remove the lower control arm from the vehicle.

12. Using SST 09922–10010 or equivalent, remove the No. 1 and No. 2 spring bumpers from the lower control arm.

To install:

13. Using a torque wrench with a fulcrum length of 13.6 in. (34.5 Nm), install the No. 1 and No. 2 spring bumpers to the lower control arm. Torque the bumpers to 17 ft. lbs. (23 Nm).

14. Install the lower control arm to the vehicle.

15. Install and align the adjusting cams and nuts to the lower control arm. Torque the nuts to 127 ft. lbs. (170 Nm).

16. Connect the lower control arm to the lower ball joint. Install the nut and torque the nut to 112 ft. lbs. (152 Nm). Install a new cotter pin to the ball joint.

17. Remove the support for the upper control arm and steering knuckle.

18. Connect the strut to the lower control arm by installing the bolt and nut. Torque to 101 ft. lbs. (135 Nm).

19. Connect the sway bar to the lower control arm. Torque the nuts for the sway bar to 51 ft. lbs. (69 Nm).

20. Install the rack and pinion steering assembly to the vehicle.

21. Install the front wheels and lower the vehicle.

22. Check the front wheel alignment.

Sway Bar

REMOVAL AND INSTALLATION

1993–94 Models and 1995 4Runner

1. Raise and safely support the vehicle.

2. Using a hexagon wrench, remove the nut and disconnect the sway bar from the sway bar link.

3. Hold the sway bar link with a wrench and remove the nut, retainers, cushions and the sway bar link lower control arm.

4. Remove the sway bar bracket bolts, brackets, and cushions from the sway bar.

5. Remove the sway bar from the vehicle.

To install:

6. Place the sway bar in position and install both sway bar cushions and brackets to the frame. Hand tighten the bolts.

7. Hold the sway bar link with a wrench and install the link onto the lower suspension arm with a new nut. Torque the nut to 19 ft. lbs. (25 Nm).

8. Using a hexagon wrench, connect the sway bar on both sides to the links with new nuts. Torque the nuts to 70 ft. lbs. (95 Nm)

9. Torque the bracket bolts to 22 ft. lbs. (29 Nm)

10. Tighten the nuts and bolts with the full weight of the vehicle on the ground.

11. Test drive the vehicle and check the front end alignment.

T-100, Tacoma and 1995 Pick-up

2-Wheel Drive Models

1. Raise and safely support the vehicle.

2. Remove the front wheels.

3. Disconnect the sway bar links from the lower control arms by removing the nuts, retainers, and cushions.

4. Disconnect the sway bar links from the sway bar by removing the nuts, retainers, and cushions.

5. Remove the sway bar links from the suspension.

6. Remove the sway bar bracket bolts and remove the brackets and bushings from the sway bar.

7. Remove the sway bar from the vehicle.

To install:

8. Place the sway bar in position and install both sway bar cushions and brackets to the frame. Hand tighten the bolts.

9. Install and connect the sway bar links on both sides to the sway bar with new nuts. Torque the nuts to 9 ft. lbs. (13 Nm)

10. Hold the sway bar links with a wrench and connect the sway bar links to the lower control arms with new nuts. Torque the nuts to 18 ft. lbs. (25 Nm) for T-100 and Pick-up, or to 29 ft. lbs. (35 Nm) for Tacoma.

11. Torque the bracket bolts to 22 ft. lbs. (29 Nm)

12. Install the wheels and lower the vehicle.

13. Test drive the vehicle and check the front end alignment.

4-Wheel Drive Models

1. Raise and safely support the vehicle.

2. Hold the sway bar link bolts with a wrench and remove the nuts, retainers, cushions and the links.

3. Remove the sway bracket bolts, brackets and cushions.

4. Remove the sway bar from the suspension.

To install:

5. Install the sway bar link, retainers and cushions to the sway bar with a new nut. Torque the nut to 22 ft. lbs. (29 Nm).

6. Place the sway bar in position and install both sway bar cushions and brackets to the frame. Tighten the bolts to 19 ft. lbs. (25 Nm).

7. Install the sway bar links, cushions and retainers. Connect the sway bar links to the lower control arms with new nuts. Torque the nuts to 19 ft. lbs. (25 Nm) for T-100 and Pickup, or to 51 ft. lbs. (69 Nm) for Tacoma models.

8. Install the wheel and lower the vehicle.

1996–97 4Runner

1. Raise and safely support the vehicle.

2. Remove the two nuts and disconnect the sway bar links from the lower control arm.

3. Remove the bracket bolts, brackets, and bushings with the sway bar.

4. Remove the sway bar from the vehicle.

To install:

5. Place the sway bar in position and install both sway bar cushions and brackets to the frame. Hand tighten the bolts.

NOTE: Tighten the nuts and bolts with the full weight of the vehicle on the ground.

6. Install the link onto the lower control arm and torque the nut to 51 ft. lbs. (69 Nm)

7. Torque the bracket bolts to 19 ft. lbs. (25 Nm)

8. Test drive the vehicle and check the front end alignment.

Front Wheel Bearings

ADJUSTMENT

1996–97 4Runner and 4-Wheel Drive Tacoma

The wheel bearing is not adjustable and must be replaced if a problem is found.

Except 1996–97 4Runner and 4-Wheel Drive Tacoma

The front wheel bearing adjustment is included in the front wheel bearing removal and installation procedure.

REMOVAL AND INSTALLATION

1993–95 4Runner, Pick-up and 1993–97 T-100

2-Wheel Drive Models

1. Raise and safely support the vehicle.

2. Remove the wheel and tire.

3. Remove the disc brake caliper support bracket by removing the two bolts and wire the caliper aside.

4. Remove the cap, cotter pin, lock cap, and the nut from the spindle.

5. Remove the hub and disc together with the outer bearing and thrust washer.

6. Using a small prybar, pry the grease seal from the disc/hub assembly, then remove the inner bearing from the assembly.

7. Using a shop cloth, wipe the grease from inside the disc/hub assembly.

8. Using a brass drift, drive the outer bearing races from each side of the disc/hub assembly.

9. Place matchmarks on the disc and axle hub.

10. Remove the six bolts and separate the disc and axle hub.

11. Using solvent, clean all of the parts.

To install:

12. Align the marks on the disc and axle hub. Install the six bolts and torque the bolts to 47 ft. lbs. (64 Nm).

13. Using a bearing installation tool, drive the outer races into the disc/hub assembly until they seat against the shoulder.

14. Using multi-purpose grease, coat the area between the races and pack the bearings.

15. Place the inner bearing into the rear of the disc/hub assembly. Using a bearing installation tool, drive a new grease seal into the rear of the disc/hub assembly until it is flush with the housing.

16. Install the disc/hub assembly onto the axle shaft, the outer bearing, the thrust washer and the adjusting nut.

17. To adjust the bearing preload, perform the following:

   a. Torque the adjusting nut to 26 ft. lbs. (35 Nm).

   b. Turn the disc/hub assembly 2–3 times, from the left to the right.

   c. Loosen the adjusting nut until it can be turned by hand.

   d. Attach a spring tension gauge to 1 lug on the hub assembly. Pull on the gauge and measure the frictional force. Frictional force = 1.1–31. ft. lbs. (5.0–14.0 N).

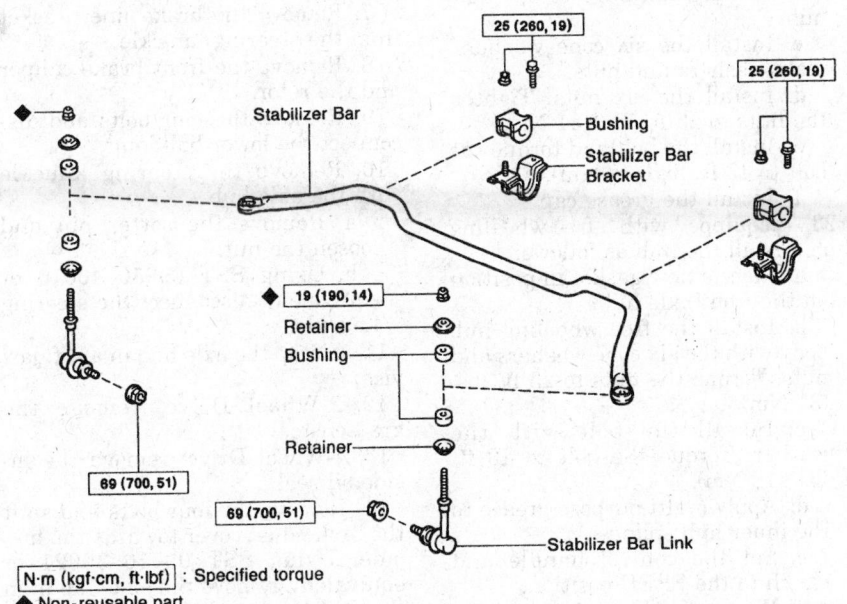

25 (260, 19)

25 (260, 19)

Stabilizer Bar

Bushing

Stabilizer Bar Bracket

◆ 19 (190, 14)

Retainer

Bushing

Retainer

69 (700, 51)

69 (700, 51)

Stabilizer Bar Link

N·m (kgf·cm, ft·lbf) : Specified torque
◆ Non-reusable part

310056

Sway bar component assembly — 1996–97 4Runner and Tacoma shown, others similar

e. Adjust the preload by tightening the nut.

18. Measure the hub axial play. The limit is 0.0020 in. (0.05mm).

19. Install the locknut, cotter pin, and the grease cap.

20. Install the disc brake caliper. Install the wheel and tire assembly.

21. Lower the vehicle.

4-Wheel Drive Models

1. Raise and safely support the vehicle.

2. Remove the wheel and tire assembly.

3. If equipped with ABS brakes, disconnect the ABS speed sensor from the steering knuckle.

4. Remove the two brake caliper support bracket bolts and wire the caliper to the side. Do not allow the caliper to hang from the brake hose.

5. Remove the free-wheel hub:

a. Set the control handle to FREE.

b. Remove the cover bolts and pull off the cover.

c. Remove the center bolt with washer.

d. Remove the mounting nuts and washer to the hub body.

e. Using a brass bar and hammer, tap on the bolts head and remove the cone washer.

f. Pull off the free wheel hub body and gasket.

6. If equipped without a free wheeling hub, remove the flange for the axle hub as follows:

a. Remove the grease cap from the flange.

b. Remove the bolt from the flange.

c. Remove the six mounting nuts to the flange.

d. Using a brass bar and hammer, tap on the bolts head and remove the six cone washers.

e. Install two bolts to the flange. Tighten the bolts to remove the flange.

f. Remove the gasket for the flange.

7. Using a screwdriver, release the taps on the lockwasher.

8. Using SST 09607–60020 or equivalent, remove the locknut.

9. Remove the lock washer and adjusting nut.

10. Remove the claw washer.

11. Remove the axle hub and rotor as an assembly. Remove the outer bearing with the hub and disc.

12. Remove the oil seal and inner bearing, using a suitable puller.

13. If replacing bearing outer race, drive out the outer bearing race using a brass bar and hammer.

To install:

14. If removed, drive in a new bearing outer race using a suitable installer.

15. Pack the bearings with MP grease. Coat the inside of the hub and cap with MP grease.

16. Install the inner bearing and oil seal. Coat the oil seal with MP grease.

17. Install the hub on the spindle. Install the outer bearing and claw washer.

18. Adjust the preload:

a. Torque the bearing adjusting nut to 43 ft. lbs. (59 Nm).

b. Turn the hub right and left 2 or 3 times and re-torque.

c. Loosen the nut until it can be turned by hand.

d. Retighten the nut to 18 ft. lbs. (25 Nm).

e. Check the bearing preload with a spring scale. The preload should be 6.4–12.6 ft. lbs. (28–56 N).

19. Install the lock washer and nut. Tighten to 35 ft. lbs. (47 Nm).

20. Check that there is no bearing end-play.

21. Secure the locknut by bending one lock washer tooth inward and another outward.

22. If equipped without a free wheeling hub, install the flange as follows:

a. Place a new gasket in position on the axle hub.

b. Install the flange to the axle hub.

c. Install the six cone washers, plate washers and nuts.

d. Install the six nuts. Tighten the nuts to 23 ft. lbs. (31 Nm)

e. Install the bolt and torque the bolt to 13 ft. lbs. (18 Nm).

f. Install the grease cap.

23. If equipped with a free wheeling hub, install the hub as follows:

a. Place a new gasket in position on the front axle hub.

b. Install the free wheeling hub body with the six cone washers and nuts. Torque the nuts to 23 ft. lbs. (31 Nm).

c. Install the bolt with the washer. Torque the bolt to 13 ft. lbs. (18 Nm).

d. Apply multi purpose grease to the inner hub splines.

e. Set the control handle and clutch to the FREE position.

f. Place a new gasket in position on the cover.

g. Install the cover to the hub body with the follower pawl tabs aligned with the non-toothed portions of the hub body.

h. Install the mounting bolts and tighten the cover bolts to 7 ft. lbs. (10 Nm).

24. Install the brake caliper support bracket to the steering knuckle and torque the two bolts to 90 ft. lbs. (123 Nm).

25. Clean the threads of the bolts and steering knuckle.

26. Apply sealant to the bolt threads and connect the knuckle arm with the brake line bracket to the steering knuckle. Torque the bolts to 135 ft. lbs. (183 Nm).

27. If equipped with ABS brakes, connect the ABS speed sensor to the steering knuckle.

28. Install the front wheel and lower the vehicle.

29. Check the ABS speed sensor signal.

1996–97 4Runner

1. Raise and safely support the vehicle.

2. Remove the front wheels.

3. Remove the shock absorber.

4. Remove the grease cap.

5. 4-Wheel Drive: Disconnect the halfshaft.

a. Remove the cotter pin and lock cap.

b. While applying the brakes, remove the locknut.

6. With ABS: Remove the ABS speed sensor and wire harness clamp from the steering knuckle.

7. Remove the brake line bracket from the steering knuckle.

8. Remove the front brake caliper and the rotor.

9. Remove the four bolts and disconnect the lower ball joint.

10. Remove the steering knuckle with the axle hub.

a. Remove the cotter pin and loosen the nut.

b. Using SST 09950–40010 or equivalent, disconnect the steering knuckle.

11. Clamp the axle hub in a soft jaw vise.

12. 2-Wheel Drive: remove the grease cap.

13. 4-Wheel Drive: remove the inside oil seal.

14. Remove the four bolts and shift the brake dust cover towards the hub side. Using SST 09710–30021 or equivalent, remove the axle hub from the steering knuckle.

15. Remove the bearing spacer and ABS speed sensor rotor/spacer.

16. Using a flat prying tool, remove the oil seal (outside) from the steering knuckle.

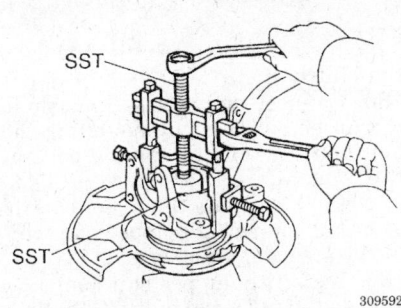

Removing the axle hub from the steering knuckle — 1996–97 4Runner

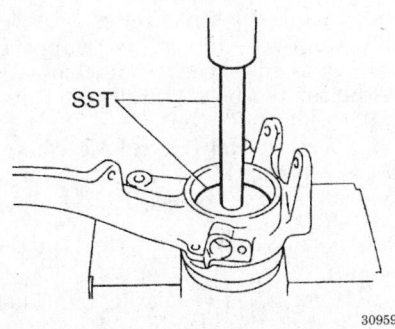

Removing the bearing from the steering knuckle — 1996–97 4Runner

17. Remove the bearing from the steering knuckle.

a. Remove the snapring.

b. Using SST 09950–60020 and 09950–70010 or equivalent, and a press, remove the bearing from the steering knuckle.

To install:

18. Install a new bearing.

a. Using SST 09527–17011 and 09950–60020 or equivalent, and a press, install a new bearing to the steering knuckle.

b. Install a new snapring.

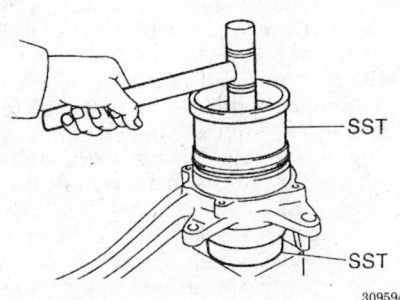

Installing a new oil seal — 1996–97 4Runner

19. Using SST 09223–15030 or equivalent, and a plastic hammer, install a new outside oil seal.

a. Coat MP grease to the oil seal lip.

20. Install the brake dust cover to the steering knuckle with the four bolts and torque to 13 ft. lbs. (18 Nm).

21. Using a press, install the axle hub to the steering knuckle.

22. Install the ABS speed sensor rotor/spacer. Be careful not to scratch the serration of the speed sensor rotor.

23. Using a press, install the bearing spacer.

24. If removed, install the grease cap.

25. If removed, install a new inside oil seal. Using SST 09527–17011 or equivalent, and a plastic hammer, strike the circumference evenly.

26. Install the steering knuckle with the axle hub. Torque the nut to 80 ft. lbs. (108 Nm). Install a new cotter pin.

27. Install the lower ball joint and torque the four bolts to 59 ft. lbs. (80 Nm).

28. Install the rotor and the caliper. Torque the caliper bolts to 90 ft. lbs. (123 Nm).

29. Install the brake line bracket to steering knuckle and torque to 21 ft. lbs. (28 Nm).

30. If removed, install the ABS speed sensor and wire harness clamp to the steering knuckle. Torque the bolts to 6 ft. lbs. (8 Nm).

31. If disconnected, install the half-shaft and torque the nut to 174 ft. lbs. (235 Nm).

32. Install the grease cap.

33. Install the shock absorber.

34. Install the front wheel.

35. Connect the negative battery cable.

Tacoma

2-Wheel Drive Models

1. Raise and safely support the front of the vehicle. Place the jackstands under the frame of the vehicle.

2. Remove the brake caliper support bracket by removing the two bolts. Support the brake caliper with a piece of wire. Do not allow the caliper to hang from the brake hose.

3. Remove the cotter pin, lock cap, nut, and the claw washer from the axle hub and disc.

4. Remove the axle hub with the disc from the steering knuckle. Do not drop the outer bearing when removing the hub.

5. Using a suitable tool, remove the inner oil seal.

6. Remove the inner bearing.

7. Using SST 09527–17011, or equivalent, a brass bar, and a hammer, drive out the bearing outer races.

8. If it is necessary to separate the hub and rotor, place matchmarks on the hub and rotor. Remove the five bolts and remove the hub from the rotor.

To install:

9. Using SST 09527–17011 and a press, install new bearing races.

10. Connect the hub to the rotor and install the five bolts. Torque the five bolts to 47 ft. lbs. (64 Nm).

11. Clean all parts.

12. Repack the bearings with multi purpose grease and apply the same grease to the outer bearings.

13. Install the inner bearing and seal to the hub. Coat the inner seal with multi purpose grease.

14. Install the outer bearing to the hub.

15. Install the hub to the steering knuckle.

16. Install the claw washer and nut to holding the axle hub to the steering knuckle.

17. To adjust the bearing preload, perform the following:

a. Torque the adjusting nut to 25 ft. lbs. (34 Nm).

b. Turn the disc/hub assembly 2–3 times, from the left to the right.

c. Loosen the adjusting nut until it can be turned by hand.

d. Attach a spring tension gauge to 1 lug on the hub assembly. Pull on the gauge and measure the frictional force, which should be 1.3–4.0 ft. lbs. (6.0–18.0 N).

e. Adjust the preload by tightening the nut.

18. Measure the hub axial play. Limit 0.0020 in. (0.05mm).

19. Install the locknut, cotter pin, and the grease cap.

20. Install the disc brake caliper by installing the two bolts. Torque the two bolts to 80 ft. lbs. (108 Nm).

21. Install the wheel and tire assembly.

22. Lower the vehicle.

4-Wheel Drive Models

1. Raise and safely support the vehicle.

2. Remove the wheel and tire assembly.

3. If not equipped with a FREE wheeling hub, disconnect the half-shaft from the steering knuckle as follows:

a. Remove the grease cap.

b. Remove the cotter pin and lock cap from the halfshaft.

c. While having an assistant apply the brakes, remove the locknut from the halfshaft.

4. If equipped with FREE wheeling hub, remove the free wheel hub as follows:

a. Set the control handle to FREE.

b. Remove the cover bolts and pull off the cover.

c. Remove the center bolt with washer.

d. Remove the mounting nuts and washer to the hub body.

e. Using a brass bar and hammer, tap on the bolts head and remove the cone washer.

f. Pull off the free wheel hub body and gasket.

g. Using a snapring expander, remove the snapring from the end of the halfshaft.

5. If equipped with ABS brakes, disconnect the ABS speed sensor from the steering knuckle.

6. Disconnect the brake hose from the steering knuckle by removing the bolt.

7. Remove the two brake caliper support bracket bolts and wire the caliper to the side. Do not allow the caliper to hang from the brake hose.

8. Remove the rotor from the steering knuckle.

9. Remove the four bolts and disconnect the lower ball joint from the steering knuckle.

10. Remove the cotter pin and nut to the upper control arm.

11. Using SST 09950–40010 or equivalent, disconnect the steering knuckle from the upper control arm.

12. Remove the steering knuckle from the vehicle.

NOTE: If it is difficult to remove the halfshaft from the steering knuckle, use a rubber hammer and tap the halfshaft from the steering knuckle.

13. Place the axle hub in a soft jaw vise.

14. Using a pry tool, remove the inside oil seal.

15. If equipped with free wheeling hubs, remove the locknut and ABS speed sensor rotor/spacer as follows:

a. Using a hammer and chisel, loosen the staked part of the locknut.

b. Using SST 09318–12010 or equivalent, remove the locknut from the hub.

c. Remove the ABS speed sensor rotor/spacer. Take care not to scratch the serration of the speed sensor rotor.

16. Remove the axle hub from the steering knuckle as follows:

a. Remove the four bolts for the backing plate and shift the plate towards the hub side.

b. Using the proper tools, press the axle hub from the steering knuckle.

c. If the vehicle is not equipped with free wheeling hubs, remove the bearing spacer and ABS speed sensor rotor/spacer.

17. Using a pry tool, remove the outside oil seal from the steering knuckle.

18. Using a snapring pliers, remove the snapring from the hub.

19. Using SST 09608–35014 or equivalent and a press, remove the bearing from the steering knuckle.

To install:

20. Using SST 09527–17011 and a press, install a new bearing to the steering knuckle.

21. Using a snapring pliers, install the snapring to the hub.

22. Using SST 09223–15030, 09527–17011 and a plastic hammer, install a new outside oil seal. Coat the multi purpose grease to the oil seal lip.

23. Turn the backing plate back into place and install the four bolts. Torque the bolts to 13 ft. lbs. (18 Nm).

24. Using SST 09649–17010 or equivalent and a press, install the axle hub to the steering knuckle.

25. Install the ABS speed sensor rotor/spacer.

26. If equipped with free wheeling hubs, install a new locknut to the hub and torque the nut to 203 ft. lbs. (274 Nm). Stake the nut with a chisel and hammer.

27. If the vehicle is not equipped with free wheeling hubs, install the bearing spacer with SST 09950–60010 and a press.

28. Using SST 09527–17011 or equivalent and a plastic hammer, install a new inside oil seal. Coat the

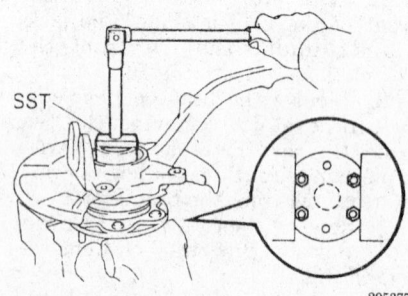

295377

Removing the hub nut — Tacoma 4-Wheel Drive models

multi purpose grease to the oil seal lip.

29. Install the steering knuckle to the halfshaft.

30. Connect the steering knuckle to the lower ball joint by installing the four bolts. Do not torque the bolts at this time.

31. Push down on the upper control arm and connect the upper ball joint to the steering knuckle.

32. Install the upper ball joint nut. Torque the nut to 80 ft. lbs. (105 Nm). Install a new cotter pin.

33. Torque the lower ball joint to steering knuckle bolts to 59 ft. lbs. (80 Nm).

34. Install the brake rotor.

35. Connect the caliper support bracket to the steering knuckle and install the two bolts. Torque the bolts to 90 ft. lbs. (123 Nm).

36. Connect the brake hose clamp to the steering knuckle by installing the bolt. Torque the bolt to 13 ft. lbs. (18 Nm).

37. If equipped with ABS brakes, connect the ABS speed sensor and wire harness to the steering knuckle.

38. Install the spacer and using a snapring expander, install the snapring to the halfshaft.

39. If equipped with a free wheeling hub, install the hub as follows:

a. Place a new gasket in position on the front axle hub.

b. Install the free wheeling hub body with the six cone washers and nuts. Torque the nuts to 23 ft. lbs. (31 Nm).

c. Install the bolt with the washer. Torque the bolt to 13 ft. lbs. (18 Nm).

d. Apply multi purpose grease to the inner hub splines.

e. Set the control handle and clutch to the FREE position.

f. Place a new gasket in position on the cover.

g. Install the cover to the hub body with the follower pawl tabs aligned with the non-toothed portions of the hub body.

h. Install the mounting bolts and tighten the cover bolts to 7 ft. lbs. (10 Nm).

40. If equipped without a free wheeling hub, install the halfshaft to the steering knuckle as follows:

a. Install the locknut to the halfshaft. Torque the nut to 174 ft. lbs. (235 Nm).

b. Install the lock cap and cotter pin to the halfshaft.

c. Install the grease cab to the hub.

41. Install the strut to the vehicle. Torque the nut holding the strut to

the lower control arm to 101 ft. lbs. (135 Nm). Torque the upper three nuts to 47 ft. lbs. (64 Nm)

42. Install the front wheels and lower the vehicle.

REAR SUSPENSION

Shock Absorber

REMOVAL AND INSTALLATION

All Models

1. Raise and safely support the frame with stands.
2. Support the axle housing with a floor jack.
3. Remove the wheel and tire assemblies.
4. Lower the floor jack to take tension off of the spring.
5. Disconnect the shock absorber from the rear axle housing by removing the bolt.
6. Remove the nut, retainers, and the cushions holding the shock absorber to the frame.

7. Remove the shock absorber from the vehicle with the washers and bushings.

To install:

8. Install the shock absorber to the frame with the washers and bushings.
9. Torque the nut to hold the shock absorber to the frame to the following values:
   • All 1993–94 models — 47 ft. lbs. (64 Nm)
   • 1995 4Runner models — 18 ft. lbs. (25 Nm)
   • 1996–97 4Runner models — 14 ft. lbs. (20 Nm)
   • 1995 Pick-up models with 2WD — 19 ft. lbs. (25 Nm)
   • 1995 Pick-up models with 4WD — 53 ft. lbs. (72 Nm)
   • 1995–97 T-100 models — 19 ft. lbs. (25 Nm)
   • Tacoma models with 2WD — 19 ft. lbs. (25 Nm)
   • Tacoma models with 4WD — 53 ft. lbs. (72 Nm)
10. Connect the shock absorber to the rear axle housing and torque the bolt to the following specifications:
   • All 1993–94 models — 27 ft. lbs. (37 Nm)
   • 1995–97 4Runner models — 47 ft. lbs. (64 Nm)
   • 1995 Pick-up models with 2WD — 19 ft. lbs. (25 Nm)

   • 1995 Pick-up models with 4WD — 53 ft. lbs. (72 Nm)
   • 1995–97 T-100 models — 19 ft. lbs. (25 Nm)
   • Tacoma models with 2WD — 19 ft. lbs. (25 Nm)
   • Tacoma models with 4WD — 53 ft. lbs. (72 Nm)
11. Install the wheels.
12. Lower the vehicle to the ground.

Coil Spring

REMOVAL AND INSTALLATION

4Runner

1993–95 Models

——— **CAUTION** ———
The spring on the axle carrier is under high pressure and can cause serious injury if not properly removed and installed.

1. Raise and safely support the vehicle at the frame.
2. Support the axle housing with a floor jack.
3. Remove the wheel and tire assembly.
4. Remove the shock absorber to axle housing bolt. Disconnect both shock absorbers from the axle housing.
5. Disconnect the sway bar brackets from the axle housing.
6. Remove the two bolts and disconnect the shackle bracket from the lateral control rod.
7. Remove the nut, washer, and the bolt from the frame and disconnect the lateral control rod.
8. If necessary, disconnect the brake hose as follows:
   a. Disconnect the brake line from the brake hose at the body bracket.
   b. Remove the clip and disconnect the brake hose from the body bracket.
9. Lower the floor jack, then remove the coil spring(s) and the insulators.

To install:

10. Place the coil spring insulators into place.
11. Place the springs into position and raise the axle housing. Make sure to fit the lower end of the coil spring into the gap of the spring seat on the lower control arm.
12. Raise the rear axle housing.
13. Connect the lateral control rod to the frame with the bolt, washer and nut. Install the bolt from the front of the vehicle but do not torque at this time.

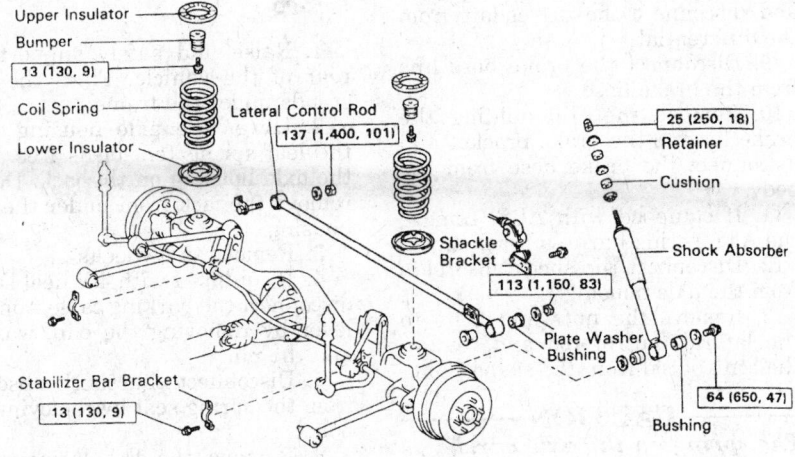

Upper Insulator
Bumper
13 (130, 9)
Coil Spring
Lateral Control Rod
137 (1,400, 101)
Lower Insulator
25 (250, 18)
Retainer
Cushion
Shackle Bracket
Shock Absorber
113 (1,150, 83)
Plate Washer Bushing
Stabilizer Bar Bracket
13 (130, 9)
64 (650, 47)
Bushing

N·m (kgf·cm, ft·lbf) : Specified torque

253097

Rear suspension component identification — 4Runner

spring into the gap of the spring seat on the lower control arm.

16. Install the lateral control rod to the suspension and torque the bolts and nuts to 64 ft. lbs. (86 Nm).

17. Connect the shock absorbers to the axle housing. Torque the bolt to 47 ft. lbs. (64 Nm).

18. If equipped, install the ABS wiring harness bracket.

19. Connect the brake hose to the bracket and install the clip.

20. Connect the brake line to the brake hose and tighten the tube.

21. Connect the parking brake cable bracket to the axle housing and torque to 9 ft. lbs. (13 Nm).

22. Connect the parking brake cable to the brake shoes.

23. Align the matchmarks and connect the driveshaft to the differential. Install the bolts and nuts and torque them to 54 ft. lbs. (73 Nm).

24. Fill the differential with the proper amount and type of oil.

25. Install the brake drum.

26. Bleed the brake system.

27. Install the wheel assemblies.

28. Lower the vehicle and bounce the vehicle several times to stabilize the suspension.

29. Torque the lower control arm to 107 ft. lbs. (145 Nm).

Leaf Spring

REMOVAL AND INSTALLATION

Pick-up

1. Raise and safely support the rear of the vehicle. Place the jack stands under the frame.

2. Lower the axle housing until the leaf spring tension is free. Rest the axle housing on the jack. Do not remove the jack from under the axle housing.

3. Remove the wheel(s).

4. If equipped with 4-Wheel Drive, disconnect the parking cable from the lever by removing the clip, washer, and the pin.

5. Disconnect the shock absorber from the spring seat by removing the bolt.

6. Remove the U-bolt mounting nuts and spring seat from the leaf spring.

7. If equipped with 2-Wheel Drive or 4-Wheel Drive regular cab, remove the pads, pad retainer, and the U-bolts.

8. If equipped with 4-Wheel Drive (regular or extra cab), remove the spring bumper when removing the U-bolts.

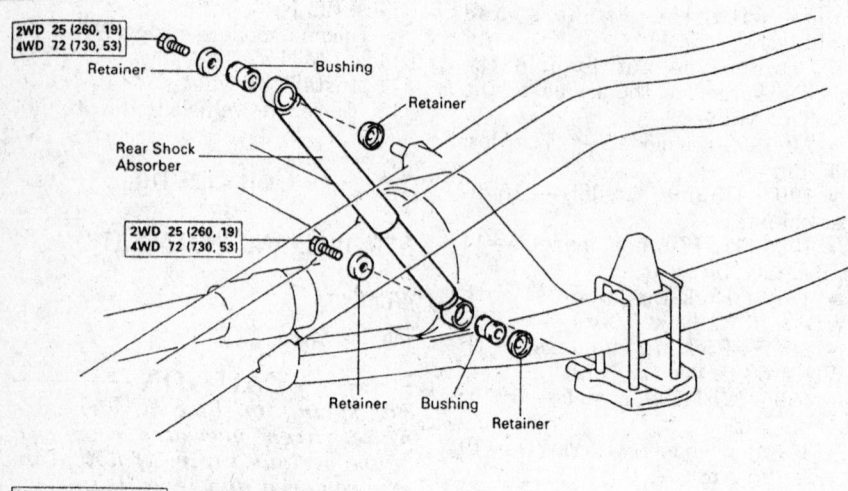

2WD 25 (260, 19)
4WD 72 (730, 53)
Retainer — Bushing
Retainer
Rear Shock Absorber
2WD 25 (260, 19)
4WD 72 (730, 53)
Retainer   Bushing   Retainer

N·m (kgf·cm, ft·lbf) : Specified torque

253085

Shock absorber removal and installation — 1995 Pick-up, 1995–97 T-100 and Tacoma

14. Connect the shackle bracket to the lateral control rod with the two bolts.

15. Connect the shock absorber to the lower control arm and install the nut. Torque the nut to 47 ft. lbs. (64 Nm).

16. If necessary, install the brake hose as follows:

a. Connect the brake hose to the bracket and install the clip.

b. Connect the brake tube to the brake hose and tighten the tube.

17. Install the sway bar brackets to the rear axle housing and torque the bolts to 9 ft. lbs. (13 Nm).

18. Bleed the brake system.

19. Install the wheels and lower the vehicle.

20. Stabilize the suspension and torque the lateral control rod nut to 101 ft. lbs. (137 Nm).

1996–97 Models

1. Raise and safely support the vehicle.

2. Remove the wheel assemblies.

3. Support the axle housing with a floor jack.

4. Remove the brake drum from the axle housing.

5. Disconnect the parking brake cable from the brake shoe.

6. Remove the bolt and disconnect the parking brake cable from the axle housing.

7. Place matchmarks on the flanges for the driveshaft and differential.

8. Remove the four bolt and nuts and disconnect the driveshaft from the differential.

9. Disconnect the brake hose line from the brake hose.

10. Remove the clip holding the brake hose to the brake bracket and disconnect the brake hose from the body.

11. If equipped with ABS, remove the ABS wiring harness bracket.

12. Disconnect the shock absorbers from the axle housing.

13. Remove the nuts and bolts to the lateral control rod and remove the control rod from the suspension.

—— CAUTION ——
The spring on the axle carrier is under high pressure and can cause serious injury if not properly removed and installed.

14. Slowly lower the rear axle housing and remove the coil spring.

To install:

15. Place the springs into position and raise the axle housing. Make sure to fit the lower end of the coil

9. Remove the bolt and nut holding the front of the leaf spring to the frame of the vehicle.

10. Remove the shackle pin mounting nuts from the rear of the spring.

11. Remove the shackle pin and plate holding the rear of the leaf spring to the frame of the vehicle.

12. Remove the leaf spring from vehicle.

To install:

13. Place the leaf spring in place and install the bolt and nut to hold the front of the leaf spring to the frame. Do not torque the bolt and nut at this time.

14. Place the rear end of the leaf spring in place and install the shackle pin. Install the plate and nuts to the shackle pin. Do not torque the nuts at this time.

15. Install the U-bolts as follows:

   a. If equipped with 2-Wheel Drive or 4-Wheel Drive regular cab, install the pads and pad retainer on the leaf spring.

   b. If equipped with 4-Wheel Drive (regular or extra cab), install the spring bumper.

   c. Install the spring seat, U-bolts, washers, and the nuts.

   d. Torque the nuts for the U-bolts as follows:

   • If equipped with 2-Wheel Drive, torque the nuts to 108 ft. lbs. (147 Nm)

   • If equipped with 4-Wheel Drive and regular cab, torque the nuts to 90 ft. lbs. (123 Nm)

   • If equipped with 4-Wheel Drive and extra cab, torque the nuts to 108 ft. lbs. (147 Nm)

NOTE: When tightening the U-bolt nuts, tighten the nuts so the lengths of the U-bolts under the spring seat are the same.

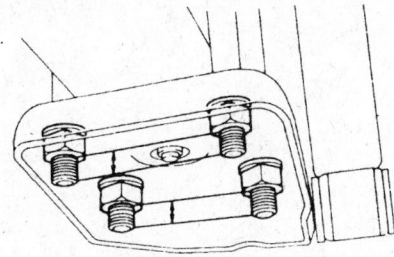

All same length

253156

Ensure that the U-bolts all the same length when installing the nuts — 1995 Pick-up

16. Connect the shock absorber to the spring seat and torque the bolt as follows:

   • 2-Wheel Drive vehicle to 19 ft. lbs. (25 Nm)

   • 4-Wheel Drive vehicle to 53 ft. lbs. (72 Nm)

17. Connect the parking brake cable to the lever and install the pin, washer, and the clip.

18. Install the wheels and lower the vehicle.

19. Bounce the vehicle up and down to stabilize the vehicle.

20. Torque the nut and bolt holding the front of the spring to the frame to 116 ft. lbs. (157 Nm).

21. Torque the shackle nuts to 67 ft. lbs. (91 Nm).

T-100

1. Raise and safely support the rear of the vehicle. Place the jack stands under the frame.

2. Lower the axle housing until the leaf spring tension is free. Rest the axle housing on the jack. Do not remove the jack from under the axle housing.

3. Remove the rear wheel(s).

4. If equipped with 4-Wheel Drive, disconnect the parking cable from the lever by removing the clip, washer, and the pin.

5. Disconnect the shock absorber from the spring seat by removing the bolt.

6. Remove the U-bolt mounting nuts and spring seat from the leaf spring.

7. Remove the pads, pad retainer the U-bolts.

8. If equipped with 4-Wheel Drive, remove the spring bumper when removing the U-bolts.

9. If the vehicle is a 2-Wheel Drive half ton or a 4-Wheel Drive (one or half ton), remove the bolt and nut holding the front of the leaf spring to the frame of the vehicle.

10. If the vehicle is a 2-Wheel Drive one ton, remove the nut holding the lock pin bolt to the leaf spring. Remove the two hanger pin lock bolts and then remove the lock pin holding the front of the leaf spring to the frame.

11. Remove the shackle pin mounting nuts from the rear of the spring.

12. Remove the shackle pin and plate holding the rear of the leaf spring to the frame of the vehicle.

13. Remove the leaf spring from the vehicle.

To install:

14. Place the leaf spring in place and install the bolt and nut to hold

the front of the leaf spring to the frame. Do not torque the bolt and nut at this time. If equipped with a 2-Wheel Drive one ton, install the two hanger lock bolts. Torque the lock bolts to 19 ft. lbs. (26 Nm).

15. Place the rear end of the leaf spring in place and install the shackle pin. Install the plate and nuts to the shackle pin. Do not torque the nuts at this time.

16. Install the U-bolts as follows:

   a. Install the pads and pad retainer on the leaf spring.

   b. If equipped with 4-Wheel Drive, install the spring bumper.

   c. Install the spring seat, U-bolts, washers, and the nuts.

   d. Torque the nuts for the U-bolts to 97 ft. lbs. (132 Nm).

NOTE: When tightening the U-bolt nuts, tighten the nuts so the lengths of the U-bolts under the spring seat are the same.

17. Connect the shock absorber to the spring seat and torque the bolt to 19 ft. lbs. (26 Nm).

18. If equipped with 4-Wheel Drive, connect the parking brake cable to the lever and install the pin, washer, and the clip.

19. Install the wheels and lower the vehicle.

20. Bounce the vehicle up and down to stabilize the vehicle.

21. Torque the nut and bolt holding the front of the spring to the frame to:

   • 1995–97 2-Wheel Drive 0.5 ton and 4-Wheel Drive: 144 ft. lbs. (196 Nm)

   • All others: 67 ft. lbs. (91 Nm)

22. Torque the shackle nuts to 67 ft. lbs. (91 Nm).

Tacoma

1. Raise and safely support the rear of the vehicle. Place the jack stands under the frame.

2. Lower the axle housing until the leaf spring tension is free. Rest the axle housing on the jack. Do not remove the jack from under the axle housing.

3. Remove the rear wheel(s).

4. If equipped with 4-Wheel Drive, disconnect the parking cable from the lever by removing the clip, washer, and the pin.

5. Disconnect the shock absorber from the spring seat by removing the bolt.

6. Remove the U-bolt mounting nuts, spring seat, pads, and the pad retainer from the leaf spring.

7. If equipped with 4-Wheel Drive, remove the spring bumper when removing the U-bolts.

8. Remove the bolt and nut holding the front of the leaf spring to the frame of the vehicle.

9. Remove the shackle pin mounting nuts from the rear of the spring.

10. Remove the shackle pin and plate holding the rear of the leaf spring to the frame of the vehicle.

11. Remove the leaf spring from the vehicle.

To install:

12. Place the leaf spring in place and install the bolt and nut to hold the front of the leaf spring to the frame. Do not torque the bolt and nut at this time.

13. Place the rear end of the leaf spring in place and install the shackle pin. Install the plate and nuts to the shackle pin. Do not torque the nuts at this time.

14. Install the U-bolts as follows:

  a. Install the pads and pad retainer on the leaf spring.

  b. If equipped with 4-Wheel Drive, install the spring bumper.

  c. Install the spring seat, U-bolts, washers, and the nuts.

  d. Torque the nuts for the U-bolts to 90 ft. lbs. (123 Nm).

NOTE: When tightening the U-bolt nuts, tighten the nuts so the lengths of the U-bolts under the spring seat are the same.

15. Connect the shock absorber to the spring seat and torque the bolt as follows:

• 2-Wheel Drive vehicle to 19 ft. lbs. (25 Nm)

• 4-Wheel Drive vehicle to 53 ft. lbs. (72 Nm)

16. If equipped with 4-Wheel Drive, connect the parking brake cable to the lever and install the pin, washer, and the clip.

17. Install the wheels and lower the vehicle.

18. Bounce the vehicle up and down to stabilize the vehicle.

19. Torque the nut and bolt holding the front of the spring to the frame to 115 ft. lbs. (155 Nm).

20. Torque the shackle nuts to 67 ft. lbs. (91 Nm).

Upper Control Arms

REMOVAL AND INSTALLATION

4Runner

1. Raise and safely support the rear of the vehicle. Place the jack stands under the frame of the vehicle.

2. Hold the rear axle housing with a jack.

3. Disconnect the upper control arm from the frame by removing the nut, washer, and the bolt.

4. Disconnect the upper control arm from the axle housing by removing the nut, washer, and the bolt.

5. Remove the upper control arm from the vehicle.

To install:

6. Install the upper control arm to the vehicle.

7. Connect the upper control arm to the axle housing by installing the bolt, washer, and the nut. Install the bolt from the outside of the vehicle. Do not torque at this time.

8. Connect the upper control arm to the frame by installing the bolt, washer, and the nut. Install the bolt from the outside of the vehicle. Do not torque at this time.

9. Lower the vehicle and stabilize the suspension.

10. Jack up the axle housing and support the housing with jack stands.

11. Torque the bolt holding the upper control arm to the frame to 148 ft. lbs. (201 Nm) for 1993–95 models, to 107 ft. lbs. (145 Nm) for 1996 models, or to 64 ft. lbs. (86 Nm) for 1997 models.

12. Torque the bolt and nut holding the upper control arm to the axle housing to 148 ft. lbs. (201 Nm) for 1993–95 models, to 107 ft. lbs. (145 Nm) for 1996 models, or to 64 ft. lbs. (86 Nm) for 1997 models.

13. Lower the vehicle.

Lower Control Arms

REMOVAL AND INSTALLATION

4Runner

1. Raise and safely support the rear of the vehicle. Place the jack stands under the frame of the vehicle.

2. Support the axle housing with a floor jack.

3. Remove the sway bar links as follows:

  a. Remove the nuts and disconnect the sway bars from the sway bar links.

  b. Hold the sway bar link with a wrench and disconnect the sway bar link from the vehicle by removing the nut, retainer and cushions. Remove both sway bar links.

  c. Remove the retainers and cushions from the sway bar links.

4. Remove the bolts and brackets holding the sway bar to the axle housing.

5. Remove the sway bar from the vehicle and remove the bushings.

To install:

6. Install the sway bar to the axle housing and install the cushions, brackets and bolts. Torque the bolts to 9 ft. lbs. (13 Nm).

7. Install both sway bar links as follows:

  a. Install the retainers and cushions to the sway bar links.

  b. Install the sway bar links, cushions and retainers onto the frame. Torque the sway bar nut to 11–13 ft. lbs. (15–17 Nm).

  c. Connect the sway bar links to the sway bar and install the nuts. Torque the nuts to 70 ft. lbs. (95 Nm) for 1993–95 models, or to 70 ft. lbs. (95 Nm) for 1996–97 models.

8. Lower the vehicle.

FIRING ORDERS

NOTE: To avoid confusion, always replace spark plug wires one at a time.

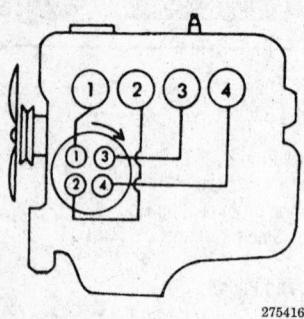

2.4L (VIN A and VIN T) Engines
Engine firing order: 1-3-4-2
Distributor rotation: clockwise

275416

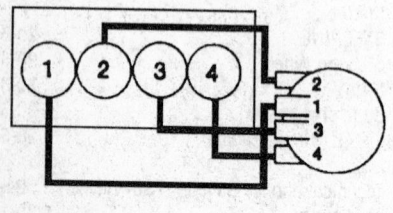

2.0L (VIN P) Engine
Engine firing order: 1-3-4-2
Distributor rotation: clockwise

353572

ENGINE ELECTRICAL

NOTE: Disconnecting the negative battery cable on some vehicles may interfere with the functions of the on board computer systems and may require the computer to undergo a relearning process, once the negative battery cable is reconnected.

Distributor

REMOVAL AND INSTALLATION

2.0L (VIN P) Engine

UNDISTURBED ENGINE

1. Disconnect the negative battery cable.

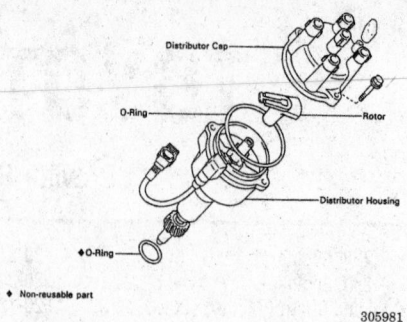

Distributor assembly — 2.0L (VIN P) engine

305981

CAUTION

Wait 90 seconds from the time the key is turned to LOCK and the negative battery cable is disconnected to begin work. This allows the SRS capacitor to discharge and prevent deployment of the air bag(s).

2. Remove the air cleaner cap assembly.
3. Disconnect the electrical connector to the distributor.
4. Remove the high tension cable from the coil.
5. Mark and remove the spark plug wires from the distributor. Remove the distributor cap bolts and then remove the cap.
6. Matchmark the rotor to the distributor housing and the distributor housing to the engine block. This will aid in correct positioning of the distributor during installation.
7. Remove the distributor hold-down clamp bolt and the distributor from the engine.

To install:

8. Install a new O-ring to the distributor and lubricate it with engine oil.
9. Insert the distributor into the engine block by aligning the matchmarks made during removal.
10. Engage the distributor drive with the slit in the intake camshaft.
11. Install the distributor hold-down clamp, the cap, the high tension wire, the spark plug wires and the electrical connector.
12. Install the air cleaner cap assembly.
13. Connect the negative battery cable.
14. Start the engine. Check and adjust the ignition timing.

DISTURBED ENGINE

1. Disconnect the negative battery cable.

CAUTION

Wait 90 seconds from the time the key is turned to LOCK and the negative battery cable is disconnected to begin work. This allows the SRS capacitor to discharge and prevent deployment of the air bag(s).

2. Remove the air cleaner cap assembly.
3. Disconnect the distributor connector.
4. Remove the high tension cable from the coil.
5. Mark and remove the spark plug wires from the distributor. Remove the distributor cap bolts and then remove the cap.
6. Remove the distributor hold-down bolt and pull the distributor from the cylinder block.

To install:

7. Install a new O-ring to the distributor and lubricate it with engine oil.
8. Remove the No. 1 cylinder spark plug.
   a. Place a finger or compression gauge over the spark plug hole.
   b. Turn the crankshaft until compression starts to build up. Continue turning the crankshaft until the crankshaft pulley groove align with the timing mark 0 of the timing chain. The position of the slit of the intake camshaft should be as shown.
9. On the distributor, align the cutout of the coupling with the line on the housing.
10. Insert the distributor, aligning the center of the flange with that of the bolt hole on the cylinder head.
11. Torque the hold-down bolt to 14 ft. lbs. (19 Nm).
12. Install the distributor cap.
13. Install the spark plug wires and the electrical connector.
14. Install the No. 1 cylinder spark plug. Install the high tension wire on the coil.
15. Connect the negative battery cable.
16. Start the engine and inspect the timing.

2.4L (VIN A and T) Engine

UNDISTURBED ENGINE

1. Disconnect the negative battery cable.
2. Label and disconnect the spark plug wires.
3. Disconnect the distributor connector and ventilation hoses.
4. Remove the cap and packing.

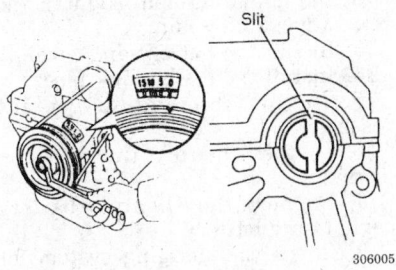

Crankshaft TDC mark and intake camshaft slit position — 2.0L (VIN P) engine

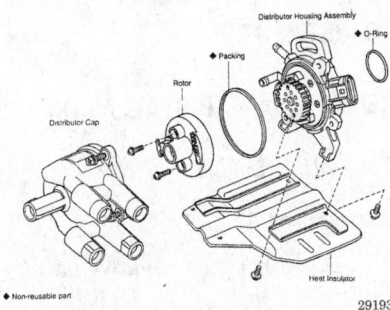

Distributor component assembly — 2.4L (VIN A and T) engine

5. Set the No. 1 cylinder to TDC of the compression stroke. Install the service bolt and nut into the equipment driveshaft to turn the crankshaft pulley. Turn the crankshaft one turn if the rotor is not facing No. 1 spark plug wire.

6. Remove the distributor hold-down and pull the distributor out of the engine.

To install:

7. Install a new O-ring to the distributor and lubricate with engine oil.

8. Insert the distributor, aligning the center of the distributor housing flange with the bolt hole on the cylinder head.

9. Engage the distributor drive with the oil pump drive shaft.

10. Install the distributor hold-down clamp, the cap, the high tension wire, the primary wire or the electrical connector and the vacuum line(s).

11. Install the spark plugs cables.

12. Connect the negative battery cable.

Engine Disturbed

1. Disconnect the negative battery cable.

2. Label and disconnect the spark plug wires.

3. Disconnect the distributor connector and ventilation hoses.

4. Remove the cap and packing.

5. Set the No. 1 cylinder to TDC of the compression stroke. Install the service bolt and nut into the equipment driveshaft to turn the crankshaft pulley. Turn the crankshaft one turn if the rotor is not facing No. 1 spark plug wire.

6. Remove the distributor hold-down and pull the distributor out of the engine.

To install:

7. If the engine was disturbed while the distributor was removed, continue as follows:

8. Remove the No. 1 spark plug, place a finger over the opening and rotate the equipment driveshaft, using a turning tool, in the clockwise direction, until pressure is felt, then replace the spark plug.

NOTE: Make sure the slit in the exhaust camshaft is in the proper position.

9. Remove the service bolt and nut.

10. Align the cut out portion of the coupling with the groove on the housing.

11. Install the distributor and align the center of the flange with the bolt hole on the cylinder head.

12. Install the hold-down bolt loosely.

13. Install the seal packing, distributor cap and connect the wiring.

14. Adjust the timing to specifications and torque the hold-down bolt to 14 ft. lbs. (19 Nm).

Ignition Timing

ADJUSTMENT

1993 Engines

1. Place the transaxle lever in **P** position.

2. Firmly apply the parking brake and block the drive wheels.

3. Connect a tachometer to the engine. Connect the tachometer positive terminal to the IG terminal of the check connector.

NOTE: Some tachometers are not compatible with this ignition system. Confirm the compatibility of your unit before using.

4. Connect a timing light to the engine, following the manufacturer's instructions.

5. Connect a jumper between terminals TE_1 and E_1 of the check connector.

6. Start the engine and allow it to idle until normal operating temperature is reached. Aim the timing light at the crankshaft pulley and check the ignition timing. Ignition timing: 5 degrees BTDC @ idle.

7. If not within specifications, loosen the distributor hold-down bolts and slowly turn the distributor until the timing mark on the crankshaft pulley is aligned with the 5 degree mark. Tighten the distributor hold-down bolts to 14 ft. lbs. (19 Nm) and recheck the ignition timing.

8. Remove the jumper from the check connector.

9. The ignition timing should now be about 12 degrees BTDC @ idle.

10. Remove the tachometer and timing light.

1994–97 Engines

Ignition timing is controlled by the ECM and is not adjustable.

Alternator

REMOVAL AND INSTALLATION

2.0L (VIN P) Engine

1. Disconnect the negative battery cable.

———— **CAUTION** ————
Failure to disconnect the battery can cause personal injury and/or damage to the vehicle.

2. Disconnect the electrical wiring from the alternator.

3. Loosen the adjusting lock bolt and the pivot bolt.

4. If equipped with A/C, loosen the adjusting bolt to relieve tension on the drive belt.

5. Remove the drive belt. It may be necessary to remove other belts for access.

6. Remove the pivot bolt first, support the alternator and remove the adjusting lock bolt.

7. Remove the alternator from the vehicle.

To install:

8. Install the alternator.

9. Leave the bolts finger tight so that the belt may be adjusted to the correct tension.

10. Make sure that the electrical plugs and connectors are properly seated and secure in their mounts. Adjust belt tension and tighten all necessary hardware. If equipped with A/C, tighten the adjusting bolt to apply tension to the belt.

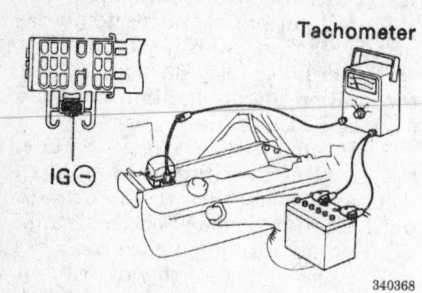

Connect the tachometer positive (+) to the IG-terminal of the check connector

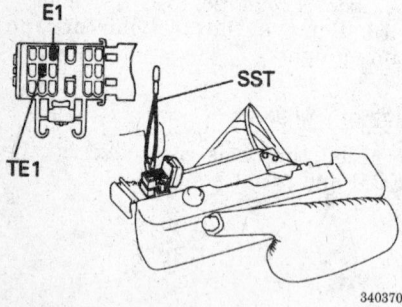

Adjust ignition timing. Connect terminals TE1 and E1 of the check connector

11. Torque the pivot bolt to 38 ft. lbs. (52 Nm) and the adjusting bolt to 13 ft. lbs. (18 Nm).
12. Connect the negative battery cable.
13. Verify that the charging system is functioning properly.

1993–95 2.4L (VIN T) Engine

1. Disconnect the negative battery cable from the battery.
2. Raise and safely support the vehicle.
3. Remove the No. 1 engine undercover.
4. Disconnect the alternator wire from the alternator adjusting bar.
5. Disconnect the alternator wire from the alternator by removing the nut.
6. Disconnect the alternator electrical connector.
7. Loosen the lock, adjusting, and pivot bolts to the alternator. Push the alternator towards the engine and remove the drive belt.
8. Remove the pivot bolt and lock bolt to the alternator.
9. Remove the alternator from the vehicle.
To install:
10. Install the alternator to the engine and install the pivot and lock

bolts. Do not torque the bolts at this time.
11. Install and adjust the drive belt with the adjusting bolt.
12. Torque the lock bolt to 13 ft. lbs. (18 Nm) and the pivot bolt to 37 ft. lbs. (50 Nm).
13. Connect the alternator connector.
14. Connect the alternator wire with the nut.
15. Connect the alternator wire to the adjusting bar.
16. Install the No. 1 engine under cover.
17. Connect the negative battery cable to the battery.
18. Check the charging system for proper operation.

1994–97 2.4L (VIN A) Engine

1. Disconnect the negative battery cable from the battery.
2. Remove the engine coolant reservoir tank.
3. Disconnect the fusible link block from the battery bracket.
4. Remove the battery cover and the battery from the engine compartment.
5. Loosen the No. 1 idler pulley nut and adjusting bolt. Remove the drive belt from the engine.
6. Disconnect the alternator wire from the alternator by removing the nut.
7. Disconnect the alternator electrical connector.
8. Remove the three bolts and remove the alternator from the engine.
To install:
9. Install the alternator to the engine and install the three bolts. Torque the bolt holding the bottom of the alternator to 37 ft. lbs. (50 Nm). Torque the alternator stay bolts to 20 ft. lbs. (26 Nm).
10. Connect the alternator connector.
11. Connect the alternator wire with the nut.

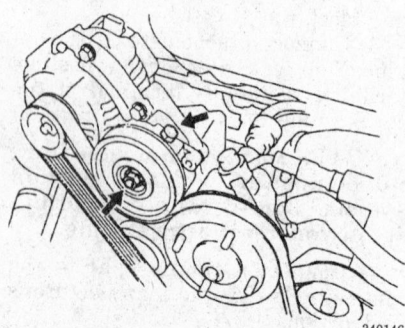

Adjusting bolt and No. 1 idler pulley nut — 1994–97 2.4L (VIN A) engine

12. Install and adjust the drive belt with the adjusting bolt. Tighten the No. 1 idler pulley nut.
13. Install the battery and cover.
14. Install the fusible link block.
15. Install the engine coolant reservoir.
16. Install the air duct to the engine.
17. Connect the negative battery cable to the battery.
18. Check the charging system for proper operation.

Drive Belt

ADJUSTMENT

REMOVAL AND INSTALLATION

2.0L (VIN P) Engine

Alternator and A/C Belt

1. Disconnect the negative battery cable.
2. Loosen the lock bolt and pivot bolt.
3. Loosen the belt by loosening the adjusting bolt at the alternator.
4. Remove the drive belt from its driven alternator and A/C units.
5. Visually check the belt for separation of the ribs, torn or worn ribs, or cracks in the inner ridges of the ribs. If necessary, replace the drive belt.
To install:
6. Position the drive belt over the component pulleys and drive pulley.
7. Tighten the adjusting bolt to tighten the belt.
8. Check the drive belt tension using a belt tension gauge. Tighten the belt to the following tension:
• With A/C – New belt to 165±25 lb.; Used belt to 110±20 lb.
• Without A/C – New belt to 125±25 lb.; Used belt to 95±20 lb.
9. Torque the pivot bolt to 38 ft. lbs. (52 Nm) and the lock bolt to 13 ft. lbs. (18 Nm).

NOTE: After installing the drive belt(s), check that the belt does not touch the bottom of the pulley groove (conventional type belts) or that it fits properly in the ribbed grooves (V-ribbed type belts).

10. Connect the negative battery cable.
11. If a new belt was installed, run the engine for approximately 5 minutes and then recheck the tension.

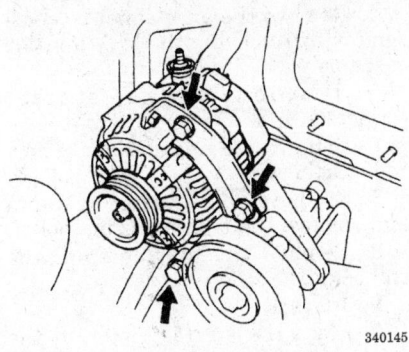

Alternator bolts — 1994–97 2.4L (VIN A) engine

Power Steering Belt

1. Disconnect the negative battery cable from the battery.
2. Remove the alternator–A/C belt from the engine.
3. Loosen the pivot bolt and adjusting bracket bolt for the power steering unit.
4. Remove the drive belt from the engine by pivoting the power steering unit.

To install:
5. Install the belt to the engine and tighten the belt by pulling back on the power steering unit.

6. Tighten the drive belt to the following specifications:
   • New belt to 95–145 ft. lbf
   • Old belt to 60–100 ft. lb.
7. Torque the pivot and adjusting bracket bolt to 32 ft. lbs. (43 Nm).
8. Connect the negative battery cable to the battery.

1993–95 2.4L (VIN A and T) Engine

Alternator/Power Steering Drive Belt

1. Disconnect the negative battery cable to the battery.
2. Raise and safely support the vehicle.
3. Remove the engine undercover.
4. Loosen the lock, adjusting, and pivot bolts to the alternator. Push the alternator towards the engine and remove the drive belt.
5. Remove the drive belt from the engine.

To install:
6. Install and adjust the drive belt with the adjusting bolt. Adjust the belt to the following specifications:
   • New belt: 135–185 lbs.
   • Old belt: 80–120 lbs.
7. Torque the lock bolt to 13 ft. lbs. (18 Nm) and the pivot bolt to 37 ft. lbs. (50 Nm).
8. Install the engine undercover.
9. Lower the vehicle and connect the negative battery cable.

A/C Compressor Drive Belt

1. Remove the engine under cover.
2. Remove the alternator/power steering drive belt.
3. Raise and safely support the vehicle.
4. Loosen the idler pulley lock nut.
5. Loosen the adjusting bolt to the idler pulley and remove the compressor drive belt.

To install:
6. Install the compressor drive belt to the engine and tighten the adjusting bolt to the idler pulley. Drive belt tension is as follows:
   • New belt: 165±26 lbs.
   • Old belt: 88±22 lbs.
7. Torque the idler pulley lock nut to 29 ft. lbs. (39 Nm).
8. Install the alternator/power steering drive belt.
9. Recheck belt tension.
10. Install the engine undercover.
11. Lower the vehicle.

1994–97 2.4L (VIN A) Engine

Alternator/Power Steering Drive Belt

1. Disconnect the negative battery cable to the battery.
2. Remove the air duct.
3. Loosen the No. 1 idler pulley nut and adjusting bolt.
4. Remove the drive belt from the engine.

To install:
5. Install the drive belt and adjust the belt with the adjusting bolt. Adjust the drive belt to the following specifications:
   • New belt: 170±10 lbs.
   • Old belt: 125±10 lbs.
6. Tighten the No. 1 idler pulley nut.
7. Install the air duct.
8. Connect the negative battery cable to the battery.

Supercharger Drive Belt

1. Disconnect the negative battery cable from the battery.
2. Remove the alternator/power steering drive belt from the engine.
3. Loosen the No. 2 idler pulley nut and adjusting bolt.
4. Remove the drive belt from the supercharger.

To install:
5. Install the drive belt to the supercharger.
6. Tighten the adjusting bolt and adjust the drive belt to the following specifications:
   • New belt: 170±10 lbs.
   • Old belt: 125±10 lbs.
7. Install the alternator/power steering drive belt to the engine.

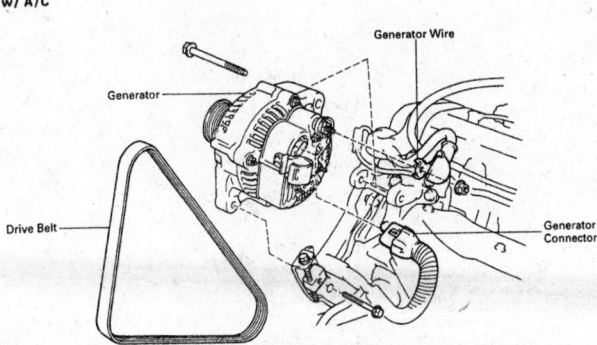

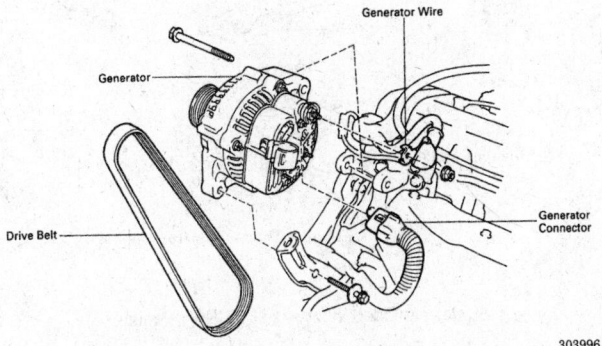

Alternator and A/C drive belt — 2.0L (VIN P) Engine

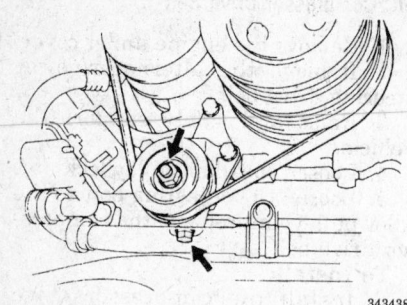

A/C compressor idler pulley — 2.4L (VIN A and T) Engine

343438

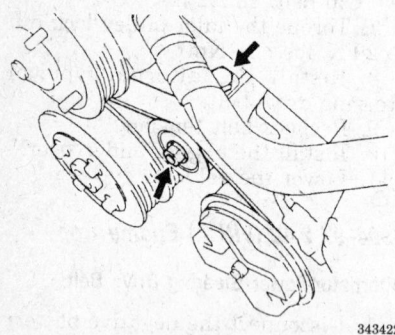

Supercharger idler pulley

343422

8. Connect the negative battery cable to the battery.

A/C Compressor Drive Belt

1. Remove the alternator/power steering drive belt.

2. Remove the supercharger drive belt.

3. Raise and safely support the vehicle.

4. Loosen the idler pulley lock nut.

5. Loosen the adjusting bolt to the idler pulley and remove the compressor drive belt.

To install:

6. Install the compressor drive belt to the engine and tighten the adjusting bolt to the idler pulley. Drive belt tension is as follows:
- New belt: 165±20 lbs.
- Old belt: 130±20 lbs.

7. Torque the idler pulley lock nut to 29 ft. lbs. (39 Nm).

8. Install the supercharger drive belt.

9. Install the alternator/power steering drive belt.

10. Connect the negative battery cable to the battery.

11. Recheck belt tension.

Starter

REMOVAL AND INSTALLATION

2.0L (VIN P) Engine

1. Disconnect the negative and positive battery cables.

---— CAUTION —

On models with an air bag, wait at least 90 seconds from the time that the ignition switch is turned to the LOCK position and the battery is disconnected before performing any further work.

---

2. Remove the engine coolant reservoir.

3. Disconnect the air cleaner cap assembly as follows:

a. Skid control relay connectors.

b. Disconnect the IAT sensor connector.

c. Disconnect the coil wire from the air cleaner hose.

d. Disconnect the PCV hose from the air cleaner hose.

e. Disconnect the air hose from the air cleaner hose.

f. Disconnect the four clamps and disconnect the air cleaner hose from the throttle body.

g. Remove the air cleaner cap assembly.

4. Remove the air cleaner case assembly by removing the VSV and the three bolts.

5. Disconnect the starter connector.

6. Remove the nut and disconnect the starter wire.

7. Support the starter by hand and remove the two mounting bolts.

8. Remove the starter from the transaxle.

To install:

9. Place the starter motor in the transaxle and support it by hand.

10. Install the two mounting bolts and torque them to 29 ft. lbs. (39 Nm)

11. Place the starter wire into position and tighten the nut.

12. Connect the electrical connector to the starter.

13. Install the air cleaner case assembly and install the three bolts and VSV.

14. Connect the air cleaner cap assembly as follows:

a. Install the air cleaner cap assembly.

b. Connect the air cleaner hose to the throttle body and clamp the four clamps for the cap assembly.

c. Connect the air hose to the air cleaner hose.

d. Connect the PCV hose to the air cleaner hose.

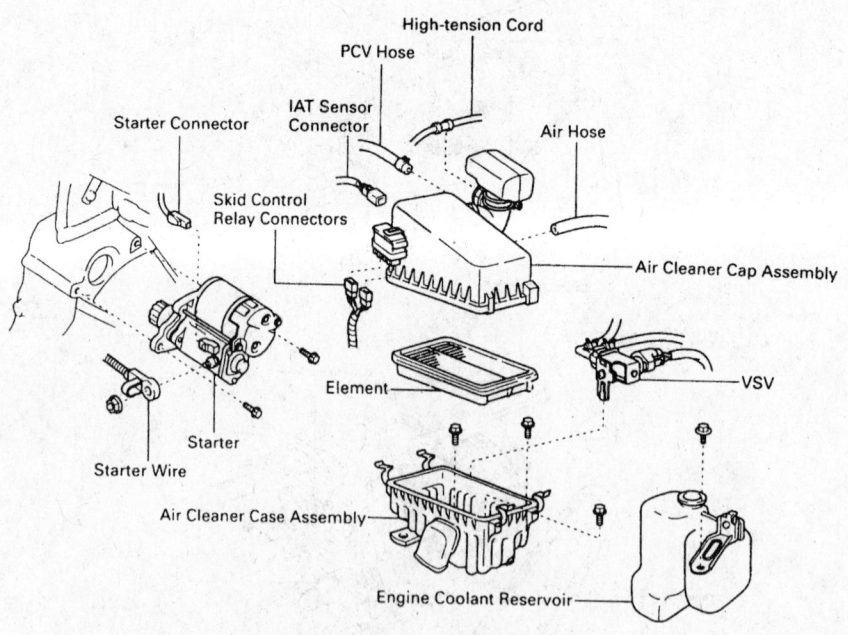

Starter exploded view — 2.0L (VIN P) Engine

303784

e. Connect the coil wire to the air cleaner hose.

f. Connect the IAT sensor connector.

g. Connect the skid control relay connectors.

15. Install the engine coolant reservoir.

16. Connect the negative and positive battery cables.

17. Verify proper starter operation.

2.4L (VIN A and T) Engine

1. Disconnect the negative battery cable from the battery.

2. Raise and safely support the vehicle.

3. If equipped with 4WD, remove the front driveshaft.

4. Remove the nut and disconnect the starter wire.

5. Disconnect the starter connector.

6. Remove the bolt holding the starter stay to the upper stiffener plate.

7. If equipped with 2WD, remove the starter by removing the nut and three bolts.

8. If equipped with 4WD, remove the starter by removing the nut, four bolts, and the center bracket.

To install:

9. If equipped with 4WD, install the starter, the center support bracket, nut, and the four bolts. Torque the bolts as follows:
 • Bolt A to 41 ft. lbs. (56 Nm)
 • Bolt B to 30 ft. lbs. (41 Nm)

10. If equipped with 2WD, install the starter, nut and three bolts. Torque the bolts as follows:
 • Bolt A to 41 ft. lbs. (56 Nm)
 • Bolt B to 30 ft. lbs. (41 Nm)

11. Install the bolt to hold the starter stay to the upper stiffener plate. Torque the bolt to 43 inch lbs. (4.9 Nm).

12. Install the starter connector.

13. Connect the starter wire with the nut. Torque the nut to 78 inch lbs. (8.8 Nm).

14. If equipped with 4WD, install the front driveshaft.

15. Connect the negative battery cable to the battery.

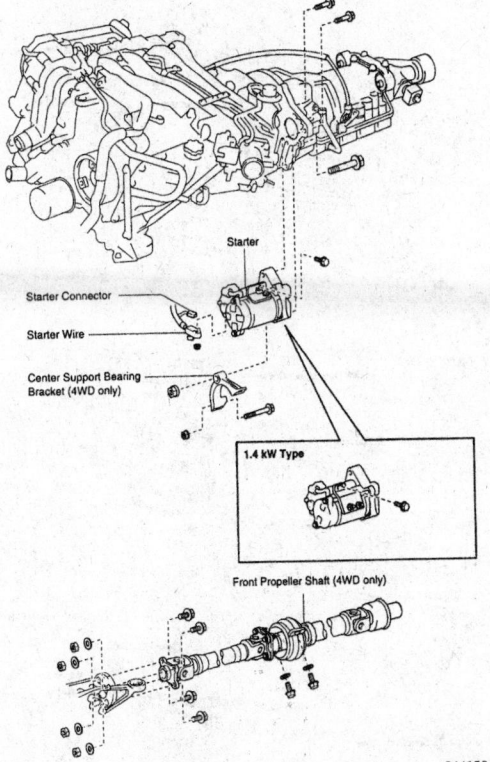

Starter

Starter Connector

Starter Wire

Center Support Bearing Bracket (4WD only)

1.4 kW Type

Front Propeller Shaft (4WD only)

344156

Starter component assembly — 2.4L (VIN A and T) Engine

CHASSIS ELECTRICAL

Blower Motor

REMOVAL AND INSTALLATION

1993–94 Previa

1. Disconnect the negative battery cable.

2. Remove the air duct, if equipped.

3. Disconnect the electrical connectors from the motor.

4. Remove the blower motor-to-case screws and lift the motor from the case.

5. To install, reverse the removal procedure. Check to ensure the seal around the motor flange is in good condition.

RAV4

1. Disconnect the negative battery cable from the battery.

2. Remove the glove box assembly.

3. Disconnect the connector from the blower motor.

4. Remove the three screws and the blower motor.

To install:

5. Install the blower motor with the three screws.

6. Connect the connector to the blower motor.

7. Install the glove box assembly.

8. Connect the negative battery cable and check the operation of the blower motor.

Wiper Motor

REMOVAL AND INSTALLATION

1993–94 Previa

NOTE: The wiper motor is removed with the linkage assembly.

1. Disconnect the negative battery cable.

2. Remove the wiper arm retaining nuts, then, the wiper arm/blade assemblies.

3. Remove both wiper arm pivot covers and the pivot-to-cowl attaching screws.

4. Remove the service hole covers from the cowl area of the engine compartment.

5. Disconnect the wiring from the wiper motor.

6. From the engine compartment, remove the wiper motor plate-to-cowl screws. Withdraw the wiper motor and the linkage from the cowl panel as an assembly.

7. Pry the linkage from of the wiper motor.

To install:

8. Attach the linkage to the wiper motor. Place the wiper motor into position and install the mounting screws.

9. Connect the wiper motor connector and install the service hole covers.

10. Install the wiper arm pivot covers.

11. Install the wiper arm/blade assemblies.

1995–97 Previa

Front Wiper Motor

1. Disconnect the negative battery cable.

2. If necessary, remove the wiper arms and the cowl louver to access the wiper motor.

3. Disconnect the wiring from the wiper motor.

4. Remove the nut and pry the wiper link from the crank arm.

5. Remove the motor.

To install:

6. Install the wiper motor.

7. Connect the wiper link to the crank arm and install the nut.

8. Connect the wiring to the wiper motor.

9. If removed, install the cowl louver and the wiper arms.

10. Connect the negative battery cable.

11. Check the wiper motor for proper operation.

Rear Wiper Motor

1. Disconnect the negative battery cable.

2. At the rear of the vehicle, remove the wiper motor cover panel.

3. Remove the wiper arm from the wiper motor.

4. Disconnect the electrical connector from the wiper motor.

5. Remove the wiper motor bolts and the motor from the vehicle.

To install:

6. Install the wiper motor.

7. Connect the electrical connector to the wiper motor.

8. Install the wiper arm to the wiper motor.

9. Install the wiper motor cover panel.

10. Connect the negative battery cable.

11. Check the wiper motor for proper operation.

RAV4

Front Wiper Motor

1. Disconnect the negative battery cable to the battery.

2. Remove the wiper arm covers and remove the nuts.

3. Remove the wiper arms from the vehicle.

4. Remove the cowl top ventilator louver by removing the clips and screws.

5. Disconnect the connector to the wiper motor.

6. Remove the four bolts to the wiper motor and remove the wiper motor with the wiper link from the vehicle.

To install:

7. Install the front wiper motor with the wiper link.

8. Install the cowl top ventilator louver.

9. Install the wiper arms with the nuts. Do not torque the nuts at this time.

10. Install the wiper arm covers.

11. Connect the negative battery cable to the battery.

12. Turn the wiper motor on once and then turn the wiper switch OFF. Adjust the installation positions of the wiper arms to approximately 0.98 inch (25mm) from the bottom of the windshield to the end of the wiper blade. Torque the nuts to 15 ft. lbs. (20 Nm).

13. Test the wiper motor operation.

Rear Wiper Motor

1. Disconnect the negative battery cable from the battery.

2. Remove the wiper arm by removing the nut.

3. Remove the back door upper cover by removing the clips.

4. Remove the door trim board.

5. Remove the service hole cover.

6. Disconnect the connector to the wiper motor.

7. Remove the three bolts to the wiper motor and remove the motor from the rear door.

To install:

8. Install the wiper motor to the rear door and install the three bolts.

9. Connect the electrical connector.

10. Install the service hole cover.

11. Install the door trim board.

12. Install the door upper cover.

13. Install the wiper blade with the nut. Torque the nut to 48 inch lbs. (5.4 Nm).

14. Connect the negative battery cable to the battery.

15. Test the operation of the wiper motor.

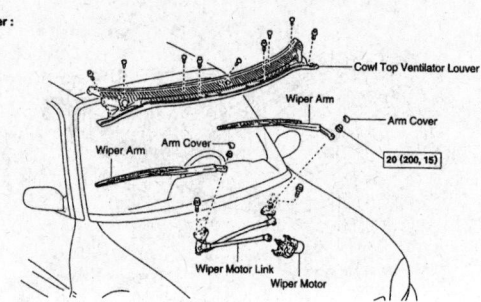

Front Wiper:

Cowl Top Ventilator Louver
Wiper Arm
Arm Cover
20 (200, 15)
Wiper Arm
Arm Cover
Wiper Motor Link
Wiper Motor

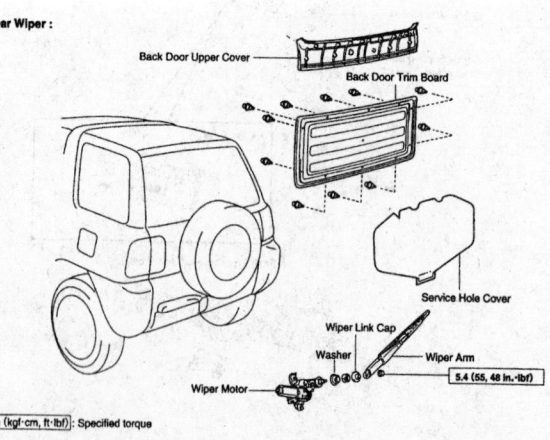

Rear Wiper:

Back Door Upper Cover
Back Door Trim Board
Service Hole Cover
Wiper Link Cap
Washer
Wiper Arm
5.4 (55, 48 in.·lbf)
Wiper Motor

N·m (kgf·cm, ft·lbf): Specified torque

304480

Wiper motor exploded view — RAV4

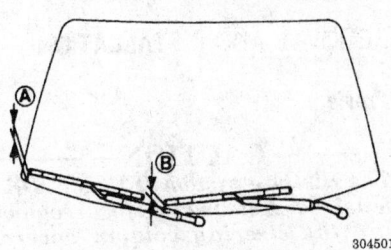

Adjustment of the wiper motor blades — RAV4

Combination Switch

REMOVAL AND INSTALLATION

1993–94 Previa

The combination switch is composed of the turn signal, the headlight control, the dimmer, the hazard, the wiper and the washer switch

1. Disconnect the negative battery cable. Remove the steering wheel.

2. Remove the upper and lower steering column shroud screws and the shrouds.

3. Remove the combination switch screws and the switch from the column.

4. Disconnect the electrical connector from the combination switch. To remove the wires from the electrical connector, perform the following procedures.

   a. Using a small prybar, insert it into the open end between the locking lugs and the terminal.

   b. Pry the locking lugs upward and pull the terminal out from the rear.

To install:

5. Push the wires into the connector until they lock securely in place.

6. Install the switch into the steering column.

7. Install the upper and lower steering column shroud.

8. Install the steering wheel.

9. Connect the negative battery cable.

1995–97 Previa

The combination switch incorporates the headlight switch, turn signal switch, dimmer switch and the windshield wiper switch.

1. Disconnect the negative battery cable.

2. Position the front wheels in a straight-ahead position.

CAUTION

Work must be started after approximately 90 seconds or longer from the time the ignition switch is turned to the LOCK position and the negative battery cable is disconnected from the battery. If the wiring connector of the air bag system is disconnected with the ignition switch at ON or ACC, diagnostic coded will be recorded.

3. Disable the air bag system.

4. Remove the steering wheel center pad.

NOTE: When removing the wheel pad, take care not to pull the air bag harness connector. When storing the wheel pad, keep the upper surface of the pad facing upward.

5. Remove the steering wheel assembly.

6. Remove the steering column covers by removing the screws.

7. Remove the combination switch with the spiral cable as follows:

   a. Remove the six screws.

   b. Disconnect the three electrical connectors.

   c. Disconnect the air bag connector.

   d. Remove the combination switch from the vehicle.

---

8. Remove the spiral cable from the combination switch by removing the screws.

9. Loosen the mounting screws and remove the dimmer switch.

10. Remove the mounting screws and disconnect the headlight/turn signal switch from the combination switch body.

11. Separate the bracket from the wiper switch body by removing the screws. Remove the wiper switch from the switch body.

To install:

12. Install the wiper switch to the switch body and connect the mounting bracket.

13. Install the dimmer and headlight/turn signal switch to the switch body and tighten the mounting screws.

14. Install the spiral cable to the combination switch by installing the screws.

15. Install the combination switch assembly to the steering column and tighten the mounting screws.

16. Connect the electrical connectors. Push in the terminals until they are securely locked in the connector lug.

17. Install both steering column covers and install the screws.

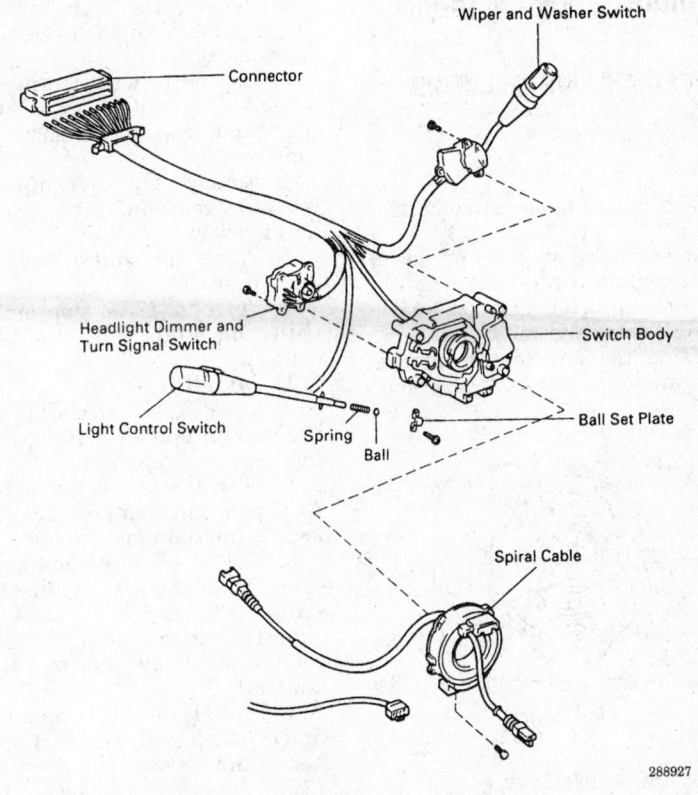

Combination switch assembly — 1995–97 Previa

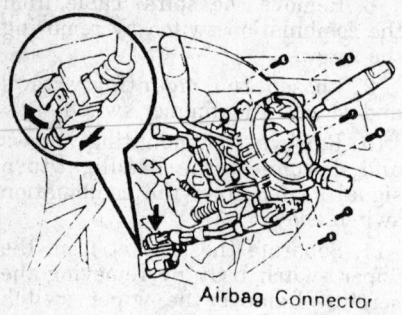

Combination switch screws — 1995–97 Previa

288926

18. Center the spiral cable as follows:
   a. Check that the front wheels are facing straight ahead.
   b. Turn the cable counterclockwise by hand until it becomes harder to turn the cable.
   c. Rotate the cable clockwise about three turns to align the red mark.
19. Install the steering wheel to the steering column.
20. Install the steering wheel center pad.
21. Connect the negative battery cable.
22. Enable the air bag system.

Ignition Lock Cylinder

REMOVAL AND INSTALLATION

Previa

1. Disconnect the negative battery terminal.
2. Remove the upper and lower steering column covers.
3. Disconnect the ignition switch from the electrical connector.
4. Using the key in the ignition switch, turn it to the ACC position.

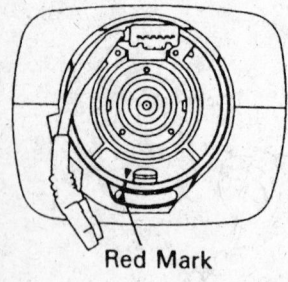

Red Mark

288924

Spiral cable red mark

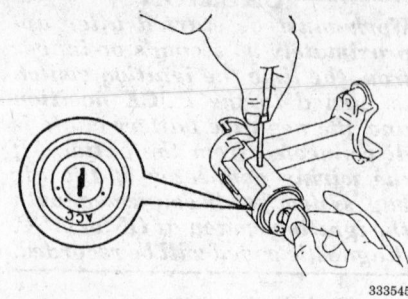

333545

Push the stop pin to remove the lock cylinder — Previa

5. Using a thin rod, place it into the hole of the cylinder lock housing. Pushing down on the thin rod, pull out the cylinder lock.
To install:
6. Using the key, install the cylinder lock into the housing until the retaining tab locks it in place.
7. Connect the ignition switch electrical connector.
8. Install the upper and lower steering column covers.
9. Connect the negative battery terminal.

RAV4

1. Disconnect the negative battery cable to the battery.
2. Remove the instrument lower finish panel from the drivers side of the vehicle.
3. Loosen the four steering column assembly set nuts. Lower the steering column enough to remove the column covers.
4. Remove the steering column covers by removing the screws in the lower cover.
5. Place the ignition key at the ACC position.
6. Push down the stop pin with a thin rod and pull out the key cylinder.
To install:
7. Install the ignition key lock to the column upper bracket by installing the one screw.
8. With the ignition switch is the ACC position, install the ignition key lock to the column upper bracket.
9. Install the upper and lower column cover and then install the screws.
10. Tighten and torque the steering column assembly nuts to 19 ft. lbs. (25 Nm).
11. Install the instrument lower finish panel to the drivers side and install the screws.
12. Connect the negative battery cable to the battery

Ignition Switch

REMOVAL AND INSTALLATION

Previa

—————— CAUTION ——————
The air bag system (SRS or SIR) must be disarmed before removing the steering column covers. Failure to do so may cause accidental deployment, property damage or personal injury.
————————————————————

1. Disconnect the negative battery terminal.
2. Remove the upper and lower steering column covers.
3. Disconnect the ignition switch from the electrical connector.
4. With the key in the ignition switch, turn the lock cylinder to the ACC position.
5. Using a thin rod, press in the stop pin and pull out the lock cylinder.
6. Remove the key warning switch from the side of the cylinder housing.
7. Remove the ignition switch from the end of the cylinder housing.
To install:
8. Position the ignition switch on the end of the housing and install the screw.
9. Using the key, install the lock cylinder into the housing until the stop pin locks it in place.
10. Install the key warning switch on the side of the cylinder housing.
11. Connect the harness connector to the ignition switch and install the steering column covers.
12. Connect the negative battery cable.

RAV4

—————— CAUTION ——————
The Supplemental Inflatable Restraint (SIR) system must be disarmed before performing service around SIR components or SIR wiring. Failure to do so may cause accidental deployment of the air bag, resulting in unnecessary SIR system repairs and/or personal injury.
————————————————————

1. Disconnect the negative battery cable.
2. If equipped, properly disarm the SRS system.
3. Remove the upper and lower steering column covers.
4. Disconnect the electrical connector to the ignition switch.

5. Remove the two screws and re-move the ignition switch from the col-umn upper bracket.

To install:

6. Install the ignition switch to the upper column bracket and install the two screws.

7. Connect the electrical connector to the ignition switch.

8. Install the upper and lower steering column covers.

9. Connect the negative battery cable to the battery.

Park/Neutral Switch

REMOVAL AND INSTALLATION

Previa

1. Disconnect the negative battery cable to the battery.

2. Apply the parking brake fully and block the rear wheels.

3. Raise and safely support the vehicle.

4. Disconnect the electrical con-nector from the park/neutral position switch.

5. Using a flat bladed tool, bend the lock washer back away from the nut.

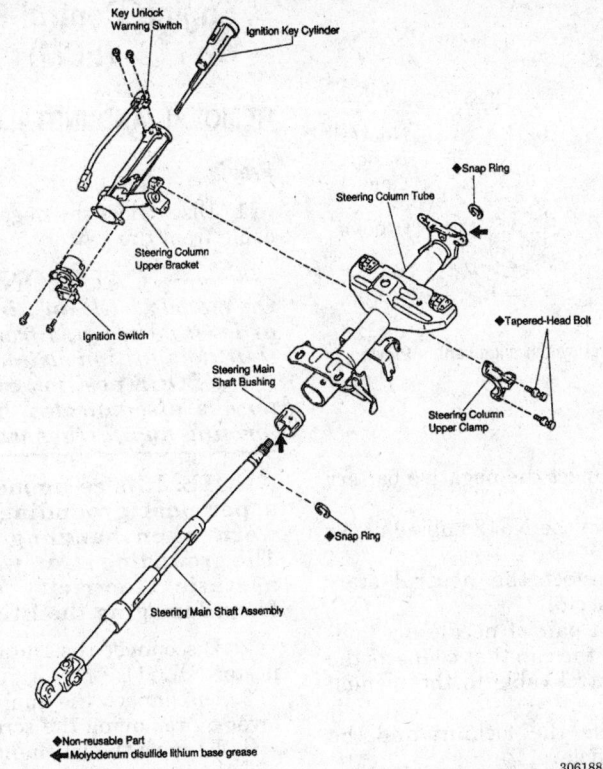

Ignition switch components (non-tilt steering) — RAV4

6. Remove the nut and lock washer from the park/neutral switch.

7. Remove the bolt and pull out the park/neutral position switch.

To install:

8. Install the park/neutral posi-tion switch to the manual valve shaft. Leave the bolt loose at this time.

9. Install the lock washer and nut. Torque the nut to 61 inch lbs. (6.9 Nm).

10. Using a flat bladed tool, bend the lock washer over the nut to se-cure the nut in place.

11. Adjust the park/neutral as follows:

    a. Turn the park/neutral switch and set the lever to the neutral (N) position.

    b. Align the groove and the neu-tral basic line on the park/neutral switch.

    c. Hold the park/neutral switch in this position and tighten the bolt to 48 inch lbs. (5.4 Nm).

12. Connect the electrical connector to the park/neutral switch.

13. Lower the vehicle and check the vehicles shifting positions.

14. Connect the negative battery cable to the battery.

15. Verify that the vehicle only starts in PARK or NEUTRAL.

Ignition switch components (tilt steering) — RAV4

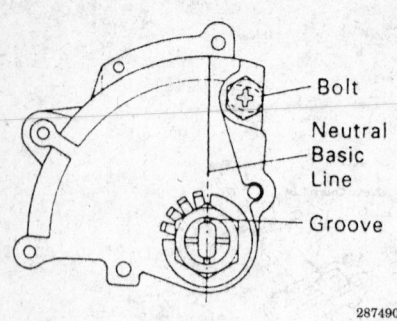

Park/neutral switch alignment — Previa

RAV4

1. Disconnect the negative battery cable.

2. Remove the No. 2 engine under cover.

3. Disconnect the neutral start switch connector.

4. With a pair of needle nose pliers, remove the clip that connects the manual control cable to the manual shift lever.

5. Remove the locknut and the manual shift lever.

6. Remove the neutral start switch with the seal gasket.

To install:

7. Install the neutral start switch making sure that the lip of the seal gasket is facing inward.

8. Install the manual shift lever.

9. Install the locknut and torque to 61 inch lbs. (6.9 Nm). Stake the nut with the locking plate.

10. Connect the switch connector.

11. Adjust the park/neutral switch.

12. Connect the transaxle shift cable and install the clip.

13. Install the No. 2 engine under cover.

14. Connect the negative battery cable.

15. Verify that the vehicle will only start in **PARK (P)** or **NEUTRAL (N)**.

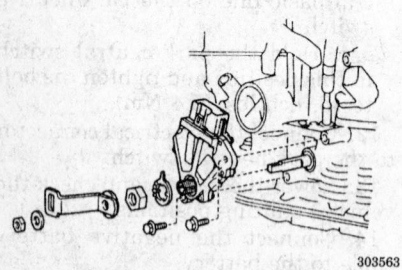

Park/Neutral switch — RAV4

Engine Control Module (ECM)

REMOVAL AND INSTALLATION

Previa

1. Disconnect the negative battery cable from the battery.

———— **CAUTION** ————

On models with an air bag, wait at least 90 seconds from the time that the ignition switch is turned to the LOCK position and the battery is disconnected before performing any further work.

NOTE: It is recommended that a personal grounding strap be worn when handling the ECM. The grounding strap will prevent a static electricity discharge from damaging the ECM.

2. Disconnect the Data Link Connector (DLC1).

3. Disconnect the fuel lid opener lever by removing the screw.

4. Remove the left-hand front seat leg.

5. Disconnect the electrical connectors from the ECM.

6. Remove the ECM from the vehicle by removing the three bolts.

To install:

7. Install the ECM to the vehicle and install the three bolts.

8. Install the electrical connectors to the ECM.

9. Install the left-hand front seat leg.

10. Install the fuel lid opener lever and install the screw.

11. Connect the Data Link Connector.

12. Connect the negative battery cable to the battery.

RAV4

———— **CAUTION** ————

If the negative (-) battery cable is disconnected, the preset AM, FM 1, and FM 2 stations stored in memory are erased. Be sure to note the stations and reset them after the battery terminal is reconnected. If the negative (-) battery cable is disconnected from the battery, the ANTI-THEFT SYSTEM will operate when the cable is reconnected, but the radio, tape player, and CD player will not operate. Be sure to input the correct ID number so that the radio, tape player, and CD player can be operated again.

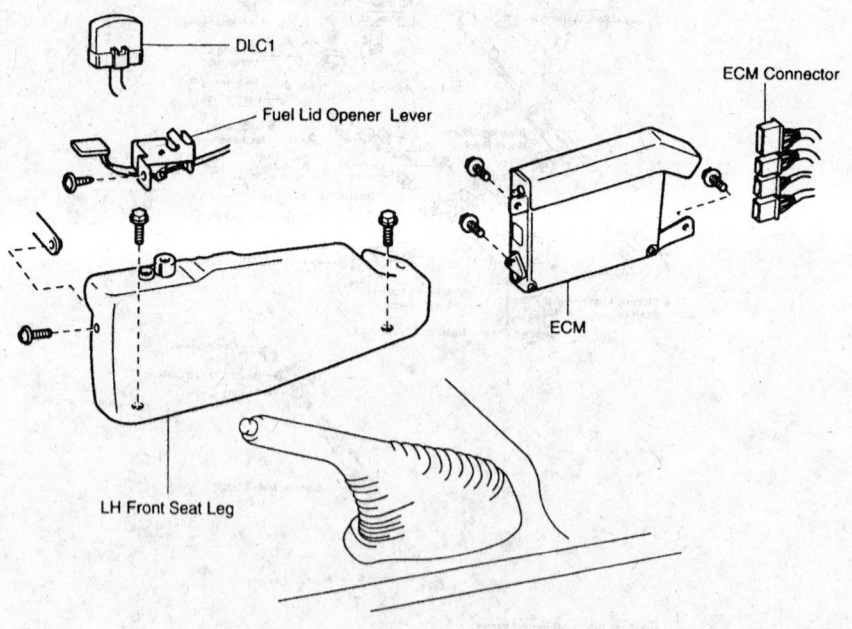

Engine control module and related components — Previa

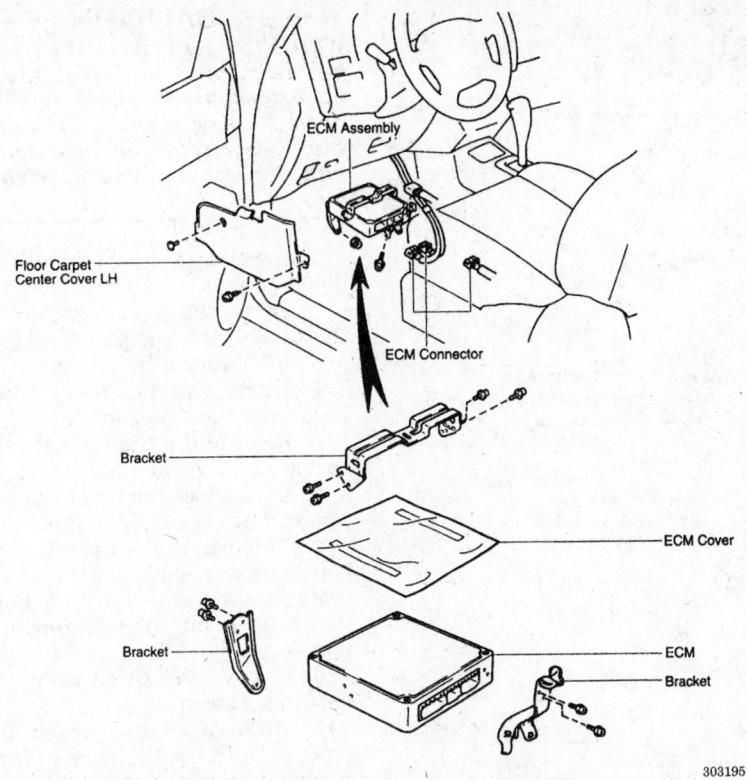

ECM exploded view — RAV4

303195

1. Disconnect the negative battery cable.

2. Remove the left hand floor carpet center cover by removing the clip and bolt.

─────── **WARNING** ───────

Due to the possibility of a static electrical discharge, a grounding strap should be worn by the technician whenever the ECM is handled.

3. Disconnect the three connectors from the ECM.

4. Remove the mounting nut and bolt and remove the ECM.

To install:

5. Install the ECM with the attaching nut and bolt.

6. Reconnect the three electrical connectors.

7. Install the left hand floor carpet center cover.

8. Install the floor mat bracket and reconnect the negative battery cable.

─────────────────────

ENGINE COOLING

Radiator

REMOVAL AND INSTALLATION

Previa

1. Disconnect the negative battery cable from the battery.

2. Raise and safely support the vehicle.

3. Remove the air intake duct.

4. Remove the No. 1 engine under cover.

5. Drain the cooling system.

6. Disconnect the radiator hoses and coolant reservoir hose from the radiator.

7. On 1995–97 vehicles, disconnect the power steering reservoir by removing the two bolts.

8. On 1995–97 vehicles, disconnect the water bypass hose for the throttle body.

9. If equipped with an automatic transaxle, disconnect the transaxle cooling lines.

10. Remove the No. 2 fan shroud by removing the four bolts.

11. Disconnect the No. 1 fan shroud from the radiator by removing the two bolts. Leave the No. 1 fan shroud in the vehicle.

12. Remove the radiator supports by removing the two bolts.

13. Remove the radiator from the vehicle.

To install:

14. Install the radiator into the vehicle. Torque the support bolts to 13 ft. lbs (18 Nm).

15. Install the fan shrouds and bolts.

16. Connect the radiator hoses.

17. On 1995–97 vehicles, connect the water bypass hose from the throttle body.

18. If equipped with an automatic transaxle, connect the transaxle cooler lines.

19. On 1995–97 vehicles, install the power steering reservoir and torque the bolts to 9 ft. lbs. (13 Nm).

20. Install the engine under cover.

21. Install the air intake duct.

22. Lower the vehicle.

23. Connect the negative battery cable.

24. Refill the cooling system. Start the engine, bleed the cooling system, and check for leaks.

RAV4

1. Disconnect the negative battery cable.

2. Remove the engine undercovers.

3. Drain the cooling system.

4. If equipped with A/C, remove the A/C condenser from the engine compartment as follows:

   a. Discharge the refrigerant from the A/C system into a recycling machine.

   b. Disconnect the liquid tube and suction tube from the vehicle.

   c. Remove the condenser by removing the two bolts.

5. Disconnect the following connectors:

   • No. 1 cooling fan connector.

   • If equipped with A/C, disconnect the No. 2 cooling fan connector

   • ECT switch connector for the electric cooling fan

   • Engine wire clamp from the No. 1 cooling fan shroud

6. Disconnect the following hoses:

   • Upper radiator hose from the radiator

   • Lower radiator hose from the radiator

   • Radiator reservoir hose from the radiator

   • If equipped with automatic transaxle, disconnect the two A/T oil cooler hoses from the oil cooler pipes

7. Remove the two bolts and two upper radiator supports.

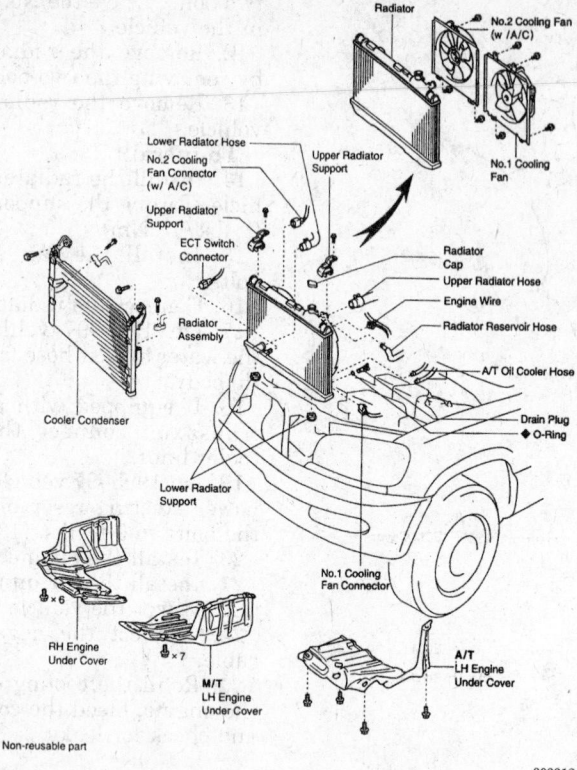

Radiator exploded view — RAV4

303312

— CAUTION —

CAUTION

Wait 90 seconds from the time the key is turned to LOCK and the negative battery cable is disconnected to begin work. This allows the SRS capacitor to discharge and prevent deployment of the air bag(s).

2. Remove the right hand engine under cover.

3. Drain the engine coolant from the radiator and engine.

4. Remove the timing belt.

5. Disconnect the lower radiator hose from the water inlet.

6. Remove the timing belt tension spring and the No. 2 idler pulley.

7. Disconnect the crankshaft position sensor connector clamp.

8. Remove the alternator drive belt adjusting bar.

9. Remove the two nuts holding the water pump to the water bypass pipe.

10. Remove the three bolts in the sequence shown.

11. Disconnect the water pump cover from the water bypass pipe and remove the water pump and water pump cover assembly.

12. Remove the gasket and two O-rings from the water pump and water bypass pipe.

13. Remove the water pump from the water pump cover by removing the three bolts, water pump and gasket.

To install:

14. Using a new gasket, install the water pump to the water pump cover. Install the three bolts and torque the bolts to 78 inch lbs. (8.8 Nm).

15. Install a new O-ring and gasket to the water pump cover.

16. Install a new O-ring to the water bypass pipe.

17. Apply soapy water to the O-ring on the water bypass pipe.

18. Connect the water pump cover to the water bypass pipe. Do not install the nuts at this time.

19. Install the water pump with the three bolts. Tighten the bolts in sequence shown. Torque the bolts to 78 inch lbs. (8.8 Nm).

20. Install the two nuts holding the water pump cover to the water pump pipe. Torque the two bolts to 82 inch lbs. (9.3 Nm).

21. Install the alternator drive belt adjusting bar. Torque the bolt to 20 ft. lbs. (27 Nm).

22. Connect the crankshaft position sensor connector clamp.

23. Install the No. 2 idler pulley and timing belt tension spring.

---

8. Remove the radiator assembly.

9. Remove the two lower radiator supports.

10. If equipped with A/C, remove the No. 2 cooling fan from the radiator by removing the three bolts.

11. Remove the No. 1 cooling fan from the radiator by removing the four bolts.

To install:

12. Install the No. 1 cooling fan to the radiator by installing the four bolts. Torque the bolts to 43 inch lbs. (5.0 Nm).

13. If equipped with A/C, install the No. 2 cooling fan to the radiator by installing the three bolts. Torque the bolts to 43 inch lbs. (5.0 Nm).

14. Install the two lower radiator supports.

15. Install the radiator assembly.

16. Install the upper radiator supports with the two bolts. Torque the bolts to 9 ft. lbs. (13 Nm)..

17. Connect the following hoses:

• If equipped with A/T, connect the oil cooler hoses to the oil cooler pipes.

• Radiator reservoir hose to the radiator.

• Lower radiator hose to the radiator.

• Upper radiator hose to the radiator.

18. Connect the following connectors and wire:

• Engine wire clamp to the No. 1 cooling fan shroud.

• ECT switch connector for the electric cooling fan.

• If equipped with A/C, connect the No. 2 cooling fan connector

• No. 1 cooling fan connector

19. Install the A/C condenser as follows:

a. Install the condenser with the two bolts.

b. Apply compressor oil to new O-rings and connect the liquid tube and suction tube to the condenser. Torque the bolts to 7 ft. lbs. (10 Nm).

c. Charge the refrigerant to the A/C system.

20. Fill the engine and radiator with engine coolant.

21. Install the engine undercovers.

22. Connect the negative battery cable to the battery.

Water Pump

REMOVAL AND INSTALLATION

2.0L (VIN P) Engine

1. Disconnect the negative battery cable.

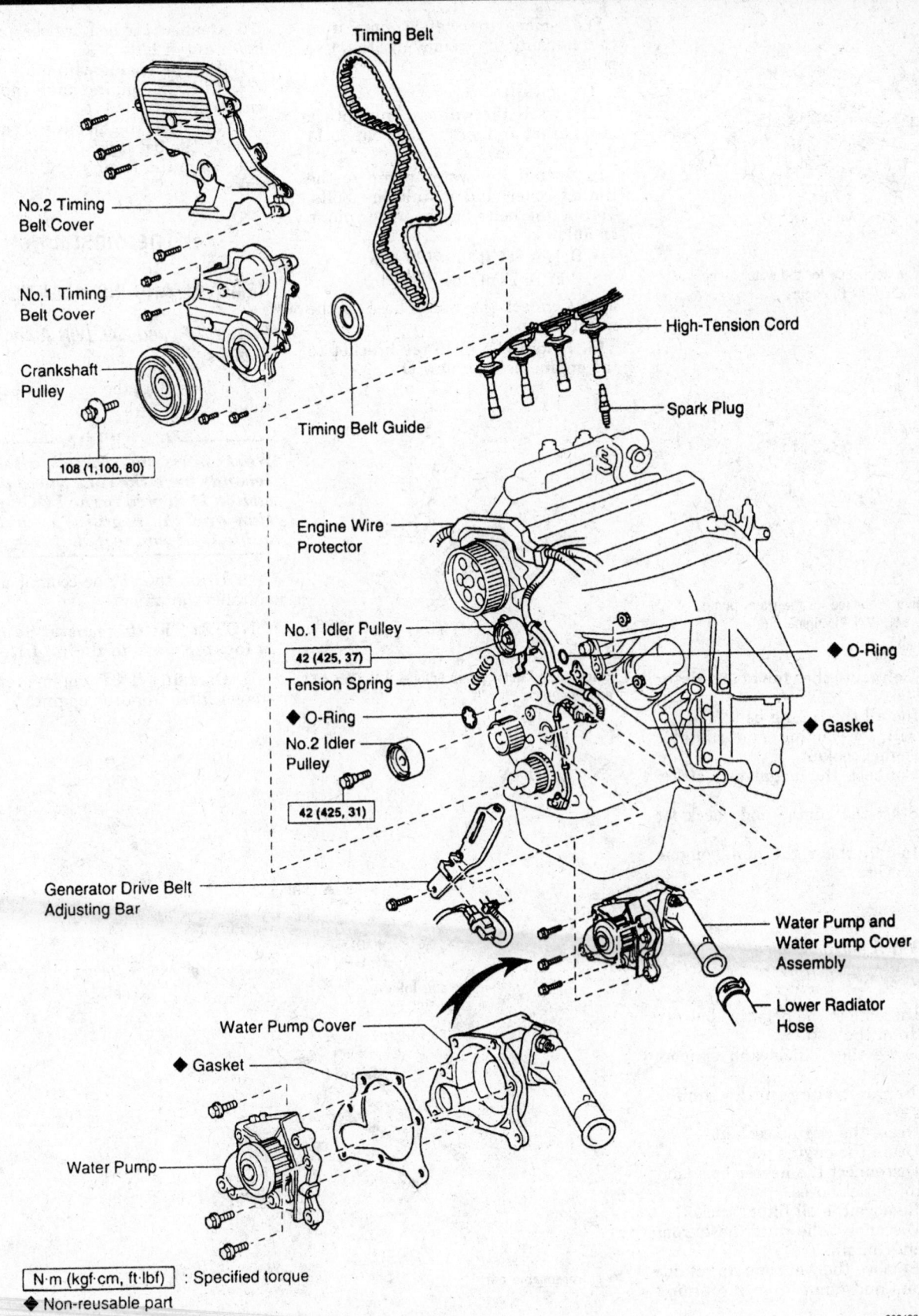

Timing Belt

No.2 Timing Belt Cover

No.1 Timing Belt Cover

Crankshaft Pulley

108 (1,100, 80)

Timing Belt Guide

High-Tension Cord

Spark Plug

Engine Wire Protector

No.1 Idler Pulley

42 (425, 37)

Tension Spring

◆ O-Ring

No.2 Idler Pulley

42 (425, 31)

◆ O-Ring

◆ Gasket

Generator Drive Belt Adjusting Bar

Water Pump and Water Pump Cover Assembly

Lower Radiator Hose

Water Pump Cover

◆ Gasket

Water Pump

N·m (kgf·cm, ft·lbf) : Specified torque

◆ Non-reusable part

303493

Water pump and related components — 2.0L (VIN P) engine

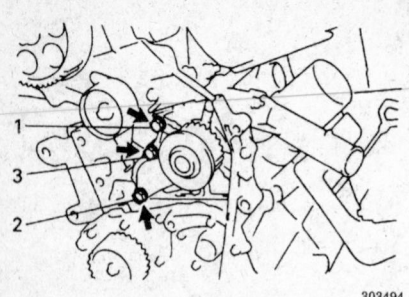

Loosening sequence for the water pump bolts — 2.0L (VIN P) engine

Tightening sequence for the water pump bolts — 2.0L (VIN P) engine

24. Connect the lower radiator hose.
25. Install the timing belt.
26. Fill the engine and radiator with engine coolant.
27. Connect the negative battery cable.
28. Start the engine and check for leaks.
29. Install the right hand engine under cover.

2.4L (VIN A and T) Engine

1. Disconnect the negative battery cable from the battery.
2. Raise the vehicle and support safely.
3. Remove the engine under covers.
4. Drain the engine coolant.
5. Drain the engine oil.
6. Disconnect the heater hose and radiator outlet hoses.
7. Remove the oil filter bracket.
8. Disconnect the water hose from the water pump.
9. Remove the water pump retaining bolts and pump from the timing cover.
10. Remove the O-ring from the water pump.

11. Remove the water pump from the housing by removing the two bolts.

To install:

12. Install the water pump with a new gasket and torque the bolts to 14 ft. lbs. (20 Nm).
13. Install the water pump to the timing cover and install the bolts. Torque the bolts for the water pump as follows:
   • Bolt A: 14 ft. lbs. (20 Nm)
   • Bolt B: 21 ft. lbs. (28 Nm)
14. Connect the water hose to the water pump.
15. Install the oil filter bracket to the engine using a new O-ring.

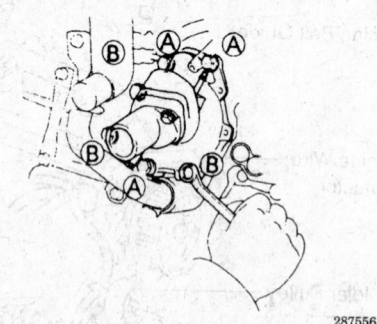

Torquing the water pump bolts — 2.4L (VIN A and T) engine

16. Connect the heater hose and radiator outlet hose.
17. Fill the engine with oil.
18. Fill the engine and radiator with coolant.
19. Connect the negative battery cable to the battery.
20. Start the engine and check for leaks.

Thermostat

REMOVAL AND INSTALLATION

2.0L (VIN P) and 2.4L (VIN A and T) Engines

1. Disconnect the negative battery cable.

— CAUTION —
Work must be started after 90 seconds from the time the ignition switch is turned to the LOCK position and the negative (-) battery cable is disconnected.

2. Drain the engine coolant into a suitable container.

NOTE: The thermostat housing is located next to the oil filter

3. On 2.0L (VIN P) engine, remove the oil filter from the engine.

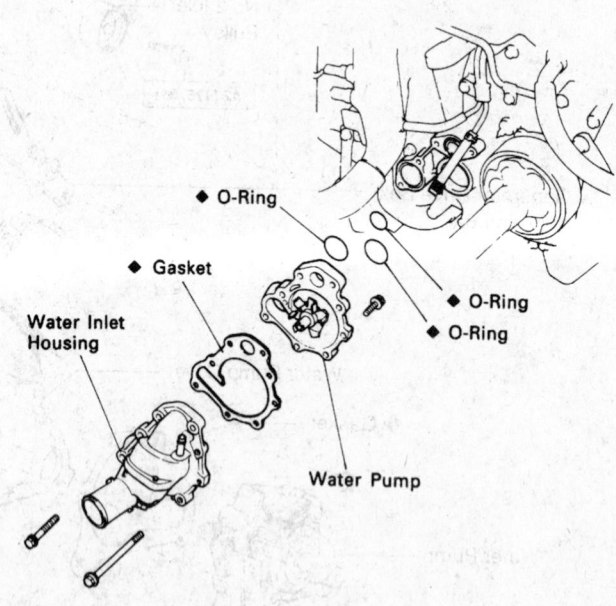

◆ O-Ring
◆ Gasket
◆ O-Ring
◆ O-Ring
Water Inlet Housing
Water Pump

◆ Non-reusable part

Water pump components — 2.4L (VIN A and T) engine

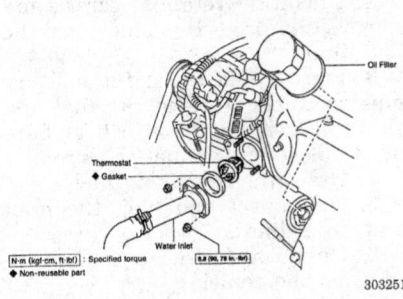

Thermostat removal and installation — 2.0L (VIN P) engine

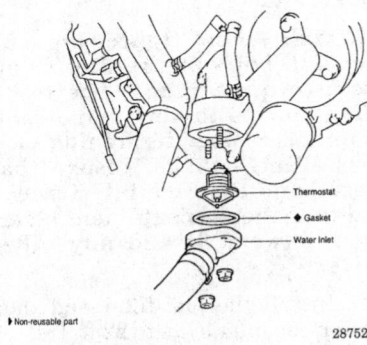

Thermostat component assembly — 2.4L (VIN A and T) engine

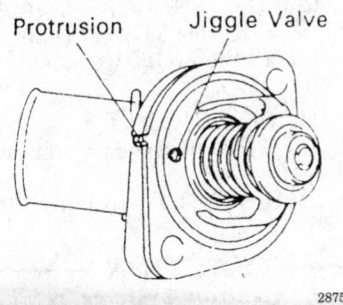

Aligning the thermostat jiggle valve — 2.4L (VIN A and T) engine

4. On 2.4L (VIN A and T) engine, disconnect the lower radiator hose from the water inlet housing.

5. Remove the two water inlet nuts and remove the water inlet from the water pump cover.

6. Remove the thermostat and gasket.

To install:

7. Install a new gasket to the thermostat.

8. On 2.0L (VIN P) engine, align the jiggle valve of the thermostat with the protrusion of the water in-

let. The jiggle valve may be installed within 5° of either side of the protrusion.

9. On 2.4L (VIN A and T) engine, align the jiggle valve with the protrusion and insert the thermostat to the water inlet.

10. Install the water inlet to the water pump cover and torque the two nuts to 7 ft. lbs. (9 Nm) for 2.0L (VIN P) engine or 14 ft. lbs (19 Nm) for 2.4L (VIN A and T) engine.

11. On 2.0L (VIN P) engine, install the oil filter to the engine.

12. Fill the engine with coolant and connect the negative battery cable.

13. Check the oil level.

14. Start the engine, bleed the cooling system, check for leaks, and verify proper operation of the thermostat.

Cooling Fan

REMOVAL AND INSTALLATION

Previa

1. Remove the air duct.
2. Remove the No. 2 fan shroud.
3. Hold the drive belt down and loosen the four nuts for the fan and fan clutch.
4. Remove the nuts and the fan clutch with the cooling fan.

To install:

5. Connect the fan clutch with the fan to the engine.
6. Install the fan nuts.
7. While holding the drive belt, tighten the nuts. Torque the nuts to 10 ft. lbs. (13 Nm).
8. Install the No. 2 fan shroud.
9. Install the air duct.

RAV4

1. Disconnect the negative battery cable.

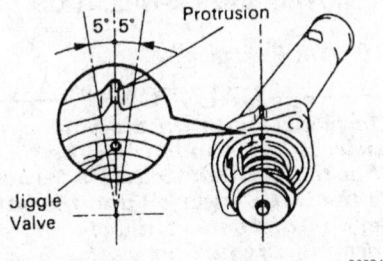

Thermostat installation - aligning the jiggle valve — 2.0L (VIN P) engine

2. Remove the engine undercovers.

3. Drain the cooling system.

4. If equipped with A/C, remove the A/C condenser from the engine compartment as follows:

a. Discharge the refrigerant from the A/C system into a recycling machine.

b. Disconnect the liquid tube and suction tube from the vehicle.

c. Remove the condenser by removing the two bolts.

5. Disconnect the following connectors:

• No. 1 cooling fan connector.

• If equipped with A/C, disconnect the No. 2 cooling fan connector

• ECT switch connector for the electric cooling fan

• Engine wire clamp from the No. 1 cooling fan shroud

6. Disconnect the following hoses:

• Upper radiator hose from the radiator

• Lower radiator hose from the radiator

• Radiator reservoir hose from the radiator

• If equipped with automatic transaxle, disconnect the two A/T oil cooler hoses from the oil cooler pipes

7. Remove the two bolts and two upper radiator supports.

8. Remove the radiator assembly.

9. Remove the two lower radiator supports.

10. If equipped with A/C, remove the No. 2 cooling fan from the radiator by removing the three bolts.

11. Remove the No. 1 cooling fan from the radiator by removing the four bolts.

To install:

12. Install the No. 1 cooling fan to the radiator by installing the four bolts. Torque the bolts to 43 inch lbs. (5.0 Nm).

13. If equipped with A/C, install the No. 2 cooling fan to the radiator by installing the three bolts. Torque the bolts to 43 inch lbs. (5.0 Nm).

14. Install the two lower radiator supports.

15. Install the radiator assembly.

16. Install the upper radiator supports with the two bolts. Torque the bolts to 9 ft. lbs. (13 Nm)..

17. Connect the following hoses:

• If equipped with A/T, connect the oil cooler hoses to the oil cooler pipes.

• Radiator reservoir hose to the radiator.

• Lower radiator hose to the radiator.

• Upper radiator hose to the radiator.

18. Connect the following connectors and wire:

• Engine wire clamp to the No. 1 cooling fan shroud.

• ECT switch connector for the electric cooling fan.

• If equipped with A/C, connect the No. 2 cooling fan connector

• No. 1 cooling fan connector

19. Install the A/C condenser as follows:

a. Install the condenser with the two bolts.

b. Apply compressor oil to new O-rings and connect the liquid tube and suction tube to the condenser. Torque the bolts to 7 ft. lbs. (10 Nm).

c. Charge the refrigerant to the A/C system.

20. Fill the engine and radiator with engine coolant.

21. Install the engine undercovers.

22. Connect the negative battery cable to the battery.

Cooling System

BLEEDING

After working on the cooling system, even to replace the thermostat, it must be bled. Air trapped in the system will prevent proper filling and leave the radiator coolant level low, causing a risk of overheating.

To bleed the system, start with the system cool, the radiator (pressure) cap off and the cooling system filled to about an inch below the filler neck.

1. Start the engine and run it at slightly above normal idle speed. This will insure adequate circulation. If air bubbles appear and the coolant level drops, fill the system with an antifreeze/water mixture to bring the level back to the proper level.

2. Run the engine this way until the thermostat opens. When this happens, coolant will move abruptly across the top of the radiator and the temperature of the radiator will suddenly rise.

3. At this point, air is often expelled and the level may drop quite a bit. Keep refilling the system until full and install the pressure cap. Fill the coolant reservoir to the full mark.

4. Run the engine until it reaches normal operating temperature. Turn the engine off and allow it to cool nearly completely. After the engine has cooled, fill the reservoir to the full mark if needed.

FUEL SYSTEM

Fuel System Pressure

RELIEVING

——— CAUTION ———

Fuel injection systems remain under pressure after the engine has been turned OFF. Properly relieve fuel pressure before disconnecting any fuel lines. Failure to do so may result in fire or personal injury.

1. Disconnect the negative battery terminal.

2. Place a catch-pan under the joint to be disconnected. A large quantity of fuel may be released when the joint is opened.

3. Wear eye or full face protection.

4. Place a shop towel over the area and slowly loosen the joint using a wrench of the correct size. Use a back-up wrench if needed.

5. Allow the fuel left in the line to bleed off slowly before fully disconnecting the joint.

6. Plug the opened lines immediately to prevent fuel spillage or the entry of dirt.

7. Dispose of the released fuel properly.

8. After connecting fuel lines, connect the negative battery cable and start the engine.

9. Check for leaks and repair as needed.

ADJUSTMENTS

Idle Speed and Mixture

Air/Fuel mixture is controlled by the ECU computer and is not adjustable.

Fuel Filter

REMOVAL AND INSTALLATION

2.0L (VIN P) Engine

——— CAUTION ———

On models with an air bag, wait at least 90 seconds from the time that the ignition switch is turned to the LOCK position and the battery is disconnected before performing any further work.

1. Properly release fuel system pressure and disconnect the negative battery cable.

2. Unbolt the retaining screws and remove the protective shield for the fuel filter.

3. Place a pan under the delivery pipe to catch the dripping fuel and slowly loosen the union bolt or flare nut to bleed off the fuel pressure.

4. Drain the remaining fuel.

5. Disconnect and plug the inlet and outlet lines.

6. Unbolt and remove the fuel filter from the vehicle.

To install:

7. Coat the flare nut, union nut and bolt threads with engine oil.

8. Hand tighten the inlet line to the fuel filter.

NOTE: When tightening the fuel line bolts to the fuel filter, use a torque wrench. The tightening torque is very important, as under or over tightening may cause fuel leakage. Insure that there is no fuel line interference and that there is sufficient clearance between it and any other parts.

9. Install the fuel filter and then tighten the inlet bolts to 22 ft. lbs. (30 Nm).

10. Reconnect the delivery pipe using new gaskets and then tighten the union bolt to 22 ft. lbs. (30 Nm).

11. Run the engine for a few minutes and check for any fuel leaks.

12. Install the protective shield.

2.4L (VIN A and T) Engine

1. Disconnect the negative battery cable.

2. Relieve the fuel system pressure.

——— CAUTION ———

Fuel injection systems remain under pressure after the engine has been turned OFF. Properly relieve fuel pressure before disconnecting any fuel lines. Failure to do so may result in fire or personal injury.

NOTE: The fuel filter is located in the engine compartment, at the inlet line to the fuel rail.

3. Disconnect and plug the inlet and outlet lines from the filter.

4. Remove the fuel filter retaining bolts and remove the filter.

To install:

5. Install the fuel filter.

6. Use new O-rings and tighten the lines to 22 ft. lbs. (29 Nm).

7. Connect the negative battery cable.

8. Start the engine and check for leaks.

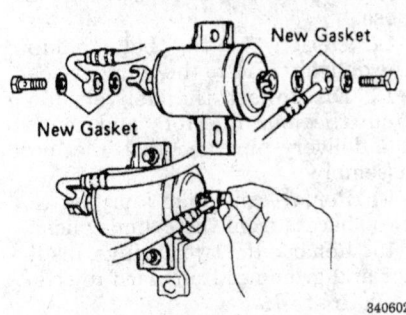

Fuel filter — 2.4L (VIN A and T) engine

Fuel Pump

REMOVAL AND INSTALLATION

2.0L (VIN P) Engine

1. Relieve the fuel system pressure.

— CAUTION —
Observe all applicable safety precautions when working around fuel. Do not allow fuel spray or fuel vapors to come into contact with a spark or open flame. Keep a dry chemical (Class B) fire extinguisher near the work area. Never drain or store fuel in an open container due to the possibility of fire or explosion.

2. Disconnect the negative battery cable from the battery.
3. Remove the left hand rear seat assembly.
4. Remove the floor service hole by pulling back the carpet and then removing the four screws.
5. Disconnect the fuel pump and sender gauge connector.

NOTE: Loosen the fuel cap to relieve any fuel pressure within the tank.

6. Remove the union bolt and two gaskets to the fuel pipe. Disconnect the outlet pipe from the fuel pump.
7. Disconnect the return vent hose from the fuel pump.
8. Remove the eight bolts to the fuel pump and remove the pump assembly from the tank.
To install:
9. Install the fuel pump to the fuel tank and install the eight bolts. Torque the bolts to 31 inch lbs. (3.5 Nm).
10. Connect the return vent hose to the fuel pump.

11. Connect the outlet pipe to the fuel pump. Using new gaskets, torque the union bolts to 22 ft. lbs. (29 Nm).
12. Connect the fuel pump and sender gauge connector.
13. Install the floor hole cover with the four screws. Replace the carpet to its original position.
14. Install the left rear seat assembly.
15. Connect the negative battery cable to the battery.
16. Tighten the fuel cap and start the vehicle. Check for leaks.

2.4L (VIN A and T) Engine

— CAUTION —
Fuel injection systems remain under pressure after the engine has been turned OFF. Properly relieve fuel pressure before disconnecting any fuel lines. Failure to do so may result in fire or personal injury.

1. Relieve the fuel pressure.
2. Disconnect the negative battery cable.
3. Drain the fuel from the fuel tank.

4. Remove the fuel tank from the vehicle.
5. Remove the access plate bolts, then pull out the fuel pump assembly.
6. Disconnect the electrical connectors from the fuel pump. Pull the bracket from the lower side of the fuel pump and remove the fuel pump from the fuel hose.
7. Remove the rubber cushion, the clip and the fuel filter from the bottom of the fuel pump.
To install:
8. Install the fuel pump filter to the fuel pump with a new clip.
9. Install the fuel pump to the fuel pump bracket and use new gaskets.
10. Connect the fuel hose to the outlet port of the fuel pump.
11. Install the fuel pump bracket. Torque the bolts to 2.9 ft. lbs. (3.9 Nm).
12. Install the fuel tank and connect all electrical and fuel connections.
13. Connect the negative battery cable.
14. Refill the fuel tank and check for leaks.

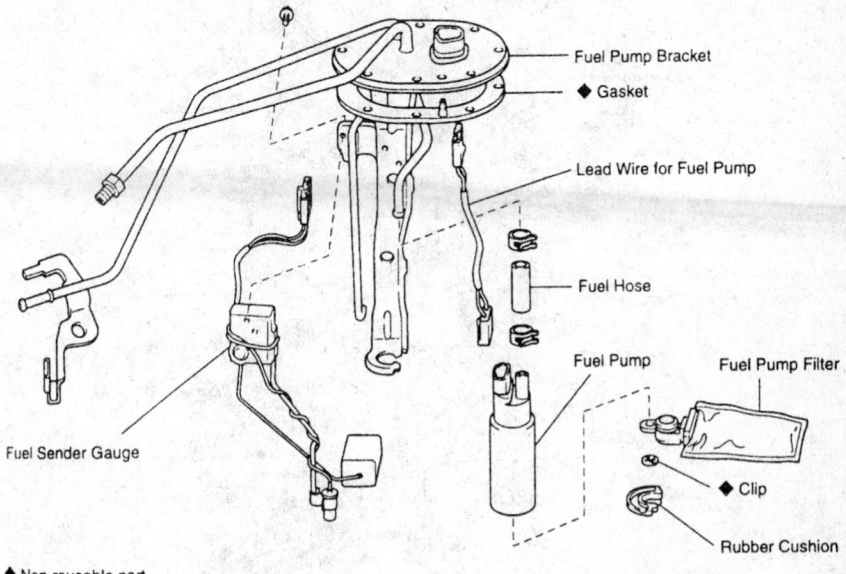

◆ Non-reusable part

Fuel pump and related components — 2.4L (VIN A and T) Engine; 2.0L engine is similar

Fuel Injector

REMOVAL AND INSTALLATION

2.0L (VIN P) Engine

CAUTION

Wait 90 seconds from the time the key is turned to LOCK and the negative battery cable is disconnected to begin work. This allows the SRS capacitor to discharge and prevent deployment of the air bag(s).

1. Relieve the fuel pressure from the fuel system.
2. Disconnect the negative battery cable from the battery.
3. Remove the air cleaner assembly.
4. Remove the cylinder head cover.
5. Disconnect the throttle body from the intake manifold.
6. Remove the distributor from the vehicle.
7. Disconnect the engine wire from the intake manifold as follows:
   a. Disconnect the four injector connectors.
   b. Disconnect the two engine wire clamps from the wire brackets on the intake manifold.

c. Disconnect the engine wire protector from the right hand side of the intake manifold by removing the bolt.
d. Disconnect the clamp of the engine wire from the wire clamp.
8. Remove the EGR valve and pipe as follows:
   a. Disconnect the vacuum hose from port E of the vacuum switching valve.
   b. Disconnect the EGR hose from the vacuum modulator.
   c. Loosen the cylinder head side of the EGR pipe union nut.
   d. Remove the two nuts, the EGR valve, pipe assembly and gasket.
9. Disconnect the engine compartment No. 2 relay block.

CAUTION

Fuel injection systems remain under pressure after the engine has been turned OFF. Properly relieve fuel pressure before disconnecting any fuel lines. Failure to do so may result in fire or personal injury.

10. Remove the union bolt and two gaskets and disconnect the fuel inlet hose from the fuel filter outlet.
11. Disconnect the air hose (for the air assist system) from the intake

manifold port and remove the air hose.
12. Loosen the two bolts holding the delivery pipe to the cylinder head.
13. Disconnect the delivery pipe from the four injectors and remove the delivery pipe and fuel inlet hose assembly.
14. Remove the four injectors and two spacers from the cylinder head.
15. Remove the two O-rings, insulator and grommet from each injector.

To install:

16. Install new insulator and grommet to each injector.
17. Apply a light coat of gasoline onto two new O-rings and install the O-rings to each injector.
18. Place the two spacers into position on the cylinder head.
19. Push each injector into the cylinder head.
20. Place the delivery pipe together with the two bolts between the intake manifold and cylinder head. Leave the bolts loose.
21. While turning the injector left and right, push each injector onto the delivery pipe.
22. Temporarily snug the two bolts holding the delivery pipe to the cylinder head. Do not torque the bolts at this time.
23. Check that the injectors rotate smoothly and position the injector connector upward.
24. Torque the two delivery pipe bolts to 9 ft. lbs. (13 Nm).
25. Install the air hose for the air assist system to the intake manifold port.
26. Connect the fuel inlet hose to the fuel filter with two new gaskets and the union nut. Torque the union nut to 22 ft. lbs. (29 Nm).
27. Install the No. 2 relay block.
28. Install the EGR valve and pipe. Torque the two nuts to 9 ft. lbs. (13 Nm) and the union nut to 43 ft. lbs. (59 Nm).
29. Connect the EGR hose to the vacuum modulator.
30. Connect the vacuum hose to port E on the VSV.
31. Connect the engine wire and injectors.

NOTE: The No. 1 and No. 3 injector connectors are brown, and the No. 2 and No. 4 injector connectors are gray.

32. Install the distributor to the vehicle.
33. Connect the throttle body to the intake manifold.
34. Install the cylinder head cover.
35. Install the air cleaner assembly.

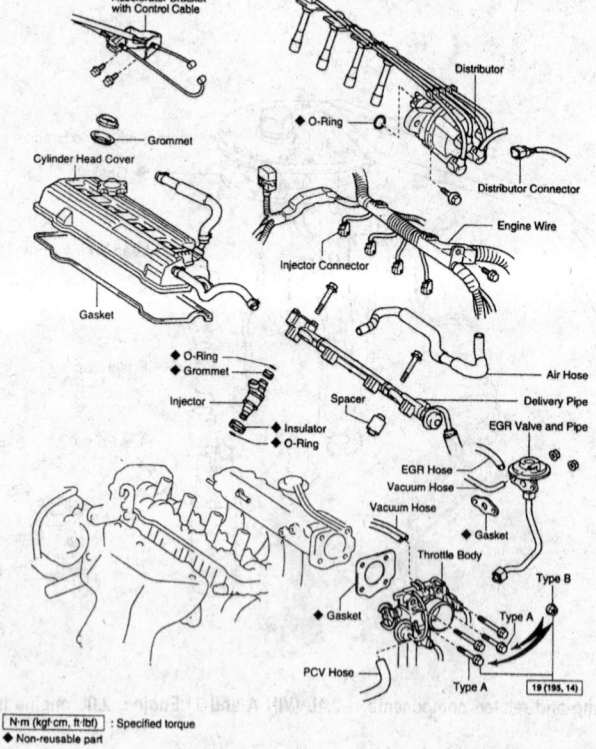

Accelerator Bracket with Control Cable

Grommet

Cylinder Head Cover

Gasket

Distributor

◆ O-Ring

Distributor Connector

Engine Wire

Injector Connector

◆ O-Ring
◆ Grommet

Injector

Spacer

◆ Insulator
◆ O-Ring

Air Hose

Delivery Pipe

EGR Valve and Pipe

EGR Hose

Vacuum Hose

Vacuum Hose

◆ Gasket

Throttle Body

Type B

◆ Gasket

Type A

PCV Hose

Type A

19 (195, 14)

N·m (kgf·cm, ft·lbf) : Specified torque
◆ Non-reusable part

303754

Fuel injector exploded view — 2.0L (VIN P) Engine

2.4L (VIN A and T) Engine

1. Disconnect the negative battery cable.
2. Relieve the fuel pressure.

— **CAUTION** —

Fuel injection systems remain under pressure after the engine has been turned OFF. Properly relieve fuel pressure before disconnecting any fuel lines. Failure to do so may result in fire or personal injury.

3. Remove the right engine service hole cover:
   a. Remove the three screws and scuff plate.
   b. Remove the bolt and disconnect the right hand seat belt from the front floor panel.
   c. Remove the four bolts and right front seat.
   d. Remove the right seat leg by removing the two bolts.
   e. Remove the jack and tool bag.
   f. Remove the jack holder by removing the two bolts.
   g. Remove the right hand engine service hole cover by removing the nine bolts.
4. Disconnect the PCV hose.

5. Disconnect the vacuum hose and fuel return hose from the pressure regulator.
6. To disconnect the engine wiring harness:
   a. Disconnect the engine wire from the intake manifold and delivery pipe by removing the four bolts.
   b. Disconnect the four injector connectors.
7. Disconnect the fuel pipe by removing the union bolt and two gaskets.
8. Remove the delivery pipe, injectors and two spacers by removing the two bolts.
9. Remove the four insulators from the delivery pipe.
10. Remove the injector covers.
11. Remove the delivery pipe with injectors.
12. Apply gasoline between the delivery pipe and injectors.
13. Place SST 09268–74010 or equivalent on the injector nozzle and push down on the delivery pipe to press out the injector.
14. Remove the four injectors from the delivery pipe.
15. Remove the insulator and two O-rings from each injector.

To install:

16. Apply a light coat of gasoline to two new O-rings.

17. Install the two O-rings and new insulator to each injector.
18. Push in the injector so that the injector connectors are in position.
19. Install the four insulators to the delivery pipe.
20. Install the two spacers, delivery pipe and two bolts. Torque the bolts to 14 ft. lbs. (20 Nm).
21. Connect the fuel pipe with the gaskets and the union bolt. Torque the bolts to 20 ft. lbs. (29 Nm).
22. Connect the four injector connectors.
23. Connect the engine wire to the intake manifold and delivery pipe by installing the four bolts.
24. Connect the vacuum hose and fuel return hose to the pressure regulator.
25. Install the PCV hose.
26. Install the right hand engine service hole cover as follows:
   a. Install the right hand engine service hole cover with the nine bolts. Torque the bolts to 10 ft. lbs. (14 Nm).
   b. Install the jack holder with the two bolts. Torque the bolts to 10 ft. lbs. (14 Nm).
   c. Install the jack and tool bag.
   d. Install the right hand front seat leg with the two bolts. Torque the bolts to 29 ft. lbs. (39 Nm).
   e. Install the right hand from seat with the four bolts. Torque the bolts to 29 ft. lbs. (39 Nm).
   f. Connect the right seat belt to the front floor panel and install the bolt. Torque the bolt to 31 ft. lbs. (42 Nm).
   g. Install the scuff plate and install the three screws.
27. Connect the negative battery cable to the battery.
28. Start the engine and check for leaks.

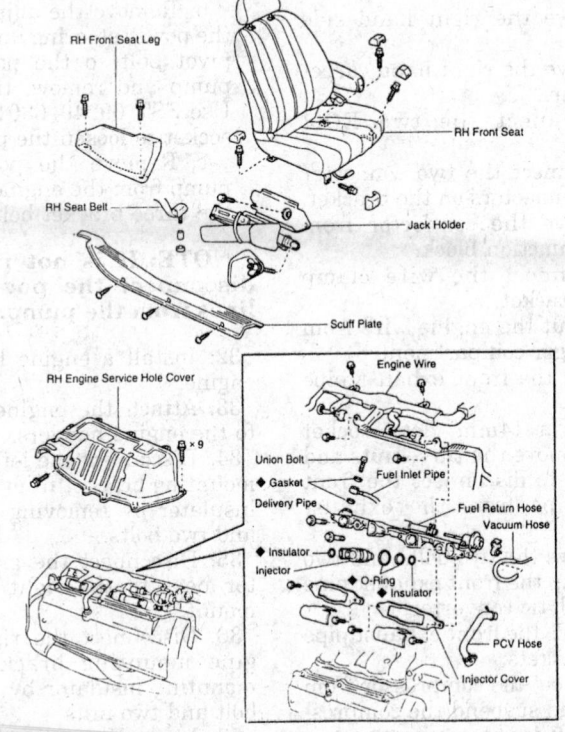

Fuel injector and related components — 2.4L (VIN A and T) engine

ENGINE MECHANICAL

Engine Assembly

REMOVAL AND INSTALLATION

2.0L (VIN P) Engine

1. Relieve the fuel system pressure.
2. Disconnect the negative battery cable.

CAUTION

On models with an air bag, wait at least 90 seconds from the time that the ignition switch is turned to the LOCK position and the battery is disconnected before performing any further work.

3. Remove the battery.

4. Remove the hood.

5. Remove the engine under cover and then drain the engine coolant and oil.

6. Drain the transaxle assembly.

7. Remove the air cleaner and case.

8. Disconnect the accelerator cable from the throttle body, bracket and clamps.

9. Disconnect the engine wire from the No.2 relay block as follows:

   a. Disconnect the No. 2 relay block from the body by removing the two bolts.

   b. Remove the upper cover to the relay block.

   c. Disconnect the connectors.

   d. Remove the engine wire by removing the two nuts.

10. Remove the charcoal canister.

11. Remove the alternator as follows:

   a. Disconnect the electrical wiring from the alternator.

   b. Loosen the adjusting lock bolt and the pivot bolt.

   c. If equipped with A/C, loosen the adjusting bolt to relieve tension on the drive belt.

   d. Remove the drive belt. It may be necessary to remove other belts for access.

   e. Remove the pivot bolt first, support the alternator and remove the adjusting lock bolt.

   f. Remove the alternator from the vehicle.

12. Remove the upper and lower radiator hoses.

13. Remove the water inlet from the engine by removing the two nuts.

14. Disconnect the heater hoses.

CAUTION

Fuel injection systems remain under pressure after the engine has been turned OFF. Properly relieve fuel pressure before disconnecting any fuel lines. Failure to do so may result in fire or personal injury.

15. Place a rag under the fuel inlet hose and disconnect the hose.

16. For vehicles with M/T, remove the starter by disconnecting the electrical connectors and two bolts.

17. Disconnect the ground cable from the transaxle by removing the bolt.

18. For vehicles with M/T, remove the clutch release cylinder from the transaxle.

19. Disconnect the transaxle control cables (two cable for M/T and one for A/T) from the transaxle.

20. For vehicles with A/T disconnect the transaxle cable from the front suspension crossmember and engine mounting center member by removing the two bolts.

21. For vehicles with A/T or 4WD with M/T, disconnect the transaxle oil cooler hoses.

22. Disconnect the following:

- Vapor pressure sensor connector
- Igniter connector
- Ignition coil connector
- Noise filter connector
- Ignition coil wire
- MAP sensor connector
- MAP sensor vacuum hose from the gas filter on the intake manifold
- Brake booster hose from the intake manifold
- If equipped with 4WD M/T, disconnect the differential lock control solenoid connector
- Ground strap from cowl

23. Disconnect the engine wire from the passenger compartment as follows:

   a. Remove the right hand scuff plate.

   b. Remove the right hand side trim.

   c. Remove the right hand carpet center cover.

   d. Disconnect the two ECM connectors.

   e. Disconnect the two connector from the connectors on the bracket.

   f. Remove the connector from the No. 4 junction block.

   g. Disconnect the wire clamp from the bracket.

   h. Pull out the engine wire from the passenger compartment.

24. Remove the front exhaust pipe as follows:

   a. Using a 14mm. deep socket wrench, remove the three nuts and the gasket to disconnect the front exhaust pipe from the exhaust manifold.

   b. Remove the two bolts and two nuts holding the front exhaust pipe to the catalytic converter.

   c. Remove the front exhaust pipe and two gaskets.

25. Disconnect the compressor from the engine and suspend the compressor securely. It is not necessary to remove the A/C compressor lines in order to remove the engine.

26. If equipped with 4WD, remove the drive shaft.

27. Remove the halfshaft from the vehicle.

28. Remove the sway bar.

29. Remove the front suspension crossmember assembly as follows:

   a. Remove the two centermember set nuts holding the centermember to the middle of the crossmember.

   b. Remove the two rack and pinion assembly set bolts and nuts from the crossmember. Securely suspend the steering gear assembly.

   c. Disconnect the catalytic converter with pipe from the ring.

   d. Support the suspension crossmember with a jack.

   e. Remove the six bolts from the suspension crossmember.

   f. Remove the suspension crossmember with the lower suspension arms.

30. Remove the engine mounting center member as follows:

   a. Remove the two bolts holding the center member to the front engine mounting insulator.

   b. Remove the two bolts holding the center member to the body and remove the center member.

31. Disconnect the power steering pump from the engine as follows:

   a. Disconnect the two vacuum hoses from the steering pump.

   b. Remove the adjusting bolt for the power steering unit. Loosen the pivot bolt to the power steering pump and remove the drive belt. Use SST 09249–63010 and a deep socket to loosen the pivot bolt.

   c. Remove the power steering pump from the engine by removing the three bracket bolts.

NOTE: It is not necessary to disconnect the power steering lines from the pump.

32. Install a engine hanger to the engine.

33. Attach the engine sling device to the engine hangers.

34. Disconnect the left hand engine mounting bracket from the mounting insulator by removing the two nuts and two bolts.

35. Disconnect the ground connector next to the right hand engine mount.

36. Disconnect the right hand engine mounting bracket from the mounting insulator by removing the bolt and two nuts.

37. Lower the engine and transaxle out of the vehicle slowly and carefully. At the same time, raise the ve-

hicle to gain clearance to the remove the engine.

38. Place the assembly on a stand and separate the engine from the transaxle.

To install:

39. Install the engine and transaxle in the vehicle.

40. Attach the left hand engine mounting bracket to the mounting insulator and install the two nuts and two bolts. Torque the bolts and nuts to 47 ft. lbs. (64 Nm).

41. Install the bolt and two nuts to hold the right hand engine mounting bracket to the mounting insulator. Torque the bolt to 27 ft. lbs. (37 Nm) and the two nuts to 38 ft. lbs. (52 Nm).

42. Connect the ground connector next to the right hand engine mount.

43. Remove the engine sling and hanger.

44. Install the power steering pump as follows:

   a. Install the pump with the bracket. Install the three bolts and torque the bolts to 32 ft. lbs. (43 Nm).

   b. Install the pivot bolt and torque the bolt to 32 ft. lbs. (43 Nm). Torque the adjusting bolt to 29 ft. lbs. (39 Nm).

   c. Install the drive belt and adjust the tension.

   d. Connect the two air hoses to the power steering pump.

45. Install the engine mounting center member to the body and install the four bolts. Do not torque the bolts at this time.

46. Install the front crossmember as follows:

   a. Raise the suspension crossmember with the lower control arms. Install the two bolts to hold the crossmember to the vehicle. Torque the bolts to 152 ft. lbs. (206 Nm).

   b. Connect the rack and pinion and install the two bolts and nuts. Torque to 83 ft. lbs. (113 Nm).

   c. Connect the centermember to the crossmember and install the two set nuts. Torque the nuts to 82 ft. lbs. (112 Nm).

   d. Torque the lower control arm rear brackets to 101 ft. lbs. (137 Nm).

   e. Torque the two bolts to hold the engine mounting center member to the front engine mounting insulator. Torque the bolts to 59 ft. lbs. (80 Nm).

   f. Torque the two bolts to hold the engine mounting center member to the body. Torque to 26 ft. lbs. (35 Nm).

47. Install the sway bar.

48. Install the halfshafts.

49. If equipped with 4WD, install the driveshaft.

50. Install the A/C compressor with the two bolts. stud bolt and nut. Torque as follows:

   • Stud bolt to 34 ft. lbs. (47 Nm)
   • Bolt to 27 ft. lbs. (37 Nm)
   • Nut to 20 ft. lbs. (27 Nm)

51. Connect the A/C compressor connector.

52. Install the front exhaust pipe with new gaskets. Torque the three front nuts to 46 ft. lbs. (62 Nm) and the two bolts to 35 ft. lbs. (48 Nm).

53. Connect the engine wire to the passenger compartment as follows:

   a. Push in the engine wire through the cowl panel.

   b. Install the wire clamp to the bracket.

   c. Connect the two ECM connectors.

   d. Connect the two connector to the connectors on the bracket.

   e. Connect the No. 4 junction block.

   f. Install the right hand floor carpet center cover.

   g. Install the right hand cowl side trim.

   h. Install the right hand scuff plate

54. Connect the following:

   • Vapor pressure sensor connector
   • Igniter connector
   • Ignition coil connector
   • Noise filter connector
   • Ignition coil wire
   • MAP sensor connector
   • MAP sensor vacuum hose to the gas filter on the intake manifold
   • Brake booster hose to the intake manifold
   • If equipped with 4WD M/T, connect the differential lock control solenoid connector
   • Ground strap from cowl

55. For A/T and 4WD with M/T, connect the transaxle oil cooler hoses.

56. Connect the transaxle control cable(s) to the transaxle.

57. For A/T, install the transaxle control cable to the front crossmember and engine mounting center member.

58. For M/T, install the clutch release cylinder. Torque the two bolts to 9 ft. lbs. (12 Nm).

59. Connect the ground cable to the transaxle.

60. For M/T, install the starter.

61. Using new gaskets, connect the fuel inlet hose to the fuel filter. Torque the union bolt to 22 ft. lbs. (29 Nm).

62. Connect the heater hoses.

63. Connect the water inlet to the engine with the two nuts. Torque the nuts to 78 inch lbs. (8.8 Nm).

64. Install the radiator upper and lower hoses.

65. Install the alternator to the engine.

66. Install the charcoal canister.

67. Connect the engine wire to the No. 2 relay box as follows:

   a. Connect the engine wire to the No. 2 relay block with the two nuts.

   b. Connect the connector.

   c. Install the upper cover.

   d. Connect the No. 2 relay block to the body with the two bolts.

68. Install the accelerator cable to the throttle body, cable bracket and clamps.

69. Install the air cleaner case and cap.

70. Install the battery.

71. Fill the transaxle with oil.

72. Fill the engine with oil.

73. Fill the engine coolant.

74. Connect the battery cables.

75. Start the engine and check for leaks.

76. Check the front wheel alignment.

77. Install the engine under covers.

78. Install the hood.

79. Recheck all fluid levels.

2.4L (VIN A and T) Engine

— **CAUTION** —

Fuel injection systems remain under pressure after the engine has been turned OFF. Properly relieve fuel pressure before disconnecting any fuel lines. Failure to do so may result in fire or personal injury.

1. Disconnect the negative battery cable and drain the engine coolant and oil.

2. Relieve the fuel system pressure.

3. Raise the vehicle and support safely. Remove the engine under covers.

4. Drain the engine oil and cooling system.

5. On 4WD vehicles, disconnect the front driveshaft.

6. Remove the rear driveshaft.

7. Remove the air duct.

8. Matchmark and disconnect the equipment (separated accessory drive system) driveshaft from the crankshaft pulley.

9. Disconnect the A/T shift cable.

10. Remove the air intake connector.

11. Disconnect the ground strap from the left hand front engine mounting.

12. Disconnect the starter wire.

13. Disconnect the following hoses:
- No. 4 radiator hose from the water inlet
- No. 1 radiator hose from the water inlet
- Heater hose from the water pump
- Oil auto feeder hose from the No. 1 oil return pipe
- On VIN T engine, disconnect the A/C idle up air hose from the union under the intake manifold
- On (VIN A) engine, disconnect the two vacuum hoses for the EGR from the transmitting pipe on the No. 2 air inlet duct.
- On VIN T engine, disconnect the P/S idle up air hose from the union under the intake manifold
- On VIN T engine, disconnect the water bypass hose from the floor pipe
- On VIN T engine, disconnect the brake booster hose from the floor pipe
- On VIN T engine, disconnect the two vacuum hoses for the fuel pressure control VSV from the engine wire and vacuum transmitting pipe on the throttle body
- Air hose for the distributor ventilation from the water bypass pipe under the intake manifold
- On VIN T engine, disconnect the vacuum hose for the EVAP from the charcoal canister

NOTE: The following steps up to Step 15 apply only to (VIN A) engine.

- On (VIN A) engine, disconnect the fuel vapor feed hose from the No.2 air inlet duct
- Water bypass hose for the EGR from the ventilation tube
- No. 2 ventilation hose from the ventilation tube
- Water bypass hose for the engine coolant reservoir from the water bypass pipe under the front floor panel

14. If equipped with automatic transaxle, disconnect the shift cable.

15. Remove the intake pipe hose.

NOTE: The following steps apply to (VIN A) and VIN T engines unless otherwise specified.

16. On VIN T engine, disconnect the accelerator cable from the throttle body.

17. On VIN T engine, disconnect the VSV connector for the fuel pressure control.

18. Disconnect the engine wire from the engine left side as follows:
   a. Disconnect the igniter connector.

b. Disconnect the two ECM connectors.

c. Disconnect the four connectors from the cowl wire on the front floor panel.

d. Disconnect the engine wire from the front floor panel by removing the bolt and disconnecting the three clamps.

e. Pull out the engine wire from the front floor panel hose.

19. Remove the A/T oil dipstick.

20. Disconnect the fuel inlet and return hoses.

21. Remove the front exhaust pipe as follows:
   a. Disconnect the two oxygen sensor connectors and clamp.
   b. Remove the two bolts, three nuts and exhaust bracket.
   c. Disconnect the three exhaust O-rings from the vehicle body.
   d. Remove the front exhaust pipe and three gaskets.

22. Remove the exhaust pipe heat insulator and ground strap by removing the four bolts.

23. Disconnect the two A/T oil cooler hoses.

24. Remove the ignition coil as follows:
   a. Disconnect the ignition coil connector.
   b. Disconnect the coil wire from the distributor.
   c. Remove the two bolts and ignition coil and disconnect the ground strap for the engine.

25. Disconnect the condenser connector.

26. Disconnect the four clamps and engine wire.

27. Disconnect the A/T and park/neutral position switch connectors.

28. Remove the engine with the transaxle as follows:
   a. Support the engine and transaxle with a supporting device.
   b. Lower the vehicle while supporting the engine and transaxle with the engine lifter.
   c. Remove the two bolts, two nuts and two plate washers holding the right and left engine mountings to the engine front support member.
   d. Remove the four through bolts, four plate washers and four nuts holding the rear mounting to the No. 2 rear engine mounting bracket.
   e. Make sure the engine and transaxle are clear of all wiring, hoses and cables.
   f. Lower the engine and transaxle to the floor.

To install:

29. Raise the engine and install the two bolts, two plate washers and two nuts to hold the right and left engine front mountings to the engine front support member. Torque the bolts and nuts to 27 ft. lbs. (37 Nm).

30. Install the two through bolts, four washers and four nuts to hold the engine rear mounting to the No. 2 rear engine mounting bracket. Torque the bolts and nuts to 31 ft. lbs. (42 Nm) on VIN T engine or 50 ft. lbs. (67 Nm) for (VIN A) engine.

31. Connect the A/T and park/neutral position switch connectors.

32. Connect the engine wire to the engine right rear side with the four clamps.

33. Connect the condenser connector.

34. Install the ignition coil as follows:
   a. Install the ignition coil and ground strap for the engine with the two bolts. Torque the two bolts to 9 ft. lbs. (12 Nm).
   b. Connect the ignition coil connector.
   c. Connect the spark plug wires to the distributor.

35. Connect the two A/T oil cooler hoses.

36. Install the heat insulator and ground strap for the transaxle with the three bolts.

37. Install the front exhaust pipe as follows:
   a. Install three new gaskets and the front exhaust pipe.
   b. Install the thee O-rings.
   c. Install and torque the two bolts and three nuts. Torque the bolts to 32 ft. lbs. (43 Nm) and the nuts to 46 ft. lbs. (62 Nm).
   d. Connect the two oxygen sensor connectors and clamp.

38. Connect the fuel inlet and return hoses.

39. On (VIN A) engine, install the A/T oil dipstick.

40. Connect the engine wire to the engine left side as follows:
   a. Push in the engine wire through the front floor panel hole.
   b. Connect the engine wire with the bolt and three clamps to the front floor panel.
   c. Connect the following connectors:
   - Igniter connector
   - Two ECM connectors
   - Four connectors to the cowl wire on the front floor panel

41. On VIN T engine, connect the VSV connector for the fuel pressure control.

42. On VIN T engine, connect the accelerator cable to the throttle body.

43. On (VIN A) engine, connect No. 2 throttle cable to the No. 2 air inlet duct.

44. On (VIN A) engine, install the intake pipe hose.

45. On (VIN A) engine, connect the A/T shift cable.

46. Connect the following hoses:
- No. 4 radiator hose from the water inlet
- No. 1 radiator hose from the water inlet
- Heater hose from the water pump
- Oil auto feeder hose from the No. 1 oil return pipe

NOTE: The following steps apply to VIN T engine.

- A/C idle up air hose from the union under the intake manifold
- P/S idle up air hose from the union under the intake manifold
- Water bypass hose from the floor pipe
- Brake booster hose from the floor pipe
- Two vacuum hoses for the fuel pressure control VSV from the engine wire and vacuum transmitting pipe on the throttle body
- Air hose for the distributor ventilation to the water bypass pipe under the intake manifold
- Vacuum hose for the EVAP from the charcoal canister

NOTE: The following steps apply to (VIN A) engine.

- Two vacuum hoses for the EGR from the transmitting pipe on the No. 2 air inlet duct.
- Air hose and vacuum hose for the distributor ventilation from the water bypass pipe under the intake manifold
- Fuel vapor feed hose from the No.2 air inlet duct
- Water bypass hose for the EGR from the ventilation tube
- No. 2 ventilation hose from the ventilation tube
- Water bypass hose for the engine coolant reservoir from the water bypass pipe under the front floor panel

NOTE: The following steps apply to (VIN A) and VIN T engines unless otherwise specified.

47. Connect the starter wire.

48. Connect the ground strap to the left front engine mounting.

49. Install the air intake connector.

50. Connect the A/T shift cable.

51. Connect the equipment drive shaft to the crankshaft pulley.

52. Install the air duct.

53. Install the rear driveshaft.

54. If equipped with 4WD, install the front driveshaft.

55. Fill the engine with oil.

56. Fill the engine with engine coolant.

57. Install the engine undercovers.

58. Start the engine and check for leaks.

Engine Mount

REMOVAL AND INSTALLATION

2.0L (VIN P) Engine

Right Hand Mounting Insulator

1. Disconnect the negative battery cable.

2. Raise and safely support the vehicle.

—————— **WARNING** ——————
Make certain vehicle is properly supported when raising the engine.

3. Raise the engine slightly to take the tension off the motor engine mounts.

NOTE: When raising the engine to replace the engine mounts, DO NOT place the jack directly under the oil pan or the crankshaft torsional damper. Use a block of wood between the engine and the jack to prevent damage at the jacking point.

4. Remove the bolt and two nuts and disconnect the right hand engine mounting bracket from the mounting insulator.

5. Remove the right hand side engine mounting insulator from the vehicle by removing the three bolts.

To Install:

6. Install the engine mounting insulator to the vehicle and install the

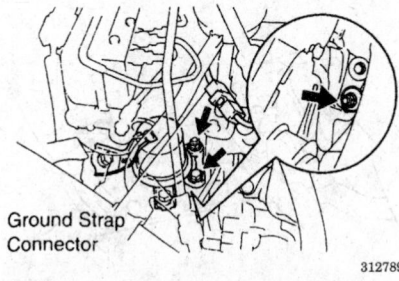

Ground Strap
Connector

312789

Right hand engine mounting — 2.0L (VIN P) engine

three bolts. Torque the bolts to 47 ft. lbs. (64 Nm).

7. Connect the right hand engine mounting insulator to the engine bracket. Torque the bolt to 27 ft. lbs. (37 Nm) and the nuts to 38 ft. lbs. (52 Nm).

8. Lower the engine and connect the negative battery cable.

Left Hand Mounting insulator

1. Disconnect the negative battery cable from the battery.

2. Raise and safely support the vehicle.

—————— **WARNING** ——————
Make certain vehicle is properly supported when raising the engine.

3. Raise the engine slightly to take the tension off the motor engine mounts.

NOTE: When raising the engine to replace the engine mounts, DO NOT place the jack directly under the oil pan or the crankshaft torsional damper. Use a block of wood between the engine and the jack to prevent damage at the jacking point.

4. Remove the two bolts and two nuts holding the left mounting insulator the vehicle.

5. Remove the nut and through bolt holding the left mounting insulator to the transaxle bracket.

6. Remove the left hand mounting insulator from the vehicle.

To install:

7. Install the left hand mounting insulator to the vehicle.

8. Install the through bolt and nut to the transaxle bracket. Do not torque the nut and bolt at this time.

9. Install the two nuts and two bolts to the mounting insulator. Torque the nuts and bolts to 47 ft. lbs. (64 Nm).

10. Torque the through bolt and nut to 54 ft. lbs. (73 Nm).

11. Lower the engine and connect the negative battery cable.

Front and Rear Mounting Insulators

1. Disconnect the negative battery cable.

2. Raise the engine slightly to take the tension off the motor engine mounts.

—————— **WARNING** ——————
Make certain vehicle is properly supported when raising the engine.

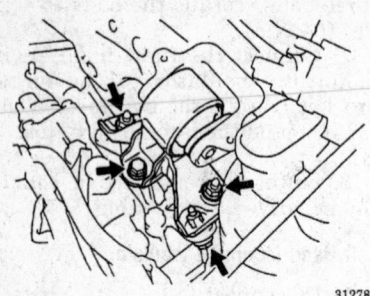

Left hand engine mounting — 2.0L (VIN P) Engine

NOTE: When raising the engine to replace the engine mounts, DO NOT place the jack directly under the oil pan or the crankshaft torsional damper. Use a block of wood between the engine and the jack to prevent damage at the jacking point.

3. Remove the through bolt for the engine mount.
4. Remove the two bolts holding the mount insulator to the engine mounting center member.
5. Remove the rear engine mount insulator from the vehicle.
To install:
6. Install the rear engine mount insulator to the vehicle and install the two bolts. Torque the bolts to 47 ft. lbs. (87 Nm).
7. Install the through bolt for the mounting insulator and torque the bolt to 64 ft. lbs. (87 Nm).
8. Lower the engine, lower the vehicle, and connect the negative battery cable to the battery.

2.4L (VIN A and T) Engine

1. Disconnect the negative battery cable.
2. Raise and safely support the vehicle.

WARNING

Make certain vehicle is properly supported when raising the engine.

3. Raise the engine slightly to take the tension off the motor engine mounts.

NOTE: When raising the engine to replace the engine mounts, DO NOT place the jack directly under the oil pan or the crankshaft torsional damper. Use a block of wood between the engine and the jack to prevent damage at the jacking point.

4. Remove the through bolt nuts and the through bolts.
5. Remove the engine mount bolts.
6. Raise the engine enough to remove the engine mounts.

NOTE: Only raise the engine enough to remove the mounts. Raising the engine too far could result in damage to some of the engine components.

To install:
7. Install the left and right engine mounts and the engine bolts. Torque the bolts to:
- 1993–94: 30 ft. lbs. (41 Nm).
- 1995–97: 27 ft. lbs. (37 Nm)
8. Install the rear engine mount and torque the bolts to:
- 1993–94: 22 ft. lbs. (29 Nm)
- 1995: 31 ft. lbs. (42 Nm)
- 1996: 50 ft. lbs. (67 Nm)
9. Lower the vehicle and connect the negative battery cable.

Cylinder Head

REMOVAL AND INSTALLATION

2.0L (VIN P) Engine

1. Release the fuel system pressure.

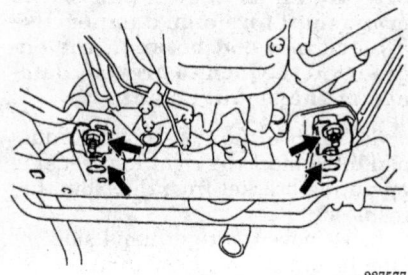

Left and right engine mounts — 2.4L (VIN T) engine

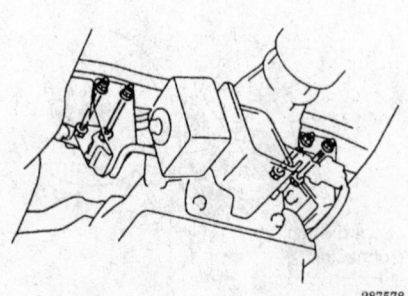

Rear engine mount — 2.4L (VIN T) engine

2. Disconnect the negative battery cable.

CAUTION

Wait 90 seconds from the time the key is turned to LOCK and the negative battery cable is disconnected to begin work. This allows the SRS capacitor to discharge and prevent deployment of the air bag(s).

3. Remove the RH engine under cover.
4. Drain the engine coolant into a suitable container.
5. See the procedure under Camshaft Removal and Installation and remove the camshafts.
6. Uniformly loosen and remove the cylinder head bolts in several passes and in the reverse of the removal sequence. Lift the cylinder head with the intake manifold from the cylinder block. Disengage the cylinder head from the block dowel pins.
7. Remove the air hose from the intake manifold. Remove the two bolts and the air tube.
8. Remove the six bolts, two nuts, and the intake manifold and gasket.
9. Disconnect the air hose from the cylinder head port and remove the air hose.
10. Remove the fuel delivery pipe and the injectors.
11. Remove the oil pressure switch.
To install:
12. Install the oil pressure switch.
13. Install the fuel injectors and the delivery pipe.
14. Install the air hose to the cylinder head port.
15. Install the intake manifold with new gaskets. Install the six bolts and two nuts, and torque the intake manifold to 14 ft. lbs. (19 Nm)
16. Install the air tube with the two bolts and connect the air hose to the intake manifold.
17. Clean the gasket mating surfaces using care not to damage the aluminum components, replace the gasket, then lower the cylinder head onto the engine. Make sure the dowel pins are aligned and no hoses or wires are between the head and cylinder block.
18. The cylinder head bolts are tightened in two progressive steps. Apply a light coat of engine oil to the cylinder head bolts. Uniformly tighten the 10 cylinder head bolts in several passes and in sequence. The torque for the head bolts is 36 ft. lbs. Mark the front of the cylinder head bolt with paint. Retighten the cylinder head bolts by 90° in sequence. Retighten an additional 90° and make

Rear engine mount — 2.4L (VIN A) engine

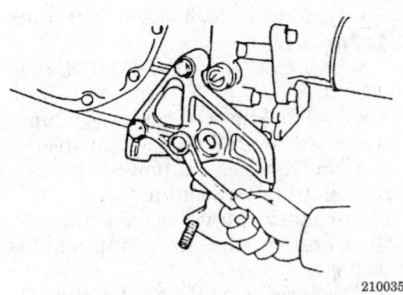

LH engine mount (VIN A) engine

sure that the paint mark is now positioned toward the rear.

19. Install the intake and exhaust camshafts and remaining parts as described in the Camshaft Removal and Installation procedure.

20. Connect the negative battery cable, fill the engine with coolant, start the engine, warm up, and check for leaks. Bleed the cooling system and top off coolant as necessary.

21. Install the RH engine under cover, check ignition timing, and road test the vehicle for proper operation.

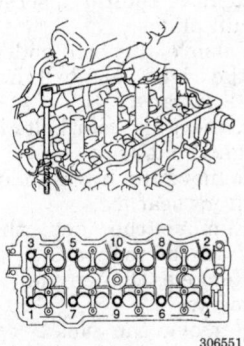

Cylinder head bolts installation sequence — 2.0L (VIN P) Engine

2.4L (VIN A and T) Engine

1. Disconnect the negative battery cable from the battery.

2. Relieve the fuel system pressure.

CAUTION

Fuel injection systems remain under pressure after the engine has been turned OFF. Properly relieve fuel pressure before disconnecting any fuel lines. Failure to do so may result in fire or personal injury.

---

3. Remove the engine/transaxle assembly from the vehicle.

4. Remove the engine wiring from the engine and move it aside.

5. See the procedure under Camshaft Removal and Installation and remove the camshafts.

6. Remove the 2 bolts in front of the head before the other head bolts are removed.

7. Using a 12 sided socket wrench, remove the 10 cylinder head retaining bolts in the proper sequence.

8. Remove the cylinder head from the block as follows:

a. Remove the two bolts in front of the head before the other head bolts are removed. The two bolts are located in front of the timing chain.

b. Uniformly remove the 10 head bolts in the reverse of the torque sequence.

c. Lift the cylinder head from the dowels on the cylinder block.

d. Remove the cylinder head gasket.

To install:

9. Clean the gasket mating surfaces and check for warpage.

10. Apply seal packing to two locations on the cylinder block as shown.

11. Install the head gasket and install the cylinder head.

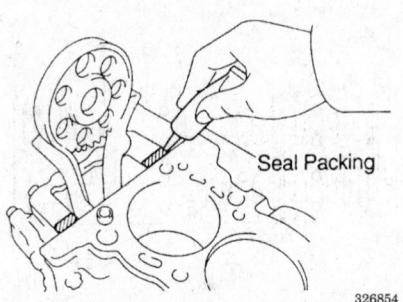

Seal Packing

Applying seal packing to the cylinder head — 2.4L (VIN A and T) Engine

12. Oil the bolts and using the proper sequence, torque the bolts in 3 steps.

a. Uniformly tighten the head bolts to 29 ft. lbs. (39 Nm).

b. Mark the front of the cylinder head bolt with paint.

c. Retighten the cylinder head bolts 90° in the numerical order as shown.

d. Check that the painted mark is now facing rearward.

e. Retighten the cylinder head bolts an additional 90°.

f. Check that the painted mark is now facing rearward.

13. See the procedure under Camshaft Removal and Installation and install the camshafts.

14. Install the engine/transaxle assembly into the vehicle.

15. Fill the cooling system and fill the engine with oil.

16. Connect the battery cable, start the engine, and check for leaks.

17. Road test the vehicle for proper operation and recheck all fluid levels.

Valve Lifters

REMOVAL AND INSTALLATION

2.0L (VIN P) Engine

1. See the procedure under Camshaft Removal and Installation and remove the camshafts.

2. Remove the valve adjusting shims from the engine. Make sure to replace the shims to their original location.

3. Remove the valve lifters from their bores in the cylinder head. Note the positions of the lifters for reinstallation.

4. Inspect the lifters and bores for signs of wear and damage. Replace any worn parts.

To install:

5. Install the valve lifters into their bores.

6. Install the valve adjusting shims to the engine.

7. See the procedures under Camshaft Removal and Installation and install the camshafts.

2.4L (VIN A and T) Engine

1. See the procedure under Camshaft Removal and Installation and remove the camshafts.

2. Remove the valve lifters from the engine. Keep the shims and the lifters together. The lifters and the shims must be reinstalled in their original location.

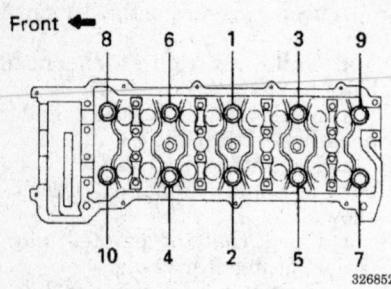

Cylinder head bolts tightening sequence — 2.4L (VIN A and T) Engine

To install:

NOTE: If any of the bolts break, deform or do not meet the torque specification replace them.

3. Install the valve lifters and shims.

4. See the procedure under Camshaft Removal and Installation and install the camshafts.

Valves

ADJUSTMENT

2.0L (VIN P) Engine

NOTE: Adjust the valve clearance when the engine is cold.

1. Disconnect the negative battery cable.

2. Disconnect the power steering reservoir.

3. Remove the cylinder head cover.

4. Disconnect the following cables and hoses:
- Accelerator cable from the throttle body
- Accelerator cable from the clamp on the intake manifold
- Accelerator cable from the clamp on the alternator bracket
- A/T throttle control cable from the throttle body
- PCV hose from the air cleaner hose and the intake manifold

5. Remove the spark plug wires.

6. Turn the crankshaft pulley and align its groove with the timing mark **0** of the No. 1 timing cover.

7. Check that the valve lifters on the No. 1 cylinder are loose and the valve lifters on the No. 4 are tight. If not, turn the crankshaft 1 complete revolution (360° degrees).

8. Measure the clearance between the valve lifter and the camshaft. Record the measurements on the intake

valves No. 1 and 2. Measure the exhaust valves at 1 and 3.

a. The intake valve clearance cold is 0.007–0.011 in. (0.19–0.29mm).

b. The exhaust valve clearance cold is 0.011–0.015 in. (0.28–0.38mm).

9. Turn the crankshaft pulley one revolution (360°) and align the groove with the timing mark **0** of the No.1 timing belt cover.

10. Measure the clearance between the valve lifter and the camshaft. Record the measurements on the intake valves No. 3 and 4. Measure the exhaust valves at 2 and 4.

a. The intake valve clearance cold is 0.007–0.011 in. (0.19–0.29mm).

b. The exhaust valve clearance cold is 0.011–0.015 in. (0.29–0.38mm).

11. To adjust the valve clearance:

a. Turn the crankshaft to position the cam lobe of the camshaft on the valve to be adjusted, upward.

b. Turn the valve lifter so that the notch is perpendicular to the camshaft and facing the spark plug side.

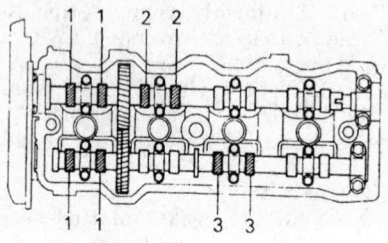

Intake valves (1 and 2) and exhaust valves (1 and 3) — 2.0L (VIN P) Engine

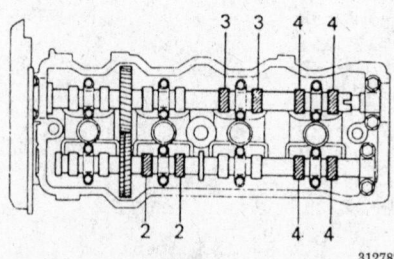

Intake valves (3 and 4) and exhaust valves (2 and 4) — 2.0L (VIN P) Engine

c. Using SST 09248–55040 (valve lifter press) or equivalent, hold the camshaft in place.

d. Using SST 09248–55040 (valve lifter press) or equivalent, press down the valve lifter and place SST 09248–05420 (valve lifter stopper) or equivalent between the camshaft and valve lifter.

e. Remove the SST 09248–44040 tool.

f. Using a small screwdriver and a magnetic finger, remove the adjusting shim.

12. Determine the replacement adjusting shim size by either using the chart or the following formula:
- Intake – N=T+(A-0.009 inch (24mm)
- Exhaust – N=T+(A-0.013 inch (0.33mm)
  - T=Thickness of removed shim
  - A=measured valve clearance
  - N=Thickness of new shim

13. Install a new shim.

14. Recheck the valve clearance.

15. Connect the following cables and hoses:
- Accelerator cable to the throttle body
- Accelerator cable to the clamp on the intake manifold
- Accelerator cable to the clamp on the alternator bracket
- A/T throttle control cable to the throttle body
- PCV hose to the air cleaner hose and the intake manifold

16. Install the cylinder head covers.

17. Install the spark plug wires.

18. Connect the p/s reservoir.

19. Connect the negative battery cable.

2.4L (VIN A and T) Engine

1. Disconnect the negative battery cable.

2. Remove the right side engine service hole cover.

a. Remove the three screws and the scuff plate.

b. Remove the bolt and disconnect the right seat belt from the front floor panel.

c. Remove the four bolts holding the right front seat.

d. Remove the two bolts and the right front seat leg.

e. Remove the jack, the jack stand and the tool bag.

f. Remove the engine service hole cover.

g. Remove the No. 2 cylinder head cover with the gasket.

h. Remove the PCV hose and disconnect the four spark plug wires from the spark plugs.

Adjusting Shim Selection Chart (Exhaust)

Installed shim thickness (mm (in.)) across the top; Measured clearance (mm (in.)) down the left. The chart cross-references these to a shim number (1–17).

Column headers (Installed shim thickness mm (in.)):
2.500 (0.0984), 2.520 (0.0992), 2.540 (0.1000), 2.550 (0.1004), 2.560 (0.1008), 2.580 (0.1016), 2.600 (0.1024), 2.620 (0.1031), 2.640 (0.1039), 2.650 (0.1043), 2.660 (0.1047), 2.680 (0.1051), 2.690 (0.1059), 2.700 (0.1063), 2.710 (0.1067), 2.720 (0.1071), 2.730 (0.1075), 2.740 (0.1079), 2.750 (0.1083), 2.760 (0.1087), 2.770 (0.1091), 2.780 (0.1094), 2.790 (0.1098), 2.800 (0.1102), 2.810 (0.1106), 2.820 (0.1110), 2.840 (0.1118), 2.850 (0.1122), 2.860 (0.1126), 2.870 (0.1130), 2.880 (0.1134), 2.890 (0.1138), 2.900 (0.1142), 2.910 (0.1146), 2.920 (0.1150), 2.930 (0.1154), 2.940 (0.1157), 2.950 (0.1161), 2.960 (0.1165), 2.970 (0.1169), 2.980 (0.1173), 2.990 (0.1177), 3.000 (0.1181), 3.010 (0.1185), 3.020 (0.1189), 3.030 (0.1193), 3.040 (0.1197), 3.050 (0.1201), 3.060 (0.1205), 3.080 (0.1213), 3.100 (0.1220), 3.120 (0.1228), 3.140 (0.1236), 3.150 (0.1240), 3.160 (0.1244), 3.180 (0.1252), 3.200 (0.1260), 3.220 (0.1268), 3.240 (0.1276), 3.250 (0.1280), 3.260 (0.1283), 3.280 (0.1291), 3.300 (0.1299)

Row labels (Measured clearance mm (in.)):
0.000 – 0.020 (0.0000 – 0.0008), 0.021 – 0.040 (0.0008 – 0.0016), 0.041 – 0.060 (0.0016 – 0.0024), 0.061 – 0.080 (0.0024 – 0.0031), 0.081 – 0.100 (0.0032 – 0.0039), 0.101 – 0.120 (0.0040 – 0.0047), 0.121 – 0.140 (0.0048 – 0.0055), 0.141 – 0.160 (0.0056 – 0.0063), 0.161 – 0.180 (0.0063 – 0.0071), 0.181 – 0.200 (0.0071 – 0.0079), 0.201 – 0.220 (0.0079 – 0.0087), 0.221 – 0.240 (0.0087 – 0.0094), 0.241 – 0.260 (0.0095 – 0.0102), 0.261 – 0.279 (0.0103 – 0.0110), 0.280 – 0.380 (0.0110 – 0.0150), 0.381 – 0.400 (0.0150 – 0.0157), 0.401 – 0.420 (0.0158 – 0.0165), 0.421 – 0.440 (0.0166 – 0.0173), 0.441 – 0.460 (0.0174 – 0.0181), 0.461 – 0.480 (0.0181 – 0.0189), 0.481 – 0.500 (0.0189 – 0.0197), 0.501 – 0.520 (0.0197 – 0.0205), 0.521 – 0.540 (0.0205 – 0.0213), 0.541 – 0.560 (0.0213 – 0.0220), 0.561 – 0.580 (0.0221 – 0.0228), 0.581 – 0.600 (0.0229 – 0.0236), 0.601 – 0.620 (0.0237 – 0.0244), 0.621 – 0.640 (0.0244 – 0.0252), 0.641 – 0.660 (0.0252 – 0.0260), 0.661 – 0.680 (0.0260 – 0.0268), 0.681 – 0.700 (0.0268 – 0.0276), 0.701 – 0.720 (0.0276 – 0.0283), 0.721 – 0.740 (0.0284 – 0.0291), 0.741 – 0.760 (0.0292 – 0.0299), 0.761 – 0.780 (0.0300 – 0.0307), 0.781 – 0.800 (0.0307 – 0.0315), 0.801 – 0.820 (0.0315 – 0.0323), 0.821 – 0.840 (0.0323 – 0.0331), 0.841 – 0.860 (0.0331 – 0.0339), 0.861 – 0.880 (0.0339 – 0.0346), 0.881 – 0.900 (0.0347 – 0.0354), 0.901 – 0.920 (0.0355 – 0.0362), 0.921 – 0.940 (0.0363 – 0.0370), 0.941 – 0.960 (0.0370 – 0.0378), 0.961 – 0.980 (0.0378 – 0.0386), 0.981 – 1.000 (0.0386 – 0.0394), 1.001 – 1.020 (0.0394 – 0.0402), 1.021 – 1.040 (0.0402 – 0.0409), 1.041 – 1.060 (0.0410 – 0.0417), 1.061 – 1.080 (0.0418 – 0.0425), 1.081 – 1.100 (0.0426 – 0.0433), 1.101 – 1.120 (0.0433 – 0.0441), 1.121 – 1.140 (0.0441 – 0.0449), 1.141 – 1.160 (0.0449 – 0.0457), 1.161 – 1.180 (0.0457 – 0.0465)

Exhaust valve clearance (Cold):
0.28 – 0.38 mm (0.011 – 0.015 in.)

EXAMPLE: The 2.800 mm (0.1102 in.) shim is installed, and the measured clearance is 0.450 mm (0.0177 in.). Replace the 2.800 mm (0.1102 in.) shim with a new No.9 shim.

New shim thickness mm (in.)

Shim No.	Thickness	Shim No.	Thickness
1	2.500 (0.0984)	10	2.950 (0.1161)
2	2.550 (0.1004)	11	3.000 (0.1181)
3	2.600 (0.1024)	12	3.050 (0.1201)
4	2.650 (0.1043)	13	3.100 (0.1220)
5	2.700 (0.1063)	14	3.150 (0.1240)
6	2.750 (0.1083)	15	3.200 (0.1260)
7	2.800 (0.1102)	16	3.250 (0.1280)
8	2.850 (0.1122)	17	3.300 (0.1299)
9	2.900 (0.1142)		

HINT: New shims have the thickness in millimeters imprinted on the face.

312784

Adjusting shim chart (intake) — 2.0L (VIN P) Engine

i. Remove the No. 2 cord clamp support plate.

j. Remove the No. 1 cylinder head cover and the gasket.

3. Install a bolt and nut to the equipment drive shaft.

4. Set the No. 1 cylinder to TDC in the compression stroke.

a. Turn the equipment drive shaft with a wrench to align the timing marks at TDC. Set the groove on the crankshaft pulley to the **0** position.

b. Check that the valve lifters on the No. 1 cylinder are loose and the valve lifters on the No. 4 are tight. If not, turn the equipment drive shaft one complete revolution and align the marks as above.

5. Inspect the valve clearance.

a. Measure the clearance between the valve lifter and the camshaft. Measure the first and second intake and the first and third exhaust valves.

b. Turn the equipment drive shaft one revolution (360°) and align the marks as above. Measure the third and fourth intake and the second and fourth exhaust valves.

6. Valve clearance cold should be:
- Intake: 0.006–0.010 in. (0.15–0.25mm)
- Exhaust: 0.010–0.014 in. (0.25–0.35mm)

7. Adjust the valve clearance by using adjusting shims.

a. Turn the equipment drive shaft so that the camlobe for the valve to be adjusted faces up.

b. Using SST 09248–55040 or equivalent, press down the valve lifter and place SST 09248–05420 or equivalent, between the camshaft and the valve lifter. Remove SST 09248–55040.

c. Remove the adjusting shim with a small flat prying tool and a magnetic finger.

d. Determine the replacement adjusting shim size according to the following formula, or use the adjusting shim charts.

Adjusting Shim Selection Chart (Intake)

Installed shim thickness mm (in.) / Measured clearance mm (in.)

(Large shim selection chart — installed shim thickness across the top from 2.500 (0.0984) to 3.300 (0.1299); measured clearance down the left side from 0.000–0.020 (0.0000–0.0008) to 1.081–1.090 (0.0426–0.0429). Grid values indicate new shim number.)

Intake valve clearance (Cold):
0.19 – 0.29 mm (0.007 – 0.011 in.)

EXAMPLE: The 2.800 mm (0.1102 in.) shim is installed, and the measured clearance is 0.450 mm (0.0177 in.). Replace the 2.800 mm (0.1102 in.) shim with a new No. 11 shim.

Adjusting shim chart (exhaust) — 2.0L (VIN P) Engine

New shim thickness mm (in.)

Shim No.	Thickness	Shim No.	Thickness
1	2.500 (0.0984)	10	2.950 (0.1161)
2	2.550 (0.1004)	11	3.000 (0.1181)
3	2.600 (0.1024)	12	3.050 (0.1201)
4	2.650 (0.1043)	13	3.100 (0.1220)
5	2.700 (0.1063)	14	3.150 (0.1240)
6	2.750 (0.1083)	15	3.200 (0.1260)
7	2.800 (0.1102)	16	3.250 (0.1280)
8	2.850 (0.1122)	17	3.300 (0.1299)
9	2.900 (0.1142)		

HINT: New shims have the thickness in millimeters imprinted on the face.

312785

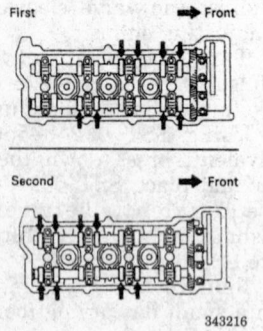

Order of valve adjustment — 2.4L (VIN A and T) Engine

343216

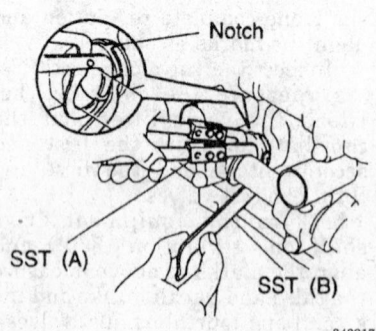

Removing the adjusting shim — 2.4L (VIN A and T) Engine

343217

e. Using a micrometer, measure the thickness of the removed shim. Calculate the thickness of a new shim so that the valve clearance comes within the specified value.

- T: Thickness of the removed shim
- A: Measured valve clearance
- N: Thickness of the new shim

f. Intake: $N = T + (A - 0.008\ \text{in.}\ (0.20\ \text{mm})$

g. Exhaust: $N = T + (A - 0.012\ \text{in.}\ (0.30\ \text{mm})$

h. Install a new adjusting shim. Place it on the valve lifter. Using the SST 09248–55040, press down the valve lifter and remove SST 09248–05420.

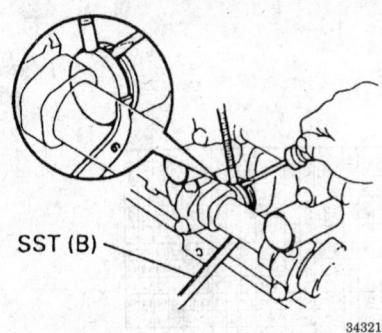

SST (B)

343218

**Pressing down on the valve lifter to remove
the shim — 2.4L (VIN A and T) Engine**

i. Recheck the valve clearance.

8. After adjustments have been made, remove the equipment drive shaft bolt and nut. If not removed, the bolt head will hit and damage the cooling fan.

9. Install the No. 1 cylinder head cover bolts in the sequence. Torque to 6 ft. lbs. (8 Nm).

10. Install the No. 2 cord clamp support plate and torque the bolt to 3.7 ft. lbs. (5 Nm).

11. Connect the spark plug wires. Install the PCV hose.

12. Install the No. 2 cylinder head cover and torque the bolts to 4 ft. lbs. (5.5 Nm).

13. Install the right engine service hole cover and torque the bolts to 10 ft. lbs. (14 Nm).

14. Install the jack holder, the jack and the tool bag.

15. Install the bolts holding the right front seat leg and torque to 29 ft. lbs. (39 Nm).

16. Install the right front seat and torque the bolts to 29 ft. lbs. (39 Nm).

17. Install the bolt holding the seat belt to the front floor panel and torque to 31 ft. lbs. (42 Nm).

18. Install the scuff plate.

Intake Manifold

REMOVAL AND INSTALLATION

2.0L (VIN P) Engine

1. Disconnect the negative battery cable to the battery.

2. Remove the air cleaner assembly.

3. Remove the throttle body from the intake manifold.

4. Disconnect the engine wire from the intake manifold:

a. Disconnect the four injector connectors.

b. Disconnect the two engine wire clamps from the wire brackets on the intake manifold.

c. Disconnect the engine wire protector from the right hand side of the intake manifold by removing the bolt.

d. Disconnect the clamp of the engine wire from the wire clamp.

5. Remove the EGR valve, EGR pipe and modulator as follows:

a. Disconnect the two vacuum hoses from the VSV for the EGR.

b. Disconnect the vacuum modulator from the clamp on the intake manifold.

c. Loosen the cylinder head side of the EGR pipe union nut.

d. Remove the two nuts, the EGR valve, pipe assembly and gasket. Remove the vacuum modulator.

6. Disconnect the following hoses:

• Fuel filter vacuum sensor hose on the intake manifold

• Brake booster vacuum hose from the intake manifold

• Ground strap from the intake manifold

7. Remove the intake manifold stay by removing the two bolts.

8. If equipped with A/T, disconnect the control cable from the clamp on the rear side of the intake manifold.

9. Disconnect the air hose from the intake manifold.

10. Remove the air tube from the intake manifold by removing the two bolts.

11. Remove the six bolts and two nuts from the intake manifold.

12. Remove the intake manifold from the vehicle.

To Install:

13. Install the intake manifold to the engine and install the six bolts and two nuts. Torque the bolts and nuts to 14 ft. lbs. (19 Nm).

14. Install the air tube with the two bolts.

15. Connect the air hose to the intake manifold.

16. If equipped with A/T, connect the control cable to the clamp on the rear side of the intake manifold.

17. Install the intake manifold stay with the two bolts. Torque the bolts to 31 ft. lbs. (42 Nm).

18. Connect the following hoses:

• Ground strap to the intake manifold

• Brake booster vacuum hose to the intake manifold

• Fuel filter vacuum sensor hose to the intake manifold

19. Install the EGR valve, EGR pipe and the vacuum modulator as follows:

a. Install the vacuum modulator.

b. Install the EGR valve and pipe. Torque the two nuts to 9 ft.

lbs. (13 Nm) and the union nut to 43 ft. lbs. (59 Nm).

c. Connect all the vacuum hoses.

20. Connect the engine wire and injectors.

NOTE: The No. 1 and No. 3 injector connectors are brown, and the No. 2 and No. 4 injector connectors are gray.

21. Install the throttle body to the intake manifold.

22. Install the air cleaner assembly.

23. Connect the negative battery cable to the battery.

2.4L (VIN A and T) Engines

1. Disconnect the negative battery cable.

2. Remove the air intake connector.

3. Disconnect and tag all wires, connectors, coolant and vacuum hoses from the intake manifold.

4. Disconnect the shift and accelerator cables.

5. Remove the fuel pipes.

6. Remove the distributor and EGR valve.

7. Remove the water outlet, by-pass pipe and gasket from the manifold.

8. Disconnect the water hose form the water pump and remove the bolt holding the water by-pass pipe and timing chain case.

9. Remove the intake manifold stays.

10. Remove the 2 nuts and 4 bolts and remove the intake manifold and gasket. Remove the cylinder block insulators.

To install:

11. Position the cylinder block insulator on the cylinder head.

12. Position a new gasket on the cylinder head and install the intake manifold with the 2 nuts and 4 bolts. Torque the fasteners to 15 ft. lbs (21 Nm).

13. Install the intake manifold stays.

14. Install the bolt holding the water by-pass pipe and timing chain case. Connect the water hose to the water pump.

15. Install the delivery pipe, water outlet and EGR valve.

16. Install the distributor.

17. Connect all wires, connectors, coolant and vacuum hoses from the intake manifold.

18. Connect the shift and accelerator cables.

19. Install the air intake connector.

20. Fill and bleed the cooling system.

Adjusting Shim Selection Chart (Intake)

Installed shim thickness mm (in.) column headers:

2.500 (0.0984), 2.520 (0.0992), 2.540 (0.1000), 2.550 (0.1004), 2.560 (0.1008), 2.580 (0.1016), 2.600 (0.1024), 2.620 (0.1031), 2.640 (0.1039), 2.650 (0.1043), 2.660 (0.1047), 2.670 (0.1051), 2.680 (0.1055), 2.690 (0.1059), 2.700 (0.1063), 2.710 (0.1067), 2.720 (0.1071), 2.730 (0.1075), 2.740 (0.1079), 2.750 (0.1083), 2.760 (0.1087), 2.770 (0.1091), 2.780 (0.1094), 2.790 (0.1098), 2.800 (0.1102), 2.810 (0.1106), 2.820 (0.1110), 2.830 (0.1114), 2.840 (0.1118), 2.850 (0.1122), 2.860 (0.1126), 2.870 (0.1130), 2.880 (0.1134), 2.890 (0.1138), 2.900 (0.1142), 2.910 (0.1146), 2.920 (0.1150), 2.930 (0.1154), 2.940 (0.1157), 2.950 (0.1161), 2.960 (0.1165), 2.970 (0.1169), 2.980 (0.1173), 2.990 (0.1177), 3.000 (0.1181), 3.010 (0.1185), 3.020 (0.1189), 3.030 (0.1193), 3.040 (0.1197), 3.050 (0.1201), 3.060 (0.1205), 3.080 (0.1213), 3.100 (0.1220), 3.120 (0.1228), 3.140 (0.1240), 3.150 (0.1240), 3.180 (0.1244), 3.200 (0.1252), 3.220 (0.1260), 3.240 (0.1276), 3.250 (0.1280), 3.260 (0.1283), 3.280 (0.1291), 3.300 (0.1299)

Measured clearance mm (in.) row labels:

0.000 – 0.020 (0.0000 – 0.0008)
0.021 – 0.040 (0.0008 – 0.0016)
0.041 – 0.060 (0.0016 – 0.0024)
0.061 – 0.080 (0.0024 – 0.0031)
0.081 – 0.100 (0.0032 – 0.0039)
0.101 – 0.120 (0.0040 – 0.0047)
0.121 – 0.140 (0.0048 – 0.0055)
0.141 – 0.149 (0.0056 – 0.0059)
0.150 – 0.250 (0.0059 – 0.0098)
0.251 – 0.260 (0.0099 – 0.0102)
0.261 – 0.280 (0.0103 – 0.0110)
0.281 – 0.300 (0.0111 – 0.0118)
0.301 – 0.320 (0.0119 – 0.0126)
0.321 – 0.340 (0.0126 – 0.0134)
0.341 – 0.360 (0.0134 – 0.0142)
0.361 – 0.380 (0.0142 – 0.0150)
0.381 – 0.400 (0.0150 – 0.0157)
0.401 – 0.420 (0.0158 – 0.0165)
0.421 – 0.440 (0.0166 – 0.0173)
0.441 – 0.460 (0.0174 – 0.0181)
0.461 – 0.480 (0.0181 – 0.0189)
0.481 – 0.500 (0.0189 – 0.0197)
0.501 – 0.520 (0.0197 – 0.0205)
0.521 – 0.540 (0.0205 – 0.0213)
0.541 – 0.560 (0.0213 – 0.0220)
0.561 – 0.580 (0.0221 – 0.0228)
0.581 – 0.600 (0.0229 – 0.0236)
0.601 – 0.620 (0.0237 – 0.0244)
0.621 – 0.640 (0.0244 – 0.0252)
0.641 – 0.660 (0.0252 – 0.0260)
0.661 – 0.680 (0.0260 – 0.0268)
0.681 – 0.700 (0.0268 – 0.0276)
0.701 – 0.720 (0.0276 – 0.0283)
0.721 – 0.740 (0.0284 – 0.0291)
0.741 – 0.760 (0.0292 – 0.0299)
0.761 – 0.780 (0.0300 – 0.0307)
0.781 – 0.800 (0.0307 – 0.0315)
0.801 – 0.820 (0.0315 – 0.0323)
0.821 – 0.840 (0.0323 – 0.0331)
0.841 – 0.860 (0.0331 – 0.0339)
0.861 – 0.880 (0.0339 – 0.0346)
0.881 – 0.900 (0.0347 – 0.0354)
0.901 – 0.920 (0.0355 – 0.0362)
0.921 – 0.940 (0.0363 – 0.0370)
0.941 – 0.960 (0.0370 – 0.0378)
0.961 – 0.980 (0.0378 – 0.0386)
0.981 – 1.000 (0.0386 – 0.0394)
1.001 – 1.020 (0.0394 – 0.0402)
1.021 – 1.040 (0.0402 – 0.0409)
1.041 – 1.050 (0.0410 – 0.0413)

Intake valve clearance (Cold):
0.15 – 0.25 mm (0.006 – 0.010 in.)

EXAMPLE: The 2.800 mm (0.1102 in.) shim is installed, and the measured clearance is 0.450 mm (0.0177 in.). Replace the 2.800 mm (0.1102 in.) shim with a new No.12 shim.

New shim thickness mm (in.)

Shim No.	Thickness	Shim No.	Thickness
1	2.500 (0.0984)	10	2.950 (0.1161)
2	2.550 (0.1004)	11	3.000 (0.1181)
3	2.600 (0.1024)	12	3.050 (0.1201)
4	2.650 (0.1043)	13	3.100 (0.1220)
5	2.700 (0.1063)	14	3.150 (0.1240)
6	2.750 (0.1083)	15	3.200 (0.1260)
7	2.800 (0.1102)	16	3.250 (0.1280)
8	2.850 (0.1122)	17	3.300 (0.1299)
9	2.900 (0.1142)		

HINT: New shims have the thickness in millimeters imprinted on the face.

343220

Intake adjusting shim selection chart — 2.4L (VIN A and T) Engine

---

21. Connect the negative battery cable. Start the engine and check for leaks.

Exhaust Manifold

REMOVAL AND INSTALLATION

2.0L (VIN P) Engine

1. Disconnect the negative battery cable.

— CAUTION —
Wait 90 seconds from the time the key is turned to LOCK and the negative battery cable is disconnected to begin work. This allows the SRS capacitor to discharge and prevent deployment of the air bag(s).

2. Raise and safely support the vehicle.

3. Using a 14mm. deep socket wrench, remove the three nuts and the gasket to disconnect the front exhaust pipe from the exhaust manifold.

4. Disconnect the main oxygen sensor and the sub oxygen sensor connectors.

5. Remove the six bolts and the upper manifold heat insulator.

6. Remove the two bolts holding the right hand exhaust manifold stay to the cylinder block.

7. Remove the six nuts, the exhaust manifold, and the three–way catalytic convertor assembly.

8. Separate the exhaust manifold and front catalytic converter.

To install:

9. Install the catalytic converter to the exhaust manifold. Torque the three bolts and two nuts to 22 ft. lbs. (29 Nm).

10. Install a new gasket, the exhaust manifold, and the front TWC assembly with the six nuts. Uniformly tighten the nuts in several

Adjusting Shim Selection Chart (Exhaust)

Installed shim thickness mm (in.): 2.500 (0.0984), 2.520 (0.0992), 2.540 (0.1000), 2.560 (0.1004), 2.580 (0.1008), 2.600 (0.1016), 2.620 (0.1024), 2.640 (0.1031), 2.650 (0.1039), 2.660 (0.1043), 2.670 (0.1047), 2.680 (0.1051), 2.690 (0.1055), 2.700 (0.1059), 2.710 (0.1063), 2.720 (0.1067), 2.730 (0.1071), 2.740 (0.1075), 2.750 (0.1079), 2.760 (0.1083), 2.770 (0.1087), 2.780 (0.1091), 2.790 (0.1094), 2.800 (0.1098), 2.810 (0.1102), 2.820 (0.1106), 2.830 (0.1110), 2.840 (0.1114), 2.850 (0.1118), 2.860 (0.1122), 2.870 (0.1126), 2.880 (0.1130), 2.890 (0.1134), 2.900 (0.1138), 2.910 (0.1142), 2.920 (0.1146), 2.930 (0.1150), 2.940 (0.1154), 2.950 (0.1157), 2.960 (0.1161), 2.970 (0.1165), 2.980 (0.1169), 2.990 (0.1173), 3.000 (0.1177), 3.010 (0.1181), 3.020 (0.1185), 3.030 (0.1189), 3.040 (0.1193), 3.050 (0.1197), 3.060 (0.1201), 3.070 (0.1205), 3.080 (0.1209), 3.090 (0.1213), 3.100 (0.1217), 3.110 (0.1220), 3.120 (0.1224), 3.130 (0.1228), 3.140 (0.1232), 3.150 (0.1236), 3.160 (0.1240), 3.170 (0.1244), 3.180 (0.1248), 3.190 (0.1252), 3.200 (0.1256), 3.210 (0.1260), 3.220 (0.1264), 3.230 (0.1268), 3.240 (0.1272), 3.250 (0.1276), 3.260 (0.1280), 3.270 (0.1283), 3.280 (0.1287), 3.290 (0.1291), 3.300 (0.1299)

Measured clearance mm (in.):

- 0.000 – 0.020 (0.0000 – 0.0008)
- 0.021 – 0.040 (0.0008 – 0.0016)
- 0.041 – 0.060 (0.0016 – 0.0024)
- 0.061 – 0.080 (0.0024 – 0.0031)
- 0.081 – 0.100 (0.0032 – 0.0039)
- 0.101 – 0.120 (0.0040 – 0.0047)
- 0.121 – 0.140 (0.0048 – 0.0055)
- 0.141 – 0.160 (0.0056 – 0.0063)
- 0.161 – 0.180 (0.0063 – 0.0071)
- 0.181 – 0.200 (0.0071 – 0.0079)
- 0.201 – 0.220 (0.0079 – 0.0087)
- 0.221 – 0.240 (0.0087 – 0.0094)
- 0.241 – 0.249 (0.0095 – 0.0098)
- 0.250 – 0.350 (0.0098 – 0.0138)
- 0.351 – 0.360 (0.0138 – 0.0142)
- 0.361 – 0.380 (0.0142 – 0.0150)
- 0.381 – 0.400 (0.0150 – 0.0157)
- 0.401 – 0.420 (0.0158 – 0.0165)
- 0.421 – 0.440 (0.0166 – 0.0173)
- 0.441 – 0.460 (0.0174 – 0.0181)
- 0.461 – 0.480 (0.0181 – 0.0189)
- 0.481 – 0.500 (0.0189 – 0.0197)
- 0.501 – 0.520 (0.0197 – 0.0205)
- 0.521 – 0.540 (0.0205 – 0.0213)
- 0.541 – 0.560 (0.0213 – 0.0220)
- 0.561 – 0.580 (0.0221 – 0.0228)
- 0.581 – 0.600 (0.0229 – 0.0236)
- 0.601 – 0.620 (0.0237 – 0.0244)
- 0.621 – 0.640 (0.0244 – 0.0252)
- 0.641 – 0.660 (0.0252 – 0.0260)
- 0.661 – 0.680 (0.0260 – 0.0268)
- 0.681 – 0.700 (0.0268 – 0.0276)
- 0.701 – 0.720 (0.0276 – 0.0283)
- 0.721 – 0.740 (0.0284 – 0.0291)
- 0.741 – 0.760 (0.0292 – 0.0299)
- 0.761 – 0.780 (0.0300 – 0.0307)
- 0.781 – 0.800 (0.0307 – 0.0315)
- 0.801 – 0.820 (0.0315 – 0.0323)
- 0.821 – 0.840 (0.0323 – 0.0331)
- 0.841 – 0.860 (0.0331 – 0.0339)
- 0.861 – 0.880 (0.0339 – 0.0346)
- 0.881 – 0.900 (0.0347 – 0.0354)
- 0.901 – 0.920 (0.0355 – 0.0362)
- 0.921 – 0.940 (0.0363 – 0.0370)
- 0.941 – 0.960 (0.0370 – 0.0378)
- 0.961 – 0.980 (0.0378 – 0.0386)
- 0.981 – 1.000 (0.0386 – 0.0394)
- 1.001 – 1.020 (0.0394 – 0.0402)
- 1.021 – 1.040 (0.0402 – 0.0409)
- 1.041 – 1.060 (0.0410 – 0.0417)
- 1.061 – 1.080 (0.0418 – 0.0425)
- 1.081 – 1.100 (0.0426 – 0.0433)
- 1.101 – 1.120 (0.0433 – 0.0441)
- 1.121 – 1.140 (0.0441 – 0.0449)
- 1.141 – 1.150 (0.0449 – 0.0453)

Exhaust valve clearance (Cold):
0.25 – 0.35 mm (0.010 – 0.014 in.)

EXAMPLE: The 2.800 mm (0.1102 in.) shim is installed, and the measured clearance is 0.450 mm (0.0177 in.).
Replace the 2.800 mm (0.1102 in.) shim with a new No.10 shim.

New shim thickness mm (in.)

Shim No.	Thickness	Shim No.	Thickness
1	2.500 (0.0984)	10	2.950 (0.1161)
2	2.550 (0.1004)	11	3.000 (0.1181)
3	2.600 (0.1024)	12	3.050 (0.1201)
4	2.650 (0.1043)	13	3.100 (0.1220)
5	2.700 (0.1063)	14	3.150 (0.1240)
6	2.750 (0.1083)	15	3.200 (0.1260)
7	2.800 (0.1102)	16	3.250 (0.1280)
8	2.850 (0.1122)	17	3.300 (0.1299)
9	2.900 (0.1142)		

HINT: New shims have the thickness in millimeters imprinted on the face.

343221

Exhaust adjusting shim selection chart — 2.4L (VIN A and T) Engine

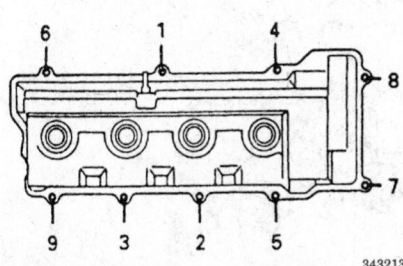

Tightening order for No. 1 cylinder head cover bolts — 2.4L (VIN A and T) Engine

343213

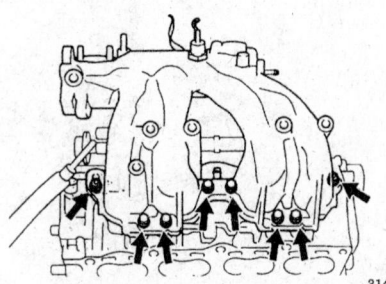

Intake manifold bolts and nuts — 2.0L (VIN P) Engine

314643

passes. Torque the nuts to 36 ft. lbs. (49 Nm).

11. Install the right hand manifold stay with the two bolts and torque to 31 ft. lbs. (42 Nm).

12. Install the manifold upper heat insulator with the six bolts and connect the main oxygen and the sub oxygen sensor connectors.

13. Install the front exhaust pipe with a new gasket to the TWC. Install the three nuts using a 14mm. deep socket wrench. Torque the nuts to 46 ft. lbs. (62 Nm).

14. Lower the vehicle safely and connect the negative battery cable.

15. Start the engine and make sure that t1here are no exhaust leaks.

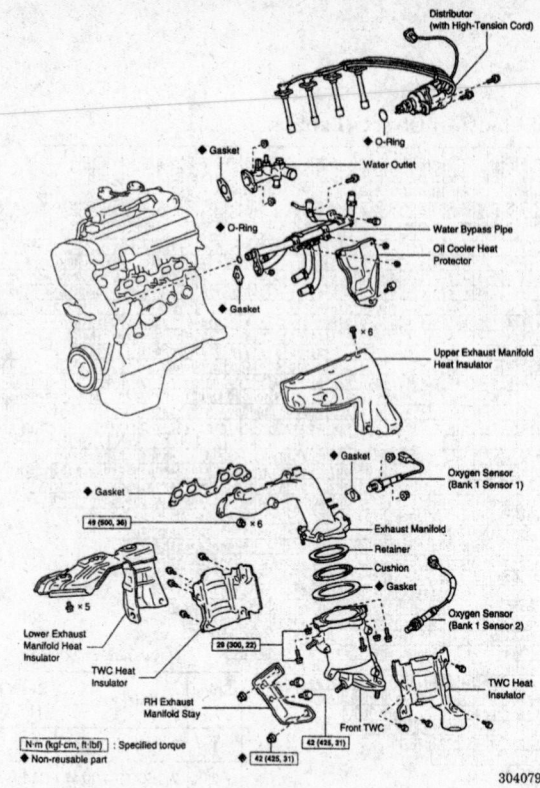

Exhaust manifold and components — 2.0L (VIN P) Engine

2.4L (VIN A and T) Engine

1. Disconnect the negative battery cable.

2. Raise and safely support the vehicle.

3. Disconnect the front exhaust pipe from the exhaust manifold by removing the three nuts.

4. Remove the 5 nuts and remove the exhaust manifold and gasket.

   To install:

5. Install the exhaust manifold with a new gasket. Install and torque the exhaust manifold nuts to 36 Nm (41 Nm).

6. Connect the front exhaust pipe to the exhaust manifold by installing new gaskets and three nuts. Torque the nuts to 46 ft. lbs. (62 Nm).

7. Lower the vehicle and connect the negative battery cable.

8. Start the engine and check for leaks.

Supercharger

REMOVAL AND INSTALLATION

2.4L (VIN A) Engine

1. Disconnect the negative battery cable from the battery.

2. Drain the engine coolant from the radiator.

3. Remove the engine coolant reservoir tank and bracket.

4. Remove the air damper case.

5. Remove the supercharger blower.

6. Remove the radiator as follows:

   a. Disconnect the radiator hoses and coolant reservoir hose from the radiator.

   b. Disconnect the power steering reservoir by removing the two bolts.

   c. If equipped with an automatic transaxle, disconnect the transaxle cooling lines.

   d. Disconnect the water bypass hose for the throttle body.

   e. Remove the No. 2 fan shroud by removing the four bolts.

   f. Stretch the alternator drive belt and remove the four nuts of the fluid coupling.

   g. Pull out the fan with the fluid coupling.

   h. Disconnect the ACV connector.

   i. Disconnect the three air hoses from the ACV.

   j. Disconnect the No. 1 fan shroud from the radiator by removing the two bolts. Leave the No. 1 fan shroud in the vehicle.

   k. Remove the radiator supports by removing the two bolts.

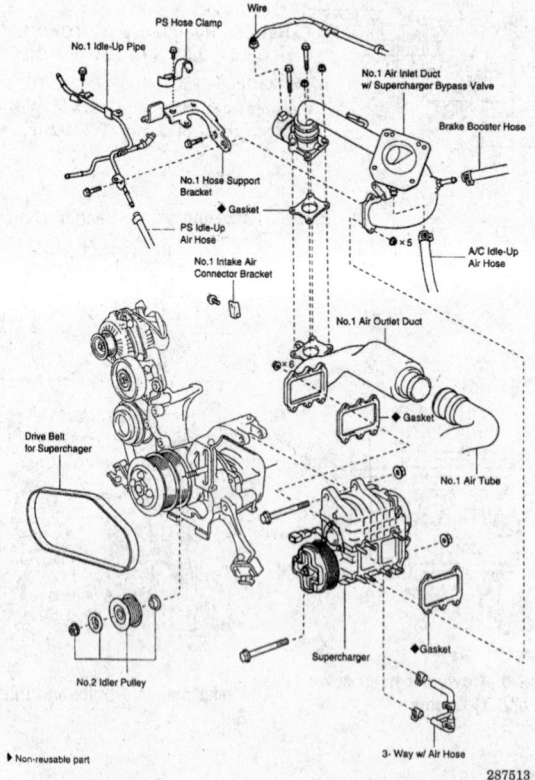

Supercharger components — 2.4L (VIN A) Engine

l. Remove the radiator from the vehicle.

7. Remove the throttle body.

8. Remove the alternator/power steering drive belt.

9. Remove the power steering pump as follows:

a. Disconnect the power steering idle up hoses.

b. Remove the bolt and power steering hose clamp.

c. Remove the five bolts, pump stay and power steering pump with the hoses attached.

10. Remove the drive belt for the supercharger.

11. Remove the No. 2 idler pulley by removing the nut, plate, and the spacer.

12. Remove the No. 1 air inlet duct with the supercharger bypass valve as follows:

a. Disconnect the supercharger bypass valve connector.

b. Disconnect the brake booster hose.

c. Disconnect the A/C idle up air hose.

d. Disconnect the supercharger magnetic clutch connector.

e. Disconnect the supercharger magnetic clutch connector from the No.1 hose support bracket.

f. Remove the air hoses and three way.

g. Remove the two bolts and two nuts holding the supercharger bypass valve to the No. 1 air outlet duct.

h. Remove the five nuts and the No. 1 air inlet duct with the supercharger bypass valve.

i. Remove the supercharger bypass valve and No. 1 air inlet duct gaskets.

13. Remove the No. 1 idle up pipe by removing the bolt and air hose.

14. Disconnect the No. 1 air tube.

15. Remove the No. 1 intake air connector bracket by removing the two bolts.

16. Remove the supercharger as follows:

a. Remove the two bolts and two nuts holding the supercharger to the equipment drive housing.

b. Remove the six nuts holding the supercharger to the No. 1 air outlet duct.

c. Separate the supercharger and No. 1 air outlet duct and remove the gasket.

d. Remove the supercharger and No. 1 air outlet duct from the vehicle.

To install

17. Install the supercharger as follows:

a. Install the supercharger and No. 1 air outlet duct to the vehicle.

b. Connect the supercharger and No. 1 air outlet duct with a new gasket.

c. Install the six nuts to hold the supercharger to the No. 1 air outlet duct. Torque the nuts to 82 inch lbs. (9.3 Nm).

d. Install the two bolts and two nuts to hold the supercharger to the equipment drive housing. Torque the bolts and nuts to 27 ft. lbs. (37 Nm).

18. Install the No. 1 intake air connector bracket with the two bolts. Torque the bolts to 13 ft. lbs. (18 Nm).

19. Connect the No.1 air tube.

20. Install the No. 1 idle up pipe by connecting the air hose and installing the bolt. Torque the bolts to 69 inch lbs. (7.5 Nm).

21. Install the No. 1 air inlet duct with the supercharger bypass valve as follows:

a. Connect the supercharger bypass valve to the No. 1 air outlet duct. Install the two bolts and two nuts and torque the bolts and nuts to 48 inch lbs. (5.4 Nm).

b. Install the air hoses and three way valve.

c. Install the No. 1 air hose support bracket and two bolts. Torque the bolts to 69 inch lbs. (7.5 Nm).

d. Connect the supercharger magnetic clutch connector to the No. 1 hose support bracket.

e. Connect the A/C idle up air hose.

f. Connect the supercharger magnetic clutch connector.

g. Connect the brake booster hose.

h. connect the supercharger bypass valve connector.

22. Install the No. 2 idler pulley with the spacer, plate, and the nut.

23. Install and adjust the drive belt for the supercharger.

24. Install the power steering pump as follows:

a. Install the power steering pump with the pump stay and five bolts. Torque the long bolts to 35 ft. lbs. (48 Nm) and the short bolts to 27 ft. lbs. (36 Nm).

b. Install the power steering hose clamp and bolt. Torque the bolt to 9 ft. lbs. (11.5 Nm).

c. Connect the power steering idle up hoses.

25. Install and adjust the alternator/power steering drive belts.

26. Install the throttle body.

27. Install the radiator. Torque the radiator bolts to 13 ft. lbs (18 Nm). Connect all connectors, hoses and shrouds.

28. Connect the power steering reservoir with the two bolts. Torque the bolts to 9 ft. lbs. (12.5 Nm).

29. Install the supercharger blower.

30. Install the air damper case.

31. Install the engine coolant reservoir and bracket.

32. Install the air duct.

33. Fill the radiator with engine coolant.

34. Connect the battery cable to the battery.

35. Check all fluids.

Front Cover Seal

REMOVAL AND INSTALLATION

2.4L (VIN A and T) Engine

NOTE: There are 2 methods to replace the oil seal; it may be done with the cover on or off the engine.

Timing Chain Cover Installed

1. Disconnect the negative battery cable.

2. Remove the engine drive belts and crankshaft pulley.

3. Using a suitable tool, pry out the oil seal. Be careful not to damage the crankshaft sealing surface.

To install:
Removing the oil seal (timing chain cover installed)1 — 2.4L (VIN A and T) Engine

340577

1. Apply MP grease to the new oil seal lip.

2. Using tool SST 09309-36010 or equivalent and a mallet, tap in the new oil seal until its surface is flush with the oil pump cover edge.

3. Install the crankshaft pulley and engine drive belts.

Timing Chain Cover Removed

1. Carefully tap the oil seal out of the timing chain cover.

2. Apply MP grease to the new oil seal lip.

3. Using tool SST 09309-36010 or equivalent and a mallet, tap in the new oil seal until its surface is flush with the oil pump cover edge.

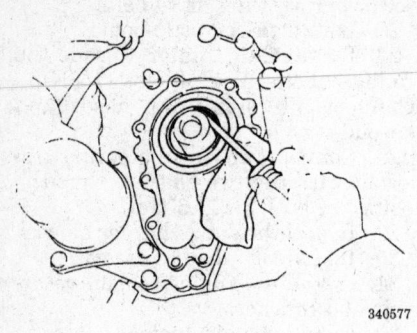

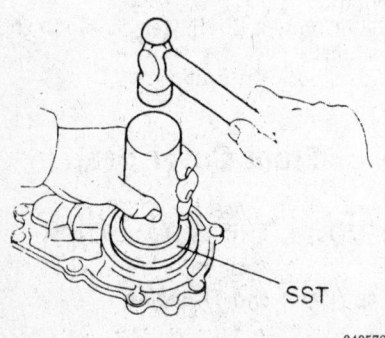

Installing the oil seal (timing chain cover removed) — 2.4L (VIN T) Engine

Front Crankshaft Seal

REMOVAL AND INSTALLATION

2.0L (VIN P) Engine

———— CAUTION ————
On models with an air bag, wait at least 90 seconds from the time that the ignition switch is turned to the LOCK position and the battery is disconnected before performing any further work.

NOTE: The front oil seal can be removed from the engine without removing the oil pump.

1. Disconnect the negative battery cable from the battery.
2. Remove the timing belt covers and the timing belt from the engine.
3. Using SST tool 09950–50010 or equivalent (crankshaft gear puller), remove the front crankshaft gear from the crankshaft. Make sure not to damage any part of the crankshaft.
4. Using a knife, cut off the oil seal lip.
5. Using a suitable tool, pry out the oil seal. Wrap the edge of the tool with a rag or tape to prevent damaging the crankshaft. Be careful not to damage the crankshaft.

To install:
6. Using a new seal, apply a thin layer of liquid sealer to the outside of the seal.
7. Apply multi purpose grease to the new oil seal lip.
8. Using SST tool 09226–00010 or equivalent (oil seal installer) and a hammer, tap in the oil seal until its surface is flush with the oil pump body edge.
9. Install the timing belt and the timing belt covers.
10. Install all other components and then connect the negative battery cable to the battery.
11. Start the engine and check for leaks.

Timing Chain Front Cover

REMOVAL AND INSTALLATION

2.4L (VIN A and T) Engine

1. Disconnect the negative battery cable.
2. Remove the engine from the vehicle.
3. Remove the cylinder head from the engine.
4. Remove the crankshaft pulley and damper.
5. Remove the left engine mounting.
6. Remove the oil pressure switch and engine ventilation case.
7. Remove the oil pan using a pan and the oil baffle.
8. Remove the 3 bolts and oil filter bracket from the timing cover.
9. Remove the 12 bolts, 2 nuts and timing cover.

NOTE: There are 3 bolts located in back of the cover. Be careful not to damage the mating surfaces during removal.

To install:
10. Before installing, clean all gasket surfaces and check for warpage.

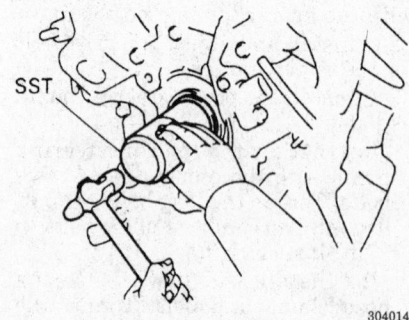

Installing the front crankshaft oil seal — 2.4L (VIN A) engines

11. Install the timing chain cover using a new gasket. Torque the (A) bolts to 14 ft. lbs. (21 Nm), (B) bolts to 21 ft. lbs. (28 Nm) and the (C) bolts to 32 ft. lbs. (43 Nm).
12. Install the oil filter bracket and torque to 14 ft. lbs. (21 Nm).
13. Install the crankcase baffle plate, oil pan and ventilation case.
14. Install the oil pressure switch and left engine mount.
15. Install the crankshaft pulley and torque the bolt to 192 ft. lbs. (260 Nm).
16. Install the cylinder head and engine assembly into the vehicle.
17. Install the remaining components and check for leaks.

Timing Chain

REMOVAL AND INSTALLATION

2.4L (VIN A and T) Engine

1. Remove the engine from the vehicle.
2. Separate the engine and transaxle.
3. Remove the cylinder head from the engine block.
4. Remove the crankshaft pulley as follows:
   a. Using SST 09213–58012 and 09330–00021, or equivalent, loosen the pulley bolt for the crankshaft pulley.
   b. Remove the SST tool and pulley bolt.
   c. Using SST 09950–40010 or equivalent (crankshaft pulley puller), remove the crankshaft pulley.
5. Remove the left engine mounting as follows:
   a. Remove the two bolts and the left hand mounting stay
   b. Remove the three bolts and the left mounting.
6. Remove the oil pressure switch.
7. Remove the engine oil dipstick.
8. Remove the No. 2 engine hanger by removing the four bolts.
9. Remove the ventilation case by removing the three bolts. Remove the gaskets.
10. Remove the No. 1 oil dipstick guide and gasket.
11. On A–2TZFE engine, remove the crankshaft position sensor by removing the two bolts.
12. Remove the crankcase as follows:
   a. Remove the 16 bolts and 2 nuts.
   b. Using SST 09032–00100, or equivalent and a brass bar, sepa-

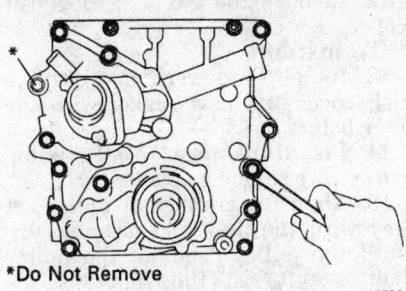

*Do Not Remove

115904

Removing timing chain cover — 2.4L (VIN A and T) engine

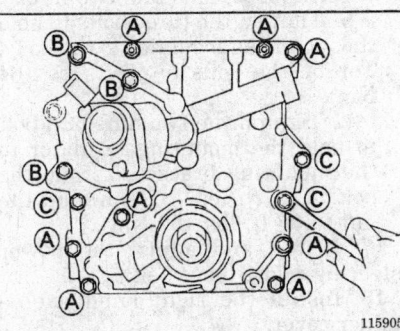

115905

Installing timing chain cover — 2.4L (VIN A and T) engine

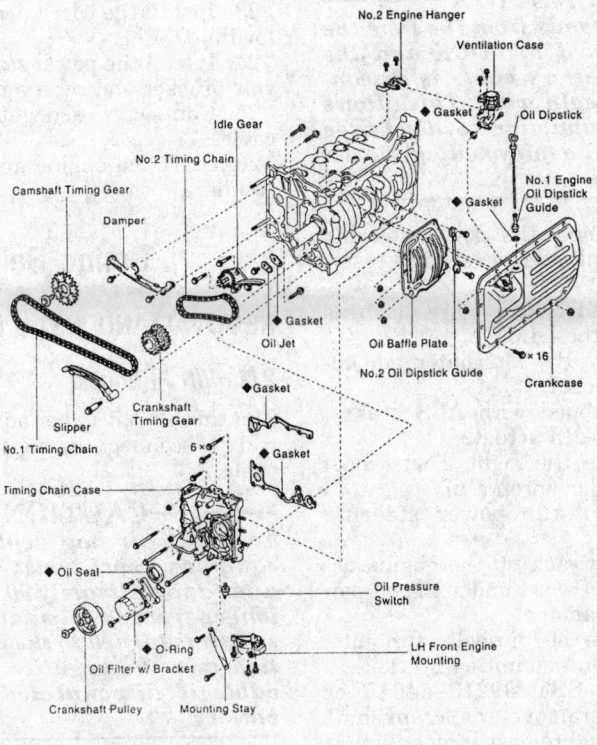

No.2 Engine Hanger
Ventilation Case
Idle Gear
No.2 Timing Chain
Camshaft Timing Gear
Damper
◆ Gasket
Oil Dipstick
No.1 Engine Oil Dipstick Guide
◆ Gasket
◆ Gasket
Oil Jet
Oil Baffle Plate
No.2 Oil Dipstick Guide
Crankcase
Slipper
Crankshaft Timing Gear
No.1 Timing Chain
6 ×
Timing Chain Case
◆ Gasket
◆ Gasket
◆ Oil Seal
Oil Pressure Switch
◆ O-Ring
Oil Filter w/ Bracket
LH Front Engine Mounting
Crankshaft Pulley
Mounting Stay

◆ Non-reusable part

343540

Timing chain and related parts — 2.4L (VIN A and T) engine

rate the crankcase from the cylinder block.

13. Remove the No. 2 oil dipstick guide and oil baffle plate by removing the two bolts and three nuts.

14. Remove the oil filter bracket with the oil filter by removing the three bolts. Remove the O-ring from the timing chain case.

15. Remove the timing chain case as follows:

a. Remove the three bolts from the rear of the timing chain cover.

b. Remove the 12 bolts and two nuts from the front of the chain cover.

c. Using a plastic faced hammer, tap the chain case and remove the timing chain case and two gaskets.

16. Remove the No. 1 timing chain and camshaft timing gear.

17. Remove the chain slipper and damper by removing the three bolts.

18. Remove the oil jet by removing the bolt.

19. Remove the No. 2 timing chain and idle gear as follows:

a. Loosen the two bolts to the idle gear chain guide.

b. Tighten the lower bolt while pushing the idle gear chain guide to the left with your finger.

c. Remove the two bolts and remove the chain and idle gear as an assembly.

20. Remove the crankshaft timing gear.

To install:

21. Install the crankshaft timing gear as follows:

a. Turn the crankshaft until the shaft key is on the top.

b. Slide the gear over the key on the crankshaft.

22. Install the No. 2 timing chain and idle gear as follows:

a. Place the No. 2 timing chain on the idle gear.

b. Place the No. 2 timing chain on the crankshaft gear.

c. Install and torque the two bolts to 14 ft. lbs. (20 Nm).

d. Loosen the lower bolt so that the chain guide presses against the chain.

e. Check that the spring is operating normally against the chain guide by pressing on the chain with your finger and then releasing your finger.

f. With the chain guide pressing against the chain, torque the bolts to hold the chain guide in place. Torque the bolts to 14 ft. lbs. (20 Nm).

23. Install the oil jet with a new gasket. Torque the bolt to 13 ft. lbs. (18 Nm).

24. Install the chain damper and slipper by installing the three bolts. Torque the chain damper bolts to 13 ft. lbs. (18 Nm) and the chain slipper bolt to 20 ft. lbs. (27 Nm).

25. Place the No. 1 timing chain and camshaft timing gear as follows:

a. Place the timing chain on the camshaft timing gear so that the timing mark is between the two bright chain links.

b. Place the timing chain on the crankshaft timing gear with the single bright link aligned with the timing mark on the crankshaft timing gear.

c. Make sure the timing chain is positioned between the damper and slipper.

d. Turn the camshaft timing gear counterclockwise to take the slack out of the chain.

e. Tie the timing chain with a cord and make sure it doesn't come loose.

26. Install the timing chain case as follows:

a. Clean the gasket surface for the timing chain case.

b. Install two new gaskets over the dowels.

c. Slide on the chain case over the dowels.

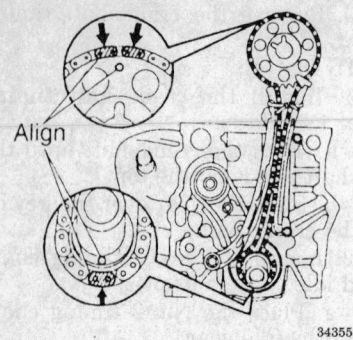

Timing chain alignment marks — 2.4L (VIN A and T) engine

d. Install the bolts and nuts and torque the bolts as follows:
- A to 14 ft. lbs. (20 Nm)
- B to 21 ft. lbs. (28 Nm)
- C to 32 ft. lbs. (44 Nm)

e. Install and torque the three chain case bolts (rear) to 13 ft. lbs. (18 Nm).

27. Using a new O-ring, Install the oil filter bracket with the oil filter. Torque the three bolts to 14 ft. lbs. (20 Nm).

28. Install the baffle plate and the No. 2 oil dipstick guide. Torque the three baffle plate nuts to 43 inch lbs. (5 Nm) and the two No. 2 oil dipstick guide bolts to 13 ft. lbs. (18 Nm).

29. Install the crankcase to the engine. Torque the 16 bolts and two nuts to 9 ft. lbs. (12 Nm).

30. On VIN A engines, install the crankshaft position sensor by installing the two bolts. Torque the bolts to 74 inch lbs. (8.5 Nm).

31. Install the ventilation case with a new gasket. Torque the three bolts to 69 inch lbs. (7.5 Nm).

32. Install the No. 1 oil dipstick guide with a new gasket. Torque the nut to 22 ft. lbs. (29 Nm).

33. Install the No. 2 engine hanger with the four bolts. Torque the bolts to the cylinder head to 27 ft. lbs. (37 Nm) and the bolts to the ventilation side to 69 inch lbs. (7.5 Nm).

34. Install the engine oil dipstick.

35. Install the oil pressure switch as follows:

a. Clean the threads of adhesive and foreign material.

b. Apply adhesive to two or three threads of the switch end.

c. Install the oil pressure switch. Torque the switch to 11 ft. lbs. (15 Nm).

36. Install the left hand engine mounting and stay. Torque the bolts for the mounting to 30 ft. lbs. (41 Nm) and the stay bolts to 27 ft. lbs. (37 Nm).

37. Install the crankshaft pulley as follows:

a. Install the crankshaft pulley to the crankshaft with the spline teeth of the crankshaft pulley engaged with the large teeth of the oil pump.

b. Rotate the crankshaft pulley to the left and right and check that the key groove of the crankshaft pulley correctly fits the crankshaft key.

c. Install the crankshaft pulley bolt.

d. Using SST 09213–58012 and 09330–00021, or equivalent, torque the bolt to 192 ft. lbs. (260 Nm).

38. Remove the cord from the timing chain.

39. Install the cylinder head.

40. Connect the engine and transaxle.

41. Install the engine and transaxle to the vehicle.

Timing Belt Front Cover

REMOVAL AND INSTALLATION

2.0L (VIN P) Engine

1. Disconnect the negative battery cable.

— CAUTION —

Wait 90 seconds from the time the key is turned to LOCK and the negative battery cable is disconnected to begin work. This allows the SRS capacitor to discharge and prevent deployment of the air bag(s).

2. Disconnect the power steering reservoir tank and remove the reservoir bracket.

3. Disconnect the wire harness bracket for the DLC1.

4. Remove the alternator and alternator bracket.

5. If equipped with ABS brakes, remove the ABS actuator.

6. Remove the right front wheel and the fender apron seal.

7. Remove the power steering drive belt.

8. Slightly jack up the engine using a block of wood under the oil pan to prevent damage.

9. Remove the four bolts, two nuts, and right hand mounting bracket.

10. Using SST 09213–54015 or equivalent, remove the crankshaft pulley bolt and remove it by pulling it straight off the crankshaft.

11. Remove the No. 2 timing belt cover.

12. Remove the No. 1 timing belt cover.

To install:

13. Install the lower (No. 1) timing belt cover and new gasket with the four bolts.

14. Install the upper (No. 2) timing cover with a new gasket(s).

15. Align the crankshaft pulley set key with the pulley key groove. Install the pulley. Tighten the pulley bolt to 80 ft. lbs. (108 Nm).

16. Install the right hand mounting insulator as follows:

a. Attach the mounting insulator to the body and mounting bracket with the four bolts and two nuts.

b. Tighten the three bolts to hold the mounting insulator to the body. Torque the bolts to 47 ft. lbs. (64 Nm).

c. Tighten the two nuts and bolt to hold the mounting insulator to the mounting bracket. Torque the bolt to 27 ft. lbs. (37 Nm) and the nut to 38 ft. lbs. (52 Nm).

17. Install and adjust the power steering pump drive belt.

18. Install the right hand engine under cover.

19. Install the right front wheel.

20. Lower the engine.

21. If equipped, install the ABS actuator.

22. Install the alternator and alternator bracket.

23. Install the wire harness bracket for the DLC 1.

24. Install the power steering reservoir bracket and reservoir.

25. Connect the negative battery cable.

26. Start the engine and check the timing.

Timing Belt

REMOVAL AND INSTALLATION

2.0L (VIN P) Engine

The timing belt is not adjustable.

1. Disconnect the negative battery cable.

———— CAUTION ————

To avoid air bag deployment, if equipped, work must be started after approximately 90 seconds or longer from the time the ignition switch is turned to the LOCK position and the negative (-) battery cable is disconnected from the battery.

2. Disconnect the power steering reservoir tank and remove the reservoir bracket.

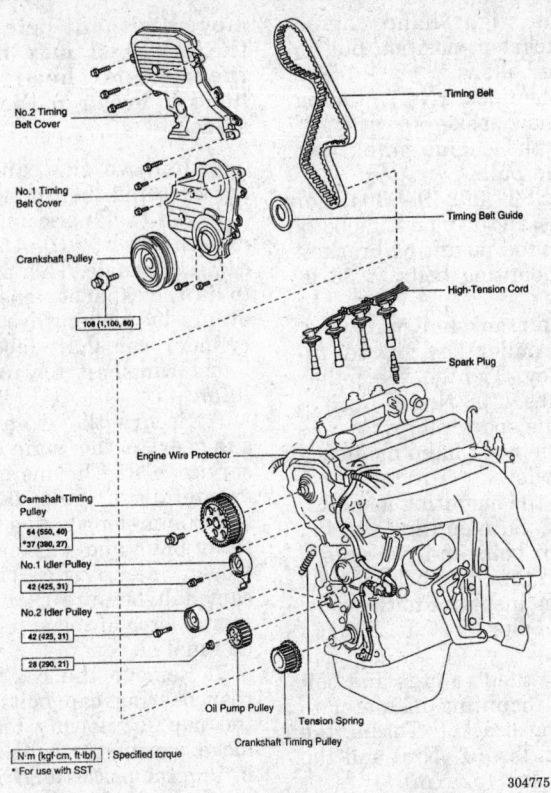

N·m (kgf·cm, ft·lbf) : Specified torque
* For use with SST

304775

Timing belt components exploded view — 2.0L (VIN P) engine

3. Disconnect the wire harness bracket for the DLC1.

4. Remove the alternator and alternator bracket.

5. If equipped with ABS brakes, remove the ABS actuator.

6. Remove the right front wheel and the fender apron seal.

7. Remove the power steering drive belt.

8. Slightly jack up the engine using a block of wood under the oil pan to prevent damage.

9. Remove the four bolts, two nuts, and right hand mounting bracket.

10. Remove the spark plugs.

11. Using SST 09213–54015 or equivalent, remove the crankshaft pulley bolt and remove it by pulling it straight off the crankshaft.

12. Using SST 09249–63010 or equivalent, remove the retaining bolts and remove the right engine mounting bracket.

13. Remove the upper (No. 2) timing belt cover.

14. Install the crankshaft pulley to the crankshaft and temporarily install the crankshaft bolt.

15. Turn the crankshaft pulley and align its groove with the timing mark **0** of the No. 1 timing belt cover. Check that the hole of the camshaft timing pulley is aligned with the timing mark of the bearing cap. If not,

turn the crankshaft 360° and align the marks.

16. Remove the timing belt from the camshaft timing pulley.

NOTE: If the timing belt is to be reused, matchmark the timing belt to the timing pulleys and timing belt covers so the belt can be reinstalled in its original position. Also, be sure to mark an arrow on the belt to indicate which direction it was turning.

17. Hold the camshaft sprocket with a spanner wrench and remove the mounting bolt. Remove the camshaft pulley.

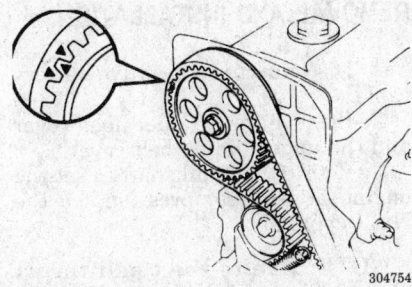

304754

Matchmarking the timing belt to the camshaft sprocket — 2.0L (VIN P) engine

18. Remove the crankshaft pulley bolt and remove the crankshaft pulley.

19. Remove the No. 1 timing belt cover.

20. Remove the timing belt guide and the timing belt.

21. Remove the No. 1 idler pulley and tension spring.

22. Remove the No. 2 idler pulley.

23. Remove the crankshaft timing pulley.

24. Support the oil pump sprocket with a spanner wrench and remove the mounting bolt and remove the sprocket.

To install:

25. Install the oil pump pulley. Torque the nut to 18 ft. lbs. (24 Nm).

26. Install the crankshaft timing pulley. Align the pulley set key with the key groove of the pulley. Slide on the pulley facing the flange side inward.

27. Install the No. 2 idler pulley and tighten the mounting bolt to 31 ft. lbs. (42 Nm). Make sure that the pulley moves smoothly.

28. Install the No. 1 idler pulley with the bolt and the tension spring. Pry the pulley toward the left as far as it will go and tighten the bolt. Make sure that the pulley moves smoothly.

29. Temporarily install the timing belt. Using the crankshaft pulley bolt, turn the crankshaft and position the key groove of the crankshaft timing pulley upward. If re–using the timing belt, align the points marked during removal.

30. Install the timing belt on the crankshaft timing pulley, oil pump pulley, No. 1 idler pulley, water pump pulley and the No. 2 idler pulley.

31. Install the timing belt guide.

NOTE: If the old timing belt is being reinstalled, make sure the directional arrow is facing in the original direction and that the belt and sprocket/cover matchmarks are properly aligned.

32. Install the lower (No. 1) timing belt cover and new gasket with the four bolts.

33. Align the crankshaft pulley set key with the pulley key groove. Temporarily install the crankshaft pulley and bolt.

34. Align the camshaft knock pin with the groove of the pulley, and slide the timing pulley onto the camshaft with the plate washer and set bolt.

35. Tighten the pulley set bolt to 40 ft. lbs. (54 Nm).

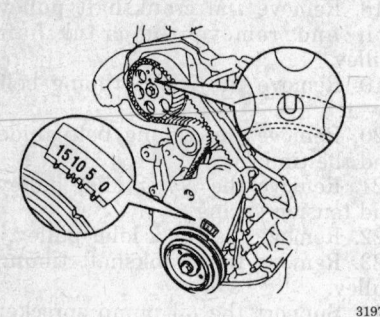

319788

Aligning the camshaft timing mark — 2.0L (VIN P) engine

36. Turn the crankshaft pulley and align the **0** mark on the lower (No. 1) timing belt cover.

37. Finish installing the timing belt and check the valve timing as follows:

 a. If re–using the old timing belt, align the matchmarks that you made previously, and install the timing belt onto the camshaft pulley.

 b. Align the marks on the timing belt with the marks on the camshaft pulley.

 c. Loosen the No. 1 idler pulley set bolt ½ turn.

 d. Turn the crankshaft pulley two complete revolutions TDC to TDC. ALWAYS turn the crankshaft CLOCKWISE. Check that the pulleys are still in alignment with the timing marks.

 e. If the No. 1 idler pulley uses a green tension spring, slowly turn the crankshaft pulley 1⅞ revolutions, and align its groove with the mark at 45° BTDC (for the No. 1 cylinder) of the No. 1 timing belt cover.

 f. Tighten the No. 1 idler pulley set bolt to 31 ft. lbs. (42 Nm).

 g. Make sure there is belt tension between the crankshaft and camshaft timing pulleys.

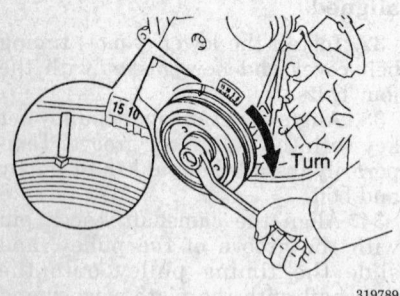

319789

Aligning the crankshaft timing mark — 2.0L (VIN P) engine

38. Place the right hand engine mounting bracket in position but do not install the bolts.

39. Install the upper (No. 2) timing cover with a new gasket(s).

40. Remove the engine crankshaft pulley bolt and pulley.

41. Using SST 09249–63010 or equivalent, install the mounting bolts for the right hand mounting bracket. Torque the mounting bolts to 38 ft. lbs. (52 Nm).

42. Align the crankshaft pulley set key with the pulley key groove. Install the pulley. Tighten the pulley bolt to 80 ft. lbs. (108 Nm).

43. Install the spark plugs.

44. Install the right hand mounting insulator as follows:

 a. Attach the mounting insulator to the body and mounting bracket with the four bolts and two nuts.

 b. Tighten the three bolts to hold the mounting insulator to the body. Torque the bolts to 47 ft. lbs. (64 Nm).

 c. Tighten the two nuts and bolt to hold the mounting insulator to the mounting bracket. Torque the bolt to 27 ft. lbs. (37 Nm) and the nut to 38 ft. lbs. (52 Nm).

45. Install and adjust the power steering pump drive belt.

46. Install the right hand engine under cover.

47. Install the right front wheel.

48. Lower the engine.

49. If equipped, install the ABS actuator.

50. Install the alternator and alternator bracket.

51. Install the wire harness bracket for the DLC 1.

52. Install the power steering reservoir bracket and reservoir.

53. Connect the negative battery cable.

54. Start the engine and check the timing.

Camshaft

REMOVAL AND INSTALLATION

1. Disconnect the negative battery cable.

2. Remove the cylinder head cover and the upper timing belt cover.

3. Rotate the crankshaft to set the engine at TDC/compression for the No. 1 cylinder.

NOTE: Due to the small thrust clearance on both the intake and exhaust camshafts, the camshafts must be kept level during removal. If the camshafts are re-

moved without being kept level, the camshaft may be caught in the cylinder head causing the head to break or the camshaft to seize.

4. Remove the camshaft timing sprocket and the timing belt.

5. Set the knock pin of the intake camshaft at 10–45° BTDC of camshaft angle. This angle will help to lift the exhaust camshaft level and evenly by pushing No. 2 and No. 4 cylinder camshaft lobes of the exhaust camshaft toward their valve lifters.

6. Secure the exhaust camshaft sub-gear to the main gear using a service bolt. The manufacturer recommends a bolt 0.63–0.79 in. (16–20mm) long with a thread diameter of 6mm and a 1mm thread pitch. When removing the exhaust camshaft be sure that the torsional spring force of the sub-gear has been eliminated.

7. Remove the No. 1 and No. 2 rear bearing cap bolts and remove the cap. Uniformly loosen and remove bearing cap bolts No. 3 to No. 8 in several passes and in the proper sequence. Do not remove bearing cap bolts No. 9 and 10 at this time. Remove the No. 1, 2, and 4 bearing caps.

8. Alternately loosen and remove bearing cap bolts No. 9 and 10. As these bolts are loosened check to see that the camshaft is being lifted out straight and level.

NOTE: If the camshaft is not lifting out straight and level retighten No. 9 and 10 bearing cap bolts. Reverse the order of steps 5 through 7 and reset the intake camshaft knock pin to 10–45 degrees BTDC and repeat steps 5 through 7 again. Do not attempt to pry the camshaft from its mounting.

9. Remove the No. 3 bearing cap and exhaust camshaft from the engine.

10. Set the knock pin of the intake camshaft at 80–115° BTDC of camshaft angle. This angle will help to lift the intake camshaft level and evenly by pushing No. 1 and No. 3 cylinder camshaft lobes of the intake camshaft toward their valve lifters.

11. Remove the No. 1 and No. 2 front bearing cap bolts and remove the front bearing cap and oil seal. If the cap will not come apart easily, leave it in place without the bolts.

12. Uniformly loosen and remove bearing cap bolts No. 3 to No. 8 in several phases and in the proper sequence. Do not remove bearing cap

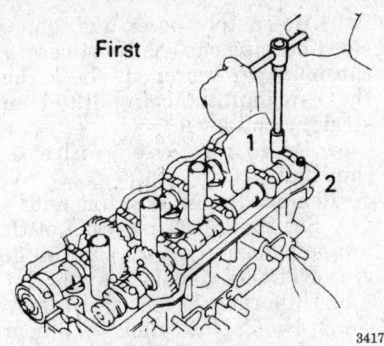

Exhaust camshaft bolt removal - step 1 — 2.0L (VIN P) and 2.4L (VIN A and T) Engine

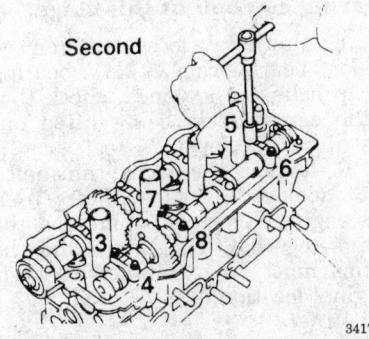

Exhaust camshaft bolt removal - step 2 — 2.0L (VIN P) and 2.4L (VIN A and T) Engine

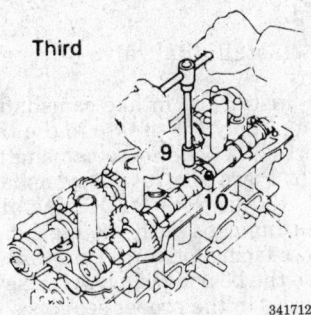

Exhaust camshaft bolt removal - step 3 — 2.0L (VIN P) and 2.4L (VIN A and T) Engine

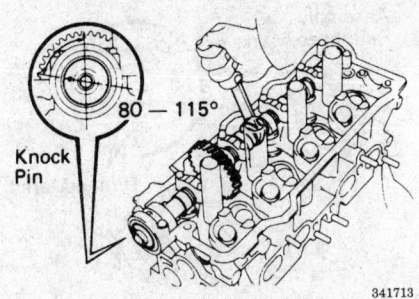

Intake camshaft knock pin alignment — 2.0L (VIN P) and 2.4L (VIN A and T) Engine

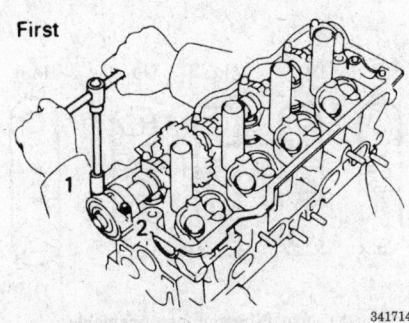

Intake camshaft bolt removal - step 1 — 2.0L (VIN P) and 2.4L (VIN A and T) Engine

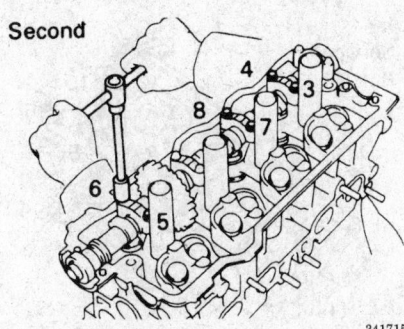

Intake camshaft bolt removal - step 2 — 2.0L (VIN P) and 2.4L (VIN A and T) Engine

bolts No. 9 and 10 at this time. Remove No. 1, 3, and 4 bearing caps.

13. Alternately loosen and remove bearing cap bolts No. 9 and 10. As these bolts are loosened and after breaking the adhesion on the front bearing cap, check to see that the camshaft is being lifted out straight and level.

NOTE: If the camshaft is not lifting out straight and level retighten No. 9 and 10 bearing cap bolts. Reverse steps 10 through

12, than start over from step 10. Do not attempt to pry the camshaft from its mounting.

14. Remove the No. 2 bearing cap with the intake camshaft from the engine.

15. Remove the valve adjusting shims from the engine. Make sure to replace the shims to their original location.

To install:

16. Install the valve adjusting shims to the engine.

17. Before installing the intake camshaft, apply multi-purpose

grease to the thrust portion of the camshaft.

18. Position the camshaft at 80–115° BTDC of camshaft angle on the cylinder head.

19. Apply sealant to the front bearing cap.

20. Coat the bearing cap bolts with clean engine oil.

21. Tighten the camshaft bearing caps evenly and in several passes to 14 ft. lbs. (19 Nm) in the proper sequence.

22. Set the knock pin of the camshaft at 10–45° BTDC of camshaft angle.

23. Apply multipurpose grease to the thrust portion of the camshaft.

24. Position the exhaust camshaft gear with the intake camshaft gear so that the timing marks are in alignment with one another. Be sure to use the proper alignment marks on the gears. Do not use the assembly reference marks.

25. Turn the intake camshaft clockwise or counterclockwise little by little until the exhaust camshaft sits in the bearing journals evenly without rocking the camshaft on the bearing journals.

26. Coat the bearing cap bolts with clean engine oil.

27. Tighten the camshaft bearing caps evenly and in several passes to 14 ft. lbs. (19 Nm). Remove the service bolt from the assembly.

28. Install the camshaft timing pulleys and the timing belt.

29. Adjust the valve clearance.

30. Install the head cover and the upper timing cover. Reconnect the negative battery cable.

31. Start the engine and check for leaks.

32. Check and adjust the ignition timing.

33. Disconnect the negative battery cable from the battery.

34. Relieve the fuel system pressure.

— **CAUTION** —
Fuel injection systems remain under pressure after the engine has been turned OFF. Properly relieve fuel pressure before disconnecting any fuel lines. Failure to do so may result in fire or personal injury.

35. Remove the engine/transaxle assembly from the vehicle.

36. Remove the engine wiring from the engine and move it aside.

37. Remove the No. 2 valve cover.

38. Mark the spark plug wires and disconnect. Matchmark the distributor, rotor and cylinder head. Remove

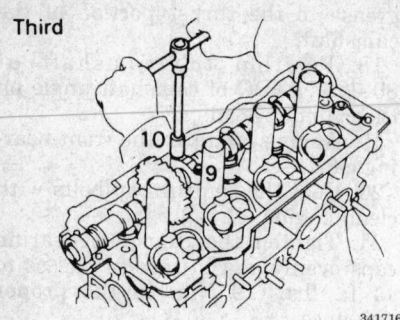

Intake camshaft bolt removal - step 3 — 2.0L (VIN P) and 2.4L (VIN A and T) Engine

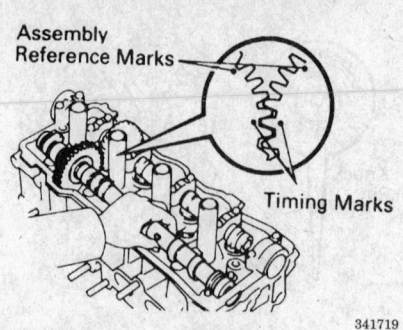

Camshaft timing mark alignment — 2.0L (VIN P) and 2.4L (VIN A and T) Engine

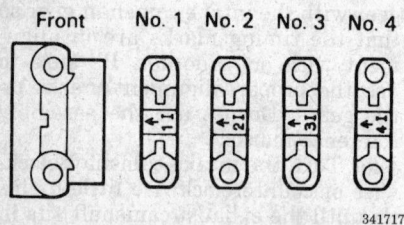

Intake camshaft bearing cap positioning — 2.0L (VIN P) and 2.4L (VIN A and T) Engine

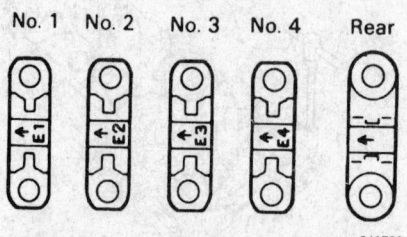

Exhaust camshaft bearing cap positioning — 2.0L (VIN P) and 2.4L (VIN A and T) Engine

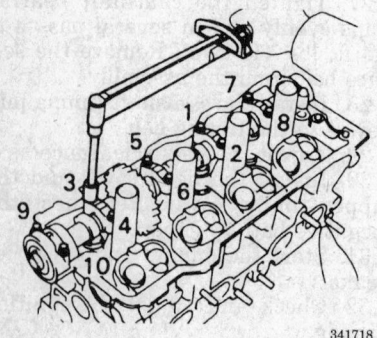

Intake camshaft bolt tightening sequence — 2.0L (VIN P) and 2.4L (VIN A and T) Engine

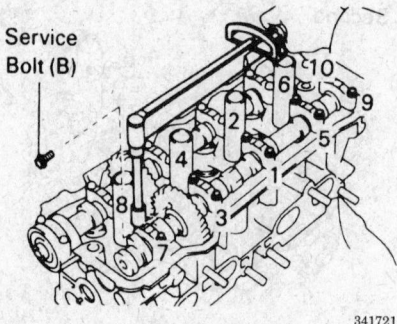

Exhaust camshaft bolt tightening sequence — 2.0L (VIN P) and 2.4L (VIN A and T) Engine

the distributor and disconnect the wiring.

39. Remove the PCV hose.

40. Remove the No. 1 valve cover and two half circular plugs.

41. Remove the chain tensioner and gasket.

42. Place matchmarks on the timing sprocket and chain. Hold the camshaft with a wrench and remove the sprocket bolt.

43. Remove the No. 6 camshaft bearing cap.

44. Remove exhaust camshaft:

a. Set the knock pin hole of the exhaust camshaft at the 5–30 de-

gree BTDC of camshaft angle. Hint: The above angle helps to lift the exhaust camshaft level and evenly by pushing No. 2 and No. 4 cylinder cam lobes of the exhaust camshaft to their valve lifters.

b. Secure the exhaust camshaft sub-gear to main gear with a service bolt.

c. Uniformly loosen and remove No. 1, No. 2, No. 3 and No. 5 bearing caps in several passes in the proper sequence.

NOTE: Do not remove No. 4 bearing cap bolt at this stage.

d. Alternately loosen and remove No. 4 bearing cap. As No. 4 bearing cap bolts are loosened, check that the camshaft is being lifted out straight and level.

e. Remove the exhaust camshaft.

45. Remove the intake camshaft:

a. Set the knock pin hole of the intake camshaft at the 75–100 degree BTDC of camshaft angle.

b. Uniformly loosen and remove No. 1, No. 2, No. 4 and No. 5 bearing caps in several passes in the proper sequence.

NOTE: Do not remove No. 3 bearing cap bolt at this stage.

c. Alternately loosen and remove No. 3 bearing cap. As No. 3 bearing cap bolts are loosened, check that the camshaft is being lifted out straight and level.

d. Remove the intake camshaft.

46. Remove the valve lifters from the engine. Keep the shims and the lifters together. The lifters and the shims must be reinstalled in their original location.

To install:

NOTE: If any of the bolts break, deform or do not meet the torque specification replace them.

47. Install the valve lifters and shims.

48. Install the intake camshaft:

a. Apply MP grease to the thrust portion of the intake camshaft.

b. Place the intake camshaft at 75–100 degrees BTDC. Install the bearing caps with the marking arrows facing forward. Uniformly torque the bearing cap bolts in several passes in the proper sequence to 12 ft. lbs. (16 Nm).

49. Install the exhaust camshaft:

a. Set the knock pin of the intake camshaft at 5–30 degrees BTDC of camshaft angle.

b. Apply MP grease to the thrust portion of the exhaust camshaft.

c. Engage the exhaust camshaft gear to the intake camshaft gear by matching the installation marks (timing marks).

d. Roll down the exhaust camshaft onto the bearing journals while engaging the gears with each other. Make sure the exhaust and intake camshaft gear alignment marks are facing each other. The one gear has 2 dots and the other has 1 dot.

e. Install the bearing caps with the marking arrows facing forward.

f. Uniformly torque the bearing cap bolts in several passes in the

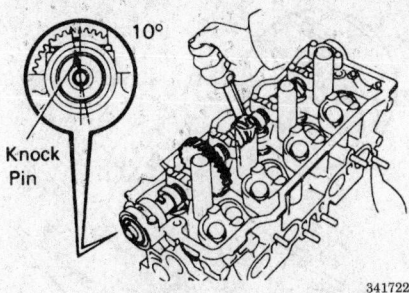

Exhaust camshaft knock pin alignment — 2.0L (VIN P) and 2.4L (VIN A and T) Engine

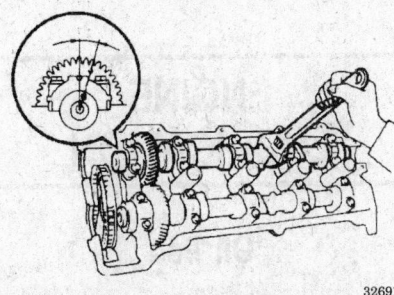

Set knock pin hole of exhaust camshaft at 5-30 degrees BTDC of camshaft angle — 2.0L (VIN P) and 2.4L (VIN A and T) Engine

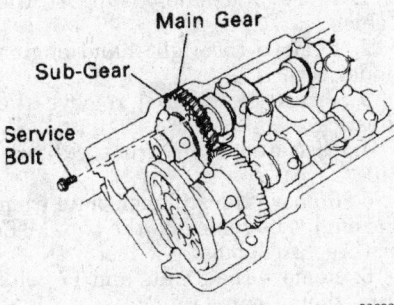

Secure exhaust camshaft sub-gear to main gear with service bolt — 2.0L (VIN P) and 2.4L (VIN A and T) Engine

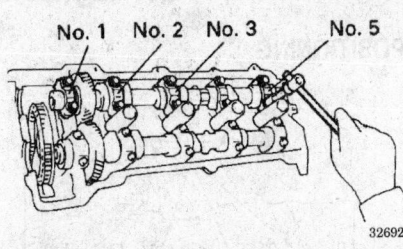

Remove No. 1, 2, 3 and 5 bearing caps (exhaust camshaft) in sequence shown — 2.0L (VIN P) and 2.4L (VIN A and T) Engine

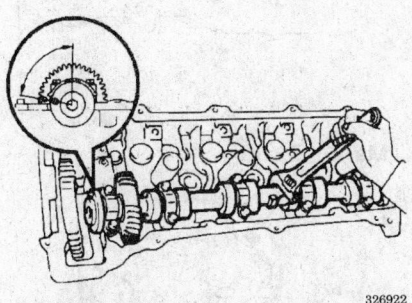

Set knock pin of intake camshaft at 75-100 degrees BTDC of camshaft angle — 2.0L (VIN P) and 2.4L (VIN A and T) Engine

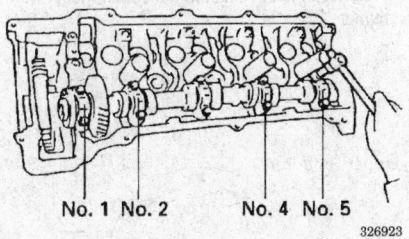

Remove No. 1, 2, 4 and 5 bearing caps (intake camshaft) in sequence shown — 2.0L (VIN P) and 2.4L (VIN A and T) Engine

proper sequence to 12 ft. lbs. (16 Nm).

50. Apply sealer to the bottom of No. 6 bearing cap and install. Torque the cap to 12 ft. lbs. (16 Nm).

51. Install the camshaft sprocket and chain. Torque the bolt to 54 ft. lbs. (74 Nm).

52. Release the chain tensioner ratchet pawl. Fully push in the plunger and apply the hook to the pin so the plunger can not spring out and install the tensioner. Torque the bolts to 15 ft. lbs. (21 Nm).

53. Set the tensioner: Turn the crankshaft to the left so the hook of the tensioner is released from the pin. If it does not spring out, press the slipper into the tensioner to release the hook.

54. Adjust the valve clearance.

55. Apply sealant (P/N 08826–00080 or equivalent) to the cylinder head. Install the 2 half-circular plugs to the cylinder head and install the valve cover. Torque the bolts to 69 inch lbs. (7.8 Nm).

56. Install the PCV hose.

57. Install the distributor.

58. Install the No. 2 cylinder head cover.

59. Install the engine wire and connect all connectors.

60. Install the engine/transaxle assembly into the vehicle.

61. Fill the cooling system. Fill the engine with oil.

62. Connect the battery cable. Start the engine and check for leaks.

63. Road test the vehicle for proper operation. Recheck all fluid levels.

Piston and Connecting Rod

POSITIONING

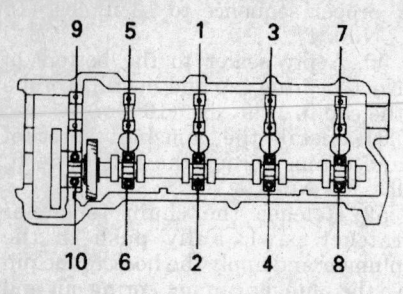

Camshaft bearing cap bolts tightening sequence — 2.0L (VIN P) and 2.4L (VIN A and T) Engines

326924

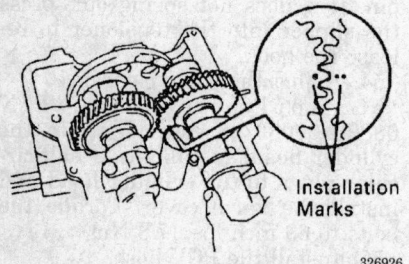

Installation Marks

326926

Engage the exhaust and intake camshafts by matching the installation marks to each gear — 2.0L (VIN P) and 2.4L (VIN A and T) Engine

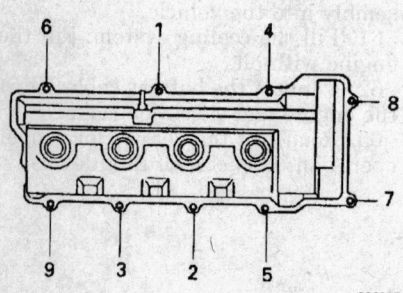

326925

Cylinder head cover tightening sequence — 2.0L (VIN P) and 2.4L (VIN A and T) Engine

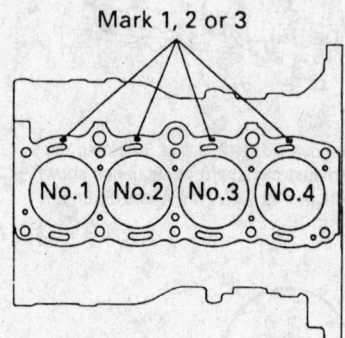

Mark 1, 2 or 3

No.1 No.2 No.3 No.4

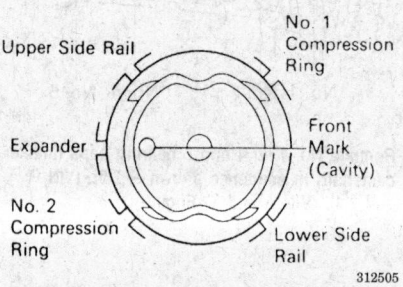

Mark 1, 2 or 3

312506

Piston diameter measurement identification marks

Upper Side Rail

No. 1 Compression Ring

Expander

Front Mark (Cavity)

No. 2 Compression Ring

Lower Side Rail

312505

Piston ring end gap positioning

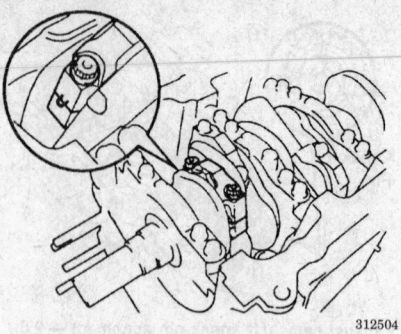

312504

Connecting rod and cap matchmarks

ENGINE LUBRICATION

Oil Pan

REMOVAL AND INSTALLATION

2.0L (VIN P) Engine

1. Raise and safely support the vehicle.

2. Remove the right hand engine under cover.

3. Drain the oil and remove the dipstick.

4. Disconnect the front exhaust pipe.

5. Remove the stiffener plate from the engine by removing the two (M/T) or three (A/T) bolts.

6. Remove the 2 nuts and 17 bolts from the oil pan.

7. Separate the oil pan from the cylinder block.

8. Remove the oil pan from the vehicle.

To install:

9. Clean all gasket surfaces completely.

10. Apply a thin bead of sealer to the oil pan mounting surfaces.

NOTE: Avoid applying too much sealant to the oil pan.

11. Place the oil pan against the block and install the bolts and nuts. Torque the nuts and bolts to 4 ft. lbs. (5.4 Nm)

12. Install the stiffener plate and torque the mounting bolts to 27 ft. lbs. (37 Nm).

13. Install the front exhaust pipe.

14. Fill the engine with oil to the proper level.

15. Lower the vehicle.

16. Start the engine and check for leaks. Recheck the engine oil level.

17. Install the right engine cover.

2.4L (VIN A and T) Engine

NOTE: This engine has 2 oil pans. If the crankshaft is going to be serviced, the side crankcase pan has to be removed. If the oil pump sump is going to be serviced, the bottom oil pan has to be removed.

1. Disconnect the negative battery cable.

2. Drain the engine oil.

3. Remove the oil level sensor and gasket. Be careful not to drop the sensor when removing.

4. Remove the 14 bolts and 2 nuts. Carefully pry the pan from the engine, being careful not to damage the flange.

To install:

5. Before installing, thoroughly clean the gasket mating surfaces. Apply gasket sealer No. 0882600080 or equivalent, to the pan and assembly within 5 minutes.

6. Install the pan and torque the bolts and nuts to 48 inch lbs. (5.4 Nm).

7. Install the gasket, oil sensor and torque to 9 ft. lbs. (13 Nm).

8. Install the remaining components.

9. Refill the engine with oil.

10. Connect the negative battery cable. Start the engine and check for leaks.

Oil Pump

REMOVAL AND INSTALLATION

2.0L (VIN P) Engine

1. Disconnect the negative battery cable.

──────── **CAUTION** ────────
Wait 90 seconds from the time the key is turned to LOCK and the negative battery cable is disconnected to begin work. This allows the SRS capacitor to discharge and prevent deployment of the air bag(s).

2. Remove the hood.

3. Raise and safely support the vehicle.

4. Remove the right hand engine under cover.

5. Drain the engine oil.

6. Remove the front exhaust pipe.

7. Remove the rear end stiffener plate.

8. Remove the oil dipstick.

9. Remove the 17 bolts and 2 nuts from the oil pan.

10. Insert the blade of the SST 09032–00100 tool between the oil pan and the cylinder block, and cut off the applied sealer and remove the oil pan.

NOTE: Do not use the tool for the oil pump body side and rear oil seal retainer.

11. Remove the bolts, nuts, oil strainer and gasket.

12. Carefully suspend the engine with a sling device or equivalent..

13. Remove the timing belt.

14. Remove the No. 2 idler pulley and crankshaft timing pulley.

15. Using SST 09960–10010 or equivalent, remove the nut and pulley to the oil pump.

16. Remove the crankshaft position sensor.

17. Remove the 12 mounting bolts, the oil pump, and the gasket.

To install:

18. Install a new gasket and the oil pump with the 12 bolts. Torque the bolts to 82 in. lbs. (9 Nm).

NOTE: The long bolts are 1.38 in. and all the others are 0.98 in.

19. Install the crankshaft position sensor.

20. Install the oil pump pulley and install the nut. Torque the nut to 18 ft. lbs. (24 Nm).

21. Install the crankshaft timing pulley and No. 2 idler pulley.

22. Install the timing belt.

23. Remove the engine sling.

24. Install the oil strainer with a new gasket, bolt, and nuts. Torque to 48 in. lbs. (5 Nm).

25. Remove any old sealant from the oil pan flange and thoroughly clean both sealing surfaces.

26. Apply a 3–5mm bead of sealant to the oil pan flange.

NOTE: The pan must be installed within five minutes of sealant application or the procedure will have to be repeated.

27. Install the oil pan with the 17 bolts and 2 nuts. Uniformly tighten the bolts and nuts in several passes. Torque the bolts and nuts to 48 in. lbs. (5 Nm) and install the dipstick.

28. Install the rear end stiffener plate and torque the bolts to 27 ft. lbs. (37 Nm).

29. Install the front exhaust pipe.

30. Lower the vehicle and fill the engine with oil.

Oil Level Sensor
◆ Gasket
Oil Pan
Drain Plug
◆ Gasket
× 14
Oil Strainer
★ Engine Coolant Drain Cock
◆ Gasket
◆ Non-reusable part
★ Precoated part
Knock Sensor

287486

Oil pan component assembly — 2.4L (VIN A and T) Engine

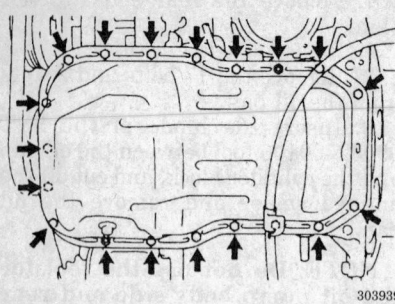

The oil pan mounting bolts and nuts — 2.0L (VIN P) Engine

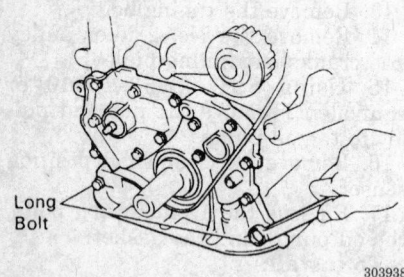

Oil pump mounting bolts — 2.0L (VIN P) Engine

— **WARNING** —

Be sure to prime the oil pump prior to initial engine start-up or engine damage may occur because of low oil pressure.

31. Connect the negative battery cable, start the engine, and check for leaks.

32. Recheck the engine oil level and install the hood.

33. Install the right hand engine undercover.

2.4L (VIN A and T) Engine

— **WARNING** —

To avoid possible air bag deployment, do not start working on the vehicle until 90 seconds has elapse form the time the ignition switch is turned OFF and the negative battery terminal is disconnected.

1. Disconnect the negative battery cable.

2. Disconnect the equipment driveshaft.

3. Remove the crankshaft pulley.

4. Remove the oil pump cover screws and cover. Remove the O-ring.

5. Remove the timing chain case as follows:

a. Remove the three bolts from the rear of the timing chain cover.

b. Remove the 12 bolts and two nuts from the front of the chain cover.

c. Using a plastic faced hammer, tap the chain case and remove the timing chain case and two gaskets.

To install:

6. Install the timing chain case as follows:

a. Clean the gasket surface for the timing chain case.

b. Install two new gaskets over the dowels.

c. Slide on the chain case over the dowels.

d. Install the bolts and nuts and torque the bolts as follows:

- A to 14 ft. lbs. (20 Nm)
- B to 21 ft. lbs. (28 Nm)
- C to 32 ft. lbs. (44 Nm)

e. Install and torque the three chain case bolts (rear) to 13 ft. lbs. (18 Nm).

7. Place a new O-ring into the groove of the timing chain case.

8. Install the oil pump cover. Torque the screws to 8 ft. lbs (10 Nm).

9. Install the crankshaft pulley and equipment driveshaft.

10. Connect the negative battery cable. Road test the vehicle for proper operation.

TRANSAXLE

Manual Tansaxle Assembly

REMOVAL AND INSTALLATION

Previa

1. Disconnect the negative battery cable and drain the transaxle fluid.

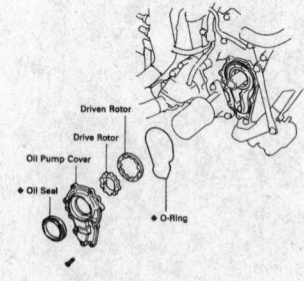

Oil pump components — 2.4L (VIN A and T) Engine

2. Raise the vehicle and support safely.

3. Remove the starter motor. Matchmark the driveshafts-to-flange and remove the front (4WD) and rear driveshafts.

4. Remove the clutch release cylinder, hose and bracket.

5. Remove the exhaust pipe and bracket.

6. Disconnect the control cables/bracket and speed sensor connector.

7. Remove the engine-to-transaxle stiffener plate.

8. Place a suitable transaxle jack under the transaxle.

9. Remove the engine rear mounting bolts and raise the rear side of the engine.

10. Remove the engine-to-transaxle bolts, pull the transaxle toward the rear and remove.

To install:

11. Align the input shaft with the clutch disc and push the transaxle fully into position.

12. Install the transaxle bolts and torque to 53 ft. lbs. (72 Nm).

13. Install the rear engine mounts and stiffener plate. Torque the bolt to 27 ft. lbs. (37 Nm).

14. Connect the speed sensor and control cables.

15. Install the exhaust pipe and torque the bracket to 37 ft. lbs. (51 Nm).

16. Install the clutch release cylinder, starter and driveshafts. Torque the starter to 41 ft. lbs. (56 Nm) and driveshaft bolts to 20 ft. lbs. (25 Nm).

17. Lower the vehicle.

18. Connect the battery cable and refill with transaxle fluid.

RAV4

2-Wheel Drive

1. Disconnect the negative battery cable.

2. Remove the air cleaner case assembly with hose.

3. Remove the engine coolant reservoir tank.

4. Remove the set nut of the engine wire clamp.

5. Remove the starter by removing the electrical connectors and the two set bolts.

6. Remove the clutch release cylinder as follows:

a. Remove the set bolts holding the clutch line bracket to the transaxle.

b. Remove the two bolts, release cylinder and line.

7. Disconnect the ground cable from the transaxle by removing the bolt.

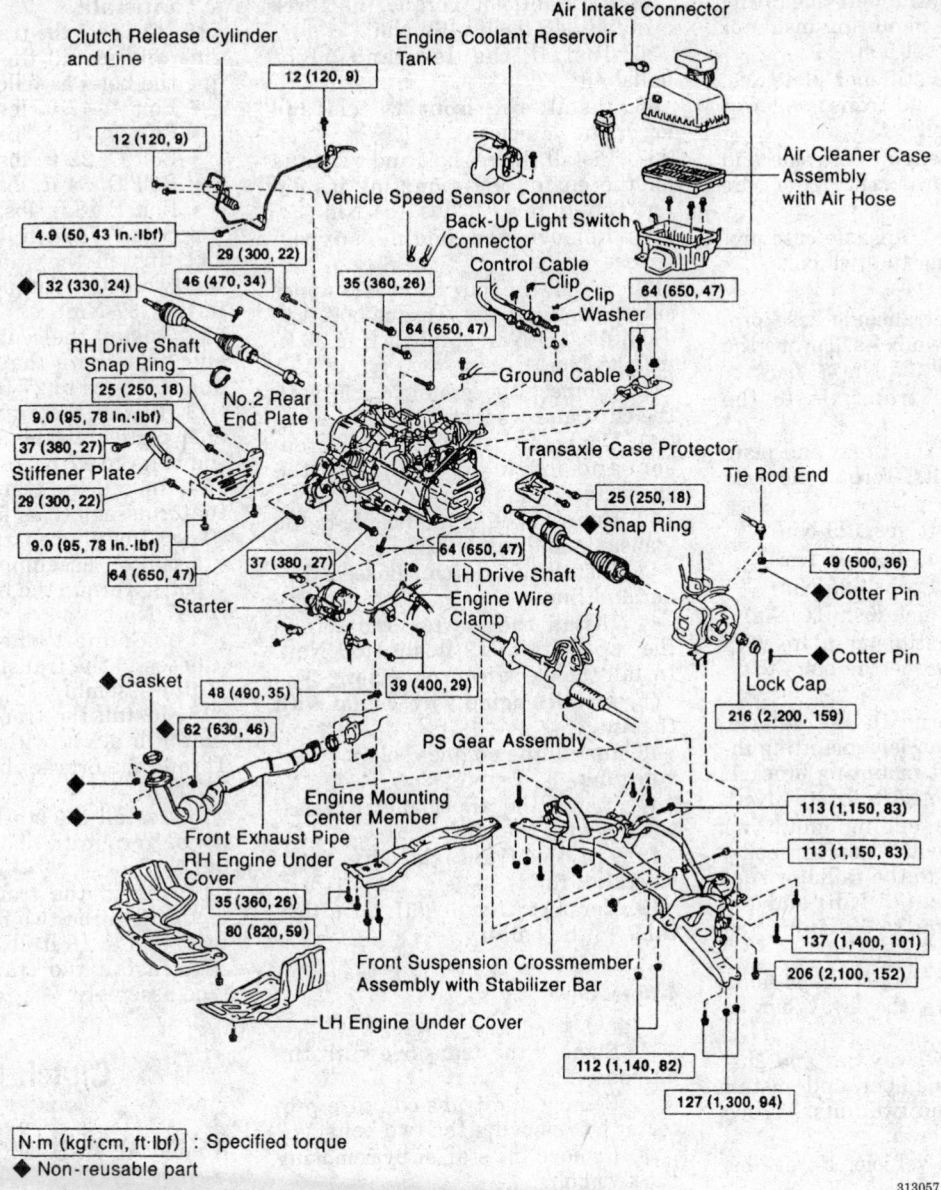

Air Intake Connector

Clutch Release Cylinder and Line

Engine Coolant Reservoir Tank

12 (120, 9)

12 (120, 9)

Air Cleaner Case Assembly with Air Hose

Vehicle Speed Sensor Connector

Back-Up Light Switch Connector

4.9 (50, 43 In.·lbf)

29 (300, 22)

Control Cable Clip

Clip

Washer

64 (650, 47)

46 (470, 34)

35 (360, 26)

32 (330, 24)

64 (650, 47)

Ground Cable

RH Drive Shaft Snap Ring

25 (250, 18)

No.2 Rear End Plate

Transaxle Case Protector

Tie Rod End

9.0 (95, 78 In.·lbf)

37 (380, 27)

◆ Snap Ring

25 (250, 18)

Stiffener Plate

29 (300, 22)

49 (500, 36)

◆ Cotter Pin

9.0 (95, 78 In.·lbf)

64 (650, 47)

◆ Cotter Pin

64 (650, 47)

37 (380, 27)

LH Drive Shaft

Lock Cap

Starter

Engine Wire Clamp

216 (2,200, 159)

◆ Gasket

48 (490, 35)

39 (400, 29)

◆ 62 (630, 46)

PS Gear Assembly

Engine Mounting Center Member

113 (1,150, 83)

Front Exhaust Pipe

113 (1,150, 83)

RH Engine Under Cover

35 (360, 26)

80 (820, 59)

137 (1,400, 101)

206 (2,100, 152)

Front Suspension Crossmember Assembly with Stabilizer Bar

LH Engine Under Cover

112 (1,140, 82)

127 (1,300, 94)

N·m (kgf·cm, ft·lbf) : Specified torque

◆ Non-reusable part

313057

Transaxle exploded view — 2WD RAV4

8. Disconnect the vehicle speed sensor and back up light switch connector.

9. Disconnect the control cable by removing the four clips and washers.

10. Remove the four upper side transaxle bolts connecting the transaxle to the engine.

11. Remove the bolt and two nuts holding the left mounting insulator to the vehicle.

12. Install a engine support to the engine.

13. Support rack and pinion to the engine support fixture with a rope.

14. Raise and safely support the front of the vehicle.

15. Remove the front wheels.

16. Remove the left and right hand engine undercovers.

17. Drain the transaxle oil.

18. Remove the left and right halfshafts.

19. Remove the front exhaust pipe as follows:

a. Remove the three nuts and gasket from the exhaust manifold.

b. Remove the two bolts holding the exhaust pipe to the center exhaust pipe.

c. Disconnect the exhaust pipe from the vehicle.

20. Remove the front suspension crossmember assembly with the sway bar as follows:

a. Support the front suspension crossmember with a jack.

b. Disconnect the ring from the center exhaust pipe.

c. Remove the two set bolts and nuts of the power steering rack and pinion assembly.

d. Remove the suspension crossmember assembly with the sway bar by removing the two nuts and six bolts.

21. Remove the engine mounting center member by removing the four bolts.

22. Jack up the transaxle slightly.

20-47

23. Disconnect the left mounting bracket from the mounting insulator by removing the set bolt.

24. Remove the stiffener plate, No. 2 rear end plate and transaxle lower side mounting bolt.

25. Lower the engine left side and remove the transaxle from the engine.

26. Remove the transaxle case protector by removing the two bolts.

To install:

27. Install the transaxle case protector with the two bolts. Torque the bolts to 18 ft. lbs. (25 Nm).

28. Install the transaxle to the engine.

29. Install the No. 2 rear end plate and transaxle bolts. Torque the bolts as follows:

- Bolt C to 22 ft. lbs. (29 Nm)
- Bolt D to 34 ft. lbs. (46 Nm)
- Bolt E to 18 ft. lbs. (25 Nm)
- Bolt F to 78 inch lbs. (9.0 Nm)

30. Install the stiffener plate with the two bolts. Torque the bolts to 27 ft. lbs. (37 Nm).

31. From underneath the vehicle, connect the engine left mounting insulator to the left mounting bracket. Torque the bolt to 47 ft. lbs. (64 Nm).

32. Install the engine mounting center member with the four bolts. Torque the bolts to the radiator support to 26 ft. lbs. (35 Nm) and the mounting insulator to 59 ft. lbs. (80 Nm).

33. Install the front suspension crossmember with the sway bar as follows:

a. Install the sway bar and suspension crossmember and install the six bolts and two nuts. Torque the bolts as follows:

- Bolt A to vehicle: 152 ft. lbs. (206 Nm)
- Bolt B to lower control arm bracket: (101 ft. lbs. (137 Nm)
- Bolt C to rear mounting bracket: 82 ft. lbs. (112 Nm)

b. Connect the rack and pinion to the crossmember with the two bolts and two nuts. Torque to 83 ft. lbs. (113 Nm).

c. Connect the ring for the center exhaust pipe.

34. Install the front exhaust pipe as follow:

a. Install the pipe with new gaskets.

b. Connect the front pipe to the center exhaust pipe with the two bolts. Torque the bolts to 35 ft. lbs. (48 Nm).

c. Install the three nuts to hold the front exhaust pipe to the ex-

haust manifold. Torque the three nuts to 46 ft. lbs. (62 Nm).

35. Install the left and right halfshafts.

36. Install the front wheels and lower the vehicle.

37. Install the set bolt and two nuts for the engine left mounting insulator. Torque to 47 ft. lbs. (64 Nm).

38. Remove the engine support fixture.

39. Install the four transaxle upper side mounting bolts. Torque bolt A to 47 ft. lbs. (64 Nm) and bolt B to 26 ft. lbs. (35 Nm).

40. Connect the ground cable with the clips and washers.

41. Connect the vehicle speed sensor and back up light switch connector.

42. Connect the ground cable to the transaxle and install the bolt.

43. Install the clutch release cylinder and line.

44. Install the starter and torque the two bolts to 29 ft. lbs. (39 Nm). Install the electrical connectors.

45. Install engine wire clamp with the nut..

46. Install the engine coolant reservoir tank.

47. Install the air cleaner case assembly with the air hose.

48. Fill the transaxle with fluid. Check all fluids.

49. Connect the negative battery cable to the battery.

4-Wheel Drive

1. Remove the transaxle with the engine.

2. Remove the transaxle case protector by removing the two bolts.

3. Remove the starter by removing the two bolts.

4. Remove the transfer vacuum actuator bracket by removing the four bolts.

5. Remove the transfer vacuum actuator assembly as follows:

a. Disconnect the four solenoid hoses from the transfer vacuum actuator assembly.

b. Remove the transfer vacuum actuator assembly by removing the two bolts.

6. Remove the right transfer stiffener plate by removing the five bolts.

7. Remove the center transfer stiffener plate by removing the three bolts.

8. Remove the stiffener plate by removing the two bolts.

9. Remove the transaxle from the engine by removing the nine transaxle mounting bolts.

To install:

10. Connect the transaxle to the engine and install the nine bolts. Torque the bolts as follows:

- Bolt A: 47 ft. lbs. (64 Nm)
- Bolt B: 26 ft. lbs. (35 Nm)
- Bolt C: 22 ft. lbs. (29 Nm)
- Bolt D: 34 ft. lbs. (46 Nm)
- Bolt E 18 ft. lbs. (25 Nm)
- Bolt F 78 inch lbs. (9.0 Nm)

11. Install the stiffener plate with the two bolts. Torque the bolts to 27 ft. lbs. (37 Nm).

12. Install the center transfer stiffener plate with the three bolts. Torque the bolts to 27 ft. lbs. (37 Nm).

13. Install the right transfer stiffener plate with the five bolts. Torque the bolts to 27 ft. lbs. (37 Nm).

14. Install the transfer vacuum actuator assembly as follows:

a. Install the transfer vacuum actuator assembly with the two bolts. Torque the bolts to 27 ft. lbs. 937 Nm).

b. Connect the four solenoid hoses to the transfer vacuum actuator assembly.

15. Install the transfer vacuum actuator bracket with the four bolts. Torque the bracket bolts to 27 ft. lbs. (37 Nm).

16. Install the starter with the two bolts. Torque the bolts to 29 ft. lbs. (39 Nm).

17. Install the transaxle case protector with the two bolts, Torque the two bolts to 18 ft. lbs. (25 Nm).

18. Install the transaxle with engine assembly.

Clutch Disc

REMOVAL AND INSTALLATION

——— **CAUTION** ———
To avoid personal injury and accidental deployment of the air bag, work must be started after about 90 seconds or longer from the time the ignition switch is turned to the LOCK position and the battery cable is disconnected from the battery.

1. Disconnect the negative battery cable from the battery.

2. Remove the transaxle assembly from the vehicle.

3. Matchmark the clutch cover to the flywheel.

4. Remove the clutch pressure plate retaining bolts in small amounts and in a crisscross pattern to relieve the clutch disc spring tension.

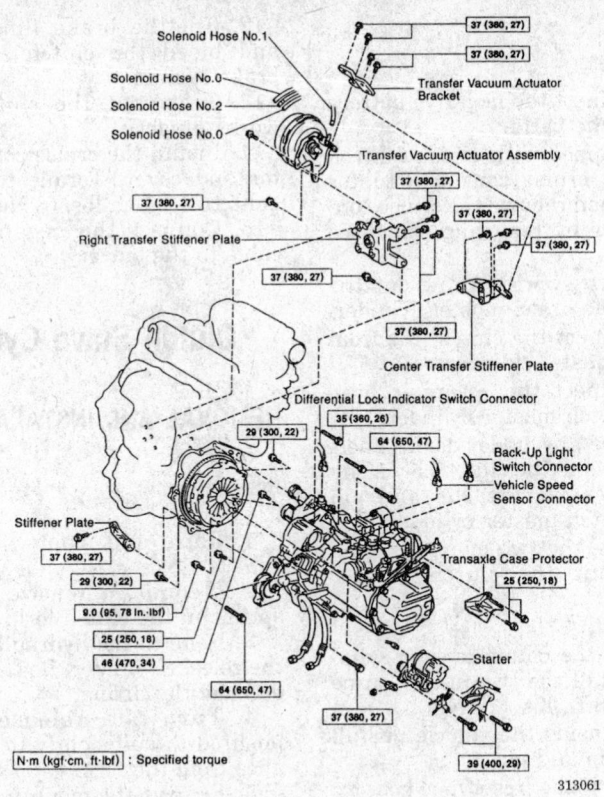

Transaxle exploded view — 4WD RAV4

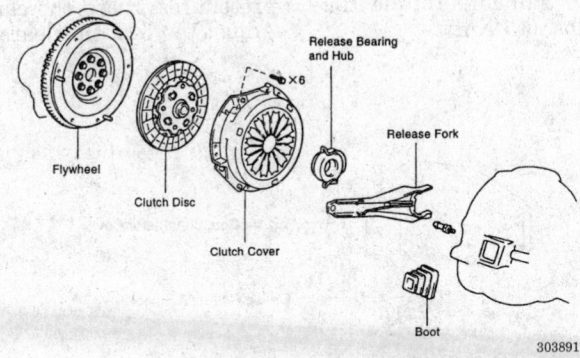

Clutch component assembly

5. At the clutch cover, loosen each bolt one turn until spring tension is released.

6. On RAV4, remove the set bolts to the clutch cover and pull off the clutch cover with the clutch disc.

7. On Previa, remove the clutch cover-to-flywheel bolts. Remove the clutch cover and the clutch disc.

8. If the clutch release bearing is to be replaced, perform the following:

a. Remove the bearing retaining clip(s), the bearing and hub.

b. Remove the release fork and the boot.

c. The bearing is press fitted to the hub.

d. Clean all parts and lightly grease the input shaft splines and all of the contact points.

e. Install the bearing/hub assembly, the fork, the boot and the retaining clip(s) in their original locations.

To install:

9. Inspect the flywheel surface for cracks, heat scoring (blue marks) and warpage. Replace or resurface the flywheel, if any damage is present.

NOTE: Before installing any new parts, make sure they are clean. During installation, do not get grease or oil on any of the components, as this will shorten clutch life considerably.

10. Using an clutch alignment tool, position the clutch disc against the flywheel. The raised center section of the disc faces the transaxle.

11. Position the clutch cover onto the flywheel and align the matchmarks.

12. Install the clutch cover retaining bolts. Torque the bolts in a crisscross pattern to 14 ft. lbs. (19 Nm).

13. Lubricate the release fork pivot and contact points, release bearing, bearing hub and input shaft spline surfaces with a suitable molybdenum disulfide lithium based or multi-purpose grease.

14. Install the boot, release fork, hub, and the bearing assemblies.

15. Install the transaxle to the vehicle.

16. Connect the negative battery cable to the battery.

Clutch Free-Play

ADJUSTMENT

RAV4

1. Check that the pedal height is correct. Pedal height from the floor panel should be: 6.889–7.283 in. (175–185mm)

2. If necessary to adjust the pedal height, loosen the lock nut and turn the stopper bolt until the height is correct. Tighten the lock nut.

3. Push in on the pedal until the beginning of the clutch resistance is felt. Free-play should be 0.197–0.591 in. (5–15mm).

4. Gently push on the pedal until the resistance begins to increase a little. Push rod play at the pedal top should be 0.039–0.197 in. (1–5mm).

5. If necessary to adjust the pedal free-play and the push rod play,

a. Loosen the lock nut and turn the push the rod until the free-play and push rod play are correct.

b. Tighten the locknut.

Clutch Master Cylinder

REMOVAL AND INSTALLATION

Previa

1. Disconnect the negative battery cable.

2. Remove the instrument panel lower finish panel and steering column cover.

3. Remove the clip and clevis pin.

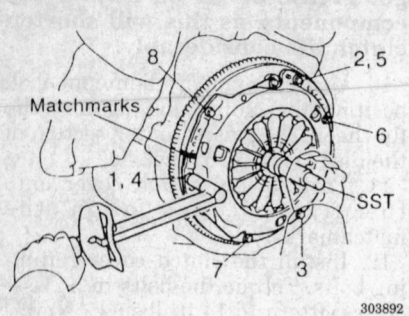

Torque sequence for the clutch cover

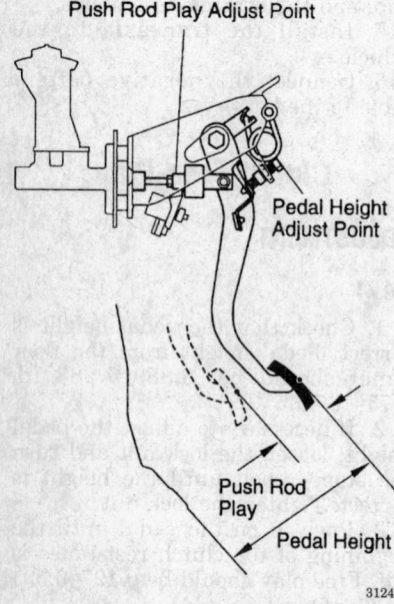

Clutch pedal — RAV4

4. Disconnect the reservoir hose and clutch union line.

5. Remove the mounting bolts and pull out the master cylinder.

6. Installation is the reverse of removal. Torque the mounting bolts to 9 ft. lbs. (12 Nm).

7. Bleed the clutch hydraulic system.

8. Connect the battery cable and check operation.

RAV4

1. Disconnect the negative battery cable from the battery.

2. If equipped with cruise control, remove the cruise control actuator cover and then remove the cruise control actuator by removing the three bolts.

3. Using a syringe, draw out the fluid from the brake master cylinder.

4. Disconnect the clutch line from the clutch master cylinder.

5. Disconnect the reservoir hose from the clutch master cylinder.

6. Remove any under-dash panels to gain access to the clutch clevis clip and pin. Remove the clip and pin from the clutch master cylinder.

7. Remove the two mounting nuts and pull out the clutch master cylinder.

To install:

8. Install the clutch master cylinder and install the two nuts. Torque the nuts to 8 ft. lbs. (12 Nm).

9. From inside the vehicle, install the clevis clip and pin.

10. Connect the reservoir hose to the master cylinder.

11. Connect the clutch line to the clutch master cylinder. Torque the line to 11 ft. lbs. (15 Nm).

12. Fill the brake fluid reservoir and bleed the clutch system and brake system.

13. Test that the clutch system works properly.

14. Install the cruise control actuator and cover. Torque the actuator bolts to 52 inch lbs. (6 Nm).

15. Connect the negative battery cable to the battery.

Clutch Slave Cylinder

REMOVAL AND INSTALATION

Previa

1. Raise and safely support the vehicle.

2. If equipped, remove the tension spring on the clutch fork.

3. Remove the hydraulic line from the release cylinder. Be careful not to damage the fitting.

4. Turn the release cylinder pushrod in sufficiently to gain clearance from the fork.

5. Remove the mounting bolts and withdraw the slave cylinder.

6. To install, reverse the removal procedures. Bleed the clutch system. Adjust the fork tip clearance.

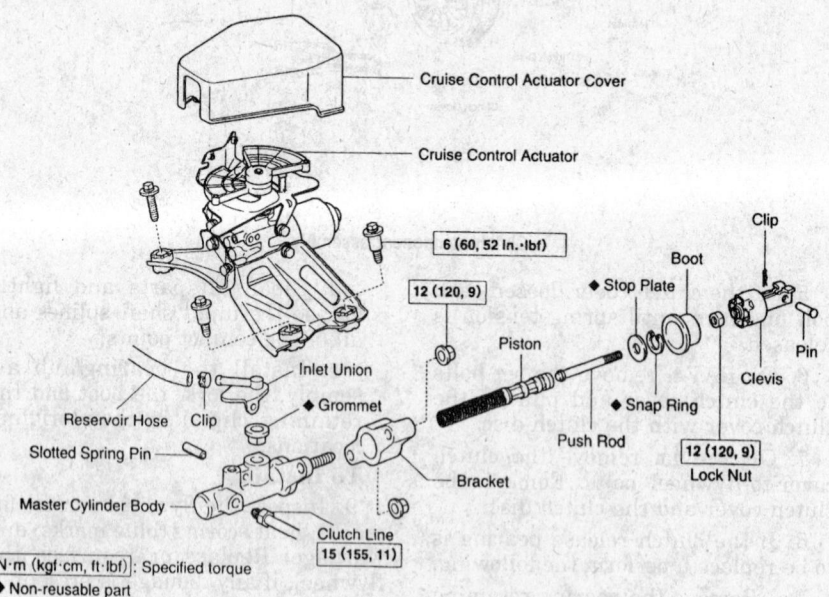

Clutch master cylinder component assembly

RAV4

---- **CAUTION** ----

To avoid personal injury and accidental deployment of the air bag, work must be started after about 90 seconds or longer from the time the ignition switch is turned to the LOCK position and the battery cable is disconnected from the battery.

---

1. Disconnect the negative battery cable from the battery.
2. Disconnect the fluid line with a line wrench. Plug the line to prevent fluid from dripping out. Dispose of fluid properly.
3. Remove the two slave cylinder retaining bolts.
4. Remove the clutch slave cylinder from the vehicle.

To Install:

5. Place the clutch slave cylinder in place and install the two bolts. Torque the bolts to 9 ft. lbs. (12 Nm).
6. Connect the fluid line with a line wrench to the slave cylinder.
7. Bleed the slave cylinder to remove any air present in the line.
8. Connect the negative battery cable to the battery.

Clutch Switch

ADJUSTMENT

RAV4

The clutch switch should be adjusted so that the engine does not start when the clutch pedal is released and the engine will start when the pedal is fully depressed. If the switch does not perform as such, replacing it may be necessary.

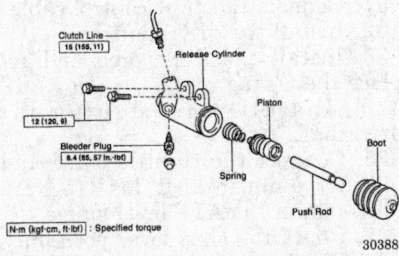

Clutch slave cylinder removal

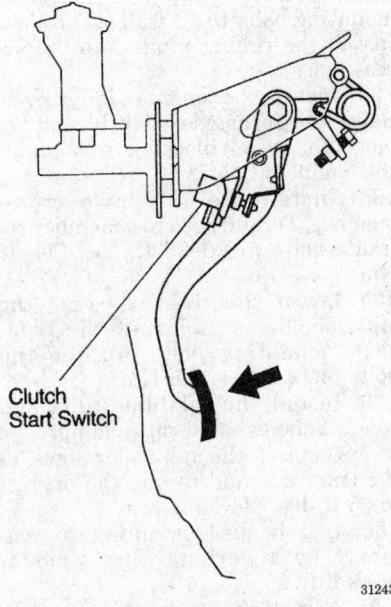

Clutch start switch — RAV4

Hydraulic System

BLEEDING

NOTE: If any maintenance on the clutch system was performed or the system is suspected of containing air, bleed the system. Use care; brake fluid will remove the paint from any surface. If the brake fluid spills onto any painted surface, wash it off immediately with soap and water.

1. Fill the clutch reservoir with brake fluid. Check the reservoir level frequently and add fluid as needed.
2. Connect one end of a vinyl tube to the bleeder plug on the slave cylinder and submerge the other end into a clear container half-filled with brake fluid.
3. Slowly pump the clutch pedal several times.
4. Have an assistant hold the clutch pedal down and loosen the bleeder plug until fluid and/or air starts to run out of the bleeder plug. Close the bleeder plug while the pedal is held to the floor.

NOTE: Do not allow the pedal to rise back up while the bleeder is still open. If this happens, it will allow air to re-enter the

slave cylinder and cause the clutch system not to work properly.

5. Repeat Steps 2 and 3 until all the air bubbles are removed from the system.
6. Tighten the bleeder plug when all the air is gone.
7. Refill the master cylinder to the proper level as required.
8. Check the system for leaks.

Automatic Transaxle

REMOVAL AND INSTALLATION

Previa

4-Speed

1. Disconnect the negative battery terminal from the battery.
2. If required, remove the air cleaner assembly.
3. Disconnect the transaxle throttle cable from the carburetor linkage or the throttle body.
4. Raise and safely support the vehicle.
5. Drain the transaxle fluid.
6. Disconnect the wiring connectors (near the starter) for the neutral start switch and the back-up light switch. If equipped, disconnect the solenoid (overdrive) switch wiring at the same location.
7. If equipped, disconnect the oil level gauge.
8. Disconnect the starter wiring at the starter. Remove the mounting bolts and the starter from the engine.
9. Make matchmarks on the rear driveshaft flange and the differential pinion flange. These marks must be aligned during installation.
10. Unbolt the rear driveshaft flange. If the vehicle has a 2 piece driveshaft, remove the center bearing bracket-to-frame bolts. Remove the driveshaft from the vehicle.
11. Disconnect the speedometer cable (tie it aside). Disconnect the shift linkage from the transaxle.
12. Disconnect the transaxle oil cooler lines at the transaxle.
13. Disconnect the exhaust pipe clamp and remove the oil filler tube, as required.
14. Support the transaxle, using a jack with a wooden block placed between the jack and the transaxle pan. Raise the transaxle, just enough to take the weight off of the rear mount.
15. Remove the rear engine mount with the bracket and the engine under cover, to gain access to the engine crankshaft pulley.

16. Remove the stiffener plates, if equipped.

17. Place a wooden block (or blocks) between the engine oil pan and the front frame crossmember.

18. Slowly, lower the transaxle until the engine rests on the wooden block.

19. Remove the rubber plug(s) from the service holes located at the rear of the engine in order to gain access to the torque convertor bolts.

20. Rotate the crankshaft (to remove the torque convertor bolts) to access the bolts through the service holes.

21. Obtain a bolt of the same dimensions as the torque convertor bolts. Cut the head off of the bolt and hacksaw a slot in the bolt opposite the threaded end.

NOTE: This modified bolt is used as a guide pin. Two guides pins are needed to properly install the transaxle.

22. Thread the guide pin into one of the torque convertor bolt holes. The guide pin will help keep the convertor with the transaxle.

23. Remove the stiffener plates from the transaxle.

24. Remove the transaxle-to-engine bolts, then carefully move the transaxle rearward by prying on the guide pin through the service hole.

25. Pull the transaxle rearward and lower it (front end down) out of the vehicle.

To install:

26. Apply a coat of multi-purpose grease to the torque convertor stub shaft and the corresponding pilot hole in the flexplate.

27. Install the torque convertor into the front of the transaxle. Push inward on the torque convertor while rotating it to completely couple the torque convertor to the transaxle.

28. To make sure the convertor is properly installed, measure the distance between the torque convertor mounting lugs and the front mounting face of the transaxle. The proper distance is 0.079 inch (20mm).

29. Install guide pins into 2 opposite mounting lugs of the torque converter.

30. Raise the transaxle to the engine, align the transaxle with the engine alignment dowels and position the convertor guide pins into the mounting holes of the flexplate.

31. Install and tighten the transaxle-to-engine mounting bolts. Torque the bolts to 47 ft. lbs. (63 Nm).

32. Remove the convertor guide pins and install the convertor mount-ing bolts. Rotate the crankshaft as necessary to gain access to the guide pins and bolts through the service holes. Evenly, tighten the convertor mounting bolts to 13 ft. lbs. (17 Nm). Install the rubber plugs into the access holes.

33. Install the engine undercover. Raise the transaxle slightly and remove the wood block(s) from under the engine oil pan.

34. Install the transaxle crossmember. Torque the crossmember-to-frame bolts to 26–36 ft. lbs. (34–48 Nm).

35. Lower the transaxle onto the crossmember and install the transaxle mounting bolts. Torque the bolts to 19 ft. lbs. (26 Nm).

36. Install the oil filler tube and connect the exhaust pipe clamp.

37. Connect the oil cooler lines to the transaxle and torque the fittings to 25 ft. lbs. (34 Nm).

38. Install the remaining components by reversing the removal procedure.

39. Adjust the transaxle throttle cable.

40. Refill the transaxle.

41. Connect the negative battery cable. Start the engine and check for leaks.

42. Road test the vehicle for proper operation.

43. Recheck all fluid levels.

3-Speed

1. Disconnect the negative battery cable.

2. Remove the ATF level gauge.

3. Loosen the nut and disconnect the throttle cable.

4. Remove the equipment driveshaft.

5. Raise and safely support the vehicle.

6. Remove the filler pipe.

7. Remove the driveshaft.

8. Disconnect the control cable.

   a. Remove the nut and two washers.

   b. Disconnect the shift control cable.

   c. Remove the clip and disconnect the shift control cable.

   d. Remove the two bolts and the bracket.

9. Disconnect the No. 1 and No. 2 vehicle speed sensor connectors.

10. Disconnect the solenoid connector.

11. Disconnect the park/neutral position switch connector.

12. Remove the bolt and disconnect the oil pipe clamp. Disconnect the oil cooler pipes. On 4WD disconnect the A/T fluid temperature switch connector.

13. Remove the starter.

14. Using a jack, support the transaxle.

15. Remove the stiffener plate.

   a. On 2WD: remove the four bolts and lower the stiffener plate.

   b. On 4WD: remove the through bolt and nut, then remove the three bolts and the upper stiffener plate.

16. Remove the torque converter clutch cover and turn the crankshaft to gain access and remove the six bolts.

17. Remove the exhaust pipe bracket.

18. Remove the rear mounting bolts.

19. Remove the transaxle.

   a. Remove the four bolts.

   b. Detach the wire harness, and remove the transaxle.

To install:

20. Connect wire harness and install the transaxle. Torque the bolts to 53 ft. lbs. (72 Nm).

21. Install the rear mounting bolts and torque to 50 ft. lbs. (67 Nm).

22. Install the exhaust pipe bracket and torque the bolts to:
   • Transaxle: 37 ft. lbs. (51 Nm)
   • Exhaust manifold: 32 ft. lbs. (43 Nm)

23. Install the torque converter clutch mounting bolts and torque to 30 ft. lbs. (41 Nm). Install the torque converter clutch cover.

24. Install the stiffener plate.

   a. 2WD: torque the four bolts to 27 ft. lbs. (37 Nm)

   b. 4WD: install the three bolts and torque to 27 ft. lbs. (37 Nm). Install the through bolt and nut and torque to 27 ft. lbs. (37 Nm).

25. Remove the transaxle jack.

26. Install the starter.

27. Connect the cooler pipes and torque to 25 ft. lbs. (34 Nm).

28. Connect the park/neutral position switch connector.

29. Connect the solenoid connector.

30. Connect the No. 1 and the No. 2 vehicle speed sensor connectors.

31. Connect the shift control cable.

32. Install the driveshaft.

33. Install the filler pipe and replace the O-ring.

34. Install the equipment driveshaft.

35. Connect the throttle cable and torque the nut to 11 ft. lbs. (15 Nm).

36. Install the ATF level gauge.

37. Check the shift lever position.

38. Check the fluid level and fill if necessary.

39. Connect the negative battery cable.

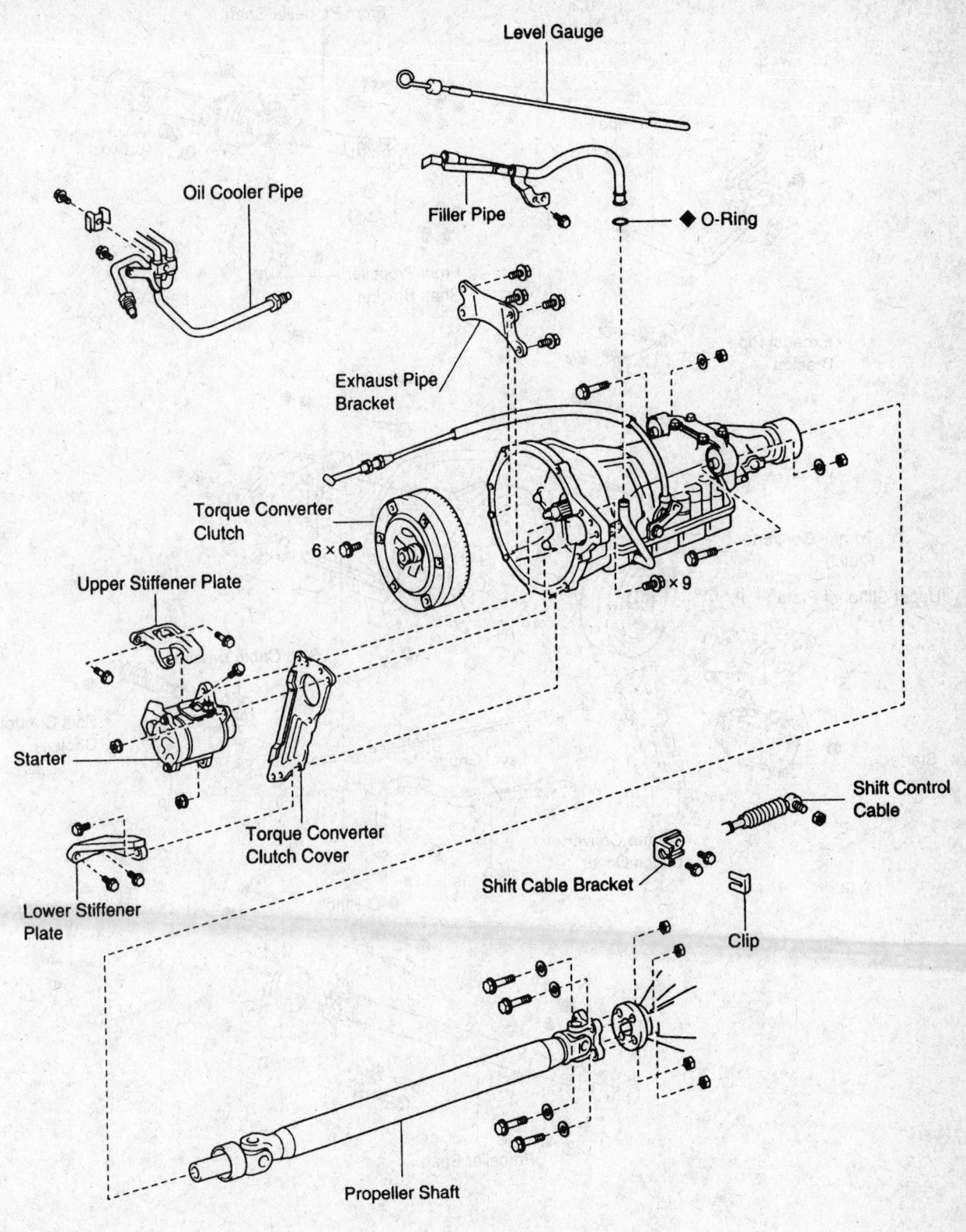

Level Gauge

Oil Cooler Pipe

Filler Pipe

◆ O-Ring

Exhaust Pipe
Bracket

Torque Converter
Clutch

6 ×

Upper Stiffener Plate

Starter

Torque Converter
Clutch Cover

Lower Stiffener
Plate

× 9

Shift Control
Cable

Shift Cable Bracket

Clip

Propeller Shaft

◆ Non-reusable part

288849

3-speed automatic transaxle assembly — 2WD Previa

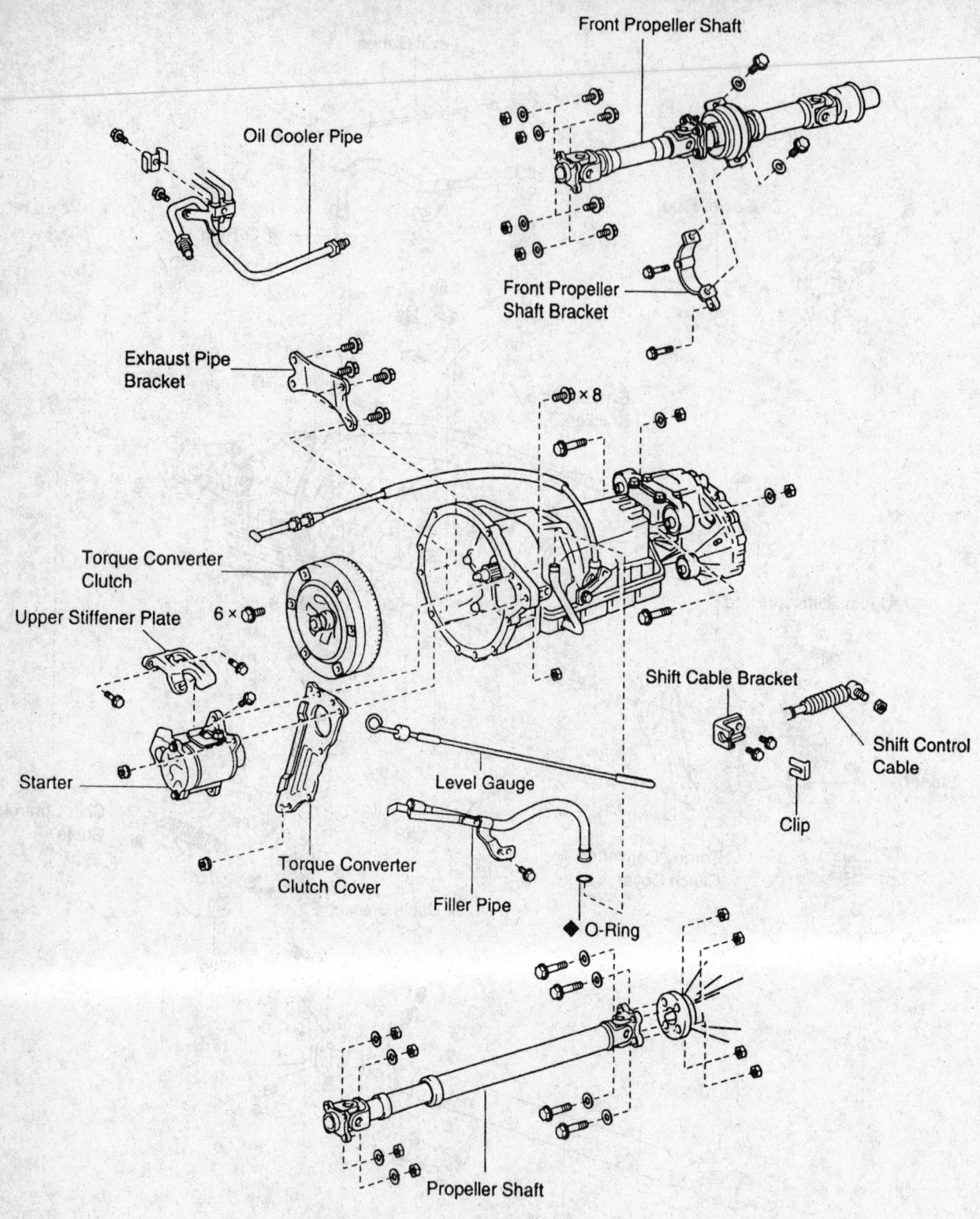

Front Propeller Shaft

Oil Cooler Pipe

Front Propeller
Shaft Bracket

Exhaust Pipe
Bracket

× 8

Torque Converter
Clutch

Upper Stiffener Plate

6 ×

Shift Cable Bracket

Shift Control
Cable

Clip

Starter

Torque Converter
Clutch Cover

Level Gauge

Filler Pipe

◆ O-Ring

Propeller Shaft

◆ Non-reusable part

3-speed automatic transaxle assembly — 4WD Previa

288850

RAV4

4-Wheel Drive

1. Disconnect the negative battery cable from the battery.

2. Remove the engine and transaxle assembly from the vehicle.

3. Remove the starter by removing the two bolts.

4. Remove the stiffener plate by removing the three bolts.

5. Remove the rear end plate by removing the four bolts.

6. Remove the six torque converter clutch mounting bolts.

7. Disconnect the connectors and wire harness from the transaxle.

8. Remove the center stiffener plate by removing the four bolts.

9. Remove the transaxle with the transfer assembly as follows:

   a. Remove the two bolts.

   b. Remove the five transaxle mounting bolts.

   c. Separate the transaxle assembly from the engine.

To install:

10. Install the transaxle to the engine.

11. Install the five transaxle mounting bolts and torque the bolts as follows:
   • 14mm head bolts to 47 ft. lbs. (64 Nm)
   • 12mm head bolts to 34 ft. lbs. (46 Nm)

12. Install the two bolts and torque the bolts to 27 ft. lbs. (37 Nm).

13. Install the center stiffener plate with the four bolts and torque the bolts to 27 ft. lbs. (37 Nm).

14. Connect the connectors and the wire harness to the transaxle.

15. Install the torque converter clutch mounting bolts. Install and

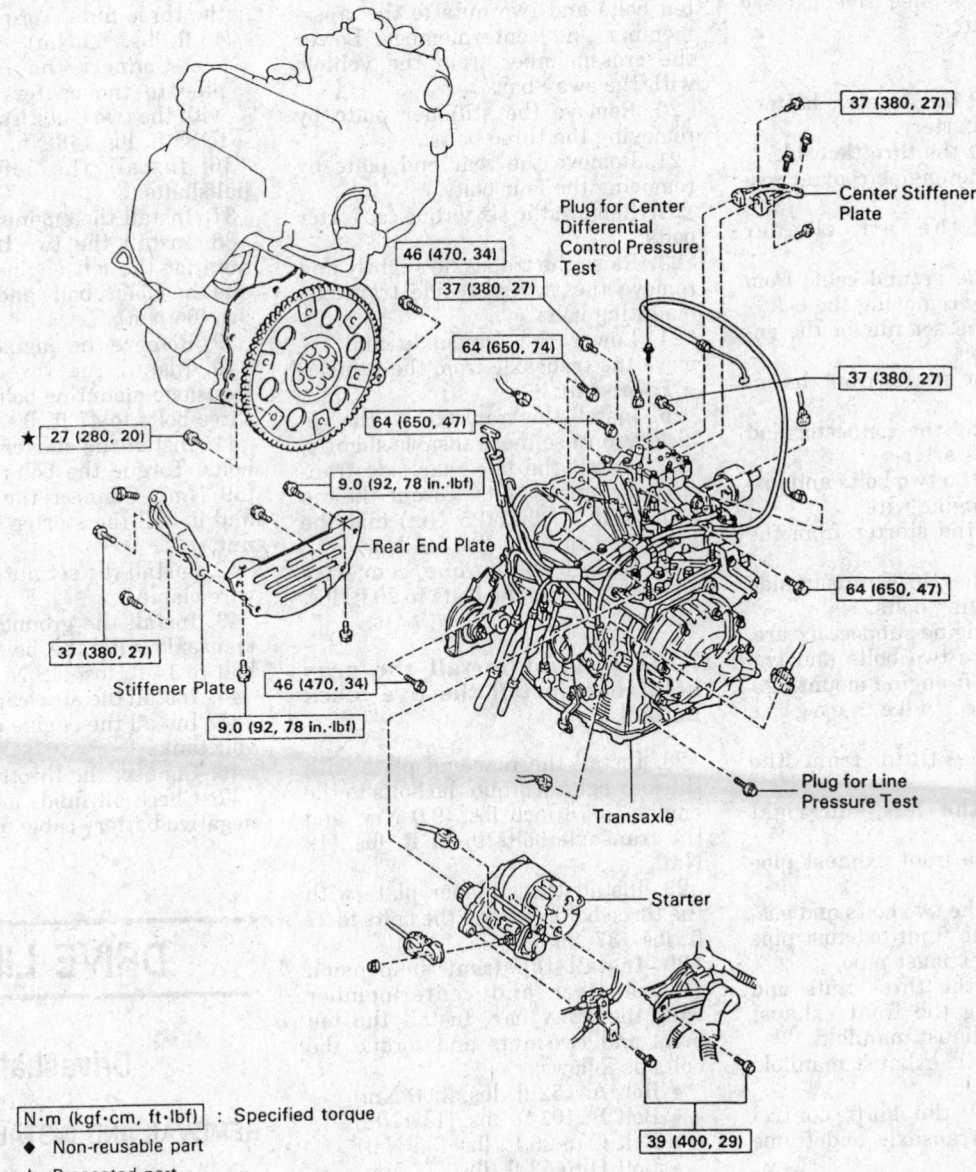

Plug for Center Differential Control Pressure Test

Center Stiffener Plate

37 (380, 27)

46 (470, 34)

37 (380, 27)

64 (650, 74)

37 (380, 27)

64 (650, 47)

★ 27 (280, 20)

9.0 (92, 78 in.·lbf)

Rear End Plate

64 (650, 47)

37 (380, 27)

Stiffener Plate

46 (470, 34)

9.0 (92, 78 in.·lbf)

Transaxle

Plug for Line Pressure Test

Starter

39 (400, 29)

N·m (kgf·cm, ft·lbf) : Specified torque
◆ Non-reusable part
★ Precoated part

314265

Transaxle exploded view — 4WD RAV4

tighten all the bolts evenly. Torque each bolt to 20 ft. lbs. (27 Nm).

NOTE: Coat the threads of the bolts with an approved locking compound.

16. Install the rear end plate with the four bolts. Torque the bolts to 80 inch lbs. (9.0 Nm).

17. Install the stiffener plate with the three bolts. Torque the bolts to 27 ft. lbs. (37 Nm).

18. Install the starter with the two bolts. Torque the bolts to 29 ft. lbs. (39 Nm).

19. Install the engine and transaxle assembly to the vehicle.

20. Connect the negative battery cable to the battery.

2-Wheel Drive

1. Disconnect the negative battery cable from the battery.

2. Disconnect the throttle cable.

3. Remove the engine coolant reservoir tank.

4. Remove the air cleaner assembly.

5. Remove the ground cable from the transaxle by removing the bolt.

6. Remove the set nut of the engine wire clamp.

7. Remove the starter from the engine as follows:

   a. Disconnect the connector and nut from the starter.

   b. Remove the two bolts and disconnect the engine wire.

   c. Remove the starter from the engine.

8. Remove the three upper side transaxle mounting bolts.

9. Install a engine support fixture.

10. Remove the two bolts and two nuts from the left engine mounting.

11. Remove the engine undercovers.

12. Drain the fluid from the transaxle.

13. Remove the left and right halfshafts.

14. Remove the front exhaust pipe as follows:

   a. Remove the two bolts and gasket holding the front exhaust pipe to the center exhaust pipe.

   b. Remove the three nuts and gasket holding the front exhaust pipe to the exhaust manifold.

   c. Remove the exhaust manifold from the vehicle.

15. Disconnect the shift control cable from the transaxle and frame as follows:

   a. Remove the nut from the control shaft lever.

   b. Remove the clip and disconnect the control cable from the transaxle.

   c. Remove the two bolts holding the shift control cable to the centermember and crossmember.

16. Disconnect the following connectors:

   • Shift solenoid valve connector
   • Park/neutral position switch connector
   • Vehicle speed sensor connector

17. Disconnect the oil cooler hoses from the transaxle.

18. Disconnect the rack and pinion from the crossmember by removing the two nuts and two bolts. Support the rack and pinion.

19. Support the suspension crossmember with a floor jack. Remove the ten bolts and two nuts to the crossmember and centermember. Lower the crossmember from the vehicle with the sway bar.

20. Remove the stiffener plate by removing the three bolts.

21. Remove the rear end plate by removing the four bolts.

22. Remove the six torque converter bolts.

23. Raise the transaxle slightly and remove the two rear side transaxle mounting bolts.

24. Lower the transaxle and remove the transaxle from the vehicle.

To install:

25. Install the transaxle to the vehicle and raise the transaxle slightly.

26. Install the two rear side transaxle mounting bolts. Torque the top bolt to 18 ft. lbs. (25 Nm) and the lower bolt to 34 ft. lbs. (46 Nm).

27. Install the torque converter bolts and torque the bolts to 20 ft. lbs. (27 Nm).

NOTE: First install the gray bolt, then install the five black bolts.

28. Install the rear end plate with the four bolts. Torque the bolts to the engine to 78 inch lbs. (9.0 Nm) and the transaxle bolts to 14 ft. lbs. (19 Nm).

29. Install the stiffener plate with the three bolts. Torque the bolts to 27 ft. lbs. (37 Nm).

30. Install the front suspension crossmember and centermember with the sway bar. Install the ten bolts and two nuts and torque the bolts as follows:

   • Bolt A: 152 ft. lbs. (206 Nm)
   • Bolt B: 101 ft. lbs. (137 Nm)
   • Bolt C to 26 ft. lbs. (35 Nm)
   • Bolt D to 53 ft. lbs. (72 Nm)
   • Nut to 54 ft. lbs. (73 Nm)

31. Connect the rack and pinion to the crossmember by installing the two bolts and two nuts. Torque the nuts to 83 ft. lbs. (113 Nm).

32. Connect the oil cooler hoses with the two clips.

33. Connect the following connectors:

   • Shift solenoid valve connector
   • Park/neutral switch connector
   • Vehicle speed sensor connector

34. Connect the shift control cable. Torque the nut to hold the control cable to the transaxle to 10 ft. lbs. (13 Nm).

35. Install the front exhaust pipe as follows:

   a. Using new gaskets, install the front exhaust pipe.

   b. Connect the front exhaust pipe to the exhaust manifold with the three nuts. Torque the nuts to 46 ft. lbs. (62 Nm).

   c. Connect the front exhaust pipe to the center exhaust pipe with the two bolts. Torque the bolts to 35 ft. lbs. (48 Nm).

36. Install the left and right halfshafts.

37. Install the engine undercovers.

38. Install the two bolts and two nuts for the left engine mount. Torque the mount bolts and nuts to 47 ft. lbs. (64 Nm).

39. Remove the engine fixture.

40. Install the three upper side transaxle mounting bolts. Torque the three bolts to 47 ft. lbs. (64 Nm).

41. Install the starter with the two bolts. Torque the bolts to 29 ft. lbs. (39 Nm). Connect the engine wire and install the starter wire with the nut.

42. Install the set nut of the engine wire clamp.

43. Install the ground cable to the transaxle with the bolt. Torque the bolt to 14 ft. lbs. (19 Nm).

44. Install the air cleaner assembly.

45. Install the engine coolant reservoir tank.

46. Connect the throttle cable.

47. Check all fluids and install the negative battery cable to the battery.

DRIVE LINE

Driveshaft

REMOVAL AND INSTALLATION

Previa

Rear Driveshaft

1. Disconnect the negative battery cable from the battery.

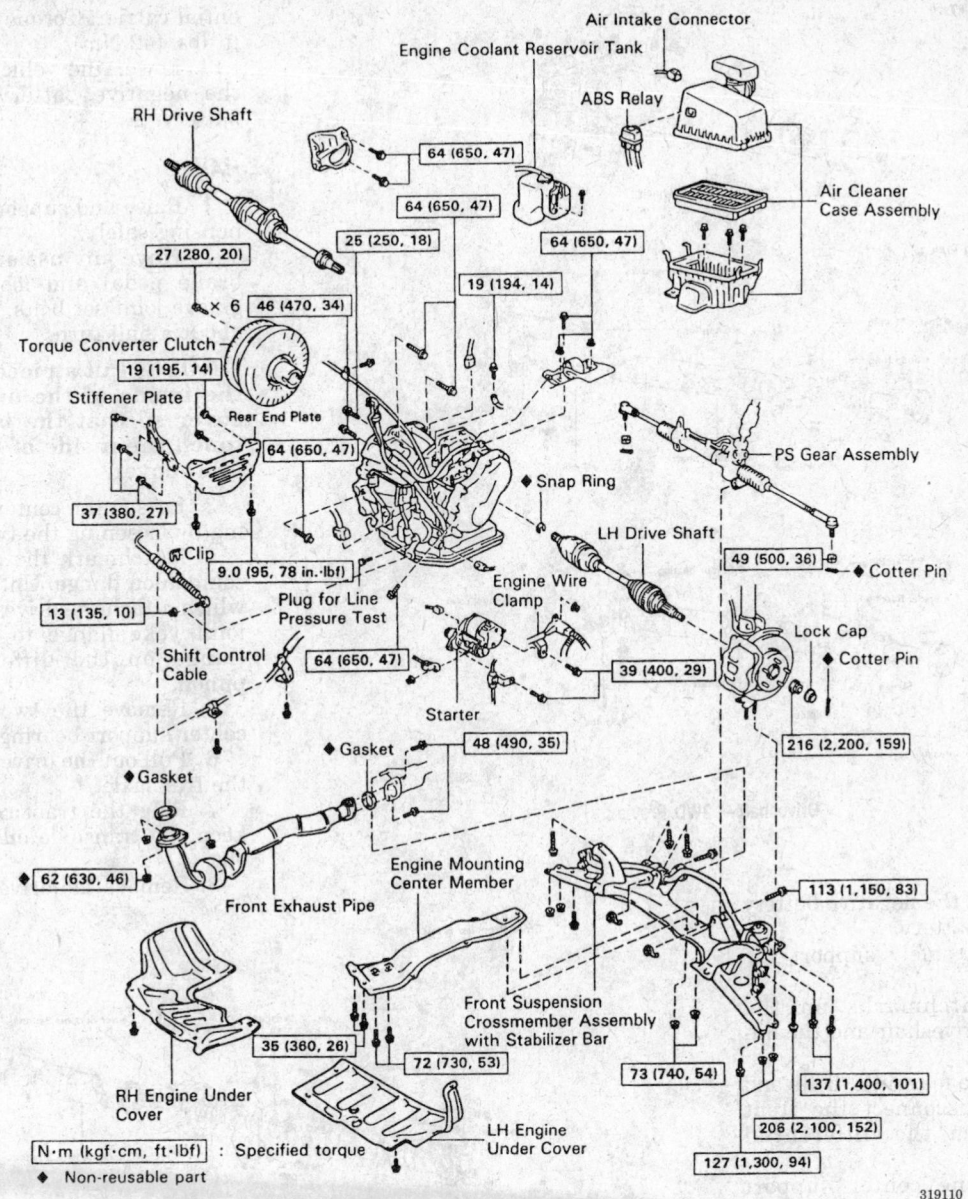

Air Intake Connector
Engine Coolant Reservoir Tank
ABS Relay
Air Cleaner Case Assembly
RH Drive Shaft
64 (650, 47)
64 (650, 47)
25 (250, 18)
64 (650, 47)
27 (280, 20)
19 (194, 14)
× 6 46 (470, 34)
Torque Converter Clutch
19 (195, 14)
Stiffener Plate
Rear End Plate
64 (650, 47)
PS Gear Assembly
37 (380, 27)
♦ Snap Ring
LH Drive Shaft
49 (500, 36) ♦ Cotter Pin
♦ Clip
9.0 (95, 78 in.·lbf)
Engine Wire Clamp
Lock Cap
13 (135, 10)
♦ Cotter Pin
Plug for Line Pressure Test
Shift Control Cable
64 (650, 47)
39 (400, 29)
216 (2,200, 159)
Starter
♦ Gasket
48 (490, 35)
♦ Gasket
♦ Gasket
Engine Mounting Center Member
113 (1,150, 83)
♦ 62 (630, 46)
Front Exhaust Pipe
Front Suspension Crossmember Assembly with Stabilizer Bar
35 (360, 26)
72 (730, 53)
73 (740, 54)
137 (1,400, 101)
RH Engine Under Cover
206 (2,100, 152)
LH Engine Under Cover
127 (1,300, 94)

N·m (kgf·cm, ft·lbf) : Specified torque
♦ Non-reusable part

319110

Transaxle exploded view — 2WD RAV4

2. Raise and safely support the vehicle.

3. Place matchmarks on the flanges of the driveshaft and the differential carrier.

4. Remove the four nuts, bolts and washers and disconnect the driveshaft from the differential housing.

5. On 2WD, pull out the driveshaft yoke from the transaxle.

6. On 2WD, install a transaxle output shaft plug into the transaxle to prevent fluid loss.

7. On 4WD, disconnect the driveshaft from the transfer case by removing the four nuts and washers.

8. On 4WD, place matchmarks on the flanges of the driveshaft and the transfer case.

9. On 4WD, remove the driveshaft from the vehicle.

To install:

10. On 2WD, remove the transaxle output shaft plug.

11. On 2WD, slide the driveshaft yoke into the transaxle.

12. On 2WD, align the matchmarks on the driveshaft and differential flanges.

13. On 2WD, install the four washers, bolts and nuts to the driveshaft and differential carrier. Torque the bolts to 54 ft. lbs. (74 Nm).

14. On 4WD, align the matchmarks on the transfer case and driveshaft.

15. On 4WD, Install the four washers and nuts to hold the driveshaft to the transfer case. Torque the nuts to 54 ft. lbs. (74 Nm).

16. On 4WD, align the matchmarks on the driveshaft and differential flanges.

17. On 4WD, install the four washers, bolts and nuts to the driveshaft and differential carrier. Torque the bolts to 54 ft. lbs. (74 Nm).

18. Lower the vehicle and connect the negative battery cable to the battery.

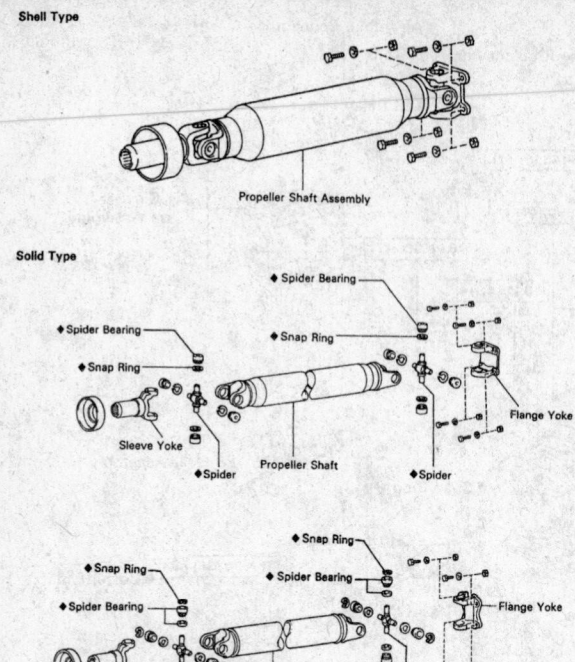

Shell Type

Solid Type

◆ Spider Bearing
◆ Spider Bearing
◆ Snap Ring
◆ Snap Ring
◆ Spider Bearing
Sleeve Yoke
Propeller Shaft
◆ Spider
◆ Spider
Flange Yoke

◆ Snap Ring
◆ Spider Bearing
◆ Snap Ring
◆ Spider Bearing
Sleeve Yoke
Propeller Shaft
◆ Spider
Flange Yoke
◆ Spider

◆ Non-reusable part

Propeller Shaft Assembly

288234

Driveshaft — 2WD Previa

Front Driveshaft

1. Disconnect the negative battery cable from the battery.

2. Raise and safely support the vehicle.

3. Place matchmarks on the flanges of the driveshaft and the differential carrier.

4. Remove the four nuts, bolts and washers and disconnect the front driveshaft from the differential housing.

5. Remove the center support bearing by removing the two bolts.

6. Pull out the driveshaft yoke from the transfer case.

7. Remove the driveshaft from the vehicle.

8. Install a transfer case output shaft plug into the transfer case to prevent fluid loss.

To install:

9. Remove the transfer case output shaft plug.

10. Slide the driveshaft into the transfer case.

11. Connect the center support bearing by installing the two bolts. Torque the bolts to 27 ft. lbs. (36 Nm).

12. Align the matchmarks on the driveshaft and differential flanges.

13. Install the four washers, bolts and nuts to the driveshaft and differ-

ential carrier. Torque the bolts to 31 ft. lbs. (42 Nm).

14. Lower the vehicle and connect the negative battery cable to the battery.

RAV4

1. Raise and support the rear axle housing safely.

2. Have an assistant hold the brake pedal and loosen the cross groove joint set bolts. Turn the bolts about a half turn.

NOTE: Put a piece of cloth into the inside of the universal joint cover so that the boot does not touch the inside of the universal joint cover.

3. Loosen the center support bearing by loosening the two bolts.

4. Matchmark the driveshaft and companion flange. Unfasten the bolts which attach the driveshaft universal joint yoke flange to the mounting flange on the differential drive pinion.

5. Remove the two bolt and the center support bearing.

6. Pull out the driveshaft end from the transaxle.

7. Plug the transaxle opening to keep the transaxle oil from running out.

8. Remove the driveshaft.

Front Propeller Shaft

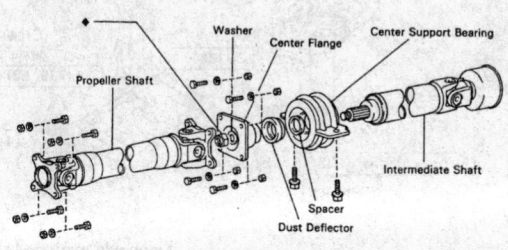

Propeller Shaft
Washer
Center Flange
Center Support Bearing
Intermediate Shaft
Spacer
Dust Deflector

Rear Propeller Shaft

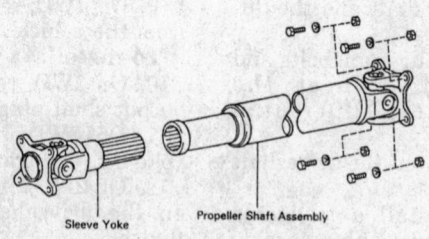

Sleeve Yoke
Propeller Shaft Assembly

◆ Non-reusable part

288235

Driveshafts — 4WD Previa

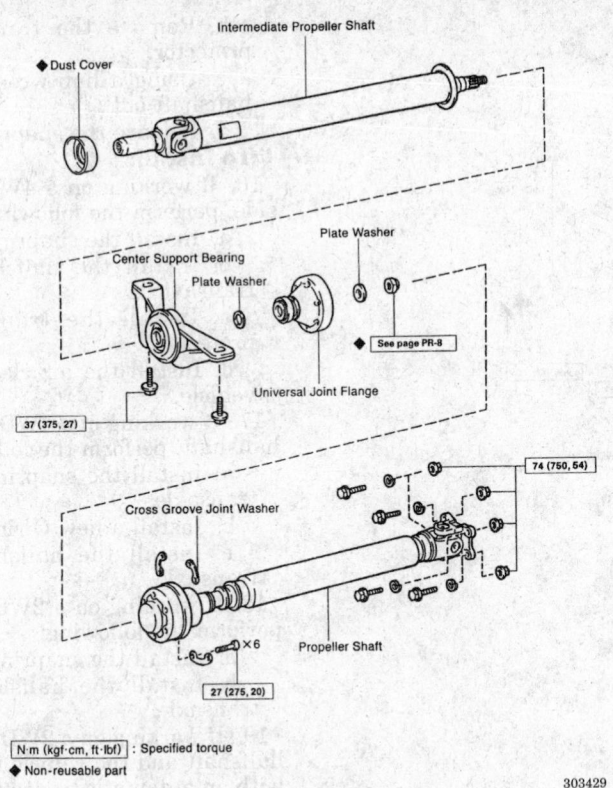

Driveshaft exploded view — RAV4

37 (375, 27)

74 (750, 54)

27 (275, 20)

Cross Groove Joint Washer

Propeller Shaft

Dust Cover

Intermediate Propeller Shaft

Center Support Bearing

Plate Washer

Plate Washer

Universal Joint Flange

◆ See page PR-8

× 6

N·m (kgf·cm, ft·lbf) : Specified torque
◆ Non-reusable part

303429

To install:

9. Apply multi-purpose grease on the section of the U-joint sleeve which is to be inserted into the transaxle.

10. Insert the driveshaft sleeve into the transaxle.

NOTE: Be careful not to damage any of the seals.

11. Install the center support bearing temporarily.

12. Align the matchmarks. Secure the flange to the differential pinion flange with the mounting bolts. Torque the nuts and bolts to 54 ft. lbs. (74 Nm).

13. Having an assistant depress the brake pedal, tighten the cross groove set bolts and torque the bolts to 20 ft. lbs. (27 Nm).

14. With the vehicle in the unladen condition, adjust the dimension between the rear side of the cover and the shaft to 2.5787–2.7756 inchs (65.5–70.5mm).

15. Under the same condition, adjust the dimension between the rear side of the center bearing housing and the rear side of the cushion at 0.4528–0.5315 inch (11.5–13.5mm).

16. Torque the center support bearing bolts to 27 ft. lbs. (37 Nm).

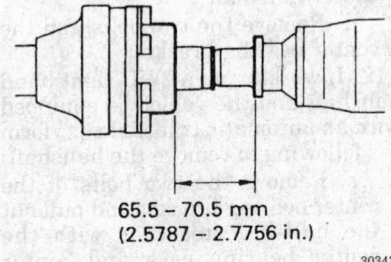

65.5 – 70.5 mm
(2.5787 – 2.7756 in.)

303430

Dimension between the rear side of the cover and the shaft — RAV4

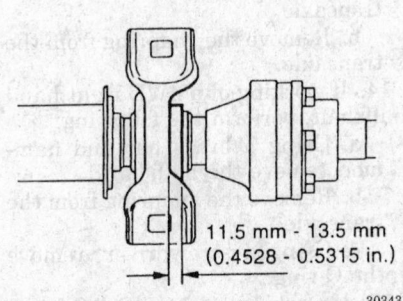

11.5 mm - 13.5 mm
(0.4528 - 0.5315 in.)

303431

Dimension between the rear side of the center bearing and the rear side of cushion — RAV4

17. Check that the center line of the bracket is at right angles at the shaft axial direction.

18. Remove the axle housing supports and lower the vehicle.

Halfshaft

REMOVAL AND INSTALLATION

Previa

1. Raise and safely support the vehicle.

2. Remove the wheel and tire assembly.

3. Remove the cotter pin and lock cap from the halfshaft.

4. Remove the locknut from the halfshaft.

5. Remove the cotter pin and locknut to the tie rod end and disconnect the tie rod end from the knuckle.

6. Remove the lower ball joint from the steering knuckle by removing the two mounting bolts.

7. Place matchmarks on the halfshaft and side gear.

8. Remove the six bolts from the inner halfshaft joint and disconnect the halfshaft from the side gear.

9. Remove the halfshaft by pulling the knuckle outward and remove the halfshaft from the wheel hub.

NOTE: If the outer shaft will not come out of the hub, soak the splines with penetrating lube, install the nut and tap on the halfshaft with a rubber hammer. Be careful not to damage the shaft threads.

To install:

NOTE: Coat the halfshaft splines with anti-seize compound to prevent spline seizure. This will help for future halfshaft removal.

10. Connect the halfshaft to the steering knuckle.

11. Connect the inner halfshaft joint to the side gear by installing the six bolts. Torque the bolts to 51 ft. lbs. (61 Nm).

12. Connect the lower ball joint and torque the bolts to 94 ft. lbs. (127 Nm).

13. Install the tie rod end, torque the nut to 36 ft. lbs. (49 Nm) and install a new cotter pin.

14. Install the halfshaft nut and torque to 152 ft. lbs. (206 Nm).

15. Install the lock cap and a new cotter pin to the halfshaft.

16. Install the wheel and lower the vehicle.

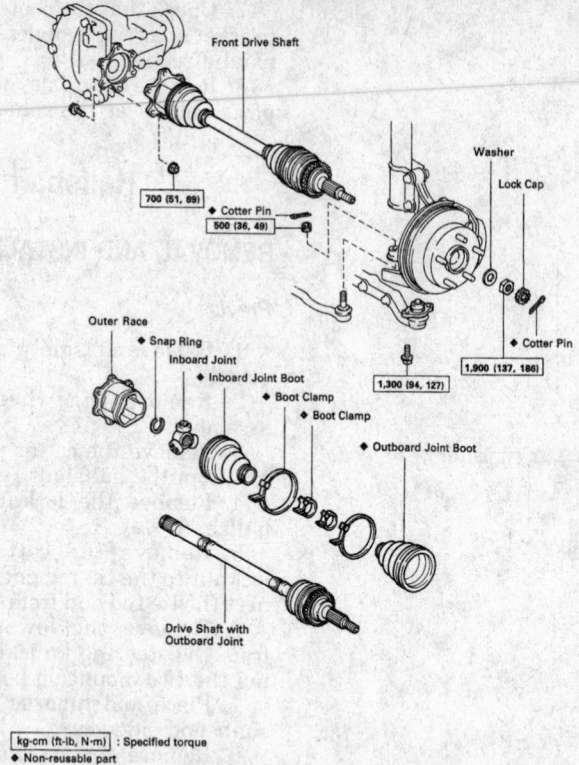

700 (51, 69)

Cotter Pin

500 (36, 49)

Front Drive Shaft

Washer

Lock Cap

Cotter Pin

1,900 (137, 186)

1,300 (94, 127)

Outer Race

Snap Ring

Inboard Joint

Inboard Joint Boot

Boot Clamp

Boot Clamp

Outboard Joint Boot

Drive Shaft with
Outboard Joint

kg-cm (ft-lb, N·m) : Specified torque
Non-reusable part

288720

Halfshaft components, exploded view — Previa

17. Check the front wheel alignment.

RAV4

Front Halfshaft

1. Disconnect the negative battery cable from the battery.
2. Raise and safely support the vehicle.
3. Remove the engine under cover.
4. Drain the transaxle.
5. If equipped with ABS brakes, disconnect the ABS sensor by removing the bolt.
6. Remove the cotter pin, lock cap, and the locknut holding the halfshaft to the steering knuckle.
7. Disconnect the tie rod ends from the steering knuckle.
8. Disconnect the sway bar link from the lower control arm.
9. Disconnect the lower ball joint from the lower control arm.
10. Using a plastic hammer, disconnect the halfshaft from the axle hub.
11. If working on a 2WD right hand halfshaft and the vehicle is equipped with a manual transaxle, perform the following to remove the halfshaft:
   a. Using a screwdriver and hammer, remove the snapring from the center bearing bracket.
   b. Remove the bolt and the center bearing bracket.

c. Remove the halfshaft with the center halfshaft.
   d. Remove the two bolts and the center bearing bracket.
12. If working on a 2WD right hand halfshaft and the vehicle is equipped with an automatic transaxle, perform the following to remove the halfshaft:
   a. Remove the two bolts of the center bearing bracket and pull out the halfshaft together with the center bearing case and center halfshaft.
   b. Remove the three bolts and the center bearing bracket.
13. If working on a 2WD left hand, perform the following:
   a. Using a brass bar and hammer, remove the halfshaft from the transaxle.
   b. Remove the snapring from the transaxle.
14. If working of a 4WD right hand halfshaft, perform the following:
   a. Using a brass bar and hammer, remove the halfshaft.
   b. Remove the snapring from the transaxle.
   c. Using a screwdriver, remove the O-ring.
15. If working on a 4WD left hand side, perform the following:
   a. Remove the air cleaner from the vehicle.

b. Remove the transaxle case protector.
   c. Using a hub wrench, pry the halfshaft out.
   d. Remove the snapring.
To install:
16. If working on a 4WD left hand side, perform the following:
   a. Install the snapring.
   b. Install the halfshaft to the transaxle.
   c. Install the transaxle case protector.
   d. Install the air cleaner to the vehicle.
17. If working of a 4WD right hand halfshaft, perform the following:
   a. Install the snapring from the transaxle.
   b. Install a new O-ring.
   c. Install the halfshaft to the transaxle.
18. If working on a 2WD left hand, perform the following:
   a. Install the snapring.
   b. Install the halfshaft to the transaxle.
19. If working on a 2WD right hand halfshaft and the vehicle is equipped with an automatic transaxle, perform the following to remove the halfshaft:
   a. Install the center bearing bracket and the three bolts. Torque the bolts to 47 ft. lbs. (64 Nm).
   b. Install the halfshaft together with the center bearing case and center halfshaft. Install the two bolts and torque the two bolts to 47 ft. lbs. (64 Nm).
20. If working on a 2WD right hand halfshaft and the vehicle is equipped with a manual transaxle, perform the following to remove the halfshaft:
   a. Install the center bearing bracket with the two bolts.
   b. Install the halfshaft with the center halfshaft.
   c. Install the bolt and the center bearing bracket. Torque the bolt to 24 ft. lbs. (32 Nm).
   d. Install the snapring to the center bearing bracket.
21. Connect the halfshaft to the axle hub.
22. Connect the lower ball joint to the lower control arm. Torque the bolt and two nuts to 94 ft. lbs. (127 Nm).
23. Connect the sway bar link to the lower control arm and torque the nut as follows:
   • 3 door vehicles to 47 ft. lbs. (64 Nm).
   • 5 door vehicles to 83 ft. lbs. (113 Nm).
24. Connect the tie rod end to the steering knuckle. Torque the nut to

2WD M/T

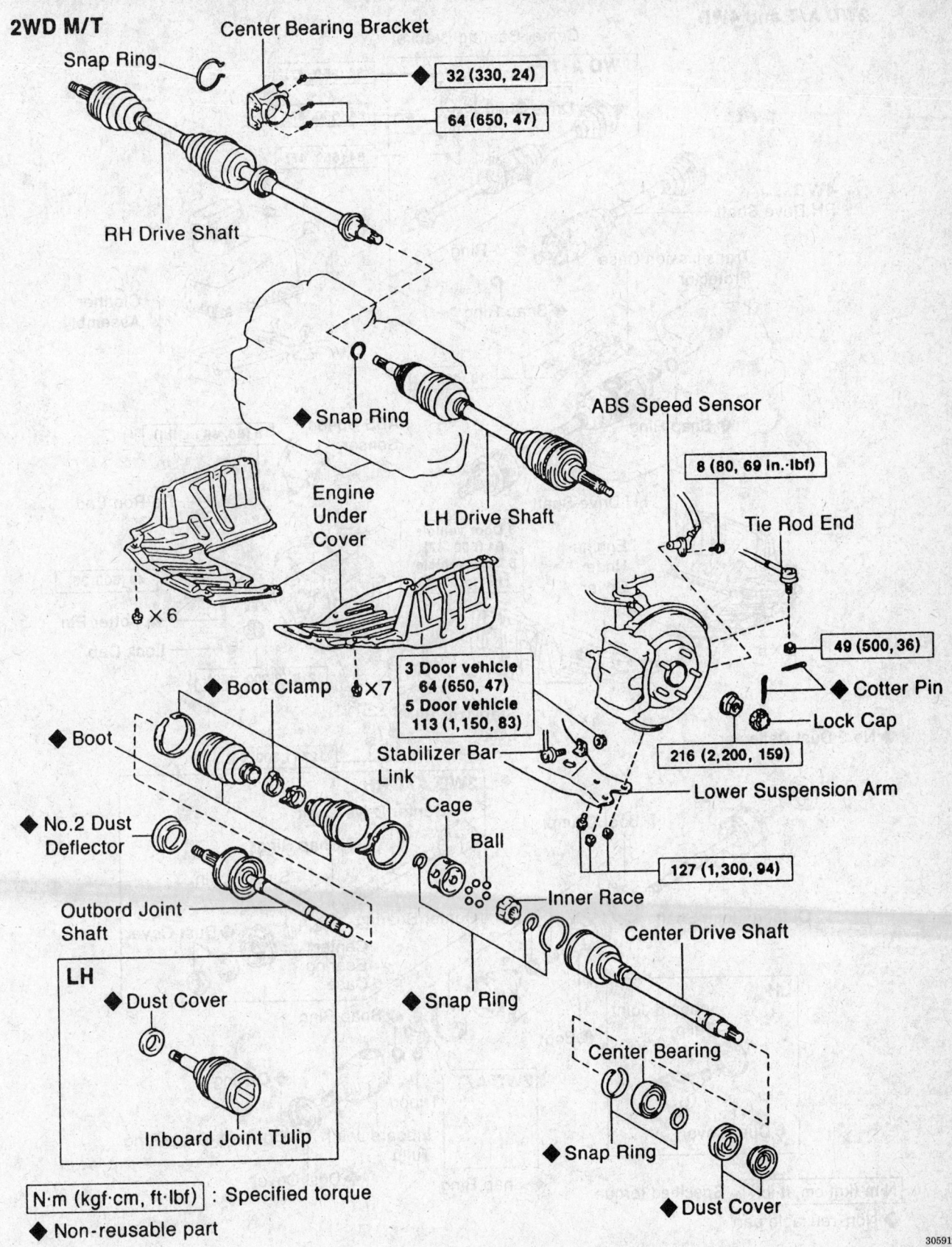

Snap Ring

Center Bearing Bracket

◆ 32 (330, 24)

64 (650, 47)

RH Drive Shaft

◆ Snap Ring

Engine Under Cover

LH Drive Shaft

ABS Speed Sensor

8 (80, 69 in.·lbf)

Tie Rod End

49 (500, 36)

◆ Cotter Pin

Lock Cap

216 (2,200, 159)

Lower Suspension Arm

◆ ×6

◆ ×7

◆ Boot Clamp

3 Door vehicle 64 (650, 47)
5 Door vehicle 113 (1,150, 83)

Stabilizer Bar Link

◆ Boot

Cage

Ball

Inner Race

◆ No.2 Dust Deflector

127 (1,300, 94)

Outboard Joint Shaft

Center Drive Shaft

LH

◆ Dust Cover

◆ Snap Ring

Inboard Joint Tulip

Center Bearing

◆ Snap Ring

◆ Dust Cover

N·m (kgf·cm, ft·lbf) : Specified torque

◆ Non-reusable part

305919

Front halfshaft exploded view (2WD with M/T only) — RAV4

2WD A/T and 4WD

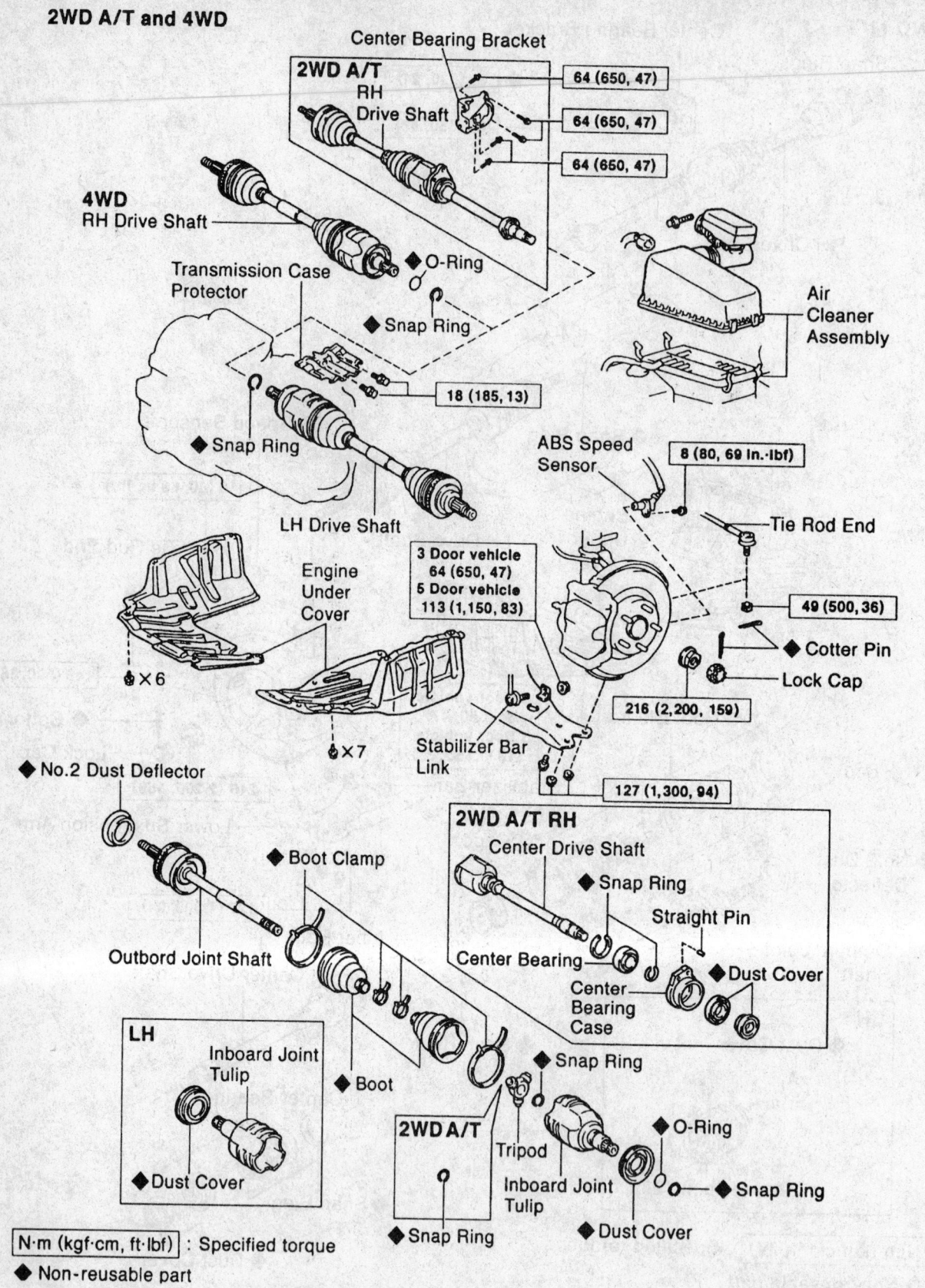

Center Bearing Bracket

2WD A/T RH Drive Shaft

64 (650, 47)

64 (650, 47)

64 (650, 47)

4WD RH Drive Shaft

Transmission Case Protector

◆ O-Ring

◆ Snap Ring

Air Cleaner Assembly

◆ Snap Ring

18 (185, 13)

LH Drive Shaft

ABS Speed Sensor

8 (80, 69 In.·lbf)

Tie Rod End

Engine Under Cover

3 Door vehicle 64 (650, 47)
5 Door vehicle 113 (1,150, 83)

49 (500, 36)

◆ Cotter Pin

Lock Cap

216 (2,200, 159)

◆×6

◆×7

Stabilizer Bar Link

127 (1,300, 94)

◆ No.2 Dust Deflector

2WD A/T RH

Center Drive Shaft

◆ Snap Ring

Straight Pin

◆ Boot Clamp

Center Bearing

◆ Dust Cover

Outbord Joint Shaft

Center Bearing Case

LH

Inboard Joint Tulip

◆ Snap Ring

◆ Boot

◆ Dust Cover

2WD A/T

Tripod

Inboard Joint Tulip

◆ O-Ring

◆ Snap Ring

◆ Dust Cover

◆ Snap Ring

N·m (kgf·cm, ft·lbf) : Specified torque

◆ Non-reusable part

Front halfshaft exploded view (2WD with A/T and 4WD) — RAV4

305920

36 ft. lbs. (49 Nm) and install a new cotter pin.

25. Install the cotter pin, lock cap, and the locknut to hold the halfshaft to the axle hub. Torque the nut to 159 ft. lbs. (216 Nm).

26. If equipped with ABS, install the ABS speed sensor with the bolt.

27. Fill the transaxle with gear oil (M/T) or ATF (A/T).

28. Install the engine under cover.

29. Install the wheels and lower the vehicle.

30. Connect the negative battery cable to the battery.

31. Check the ABS sensor signal.

Rear halfshaft

1. Disconnect the negative battery cable from the battery.

2. Raise and safely support the rear of the vehicle.

3. Remove the rear wheels.

4. If equipped with ABS brakes, remove the ABS speed sensor from the axle assembly by removing the bolt.

5. Remove the cotter pin, lock cap, and the nut holding the halfshaft to the axle carrier.

6. Place matchmarks on the halfshaft and side gear shaft.

7. Disconnect the halfshaft from the differential side gear shaft by removing the four nuts and washers.

8. Using a plastic hammer, disconnect the halfshaft from the axle carrier.

To install:

9. Install the halfshaft to the axle carrier.

10. Aligning the marks, connect the halfshaft to the differential side gear shaft with the four nuts. Torque the nuts to 41 ft. lbs. (56 Nm).

11. Install the nut, lock cap, and the cotter pin to hold the halfshaft to the axle carrier. Torque the nut to 152 ft. lbs. (206 Nm).

12. Install the ABS sensor with the bolt. Torque the bolt to 69 inch lbs. (8 Nm).

13. Install the rear wheels and lower the vehicle.

14. Connect the negative battery cable to the battery.

15. Check the ABS sensor signal.

CV-Joint

REMOVAL AND INSTALLATION

Previa

4WD Vehicles

1. Raise and safely support the vehicle.

2. Remove the wheel and tire assembly.

3. Remove the cotter pin and lock cap from the halfshaft.

4. Remove the locknut from the halfshaft.

5. Remove the cotter pin and locknut to the tie rod end and disconnect the tie rod end from the knuckle.

6. Remove the lower ball joint from the steering knuckle by removing the two mounting bolts.

7. Place matchmarks on the halfshaft and side gear.

8. Remove the six bolts from the inner halfshaft joint and disconnect the halfshaft from the side gear.

9. Remove the halfshaft by pulling the knuckle outward and remove the halfshaft from the wheel hub.

NOTE: If the outer shaft will not come out of the hub, soak the splines with penetrating lube, install the nut and tap on the halfshaft with a rubber hammer. Be careful not to damage the shaft threads.

10. Remove the inboard joint boot clamps.

11. Slide the inboard joint boot toward the outboard joint.

12. Place matchmarks on the inboard joint tulip and the shaft. Remove the inboard joint tulip from the drive shaft.

13. Using a snap ring expander remove the snap ring and disassemble the tripod joint.

14. Using a brass bar and hammer, remove the tripod joint from the drive shaft. Do not punch the roller.

15. Remove the inboard joint boot.

16. Remove the outboard joint boot clamps and boot. Do not disassemble the outboard joint.

To install:

17. Temporarily install the new boot and new boot clamps to the outboard joint.

NOTE: Before installing the boot, wrap vinyl tape around the spline of the shaft to prevent damaging the boot.

18. Temporarily install the new boot and the new boot clamps for the inboard joint to the drive shaft.

19. Assemble the tripod joint.

 a. Place the beveled side of the tripod axial spline toward the outboard joint.

 b. Align the matchmarks placed before disassembly.

 c. Using a brass bar and hammer, tap in the tripod joint onto the drive shaft. Do not punch the roller.

20. Using a snap ring expander, install a new snap ring.

21. Before assembling the boot to the outboard joint, pack the boot with grease.

22. Assemble the inboard joint to the inboard joint tulip. Pack in grease to the inboard tulip and the boot.

23. Assemble the boot clamps to both boots. Be sure that the boot is on the shaft groove and that it is not stretched or contracted.

NOTE: Coat the halfshaft splines with anti-seize compound to prevent spline seizure. This will help for future halfshaft removal.

24. Connect the halfshaft to the steering knuckle.

25. Connect the inner halfshaft joint to the side gear by installing the six bolts. Torque the bolts to 51 ft. lbs. (61 Nm).

26. Connect the lower ball joint and torque the bolts to 94 ft. lbs. (127 Nm).

27. Install the tie rod end, torque the nut to 36 ft. lbs. (49 Nm) and install a new cotter pin.

28. Install the halfshaft nut and torque to 152 ft. lbs. (206 Nm).

29. Install the lock cap and a new cotter pin to the halfshaft.

30. Install the wheel and lower the vehicle.

31. Check the front wheel alignment.

2WD with Manual Transaxle

1. Remove the halfshaft(s) and secure to a suitable holding device.

2. Using a flat prying tool, remove the two inboard joint boot clamps.

3. Using a side cutter, cut the small outboard joint boot clamp and remove them.

4. Using pliers, draw the hooks together and remove the large outboard joint boot clamp.

5. Place matchmarks on the inboard joint tulip or center driveshaft and the halfshaft. Do not punch the marks. Remove the inboard tulip from the halfshaft.

6. Place matchmarks on the halfshaft, the inner race, and the cage.

7. Remove the six balls and the cage.

8. Using a snapring expander, remove the snapring.

9. Remove the inner race by using a brass bar and hammer.

10. Remove the snapring. Be careful not to damage the outboard joint.

11. Slide out the inboard and outboard joint boots.

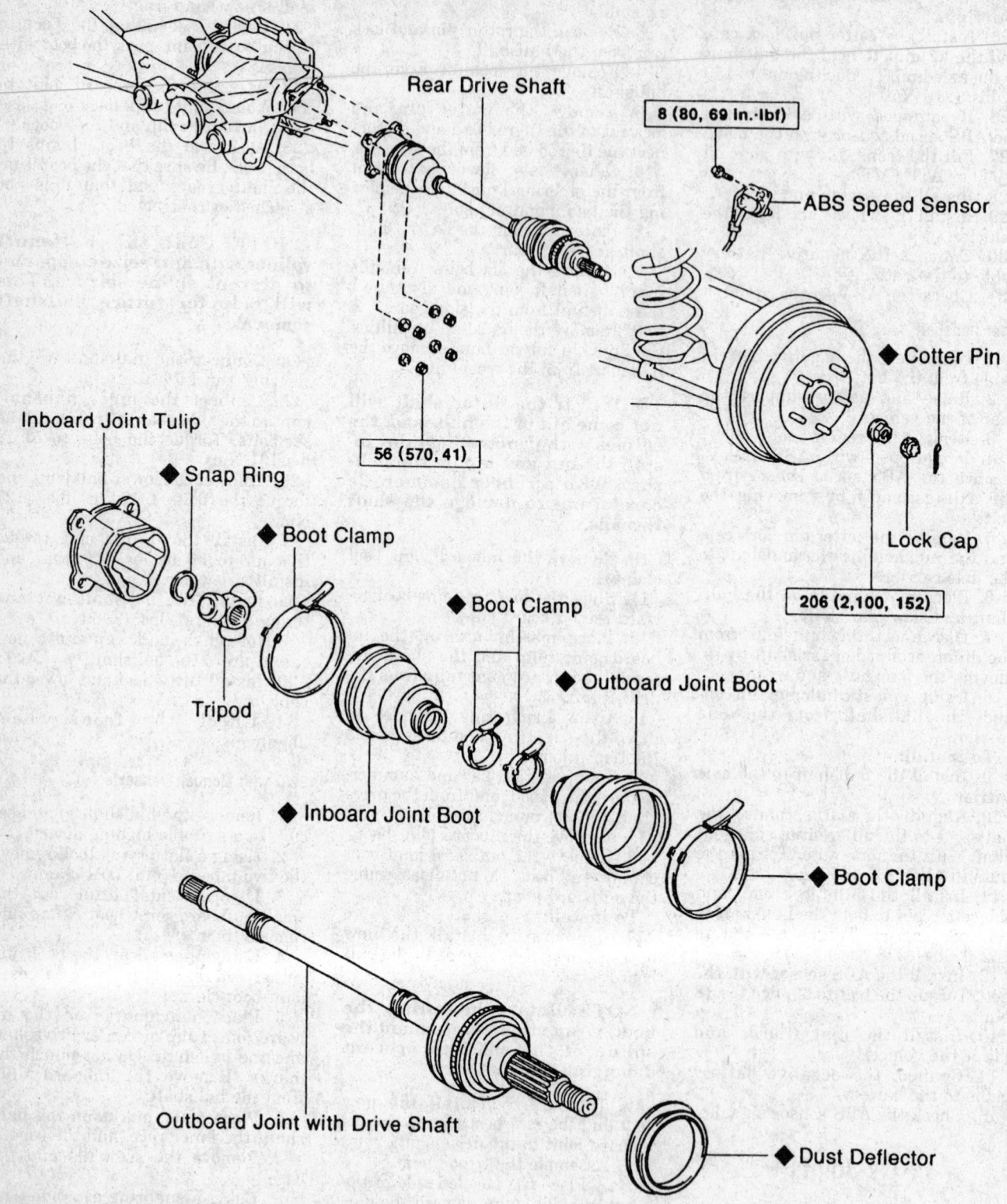

Rear Drive Shaft

8 (80, 69 in.·lbf)

ABS Speed Sensor

Cotter Pin

Inboard Joint Tulip

Snap Ring

Boot Clamp

Boot Clamp

56 (570, 41)

Outboard Joint Boot

Tripod

Lock Cap

206 (2,100, 152)

Inboard Joint Boot

Boot Clamp

Outboard Joint with Drive Shaft

Dust Deflector

N·m (kgf·cm, ft·lbf) : Specified torque

◆ Non-reusable part

Rear halfshaft removal and installation (4WD only) — RAV4

305972

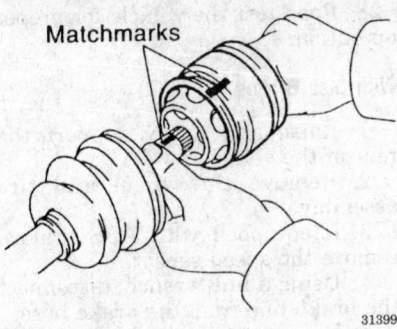

Matchmarks on the inboard tulip joint (2WD M/T) — RAV4

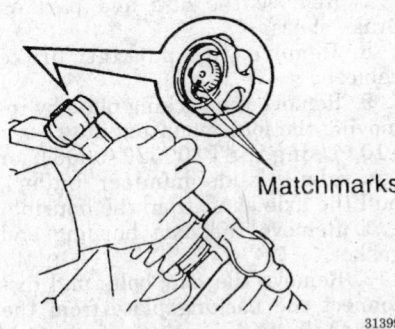

Removing the inner race (2WD M/T) — RAV4

To install:

12. Before install the two boots, wrap vinyl tape around the spline of the halfshaft to prevent damaging the boots.

13. Place a new clamp on the small ends of the new outboard joint boot and install the boot to the halfshaft.

14. Temporarily install a new inboard joint boot to the halfshaft.

15. Install a new snapring and install the cage to the halfshaft.

NOTE: The side with the smaller diameter must face outboard.

Installing the boot (2WD M/T) — RAV4

16. Align the matchmarks and using a brass bar and hammer, tap in the inner race to the halfshaft. Install a new snapring.

17. Install the outboard boot to the joint and pack with grease.

18. Pack the boot and the inboard tulip joint with grease.

19. Align the matchmarks placed before removal and install the cage to the inner race.

20. Apply grease to the six balls and install them.

21. Align the matchmarks and install the inboard joint tulip or the center drive shaft to the halfshaft.

22. Install a new snapring and install the boot to the inboard tulip joint.

23. Make sure that the two boots are on the shaft groove.

24. Place two new boot clamps to the inboard joint boot.

25. Bend the band and lock it. Holding the clamp near the closing hooks, position the holes in the clamp's free end over the closing hooks.

26. Secure the clamps by drawing the closing hooks together. Secure the clamp onto the boot.

27. Place SST 09521–24010 or equivalent onto the clamp. Tighten the SST so that the clamp is pinched.

28. Using SST 09240–00020 or equivalent, adjust the clearance of the clamp. Clearance should be 0.075 in. or less.

2WD with Automatic Transaxle and 4WD

1. Remove the halfshaft(s) and secure to a suitable holding device.

2. Using a flat prying tool, remove the inboard and the outboard joint boot clamps.

3. Place matchmarks on the tripod, the inboard tulip joint or center driveshaft and the halfshaft.

4. Remove the inboard tulip joint from the driveshaft.

5. Remove the tripod.

   a. Remove the snapring.

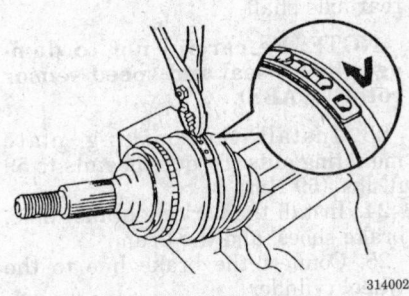

Installing the boot clamp (2WD M/T) — RAV4

   b. 2WD A/T: Slide the snapring toward the outboard joint side.

   c. Place matchmarks on the driveshaft and the tripod.

   d. Using a brass bar and hammer, remove the tripod from the driveshaft.

   e. 2WD A/T: Remove the snapring.

6. Remove the inboard and outboard joint boots.

To install:

7. Before installing the two boots, wrap vinyl tape around the spline of the halfshaft to prevent damaging the boots.

8. Temporarily install a new outboard joint boot to the halfshaft.

9. Temporarily install a new inboard joint boot to the halfshaft.

10. Install the tripod.

   a. 2WD A/T: Install a new snapring.

   b. Place the beveled side of the tripod axial spline toward the outboard joint.

   c. Align the matchmarks placed before removal.

   d. Using a brass bar and hammer, tap in the tripod to the halfshaft. Install a new snapring.

11. Install the outboard joint boot and pack with grease.

12. Pack the boot and the inboard tulip joint with grease. Align the matchmarks and install the inboard tulip joint to the center driveshaft or the halfshaft.

13. Make sure that the two boots are on the shaft groove. Place four new boot clamps to the two boots and bend the band to lock it.

Rear Axle Shaft, Bearing and Seal

REMOVAL AND INSTALLATION

Previa

With Drum Brakes

1. Raise and safely support the rear of the vehicle.

2. Remove the wheel and tire assembly.

3. Remove the brake drum.

4. If equipped with ABS, remove the speed sensor.

5. Using a line wrench, disconnect the brake line from the wheel cylinder.

6. Remove the brake shoes from the vehicle.

7. Remove the two bolts and remove the parking brake cable from the backing plate.

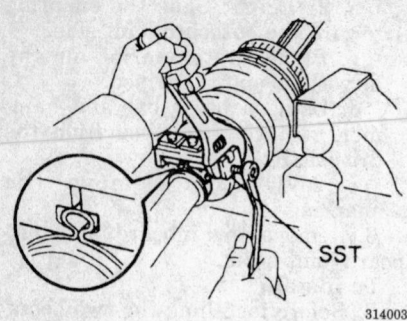

Tightening the SST to pinch the clamp (2WD M/T) — RAV4

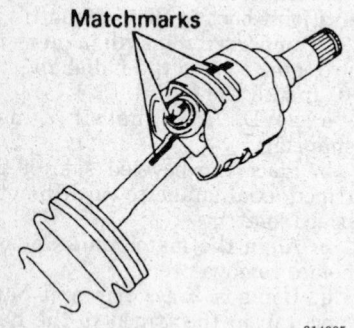

Matchmarks

Matchmarks on the tripod (2WD A/T and 4WD) — RAV4

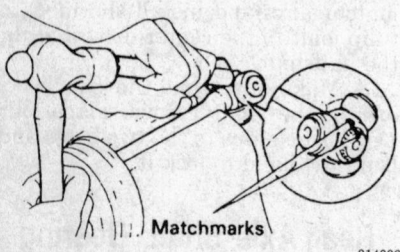

Matchmarks

Removing the tripod from the drive shaft (2WD A/T and 4WD) — RAV4

8. Working through the hole in the axle flange, remove the four backing plate mounting nuts.
9. Using SST 09520–00031 or equivalent (slide hammer puller), pull the axle shaft from the housing.
10. Remove the backing plate.
11. Remove the end gasket to the axle housing.
12. If equipped with ABS, press the seal and speed sensor rotor from the axle shaft.
13. Using a grinder, grind down the inner bearing retainer on the axle shaft. Using a chisel and a hammer,

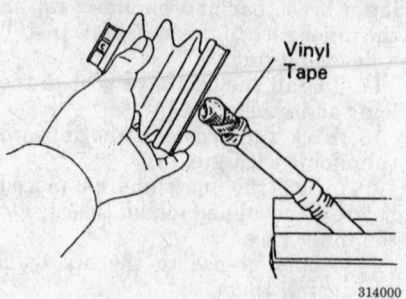

Vinyl Tape

Installing the boot (2WD A/T and 4WD) — RAV4

cut off the retainer and remove it from the shaft.

———— WARNING ————
When removing the bearing, be careful not to damage the axle shaft.

14. Using a press, press the bearing from the axle shaft.
15. Remove the bearing outer retainer.
   To install:
16. Install the bearing outer retainer to the axle shaft.
17. Using SST 09506–30012 or equivalent and a press, install a new bearing.
18. Heat the new inner retainer to approximately 302°F (150°C) in an oil bath. Using a suitable installer and a press, install the inner retainer to the axle shaft while the retainer is still hot.

NOTE: Face the non-beveled side of the inner retainer toward the bearing.

19. If equipped with ABS, carefully install the speed sensor rotor. Using a suitable installer and a press, install the new oil seal.
20. Apply liquid sealant on a new axle housing gasket and install the end gasket on the rear axle housing.
21. Install the backing plate.
22. Using a suitable tool, install the rear axle shaft.

NOTE: Be careful not to damage the oil seal and speed sensor rotor (w/ABS).

23. Install the backing plate mounting nuts. Torque the nuts to 59 ft. lbs. (59 Nm).
24. Install the parking brake cable, brake shoes, and the drum.
25. Connect the brake line to the wheel cylinder.
26. Bleed the brake system.
27. Install the rear wheel and lower the vehicle.

28. Road test the vehicle for proper operation.

With Disc Brakes

1. Raise and safely support the rear of the vehicle.
2. Remove the wheel and tire assembly.
3. If equipped with ABS brakes, remove the speed sensor.
4. Using a line wrench, disconnect the brake line from the brake hose.
5. Disconnect the brake hose from the axle bracket by removing the clip.
6. Remove the brake caliper support by removing the two bolts.
7. Remove the disc and parking brake shoes.
8. Remove the parking brake cable.
9. Remove the backing plate by removing the four mounting nuts.
10. Using SST 09520–00031 or equivalent (slide hammer puller), pull the axle shaft from the housing.
11. Remove the axle housing end gasket.
12. Remove the four bolts and disconnect the backing plate from the axle shaft.
13. If equipped with ABS, press the seal and speed sensor rotor from the axle shaft.
14. Using a grinder, grind down the inner bearing retainer on the axle shaft. Using a chisel and a hammer, cut off the retainer and remove it from the shaft.

———— WARNING ————
When removing the bearing, be careful not to damage the axle shaft.

15. Using a press, press the bearing from the axle shaft.
16. Remove the bearing outer retainer.
   To install:
17. Place a new retainer gasket and bearing on the backing plate. Using a socket wrench and hammer, install the four bolts for the backing plate.
18. Install the backing plate to the axle shaft.
19. Using SST 09506–30012 or equivalent and a press, install a new bearing.
20. Heat the new inner retainer to approximately 302°F (150°C) in an oil bath. Using a suitable installer and a press, install the inner retainer to the axle shaft while the retainer is still hot.

NOTE: Face the non-beveled side of the inner retainer toward the bearing.

[DRUM BRAKE]

[DISC BRAKE]

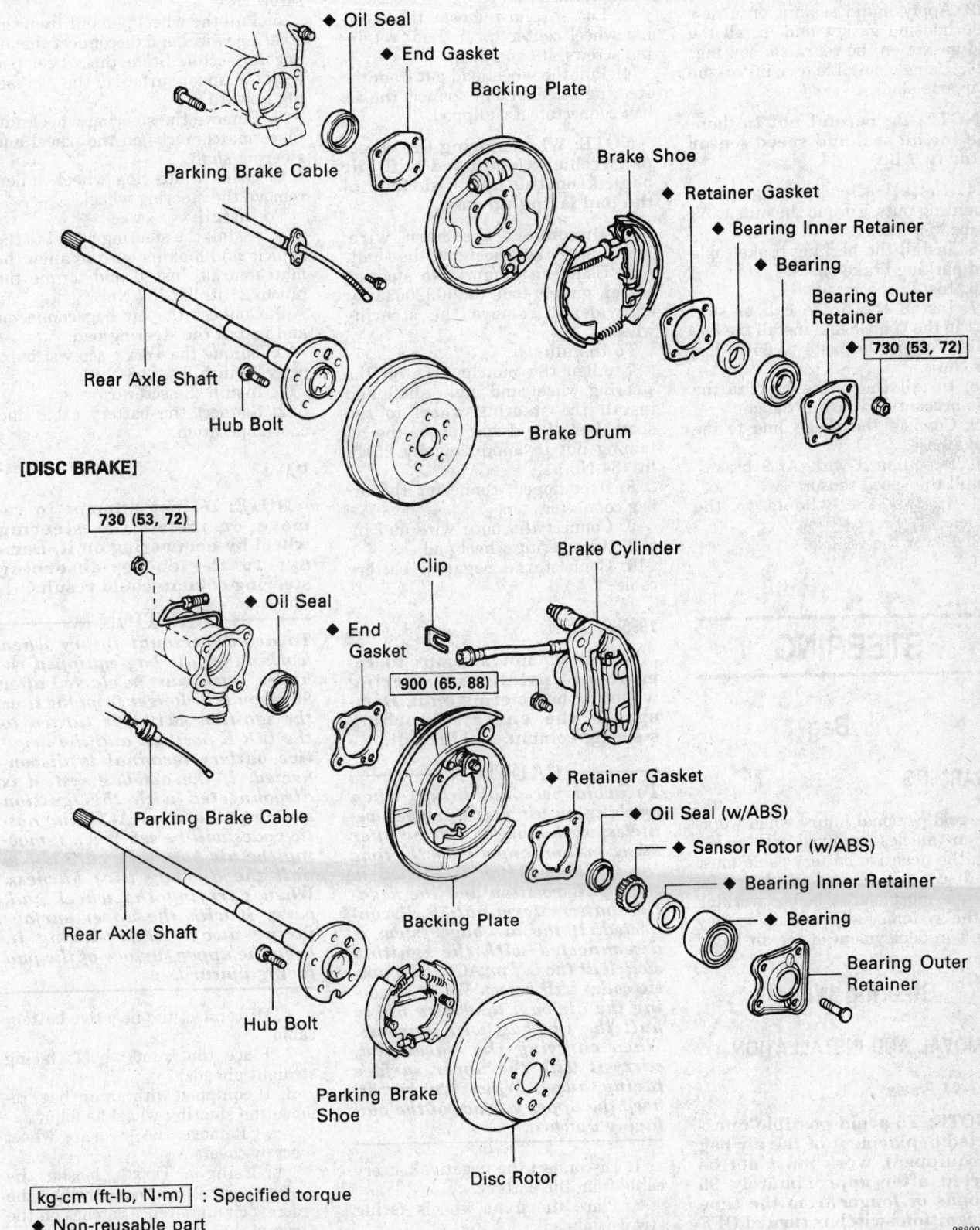

kg-cm (ft-lb, N·m) : Specified torque

◆ Non-reusable part

288005

Rear axle shaft components, exploded view — Previa with drum brakes

21. If equipped with ABS, carefully install the speed sensor rotor. Using a suitable installer and a press, install the new oil seal.

22. Apply liquid sealant on a new axle housing gasket and install the end gasket on the rear axle housing.

23. Using a suitable tool, install the rear axle shaft.

NOTE: Be careful not to damage the oil seal and speed sensor rotor (w/ABS).

24. Install the backing plate mounting nuts. Torque the nuts to 59 ft. lbs. (59 Nm).

25. Install the parking brake cable and parking brake shoes.

26. Install the rotor.

27. Install the brake caliper support to the vehicle and install the two bolts. Torque the bolts to 65 ft. lbs. (88 Nm).

28. Install the brake hose to the axle bracket and install the clip.

29. Connect the brake line to the brake hose.

30. If equipped with ABS brakes, install the speed sensor.

31. Install the wheels to the vehicle.

32. Lower the vehicle.

STEERING

Air Bag

DISARMING

To avoid personal injury when working on vehicles equipped with an air bag, the negative battery cable must be disconnected and at least 90 seconds must elapse before working on the system. Failure to do so may result in deployment of the air bag.

Steering Wheel

REMOVAL AND INSTALLATION

1993–94 Previa

NOTE: To avoid possible unexpected deployment of the air bag (if equipped), work must not be started after approximately 90 seconds or longer from the time the ignition switch is turned OFF and the negative battery cable disconnected.

1. Disconnect the negative battery cable.

2. Position the wheels in a straight ahead position.

3. Loosen and remover the steering wheel center cover (pad) retaining screws, if equipped.

4. Pull the wheel pad out from the steering wheel and disconnect the air bag connector, if equipped.

NOTE: When storing the wheel pad (vehicles equipped with air bag), keep the upper surface of the pad facing upward.

5. Disconnect the horn wire. Matchmark the wheel and the shaft.

6. Using an appropriate steering wheel puller tool (0960920011 or equivalent), remove the steering wheel.

To install:

7. Align the matchmarks on the steering wheel and main shaft and install the steering wheel to the shaft. Install and and torque the retaining nut to approximately 25 ft. lbs (34 Nm).

8. If equipped, connector the air bag connector.

9. Connect the horn wire and install the steering wheel pad.

10. Connect the negative battery cable.

1995–97 Previa

NOTE: Do not attempt to remove or install the steering wheel by hammering on it. Damage to the energy-absorbing steering column could result.

--- CAUTION ---

To avoid personal injury when working on air bag equipped vehicles, work must be started after 90 seconds or longer from the time the ignition switch is turned to the LOCK position and the negative battery terminal is disconnected. If the air bag system is disconnected with the ignition switch at the ON or ACC, diagnostic codes will be set. When removing the air bag, take care not to pull the air bag wire harness. When carrying the wheel pad, carry it with the upper surface facing away. When storing it, keep the upper surface of the pad facing upward.

1. Disconnect the negative battery cable from the battery.

2. Place the front wheels facing straight ahead.

3. Remove the steering wheel screw covers.

4. Using a Torx®, loosen the screws until the groove trailing the screw circumference catches on the screw case.

5. Pull the wheel pad out from the steering wheel and disconnect the air bag connector. Store the wheel pad with the upper surface of the pad facing upward.

6. Remove the steering wheel nut. Place matchmarks on the wheel and steering shaft.

7. Using a steering wheel puller, remove the steering wheel.

To install:

8. Install the steering wheel to the vehicle and making sure to align the matchmarks. Install and torque the nut to 25 ft. lbs. (34 Nm).

9. Connect the air bag connector and install the steering pad.

10. Torque the Torx® screws to exactly 78 inch lbs. (8.8 Nm).

11. Install the screw covers.

12. Connect the battery cable and check operation.

RAV4

NOTE: Do not attempt to remove or install the steering wheel by hammering on it. Damage to the energy-absorbing steering column could result.

--- CAUTION ---

To avoid personal injury when working on air bag equipped vehicles, work must be started after 90 seconds or longer from the time the ignition switch is turned to the LOCK position and the negative battery terminal is disconnected. If the air bag system is disconnected with the ignition switch at the ON or ACC, diagnostic codes will be set. When removing the air bag, take care not to pull the air bag wire harness. When carrying the wheel pad, carry it with the upper surface facing away. When storing it, keep the upper surface of the pad facing upward.

1. Disconnect the negative battery cable.

2. Place the front wheels facing straight ahead.

3. If equipped with an air bag, remove the steering wheel as follows;

   a. Remove the steering wheel screw covers.

   b. Using a Torx®, loosen the screws until the groove trailing the screw circumference catches on the screw case.

   c. Pull the wheel pad out from the steering wheel and disconnect the air bag connector. Store the wheel pad with the upper surface of the pad facing upward.

d. Remove the steering wheel nut. Place matchmarks on the wheel and steering shaft.

e. Using a steering wheel puller, remove the steering wheel.

4. If equipped with no air bag, remove the steering wheel as follows:

a. Remove the screw located at the bottom of the steering pad.

b. Remove the three clips.

c. Remove the steering pad from the steering wheel.

d. Remove the steering wheel nut.

e. Using a steering wheel puller, remove the steering wheel.

To install:

5. If equipped with an air bag, install the steering wheel as follows:

a. Install the steering wheel to the vehicle and making sure to align the matchmarks. Install and torque the nut to 25 ft. lbs. (34 Nm).

b. Connect the air bag connector and install the steering pad.

c. Torque the Torx® screws to exactly 78 inch lbs. (8.8 Nm).

d. Install the screw covers.

6. If equipped with no air bag, install the steering wheel as follows:

a. Install the steering wheel and nut. Torque the nut to 25 ft. lbs. (34 Nm).

b. Install the steering wheel pad and connect the three clips and screw.

7. Connect the battery cable and check operation.

Tie Rod Ends

REMOVAL AND INSTALLATION

Previa

1. Raise and safely support the vehicle. Remove the front wheels.

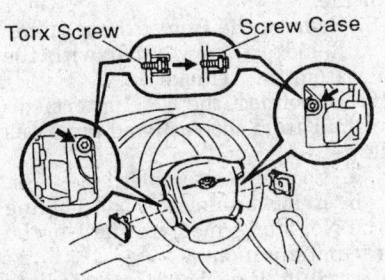

Steering pad screws (with air bag) — RAV4

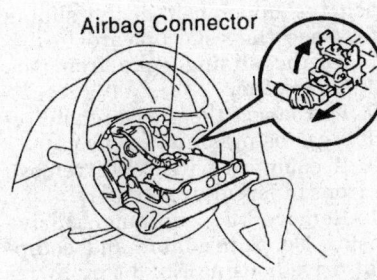

Storing the air bag — RAV4

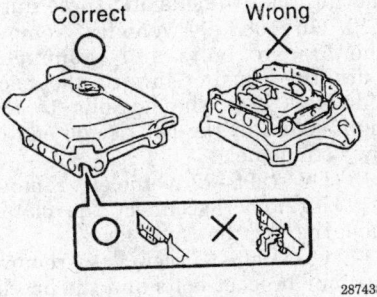

Removing the steering wheel — RAV4

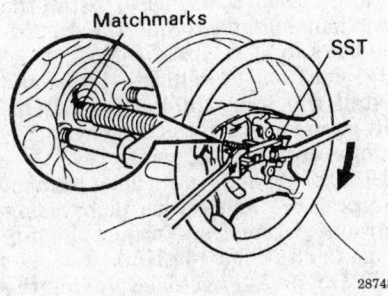

Removing the tie rod — Previa

2. Working at the steering knuckle arm, pull out the cotter pin and then remove the nut to the tie rod.

3. Using a tie rod end puller, disconnect the tie rod from the steering knuckle arm.

4. Loosen the locknut for the tie rod. Turn the tie rod in the opposite direction of the lock nut until it is removed from steering rack.

NOTE: When removing the tie rods, count the amount of turns needed to remove the tie rod. This will help with getting the steering alignment close when installing the tie rod.

To install:

5. Install the tie rod end.

6. Install the tie rod the same amount of turns that were needed to remove the tie rod.

7. Tighten the tie rod end to steering knuckle nut to 36 ft. lbs. (49 Nm). Install a new cotter pin.

NOTE: If the hole on the tie rod does not line up with the nut, always tighten the nut until the hole lines up.

8. Install the front wheels and lower the vehicle. Check the front end alignment.

9. Torque the tie rod locknut to 67 ft. lb (91 Nm).

10. Check steering wheel center point and front end alignment. Adjust the alignment as needed.

RAV4

1. Raise and safely support the vehicle. Remove the wheel.

2. Remove the cotter pin and nut holding the tie rod to the steering knuckle.

3. Using a tie rod separator, press the tie rod out of the knuckle.

NOTE: Use only the correct tool to separate the tie rod joint. Replace the boot if the rubber is cracked or ripped.

4. Matchmark the inner end of the tie rod to the end of the steering rack.

5. Loosen the locknut and remove the tie rod from the steering rack. Count the amount of turns it takes to remove the tie rod. This will help in getting the alignment close when reinstalling.

To install:

6. Install the tie rod ends (the same number of turns as required to remove it) onto the rack ends and align the matchmarks made earlier.

7. Tighten the locknuts to 41 ft. lbs. (56 Nm).

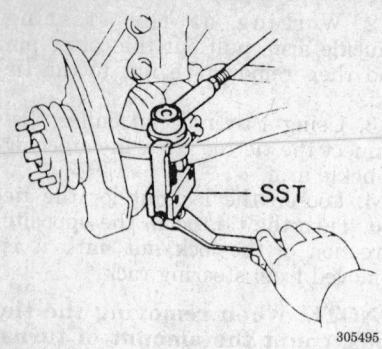

Separating the outer tie rod end from the steering knuckle — RAV4

305495

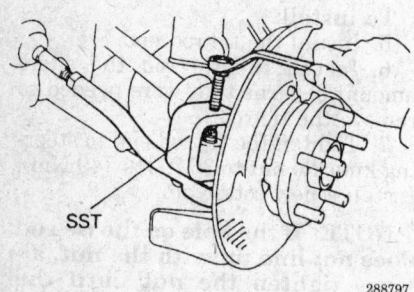

Removing the tie rod — RAV4

288797

8. Connect the tie rod joint to the knuckle. Tighten the nut to 36 ft. lbs. (49 Nm) and install a new cotter pin.

NOTE: If the hole on the tie rod does not line up with the nut, always tighten the nut until the hole lines up.

9. Install the wheel.
10. Lower the vehicle.
11. Check and adjust the front toe as necessary.

Rack and Pinion

REMOVAL AND INSTALLATION

Previa

1. Remove the battery from the vehicle.
2. Raise the vehicle and support safely.
3. Remove the front wheels.
4. Remove the engine undercovers.
5. Remove the cotter pins and the nuts to the tie rod ends.
6. Using a separator tool (SST 0961112010 or equivalent), disconnect the tie rod ends from the steering knuckle.

7. Place matchmarks on the universal joint and steering column shaft.
8. Remove the lower bolt and loosen the upper bolt to the sliding yoke. Slide the yoke upward to disconnect the sliding yoke from the rack and pinion.
9. Disconnect the pressure and return pipes using a line wrench.
10. If equipped with 4WD, remove the front differential assembly.
11. Remove the equipment drive housing No. 2 insulator and equipment drive housing No. 3 stay by removing the two bolts and three nuts.
12. On 1993–94 vehicles, remove the bracket bolts and grommets. Turn the housing toward the back side and slide the housing to the right side. Put the left tie rod end in the body panel.
13. On 1995–97 vehicles, remove the bolt and disconnect the clamp from the rack housing.
14. On 1995–97 vehicles, remove the four bracket bolts and the brackets from the steering rack.
15. Pull the housing out through the opening in the left side of the body.
16. On 1995–97 vehicles, remove the grommets from the steering rack.

To install:

17. On 1993–94 vehicles, install the gear housing and torque the mounting bolts to 56 ft. lbs. (76 Nm).
18. On 1993–94 vehicles, 4WD only: install the front differential assembly, equipment driveshaft housing insulator and drive housing stay.
19. On 1993–94 vehicles, connect the pressure and return pipes using flarenut wrenches. Torque the fittings to 33 ft. lbs. (44 Nm).
20. On 1995–97 vehicles, install the grommets to the rack and pinion.
21. On 1995–97 vehicles, install the gear housing and torque the mounting bolts to 70 ft. lbs. (95 Nm).
22. On 1995–97 vehicles, connect the power steering lines to the power steering rack and pinion. Torque the steering lines to 27 ft. lbs. (36 Nm).
23. On 1995–97 vehicles, install the bolt to hold the clamp and power steering lines to the steering rack.
24. On 1995–97 vehicles, install the equipment drive housing No. 2 insulator and equipment drive housing No. 3 stay by installing the two bolts and three nuts. Torque the bolts to 13 ft. lbs. (18 Nm) and the nuts to 18 ft. lbs. (25 Nm).

NOTE: The following steps are for all vehicles unless otherwise specified.

25. Install the lower bolt to the sliding yoke. Torque the upper and lower bolts to 26 ft. lbs. (35 Nm).
26. On 1995–97 vehicles, 4WD only: install the front differential assembly.
27. Connect the tie rod ends to the steering knuckle and torque the nuts to 36 ft. lbs. (49 Nm). Install a new cotter pin.
28. Install the engine undercovers.
29. Install the battery to the vehicle.
30. Install the front wheels and lower the vehicle.
31. Perform a front end alignment.

RAV4

1. Disconnect the negative battery cable from the battery.

─────── **CAUTION** ───────
To avoid personal injury when working on air bag equipped vehicles, work must be started after 90 seconds or longer from the time the ignition switch is turned to the LOCK position and the negative battery terminal is disconnected. If the air bag system is disconnected with the ignition switch at the ON or ACC, diagnostic codes will be set. When removing the air bag, take care not to pull the air bag wire harness. When carrying the wheel pad, carry it with the upper surface facing away. When storing it, keep the upper surface of the pad facing upward.

2. Turn the key to the lock position and lock the steering wheel in place.
3. Place a drain pan under the steering rack.
4. Raise and safely support the front of the vehicle.
5. Remove the front wheels.
6. Remove the right and left hand engine under covers.
7. Disconnect the right and left hand tie rod ends from the steering knuckle.
8. Remove the front exhaust pipe.
9. Remove the sway bar with the links from the vehicle.
10. Disconnect the No. 2 intermediate shaft from the rack and pinion as follows:
   a. Loosen the top bolt.
   b. Remove the lower bolt holding the No. 2 intermediate shaft to the rack and pinion.
   c. Shift the No. 2 intermediate shaft and place matchmarks on the control valve shaft and the No. 2 intermediate shaft.

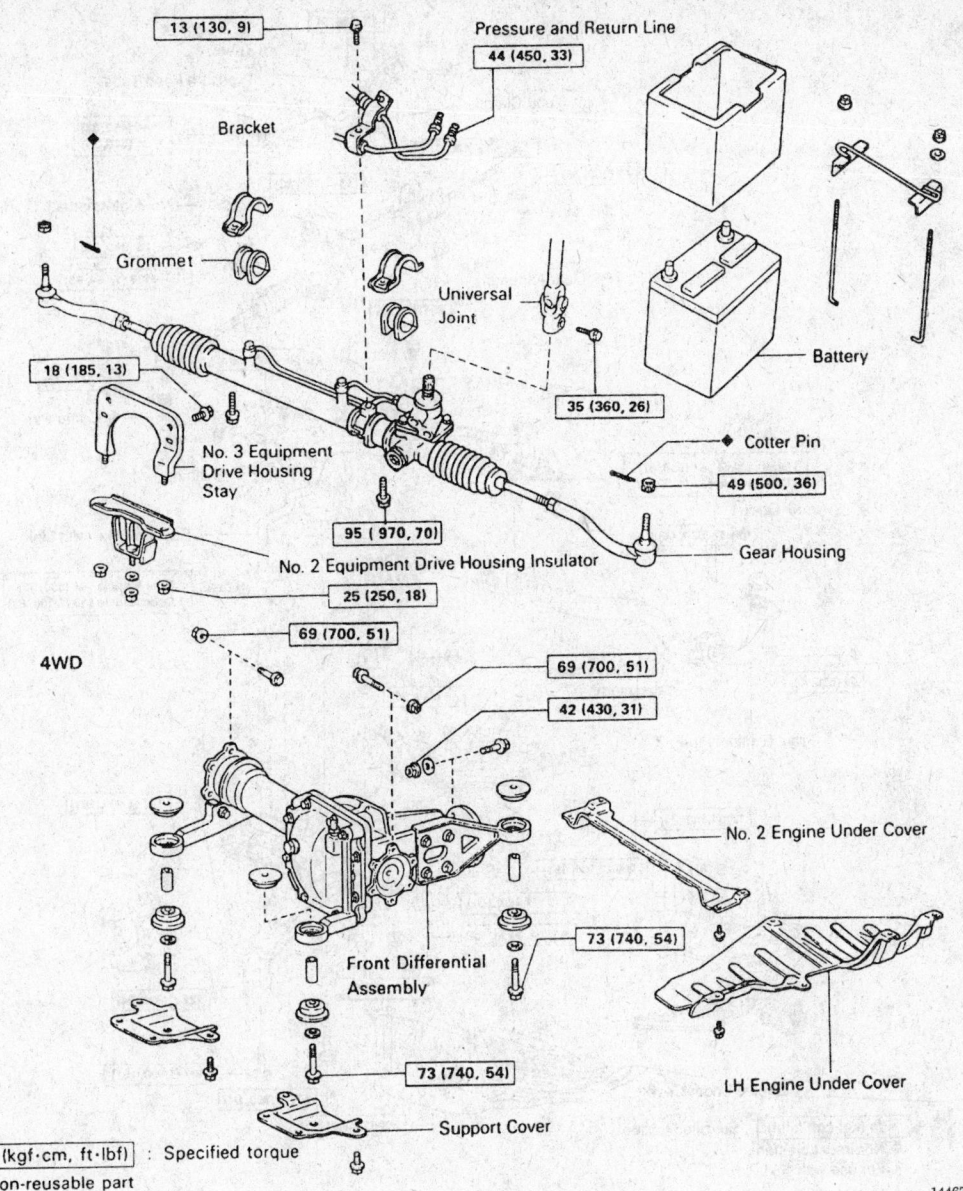

13 (130, 9)

Pressure and Return Line
44 (450, 33)

Bracket

Grommet

Universal Joint

18 (185, 13)

No. 3 Equipment Drive Housing Stay

35 (360, 26)

Battery

Cotter Pin
49 (500, 36)

95 (970, 70)

No. 2 Equipment Drive Housing Insulator

Gear Housing

25 (250, 18)

69 (700, 51)

4WD

69 (700, 51)

42 (430, 31)

No. 2 Engine Under Cover

73 (740, 54)

Front Differential Assembly

73 (740, 54)

LH Engine Under Cover

Support Cover

N·m (kgf·cm, ft·lbf) : Specified torque

◆ Non-reusable part

144677

Rack and pinion removal and installation — Previa

d. Disconnect the No. 2 shaft from the rack and pinion.

11. Using a line wrench, disconnect the pressure feed and return tubes from the rack and pinion.

12. Disconnect the pressure feed and return tube clamps by removing the bolt.

13. Disconnect the right and left lower control arms from the steering knuckle.

14. Remove the front suspension crossmember assembly as follows:

a. Remove the two centermember set nuts holding the centermember to the middle of the crossmember.

b. Remove the two rack and pinion assembly set bolts and nuts from the crossmember. Securely suspend the steering gear assembly.

c. Support the suspension crossmember with a jack.

d. Remove the two bolts from the suspension crossmember.

e. Remove the suspension crossmember with the lower suspension arms.

15. Remove the rack and pinion from the vehicle.

To install:

16. Install the rack and pinion to the vehicle.

17. Install the crossmember to the vehicle as follows:

a. Raise the suspension crossmember with the lower control arms. Install the two bolts to hold the crossmember to the vehicle. Torque the bolts to 152 ft. lbs. (206 Nm).

b. Connect the rack and pinion and install the two bolts and nuts. Torque to 83 ft. lbs. (113 Nm).

c. Connect the centermember to the crossmember and install the two set nuts. Torque the nuts to 82 ft. lbs. (112 Nm).

18. Connect the right and left lower control arms.

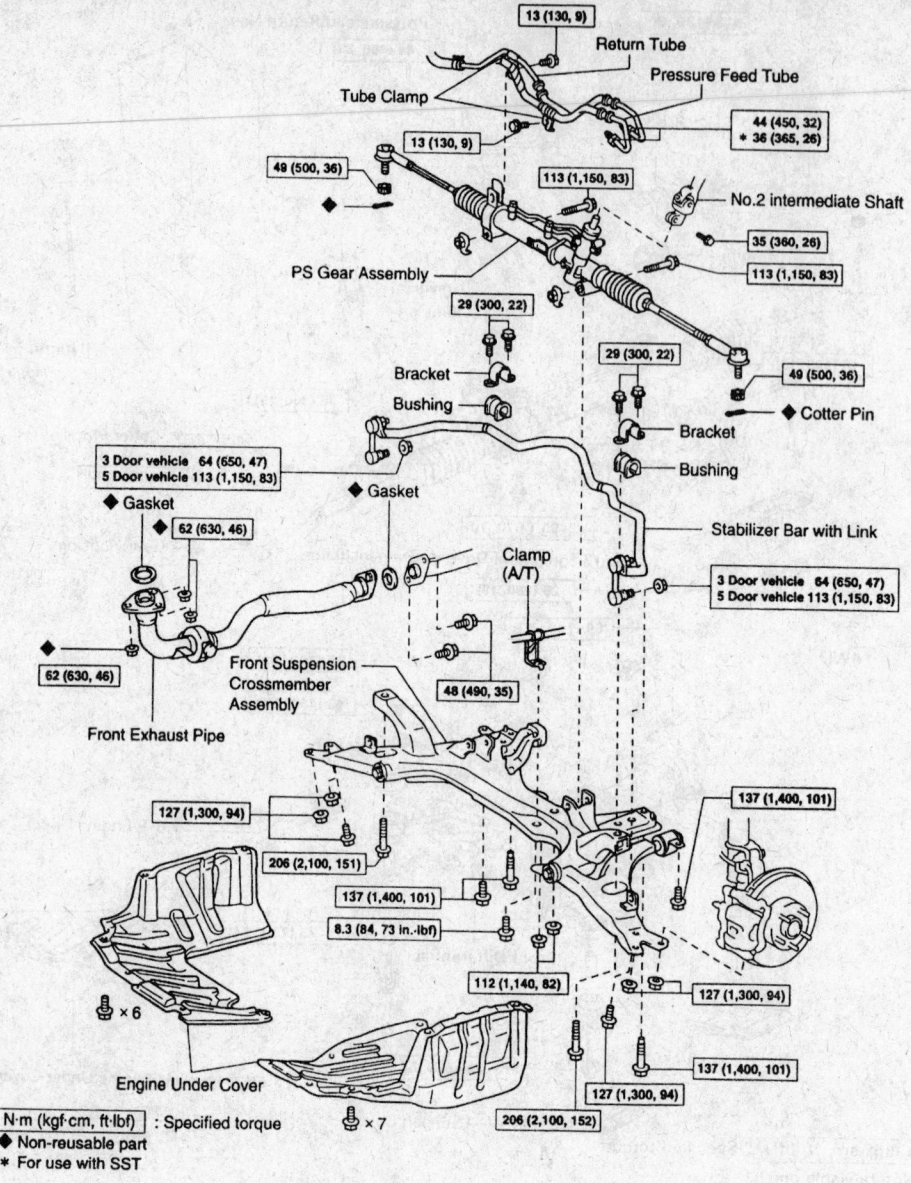

13 (130, 9)   Return Tube

Tube Clamp   Pressure Feed Tube

44 (450, 32)
* 36 (365, 26)

13 (130, 9)

49 (500, 36)

113 (1,150, 83)   No.2 intermediate Shaft

35 (360, 26)

PS Gear Assembly   113 (1,150, 83)

29 (300, 22)

29 (300, 22)

Bracket   49 (500, 36)

Bushing   Bracket   Cotter Pin

Bushing

3 Door vehicle 64 (650, 47)
5 Door vehicle 113 (1,150, 83)

Stabilizer Bar with Link

◆ Gasket   ◆ Gasket

◆ 62 (630, 46)   3 Door vehicle 64 (650, 47)
5 Door vehicle 113 (1,150, 83)

Clamp
(A/T)

Front Suspension
Crossmember
Assembly

48 (490, 35)

62 (630, 46)

137 (1,400, 101)

Front Exhaust Pipe

127 (1,300, 94)

206 (2,100, 151)

137 (1,400, 101)

8.3 (84, 73 in.·lbf)

112 (1,140, 82)   127 (1,300, 94)

× 6   137 (1,400, 101)

Engine Under Cover   127 (1,300, 94)

N·m (kgf·cm, ft·lbf) : Specified torque   × 7   206 (2,100, 152)
◆ Non-reusable part
* For use with SST

305216

Rack and pinion exploded view — RAV4

19. Connect the pressure feed and return tubes clamps.
20. Connect the pressure feed and return tubes to the rack and pinion. Torque the tubes to 26 ft. lbs. (36 Nm) using a torque wrench with a fulcrum length of 11.81 inches (300mm).
21. Connect the steering column No. 2 intermediate shaft to the rack and pinion. Align the marks and torque the upper and lower pinch bolts to 26 ft. lbs. (35 Nm).
22. Connect the stabilizer bar links. Torque the nuts to 22 ft. lbs. (29 Nm).
23. Install the front exhaust pipe with the two bolts, three nuts and two gaskets. Torque the bolts to 35 ft.

lbs. (48 Nm) and the nuts to 46 ft. lbs. (62 Nm).
24. Connect the right and left hand tie rod ends to the steering knuckle. Torque the nuts to 36 ft. lbs. (49 Nm) and install new cotter pins.
25. Install the right and left hand engine under covers.
26. Fill the power steering unit and bleed the system. Check for leaks.
27. Install the front wheels and lower the vehicle. Check the front wheel alignment.

Power Steering Pump

REMOVAL AND INSTALLATION

Previa

1. Disconnect the negative battery cable from the battery.
2. Remove the air duct by removing the six screws.
3. Remove the upper fan shroud by removing the four bolts.
4. Remove the engine cooling fan with coupling by removing the four nuts.
5. Place a suitable drain pan into position and disconnect the return hose from the power steering pump.

Drain the fluid from the power steering pump.

6. On 1995 non-supercharged engines: loosen the bolt and nut to the alternator and remove the drive belt from the engine.

7. On 1995 supercharged and 1996 engines: disconnect the pressure feed tube bracket by removing the bolt.

8. Disconnect the pressure tube from the power steering pump.

9. On 1995 supercharged and 1996 engines: loosen the adjusting bolt and the No. 1 idler pulley nut and remove the drive belt from the engine.

10. Remove the pump mounting bolts and remove the power steering pump from the vehicle.

To install:

11. Install the power steering pump to the engine and install the four bolts. Torque the bolts as follows:

• Bolt A: 35 ft. lbs. (48 Nm)
• Bolt B: 27 ft. lbs. (36 Nm)

12. On 1995 non-supercharged engines: install the drive belt and adjust the tension by pulling back on the alternator. Torque the bolt for the alternator to 37 ft. lbs. (50 Nm) and the nut to 13 ft. lbs. (18 Nm).

13. On 1995 supercharged and 1996 engines: install the drive belt

and tighten the adjusting bolt and No. 1 idler pulley nut.

14. Connect the return hose to the power steering pump.

15. Connect the pressure feed line by installing the two gaskets and union bolt. Torque the bolt to 36 ft. lbs. (49 Nm).

16. On 1995 supercharged and 1996 engines: connect the pressure feed tube bracket by install the bolt. Torque the bolt to 10 ft. lbs. (13 Nm).

17. Install the engine cooling fan with coupling by installing the four nuts. Torque the nuts to 10 ft. lbs. (13 Nm).

18. Install the upper fan shroud with the four bolts.

19. Install the air duct with the four screws.

20. Fill and bleed the power steering system.

21. Connect the negative battery cable to the battery.

RAV4

1. Disconnect the negative battery cable from the battery.

2. Place a drain pan under the power steering pump.

3. Raise and safely support the front of the vehicle.

4. Remove the front wheels.

5. Remove the right and left hand engine under covers.

6. Remove the front exhaust pipe.

7. Remove the sway bar with the links from the vehicle.

8. Disconnect the right and left lower control arms from the steering knuckle.

9. Remove the front suspension crossmember assembly as follows:

a. Remove the two centermember set nuts holding the centermember to the middle of the crossmember.

b. Remove the two rack and pinion assembly set bolts and nuts from the crossmember. Securely suspend the steering gear assembly.

c. Support the suspension crossmember with a jack.

d. Remove the two bolts from the suspension crossmember.

e. Remove the suspension crossmember with the lower suspension arms.

10. Disconnect the tube clamps by removing the bolt.

11. Disconnect the two vacuum hoses from the steering pump.

12. Disconnect the return hose from the steering pump.

13. Loosen the two bolts to the power steering pump and remove the drive belt.

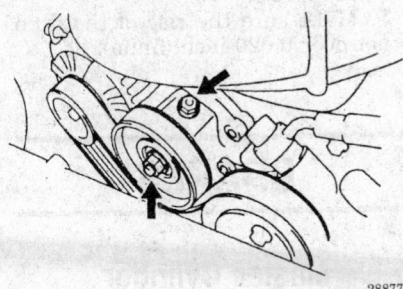

288770

Drive belt tensioner bolt and nut — Previa

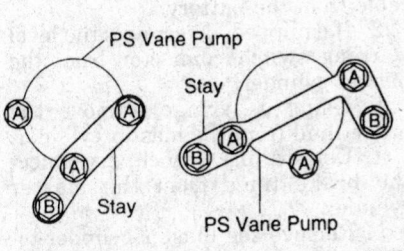

250723

Torquing the power steering bolts — Previa

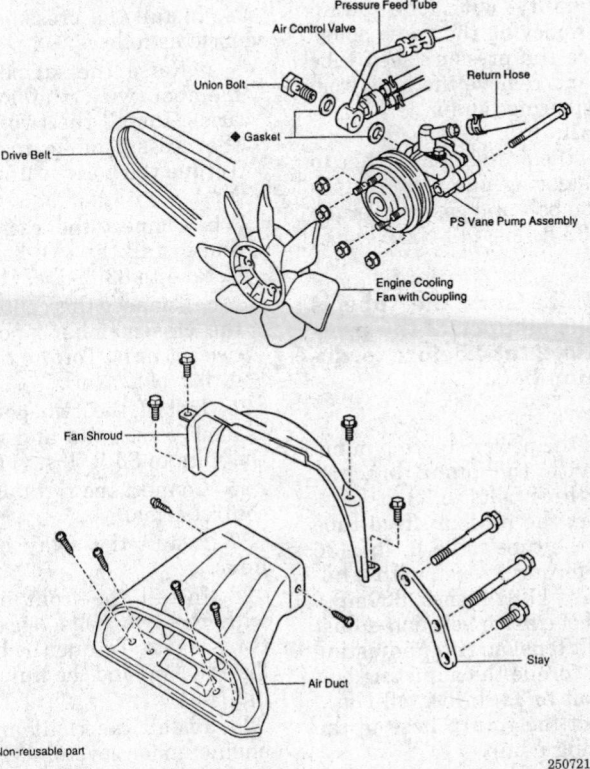

◆ Non-reusable part

250721

Steering pump component assembly (non-supercharged engine) — Previa

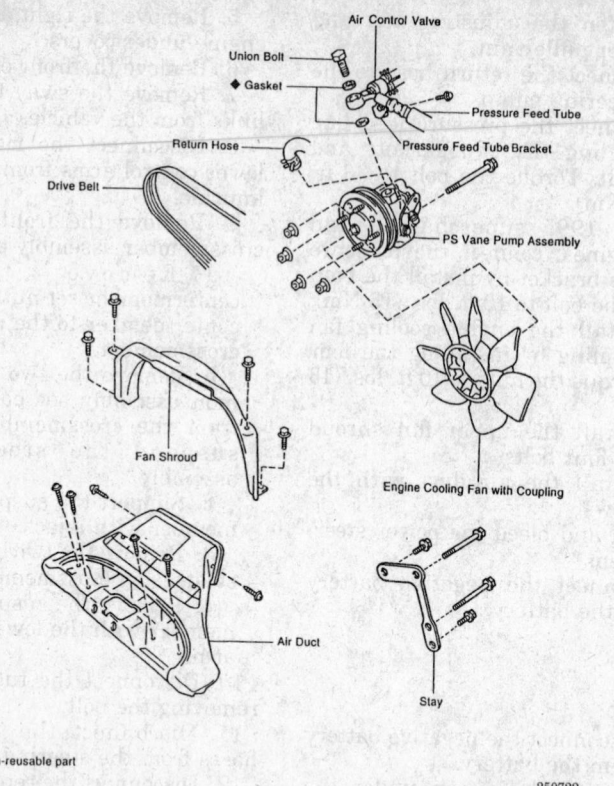

Steering pump component assembly (supercharged engine) — Previa

14. Disconnect the pressure feed tube to the power steering pump.

15. Remove the power steering pump assembly with the pump bracket by removing the three bolts.

16. Remove the pressure feed tube union bolt and remove the tube from the power steering pump.

To install:

17. Install the pressure feed tube to the power steering pump by installing the union bolt and gasket. Torque the bolt to 32 ft. lbs. (43 Nm).

NOTE: Make sure the tube is touching the stopper of the pressure feed No. 2 tube before torquing the union bolt.

18. Install the power steering pump assembly with the pump bracket. Torque the three bolts to 32 ft. lbs.

19. Connect the pressure feed tube and torque the tube to 26 ft. lbs. (36 Nm). Use a torque wrench with a fulcrum length of 11.81 inches (300mm).

20. Install the drive belt and adjust the drive belt tension. After adjusting the tension, torque the adjusting bolt and pivot bolt to 32 ft. lbs. (43 Nm).

21. Connect the return hose to the power steering pump.

22. Connect the two vacuum hoses to the power steering pump.

23. Connect the two tube clamps and torque the bolt to 9 ft. lbs. (13 Nm).

24. Install the crossmember to the vehicle as follows:

a. Raise the suspension crossmember with the lower control arms. Install the two bolts to hold the crossmember to the vehicle. Torque the bolts to 152 ft. lbs. (206 Nm).

b. Connect the rack and pinion and install the two bolts and nuts. Torque to 83 ft. lbs. (113 Nm).

c. Connect the centermember to the crossmember and install the two set nuts. Torque the nuts to 82 ft. lbs. (112 Nm).

25. Install the two power steering assembly set bolts and nuts. Torque the nuts to 83 ft. lbs. (113 Nm).

26. Connect the right and left lower control arms.

27. Install the sway bar with the links.

28. Install the front exhaust pipe with the two bolts, three nuts and two gaskets. Torque the bolts to 35 ft. lbs. (48 Nm) and the nuts to 46 ft. lbs. (62 Nm).

29. Install the right and left hand engine under covers.

30. Fill and bleed the power steering system.

31. Install the front wheels and lower the vehicle.

POWER STEERING SYSTEM BLEEDING

1993–94 Previa

1. Raise and safely support the vehicle.

2. Fill the pump reservoir with power steering fluid.

3. With the engine running, rotate the steering wheel from lock to lock several times. Add fluid as necessary.

NOTE: Perform the bleeding procedure until all of the air is bled from the system.

4. The fluid level should not have risen more than 0.20 inch (5mm); if it does, check the pump.

1995–97 Previa and RAV4

1. Check that the fluid level in the reservoir tank is at the maximum level. Fill with Dexron®II or equivalent.

2. Start the engine.

3. With the engine idle below 1000 rpm, turn the steering wheel from lock to lock. Hold the steering wheel at each lock for 2–3 seconds. Repeat this procedure until all the air bubbles are removed from the fluid.

4. Stop the engine and measure the fluid level.

5. Make sure the rise of the fluid is not over 0.020 inch (5mm).

BRAKES

Master Cylinder

REMOVAL AND INSTALLATION

Previa

1. Disconnect the negative battery cable from the battery.

2. If equipped, disconnect the level warning switch connector from the master cylinder.

3. Using a syringe, remove the brake fluid from the master cylinder.

4. Using a line wrench, disconnect the brake lines from the master cylinder.

5. Remove the master cylinder-to-power booster mounting nuts. Remove the master cylinder assembly and gasket from the power brake unit.

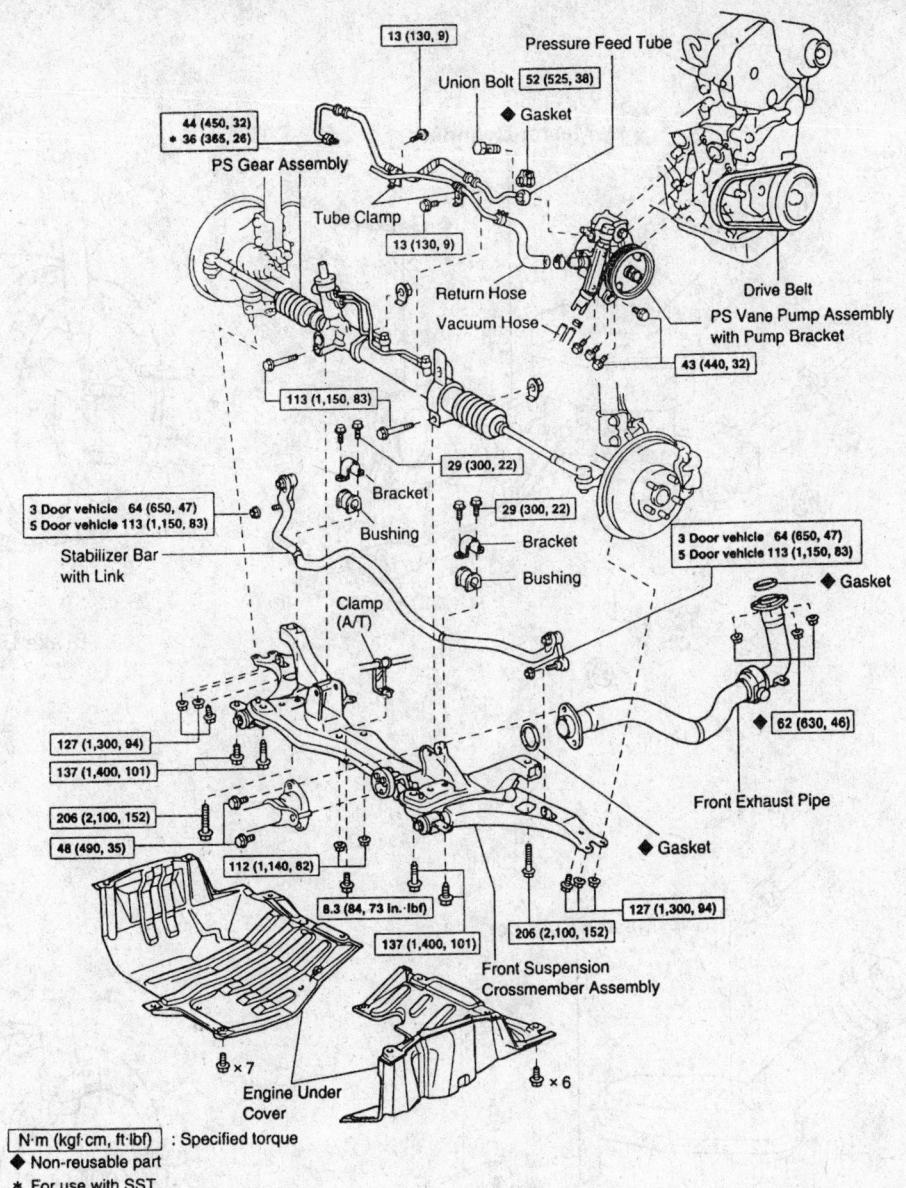

13 (130, 9)

Pressure Feed Tube

Union Bolt 52 (525, 38)

◆ Gasket

44 (450, 32)
* 36 (365, 26)

PS Gear Assembly

Tube Clamp

13 (130, 9)

Return Hose

Vacuum Hose

Drive Belt

PS Vane Pump Assembly
with Pump Bracket

43 (440, 32)

113 (1,150, 83)

29 (300, 22)

Bracket

Bushing

29 (300, 22)

Bracket

Bushing

3 Door vehicle 64 (650, 47)
5 Door vehicle 113 (1,150, 83)

Stabilizer Bar
with Link

Clamp
(A/T)

3 Door vehicle 64 (650, 47)
5 Door vehicle 113 (1,150, 83)

◆ Gasket

62 (630, 46)

127 (1,300, 94)

137 (1,400, 101)

206 (2,100, 152)

48 (490, 35)

112 (1,140, 82)

8.3 (84, 73 in.·lbf)

137 (1,400, 101)

127 (1,300, 94)

206 (2,100, 152)

◆ Gasket

Front Exhaust Pipe

Front Suspension
Crossmember Assembly

×7

Engine Under
Cover

×6

N·m (kgf·cm, ft·lbf) : Specified torque
◆ Non-reusable part
* For use with SST

305472

Steering system exploded view — RAV4

To install:

6. Adjust the length of the brake booster push rod as follows:

   a. Install a new gasket on the master cylinder.

   b. Install SST 09737–00010, or equivalent, on the master cylinder gasket and lower the pin until its tip slightly touches the piston.

   c. Turn the tool upside down and position it on the brake booster.

   d. Measure the clearance between the booster push rod and the pin head on the SST tool.

   e. Clearance should be 0 inch (0mm). If not, adjust the booster

push rod length until the push rod lightly touches the pin head.

NOTE: When adjusting the push rod, depress the brake pedal so that the push rod sticks out.

7. Remove the SST tool and install the master cylinder to the brake booster using a new gasket.

8. Install the master cylinder four nuts and torque the nuts to 9 ft. lbs. (13 Nm).

9. Using a line wrench, connect the two brake lines to the master cylinder. Torque the union nuts to 11 ft. lbs. (15 Nm).

10. Connect the level warning switch connector.

11. Fill the brake reservoir with brake fluid and bleed the brake system.

12. Check and adjust brake pedal.

13. Connect the negative battery cable to the battery.

RAV4

1. Disconnect the brake fluid level warning switch connector.

2. Remove and safely dispose of the brake fluid.

NOTE: Do not let brake fluid remain on a painted surface. Wash it off immediately.

w/o ABS

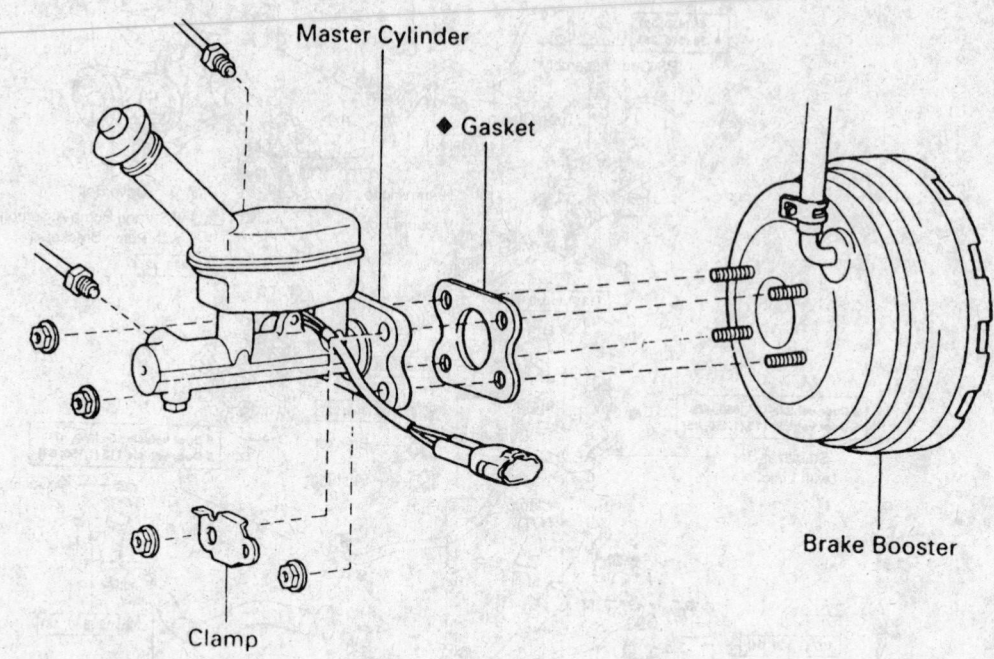

Master Cylinder

◆ Gasket

Brake Booster

Clamp

w/ ABS

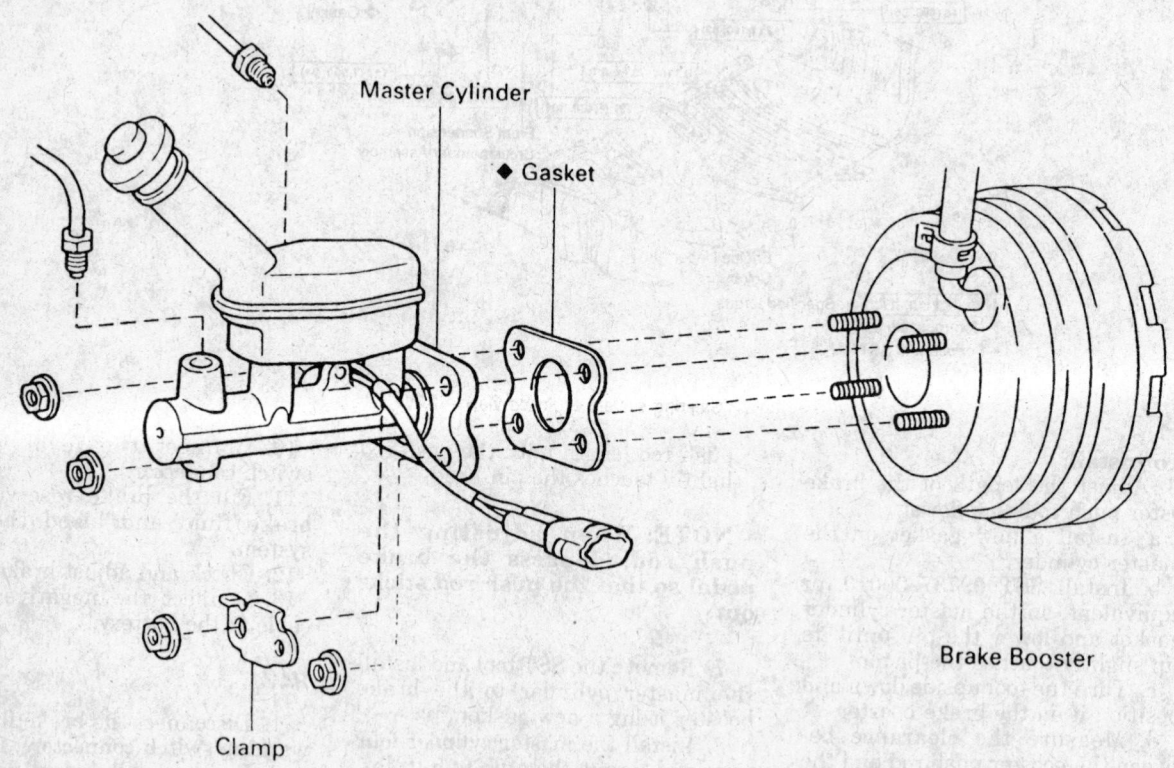

Master Cylinder

◆ Gasket

Brake Booster

Clamp

▶ Non-reusable part

Master cylinder component assembly — Previa

287819

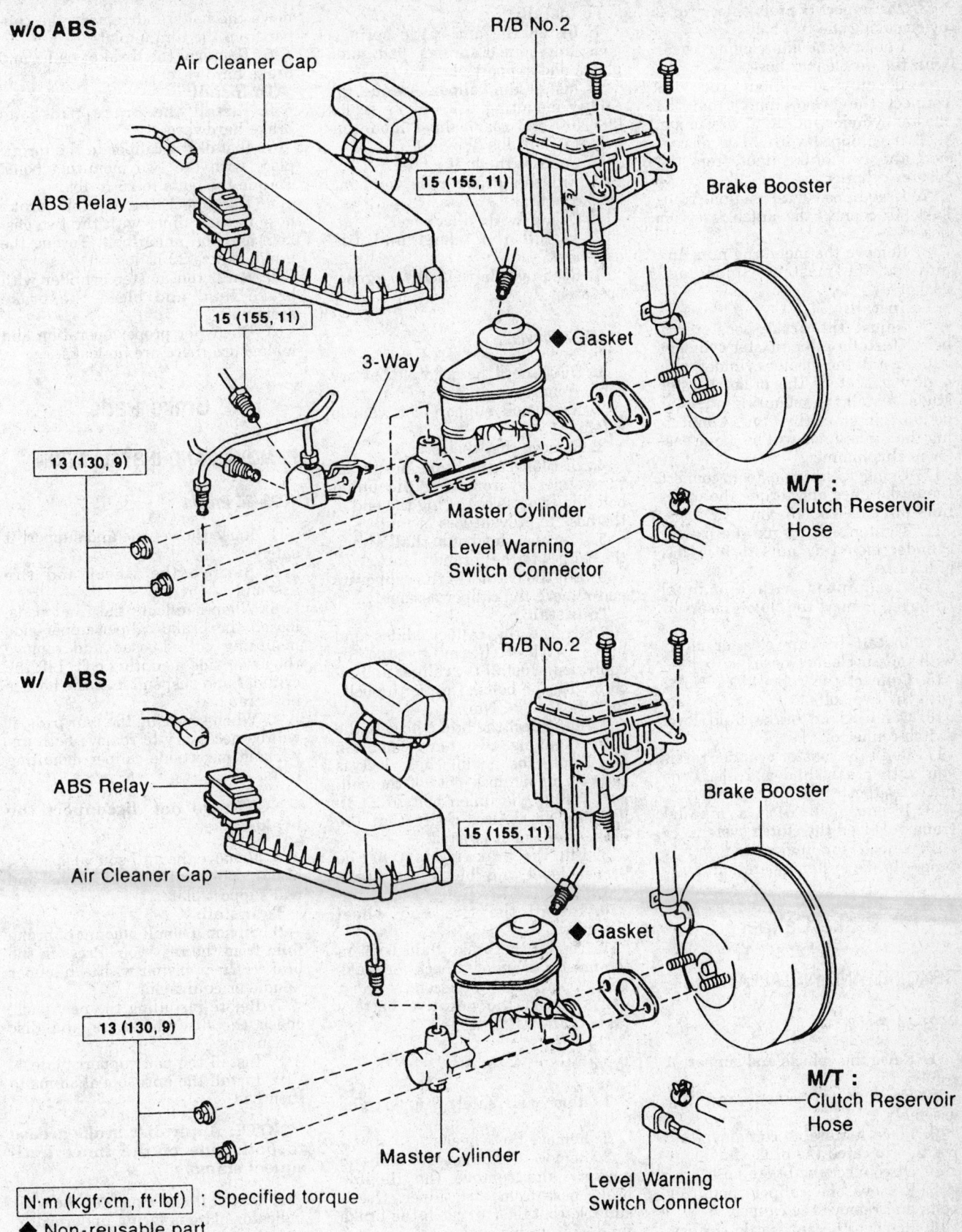

w/o ABS

Air Cleaner Cap

R/B No.2

ABS Relay

15 (155, 11)

Brake Booster

15 (155, 11)

3-Way

◆ Gasket

13 (130, 9)

Master Cylinder

Level Warning
Switch Connector

M/T :
Clutch Reservoir
Hose

w/ ABS

R/B No.2

ABS Relay

15 (155, 11)

Brake Booster

Air Cleaner Cap

◆ Gasket

13 (130, 9)

Master Cylinder

Level Warning
Switch Connector

M/T :
Clutch Reservoir
Hose

N·m (kgf·cm, ft·lbf) : Specified torque

◆ Non-reusable part

301244

Master cylinder and 3-way union — RAV4

3. Disconnect the relay block No. 2 by removing the two bolts.

4. Remove the air cleaner cover with the air cleaner hose.

5. If equipped without ABS, disconnect the brake lines from the master cylinder and the 3–way union.

6. If equipped with ABS, disconnect the two brake lines from the master cylinder.

7. If equipped with manual transaxle, disconnect the clutch reservoir hose.

8. Remove the mounting nuts and pull out the master cylinder and gasket.

To install:

9. Adjust the brake booster rod before installing the master cylinder.

10. Install the master cylinder with a new gasket to the brake booster studs. Install the mounting nuts but do not tighten at this time. Connecting the brake lines will be made easier in this manner.

11. Using a suitable tool, connect the brake lines and torque the union nuts to 11 ft. lbs. (15 Nm).

12. Tighten and torque the master cylinder mounting nuts to 9 ft. lbs. (13 Nm).

13. If equipped with a manual tranaxle, connect the clutch reservoir hose.

14. Install the air cleaner cover with the air cleaner hose.

15. Connect the relay block No. 2 with the two bolts.

16. Connect the brake fluid level switch connector.

17. Refill the master cylinder reservoir with brake fluid and bleed the brake system.

18. If equipped with a manual tranaxle, bleed the clutch system.

19. Check for leaks and verify proper brake system operation.

Brake Caliper

REMOVAL AND INSTALLATION

1993–94 Previa

1. Raise the vehicle and support it safely.

2. Remove the wheel and tire assembly.

3. Place a suitable container into position to catch the brake fluid.

4. Disconnect the brake hose.

5. Remove the caliper mounting bolts and remove the caliper.

6. Remove the anti-rattle springs, pads, anti-squeal springs, pad guide plates and support plate.

To install:

7. Install the anti-rattle springs, pads, anti-squeal springs, pad guide plates and support plate.

8. Install the caliper. Torque the caliper mounting bolts to 27 ft. lbs (36 Nm) and rear caliper mounting bolts to 18 ft. lbs (25 Nm).

9. Connect the brake hose.

10. Fill the brake reservoir and bleed the brake system.

11. Check for fluid leakage.

12. Install the wheel and tire assembly.

13. Road test the vehicle for proper operation.

1995–97 Previa

1. Disconnect the negative battery cable from the battery.

2. Raise and support the vehicle safely.

3. Remove the wheels.

4. Disconnect the brake hose from the caliper by removing the union bolt and two gaskets. Plug the end of the hose to prevent loss of fluid.

5. Remove the bolts that attach the caliper to the torque plate.

6. Lift the bottom of the caliper up and remove the caliper assembly.

To install:

7. Grease the caliper slides and bolts with lithium grease or equivalent. Install the caliper and secure with the bolts. Torque the bolts to 27 ft. lbs. (36 Nm).

8. Reconnect the brake hose to the caliper, using two new washers. Make sure the flexible hose lock is securely in the lock hole of the caliper. Torque the union bolt to 22 ft. lbs. (30 Nm). Also, verify that the brake hose is not twisted.

9. Fill the brake system to the proper level and bleed the brake system.

10. Install the tire and wheel assembly.

11. Top off the brake fluid level in the master cylinder. Check for leaks and proper brake operation.

12. Connect the negative battery cable to the battery.

RAV4

1. Raise and safely support the vehicle.

2. Remove the wheel(s).

3. Remove the union bolt and two gaskets and remove the flexible brake hose from the caliper. Use a suitable container to catch the brake fluid as it drains out.

4. Hold the sliding pin and loosen the two caliper mounting bolts. Re-move the bolts and remove the caliper from the torque plate.

5. Remove the brake pads and brake hardware.

To install:

6. Install the brake pads and brake hardware.

7. Install the caliper to the torque plate with the two mounting bolts. Torque the bolts to 25 ft. lbs.

8. Reconnect the flexible brake hose to the caliper with the two gaskets and the union bolt. Torque the union bolt to 22 ft. lbs.

9. Refill the master cylinder with brake fluid and bleed the brake system.

10. Check for proper operation and make sure there are no leaks.

Brake Pads

REMOVAL AND INSTALLATION

1993–94 Previa

1. Raise the vehicle and support it safely.

2. Remove the wheel and tire assembly.

3. When servicing the front pads, loosen the brake caliper upper side mounting bolt. Loosen and remove the lower side mounting bolt. Lift the cylinder and suspend it so the hose is not stretched.

4. When servicing the rear pads, it will be necessary to remove both upper and lower side caliper mounting bolts.

NOTE: Do not disconnect the brake hose.

5. Remove the anti-squeal springs, brake pads, anti-squeal shims and pad support plates.

To install:

6. Siphon a small amount of brake fluid from the reservoir. Press in the brake caliper piston with a hammer handle or equivalent.

7. Before installing the new pads, check the disc thickness and disc runout.

8. Install the pad support plates.

9. Install the anti-squeal shims to each pad.

NOTE: Apply disc brake grease to both side of the inner anti-squeal shims.

10. Install the disc pads so the wear indicator plate is facing underneath.

11. Install the anti-squeal springs.

12. Carefully install the brake caliper so the boot is not wedged. Torque

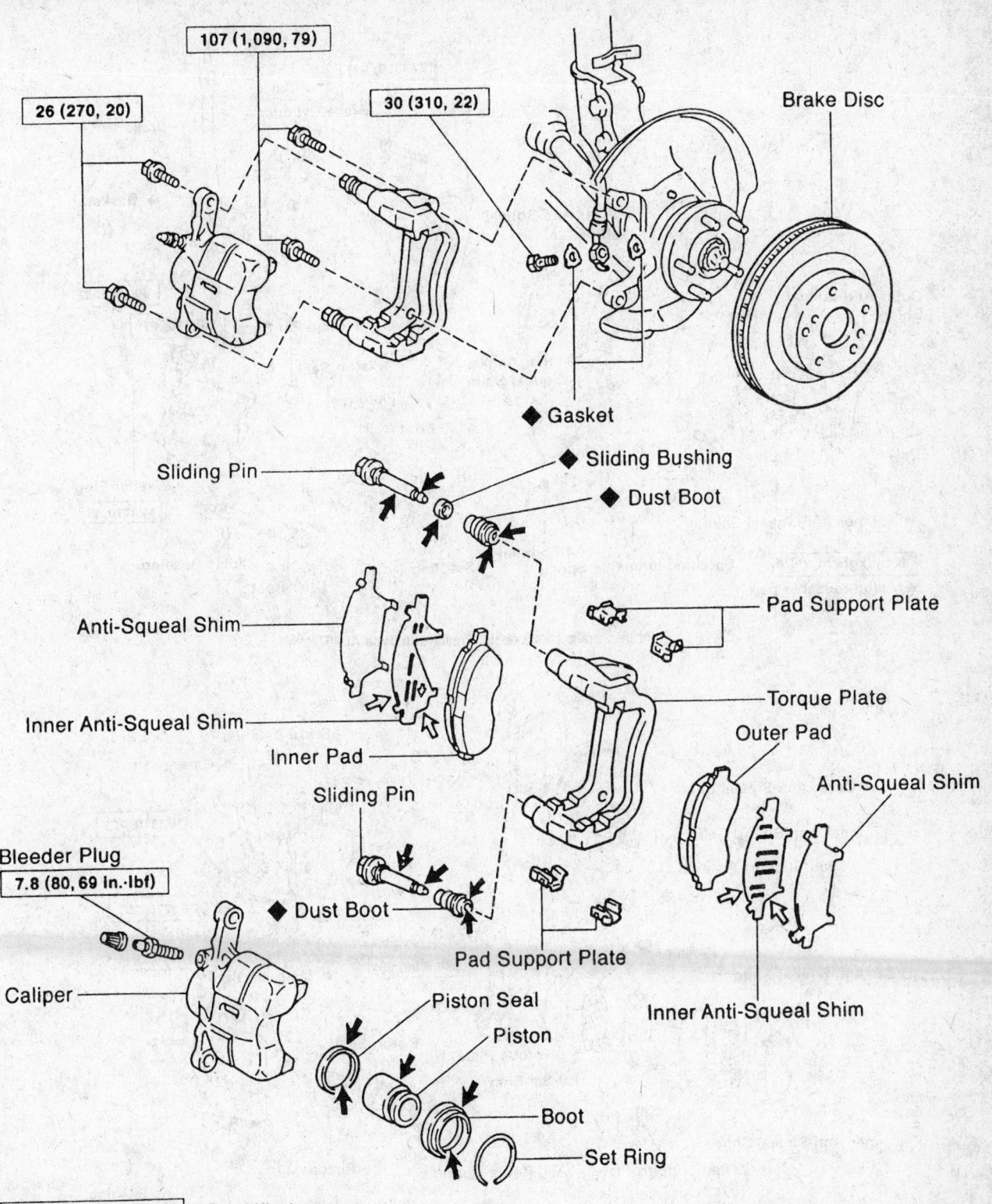

107 (1,090, 79)

26 (270, 20)

30 (310, 22)

Brake Disc

◆ Gasket

Sliding Pin

◆ Sliding Bushing

◆ Dust Boot

Anti-Squeal Shim

Pad Support Plate

Inner Anti-Squeal Shim

Torque Plate

Inner Pad

Outer Pad

Sliding Pin

Anti-Squeal Shim

Bleeder Plug

7.8 (80, 69 in.·lbf)

◆ Dust Boot

Pad Support Plate

Inner Anti-Squeal Shim

Caliper

Piston Seal

Piston

Boot

Set Ring

N·m (kgf·cm, ft·lbf) : Specified torque

◆ Non-reusable part

➤ Lithium soap base glycol grease

⇨ Disc brake grease

310152

Front brake components — RAV4

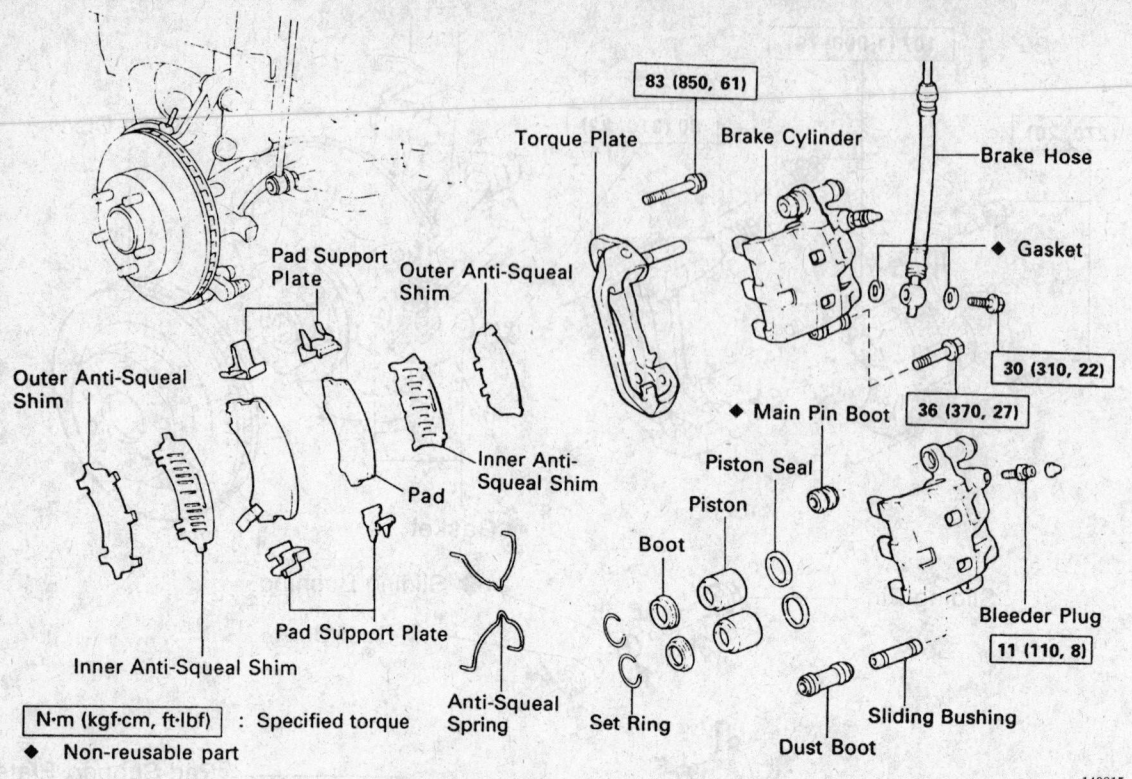

83 (850, 61)

Torque Plate   Brake Cylinder   Brake Hose

Pad Support Plate

Outer Anti-Squeal Shim

◆ Gasket

Outer Anti-Squeal Shim

Inner Anti-Squeal Shim

Pad

30 (310, 22)

◆ Main Pin Boot

36 (370, 27)

Piston Seal

Piston

Boot

Pad Support Plate

Bleeder Plug

11 (110, 8)

Inner Anti-Squeal Shim

Anti-Squeal Spring

Set Ring

Sliding Bushing

Dust Boot

N·m (kgf·cm, ft·lbf) : Specified torque
◆ Non-reusable part

140815

Front disc brake components, Previa with Type AD45T

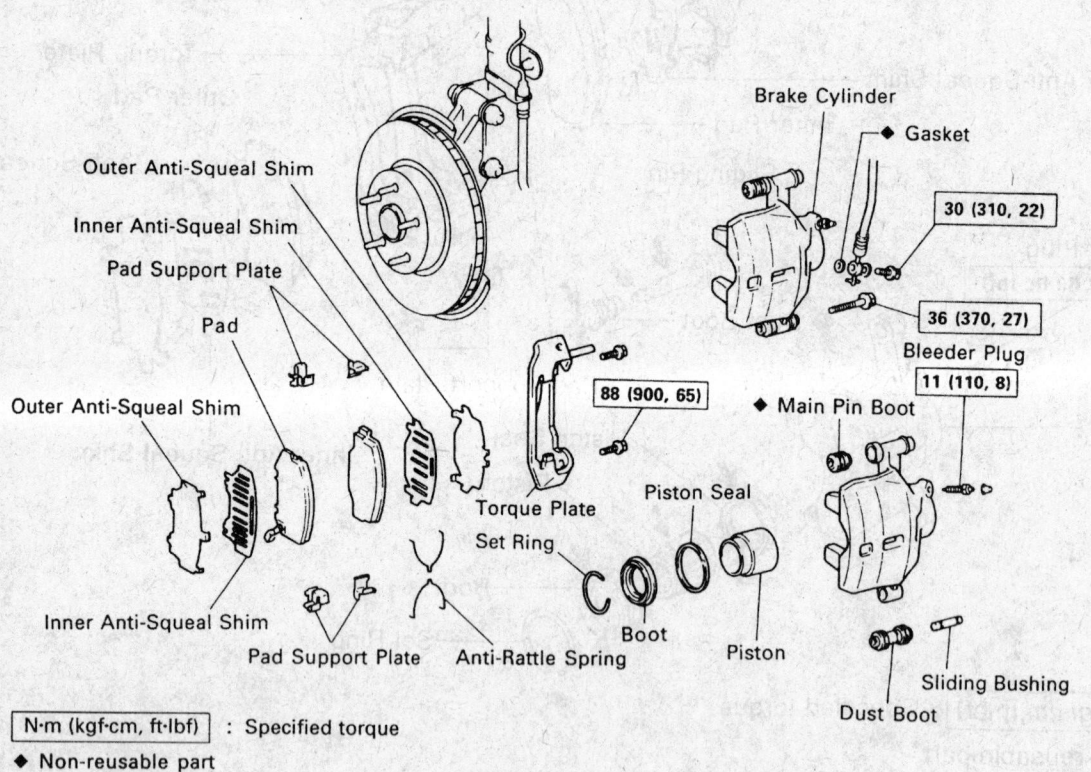

Brake Cylinder

◆ Gasket

Outer Anti-Squeal Shim

Inner Anti-Squeal Shim

Pad Support Plate

30 (310, 22)

Pad

36 (370, 27)

Outer Anti-Squeal Shim

Bleeder Plug

11 (110, 8)

◆ Main Pin Boot

88 (900, 65)

Torque Plate

Piston Seal

Set Ring

Boot

Inner Anti-Squeal Shim

Piston

Pad Support Plate   Anti-Rattle Spring

Sliding Bushing

Dust Boot

N·m (kgf·cm, ft·lbf) : Specified torque
◆ Non-reusable part

140876

Front disc brake components, Previa with Type AD60

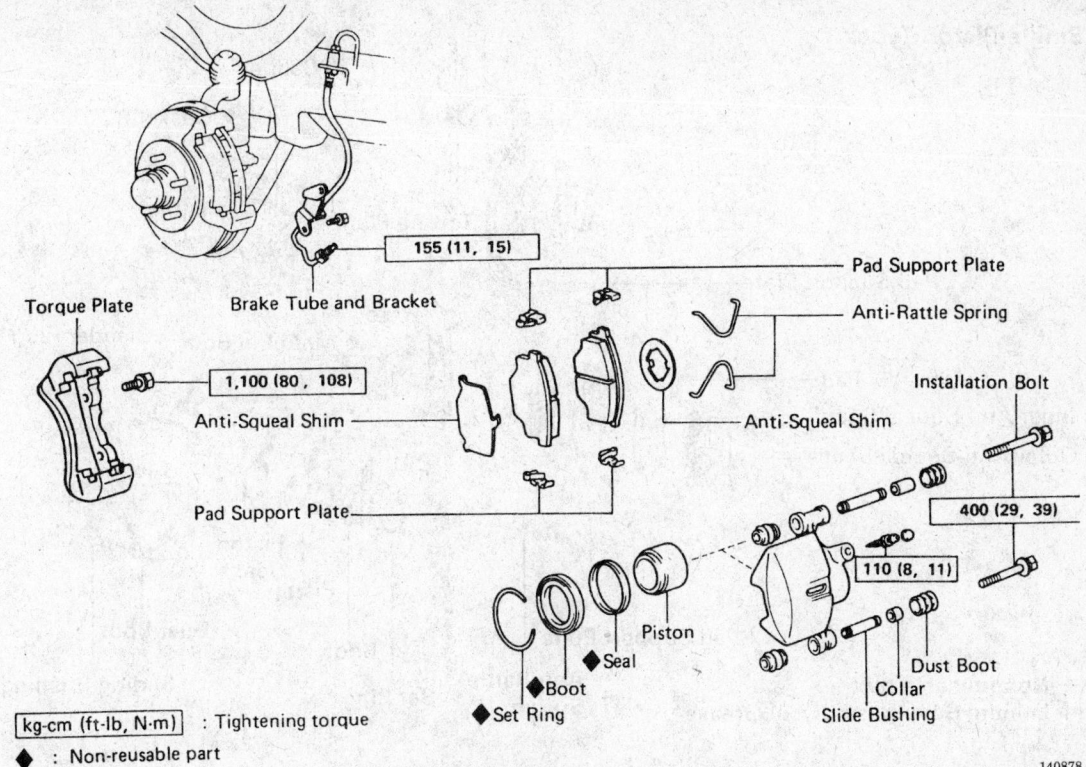

155 (11, 15)

Torque Plate

Brake Tube and Bracket

Pad Support Plate

Anti-Rattle Spring

1,100 (80, 108)

Installation Bolt

Anti-Squeal Shim

Anti-Squeal Shim

400 (29, 39)

Pad Support Plate

110 (8, 11)

Piston

Dust Boot

◆ Seal

Collar

◆ Boot

Slide Bushing

◆ Set Ring

kg-cm (ft-lb, N·m) : Tightening torque

◆ : Non-reusable part

140878

Front disc brake components, Previa with Type PD60

the caliper mounting bolts, as follows:

• Front caliper mounting bolts to: 27 ft. lbs (36 Nm)

• Rear caliper mounting bolts to: 18 ft. lbs (25 Nm).

13. Install the wheel and tire assembly.

14. Check and adjust the fluid level. Apply the brake pedal several times.

15. Road test the vehicle for proper operation.

1995–97 Previa

Front Disc Brake Pads

—— **CAUTION** ——

Brake shoes contain asbestos, which has been determined to be a cancer causing agent. Never clean the brake surfaces with compressed air! Avoid inhaling any dust from any brake surfaces. When cleaning brake surfaces, use a commercially available brake cleaning fluid.

1. Raise and safely support the front of the vehicle.

2. Remove the front wheels and temporarily fasten the rotor disc with the hub nuts.

NOTE: Always replace the front disc pads as a set.

3. Hold the sliding pin on the bottom of the caliper and loosen the installation bolt.

4. Remove the lower installation bolt.

5. Lift up the caliper and suspend it securely. Do not remove the upper installation bolt.

6. Remove the following parts:
   a. The 2 anti squeal springs.
   b. The 2 brake pads.
   c. The 4 anti squeal shims.
   d. The 4 pad support plates.

To install:

7. Install the pad support plates.

8. Install a pad wear indicator plate to the pad. Install the anti-squeal shims and support plates to each pad.

NOTE: It recommended that a suitable anti-squeal compound (available at your local parts house) be applied to both sides of the inner anti-squeal shim.

9. Draw out a small amount of brake fluid from the brake reservoir. Press in the caliper piston with a suitable tool.

10. Press the brake piston in carefully so the boot will not become wedged.

11. Install the two pads so that the wear indicator plate is facing upward. Do not allow oil or grease to get in the rubbing face of the pads.

12. Lower and install the caliper. Torque the sliding main pin to 27 ft. lbs. (36 Nm).

NOTE: When installing the sliding main pin, be careful that the plug installed in the torque plate does not come loose.

13. Install the front wheels and lower the vehicle.

14. Check the fluid level in the master cylinder and add as necessary. Be sure to pump the brake pedal a few times before road testing the vehicle.

Rear Disc Brake Pads

—— **CAUTION** ——

Brake shoes contain asbestos, which has been determined to be a cancer causing agent. Never clean the brake surfaces with compressed air! Avoid inhaling any dust from any brake surfaces. When cleaning brake surfaces, use a commercially available brake cleaning fluid.

Single-Piston Type:

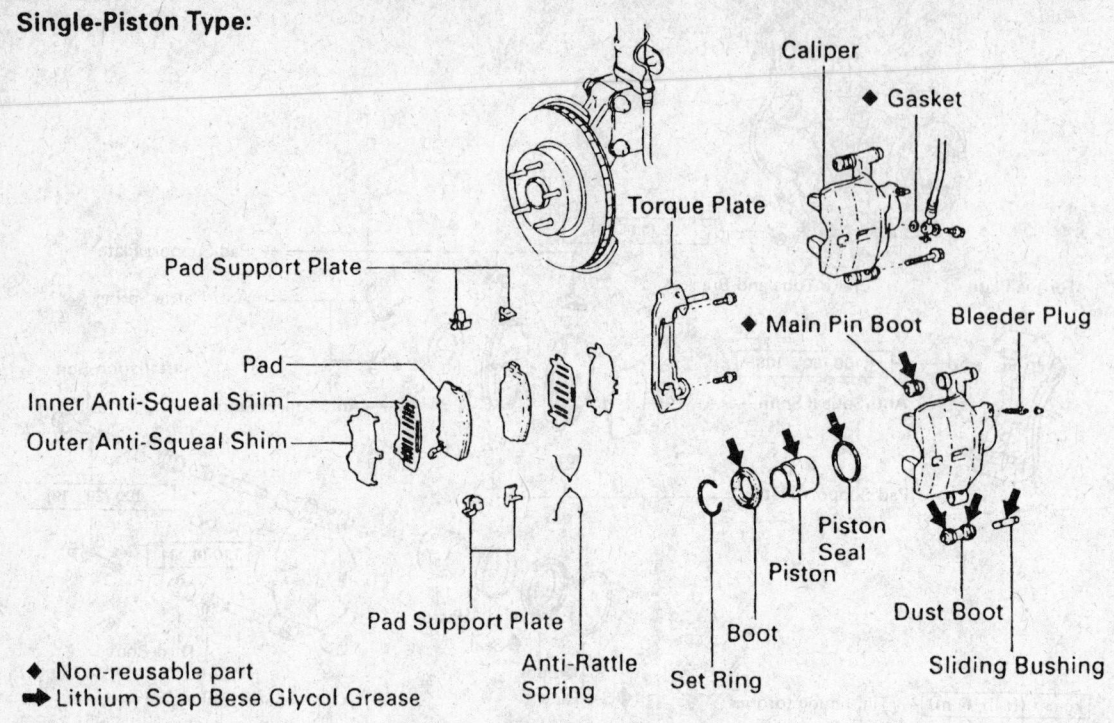

Rear disc brake pads (single piston type)

◆ Non-reusable part
➡ Lithium Soap Bese Glycol Grease

287810

1. Raise and safely support the rear of the vehicle.

2. Remove the rear wheels and temporarily fasten the rotor disc with the hub nuts.

NOTE: Always replace the rear disc pads as a set.

3. Hold the sliding pin on the bottom of the caliper and loosen the installation bolt.

4. Remove the lower installation bolt.

5. Lift up the caliper and suspend it securely. Do not remove the upper installation bolt.

6. Remove the following parts:
 a. The 2 anti squeal springs.
 b. The 2 brake pads.
 c. The 4 anti squeal shims.
 d. The 4 pad support plates.

To install:

7. Install the pad support plates.

8. Install a pad wear indicator plate to the pad. Install the anti-squeal shims and support plates to each pad.

NOTE: It recommended that a suitable anti-squeal compound (available at your local parts house) be applied to both sides of the inner anti-squeal shim.

9. Draw out a small amount of brake fluid from the brake reservoir.

Press in the caliper piston with a suitable tool.

10. Press the brake piston in carefully so the boot will not become wedged.

11. Install the two pads so that the wear indicator plate is facing upward. Do not allow oil or grease to get in the rubbing face of the pads.

12. Lower and install the caliper. Torque the sliding main pin to 25 ft. lbs. (34 Nm).

NOTE: When installing the sliding main pin, be careful that the plug installed in the torque plate does not come loose.

13. Install the rear wheels and lower the vehicle.

14. Check the fluid level in the master cylinder and add as necessary. Be sure to pump the brake pedal a few times before road testing the vehicle.

RAV4

1. Raise and safely support the vehicle.

2. Remove the wheel(s).

3. Temporarily install two wheel stud nuts to hold the brake rotor in place.

4. If necessary, siphon a sufficient quantity of brake fluid from the

master cylinder reservoir to prevent any brake fluid from overflowing the master cylinder when removing or installing new pads. This may be necessary as the piston must be forced into the caliper bore to provide sufficient clearance when installing the pads.

5. Grasp the caliper from behind and carefully pull it towards you. This will start to seat the piston(s) in its bore. Using a C-clamp or other suitable tool, press the piston the remaining way into the caliper. Be careful not to cock the piston in the bore. Also, do not force the piston or the caliper and piston may be damaged.

6. Hold the sliding pin and loosen the two caliper mounting bolts. Remove the bolts and remove the caliper from the torque plate.

7. Secure the caliper assembly out of the way with a wire; so as not to stress the flexible hose.

——— **CAUTION** ———

Brake pads contain asbestos, which has been determined to be a cancer causing agent. Never clean the brake surfaces with compressed air! Avoid inhaling any dust from any brake surfaces. When cleaning brake surfaces, use a commercially available brake cleaning fluid.

8. Slide out the old brake pads along with any anti-squeal shims, springs, pad wear indicators and pad support plates. Make sure to note the position of all assorted pad hardware.

To install:

9. Check the brake disc (rotor) for thickness and run-out. Inspect the caliper and piston assembly for breaks, cracks, fluid seepage or other damage. Overhaul or replace as necessary.

10. Install the pad support plates into the torque plate.

11. Install the pad wear indicators onto the pads. Be sure the arrow on the indicator plate is pointing in the direction of rotation.

12. Install the anti-squeal shims on the outside of each pad and then install the pad assemblies into the torque plate.

13. Install the caliper to the torque plate with the two mounting bolts. Torque the bolts to 20 ft. lbs. (26 Nm).

14. Remove the two temporary wheel stud nuts and check that the rotor turns freely.

15. Reinstall the wheel(s), safely lower the vehicle, and road test for proper brake operation.

16. Be sure to pump the brakes several times prior to moving the vehicle.

Brake Rotor

REMOVAL AND INSTALLATION

1993–94 Previa

1. Raise the vehicle and support it safely.

2. Remove the wheel and tire assembly.

3. Remove the brake caliper.

4. On 2WD vehicles, remove the grease cap, cotter pin and nut from the hub. Remove the rotor from the vehicle.

5. On 4WD vehicles, remove the locknut and washer from the free wheeling hub. Remove the adjusting nut and thrust washer. Remove the hub with disc together with the outer bearing.

To install:

6. Before installing the rotor, thoroughly clean and repack the wheel bearings, using MP grease. Coat inside the hub and cap with MP grease.

7. Install a new bearing seal. Coat the seal lip with MP grease.

8. On 2WD vehicles, install the rotor, outer bearing and thrust washer. Adjust the bearing preload.

9. On 4WD vehicles, install the free wheeling hub locknut and washer. Adjust the preload and check that the bearing has no play. Secure the lock nut by bending one of the lock washer teeth inward and the other lock washer teeth outward.

10. On 2WD vehicles, install the locknut, cotter pin and hub grease cap.

11. On 4WD vehicles, install the free wheeling hub or automatic locking hub.

12. Install the brake caliper.

13. Install the wheel and tire assembly.

14. Check and adjust the brake fluid level.

15. Road test the vehicle for proper operation.

1995–97 Previa

1. Disconnect the negative battery cable to the battery.

2. Loosen the wheel lugs slightly, then raise and safely support the vehicle.

3. Remove the wheel(s) and temporarily install two of the wheel lug nuts.

4. Hold the sliding pin on the bottom with a wrench and loosen the installation bolt.

5. Remove the two installation bolts holding the caliper to the torque plate.

6. Remove the caliper from the torque plate and hang the caliper from a piece of wire. Do not disconnect the brake hose.

———— **WARNING** ————
Do not allow the caliper to hang freely from the vehicle. Always support the caliper with a wire from the vehicle.

7. Unbolt and remove the torque plate.

8. Remove the two wheel nuts and pull the disc from the axle hub.

To install:

9. Position the new rotor disc onto the hub and reinstall the two wheel nuts temporarily.

10. Install the torque plate onto the vehicle. Torque the two torque plate bolts to 65 ft. lbs. (88 Nm). Make sure the brake pads are seated correctly within the torque plate.

11. Install the caliper to the torque plate. Torque the installation bolts to 27 ft. lbs. (36 Nm).

12. Remove the wheel lug nuts and install the wheels. Secure the wheel lugs.

13. Lower the vehicle and tighten the lug nuts. Before moving the vehicle, make sure to pump the brake pedal to seat the brake pads against the rotors.

14. Connect the negative battery cable to the battery.

RAV4

1. Raise and safely support the vehicle.

2. Remove the wheel(s).

3. Hold the sliding pin and loosen the two caliper mounting bolts. Remove the bolts and remove the caliper from the torque plate.

4. Secure the caliper assembly out of the way with a wire so as not to stress the flexible hose.

5. Remove the disc brake pads.

6. Remove the two torque plate mounting bolts and remove the torque plate from the steering knuckle.

7. Remove the brake rotor.

To install:

8. Install the brake rotor and temporarily install two wheel stud nuts to hold the rotor in place.

9. Install the torque plate to the steering knuckle with the two mounting bolts. Torque the bolts to 79 ft. lbs. (107 Nm).

10. Install the disc brake pads.

11. Install the caliper to the torque plate with the two mounting bolts. Torque the bolts to 20 ft. lbs. (26 Nm).

12. Remove the two temporary wheel stud nuts and check that the rotor turns freely.

13. Reinstall the wheel(s), safely lower the vehicle, and road test for proper brake operation.

NOTE: Pump the brakes several times and verify that the brake pedal is firm prior to moving the vehicle.

Brake Drum

REMOVAL AND INSTALLATION

Previa

1. Raise and safely support the vehicle.

2. Remove the rear wheel(s).

3. On 1993–94 vehicles, remove the brake drum retaining screws, if equipped. Remove the brake drum.

4. On 1995–97 vehicles, remove the brake drum from the axle hub.

5. If equipped with double tire, remove the rear axle shaft and remove the drum with axle hub.

NOTE: If the brake drum cannot be removed easily, insert a screwdriver through the hole in the backing plate. Hold the adjuster lever away from the adjuster and reduce the brake shoe clearance by turning the adjuster wheel using a brake adjuster tool.

To install:

6. Install the brake drum and pull the parking brake lever all the way up until a clicking sound can no longer be heard.

7. Verify that the rear wheels will not turn. If the rear wheels turn, adjust the parking brake cable as necessary.

8. Release the parking brake and remove the brake drum. Measure the brake drum inside diameter and diameter of the brake shoes. Check that the difference between the diameters is the correct shoe clearance. Clearance is: 0.024 in. (6mm).

9. If the brake shoe clearance is not correct, adjust the brake shoes until the clearance is correct.

10. Reinstall the brake drum, install the wheel(s), and safely lower the vehicle.

11. Road test the vehicle for proper brake operation.

RAV4

1. Raise and safely support the vehicle.

2. Remove the rear wheel(s).

3. Remove the brake drum from the axle hub. If there is difficulty in removing the drum, insert a suitable tool through the hole in the brake drum, and hold the automatic adjusting lever away from the adjuster. Using another suitable tool at the same time, reduce the brake shoe adjuster by turning the adjusting wheel.

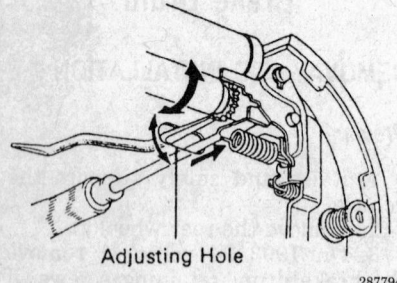

Adjusting Hole

287794

Brake shoes adjusting hole

To install:

4. Install the brake drum and pull the parking brake lever all the way up until a clicking sound can no longer be heard.

5. Verify that the rear wheels will not turn. If the rear wheels turn, adjust the parking brake cable as necessary.

6. Release the parking brake and remove the brake drum. Measure the brake drum inside diameter and diameter of the brake shoes. Check that the difference between the diameters is the correct shoe clearance. Clearance is: 0.024 in. (0.6mm).

7. If the brake shoe clearance is not correct, adjust the brake shoes until the clearance is correct.

8. Reinstall the brake drum, replace the wheel(s), and safely lower the vehicle.

9. Road test the vehicle for proper brake operation.

Wheel Cylinder

REMOVAL AND INSTALLATION

1. Raise and safely support the vehicle.

2. Remove the rear wheel(s).

3. Remove the brake drum.

4. Remove the upper brake shoe return springs.

5. Working from behind the backing plate, disconnect the hydraulic line from the wheel cylinder and plug the opening of the brake line.

6. Unfasten the two bolts holding the wheel cylinder to the backing plate and remove the cylinder.

To install:

7. Insert the wheel cylinder into position and secure with the two retaining bolts.

NOTE: Start the brake line fitting into the wheel cylinder body before tightening the wheel cylinder mounting bolts.

8. Connect the brake line to the wheel cylinder.

9. Torque the bolts to 7 ft. lbs. (10 Nm) and the line to 11 ft. lbs. (15 Nm).

10. Install the upper brake shoe return springs. Make sure that the brake shoes are seated correctly.

11. Install the brake drum and replace the rear wheel(s).

12. Bleed the brake system and adjust the rear brake shoes.

13. Road test the vehicle for proper brake operation.

Brake Shoes

REMOVAL AND INSTALLATION

1. Disconnect the negative battery cable from the battery.

2. Loosen the rear wheel lug nuts slightly. Release the parking brake.

3. Block the front wheels, raise the rear of the vehicle, and safely support it with jackstands.

4. Remove the wheel lug nuts and the wheel.

5. Remove the brake drum retaining screws, if equipped. Remove the brake drum.

6. If the drum is difficult to remove, perform the following:

   a. Insert the end of a bent wire (a coat hanger will do nicely) through the hole in the brake drum and hold the automatic adjusting lever away from the adjuster.

   b. Reduce the brake shoe adjustment by turning the adjuster bolt with a brake tool.

   c. The drum should now be loose enough to remove without much effort.

NOTE: Do not depress the brake pedal once the brake drum has been removed.

7. Carefully unhook the return spring from the leading (front) brake shoe.

8. Press the hold down spring retainer in and turn the pin on the front brake shoe.

9. Remove the hold down spring, retainers and the pin for the front brake shoe.

10. Pull out the brake shoe and unhook the anchor spring from the lower edge.

11. Remove the hold down spring from the trailing (rear) shoe. Pull the shoe out with the adjuster, automatic adjuster assembly and springs attached. Disconnect the parking brake cable. Remove the tension/return and anchor springs from the rear shoe.

12. Unhook the adjusting lever spring from the rear shoe and then remove the automatic adjuster assembly.

To install:

13. Inspect the shoes for signs of unusual wear or scoring.

14. Check the wheel cylinder for any sign of fluid seepage or frozen pistons.

15. Clean and inspect the brake backing plate and all other components. Check that the brake drum inner diameter is within specified limits. Lubricate the backing plate at the positions the brakes come in contact

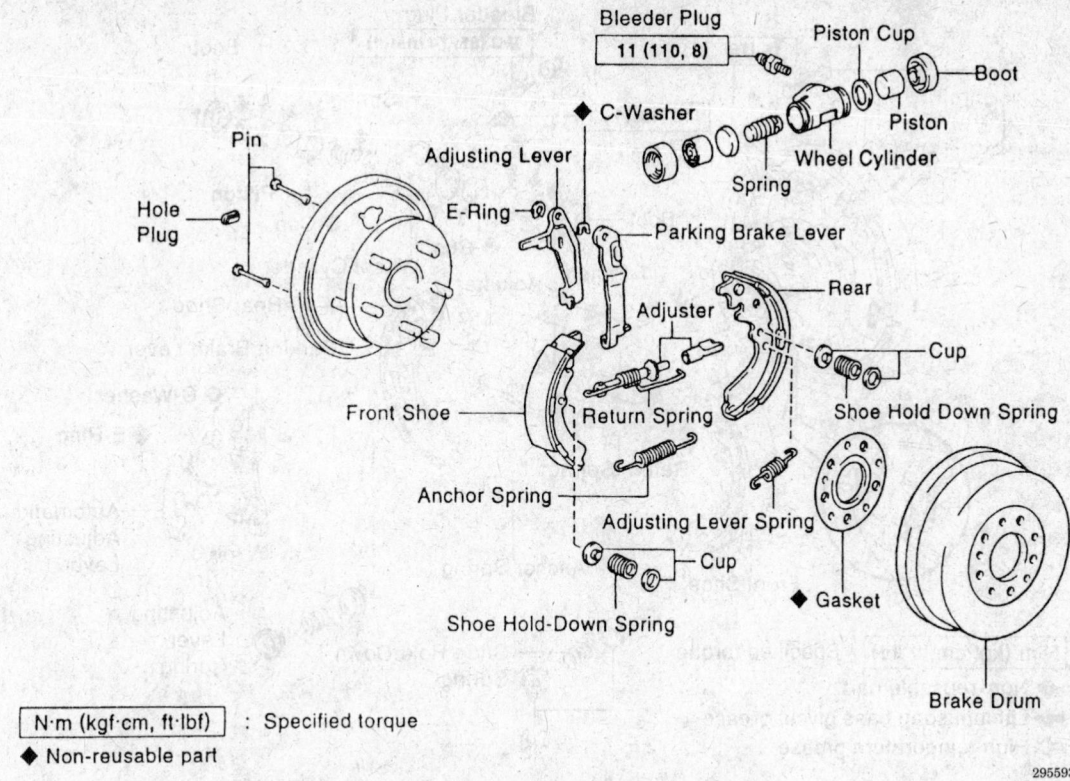

N·m (kgf·cm, ft·lbf) : Specified torque
◆ Non-reusable part

295593

Rear brake drums and related components — 2WD Previa

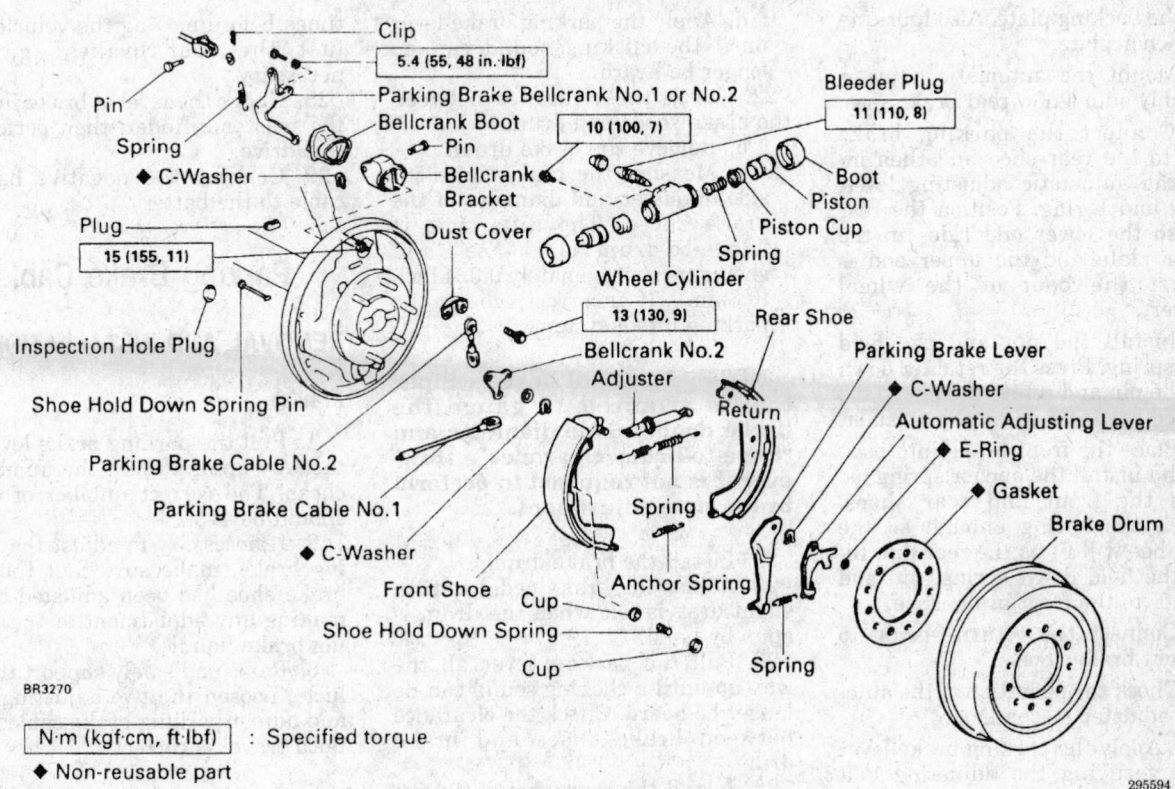

BR3270

N·m (kgf·cm, ft·lbf) : Specified torque
◆ Non-reusable part

295594

Rear brake drums and related components — 4WD Previa

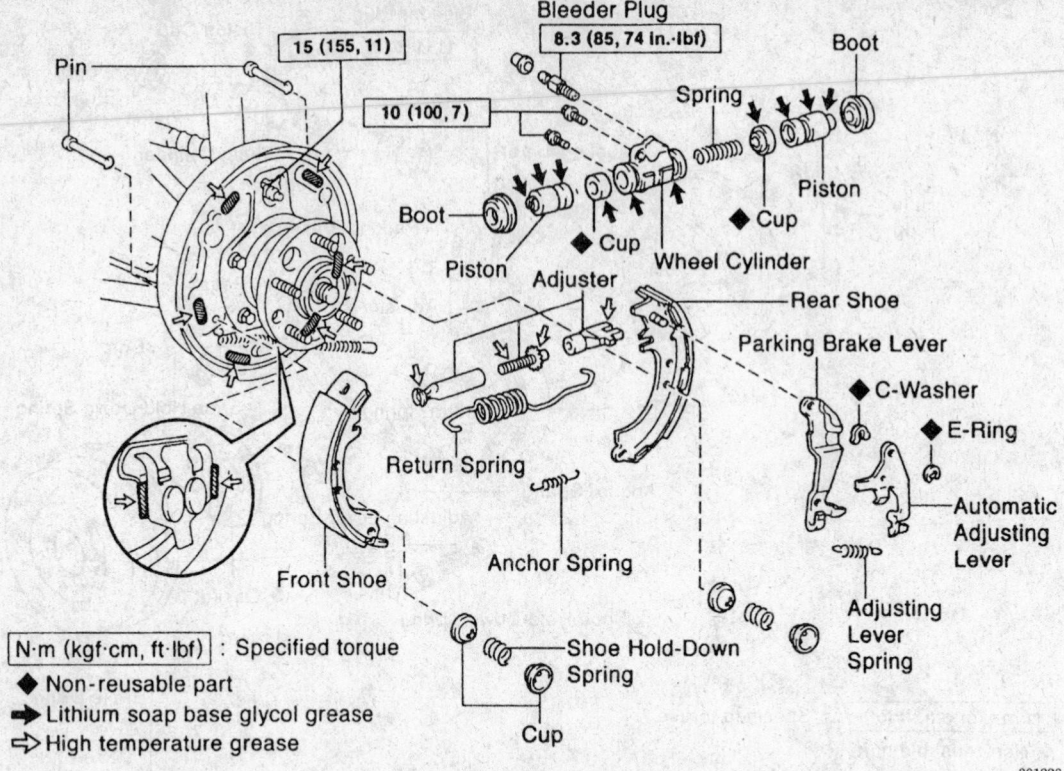

Bleeder Plug
8.3 (85, 74 in.·lbf)

15 (155, 11)

10 (100, 7)

Pin

Boot

Piston

Cup

Adjuster

Return Spring

Front Shoe

Anchor Spring

Shoe Hold-Down Spring

Cup

Boot

Spring

Piston

Cup

Wheel Cylinder

Rear Shoe

Parking Brake Lever

◆ C-Washer

◆ E-Ring

Automatic Adjusting Lever

Adjusting Lever Spring

N·m (kgf·cm, ft·lbf) : Specified torque
◆ Non-reusable part
➡ Lithium soap base glycol grease
⇨ High temperature grease

301228

Rear brake shoe components — RAV4

with the backing plate. Also lubricate the anchor plate.

16. Mount the automatic adjuster assembly onto a new rear brake shoe.

17. Connect the parking brake cable to the rear shoe and then install the automatic adjusting lever, spring and E-ring. Position the rear shoe so the lower end rides in the anchor plate and the upper end is against the boot of the wheel cylinder.

18. Install the pin and the hold down spring. Press the retainer down over the pin and rotate the pin so the crimped edge is held by the retainer.

19. Place the front brake into position and install the anchor spring between the front and rear shoes. Stretch the spring enough so the front shoe will fit as the rear did. Install the hold down spring, pin and retainer to the front brake shoe.

20. Connect the return spring to the front brake shoe.

21. Check the operation of the automatic adjuster mechanism:

　a. Apply the parking brake lever and verifying the adjusting bolt turns.

　b. Adjust the strut to where it is the shortest possible length.

　c. Install the brake drum.

　d. Apply the parking brake lever until the clicking sound can no longer be heard.

22. Check the clearance between the brake shoes and drum:

　a. Remove the brake drum.

　b. Measure the brake drum inside diameter and diameter of the brake shoes. The difference is "Shoe-to-drum clearance" and should be approximately 0.024 inch (0.6mm). If incorrect, check the parking brake system.

　NOTE: A special brake caliper tool is required to gauge the brake drum inside diameter and "Shoe-to-drum clearance". However it is not required to perform brake shoe adjustment.

23. Install the brake drum.
24. Adjust the brake pedal until a slight drag is felt when the drum is spun by hand.
25. Pull the parking lever all the way up until a clicking sound can no longer be heard. Check the clearance between brake shoes and brake drum.
26. Install the rear wheels, tighten the wheel lug nuts and lower the vehicle.
27. Retighten the wheel lug nuts and pump the brake pedal a few

times before moving the vehicle. Adjust the rear brakes again if necessary.

28. Check the level of brake fluid in the master cylinder, then perform a test drive.

29. Connect the negative battery cable to the battery.

Parking Brake Cable

REMOVAL AND INSTALLATION

Previa

1. Pull the parking brake lever all the way up and count the number of clicks. The correct number of clicks should be 4–5.

2. If necessary to adjust the parking brake, make sure that the rear brake shoe has been adjusted before making any adjustment at the parking brake handle.

3. Raise and safely support the vehicle. Loosen the two adjusting nuts and adjust parking brake cable No. 1 until travel is correct.

RAV4

1. Pull the parking brake lever all the way up and count the number of clicks. The parking brake lever

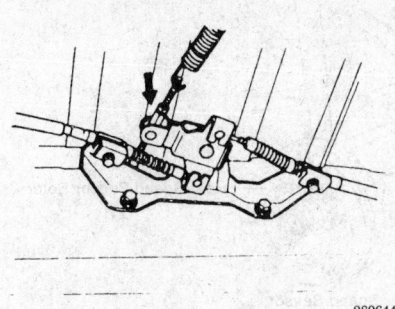

289644

Parking brake cable — Previa

should travel approximately 5–8 clicks at 44 lbs (196 N) of force.

NOTE: Before adjusting the parking brake, make sure that the rear brake shoe clearance has been adjusted.

2. Remove the rear console box.
3. Loosen the lock nut and turn the adjusting nut (at the parking brake lever) until the lever travel is correct.
4. Apply the parking brake several times and again check that there is no drag with the brake released.
5. Tighten the lock nut to 48 in. lbs. (5.4 Nm).
6. Install the rear console box.

Hydraulic System

BRAKE SYSTEM BLEEDING

Start the bleeding procedure at the caliper or wheel cylinder the furthest from the master cylinder.

1. Connect a vinyl tube to the bleeder screw on the brake cylinder and submerge the other end of the tube in a transparent container half filled with clean brake fluid.
2. Pump the brake pedal several times and loosen the bleeder screw with the pedal held down.

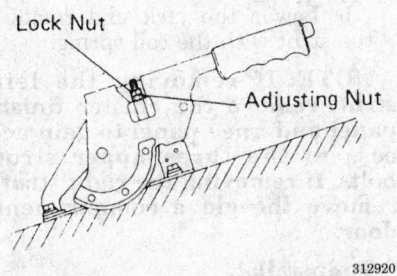

312920

Adjusting nut at the brake handle — RAV4

3. When brake fluid stops coming out of the tube with the brake pedal held to the floor, tighten the bleeder screw and release the brake pedal.
4. Repeat Steps 2 and 3 until no air bubbles can be seen in the container.
5. Repeat the procedure for each wheel.
6. Check the level in the master cylinder. Add fluid as necessary.

Front Wheel Speed Sensor

REMOVAL AND INSTALLATION

Previa

1. Disconnect the negative battery cable.
2. Raise and safely support the vehicle.
3. Disconnect the speed sensor connector.
4. Remove the speed sensor.
• Remove the clamp bolts holding the sensor harness from the frame, upper arm and the steering knuckle.
• Remove the speed sensor from the steering knuckle.
To install:
5. Install the speed sensor.

NOTE: Never force the speed sensor into place, always install it by hand.

6. Torque the clamp bolts holding the sensor harness to the frame, upper arm and the steering knuckle to 3.7 ft. lbs. (5 Nm).
7. Install the speed sensor to the steering knuckle and torque to 69 inch. lbs. (7.8 Nm).
8. Connect the speed sensor connector.
9. Connect the negative battery cable.
10. After installation, check speed sensor signal.

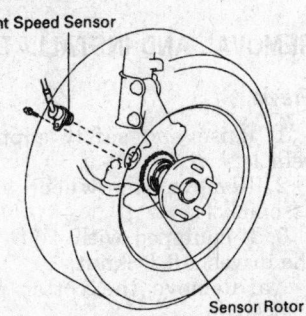

2WD Model

Front Speed Sensor

Sensor Rotor

RAV4

1. Disconnect the negative battery cable.
2. Raise and safely support the vehicle.
3. Remove the fender liner from the front fender.
4. Disconnect the speed sensor connector.
5. Remove the four clamp bolts holding the sensor harness to the frame, strut and steering knuckle.
6. Remove the bolt holding the speed sensor to the steering knuckle. Remove the speed sensor from the vehicle.
To install:
7. Install the speed sensor to the steering knuckle and install the bolt. Torque the bolt to 71 inch lbs. (8 Nm).
8. Torque the clamp bolts holding the sensor harness to the frame, steering knuckle and strut to 48 inch lbs. (5 Nm).
9. Connect the speed sensor connector.
10. Install the fender liner and install the wheels.
11. Lower the vehicle.
12. Connect the negative battery cable.
13. After installation, check speed sensor signal.

Rear Wheel Speed Sensor

REMOVAL AND INSTALLATION

Previa

1. Disconnect the negative battery cable.
2. Raise and safely support the vehicle.
3. Disconnect the sensor connector.
4. Remove the clamp bolts holding the sensor wire harness to the body and lower control arm.

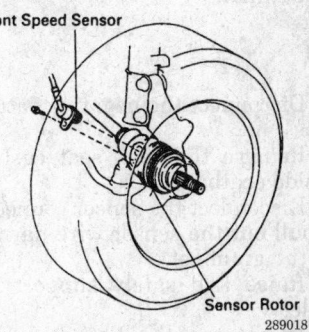

4WD Model

Front Speed Sensor

Sensor Rotor

289018

Front speed sensor location — Previa

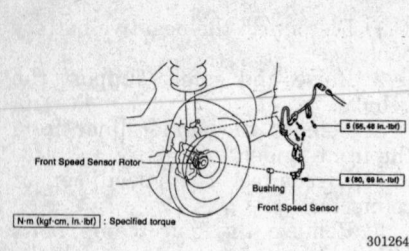

Front speed sensor — RAV4

301264

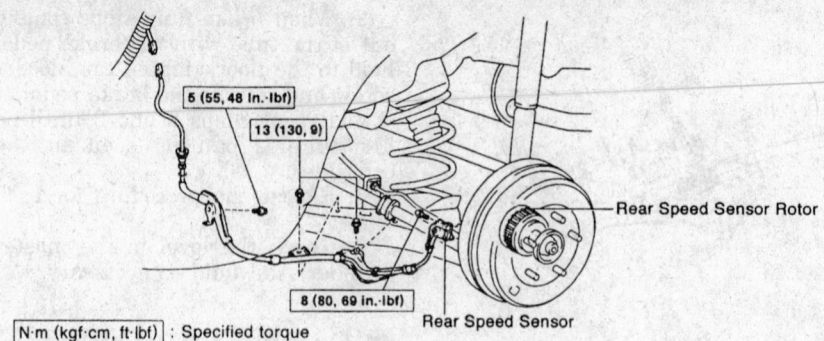

Rear speed sensor — RAV4

301265

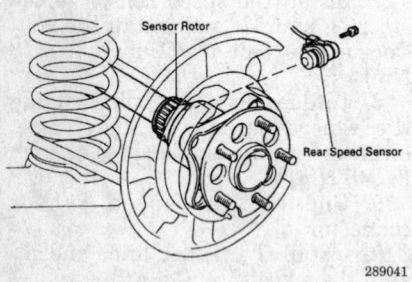

Rear speed sensor location — Previa

289041

5. Remove the lock bolt from the axle carrier and remove the speed sensor.

To install:

6. Install the speed sensor and tighten the lock bolt to 69 inch lbs. (7.8 Nm).

NOTE: Never force the speed sensor into place, always install it by hand.

7. Install the clamp bolts to hold the sensor wire harness to the body and lower control arm.

8. Connect the sensor connector.

9. Lower the vehicle.

10. Connect the negative battery cable.

11. After installation, check speed sensor signal.

RAV4

1. Disconnect the negative battery cable.

2. Remove the back seat cushion and side seatback.

3. Disconnect the sensor connector and pull out the sensor wire harness with the grommet.

4. Raise and safely support the vehicle.

5. Disconnect the parking brake cable from the lower control arm.

6. Remove the three clamp bolts holding the sensor wire harness to the body and lower control arm.

7. Remove the lock bolt from the axle carrier and remove the speed sensor.

To install:

8. Install the speed sensor and tighten the lock bolt to 71 inch lbs. (7.8 Nm).

9. Install the three clamp bolts to hold the sensor wire harness to the body and lower control arm. Torque the bolts to the lower control arm 9 ft. lbs. (13 Nm). Torque the body bolts to 48 inch lbs. (5 Nm).

10. Connect the parking brake cable to the lower control arm.

11. Connect the speed sensor connector and install the grommet.

12. Install the quarter trim panel.

13. Install the seat cushion and seatback.

14. Lower the vehicle.

15. Connect the negative battery cable.

16. After installation, check speed sensor signal.

FRONT SUSPENSION

Strut Front

REMOVAL AND INSTALLATION

Previa

1. Raise and safely support the vehicle.

2. Remove the wheel and tire assembly.

3. If equipped with 4WD, remove the driveshaft locknut:

   a. Remove the cotter pin and lock cap.

   b. While applying the brake, remove the locknut.

c. Remove the washer.

4. Disconnect the sway bar link from the strut by removing the nut.

5. If equipped with ABS, remove the speed sensor.

6. Using a line wrench, disconnect the brake line at the strut.

7. Remove the clips and disconnect the brake hose from the strut bracket.

8. Loosen the two nuts on the lower side of the strut. Do not remove the bolts.

9. Loosen the bolts on the lower ball joint, but do not remove the bolts.

10. Remove the cotter pin and nut from the tie rod end.

11. Using SST09628–10011, or equivalent, disconnect the tie rod from the steering knuckle.

12. Remove the lower ball joint bolts, two nuts, and the bolts on the lower side of the strut and remove the steering knuckle.

NOTE: If equipped with four wheel drive, it will be necessary to use a rubber hammer to disconnect the halfshaft from the steering knuckle. Be careful not to damage the oil seal, driveshaft boot, and/or the speed sensor rotor.

13. Place a service jack underneath the strut to support it.

   a. Remove the three nuts on the upper side of the strut.

   b. Lower the jack and remove the strut with the coil spring.

NOTE: If removing the left strut, remove the cluster finish panel and knee panel to gain access to the three upper strut bolts. If removing the right strut, remove the glove compartment door.

To install:

14. Place the strut with coil spring into position and support it with a jack. Install the three upper side nuts

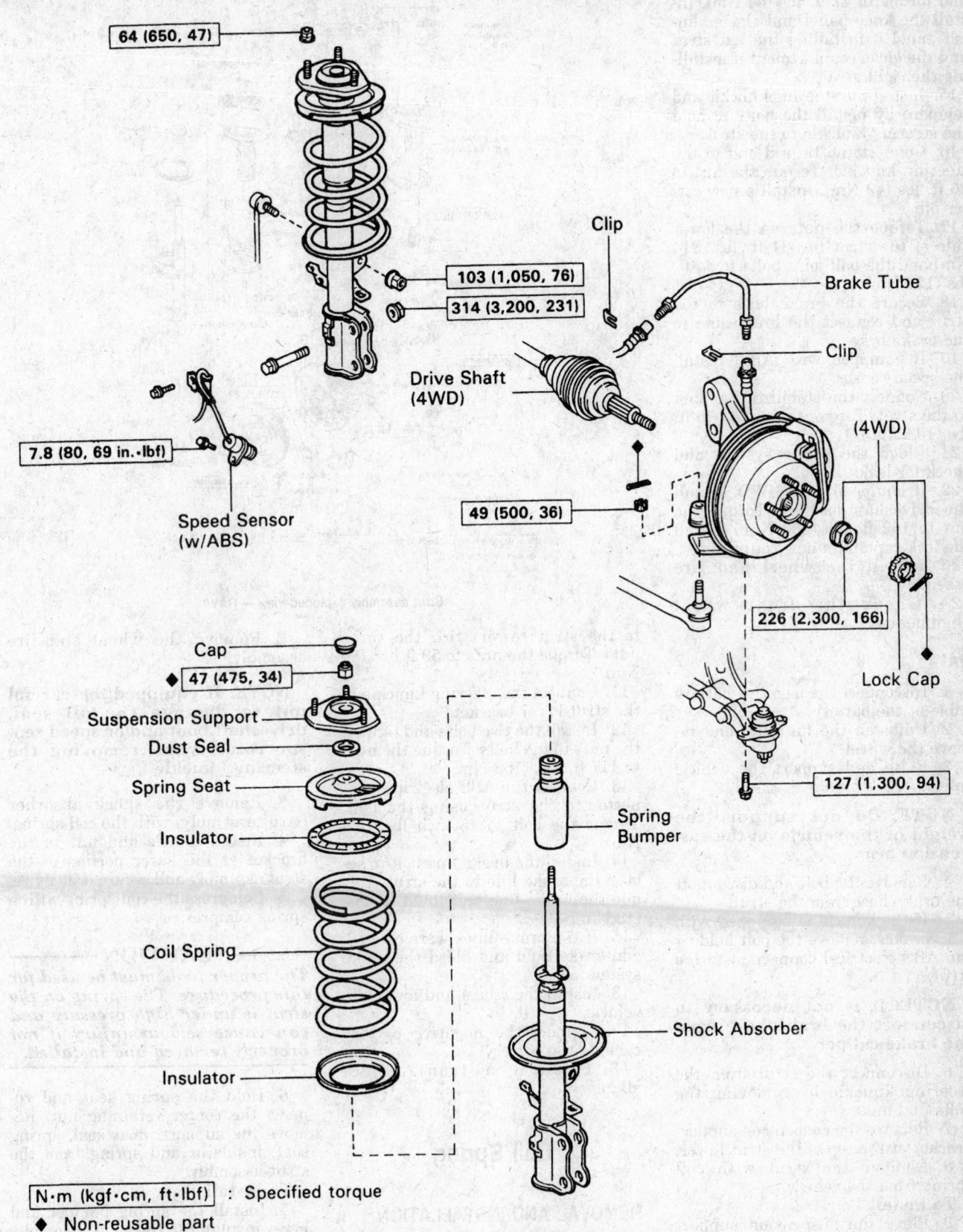

64 (650, 47)

103 (1,050, 76)

314 (3,200, 231)

Clip

Brake Tube

Clip

Drive Shaft (4WD)

(4WD)

7.8 (80, 69 in.·lbf)

Speed Sensor (w/ABS)

49 (500, 36)

226 (2,300, 166)

Lock Cap

Cap

◆ 47 (475, 34)

Suspension Support

Dust Seal

Spring Seat

Insulator

Spring Bumper

127 (1,300, 94)

Coil Spring

Shock Absorber

Insulator

N·m (kgf·cm, ft·lbf) : Specified torque

◆ Non-reusable part

287738

Front strut exploded view — Previa

and torque to 47 ft. lbs (64 Nm). Install the knee panel and cluster finish panel if installing the left strut and the glove compartment if installing the right strut.

15. Install the steering knuckle and temporarily install the bolts to hold the steering knuckle to the strut.

16. Connect the tie rod end to the steering knuckle. Torque the nut to 36 ft. lbs (49 Nm). Install a new cotter pin.

17. Torque the nuts on the lower side of the strut to 231 ft. lbs (314 Nm) and the ball joint bolts to 94 ft. lbs (127 Nm).

18. Secure the brake hose to the strut and connect the brake line to the brake hose.

19. If equipped with ABS, install the speed sensor.

20. Connect the stabilizer bar link to the strut. Torque the nut to 76 ft. lbs (103 Nm).

21. Bleed the brake system and check for leaks.

22. If equipped with 4WD, install the driveshaft locknut. Torque the nut to 152 ft. lbs (206 Nm). Install the lock cap and a new cotter pin.

23. Install the wheel and tire assembly.

24. Check the front wheel alignment.

RAV4

1. Disconnect the negative battery cable at the battery

2. Unfasten the lug nuts and remove the wheel.

3. Raise and support the vehicle safely.

NOTE: Do not support the weight of the vehicle on the suspension arm.

4. Remove the bolt and disconnect the brake hose from the strut.

5. If the vehicle is equipped with ABS brakes, remove the bolt holding the ABS electrical connection to the strut.

NOTE: It is not necessary to disconnect the brake hose from the brake caliper.

6. Disconnect the strut from the steering knuckle by removing the bolts and nuts.

7. Remove the suspension support bracket at the top of the strut tower.

8. Remove the strut with coil spring from the vehicle.

To install:

9. Place the suspension support bracket at the top of the strut tower.

10. Align the strut to the strut tower bolt holes and secure the strut

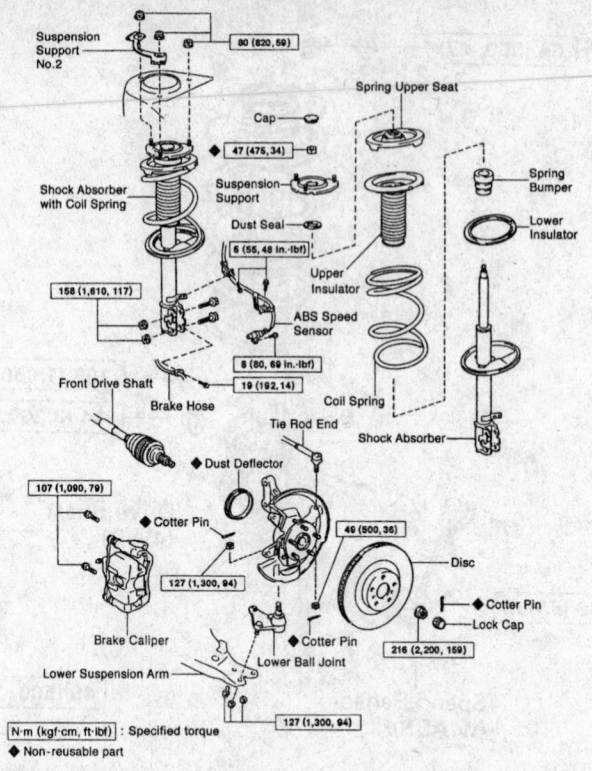

to the strut tower with the three nuts. Torque the nuts to 59 ft lbs. (80 Nm).

11. Connect the steering knuckle to the strut lower bracket.

12. Install the two bolts and tighten the nuts to the bolts. Torque the nuts to 117 ft. lbs. (158 Nm).

13. Connect the ABS electrical connector to the strut using the bolt. Torque the bolt to 48 inch lbs. (5.4 Nm).

14. Install the brake line bolt to secure the brake line to the strut. Torque the brake line bolt to 14 ft. lbs. (19 Nm).

15. If the brake lines were opened, add brake fluid and bleed the brake system.

16. Install the wheel and lower the vehicle.

17. Connect the negative battery cable to the battery.

18. Perform a front wheel alignment.

Coil Spring

REMOVAL AND INSTALLATION

1. Raise and safely support the vehicle.

2. Remove the wheel and tire assembly.

NOTE: If equipped, be careful not to damage the oil seal, driveshaft boot and/or speed sensor rotor when removing the steering knuckle.

3. Remove the shock absorber (strut assembly) with the coil spring.

4. Install a bolt and nut to the bracket at the lower portion of the strut assembly and secure it in a vice.

5. Compress the coil spring with a spring compressor.

—— CAUTION ——
The proper tools must be used for this procedure. The spring on the strut is under high pressure and can cause serious injury if not properly removed and installed.

6. Hold the spring seat and remove the center retaining nut. Remove the support, dust seal, spring seat, insulator and spring from the strut assembly.

To install:

7. Install the spring bumper and lower insulator to the strut assembly.

8. Compress the coil spring and fit the lower end of the spring into the gap of the spring seat.

Strut assembly exploded view — RAV4

304288

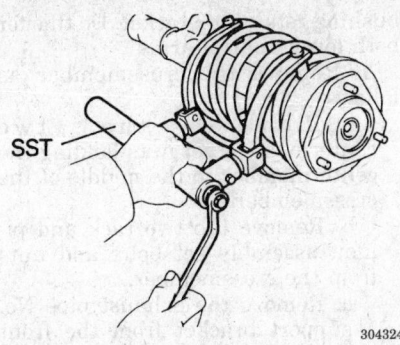

Compress the coil spring with a suitable compressor

9. Install the upper insulator, spring seat, dust seal, support and spring seat.

10. Install a new retaining nut and torque to 34 ft. lbs (47 Nm).

11. Rotate the spring seat so that the OUT mark of the spring seat faces the outside of the vehicle.

12. Install the strut assembly with coil spring.

13. If required, bleed the brake system and check for leaks.

14. Install the wheel and tire assembly.

15. Check the front wheel alignment.

Lower Ball Joint

REMOVAL AND INSTALLATION

Previa

1. Raise and safely support the vehicle.

2. Remove the wheel and tire assembly.

3. Remove the cotter pin and nut from the ball joint.

4. Disconnect the ball joint from the steering knuckle by removing the two bolts.

5. Using a suitable ball joint separator tool (SST 09628–62011), remove the lower ball joint.

To install:

6. Install the lower ball joint with the nut and bolts. Torque the nut to 76 ft. lbs (103 Nm) and the two bolts to 94 ft. lbs (127 Nm).

7. Install a new cotter pin to the ball joint.

8. Install the wheel and tire assembly and lower the vehicle.

RAV4

────── **CAUTION** ──────
The Supplemental Inflatable Restraint (SIR) system must be disarmed before removing the ball

joint. Failure to do so may cause accidental deployment of the air bag, resulting in unnecessary SIR system repairs and/or personal injury.

1. Disconnect the negative battery cable.

2. Raise the front of the vehicle and support it safely.

3. Remove the front wheel(s).

4. Remove the steering knuckle with the axle hub, from the vehicle.

5. Pry the dust deflector from the knuckle.

6. Remove the cotter pin and the nut from the ball joint stud.

7. Using SST 09628–62011 or equivalent, remove the lower ball joint from the steering knuckle.

To install:

8. Install the lower ball joint onto the steering knuckle and tighten nut to 94 ft. lbs. (127 Nm). Install new cotter pin.

9. Align the hole in the dust deflector with the ABS speed sensor. Using the appropriate driver, install a new dust deflector.

10. Install the steering knuckle and hub onto the vehicle.

11. Install the front wheel(s).

12. Connect the negative battery cable.

Lower Control Arm

REMOVAL AND INSTALLATION

Previa

1. Raise and safely support the vehicle. Place jackstands under the vehicle frame.

2. Remove the wheel and tire assembly.

3. Remove the engine under cover.

4. Disconnect the lower control arm from the steering knuckle by removing the two bolts.

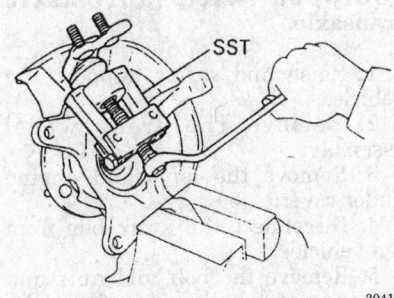

Removing the ball joint from the knuckle — RAV4

5. Remove the two bolts and brackets holding the rear of the lower control arm to the vehicle frame.

6. Remove the through bolt holding the front of the lower control arm to the frame.

7. Remove the lower arm from the vehicle.

8. Remove the cotter pin and nut from the ball joint.

9. Using SST 09628–62011 or equivalent, separate the ball joint from the lower control am.

10. Remove the rear bushing from the No. 2 lower control arm by removing the end nut.

11. Remove the No. 2 lower control arm from the No. 1 lower control arm by removing the two bolts.

To install:

12. Connect the No. 2 lower control arm to the No. 1 lower control arm by installing the two bolts. Torque the bolts to 136 ft. lbs. (184 Nm).

13. Install a new bushing to the No. 2 lower control arm and install the end nut. Torque the nut to 80 ft. lbs. (109 Nm).

14. Install the ball joint to the lower control arm and install the nut. Torque the nut to 76 ft. lbs. (103 Nm). Install a new cotter pin to the ball joint.

15. Place the lower control arm into position and the through bolt and nut. Do not torque the bolts at this time.

16. Install the lower control arm bracket and install the two bolts. Do not torque the bolts at this time.

17. Connect the lower arm to the steering knuckle and torque to 94 ft. lbs (127 Nm).

18. Install the wheel and tire assembly and lower the vehicle.

19. Bounce the vehicle several times to stabilize the suspension.

20. Raise the vehicle and torque the lower control arm through bolt to 121 ft. lbs (164 Nm). Torque the lower control arm bracket bolts to to 105 ft. lbs (142 Nm).

21. Install the under cover.

22. Check the front wheel alignment.

RAV4

Except Left Side in Vehicles with Automatic Transaxle

1. Raise and safely support the vehicle.

2. Remove the front wheel assembly.

3. Disconnect the sway bar link from the lower control arm by removing the nut.

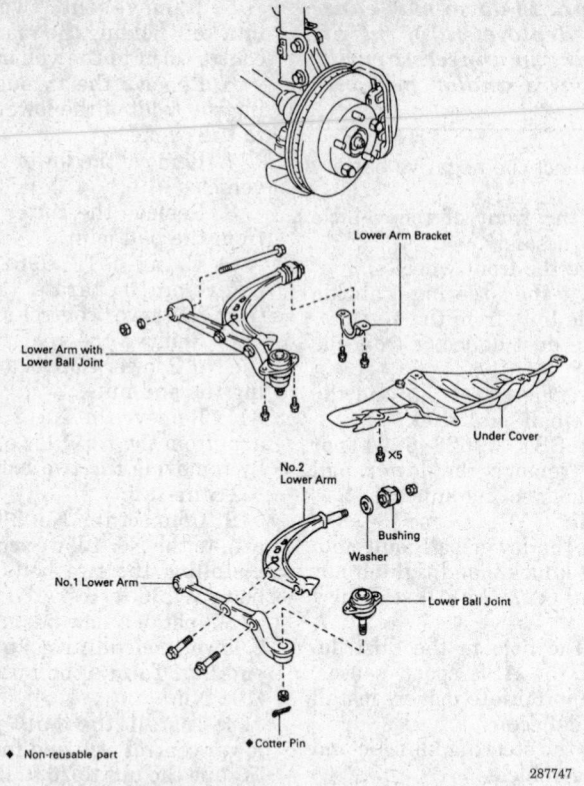

Lower control arm assembly — Previa

287747

◆ Non-reusable part
◆ Cotter Pin
Lower Arm Bracket
Lower Arm with Lower Ball Joint
Under Cover
×5
No.2 Lower Arm
Bushing
Washer
No.1 Lower Arm
Lower Ball Joint

4. Remove the bolt and nuts and separate the ball joints from the lower control arm.

5. Remove the nut and two bolts and remove the lower control arm bushing retaining bracket.

6. Remove the control arm to crossmember mounting bolt and remove the control arm from the crossmember.

To install:

7. Install the control arm to the suspension crossmember and install the front bolt. Do not torque the bolt at this time.

8. Install the lower control arm bushing retaining bracket and install the nut and two bolts. Do not torque the bolts or nut at this time.

9. Connect the lower control arm to the lower ball joint by installing the bolt and two nuts. Torque the bolt and nuts to 94 ft. lbs. (127 Nm).

10. Connect the sway bar link to the lower control arm and torque the nut as follows:

• 3 door vehicles: 47 ft. lbs. (64 Nm)

• 5 door vehicles: 83 ft. lbs. (113 Nm)

11. Install the wheels and lower the vehicle to the ground.

12. Stabilize the suspension by pushing up and down the the vehicle.

13. Torque the front lower control arm bolt to 126 ft. lbs. (172 Nm).

14. Torque the rear lower control arm bracket bolts and nuts as follows:

• Bolt: Control arm bracket bolts to 101 ft. lbs. (137 Nm)

• Nut: Bracket nut to 21 ft. lbs. (28 Nm)

15. Check wheel alignment.

Left Side Control Arms in Vehicles with Automatic Transaxle

NOTE: Both lower control arms and the suspension crossmember must be removed as as unit when replacing the left hand control arm on a vehicle equipped with automatic transaxle.

1. Raise and safely support the vehicle.

2. Remove the front wheel assembly.

3. Remove the left hand engine under cover.

4. Disconnect the sway bar from the vehicle.

5. Remove the bolt and nuts and separate the ball joints from the lower control arms.

6. Remove the nut and two bolts and remove the rear control arm bushing retaining bracket. Do this for both lower control arms.

7. Remove the crossmember as follows:

a. Remove the two centermember set nuts holding the centermember to the middle of the crossmember.

b. Remove the two rack and pinion assembly set bolts and nuts from the crossmember.

c. Remove the exhaust pipe No. 1 support bracket from the front exhaust pipe.

d. Remove the bolt and clamp.

e. Support the suspension crossmember with a jack.

f. Remove the two bolts from the suspension crossmember.

g. Remove the suspension crossmember with the lower suspension arms.

8. Remove the bolt holding the front of the lower control arm to the suspension crossmember.

9. Disconnect the lower control arm from the suspension crossmember.

To install:

10. Install the lower control arm to the suspension crossmember.

11. Install the bolt to hold the lower control arm to the crossmember. Do not torque the bolt at this time.

12. Install the crossmember as follows:

a. Raise the suspension crossmember with the lower control arms. Install the two bolts to hold the crossmember to the vehicle. Torque the bolts to 152 ft. lbs. (206 Nm).

b. Install the bolt and clamp.

c. Install the exhaust pipe No. 1 support bracket to the front exhaust pipe.

d. Connect the rack and pinion and install the two bolts and nuts. Torque to 83 ft. lbs. (113 Nm).

e. Connect the centermember to the crossmember and install the two set nuts. Torque the nuts to 82 ft. lbs. (112 Nm).

13. Install the lower control arm bushing retaining bracket and install the nut and two bolts. Do not torque the bolts or nut at this time.

14. Connect the lower control arm to the lower ball joint by installing the bolt and two nuts. Torque the bolt and nuts to 94 ft. lbs. (127 Nm). Connect both lower control arms to the ball joints.

15. Install the sway bar.

16. Install the wheels and lower the vehicle to the ground.

17. Stabilize the suspension by pushing up and down the the vehicle.

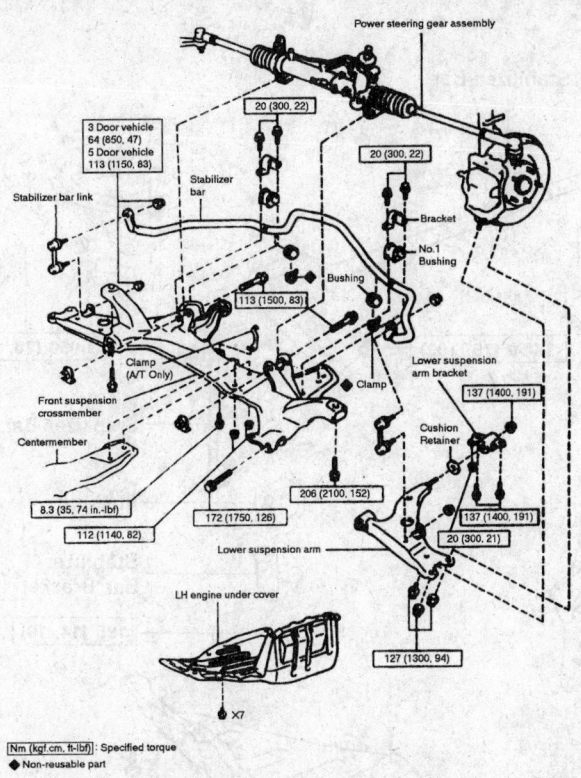

Exploded view of the lower suspension — RAV4

18. Torque the front lower control arm bolt to 126 ft. lbs. (172 Nm).
19. Torque the rear lower control arm bracket bolts and nuts as follows:
   • Bolt: Control arm bracket bolts to 101 ft. lbs. (137 Nm)
   • Nut: Bracket nut to 21 ft. lbs. (28 Nm)
20. Install the left hand engine under cover.
21. Check the wheel alignment.

Sway Bar

REMOVAL AND INSTALLATION

Previa

1. Raise and support the vehicle safely.
2. Remove the wheels and the engine under covers.
3. Remove the sway bar links from the struts and sway bar.
4. Remove the sway bar brackets by removing the four bolts.
5. Remove the sway bar from the vehicle.

To install:
6. Install the sway bar.

NOTE: Make sure that the full weight of the vehicle is on the ground before torquing the nuts and bolts.

7. Install the sway bar brackets. Torque the bolts to 14 ft. lbs. (19 Nm).
8. Install the sway bar links and torque the bolts to 76 ft. lbs. (103 Nm)
9. Install the under covers and the wheels.
10. Test drive the vehicle and check the front end alignment.

RAV4

1. Raise and safely support the front of the vehicle.
2. Remove the front wheels.
3. Remove the four nuts and remove the two sway bar links from the sway bar and lower control arms.

NOTE: If the ball joint turns together with the nut, use a 5mm hexagon wrench to hold the stud.

4. Remove the sway bar brackets and bushings by removing the four bolts.
5. Remove the sway bar from the right hand side of the vehicle.

To install:
6. Install the sway bar to the vehicle.
7. Install the bushings and brackets. Install the bolts and torque the bolts to 22 ft. lbs. (29 Nm).
8. Connect the sway bar links to the sway bar and lower control arms. Install the nuts and torque the nuts as follows:
   • 3 door vehicle: 47 ft. lbs. (64 Nm)
   • 5 door vehicle: 83 ft. lbs. (113 Nm)
9. Install the front wheels and lower the vehicle.

Wheel Bearings

ADJUSTMENT

Check the bearing play in the axial direction and also check the axle hub deviation. The maximum play for both checks should be 0.0020 in. (0.05mm). If greater than the specified maximum, replace the bearing. The wheel bearing is not adjustable.

REMOVAL AND INSTALLATION

Previa

1. Raise and safely support the vehicle.
2. Remove the wheel and tire assembly.
3. Remove the steering knuckle from the vehicle.
4. On 2WD vehicles, pry off the grease cap using a small prybar.
   a. Using a chisel and hammer, release the nut calking.
   b. Remove the locknut from the hub.
   c. Remove the spacer (w/o ABS) or speed sensor rotor (w/ABS).
5. Using a suitable puller (SST 09520–00031), remove the wheel hub from the knuckle.
6. Remove the bearing from the hub using a press and suitable bearing separator tool.
7. Remove the oil seal from the axle hub.
8. Remove the backing plate by removing the three bolts.
9. If equipped with 4WD, remove the dust deflector and oil seal..
10. Remove the bearing snapring from the knuckle.
11. Press the inner bearing from the knuckle.

To install:
12. Using a press and arbor tool 09608–10010 or equivalent, press the inner bearing into the steering knuckle.

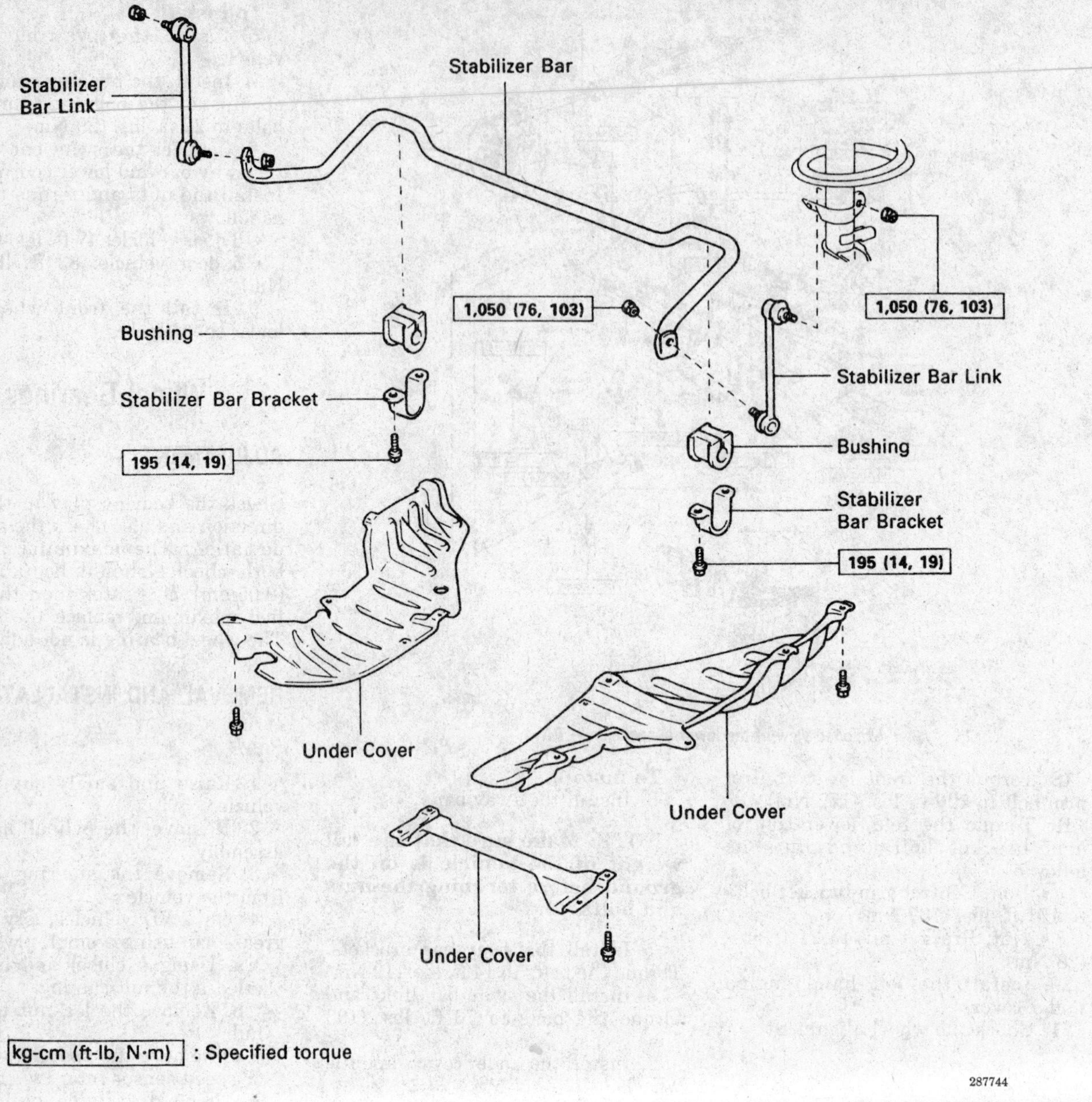

Stabilizer Bar Link

Stabilizer Bar

Bushing

Stabilizer Bar Bracket

1,050 (76, 103)

1,050 (76, 103)

Stabilizer Bar Link

Bushing

Stabilizer Bar Bracket

195 (14, 19)

195 (14, 19)

Under Cover

Under Cover

Under Cover

kg-cm (ft-lb, N·m) : Specified torque

287744

Stabilizer bar — Previa

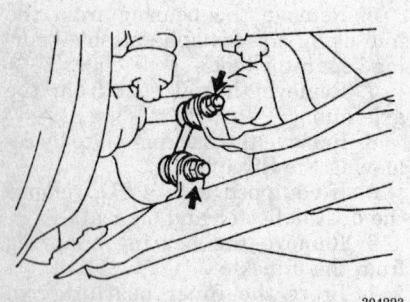

304223

Sway bar link nuts — RAV4

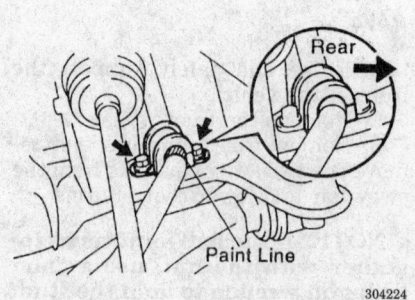

304224

Sway bar bracket bolts — RAV4

Rear

Paint Line

13. Install the snapring.
14. Install the outer bearing.
15. Install the outer oil seal. The seal should be flush with the end surface of the steering knuckle.
16. Install the dust deflector with the three bolts..
17. Press the axle hub onto the steering knuckle.
18. On 2WD vehicles, install the spacer (w/o ABS) or speed sensor rotor (w/ABS).

   a. Install a new nut to the hub and torque to 147 ft. lbs (199 Nm). Caulk the nut.

   b. Install the grease cap.

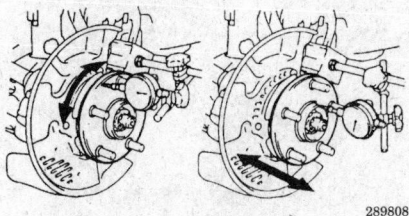

Checking wheel bearings

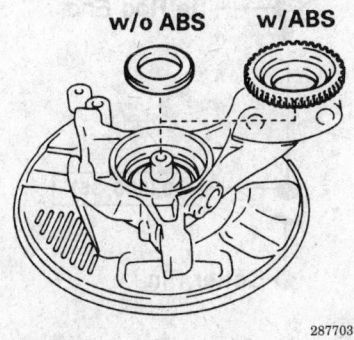

w/o ABS   w/ABS

Spacer or speed sensor rotor components — 2WD Previa

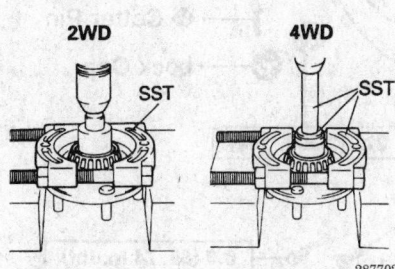

2WD   4WD

SST   SST

Removing bearing from axle hub — Previa

19. On 4WD vehicles, install a new inner oil seal. Install the dust deflector.

20. Install the steering knuckle assembly onto the vehicle.

21. Install the tie rod end to the steering knuckle.

22. Install the wheel. Lower the vehicle.

23. Check the vehicle's front end alignment.

RAV4

1. Disconnect the negative battery cable.

2. Raise the vehicle and support safely.

3. Remove the front wheels.

4. Remove the cotter pin and lock cap from the end of the halfshaft.

5. While applying the front brakes, remove the halfshaft locknut.

6. Remove the brake caliper and use a wire to support it out of the way.

NOTE: Never allow the caliper to hang freely from the brake hose.

7. Matchmark the rotor to the hub and remove the rotor.

8. If equipped with ABS brakes, remove the ABS speed sensor from the steering knuckle.

9. Loosen the nuts on the lower end of the strut.

10. Disconnect and separate the tie rod end from the steering knuckle.

11. Disconnect the lower control arm from the ball joint by removing the bolt and two nuts.

12. Remove the halfshaft from the axle hub. Secure the shaft out of the way using a wire. Be careful not to damage the shaft boot or ABS sensor rotor.

13. Remove the two nuts on the lower end of the strut and remove the steering knuckle.

14. Clamp the steering knuckle in a vise with soft jaws to protect the knuckle.

15. Carefully pry the dust deflector from the hub.

16. Remove the ball joint from the steering knuckle.

17. Using SST 09520–00031 or equivalent, remove the hub from the knuckle.

18. Using SST 09950–00020 or equivalent, remove the inner race from the hub.

19. Remove the four bolts to the dust cover and then remove dust cover.

20. Using SST 09308–00010 or equivalent, remove the inner oil seal.

21. Using SST 09308–00010 or equivalent, remove the outer oil seal.

22. Using snapring pliers, remove the snapring.

23. Take the inner race (removed from the hub) and install it on the outside of the bearing.

24. Using a bearing driver, drive the bearing from the steering knuckle.

To install:

25. Clean bearing seating surfaces with a clean, dry rag.

26. Using a press and SST 09608–32010 or equivalent, install the bearing into the knuckle.

27. Install the snapring.

28. Install the dust cover. Torque the four bolts to 74 in. lbs. (8.3 Nm).

29. Using SST 09608–32010 or equivalent, install a new outer oil seal. Apply multi–purpose grease to the oil seal lip.

30. Press the hub into the steering knuckle.

31. Using SST 09608–32010 or equivalent, install a new inner oil seal. Apply multi–purpose grease to the oil seal lip.

32. Install the lower ball joint to the steering knuckle. Torque the nut to 94 ft. lbs. (127 Nm) and install a new cotter pin.

33. Align the hole in the dust deflector and the hole for the ABS speed sensor and install the dust deflector.

34. Position the knuckle to the lower strut and install the bolts.

35. Install the lower ball joint to the lower arm. Torque the bolts to 94 ft. lbs. (127 Nm).

36. Connect the tie rod end to the steering knuckle. Torque the nut to 36 ft. lbs. (49 Nm).

37. Install the halfshaft the hub and knuckle.

38. Install and torque the nuts on the lower strut to 117 ft. lbs. (158 Nm).

39. Install the ABS speed sensor. Torque the mounting bolt to 69 in. lbs. (7.8 Nm).

40. Align the matchmark and install the rotor on the hub. Install the brake caliper. Torque the mounting bolts to 79 ft. lbs. (107 Nm).

41. Have a helper apply the brakes and install the axle locknut. Torque the nut to 159 ft. lbs (216 Nm). Install the lock cap and a new cotter pin.

42. Install the wheel.

43. Turn the wheel by hand, verify that the wheel turns without noise and without binding.

44. Lower the vehicle.

45. Connect the negative battery cable to the battery and check the signal from the ABS sensor.

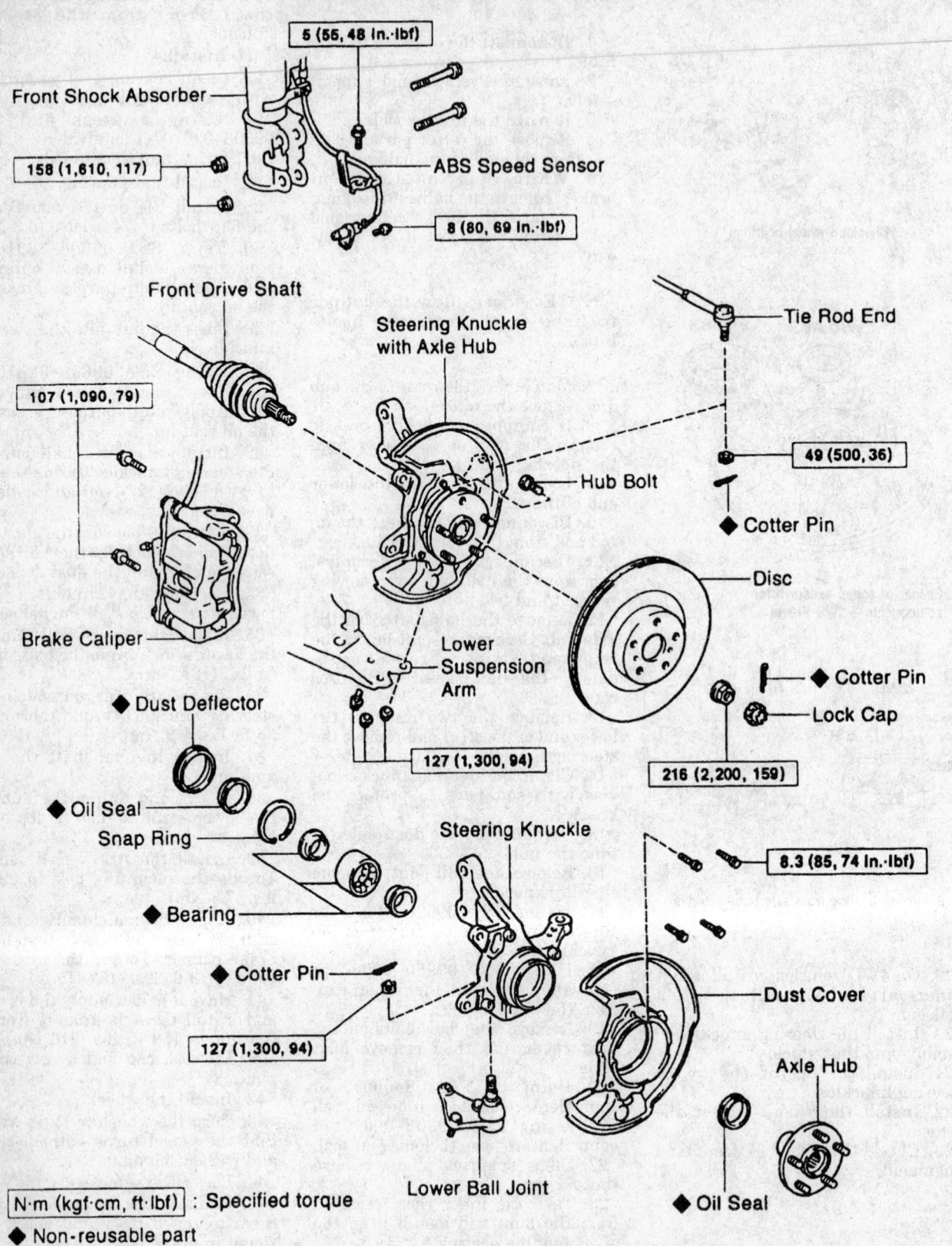

5 (55, 48 In.·lbf)

Front Shock Absorber

158 (1,610, 117)

ABS Speed Sensor

8 (80, 69 In.·lbf)

Front Drive Shaft

Steering Knuckle
with Axle Hub

Tie Rod End

107 (1,090, 79)

49 (500, 36)

Hub Bolt

Cotter Pin

Disc

Brake Caliper

Lower
Suspension
Arm

Cotter Pin

Lock Cap

◆ Dust Deflector

127 (1,300, 94)

216 (2,200, 159)

◆ Oil Seal

Snap Ring

Steering Knuckle

8.3 (85, 74 In.·lbf)

◆ Bearing

◆ Cotter Pin

Dust Cover

127 (1,300, 94)

Axle Hub

Lower Ball Joint

◆ Oil Seal

N·m (kgf·cm, ft·lbf) : Specified torque

◆ Non-reusable part

304128

Exploded view of the front steering knuckle and components — RAV4

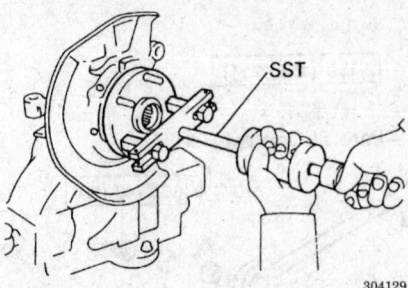

Removing the axle hub from the steering knuckle — RAV4

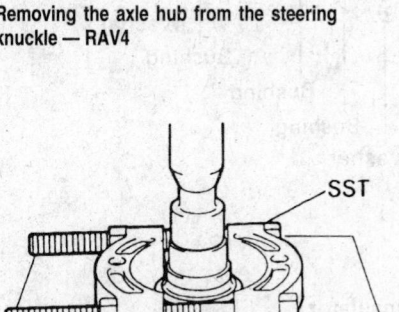

Removing the inner race from the hub — RAV4

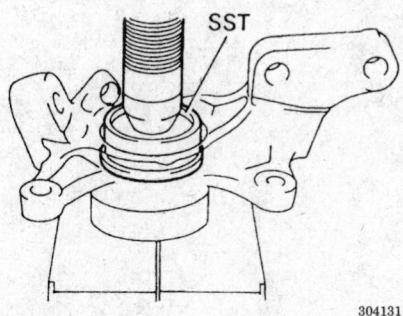

Installing a new bearing in the steering knuckle — RAV4

REAR SUSPENSION

Shock Absorber

REMOVAL AND INSTALLATION

Previa

1. Raise and safely support the rear of the vehicle. Place the jack stands under the frame.

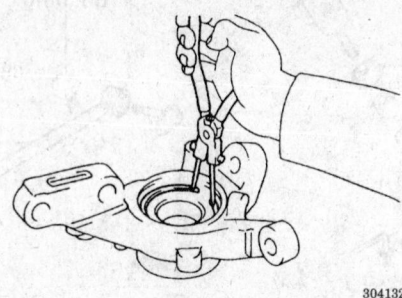

Installing a new snapring in the steering knuckle — RAV4

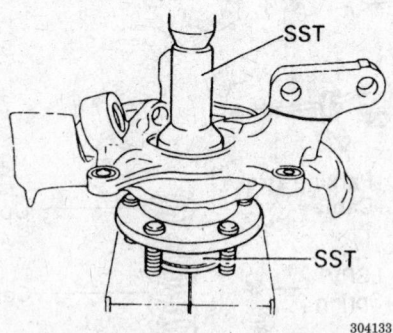

Pressing the axle hub into the steering knuckle — RAV4

2. Support the rear differential with a jack.

— CAUTION —
The coil spring is under high tension. Do not remove the shock absorber without first supporting the differential.

3. Disconnect the shock absorber from the lower control arm by removing the nut, washer and bushing.
4. Remove the shock absorber from the vehicle body by removing the bolt.
5. Remove the shock absorber from the vehicle.
To install:

NOTE: When installing the shock absorber, make sure to install all washers and bushings.

6. Install the shock absorber to the vehicle.
7. Connect the shock absorber to the body by installing the bolt. Torque the bolt to 27 ft. lbs. (37 Nm).
8. Connect the shock absorber to the lower control arm and install the nut. Tighten the nut until the bolt protrudes 0.0059 inch (1.5mm) or more.

9. With the shock absorbers installed, remove the jacks from the vehicle.
10. Lower the vehicle to the ground.

RAV4

1. Raise and safely support the rear of the vehicle.
2. Remove the rear wheel and support the No. 1 control arm with a floor jack.
3. Remove the suspension cap from inside the vehicle.
4. Remove the two nuts from the top of the shock absorber and remove the two retainers and cushion.
5. Disconnect the shock absorber from the lower control arm by removing the bolt and two retainers.
6. Remove the shock absorber from the vehicle.
To install:
7. Install the shock absorber to the vehicle.
8. Install the two retainers and bolt to hold the shock absorber to the lower control arm. Torque the bolt to 27 ft. lbs. (37 Nm).
9. Install the cushion, two retainers and two nuts to hold the shock absorber to the body. Torque the nuts to 18 ft. lbs. (25 Nm).
10. Install the suspension cap.
11. Install the wheel and remove the floor jack.
12. Lower the vehicle.

Coil Spring

REMOVAL AND INSTALLATION

Previa

— CAUTION —
The spring on the axle carrier is under high pressure and can cause serious injury if not properly removed and installed.

1. Raise and safely support the vehicle at the frame.
2. Support the axle housing with a floor jack.
3. Remove the wheel and tire assembly.
4. Remove the shock absorber to axle housing bolt. Disconnect both shock absorbers from the axle housing.
5. Disconnect the lateral control arm from axle housing by removing the nut.
6. Disconnect the LSPV spring from the lower control arm by removing the nut.
7. Disconnect the brake line from the brake hose at the body bracket.

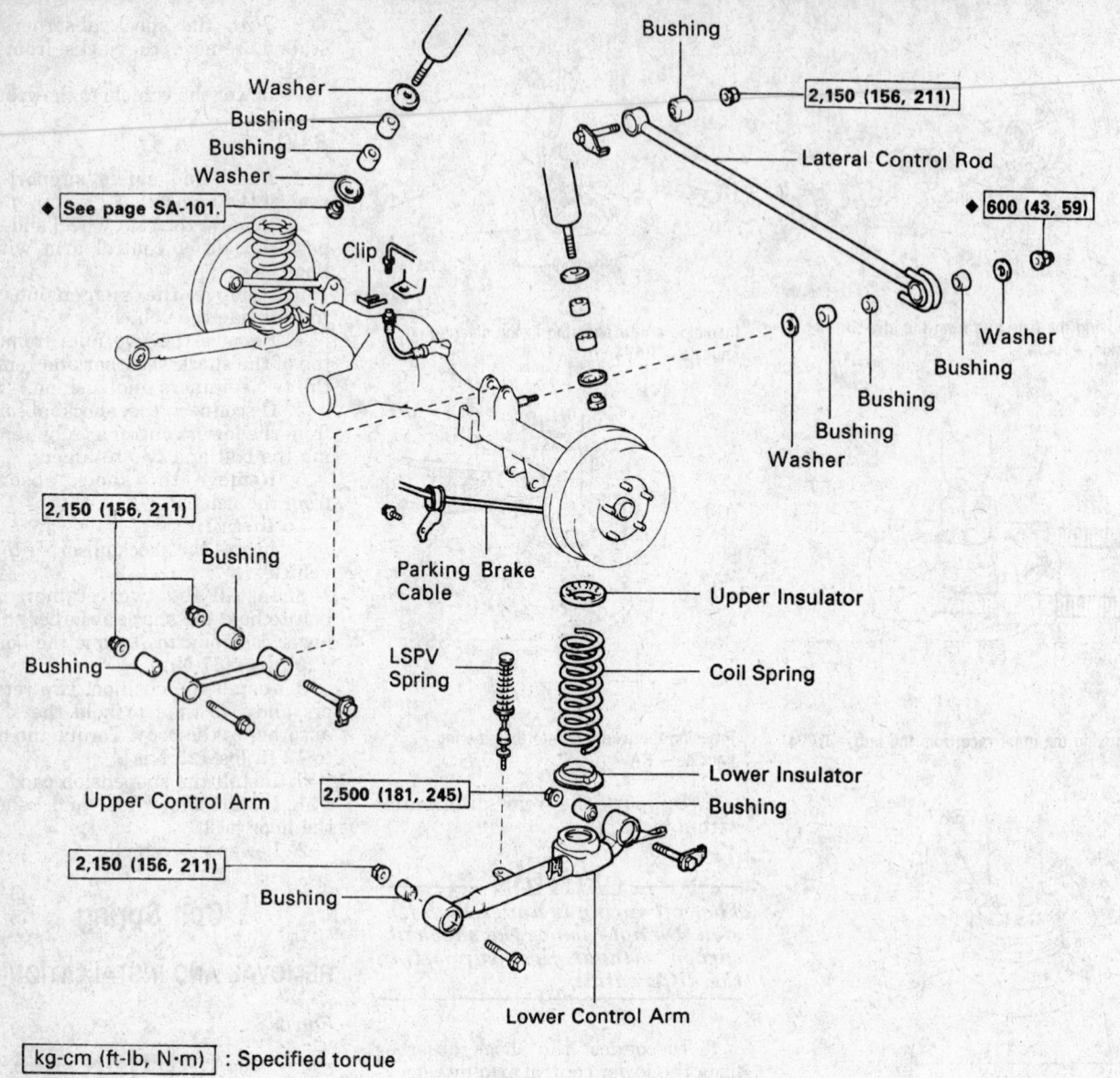

Washer
Bushing
Bushing
Washer
◆ See page SA-101.
Clip
Bushing
2,150 (156, 211)
Lateral Control Rod
◆ 600 (43, 59)
Washer
Bushing
Bushing
Bushing
Washer
2,150 (156, 211)
Bushing
Bushing
Parking Brake Cable
Upper Insulator
LSPV Spring
Coil Spring
Upper Control Arm
2,500 (181, 245)
Lower Insulator
Bushing
2,150 (156, 211)
Bushing
Lower Control Arm

kg-cm (ft-lb, N·m) : Specified torque

◆ Non-reusable part

287712

Rear suspension components, exploded view — Previa

8. Remove the clip and disconnect the brake hose from the body bracket.

9. If equipped with ABS brakes, remove the ABS wiring harness bracket.

10. Disconnect the parking brake cable from the lower control arm.

11. Lower the floor jack, then remove the coil spring(s) and the insulators.

To install:

12. Place the springs into position and raise the axle housing. Make sure to fit the lower end of the coil spring into the gap of the spring seat on the lower control arm.

13. Raise the rear axle housing.

14. Connect the parking brake cable to the lower control arm.

15. Install the lateral control rod to the suspension and torque the bolts and nuts as follows:

• Body side bolt to 156 ft. lbs. (211 Nm).

• Axle housing side to 43 ft. lbs. (59 Nm)

16. Connect the shock absorber to the lower control arm and install the nut.

17. Connect the LSPV spring to the lower control arm and install the nut. Torque the nut to 9 ft. lbs. (13 Nm).

18. If equipped, install the ABS wiring harness bracket.

19. Connect the brake hose to the bracket and install the clip.

20. Connect the brake tube to the brake hose and tighten the tube.

21. Bleed the brake system.

22. Install the wheels and lower the vehicle.

RAV4

1. Disconnect the negative battery cable from the battery.

2. Raise and safely support the rear of the vehicle.

3. If equipped with 2WD, remove the axle shaft.

4. If equipped with 4WD, remove the halfshaft.

5. Remove the brake drum.

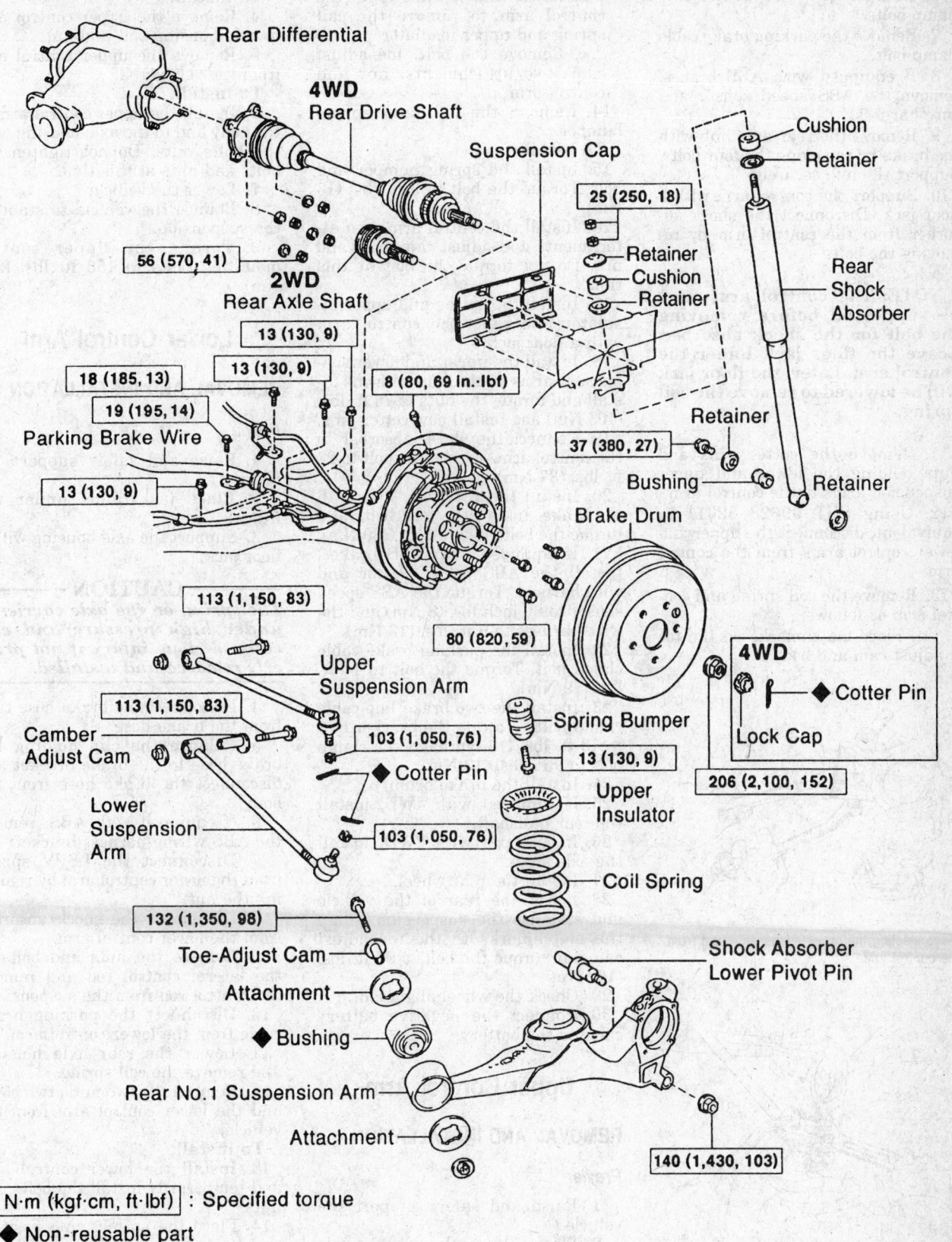

Rear Differential

4WD
Rear Drive Shaft

Suspension Cap

Cushion

Retainer

25 (250, 18)

56 (570, 41)

2WD
Rear Axle Shaft

Retainer

Cushion

Retainer

Rear Shock Absorber

13 (130, 9)

13 (130, 9)

18 (185, 13)

8 (80, 69 In.·lbf)

Retainer

19 (195, 14)

Parking Brake Wire

37 (380, 27)

Bushing

13 (130, 9)

Brake Drum

Retainer

113 (1,150, 83)

Upper Suspension Arm

80 (820, 59)

4WD

113 (1,150, 83)

Camber Adjust Cam

103 (1,050, 76)

Spring Bumper

Cotter Pin

Lock Cap

Lower Suspension Arm

Cotter Pin

13 (130, 9)

206 (2,100, 152)

103 (1,050, 76)

Upper Insulator

Coil Spring

132 (1,350, 98)

Shock Absorber Lower Pivot Pin

Toe-Adjust Cam

Attachment

Bushing

Rear No.1 Suspension Arm

Attachment

140 (1,430, 103)

N·m (kgf·cm, ft·lbf) : Specified torque

◆ Non-reusable part

307274

Rear suspension exploded view — RAV4

6. Remove the two brake line clamp bolts.

7. Remove the parking brake cable clamp bolt.

8. If equipped with ABS brakes, remove the ABS speed sensor and wire harness.

9. Remove the rear axle hub with the brake by removing the four bolts. Support the hub securely.

10. Support the control arm with a floor jack. Disconnect the shock absorber from the control arm by removing the bolt.

NOTE: The control arm must be supported before removing the bolt for the shock absorber. Leave the floor jack under the control arm. Later, the floor jack will be lowered to remove the coil spring.

11. Remove the cotter pins and nuts holding the lower and upper suspension arms to the control arm.

12. Using SST 09628–62011 or equivalent, disconnect the upper and lower control arms from the control arm.

13. Remove the coil spring and control arm as follows:

    a. Place matchmarks on the toe adjust cam and body.

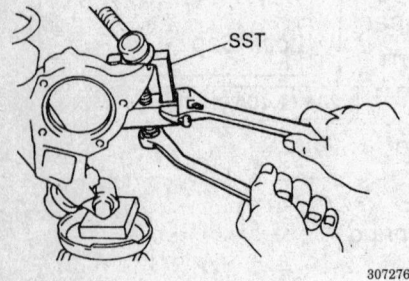

307276

Disconnecting the upper suspension arm from the control arm

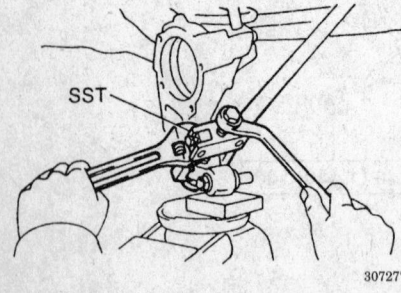

307277

Disconnecting the lower suspension arm from the control arm

    b. Loosen the bolt and lower the control arm to remove the coil spring and upper insulator.

    c. Remove the bolt, toe–adjust cam, two attachments, nut and control arm.

14. Remove the bolt and spring bumper.

To install:

15. Install the spring bumper and bolt. Torque the bolt to 9 ft. lbs. (13 Nm).

16. Install the control arm, two attachments, toe–adjust cam, bolt, and nut. Do not torque the bolt at this time.

17. Install the spring and upper insulator and raise the control arm with a floor jack.

18. Install the upper and lower suspension arms to the control arm. Install and torque the nuts to 76 ft. lbs. (103 Nm) and install new cotter pins.

19. Connect the shock absorber to the control arm. Torque the bolt to 27 ft. lbs. (37 Nm).

20. Install the rear axle hub with the brake. Install the four bolts and torque the bolts to 59 ft. lbs. (80 Nm).

21. If equipped with ABS brakes, install the ABS speed sensor and wire harness. Torque the ABS speed sensor to 69 inch lbs. (8 Nm) and the wire harness to 9 ft. lbs. (13 Nm).

22. Install the parking brake cable clamp bolt. Torque the bolt to 14 ft. lbs. (19 Nm).

23. Install the two brake line cable clamp bolts. Torque the bracket bolt to 13 ft. lbs. (18 Nm) and the clamp bolt to 9 ft. lbs. (13 Nm).

24. Install the brake drum.

25. If equipped with 4WD, install the rear halfshaft.

26. If equipped with 2WD, install the axle shaft.

27. Install the rear wheel.

28. Lower the rear of the vehicle and stabilize the suspension. Align the matchmarks to the toe–adjust cam and torque the bolt to 98 ft. lbs. (132 Nm).

29. Check the wheel alignment.

30. Connect the negative battery cable to the battery.

Upper Control Arm

REMOVAL AND INSTALLATION

Previa

1. Raise and safely support the vehicle.

2. Place a floor jack under the axle housing to support it.

3. Remove the upper control arm to body bolt and nut.

To install:

4. Remove the upper control arm to axle housing bolt and nut.

5. Remove the upper control arm from the vehicle.

To install:

6. Install the upper control arm to the body and to the axle housing with the bolts nuts. Do not tighten the bolts and nuts at this time.

7. Lower the vehicle.

8. Bounce the vehicle to stabilize the suspension.

9. Torque the upper control mounting bolts to 156 ft. lbs. (211 Nm).

Lower Control Arm

REMOVAL AND INSTALLATION

Previa

1. Raise and safely support the vehicle.

2. Place jackstands under the frame.

3. Support the axle housing with a floor jack.

— CAUTION —

The spring on the axle carrier is under high pressure and can cause serious injury if not properly removed and installed.

4. Disconnect the brake hose tube from the brake hose.

5. Remove the clip holding the brake hose to the brake bracket and disconnect the brake hose from the body.

6. If equipped with ABS, remove the ABS wiring harness bracket.

7. Disconnect the LSPV spring from the lower control arm by removing the nut.

8. Disconnect the shock absorber from the lower control arm.

9. Remove the nuts and bolts to the lateral control rod and remove the control rod from the suspension.

10. Disconnect the parking brake cable from the lower control arm.

11. Lower the rear axle housing and remove the coil spring.

12. Remove the two nuts, two bolts, and the lower control arm from the vehicle.

To install:

13. Install the lower control arm and temporarily install the bolts and nuts.

14. Place the springs into position and raise the axle housing. Make sure to fit the lower end of the coil spring into the gap of the spring seat on the lower control arm.

15. Raise the rear axle housing.

16. Connect the parking brake cable to the lower control arm.

17. Install the lateral control rod to the suspension and torque the bolts and nuts as follows:

• Body side bolt to 156 ft. lbs. (211 Nm).

• Axle housing side to 43 ft. lbs. (59 Nm)

18. Connect the shock absorber to the lower control arm and install the nut.

19. Connect the LSPV spring to the lower control arm and install the nut. Torque the nut to 9 ft. lbs. (13 Nm).

20. If equipped, install the ABS wiring harness bracket.

21. Connect the brake hose to the bracket and install the clip.

22. Connect the brake tube to the brake hose and tighten the tube.

23. Bleed the brake system.

24. Lower the vehicle and bounce the vehicle several times to stabilize the suspension.

25. Torque the lower control arm to body nuts to 156 ft. lbs (211 Nm) and control arm to axle housing nuts to 181 ft. lbs (245 Nm).

RAV4

1. Raise and safely support the rear of the vehicle.

2. Remove the wheel.

3. Remove the cotter pin and nut holding the lower control arm to the control arm.

4. Using SST 09629–62011 or equivalent, disconnect the lower control arm from the control arm.

5. Place matchmarks on the camber adjust cam and suspension member.

6. Remove the nut, bolt, and camber adjust cam holding the lower control arm to the suspension member.

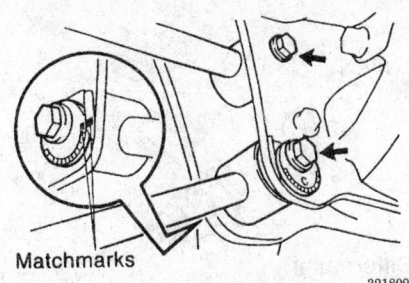

Matchmarks

321809

Lower control arm bolt and camber adjusting cam — RAV4

7. Remove the lower control arm from the vehicle.

To install:

8. Install the lower control arm to the vehicle.

9. Install the camber adjust cam, bolt and nut to hold the lower control arm to the suspension member. Do not torque the bolt and nut at this time.

10. Connect the lower control arm to the control arm. Install the nut and torque to 76 ft. lbs. (103 Nm). Install a new cotter pin.

11. Install the wheel and lower the vehicle.

12. Stabilize the suspension and torque the camber adjust cam bolt to 83 ft. lbs. (113 Nm).

Wheel Bearing

REMOVAL AND INSTALLATION

RAV4

1. Raise and safely support the rear of the vehicle.

2. Remove the wheel.

3. Remove the brake drum.

4. If equipped with ABS brakes, remove the bolt and the ABS speed sensor.

5. If equipped with 2WD, remove the cotter pin and lock cap to the axle shaft. Remove the nut to the axle shaft and remove the shaft from the hub.

6. If equipped with 4WD, remove the rear halfshaft from the hub.

7. Disassembly the rear brake components.

8. Disconnect the brake line from the wheel cylinder.

9. Disconnect the parking brake cable from the backing plate by removing the two bolts.

10. Remove the rear axle hub with the backing plate by removing the four bolts.

11. Using the proper tools, press out the axle hub from the bearing.

12. Using the proper tools, remove the inner race (outside) from the axle hub.

To install:

13. Using the proper tools, install the axle hub to a new bearing.

14. Install the axle hub with the backing plate. Install the four bolts and torque to 59 ft. lbs. (80 Nm).

15. Connect the parking brake cable with the two bolts. Torque the bolts to 69 inch lbs. (80 Nm).

16. Connect the brake line to the wheel cylinder.

17. Assembly the brake assembly.

18. If equipped with 4WD, install the halfshaft.

19. If equipped with 2WD, install the axle shaft and nut. Torque the nut to 152 ft. lbs. (206 Nm). Install the lock cap and a new cotter pin.

20. If equipped with ABS, connect the ABS speed sensor and bolt.

21. Install the brake drum.

22. Bleed the brake system.

23. Install the wheel and lower the vehicle.

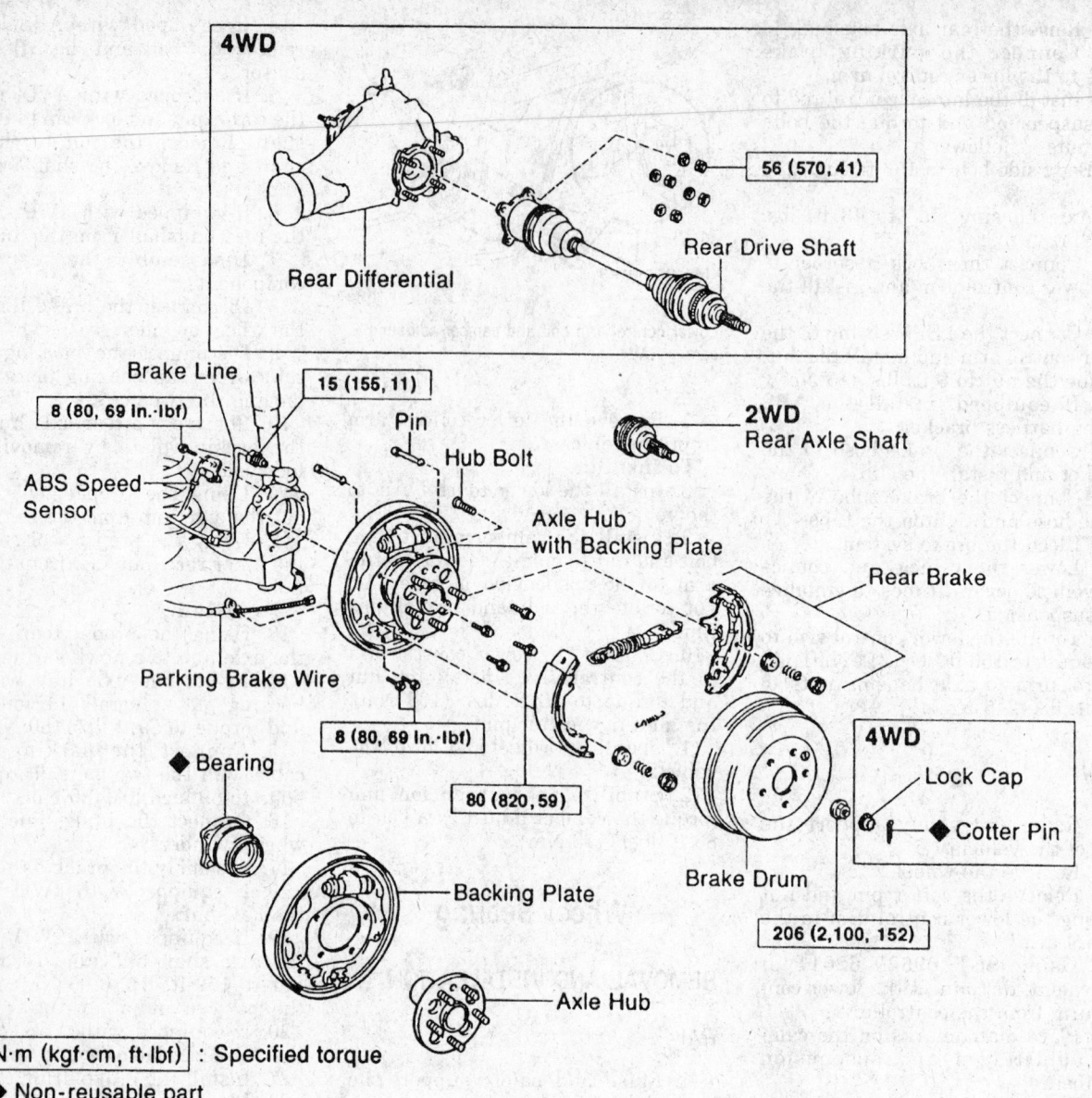

4WD

56 (570, 41)

Rear Drive Shaft

Rear Differential

Brake Line

8 (80, 69 In.·lbf)

15 (155, 11)

Pin

Hub Bolt

2WD
Rear Axle Shaft

ABS Speed
Sensor

Axle Hub
with Backing Plate

Rear Brake

Parking Brake Wire

◆ Bearing

8 (80, 69 In.·lbf)

80 (820, 59)

4WD

Lock Cap

◆ Cotter Pin

Brake Drum

206 (2,100, 152)

Backing Plate

Axle Hub

N·m (kgf·cm, ft·lbf) : Specified torque

◆ Non-reusable part

313770

Hub and wheel bearing removal and installation — RAV4

TOOLS AND EQUIPMENT 21

WHAT TOOLS WILL BE NEEDED?

Analyzing Specific Needs

Nearly everybody needs some tools, whether they are fixing a kitchen sink, or overhauling the engine in the family truck. As far as truck repairs go, pliers and a can of oil will not get one very far down the path of do-it-yourself service. But, a do-it-yourselfer's garage does not have to be equipped like the local service station either. Somewhere between these two extremes is a level that suits the average do-it-yourselfer. Just where that point is depends on the home mechanic's ability and level of interest. The strategy is to match the tools and equipment to the tasks which are to be tackled.

First, things should be sorted out in an orderly manner. The do-it-yourselfer should think about his/her repair work in three levels: Basic, average and advanced. Before purchasing any tools, he/she should sit down and determine the level of the expertise and cost needed to accomplish the job at hand. Knowing what repairs can, or are needed to be performed is the most important step. Obviously, if all that is intended is changing the oil and spark plugs, many tools will not be needed. If fairly extensive repair work is planned, the do-it-yourselfer will end up with a pretty complete collection of tool. Many expensive tools can be rented from automotive parts jobbers or tool rental centers. This allows many of home mechanics to do special repairs on an occasional basis.

Basic Automotive Tools

Naturally, without the proper tools it is impossible to properly service a truck. It would be impossible to catalog each tool which would be needed to perform every operation in this book. It would also be unwise for the amateur to rush out and buy an expensive set of tools, on the theory that one or more may be needed at sometime.

The best approach is to proceed slowly, gathering together a good quality set of those tools that are used most frequently. Don't be misled by the low cost of bargain tools. It is far better to spend a little extra money for better quality tools. Forged wrenches, 6-point sockets and fine tooth ratchets are preferable to their less expensive counterparts. As any good mechanic can concur, there are few worse experiences than trying to work on a truck with bad tools. Any monetary savings will be far outweighed by frustration and mangled knuckles.

Certain tools, plus a basic ability to handle them, are required to get started. A basic tool set and a torque wrench, are good for a start. Begin by accumulating those tools that are used most frequently (tools associated with routine maintenance/tune-up and engine repair). In addition to the normal assortment of screwdrivers and pliers, the following tools should be acquired for general routine maintenance:

• Metric and standard wrenches, sockets and combination open end/box end wrenches in sizes from 3–19mm and $\frac{1}{4}$–$\frac{7}{8}$ in., and a spark plug socket ($\frac{5}{8}$ inch or 16mm). If possible, buy various length socket drive extensions. One advantage in this area is that the metric sockets available in the United States will fit SAE ratchet handles and extensions one may already have ($\frac{1}{4}$ in., $\frac{3}{8}$ in., and $\frac{1}{2}$ in. drive).

• Jackstands for support.
• Oil filter wrench.
• Oil filler spout or funnel.
• Grease gun for chassis lubrication.
• Hydrometer or battery tester for checking the battery.
• A low flat pan for draining oil.
• Lots of rags for wiping up the inevitable mess.

In addition to the above items, there are several other tools which, although not absolutely necessary, are handy to have around. These include oil-dry, a transmission fluid funnel and the usual supply of lubricants and fluids, all of which can be purchased on a need basis. This is a basic list for routine maintenance, but only personal needs and desires can accurately determine the final list of tools necessary.

The second list of tools is for tune-ups. While these tools are slightly more sophisticated, they need not be

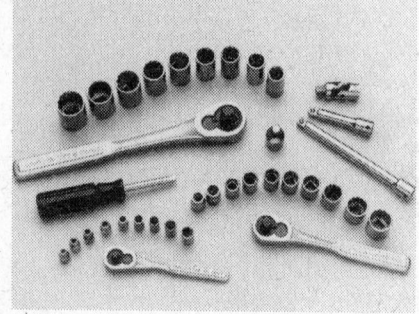

All but the most basic procedure will require an assortment of ratchets and sockets

TCCS1200

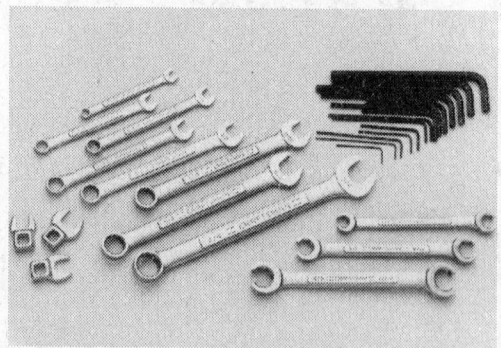

In addition to ratchets, a good set of wrenches and hex keys will be necessary

TCCS1201

Here again, it is best to be guided by particular needs. In addition to these basic tools, there are several other tools and gauges which may be useful. These include:

• A compression gauge. The screw-in type is slower to use, but eliminates the possibility of a faulty reading due to escaping pressure.

• A manifold vacuum gauge.

• A test light.

• An Digital Volt-Ohmmeter (DVOM). This meter allows direct testing of electrical components and grounds.

As a final note, a torque wrench will be necessary for most work. The beam type models are perfectly adequate, although the newer click (break-away) type is more precise, and does not require one to crane one's neck to see a torque reading in awkward situations. The break-away torque wrenches are more expensive and should be recalibrated periodically.

Tightening bolts to the correct torque value is extremely important on today's trucks. The torque specification for each fastener will be given in the procedure whenever a specific torque value is required. An example of torque specifications are given (the following values are only a guide), based upon fastener size:

Bolts marked 6T

6mm bolt/nut: 5–7 ft. lbs. (7–10 Nm)

8mm bolt/nut: 12–17 ft. lbs. (16–23 Nm)

10mm bolt/nut: 23–34 ft. lbs. (31–46 Nm)

12mm bolt/nut: 41–59 ft. lbs. (56–80 Nm)

14mm bolt/nut: 56–76 ft. lbs. (76–103 Nm)

Bolts marked 8T

6mm bolt/nut: 6–9 ft. lbs. (8–12 Nm)

8mm bolt/nut: 13–20 ft. lbs. (18–27 Nm)

10mm bolt/nut: 27–40 ft. lbs. (37–54 Nm)

12mm bolt/nut: 46–69 ft. lbs. (62–94 Nm)

14mm bolt/nut: 75–101 ft. lbs. (102–137 Nm)

Special Tools

Normally, special factory tools are avoided for repair procedures, since these many not be readily available for the do-it-yourself mechanic. When it is possible to perform the job with more commonly available tools, it will be pointed out, but occasionally, a special tool was designed to per-

TCCS1202

A hydraulic floor jack and a set of jackstands are essential for lifting and supporting the vehicle

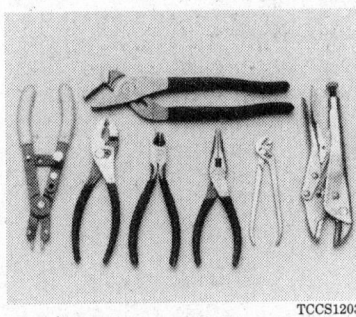

TCCS1203

An assortment of pliers will be handy, especially for old rusted parts and stripped bolt heads

outrageously expensive. There are several inexpensive tach/dwell meters on the market that are every bit as good for the average mechanic as a costly professional model. Just be sure the tach/dwell meter measures to at least 1200–1500 rpm on the tach scale, and that it works on 4, 6 and 8-cylinder engines. A basic list of tune-up equipment could include:

• Tach/dwell meter.

• Spark plug wrench.

• Timing light (a DC light that works from the vehicle's battery is best).

• Wire spark plug gauge/adjusting tools.

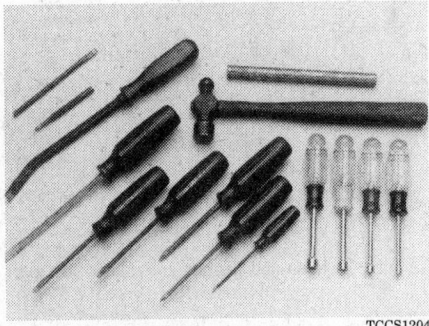

TCCS1204

Various screwdrivers, a hammer, chisels and prybars are necessary to have in one's tool box

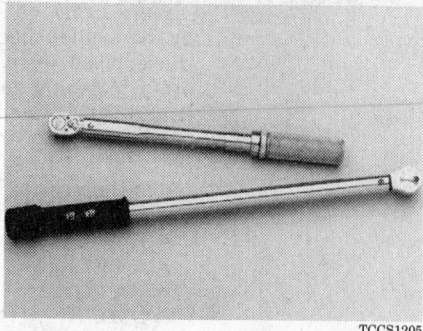

Many repairs will require the use of a torque wrench to assure the components are properly fastened

TCCS1205

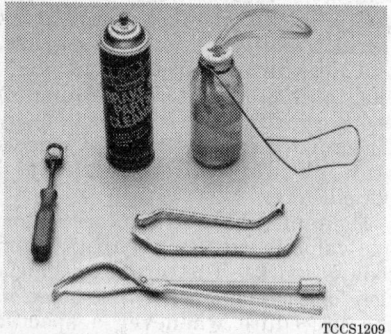

Although not always necessary, using specialized brake tools will save time

TCCS1209

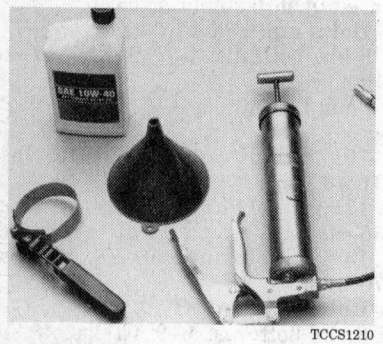

A few inexpensive lubrication tools will make regular service easier

TCCS1210

Among other features, scan tools combine many standard testers into a single device for quick and accurate circuit diagnosis. For many tests, a multi-meter, test light, or other general test equipment can be substituted but the technician must be aware of the risk involved. The general test equipment may not be capable of safely testing the system or may generate incomplete or inaccurate test results. Some tests require activating system components and often this can only be done with scan tools or other special equipment.

Most test equipment is available through aftermarket tool manufacturers, but some can only be obtained through the vehicle manufacturer. Care should be taken that all test equipment being used is designed to diagnose that particular system accurately without damaging control modules or other components.

NOTE: When using special test equipment, the manufacturer's instructions provided with the tester should be read and clearly understood before attempting any test procedures.

SPECIAL DIAGNOSTIC TOOLS

Frequent references to specific test equipment will be found in the text and in the diagnostic charts. This usually refers to scan tools used to communicate with electronic control units or special electronic testers.

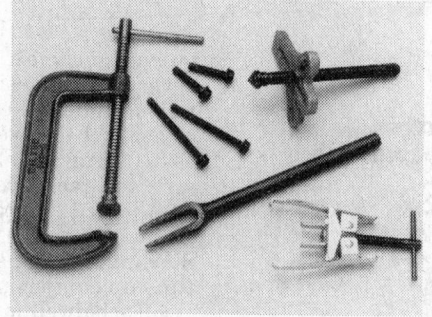

Various pullers, clamps and separator tools are needed for the repair of many components

TCCS1211

form a specific function and should be used. Before substituting another tool, one should be convinced that neither one's safety nor the performance of the vehicle will be compromised.

Some special tools are available commercially from major tool manufacturers. Others can be purchased from a truck dealership or automotive parts store.

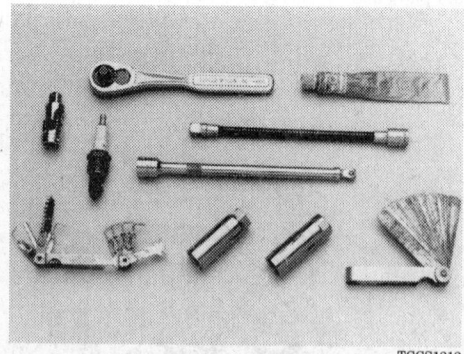

A variety of tools and gauges are needed for spark plug service

TCCS1212

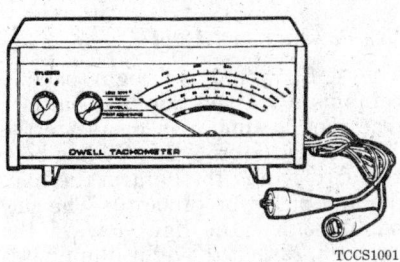

TCCS1001

Dwell/tachometer unit (typical)

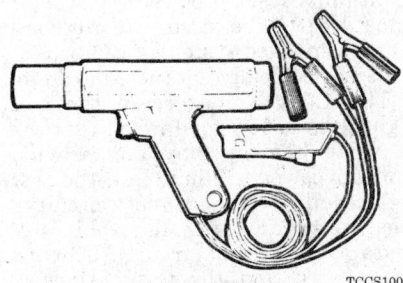

TCCS1002

Inductive type timing light

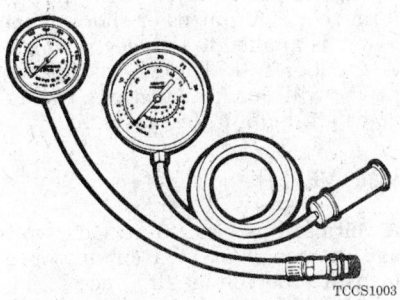

TCCS1003

Compression gauge and a combination
vacuum/fuel pressure test gauge

Electrical Test Tools

ORGANIZED TROUBLESHOOTING

When diagnosing a specific problem, there are certain troubleshooting techniques that are standard:

1. Establish when the problem occurs. Does the problem appear only under certain conditions? Were there any noises, odors, or other unusual symptoms? Make notes on any symp-toms found, including warning lights and trouble codes, if applicable.

2. Isolate the problem area. To do this, make some simple tests and observations; then eliminate the systems that are working properly. Check for obvious problems such as broken wires, split or disconnected vacuum hoses. Always check the obvious before assuming something complicated is the cause. Be suspicious of fuses, switches and connectors; wiring itself rarely fails.

3. Test for problems systematically to determine the cause once the problem area is isolated. Are all the components functioning properly? Is there power going to electrical switches and motors? Is there vacuum at vacuum switches and/or actuators? Doing careful, systematic checks will often turn up most causes on the first inspection without wasting time checking components that have little or no relationship to the problem.

4. Test all repairs after the work is done to make sure that the problem is fixed. Some causes can be traced to more than 1 component, so a careful verification of repair work is important to pick up additional malfunctions that may cause a problem to reappear or a different problem to arise. A blown fuse, for example, is a simple problem which may require more than another fuse to repair.

The diagnostic tree charts are designed to help solve problems by leading the user through closely defined conditions and tests. Only the most likely components, vacuum and electrical circuits are checked for proper operation when troubleshooting a particular malfunction. By using the diagnostic trees to eliminate those systems and components which normally will not cause the condition described, a problem can be isolated within 1 or more systems or circuits without wasting time on unnecessary testing.

Experience has shown that most problems tend to be the result of a fairly simple and obvious cause, such as loose or corroded connectors; making careful inspection of components during testing is essential to quick and accurate troubleshooting. Frequent references to special test equipment will be found in the text and in the diagnosis charts. These devices or a compatible equivalent are necessary to perform some of the more complicated test procedures listed. Testers are available from a variety of aftermarket sources as well as from the vehicle manufacturer. Care should be taken that any test equipment being used is designed to diagnose that particular system accurately without damaging the computer control modules or components being tested.

NOTE: Pinpointing the exact cause of trouble in an electrical system can sometimes be accomplished only by the use of special test equipment. In addition to the information covered in this section, the manufacturer's instructions booklet provided with the tester should be read and clearly understood before attempting any test procedures.

Testers and Equipment

JUMPER WIRES

Jumper wires are simple, yet extremely valuable, pieces of test equipment. Jumper wires are merely wires that are used to bypass sections of a circuit. The simplest type of jumper wire is a length of multi-strand wire with an alligator clip at each end. Jumper wires are usually fabricated from lengths of standard automotive wire and whatever type of connector (alligator clip, spade connector or pin connector) is required for the vehicle being tested. Some jumper wires are made with 3 or more terminals coming from a common splice for special-purpose testing. In cramped, hard-to-reach areas it is advisable to have insulated boots over the jumper wire terminals in order to prevent accidental grounding and possible system damage.

Jumper wires are used primarily to locate open electrical circuits, on either the ground (-) side of the circuit or on the hot (+) side. If an electrical component fails to operate, connect the jumper wire between the component and a good ground. If the component operates only with the jumper installed, the ground circuit is open. If the ground circuit is good, but the component does not operate, the circuit between the power feed and component is open. Sometimes a fused jumper wire is connected directly from the battery to the hot terminal of the component, but first make sure the component uses 12 volts in operation.

By inserting an in-line fuse between a set of test leads, a fused jumper wire is created. A fused jumper wire can be used for bypas-

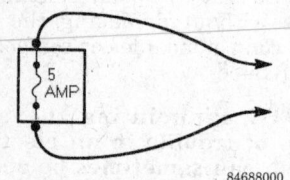

Schematic of a fused jumper wire

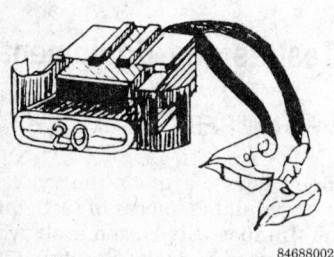

Fused jumper wire

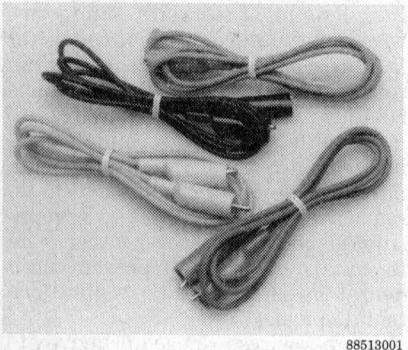

Jumper wires come in different gauges

sing open circuits. Use a 5 amp fuse to provide circuit protection.

NOTE: Never use jumpers made from wire that is of lighter gauge (smaller diameter) than used in the circuit under test. If the jumper wire is too small, it may overheat and possibly melt. Never use jumpers to bypass high-resistance loads (such as motors) in a circuit. Bypassing resistances, in effect, creates a short circuit which may cause damage and fire. Never use a jumper for anything other than

temporary bypassing of components in a circuit, otherwise damage or fire could result.

TEST LIGHTS

12 Volt Test Light

The 12 volt test light is used to check circuits and components while electrical current is flowing through them. It is used for voltage and ground tests. Twelve volt test lights come in different styles, but all have 3 main parts; a ground clip, a probe and a light.

NOTE: Avoid piercing the insulation of any wire. While most probes are designed to pierce insulation, this can lead to corrosion or broken conductors within the wire. Trace the wire to a terminal that can be probed before piercing the insulation.

The most commonly used 12 volt test lights have pick-type probes. To use a 12 volt test light, connect the ground clip to a good ground and probe wherever necessary with the pick.

The wrap-around light is handy in hard to reach areas or where it is difficult to support a wire to push a probe pick into it. To use the wrap around light, hook the wire to be probed with the hook and pull the trigger. A small pick will be forced through the wire insulation into the wire core. Only use this type of test light as a last resort and do not use it on SRS (air bag systems) or computer data lines.

NOTE: Never use a pick-type test light to probe wiring on computer controlled systems unless specifically instructed to do so. Any wire insulation that is pierced by the test light probe should be taped and sealed with silicone after testing.

The test light does not detect specific voltage amounts; it only detects that voltage is present. It is advisable before using the test light to touch its terminals across the battery posts to make sure the light is operating properly. Do not attempt to determine voltage by how brightly the tester glows; use a voltmeter if an exact reading is needed.

Use of a LED type test light is recommended for computer controlled circuits. A standard incandescent bulb test light can load the circuit causing a high current to flow and damage the components. An LED type test light will not load the circuit

and is safer to use in a computer controlled circuit.

Self-Powered Test Light

The self-powered test light usually contains a 1.5 volt penlight battery. One type is similar in design to the 12 volt test light. This type has both the battery and the light in the handle and pick-type probe tip. The second type has the light toward the open tip, so that the light illuminates the contact point. The self-powered test light is a dual-purpose piece of equipment. It can be used to test for either open or short circuits when power is isolated from the circuit (continuity test). A powered test light should never be used on any computer controlled system or component unless specifically instructed to do so.

The 1.5 volt battery in the test light does not provide much current. A weak battery may not provide enough power to illuminate the test light even when a complete circuit is made (especially if there are high resistances in the circuit). Always make sure that the test battery is strong. To check the battery, briefly touch the ground clip to the probe; if the light glows brightly, the battery is strong enough for testing. Never use a self-powered test light to perform checks for opens or shorts when power is applied to the electrical system under test. The 12 volt vehicle power will quickly burn out the 1.5 volt light bulb in the test light.

VOLTMETER

A voltmeter is used to measure voltage at any point in a circuit, or to measure the voltage drop across any part of a circuit. Voltmeters usually have various scales on the meter dial and a selector switch to allow the selection of different test ranges. The voltmeter has a positive and a negative lead. To avoid damage to the meter, connect the negative lead, usually black, to the negative (-) side of circuit or to ground. Connect the positive lead, usually red, to the positive (+) or power side of the circuit.

A voltmeter can be connected either in parallel or in series with a circuit and has a very high resistance to current flow. When connected in parallel, only a small amount of current will flow through the voltmeter current path; the rest will flow through the normal current path and the circuit will work normally. When the voltmeter is connected in series with a circuit, only a small amount of current can flow through the circuit.

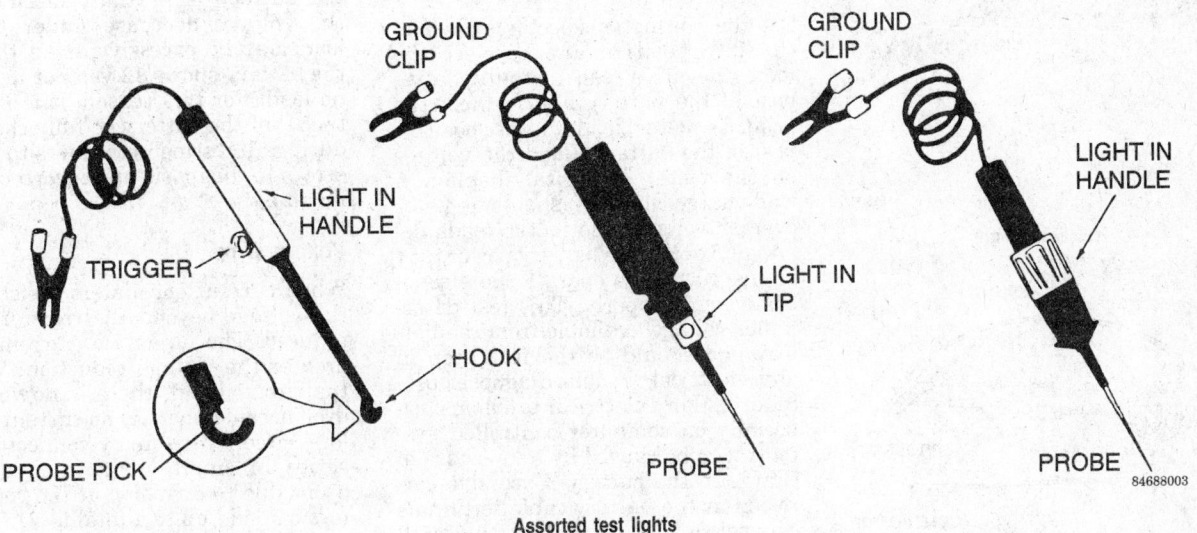

GROUND CLIP

LIGHT IN HANDLE

TRIGGER

PROBE PICK

HOOK

GROUND CLIP

LIGHT IN TIP

PROBE

GROUND CLIP

LIGHT IN HANDLE

PROBE

84688003

Assorted test lights

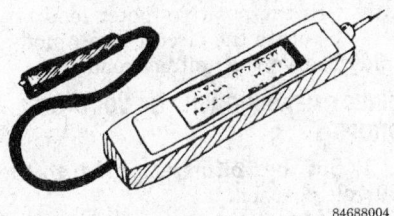

84688004

Logic probe type tester

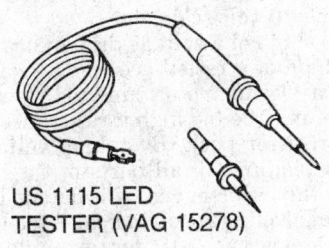

US 1115 LED TESTER (VAG 15278)

84688005

LED type test light for use on computer circuits

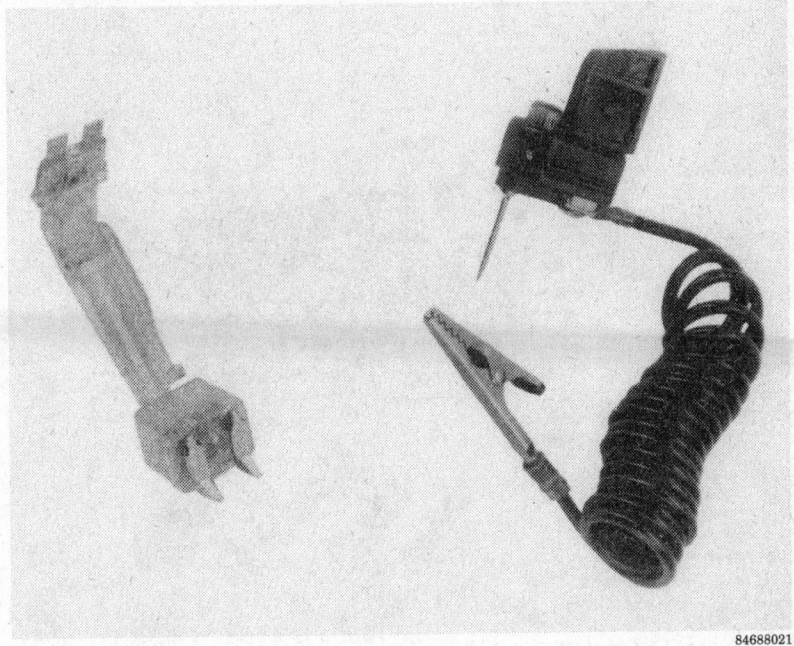

84688021

The device on the left is a fuse checker and the test light on the right is a LED type for use on computer circuits

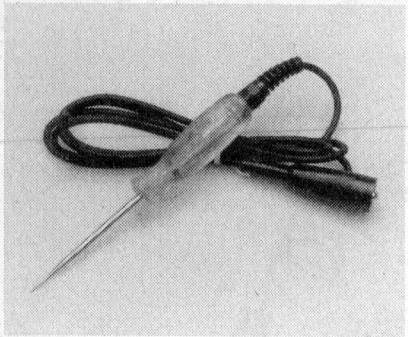

Typical test light

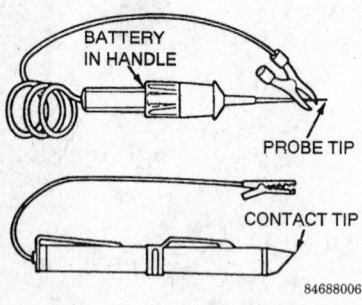

BATTERY IN HANDLE

PROBE TIP

CONTACT TIP

Types of self-powered test lights

The circuit will not work properly, but the voltmeter reading will show if the circuit is complete or not.

Available Voltage Measurement

Set the voltmeter selector switch to the 20V position and connect the meter negative lead to the negative post of the battery and connect the positive meter lead to the positive post of the battery. Read the voltage on the meter or digital display. A well-charged battery should register over 12 volts. If the meter reads below 11.5 volts, the battery power may be insufficient to operate the electrical system properly. This test determines voltage available from the battery and should be the first step in any electrical trouble diagnosis procedure. Many electrical problems, especially on computer controlled systems, can be caused by a low state of charge in the battery. Excessive corrosion at the battery cable terminals can cause a poor contact that will prevent proper charging and full battery current flow.

Nominal battery voltage is 12 volts but, when fully charged, should be about 13.2 volts. When the battery is supplying current to 1 or more circuits it is said to be under load. When everything is **OFF** the electrical sys-tem is under a no-load condition. A fully charged battery showing about 12.5 volts at no load may drop to 12 volts under medium load and will drop even lower under heavy load. If the battery is partially discharged, the voltage decrease under heavy load may be excessive, even though the battery shows 12 volts or more at no load. For this reason, it is important that the battery be fully charged during all testing procedures to avoid errors in diagnosis and incorrect test results.

Voltage Drop

When current encounters resistance, the voltage beyond the resistance is reduced. The larger the current, the greater the voltage reduction. When the circuit is off, there is no voltage drop because there is no current. In a long circuit with many connectors, a series of small, unwanted voltage drops due to corrosion at the connectors can add up to a total loss of voltage which impairs the operation of the loads in the circuit.

INDIRECT COMPUTATION OF VOLTAGE DROPS

1. Set the voltmeter selector switch to the 20 volt position.
2. Connect the meter negative lead to a good ground.
3. Probe all resistances in the circuit with the positive meter lead.
4. Operate the circuit in all modes and observe the voltage readings.

DIRECT MEASUREMENT OF VOLTAGE DROPS

1. Set the voltmeter switch to the 20 volt position.
2. Connect the voltmeter negative lead to the ground side of the resistance load to be measured.
3. Connect the positive lead to the positive side of the resistance or load to be measured.
4. Read the voltage drop directly on the 20 volt scale.

Too high of a voltage drop indicates too high of a resistance. If, for example, a blower motor runs too slowly, there may be too high a resistance in the resistor pack. By taking voltage drop readings in all parts of the circuit, the problem can be isolated. Too low of a voltage drop indicates too low of a resistance. If, for example, a blower motor runs too fast in the **MED** and/or **LOW** position, the problem can be isolated to the resistor pack by taking voltage drop readings in all parts of the circuit to locate a possibly shorted resistor. The maximum allowable voltage drop under

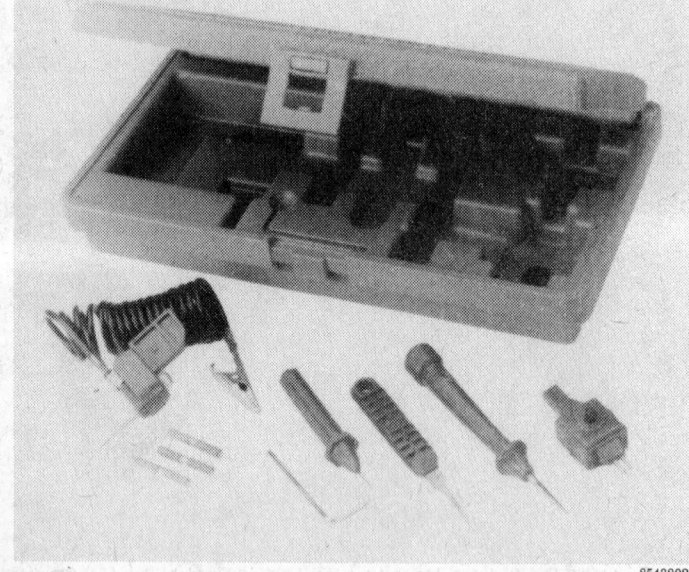

This computer circuit testing kit includes LED test lights that are safe for use on electronic circuits.

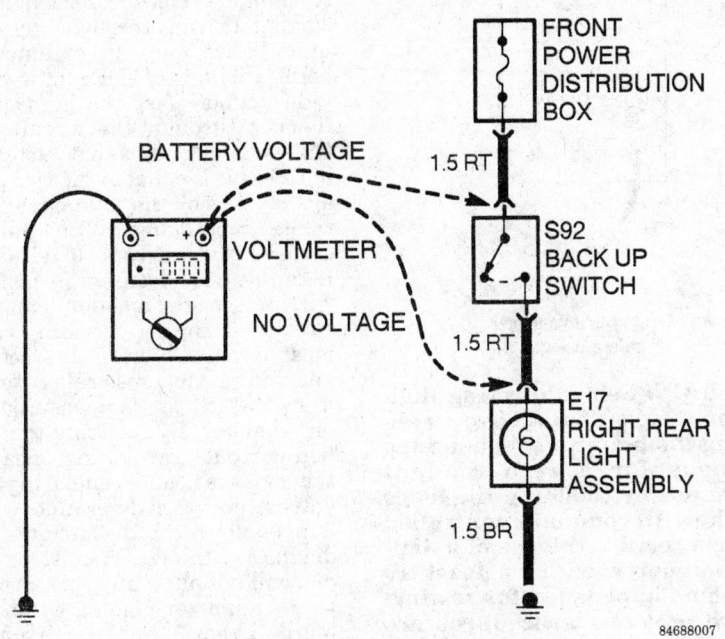

Measuring voltage at different points in the circuit

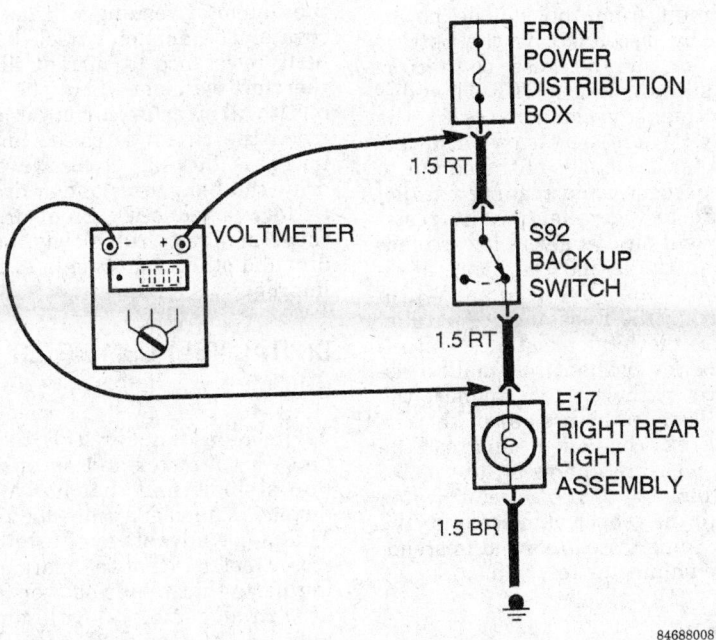

Checking for the voltage drop across a component in the circuit

load is critical, especially if there is more than one high resistance problem in a circuit; all voltage drops are cumulative. A small drop is normal due to the resistance of the conductors.

High Resistance Testing

1. Set the voltmeter selector switch to the 2 volt position.
2. Connect the voltmeter positive lead to the positive post of the battery.
3. Turn **ON** the headlights and heater blower to provide a load.
4. Probe various points in the circuit with the negative voltmeter lead.
5. Read the voltage drop. Some average maximum allowable voltage drops are:

    Fuse panel — 0.7 volts
    Ignition switch — 0.5 volts
    Headlight switch — 0.7 volts
    Ignition coil (+) — 0.5 volts
    Any other load — 0.5–1.3 volts

NOTE: Voltage drops are all measured while a load is operating; without current flow, there will be no voltage drop.

OHMMETER

The ohmmeter is designed to read resistance (ohms) in a circuit or component. Although there are several different styles of ohmmeters, all will usually have a selector switch which permits the measurement of different ranges of resistance. Usually the selector switch allows the multiplication of the meter reading by 10, 100, 1000 or 10,000. A calibration knob allows the meter to be set at zero for accurate measurement. Since all ohmmeters are powered by an internal battery (usually 9 volts), the ohmmeter can be used as a self-powered test light. When the ohmmeter is connected, current from the ohmmeter flows through the circuit or component being tested. Since the ohmmeter's internal resistance and voltage are known values, the amount of current flow through the meter depends on the resistance of the circuit or component being tested.

The ohmmeter can be used to perform continuity tests for opens or shorts and to read actual resistance in a circuit. It should be noted that the ohmmeter is used to check the resistance of a component or wire while there is no voltage applied to the circuit. Current flow from an outside voltage source (such as the vehicle battery) can damage the ohmmeter, so the circuit or component should be isolated from the vehicle electrical system before any testing is done. Since the ohmmeter uses its own voltage source, either lead can be connected to any test point.

NOTE: When checking diodes or other solid state components, the ohmmeter leads can only be connected one way in order to measure current flow in a single direction. Make sure the positive (+) and negative (-) terminal connections are as described in the test procedures to verify the one-way diode operation.

When using the meter for continuity checks, do not be concerned with the actual resistance readings. Zero resistance, or any resistance reading, indicates continuity in the circuit. Infinite resistance indicates an open in the circuit. A high resistance reading where there should be none indicates a problem in the circuit. Checks for short circuits are made in the same manner as checks for open circuits except that the circuit must be isolated from both power and normal ground. Infinite resistance indicates no continuity to ground, while zero resistance indicates a dead short to ground.

Resistance Measurement

The batteries in an ohmmeter may be affected by temperature and will weaken with age. The ohmmeter must be calibrated or "zeroed" before taking measurements. To zero the meter, place the selector switch in its lowest range and touch the 2 leads together. Turn the calibration knob until the meter needle is exactly on zero.

NOTE: All analog (needle) type ohmmeters must be zeroed before use, but some digital ohmmeter models are automatically calibrated when the switch is

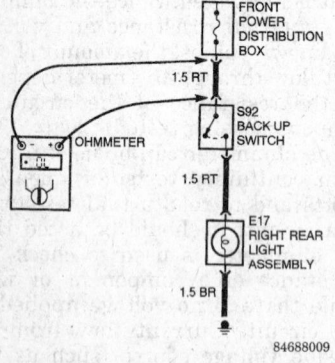

Using an ohmmeter to do a continuity test

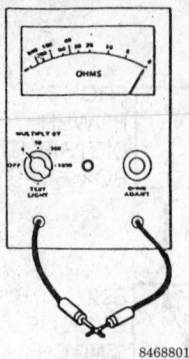

Zeroing the ohmmeter before using it

turned ON. Self-calibrating digital ohmmeters do not have an adjusting knob, but it's a good idea to check for a zero readout before use by touching the leads together. All computer controlled systems require the use of a digital ohmmeter with at least 10 megohms impedance for testing. Before any test procedures are attempted, make sure the ohmmeter used is compatible with the electrical system, or damage to the on-board computer could result.

To measure resistance, first isolate the circuit from the vehicle power source by disconnecting the battery cables or the harness connector. Make sure the key is **OFF** when disconnecting any components or the battery. Where necessary, also isolate at least one side of the circuit to be checked to avoid reading parallel resistances. Parallel circuit resistances will always give a lower reading than the actual resistance of either of the branches. When measuring the resistance of parallel circuits, the total resistance will always be lower than the smallest resistance in the circuit. Connect the meter leads to both sides of the circuit (wire or component) and read the actual measured ohms on the meter scale. Make sure the selector switch is set to the proper ohm scale for the circuit being tested to avoid misreading the ohmmeter test value.

NOTE: Never use an ohmmeter with power applied to the circuit. Like the self-powered test light, the ohmmeter is designed to operate on its own power supply. The normal 12 volt automotive system could damage the meter.

AMMETERS

An ammeter measures the amount of current flowing through a circuit in units called amperes or amps. Amperes are units of electron flow which indicate how fast the electrons are flowing through the circuit. Since Ohm's Law dictates that current flow in a circuit is equal to the circuit voltage divided by the total circuit resistance, increasing voltage also increases the current level (amps). Likewise, any decrease in resistance will increase the amount of amps in a circuit. At normal operating voltage, most circuits have a characteristic amount of amperes, called "current draw'" which can be measured using an ammeter. By referring to a specified current draw rating, measuring the amperes, and comparing the 2 values, one can determine what is happening within the circuit to aid in diagnosis. An open circuit, for example, will not allow any current to flow so the ammeter reading will be zero. More current flows through a heavily loaded circuit or when the charging system is operating.

An ammeter is always connected in series with the circuit being tested. All of the current that normally flows through the circuit must also flow through the ammeter; if there is any other path for the current to follow, the ammeter reading will not be accurate. The ammeter itself has very little resistance to current flow and therefore will not affect the circuit, but it will measure current draw only when the circuit is closed and electricity is flowing. Excessive current draw can blow fuses and/or drain the battery; a reduced current draw can cause motors to run slowly, lights to dim and other components to operate improperly.

DIGITAL VOLT-OHM METER (DVOM)

As its name implies, this tool combines a voltmeter and an ohmmeter into a single unit that has a digital display instead of a scale and pointer. The major advantage of a fully electronic meter is that there are no moving parts that require power to operate. Analog meters with an ultra light weight needle still require some power to move the needle. This limits the range and the features that can be built into the meter. Even the most basic DVOM can read a much greater range of voltage and resistance without imposing any load on the circuit being tested. It is usually

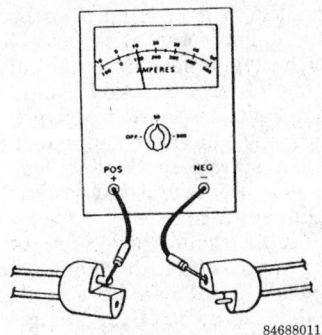

84688011

The ammeter is placed in line with the circuit to be tested

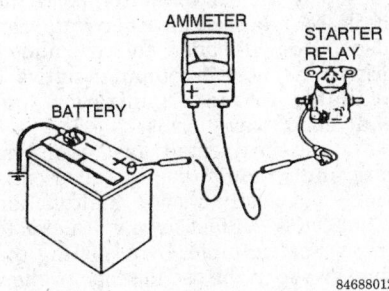

84688012

Checking the draw of the starter relay with an ammeter

88513005

Different styles of multi-meters allow a choice of meter functions

matically turns the unit on when needed. Its range of 10–5000 Hz makes it useful for checking rpm sensors, mass air flow sensors, Hall effect sensors and more. Instructions provided with the processor show how to interpret the readings.

BREAK-OUT BOX

The electronic Break-Out Box (BOB) is used to tap into the wiring of a control unit. The main connector to the electronic control unit is connected to the break-out box and another wire harness is connected from the box to the control unit. The break-out box then allows the technician to access each circuit while it is operating without piercing the wire or causing damage to the connectors. All testing with the DVOM can be done safely at these terminals, eliminating the risk of damage due to backprobing at the control unit. Many times a break-out box is the only way to test a control unit function.

An Intelligent Break-Out Box (IBOB) connects to the vehicle diagnostic connector and has connector ports for a scan tool and/or a computer. On earlier electronic control units that do not generate a data

the only equipment suitable for testing computer controlled circuits and is often the only test equipment needed.

Several additional features can be built into the same unit, such as circuitry for testing diodes and measuring AC voltage, AC and DC current, temperature, duty cycle, frequency, pulse width, dwell, and rpm. Some of the more sophisticated units also have storage capability, bar graph display, automatic shut-off, and can display the difference between two readings. A top-of-the-line DVOM designed for automotive testing is probably the most useful and cost effective diagnostic tool available. Be sure to buy a unit with a high impedance, usually 10 megohms or higher.

Specialty Testers

FREQUENCY PROCESSOR

Some older DVOMs are not equipped to read frequency. There is at least one unit on the market that converts frequency signals to a millivolt signal that any DVOM can read. It is a simple box with input and output jacks and a "wake-up" circuit that auto-

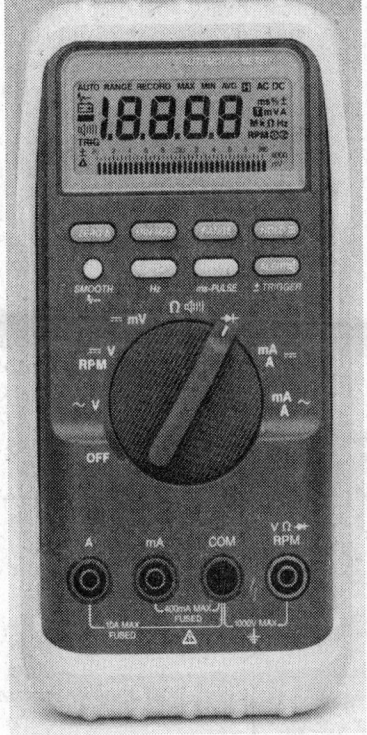

88513006

A good quality DVOM designed for automotive testing is the most useful diagnostic tool available.

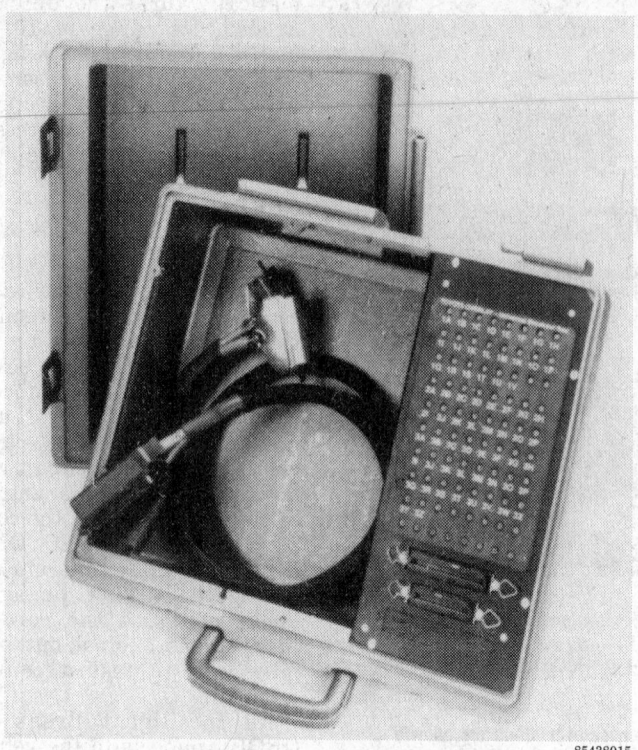

85438015

A break-out box makes it possible to tee into control unit circuits.

stream, an IBOB will collect input/output data while the engine is running and present it to a scan tool or PC. Additionally, some manufacturers provide plastic overlays for the break-out box. This allows the box to be used on a variety of models; different overlays identify the changes in wire use or labeling. With the proper cable adaptors, an IBOB can be used with any engine, body or ABS control unit on any vehicle.

OSCILLOSCOPE

An oscilloscope is a voltmeter that presents a graphic picture of the volt-

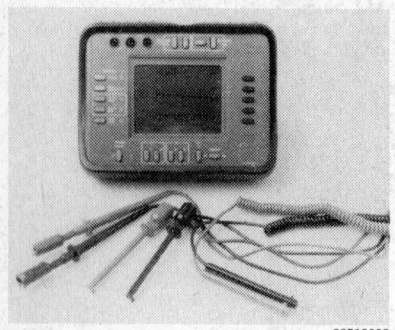

88513003

An oscilloscope shown with related testing probes

age reading over time. Unlike a DVOM, it can show a voltage that exists for only a fraction of a second or occurs only at a specific time. Ignition oscilloscopes have been around for many years, but the latest generation of service bay oscilloscopes are more like those found in electronics labs. They can read voltages as small as one millivolt and can show a spike that occurs for as little as 10 nanoseconds (1 ns = of a second). Both the voltage and time scales are adjustable, so the same tool can be used to measure the fast, high voltage signal of the secondary ignition system and slow stable signals such as a temperature sensor. Another major feature of all oscilloscopes is an extremely high input impedance, meaning the oscilloscope imposes negligible current draw on the circuit being measured that might influence that measurement. Many times an oscilloscope is the only tool that can be used to measure low voltage, frequency, or duty cycle signals.

Like a timing light, an oscilloscope must be triggered. The trigger can be internal (automatic) or can come from an external source. On a multi-channel oscilloscope, displaying the external trigger signal can show the timing of two events. For example, by

taking the trigger from a suspected faulty fuel injector, it is possible to see the oxygen sensor signal only at the time of that injection event. The voltage level required to trigger the oscilloscope can also be adjusted, providing a simple method to look for low level or intermittent faults that may not set a code.

A digital oscilloscope converts the analog input signal to a digital form. A digital signal can be stored and played back by itself or along with another trace. Some units can also display the signal as numbers, min/max values, change value and average value. If the oscilloscope is equipped with a computer port, the digitized traces and other data can also be down-loaded to save and/or print out. There is computer software available to aid organization and analysis of wave forms.

With its extremely fast sampling rate and graphic display, an oscilloscope can easily show a malfunction that occurs too fast for a voltmeter to show. For example, by adjusting the time sweep of the oscilloscope to show the full up-and-down stroke of a throttle position sensor, an intermittent fault in the signal can be clearly shown. A voltmeter may also detect the fault but cannot change the display fast enough to show the resistance spike. A storage oscilloscope with minimum/maximum value capability can locate intermittent faults that even other oscilloscopes cannot.

Any oscilloscope used for automotive testing must be designed for use with trucks. Standard lab oscilloscopes are usually not able to cope with the relatively harsh automotive electronic environment. Automotive oscilloscopes are available in a variety of types with a variety of features. Some are portable hand held models that operate on batteries or vehicle power. Even though they are small, the newest portable units include multi-trace and storage capabilities and are rugged enough to be used under the hood or on road tests. Larger more powerful models mounted in a console are often part of a top-of-the-line engine analyzer package. Most of the major diagnostic tool manufacturers produce at least one oscilloscope model.

As vehicles become more sophisticated and electronic controls become more powerful, an oscilloscope is fast becoming a necessary diagnostic tool. When the technician becomes proficient with an oscilloscope, many other diagnostic tools become unnecessary.

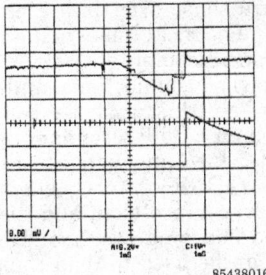

The oxygen sensor trace (top) shows a delayed cross-over coinciding with an injector pulse (bottom trace).

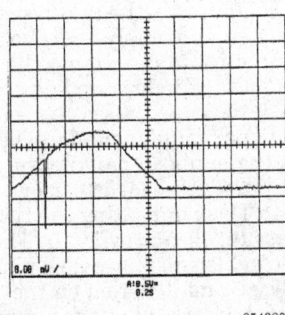

An intermittent fault in the throttle position sensor shows clearly on the scope trace.

SCAN TOOL

This is the generic name for portable diagnostic equipment that communicates directly with an electronic control unit. The major vehicle manufacturers each have their own scan tool that is used by dealership technicians. Some of these are available through the dealer parts network or are sold outside the network under another name. Others are available only to authorized dealerships.

Scan tools are used to read and erase trouble codes stored in the con-

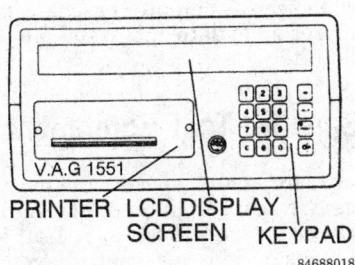

PRINTER LCD DISPLAY
SCREEN KEYPAD

Scan tools allow the testing of electronic control units

Aftermarket scan tool with a program module installed

trol unit memory and to provide a direct data transfer link with the control unit's On Board Diagnostic (OBD) system. Reading the control unit memory through the scan tool is more complete than reading codes with the flashing light on the instrument panel. Some information is only available through the scan tool, such as the number of engine starts since the fault first appeared. Data transfer provides a real time display of control unit input/output signals. Data such as the oxygen sensor reading or idle control motor duty cycle can be displayed while the engine is running. The scan tool can also be used as a volt/ohmmeter to check selected circuits without disconnecting them.

Some scan tools are designed to simulate sensor inputs to test the sensor circuit, the control unit and the output device. On many vehicles, the scan tool can communicate with every control unit on the vehicle through a single diagnostic connector. Some of the more advanced scan tools are equipped with a data memory to store test data during a road test. The test data is down loaded into a PC through an RS232 computer port, greatly increasing the processing power and expanding the

amount of memory and information available.

Aftermarket scan tools require adaptors to match the different vehicle diagnostic connectors. A cartridge that plugs into the scan tool contains software needed for communicating with the different control units. The software is the tool's real power and is continuously evolving to enhance its capabilities. As the tool manufacturer's data base has grown, software now includes VIN-specific information that addresses some of the most common trouble codes and driveability problems that don't always generate codes. Tests are menu-driven and many of the specifications are included right there in the program. Depending on the vehicle and the amount of computer control used, systems which may be viewed or investigated with a scan tool include:

1. Engine controllers/ECUs
2. Fuel/ignition systems
3. Electronic transmission control
4. Charging system
5. Suspension control functions
6. Anti-Lock brake system.
7. Passive restraint system
8. Anti-theft system
9. Climate Control Systems
10. Body electrical systems — including power locks, entertainment systems, sunroofs, and defoggers.

Aftermarket scan tools work well with most vehicles, but no scan tool can be used on all models and all are limited in their ability to communicate with European models. The Federal On-Board Diagnostic (OBD) specification requires all vehicles to have the same diagnostic connector and diagnostic trouble codes and to use the same data transfer language. This makes it possible to use a single scan tool to communicate with all engine control units from all manufacturers. Some manufacturers began production of OBD vehicles for the 1994 model year and all new vehicles must comply by the 1998 model year. As vehicle control units and scan tools become more powerful, data acquisition and test capabilities will also improve dramatically with each new generation of control unit and scan tool software.

EXHAUST GAS ANALYZERS

Exhaust gas analysis has long been an extremely valuable and versatile diagnostic tool. It can be used to troubleshoot fuel and ignition systems, locate vacuum leaks or EGR malfunctions, even diagnose mechanical problems such as worn valve

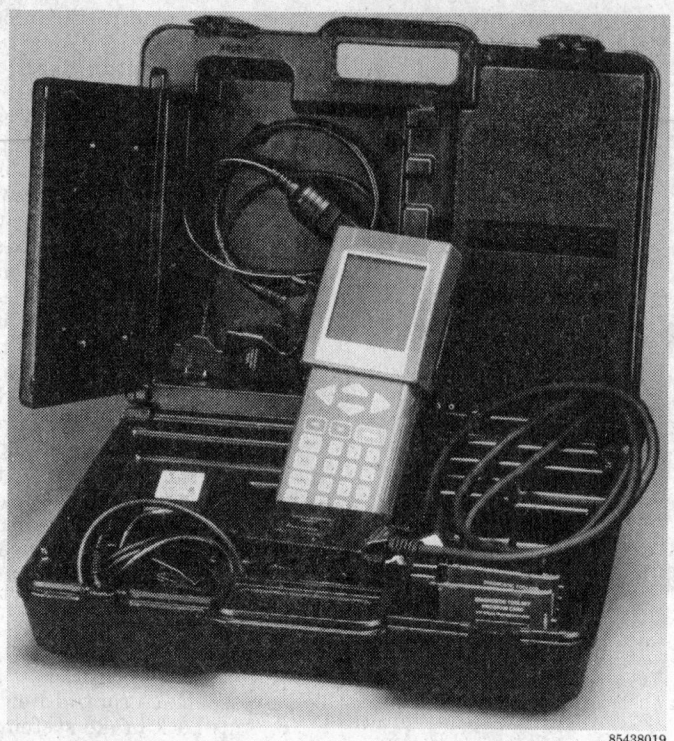

85438019

This type scan tool, used by dealer technicians, is available outside the dealer network and can scan all OBD II-specific vehicles — this one includes a built-in oscilloscope

guides. On most vehicles, it is the only way to accurately check the air/fuel mixture.

The federal government regulates three exhaust gas components: hydrocarbons (HC), carbon monoxide (CO), and oxides of nitrogen (NOx). HC and CO are relatively easy to measure and have long been tested in states that require emissions inspections. Measuring carbon dioxide (CO_2) and oxygen (O_2) are also valuable diagnostic aids but NOx cannot be accurately measured at idle or no-load conditions. However as states enact tighter Inspection and Maintenance programs, service bay NOx analyzers and test procedures are being developed. For complete diagnostic and certification testing, a five-gas analyzer is required.

Most gas analyzers include a single tailpipe sample probe, sample pump and filtration system and a detector cell for each of the gasses being measured. Most stand-alone four-gas analyzers used for emissions inspections are equipped with a small microprocessor that includes testing and calibration programs, a self diagnostic program and a built-in printer. Other units are designed as part of a complete diagnostic station and are connected to a PC based computer, printer and monitor screen. There is

at least one portable four-gas analyzer that is used along with the same manufacturer's scan tool. The scan tool's software guides the user through test procedures based on the gas and sensor readings.

There is also a series of small, hand-held oxygen and CO monitors available that do not require a sample pump and filter system. The measuring cell is built into the tailpipe probe and the monitor is battery operated. Since there is only a wire between the probe and the monitor, these units can easily be used on a road test. They are also equipped with a memory that can record three minutes of test data. The CO monitor can be particularly useful for routine air/fuel adjustments. Each unit is available with its own display, with voltage outputs to use a DVOM display, or with computer ports and PC based software that includes test procedures.

All gas analyzers must be calibrated at least once per day. Industry standard calibration gasses are usually available through parts stores or tool outlets.

ENGINE ANALYZERS

A large, fully-equipped engine analyzer usually includes an oscilloscope,

exhaust gas analyzer, vacuum and pressure sensors, a timing light and probe, electrical measuring equipment, and a computer. With a variety of electrical connections and a tail pipe probe, this analyzer can check the primary and secondary ignition systems, fuel injection controls and injectors, EGR systems, engine vacuum and compression, and the starting and charging systems. The computer can be used to read the engine control unit's data stream through the vehicle diagnostic connector. Even on earlier vehicles with no sensors or data stream, an engine analyzer is still a powerful diagnostic tool.

The computer is the real power behind an engine analyzer. The computer's ability to determine the ignition pulse at cylinder number one can be used to index all other engine events to particular cylinders. For example, the analyzer can measure the starter current needed to move each piston to TDC, indicating the relative compression of each cylinder. The analyzer can also display spark plug firing voltage and duration on the oscilloscope. A spark plug that requires more voltage with less duration could indicate a faulty injector. The analyzer can detect and clearly show all the differences between cylinders. The computer can help the technician diagnose the data, determine the necessary repairs and even provide a print-out to present a clear explanation to the vehicle owner.

These analyzers are a major investment and are well supported by the manufacturer. They are frequently updated with a computer floppy disc that includes new vehicle information and test procedures. Some machines include CD-ROM equipment to read service manuals that are available on disc. They may also include a modem to communicate with the manufacturer or other computers via telephone. As vehicles and other shop equipment become more sophisticated, it should be possible to keep a computer based engine analyzer up to date and useful almost indefinitely.

Specific Test Equipment

There are many special diagnostic tools for testing individual components or systems, such as a Hall effect sensor, idle air control motor, fuel injectors, secondary ignition systems, and others. Most are designed for use on as many vehicles as possible. Some are designed to test parts or systems on specific vehicles, such as the ABS tester for Mitsubishi trucks. Generally these devices allow the technician to quickly test components or sub-systems without going through a long diagnostic procedure. However there is a risk of incorrect diagnosis. These tools can only be dependable if the technician is familiar with their use and understands what the test results really mean. A simple vacuum leak or loose connection may produce the same test result as a faulty component.

LEAK DETECTORS

A battery powered, hand-held vacuum leak detector uses a microphone and amplifier that detects noise in the ultrasonic range. Air moving through a vacuum leak will generate sound waves in the 40 kHz range, well above the range of human hearing. The detector will sound a beeper when a leak is found. Because of the high frequency sensed by the detector, it is not generally affected by normal engine or shop noises.

Leak detectors for air conditioning systems have a vacuum pump and probe to draw an air sample into the detecting cell. The cell detects halogen gas that is common to all air con-

ditioning refrigerants. Most are capable of indicating the type of freon in the system, as well as the rate of leakage. There are battery powered hand-held models and larger AC powered units suitable for mounting on an air conditioning service cart. The newest models are capable of detecting R-134a and the sensitivity can be adjusted for possible background interference.

A combustible gas leak detector reacts to hydrocarbons present in fuels, exhaust gases, coolants, and lubricants. Models with adjustable sensitivity are typically used to look for fuel vapor leaks, head gasket leaks, and to measure the amount of exhaust leaking into the interior of a vehicle. With some imagination, this can be an extremely useful tool.

PYROMETER

A pyrometer measures a wide range of temperatures with a probe that only needs to touch the item being measured. As a general diagnostic tool, a hand held pyrometer can quickly locate hot or cold spots in a cooling system, a seized brake caliper, a dry bearing, test heater and A/C performance or even find a weak cylinder by measuring exhaust mani-

fold runner temperatures. Most pyrometers are available with special probes for penetration and for measuring tire temperatures. There are even optical infrared non-contact pyrometers that measure temperature by the heat emission of a surface. This is useful as the surface to be measured does not have to actually touched by a probe. They can be calibrated quickly, have a very wide range and can usually be switched to display temperature in either Fahrenheit or Celsius.

IDLE AIR CONTROL TESTER

This is a kit used to isolate and test idle air control solenoids, motors, and signals. Some are made for use with a specific system, others include adaptors for use with many different vehicles. The device can activate solenoid valves and control motors to test the full range of motion with the engine not running. It can also be used to control idle speed for timing adjustment or other engine tests. Some can also check the control unit output signal to the idle air control motor. Although these functions can also be accomplished with a scan tool, this tester can be faster and easier to use for some tests.

FUEL INJECTOR TESTER

This device can quickly check the coil resistance and current draw of an electric fuel injector while it is under load. Each injector is tested individually and the results are reported on a DVOM or oscilloscope. This information makes it possible to electronically check injector balance and detect intermittent faults. When used with equipment that measures fuel pressure and injection quantity, every function of the fuel system can be tested.

OXYGEN SENSOR TESTER

This kit usually includes a propane enrichment control valve, special connectors and test instructions, and the hose and fittings needed for connecting the valve to an intake manifold. The kit allows the technician to control air/fuel mixture and check the oxygen sensor response time. When the oxygen sensor is disconnected, forcing the control unit into open loop, sensor output voltage or resistance can be read with a DVOM. The instructions also include procedures for testing the control unit's response to the oxygen sensor signal.

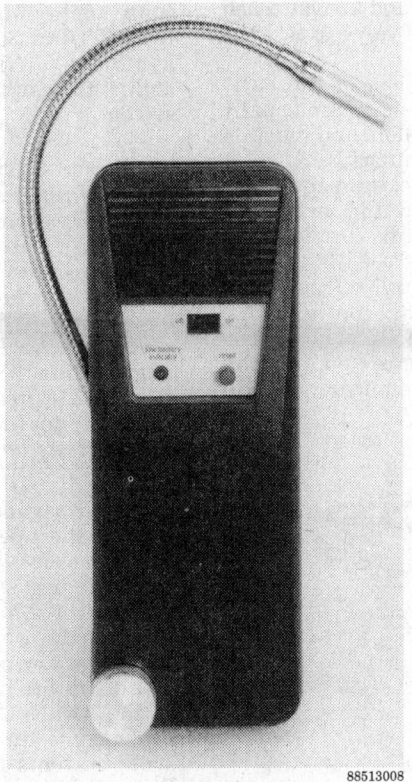

88513008

An ultrasonic vacuum leak detector is unaffected by engine or shop noise.

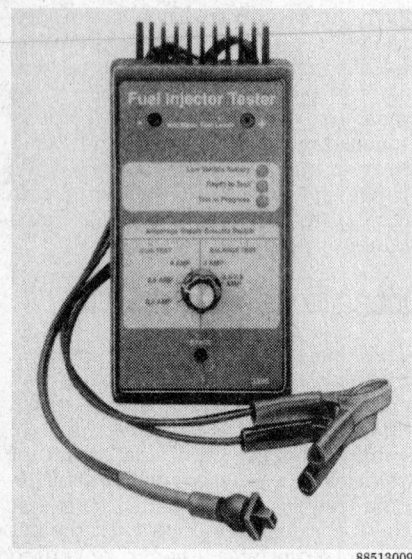

88513009

A fuel injector tester measures voltage drop while the injector is being activated.

SENSOR SIMULATOR

This device is used to take a sensor "out of the loop" and simulate its input signals to the control unit. It can simulate every type of voltage, resistance, and frequency signal one at a time to test the control unit's response to the input. The simulator can also measure any sensor output signal by back-probing the sensor connector. In addition to displaying the reading directly, some units can also output the reading to an oscilloscope, scan tool, or other diagnostic equipment.

POWER STEERING TESTER

A power steering tester can quickly confirm that the hydraulic system is functioning properly and indicate excess loads on the system due to mechanical malfunction in the suspension or steering linkage. The mechanical tester consists of a gauge, a heavy duty hose, and adapters for various models. Newer testers use an electronic pressure transducer that converts the pressure to an electrical signal. This allows the technician to road test variable effort power steering systems with a DVOM.

ELECTRONIC SIGHTGLASS

This device, used for troubleshooting air conditioning systems, includes two transducers and a battery operated meter. The transducers attach to the outside of the metal air conditioning lines without disconnecting the lines. While the A/C system is operating, the device ultrasonically detects bubbles in the refrigerant and uses an LED display to simulate a sightglass, allowing the technician to actually "see" the bubbles in the system. The transducers can be fitted to any metal line at almost any point in the system, making it possible to test expansion valves and capillary tubes, or find other undesired restrictions.

ANTI-LOCK BRAKE TESTER

Anti-lock brake system control units are equipped with a self-diagnostic program that checks the system at engine start-up and de-activates the system if a fault is detected. Most control units also store diagnostic trouble codes. An ABS tester is basically a scan tool used to read and erase trouble codes and to provide a data link with the ABS control unit. The data link allows the tester to check sensor inputs and activate each solenoid and control valve to test the output system.

The testers for some of the early anti-lock braking systems are usually available only to dealers and function only with that manufacturer's vehicles. The aftermarket scan tool makers have developed the necessary adapters and software cartridges so that most scan tools used to communicate with engine control units can also be used on ABS control units. Unfortunately there are major differences in ABS designs and no scan tool is able to communicate with all ABS control units.

85438023

A power steering system testing gauge can quickly isolate hydraulic or linkage problems.

The most complete ABS test equipment is a software package, connector and interface pod that establishes a data link between the control unit and a PC-based computer in an engine analyzer or alignment station. This software uses the power of the computer to completely check the control unit and "bench test" every component of the system in about one minute. It is suitable for use with every anti-lock braking system on the market and the software can be updated as required to keep the system current.

Even without a scan tool, all of the system's sensors, solenoids, and actuators can still be individually checked with a DVOM and an oscilloscope. The trouble codes and other control unit functions can only be accessed with the correct scan tool or tester.

PASSIVE RESTRAINT TESTER

While there are some differences, the air bag system in most vehicles operates basically the same way. This has made it possible to develop a unit that will test the circuits while they are fully connected and operating. It uses LEDs to read out circuit continuity, power supply, switch state and output device state. On some scan tools, such as the Nissan CONSULT or Subaru Select Monitor, a software cartridge gives the tool the ability to test the passive restraint system.

—— CAUTION ——

If not familiar with disarming procedures and air bag system operation, do not attempt air bag system service. The air bag system must be disabled for some vehicle tests and before removing the steering wheel or dashboard air bag module. Failure to follow air bag safety procedures could result in accidental deployment and serious or fatal injury.

The air bag must function in an extremely short period of time after the crash sensor switches have closed. Most air bag systems include their own power supply. The squib that fires the gas generator charge must operate at low power levels to assure it will still fire if vehicle power is lost in the accident. This makes an air bag module quite sensitive to even small currents like static electricity. Even with such a "hair trigger", it is not difficult to test the system or handle an air bag module safely.

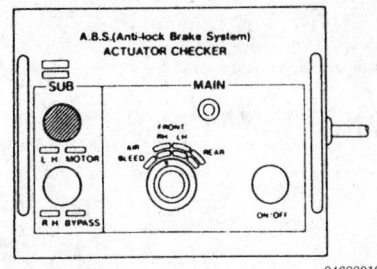

An ABS checker allows the testing of the anti-lock system

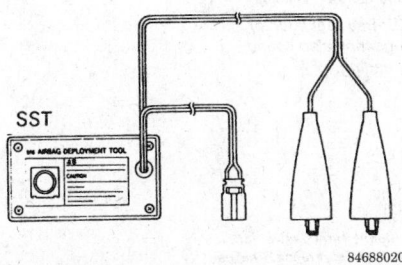

An air bag deployment tool used when disposing of an air bag

Air Bag Service Precautions

——— CAUTION ———

An air bag is an explosive device. Before beginning any air bag system service, disconnect and isolate the negative battery cable and backup power supply. Follow the procedures for disarming the air bag system exactly. Do not use any type of computer memory saver device. Do not use any self powered test equipment or test lights until the system is properly disabled. Failure to follow these precautions could result in accidental deployment of the air bag and possible serious or fatal injury.

• Disconnect and isolate the battery cable and any backup power supply when servicing the air bag system.
• Do not use a memory saver.
• When re-activating an air bag system, connect the power last and make sure no one is in the vehicle.
• Do not attempt to measure the resistance across the air bag module connectors with any type of ohmmeter. The ohmmeter battery can fire the air bag module.
• When working on air bag components in the passenger compartment, try to work away from where an air bag would deploy. Accidental deployment of the air bag against a body that is not in the proper position could result in severe or fatal injury.
• A removed air bag module must be placed away from sources of heat, sparks, or electricity, including static electricity.
• A removed air bag module must be placed with the cover pad facing up, so that accidental deployment will not launch the module into the air. Also be sure to carry the module with the cover pad facing away from the body.
• When the air bag module is removed, place it away from loose objects that would be thrown in the event of accidental deployment.
• When removing a steering column with the steering wheel attached, pay attention to where the air bag is aimed. Never stand the column on the face of the steering wheel. Lock the column to avoid damage to the clockspring.
• When handling an air bag which has been deployed, there may be a powdery residue which, though mostly talc, may irritate skin, eyes, and breathing passages. Wear gloves, glasses, and a dust mask while wrapping the deployed air bag in a plastic bag for disposal.
• Sensor positioning is critical for proper system operation. If the sensor is in an area of vehicle damage, replace the sensor whether or not the air bag deployed. The proper torquing of the sensor retainers is critical.
• Any part of the air bag system found to be faulty must be replaced. No part can be repaired, not even the wiring.
• All diagnostic work is to be done with the air bag module(s) removed. Air bag simulators are available and can be installed for a full functional test of the control unit.

Mechanical Test Equipment

VACUUM GAUGE

Intake manifold vacuum is used to operate various systems and devices on all trucks. To correctly diagnose and solve problems in vacuum control systems, a vacuum source is necessary for testing. In some cases, vacuum can be taken from the intake manifold when the engine is running, but vacuum is normally provided by a hand vacuum pump.

Most gauges are graduated in inches of Mercury (in. Hg) or kilopascals (kPa), although a device called a manometer reads vacuum in inches of water (in. H₂O). The vacuum reading usually varies between 18–22 in. Hg (60–74 kPa) at sea level. To test engine vacuum, the vacuum gauge must be connected to a source of manifold vacuum. Many engines have a plug in the intake manifold which can be removed and replaced with an adapter fitting. Connect the vacuum gauge to the fitting with a suitable rubber hose or, if no manifold plug is available, connect the vacuum gauge to any device using manifold vacuum, such as EGR valves. The vacuum gauge can be used to determine the amount of vacuum reaching a component.

HAND VACUUM PUMP

Small, hand-held vacuum pumps come in a variety of designs and provide a source of vacuum for testing components without the engine operating. Most have a built-in vacuum gauge and allow a component to be tested without removing it from the vehicle. Operate the pump lever or plunger, applying the correct amount of vacuum required for the test. The level of vacuum in inches of Mercury (in. Hg) or kilopascals (kPa) is indicated on the pump gauge. For some testing, an additional vacuum gauge may be necessary.

COMPRESSION GAUGE

A compression gauge measures the amount of pressure in pounds per square inch (psi) or kilopascals (kPa) that a cylinder is producing. Some gauges have a hose that screws into the spark plug hole while others have a tapered rubber tip which is held by hand in the spark plug hole. Engine compression depends on the sealing ability of the rings, valves, head gasket and spark plug gaskets. If any of these parts are not sealing properly, compression will be lost and the power output of the engine will be reduced. The compression in each cylinder should be measured and the variation between cylinders should be noted. The engine should be cranked through 5 or 6 compression strokes while warm, with all plugs removed, ignition disabled and throttle valves wide open.

USING A VACUUM GAUGE

White needle = steady needle   Dark needle = drifting needle

The vacuum gauge is one of the most useful and easy-to-use diagnostic tools. It is inexpensive, easy to hook up, and provides valuable information about the condition of your engine.

Indication: Normal engine in good condition

Gauge reading: Steady, from 17–22 in./Hg.

Indication: Sticking valve or ignition miss

Gauge reading: Needle fluctuates from 15–20 in./Hg. at idle

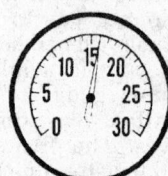

Indication: Late ignition or valve timing, low compression, stuck throttle valve, leaking carburetor or manifold gasket.

Gauge reading: Low (15–20 in./Hg.) but steady

Indication: Improper carburetor adjustment, or minor intake leak at carburetor or manifold

NOTE: Bad fuel injector O-rings may also cause this reading.

Gauge reading: Drifting needle

Indication: Weak valve springs, worn valve stem guides, or leaky cylinder head gasket (vibrating excessively at all speeds).

NOTE: A plugged catalytic converter may also cause this reading.

Gauge reading: Needle fluctuates as engine speed increases

Indication: Burnt valve or improper valve clearance. The needle will drop when the defective valve operates.

Gauge reading: Steady needle, but drops regularly

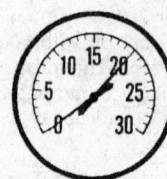

Indication: Choked muffler or obstruction in system. Speed up the engine. Choked muffler will exhibit a slow drop of vacuum to zero.

Gauge reading: Gradual drop in reading at idle

Indication: Worn valve guides

Gauge reading: Needle vibrates excessively at idle, but steadies as engine speed increases

88513010

A vacuum gauge is a good tool for diagnosing the general condition of an engine

FUEL PRESSURE GAUGE

A fuel pressure gauge is required to test the operation of the fuel delivery and injection systems. Some systems also need a 3-way valve to check the fuel pressure in various modes of operation. Gauges may require special adapters for making fuel connections. Always observe fuel system cautions when working around any pressurized fuel system.

BASIC MAINTENANCE AND TROUBLESHOOTING

WHERE DO I START?

Logical Diagnostic Procedures

Diagnosis of a driveability problem requires attention to detail and following the diagnostic procedures in the correct order. Resist the temptation to begin extensive testing before completing the preliminary diagnostic steps. The preliminary or visual inspection must be completed in detail before diagnosis begins. In many cases this will shorten diagnostic time and often cure the problem without the need for involved electronic testing.

There are two basic ways to check the vehicle's engine for electronic problems. These are by symptom diagnosis and by the on-board computer self-diagnostic system. The first place to start is always the preliminary inspection. Intermittent problems are the most difficult to locate. If the problem is not present at the time of testing, the fault may be able to be located.

PRELIMINARY INSPECTION

The visual inspection of all components is possibly the most critical step of diagnosis. A detailed examination of connectors, wiring and vacuum hoses can often lead to a repair without further diagnosis. Also, take into consideration if the vehicle has been serviced recently. Sometimes things get reconnected in the wrong place, or not at all. A careful inspector will check the undersides of hoses as well as the integrity of hard-to-reach hoses blocked by the air

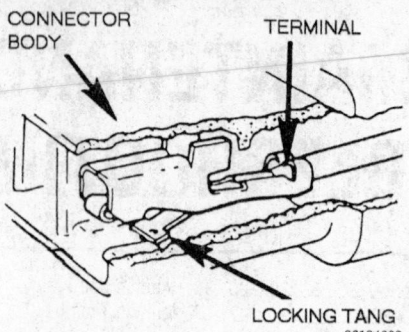

Check the individual terminals and wiring connectors for damage

cleaner or other components. Correct routing for vacuum hoses can be obtained from the specific vehicle service manual, or Vehicle Emission Control Information (VECI) label in the engine compartment of the vehicle. Wiring should be checked carefully for any sign of strain, burning, crimping or terminals pulled out from a connector.

Checking connectors at components or in harnesses is required; usually, pushing them together will reveal a loose fit. Also, check electrical connectors for corroded, bent, damaged, improperly seated pins, and bad wire crimps to terminals.

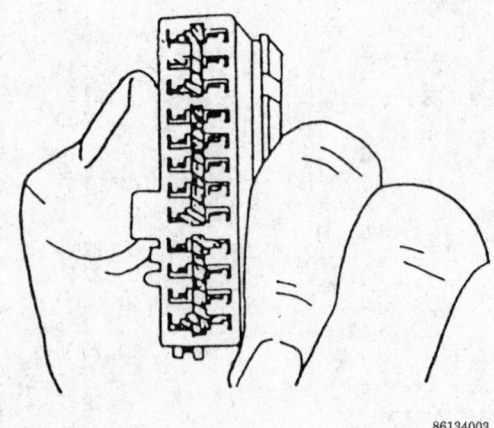

Inspect the connector terminals for damage

Perform underhood inspection of all wiring and hoses

Check for damaged or broken wires

Pay particular attention to ground circuits, making sure they are not loose or corroded. Remember to inspect connectors and hose fittings at components not mounted on the engine, such as the evaporative canister or relays mounted on the fender aprons. Any component or wiring in the vicinity of a fluid leak or spillage should be given extra attention during inspection.

Additionally, inspect maintenance items such as belt condition and tension, battery charge and condition and the radiator cap carefully. Any of these very simple items may affect the system enough to set a fault code.

DIAGNOSIS BY SYMPTOM

Before the advent of the self-diagnostic system, diagnosis by symptom was the only method for investigation of an automotive problem. An attempt was made to solve problems by reviewing the symptoms and performing tests on suspected components until a defective component was located. The problem was then corrected and the vehicle checked for any other problems. This method is still used frequently today when a driveability complaint is made, but no code is set in the electronic control unit's memory.

When diagnosing by symptom the first step is to find out if the problem really exists. This may sound like a waste of time but the problem must be recreated before testing is begun. This is called an ""operational check". Each operational check will give either a positive or negative answer (symptom). A positive answer is found when the check gives a positive result (the horn blows when the horn button is pressed). A negative answer is found when the check gives a nega-

tive result (the radio does not play when the knob is turned on). After performing several operational checks, a pattern may develop. This pattern is used in the next step of diagnosis to determine related symptoms.

In order to determine related symptoms, perform operational checks on circuits related to the problem circuit (the radio does not work and the dash lights do not go on). These checks can be made without the use of any test equipment. Simply follow the wires in the wiring harness or, if available, obtain a copy of the vehicle's specific wiring diagram. If the radio and the dash lights are on the same circuit, first check the radio to see if it works. Then check the dash lights. If the neither the radio or dash lights work, this indicates that there is a problem in that circuit. Perform additional operational checks on that circuit and compile a list of symptoms.

When analyzing the answers, a defect will always lie between a check which gave a positive answer and one which gave a negative answer. Look at the list of symptoms and try to determine probable areas to test. If there are negative answers on related circuits, then maybe the problem is at the common junction. After you have determined what the symptoms are and where to look for defects, develop a plan for isolating the trouble. Ask a knowledgable automotive person which components frequently fail on the vehicle. Also notice which parts or components are easiest to reach and how the most can be accomplished by doing the least amount of checks.

A common way of diagnosis is to use the split-in-half technique. Each test that is made essentially splits the trouble area in half. By perform-

ing this technique several times the area where a problem is located becomes smaller and smaller until the problem can be isolated in a single wire or component. This area is most commonly between the two closest check points that produced a negative answer and a positive answer.

After the problem is located, perform the repair procedure. This may involve replacing a component, repairing a component or damaged wire, or making an adjustment.

NOTE: Never assume a component is defective until it has been thoroughly tested.

The final step is to make sure the complaint is corrected. Remember, the symptoms that are uncovered may lead to several problems that require separate repairs. Repeat the diagnosis and test procedures again and again until all negative symptoms are corrected.

DIAGNOSIS BY THE STORED TROUBLE CODE

When a fault code is detected, it appears as a flash of the CHECK ENGINE light on the instrument panel. This indicates that an abnormal signal in the system has been recognized by the ECM.

When diagnosing by code, the first step is to read any fault codes from the ECM. These codes will identify the area to perform more in-depth testing. After the fault codes have been read, proceed to test each of the components and component circuits indicated. Continue performing individual component tests until the failed component is located. Remember, fault codes do indicate the presence of a failure, but they do not identify the failed component directly.

DIAGNOSIS BY SYMPTOM — QUICK REFERENCE CHART

	Throttle Position Sensor	Coolant Temperature Sensor	MAP or MAF sensor	Air Temperature Sensor	Ignition Coil	Distributor	Spark Plug Wires	Fuel Filter	Air Filter	Vacuum Leak	Engine Mechanical	Knock Sensor / Spark Control	EGR System	Idle Control System	Camshaft sensor/Dist. pick-up	Oxygen Sensor	Ignition Module / Engine Computer	Torque Converter Clutch	PCV System
No Start	u	u	u		•	•		•			u	•			u	•		•	
Hard Start	•	•	•		•	•	•												u
Hesitation	•		•		•	•	•	•	•	•		•							
Stalling	•				•	•	•	•	•	•	•			•				•	u
Poor Idle	•				•	•	•			•				•					•
Dieseling								•			•		•						
Engine Lamp ON	•	•	•	•								•	•		•	•	•		
Knocks or Pings							•					•	•				•	•	•
High Hydrocarbons	u	u	u		•	•		•	u	•	u	•		u					
Black Smoke	u	•	•	•						•						•			
Blue Smoke											•								
Poor Fuel Mileage	•	•	•	•						•				•		•			u
Lack of Power	•	•	•			•	•	•	•	•	•							•	u
Back fires		•	•			•	•	•	•	•							•		u
Runs Poor Cold	•				•	•	•	•	•	•	•	u		•			•	u	u
Runs Poor Hot	•	•	•							•	•	•	•	•	•	•	u	u	u
High speed surging		•	•					•										•	u

u: Although possible it is unlikely this component is at fault. A totally open or shorted circuit, or severe
component fault may cause this condition.

8736MW01

SAFE VEHICLE SERVICING TIPS

It is virtually impossible to anticipate all of the hazards involved with automotive service, but care and common sense should prevent most accidents.

The rules of safety for mechanics range from "don't smoke around gasoline" to "use the proper tool for the job." The trick to avoiding injuries is to develop safe work habits and take every possible precaution.

Do's

• Do keep a fire extinguisher and first aid kit handy.

• Do wear safety glasses or safety goggles when cutting, drilling, grinding or prying. If you wear glasses for the sake of vision, wear safety goggles over your regular glasses.

• Do shield the eyes whenever you work around the battery. Batteries contain sulfuric acid. In case of contact with the eyes or skin, flush the area with water or a mixture of water and baking soda, then get medical attention immediately.

• Do use safety stands for any under-vehicle service. Jacks are for raising vehicles; jackstands are for making sure the vehicle stays raised until you want it to come down. Whenever the vehicle is raised, block the wheels remaining on the ground and set the parking brake.

• Do use adequate ventilation when working with chemicals. Asbestos dust from some worn brake linings can cause cancer.

• Do disconnect the negative battery cable when working on the vehicle.

• Do follow manufacturer's directions whenever working with potentially hazardous materials. Both brake fluid and most types of anti-

...if taken
...tain tools.
...ushroomed
...ed or poorly
...excessively
...d wrenches
...ts, slipping
...ght sockets

...e and type

...pull on a
...n push on
...vent a fall.
...djustable
...nut or bolt
...n the fixed

...socket that
fits the nut or bolt. The wrench or socket should sit straight, not cocked.

• Do strike squarely with a hammer. Avoid glancing blows.

• Do set the parking brake and block the wheels if work requires that the engine be running.

Don'ts

• Don't run an engine in a garage or anywhere else without proper ventilation — EVER! Carbon monoxide is poisonous and it is absorbed by the body faster than oxygen. It takes a long time to leave the human body, and a deadly supply of it can be built up it in your system by simply breathing in a little every day. You may not realize you are slowly poisoning yourself. Always use power vents, windows, fans or open the garage.

• Don't work around moving parts while wearing loose clothing. Short sleeves are much safer than long, loose sleeves. Hard-toed shoes with neoprene soles protect your toes and give a better grip on slippery surfaces. Jewelry such as watches, fancy belt buckles, beads, or body adornment of any kind is not safe while working around a vehicle. Long hair should be hidden under a hat or cap.

• Don't use pockets for toolboxes. A fall or bump can drive a screwdriver deep into you body. Even a wiping cloth hanging from the back pocket can wrap around a spinning shaft or fan.

• Don't smoke around gasoline, solvent or any flammable material.

• Don't smoke around the battery. When the battery is being charged, it gives off explosive hydrogen gas.

• Don't use gasoline to wash the hands. There are excellent soaps available. Gasoline contains compounds which are hazardous to your health and it removes natural oils from the skin so that bone dry hands will suck up oil and grease.

• Don't service the air conditioning system unless equipped with the necessary tools and training. The refrigerant is extremely cold, and when exposed to the air, will instantly freeze any surface it comes in contact with, including your eyes. Keep refrigerant away from open flames; poisonous gas will be produced if refrigerant burns.

SERIAL NUMBER IDENTIFICATION

Vehicle Identification Number

The Vehicle Identification Number (VIN) number is somewhat like a Social Security Number in an automotive sense. The VIN is a standardized 17 digit number. Each digit of this number has a specific meaning or designation. Example: the 8th digit designates the engine code, the 10th digit designates the model year of the vehicle etc. The vehicle serial number is stamped on a plate fastened to the driver's side door pillar.

This number is usually located on the one of the fender aprons in the engine compartment (behind the wheel arch).

All models have the vehicle identification number stamped on a plate attached to the left side of the instrument panel. The plate is visible through the windshield.

The vehicle identification (model variation codes) may be interpreted as follows:

Using a Nissan vehicle as an illustration, look at the VIN number (JN6H D2 1S*MW 000001), all models use a four letter prefix followed by the model designation (D21), then a five letter suffix, as shown in the illustration.

The serial number on all models is the new 17-digit format. The first three digits are the World Manufacturer Identification number. The next five digits are the Vehicle Description Section (same as the series identification number). The remaining nine digits are the production numbers.

NOTE: For specific identification of the vehicle see the Vehicle Identification Label on the vehicle you are working with. If the vehicle has been altered in some way or the engine or transmission has been changed the VIN may not coincide with the change that has been made. It a case like this, look for the serial number on the component to be sure the correct ordering of parts are made.

Vehicle Identification Number — visible through the windshield

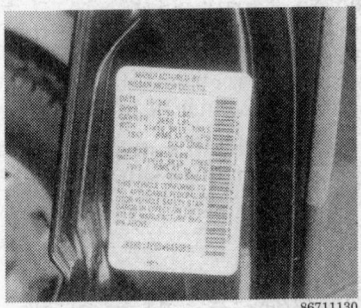

Manufacturer's label located in the door pillar area — note that the build date of vehicle is at the top

Vehicle identification number is on the firewall-mounted label in the engine compartment

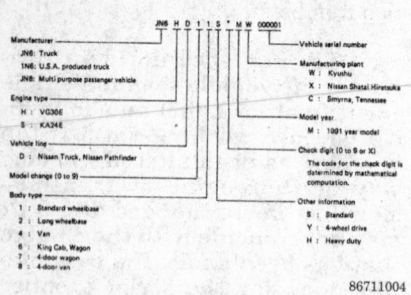

Vehicle Identification Number (VIN) translation

Engine Serial Number

The engine serial number consists of an engine series identification number followed by a six-digit production number. The number may be found in various places, depending upon the particular engine. Observe the following examples:

- Z24i Engine — the serial number is stamped on the left side of the cylinder block, below the No. 3 and No. 4 spark plugs.
- KA24E Engine — the serial number is stamped on the left side of the cylinder block, below the No. 2 and No. 3 spark plugs.
- VG30i and VG30E Engines — the serial number is stamped on the cylinder block, below the rear of the right side cylinder head.

NOTE: The illustrations given here are for a Nissan.

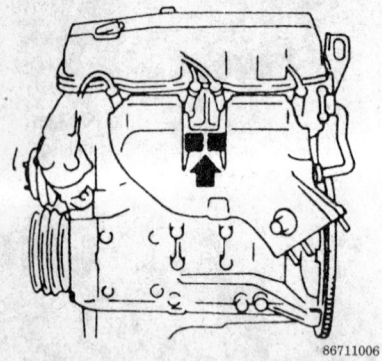

Typical engine serial number location

Transmission Serial Number

The transmission serial number is stamped on the front upper face of the transmission case on manual transmissions, or on the right side of

the transmission case on automatic transmissions.

NOTE: The illustrations given here are for a Nissan.

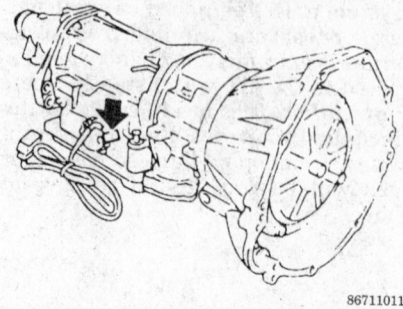

Typical automatic transmission serial number location

Transfer Case Serial Number

The transfer case serial number is stamped on the front upper face of the transfer case.

NOTE: The illustrations given here are a generalization and may not apply to the vehicle you are working on.

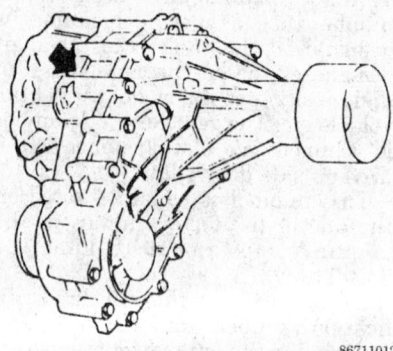

Typical transfer case serial number location

Vehicle Emissions Control Information (VECI) label

The Vehicle Emissions Control Information (VECI) label provides a wealth of information pertaining to the engine's Emission Control System. First, it identifies the engine's Cubic Inch Displacement (CID) and size in Liter(s). It provides information for tune-up, such as spark plug gap, ignition timing, idle/mixture and valve lash specifications. In some cases, it will provide a specific adjustment procedure as required. Some la-

bels will also incorporate the vacuum routing of the engine's emission control system. Although the VECI label is very helpful to the person working on the vehicle, it should not be used as the main source for repair information. However, if there have not been any alterations to the engine and the Manual and sticker do not agree, use the Vehicle Emissions Control Information (VECI) sticker information, as it often reflects changes made during the production run.

Always keep in mind that the vehicle you are working with may have had an engine change. If this is the case, the year and engine code will have to be identified. If the vehicle is missing the Emission Control Information label, a new one can be ordered from a local dealer.

The Vehicle Emissions Control Information (VECI) label is usually located in the engine compartment. On some vehicles it will be found directly under the hood. Others may have it on the strut tower or radiator support.

Vehicle Emission Control Information (VECI) label located in the engine compartment

JUMP STARTING

Jump Starting a Dead Battery

Whenever a vehicle must be jump started, precautions must be followed in order to prevent the possibility of personal injury. Remember that batteries contain a small amount of explosive hydrogen gas which is a by-product of battery charging. Sparks should always be avoided when working around batteries, especially when attaching jumper cables. To minimize the possibility of accidental sparks, follow the procedure carefully.

---- **WARNING** ----

NEVER hook the batteries up in a series circuit or the entire electrical system will go up in smoke, especially the starter!

Vehicles equipped with a diesel engine utilize two 12 volt batteries, one on either side of the engine compartment. The batteries are connected in a parallel circuit (positive terminal to positive terminal, negative terminal to negative terminal). Hooking the batteries up in parallel circuit increases battery cranking power without increasing total battery voltage output. Output remains at 12 volts. On the other hand, hooking two 12 volt batteries up in a series circuit (positive terminal to negative terminal, positive terminal to negative terminal) increases total battery output to 24 volts (12 volts plus 12 volts).

Jump Starting Precautions

1. Be sure that both batteries are of the same voltage. All vehicles covered by this manual and most vehicles on the road today utilize a 12 volt charging system.
2. Be sure that both batteries are of the same polarity (have the same grounded terminal; in most cases NEGATIVE).
3. Be sure that the vehicles are not touching or a short circuit could occur.
4. On serviceable batteries, be sure the vent cap holes are not obstructed.
5. Do not smoke or allow sparks anywhere near the batteries.
6. In cold weather, make sure the battery electrolyte is not frozen. This can occur more readily in a battery that has been in a state of discharge.
7. Do not allow electrolyte to contact skin or clothing.

Jump Starting Procedure

1. Make sure that the voltages of the 2 batteries are the same. Most batteries and charging systems are of the 12 volt variety.
2. Pull the jumping vehicle (with the good battery) into a position so the jumper cables can reach the dead battery and that vehicle's engine. Make sure that the vehicles do NOT touch.
3. Place the transmissions of both vehicles in NEUTRAL or PARK, as

```
MAKE CONNECTIONS IN NUMERICAL ORDER

                    ① FIRST JUMPER CABLE
DO NOT ALLOW
VEHICLES TO TOUCH       DISCHARGED
                        BATTERY

                    SECOND JUMPER CABLE

                    MAKE LAST
                    CONNECTION ON
                    ENGINE, AWAY
                    FROM BATTERY

                    BATTERY IN VEHICLE
                    WITH CHARGED BATTERY

                                    TCCS1080
```

Connect the jumper cables to the batteries and engine in the order shown

applicable, then firmly set their parking brakes.

NOTE: If necessary for safety reasons, both vehicle's hazard lights may be operated throughout the entire procedure without significantly increasing the difficulty of jump starting the dead battery.

4. Turn all lights and accessories **OFF** on both vehicles. Make sure the ignition switches on both vehicles are turned to the **OFF** position.
5. Cover the battery cell caps with a rag, but do not cover the terminals.
6. Make sure the terminals on both batteries are clean and free of corrosion or proper electrical connection will be impeded. If necessary, clean the battery terminals before proceeding.
7. Identify the positive (+) and negative (-) terminals on both batteries.
8. Connect the first jumper cable to the positive (+) terminal of the dead battery, then connect the other end of that cable to the positive (+) terminal of the booster (good) battery.
9. Connect one end of the other jumper cable to the negative (-) terminal of the booster battery and the other cable clamp to an engine bolt head, alternator bracket or other solid, metallic point on the dead battery's engine. Try to pick a ground on the engine that is positioned away from the battery, in order to minimize the possibility of the 2 clamps touching should one loosen during the procedure. DO NOT connect this clamp to the negative (-) terminal of the bad battery.

---- **CAUTION** ----

Be very careful to keep the jumper cables away from moving parts (cooling fan, belts, etc.) on both engines.

10. Check to make sure that the cables are routed away from any moving parts, then start the donor vehicle's engine. Run the engine at moderate speed for several minutes to allow the dead battery a chance to receive some initial charge.
11. With the donor vehicle's engine still running slightly above idle, try to start the vehicle with the dead battery. Crank the engine for no more than 10 seconds at a time and let the starter cool for at least 20 seconds between tries. If the vehicle does not start within 3 tries, it is likely that something else is also wrong.
12. Once the vehicle is started, allow it to run at idle for a few seconds to make sure that it is properly operating.
13. Turn on the headlights, heater blower and, if equipped, the rear defroster of both vehicles in order to reduce the severity of voltage spikes and subsequent risk of damage to the vehicles' electrical systems when the cables are disconnected.
14. Carefully disconnect the cables in the reverse order of connection. Start with the negative cable that is attached to the engine ground, then the negative cable on the donor battery. Disconnect the positive cable from the donor battery, then disconnect the positive cable from the formerly dead battery. Be careful when disconnecting the cables from the positive terminals not to allow the alligator clips to touch any metal on either vehicle or a short circuit and sparks will occur.

CHARGING SYSTEM

Alternator

The alternator converts the mechanical energy supplied by the drive belt into electrical energy by a process of electromagnetic induction. When the ignition switch is turned **ON**, current flows from the battery through the charging system light (or ammeter) to the voltage regulator, and finally to the alternator. When the engine is started, the drive belt turns the rotating field (rotor) in the stationary windings (stator), inducing alternating current. This alternating current is converted into usable direct current by the diode rectifier. Most of this current is used to charge the battery and to supply power for the vehicle's electrical accessories. A small part of this current is returned to the field windings of the alternator, enabling it to increase its power output. When the current in the field windings reaches a predetermined level, the voltage regulator grounds the circuit preventing any further increase. The cycle is continued so that the voltage supply remains constant.

All models use a 12-volt alternator. Amperage ratings vary according to the year and model. All models have an electronic, nonadjustable regulator, integral with the alternator.

ALTERNATOR PRECAUTIONS

To prevent damage to the alternator and regulator, the following precautionary measures must be taken when working with the electrical system:

• Never reverse the battery connections. Always visually check the battery polarity before any connections are made, to ensure that all connections correspond to the vehicle's battery ground polarity.

• Booster batteries must be connected properly. Make sure the positive cable of the booster battery is connected to the positive terminal of the battery which is getting the boost.

• Disconnect the battery cables before using a fast charger; the charger has a tendency to force current through the diodes in the opposite direction for which they were designed.

• Never use a fast charger as a booster for starting the vehicle.

• Never disconnect the voltage regulator while the engine is running, unless as noted for testing purposes.

• Do not ground the alternator output terminal.

• Do not operate the alternator on an open circuit with the field energized.

• Do not attempt to polarize the alternator.

• Disconnect the battery cables and remove the alternator before using an electric arc welder on the vehicle.

• Protect the alternator from excessive moisture. If the engine is to be steam cleaned, cover or remove the alternator.

REMOVAL & INSTALLATION

The procedure below is just a general procedure.

NOTE: On some models, the alternator is mounted very low on the engine. On these models, it may be necessary to remove the gravel shield and work from beneath the vehicle in order to gain access to the alternator.

1. Disconnect the negative battery cable.

2. On vehicles where the alternator can only be accessed from underneath the vehicle, raise the vehicle and support it safely with jackstands. Make sure the jackstands are at proper locations.

3. Remove the alternator pivot bolt. Push the alternator inward and remove the drive belt.

4. Pull back the rubber boots and disconnect the wiring from the back of the alternator.

5. Remove the alternator mounting bolt, then withdraw the alternator from its bracket.

To install:

6. Position the alternator in its mounting bracket, then lightly tighten the mounting and adjusting bolts.

7. Connect the electrical leads at the rear of the alternator, and return rubber boots to the proper position.

8. Adjust the belt tension.

9. Connect the negative battery cable.

10. Start the engine and perform a charging system voltage test to in-

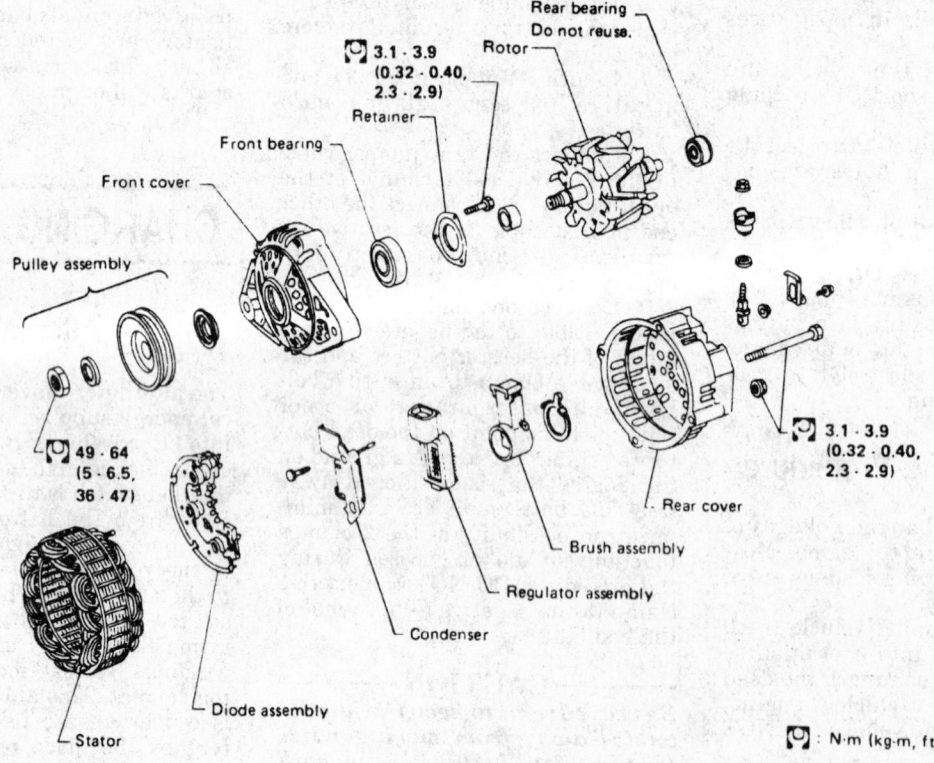

Exploded view of a typical alternator assembly

sure the charging system is putting out adequately.

Make a basic check to see if the charging system is charging by using a voltmeter. Connect the voltmeter to the battery, battery voltage is approximately 12.6 volts. Start the engine and observe the voltmeter reading, it should read between 13.2–14 volts. If it remains at 12.6 volts the system is not charging.

Regulator

Regulators may be located internally to the the alternator or externally mounted on the firewall or inner fender panel depending on the vehicle. If faulty, it must be replaced; there are no adjustments which can be made.

REMOVAL & INSTALLATION

Internal Regulator

The electronic regulator is located inside the alternator. On some alternators the regulator is a simple bolt-on procedure, others may be soldered to the brush assembly. With a little knowledge of soldering the job should be able to be accomplished. The following procedure is for a regulator that is soldered to the brush assembly. The regulator is non-adjustable, and must be replaced together with the brush assembly, if faulty.

1. Remove the alternator.
2. Remove the thru-bolts and separate the front cover from the stator housing.
3. Unsolder the wire connecting the diode plate to the brush at the brush terminal.
4. Remove the bolt retaining the diode plate to the rear cover.
5. Remove the nut securing the battery terminal bolt.
6. Lift the stator slightly, together with the diode plate, to gain access to the diode plate screw. Remove the screw.
7. Separate the stator and diode, then remove the brush and regulator assembly.
8. On assembly, apply soldering heat sparingly, carrying out the operation as quickly as possible, to avoid damage to the transistors and diodes. Before assembling the alternator halves, bend a piece of wire into an L-shape and slip it through the rear cover, next to the brushes. Use the wire to hold the brushes in a retracted position until the case halves are assembled. Remove the wire carefully, to prevent damage to the slip rings.

9. Install the alternator.
10. Start the engine and perform a charging system voltage test to insure charging system is putting out adequately.

Make a basic check to see if the charging system is charging by using a voltmeter. Connect the voltmeter to the battery, battery voltage is approximately 12.6 volts. Start the engine and observe the voltmeter reading, it should read between 13.2–14 volts. If it remains at 12.6 volts the system is not charging.

External Regulator

Depending on the vehicle, the external regulator may be mounted on the firewall or inner fender panel. These regulators are much simpler to replace.

1. Disconnect the negative battery cable.
2. Locate the regulator.
3. Disconnect the harness connector from the regulator.
4. Remove the bolts holding the regulator to its mounting base and remove the regulator.

To install:

5. Install the regulator to the mounting base and secure it in place with the mounting screws.
6. Apply a coating of dielectric grease on the harness electrical connectors and connect.
7. Connect the negative battery cable.
8. Start the engine a perform a charging system voltage test to insure charging system is putting out adequately.

Make a basic check to see if the charging system is charging by using a voltmeter. Connect the voltmeter to the battery, battery voltage is approximately 12.6 volts. Start the engine and observe the voltmeter reading, it should read between 13.2–14 volts. If it remains at 12.6 volts the system is not charging.

ADJUSTMENT

Voltage regulators on modern vehicles are not adjustable, most are electronic or even computer controlled.

STARTING SYSTEM

Starter

REMOVAL & INSTALLATION

The procedure below is just a general procedure.

1. Disconnect the negative battery cable at the battery, then disconnect the positive battery cable at the starter.
2. On vehicles where the starter can only be accessed from underneath the vehicle, observe the following caution:

——— CAUTION ———
Raise the vehicle a support it safely with suitable jackstands. Be sure to position the jackstands at proper frame locations.

3. On some 4wd vehicles it may be necessary to perform the following procedure:
   a. Remove the front gravel shield.
   b. Detach the oil pressure switch connector.
   c. Drain the engine oil and remove the oil filter.

——— CAUTION ———
The EPA warns that prolonged contact with used engine oil may cause a number of skin disorders, including cancer! Make every effort to minimize exposure to used engine oil. Protective gloves should be worn when changing the oil. Wash hands and any other exposed skin areas as soon as possible after exposure to used engine oil. Soap and water, or waterless hand cleaner should be used.

   d. Remove the exhaust manifold heat insulator.
   e. Remove the fuel tube retainer bolt.
4. On some vehicles it may be necessary to perform the following procedure:
   a. Remove the front right wheel.
   b. Remove the front gravel shield.
   c. Remove the exhaust manifold heat insulator.
   d. Remove the exhaust manifold.
   e. Detach the oil pressure switch connector.
5. Unfasten the remaining electrical connections at the starter solenoid.

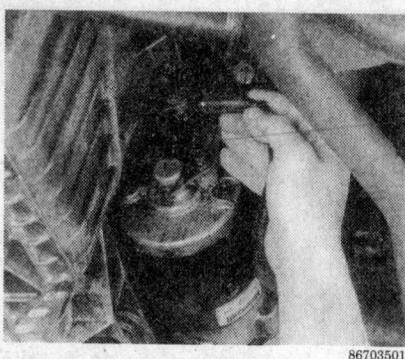

Disconnect the cable attached the starter

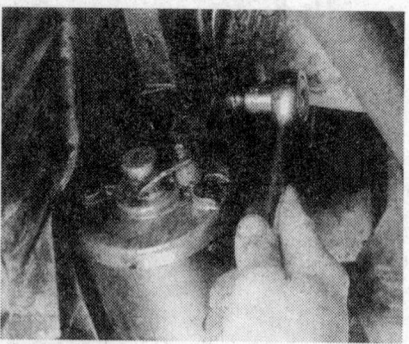

Remove the starter bracket, if equipped

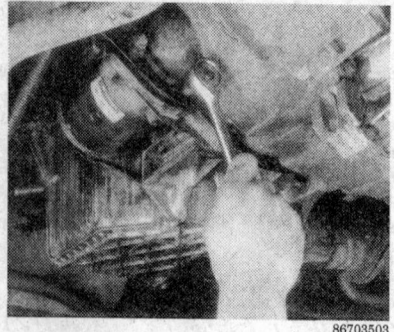

Loosen the holding the nose bracket, if equipped

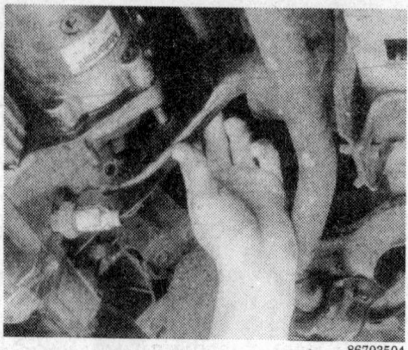

Remove the nose bracket, if equipped

Remove the bolts holding the starter in place

Remove the the starter from the engine

6. Remove the two nuts holding the starter to the bell housing, then pull the starter toward the front of the vehicle and out.

To install:

7. Insert the starter into the bell housing, being sure that the starter drive is not jammed against the flywheel.

8. Tighten the attaching nuts and secure all electrical connections to the starter assembly.

9. Install all remaining components in reverse order of removal.

10. Reconnect the battery cables. If applicable, refill and check the oil level. Check the starter assembly for proper operation.

OVERHAUL

The procedure below is a general overhaul procedure.

Solenoid Replacement

1. Remove the starter.
2. Unscrew the two solenoid switch (magnetic switch) retaining screws.

3. Remove the solenoid. In order to unhook the solenoid from the starter drive lever, lift it up at the same time that you are pulling it out of the starter housing.

4. Installation is in the reverse order of removal. Make sure that the solenoid switch is properly engaged with the drive lever before tightening the mounting screws.

Brush Replacement

NON-REDUCTION GEAR TYPE

1. Remove the starter.
2. Remove the solenoid (magnetic switch).
3. Unfasten the two end frame cap mounting bolts and remove the end frame cap.
4. Remove the O-ring and lock plate from the armature shaft groove, then slide the shims off the shaft.
5. Unfasten the two long housing screws (at the front of the starter) and carefully pull off the end plate.
6. Using a screwdriver, separate the brushes from the brush holder.
7. Slide the brush holder off of the armature shaft.
8. Crush the old brushes off of the copper braid and file away any remaining solder.
9. Fit the new brushes to the braid and spread the braid slightly.

NOTE: Use a soldering iron of at least 250 watts.

10. Using a light-grade solder, solder the brush to the braid. Grip the copper braid with flat pliers to prevent the solder from flowing down its length.
11. File off any extra solder and then repeat the procedure for the remaining three brushes.
12. Installation is in the reverse order of removal.

NOTE: When installing the brush holder, make sure that the brushes line up properly.

REDUCTION GEAR TYPE

1. Remove the starter and the solenoid.
2. Remove the through-bolts and the rear cover. The rear cover can be pried off with a small pry tool, but be careful not to damage the O-ring.
3. Separate the starter housing, armature, and brush holder from the center housing. They can be removed as an assembly.
4. Remove the positive side brush from its holder. The positive brush is insulated from the brush holder, and its lead wire is connected to the field coil.

5. Carefully lift the negative brush from the commutator and remove it from the holder.

6. Installation is in the reverse order of removal.

Starter Drive Replacement

NON-REDUCTION GEAR TYPE

1. With the starter motor removed from the vehicle, separate the solenoid from the starter.

2. Remove the two through-bolts and separate the gear case from the yoke housing.

3. Remove the pinion stopper clip and the pinion stopper.

4. Slide the starter drive off the armature shaft.

5. Install the starter drive and reassemble the starter in the reverse order of removal.

REDUCTION GEAR TYPE

1. Remove the starter.

2. Unfasten the solenoid and the shift lever.

3. Remove the bolts securing the center housing to the front cover and separate the parts.

4. Remove the gears and starter drive.

5. Installation is in the reverse order of removal.

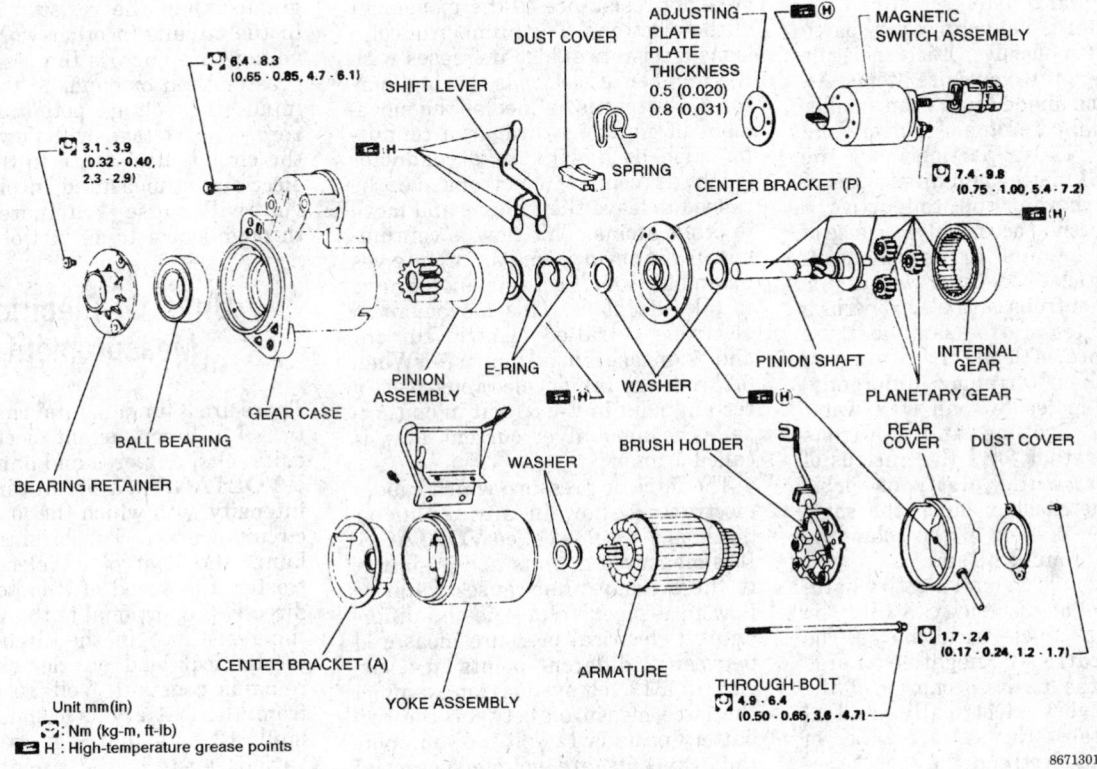

Unit mm(in)

: Nm (kg-m, ft-lb)

H : High-temperature grease points

86713017

Exploded view of a reduction gear type starter — typical

FUNDAMENTALS OF ELECTRICITY

A good understanding of basic electrical theory and how circuits work is necessary to successfully perform the service and testing outlined in this manual. Therefore, this section should be read before attempting any diagnosis and repair.

All matter is made up of tiny particles called molecules. Each molecule is made up of two or more atoms. Atoms may be divided into even smaller particles called protons, neutrons and electrons. These particles are the same in all matter and differences in materials (hard or soft, conductive or non-conductive) occur only because of the number and arrangement of these particles. In other words, the protons, neutrons and electrons in a drop of water are the same as those in an ounce of lead, there are just more of them (arranged differently) in a lead molecule than in a water molecule. Protons and neutrons packed together form the nucleus of the atom, while electrons orbit around the nucleus much the same way as the planets of the solar system orbit around the sun.

The proton is a small positive natural charge of electricity, while the neutron has no electrical charge. The electron carries a negative charge equal to the positive charge of the proton. Every electrically neutral atom contains the same number of protons and electrons, the exact number of which determines the element. The only difference between a conductor and an insulator is that a conductor possesses free electrons in large quantities, while an insulator has only a few. An element must have very few free electrons to be a

good insulator and vice-versa. When we speak of electricity, we're talking about these free electrons.

In a conductor, the movement of the free electrons is hindered by collisions with the adjoining atoms of the element (matter). This hindrance to movement is called **RESISTANCE** and it varies with different materials and temperatures. As temperature increases, the movement of the free electrons increases, causing more frequent collisions and therefore increasing resistance to the movement of the electrons. The number of collisions (resistance) also increases with the number of electrons flowing (current). Current is defined as the movement of electrons through a conductor such as a wire. In a conductor (such as copper) electrons can be caused to leave their atoms and move to other atoms. This flow is continuous in that every time an atom gives up an electron, it collects another one to take its place. This movement of electrons is called electric current and is measured in amperes. When 6.28 billion, billion electrons pass a certain point in the circuit in one second, the amount of current flow is called 1 ampere.

The force or pressure which causes electrons to flow in any conductor (such as a wire) is called **VOLTAGE**. It is measured in volts and is similar to the pressure that causes water to flow in a pipe. Voltage is the difference in electrical pressure measured between 2 different points in a circuit. In a 12 volt system, for example, the force measured between the two battery posts is 12 volts. Two important concepts are voltage potential and polarity. Voltage potential is the amount of voltage or electrical pressure at a certain point in the circuit with respect to another point. For example, if the voltage potential at one post of the 12 volt battery is 0, the voltage potential at the other post is 12 volts with respect to the first post.

One post of the battery is said to be positive (+); the other post is negative (-) and the conventional direction of current flow is from positive to negative in an electrical circuit. It should be noted that the electron flow in the wire is opposite the current flow. In other words, when the circuit is energized, the current flows from positive to negative, but the electrons actually move from negative to positive. The voltage or pressure needed to produce a current flow in a circuit must be greater than the resistance present in the circuit. In other words, if the voltage drop across the resistance is greater than or equal to the voltage input, the voltage potential will be zero — no voltage will flow through the circuit. Resistance to the flow of electrons is measured in ohms. One volt will cause 1 ampere to flow through a resistance of 1 ohm.

Units Of Electrical Measurement

There are 3 fundamental characteristics of a direct-current electrical circuit: volts, amperes and ohms.

VOLTAGE in a circuit controls the intensity with which the loads in the circuit operate. The brightness of a lamp, the heat of an electrical defroster, the speed of a motor are all directly proportional to the voltage, if the resistance in the circuit and/or mechanical load on electric motors remains constant. Voltage available from the battery is constant (normally 12 volts), but as it operates the various loads in the circuit, voltage decreases (drops).

AMPERE is the unit of measurement of current in an electrical circuit. One ampere is the quantity of current that will flow through a resistance of 1 ohm at a pressure of 1 volt. The amount of current that flows in a circuit is controlled by the voltage and the resistance in the circuit. Current flow is directly proportional to resistance. Thus, as voltage is increased or decreased, current is increased or decreased accordingly. Current is decreased as resistance is increased. However, current is also increased as resistance is decreased. With little or no resistance in a circuit, current is high.

OHM is the unit of measurement of resistance, represented by the Greek letter Omega (ω). One ohm is the resistance of a conductor through which a current of one ampere will flow at a pressure of one volt. Electrical resistance can be measured on an

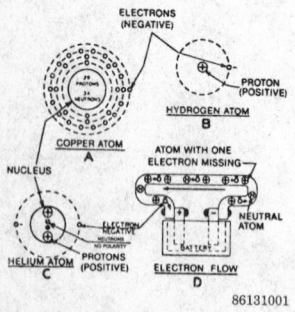

ATOMS AND ELECTRONS

86131001

Typical atoms of Copper (A), Hydrogen (B) and Helium (C). Electron flow in a battery circuit

86131002

Electrical resistance can be compared to water flow through a pipe. The smaller the wire (pipe), the more resistance the flow of electrons (water)

instrument called an ohmmeter. The loads (electrical devices) are the primary resistances in a circuit. Loads such as lamps, solenoids and electric heaters have a resistance that is essentially fixed; at a normal fixed voltage, they will draw a fixed current. Motors, on the other hand, do not have a fixed resistance. Increasing the mechanical load on a motor (such as might be caused by a misadjusted track in a power window system) will decrease the motor speed. The drop in motor rpm has the effect of reducing the internal resistance of the motor because the current draw of the motor varies directly with the mechanical load on the motor, although its actual resistance is unchanged. Thus, as the motor load increases, the current draw of the motor increases, and may increase up to the point where the motor stalls (cannot move the mechanical load).

Circuits are designed with the total resistance of the circuit taken into account. Troubles can arise when unwanted resistances enter into a circuit. If corrosion, dirt, grease, or any other contaminant occurs in places like switches, connectors and grounds, or if loose connections occur, resistances will develop in these areas. These resistances act like additional loads in the circuit and cause problems.

OHM'S LAW

Ohm's law is a statement of the relationship between the 3 fundamental characteristics of an electrical circuit. These rules apply to direct current (DC) only.

Ohm's law provides a means to make an accurate circuit analysis without actually seeing the circuit. If, for example, one wanted to check the condition of the rotor winding in a alternator whose specifications indicate that the field (rotor) current draw is normally 2.5 amperes at 12 volts, simply connect the rotor to a 12 volt battery and measure the current with an ammeter. If it measures about 2.5 amperes, the rotor winding can be assumed good.

An ohmmeter can be used to test components that have been removed from the vehicle in much the same manner as an ammeter. Since the voltage and the current of the rotor windings used as an earlier example are known, the resistance can be calculated using Ohms law. The formula would be ohms equals volts divided by amperes.

If the rotor resistance measures about 4.8 ohms when checked with an ohmmeter, the winding can be assumed good. By plugging in different specifications, additional circuit information can be determined such as current draw, etc.

$$I = \frac{E}{R} \quad \text{or} \quad AMPERES = \frac{VOLTS}{OHMS}$$

$$R = \frac{E}{I} \quad \text{or} \quad OHMS = \frac{VOLTS}{AMPERES}$$

$$E = I \times R \quad \text{or} \quad VOLTS = AMPERES \times OHMS$$

86131003

Ohms Law is the basis for all electrical measurement. By simply plugging in two values, the third can be calculated using this formula

Electrical Circuits

An electrical circuit must start from a source of electrical supply and return to that source through a continuous path. Circuits are designed to handle a certain maximum current flow. The maximum allowable current flow is designed higher than the normal current requirements of all the loads in the circuit. Wire size, connections, insulation, etc., are designed to prevent undesirable voltage drop, overheating of conductors, arcing of contacts and other adverse effects. If the safe maximum current flow level is exceeded, damage to the circuit components will result; it is this condition that circuit protection devices are designed to prevent.

Protection devices are fuses, fusible links or circuit breakers designed to open or break the circuit quickly whenever an overload, such as a short circuit, occurs. By opening the circuit quickly, the circuit protection

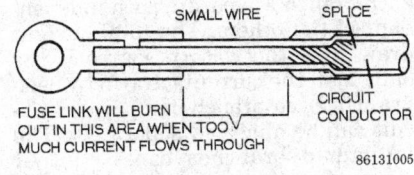

Typical fusible link wire

device prevents damage to the wiring, battery and other circuit components. Fuses and fusible links are designed to carry a preset maximum amount of current and to melt when that maximum is exceeded, while circuit breakers merely break the connection and may be manually reset. The maximum amperage rating of each fuse is marked on the fuse body and all contain a see-through portion that shows the break in the fuse element when blown. Fusible link maximum amperage rating is indicated by gauge or thickness of the wire. Never replace a blown fuse or fusible link with one of a higher amperage rating.

CAUTION

Resistance wires, like fusible links, are also spliced into conductors in some areas. Do not make the mistake of replacing a fusible link with a resistance wire. Resistance wires are longer than fusible links and are stamped "RESISTOR-DO NOT CUT OR SPLICE."

Circuit breakers consist of 2 strips of metal which have different coefficients of expansion. As an overload or current flows through the bimetallic strip, the high-expansion metal will elongate due to heat and break the contact. With the circuit open, the bi-metal strip cools and shrinks, drawing the strip down until contact is re-established and current flows once again. In actual operation, the contact is broken very quickly if the overload is continuous and the circuit will be repeatedly broken and remade until the source of the overload is corrected.

The self-resetting type of circuit breaker is the one most generally used in automotive electrical systems. On manually reset circuit breakers, a button will pop up on the circuit breaker case. This button must be pushed in to reset the circuit breaker and restore power to the circuit. Always repair the source of the overload before resetting a circuit breaker or replacing a fuse or fusible link. When searching for overloads, keep in mind that the circuit protection devices protect only against overloads between the protection device and ground.

There are 2 basic types of circuit; Series and Parallel. In a series circuit, all of the elements are connected in chain fashion with the same amount of current passing through each element or load. No matter where an ammeter is connected in a series circuit, it will always read the

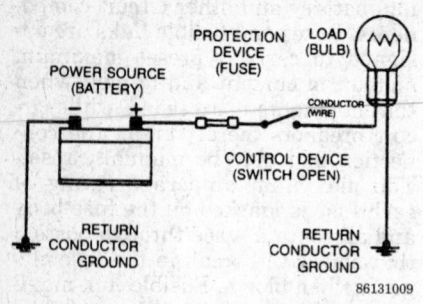

Typical circuit with all essential components

86131009

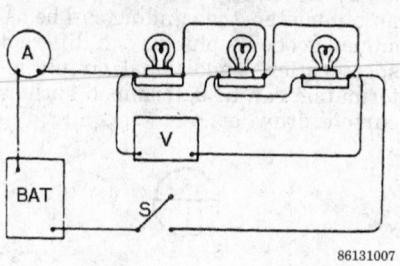

Example of a parallel circuit

86131007

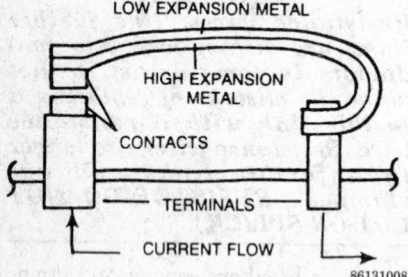

Typical circuit breaker construction

86131008

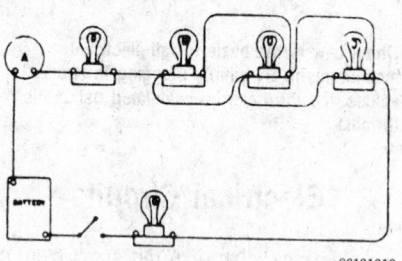

Example of a series-parallel circuit

86131010

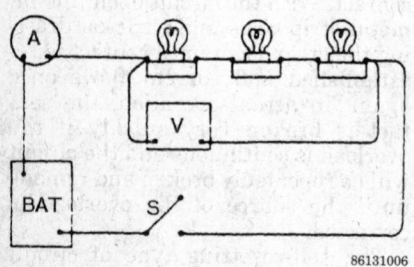

Example of a series circuit

86131006

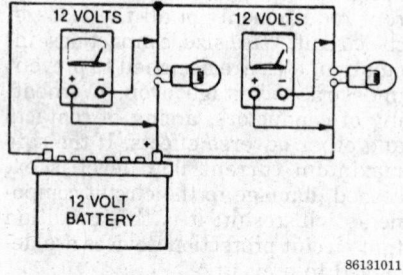

Voltage drop in a parallel circuit. Voltage drop across each lamp is 12 volts

86131011

just as it applies to series circuits, by considering each branch independently of the others. The most important thing to remember is that the voltage across each branch is the same as the source voltage. The current in any branch is that voltage divided by the resistance of the branch. A practical method of determining the resistance of a parallel circuit is to divide the product of the 2 resistances by the sum of 2 resistances at a time. Amperes through a parallel circuit is the sum of the amperes through the separate branches. Voltage across a parallel circuit is the same as the voltage across each branch.

By measuring the voltage drops, the resistance of each element within the circuit is being measured. The greater the voltage drop, the greater the resistance. Voltage drop measurements are a common way of checking circuit resistances in automotive electrical systems. When part of a circuit develops excessive resistance (due to a bad connection) the element will show a higher than normal voltage drop. Normally, automotive wiring is selected to limit voltage drops to a few tenths of a volt. In parallel circuits, the total resistance is less than the sum of the individual resistances; because the current has 2 paths to take, the total resistance is lower.

Magnetism and Electromagnets

Electricity and magnetism are very closely associated because when electric current passes through a wire, a magnetic field is created around the wire. When a wire carrying electric current is wound into a coil, a magnetic field with North and South poles is created just like in a bar magnet. If an iron core is placed within the coil, the magnetic field becomes stronger because iron conducts magnetic lines much easier than air. This arrangement is called an electromagnet and is the basic principle behind the operation of such components as relays, buzzers and solenoids.

A relay is basically just a remote-controlled switch that uses a small amount of current to control the flow of a large amount of current. The simplest relay contains an electromagnetic coil in series with a voltage source (battery) and a switch. A movable armature made of some magnetic material pivots at one end and

same. The most important fact to remember about a series circuit is that the sum of the voltages across each element equals the source voltage. The total resistance of a series circuit is equal to the sum of the individual resistances within each element of the circuit. Using ohms law, one can determine the voltage drop across each element in the circuit. If the total resistance and source voltage is known, the amount of current can be calculated. Once the amount of current (amperes) is known, values can be substituted in the Ohms law formula to calculate the voltage drop

across each individual element in the series circuit. The individual voltage drops must add up to the same value as the source voltage.

A parallel circuit, unlike a series circuit, contains 2 or more branches, each branch a separate path independent of the others. The total current draw from the voltage source is the sum of all the currents drawn by each branch. Each branch of a parallel circuit can be analyzed separately. The individual branches can be either simple circuits, series circuits or combinations of series-parallel circuits. Ohms law applies to parallel circuits

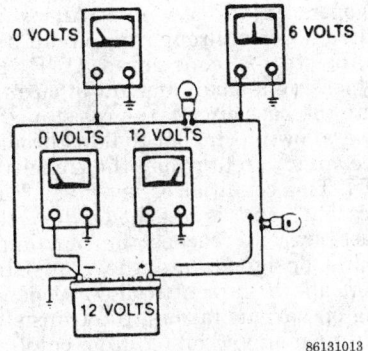

Total current in a parallel circuit: 4 + 6 + 12 = 22 amps

Voltage drop in a series circuit

ELECTRO-MAGNETS

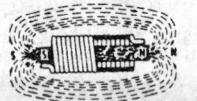

FORCE FIELD SURROUNDING A CURRENT CARRYING COIL
(WITHOUT IRON CORE)
ALL FORCE LINES ARE COMPLETE LOOPS

FORCE FIELD WITH SOFT IRON CORE
NOTE CONCENTRATION OF LINES IN IRON CORE
86131014

Magnetic field surrounding an electromagnet

is held a small distance away from the electromagnet by a spring or the spring steel of the armature itself. A contact point, made of a good conductor, is attached to the free end of the armature with another contact point a small distance away. When the relay is switched on (energized), the magnetic field created by the current flow attracts the armature, bending it until the contact points meet, closing a circuit and allowing current to flow in the second circuit through the relay to the load the circuit operates. When the relay is switched off (de-

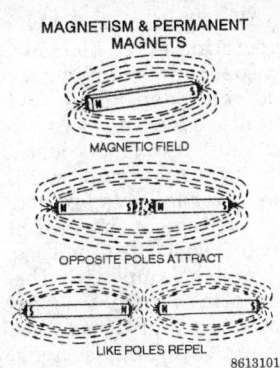

MAGNETISM & PERMANENT MAGNETS

MAGNETIC FIELD

OPPOSITE POLES ATTRACT

LIKE POLES REPEL

86131015

Magnetic field surrounding a bar magnet

energized), the armature springs back and opens the contact points, cutting off the current flow in the secondary, or controlled, circuit. Relays can be designed to be either open or closed when energized, depending on the type of circuit control a manufacturer requires.

A buzzer is similar to a relay, but its internal connections are different. When the switch is closed, the current flows through the normally closed contacts and energizes the coil. When the coil core becomes magnetized, it bends the armature down and breaks the circuit. As soon as the circuit is broken, the spring-loaded

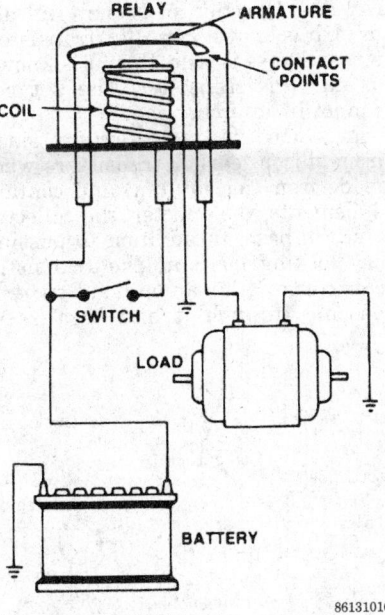

Typical relay circuit with basic components

armature remakes the circuit and again energizes the coil. This cycle repeats rapidly to cause the buzzing sound.

A solenoid is constructed like a relay, except that its core is allowed to move, providing mechanical motion that can be used to actuate mechanical linkage to operate a door or trunk lock or control any other mechanical function. When the switch is closed, the coil is energized and the movable core is drawn into the coil. When the switch is opened, the coil is de-energized and spring pressure returns the core to its original position.

Basic Solid State

The term "solid state" refers to devices utilizing transistors, diodes and other components which are made from materials known as semiconductors. A semiconductor is a material that is neither a good insulator nor a good conductor; principally silicon and germanium. The semiconductor material is specially treated to give it certain qualities that enhance its function, therefore becoming either P-type (positive) or N-type (negative) material. Most semiconductors are constructed of silicon and can be designed to function either as an insulator or conductor.

DIODES

The simplest semiconductor function is that of the diode or rectifier (the 2 terms mean the same thing). A diode will pass current in one direction only, like a one-way valve, because it has low resistance in one direction and high resistance on the other. Whether the diode conducts or not depends on the polarity of the voltage applied to it. A diode has 2 electrodes, an anode and a cathode. When the anode receives positive (+) voltage and the cathode receives negative (-) voltage, current can flow easily through the diode. When the voltage is reversed, the diode becomes nonconducting and only allows a very slight amount of current to flow in the circuit. Because the semiconductor is not a perfect insulator, a small amount of reverse current leakage will occur, but the amount is usually too small to consider. The application of voltage to maintain the current flow described is called "forward bias."

A light-emitting diode (LED) is made of a particular type of crystal that glows when current is passed through it. LED's are used in display faces of many digital or electronic instrument clusters. LED's are usually arranged to display numbers (digital readout), but can be used to illuminate a variety of electronic graphic displays.

Like any other electrical device, diodes have certain ratings that must be observed and should not be exceeded. The forward current rating (or bias) indicates how much current can safely pass through the diode without causing damage or destroying it. Forward current rating is usually given in either amperes or milliamperes. The voltage drop across a diode remains constant regardless of the current flowing through it. Small diodes designed to carry low amounts of current need no special provision for dissipating the heat generated in any electrical device, but large current carrying diodes are usually mounted on heat sinks to keep the internal temperature from rising to the point where the silicon will melt and destroy the diode. When diodes are operated in a high ambient temperature environment, they must be de-rated to prevent failure.

Another diode specification is its peak inverse voltage rating. This value is the maximum amount of

voltage the diode can safely handle when operating in the blocking mode. This value can be anywhere from 50–1000 volts, depending on the diode. If voltage amount is exceeded, it will damage the diode just as too much forward current will. Most semiconductor failures are caused by excessive voltage or internal heat.

One can test a diode with a small battery and a lamp with the same voltage rating. With this arrangement one can find a bad diode and determine the polarity of a good one. A diode can fail and cause either a short or open circuit, but in either case it fails to function as a diode. Testing is simply a matter of connecting the test bulb first in one direction and then the other and making sure that current flows in one direction only. If the diode is shorted, the test bulb will remain on no matter how the light is connected.

TRANSISTORS

The transistor is an electrical device used to control voltage within a circuit. A transistor can be considered a "controllable diode" in that, in addition to passing or blocking current, the transistor can control the amount of current passing through it. Simple transistors are composed of 3 pieces of semiconductor material, P and N type, joined together and enclosed in a container. If 2 sections of P material and 1 section of N material are used, it is known as a PNP transistor; if the reverse is true, then it is known as an NPN transistor. The 2 types cannot be interchanged.

Most modern transistors are made from silicon (earlier transistors were made from germanium) and contain 3 elements; the emitter, the collector and the base. In addition to passing or blocking current, the transistor can control the amount of current passing through it and because of

this can function as an amplifier or a switch. The collector and emitter form the main current-carrying circuit of the transistor. The amount of current that flows through the collector-emitter junction is controlled by the amount of current in the base circuit. Only a small amount of base-emitter current is necessary to control a large amount of collector-emitter current (the amplifier effect). In automotive applications, however, the transistor is used primarily as a switch.

When no current flows in the base-emitter junction, the collector-emitter circuit has a high resistance, like to open contacts of a relay. Almost no current flows through the circuit and transistor is considered **OFF**. By bypassing a small amount of current into the base circuit, the resistance is low, allowing current to flow through the circuit and turning the transistor **ON**. This condition is known as "saturation" and is reached when the base current reaches the maximum value designed into the transistor that allows current to flow. Depending on various factors, the transistor can turn on and off (go from cutoff to saturation) in less than one millionth of a second.

Much of what was said about ratings for diodes applies to transistors, since they are constructed of the same materials. When transistors are required to handle relatively high currents, such as in voltage regulators or ignition systems, they are generally mounted on heat sinks in the same manner as diodes. They can be damaged or destroyed in the same manner if their voltage ratings are exceeded. A transistor can be checked for proper operation by measuring the resistance with an ohmmeter between the base-emitter terminals and then between the base-collector terminals. The forward resistance should be small, while the reverse resistance should be large. Compare the readings with those from a known good transistor. As a final check, measure the forward and reverse resistance between the collector and emitter terminals.

INTEGRATED CIRCUITS

The integrated circuit (IC) is an extremely sophisticated solid state device that consists of a silicone wafer (or chip) which has been doped, insulated and etched many times so that it contains an entire electrical circuit with transistors, diodes, conductors and capacitors miniaturized within each tiny chip. Integrated circuits are

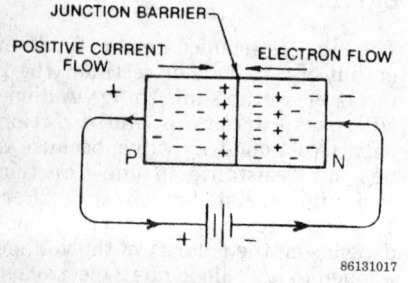

Diode with forward bias

Diode with reverse bias

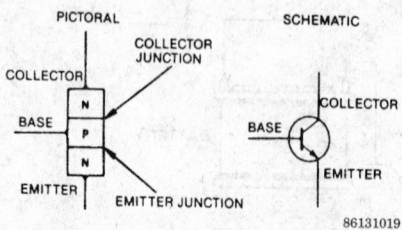

NPN transistor illustration (pictorial and schematic)

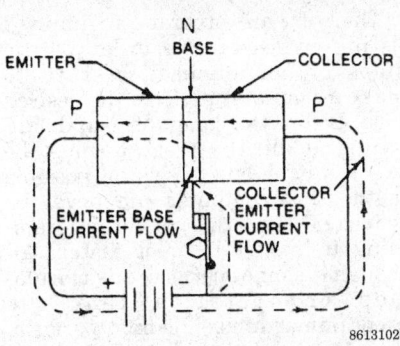

PNP transistor with base switch closed (base emitter and collector emitter current (flow)

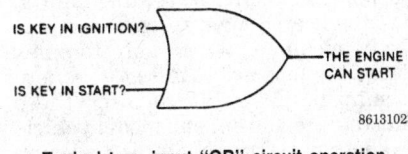

Typical two- input "OR" circuit operation

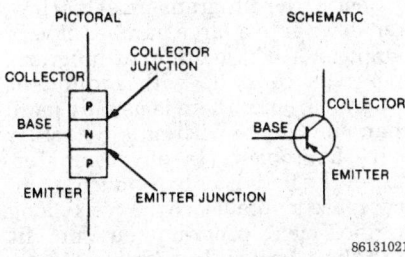

PNP transistor illustrations (pictorial and schematic)

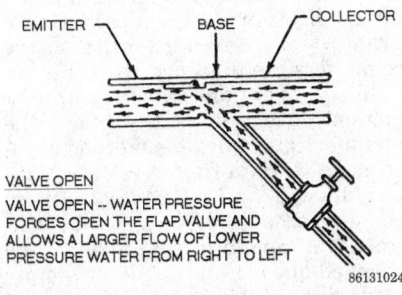

VALVE OPEN -- WATER PRESSURE FORCES OPEN THE FLAP VALVE AND ALLOWS A LARGER FLOW OF LOWER PRESSURE WATER FROM RIGHT TO LEFT

Hydraulic analogy to transistor function is shown with the base circuit energized

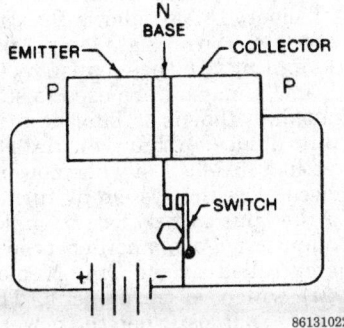

PNP transistor with base switch open (no current)

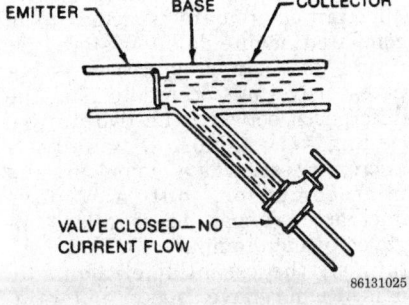

VALVE CLOSED—NO CURRENT FLOW

Hydraulic analogy to transistor function is shown with the base circuit off

often referred to as "computers on a chip" and are largely responsible for the current boom in electronic control technology.

Microprocessors, Computers and Logic Systems

Mechanical or electromechanical control devices lack the precision necessary to meet the requirements of modern control standards. They do not have the ability to respond to a variety of input conditions common to antilock brakes, climate control and electronic suspension operation. To meet these requirements, manufacturers have gone to solid state logic systems and microprocessors to control the basic functions of suspension, brake and temperature control, as well as other systems and accessories.

One of the more vital roles of microprocessor-based systems is their ability to perform logic functions and make decisions. Logic designers use a shorthand notation to indicate whether a voltage is present in a circuit (the number 1) or not present (the number 0). Their systems are designed to respond in different ways depending on the output signal (or the lack of it) from various control devices.

There are 3 basic logic functions or "gates" used to construct a microprocessor control system: the AND gate, the OR gate or the NOT gate. Stated simply, the AND gate works when voltage is present in 2 or more circuits which then energize a third (A and B energize C). The OR gate works when voltage is present at either circuit A or circuit B which then energizes circuit C. The NOT function is performed by a solid state device called an "inverter" which reverses the input from a circuit so that, if voltage is going in, no voltage comes out and vice versa. With these three basic building blocks, a logic designer can create complex systems easily. In actual use, a logic or decision making system may employ many logic gates and receive inputs from a number of sources (sensors), but for the most part, all utilize the basic logic gates discussed above.

Stripped to its bare essentials, a computerized decision-making system is made up of 3 subsystems:

- Input devices (sensors or switches)
- Logic circuits (computer control unit)
- Output devices (actuators or controls)

The input devices are usually nothing more than switches or sensors that provide a voltage signal to the control unit logic circuits that is read as a 1 or 0 (on or off) by the logic circuits. The output devices are anything from a warning light to solenoid-operated valves, motors, linkage, etc. In most cases, the logic circuits themselves lack sufficient output power to operate these devices directly. Instead, they operate some intermediate device such as a relay or power transistor which in turn operates the appropriate device or control. Many problems diagnosed as computer failures are really the result of a malfunctioning intermediate device like a relay. This must be kept in mind whenever troubleshooting any microprocessor-based control system.

As computer capacity is improved by the manufacturers, so does sensor technology. A few years ago, the on-board computer would receive a message from an engine sensor in a "go or no-go" form; for example the coolant

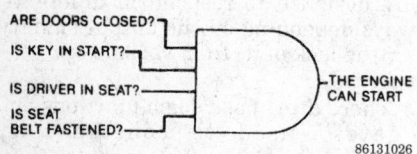

Multiple inputs "AND" operation in a typical automotive starting circuit

86131026

temperature would either be above or below 150°F. Today's systems allow the same sensor to to pass progressively more voltage as the engine warms up. The engine computer now knows exactly what temperature the coolant is at all times. With this information the the computer can react to the changing voltage signal from the sensor instantly and control other engine functions based on engine warm-up or over heating conditions.

The logic systems discussed above are called "hardware" systems, because they consist only of the physical electronic components (gates, resistors, transistors, etc.). Hardware

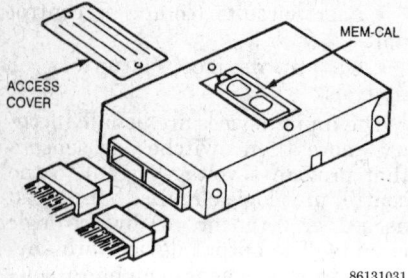

Typical General Motors engine control computer

86131031

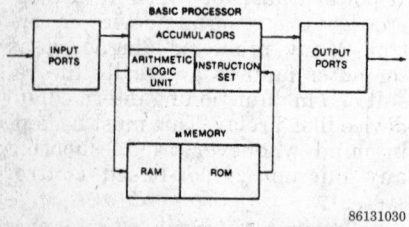

Schematic of typical microprocessor based on-board computer, showing essential common

86131030

systems do not contain a program and are designed to perform specific or "dedicated" functions which cannot readily be changed. For many simple automotive control requirements, such dedicated logic systems are perfectly adequate. When more complex logic functions are required, or where it may be desirable to alter these functions (e.g. from one model vehicle to another) a true computer system is used. A computer can be programmed through its software to perform many different functions and, if that program is stored on a separate integrated circuit chip called a ROM (Read Only Memory), it can be easily changed simply by plugging in a different ROM with the desired program. Most on-board automotive computers are designed with this capability. The on-board computer method of engine control offers the manufacturer a flexible method of responding to data from a variety of input devices and of controlling an equally large variety of output controls. The computer response can be changed quickly and easily by simply modifying its software program.

The microprocessor is the heart of the microcomputer. It is the thinking part of the computer system through which all the data from the various sensors passes. Within the microprocessor, data is acted upon, compared, manipulated or stored for future use. A microprocessor is not necessarily a microcomputer, but the differences between the two are becoming very minor. Originally, a microprocessor was a major part of a microcomputer, but nowadays microprocessors are being called "single-chip microcomputers". They contain all the essential elements to make them behave as a computer, including the most important ingredient–the program.

All computers require a program. In a general purpose computer, the program can be easily changed to allow different tasks to be performed. In a "dedicated" computer, such as most on-board automotive computers, the program isn't quite so easily altered. These automotive computers are designed to perform one or several specific tasks, such as maintaining the passenger compartment temperature at a specific, predetermined level. A program is what makes a computer smart; without a program a computer can do absolutely nothing. The term "software" refers to the computer's program that makes the hardware preform the function needed.

The software program is simply a listing in sequential order of the steps or commands necessary to make a computer perform the desired task. Before the computer can do anything at all, the program must be fed into it by one of several possible methods. A computer can never be "smarter" than the person programming it, but it is a lot faster. Although it cannot perform any calculation or operation that the programmer himself cannot perform, its processing time is measured in millionths of a second.

Because a computer is limited to performing only those operations (instructions) programmed into its memory, the program must be broken down into a large number of very simple steps. Two different programmers can come up with 2 different programs, since there is usually more than one way to perform any task or solve a problem. In any computer, however, there is only so much memory space available, so an overly long or inefficient program may not fit into the memory. In addition to performing arithmetic functions (such as with a trip computer), a computer can also store data, look up data in a table and perform the logic functions previously discussed. A Random Access Memory (RAM) allows the computer to store bits of data temporarily while waiting to be acted upon by the program. It may also be used to store output data that is to be sent to an output device. Whatever data is stored in a RAM is lost when power is removed from the system by turning **OFF** the ignition key, for example.

Computers have another type of memory called a Read Only Memory (ROM) which is permanent. This memory is not lost when the power is removed from the system. Most programs for automotive computers are stored on a ROM memory chip. Data is usually in the form of a look-up table that saves computing time and program steps. For example, a computer designed to control the amount of distributor advance can have this information stored in a table. The information that determines distributor advance (engine rpm, manifold vacuum and temperature) is coded to produce the correct amount of distributor advance over a wide range of engine operating conditions. Instead of the computer computing the required advance, it simply looks it up in a pre-programmed table. However, not all electronic control functions can be handled in this manner; some must be computed. On an antilock

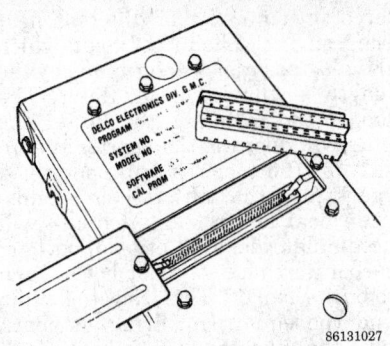

Electronic control module with Mem-Cal, used in General Motors vehicles

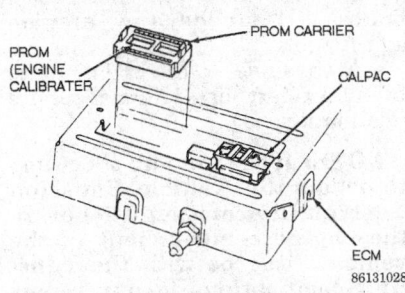

Electronic control module with PROM and CALPAK, used in General Motors vehicles

brake system, for example, the computer must measure the rotation of each separate wheel and then calculate how much brake pressure to apply in order to prevent one wheel from locking up and causing a loss of control.

There are several ways of programming a ROM, but once programmed the ROM cannot be changed. If the ROM is made on the same chip that contains the microprocessor, the whole computer must be altered if a program change is needed. For this reason, a ROM is usually placed on a separate chip. Another type of memory is the Programmable Read Only Memory (PROM) that has the program "burned in" with the appropriate programming machine. Like the ROM, once a PROM has been programmed, it cannot be changed. The advantage of the PROM is that it can be produced in small quantities economically, since it is manufactured with a blank memory. Program changes for various vehicles can be made readily. There is still another type of memory called an EPROM (Erasable PROM) which can be erased and programmed many times. EPROM's are used only in research and development work, not on production vehicles.

General Motors refers to the engine controlling computer as an Electronic Control Module (ECM). The ECM contains the PROM necessary for all engine functions, it also contains a device called a CalPak. This allows the fuel delivery function should other parts of the ECM become damaged. It has an access door in the ECM, like the PROM has. There is a third type control module used in some ECMs called a Mem-Cal. The Mem-Cal contains the function of PROM, CalPak and Electronic Spark Control (EST) module. Like the PROM, it contains the calibrations needed for a specific vehicle, as well as the back-up fuel control circuitry required if the rest of the ECM should become damaged and the spark control. An ECM containing a PROM and CalPak can be identified by the 2 connector harnesses, while the ECM containing the Mem-Cal has 3 connector harnesses attached to it.

Engines coupled to electronically controlled transmissions employ a Powertrain Control Module (PCM) to oversee both engine and transmission operation. This unit may be referred to as the PCM, the ECM/PCM or the PCM/TCM (Transmission Control Module). The integrated functions of engine and transmission control allow accurate gear selection and improved fuel economy.

For engine diagnostics, the PCM may be considered identical to an ECM system, although the combined unit will display additional codes relating to transmission function and components.

NOTE: When the term Powertrain Control Module (PCM) is used in this manual it will refer to the engine control computer regardless that it may be a Powertrain Control Module (PCM) or Electronic Control Module (ECM).

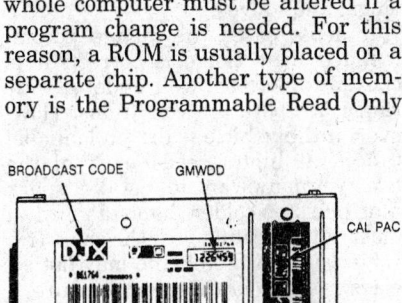

Identification of a General Motors Powertrain Control Module (PCM)

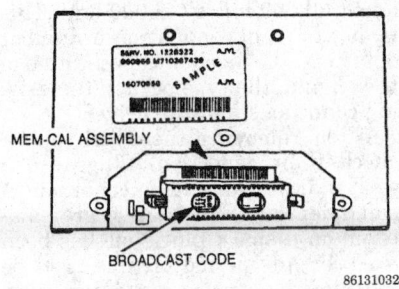

Typical General Motors Mem-Cal identification

BASIC TUNE-UP PROCEDURES

In order to extract the best performance and economy from an engine, it is essential that it be properly tuned at regular intervals. A regular tune-up/inspection will keep the engine running smoothly and will prevent the annoying minor breakdowns and poor performance associated with an untuned engine.

A complete tune-up/inspection should generally be performed every 30,000 miles (48,000 km) or 24 months, whichever comes first. This interval should be halved (as a general rule of thumb) if the vehicle is operated under severe conditions, such as trailer towing, prolonged idling, continual stop and start driving, or if starting or running problems are noticed. It is assumed that the routine maintenance has been kept up, as this will have a decided effect on the results of a tune-up.

Some 1994 and newer vehicles are specified to go up to 100,000 miles between engine tune-ups, you can always refer the owners manual. A tune-up/inspection for all models should consists of the following items

- Inspect the drive belts.
- If necessary, check and adjust valve clearance.
- Clean the air filter housing and replacing the air filter element.
- If equipped, replace the PCV filter and Pulsed secondary air injection filter.
- Inspect or replace the fuel filter assembly.
- Inspect all fuel and vapor lines.
- Check or replace the distributor cap, rotor and ignition wires.

86712038

Because of the tangle of underhood wiring, ALWAYS tag wires before removal

- Replace the spark plugs and make all necessary engine adjustments.
- Always refer to the Maintenance Interval Chart for additional service information.

NOTE: If the tune-up specifications on the Vehicle Emission Control Information sticker in the engine compartment of the vehicle disagree with the Tune-Up Specifications in any repair manual that may be used, the figures on the sticker must be used. The sticker often reflects changes made during the production run.

Spark Plugs

NOTE: The platinum type spark plug is not recommended by all manufacturers. If the vehicle has an aftermarket type platinum plug installed, these plugs are usually marked and are not to be cleaned or regapped.

Spark plugs ignite the air and fuel mixture in the cylinder as the piston reaches the top of the compression stroke. The controlled explosion that results forces the piston down, turning the crankshaft and the rest of the drive train.

The average life of a spark plug is about 30,000 miles (48,000 km). This is, however, dependent on a number of factors: the mechanical condition of the engine, the type of fuel, the driving conditions and the driver.

When removing the spark plugs, check their condition. Plugs are a good indicator of engine condition. A small deposit of light tan or gray material on a spark plug that has been used for any period of time is to be considered normal. Any other color, or abnormal amounts of deposit, indicates that there may be something wrong in the engine.

When a spark plug is functioning normally or, more accurately, when the plug is installed in an engine that is functioning properly, the plugs can be taken out, cleaned, regapped, and reinstalled in the engine without causing the engine any harm.

When, and if, a plug fouls and begins to misfire, you will have to investigate, correct the cause of the fouling, and either clean or replace the plug. There are several reasons why a spark plug will foul and the fault can be learned by just looking at the plug.

There are many spark plugs suitable for use in the engine and are of-fered in a number of different heat ranges. The amount of heat which the plug absorbs is determined by the length of the lower insulator. The longer the insulator the hotter the plug will operate; the shorter the insulator, the cooler it will operate. A spark plug that absorbs (or retains) little heat and remains too cool will accumulate deposits of lead, oil, and carbon, because it is not hot enough to burn them off. This leads to fouling and consequent misfiring. A spark plug that absorbs too much heat will have no deposits, but the electrodes will burn away quickly and, in some cases, pre-ignition may result. Pre-ignition occurs when the spark plug tips get so hot that they ignite the air/fuel mixture before the actual spark fires. This premature ignition will usually cause a pinging sound under conditions of low speed and heavy load. In severe cases, the heat may become high enough to start the air/fuel mixture burning throughout the combustion chamber rather than just to the front of the plug. In this case, the resultant explosion could be strong enough to damage pistons, rings, and valves.

In most cases the factory recommended heat range is correct; it is chosen to perform well under a wide range of operating conditions. However, if the vehicle is driven long distances at high speeds most of the time you may want to install a spark plug one step colder than standard. If most of the driving is of the short trip variety, when the engine may not always reach operating temperature, a hotter plug may help burn off the deposits normally accumulated under those conditions.

REMOVAL

NOTE: Some engines use two spark plugs per cylinder, be sure to replace them all.

1. Disconnect the negative battery cable.

NOTE: Always keep track of the spark plug cable routing and plug wire bracket locations.

2. Number the spark plug wires so that they won't get crossed when they are reconnected.

3. Remove the wire from the end of the spark plug by grasping the rubber boot. If the boot sticks to the plug, remove it by twisting and pulling at the same time. DO NOT pull wire itself or damage to the core will occur.

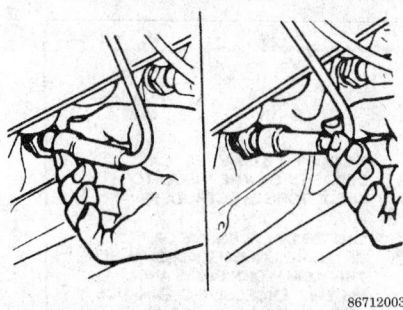

86712003

Twist and pull on the rubber boot to remove the spark plug wires; NEVER pull on the wire itself

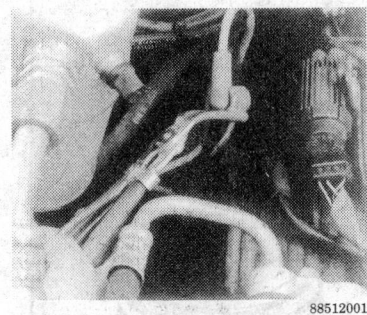

88512001

Using this special tool to remove the spark plug wire makes the job easier; NEVER pull on the wire itself

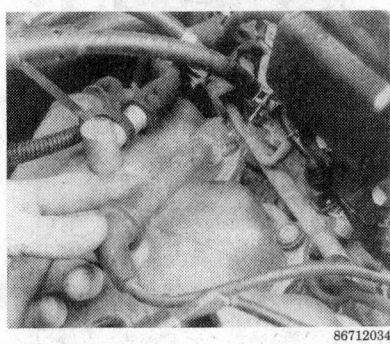

86712034

Mark and remove the spark plug wires one at a time to avoid a mix-up

4. Use a spark plug socket to loosen all of the plugs about two turns.

NOTE: Remove the spark plugs when the engine is cold, if possible, to prevent damage to the threads. If removal of the plugs is difficult, apply a few drops of penetrating oil or silicone spray to the area around the base of the plug, and allow it a few minutes to work.

5. If compressed air is available, apply it to the area around the spark plug holes. Otherwise, use a rag or a

86712035

Carefully unthread the spark plug from the cylinder head using the proper tools

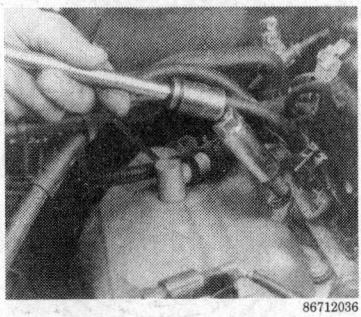

86712036

A U-joint can make removal easier — but use caution with spark plugs; a shear force is applied if not supported

brush to clean the area. Be careful not to allow any foreign material to drop into the spark plug holes.

6. Remove the plugs by unscrewing them the rest of the way from the engine.

INSPECTION

Check the plugs for deposits and wear. If they are not going to be replaced, clean the plugs thoroughly. Remember that any kind of deposit will decrease the efficiency of the plug. Plugs can be cleaned on a spark plug cleaning machine, which can sometimes be found in service stations, or an acceptable job of cleaning can be done with a stiff brush. If the plugs are cleaned, the electrodes must be filed flat. Use an ignition points file, not an emery board or the like, which will leave deposits. The electrodes must be filed perfectly flat with sharp edges; rounded edges reduce the spark plug voltage by as much as 50%.

Check spark plug gap before installation. The ground electrode (the L-shaped one connected to the body of the plug) must be parallel to the center electrode and the specified size wire gauge (please refer to the Tune-

Up Specifications chart for details) must pass between the electrodes with a slight drag.

NOTE: NEVER adjust the gap on a used platinum type spark plug.

Always check the gap on new plugs as they are not always set correctly at the factory. Do not use a flat feeler gauge when measuring the gap on a used plug, because the reading may be inaccurate. A wire type gapping tool is the best way to check the gap. Wire gapping tools usually have a bending tool attached. Use that to adjust the side electrode until the proper distance is obtained. Absolutely never attempt to bend the center electrode. Also, be careful not to bend the side electrode too far or too often as it may weaken and break off within the engine, requiring removal of the cylinder head to retrieve it.

INSTALLATION

1. Lubricate the threads of the spark plugs with a drop of oil. Install the plugs and tighten them by hand first. Take care not to cross-thread them.

2. Tighten the spark plugs with a plug socket. Do not apply the same amount of force you would use for a bolt; just snug them in. If a torque wrench is available, tighten to specific specifications for the vehicle you are working with.

3. Install the wires on their respective plugs. Make sure the wires are firmly connected. They should be felt to click into place. Check the spark plug cable routing and always make sure the plug wires are in the correct plug wire bracket.

4. Connect the negative battery cable.

Spark Plug Wires

CHECKING & REPLACEMENT

At every tune-up/inspection, visually inspect the spark plug cables for burns cuts, or breaks in the insulation. Check the boots and the nipples on the distributor cap and coil. Replace any damaged wiring.

Every 50,000 miles (80,000 Km) or 60 months, the resistance of the wires should be checked with an ohmmeter. Wires with excessive resistance will cause misfiring, and may make the engine difficult to start in damp weather.

GAP BRIDGED

IDENTIFIED BY DEPOSIT BUILD-UP CLOSING GAP BETWEEN ELECTRODES.

CAUSED BY OIL OR CARBON FOULING. REPLACE PLUG, OR, IF DEPOSITS ARE NOT EXCESSIVE THE PLUG CAN BE CLEANED.

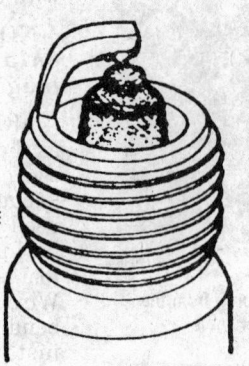

OIL FOULED

IDENTIFIED BY WET BLACK DEPOSITS ON THE INSULATOR SHELL BORE ELECTRODES.

CAUSED BY EXCESSIVE OIL ENTERING COMBUSTION CHAMBER THROUGH WORN RINGS AND PISTONS, EXCESSIVE CLEARANCE BETWEEN VALVE GUIDES AND STEMS, OR WORN OR LOOSE BEARINGS. CORRECT OIL PROBLEM. REPLACE THE PLUG.

CARBON FOULED

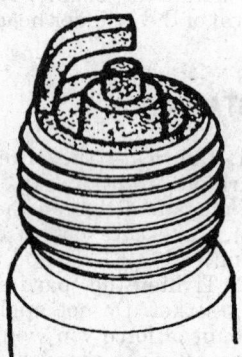

IDENTIFIED BY BLACK, DRY FLUFFY CARBON DEPOSITS ON INSULATOR TIPS, EXPOSED SHELL SURFACES AND ELECTRODES.

CAUSED BY TOO COLD A PLUG, WEAK IGNITION, DIRTY AIR CLEANER, DEFECTIVE FUEL PUMP, TOO RICH A FUEL MIXTURE, IMPROPERLY OPERATING HEAT RISER OR EXCESSIVE IDLING. CAN BE CLEANED.

NORMAL

IDENTIFIED BY LIGHT TAN OR GRAY DEPOSITS ON THE FIRING TIP.

PRE-IGNITION

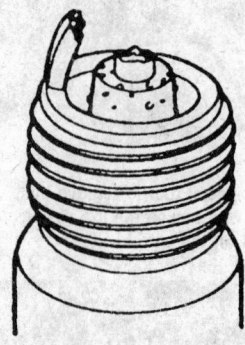

IDENTIFIED BY MELTED ELECTRODES AND POSSIBLY BLISTERED INSULATOR. METALIC DEPOSITS ON INSULATOR INDICATE ENGINE DAMAGE.

CAUSED BY WRONG TYPE OF FUEL, INCORRECT IGNITION TIMING OR ADVANCE, TOO HOT A PLUG, BURNT VALVES OR ENGINE OVERHEATING. REPLACE THE PLUG.

OVERHEATING

IDENTIFIED BY A WHITE OR LIGHT GRAY INSULATOR WITH SMALL BLACK OR GRAY BROWN SPOTS AND WITH BLUISH-BURNT APPEARANCE OF ELECTRODES.

CAUSED BY ENGINE OVER-HEATING, WRONG TYPE OF FUEL, LOOSE SPARK PLUGS, TOO HOT A PLUG, LOW FUEL PUMP PRESSURE OR INCORRECT IGNITION TIMING. REPLACE THE PLUG.

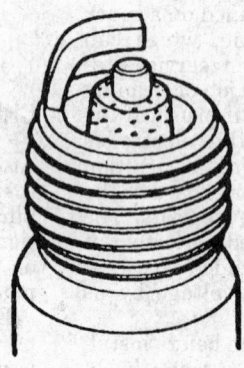

FUSED SPOT DEPOSIT

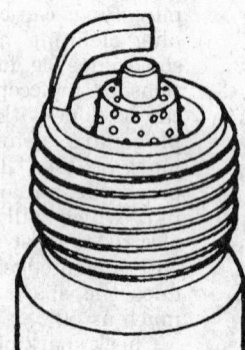

IDENTIFIED BY MELTED OR SPOTTY DEPOSITS RESEMBLING BUBBLES OR BLISTERS.

CAUSED BY SUDDEN ACCELERATION. CAN BE CLEANED IF NOT EXCESSIVE, OTHERWISE REPLACE PLUG.

TCCS2002

Inspect the spark plug to determine engine running conditions

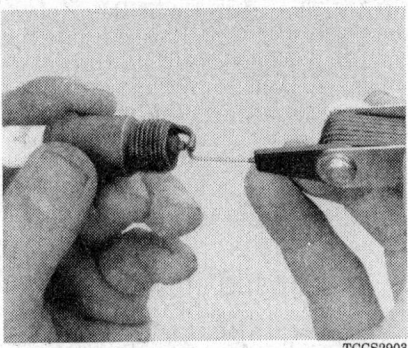

TCCS2903

Check the spark plugs with a wire feeler gauge

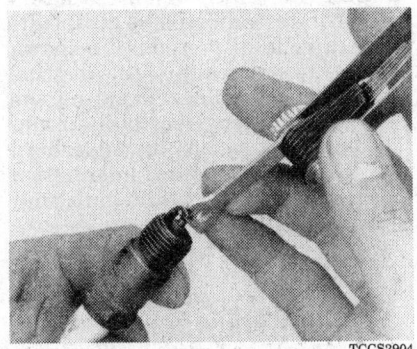

TCCS2904

Bend the side electrode to adjust the gap

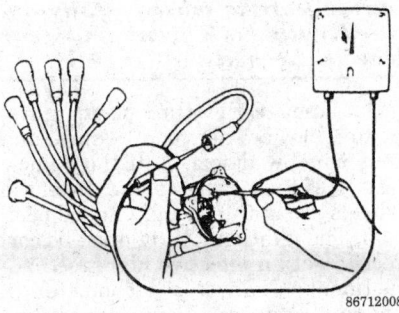

86712008

Checking plug wire resistance through the distributor cap with an ohmmeter

To check resistance, remove the distributor cap, leaving the wires attached. Connect one lead of an ohmmeter to an electrode within the cap; connect the other lead to the corresponding spark plug terminal (remove it from the plug for this test). Replace any wire which shows a resistance over 30,000 ohms. Test the high tension lead from the coil by connecting the ohmmeter between the center contact in the distributor cap and either of the primary terminals of the coil. If resistance is more than 25,000 ohms, remove the cable from the coil and check the resistance

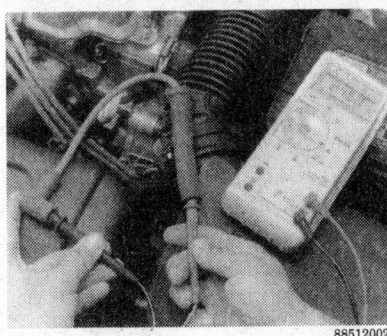

88512002

Checking individual plug wire resistance with an digital ohmmeter

of the cable alone. Anything over 15,000 ohms is cause for replacement. It should be remembered that resistance is also a function of length; the longer the cable, the greater the resistance. Thus, if the cables on the vehicle are longer than the factory originals, resistance will be higher, and quite possibly outside these limits.

When installing new cables, replace them one at a time to avoid mix-ups. Start by replacing the longest one first. Install the boot firmly over the spark plug. Route the wire over the same path as the original. Insert the nipple firmly into the tower on the cap or the coil. Check the spark plug cable routing and always make sure the plug wires are in the correct plug wire bracket.

Ignition Timing

Ignition timing is the measurement in degrees of crankshaft rotation of the instant the spark plug fires, in relation to the location of the piston (while the piston is on its compression stroke).

Although no periodic service is necessary, ignition timing can be adjusted by loosening the distributor locking device and turning the distributor in the engine.

Ideally, the air/fuel mixture in the cylinder will be ignited (by the spark plug) and just begin its rapid expansion as the piston passes top dead center (TDC) of the compression stroke. If this happens, the piston will be beginning the power stroke just as the compressed (by the movement of the piston) air/fuel mixture starts to expand. The expansion of the air/fuel mixture will then force the piston down on the power stroke and turn the crankshaft.

It takes a fraction of a second for the spark from the plug to completely ignite the mixture in the cylinder.

Because of this, the spark plug must fire before the piston reaches TDC, if the mixture is to be completely ignited as the piston passes TDC. This measurement is given in degrees (of crankshaft rotation) Before the piston reaches Top Dead Center (BTDC). For example: if the ignition timing setting for an engine is seven (7°) BTDC, this means that the spark plug must fire at a time when the piston for that cylinder is 7° before top dead center of its compression stroke. However, this only holds true while the engine is at idle speed.

As the engine accelerates from idle, the speed of the engine, in revolutions per minute (rpm), increases. The increase in rpm means that the pistons are now traveling up and down much faster. Because of this, the spark plugs will have to fire even sooner if the mixture is to be completely ignited as the piston passes TDC. To accomplish this, the ECU unit incorporates means to advance the timing of the spark as engine speed increases.

If ignition timing is set too far advanced (too far BTDC), the ignition and expansion of the air/fuel mixture in the cylinder will try to force the piston down the cylinder while it is still traveling upward. This causes engine "ping", a sound which resembles marbles being dropped into an empty tin can. If the ignition timing is too far retarded (after, or ATDC), the piston will have already started down on the power stroke when the air/fuel mixture ignites and expands. This will cause the piston to be forced down only a portion of its travel, resulting in poor engine performance and lack of power.

Ignition timing adjustment is checked with a timing light. This instrument is connected to the Number One (No. 1) spark plug of the engine. The timing light flashes every time an electrical current is sent from the distributor, through the No. 1 spark plug wire, to the spark plug. The crankshaft pulley and the front cover of the engine are marked with a timing pointer and a timing scale. When the timing pointer is aligned with the **0** mark on the timing scale, the piston for the No. 1 cylinder is at TDC of its compression stroke. With the engine running, and the timing light aimed at the timing pointer/scale, the flashes from the timing light will allow you to check the ignition timing. The timing light flashes every time the spark plug in the No. 1 cylinder of the engine fires. Since the flash from the timing light makes the crank-

Checking and adjusting the ignition timing to specifications

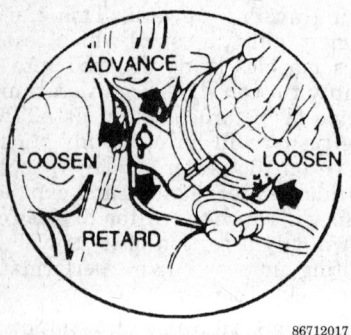

Adjust the ignition timing by rotating the distributor

Timing marks on a 6 cylinder engine — note the fan was removed for a better view

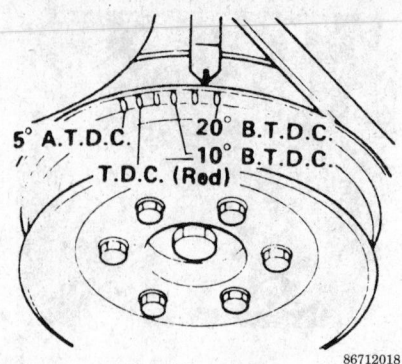

Typical timing marks

shaft pulley seem stationary for a moment you will be able to read the exact position of the piston in the No. 1 cylinder on the timing scale.

There are three basic types of timing lights available. The first is a simple neon bulb with two wire connections (one for the spark plug and one for the plug wire, connecting the light in series). This type of light is quite dim, and must be held closely to the marks to be seen, but it is inexpensive. The second type of light operates from the battery. Two alligator clips connect to the battery terminals, while a third wire connects to the spark plug with an adapter. This type of light is more expensive, but the xenon bulb provides a nice bright flash which can even be seen in sunlight. The third type replaces the battery source with 110 volt house current. Some timing lights have other functions built into them, such as dwell meters, tachometers, or remote starting switches. These are convenient, in that they reduce the tangle of wires under the hood, but may duplicate the functions of tools you already have.

For most vehicles, it is best to use a timing light with an inductive pickup. This pickup simply clamps onto the No. 1 plug wire, eliminating the adapter. It is not susceptible to crossfiring or false triggering, which may occur with a conventional light, due to the greater voltages produced by electronic ignition.

INSPECTION AND ADJUSTMENT

Idle Mixture Adjustment

Most vehicles today use a rather complex electronic fuel injection system which is regulated by a series of temperature, altitude (for California) and air flow sensors which feed information into an Electronic Control Unit

(ECU). The control unit then relays an electronic signal to the injector nozzle(s), which allow(s) a predetermined amount of fuel into the combustion chamber. In this way all mixture control adjustments are regulated by the ECU, therefore on these vehicles no manual adjustments are necessary or possible.

Idle Speed Adjustment

Because of ECU control used on many vehicles today, no periodic service adjustments are necessary. If however the vehicle you are working with requires adjustment or an idle check, a general procedure is shown below. Always refer to the instructions or specifications found on the Vehicle Emission Control Information (VECI) label found underhood for additional or updated information which is applicable to the particular vehicle.

CAUTION

For manual transmission models, set parking brake and check idle speed in N position. For automatic transmission equipped models, shifted into D for idle speed checks. When in Drive, the parking brake must be fully applied with both front and rear wheels chocked.

---

1. Turn **OFF** the: headlights, heater blower, air conditioning, and rear window defogger. If the vehicle has power steering, make sure the wheels are in the straight ahead position. The ignition timing must be correct to get an effective idle speed adjustment. Connect a tachometer (a special adapter harness may be needed) according to the instrument manufacturer's directions.

2. Start the engine and warm the engine so it reaches normal operating temperature. The water temperature indicator should be in the middle of the gauge.

CAUTION

NEVER run the engine in a closed garage. Always make sure there is proper ventilation to prevent carbon monoxide poisoning.

---

3. Run engine at 2000 rpm for about 2 minutes under no load.

4. Race the engine to 2000–3000 rpm a few times under no load and then allow it to return to idle speed.

5. Apply the parking brake securely. If equipped with an automatic, put the transmission into **D**.

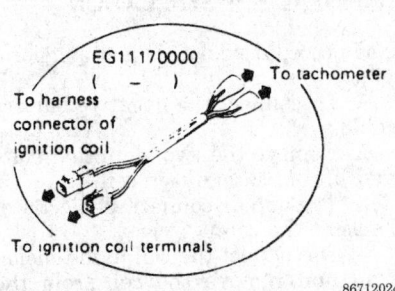

EG11170000

To harness connector of ignition coil

To tachometer

To ignition coil terminals

86712024

Some vehicles require a special harness for the tachometer connection to check idle speed

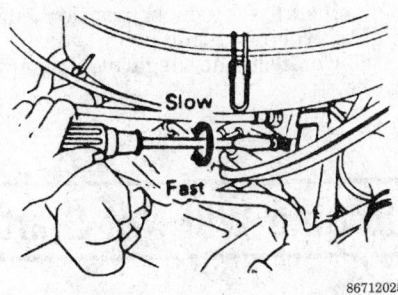

Slow

Fast

86712025

Idle speed adjustment — Carbureted engine

6. Adjust the idle speed by turning the idle speed adjusting screw.

7. Turn the engine **OFF** and remove the tachometer. Road test for proper operation.

Distributor Cap

CHECKING & REPLACEMENT

Disconnect the negative battery cable. Individually disconnect each ignition wire (one at a time) from the distributor cap and inspect the cap towers for corrosion build-up. Note,

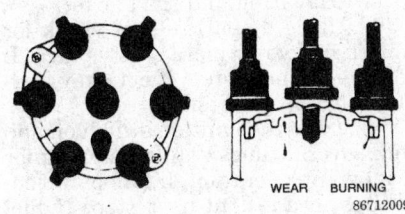

WEAR   BURNING
86712009

Check under the distributor cap for cracks; check the cable ends for wear

do not remove all of the spark plug wires from the cap, do this removing one wire, inspect the tower and wire contact then plug it back in. This will avoid getting the wires mixed up and out of the correct firing order. If all towers and spark plug wire contacts look good, make sure each wire is securely plug in its correct tower. Unfasten the retaining clips or unscrew the caps retaining screws and lift the cap off with the wires still attached. Inspect the underside of the distributor cap for cracks or carbon streaking between the contacts. Inspect the contacts for corrosion or wear. Replace the cap if any of the signs exists.

Some times water or condensation under the distributor cap will cause electrical current to short out between the contacts of the distributor cap or even wet ignition wires. If you have ever gone through a deep puddle of water and the engine stalls, chances are the ignition wires have become wet or the under the distributor cap. In this case it is possible to get the engine started by using a dry cloth and thoroughly drying the under side of the cap and the wires as best as possible.

Distributor

REMOVAL

The procedure given below is a generalization and may not apply to the vehicle you are working with.

1. Disconnect the negative battery cable.

2. Unfasten the retaining clips (only remove the coil wire if necessary) and lift the distributor cap straight off. It will be easier to install the distributor if the spark plug wiring is left connected to the cap. If the plug wires must be removed from the cap, mark their positions to aid in installation.

3. Remove the dust cover and mark the position of the rotor relative to the distributor body; then mark the position of the distributor body relative to the engine block.

4. Detach the harness assembly connector.

5. Remove the pinch-bolt and lift the distributor straight out, away from the engine. The rotor and body are marked so that they can be returned to the position from which they were removed. Do not turn or disturb the engine (unless absolutely necessary) after the distributor assembly has been removed.

INSTALLATION

Timing Not Disturbed

1. Insert the distributor in the block and align ALL matchmarks made during removal.

2. Engage the distributor driven gear with the distributor drive.

3. Install the distributor clamp and secure it with the pinch-bolt.

4. Install the distributor cap and fasten the harness electrical connector.

5. If necessary, install the spark plug wires and coil wire.

6. Start the engine. Check the timing and adjust it if necessary.

Timing Disturbed

This procedure gives a basic and simple way to install the distributor correctly if the engine was cranked while the distributor was out of the engine. Another reason this procedure may be helpful is if when the distributor was removed from the engine it was not marked for installation position as mentioned in the above procedure.

1. It is necessary to place the No. 1 cylinder in the firing position to correctly install the distributor. To locate this position, the ignition timing marks on the crankshaft front pulley can be used.

2. Remove the No. 1 cylinder spark plug. Turn the crankshaft until the piston in the No. 1 cylinder is moving up on the compression stroke. This can be determined by placing a thumb over the spark plug hole and feeling the air being forced out of the cylinder. Stop turning the crankshaft when the timing marks indicate **TDC** or **0**.

3. Oil the distributor housing lightly where the distributor bears on the cylinder block.

4. Install the distributor so that the rotor, which is mounted on the shaft, points toward the No. 1 spark plug terminal tower position when the cap is installed. Of course, you won't be able to see the direction in which the rotor is pointing if the cap is installed, so lay the cap on the top of the distributor and make a mark on the side of the distributor housing just below the No. 1 spark plug terminal. Make sure that the rotor points toward that mark when the distributor is installed.

NOTE: Some engines may have an alignment mark on the distributor shaft which should be aligned with the protruding mark on the distributor housing.

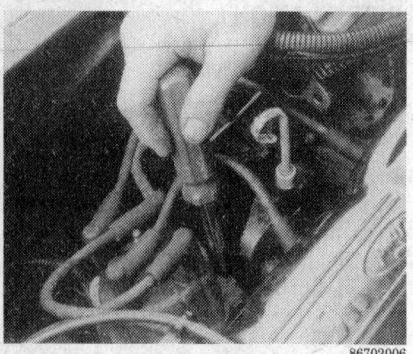

Unscrew the distributor cap retaining screws

Place the distributor cap and wires aside

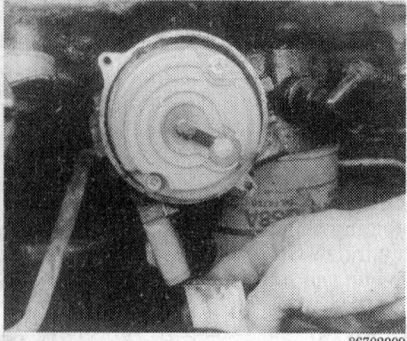

Unplug the distributor harness connector

Paint alignment marks on both the rotor cap and engine block

Remove the distributor retaining bolt

Carefully remove the distributor

5. When the distributor shaft has reached the bottom of the hole, gently move the rotor back and forth slightly until the driving lug on the end of the shaft enters the slots cut in the end of the oil pump shaft and the distributor assembly slides down into place.

6. Fasten the distributor hold-down bolt.

7. Install the spark plug into the No. 1 spark plug hole.

8. Install the distributor cap and engage the harness electrical connector.

9. If necessary, attach the spark plug wires and coil wire.

10. Start the engine. Check the timing and adjust it if necessary.

There are many variations depending on which vehicle you are working with.

Ignition Coil

Ignition coils may be externally mounted or internally located within the distributor. The procedure given below is a basic procedure for an externally mounted ignition coil.

REMOVAL & INSTALLATION

The procedure below is just a general procedure.

1. Disconnect the negative battery cable.

2. Remove the two mounting bolts and lift off the ignition coil.

3. Tag and disconnect all electrical leads at the coil.

4. Disconnect the coil high tension lead and remove the coil from the engine.

To install:

5. Connect all electrical leads to the coil as tagged.

6. Plug the high tension coil wire into the coil tower.

7. Install the coil in position and tighten the mounting bolts.

8. Connect the negative battery cable.

Engine Will Not Start

The procedure below is just a general procedure.

No Start Testing

1. Connect a voltmeter across the battery terminals. If battery voltage is not at least 12 volts, charge and test the battery before proceeding.

2. Turn the key to the **START** position and observe the voltmeter. If the engine turned over and battery voltage remained above 9.6 volts, go to next step. If the engine failed to crank and/or voltage was below 9.6 volts, proceed as follows:

a. If the instrument panel lights dim, load test the battery, check the battery terminals and cables, test the starter motor and verify the engine turns.

b. If the instrument panel lights do not dim, check the battery terminal connections, the ignition switch/wiring and the starter.

3. Using a spark tester, check for spark at two or more spark plugs. If okay go to next step, if not okay, perform No Spark Testing.

4. Cycle the ignition switch on and off, several times, while listening for fuel pump operation. If fuel pump operates, proceed to next step. If fuel pump does not operate begin testing of the fuel pump circuit.

5. Verify adequate fuel in the tank, then connect a fuel pressure gauge and check fuel pressure. If fuel

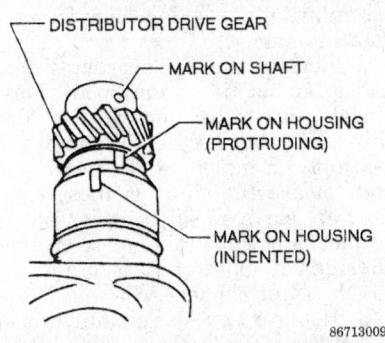

- DISTRIBUTOR DRIVE GEAR
- MARK ON SHAFT
- MARK ON HOUSING (PROTRUDING)
- MARK ON HOUSING (INDENTED)

86713009

Align the mark on the housing with the mark on the shaft

86713217

Remove the coil assembly mounting bolts

pressure is within specifications, proceed to next step, if not okay continue on checking the fuel pump and supply system.

6. Disconnect the fuel injector connector and connect a noid light to the wiring harness. Crank the engine, while watching the light. Perform this test on at least two injectors before proceeding. If the light does not flash, go to the next step. If the light flashes, check the engine valve timing and overall mechanical condition of the engine. If okay, items such as; poor fuel quality, faulty injectors and computer controlled devices

should be checked. Although these items are less likely, a shorted TPS or faulty coolant temperature sensor, are possibilities.

7. Check and verify the Malfunction Indicator Lamp (MIL) is operating properly. If the light does not operate, check the ECM and related wiring. If the MIL lamp is operational, check the injector wiring and circuitry.

No Spark Testing

1. Check for spark at two or more spark plugs. If spark does not exist go

to next step, if spark is okay, check spark plugs, fuel system and engine mechanical condition.

2. Check for spark from the ignition coil wire. If spark does not exist go to next step, if spark is okay, check distributor cap, rotor and ignition wires.

3. Check the ignition coil wire with an ohmmeter. Resistance should not exceed 1000 ohms per inch of cable. If wire resistance exceeds specification, replace the wire and retest. If wire is okay, proceed to the next step.

4. Connect a test light to the negative side of the ignition coil. Turn the key to the **ON** position and observe the test light. If the light remains brightly lit, proceed to the next step. If the light did not light, or glowed dim, check the ignition switch and power supply circuit.

5. Observe the light while cranking the engine. If the light was flashing during cranking, check the ignition coil. If the light did not flash, verify the distributor rotates smoothly, then test the ignition module, pick-up coil or hall effect switch.

RELIEVING FUEL PRESSURE

The procedure is a generic procedure for most fuel injected vehicles. Carbureted vehicles may not use an electric fuel pump. Carbureted systems are lower pressure.

1. Disable the fuel pump by one of the following methods:
- Remove the fuel pump fuse.
- Remove the fuel pump relay.
- Locate and disconnect the fuel pump wiring.

NOTE: When removing the fuel pump fuse or relay to disable the fuel pump, it is important to make certain that the fuel injectors are not part of this circuit. If the injectors do not operate the residual fuel system pressure will not be relieved.

2. Start the engine and operate it until it stalls. Once the engine has stalled, crank the starter for an additional 10 seconds.

3. Place a rag over the connection that is to be disconnected and carefully separate the connections. Use the rag to absorb any remaining fuel.

- CAP
- CARBON POINT
- ROTOR HEAD
- METAL SUPPORT
- HARNESS ASSEMBLY
- O-RING

86713008

Exploded view of the distributor assembly — typical 6 cylinder engine

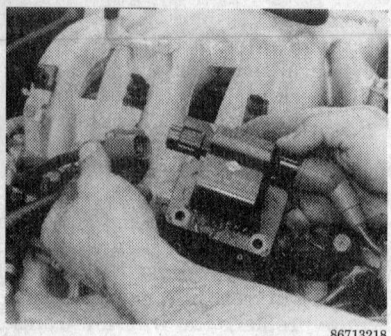

Disconnect the electrical connection from the coil assembly

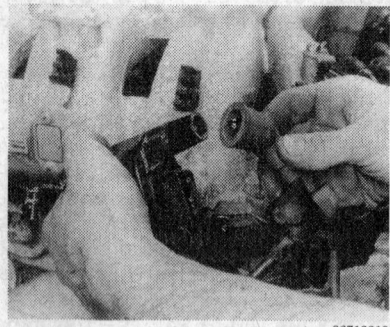

Remove the coil ignition wire from the coil assembly

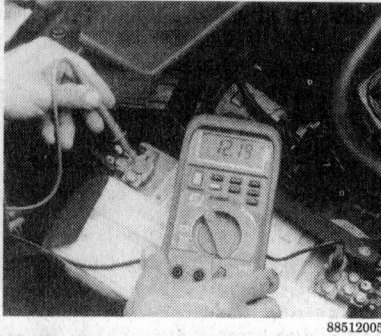

Battery voltage should remain over 9.6 volts, while the engine is cranking

PREVENTIVE MAINTENANCE

Air Cleaner

The air cleaner element should be replaced at the recommended maintenance intervals. If the vehicle is oper-

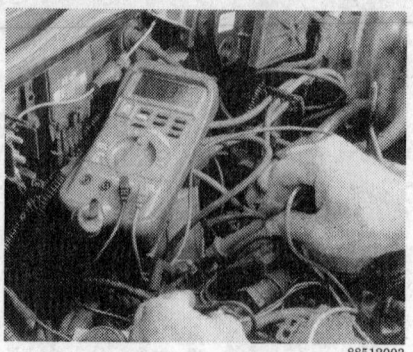

Testing the ignition coil resistance

ated under severely dusty conditions or severe operating conditions, more frequent changes will certainly be necessary. Inspect the element at least twice a year. Early spring and early fall are good times for an inspection. Remove the element and check for any perforations or tears in the filter. Check the cleaner housing for signs of dirt or dust that may have leaked through the filter element or in through the snorkel tube. Position a droplight on one side of the element and look through the filter at the light. If no glow of light can be seen through the element material, replace the filter. If holes in the filter element are apparent, or signs of dirt seepage through the filter are evident, replace the filter.

REMOVAL & INSTALLATION

Air cleaners come a wide selection of shapes and sizes. Most common are either round or rectangular. In any event, air filter element replacement is usually pretty simple.

If the vehicle is equipped with a round type air cleaner, it probably has one or two wing nuts and/or some clips holding the air cleaner lid in place. Just remove the wing nuts and unclip the retaining clips. Lift the lid off and remove the element.

If the vehicle s equipped with a rectangular type air cleaner, unclip the retaining clips and lift off the air cleaner lid. remove the element from the assembly. Some these units may not be as easily accessible and may require removing a hose or two, but all in all it's pretty simple to do.

Air Cleaner Assembly (Housing)

1. Disconnect all hoses, ducts and vacuum tubes from the air cleaner

assembly, after tagging them for easy identification.

2. Remove the top cover wing nuts and grommet (if so equipped). Most models also utilize four to five side clips to further secure the top of the assembly. Simply pull the wire tab and release the clip. On most vehicles, air cleaners are secured solely by means of clips (air box-to-cleaner housing). Remove the cover and lift out the filter element.

3. Remove any side mount brackets and/or retaining bolts, then lift off the air cleaner assembly.

4. Clean or replace the filter element as detailed previously. Wipe clean all surfaces of the air cleaner housing and cover. Check the condition of the mounting gasket and replace if it appears worn or broken.

5. Reposition the air cleaner assembly, then install the mounting bracket and/or bolts.

6. Reposition the filter element in the case and install the cover being careful not to overtighten the wingnut(s). On round-style cleaners, be certain that the arrows on the cover lid and the snorkel match up properly.

NOTE: Filter elements on many engines have a TOP and BOTTOM side; be sure they are inserted correctly.

7. Reconnect all hoses, ductwork and vacuum lines.

NOTE: Never operate the engine without the air filter element in place.

Air Cleaner Element

The air cleaner element can be replaced by removing the wingnut(s) and/or side clips, then removing the top cover as previously detailed.

Crankcase Ventilation Filter

Certain models may also utilize an air cleaner-mounted crankcase ventilation filter. If so, it should also be cleaned or replaced at the same time as the regular air filter element. To replace the filter, remove the air cleaner top cover and pull the filter from its housing on the side of the air cleaner assembly. Push a new filter into the housing and reinstall the cover. If the filter and plastic holder need replacement, remove the clip mounting the feeder tube to the air cleaner housing, then remove the assembly from the air cleaner.

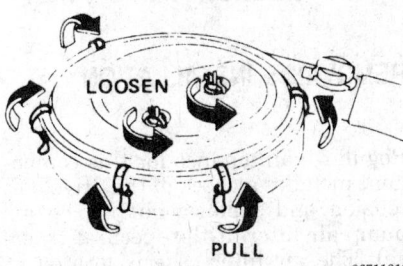

Removing the air cleaner filter element — round type

The air filter element may be cleaned with low pressure compressed air

Removing the air cleaner filter element — rectangular

Loosen the air intake hose clamp before removing the filter element

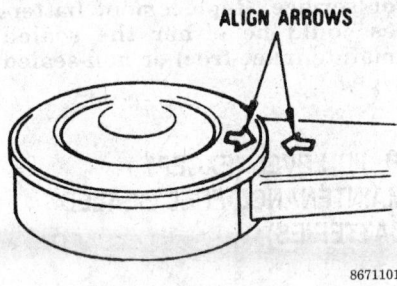

Many air cleaner assemblies have arrows on the housing and lid — always make sure they align

Fuel Filter

REMOVAL & INSTALLATION

——— CAUTION ———
NEVER SMOKE WHEN WORKING AROUND OR NEAR GASOLINE! MAKE SURE THAT THERE IS NO ACTIVE IGNITION SOURCE NEAR THE WORK AREA!

Disconnect the air intake hose, being careful not to lose the retaining clamp

——— WARNING ———
Never attempt to remove the fuel filter without first relieving the fuel system pressure!

The procedure below is just a general procedure, which may differ slightly depending upon what kind of vehicle you are working on.

1. Release the fuel pressure from the fuel line as follows:

   a. Remove the fuel pump fuse at the fuse box.

   b. Start the engine.

   c. After the engine stalls, crank the engine two or three times to make sure that the fuel pressure is released.

   d. Turn the ignition switch **OFF** and reinstall the fuel pump fuse.

2. Loosen the hose clamps at the fuel inlet and outlet lines. Wrap a shop towel or absorbent rag around the filter, then slide each line off the filter nipples.

3. Remove the fuel filter and old hose clamps.

To install:

4. Place new hose clamps on the fuel inlet and outlet lines.

5. Connect the fuel filter, being careful to observe the correct direction of flow, then tighten the hose clamps.

6. Start the engine and check for fuel leaks.

NOTE: Always use a high pressure-type fuel filter assembly. Do not use a synthetic resinous fuel filter.

PCV Valve

The PCV valve regulates crankcase ventilation during various engine operating conditions. At high vacuum (idle speed and partial load range) it will open slightly, and at low vacuum (full throttle) it will open fully. This causes vapor to be removed from the crankcase by the engine vacuum and then be sucked into the combustion chamber where it is burned.

NOTE: The PCV system will not function properly unless the oil filler cap is tightly sealed. Check the gasket on the cap and be certain it is not leaking. Replace the cap and/or gasket, if necessary, to ensure proper sealing.

TESTING

1. Check the ventilation hoses and lines for leaks or clogging. Clean or replace as necessary.

2. With the engine running at idle, locate the PCV valve in the cylinder head cover or intake manifold and remove the ventilation hose from the valve; a strong hissing sound should be heard as air passes through the valve.

3. With the engine still idling, place a finger over the valve; a strong vacuum should be felt.

4. If the PCV valve failed either of the preceding two checks (and the ventilation hose is not clogged or bro-

Unfasten the side retaining clamps so that the air filter housing can be opened

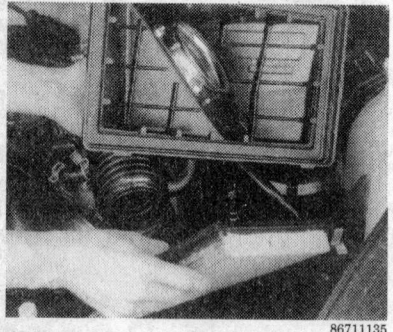

Remove the air filter element from the air filter housing

View of the air filter element. Make sure that the element is installed in the housing properly before fastening the side clamps

ken), the valve will require replacement.

REMOVAL & INSTALLATION

1. If not already done, detach the ventilation hose from the PCV valve.
2. Remove the PCV valve. If its base is threaded, unscrew the valve;

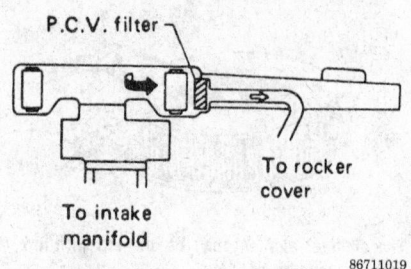

P.C.V. filter

To rocker cover

To intake manifold

Crankcase ventilation filter replacement

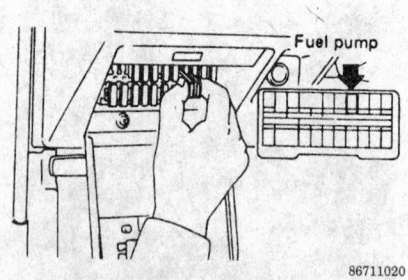

Fuel pump

Remove the fuel pump fuse when releasing the fuel pressure — the fuse's location may vary in the box

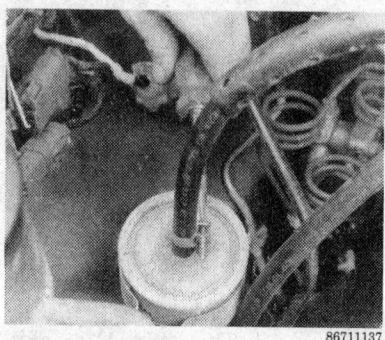

Remove the fuel line hose clamp after releasing the fuel pressure

otherwise, simply pull the valve from its retaining grommet.
   To install:
3. Depending on the type of valve, either screw in the replacement PCV valve or push it into its retaining grommet.
4. Slide the ventilation hose onto the end of the PCV valve.

Air Induction Valve Filter

REMOVAL & INSTALLATION

Regular maintenance for this component includes a check of the drive belt tension and replacement of the air pump air filter at the specified interval. The air filter case is located in the left front of the engine compartment on most models. To replace the air filter, simply unscrew the wing nut(s) securing the cover to the case, withdraw the old filter, install the new one, and reinstall the case.

Battery

NOTE: On a maintenance-free sealed battery, a built-in hydrometer or "eye" is used for checking the fluid level and specific gravity readings. If the battery is equipped with an eye, use it for checking the condition of the battery by observing the color of the eye. A green colored eye indicates good condition and a dark colored eye indicates the need for service. Replacement batteries could be either the sealed (maintenance-free) or non-sealed type.

FLUID LEVEL (EXCEPT MAINTENANCE-FREE SEALED BATTERIES)

Check the battery electrolyte level at least once a month, or more often in hot weather or during periods of extended operation. The level can be checked through the case on translucent polypropylene batteries; the cell caps must be removed on other models. The electrolyte level in each cell should be kept filled to the bottom of the split ring inside, or to the line marked on the outside of the case.

If the level is low, add only distilled water, or colorless, odorless drinking water, through the opening until the level is correct. Each cell is completely separate from the others, so each must be checked and filled individually.

If water is added in freezing weather, the vehicle should be driven several miles to allow the water to mix with the electrolyte. Otherwise, the battery could freeze.

When disconnecting fuel lines, have a rag in position to catch any fuel that may spill

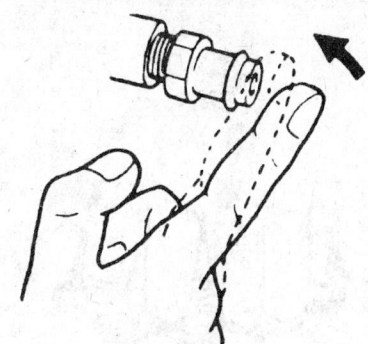

Checking the PCV valve

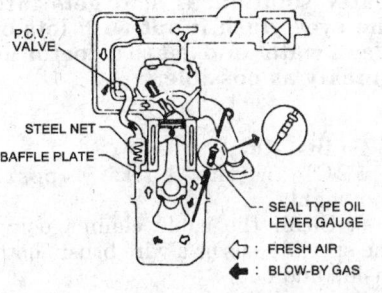

PCV valve location — typical 4 cylinder engine

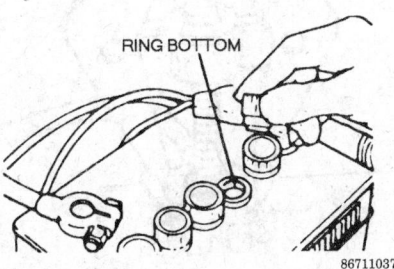

Fill each battery cell to the bottom of the split ring with distilled water

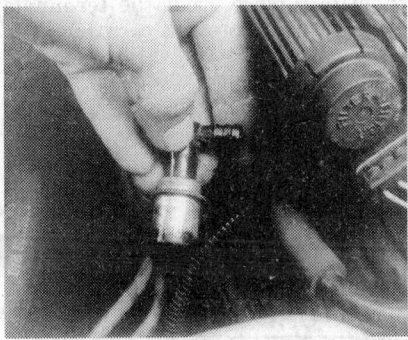

Removing the PCV valve

SPECIFIC GRAVITY (EXCEPT MAINTENANCE-FREE BATTERIES)

NOTE: On a maintenance-free sealed battery, a built-in eye is used for checking the specific gravity readings. Refer to the battery case for further instructions.

At least once a year, check the specific gravity of the battery. It should be 1.26–1.28 at room temperature.

The specific gravity can be checked with the use of a hydrometer, an inexpensive instrument available from many sources, including auto parts stores. The hydrometer has a squeeze bulb at one end and a nozzle at the other. Battery electrolyte is sucked into the hydrometer until the float is lifted from its seat. The specific gravity is then read by noting the position of the float. Generally, if after charging, the specific gravity of any two cells varies more than 50 points (0.50), the battery is bad and should be replaced.

It is not possible to check the specific gravity in this manner on sealed (maintenance-free) batteries. Instead, the indicator built into the top of the case must be relied on to display any signs of battery deterioration. On most batteries if the indicator is a light color, the battery can be assumed to be OK. If the indicator is a dark color, the specific gravity is low, and the battery should be charged or replaced. There should be specific notations on the battery as to what color the indicator will be depending on the batteries state of charge.

CABLES AND CLAMPS

Once a year, the battery terminals and the cable clamps should be checked and cleaned, if necessary. Make sure that the ignition switch is turned to the **OFF** position. Loosen the clamps and remove the cables, negative cable first. On batteries with posts on top, the use of a puller specially made for this purpose is recommended. These are inexpensive, and available in most auto parts stores. Side terminal battery cables are secured with a bolt.

Clean the cable clamps and the battery terminal with a wire brush, until all corrosion, grease, etc., is removed and the metal is shiny. It is especially important to clean the inside of the clamp thoroughly, since a small deposit of foreign material or oxidation there will prevent a sound electrical connection and inhibit either starting or charging. Special tools are available for cleaning these parts, one type for top post batteries and another type for side terminal batteries.

Before installing the cables, loosen the battery hold-down clamp or strap, remove the battery and check the battery tray. Clear it of any debris, and check it for soundness. Rust should be wire brushed away, and the metal given a coat of anti-rust paint. Install the battery and tighten the hold-down clamp or strap securely, but be careful not to overtighten, as doing so may crack the battery case.

After the clamps and terminals are clean, reinstall the cables, negative cable last; do not hammer on the clamps to install. Tighten the clamps securely, but do not distort them. Give the clamps and terminals a thin external coat of grease after installation, to retard corrosion.

Check the cables at the same time that the terminals are cleaned. If the cable insulation is cracked or broken, or if the ends are frayed, the cable should be replaced with a new cable of the same length and gauge.

— CAUTION —
Keep flame or sparks away from the battery; it gives off explosive hydrogen gas! Battery electrolyte contains sulfuric acid! If you should splash any on your skin or in your eyes, flush the affected area with plenty of clear water. If it lands in your eyes, get medical help immediately!

REPLACEMENT

When it becomes necessary to replace the battery, be sure to select a new battery with a cold cranking power

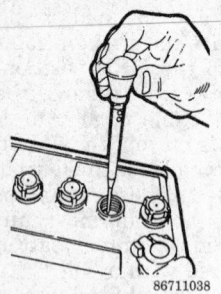

The specific gravity of the battery can be checked with a simple float-type hydrometer

rating equal to or greater than the battery originally installed. Deterioration, embrittlement and just plain aging of the battery cables, starter motor and associated wires makes the battery's job all the more difficult in successive years. The slow increase in electrical resistance over time makes it prudent to install a new battery with a greater capacity than the old.

REMOVAL

1. Make sure the ignition switch is turned **OFF**.
2. Disconnect the negative battery cable from the terminal, then discon-

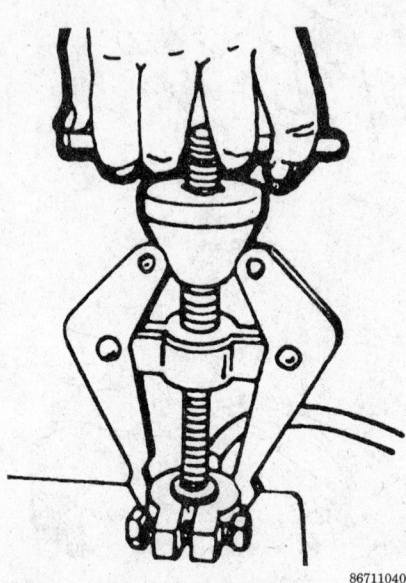

Special pullers are available to remove cable clamps

nect the positive cable. Special pullers are available to remove the clamps.

NOTE: To avoid sparks, always disconnect the negative cable first and reconnect it last.

3. Unscrew and remove the battery hold-down clamp.
4. Remove the battery, being careful not to spill any of the acid.

NOTE: Spilled acid can be neutralized with a baking soda and water solution. If acid gets into the eyes, flush it out with lots of clean water and get to a doctor as quickly as possible.

To install:
5. Clean the battery posts thoroughly.
6. Clean the cable clamps using the special tools or a wire brush, both inside and out.
7. Install the battery, then fasten the hold-down clamp.
8. Connect the positive and then the negative cable. Do not hammer them into place. Coat the terminals with grease to prevent corrosion.

BATTERY STATE OF CHARGE AT ROOM TEMPERATURE

Specific Gravity Reading	Charged Condition
1.260–1.280	Fully Charged
1.230–1.250	3/4 Charged
1.200–1.220	1/2 Charged
1.170–1.190	1/4 Charged
1.140–1.160	Almost no Charge
1.110–1.130	No Charge

Battery state of charge at room temperature — Generalized Specifications

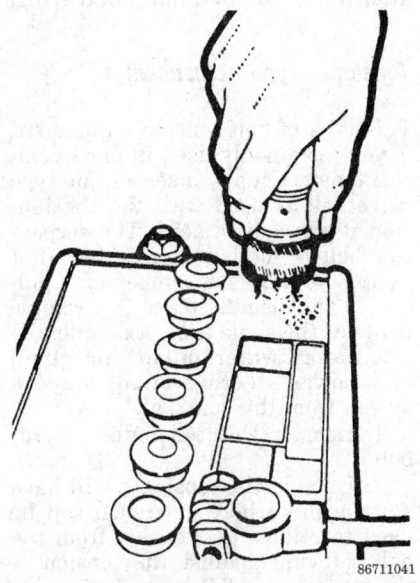

Clean the battery posts with a wire brush or the special tool shown

Clean the inside of the clamps with a wire brush or the special tool

CAUTION

Make absolutely sure that the battery is connected properly before the ignition switch is turned on. Reversed polarity can burn out the alternator and regulator in a matter of seconds.

Drive Belts

INSPECTION

Check the condition of the drive belts, and check the belt tension at least

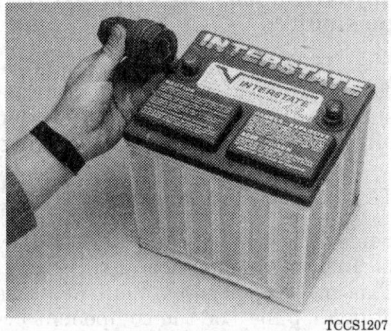

The underside of this special battery tool has a wire brush to clean post terminals

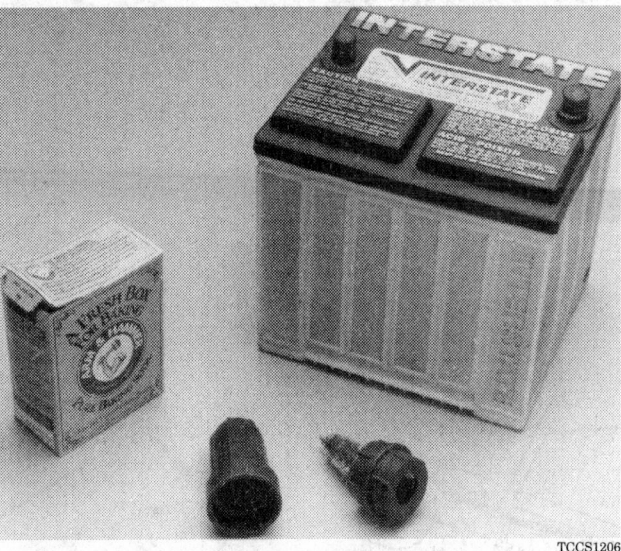

Battery maintenance may be accomplished with household items (such as baking soda to neutralize spilled acid) or with special tools such as this post and terminal cleaner

every 30,000 miles (48,000 km) or every 24 months.

Periodic inspection of the drive belts is important because of the following reasons; first of all, the drive belts drive various components such as the engine water pump, alternator, power steering pump and emission pump, etc.

Two of the components mentioned above play a vital part in keeping the engine running. They are the alternator and water pump. To give an example of how important drive belt inspection is picture this; suppose the alternator belt were to break due to wear or cracking, the alternator would be completely disabled and the battery would eventually go dead.

In case a like this you just may find yourself sitting on the side of the road seeking someone to give the battery a jump to get started. Not to mention, a possible tow job, battery charge and replacement of that drive belt that could have been detected during the inspection.

The water pump drive belt could cause even more severe complications, such as an excessively overheated engine. This could result in a very expensive engine repair, like a head gasket replacement, etc. So be

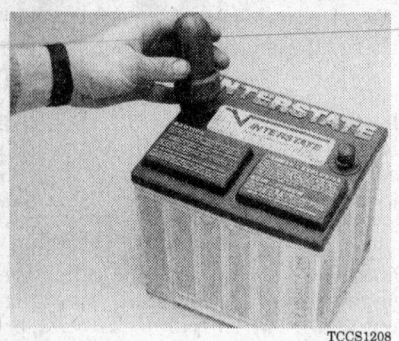

Place the tool over the terminals and twist to clean the post

TCCS1208

sure to keep a good maintenance check on the drive belts.

1. Inspect the belts for signs of glazing or cracking. A glazed belt will be perfectly smooth from slippage, while a good belt will have a slight texture of fabric visible. Cracks will generally start at the inner edge of the belt and run outward. Replace the belt at the first sign of cracking or if the glazing is severe.

2. By placing a thumb midway between the two pulleys, it should be possible to depress the belt ¼–½ in. (6–13mm). If any of the belts can be depressed more than this, or cannot be depressed this much, adjust the tension. Inadequate tension will result in slippage or wear, while excessive tension will damage pulley bearings and cause belts to fray and crack.

3. It's not a bad idea to replace all drive belts at 60,000 miles (96,000 km) or 48 months, regardless of their condition.

ADJUSTMENT

Pivot Type Adjustment

This type of belt tension adjustment is commonly used in most vehicles today.

1. Loosen the pivot and mounting bolts on the alternator.

2. Using a wooden hammer handle or broomstick, move the alternator one way or the other until the tension is within acceptable limits.

NOTE: Never use a screwdriver or any other metal device, such as a pry bar, as a lever when adjusting the alternator belt tension!

3. Tighten the mounting bolts securely. If a new belt has been installed, always recheck the tension after a few hundred miles of driving.

Tension Bolt Type Adjustment

Some belt tensions are adjusted by means of a tension adjusting bolt. This method of adjustment may use an idler pulley or the component being moved may slide on a bracket to increase or decrease belt tension.

1. Loosen the pivot bolt, then turn the adjusting bolt until proper tension is achieved.

2. Tighten the mounting bolts securely. If a new belt has been installed, always recheck the tension after a few hundred miles of driving.

Tensioner Type Adjustment

This type of belt tension adjustment is very commonly used in most vehicles today. Usually a serpentine type drive belt is used with the the tensioner type adjustment. The serpentine belt is one large single belt that wraps around each component's pulley. It will usually drive 3–4 components at the same time. Example: the alternator, water pump, air pump and power steering pump may be driven from this one belt.

1. Loosen the tensioner's pivot bolt.

2. Usually the tensioner will have a large nut where a wrench can be used to relieve the tension from the belt. Moving against the tension of the tensioner will relieve the tension on the the belt. Allowing the tension of the tensioner to release will increase the tension on the belt within acceptable limits.

3. Tighten the pivot bolt securely. If a new belt has been installed, always recheck the tension after a few hundred miles of driving.

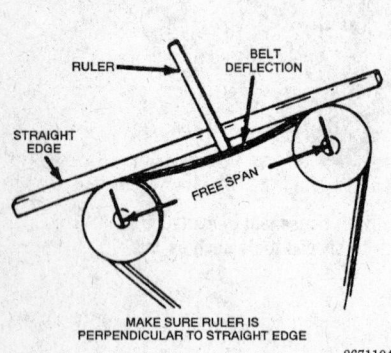

Measuring belt deflection with a straightedge and ruler

86711044

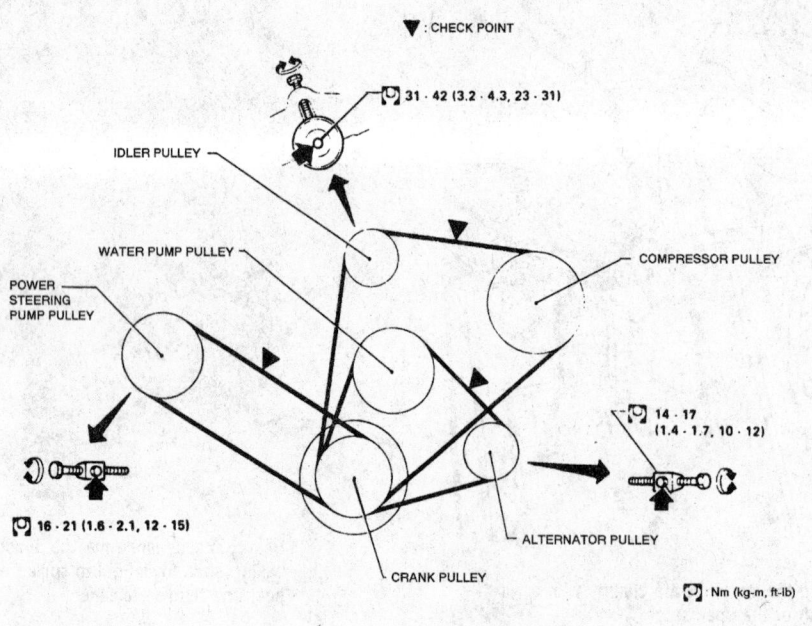

Drive belt tension inspection and adjustment points — typical

86711045

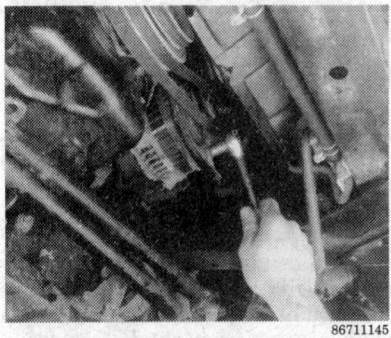

On some vehicles it is easier to access a component from underneath the vehicle

Loosen the alternator pivot bolt with a box wrench or a ratchet and socket

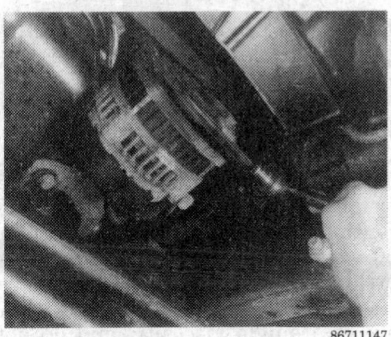

Use the adjusting bolt to vary tension on the belt

Timing Belt

INSPECTION

The timing belt is a bit more involved and a more critical service procedure. Although the timing belt can be serviced, remember that the correct procedures must be followed exactly. Be sure to have the correct repair manual for the vehicle. If you can enlist the aid of someone who is exper-

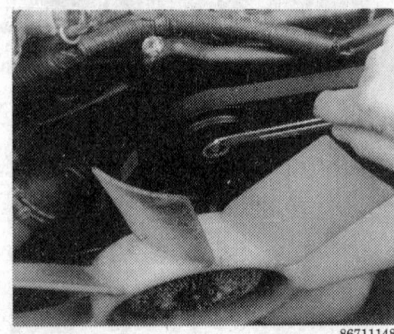

Loosen the locknut on the idler pulley before adjusting the belt

Turn the adjusting bolt until the correct belt tension is achieved

ienced with timing belt replacement, it would be helpful.

NOTE: Do not bend or twist the timing belt. If the timing belt breaks while driving, or the crankshaft and/or camshaft are turned separately after the timing belt is removed, valves may strike the piston heads, causing engine damage. Make sure the timing belt and tensioner are clean and free from oil and water.

As a average rule, replace the timing belt at 60,000 miles (96,000 km).

Evaporative Canister

SERVICING

Check the evaporation control system, if so equipped, every 15,000 miles (24,000 km) or every 12 months. Check the fuel and vapor lines/hoses for proper connections, correct routing, and condition. Replace damaged or deteriorated parts as necessary.

To check the operation of the carbon canister purge control valve, disconnect the rubber hose between the

canister control valve and the T-fitting at the T-fitting. Apply vacuum to the hose leading to the control valve. The vacuum condition should be maintained indefinitely. If the control valve leaks, remove the top cover of the valve and check for a dislocated or cracked diaphragm. If the diaphragm is damaged, a repair kit containing a new diaphragm, retainer, and spring is available and should be installed.

The carbon canister has a replaceable air filter in the bottom of the canister. The filter element should be checked once a year or every 15,000 miles (24,000 km); more frequently if the vehicle is operated in dusty areas. Replace the filter by pulling it out of the bottom of the canister and installing a new one.

Hoses

INSPECTION

Inspect the condition of the radiator hoses, heater hoses and clamps periodically. Early spring and late fall are often good times to perform this, as well as other routine maintenance. Make sure the engine and cooling system are cold. Visually inspect for cracked, rotted or collapsed hoses, and replace as necessary. Run a hand along the length of the hose. If a weak or swollen spot is noted when squeezing the hose wall, replace the hose.

REPLACEMENT

1. Drain the coolant into a suitable container (if the coolant is to be reused).

———— CAUTION ————
When draining the coolant, keep in mind that cats and dogs are attracted by ethylene glycol antifreeze, and are quite likely to drink any that is left in an uncovered container or in puddles on the ground. This will prove fatal in sufficient quantity. Always drain the coolant into a sealable container. Coolant should be reused unless it is contaminated or several years old.

2. Loosen the hose clamps at each end of the hose that requires replacement.

3. Twist, pull and slide the hose off the radiator, water pump, thermostat housing or heater connection.

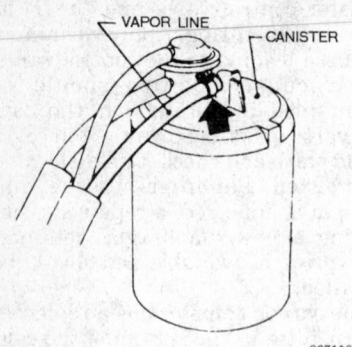

Checking the evaporative canister

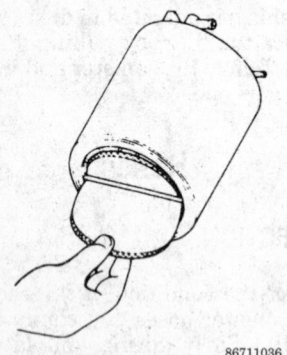

Replacing the evaporative canister filter

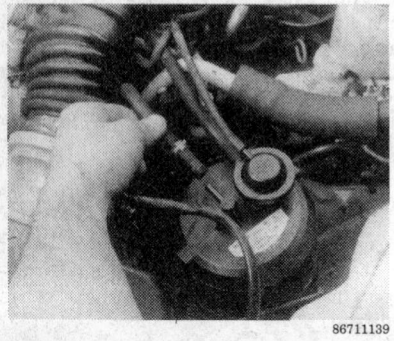

Remove the lines to the evaporative canister assembly before removing the canister

4. Clean the hose mounting connections. Inspect the hose clamps and replace any which are rusted or worn.

5. Position the hose clamps on the new hose.

6. Coat the connection surfaces with a water resistant sealer or equivalent and slide the hose into position. Make sure the hose clamps are located beyond the raised bead of the connector (if equipped) and centered in the clamping area of the connection.

7. Tighten the clamps evenly. Do not overtighten.

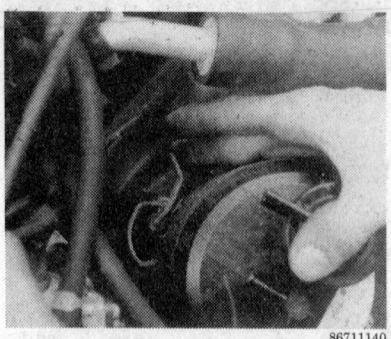

Unfasten the evaporative canister assembly retaining clamp

Remove the evaporative canister assembly from the vehicle

8. Refill the cooling system.

9. Start the engine and allow it to reach normal operating temperature. Check for coolant leaks, then top off the coolant level as necessary.

ADDITIONAL PREVENTIVE MAINTENANCE CHECKS

Antifreeze

In order to prevent heater core freeze-up during A/C operation, it is necessary to maintain permanent-type antifreeze protection of +15°F (-9°C) or lower. A reading of -15°F (-26°C) is ideal since this protection also supplies sufficient corrosion inhibitors for the protection of the engine cooling system.

NOTE: The same antifreeze should not be used longer than the manufacturer specifies.

Radiator Cap

For efficient operation of the vehicle's cooling system, the radiator cap should have a holding pressure which meets manufacturer's specifications. A cap which fails to hold the specified pressure should be replaced.

FLUIDS AND LUBRICANTS

Fluid Disposal

Used fluids such as engine oil, transmission fluid, antifreeze and brake fluid are hazardous wastes and must be disposed of properly. Before draining any fluids, consult with the local authorities; in many areas, waste oil, antifreeze, etc. are being accepted as a part of recycling programs. A number of service stations and auto parts stores are also accepting waste fluids for recycling.

Be sure of the recycling center's policies before draining any fluids, as many will not accept different fluids that have been mixed together, such as oil and antifreeze.

Oil and Fuel Recommendations

ENGINE OIL

The SAE (Society of Automotive Engineers) grade number indicates the viscosity of the engine oil (its resistance to flow at a given temperature). The lower the SAE grade number, the lighter the oil. For example, the mono-grade oils begin with SAE 5 weight, which is a thin, light oil, and continue in viscosity up to SAE 80 or 90 weight, which are heavy gear lubricants. These oils are also known as "straight weight," meaning they are of a single viscosity, and do not vary with engine temperature.

Multi-viscosity oils offer the important advantage of being adaptable to temperature extremes. These oils have designations such as 10W-40, 20W-50, etc. For example, 10W-40 means that in winter (the "W" in the designation) the oil acts like a thin 10 weight oil, allowing the engine to spin easily when cold and offering rapid lubrication. Once the engine has warmed up, however, the oil acts like a straight 40 weight, maintaining good lubrication and protection for the engine's internal components. A 20W-50 oil would therefore be slightly heavier than, and not as ideal, in cold weather as the 10W-40, but would offer better protection at higher rpm and temperatures because, when warm, it acts like a 50 weight oil. Whichever oil viscosity

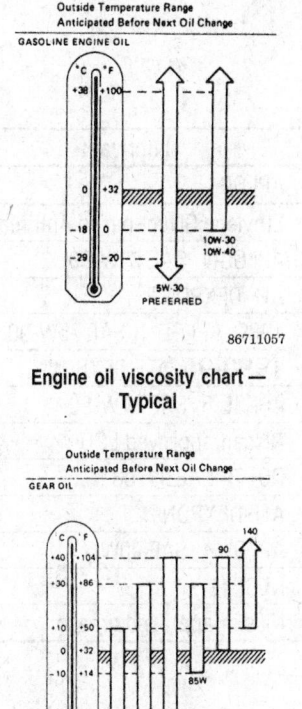

86711057

**Engine oil viscosity chart —
Typical**

86711058

**Gear oil viscosity chart —
Typical**

you choose when changing the oil and filter, you are anticipating the temperatures your engine will be operating in until the oil is changed again. Refer to the oil viscosity chart that applies to your specific vehicle for oil recommendations according to temperature.

The API (American Petroleum Institute) designation indicates the classification of engine oil used under certain given operating conditions. Only oils designated for use "Service SH" or latest superceding grade should be used. Oils of the SH type perform a variety of functions inside the engine in addition to the basic function as a lubricant. Through a balanced system of metallic detergents and polymeric dispersants, the oil prevents the formation of high and low temperature deposits, and also keeps sludge and dirt particles in suspension. Acids, particularly sulfuric acid, as well as other by-products of combustion, are neutralized. Both the SAE grade number and the API

designation can be found on the oil container.

Synthetic Oil

There are many excellent synthetic and fuel-efficient oils currently available that can provide better gas mileage, longer service life and, in some cases, better engine protection. These benefits do not come without a few hitches, however, the main one being the price of synthetic oils, which is three or four times the price per quart of conventional oil.

Synthetic oil is not for every vehicle and every type of driving, so you should consider the engine's condition and type of driving. Also, check the vehicle's warranty conditions regarding the use of synthetic oils.

Brand new engines and older, high mileage engines are not good candidates for synthetic oil. The synthetic oils are so slippery that they can prevent the proper break-in of new engines; most manufacturers recommend that you wait until the engine is properly broken in (3000 miles) before using synthetic oil. Older engines with wear have a different problem with synthetics: they "use" (consume during operation) more oil as they age. Slippery synthetic oils get past these worn parts easily. If the engine is using conventional oil, it will use synthetics much faster. Also, if the vehicle is leaking oil past old seals, there will be a much greater leak problem with synthetics.

Consider your type of driving. If most of your accumulated mileage is high speed, highway type driving, the more expensive synthetic oils may be a benefit. Extended highway driving gives the engine a chance to warm up, accumulating fewer acids in the oil, and putting less stress on the engine over the long run. Under these conditions, the oil change interval can be extended (as long as your oil filter can last the extended life of the oil) up to the advertised mileage claims of the synthetics. Vehicles with synthetic oils may show increased fuel economy in highway driving, due to less internal friction. However, many automotive experts agree that 50,000 miles (80,000 km) is too long to keep any oil in your engine.

Vehicles used under harsher circumstances, such as stop-and-go, city type driving, short trips, or extended idling, should be serviced more fre-

quently. For the engines in these vehicles, the much greater cost of synthetic or fuel-efficient oils may not be worth the investment. Internal wear increases much quicker on these vehicles, causing greater oil consumption and leakage.

NOTE: The mixing of conventional and synthetic oils is possible but not recommended. Nondetergent or straight mineral oils must never be used in the engine.

FUEL

It is important to use fuel of the proper octane rating in the vehicle. Octane rating is based on the quantity of anti-knock compounds added to the fuel, and also reflects the speed at which the gas will burn. The lower the octane rating, the faster it burns. The higher the octane, the slower the fuel will burn, and the greater the percentage of compounds in the fuel to prevent spark ping (knock), detonation and preignition (dieseling).

As the temperature of the engine increases, the air/fuel mixture exhibits a tendency to ignite before the spark plug is fired. If fuel of an octane rating too low for the engine is used, this will allow combustion to occur before the piston has completed its compression stroke, thereby creating a very high pressure very rapidly.

Fuel of the proper octane rating, for the compression ratio and ignition timing of the vehicle, will slow the combustion process sufficiently to allow the spark plug enough time to ignite the mixture completely and smoothly. The use of super-premium fuel is no substitution for a properly tuned and maintained engine.

Light spark knock may be noticed when accelerating or driving up hills. The slight knocking may be considered normal (with 87 octane) because the maximum fuel economy is obtained under condition of occasional light spark knock. Gasoline with an octane rating higher than 87 may be used, but it is not necessary (in most cases) for proper operation.

NOTE: An engine's fuel requirement can change with time, mainly due to carbon buildup, which changes the compression ratio. If the engine pings, knocks or runs on, switch to a higher grade of fuel. Sometimes just changing brands may cure the problem.

RECOMMENDED LUBRICANTS

Component	Lubricant
Engine oil	API SG
Coolant	Ethylene Glycol-based Antifreeze
Manual Transmission	API GL-4, SAE 75W-90
Automatic Transmission	ATF DEXRON®
Transfer Case	1989: API GL-4, SAE 75W-90
	1990-95: ATF DEXRON®
Differentials	API GL-5, SAE 80W-90
Limited Slip	Nissan-approved LSD
Master Cylinder	DOT 3, SAE J1703
Power Steering	ATF DEXRON®
Manual Steering	API GL-4, SAE 90W
Multi-Purpose Grease	NLGI #2
Free-Running Hub	Nissan-approved grease

86711059

Example of a Recommended lubricants chart

OIL LEVEL CHECK

Check the engine oil as follows:

1. Park the vehicle on level ground.

NOTE: Although it is best for the engine to be at operating temperature, checking the oil immediately after stopping will lead to a false reading. Wait a few minutes after turning off the engine to allow the oil to drain back into the crankcase.

2. Open the hood and locate the dipstick. Pull the dipstick from its tube, wipe it clean and reinsert it.

NOTE: Keep in mind that this is a generalized procedure. The actual markings on the vehicle's dipstick may vary from those described here.

3. Pull the dipstick out again, and holding it horizontally, read the oil level. The oil should be between the **H** and **L** marks on the dipstick. If the oil is below the **L** mark, add oil of the proper viscosity and classification through the capped opening on top of the cylinder head cover.

4. Insert the dipstick and check the oil level again after adding any oil. Be careful not to overfill the crankcase. Approximately one quart

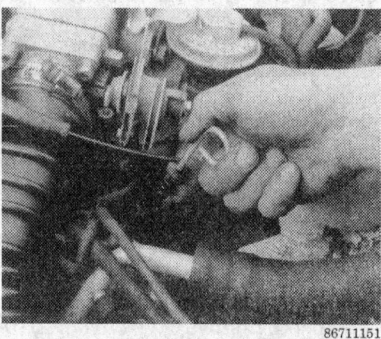

86711151

The oil dipstick in the engine compartment may be painted yellow on newer models

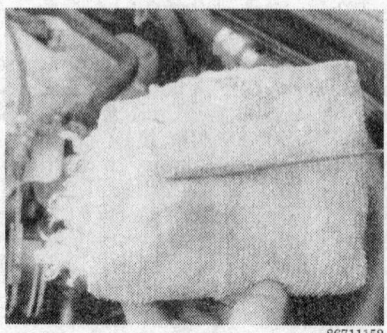

86711152

Check the oil dipstick for the correct level of engine oil — never overfill the engine oil

of oil will raise the level from the **L** mark to the **H** mark. Excess oil will generally be consumed at an accelerated rate.

OIL AND FILTER CHANGE

NOTE: It may be a good idea to look under the vehicle, before starting any service procedure, in order to be familiar with the necessary components and locations.

The oil should be changed at least every 7500 miles (12,000 km) or every 6 months. Some manufacturers recommend changing the oil filter with every other oil change; we suggest that the filter be changed with **every** oil change. There is approximately 1 quart of dirty oil remaining in the old oil filter if it is not changed! A few dollars more every year seems a small price to pay for extended engine life — so change the filter every time the oil is changed!

—— CAUTION ——
Prolonged and repeated skin contact with used engine oil, with no effort to remove the oil, may be harmful. Always follow these simple precautions when handling used motor oil.

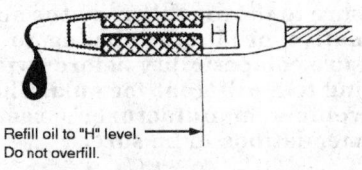

Refill oil to "H" level.
Do not overfill.

86711064

The engine oil level should be maintained between the L and H marks

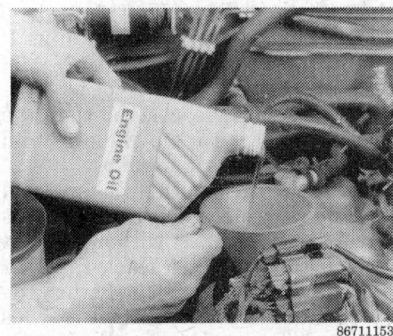

86711153

If the engine oil level is low, add engine oil, but do not overfill

• Avoid prolonged skin contract with used motor oil
• Remove oil from skin by washing thoroughly with soap and water, or waterless hand cleaner. Do not use gasoline, thinners or other solvents
• Avoid prolonged skin contact with oil-soaked clothing

The mileage figures given are sample recommended intervals assuming normal driving and conditions. If the vehicle is being used under dusty, polluted or off-road conditions, change the oil and filter more frequently than specified. The same goes for vehicles driven in stop-and-go traffic or only for short distances. Always drain the oil after the engine has been running long enough to bring it to normal operating temperature. Hot oil will flow easier and more contaminants will be removed along with the oil than if it were drained cold. To change the oil and filter:

CAUTION

The EPA warns that prolonged contact with used engine oil may cause a number of skin disorders, including cancer! Make every effort to minimize exposure to used engine oil. Protective gloves

should be worn when changing the oil. Wash hands and any other exposed skin areas as soon as possible after exposure to used engine oil. Soap and water, or waterless hand cleaner should be used.

1. Run the engine until it reaches normal operating temperature.
2. Jack up the front of the vehicle and support it on safety stands.
3. Slide a drain pan of at least 6 quarts capacity under the oil pan.
4. Loosen the drain plug. Turn the plug out by hand. By keeping inward pressure on the plug while unscrewing it, oil won't escape past the threads, and it is possible to remove it without being burned by hot oil.

CAUTION

The oil will be HOT! Be careful when removing the plug, so that you don't take a bath in hot engine oil.

5. Allow the oil to drain completely. Clean and inspect the drain plug and oil pan sealing surface. If the plug is equipped with a removable gasket, also clean and inspect it.
6. Using a new plug gasket, if necessary, install the drain plug and tighten to correct specifications. Don't overtighten the plug; otherwise, you'll be buying a new pan or a replacement plug for stripped threads.
7. Some engines will require the use of a oil filter strap wrench to remove the oil filter. Others may require a cap-type filter removal tool. Keep in mind that the filter is holding about one quart of dirty, hot oil.

NOTE: If the oil filter cannot be loosened by conventional methods, punch a hole through both sides near the mounting

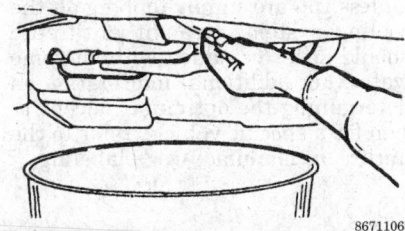

86711065

By keeping inward pressure on the drain plug as you unscrew it, oil won't escape past the threads

base of the filter and insert a punch, then turn to loosen the oil filter. After the oil filter is loosened, remove it from the engine with an oil filter wrench or equivalent.

8. Empty the old filter into the drain pan and properly dispose of the filter.
9. Using a clean rag, wipe off the filter adapter on the engine block. Be sure that the rag doesn't leave any lint which could clog an oil passage.
10. Coat the rubber gasket on the filter with fresh oil. Spin it onto the engine by hand; when the gasket touches the adapter surface, give it another $1/2-3/4$ turn. Do not overtighten, or the gasket will be distorted and it will leak.
11. Refill the engine with the correct amount of fresh oil.
12. Check the oil level on the dipstick. It is normal for the level to be a bit above the full mark. Start the engine and allow it to idle for a few minutes.

NOTE: Do not run the engine above idle speed until it has built up oil pressure, as indicated when the oil light goes out.

13. Shut off the engine and allow the oil to drain into the crankcase for a few minutes, then check the oil level. Check around the filter and drain plug for any leaks and correct as necessary.

Power Steering Pump

Check the power steering fluid level every 6 months or 6000 miles (9600 km).

1. Park the vehicle on a level surface. Run the engine until normal operating temperature is reached.
2. Turn the steering all the way to the left and then all the way to the right several times. Center the steering wheel and shut off the engine.
3. Open the hood and check the power steering reservoir fluid level.
4. Remove the filler cap and wipe the attached dipstick clean.
5. Reinsert the dipstick and tighten the cap. Remove the dipstick and note the fluid level indicated on the dipstick.
6. The level should be at any point below the upper hash mark, but not below the lower hash mark (in the HOT or COLD ranges).
7. Add fluid as necessary, but do not overfill.

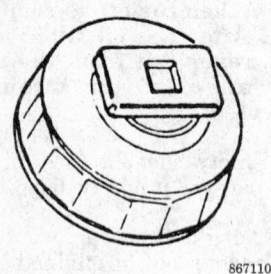

On some models, a cap-type oil filter removal tool works best

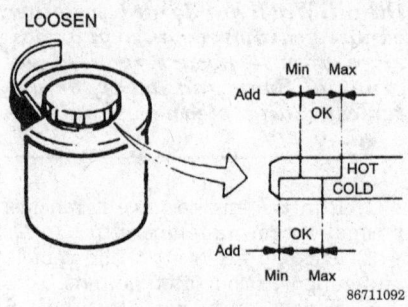

Checking the power steering fluid level

Lubricate the gasket on the new filter with clean engine oil. A dry gasket may not make as good a seal, and could allow the filter to leak

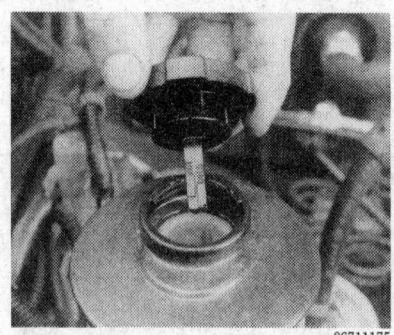

Remove the power steering cap to check the fluid level

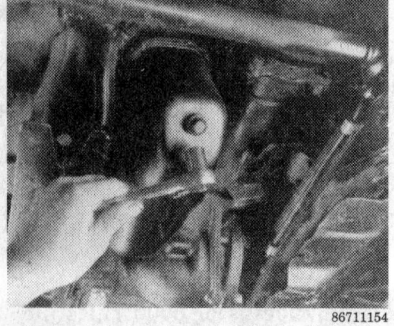

Removing the oil drain plug — do not over torque this drain plug upon installation

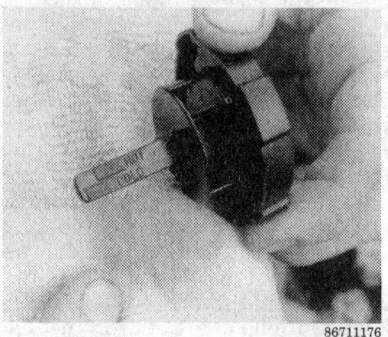

View of the power steering cap dipstick — note the hot and cold marks

Cooling System

FLUID RECOMMENDATIONS

When additional coolant is required to maintain the proper level, always add a mixture of aluminum-compatible antifreeze/coolant and water. Typically, a 50/50 mixture of antifreeze and water is recommended (even for vehicles which are not exposed to cold winter temperatures), since this mixture also imparts the necessary corrosion inhibition. A greater concentration of antifreeze may be used, but the coolant mixture's level of protection actually lessens if too much antifreeze is used. Unless you are simply topping off the cooling system, straight antifreeze should never be added without some water. For additional information on determining the optimum concentration for a specific vehicle, refer to the antifreeze manufacturer's labeling.

NOTE: Although most manufacturers recommend ethylene glycol-based antifreeze (which has long been the prevalent type on the market), other types (such as propylene glycol) may also be suitable for use in the vehicle. Be sure to thoroughly read the alternative product's labeling to ensure compatibility before switching to a different formula. Check vehicle manufacturer's recommendations to be sure.

FLUID LEVEL CHECK

Dealing with the cooling system can be a tricky matter unless the proper precautions are observed. It is best to check the coolant level in the radiator when the engine is cold. This is done by checking the expansion tank. If coolant is visible above the **MIN** mark on the tank, the level is satisfactory. Always be certain that the filler caps on both the radiator and the reservoir are tightly closed.

In the event that the coolant level must be checked when the engine is warm or on engines without an expansion tank, place a thick rag over the radiator cap, then slowly turn the cap counterclockwise until it reaches the first detent. Allow all the hot steam to escape. This will allow the pressure in the system to drop gradually, preventing an explosion of hot coolant. When the hissing noise stops, remove the cap the rest of the way.

It's a good idea to check the coolant every time that you stop for fuel. If the coolant level is low, add equal amounts of suitable antifreeze and clean water. Fill the expansion tank to the **MAX** level. On models without an expansion tank, add coolant through the radiator filler neck.

NOTE: Never add cold coolant to a hot engine unless the engine is running, to avoid cracking the engine block.

Avoid using water that is known to have a high alkaline content or is very hard, except in emergency situations. Drain and flush the cooling system as soon as possible after using such water.

The radiator hoses and clamps and the radiator cap should be checked at the same time as the coolant level. Hoses which are brittle, cracked, or swollen should be replaced. Clamps should be checked for tightness (screwdriver-tight only)! Do not allow the clamp to cut into the hose or crush the fitting. The radiator cap gasket should be checked for any tears, cracks, swelling, or any signs of incorrect seating in the radiator neck.

Adding power steering fluid — use a funnel to avoid spills

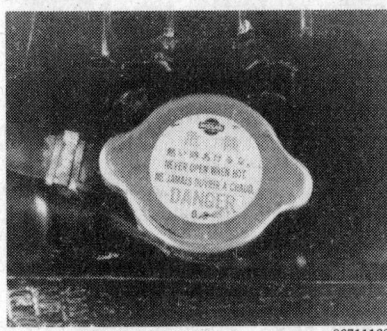

View of the radiator cap installed — never open when the engine is hot!

Add engine coolant to the radiator with a funnel to avoid spills

DRAIN, REFILL AND FLUSH

CAUTION

When draining the coolant, keep in mind that cats and dogs are attracted by ethylene glycol antifreeze, and are quite likely to drink any that is left in an uncovered container or in puddles on the ground. This will prove fatal in sufficient quantity. Always drain the coolant into a sealable container. Coolant should be reused unless it is contaminated or two years old.

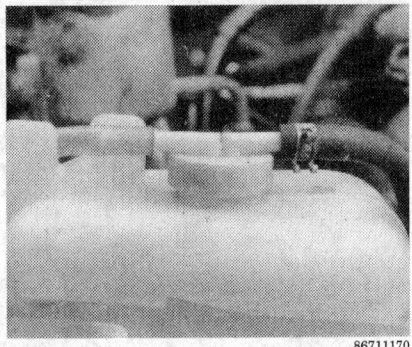

View of the coolant expansion tank

Remove the cap on the coolant expansion tank and add coolant to the proper level

Fluid level marks on the coolant recovery tank

Complete draining and refilling of the cooling system at least once every two years will remove accumulated rust, scale and other deposits.

NOTE: Use a good quality antifreeze with water pump lubricants, rust inhibitors and other corrosion inhibitors along with acid neutralizers.

1. Drain the existing coolant as follows: Position suitable drain pans beneath the radiator and engine block. Open the radiator petcock and engine drain plug(s); there may be 1

or 2 drain plugs on the engine block depending on the type of engine. Another method of draining coolant is to disconnect the bottom radiator hose at the radiator outlet.

NOTE: If it is rusted or difficult to open, spray the radiator petcock with some penetrating lubricant.

2. Set the heater temperature controls to the full HOT position.
3. Close the petcock and tighten the drain plug(s) to correct specifications or reconnect the lower hose. Open the air relief plug, if so equipped, then fill the system with water.
4. Add a can of quality radiator flush. Be sure the flush is safe to use in engines having aluminum components.
5. Idle the engine until the upper radiator hose gets hot. Race it 2 or 3 times, then shut it **OFF**. Let the engine cool down.
6. Drain the system again.
7. Repeat this process until the drained water is clear and free of scale.
8. Close the petcock and drain plug(s) or, if applicable, connect the radiator hose.
9. If equipped with a coolant recovery system, flush the reservoir with water and leave empty.

NOTE: Always open the air relief plug before filling the cooling system, in order to bleed the trapped air. Only when the cooling system is bled properly can the correct amount of coolant be added to the system.

10. Determine the capacity of the cooling system. Add the appropriate ratio of quality aluminum-compatible antifreeze and water (normally a 50/50 mix) to provide the desired protection. With the air relief plug open, add the coolant mixture through the radiator filler neck until full, then close the bleeder plug and radiator cap.
11. Using the same concentration of clean antifreeze and water, fill the expansion tank to the **MAX** line, then cap the tank.

SYSTEM INSPECTION

Most permanent antifreeze/coolants have a colored dye added which makes the solution an excellent leak

Add engine coolant to the expansion tank with a funnel to avoid spills

detector. When servicing the cooling system, check for leakage at:

- All hoses and hose connections
- Radiator seams, radiator core, and radiator draincock
- All engine block and cylinder head freeze (core) plugs, and drain plugs
- Edges of all cooling system gaskets (head gaskets, thermostat gasket)
- Transmission fluid cooler
- Heating system components
- Water pump

In addition, check the engine oil dipstick for signs of coolant in the oil; also, check the coolant in the radiator for signs of oil. Investigate and correct any indication of coolant leakage.

Check the Radiator Cap

While checking the coolant level, check the radiator cap for a worn or cracked gasket. If the cap doesn't seal properly, fluid will be lost and the engine will overheat. A worn cap should be replaced with a new one. The radiator cap must maintain pressure when the engine is running, or the cooling system will "boil over". The radiator cap also has a 2-way valve design to allow coolant to be drawn into the radiator from the coolant overflow tank. If this valve is not functioning properly, the vacuum

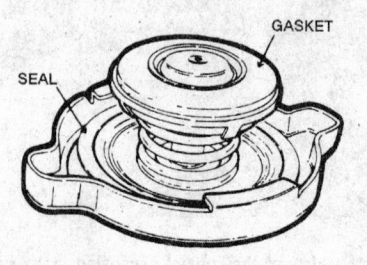

Check the radiator cap seal and gasket condition

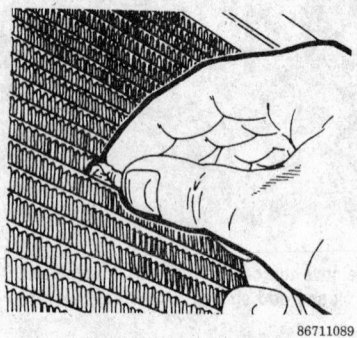

Clean the radiator fins of any debris which impedes air flow

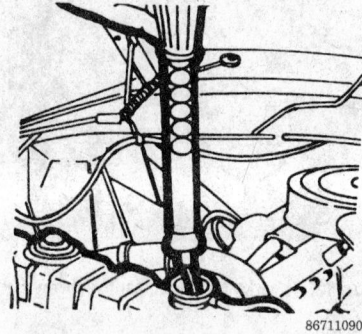

The freeze protection rating can be checked with an antifreeze tester

cause in the system as the engine cools down can collapse and damage the hoses.

Clean Radiator of Debris

Periodically clean any debris such as leaves, paper, insects, etc., from the radiator fins. Pick the large pieces off by hand. The smaller pieces can be washed away with water pressure from a hose.

Carefully straighten any bent radiator fins with a pair of needlenose pliers. Be careful, the fins are very soft. Don't wiggle the fins back and forth too much. Straighten them once and try not to move them again.

CHECKING SYSTEM PROTECTION

A 50/50 mix of antifreeze/coolant concentrate and water will usually provide the necessary protection. Freeze protection may be checked by using a cooling system hydrometer. Inexpensive hydrometers (floating ball types) may be obtained from a local department store (automotive section) or an auto supply store. Follow the directions packaged with the coolant hydrometer when checking protection.

SPECIFICATIONS

CHRYSLER CORPORATION

B Series (Van) • Dakota • DW Series (Pick Up) • Ramcharger

VEHICLE IDENTIFICATION CHART

Engine Code						Model Year	
Code	Liters	Cu. In. (cc)	Cyl.	Fuel Sys.	Eng. Mfg.	Code	Year
5	5.9	360 (5899)	V8	TFI	Chrysler	P	1993
8	5.9	359 (5882)	I6	DSL	Cummins	R	1994
A	5.9	360 (5899)	V8	MFI	Chrysler	S	1995
C	5.9	359 (5882)	I6	DSL	Cummins	T	1996
G	2.5	153 (2507)	I4	TFI	Chrysler	V	1997
P	2.5	153 (2507)	I4	MFI	Chrysler		
K	2.5	153 (2507)	I4	TFI	Chrysler		
W	8.0	488 (7997)	V10	MFI	Chrysler		
X	3.9	238 (3916)	V6	MFI	Chrysler		
Y	5.2	318 (5211)	V8	MFI	Chrysler		
Z	5.9	360 (5899)	V8	TFI	Chrysler		

DSL - Diesel   MFI - Multiport fuel injection   TFI - Throttle body fuel injection

ENGINE IDENTIFICATION

Year	Model	Engine Displacement Liters (cc)	Engine Series (ID/VIN)	Fuel System	No. of Cylinders	Engine Type
1993	Dakota	2.5 (2507)	K	TFI	4	SOHC
	Dakota	3.9 (3916)	X	MFI	6	OHV
	Dakota	5.2 (5211)	Y	MFI	8	OHV
	B150 Van	3.9 (3916)	X	MFI	6	OHV
	B150 Van	5.2 (5211)	Y	MFI	8	OHV
	B250 Van	3.9 (3916)	X	MFI	6	OHV
	B250 Van	5.2 (5211)	Y	MFI	8	OHV
	B250 Van	5.9 (5899)	Z	MFI	8	OHV
	B350 Van	5.2 (5211)	Y	MFI	8	OHV
	B350 Van	5.9 (5899)	Z	MFI	8	OHV
	D/W 150 Pick-up	3.9 (3916)	X	MFI	6	OHV
	D/W 150 Pick-up	5.2 (5211)	Y	MFI	8	OHV
	D/W 150 Pick-up	5.9 (5899)	Z	MFI	8	OHV
	D 250 Pick-up	3.9 (3916)	X	MFI	6	OHV
	D/W 250 Pick-up	5.2 (5211)	Y	MFI	8	OHV
	D/W 250 Pick-up	5.9 (5882)	8	DSL	6	OHV
	D/W 250 Pick-up	5.9 (5899)	Z	MFI	8	OHV
	D/W 350 Pick-up	5.9 (5882)	8	DSL	6	OHV
	D/W 350 Pick-up	5.9 (5899)	5	MFI	8	OHV
	Ramcharger	5.2 (5211)	Y	MFI	8	OHV
	Ramcharger	5.9 (5899)	Z	MFI	8	OHV

ENGINE IDENTIFICATION

Year	Model	Engine Displacement Liters (cc)	Engine Series (ID/VIN)	Fuel System	No. of Cylinders	Engine Type
1994	Dakota	2.5 (2507)	K	TFI	4	SOHC
	Dakota	3.9 (3916)	X	MFI	6	OHV
	Dakota	5.2 (5211)	Y	MFI	8	OHV
	B150 Van	3.9 (3916)	X	MFI	6	OHV
	B150 Van	5.2 (5211)	Y	MFI	8	OHV
	B250 Van	3.9 (3916)	X	MFI	6	OHV
	B250 Van	5.2 (5211)	Y	MFI	8	OHV
	B250 Van	5.9 (5899)	A	MFI	8	OHV
	B350 Van	5.2 (5211)	Y	MFI	8	OHV
	B350 Van	5.9 (5899)	A	MFI	8	OHV
	C 1500 Pick-up	3.9 (3916)	X	MFI	6	OHV
	C/F 1500 Pick-up	5.2 (5211)	Y	MFI	8	OHV
	C/F 1500 Pick-up	5.9 (5899)	Z	MFI	8	OHV
	C/F 2500 Pick-up	5.2 (5211)	Y	MFI	8	OHV
	C/F 2500 Pick-up	5.9 (5882)	C	DSL	6	OHV
	C/F 2500 Pick-up	5.9 (5899)	Z	MFI	8	OHV
	C/F 2500 Pick-up	8.0 (7997)	W	MFI	10	OHV
	C/F 3500 Pick-up	5.9 (5882)	C	DSL	6	OHV
	C/F 3500 Pick-up	5.9 (5899)	5	MFI	8	OHV
	C/F 3500 Pick-up	8.0 (7997)	W	MFI	10	OHV
1995	Dakota	2.5 (2507)	G	TFI	4	SOHC
	Dakota	3.9 (3916)	X	MFI	6	OHV
	Dakota	5.2 (5211)	Y	MFI	8	OHV
	B150 Van	3.9 (3916)	X	MFI	6	OHV
	B150 Van	5.2 (5211)	Y	MFI	8	OHV
	B250 Van	3.9 (3916)	X	MFI	6	OHV
	B250 Van	5.2 (5211)	Y	MFI	8	OHV
	B250 Van	5.9 (5899)	Z	MFI	8	OHV
	B350 Van	5.2 (5211)	Y	MFI	8	OHV
	B350 Van	5.9 (5899)	Z	MFI	8	OHV
	C 1500 Pick-up	3.9 (3916)	X	MFI	6	OHV
	C/F 1500 Pick-up	5.2 (5211)	Y	MFI	8	OHV
	C/F 1500 Pick-up	5.9 (5899)	Z	MFI	8	OHV
	C/F 2500 Pick-up	5.2 (5211)	Y	MFI	8	OHV
	C/F 2500 Pick-up	5.9 (5882)	C	DSL	6	OHV
	C/F 2500 Pick-up	5.9 (5899)	Z	MFI	8	OHV
	C/F 2500 Pick-up	8.0 (7997)	W	MFI	10	OHV
	C/F 3500 Pick-up	5.9 (5882)	C	DSL	6	OHV
	C/F 3500 Pick-up	5.9 (5899)	5	MFI	8	OHV
	C/F 3500 Pick-up	8.0 (7997)	W	MFI	10	OHV
1996-97	Dakota	2.5 (2458)	P	MFI	4	OHV
	Dakota	3.9 (3916)	X	MFI	6	OHV
	Dakota	5.2 (5211)	Y	MFI	8	OHV
	B1500 Van	3.9 (3916)	X	MFI	6	OHV
	B1500 Van	5.2 (5211)	Y	MFI	8	OHV
	B2500 Van	3.9 (3916)	X	MFI	6	OHV
	B2500 Van	5.2 (5211)	Y	MFI	8	OHV
	B2500 Van	5.9 (5899)	Z	MFI	8	OHV
	B3500 Van	5.2 (5211)	Y	MFI	8	OHV
	B3500 Van	5.9 (5899)	Z	MFI	8	OHV

ENGINE IDENTIFICATION

Year	Model	Engine Displacement Liters (cc)	Engine Series (ID/VIN)	Fuel System	No. of Cylinders	Engine Type
1996-97	C/F 1500 Pick-up	3.9 (3916)	X	MFI	6	OHV
	C/F 1500 Pick-up	5.2 (5211)	Y	MFI	8	OHV
	C/F 1500 Pick-up	5.9 (5899)	Z	MFI	8	OHV
	C/F 2500 Pick-up	5.2 (5211)	Y	MFI	8	OHV
	C/F 2500 Pick-up	5.9 (5882)	C	DSL	6	OHV
	C/F 2500 Pick-up	5.9 (5899)	Z	MFI	8	OHV
	C/F 2500 Pick-up	8.0 (7997)	W	MFI	10	OHV
	C/F 3500 Pick-up	5.9 (5882)	C	DSL	6	OHV
	C/F 3500 Pick-up	5.9 (5899)	5	MFI	8	OHV
	C/F 3500 Pick-up	8.0 (7997)	W	MFI	10	OHV

DSL - Diesel
MFI - Multiport fuel injection

TFI - Throttle body fuel injection
OHV - Overhead valve

SOHC - Single overhead camshaft
DOHC - Double overhead camshaft

GENERAL ENGINE SPECIFICATIONS

Year	Engine ID/VIN	Engine Displacement Liters (cc)	Fuel System Type	Net Horsepower @ rpm	Net Torque @ rpm (ft. lbs.)	Bore x Stroke (in.)	Compression Ratio	Oil Pressure @ rpm
1993	K	2.5 (2507)	TFI	100@4800	135@2800	3.45x4.09	8.9:1	35-65@2000
	X	3.9 (3916)	MFI	175@4800	220@3200	3.91x3.31	9.1:1	30-80@3000
	Y	5.2 (5211)	MFI	230@4800	280@3200	3.91x3.31	9.1:1	30-80@3000
	8	5.9 (5882)	DSL	160@2500	400@1750	4.02x4.72	17.5:1	30-70@2500
	Z	5.9 (5899)	MFI	230@4000	325@2500	4.00x3.58	8.9:1	30-80@3000
	5	5.9 (5899)	MFI	230@4000	230@2800	4.00x3.58	8.9:1	30-80@3000
1994	K	2.5 (2507)	TFI	100@4800	135@2800	3.45x4.09	8.9:1	25-80@3000
	X	3.9 (3916)	MFI	175@4800	220@3200	3.91x3.31	9.1:1	30-80@3000
	Y	5.2 (5211)	MFI	220@4400	300@3200	3.91x3.31	9.1:1	30-80@3000
	C	5.9 (5882)	DSL	160@2500 ①	400@1600 ②	4.02x4.72	17.5:1	30@2500
	A	5.9 (5899)	MFI	230@4000	330@3200	4.00x3.58	8.9:1	30-80@3000
	5	5.9 (5899)	MFI	230@4000	330@2800	4.00x3.58	8.9:1	30-80@3000
	Z	5.9 (5899)	MFI	230@4000	330@3200	4.00x3.58	8.9:1	30-80@3000
	W	8.0 (7997)	MFI	300@4000	450@2400	4.00x3.88	8.6:1	50-60@3000
1995	G	2.5 (2507)	TFI	100@4800	135@2800	3.45x4.09	8.9:1	25-80@3000
	Y	5.2 (5211)	MFI	220@4400	300@3200	3.91x3.31	9.1:1	30-80@3000
	C	5.9 (5882)	DSL	160@2500 ①	400@1600 ②	4.02x4.72	17.5:1	30@2500
	5	5.9 (5899)	MFI	230@4000	330@2800	4.00x3.58	8.9:1	30-80@3000
	Z	5.9 (5899)	MFI	230@4000	330@3200	4.00x3.58	8.9:1	30-80@3000
	W	8.0 (7997)	MFI	300@4000	450@2400	4.00x3.88	8.6:1	50-60@3000
1996-97	B	2.4 (2429)	MFI	150@5200	167@4000	3.44x3.98	9.4:1	25-80@3000
	P	2.5 (2458)	MFI	120@5200	145@3400	3.88x3.19	9.2:1	25-80@3000
	X	3.9 (3916)	MFI	175@4800	220@3200	3.91x3.31	9.1:1	30-80@3000
	Y	5.2 (5211)	MFI	220@4400 ③	300@3200 ④	3.91x3.31	9.1:1	30-80@3000
	C	5.9 (5882)	DSL	180@2500 ⑤	420@1600 ⑥	4.02x4.72	17.5:1	30@2500
	5	5.9 (5899)	MFI	230@4000	330@2800	4.00x3.58	8.9:1	30-80@3000
	Z	5.9 (5899)	MFI	230@4000	330@3250	4.00x3.58	9.1:1	30-80@3000
	W	8.0 (7997)	MFI	300@4000	450@2400	4.00x3.58	8.4:1	50-60@3000

MFI - Multiport fuel injection
TFI - Throttle body fuel injection
DSL - Diesel

① Manual transmission: 430@1600
② Manual transmission: 175@2500
③ Compressed natural gas engine: 200@4400

④ Compressed natural gas engine: 250@3600
⑤ Manual transmission: 200@2500
⑥ Manual transmission: 440@1600

GASOLINE ENGINE TUNE-UP SPECIFICATIONS

Year	Engine ID/VIN	Engine Displacement Liters (cc)	Spark Plugs Gap (in.)	Ignition Timing (deg.) MT	Ignition Timing (deg.) AT	Fuel Pump (psi)	Idle Speed (rpm) MT	Idle Speed (rpm) AT	Valve Clearance In.	Valve Clearance Ex.
1993	K	2.5 (2507)	0.035	12B	12B	37-41	850	850	HYD	HYD
	X	3.9 (3916)	0.035	①	①	37-41	750	750	HYD	HYD
	Y	5.2 (5211)	0.035	①	①	37-41	700	700	HYD	HYD
	5	5.9 (5899)	0.035	①	①	35-45	700	700	HYD	HYD
	Z	5.9 (5899)	0.035	①	①	35-45	700	700	HYD	HYD
1994	K	2.5 (2507)	0.035	12B	12B	37-41	②	②	HYD	HYD
	X	3.9 (3916)	0.035	①	①	35-45	②	②	HYD	HYD
	Y	5.2 (5211)	0.035	①	①	35-45	②	②	HYD	HYD
	A	5.9 (5899)	0.035	①	①	35-45	②	②	HYD	HYD
	5	5.9 (5899)	0.035	①	①	35-45	②	②	HYD	HYD
	Z	5.9 (5899)	0.035	①	①	35-45	②	②	HYD	HYD
	W	8.0 (7997)	0.035	①	①	35-45	②	②	HYD	HYD
1995	G	2.5 (2507)	0.035	①	①	14.5	②	②	HYD	HYD
	X	3.9 (3916)	0.030	①	①	35-45	②	②	HYD	HYD
	Y	5.2 (5211)	0.030	①	①	35-45	②	②	HYD	HYD
	5	5.9 (5899)	0.030	①	①	35-45	②	②	HYD	HYD
	Z	5.9 (5899)	0.030	①	①	35-45	②	②	HYD	HYD
	W	8.0 (7997)	0.030	①	①	35-45	②	②	HYD	HYD
1996-97	P	2.5 (2458)	0.035	-	③	49.2	②	②	HYD	HYD
	X	3.9 (3916)	0.035	①	①	49.2	②	②	HYD	HYD
	Y	5.2 (5211)	0.035	①	①	49.2	②	②	HYD	HYD
	5	5.9 (5899)	0.035	①	①	49.2	②	②	HYD	HYD
	Z	5.9 (5899)	0.035	①	①	49.2	②	②	HYD	HYD
	W	8.0 (7997)	0.045	①	①	49.2	②	②	HYD	HYD

NOTE: The Vehicle Emission Control Information label often reflects specification changes made during production. The label figures must be used if they differ from those in this chart.

B - Before top dead center

HYD - Hydraulic

① Ignition timing cannot be adjusted. Base engine timing is set at TDC during assembly

② Refer to the Vehicle Emission Control Information (VECI) label for correct specification

③ Refer to the Vehicle Emission Control Information label for correct timing specification with a range of +/- 2.

DIESEL ENGINE TUNE-UP SPECIFICATIONS

Year	Engine ID/VIN	Engine Displacement cu. in. (cc)	Valve Clearance Intake (in.)	Valve Clearance Exhaust (in.)	Intake Valve Opens (deg.)	Injection Pump Setting (deg.)	Injection Nozzle Pressure (psi) New	Injection Nozzle Pressure (psi) Used	Idle Speed (rpm)	Cranking Compression Pressure (psi)
1993	8	5.9 (5882)	0.010	0.020	NA	①	3550	NA	②	NA
1994	C	5.9 (5882)	0.010	0.020	NA	①	3822	NA	③	NA
1995	C	5.9 (5882)	0.010	0.020	NA	①	3822	NA	③	NA
1996-97	C	5.9 (5882)	0.010	0.020	NA	①	3822	NA	③	NA

NOTE: The Vehicle Emission Control Information label often reflects specification changes made during production. The label figures must be used if they differ from those in this chart

NA - Not Available

① Align marks on pump flange and gear housing

② Automatic transmission with A/C - 700 rpm
Manual transmission with A/C: 750 rpm

③ Automatic transmission with A/C: 750-800 rpm
Manual transmission with A/C: 780 rpm

CAPACITIES

Year	Model	Engine ID/VIN	Engine Displacement Liters (cc)	Engine Oil with Filter (qts.)	Transmission (pts.)			Transfer Case (pts.)	Drive Axle		Fuel Tank (gal.)	Cooling System (qts.)
					4-Spd	5-Spd	Auto.		Front (pts.)	Rear (pts.)		
1993	Dakota	K	2.5 (2507)	4.5	-	6.6	10.4	4.5	2.6	[1]	15.0 [2]	9.8
	Dakota	X	3.9 (3916)	4.5	-	6.6	10.4	4.5	2.6	[1]	15.0 [2]	14.0
	Dakota	Y	5.2 (5211)	5.0	-	6.6	10.4	4.5	2.6	[1]	15.0 [2]	14.3
	B150 Van	X	3.9 (3916)	4.5	-	6.8	[3]	-	-	[4]	22.0 [5]	14.6 [6]
	B150 Van	Y	5.2 (5211)	4.5	-	6.8	[3]	-	-	[4]	22.0 [5]	16.5 [6]
	B250 Van	X	3.9 (3916)	4.5	-	6.8	[3]	-	-	[4]	22.0 [5]	14.6 [6]
	B250 Van	Y	5.2 (5211)	4.5	-	6.8	[3]	-	-	[4]	22.0 [5]	16.5 [6]
	B250 Van	Z	5.9 (5899)	4.5	-	6.8	[3]	-	-	[4]	35.0	15.0 [6]
	B350 Van	Y	5.2 (5211)	4.5	-	-	[3]	-	-	[4]	22.0 [5]	11.5 [6]
	B350 Van	Z	5.9 (5899)	4.5	-	-	[3]	-	-	[4]	35.0	15.0 [6]
	D/W 150 Pick-up	X	3.9 (3916)	4.5	-	8.0	[3]	[7]	[8]	[9]	22.0 [10]	15.1
	D/W 150 Pick-up	Y	5.2 (5211)	4.5	-	8.0	[3]	[7]	[8]	[9]	22.0 [10]	17.0
	D/W 150 Pick-up	Z	5.9 (5899)	4.5	-	8.0	[3]	[7]	[8]	[9]	30.0 [10]	15.5
	D250 Pick-up	X	3.9 (3916)	4.5	-	8.0	[3]	-	-	[9]	22.0 [10]	15.1
	D/W 250 Pick-up	Y	5.2 (5211)	4.5	-	8.0	[3]	[7]	[8]	[9]	22.0 [10]	17.0
	D/W 250 Pick-up	8	5.9 (5882)	12.5	-	7.0	22.0	[7]	6.5	[9]	30.0	[11]
	D/W 250 Pick-up	Z	5.9 (5899)	4.5	-	8.0	[3]	[7]	[8]	[9]	30.0	15.5
	D/W 350 Pick-up	8	5.9 (5882)	12.5	-	7.0	22.0	[7]	6.5	[9]	30.0	[11]
	D/W 350 Pick-up	5	5.9 (5899)	4.5	-	8.0	[3]	[7]	[8]	[9]	30.0	15.5
	Ramcharger	Y	5.2 (5211)	4.5	-	8.0	[3]	[7]	[8]	[9]	34.0	17.0
	Ramcharger	Z	5.9 (5899)	4.5	-	8.0	[3]	[7]	[8]	[9]	34.0	15.5
1994	Dakota	K	2.5 (2507)	4.5	-	[12]	[13]	2.5	3.0	[14]	15.0 [2]	9.8
	Dakota	X	3.9 (3916)	4.5	-	[12]	[13]	2.5	3.0	[14]	15.0 [2]	14.0
	Dakota	Y	5.2 (5211)	5.0	-	[12]	[13]	2.5	3.0	[14]	15.0 [2]	14.3
	B150 Van	X	3.9 (3916)	4.0	-	-	[15]	-	-	[16]	22.0 [5]	14.6
	B150 Van	Y	5.2 (5211)	4.5	-	-	[15]	-	-	[16]	22.0 [5]	16.5
	B250 Van	X	3.9 (3916)	4.5	-	-	[15]	-	-	[16]	22.0 [5]	14.6
	B250 Van	Y	5.2 (5211)	4.5	-	-	[15]	-	-	[16]	22.0 [5]	16.5
	B250 Van	A	5.9 (5899)	4.5	-	-	[15]	-	-	[16]	35.0	15.0 [17]
	B350 Van	Y	5.2 (5211)	4.5	-	-	[15]	-	-	[16]	22.0 [5]	16.5
	B350 Van	A	5.9 (5899)	4.5	-	-	[15]	-	-	[16]	35.0	15.0 [17]
	C 1500 Pick-up	X	3.9 (3916)	4.0	-	[12]	[13]	-	-	[18]	26.0 [5]	20.0
	C/F 1500 Pick-up	Y	5.2 (5211)	5.0	-	[12]	[13]	[19]	[20]	[18]	26.0 [5]	20.0
	C/F 1500 Pick-up	Z	5.9 (5899)	5.0	-	[12]	[13]	[19]	[20]	[18]	26.0 [5]	20.0
	C/F 2500 Pick-up	Y	5.2 (5211)	5.0	-	[12]	[13]	[19]	[20]	[18]	26.0 [5]	20.0
	C/F 2500 Pick-up	C	5.9 (5882)	11.0	-	[12]	[13]	[19]	[20]	[18]	26.0 [5]	26.0
	C/F 2500 Pick-up	Z	5.9 (5899)	5.0	-	[12]	[13]	[19]	[20]	[18]	26.0 [5]	20.0
	C/F 2500 Pick-up	W	8.0 (7997)	7.0	-	[12]	[13]	[19]	[20]	[18]	26.0 [5]	24.0
	C/F 3500 Pick-up	C	5.9 (5882)	11.0	-	[12]	[13]	[19]	[20]	[18]	26.0 [5]	26.0
	C/F 3500 Pick-up	5	5.9 (5899)	5.0	-	[12]	[13]	[19]	[20]	[18]	26.0 [5]	20.0
	C/F 3500 Pick-up	W	8.0 (7997)	7.0	-	[12]	[13]	[19]	[20]	[18]	26.0 [5]	24.0
1995	Dakota	G	2.5 (2507)	4.5	-	[12]	[13]	2.5	3.0	[14]	15.0 [2]	9.8
	Dakota	X	3.9 (3916)	4.5	-	[12]	[13]	2.5	3.0	[14]	15.0 [2]	14.0
	Dakota	Y	5.2 (5211)	5.0	-	[12]	[13]	2.5	3.0	[14]	15.0 [2]	14.3
	B150 Van	X	3.9 (3916)	4.0	-	-	[13]	-	-	[16]	22.0 [5]	14.6
	B150 Van	Y	5.2 (5211)	5.0	-	-	[13]	-	-	[16]	22.0 [5]	16.5
	B250 Van	X	3.9 (3916)	4.0	-	-	[13]	-	-	[16]	22.0 [5]	14.6
	B250 Van	Y	5.2 (5211)	5.0	-	-	[13]	-	-	[16]	22.0 [5]	16.5

CAPACITIES

Year	Model	Engine ID/VIN	Engine Displacement Liters (cc)	Engine Oil with Filter (qts.)	Transmission (pts.)			Transfer Case (pts.)	Drive Axle		Fuel Tank (gal.)	Cooling System (qts.)
					4-Spd	5-Spd	Auto.		Front (pts.)	Rear (pts.)		
1995	B250 Van	Z	5.9 (5899)	5.0	-	-	(13)	-	-	(16)	35.0	15.0 (17)
	B350 Van	Y	5.2 (5211)	5.0	-	-	(13)	-	-	(16)	22.0 (5)	16.5
	B350 Van	Z	5.9 (5899)	5.0	-	-	(13)	-	-	(16)	35.0	15.0 (17)
	C 1500 Pick-up	X	3.9 (3916)	4.0	-	(12)	(13)	-	-	(16)	26.0 (5)	20.0
	C/F 1500 Pick-up	Y	5.2 (5211)	5.0	-	(12)	(13)	(19)	(20)	(16)	26.0 (5)	20.0
	C/F 1500 Pick-up	Z	5.9 (5899)	5.0	-	(12)	(13)	(19)	(20)	(16)	26.0 (5)	20.0
	C/F 2500 Pick-up	Y	5.2 (5211)	5.0	-	(12)	(13)	(19)	(20)	(16)	26.0 (5)	20.0
	C/F 2500 Pick-up	C	5.9 (5882)	11.0	-	(12)	(13)	(19)	(20)	(16)	26.0 (5)	26.0
	C/F 2500 Pick-up	Z	5.9 (5899)	5.0	-	(12)	(13)	(19)	(20)	(16)	26.0 (5)	20.0
	C/F 2500 Pick-up	W	8.0 (7997)	7.0	-	(12)	(13)	(19)	(20)	(16)	26.0 (5)	24.0
	C/F 3500 Pick-up	C	5.9 (5882)	11.0	-	(12)	(13)	(19)	(20)	(16)	26.0 (5)	26.0
	C/F 3500 Pick-up	5	5.9 (5899)	5.0	-	(12)	(13)	(19)	(20)	(16)	26.0 (5)	20.0
	C/F 3500 Pick-up	W	8.0 (7997)	7.0	-	(12)	(13)	(19)	(20)	(16)	26.0 (5)	24.0
1996-97	Dakota	P	2.5 (2458)	4.5	-	(12)	(13)	2.5	3.0	(14)	15.0 (12)	9.8
	Dakota	X	3.9 (3916)	4.5	-	(12)	(13)	2.5	3.0	(14)	15.0 (12)	14.0
	Dakota	Y	5.2 (5211)	5.0	-	(12)	(13)	2.5	3.0	(14)	15.0 (12)	14.3
	C/F 1500 Pick-up	X	3.9 (3916)	4.0	-	(12)	(13)	-	-	(18)	26.0 (5)	20.0
	C/F 1500 Pick-up	Y	5.2 (5211)	5.0	-	(12)	(13)	(21)	(20)	(18)	26.0 (5)	20.0
	C/F 1500 Pick-up	Z	5.9 (5899)	5.0	-	(12)	(13)	(21)	(20)	(18)	26.0 (5)	20.0
	C/F 2500 Pick-up	Y	5.2 (5211)	5.0	-	(12)	(13)	(21)	(20)	(18)	26.0 (5)	20.0
	C/F 2500 Pick-up	C	5.9 (5882)	11.0	-	(12)	(13)	(21)	(20)	(18)	26.0 (5)	26.0
	C/F 2500 Pick-up	Z	5.9 (5899)	5.0	-	(12)	(13)	(21)	(20)	(18)	26.0 (5)	20.0
	C/F 2500 Pick-up	W	8.0 (7997)	7.0	-	(12)	(13)	(21)	(20)	(18)	26.0 (5)	24.0
	C/F 3500 Pick-up	C	5.9 (5882)	11.0	-	(12)	(13)	(21)	(20)	(18)	26.0 (5)	26.0
	C/F 3500 Pick-up	5	5.9 (5899)	5.0	-	(12)	(13)	(21)	(20)	(18)	26.0 (5)	20.0
	C/F 3500 Pick-up	W	8.0 (7997)	7.0	-	(12)	(13)	(21)	(20)	(18)	26.0 (5)	24.0
	B1500 Van	X	3.9 (3916)	4.0	-	-	(13)	-	-	(16)	22.0 (5)	14.6
	B1500 Van	Y	5.2 (5211)	5.0	-	-	(13)	-	-	(16)	22.0 (5)	16.5
	B2500 Van	X	3.9 (3916)	4.0	-	-	(13)	-	-	(16)	22.0 (5)	14.6
	B2500 Van	Y	5.2 (5211)	5.0	-	-	(13)	-	-	(16)	22.0 (5)	16.5
	B2500 Van	Z	5.9 (5899)	5.0	-	-	(13)	-	-	(16)	35.0	15.0 (17)
	B3500 Van	Y	5.2 (5211)	5.0	-	-	(13)	-	-	(16)	22.0 (5)	16.5
	B3500 Van	Z	5.9 (5899)	5.0	-	-	(13)	-	-	(16)	35.0	15.0 (17)

① With 7.25 in. rear: 3.0 pts.
  With 8.25 in. rear: 4.4 pts.
② Optional fuel tank: 22 gals.
③ A998/A999 and A727: 17.1 pts.
  A500/A518: 20.4 pts.
④ Chrysler: 4.5 pts.
  Dana 60: 6.25 pts.
⑤ Optional fuel tank: 35 gals.
⑥ With HD cooling or A/C, add one quart
⑦ (4W/D ONLY)
  NP205: 4.5 pts.
  NP241: 6.0 pts.
⑧ (4W/D ONLY)
  Dana 44: 5.6 pts.
  Dana 60: 6.5 pts.
⑨ Chrysler 8.25 in. and 9.25 in.: 4.5 pts.
  Spicer and Dana 60: 6.0 pts.
  Dana 70: 7.0 pts.

⑩ Optional fuel tank: 30 gals.
⑪ Manual transmission: 15.5 qts.
  Automatic transmission: 16.5 qts.
⑫ NV3500: 4.2 pts.
  NV4500: 8.0 pts.
  AX15: 6.6 pts.
  Getrag: 7.0 pts.
⑬ 32RH: 17.0 pts.
  36RH: 16.6 pts.
  42RH: 20.2 pts.
  32RH: 17.0 pts.
  36RH: 16.6 pts.
  42RH: 20.2 pts.
⑭ 7.25 in.: 2.9 pts.
  8.25 in.: 4.4 pts.
⑮ A998/A999 and A727: 17.2 pts.
  A500: 20.4 pts.
  A518: 21.4 pts.

⑯ Chrysler 8.25 in.: 4.4 pts.
  Chrysler 9.25 in.: 4.8 pts.
  Dana 60: 6.3 pts.
⑰ With rear heater: 16.0 qts.
⑱ Chrysler 7.25 in.: 3 pts.
  Chrysler 8.25 in. and 9.25 in.: 4.8 pts.
  Spicer and Dana 60: 6.0 pts.
  Dana 70 and 80: 7.0 pts.
⑲ NP231HD: 2.5 pts.
  NP241: 4.7 pts.
  NP241HD: 13 pts.
⑳ 7.25 in.: 3 pts.
  Dana 44: 5.6 pts.
  Dana 60: 6.5 pts.
㉑ NP231HD: 2.5 pts.
  NP241: 4.7 pts.
  NP241HD: 6.5 pts.

VALVE SPECIFICATIONS

Year	Engine ID/VIN	Engine Displacement Liters (cc)	Seat Angle (deg.)	Face Angle (deg.)	Spring Test Pressure (lbs. @ in.)	Spring Installed Height (in.)	Stem-to-Guide Clearance (in.)		Stem Diameter (in.)	
							Intake	Exhaust	Intake	Exhaust
1993	K	2.5 (2507)	45	45	115@1.65	1.65	0.001-0.005	0.003-0.005	0.3124	0.3103
	X	3.9 (3916)	44.25-44.75	43.25-43.75	85@1.64	1.64	0.001-0.003	0.001-0.003	0.3110-0.3120	0.3110-0.3120
	Y	5.2 (5211)	44.25-44.75	43.25-43.75	85@1.64	1.64	0.001-0.003	0.001-0.003	0.3110-0.3120	0.3110-0.3120
	5	5.9 (5899)	44.25-44.75	43.25-43.75	85@1.64	1.64	0.001-0.003	0.002-0.004	0.3720-0.3730	0.3710-0.3720
	Z	5.9 (5899)	44.25-44.75	43.25-43.75	85@1.64	1.64	0.001-0.003	0.002-0.004	0.3720-0.3730	0.3710-0.3720
	8	5.9 (5882)	①	①	65@1.94 ②	2.19	0.002-0.006	0.002-0.006	0.3130-0.3140	0.3130-0.3140
1994	K	2.5 (2507)	45	45	115@1.65	1.65	0.001-0.005	0.003-0.005	0.3124	0.3103
	X	3.9 (3916)	44.25-44.75	43.25-43.75	85@1.64	1.64	0.001-0.003	0.001-0.003	0.3110-0.3120	0.3110-0.3120
	Y	5.2 (5211)	44.25-44.75	43.25-43.75	85@1.64	1.64	0.001-0.003	0.001-0.003	0.3110-0.3120	0.3110-0.3120
	C	5.9 (5882)	①	①	81@1.94	2.36	0.002-0.006	0.002-0.006	0.3130-0.3140	0.3130-0.3140
	5	5.9 (5899)	44.25-44.75	43.25-43.75	85@1.64	1.64	0.001-0.003	0.002-0.004	0.3720-0.3730	0.3710-0.3720
	A	5.9 (5899)	44.25-44.75	43.25-43.75	85@1.64	1.64	0.001-0.003	0.002-0.004	0.3720-0.3730	0.3710-0.3720
	Z	5.9 (5899)	44.25-44.75	43.25-43.75	85@1.64	1.64	0.001-0.003	0.002-0.004	0.3720-0.3730	0.3710-0.3720
	W	8.0 (7997)	44.5	45	81-89@1.64	1.64	0.001-0.003	0.001-0.003	0.3110-0.3120	0.3110-0.3120
1995	G	2.5 (2507)	45	45	115@1.65	1.65	0.001-0.003	0.003-0.005	0.3124	0.3103
	X	3.9 (3916)	44.25-44.75	43.25-43.75	85@1.64	1.64	0.001-0.003	0.001-0.003	0.3110-0.3120	0.3110-0.3120
	Y	5.2 (5211)	44.25-44.75	43.25-43.75	85@1.64	1.64	0.001-0.003	0.001-0.003	0.3110-0.3120	0.3110-0.3120
	C	5.9 (5882)	①	①	81@1.94	2.36	0.002-0.006	0.002-0.006	0.3130-0.3140	0.3130-0.3140
	5	5.9 (5899)	44.25-44.75	43.25-43.75	85@1.64	1.64	0.001-0.003	0.002-0.004	0.3720-0.3730	0.3710-0.3720
	Z	5.9 (5899)	44.25-44.75	43.25-43.75	85@1.64	1.64	0.001-0.003	0.002-0.004	0.3720-0.3730	0.3710-0.3720
	W	8.0 (7997)	44.5	45	81-89@1.64	1.64	0.001-0.003	0.001-0.003	0.3110-0.3120	0.3110-0.3120
1996-97	P	2.5 (2507)	45	45	184-196@1.22	1.65	0.001-0.003	0.001-0.003	0.311-0.312	0.311-0.312
	X	3.9 (3916)	44.25-44.75	43.25-43.75	200@1.21	1.64	0.001-0.003	0.001-0.003	0.311-0.312	0.311-0.312
	Y	5.2 (5211)	44.25-44.75	43.25-43.75	200@1.21	1.64	0.001-0.003	0.001-0.003	0.311-0.312	0.311-0.312

VALVE SPECIFICATIONS

Year	Engine ID/VIN	Engine Displacement Liters (cc)	Seat Angle (deg.)	Face Angle (deg.)	Spring Test Pressure (lbs. @ in.)	Spring Installed Height (in.)	Stem-to-Guide Clearance (in.)		Stem Diameter (in.)	
							Intake	Exhaust	Intake	Exhaust
1996-97	C	5.9 (5882)	③	③	81@1.94	2.36	0.002-0.006	0.002-0.006	0.313-0.314	0.313-0.314
	5	5.9 (5899)	44.25-44.75	43.25-43.75	200@1.21	1.64	0.001-0.003	0.002-0.004	0.372-0.373	0.371-0.372
	Z	5.9 (5899)	44.25-44.75	43.25-43.75	200@1.21	1.64	0.001-0.003	0.002-0.004	0.372-0.373	0.371-0.372
	W	8.0 (7997)	44.5	45	190-210@1.22	1.64	0.001-0.003	0.001-0.003	0.311-0.312	0.311-0.312

① Intake: 30
Exhaust: 45
Exhaust: 80-90 at 1.20 in.

② Minimum acceptable specification

③ Intake: 30 degrees
Exhaust: 45 degrees

TORQUE SPECIFICATIONS
All readings in ft. lbs.

Year	Engine ID/VIN	Engine Displacement Liters (cc)	Cylinder Head Bolts	Main Bearing Bolts	Rod Bearing Bolts	Crankshaft Damper Bolts	Flywheel Bolts	Manifold		Spark Plugs	Lug Nut
								Intake *	Exhaust		
1993	K	2.5 (2507)	①	②	③	50	70	17	17	26	95
	X	3.9 (3916)	105	85	45	135	55	45	⑥	30	⑦
	Y	5.2 (5211)	105	85	45	135	55	40	⑥	30	⑦
	8	5.9 (5882)	⑧	⑨	⑩	92	101	-	32	-	⑦
	5	5.9 (5899)	105	85	45	135	55	40	⑥	30	⑦
	Z	5.9 (5899)	105	85	45	135	55	40	⑥	30	⑦
1994	K	2.5 (2507)	①	②	③	85	70	17	17	20	95
	X	3.9 (3916)	⑯	85	45	135	55	⑫	25	30	⑬
	Y	5.2 (5211)	⑯	85	45	135	55	⑫	25	30	⑬
	C	5.9 (5882)	⑧	⑨	⑩	92	101	⑭	32	-	⑬
	5	5.9 (5899)	⑯	85	45	135	55	⑫	25	30	⑬
	A	5.9 (5899)	⑯	85	45	135	55	⑫	25	30	⑬
	Z	5.9 (5899)	⑯	85	45	135	55	⑫	25	30	⑬
	W	8.0 (7997)	⑯	85	45	135	55	⑯	16	30	⑬
1995	G	2.5 (2507)	①	②	③	85	70	17	17	20	95
	L	3.8 (3785)	①	②	③	40	70	17	17	20	95
	X	3.9 (3916)	⑯	85	45	135	55	⑫	25	30	⑬
	Y	5.2 (5211)	⑯	85	45	135	55	⑫	25	30	⑬
	C	5.9 (5882)	⑧	⑨	⑩	92	101	⑭	32	-	⑬
	5	5.9 (5899)	⑯	85	45	135	55	⑫	25	30	⑬
	Z	5.9 (5899)	⑯	85	45	135	55	⑫	25	30	⑬
	W	8.0 (7997)	⑯	85	45	135	55	⑯	16	30	⑬
1996-97	P	2.5 (2507)	⑱	80	33	80	105	17	17	27	95
	X	3.9 (3916)	⑮	85	45	135	55	⑫	25	30	⑬
	Y	5.2 (5211)	⑮	85	45	135	55	⑫	25	30	⑬
	C	5.9 (5882)	⑧	⑨	⑩	92	101	⑭	32	-	⑬

TORQUE SPECIFICATIONS
All readings in ft. lbs.

Year	Engine ID/VIN	Engine Displacement Liters (cc)	Cylinder Head Bolts	Main Bearing Bolts	Rod Bearing Bolts	Crankshaft Damper Bolts	Flywheel Bolts	Manifold		Spark Plugs	Lug Nut
								Intake	Exhaust		
1996-97	5	5.9 (5899)	⑮	85	45	135	55	⑫	25	30	⑬
	Z	5.9 (5899)	⑮	85	45	135	55	⑫	25	30	⑬
	W	8.0 (7997)	⑮	85	45	135	55	⑯	16	30	⑬

* NOTE: Applies to Lower Manifold only.

① Step 1: 45 ft. lbs.
Step 2: 65 ft. lbs.
Step 3: 65 ft. lbs.
Step 4: Plus 1/4 turn
② Step 1: 30 ft. lbs.
Step 2: Plus 1/4 turn
③ Step 1: 40 ft. lbs.
Step 2: Plus 1/4 turn
④ Step 1: 45 ft. lbs.
Step 2: 65 ft. lbs.
Step 3: 65 ft. lbs.
Step 4: Plus 1/4 turn
Torque small bolt in rear of head to 25 ft. lbs.
⑤ Town & Country
⑥ Bolts: 20 ft. lbs.
Nuts: 15 ft. lbs.
⑦ 1/2x20 stud with cone nut: 85-110 ft. lbs.
5/8x18 stud with cone nut: 175-225 ft. lbs.
5/8x18 stud with flanged nut: 300-350 ft. lbs.

⑧ All bolts: 66 ft. lbs.
Long bolts: 89 ft. lbs.
All bolts and additional 1/4 turn
⑨ Step 1: 45 ft. lbs.
Step 2: 88 ft. lbs.
Step 3: 129 ft. lbs.
⑩ Step 1: 26 ft. lbs.
Step 2: 51 ft. lbs.
Step 3: 73 ft. lbs.
⑪ Voyager
⑫ Step 1: Bolts 1-12: 6 ft. lbs. in sequence
Step 2: Torque all bolts to 12 in. lbs. in alternating steps
Step 3: Check that all bolts are tightened to 6 ft. lbs.
Step 4: Tighten all bolts in sequence to 12 ft. lbs.
Step 5: Check that all bolts are tightened to 12 ft. lbs.
⑬ 5 stud wheel: 95 ft. lbs.
8 stud wheel: 135 ft. lbs.
8 stud dual wheel: 145 ft. lbs.

⑭ Intake manifold cover bolts: 18 ft. lbs.
⑮ 50 ft. lbs. then 105 ft. lbs., in sequence
⑯ Lower intake: 40 ft. lbs.
Upper intake: 16 ft. lbs.
⑰ Step 1: 25 ft. lbs.
Step 2: 50 ft. lbs.
Step 3: 50 ft. lbs.
Step 4: Plus 1/4 turn
⑱ Bolts 1-10 and 12-14: 110 ft. lbs.
Bolt 11: 100 ft. lbs.
⑲ M8 bolts: 250 in. lbs. plus 1/4 turn
M11 bolts: 30 ft. lbs. plus 1/4 turn
⑳ Step 1: 20 ft. lbs.
Step 2: Plus 1/4 turn
㉑ Lower Manifold only

BRAKE SPECIFICATIONS
All measurements in inches unless noted

Year	Model	Master Cylinder Bore	Brake Disc			Brake Drum Diameter			Minimum Lining Thickness	
			Original Thickness	Minimum Thickness	Maximum Runout	Original Inside Diameter	Max. Wear Limit	Maximum Machine Diameter	Front	Rear
1993	B150 Van	1.125	1.240	1.180	0.004	11.00	11.09	11.06	0.062	0.062
	B250 Van	1.125	1.240	1.180	0.004	11.00	11.09	11.06	0.062	0.062
	B350 Van	1.125	①	②	0.004	12.00	12.09	12.06	0.062	0.062
	Dakota	NA	0.861	0.810	0.004	9.00	9.09	9.06	0.060	0.060
	Dakota	NA	0.861	0.810	0.004	10.00	10.09	10.06	0.060	0.060
	D150 Pick-up	1.125	1.240	1.180	0.004	11.00	11.09	11.06	0.062	0.062
	D250 Pick-up	1.125	1.240	1.180	0.005	12.00	12.09	11.06	0.062	0.062
	D350 Pick-up	1.125	1.18	1.125	0.005	12.00	12.09	12.06	0.062	0.062
	Ramcharger	1.125	1.240	1.180	0.004	11.00	11.09	11.06	0.062	0.062
	W150 Pick-up	1.125	1.240	1.180	0.004	11.00	11.09	11.06	0.062	0.062
	W250 Pick-up	1.125	①	②	0.005	12.00	12.09	12.06	0.062	0.062
	W350 Pick-up	1.125	1.180	1.125	0.005	12.00	12.09	12.06	0.062	0.062
1994	B150 Van	1.125	-	③	0.004	11.00	11.09	11.06	0.125	0.062 ④
	B250 Van	1.125	-	③	0.004	11.00	11.09	11.06	0.125	0.062 ④
	B350 Van	1.125	-	③	0.004	12.00	12.09	12.06	0.125	0.062 ④
	Dakota	NA	0.861	0.810	0.004	9.00	9.09	9.06	0.060	0.060 ④

BRAKE SPECIFICATIONS

All measurements in inches unless noted

Year	Model	Master Cylinder Bore	Brake Disc Original Thickness	Brake Disc Minimum Thickness	Maximum Runout	Brake Drum Diameter Original Inside Diameter	Brake Drum Diameter Max. Wear Limit	Brake Drum Diameter Maximum Machine Diameter	Minimum Lining Thickness Front	Minimum Lining Thickness Rear
1994	Dakota	NA	0.861	0.810	0.004	10.00	10.09	10.06	0.060	0.060 ④
	D1500 Pick-up	1.125	1.260	③	0.004	11.00	11.09	11.06	0.062	0.062 ④
	D2500 Pick-up	1.250	1.500	③	0.005	13.00	13.09	13.06	0.062	0.062 ④
	D3500 Pick-up	1.250	1.500	③	0.005	13.00	13.09	13.06	0.062	0.062 ④
	W1500 Pick-up	1.125	1.260	③	0.004	11.00	11.09	11.06	0.062	0.062 ④
	W2500 Pick-up	1.250	1.500	③	0.005	13.00	13.09	13.06	0.062	0.062 ④
	W3500 Pick-up	1.250	1.500	③	0.005	13.00	13.09	13.06	0.062	0.062 ④
1995	B150 Van	1.125	-	③	0.004	11.00	11.09	11.06	0.125	0.062 ④
	B250 Van	1.125	-	③	0.004	11.00	11.09	11.06	0.125	0.062 ④
	B350 Van	1.125	-	③	0.004	12.00	12.09	12.06	0.125	0.062 ④
	Dakota	NA	0.861	0.810	0.004	9.00	9.09	9.06	0.060	0.060 ④
	Dakota	NA	0.861	0.810	0.004	10.00	10.09	10.06	0.060	0.060 ④
	D1500 Pick-up	1.125	1.260	③	0.004	11.00	11.09	11.06	0.062	0.062 ④
	D2500 Pick-up	1.250	1.500	③	0.005	13.00	13.09	13.06	0.062	0.062 ④
	D3500 Pick-up	1.250	1.500	③	0.005	13.00	13.09	13.06	0.062	0.062 ④
	W1500 Pick-up	1.125	1.260	③	0.004	11.00	11.09	11.06	0.062	0.062 ④
	W2500 Pick-up	1.250	1.500	③	0.005	13.00	13.09	13.06	0.062	0.062 ④
	W3500 Pick-up	1.250	1.500	③	0.005	13.00	13.09	13.06	0.062	0.062 ④
1996-97	Ram 1500 Pick-up	1.125	1.260	③	0.004	11.00	11.09	11.06	0.062	0.062 ④
	Ram 2500 Pick-up	1.250	1.500	③	0.005	13.00	13.09	13.06	0.062	0.062 ④
	Ram 3500 Pick-up	1.250	1.500	③	0.005	13.00	13.09	13.06	0.062	0.062 ④
	B1500 Van	1.125	-	③	0.004	11.00	11.09	11.06	0.125	0.062 ④
	B2500 Van	1.125	-	③	0.004	11.00	11.09	11.06	0.125	0.062 ④
	B3500 Van	1.125	-	③	0.004	12.00	12.09	12.06	0.125	0.062 ④
	Dakota ⑤	NA	0.861	0.810	0.004	9.00	9.09	9.06	0.060	0.060 ④
	Dakota ⑥	NA	0.861	0.810	0.004	10.00	10.09	10.06	0.060	0.060 ④

① Except 4000 lb. rear axle: 1.240 in.
  With 4000 lb. rear axle: 1.180 in.
② With 3300 lb. or 3600 lb. rear axle: 1.180 in.
  With 4000 lb. rear axle: 1.125 in.
③ Minimum thickness indicated on rotor hub
④ For riveted brake shoes: 0.031
⑤ With 9 inch rear brakes
⑥ With 10 inch rear brakes

FREQUENT MAINTENANCE LABOR
DAKOTA

The following should be used as a guide when determining the amount of work required for a particular service if taken to a repair shop. In estimating how long a particular Frequent Maintenance Service item should take, please observe the following:
- **Factory Time** is time that is generated by the vehicle manufacturer.
- **Chilton Time** is time that is based on field research and data supplied by the vehicle manufacturer.
- All labor time operations are given in hours and tenths of an hour.
- All labor operations, are to be used as a **guide**.

COOLING

(G) Winterize Cooling System
Includes: Run engine to check for leaks, tighten all hose connections. Test radiator and pressure cap, drain radiator and engine block. Add antifreeze and refill system.
1993-97 5

(G) Belt, Drive, Renew
V belt
1993-97
  Alternator (.2) 3
w/PS add 2
w/AC add 1
w/AIR add 1
  Power Steering (.2) 3
w/AC add 1
w/AIR add 1
  AC (.2) 3
w/PS add 1
w/AIR add 1
  Air Pump (.2) 3
Serpentine belt
1993-97 (.2) 3

(G) Belt, Drive, Adjust
1993-97
  one (.2) 3
  each adtnl. 1

(G) Hoses, Radiator, Renew
Includes: Drain and refill cooling system as required.
1993-97
  upper or lower, each (.4) 6
1993-97 by-pass (.5) 7
w/AIR add 1
w/AC add 1

(G) Thermostat, Coolant, Renew
1993-97
  4 cyl., V6 (.4) 6
  V8 (.8) 1.0
w/AC add 1

FUEL

(M) Air Cleaner, Service
1993-97 2

(G) Filter, Fuel, Renew
1993-97
  standard filter assy.
    in line (.2) 3
    in tank (.9) 1.3
  filter regulator assy.
    Vans (1.4) 1.9
    Pickups (1.1) 1.5

BRAKES

(G) Bleed Brakes (Four Wheels)
Includes: Add fluid.
1993-97 (.4) 6

(G) Brakes, Adjust (Minor)
Includes: Adjust brakes, fill master cylinder.
1993-97, two wheels 5

(M) Parking Brake, Adjust
1993-97 (.3) 5

LUBRICATION SERVICES

(M) Engine Oil & Filter, Renew
Includes: Inspect and correct all fluid levels.
1993-97 3

(M) Lubricate Chassis
Includes: Inspect and correct all fluid levels.
1993-97 4
Install grease fittings, add 1

ELECTRICAL

(G) Headlamps, Aim
1993-97
  two 4
  four 6

(G) Parking Lamp Lens or Bulb, Renew
1993-97, one (.2) 3

(G) Horn, Renew
1993-97, one (.2) 3
w/AC add 3
w/aux. trans oil cooler, add 3

(M) Terminals, Battery, Clean
1993-97 3

FREQUENT MAINTENANCE LABOR
RAM VANS (B SERIES), RAM PICKUPS (D, W SERIES)

The following should be used as a guide when determining the amount of work required for a particular service if taken to a repair shop. In estimating how long a particular Frequent Maintenance Service item should take, please observe the following:

- **Factory Time** is time that is generated by the vehicle manufacturer.
- **Chilton Time** is time that is based on field research and data supplied by the vehicle manufacturer.
- All labor time operations are given in hours and tenths of an hour.
- All labor operations, are to be used as a **guide**.

	(Factory Time)	Chilton Time

COOLING

(G) Winterize Cooling System
Includes: Run engine to check for leaks, tighten all hose connections. Test radiator and pressure cap, drain radiator and engine block. Add antifreeze and refill system.

1993-97		.5

(G) Belt, Fan Drive, Renew

1993-97		
V belt (.3)		.6
Serpentine belt (.2)		.4
w/AIR add		.1
w/PS add		.2

(G) Hoses, Radiator, Renew
Includes: Drain and refill cooling system as required.

Upper
1993-97 Vans (.5)		.7
1993-97 Pickups (.4)		.6

Lower
1993-97 (.4)		.6

By-pass
1993-97 Vans (.5)		.7
1993-97 Pickups-V6, V8, V10 (.3)		.5
w/AC add		.2

(G) Thermostat, Coolant, Renew
Vans
1993-97 (.9)		1.2

Pickups
1993-97		
gas (.4)		.6
diesel (.6)		1.0
w/AC add		.1

FUEL

(M) Air Cleaner, Service
1993-97		.3

(G) Filter, Fuel, Renew
Vans
1993-97		
gas		
in line (.2)		.3
in tank (.9)		1.3
CNG (.4)		.6

Pickups
1993		
gas		
in line (.2)		.3
in tank (.9)		1.3
diesel (.2)		.3
1994-97		
gas (1.3)		1.7
diesel (.2)		.3
CNG (.4)		.6

BRAKES

(G) Bleed Brakes (Four Wheels)
Includes: Add fluid.
1993-97 (.4)		.6

(G) Brakes, Adjust (Minor)
Includes: Adjust brakes, fill master cylinder.
1993-97		
two wheels		.5
four wheels		.8

(M) Parking Brake, Adjust
1993-97 (.3)		.5

LUBRICATION SERVICES

(M) Engine Oil & Filter, Renew
Includes: Inspect and correct all fluid levels.
1993-97 (.2)		.3

ELECTRICAL

(G) Headlamps, Aim
1993-97		
two		.4
four		.6

(G) Park & Turn Signal Lamp Bulb or Lens, Renew
1993-97, one (.2)		.3

(G) Tail Lamp Lens or Bulb, Renew
1993-97, one (.2)		.3

(G) Horn, Renew
1993-97		
Vans (.3)		.4
Pickups (.2)		.3
w/AC add		.3
w/auxiliary oil cooler, add		.3

SCHEDULED MAINTENANCE INTERVALS
(CHRYSLER B SERIES (VAN), DAKOTA, D, W SERIES (PICK UP), RAMCHARGER & RAM TRUCK (LIGHT DUTY))

TO BE SERVICED	TYPE OF SERVICE	VEHICLE MILEAGE INTERVAL (x1000)												
		7.5	15	22.5	30	37.5	45	52.5	60	67.5	75	82.5	90	97.5
Engine oil & filter	R	✓	✓	✓	✓	✓	✓	✓	✓	✓	✓	✓	✓	✓
Exhaust system	S/I	✓	✓	✓	✓	✓	✓	✓	✓	✓	✓	✓	✓	✓
Engine coolant level, hoses & clamps	S/I	✓	✓	✓	✓	✓	✓	✓	✓	✓	✓	✓	✓	✓
Rotate tires	S/I	✓		✓		✓		✓		✓		✓		✓
Drive belts	S/I		✓		✓		✓		✓		✓		✓	
Brake booster bellcrank pivot (B Series)	S/I		✓		✓		✓		✓		✓		✓	
Steering linkage (B Series)	S/I		✓		✓		✓		✓		✓		✓	
Brake hoses & linings	S/I			✓			✓			✓			✓	
Front suspension ball joints	S/I			✓			✓			✓			✓	
Front wheel bearings	S/I			✓			✓			✓			✓	
Steering linkage (except B series)	S/I			✓			✓			✓			✓	
Spark plugs	R				✓				✓				✓	
Engine air cleaner element	R				✓				✓				✓	
Automatic transmission fluid, filter & adjust bands	S/I					✓					✓			
Manual transmission fluid	R					✓					✓			
Transfer case fluid	R					✓					✓			
Engine Coolant	R						✓				✓			
PCV valve	R								✓				✓	
Battery①	R								✓					
Fuel filter	R								✓					
Ignition cables, distributor cap & rotor	R								✓					
Timing belt (2.5L)	R								✓					

① Replace at 60,000 miles, if not replaced previously.
R – Replace   S/I – Service or Inspect

SCHEDULED MAINTENANCE INTERVALS
(CHRYSLER B SERIES (VAN), DAKOTA, D, W SERIES (PICK UP), RAMCHARGER & RAM TRUCK (LIGHT DUTY)) (Cont.)

FREQUENT OPERATION MAINTENANCE (SEVERE SERVICE)

If a vehicle is operated under any of the following conditions it is considered severe service:
- Extremely dusty areas.
- 50% or more of the vehicle operation is in 32°C (90°F) or higher temperatures, or constant operation in temperatures below 0°C (32°F).
- Prolonged idling (vehicle operation in stop and go traffic).
- Frequent short running periods (engine does not warm to normal operating temperatures).
- Police, taxi, delivery usage or trailer towing usage.

Oil & oil filter change – change every 3000 miles.
Air filter/air pump air filter – change every 24,000 miles.
Engine coolant level, hoses & clamps - check every 6000 miles.
Exhaust system - check every 6000 miles.
Drive belts - check every 18,000 miles; replace every 24,000 miles.
Crankcase inlet air filter (6 & 8 cyl.) - clean every 24,000 miles.
Ignition timing (1993-95 2.5L) - adjust every 60,000 miles.
Oxygen sensor - replace every 82,500 miles.
Automatic transmission fluid, filter & bands - change & adjust every 12,000 miles.
Steering linkage - lubricate every 6000 miles.
Rear axle fluid (B Series) - change every 12,000 miles.

SCHEDULED MAINTENANCE INTERVALS
(CHRYSLER RAM TRUCK (GASOLINE MEDIUM & HEAVY DUTY))

TO BE SERVICED	TYPE OF SERVICE	VEHICLE MILEAGE INTERVAL (x1000)												
		6	12	18	24	30	36	42	48	54	60	66	72	78
Engine oil & filter	R	✓	✓	✓	✓	✓	✓	✓	✓	✓	✓	✓	✓	✓
Exhaust system	S/I	✓	✓	✓	✓	✓	✓	✓	✓	✓	✓	✓	✓	✓
Engine coolant level, hoses & clamps	S/I	✓	✓	✓	✓	✓	✓	✓	✓	✓	✓	✓	✓	✓
Rotate tires	S/I	✓		✓		✓		✓		✓		✓		✓
Drive belts	S/I		✓		✓		✓		✓		✓		✓	
Brake hoses & linings	S/I			✓			✓			✓			✓	
Engine air cleaner element & air pump filter (heavy duty)	R				✓				✓				✓	
Automatic transmission fluid, filter & adjust bands	S/I				✓				✓				✓	
Crankcase inlet air filter (5.9L) (heavy duty)	S/I				✓				✓				✓	
Front wheel bearings (4x2)	S/I				✓				✓				✓	

SCHEDULED MAINTENANCE INTERVALS
(CHRYSLER RAM TRUCK (GASOLINE MEDIUM & HEAVY DUTY)) (Cont.)

TO BE SERVICED	TYPE OF SERVICE	VEHICLE MILEAGE INTERVAL (x1000)												
		6	12	18	24	30	36	42	48	54	60	66	72	78
Engine air cleaner element (medium duty)	R					✓					✓			
Engine Coolant	R						✓					✓		
Spark plugs	R						✓				✓			
Transfer case fluid	R						✓						✓	
Battery①	R										✓			
Distributor cap & rotor (5.9L) (heavy duty)	R										✓			
EGR valve (5.9L) (heavy duty)	R										✓			
Ignition cables	R										✓			
Oxygen sensor (5.9L) (heavy duty)	R										✓			
PCV valve (5.9L) (heavy duty)	R										✓			
EGR passages (5.9L) (heavy duty)	S/I										✓			

① Replace at 60,000 miles, if not replaced previously.
R – Replace   S/I – Service or Inspect

FREQUENT OPERATION MAINTENANCE (SEVERE SERVICE)

If a vehicle is operated under any of the following conditions it is considered severe service:
- Extremely dusty areas.
- 50% or more of the vehicle operation is in 32°C (90°F) or higher temperatures, or constant operation in temperatures below 0°C (32°F).
- Prolonged idling (vehicle operation in stop and go traffic).
- Frequent short running periods (engine does not warm to normal operating temperatures).
- Police, taxi, delivery usage or trailer towing usage.

Oil & oil filter change – change every 3000 miles.
Automatic transmission fluid, filter & bands - change & adjust every 12,000 miles.
Rear axle fluid - change every 12,000 miles.
Brake hoses & linings - check every 12,000 miles.
Front axle fluid (4x4) - change every 24,000 miles.
Engine air cleaner element & air pump filter - replace every 12,000 miles.
Crankcase inlet air filter (5.9L) (heavy duty) - clean & relubricate every 12,000 miles.
PCV valve (5.9L) (heavy duty) - check every 30,000 miles.

SCHEDULED MAINTENANCE INTERVALS
(CHRYSLER RAM TRUCK (DIESEL))

TO BE SERVICED	TYPE OF SERVICE	VEHICLE MILEAGE INTERVAL (x1000)												
		6	12	18	24	30	36	42	48	54	60	66	72	78
Engine oil & filter	R	✓	✓	✓	✓	✓	✓	✓	✓	✓	✓	✓	✓	✓
Brake hoses	S/I	✓	✓	✓	✓	✓	✓	✓	✓	✓	✓	✓	✓	✓
Exhaust system	S/I	✓	✓	✓	✓	✓	✓	✓	✓	✓	✓	✓	✓	✓
Engine coolant level, hoses & clamps	S/I	✓	✓	✓	✓	✓	✓	✓	✓	✓	✓	✓	✓	✓
Steering linkage	S/I	✓	✓	✓	✓	✓	✓	✓	✓	✓	✓	✓	✓	✓
Rotate tires	S/I	✓		✓		✓		✓		✓		✓		✓
Fuel filter	R		✓		✓		✓		✓		✓		✓	
Water pump weep hole	S/I		✓		✓		✓		✓		✓		✓	
Drive belts	S/I			✓			✓			✓			✓	
Brake linings	S/I			✓			✓			✓			✓	
Automatic transmission fluid, filter & adjust bands	S/I				✓				✓				✓	
Damper	S/I				✓				✓				✓	
Fan hub	S/I				✓				✓				✓	
Front wheel bearings	S/I				✓				✓				✓	
Valve lash clearance	S/I				✓				✓				✓	
Air filter	R					✓					✓			
Engine Coolant	R						✓					✓		
Transfer case fluid	R						✓						✓	

R – Replace   S/I – Service or Inspect

FREQUENT OPERATION MAINTENANCE (SEVERE SERVICE)

If a vehicle is operated under any of the following conditions it is considered severe service:
- Extremely dusty areas.
- 50% or more of the vehicle operation is in 32°C (90°F) or higher temperatures, or constant operation in temperatures below 0°C (32°F).
- Prolonged idling (vehicle operation in stop and go traffic).
- Frequent short running periods (engine does not warm to normal operating temperatures).
- Police, taxi, delivery usage or trailer towing usage.

Oil & oil filter change – change every 3000 miles.
Automatic transmission fluid, filter & adjust bands - change & adjust every 12,000 miles.
Rear axle fluid - change every 12,000 miles.
Brake linings - check every 12,000 miles.
Front axle fluid (4x4) - change every 24,000 miles.

CHRYSLER CORPORATION
Caravan • Town & Country • Voyager

VEHICLE IDENTIFICATION CHART

Engine Code						Model Year	
Code	Liters	Cu. In. (cc)	Cyl.	Fuel Sys.	Eng. Mfg.	Code	Year
B	2.4	148 (2429)	I4	MFI	Chrysler	P	1993
K	2.5	153 (2507)	I4	TFI	Chrysler	R	1994
3	3.0	181 (2972)	V6	MFI	Mitsubishi	S	1995
R	3.3	201 (3300)	V6	MFI	Chrysler	T	1996
L	3.8	231 (3785)	V6	MFI	Chrysler	V	1997

MFI - Multiport fuel injection
TFI - Throttle body fuel injection

ENGINE IDENTIFICATION

Year	Model	Engine Displacement Liters (cc)	Engine Series (ID/VIN)	Fuel System	No. of Cylinders	Engine Type
1993	Caravan	2.5 (2507)	K	TFI	4	SOHC
	Caravan	3.0 (2972)	3	MFI	6	SOHC
	Caravan	3.3 (3300)	R	MFI	6	OHV
	Town & Country	3.3 (3300)	R	MFI	6	OHV
	Voyager	2.5 (2507)	K	TFI	4	SOHC
	Voyager	3.0 (2972)	3	MFI	6	SOHC
	Voyager	3.3 (3300)	R	MFI	6	OHV
1994	Caravan	2.5 (2507)	K	TFI	4	SOHC
	Caravan	3.0 (2972)	3	MFI	6	SOHC
	Caravan	3.3 (3300)	R	MFI	6	OHV
	Caravan	3.8 (3785)	L	MFI	6	OHV
	Town & Country	3.8 (3785)	L	MFI	6	OHV
	Voyager	2.5 (2507)	K	TFI	4	SOHC
	Voyager	3.0 (2972)	3	MFI	6	SOHC
	Voyager	3.3 (3300)	R	MFI	6	OHV
	Voyager	3.8 (3785)	L	MFI	6	OHV
1995	Caravan	3.0 (2972)	3	MFI	6	SOHC
	Caravan	3.3 (3300)	R	MFI	6	OHV
	Caravan	3.8 (3785)	L	MFI	6	OHV
	Town & Country	3.8 (3785)	L	MFI	6	OHV
	Voyager	2.5 (2507)	K	TFI	4	SOHC
	Voyager	3.0 (2972)	3	MFI	6	SOHC
	Voyager	3.3 (3300)	R	MFI	6	OHV
	Voyager	3.8 (3785)	L	MFI	6	OHV

ENGINE IDENTIFICATION

Year	Model	Engine Displacement Liters (cc)	Engine Series (ID/VIN)	Fuel System	No. of Cylinders	Engine Type
1996-97	Caravan	2.4 (2429)	B	MFI	4	DOHC
	Caravan	3.0 (2972)	3	MFI	6	SOHC
	Caravan	3.3 (3300)	R	MFI	6	OHV
	Caravan	3.8 (3785)	L	MFI	6	OHV
	Town & Country	3.3 (3301)	R	MFI	6	OHV
	Town & Country	3.8 (3778)	L	MFI	6	OHV
	Voyager	2.4 (2429)	B	MFI	4	DOHC
	Voyager	3.0 (2972)	3	MFI	6	SOHC
	Voyager	3.3 (3301)	R	MFI	6	OHV
	Voyager	3.8 (3785)	L	MFI	6	OHV

DSL - Diesel
MFI - Multiport fuel injection
TFI - Throttle body fuel injection
OHV - Overhead valve
SOHC - Single overhead camshaft
DOHC - Double overhead camshaft

GENERAL ENGINE SPECIFICATIONS

Year	Engine ID/VIN	Engine Displacement Liters (cc)	Fuel System Type	Net Horsepower @ rpm	Net Torque @ rpm (ft. lbs.)	Bore x Stroke (in.)	Compression Ratio	Oil Pressure @ rpm
1993	K	2.5 (2507)	TFI	100@4800	135@2800	3.45x4.09	8.9:1	35-65@2000
	3	3.0 (2972)	MFI	143@5000	168@2500	3.59x2.99	8.9:1	30-80@3000
	R	3.3 (3300)	MFI	150@4800	185@3600	3.66x3.19	8.9:1	30-80@3000
1994	K	2.5 (2507)	TFI	100@4800	135@2800	3.45x4.09	8.9:1	25-80@3000
	3	3.0 (2972)	MFI	143@5000	168@2500	3.59x2.99	8.9:1	30-80@3000
	R	3.3 (3300)	MFI	150@4800	185@3600	3.66x3.19	8.9:1	30-80@3000
	L	3.8 (3785)	MFI	162@4400	213@3300	3.78x3.43	9.0:1	30-80@3000
1995	K	2.5 (2507)	TFI	100@4800	135@2800	3.44x4.09	8.9:1	25-80@3000
	3	3.0 (2972)	MFI	143@5000	170@2800	3.59x2.99	8.9:1	30-80@3000
	R	3.3 (3300)	MFI	162@4800	194@3600	3.66x3.19	8.9:1	30-80@3000
	L	3.8 (3785)	MFI	162@4400	213@3300	3.78x3.43	9.0:1	30-80@3000
1996-97	B	2.4 (2429)	MFI	150@5200	167@4000	3.44x3.98	9.4:1	25-80@3000
	3	3.0 (2972)	MFI	143@5000	170@2800	3.59x2.99	8.9:1	30-80@3000
	R	3.3 (3300)	MFI	162@4800	194@3600	3.66x3.19	8.9:1	30-80@3000
	L	3.8 (3785)	MFI	162@4400	213@3300	3.78x3.43	9.0:1	30-80@3000

MFI - Multiport fuel injection
TFI - Throttle body fuel injection

GASOLINE ENGINE TUNE-UP SPECIFICATIONS

Year	Engine ID/VIN	Engine Displacement Liters (cc)	Spark Plugs Gap (in.)	Ignition Timing (deg.) MT	Ignition Timing (deg.) AT	Fuel Pump (psi)	Idle Speed (rpm) MT	Idle Speed (rpm) AT	Valve Clearance In.	Valve Clearance Ex.
1993	K	2.5 (2507)	0.035	12B	12B	37-41	850	850	HYD	HYD
	3	3.0 (2972)	0.039-0.043	12B	12B	46-50	800	800	HYD	HYD
	R	3.3 (3300)	0.048-0.053	①	①	46-50	750	750	HYD	HYD
1994	K	2.5 (2507)	0.035	12B	12B	37-41	②	②	HYD	HYD
	3	3.0 (2972)	0.039-0.043	12B	12B	46-50	②	②	HYD	HYD
	R	3.3 (3300)	0.048-0.053	①	①	46-50	②	②	HYD	HYD
	L	3.8 (3785)	0.048-0.053	①	①	46-50	②	②	HYD	HYD

GASOLINE ENGINE TUNE-UP SPECIFICATIONS

Year	Engine ID/VIN	Engine Displacement Liters (cc)	Spark Plugs Gap (in.)	Ignition Timing (deg.) MT	Ignition Timing (deg.) AT	Fuel Pump (psi)	Idle Speed (rpm) MT	Idle Speed (rpm) AT	Valve Clearance In.	Valve Clearance Ex.
1995	K	2.5 (2507)	0.035	-	③	39	-	②	HYD	HYD
	3	3.0 (2972)	0.035	-	③	48	-	②	HYD	HYD
	R	3.3 (3300)	0.050	-	①	48	-	②	HYD	HYD
	L	3.8 (3785)	0.050	-	①	48	-	②	HYD	HYD
1996-97	B	2.4 (2429)	0.050	③	③	49	②	②	HYD	HYD
	3	3.0 (2972)	0.035	-	③	48	-	②	HYD	HYD
	R	3.3 (3300)	0.050	-	①	49	-	②	HYD	HYD
	L	3.8 (3785)	0.050	-	①	49	-	②	HYD	HYD

NOTE: The Vehicle Emission Control Information label often reflects specification changes made during production. The label figures must be used if they differ from those in this chart.

B - Before top dead center

HYD - Hydraulic

① Ignition timing cannot be adjusted. Base engine timing is set at TDC during assembly

② Refer to the Vehicle Emission Control Information (VECI) label for correct specification

③ Refer to the Vehicle Emission Control Information label for correct timing specification with a range of +/- 2.

CAPACITIES

Year	Model	Engine ID/VIN	Engine Displacement Liters (cc)	Engine Oil with Filter (qts.)	Transmission (pts.) 4-Spd	Transmission (pts.) 5-Spd	Transmission (pts.) Auto.	Transfer Case (pts.)	Drive Axle Front (pts.)	Drive Axle Rear (pts.)	Fuel Tank (gal.)	Cooling System (qts.)
1993	Caravan	K	2.5 (2507)	4.5	-	4.8	①	2.4	-	4.0 ②	③	9.5
	Caravan	3	3.0 (2972)	4.5	-	4.8	①	2.4	-	4.0 ②	③	10.0
	Caravan	R	3.3 (3300)	4.5	-	4.8	①	2.4	-	4.0 ②	③	10.0
	Town & Country	R	3.3 (3300)	4.5	-	-	18.0	2.4	4.0	4.0	②	10.5
	Voyager	K	2.5 (2507)	4.5	-	4.8	①	2.4	-	4.0 ②	③	9.5
	Voyager	3	3.0 (2972)	4.5	-	4.8	①	2.4	-	4.0 ②	③	10.0
	Voyager	R	3.3 (3300)	4.5	-	4.8	①	2.4	-	4.0 ②	③	10.0
1994	Caravan	R	3.3 (3300)	4.5	-	4.6	18.0	2.4	-	4.0 ⑤	③	10.5
	Caravan	L	3.8 (3785)	4.5	-	4.6	18.0	2.4	-	4.0 ⑤	③	10.5
	Town & Country	L	3.8 (3785)	4.5	-	-	③	2.4	4.0	4.0 ④	②	10.5
	Voyager	K	2.5 (2507)	4.5	-	4.6	18.0	-	-	-	18.0	9.5
	Voyager	3	3.0 (2972)	4.5	-	4.6	18.0	-	-	-	18.0	10.5
	Voyager	R	3.3 (3300)	4.5	-	4.6	18.0	2.4	-	4.0 ②	③	10.5
	Voyager	L	3.8 (3785)	4.5	-	4.6	18.0	2.4	-	4.0 ②	③	10.5
1995	Caravan	R	3.3 (3300)	4.5	-	-	18.0 ⑥	2.4	-	4.0 ⑤	③	10.5
	Caravan	L	3.8 (3785)	4.5	-	-	18.0 ⑥	2.4	-	4.0 ⑤	③	10.5
	Town & Country	L	3.8 (3785)	4.5	-	-	③	2.4	4.0	4.0 ④	②	10.5
	Voyager	K	2.5 (2507)	4.5	-	-	①	-	-	-	20.0	9.5
	Voyager	3	3.0 (2972)	4.5	-	-	①	-	-	-	20.0	10.5
	Voyager	R	3.3 (3300)	4.5	-	-	①	2.4	-	4.0 ②	③	10.5
	Voyager	L	3.8 (3785)	4.5	-	-	①	2.4	-	4.0 ②	③	10.5
1996-97	Caravan	B	2.4 (2429)	4.5	-	-	18.0 ⑥	-	-	-	20.0	9.5
	Caravan	3	3.0 (2972)	4.5	-	-	18.0 ⑥	-	-	-	20.0	10.5
	Caravan	R	3.3 (3300)	4.5	-	-	18.0 ⑥	2.4	-	4.0 ⑤	③	10.5
	Caravan	L	3.8 (3785)	4.5	-	-	18.0 ⑥	2.4	-	4.0 ⑤	③	10.5
	Town & Country	R	3.3 (3301)	4.5	-	-	18.2 ⑥	NA	NA	NA	20.0	10.5
	Town & Country	L	3.8 (3778)	4.5	-	-	18.2 ⑥	NA	NA	NA	20.0	10.5
	Voyager	B	2.4 (2429)	4.0	-	-	8.0 ⑦	-	-	-	20.0	9.5

CAPACITIES

Year	Model	Engine ID/VIN	Engine Displacement Liters (cc)	Engine Oil with Filter (qts.)	Transmission (pts.) 4-Spd	5-Spd	Auto.	Transfer Case (pts.)	Drive Axle Front (pts.)	Rear (pts.)	Fuel Tank (gal.)	Cooling System (qts.)
1996-97	Voyager	3	3.0 (2972)	4.5	-	-	8.0 ⑦	-	-	-	20.0	10.5
	Voyager	R	3.3 (3301)	4.5	-	-	8.0 ⑦	-	-	-	20.0	10.5
	Voyager	L	3.8 (3785)	4.5	-	-	8.0 ⑦	-	-	4.0 ⑤	20.0	10.5

① Fleet vehicles: 19.0 pts.
   Non-fleet vehicles: 18.0 pts.
② Overrunning clutch: 0.78 pts.

③ FWD: 20.0 gals.
   AWD: 18.0 gals.
④ With 7.25 in. rear: 2.5 pts.
   With 8.25 in. rear: 4.4 pts.

⑤ Overrunning clutch: 0.75 pts.
⑥ Overhaul fill capacity with torque converter empty
⑦ 31TH overhaul fill capacity with torque converter empty: 17.0
   41TE overhaul fill capacity with torque converter empty: 18.2

VALVE SPECIFICATIONS

Year	Engine ID/VIN	Engine Displacement Liters (cc)	Seat Angle (deg.)	Face Angle (deg.)	Spring Test Pressure (lbs. @ in.)	Spring Installed Height (in.)	Stem-to-Guide Clearance (in.) Intake	Exhaust	Stem Diameter (in.) Intake	Exhaust
1993	K	2.5 (2507)	45	45	115@1.65	1.65	0.001-0.005	0.003-0.005	0.3124	0.3103
	3	3.0 (2972)	44.5	45.5	73@1.59	1.59	0.001-0.002	0.002-0.003	0.3130-0.3140	0.3120-0.3130
	R	3.3 (3300)	45	44.5	95@1.57	1.62-1.68	0.001-0.003	0.002-0.006	0.3120-0.3130	0.3110-0.3120
	R ①	3.3 (3300)	45	44.5	60@1.56	1.56	0.001-0.003	0.002-0.006	0.3110-0.3120	0.3110-0.3120
1994	K	2.5 (2507)	45	45	115@1.65	1.65	0.001-0.005	0.003-0.005	0.3124	0.3103
	3	3.0 (2972)	44.5	45.5	73@1.59	1.59	0.001-0.002	0.002-0.003	0.3130-0.3140	0.3120-0.3130
	R	3.3 (3300)	45	44.5	95@1.57	1.62-1.68	0.001-0.003	0.002-0.006	0.3120-0.3130	0.3110-0.3120
	L	3.8 (3785)	45	44.5	95@1.57	1.62-1.68	0.001-0.003	0.002-0.006	0.3120-0.3130	0.3110-0.3120
	L ①	3.8 (3785)	45	44.5	60@1.56	1.56	0.001-0.003	0.003-0.005	0.3120-0.3130	0.3110-0.3120
1995	K	2.5 (2507)	45	45	115@1.65	1.65	0.001-0.003	0.003-0.005	0.3124	0.3103
	3	3.0 (2972)	44.5	45.5	73@1.59	1.59	0.001-0.002	0.002-0.003	0.3130-0.3140	0.3120-0.3130
	R	3.3 (3300)	45	44.5	95@1.57	1.62-1.68	0.001-0.003	0.002-0.006	0.3120-0.3130	0.3110-0.3120
	L	3.8 (3785)	45	44.5	95@1.57	1.62-1.68	0.001-0.003	0.002-0.006	0.3120-0.3130	0.3110-0.3120
1996-97	B	2.4 (2429)	45	44.5-45	121-143@1.17	1.50	0.002-0.003	0.003-0.004	0.234	0.233
	3	3.0 (2972)	44.5	45.5	73@1.59	1.59	0.001-0.002	0.002-0.003	0.313-0.314	0.312-0.313
	R	3.3 (3300)	45	44.5	207-229@1.17	1.62-1.68	0.001-0.003	0.002-0.006	0.312-0.313	0.311-0.312
	L	3.8 (3785)	45	44.5	207-229@1.17	1.62-1.68	0.001-0.003	0.002-0.006	0.312-0.313	0.311-0.312

① Town & Country

TORQUE SPECIFICATIONS
All readings in ft. lbs.

Year	Engine ID/VIN	Engine Displacement Liters (cc)	Cylinder Head Bolts	Main Bearing Bolts	Rod Bearing Bolts	Crankshaft Damper Bolts	Flywheel Bolts	Manifold Intake *	Manifold Exhaust	Spark Plugs	Lug Nut
1993	K	2.5 (2507)	①	②	③	50	70	17	17	26	95
	3	3.0 (2972)	80	60	38	110	70	17	17	20	95
	R	3.3 (3300)	④	②	③	40	70	17	17	30	95
	R ⑤	3.3 (3300)	④	②	③	50	70	17	17	26	95
1994	K	2.5 (2507)	①	②	③	85	70	17	17	20	95
	3 ⑥	3.0 (2972)	80	60	38	112	70	17	17	20	95
	R	3.3 (3300)	①	②	③	40	70	17	17	20	95
	L	3.8 (3785)	①	②	③	40	70	17	17	20	95
1995	K	2.5 (2507)	①	②	③	85	70	17	17	20	95
	3	3.0 (2972)	80	60	38	112	70	15	16	20	95
	R	3.3 (3300)	①	②	③	40	70	17	17	20	95
	L	3.8 (3785)	①	②	③	40	70	17	17	20	95
1996-97	B	2.4 (2429)	⑦	⑧	⑨	100	70	20	17	20	95
	3	3.0 (2972)	80	60	38	112	70	15	16	20	95
	R	3.3 (3300)	①	②	③	40	70	17	17	20	95
	L	3.8 (3785)	①	②	③	40	70	17	17	20	95

* NOTE: Applies to Lower Manifold only.

① Step 1: 45 ft. lbs.
   Step 2: 65 ft. lbs.
   Step 3: 65 ft. lbs.
   Step 4: Plus 1/4 turn
② Step 1: 30 ft. lbs.
   Step 2: Plus 1/4 turn

③ Step 1: 40 ft. lbs.
   Step 2: Plus 1/4 turn
④ Step 1: 45 ft. lbs.
   Step 2: 65 ft. lbs.
   Step 3: 65 ft. lbs.
   Step 4: Plus 1/4 turn
   Torque small bolt in rear of head to 25 ft. lbs.

⑤ Town & Country
⑥ Step 1: 26 ft. lbs.
   Step 2: 51 ft. lbs.
   Step 3: 73 ft. lbs.

⑦ Step 1: 25 ft. lbs.
   Step 2: 50 ft. lbs.
   Step 3: 50 ft. lbs.
   Step 4: Plus 1/4 turn

⑧ M8 bolts: 250 in. lbs. plus 1/4 turn
   M11 bolts: 30 ft. lbs. plus 1/4 turn
⑨ Step 1: 20 ft. lbs.
   Step 2: Plus 1/4 turn

BRAKE SPECIFICATIONS
All measurements in inches unless noted

Year	Model	Master Cylinder Bore	Brake Disc Original Thickness	Brake Disc Minimum Thickness	Brake Disc Maximum Runout	Brake Drum Diameter Original Inside Diameter	Brake Drum Diameter Max. Wear Limit	Brake Drum Diameter Maximum Machine Diameter	Minimum Lining Thickness Front	Minimum Lining Thickness Rear
1993	Caravan	0.940	0.940	0.880	0.005	9.00	9.09	9.06	0.060	0.060
	Town & Country	0.940	0.861	0.800	0.005	9.00	9.09	9.06	0.062	0.062
	Voyager	0.940	0.940	0.880	0.005	9.00	9.09	9.06	0.060	0.060
1994	Caravan	0.940	0.940	0.880	0.005	9.00	9.09	9.06	0.060	0.060 ④
	Town & Country	0.940	0.940	0.880	0.005	9.00	9.09	9.06	0.062	0.062 ④
	Voyager	0.940	0.940	0.880	0.005	9.00	9.09	9.06	0.060	0.060 ④
1995	Caravan	0.940	0.940	0.880	0.005	9.00	9.09	9.06	0.060	0.060 ④
	Town & Country	0.940	0.940	0.880	0.005	9.00	9.09	9.06	0.062	0.062 ④
	Voyager	0.940	0.940	0.880	0.005	9.00	9.09	9.06	0.060	0.060 ④
1996-97	Caravan	0.940	0.940	0.880	0.005	9.00	9.09	9.06	0.060	0.060 ④
	Town & Country	0.937	0.940	0.881	0.005	9.84	9.93	9.90	0.062	0.062 ④
	Voyager	0.937	0.939	0.881	0.005	9.84	9.93	9.90	0.060	0.060 ④

① Except 4000 lb. rear axle: 1.240 in.
   With 4000 lb. rear axle: 1.180 in.
② With 3300 lb. or 3600 lb. rear axle: 1.180 in.
   With 4000 lb. rear axle: 1.125 in.
③ Minimum thickness indicated on rotor hub
④ For riveted brake shoes: 0.031

FREQUENT MAINTENANCE LABOR
CARAVAN, VOYAGER, TOWN & COUNTRY

The following should be used as a guide when determining the amount of work required for a particular service if taken to a repair shop. In estimating how long a particular Frequent Maintenance Service item should take, please observe the following:
- **Factory Time** is time that is generated by the vehicle manufacturer.
- **Chilton Time** is time that is based on field research and data supplied by the vehicle manufacturer.
- All labor time operations are given in hours and tenths of an hour.
- All labor operations, are to be used as a **guide**.

	(Factory Time)	Chilton Time

COOLING

(G) Winterize Cooling System
Includes: Run engine to check for leaks, tighten all hose connections. Test radiator and pressure cap, drain radiator and engine block. Add antifreeze and refill system.
1993-97 5

(G) Belt, Serpentine Drive, Renew
1993-97 (.2) 4
w/AC add 1

(G) Belt, Drive, Renew
1993-97
 fan & alternator (.2) 3
w/AC add 1
 power steering (.4) 6
w/AC add 1
 AC (.2) 3

(G) Belt, Drive, Adjust
1993-97
 one .2
 each adtnl. 1

(G) Hoses, Radiator, Renew
Includes: Drain and refill cooling system as required.
1993-97
 upper (.4) 6
 lower (.5) 8

(G) Thermostat, Coolant, Renew
1993-97 (.4) 6

FUEL

(M) Air Cleaner, Service
1993-97 3

(G) Filter, Fuel, Renew
1993-97
 in line (.3) 4
 in tank (1.0) 1.6
w/AWD add 1

BRAKES

(G) Bleed Brakes (Four Wheels)
Includes: Add fluid.
1993-95 (.4) 6
w/antilock 4 add 4
Bleed modulator, add 2.2
1996-97 (.4) 6
Bleed modulator, add 4

(G) Brakes, Adjust (Minor)
Includes: Adjust brakes, fill master cylinder.
1993-97, two wheels 4

(M) Parking Brake, Adjust
1993-97 (.3) 4

LUBRICATION SERVICES

(M) Engine Oil & Filter, Renew
Includes: Inspect and correct all fluid levels.
1993-97 (.2) 3

(M) Lubricate Chassis
Includes: Inspect and correct all fluid levels.
1993-97 4
Install grease fittings, add 1

WHEELS

(G) Wheel, Renew (One)
1993-97, one 5

(G) Wheel, Balance
1993-97
 one .3
 each adtnl. 2

(G) Wheels, Rotate (All)
1993-97 5

ELECTRICAL

(G) Headlamps, Aim
1993-97 4

(G) Halogen Headlamp Bulb, Renew
1993-97, each (.2) 3

(G) License Lamp Assy., Renew
1993-95 (.2) 3
1996-97 (.5) 8

(G) Tail Lamp Assy., Renew
1993-97 (.2) 4
w/dual rear doors, add 2

(G) Turn Signal & Parking Lamp Assy., Renew
1993-97 (.2) 3

(G) Horn, Renew
1993-97, one (.5) 7

(M) Terminals, Battery, Clean
1993-97 3

SCHEDULED MAINTENANCE INTERVALS
(CHRYSLER CARAVAN, TOWN & COUNTRY, VOYAGER)

TO BE SERVICED	TYPE OF SERVICE	VEHICLE MILEAGE INTERVAL (x1000)												
		7.5	15	22.5	30	37.5	45	52.5	60	67.5	75	82.5	90	97.5
Engine oil & filter	R	✓	✓	✓	✓	✓	✓	✓	✓	✓	✓	✓	✓	✓
Driveshaft boots	S/I	✓	✓	✓	✓	✓	✓	✓	✓	✓	✓	✓	✓	✓
Exhaust system	S/I	✓	✓	✓	✓	✓	✓	✓	✓	✓	✓	✓	✓	✓
Engine coolant level, hoses & clamps	S/I	✓	✓	✓	✓	✓	✓	✓	✓	✓	✓	✓	✓	✓
Rotate tires	S/I	✓	✓	✓	✓	✓	✓	✓	✓	✓	✓	✓	✓	✓
Drive belts	S/I		✓		✓		✓		✓		✓		✓	
Brake hoses & linings	S/I			✓			✓			✓			✓	
Automatic transaxle fluid & filter	R				✓				✓				✓	
Air filter	R				✓				✓				✓	
Spark plugs① (2.5L & 3.0L)	R				✓				✓				✓	
Serpentine belts (3.0L & 3.3L)	S/I								✓		✓		✓	
Lubricate tie rod ends	S/I				✓				✓				✓	
PCV valve	S/I				✓				✓				✓	
Engine Coolant	R							✓				✓		
Timing belt (2.5L)	R												✓	
Timing belt (3.0L)	R								✓					
Distributor cap & rotor	R							✓						
Ignition cables (2.5L & 3.0L)	R								✓					
Ignition timing	S/I								✓					

① Platinum tip spark plugs & ignition cables (3.3L & 3.8L) - replace every 100,000 miles.

R – Replace   S/I – Service or Inspect

FREQUENT OPERATION MAINTENANCE (SEVERE SERVICE)

If a vehicle is operated under any of the following conditions it is considered severe service:
- **Extremely dusty areas.**
- **50% or more of the vehicle operation is in 32°C (90°F) or higher temperatures, or constant operation in temperatures below 0°C (32°F).**
- **Prolonged idling (vehicle operation in stop and go traffic).**
- **Frequent short running periods (engine does not warm to normal operating temperatures).**
- **Police, taxi, delivery usage or trailer towing usage.**

Oil & oil filter change – change every 3000 miles.
Automatic transaxle fluid & filter - change every 15,000 miles.
Brake hoses & linings - check every 9000 miles.
CV joints & front suspension ball joints - check every 3000 miles.
Tie rod ends & steering linkage - check every 15,000 miles.
Air filter - change every 15,000 miles.

FORD MOTOR COMPANY
E Series (Van) • F Series (Pick Up) • Bronco • Expedition
Aerostar • Ranger • Explorer • Mountaineer

VEHICLE IDENTIFICATION CHART

Engine Code							Model Year	
Code	Liters	Cu. In. (cc)	Cyl.	Fuel Sys.	Eng. Mfg.		Code	Year
2	4.2	256 (4195)	6	MFI	Ford		P	1993
6	4.6	280 (4588)	8	MFI	Ford		R	1994
9	4.6	280 (4588)	8	NA	Ford		S	1995
A	2.3	140 (2294)	4	MFI	Ford		T	1996
F	7.3	445 (7292)	8	DI	Navistar		V	1997
G	7.5	460 (7538)	8	MFI	Ford			
H	5.8	351 (5752)	8	MFI	Ford			
K	7.3	445 (7292)	8	DSL	Navistar			
M	7.3	445 (7292)	8	DSL	Ford			
N	5.0	302 (4949)	8	MFI	Ford			
P	5.0	302 (4949)	8	MFI	Ford			
R	5.8	351 (5752)	8	MFI	Ford			
U	3.0	183 (2999)	6	MFI	Ford			
W	4.6	280 (4588)	8	MFI	Ford			
X	4.0	244 (3998)	6	MFI	Ford			
Y	4.9	300 (4916)	6	MFI	Ford			

MFI - Multiport fuel injection   DSL - Diesel   DI - Direct injection Turbo

ENGINE IDENTIFICATION

Year	Model	Engine Displacement Liters (cc)	Engine Series (ID/VIN)	Fuel System	No. of Cylinders	Engine Type
1993	Aerostar	3.0 (2999)	U	MFI	6	OHV
	Aerostar	4.0 (3998)	X	MFI	6	OHV
	Bronco	4.9 (4916)	Y	MFI	6	OHV
	Bronco	5.0 (4949)	N	MFI	8	OHV
	Bronco	5.8 (5752)	H	MFI	8	OHV
	E-150	4.9 (4916)	Y	MFI	6	OHV
	E-150	5.0 (4949)	N	MFI	8	OHV
	E-150	5.8 (5752)	H	MFI	8	OHV
	E-250	4.9 (4916)	Y	MFI	6	OHV
	E-250	5.0 (4949)	N	MFI	8	OHV
	E-250	5.8 (5752)	H	MFI	8	OHV
	E-250	7.3 (7292)	M	IDI	8	OHV
	E-250	7.5 (7538)	G	MFI	8	OHV
	E-350	4.9 (4916)	Y	MFI	6	OHV
	E-350	5.8 (5752)	H	MFI	8	OHV
	E-350	7.3 (7292)	M	IDI	8	OHV

ENGINE IDENTIFICATION

Year	Model	Engine Displacement Liters (cc)	Engine Series (ID/VIN)	Fuel System	No. of Cylinders	Engine Type
1993	E-350	7.5 (7538)	G	MFI	8	OHV
	Explorer	4.0 (3998)	X	MFI	6	OHV
	F-150	4.9 (4916)	Y	MFI	6	OHV
	F-150	5.0 (4949)	N	MFI	8	OHV
	F-150	5.8 (5752)	H	MFI	8	OHV
	F-250	4.9 (4916)	Y	MFI	6	OHV
	F-250	5.0 (4949)	N	MFI	8	OHV
	F-250	5.8 (5752)	H	EFI	8	OHV
	F-250	7.3 (7292)	M	IDI	8	OHV
	F-250	7.5 (7538)	G	EFI	8	OHV
	F-350	4.9 (4916)	Y	EFI	6	OHV
	F-350	5.8 (5752)	H	EFI	8	OHV
	F-350	7.3 (7292)	M	DDI	8	OHV
	F-350	7.5 (7538)	G	EFI	8	OHV
	F-Super Duty	7.3 (7292)	M	DDI	8	OHV
	F-Super Duty	7.5 (7538)	G	EFI	8	OHV
	Lightning Pick-up	5.8 (5752)	R	EFI	8	OHV
	Ranger	2.3 (2294)	A	MFI	4	SOHC
	Ranger	3.0 (2999)	U	MFI	6	OHV
	Ranger	4.0 (3998)	X	MFI	6	OHV
1994	Aerostar	3.0 (2999)	U	MFI	6	OHV
	Aerostar	4.0 (3998)	X	MFI	6	OHV
	Bronco	5.0 (4949)	N	MFI	8	OHV
	Bronco	5.8 (5752)	H	MFI	8	OHV
	E-150	4.9 (4916)	Y	MFI	6	OHV
	E-150	5.0 (4949)	N	MFI	8	OHV
	E-150	5.8 (5752)	H	MFI	8	OHV
	E-250	4.9 (4916)	Y	MFI	6	OHV
	E-250	5.0 (4949)	N	MFI	8	OHV
	E-250	5.8 (5752)	H	MFI	8	OHV
	E-350	4.9 (4916)	Y	MFI	6	OHV
	E-350	5.8 (5752)	H	MFI	8	OHV
	E-350	7.3 (7292)	F	DI	8	OHV
	E-350	7.3 (7292)	K	IDI	8	OHV
	E-350	7.3 (7292)	M	IDI	8	OHV
	E-350	7.5 (7538)	G	MFI	8	OHV
	Explorer	4.0 (3998)	X	MFI	6	OHV
	F-150	4.9 (4916)	Y	MFI	6	OHV
	F-150	5.0 (4949)	N	MFI	8	OHV
	F-150	5.8 (5752)	H	MFI	8	OHV
	F-250	4.9 (4916)	Y	MFI	6	OHV
	F-250	5.0 (4949)	N	MFI	8	OHV
	F-250	5.8 (5752)	H	EFI	8	OHV
	F-250	7.3 (7292)	M	IDI	8	OHV
	F-250	7.5 (7538)	G	EFI	8	OHV
	F-350	4.9 (4916)	Y	EFI	6	OHV
	F-350	5.8 (5752)	H	EFI	8	OHV
	F-350	7.3 (7292)	F	DI	8	OHV
	F-350	7.3 (7292)	K	DDI	8	OHV
	F-350	7.3 (7292)	M	DDI	8	OHV
	F-350	7.5 (7538)	G	EFI	8	OHV
	F-Super Duty	7.3 (7292)	F	DI	8	OHV

ENGINE IDENTIFICATION

Year	Model	Engine Displacement Liters (cc)	Engine Series (ID/VIN)	Fuel System	No. of Cylinders	Engine Type
1994	F-Super Duty	7.3 (7292)	K	IDI	8	OHV
	F-Super Duty	7.3 (7292)	M	DDI	8	OHV
	F-Super Duty	7.5 (7538)	G	EFI	8	OHV
	Lightning Pick-up	5.8 (5752)	R	EFI	8	OHV
	Ranger	2.3 (2294)	A	MFI	4	SOHC
	Ranger	3.0 (2999)	U	MFI	6	OHV
	Ranger	4.0 (3998)	X	MFI	6	OHV
1995	Aerostar	3.0 (2999)	U	MFI	6	OHV
	Aerostar	4.0 (3998)	X	MFI	6	OHV
	Bronco	5.0 (4949)	N	MFI	8	OHV
	Bronco	5.8 (5752)	H	MFI	8	OHV
	E-150	4.9 (4916)	Y	MFI	6	OHV
	E-150	5.0 (4949)	N	MFI	8	OHV
	E-150	5.8 (5752)	H	MFI	8	OHV
	E-250	4.9 (4916)	Y	MFI	6	OHV
	E-250	5.0 (4949)	N	MFI	8	OHV
	E-250	5.8 (5752)	H	MFI	8	OHV
	E-350	4.9 (4916)	Y	MFI	6	OHV
	E-350	5.8 (5752)	H	MFI	8	OHV
	E-350	7.3 (7292)	F	DI	8	OHV
	E-350	7.3 (7292)	K	IDI	8	OHV
	E-350	7.3 (7292)	M	IDI	8	OHV
	E-350	7.5 (7538)	G	MFI	8	OHV
	Explorer	4.0 (3998)	X	MFI	6	OHV
	F-150	4.9 (4916)	Y	MFI	6	OHV
	F-150	5.0 (4949)	N	MFI	8	OHV
	F-150	5.8 (5752)	H	MFI	8	OHV
	F-250	4.9 (4916)	Y	MFI	6	OHV
	F-250	5.0 (4949)	N	MFI	8	OHV
	F-250	5.8 (5752)	H	MFI	8	OHV
	F-250	7.3 (7292)	M	IDI	8	OHV
	F-250	7.5 (7538)	G	MFI	8	OHV
	F-350	4.9 (4916)	Y	MFI	6	OHV
	F-350	5.8 (5752)	H	MFI	8	OHV
	F-350	7.3 (7292)	K	IDI	8	OHV
	F-350	7.3 (7292)	F	DI	8	OHV
	F-350	7.3 (7292)	M	IDI	8	OHV
	F-350	7.5 (7538)	G	MFI	8	OHV
	F-Super Duty	7.3 (7292)	F	DI	8	OHV
	F-Super Duty	7.3 (7292)	M	DDI	8	OHV
	F-Super Duty	7.3 (7292)	K	IDI	8	OHV
	F-Super Duty	7.5 (7538)	G	EFI	8	OHV
	Lightning Pick-up	5.8 (5752)	R	EFI	8	OHV
	Ranger	2.3 (2294)	A	MFI	4	SOHC
	Ranger	3.0 (2999)	U	MFI	6	OHV
	Ranger	4.0 (3998)	X	MFI	6	OHV
1996-97	Aerostar	3.0 (2982)	U	MFI	6	OHV
	Aerostar	4.0 (3950)	X	MFI	6	OHV
	Bronco	5.0 (4949)	N	MFI	8	OHV
	Bronco	5.8 (5752)	H	MFI	8	OHV
	E-150	4.9 (4916)	Y	MFI	6	OHV
	E-150	5.0 (4949)	N	MFI	8	OHV
	E-150	5.8 (5752)	H	MFI	8	OHV

ENGINE IDENTIFICATION

Year	Model	Engine Displacement Liters (cc)	Engine Series (ID/VIN)	Fuel System	No. of Cylinders	Engine Type
1996-97	E-250	4.9 (4916)	Y	MFI	6	OHV
	E-250	5.0 (4949)	N	MFI	8	OHV
	E-250	5.8 (5752)	H	MFI	8	OHV
	E-350	4.9 (4916)	Y	MFI	6	OHV
	E-350	5.8 (5752)	H	MFI	8	OHV
	E-350	7.3 (7292)	F	DI	8	OHV
	E-350	7.5 (7538)	G	MFI	8	OHV
	Explorer	4.0 (3950)	X	MFI	6	OHV
	Explorer	5.0 (4949)	P	MFI	8	OHV
	F-150	4.2 (4195)	2	MFI	6	OHV
	F-150	4.6 (4588)	W	MFI	8	SOHC
	F-150	4.6 (4588)	6	MFI	8	SOHC
	F-150	4.6 (4588)	9	NA	8	SOHC
	F-150	4.9 (4916)	Y	MFI	6	OHV
	F-150	5.0 (4949)	N	MFI	8	OHV
	F-150	5.8 (5752)	H	MFI	8	OHV
	F-250	4.9 (4916)	Y	MFI	6	OHV
	F-250	5.0 (4949)	N	MFI	8	OHV
	F-250	5.8 (5752)	H	MFI	8	OHV
	F-250	7.3 (7292)	F	DI	8	OHV
	F-250	7.5 (7538)	G	MFI	8	OHV
	F-350	4.9 (4916)	Y	MFI	6	OHV
	F-350	5.8 (5752)	H	MFI	8	OHV
	F-350	7.3 (7292)	F	DI	8	OHV
	F-350	7.5 (7538)	G	MFI	8	OHV
	F-Super Duty	7.3 (7292)	F	DI	8	OHV
	F-Super Duty	7.5 (7538)	G	MFI	8	OHV
	Mountaineer	5.0 (4949)	P	MFI	8	OHV
	Ranger	2.3 (2300)	A	MFI	4	SOHC
	Ranger	3.0 (2982)	U	MFI	6	OHV
	Ranger	4.0 (3950)	X	MFI	6	OHV

MFI - Multiport fuel injection
DDI - Direct diesel injection
OHV - Overhead valve
SOHC - Single overhead camshaft
IDI - Indirect diesel injection
EFI - Electronic fuel injection
DI - Direct injection Turbo

GENERAL ENGINE SPECIFICATIONS

Year	Engine ID/VIN	Engine Displacement Liters (cc)	Fuel System Type	Net Horsepower @ rpm	Net Torque @ rpm (ft. lbs.)	Bore x Stroke (in.)	Compression Ratio	Oil Pressure @ rpm
1993	A	2.3 (2294)	EFI	100@4600	133@2600	3.78x3.13	9.2:1	40-60@2000
	C	7.3 (7292)	IDI	190@3000	395@1400	4.11x4.18	21.5:1	40-70@2000
	G	7.5 (7538)	EFI	245@4000	400@2200	4.36x3.85	8.5:1	40-65@2000
	H	5.8 (5752)	EFI	200@3800	310@2800	4.00x3.50	8.8:1	40-65@2000
	M	7.3 (7292)	IDI	185@3000 ③	360@400 ④	4.11x4.18	21.5:1	40-70@2000
	N	5.0 (4949)	EFI	185@3800	270@2400	4.00x3.98	9.0:1	40-60@2000
	R	5.8 (5752)	EFI	240@4200	340@3200	4.00x3.50	8.8:1	40-65@2000
	U	3.0 (2999)	EFI	145@4800	165@3600	3.50x3.14	9.3:1	40-60@2500
	X	4.0 (3998)	EFI	①	②	3.95x3.32	9.0:1	40-60@2000
	Y	4.9 (4916)	EFI	①	②	4.00x3.98	8.8:1	40-60@2000

GENERAL ENGINE SPECIFICATIONS

Year	Engine ID/VIN	Engine Displacement Liters (cc)	Fuel System Type	Net Horsepower @ rpm	Net Torque @ rpm (ft. lbs.)	Bore x Stroke (in.)	Compression Ratio	Oil Pressure @ rpm
1994	A	2.3 (2294)	EFI	100@4600	133@2600	3.78x3.13	9.2:1	40-60@2000
	F	7.3 (7292)	DI	210@3000	425@2000	4.11x4.18	17.5:1	40-70@3000
	G	7.5 (7538)	EFI	245@4000	400@2200	4.36x3.85	8.5:1	40-88@2000
	H	5.8 (5752)	EFI	210@3600	325@2800	4.00x3.50	8.8:1	40-65@2000
	K	7.3 (7292)	IDI	190@3000	395@1400	4.11x4.18	21.5:1	40-70@2000
	M	7.3 (7292)	IDI	185@3000 ③	360@1400 ④	4.11x4.18	21.5:1	40-70@2000
	N	5.0 (4949)	EFI	205@4000	275@3000	4.00x3.00	9.0:1	40-60@2000
	R	5.8 (5752)	EFI	240@4200	340@3200	4.00x3.50	8.8:1	40-65@2000
	U	3.0 (2999)	MFI	135@4600	160@2800	3.50x3.14	9.3:1	40-60@2500
	X	4.0 (3998)	EFI	160@4000	225@2500	3.95x3.32	9.0:1	40-60@2000
	Y	4.9 (4916)	EFI	①	②	4.00x3.98	8.8:1	40-60@2000
1995	A	2.3 (2294)	EFI	100@4600	133@2600	3.78x3.13	9.2:1	40-60@2000
	F	7.3 (7292)	DI	210@3000	425@2000	4.11x4.18	17.5:1	40-70@3000
	G	7.5 (7538)	EFI	245@4000	400@2200	4.36x3.85	8.5:1	40-88@2000
	H	5.8 (5752)	EFI	210@3600	325@2800	4.00x3.50	8.8:1	40-65@2000
	M	7.3 (7292)	IDI	185@3000 ③	360@1400 ④	4.11x4.18	21.5:1	40-70@2000
	N	5.0 (4949)	EFI	205@4000	275@3000	4.00x3.00	9.0:1	40-60@2000
	R	5.8 (5752)	EFI	240@4200	340@3200	4.00x3.50	8.8:1	40-65@2000
	U	3.0 (2999)	MFI	135@4600	160@2800	3.50x3.14	9.3:1	40-60@2500
	X	4.0 (3998)	EFI	160@4000	225@2500	3.81x3.39	9.0:1	40-60@2000
	Y	4.9 (4916)	EFI	145@3400 ⑤	265@2000 ⑤	4.00x3.98	8.8:1	40-60@2000
1996-97	2	4.2 (4195)	MFI	205@4400	255@3000	3.81x3.74	9.3:1	50@2000
	6	4.6 (4588)	MFI	210@4400	290@3250	3.55x3.54	9.0:1	20-45@1500
	9	4.6 (4588)	NA	NA	NA	3.55x3.54	9.0:1	20-45@1500
	A	2.3 (2300)	MFI	112@4800	135@2400	3.78x3.13	9.4:1	40-60@2000
	F	7.3 (7292)	DI	210@3000	425@2000	4.11x4.18	17.5:1	40-70@3000
	G	7.5 (7538)	MFI	245@4000	400@2200	4.36x3.85	8.5:1	40-88@2000
	H	5.8 (5752)	MFI	210@3600	325@2800	4.00x3.50	8.8:1	40-65@2000
	N	5.0 (4949)	MFI	199@4200	270@2400	4.00x3.00	9.0:1	40-60@2000
	P	5.0 (4949)	MFI	210@4500	280@3500	4.00x3.00	9.0:1	40-60@2500
	U	3.0 (2982)	MFI	147@5000	162@3250	3.50x3.14	9.3:1	40-60@2500
	W	4.6 (4588)	MFI	210@4400	290@3250	3.55x3.54	9.0:1	20-45@1500
	X	4.0 (3950)	MFI	160@4000	225@2500	3.81x3.39	9.0:1	40-60@2000
	Y	4.9 (4916)	MFI	145@3400 ⑤	265@2000 ⑤	4.00x3.98	8.8:1	40-60@2000

NA - Not Available
MFI - Multiport fuel injection
EFI - Electronic fuel injection
IDI - Indirect diesel injection
DI - Direct injection Turbo
① E-Series 3 spd automatic: 150@3400
   E-Series 4 spd OD automatic: 145@3400
   F-Series 5 spd manual or 4 spd OD automatic
   and 2.73 rear axle ratio: 145@3400
   F-Series 5 spd HD or 3 spd automatic: 150@3400

② 3 spd automatic: 260@2000
   4 spd OD automatic: 265@2000
   5 spd manual OD or 4 spd automatic OD: 265@2000
   5 spd manual HD or 3 spd automatic: 260@2000
③ High altitude: 165@3000
④ High altitude: 325@1600
⑤ Ratings are for E150-250 Van and regular Wagon with
   4 spd automatic OD (E40D). Use 150hp@3400 rpm and
   260 ft. lbs. @2000 rpm for all other applications

GASOLINE ENGINE TUNE-UP SPECIFICATIONS

Year	Engine ID/VIN	Engine Displacement Liters (cc)	Spark Plugs Gap (in.)	Ignition Timing (deg.) MT	Ignition Timing (deg.) AT	Fuel Pump (psi)	Idle Speed (rpm) MT	Idle Speed (rpm) AT	Valve Clearance In.	Valve Clearance Ex.
1993	A	2.3 (2294)	0.044	10B	10B	35-45	725	675	HYD	HYD
	G	7.5 (7538)	0.044	10B	10B	35-45	775	675	HYD	HYD
	H	5.8 (5752)	0.044	10B	10B	35-45	775	675	HYD	HYD
	N	5.0 (4949)	0.044	10B	10B	35-45	775	675	HYD	HYD
	R	5.8 (5752)	0.044	10B	10B	35-45	775	675	HYD	HYD
	U	3.0 (2999)	0.044	10B	10B	35-45	①	①	HYD	HYD
	X	4.0 (3998)	0.054	10B	10B	35-45	①	①	HYD	HYD
	Y	4.9 (4916)	0.044	10B	10B	50-60	700	575	HYD	HYD
1994	A	2.3 (2294)	0.044	10B	10B	35-45	725	675	HYD	HYD
	G	7.5 (7538)	0.044	10B	10B	35-45	775	675	HYD	HYD
	H	5.8 (5752)	0.044	10B	10B	35-45	775	675	HYD	HYD
	N	5.0 (4949)	0.044	10B	10B	35-45	775	675	HYD	HYD
	R	5.8 (5752)	0.044	10B	10B	35-45	775	675	HYD	HYD
	U	3.0 (2999)	0.044	10B	10B	35-45	①	①	HYD	HYD
	X	4.0 (3998)	0.054	10B	10B	35-45	①	①	HYD	HYD
	Y	4.9 (4916)	0.044	10B	10B	50-60	700	575	HYD	HYD
1995	A	2.3 (2294)	0.044	10B	10B	35-45	725	675	HYD	HYD
	G	7.5 (7538)	0.044	10B	10B	35-45	775	675	HYD	HYD
	H	5.8 (5752)	0.044	10B	10B	35-45	775	675	HYD	HYD
	N	5.0 (4949)	0.044	10B	10B	35-45	775	675	HYD	HYD
	R	5.8 (5752)	0.044	10B	10B	35-45	775	675	HYD	HYD
	U	3.0 (2999)	0.044	10B	10B	35-45	①	①	HYD	HYD
	X	4.0 (3998)	0.054	10B	10B	35-45	①	①	HYD	HYD
	Y	4.9 (4916)	0.044	10B	10B	50-60	700	575	HYD	HYD
1996-97	A	2.3 (2300)	0.044	10B	10B	35-45	725	675	HYD	HYD
	U	3.0 (2982)	0.044	10B	10B	35-45	①	①	HYD	HYD
	2	4.2 (4195)	0.052-0.056	10B ②	10B ②	30-45 ③	NA	NA	HYD	HYD
	W	4.6 (4588)	0.052-0.056	8-12B ②	8-12B ②	30-45 ③	NA	NA	HYD	HYD
	6	4.6 (4588)	0.052-0.056	8-12B ②	8-12B ②	30-45 ③	NA	NA	HYD	HYD
	9	4.6 (4588)	NA	NA	NA	30-45 ③	NA	NA	HYD	HYD
	X	4.0 (3950)	0.054	10B	10B	35-45	①	①	HYD	HYD
	P	5.0 (4949)	0.044	-	10B	35-45	-	①	HYD	HYD
	Y	4.9 (4916)	0.044	10B	10B	50-60	700	575	HYD	HYD
	N	5.0 (4949)	0.044	10B	10B	35-45	775	675	HYD	HYD
	H	5.8 (5752)	0.044	10B	10B	35-45	775	675	HYD	HYD
	G	7.5 (7538)	0.044	10B	10B	35-45	775	675	HYD	HYD

NOTE: The Vehicle Emission Control Information label often reflects specification changes changes made during production. The label figures must be used if they differ from those in this chart.
B - Before top dead center
HYD - Hydraulic
NA - Not Available
① Idle speed is electronically controlled and cannot be adjusted
② Ignition timing is preset and cannot be adjusted
③ With engine running

DIESEL ENGINE TUNE-UP SPECIFICATIONS

Year	Engine ID/VIN	Engine Displacement cu. in. (cc)	Valve Clearance Intake (in.)	Exhaust (in.)	Intake Valve Opens (deg.)	Injection Pump Setting (deg.)	Injection Nozzle Pressure (psi) New	Used	Idle Speed (rpm)	Cranking Compression Pressure (psi)
1993	M	7.3 (7292)	HYD	HYD	-	8.5B ①	1875	1425	②	④
1994	M	7.3 (7292)	HYD	HYD	-	8.5B ①	1875	1425	②	④
	F	7.3 (7292)	HYD	HYD	-	③	NA	NA	②	④
	K	7.3 (7292)	HYD	HYD	-	③	NA	NA	②	④
1995	F	7.3 (7292)	HYD	HYD	-	③	1875	1425	②	④
	M	7.3 (7292)	HYD	HYD	-	8.5B ①	NA	NA	②	④
1996-97	F	7.3 (7292)	HYD	HYD	-	③	1875	1425	②	④

NOTE: The Vehicle Emission Control Information label often reflects specification changes made during production. The label figures must be used if they differ from those in this chart

HYD - Hydraulic
B - Before top dead center
NA - Not Available

① At 2000 rpm
② See underhood emission label
③ PCM controlled

④ Compression pressure in the lowest cylinder must be at least 75% of the highest cylinder
Minimum pressure: 195 psi
Maximum pressure: 440 psi

CAPACITIES

Year	Model	Engine ID/VIN	Engine Displacement Liters (cc)	Engine Oil with Filter (qts.)	Transmission (pts.) 4-Spd	5-Spd	Auto.	Transfer Case (pts.)	Drive Axle Front (pts.)	Rear (pts.)	Fuel Tank (gal.)	Cooling System (qts.)
1993	Aerostar	U	3.0 (2999)	5.0	-	5.6	02.0	2.5	3.0	⑭	21.0	11.8
	Aerostar	X	4.0 (3998)	5.0	-	5.6	19.0	2.5	3.0	⑭	21.0	8.5
	Bronco	Y	4.9 (4916)	6.0	7.0	7.0	24.0	①	5.5	5.5	32.0	14.0
	Bronco	N	5.0 (4949)	6.0	7.0 ②	7.0	24.0	①	5.5	5.5	32.0	14.0
	Bronco	H	5.8 (5752)	6.0	7.0 ②	7.0	24.0	①	5.5	5.5	32.0	15.0
	E-150	Y	4.9 (4916)	6.0	7.0	7.0	24.0	-	-	6.0 ③	④	17.5
	E-150	N	5.0 (4949)	6.0	7.0 ②	7.0	24.0	-	-	6.0 ③	④	⑤
	E-150	H	5.8 (5752)	6.0	7.0 ②	7.0	24.0	-	-	6.0 ③	④	⑥
	E-250	Y	4.9 (4916)	6.0	7.0 ②	7.0	24.0	-	-	6.0 ③	④	17.5
	E-250	N	5.0 (4949)	6.0	7.0 ②	7.0	24.0	-	-	6.0 ③	④	⑤
	E-250	H	5.8 (5752)	6.0	7.0 ②	7.0	24.0	-	-	6.0 ③	④	⑥
	E-250	M	7.3 (7292)	10.0	7.0 ②	7.0	24.0	-	-	6.0 ③	④	31.0
	E-250	L	7.5 (7538)	6.0	7.0 ②	7.0	24.0	-	-	6.0 ③	④	28.0
	E-350	Y	4.9 (4916)	6.0	7.0 ②	7.0	24.0	-	-	6.0 ③	④	17.5
	E-350	H	5.8 (5752)	6.0	7.0 ②	7.0	24.0	-	-	6.0 ③	④	⑥
	E-350	C	7.3 (7292)	10.0	7.0 ②	7.0	24.0	-	-	6.0 ③	④	31.0
	E-350	M	7.3 (7292)	10.0	7.0 ②	7.0	24.0	-	-	6.0 ③	④	31.0
	E-350	L	7.5 (7538)	6.0	7.0 ②	7.0	24.0	-	-	6.0 ③	④	28.0
	Explorer	X	4.0 (3998)	5.0	-	5.6	19.0	3.0	⑭	⑭	21.0	8.5
	F-150	Y	4.9 (4916)	6.0	7.0	7.0	24.0 ⑬	①	6.0	6.0 ③	19.0	14.0
	F-150	N	5.0 (4949)	6.0	7.0 ②	7.0	24.0 ⑬	①	6.0	6.0 ③	19.0	14.0
	F-150	H	5.8 (5752)	6.0	7.0 ②	7.0	24.0 ⑬	①	6.0	6.0 ③	19.0	15.0
	F-250	Y	4.9 (4916)	6.0	7.0 ②	7.0	24.0 ⑬	①	6.0	6.0 ③	19.0	14.0
	F-250	N	5.0 (4949)	6.0	7.0 ②	7.0	24.0 ⑬	①	6.0	6.0 ③	19.0	14.0
	F-250	H	5.8 (5752)	6.0	7.0 ②	7.0	24.0 ⑬	①	6.0	6.0 ③	19.0	15.0
	F-250	C	7.3 (7292)	10.0	7.0 ②	7.0	24.0 ⑬	①	6.0	6.0 ③	19.0	31.0
	F-250	M	7.3 (7292)	10.0	7.0 ②	7.0	24.0 ⑬	①	6.0	6.0 ③	19.0	31.0
	F-250	L	7.5 (7538)	6.0	7.0 ②	7.0	24.0 ⑬	①	6.0	6.0 ③	19.0	16.0

CAPACITIES

Year	Model	Engine ID/VIN	Engine Displacement Liters (cc)	Engine Oil with Filter (qts.)	Transmission (pts.) 4-Spd	Transmission (pts.) 5-Spd	Transmission (pts.) Auto.	Transfer Case (pts.)	Drive Axle Front (pts.)	Drive Axle Rear (pts.)	Fuel Tank (gal.)	Cooling System (qts.)
1993	F-350	Y	4.9 (4916)	6.0	7.0 ②	7.0	24.0 ⑬	①	6.0	6.0 ③	19.0	17.5
	F-350	H	5.8 (5752)	6.0	7.0 ②	7.0	24.0 ⑬	①	6.0	6.0 ③	19.0	15.0
	F-350	C	7.3 (7292)	10.0	7.0 ②	7.0	24.0 ⑬	①	6.0	6.0 ③	19.0	31.0
	F-350	M	7.3 (7292)	10.0	7.0 ②	7.0	24.0 ⑬	①	6.0	6.0 ③	19.0	31.0
	F-350	L	7.5 (7538)	6.0	7.0 ②	7.0	24.0 ⑬	①	6.0	6.0 ③	19.0	16.0
	F-Super Duty	C	7.3 (7292)	10.0	7.0 ②	7.0	24.0 ⑬	①	6.0	6.0 ③	19.0	31.0
	Ranger	A	2.3 (2294)	5.0	-	⑮	19.4	3.0	⑭	5.5	⑯	7.2
	Ranger	U	3.0 (2999)	5.0	-	3.0	⑰	⑱	⑲	⑲	⑯	11.8
	Ranger	X	4.0 (3998)	5.0	-	3.0	⑰	⑱	⑲	⑲	⑯	8.5
1994	Aerostar	U	3.0 (2999)	4.5	-	5.6	19.0	2.5	3.0	⑭	21.0	11.8
	Aerostar	X	4.0 (3998)	5.0	-	5.6	19.0	2.5	3.0	⑭	21.0	12.6
	Bronco	Y	4.9 (4916)	6.0	7.0 ②	7.0	24.0	①	5.5	5.5	32.0	14.0
	Bronco	N	5.0 (4949)	6.0	7.0 ②	7.0	24.0	①	5.5	5.5	32.0	14.0
	Bronco	H	5.8 (5752)	6.0	7.0 ②	7.0	24.0	①	5.5	5.5	32.0	15.0
	E-150	Y	4.9 (4916)	6.0	7.0 ②	7.0	24.0	-	-	6.0 ③	④	14.0
	E-150	H	5.0 (4949)	6.0	7.0 ②	7.0	24.0	-	-	6.0 ③	④	15.0
	E-150	N	5.8 (5752)	6.0	7.0 ②	7.0	24.0	-	-	6.0 ③	④	14.0
	E-250	Y	4.9 (4916)	6.0	7.0 ②	7.0	24.0	-	-	6.0 ③	④	14.0
	E-250	N	5.0 (4949)	6.0	7.0 ②	7.0	24.0	-	-	6.0 ③	④	15.0
	E-250	H	5.8 (5752)	6.0	7.0 ②	7.0	24.0	-	-	6.0 ③	④	15.0
	E-250	M	7.3 (7292)	10.0	7.0 ②	7.0	24.0	-	-	6.0 ③	④	20.0
	E-250	L	7.5 (7538)	6.0	7.0 ②	7.0	24.0	-	-	6.0 ③	④	19.8
	E-350	Y	4.9 (4916)	6.0	7.0 ②	7.0	24.0	-	-	6.0 ③	④	17.5
	E-350	H	5.8 (5752)	6.0	7.0 ②	7.0	24.0	-	-	6.0 ③	④	15.0
	E-350	C	7.3 (7292)	10.0	7.0 ②	7.0	24.0	-	-	6.0 ③	④	20.0
	E-350	M	7.3 (7292)	10.0	7.0 ②	7.0	24.0	-	-	6.0 ③	④	20.0
	E-350	L	7.5 (7538)	6.0	7.0 ②	7.0	24.0	-	-	6.0 ③	④	19.8
	Explorer	X	4.0 (3998)	5.0	-	5.6	⑰	3.0	⑭	⑭	19.3	㉑
	F-150	Y	4.9 (4916)	6.0	7.0 ②	7.0	24.0 ⑬	①	6.0	6.0 ③	④	⑧
	F-150	N	5.0 (4949)	6.0	7.0 ②	7.0	24.0 ⑬	①	6.0	6.0 ③	④	⑨
	F-150	H	5.8 (5752)	6.0	7.0 ②	7.0	24.0 ⑬	①	6.0	6.0 ③	④	⑩
	F-250	Y	4.9 (4916)	6.0	7.0 ②	7.0	24.0 ⑬	①	6.0	6.0 ③	④	⑧
	F-250	N	5.0 (4949)	6.0	7.0 ②	7.0	24.0 ⑬	①	6.0	6.0 ③	④	⑧
	F-250	H	5.8 (5752)	6.0	7.0 ②	7.0	24.0 ⑬	①	6.0	6.0 ③	④	⑩
	F-250	C	7.3 (7292)	10.0	7.0 ②	7.0	24.0	-	-	6.0 ③	④	20.0
	F-250	L	7.5 (7538)	6.0	7.0 ②	7.0	24.0 ⑬	①	6.0	6.0 ③	19.0	19.8
	F-350	Y	4.9 (4916)	6.0	7.0 ②	7.0	24.0 ⑬	①	6.0	6.0 ③	19.0	⑧
	F-350	H	5.8 (5752)	6.0	7.0 ②	7.0	24.0 ⑬	①	6.0	6.0 ③	19.0	⑩
	F-350	C	7.3 (7292)	10.0	7.0 ②	7.0	24.0 ⑬	①	6.0	6.0 ③	19.0	20.0
	F-350	M	7.3 (7292)	10.0	7.0 ②	7.0	24.0 ⑬	①	6.0	6.0 ③	19.0	20.0
	F-350	L	7.5 (7538)	6.0	7.0 ②	7.0	24.0 ⑬	①	6.0	6.0 ③	19.0	19.8
	F-Super Duty	C	7.3 (7292)	10.0	7.0 ②	7.0	24.0 ⑬	①	6.0	6.0 ③	19.0	20.0
	Ranger	A	2.3 (2294)	5.0	-	⑮	⑰	3.0	⑭	5.5	⑳	㉒
	Ranger	U	3.0 (2999)	4.5	-	5.6	⑰	⑱	⑲	⑲	⑳	㉓
	Ranger	X	4.0 (3998)	5.0	-	5.6	⑰	⑱	⑲	⑲	⑳	㉑
1995	Aerostar	U	3.0 (2999)	4.5	-	5.6	19.0	2.5	3.0	⑭	21.0	11.8
	Aerostar	X	4.0 (3998)	5.0	-	5.6	19.0	2.5	3.0	⑭	21.0	12.6
	Bronco	Y	4.9 (4916)	6.0	7.0	7.0	24.0	①	5.5	5.5	32.0	14.0
	Bronco	N	5.0 (4949)	6.0	7.0 ②	7.0	24.0	①	5.5	5.5	32.0	14.0
	Bronco	H	5.8 (5752)	6.0	7.0 ②	7.0	24.0	①	5.5	5.5	32.0	15.0
	E-150	Y	4.9 (4916)	6.0	7.0 ②	7.0	24.0	-	-	6.0 ③	④	14.0

CAPACITIES

Year	Model	Engine ID/VIN	Engine Displacement Liters (cc)	Engine Oil with Filter (qts.)	Transmission 4-Spd	5-Spd	Auto.	Transfer Case (pts.)	Drive Axle Front (pts.)	Drive Axle Rear (pts.)	Fuel Tank (gal.)	Cooling System (qts.)
1995	E-150	N	5.0 (4949)	6.0	7.0 [2]	7.0	24.0	-	-	6.0 [3]	[4]	15.0
	E-150	H	5.8 (5752)	6.0	7.0 [2]	7.0	24.0	-	-	6.0 [3]	[4]	14.0
	E-250	Y	4.9 (4916)	6.0	7.0 [2]	7.0	24.0	-	-	6.0 [3]	[4]	14.0
	E-250	N	5.0 (4949)	6.0	7.0 [2]	7.0	24.0	-	-	6.0 [3]	[4]	15.0
	E-250	H	5.8 (5752)	6.0	7.0 [2]	7.0	24.0	-	-	6.0 [3]	[4]	15.0
	E-250	M	7.3 (7292)	10.0	7.0 [2]	7.0	24.0	-	-	6.0 [3]	[4]	20.0
	E-250	L	7.5 (7538)	6.0	7.0 [2]	7.0	24.0	-	-	6.0 [3]	[4]	10.9
	E-350	Y	4.9 (4916)	6.0	7.0 [2]	7.0	24.0	-	-	6.0 [3]	[4]	17.5
	E-350	H	5.8 (5752)	6.0	7.0 [2]	7.0	24.0	-	-	6.0 [3]	[4]	15.0
	E-350	M	7.3 (7292)	10.0	7.0 [2]	7.0	24.0	-	-	6.0 [3]	[4]	20.0
	E-350	L	7.5 (7538)	6.0	7.0 [2]	7.0	24.0	-	-	6.0 [3]	[4]	19.8
	Explorer	X	4.0 (3998)	5.0	-	5.6	[17]	3.0	[14]	[14]	19.3	[21]
	F-150	Y	4.9 (4916)	6.0	7.0 [2]	7.0	24.0 [13]	[1]	6.0	6.0 [3]	[4]	[8]
	F-150	N	5.0 (4949)	6.0	7.0 [2]	7.0	24.0 [13]	[1]	6.0	6.0 [3]	[4]	[9]
	F-150	H	5.8 (5752)	6.0	7.0 [2]	7.0	24.0 [13]	[1]	6.0	6.0 [3]	[4]	[10]
	F-250	Y	4.9 (4916)	6.0	7.0 [2]	7.0	24.0 [13]	[1]	6.0	6.0 [3]	[4]	[8]
	F-250	N	5.0 (4949)	6.0	7.0 [2]	7.0	24.0 [13]	[1]	6.0	6.0 [3]	[4]	[8]
	F-250	H	5.8 (5752)	6.0	7.0 [2]	7.0	24.0 [13]	[1]	6.0	6.0 [3]	[4]	[10]
	F-250	L	7.5 (7538)	6.0	7.0 [2]	7.0	24.0 [13]	[1]	6.0	6.0 [3]	19.0	19.8
	F-350	Y	4.9 (4916)	6.0	7.0 [2]	7.0	24.0 [13]	[1]	6.0	6.0 [3]	19.0	[8]
	F-350	H	5.8 (5752)	6.0	7.0 [2]	7.0	24.0 [13]	[1]	6.0	6.0 [3]	19.0	[10]
	F-350	M	7.3 (7292)	10.0	7.0 [2]	7.0	24.0 [13]	[1]	6.0	6.0 [3]	19.0	20.0
	F-350	L	7.5 (7538)	6.0	7.0 [2]	7.0	24.0 [13]	[1]	6.0	6.0 [3]	19.0	19.8
	F-Super Duty	C	7.3 (7292)	10.0	7.0 [2]	7.0	24.0 [13]	[1]	6.0	6.0 [3]	19.0	20.0
	Ranger	A	2.3 (2294)	5.0	-	[15]	[17]	3.0	[14]	5.5	[20]	[22]
	Ranger	U	3.0 (2999)	4.5	-	3.0	[17]	[18]	[19]	[19]	[20]	[23]
	Ranger	X	4.0 (3998)	5.0	-	3.0	[17]	[18]	[19]	[19]	[20]	[21]
1996-97	Aerostar	U	3.0 (2982)	5.0	-	5.6	19.0	2.5	3.0	[14]	21.0	11.8
	Aerostar	X	4.0 (3950)	5.0	-	5.6	19.0	2.5	3.0	[14]	21.0	12.6
	Bronco	Y	4.9 (4916)	6.0	[12]	7.6	24.0 [13]	[11]	5.5	5.5	32.0	14.0
	Bronco	N	5.0 (4949)	6.0	[12]	7.6	24.0	[11]	5.5	5.5	32.0	14.0
	Bronco	H	5.8 (5752)	6.0	[12]	7.6	24.0 [12]	[11]	5.5	5.5	32.0	15.0
	E-150	Y	4.9 (4916)	6.0	[12]	7.6	24.0 [12]	-	-	6.0 [3]	[4]	14.0
	E-150	N	5.0 (4949)	6.0	[12]	7.6	24.0 [12]	-	-	6.0 [3]	[4]	15.0
	E-150	H	5.8 (5752)	6.0	[12]	7.6	24.0 [12]	-	-	6.0 [3]	[4]	14.0
	E-250	Y	4.9 (4916)	6.0	[12]	7.6	24.0 [12]	-	-	6.0 [3]	[4]	14.0
	E-250	N	5.0 (4949)	6.0	[12]	7.6	24.0 [12]	-	-	6.0 [3]	[4]	15.0
	E-250	H	5.8 (5752)	6.0	[12]	7.6	24.0 [12]	-	-	6.0 [3]	[4]	15.0
	E-250	L	7.5 (7538)	6.0	[12]	7.6	24.0 [12]	-	-	6.0 [3]	[4]	10.9
	E-350	Y	4.9 (4916)	6.0	[12]	7.6	24.0 [12]	-	-	6.0 [3]	[4]	17.5
	E-350	H	5.8 (5752)	6.0	[12]	7.6	24.0 [12]	-	-	6.0 [3]	[4]	15.0
	E-350	F	7.3 (7292)	14.0	[12]	7.6	24.0 [12]	-	-	6.0 [3]	[4]	23.0
	E-350	L	7.5 (7538)	6.0	[12]	7.6	24.0 [12]	-	-	6.0 [3]	[4]	19.8
	Explorer	X	4.0 (3950)	5.0	-	5.6	[17]	3.0	[14]	[14]	19.3	[21]
	Explorer	P	5.0 (4949)	5.0	-	-	13.9	-	-	5.5	19.0	12.8
	F-150	2	4.2 (4195)	6.0	-	7.6	26.0	4.0	3.7	5.5	24.5 [24]	15.7 [25]
	F-150	W	4.6 (4588)	6.0	-	7.6	26.0	4.0	3.7	5.5	24.5 [24]	17.9
	F-150	6	4.6 (4588)	6.0	-	7.6	26.0	4.0	3.7	5.5	24.5 [24]	17.9
	F-150	9	4.6 (4588)	6.0	-	7.6	26.0	4.0	3.7	5.5	24.5 [24]	17.9
	F-150	Y	4.9 (4916)	6.0	[12]	7.6	24.0 [12]	[11]	6.0	6.0 [3]	[13]	[8]

CAPACITIES

Year	Model	Engine ID/VIN	Engine Displacement Liters (cc)	Engine Oil with Filter (qts.)	Transmission (pts.)			Transfer Case (pts.)	Drive Axle		Fuel Tank (gal.)	Cooling System (qts.)
					4-Spd	5-Spd	Auto.		Front (pts.)	Rear (pts.)		
1996-97	F-150	N	5.0 (4949)	6.0	(12)	7.6	24.0 (12)	(11)	6.0	6.0 (3)	(13)	(9)
	F-150	H	5.8 (5752)	6.0	(12)	7.6	24.0 (12)	(11)	6.0	6.0 (3)	(13)	(10)
	F-250	Y	4.9 (4916)	6.0	(12)	7.6	24.0 (12)	(11)	6.0	6.0 (3)	(13)	(8)
	F-250	N	5.0 (4949)	6.0	(12)	7.6	24.0 (12)	(11)	6.0	6.0 (3)	(13)	(9)
	F-250	H	5.8 (5752)	6.0	(12)	7.6	24.0 (12)	(11)	6.0	6.0 (3)	(13)	(10)
	F-250	F	7.3 (7292)	14.0	(12)	7.6	24.0 (12)	(11)	6.0	6.0 (3)	(13)	23.0
	F-250	L	7.5 (7538)	6.0	(12)	7.6	24.0 (12)	(11)	6.0	6.0 (3)	(13)	19.8
	F-350	Y	4.9 (4916)	6.0	(12)	7.6	24.0 (12)	(11)	6.0	6.0 (3)	(13)	(8)
	F-350	H	5.8 (5752)	6.0	(12)	7.6	24.0 (12)	(11)	6.0	6.0 (3)	(13)	(10)
	F-350	F	7.3 (7292)	14.0	-	7.6	24.0 (12)	(11)	6.0	6.0 (3)	(13)	23.0
	F-350	L	7.5 (7538)	6.0	(12)	7.6	24.0 (12)	(11)	6.0	6.0 (3)	(13)	19.8
	F-Super Duty	C	7.3 (7292)	10.0	7.0 (2)	7.0	24.0 (13)	(1)	6.0	6.0 (3)	19.0	20.0
	Mountaineer	P	5.0 (4949)	5.0	-	-	13.9	-	-	5.5		12.8
	Ranger	A	2.3 (2300)	5.0	-	(15)	(17)	3.0	(14)	5.5	(20)	(22)
	Ranger	U	3.0 (2982)	4.5	-	3.0	(17)	(18)	(19)	(19)	(20)	(23)
	Ranger	X	4.0 (3950)	5.0	-	3.0	(17)	(18)	(19)	(19)	(20)	(21)

① New Process: 9 pts. Dextron II
   BW 1345: 6.5 pts. Dextron II
   BW 1356: 4 pts. Mercon

② With OD: 4.5 pts.

③ Heavy duty: 7.5 pts.

④ 124" Wheelbase: 18 gals.
   138", 158" and 176" Wheelbase and front-mounted tank: 22 gals.
   138", 158" and 176" Wheelbase and rear-mounted tank: 16 gals.

⑤ Manual trans.: 17.5 pts.
   Automatic trans.: 18.5 pts.

⑥ Manual trans.: 15 pts.
   Automatic trans.: 21 pts.

⑦ 4WD: 27 pts.

⑧ 4.9L without AC: 13.0 qts.
   4.9L with AC or supercooling: 14.0 qts.
   4.9L with AC and supercooling: 15.6 qts.

⑨ 5.0L with manual trans. and standard cooling system: 15.7 qts.
   5.0L with automatic trans. and standard cooling: 16.5 qts.
   5.0L with manual/automatic trans. and AC: 16.4 qts.
   5.0L with manual/automatic trans. with supercooling or supercooling and AC: 18.3 qts.

⑩ Manual trans. with standard cooling: 15.7 qts.
   Automatic trans. with standard cooling: 16.4 qts.
   Manual/automatic trans. with AC: 16.4 qts.
   Manual/automatic trans. with supercooling and AC: 18.0 qts.

⑪ Without PTO: 4.2 pts.
   With PTO: 12.0 pts.

⑫ With 4R70W: 28 pts.
   With E40D: 32.0 pts.

⑬ 4WD: 27 pts.

⑭ Front axle Dana 28: 1.1 pts.
   Front axle Dana 35: 3.5 pts.
   Rear axle: 5.5 pts.

⑮ Mazda trans.: 3 pts.
   Mitsubishi trans.: 4.8 pts.

⑯ Short wheelbase: 17 gals.
   Long wheelbase: 17 or 21 gals.
   Ranger Supercab: 17 or 21 gals.

⑰ 2WD: 19.4 pts.
   4WD: 20.0 pts.

⑱ BW 13-50 manual shift: 3.0 pts. Dextron II
   BW 13-50 electric shift: 6.5 pts. Dextron II
   BW 13-54 mechanical shift: 3.0 pts. Dextron II
   BW 13-50 contains no lubricant and none should be added

⑲ 6.75" ring gear: 3 pts.
   7.50" ring gear: 5 pts.

⑳ Short wheelbase: 16.3
   Long wheelbase: 19.6
   Ranger Supercab: 19.6

㉑ 4.0L without AC: 7.8 qts.
   4.0L with AC: 8.6 qts.

㉒ 2.3L without AC: 6.5 qts.
   2.3L with AC: 7.2 qts.

㉓ 3.0L without AC: 9.5 qts.
   3.0L with AC: 10.2 qts.

㉔ Also available with a 30 gallon tank with 8 ft. box

㉕ Includes radiator coolant recovery reservoir fluid level between the "coolant level" lines

VALVE SPECIFICATIONS

Year	Engine ID/VIN	Engine Displacement Liters (cc)	Seat Angle (deg.)	Face Angle (deg.)	Spring Test Pressure (lbs. @ in.)	Spring Installed Height (in.)	Stem-to-Guide Clearance (in.)		Stem Diameter (in.)	
							Intake	Exhaust	Intake	Exhaust
1993	A	2.3 (2300)	45	44	149@1.12	1.53-1.59	0.0010-0.0027	0.0015-0.0032	0.3416-0.3423	0.3411-0.3418
	U	3.0 (2999)	45	44	185@1.16	1.58-1.61	0.0010-0.0027	0.0015-0.0032	0.3126-0.3134	0.3121-0.3129
	X	4.0 (3998)	45	44	138@1.22	1.58-1.61	0.0008-0.0025	0.0018-0.0035	0.3159-0.3167	0.3149-0.3156
	Y	4.9 (4916)	45	44	192@1.18	①	0.0010-0.0027	0.0010-0.0027	0.3415-0.3420	0.3415-0.3420
	N	5.0 (4949)	45	44	200@1.35	②	0.0010-0.0027	0.0015-0.0032	0.3415-0.3420	0.3415-0.3420
	H	5.8 (5752)	45	44	200@1.20	③	0.0010-0.0027	0.0015-0.0032	0.3415-0.3420	0.3415-0.3420
	R	5.8 (5752)	45	44	200@1.20	③	0.0010-0.0027	0.0015-0.0032	0.3415-0.3420	0.3415-0.3420
	C	7.3 (7292)	④	④	80@1.83	⑤	0.0055	0.0055	0.3716-0.3723	0.3716-0.3723
	M	7.3 (7292)	④	④	80@1.83	⑤	0.0055	0.0055	0.3716-0.3723	0.3716-0.3723
	G	7.5 (7538)	45	44	220@1.33	1.83	0.0010-0.0027	0.0010-0.0027	0.3415-0.3420	0.3415-0.3420
1994	A	2.3 (2300)	45	44	126-142@1.12	1.53-1.59	0.0010-0.0027	0.0015-0.0032	0.3416-0.3423	0.3411-0.3418
	U	3.0 (2999)	45	44	185@1.16	1.58-1.61	0.0010-0.0027	0.0015-0.0032	0.3126-0.3134	0.3121-0.3129
	X	4.0 (3998)	45	44	138@1.22	1.58-1.61	0.0008-0.0025	0.0018-0.0035	0.3159-0.3167	0.3149-0.3156
	Y	4.9 (4916)	45	44	⑦	①	0.0010-0.0027	0.0010-0.0027	0.3415-0.3423	0.3415-0.3423
	N	5.0 (4949)	45	44	200@1.20	⑥	0.0010-0.0027	0.0015-0.0032	0.3415-0.3423	0.3415-0.3418
	H	5.8 (5752)	45	44	200@1.20	③	0.0010-0.0027	0.0010-0.0027	0.3415-0.3420	0.3415-0.3420
	R	5.8 (5752)	45	44	200@1.20	③	0.0010-0.0027	0.0010-0.0027	0.3415-0.3420	0.3415-0.3420
	F	7.3 (7292)	④	④	200@1.38	⑤	0.0055	0.0055	0.3119-0.3126	0.3119-0.3126
	K	7.3 (7292)	④	④	200@1.38	⑤	0.0055	0.0055	0.3717-0.3724	0.3717-0.3724
	M	7.3 (7292)	④	④	200@1.40	⑤	0.0055	0.0055	0.3716-0.3723	0.3716-0.3723
	G	7.5 (7538)	45	44	220@1.33	1.83	0.0010-0.0027	0.0010-0.0027	0.3415-0.3423	0.3415-0.3423
1995	A	2.3 (2300)	45	44	126-142@1.12	1.53-1.59	0.0010-0.0027	0.0015-0.0032	0.3416-0.3423	0.3411-0.3418
	U	3.0 (2999)	45	44	185@1.16	1.58-1.61	0.0010-0.0027	0.0015-0.0032	0.3126-0.3134	0.3121-0.3129
	X	4.0 (3998)	45	44	138@1.22	1.58-1.61	0.0008-0.0025	0.0018-0.0035	0.3159-0.3167	0.3149-0.3156

VALVE SPECIFICATIONS

Year	Engine ID/VIN	Engine Displacement Liters (cc)	Seat Angle (deg.)	Face Angle (deg.)	Spring Test Pressure (lbs. @ in.)	Spring Installed Height (in.)	Stem-to-Guide Clearance (in.) Intake	Exhaust	Stem Diameter (in.) Intake	Exhaust
1995	Y	4.9 (4916)	45	44	⑧	①	0.0010-0.0027	0.0010-0.0027	0.3415-0.3423	0.3415-0.3423
	N	5.0 (4949)	45	44	200@1.20	⑥	0.0010-0.0027	0.0015-0.0032	0.3415-0.3423	0.3415-0.3423
	H	5.8 (5752)	45	44	200@1.20	③	0.0010-0.0027	0.0010-0.0027	0.3415-0.3420	0.3415-0.3420
	R	5.8 (5752)	45	44	200@1.20	③	0.0010-0.0027	0.0010-0.0027	0.3415-0.3420	0.3415-0.3420
	F	7.3 (7292)	④	④	200@1.38	⑤	0.0055	0.0055	0.3119-0.3126	0.3119-0.3126
	M	7.3 (7292)	④	④	200@1.40	⑤	0.0055	0.0055	0.3716-0.3723	0.3716-0.3723
	G	7.5 (7538)	45	44	220@1.33	1.83	0.0010-0.0027	0.0010-0.0027	0.3415-0.3423	0.3415-0.3423
1996-97	A	2.3 (2300)	45	44	126-142@1.12	1.53-1.59	0.0010-0.0027	0.0015-0.0032	0.3416-0.3423	0.3411-0.3418
	U	3.0 (2982)	45	44	185@1.16	1.58-1,61	0.0010-0.0027	0.0015-0.0032	0.3126-0.3134	0.3121-0.3129
	X	4.0 (3950)	45	44	138@1.22	1.58-1.61	0.0008-0.0025	0.0018-0.0035	0.3159-0.3167	0.3149-0.3156
	2	4.2 (4195)	44.75	NA	NA	1.566-1.637	0.0008-0.0027	0.0018-0.0037	0.3423-0.3415	0.3418-0.3410
	W	4.6 (4588)	45	45.5	132@1.100	1.570	0.0008-0.0027	0.0018-0.0037	0.2750-0.2746	0.2740-0.2736
	6	4.6 (4588)	45	45.5	132@1.100	1.570	0.0008-0.0027	0.0018-0.0037	0.2750-0.2746	0.2740-0.2736
	9	4.6 (4588)	45	45.5	132@1.100	1.570	0.0008-0.0027	0.0018-0.0037	0.2750-0.2746	0.2740-0.2736
	Y	4.9 (4916)	45	44	⑧	①	0.0010-0.0027	0.0010-0.0027	0.3415-0.3423	0.3415-0.3423
	N	5.0 (4949)	45	44	200@1.20	⑥	0.0010-0.0027	0.0015-0.0032	0.3415-0.3423	0.3415-0.3423
	P	5.0 (4949)	45	44	200@1.20	⑥	0.0010-0.0027	0.0015-0.0032	0.3415-0.3423	0.3410-0.3418
	H	5.8 (5752)	45	44	200@1.20	③	0.0010-0.0027	0.0010-0.0027	0.3415-0.3420	0.3415-0.3420
	F	7.3 (7292)	④	④	200@1.38	⑤	0.0055	0.0055	0.3119-0.3126	0.3119-0.3126
	G	7.5 (7538)	45	44	220@1.33	1.83	0.0010-0.0027	0.0010-0.0027	0.3415-0.3423	0.3415-0.3423

① Intake: 1.64 in.
  Exhaust: 1.47 in.
② Intake: 1.68 in.
  Exhaust: 1.59 in.
③ Intake: 1.78 in.
  Exhaust: 1.59 in.
④ Intake: 30
  Exhaust: 37.5
⑤ Intake: 1.767 in.
  Exhaust: 1.833 in.
⑥ Intake: 1.75-1.81 in.
  Exhaust: 1.59 in.
⑦ Intake: 166@1.240
  Exhaust: 166@1.070
⑧ Intake: 166-184@1.240
  Exhaust: 1.66-184@1.070

TORQUE SPECIFICATIONS
All readings in ft. lbs.

	Engine ID/VIN	Engine Displacement Liters (cc)	Cylinder Head Bolts	Main Bearing Bolts	Rod Bearing Bolts	Crankshaft Damper Bolts	Flywheel Bolts	Manifold Intake *	Manifold Exhaust	Spark Plugs	Lug Nut
1993	A	2.3 (2294)	(16)	(16)	(17)	100-120	56-64	14-21	14-21	5-10	(18)
	U	3.0 (2999)	(19)	65-81	(20)	141-169	54-64	18	25	8-10	(18)
	X	4.0 (3998)	(21)	66-77	19-24	(22)	59	(23)	19	10-15	(18)
	Y	4.9 (4916)	(1)	60-70	40-45	130-150	75-85	22-32	22-32	10-15	(2)
	N	5.0 (4949)	(3)	95-105	19-24	70-90	75-85	23-25	18-24	10-15	(2)
	H	5.8 (5752)	(3)	60-70	19-24	70-90	75-90	23-25	20-24	10-15	(2)
	R	5.8 (5752)	(3)	60-70	19-24	70-90	75-90	23-25	20-24	10-15	(2)
	C	7.3 (7292)	(4)	(5)	(6)	90	47	23-25	20-24	-	(2)
	M	7.3 (7292)	(4)	(5)	(6)	90	47	23-25	20-24	-	(2)
	G	7.5 (7538)	(7)	95-105	95-105	70-90	75-85	22-32	28-33	5-10	(2)
1994	A	2.3 (2294)	51	75-85	30-36	103-133	56-64	19-28	14-21	5-10	(18)
	U	3.0 (2999)	(24)	60	26	107	54-64	24	25	8-10	(18)
	X	4.0 (3998)	(21)	66-77	19-24	(22)	59	(23)	19	10-15	(18)
	Y	4.9 (4916)	(1)	60-70	40-45	130-150	75-85	22-32	22-32	10-15	(2)
	N	5.0 (4949)	(9)	60-70	19-24	70-90	75-85	23-25	18-24	10-15	(2)
	H	5.8 (5752)	(10)	95-105	40-45	70-90	75-90	23-25	20-24	10-15	(2)
	R	5.8 (5752)	(10)	95-105	40-45	70-90	75-90	23-25	20-24	10-15	(2)
	F	7.3 (7292)	(11)	95	70	90	89	18	45	-	-
	K	7.3 (7292)	(12)	(13)	(14)	90	(8)	24	-	-	-
	M	7.3 (7292)	(11)	(13)	(14)	90	47	23-25	20-24	-	(2)
	G	7.5 (7538)	(5)	95-105	45-50	70-90	(8)	22-32	28-33	5-10	(2)
1995	A	2.3 (2294)	51	75-85	30-36	103-133	54-64	19-28	14-21	5-10	(18)
	U	3.0 (2999)	(24)	60	26	107	54-64	24	25	8-10	(18)
	X	4.0 (3998)	(21)	66-77	19-24	(22)	59	(23)	19	10-15	(18)
	Y	4.9 (4916)	(1)	60-70	40-45	130-150	75-85	22-32	22-32	10-15	(2)
	N	5.0 (4949)	(9)	60-70	19-24	70-90	75-85	23-25	18-24	10-15	(2)
	H	5.8 (5752)	(10)	95-105	40-45	70-90	75-90	23-25	20-24	10-15	(2)
	R	5.8 (5752)	(10)	95-105	40-45	70-90	75-90	23-25	20-24	10-15	(2)
	F	7.3 (7292)	(11)	95	70	90	89	18	45	-	-
	M	7.3 (7292)	(11)	13	14	90	47	23-25	20-24	-	(2)
	G	7.5 (7538)	(15)	95-105	45-40	70-90	75-85	22-32	28-33	5-10	(2)
1996-97	A	2.3 (2300)	51	75-85	30-36	103-133	54-64	19-28	14-21	5-10	100
	U	3.0 (2982)	(24)	60	26	107	54-64	24	25	8-10	100
	X	4.0 (3950)	(25)	66-77	19-24	(22)	59	(23)	19	10-15	100
	2	4.2 (4195)	(26)	81-88	(27)	103-117	54-63	(28)	15-22	8-14	83-113
	W	4.6 (4588)	(29)	(30)	29-33	(31)	54-64	(32)	15	7-14	83-113
	6	4.6 (4588)	(29)	(33)	29-33	(31)	54-64	(32)	15	7-14	83-113
	9	4.6 (4588)	(29)	NA	29-33	(31)	54-64	(32)	15	7-14	83-113
	P	5.0 (4949)	(9)	60-70	19-24	110-130	75-85	12-18	26-32	7-15	100
	Y	4.9 (4916)	(1)	60-70	40-45	130-150	75-85	22-32	22-32	10-15	(2)
	N	5.0 (4949)	(9)	60-70	19-24	70-90	75-85	23-25	18-24	10-15	(2)

TORQUE SPECIFICATIONS
All readings in ft. lbs.

	Engine ID/VIN	Engine Displacement Liters (cc)	Cylinder Head Bolts	Main Bearing Bolts	Rod Bearing Bolts	Crankshaft Damper Bolts	Flywheel Bolts	Manifold		Spark Plugs	Lug Nut
								Intake *	Exhaust		
1996-97	H	5.8 (5752)	⑩	95-105	40-45	70-90	75-90	23-25	20-24	10-15	②
	F	7.3 (7292)	⑪	95	70	90	89	18	45	-	-
	G	7.5 (7538)	⑮	95-105	40-45	70-90	75-85	22-32	28-33	5-10	

* NOTE: Applies to Lower Manifold only.

① Step 1: 55 ft. lbs.
  Step 2: 65 ft. lbs.
  Step 3: 85 ft. lbs.
② E-F100, E-F150, E-F250: 90 ft. lbs.
  E-F350 with single rear wheels: 135 ft. lbs.
  F350 with dual rear wheels: 210 ft. lbs.
③ Step 1: 55-65 ft. lbs.
  Step 2: 65-72 ft. lbs.
④ Step 1: 65 ft. lbs.
  Step 2: 90 ft. lbs.
  Step 3: 100 ft. lbs.
⑤ Step 1: 75 ft. lbs.
  Step 2: 95 ft. lbs.
⑥ Step 1: 38 ft. lbs.
  Step 2: 50-55 ft. lbs.
⑦ Step 1: 80 ft. lbs.
  Step 2: 110 ft. lbs.
  Step 3: 130-140 ft. lbs.
⑧ Step 1: 47 ft. lbs.
  Step 2: Plus 45 degrees
⑨ With flanged head bolts:
  Step 1: 25-35 ft. lbs.
  Step 2: 40-55 ft. lbs.
  Step 3: Turn an additional 1/4 turn
  With hex head bolts:
  Step 1: 55-65 ft. lbs.
  Step 2: 65-72 ft. lbs.
⑩ Step 1: 95-105 ft. lbs.
  Step 2: 105-112 ft. lbs.
⑪ Step 1: 65 ft. lbs.
  Step 2: 85 ft. lbs.
  Step 3: 105 ft. lbs.
⑫ Step 1: 65 ft. lbs.
  Step 2: 90 ft. lbs.
  Step 3: 110 ft. lbs.

⑬ Step 1: 95 ft. lbs.
  Step 2: Plus 45 degrees
⑭ Step 1: 38 ft. lbs.
  Step 2: 51 ft. lbs.
⑮ Step 1: 70-80 ft. lbs.
  Step 2: 100-110 ft. lbs.
  Step 3: 130-140 ft. lbs.
⑯ Step 1: 50-60 ft. lbs.
  Step 2: 80-90 ft. lbs.
⑰ Step 1: 25-30 ft. lbs.
  Step 2: 30-36 ft. lbs.
⑱ Aerostar, Explorer and Ranger: 100 ft. lbs.
⑲ Step 1: 48-54 ft. lbs.
  Step 2: 63-80 ft. lbs.
⑳ Step 1: 20-28 ft. lbs.
  Step 2: Back off a minimum of two turns
  Step 3: 20-25 ft. lbs.
㉑ Tighten cylinder head bolts to 44 ft. lbs.
  Tighten intake manifold bolts to 3-6 ft. lbs.
  Tighten cylinder head bolts to 59 ft. lbs.
  Tighten intake manifold bolts to 6-11 ft. lbs.
  Tighten cylinder head bolts 85 degrees
  Tighten intake manifold bolts to 11-15 ft. lbs.
  Tighten intake manifold bolts to 15-18 ft. lbs.
㉒ Step 1: 30-37 ft. lbs.
  Step 2: Turn 90 degrees
㉓ Step 1: 3-6 ft. lbs.
  Step 2: 6-11 ft. lbs.
  Step 3: 11-15 ft. lbs.
  Step 4: 15-18 ft. lbs.
㉔ Step 1: 37 ft. lbs.
  Step 2: 68 ft. lbs.

㉕ Cylinder head bolts:
  Step 1: 44 ft. lbs.
  Step 2: 59 ft. lbs.
  Step 3: Plus 85 degrees
  Intake manifold bolts:
  Step 1: 3-6 ft. lbs.
  Step 2: 6-11 ft. lbs.
  Step 3: 11-15 ft. lbs.
  Step 4: 15-18 ft. lbs.
㉖ Step 1: 29 ft. lbs.
  Step 2: 36 ft. lbs.
  Step 3: Loosen each bolt and tighten one at a time
    Short bolts to 32 ft. lbs.
    Long bolts to 36 ft. lbs.
  Step 4: Turn each bolt an additional 135 degrees
㉗ Step 1: 29 ft. lbs.
  Step 2: Plus 90 degrees
㉘ Tighten bolts in sequence to 71-101 in. lbs.
㉙ Step 1: 31 ft. lbs.
  Step 2: Plus 90 degrees
  Step 3: Plus 90 degrees
㉚ Tighten main bearing jack screws in sequence as follows:
  Step 1: 45 in. lbs.
  Step 2: 98 in. lbs.
  Tighten cross-mounted cap bolts in sequence as follows:
  Step 1: 89 in. lbs.
  Step 2: 17 in. lbs.
㉛ Step 1: 88 ft. lbs.
  Step 2: Loosen bolt
  Step 3: 39 ft. lbs.
  Step 4: Plus 90 degrees
㉜ Step 1: 18 in. lbs.
  Step 2: 8 ft. lbs.
㉝ Tighten jack screws in sequence as follows:
  Step 1: 45 in. lbs.
  Step 2: 98 in. lbs.
  Tighten cross-mounted cap bolts in sequence as follows:
  Step 1: 24 ft. lbs.
  Step 2: Plus 90 degrees

BRAKE SPECIFICATIONS
All measurements in inches unless noted

Year	Model		Master Cylinder Bore	Brake Disc Original Thickness	Brake Disc Minimum Thickness	Brake Disc Maximum Runout	Brake Drum Diameter Original Inside Diameter	Brake Drum Diameter Max. Wear Limit	Brake Drum Diameter Maximum Machine Diameter	Minimum Lining Thickness Front	Minimum Lining Thickness Rear
1993	Aerostar	④	NA	0.850	0.810	0.003	9.00	9.09	9.06	0.030	0.030
		⑤	NA	0.850	0.810	0.003	10.00	10.09	10.06	0.030	0.030
	Bronco		NA	1.160	1.120	0.003	11.03	11.09	11.06	0.030	0.030
	E-150		NA	1.160	1.120	0.003	11.03	11.09	11.06	0.030	0.030
	E-250		NA	1.220	1.180	0.003	12.00	12.09	12.06	0.030	0.030
	E-350		NA	1.220	1.180	0.003	12.00	12.09	12.06	0.030	0.030
	Explorer	④	0.938	0.850	0.810	0.003	9.00	9.09	9.06	0.030	0.030
		⑥	0.938	0.850	0.810	0.003	10.00	10.09	10.06	0.030	0.030
		⑦	0.975	0.850	0.810	0.003	10.00	10.09	10.06	0.030	0.030
	F-150		NA	1.160	1.120	0.003	11.03	11.09	11.06	0.030	0.030
	F-250		NA	1.220	1.180	0.003	12.00	12.09	12.06	0.030	0.030
	F-350		NA	1.220	1.180	0.003	12.00	12.09	12.06	0.030	0.030
	F-Super Duty		NA	1.220	1.180	0.008	12.00	12.09	12.06	0.030	0.030
	Ranger	④	0.938	0.850	0.810	0.003	9.00	9.09	9.06	0.030	0.030
		⑥	0.938	0.850	0.810	0.003	10.00	10.09	10.06	0.030	0.030
		⑦	0.975	0.850	0.810	0.003	10.00	10.09	10.06	0.030	0.030
1994	Aerostar	④	0.938	0.850	0.810	0.003	9.00	9.09	9.06	0.030	0.030
		⑤	0.938	0.850	0.810	0.003	10.00	10.09	10.06	0.030	0.030
	Bronco		1.000	1.160	1.120	0.003	11.03	11.09	11.06	0.030	0.030
	E-150		1.000	1.160	1.120	0.003	11.03	11.09	11.06	0.030	0.030
	E-250		①	1.220	1.180	0.003	12.00	12.09	12.06	0.030	0.030
	E-350		NA	1.220	1.180	0.003	12.00	12.09	12.06	0.030	0.030
	Explorer	④	0.938	0.850	0.810	0.003	9.00	9.09	9.06	0.030	0.030
		⑥	0.938	0.850	0.810	0.003	10.00	10.09	10.06	0.030	0.030
		⑦	0.938	0.850	0.810	0.003	10.00	10.09	10.06	0.030	0.030
	F-150		1.000	1.160	1.120	0.003	11.03	11.09	11.06	0.030	0.030
	F-250		①	1.220	1.180	0.003	12.00	12.09	12.06	0.030	0.030
	F-350		1.125	1.220	1.180	0.003	12.00	12.09	12.06	0.030	0.030
	F-Super Duty	F	NA	1.220	1.180	0.008	-	-	-	0.030	-
		R	-	NA	1.430	0.008	-	-	-	-	0.030
	Ranger	④	0.938	0.850	0.810	0.003	9.00	9.09	9.06	0.030	0.030
		⑥	0.938	0.850	0.810	0.003	10.00	10.09	10.06	0.030	0.030
		⑦	0.938	0.850	0.810	0.003	10.00	10.09	10.06	0.030	0.030
1995	Aerostar	④	0.938	0.850	0.810	0.003	9.00	9.09	9.06	0.030	0.030
		⑤	0.938	0.850	0.810	0.003	10.00	10.09	10.06	0.030	0.030
	Bronco		1.000	1.160	0.960	0.003	11.03	11.09	11.06	0.030	0.030
	E-150		1.000	1.160	1.120	0.003	11.03	11.09	11.06	0.030	0.030
	E-250		①	1.220	1.180	0.003	12.00	12.09	12.06	0.030	0.030
	E-350		NA	1.220	1.180	0.003	12.00	12.09	12.06	0.030	0.030
	Explorer	④	0.938	0.850	0.810	0.003	9.00	9.09	9.06	0.030	0.030
		⑥	0.938	0.850	0.810	0.003	10.00	10.09	10.06	0.030	0.030
		⑦	0.938	0.850	0.810	0.003	10.00	10.09	10.06	0.030	0.030
	F-150		1.000	1.160	0.960	0.003	11.03	11.09	11.06	0.030	0.030
	F-250		①	1.220	②	0.003	12.00	12.09	12.06	0.030	0.030
	F-350		1.125	1.220	②	③	12.00	12.09	12.06	0.030	0.030
	F-Super Duty	F	NA	1.220	1.180	0.008	-	-	-	0.030	-
		R	-	NA	1.430	0.008	-	-	-	-	0.030

BRAKE SPECIFICATIONS
All measurements in inches unless noted

Year	Model			Master Cylinder Bore	Brake Disc			Brake Drum Diameter			Minimum Lining Thickness	
					Original Thickness	Minimum Thickness	Maximum Runout	Original Inside Diameter	Max. Wear Limit	Maximum Machine Diameter	Front	Rear
1995	Ranger	④		0.938	0.850	0.810	0.003	9.00	9.09	9.06	0.030	0.030
		⑥		0.938	0.850	0.810	0.003	10.00	10.09	10.06	0.030	0.030
		⑦		0.938	0.850	0.810	0.003	10.00	10.09	10.06	0.030	0.030
	Windstar			NA	1.020	0.097	0.003	9.84	NA	9.90	0.040	0.590
1996-97	Aerostar	④		0.938	0.850	0.810	0.003	9.00	9.09	9.06	0.030	0.030
		⑤		0.938	0.850	0.810	0.003	10.00	10.09	10.06	0.030	0.030
	Bronco			1.000	1.160	0.960	0.003	11.03	11.09	11.06	0.030	0.030
	E-150			1.000	1.160	1.120	0.003	11.03	11.09	11.06	0.030	0.030
	E-250			①	1.220	1.180	0.003	12.00	12.09	12.06	0.030	0.030
	E-350			NA	1.220	1.180	0.003	12.00	12.09	12.06	0.030	0.030
	Explorer	④		0.938	0.850	0.810	0.003	9.00	9.09	9.06	0.030	0.030
		⑥		0.938	0.850	0.810	0.003	10.00	10.09	10.06	0.030	0.030
		⑦		0.938	0.850	0.810	0.003	10.00	10.09	10.06	0.030	0.030
	F-150	⑧		1.062	NA	0.972	NA	11.03	11.12	NA	0.156	0.030
	F-150			1.000	1.160	0.960	0.003	11.03	11.09	11.06	0.030	0.030
	F-250			①	1.220	②	0.003	12.00	12.09	12.06	0.030	0.030
	F-350			1.125	1.220	②	③	12.00	12.09	12.06	0.030	0.030
	Mountaineer	④		0.938	0.850	0.810	0.003	9.00	9.09	9.06	0.030	0.030
		⑥		0.938	0.850	0.810	0.003	10.00	10.09	10.06	0.030	0.030
		⑦		0.938	0.850	0.810	0.003	10.00	10.09	10.06	0.030	0.030
	Ranger	④		0.938	0.850	0.810	0.003	9.00	9.09	9.06	0.030	0.030
		⑥		0.938	0.850	0.810	0.003	10.00	10.09	10.06	0.030	0.030
		⑦		0.938	0.850	0.810	0.003	10.00	10.09	10.06	0.030	0.030
	F-Super Duty	F		NA	1.220	1.180	0.008	-	-	-	0.030	-
		R		-	NA	1.430	0.008	-	-	-	-	0.030

NOTE: Due to changes made during production, refer to manufacturer's specifications if they differ from those in this chart

NA - Not Available

F - Front

R - Rear

① Under 6900 GVW: 1.062
Over 6900 GVW: 1.125

② 4x2: 1.100
4x4: 1.120

③ Except F-350 4x2 with dual rear wheel and 2-piece rotor/hub:0.003 in.
F-350 4x2 with dual rear wheel and 2-piece rotor/hub: 0.010 in.

④ With 9 inch brakes

⑤ With 10 inch brakes

⑥ 4x2 with 10 inch brakes

⑦ 4x4 with 10 inch brakes

⑧ 1997 only

FREQUENT MAINTENANCE LABOR
EXPLORER

The following should be used as a guide when determining the amount of work required for a particular service if taken to a repair shop. In estimating how long a particular Frequent Maintenance Service item should take, please observe the following:
- **Factory Time** is time that is generated by the vehicle manufacturer.
- **Chilton Time** is time that is based on field research and data supplied by the vehicle manufacturer.
- All labor time operations are given in hours and tenths of an hour.
- All labor operations, are to be used as a **guide**.

COOLING

(G) Winterize Cooling System
Includes: Run engine to check for leaks, tighten all hose connections. Test radiator and pressure cap, drain radiator and engine block. Add antifreeze and refill system.
1993-97 5

(G) Belt, Drive, Renew
1993-97
Serpentine belt
4.0L (.5) 6
5.0L (.3) 4

(G) Hoses, Radiator, Renew
Includes: Drain and refill cooling system as required.
1993-97
upper (.3) 4
lower (.4) 5
both (.5) 6

(G) Thermostat, Coolant, Renew
1993-97
4.0L (.7) 9
5.0L (.5) 7

FUEL

(M) Air Cleaner, Service
1993-97 (.3) 5

(G) Filter, Fuel, Renew
1993-97 (.3) 4

BRAKES

(G) Bleed Brakes (Four Wheels)
Includes: Add fluid.
1993-97 5

(G) Brakes, Adjust (Minor)
Includes: Adjust brakes, fill master cylinder.
1993-94, two wheels 4

(M) Parking Brake, Adjust
1993-97 (.2) 3

LUBRICATION SERVICES

(M) Engine Oil & Filter, Renew
Includes: Inspect and correct all fluid levels.
1993-97 3

(M) Lubricate Chassis, Change Oil & Filter
Includes: Inspect and correct all fluid levels.
1993-97 6
Install grease fittings add 1

(M) Lubricate Chassis
Includes: Inspect and correct all fluid levels.
1993-97 4
Install grease fittings add 1

ELECTRICAL

(G) Headlamps, Aim
1993-97
two 4
four 6

(G) Halogen Headlamp Bulb, Renew
1993-97, each (.3) 3

(G) License Lamp Assy., Renew
1993-97 (.2) 3

(G) Tail Lamp Assy., Renew
1993-97 (.2) 3

(G) Horn, Renew
1993-97
one (.3) 4
each adtnl. 1

(M) Terminals, Battery, Clean
1993-97 3

FREQUENT MAINTENANCE LABOR
F SERIES, ECONOLINE, BRONCO, AEROSTAR, RANGER

The following should be used as a guide when determining the amount of work required for a particular service if taken to a repair shop. In estimating how long a particular Frequent Maintenance Service item should take, please observe the following:
- **Factory Time** is time that is generated by the vehicle manufacturer.
- **Chilton Time** is time that is based on field research and data supplied by the vehicle manufacturer.
- All labor time operations are given in hours and tenths of an hour.
- All labor operations, are to be used as a **guide**.

COOLING

(G) Winterize Cooling System
Includes: Run engine to check for leaks, tighten all hose connections. Test radiator and pressure cap, drain radiator and engine block. Add antifreeze and refill system.
1993-97 5

(G) Belt, Serpentine Drive, Renew
1993-97
  2.3L (.4) 5
  3.0L, 4.0L (.5) 6
  6 cyl., V8
    Econoline (.4) 5
    F Series, Bronco (.3) 4
  diesel (.4) 5

(G) Belt, Drive, Renew
1993-97 (.3) 4

(G) Belt, Drive, Adjust
1993-97
  one (.3) 4
  each adtnl. (.2) 3

(G) Hoses, Radiator, Renew
Includes: Drain and refill cooling system as required.
1993-97
  Aerostar
    upper or lower (.4) 5
    both (.5) 7
  Bronco, Econoline, F Series
    upper or lower (.4) 5
    both (.5) 7
  Ranger
    upper (.3) 4
    lower (.4) 5
    both (.5) 6

(G) Thermostat, Coolant, Renew
1993-97
  gas
    4 cyl. (.5) 6
    6 cyl. (.4) 6
    V6 (.7) 9
    V8 (.9) 1.0
  diesel
    V8 (1.1) 1.4
  Econoline, add 5
  1997 (.4) 6

FUEL

(G) Filter, Fuel, Renew
Gasoline Engines
Aerostar
1993-97
  V6
    3.0L (.6) 8
    4.0L (.3) 4
Econoline, F Series
1993-97
  6 cyl. (.4) 5
  V6 (.4) 6
  V8 (.4) 6
Ranger, Bronco
1993-97 (.3) 4
Diesel Engines
1993-97
  V8
    6.9L, 7.3L (.3) 4
    7.3L IDI Turbo (.5) 8
    7.3L DI Turbo (.4) 6

BRAKES

(G) Bleed Brakes (Four Wheels)
Includes: Add fluid.
1993-97 (.3) 6

(G) Brakes, Adjust (Minor)
Includes: Adjust brakes, fill master cylinder.
1993-97, two wheels 4

(M) Parking Brake, Adjust
1993-97 (.3) 5

LUBRICATION SERVICES

(M) Lubricate Chassis, Change Oil & Filter
Includes: Inspect and correct all fluid levels.
1993-97 6
Install grease fittings, add 1

(M) Lubricate Chassis
Includes: Inspect and correct all fluid levels.
1993-97 4
Install grease fittings, add 1

ELECTRICAL

(G) Headlamps, Aim
1993-97
  two .4
  four .6

(G) Halogen Headlamp Bulb, Renew
1993-97 one (.3) 3

(G) License Lamp Assy., Renew
1993-97
  Aerostar (.4) 5
  Bronco, F Series (.2) 3
  Econoline (.3) 4
  Ranger (.2) 4

(G) Horn, Renew
1993-97
  one (.3) 4
  each adtnl. (.1) 1

(M) Terminals, Battery, Clean
1993-97 3

SCHEDULED MAINTENANCE INTERVALS
(FORD F-150/250 & BRONCO/E-150/250 & CLUB WAGON (LIGHT DUTY EMISSIONS UNDER 8500 LBS GVWR) (1993))

TO BE SERVICED	TYPE OF SERVICE	VEHICLE MILEAGE INTERVAL (x1000)												
		7.5	15	22.5	30	37.5	45	52.5	60	67.5	75	82.5	90	97.5
Engine oil & filter	R	✓	✓	✓	✓	✓	✓	✓	✓	✓	✓	✓	✓	✓
Automatic transmission fluid & filter	S/I	✓	✓	✓	✓	✓	✓	✓	✓	✓	✓	✓	✓	✓
Automatic transmission shift linkage (Bell crank system)	S/I	✓	✓	✓	✓	✓	✓	✓	✓	✓	✓	✓	✓	✓
Clutch reservoir fluid level	S/I	✓	✓	✓	✓	✓	✓	✓	✓	✓	✓	✓	✓	✓
Engine coolant level, hoses & clamps	S/I	✓	✓	✓	✓	✓	✓	✓	✓	✓	✓	✓	✓	✓
Exhaust system & heat shields	S/I	✓	✓	✓	✓	✓	✓	✓	✓	✓	✓	✓	✓	✓
Steering linkage & driveshaft slip yoke	S/I	✓	✓	✓	✓	✓	✓	✓	✓	✓	✓	✓	✓	✓
Wheel lug nut torque	S/I	✓	✓	✓	✓	✓	✓	✓	✓	✓	✓	✓	✓	✓
Rotate tires	S/I	✓		✓		✓		✓		✓		✓		✓
Fuel filter	R		✓		✓		✓		✓		✓		✓	
Disc brake system & caliper slide rails (F-150/250 & Bronco)	S/I		✓		✓		✓		✓		✓		✓	
Drum brake systems, hoses & lines	S/I		✓		✓		✓		✓		✓		✓	
Transfer case shift lever pivot bolt & control rod connecting pins	S/I		✓		✓		✓		✓		✓		✓	
Air filter	R				✓				✓				✓	
Disc brake system & caliper slide rails & knuckle top & bottom inner pad slots (E-150/250 & Club Wagon)	S/I		✓		✓		✓		✓		✓		✓	
Crankcase emission air filter	R				✓				✓				✓	
Engine Coolant	R				✓				✓				✓	
PCV valve	R				✓				✓				✓	
Spark plugs	R				✓				✓				✓	
Front wheel bearings	S/I				✓				✓				✓	
Hub lock (4x4)	S/I				✓				✓				✓	

SCHEDULED MAINTENANCE INTERVALS
(FORD F-150/250 & BRONCO/E-150/250 & CLUB WAGON (LIGHT DUTY EMISSIONS UNDER 8500 LBS GVWR) (1993)) (Cont.)

TO BE SERVICED	TYPE OF SERVICE	VEHICLE MILEAGE INTERVAL (x1000)												
		7.5	15	22.5	30	37.5	45	52.5	60	67.5	75	82.5	90	97.5
Parking brake system	S/I				✓				✓				✓	
Spindle needle bearing	S/I				✓				✓				✓	
Throttle & kickdown lever ball studs	S/I				✓				✓				✓	
Manual transmission oil	R								✓					
Rear axle oil	R													✓
Transfer case oil	R								✓					
Drive belts	S/I								✓					
Front axle R.H. axle shaft slip yoke	S/I								✓					
Secondary air injection hoses & clamps	S/I								✓					

R – Replace   S/I – Service or Inspect

FREQUENT OPERATION MAINTENANCE (SEVERE SERVICE)

If a vehicle is operated under any of the following conditions it is considered severe service:
- **Extremely dusty areas.**
- **50% or more of the vehicle operation is in 32°C (90°F) or higher temperatures, or constant operation in temperatures below 0°C (32°F).**
- **Prolonged idling (vehicle operation in stop and go traffic).**
- **Frequent short running periods (engine does not warm to normal operating temperatures).**
- **Police, taxi, delivery usage or trailer towing usage.**

Oil & oil filter change – change every 3000 miles.
Automatic transmission fluid & filter - change every 30,000 miles.
Manual transmission oil - change every 30,000 miles.

Additional items for vehicles operated off-highway:
Front axle spindle pins, steering & clutch linkages, axle & driveshaft U-joints & slip yoke - lubricate every 1000 miles.
Front wheel bearings - check every 1000 miles.
Disc brake system, caliper slide rails - check every 1000 miles.
Drum brake system, hoses & lines - check every 1000 miles.
Exhaust system - check every 1000 miles.
Clutch release lever pivot (7.3L diesel & 7.5L) - lubricate every 1000 miles.

SCHEDULED MAINTENANCE INTERVALS
(FORD F-250 HD/350 & SUPER DUTY/E-250/350 & CLUB WAGON (4.9L,5.8L,7.5L EFI/MFI)
(HEAVY DUTY EMISSIONS OVER 8500 LBS GVWR) (1993))

TO BE SERVICED	TYPE OF SERVICE	VEHICLE MILEAGE INTERVAL (x1000)												
		5	10	15	20	25	30	35	40	45	50	55	60	65
Engine oil & filter	R	✓	✓	✓	✓	✓	✓	✓	✓	✓	✓	✓	✓	✓
Automatic transmission fluid & filter	S/I	✓	✓	✓	✓	✓	✓	✓	✓	✓	✓	✓	✓	✓
Automatic transmission shift linkage (Bell crank system)	S/I	✓	✓	✓	✓	✓	✓	✓	✓	✓	✓	✓	✓	✓
Clutch release lever (7.5L)	S/I	✓	✓	✓	✓	✓	✓	✓	✓	✓	✓	✓	✓	✓
Clutch reservoir fluid level	S/I	✓	✓	✓	✓	✓	✓	✓	✓	✓	✓	✓	✓	✓
Engine coolant level, hoses & clamps	S/I	✓	✓	✓	✓	✓	✓	✓	✓	✓	✓	✓	✓	✓
Exhaust system & heat shields	S/I	✓	✓	✓	✓	✓	✓	✓	✓	✓	✓	✓	✓	✓
Front axle spindle pins, steering linkage, driveshaft slip yoke	S/I	✓	✓	✓	✓	✓	✓	✓	✓	✓	✓	✓	✓	✓
Wheel lug nut torque	S/I	✓	✓	✓	✓	✓	✓	✓	✓	✓	✓	✓	✓	✓
Front & rear spring U bolts (F-Super Duty commercial & motorhome)	S/I	✓		✓			✓			✓			✓	
Rotate tires	S/I	✓		✓			✓			✓			✓	
Fuel filter	R			✓			✓			✓			✓	
Disc brake system & caliper slide rails & knuckle top & bottom inner pad slots	S/I			✓			✓			✓			✓	
Drive belts	S/I			✓			✓			✓			✓	
Drum brake systems, hoses & lines	S/I			✓			✓			✓			✓	
Front axle spindle pins (F-Super Duty)	S/I			✓			✓			✓			✓	
Transfer case shift lever pivot bolt & control rod connecting pins	S/I			✓			✓			✓			✓	
Parking brake fluid level (F-Super Duty)	S/I			✓			✓			✓			✓	
Air filter	R						✓						✓	

SCHEDULED MAINTENANCE INTERVALS
(FORD F-250 HD/350 & SUPER DUTY/E-250/350 & CLUB WAGON (4.9L,5.8L,7.5L EFI/MFI)
(HEAVY DUTY EMISSIONS OVER 8500 LBS GVWR) (1993)) (Cont.)

TO BE SERVICED	TYPE OF SERVICE	VEHICLE MILEAGE INTERVAL (x1000)												
		5	10	15	20	25	30	35	40	45	50	55	60	65
Crankcase emission air filter	R						✓						✓	
Engine Coolant	R						✓						✓	
Spark plugs	R						✓						✓	
Transfer case oil	R						✓						✓	
Engine air induction system (E-350 over 10,000 lbs GVWR)	S/I						✓						✓	
Fan & fan shroud (E-350 over 10,000 lbs (GVWR)	S/I						✓						✓	
Front drive axle R.H. axle slip yoke (4x4) (F-250)	S/I						✓						✓	
Front wheel bearings	S/I						✓						✓	
Hub lock (4x4)	S/I						✓						✓	
Parking brake system	S/I						✓						✓	
Spindle needle bearing (4x4)	S/I						✓						✓	
Throttle & kickdown lever ball studs	S/I						✓						✓	
Ignition wires	R												✓	
Manual transmission & rear axle oil①	R												✓	
PCV valve	R												✓	
Secondary air injection hoses & clamps	S/I												✓	

① Rear axle oil - change every 100,000 miles.
R – Replace   S/I – Service or Inspect

FREQUENT OPERATION MAINTENANCE (SEVERE SERVICE)
If a vehicle is operated under any of the following conditions it is considered severe service:
- Extremely dusty areas.
- 50% or more of the vehicle operation is in 32°C (90°F) or higher temperatures, or constant operation in temperatures below 0°C (32°F).
- Prolonged idling (vehicle operation in stop and go traffic).
- Frequent short running periods (engine does not warm to normal operating temperatures).
- Police, taxi, delivery usage or trailer towing usage.
Oil & oil filter change – change every 3000 miles.
Manual/automatic transmission fluid & filter - change every 30,000 miles.

Additional is for vehicles operated off-highway:
Front wheel bearings, front axle spindle pins, steering & clutch linkages, axle & driveshaft U-joints & slip yoke - lubricate every 1000 miles.
Disc brake system, caliper slide rails, drum brake system, hoses & lines - check every 1000 miles.
Exhaust system - check every 1000 miles.
Clutch release lever pivot (7.5L) - lubricate every 1000 miles.

SCHEDULED MAINTENANCE INTERVALS
(FORD F-250 HD/350 & SUPER DUTY/E-250/350 & CLUB WAGON (7.3L DIESEL) (HEAVY DUTY EMISSIONS OVER 8500 LBS GVWR) (1993))

TO BE SERVICED	TYPE OF SERVICE	VEHICLE MILEAGE INTERVAL (x1000)												
		5	10	15	20	25	30	35	40	45	50	55	60	65
Engine oil & filter	R	✓	✓	✓	✓	✓	✓	✓	✓	✓	✓	✓	✓	✓
Automatic transmission fluid & filter	S/I	✓	✓	✓	✓	✓	✓	✓	✓	✓	✓	✓	✓	✓
Automatic transmission shift linkage (Bell crank system)	S/I	✓	✓	✓	✓	✓	✓	✓	✓	✓	✓	✓	✓	✓
Clutch reservoir fluid level	S/I	✓	✓	✓	✓	✓	✓	✓	✓	✓	✓	✓	✓	✓
Drain water from fuel/filter bowl	S/I	✓	✓	✓	✓	✓	✓	✓	✓	✓	✓	✓	✓	✓
Exhaust system & heat shields	S/I	✓	✓	✓	✓	✓	✓	✓	✓	✓	✓	✓	✓	✓
Front axle spindle pins, steering linkage, driveshaft slip yoke	S/I	✓	✓	✓	✓	✓	✓	✓	✓	✓	✓	✓	✓	✓
Wheel lug nut torque	S/I	✓	✓	✓	✓	✓	✓	✓	✓	✓	✓	✓	✓	✓
Engine idle speed	S/I		✓			✓		✓		✓		✓		
Front & rear spring U bolts (F-Super Duty commercial & motorhome)	S/I	✓		✓			✓			✓			✓	
Rotate tires	S/I	✓		✓			✓			✓			✓	
Throttle operation & idle return spring	S/I	✓		✓			✓			✓			✓	
Disc brake system & caliper slide rails & knuckle top & bottom inner pad slots	S/I			✓			✓			✓			✓	
Drum brake systems, hoses & lines	S/I			✓			✓			✓			✓	
Engine coolant level, hoses & clamps	S/I	✓	✓	✓	✓	✓	✓	✓	✓	✓	✓	✓	✓	✓
Fan & fan shroud (E- & F-350 over 10,000 lbs GVWR)	S/I			✓			✓			✓			✓	
Front axle spindle pins (F-Super Duty)	S/I			✓			✓			✓			✓	
Transfer case shift lever pivot bolt & control rod connecting pins	S/I			✓			✓			✓			✓	

SCHEDULED MAINTENANCE INTERVALS
(FORD F-250 HD/350 & SUPER DUTY/E-250/350 & CLUB WAGON (7.3L DIESEL) (HEAVY DUTY EMISSIONS OVER 8500 LBS GVWR) (1993)) (Cont.)

TO BE SERVICED	TYPE OF SERVICE	VEHICLE MILEAGE INTERVAL (x1000)												
		5	10	15	20	25	30	35	40	45	50	55	60	65
Parking brake system	S/I			✓			✓			✓			✓	
Parking brake fluid level (F-Super Duty)	S/I			✓			✓			✓			✓	
Air filter	R						✓						✓	
Crankcase emission air filter	R						✓						✓	
Engine Coolant	R						✓						✓	
Brake master cylinder fluid level	S/I						✓						✓	
Drive belts	S/I						✓						✓	
Engine air induction system (E-350 over 10,000 lbs GVWR)	S/I						✓						✓	
Front drive axle R.H. axle slip yoke (4x4) (F-250)	S/I						✓						✓	
Front wheel bearings	S/I						✓						✓	
Hub lock (4x4)	S/I						✓						✓	
Spindle needle bearing (4x4)	S/I						✓						✓	
Throttle ball stud	S/I						✓						✓	
Fuel filter	R												✓	
Manual transmission & transfer case oil	R												✓	
PCV valve	R												✓	
Rear axle oil①	R													

① Rear axle oil - change every 100,000 miles.

R – Replace   S/I – Service or Inspect

FREQUENT OPERATION MAINTENANCE (SEVERE SERVICE)
If a vehicle is operated under any of the following conditions it is considered severe service:
- Extremely dusty areas.
- 50% or more of the vehicle operation is in 32°C (90°F) or higher temperatures, or constant operation in temperatures below 0°C (32°F).
- Prolonged idling (vehicle operation in stop and go traffic).
- Frequent short running periods (engine does not warm to normal operating temperatures).
- Police, taxi, delivery usage or trailer towing usage.

Oil & oil filter change – change every 3000 miles.

Manual/automatic transmission fluid & filter - change every 30,000 miles.

Additional items for vehicles operated off-highway:

Front wheel bearings, front axle spindle pins, steering & clutch linkages, axle & driveshaft U-joints & slip yoke - lubricate every 1000 miles.

Disc brake system, caliper slide rails, drum brake system, hoses & lines - check every 1000 miles.

Exhaust system - check every 1000 miles.

Clutch release lever pivot - lubricate every 1000 miles.

SCHEDULED MAINTENANCE INTERVALS
(FORD F-150/250/250HD/350 & SUPER DUTY/E-150/250/350 & CLUB WAGON (1994-97))

TO BE SERVICED	TYPE OF SERVICE	VEHICLE MILEAGE INTERVAL (x1000)												
		5	10	15	20	25	30	35	40	45	50	55	60	65
Engine oil & filter	R	✓	✓	✓	✓	✓	✓	✓	✓	✓	✓	✓	✓	✓
Automatic transmission shift linkage	S/I	✓		✓		✓		✓		✓		✓		✓
Clutch reservoir fluid level	S/I	✓		✓		✓		✓		✓		✓		✓
Exhaust system & heat shields	S/I	✓		✓		✓		✓		✓		✓		✓
Rotate tires⑥	S/I	✓		✓		✓		✓		✓		✓		✓
Steering linkage suspension, driveshaft U-joint, & slip-yoke (if equipped)	S/I	✓		✓		✓		✓		✓		✓		✓
Clutch release lever (7.3L diesel & 7.5L)	S/I	✓			✓			✓			✓			✓
Fuel filter③	R			✓			✓			✓			✓	
Disc brake system & caliper slide rails	S/I			✓			✓			✓			✓	
Drum brake systems, hoses & lines	S/I			✓			✓			✓			✓	
Parking brake fluid level (F-Super Duty)	S/I			✓			✓			✓			✓	
Spring U bolts (F-Super Duty)	S/I			✓			✓			✓			✓	
Transfer case shift lever pivot bolt & control rod connecting pins (4x4)	S/I			✓			✓			✓			✓	
Engine coolant strength, hoses & clamps	S/I			✓			✓			✓			✓	
Air cleaner filter④	R						✓						✓	
Automatic transmission fluid & filter⑤	R						✓						✓	
Front axle RH axle slip yoke (4x4)	S/I						✓						✓	
Front & rear driveshaft slip-yoke (1997 F-150 4x4)	S/I						✓						✓	
Front wheel bearings	S/I						✓						✓	
Hub lock (4x4)	S/I						✓						✓	

SCHEDULED MAINTENANCE INTERVALS
(FORD F-150/250/250HD/350 & SUPER DUTY/E-150/250/350 & CLUB WAGON(Cont.)

TO BE SERVICED	TYPE OF SERVICE	VEHICLE MILEAGE INTERVAL (x1000)												
		5	10	15	20	25	30	35	40	45	50	55	60	65
Parking brake system	S/I						✓						✓	
Spindle needle bearing (4x4)	S/I						✓						✓	
Throttle & TV lever ball studs	S/I						✓						✓	
Crankcase emission air filter	R						✓						✓	
Engine coolant②	R										✓			
Manual transmission & rear axle oil①	R												✓	
PCV valve	R												✓	
Spark plugs	R												✓	
Transfer case oil	R												✓	
Drive belts	S/I												✓	
Thermactor hoses & clamps	S/I												✓	

① Rear axle oil - change every 100,000 miles.
② Engine coolant - change at 50,000 miles:
      Gasoline - change every 30,000 miles thereafter.
      Diesel - add 8-10 oz. FW-15 every 15,000 miles & 4 pints FW-15 every time coolant is changed.
③ Fuel filter (7.3L diesel) - change filter at 15,000 miles & when ever fuel restriction lamp is illuminated.
④ Air cleaner filter (7.3L diesel) - change filter when restriction gauge is in red zone.
⑤ Automatic transmission fluid & filter - C6 & E40D transmissions do not require regular fluid changes under normal operating conditions.
⑥ Rotate front tires for dual rear wheel vehicles from side to side only.
R – Replace   S/I – Service or Inspect

FREQUENT OPERATION MAINTENANCE (SEVERE SERVICE)
If a vehicle is operated under any of the following conditions it is considered severe service:
- Extremely dusty areas.
- 50% or more of the vehicle operation is in 32°C (90°F) or higher temperatures, or constant operation in temperatures below 0°C (32°F).
- Prolonged idling (vehicle operation in stop and go traffic).
- Frequent short running periods (engine does not warm to normal operating temperatures).
- Police, taxi, delivery usage or trailer towing usage.
Air cleaner filter - check every 3000 miles.
Oil & oil filter change – change every 3000 miles.
Rear axle oil (F-Super Duty) - change every 3000 miles.
Automatic transmission shift linkage - lubricate every 6000 miles.
Clutch reservoir fluid level - check every 6000 miles.
Exhaust system - check every 6000 miles.
Steering linkage suspension, driveshaft U-joint & slip-yoke (if equipped) - lubricate every 6000 miles.
Rotate tires every 9000 miles. (City delivery vehicles & other unique applications that require constant turning, may need more frequent tire rotation.)
Clutch release lever (7.3L diesel & 7.5L) - lubricate every 15,000 miles.
Automatic transmission fluid & filter - change every 21,000 miles.

SCHEDULED MAINTENANCE INTERVALS
(FORD AEROSTAR, RANGER & EXPLORER (1993))

TO BE SERVICED	TYPE OF SERVICE	VEHICLE MILEAGE INTERVAL (x1000)												
		7.5	15	22.5	30	37.5	45	52.5	60	67.5	75	82.5	90	97.5
Engine oil & filter	R	✓	✓	✓	✓	✓	✓	✓	✓	✓	✓	✓	✓	✓
Automatic transmission shift linkage (Cable system)	S/I	✓	✓	✓	✓	✓	✓	✓	✓	✓	✓	✓	✓	✓
Clutch reservoir fluid level	S/I	✓	✓	✓	✓	✓	✓	✓	✓	✓	✓	✓	✓	✓
Driveshaft (if equipped with fittings)	S/I	✓	✓	✓	✓	✓	✓	✓	✓	✓	✓	✓	✓	✓
Exhaust system & heat shields	S/I	✓	✓	✓	✓	✓	✓	✓	✓	✓	✓	✓	✓	✓
Rear driveshaft double cardan joint centering ball (Ranger SWB 4x4 & Explorer)	S/I	✓	✓	✓	✓	✓	✓	✓	✓	✓	✓	✓	✓	✓
Steering linkage joints (Ranger/Explorer)	S/I	✓	✓	✓	✓	✓	✓	✓	✓	✓	✓	✓	✓	✓
Wheel lug nut torque	S/I	✓	✓	✓	✓	✓	✓	✓	✓	✓	✓	✓	✓	✓
Rotate tires	S/I	✓		✓		✓		✓		✓		✓		✓
Disc brake system & caliper slide rails	S/I		✓		✓		✓		✓		✓		✓	
Drum brake systems, hoses & lines	S/I		✓		✓		✓		✓		✓		✓	
Air cleaner filter	R				✓				✓				✓	
Crankcase emission air filter (Aerostar 4.0L)	R				✓				✓				✓	
Engine Coolant	R				✓				✓				✓	
Spark plugs①	R				✓				✓				✓	
Automatic transmission fluid & filter	S/I				✓				✓				✓	
Engine coolant level, hoses & clamps	S/I		✓		✓				✓				✓	
Front suspension ball joints (Aerostar)	S/I				✓				✓				✓	
Front wheel bearings	S/I				✓				✓				✓	
Hub lock (Ranger/Explorer 4x4)	S/I				✓				✓				✓	
Parking brake system	S/I				✓				✓				✓	

SCHEDULED MAINTENANCE INTERVALS
(FORD AEROSTAR, RANGER & EXPLORER (1993)) (Cont.)

TO BE SERVICED	TYPE OF SERVICE	VEHICLE MILEAGE INTERVAL (x1000)												
		7.5	15	22.5	30	37.5	45	52.5	60	67.5	75	82.5	90	97.5
Spindle needle bearing spindle thrust bearing (Ranger/Explorer 4x4)	S/I				✓				✓				✓	
Suspension bushings, arms, springs & rear jounce bumpers (Aerostar)	S/I				✓				✓				✓	
Throttle & kickdown cable ball studs	S/I				✓				✓				✓	
Front drive axle RH axle shaft slip yoke (Ranger/Explorer 4x4)	S/I				✓				✓				✓	
Manual transmission oil	S/I	✓			✓				✓				✓	
PCV valve	R								✓					
Transfer case oil	R								✓					
Drive belts	S/I								✓					
Rear axle oil②	R													

① Platinum tip spark plugs - replace every 60,000 miles.
② Rear axle oil - replace every 100,000 miles.
R – Replace   S/I – Service or Inspect

FREQUENT OPERATION MAINTENANCE (SEVERE SERVICE)

If a vehicle is operated under any of the following conditions it is considered severe service:
- Extremely dusty areas.
- 50% or more of the vehicle operation is in 32°C (90°F) or higher temperatures, or constant operation in temperatures below 0°C (32°F).
- Prolonged idling (vehicle operation in stop and go traffic).
- Frequent short running periods (engine does not warm to normal operating temperatures).
- Police, taxi, delivery usage or trailer towing usage.
Oil & oil filter change – change every 3000 miles.
Automatic transmission fluid & filter - change every 30,000 miles.

Additional items for vehicles operated off-highway:
Front axle, steering & clutch linkages, axle & driveshaft U-joints & slip yoke - lubricate every 1000 miles.
Front wheel bearings - check every 1000 miles.
Disc brake system, caliper slide rails - check every 1000 miles.
Drum brake system, hoses & lines - check every 1000 miles.
Exhaust system - check every 1000 miles.
Automatic transmission shift linkage (cable system) - lubricate every 1000 miles.
Driveshaft (if equipped with fittings) - lubricate every 1000 miles.

SCHEDULED MAINTENANCE INTERVALS
(FORD AEROSTAR, RANGER, EXPLORER, & MOUNTAINEER) (1994-97))

TO BE SERVICED	TYPE OF SERVICE	VEHICLE MILEAGE INTERVAL (x1000)												
		5	10	15	20	25	30	35	40	45	50	55	60	65
Engine oil & filter	R	✓	✓	✓	✓	✓	✓	✓	✓	✓	✓	✓	✓	✓
Automatic transmission shift linkage (Bell crank system)	S/I	✓	✓	✓	✓	✓	✓	✓	✓	✓	✓	✓	✓	✓
Exhaust system & heat shields	S/I	✓		✓		✓		✓		✓		✓		✓
Rotate tires	S/I	✓		✓		✓		✓		✓		✓		✓
Steering linkage & driveshaft U-joint (if equipped with fitting)	S/I	✓		✓		✓		✓		✓		✓		✓
Clutch reservoir fluid level (Ranger, Explorer & Mountaineer)	S/I	✓		✓		✓		✓		✓		✓		✓
Rear driveshaft double cardan joint centering ball (Ranger, Explorer & Mountaineer SWB 4x4)	S/I	✓		✓		✓		✓		✓		✓		✓
Disc brake system & caliper slide rails	S/I			✓			✓			✓			✓	
Drum brake systems, hoses & lines	S/I			✓			✓			✓			✓	
Engine coolant strength, hoses & clamps	S/I			✓			✓			✓			✓	
Transfer case shift lever pivot bolt & control rod connecting pins (Ranger, Explorer & Mountaineer 4x4)	S/I			✓			✓			✓			✓	
Air cleaner filter	R						✓						✓	
Automatic transmission fluid & filter	R						✓						✓	
Engine Coolant①	R						✓						✓	
Fuel filter	R						✓						✓	
Front axle RH axle shaft slip yoke (Ranger, Explorer & Mountaineer 4x4)	S/I						✓						✓	

SCHEDULED MAINTENANCE INTERVALS
(FORD AEROSTAR, RANGER, EXPLORER, & MOUNTAINEER) (1994-97)) (Cont.)

TO BE SERVICED	TYPE OF SERVICE	VEHICLE MILEAGE INTERVAL (x1000)												
		5	10	15	20	25	30	35	40	45	50	55	60	65
Front suspension ball joints, bushings, arms, springs & rear jounce bumpers (Aerostar)	S/I						✓						✓	
Front wheel bearings (4x2)	S/I						✓						✓	
Hub lock (Ranger, Explorer & Mountaineer 4x4)	S/I						✓						✓	
Parking brake system	S/I						✓						✓	
Spindle needle bearing (Ranger, Explorer & Mountaineer 4x4)	S/I						✓						✓	
Throttle or TV lever ball studs	S/I						✓						✓	
Front axle & transfer case oil (E-4WD)	R												✓	
Manual transmission	R												✓	
PCV valve	R												✓	
Spark plugs (2.3L)②	R												✓	
Transfer case oil (Ranger, Explorer & Mountaineer 4x4)	R												✓	
Drive belts	S/I						✓						✓	

① Engine coolant (1995-97) - change initially at 50,000 miles and every 30,000 miles thereafter.
② Spark plugs (3.0L & 4.0L) - replace every 100,000 miles.
R – Replace   S/I – Service or Inspect

FREQUENT OPERATION MAINTENANCE (SEVERE SERVICE)

If a vehicle is operated under any of the following conditions it is considered severe service:
- **Extremely dusty areas.**
- **50% or more of the vehicle operation is in 32°C (90°F) or higher temperatures, or constant operation in temperatures below 0°C (32°F).**
- **Prolonged idling (vehicle operation in stop and go traffic).**
- **Frequent short running periods (engine does not warm to normal operating temperatures).**
- **Police, taxi, delivery usage or trailer towing usage.**
Oil & oil filter change – change every 3000 miles.
Air cleaner filter - service or inspect every 6000 miles.
Exhaust system - check every 6000 miles.
Rotate tires every 9000 miles. (City delivery vehicles & other unique applications that require constant turning may need frequent tire rotation.)
Automatic transmission fluid & filter - change every 21,000 miles.

FORD MOTOR COMPANY
Windstar • Villager

ENGINE IDENTIFICATION

Year	Model	Engine Displacement Liters (cc)	Engine Series (ID/VIN)	Fuel System	No. of Cylinders	Engine Type
1993	Villager	3.0 (2966)	W	MFI	6	SOHC
1994	Villager	3.0 (2966)	W	MFI	6	SOHC
1995	Windstar	3.8 (3802)	4	MFI	6	OHV
1996-97	Windstar	3.0 (2982)	U	MFI	6	OHV
	Windstar	3.8 (3802)	4	SPI	6	OHV
	Villager	3.0 (2966)	W	MFI	6	SOHC

MFI - Multiport fuel injection
SPI - Split port injection
OHV - Overhead valve
SOHC - Single overhead camshaft

7914gc02 not found

VEHICLE IDENTIFICATION CHART

Code	Liters	Cu. In. (cc)	Cyl.	Fuel Sys.	Eng. Mfg.
U	3.0	183 (2999)	6	MFI	Ford
W	3.0	183 (2999)	6	MFI	Ford
4	3.8	231 (3785)	6	MFI	Ford

MFI - Multiport fuel injection

Code	Year
P	1993
R	1994
S	1995
T	1996
V	1997

GENERAL ENGINE SPECIFICATIONS

Year	Engine ID/VIN	Engine Displacement Liters (cc)	Fuel System Type	Net Horsepower @ rpm	Net Torque @ rpm (ft. lbs.)	Bore x Stroke (in.)	Compression Ratio	Oil Pressure @ rpm
1993	W	3.0 (2966)	MFI	151@4800	174@4400	3.43x3.27	9.0:1	40-60@2000
1994	W	3.0 (2966)	MFI	151@4800	174@4400	3.43x3.27	9.0:1	40-60@2000
1995	4	3.8 (3802)	MFI	140@3800	215@2400	3.81x3.39	9.0:1	40-60@2500
1996-97	U	3.0 (2982)	MFI	147@5000	162@3250	3.50x3.14	9.3:1	40-60@2500
	W	3.0 (2966)	MFI	151@4800	174@4400	3.43x3.27	9.0:1	40-60@2500
	4	3.8 (3802)	MFI	200@5000	230@3000	3.81x3.39	9.3:1	40-60@2500

MFI - Multiport fuel injection

GASOLINE ENGINE TUNE-UP SPECIFICATIONS

Year	Engine ID/VIN	Engine Displacement Liters (cc)	Spark Plugs Gap (in.)	Ignition Timing (deg.) MT	Ignition Timing (deg.) AT	Fuel Pump (psi)	Idle Speed (rpm) MT	Idle Speed (rpm) AT	Valve Clearance In.	Valve Clearance Ex.
1993	W	3.0 (2966)	0.033	-	15B	36-38 ②	-	①	HYD	HYD
1994	W	3.0 (2966)	0.033	-	15B	36-38 ②	-	①	HYD	HYD
1995	4	3.8 (3802)	0.054	-	10B	30-45	①	①	HYD	HYD
1996-97	U	3.0 (2982)	0.044	-	10B	35-45	①	①	HYD	HYD
	W	3.0 (2966)	0.033	-	15B	36-38 ②	-	①	HYD	HYD
	4	3.8 (3802)	0.054	-	10B	30-45	①	①	HYD	HYD

NOTE: The Vehicle Emission Control Information label often reflects specification changes changes made during production. The label figures must be used if they differ from those in this chart.

B - Before top dead center

HYD - Hydraulic

① Idle speed is electronically controlled and cannot be adjusted

② Fuel pressure is with engine running, pressure regulator vacuum hose disconnected

CAPACITIES

Year	Model	Engine ID/VIN	Engine Displacement Liters (cc)	Engine Oil with Filter (qts.)	Transmission (pts.) 4-Spd	Transmission (pts.) 5-Spd	Transmission (pts.) Auto.	Transfer Case (pts.)	Drive Axle Front (pts.)	Drive Axle Rear (pts.)	Fuel Tank (gal.)	Cooling System (qts.)
1993	Villager	W	3.0 (2966)	4.2	-	-	17.4	-	-	①	20.0	11.6
1994	Villager	W	3.0 (2966)	4.2	-	-	17.4 ④	-	①	①	20.0	11.6
1995	Windstar	4	3.8 (3802)	4.5	-	-	24.5	-	-	-	20.0	12.1
1996-97	Windstar	U	3.0 (2966)	4.2	-	-	16.5	-	①	-	20.0	②
	Windstar	4	3.8 (3802)	4.5	-	-	24.5	-	-	-	20.0 ③	12.1
	Villager	W	3.0 (2966)	4.2	-	-	16.5	-	①	-	20.0	②

① Included in transaxle capacity
② With coolant recovery reservoir: 7.9
   Without coolant recovery reservoir: 5.8
③ Optional: 26 gals.
④ Includes torque converter

VALVE SPECIFICATIONS

Year	Engine ID/VIN	Engine Displacement Liters (cc)	Seat Angle (deg.)	Face Angle (deg.)	Spring Test Pressure (lbs. @ in.)	Spring Installed Height (in.)	Stem-to-Guide Clearance (in.) Intake	Stem-to-Guide Clearance (in.) Exhaust	Stem Diameter (in.) Intake	Stem Diameter (in.) Exhaust
1993	W	3.0 (2966)	45	45	①	②	0.0008-0.0021	0.0016-0.0029	0.2742-0.2748	0.3136-0.3138
1994	W	3.0 (2966)	45	45	①	②	0.0008-0.0021	0.0016-0.0029	0.2742-0.2748	0.3136-0.3138
1995	4	3.8 (3802)	44.5	45.8	220@1.18	1.97	0.0010-0.0028	0.0015-0.0033	0.3423-0.3415	0.3418-0.3410
1996-97	U	3.0 (2982)	45	44	185@1.16	1.58-1.61	0.0010-0.0027	0.0015-0.0032	0.3126-0.3134	0.3121-0.3129
	W	3.0 (2966)	45	45	①	②	0.0008-0.0021	0.0016-0.0029	0.2742-0.2748	0.3136-0.3138
	4	3.8 (3802)	44.75	45.8	220@1.18	1.97	0.0010-0.0028	0.0015-0.0033	0.3423-0.3415	0.3410-0.3418

① Outer spring: 118@1.81
   Inner spring: 57.3@0.984
② Spring height measured unloaded
   Minimum length, outer spring: 2.016
   Minimum length, inner spring: 1.736

TORQUE SPECIFICATIONS
All readings in ft. lbs.

	Engine ID/VIN	Engine Displacement Liters (cc)	Cylinder Head Bolts	Main Bearing Bolts	Rod Bearing Bolts	Crankshaft Damper Bolts	Flywheel Bolts	Manifold		Spark Plugs	Lug Nut
								Intake	Exhaust		
1993	W	3.0 (2966)	④	67-74	⑤	141-156	61-69	⑥	13-16	14-22	80
1994	W	3.0 (2966)	④	67-74	⑤	141-156	61-69	⑥	13-16	14-22	80
1995	4	3.8 (3802)	①	65-81	31-36	103-132	54-64	②	19	8-10	85-105
1996-97	U	3.0 (2982)	③	60	26	107	54-64	24	25	8-10	100
	W	3.0 (2966)	④	67-74	⑤	141-156	61-69	⑥	13-16	14-22	80
	4	3.8 (3802)	①	65-81	31-36	103-132	54-64	②	19	8-10	85-105

① Step 1: 15 ft. lbs.
   Step 2: 29 ft. lbs.
   Step 3: 37 ft. lbs.
   Step 4: Loosen each bolt one at a time
   Step 5: Long bolts to 11-19 ft. lbs. plus 1/4 turn
   Step 6: Short bolts to 7-15 ft. lbs. plus 1/4 turn
② Lower intake manifold:
   Step 1: 13 ft. lbs.
   Step 2: 16 ft. lbs.
   Upper intake manifold:
   Step 1: 8 ft. lbs.
   Step 2: 15 ft. lbs.
   Step 3: 24 ft. lbs.

③ Step 1: 37 ft. lbs.
   Step 2: 68 ft. lbs.
④ Step 1: 22 ft. lbs.
   Step 2: 43 ft. lbs.
   Step 3: Loosen bolts one turn
   Step 4: 22 ft. lbs.
   Step 5: Rotate 60-65 degrees or 40-47 ft. lbs.
   Step 6: Small cylinder head bolt located outside of valve cover: 6 ft. lbs.
⑤ Step 1: 10-12 ft. lbs.
   Step 2: 28-33 ft. lbs.
⑥ Step 1: Nuts and bolts: 3 ft. lbs.
   Step 2: Nuts: 17-20 ft. lbs., Bolts: 12-14 ft. lbs.
   Step 3: Repeat Step 2

BRAKE SPECIFICATIONS
All measurements in inches unless noted

Year	Model		Master Cylinder Bore	Brake Disc			Brake Drum Diameter			Minimum Lining Thickness	
				Original Thickness	Minimum Thickness	Maximum Runout	Original Inside Diameter	Max. Wear Limit	Maximum Machine Diameter	Front	Rear
1993	Villager		NA	1.024	0.974	0.003	9.84	9.89	9.86	0.080	0.030
1994	Villager		NA	1.024	0.974	0.003	9.84	9.90	9.86	0.080	0.059
1995	Windstar		NA	1.020	0.097	0.003	9.84	NA	9.90	0.040	0.590
1996-97	Windstar	F	NA	1.020	0.907	0.003	-	-	-	0.125	-
		R	NA	0.472	0.409	0.003	9.84	NA	9.90	-	①
	Villager	F	1.000	1.005	0.945	0.003	-	-	-	0.079	-
		R	-	-	-	-	9.84	9.90	9.86	-	0.059

NOTE: Due to changes made during production, refer to manufacturer's specifications if they differ from those in this chart
NA - Not Available
F - Front
R - Rear
① With drum brakes: 0.030
   With disc brakes: 0.125

FREQUENT MAINTENANCE LABOR
FORD WINDSTAR

The following should be used as a guide when determining the amount of work required for a particular service if taken to a repair shop. In estimating how long a particular Frequent Maintenance Service item should take, please observe the following:

- **Factory Time** is time that is generated by the vehicle manufacturer.
- **Chilton Time** is time that is based on field research and data supplied by the vehicle manufacturer.
- All labor time operations are given in hours and tenths of an hour.
- All labor operations, are to be used as a **guide**.

COOLING

(G) Winterize Cooling System
Includes: Run engine to check for leaks, tighten all hose connections. Test radiator and pressure cap, drain radiator and engine block. Add antifreeze and refill system.
1995-97 5

(G) Belt, Serpentine Drive, Renew
1995-97 (.3) 4

(G) Hoses, Radiator, Renew
Includes: Drain and refill cooling system as required.
1995-97
  upper (.5) 6
  lower (.6) 7
  both (.7) 1.0

(G) Thermostat, Coolant, Renew
1995-97 (.4) 6

FUEL

(M) Air Cleaner, Service
1995-97 (.3) 4

(G) Filter, Fuel, Renew
1995-97 (.3) 4

BRAKES

(G) Bleed Brakes (Four Wheels)
Includes: Add fluid.
1995-97 (.3) 4

LUBRICATION SERVICES

(M) Engine Oil & Filter, Renew
Includes: Inspect and correct all fluid levels.
1995-97 3

(M) Lubricate Chassis, Change Oil & Filter
Includes: Inspect and correct all fluid levels.
1995-97 6
Install grease fittings add 1

(M) Lubricate Chassis
Includes: Inspect and correct all fluid levels.
1995-97 4
Install grease fittings add 1

WHEELS

(G) Wheel, Renew (One)
1995-97 5

(G) Wheel, Balance
1995-97
  one .3
  each adtnl. 2

(G) Wheels, Rotate (All)
1995-97 5

ELECTRICAL

(G) Headlamps, Aim
1995-97
  one .4
  all 6

(G) Halogen Headlamp Bulb, Renew
1995-97, one (.3) 4

(G) License Lamp Assy., Renew
1995-97
  one (.3) 4
  both (.4) 5

(G) Rear Combination Lamp Assy., Renew
1995-97
  one (.3) 4
  both (.4) 6

(G) Horn, Renew
1995-97 (.3) 4

(M) Terminals, Battery, Clean
1995-97 3

FREQUENT MAINTENANCE LABOR
MERCURY VILLAGER

The following should be used as a guide when determining the amount of work required for a particular service if taken to a repair shop. In estimating how long a particular Frequent Maintenance Service item should take, please observe the following:
- **Factory Time** is time that is generated by the vehicle manufacturer.
- **Chilton Time** is time that is based on field research and data supplied by the vehicle manufacturer.
- All labor time operations are given in hours and tenths of an hour.
- All labor operations, are to be used as a **guide**.

COOLING

(G) Winterize Cooling System
Includes: Run engine to check for leaks, tighten all hose connections. Test radiator and pressure cap, drain radiator and engine block. Add antifreeze and refill system.
1993-97 5

(G) Belt, Drive, Renew
1993-97
AC (.5) 7
PS (.7) 1.0
Alternator (.6) 8

(G) Belt, Drive, Adjust
1993-97
one (.2) 3
each adtnl. 1

(G) Hoses, Radiator, Renew
Includes: Drain and refill cooling system as required.
1993-97
upper (.5) 7
lower (.6) 8
both (.7) 9

(G) Thermostat, Coolant, Renew
1993-97 (.9) 1.3

FUEL

(M) Air Cleaner, Service
1993-97 3

(G) Filter, Fuel, Renew
1993-97 (.4) 6

BRAKES

(G) Bleed Brakes (Four Wheels)
Includes: Add fluid.
1993-97 (.3) 4

(M) Parking Brake, Adjust
1993-97 (.3) 4

LUBRICATION SERVICES

(M) Engine Oil & Filter, Renew
Includes: Inspect and correct all fluid levels.
1993-97 3

(M) Lubricate Chassis, Change Oil & Filter
Includes: Inspect and correct all fluid levels.
1993-97 6
Install grease fittings add 1

(M) Lubricate Chassis
Includes: Inspect and correct all fluid levels.
1993-97 4
Install grease fittings add 1

WHEELS

(G) Wheel, Renew (One)
1993-97 5

(G) Wheel, Balance
1993-97
one .3
each adtnl. 2

(G) Wheels, Rotate (All)
1993-97 5

ELECTRICAL

(G) Headlamps, Aim
1993-97 (.5) 6

(G) Halogen Headlamp Bulb, Renew
1993-97 (.3) 4

(G) License Lamp Assy., Renew
1993-97, each (.3) 4

(G) Rear Combination Lamp Assy., Renew
1993-97
one (.3) 4
both (.4) 6

(G) Horn, Renew
1993-97, one (.4) 5
each adtnl. 1

(M) Terminals, Battery, Clean
1993-97 3

Factory Time / *Chilton Time*

SCHEDULED MAINTENANCE INTERVALS
(MERCURY VILLAGER (1993))

TO BE SERVICED	TYPE OF SERVICE	VEHICLE MILEAGE INTERVAL (x1000)												
		7.5	15	22.5	30	37.5	45	52.5	60	67.5	75	82.5	90	97.5
Engine oil & filter	R	✓	✓	✓	✓	✓	✓	✓	✓	✓	✓	✓	✓	✓
Rotate tires	S/I	✓		✓		✓		✓		✓		✓		✓
Engine coolant strength, hoses & clamps	S/I			✓				✓		✓			✓	
Air cleaner filter	R				✓				✓				✓	
Engine Coolant	R				✓				✓				✓	
Spark plugs	R				✓				✓				✓	
Automatic transaxle fluid & filter	S/I				✓				✓				✓	
Drive belts	S/I				✓				✓				✓	
Exhaust system & heat shields	S/I				✓				✓				✓	
Front & rear brakes	S/I				✓				✓				✓	
Cam belt	R								✓					
PCV valve	R								✓					

R – Replace   S/I – Service or Inspect

FREQUENT OPERATION MAINTENANCE (SEVERE SERVICE)

If a vehicle is operated under any of the following conditions it is considered severe service:
- Extremely dusty areas.
- 50% or more of the vehicle operation is in 32°C (90°F) or higher temperatures, or constant operation in temperatures below 0°C (32°F).
- Prolonged idling (vehicle operation in stop and go traffic).
- Frequent short running periods (engine does not warm to normal operating temperatures).
- Police, taxi, delivery usage or trailer towing usage.

Oil & oil filter change – change every 3000 miles.
Automatic transaxle fluid & filter - change every 21,000 miles.
Rotate tires every 12,000 miles.
Exhaust system - check every 15,000 miles.
Air cleaner filter - service or inspect every 15,000 miles.
Engine coolant strength, hoses & clamps - check every 15,000 miles.

SCHEDULED MAINTENANCE INTERVALS
(FORD WINDSTAR, MERCURY VILLAGER (1994-97))

TO BE SERVICED	TYPE OF SERVICE	VEHICLE MILEAGE INTERVAL (x1000)												
		5	10	15	20	25	30	35	40	45	50	55	60	65
Engine oil & filter	R	✓	✓	✓	✓	✓	✓	✓	✓	✓	✓	✓	✓	✓
Rotate tires	S/I	✓		✓		✓		✓		✓		✓		✓
Engine coolant strength, hoses & clamps	S/I			✓			✓			✓			✓	
Air cleaner filter	R						✓						✓	
Automatic transmission fluid & filter	R						✓						✓	
Engine Coolant①	R						✓						✓	
PCV valve	R												✓	
Spark plugs②	R						✓						✓	
Drive belts	S/I						✓						✓	
Exhaust system & heat shields	S/I						✓						✓	
Front & rear brakes	S/I						✓						✓	

① Engine coolant (1995-97) - change initially at 50,000 miles and every 30,000 miles thereafter.
② Spark plugs (Windstar) - replace every 100,000 miles.
R – Replace   S/I – Service or Inspect

FREQUENT OPERATION MAINTENANCE (SEVERE SERVICE)
If a vehicle is operated under any of the following conditions it is considered severe service:
- Extremely dusty areas.
- 50% or more of the vehicle operation is in 32°C (90°F) or higher temperatures, or constant operation in temperatures below 0°C (32°F).
- Prolonged idling (vehicle operation in stop and go traffic).
- Frequent short running periods (engine does not warm to normal operating temperatures).
- Police, taxi, delivery usage or trailer towing usage.
Oil & oil filter change – change every 3000 miles.
Rotate tires initially at 6000 miles, and every 9000 miles thereafter.
Air cleaner filter - change every 15,000 miles.
Engine coolant strength, hoses & clamps - check every 15,000 miles.
Exhaust system - check every 15,000 miles.
Automatic transmission fluid & filter - change every 21,000 miles.

GENERAL MOTORS CORPORATION
G Series (Vans) • C/K Series (Pick Up)
Blazer • Jimmy • Tahoe • Yukon • Suburban
Astro • Safari • Blazer (S10) • Jimmy (S10) • Bravada
Sonoma • S Series (Pick Up) • Typhoon

VEHICLE IDENTIFICATION CHART

		Engine Code					Model Year	
Code	Liters	Cu. In. (cc)	Cyl.	Fuel Sys.	Eng. Mfg.		Code	Year
4	2.2	134 (2189)	4	MFI	CPC		P	1993
A	2.5	151 (2474)	4	TFI	CPC		R	1994
C	6.2	379 (6210)	8	DSL	CPC		S	1995
F	6.5	395 (6473)	8	DSL	CPC		T	1996
H	5.0	305 (4999)	8	TFI	CPC		V	1997
J	6.2	379 (6210)	8	DSL	CPC			
J	7.4	454 (7440)	8	MFI	CPC			
K	5.7	350 (5735)	8	TFI	CPC			
M	5.0	305 (4999)	8	MFI	CPC			
N	7.4	454 (7440)	8	TFI	CPC			
P	6.5	395 (6473)	8	DSL	CPC			
R	2.8	173 (2835)	6	TFI	CPC			
R	5.7	350 (5735)	8	MFI	CPC			
S	6.5	395 (6473)	8	DSL	CPC			
W	4.3	263 (4293)	6	TFI	CPC			
W	4.3	263 (4293)	6	MFI	CPC			
X	4.3	263 (4293)	6	MFI	CPC			
Y	6.5	395 (6473)	8	DSL	CPC			
Z	4.3	263 (4293)	6	TFI	CPC			
Z	4.3	263 (4293)	6	MFI	CPC			

TFI - Throttle body fuel injection   DSL - Diesel
MFI - Multiport fuel injection   CPC - Chevrolet/Pontiac/Canada

ENGINE IDENTIFICATION

Year	Model	Engine Displacement Liters (cc)	Engine Series (ID/VIN)	Fuel System	No. of Cylinders	Engine Type
1993	Astro/Safari	4.3 (4293)	W	MFI	6	OHV
	Astro/Safari	4.3 (4293)	Z	TFI	6	OHV
	Blazer/Yukon	5.7 (5735)	K	TFI	8	OHV
	C1500	4.3 (4293)	Z	TFI	6	OHV
	C1500	5.0 (4999)	H	TFI	8	OHV
	C1500	5.7 (5735)	K	TFI	8	OHV
	C1500	6.2 (6210)	C	DSL	8	OHV

ENGINE IDENTIFICATION
All measurements are given in inches.

Year	Model	Engine Displacement Liters (cc)	Engine Series (ID/VIN)	Fuel System	No. of Cylinders	Engine Type
1993	C1500	6.2 (6210)	J	DSL	8	OHV
	C1500	7.4 (7440)	N	TFI	8	OHV
	C2500	4.3 (4293)	Z	TFI	6	OHV
	C2500	5.0 (4999)	H	TFI	8	OHV
	C2500	5.7 (5735)	K	TFI	8	OHV
	C2500	6.2 (6210)	C	DSL	8	OHV
	C2500	6.2 (6210)	J	DSL	8	OHV
	C2500	6.5 (6505)	F	DSL	8	OHV
	C2500	7.4 (7440)	N	TFI	8	OHV
	C3500	4.3 (4293)	Z	TFI	6	OHV
	C3500	5.0 (4999)	H	TFI	8	OHV
	C3500	5.7 (5735)	K	TFI	8	OHV
	C3500	6.2 (6210)	C	DSL	8	OHV
	C3500	6.2 (6210)	J	DSL	8	OHV
	C3500	6.5 (6505)	F	DSL	8	OHV
	C3500	7.4 (7440)	N	TFI	8	OHV
	G10	4.3 (4293)	Z	TFI	6	OHV
	G10	5.0 (4999)	H	TFI	8	OHV
	G10	5.7 (5735)	K	TFI	8	OHV
	G10	6.2 (6210)	C	DSL	8	OHV
	G10	6.2 (6210)	J	DSL	8	OHV
	G10	7.4 (7440)	N	TFI	8	OHV
	G20	4.3 (4293)	Z	TFI	6	OHV
	G20	5.0 (4999)	H	TFI	8	OHV
	G20	5.7 (5735)	K	TFI	8	OHV
	G20	6.2 (6210)	C	DSL	8	OHV
	G20	6.2 (6210)	J	DSL	8	OHV
	G20	7.4 (7440)	N	TFI	8	OHV
	G30	4.3 (4293)	Z	TFI	6	OHV
	G30	5.0 (4999)	H	TFI	8	OHV
	G30	5.7 (5735)	K	TFI	8	OHV
	G30	6.2 (6210)	C	DSL	8	OHV
	G30	6.2 (6210)	J	DSL	8	OHV
	G30	7.4 (7440)	N	TFI	8	OHV
	K1500	4.3 (4293)	Z	TFI	6	OHV
	K1500	5.0 (4999)	H	TFI	8	OHV
	K1500	5.7 (5735)	K	TFI	8	OHV
	K1500	6.2 (6210)	C	DSL	8	OHV
	K1500	6.2 (6210)	J	DSL	8	OHV
	K1500	7.4 (7440)	N	TFI	8	OHV
	K2500	4.3 (4293)	Z	TFI	6	OHV
	K2500	5.0 (4999)	H	TFI	8	OHV
	K2500	5.7 (5735)	K	TFI	8	OHV
	K2500	6.2 (6210)	C	DSL	8	OHV
	K2500	6.2 (6210)	J	DSL	8	OHV
	K2500	6.5 (6505)	F	DSL	8	OHV
	K2500	7.4 (7440)	N	TFI	8	OHV
	K3500	4.3 (4293)	Z	TFI	6	OHV
	K3500	5.0 (4999)	H	TFI	8	OHV
	K3500	5.7 (5735)	K	TFI	8	OHV

ENGINE IDENTIFICATION

Year	Model	Engine Displacement Liters (cc)	Engine Series (ID/VIN)	Fuel System	No. of Cylinders	Engine Type
1993	K3500	6.2 (6210)	C	DSL	8	OHV
	K3500	6.2 (6210)	J	DSL	8	OHV
	K3500	6.5 (6505)	F	DSL	8	OHV
	K3500	7.4 (7441)	N	TFI	8	OHV
	S10 Blazer/Jimmy/Bravada	4.3 (4293)	W	MFI	6	OHV
	S10 Blazer/Jimmy/Bravada/ Typhoon	4.3 (4293)	Z	TFI	6	OHV
	S10/S15 Pick-up/Sonoma	2.5 (2474)	A	TFI	4	OHV
	S10/S15 Pick-up/Sonoma	2.8 (2835)	R	TFI	6	OHV
	S10/S15 Pick-up/Sonoma	4.3 (4293)	W	MFI	6	OHV
	Suburban	5.7 (5735)	K	TFI	8	OHV
	Suburban	7.4 (7440)	N	TFI	8	OHV
1994	Astro/Safari	4.3 (4293)	W	TFI	6	OHV
	Astro/Safari	4.3 (4293)	Z	TFI	6	OHV
	Blazer/Yukon	5.7 (5735)	K	MFI	8	OHV
	Blazer/Yukon	6.5 (6505)	S	DSL	8	OHV
	C1500	4.3 (4293)	Z	TFI	6	OHV
	C1500	5.0 (4999)	H	TFI	8	OHV
	C1500	5.7 (5735)	K	TFI	8	OHV
	C1500	6.5 (6505)	P	DSL	8	OHV
	C1500	6.5 (6505)	S	DSL	8	OHV
	C2500	4.3 (4293)	Z	TFI	6	OHV
	C2500	5.0 (4999)	H	TFI	8	OHV
	C2500	5.7 (5735)	K	TFI	8	OHV
	C2500	6.5 (6505)	F	DSL	8	OHV
	C2500	6.5 (6505)	P	DSL	8	OHV
	C2500	6.5 (6505)	S	DSL	8	OHV
	C2500	7.4 (7440)	N	TFI	8	OHV
	C3500	5.7 (5735)	K	TFI	8	OHV
	C3500	6.5 (6505)	F	DSL	8	OHV
	C3500	6.5 (6505)	S	DSL	8	OHV
	G20	4.3 (4293)	Z	TFI	6	OHV
	G20	5.0 (4999)	H	TFI	8	OHV
	G20	5.7 (5735)	K	TFI	8	OHV
	G20	6.5 (6505)	P	DSL	8	OHV
	G30	4.3 (4293)	Z	TFI	6	OHV
	G30	5.7 (5735)	K	TFI	8	OHV
	G30	6.5 (6505)	Y	DSL	8	OHV
	G30	7.4 (7440)	N	TFI	8	OHV
	K1500	4.3 (4293)	Z	TFI	6	OHV
	K1500	5.0 (4999)	H	TFI	8	OHV
	K1500	5.7 (5735)	K	TFI	8	OHV
	K1500	6.5 (6505)	P	DSL	8	OHV
	K1500	6.5 (6505)	S	DSL	8	OHV
	K1500	7.4 (7440)	N	TFI	8	OHV
	K2500	4.3 (4293)	Z	TFI	6	OHV
	K2500	5.0 (4999)	H	TFI	8	OHV
	K2500	5.7 (5735)	K	TFI	8	OHV
	K2500	6.5 (6505)	F	DSL	8	OHV
	K2500	6.5 (6505)	P	DSL	8	OHV
	K2500	6.5 (6505)	S	DSL	8	OHV

ENGINE IDENTIFICATION

Year	Model	Engine Displacement Liters (cc)	Engine Series (ID/VIN)	Fuel System	No. of Cylinders	Engine Type
1994	K2500	7.4 (7440)	N	TFI	8	OHV
	K3500	5.7 (5735)	K	TFI	8	OHV
	K3500	6.5 (6505)	F	DSL	8	OHV
	K3500	7.4 (7440)	N	TFI	8	OHV
	S10 Blazer/Jimmy/Bravada	4.3 (4293)	W	MFI	6	OHV
	S10 Blazer/Jimmy/Bravada	4.3 (4293)	Z	TFI	6	OHV
	S10/S15 Pick-up/Sonoma	2.2 (2189)	4	MFI	4	OHV
	S10/S15 Pick-up/Sonoma	4.3 (4293)	W	MFI	6	OHV
	S10/S15 Pick-up/Sonoma	4.3 (4293)	Z	MFI	6	OHV
	Suburban	5.7 (5735)	K	TFI	8	OHV
	Suburban	6.5 (6505)	F	DSL	8	OHV
	Suburban	7.4 (7440)	N	TFI	8	OHV
1995	Astro/Safari	4.3 (4293)	W	TFI	6	OHV
	C1500	4.3 (4293)	Z	TFI	6	OHV
	C1500	5.0 (4999)	H	TFI	8	OHV
	C1500	5.7 (5735)	K	TFI	8	OHV
	C1500	6.5 (6505)	P	DSL	8	OHV
	C1500	6.5 (6505)	S	DSL	8	OHV
	C2500	4.3 (4293)	Z	TFI	6	OHV
	C2500	5.0 (4999)	H	TFI	8	OHV
	C2500	5.7 (5735)	K	TFI	8	OHV
	C2500	6.5 (6505)	F	DSL	8	OHV
	C2500	6.5 (6505)	P	DSL	8	OHV
	C2500	6.5 (6505)	S	DSL	8	OHV
	C2500	7.4 (7440)	N	TFI	8	OHV
	C3500	5.7 (5735)	K	TFI	8	OHV
	C3500	6.5 (6505)	F	DSL	8	OHV
	C3500	6.5 (6505)	S	DSL	8	OHV
	G20	4.3 (4293)	Z	TFI	6	OHV
	G20	5.0 (4999)	H	TFI	8	OHV
	G20	5.7 (5735)	K	TFI	8	OHV
	G20	6.5 (6505)	P	DSL	8	OHV
	G30	4.3 (4293)	Z	TFI	6	OHV
	G30	5.7 (5735)	K	TFI	8	OHV
	G30	6.5 (6505)	Y	DSL	8	OHV
	G30	7.4 (7440)	N	TFI	8	OHV
	K1500	4.3 (4293)	Z	TFI	6	OHV
	K1500	5.0 (4999)	H	TFI	8	OHV
	K1500	5.7 (5735)	K	TFI	8	OHV
	K1500	6.5 (6505)	P	DSL	8	OHV
	K1500	6.5 (6505)	S	DSL	8	OHV
	K1500	7.4 (7440)	N	TFI	8	OHV
	K2500	4.3 (4293)	Z	TFI	6	OHV
	K2500	5.0 (4999)	H	TFI	8	OHV
	K2500	5.7 (5735)	K	TFI	8	OHV
	K2500	6.5 (6505)	F	DSL	8	OHV
	K2500	6.5 (6505)	P	DSL	8	OHV
	K2500	6.5 (6505)	S	DSL	8	OHV
	K2500	7.4 (7440)	N	TFI	8	OHV
	K3500	5.7 (5735)	K	TFI	8	OHV

ENGINE IDENTIFICATION

Year	Model	Engine Displacement Liters (cc)	Engine Series (ID/VIN)	Fuel System	No. of Cylinders	Engine Type
1995	K3500	6.5 (6505)	F	DSL	8	OHV
	K3500	7.4 (7440)	N	TFI	8	OHV
	S10 Blazer/Jimmy/Bravada	4.3 (4293)	W	MFI	6	OHV
	S10/S15 Pick-up/Sonoma	2.2 (2189)	4	MFI	4	OHV
	S10/S15 Pick-up/Sonoma	4.3 (4293)	W	MFI	6	OHV
	S10/S15 Pick-up/Sonoma	4.3 (4293)	Z	MFI	6	OHV
	Suburban	5.7 (5735)	K	TFI	8	OHV
	Suburban	6.5 (6505)	F	DSL	8	OHV
	Suburban	7.4 (7440)	N	TFI	8	OHV
	Tahoe/Yukon	5.7 (5735)	K	MFI	8	OHV
	Tahoe/Yukon	6.5 (6505)	S	DSL	8	OHV
1996-97	Astro/Safari	4.3 (4293)	W	MFI	6	OHV
	C1500	4.3 (4293)	W	MFI	6	OHV
	C1500	5.0 (4999)	M	MFI	8	OHV
	C1500	5.7 (5735)	R	MFI	8	OHV
	C2500	4.3 (4293)	W	MFI	6	OHV
	C2500	5.0 (4999)	M	MFI	8	OHV
	C2500	5.7 (5735)	R	MFI	8	OHV
	C2500	6.5 (6374)	F	DSL	8	OHV
	C3500	6.5 (6374)	F	DSL	8	OHV
	C3500	7.4 (7440)	J	MFI	8	OHV
	G1500	4.3 (4293)	W	MFI	6	OHV
	G1500	5.0 (4999)	M	MFI	8	OHV
	G1500	5.7 (5735)	R	MFI	8	OHV
	G2500	5.0 (4999)	M	MFI	8	OHV
	G2500	5.7 (5735)	R	MFI	8	OHV
	G3500	6.5 (6374)	F	DSL	8	OHV
	G3500	7.4 (7440)	J	MFI	8	OHV
	K1500	4.3 (4293)	W	MFI	6	OHV
	K1500	5.0 (4999)	M	MFI	8	OHV
	K1500	5.7 (5735)	R	MFI	8	OHV
	K1500	6.5 (6374)	F	DSL	8	OHV
	K2500	5.0 (4999)	M	MFI	8	OHV
	K2500	5.7 (5735)	R	MFI	8	OHV
	K2500	6.5 (6374)	F	DSL	8	OHV
	K3500	5.7 (5735)	R	MFI	8	OHV
	K3500	6.5 (6374)	F	DSL	8	OHV
	K3500	7.4 (7440)	J	MFI	8	OHV
	S10 Blazer/Jimmy/Bravada	4.3 (4293)	W	MFI	6	OHV
	S10 Blazer/Jimmy/Bravada	4.3 (4293)	X	MFI	6	OHV
	S10/S15 Pick-up/Sonoma	2.2 (2189)	4	MFI	4	OHV
	S10/S15 Pick-up/Sonoma	4.3 (4293)	W	MFI	6	OHV
	S10/S15 Pick-up/Sonoma	4.3 (4293)	X	MFI	6	OHV
	Suburban	5.7 (5735)	R	MFI	8	OHV
	Suburban	7.4 (7440)	J	MFI	8	OHV
	Tahoe/Yukon	6.5 (6374)	S	DSL	8	OHV
	Tahoe/Yukon	5.7 (5735)	R	MFI	8	OHV

TFI - Throttle body fuel injection
DSL - Diesel

MFI - Multiport fuel injection
OHV - Overhead valve

GENERAL ENGINE SPECIFICATIONS

Year	Engine ID/VIN	Engine Displacement Liters (cc)	Fuel System Type	Net Horsepower @ rpm	Net Torque @ rpm (ft. lbs.)	Bore x Stroke (in.)	Compression Ratio	Oil Pressure @ rpm
1993	A	2.5 (2474)	TFI	105@4800	135@3200	4.00x3.00	8.3:1	41@2000
	R	2.8 (2835)	TFI	125@2800	150@2200	3.56x3.04	8.5:1	50@2000
	W	4.3 (4293)	TFI	200@4500	260@3600	4.00x3.48	9.1:1	18@2000
	Z	4.3 (4293)	TFI	①	②	4.00x3.48	9.3:1	18@2000
	H	5.0 (4999)	TFI	165@4400	240@2000	3.74x3.48	9.0:1	18@2000
	K	5.7 (5735)	TFI	③	④	4.00x3.48	8.5:1	18@2000
	C	6.2 (6210)	DSL	130@3600	240@2000	3.98x3.80	21.3:1	35@2000
	J	6.2 (6210)	DSL	135@3600	240@2000	3.98x3.80	21.3:1	35@2000
	F	6.5 (6473)	DSL	⑤	⑥	4.05x3.80	21.0:1	40-45@2000
	N	7.4 (7440)	TFI	230@3600	385@1600	4.25x4.00	8.0:1	40@2000
1994	4	2.2 (2189)	MFI	118@5200	130@2800	3.50x3.46	9.0:1	56@3000
	W	4.3 (4293)	TFI	200@4500	260@3600	4.00x3.48	9.1:1	18@2000
	Z	4.3 (4293)	MFI	①	②	4.00x3.48	9.1:1	18@2000
	Z	4.3 (4293)	TFI	①	②	4.00x3.48	9.1:1	18@2000
	H	5.0 (4999)	TFI	175@4200	265@2800	3.74x3.48	9.0:1	18@2000
	K	5.7 (5735)	MFI	③	④	4.00x3.48	9.1:1	18@2000
	K	5.7 (5735)	TFI	③	④	4.00x3.48	9.1:1	18@2000
	F	6.5 (6473)	DSL	⑤	⑥	4.05x3.80	21.5:1	40-45@2000
	P	6.5 (6473)	DSL	190@3400	385@1700	4.06x3.82	21.5:1	40-45@2000
	S	6.5 (6473)	DSL	190@3400	275@1700	4.06x3.82	21.5:1	40-45@2000
	Y	6.5 (6473)	DSL	⑦	⑧	4.06x3.82	21.5:1	40-45@2000
	N	7.4 (7440)	TFI	230@3600	385@1600	4.25x4.00	8.0:1	40@2000
1995	4	2.2 (2189)	MFI	118@5200	130@2800	3.50x3.46	9.0:1	56@3000
	W	4.3 (4293)	TFI	191@4500	260@3600	4.00x3.48	9.1:1	18@2000
	Z	4.3 (4293)	MFI	①	②	4.00x3.48	9.1:1	18@2000
	Z	4.3 (4293)	TFI	①	②	4.00x3.48	9.1:1	18@2000
	H	5.0 (4999)	TFI	175@4200	265@2800	3.74x3.48	9.1:1	18@2000
	K	5.7 (5735)	MFI	③	⑨	4.00x3.48	9.1:1	18@2000
	K	5.7 (5735)	TFI	③	⑨	4.00x3.48	9.1:1	18@2000
	F	6.5 (6473)	DSL	190@3400	⑥	4.06x3.82	21.5:1	40-45@2000
	P	6.5 (6473)	DSL	155@3600	275@1700	4.06x3.82	21.5:1	40-45@2000
	S	6.5 (6473)	DSL	155@3600	275@1700	4.06x3.82	21.5:1	40-45@2000
	Y	6.5 (6473)	DSL	⑦	⑧	4.06x3.82	21.5:1	40-45@2000
	N	7.4 (7440)	TFI	230@3600	385@1600	4.25x4.00	7.9:1	25@2000
1996-97	4	2.2 (2189)	MFI	118@5200	130@2800	3.50x3.46	9.0:1	56@3000
	W	4.3 (4293)	MFI	⑩	⑪	4.00x3.48	9.2:1	18@2000
	X	4.3 (4293)	MFI	⑫	⑬	4.00x3.48	9.2:1	18@2000
	M	5.0 (4999)	MFI	220@4600	285@2800	3.74x3.48	9.4:1	18@2000
	R	5.7 (5735)	MFI	250@4600	335@2800	4.00x3.48	9.4:1	18@2000
	F	6.5 (6473)	DSL	⑤	⑥	4.05x3.80	21.5:1	40-45@2000
	S	6.5 (6473)	DSL	180@3400	360@1700	4.06x3.82	21.5:1	40-45@2000
	J	7.4 (7440)	MFI	290@4200	410@3200	4.25x4.00	9.0:1	40@2000

TFI - Throttle body fuel injection
MFI - Multiport fuel injection
DSL - Diesel

① S10: 155@4000
  C/K: 160@4000
  C/K HD: 155@4000
  G Van: 165@4000

② S10: 230@2800
  C/K Pick-up: 235@2400
  C/K HD Pick-up and G-Van: 230@2400

③ Below 8500 GVWR: 210@4000
  Above 8500 GVWR: 190@4000

④ Below 8500 GVWR: 300@2800
  Above 8500 GVWR: 300@2400

⑤ Below 15,000 GVWR: 180@3400
  Above 15,000 GVWR: 190@3400

⑥ Below 15,000 GVWR: 385@1700
  Above 15,000 GVWR: 380@1700

⑦ Below 8500 GVWR: 155@3600
  Above 8500 GVWR: 160@3600

⑧ Below 8500 GVWR: 275@1700
  Above 8500 GVWR: 290@1700

⑨ Below 8500 GVWR: 385@1700
  Above 8500 GVWR: 300@2400

⑩ 2WD: 180@4400
  4WD: 190@4400

⑪ 2WD: 245@2800
  4WD: 250@2800

⑫ 2WD: 170@4400
  4WD: 180@4400

⑬ 2WD: 235@2800
  4WD: 240@2800

GASOLINE ENGINE TUNE-UP SPECIFICATIONS

Year	Engine ID/VIN	Engine Displacement Liters (cc)	Spark Plugs Gap (in.)	Ignition Timing (deg.) MT	Ignition Timing (deg.) AT	Fuel Pump (psi)	Idle Speed (rpm) MT	Idle Speed (rpm) AT	Valve Clearance In.	Valve Clearance Ex.
1993	A	2.5 (2474)	0.060	①	①	9-13	①	①	HYD	HYD
	R	2.8 (2835)	0.040	①	①	9-13	①	①	HYD	HYD
	W	4.3 (4293)	0.035	①	①	55-61 ②	①	①	HYD	HYD
	Z	4.3 (4293)	0.035	①	①	9-13	①	①	HYD	HYD
	H	5.0 (4999)	0.045	①	①	9-13	①	①	HYD	HYD
	K	5.7 (5735)	0.045	①	①	9-13	①	①	HYD	HYD
	N	7.4 (7440)	0.045	①	①	9-13	①	①	HYD	HYD
1994	4	2.2 (2189)	0.060	①	①	9-13	①	①	HYD	HYD
	W	4.3 (4293)	0.035	①	①	55-61 ②	①	①	HYD	HYD
	Z	4.3 (4293)	0.035	①	①	9-13	①	①	HYD	HYD
	H	5.0 (4999)	0.045	①	①	9-13	①	①	HYD	HYD
	K	5.7 (5735)	0.045	①	①	9-13	①	①	HYD	HYD
	N	7.4 (7440)	0.045	①	①	9-13	①	①	HYD	HYD
1995	4	2.2 (2189)	0.060	③	③	41-47 ②	950	890	HYD	HYD
	W	4.3 (4293)	0.035	④	④	55-61 ②		625	HYD	HYD
	Z	4.3 (4293)	0.035	⑤	⑤	9-13	650	725	HYD	HYD
	H	5.0 (4999)	0.035	①	①	9-13	①	①	HYD	HYD
	K	5.7 (5735)	0.035	①	①	9-13	①	①	HYD	HYD
	N	7.4 (7440)	0.035	①	①	26-32	①	①	HYD	HYD
1996-97	4	2.2 (2189)	0.060	③	③	41-47	⑥	⑥	HYD	HYD
	W	4.3 (4293)	0.060	④	④	58-64 ②	600	625	HYD	HYD
	X	4.3 (4293)	0.045	③	③	41-47	⑥	⑥	HYD	HYD
	M	5.0 (4999)	0.060	④	④	60-66 ②	650 ⑦	550	HYD	HYD
	R	5.7 (5735)	0.060	④	④	60-66 ②	660 ⑧	525	HYD	HYD
	J	7.4 (7440)	0.060	④	④	60-66 ②	750 ⑨	675 ⑨	HYD	HYD

NOTE: The Vehicle Emission Control Information label often reflects specification changes made during production. The label figures must be used if they differ from those in this chart.

HYD - Hydraulic

① Refer to underhood label for exact setting
② With key on and engine off
③ Distributorless ignition
④ Ignition timing is preset and cannot be adjusted
⑤ 0°, disconnect set timing connector (tan with black striped wire taped to engine harness near distributor)

⑥ Idle speed is maintained by the PCM
⑦ Under 8500 GVW
⑧ Over 8500:
Manual: 565-615
Automatic: 525-575
⑨ Over 8500 GVW

DIESEL ENGINE TUNE-UP SPECIFICATIONS

Year	Engine ID/VIN	Engine Displacement cu. in. (cc)	Valve Clearance Intake (in.)	Valve Clearance Exhaust (in.)	Intake Valve Opens (deg.)	Injection Pump Setting (deg.)	Injection Nozzle Pressure (psi) New	Injection Nozzle Pressure (psi) Used	Idle Speed (rpm)	Cranking Compression Pressure (psi)
1993	C	6.2 (6210)	HYD	HYD	①	②	1600	1500	①	NA
	J	6.2 (6210)	HYD	HYD	①	②	1600	1500	①	NA
	F	6.5 (6473)	HYD	HYD	①	①	1600	1500	①	NA
1994	F	6.5 (6473)	HYD	HYD	①	①	1600	1500	①	NA
	P	6.5 (6473)	HYD	HYD	①	①	1800	1700	①	NA
	S	6.5 (6473)	HYD	HYD	①	①	1800	1700	①	NA
	Y	6.5 (6473)	HYD	HYD	①	①	1600	1500	①	NA
1995	F	6.5 (6473)	HYD	HYD	①	①	1600	1500	①	NA
	P	6.5 (6473)	HYD	HYD	①	①	1800	1700	①	NA
	S	6.5 (6473)	HYD	HYD	①	①	1800	1700	①	NA
	Y	6.5 (6473)	HYD	HYD	①	①	1600	1500	①	NA

DIESEL ENGINE TUNE-UP SPECIFICATIONS

Year	Engine ID/VIN	Engine Displacement cu. in. (cc)	Valve Clearance Intake (in.)	Valve Clearance Exhaust (in.)	Intake Valve Opens (deg.)	Injection Pump Setting (deg.)	Injection Nozzle Pressure (psi) New	Injection Nozzle Pressure (psi) Used	Idle Speed (rpm)	Cranking Compression Pressure (psi)
1996-97	F	6.5 (6473)	HYD	HYD	①	①	1800	1700	①	NA
	S	6.5 (6473)	HYD	HYD	①	①	1800	1700	①	NA

NOTE: The Vehicle Emission Control Information label often reflects specification changes made during production. The label figures must be used if they differ from those in this chart

HYD - Hydraulic

NA - Not Available

① Refer to Vehicle Emission Control Information label

② Set by aligning marks on top of engine front cover and injection pump flange

CAPACITIES

Year	Model	Engine ID/VIN	Engine Displacement Liters (cc)	Engine Oil with Filter (qts.)	Transmission (pts.) 5-Spd	Transmission (pts.) Auto.	Transfer Case (pts.)	Drive Axle Front (pts.)	Drive Axle Rear (pts.)	Fuel Tank (gal.)	Cooling System (qts.)
1993	Astro/Safari	W	4.3 (4293)	5.0	4.4	10.0	-	-	3.8	27.0	13.5 ③
	Astro/Safari	Z	4.3 (4293)	5.0	4.4	10.0	-	-	3.8	27.0	13.5 ③
	Blazer/Yukon	K	5.7 (5735)	5.0	-	②	10.0	4.0	④	25.0 ⑤	18.0
	C1500	Z	4.3 (4293)	5.0	①	②	-	-	④	⑥	11.0
	C1500	H	5.0 (4999)	5.0	①	②	-	-	④	⑥	18.0
	C1500	K	5.7 (5735)	5.0	①	②	-	-	④	⑥	18.0
	C1500	C	6.2 (6210)	7.0	①	②	-	-	④	⑥	25.0
	C1500	J	6.2 (6210)	7.0	①	②	-	-	④	⑥	25.0
	C1500	N	7.4 (7440)	6.0	①	②	-	-	④	⑥	25.0
	C2500	Z	4.3 (4293)	5.0	①	②	-	-	④	⑥	11.0
	C2500	H	5.0 (4999)	5.0	①	②	-	-	④	⑥	18.0
	C2500	K	5.7 (5735)	5.0	①	②	-	-	④	⑥	18.0
	C2500	C	6.2 (6210)	7.0	①	②	-	-	④	⑥	25.0
	C2500	J	6.2 (6210)	7.0	①	②	-	-	④	⑥	25.0
	C2500	F	6.5 (6473)	7.0	①	②	-	-	④	⑥	26.5
	C3500	H	5.0 (4999)	5.0	①	②	-	-	④	⑥	18.0
	C3500	K	5.7 (5735)	5.0	①	②	-	-	④	⑥	18.0
	C3500	C	6.2 (6210)	7.0	①	②	-	-	④	⑥	25.0
	C3500	J	6.2 (6210)	7.0	①	②	-	-	④	⑥	25.0
	C3500	F	6.5 (6473)	7.0	①	②	-	-	④	⑥	26.5
	C3500	N	7.4 (7440)	6.0	①	②	-	-	④	⑥	25.0 ⑦
	G10	Z	4.3 (4293)	5.0		②	-	-	④	22.0 ⑧	11.0 ⑨
	G10	H	5.0 (4999)	5.0		②	-	-	④	22.0 ⑧	17.0 ⑨
	G10	K	5.7 (5735)	5.0		②	-	-	④	⑩	18.0 ⑨
	G10	C	6.2 (6210)	7.0		②	-	-	④	22.0 ⑧	24.0 ⑨
	G10	J	6.2 (6210)	7.0		②	-	-	④	22.0 ⑧	24.0 ⑨
	G20	Z	4.3 (4293)	5.0		②	-	-	④	22.0 ⑧	11.0 ⑨
	G20	H	5.0 (4999)	5.0		②	-	-	④	22.0 ⑧	17.0 ⑨
	G20	K	5.7 (5735)	5.0		②	-	-	④	⑩	18.0 ⑨
	G20	C	6.2 (6210)	7.0		②	-	-	④	22.0 ⑧	24.0 ⑨
	G20	J	6.2 (6210)	7.0		②	-	-	④	22.0 ⑧	24.0 ⑨
	G30	Z	4.3 (4293)	5.0		②	-	-	④	22.0 ⑧	11.0 ⑨
	G30	K	5.7 (5735)	5.0		②	-	-	④	⑩	18.0 ⑨
	G30	C	6.2 (6210)	7.0		②	-	-	④	22.0 ⑧	24.0 ⑨

CAPACITIES

Year	Model	Engine ID/VIN	Engine Displacement Liters (cc)	Engine Oil with Filter (qts.)	Transmission (pts.)		Transfer Case (pts.)	Drive Axle		Fuel Tank (gal.)	Cooling System (qts.)
					5-Spd	Auto.		Front (pts.)	Rear (pts.)		
1993	G30	J	6.2 (6210)	7.0		②	-	-	④	22.0 ⑧	24.0 ⑨
	G30	N	7.4 (7440)	6.0		②	-	-	⑪	⑫	24.5 ⑨
	K1500	Z	4.3 (4293)	5.0	①	②	-	-	④	⑥	11.0
	K1500	H	5.0 (4999)	5.0	①	②	-	-	④	⑥	18.0
	K1500	K	5.7 (5735)	5.0	①	②	-	-	④	⑥	18.0
	K1500	C	6.2 (6210)	7.0	①	②	-	-	④	⑥	25.0
	K1500	N	7.4 (7440)	6.0	①	②	-	-	④	⑥	25.0 ⑦
	K2500	Z	4.3 (4293)	5.0	①	②	-	-	④	⑥	11.0
	K2500	H	5.0 (4999)	5.0	①	②	-	-	④	⑥	18.0
	K2500	K	5.7 (5735)	5.0	①	②	-	-	④	⑥	18.0
	K2500	C	6.2 (6210)	7.0	①	②	-	-	④	⑥	25.0
	K2500	J	6.2 (6210)	7.0	①	②	-	-	④	⑥	25.0
	K2500	F	6.5 (6473)	7.0	①	②	-	-	④	⑥	26.5
	K2500	N	7.4 (7440)	6.0	①	②	-	-	④	⑥	25.0 ⑦
	K2500	H	5.0 (4999)	5.0	①	②	-	-	④	⑥	18.0
	K3500	K	5.7 (5735)	5.0	①	②	-	-	④	⑥	18.0
	K3500	C	6.2 (6210)	7.0	①	②	-	-	④	⑥	25.0
	K3500	J	6.2 (6210)	7.0	①	②	-	-	④	⑥	25.0
	K3500	F	6.5 (6473)	7.0	①	②	-	-	④	⑥	26.5
	K3500	N	7.4 (7440)	6.0	①	②	-	-	④	⑥	25.0 ⑦
	S10 Blazer/Jimmy/Bravada	W	4.3 (4293)	4.5	-	10.0	-	3.5	3.5	20.0	12.0
	S10 Blazer/Jimmy/Bravada/ Typhoon	Z	4.3 (4293)	5.0	4.4	10.0	4.6	3.5	3.5	20.0 ⑬	12.1
	S10/S15 Pick-up/Sonoma	A	2.5 (2474)	4.0	4.4	10.0	4.6	3.5	3.9	13.0 ⑭	11.5
	S10/S15 Pick-up/Sonoma	R	2.8 (2835)	4.0	4.5	10.0	4.6	3.5	3.9	13.0 ⑭	10.5
	S10/S15 Pick-up/Sonoma	W	4.3 (4293)	4.5	4.4	10.0	4.4	2.6	3.9	20.0 ⑬	12.1
1994	Astro/Safari	W	4.3 (4293)	5.0	4.4	10.0	-	-	3.8	27.0	13.5 ③
	Astro/Safari	Z	4.3 (4293)	5.0	4.4	10.0	-	-	3.8	27.0	13.5 ③
	Blazer/Yukon	K	5.7 (5735)	5.0	-	②	10.0	4.0	④	25.0 ⑤	18.0
	Blazer/Yukon	F	6.5 (6473)	7.0	①	②	-	-	④	⑥	26.5
	C1500	Z	4.3 (4293)	5.0	①	②	-	-	④	⑥	11.0
	C1500	H	5.0 (4999)	5.0	①	②	-	-	④	⑥	18.0
	C1500	K	5.7 (5735)	5.0	①	②	-	-	④	⑥	18.0
	C1500	F	6.5 (6473)	7.0	①	②	-	-	④	⑥	26.5
	C1500	N	7.4 (7440)	6.0	①	②	-	-	④	⑥	25.0
	C2500	Z	4.3 (4293)	5.0	①	②	-	-	④	⑥	11.0
	C2500	H	5.0 (4999)	5.0	①	②	-	-	④	⑥	18.0
	C2500	K	5.7 (5735)	5.0	①	②	-	-	④	⑥	18.0
	C2500	P	6.5 (6473)	7.0	①	②	-	-	④	⑥	26.5
	C2500	S	6.5 (6473)	7.0	①	②	-	-	④	⑥	26.5
	C3500	H	5.0 (4999)	5.0	①	②	-	-	④	⑥	18.0
	C3500	K	5.7 (5735)	5.0	①	②	-	-	④	⑥	18.0
	C3500	F	6.5 (6473)	7.0	①	②	-	-	④	⑥	26.5
	C3500	P	6.5 (6473)	7.0	①	②	-	-	④	⑥	26.5
	C3500	S	6.5 (6473)	7.0	①	②	-	-	④	⑥	26.5
	C3500	N	7.4 (7440)	6.0	①	②	-	-	④	⑥	25.0 ⑦
	G10	Z	4.3 (4293)	5.0	①	②	-	-	④	22.0 ⑧	11.0 ⑨
	G10	H	5.0 (4999)	5.0	①	②	-	-	④	22.0 ⑧	17.0 ⑨
	G10	K	5.7 (5735)	5.0	①	②	-	-	④	⑩	18.0 ⑨
	G10	P	6.5 (6505)	7.0	①	②	-	-	④	22.0 ⑧	24.0 ⑨
	G10	S	6.5 (6505)	7.0	①	②	-	-	④	22.0 ⑧	24.0 ⑨

CAPACITIES

Year	Model	Engine ID/VIN	Engine Displacement Liters (cc)	Engine Oil with Filter (qts.)	Transmission (pts.) 5-Spd	Transmission (pts.) Auto.	Transfer Case (pts.)	Drive Axle Front (pts.)	Drive Axle Rear (pts.)	Fuel Tank (gal.)	Cooling System (qts.)
1994	G20	Z	4.3 (4293)	5.0	①	②	-	-	④	22.0 ⑧	11.0 ⑨
	G20	H	5.0 (4999)	5.0	①	②	-	-	④	22.0 ⑧	17.0 ⑨
	G20	K	5.7 (5735)	5.0	①	②	-	-	④	⑩	18.0 ⑨
	G20	P	6.5 (6505)	7.0	①	②	-	-	④	22.0 ⑧	24.0 ⑨
	G20	Y	6.5 (6505)	7.0	①	②	-	-	④	22.0 ⑧	24.0 ⑨
	G30	Z	4.3 (4293)	5.0	①	②	-	-	④	22.0 ⑧	11.0 ⑨
	G30	K	5.7 (5735)	5.0	①	②	-	-	④	⑩	18.0 ⑨
	G30	P	6.5 (6505)	7.0	①	②	-	-	④	22.0 ⑧	24.0 ⑨
	G30	Y	6.5 (6505)	7.0	①	②	-	-	④	22.0 ⑧	24.0 ⑨
	G30	N	7.4 (7440)	6.0	①	②	-	-	④	⑫	24.5 ⑨
	K1500	Z	4.3 (4293)	5.0	①	②	-	-	④	⑥	11.0
	K1500	H	5.0 (4999)	5.0	①	②	-	-	④	⑥	18.0
	K1500	K	5.7 (5735)	5.0	①	②	-	-	④	⑥	18.0
	K1500	F	6.5 (6505)	7.0	①	②	-	-	④	⑥	25.0
	K1500	N	7.4 (7440)	6.0	①	②	-	-	④	⑥	25.0 ⑦
	K2500	Z	4.3 (4293)	5.0	①	②	-	-	④	⑥	11.0
	K2500	H	5.0 (4999)	5.0	①	②	-	-	④	⑥	18.0
	K2500	K	5.7 (5735)	5.0	①	②	-	-	④	⑥	18.0
	K2500	P	6.5 (6505)	7.0	①	②	-	-	④	⑥	25.0
	K2500	S	6.5 (6505)	7.0	①	②	-	-	④	⑥	25.0
	K2500	F	6.5 (6473)	7.0	①	②	-	-	④	⑥	26.5
	K2500	N	7.4 (7440)	6.0	①	②	-	-	④	⑥	25.0 ⑦
	K3500	H	5.0 (4999)	5.0	①	②	-	-	④	⑥	18.0
	K3500	K	5.7 (5735)	5.0	①	②	-	-	④	⑥	18.0
	K3500	P	6.5 (6505)	7.0	①	②	-	-	④	⑥	25.0
	K3500	S	6.5 (6505)	7.0	①	②	-	-	④	⑥	25.0
	K3500	F	6.5 (6473)	7.0	①	②	-	-	④	⑥	26.5
	K3500	N	7.4 (7440)	6.0	①	②	-	-	④	⑥	25.0 ⑦
	S10 Blazer/Jimmy/Bravada	W	4.3 (4293)	4.5	-	10.0	-	3.5	3.5	20.0	12.0
	S10 Blazer/Jimmy/Bravada	Z	4.3 (4293)	5.0	4.4	10.0	4.6	3.5	3.5	20.0	12.1
	S10/S15 Pick-up/Sonoma	4	2.2 (2189)	4.0	4.4	10.0	4.6	3.5	3.5	13.0 ⑬	11.5
	S10/S15 Pick-up/Sonoma	W	4.3 (4293)	4.5	4.4	10.0	4.6	3.5	3.5	20.0	12.0
	S10/S15 Pick-up/Sonoma	Z	4.3 (4293)	5.0	4.4	10.0	4.6	3.5	3.5	20.0	12.0
	Suburban	K	5.7 (5735)	5.0	-	②	10.0	4.0	④	25.0 ⑭	18.0
	Suburban	P	6.5 (6505)	7.0	①	②	-	-	④	⑥	25.0
	Suburban	S	6.5 (6505)	7.0	①	②	-	-	④	⑥	25.0
	Suburban	N	7.4 (7440)	6.0	-	②	10.0	4.0	④	25.0 ⑭	24.5
1995	Astro/Safari	W	4.3 (4293)	5.0	4.4	10.0	3.0	2.6	3.8	27.0	12.8 ③
	Astro/Safari	Z	4.3 (4293)	5.0	4.4	10.0	-	-	3.8	27.0	12.8 ③
	C1500	Z	4.3 (4293)	5.0	①	②	-	-	④	⑥	11.0
	C1500	H	5.0 (4999)	5.0	①	②	-	-	④	⑥	18.0
	C1500	K	5.7 (5735)	5.0	①	②	-	-	④	⑥	18.0
	C1500	F	6.5 (6473)	7.0	①	②	-	-	④	⑥	26.5
	C1500	N	7.4 (7440)	7.0	①	②	-	-	④	⑥	25.0
	C2500	Z	4.3 (4293)	5.0	①	②	-	-	④	⑥	11.0
	C2500	H	5.0 (4999)	5.0	①	②	-	-	④	⑥	18.0
	C2500	K	5.7 (5735)	5.0	①	②	-	-	④	⑥	18.0
	C2500	P	6.5 (6473)	7.0	①	②	-	-	④	⑥	26.5
	C2500	S	6.5 (6473)	7.0	①	②	-	-	④	⑥	26.5
	C3500	H	5.0 (4999)	5.0	①	②	-	-	④	⑥	18.0
	C3500	K	5.7 (5735)	5.0	①	②	-	-	④	⑥	18.0 ⑮

CAPACITIES

Year	Model	Engine ID/VIN	Engine Displacement Liters (cc)	Engine Oil with Filter (qts.)	Transmission (pts.) 5-Spd	Transmission (pts.) Auto.	Transfer Case (pts.)	Drive Axle Front (pts.)	Drive Axle Rear (pts.)	Fuel Tank (gal.)	Cooling System (qts.)
1995	C3500	F	6.5 (6473)	7.0	①	②	-	-	④	⑥	26.5
	C3500	P	6.5 (6473)	7.0	①	②	-	-	④	⑥	26.5
	C3500	S	6.5 (6473)	7.0	①	②	-	-	④	⑥	26.5
	C3500	N	7.4 (7440)	6.0	①	②	-	-	④	⑥	25.0 ⑦
	G10	Z	4.3 (4293)	5.0	①	②	-	-	④	22.0 ⑧	11.0 ⑨
	G10	H	5.0 (4999)	5.0	①	②	-	-	④	22.0 ⑧	17.0 ⑨
	G10	K	5.7 (5735)	5.0	①	②	-	-	④	⑲	18.0 ⑨
	G10	P	6.5 (6505)	7.0	①	②	-	-	④	22.0 ⑧	24.0 ⑨
	G20	Z	4.3 (4293)	5.0	①	②	-	-	④	22.0 ⑧	11.0 ⑨
	G20	H	5.0 (4999)	5.0	①	②	-	-	④	22.0 ⑧	17.0 ⑨
	G20	K	5.7 (5735)	5.0	①	②	-	-	④	⑲	18.0 ⑨
	G20	P	6.5 (6505)	7.0	①	②	-	-	④	22.0 ⑧	24.0 ⑨
	G20	Y	6.5 (6505)	7.0	①	②	-	-	④	22.0 ⑧	24.0 ⑨
	G30	Z	4.3 (4293)	5.0	①	②	-	-	④	22.0 ⑧	11.0 ⑨
	G30	K	5.7 (5735)	5.0	①	②	-	-	④	⑲	18.0 ⑨
	G30	P	6.5 (6505)	7.0	①	②	-	-	④	22.0 ⑧	24.0 ⑨
	G30	Y	6.5 (6505)	7.0	①	②	-	-	④	22.0 ⑧	24.0 ⑨
	G30	N	7.4 (7440)	6.0	①	②	-	-	④	⑫	24.5 ⑨
	K1500	Z	4.3 (4293)	5.0	①	②	⑯	⑰	④	⑥	11.0
	K1500	H	5.0 (4999)	5.0	①	②	⑯	⑰	④	⑥	18.0
	K1500	K	5.7 (5735)	5.0	①	②	⑯	⑰	④	⑥	18.0
	K1500	F	6.5 (6505)	7.0	①	②	⑯	⑰	④	⑥	25.0
	K1500	N	7.4 (7440)	7.0	①	②	⑯	⑰	④	⑥	25.0 ⑦
	K2500	Z	4.3 (4293)	5.0	①	②	⑯	⑰	④	⑥	11.0
	K2500	H	5.0 (4999)	5.0	①	②	⑯	⑰	④	⑥	18.0
	K2500	K	5.7 (5735)	5.0	①	②	⑯	⑰	④	⑥	18.0
	K2500	P	6.5 (6505)	7.0	①	②	⑯	⑰	④	⑥	25.0
	K2500	S	6.5 (6505)	7.0	①	②	⑯	⑰	④	⑥	25.0
	K2500	F	6.5 (6473)	7.0	①	②	⑯	⑰	④	⑥	26.5
	K2500	N	7.4 (7440)	7.0	①	②	⑯	⑰	④	⑥	25.0 ⑦
	K3500	H	5.0 (4999)	5.0	①	②	⑯	⑰	④	⑥	18.0
	K3500	K	5.7 (5735)	5.0	①	②	⑯	⑰	④	⑥	18.0
	K3500	P	6.5 (6505)	7.0	①	②	⑯	⑰	④	⑥	25.0
	K3500	S	6.5 (6505)	7.0	①	②	⑯	⑰	④	⑥	25.0
	K3500	F	6.5 (6473)	7.0	①	②	⑯	⑰	④	⑥	26.5
	K3500	N	7.4 (7440)	7.0	①	②	⑯	⑰	④	⑥	25.0
	S10 Blazer/Jimmy/Bravada	W	4.3 (4293)	4.5	4.4	10.0	-	3.5	3.5	20.0	12.1
	S10/S15 Pick-up/Sonoma	4	2.2 (2189)	4.0	4.4	10.0	4.6	2.6	4.0	13.0 ⑭	11.5
	S10/S15 Pick-up/Sonoma	W	4.3 (4293)	4.5	4.4	10.0	4.6	2.6	4.0	20.0	12.0
	S10/S15 Pick-up/Sonoma	Z	4.3 (4293)	5.0	4.4	10.0	4.6	2.6	4.0	20.0	12.0
	Suburban	K	5.7 (5735)	5.0	-	②	-	-	④	⑥	18.0
	Suburban	P	6.5 (6505)	7.0	①	②	-	-	④	⑥	25.0
	Suburban	S	6.5 (6505)	7.0	①	②	-	-	④	⑥	25.0
	Suburban	N	7.4 (7440)	6.0	-	②	10.0	4.0	④	25.0 ⑬	24.5
	Tahoe/Yukon	K	5.7 (5735)	5.0	-	②	10.0	4.0	④	25.0 ⑤	18.0
	Tahoe/Yukon	F	6.5 (6473)	7.0	①	②	-	-	④	⑥	26.5

CAPACITIES

Year	Model	Engine ID/VIN	Engine Displacement Liters (cc)	Engine Oil with Filter (qts.)	Transmission (pts.) 5-Spd	Transmission (pts.) Auto.	Transfer Case (pts.)	Drive Axle Front (pts.)	Drive Axle Rear (pts.)	Fuel Tank (gal.)	Cooling System (qts.)
1996-97	Astro/Safari	W	4.3 (4293)	5.0	-	10.0	3.0	2.6	3.8	27.0	㉑
	C1500	W	4.3 (4293)	5.0	①	㉒	-	-	④	⑥	13.0
	C1500	M	5.0 (4999)	5.0	①	㉒	-	-	④	⑥	18.0
	C1500	R	5.7 (5735)	5.0	①	㉒	-	-	④	⑥	18.0
	C2500	W	4.3 (4293)	5.0	①	㉒	-	-	④	⑥	13.0
	C2500	M	5.0 (4999)	5.0	①	㉒	-	-	④	⑥	18.0
	C2500	R	5.7 (5735)	5.0	①	㉒	-	-	④	⑥	18.0
	C2500	F	6.5 (6473)	7.0	①	㉒	-	-	④	⑥	27.5
	C2500	S	6.5 (6473)	7.0	①	㉒	-	-	④	⑥	27.5
	C3500	F	6.5 (6473)	7.0	①	㉒	-	-	④	⑥	27.5
	C3500	J	7.4 (7440)	6.0	①	㉒	-	-	④	⑥	25.0 ⑦
	G1500	W	4.3 (4293)	5.0	①	㉒	-	-	④	22.0 ⑬	13.0 ⑨
	G1500	M	5.0 (4999)	5.0	①	㉒	-	-	④	22.0 ⑬	17.0 ⑨
	G1500	R	5.7 (5735)	5.0	①	㉒	-	-	④	⑬	18.0 ⑨
	G2500	M	5.0 (4999)	5.0	①	㉒	-	-	④	22.0 ⑬	17.0 ⑨
	G2500	R	5.7 (5735)	5.0	①	㉒	-	-	④	⑬	18.0 ⑨
	G3500	F	6.5 (6473)	7.0	①	㉒	-	-	④	22.0 ⑬	27.5 ⑨
	G3500	J	7.4 (7440)	6.0	①	㉒	-	-	④	⑫	24.5 ⑨
	K1500	W	4.3 (4293)	5.0	①	㉒	㉓	⑰	④	⑥	11.0
	K1500	M	5.0 (4999)	5.0	①	㉒	㉓	⑰	④	⑥	18.0
	K1500	R	5.7 (5735)	5.0	①	㉒	㉓	⑰	④	⑥	18.0
	K1500	F	6.5 (6473)	7.0	①	㉒	㉓	⑰	④	⑥	27.5
	K2500	M	5.0 (4999)	5.0	①	㉒	㉓	⑰	④	⑥	18.0
	K2500	R	5.7 (5735)	5.0	①	㉒	㉓	⑰	④	⑥	18.0
	K2500	F	6.5 (6473)	7.0	①	㉒	㉓	⑰	④	⑥	27.5
	K3500	F	6.5 (6473)	7.0	①	㉒	㉓	⑰	④	⑥	27.5
	K3500	J	7.4 (7440)	6.0	①	㉒	㉓	⑰	④	⑥	25.0
	S10 Blazer/Jimmy/Bravada	W	4.3 (4293)	5.0	4.4	10.0	2.6	2.6	3.9	20.0	11.9
	S10 Blazer/Jimmy/Bravada	X	4.3 (4293)	5.0	4.4	10.0	2.6	2.6	3.9	20.0	11.9
	S10/S15 Pick-up/Sonoma	4	2.2 (2189)	4.0	4.4	10.0	2.6	2.6	3.9	13.0 ⑭	11.5
	S10/S15 Pick-up/Sonoma	W	4.3 (4293)	5.0	4.4	10.0	2.6	2.6	3.9	20.0	11.9
	S10/S15 Pick-up/Sonoma	X	4.3 (4293)	5.0	4.4	10.0	2.6	2.6	3.9	20.0	11.9
	Suburban	R	5.7 (5735)	5.0	-	㉒	-	-	④	⑥	18.0
	Suburban	F	7.4 (7440)	6.0	-	㉒	㉓	⑰	④	25.0 ⑬	24.5
	Tahoe/Yukon	R	5.7 (5735)	5.0	-	㉒	㉓	⑰	④	⑲	18.0
	Tahoe/Yukon	S	6.5 (6473)	7.0	-	㉒	-	-	④	⑲	23.8

① New Venture gear 4500: 8.0 pts.
  New Venture gear 5LM60: 4.4 pts.
② 350C trans.: 6.3 pts.
  THM400 and 4L80 trans.: 9.0 pts.
  THM700 R4 and 4L60 trans.: 10.0 pts.
  THM700 R4 and 4L80-E trans.: 14.3 pts.
③ 16.5 qts. with rear heater
④ 8.5" ring gear: 4.2 pts.
  9.5" ring gear: 6.5 pts.
  9.75" ring gear: 6.0 pts.
  10.5" ring gear: 6.5 pts.
⑤ Available with optional 31 gallon tank
⑥ Std. available with 25 and 34 gallon tanks
  Chassis cab available with 22, 30 and 34 gallon tanks

⑦ 3500HD: 28.5 qts. capacity
⑧ Available 32 and 41 gallon tanks
⑨ Add three qts. with rear heater
⑩ Short bed: 20 gals.
  Long bed: 34 gals.
⑪ 8.5" ring gear: 4.2 pts.
  9.5" ring gear: 6.5 pts.
  Chevrolet 10.5" ring gear: 6.5 pts.
  Dana 9.75" ring gear: 6.0 pts.
  Rockwell 12" ring gear: 12.5 pts.
⑫ Available with a variety of fuel tanks
⑬ Available 31 and 40 gallon tanks
⑭ Available with 20 gallon tank
⑮ HD: 27.0 qts.

⑯ Fill to bottom of filler plug hole
⑰ K2 models: 1.75 qts.
  K3 models: 2.25 qts.
⑱ Saginaw 10.5" ring gear: 7.2 pts.
  Dana 80 11" ring gear: 8.2 pts.
⑲ Short bed: 26 gals.
  Long bed: 34 gals.
⑳ With rear heater: 13.5 qts.
  Without rear heater: 11.8 qts.
㉑ With rear heater: 16.5 qts.
  Without rear heater: 13.5 qts.
㉒ 4L60E trans.: 10.0 pts.
  4L80E trans.: 14.5 pts.
㉓ NV241 and NV243: 4.5 pts.
  4401 and 4470: 6.6 pts.

VALVE SPECIFICATIONS

Year	Engine ID/VIN	Engine Displacement Liters (cc)	Seat Angle (deg.)	Face Angle (deg.)	Spring Test Pressure (lbs. @ in.)	Spring Installed Height (in.)	Stem-to-Guide Clearance (in.)		Stem Diameter (in.)	
							Intake	Exhaust	Intake	Exhaust
1993	A	2.5 (2474)	46	45	1.58-1.70@ 1.04	1.44	0.0010-0.0025	0.0013-0.0030	0.3133-0.3138	0.3128-0.3135
	R	2.8 (2835)	46	45	88@1.57	1.57	0.0010-0.0027	0.0010-0.0027	0.3410-0.3417	0.3410-0.3417
	W	4.3 (4293)	46	45	194-206@1.25	1.69-1.71	0.0011-0.0027	0.0011-0.0027	NA	NA
	Z	4.3 (4293)	46	45	194-206@1.25	1.72	0.0010-0.0027	0.0010-0.0027	NA	NA
	H	5.0 (4999)	46	45	76-84@1.70	1.72	0.0010-0.0027	0.0010-0.0027	NA	NA
	K	5.7 (5735)	46	45	76-84@1.70	1.72	0.0010-0.0027	0.0010-0.0027	NA	NA
	C	6.2 (6210)	46	45	230@1.39	1.81	0.0010-0.0027	0.0010-0.0027	NA	NA
	J	6.2 (6210)	46	45	230@1.39	1.81	0.0010-0.0027	0.0010-0.0027	NA	NA
	F	6.5 (6473)	46	45	230@1.39	1.81	0.0010-0.0027	0.0010-0.0027	NA	NA
	N	7.4 (7440)	46	45	74-86@1.80	1.80	0.0010-0.0027	0.0012-0.0029	NA	NA
1994	4	2.2 (2189)	46	45	228@1.28	1.71	0.0010-0.0020	0.0010-0.0030	NA	NA
	W	4.3 (4293)	46	45	194-206@1.25	1.69-1.71	0.0011-0.0027	0.0011-0.0027	NA	NA
	Z	4.3 (4293)	46	45	194-206@1.25	1.72	0.0010-0.0027	0.0010-0.0027	NA	NA
	H	5.0 (4999)	46	45	76-84@1.70	1.72	0.0010-0.0027	0.0010-0.0027	NA	NA
	K	5.7 (5735)	46	45	76-84@1.70	1.72	0.0010-0.0027	0.0010-0.0027	NA	NA
	F	6.5 (6473)	46	45	230@1.39	1.81	0.0010-0.0027	0.0010-0.0027	NA	NA
	P	6.5 (6473)	46	45	230@1.39	1.81	0.0010-0.0027	0.0010-0.0027	NA	NA
	S	6.5 (6473)	46	45	230@1.39	1.81	0.0010-0.0027	0.0010-0.0027	NA	NA
	Y	6.5 (6473)	46	45	230@1.39	1.81	0.0010-0.0027	0.0010-0.0027	NA	NA
	N	7.4 (7440)	46	45	74-86@1.80	1.80	0.0010-0.0027	0.0012-0.0029	NA	NA
1995	4	2.2 (2189)	46	45	228@1.27	1.71	0.0010-0.0020	0.0010-0.0030	NA	NA
	W	4.3 (4293)	46	45	194-206@1.25	1.69-1.71	0.0011-0.0027	0.0011-0.0027	NA	NA
	Z	4.3 (4293)	46	45	194-206@1.25	1.69-1.71	0.0010-0.0027	0.0010-0.0027	NA	NA
	H	5.0 (4999)	46	45	76-84@1.70	1.72	0.0010-0.0027	0.0010-0.0027	NA	NA

VALVE SPECIFICATIONS

Year	Engine ID/VIN	Engine Displacement Liters (cc)	Seat Angle (deg.)	Face Angle (deg.)	Spring Test Pressure (lbs. @ in.)	Spring Installed Height (in.)	Stem-to-Guide Clearance (in.) Intake	Stem-to-Guide Clearance (in.) Exhaust	Stem Diameter (in.) Intake	Stem Diameter (in.) Exhaust
1995	K	5.7 (5735)	46	45	76-84@1.70	1.72	0.0010-0.0027	0.0010-0.0027	NA	NA
	F	6.5 (6473)	46	45	230@1.39	1.81	0.0010-0.0027	0.0010-0.0027	NA	NA
	P	6.5 (6473)	46	45	230@1.39	1.81	0.0010-0.0027	0.0010-0.0027	NA	NA
	S	6.5 (6473)	46	45	230@1.39	1.81	0.0010-0.0027	0.0010-0.0027	NA	NA
	Y	6.5 (6473)	46	45	230@1.39	1.81	0.0010-0.0027	0.0010-0.0027	NA	NA
	N	7.4 (7440)	46	45	205-225@1.40	1.80	0.0010-0.0027	0.0012-0.0029	NA	NA
1996-97	4	2.2 (2189)	46	45	228@1.28	1.71	0.0010-0.0020	0.0010-0.0030	NA	NA
	W	4.3 (4293)	46	45	187-203@1.27	1.69-1.71	0.0010	0.0020	NA	NA
	X	4.3 (4293)	46	45	187-203@1.27	1.69-1.71	0.0010	0.0020	NA	NA
	M	5.0 (4999)	46	45	187-203@1.27	1.69-1.71	0.0010-0.0027	0.0010-0.0027	NA	NA
	R	5.7 (5735)	46	45	187-203@1.27	1.69-1.71	0.0010-0.0027	0.0010-0.0027	NA	NA
	F	6.5 (6473)	46	45	230@1.40	1.80	0.0010-0.0027	0.0010-0.0027	NA	NA
	S	6.5 (6473)	46	45	230@1.40	1.80	0.0010-0.0027	0.0010-0.0027	NA	NA
	J	7.4 (7440)	46	45	238-262@1.34	1.83	0.0010-0.0029 ②	0.0012-0.0031 ②	NA	NA

NA - Not Available
① Upper: 0.3129-0.3137
   Lower: 0.3118-0.3126

② Service limit:
   Intake: 0.0037 MAX
   Exhaust: 0.0049 MAX

TORQUE SPECIFICATIONS
All readings in ft. lbs.

Year	Engine ID/VIN	Engine Displacement Liters (cc)	Cylinder Head Bolts	Main Bearing Bolts	Rod Bearing Bolts	Crankshaft Damper Bolts	Flywheel Bolts	Manifold Intake *	Manifold Exhaust	Spark Plugs	Lug Nut
1993	A	2.5 (2474)	①	70	32	160	②	25	③	7-15	90
	R	2.8 (2835)	70	70	39	70	52	15	25	22	90
	W	4.3 (4293)	65	75	⑬	70	75	35	⑭	22	90
	Z	4.3 (4293)	65	75	⑬	70	75	35	⑭	22	90
	H	5.0 (4999)	65	⑯	45	70	75	35	⑭	15	⑮
	K	5.7 (5735)	65	⑯	45	70	75	35	⑭	15	⑮
	C	6.2 (6210)	⑰	⑱	48	200	66	31	26	-	⑮
	J	6.2 (6210)	⑰	⑱	48	200	66	31	26	-	⑮
	F	6.5 (6473)	⑰	⑱	48	200	66	31	26	-	⑮
	N	7.4 (7440)	80	100	48	85	65	40	40	22	⑮
1994	4	2.2 (2189)	⑲	70	38	77	55	24	10	⑦	100
	W	4.3 (4293)	65	75	⑬	70	75	35	⑭	22	90
	Z	4.3 (4293)	65	75	⑬	70	75	35	⑭	11	90
	H	5.0 (4999)	65	⑯	45	70	75	35	⑭	15	⑮
	K	5.7 (5735)	65	⑯	45	70	75	35	⑭	15	⑮
	F	6.5 (6473)	㉑	⑱	48	200	66	31	26	-	⑮
	P	6.5 (6473)	㉑	⑱	48	200	66	31	26	-	⑮
	S	6.5 (6473)	㉑	⑱	48	200	66	31	26	-	⑮
	Y	6.5 (6473)	㉑	⑱	48	200	66	31	26	-	⑮
	N	7.4 (7440)	80	100	48	85	65	35	40	22	⑮
1995	4	2.2 (2189)	⑲	70	38	77	55	㉒	10	⑦	100
	W	4.3 (4293)	65	81	⑬	70	74	35	⑭	11	90
	Z	4.3 (4293)	65	81	⑬	70	74	35	⑭	11	90
	H	5.0 (4999)	65	⑯	45	70	75	35	⑭	15	⑮
	K	5.7 (5735)	65	⑯	45	70	75	35	⑭	15	⑮
	F	6.5 (6473)	㉓	⑱	48	200	66	31	26	-	⑮
	P	6.5 (6473)	㉓	⑱	48	200	66	31	26	-	⑮
	S	6.5 (6473)	㉓	⑱	48	200	66	31	26	-	⑮
	Y	6.5 (6473)	㉓	⑱	48	200	66	31	26	-	⑮
	N	7.4 (7440)	80	100	48	85	65	35	40	22	⑮
1996-97	4	2.2 (2189)	⑲	70	38	77	55	㉒	10	11	100
	W	4.3 (4293)	㉖	77	⑬	74	74	㉗	⑥	11	90
	X	4.3 (4293)	㉖	77	⑬	74	74	㉗	⑥	11	90
	M	5.0 (4999)	㉘	㉙	㉑	74	74	㉗	⑥	15	⑮
	R	5.7 (5735)	㉘	㉙	㉑	74	74	㉗	⑥	15	⑮

TORQUE SPECIFICATIONS
All readings in ft. lbs.

Year	Engine ID/VIN	Engine Displacement Liters (cc)	Cylinder Head Bolts	Main Bearing Bolts	Rod Bearing Bolts	Crankshaft Damper Bolts	Flywheel Bolts	Manifold Intake *	Manifold Exhaust	Spark Plugs	Lug Nut
1996-97	F	6.5 (6473)	㉓	⑱	48	200	65	31	26	-	⑮
	S	6.5 (6473)	㉓	⑱	48	200	65	31	26	-	⑮
	J	7.4 (7440)	85	100	45	110	67	30	22	15	⑮

* NOTE: Applies to Lower Manifold only.

* NOTE: Applies to Lower Manifold only.
① Step 1: Tighten all head bolts to 18 ft. lbs.
  Step 2: Tighten all bolts to 26 ft. lbs. except No. 9
  Retorque No. 9 to 18 ft. lbs.
  Step 3: Tighten all bolts an additional 90 degrees
② Automatic trans.: 55 ft. lbs.
  Manual trans.: 65 ft. lbs.
③ Center bolts: 36 ft. lbs.
  Outer bolts: 28 ft. lbs.
④ Not Used
⑤ Not Used
⑥ Tighten bolts to 12 ft. lbs.
  Retorque to 22 ft. lbs.
⑦ 1st-time installation (new head): 22 ft. lbs.
  All other installations: 12 ft. lbs.
⑧ Not Used
⑨ Not Used
⑩ Not Used
⑪ Not Used
⑫ Not Used
⑬ 20 ft. lbs. plus 70 degrees
⑭ Two center bolts: 26 ft. lbs.
  All others: 20 ft. lbs.
⑮ All 5 & 6 stud single rear wheels: 110 ft. lbs.
  All 8 stud single rear wheels: 120 ft. lbs.
  All 8 stud dual rear wheels: 140 ft. lbs.
  All 10 stud dual wheels: 175 ft. lbs.
⑯ Outer bolts on caps 2-4: 70 ft. lbs
  All others: 80 ft. lbs.
⑰ Coat threads with sealant
  Tighten all bolts to to 20 ft. lbs.
  Tighten all bolts an additional 90 degrees (1/4 turn)
⑱ Outer bolts: 100 ft. lbs.
  Inner bolts: 111 ft. lbs.
⑲ Short bolts: 43 ft. lbs. plus 90 degrees
  Long bolts: 46 ft. lbs. plus 90 degrees

⑳ Not Used
㉑ Coat threads with sealant
  Tighten all bolts to 20 ft. lbs.
  Retorque to 50 ft. lbs.
㉒ Lower intake manifold nuts: 24 ft. lbs.
  Lower intake manifold studs: 22 ft. lbs.
  Upper intake manifold bolts: 22 ft. lbs.
㉓ Apply sealer
  Step 1: 20 ft. lbs.
  Step 2: 50 ft. lbs.
  Step 3: 50 ft. lbs.
  Step 4: Plus 90-100 degrees
㉔ Not Used
㉕ Not Used
㉖ 1st pass: 22 ft. lbs.
  2nd pass:
    Short bolt: Plus 55 degrees
    Medium bolt: Plus 65 degrees
    Long bolt: Plus 75 degrees
㉗ Lower intake manifold:
  1st pass: 27 in. lbs.
  2nd pass: 106 in. lbs.
  Final pass: 11 ft. lbs.
  Upper manifold bolts:
  1st pass: 44 in. lbs.
  2nd pass: 88 in. lbs.
㉘ Step 1: 22 ft. lbs.
  Step 2:
    Short bolt: Plus 55 degrees
    Medium bolt: Plus 65 degrees
    Long bolt: Plus 75 degrees
㉙ Outer bolts on caps 2-4: 67 ft. lbs.
  All others: 74 ft. lbs.

BRAKE SPECIFICATIONS
All measurements in inches unless noted

Year	Model	Master Cylinder Bore	Brake Disc Original Thickness	Brake Disc Minimum Thickness	Maximum Runout	Brake Drum Diameter Original Inside Diameter	Brake Drum Diameter Max. Wear Limit	Brake Drum Diameter Maximum Machine Diameter	Minimum Lining Thickness Front	Minimum Lining Thickness Rear
1993	Astro/Safari	NA	①	②	0.004	9.50	9.59	9.56	0.030	0.030
	Blazer/Yukon	NA	1.500	1.480	0.004	⑤	⑥	⑦	0.030	0.030
	C1500	NA	1.250	1.230	0.004	⑤	⑥	⑦	0.030	0.030
	C2500	NA	1.500	1.480	0.004	⑤	⑥	⑦	0.030	0.030
	C3500	NA	1.500	1.480	0.004	⑤	⑥	⑦	0.030	0.030
	G10	NA	③	④	0.004	⑤	⑥	⑦	0.030	0.030
	G20	NA	③	④	0.004	⑤	⑥	⑦	0.030	0.030
	G30	NA	③	④	0.004	⑤	⑥	⑦	0.030	0.030
	K1500	NA	1.500	1.480	0.004	⑤	⑥	⑦	0.030	0.030
	K2500	NA	1.500	1.480	0.004	⑤	⑥	⑦	0.030	0.030
	K3500	NA	1.500	1.480	0.004	⑤	⑥	⑦	0.030	0.030
	S10 Blazer/Jimmy/Bravada	NA	1.040	0.980	0.004	9.50	9.59	9.56	0.030	0.030
	S10/S15 Pick-up/Sonoma	NA	1.040	0.980	0.004	9.50	9.59	9.56	0.030	0.030
	Suburban	NA	1.500	1.480	0.004	⑤	⑥	⑦	0.030	0.030
1994	Astro/Safari	NA	①	②	0.004	9.50	9.59	9.56	0.030	0.030
	Blazer/Yukon	NA	1.500	1.480	0.004	⑤	⑥	⑦	0.030	0.030
	C1500	NA	1.250	1.230	0.004	⑤	⑥	⑦	0.030	0.030
	C2500	NA	1.500	1.480	0.004	⑤	⑥	⑦	0.030	0.030
	C3500	NA	1.500	1.480	0.004	⑤	⑥	⑦	0.030	0.030
	G20	NA	③	④	0.004	⑤	⑥	⑦	0.030	0.030
	G30	NA	③	④	0.004	⑤	⑥	⑦	0.030	0.030
	K1500	NA	1.500	1.480	0.004	⑤	⑥	⑦	0.030	0.030
	K2500	NA	1.500	1.480	0.004	⑤	⑥	⑦	0.030	0.030
	K3500	NA	1.500	1.480	0.004	⑤	⑥	⑦	0.030	0.030
	S10 Blazer/Jimmy/Bravada	NA	1.040	0.980	0.004	9.50	9.59	9.56	0.030	0.030
	S10/S15 Pick-up/Sonoma	NA	1.040	0.980	0.004	9.50	9.59	9.56	0.030	0.030
	Suburban	NA	1.500	1.480	0.004	⑤	⑥	⑦	0.030	0.030
1995	Astro/Safari	NA	①	②	0.004	9.50	9.59	9.56	0.030	0.030
	C1500	NA	1.250	1.230	0.004	⑤	⑥	⑦	0.030	0.030
	C2500	NA	1.500	1.480	0.004	⑤	⑥	⑦	0.030	0.030
	C3500	NA	1.500	1.480	0.004	⑤	⑥	⑦	0.030	0.030
	G20	NA	③	④	0.004	⑤	⑥	⑦	0.030	0.030
	G30	NA	③	④	0.004	⑤	⑥	⑦	0.030	0.030
	K1500	NA	1.500	1.480	0.004	⑤	⑥	⑦	0.030	0.030
	K2500	NA	1.500	1.480	0.004	⑤	⑥	⑦	0.030	0.030
	K3500	NA	1.500	1.480	0.004	⑤	⑥	⑦	0.030	0.030
	S10 Blazer/Jimmy/Bravada	NA	1.030	0.980	0.002	9.50	9.59	9.56	0.030	0.030
	S10/S15 Pick-up/Sonoma	NA	1.030	0.980	0.002	9.50	9.59	9.56	0.030	0.030
	Suburban	NA	1.500	1.480	0.004	⑤	⑥	⑦	0.030	0.030
	Tahoe/Yukon	NA	1.500	1.480	0.004	⑤	⑥	⑦	0.030	0.030
1996-97	Astro/Safari	NA	①	②	0.004	9.50	9.59	9.56	0.030	0.030
	C1500	NA	1.250	1.230	0.004	⑤	⑥	⑦	0.030	0.030
	C2500	NA	1.500	1.480	0.004	⑤	⑥	⑦	0.030	0.030
	C3500	NA	1.500	1.480	0.004	⑤	⑥	⑦	0.030	0.030
	G1500	NA	③	④	0.004	⑤	⑥	⑦	0.030	0.030
	G2500	NA	③	④	0.004	⑤	⑥	⑦	0.030	0.030
	G3500	NA	③	④	0.004	⑤	⑥	⑦	0.030	0.030
	K1500	NA	1.500	1.480	0.004	⑤	⑥	⑦	0.030	0.030

BRAKE SPECIFICATIONS
All measurements in inches unless noted

Year	Model	Master Cylinder Bore	Brake Disc			Brake Drum Diameter			Minimum Lining Thickness	
			Original Thickness	Minimum Thickness	Maximum Runout	Original Inside Diameter	Max. Wear Limit	Maximum Machine Diameter	Front	Rear
1996-97	K2500	NA	1.500	1.480	0.004	⑤	⑥	⑦	0.030	0.030
	K3500	NA	1.500	1.480	0.004	⑤	⑥	⑦	0.030	0.030
	S10 Blazer/Jimmy/Bravada	NA	1.030	0.965	0.003	9.50	9.59	9.56	0.030	0.030
	S10/S15 Pick-up/Sonoma	NA	1.030	0.965	0.003	9.50	9.59	9.56	0.030	0.030
	Suburban	NA	1.500	1.480	0.004	⑤	⑥	⑦	0.030	0.030
	Tahoe/Yukon	NA	1.500	1.480	0.004	⑤	⑥	⑦	0.030	0.030

NA - Not Available
① Available with 1.040" and 1.250" rotors
② 1.040" rotors: 0.980
  1.250" rotors" 1.230
③ Available with 1.280 and 1.540 discs
④ 1.28" disc: 1.230
  1.54" disc: 1.480
⑤ Available with 10", 11.15" and 13" drums
⑥ 10" drum: 10.05
  11.15" drum: 11.24
  13" drum: 13.09
⑦ 10" drum: 10.09
  11.15" drum: 11.21
  13" drum: 13.06

FREQUENT MAINTENANCE LABOR
VANS (G SERIES), PICK UPS (C/K SERIES), BLAZER/JIMMY TAHOE/YUKON, SUBURBAN, ASTRO/SAFARI

The following should be used as a guide when determining the amount of work required for a particular service if taken to a repair shop. In estimating how long a particular Frequent Maintenance Service item should take, please observe the following:
- **Factory Time** is time that is generated by the vehicle manufacturer.
- **Chilton Time** is time that is based on field research and data supplied by the vehicle manufacturer.
- All labor time operations are given in hours and tenths of an hour.
- All labor operations, are to be used as a **guide**.

COOLING

(G) Winterize Cooling System
Includes: Run engine to check for leaks, tighten all hose connections. Test radiator and pressure cap, drain radiator and engine block. Add antifreeze and refill system.
1993-97 .... (Chilton) .8

(G) Belt, Drive, Renew
1993-97
V6, Serpentine (.2) 3
1993 (.2) 3
1994-97
code Z (.2) 3
code W (.4) 5
Renew tensioner add 1

V8 305, 350
AC
P Vans (.3) 4
AIR
P Vans (.2) 3
Fan
P Vans (.2) 3
PS
P Vans (.4) 6
w/AC add 3
Serpentine (.2) 4
V8 454
AC (.3) 4
AIR & Fan (.2) 3
PS (.5) 6
Serpentine (.2) 4

V8, Diesel
Serpentine (.2) 4

(G) Belt, Drive, Adjust
1993
Six, V8, one (.2) 3
each adtnl. (.1) 1

(G) Hoses, Radiator, Renew
Includes: Drain and refill cooling system as required.
1993-97
upper (.4) 5
lower (.5) 6
both (.6) 9

FREQUENT MAINTENANCE LABOR (cont.)
VANS (G SERIES), PICK UPS (C/K SERIES), BLAZER/JIMMY
TAHOE/YUKON, SUBURBAN, ASTRO/SAFARI

	(Factory Time)	Chilton Time
(G) Thermostat, Coolant, Renew		
1993-97		
C, K series		
Gas (.5)		.7
Diesel (.5)		.6
Vans		
V6 (.8)		1.0
V8 (.8)		1.0
w/AC add		.5
Diesel (.8)		1.0
Express, Van, Savana (.8)		1.0
w/AC add		.2
Astro, Safari		
V6		
code Z (.6)		.9
code W (.8)		1.1

FUEL

	(Factory Time)	Chilton Time
(M) Air Cleaner, Service		
1993-97		.2
(G) Filter, Fuel, Renew		
1993-97 (.3)		.4

BRAKES

(G) Bleed Brakes (Four Wheels)
Includes: Add fluid.

	(Factory Time)	Chilton Time
1993-97		
wo/Antilock (.4)		.6
w/Two wheel antilock (1.0)		1.3
w/Four wheel antilock (1.4)		1.8

(G) Brakes, Adjust (Minor)
Includes: Adjust brakes, fill master cylinder.

	(Factory Time)	Chilton Time
1993-97, two wheels		.4

(M) Parking Brake, Adjust

	(Factory Time)	Chilton Time
1993-97		
C, K, R, V series		
1500, 2500 (.8)		1.0
3500HD (.3)		.4
G, P Vans		
Manual (.3)		.4
Auto (.6)		.8
Astro, Safari (.3)		.4
Express, Van, Savana (.3)		.4

LUBRICATION SERVICES

(M) Engine Oil & Filter, Renew
Includes: Inspect and correct all fluid levels.

	(Factory Time)	Chilton Time
1993-97		.3

(M) Lubricate Chassis, Change Oil & Filter
Includes: Inspect and correct all fluid levels.

	(Factory Time)	Chilton Time
1993-97		.6
Install grease fittings add		.1

(M) Lubricate Chassis
Includes: Inspect and correct all fluid levels.

	(Factory Time)	Chilton Time
1993-97		.4
Install grease fittings add		.1

ELECTRICAL

	(Factory Time)	Chilton Time
(G) Headlamps, Aim		
1993-97		
two		.4
four		.6
(G) License Lamp Assy., Renew		
1993-97, one or both (.2)		.3
(G) Turn Signal & Parking Lamp Assy., Renew		
1993-97, each		
P Vans, R, V series (.4)		.5
C, K series, G Vans (.2)		.4
Astro, Safari (.2)		.3
Express, Van, Savana (.2)		.3
(G) Horn, Renew		
1993		
Vans, C, K series (.2)		.4
Astro, Safari (.7)		1.0
1994-97		
Vans (.2)		.4
Astro, Safari (.7)		1.0
C, K series		
left (.2)		.3
center (.5)		.7
Express, Van, Savana (.3)		.5
(M) Terminals, Battery, Clean		
1993-97		.3

FREQUENT MAINTENANCE LABOR
S SERIES, BRAVADA

The following should be used as a guide when determining the amount of work required for a particular service if taken to a repair shop. In estimating how long a particular Frequent Maintenance Service item should take, please observe the following:
- **Factory Time** is time that is generated by the vehicle manufacturer.
- **Chilton Time** is time that is based on field research and data supplied by the vehicle manufacturer.
- All labor time operations are given in hours and tenths of an hour.
- All labor operations, are to be used as a **guide**.

COOLING

(G) Winterize Cooling System
Includes: Run engine to check for leaks, tighten all hose connections. Test radiator and pressure cap, drain radiator and engine block. Add antifreeze and refill system.
1993-97	.5

(G) Belt, Drive, Renew
1993-97, Serpentine	.3
Renew tensioner add	.1

(G) Hoses, Radiator, Renew
Includes: Drain and refill cooling system as required.
1993-97
upper	(.4)	.5
lower	(.5)	.6
both	(.6)	.9
by-pass	(.3)	.5
w/4.3L code W add	(.1)	.1

(G) Thermostat, Coolant, Renew
1993-97
4 cyl.	(.3)	.5
V6	(.4)	.6

FUEL

(M) Air Cleaner, Service
1993-97	.2

(G) Filter, Fuel, Renew
1993-97	(.4)	.5

BRAKES

(G) Bleed Brakes (Four Wheels)
Includes: Add fluid.
1993-97
w/2 wheel ABS	(1.0)	1.3
w/4 wheel ABS	(1.2)	1.5

(G) Brakes, Adjust (Minor)
Includes: Adjust brakes, fill master cylinder.
1993-97, two wheels	.4

(M) Parking Brake, Adjust
1993-97	(.3)	.4

LUBRICATION SERVICES

(M) Engine Oil & Filter, Renew
Includes: Inspect and correct all fluid levels.
1993-97	.3
w/Turbo add	.1

(M) Lubricate Chassis, Change Oil & Filter
Includes: Inspect and correct all fluid levels.
1993-97	.6
Install grease fittings add	.1

(M) Lubricate Chassis
Includes: Inspect and correct all fluid levels.
1993-97	.4
Install grease fittings add	.1

ELECTRICAL

(G) Headlamps, Aim
1993-97
two	.4
four	.6

(G) Halogen Headlamp Bulb, Renew
1993-97
one	.3
each adtnl.	.1

(G) License Lamp Assy., Renew
1993-97, one or both	.3

(G) Park & Turn Signal Lamp Assy., Renew
1993, one	(.2)	.3
1994-97, one	(.6)	.8

(G) Rear Combination Lamp Assy., Renew
1993-97, one	(.2)	.3

(G) Horn, Renew
1993	(.3)	.4
1994-97	(.2)	.3
w/Remote air cleaner add	(.1)	.1

(M) Terminals, Battery, Clean
1993-97	.3

SCHEDULED MAINTENANCE INTERVALS
(GENERAL MOTORS G SERIES (VANS), C/K SERIES (PICK UP) BLAZER, JIMMY, TAHOE, YUKON & SUBURBAN (GASOLINE 1993-95 LIGHT DUTY EMISSIONS & 1996-97))

TO BE SERVICED	TYPE OF SERVICE	VEHICLE MILEAGE INTERVAL (x1000)												
		7.5	15	22.5	30	37.5	45	52.5	60	67.5	75	82.5	90	97.5
Engine oil & filter	R	✓	✓	✓	✓	✓	✓	✓	✓	✓	✓	✓	✓	✓
Chassis lubrication	S/I	✓	✓	✓	✓	✓	✓	✓	✓	✓	✓	✓	✓	✓
CV joints & axle seals	S/I	✓	✓	✓	✓	✓	✓	✓	✓	✓	✓	✓	✓	✓
Front axle propshaft splines	S/I	✓	✓	✓	✓	✓	✓	✓	✓	✓	✓	✓	✓	✓
Front suspension	S/I	✓	✓	✓	✓	✓	✓	✓	✓	✓	✓	✓	✓	✓
Kingpin bushings	S/I	✓	✓	✓	✓	✓	✓	✓	✓	✓	✓	✓	✓	✓
Parking brake cable guides	S/I	✓	✓	✓	✓	✓	✓	✓	✓	✓	✓	✓	✓	✓
Rear driveline center splines	S/I	✓	✓	✓	✓	✓	✓	✓	✓	✓	✓	✓	✓	✓
Rear axle fluid level	S/I	✓	✓	✓	✓	✓	✓	✓	✓	✓	✓	✓	✓	✓
Steering linkage	S/I	✓	✓	✓	✓	✓	✓	✓	✓	✓	✓	✓	✓	✓
Transfer case shift linkage (4WD)	S/I	✓	✓	✓	✓	✓	✓	✓	✓	✓	✓	✓	✓	✓
Transmission shift linkage	S/I	✓	✓	✓	✓	✓	✓	✓	✓	✓	✓	✓	✓	✓
Rotate tires	S/I	✓		✓		✓		✓		✓		✓		✓
Engine coolant hoses, ducts & valves①	S/I		✓		✓		✓		✓		✓		✓	
Air cleaner filter	R				✓				✓				✓	
Automatic transmission fluid & filter	R②				✓				✓				✓	
Engine Coolant③	R				✓				✓				✓	
Fuel filter	R				✓				✓				✓	
Spark plugs④	R				✓				✓				✓	
Clutch fork ball stud	S/I				✓				✓				✓	
Exhaust system & shields	S/I				✓				✓				✓	
Front wheel bearings (2 wheel drive)	S/I				✓				✓				✓	
Serpentine belt	S/I				✓				✓					
EGR system	S/I								✓					
Engine timing check	S/I								✓					
EVAP system	S/I								✓					

SCHEDULED MAINTENANCE INTERVALS
(GENERAL MOTORS G SERIES (VANS), C/K SERIES (PICK UP) BLAZER, JIMMY, TAHOE, YUKON & SUBURBAN (GASOLINE 1993-95 LIGHT DUTY EMISSIONS & 1996-97)) (Cont.)

TO BE SERVICED	TYPE OF SERVICE	VEHICLE MILEAGE INTERVAL (x1000)												
		7.5	15	22.5	30	37.5	45	52.5	60	67.5	75	82.5	90	97.5
Fuel tank, cap & lines	S/I								✓					

① Only for thermostatically controlled cooling fan.
② Under 8600 GVWR shown; over 8600 GVWR change every 24,000 miles.
③ Engine coolant (1996-97) - replace every 100,000 miles. Use O.E. specified (DEX-COOL™) coolant only. If any silicate coolant is used, the service interval is every 30,000 miles.
④ Spark plugs (1996-97) - replace every 100,000 miles.
R – Replace   S/I – Service or Inspect

FREQUENT OPERATION MAINTENANCE (SEVERE SERVICE)

If a vehicle is operated under any of the following conditions it is considered severe service:
- Extremely dusty areas.
- 50% or more of the vehicle operation is in 32°C (90°F) or higher temperatures, or constant operation in temperatures below 0°C (32°F).
- Prolonged idling (vehicle operation in stop and go traffic).
- Frequent short running periods (engine does not warm to normal operating temperatures).
- Police, taxi, delivery usage or trailer towing usage.

Oil & oil filter change – change every 3000 miles.
Lubricate chassis every 3000 miles.
Drive axle - check every 3000 miles.
Rotate tires every 6000 miles.
Automatic transmission fluid & filter (1993-95 over 8600 GVWR) - change every 12,000 miles.
Automatic transmission fluid & filter (1993-95 under 8600 GVWR) - change every 15,000 miles.
Exhaust system & shields - check every 15,000 miles.
Front wheel bearings - repack every 15,000 miles.
Air cleaner filter - change every 24,000 miles.

SCHEDULED MAINTENANCE INTERVALS
(GENERAL MOTORS G SERIES (VANS), C/K SERIES (PICK UP) BLAZER, JIMMY, TAHOE, YUKON & SUBURBAN (GASOLINE 1993-95 HEAVY DUTY EMISSIONS))

TO BE SERVICED	TYPE OF SERVICE	VEHICLE MILEAGE INTERVAL (x1000)												
		6	12	18	24	30	36	42	48	54	60	66	72	78
Engine oil & filter	R	✓	✓	✓	✓	✓	✓	✓	✓	✓	✓	✓	✓	✓
Chassis lubrication	S/I	✓	✓	✓	✓	✓	✓	✓	✓	✓	✓	✓	✓	✓
Drive axle	S/I	✓	✓	✓	✓	✓	✓	✓	✓	✓	✓	✓	✓	✓
Rotate tires	S/I	✓	✓	✓	✓	✓	✓	✓	✓	✓	✓	✓	✓	✓
Engine coolant hoses, ducts & valves	S/I		✓		✓		✓		✓		✓		✓	
Exhaust system & shields	S/I		✓		✓		✓		✓		✓		✓	
Serpentine belt	S/I		✓		✓		✓		✓		✓		✓	

SCHEDULED MAINTENANCE INTERVALS
(GENERAL MOTORS G SERIES (VANS), C/K SERIES (PICK UP) BLAZER, JIMMY, TAHOE, YUKON & SUBURBAN (GASOLINE 1993-95 HEAVY DUTY EMISSIONS)) (Cont.)

TO BE SERVICED	TYPE OF SERVICE	VEHICLE MILEAGE INTERVAL (x1000)												
		6	12	18	24	30	36	42	48	54	60	66	72	78
Air cleaner filter	R				✓				✓				✓	
Engine Coolant	R				✓				✓				✓	
Fuel filter	R				✓				✓				✓	
Spark plugs	R				✓				✓				✓	
Front wheel bearings (2 wheel drive)	S/I				✓				✓				✓	
Front wheel bearings (2 wheel drive)	S/I				✓				✓				✓	
Thermostatically controlled air cleaner	S/I				✓				✓				✓	
Automatic transmission fluid & filter	R②					✓					✓			
Clutch fork ball stud	S/I					✓					✓			
EGR system	S/I										✓			
Engine timing check	S/I										✓			
EVAP system	S/I										✓			
EVRV system	S/I										✓			
Fuel tank, cap & lines	S/I										✓			
Ignition wires	S/I										✓			

① Only for thermostatically controlled cooling fan.
② Under 8600 GVWR shown; over 8600 GVWR change every 24,000 miles.
R – Replace   S/I – Service or Inspect

FREQUENT OPERATION MAINTENANCE (SEVERE SERVICE)
If a vehicle is operated under any of the following conditions it is considered severe service:
- Extremely dusty areas.
- 50% or more of the vehicle operation is in 32°C (90°F) or higher temperatures, or constant operation in temperatures below 0°C (32°F).
- Prolonged idling (vehicle operation in stop and go traffic).
- Frequent short running periods (engine does not warm to normal operating temperatures).
- Police, taxi, delivery usage or trailer towing usage.
Oil & oil filter change – change every 3000 miles.
Lubricate chassis every 3000 miles.
Drive axle - check every 3000 miles.
Automatic transmission fluid & filter (over 8600 GVWR) - change every 12,000 miles.
Automatic transmission fluid & filter (under 8600 GVWR) - change every 15,000 miles.

SCHEDULED MAINTENANCE INTERVALS
(GENERAL MOTORS G SERIES (VANS), C/K SERIES (PICK UP) BLAZER, JIMMY, TAHOE, YUKON & SUBURBAN (DIESEL)

TO BE SERVICED	TYPE OF SERVICE	VEHICLE MILEAGE INTERVAL (x1000)												
		5	10	15	20	25	30	35	40	45	50	55	60	65
Engine oil & filter	R	✓	✓	✓	✓	✓	✓	✓	✓	✓	✓	✓	✓	✓
Brake pedal springs	S/I	✓	✓	✓	✓	✓	✓	✓	✓	✓	✓	✓	✓	✓
Chassis lubrication	S/I	✓	✓	✓	✓	✓	✓	✓	✓	✓	✓	✓	✓	✓
CV joints & axle seals	S/I	✓	✓	✓	✓	✓	✓	✓	✓	✓	✓	✓	✓	✓
Front axle propshaft splines	S/I	✓	✓	✓	✓	✓	✓	✓	✓	✓	✓	✓	✓	✓
Front suspension	S/I	✓	✓	✓	✓	✓	✓	✓	✓	✓	✓	✓	✓	✓
Kingpin bushings	S/I	✓	✓	✓	✓	✓	✓	✓	✓	✓	✓	✓	✓	✓
Parking brake cable guides	S/I	✓	✓	✓	✓	✓	✓	✓	✓	✓	✓	✓	✓	✓
Rear driveline center splines	S/I	✓	✓	✓	✓	✓	✓	✓	✓	✓	✓	✓	✓	✓
Rear axle fluid level	S/I	✓	✓	✓	✓	✓	✓	✓	✓	✓	✓	✓	✓	✓
Steering linkage	S/I	✓	✓	✓	✓	✓	✓	✓	✓	✓	✓	✓	✓	✓
Transfer case shift linkage (4WD)	S/I	✓	✓	✓	✓	✓	✓	✓	✓	✓	✓	✓	✓	✓
Transmission shift linkage	S/I	✓	✓	✓	✓	✓	✓	✓	✓	✓	✓	✓	✓	✓
Rotate tires	S/I	✓		✓		✓		✓		✓		✓		✓
Air intake system	S/I		✓		✓		✓		✓		✓		✓	
Engine coolant hoses, ducts & valves①	S/I		✓		✓		✓		✓		✓		✓	
Exhaust system & shields	S/I		✓		✓		✓		✓		✓		✓	
Clutch fork ball stud	S/I				✓				✓				✓	
Front wheel bearings (2 wheel drive)	S/I				✓				✓				✓	
Air cleaner filter	R						✓						✓	
Engine idle speed adjustment	S/I						✓						✓	
Fuel filter	R						✓						✓	
Automatic transmission fluid & filter②	S/I						✓						✓	
Crankcase depression regulator valve	S/I												✓	
EGR system	S/I												✓	

SCHEDULED MAINTENANCE INTERVALS
(GENERAL MOTORS G SERIES (VANS), C/K SERIES (PICK UP) BLAZER, JIMMY, TAHOE, YUKON & SUBURBAN (DIESEL) (Cont.)

TO BE SERVICED	TYPE OF SERVICE	VEHICLE MILEAGE INTERVAL (x1000)												
		5	10	15	20	25	30	35	40	45	50	55	60	65
Fuel tank, cap & lines	S/I												✓	
Serpentine belt	S/I												✓	
Engine Coolant③	R													

① Only for thermostatically controlled cooling fan.
② Under 8600 GVWR shown; over 8600 GVWR change every 24,000 miles.
③ Engine coolant (1996-97) - replace every 100,000 miles. Use O.E. specified (DEX-COOL™) coolant only. If any silicate coolant is used, the service interval is every 30,000 miles.

R – Replace   S/I – Service or Inspect

FREQUENT OPERATION MAINTENANCE (SEVERE SERVICE)

If a vehicle is operated under any of the following conditions it is considered severe service:
- Extremely dusty areas.
- 50% or more of the vehicle operation is in 32°C (90°F) or higher temperatures, or constant operation in temperatures below 0°C (32°F).
- Prolonged idling (vehicle operation in stop and go traffic).
- Frequent short running periods (engine does not warm to normal operating temperatures).
- Police, taxi, delivery usage or trailer towing usage.

Oil & oil filter change – change every 2500 miles.
Lubricate chassis every 2500 miles.
Drive axle - check every 2500 miles.
Rotate tires every 7500 miles.
Exhaust system & shields - check every 10,000 miles.
Automatic transmission fluid & filter (1993-95 over 8600 GVWR) - change every 12,000 miles.
Automatic transmission fluid & filter (1993-95 under 8600 GVWR) - change every 15,000 miles.
Air cleaner filter - change every 15,000 miles.
Front wheel bearings - repack every 15,000 miles.

SCHEDULED MAINTENANCE INTERVALS
(GENERAL MOTORS ASTRO, SAFARI, BLAZER, JIMMY, BRAVADA, SONOMA, S SERIES (PICK-UP) & TYPHOON)

TO BE SERVICED	TYPE OF SERVICE	VEHICLE MILEAGE INTERVAL (x1000)												
		7.5	15	22.5	30	37.5	45	52.5	60	67.5	75	82.5	90	97.5
Engine oil & filter	R	✓	✓	✓	✓	✓	✓	✓	✓	✓	✓	✓	✓	✓
Brake/clutch pedal springs & parking brake cable guides	S/I	✓	✓	✓	✓	✓	✓	✓	✓	✓	✓	✓	✓	✓
Chassis lubrication & CV joints, ball joints & axle seals	S/I	✓	✓	✓	✓	✓	✓	✓	✓	✓	✓	✓	✓	✓
Drive axle, steering linkage & front suspension	S/I	✓	✓	✓	✓	✓	✓	✓	✓	✓	✓	✓	✓	✓
Front/rear axle fluid & transfer case shift linkage (4WD)	S/I	✓	✓	✓	✓	✓	✓	✓	✓	✓	✓	✓	✓	✓
Rotate tires	S/I	✓		✓		✓		✓		✓		✓		✓
Engine coolant strength, hoses & clamps	S/I		✓		✓		✓		✓		✓		✓	
Air cleaner filter & fuel filter	R				✓				✓				✓	
Automatic transmission fluid & filter (Astro & Safari)	R				✓				✓				✓	
Automatic transmission fluid & filter (1993-95 Blazer, Jimmy, Bravada, Sonoma, S Series (Pick-up) & Typhoon②	R				✓				✓				✓	
Engine coolant③	R				✓				✓				✓	
Spark plugs①	R				✓				✓				✓	
Accessory drive belt	S/I				✓				✓				✓	
Exhaust system	S/I				✓				✓				✓	
Front wheel bearings	S/I				✓				✓				✓	
Fuel tank, cap & lines	S/I								✓				✓	
Engine timing check	S/I								✓					

① Check ignition wires. Spark plugs (1996-97) - replace every 100,000 miles.
② 1993-95 Blazer, Jimmy, Bravada, Sonoma, S Series (Pick-up) & Typhoon - change every 30,000 miles. 1996-97 Blazer, Jimmy, Bravada, Sonoma, S Series (Pick-up) & Typhoon - change only when necessary.
③ Engine coolant (1996-97) - replace every 100,000 miles. Use O.E. specified (DEX-COOL™) coolant only. If any silicate coolant is used, the service interval is every 30,000 miles.
R – Replace   S/I – Service or Inspect

SCHEDULED MAINTENANCE INTERVALS
(GENERAL MOTORS ASTRO, SAFARI, BLAZER, JIMMY, BRAVADA, SONOMA, S SERIES (PICK-UP) & TYPHOON) (Cont.)

FREQUENT OPERATION MAINTENANCE (SEVERE SERVICE)
If a vehicle is operated under any of the following conditions it is considered severe service:
- Extremely dusty areas.
- 50% or more of the vehicle operation is in 32°C (90°F) or higher temperatures, or constant operation in temperatures below 0°C (32°F).
- Prolonged idling (vehicle operation in stop and go traffic).
- Frequent short running periods (engine does not warm to normal operating temperatures).
- Police, taxi, delivery usage or trailer towing usage.

Oil & oil filter change – change every 3000 miles.
Lubricate chassis every 3000 miles.
Rotate tires at 6000 miles, then every 15,000 miles.
Drive axle - check every 15,000 miles.
Automatic transmission fluid & filter (Astro & Safari) - change every 15,000 miles.
Automatic transmission fluid & filter (1993-95 Blazer, Jimmy, Bravada, Sonoma, S Series (Pick-up) & Typhoon - change every 15,000 miles.
Automatic transmission fluid & filter (1996-97 Blazer, Jimmy, Bravada, Sonoma, S Series (Pick-up) & Typhoon) - change every 50,000 miles.
Air cleaner filter - change every 15,000 miles.

GENERAL MOTORS CORPORATION
Lumina APV • Silhouette • Trans Sport

VEHICLE IDENTIFICATION CHART

Engine Code						Model Year	
Code	Liters	Cu. In. (cc)	Cyl.	Fuel Sys.	Eng. Mfg.	Code	Year
D	3.1	191 (3130)	6	TFI	CPC	P	1993
E	3.4	207 (3350)	6	MFI	CPC	R	1994
L	3.8	231 (3785)	6	MFI	CPC	S	1995
						T	1996
						V	1997

TFI - Throttle body fuel injection
MFI - Multiport fuel injection
CPC - Chevrolet/Pontiac/Canada

ENGINE IDENTIFICATION
All measurements are given in inches.

Year	Model	Engine Displacement Liters (cc)	Engine Series (ID/VIN)	Fuel System	No. of Cylinders	Engine Type
1993	Lumina APV	3.1 (3097)	D	TFI	6	OHV
	Silhouette	3.1 (3097)	D	TFI	6	OHV
	Trans Sport	3.1 (3097)	D	TFI	6	OHV
	Lumina APV	3.8 (3785)	L	MFI	6	OHV
	Silhouette	3.8 (3785)	L	MFI	6	OHV
	Trans Sport	3.8 (3785)	L	MFI	6	OHV
1994	Lumina APV	3.1 (3097)	D	TFI	6	OHV
	Silhouette	3.1 (3097)	D	TFI	6	OHV
	Trans Sport	3.1 (3097)	D	TFI	6	OHV
	Lumina APV	3.8 (3785)	L	MFI	6	OHV
	Silhouette	3.8 (3785)	L	MFI	6	OHV
	Trans Sport	3.8 (3785)	L	MFI	6	OHV
1995	Lumina APV	3.1 (3097)	D	TFI	6	OHV
	Silhouette	3.1 (3097)	D	TFI	6	OHV
	Trans Sport	3.1 (3097)	D	TFI	6	OHV
	Lumina APV	3.8 (3785)	L	MFI	6	OHV
	Silhouette	3.8 (3785)	L	MFI	7	OHV
	Trans Sport	3.8 (3785)	L	MFI	8	OHV
1996-97	Lumina APV	3.4 (3350)	E	MFI	6	OHV
	Silhouette	3.4 (3350)	E	MFI	6	OHV
	Trans Sport	3.4 (3350)	E	MFI	6	OHV

TFI - Throttle body fuel injection   MFI - Multiport fuel injection

GENERAL ENGINE SPECIFICATIONS

Year	Engine ID/VIN	Engine Displacement Liters (cc)	Fuel System Type	Net Horsepower @ rpm	Net Torque @ rpm (ft. lbs.)	Bore x Stroke (in.)	Compression Ratio	Oil Pressure @ rpm
1993	D	3.1 (3130)	TFI	120@4400	175@2200	3.50x3.30	8.9:1	15@1100
	L	3.8 (3785)	MFI	165@4300	220@3200	3.80x3.40	8.5:1	60@1850
1994	D	3.1 (3130)	TFI	120@4400	175@2200	3.50x3.30	8.9:1	15@1100
	L	3.8 (3785)	MFI	165@4300	220@3200	3.80x3.40	8.5:1	60@1850
1995	D	3.1 (3130)	TFI	120@4400	175@2200	3.50x3.31	8.5:1	15@1100
	L	3.8 (3785)	MFI	170@4300	225@3200	3.80x3.40	9.0:1	60@1850
1996-97	E	3.4 (3350)	MFI	180@5200	205@4000	3.62x3.31	9.5:1	15@1100

GASOLINE ENGINE TUNE-UP SPECIFICATIONS

Year	Engine ID/VIN	Engine Displacement Liters (cc)	Spark Plugs Gap (in.)	Ignition Timing (deg.) MT	Ignition Timing (deg.) AT	Fuel Pump (psi)	Idle Speed (rpm) MT	Idle Speed (rpm) AT	Valve Clearance In.	Valve Clearance Ex.
1993	D	3.1 (3130)	0.045	①	①	9-13	-	①	HYD	HYD
	L	3.8 (3785)	0.060	①	①	41-47 ②	-	①	HYD	HYD
1994	D	3.1 (3130)	0.045	①	①	9-13	-	①	HYD	HYD
	L	3.8 (3785)	0.060	①	①	41-47 ②	-	①	HYD	HYD

GASOLINE ENGINE TUNE-UP SPECIFICATIONS

Year	Engine ID/VIN	Engine Displacement Liters (cc)	Spark Plugs Gap (in.)	Ignition Timing (deg.)		Fuel Pump (psi)	Idle Speed (rpm)		Valve Clearance	
				MT	AT		MT	AT	In.	Ex.
1995	D	3.1 (3130)	0.045	①	①	9-13	-	725	HYD	HYD
	L	3.8 (3785)	0.060	①	①	41-47 ②	-	725	HYD	HYD
1996-97	E	3.4 (3350)	0.060	①	①	41-47	-	③	HYD	HYD

NOTE: The Vehicle Emission Control Information label often reflects specification changes made during production. The label figures must be used if they differ from those in this chart.

HYD - Hydraulic

① Refer to underhood label for exact setting
② With key on and engine off
③ Idle speed is maintained by the PCM

CAPACITIES

Year	Model	Engine ID/VIN	Engine Displacement Liters (cc)	Engine Oil with Filter (qts.)	Transmission (pts.)			Fuel Tank (gal.)	Cooling System (qts.)
					4-Spd	5-Spd	Auto.		
1993	Lumina APV	D	3.1 (3130)	4.5	-	-	8.0	20.0	13.4
	Silhouette	D	3.1 (3130)	4.5	-	-	8.0	20.0	13.4
	Trans Sport	D	3.1 (3130)	4.5	-	-	8.0	20.0	13.4
	Lumina APV	L	3.8 (3785)	4.5	-	-	12.0	20.0	13.4
	Silhouette	L	3.8 (3785)	4.5	-	-	12.0	20.0	13.4
	Trans Sport	L	3.8 (3785)	4.5	-	-	12.0	20.0	13.4
1994	Lumina APV	D	3.1 (3130)	4.5	-	-	8.0	20.0	13.4
	Silhouette	D	3.1 (3130)	4.5	-	-	8.0	20.0	13.4
	Trans Sport	D	3.1 (3130)	4.5	-	-	8.0	20.0	13.4
	Lumina APV	L	3.8 (3785)	4.5	-	-	12.0	20.0	13.4
	Silhouette	L	3.8 (3785)	4.5	-	-	12.0	20.0	13.4
	Trans Sport	L	3.8 (3785)	4.5	-	-	12.0	20.0	13.4
1995	Lumina APV	D	3.1 (3130)	4.5	-	-	8.0	20.0	13.4
	Silhouette	D	3.1 (3130)	4.5	-	-	8.0	20.0	13.4
	Trans Sport	D	3.1 (3130)	4.5	-	-	8.0	20.0	13.4
	Lumina APV	L	3.8 (3785)	4.5	-	-	12.0	20.0	11.4
	Silhouette	L	3.8 (3785)	4.5	-	-	12.0	20.0	11.4
	Trans Sport	L	3.8 (3785)	4.5	-	-	12.0	20.0	11.4
1996-97	Lumina APV	E	3.4 (3350)	4.5	-	-	12.0	20.0	①
	Silhouette	E	3.4 (3350)	4.5	-	-	12.0	20.0	①
	Trans Sport	E	3.4 (3350)	4.5	-	-	12.0	20.0	①

① With rear heater: 13.5 qts.
　 Without rear heater: 11.8 qts.

VALVE SPECIFICATIONS

Year	Engine ID/VIN	Engine Displacement Liters (cc)	Seat Angle (deg.)	Face Angle (deg.)	Spring Test Pressure (lbs. @ in.)	Spring Installed Height (in.)	Stem-to-Guide Clearance (in.)		Stem Diameter (in.)	
							Intake	Exhaust	Intake	Exhaust
1993	D	3.1 (3130)	46	45	82@1.58	1.57	0.0010-0.0027	0.0010-0.0027	NA	NA
	L	3.8 (3785)	46	45	210@1.32	1.69-1.72	0.0015-0.0035	0.0015-0.0032	NA	①

VALVE SPECIFICATIONS

Year	Engine ID/VIN	Engine Displacement Liters (cc)	Seat Angle (deg.)	Face Angle (deg.)	Spring Test Pressure (lbs. @ in.)	Spring Installed Height (in.)	Stem-to-Guide Clearance (in.)		Stem Diameter (in.)	
							Intake	Exhaust	Intake	Exhaust
1994	D	3.1 (3130)	46	45	82@1.58	1.57	0.0010-0.0027	0.0010-0.0027	NA	NA
	L	3.8 (3785)	46	45	210@1.32	1.69-1.72	0.0015-0.0035	0.0015-0.0032	NA	①
1995	D	3.1 (3130)	46	45	191@1.18	1.58	0.0010-0.0027	0.0010-0.0027	NA	NA
	L	3.8 (3785)	45	45	210@1.32	1.69-1.72	0.0015-0.0035	0.0015-0.0032	NA	NA
1996-97	E	3.4 (3350)	46	45	230@1.26	1.70	0.0010-0.0027	0.0010-0.0027	NA	NA

① Upper: 0.3129-0.3137
  Lower: 0.3118-0.3126

TORQUE SPECIFICATIONS
All readings in ft. lbs.

Year	Engine ID/VIN	Engine Displacement Liters (cc)	Cylinder Head Bolts	Main Bearing Bolts	Rod Bearing Bolts	Crankshaft Damper Bolts	Flywheel Bolts	Manifold Intake	Manifold Exhaust	Spark Plugs	Lug Nut
1993	D	3.1 (3130)	①	②	37	76	52	③	24	④	100
	L	3.8 (3785)	⑤	⑥	⑦	⑧	⑨	7	38	12	100
1994	D	3.1 (3130)	⑩	②	37	76	52	③	25	④	100
	L	3.8 (3785)	⑤	⑥	⑦	⑧	⑨	7	38	11	100
1995	D	3.1 (3130)	⑩	78	37	76	57	③	25	④	100
	L	3.8 (3785)	⑤	⑥	⑦	⑧	⑨	7	38	11	100
1996-97	E	3.4 (3350)	⑩	②	⑪	76	61	⑫	12	11	100

① Coat threads with sealer
  Tighten all bolts 40 ft. lbs.
  Tighten all an additional 90 degrees (1/4 turn)
② 37 ft. lbs. plus 77 degrees
③ Tighten bolts to 12 ft. lbs.
  Retorque to 22 ft. lbs.
④ 1st-time installation (new head): 22 ft. lbs.
  All other installations: 12 ft. lbs.

⑤ Tighten all bolts to 35 ft. lbs. then rotate 130 degrees
  Tighten four center bolts an additional 30 degrees
⑥ 26 ft. lbs. plus 50 degrees
⑦ 20 ft. lbs. plus 50 degrees
⑧ 110 ft. lbs. plus 76 degrees
⑨ 11 ft. lbs. plus 50 degrees

⑩ Coat threads with sealer
  Tighten all bolts to 33 ft. lbs.
  Tighten all an additional 90 degrees (1/4 turn)
⑪ Step 1: 15 ft. lbs.
  Step 2: Plus 75 degrees
⑫ Lower manifold: 10 ft. lbs.
  Upper manifold: 18 ft. lbs.

BRAKE SPECIFICATIONS
All measurements in inches unless noted

Year	Model	Master Cylinder Bore	Brake Disc Original Thickness	Brake Disc Minimum Thickness	Brake Disc Maximum Runout	Brake Drum Diameter Original Inside Diameter	Max. Wear Limit	Maximum Machine Diameter	Minimum Lining Thickness Front	Minimum Lining Thickness Rear
1993	Lumina APV	0.944	1.043	0.972	0.004	8.86	8.91	8.88	0.030	0.030
	Silhouette	0.944	1.043	0.972	0.004	8.86	8.91	8.88	0.030	0.030
	Trans Sport	0.944	1.043	0.972	0.004	8.86	8.91	8.88	0.030	0.030
1994	Lumina APV	0.944	1.043	0.972	0.004	8.86	8.91	8.88	0.030	0.030
	Silhouette	0.944	1.043	0.972	0.004	8.86	8.91	8.88	0.030	0.030
	Trans Sport	0.944	1.043	0.972	0.004	8.86	8.91	8.88	0.030	0.030
1995	Lumina APV	0.944	1.260	1.209	0.002	8.86	8.92	8.90	0.030	0.030
	Silhouette	0.944	1.260	1.209	0.002	8.86	8.92	8.91	0.030	0.030
	Trans Sport	0.944	1.260	1.209	0.002	8.86	8.92	8.92	0.030	0.030

BRAKE SPECIFICATIONS
All measurements in inches unless noted

Year	Model	Master Cylinder Bore	Brake Disc			Brake Drum Diameter			Minimum Lining Thickness	
			Original Thickness	Minimum Thickness	Maximum Runout	Original Inside Diameter	Max. Wear Limit	Maximum Machine Diameter	Front	Rear
1996-97	Lumina APV	0.944	1.260	1.209	0.002	8.86	8.92	8.90	0.030	0.030
	Silhouette	0.944	1.260	1.209	0.002	8.86	8.92	8.91	0.030	0.030
	Trans Sport	0.944	1.260	1.209	0.002	8.86	8.92	8.92	0.030	0.030

FREQUENT MAINTENANCE LABOR
CHEVROLET LUMINA/VENTURE, OLDSMOBILE SILHOUETTE PONTIAC TRANS SPORT

The following should be used as a guide when determining the amount of work required for a particular service if taken to a repair shop. In estimating how long a particular Frequent Maintenance Service item should take, please observe the following:
- **Factory Time** is time that is generated by the vehicle manufacturer.
- **Chilton Time** is time that is based on field research and data supplied by the vehicle manufacturer.
- All labor time operations are given in hours and tenths of an hour.
- All labor operations, are to be used as a **guide**.

(Factory Time) / Chilton Time

COOLING

(G) Winterize Cooling System
Includes: Run engine to check for leaks, tighten all hose connections. Test radiator and pressure cap, drain radiator and engine block. Add antifreeze and refill system.
1993-97 5

(G) Belt, Serpentine Drive, Renew
1993-97 (.2) 4

(G) Hoses, Radiator, Renew
Includes: Drain and refill cooling system as required.
1993-97
  upper or lower (.3) 4
  both (.4) 5

(G) Thermostat, Coolant, Renew
1993 (.4) 6
1994-97
  3.1L code D (.4) 6
  3.4L code E (.4) 6
  3.8L code L (.5) 7

FUEL

(M) Air Cleaner, Service
1993-97 2

BRAKES

(G) Bleed Brakes (Four Wheels)
Includes: Add fluid.
1993-97 6

(M) Parking Brake, Adjust
1993-97 (.3) 4

LUBRICATION SERVICES

(M) Engine Oil & Filter, Renew
Includes: Inspect and correct all fluid levels.
1993-97 3

(M) Lubricate Chassis, Change Oil & Filter
Includes: Inspect and correct all fluid levels.
1993-97 4
Install grease fittings add 1

(M) Lubricate Chassis
Includes: Inspect and correct all fluid levels.
1993-97 4
Install grease fittings add 1

WHEELS

(G) Wheel, Renew (One)
1993-97 5

(G) Wheel, Balance
1993-97
  one 3
  each adtnl. 2

(G) Wheels, Rotate (All)
1993-97 5

ELECTRICAL

(G) Headlamps, Aim
1993-97
  two 4
  four 6

(G) License Lamp Bulb, Renew
1993-97, one or all (.3) 4

(G) Park & Turn Signal Lamp Assy., Renew
1993-97, each (.2) 3

(G) Park & Turn Signal Lamp Bulb or Lens, Renew
1993-97, one 3

(G) Stop, Tail & Turn Signal Lamp Bulb, Renew
1993-97
  one 3
  each adtnl. 1

(G) Horn, Renew
1993
  one (.3) 4
  each adtnl. 1
1994-97 (.4) 5

(M) Terminals, Battery, Clean
1993-97 3

SCHEDULED MAINTENANCE INTERVALS
(GENERAL MOTORS CHEVROLET LUMINA, VENTURE, OLDSMOBILE SILHOUETTE & PONTIAC TRANS SPORT)

TO BE SERVICED	TYPE OF SERVICE	VEHICLE MILEAGE INTERVAL (x1000)												
		7.5	15	22.5	30	37.5	45	52.5	60	67.5	75	82.5	90	97.5
Engine oil & filter	R	✓	✓	✓	✓	✓	✓	✓	✓	✓	✓	✓	✓	✓
Automatic transaxle fluid & filter③	S/I	✓	✓	✓	✓	✓	✓	✓	✓	✓	✓	✓	✓	✓
Chassis lubrication	S/I	✓	✓	✓	✓	✓	✓	✓	✓	✓	✓	✓	✓	✓
Throttle body mount bolt torque	S/I	✓												
Disc brake pads, rear drum brake linings, wheel cylinders & parking brake	S/I	✓		✓		✓		✓		✓		✓		✓
Rotate tires	S/I	✓		✓		✓		✓		✓		✓		✓
Air cleaner filter & PCV valve & filter	R				✓				✓				✓	
Engine coolant②	R				✓				✓				✓	
Spark plugs①	R				✓				✓				✓	
Accessory drive belt	S/I				✓				✓				✓	
Engine coolant strength, hoses & clamps	S/I				✓				✓				✓	
EGR system	S/I				✓				✓				✓	
Exhaust system & shields	S/I				✓				✓				✓	
Ignition wires	S/I				✓				✓				✓	
Fuel tank, cap & lines	S/I								✓				✓	

① Replace platinum tip spark plugs every 100,000 miles.
② Engine coolant (1996-97) - replace every 100,000 miles. Use O.E. specified (DEX-COOL™) coolant only. If any silicate coolant is used, the service interval is every 30,000 miles.
③ (1993-95) - change every 100,000 miles.
R – Replace   S/I – Service or Inspect

FREQUENT OPERATION MAINTENANCE (SEVERE SERVICE)
If a vehicle is operated under any of the following conditions it is considered severe service:
- Extremely dusty areas.
- 50% or more of the vehicle operation is in 32°C (90°F) or higher temperatures, or constant operation in temperatures below 0°C (32°F).
- Prolonged idling (vehicle operation in stop and go traffic).
- Frequent short running periods (engine does not warm to normal operating temperatures).
- Police, taxi, delivery usage or trailer towing usage.

Oil & oil filter change – change every 3000 miles.
Lubricate parking brake cable guides, underbody contact points & linkage every 6000 miles.
Rotate tires at 6000 miles, then every 15,000 miles (1993-95) or every 12,000 miles (1996-97).
Throttle body mounting torque - check at 6000 miles.
Air cleaner filter - service or inspect every 15,000 miles.
Automatic transaxle - change fluid & filter every 15,000 miles (1993-95) or every 50,000 miles (1996-97).

JEEP
Cherokee • Grand Cherokee • Grand Wagoneer • Wrangler

VEHICLE IDENTIFICATION CHART

		Engine Code					Model Year	
Code	Liters	Cu. In. (cc)	Cyl.	Fuel Sys.	Eng. Mfg.		Code	Year
P	2.5	150 (2458)	4	MFI	Chrysler		P	1993
S	4.0	242 (3966)	6	MFI	Chrysler		R	1994
Y	5.2	318 (5211)	8	MFI	Chrysler		S	1995
							T	1996
							V	1997

MFI - Multiport fuel injection

ENGINE IDENTIFICATION

Year	Model	Engine Displacement Liters (cc)	Engine Series (ID/VIN)	Fuel System	No. of Cylinders	Engine Type
1993	Cherokee	2.5 (2458)	P	MFI	4	OHV
	Cherokee	4.0 (3966)	S	MFI	6	OHV
	Wrangler	2.5 (2458)	P	MFI	4	OHV
	Wrangler	4.0 (3966)	S	MFI	6	OHV
	Grand Cherokee	4.0 (3966)	S	MFI	6	OHV
	Grand Cherokee	5.2 (5211)	Y	MFI	8	OHV
	Grand Wagoneer	5.2 (5211)	Y	MFI	8	OHV
1994	Cherokee	2.5 (2458)	P	MFI	4	OHV
	Cherokee	4.0 (3966)	S	MFI	6	OHV
	Wrangler	2.5 (2458)	P	MFI	4	OHV
	Wrangler	4.0 (3966)	S	MFI	6	OHV
	Grand Cherokee	4.0 (3966)	S	MFI	6	OHV
	Grand Cherokee	5.2 (5211)	Y	MFI	8	OHV
1995	Cherokee	2.5 (2458)	P	MFI	4	OHV
	Cherokee	4.0 (3966)	S	MFI	6	OHV
	Wrangler	2.5 (2458)	P	MFI	4	OHV
	Wrangler	4.0 (3966)	S	MFI	6	OHV
	Grand Cherokee	4.0 (3966)	S	MFI	6	OHV
	Grand Cherokee	5.2 (5211)	Y	MFI	8	OHV
1996-97	Cherokee	2.5 (2458)	P	MFI	4	OHV
	Cherokee	4.0 (3966)	S	MFI	6	OHV
	Grand Cherokee	4.0 (3966)	S	MFI	6	OHV
	Grand Cherokee	5.2 (5211)	Y	MFI	8	OHV
	Wrangler	2.5 (2464)	P	MFI	4	OHV
	Wrangler	4.0 (3958)	S	MFI	6	OHV

MFI - Multiport fuel injection
OHV - Overhead Valve

GENERAL ENGINE SPECIFICATIONS

Year	Engine ID/VIN	Engine Displacement Liters (cc)	Fuel System Type	Net Horsepower @ rpm	Net Torque @ rpm (ft. lbs.)	Bore x Stroke (in.)	Compression Ratio	Oil Pressure @ rpm
1993	P	2.5 (2458)	MFI	130@5250	139@3250 ①	3.88x3.19	9.1:1	37@1600 ③
	S	4.0 (3966)	MFI	180@4750	220@2500 ②	3.88x3.44	8.8:1	37@1600 ③
	Y	5.2 (5211)	MFI	220@4800	285@3600	3.91x3.31	9.1:1	30@3000 ③
1994	P	2.5 (2458)	MFI	130@5250	139@3250 ④	3.88x3.19	9.1:1	37@1600 ③
	S	4.0 (3966)	MFI	180@4750	220@4000 ⑤	3.88x3.44	8.8:1	37@1600 ③
	Y	5.2 (5211)	MFI	220@4800	285@3600	3.91x3.31	9.1:1	30@3000 ③
1995	P	2.5 (2458)	MFI	130@5250	139@3250 ④	3.88x3.19	9.1:1	37@1600 ⑦
	S	4.0 (3966)	MFI	180@4750 ⑥	220@4000 ⑤	3.88x3.44	8.7:1	37@1600 ⑦
	Y	5.2 (5211)	MFI	220@4400	285@3600	3.91x3.31	9.1:1	30@3000 ⑦
1996-97	P	2.5 (2458)	MFI	130@5250	149@3250	3.88x3.19	9.1:1	37@1600 ⑦
	S	4.0 (3966)	MFI	190@4750	225@4000	3.88x3.44	8.7:1	37@1600 ⑦
	Y	5.2 (5211)	MFI	220@4400	285@3600	3.91x3.31	9.1:1	30@3000 ⑦

MFI - Multiport fuel injection
① Cherokee: 149@3250
② Cherokee and Grand Wagoneer :225@4000
③ Above 3000 rpm, pressure can vary to a maximum of 80 psi.
④ Cherokee: 149@3250
⑤ Cherokee and Grand Wagoneer: 225@4000
⑥ Cherokee and Grand Cherokee: 190@4750
⑦ Above 3000 rpm, pressure can vary to a maximum of 75 psi, except on 5.2L.
On 5.2L, pressure can vary to 80 psi maximum

GASOLINE ENGINE TUNE-UP SPECIFICATIONS

Year	Engine ID/VIN	Engine Displacement Liters (cc)	Spark Plugs Gap (in.)	Ignition Timing (deg.) MT	Ignition Timing (deg.) AT	Fuel Pump (psi)	Idle Speed (rpm) MT	Idle Speed (rpm) AT	Valve Clearance In.	Valve Clearance Ex.
1993	P	2.5 (2458)	0.035	①	①	39-41	①	①	HYD	HYD
	S	4.0 (3966)	0.035	①	①	39-41	①	①	HYD	HYD
	Y	5.2 (5211)	0.035	①	①	39-41	①	①	HYD	HYD
1994	P	2.5 (2458)	0.035	①	①	39-41	①	①	HYD	HYD
	S	4.0 (3966)	0.035	①	①	39-41	①	①	HYD	HYD
	Y	5.2 (5211)	0.035	①	①	39-41	①	①	HYD	HYD
1995	P	2.5 (2458)	0.035	①	①	39-41 ②	①	①	HYD	HYD
	S	4.0 (3966)	0.035	①	①	39-41 ②	①	①	HYD	HYD
	Y	5.2 (5211)	0.035	①	①	39-41 ②	①	①	HYD	HYD
1996-97	P	2.5 (2458)	0.035	①	①	47-51 ②	①	①	HYD	HYD
	S	4.0 (3966)	0.035	①	①	47-51 ②	①	①	HYD	HYD
	Y	5.2 (5211)	0.035	①	①	47-51 ②	①	①	HYD	HYD

NOTE: The Vehicle Emission Control Information label often reflects specification changes made during production. The label figures must be used if they differ from those in this chart.
HYD - Hydraulic
① Not adjustable
② With the vacuum line disconnected from the fuel pressure regulator (if equipped).
Fuel pressure is measured at the test port pressure fitting on fuel rail.

CAPACITIES

Year	Model	Engine ID/VIN	Engine Displacement Liters (cc)	Engine Oil with Filter	Transmission (pts.) 4-Spd	Transmission (pts.) 5-Spd	Transmission (pts.) Auto.	Transfer Case (pts.)	Drive Axle Front (pts.)	Drive Axle Rear (pts.)	Fuel Tank (gal.)	Cooling System (qts.)
1993	Cherokee	P	2.5 (2458)	4.0	-	7.4 ①	17.0	3.0 ②	2.5	2.5 ③	20.2	10.0
	Cherokee	S	4.0 (3966)	6.0	-	6.7	17.0	3.0 ②	2.5	2.5 ③	20.2	12.0
	Wrangler	P	2.5 (2458)	4.0	-	7.4 ①	16.0	3.3	2.5	2.5 ③	15.0 ⑤	9.0
	Wrangler	S	4.0 (3966)	6.0	-	6.7	16.0	3.3	2.5	2.5 ③	15.0 ⑤	10.5
	Grand Cherokee	S	4.0 (3966)	6.0	-	6.5	17.0	3.2 ⑥	3.1	3.4	23.0	9.3
	Grand Cherokee	Y	5.2 (5211)	5.0	-	6.5	17.0	3.2 ⑥	3.1	3.4	23.0	14.9
	Grand Wagoneer	Y	5.2 (5211)	5.0	-	-	17.0	3.2 ⑥	3.1	3.4	23.0	14.9
1994	Cherokee	P	2.5 (2458)	4.0	-	7.4 ①	17.0	3.0 ②	2.5	2.5 ③	20.2	10.0
	Cherokee	S	4.0 (3966)	6.0	-	6.7	17.0	3.0 ②	2.5	2.5 ③	20.2	12.0
	Wrangler	P	2.5 (2458)	4.0	-	7.4 ①	16.0	3.3	2.5	2.5 ③	15.0 ⑤	9.0
	Wrangler	S	4.0 (3966)	6.0	-	6.7	16.0	3.3	2.5	2.5 ③	15.0 ⑤	10.5
	Grand Cherokee	S	4.0 (3966)	6.0	-	6.5	17.0	3.2 ⑥	3.1	3.4	23.0	12.0
	Grand Cherokee	Y	5.2 (5211)	5.0	-	6.5	17.0	3.2 ⑥	3.1	3.4	23.0	14.9
1995	Cherokee	P	2.5 (2468)	4.0	-	6.6 ⑦	17.0	3.0 ②	3.1	3.5 ⑧	20.2	10.0
	Cherokee	S	4.0 (3966)	6.0	-	6.6	17.0	3.0 ②	3.1	3.5 ⑧	20.2	12.0
	Wrangler	P	2.5 (2468)	4.0	-	6.6	17.5	⑥	3.7	3.5 ⑧	15.0 ⑤	9.0
	Wrangler	S	4.0 (3966)	6.0	-	6.6	17.5	⑥	3.7	3.5 ⑧	15.0 ⑤	10.5
	Grand Cherokee	S	4.0 (3966)	6.0	-	6.5	17.0	3.2 ⑩	3.1	3.4	23.0	12.0
	Grand Cherokee	Y	5.2 (5211)	5.0	-	6.5	19.5	3.2 ⑩	3.1	3.4	23.0	14.9
1996-97	Cherokee	P	2.5 (2468)	4.0	-	6.6 ⑪	17.0	3.0 ②	3.1	3.5 ⑧	20.2	10.0
	Cherokee	S	4.0 (3966)	6.0	-	6.6	17.0	3.0 ②	3.1	3.5 ⑧	20.2	12.0
	Grand Cherokee	S	4.0 (3966)	6.0	-	6.5	17.0	3.2 ⑫	3.1	3.4	23.0	12.0
	Grand Cherokee	Y	5.2 (5211)	5.0	-	6.5	19.5	3.2 ⑫	3.1	3.4	23.0	14.9
	Wrangler	P	2.5 (2464)	4.0	-	6.6	17.5	⑨	3.7	3.5	⑬	9.0
	Wrangler	S	4.0 (3958)	6.0	-	6.6	17.5	⑨	3.7	3.5	⑬	10.5

① 2WD - 7.4 pts.
② Command-Trac - 2.2 pts.
③ Heavy Duty - 3.0 pts.
④ Long Bed - 23.5 gals.
⑤ Optional - 20 gal.

⑥ NP242: 2.9
NP249: 3.0
⑦ 2WD: 7.0 pts.
⑧ 8 1/4 axle: 4.4 pts.
When equipped with TRAC-LOK,
include 2 oz. of friction modifier additive

⑨ Command-Trac:
Automatic: 2.2 pts.
Manual: 3.3 pts.
⑩ NP242: 2.9
NP249: 2.5

⑪ 2WD: 7.0 pts.
⑫ NP242: 2.9 pts.
NP249: 2.5 pts.
⑬ Standard: 15.0 gals.
Optional: 19.6 gals.

VALVE SPECIFICATIONS

Year	Engine ID/VIN	Engine Displacement Liters (cc)	Seat Angle (deg.)	Face Angle (deg.)	Spring Test Pressure (lbs. @ in.)	Spring Installed Height (in.)	Stem-to-Guide Clearance (in.) Intake	Stem-to-Guide Clearance (in.) Exhaust	Stem Diameter (in.) Intake	Stem Diameter (in.) Exhaust
1993	P	2.5 (2458)	44.5	45	200@1.216	1.640	0.0010-0.0030	0.0010-0.0030	0.3110-0.3120	0.3110-0.3120
	S	4.0 (3966)	44.5	45	210@1.200	1.625	0.0010-0.0030	0.0010-0.0030	0.3110-0.3120	0.3110-0.3120
	Y	5.2 (5211)	44.25-44.75	43.25-43.75	200@1.212	1.640	0.0010-0.0030	0.0010-0.0030	0.3110-0.3120	0.3110-0.3120

VALVE SPECIFICATIONS

Year	Engine ID/VIN	Engine Displacement Liters (cc)	Seat Angle (deg.)	Face Angle (deg.)	Spring Test Pressure (lbs. @ in.)	Spring Installed Height (in.)	Stem-to-Guide Clearance (in.) Intake	Stem-to-Guide Clearance (in.) Exhaust	Stem Diameter (in.) Intake	Stem Diameter (in.) Exhaust
1994	P	2.5 (2458)	44.5	45	200@1.216	1.640	0.0010-0.0030	0.0010-0.0030	0.3110-0.3120	0.3110-0.3120
	S	4.0 (3966)	44.5	45	200@1.216	1.640	0.0010-0.0030	0.0010-0.0030	0.3110-0.3120	0.3110-0.3120
	Y	5.2 (5211)	44.25-44.75	43.25-43.75	200@1.212	1.640	0.0010-0.0030	0.0010-0.0030	0.3110-0.3120	0.3110-0.3120
1995	P	2.5 (2458)	44.5	45	200@1.216	1.640	0.0010-0.0030	0.0010-0.0030	0.3110-0.3120	0.3110-0.3120
	S	4.0 (3966)	44.5	45	200@1.216	1.640	0.0010-0.0030	0.0010-0.0030	0.3110-0.3120	0.3110-0.3120
	Y	5.2 (5211)	44.25-44.75	43.25-43.75	200@1.212	1.640	0.0010-0.0030	0.0010-0.0030	0.3110-0.3120	0.3110-0.3120
1996-97	P	2.5 (2458)	44.5	45	184-196@1.216	1.640	0.0010-0.0030	0.0010-0.0030	0.3110-0.3120	0.3110-0.3120
	S	4.0 (3966)	44.5	45	184-196@1.216	1.640	0.0010-0.0030	0.0010-0.0030	0.3110-0.3120	0.3110-0.3120
	Y	5.2 (5211)	44.25-44.75	43.25-43.75	200@1.212	1.640	0.0010-0.0030	0.0010-0.0030	0.3110-0.3120	0.3110-0.3120

TORQUE SPECIFICATIONS
All readings in ft. lbs.

Year	Engine ID/VIN	Engine Displacement Liters (cc)	Cylinder Head Bolts	Main Bearing Bolts	Rod Bearing Bolts	Crankshaft Damper Bolts	Flywheel Bolts	Manifold Intake *	Manifold Exhaust	Spark Plugs	Lug Nut
1993	P	2.5 (2458)	①	80	33	80	③	④	30	27	80-110
	S	4.0 (3966)	②	80	33	80	105	⑤	⑥	27	80-110
	Y	5.2 (5211)	⑥	85	45	135	105	⑦	20	30	80-110
1994	P	2.5 (2458)	①	80	33	80	105	⑤	⑤	27	80-110
	S	4.0 (3966)	②	80	33	80	105	⑥	⑥	27	80-110
	Y	5.2 (5211)	⑥	85	45	135	105	⑦	20	30	80-110
1995	P	2.5 (2458)	①	80	33	80	105	⑤	⑤	27	80-110
	S	4.0 (3966)	②	80	33	80	105	⑧	8	27	80-110
	Y	5.2 (5211)	⑥	85	45	135	105	⑦	20	30	80-110
1996-97	P ⑩	2.5 (2458)	①	80	33	80	105	④	④	27	80-110
	S ⑩	4.0 (3966)	②	80	33	80	105	⑧	⑧	27	80-110
	P ⑪	2.5 (2464)	⑨	80	33	80	105	④	20	27	95
	S ⑪	4.0 (3958)	⑨	80	33	80	105	⑧	20	27	95
	Y	5.2 (5211)	⑥	85	45	135	105	⑦	20	30	80-110

* NOTE: Applies to Lower Manifold only.

① Step 1: 22 ft. lbs.
  Step 2: 45 ft. lbs.
  Step 3: Bolts 1-6: 110 ft. lbs.
  Step 4: Bolt 7: 100 ft. lbs.
② Step 5: Bolts 8-10: 110 ft. lbs.
  Step 1: 22 ft. lbs.
  Step 2: 45 ft. lbs.
  Step 3: Bolts 1-10 and 12-14: 110 ft. lbs.
③ Step 4: Bolt 11: 100 ft. lbs.
  Step 1: 50 ft. lbs.

④ Step 2: Plus 60 degrees
  Bolts 1, 6 & 7: 30 ft. lbs.
  Bolts 2-5: 23 ft. lbs.
⑤ Bolt 1: 30 ft. lbs.
  Bolts 2-7: 23 ft. lbs.
⑥ Step 1: 50 ft. lbs.
  Step 2: 105 ft. lbs.
⑦ Step 1: Torque bolts 1-4 to 72 in. lbs.
  Step 2: Torque bolts 5-12 to 72 in. lbs.
  Step 3: Torque all bolts to 12 ft. lbs.

⑧ Bolts 1-5 and 8-11: 24 ft. lbs.
  Bolts 6-7: 23 ft. lbs.
⑨ Bolts 1-10 and 12-14: 110 ft. lbs.
  Bolt 11: 100 ft. lbs.
⑩ Cherokee and Grand Cherokee
⑪ Wrangler

BRAKE SPECIFICATIONS

All measurements in inches unless noted

Year	Model	Master Cylinder Bore	Brake Disc Original Thickness	Brake Disc Minimum Thickness	Maximum Runout	Brake Drum Diameter Original Inside Diameter	Brake Drum Diameter Max. Wear Limit	Brake Drum Diameter Maximum Machine Diameter	Minimum Lining Thickness Front	Minimum Lining Thickness Rear
1993	Wrangler	NA	NA	0.890	0.005	NA	①	NA	NA	NA
	Cherokee	NA	NA	0.890 ②	0.005	NA	①	NA	NA	NA
	Grand Cherokee	NA	NA	0.890 ③	0.005	-	-	-	NA	NA
	Grand Wagoneer	NA	NA	0.890 ③	0.005	-	-	-	NA	NA
1994	Wrangler	NA	0.940	0.890	0.005	9.00	①	9.06	0.030	0.030
	Cherokee	NA	0.940	0.890	0.005	9.00	①	9.06	0.030	0.030
	Grand Cherokee	0.990	0.940 ④	0.890 ③	0.005	-	-	-	0.030	0.030
1995	Wrangler	NA	0.940	0.890	0.005	9.00	①	9.06	0.030	0.030
	Cherokee	NA	0.940	0.890	0.005	9.00	①	9.06	0.030	0.030
	Grand Cherokee	0.990	0.940 ④	0.890 ⑤	0.005	-	-	-	0.030	0.030
1996-97	Wrangler	NA	0.940	0.890	0.005	9.00	①	9.06	0.030	0.030
	Cherokee	NA	0.940	0.890	0.005	9.00	①	9.06	0.030	0.030
	Grand Cherokee	0.990	0.940 ④	0.890 ⑥	0.005	-	-	-	0.030	0.030

NA - Not Available

① Maximum diameter is listed on outside of drum

② 2WD - 0.866 inch

③ Rear rotors have minimum allowable thickness listed on edge of parking brake drum

④ Rear rotor original thickness: 0.440

⑤ Rear rotors have minimum allowable thickness listed 0.370

⑥ Rear rotors have minimum allowable thickness listed: 0.370

FREQUENT MAINTENANCE LABOR
JEEP, CHEROKEE, GRAND CHEROKEE, GRAND WAGONEER, WRANGLER

The following should be used as a guide when determining the amount of work required for a particular service if taken to a repair shop. In estimating how long a particular Frequent Maintenance Service item should take, please observe the following:
- **Factory Time** is time that is generated by the vehicle manufacturer.
- **Chilton Time** is time that is based on field research and data supplied by the vehicle manufacturer.
- All labor time operations are given in hours and tenths of an hour.
- All labor operations, are to be used as a **guide**.

	(Factory Time)	Chilton Time

COOLING

(G) Winterize Cooling System
Includes: Run engine to check for leaks, tighten all hose connections. Test radiator and pressure cap, drain radiator and engine block. Add antifreeze and refill system.
1993-97 5

(G) Belt, Drive, Renew
1993-97
V-belt
one (.2) 4
each adtnl. 1
Serpentine (.4) 6

(G) Belt, Drive, Adjust
1993-97
V-belt
one 3
each adtnl. 1
Serpentine 4

(G) Hoses, Radiator, Renew
Includes: Drain and refill cooling system as required.
1993-97
upper (.4) 5
lower (.4) 6

(G) Thermostat, Coolant, Renew
1993-94 (.5) 8
w/AC add 5
1995-97
Cherokee (.3) 5
Grand Cherokee
4 cyl. (.5) 8
6 cyl. (.5) 8
V8 (.8) 1.2
Wrangler (.4) 7

FUEL

(M) Air Cleaner, Service
1993-97 2

(G) Filter, Fuel, Renew
In line
1993-97 (.3) 4
In tank
1993-97 Cherokee (.6) 9
1993-95 Wrangler (1.2) 1.5
1997 Wrangler (1.2) 1.5
1993-97 Grand Cherokee (1.4) . . 2.0
w/skid plate add 3

BRAKES

(G) Bleed Brakes (Four Wheels)
Includes: Add fluid.
1993-97 (.4) 5
w/ABS add 5

(M) Parking Brake, Adjust
1993-97 (.3) 4

LUBRICATION SERVICES

(M) Engine Oil & Filter, Renew
Includes: Inspect and correct all fluid levels.
1993-97 (.2) 3

(M) Lubricate Chassis, Change Oil & Filter
Includes: Inspect and correct all fluid levels.
1993-97 6
Install grease fittings, add 1

(M) Lubricate Chassis
Includes: Inspect and correct all fluid levels.
1993-97 4
Install grease fittings, add 1

ELECTRICAL

(G) Headlamps, Aim
1993-97
two 4
four 6

(G) Halogen Headlamp Bulb, Renew
1993-97, one (.2) 3

(G) License Lamp Assy., Renew
1993-97, one (.2) 3

(G) Park & Turn Signal Lamp Assy., Renew
1993-97, one (.2) 3

(G) Park & Turn Signal Lamp Bulb or Lens, Renew
1993-97, one (.2) 3

(G) Tail Lamp Assy., Renew
1993-97, one (.2) 4

(G) Tail Lamp Lens or Bulb, Renew
1993-97, one (.3) 4

(G) Horn, Renew
1993-97
one (.3) 4
each adtnl. 2

(M) Terminals, Battery, Clean
1993-97 3

SCHEDULED MAINTENANCE INTERVALS
(JEEP CHEROKEE, GRAND CHEROKEE, GRAND WAGONEER, & WRANGLER)

TO BE SERVICED	TYPE OF SERVICE	VEHICLE MILEAGE INTERVAL (x1000)												
		7.5	15	22.5	30	37.5	45	52.5	60	67.5	75	82.5	90	97.5
Engine oil & filter	R	✓	✓	✓	✓	✓	✓	✓	✓	✓	✓	✓	✓	✓
Brake hoses & linings	S/I	✓	✓	✓	✓	✓	✓	✓	✓	✓	✓	✓	✓	✓
Engine coolant level, hoses & clamps	S/I	✓	✓	✓	✓	✓	✓	✓	✓	✓	✓	✓	✓	✓
Exhaust system	S/I	✓	✓	✓	✓	✓	✓	✓	✓	✓	✓	✓	✓	✓
Lubricate steering linkage (4x2)	S/I	✓	✓	✓	✓	✓	✓	✓	✓	✓	✓	✓	✓	✓
Lubricate steering linkage (4x4)	S/I	✓		✓		✓		✓		✓		✓		✓
Air filter	R				✓				✓				✓	
Automatic transmission fluid & filter	R				✓				✓				✓	
Spark plugs	R				✓				✓				✓	
Transfer case oil	R				✓				✓				✓	
Drive belts	S/I				✓				✓				✓	
Front & rear axle oil	R				✓				✓				✓	
Prop shaft universal joints	S/I				✓				✓				✓	
Rotate tires	S/I				✓				✓				✓	
Engine Coolant	R					✓					✓			
Manual transmission fluid	R					✓					✓			
Distributor cap & rotor	R								✓					
Fuel filter	R								✓					
Ignition cables	R								✓					

R – Replace   S/I – Service or Inspect

FREQUENT OPERATION MAINTENANCE (SEVERE SERVICE)

If a vehicle is operated under any of the following conditions it is considered severe service:
- Extremely dusty areas.
- 50% or more of the vehicle operation is in 32°C (90°F) or higher temperatures, or constant operation in temperatures below 0°C (32°F).
- Prolonged idling (vehicle operation in stop and go traffic).
- Frequent short running periods (engine does not warm to normal operating temperatures).
- Police, taxi, delivery usage or trailer towing usage.

Oil & oil filter change – change every 3000 miles.
Automatic transmission fluid & filter - change every 12,000 miles.
Brake hoses & linings - check every 12,000 miles.
Lubricate steering linkage - check every 3000 miles.
Manual transmission fluid - change every 18,000 miles.
Prop shaft universal joints - lubricate every 3000 miles.
Front & rear axle oil - change every 12,000 miles.

ACURA
SLX

VEHICLE IDENTIFICATION CHART

		Engine Code					Model Year	
Code	Liters	Cu. In. (cc)	Cyl.	Fuel Sys.	Eng. Mfg.		Code	Year
6VD1	3.2	193 (3165)	6	MFI	Isuzu		T	1996
							V	1997

MFI - Multiport fuel injection

ENGINE IDENTIFICATION

Year	Model	Engine Displacement Liters (cc)	Engine Series (ID/VIN)	Fuel System	No. of Cylinders	Engine Type
1996-97	SLX	3.2 (3165)	6VD1	MFI	6	SOHC

MFI - Multiport fuel injection
SOHC - Single overhead camshaft

GENERAL ENGINE SPECIFICATIONS

Year	Engine ID/VIN	Engine Displacement Liters (cc)	Fuel System Type	Net Horsepower @ rpm	Net Torque @ rpm (ft. lbs.)	Bore x Stroke (in.)	Compression Ratio	Oil Pressure @ rpm
1996-97	6VD1	3.2 (3165)	MFI	190@5600	188@4000	3.68x3.03	9.1:1	57-80@3000

MFI - Multiport fuel injection

GASOLINE ENGINE TUNE-UP SPECIFICATIONS

Year	Engine ID/VIN	Engine Displacement Liters (cc)	Spark Plugs Gap (in.)	Ignition Timing (deg.) MT	AT	Fuel Pump (psi)	Idle Speed (rpm) MT	AT	Valve Clearance In.	Ex.
1996-97	6VD1	3.2 (3165)	0.040	5B ①	5B ①	41-46	750 ①	750 ①	HYD	HYD

NOTE: The Vehicle Emission Control Information label often reflects specification changes made during production. The label figures must be used if they differ from those in this chart.
B - Before top dead center
HYD - Hydraulic
① Controlled by the ECM and is not adjustable

CAPACITIES

| Year | Model | Engine ID/VIN | Engine Displacement Liters (cc) | Engine Oil with Filter (qts.) | Transmission (pts.) | | | Transfer Case (pts.) | Drive Axle | | Fuel Tank (gal.) | Cooling System (qts.) |
					4-Spd	5-Spd	Auto.		Front (pts.)	Rear (pts.)		
1996-97	SLX	6VD1	3.2 (3165)	5.7	-	6.2	18.2	3.0	3.2	3.8	22.5	①

① Manual transmission: 9.3
  Automatic transmission: 9.0

VALVE SPECIFICATIONS

| Year | Engine ID/VIN | Engine Displacement Liters (cc) | Seat Angle (deg.) | Face Angle (deg.) | Spring Test Pressure (lbs. @ in.) | Spring Installed Height (in.) | Stem-to-Guide Clearance (in.) | | Stem Diameter (in.) | |
							Intake	Exhaust	Intake	Exhaust
1996-97	6VD1	3.2 (3165)	45	45	45-55@1.54	1.54	0.0009-0.0079	0.0012-0.0079	0.2323-0.2353	0.2323-0.2350

TORQUE SPECIFICATIONS
All readings in ft. lbs.

| Year | Engine ID/VIN | Engine Displacement Liters (cc) | Cylinder Head Bolts | Main Bearing Bolts | Rod Bearing Bolts | Crankshaft Damper Bolts | Flywheel Bolts | Manifold | | Spark Plugs | Lug Nut |
								Intake	Exhaust		
1996-97	6VD1	3.2 (3165)	①	②	40	123	40	17	42	13	87

① 8x1.25 bolts: 15 ft. lbs.
  11x1.5 bolts: 47 ft. lbs.

② Main bearing cap bolts: 29 ft. lbs.
  Oil gallery bolts: 29 ft. lbs. plus 55-65 degrees
  Buttress bolts: 29 ft. lbs.

BRAKE SPECIFICATIONS
All measurements in inches unless noted

| Year | Model | | Master Cylinder Bore | Brake Disc | | | Brake Drum Diameter | | | Minimum Lining Thickness | |
				Original Thickness	Minimum Thickness	Maximum Runout	Original Inside Diameter	Max. Wear Limit	Maximum Machine Diameter	Front	Rear
1996-97	SLX	F	1.000	1.024	0.969 ①	0.005	-	-	-	0.039	-
		R	-	0.710	0.654 ②	0.005	8.27 ③	8.32 ③	8.32 ③	-	0.039 ④

① Minimum machine diameter: 0.983
② Minimum machine diameter: 0.668

③ Emergency brake drum surface
④ Specification includes disc pads and parking brake shoes

SCHEDULED MAINTENANCE INTERVALS
(ACURA SLX)

TO BE SERVICED	TYPE OF SERVICE	VEHICLE MILEAGE INTERVAL (x1000)												
		7.5	15	22.5	30	37.5	45	52.5	60	67.5	75	82.5	90	97.5
Engine oil & filter	R	✓	✓	✓	✓	✓	✓	✓	✓	✓	✓	✓	✓	✓
Battery fluid level	S/I	✓	✓	✓	✓	✓	✓	✓	✓	✓	✓	✓	✓	✓
Body & chassis lubrication	S/I	✓	✓	✓	✓	✓	✓	✓	✓	✓	✓	✓	✓	✓
Brake & clutch fluid level	S/I	✓	✓	✓	✓	✓	✓	✓	✓	✓	✓	✓	✓	✓
Brake lines & hoses	S/I	✓	✓	✓	✓	✓	✓	✓	✓	✓	✓	✓	✓	✓
Check & rotate tires	S/I	✓	✓	✓	✓	✓	✓	✓	✓	✓	✓	✓	✓	✓
Engine coolant	S/I	✓	✓	✓	✓	✓	✓	✓	✓	✓	✓	✓	✓	✓
Exhaust system	S/I	✓	✓	✓	✓	✓	✓	✓	✓	✓	✓	✓	✓	✓
Lubricate accelerator linkage	S/I	✓	✓	✓	✓	✓	✓	✓	✓	✓	✓	✓	✓	✓
Starter safety switch	S/I	✓	✓	✓	✓	✓	✓	✓	✓	✓	✓	✓	✓	✓
Steering operation	S/I	✓	✓	✓	✓	✓	✓	✓	✓	✓	✓	✓	✓	✓
Lubricate front & rear propeller shaft	S/I	✓		✓		✓		✓		✓		✓		✓
Propeller shaft flange torque 46 lb. ft. (63 Nm)	S/I	✓		✓					✓			✓		✓
Accelerator linkage	S/I		✓		✓		✓		✓		✓		✓	
Auto cruise control linkage & hose	S/I		✓		✓		✓		✓		✓		✓	
Brake pedal play	S/I		✓		✓		✓		✓		✓		✓	
Clutch pedal play	S/I		✓		✓		✓		✓		✓		✓	
Cooling & heater hoses	S/I		✓		✓		✓		✓		✓		✓	
Disc brakes	S/I		✓		✓		✓		✓		✓		✓	
Front & rear axle oil	R		✓		✓				✓				✓	
Lubricate clutch pedal spring, bushing & clevis pin	S/I		✓		✓		✓		✓		✓		✓	
Lubricate key lock cylinder	S/I		✓		✓		✓		✓		✓		✓	
Parking brake	S/I		✓		✓		✓		✓		✓		✓	
Suspension & steering	S/I		✓		✓		✓		✓		✓		✓	
Manual transmission & transfer case oil	R		✓		✓				✓				✓	
Air cleaner filter	R				✓				✓				✓	

SCHEDULED MAINTENANCE INTERVALS
(ACURA SLX) (Cont.)

TO BE SERVICED	TYPE OF SERVICE	VEHICLE MILEAGE INTERVAL (x1000)												
		7.5	15	22.5	30	37.5	45	52.5	60	67.5	75	82.5	90	97.5
Engine coolant	R				✓				✓				✓	
Power steering fluid	R				✓				✓				✓	
Clutch lines & hose	S/I				✓				✓				✓	
Engine drive belt	S/I				✓				✓				✓	
Front wheel bearings	S/I				✓				✓				✓	
Clean radiator core & A/C condenser	S/I								✓				✓	
Spark plugs	R								✓					
Timing belt	R								✓					
Fuel tank, cap & lines	S/I								✓					

R – Replace   S/I – Service or Inspect

FREQUENT OPERATION MAINTENANCE (SEVERE SERVICE)

If a vehicle is operated under any of the following conditions it is considered severe service:
- **Extremely dusty areas.**
- **50% or more of the vehicle operation is in 32°C (90°F) or higher temperatures, or constant operation in temperatures below 0°C (32°F).**
- **Prolonged idling (vehicle operation in stop and go traffic).**
- **Frequent short running periods (engine does not warm to normal operating temperatures).**
- **Police, taxi, delivery usage or trailer towing usage.**

Oil & oil filter change – change every 3000 miles.
Front & rear axle oil - change every 15,000 miles.
Automatic transmission fluid & filter - change every 20,000 miles.

GEO
Tracker

ENGINE IDENTIFICATION

Year	Model		Engine Displacement Liters (cc)	Engine Series (ID/VIN)	Fuel System	No. of Cylinders	Engine Type
1993	Tracker		1.6 (1590)	U	TFI	4	SOHC
1994	Tracker		1.6 (1590)	U	TFI	4	SOHC
	Tracker	①	1.6 (1590)	U	MFI	4	SOHC
1995	Tracker		1.6 (1590)	U	TFI	4	SOHC
	Tracker	①	1.6 (1590)	6	MFI	4	SOHC
1996-97	Tracker		1.6 (1590)	6	MFI	4	SOHC

TFI - Throttle body fuel injection
MFI - Multiport fuel injection
SOHC - Single overhead camshaft
① California and New York models

GENERAL ENGINE SPECIFICATIONS

Year	Engine ID/VIN		Engine Displacement Liters (cc)	Fuel System Type	Net Horsepower @ rpm	Net Torque @ rpm (ft. lbs.)	Bore x Stroke (in.)	Compression Ratio	Oil Pressure @ rpm
1993	U		1.6 (1590)	TFI	80@5400	94@3000	2.95x3.54	8.9:1	51-62@3000
1994	U		1.6 (1590)	TFI	80@5400	94@3000	2.95x3.54	8.9:1	47-61@3000
	U	①	1.6 (1590)	MFI	95@5600	98@4000	2.95x3.54	8.9:1	47-61@3000
1995	U		1.6 (1590)	TFI	80@5400	94@3000	2.95x3.54	8.9:1	47-61@4000
	6	①	1.6 (1590)	MFI	95@5600	98@4000	2.95x3.54	9.5:1	47-61@4000
1996-97	6		1.6 (1590)	MFI	95@5600	98@4000	2.95x3.54	9.5:1	47-61@4000

TFI - Throttle body fuel injection
MFI - Multiport fuel injection
① California and New York models

GASOLINE ENGINE TUNE-UP SPECIFICATIONS

Year	Engine ID/VIN		Engine Displacement Liters (cc)	Spark Plugs Gap (in.)	Ignition Timing (deg.) MT	AT	Fuel Pump (psi)	Idle Speed (rpm) MT	AT	Valve Clearance In.		Ex.
1993	U		1.6 (1590)	0.030	8B	8B	34-40	800	800	0.0090-0.0110	①	0.0102-0.0118
1994	U		1.6 (1590)	0.030	8B	8B	34-40	800	800	0.0090-0.0110	①	0.0102-0.0118
	U	②	1.6 (1590)	0.030	8B	8B	36-43	800	800	0.0050-0.0070		0.0050-0.0070

GASOLINE ENGINE TUNE-UP SPECIFICATIONS

Year	Engine ID/VIN	Engine Displacement Liters (cc)	Spark Plugs Gap (in.)	Ignition Timing (deg.) MT	Ignition Timing (deg.) AT	Fuel Pump (psi)	Idle Speed (rpm) MT	Idle Speed (rpm) AT	Valve Clearance In.	Valve Clearance Ex.
1995	U	1.6 (1590)	0.030	8B ⑤	8B ⑤	34-41	800-850	800-850	0.0090- 0.0110 ①	0.0102- 0.0114 ①
	6 ②	1.6 (1590)	0.030	8B ⑤	8B ⑤	30-37	800-850	800-850	0.0050- 0.0070	0.0050- 0.0070
1996-97	6	1.6 (1590)	0.030	5B ③	5B ③	30-37	800-850	800-850	0.0050- 0.0070 ④	0.0050- 0.0070 ④

NOTE: The Vehicle Emission Control Information label often reflects specification changes made during production. The label figures must be used if they differ from those in this chart.

B - Before top dead center

HYD - Hydraulic

① Specifications for hot engine.Cold adjustment set valve lash:
   Intake: 0.0051-0.0067
   Exhaust: 0.0063-0.0073

② California and New York models

③ Connect fused jumper from Duty Check Cavity 4 to cavity 5 for fixed timing
   (DLC connector located at left strut tower)

④ Cold settings

⑤ Connect jumper wire from Duty Check DLC cavity 3 to 4
   (DLC connector located next to battery)

CAPACITIES

Year	Model	Engine ID/VIN	Engine Displacement Liters (cc)	Engine Oil with Filter (qts.)	Transmission (pts.) 4-Spd	Transmission (pts.) 5-Spd	Transmission (pts.) Auto.	Transfer Case (pts.)	Drive Axle Front (pts.)	Drive Axle Rear (pts.)	Fuel Tank (gal.)	Cooling System (qts.)
1993	Tracker	U	1.6 (1590)	4.5	-	3.2	10.2 ①	3.6	3.6	2.2	11.1	②
1994	Tracker	U	1.6 (1590)	4.5	-	3.2	10.2 ①	3.6	3.6	2.2	11.1	②
1995	Tracker	U	1.6 (1590)	4.5	-	3.2	9.2 ①	1.8	2.4	2.2	11.1	②
	Tracker ③	6	1.6 (1590)	4.5	-	3.2	9.2 ①	3.6	2.4	2.2	11.1	②
1996-97	Tracker	6	1.6 (1590)	4.5	-	3.2	10.6	3.6	2.4	4.6	11.0	②

① Automatic transmission - Specification is after complete overhaul. Drain and fill will be less

② Manual transmission: 5.6 qts.
   Automatic transmission: 5.5 qts.

③ California and New York

VALVE SPECIFICATIONS

Year	Engine ID/VIN	Engine Displacement Liters (cc)	Seat Angle (deg.)	Face Angle (deg.)	Spring Test Pressure (lbs. @ in.)	Spring Installed Height (in.)	Stem-to-Guide Clearance (in.) Intake	Stem-to-Guide Clearance (in.) Exhaust	Stem Diameter (in.) Intake	Stem Diameter (in.) Exhaust
1993	U	1.6 (1590)	45	45	54.7-64.3@ 1.63	1.91	0.0008- 0.0019	0.0014- 0.0025	0.2742- 0.2748	0.2737- 0.2742
1994	U	1.6 (1590)	45	45	50.2-64.3@ 1.63	1.63	0.0008- 0.0019	0.0014- 0.0025	0.2742- 0.2748	0.2737- 0.2742
	U ①	1.6 (1590)	45	45	23.6-27.5@ 1.63	1.24	0.0008- 0.0018	0.0018- 0.0028	0.2152- 0.2157	0.2142- 0.2148
1995	U	1.6 (1590)	45	45	50.2-64.3@ 1.63	1.63	0.0008- 0.0019	0.0014- 0.0025	0.2742- 0.2748	0.2737- 0.2742
	6 ①	1.6 (1590)	45	45	23.6-27.5@ 1.24	1.24	0.0008- 0.0018	0.0018- 0.0028	0.2152- 0.2157	0.2142- 0.2148
1996-97	6	1.6 (1590)	45	45	23.6-27.5@ 1.24	1.24	0.0008- 0.0018	0.0018- 0.0028	0.2152- 0.2157	0.2142- 0.2148

① California and New York models

TORQUE SPECIFICATIONS
All readings in ft. lbs.

Year	Engine ID/VIN	Engine Displacement Liters (cc)	Cylinder Head Bolts	Main Bearing Bolts	Rod Bearing Bolts	Crankshaft Damper Bolts	Flywheel Bolts	Manifold Intake	Manifold Exhaust	Spark Plugs	Lug Nut
1993	U	1.6 (1590)	54	39	26	81 ①	58	17	17	18	87
1994	U	1.6 (1590)	52 ②	40	26	81 ①	58	17	17	18	87
1995	6 ③	1.8 (1803)	④	44	⑥	87 ①	⑤	14	25	21	76
	U	1.6 (1590)	②	40	40	81 ①	58	17	17	21	70
1996-97	6	1.6 (1590)	②	40	26	44	58	17	17	21	70

① Crankshaft timing belt sprocket
② Step 1: 26 ft. lbs.
   Step 2: 41 ft. lbs.
   Step 3: 52 ft. lbs.

③ California and New York
④ Step 1: 22 ft. lbs.
   Step 2: Plus two additional steps of 90 degrees

⑤ Manual transaxle: 58 ft. lbs.
   Automatic transaxle: 47 ft. lbs.
⑥ Step 1: 18 ft. lbs.
   Step 2: Plus 90 degrees

BRAKE SPECIFICATIONS
All measurements in inches unless noted

Year	Model	Master Cylinder Bore	Brake Disc Original Thickness	Brake Disc Minimum Thickness	Brake Disc Maximum Runout	Brake Drum Diameter Original Inside Diameter	Max. Wear Limit	Maximum Machine Diameter	Minimum Lining Thickness Front	Minimum Lining Thickness Rear
1993	Tracker	NA	0.394	0.315	0.006	8.66	8.74	8.74	0.236 ①	0.210 ①
1994	Tracker	NA	0.394	0.315	0.006	8.66	8.74	8.74	0.236 ①	0.210 ①
1995	Tracker	NA	0.394	0.315	0.006	8.66	8.74	8.74	0.236 ①	0.210 ①
1996-97	Tracker	NA	②	0.315	0.006	③	8.74	8.74	0.236 ③	0.210 ①

NA - Not Available
① Minimum lining thickness includes pad/shoe backing
② 2 door: 0.394
   4 door: 0.670 (service limit 0.590)
③ 2 door: 8.66 (service limit 0.874)
   4 door: 10.00 (service limit 10.07)

FREQUENT MAINTENANCE LABOR
GEO TRACKER

The following should be used as a guide when determining the amount of work required for a particular service if taken to a repair shop. In estimating how long a particular Frequent Maintenance Service item should take, please observe the following:
- **Chilton Time** is time that is based on field research and data supplied by the vehicle manufacturer.
- All labor time operations are given in hours and tenths of an hour.
- All labor operations, are to be used as a **guide**.

	(Factory Time)	Chilton Time

COOLING

(G) Winterize Cooling System
Includes: Run engine to check for leaks, tighten all hose connections. Test radiator and pressure cap, drain radiator and engine block. Add antifreeze and refill system.
1993-97 5

(G) Belt, Drive, Renew
1993-97
code U
AC .3
Fan or Alternator 5
PS .3
code 6
one .3
each adtnl. 1

(G) Belt, Drive, Adjust
1993-97
one .3
all 4

(G) Hoses, Radiator, Renew
Includes: Drain and refill cooling system as required.
1993-97
upper 4
lower
radiator to outlet pipe 5
outlet to inlet pipe 5
both 7

(G) Thermostat, Coolant, Renew
1993-97 4

FUEL

(M) Air Cleaner, Service
1993-97 2

(G) Filter, Fuel, Renew
1993-97 4

BRAKES

(G) Bleed Brakes (Four Wheels)
Includes: Add fluid.
1993-97 5

(G) Brakes, Adjust (Minor)
Includes: Adjust brakes, fill master cylinder.
1993-97, two wheels 4
Remove knock out plugs add,
each 1

(M) Parking Brake, Adjust
1993-97 4

LUBRICATION SERVICES

(M) Engine Oil & Filter, Renew
Includes: Inspect and correct all fluid levels.
1993-97 3

(M) Lubricate Chassis, Change Oil & Filter
Includes: Inspect and correct all fluid levels.
1993-97 6
Install grease fittings add 1

(M) Lubricate Chassis
Includes: Inspect and correct all fluid levels.
1993-97 4
Install grease fittings add 1

ELECTRICAL

(G) Headlamps, Aim
1993-97
two .4
four 6

(G) High Mount Stop Lamp Bulb, Renew
1993-97 3

(G) License Lamp Bulb, Renew
1993-97, one or all 2

(G) Park & Turn Signal Lamp Assy., Renew
1993-97, each 5

(G) Park & Turn Signal Lamp Bulb or Lens, Renew
1993-97
one .3
all 4

(G) Rear Combination Lamp Assy., Renew
1993-97, each 5

(G) Stop, Tail & Turn Signal Lamp Bulb, Renew
1993-97
one .2
each adtnl. 1

(G) Horn, Renew
1993-97 3

(M) Terminals, Battery, Clean
1993-97 3

SCHEDULED MAINTENANCE INTERVALS
(GEO TRACKER)

TO BE SERVICED	TYPE OF SERVICE	VEHICLE MILEAGE INTERVAL (x1000)												
		7.5	15	22.5	30	37.5	45	52.5	60	67.5	75	82.5	90	97.5
Engine oil & filter	R	✓	✓	✓	✓	✓	✓	✓	✓	✓	✓	✓	✓	✓
Automatic transmission fluid⑤	S/I	✓	✓	✓	✓	✓	✓	✓	✓	✓	✓	✓	✓	✓
Disc brake pads, rotors, drum brake linings, drums, wheel cylinders & parking brake	S/I	✓	✓	✓	✓	✓	✓	✓	✓	✓	✓	✓	✓	✓
Exhaust system	S/I	✓	✓	✓	✓	✓	✓	✓	✓	✓	✓	✓	✓	✓
Free-wheeling hubs	S/I	✓	✓	✓	✓	✓	✓	✓	✓	✓	✓	✓	✓	✓
Locking front hubs	S/I	✓	✓	✓	✓	✓	✓	✓	✓	✓	✓	✓	✓	✓
Manual transmission/transfer case fluids①	S/I	✓	✓	✓	✓	✓	✓	✓	✓	✓	✓	✓	✓	✓
Rotate tires	S/I	✓	✓	✓	✓	✓	✓	✓	✓	✓	✓	✓	✓	✓
Steering & suspension	S/I	✓	✓	✓	✓	✓	✓	✓	✓	✓	✓	✓	✓	✓
Throttle linkage	S/I	✓	✓	✓	✓	✓	✓	✓	✓	✓	✓	✓	✓	✓
Adjust valve lash	S/I		✓		✓		✓		✓		✓		✓	
Engine idle speed	S/I		✓		✓		✓		✓		✓		✓	
Propeller shafts & U-joints	S/I		✓		✓		✓		✓		✓		✓	
Air cleaner filter	R				✓				✓				✓	
Camshaft timing belt	R								✓					
Fuel filter	R				✓				✓				✓	
Spark plugs	R				✓				✓				✓	
Engine coolant	R				✓				✓				✓	
Engine accessory drive belt③	S/I				✓				✓				✓	
Front wheel bearings	S/I				✓				✓				✓	
Fuel tank, cap & lines	S/I				✓				✓				✓	
Brake fluid	R								✓					
Fuel tank cap gasket	R								✓					
Ignition wires	R								✓					
Emission system hoses	S/I								✓					
Engine timing	S/I								✓					

SCHEDULED MAINTENANCE INTERVALS
(GEO TRACKER) (Cont.)

TO BE SERVICED	TYPE OF SERVICE	VEHICLE MILEAGE INTERVAL (x1000)												
		7.5	15	22.5	30	37.5	45	52.5	60	67.5	75	82.5	90	97.5
EVAP canister④	R													
PCV valve②	R													
Fuel injectors⑥	S/I													

① Replace every 30,000 miles.
② Replace every 50,000 miles.
③ Replace every 60,000 miles.
③ Replace every 100,000 miles.
⑤ Replace every 100,000 miles (1993-94)
⑥ Service or inspect every 100,000 miles.
R – Replace   S/I – Service or Inspect

FREQUENT OPERATION MAINTENANCE (SEVERE SERVICE)
  If a vehicle is operated under any of the following conditions it is considered severe service:
- Extremely dusty areas.
- 50% or more of the vehicle operation is in 32°C (90°F) or higher temperatures, or constant operation in temperatures below 0°C (32°F).
- Prolonged idling (vehicle operation in stop and go traffic).
- Frequent short running periods (engine does not warm to normal operating temperatures).
- Police, taxi, delivery usage or trailer towing usage.
Oil & oil filter change – change every 3000 miles.
Free-wheeling hubs - service or inspect every 3000 miles.
Rotate tires every 6000 miles.
Air cleaner filter - service or inspect every 15,000 miles.
Repack front wheel bearings every 15,000 miles.
Manual transmission fluid - change every 15,000 miles.
Engine idle speed - check every 15,000 miles.
Propeller shafts & U-joints - service or inspect every 15,000 miles.
Automatic transmission fluid & filter - change every 15,000 miles (1993-94) or every 50,000 miles (1995-97).

HONDA
Passport • Odyssey

ENGINE IDENTIFICATION
All measurements are given in inches.

Year	Model	Engine Displacement Liters (cc)	Engine Series (ID/VIN)	Fuel System	No. of Cylinders	Engine Type
1995	Odyssey	2.2 (2156)	F22B6	MFI	4	SOHC 16V
	Passport	2.6 (2559)	4ZE1/E	MFI	4	SOHC 8V
	Passport	3.2 (3165)	6VD1/V	MFI	6	SOHC 24V
1996-97	Odyssey	2.2 (2156)	F22B6	MFI	4	SOHC 16V
	Passport	2.6 (2559)	4ZE1/E	MFI	4	SOHC 8V
	Passport	3.2 (3165)	6VD1/V	MFI	6	SOHC 24V

MFI - Multipoint fuel injection
SOHC - Single overhead camshaft
DOHC - Double overhead camshaft

GENERAL ENGINE SPECIFICATIONS

Year	Engine ID/VIN	Engine Displacement Liters (cc)	Fuel System Type	Net Horsepower @ rpm	Net Torque @ rpm (ft. lbs.)	Bore x Stroke (in.)	Compression Ratio	Oil Pressure @ rpm
1995	F22B6	2.2 (2156)	MFI	140@5600	145@4600	3.35x3.74	8.8:1	50@3000
	4ZE1	2.6 (2559)	MFI	120@4600	150@2600	3.65x3.74	8.6:1	57-71@4000
	6VD1	3.2 (3165)	MFI	175@5200	188@4000	3.68x3.03	9.3:1	57-80@3000
1996-97	F22B6	2.2 (2156)	MFI	140@5600	145@4600	3.35x3.74	8.8:1	50@3000
	4ZE1	2.6 (2559)	MFI	120@4600	150@2600	3.65x3.74	8.6:1	57-80@3000
	6VD1	3.2 (3165)	MFI	190@5600	188@4000	3.68x3.03	9.0:1	57-80@3000

MFI - Multiport fuel injection

GASOLINE ENGINE TUNE-UP SPECIFICATIONS

Year	Engine ID/VIN	Engine Displacement Liters (cc)	Spark Plugs Gap (in.)	Ignition Timing (deg.) MT	Ignition Timing (deg.) AT	Fuel Pump (psi)	Idle Speed (rpm) MT	Idle Speed (rpm) AT	Valve Clearance In.	Valve Clearance Ex.
1995	F22B6	2.2 (2156)	0.039-0.043	-	15B	36-43	-	650-750	0.009-0.010	0.011-0.013
	4ZE1	2.6 (2559)	0.039-0.043	①	-	35	850-950	-	0.006	0.001
	6VD1	3.2 (3156)	0.040	NA	NA	25-30	750	750	NA	NA

23 SPECIFICATIONS

GASOLINE ENGINE TUNE-UP SPECIFICATIONS

Year	Engine ID/VIN	Engine Displacement Liters (cc)	Spark Plugs Gap (in.)	Ignition Timing (deg.) MT	Ignition Timing (deg.) AT	Fuel Pump (psi)	Idle Speed (rpm) MT	Idle Speed (rpm) AT	Valve Clearance in.	Valve Clearance Ex.
1996-97	F22B6	2.2 (2156)	0.039-0.043	-	15B	30-37	-	650-750	0.009-0.011	0.011-0.013
	4ZE1	2.6 (2559)	0.040	①	-	35	900 ②	-	0.006	0.010
	6VD1	3.2 (3156)	0.040	5B ②	5B ②	41-46	750 ②	750 ②	HLA	HLA

NOTE: The Vehicle Emission Control Information label often reflects specification changes made during production. The label figures must be used if they differ from those in this chart.
NOTE: The fuel pressure readings are given with the vacuum hose connected to the regulator and the engine running
B - Before to
HLA - Hydraulic Lash Adjuster
① 12B at 900 rpm
② Controlled by ECM and is not adjustable

CAPACITIES

Year	Model	Engine ID/VIN	Engine Displacement Liters (cc)	Engine Oil with Filter (qts.)	Transmission (pts.) 4-Spd	Transmission (pts.) 5-Spd	Transmission (pts.) Auto.	Transfer Case (pts.)	Drive Axle Front (pts.)	Drive Axle Rear (pts.)	Fuel Tank (gal.)	Cooling System (qts.)
1995	Odyssey	F22B6	2.2 (2156)	4.0	-	-	5.0	-	-	-	17.2	6.7
	Passport	4ZE1	2.6 (2559)	4.4	-	4.5	-	3.0	3.2	①	21.9	9.5
	Passport	6VD1	3.2 (3165)	5.7	-	5.9	9.1	3.0	3.2	①	21.9	②
1996-97	Odyssey	F22B6	2.2 (2156)	4.0	-	-	5.0	-	-	-	17.2	6.7
	Passport	4ZE1	2.6 (2559)	4.4	-	4.5	-	3.0	3.2	①	21.9	9.5
	Passport	6VD1	3.2 (3165)	5.7	-	5.9	18.2	3.0	3.2	①	21.9	②

NOTE: Capacities given are service, not overhaul capacities
① Saginaw: 4.0
   Dana: 3.8
② Automatic transmission: 9.3
   Manual transmission: 9.7

VALVE SPECIFICATIONS

Year	Engine ID/VIN	Engine Displacement Liters (cc)	Seat Angle (deg.)	Face Angle (deg.)	Spring Test Pressure (lbs. @ in.)	Spring Installed Height (in.)	Stem-to-Guide Clearance (in.) Intake	Stem-to-Guide Clearance (in.) Exhaust	Stem Diameter (in.) Intake	Stem Diameter (in.) Exhaust
1995	F22B6	2.2 (2156)	45	45	NA	NA	0.0008-0.0030	0.0022-0.0050	0.2148-0.2163	0.2134-0.2150
	4ZE1	2.6 (2559)	45	45	45-55@1.61	1.610	0.0009-0.0079	0.0015-0.0098	0.3120-0.3134	0.3091-0.3124
	6VD1	3.2 (3165)	45	45	45-55@1.54	1.540	0.0009-0.0079	0.0012-0.0079	0.2323-0.2353	0.2323-0.2350
1996-97	F22B6	2.2 (2156)	45	45	NA	NA	0.0008-0.0018	0.0022-0.0050	0.2159-0.2163	0.2146-0.2150
	4ZE1	2.6 (2559)	45	45	45-55@1.61	1.610	0.0009-0.0079	0.0015-0.0098	0.3120-0.3134	0.3091-0.3124
	6VD1	3.2 (3165)	45	45	45-55@1.54	1.540	0.0009-0.0079	0.0012-0.0079	0.2323-0.2353	0.2323-0.2350

NA - Not Available

TORQUE SPECIFICATIONS
All readings in ft. lbs.

Year	Engine ID/VIN	Engine Displacement Liters (cc)	Cylinder Head Bolts	Main Bearing Bolts	Rod Bearing Bolts	Crankshaft Damper Bolts	Flywheel Bolts	Manifold Intake *	Manifold Exhaust	Spark Plugs	Lug Nut
1995	F22B6	2.2 (2156)	①	②	34	181	54	16	23	13	80
	4ZE1/E	2.6 (2559)	③	72	43	87	40	16	33	14	④
	6VD1/V	3.2 (3165)	⑤	29	49	123	40	17	42	13	④
1996-97	F22B6	2.2 (2156)	①	②	34	181	54	16	23	13	80
	4ZE1/E	2.6 (2559)	③	72	43	87	40	16	33	14	④
	6VD1/V	3.2 (3165)	⑤	⑥	40	123	40	17	42	13	87

NOTE: Dip main bearing bolts and crankshaft damper bolt in clean engine oil

* NOTE: Applies to Lower Manifold only.

① Step 1: 29 ft. lbs.
    Step 2: 51 ft. lbs.
    Step 3: 72 ft. lbs.

② Step 1: 22 ft. lbs.
    Step 2: 54 ft. lbs.

③ Step 1: 58 ft. lbs.
    Step 2: 72 ft. lbs.

④ Steel wheels: 66 ft. lbs.
    Aluminum wheels: 87 ft. lbs.

⑤ M8x1.25 bolts: 15 ft. lbs.
    M11x1.50 bolts: 47 ft. lbs.

⑥ Main bearing cap bolts: 29 ft. lbs.
    Oil gallery bolts: 29 ft. lbs. plus 55-65 degrees
    Buttress bolts: 29 ft. lbs.

BRAKE SPECIFICATIONS
All measurements in inches unless noted

Year	Model		Master Cylinder Bore	Brake Disc Original Thickness	Brake Disc Minimum Thickness	Maximum Runout	Brake Drum Diameter Original Inside Diameter	Brake Drum Diameter Max. Wear Limit	Brake Drum Diameter Maximum Machine Diameter	Minimum Lining Thickness Front	Minimum Lining Thickness Rear
1995	Odyssey	F	NA	0.930	0.830	0.004	-	-	-	0.060	-
		R	-	0.350	0.300	0.004	6.69	6.73	6.73	-	0.040 ①
	Passport	F	1.000	②	③	0.005	-	-	-	0.039	-
		R	-	0.710	0.654	0.005	10.00 ④	10.59 ④	10.059	-	0.039 ⑤
1996-97	Odyssey	F	NA	0.937	0.830	0.004	-	-	-	0.060	-
		R	-	0.358	0.300	0.004	6.69	6.73	6.73	-	0.040
	Passport	F	1.000	1.020	0.983	0.005	-	-	-	0.039	-
		R	-	0.710	0.654	0.005	10.00 ④	10.06 ④	10.06	-	0.039 ⑤

NA - Not Available
F - Front
R - Rear

① Rear disc brakes: 0.060

② 3.2L engine: 1.020
    2.6L engine: 0.866

③ 3.2L engine: 0.983
    2.6L engine: 0.811

④ Parking brake drum:
    Original inside diameter: 8.27
    Maximum wear limit: 8.32

⑤ Specifications include disc pads and parking brake shoes

FREQUENT MAINTENANCE LABOR
ODYSSEY

The following should be used as a guide when determining the amount of work required for a particular service if taken to a repair shop. In estimating how long a particular Frequent Maintenance Service item should take, please observe the following:
- **Chilton Time** is time that is based on field research and data supplied by the vehicle manufacturer.
- All labor time operations are given in hours and tenths of an hour.
- All labor operations, are to be used as a **guide**.

COOLING

(G) Winterize Cooling System
Includes: Run engine to check for leaks, tighten all hose connections. Test radiator and pressure cap, drain radiator and engine block. Add antifreeze and refill system.
1994-97 5

(G) Hoses, Radiator, Renew
Includes: Drain and refill cooling system as required.
1994-97
upper 5
lower 6
by-pass 6

(G) Thermostat, Coolant, Renew
1995-97 Odyssey 8

FUEL

(G) Filter, Fuel, Renew
1995-97 Odyssey 5

BRAKES

(G) Bleed Brakes (Four Wheels)
Includes: Add fluid.
1994-97 5

(G) Brakes, Adjust (Minor)
Includes: Adjust brakes, fill master cylinder.
1994-97 4

(M) Parking Brake, Adjust
1994-97 4

LUBRICATION SERVICES

(M) Engine Oil & Filter, Renew
Includes: Inspect and correct all fluid levels.
1994-97 3

(M) Lubricate Chassis, Change Oil & Filter
Includes: Inspect and correct all fluid levels.
1994-97 6
Install grease fittings add 1

(M) Lubricate Chassis
Includes: Inspect and correct all fluid levels.
1994-97 4
Install grease fittings add 1

WHEELS

(G) Wheel, Renew (One)
1994-97 5

(G) Wheel, Balance
1994-97
one 3
each adtnl. 2

(G) Wheels, Rotate (All)
1994-97 5

ELECTRICAL

(G) Headlamps, Aim
1995-97 Odyssey 4

(G) Halogen Headlamp Bulb, Renew
1994-97, each 4

(G) Parking Lamp Lens or Bulb, Renew
1994-97, one 3

(G) Tail Lamp Lens or Bulb, Renew
1995-97 Odyssey 2

(G) Horn, Renew
1995-97 Odyssey 3

(M) Terminals, Battery, Clean
1994-97 3

FREQUENT MAINTENANCE LABOR
PASSPORT

The following should be used as a guide when determining the amount of work required for a particular service if taken to a repair shop. In estimating how long a particular Frequent Maintenance Service item should take, please observe the following:
- **Chilton Time** is time that is based on field research and data supplied by the vehicle manufacturer.
- All labor time operations are given in hours and tenths of an hour.
- All labor operations, are to be used as a **guide**.

COOLING

Chilton Time

(G) Winterize Cooling System
Includes: Run engine to check for leaks, tighten all hose connections. Test radiator and pressure cap, drain radiator and engine block. Add antifreeze and refill system.
1994-97 5

(G) Hoses, Radiator, Renew
Includes: Drain and refill cooling system as required.
1994-97
  upper or lower, one 4
  both 8

(G) Thermostat, Coolant, Renew
1994-97 5

FUEL

(G) Filter, Fuel, Renew
1994-97 3

BRAKES

Chilton Time

(G) Bleed Brakes (Four Wheels)
Includes: Add fluid.
1994-97 5

(G) Brakes, Adjust (Minor)
Includes: Adjust brakes, fill master cylinder.
1994-97 4

(M) Parking Brake, Adjust
1994-97
  drum brakes 5
  disc brakes 3

LUBRICATION SERVICES

(M) Engine Oil & Filter, Renew
Includes: Inspect and correct all fluid levels.
1994-97 3

ELECTRICAL

Chilton Time

(G) Headlamps, Aim
1994-97
  one side 4
  both sides 6

(G) Halogen Headlamp Bulb, Renew
1994-97 3

(G) High Mount Stop Lamp Bulb, Renew
1994-97 3

(G) Park & Turn Signal Lamp Bulb or Lens, Renew
1994-97
  one 3
  each adtnl. 2

(G) Horn, Renew
1994-97, one or all 6

(M) Terminals, Battery, Clean
1994-97 3

SCHEDULED MAINTENANCE INTERVALS
(HONDA PASSPORT & ODYSSEY)

TO BE SERVICED	TYPE OF SERVICE	VEHICLE MILEAGE INTERVAL (x1000)												
		7.5	15	22.5	30	37.5	45	52.5	60	67.5	75	82.5	90	97.5
Engine oil & filter	R	✓	✓	✓	✓	✓	✓	✓	✓	✓	✓	✓	✓	✓
Suspension & steering	S/I	✓	✓	✓	✓	✓	✓	✓	✓	✓	✓	✓	✓	✓
Automatic transmission fluid	S/I	✓	✓	✓	✓	✓	✓	✓	✓	✓	✓	✓	✓	✓
Body & chassis lubrication	S/I	✓	✓	✓	✓	✓	✓	✓	✓	✓	✓	✓	✓	✓
Brake & clutch fluid level	S/I	✓	✓	✓	✓	✓	✓	✓	✓	✓	✓	✓	✓	✓
Brake lines & hoses	S/I	✓	✓	✓	✓	✓	✓	✓	✓	✓	✓	✓	✓	✓
Check & rotate tires	S/I	✓	✓	✓	✓	✓	✓	✓	✓	✓	✓	✓	✓	✓
Driveshaft boots	S/I	✓	✓	✓	✓	✓	✓	✓	✓	✓	✓	✓	✓	✓
Engine coolant	S/I	✓	✓	✓	✓	✓	✓	✓	✓	✓	✓	✓	✓	✓
Exhaust system	S/I	✓	✓	✓	✓	✓	✓	✓	✓	✓	✓	✓	✓	✓
Lubricate accelerator linkage	S/I	✓	✓	✓	✓	✓	✓	✓	✓	✓	✓	✓	✓	✓
Starter safety switch	S/I	✓	✓	✓	✓	✓	✓	✓	✓	✓	✓	✓	✓	✓
Parking brake	S/I	✓	✓	✓	✓	✓	✓	✓	✓	✓	✓	✓	✓	✓
Lubricate front & rear propeller shaft	S/I	✓		✓		✓		✓		✓		✓		✓
Propeller shaft flange torque 46 lb. ft. (63 Nm)	S/I	✓		✓		✓		✓		✓		✓		✓
Auto cruise control linkage & hose	S/I		✓		✓		✓		✓		✓		✓	
Brake & clutch pedal play	S/I		✓		✓		✓		✓		✓		✓	
Cooling & heater hoses	S/I		✓		✓		✓		✓		✓		✓	
Disc & drum brakes	S/I		✓		✓		✓		✓		✓		✓	
Lubricate clutch pedal spring, bushing & clevis pin	S/I		✓		✓		✓		✓		✓		✓	
Shift on-the-fly system gear fluid	S/I		✓		✓		✓		✓		✓		✓	
Throttle linkage	S/I		✓		✓		✓		✓		✓		✓	
Valve clearance (2.2L & 2.6L)	S/I		✓		✓		✓		✓		✓		✓	
Automatic transmission fluid & filter (Odyssey)	R				✓				✓				✓	

SCHEDULED MAINTENANCE INTERVALS
(HONDA PASSPORT & ODYSSEY) (Cont.)

TO BE SERVICED	TYPE OF SERVICE	VEHICLE MILEAGE INTERVAL (x1000)												
		7.5	15	22.5	30	37.5	45	52.5	60	67.5	75	82.5	90	97.5
Manual transmission & transfer case oil	R		✓		✓				✓					
Air cleaner filter	R				✓				✓				✓	
Power steering fluid	R				✓				✓				✓	
Clutch lines & hose	S/I				✓				✓				✓	
Engine drive belts	S/I				✓				✓				✓	
Engine idle speed (2.6L)	S/I	✓			✓				✓				✓	
Front wheel bearings & free wheeling hubs	S/I				✓				✓				✓	
Front & rear axle oil	R		✓		✓				✓					
Steering gear play	S/I				✓				✓				✓	
Brake fluid (include. ABS) (Odyssey)	R				✓				✓				✓	
Spark plugs (2.2L & 2.6L)	R				✓				✓				✓	
Spark plugs (3.2L)	R								✓					
Engine coolant	R				✓				✓				✓	
Distributor cap, rotor & ignition wires (2.2L & 2.6L)	S/I								✓				✓	
Oxygen sensor (1993-95)	R												✓	
Timing belt①	R													
PCV valve (Odyssey)	S/I								✓					
Clean radiator core & A/C condenser	S/I								✓					
Fuel tank, cap & lines	S/I								✓					

① Replace timing belt for Odyssey at 90,000 miles or for Passport at 60,000 miles.

R – Replace   S/I – Service or Inspect

FREQUENT OPERATION MAINTENANCE (SEVERE SERVICE)

If a vehicle is operated under any of the following conditions it is considered severe service:
- Extremely dusty areas.
- 50% or more of the vehicle operation is in 32°C (90°F) or higher temperatures, or constant operation in temperatures below 0°C (32°F).
- Prolonged idling (vehicle operation in stop and go traffic).
- Frequent short running periods (engine does not warm to normal operating temperatures).
- Police, taxi, delivery usage or trailer towing usage.

Oil & oil filter change – change every 3000 miles.
Front & rear axle oil - change every 15,000 miles.
Automatic transmission fluid & filter - change every 20,000 miles.

ISUZU

Amigo • Hombre • Oasis • Pick Up • Rodeo • Trooper

VEHICLE IDENTIFICATION CHART

Engine Code							Model Year	
Code	Liters	Cu. In. (cc)	Cyl.	Fuel Sys.		Eng. Mfg.	Code	Year
E	2.6	156 (2559)	4	MFI		Isuzu	P	1993
F22B6	2.2	134 (2156)	4	MFI		Isuzu	R	1994
L	2.3	137 (2254)	4	MFI	①	Isuzu	S	1995
V	3.2	193 (3165)	6	MFI		Isuzu	T	1996
W	2.2	134 (2156)	4	MFI		CPC	V	1997
W	3.2	194 (3165)	6	MFI		Isuzu		
Z	3.1	189 (3098)	6	TFI		Isuzu		

MFI - Multiport fuel injection
TFI - Throttle body fuel injection
CPC - Chevrolet/Pontiac/Canada

① For 1994 vehicles:
    Federal: 2 barrel carburetor
    California: Multiport fuel injection

ENGINE IDENTIFICATION

Year	Model	Engine Displacement Liters (cc)	Engine Series (ID/VIN)	Fuel System	No. of Cylinders	Engine Type
1993	Amigo	2.3 (2254)	L	2BC	4	SOHC
	Amigo	2.6 (2559)	E	MFI	4	SOHC
	Pick-up	2.3 (2254)	L	2BC	4	SOHC
	Pick-up	2.6 (2559)	E	MFI	4	SOHC
	Pick-up	3.1 (3098)	Z	TFI	6	OHV
	Rodeo	2.6 (2559)	E	MFI	4	SOHC
	Rodeo	3.2 (3165)	V	MFI	6	SOHC
	Trooper	3.2 (3165)	V	MFI	6	SOHC
	Trooper	3.2 (3165)	V	MFI	6	DOHC
1994	Amigo	2.6 (2559)	E	MFI	4	SOHC
	Pick-up	2.3 (2254)	L	2BC	4	SOHC
	Pick-up	2.3 (2254)	L	MFI	4	SOHC
	Pick-up	2.6 (2559)	E	MFI	4	SOHC
	Pick-up	3.1 (3098)	Z	TFI	6	OHV
	Rodeo	2.6 (2559)	E	MFI	4	SOHC
	Rodeo	3.2 (3165)	V	MFI	6	SOHC
	Trooper	3.2 (3165)	V	MFI	6	SOHC
	Trooper	3.2 (3165)	V	MFI	6	DOHC
1995	Pick-up	2.3 (2254)	L	MFI	4	SOHC
	Pick-up	2.6 (2559)	E	MFI	4	SOHC
	Rodeo	2.6 (2559)	E	MFI	4	SOHC
	Rodeo	3.2 (3165)	V	MFI	6	SOHC
	Trooper	3.2 (3165)	V	MFI	6	SOHC
	Trooper	3.2 (3165)	V	MFI	6	DOHC

ENGINE IDENTIFICATION

Year	Model	Engine Displacement Liters (cc)	Engine Series (ID/VIN)	Fuel System	No. of Cylinders	Engine Type
1996-97	Oasis	2.2 (2156)	F22B6	MFI	4	SOHC
	Hombre	2.2 (2189)	W	MFI	4	SOHC
	Rodeo	2.6 (2559)	E	MFI	4	SOHC
	Rodeo	3.2 (3165)	V	MFI	6	SOHC
	Trooper	3.2 (3165)	V	MFI	6	SOHC

MFI - Multiport fuel injection
BC - Barrel carburetor
TFI - Throttle body fuel injection

DOHC - Double overhead camshaft
SOHC - Single overhead camshaft
OHV - Overhead valve

GENERAL ENGINE SPECIFICATIONS

Year	Engine ID/VIN		Engine Displacement Liters (cc)	Fuel System Type	Net Horsepower @ rpm	Net Torque @ rpm (ft. lbs.)	Bore x Stroke (in.)	Compression Ratio	Oil Pressure @ rpm
1993	L		2.3 (2243)	2BC	96@4600	123@2600	3.52x3.54	8.3:1	57@3000
	E		2.6 (2559)	MFI	120@4600	150@2600	3.65x3.74	8.6:1	57-71@4000
	Z		3.1 (3098)	TFI	120@4400	165@2600	3.50x3..31	8.5:1	30-55@2000
	V	①	3.2 (3165)	MFI	175@5200	188@4000	3.67x3.03	9.3:1	57-80@3000
	V	②	3.2 (3165)	MFI	190@5600	195@3800	3.67x3.03	9.3:1	57-80@3000
1994	L		2.3 (2254)	2BC	96@4600	123@2600	3.52x3.54	8.3:1	57@3000
	L		2.3 (2254)	MFI	100@4600	125@2600	3.52x3.54	8.3:1	57@3000
	E		2.6 (2559)	MFI	120@4600	150@2600	3.65x3.74	8.6:1	57-71@4000
	Z		3.1 (3098)	TFI	120@4400	165@2800	3.50x3.31	8.5:1	30-55@2000
	V	①	3.2 (3165)	MFI	175@5200	188@4000	3.67x3.03	9.3:1	57-80@3000
	V	②	3.2 (3165)	MFI	190@5600	195@3800	3.67x3.03	9.3:1	57-80@3000
1995	L		2.3 (2254)	MFI	100@4600	125@2600	3.52x3.54	8.3:1	57@3000
	E		2.6 (2559)	MFI	120@4600	150@2600	3.65x3.74	8.6:1	57-71@4000
	V	①	3.2 (3165)	MFI	175@5200	188@4000	3.67x3.03	9.3:1	57-80@3000
	V	②	3.2 (3165)	MFI	190@5600	195@3800	3.67x3.03	9.8:1	57-80@3000
1996-97	F22B6		2.2 (2156)	MFI	140@5600	145@4500	3.35x3.74	8.8:1	50@3000
	W		2.2 (2189)	MFI	118@5200	130@2800	3.50x3.46	9.0:1	56@3000
	E		2.6 (2559)	MFI	120@4600	150@2600	3.65x3.74	8.6:1	57-80@3000
	V		3.2 (3165)	MFI	190@5600	188@4000	3.68x3.03	9.1:1	57-80@3000

BC - Barrel carburetor
MFI - Multiport fuel injection
TFI - Throttle body fuel injection

① Single overhead camshaft
② Double overhead camshaft

GASOLINE ENGINE TUNE-UP SPECIFICATIONS

Year	Engine ID/VIN	Engine Displacement Liters (cc)	Spark Plugs Gap (in.)	Ignition Timing (deg.) MT	AT	Fuel Pump (psi)	Idle Speed (rpm) MT	AT	Valve Clearance In.	Ex.
1993	L	2.3 (2254)	0.040	6B	6B	3.5	850	950	0.006	0.010
	E	2.6 (2559)	0.040	12B	12B	35	850	950	0.008	0.008
	Z	3.1 (3098)	0.040	10B	10B	9-13	800	800	①	①
	V	3.2 (3165)	0.040-0.043	5B	5B	41-46	750	750	NA	NA

GASOLINE ENGINE TUNE-UP SPECIFICATIONS

Year	Engine ID/VIN	Engine Displacement Liters (cc)	Spark Plugs Gap (in.)	Ignition Timing (deg.) MT	AT	Fuel Pump (psi)	Idle Speed (rpm) MT	AT	Valve Clearance In.	Ex.
1994	L	2.3 (2254)	0.040	②	②	③	850	950	④	④
	E	2.6 (2559)	0.040	12B	12B	35	850	950	0.008	0.008
	Z	3.1 (3098)	0.040	10B	10B	9-13	800	800	①	①
	V	3.2 (3165)	0.040-0.043	5B	5B	41-46	750	750	NA	NA
1995	L	2.3 (2254)	0.040	12B	-	35	850	950	0.008	0.008
	E	2.6 (2559)	0.040	12B	12B	35	850	950	0.008	0.008
	V	3.2 (3165)	0.040-0.043	5B	5B	41-46	750	750	NA	NA
1996-97	F22B6	2.2 (2156)	0.039-0.043	-	15B	38-46	-	650-750	0.009-0.011	0.011-0.013
	W	2.2 (2189)	0.060	NA	-	41-47	NA	-	NA	NA
	E	2.6 (2559)	0.040	12B ⑤	12B ⑤	35	900 ⑤	900 ⑤	0.006	0.010
	V	3.2 (3165)	0.040	5B ⑤	5B ⑤	41-46	750 ⑤	750 ⑤	HLA	HLA

NOTE: The Vehicle Emission Control Information label often reflects specification changes made during production. The label figures must be used if they differ from those in this chart.

B - Before top dead center
HYD - Hydraulic
NA - Non-adjustable

① Zero lash, plus 1 1/4 turns
② Carbureted: 6B
   Fuel-injected: 12B
③ Carbureted: 3.5 psi
   Fuel-injected: 35 psi
④ Carbureted, Intake: 0.006
   Carbureted, Exhaust: 0.010
   Fuel-injected: 0.008
⑤ Controlled by the ECM and is not adjustable

CAPACITIES

Year	Model	Engine ID/VIN	Engine Displacement Liters (cc)	Engine Oil with Filter (qts.)	Transmission (pts.) 4-Spd	5-Spd	Auto.	Transfer Case (pts.)	Drive Axle Front (pts.)	Rear (pts.)	Fuel Tank (gal.)	Cooling System (qts.)
1993	Amigo	L	2.3 (2254)	4.2	-	3.2	-	-	-	3.2	21.9	9.5
	Amigo	E	2.6 (2559)	5.2	-	6.2	13.8	3.0	3.2	3.8	21.9	9.5
	Pick-up	L	2.3 (2254)	4.2	-	3.2	13.8	-	-	3.2	①	9.5
	Pick-up	E	2.6 (2559)	5.2	-	6.2	13.8	3.0	3.2	3.8	①	9.5
	Pick-up	Z	3.1 (3098)	4.5	-	6.2	-	3.0	3.2	3.8	21.9	11.4
	Rodeo	E	2.6 (2559)	5.8	-	②	19.0	3.0	3.2	3.9	21.9	9.5
	Rodeo	V	3.2 (3165)	6.2	-	2	18.2	3.0	3.2	3.9	21.9	③
	Trooper	V	3.2 (3165)	6.3	-	6.2	18.2	3.0	3.2	3.9	22.5	④
	Trooper	V	3.2 (3165)	6.3	-	6.2	18.2	3.0	3.2	3.9	22.5	④
1994	Amigo	E	2.6 (2559)	5.2	-	6.2	13.8	3.0	3.2	3.8	21.9	9.5
	Pick-up	L	2.3 (2254)	4.2	-	3.2	-	-	-	3.2	①	9.5
	Pick-up	E	2.6 (2559)	5.2	-	6.2	13.8	3.0	3.2	3.2	①	9.5
	Pick-up	Z	3.1 (3098)	4.5	-	6.2	-	3.0	3.2	3.2	21.9	11.4
	Rodeo	E	2.6 (2559)	5.8	-	②	19.0	3.0	3.2	3.9	21.9	9.5
	Rodeo	V	3.2 (3165)	6.2	-	②	18.2	3.0	3.2	3.9	21.9	③
	Trooper	V	3.2 (3165)	6.3	-	6.2	18.2	3.0	3.2	3.9	22.5	④
	Trooper	V	3.2 (3165)	6.3	-	6.2	18.2	3.0	3.2	3.9	22.5	④
1995	Pick-up	L	2.3 (2254)	3.7	-	3.2	-	-	-	3.2	①	9.5
	Pick-up	E	2.6 (2559)	4.4	-	6.2	-	3.0	3.2	3.8	①	9.5
	Rodeo	E	2.6 (2559)	4.4	-	②	-	-	-	⑤	21.9	9.5
	Rodeo	V	3.2 (3165)	6.2	-	②	18.2	3.0	3.2	⑤	21.9	③
	Trooper	V	3.2 (3165)	5.7	-	6.2	18.2	3.0	3.2	3.8	22.5	④
	Trooper	V	3.2 (3165)	5.7	-	6.2	18.2	3.0	3.2	3.8	22.6	④

CAPACITIES

Year	Model	Engine ID/VIN	Engine Displacement Liters (cc)	Engine Oil with Filter (qts.)	Transmission (pts.) 4-Spd	5-Spd	Auto.	Transfer Case (pts.)	Drive Axle Front (pts.)	Rear (pts.)	Fuel Tank (gal.)	Cooling System (qts.)
1996-97	Oasis	F22B6	2.2 (2156)	3.3	-	-	10.6	-	-	-	14.3	6.9
	Hombre	W	2.2 (2189)	4.5	-	4.4	-	-	-	3.9	19.0	11.5
	Rodeo	E	2.6 (2559)	4.4	-	②	-	-	-	⑤	21.9	9.5
	Rodeo	V	3.2 (3165)	5.7	-	②	18.2	3.0	3.2	⑤	21.9	③
	Trooper	V	3.2 (3165)	5.7	-	6.2	18.2	3.0	3.2	3.8	22.5	④

① Standard bed: 14.0
　Spacecab and long bed: 19.8
② MUA transmission: 6.2
　Borg-Warner transmission: 4.8
③ Manual transmission: 9.7
　Automatic transmission: 9.3
④ Manual transmission: 9.3
　Automatic transmission: 9.0
⑤ Saginaw: 4.0
　Dana: 3.8

VALVE SPECIFICATIONS

Year	Engine ID/VIN	Engine Displacement Liters (cc)	Seat Angle (deg.)	Face Angle (deg.)	Spring Test Pressure (lbs. @ in.)	Spring Installed Height (in.)	Stem-to-Guide Clearance (in.) Intake	Exhaust	Stem Diameter (in.) Intake	Exhaust
1993	L	2.3 (2254)	45	45	49-56@1.61	1.61	0.0009-0.0080	0.0015-0.0098	0.3102-0.3134	0.3091-0.3124
	E	2.6 (2559)	45	45	49-56@1.61	1.61	0.0009-0.0080	0.0015-0.0098	0.3102-0.3134	0.3191-0.3124
	Z	3.1 (3098)	46	45	82@1.58	1.58	0.0010-0.0027	0.0010-0.0027	0.3410-0.3420	0.3410-0.3420
	V	3.2 (3165)	45	45	45-55@1.54	1.54	0.0009-0.0078	0.0012-0.0078	0.2323-0.2346	0.2323-0.2350
1994	L	2.3 (2254)	45	45	49-56@1.61	1.61	0.0009-0.0080	0.0015-0.0098	0.3102-0.3134	0.3091-0.3124
	E	2.6 (2559)	45	45	49-56@1.61	1.61	0.0009-0.0080	0.0015-0.0098	0.3102-0.3134	0.3191-0.3124
	Z	3.1 (3098)	46	45	82@1.58	1.58	0.0010-0.0027	0.0010-0.0027	0.3410-0.3420	0.3410-0.3420
	V	3.2 (3165)	45	45	45-55@1.54	1.54	0.0009-0.0078	0.0012-0.0078	0.2323-0.2346	0.2323-0.2350
1995-96	L	2.3 (2254)	45	45	49-56@1.61	1.61	0.0009-0.0080	0.0015-0.0098	0.3102-0.3134	0.3091-0.3124
	E	2.6 (2559)	45	45	45-55@1.61	1.61	0.0009-0.0080	0.0015-0.0098	0.3102-0.3134	0.3091-0.3124
	V	3.2 (3165)	45	45	45-55@1.54	1.54	0.0009-0.0078	0.0012-0.0078	0.2323-0.2353	0.2323-0.2350
1996-97	F22B6	2.2 (2156)	45	45	NA	NA	0.0008-0.0018	0.0022-0.0031	0.2159-0.2163	0.2146-0.2150
	W	2.2 (2189)	45	45	75-81@1.71	1.71	0.0010-0.0027	0.0014-0.0031	-	-
	E	2.6 (2559)	45	45	45-55@1.61	1.61	0.0009-0.0079	0.0015-0.0098	0.3102-0.3134	0.3091-0.3124
	V	3.2 (3165)	45	45	45-55@1.54	1.54	0.0009-0.0079	0.0012-0.0079	0.2323-0.2353	0.2323-0.2350

NA - Not Available

TORQUE SPECIFICATIONS
All readings in ft. lbs.

Year	Engine ID/VIN	Engine Displacement Liters (cc)	Cylinder Head Bolts	Main Bearing Bolts	Rod Bearing Bolts	Crankshaft Damper Bolts	Flywheel Bolts	Manifold Intake	Manifold Exhaust	Spark Plugs	Lug Nut
1993	L	2.3 (2254)	①	72	43	87	40	16	16	14	②
	E	2.6 (2559)	①	72	43	87	40	16	16	14	②
	Z	3.1 (3098)	③	72	39	70	52	19	25	14	②
	V	3.2 (3165)	④	29	49	123	40	17	42	13	87
1994	L	2.3 (2254)	①	72	43	87	40	16	16	14	②
	E	2.6 (2559)	①	72	43	87	40	16	16	14	②
	Z	3.1 (3098)	③	72	39	70	52	19	25	14	②
	V	3.2 (3165)	④	29	49	123	40	17	42	13	87
1995	L	2.3 (2254)	①	72	43	87	40	16	16	14	②
	E	2.6 (2559)	①	72	43	87	40	16	16	14	②
	V	3.2 (3165)	④	29	40	123	40	17	42	13	87
1996-97	F22B6	2.2 (2156)	⑤	54	34	181	54	16	23	13	80
	W	2.2 (2189)	⑥	70	38	77	55	22	10	13	95
	E	2.6 (2559)	①	72	43	87	40	16	33	14	②
	V	3.2 (3165)	④	⑦	40	123	40	17	42	13	87

① Step 1: 58 ft. lbs.
  Step 2: 72 ft. lbs.
② Steel wheels: 58-72 ft. lbs.
  Aluminum wheels: 80-94 ft. lbs.
③ Step 1: 41 ft. lbs.
  Step 2: Turn an additional 90 degrees
④ 8x1.25 bolts: 15 ft. lbs.
  11x1.5 bolts: 47 ft. lbs.
⑤ Step 1: 29 ft. lbs.
  Step 2: 51 ft. lbs.
  Step 3: 72 ft. lbs.
⑥ Long bolts: 46 ft. lbs.
  Short bolts: 43 ft. lbs.
⑦ Main bearing cap bolts: 29 ft. lbs.
  Oil gallery bolts: 29 ft. lbs. plus 55-65 degrees
  Buttress bolts: 29 ft. lbs.

BRAKE SPECIFICATIONS
All measurements in inches unless noted

Year	Model		Master Cylinder Bore	Brake Disc Original Thickness	Brake Disc Minimum Thickness	Maximum Runout	Brake Drum Diameter Original Inside Diameter	Brake Drum Diameter Max. Wear Limit	Brake Drum Diameter Maximum Machine Diameter	Minimum Lining Thickness Front	Minimum Lining Thickness Rear
1993	Amigo		1.000	③	④	0.005	8.27 ⑤	8.32 ⑤	NA	0.039	0.039 ⑥
	Pick-up		0.938	⑦	⑧	⑨	10.01	10.06	NA	0.039	⑩
	Rodeo ①		1.000	0.866	⑪	0.005	10.00	10.06	NA	0.039	0.039
	Rodeo ②		1.000	⑫	⑬	0.005	8.27 ⑤	8.32 ⑤	NA	0.039	0.039 ⑥
	Trooper		1.000	⑫	⑬	0.005	8.27 ⑤	8.27 ⑤	NA	0.039	0.039 ⑥
1994	Amigo		1.000	③	④	0.005	8.27 ⑤	8.32 ⑤	NA	0.039	0.039 ⑥
	Pick-up		0.938	⑦	⑧	⑨	10.01	10.06	NA	0.039	⑩
	Rodeo ①		1.000	0.866	⑪	0.005	10.00	10.06	NA	0.039	0.039
	Rodeo ②		1.000	⑫	⑬	0.005	8.27 ⑤	8.32 ⑤	NA	0.039	0.039 ⑥
	Trooper		1.000	⑫	⑬	0.005	8.27 ⑤	8.27 ⑤	NA	0.039	0.039 ⑥
1995	Pick-up		0.938	⑦	⑧	⑨	10.01	10.06	10.06	0.039	⑩
	Rodeo ①		1.000	0.866	⑥	0.005	10.00	10.06	10.06	0.039	0.039
	Rodeo ②		1.000	③	⑬	0.005	8.27 ⑤	8.32 ⑤	8.32 ⑤	0.039	0.039 ⑥
	Trooper		1.000	③	⑬	0.005	8.27 ⑤	8.32 ⑤	8.32 ⑤	0.039	0.039 ⑥
1996-97	Hombre		NA	1.040	0.980	0.004	9.50	9.59	9.56	0.030	0.030
	Oasis	F	NA	0.930	0.830	0.004	-	-	-	0.060	-
		R	-	0.350	0.300	0.004	6.69 ⑤	6.73 ⑤	6.73 ⑤	-	0.040
	Rodeo ①	F	1.000	1.020	0.983	0.005	-	-	-	0.039	-
		R	-	-	-	0.005	10.00	10.06	10.06	-	0.039

BRAKE SPECIFICATIONS
All measurements in inches unless noted

Year	Model			Master Cylinder Bore	Brake Disc			Brake Drum Diameter			Minimum Lining Thickness	
					Original Thickness	Minimum Thickness	Maximum Runout	Original Inside Diameter	Max. Wear Limit	Maximum Machine Diameter	Front	Rear
1996-97	Rodeo ②	F		1.000	1.024	0.969 ⑭	0.005	-	-	-	0.039	-
		R		-	0.710	0.654 ⑮	0.005	8.27 ⑤	8.32 ⑤	8.32 ⑤	-	0.039 ⑥
	Trooper	F		1.000	1.024	0.969 ⑭	0.005	-	-	-	0.039	-
		R		-	0.710	0.654 ⑮	0.005	8.27 ⑤	8.32 ⑤	8.32 ⑤	-	0.039 ⑥

NA - Not Available

① 2.6L engine
② 3.2L engine
③ Front: 1.026
  Rear: 0.709
④ Front: 0.970
  Rear: 0.654 (Minimum machine diameter: 0.668)
⑤ Emergency brake drum surface
⑥ Specification includes disc pads and parking brake shoes

⑦ Front: 0.886
  Rear: 0.472
⑧ Front: 0.811
  Rear: 0.417 (Minimum machine diameter: 0.417)
⑨ Front: 0.0050
  Rear: 0.0051
⑩ Disc: 0.040
  Drum: 0.039

⑪ 0.811 (Minimum machine diameter: 0.826)
⑫ Front: 1.020
  Rear: 0.710
⑬ Front: 0.969 (Minimum machine diameter: 0.983)
  Rear: 0.654 (Minimum machine diameter: 0.668)
⑭ Minimum machine diameter: 0.983
⑮ Minimum machine diameter: 0.668

FREQUENT MAINTENANCE LABOR
AMIGO, HOMBRE, OASIS, PICKUP, RODEO, TROOPER

The following should be used as a guide when determining the amount of work required for a particular service if taken to a repair shop. In estimating how long a particular Frequent Maintenance Service item should take, please observe the following:
- **Chilton Time** is time that is based on field research and data supplied by the vehicle manufacturer.
- All labor time operations are given in hours and tenths of an hour.
- All labor operations, are to be used as a **guide**.

COOLING

(G) Winterize Cooling System
Includes: Run engine to check for leaks, tighten all hose connections. Test radiator and pressure cap, drain radiator and engine block. Add antifreeze and refill system.
1993-97 5

(G) Belt, Drive, Renew
1993-97 4 cyl.
  one 5
  each adtnl. 1
1993-97 V6
  one 3
  each adtnl. 1

(G) Belt, Drive, Adjust
1993-97
  one 2
  each adtnl. 1

(G) Hoses, Radiator, Renew
Includes: Drain and refill cooling system as required.
1993-97, each 4

(G) Thermostat, Coolant, Renew
1993-97 4 cyl.
  gas 5
1993-97 V6
  2.8L, 3.1L 7
  3.2L 4

BRAKES

(G) Bleed Brakes (Four Wheels)
Includes: Add fluid.
1993-97 .6

(G) Brakes, Adjust (Minor)
Includes: Adjust brakes, fill master cylinder.
1993-97, two wheels 4

(M) Parking Brake, Adjust
1993-97 .5

LUBRICATION SERVICES

(M) Engine Oil & Filter, Renew
Includes: Inspect and correct all fluid levels.
1993-97 .4

ELECTRICAL

(G) Headlamps, Aim
1993-97 .4

(G) Halogen Headlamp Bulb, Renew
1993-97 .3

(G) High Mount Stop Lamp Bulb, Renew
1993-97 Pickup, Rodeo 3

(G) License Lamp Assy., Renew
1993-97 Pickup 3
1993-97 Trooper/Trooper II 3
1993-94 Amigo 6
1993-97 Rodeo 3

(G) Stop & Tail Lamp Bulb, Renew
1993-97 .3

(G) Horn, Renew
1993-97 Pickup 3
1993-97 Trooper 5
1993-94 Amigo 3
1993-97 Rodeo 5
For each adtnl., add 1

(M) Terminals, Battery, Clean
1993-97 .3

SCHEDULED MAINTENANCE INTERVALS
(ISUZU AMIGO, HOMBRE, OASIS, PICK UP, RODEO & TROOPER)

TO BE SERVICED	TYPE OF SERVICE	VEHICLE MILEAGE INTERVAL (x1000)												
		7.5	15	22.5	30	37.5	45	52.5	60	67.5	75	82.5	90	97.5
Engine oil & filter	R	✓	✓	✓	✓	✓	✓	✓	✓	✓	✓	✓	✓	✓
Automatic transmission fluid	S/I	✓	✓	✓	✓	✓	✓	✓	✓	✓	✓	✓	✓	✓
Battery fluid level	S/I	✓	✓	✓	✓	✓	✓	✓	✓	✓	✓	✓	✓	✓
Body & chassis lubrication	S/I	✓	✓	✓	✓	✓	✓	✓	✓	✓	✓	✓	✓	✓
Brake & clutch fluid level	S/I	✓	✓	✓	✓	✓	✓	✓	✓	✓	✓	✓	✓	✓
Brake lines & hoses	S/I	✓	✓	✓	✓	✓	✓	✓	✓	✓	✓	✓	✓	✓
Check & rotate tires	S/I	✓	✓	✓	✓	✓	✓	✓	✓	✓	✓	✓	✓	✓
Engine coolant strength, hoses & clamps	S/I	✓	✓	✓	✓	✓	✓	✓	✓	✓	✓	✓	✓	✓
Exhaust system	S/I	✓	✓	✓	✓	✓	✓	✓	✓	✓	✓	✓	✓	✓
Front suspension, ball joints, steering linkage, parking brake cable guides, propeller shaft splines, universal joints, brake & clutch pedal springs (Hombre)	S/I	✓	✓	✓	✓	✓	✓	✓	✓	✓	✓	✓	✓	✓
Lubricate accelerator linkage (except Rodeo)	S/I	✓	✓	✓	✓	✓	✓	✓	✓	✓	✓	✓	✓	✓
Rear axle seals (Hombre)	S/I	✓	✓	✓	✓	✓	✓	✓	✓	✓	✓	✓	✓	✓
Starter safety switch	S/I	✓	✓	✓	✓	✓	✓	✓	✓	✓	✓	✓	✓	✓
Suspension & steering (Rodeo)	S/I	✓	✓	✓	✓	✓	✓	✓	✓	✓	✓	✓	✓	✓
Suspension & steering (Trooper)	S/I		✓		✓		✓		✓		✓		✓	
Lubricate front & rear propeller shaft	S/I	✓		✓		✓		✓		✓		✓		✓
Propeller shaft flange torque 46 lb. ft. (63 Nm)	S/I	✓		✓		✓		✓		✓		✓		✓
Auto cruise control linkage & hose	S/I		✓		✓		✓		✓		✓		✓	
Brake & clutch pedal play	S/I		✓		✓		✓		✓		✓			
Cooling & heater hoses	S/I		✓		✓		✓		✓			✓	✓	

SCHEDULED MAINTENANCE INTERVALS
(ISUZU AMIGO, HOMBRE, OASIS, PICK UP, RODEO & TROOPER) (Cont.)

TO BE SERVICED	TYPE OF SERVICE	VEHICLE MILEAGE INTERVAL (x1000)												
		7.5	15	22.5	30	37.5	45	52.5	60	67.5	75	82.5	90	97.5
Disc brakes (Trooper)	S/I		✓		✓		✓		✓		✓		✓	
Disc & drum brakes (Amigo, Pick Up & Rodeo)	S/I		✓		✓		✓		✓		✓		✓	
Lubricate clutch pedal spring, bushing & clevis pin	S/I		✓		✓		✓		✓		✓		✓	
Lubricate key lock cylinder	S/I		✓		✓		✓		✓		✓		✓	
Parking brake	S/I		✓		✓		✓		✓		✓		✓	
Shift on-the-fly system gear fluid (Rodeo)	S/I		✓		✓		✓		✓		✓		✓	
Clutch control cable (2.3L)	S/I		✓		✓		✓		✓		✓		✓	
Throttle linkage (Rodeo)	S/I		✓		✓		✓		✓		✓		✓	
Valve clearance (Rodeo 2.6L)	S/I		✓		✓		✓		✓		✓		✓	
Manual transmission & transfer case oil	R		✓		✓				✓				✓	
Front & rear axle oil (Trooper)	R		✓		✓				✓				✓	
Engine idle speed (Amigo & Pick Up 2.3L & 2.6L)	S/I	✓							✓		✓		✓	
Air cleaner filter	R				✓				✓				✓	
Automatic transmission fluid & filter (Oasis)	R				✓				✓				✓	
Brake fluid (include ABS) (Oasis)	R				✓				✓				✓	
Engine coolant (Trooper)	R				✓				✓				✓	
Engine coolant (Rodeo)	R				✓								✓	
Front & rear axle oil	R		✓		✓				✓					
Fuel filter (Hombre)	R				✓				✓				✓	
Power steering fluid	R				✓				✓				✓	
Spark plugs (Rodeo 2.6L, Amigo & Pick Up)①	R				✓				✓				✓	

SCHEDULED MAINTENANCE INTERVALS
(ISUZU AMIGO, HOMBRE, OASIS, PICK UP, RODEO & TROOPER) (Cont.)

TO BE SERVICED	TYPE OF SERVICE	VEHICLE MILEAGE INTERVAL (x1000)												
		7.5	15	22.5	30	37.5	45	52.5	60	67.5	75	82.5	90	97.5
Spark plugs (Rodeo 3.2L & Trooper)①	R								✓					
Carburetor choke (2.3L)	S/I				✓				✓				✓	
Clutch lines & hose	S/I				✓				✓				✓	
Engine drive belts	S/I				✓				✓				✓	
Engine idle speed (Rodeo 2.6L)	S/I	✓							✓				✓	
Front wheel bearings & free wheeling hubs	S/I				✓				✓				✓	
Steering gear play (Amigo & Pick Up)	S/I				✓				✓				✓	
Thermostatically controlled air cleaner (3.1L)	S/I				✓				✓				✓	
Clean radiator core & A/C condenser	S/I				✓				✓				✓	
Ignition wires (Amigo, Hombre, Pick Up & Rodeo 2.6L)	S/I								✓				✓	
Oxygen sensor (Amigo, Pick Up & 1993-95 Rodeo)	R												✓	
Timing belt (Oasis)	R												✓	
Timing belt (Amigo, Rodeo, Pick Up 2.3L & 2.6L)	R								✓					
Engine timing (Amigo, Hombre & Pick Up)	S/I								✓					
Fuel tank, cap & lines	S/I								✓					
PCV valve (Oasis)	S/I								✓					

① Spark plugs (Hombre) - replace every 100,000 miles.
R – Replace   S/I – Service or Inspect

FREQUENT OPERATION MAINTENANCE (SEVERE SERVICE)

If a vehicle is operated under any of the following conditions it is considered severe service:
- Extremely dusty areas.
- 50% or more of the vehicle operation is in 32°C (90°F) or higher temperatures, or constant operation in temperatures below 0°C (32°F).
- Prolonged idling (vehicle operation in stop and go traffic).
- Frequent short running periods (engine does not warm to normal operating temperatures).
- Police, taxi, delivery usage or trailer towing usage.

Oil & oil filter change, body & chassis lubrication – every 3000 miles.
Rotate tires every 6000 miles.
Front & rear axle oil - change every 15,000 miles. Automatic transmission fluid & filter - change every 20,000 miles.

LEXUS
LX450

ENGINE IDENTIFICATION

Year	Model	Engine Displacement Liters (cc)	Engine Series (ID/VIN)	Fuel System	No. of Cylinders	Engine Type
1996-97	LX450	4.5 (4477)	1FZ-FE	EFI	6	DOHC

EFI - Electronic fuel injection
DOHC - Double overhead camshaft

GENERAL ENGINE SPECIFICATIONS

Year	Engine ID/VIN	Engine Displacement Liters (cc)	Fuel System Type	Net Horsepower @ rpm	Net Torque @ rpm (ft. lbs.)	Bore x Stroke (in.)	Compression Ratio	Oil Pressure @ rpm
1996-97	1FZ-FE	4.5 (4477)	EFI	212@4600	275@3200	3.94x3.74	9.0:1	36-71@3000

EFI - Electronic fuel injection

GASOLINE ENGINE TUNE-UP SPECIFICATIONS

Year	Engine ID/VIN	Engine Displacement Liters (cc)	Spark Plugs Gap (in.)	Ignition Timing (deg.) MT	Ignition Timing (deg.) AT	Fuel Pump (psi)	Idle Speed (rpm) MT	Idle Speed (rpm) AT	Valve Clearance In.	Valve Clearance Ex.
1996-97	1FZ-FE	4.5 (4477)	0.031	-	3B ①	38-44	-	600-700	0.006-0.010	0.010-0.014

NOTE: The Vehicle Emission Control Information label often reflects specification changes
made during production. The label figures must be used if they differ from those in this chart.
B - Before top dead center
① Terminals TE1 and E1 check connector must be connected

CAPACITIES

Year	Model	Engine ID/VIN	Engine Displacement Liters (cc)	Engine Oil with Filter (qts.)	Transmission (pts.) 4-Spd	Transmission (pts.) 5-Spd	Transmission (pts.) Auto.	Transfer Case (pts.)	Drive Axle Front (pts.)	Drive Axle Rear (pts.)	Fuel Tank (gal.)	Cooling System (qts.)
1996-97	LX450	1FZ-FE	4.5 (4477)	7.8	-	-	4.0	3.6	①	6.8	25.1	②

① With differential lock: 5.6
　Without differential lock: 5.8

② With rear heater: 14.2
　Without rear heater: 13.2

VALVE SPECIFICATIONS

Year	Engine ID/VIN	Engine Displacement Liters (cc)	Seat Angle (deg.)	Face Angle (deg.)	Spring Test Pressure (lbs. @ in.)	Spring Installed Height (in.)	Stem-to-Guide Clearance (in.)		Stem Diameter (in.)	
							Intake	Exhaust	Intake	Exhaust
1996-97	1FZ-FE	4.5 (4477)	45	44.5	48.1-53.4@ 1.437	1.437	0.0010-0.0024	0.0012-0.0026	0.2744-0.2750	0.2742-0.2748

TORQUE SPECIFICATIONS
All readings in ft. lbs.

Year	Engine ID/VIN	Engine Displacement Liters (cc)	Cylinder Head Bolts	Main Bearing Bolts	Rod Bearing Bolts	Crankshaft Damper Bolts	Flywheel Bolts	Manifold		Spark Plugs	Lug Nut
								Intake	Exhaust		
1996-97	1FZ-FE	4.5 (4477)	①	②	③	304	74	15	29	14	④

① Step 1: 29 ft. lbs.
   Step 2: Plus 90 degrees
   Step 3: Plus 90 degrees

② Step 1: 54 ft. lbs.
   Step 2: Plus 90 degrees

③ Step 1: 35 ft. lbs.
   Step 2: Plus 90 degrees

④ Steel wheels: 109 ft. lbs.
   Aluminum wheels: 76 ft. lbs.

BRAKE SPECIFICATIONS
All measurements in inches unless noted

Year	Model		Master Cylinder Bore	Brake Disc			Brake Drum Diameter			Minimum Lining Thickness	
				Original Thickness	Minimum Thickness	Maximum Runout	Original Inside Diameter	Max. Wear Limit	Maximum Machine Diameter	Front	Rear
1996-97	LX450	F	NA	1.260	1.181	0.0059	-	-	-	0.039	-
		R	NA	0.709	0.630	0.0059	-	-	-	-	0.039

NA - Not Available
F - Front
R - Rear

SCHEDULED MAINTENANCE INTERVALS
(LEXUS LX450)

TO BE SERVICED	TYPE OF SERVICE	VEHICLE MILEAGE INTERVAL (x1000)												
		7.5	15	22.5	30	37.5	45	52.5	60	67.5	75	82.5	90	97.5
Engine oil & filter	R	✓	✓	✓	✓	✓	✓	✓	✓	✓	✓	✓	✓	✓
Automatic transmission fluid & filter	S/I		✓		✓		✓		✓		✓		✓	
Ball joints & dust covers	S/I		✓		✓		✓		✓		✓		✓	
Bolts & nuts on chassis & body	S/I		✓		✓		✓		✓		✓		✓	

SCHEDULED MAINTENANCE INTERVALS
(LEXUS LX450) (Cont.)

TO BE SERVICED	TYPE OF SERVICE	VEHICLE MILEAGE INTERVAL (x1000)												
		7.5	15	22.5	30	37.5	45	52.5	60	67.5	75	82.5	90	97.5
Brake linings & drums	S/I		✓		✓		✓		✓		✓		✓	
Brake line pipes & hoses	S/I		✓		✓		✓		✓		✓		✓	
Brake pads discs (front & rear)	S/I		✓		✓		✓		✓		✓		✓	
Propeller shaft grease	S/I		✓		✓		✓		✓		✓		✓	
Steering knuckle & chassis grease	S/I		✓		✓		✓		✓		✓		✓	
Steering linkage	S/I		✓		✓		✓		✓		✓		✓	
Transfer, differential & steering gear box oil	S/I		✓		✓		✓		✓		✓		✓	
Air cleaner filter	R				✓				✓				✓	
Front wheel bearing & thrust bush grease	R				✓				✓				✓	
Spark plugs	R				✓				✓				✓	
Drive belts	S/I				✓				✓				✓	
Exhaust pipes & mountings	S/I				✓				✓				✓	
Fuel lines & connections	S/I				✓				✓				✓	
Engine Coolant	R					✓					✓			
Charcoal canister (Calif.)	R								✓					
Fuel tank cap gasket	R								✓					
Heated oxygen sensors (except Calif.)①	R													

① Heated oxygen sensors (except Calif.) - replace every 80,000 miles.
R – Replace   S/I – Service or Inspect

FREQUENT OPERATION MAINTENANCE (SEVERE SERVICE)

If a vehicle is operated under any of the following conditions it is considered severe service:
- Extremely dusty areas.
- 50% or more of the vehicle operation is in 32°C (90°F) or higher temperatures, or constant operation in temperatures below 0°C (32°F).
- Prolonged idling (vehicle operation in stop and go traffic).
- Frequent short running periods (engine does not warm to normal operating temperatures).
- Police, taxi, delivery usage or trailer towing usage.

Air cleaner filter - service or inspect every 3750 miles.
Oil & oil filter change – change every 3750 miles.
Ball joints & dust covers - service or inspect every 7500 miles.
Bolts & nuts on body & chassis - service or inspect every 7500 miles.
Brake linings & drums - service or inspect every 7500 miles.
Brake pads & discs (front & rear) - service or inspect every 7500 miles.
Steering knuckle & chassis grease - service or inspect every 7500 miles.
Steering linkage - service or inspect every 7500 miles.
Propeller shaft grease - service or inspect every 7500 miles.
Exhaust pipes & mountings - service or inspect every 15,000 miles.

MAZDA
B Series • MPV • Navajo

ENGINE IDENTIFICATION

Year	Model	Engine Displacement Liters (cc)	Engine Series (ID/VIN)	Fuel System	No. of Cylinders	Engine Type
1993	B2200	2.2 (2184)	F2	2BC	4	SOHC
	B2200	2.2 (2184)	F2	EFI	4	SOHC
	B2600i	2.6 (2606)	G6	EFI	4	SOHC
	MPV	2.6 (2606)	G6	EFI	4	SOHC
	MPV	3.0 (2954)	JE	EFI	6	SOHC
	Navajo	4.0 (4016)	X	EFI	6	OHV
1994	B2300	2.3 (2298)	A	EFI	4	SOHC
	B3000	3.0 (2968)	U	EFI	6	OHV
	B4000	4.0 (4016)	X	EFI	6	OHV
	MPV	2.6 (2606)	G6	EFI	4	SOHC
	MPV	3.0 (2954)	JE	EFI	6	SOHC
	Navajo	4.0 (4016)	X	EFI	6	OHV
1995	B2300	2.3 (2298)	A	EFI	4	SOHC
	B3000	3.0 (2968)	U	EFI	6	OHV
	B4000	4.0 (4016)	X	EFI	6	OHV
	MPV	3.0 (2954)	JE	EFI	6	SOHC
1996-97	B2300	2.3 (2298)	A	EFI	4	SOHC
	B3000	3.0 (2968)	U	EFI	6	OHV
	B4000	4.0 (4016)	X	EFI	6	OHV
	MPV	3.0 (2954)	JE	EFI	6	SOHC

EFI - Electronic fuel injection
BC - Barrel carburetor
SOHC - Single overhead camshaft
OHV - Overhead valve

GENERAL ENGINE SPECIFICATIONS

Year	Engine ID/VIN	Engine Displacement Liters (cc)	Fuel System Type	Net Horsepower @ rpm	Net Torque @ rpm (ft. lbs.)	Bore x Stroke (in.)	Compression Ratio	Oil Pressure @ rpm
1993	F2	2.2 (2184)	EFI	145@4300	190@3500	3.39x3.70	7.8:1	43-57@3000
	F2	2.2 (2184)	2BC	110@4700	130@3000	3.39x3.70	8.6:1	43-56@3000
	G6	2.6 (2606)	EFI	121@4600	149@3500	3.62x3.86	8.4:1	45-58@3000
	X	4.0 (4016)	EFI	160@4500	①	3.94x3.31	9.0:1	40-60@2000
	JE	3.0 (2954)	EFI	150@5000	165@4000	3.54x3.05	8.5:1	53-75@3000
1994	A	2.3 (2298)	EFI	98@4600	130@2600	3.78x3.13	9.2:1	40-60@2000
	G6	2.6 (2606)	EFI	121@4600	149@3500	3.60x3.90	8.4:1	45-58@3000
	JE	3.0 (2954)	EFI	155@5000	169@4000	3.50x3.00	8.5:1	53-75@3000
	U	3.0 (2968)	EFI	140@4800	160@3000	3.50x3.14	9.3:1	40-60@2500
	X	4.0 (4016)	EFI	160@4500	②	3.95x3.32	9.1:1	40-60@2000
1995	A	2.3 (2298)	EFI	112@4800	135@2400	3.78x3.13	9.2:1	36-71@3000
	JE	3.0 (2954)	EFI	155@5000	169@4000	3.50x3.00	8.5:1	53-75@3000
	U	3.0 (2968)	EFI	145@4800	165@3000	3.50x3.14	9.3:1	36-71@3000
	X	4.0 (4016)	EFI	160@4200	220@3000	3.95x3.32	9.0:1	36-71@3000

GENERAL ENGINE SPECIFICATIONS

Year	Engine ID/VIN	Engine Displacement Liters (cc)	Fuel System Type	Net Horsepower @ rpm	Net Torque @ rpm (ft. lbs.)	Bore x Stroke (in.)	Compression Ratio	Oil Pressure @ rpm
1996-97	A	2.3 (2298)	EFI	112@4800	135@2400	3.78x3.13	9.2:1	40-60@2000
	JE	3.0 (2954)	EFI	155@5000	169@4000	3.54x3.05	8.5:1	53-75@3000
	U	3.0 (2968)	EFI	145@4800	165@3000	3.50x3.14	9.3:1	40-60@2500
	X	4.0 (4016)	EFI	160@4200	220@3000	3.94x3.31	9.0:1	40-60@2000

EFI - Electronic fuel injection
BC - Barrel carburetor
① Manual transmission: 220@2500
　 Automatic transmission: 220@2200
② Automatic transmission: 220@2800
　 Manual transmission: 225@2500

GASOLINE ENGINE TUNE-UP SPECIFICATIONS

Year	Engine ID/VIN	Engine Displacement Liters (cc)	Spark Plugs Gap (in.)	Ignition Timing (deg.) MT	Ignition Timing (deg.) AT	Fuel Pump (psi)	Idle Speed (rpm) MT	Idle Speed (rpm) AT	Valve Clearance In.	Valve Clearance Ex.
1993	F2	2.2 (2184)	0.039-0.043 ⑪	5-7B ①	5-7B ①	30-38 ⑫	730-770 ⑩	750-790 ⑩	HYD	HYD
	G6	2.6 (2606)	0.039-0.043	4-6B ①	4-6B ①	30-38 ②	730-770 ①	750-790 ①	HYD	HYD
	JE	3.0 (2954)	0.039-0.043	10-12B ①	10-12B ①	30-38 ②	780-820	780-820	HYD	HYD
	X	4.0 (4016)	0.052-0.056	10B ①	10B ①	35-45 ②	750-850	750-850	HYD	HYD
1994	A	2.3 (2298)	0.042-0.046	8-12B ⑧	8-12B ⑧	30-45 ②	475-575	475-575	HYD	HYD
	G6	2.6 (2606)	0.039-0.043	4-6B ①	4-6B ①	30-38 ②	750-800	750-800	HYD	HYD
	JE	3.0 (2954)	0.039-0.043	10-12B ①	10-12B ①	30-38 ②	780-820	780-820	HYD	HYD
	U	3.0 (2968)	0.042-0.046	8-12B ⑧	8-12B ⑧	30-45 ②	⑥	⑥	HYD	HYD
	X	4.0 (4016)	0.052-0.056	8-12B ⑨	8-12B ⑨	30-45 ②	⑥	⑥	HYD	HYD
1995	A	2.3 (2298)	0.042-0.046	8-12B ⑦	8-12B ⑦	35-45 ②	475-575	475-575	HYD	HYD
	JE	3.0 (2954)	0.039-0.043	-	10-12B	30-37 ②	-	780-820 ⑤	HYD	HYD
	U	3.0 (2968)	0.042-0.046	8-12B ⑧	8-12B ⑧	35-45 ②	④	④	HYD	HYD
	X	4.0 (4016)	0.052-0.056	8-12B ⑦	8-12B ⑦	35-45 ②	④	④	HYD	HYD
1996-97	A	2.3 (2298)	③	8-12B ⑨	8-12B ⑨	35-45 ②	475-575	475-575	HYD	HYD
	JE	3.0 (2954)	0.040-0.043	-	10-12B	30-37 ②	-	780-820	HYD	HYD
	U	3.0 (2968)	③	8-12B ⑨	8-12B ⑨	35-45 ②	④	④	HYD	HYD
	X	4.0 (4016)	③	8-12B ⑨	8-12B ⑨	35-45 ②	④	④	HYD	HYD

NOTE: The Vehicle Emission Control Information label often reflects specification changes made during production. The label figures must be used if they differ from those in this chart.
B - Before top dead center
HYD - Hydraulic
① Data link connector terminal 10 grounded
② Pressure indicated is with gauge in-line, regulator vacuum hose connected and engine idling
③ Refer to Vehicle's Emission Control Information label
④ Not adjustable
⑤ Data link connector terminal 10 grounded and transmission in park

⑥ Automatically adjusted
⑦ With "SPOUT" shorting bar disconnected
⑧ With "SPOUT" shorting bar disconnected
⑨ Base timing, not adjustable
⑩ Carbureted models: 800-850 rpm
⑪ Carbureted models: 0.028-0.033
⑫ Pressure indicated is with gauge in-line, regulator vacuum hose connected and engine idling
　 Carbureted models with manual transmission: 3.7 to 4.7
　 Carbureted models with automatic transmission: 2.8 to 3.6

CAPACITIES

Year	Model	Engine ID/VIN	Engine Displacement Liters (cc)	Engine Oil with Filter (qts.)	Transmission (pts.)			Transfer Case (pts.)	Drive Axle		Fuel Tank (gal.)	Cooling System (qts.)
					4-Spd	5-Spd	Auto.		Front (pts.)	Rear (pts.)		
1993	B2200	F2	2.2 (2184)	4.3	-	4.2	15.8 ①	-	-	2.6	14.8 ②	7.9
	B2600i	G6	2.6 (2606)	5.0	-	③	15.8 ①	4.2	3.2	3.6	14.8 ②	7.9
	MPV	G6	2.6 (2606)	5.0	-	-	15.8 ④	-	-	3.2	15.9	7.6
	MPV	JE	3.0 (2954)	5.0	-	-	15.8 ④	3.2	3.6	3.2	19.6 ⑥	10.3
	Navajo	X	4.0 (4016)	5.0	-	5.6	20.0	2.5	3.5	5.3	19.3	7.8 ④
1994	B2300	A	2.3 (2298)	5.0	-	3.6	19.4	-	-	5.0 ⑧	16.3 ⑨	7.2
	B3000	U	3.0 (2968)	4.5	-	3.6	⑩	2.5	5.0	5.0 ⑧	16.3 ⑨	11.8
	B4000	X	4.0 (4016)	5.0	-	3.6	⑩	2.5	5.0	5.0 ⑧	16.3 ⑨	8.1 ⑪
	MPV	G6	2.6 (2606)	5.0	-	-	15.8 ④	-	-	3.2	19.6	7.6
	MPV	JE	3.0 (2954)	5.0	-	-	15.8 ④	3.2	3.6	3.2	19.6 ⑥	10.3
	Navajo	X	4.0 (4016)	5.0	-	5.6	20.0	2.5	3.5	5.3	19.3	8.0
1995	B2300	A	2.3 (2298)	5.0	-	5.6	19.4	2.5	3.5	5.0	16.3 ⑫	⑬
	B3000	U	3.0 (2968)	5.0	-	5.6	⑩	2.5	3.5	5.0	16.3 ⑫	⑭
	B4000	X	4.0 (4016)	5.0	-	5.6	⑩	2.5	3.5	5.0	16.3 ⑫	⑮
	MPV	JE	3.0 (2954)	5.0	-	-	18.2	3.2	3.6	3.2	⑯	10.3
1996-97	B2300	A	2.3 (2298)	5.0	-	5.6	19.0	2.5	⑰	5.0	16.3 ⑫	⑤
	B3000	U	3.0 (2968)	4.5	-	5.6	⑩	2.5	⑰	5.0	16.3 ⑫	⑦
	B4000	X	4.0 (4016)	5.0	-	5.6	⑩	2.5	⑰	5.0	16.3 ⑫	⑮
	MPV	JE	3.0 (2954)	5.0	-	-	18.2	3.2	3.6	3.2	⑯	10.3

① Electronically-controlled transmission: 18.2 pts.
② Long bed: 17.4 gals.
③ 2WD: 6.0 pts.
  4WD: 6.8 pts.
④ With AC: 8.6 qts.
⑤ Without A/C: 6.5 qts.
  With A/C: 7.2 qts.
⑥ With 4 wheel drive: 19.8 gals.
⑦ Without A/C: 9.5 qts.
  With A/C: 10.2 qts.
⑧ Limited slip differential: 5.0 to 5.3 pts. plus four oz. of friction modifier
⑨ Long bed and cab plus: 19.6 gals.
⑩ 2WD: 19.4 pts.
  4WD: 20.0 pts.
⑪ With AC super cool and automatic transmission: 8.5 qts.
⑫ Long bed and Supercab: 19.6 gals.
⑬ Without A/C: 6.5 qts.
  With A/C: 7.2 qts.
⑭ Without A/C: 9.5 qts.
  With A/C: 10.2 qts.
⑮ Without A/C: 7.8 qts.
  Wtih A/C: 8.6 qts.
⑯ 2WD: 19.6 gals.
  4WD: 19.8 gals.
⑰ Dana 28: 3.0 pts.
  Dana 35: 3.5 pts.

VALVE SPECIFICATIONS

Year	Engine ID/VIN	Engine Displacement Liters (cc)	Seat Angle (deg.)	Face Angle (deg.)	Spring Test Pressure (lbs. @ in.)	Spring Installed Height (in.)	Stem-to-Guide Clearance (in.)		Stem Diameter (in.)	
							Intake	Exhaust	Intake	Exhaust
1993	F2	2.2 (2184)	45	45	NA	①	0.0010-0.0024	0.0012-0.0026	0.3162-0.3167	0.3160-0.3165
	G6	2.6 (2606)	45	45	NA	1.963-1.970	0.0010-0.0023	0.0012-0.0025	0.2744-0.2749	0.2743-0.2748
	JE	3.0 (2954)	45	45	②	③	0.0010-0.0023	0.0012-0.0025	0.2745-0.2750	0.3160-0.3165
	X	4.0 (4016)	45	44	60-68@1.59	1.641-1.729	0.0008-0.0025	0.0018-0.0035	0.3159-0.3167	0.3149-0.3156

VALVE SPECIFICATIONS

Year	Engine ID/VIN	Engine Displacement Liters (cc)	Seat Angle (deg.)	Face Angle (deg.)	Spring Test Pressure (lbs. @ in.)	Spring Installed Height (in.)	Stem-to-Guide Clearance (in.)		Stem Diameter (in.)	
							Intake	Exhaust	Intake	Exhaust
1994	A	2.3 (2299)	45	44	71-79@ 1.52	1.490- 1.550	0.0010- 0.0055	0.0015- 0.0055	0.3416- 0.3423	0.3411- 0.3418
	G6	2.6 (2606)	45	45	NA	1.955	0.0010- 0.0023	0.0012- 0.0025	0.2744- 0.2749	0.2743- 0.2748
	JE	3.0 (2954)	45	45	②	③	0.0010- 0.0023	0.0012- 0.0025	0.2745- 0.2750	0.3160- 0.3165
	U	3.0 (2968)	44	44	65@1.58	1.580	0.0010- 0.0027	0.0015- 0.0032	0.3134- 0.3126	0.3129- 0.3121
	X	4.0 (4016)	45	44	60-68@ 1.59	1.641- 1.729	0.0008- 0.0025	0.0018- 0.0035	0.3159- 0.3167	0.3149- 0.3156
1995	A	2.3 (2299)	45	44	57-63@ 1.56	1.540- 1.580	0.0010- 0.0027	0.0015- 0.0032	0.2746- 0.2754	0.2736- 0.2744
	JE	3.0 (2954)	45	45	②	③	0.0010- 0.0023	0.0012- 0.0025	0.2745- 0.2750	0.3160- 0.3165
	U	3.0 (2968)	45	44	65@1.58	1.736- 1.650	0.0010- 0.0027	0.0015- 0.0032	0.3134- 0.3126	0.3129- 0.3121
	X	4.0 (4016)	45	44	60-68@ 1.59	1.641- 1.729	0.0008- 0.0025	0.0018- 0.0035	0.3159- 0.3167	0.3149- 0.3156
1996-97	A	2.3 (2299)	45	44	57-63@1.56	1.540- 1.580	0.0010- 0.0027	0.0015- 0.0032	0.2746- 0.2754	0.2736- 0.2744
	JE	3.0 (2954)	45	45	④	⑤	0.0010- 0.0023	0.0012- 0.0025	0.2745- 0.2750	0.3160- 0.3165
	U	3.0 (2968)	45	44	⑥	1.736- 1.650	0.0010- 0.0017	0.0015- 0.0032	0.3134- 0.3126	0.3129- 0.3121
	X	4.0 (4016)	45	44	60-68@1.59	1.910 ⑦	0.0008- 0.0025	0.0018- 0.0035	0.3159- 0.3167	0.3149- 0.3156

NA - Not Available
① Free length:
Inner: 1.732
Outer: 2.047

② Intake:
Inner: 21-22@1.56
Outer: 31-33@1.73
Exhaust:
Inner: 33-37@1.59
Outer: 52-58@1.77

③ Intake:
Inner: 1.555
Outer: 1.732
Exhaust:
Inner: 1.594
Outer: 1.772

④ Intake:
Inner: 21-22@1.77
Outer: 31-33@1.73
Exhaust:
Inner: 33-37@1.59
Outer: 21-22@1.56

⑤ Intake:
Inner: 1.840
Outer: 2.004
Exhaust:
Inner: 2.092
Outer: 2.296

⑥ Loaded: 180@1.16
Unloaded: 65@1.58
⑦ Free length only

TORQUE SPECIFICATIONS
All readings in ft. lbs.

Year	Engine ID/VIN	Engine Displacement Liters (cc)	Cylinder Head Bolts	Main Bearing Bolts	Rod Bearing Bolts	Crankshaft Damper Bolts	Flywheel Bolts	Manifold		Spark Plugs	Lug Nut
								Intake	Exhaust		
1993	F2	2.2 (2184)	59-64	61-65	48-51	116-123	71-76	14-22	25-36	11-17	65-87
	G6	2.6 (2606)	①	61-65	48-51	130-145	67-72	14-19	16-21	11-17	65-87
	JE	3.0 (2954)	②	③	④	116-123	76-81	14-19	16-21	10-13	65-87
	X	4.0 (4016)	⑤	66-77	18-24	30-37	59	⑤	18	11-17	100
1994	A	2.3 (2298)	⑥	⑦	30-36	15-22	60	19-28	⑧	7-15	100
	G6	2.6 (2606)	①	61-65	48-51	130-145	67-72	14-19	16-21	11-17	65-87
	JE	3.0 (2954)	②	③	④	116-123	76-81	14-19	16-21	10-13	65-87
	U	3.0 (2968)	⑨	59	23-28	24	59	⑩	18	7-15	100
	X	4.0 (4016)	⑤	66-77	18-24	30-37	59	⑤	18	11-17	100

TORQUE SPECIFICATIONS
All readings in ft. lbs.

Year	Engine ID/VIN	Engine Displacement Liters (cc)	Cylinder Head Bolts	Main Bearing Bolts	Rod Bearing Bolts	Crankshaft Damper Bolts	Flywheel Bolts	Manifold Intake	Manifold Exhaust	Spark Plugs	Lug Nut
1995	A	2.3 (2298)	⑪	⑧	30-36	103-133	56-64	19-28	⑦	7-15	100
	JE	3.0 (2954)	⑫	⑬	⑭	116-122	76-81	14-19	16-21	10-13	65-87
	U	3.0 (2968)	⑮	55-62	23-28	92-122	54-64	⑯	15-22	7-14	100
	X	4.0 (4016)	⑰	66-77	18-24	30-37	⑱	⑲	18	7-15	100
1996-97	A	2.3 (2298)	㉑	⑧	⑳	92-121	56-64	19-28	⑦	7-15	100
	JE	3.0 (2954)	⑫	⑬	⑭	116-122	76-81	14-19	16-21	11-16	65-87
	U	3.0 (2968)	㉒	55-62	23-28	92-122	54-64	⑯	15-22	7-14	100
	X	4.0 (4016)	⑰	66-77	18-24	30-37	60	㉓	18	7-15	100

① Step 1: 59-64 ft. lbs.
Step 2: Tighten two bolts nearest gear 12-17 ft. lbs.

② Step 1: 14 ft. lbs.
Step 2: Turn each bolt 90 degrees
Step 3: Repeat Step 2

③ Step 1: 14 ft. lbs.
Step 2: Plus 90 degrees
Step 3: Plus 45 degrees

④ Step 1: 22 ft. lbs.
Step 2: Turn each nut 40 degrees

⑤ Step 1: Tighten cylinder head to 44 ft. lbs.
Step 2: Tighten intake manifold to 3-6 ft. lbs.
Step 3: Tighten cylinder head to 59 ft. lbs.
Step 4: Tighten intake to 6-11 ft. lbs.
Step 5: Tighten cylinder head 80 to 85 degrees
Step 6: Tighten intake manifold 11-15 ft. lbs.,
then 15-18 ft. lbs.

⑥ Step 1: 51-59 ft. lbs.
Step 2: 80-89 ft. lbs.

⑦ Step 1: 15-22 ft. lbs.
Step 2: 45-59 ft. lbs.

⑧ Step 1: Tighten by hand until seated
Step 2: 50-60 ft. lbs.
Step 3: 75-85 ft. lbs.

⑨ Step 1: 59 ft. lbs.
Step 2: Back off one full turn
Step 3: 37 ft. lbs.
Step 4: 68 ft. lbs.

⑩ Step 1: 11 ft. lbs.
Step 2: 19 ft. lbs.

⑪ Step 1: 51 ft. lbs.
Step 2: Plus 90-100 degrees

⑫ Step 1: 12.7-16.2 ft. lbs.
Step 2: Turn each bolt, in order, 90 degrees
Step 3: Repeat Step 2

⑬ Step 1: 12.7-16.2 ft. lbs.
Step 2: Turn each bolt, in order, 90 degrees
Step 3: Turn each bolt, in order, 45 degrees

⑭ Step 1: 20-23.5 ft. lbs.
Step 2: Plus 90 degrees

⑮ Step 1: 33-41 ft. lbs.
Step 2: 63-73 ft. lbs.

⑯ Step 1: 11 ft. lbs.
Step 2: 19-24 ft. lbs.

⑰ Step 1: 22-26 ft. lbs.
Step 2: 52-56 ft. lbs.
Step 3: Plus 90 degrees

⑱ Step 1: 9-11 ft. lbs.
Step 2: 50-55 ft. lbs.

⑲ Step 1: 6 ft. lbs.
Step 2: 11 ft. lbs.
Step 3: 16 ft. lbs.

⑳ Step 1: 25-30 ft. lbs.
Step 2: 30-36 ft. lbs.

㉑ Step 1: 52 ft. lbs.
Step 2: 52 ft. lbs.
Step 3: Plus 90-100 degrees

㉒ Step 1: 59 ft. lbs.
Step 2: Loosen all bolts 1 turn
Step 3: 33-41 ft. lbs.
Step 4: 63-73 ft. lbs.

㉓ Step 1: 6 ft. lbs.
Step 2: 11 ft. lbs.

BRAKE SPECIFICATIONS
All measurements in inches unless noted

Year	Model	Master Cylinder Bore	Brake Disc Original Thickness	Brake Disc Minimum Thickness	Brake Disc Maximum Runout	Brake Drum Diameter Original Inside Diameter	Brake Drum Diameter Max. Wear Limit	Brake Drum Diameter Maximum Machine Diameter	Minimum Lining Thickness Front	Minimum Lining Thickness Rear
1993	MPV	0.940	③	④	0.004	10.24	10.3	NA	0.080	0.040
	B2200	0.875	①	②	0.006	10.24	10.3	NA	0.118	0.040
	B2600	0.875	①	②	0.006	10.24	10.3	NA	0.118	0.040
	Navajo	NA	NA	0.810	0.010	NA	⑤	0.06	NA	NA
1994	MPV	0.940	③	④	0.004	10.24	10.3	NA	0.080	0.040
	B2300	NA	NA	0.810	0.003	NA	⑤	0.003	0.012	0.003
	B3000	NA	NA	0.810	0.003	NA	⑤	0.003	0.012	0.003
	B4000	NA	NA	0.810	0.003	NA	⑤	0.003	0.012	0.003
	Navajo	NA	NA	0.810	0.010	NA	⑤	0.060	NA	NA
1995	B2300	NA	NA	0.810	0.003	NA	⑤	0.003	0.012	0.003
	B3000	NA	NA	0.810	0.003	NA	⑤	0.003	0.012	0.003
	B4000	NA	NA	0.810	0.003	NA	⑤	0.003	0.012	0.003
	MPV	0.940	⑥	⑦	0.004	NA	NA	NA	0.080	0.080

BRAKE SPECIFICATIONS
All measurements in inches unless noted

Year	Model		Master Cylinder Bore	Brake Disc			Brake Drum Diameter			Minimum Lining Thickness	
				Original Thickness	Minimum Thickness	Maximum Runout	Original Inside Diameter	Max. Wear Limit	Maximum Machine Diameter	Front	Rear
1996-97	B2300		NA	NA	0.810	0.003	NA	⑤	0.003	0.012	0.003
	B3000		NA	NA	0.810	0.003	NA	⑤	0.003	0.012	0.003
	B4000		NA	NA	0.810	0.003	NA	⑤	0.003	0.012	0.003
	MPV	F	0.940	1.100	1.020	0.004	NA	NA	NA	0.080	NA
		R	NA	0.710	0.630	0.004	NA	NA	NA	NA	0.040

NA - Not Available

① 4x2: 0.790
   4x4: 0.870
② 4x2: 0.710
   4x4: 0.790

③ 4x2: 1.180
   4x4: 1.100
④ 4x2: 1.100
   4x4: 1.020

⑤ Refer to the maximum diameter stamped on drum
⑥ Front 4x2: 1.180
   Front 4x4: 1.100
   Rear: 0.710

⑦ Front 4x2: 1.100
   Front 4x4: 1.020
   Rear: 0.630

FREQUENT MAINTENANCE LABOR
MPV, B SERIES, NAVAJO

The following should be used as a guide when determining the amount of work required for a particular service if taken to a repair shop. In estimating how long a particular Frequent Maintenance Service item should take, please observe the following:
- **Chilton Time** is time that is based on field research and data supplied by the vehicle manufacturer.
- All labor time operations are given in hours and tenths of an hour.
- All labor operations, are to be used as a **guide**.

COOLING
Chilton Time

(G) Winterize Cooling System
Includes: Run engine to check for leaks, tighten all hose connections. Test radiator and pressure cap, drain radiator and engine block. Add antifreeze and refill system.
1993-97 5

(G) Belt, Fan Drive, Renew
1993-97 3

(G) Belt, Serpentine Drive, Renew
1994-97 B Series
  4 cyl. 6
  V6 8
1993-94 Navajo 6

(G) Hoses, Radiator, Renew
Includes: Drain and refill cooling system as required.
1993-97
  upper 3
  lower 5

(G) Thermostat, Coolant, Renew
1993 B2200, B2600 7
1994-97 B Series
  4 cyl. 8
  V6 . . . . 1.0
1993-97 MPV 8
1993-94 Navajo 9

FUEL
Chilton Time

(M) Air Cleaner, Service
1993-97 2

(G) Filter, Fuel, Renew
1993-97 3

(G) Carburetor, Adjust (On Car)
1993 . . . . 1.7

BRAKES

(G) Bleed Brakes (Four Wheels)
Includes: Add fluid.
1993-97 6

(G) Brakes, Adjust (Minor)
Includes: Adjust brakes, fill master cylinder.
1993-97, two wheels 4

(M) Parking Brake, Adjust
1993-97 3

LUBRICATION SERVICES

(M) Engine Oil & Filter, Renew
Includes: Inspect and correct all fluid levels.
1993-97 3

ELECTRICAL
Chilton Time

(G) Headlamps, Aim
1993-97
  one side 4
  both sides 6

(G) Halogen Headlamp Bulb, Renew
1993-97 each 2

(G) License Lamp Assy., Renew
1993-97, each 3

(G) Rear Combination Lamp Bulb, Renew
1993 B Series
  one side 2
  both sides 3
1994-97 B Series
  one side 4
  both sides 6
1993-97 MPV
  one side 3
  both sides 5
1993-94 Navajo
  one .4
  each adtnl. 2

(G) Horn, Renew
1993-97
  one .3
  both 4

(M) Terminals, Battery, Clean
1993-97 3

SCHEDULED MAINTENANCE INTERVALS
(MAZDA MPV)

TO BE SERVICED	TYPE OF SERVICE	VEHICLE MILEAGE INTERVAL (x1000)												
		7.5	15	22.5	30	37.5	45	52.5	60	67.5	75	82.5	90	97.5
Engine oil & filter	R	✓	✓	✓	✓	✓	✓	✓	✓	✓	✓	✓	✓	✓
Air cleaner filter	R				✓				✓				✓	
Brake fluid	R				✓				✓				✓	
Spark plugs	R				✓				✓				✓	
Bolts & nuts on chassis & body	S/I				✓				✓				✓	
Cooling system	S/I				✓				✓				✓	
Disc brakes, brake lines, hoses & connections	S/I				✓				✓				✓	
Drive belt(s)	S/I				✓				✓				✓	
Drive shaft dust boots (4WD)	S/I				✓				✓				✓	
Exhaust system heat shields	S/I				✓				✓				✓	
Front suspension ball joints	S/I				✓				✓				✓	
Fuel lines & hoses	S/I				✓				✓				✓	
Idle speed	S/I		✓				✓				✓			
Steering operation & linkages	S/I				✓				✓				✓	
Engine coolant	R						✓				✓			
Timing belt (except Calif.)	R								✓					
Timing belt (Calif.)①	S/I								✓				✓	
Automatic transmission fluid & filter	R								✓					
Front & rear axle oil	R								✓					
Fuel filter & PCV valve	R								✓					
Transfer case oil (4WD)	R								✓					
Emission hoses & tubes②	S/I								✓					
Ignition timing	S/I								✓					

① Timing belt (Calif.) - replace at 105,000 miles, unless previously replaced.
② Emission hoses & tubes - replace at 80,000 miles.
R – Replace   S/I – Service or Inspect

SCHEDULED MAINTENANCE INTERVALS
(MAZDA MPV) (Cont.)

FREQUENT OPERATION MAINTENANCE (SEVERE SERVICE)

   If a vehicle is operated under any of the following conditions it is considered severe service:
- Extremely dusty areas.
- 50% or more of the vehicle operation is in 32°C (90°F) or higher temperatures, or constant operation in temperatures below 0°C (32°F).
- Prolonged idling (vehicle operation in stop and go traffic).
- Frequent short running periods (engine does not warm to normal operating temperatures).
- Police, taxi, delivery usage or trailer towing usage.

Oil & oil filter - change every 5000 miles.
Air cleaner filter - service or inspect every 15,000 miles.
Bolts & nuts on chassis & body - tighten every 15,000 miles.
Spark plugs - replace every 15,000 miles.
Automatic transmission fluid & filter - replace every 30,000 miles.
Front & rear axle oil - replace every 30,000 miles.
Transfer case oil (4WD) - replace every 30,000 miles.

SCHEDULED MAINTENANCE INTERVALS
(MAZDA B SERIES & NAVAJO)

TO BE SERVICED	TYPE OF SERVICE	VEHICLE MILEAGE INTERVAL (x1000)												
		5	10	15	20	25	30	35	40	45	50	55	60	65
Engine oil & filter	R	✓	✓	✓	✓	✓	✓	✓	✓	✓	✓	✓	✓	✓
Automatic transmission shift linkage	S/I	✓		✓		✓		✓		✓		✓		✓
Clutch reservoir fluid	S/I	✓		✓		✓		✓		✓		✓		✓
Exhaust system	S/I	✓		✓		✓		✓		✓		✓		✓
Propeller shaft slip yoke (B Series)	S/I	✓		✓		✓		✓		✓		✓		✓
Propeller shaft U-joints	S/I	✓		✓		✓		✓		✓		✓		✓
Rear propeller shaft double cardan joint centering ball (B Series short bed 4x4)	S/I	✓		✓		✓		✓		✓		✓		✓
Rotate tires	S/I	✓		✓		✓		✓		✓		✓		✓
Slip yoke (Navajo) (if equipped)	S/I	✓		✓		✓		✓		✓		✓		✓
Steering linkage suspension	S/I	✓		✓		✓		✓		✓		✓		✓
Disc brake system & caliper slide rails	S/I			✓			✓			✓			✓	
Drum brake linings, lines & hoses	S/I			✓			✓			✓			✓	
Engine cooling hoses, clamps & coolant condition	S/I			✓			✓			✓			✓	
Transfer case shift lever pivot bolt & control rod connecting pins (4x4)	S/I			✓			✓			✓			✓	
Air cleaner filter	R						✓						✓	
Automatic transmission fluid & filter (B Series)	R						✓						✓	
Engine coolant (Navajo)	R						✓						✓	
Engine coolant (B Series)③	R										✓			
Fuel filter	R						✓						✓	
Accessory drive belts	S/I						✓						✓	
Front axle R.H. axle - shaft slip yoke (4x4)	S/I						✓						✓	

SCHEDULED MAINTENANCE INTERVALS
(MAZDA B SERIES & NAVAJO) (Cont.)

TO BE SERVICED	TYPE OF SERVICE	VEHICLE MILEAGE INTERVAL (x1000)												
		5	10	15	20	25	30	35	40	45	50	55	60	65
Front wheel bearings	S/I						✓						✓	
Hub lock (4x4)	S/I						✓						✓	
Parking brake system	S/I						✓						✓	
Spindle needle bearing spindle thrust bearing (4x4)	S/I						✓						✓	
Throttle linkage & kick down cable ball studs (Navajo)	S/I						✓						✓	
Manual transmission oil	R												✓	
PCV valve	R												✓	
Spark plugs (1993-95)	R						✓						✓	
Spark plugs (1996-97 Calif.)	R												✓	
Spark plugs (1996-97) (exc. Calif.)①	R													
Timing belt (B Series 2.2L) ②	R												✓	
Timing belt (B Series 2.3L) ②	S/I													
Transfer case oil (4x4)	R												✓	

① Replace every 100,000 miles.
② Timing belt (B Series 2.3L) - service or inspect at 120,000 miles.
③ Engine coolant (B Series) - replace initially at 50,000 miles, and every 30,000 miles thereafter.

R – Replace   S/I – Service or Inspect

FREQUENT OPERATION MAINTENANCE (SEVERE SERVICE)

If a vehicle is operated under any of the following conditions it is considered severe service:
- Extremely dusty areas.
- 50% or more of the vehicle operation is in 32°C (90°F) or higher temperatures, or constant operation in temperatures below 0°C (32°F).
- Prolonged idling (vehicle operation in stop and go traffic).
- Frequent short running periods (engine does not warm to normal operating temperatures).
- Police, taxi, delivery usage or trailer towing usage.

Oil & oil filter - replace every 3000 miles.
Automatic transmission shift linkage - lubricate every 6000 miles.
Exhaust system - service or inspect every 6000 miles.
Clutch reservoir fluid - service or inspect every 6000 miles.
Propeller shaft slip yoke (B Series) - lubricate every 6000 miles.
Propeller shaft U-joints - lubricate every 6000 miles.
Rear propeller shaft double cardan joint centering ball (B Series short bed 4x4) - lubricate every 6000 miles.
Rotate tires (B Series) - rotate every 6000 miles.
Slip yoke (Navajo if equipped) - lubricate every 6000 miles.
Steering linkage suspension - lubricate every 6000 miles.
Automatic transmission fluid & filter (B Series) - replace every 21,000 miles.
Automatic transmission fluid & filter (Navajo) - replace every 24,000 miles.
Throttle linkage & kick down cable ball (Navajo) - lubricate every 24,000 miles.
Manual transmission oil - replace every 30,000 miles.
Rear axle oil - replace at 99,000 miles.

MITSUBISHI
Montero • Pick Up

ENGINE IDENTIFICATION

Year	Model	Engine Displacement Liters (cc)	Engine Series (ID/VIN)	Fuel System	No. of Cylinders	Engine Type
1993	Mighty Max	2.4 (2350)	G	MFI	4	SOHC
	Mighty Max	3.0 (2972)	H	MFI	6	SOHC
	Montero	3.0 (2972)	H	MFI	6	SOHC
1994	Mighty Max	2.4 (2350)	G	MFI	4	SOHC
	Mighty Max	3.0 (2972)	A	MFI	6	SOHC
	Montero	3.0 (2972)	H	MFI	6	SOHC
	Montero	3.5 (3497)	M	MFI	6	DOHC
1995	Mighty Max	2.4 (2350)	G	MFI	4	SOHC
	Montero	3.0 (2972)	H	MFI	6	SOHC
	Montero	3.5 (3497)	M	MFI	6	DOHC
1996-97	Mighty Max	2.4 (2350)	G	MFI	4	SOHC
	Montero	3.0 (2972)	H	MFI	6	SOHC
	Montero	3.5 (3497)	M	MFI	6	DOHC

MFI - Multiport fuel injection   SOHC - Single overhead camshaft   DOHC - Double overhead camshaft

GENERAL ENGINE SPECIFICATIONS

Year	Engine ID/VIN	Engine Displacement Liters (cc)	Fuel System Type	Net Horsepower @ rpm	Net Torque @ rpm (ft. lbs.)	Bore x Stroke (in.)	Compression Ratio	Oil Pressure @ rpm
1993	G	2.4 (2350)	MFI	116@5000	136@3500	3.41x3.94	8.5:1	41@2000
	H	3.0 (2972)	MFI	151@5000	174@4000	3.59x2.99	8.9:1	30-80@2000
1994	G	2.4 (2350)	MFI	116@5000	136@3500	3.41x3.94	8.5:1	41@2000
	H	3.0 (2972)	MFI	151@5000	174@4000	3.59x2.99	8.9:1	30-80@2000
	M	3.5 (3496)	MFI	215@5500	228@3000	3.66x3.38	9.5:1	30-80@2000
1995	G	2.4 (2350)	MFI	①	148@3000	3.41x3.94	9.5:1	41@2000
	H	3.0 (2972)	MFI	②	③	3.59x2.99	9.0:1	30-80@2000
	M	3.5 (3497)	MFI	214@5000	228@3000	3.66x3.38	9.5:1	30-80@2000
1996-97	G	2.4 (2350)	MFI	①	148@3000	3.41x3.94	9.5:1	41@2000
	H	3.0 (2972)	MFI	②	③	3.59x2.99	9.0:1	30-80@2000
	M	3.5 (3497)	MFI	214@5000	228@3000	3.66x3.38	9.5:1	30-80@2000

MFI - Multiport fuel injection
① California: 138@5500   ② California: 168@5500   ③ California: 183@4500
Except California: 141@5500   Except California: 177@5500   Except California: 188@4500

GASOLINE ENGINE TUNE-UP SPECIFICATIONS

Year	Engine ID/VIN	Engine Displacement Liters (cc)	Spark Plugs Gap (in.)	Ignition Timing (deg.) MT	Ignition Timing (deg.) AT	Fuel Pump (psi)	Idle Speed (rpm) MT	Idle Speed (rpm) AT	Valve Clearance In.	Valve Clearance Ex.
1993	G	2.4 (2350)	0.039-0.043	5B	5B	38	750	750	HYD	HYD
	H	3.0 (2972)	0.039-0.043	5B	5B	38	700	700	HYD	HYD
1994	G	2.4 (2350)	0.039-0.043	5B	5B	38	750	750	HYD	HYD
	H	3.0 (2972)	0.039-0.043	5B	5B	38	700	700	HYD	HYD
	M	3.5 (3496)	0.039-0.043	5B	5B	38	700	700	HYD	HYD
1995	G	2.4 (2350)	0.039-0.043	5B	5B	38	800	800	HYD	HYD
	H	3.0 (2972)	0.039-0.043	5B	5B	38	700	700	HYD	HYD
	M	3.5 (3497)	0.039-0.043	5B	5B	38	700	700	HYD	HYD
1996-97	G	2.4 (2350)	0.039-0.043	5B	5B	38 ①	650-850	650-850	HYD	HYD
	H	3.0 (2972)	0.039-0.043	5B	5B	38 ①	600-800	600-800	HYD	HYD
	M	3.5 (3497)	0.039-0.043	5B	5B	47-53 ②	600-800	600-800	HYD	HYD

NOTE: The Vehicle Emission Control Information label often reflects specification changes made during production. The label figures must be used if they differ from those in this chart.

B - Before top dead center

HYD - Hydraulic

① With vacuum hose connected

② With vacuum hose disconnected

CAPACITIES

Year	Model	Engine ID/VIN	Engine Displacement Liters (cc)	Engine Oil with Filter (qts.)	Transmission (pts.) 4-Spd	Transmission (pts.) 5-Spd	Transmission (pts.) Auto.	Transfer Case (pts.)	Drive Axle Front (pts.)	Drive Axle Rear (pts.)	Fuel Tank (gal.)	Cooling System (qts.)
1993	Mighty Max	G	2.4 (2350)	4.2	-	4.9	14.8	-	-	3.2	①	②
	Mighty Max	H	3.0 (2972)	5.0	-	5.3	-	4.7	2.4	5.5	15.9	8.9
	Montero	H	3.0 (2972)	5.5	-	5.3	15.2	4.8	2.6	5.5	24.3	10.0
1994	Mighty Max	G	2.4 (2350)	4.2	-	4.9	14.8	-	-	3.2	①	②
	Mighty Max	H	3.0 (2972)	5.0	-	5.3	-	4.7	2.4	5.5	15.9	8.9
	Montero	H	3.0 (2972)	5.5	-	5.3	15.2	4.8	2.6	5.5	24.3	10.0
	Montero	M	3.5 (3497)	5.5	-	5.3	15.2	5.2	2.6	5.5	24.3	10.0
1995	Mighty Max	G	2.4 (2350)	4.2	-	4.9	14.8	NA	NA	3.2	13.7	6.3
	Montero	H	3.0 (2972)	5.1	-	5.3	15.2	4.8	2.6	5.5	24.3	10.0
	Montero	M	3.5 (3497)	5.1	-	5.3	15.2	5.2	2.6	5.5	24.3	10.0
1996-97	Mighty Max	G	2.4 (2350)	4.2	-	4.9	14.8	NA	NA	3.2	①	6.3
	Montero	H	3.0 (2972)	5.0	-	5.3	15.2	4.8	2.6	5.5	24.3	10.0
	Montero	M	3.5 (3497)	5.0	-	5.3	15.2	5.2	2.5	5.5	24.3	10.0

NA - Not Available

① Std. body: 13.7 gals.
   Long body: 18.2 gals.

② Manual transmission: 6.3 qts.
   Automatic transmission: 6.4 qts.

VALVE SPECIFICATIONS

Year	Engine ID/VIN	Engine Displacement Liters (cc)	Seat Angle (deg.)	Face Angle (deg.)	Spring Test Pressure (lbs. @ in.)	Spring Installed Height (in.)	Stem-to-Guide Clearance (in.) Intake	Stem-to-Guide Clearance (in.) Exhaust	Stem Diameter (in.) Intake	Stem Diameter (in.) Exhaust
1993	G	2.4 (2350)	44-44.5	45-45.5	73@1.59	1.59	0.0010-0.0020	0.0020-0.0040	0.315	0.311
	H	3.0 (2972)	44-44.5	45-45.5	72.5@1.59	1.59	0.0010-0.0020	0.0020-0.0040	0.315	0.311

VALVE SPECIFICATIONS

Year	Engine ID/VIN	Engine Displacement Liters (cc)	Seat Angle (deg.)	Face Angle (deg.)	Spring Test Pressure (lbs. @ in.)	Spring Installed Height (in.)	Stem-to-Guide Clearance (in.)		Stem Diameter (in.)	
							Intake	Exhaust	Intake	Exhaust
1994	G	2.4 (2350)	44-44.5	45-45.5	73@1.59	1.59	0.0010-0.0020	0.0020-0.0040	0.315	0.311
	H	3.0 (2972)	44-44.5	45-45.5	72.5@1.59	1.59	0.0010-0.0020	0.0020-0.0040	0.315	0.311
	M	3.5 (3497)	44-44.5	45-45.5	52.9@1.49	1.49	0.0010-0.0020	0.0020-0.0040	0.260	0.256
1995	G	2.4 (2350)	44-44.5	45-45.5	73@1.59	1.59	0.0010-0.0020	0.0020-0.0035	0.315	0.311
	H	3.0 (2972)	44-44.5	45-45.5	72.5@1.59	1.59	0.0010-0.0020	0.0020-0.0035	0.315	0.311
	M	3.5 (3497)	44-44.5	45-45.5	52.9@1.49	1.49	0.0010-0.0020	0.0020-0.0035	0.260	0.256
1996-97	G	2.4 (2350)	44-44.5	45-45.5	73@1.59	1.59	0.0008-0.0020	0.0020-0.0035	0.315	0.311
	H	3.0 (2972)	44-44.5	45-45.5	72.5@1.59	1.59	0.0012-0.0024	0.0020-0.0035	0.315	0.311
	M	3.5 (3497)	44-44.5	45-45.5	52.9@1.49	1.49	0.0008-0.0020	0.0020-0.0035	0.260	0.256

TORQUE SPECIFICATIONS
All readings in ft. lbs.

Year	Engine ID/VIN	Engine Displacement Liters (cc)	Cylinder Head Bolts	Main Bearing Bolts	Rod Bearing Bolts	Crankshaft Damper Bolts	Flywheel Bolts	Manifold Intake *	Exhaust	Spark Plugs	Lug Nut
1993	G	2.4 (2350)	①	②	③	87	98	13	13	18	65-80
	H	3.0 (2972)	80	57	38	136	54	10	14	18	⑤
1994	G	2.4 (2350)	①	②	③	87	98	13	13	18	65-80
	H	3.0 (2972)	80	57	38	136	54	10	14	18	⑤
	M	3.5 (3496)	80	54	38	136	54	10	14	18	72-87
1995	G	2.4 (2350)	①	②	③	87	98	13	13	18	87-101
	H	3.0 (2972)	80	47	38	136	54	10	14	18	72-87
	M	3.5 (3497)	80	54	NA	134	54	10	33	18	72-87
1996-97	G	2.4 (2350)	①	③	③	87	98	13	⑤	18	87-101
	H	3.0 (2972)	80	57	38	136	54	10	14	18	72-87
	M	3.5 (3497)	80	54	25 ④	134	54	10	33	18	72-87

* NOTE: Applies to Lower Manifold only.
① Step 1: 54 ft. lbs.
  Step 2: 14.5 ft. lbs. plus 1/4 turn
  Step 3: Plus an additional 1/4 turn
② Step 1: 18 ft. lbs.
  Step 2: Plus 1/4 turn
③ Step 1: 14.5 ft. lbs.
  Step 2: Plus 1/4 turn
④ Torque to value plus an additional 1/4 turn
⑤ Mighty Max: 87-101 ft. lbs.
  Montero: 72-87 ft. lbs.

BRAKE SPECIFICATIONS
All measurements in inches unless noted

Year	Model	Master Cylinder Bore	Brake Disc Original Thickness	Minimum Thickness	Maximum Runout	Brake Drum Diameter Original Inside Diameter	Max. Wear Limit	Maximum Machine Diameter	Minimum Lining Thickness Front	Rear
1993	Mighty Max	0.938	0.866	0.866	0.006	10.00	10.08	-	0.079	0.039
	Montero	0.938	③	①	0.003	-	7.80	-	0.079	0.079 ④
1994	Mighty Max	0.938	0.866	0.803	0.006	10.00	10.08	-	0.079	0.039
	Montero	0.938	③	①	0.003	-	7.80	-	0.079	0.079 ④

BRAKE SPECIFICATIONS
All measurements in inches unless noted

Year	Model		Master Cylinder Bore	Brake Disc			Brake Drum Diameter			Minimum Lining Thickness	
				Original Thickness	Minimum Thickness	Maximum Runout	Original Inside Diameter	Max. Wear Limit	Maximum Machine Diameter	Front	Rear
1995	Mighty Max		0.938	0.866	0.803	0.006	10.00	10.08	-	0.079	0.039 ②
	Montero		0.938	③	①	0.003	-	7.80	-	0.079	0.079 ②
1996-97	Mighty Max		0.938	0.866	0.803	0.004	10.00	-	10.08	0.079	0.039
	Montero	F	0.938	⑤	⑥	0.004	-	-	-	0.079	-
		R	-	0.710	0.646	0.003	-	-	-	-	0.040

① Front: 0.880
   Rear: 0.330

② Drum shoe: 0.040

③ Front: 0.940
   Rear: 0.710

④ Drum shoe: 0.177

⑤ 3.0L engine: 0.940
   3.5L engine: 1.060

⑥ 3.0L engine: 0.880
   3.5L engine: 1.000

FREQUENT MAINTENANCE LABOR
MONTERO, PICKUP

The following should be used as a guide when determining the amount of work required for a particular service if taken to a repair shop. In estimating how long a particular Frequent Maintenance Service item should take, please observe the following:

- **Chilton Time** is time that is based on field research and data supplied by the vehicle manufacturer.
- All labor time operations are given in hours and tenths of an hour.
- All labor operations, are to be used as a **guide**.

COOLING

Chilton Time

(G) Winterize Cooling System
Includes: Run engine to check for leaks, tighten all hose connections. Test radiator and pressure cap, drain radiator and engine block. Add antifreeze and refill system.
1993-97 5

(G) Belt, Drive, Renew
1993-97 V belt 4
1993-97 Serpentine belt 6

(G) Belt, Drive, Adjust
1993-97 3

(G) Hoses, Radiator, Renew
Includes: Drain and refill cooling system as required.
1993-97 Truck
  upper 4
  lower 6
1993-97 Montero
  upper 4
  lower 8
1993-95 Expo, LRV
  upper 6
  lower 1.0

(G) Thermostat, Coolant, Renew
1993-97 6

FUEL

Chilton Time

(G) Filter, Fuel, Renew
1993-97 Montero
  in line or in tank 7
1993-97 Truck
  in line 5
  in tank 1.1
1993-95 Expo, LRV
  in line 5
  in tank
    2 WD 1.6
    4 WD 2.0

BRAKES

(G) Bleed Brakes (Four Wheels)
Includes: Add fluid.
1993-97 5

(G) Brakes, Adjust (Minor)
Includes: Adjust brakes, fill master cylinder.
1993-97, two wheels 4

(M) Parking Brake, Adjust
1993-97 3

LUBRICATION SERVICES

(M) Engine Oil & Filter, Renew
Includes: Inspect and correct all fluid levels.
1993-97 3

ELECTRICAL

Chilton Time

(G) Headlamps, Aim
1993-97
  two 4
  four 6

(G) License Lamp Assy., Renew
1993-95 Expo, LRV 9
1993-97 Montero 7
1993-97 Truck 4

(G) Park & Turn Signal Lamp Assy., Renew
1993-97 4

(G) Horn, Renew
1993-97 Truck 4
1993-97 Montero 4
1993-95 Expo, LRV 5

(M) Terminals, Battery, Clean
1993-97 3

SCHEDULED MAINTENANCE INTERVALS
(MITSUBISHI MONTERO & PICK UP)

TO BE SERVICED	TYPE OF SERVICE	VEHICLE MILEAGE INTERVAL (x1000)												
		7.5	15	22.5	30	37.5	45	52.5	60	67.5	75	82.5	90	97.5
Engine oil & filter	R	✓	✓	✓	✓	✓	✓	✓	✓	✓	✓	✓	✓	✓
Automatic transmission & transfer oil	S/I		✓		✓		✓		✓		✓		✓	
Brake hoses	S/I		✓		✓		✓		✓		✓		✓	
Disc brake pads & rotors	S/I		✓		✓		✓		✓		✓		✓	
Drive shaft boots	S/I		✓		✓		✓		✓		✓		✓	
Air cleaner filter	R				✓				✓				✓	
Automatic transmission & transfer oil (4WD)	R				✓				✓				✓	
Engine coolant	R								✓				✓	
Ball joints & steering linkage seals	S/I				✓				✓				✓	
Drive belt(s)	S/I				✓				✓				✓	
Drum brake linings & wheel cylinders	S/I				✓				✓				✓	
Exhaust system	S/I				✓				✓				✓	
Front & rear axle	S/I				✓				✓				✓	
Fuel hoses	S/I				✓				✓				✓	
Manual transmission & transfer oil (4WD)	S/I				✓				✓				✓	
Propeller shaft joint	S/I				✓				✓				✓	
Spark plugs (Montero)	R				✓				✓				✓	
Spark plugs (Pick Up or Montero w/platinum tip)	R								✓					
Ignition cables	R								✓					
Timing belt	R								✓					
Distributor cap & rotor	S/I								✓					
EVAP system (except EVAP canister)	S/I								✓					
Fuel system (tank, pipe line connection & fuel tank filler tube cap)	S/I								✓					

SCHEDULED MAINTENANCE INTERVALS
(MITSUBISHI MONTERO & PICK UP) (Cont.)

TO BE SERVICED	TYPE OF SERVICE	VEHICLE MILEAGE INTERVAL (x1000)												
		7.5	15	22.5	30	37.5	45	52.5	60	67.5	75	82.5	90	97.5
EGR valve②	S/I													
EVAP canister②	S/I													
PCV system①	S/I													

① PCV system (except EVAP canister) - service or inspect at 100,000 miles.
② Replace at 100,000 miles.
R – Replace　　　S/I – Service or Inspect

FREQUENT OPERATION MAINTENANCE (SEVERE SERVICE)

If a vehicle is operated under any of the following conditions it is considered severe service:
- **Extremely dusty areas.**
- **50% or more of the vehicle operation is in 32°C (90°F) or higher temperatures, or constant operation in temperatures below 0°C (32°F).**
- **Prolonged idling (vehicle operation in stop and go traffic).**
- **Frequent short running periods (engine does not warm to normal operating temperatures).**
- **Police, taxi, delivery usage or trailer towing usage.**

Oil & oil filter - change every 3000 miles.
Front disc brake pads (dusty or salty conditions) - service or inspect every 6000 miles.
Front disc brake pads - service or inspect every 7500 miles.
Air cleaner filter - service or inspect every 15,000 miles.
Rear drum brake linings & rear wheel cylinders - service or inspect every 15,000 miles.
Spark plugs (except platinum tip) - replace every 15,000 miles.
PCV system - service or inspect every 60,000 miles.

NISSAN
Pathfinder • Pick Up • Quest

ENGINE IDENTIFICATION

Year	Model	Engine Displacement Liters (cc)	Engine Series (ID/VIN)	Fuel System	No. of Cylinders	Engine Type
1993	Pick-up	2.4 (2389)	KA24E	MFI	4	SOHC
	Pick-up	3.0 (2960)	VG30E	MFI	6	SOHC
	Quest	3.0 (2960)	VG30E	MFI	6	SOHC
	Pathfinder	3.0 (2960)	VG30E	MFI	6	SOHC
1994	Pick-up	2.4 (2389)	KA24E	MFI	4	SOHC
	Pick-up	3.0 (2960)	VG30E	MFI	6	SOHC
	Quest	3.0 (2960)	VG30E	MFI	6	SOHC
	Pathfinder	3.0 (2960)	VG30E	MFI	6	SOHC
1995	Pick-up	2.4 (2389)	KA24E (S)	MFI	4	SOHC
	Pick-up	3.0 (2960)	VG30E (H)	MFI	6	SOHC
	Quest	3.0 (2960)	VG30E (W)	MFI	6	SOHC
	Pathfinder	3.0 (2960)	VG30E (H)	MFI	6	SOHC

ENGINE IDENTIFICATION

Year	Model	Engine Displacement Liters (cc)	Engine Series (ID/VIN)	Fuel System	No. of Cylinders	Engine Type
1996-97	Pick-up	2.4 (2389)	KA24E (S)	MFI	4	SOHC
	Pick-up	3.0 (2960)	VG30E (H)	MFI	6	SOHC
	Quest	3.0 (2960)	VG30E (W)	MFI	6	SOHC
	Pathfinder	3.3 (3277)	VG33E	MFI	6	SOHC

MFI - Multiport fuel injection
SOHC - Single overhead camshaft

GENERAL ENGINE SPECIFICATIONS

Year	Engine ID/VIN	Engine Displacement Liters (cc)	Fuel System Type	Net Horsepower @ rpm	Net Torque @ rpm (ft. lbs.)	Bore x Stroke (in.)	Compression Ratio	Oil Pressure @ rpm
1993	KA24E	2.4 (2389)	MFI	134@5200	154@3600	3.50x3.78	8.6:1	60@3000
	VG30E	3.0 (2960)	MFI	153@4800	180@4000	3.43x3.27	9.0:1	53@3200
1994	KA24E	2.4 (2389)	MFI	134@5200	154@3600	3.50x3.78	8.6:1	60@3000
	VG30E	3.0 (2960)	MFI	153@4800	180@4000	3.43x3.27	9.0:1	53@3200
1995	KA24E	2.4 (2389)	MFI	134@5200	154@3600	3.50X3.78	8.6:1	60@3000
	VG30E	3.0 (2960)	MFI	153@4800	180@4000	3.43X3.27	9.0:1	53@3200
1996-97	KA24E	2.4 (2389)	MFI	134@5200	154@3600	3.50x3.78	8.6:1	60@3000
	VG30E	3.0 (2960)	MFI	①	②	3.43x3.27	9.0:1	53@3200
	VG33E	3.3 (3277)	MFI	168@4800	196@2800	3.60x3.27	8.9:1	53@3200

NA - Not Available
① Quest: 151@4800
   Pick-up and Pathfinder: 153@4800
② Quest: 174@4400
   Pick-up and Pathfinder: 180@4000

GASOLINE ENGINE TUNE-UP SPECIFICATIONS

Year	Engine ID/VIN	Engine Displacement Liters (cc)	Spark Plugs Gap (in.)	Ignition Timing (deg.) MT	Ignition Timing (deg.) AT	Fuel Pump (psi)	Idle Speed (rpm) MT	Idle Speed (rpm) AT	Valve Clearance In.	Valve Clearance Ex.
1993	KA24E	2.4 (2389)	0.033	10B	10B	33 ①	800	800 ②	HYD	HYD
	VG30E	3.0 (2960)	0.041	15B	15B	33 ①	750	750 ②	HYD	HYD
1994	KA24E	2.4 (2389)	0.033	10B	10B	33 ①	800	800 ②	HYD	HYD
	VG30E	3.0 (2960)	0.041	15B	15B	33 ①	750	750 ②	HYD	HYD
1995	KA24E	2.4 (2389)	0.033	10B	10B	36 ①	800	800 ②	HYD	HYD
	VG30E	3.0 (2960)	③	15B	15B	34 ①	750	750 ②	HYD	HYD
1996-97	KA24E	2.4 (2389)	0.033	10B	10B	36 ①	800	800 ②	HYD	HYD
	VG30E	3.0 (2960)	③	15B	15B	34 ①	700	700 ②	HYD	HYD
	VG33E	3.3 (3277)	0.041	15B	15B	34 ①	750	750 ②	HYD	HYD

NOTE: The Vehicle Emission Control Information label often reflects specification changes made during production. The label figures must be used if they differ from those in this chart.
B - Before top dead center
HYD - Hydraulic
① System pressure at idle with vacuum hose connected
   Should increase to 43 psi when disconnected
② Automatic transmission in neutral
③ Quest: 0.033
   Pick-up and Pathfinder: 0.041

CAPACITIES

Year	Model	Engine ID/VIN	Engine Displacement Liters (cc)	Engine Oil with Filter (qts.)	Transmission (pts.) 4-Spd	5-Spd	Auto.	Transfer Case (pts.)	Drive Axle Front (pts.)	Rear (pts.)	Fuel Tank (gal.)	Cooling System (qts.)
1993	Pick-up	KA24E	2.4 (2389)	①	-	②	③	-	④	⑤	16.0	⑥
	Pick-up	VG30E	3.0 (2960)	⑦	-	⑧	③	-	⑨	5.8	21.0 ⑩	⑪
	Pathfinder	VG30E	3.0 (2960)	⑦	-	⑧	③	-	⑨	5.8	20.0	⑪
	Quest	VG30E	3.0 (2960)	4.0	-	-	20.0	-	-	-	20.0	⑫
1994	Pick-up	KA24E	2.4 (2389)	①	-	②	16.8	-	④	⑤	16.0	⑥
	Pick-up	VG30E	3.0 (2960)	⑦	-	⑧	③	-	11.0	5.8	21.0 ⑩	⑪
	Pathfinder	VG30E	3.0 (2960)	⑦	-	⑧	③	-	11.0	5.8	20.0	⑫
	Quest	VG30E	3.0 (2960)	4.0	-	-	20.0	-	-	-	20.0	⑪
1995	Pick-up	KA24E	2.4 (2389)	①	-	⑬	-	-	⑭	⑮	15.9	⑥
	Pick-up	VG30E	3.0 (2960)	⑦	-	⑬	⑰	2.4	⑭	⑮	21.1	⑯
	Pathfinder	VG30E	3.0 (2960)	⑦	-	⑬	⑰	2.4	⑭	⑮	20.4	⑯
	Quest	VG30E	3.0 (2960)	4.3	-	-	20.0	-	-	-	20.0	⑫
1996-97	Pick-up	KA24E	2.4 (2389)	①	-	⑬	-	-	⑭	⑮	15.9	⑥
	Pick-up	VG30E	3.0 (2960)	⑦	-	⑬	⑪	2.4	⑭	⑮	21.1	⑯
	Pathfinder	VG33E	3.3 (3277)	3.8	-	⑱	⑲	2.4	4.4	5.9	21.1	10
	Quest	VG30E	3.0 (2960)	4.3	-	-	20.0	-	-	-	20.0	⑫

① 2WD: 4.1 qts.; 4WD: 3.5 qts.
② 2WD: 4.25 pts.; 4WD: 8.50 pts.
③ 2WD: 16.75 pts.; 4WD: 18.0 pts.
④ Front differential: 2.75 pts.
　Transfer case: 4.62 pts.
⑤ 2WD: 3.12 pts.; 4WD: 2.75 pts.
⑥ 2WD: 8.6 qts.; 4WD: 9.5 qts.
⑦ 2WD: 4.25 qts.; 4WD: 3.60 qts.
⑧ 2WD: 5.1 pts.; 4WD: 7.6 pts.

⑨ Front differential: 3.12 pts.
　Transfer case: 4.62 pts.
⑩ SE models: 16 gals.
⑪ 2WD: 11.4 qts.; 4WD: 12.4 qts.
⑫ With rear heater: 12.75 qts.
　Without rear heater: 11.3 qts.
⑬ 2WD KA24DE: 4.3 pts.
　4WD KA24DE: 8.5 pts.
　2WD VG30DE: 5.1 pts.
　4WD VG30DE: 7.6 pts.

⑭ R180A: 2.75 pts.
　R200A: 3.1 pts.
⑮ H190A: 3.1 pts.
　C200: 2.75 pts.
　H233B: 5.9 pts.
⑯ 2WD: 11.4 qts.; 4WD: 12.4 qts.
⑰ 2WD: 8.4 qts.; 4WD: 9.0 qts.
⑱ 2WD: 5.2 qts.; 4WD: 10.8 qts.
⑲ 2WD: 5.2 qts.; 4WD: 10.8 qts.

VALVE SPECIFICATIONS

Year	Engine ID/VIN	Engine Displacement Liters (cc)	Seat Angle (deg.)	Face Angle (deg.)	Spring Test Pressure (lbs. @ in.)	Spring Installed Height (in.)	Stem-to-Guide Clearance (in.) Intake	Exhaust	Stem Diameter (in.) Intake	Exhaust
1993	KA24E	2.4 (2389)	45	45.5	①	NA	0.0008-0.0021	0.0016-0.0028	0.2742-0.2748	0.3129-0.3134
	VG30E	3.0 (2960)	45	45.25-45.75	②	NA	0.0008-0.0021	0.0016-0.0029	0.2742-0.2748	0.3136-0.3138
1994	KA24E	2.4 (2389)	45	45.5	106@1.026	NA	0.0008-0.0021	0.0016-0.0028	0.2742-0.2748	0.3129-0.3134
	VG30E	3.0 (2960)	45	45.25-45.75	①	NA	0.0008-0.0021	0.0016-0.0029	0.2742-0.2748	0.3136-0.3138
1995	KA24E	2.4 (2389)	45	45.5	①	NA	0.0008-0.0021	0.0016-0.0028	0.2742-0.2748	0.3129-0.3134
	VG30E	3.0 (2960)	45	45.25-45.75	②	NA	0.0008-0.0021	0.0016-0.0029	0.2742-0.2748	0.3136-0.3138

VALVE SPECIFICATIONS

Year	Engine ID/VIN	Engine Displacement Liters (cc)	Seat Angle (deg.)	Face Angle (deg.)	Spring Test Pressure (lbs. @ in.)	Spring Installed Height (in.)	Stem-to-Guide Clearance (in.)		Stem Diameter (in.)	
							Intake	Exhaust	Intake	Exhaust
1996-97	KA24E	2.4 (2389)	45	45.5	①	NA	0.0008-0.0021	0.0016-0.0029	0.2742-0.2748	0.3129-0.3134
	VG30E	3.0 (2960)	45	45.25-45.75	②	NA	0.0008-0.0021	0.0016-0.0029	0.2742-0.2748	0.3136-0.3138
	VG33E	3.3 (3277)	45	45.25-46.75	②	NA	0.0008-0.0021	0.0016-0.0029	0.2742-0.2748	0.3136-0.3138

NA - Not Available

① Intake:
Inner: 63.9 @ 1.28
Outer: 135.2 @ 1.48
Exhaust:
Inner: 74 @ 1.15
Outer: 144 @ 1.34

② Inner: 57.3 @ 0.984
Outer: 117.7 @ 1.181

TORQUE SPECIFICATIONS
All readings in ft. lbs.

Year	Engine ID/VIN	Engine Displacement Liters (cc)	Cylinder Head Bolts	Main Bearing Bolts	Rod Bearing Bolts	Crankshaft Damper Bolts	Flywheel Bolts	Manifold		Spark Plugs	Lug Nut
								Intake	Exhaust		
1993	KA24E	2.4 (2389)	①	34-38	②	105-112	③	14	14	18	④
	VG30E	3.0 (2960)	⑤	67-74	②	90-98	72-80	⑥	15	18	④
1994	KA24E	2.4 (2389)	①	34-38	②	105-112	③	14	14	18	⑦
	VG30E	3.0 (2960)	⑤	67-74	②	90-98	72-80	⑥	15	18	⑦
1995	KA24E	2.4 (2389)	①	34-38	⑧	87-116	③	14	14	18	⑨
	VG30E	3.0 (2960)	⑤	67-74	⑧	90-98	72-80	⑩	15	18	⑨
1996-97	KA24E	2.4 (2389)	①	34-38	⑧	87-116	③	14	14	18	⑨
	VG30E	3.0 (2960)	⑤	67-74	⑧	90-98	72-80	⑩	15	18	⑨
	VG33E	3.3 (3277)	⑤	67-74	⑧	141-156	61-69	⑩	21-25	18	⑨

① Step 1: 22 ft. lbs.
Step 2: 58 ft. lbs.
Step 3: Loosen completely then retorque to 22 ft. lbs.
Step 4: 58 ft. lbs. or an additional 80-85 degrees
② 12 ft. lbs. then an additional 60-65 degrees
③ Manual transmission: 105-112 ft. lbs.
Automatic transmission: 69-76 ft. lbs.
④ Pick-up and Pathfinder with single wheel: 87-108 ft. lbs.
Dual wheel: 166-203 ft. lbs.
⑤ Step 1: 22 ft. lbs.
Step 2: 43 ft. lbs.
Step 3: Loosen completely then retorque to 22 ft. lbs.
Step 4: 40-47 ft. lbs. or an additional 60-65 degrees
⑥ Nuts: 18 ft. lbs. in two steps
Bolts: 13 ft. lbs. in two steps

⑦ Pick-up with single wheel: 87-108 ft. lbs.
Pick-up with dual wheel: 166-203 ft. lbs.
⑧ 10-12 ft. lbs. plus 60-65 degrees or 28-33 ft. lbs.
⑨ Quest: 80 ft. lbs.
Pick-up with single wheels: 87-108 ft. lbs.
Pick-up with dual wheels: 166-203 ft. lbs.
⑩ Step 1: Tighten nuts and bolts to 3 ft. lbs.
Step 2: Tighten bolts to 12-14 ft. lbs.; nuts to 17-20 ft. lbs.
Step 3: Repeat Step 2

BRAKE SPECIFICATIONS
All measurements in inches unless noted

Year	Model	Master Cylinder Bore	Brake Disc			Brake Drum Diameter			Minimum Lining Thickness	
			Original Thickness	Minimum Thickness	Maximum Runout	Original Inside Diameter	Max. Wear Limit	Maximum Machine Diameter	Front	Rear
1993	Quest	1.000 ①	NA	0.945	0.003	NA	-	9.90	0.079	0.079
	Pick-up	1.000	NA	②	0.003	NA	-	③	0.079	④
	Pathfinder	1.000	NA	⑤	0.003	NA	-	⑦	0.079	④
1994	Quest	1.000	NA	0.945	0.003	NA	-	9.90	0.079	0.079
	Pick-up	1.000	NA	②	0.003	NA	-	③	0.079	⑧
	Pathfinder	1.000	NA	⑤	0.003	NA	-	⑦	0.079	⑧
1995	Quest	1.000	1.020	0.945	0.003	9.84	NA	9.90	0.079	0.079
	Pick-up	1.000	⑨	⑩	0.003	⑩	NA	③	0.079	0.059
	Pathfinder	1.000	⑫	⑤	0.003	⑬	NA	⑦	0.079	0.059
1996-97	Quest	1.000	1.020	0.945	0.003	9.84	NA	9.90	0.079	0.079
	Pick-up	1.000	⑨	⑩	0.003	⑩	NA	③	0.079	0.059
	Pathfinder	1.000	⑫	⑤	0.003	⑬	NA	⑦	0.079	0.059

NA - Not Available
① With ABS: 1.000
   Without ABS: 0.938
② 2WD, KA24 engine:
   Front: 0.787; Rear: 0.630
   VG30E engine:
   Front: 0.945; Rear: 0.630
③ 2WD: 10.30
   4WD: 11.67
   Parking brake drum: 7.52
④ Disc brake: 0.079
   Drum brake: 0.059

⑤ Front: 0.945
   Rear: 0.630
⑥ Rear disc parking brake drum: 7.52
⑦ Rear drum brake: 10.30
   Rear disc parking brake drum: 7.52
⑧ Disc brake: 0.079
   Drum brake: 0.059
⑨ 2WD KA24E: 0.870
   2WD VG30E: 1.020
   4WD: 1.020
⑩ 2WD KA24E: 0.787
   2WD/4WD VG30E: 0.945

⑪ 2WD: 10.24
   4WD: 11.61
⑫ Front: 1.020
   Rear: 0.710
⑬ Rear drum: 10.24
   Parking drum: 7.48

FREQUENT MAINTENANCE LABOR
PATHFINDER, QUEST, PICKUP

The following should be used as a guide when determining the amount of work required for a particular service if taken to a repair shop. In estimating how long a particular Frequent Maintenance Service item should take, please observe the following:

- **Chilton Time** is time that is based on field research and data supplied by the vehicle manufacturer.
- All labor time operations are given in hours and tenths of an hour.
- All labor operations, are to be used as a **guide**.

COOLING

(G) Winterize Cooling System
Includes: Run engine to check for leaks, tighten all hose connections. Test radiator and pressure cap, drain radiator and engine block. Add antifreeze and refill system.
1993-97 5

(G) Belt, Drive, Renew
1993-97 Pickup, Pathfinder
4 cyl.
KA 4
V6 7
1993-97 Quest 6
Renew each adtnl. belt add 1

(G) Belt, Drive, Adjust
1993-97 4

(G) Hoses, Radiator, Renew
Includes: Drain and refill cooling system as required.
1993-97 Pickup, Pathfinder
upper 7
lower
4 cyl.
KA 1.0
V6 8

1993-97 Quest
upper 5
lower 4

(G) Thermostat, Coolant, Renew
1993-97 Pickup, Pathfinder
4 cyl.
KA 1.4
V6 . . . . 1.8
1993-97 Quest 1.3

BRAKES

(G) Bleed Brakes (Four Wheels)
Includes: Add fluid.
1993-97 . . . . 1.0

(M) Parking Brake, Adjust
1993-97 4

LUBRICATION SERVICES

(M) Engine Oil & Filter, Renew
Includes: Inspect and correct all fluid levels.
1993-97 Pickup, Pathfinder 6
1993-97 Quest 8

WHEELS

(G) Wheel, Renew (One)
1993-97 Quest 5

(G) Wheel, Balance
1993-97 Quest
one 3
each adtnl. 2

(G) Wheels, Rotate (All)
1993-97 Quest 5

ELECTRICAL

(G) Headlamps, Aim
1993-97
two 4
four 6

(G) Halogen Headlamp Bulb, Renew
1993-97 4

(G) License Lamp Assy., Renew
1993-97 4

(G) Rear Combination Lamp Assy., Renew
1993-97, each 5

(G) Horn, Renew
1993-97 4

SCHEDULED MAINTENANCE INTERVALS
(NISSAN PATHFINDER, PICK UP & QUEST)

TO BE SERVICED	TYPE OF SERVICE	VEHICLE MILEAGE INTERVAL (x1000)												
		7.5	15	22.5	30	37.5	45	52.5	60	67.5	75	82.5	90	97.5
Engine oil & filter	R	✓	✓	✓	✓	✓	✓	✓	✓	✓	✓	✓	✓	✓
Brake lines & cables	S/I		✓		✓		✓		✓		✓		✓	
Brake pads, discs, drums & linings	S/I		✓		✓		✓		✓		✓		✓	
Drive shaft boots (Quest & 1993 Pathfinder & Pick Up)	S/I		✓		✓		✓		✓		✓		✓	
Drive shaft boots & propeller shaft (1994-97 4x4 Pathfinder & Pick Up)	S/I		✓		✓		✓		✓		✓		✓	
Front wheel bearings (4x2 Pathfinder & Pick Up)	S/I				✓				✓				✓	
Automatic transaxle oil (Quest)	S/I		✓		✓		✓		✓		✓		✓	
Automatic & manual transmission, transfer & differential gear oil (Pathfinder & Pick Up)②	S/I		✓		✓		✓		✓		✓		✓	
Front wheel bearings (1996-97 4x4 Pathfinder)	S/I		✓		✓		✓		✓		✓		✓	
Front wheel bearings & free running hubs (4x4 Pick Up & 1993-95 4x4 Pathfinder)	S/I		✓		✓		✓		✓		✓		✓	
Propeller shaft (1996-97 Pathfinder)	S/I		✓		✓		✓		✓		✓		✓	
Air cleaner filter	R				✓				✓				✓	
Engine coolant (Quest)	R				✓				✓				✓	
Engine coolant (Pathfinder & Pick Up)	R								✓					
PCV filter (Pathfinder & Pick Up KA24E)	R				✓				✓				✓	
Spark plugs	R				✓				✓				✓	
Drive belt(s) (Pathfinder & Pick Up)	S/I				✓				✓				✓	
Exhaust system	S/I				✓				✓				✓	

SCHEDULED MAINTENANCE INTERVALS
(NISSAN PATHFINDER, PICK UP & QUEST) (Cont.)

TO BE SERVICED	TYPE OF SERVICE	VEHICLE MILEAGE INTERVAL (x1000)												
		7.5	15	22.5	30	37.5	45	52.5	60	67.5	75	82.5	90	97.5
Drive belt(s) (Quest)	S/I								✓		✓		✓	
Fuel lines	S/I				✓				✓				✓	
Steering gear (box) & linkage, (steering damper-4x4), axle & suspension parts (Pathfinder & Pick Up)	S/I				✓				✓				✓	
Steering gear linkage, axle & suspension parts (Quest)	S/I				✓				✓				✓	
Vapor lines	S/I				✓				✓				✓	
Steering linkage ball joints & front suspension ball joints (Pathfinder & Pick Up)	S/I								✓					
Timing belt (1993 VG30E)	R								✓					
Timing belt (1994-97)①	R													

① Timing belt - replace at 105,000 miles.
② Differential (w/limited-slip differential) oil - replace oil every 30,000 miles.
R – Replace   S/I – Service or Inspect

FREQUENT OPERATION MAINTENANCE (SEVERE SERVICE)
If a vehicle is operated under any of the following conditions it is considered severe service:
- Extremely dusty areas.
- 50% or more of the vehicle operation is in 32°C (90°F) or higher temperatures, or constant operation in temperatures below 0°C (32°F).
- Prolonged idling (vehicle operation in stop and go traffic).
- Frequent short running periods (engine does not warm to normal operating temperatures).
- Police, taxi, delivery usage or trailer towing usage.
Oil & oil filter change – change every 3750 miles.
Brake pads, discs, drums & linings - service or inspect every 7500 miles.
Drive shaft boots (Quest) - service or inspect every 7500 miles.
Drive shaft boots & propeller shaft (Pathfinder & Pick Up) - service or inspect every 7500 miles.
Exhaust system - service or inspect every 7500 miles.
Propeller shaft (1996-97 Pathfinder) - service or inspect every 7500 miles. (if immersed in water, grease daily.)
Steering gear (box) & linkage, (steering damper-4x4), axle & suspension parts (Pathfinder & Pick Up) - service or inspect every 7500 miles.
Steering gear linkage, axle & suspension parts (Quest) - service or inspect every 7500 miles.
Steering linkage ball joints & front suspension ball joints - service or inspect every 7500 miles.

SUZUKI
Samurai • Sidekick • X90

ENGINE IDENTIFICATION

Year	Model		Engine Displacement Liters (cc)	Engine Series (ID/VIN)	Fuel System	No. of Cylinders	Engine Type
1993	Samurai		1.3 (1298)	3	EFI	4	SOHC
	Sidekick		1.6 (1590)	0	EFI	4	SOHC
	Sidekick	①	1.6 (1590)	0	MFI	4	SOHC
1994	Samurai		1.3 (1298)	3	TFI	4	SOHC
	Sidekick		1.6 (1590)	0	TFI	4	SOHC
	Sidekick	①	1.6 (1590)	0	MFI	4	SOHC
1995	Samurai		1.3 (1298)	3	TFI	4	SOHC
	Sidekick		1.6 (1590)	0	TFI	4	SOHC
	Sidekick	①	1.6 (1590)	0	MFI	4	SOHC
1996-97	X90		1.6 (1590)	0	MFI	4	SOHC
	Sidekick		1.6 (1590)	0	MFI	4	SOHC
	Sidekick		1.8 (1843)	2	MFI	4	DOHC

EFI - Electronic fuel injection
TFI - Throttle body fuel injection
MFI - Multiport fuel injection

SOHC - Single overhead camshaft
DOHC - Double overhead camshaft
① 16 valve engine

GENERAL ENGINE SPECIFICATIONS

Year	Engine ID/VIN		Engine Displacement Liters (cc)	Fuel System Type	Net Horsepower @ rpm	Net Torque @ rpm (ft. lbs.)	Bore x Stroke (in.)	Compression Ratio	Oil Pressure @ rpm
1993	3	①	1.3 (1298)	EFI	66@6000	76@3500	2.91x2.97	9.5:1	43-60@3000
	0		1.6 (1590)	EFI	80@5400	94@3000	2.95x3.54	8.9:1	51-63@3000
	0	③	1.6 (1590)	MFI	95@5600	98@4000	2.95x3.54	9.5:1	47-61@3000
1994	3	①	1.3 (1298)	TFI	66@6000	76@3500	2.91x2.97	9.5:1	43-60@3000
	0		1.6 (1590)	TFI	80@5400	94@3000	2.95x3.54	8.9:1	51-63@3000
	0	③	1.6 (1590)	MFI	95@5600	98@4000	2.95x3.54	9.5:1	47-61@3000
1995	3	①	1.3 (1298)	TFI	66@6000	76@3500	2.91x2.97	9.5:1	43-60@3000
	0		1.6 (1590)	TFI	80@5400	94@3000	2.95x3.54	8.9:1	51-63@3000
	0	③	1.6 (1590)	MFI	95@5600	98@4000	2.95x3.54	9.5:1	47-61@3000
1996-97	0	②	1.6 (1590)	MFI	95@5600	98@4000	2.95x3.54	9.5:1	47-61@3000
	2		1.8 (1843)	MFI	120@6500	114@3500	3.31x3.27	9.8:1	55-67@4000

EFI - Electronic fuel injection
MFI - Multiport fuel injection
TFI - Throttle body fuel injection

① Samurai
② X90 and Sidekick
③ Sidekick 16 valve engine

GASOLINE ENGINE TUNE-UP SPECIFICATIONS

Year	Engine ID/VIN	Engine Displacement Liters (cc)	Spark Plugs Gap (in.)	Ignition Timing (deg.) MT	Ignition Timing (deg.) AT	Fuel Pump (psi)	Idle Speed (rpm) MT	Idle Speed (rpm) AT	Valve Clearance In.	Valve Clearance Ex.
1993	3 ①	1.3 (1298)	0.029	8B	NA	34-40	800	NA	0.009-0.011 ④	0.010-0.012 ④
	0	1.6 (1590)	0.029	8B	8B	34-40	800	800	0.009-0.011 ④	0.010-0.011 ④
1994	3 ①	1.3 (1298)	0.029	8B	NA	34-40	800	NA	0.009-0.011 ④	0.010-0.012 ④
	0	1.6 (1590)	0.029	8B	8B	34-40	800	800	0.009-0.011 ④	0.010-0.012 ④
1995	3 ①	1.3 (1298)	0.029	8B	NA	34-40 ②	800	NA	0.009-0.011 ④	0.010-0.012 ④
	0 ⑤	1.6 (1590)	0.029	5B	5B	34-40 ②	800	800	⑥	⑥
	0 ⑦	1.6 (1590)	0.029	8B	8B	30-37 ②	800	800	⑧	⑨
1996-97	0 ③	1.6 (1590)	0.029	5B	5B	28-37 ②	800	800	⑩	⑩
	2	1.8 (1843)	0.029	5B	5B	31-37 ②	750-800	750-800	HYD	HYD

NOTE: The Vehicle Emission Control Information label often reflects specification changes made during production. The label figures must be used if they differ from those in this chart.

B - Before top dead center
NA - Not Applicable
HYD - Hydraulic
① Samurai
② At idle
③ X90 and Sidekick
④ Specifications for hot engine
Cold engine: Intake - 0.005-0.007; Exhaust: 0.006-0.008
⑤ Sidekick 16 valve engine

⑥ When cold: 0.005-0.007
When hot: 0.007-0.008
⑦ Sidekick 8 valve engine
⑧ When cold: 0.005-0.007
When hot: 0.009-0.011
⑨ When cold: 0.006-0.008
When hot: 0.010-0.012
⑩ When cold: 0.005-0.007 in.
When hot: 0.007-0.008 in.

CAPACITIES

Year	Model	Engine ID/VIN	Engine Displacement Liters (cc)	Engine Oil with Filter (qts.)	Transmission (pts.) 4-Spd	Transmission (pts.) 5-Spd	Transmission (pts.) Auto.	Transfer Case (pts.)	Drive Axle Front (pts.)	Drive Axle Rear (pts.)	Fuel Tank (gal.)	Cooling System (qts.)
1993	Samurai	3	1.3 (1298)	3.7	-	2.7	-	1.7	4.2	3.2	10.6	5.1
	Sidekick	0	1.6 (1590)	4.5	-	3.2	10.8	3.6	2.1	4.6	11.1	⑤
1994	Samurai	3	1.3 (1298)	3.7	-	2.8	-	1.7	4.2	3.2	10.6	5.1
	Sidekick	0	1.6 (1590)	4.4	-	②	③	3.6	2.2	4.6	④	⑤
1995	Samurai	3	1.3 (1298)	3.7	-	2.8	-	1.7	4.2	3.2	10.6	5.1
	Sidekick	0	1.6 (1590)	4.4	-	②	③	3.6	2.2	4.6	④	⑤
1996-97	X90	0	1.6 (1590)	4.3	-	②	13.0 ①	3.6	2.2	4.6	11.1	⑥
	Sidekick	0	1.6 (1590)	4.3	-	②	③	3.6	2.2	4.6	④	5.8
	Sidekick	2	1.8 (1843)	5.3	-	3.2	13.0 ①	3.6	2.2	4.6	18.5	6.9

① Specification for automatic transaxle is after complete overhaul.
Drain and fill will be less.
② 2WD: 4.0 pts.
4WD: 3.2 pts.
③ 2 door: 6.0 pts.
4 door: 5.3 pts.

④ 2 door: 11.1 gals.
4 door: 14.5 gals.
⑤ With automatic transmission: 5.5 qts.
With manual transmission: 5.6 qts.
⑥ MTX 5.6 qts.
ATX 5.5 qts.

VALVE SPECIFICATIONS

Year	Engine ID/VIN	Engine Displacement Liters (cc)	Seat Angle (deg.)	Face Angle (deg.)	Spring Test Pressure (lbs. @ in.)	Spring Installed Height (in.)	Stem-to-Guide Clearance (in.) Intake	Exhaust	Stem Diameter (in.) Intake	Exhaust
1993	3 ①	1.3 (1298)	45	45	55-64@1.63	1.9409	0.0008-0.0019	0.0014-0.0025	0.2742-0.2748	0.2737-0.2742
	0	1.6 (1590)	45	45	55-64@1.63	1.9866	0.0008-0.0019	0.0014-0.0025	0.2742-0.2748	0.2737-0.2742
	0 ④	1.6 (1590)	45	45	50-57@1.28	1.4500	0.0008-0.0018	0.0018-0.0028	0.2152-0.2157	0.2142-0.2148
1994	3 ①	1.3 (1298)	45	45	55-64@1.63	1.9409	0.0008-0.0019	0.0014-0.0025	0.2742-0.2748	0.2737-0.2742
	0	1.6 (1590)	45	45	55-64@1.63	1.9866	0.0008-0.0019	0.0014-0.0025	0.2742-0.2748	0.2737-0.2742
	0 ④	1.6 (1590)	45	45	50-57@1.28	1.4500	0.0008-0.0018	0.0018-0.0028	0.2152-0.2157	0.2142-0.2148
1995	3 ①	1.3 (1298)	45	45	61-74@1.67	2.008	0.0008-0.0018	0.0014-0.0024	0.2152-0.2157	0.2146-0.2151
	0 ③	1.6 (1590)	45	45	55-64@1.63	1.986	0.0008-0.0019	0.0014-0.0025	0.2742-0.2748	0.2737-0.2742
	0 ④	1.6 (1590)	45	45	50-57@1.28	1.450	0.0008-0.0018	0.0018-0.0028	0.2152-0.2157	0.2142-0.2148
1996-97	0 ②	1.6 (1590)	45	45	24-28@1.24	1.245	0.0008-0.0018	0.0018-0.0028	0.2152-0.2157	0.2142-0.2148
	2	1.8 (1843)	45	45	50-57@1.28	1.280	0.0008-0.0018	0.0018-0.0028	0.2348-0.2354	0.2339-0.2344

① Samurai
② X90 and Sidekick
③ Sidekick 8 valve engine
④ Sidekick 16 valve engine

TORQUE SPECIFICATIONS
All readings in ft. lbs.

Year	Engine ID/VIN	Engine Displacement Liters (cc)	Cylinder Head Bolts	Main Bearing Bolts	Rod Bearing Bolts	Crankshaft Damper Bolts	Flywheel Bolts	Manifold Intake	Exhaust	Spark Plugs	Lug Nut
1993	3 ②	1.3 (1298)	51-54	36-41	24-26	76-83 ①	41-47	13-20	13-20	14-21	58-80
	0	1.6 (1590)	51-54	36-41	24-26	76-83 ①	55-58	13-20	13-20	14-21	58-80
1994	3 ②	1.3 (1298)	51-54	36-41	24-26	76-83 ①	41-47	13-20	13-20	14-21	58-80
	0	1.6 (1590)	48-51	36-41	24-26	76-83 ①	55-58	13-20	13-20	14-21	58-80
1995	3 ②	1.3 (1298)	51-54	36-41	24-26	76-83 ①	50-52	13-20	13-20	15-22	36-50
	0 ⑤	1.6 (1590)	48-51	36-41	24-26	76-83 ①	55-58	13-20	13-20	14-21	58-80
	0 ⑥	1.6 (1590)	51-54	36-41	24-26	76-83 ①	55-58	13-20	13-20	14-21	58-80
1996-97	0 ③	1.6 (1590)	48-51	36-41	24-26	76-83 ①	55-58	13-20	13-20	14-21	58-80
	2	1.8 (1843)	④	⑦	33	108.5	51	13-20	13-20	14-21	58-80

① Specification shown is for crankshaft timing sprocket bolt
② Samurai
③ X90 and Sidekick
④ M10: 76 ft. lbs.
   M6: 8 ft. lbs.
⑤ Sidekick 8 valve engine
⑥ Sidekick 16 valve engine
⑦ 10mm threaded diameter: 42 ft. lbs.
   8mm threaded diameter: 19.5 ft. lbs.

BRAKE SPECIFICATIONS
All measurements in inches unless noted

Year	Model		Master Cylinder Bore	Brake Disc			Brake Drum Diameter			Minimum Lining Thickness	
				Original Thickness	Minimum Thickness	Maximum Runout	Original Inside Diameter	Max. Wear Limit	Maximum Machine Diameter	Front	Rear
1993	Samurai		NA	0.394	0.334	0.006	8.66	8.74	8.74	0.236 ①	0.120 ①
	Sidekick		NA	0.394	0.315	0.006	8.66	8.74	8.74	0.315 ①	0.120 ①
	Sidekick	②	NA	0.669	0.591	0.006	10.00	10.07	10.07	0.315 ①	0.120 ①
1994	Samurai		NA	0.394	0.334	0.006	8.66	8.74	8.74	0.236 ①	0.120 ①
	Sidekick		NA	0.669	0.591	0.006	10.00	10.07	10.07	0.315 ①	0.120 ①
1995	Samurai		NA	0.394	0.334	0.006	8.66	8.74	8.74	0.236 ①	0.120 ①
	Sidekick	②	NA	0.669	0.591	0.006	10.00	10.07	10.07	0.315 ①	0.120 ①
	Sidekick	③	NA	0.394	0.315	0.006	8.66	8.74	8.74	0.315 ①	0.120 ①
1996-97	X90		NA	0.394	0.315	0.006	8.66	8.74	8.74	0.240 ①	0.120 ①
	Sidekick	③	NA	0.394	0.315	0.006	8.66	8.74	8.74	0.240 ①	0.120 ①
	Sidekick	④	NA	0.669	0.591	0.006	10.00	10.07	10.07	0.315 ①	0.120 ①
	Sidekick	⑤	NA	0.866	0.787	0.006	10.00	10.07	10.07	0.275 ①	0.120 ①

NA - Not Available
① Measurement is for lining and backing
② 4 door model
③ 2 door model
④ 4 door model with 1.6L engine
⑤ 4 door model with 1.8L engine

FREQUENT MAINTENANCE LABOR
SAMURAI, SIDEKICK, X90

The following should be used as a guide when determining the amount of work required for a particular service if taken to a repair shop. In estimating how long a particular Frequent Maintenance Service item should take, please observe the following:
- **Chilton Time** is time that is based on field research and data supplied by the vehicle manufacturer.
- All labor time operations are given in hours and tenths of an hour.
- All labor operations, are to be used as a **guide**.

COOLING

(G) Winterize Cooling System
Includes: Run engine to check for leaks, tighten all hose connections. Test radiator and pressure cap, drain radiator and engine block. Add antifreeze and refill system.
1993-97 5

(G) Hoses, Radiator, Renew
Includes: Drain and refill cooling system as required.
1993-97
upper 5
lower 7

(G) Thermostat, Coolant, Renew
1993-95 Samurai 4
1993-97 Sidekick, X-90 5

FUEL

(G) Filter, Fuel, Renew
w/Fuel Injection
1993-97 Samurai, Sidekick, X-90 7

BRAKES

(G) Bleed Brakes (Four Wheels)
Includes: Add fluid.
1993-97 8
w/ABS add 2

(G) Brakes, Adjust (Minor)
Includes: Adjust brakes, fill master cylinder.
1993-97 4

(M) Parking Brake, Adjust
1993-97 5

LUBRICATION SERVICES

(M) Engine Oil & Filter, Renew
Includes: Inspect and correct all fluid levels.
1993-97 3

WHEELS

(G) Wheel, Renew (One)
1993-97 5

(G) Wheel, Balance
1993-97
one 3
each adtnl. 2

(G) Wheels, Rotate (All)
1993-97 5

ELECTRICAL

(G) Headlamps, Aim
1993-97 4

(G) Halogen Headlamp Bulb, Renew
1993-97 3

(G) License Lamp Assy., Renew
1993-95 Samurai 5
1993-97 Sidekick, X-90 4

(G) Rear Combination Lamp Assy., Renew
1993-95 Samurai 4
1993-97 Sidekick, X-90 3

(G) Tail Lamp Lens or Bulb, Renew
1993-97 3

(G) Horn, Renew
1993-97 3

(M) Terminals, Battery, Clean
1993-97 3

SCHEDULED MAINTENANCE INTERVALS
(SUZUKI SAMURAI, SIDEKICK & X90)

TO BE SERVICED	TYPE OF SERVICE	VEHICLE MILEAGE INTERVAL (x1000)												
		7.5	15	22.5	30	37.5	45	52.5	60	67.5	75	82.5	90	97.5
Engine oil & filter	R	✓	✓	✓	✓	✓	✓	✓	✓	✓	✓	✓	✓	✓
Automatic transmission fluid (Sidekick) ④	S/I	✓	✓	✓	✓	✓	✓	✓	✓	✓	✓	✓	✓	✓
Manual transmission oil (Sidekick) ⑧	S/I	✓	✓	✓	✓	✓	✓	✓	✓	✓	✓	✓	✓	✓
Power steering system (Sidekick)	S/I	✓	✓	✓	✓	✓	✓	✓	✓	✓	✓	✓	✓	✓
Steering system	S/I	✓	✓	✓	✓	✓	✓	✓	✓	✓	✓	✓	✓	✓
Transfer & differential oil (Sidekick) ⑧	S/I	✓	✓	✓	✓	✓	✓	✓	✓	✓	✓	✓	✓	✓
Transmission, transfer & differential oil (Samurai) ⑦	S/I	✓	✓	✓	✓	✓	✓	✓	✓	✓	✓	✓	✓	✓
Wheel discs & free wheeling hubs	S/I	✓	✓	✓	✓	✓	✓	✓	✓	✓	✓	✓	✓	✓
Shock absorbers (Samurai)	S/I	✓	✓		✓		✓		✓		✓		✓	
Suspension bolts & nuts (Samurai)	S/I	✓	✓		✓		✓		✓		✓		✓	
Suspension system (Sidekick)	S/I	✓	✓		✓		✓		✓		✓		✓	
Brake discs & pads (front)	S/I		✓		✓		✓		✓		✓		✓	
Brake drums & shoes (rear)	S/I		✓		✓		✓		✓		✓		✓	
Brake fluid ⑥	S/I		✓		✓		✓		✓		✓		✓	
Brake hoses & pipes	S/I		✓		✓		✓		✓		✓		✓	
Brake pedal	S/I		✓		✓		✓		✓		✓		✓	
Brake lever & cable	S/I		✓		✓		✓		✓		✓		✓	
Clutch	S/I		✓		✓		✓		✓		✓		✓	
Idle speed	S/I		✓		✓		✓		✓		✓		✓	
Propeller shafts	S/I		✓		✓		✓		✓		✓		✓	
Valve lash (clearance)	S/I		✓		✓		✓		✓		✓		✓	
Wheel bearings	S/I		✓		✓		✓		✓		✓		✓	
Steering knuckle oil seals (Samurai)	R			✓			✓			✓			✓	

SCHEDULED MAINTENANCE INTERVALS
(SUZUKI SAMURAI, SIDEKICK & X90) (Cont.)

TO BE SERVICED	TYPE OF SERVICE	VEHICLE MILEAGE INTERVAL (x1000)												
		7.5	15	22.5	30	37.5	45	52.5	60	67.5	75	82.5	90	97.5
Air cleaner filter element	R				✓				✓				✓	
Engine Coolant	R				✓				✓				✓	
Fuel filter	R				✓				✓				✓	
Spark plugs	R				✓				✓				✓	
Cooling system hoses & connections	S/I				✓				✓				✓	
Drive belt(s)	S/I				✓				✓				✓	
Exhaust pipes & mountings	S/I				✓				✓				✓	
Fuel lines & connections	S/I				✓				✓				✓	
Fuel tank cap gasket	S/I				✓				✓				✓	
Leaf springs (Samurai)	S/I				✓				✓				✓	
Camshaft timing belt	R								✓				✓	
Distributor cap & rotor	S/I								✓					
Emission-related hoses & tubes	S/I								✓					
Oxygen sensor or heated oxygen sensor②	R													
EVAP canister④	R													
PCV valve①	R													
EGR system⑤	S/I													
Fuel Injectors③	S/I													
TWC converter③	S/I													

① PCV valve - replace every 50,000 miles.
② Oxygen sensor or heated oxygen sensor - service or inspect at 80,000 miles.
③ Service or inspect at 100,000 miles.
④ Replace at 100,000 miles.
⑤ EGR system - service or inspect every 50,000 miles.
⑥ Replace every 60,000 miles.
⑦ Replace oil at 7500 miles and every 30,000 miles thereafter.
⑧ Replace oil every 30,000 miles.
R – Replace   S/I – Service or Inspect

SCHEDULED MAINTENANCE INTERVALS
(SUZUKI SAMURAI, SIDEKICK & X90) (Cont.)

FREQUENT OPERATION MAINTENANCE (SEVERE SERVICE)

If a vehicle is operated under any of the following conditions it is considered severe service:

- Extremely dusty areas.
- 50% or more of the vehicle operation is in 32°C (90°F) or higher temperatures, or constant operation in temperatures below 0°C (32°F).
- Prolonged idling (vehicle operation in stop and go traffic).
- Frequent short running periods (engine does not warm to normal operating temperatures).
- Police, taxi, delivery usage or trailer towing usage.

Oil & oil filter change – change every 3000 miles.
Air cleaner filter element - service or inspect every 3000 miles & replace every 15,000 miles.
Steering wheel free play, gear box oil & linkage - service or inspect every 3000 miles.
Bolts & nuts on chassis - tighten every 6000 miles.
Brake discs & pads (front) - service or inspect every 6000 miles.
Brake drums & shoes (rear) - service or inspect every 6000 miles.
Exhaust pipes & mountings - tighten every 6000 miles.
Propeller shafts - service or inspect every 6000 miles.
Automatic transmission fluid & filter - replace every 15,000 miles.
Distributor cap & ignition wires - service or inspect every 15,000 miles.
Drive belt(s) - service or inspect every 15,000 miles.
Manual transmission oil - replace every 15,000 miles.
Transfer & differential oil - replace every 15,000 miles.

TOYOTA
Land Cruiser • Pick Up • Previa • RAV4 • T100 • Tacoma • 4Runner

ENGINE IDENTIFICATION

Year	Model	Engine Displacement Liters (cc)	Engine Series (ID/VIN)	Fuel System	No. of Cylinders	Engine Type
1993	Previa ①	2.4 (2438)	2TZ-FE	EFI	4	DOHC
	4Runner	2.4 (2366)	22R-E	EFI	4	SOHC
	4Runner	3.0 (2959)	3VZ-E	EFI	6	SOHC
	Pick-up	2.4 (2366)	22R-E	EFI	4	SOHC
	Pick-up	3.0 (2959)	3VZ-E	EFI	6	SOHC
	T100	3.0 (2959)	3VZ-E	EFI	6	SOHC
	Land Cruiser	4.5 (4477)	1FZ-FE	EFI	6	DOHC
1994	Previa ①	2.4 (2438)	2TZ-FE	EFI	4	DOHC
	4Runner	2.4 (2366)	22R-E	EFI	4	SOHC
	4Runner	3.0 (2959)	3VZ-E	EFI	6	SOHC
	Pick-up	2.4 (2366)	22R-E	EFI	4	SOHC
	Pick-up	3.0 (2959)	3VZ-E	EFI	6	SOHC
	T100	2.7 (2693)	3RZ-FE	EFI	4	DOHC
	T100	3.0 (2959)	3VZ-E	EFI	6	SOHC
	Land Cruiser	4.5 (4477)	1FZ-FE	EFI	6	DOHC

ENGINE IDENTIFICATION

Year	Model		Engine Displacement Liters (cc)	Engine Series (ID/VIN)	Fuel System	No. of Cylinders	Engine Type
1995	Previa		2.4 (2438)	2TZ-FE	EFI	4	DOHC
	Previa	①	2.4 (2438)	2TZ-FE	EFI	4	DOHC
	4Runner		2.4 (2366)	22R-E	EFI	4	SOHC
	4Runner		3.0 (2959)	3VZ-E	EFI	6	SOHC
	Pick-up		2.4 (2366)	22R-E	EFI	4	SOHC
	Pick-up		3.0 (2959)	3VZ-E	EFI	6	SOHC
	Tacoma		2.4 (2438)	2RZ-FE	EFI	4	DOHC
	Tacoma		2.7 (2693)	3RZ-FE	EFI	4	DOHC
	Tacoma		3.4 (3378)	5VZ-FE	EFI	6	DOHC
	T100		2.7 (2693)	3RZ-FE	EFI	4	DOHC
	T100		3.4 (3378)	5VZ-FE	EFI	6	DOHC
	Land Cruiser		4.5 (4477)	1FZ-FE	EFI	6	DOHC
1996-97	RAV4		2.0 (1998)	3S-FE	EFI	4	DOHC
	Previa	①	2.4 (2438)	2TZ-FZE	EFI	4	DOHC
	4Runner		2.7 (2693)	3RZ-FE	EFI	4	DOHC
	4Runner		3.4 (3378)	5VZ-FE	EFI	6	DOHC
	Tacoma		2.4 (2438)	2RZ-FE	EFI	4	DOHC
	Tacoma		2.7 (2693)	3RZ-FE	EFI	4	DOHC
	Tacoma		3.4 (3378)	5VZ-FE	EFI	4	DOHC
	T100		2.7 (2693)	3RZ-FE	EFI	4	DOHC
	T100		3.4 (3378)	5VZ-FE	EFI	6	DOHC
	Land Cruiser		4.5 (4477)	1FZ-FE	EFI	6	DOHC

EFI - Electronic fuel injection
SOHC - Single overhead camshaft
DOHC - Double overhead camshaft

OHV - Overhead valve
① Supercharged

GENERAL ENGINE SPECIFICATIONS

Year	Engine ID/VIN	Engine Displacement Liters (cc)	Fuel System Type	Net Horsepower @ rpm	Net Torque @ rpm (ft. lbs.)	Bore x Stroke (in.)	Compression Ratio	Oil Pressure @ rpm
1993	2TZ-FE	2.4 (2438)	EFI	138@5000	154@4000	3.74x3.39	9.1:1	36-71@3000
	22R-E	2.4 (2366)	EFI	116@4800	140@2800	3.62x3.50	9.3:1	36-71@3000
	3VZ-E	3.0 (2959)	EFI	150@4800	180@3400	3.44x3.23	9.0:1	36-71@4000
	1FZ-FE	4.5 (4477)	EFI	212@4600	275@3200	3.94x3.74	9.0:1	36-71@3000
1994	2TZ-FE	2.4 (2438)	EFI	161@5000	201@3000	3.74x3.39	8.9:1	36-71@3000
	2TZ-FE	2.4 (2438)	EFI	138@5000	154@4000	3.74x3.39	9.1:1	36-71@3000
	22R-E	2.4 (2366)	EFI	116@4800	140@2800	3.62x3.50	9.3:1	36-71@3000
	3RZ-FE	2.7 (2693)	EFI	150@4800	177@4000	3.74x3.74	9.5:1	36-71@3000
	3VZ-E	3.0 (2959)	EFI	150@4800	180@3400	3.44x3.23	9.0:1	36-71@4000
	1FZ-FE	4.5 (4477)	EFI	212@4600	275@3200	3.94x3.64	9.0:1	36-71@3000
1995	2TZ-FE	2.4 (2438)	EFI	161@5000	201@3000	3.74x3.39	8.9:1	36-71@3000
	2TZ-FE	2.4 (2438)	EFI	138@5000	154@4000	3.74x3.39	9.1:1	36-71@3000
	22R-E	2.4 (2366)	EFI	116@4800	140@2800	3.62x3.50	9.3:1	36-71@3000
	2RZ-FE	2.4 (2438)	EFI	142@5000	160@4000	3.74x3.38	9.5:1	36-71@3000
	3RZ-FE	2.7 (2693)	EFI	150@4800	177@4000	3.74x3.74	9.5:1	36-71@3000
	3VZ-E	3.0 (2959)	EFI	150@4800	180@3400	3.44x3.23	9.0:1	36-71@3000
	5VZ-FE	3.4 (3378)	EFI	190@4800	220@3400	3.68x3.23	9.6:1	NA
	1FZ-FE	4.5 (4477)	EFI	212@4600	275@3200	3.94x3.64	9.0:1	36-71@3000

GENERAL ENGINE SPECIFICATIONS

Year	Engine ID/VIN	Engine Displacement Liters (cc)	Fuel System Type	Net Horsepower @ rpm	Net Torque @ rpm (ft. lbs.)	Bore x Stroke (in.)	Com-pression Ratio	Oil Pressure @ rpm
1996-97	3S-FE	2.0 (1998)	EFI	120@5400	125@4600	3.40x3.40	9.5:1	NA
	2TZ-FZE	2.4 (2438)	EFI	161@5000	201@3600	3.74x3.39	8.9:1	36@3000
	2RZ-FE	2.4 (2438)	EFI	142@5000	160@4000	3.74x3.38	9.5:1	36-71@3000
	3RZ-FE	2.7 (2693)	EFI	150@4800	177@4000	3.74x3.74	9.5:1	36-71@3000
	5VZ-FE	3.4 (3378)	EFI	190@4800	220@3600	3.68x3.23	9.6:1	NA
	1FZ-FE	4.5 (4477)	EFI	212@4600	275@3200	3.94x3.74	9.0:1	36-71@3000

EFI - Electronic fuel injection

NA - Not Available

GASOLINE ENGINE TUNE-UP SPECIFICATIONS

Year	Engine ID/VIN	Engine Displacement Liters (cc)	Spark Plugs Gap (in.)	Ignition Timing (deg.) MT	Ignition Timing (deg.) AT	Fuel Pump (psi)	Idle Speed (rpm) MT	Idle Speed (rpm) AT	Valve Clearance In.	Valve Clearance Ex.
1993	22R-E	2.4 (2366)	0.031	5B	5B	38-44	750	850	0.008	0.012
	2TZ-FE	2.4 (2438)	0.043	5B	5B	38-44	700	750	①	②
	3VZ-E	3.0 (2959)	0.041	10B	10B	38-44	800	800	③	④
	1FZ-FE	4.5 (4477)	0.031	-	3B	38-44	-	600-700	0.006-0.010	0.010-0.014
1994	22R-E	2.4 (2366)	0.031	5B	5B	38-44	750	850	0.008	0.012
	2TZ-FE	2.4 (2438)	0.043	5B	5B	38-44	-	750	①	②
	3RZ-FE	2.7 (2693)	0.031	5B	-	38-44	750	-	0.008	0.012
	3VZ-E	3.0 (2959)	0.031	10B	10B	38-44	800	800	③	④
	1FZ-FE	4.5 (4477)	0.031	-	3B	38-44	-	600-700	0.006-0.010	0.010-0.014
1995	22R-E	2.4 (2366)	0.031	5B	5B	38-44	750	850	0.008	0.012
	2TZ-FE	2.4 (2438)	0.043	5B	5B	38-44	-	750	①	②
	2RZ-FE	2.4 (2438)	0.031	⑤	⑤	38-44	650-750	650-750	0.006-0.010	0.010-0.014
	3RZ-FE	2.7 (2693)	0.031	⑤	⑤	38-44	750	-	0.008	0.012
	3VZ-E	3.0 (2959)	0.031	10B	10B	38-44	800	800	③	④
	5VZ-FE	3.4 (3378)	0.043	⑥	⑥	38-44	650-750	650-750	0.006-0.009	0.011-0.014
	1FZ-FE	4.5 (4477)	0.031	-	3B	38-44	-	600-700	0.006-	0.010-
1996-97	3S-FE	2.0 (1998)	0.043	10B ⑤	5B ⑤	44-50	700-800	700-800	0.007-0.011	0.011-0.015
	2TZ-FZE	2.4 (2438)	0.043	-	5B ⑥	38-44	700-800	700-800	0.006-0.010	0.010-0.014
	2RZ-FE	2.4 (2438)	0.031	5B ⑥	5B ⑥	38-44	650-750	-	0.006-0.010	0.010-0.014
	3RZ-FE	2.7 (2693)	0.031	5B ⑥	5B ⑥	38-44	650-750	650-750	0.006-0.010	0.010-0.014
	5VZ-FE	3.4 (3378)	0.043	10B ⑤	10B ⑤	38-44	650-750	650-750	0.006-0.009	0.011-0.014
	1FZ-FE	4.5 (4477)	0.031	-	3B	38-44	-	600-700	0.006-0.010	0.010-0.014

NOTE: The Vehicle Emission Control Information label often reflects specification changes made during production. The label figures must be used if they differ from those in this chart.

B - Before top dead center

① Intake: 0.006-0.010 (cold)

② Exhaust: 0.010-0.014 (cold)

③ Intake: 0.007-0.011 (cold)

④ Exhaust: 0.009-0.013 (cold

⑤ 5B at idle, with terminal TE1 and E1 connected of DLC1

⑥ 10B at idle, with terminal TE1 and E1 connected of DLC1

23 SPECIFICATIONS

CAPACITIES

Year	Model	Engine ID/VIN	Engine Displacement Liters (cc)	Engine Oil with Filter (qts.)	Transmission (pts.) 4-Spd	Transmission (pts.) 5-Spd	Transmission (pts.) Auto.	Transfer Case (pts.)	Drive Axle Front (pts.)	Drive Axle Rear (pts.)	Fuel Tank (gal.)	Cooling System (qts.)
1993	Previa	2TZ-FE	2.4 (2438)	6.1	-	[1]	5.0	3.0	2.2	3.2	19.8	12.3
	Pick-up	22R-E	2.4 (2366)	4.5	-	[12]	[13]	[14]	[5]	[15]	[16]	[17]
	Pick-up	3VZ-E	3.0 (2959)	4.8	-	[12]	[13]	[14]	[5]	[15]	[16]	[18]
	4Runner	22R-E	2.4 (2366)	4.5	-	[12]	[13]	[14]	[5]	[15]	[16]	[17]
	4Runner	3VZ-E	3.0 (2959)	4.8	-	[12]	[13]	[14]	[5]	[15]	[16]	[18]
	T100	3VZ-E	3.0 (2959)	4.8	-	6.4	[13]	2.4	[5]	[15]	[16]	[18]
	Land Cruiser	1FZ-FE	4.5 (4477)	7.8	-	-	12.6	[19]	[20]	6.8	25.1	14.8
1994	Previa	2TZ-FE	2.4 (2438)	6.1	-	-	5.0	3.0	2.2	3.2	19.8	13.0
	Previa	2TZ-FZE	2.4 (2438)	6.1	-	-	5.0	3.0	2.2	3.2	19.8	13.0
	Pick-up	22R-E	2.4 (2366)	4.5	-	[12]	[13]	[14]	[5]	[15]	[16]	[17]
	Pick-up	3VZ-E	3.0 (2959)	4.8	-	[12]	[13]	[14]	[5]	[15]	[16]	[18]
	4Runner	22R-E	2.4 (2366)	4.5	-	[12]	[13]	[14]	[5]	[15]	[16]	[17]
	4Runner	3VZ-E	3.0 (2959)	4.8	-	[12]	[13]	[14]	[5]	[15]	[16]	[18]
	T100	3RZ-FE	2.7 (2693)	5.6	-	2.7	-	-		3.8	19.8	9.2
	T100	3VZ-E	3.0 (2959)	4.8	-	6.4	[13]	2.4	[5]	[15]	[16]	[18]
	Land Cruiser	1FZ-FE	4.5 (4477)	7.8	-	-	12.6	[19]	[20]	6.8	25.1	14.8
1995	Previa	2TZ-FE	2.4 (2438)	6.1	-	-	5.0	3.0	2.2	3.2	19.8	13.0
	Previa	2TZ-FZE	2.4 (2438)	6.1	-	-	5.0	3.0	2.2	3.2	19.8	13.0
	Pick-up	22R-E	2.4 (2366)	4.5	-	[12]	[21]	[14]	[5]	[22]	[16]	[17]
	Pick-up	3VZ-E	3.0 (2959)	4.8	-	[12]	[21]	[14]	[5]	[22]	[16]	[23]
	4Runner	22R-E	2.4 (2366)	4.5	-	[12]	[21]	[14]	[5]	[22]	[16]	[23]
	4Runner	3VZ-E	3.0 (2959)	4.8	-	[12]	[21]	[14]	[5]	[22]	[16]	[23]
	Tacoma	2RZ-FE	2.4 (2438)	5.8	-	[24]	[25]	2.2	[26]	2.9	15.1	[28]
	Tacoma	3RZ-FE	2.7 (2693)	5.8	-	[24]	[25]	2.2	[26]	[27]	18.0	[28]
	Tacoma	5VZ-FE	3.4 (3378)	[29]	-	[24]	[25]	2.2	[26]	[27]	18.0	[9]
	T100	3RZ-FE	2.7 (2693)	5.6	-	2.7	-	-		3.8	19.8	9.2
	T100	5VZ-FE	3.4 (3378)	[10]	[11]	[21]	2.4	3.9	[22]		24.0	[23]
	Land Cruiser	1FZ-FE	4.5 (4477)	7.8	-	-	13	[19]	[20]	6.8	25.1	14.8
1996-97	RAV4	3S-FE	2.0 (1998)	4.1	-	[2]	7.0	-	-	-	15.3	[3]
	Previa	2TZ-FZE	2.4 (2438)	6.1	-	-	3.4	2.8	[4]	3.2	19.8	13.0
	4Runner	3RZ-FE	2.7 (2693)	5.8	-	[24]	[25]	2.4	[26]	[6]	18.0	[7]
	4Runner	5VZ-FE	3.4 (3378)	5.5	-	[24]	[25]	2.4	[26]	[6]	18.0	[8]
	Tacoma	2RZ-FE	2.4 (2438)	5.8	-	[24]	[25]	2.4	[26]	2.9	15.1	[28]
	Tacoma	3RZ-FE	2.7 (2693)	5.8	-	[24]	[25]	2.4	[26]	[27]	18.0	[28]
	Tacoma	5VZ-FE	3.4 (3378)	[29]	-	[24]	[25]	2.4	[26]	[27]	18.0	[9]
	T100	3RZ-FE	2.7 (2693)	5.8	-	2.7	-	-		3.8	24.0	9.2
	T100	5VZ-FE	3.4 (3378)	5.8	[10]	[10]	[21]	2.4	3.9	[22]	-	24.0	[23]
	Land Cruiser	1FZ-FE	4.5 (4477)	7.8	-	-	4.0	3.6	[20]	6.8	25.1	38.0

① 2WD: 4.6; 4WD: 5.4
② 2WD: 8.2 / 4WD: 10.6
③ M/T: 8.5 / A/T: 8.1
④ 2WD: 3.2 / 2WD: 2.2
⑤ Standard: 3.4; ADD: 4.0
⑥ 2WD: 5.8 / 4WD with differential locks: 5.8 / 4WD without differential locks: 5.2
⑦ With rear heater: 11.6 / Without rear heater: 10.6
⑧ With rear heater: 9.5 / Without rear heater: 8.5

⑨ M/T: 10.7 / A/T: 10.5
⑩ 2WD: 5.5 / 4WD: 5.0
⑪ 2WD: 5.4 / 4WD: 4.6 / All others: 5.4
⑫ G58: 8.2 / R150F: 6.4
⑬ 2WD: 3.4 / 4WD with A340H: 9.6 / 4WD with A340F: 4.2
⑭ Except 3VZ-E AT (VF1A type): 2.4 / 3VZ-E AT (A340H): 1.6
⑮ 2WD: 3.8; 4WD: 4.6

⑯ With standard tires: 17.2 / With optional 31x10.5 tires: 18.8
⑰ With rear heater: 9.2 / Without rear heater: 8.9
⑱ 2WD M/T: 11.0 / 2WD A/T: 10.8 / 4WD M/T: 11.1 / 4WD A/T: 10.9
⑲ With ABS: 2.8 / Without ABS: 3.6
⑳ With differential lock: 5.6 / Without differential lock: 5.8
㉑ Drain and refill: A340E: 3.4 / A340F: 4.2

㉒ 2WD: 4.4 / 4WD: 4.3
㉓ 2WD M/T: 10.6 / 2WD A/T: 10.5 / 4WD M/T: 10.6 / 4WD A/T: 10.8
㉔ W59: 2WD: 5.4 / 4WD: 5.2 / R150, R150F: 2WD: 5.4 / 4WD: 4.6
㉕ A43D: 5.0 / A340E: 3.4 / A340F: 4.2

㉖ Without ADD: 2.32 / With ADD: 2.44
㉗ Extra long: 4.4 / All others: 5.4
㉘ 2WD M/T: 8.5 / 2WD A/T: 8.2
㉙ 2WD: 5.7 / 4WD: 5.5

VALVE SPECIFICATIONS

Year	Engine ID/VIN	Engine Displacement Liters (cc)	Seat Angle (deg.)		Face Angle (deg.)	Spring Test Pressure (lbs. @ in.)	Spring Installed Height (in.)	Stem-to-Guide Clearance (in.)		Stem Diameter (in.)	
								Intake	Exhaust	Intake	Exhaust
1993	2TZ-FE	2.4 (2438)	45	①	44.5	57-63	1.406	0.0010-0.0024	0.0012-0.0026	0.2350-0.2356	0.2348-0.2354
	22R-E	2.4 (2366)	45	①	44.5	66.1	1.594	0.0010-0.0024	0.0012-0.0026	0.3138-0.3144	0.3136-0.3142
	3VZ-E	3.0 (2959)	45	①	44.5	54-57	1.575	0.0010-0.0024	0.0012-0.0026	0.3138-0.3144	0.3136-0.3142
	1FZ-FE	4.5 (4477)	45		44.5	53.4	1.437	0.0010-0.0024	0.0012-0.0026	0.2744-0.2750	0.2742-0.2748
1994	2TZ-FE	2.4 (2438)	45	①	44.5	57-63	1.406	0.0010-0.0024	0.0012-0.0026	0.2350-0.2356	0.2348-0.2354
	22R-E	2.4 (2366)	45	①	44.5	66.1	1.594	0.0010-0.0024	0.0012-0.0026	0.3138-0.3144	0.3136-0.3142
	3RZ-FE	2.7 (2693)	45	①	44.5	57-63	1.406	0.0010-0.0024	0.0012-0.0026	0.2350-0.2356	0.2348-0.2354
	3VZ-E	3.0 (2959)	45	①	44.5	54-57	1.575	0.0010-0.0024	0.0012-0.0026	0.3138-0.3144	0.3136-0.3142
	1FZ-FE	4.5 (4477)	45		44.5	53.4	1.437	0.0010-0.0024	0.0012-0.0026	0.2744-0.2750	0.2742-0.2748
1995	2TZ-FE	2.4 (2438)	45	①	44.5	57-63@1.406	1.406	0.0010-0.0024	0.0012-0.0026	0.2350-0.2356	0.2348-0.2354
	22R-E	2.4 (2366)	45	①	44.5	66.1@1.594	1.594	0.0010-0.0024	0.0012-0.0026	0.3138-0.3144	0.3136-0.3142
	2RZ-FE	2.4 (2438)	45	①	44.5	40-46@1.406	1.406	0.0010-0.0024	0.0012-0.0026	0.2350-0.2356	0.2348-0.2354
	3RZ-FE	2.7 (2693)	45	①	44.5	57-63@1.406	1.406	0.0010-0.0024	0.0012-0.0026	0.2350-0.2356	0.2348-0.2354
	3VZ-E	3.0 (2959)	45	①	44.5	54-57@1.575	1.575	0.0010-0.0024	0.0012-0.0026	0.3138-0.3144	0.3136-0.3142
	5VZ-FE	3.4 (3378)	45		44.5	41.9-46.3@1.311	1.311	0.0010-0.0024	0.0012-0.0026	0.2350-0.2356	0.2348-0.2354
	1FZ-FE	4.5 (4477)	45		44.5	53.4@1.437	1.437	0.0010-0.0024	0.0012-0.0026	0.2744-0.2750	0.2742-0.2748
1996-97	3S-FE	2.0 (1998)	45		44.5	36.8-42.5@1.366	1.366	0.0010-0.0024	0.0012-0.0026	0.2350-0.2356	0.2348-0.2354
	2RZ-FE	2.4 (2438)	45		44.5	40-46@1.406	1.406	0.0010-0.0024	0.0012-0.0026	0.2350-0.2356	0.2348-0.2354
	2TZ-FZE	2.4 (2438)	45		44.5	38.7-42.8@1.594	1.406	0.0010-0.0024	0.0012-0.0026	0.2350-0.2356	0.2348-0.2354
	3RZ-FE	2.7 (2693)	45		44.5	40-46@1.406	1.406	0.0010-0.0024	0.0012-0.0026	0.2350-0.2356	0.2348-0.2354
	5VZ-FE	3.4 (3378)	45		44.5	41.9-46.3@1.311	1.311	0.0010-0.0024	0.0012-0.0026	0.2350-0.2356	0.2348-0.2354
	1FZ-FE	4.5 (4477)	45		44.5	48.1-53.4@1.437	1.437	0.0010-0.0024	0.0012-0.0026	0.2744-0.2750	0.2742-0.2748

① Blend seat with 30 and 60 degree cutters to center the 45 degree portion on valve face

TORQUE SPECIFICATIONS
All readings in ft. lbs.

Year	Engine ID/VIN	Engine Displacement Liters (cc)	Cylinder Head Bolts	Main Bearing Bolts	Rod Bearing Bolts	Crankshaft Damper Bolts	Flywheel Bolts	Manifold Intake	Manifold Exhaust	Spark Plugs	Lug Nut
1993	2TZ-FE	2.4 (2438)	①	②	③	192	④	15	36	11-15	-
	22R-E	2.4 (2366)	53-63	69-83	40-47	120-130	73-86	13-19	26-36	11-15	-
	3VZ-E	3.0 (2959)	⑤	⑥	⑦	176-186	63-67	11-15	25-33	11-15	-
	1FZ-FE	4.5 (4477)	⑩	⑩	⑪	304	74	15	29	15	-
1994	2TZ-FE	2.4 (2438)	①	②	③	192	④	15	36	11-15	-
	22R-E	2.4 (2366)	53-63	69-83	40-47	120-130	73-86	13-19	26-36	11-15	-
	3RZ-FE	2.7 (2693)	⑩	⑩	⑭	192	⑥	22	36	14	-
	3VZ-E	3.0 (2959)	⑤	⑥	⑦	176-186	63-67	11-15	25-33	11-15	-
	1FZ-FE	4.5 (4477)	⑩	⑪	⑫	304	74	15	29	15	-
1995	2TZ-FE	2.4 (2438)	①	②	③	192	④	15	36	11-15	-
	22R-E	2.4 (2366)	53-63	69-83	40-47	120-130	73-86	13-19	26-36	11-15	-
	2RZ-FE	2.4 (2438)	①	⑬	⑭	193	65	22	36	14	83
	3RZ-FE	2.7 (2693)	①	⑬	⑭	⑮	⑮	22	36	14	-
	3VZ-E	3.0 (2959)	⑤	⑥	⑦	176-186	63-67	11-15	25-33	11-15	-
	5VZ-FE	3.4 (3378)	⑤	⑦	⑮	176-186	63-67	11-15	25-33	11-15	76
	1FZ-FE	4.5 (4477)	⑩	⑪	⑫	304	74	15	29	15	-
1996-97	3S-FE	2.0 (1998)	⑱	43	⑰	80	⑲	14	36	13	76
	2RZ-FE	2.4 (2438)	⑧	⑨	⑯	193	⑳	22	36	14	83
	2TZ-FZE	2.4 (2438)	⑧	⑨	⑯	192	54	15	36	14	76
	3RZ-FE	2.7 (2693)	⑧	⑨	⑯	⑰	㉑	22	36	14	83
	5VZ-FE	3.4 (3378)	㉒	㉓	⑰	184	63-67	13	30	13	76
	1FZ-FE	4.5 (4477)	⑧	㉔	㉕	304	74	15	29	14	㉖

① Step 1: 29 ft. lbs.
　Step 2: 90 degree turn
　Step 3: 90 degree turn
② Step 1: 20 ft. lbs.
　Step 2: 35 ft. lbs.
　Step 3: 58 ft. lbs.
③ Step 1: 22 ft. lbs.
　Step 2: 90 degree turn
④ Manual transmission: 65 ft. lbs.
　Automatic transmission: 54 ft. lbs.
⑤ Step 1: 27 ft. lbs.
　Step 2: 33 ft. lbs.
　Step 3: 90 degree turn
　Step 4: 90 degree turn
⑥ Step 1: 18 ft. lbs.
　Step 2: 90 degree turn
⑦ Step 1: 45 ft. lbs.
　Step 2: 90 degree turn

⑧ Step 1: 29 ft. lbs.
　Step 2: 90 degree turn
　Step 3: 90 degree turn
⑨ Step 1: 29 ft. lbs.
　Step 2: 90 degree turn
⑩ Step 1: 27 ft. lbs.
　Step 2: 90 degree turn
　Step 3: 90 degree turn
⑪ Step 1: 54 ft. lbs.
　Step 2: 90 degree turn
⑫ Step 1: 35 ft. lbs.
　Step 2: 90 degree turn
⑬ Step 1: 29 ft. lbs.
　Step 2: 90 degree turn
⑭ Step 1: 33 ft. lbs.
　Step 2: 90 degree turn
⑮ Step 1: 19 ft. lbs.
　Step 2: 90 degree turn

⑯ Step 1: 33 ft. lbs.
　Step 2: 90 degree turn
⑰ Step 1: 18 ft. lbs.
　Step 2: 90 degree turn
⑱ Step 1: 36 ft. lbs.
　Step 2: 90 degree turn
⑲ Manual transmission: 65 ft. lbs.
　Automatic transmission: 61 ft. lbs.
⑳ Manual trans.: 65 ft. lbs.
　Automatic trans.: 54 ft. lbs.
㉑ Manual trans.: 19 ft. lbs. + 90 degrees
　Automatic trans.: 54 ft. lbs.
㉒ Step 1: 25 ft. lbs.
　Step 2: 90 degree turn
　Recessed head: 13 ft. lbs.
㉓ Step 1: 45 ft. lbs.
　Step 2: 90 degree turn

㉔ Step 1: 54 ft. lbs.
　Step 2: 90 degree turn
㉕ Step 1: 35 ft. lbs.
　Step 2: 90 degree turn
㉖ Steel wheel: 109 ft. lbs.
　Aluminum wheel: 76 ft. lbs.

BRAKE SPECIFICATIONS
All measurements in inches unless noted

Year	Model		Master Cylinder Bore	Brake Disc Original Thickness	Brake Disc Minimum Thickness	Maximum Runout	Brake Drum Diameter Original Inside Diameter	Max. Wear Limit	Maximum Machine Diameter	Minimum Lining Thickness Front	Minimum Lining Thickness Rear
1993	Previa		NA	①	②	0.0028	10.00	-	10.08	0.039	0.039
	Pick-up		NA	③	④	⑦	⑨	-	⑩	0.039	0.039
	4Runner		NA	0.984	0.906	0.0035	11.61	-	11.69	0.059	0.039
	Land Cruiser		NA	1.260 ⑫	1.181 ⑬	0.0059	11.61	-	11.69	0.059	⑮
1994	Previa		NA	①	②	0.0028	10.00	-	10.08	0.039	0.039
	Pick-up		NA	⑪	⑬	0.0035	⑨	-	⑩	⑭	0.039
	4Runner		NA	0.984	0.906	0.0035	11.61	-	11.69	0.059	0.039
	Land Cruiser		NA	1.260 ⑫	1.181 ⑬	0.0059	11.61	-	11.69	0.059	⑮
1995	Previa		NA	①	②	0.0028	10.00	-	10.08	0.039	0.039
	Pick-up		NA	⑪	⑫	0.0035	⑨	-	⑩	⑭	0.039
	4Runner		NA	0.984	0.906	0.0035	11.61	-	11.69	0.059	0.039
	Land Cruiser		NA	1.260 ⑫	1.181 ⑬	0.0059	11.61	-	11.69	0.059	⑮
	Tacoma		NA	0.866	0.787	0.0028	⑨	-	⑩	0.039	0.039
1996-97	Previa	F	NA	⑤	⑥	0.0028	-	-	-	0.039	-
		R	NA	0.709	0.630	0.0039	10.00	-	10.18	-	0.039
	4Runner		NA	0.866	0.787	0.0028	11.61	-	11.69	0.039	0.039
	RAV4		NA	0.709	0.630	0.0020	9.00	-	9.08	0.039	0.039
	Land Cruiser	F	NA	1.260	1.181	0.0059	-	-	-	0.039	-
		R	NA	0.709	0.630	-	-	-	-	-	⑮
	T100 2WD		NA	0.984	0.906	⑧	11.61	-	11.69	0.039	0.039
	T100 4WD		NA	0.984	0.984	0.0028	11.61	-	11.69	0.039	0.039
	Tacoma 2WD		NA	0.866	0.787	0.0028	10.00	-	10.08	0.039	-
	Tacoma 4WD		NA	0.866	0.787	0.0028	11.61	-	11.69	-	0.039

NA - Not Available

① Front with rear drum brake: 0.984
   Front with rear disc brake: 0.866
   Rear disc: 0.709
② Front with rear drum brake: 0.906
   Front with rear disc brake: 0.787
   Rear disc: 0.669
③ 2WD (PD60 type disc): 0.984
   2WD (PD66 type disc): 1.181
   2WD (FS17, 18 type disc): 0.866
   4WD (S12+12 type disc): 0.787
④ 2WD (PD60 type disc): 0.906
   2WD (PD66 type disc): 1.102
   2WD (FS17, 18 type disc): 0.787
   4WD (S12+12 type disc): 0.709

⑤ With rear drum brake: 0.985
   With rear disc brake: 0.866
⑥ With rear drum brake: 0.906
   With rear disc brake: 0.787
⑦ PD60, FS17, FS18, S12+12 type discs: 0.0035
⑧ 1 ton: 0.0035
   1/2 ton: 0.0028
⑨ 2WD: 10.00; 4WD: 11.61
⑩ 2WD: 10.08; 4WD: 11.69
⑪ 2WD: 0.866; 4WD: 0.787
⑫ Rear disc: 0.630
⑬ Rear disc: 0.709
⑭ 2WD: 0.059; 4WD: 0.039
⑮ Brake shoe lining: 0.059
   Disc pad lining: 0.039

FREQUENT MAINTENANCE LABOR
LAND CRUISER, PICKUP, PREVIA, RAV4, T100, TACOMA, 4RUNNER

The following should be used as a guide when determining the amount of work required for a particular service if taken to a repair shop. In estimating how long a particular Frequent Maintenance Service item should take, please observe the following:
- **Chilton Time** is time that is based on field research and data supplied by the vehicle manufacturer.
- All labor time operations are given in hours and tenths of an hour.
- All labor operations, are to be used as a **guide**.

COOLING

(G) Winterize Cooling System
Includes: Run engine to check for leaks, tighten all hose connections. Test radiator and pressure cap, drain radiator and engine block. Add antifreeze and refill system.

	Chilton Time
1993-97	.5

(G) Belt, Fan Drive, Renew

1993-97 Previa	.6
1993-97 T100	.4
1993-95 Pickup	.4
1993-95 4Runner	.4
1993-97 Land Cruiser	.4
1995-97 Tacoma	.4

(G) Hoses, Radiator, Renew
Includes: Drain and refill cooling system as required.

1993-97	
upper	.8
lower	1.0

(G) Thermostat, Coolant, Renew

Four	
1993-97 Previa	1.0
1995-97 T100	1.0
1993-95 Pickup	.9
1993-95 4Runner	.9
1995-97 Tacoma	1.0

Six	
1993-97 Land Cruiser	1.1

V6	
1993-95 Pickup, 4Runner	1.0
1993-97 T100	1.0
1995-97 Tacoma	1.0

FUEL

(G) Filter, Fuel, Renew

	Chilton Time
1993-95 Pickup, 4Runner	1.0
1993-97 T100	1.0
1993-97 Previa	1.0
1993-97 Land Cruiser	.9
1995-97 Tacoma	1.0

BRAKES

(G) Bleed Brakes (Four Wheels)
Includes: Add fluid.

1993-97	
two wheels	.6
four wheels	1.0

(G) Brakes, Adjust (Minor)
Includes: Adjust brakes, fill master cylinder.

1993-97, two wheels	.4

(M) Parking Brake, Adjust

1993-97 Previa	.6
1993-97 T100	.4

	Chilton Time
1993-95 Pickup	.4
1993-95 4Runner	.4
1993-97 Land Cruiser	.6
1995-97 Tacoma	.4

LUBRICATION SERVICES

(M) Engine Oil & Filter, Renew
Includes: Inspect and correct all fluid levels.

1993-97	.3

ELECTRICAL

(G) Headlamps, Aim

1993-97	
two	.4
four	.6

(G) Halogen Headlamp Bulb, Renew

1993-97	.3

(G) Parking Lamp Lens or Bulb, Renew

1993-97	.3

(G) Horn, Renew

1993-97	.4

(M) Terminals, Battery, Clean

1993-97	.3

SCHEDULED MAINTENANCE INTERVALS
(TOYOTA LAND CRUISER, PICK UP, PREVIA, RAV4, T100, TACOMA & 4RUNNER)

TO BE SERVICED	TYPE OF SERVICE	VEHICLE MILEAGE INTERVAL (x1000)												
		7.5	15	22.5	30	37.5	45	52.5	60	67.5	75	82.5	90	97.5
Engine oil & filter	R	✓	✓	✓	✓	✓	✓	✓	✓	✓	✓	✓	✓	✓
Rotate tires②	S/I	✓	✓	✓	✓	✓	✓	✓	✓	✓	✓	✓	✓	✓
Driveshaft boots (Pick Up 4WD)	S/I	✓	✓	✓	✓	✓	✓	✓	✓	✓	✓	✓	✓	✓
Driveshaft boots (T100 & Tacoma 4WD)	S/I		✓			✓		✓		✓		✓	✓	
Idle speed (T100, 4Runner & 1995-97 Pick Up)	S/I	✓		✓		✓		✓		✓		✓		✓
Drive belts (T100, 4Runner & 1995-97 Pick Up)	S/I					✓			✓	✓	✓	✓	✓	✓
Drive belts (Previa & Tacoma)	S/I								✓	✓	✓	✓	✓	✓
Drive belts (Land Cruiser & 1993-94 Pick Up)	S/I					✓			✓				✓	
Automatic transmission fluid & filter	S/I		✓		✓		✓		✓		✓		✓	
Ball joints & dust covers	S/I		✓		✓		✓		✓		✓		✓	
Bolts & nuts on chassis & body	S/I		✓		✓		✓		✓		✓		✓	
Brake linings & drums	S/I		✓		✓		✓		✓		✓		✓	
Brake line pipes & hoses	S/I		✓		✓		✓		✓		✓		✓	
Brake pads discs (front & rear)	S/I		✓		✓		✓		✓		✓		✓	
Manual transmission fluid (Tacoma, 4Runner & Pick Up)	S/I		✓		✓		✓		✓		✓		✓	
Propeller shaft grease	S/I		✓		✓		✓		✓		✓		✓	
Steering knuckle & chassis grease	S/I		✓		✓		✓		✓		✓		✓	
Steering linkage	S/I		✓		✓		✓		✓		✓		✓	
Transfer, differential & steering gear box oil	S/I		✓		✓		✓		✓		✓		✓	
Air cleaner filter	R				✓				✓				✓	

SCHEDULED MAINTENANCE INTERVALS
(TOYOTA LAND CRUISER, PICK UP, PREVIA, RAV4, T100, TACOMA & 4RUNNER) (Cont.)

TO BE SERVICED	TYPE OF SERVICE	VEHICLE MILEAGE INTERVAL (x1000)												
		7.5	15	22.5	30	37.5	45	52.5	60	67.5	75	82.5	90	97.5
Front wheel bearing & thrust bushing grease	R				✓				✓				✓	
Spark plugs (except Previa)	R				✓				✓				✓	
Spark plugs (Previa)	R								✓					
Exhaust pipes & mountings	S/I				✓				✓				✓	
Fuel lines & connections	S/I				✓				✓				✓	
Valve clearance (4Runner & Pick Up 22R-E)	S/I				✓				✓				✓	
Valve clearance (Previa, Tacoma, T100, 4Runner & Pick Up 3VZ-E)	S/I								✓					
Engine Coolant	R						✓				✓			
Charcoal canister (Calif.)	R								✓					
Fuel tank cap gasket	R								✓					
Oxygen sensor or heated oxygen sensors①	R													

① Oxygen sensor or heated oxygen sensors - replace every 80,000 miles.
② 4WD vehicles - rotate tires every 5000 miles.
R – Replace   S/I – Service or Inspect

FREQUENT OPERATION MAINTENANCE (SEVERE SERVICE)
   If a vehicle is operated under any of the following conditions it is considered severe service:
- Extremely dusty areas.
- 50% or more of the vehicle operation is in 32°C (90°F) or higher temperatures, or constant operation in temperatures below 0°C (32°F).
- Prolonged idling (vehicle operation in stop and go traffic).
- Frequent short running periods (engine does not warm to normal operating temperatures).
- Police, taxi, delivery usage or trailer towing usage.
Air cleaner filter - service or inspect every 3750 miles.
Oil & oil filter change – change every 3750 miles.
Ball joints & dust covers - service or inspect every 7500 miles.
Bolts & nuts on body & chassis - service or inspect every 7500 miles.
Brake linings & drums - service or inspect every 7500 miles.
Brake pads & discs (front & rear) - service or inspect every 7500 miles.
Drivebelts (T100 & Tacoma) - service or inspect initially at 45,000 miles and every 7500 miles thereafter.
Steering knuckle & chassis grease - service or inspect every 7500 miles.
Steering linkage - service or inspect every 7500 miles.
Propeller shaft grease - service or inspect every 7500 miles.
Driveshaft boots (Previa, T100 & Tacoma 4WD) - service or inspect every 7500 miles.
Driveshaft boots (4Runner 4WD) - service or inspect every 15,000 miles.
Exhaust pipes & mountings - service or inspect every 15,000 miles.
Timing belt (T100, Tacoma 5VZ-FE & Pick Up 3VZ-E) - replace every 60,000 miles.

VOLKSWAGEN
Eurovan

ENGINE IDENTIFICATION

Year	Model	Engine Displacement Liters (cc)	Engine Series (ID/VIN)	Fuel System	No. of Cylinders	Engine Type
1993	Eurovan	2.5 (2461)	AAF	Digifant	5	SOHC
1994	Eurovan	2.5 (2461)	AAF	Digifant	5	SOHC
1995	Eurovan	2.5 (2461)	AAF	Digifant	5	SOHC
1996-97	Eurovan	2.5 (2461)	AAF	Digifant	5	SOHC

SOHC - Single overhead camshaft

GENERAL ENGINE SPECIFICATIONS

Year	Engine ID/VIN	Engine Displacement Liters (cc)	Fuel System Type	Net Horsepower @ rpm	Net Torque @ rpm (ft. lbs.)	Bore x Stroke (in.)	Compression Ratio	Oil Pressure @ rpm
1993	AAF	2.5 (2461)	Digifant	109@4500	140@2200	3.19x3.76	8.5:1	29@2000
1994	AAF	2.5 (2461)	Digifant	109@4500	140@2200	3.19x3.76	8.5:1	29@2000
1995	AAF	2.5 (2461)	Digifant	109@4500	140@2200	3.19x3.76	8.5:1	29@2000
1996-97	AAF	2.5 (2461)	Digifant	109@4500	140@2200	3.19x3.76	8.5:1	29@2000

GASOLINE ENGINE TUNE-UP SPECIFICATIONS

Year	Engine ID/VIN	Engine Displacement Liters (cc)	Spark Plugs Gap (in.)	Ignition Timing (deg.) MT	Ignition Timing (deg.) AT	Fuel Pump (psi)	Idle Speed (rpm) MT	Idle Speed (rpm) AT	Valve Clearance In.	Valve Clearance Ex.
1993	AAF	2.5 (2461)	0.030	5-7B	5-7B	43.5	775-825	775-825	HYD	HYD
1994	AAF	2.5 (2461)	0.030	5-7B	5-7B	43.5	775-825	775-825	HYD	HYD
1995	AAF	2.5 (2461)	0.030	5-7B	5-7B	43.5	775-825	775-825	HYD	HYD
1996-97	AAF	2.5 (2461)	0.030	5-7B	5-7B	43.5	775-825	775-825	HYD	HYD

NOTE: The Vehicle Emission Control Information label often reflects specification changes made during production. The label figures must be used if they differ from those in this chart.
B - Before top dead center
HYD - Hydraulic

CAPACITIES

Year	Model	Engine ID/VIN	Engine Displacement Liters (cc)	Engine Oil with Filter (qts.)	Transmission (pts.)			Transfer Case (pts.)	Drive Axle		Fuel Tank (gal.)	Cooling System (qts.)
					4-Spd	5-Spd	Auto.		Front (pts.)	Rear (pts.)		
1993	Eurovan	AAF	2.5 (2461)	5.9	-	5.4	13	-	-	-	21.1	12.2
1994	Eurovan	AAF	2.5 (2461)	5.9	-	5.4	11.5	-	-	-	21.1	12.2
1995	Eurovan	AAF	2.5 (2461)	5.9	-	5.4	11.5	-	-	-	21.1	12.2
1996-97	Eurovan	AAF	2.5 (2461)	5.9	-	5.4	11.5	-	-	-	21.1	12.2

VALVE SPECIFICATIONS

Year	Engine ID/VIN	Engine Displacement Liters (cc)	Seat Angle (deg.)	Face Angle (deg.)	Spring Test Pressure (lbs. @ in.)	Spring Installed Height (in.)	Stem-to-Guide Clearance (in.)		Stem Diameter (in.)	
							Intake	Exhaust	Intake	Exhaust
1993	AAF	2.5 (2461)	NA	NA	NA	NA	0.039	0.051	0.3138	0.3130
1994	AAF	2.5 (2461)	45	45	NA	NA	0.039	0.051	0.3138	0.3130
1995	AAF	2.5 (2461)	45	45	NA	NA	0.039	0.051	0.3138	0.3130
1996-97	AAF	2.5 (2461)	45	45	NA	NA	0.039	0.051	0.3138	0.3130

NA - Not Available

TORQUE SPECIFICATIONS
All readings in ft. lbs.

Year	Engine ID/VIN	Engine Displacement Liters (cc)	Cylinder Head Bolts	Main Bearing Bolts	Rod Bearing Bolts	Crankshaft Damper Bolts	Flywheel Bolts	Manifold		Spark Plugs	Lug Nut
								Intake	Exhaust		
1993	AAF	2.5 (2461)	①	48	②	340	③	22	④	18	80
1994	AAF	2.5 (2461)	①	48	②	340	③	22	④	18	80
1995	AAF	2.5 (2461)	①	48	②	340	③	22	④	18	80
1996-97	AAF	2.5 (2461)	①	48	②	340	③	22	④	18	80

① Step 1: 30 ft. lbs.
   Step 2: 44 ft. lbs.
   Step 3: Plus 90 degrees
   Step 4: Plus 90 degrees
② Step 1: 22 ft. lbs.
   Step 2: Plus 90 degrees

③ Step 1: 44 ft. lbs.
   Step 2: Plus 90 degrees
④ M6 bolts: 7 ft. lbs.
   M8 bolts: 18 ft. lbs.
   M10 bolts: 30 ft. lbs.

BRAKE SPECIFICATIONS
All measurements in inches unless noted

Year	Model	Master Cylinder Bore	Brake Disc			Brake Drum Diameter			Minimum Lining Thickness	
			Original Thickness	Minimum Thickness	Maximum Runout	Original Inside Diameter	Max. Wear Limit	Maximum Machine Diameter	Front	Rear
1993	Eurovan	0.874	0.945	0.787	0.003	10.55	10.61	NA	0.079	0.040
1994	Eurovan	0.874	0.945	0.787	0.003	10.55	10.61	NA	0.079	0.040
1995	Eurovan	0.874	0.945	0.787	0.003	10.55	10.61	NA	0.079	0.040
1996-97	Eurovan	0.874	0.945	0.787	0.003	10.55	10.61	NA	0.079	0.040

NA - Not Available

FREQUENT MAINTENANCE LABOR
EUROVAN

The following should be used as a guide when determining the amount of work required for a particular service if taken to a repair shop. In estimating how long a particular Frequent Maintenance Service item should take, please observe the following:
- **Chilton Time** is time that is based on field research and data supplied by the vehicle manufacturer.
- All labor time operations are given in hours and tenths of an hour.
- All labor operations, are to be used as a **guide**.

COOLING

	Chilton Time
(G) Belt, Drive, Renew	
1993-94 Eurovan	.8
(G) Thermostat, Coolant, Renew	
1993-94 Eurovan	1.1

FUEL

(M) Air Cleaner, Service	
1993-94	
Gas	.2
(G) Filter, Fuel, Renew	
1993-94	
Gas	.3

BRAKES

(G) Bleed Brakes (Four Wheels)	
Includes: Add fluid.	
1993-94	.6

	Chilton Time
(G) Brakes, Adjust (Minor)	
Includes: Adjust brakes, fill master cylinder.	
1993-94, two wheels	.4
(M) Parking Brake, Adjust	
1993-94	.5

LUBRICATION SERVICES

(M) Engine Oil & Filter, Renew	
Includes: Inspect and correct all fluid levels.	
1993-94	.3

WHEELS

(G) Wheel, Renew (One)	
1993-94 Eurovan	.5
(G) Wheel, Balance	
1993-94 Eurovan	
one	.3
each adtnl.	.2

	Chilton Time
(G) Wheels, Rotate (All)	
1993-94 Eurovan	.5

ELECTRICAL

(G) Headlamps, Aim	
1993-94	.4
(G) High Mount Stop Lamp Bulb, Renew	
1993-94 Eurovan	.3
(G) License Lamp Bulb, Renew	
1993-94 Eurovan	
one or all	.3
(G) Parking Lamp Lens or Bulb, Renew	
1993-94, each	.3
(G) Stop & Tail Lamp Bulb, Renew	
1993-94	.3
(G) Horn, Renew	
1993-94 Eurovan	.4

SCHEDULED MAINTENANCE INTERVALS
(VOLKSWAGEN EUROVAN)

TO BE SERVICED	TYPE OF SERVICE	VEHICLE MILEAGE INTERVAL (x1000)												
		7.5	15	22.5	30	37.5	45	52.5	60	67.5	75	82.5	90	97.5
Engine oil & filter	R	✓	✓	✓	✓	✓	✓	✓	✓	✓	✓	✓	✓	✓
Brake pads & fluid level	S/I	✓	✓	✓	✓	✓	✓	✓	✓	✓	✓	✓	✓	✓
Automatic transmission fluid & filter	S/I	✓		✓		✓		✓		✓		✓		✓
Drive shaft boots	S/I	✓		✓		✓		✓		✓		✓		✓
Engine coolant level, hoses & clamps	S/I	✓		✓		✓		✓		✓		✓		✓
Engine idle speed (gasoline except Calif.)	S/I	✓		✓		✓		✓		✓		✓		✓
Engine idle speed (diesel)	S/I				✓				✓				✓	
Exhaust system	S/I	✓		✓		✓		✓		✓		✓		✓
Power steering fluid	S/I	✓		✓		✓		✓		✓		✓		✓
Rotate tires①	S/I	✓		✓		✓		✓		✓		✓		✓
Timing belt	S/I				✓				✓				✓	
Water separator (diesel)	S/I	✓		✓		✓		✓		✓		✓		✓
Air cleaner filter & fuel filter (diesel)	R				✓				✓				✓	
Drive belts (except auto tensioner)	S/I				✓				✓				✓	
Drive belts (auto tensioner)	S/I								✓					
Front axle, dust seals on ball joints, tie rod ends & tie rods	S/I				✓				✓				✓	
Spark plugs	R				✓				✓				✓	
Brake fluid②	R								✓					

① Rotate tires front to rear only.
② Replace every 2 years regardless of mileage.
R – Replace   S/I – Service or Inspect

FREQUENT OPERATION MAINTENANCE (SEVERE SERVICE)

If a vehicle is operated under any of the following conditions it is considered severe service:
- Extremely dusty areas.
- 50% or more of the vehicle operation is in 32°C (90°F) or higher temperatures, or constant operation in temperatures below 0°C (32°F).
- Prolonged idling (vehicle operation in stop and go traffic).
- Frequent short running periods (engine does not warm to normal operating temperatures).
- Police, taxi, delivery usage or trailer towing usage.

Oil & oil filter change – change every 3750 miles.
Rotate tires - rotate every 7500 miles.
Automatic transmission fluid & filter - change every 30,000 miles.
Air cleaner filter - service or inspect every 15,000 miles.

TECHNICAL SERVICE BULLETINS 24

TECHNICAL SERVICE BULLETINS

What is a TSB?

All vehicle manufacturers experience occasional problems with one or more of their model lines, requiring that fixes be made after the vehicle is sold to the customer. Manufacturers therefore issue Technical Service Bulletins (TSBs) to inform and to suggest certain repairs or component replacements. These fixes may cover a variety of issues including: safety, general maintenance, part replacement, engine driveability improvements or general repairs. If the item at issue is a noted safety related problem, it is likely that the manufacturer will also issue SAFETY RECALL CAMPAIGN notices.

NOTE: The major difference between a TSB and a Recall Campaign is that the manufacturer wants the repairs performed to ALL vehicles affected by a recall, in order to prevent a problem (often safety related) from occurring. TSBs, on the other hand, are issued to help the dealership service facility cope with a problem (usually non-safety related) that may occur to SOME vehicles. The repair or change may not be necessary if the component or system in question never develops a problem.

The TSBs and Recall Campaigns notices are sent directly to the dealership repair facility. Safety Recall Campaign notices are also sent to the vehicle owners, but in case of a sale, transfer of title or other circumstances, the owner may not receive the actual notice.

All of these factory notifications provide a description of the problem, the vehicles which are affected by the problem, how the problem is fixed and whether or not the problem is warranty-related.

Federal law also requires that the general public have access to TSBs and Recalls. You can obtain copies of any authorized bulletins from various sources including the manufacturer service information groups or distributors, federal or state government agencies dealing in transportation or publication, electronic based professional information systems such as Chilton On Disc, or even a cooperative dealer service department.

We have provided the following examples of bulletins so that you can see the kind of information that a TSB might provide. You will also note that each bulletin is numbered, providing a valuable way to access this information. See the index for the list of bulletins applying to your vehicle which was current at time of publication.

Using the TSB Index

The TSB index in this manual is divided into groups of charts covering each manufacturer. Each group contains separate charts for the various vehicle sub-systems or service categories.

The charts we have provided contain 4 columns. The first column, **MODELS**, provides you with a coded listing of what actual vehicle nameplates from that manufacturer are affected by the bulletin. The model listing is a numerical code, explained in the footer (the bottom) of each chart.

The second column, **YEAR**, lists the model year(s) of those nameplates which are affected. For example, 92–94 would include all 1992, 1993 and 1994 models carrying the nameplate listed in the first column. Similarly, 94–94 would only affect vehicles built for the 1994 model year.

The third column, **TSB#**, was the last revised (if more than one was published) part number or code to retrieve that particular service bulletin.

Finally, the fourth column, **DESCRIPTION**, provides a brief idea of what the bulletin is about (new components, procedures, specifications, or possibly just dealer network service policy revisions).

To determine if there are any bulletins for your vehicle within a certain category:

1. Locate the group containing charts for your vehicle's manufacturer.

2. Next, turn to the page(s) covering the specific system you are curious about (such as Heating, Air Conditioning, Ventilation, Defogger or Lighting, Horns, Turn Signals, Steering Column).

3. Once you have reached the chart for the category, the next step is to determine if any bulletins apply to the particular model on which you are working. Determine what number represents your model using the footer at the bottom of the chart. Scan the **MODELS** column for that number.

4. Whatever matches you find are service bulletins that *could* apply to your vehicle. If the **YEAR** and **DESCRIPTION** columns also match your model and problem, then you can use the **TSB#** to help attain the bulletin in question.

5. If you do not find a number match, check the chart footer again to see if any codes such as 1=All or 2=Most are used for that manufacturer. Sometimes components or procedures are shared by all or the majority of a manufacturer's model lines. In this case, a bulletin may apply to your model as well, but only if the **YEAR** and **DESCRIPTIONS** still match.

NOTE: Keep in mind that even if your model and year match a bulletin, it does not necessarily mean that the repair is required for your vehicle. A TSB (not a Safety Recall) repair should be performed ONLY if the problem (not just the symptom) exists on your vehicle. Remember that a TSB lists the probable cause of a symptom, but it is often not the only possible cause.

Service Bulletin

File In Section: 1 - HVAC

Bulletin No.: 66-11-03

Date: July, 1996

Subject: HVAC Fan Only Operates on High Speed
(Replace Blower Motor Resistor Assembly)

Models: 1995-96 Chevrolet and GMC C/K Models
Built before the following VIN breakpoints:

Make	Plant	VIN
Chevrolet	Flint	TF013550
GMC	Flint	TF013559
Chevrolet	Ft. Wayne	TZ167736
GMC	Ft. Wayne	TZ523792
Chevrolet	Janesville	TJ360493
GMC	Janesville	TJ725741
Chevrolet	Oshawa	T1191428
GMC	Oshawa	T1534201
Chevrolet	Pontiac	TE197119
GMC	Pontiac	TE532000
Chevrolet	Silao	TG133568
GMC	Silao	TG511601

Condition

Some owners may comment that the HVAC fan will operate only on high speed setting.

Cause

The blower motor resistor fails open.

Correction

Replace blower motor resistor assembly with new part, following the service procedure in Section 1B of the appropriate C/K Truck Service Manual.

Parts Information

P/N	Description	Qty
15024815	Resistor Asm - Blower Motor	1

Parts are currently available from GMSPO.

Warranty Information

For vehicles repaired under warranty, use:

Labor Operation	Description	Labor Time
D1002	Resistor, Blower Motor - Replace	Use Published Labor Operation Time

Important: Labor operation is coded to base vehicle coverage in the warranty system.

**WE SUPPORT
VOLUNTARY TECHNICIAN
CERTIFICATION**

7920tsb3

A typical GM TSB for the heater and A/C blower motor

CHRYSLER CORPORATION

Lighting, Horns, Turn Signals, Steering Column

Models	Year	TSB#	Description
12,14	93 - 93	8/31/93	LAMP (TAIL/STOP/BACK UP OR SIDE MARKER) INOP - PROC
10,14	93 - 93	8/31/93	LAMP (TAIL/STOP/BACK UP OR SIDE MARKER) INOP - PROC
5	95 - 95	08-54-94	MAP/COURTESY LAMP OPS - SH WHL BASE W/O O/H CONSOLE
11	95 - 95	08-54-94	MAP/COURTESY LAMP OPS - SH WHL BASE W/O O/H CONSOLE
10,14	93 - 93	08-32-93	TAIL LAMP - HOUSING COLLECTS WATER - GASKET/PROC
12,14	93 - 93	08-32-93	TAIL LAMP - HOUSING COLLECTS WATER - GASKET/PROC
7	94 - 94	8/8/94	WEAK SOUNDING HORN - DUAL NOTE UPGRADE
7	94 - 94	8/8/94	WEAK SOUNDING HORN - DUAL NOTE UPGRADE - REVISED
12	94 - 94	8/8/94	WEAK SOUNDING HORN - DUAL NOTE UPGRADE - REVISED

Model Legend: 1= All 2= Most 3= Town & Country 4= D-50 Pickup 5= Caravan 6= Dakota 7= Ram Pickup 8= Ram 50 9= Ram Van-B Series 10=Grand Cherokee 11= Voyager 12= Cherokee 13= Comanchee 14= Grand Wagoneer 15= CJ, YJ, Wrangler 16= Ramcharger 17= Wagoneer

Instruments, Dash Cluster, Warning Lights, Mirrors

Models	Year	TSB#	Description
7	94 - 96	23-02-96	CREAK/TICK NOISE - RT SIDE OF IP
2	91 - 96	8/16/95	CRUISE CNTRL OVER/UNDERSHOOT DURING INITIAL SET -REV
2	89 - 96	34927	CRUISE CNTRL OVER/UNDERSHOOT DURING INITIAL SET -REV
2	89 - 96	8/16/95	CRUISE CNTRL OVER/UNDERSHOOT DURING INITIAL SET -REV
2	89 - 96	8/16/95	CRUISE CNTRL OVER/UNDERSHOOT DURING INITIAL SET -REV
2	89 - 96	34927	CRUISE CONTROL OVER/UNDERSHOOT DURING INITIAL SET
2	89 - 96	8/16/95	CRUISE CONTROL OVER/UNDERSHOOT DURING INITIAL SET
2	89 - 96	8/16/95	CRUISE CONTROL OVER/UNDERSHOOT DURING INITIAL SET
3	93 - 95	8/7/94	DIG SPEEDO FLUCTUATES AT IDLE - 3.3/3.8L W/41TE -REV
5	93 - 95	8/7/94	DIG SPEEDO FLUCTUATES AT IDLE - 3.3/3.8L W/41TE -REV
11	93 - 95	34553	DIG SPEEDO FLUCTUATES AT IDLE - 3.3/3.8L W/41TE -REV
3	93 - 95	08-07-94A	DIG SPEEDO FLUCTUATES AT IDLE -3.3/3.8L 41TE - REV B
5	93 - 95	08-07-94A	DIG SPEEDO FLUCTUATES AT IDLE -3.3/3.8L 41TE - REV B
11	93 - 95	08-07-94A	DIG SPEEDO FLUCTUATES AT IDLE -3.3/3.8L 41TE - REV B
10	96 - 96	8/7/96	ERRATIC OIL PRESSURE GAUGE READINGS - PCM
12	96 - 96	8/7/96	ERRATIC OIL PRESSURE GAUGE READINGS - PCM
9	93 - 94	8/12/94	FUEL GAUGE - EXCESSIVE MOVEMENT AS VHCL TURNS CORNER
6,7,16	91 - 94	08-58-93	FUEL GAUGE INACCURATE - PARTS & PROCEDURE
6,7,16	91 - 94	08-58-93	FUEL GAUGE INACCURATE - PARTS & PROCEDURE - REVISED
7	94 - 94	8/10/94	FUEL GAUGE STICKS - REPLACE FUEL PUMP MODULE
7	94 - 94	08-47-93	GAUGE/COOLANT TEMP - READING EXPLAINED - DIESEL
2	89 - 93	8/15/93	GAUGE/TEMP - FLUCTUATES DURING COLD WEATHER - INFO
11	89 - 93	34196	GAUGE/TEMP - FLUCTUATES DURING COLD WEATHER - INFO
4	93 - 93	23-17-93	I/P - SUNGLASS POCKET R&I - REVISED PROCEDURE
7	94 - 94	23-41-94	I/P BEZEL CREAK
6	93 - 93	8/21/93	I/P CLUSTER - PRNDL CABLE CAUTION DURING SERVICING
3	91 - 94	08-33-93	IC/ELECTRONIC - FUEL GAUGE OPERATION EXPLAINED
10,14	93 - 93	8/17/93	INSTRUMENT CLUSTER - SQUEAK - DIAGNOSIS/REPAIR PROC
12,14	93 - 93	8/17/93	INSTRUMENT CLUSTER - SQUEAK - DIAGNOSIS/REPAIR PROC
2	93 - 95	34926	INTERACTIVE CRUISE CONTROL OPS & 41TE/42LE
2	93 - 95	8/15/95	INTERACTIVE CRUISE CONTROL OPS & 41TE/42LE
2	93 - 95	8/15/95	INTERACTIVE CRUISE CONTROL OPS & 41TE/42LE
2	93 - 96	08-15-95A	INTERACTIVE CRUISE CONTROL OPS & 41TE/42LE - REV
2	93 - 96	08-15-95A	INTERACTIVE CRUISE CONTROL OPS & 41TE/42LE - REV
2	93 - 96	08-15-95A	INTERACTIVE CRUISE CONTROL OPS & 41TE/42LE - REV
7	94 - 94	23-57-93	IP CREAK - LEFT SIDE - REPAIR PROCEDURE
12,17,10	93 - 93	23-08-93	MIRROR - AVOIDING SCRATCHED GLASS - INFORMATION
2	93 - 93	23-08-93	MIRROR - AVOIDING SCRATCHED GLASS - INFORMATION
2	93 - 93	23-08-93	MIRROR - AVOIDING SCRATCHED GLASS - INFORMATION
11	93 - 93	23-08-93	MIRROR - AVOIDING SCRATCHED GLASS - INFORMATION
12,16,12	93 - 93	23-08-93	MIRROR - AVOIDING SCRATCHED GLASS - INFORMATION
3	95 - 95	23-88-94	NOISE - LEFT SIDE OF INSTRUMENT PANEL
5	95 - 95	23-88-94	NOISE - LEFT SIDE OF INSTRUMENT PANEL
11	95 - 95	23-88-94	NOISE - LEFT SIDE OF INSTRUMENT PANEL
7	94 - 95	8/25/95	POWER MIRROR VIBRATES - BUGSCREEN DEFLECTOR

Model Legend: 1= All 2= Most 3= Town & Country 4= D-50 Pickup 5= Caravan 6= Dakota 7= Ram Pickup 8= Ram 50 9= Ram Van-B Series 10=Grand Cherokee 11= Voyager
12= Cherokee 13= Comanchee 14= Grand Wagoneer 15= CJ, YJ, Wrangler 16= Ramcharger 17= Wagoneer

Instruments, Dash Cluster, Warning Lights, Mirrors

Models	Year	TSB#	Description
7	94 - 95	08-64-94	POWER REAR VIEW MIRROR VIBRATION OR BLURRED IMAGES
11	94 - 94	34562	RATTLE/LOOSE SPEED CONTROL SWITCH
5	94 - 94	8/16/94	RATTLE/LOOSE SPEED CONTROL SWITCH
12	96 - 96	8/1/96	SPEED CONTROL LIGHT FLICKERS AND/OR SERVO CLICKS
7,6,9	96 - 96	8/1/96	SPEED CONTROL LIGHT FLICKERS AND/OR SERVO CLICKS
7,16	92 - 93	8/7/93	SPEED CONTROL SURGE - REPAIR/PCM REFERENCE CHART
16,7	92 - 93	8/7/93	SPEED CONTROL SURGE - REPAIR/PCM REFERENCE CHART/REV
13,10,14	92 - 93	8/28/92	SPEED CONTROL SYS - INTERMIT DISENGAGE AT HWY SPEEDS
13,10,15	92 - 93	08-28-92A	SPEED CONTROL SYS - INTERMIT DISENGAGE AT HWY SPEEDS
2	92 - 93	8/28/92	SPEED CONTROL SYS - INTERMIT DISENGAGE AT HWY SPEEDS
2	92 - 93	8/28/92	SPEED CONTROL SYS - INTERMIT DISENGAGE AT HWY SPEEDS
2	92 - 93	8/28/92	SPEED CONTROL SYS - INTERMIT DISENGAGE AT HWY SPEEDS
2	92 - 93	8/28/92	SPEED CONTROL SYS - INTERMIT DISENGAGE AT HWY SPEEDS
11	92 - 93	33844	SPEED CONTROL SYS - INTERMIT DISENGAGE AT HWY SPEEDS
11	92 - 93	33844	SPEED CONTROL SYS - INTERMIT DISENGAGE AT HWY SPEEDS
13,12,14	92 - 93	8/28/92	SPEED CONTROL SYS - INTERMIT DISENGAGE AT HWY SPEEDS
13,12,14	92 - 93	8/28/92	SPEED CONTROL SYS - INTERMIT DISENGAGE AT HWY SPEEDS
2	92 - 93	08-28-92A	SPEED CONTROL SYS - INTERMIT DISENGAGE AT HWY SPEEDS
2	92 - 93	08-28-92A	SPEED CONTROL SYS - INTERMIT DISENGAGE AT HWY SPEEDS
11	92 - 93	08-28-92A	SPEED CONTROL SYS - INTERMIT DISENGAGE AT HWY SPEEDS
13,12,14	92 - 93	08-28-92A	SPEED CONTROL SYS - INTERMIT DISENGAGE AT HWY SPEEDS
3	94 - 95	08-77-94	SPEEDO INOP - NO VSS SIGNAL
11	94 - 95	08-77-94	SPEEDO INOP - NO VSS SIGNAL
5	94 - 95	08-77-94	SPEEDO INOP - NO VSS SIGNAL
2	88 - 93	HL-58-92	SPEEDOMETER - CHART FOR PROPER PINION SELECTION
11	89 - 93	HL-58-92	SPEEDOMETER - CHART FOR PROPER PINION SELECTION
5	89 - 93	HL-58-92	SPEEDOMETER - CHART FOR PROPER PINION SELECTION
3	93 - 94	8/7/94	SPEEDOMTR FLUCTUATES AT IDLE - LOW SPEED HARSH SHIFT
5	93 - 94	8/7/94	SPEEDOMTR FLUCTUATES AT IDLE - LOW SPEED HARSH SHIFT
11	93 - 94	34553	SPEEDOMTR FLUCTUATES AT IDLE - LOW SPEED HARSH SHIFT
3	94 - 94	08-59-93	SQUEAK/RATTLE RIGHT SIDE IP - CHECK PAB (AIR BAG)
5	94 - 94	08-59-93	SQUEAK/RATTLE RIGHT SIDE IP - CHECK PAB (AIR BAG)
11	94 - 94	08-59-93	SQUEAK/RATTLE RIGHT SIDE IP - CHECK PAB (AIR BAG)
10,14	93 - 93	8/22/93	VEHICLE INFORMATION CENTER - DIAGNOSTICS INFORMATION
12,14	93 - 93	8/22/93	VEHICLE INFORMATION CENTER - DIAGNOSTICS INFORMATION
10	93 - 93	9/16/92	VIC - OIL LEVEL LIGHT PREMATURE ON - SERVICE PROC
12	93 - 93	9/16/92	VIC - OIL LEVEL LIGHT PREMATURE ON - SERVICE PROC

Model Legend: 1= All 2= Most 3= Town & Country 4= D-50 Pickup 5= Caravan 6= Dakota 7= Ram Pickup 8= Ram 50 9= Ram Van-B Series 10=Grand Cherokee 11= Voyager 12= Cherokee 13= Comanchee 14= Grand Wagoneer 15= CJ, YJ, Wrangler 16= Ramcharger 17= Wagoneer

Chassis Electrical, Wiring Harness, Fuses-Circuit Breakers, Wipers, Window Motors

Models	Year	TSB#	Description
12,14	93 - 93	8/27/92	30-WAY DOOR HARNESS CONNECTOR - BOOT SEPARATES - FIX
10,14	93 - 93	8/27/92	30-WAY DOOR HARNESS CONNECTOR - BOOT SEPARATES - FIX
7	94 - 95	8/24/95	ACCESSORY FRAME GROUND JUMPER WIRE
7,6,9	95 - 95	08-55-94	ELECT SERV - ADL CONNECTOR FROM 6 TO 8 PIN - S/M REV
12,10,14	93 - 94	8/17/94	ERRATIC LIGHT OP/BATTERY DRAIN W/ TRAILER TOW PKG
7	94 - 94	8/17/94	ERRATIC LIGHT OP/BATTERY DRAIN W/ TRAILER TOW PKG
12,12,14	93 - 94	8/17/94	ERRATIC LIGHT OP/BATTERY DRAIN W/ TRAILER TOW PKG
3	94 - 94	8/28/94	FUEL PUMP/SENDING UNIT MODULE CONNECTOR SERVICE
11	94 - 95	34574	FUEL PUMP/SENDING UNIT MODULE CONNECTOR SERVICE
5	94 - 95	8/28/94	FUEL PUMP/SENDING UNIT MODULE CONNECTOR SERVICE
3	94 - 94	08-46-93	FUSE BLOCK CRACKED - PARTS/PROCEDURE REVISED
5	94 - 94	08-46-93	FUSE BLOCK CRACKED - PARTS/PROCEDURE REVISED
11	94 - 94	08-46-93	FUSE BLOCK CRACKED - PARTS/PROCEDURE REVISED
3	94 - 94	08-46-93	FUSE BLOCK CRACKED - REPLACE/CLEAN OIL FROM BRACKET
5	94 - 94	08-46-93	FUSE BLOCK CRACKED - REPLACE/CLEAN OIL FROM BRACKET
11	94 - 94	08-46-93	FUSE BLOCK CRACKED - REPLACE/CLEAN OIL FROM BRACKET
3	93 - 93	08-33-92	FUSIBLE LINK - REPEATED FAILURE/COOLING FAN LOSS
5	93 - 93	08-33-92	FUSIBLE LINK - REPEATED FAILURE/COOLING FAN LOSS
11	93 - 93	08-33-92	FUSIBLE LINK - REPEATED FAILURE/COOLING FAN LOSS
10	96 - 96	08-47-95	HOOD SWITCH DELETED (HOOD AJAR WARNING)
12	96 - 96	08-47-95	HOOD SWITCH DELETED (HOOD AJAR WARNING)
15	97 - 97	8/6/96	IGNITION OFF DRAW (IOD) FUSE LOCATION
3	96 - 96	8/29/95	IGNITION OFF DRAW (IOD) FUSE MUST BE FULLY SEATED
5	96 - 96	8/29/95	IGNITION OFF DRAW (IOD) FUSE MUST BE FULLY SEATED
11	96 - 96	34940	IGNITION OFF DRAW (IOD) FUSE MUST BE FULLY SEATED
9	94 - 94	08-57-94	IMPROVED WIPER BLADES - PERFORMANCE AT HWY SPEEDS
12	93 - 94	08-52-94	INOP PWR WINDOW/LOCK/MIRROR OR BLOWN 9 OR 13 FUSE
10	93 - 93	18-04-92	INTERMIT ELEC CONNECTION - POOR PERFORMANCE/NO START
10	93 - 93	18-04-92A	INTERMIT ELEC CONNECTION - POOR PERFORMANCE/NO START
2	92 - 93	18-04-92	INTERMIT ELEC CONNECTION - POOR PERFORMANCE/NO START
2	92 - 93	18-04-92	INTERMIT ELEC CONNECTION - POOR PERFORMANCE/NO START
12	93 - 93	18-04-92	INTERMIT ELEC CONNECTION - POOR PERFORMANCE/NO START
12	93 - 93	18-04-92	INTERMIT ELEC CONNECTION - POOR PERFORMANCE/NO START
3	91 - 93	08-34-93	LIFTGATE WIPER DOESN'T WIPE CLEAN- NEW ARM/BLADE ASM
5	91 - 93	08-34-93	LIFTGATE WIPER DOESN'T WIPE CLEAN- NEW ARM/BLADE ASM
11	91 - 93	08-34-93	LIFTGATE WIPER DOESN'T WIPE CLEAN- NEW ARM/BLADE ASM
12,10	96 - 96	08-41-95	LOSS OF BATTERY FEED TO PDC - ERRATIC OP
12,12	96 - 96	08-41-95	LOSS OF BATTERY FEED TO PDC - ERRATIC OP
10	96 - 96	08-48-95	O2 HARNESS CONTACTS PROPELLER SHAFT
12	96 - 96	08-48-95	O2 HARNESS CONTACTS PROPELLER SHAFT
12	94 - 94	8/9/94	POWER WINDOW SWITCH WIRING DIAGRAM - S/M REVISION
3	96 - 96	23-65-95	PWR VENT WINDOW ACTUATOR BALL & SOCKET SEPARATE
5	96 - 96	23-65-95	PWR VENT WINDOW ACTUATOR BALL & SOCKET SEPARATE
11	96 - 96	23-65-95	PWR VENT WINDOW ACTUATOR BALL & SOCKET SEPARATE
3	96 - 96	23-65-95	PWR VENT WINDOW ACTUATOR BALL & SOCKET SEPARATE -REV

Model Legend: 1= All 2= Most 3= Town & Country 4= D-50 Pickup 5= Caravan 6= Dakota 7= Ram Pickup 8= Ram 50 9= Ram Van-B Series 10=Grand Cherokee 11= Voyager 12= Cherokee 13= Comanchee 14= Grand Wagoneer 15= CJ, YJ, Wrangler 16= Ramcharger 17= Wagoneer

Chassis Electrical, Wiring Harness, Fuses-Circuit Breakers, Wipers, Window Motors

Models	Year	TSB#	Description
5	96 - 96	23-65-95	PWR VENT WINDOW ACTUATOR BALL & SOCKET SEPARATE -REV
11	96 - 96	23-65-95	PWR VENT WINDOW ACTUATOR BALL & SOCKET SEPARATE -REV
3	93 - 93	08-50-93	RELAY FAILURES - AUTO SHUTDWN/FUEL PUMP/RADIATOR FAN
5	93 - 93	08-50-93	RELAY FAILURES - AUTO SHUTDWN/FUEL PUMP/RADIATOR FAN
11	93 - 93	08-50-93	RELAY FAILURES - AUTO SHUTDWN/FUEL PUMP/RADIATOR FAN
3	91 - 94	HL-46-92	REPLACE AJAR SWITCH AFTER SLIDING DOOR ADJUSTMENT
5	91 - 94	HL-46-92	REPLACE AJAR SWITCH AFTER SLIDING DOOR ADJUSTMENT
11	91 - 94	HL-46-92	REPLACE AJAR SWITCH AFTER SLIDING DOOR ADJUSTMENT
3	91 - 94	08-62-93	SLIDING DOOR AJAR SWITCH/ADJUST (CHIME/LOCKS/LITES)
5	91 - 94	08-62-93	SLIDING DOOR AJAR SWITCH/ADJUST (CHIME/LOCKS/LITES)
11	91 - 94	08-62-93	SLIDING DOOR AJAR SWITCH/ADJUST (CHIME/LOCKS/LITES)
3	96 - 96	23-60-95	SLIDING DOOR/LIFTGATE POWER LOCK MOTOR INOP
5	96 - 96	23-60-95	SLIDING DOOR/LIFTGATE POWER LOCK MOTOR INOP
11	96 - 96	23-60-95	SLIDING DOOR/LIFTGATE POWER LOCK MOTOR INOP
10	94 - 94	08-67-93	STOP LIGHT SWITCH WIRING CONNECTOR SERVICE
1	94 - 94	08-67-93	STOP LIGHT SWITCH WIRING CONNECTOR SERVICE
2	94 - 95	08-67-93	STOP LIGHT SWITCH WIRING CONNECTOR SERVICE
2	94 - 94	08-67-93	STOP LIGHT SWITCH WIRING CONNECTOR SERVICE
12	94 - 94	08-67-93	STOP LIGHT SWITCH WIRING CONNECTOR SERVICE
7	94 - 95	08-41-94	TRAILER TOW BRAKE WIRE LOCATION
10,14	93 - 94	8/3/94	TRAILOR TOW WIRING - BYPASS LAMP OUTAGE MODULE
12,14	93 - 94	8/3/94	TRAILOR TOW WIRING - BYPASS LAMP OUTAGE MODULE
11	89 - 93	23-26-92	WINDSHIELD WASHER SYSTEM - NOZZLE FREEZE UP - SERV
9	94 - 94	23-31-94	WINDSHIELD WASHERS INOP - REVISED WASHER HOSE ELBOW
9	94 - 94	08-43-94	WINDSHIELD WIPER BLADE USAGE
3	91 - 93	23-31-93	WINDSHIELD WIPER PIVOTS REVISED - PRODUCTION CHANGE
5	91 - 93	23-31-93	WINDSHIELD WIPER PIVOTS REVISED - PRODUCTION CHANGE
11	91 - 93	23-31-93	WINDSHIELD WIPER PIVOTS REVISED - PRODUCTION CHANGE
3	96 - 96	08-36-95	WIPERS SELF-ACTIVATE WHILE DRIVING
5	96 - 96	08-36-95	WIPERS SELF-ACTIVATE WHILE DRIVING
11	96 - 96	08-36-95	WIPERS SELF-ACTIVATE WHILE DRIVING
3	96 - 96	08-36-95	WIPERS SELF-ACTIVATE WHILE DRIVING - REV
5	96 - 96	08-36-95	WIPERS SELF-ACTIVATE WHILE DRIVING - REV
11	96 - 96	08-36-95	WIPERS SELF-ACTIVATE WHILE DRIVING - REV
10,14	93 - 93	08-42-93	WIRE HARNESS ROUTING - INFORMATION - V8 WITH 46RH
12,14	93 - 93	08-42-93	WIRE HARNESS ROUTING - INFORMATION - V8 WITH 46RH
10	93 - 93	8/11/92	WIRING DIAGRAMS - SERVICE MANUAL REVISIONS
3	92 - 93	8/11/92	WIRING DIAGRAMS - SERVICE MANUAL REVISIONS
11	92 - 93	33827	WIRING DIAGRAMS - SERVICE MANUAL REVISIONS
12	93 - 93	8/11/92	WIRING DIAGRAMS - SERVICE MANUAL REVISIONS
5	92 - 93	8/11/92	WIRING DIAGRAMS - SERVICE MANUAL REVISIONS

Model Legend: 1= All 2= Most 3= Town & Country 4= D-50 Pickup 5= Caravan 6= Dakota 7= Ram Pickup 8= Ram 50 9= Ram Van-B Series 10=Grand Cherokee 11= Voyager 12= Cherokee 13= Comanchee 14= Grand Wagoneer 15 = CJ, YJ, Wrangler 16= Ramcharger 17= Wagoneer

Auxiliary Equipment, Jacks, Trailer Hitches, Towing

Models	Year	TSB#	Description
7	94 - 95	8/24/95	ACCESSORY FRAME GROUND JUMPER WIRE
5	94 - 95	23-77-94	BODY CLADDING INTERFERES WITH USE OF JACK
7	94 - 96	2/1/96	CAMPER SPECIAL SERVICE KIT - REAR SUSPENSION
7	94 - 96	2/1/96	CAMPER SPECIAL SERVICE KIT - REAR SUSPENSION - REV
2	89 - 95	8/31/94	INSTALLATION OF RADIO TRANSMITTING EQUIPMENT
2	89 - 95	34577	INSTALLATION OF RADIO TRANSMITTING EQUIPMENT
2	89 - 95	8/31/94	INSTALLATION OF RADIO TRANSMITTING EQUIPMENT
2	89 - 95	8/31/94	INSTALLATION OF RADIO TRANSMITTING EQUIPMENT - REV
2	89 - 95	34577	INSTALLATION OF RADIO TRANSMITTING EQUIPMENT - REV
2	89 - 95	8/31/94	INSTALLATION OF RADIO TRANSMITTING EQUIPMENT - REV
6	95 - 95	8/10/95	SNOW PLOW PACKAGE - TRANS TEMP SENSOR - 2 PIN TO 1
12,14	93 - 93	HL-52-92	TRAILER TOW PREP PACKAGE AVAILABLE - INFORMATION

Model Legend: 1= All 2= Most 3= Town & Country 4= D-50 Pickup 5= Caravan 6= Dakota 7= Ram Pickup 8= Ram 50 9= Ram Van-B Series 10=Grand Cherokee 11= Voyager 12= Cherokee 13= Comanchee 14= Grand Wagoneer 15 = CJ, YJ, Wrangler 16= Ramcharger 17= Wagoneer

Heating, Air Conditioning, Ventilation, Defogger

Models	Year	TSB#	Description
12	93 - 93	24-08-92	A/C & HEAT - LOW OUTPUT (DRIVER'S FLOOR) - ABS CAUSE
10	93 - 93	24-08-92	A/C & HEAT - LOW OUTPUT (DRIVER'S FLOOR) - ABS CAUSE
3	93 - 93	24-13-93	A/C - COMPRESSOR LOCK-UP AT LOW MILEAGE - DIAG/PROC
5	93 - 93	24-13-93	A/C - COMPRESSOR LOCK-UP AT LOW MILEAGE - DIAG/PROC
11	93 - 93	24-13-93	A/C - COMPRESSOR LOCK-UP AT LOW MILEAGE - DIAG/PROC
3	92 - 94	24-12-92	A/C - OFFENSIVE ODOR FROM DUCTS - REPAIR REVISED
11	92 - 94	24-12-92	A/C - OFFENSIVE ODOR FROM DUCTS - REPAIR REVISED
5	92 - 94	24-12-92	A/C - OFFENSIVE ODOR FROM DUCTS - REPAIR REVISED
2	88 - 94	24-01-95	A/C - R-12 TO R-134A ADAPTATION SERVICE PROCED - REV
2	88 - 94	24-01-95	A/C - R-12 TO R-134A ADAPTATION SERVICE PROCEDURE
10	93 - 93	HL-44-92	A/C - SYSTEM SERVICING INFORMATION (4.0L)
12	93 - 93	HL-44-92	A/C - SYSTEM SERVICING INFORMATION (4.0L)
3	91 - 93	24-08-93	A/C - TEMP CONTROL LEVER FAILS TO HOLD POSITION/FIX
5	91 - 93	24-08-93	A/C - TEMP CONTROL LEVER FAILS TO HOLD POSITION/FIX
11	91 - 93	24-08-93	A/C - TEMP CONTROL LEVER FAILS TO HOLD POSITION/FIX
10	93 - 93	24-04-92	A/C - TEMPERATURE DIFFERENTIAL CHART - S/M REVISION
12	93 - 93	24-04-92	A/C - TEMPERATURE DIFFERENTIAL CHART - S/M REVISION
10,14	93 - 93	24-11-93	A/C - WHISTLE FROM I/P W/SYS OP IN OUTSIDE AIR MODE
12,14	93 - 93	24-11-93	A/C - WHISTLE FROM I/P W/SYS OP IN OUTSIDE AIR MODE
10,14	92 - 95	24-08-95	A/C - WHITE FLAKES FROM IP AND DEFROST OUTLETS
2	92 - 95	24-08-95	A/C - WHITE FLAKES FROM IP AND DEFROST OUTLETS
12,14	92 - 95	24-08-95	A/C - WHITE FLAKES FROM IP AND DEFROST OUTLETS
10	93 - 93	24-10-92	A/C - WILL NOT COOL - DIAGNOSIS/SERVICE PROCEDURE
12	93 - 93	24-10-92	A/C - WILL NOT COOL - DIAGNOSIS/SERVICE PROCEDURE
3	91 - 93	24-17-92	A/C AIRFLOW/COOLING LOSS W/BLOWER FAN STILL OP- INFO
5	91 - 93	24-17-92	A/C AIRFLOW/COOLING LOSS W/BLOWER FAN STILL OP- INFO
11	91 - 93	24-17-92	A/C AIRFLOW/COOLING LOSS W/BLOWER FAN STILL OP- INFO
2	91 - 95	24-17-92	A/C AIRFLOW/COOLING LOSS W/BLOWER FAN STILL OP- REV
2	91 - 95	24-17-92	A/C AIRFLOW/COOLING LOSS W/BLOWER FAN STILL OP- REV
2	91 - 95	24-17-92	A/C AIRFLOW/COOLING LOSS W/BLOWER FAN STILL OP- REV
2	91 - 95	24-17-92	A/C AIRFLOW/COOLING LOSS W/BLOWER FAN STILL OP- REV
9	94 - 94	CSN#585	A/C COMPRESSOR - R134A OIL WITH R12 REFRIG - RECALL
10	96 - 96	24-03-96	A/C DOES NOT COOL - 5.2/5.9/8.0L - PCM RECALIB
12	96 - 96	24-03-96	A/C DOES NOT COOL - 5.2/5.9/8.0L - PCM RECALIB
7,6,9	96 - 96	24-03-96	A/C DOES NOT COOL - 5.2/5.9/8.0L - PCM RECALIB
9	93 - 94	24-28-93	A/C LEAK AT HIGH PRESSURE HOSE CONNECTION - WASHER
3	93 - 94	24-25-93	A/C MOAN - NEW A/C CLUTCH PLATE
5	93 - 94	24-25-93	A/C MOAN - NEW A/C CLUTCH PLATE
11	93 - 94	24-25-93	A/C MOAN - NEW A/C CLUTCH PLATE
12,15	94 - 94	24-09-94	A/C MOAN/HUM - R134A SYSTEM
12	94 - 94	08-56-94	A/C POOR/NONE - AUX COOLING FAN INOP - 25 AMP FUSE
3	96 - 96	24-13-95	A/C RELATED MOAN/WHINE NOISE
5	96 - 96	24-13-95	A/C RELATED MOAN/WHINE NOISE
11	96 - 96	24-13-95	A/C RELATED MOAN/WHINE NOISE
3	93 - 93	24-02-93	A/C SUCTION/LIQUID LINE - HARD TO CONNECT SERV EQUIP

Model Legend: 1= All 2= Most 3= Town & Country 4= D-50 Pickup 5= Caravan 6= Dakota 7= Ram Pickup 8= Ram 50 9= Ram Van-B Series 10=Grand Cherokee 11= Voyager 12= Cherokee 13= Comanchee 14= Grand Wagoneer 15 = CJ, YJ, Wrangler 16= Ramcharger 17= Wagoneer

Heating, Air Conditioning, Ventilation, Defogger

Models	Year	TSB#	Description
5	93 - 93	24-02-93	A/C SUCTION/LIQUID LINE - HARD TO CONNECT SERV EQUIP
11	93 - 93	24-02-93	A/C SUCTION/LIQUID LINE - HARD TO CONNECT SERV EQUIP
10	93 - 93	HL-22-92	A/C SYSTEM REQUIREMENTS - REFRIGERANT/LUBRICANT
12	93 - 93	HL-22-92	A/C SYSTEM REQUIREMENTS - REFRIGERANT/LUBRICANT
10,14	93 - 93	24-04-93	ATC - TEMPERATURE & BLOWER SPEED VARY - SERVICE PROC
10,14	93 - 93	24-04-93	ATC - TEMPERATURE & BLOWER SPEED VARY - SERVICE PROC
12,14	93 - 93	24-04-93	ATC - TEMPERATURE & BLOWER SPEED VARY - SERVICE PROC
10	96 - 96	24-05-96	ATC ERRONEOUS FAULT CODES - CODE 40
12	96 - 96	24-05-96	ATC ERRONEOUS FAULT CODES - CODE 40
3	96 - 96	24-10-95	BLOWER MOTOR WHINE IN LOW OR 2ND SPEED
5	96 - 96	24-10-95	BLOWER MOTOR WHINE IN LOW OR 2ND SPEED
11	96 - 96	24-10-95	BLOWER MOTOR WHINE IN LOW OR 2ND SPEED
3	96 - 96	24-04-96	COOLANT SEEP/LEAK AT AUX REAR HEATER HOSE CONNECTION
5	96 - 96	24-04-96	COOLANT SEEP/LEAK AT AUX REAR HEATER HOSE CONNECTION
11	96 - 96	24-04-96	COOLANT SEEP/LEAK AT AUX REAR HEATER HOSE CONNECTION
6	94 - 95	24-18-94	DRAIN TUBE - A/C CONDENSATE ELIMINATED
12	93 - 95	24-18-94	DRAIN TUBE - A/C CONDENSATE ELIMINATED
3	93 - 94	24-23-93	ECCS (ELECTRONIC CYCLING CLUTCH SWITCH) AVAILABLE
5	93 - 94	24-23-93	ECCS (ELECTRONIC CYCLING CLUTCH SWITCH) AVAILABLE
11	93 - 94	24-23-93	ECCS (ELECTRONIC CYCLING CLUTCH SWITCH) AVAILABLE
13,12,15	91 - 94	24-27-93	GROWL/NOISE - A/C COMPRESSOR - DIAGNOSIS & REPAIR
7,16,6	91 - 94	24-27-93	GROWL/NOISE - A/C COMPRESSOR - DIAGNOSIS & REPAIR
15,12	95 - 97	24-08-96	GROWLING NOISE FROM A/C COMPRESSOR
6	96 - 96	24-08-96	GROWLING NOISE FROM A/C COMPRESSOR - 2.5L
10,14	93 - 93	24-16-93	HEAT & A/C - BLOWER NOISE - REPAIR PROCEDURE
12,14	93 - 93	24-16-93	HEAT & A/C - BLOWER NOISE - REPAIR PROCEDURE
12,14	93 - 95	24-16-93	HEAT & A/C - BLOWER NOISE - REPAIR PROCEDURE - REV
10,14	93 - 95	24-22-94	HEAT & A/C - BLOWER NOISE - REPAIR PROCEDURE - REV
12,14	93 - 93	HL-56-92	HEAT & A/C SYS - INTERMITTENT PROBLEMS - SERV INFO
3	92 - 94	24-03-94	HEAT - POOR PERFORMANCE - ADJUST DOOR CABLE CLIP
5	92 - 94	24-03-94	HEAT - POOR PERFORMANCE - ADJUST DOOR CABLE CLIP
11	92 - 94	24-03-94	HEAT - POOR PERFORMANCE - ADJUST DOOR CABLE CLIP
3	93 - 93	24-12-93	HEAT INSUFFICIENT AT DRIVER'S RIGHT FOOT - REPAIR
5	93 - 93	24-12-93	HEAT INSUFFICIENT AT DRIVER'S RIGHT FOOT - REPAIR
11	93 - 93	24-12-93	HEAT INSUFFICIENT AT DRIVER'S RIGHT FOOT - REPAIR
7	94 - 95	23-19-95	HEATER - A/C CONTROL KNOBS PARTS REPLACEMENT
7	94 - 95	23-19-95	HEATER - A/C CONTROL KNOBS PARTS REPLACEMENT - REV
3	91 - 94	24-05-94	HEATER - POOR PERFORMANCE - DIAGNOSIS & REPAIR
5	91 - 94	24-05-94	HEATER - POOR PERFORMANCE - DIAGNOSIS & REPAIR
11	91 - 94	24-05-94	HEATER - POOR PERFORMANCE - DIAGNOSIS & REPAIR
3	91 - 95	24-05-94	HEATER - POOR PERFORMANCE - DIAGNOSIS & REPAIR - REV
5	91 - 95	24-05-94	HEATER - POOR PERFORMANCE - DIAGNOSIS & REPAIR - REV
11	91 - 95	24-05-94	HEATER - POOR PERFORMANCE - DIAGNOSIS & REPAIR - REV
6	94 - 94	24-06-94	HEATER - POOR PERFORMANCE - NO CONTROL

Model Legend: 1= All 2= Most 3= Town & Country 4= D-50 Pickup 5= Caravan 6= Dakota 7= Ram Pickup 8= Ram 50 9= Ram Van-B Series 10=Grand Cherokee 11= Voyager 12= Cherokee 13= Comanchee 14= Grand Wagoneer 15= CJ, YJ, Wrangler 16= Ramcharger 17= Wagoneer

Heating, Air Conditioning, Ventilation, Defogger

Models	Year	TSB#	Description
10	96 - 96	24-14-95	HEATER A/C BLOWER RUNS W/ IGN OFF - MANUAL SYS
12	96 - 96	24-14-95	HEATER A/C BLOWER RUNS W/ IGN OFF - MANUAL SYS
10	93 - 96	24-01-96	HEATER A/C SYS CHANGES MODE TO DEFROST ON ACCEL
7	94 - 96	24-01-96	HEATER A/C SYS CHANGES MODE TO DEFROST ON ACCEL
12	93 - 96	24-01-96	HEATER A/C SYS CHANGES MODE TO DEFROST ON ACCEL
10	95 - 95	24-19-94	HEATER A/C WHISTLE WHEN FAN ON HIGH
12	95 - 95	24-19-94	HEATER A/C WHISTLE WHEN FAN ON HIGH
3	96 - 96	24-02-96	HVAC CONTROL BUTTONS OR KNOBS STICK
5	96 - 96	24-02-96	HVAC CONTROL BUTTONS OR KNOBS STICK
11	96 - 96	24-02-96	HVAC CONTROL BUTTONS OR KNOBS STICK
3	96 - 96	24-16-95	HVAC DIAGNOSIS
5	96 - 96	24-16-95	HVAC DIAGNOSIS
11	96 - 96	24-16-95	HVAC DIAGNOSIS
12	94 - 94	18-11-94	NO/POOR A/C COOLING AND/OR ENGINE OVERHEAT
3	93 - 94	24-30-93	NOISE/WHISTLE - A/C EVAPORATOR - NEW A/C CORE - REV
5	93 - 94	24-30-93	NOISE/WHISTLE - A/C EVAPORATOR - NEW A/C CORE - REV
11	93 - 94	24-30-93	NOISE/WHISTLE - A/C EVAPORATOR - NEW A/C CORE - REV
3	93 - 94	24-30-93	NOISE/WHISTLE - A/C EVAPORATOR - REVISED A/C CORE
5	93 - 94	24-30-93	NOISE/WHISTLE - A/C EVAPORATOR - REVISED A/C CORE
11	93 - 94	24-30-93	NOISE/WHISTLE - A/C EVAPORATOR - REVISED A/C CORE
10,14	93 - 94	24-08-94	ODOR - A/C EVAPORATOR
2	93 - 94	24-08-94	ODOR - A/C EVAPORATOR
11	93 - 95	24-08-94	ODOR - A/C EVAPORATOR
12,14	93 - 94	24-08-94	ODOR - A/C EVAPORATOR
2	92 - 95	24-06-95	ODOR - A/C EVAPORATOR - REV TWO
2	92 - 95	24-08-94	ODOR - A/C EVAPORATOR - REVISED
3	89 - 94	24-11-94	POOR PERF - A/C & HEAT - BLEND AIR DOOR STICKING
5	84 - 94	24-11-94	POOR PERF - A/C & HEAT - BLEND AIR DOOR STICKING
11	84 - 94	24-11-94	POOR PERF - A/C & HEAT - BLEND AIR DOOR STICKING
10,14	93 - 94	24-20-93	R-134A A/C - REFRIGERANT LEAK DETECTION UNIT INFO
2	93 - 94	24-20-93	R-134A A/C - REFRIGERANT LEAK DETECTION UNIT INFO
2	93 - 94	24-20-93	R-134A A/C - REFRIGERANT LEAK DETECTION UNIT INFO
11	93 - 94	24-20-93	R-134A A/C - REFRIGERANT LEAK DETECTION UNIT INFO
12,14	93 - 94	24-20-93	R-134A A/C - REFRIGERANT LEAK DETECTION UNIT INFO
3	89 - 95	24-03-95	REVISED UNDERBODY REFRIG LINE GASKET - REAR A/C
5	89 - 95	24-03-95	REVISED UNDERBODY REFRIG LINE GASKET - REAR A/C
11	89 - 95	24-03-95	REVISED UNDERBODY REFRIG LINE GASKET - REAR A/C
15	97 - 97	24-10-96	WATER LEAK FROM A/C HOUSING

Model Legend: 1= All 2= Most 3= Town & Country 4= D-50 Pickup 5= Caravan 6= Dakota 7= Ram Pickup 8= Ram 50 9= Ram Van-B Series 10= Grand Cherokee 11= Voyager 12= Cherokee 13= Comanchee 14= Grand Wagoneer 15 = CJ, YJ, Wrangler 16= Ramcharger 17= Wagoneer

Entertainment Devices, Stereo, Radio, Etc.

Models	Year	TSB#	Description
7	94 - 94	08-65-94	AM RADIO - POOR OR NO RECEPTION
3	96 - 96	SA#95-38	AM/FM/CD/CASSETTE RADIO MALFUNCTIONS
5	96 - 96	SA#95-38	AM/FM/CD/CASSETTE RADIO MALFUNCTIONS
11	96 - 96	SA#95-38	AM/FM/CD/CASSETTE RADIO MALFUNCTIONS
10	94 - 94	8/13/94	CLICK/POP - INFINITY GOLD RADIO - WIPER/POWER SEAT
12	94 - 94	8/13/94	CLICK/POP - INFINITY GOLD RADIO - WIPER/POWER SEAT
3	94 - 94	08-63-93	NOISE(BSR)- RADIO REMOVE/INSTALL PROCEDURE - S/M REV
5	94 - 94	08-63-93	NOISE(BSR)- RADIO REMOVE/INSTALL PROCEDURE - S/M REV
11	94 - 94	08-63-93	NOISE(BSR)- RADIO REMOVE/INSTALL PROCEDURE - S/M REV
3	96 - 96	8/3/96	NOISE/STATIC - POOR AM RADIO RECEPTION
5	96 - 96	8/3/96	NOISE/STATIC - POOR AM RADIO RECEPTION
11	96 - 96	35280	NOISE/STATIC - POOR AM RADIO RECEPTION
7	94 - 94	8/5/94	POOR AM RADIO RECEPTION - LOOSE ANTENNA BASE
10	96 - 96	8/14/96	PREM CASSETTE - INTERMIT LOSS OF AUDIO
12	96 - 96	8/14/96	PREM CASSETTE - INTERMIT LOSS OF AUDIO
3	93 - 93	08-36-93	RADIO - IGN SYS INTERFERENCE WITH AM RECEPTION - FIX
5	93 - 93	08-36-93	RADIO - IGN SYS INTERFERENCE WITH AM RECEPTION - FIX
11	93 - 93	08-36-93	RADIO - IGN SYS INTERFERENCE WITH AM RECEPTION - FIX
9	94 - 94	08-47-94	RADIO - INOP OR POOR RECEPTION
2	93 - 93	8/29/93	RADIO - POOR RECEPTION AT FRINGE OF BROADCAST AREA
5	93 - 93	8/29/93	RADIO - POOR RECEPTION AT FRINGE OF BROADCAST AREA
11	93 - 93	34210	RADIO - POOR RECEPTION AT FRINGE OF BROADCAST AREA
10	96 - 96	8/5/96	RADIO DIAGNOSIS - TIPS
12	96 - 96	8/5/96	RADIO DIAGNOSIS - TIPS
12	93 - 93	08-52-93	RADIO INOP - CHECK FOR WATER DAMAGE/SEAL WINDSHIELD
10	94 - 94	08-60-93	RADIO INOP OR ONLY ONE SPEAKER WORKS - HARNESS
12	94 - 94	08-60-93	RADIO INOP OR ONLY ONE SPEAKER WORKS - HARNESS
7	94 - 94	8/6/94	RADIO LOOSES RIGHT SIDE SPEAKERS - RAY(TAPE W/EQUAL)
10	95 - 95	08-76-94	SLOW OR ERRATIC TAPE OPERATION - RAY RADIO
2	95 - 95	08-76-94	SLOW OR ERRATIC TAPE OPERATION - RAY RADIO
12	95 - 95	08-76-94	SLOW OR ERRATIC TAPE OPERATION - RAY RADIO
,9	95 - 95	08-76-94	SLOW OR ERRATIC TAPE OPERATION - RAY RADIO
10	94 - 94	08-45-93	STEREO RADIO LOCK UP (INOP) - PROCEDURES
2	94 - 94	08-45-93	STEREO RADIO LOCK UP (INOP) - PROCEDURES
2	94 - 94	08-45-93	STEREO RADIO LOCK UP (INOP) - PROCEDURES
11	94 - 94	08-45-93	STEREO RADIO LOCK UP (INOP) - PROCEDURES
12	94 - 94	08-45-93	STEREO RADIO LOCK UP (INOP) - PROCEDURES

Model Legend: 1= All 2= Most 3= Town & Country 4= D-50 Pickup 5= Caravan 6= Dakota 7= Ram Pickup 8= Ram 50 9= Ram Van-B Series 10=Grand Cherokee 11= Voyager 12= Cherokee 13= Comanchee 14= Grand Wagoneer 15= CJ, YJ, Wrangler 16= Ramcharger 17= Wagoneer

Seats, Belts, Interior Trim, Carpets, Air Bags

Models	Year	TSB#	Description
3	91 - 93	23-50-93	2ND/3RD SEAT OUTBOARD SEAT BELT EXTENSIONS AVAILABLE
5	91 - 93	23-50-93	2ND/3RD SEAT OUTBOARD SEAT BELT EXTENSIONS AVAILABLE
11	91 - 93	23-50-93	2ND/3RD SEAT OUTBOARD SEAT BELT EXTENSIONS AVAILABLE
3	96 - 96	23-59-95	ADDED DRIVER'S SEAT LUMBAR SUPPORT
5	96 - 96	23-59-95	ADDED DRIVER'S SEAT LUMBAR SUPPORT
11	96 - 96	23-59-95	ADDED DRIVER'S SEAT LUMBAR SUPPORT
3	96 - 96	23-59-95	ADDED DRIVER'S SEAT LUMBAR SUPPORT - REV
5	96 - 96	23-59-95	ADDED DRIVER'S SEAT LUMBAR SUPPORT - REV
11	96 - 96	23-59-95	ADDED DRIVER'S SEAT LUMBAR SUPPORT - REV
3	92 - 93	23-26-93	CARPET/LOWER DOOR TRIM PANEL - WARPED - REPAIR PROC
5	92 - 93	23-26-93	CARPET/LOWER DOOR TRIM PANEL - WARPED - REPAIR PROC
11	92 - 93	23-26-93	CARPET/LOWER DOOR TRIM PANEL - WARPED - REPAIR PROC
3	96 - 96	23-53-95	CARPETED POWER SEAT TRACK COVER FALLS DOWN
5	96 - 96	23-53-95	CARPETED POWER SEAT TRACK COVER FALLS DOWN
11	96 - 96	23-53-95	CARPETED POWER SEAT TRACK COVER FALLS DOWN
3	92 - 93	23-09-93	CHILD SEAT HEAD RESTRAINT RATTLE - DIAGNOSIS/REPAIR
5	92 - 93	23-09-93	CHILD SEAT HEAD RESTRAINT RATTLE - DIAGNOSIS/REPAIR
11	92 - 93	23-09-93	CHILD SEAT HEAD RESTRAINT RATTLE - DIAGNOSIS/REPAIR
10,14	93 - 94	23-06-94	CLUNK/KNOCK FROM REAR SEAT - REPLACE LATCH
12,14	93 - 94	23-06-94	CLUNK/KNOCK FROM REAR SEAT - REPLACE LATCH
7	94 - 96	23-18-96	CLUTCH PEDAL CUTS CARPET
10	96 - 96	23-69-95	CONSOLE LID WILL NOT OPEN
12	96 - 96	23-69-95	CONSOLE LID WILL NOT OPEN
7	94 - 95	23-29-95	CRACKED SUNVISOR SUPPORT BRACKET/RETAINER
10	93 - 93	23-16-92	DASH/KICK PANELS - LEAK WATER INTO DRIVER'S FOOTWELL
12	93 - 93	23-16-92	DASH/KICK PANELS - LEAK WATER INTO DRIVER'S FOOTWELL
12	95 - 95	23-03-95	FLOOR MATS CUT OFF AT REAR EDGE
10,14	93 - 93	23-45-93	FRONT SEAT BACK PIVOT KNOCK NOISE - INSTALL SPACERS
12,14	93 - 93	23-45-93	FRONT SEAT BACK PIVOT KNOCK NOISE - INSTALL SPACERS
10,14	93 - 94	23-70-94	FRONT SEAT BACK PIVOT KNOCK NOISE - PIVOT BOLT - REV
12,14	93 - 94	23-45-93	FRONT SEAT BACK PIVOT KNOCK NOISE - PIVOT BOLT - REV
7,16	94 - 95	23-95-94	FRONT SEAT COVER WEAR ABOVE RECLINER PIVOT
7,16	94 - 95	23-95-94	FRONT SEAT COVER WEAR ABOVE RECLINER PIVOT - REVISED
6	90 - 94	23-96-94	FRONT SEAT CUSHION COMFORT
10	93 - 95	23-90-94	HEAD REST RATTLE
12	93 - 95	23-90-94	HEAD REST RATTLE
3	93 - 93	23-12-93	HEADLINER - FISH ODOR IN DAMP/WARM WEATHER - INFO
3	93 - 93	23-12-93	HEADLINER - FISH ODOR IN DAMP/WARM WEATHER - INFO
5	93 - 93	23-12-93	HEADLINER - FISH ODOR IN DAMP/WARM WEATHER - INFO
5	93 - 93	23-12-93	HEADLINER - FISH ODOR IN DAMP/WARM WEATHER - INFO
11	93 - 93	23-12-93	HEADLINER - FISH ODOR IN DAMP/WARM WEATHER - INFO
11	93 - 93	23-12-93	HEADLINER - FISH ODOR IN DAMP/WARM WEATHER - INFO
3	93 - 93	23-12-93A	HEADLINER - FISH ODOR IN DAMP/WARM WEATHER - INFO
5	93 - 93	23-12-93A	HEADLINER - FISH ODOR IN DAMP/WARM WEATHER - INFO
11	93 - 93	23-12-93A	HEADLINER - FISH ODOR IN DAMP/WARM WEATHER - INFO

Model Legend: 1= All 2= Most 3= Town & Country 4= D-50 Pickup 5= Caravan 6= Dakota 7= Ram Pickup 8= Ram 50 9= Ram Van-B Series 10=Grand Cherokee 11= Voyager 12= Cherokee 13= Comanchee 14= Grand Wagoneer 15= CJ, YJ, Wrangler 16= Ramcharger 17= Wagoneer

Seats, Belts, Interior Trim, Carpets, Air Bags

Models	Year	TSB#	Description
3	94 - 94	23-20-94	HEADLINER FASTENER PINS - NEW STYLE
5	94 - 94	23-20-94	HEADLINER FASTENER PINS - NEW STYLE
11	94 - 94	23-20-94	HEADLINER FASTENER PINS - NEW STYLE
10,14	93 - 94	23-58-94	HEADLINER ODOR - REVISED PART
12,14	93 - 94	23-58-94	HEADLINER ODOR - REVISED PART
3	96 - 96	23-20-96	HIGH EFFORT TO UNLATCH REAR BENCH SEATS FROM FLOOR
5	96 - 96	23-20-96	HIGH EFFORT TO UNLATCH REAR BENCH SEATS FROM FLOOR
11	96 - 96	23-20-96	HIGH EFFORT TO UNLATCH REAR BENCH SEATS FROM FLOOR
3	91 - 93	23-21-93	INTERIOR TRIM RATTLES - LOOSE SCREW CONDITION
5	91 - 93	23-21-93	INTERIOR TRIM RATTLES - LOOSE SCREW CONDITION
11	91 - 93	23-21-93	INTERIOR TRIM RATTLES - LOOSE SCREW CONDITION
6	94 - 95	23-12-95	IP APPEARS WAVY OR BUMPS AT WINDSHIELD EDGE
3	96 - 96	SA#95-34	IP CUPHOLDER/CONV TRAY - INFO
5	96 - 96	SA#95-34	IP CUPHOLDER/CONV TRAY - INFO
11	96 - 96	SA#95-34	IP CUPHOLDER/CONV TRAY - INFO
10,14	93 - 95	23-23-95	KICK PANEL WATER LEAKS
12,14	93 - 95	23-23-95	KICK PANEL WATER LEAKS
3	91 - 94	23-53-94	LIFTGATE TRIM PANEL CRACKING - REVISED SCREWS
5	91 - 94	23-53-94	LIFTGATE TRIM PANEL CRACKING - REVISED SCREWS
11	91 - 94	23-53-94	LIFTGATE TRIM PANEL CRACKING - REVISED SCREWS
9	95 - 95	23-44-95	MANUAL SEAT POSITION CANNOT BE ADJUSTED
10	96 - 96	08-45-95	MEMORY SEATS SOFT STOPS - INFO
12	96 - 96	08-45-95	MEMORY SEATS SOFT STOPS - INFO
3	94 - 94	23-30-94	MIDDLE SEAT - TICKING NOISE FROM UNDER FLOOR PAN
5	94 - 94	23-30-94	MIDDLE SEAT - TICKING NOISE FROM UNDER FLOOR PAN
11	94 - 94	23-30-94	MIDDLE SEAT - TICKING NOISE FROM UNDER FLOOR PAN
9	92 - 94	23-86-94	NOISE - REAR SEAT ANCHOR
7	94 - 95	23-04-95	NOISE/RATTLE - SEAT BELT LATCH PLATE BUMPS TRIM
7	94 - 94	23-35-94	NOISE/SQUEAK - BASE OF IP/STEERING COLUMN
9	94 - 95	23-45-95	OUTSIDE EDGE OF SEAT CUSHION BOLSTER APPEARS SLANTED
3	96 - 96	23-48-95	QUAD SEATS LOOSE TO FLOOR
5	96 - 96	23-48-95	QUAD SEATS LOOSE TO FLOOR
11	96 - 96	23-48-95	QUAD SEATS LOOSE TO FLOOR
3	91 - 93	23-42-93	QUARTER TRIM PANEL/UPPER - REPLACEMENT INFORMATION
5	91 - 93	23-42-93	QUARTER TRIM PANEL/UPPER - REPLACEMENT INFORMATION
11	91 - 93	23-42-93	QUARTER TRIM PANEL/UPPER - REPLACEMENT INFORMATION
7	94 - 94	23-73-94	RATTLE - MUG HOLDER
3	96 - 96	23-57-95	RATTLE - REAR BENCH SEAT TO FLOOR ATTACHMENT
5	96 - 96	23-57-95	RATTLE - REAR BENCH SEAT TO FLOOR ATTACHMENT
11	96 - 96	23-57-95	RATTLE - REAR BENCH SEAT TO FLOOR ATTACHMENT
3	94 - 95	23-74-94	RATTLE NOISE - QUAD SEAT FLOOR ATTACHMENT
5	94 - 95	23-74-94	RATTLE NOISE - QUAD SEAT FLOOR ATTACHMENT
11	94 - 95	23-74-94	RATTLE NOISE - QUAD SEAT FLOOR ATTACHMENT
3	94 - 95	23-74-94	RATTLE NOISE - QUAD SEAT FLOOR ATTACHMENT - REV

Model Legend: 1= All 2= Most 3= Town & Country 4= D-50 Pickup 5= Caravan 6= Dakota 7= Ram Pickup 8= Ram 50 9= Ram Van-B Series 10=Grand Cherokee 11= Voyager 12= Cherokee 13= Comanchee 14= Grand Wagoneer 15= CJ, YJ, Wrangler 16= Ramcharger 17= Wagoneer

Seats, Belts, Interior Trim, Carpets, Air Bags

Models	Year	TSB#	Description
5	94 - 95	23-74-94	RATTLE NOISE - QUAD SEAT FLOOR ATTACHMENT - REV
11	94 - 95	23-74-94	RATTLE NOISE - QUAD SEAT FLOOR ATTACHMENT - REV
3	94 - 94	23-37-94	RATTLE/NOISE - REAR SEAT ADJUSTER
5	94 - 94	23-37-94	RATTLE/NOISE - REAR SEAT ADJUSTER
11	94 - 94	23-37-94	RATTLE/NOISE - REAR SEAT ADJUSTER
10,14	93 - 93	23-52-93	REAR SEAT BELT ACCESS WITH SEAT BACK FOLDED
12,14	93 - 93	23-52-93	REAR SEAT BELT ACCESS WITH SEAT BACK FOLDED
12	94 - 94	SR#596	REAR SEAT BELT ATTACHMENT BOLTS - SAFETY RECALL
3	94 - 94	23-11-94	REAR SEAT SLIDING EFFORT
5	94 - 94	23-11-94	REAR SEAT SLIDING EFFORT
11	94 - 94	23-11-94	REAR SEAT SLIDING EFFORT
3	94 - 94	23-03-94	SEAT BELT - HIGH EFFORT OR WON'T RETURN -MIDDLE SEAT
5	94 - 94	23-03-94	SEAT BELT - HIGH EFFORT OR WON'T RETURN -MIDDLE SEAT
11	94 - 94	23-03-94	SEAT BELT - HIGH EFFORT OR WON'T RETURN -MIDDLE SEAT
3	92 - 93	23-25-93	SEAT BELT/FRONT - DOESN'T FULLY RETRACT - NEW ASSY
3	92 - 93	23-25-93	SEAT BELT/FRONT - DOESN'T FULLY RETRACT - NEW ASSY
5	92 - 93	23-25-93	SEAT BELT/FRONT - DOESN'T FULLY RETRACT - NEW ASSY
5	92 - 93	23-25-93	SEAT BELT/FRONT - DOESN'T FULLY RETRACT - NEW ASSY
11	92 - 93	23-25-93	SEAT BELT/FRONT - DOESN'T FULLY RETRACT - NEW ASSY
11	92 - 93	23-25-93	SEAT BELT/FRONT - DOESN'T FULLY RETRACT - NEW ASSY
5	91 - 93	23-32-93	SEAT/FRONT - MANUAL ADJUSTER DOES NOT LATCH SMOOTHLY
11	91 - 93	23-32-93	SEAT/FRONT - MANUAL ADJUSTER DOES NOT LATCH SMOOTHLY
5	91 - 94	23-32-93	SEAT/FRONT - MANUAL ADJUSTER LATCHING - REVISED
11	91 - 94	23-32-93	SEAT/FRONT - MANUAL ADJUSTER LATCHING - REVISED
7	96 - 96	23-09-96	SEATBELT BUCKLE DIFFICULT TO ENGAGE - CLUB CAB
3	93 - 93	23-24-93	SEATS - INTERCHANGING 2 PASSENGER & QUAD SEATS/INFO
5	93 - 93	23-24-93	SEATS - INTERCHANGING 2 PASSENGER & QUAD SEATS/INFO
11	93 - 93	23-24-93	SEATS - INTERCHANGING 2 PASSENGER & QUAD SEATS/INFO
3	96 - 96	SA#95-21	SHILD SEAT BELT BUCKLE LUBE - SERVICE ACTION
5	96 - 96	SA#95-21	SHILD SEAT BELT BUCKLE LUBE - SERVICE ACTION
11	96 - 96	SA#95-21	SHILD SEAT BELT BUCKLE LUBE - SERVICE ACTION
9	95 - 95	23-35-95	SHOULDER BELT/RETRACTOR - SERVICE/REPLACE PROCEDURE
12	93 - 95	23-10-96	SLOW FRONT SEAT BELT RETRACTION
7	94 - 94	23-45-94	SNAPPING NOISE AT RIGHT SIDE OF IP
3	94 - 94	23-24-94	SQUEAK NOISE - FRONT SEAT BACK
5	94 - 94	23-24-94	SQUEAK NOISE - FRONT SEAT BACK
11	94 - 94	23-24-94	SQUEAK NOISE - FRONT SEAT BACK
3	94 - 94	08-59-93	SQUEAK/RATTLE RIGHT SIDE IP - CHECK PAB (AIR BAG)
5	94 - 94	08-59-93	SQUEAK/RATTLE RIGHT SIDE IP - CHECK PAB (AIR BAG)
11	94 - 94	08-59-93	SQUEAK/RATTLE RIGHT SIDE IP - CHECK PAB (AIR BAG)
10,14	93 - 95	23-37-95	SQUEAKS/CREAKS FROM IP/DASH PAD AREA
12,14	93 - 95	23-37-95	SQUEAKS/CREAKS FROM IP/DASH PAD AREA
4	94 - 95	8/13/95	SRS - ECU CONNECTOR LOCK CHANGE
9	96 - 96	23-12-96	THREE PASS SEAT LEFT ARM REST ELIMINATED

Model Legend: 1= All 2= Most 3= Town & Country 4= D-50 Pickup 5= Caravan 6= Dakota 7= Ram Pickup 8= Ram 50 9= Ram Van-B Series 10=Grand Cherokee 11= Voyager 12= Cherokee 13= Comanchee 14= Grand Wagoneer 15 = CJ, YJ, Wrangler 16= Ramcharger 17= Wagoneer

Seats, Belts, Interior Trim, Carpets, Air Bags

Models	Year	TSB#	Description
3	93 - 95	23-78-94	UNDERSEAT STORAGE DRAWER CRACKS
5	93 - 95	23-78-94	UNDERSEAT STORAGE DRAWER CRACKS
11	93 - 95	23-78-94	UNDERSEAT STORAGE DRAWER CRACKS
10,14	93 - 94	23-14-94	WATER LEAK - LEFT KICK PANEL AND DASH PANEL AREAS
12,14	93 - 94	23-14-94	WATER LEAK - LEFT KICK PANEL AND DASH PANEL AREAS
3	96 - 96	23-64-95	WHITE STRESS MARKS ON INTERIOR TRIM PANEL
5	96 - 96	23-64-95	WHITE STRESS MARKS ON INTERIOR TRIM PANEL
11	96 - 96	23-64-95	WHITE STRESS MARKS ON INTERIOR TRIM PANEL

Model Legend: 1= All 2= Most 3= Town & Country 4= D-50 Pickup 5= Caravan 6= Dakota 7= Ram Pickup 8= Ram 50 9= Ram Van-B Series 10=Grand Cherokee 11= Voyager 12= Cherokee 13= Comanchee 14= Grand Wagoneer 15 = CJ, YJ, Wrangler 16= Ramcharger 17= Wagoneer

Glass, Doors, Hood, Decklid, Tailgate, Liftgate, Locks

Models	Year	TSB#	Description
3	96 - 96	23-54-95	B-POST AREA RATTLE NOISE
5	96 - 96	23-54-95	B-POST AREA RATTLE NOISE
11	96 - 96	23-54-95	B-POST AREA RATTLE NOISE
3	96 - 96	23-54-95	B-POST AREA RATTLE NOISE - REV
5	96 - 96	23-54-95	B-POST AREA RATTLE NOISE - REV
10	93 - 93	23-09-92	DOOR - OUTSIDE HANDLE BUTTON INOP - REPAIR/PARTS
10,14	93 - 94	23-70-93	DOOR LOCK OR LATCH INOPERATIVE - ADJUSTMENTS
10,14	93 - 95	23-15-95	DOOR OPENING & CLOSING - HIGH EFFORT
14,10	93 - 95	23-06-95	FRONT DOOR SAG
10	93 - 93	HL-14-92	HOOD - IN-TRANSIT PROTECTIVE COVER - DEALER REMOVE
10,14	93 - 94	23-52-94	HOOD HINGE RATTLE
10	93 - 93	23-22-92	LIFTGATE - POWER LOCK SYS WON'T LOCK AND/OR OPEN
10	93 - 93	23-22-92A	LIFTGATE - POWER LOCK SYS WON'T LOCK AND/OR OPEN
10	94 - 94	23-02-94	LIFTGATE HANDLE LOCKS WHEN SLAMMED - PROCEDURE
10	93 - 93	23-22-92A	LIFTGATE POWER LOCK SYS WON'T LOCK AND/OR OPEN - REV
10,14	93 - 93	23-33-92	LIFTGATE RATTLES - DIAGNOSIS/PARTS/REPAIR PROCEDURE
10,14	93 - 93	23-44-93	REAR CARGO COVER - LABEL TO AID PROPER INSTALLATION
10	95 - 95	23-22-95	SNAP NOISE AFTER SUN ROOF IS OPENED
10	94 - 95	08-60-94	SNAP/POP NOISE WHEN DOOR IS OPENED -DOOR AJAR SWITCH
10	96 - 96	23-70-95	WATER LEAKS AT BOTTOM OF DOORS
10,14	93 - 94	23-69-94	WIND NOISE AT LOWER WINDSHIELD CORNERS
11	96 - 96	23-54-95	B-POST AREA RATTLE NOISE - REV
7,16	Up - 93	23-63-94	CRACK/POP NOISE FROM COWL AREA - COWL CRACK
3	96 - 96	23-47-95	CREAK/SQUEAK NOISE AT FRONT DOOR
5	96 - 96	23-47-95	CREAK/SQUEAK NOISE AT FRONT DOOR
11	96 - 96	23-47-95	CREAK/SQUEAK NOISE AT FRONT DOOR
12	93 - 93	23-09-92	DOOR - OUTSIDE HANDLE BUTTON INOP - REPAIR/PARTS
3	93 - 93	23-13-93	DOOR - OUTSIDE HANDLE SEPARATES FROM BODY - REPAIR
3	93 - 93	23-13-93	DOOR - OUTSIDE HANDLE SEPARATES FROM BODY - REPAIR
5	93 - 93	23-13-93	DOOR - OUTSIDE HANDLE SEPARATES FROM BODY - REPAIR
5	93 - 93	23-13-93	DOOR - OUTSIDE HANDLE SEPARATES FROM BODY - REPAIR
11	93 - 93	23-13-93	DOOR - OUTSIDE HANDLE SEPARATES FROM BODY - REPAIR
11	93 - 93	23-13-93	DOOR - OUTSIDE HANDLE SEPARATES FROM BODY - REPAIR
3	92 - 93	23-13-93A	DOOR - OUTSIDE HANDLE SEPARATES FROM BODY - REPAIR
5	92 - 93	23-13-93A	DOOR - OUTSIDE HANDLE SEPARATES FROM BODY - REPAIR
11	92 - 93	23-13-93A	DOOR - OUTSIDE HANDLE SEPARATES FROM BODY - REPAIR
7	94 - 95	23-43-95	DOOR DOES NOT OPEN SMOOTHLY OR FEELS LOOSE
7	94 - 94	23-32-94	DOOR FIT AT ROOFLINE - SPECIFICATION
7	94 - 94	23-40-94	DOOR GLASS RATTLE - BELTLINE WEATHERSTRIP
7	94 - 94	23-40-94	DOOR GLASS RATTLE - REVISED WEATHER STRIP
9	90 - 96	23-13-96	DOOR GLASS SEPARATES FROM LIFT CHANNEL
12,14	93 - 94	23-70-93	DOOR LOCK OR LATCH INOPERATIVE - ADJUSTMENTS
12,14	93 - 95	23-15-95	DOOR OPENING & CLOSING - HIGH EFFORT
12	93 - 95	23-04-96	DOOR SAG/CREAKING NOISE
9	95 - 96	23-71-95	DRIVER SEAT BELT LOCKS UP

Model Legend: 1= All 2= Most 3= Town & Country 4= D-50 Pickup 5= Caravan 6= Dakota 7= Ram Pickup 8= Ram 50 9= Ram Van-B Series 10=Grand Cherokee 11= Voyager 12= Cherokee 13= Comanchee 14= Grand Wagoneer 15= CJ, YJ, Wrangler 16= Ramcharger 17= Wagoneer

Glass, Doors, Hood, Decklid, Tailgate, Liftgate, Locks

Models	Year	TSB#	Description
9	96 - 96	23-31-96	EXCESS EFFORT TO LATCH SINGLE REAR CARGO DOOR
15	93 - 95	23-26-95	EXCESSIVE DOOR WINDOW EFFORT
14,12	93 - 95	23-06-95	FRONT DOOR SAG
3	91 - 95	23-66-94	FRONT DOOR SAG - FORWARD HEM SEPARATION
5	91 - 95	23-66-94	FRONT DOOR SAG - FORWARD HEM SEPARATION
11	91 - 95	23-66-94	FRONT DOOR SAG - FORWARD HEM SEPARATION
3	91 - 93	23-46-93	FRT WINDOW ROLLS UP OUT OF RUN WEATHERSTRIP CHAN/REV
5	91 - 93	23-46-93	FRT WINDOW ROLLS UP OUT OF RUN WEATHERSTRIP CHAN/REV
11	91 - 93	23-46-93	FRT WINDOW ROLLS UP OUT OF RUN WEATHERSTRIP CHAN/REV
3	91 - 93	23-46-93	FRT WINDOW ROLLS UP OUT OF RUN WEATHERSTRIP CHANNEL
5	91 - 93	23-46-93	FRT WINDOW ROLLS UP OUT OF RUN WEATHERSTRIP CHANNEL
11	91 - 93	23-46-93	FRT WINDOW ROLLS UP OUT OF RUN WEATHERSTRIP CHANNEL
3	91 - 93	23-46-93A	FRT WINDOW ROLLS UP OUT OF WEATHERSTRIP CHAN/REV 2
5	91 - 93	23-46-93A	FRT WINDOW ROLLS UP OUT OF WEATHERSTRIP CHAN/REV 2
11	91 - 93	23-46-93A	FRT WINDOW ROLLS UP OUT OF WEATHERSTRIP CHAN/REV 2
5	93 - 93	23-31-92	GLASS (DEEP TINT) - IDENTIFYING SOLARBAN VS PRIVACY
11	93 - 93	23-31-92	GLASS (DEEP TINT) - IDENTIFYING SOLARBAN VS PRIVACY
2	94 - 95	23-68-94	GLUE OOZES OUT AT BACKLIGHT OR WINDSHIELD MOLDING
2	94 - 95	23-68-94	GLUE OOZES OUT AT BACKLIGHT OR WINDSHIELD MOLDING
11	94 - 95	23-68-94	GLUE OOZES OUT AT BACKLIGHT OR WINDSHIELD MOLDING
12	93 - 93	HL-14-92	HOOD - IN-TRANSIT PROTECTIVE COVER - DEALER REMOVE
12,14	93 - 94	23-52-94	HOOD HINGE RATTLE
1	95 - 96	23-74-95	INTERIOR WINDOW FILM BUILD UP
1	95 - 96	23-74-95	INTERIOR WINDOW FILM BUILD UP
1	95 - 96	23-74-95	INTERIOR WINDOW FILM BUILD UP
3	96 - 96	08-40-95	INTERMIT OPS - SLIDING DOOR POWER LOCK
5	96 - 96	08-40-95	INTERMIT OPS - SLIDING DOOR POWER LOCK
11	96 - 96	08-40-95	INTERMIT OPS - SLIDING DOOR POWER LOCK
3	96 - 96	08-40-95	INTERMIT OPS - SLIDING DOOR POWER LOCK - REV
5	96 - 96	08-40-95	INTERMIT OPS - SLIDING DOOR POWER LOCK - REV
11	96 - 96	08-40-95	INTERMIT OPS - SLIDING DOOR POWER LOCK - REV
5	91 - 93	23-05-93	LIFTGATE - HOWL WIND NOISE ABOVE 40 MPH - REPAIR
11	91 - 93	23-05-93	LIFTGATE - HOWL WIND NOISE ABOVE 40 MPH - REPAIR
12	93 - 93	23-22-92	LIFTGATE - POWER LOCK SYS WON'T LOCK AND/OR OPEN
12	93 - 93	23-22-92	LIFTGATE - POWER LOCK SYS WON'T LOCK AND/OR OPEN
12	94 - 94	23-02-94	LIFTGATE HANDLE LOCKS WHEN SLAMMED - PROCEDURE
12	93 - 93	23-22-92A	LIFTGATE POWER LOCK SYS WON'T LOCK AND/OR OPEN - REV
12,14	93 - 93	23-33-92	LIFTGATE RATTLES - DIAGNOSIS/PARTS/REPAIR PROCEDURE
3	96 - 96	23-56-95	LIFTGATE TO REAR FASCIA GAP TOO SMALL
5	96 - 96	23-56-95	LIFTGATE TO REAR FASCIA GAP TOO SMALL
11	96 - 96	23-56-95	LIFTGATE TO REAR FASCIA GAP TOO SMALL
2	93 - 93	23-60-93	LOCK CYLINDER KEY BREAKAGE - REPAIR PROCEDURES
11	93 - 93	23-60-93	LOCK CYLINDER KEY BREAKAGE - REPAIR PROCEDURES
3	94 - 95	23-67-94	MIDDLE ROW SEAT INTERCHANGEABILITY - PROCEDURES

Model Legend: 1= All 2= Most 3= Town & Country 4= D-50 Pickup 5= Caravan 6= Dakota 7= Ram Pickup 8= Ram 50 9= Ram Van-B Series 10=Grand Cherokee 11= Voyager 12= Cherokee 13= Comanchee 14= Grand Wagoneer 15 = CJ, YJ, Wrangler 16= Ramcharger 17= Wagoneer

Glass, Doors, Hood, Decklid, Tailgate, Liftgate, Locks

Models	Year	TSB#	Description
5	94 - 95	23-67-94	MIDDLE ROW SEAT INTERCHANGEABILITY - PROCEDURES
11	94 - 95	23-67-94	MIDDLE ROW SEAT INTERCHANGEABILITY - PROCEDURES
3	91 - 94	23-62-93	MIDDLE SIDE GLASS CONTROL EFFORT - REVISED LATCH
5	91 - 94	23-62-93	MIDDLE SIDE GLASS CONTROL EFFORT - REVISED LATCH
11	91 - 94	23-62-93	MIDDLE SIDE GLASS CONTROL EFFORT - REVISED LATCH
6	96 - 96	23-26-96	POP/SCRATCH NOISE FROM A-PILLAR AREA
7	94 - 94	23-60-94	POP/SNAP NOISE FROM WINDSHIELD
3	93 - 93	8/14/93	POWER DOOR LOCK SWITCH AND BEZEL SERVICE - INFO/PROC
5	93 - 93	8/14/93	POWER DOOR LOCK SWITCH AND BEZEL SERVICE - INFO/PROC
11	93 - 93	34195	POWER DOOR LOCK SWITCH AND BEZEL SERVICE - INFO/PROC
12,14	93 - 93	23-44-93	REAR CARGO COVER - LABEL TO AID PROPER INSTALLATION
3	91 - 94	HL-46-92	REPLACE AJAR SWITCH AFTER SLIDING DOOR ADJUSTMENT
5	91 - 94	HL-46-92	REPLACE AJAR SWITCH AFTER SLIDING DOOR ADJUSTMENT
11	91 - 94	HL-46-92	REPLACE AJAR SWITCH AFTER SLIDING DOOR ADJUSTMENT
3	92 - 94	23-04-94	REPLACING FRONT DOOR - REPLACE IMPACT BEAM ASSEMBLY
5	92 - 94	23-04-94	REPLACING FRONT DOOR - REPLACE IMPACT BEAM ASSEMBLY
11	92 - 94	23-04-94	REPLACING FRONT DOOR - REPLACE IMPACT BEAM ASSEMBLY
3	96 - 96	23-55-95	RIGHT SIDE QUARTER TRIM PANEL WARM TO TOUCH
5	96 - 96	23-55-95	RIGHT SIDE QUARTER TRIM PANEL WARM TO TOUCH
11	96 - 96	23-55-95	RIGHT SIDE QUARTER TRIM PANEL WARM TO TOUCH
3	94 - 94	08-51-93	RKE - KEYLESS ENTRY - MUST DEPRESS BUTTON REPEATEDLY
5	94 - 94	08-51-93	RKE - KEYLESS ENTRY - MUST DEPRESS BUTTON REPEATEDLY
11	94 - 94	08-51-93	RKE - KEYLESS ENTRY - MUST DEPRESS BUTTON REPEATEDLY
3	91 - 93	23-47-93	SIDE VENT GLASS CONTACTS BODY - NOISE
5	91 - 93	23-47-93	SIDE VENT GLASS CONTACTS BODY - NOISE
11	91 - 93	23-47-93	SIDE VENT GLASS CONTACTS BODY - NOISE
3	89 - 93	23-06-93	SLIDING DOOR - REVISED CENTER AND LOWER TRACK STOPS
5	87 - 93	23-06-93	SLIDING DOOR - REVISED CENTER AND LOWER TRACK STOPS
11	87 - 93	23-06-93	SLIDING DOOR - REVISED CENTER AND LOWER TRACK STOPS
3	91 - 93	HL-46-92	SLIDING DOOR ADJUSTMENT - AJAR SWITCH RECOMMENDATION
5	91 - 93	HL-46-92	SLIDING DOOR ADJUSTMENT - AJAR SWITCH RECOMMENDATION
11	91 - 93	HL-46-92	SLIDING DOOR ADJUSTMENT - AJAR SWITCH RECOMMENDATION
3	91 - 94	08-62-93	SLIDING DOOR AJAR SWITCH/ADJUST (CHIME/LOCKS/LITES)
5	91 - 94	08-62-93	SLIDING DOOR AJAR SWITCH/ADJUST (CHIME/LOCKS/LITES)
11	91 - 94	08-62-93	SLIDING DOOR AJAR SWITCH/ADJUST (CHIME/LOCKS/LITES)
3	96 - 96	SA#95-36	SLIDING DOOR POWER LOCKS INTERMITTENT
5	96 - 96	SA#95-36	SLIDING DOOR POWER LOCKS INTERMITTENT
11	96 - 96	SA#95-36	SLIDING DOOR POWER LOCKS INTERMITTENT
3	96 - 96	23-60-95	SLIDING DOOR/LIFTGATE POWER LOCK MOTOR INOP
5	96 - 96	23-60-95	SLIDING DOOR/LIFTGATE POWER LOCK MOTOR INOP
11	96 - 96	23-60-95	SLIDING DOOR/LIFTGATE POWER LOCK MOTOR INOP
12	95 - 95	23-22-95	SNAP NOISE AFTER SUN ROOF IS OPENED
12	94 - 95	08-60-94	SNAP/POP NOISE WHEN DOOR IS OPENED -DOOR AJAR SWITCH
15	93 - 95	23-18-95	SOFT TOP APPEARS LOOSE

Model Legend: 1= All 2= Most 3= Town & Country 4= D-50 Pickup 5= Caravan 6= Dakota 7= Ram Pickup 8= Ram 50 9= Ram Van-B Series 10=Grand Cherokee 11= Voyager 12= Cherokee 13= Comanchee 14= Grand Wagoneer 15 = CJ, YJ, Wrangler 16= Ramcharger 17= Wagoneer

Glass, Doors, Hood, Decklid, Tailgate, Liftgate, Locks

Models	Year	TSB#	Description
3	94 - 94	23-27-94	SQUEAK NOISE - BASE OF WINDSHIELD - NEW CAR
5	94 - 94	23-27-94	SQUEAK NOISE - BASE OF WINDSHIELD - NEW CAR
11	94 - 94	23-27-94	SQUEAK NOISE - BASE OF WINDSHIELD - NEW CAR
7	94 - 94	23-36-94	SQUEAK/CREAK - FRONT DOOR TO WINDSHIELD MOLDING
3	96 - 96	23-50-95	SUCTION CUP MARKS ON GLASS
5	96 - 96	23-50-95	SUCTION CUP MARKS ON GLASS
11	96 - 96	23-50-95	SUCTION CUP MARKS ON GLASS
3	96 - 96	23-50-95A	SUCTION CUP MARKS ON GLASS - REV
5	96 - 96	23-50-95A	SUCTION CUP MARKS ON GLASS - REV
11	96 - 96	23-50-95A	SUCTION CUP MARKS ON GLASS - REV
7	94 - 96	23-29-96	TAILGATE CRACKING ON TOP INNER ENDS
7	94 - 96	23-21-96	TAILGATE HARD TO CLOSE IN COLD WEATHER
7,6	93 - 94	23-98-94	TAILGATE HARD TO LATCH IN COLD TEMP
15	87 - 93	23-49-93	TAILGATE HINGE CORRODES/BINDS - IMPROVED HINGE
15	87 - 95	23-68-95	TAILGATE HINGE CORRODES/BINDS - REV HINGE
7	95 - 95	23-71-94	TAILGATE LATCH HANDLE ESCUTCHEON LOOSE
7	94 - 94	23-51-94	TAILGATE RATTLE
7	94 - 94	23-64-93	TAILGATE RATTLE - INSTL OVERSLAM & ALIGNMENT BUMPERS
7	95 - 95	23-33-95	TAILGATE RATTLE/EXCESSIVE MOVEMENT
3	94 - 94	23-61-94	VENT WINDOW LATCH - RATTLE/POP OPEN/WON'T STAY OPEN
5	94 - 94	23-61-94	VENT WINDOW LATCH - RATTLE/POP OPEN/WON'T STAY OPEN
11	94 - 94	23-61-94	VENT WINDOW LATCH - RATTLE/POP OPEN/WON'T STAY OPEN
12	96 - 96	23-70-95	WATER LEAKS AT BOTTOM OF DOORS
7	94 - 94	23-67-93	WIND NOISE - FRONT DOOR/LOWER A-POST AREA - REVISED
7	94 - 94	23-67-93	WIND NOISE - FRONT DOOR/LOWER A-POST AREA - SEAL
12,14	93 - 94	23-69-94	WIND NOISE AT LOWER WINDSHIELD CORNERS
12,14,10	93 - 94	23-69-94	WIND NOISE AT LOWER WINDSHIELD CORNERS - REVISED
6	93 - 95	23-27-95	WIND NOISE FROM WINDSHIELD/IP AREA
6	96 - 96	23-34-96	WINDNOISE AROUND FRONT DOORS
3	91 - 93	23-04-93	WINDOW LATCH/MANUAL VENT - SNAP/CLICK ON ROUGH ROAD
5	91 - 93	23-04-93	WINDOW LATCH/MANUAL VENT - SNAP/CLICK ON ROUGH ROAD
11	91 - 93	23-04-93	WINDOW LATCH/MANUAL VENT - SNAP/CLICK ON ROUGH ROAD
9	94 - 94	23-13-94	WINDSHIELD AIR NOISE
7	94 - 94	23-17-94	WINDSHIELD INSTALLATION PROCEDURE - S/M REVISION
9	94 - 94	23-13-94	WINDSHIELD OR DOOR AREA AIR NOISE - REVISED

Model Legend: 1= All 2= Most 3= Town & Country 4= D-50 Pickup 5= Caravan 6= Dakota 7= Ram Pickup 8= Ram 50 9= Ram Van-B Series 10=Grand Cherokee 11= Voyager 12= Cherokee 13= Comanchee 14= Grand Wagoneer 15= CJ, YJ, Wrangler 16= Ramcharger 17= Wagoneer

Finishes, Body Structure, Frame, Bumpers

Models	Year	TSB#	Description
3	91 - 93	23-19-93	A-POST WIND NOISE - DIAGNOSIS/REPAIR/CLEANER/SEALER
5	91 - 93	23-19-93	A-POST WIND NOISE - DIAGNOSIS/REPAIR/CLEANER/SEALER
11	91 - 93	23-19-93	A-POST WIND NOISE - DIAGNOSIS/REPAIR/CLEANER/SEALER
7	94 - 95	23-101-94	ANTI-FRICTION TAPE ON A-PILLAR - DO NOT REMOVE
10	93 - 93	23-24-92	BUMPER SILL/FRONT - SQUEAK/CREAK ON STEER WHEEL TURN
10,17	93 - 93	13-01-92	FASCIA - PLASTIC BUMPER GUARD SERVICE INFORMATION
10	93 - 93	HL-21-92	HOOD - ADHESIVE RESIDUE FROM IN-TRANSIT COVER - PROC
10,14	93 - 95	23-23-95	KICK PANEL WATER LEAKS
10,14	93 - 94	23-10-94	PAINT CHIPPING AT D-PILLAR - PROCEDURE
10	96 - 96	23-03-96	POOR DOOR CLADDING FIT - FRONT DOOR TO FENDER
12,17,10,14	93 - 93	23-20-93	RATTLE AT VEHICLE REAR - LICENSE PLATE INSTALL INFO
10,14	93 - 94	11/1/94	REAR BUMPER FASCIA DISTORTS - TAILPIPE HEAT
10,14	93 - 93	23-27-93	ROOF AREA RATTLE/LOOSE ROOF RACK CROSS BOW - REPAIR
10	93 - 93	23-07-92	ROOF RACK - WIND NOISE AT LOW SPD W/CROSSWIND - PROC
10	96 - 96	23-72-95	ROOF RACK RATTLE
10,14	93 - 96	23-72-95	ROOF RACK RATTLE - REV
10,14	93 - 95	23-16-94	ROOF RACK RATTLE/LOOSE OR STICKING BUTTON - REV2
10,14	93 - 94	23-27-93	ROOF RACK RATTLE/LOOSE OR STICKING BUTTON - REVISION
10	95 - 95	23-36-95	ROOF RACK/SUNROOF WATER LEAKS
12,15,10,14	93 - 93	23-43-93	TAPE STRIPE AND MOULDING ADHESIVE REMOVAL/PRODUCT
10,14	93 - 94	23-14-94	WATER LEAK - LEFT KICK PANEL AND DASH PANEL AREAS
10,14	93 - 93	23-56-93	WAVY ROOF SHEET METAL AT THE B PILLAR - REPAIR
4	93 - 93	HL-45-92	BODY - PLASTIC WRAP PROTECTION DURING SHIPPING INFO
7,16	94 - 94	13-02-93	BODY - SHIPPING TIE-DOWN BRACKET REMOVAL
5	96 - 96	23-21-95	BODY SIDE MOLDINGS/CLADDING REV - WIDER
11	96 - 96	23-21-95	BODY SIDE MOLDINGS/CLADDING REV - WIDER
12	93 - 93	23-24-92	BUMPER SILL/FRONT - SQUEAK/CREAK ON STEER WHEEL TURN
6	93 - 95	23-38-95	CLICK/SQUEAK - LEFT FRONT WHEEL WELL AREA
1	93 - 93	23-21-92	COLOR INFORMATION & VEHICLE CODE PLATE LOCATION
1	93 - 93	23-21-92	COLOR INFORMATION & VEHICLE CODE PLATE LOCATION
1	93 - 93	23-21-92	COLOR INFORMATION & VEHICLE CODE PLATE LOCATION
2	93 - 93	23-21-92	COLOR INFORMATION & VEHICLE CODE PLATE LOCATION
3	96 - 96	23-61-95	COWL COVER FIT - ENDS UNSNAPPED
5	96 - 96	23-61-95	COWL COVER FIT - ENDS UNSNAPPED
11	96 - 96	23-61-95	COWL COVER FIT - ENDS UNSNAPPED
7	94 - 94	23-57-93	CREAK - LEFT SIDE OF IP - REPAIR PROCEDURE
3	94 - 94	23-61-93	CREAK AT LEFT B-PILLAR - REPOSITION BAFFLE
5	94 - 94	23-61-93	CREAK AT LEFT B-PILLAR - REPOSITION BAFFLE
11	94 - 94	23-61-93	CREAK AT LEFT B-PILLAR - REPOSITION BAFFLE
7	94 - 95	23-52-95	CREAK NOISE OR EXTERIOR NOISE FROM BACK OF CAB
7	94 - 95	23-52-95	CREAK NOISE OR EXTERIOR NOISE FROM BACK OF CAB - REV
3	96 - 96	23-75-95	DISCOLORED/HAZED COWL GRILL OR OUTSIDE MIRROR
5	96 - 96	23-75-95	DISCOLORED/HAZED COWL GRILL OR OUTSIDE MIRROR
11	96 - 96	23-75-95	DISCOLORED/HAZED COWL GRILL OR OUTSIDE MIRROR
3	96 - 96	23-11-96	DUST ENTERS REAR INTERIOR CARGO AREA

Model Legend: 1= All 2= Most 3= Town & Country 4= D-50 Pickup 5= Caravan 6= Dakota 7= Ram Pickup 8= Ram 50 9= Ram Van-B Series 10=Grand Cherokee 11= Voyager 12= Cherokee 13= Comanchee 14= Grand Wagoneer 15= CJ, YJ, Wrangler 16= Ramcharger 17= Wagoneer

Finishes, Body Structure, Frame, Bumpers

Models	Year	TSB#	Description
5	96 - 96	23-11-96	DUST ENTERS REAR INTERIOR CARGO AREA
11	96 - 96	23-11-96	DUST ENTERS REAR INTERIOR CARGO AREA
12,16	93 - 93	13-01-92	FASCIA - PLASTIC BUMPER GUARD SERVICE INFORMATION
12	93 - 93	HL-21-92	HOOD - ADHESIVE RESIDUE FROM IN-TRANSIT COVER - PROC
5	91 - 93	23-48-93	INTERMITTENT SQUEAK AT RR - BRAKE CABLE RUBS METAL
11	91 - 93	23-48-93	INTERMITTENT SQUEAK AT RR - BRAKE CABLE RUBS METAL
12,14	93 - 95	23-23-95	KICK PANEL WATER LEAKS
3	91 - 94	23-59-94	LEFT B-POST BODY EXHAUSTER RETENTION
5	91 - 94	23-59-94	LEFT B-POST BODY EXHAUSTER RETENTION
11	91 - 94	23-59-94	LEFT B-POST BODY EXHAUSTER RETENTION
3	94 - 94	23-42-94	PAINT - PROTECTIVE SHIPPING FILM
5	94 - 94	23-42-94	PAINT - PROTECTIVE SHIPPING FILM
11	94 - 94	23-42-94	PAINT - PROTECTIVE SHIPPING FILM
3	94 - 95	23-42-94	PAINT - PROTECTIVE SHIPPING FILM - REVISED
11	94 - 95	23-42-94	PAINT - PROTECTIVE SHIPPING FILM - REVISED
5	94 - 95	23-42-94	PAINT - PROTECTIVE SHIPPING FILM - REVISED
3	95 - 95	23-42-94A	PAINT - PROTECTIVE SHIPPING FILM - REVISED
11	95 - 95	23-42-94A	PAINT - PROTECTIVE SHIPPING FILM - REVISED
5	95 - 95	23-42-94A	PAINT - PROTECTIVE SHIPPING FILM - REVISED
12	93 - 95	23-77-95	PAINT CHAFING ON FRONT LOWER DOOR SILL
12,14	93 - 94	23-10-94	PAINT CHIPPING AT D-PILLAR - PROCEDURE
3	96 - 96	23-76-95	PAINT CHIPS AT UPPER FRONT CORNER OF SLIDING DOOR
5	96 - 96	23-76-95	PAINT CHIPS AT UPPER FRONT CORNER OF SLIDING DOOR
11	96 - 96	23-76-95	PAINT CHIPS AT UPPER FRONT CORNER OF SLIDING DOOR
2	90 - 93	23-37-93	PAINT/ANTI-CHIP (STONE GUARD) - REPAIR PROCEDURES
2	89 - 94	23-37-93	PAINT/ANTI-CHIP (STONE GUARD) - REPAIR PROCEDURES
2	89 - 94	23-37-93	PAINT/ANTI-CHIP (STONE GUARD) - REPAIR PROCEDURES
11	89 - 93	23-37-93	PAINT/ANTI-CHIP (STONE GUARD) - REPAIR PROCEDURES
7	94 - 94	23-50-94	PICK UP BOX MOUNTING BOLTS ARE SINGLE SERVICE
7	94 - 94	23-66-93	PICKUP BOX SIDE PANEL CRACKS - PROCEDURE
12	96 - 96	23-03-96	POOR DOOR CLADDING FIT - FRONT DOOR TO FENDER
7	94 - 94	23-39-94	RATTLE - PICK UP BOX FLOOR
9	94 - 94	23-64-94	RATTLE - SPARE TIRE CARRIER
11	93 - 93	23-20-93	RATTLE AT VEHICLE REAR - LICENSE PLATE INSTALL INFO
12,16,12,14	93 - 93	23-20-93	RATTLE AT VEHICLE REAR - LICENSE PLATE INSTALL INFO
5,9	93 - 93	23-20-93	RATTLE AT VEHICLE REAR - LICENSE PLATE INSTALL INFO
12,14	93 - 94	11/1/94	REAR BUMPER FASCIA DISTORTS - TAILPIPE HEAT
9	94 - 94	23-41-93	REAR QUARTER PANEL - AIR EXHAUSTER WHISTLE - REPAIR
7	94 - 96	23-01-96	REPLACEMENT CARGO BOX INFORMATION
12,14	93 - 93	23-27-93	ROOF AREA RATTLE/LOOSE ROOF RACK CROSS BOW - REPAIR
12	93 - 93	23-07-92	ROOF RACK - WIND NOISE AT LOW SPD W/CROSSWIND - PROC
12	96 - 96	23-72-95	ROOF RACK RATTLE
12,14	93 - 96	23-72-95	ROOF RACK RATTLE - REV
12,14	93 - 95	23-16-94	ROOF RACK RATTLE/LOOSE OR STICKING BUTTON - REV2

Model Legend: 1= All 2= Most 3= Town & Country 4= D-50 Pickup 5= Caravan 6= Dakota 7= Ram Pickup 8= Ram 50 9= Ram Van-B Series 10=Grand Cherokee 11= Voyager 12= Cherokee 13= Comanchee 14= Grand Wagoneer 15 = CJ, YJ, Wrangler 16= Ramcharger 17= Wagoneer

Finishes, Body Structure, Frame, Bumpers

Models	Year	TSB#	Description
12,14	93 - 94	23-27-93	ROOF RACK RATTLE/LOOSE OR STICKING BUTTON - REVISION
12	95 - 95	23-36-95	ROOF RACK/SUNROOF WATER LEAKS
6	93 - 93	23-40-93	ROOF SEAM WATERLEAK/APPEARS AT WINDSHIELD CORNERS
15	95 - 95	23-07-95	RUSTED HOOD BUMPERS (WINDSHIELD SUPPORT)
2	93 - 93	23-43-93	TAPE STRIPE AND MOULDING ADHESIVE REMOVAL/PRODUCT
2	93 - 94	23-43-93	TAPE STRIPE AND MOULDING ADHESIVE REMOVAL/PRODUCT
2	93 - 93	23-43-93	TAPE STRIPE AND MOULDING ADHESIVE REMOVAL/PRODUCT
12,15,12,14	93 - 93	23-43-93	TAPE STRIPE AND MOULDING ADHESIVE REMOVAL/PRODUCT
9	94 - 94	CSN#609	TURQUOISE PAINT DELAMINATION - RECALL
15	93 - 95	23-31-95	UNDERHOOD RATTLE
12,14	93 - 94	23-14-94	WATER LEAK - LEFT KICK PANEL AND DASH PANEL AREAS
12,14	93 - 93	23-56-93	WAVY ROOF SHEET METAL AT THE B PILLAR - REPAIR
5	93 - 94	23-58-93	WELDING PRECAUTIONS - TEVAN (ELECTRIC MINIVAN)
11	93 - 94	23-58-93	WELDING PRECAUTIONS - TEVAN (ELECTRIC MINIVAN)
7	96 - 96	23-27-96	WIND NOISE/WHISTLE AROUND GRILLE AREA
3	93 - 93	23-07-93	WOODGRAIN OVERLAY WRINKLE AT CONTACT W/SIDE APPLIQUE

Model Legend: 1= All 2= Most 3= Town & Country 4= D-50 Pickup 5= Caravan 6= Dakota 7= Ram Pickup 8= Ram 50 9= Ram Van-B Series 10=Grand Cherokee 11= Voyager 12= Cherokee 13= Comanchee 14= Grand Wagoneer 15 = CJ, YJ, Wrangler 16= Ramcharger 17= Wagoneer

FORD MOTOR COMPANY

Lighting, Horns, Turn Signals, Steering Column

Models	Year	TSB#	Description
4,8	92 - 95	95-05-21	DRL OR BACK-UP LAMPS INOP - 5.0/5.8L
4,5,8	92 - 93	93-19B-26	ELECTRIC HORN REMOVAL & INSTALL PROCEDURES - S/M REV
11	93 - 93	94-21-12	FRONT LIGHT BAR - EARLY BULB FAILURE
2	89 - 93	89-16-12	HEADLAMP - HORIZONTAL ADJUSTER DIFFICULT - ACCESS
7	93 - 94	93-18-13	HEADLAMP CIRCUITS -ELECTROCHROMATIC REAR VIEW MIRROR
7	93 - 93	93-18-13	HEADLAMP CKTS - ELECTROCHROMATIC REAR VIEW MIRROR
11	96 - 96	96-09-09	HORN CHIRP - INADVERTANT ACTIVATION
11	96 - 96	96-09-09	HORN CHIRP - INADVERTANT ACTIVATION - REV
7,9	95 - 95	95-23-11	HORN INTERMITTENT WHEN TURNING STEER WHEEL
7,9	93 - 93	93-26B-45	HORN LOCATION CLARIFIED - S/M REVISION
11	93 - 93	93-05B-42	LIGHTING DIAGNOSIS & TESTING PROCEDURES - S/M REV
7,4	93 - 96	96-08-15	MAP LAMP ERRATIC/BURNS OUT - COMPASS ERRATIC
11	93 - 93	93-23-22	SQUEAK NOISE IN STEERING WHEEL WHEN TURNED - GREASE

Model Legend: 1= All 2= Most 3= Aerostar 4= Bronco 5= Econoline 6= E Series 7= Explorer/Mountaineer 8= F Series 9= Ranger 10= Windstar 11= Villager

Instruments, Dash Cluster, Warning Lights, Mirrors

Models	Year	TSB#	Description
3	92 - 96	96-02-06	BUZZ/RATTLE AT SIDES OF IP AND/OR SIDE 180W
7	93 - 93	93-05B-29	COMPASS/OUTSIDE TEMP DISPLAY SERVICE INFO - S/M ADD
7	93 - 94	93-18-13	ELECTROCHROMATIC REAR VIEW MIRROR - FUNCTION/DIAGNOS
7	93 - 93	93-18-13	ELECTROCHROMATIC REAR VIEW MIRROR - FUNCTION/DIAGNOS
10	95 - 96	96-11-10	FUEL GAUGE DOES NOT READ OVER 3/4 OR SLOW - RISE
3,7,9	92 - 93	94-10B-60	FUEL GAUGE PINPOINT TESTS D & E - S/M REVISION
11	93 - 93	94-08-11	GLOVE BOX LOOSE OR MISSING MOUNTING SCREWS
4,8	94 - 94	94-09-12	I/P - PSOM (PROG SPEEDO/ODO) INOP - #8 FUSE BLOWN
11	94 - 94	94-10B-71	IC & PANEL LIGHTING - TEST & SECTION REF - S/M REV
5	94 - 94	94-17B-62	IC (CONV) - PIN OUT CHART - S/M REV
4,8,5	94 - 94	94-17B-58	IC (CONV) SYMP-M CHART/TEST DESIGNATIONS - S/M REV
2	83 - 94	93-24-10	IC LAMPS DIM - IGNITION SWITCH DIAGNOSIS
5	92 - 94	94-17B-64	IC PANEL REMOVAL PROCEDURES - S/M REV
11	93 - 93	93-05B-39	INSTRUMENTATION & WARN SYS DIAG/TEST PROCS - S/M REV
5	92 - 93	93-19B-29	IP - P/N & -RQUE VALUE SCREW/WASHER ASSEM - S/M REV
8	92 - 93	92-25-09	IP - POP/CRACK NOISES WITH A/C ON - DIAGNOSIS/REPAIR
11	93 - 93	93-18-11	LIFTGATE/DOOR AJAR LIGHT FLICKERS WHILE DRIVING
2	91 - 93	92-04-04	MIL/CES/SES - LIGHT ON W/NO SELF-TEST CODES - REV
9	93 - 94	94-23-13	MIRROR - DRIVER SIDE - NEW SERVICE PART - TIPS
5	92 - 94	94-18-07	MIRRORS - POWER OUTSIDE - SERVICE PARTS INFO
4	95 - 95	95-17-10	OUTSIDE TEMP GAUGE READS HIGH - EDDIE BAUER ONLY
10	95 - 95	94-22-10	RATTLE UNDER DASH - AIR BAG INTERNAL SAFING SENSOR
1	90 - 95	94-17-02	REAR VIEW MIRROR DETACHES FROM 18SHIELD - SVC TIP
4,8	94 - 94	94-17B-56	SPEED CONTRO SERVO/AMPLIFIER PIN NUMBERS - S/M REV
11	93 - 93	93-22-16	SPEED CONTROL 3-5 MPH LOSS ON HILLS - NEW MODULE
5	92 - 93	94-08-17	SPEED CONTROL INOP - DEACTIVATION SWITCH & GASKET
7,9	95 - 95	95-23-11	SPEED CONTROL LAMP FLICKERS W/TURNING STEER WHEEL
11	93 - 93	93-05B-36	SPEED CONTROL SYSTEM DIAGNOSIS & TEST PROC - S/M REV
11	93 - 93	93-05B-37	SPEED CONTROL SYSTEM TEST STEP SC22 - S/M REVISION
5	92 - 93	93-26B-02	SPEED CONTROL TEST STEPS REVISED - S/M REVISION
4,8	93 - 93	93-26B-02	SPEED CONTROL TEST STEPS REVISED - S/M REVISION
3,4,5,8	92 - 94	94-16-16	SPEEDO (PSOM) RE-CALIBRATION - SERVICE TIPS
4,8	92 - 95	95-05-21	SPEEDO INOP - 5.0/5.8L
10	95 - 95	ONP/94B47	SPEEDO INOP/POOR SHIFT - INCORRECT SPEEDO GEAR
4,5,8	93 - 93	94-10B-63	SPEEDO/ODO FORMULA FOR CONVERSION CONSTANT - S/M REV
4,5,8	93 - 93	93-13B-21	SPEEDO/ODOMETER - TEST STEP A6 INFO WRONG - S/M REV
11	93 - 93	93-05B-40	SPEEDOMETER/ODOMETER (ANALOG) - DIAG TEST - S/M REV
2	85 - 93	91-08-14	SPEEDOMETER/ODOMETER-REPLACEMENTS W/PRE-SET MI AVAIL
2	90 - 94	95-15-03	TEMP GAUGE - FLUCTUATES AND/OR FALSE HIGH READINGS
3,9,7	90 - 94	95-15-03	TEMP GAUGE - FLUCTUATES OR FALSE HIGH READINGS - REV
3,7,9	92 - 93	94-08-16	TEMP GAUGE READS HIGH
11	93 - 93	94-05B-67	TEMP GAUGE SYS PINPOINT TEST STEP TG7 - S/M REVISION

Model Legend: 1= All 2= Most 3= Aerostar 4= Bronco 5= Econoline 6= E Series 7= Explorer/Mountaineer 8= F Series 9= Ranger 10= Windstar 11= Villager

Chassis Electrical, Wiring Harness, Fuses-Circuit Breakers, Wipers, Window Motors

Models	Year	TSB#	Description
4,8	93 - 93	93-05B-27	18SHIELD WIPER MO-R CONNEC-R GRAPHICS - S/M REV
8,5	94 - 96	96-06-10	ELEC PARTS - WEATHER PACK CONN SEALS - 7.3L DI DSL
10	95 - 95	95-02-09	FALSE DOOR AJAR/CHIME/ALARM/DOME LAMP - REV PART
7	93 - 93	93-26B-44	FOG LAMP ELECT SCHEMATIC/CONNEC-RS - S/M REVISION
4,5,8	93 - 93	93-26B-42	FUEL TANK SELEC-R SWITCH CKT SCHEMATIC - S/M REV
4,8	92 - 95	95-05-21	FUSE "E" INOP - 5.0/5.8L
5	94 - 94	94-24B-78	FUSE AMPERAGES & WIRE ROUTINGS - S/M REV
1	94 - 94	94-19-07	HARNESS CONNEC-R 14401 (IGN SW) DOES NOT MATE
4,8	94 - 95	95-14-11	LAMPS/PWR MIRRORS/RKE INOP - FUSE 8 CKT 54 - PROCED
2	83 - 94	93-24-10	MALFUNCTION OF 4 PIN LOW OIL LEVEL RELAY - DIAGNOSIS
11	93 - 95	95-05-20	REAR WIPER MOTOR INOP OR INTERMITTENT - WATER ENTRY
10	95 - 95	94-11-19	RELAY BOX COVER REMOVAL PROCEDURE
2	94 - 94	94-07-08	REVISED ICM (EDIS) CONNEC-R - SEPARATING TIP
4,8	94 - 94	94-11-18	SWITCHES STICK - DOOR LOCKS & 18OWS
10	95 - 95	96-04-13	WASHER SPRAY PATTERN DOES NOT STRIKE CENTER OF W/S
7	91 - 93	92-24-12	WIPER (REAR) - ERRATIC OPERATION - DIAGNOSIS/REPAIR
2	86 - 94	95-04-06	WIPER MO-R REPLACEMENT - UPGRADED SERVICE KITS
2	86 - 94	95-04-06	WIPER MO-R REPLACEMENT - UPGRADED SERVICE KITS -REV
4,8	87 - 94	94-11-20	WIPER MO-R SERVICE REPLACEMENT - NEW MO-R - TIP
4,8	95 - 96	95-24-07	WIPER WASHER FLUID RESERVOIR LOOSE
3,4,5,8	93 - 93	93-26B-40	WIPERS - TEST E - AEROSTAR DELETED - S/M REVISION
4,5,8	93 - 93	93-26B-41	WIPERS - TEST STEP F1 REVISED - S/M REVISION
4,8	94 - 94	94-17B-57	WIPERS/WASHERS - PINPOINT TEST STEP C2 - S/M REV
4,8,5	94 - 94	94-17B-59	WIRING & CIRCUIT PROTECTION INFO - S/M REV
10	95 - 95	94-24B-84	WIRING - CONNEC-R C248 COLOR - S/M REV
10	95 - 95	94-24B-85	WIRING - REAR WIN DEFROST/HEATED MIRRORS - S/M REV
4,8	94 - 95	95-02-11	WIRING - SUMMARY OF POSSIBLE LOCATIONS OF SHORTS
5	92 - 93	93-11-09	WIRING - TRAILER -W AUX BAT FEED NOT ALWAYS HOT/FIX
7	92 - 93	93-16-10	WIRING DAMAGED BY BATTERY ACID EMISSIONS - SERV PROC
11	93 - 93	94-10B-77	WIRING DIAGRAM - ZONE 85-C - S/M REVISION
2	85 - 93	92-13-05	WIRING HARNESS - TERMINAL REPAIR KIT - REVISED
2	85 - 96	93-10-05	WIRING HARNESS - TERMINAL REPAIR KIT - REVISED
2	85 - 93	91-25-08	WIRING HARNESSES - NEW KIT/REPAIR INSTEAD OF REPLACE

Model Legend: 1= All 2= Most 3= Aerostar 4= Bronco 5= Econoline 6= E Series 7= Explorer/Mountaineer 8= F Series 9= Ranger 10= Windstar 11= Villager

Auxiliary Equipment, Jacks, Trailer Hitches, Towing

Models	Year	TSB#	Description
2	88 - 94	91-21-05	RECOMMENDATIONS WHEN -WING THIS VHCL BEHIND ANOTHER
8	94 - 96	94-02-19	SNOW PLOWING WITH SRS (AIR BAG) VEHICLE - INFO - REV
8	94 - 94	94-02-19	SNOW PLOWING WITH SRS (AIR BAG) VEHICLE -INFORMATION

Model Legend: 1= All 2= Most 3= Aerostar 4= Bronco 5= Econoline 6= E Series 7= Explorer/Mountaineer 8= F Series 9= Ranger 10= Windstar 11= Villager

Heating, Air Conditioning, Ventilation, Defogger

Models	Year	TSB#	Description
4,8	94 - 94	94-23-14	18/WHISTLE NOISE FROM PLENUM WHEN FAN IS ON
2	81 - 94	93-15-05	A/C "O" RING REMOVAL FROM SPRING LOCK COUPLING/INFO
5	92 - 93	93-08-12	A/C & HEATER - VACUUM TANK REPLACEMENT PROCEDURE
4,8	92 - 93	93-08-13	A/C & HEATER - VACUUM TANK REPLACEMENT PROCEDURE
4,8	95 - 95	95-06-15	A/C - WHISTLE NOISE WHEN ON MAX A/C
8	92 - 94	94-04-12	A/C - 18SHIELD FOG - NEW SELEC-R SWITCH ASSEMBLY
8	92 - 94	94-04-12	A/C - 18SHIELD FOG - NEW SWITCH ASSEMBLY - REVISED
1	90 - 95	95-05-12	A/C - ADDING REFRIGERANT OIL - SERVICE TIP
1	85 - 95	95-08-01	A/C - APPROVED FLUSHING PROCEDURES - SERVICE TIP
1	85 - 95	95-08-01	A/C - APPROVED FLUSHING PROCEDURES - SRVC TIP - REV
11	93 - 93	93-04-10	A/C - BLOWER MOTOR INOP IN LOW/MED LOW/MED HIGH SPDS
10	95 - 95	94-21-10	A/C - CENTER REGISTERS BLOW CLOSED WHEN OUTER CLOSED
2	82 - 94	92-25-3 & 93-9-8	A/C - FILTER REFRIG AFTER REP'L COMPRESSOR - REVISED
11	93 - 93	92-25-14	A/C - FIXED ORIFICE TUBE REPLACEMENT PROCEDURE
2	88 - 94	94-15-06	A/C - FS-10 & FX-15 COMPRESSORS - SERVICE TIPS
2	88 - 93	92-20-05	A/C - FX-15 COMPRESSOR AIR GAP SPECIFICATION REVISED
2	92 - 93	93-15-08	A/C - FX-15 COMPRESSOR SQUEAL - INSPECTION/SERV PROC
2	85 - 96	95-22-07	A/C - IDENT OF NON-APPROVED REFRIGERANTS - REV
2	85 - 96	95-22-07	A/C - IDENTIFICATION OF NON-APPROVED REFRIGERANTS
2	85 - 96	95-22-07	A/C - IDENTIFICATION OF NON-APPROVED REFRIGERANTS
4,8	94 - 95	94-19-20	A/C - INCREASE IN R-134A CHARGE - S/M REV
4,8	96 - 96	96-08-13	A/C - INSUFFICIENT COOLING
7,9	94 - 94	95-15-14	A/C - LACK OF COOLING - ORIFACE TUBE
8,6	94 - 95	95-16-16	A/C - LACK OF COOLING AT IDLE - 7.3L DI TURBO
2	80 - 93	92-25-06	A/C - MUSTY & MILDEW ODORS - NEW PRODUCT AVAIL - REV
2	80 - 93	92-25-06	A/C - MUSTY & MILDEW ODORS - NEW PRODUCT AVAILABLE
10	95 - 96	96-07-24	A/C - MUSTY ODOR
2	80 - 93	92-05-07	A/C - NEW FILTER KIT AVAILABLE - SERVICE INFO - REV
2	80 - 93	93-23-11	A/C - NO USE OF R12 REFRIG SUBSTITUTES - REVISED
2	80 - 93	93-23-11	A/C - NO USE OF R12 REFRIG SUBSTITUTES - SERVICE TIP
2	81 - 95	93-15-05	A/C - O-RING REMOVAL FROM SPRING LOCK COUPLER - TIP
2	94 - 95	94-20-06	A/C - OIL VISIBLE AT SPRING LOCK COUPLERS - TIP
2	92 - 93	93-10-04	A/C - ON-VEHICLE EVAP CORE LEAK TEST/WARRANTY - REV
2	92 - 95	94-10-07	A/C - ON-VEHICLE EVAP CORE LEAK TEST/WARRANTY - REV2
2	92 - 93	93-10-04	A/C - ON-VEHICLE EVAP CORE LEAK TEST/WARRANTY INFO
11	93 - 93	92-25-15	A/C - POOR COOLING IN HIGH AMBIENTS 17AINED
3	90 - 93	95-16-14	A/C - POOR PERFORMANCE - TIPS
2	92 - 95	94-26-06	A/C - R-134A TRACER DYE FOR LEAK DETECTION
1	94 - 96	94-26-06	A/C - R-134A TRACER DYE INSTALLED IN SYSTEM - REV
1	95 - 95	94-23-10	A/C - R-134A TRACER DYE INSTALLED IN SYSTEM - TIPS
3	92 - 94	94-04-13	A/C - RATTLE/NOISE - 3.0L - CRANKSHAFT PULLEY
2	93 - 94	94-05B-08	A/C - REFRIGERANT CAPACITIES REVISED/ADDED - S/M REV
7,9	95 - 96	95-23-03	A/C - WATER DRIPS ON FLOOR WHEN BLOWER ON HIGH
11	94 - 95	94-24B-82	A/C CLUTCH ELECT SCHEMATIC & AIR CON RELAY - S/M REV
2	88 - 93	92-18-04	A/C COMP (FX-15) - OIL RECOVERY/MEASURING PROC - REV

Model Legend: 1= All 2= Most 3= Aerostar 4= Bronco 5= Econoline 6= E Series 7= Explorer/Mountaineer 8= F Series 9= Ranger 10= Windstar 11= Villager

Heating, Air Conditioning, Ventilation, Defogger

Models	Year	TSB#	Description
2	88 - 93	92-18-04	A/C COMP (FX-15) - OIL RECOVERY/MEASURING PROCEDURE
4,8	93 - 94	94-15-17	A/C COMPRESSOR FAILURE DUE - INSUFFICIENT REFR OIL
10	93 - 95	95-16-08	A/C COMPRESSOR NOISE/MOAN - 3.8L
2	91 - 93	92-25-05	A/C COMPRESSOR/FX-15 - COIL/PULLEY APPLICATION CHART
2	92 - 93	93-15-08	A/C FX-15 COMPRESSOR SQUEAL - INSPECT/SERV PROC REV
2	92 - 93	93-23-10	A/C FX-15 COMPRESSOR SQUEAL - INSPECT/SERV PROC REV
4,8	95 - 95	95-07-16	A/C HEATER BLOWER WHEEL BREAKAGE
2	95 - 95	94-23-09	A/C NO/POOR PERFORMANCE - FS-10 COMPRESSOR
11	93 - 93	93-21-16	A/C ODOR - BLOWER NOISE - CLEAN/INSTALL INLET SCREEN
8	95 - 95	95-03-18	A/C POOR PERF - 7.3L DI DIESEL - HOSE CUT
4,8	94 - 94	94-13-14	A/C POOR PERF - ORIFACE TUBE - DEALER INSTALLED UNIT
7	91 - 93	94-22-11	A/C POOR PERFORMANCE
5	95 - 96	96-10-15	A/C REFRIGERANT OIL REFILL & CHARGE SERVICE TIP
5	94 - 94	94-21-09	A/C SUCTION HOSE KINKED - 7.3/7.5L - REV HOSE
7	91 - 93	93-24-18	A/C SUCTION HOSE ROUTING
2	80 - 93	93-20-06	A/C USE OF CORRECT FLUORESCENT DYE - SERVICE TIP
7,9	94 - 94	95-16-15	A/C WHISTLE NOISE - CONTROLS SET - OFF OR MAX A/C
11	93 - 93	93-20-10	A/C-HEAT - FOGGING/FROST SIDE 18OWS - REVISED
11	93 - 93	93-20-10	A/C-HEAT - FOGGING/FROST SIDE 18OWS WHILE DRIVING
7,9	93 - 93	93-13B-28	A/C-HEATER & BLOWER MO-R SCHEMATICS - S/M REVISIONS
5,8	92 - 93	93-16-08	A/C-HEATER - DASH MOUNTED CONTROL KNOB BREAKS - PROC
4,8	91 - 93	93-02-09	ACC - HEAT/COOL LACK - NEW TEMP DOOR CABLE ADJ PROC
10	95 - 95	95-12-07	AUX BLOWER MO-R RUNS AFTER VHCL SHUT OFF
5	94 - 94	94-05B-64	BLEND DOOR ACTUA-R/POTENTIOMETER INFO - S/M REV
3	93 - 94	95-05-11	CHIRP/SQUEAL FROM BLOWER MO-R AT LOW BLOWER SPEEDS
10	95 - 95	95-05-11	CHIRP/SQUEAL FROM BLOWER MO-R AT LOW BLOWER SPEEDS
3	93 - 94	95-05-11	CHIRP/SQUEAL FROM BLOWER MO-R AT LOW SPEEDS - REV
10	95 - 95	95-05-11	CHIRP/SQUEAL FROM BLOWER MO-R AT LOW SPEEDS - REV
3,10	93 - 94	95-06-06	CHIRP/SQUEAL FROM BLOWER MO-R AT LOW SPEEDS - REV 2
11	93 - 94	95-05-11	CHIRP/SQUEAL FROM BLOWER MOTOR AT LOW BLOWER SPEEDS
11	93 - 94	95-05-11	CHIRP/SQUEAL FROM BLOWER MOTOR AT LOW SPEEDS - REV
11	93 - 94	95-06-06	CHIRP/SQUEAL FROM BLOWER MOTOR AT LOW SPEEDS - REV 2
2	91 - 93	93-05B-03	CLIMATE CONTROL FIXED ORIFICE TUBE INFO - S/M REV
3,7,9	94 - 94	94-05B-49	CLIMATE CONTROL SYSTEM SERVICE INFORMATION - S/M REV
10	95 - 95	94-24-14	CONDENSATE LEAK AT AUX A/C UNIT - DRAIN HOSE/CLAMPS
2	85 - 94	91-05-01	DEFROSTER - HEATED BACKLITE INOPERATIVE - REVISED
2	80 - 96	95-04-05	FILTER REFRIGERANT AFTER COMPRESSOR REPL - REV FOUR
2	80 - 95	94-15-05	FILTER REFRIGERANT AFTER COMPRESSOR REPL - REV THREE
2	80 - 94	94-03-04	FILTER REFRIGERANT AFTER COMPRESSOR REPL - REV TWO
2	80 - 94	93-26-08	FILTER REFRIGERANT AFTER COMPRESSOR REPL - TIPS REV
9	93 - 93	93-05B-30	HEAT & A/C CIRCUITS 752 & 754 - EVTM REVISIONS
4,8	92 - 94	94-20-16	HEAT - POOR/NO PERF WHEN SET AT FULL HEAT
7,9	93 - 94	95-06-14	HEAT- OBJECTIONABLE AIR FLOW FROM PANEL - HVY ACCEL
11	93 - 93	93-05B-38	HEAT/DEFROST/FX-15 COMPRSSR DIAG/TEST PROC - S/M REV
5	93 - 94	94-17B-63	HEATER & A/C HEATER (VACUUM) SCHEMATICS - S/M REV

Model Legend: 1= All 2= Most 3= Aerostar 4= Bronco 5= Econoline 6= E Series 7= Explorer/Mountaineer 8= F Series 9= Ranger 10= Windstar 11= Villager

Heating, Air Conditioning, Ventilation, Defogger

Models	Year	TSB#	Description
3	93 - 93	93-05B-25	HEATER & AIR CONDITIONER/HEATER SCHEMATICS - S/M REV
4,5,8	92 - 94	94-09-13	HEATER - GURGLING NOISE FROM CORE - E40D
7,9	94 - 94	94-10B-68	HEATER SYSTEM - SERVICE INFO - S/M REVISION
2	83 - 94	93-14-04	HEATER/DEFROST - POOR OUTPUT/THERMOSTAT STUCK - REV
2	83 - 93	93-14-04	HEATER/DEFROSTER - POOR OUTPUT/THERMOSTAT STUCK OPEN
3	94 - 94	94-05B-50	HEATER/POWER VENT SYSTEM SERVICE INFO - S/M REVISION
9	91 - 94	95-08-07	KNOCK/THUMP NOISE IN PASS COMP/HEAT CORE - 3.0L -REV
9	91 - 94	95-08-07	KNOCK/THUMP NOISE IN PASS COMP/HEATER CORE - 3.0L
3	94 - 94	94-05B-51	MANUAL A/C-HEATER SERVICE INFORMATION - S/M REVISION
3	92 - 93	94-19-19	NOISE/CLUNK WHEN A/C CLUTCH ENGAGES - 3.0L
7,9	94 - 94	94-10-17	NOISE/WHISTLE IN HIGH BLOWER ON FULL HEAT
8	97 - 97	96-08-16	NOISE/WHISTLE WITH BLOWER MO-R OPERATING
5	92 - 93	93-20-12	RUMBLE NOISE FROM HEATER CASE (HEATER ONLY VEHICLE)
5	92 - 93	93-20-12	RUMBLE NOISE FROM HEATER CASE - "HEATER ONLY" VEHIC
4,8	95 - 95	95-11-10	TEMP CONTROL CABLE/BRACKET/PLENUM REVISED - INFO
4,8	88 - 94	94-21-11	VACUUM TANK REPLACEMENT TIPS
7,9	91 - 94	94-08-15	VIBRATION NOISE UNDER DASH -RECIRC DOOR VACUUM MO-R

Model Legend: 1= All 2= Most 3= Aerostar 4= Bronco 5= Econoline 6= E Series 7= Explorer/Mountaineer 8= F Series 9= Ranger 10= Windstar 11= Villager

Entertainment Devices, Stereo, Radio, Etc.

Models	Year	TSB#	Description
4,5,8	94 - 94	94-05B-59	AMPLIFIER ASSEMBLY - REMOVE/INSTALL - S/M REVISION
3,7,9	93 - 93	94-05B-55	AUDIO - GENERAL - TEST STEP F7 - S/M REVISION
7	94 - 94	94-24B-79	AUDIO SYS - SCHEMATICS/CONNEC-R FACES - S/M REV
9	92 - 94	94-17-10	RADIO - ENGINE IGNITION STATIC - 3.0L
9	92 - 94	94-17-10	RADIO - ENGINE IGNITION STATIC - 3.0L - REV
10	95 - 95	94-21-08	RADIO STATIC - IGN NOISE - AM BAND
2	85 - 94	92-06-09	RADIO/2-WAY RADIO - WHINE/BUZZ NOISE - FUEL PUMP-REV
11	93 - 93	93-05B-41	RADIO/TAPE CHASSIS DIAGNOSIS & TEST PROCS - S/M REV
11	93 - 94	94-14-12	RATTLE NOISE - 2ND ROW PASS SIDE SPEAKER/BELT AREA
2	85 - 95	93-18-04	WHINE/BUZZ IN SPEAKER - FUEL PUMP RFI NOISE - REV
2	85 - 96	95-11-03	WHINE/BUZZ IN SPEAKER - FUEL PUMP RFI NOISE - REV2
2	85 - 94	93-15-06	WHINE/BUZZ IN SPEAKER - FUEL PUMP RFI NOISE FILTER

Model Legend: 1= All 2= Most 3= Aerostar 4= Bronco 5= Econoline 6= E Series 7= Explorer/Mountaineer 8= F Series 9= Ranger 10= Windstar 11= Villager

Seats, Belts, Interior Trim, Carpets, Air Bags

Models	Year	TSB#	Description
3	93 - 94	94-17B-55	AIR BAG (SIR) DIAGNOSTIC & TESTING INFO - S/M REV
3	93 - 93	94-10B-59	AIR BAG - CODE 23 TEST STEP 23-2 - S/M REVISION
7,9,10	95 - 95	95-22-08	AIR BAG CODE 24 - INCORRECT DIAGNOSIS - TIP
11	96 - 96	96-07-25	AIR BAG IND LAMP FLASHES CONSTANTLY
11	93 - 93	93-09-13	CARPET LIFTS AT RIGHT SLIDING SEAT TRACK - PROCEDURE
11	93 - 96	93-09-13	CARPET LIFTS AT RIGHT SLIDING SEAT TRACK - REV
8	92 - 94	ONP/94B50	CHILD SEAT LOCKING CLIP - F350 CREW - REV OWNER MAN
7,9	91 - 93	93-23-19	CONSOLE ARMREST MOUNTING BRACKET BREAKS - KIT
9	91 - 93	94-17-07	DRIVER 60/40 BENCH SEAT - NO/POOR ADJUST - SUPERCAB
11	93 - 93	93-23-21	FRONT LAP BELTS RETRACT TOO QUICKLY - NICKS DOOR
3,7	92 - 94	94-04-01	HEADLINER ODOR (FISH OR MUSTY) - NEW HEADLINER - REV
3,7	92 - 94	94-04-01	HEADLINER ODOR (FISH OR MUSTY) - REVISED HEADLINER
2	89 - 94	94-24-10	LOOSE FRONT CENTER ARMREST - W/BUCKET SEAT & CONSOLE
11	93 - 93	93-18-12	MANUAL SEATS - ARMRESTS NOT HORIZONTAL
7,9	94 - 94	94-19-15	NOISE/SQUEAK FROM CONSOLE ARMREST WHEN DOWN
4,8	92 - 94	94-17-08	NOISE/SQUEAK FROM IP AT SIDES
10	95 - 96	96-07-16	ODOR - FISHY - FROM CARPET
7	93 - 94	94-05-09	PAINT WEARS OFF DOOR TRIM PANEL PULL HANDLES - REV
7	93 - 94	94-05-09	PAINT WEARS OFF INSIDE DOOR TRIM PANEL PULL HANDLES
11	93 - 93	92S57	PASSIVE SEAT BELT GUIDE RAIL BOLTS NOT TIGHT- RECALL
7	91 - 93	94-14-10	RATTLE - FRONT PASSENGER SEAT BACK - WHEN UNOCCUPIED
3,4,8,9	90 - 93	94-14-10	RATTLE - FRONT PASSENGER SEAT BACK - WHEN UNOCCUPIED
4,8	92 - 94	94-15-13	RATTLE NOISE/IP LOOSE - RIGHT SIDE OF IP
10	95 - 95	94-22-10	RATTLE UNDER DASH - AIR BAG INTERNAL SAFING SENSOR
4,8	92 - 95	95-08-05	RATTLE/SQUEAK - FRONT PASS SEAT BACK
8	94 - 94	94-19-14	REAR JUMP SEAT DIFFICULT - FOLD DOWN - SUPERCAB
4	94 - 94	94-21-07	REAR SEAT TRIM COVER - POOR APPEARANCE/LOOSE
5	92 - 93	93-23-20	SEAT BACK COVER LOOSE/BAGGY - 3RD ROW BENCH SEAT-PAD
10	95 - 95	95-20-05	SEAT BELT - SHOULDER BELT MAY TWIST AT D-RING
4	93 - 93	93-15-13	SEAT BELT D-RING ATTACH AREA SQUEAK NOISE - REPAIR
8,9	92 - 93	92-16-07	SEAT HEADREST - NEW DESIGN GUIDE ROD SLEEVE/R&I PROC
7	93 - 93	93-05B-29	SEAT/PROGRAMMABLE MEMORY - SERVICE INFO - S/M ADD
11	93 - 93	93-12-15	SEAT/REAR BENCH - SQUEAKS OR RATTLES - SERV PROC
7	91 - 93	93-03-14	SEATS/FRONT - DIFFICULT - OR WON'T ADJUST - PROC
5	93 - 93	94-05B-65	SIR - CODE 41 STEP 41-1 - S/M REVISION
10	95 - 95	94-08-10	SLIDING DOOR TRIM PANEL REMOVE/INSTALL TIPS - LX
4,8	92 - 95	95-05-17	SQUEAK/CREAK NOISE - DOOR TRIM PANEL - XLT MODELS
11	93 - 94	94-07-13	SQUEAKS/RATTLES - COMPREHENSIVE SUMMARY & PROCEDURES
8	94 - 94	94-02-19	SRS (AIR BAG) VEHICLE & SNOW PLOWING - INFORMATION
2	88 - 96	92-24-01	SRS - AIR BAG MODULE COVERS - PAINTING RESTRICT -REV
3,5	88 - 93	92-24-01	SRS - AIR BAG MODULE COVERS - PAINTING RESTRICTIONS
5	92 - 93	93-15-14	TRIM - ENG COVER CONSOLE DOOR BREAKAWAY FEATURE INFO
11	93 - 94	93-15-15	UNOCCUP SEAT (CENTER BUCKET/BENCH) SQUEAK/RATTLE-REV
11	93 - 93	93-15-15	UNOCCUPIED SEAT (CENTER BUCKET/BENCH) SQUEAK/RATTLE
5	95 - 96	95-22-12	WATER LEAK IN IP AT A-PILLAR

Model Legend: 1= All 2= Most 3= Aerostar 4= Bronco 5= Econoline 6= E Series 7= Explorer/Mountaineer 8= F Series 9= Ranger 10= Windstar 11= Villager

Glass, Doors, Hood, Decklid, Tailgate, Liftgate, Locks

Models	Year	TSB#	Description
11	93 - 94	95-15-10	18/ROAD NOISE - SUMMARY OF REPAIRS - TIPS
10	95 - 96	95-19-11	18NOISE AT FRONT DOORS
3	94 - 94	95-21-07	18NOISE NEAR DOOR IN THE A-PILLAR
3	94 - 95	95-21-07	18NOISE NEAR DOOR IN THE A-PILLAR - REV
7	91 - 94	94-06-09	AIR LEAK FROM REAR DOOR GRILLE & INSIDE DOOR HANDLES
11	93 - 96	96-03-16	BODY SIDE 18OWS WON'T STAY CLOSED
3	92 - 96	96-02-06	BUZZ/RATTLE AT SIDES OF IP AND/OR SIDE 18OW
10	95 - 96	95-19-10	CREAK OR RATTLE NOISE FROM SLIDING DOOR
9	93 - 93	93-08-10	DOOR GLASS - WON'T ROLL UP COMPLETELY - SERV PROC
4,8	92 - 93	93-19-09	DOOR GLASS SEPARATION FROM CHANNEL
3	86 - 93	94-15-12	DOOR GLASS TIPS/BINDS GOING DOWN
3	86 - 94	94-15-12	DOOR GLASS TIPS/BINDS GOING DOWN - REVISED
5	92 - 93	93-01-14	DOOR/40% SIDE - FORWARD EDGE SHEET METAL CRACKS
5	92 - 93	93-04-04	DOOR/REAR & SIDE CARGO - CHECK ARM BINDING/NOISY
5	96 - 96	96-02-05	DRIVER DOOR - SHEET MOLDED COMPOSITE - INFO
2	81 - 95	95-12-01	ESSEX 18SHIELD REPLACE PRODUCTS - SHELF LIFE TIP
11	93 - 93	ONP/94B39	FRONT DOOR GUARD BEAM REPAIR - CAMPAIGN
7	91 - 94	94-22-09	INSIDE DOOR HANDLE BREAKS WHEN OPENING DOOR
2	84 - 94	93-21-01	IRIDESCENCE OR MOTTLING IN TEMPERED GLASS 17AINED
11	93 - 93	93-22-17	KEYLESS ENTRY INOP - ELECT INTERFERENCE - REV MODULE
5,7,8	96 - 96	96-11-01	LOCKS - NEW DESIGN - SERVICE TIP
5,7,8,9	96 - 96	96-07-15	LOCKS - NEW DESIGN ON TRIAL VEHCLS - APPL & TIPS
2	91 - 93	93-02-02	LOCKS/POWER DOOR - INOP/ACTUA-R DISENGAGES - PROC
11	93 - 93	93-18-11	NO START AFTER LONG SETTING - LIFTGATE/DOOR SWITCH
11	93 - 93	94-04-09	NOISE/RATTLE - FRONT DOOR - PROCEDURES
4,8	94 - 94	94-10B-62	POWER DOOR LOCK - ELECTRICAL SCHEMATIC - S/M REV
10	95 - 96	95-23-05	POWER DOOR LOCKS - CONSTANT CYCLING
4,8	93 - 95	94-24B-75	PWR DOOR LOCK SWITCH SCHEMATICS (RH) - S/M REV
5	95 - 95	95-17-09	RATTLE IN FRONT DOOW - 18OW REG CABLE
11	93 - 94	94-20-11	RATTLE/CHUCKING NOISE - SLIDING DOOR
10	95 - 95	95-11-07	RATTLE/SQUEAK NOISE FROM LIFTGATE
3	86 - 95	95-24-04	REPLACEMENT DOOR - LOOSE FITTING LOCK CYLINDER
5,10	92 - 96	95-24-04	REPLACEMENT DOOR - LOOSE FITTING LOCK CYLINDER
4,8,9	81 - 96	95-24-04	REPLACEMENT DOOR - LOOSE FITTING LOCK CYLINDER
10	95 - 95	94-15-14	SLIDING DOOR ADJUSTMENT PROCEDURE - TIPS
3	86 - 93	95-16-12	SLIDING DOOR REPLACEMENT - REV DOOR - TIP
3	86 - 94	94-19-16	SLIDING DOOR WEATHERSTRIP WORN/DAMAGED
11	93 - 96	96-07-17	SQUAEK/RATTLE AND/OR HIGH EFFORT - SLIDING DOOR
11	93 - 94	94-07-13	SQUEAKS/RATTLES - COMPREHENSIVE SUMMARY & PROCEDURES
9	93 - 93	92-25-10	TAILGATE & PICK-UP BOX - GAPPING CONDITION - REPAIR
5	95 - 96	95-22-12	WATER LEAK IN IP AT A-PILLAR

Model Legend: 1= All 2= Most 3= Aerostar 4= Bronco 5= Econoline 6= E Series 7= Explorer/Mountaineer 8= F Series 9= Ranger 10= Windstar 11= Villager

Finishes, Body Structure, Frame, Bumpers

Models	Year	TSB#	Description
5	96 - 96	96-08-09	18NOISE AT A-PILLAR
8	91 - 94	92-13-14	BOOM NOISE FROM CAB BACK PANEL - 4.9L - E40D OR C6
7	91 - 93	93-12-13	BUMPER - TRAILER HITCH PLATE W/SAFETY CHAIN HOLES
3	93 - 93	93-05-03	BUMPER COVER/PLASTIC - MINOR SCRATCH - REPAIR INFO
2	79 - 94	93-24-01	BUMPER COVERS - IMPROVED FLEXIBLE PART REPAIR MAT'L
7,10	95 - 95	95-03-03	BUMPER COVERS - PAINT SERVICE TIP
2	80 - 94	94-08-01	BUMPER ISOLA-R & BRACKET ASSMBLY REPLCMNT GUIDELINE
4,8,5	92 - 94	94-17B-61	BUMPERS - FRONT - REMOVE/INSTALL - S/M REV
3	86 - 93	87-12-01	BUMPERS - LOOSE OR UNATTACHED ENDS - REPAIR PROC
10	95 - 95	95-24-05	CLUNK NOISE ON HEAVY ACCEL/DECEL IN FLOOR PAN AREA
4,8	87 - 93	93-13B-23	COWL GRILL R&I PROCEDURES - S/M OMISSIONS
2	94 - 94	93-24-03	FLEET PAINT CODES FOR DSO PAINT
2	80 - 93	91-21-07	FRAME - RIVET REPLACEMENT W/BOLTS - SERVICE TIPS
4,8	94 - 95	95-07-07	FRONT BUMPER - NEW -RQUE SPEC - TIP
10	95 - 95	94-15-15	FRONT BUMPER COVER HOLES MISSING FOR LICENSE PLATE
9	93 - 93	93-15-18	GUARD/FRONT BUMPER GRILLE - OBSOLETE PARTS RETURN
3	92 - 95	96-10-13	NOISE - LR TIRE RUBS ON WHEELHOUSE - BRACKET
2	93 - 93	92-20-01	PAINT - 1993 COLOR CHART/AFTERMARKET SUPPLIER INFO
1	80 - 95	95-06-01	PAINT - ACID RAIN/IRON PARTICLE REMOVAL PROCEDURES
2	83 - 93	93-08-04	PAINT - EXT CLEARCOAT HAZING/PEELING/MICROCHECKING
2	92 - 93	92-05-02	PAINT - NEW ID LABEL FOR PAINT CODE CHANGES
1	95 - 96	95-16-02	PAINT - TRANSIT COATING - RAPGARD - SERVICE TIP
1	96 - 96	95-16-02	PAINT - TRANSIT COATING - RAPGARD - SERVICE TIP -REV
1	95 - 96	95-07-02	PAINT CODES - AFTERMARKET CROSS REFERENCE LIST
2	94 - 94	93-24-04	PAINT CODES FOR REGULAR PRODUCTION - 1994 MODELS
4,8	80 - 94	91-25-18	POP/CREAK NOISE - FRAME #1 CROSSMEMBER RIVET
4,8	80 - 95	94-08-13	POP/CREAK NOISE - FRAME #1 CROSSMEMBER RIVET - REV
4,8	80 - 96	95-09-07	POP/CREAK NOISE - FRONT FRAME OR SUSP BRACKETS - REV
1	80 - 95	94-23-04	PRIMED SHEET METAL - PREP PROCEDURE & MSDS INFO
9	93 - 94	93-24-17	REAR BUMPER RUST ON INSIDE - NEW COATED BUMPER
3	86 - 94	95-06-12	ROCKER PANEL RUST PERFORATION - TIPS
4	91 - 95	92-13-13	ROOF -COSMETIC CRACKS IN FIBERGLASS SEAM SEALER -REV
7,10	91 - 96	92-15-03	ROOF PANEL/DRIP RAIL REPLACEMENT - SEALING PROC -REV
4	80 - 95	95-14-09	ROOF PANEL/SHEET METAL - CRACKS - REPAIR PROCED -REV
2	87 - 95	91-04-11	RUNNING BOARDS NOISY - INCORRECTLY INSTALLED - REV2
11	93 - 94	94-07-13	SQUEAKS/RATTLES - COMPREHENSIVE SUMMARY & PROCEDURES
4,8	90 - 95	95-01-06	TAILGATE APPLIQUE POOR APPEARANCE AROUND FORD LTRS

Model Legend: 1= All 2= Most 3= Aerostar 4= Bronco 5= Econoline 6= E Series 7= Explorer/Mountaineer 8= F Series 9= Ranger 10= Windstar 11= Villager

GENERAL MOTORS CORPORATION

Lighting, Horns, Turn Signals, Steering Column

Models	Year	TSB#	Description
7,10,5	94 - 94	468106	CRUISE CONTROL DIAGNOSIS - ADDED INFO - S/M REV
14	94 - 94	469001	CRUISE CONTROL MODULE LOCATION - S/M REV
2	Up - 96	53-82-09	DAYTIME RUNNING LAMPS RETROFIT KITS AVAILABLE
7,10,12	95 - 95	56-82-05	DOME LAMP STAYS ON - DEAD BATTERY - NO START
16	94 - 94	56-82-08	EXTERIOR LAMPS - COMPONENT LOCATION CHART - S/M REV
9,8	94 - 94	56-82-08	EXTERIOR LAMPS - COMPONENT LOCATION CHART - S/M REV
1	Up - 96	63-81-11	HEADLAMP AIMING EQUIP DISTORTS PLASTIC LENS -CAUTION
4	94 - 94	66-81-04	INTERIOR LAMPS - SCHEMATICS/DIAGN - S/M REV
7,10,5	95 - 95	56-82-02	INTERIOR/DOME LAMP ROTARY WHEEL SWITCH OPS - INFO
16	93 - 93	368103	LAMPS/INT - NEW AND/OR REV WIRING & INFO - S/M UP
1	92 - 93	333210	POP NOISE - TILT STEER COLUMN - REPLACE BEARING SEAT
2	92 - 93	333210	POP NOISE/TILT STEER COLUMN- REPL BEARING SEAT - REV
7,10,5	95 - 95	56-32-01	STEER WHEEL & COLUMN ON-VEHICLE SERVICE - S/M REV

Model Legend: 1= All 2= Most 3= Trans Sport 4= Astro/Safari 5= Blazer,C/K 6= Silhoutte 7= C/K Pickup 8= S/T-Series (Pickup, Sonoma) 9= S/T -Series (Blazer, Jimmy)
10= Suburban 11= Sierra 12= Tahoe/Yukon 13= Lumina Van 14= G Series Van 15= Tracker 16= Bravada

Instruments, Dash Cluster, Warning Lights, Mirrors

Models	Year	TSB#	Description
2	91 - 93	67138R	CEL - INTERMIT OR CONSTANT ON/TCC FLUCTUATION
2	93 - 93	377127	CRUISE - NO 3-4 SHIFT AFTER 4-3 DOWNSHIFT- WIRE PROC
9,8	95 - 95	56-81-04	CRUISE CONTROL DIAGNOSTIC PROCEDURE - S/M REV
7,10,12	95 - 95	56-81-08	CRUISE CONTROL SCHEMATICS/TABLES - S/M REV
4	89 - 94	89-144-08C	ELECT IP - IRREG/FALSE TEMP WHEN HORN ACTIVATED -REV
5,7,10,11	92 - 93	361005	ELECTRIC MIRROR WILL NOT HOLD ADJUSTMENT - LABOR REV
5,7,10,11	92 - 93	361005	ELECTRIC MIRROR WILL NOT HOLD ADJUSTMENT - RPO D48
4	93 - 95	56-82-03	ERRATIC GAUGE/MIL OPS - 4.3L W/ A/C
9,8	94 - 94	46-83-06	ERRATIC TEMP GAUGE - STD ANALOG IC
7,10,12,14	96 - 96	67-62-01	EXTREME ENG TEMP GAUGE FLUCTUATION - 4.3/5.0/5.7L
7,10,12	95 - 95	56-83-02	FUEL GAUGE STICKS/INACCURATE
16	91 - 93	268304	GAUGE/OIL PRESS - ERRATIC/INCORRECT/INOP - DIAG/FIX
1	90 - 93	268304	GAUGE/OIL PRESSURE - ERRATIC/INCORRECT READINGS/FIX
2	92 - 93	268306	GAUGE/OIL PRESSURE - HIGH READING AT START-UP - INFO
18	92 - 93	318103	GAUGE/OIL PRESSURE INDICATOR - INFO - S/M UPDATE
16	93 - 93	368103	I/P - NEW AND/OR REV WIRING & INFO - S/M UPDATE
5,7,10,11	91 - 93	238301R	I/P CUPHOLDER - BINDING/STUCK - SERVICE PROCEDURE
16	93 - 93	368301	IC BLANK & SPEEDO READS 188 WHEN HORN IS DEPRESSED
9,8	93 - 93	368301	IC BLANK & SPEEDO READS 188 WHEN HORN IS DEPRESSED
7,10,5	95 - 95	56-82-02	IC DIMMER SWITCH OPS - INFO
9,8	94 - 95	56-83-03	IC REMOVAL - CAUTIONS - DAMAGE TO HAZARD BUTTON/MFS
5,7,10,11	88 - 93	261515	MIRROR/OUTSIDE REAR VIEW - VIB - CAUSE/CORRECTION
4,9,8,14	91 - 94	47-61-49	OIL PRESSURE GAUGE READING CONCERNS - REV SEND UNIT
5,7,10,11	91 - 94	47-61-49	OIL PRESSURE GAUGE READING CONCERNS - REV SEND UNIT
9,8	82 - 94	461001	OUTSIDE REAR VIEW MIRROR VIBRATION - INSTL SET SCREW
18	92 - 94	431033	OUTSIDE REAR VIEW MIRROR VIBRATION - NEW MIRROR
6	92 - 94	431033	OUTSIDE REAR VIEW MIRROR VIBRATION - NEW MIRROR
13	92 - 94	431033	OUTSIDE REAR VIEW MIRROR VIBRATION - NEW MIRROR
4	93 - 93	266604R	SES LAMP ON/POOR ENG PERFORM/DTC 44 - SERVICE (4.3L)
4	93 - 93	46-82-03	SPEEDO INOP - POSS CODES 24/72 - A/T ERRATIC SHIFT
4	93 - 94	46-82-03	SPEEDO INOP -POSS DTC 24/72- A/T ERRATIC SHIFT - REV
9,8	94 - 94	46-83-04	SPEEDO MOMENTARY DROP W/ TURNING ON PARK/TURN/FLASH
5,7,10,11	93 - 94	46-83-04	SPEEDO MOMENTARY DROP W/ TURNING ON PARK/TURN/FLASH
2	88 - 93	268305R	SPEEDOMETER - REGISTERS WHEN VEHICLE IS STAIONARY
4	94 - 95	56-90-02	STEPPER MOTOR CRUISE CNTRL - NO RESUME OR RESET
7,10,5,9,8	94 - 95	56-90-02	STEPPER MOTOR CRUISE CNTRL - NO RESUME OR RESET
4	93 - 95	56-90-02	STEPPER MOTOR CRUISE CNTRL - NO RESUME OR RESET -REV
7,10,5,9,8	94 - 95	56-90-02	STEPPER MOTOR CRUISE CNTRL - NO RESUME OR RESET -REV
7,10,11,5	92 - 94	368305	TACHOMETER ACCURACY AT IDLE
9,8	94 - 94	478101	TEMPERATURE GAUGE FLUCTUATION - MAY BE NORMAL - 2.2L

Model Legend: 1= All 2= Most 3= Trans Sport 4= Astro/Safari 5= Blazer,C/K 6= Silhoutte 7= C/K Pickup 8= S/T-Series (Pickup, Sonoma) 9= S/T -Series (Blazer, Jimmy) 10= Suburban 11= Sierra 12= Tahoe/Yukon 13= Lumina Van 14= G Series Van 15= Tracker 16= Bravada

Chassis Electrical, Wiring Harness, Fuses-Circuit Breakers, Wipers, Window Motors

Models	Year	TSB#	Description
18	90 - 93	318108	ELEC DIAG CHART "GATE AJAR INDICATOR ON..." - S/M UP
1	Up - 94	149301	EMI/RFI DISTURBANCE - NICKEL TAPE AVAILABLE
18	90 - 95	431042	FRONT WINDOW INOP - NEW REGULATOR/MOTOR KIT - REV
6	90 - 95	431042	FRONT WINDOW INOP - NEW REGULATOR/MOTOR KIT - REV
13	90 - 95	431042	FRONT WINDOW INOP - NEW REGULATOR/MOTOR KIT - REV
18	94 - 95	43-81-25	LOCATION OF AIR BAG FUSE - S/M REVISION
6	94 - 95	43-81-25	LOCATION OF AIR BAG FUSE - S/M REVISION
13	94 - 95	43-81-25	LOCATION OF AIR BAG FUSE - S/M REVISION
14	91 - 93	368102	POWER DOOR LOCK SWITCH CHROME FINISH PEELING - PROC
18	94 - 94	418101	POWER LOCKS/KEYLESS ENTRY WIRING DIAGRAMS - S/M REV
6	94 - 94	418101	POWER LOCKS/KEYLESS ENTRY WIRING DIAGRAMS - S/M REV
13	94 - 94	418101	POWER LOCKS/KEYLESS ENTRY WIRING DIAGRAMS - S/M REV
9,8	95 - 95	95-C-33	REAR WIPER SCRATCHES ENDGATE - BLAZER - CAMPAIGN
9,8	95 - 95	95-C-33	REAR WIPER SCRATCHES ENDGATE -BLAZER - CAMPAIGN SUPP
9,8	95 - 95	56-81-03	REAR WIPER/WASHER DIAGNOSIS - S/M REV
18	90 - 93	431042	RIGHT FRONT WINDOW INOP - NEW REGULATOR/MOTOR KIT
6	90 - 93	431042	RIGHT FRONT WINDOW INOP - NEW REGULATOR/MOTOR KIT
13	90 - 93	431042	RIGHT FRONT WINDOW INOP - NEW REGULATOR/MOTOR KIT
7,10,12	95 - 95	56-82-07	STOP LAMP RELAY - UPDATED PART
18	94 - 94	418102	WARNINGS & ALARMS/RADIO WIRING DIAGRAMS - S/M REV
6	94 - 94	418102	WARNINGS & ALARMS/RADIO WIRING DIAGRAMS - S/M REV
13	94 - 94	418102	WARNINGS & ALARMS/RADIO WIRING DIAGRAMS - S/M REV
7,10,12	94 - 95	66-82-02	WASHER FLUID LEAKS OUT OF SPRAY NOZZLES
4,14	93 - 94	468201	WIPER - PHANTOM WIPE/SINGLE UNWANTED CYCLE
9,8	94 - 94	468201	WIPER - PHANTOM WIPE/SINGLE UNWANTED CYCLE
5,7,10,11	93 - 94	468201	WIPER - PHANTOM WIPE/SINGLE UNWANTED CYCLE
9,8	94 - 95	56-82-04	WIPER BLADE CHATTER
7,10,5,12	88 - 95	56-82-04	WIPER BLADE CHATTER
9,8	94 - 95	56-82-04	WIPER BLADE CHATTER - REV
7,10,5,12	88 - 95	56-82-04	WIPER BLADE CHATTER - REV
7,10,5	88 - 94	56-82-01	WIPER BLADE CHATTER - REVISED ARMS
18	90 - 94	331047	WIPER BLADE ELEMENT REPLACEMENT - S/M REVISION
6	90 - 94	331047	WIPER BLADE ELEMENT REPLACEMENT - S/M REVISION
13	90 - 94	331047	WIPER BLADE ELEMENT REPLACEMENT - S/M REVISION
7,10,12	95 - 95	56-81-10	WIPER/WASHER PULSE - CHARTS - S/M REV
13	94 - 94	377133	WIRE HARNESS & TRANSAXLE PIN LOCATION CHANGES -4T60E
16	93 - 93	377112	WIRING HARNESS - INTERMIT SHORT TO GROUND/BLOWS FUSE

Model Legend: 1= All 2= Most 3= Trans Sport 4= Astro/Safari 5= Blazer,C/K 6= Silhoutte 7= C/K Pickup 8= S/T-Series (Pickup, Sonoma) 9= S/T -Series (Blazer, Jimmy)
10= Suburban 11= Sierra 12= Tahoe/Yukon 13= Lumina Van 14= G Series Van 15= Tracker 16= Bravada

Auxiliary Equipment, Jacks, Trailer Hitches, Towing

Models	Year	TSB#	Description
9,8	94 - 94	94C30	LOW TORQUE ON PLATFORM TRAILER - 2WD - RECALL
7	92 - 93	362001	TOWING A DEDICATED NATURAL GAS PICKUP - INFO
16	91 - 96	16-01-01	TOWING/RECREATIONAL - PROCEDURES & CAUTIONS - REV
12	88 - 96	16-01-01	TOWING/RECREATIONAL - PROCEDURES & CAUTIONS - REV
5,10,9,8,7	88 - 96	16-01-01	TOWING/RECREATIONAL - PROCEDURES & CAUTIONS - REV

Model Legend: 1= All 2= Most 3= Trans Sport 4= Astro/Safari 5= Blazer,C/K 6= Silhoutte 7= C/K Pickup 8= S/T-Series (Pickup, Sonoma) 9= S/T -Series (Blazer, Jimmy)
10= Suburban 11= Sierra 12= Tahoe/Yukon 13= Lumina Van 14= G Series Van 15= Tracker 16= Bravada

Heating, Air Conditioning, Ventilation, Defogger

Models	Year	TSB#	Description
18	90 - 93	231212	A/C - FRONT EVAPORATOR PROCEDURES - S/M REVISION
10	92 - 93	361202	A/C - GROWL NOISE DURING OPERATION - SERVICE PROC
6	93 - 93	261207R	A/C - HARRISON R134A COMPRESSOR - SERV PARTS PROGRAM
4	92 - 94	46-11-03	A/C - HVAC FAN INOP AT ONE OR MORE SPEEDS
13	92 - 93	201201R	A/C - INOP OR NO COOLING - DIAGNOSTIC INFORMATION
5,7,10,11	93 - 93	371201	A/C - INOP/ACCESSORY DRIVE BELT COMES OFF - INFO
9,8	88 - 94	91-042A-01B	A/C - INSUFFICIENT COOLING IN HIGH AMBIENT TEMP -REV
16	91 - 94	91-T-232	A/C - INSUFFICIENT COOLING IN HIGH AMBIENT TEMP -REV
7,10	94 - 95	56-12-01	A/C - LOUD KNOCK FROM COMPRESSOR
6	92 - 93	236115	A/C - MOAN/GROWL/VIB - NEW ENG MOUNT BRACKET (3.8L)
6	93 - 93	331205	A/C - SYSTEM EVACUATION USING ACR4 UNIT - S/M REV
16	91 - 94	56-12-04	A/C - WATER LEAK AT RECIRCULATION DOOR
9,8	83 - 94	56-12-04	A/C - WATER LEAK AT RECIRCULATION DOOR
9,8	92 - 93	361203	A/C - WHISTLE (RT COWL VENT AREA) IN BI-LEVEL MODE
16	92 - 93	361203	A/C - WHISTLE FROM COWL VENT WHEN IN BI-LEVEL MODE
9,8	94 - 94	461203	A/C BLOWS WARM AIR - REPLACE A/C DELAY RELAY - 4.3L
18	93 - 95	51-65-56	A/C CLUTCH CONTROL DIAG CHART C-10 - 3.1L - S/M REV
6	93 - 94	51-65-56	A/C CLUTCH CONTROL DIAG CHART C-10 - 3.1L - S/M REV
13	93 - 95	51-65-56	A/C CLUTCH CONTROL DIAG CHART C-10 - 3.1L - S/M REV
9,8	94 - 94	56-81-05	A/C COMP CONTROLS - DIAGN PROCED - 4.3L - S/M REV
18	93 - 93	261207R	A/C COMPRESSOR - DIAG/REPLACEMENT/SERV PARTS PROGRAM
5,7,10,11	88 - 94	461202	A/C CONTROL HEAD FLASHES OR ERRATIC OPERATION
18	93 - 95	53-12-02	A/C CONTROL SETTINGS LOST - BATTERY RUN DOWN
6	93 - 95	53-12-02	A/C CONTROL SETTINGS LOST - BATTERY RUN DOWN
13	93 - 95	53-12-02	A/C CONTROL SETTINGS LOST - BATTERY RUN DOWN
1	93 - 96	53-12-12	A/C ODOR AT START UP IN HUMID CLIMATES
7,10,5	95 - 95	56-12-03	A/C POOR PERF OR INOP RECIRCULATION MODE
18	96 - 96	63-12-07	A/C POOR PERF WHEN USING FRONT SYS ONLY (REAR A/C)
6	96 - 96	63-12-07	A/C POOR PERF WHEN USING FRONT SYS ONLY (REAR A/C)
13	96 - 96	63-12-07	A/C POOR PERF WHEN USING FRONT SYS ONLY (REAR A/C)
13	92 - 93	236115	A/C SYS - MOAN/GROWL/VIB - INSTALL ENG MOUNT BRACKET
7,5,10	93 - 93	93C22	A/C SYS FREON OVERCHARGE - FRONT A/C ONLY - RECALL
18	93 - 93	331205	A/C SYSTEM - EVACUATION USING ACR4 - INFORMATION
10	95 - 95	95-C-41	A/C SYSTEM MISBUILD - 2WD W/ REAR A/C - CAMPAIGN
9,8	95 - 95	56-12-05	BLOWER MOTOR & RESISTOR REPLACE PROCEDURES
1	Up - 94	43-12-23	CONTAMINATED A/C REFRIGERENT - R-12 SYSTEM
1	Up - 94	53-12-05	CONTAMINATED R12 REFRIGERANT TESTING & HANDLING
18	90 - 93	231211	FRONT EVAPORATOR CORE R&R - S/M UPDATE
4	91 - 93	431207A	GUIDELINES FOR RETROFIT R-12 VHCLS TO R-134A - REV
4	91 - 93	43-12-07B	GUIDELINES FOR RETROFIT R-12 VHCLS TO R-134A - REV 2
4	91 - 93	43-12-07C	GUIDELINES FOR RETROFIT R-12 VHCLS TO R-134A - REV 4
7,10,5	94 - 94	468102	HEATER & A/C CONTROLS - TESTS F & G - S/M REV
14	90 - 94	46-11-02	HEATER - POOR PERF - 5.0/5.7L
5,7,10	93 - 94	468101	HEATER AND A/C CONTROLS - S/M REVISION
7,10,12	95 - 95	56-81-11	HEATER BLOWER MOTOR T/S CHART - S/M REV
6	90 - 93	331202	HEATER CORE & A/C EVAP (FRONT) - SERV PROC - S/M REV

Model Legend: 1= All 2= Most 3= Trans Sport 4= Astro/Safari 5= Blazer,C/K 6= Silhoutte 7= C/K Pickup 8= S/T-Series (Pickup, Sonoma) 9= S/T -Series (Blazer, Jimmy)
10= Suburban 11= Sierra 12= Tahoe/Yukon 13= Lumina Van 14= G Series Van 15= Tracker 16= Bravada

Heating, Air Conditioning, Ventilation, Defogger

Models	Year	TSB#	Description
4	95 - 95	56-12-06	HVAC - COMPRESSOR/CONDENSER REPL PROCED - S/M REV
5,7,10,11	90 - 93	261210	HVAC - CONTROL HEAD INOP/NO DISPLAY - PROCEDURE
5,7,10,11,12	95 - 96	66-11-03	HVAC - FAN ONLY OPERATES ON HIGH SPEED
7,10,12	92 - 94	56-12-02	HVAC - MODE & TEMP ACTUATORS LOCATIONS - S/M REV
18	90 - 94	451101	HVAC - REAR SYSTEM CONTENT CLARIFIED
6	90 - 94	451101	HVAC - REAR SYSTEM CONTENT CLARIFIED
13	90 - 94	451101	HVAC - REAR SYSTEM CONTENT CLARIFIED
4	95 - 95	66-81-03	HVAC BLOWER MOTOR SCHEMATIC - S/M REV
7,5,10	Up - 94	361102	HVAC ERRATIC OP/FUSE BLOWS/DIAGN INDICATOR BLINKS
5,7,10,11	94 - 94	361103	HVAC WHISTLE NOISE AT HIGH BLOWER MOTOR SPEEDS
5,7,10,11	88 - 94	461101	MANUAL TEMP CONTROL LEVER SLIPS FROM HOT TO COLD
4	96 - 96	66-11-02	NO HEAT FROM AIR DUCTS
14	93 - 93	361204	NOISE/RATTLE - A/C COMPRESSOR ON START-UP - HR6
1	93 - 95	43-12-15	R134A LEAK DETECTION WITH TRACER DYE
4	91 - 93	53-12-01	RETROFITTING OUT-OF-WARRANTY VHCLS FROM R12 TO R134A
1	Up - 94	331226	RETROFITTING R-12 VEHICLES TO R-134A
1	88 - 96	63-12-09	SERVICE ISSUES FOR R12 OR R134A A/C SYSTEMS
16	96 - 96	66-12-04	WIND RUSH NOISE FROM DASH WITH MAX A/C ON
9,8	95 - 96	66-12-04	WIND RUSH NOISE FROM DASH WITH MAX A/C ON

Model Legend: 1= All 2= Most 3= Trans Sport 4= Astro/Safari 5= Blazer,C/K 6= Silhoutte 7= C/K Pickup 8= S/T-Series (Pickup, Sonoma) 9= S/T -Series (Blazer, Jimmy) 10= Suburban 11= Sierra 12= Tahoe/Yukon 13= Lumina Van 14= G Series Van 15= Tracker 16= Bravada

Entertainment Devices, Stereo, Radio, Etc.

Models	Year	TSB#	Description
7,10,12	95 - 95	56-96-02	AM RADIO - POOR RECEPTION OR STATIC
16	96 - 96	56-96-04	ANTENNA CABLE ASSMBLY REMOVE/INSTALL - S/M REV
9,8	95 - 96	56-96-04	ANTENNA CABLE ASSMBLY REMOVE/INSTALL - S/M REV
4	85 - 93	92-280-09A	ANTENNA WIND NOISE - REPLACE WITH NEW ANTENNA
14	92 - 95	46-90-03	BENT OR BROKEN POWER ANTENNA MAST
16	92 - 94	449601	BUZZ NOISE IN AM BAND - FILTER
2	92 - 94	449601	BUZZ NOISE IN AM BAND - FILTER - TRUCKS ONLY
2	94 - 94	439601	INSTALLATION OF FIXED MAST ANTENNAS
16	96 - 96	56-96-03	MOBILE 2-WAY RADIO - ELECTRICAL INTERFERENCE
9,8	95 - 96	56-96-03	MOBILE 2-WAY RADIO - ELECTRICAL INTERFERENCE
7,10,12,4	95 - 96	56-96-03	MOBILE 2-WAY RADIO - ELECTRICAL INTERFERENCE
1	91 - 94	439604	POWER ANTENNA MAST REPLACEMENT
5,7,10,11	88 - 93	369601	RADIO - POOR RECEPTION/STATIC - CAUSE/CORRECTION
5,7,10,11	88 - 93	93-239-09A	RADIO - POOR RECEPTION/STATIC - CAUSE/CORRECTION-REV
2	94 - 96	54-90-01	RADIO DISPLAY - "CLN" APPEARS - CLEAN/RESET
7,10	95 - 96	64-90-04	RADIO DISPLAY - ERROR CODES - CAUSE/CORRECTION
13,4,14	96 - 96	64-90-04	RADIO DISPLAY - ERROR CODES - CAUSE/CORRECTION
7,10	95 - 96	64-90-04	RADIO DISPLAY - ERROR CODES - CAUSE/CORRECTION
1	Up - 93	349212R	RADIO FREQUENCY INTERFERENCE ID & DIAGNOSIS
1	Up - 96	93-254-09A	RADIO FREQUENCY INTERFERENCE ID & DIAGNOSIS - REV
5,9,8	94 - 95	56-96-01	RADIO/CASSETTE TEMP HOT AFTER EXTENDED PLAY
5,7,10,11	88 - 93	93-249-09	TAPE DECK - STATIC DURING PAUSES - PROCEDURE - REV
5,7,10,11	88 - 93	369602	TAPE DECK - STATIC DURING PROGRAM PAUSES - SERVICE

Model Legend: 1= All 2= Most 3= Trans Sport 4= Astro/Safari 5= Blazer,C/K 6= Silhoutte 7= C/K Pickup 8= S/T-Series (Pickup, Sonoma) 9= S/T -Series (Blazer, Jimmy) 10= Suburban 11= Sierra 12= Tahoe/Yukon 13= Lumina Van 14= G Series Van 15= Tracker 16= Bravada

Seats, Belts, Interior Trim, Carpets, Air Bags

Models	Year	TSB#	Description
18	94 - 94	431032	2ND/3RD ROW SEAT BACK DIFFICULT TO LOCK
6	94 - 94	431032	2ND/3RD ROW SEAT BACK DIFFICULT TO LOCK
13	94 - 94	431032	2ND/3RD ROW SEAT BACK DIFFICULT TO LOCK
9,8	95 - 95	95-C-34	CARGO SHADE RATTLES - CAMPAIGN
9,8	95 - 95	95-C-34A	CARGO SHADE RATTLES - CAMPAIGN SUPPLEMENT
2	Up - 95	43-16-06	CLEANING PROCEDURE FOR LEATHER SEAT COVERS
7,5,10,11	91 - 94	361015	DOOR TRIM PANEL ILLUSTRATION - S/M REVISION
16	91 - 94	56-16-04	EXPORT & DOMESTIC SEAT BELTS INCOMPATIBLE
9,8,4	Up - 95	56-16-04	EXPORT & DOMESTIC SEAT BELTS INCOMPATIBLE
16	96 - 96	56-16-07	FLOOR CONSOLE/SHIFT BOOT SERVICE - S/M REV
18	94 - 94	431602	FRONT SEAT BELT RETRACTORS LAZY - REVISED PART
6	94 - 94	431602	FRONT SEAT BELT RETRACTORS LAZY - REVISED PART
13	94 - 94	431602	FRONT SEAT BELT RETRACTORS LAZY - REVISED PART
18	93 - 95	63-16-04	FRONT SEAT BELTS SLOW TO RETRACT
6	93 - 95	63-16-04	FRONT SEAT BELTS SLOW TO RETRACT
13	93 - 95	63-16-04	FRONT SEAT BELTS SLOW TO RETRACT
4,13	92 - 93	361011	FRONT SEAT BELTS TOO SHORT - PARTS & PROCEDURE
18	94 - 94	94-C-02	LAP BELT RETRACTOR ASSEMBLY - RECALL CAMPAIGN
6	94 - 94	94-C-03	LAP BELT RETRACTOR ASSEMBLY - RECALL CAMPAIGN
7,10,12	95 - 95	66-81-02	LUMBAR SUPPORTS - SCHEMATICS - S/M REV
18	94 - 96	63-16-02	RATTLE IN REAR COMP AREA OF ROOF CONSOLE
6	94 - 96	63-16-02	RATTLE IN REAR COMP AREA OF ROOF CONSOLE
9,8	92 - 94	461602	RATTLE/LOOSE REAR SEAT
9,8	92 - 94	461602	RATTLE/LOOSE REAR SEAT - KIT - REVISED
7	93 - 93	93C20	REAR BENCH SEAT LATCH ASSEMBLY - CREW CAB - RECALL
7	95 - 95	56-16-05	REAR SEAT RATTLES/ROCKS - EXT CAB
4	92 - 93	92-64-10	SEAT BELT - STOP BUTTON BREAKS - REPLACEMENT KIT
4,14	Up - 94	92-183-10	SEAT BELT EXTENDER - FASTEN INFANT/CHILD SEAT - REV
9,8,7	83 - 95	56-16-01	SEAT BELT OPERATION/FEATURES/LOCATION
13	94 - 94	94C20	SEAT BELT RETRACTORS (3RD ROW) LOCK UP - RECALL
14	92 - 93	361601	SEAT BELT/DRIVER - LATCH PLATE NOT EASILY ACCESSIBLE
13	93 - 93	94C27	SEAT TO POWER SLIDE DOOR TRIM INTERFERENCE - RECALL
16	96 - 96	56-16-08	SEAT TRIM - ADD'L SERVICE INFO - S/M REV
9,8	96 - 96	56-16-08	SEAT TRIM - ADD'L SERVICE INFO - S/M REV
18	90 - 93	331003	SEAT/DRIVER - ADJUSTER SWITCH REPLACEMENT - S/M ADD
16	93 - 93	261013	SEAT/POWER - DIAGNOSIS & REPAIR INFO - S/M UPDATE
4	93 - 94	56-90-01	SIR - CHART B - WARNING LAMP COMES ON - S/M REV
16	96 - 96	56-90-07	SIR DTC 24 DIAGNOSTIC CHARTS/TABLES - S/M REV
4	94 - 95	56-90-07	SIR DTC 24 DIAGNOSTIC CHARTS/TABLES - S/M REV
9,8	96 - 96	56-90-07	SIR DTC 24 DIAGNOSTIC CHARTS/TABLES - S/M REV
7,10,12	95 - 96	56-90-07	SIR DTC 24 DIAGNOSTIC CHARTS/TABLES - S/M REV
1	92 - 94	319003	SIR INFLATOR MODULE SCRAPPING PROCEDURE
18	94 - 95	43-90-16	SIR SYSTEM DIAGNOSIS CHART B - S/M REVISION
6	94 - 95	43-90-16	SIR SYSTEM DIAGNOSIS CHART B - S/M REVISION
13	94 - 95	43-90-16	SIR SYSTEM DIAGNOSIS CHART B - S/M REVISION
18	94 - 94	316402	SIR WARNING LAMP COMES ON STEADY - S/M UPDATE

Model Legend: 1= All 2= Most 3= Trans Sport 4= Astro/Safari 5= Blazer,C/K 6= Silhoutte 7= C/K Pickup 8= S/T-Series (Pickup, Sonoma) 9= S/T -Series (Blazer, Jimmy) 10= Suburban 11= Sierra 12= Tahoe/Yukon 13= Lumina Van 14= G Series Van 15= Tracker 16= Bravada

Seats, Belts, Interior Trim, Carpets, Air Bags

Models	Year	TSB#	Description
6	94 - 94	316402	SIR WARNING LAMP COMES ON STEADY - S/M UPDATE
13	94 - 94	316402	SIR WARNING LAMP COMES ON STEADY - S/M UPDATE
18	94 - 94	316402	SIR WARNING LAMP COMES ON STEADY - S/M UPDATE - REV
6	94 - 94	316402	SIR WARNING LAMP COMES ON STEADY - S/M UPDATE - REV
13	94 - 94	316402	SIR WARNING LAMP COMES ON STEADY - S/M UPDATE - REV
18	94 - 95	43-10-54	SQUEAK/RATTLE/NOISE DIAGNOSTIC GUIDE
6	94 - 95	43-10-54	SQUEAK/RATTLE/NOISE DIAGNOSTIC GUIDE
13	94 - 95	43-10-54	SQUEAK/RATTLE/NOISE DIAGNOSTIC GUIDE

Model Legend: 1= All 2= Most 3= Trans Sport 4= Astro/Safari 5= Blazer,C/K 6= Silhoutte 7= C/K Pickup 8= S/T-Series (Pickup, Sonoma) 9= S/T -Series (Blazer, Jimmy)
10= Suburban 11= Sierra 12= Tahoe/Yukon 13= Lumina Van 14= G Series Van 15= Tracker 16= Bravada

Glass, Doors, Hood, Decklid, Tailgate, Liftgate, Locks

Models	Year	TSB#	Description
14	94 - 94	461501	DOOR HINGE COVERS FALL OFF
18	90 - 96	63-10-25	DOOR INNER PANEL REPAIRS ON SMC TYPE DOORS
6	90 - 96	63-10-25	DOOR INNER PANEL REPAIRS ON SMC TYPE DOORS
13	90 - 96	63-10-25	DOOR INNER PANEL REPAIRS ON SMC TYPE DOORS
1	Up - 93	92-041-10	DOOR LOCK CYLINDER - FREEZES/STICKS/BINDS - REVISION
5,7,10,11	92 - 93	93C24	DOOR LOCK CYLINDER FREEZE-UP - RECALL CAMPAIGN
5,7,10,11	88 - 93	361009	DOOR STRIKER RATTLE - REDESIGNED DOOR LOCK/LATCH
18	90 - 93	331023	DOOR/FUEL FILLER - BINDS/STICKS/HARD TO OPEN - SERV
6	90 - 93	331023	DOOR/FUEL FILLER - BINDS/STICKS/HARD TO OPEN - SERV
13	90 - 93	331023	DOOR/FUEL FILLER - BINDS/STICKS/HARD TO OPEN - SERV
9,8	91 - 93	92-352-10	DOOR/REAR - WATER LEAKS - PROCEDURE - REVISED
16	91 - 93	92-T-62	DOOR/REAR - WATER LEAKS - PROCEDURE - REVISED
4	92 - 93	361501	DOORS/DUTCH - WATER LEAKS - SERVICE PROCEDURE
4	92 - 93	261007	DUTCH DOOR - WATER ENTRY - CAUSES/CORRECTION
9,8	83 - 94	92-153-10	END GATE LATCH - RATTLES - DIAGNOSIS/SERVICE - REV
16	91 - 94	92-T-077	END GATE LATCH - RATTLES - DIAGNOSIS/SERVICE - REV
7,10,5,11	88 - 93	92-184-10	FRONT DOOR HINGE PIN/BUSHING WEAR - PROCEDURE - REV
7,10,12,5	88 - 95	56-15-05	FRONT DOOR/A-PILLAR WIND NOISE - SERVICE PACKAGE
7,10,12,5	88 - 95	56-15-05	FRONT DOOR/A-PILLAR WIND NOISE - SERVICE PKG - REV
16	91 - 93	261614	GLASS/FRONT DOOR SIDE - BINDING/TRACK DISENGAGEMENT
9,8	82 - 93	261614	GLASS/FRONT DOOR SIDE - BINDING/TRACK DISENGAGEMENT
9,8	95 - 95	52-15-01	HOOD HINGE RATTLE
18	93 - 95	53-81-15	HOW TO REINITIALIZE THE POWER SLIDING SIDE DOOR
6	93 - 95	53-81-15	HOW TO REINITIALIZE THE POWER SLIDING SIDE DOOR
13	93 - 95	53-81-15	HOW TO REINITIALIZE THE POWER SLIDING SIDE DOOR
18	93 - 95	53-81-15	HOW TO REINITIALIZE THE POWER SLIDING SIDE DOOR -REV
6	93 - 95	53-81-15	HOW TO REINITIALIZE THE POWER SLIDING SIDE DOOR -REV
13	93 - 95	53-81-15	HOW TO REINITIALIZE THE POWER SLIDING SIDE DOOR -REV
16	91 - 94	461601	INTERIOR DOOR HANDLE SPRING BREAKAGE
9,8	90 - 94	461601	INTERIOR DOOR HANDLE SPRING BREAKAGE
7,10,5	88 - 94	56-16-03	INTERIOR DOOR HANDLE SPRING BREAKS
4	94 - 94	94C26	LEFT FRONT QUARTER STATIONARY GLASS - RECALL
18	94 - 94	431040	LIFT GATE KEY CYLINDER LOOSE/WON'T RETURN TO NORMAL
6	94 - 94	431040	LIFT GATE KEY CYLINDER LOOSE/WON'T RETURN TO NORMAL
13	94 - 94	431040	LIFT GATE KEY CYLINDER LOOSE/WON'T RETURN TO NORMAL
18	92 - 94	331068	LIFTGATE - LOSS OF LOCKING FEATURE - NON-POWER LOCKS
6	92 - 94	331068	LIFTGATE - LOSS OF LOCKING FEATURE - NON-POWER LOCKS
13	92 - 94	331068	LIFTGATE - LOSS OF LOCKING FEATURE - NON-POWER LOCKS
18	94 - 95	53-10-12	LIFTGATE WINDOW - REMOVE/INSTALL - S/M REV
6	94 - 95	53-10-12	LIFTGATE WINDOW - REMOVE/INSTALL - S/M REV
13	94 - 95	53-10-12	LIFTGATE WINDOW - REMOVE/INSTALL - S/M REV
16	96 - 96	56-16-06	LIFTGATE WINDOW GLASS SUPPORT REPLACEMENT - S/M REV
9,8	95 - 96	56-16-06	LIFTGATE WINDOW GLASS SUPPORT REPLACEMENT - S/M REV
18	94 - 94	431038	NOISE/CHATTER/SCRAPE - FRONT DOOR GLASS TO CHANNEL
6	94 - 94	431038	NOISE/CHATTER/SCRAPE - FRONT DOOR GLASS TO CHANNEL
13	94 - 94	431038	NOISE/CHATTER/SCRAPE - FRONT DOOR GLASS TO CHANNEL

Model Legend: 1= All 2= Most 3= Trans Sport 4= Astro/Safari 5= Blazer,C/K 6= Silhoutte 7= C/K Pickup 8= S/T-Series (Pickup, Sonoma) 9= S/T -Series (Blazer, Jimmy)
10= Suburban 11= Sierra 12= Tahoe/Yukon 13= Lumina Van 14= G Series Van 15= Tracker 16= Bravada

Glass, Doors, Hood, Decklid, Tailgate, Liftgate, Locks

Models	Year	TSB#	Description
7,10,12	95 - 95	56-81-07	POWER DOOR LOCKS - NEW DIAGN CHARTS - S/M REV
18	94 - 94	418101	POWER LOCKS/KEYLESS ENTRY WIRING DIAGRAMS - S/M REV
6	94 - 94	418101	POWER LOCKS/KEYLESS ENTRY WIRING DIAGRAMS - S/M REV
13	94 - 94	418101	POWER LOCKS/KEYLESS ENTRY WIRING DIAGRAMS - S/M REV
18	93 - 95	63-10-30	POWER SLIDING SIDE DOOR WILL NOT OPEN
6	93 - 95	63-10-30	POWER SLIDING SIDE DOOR WILL NOT OPEN
13	93 - 95	63-10-30	POWER SLIDING SIDE DOOR WILL NOT OPEN
18	94 - 94	43-10-47	POWER/MANUAL SLIDING DOOR MAY NOT OPEN AT TIMES
6	94 - 94	43-10-47	POWER/MANUAL SLIDING DOOR MAY NOT OPEN AT TIMES
13	94 - 94	43-10-47	POWER/MANUAL SLIDING DOOR MAY NOT OPEN AT TIMES
18	95 - 95	63-81-07	PWR DOOR LOCKS CYCLE WHEN BRAKES APPLIED - 3.1L
13	95 - 95	63-81-07	PWR DOOR LOCKS CYCLE WHEN BRAKES APPLIED - 3.1L
18	94 - 95	53-10-25	PWR SIDE DOOR MAY NOT STAY OPEN OR DOOR AJAR LITE ON
6	94 - 95	53-10-25	PWR SIDE DOOR MAY NOT STAY OPEN OR DOOR AJAR LITE ON
13	94 - 95	53-10-25	PWR SIDE DOOR MAY NOT STAY OPEN OR DOOR AJAR LITE ON
18	90 - 94	43-10-51	RATTLE - SLIDING DOOR - NEW BRACKETS
6	90 - 94	43-10-51	RATTLE - SLIDING DOOR - NEW BRACKETS
13	90 - 94	43-10-51	RATTLE - SLIDING DOOR - NEW BRACKETS
14	85 - 94	46-15-06	RATTLE/NOISE - REAR & SIDE CARGO DOOR LATCH
10	92 - 93	361014	REAR CARGO DOOR HINGE BINDS
4	85 - 93	361014	REAR CARGO DOOR HINGE BINDS
10	92 - 94	361014	REAR CARGO DOOR HINGE BINDS - REV
4	85 - 93	361014	REAR CARGO DOOR HINGE BINDS - REV
14	89 - 93	361012	REAR CARGO DOOR INNER PANEL CRACKED OR BROKEN
14	89 - 93	361012	REAR CARGO DOOR INNER PANEL CRACKED OR BROKEN - REV
4	92 - 93	261510	REAR LIFTGATE OR LOWER DOORS WON'T OPEN - SERV PROC
5,10	92 - 93	361007	REAR LIFTGLASS DOES NOT RELEASE - REPL WEATHERSTRIP
5,10	92 - 94	36-10-07A	REAR LIFTGLASS DOES NOT RELEASE - REV TWO
5,10	92 - 94	361007	REAR LIFTGLASS DOES NOT RELEASE - REVISED
7,10	92 - 93	361010	REAR SIDE DOOR WINDOW OUTER BELT SEAL CURLS UP
7,10	92 - 93	361010	REAR SIDE DOOR WINDOW OUTER BELT SEAL CURLS UP - REV
7	94 - 95	56-15-01	REAR SIDE WINDOWS MAY NOT LATCH/OPEN - EXT CAB
7	94 - 95	56-15-01	REAR SIDE WINDOWS MAY NOT LATCH/OPEN - EXT CAB - REV
18	94 - 94	339007	REMOTE KEYLESS ENTRY - INFO ON INTERMIT'T OPERATION
6	94 - 94	339007	REMOTE KEYLESS ENTRY - INFO ON INTERMIT'T OPERATION
13	94 - 94	339007	REMOTE KEYLESS ENTRY - INFO ON INTERMIT'T OPERATION
18	94 - 94	339007	REMOTE KEYLESS ENTRY/RKE INTERMITTENT - NEW PCM
6	94 - 94	339007	REMOTE KEYLESS ENTRY/RKE INTERMITTENT - NEW PCM
13	94 - 94	339007	REMOTE KEYLESS ENTRY/RKE INTERMITTENT - NEW PCM
18	94 - 94	339007A	REMOTE KEYLESS ENTRY/RKE INTERMITTENT - NEW PCM -REV
6	94 - 94	339007A	REMOTE KEYLESS ENTRY/RKE INTERMITTENT - NEW PCM -REV
13	94 - 94	339007A	REMOTE KEYLESS ENTRY/RKE INTERMITTENT - NEW PCM -REV
14	96 - 96	66-10-02	SIDE & REAR CARGO DOOR SEALS - NEW DESIGN
9,8	94 - 94	46-15-03	SIDE DOOR WATER LEAKS - REVISED WEATHERSTRIP
4	85 - 93	91-369-10	SLIDING DOOR - HARD TO OPEN/CLOSE - DIAG & REPAIR
18	94 - 95	43-10-54	SQUEAK/RATTLE/NOISE DIAGNOSTIC GUIDE

Model Legend: 1= All 2= Most 3= Trans Sport 4= Astro/Safari 5= Blazer,C/K 6= Silhoutte 7= C/K Pickup 8= S/T-Series (Pickup, Sonoma) 9= S/T -Series (Blazer, Jimmy) 10= Suburban 11= Sierra 12= Tahoe/Yukon 13= Lumina Van 14= G Series Van 15= Tracker 16= Bravada

Glass, Doors, Hood, Decklid, Tailgate, Liftgate, Locks

Models	Year	TSB#	Description
6	94 - 95	43-10-54	SQUEAK/RATTLE/NOISE DIAGNOSTIC GUIDE
13	94 - 95	43-10-54	SQUEAK/RATTLE/NOISE DIAGNOSTIC GUIDE
2	88 - 94	431018	SUNROOF (WEBASTO SYSTEM) SERVICE HOTLINE - REVISED
14	86 - 95	56-15-03	SWING OUT DOORS HARD TO OPEN
5,7,10,11	92 - 93	261513	TAILGATE - LATCH RELEASES OVER ROUGH ROADS - REPAIR
5,7,10,11	92 - 95	93-069-10	TAILGATE - LATCH RELEASES OVER ROUGH ROADS - REVISED
4	92 - 94	46-15-02	WATER LEAK - DUTCH DOOR AT CENTER WEATHERSTRIP
14	92 - 94	46-10-06	WIND NOISE - LH VENT WINDOW
16	91 - 94	461003	WIND NOISE - REAR SIDE DOOR WINDOW
9,8	91 - 94	461003	WIND NOISE - REAR SIDE DOOR WINDOW
9,8	95 - 95	52-15-04	WIND WHISTLE NOISE FROM ROOF RACK
9,8	94 - 95	56-15-02	WIND WHISTLE NOISE OUTSIDE WINDOWS - MIIRORS
8	94 - 94	321503	WINDNOISE SWING OUT GLASS - EXTENDED CAB
4	85 - 94	89-073-10	WINDOW/SLIDING DOOR - LATCH SELF RELEASES - REV

Model Legend: 1= All 2= Most 3= Trans Sport 4= Astro/Safari 5= Blazer,C/K 6= Silhoutte 7= C/K Pickup 8= S/T-Series (Pickup, Sonoma) 9= S/T -Series (Blazer, Jimmy) 10= Suburban 11= Sierra 12= Tahoe/Yukon 13= Lumina Van 14= G Series Van 15= Tracker 16= Bravada

Finishes, Body Structure, Frame, Bumpers

Models	Year	TSB#	Description
1	Up - 94	43-10-48	ADHESIVE CAULKING KIT - REVISIONS/PRIMERS/PROCEDURES
1	90 - 95	53-6-03	ALUMINUM WHEEL REFINISHING - REVISED PROCEDURE
1	91 - 96	53-6-03	ALUMINUM WHEEL REFINISHING - REVISED PROCEDURE
1	Up - 93	93-058-10	BASE COAT/CLEAR COAT - RAIL DUST DAMAGE - REPAIR
1	84 - 93	92-033-10	BASE COAT/CLEAR COAT PAINT - INDUSTRIAL DUST DAMAGE
1	93 - 96	11602	BASE COAT/CLEAR COAT PAINT - POLISHING - REV
18	90 - 93	331001	BODY SIDE MOULDING/GAP TO FRONT OR SLIDING SIDE DOOR
1	Up - 94	43601	BUMPS OR RUST COLORED SPOTS IN PAINT - REMOVE PROCED
1	94 - 95	43601	BUMPS OR RUST COLORED SPOTS IN PAINT - REVISED
1	Up - 94	33608	CLEARCOAT DEGRADATION/CHALKING & WHITENING - PROCED
9,8	83 - 93	REF: 91-351-10	COWL - POP NOISE FROM WINDSHIELD PILLAR AREA - PROC
16	91 - 93	REF: 91-T-167	COWL - POP/TAPPING NOISE - SERVICE PROCEDURE
16	96 - 96	56-15-06	FASCIA/GRILLE/BODY CLADDING - PROCEDURES - S/M REV
5,7,10,11	88 - 94	46-20-01	FRAME CRACKS AT ENGINE MOUNT BRACKETS - 4WD
5	95 - 95	95-C-44	GMSPO RUNNING BOARDS - RECALL
7,10,12	88 - 96	66-15-05	HOOD TO GRILLE CONTACT
16	91 - 94	261501	MOULDING/FRONT DOOR ROOF - WIND NOISE - SEAL - REV
9,8	82 - 94	92-128-10	MOULDING/FRONT DOOR ROOF - WIND NOISE - SEAL - REV
14	93 - 93	361602	MOULDING/WINDSHIELD SUPPORT - FALLS DOWN - REPAIR
9,8,4	96 - 96	63-6-05	NEW COLOR PULL-AHEAD PROGRAM
10,7,12,14	96 - 96	63-6-05	NEW COLOR PULL-AHEAD PROGRAM
9,8	94 - 94	321504	NO FRONT LICENSE PLATE MOUNTING - INSTALL BRACKET
1	94 - 94	33607	PAINT INFORMATION - TRUCK MODELS ONLY
9,8	82 - 93	261514	PICKUP BOX MISALIGNED TO THE CAB - SERVICE PROCEDURE
18	94 - 95	43-10-59	POP NOISE - FRAME INSULATOR
6	94 - 95	43-10-59	POP NOISE - FRAME INSULATOR
13	94 - 95	43-10-59	POP NOISE - FRAME INSULATOR
1	94 - 94	32602	VINYL TRIM AND VINYL TOP REPAIR
18	90 - 94	53-15-21	WINDNOISE AT ROOF AREA AT HWY SPEEDS
6	90 - 94	53-15-21	WINDNOISE AT ROOF AREA AT HWY SPEEDS
13	90 - 94	53-15-21	WINDNOISE AT ROOF AREA AT HWY SPEEDS
6	94 - 94	431503	WRINKLED LIFTGATE DECAL - REPLACE HANDLE

Model Legend: 1= All 2= Most 3= Trans Sport 4= Astro/Safari 5= Blazer,C/K 6= Silhoutte 7= C/K Pickup 8= S/T-Series (Pickup, Sonoma) 9= S/T -Series (Blazer, Jimmy) 10= Suburban 11= Sierra 12= Tahoe/Yukon 13= Lumina Van 14= G Series Van 15= Tracker 16= Bravada

HONDA

Instruments, Dash Cluster, Warning Lights, Mirrors

Models	Year	TSB#	Description
3	95 - 95	95-010	BUZZ NOISE IN IP NEAR SPEEDOMETER
3	95 - 95	95-019	FUEL GAUGE READS LESS THAN FULL
4	94 - 94	SN06/94	ILLUMINATION CONTROLLER TEST - S/M CORRECTION

Model Legend: 1= All 2= Most 3= Odyssey 4= Passport

Chassis Electrical, Wiring Harness, Fuses-Circuit Breakers, Wipers, Window Motors

Models	Year	TSB#	Description
1	Up - 95	95-023	CONNECTOR TERMINAL REPLACEMENT - KIT/INSTRUCTIONS
1	Up - 96	SN05/95	ISOLATING SHORTS WITH IN-LINE FUSES - TIPS
3	95 - 95	95-012	LOOSE CIGARETTE LIGHTER SOCKET - PDI

Model Legend: 1= All 2= Most 3= Odyssey 4= Passport

Auxiliary Equipment, Jacks, Trailer Hitches, Towing

Models	Year	TSB#	Description
3	94 - 95	95-008	CELLULAR PHONE - OUT-OF-WARRANTY REPAIR
1	Up - 95	SN05/95	REPLACING DAMAGED IN-DASH PHONE LENSES
3	95 - 95	AHM/A50	ROOF RACK BASKET & SKI ATTACHMENTS - REV KIT

Model Legend: 1= All 2= Most 3= Odyssey 4= Passport

Heating, Air Conditioning, Ventilation, Defogger

Models	Year	TSB#	Description
1	Up - 94	SN02/94	A/C - R-134A SYS WON'T CHARGE - CHECK SCHRADER VALVE
1	Up - 95	SN03/95	A/C - R134A O-RING LUBRICATION
4	94 - 94	94-002	CONDENSER FAN MOUNTING SCREWS BREAK - UPDATED SCREWS
1	Up - 93	95-020	CONVERTING R-12 A/C SYSTEMS TO R-134A
3	95 - 95	95-015	TEMPERATURE LEVER IS HARD TO MOVE

Model Legend: 1= All 2= Most 3= Odyssey 4= Passport

Entertainment Devices, Stereo, Radio, Etc.

Models	Year	TSB#	Description
1	Up - 93	87-014 JUL`92	AUDIO COMPONENTS - OUT-OF-WARRANTY REPAIR INFO
1	Up - 95	89-029 FEB'93	AUDIO UNIT - EXCHANGE PROGRAM PROC - REV
1	Up - 93	92-026	CD CHANGER MAGAZINE - WON'T EJECT - CORRECTION
1	Up - 93	92-026 OCT`92	CD CHANGER MAGAZINE WON'T EJECT-REPLACEMENT P/N REV
1	Up - 95	SN08/94	PHONE ESN CHANGES EXPLAINED
1	Up - 94	SN06/94	RADIO - ENTERING ANTI-THEFT CODES - TIPS
1	82 - 94	87-014 JAN`93	RADIO - OUT-OF-WARRANTY REPAIR INFORMATION - REV
1	Up - 93	SN11/92	RADIO/1000 SERIES - POOR SOUND - CHECK FADER CONTROL
1	Up - 93	89-029 MAY`92	RADIO/TAPE/CD PLAYER - EXCHANGE PROGRAM PROC - REV

Model Legend: 1= All 2= Most 3= Odyssey 4= Passport

Seats, Belts, Interior Trim, Carpets, Air Bags

Models	Year	TSB#	Description
1	Up - 95	SN05/95	DEFOGGER GRILLE REPAIR KIT
3	95 - 95	95-016	FRONT DOOR PANELS CREAK
1	92 - 95	95-024	SRS - CABLE REEL REMOVAL - HOLDING FIXTURE/CAUTIONS
3	95 - 95	SN05/95	UNDER SEAT DRAWER WON'T FIT BENCH SEAT

Model Legend: 1= All 2= Most 3= Odyssey 4= Passport

Glass, Doors, Hood, Decklid, Tailgate, Liftgate, Locks

Models	Year	TSB#	Description
1	94 - 94	AHM/A77	PDI CHECK - MISSING TRUNK LIGHT LENSE
4	94 - 94	SN01/94	RATTLE - TAILGATE AND SPARE TIRE CARRIER
4	94 - 95	SN12/94	TAILGATE RELEASE INOP - TIPS
4	94 - 94	SN01/94	TAILGATE RELEASE OPERATION EXPLAINED

Model Legend: 1= All 2= Most 3= Odyssey 4= Passport

Finishes, Body Structure, Frame, Bumpers

Models	Year	TSB#	Description
1	93 - 93	87-006 APR`89	BODY - PDI SHIPPING WAX REMOVAL/SOLVENTS - REVISION
1	94 - 94	93-036	PAINT CODES
1	95 - 95	94-035 OCT`94	PAINT CODES - 1995 MODELS
4	95 - 95	95-009	PAINT CODES - 1995 PASSPORT
1	93 - 93	92-045	PAINT CODES AND DESCRIPTIONS
4	94 - 94	93-053	PASSPORT PAINT CODES
4	94 - 94	93-052	TRANSIT COATING REMOVAL

Model Legend: 1= All 2= Most 3= Odyssey 4= Passport

ISUZU

Instruments, Dash Cluster, Warning Lights, Mirrors

Models	Year	TSB#	Description
7	92 - 94	94-11-003	POWER DOOR MIRROR VIBRATION
2	88 - 93	93-04-004	WARNING UNIT/LOW FUEL - INSPECTION - S/M CORRECTION

Model Legend: 1= All 2= Most 3= Amigo Sport Truck 4= Pickup 5= Rodeo 6= Trooper II 7= Trooper

Chassis Electrical, Wiring Harness, Fuses-Circuit Breakers, Wipers, Window Motors

Models	Year	TSB#	Description
1	Up - 96	95-14-004	ELECT TERMINAL KIT AVAILABLE
5	94 - 94	93-11-008	WINDOW REGULATOR PROCEDURES - S/M REV

Model Legend: 1= All 2= Most 3= Amigo Sport Truck 4= Pickup 5= Rodeo 6= Trooper II 7= Trooper

Auxiliary Equipment, Jacks, Trailer Hitches, Towing

Models	Year	TSB#	Description
5	92 - 93	93-13-001	TRAILER HITCH WIRING CONVERTER BOX - AIPDN 3,500 LB

Model Legend: 1= All 2= Most 3= Amigo Sport Truck 4= Pickup 5= Rodeo 6= Trooper II 7= Trooper

Heating, Air Conditioning, Ventilation, Defogger

Models	Year	TSB#	Description
5	94 - 94	94-12-001	A/C COND FAN SHROUD BOLT - 3.2L - CAMPAIGN 94S941
3,4,5	93 - 93	93-12-003	A/C EVAP FREEZING UP - THERMOSTAT RELOCATION KIT
3,4,5,7	93 - 93	93-12-005	A/C THERMOSTAT - PROCEDURES - S/M REV
1	93 - 93	93-12-002	R-12 A/C SYSTEM WARNING LABEL
3,4,5,7	93 - 93	93-12-001	R-134A A/C S/M SUPPLEMENT - LUBRICANTS & CAUTIONS
3,4,5,7	93 - 93	93-12-004	R-134A A/C SERVICE VALVE OPERATION - S/M CORRECTION

Model Legend: 1= All 2= Most 3= Amigo Sport Truck 4= Pickup 5= Rodeo 6= Trooper II 7= Trooper

Seats, Belts, Interior Trim, Carpets, Air Bags

Models	Year	TSB#	Description
7	92 - 93	93-11-007	CENTER CONSOLE LID - POOR OPERATION
1	Up - 95	95-11-003	SEAT BELT EXTENDER PROGRAM
7	95 - 95	95-14-003	SRS DIAGNOSTIC PROCEDURE - S/M REV

Model Legend: 1= All 2= Most 3= Amigo Sport Truck 4= Pickup 5= Rodeo 6= Trooper II 7= Trooper

Glass, Doors, Hood, Decklid, Tailgate, Liftgate, Locks

Models	Year	TSB#	Description
7	92 - 94	94-11-007	DIRT/DUST ENTERS CARGO AREA
5	93 - 93	93-04-003	DOOR LOCKS/POWER - OP CLARIFICATION/SERVICE INFO
1	93 - 93	93-14-004	LABEL ADHESIVE REMOVAL/ DON'T SCRAPE GLASS - REMOVER
5	93 - 93	94-11-006	POWER DOOR LOCK OPERATION - S/M REV
7	93 - 93	93-11-004	TAILGATE DOOR ASSIST GRIP MISSING
7	92 - 94	94-11-002	WIND NOISE - FRONT DOOR - A PILLAR

Model Legend: 1= All 2= Most 3= Amigo Sport Truck 4= Pickup 5= Rodeo 6= Trooper II 7= Trooper

Finishes, Body Structure, Frame, Bumpers

Models	Year	TSB#	Description
7	92 - 93	93-11-005	BODY NOISE DIAGNOSIS AND REPAIR
7	92 - 94	93-11-006	BUMPER FINISH REPAIR PROCEDURES
7	92 - 93	93-11-002	BUMPER INSTALLATION - TORQUE SPECS - S/M CORRECTION
3,4,5,7	94 - 94	93-11-003	PAINT - COLORS AND CODES
7	94 - 94	94-11-001	PAINT CODES - SPECIAL EDITION
5,4,7	95 - 95	95-11-004	PAINT COLORS & CODES - INC 1995.5 MODELS
7,5,4	95 - 95	94-11-004	PAINT COLORS AND CODES
1	93 - 93	92-11-010	PAINT COLORS AND CODES - 1993 MODELS
7,5	95 - 95	94-14-002	PROTECTIVE PAINT FILM REMOVAL
5	93 - 93	94-11-005	ROOF RACK WIND NOISE

Model Legend: 1= All 2= Most 3= Amigo Sport Truck 4= Pickup 5= Rodeo 6= Trooper II 7= Trooper

MAZDA

Lighting, Horns, Turn Signals, Steering Column

Models	Year	TSB#	Description
6	89 - 94	T-004/95	AFTER IGN SWITCH REPLACE - DEFROST LITE FLASHES

Model Legend: 1= All 2= Most 3= B-2000 4= B-2200 5= B-2600 6= MPV (Van) 7= Navajo 8= Pickup

Instruments, Dash Cluster, Warning Lights, Mirrors

Models	Year	TSB#	Description
3,4,5	94 - 94	F-009/94	GROWL NOISE FROM DASH - DPFE SENSOR
7	93 - 94	S-033/94	REAR VIEW MIRROR FALLS OFF
7	93 - 94	S-033/94	REAR VIEW MIRROR FALLS OFF - REV

Model Legend: 1= All 2= Most 3= B-2000 4= B-2200 5= B-2600 6= MPV (Van) 7= Navajo 8= Pickup

Chassis Electrical, Wiring Harness, Fuses-Circuit Breakers, Wipers, Window Motors

Models	Year	TSB#	Description
3,4,5	95 - 95	TIPS04/95	POWER POINT (ACCES CIGAR LIGHTER) - IMPROPER OPS

Model Legend: 1= All 2= Most 3= B-2000 4= B-2200 5= B-2600 6= MPV (Van) 7= Navajo 8= Pickup

Heating, Air Conditioning, Ventilation, Defogger

Models	Year	TSB#	Description
1	93 - 94	U-002/93	A/C (R-12 CFC SYS) - WARNING LABEL INFORMATION
7	91 - 93	U-020/92	A/C - FILTER KIT USAGE WHEN REPLACING COMPRESSOR
2	88 - 93	U-016/92	A/C - O-RING INSTALLATION PROCEDURE
2	Up - 94	U-003/93	A/C - R134A SERVICE PRECAUTIONS & PARTS COMPARISON
2	Up - 94	U-003/93	A/C - R134A SERVICE PRECAUTIONS/PARTS REVISED
6	94 - 95	U-005/95	A/C - REAR COOLER PIPE COVERS DISCONTINUED
2	94 - 94	U-002/94	A/C 134A SIGHT GLASS ELIMINATED - INSPECT PROCEDURE
1	88 - 94	U-001/94	A/C O-RING REPLACEMENT - APPLICATIONS
1	88 - 94	U-001/94	A/C O-RING REPLACEMENT - APPLICATIONS - REVISED
2	Up - 95	U-001/95	A/C RECEIVER DRIER REPLACEMENT CRITERIA
6	89 - 93	U-005/93	A/C REFRIGERANT LEAKS AT CONDENSER -R12 SINGLE UNITS
3,4,5	95 - 95	TIPS04/95	A/C TOUBLESHOOTING - TRACER DYE IN SYSTEM
6	89 - 94	T-004/95	DEFROSTER DASH LITE FLASHES AFTER IGN SWITCH REPLACE
7,3,4,5	91 - 94	S-022/94	VIBRATION NOISE UNDER DASH - A/C DOOR MOTOR SPRING
7,3,4,5	94 - 94	S-007/95	WHISTLE NOISE WITH BLOWER MOTOR ON "HI"

Model Legend: 1= All 2= Most 3= B-2000 4= B-2200 5= B-2600 6= MPV (Van) 7= Navajo 8= Pickup

Electronic Devices, Computers, PROMS, Sensors

Models	Year	TSB#	Description
8,7	94 - 94	W-005/94	CAMSHAFT POSITION SENSOR ADJUSTMENT - S/M REVISION
7,3,4,5	91 - 94	F-009/95	DRIVEABILITY/MIL ON - ICM CONNECTOR
2	88 - 94	G-002/93	ECU - ENGINE CONTROL UNIT - DIAGNOSTIC PROCEDURES
3,4,5	94 - 94	F-009/94	GROWL NOISE FROM DASH - DPFE (EGR) SENSOR
3,4,5	94 - 94	F-008/95	ROUGH IDLE/STALL AFTER COLD START - PCM
3,4,5	87 - 93	W-045/93	SENSOR/BAROMETRIC PRESSURE - INFO - S/M CORRECTION
7	93 - 93	F-008/94	STALL AFTER COLD START - PCM
7	93 - 93	F-008/94	STALL AFTER COLD START - PCM - REVISED
2	88 - 94	F-001/94	THROTTLE POSITION SENSOR (TPS) INSPECT/ADJUST - REV
2	88 - 94	F-001/94	THROTTLE POSITION SENSOR (TPS) INSPECT/ADJUST PROCED
3,4,5	94 - 94	F-007/95	UNSTABLE IDLE ON COLD START - PCM

Model Legend: 1= All 2= Most 3= B-2000 4= B-2200 5= B-2600 6= MPV (Van) 7= Navajo 8= Pickup

Entertainment Devices, Stereo, Radio, Etc.

Models	Year	TSB#	Description
2	Up - 95	T-005/95	POWER ANTENNA MAST REPLACEMENT
3,4,5	94 - 94	TIPS03/95	RADIO FADES
5	94 - 94	T-005/94	RADIO NOISE - RFI ON FM BAND

Model Legend: 1= All 2= Most 3= B-2000 4= B-2200 5= B-2600 6= MPV (Van) 7= Navajo 8= Pickup

Seats, Belts, Interior Trim, Carpets, Air Bags

Models	Year	TSB#	Description
6	93 - 95	TIPS03/95	CARPET - POOR FIT
6	88 - 93	S-021/93	MAP BOX - WON'T STAY CLOSED - DOOR BUTTON STICKS
6	94 - 95	S-003/95	MAP BOX STRAP - LEFT SIDE - DISCONTINUED
7	91 - 94	S-032/92	SEAT BELT - EXTENDER AVAILABLE - REVISED
6	91 - 93	S-029/93	SUNVISOR - DEFORMED - DIAGNOSIS & REPAIR

Model Legend: 1= All 2= Most 3= B-2000 4= B-2200 5= B-2600 6= MPV (Van) 7= Navajo 8= Pickup

Glass, Doors, Hood, Decklid, Tailgate, Liftgate, Locks

Models	Year	TSB#	Description
8	94 - 94	S-015/94	DOOR GLASS ADJUSTMENT
7,8	91 - 94	S-019/94	FRONT SEAT WILL NOT MOVE - REVISED PARTS
1	Up - 95	AD-005/94	KEY REPLACEMENT INFO - CALIFORNIA ONLY
7,8	91 - 94	S-014/94	LOOSE POWER DOOR LOCK ACTUATOR BRACKET
6	93 - 93	S-014/95	PASS REAR DOOR WEATHER STRIP BREAKS
7,3,4,5	91 - 94	S-006/95	PASS SEAT BACK RATTLES WHEN UNOCCUPIED
6	93 - 93	TIPS08/94	REAR SIDE DOOR WEATHERSTRIP TEARS - TIP

Model Legend: 1= All 2= Most 3= B-2000 4= B-2200 5= B-2600 6= MPV (Van) 7= Navajo 8= Pickup

Finishes, Body Structure, Frame, Bumpers

Models	Year	TSB#	Description
7,8	91 - 94	S-016/94	CLEAR COAT PAINT CRACKING
7,3,4,5	94 - 94	S-045/93	EXTERIOR PAINT CODES
2	Up - 95	TIPS10/94	LICENSE PLATE FRAME INSTALLATION - TIP
1	92 - 93	S-026/92	TRANSIT COATING REMOVAL - SPECIAL SOLUTION
1	95 - 95	TIPS02/95	TRANSIT COATING REMOVAL TIPS

Model Legend: 1= All 2= Most 3= B-2000 4= B-2200 5= B-2600 6= MPV (Van) 7= Navajo 8= Pickup

MITSUBISHI

Lighting, Horns, Turn Signals, Steering Column

Models	Year	TSB#	Description
4	92 - 93	93-54-008	HEADLIGHT AIMING PROCEDURES - S/M REVISION
2	Up - 93	93-54-003	HORN - UNNECESSARY R&R - ADJUSTMENT PROCEDURE

Model Legend: 1= All 2= Most 3= Expo, Expo LRV 4= Montero 5= Truck

Auxiliary Equipment, Jacks, Trailer Hitches, Towing

Models	Year	TSB#	Description
4	91 - 93	SR-93-001	TRAILER HITCH - SAFETY RECALL

Model Legend: 1= All 2= Most 3= Expo, Expo LRV 4= Montero 5= Truck

Heating, Air Conditioning, Ventilation, Defogger

Models	Year	TSB#	Description
1	Up - 93	93-55-001	A/C - MUSTY ODOR EMITTED - CAUSE/CORRECTION
1	94 - 94	93-55-001	A/C COMPRESSOR OIL REQUIREMENTS - R134A SYSTEMS
1	Up - 94	94-55-001	A/C LEAK CHECKING PROCEDURES
5	90 - 93	93-55-002	A/C WIRING DIAGRAMS - S/M REVISION

Model Legend: 1= All 2= Most 3= Expo, Expo LRV 4= Montero 5= Truck

Seats, Belts, Interior Trim, Carpets, Air Bags

Models	Year	TSB#	Description
3	93 - 93	93-52A-001	PROTECTIVE COVERINGS/INTERIOR - DESCRIPTION/REMOVAL
3	89 - 93	92-52-003	SHOULDER BELT - NORMAL MOVEMENT/DIAGNOSIS INFO
2	90 - 94	94-52B-001	SRS - INSPECTION/REPLACEMENT PROCEDURE - S/M UPDATE

Model Legend: 1= All 2= Most 3= Expo, Expo LRV 4= Montero 5= Truck

Glass, Doors, Hood, Decklid, Tailgate, Liftgate, Locks

Models	Year	TSB#	Description
2	94 - 94	93-00-004	PWR DOOR LOCK/KEYLESS ENTRY "TWO TURN" SYS EXPLAINED

Model Legend: 1= All 2= Most 3= Expo, Expo LRV 4= Montero 5= Truck

Finishes, Body Structure, Frame, Bumpers

Models	Year	TSB#	Description
3,3	92 - 93	93-51-002	BUMPERS - SLIDE/GUIDE BRACKET ELIMINATION - INFO
2	93 - 93	92-51-004	PAINT - EXTERIOR COLOR CODES AND APPLICATIONS
2	94 - 94	93-51-004	PAINT - EXTERIOR COLOR CODES AND APPLICATIONS
1	95 - 95	94-51-001	PAINT CODES & APPLICATIONS - 1995 MODELS
3	93 - 93	93-52A-001	PROTECTIVE COVERINGS/EXTERIOR - DESCRIPTION/REMOVAL
3,4	92 - 93	93-51-003	ROOF RACK FADING/PAINT PEELING - CAUSE/CORRECTION
2	93 - 94	93-52A-001	VEHICLE PROTECTIVE COVERINGS - REVISED

Model Legend: 1= All 2= Most 3= Expo, Expo LRV 4= Montero 5= Truck

NISSAN

Lighting, Horns, Turn Signals, Steering Column

Models	Year	TSB#	Description
4	93 - 93	FC0008317	HEADLIGHT/LEFT - LOW BEAM INOPERATIVE - SERVICE PROC
5	91 - 93	FC0008437	HEADLIGHT/RIGHT SIDE - FUSE BLOWS W/HIGH BEAM ON
4	93 - 95	NTB95-073	INTERMIT TAIL/STOP LAMP - LOOSE SOCKET
4	93 - 93	FC0008315	LAMPS/STOP - INOP/FUSE BLOWS ON INSTALL - FIX
4	93 - 93	FC0008316	LIGHTS/TAIL - GO OFF INTERMITTENTLY - SERVICE PROC
5	91 - 93	FC0008438	PARK LIGHTS/RIGHT HEAD LIGHT - STAY ON ALL THE TIME
4	93 - 93	FC0008450	TAIL LIGHTS - IMPROPER OPERATION - WIRING/BULB INFO
5	91 - 93	FC0008470	TURN SIGNAL (RIGHT) - INOPERATIVE - WIRE REPAIR
4	93 - 95	NTB95-094	VANITY LIGHT WIRING REPAIR

Model Legend: 1= All 2= Most 3= Axxess 4= Quest 5= 720 Pickup 6= D21 Pickup 7= Pathfinder 8= Hardbody

Instruments, Dash Cluster, Warning Lights, Mirrors

Models	Year	TSB#	Description
4	93 - 93	FC0008112	CEL ON/SPEEDOMETER INTERMITTENT - DIAGNOSIS/REPAIR
4	93 - 93	FC0008487	CRUISE CONTROL (ASCD) - INOP/WARN LITE BLINKS - FIX
4	93 - 93	FC0008350	CRUISE CONTROL INOP/CRUISE LITE BLINKING - FIELD FIX
4	93 - 93	FC0008315	DASH ILLUMINATION INOP/FUSE BLOWS ON INSTALL - FIX
4	93 - 93	FC0008316	DASH LIGHTS - GO OFF INTERMITTENTLY - SERVICE PROC
7	91 - 93	FC0007827	DASH/LEFT SIDE - SQUEAKS/LOW SPDS - FENDER RUBBING
1	Up - 94	FC0008618	DIGITAL DASH INOP ON REPLACEMENT UNIT - RESET ODOM
4	93 - 93	FC0008202	GAUGE/FUEL - ALWAYS READS FULL - SERVICE PROCEDURE
4	93 - 93	FC0008310	GAUGE/FUEL - INOP/ALWAYS READS FULL - SERVICE INFO
4	93 - 93	FC0008312	GAUGE/TEMP - INTERMITTENT OP/SENDER & WIRING OK
4	93 - 93	FC0008460	GLOVE BOX RATTLE-2 CONTROL MODULES HITTING TOGETHER
4	93 - 93	FC0008100	SPEEDO AREA - CLICKING NOISE AT 5 MPH - A/C SHIELD
4	93 - 93	FC0007826	SPEEDO FLUCTUATES/STEADY SPEEDS-SPEED SENSOR WIRING
4	93 - 93	FC0008314	SPEEDOMETER - INOP/SIGNAL FROM SPEED SENSOR OK - FIX
4	93 - 93	FC0007801	SPEEDOMETER WORKS INTERMIT - CEL ON W/CODE 35 (EGR)
4	93 - 93	FC0008097	TACH/SPEEDO/TEMP & FUEL GAUGES/ASCD LIGHT - ERRATIC
4	93 - 93	NTB93-032	TACHOMETER &/OR SPEEDOMETER - INOP - SERVICE INFO
6,7	86 - 93	NTB93-186	VIBRATION - DOOR MIRROR - REVISED PARTS
4	93 - 93	FC0008486	WARN LIGHT (DOOR OPEN) - CONSTANT ON - FIELD FIX
4	93 - 93	FC0008311	WARN LITE/ABS - FLASHING CODE 12 - SERVICE INFO
5	91 - 93	FC0008471	WARN LITES ON (BRAKE/ALTERNATOR) - TURN SIGNAL INOP
7,5,6	91 - 93	FC0007692	WARNING LIGHT/ABS - INTERMIT/CONTINUOUS ON - CODE 4
4	93 - 93	FC0008505	WARNING LIGHT/LOW OIL LEVEL - FALSE - TEST PROCEDURE

Model Legend: 1= All 2= Most 3= Axxess 4= Quest 5= 720 Pickup 6= D21 Pickup 7= Pathfinder 8= Hardbody

Chassis Electrical, Wiring Harness, Fuses-Circuit Breakers, Wipers, Window Motors

Models	Year	TSB#	Description
5	93 - 93	FC0008365	10 AMP FUSE BLOWS - AFFECTS VARIOUS LAMPS/WARN CHIME
4	93 - 94	NTB94-037	DIGITAL TOUCH ENTRY HARNESS ROUTING - S/M REV
1	Up - 94	NTB93-148	ELECTRICAL SYSTEM DIAGNOSTIC PROCEDURES
5	90 - 93	FC0008132	FUSE BLOWS/GAUGES & SHIFT INTERLOCK INOP - REPAIR
1	Up - 95	NTB95-070	GREEN OR BLUE (FUEL PUMP & A/C) RELAY CAUTION
4	93 - 93	EL93-021	HARNESS - TRAILER TOW - SERVICE INFORMATION
4	93 - 93	NTB93-120	HARNESS - TRAILER TOW - SERVICE INFORMATION
4	93 - 93	FC0008646	INTERIOR LAMPS INOP - FUSE BLOWS WHEN DOOR IS OPENED
4	93 - 93	NTB93-151	PASSIVE RESTRAINT MOTOR REPLACEM'T REQ CONNECTOR MOD
4	93 - 93	NTB93-131	REAR WIPER MOTOR INOP/IMPROPER OP
1	Up - 94	NTB93-176	RELAY TYPE 1M (GREEN OR BLUE) - SERVICE CAUTIONS
4	93 - 95	NTB95-094	SRS LITE ON/INT LIGHTS INOP/FUSE BLOWN/DOOR AJAR ON
4	93 - 95	PI94-006	TOWING - WIRE HARNESS - INFO & P/N
4	94 - 95	NTB95-044	TRAILER TOW CONTROL MODULE - WIRING/DIAGN - S/M REV
4	93 - 93	FC0008432	WINDSHIELD WASHERS/FRONT - ERRATIC OPERATION
4	93 - 93	FC0008496	WIPER (REAR) - ERRATIC ON AND OFF - FIELD FIX
4	93 - 93	FC0008489	WIPER (REAR) - INOP & FUSE IS BLOWN - FIELD FIX
1	Up - 93	NTB92-083	WIPER BLADES - MAINTENANCE/SERVICE INFORMATION
4	93 - 93	FC0008491	WIPER MOTOR/REAR - CANNOT BE TURNED OFF - FIELD FIX
4	93 - 93	FC0008485	WIPERS/REAR - STAY ON/RUN FAST - DIODES MALFUNCTION
4	93 - 95	PI94-006	WIRE HARNESS - TOWING - INFO & P/N

Model Legend: 1= All 2= Most 3= Axxess 4= Quest 5= 720 Pickup 6= D21 Pickup 7= Pathfinder 8= Hardbody

Auxiliary Equipment, Jacks, Trailer Hitches, Towing

Models	Year	TSB#	Description
4	93 - 93	FC0008219	TRAILER LIGHTS INOP AFTER INSTALLING TOW HARNESS
4	93 - 93	EL93-021	TRAILER TOW HARNESS - SERVICE INFORMATION

Model Legend: 1= All 2= Most 3= Axxess 4= Quest 5= 720 Pickup 6= D21 Pickup 7= Pathfinder 8= Hardbody

Heating, Air Conditioning, Ventilation, Defogger

Models	Year	TSB#	Description
1	93 - 93	NTB93-097	A/C (R134a SYS) - SERVICE PARTS ORDERING INFORMATION
6,7	93 - 93	NTB92-055	A/C - CHANGES FOR 134a REFRIGERANT/SERVICE INFO
6,7	93 - 93	NTB92-098	A/C - CHARGE SPECIFICATION AND LABELING
4	93 - 93	FC0008558	A/C - COMPRESSOR INOP - SERVICE INFORMATION
4	93 - 94	NTB95-092	A/C - HEATER - CIRCUIT WIRING - REFERENCES
2	Up - 94	NTB94-013	A/C - LUBRICANTS & OILS
6,7	94 - 94	REF: HA93-001	A/C - R134A CHANGES FOR DEALER INSTALLED UNITS
7	94 - 94	NTB93-170	A/C - SEMI-AUTOMATIC - CHANGES FOR A994 MODELS
4	93 - 93	NTB92-110	A/C - UNIQUE COMPRESSOR OIL - SERVICE INFORMATION
4	93 - 93	FC0008574	A/C - WON'T CHANGE FROM VENT MODE - WIRING PROCEDURE
1	93 - 93	NTB93-001	A/C CHANGES FOR R134A REFRIGERANT - DETAILS
1	Up - 95	NTB95-057	A/C COMPRESSOR LEAK/NOISE DIAGNOSIS
4	93 - 93	FC0008640	A/C EVAP DRAIN - WET CARPET - EXTEND DRAIN
1	Up - 96	NTB95-106	A/C LUBRICANT/OIL - SPECS
1	Up - 95	NTB95-015	A/C PARTS - PROPER INSTALLATION PROCEDURE
7	93 - 95	NTB95-084	A/C POOR PERF IN HIGH AMBIENT TEMP - 4x4
8	93 - 94	NTB95-048	A/C POOR PERFORMANCE - COMPRESSOR BELT REPLACEMENT
6	92 - 93	NTB93-147	A/C REFR LEAK - CHECK LOW PRESSURE HOSE - PROCEDURE
8	95 - 95	NTB94-122	A/C RELAY INSTALLATION - NEW LOCATION
4	Up - 95	NTB95-054	BLOWER FAN WHEEL INSTALLATION PROCEDURE
4	93 - 93	NTB92-111	BLOWER MOTOR - OPERATES ON FAN SPD #4 ONLY - REPAIR
4	93 - 93	NTB93-153	FOG/FROST ON SIDE WINDOWS - PROCEDURES
4	93 - 93	FC0008290	HEAT - INTERMITTENT COOL AIR FROM DUCTS - PROCEDURE
1	Up - 93	FC0008303	HEATER - PERFORMANCE INFORMATION
7	95 - 95	NTB95-028	HEATER FAN START UP DELAY - LE 4X2 W/ SEMI AUTO A/C
8	95 - 95	NTB95-088	HVAC - AIR MIX DOOR 2 DELETED
6	92 - 93	HA93-015	LEAK - A/C LOW PRESSURE HOSE - NEW TOOL/REVISION
4	93 - 93	NTB93-174	NOISE FRONT BLOWER - PROCEDURES
4	93 - 93	HA93-022	NOISE FRONT BLOWER - PROCEDURES - REVISED
1	Up - 96	NTB95-068	R134A CHARGE AMOUNTS/PAG LUBRICATION CHART
8,7	93 - 95	NTB95-043	RATTLE/RUMBLE BLOWER NOISE
8,7	93 - 95	HA95-008	RATTLE/RUMBLE BLOWER NOISE - REV
1	Up - 95	NTB95-014	REFRIGERANT LEAK DETECTION PROCEDURE
2	Up - 94	NTB94-091	SRVC PROCED - R134A RETROFITTED A/C SYS (EXC QUEST)
1	Up - 95	NTB94-020	THERMOSTATS AND HEATER PERFORMANCE

Model Legend: 1= All 2= Most 3= Axxess 4= Quest 5= 720 Pickup 6= D21 Pickup 7= Pathfinder 8= Hardbody

Entertainment Devices, Stereo, Radio, Etc.

Models	Year	TSB#	Description
4	93 - 93	FC0008429	ANTENNA - INOPERATIVE - WIRING REPAIR
7	93 - 93	FC0008364	ANTENNA/POWER - INOP - MOTOR HOT/TIMER BURNT OUT
1	Up - 93	NTB93-044	ANTENNA/POWER - ROD RELACEMENT PROCEDURE
1	Up - 96	EL94-004	ANTENNA/POWER - ROD RELACEMENT PROCEDURE - REV2
1	Up - 94	EL93-009	ANTENNA/POWER - ROD RELACEMENT PROCEDURE - REVISED
2	Up - 93	NTB94-034	AUDIO HEADUNIT MOUNTING SCREWS - IMPROVED
4	93 - 94	PI94-001	AUDIO SYS DATA SHEETS - SPECIFICATIONS
7	Up - 93	NTB94-016	AUDIO UNIT MISALIGNED (AT ANGLE) - PINCHED HARNESS
1	Up - 94	NTB94-060	BOSE SPEAKER EXCHANGE - REPLACE AS UNIT WITH AMP
2	Up - 93	NTB93-017	CASSETTE TAPE PLAYER - MAINTENANCE GUIDELINES
1	Up - 93	NTB92-068	CD AUTO CHANGER/DIAGNOSING AFTERMARKET INSTALLATION
7	93 - 93	FC0008232	CD INOP AFTER INSTALLATION/UNIT OK - WIRING REPAIR
2	93 - 93	NTB92-101	CD PLAYER - CHANGE IN AUDIO UNIT HARNESS CONNECTOR
4	93 - 93	FC0008111	CD PLAYER - INOP - INSTALL INSTRUCTIONS WRONG - FIX
4	93 - 93	NTB93-072	CD PLAYER - INSTALLATION INSTRUCTIONS
1	Up - 93	NTB92-083	POWER ANTENNA - MAINTENANCE/SERVICE INFORMATION
7	93 - 93	NTB93-127	POWER ANTENNA MODIFICATION - PREVENT WATER ENTRY
1	93 - 93	NTB93-035	RADIO - IDENTIFICATION INFORMATION
4	93 - 93	FC0008362	RADIO - LOSES POWER WHILE DRIVING - LOSES MEMORY
2	95 - 95	NTB94-087	RADIO CONNECTOR MODIFIED - ADAPTER HARNESS
7	91 - 93	FC0007805	RADIO/CD COMBO - INSTALLATION INSTRUCTIONS AVAILABLE
4	93 - 93	FC0008513	RADIO/TAPE - NO SOUND/ALL OTHER FUNCTIONS OK - FIX
4	93 - 93	FC0008639	REAR SPEAKERS INOP - FRONT OK - MISSING CONNECTOR
1	Up - 96	PI95-013	REMAN BOSE SPEAKERS AVAILABLE
4	93 - 93	FC0008581	RUMBLE NOISE IN REAR SPEAKER - BODY PANEL VIBRATION
7	93 - 93	FC0008260	SPEAKER/LEFT FRONT - CRACKLING NOISE - SERVICE PROC
7	93 - 93	FC0008507	SPEAKERS/REAR - INOP/INTERMIT LOW HUM - FIELD FIX
7	93 - 93	FC0008699	STATIC NOISE/WHINE IN SPEAKER - BAD WIRES

Model Legend: 1= All 2= Most 3= Axxess 4= Quest 5= 720 Pickup 6= D21 Pickup 7= Pathfinder 8= Hardbody

Seats, Belts, Interior Trim, Carpets, Air Bags

Models	Year	TSB#	Description
4	93 - 93	NTB93-029	FLOOR CARPET LIFTING AT SEAT TRACK - INFORMATION
4	93 - 93	FC0008106	FLOOR CARPET/RIGHT FRONT - VERY WET - DIAG/REPAIR
7	91 - 93	FC0008400	FLOOR MAT WET/RIGHT FRONT-HOOD HINGE BRACKET REPAIR
4	93 - 93	FC0008636	FRONT PASS SHOULDER BELT INOP - MOTOR DRIVE STRIPPED
1	94 - 94	NTB94-039	FRONT SEAT BELT EXTENDERS - CAUTION/APPLICATIONS
4	93 - 93	NTB93-151	PASSIVE RESTRAINT MOTOR REPLACEM'T REQ CONNECTOR MOD
4	93 - 94	NTB94-084	RATTLE NOISE - 2ND SEAT BELT RETRACTER/SPEAKER AREA
4	93 - 93	NTB94-029	RATTLE/SQUEAK FROM SECOND SEAT
4	93 - 93	FC0008318	SEAT BELT - MOTOR RUNS CONSTANTLY W/IGN ON - REPAIR
4	93 - 93	NTB92-088	SEAT BELT TRACK BOLTS - TIGHTENING DEFECT - RECALL
2	Up - 93	NTB93-112	SEAT BELTS - EXTENDERS AVAILABLE - P/N INFO
4	93 - 93	NTB92-064	SEAT/SLIDING 3RD - SQUEAK/RATTLE - ADJUSTMENT PROC
1	Up - 94	NTB94-072	SRS - AIR BAG SYSTEM DIAGNOSTIC PROCEDURES
1	Up - 93	NTB92-065	SRS - CLARIFICATION OF INSPECTION/REPAIR PROCEDURES
1	Up - 94	NTB93-133	TWO POINT MOTORIZED SHOULDER BELT - LOCKUP EXPLAINED
4	93 - 93	FC0008640	WET CARPET - A/C EVAP DRAIN - EXTEND DRAIN
4	94 - 94	NTB93-113	WIRING DIAGRAMS - S/M IMPROVEMENTS

Model Legend: 1= All 2= Most 3= Axxess 4= Quest 5= 720 Pickup 6= D21 Pickup 7= Pathfinder 8= Hardbody

Glass, Doors, Hood, Decklid, Tailgate, Liftgate, Locks

Models	Year	TSB#	Description
4	93 - 94	B3013	DOOR GUARD BEAM REPAIR - CAMPAIGN/RECALL
8	95 - 95	NTB95-007	DOOR LOCK INSPECTION - RECALL
1	92 - 93	NTB92-070	DOOR LOCKS/POWER - LOGIC REFERENCE CHART
4	93 - 93	FC0008275	IGN KEY WON'T GO INTO ACCESSORY POSITION - REPAIR
4	93 - 93	NTB93-031	KEYS LOCKED IN VEHICLE - KEY CODE & SERVICE INFO
4	93 - 93	NTB93-175	LEAK - WET CARPETS - DIAGNOSTICS AND PROCEDURES
4	93 - 94	B3013	NOISE/RATTLE - DOOR GUARD BEAM - RECALL CAMPAIGN
4	93 - 93	FC0008609	PWR DOOR LOCKS NOT LOCKING SLIDING DOOR - EXPLAINED
4	93 - 93	NTB93-130	SLIDING DOOR RATTLE
4	93 - 93	FC0008681	SLIDING DOOR RATTLE - LUBE WAIST (CENTER) ROLLER
4	93 - 93	NTB93-129	SUN ROOF WATER LEAKS
4	Up - 96	NTB95-071	WATER LEAK - SUNROOF REAR DRAIN
6,7	87 - 93	REF: BF93-023	WATER LEAKS TROUBLE DIAGNOSIS
4	93 - 95	NTB95-086	WINDSHIELD REPLACEMENT WITH DIVERSITY ANTENNA

Model Legend: 1= All 2= Most 3= Axxess 4= Quest 5= 720 Pickup 6= D21 Pickup 7= Pathfinder 8= Hardbody

Finishes, Body Structure, Frame, Bumpers

Models	Year	TSB#	Description
1	Up - 93	NTB93-075	BODY - DIAGNOSING WATER LEAKS W/VAC PRESSURE DEVICE
1	Up - 95	PI94-003	BUMPER REFINISHING TECHNIQUES
8	95 - 95	PI95-001	NEW PAINT GUARD FILM
7	95 - 95	PI95-009	NEW PAINT GUARD FILM
1	Up - 93	NTB93-058	PAINT - CONTAMINATION IDENTIFICATION & REPAIR PROC
1	Up - 93	NTB93-109	PAINT - MAINTENANCE/STORAGE RECOMMENDATIONS
1	94 - 94	NTB93-139	PAINT CODE NUMBERS - 1994 MODELS
1	95 - 95	PI94-005	PAINT CODES - 1995 MODELS
1	93 - 93	NTB92-069	PAINT CODES/COLOR NAMES/REFINISH PAINT CODES INFO
1	Up - 94	NTB94-015	PAINT CONTAMINATION IDENTIFICATION AND REPAIR
2	Up - 93	NTB92-061	PAINT GUARD COATING - REMOVER/SERVICE INFORMATION
1	93 - 93	NTB93-052	PAINT GUARD COATING REMOVER - NEW FORMULA/NO AMMONIA
2	93 - 94	NTB93-103	PAINT GUARD FILM - CLEAN UP PROCEDURES DURING PDI
1	94 - 95	NTB94-042	PAINT GUARD FILM - TIMING OF REMOVAL
1	94 - 95	BF94-010	PAINT GUARD FILM - TIMING OF REMOVAL - REVISED
2	93 - 93	NTB93-041	PAINT GUARD FILM REMOVAL
8	95 - 95	PI95-010	PAINT GUARD FILM SWELLING - PROCEDURE
2	95 - 95	PI95-002	POLYURETHANE & POLYPROPYLENE BUMPER FACIA REPAIR
1	96 - 96	PI95-015	REFINISH PAINT FORMULA CODES
4	93 - 93	NTB93-028	ROOF RACK - WIND NOISE - CORRECTION
6	94 - 94	NTB93-179	UNDERBODY BLACK BITUMIN WAX DISCONTINUED
6,7	87 - 93	REF: BF93-023	WATER LEAKS TROUBLE DIAGNOSIS
1	Up - 95	PI95-007	WATERBORNE REFINISH PAINTS - APPROVED SUPPLIERS

Model Legend: 1= All 2= Most 3= Axxess 4= Quest 5= 720 Pickup 6= D21 Pickup 7= Pathfinder 8= Hardbody

SUZUKI

Lighting, Horns, Turn Signals, Steering Column

Models	Year	TSB#	Description
4	92 - 93	TS02-11B	HEADLAMP ASSY - MODIFICATION - PARTS/SERVICE INFO
4	89 - 93	TS02-12	INT DOME LIGHT/WARNING BUZZER TEST PROC - S/M REV

Model Legend: 1= All 2= Most 3= X-90 4= Sidekick 5= Samari

Instruments, Dash Cluster, Warning Lights, Mirrors

Models	Year	TSB#	Description
4,3	Up - 96	TS04-22	FUEL GAUGE ACCURACY - PARTS MODIFICATION

Model Legend: 1= All 2= Most 3= X-90 4= Sidekick 5= Samari

Chassis Electrical, Wiring Harness, Fuses-Circuit Breakers, Wipers, Window Motors

Models	Year	TSB#	Description
1	Up - 93	SA01-01/93	ELECTRICAL TERMINAL KIT - NEW - MANDATORY PARTS
1	94 - 94	TS00-07	STOP LIGHT SWITCH MODIFIED - ADJUST PROCEDURES

Model Legend: 1= All 2= Most 3= X-90 4= Sidekick 5= Samari

Heating, Air Conditioning, Ventilation, Defogger

Models	Year	TSB#	Description
4	94 - 94	TS07-06/A	A/C - MODIFICATION/CHANGE IN SYSTEM AND S/M
1	95 - 95	TS00-11	A/C - RECOMMENDED OIL FOR R134A SYSTEMS
4	91 - 93	TS07-04	A/C COMPRESSOR - MODIFICATIONS AND SPECIAL TOOLS
1	Up - 94	TS00-01	AUTHORIZED A/C REFRIGERANT - FREON R-12
1	Up - 96	TS00-01 06031	AUTHORIZED A/C REFRIGERANT - R-12 OR R134A - REV

Model Legend: 1= All 2= Most 3= X-90 4= Sidekick 5= Samari

Seats, Belts, Interior Trim, Carpets, Air Bags

Models	Year	TSB#	Description
1	95 - 95	TS00-10R	AIR BAG SYSTEM ENABLE/DISABLE PROCEDURE
1	Up - 95	TS01-10	AIR BAG SYSTEM ENABLE/DISABLE PROCEDURE
1	95 - 96	TS00-10R	AIR BAG SYSTEM ENABLE/DISABLE PROCEDURE - REV

Model Legend: 1= All 2= Most 3= X-90 4= Sidekick 5= Samari

Glass, Doors, Hood, Decklid, Tailgate, Liftgate, Locks

Models	Year	TSB#	Description
4	94 - 95	TS01-08	HOOD AND HOOD LATCH MODIFICATION
1	Up - 93	SA01-07/87	KEY CODES - AVAILABILITY & PROCEDURE TO BE FOLLOWED

Model Legend: 1= All 2= Most 3= X-90 4= Sidekick 5= Samari

Finishes, Body Structure, Frame, Bumpers

Models	Year	TSB#	Description
5	Up - 95	TS01-07	ANTI-CORROSION TREATMENT PROCEDURE - S/M REV
1	Up - 94	TS00-06	BUMPER - PRECAUTIONS WHEN PAINTING REPLACEMENT PART
5	87 - 94	TS01-01R	LOCATION OF PAINT CODE IDENTIFICATION PLATE
4	95 - 96	TS01-09	MODIFICATION OF SPLASH GUARDS
4	89 - 93	TS01-02	PAINT - CODE IDENTIFICATION PLATE LOCATION
5	86 - 93	TS01-04	PAINT - CODE IDENTIFICATION PLATE LOCATION
4	89 - 95	TS01-02R	PAINT CODES
5	86 - 95	TS01-04R	PAINT CODES
4,3	89 - 96	TS1-02	PAINT CODES
4	89 - 94	TS01-02R	PAINT CODES - IDENTIFICATION PLATE LOCATION
5	86 - 94	TS01-04R	PAINT CODES - IDENTIFICATION PLATE LOCATION
3	96 - 96	TS01-01	PAINT CODES - LOCATION OF PAINT CODE ID PLATE
4	89 - 96	TS01-02	PAINT CODES - LOCATION OF PAINT CODE ID PLATE
4	89 - 96	TS01-02R	PAINT CODES - REV
1	Up - 94	TS00-06R	REPLACEMENT BUMPER PAINTING PRECAUTIONS
1	Up - 94	TS00-08	UNDERHOOD (PRIMER) PAINT COLOR CODES
1	Up - 96	TS0-08	UNDERHOOD (PRIMER) PAINT COLOR CODES - REV
1	Up - 95	TS00-08	UNDERHOOD (PRIMER) PAINT COLOR CODES - REV
1	Up - 96	TS00-08R	UNDERHOOD (PRIMER) PAINT COLOR CODES - REV

Model Legend: 1= All 2= Most 3= X-90 4= Sidekick 5= Samari

TOYOTA

Instruments, Dash Cluster, Warning Lights, Mirrors

Models	Year	TSB#	Description
4	95 - 95	AX95-006	WOOD DASH REPAIR PROCEDURE

Model Legend: 1= All 2= Most 3= 4Runner 4= Land Cruiser 5= Pickup 6= RAV4 7= T100 8= Tacoma

Chassis Electrical, Wiring Harness, Fuses-Circuit Breakers, Wipers, Window Motors

Models	Year	TSB#	Description
3	94 - 95	BO94-012	BACK DOOR WINDOW REGULATOR RUST - MODIFICATIONS
1	Up - 96	EL95-004	WIPER MOTOR CKT BREAKER INSPECT PROCEDURE

Model Legend: 1= All 2= Most 3= 4Runner 4= Land Cruiser 5= Pickup 6= RAV4 7= T100 8= Tacoma

Auxiliary Equipment, Jacks, Trailer Hitches, Towing

Models	Year	TSB#	Description
3	89 - 95	AX94-002	FRONT MUDGUARD AVAILABLE - INSTALLATION INSTRUCTIONS
1	Up - 94	AU93-001	MOBILE COMMUNICATION EQUIP - INSTALL PRECAUTIONS
3	96 - 96	AX001-96	MUDGUARD INSTALLATION
6	96 - 96	AX002-96	MUDGUARD INSTALLATION
8	95 - 95	AX95-004	MUDGUARD INSTALLATION INSTRUCTIONS - 4X2

Model Legend: 1= All 2= Most 3= 4Runner 4= Land Cruiser 5= Pickup 6= RAV4 7= T100 8= Tacoma

Heating, Air Conditioning, Ventilation, Defogger

Models	Year	TSB#	Description
2	89 - 93	SSC#P01	A/C - EXPANSION VALVE MALFUNCTIONS - SPECIAL CMPGN
1	Up - 94	SUPP: SSC#P01	A/C - EXPANSION VALVE REPLACEMENT/WARRANTY INFO
1	Up - 93	AC93-003	A/C - INSPECTION/DIAG/REPAIR OF LOW REFRIG LEVEL
5,3	93 - 93	AC92-006	A/C - INSTALLATION CHANGES/VEHICLES W/ABS - I/M REV
1	92 - 93	AC92-004	A/C - O-RING CHANGED DUE TO R134a REFRIGERANT
7	93 - 93	AC92-005	A/C - REFRIGERANT PRESSURE CONVERSION REF CHART
3,5	92 - 93	AC92-003	A/C - SUCTION HOSE CHANGED (V6)
1	Up - 96	AC002-96	A/C COMPRESSOR MAINTENANCE FOR STORED VEHICLES
1	Up - 94	AC93-005	A/C COMPRESSOR OIL APPLICATIONS
7	Up - 93	AC93-002	A/C W/HFC134a REFRIGERANT - NEW COMPONENTS & O-RINGS
1	Up - 94	AC94-001	ALTERNATIVE REFRIGERANTS FOR R12
1	93 - 94	PG93-003	CAUTION LABELS/EPA REQUIRED - A/C SYS W/REFRIG R-12
1	Up - 95	AC95-001	REFRIGERANT LEAK DETECTION - SERVICE HINTS

Model Legend: 1= All 2= Most 3= 4Runner 4= Land Cruiser 5= Pickup 6= RAV4 7= T100 8= Tacoma

Entertainment Devices, Stereo, Radio, Etc.

Models	Year	TSB#	Description
1	92 - 93	AU92-002	CASSETTE PLAYER WON'T PLAY - DIAGNOSIS/REPAIR
2	92 - 93	AU94-001	CD CHANGER INOP/SKIP/DISPLAYS ERROR - PROCEDURE

Model Legend: 1= All 2= Most 3= 4Runner 4= Land Cruiser 5= Pickup 6= RAV4 7= T100 8= Tacoma

Seats, Belts, Interior Trim, Carpets, Air Bags

Models	Year	TSB#	Description
4	93 - 93	BO94-001	POWER SEAT BACK - PLASTIC TUBE PREVENTS CHAFING
2	93 - 93	SSC#N06	SEAT BELT - INCORRECT TONGUE PLATE SIZE - RECALL
3,5	93 - 93	SSC#N06R	SEAT BELT - RENOTIFICATION - CAMPAIGN
1	85 - 94	BO94-006	SEAT BELT EXTENDER - APPLICATIONS & PROCEDURES
1	83 - 93	BO93-006	SEAT BELTS - EXTENDERS AVAIL THRU SPECIAL PROGRAM
1	Up - 93	EL93-001	SRS - SERVICE INFORMATION PRECAUTION

Model Legend: 1= All 2= Most 3= 4Runner 4= Land Cruiser 5= Pickup 6= RAV4 7= T100 8= Tacoma

Glass, Doors, Hood, Decklid, Tailgate, Liftgate, Locks

Models	Year	TSB#	Description
8	95 - 95	BO95-011	BACK WINDOW WEATHERSTRIP ADHESION
4	96 - 96	BO007-96	FRONT DOOR BELT MOLDING WIND NOISE
8	95 - 95	BO95-007	WIND NOISE - MODIFIED DOOR WEATHERSTRIP
8	95 - 95	BO002-96	WIND NOISE - WINDSHIELD MOLDING
5	Up - 94	BO94-007	WINDSHIELD WIND NOISE REPAIR

Model Legend: 1= All 2= Most 3= 4Runner 4= Land Cruiser 5= Pickup 6= RAV4 7= T100 8= Tacoma

Finishes, Body Structure, Frame, Bumpers

Models	Year	TSB#	Description
3	91 - 95	PA002-96	FRONT BUMPER/VALANCE PANEL COLORS
7	Up - 93	BO93-007	HOOD - VIB/FLUTTER REDUCTION PROCEDURE & ADHESIVE
1	93 - 93	PA92-001	PAINT & REFINISH FORMULA CODES
1	93 - 93	BO92-005	PAINT - NEW VEHICLE WASHING SCHEDULE FOR PROTECTION
1	95 - 95	PA94-001	PAINT AND REFINISH CODES - 1995 MODELS
1	94 - 94	PA93-002	PAINT AND REFINISH CODES - OEM AND REFURBISH
1	96 - 96	PA001-96	PAINT AND REFINISH FORMULA CODES
1	Up - 96	PA95-001	PAINT REPAIRS ON POLYURETHANE BUMPER
3	94 - 94	AX93-004	RUNNING BOARD INSTALLATION PRECAUTIONS

Model Legend: 1= All 2= Most 3= 4Runner 4= Land Cruiser 5= Pickup 6= RAV4 7= T100 8= Tacoma

VOLKSWAGEN

Instruments, Dash Cluster, Warning Lights, Mirrors

Models	Year	TSB#	Description
3	92 - 93	27/93-02	CRUISE CONTROL INOPERATIVE
1	Up - 95	90/95-05	SPEEDO REPLACEMENT - COMPLY WITH REGULATIONS

Model Legend: 1= All 2= Most 3= Eurovan 4= Vanagon

Chassis Electrical, Wiring Harness, Fuses-Circuit Breakers, Wipers, Window Motors

Models	Year	TSB#	Description
1	Up - 93	91/93-02	ELECTRICAL AFTERMARKT EQUIPMENT - PRECAUTIONS
1	Up - 95	97/94-04	ELECTRICAL CONTACT ENHANCER - STABILANT 22A
1	Up - 95	97/94-04	ELECTRICAL CONTACT ENHANCER - STABILANT 22A - REV
1	90 - 94	97/93-02	HARNESS CONNECT TERMINALS MUST MATCH SENSOR (GOLD)
3	92 - 93	97/92-01A	RELAY (AUX BATTERY CUT-OUT) - INSTALLATION INFO
3	92 - 93	CCC:SW	REPLACE POWER WINDOW REGULATORS - CAMPAIGN
1	Up - 96	69/95-03	USING STABILANT 22A ON ELECT CONN OF AIR BAG SYS
1	76 - 95	92/95-02	WIPER BLADES JERK OR SKIP
3	92 - 93	92/93-01	WIPER RELAY - INFINITELY VARIABLE - NOT INSTALLED
3	92 - 93	92/92-02	WIPER/WINDSHIELD - INFINITELY VARIABLE INTERMIT OPER
3	92 - 93	SI/02	WIRING DIAGRAMS - NEW INDEX & SENSORS/COMPUTERS
3	92 - 96	SI/06	WIRING DIAGRAMS/TROUBLESHOOT/COMPONENT LOC

Model Legend: 1= All 2= Most 3= Eurovan 4= Vanagon

Heating, Air Conditioning, Ventilation, Defogger

Models	Year	TSB#	Description
3	93 - 94	87/93-05	A/C - R134a SYSTEM - HISSING SOUND - SERVICE
3	93 - 94	87/93-04	A/C COMP - LEAKS REFRIGERANT R134a - SERVICE PROC
2	93 - 95	87/94-02	A/C POOR PERFORMANCE - VARIABLE DISPLACEMENT COMP
3	93 - 94	87/93-07	A/C REAR UNIT DOES NOT COOL
3	93 - 93	87/93-01	R-134A COMPONENTS - REPLACING & CAPABILITIES
3	93 - 94	87/93-01	R-134A COMPONENTS - REPLACING & CAPABILITIES - REV

Model Legend: 1= All 2= Most 3= Eurovan 4= Vanagon

Entertainment Devices, Stereo, Radio, Etc.

Models	Year	TSB#	Description
3	92 - 93	91/93-01A	CASSETTE RADIO - DELUXE & PREMIUM - GEN INFO
1	Up - 95	91/95-03	CD CHANGER SKIPS
1	Up - 93	WAR/V-93-13	HEIDELBERG LOCKED-UP RADIOS/OUT OF WARR UNLOCK INFO
1	Up - 96	91/95-04	PANASONIC CD CHANGER NOISE - REMOVE ADD'L GROUND
1	Up - 94	91/94-02	RADIO DIAGNOSIS INSTRUCTIONS

Model Legend: 1= All 2= Most 3= Eurovan 4= Vanagon

Seats, Belts, Interior Trim, Carpets, Air Bags

Models	Year	TSB#	Description
1	Up - 95	68/94-04	AIR BAG DEPLOYMENT - REPORT INCIDENT
1	Up - 96	69/95-03	AIR BAG SYS - USE OF STABILANT 22A ON ELECT CONNECT
2	85 - 94	69/95-04	CHILD SAFETY SEATS - ON FRONT PASS SEAT - INSTALL
1	Up - 94	70/93-04	FABRIC/VINYL/PLASTIC CLEANING PROCEDURES
3	93 - 93	51/93-01	NOISE BET FR SEATS - COVER FOR FUEL GAUGE SEND UNIT
2	80 - 96	69/95-05	REAR CHILD RESTRAINT ANCHORAGES

Model Legend: 1= All 2= Most 3= Eurovan 4= Vanagon

Glass, Doors, Hood, Decklid, Tailgate, Liftgate, Locks

Models	Year	TSB#	Description
3	85 - 95	64/95-01	ADHESIVE MAT'L FOR FLUSH BONDED WINDOWS - S/M REV
3	93 - 93	55/93-01	GAS FILLED REAR HATCH STRUT MODIFIED

Model Legend: 1= All 2= Most 3= Eurovan 4= Vanagon

Finishes, Body Structure, Frame, Bumpers

Models	Year	TSB#	Description
1	Up - 94	02/92-02	INDUSTRIAL FALLOUT REMOVAL
1	95 - 95	50/94-03	PAINT - REDUCING SWIRL MARKS & FINE SCRATCHES
3	95 - 95	66/94-05	RADIATOR GRILLE REMOVE/INSTALL PROCEDURES
3	Up - 93	40/92-02	SUBFRAME - ADDITIONAL PARTS NEEDED WHEN REPLACING

Model Legend: 1= All 2= Most 3= Eurovan 4= Vanagon

—————NOTES—————

———NOTES———

————NOTES————

—NOTES—

—————NOTES—————

NOTES

—————NOTES—————